Provided
by

Measure B

which was approved by
the voters in
November, 1998

CHILTON

ASIAN

SERVICE MANUAL
2004 Edition

THOMSON
™
DELMAR LEARNING

Australia • Canada • Mexico • Singapore • Spain • United Kingdom • United States

THOMSON
★
TM

DELMAR LEARNING

Chilton
Asian Service Manual
2004 Edition

Vice President, Technology and Trades SBU:
Alar Elken

Executive Director, Professional Business Unit:
Greg Clayton

Publisher, Professional Business Unit:
David Koontz

Channel Manager:
Beth A. Lutz

Marketing Specialist:
Brian McGrath

Production Director:
Mary Ellen Black

Production Manager:
Larry Main

Production Editor:
Elizabeth Hough

Editorial Assistant:
Kristen Shenfield

Editors:
Timothy A. Crain
Thomas A. Mellon
Richard J. Rivele
Christine L. Sheeky

COPYRIGHT 2004 by Delmar Learning, a division of Thomson Learning, Inc. Thomson Learning™ is a trademark used herein under license.

Printed in the United States of America
1 2 3 4 5 6 XX 07 06 05 04 03

For more information contact
Delmar Learning
Executive Woods
5 Maxwell Drive, PO Box 8007,
Clifton Park, NY 12065-8007
Or find us on the World Wide Web at :
www.trainingbay.com
or, www.chiltonsonline.com

ISBN: 1-4018-4235-6

NOTICE TO THE READER

Publisher does not warrant or guarantee any of the products described herein or perform any independent analysis in connection with any of the product information contained herein. Publisher does not assume, and expressly disclaims, any obligation to obtain and include information other than that provided to it by the manufacturer.

The reader is expressly warned to consider and adopt all safety precautions that might be indicated by the activities herein and to avoid all potential hazards. By following the instructions contained herein, the reader willingly assumes all risks in connection with such instructions.

The publisher makes no representation or warranties of any kind, including but not limited to, the warranties of fitness for particular purpose or merchantability, nor are any such representations implied with respect to the material set forth herein, and the publisher takes no responsibility with respect to such material. The publisher shall not be liable for any special, consequential, or exemplary damages resulting, in whole or part, from the readers' use of, or reliance upon, this material.

Table of Contents

Model Index

EDITORIAL POLICY

Manufacturer and Model Coverage

This manual does not cover every make and model that is currently available on the market. Rather, the Chilton editorial staff makes judicious decisions as to which makes and models warrant coverage, based on which vehicles are serviced by most technicians. In general, this manual does not cover:

- Exotic vehicles (e.g. Rolls-Royce, Ferrari, Dodge Viper)
- Vehicle manufacturers with no current U.S. presence (e.g. Fiat and Peugeot)
- Vehicle models that have not sold enough units to be a factor in the repair market

Model Year Information

This manual is published toward the end of the year prior to the edition year. Every effort is made to gather current data from the Original Vehicle Manufacturers (OEMs) when they publish it. Different OEMs choose to release their new model information at different times of the year. Indeed, the same OEM can publish information early one season and late the next season. As a result, not all models are equally current when each edition of this manual is published.

Although information in this manual is based on industry sources and is as complete as possible at the time of publication, some vehicle manufacturers may make changes which cannot be included here. Information on very late models may not be available in some circumstances. While striving for total accuracy, the publisher cannot assume responsibility for any errors, changes, or omissions that may occur in the compilation of this data.

Safety Notice

Proper service and repair procedures are vital to the safe, reliable operation of all motor vehicles, as well as the personal safety of those performing the repairs. This manual outlines procedures for servicing and repairing vehicles using safe, effective methods. The procedures contain many NOTES, WARNINGS and CAUTIONS which should be followed along with standard safety procedures to reduce the possibility of personal injury or improper service which could damage the vehicle or compromise its safety.

Repair procedures, tools, parts, and technician skill and experience vary widely. It is not possible to anticipate all conceivable ways or conditions under which vehicles may be serviced, or to provide cautions for all possible hazards that may result. Standard and accepted safety precautions and equipment should be used when handling toxic or flammable fluids, and safety goggles or other protection should be used during cutting, grinding, chiseling, prying, or any other process that can cause material removal or projectiles.

Some procedures require the use of tools specially designed for a specific purpose. Before substituting another tool or procedure, you must be completely satisfied that neither your personal safety, nor the performance of the vehicle will be endangered.

LOCATING AND USING THE INFORMATION

Organization

To find where a particular model section is located, look in the Table of Contents. On the first page of each model section, main topics are listed with the page number on which they may be found. Following the main topics is an alphabetical listing of all of the procedures within the section and their page numbers.

Part Numbers

Part numbers listed in this book are not recommendations by the publisher for any product by brand name. They are references that can be used with interchanges manuals and aftermarket supplier catalogs to locate each brand supplier's discrete part number.

Special Tools

Special tools are recommended by the vehicle manufacturer to perform specific jobs. When necessary, special tools are referred to in the text by the part number of the tool manufacturer. These tools may be purchased, under the appropriate part number, from your local dealer or regional distributor, or an equivalent tool can be purchased locally from a tool supplier or parts outlet. Before substituting any tool for the one recommended, read the previous Safety Notice.

ACKNOWLEDGEMENT

The publisher would like to express appreciation to the following vehicle manufacturers for their assistance in producing this publication. No further reproduction or distribution of the material in this manual is allowed without the expressed written permission of the vehicle manufacturers and the publisher.

American Honda Motor Co., including Acura and Honda Division
Fuji Heavy Industries Ltd., including Subaru
Isuzu Motors Ltd.
Hyundai Group, including Hyundai and Kia Motor
Mazda Motor Corp.
Mitsubishi
Nissan North America, including Infiniti and Nissan Division
Suzuki Motor Corp.
Toyota Motor Sales USA, including Lexus and Toyota Division

ACURA

3.2TL • 3.2CL • 3.5RL • Integra • RSX • NSX

1

SPECIFICATION CHARTS

ENGINE AND VEHICLE IDENTIFICATION

		Engine						Model Year	
Code	Liters (cc)	Cu. In.	Cyl.	Fuel Sys.	Engine Type	Eng. Mfg.		Code ①	Year
B18B1	1.8 (1834)	112	4	PGM-FI	DOHC	Honda		Y	2000
B18C1	1.8 (1797)	110	4	PGM-FI	DOHC	Honda		1	2001
B18C5	1.8 (1797)	110	4	PGM-FI	DOHC	Honda		2	2002
K20A2	2.0 (1999)	122	4	PGM-FI	DOHC	Honda		3	2003
K20A3	2.0 (1999)	122	4	PGM-FI	DOHC	Honda		4	2004
C30A1	3.0 (2977)	183	6	PGM-FI	DOHC	Honda			
J30A1	3.0 (2997)	183	6	PGM-FI	SOHC	Honda			
C32A6	3.2 (3206)	196	6	PGM-FI	SOHC	Honda			
C32B1	3.2 (3206)	196	6	PGM-FI	DOHC	Honda			
J32A1	3.2 (3210)	196	6	PGM-FI	SOHC	Honda			
C35A1	3.5 (3474)	211	6	PGM-FI	SOHC	Honda			

PGM-FI: Programmed Fuel Injection

DOHC: Double Overhead Camshaft

SOHC: Single Overhead Camshaft

① 10th digit of the Vehicle Identification Number (VIN)

42356-INTE-C01

GENERAL ENGINE SPECIFICATIONS

Year	Model	Engine Displacement Liters (cc)	Engine ID	Fuel System Type	Net Horsepower @ rpm	Net Torque @ rpm (ft. lbs.)	Bore x Stroke (in.)	Compression Ratio	Oil Pressure @ rpm
2000	Integra	1.8 (1834)	B18B1/①	PGM-FI	140@6300	127@5200	3.19x3.50	9.2:1	50@3000
	Integra GSR	1.8 (1797)	B18C1/②	PGM-FI	170@7600	128@6200	3.19x3.43	10.0:1	50@3000
	NSX	3.0 (2977)	C30A1/NA1	PGM-FI	252@6600	210@5300	3.54x3.07	10.2:1	50@3000
	NSX	3.2 (3206)	C32B1/NA2	PGM-FI	290@7100	224@5500	3.66x3.07	10.2:1	50@3000
	3.2TL	3.2 (3206)	J32A1/UA5	PGM-FI	225@5500	216@5000	3.50x3.39	9.8:1	71@3000
	3.5RL	3.5 (3474)	C35A1/KA9	PGM-FI	210@5200	224@2800	3.54x3.58	9.6:1	50@3000
2001	Integra	1.8 (1834)	B18B1/①	PGM-FI	140@6300	127@5200	3.19x3.50	9.2:1	50@3000
	Integra GSR	1.8 (1797)	B18C1/②	PGM-FI	170@7600	128@6200	3.19x3.43	10.0:1	50@3000
	NSX	3.0 (2977)	C30A1/NA1	PGM-FI	252@6600	210@5300	3.54x3.07	10.2:1	50@3000
	NSX	3.2 (3206)	C32B1/NA2	PGM-FI	290@7100	224@5500	3.66x3.07	10.2:1	50@3000
	3.2TL	3.2 (3206)	J32A1/UA5	PGM-FI	225@5500	216@5000	3.50x3.39	9.8:1	71@3000
	3.2CL	3.2 (3210)	J32A2/YA4	PGM-FI	260@6100	232@3500	3.50x3.39	10.5:1	71@3000
	3.5RL	3.5 (3474)	C35A1/KA9	PGM-FI	210@5200	224@2800	3.54x3.58	9.6:1	50@3000
2002	RSX	2.0 (1999)	K20A3	PGM-FI	160@6500	141@4000	3.39x3.39	9.8:1	44@3000
	RSX ③	2.0 (1999)	K20A2	PGM-FI	200@7400	142@6000	3.39x3.39	11.0:1	44@3000
	NSX	3.0 (2977)	C30A1/NA1	PGM-FI	252@6600	210@5300	3.54x3.07	10.2:1	50@3000
	NSX	3.2 (3206)	C32B1/NA2	PGM-FI	290@7100	224@5500	3.66x3.07	10.2:1	50@3000
	3.2TL	3.2 (3206)	J32A1/UA5	PGM-FI	225@5500	216@5000	3.50x3.39	9.8:1	71@3000
	3.2CL	3.2 (3210)	J32A2/YA4	PGM-FI	260@6100	232@3500	3.50x3.39	10.5:1	71@3000
	3.5RL	3.5 (3474)	C35A1/KA9	PGM-FI	210@5200	224@2800	3.54x3.58	9.6:1	50@3000
2003-04	RSX	2.0 (1999)	K20A3	PGM-FI	160@6500	141@4000	3.39x3.39	9.8:1	44@3000
	RSX ③	2.0 (1999)	K20A2	PGM-FI	200@7400	142@6000	3.39x3.39	11.0:1	44@3000
	NSX	3.0 (2977)	C30A1/NA1	PGM-FI	252@6600	210@5300	3.54x3.07	10.2:1	50@3000
	NSX	3.2 (3206)	C32B1/NA2	PGM-FI	290@7100	224@5500	3.66x3.07	10.2:1	50@3000
	3.2TL	3.2 (3206)	J32A1/UA5	PGM-FI	225@5500	216@5000	3.50x3.39	9.8:1	71@3000
	3.2CL	3.2 (3210)	J32A2/YA4	PGM-FI	260@6100	232@3500	3.50x3.39	10.5:1	71@3000
	3.5RL	3.5 (3474)	C35A1/KA9	PGM-FI	210@5200	224@2800	3.54x3.58	9.6:1	50@3000

PGM-FI: Programmed Fuel Injection

① DB7: 4 door

DC4: 3 door

② DB8: 4 door (Except Type R)

DC2: 3 door

③ Type-S

42356-INTE-C02

ENGINE TUNE-UP SPECIFICATIONS

Year	Engine Displacement Liters (cc)	Engine ID/VIN	Spark Plug Gap (in.)	Ignition Timing (deg.)		Fuel Pump (psi)	Idle Speed (rpm)		Valve Clearance	
				MT	AT		MT	AT	In.	Ex.
2000	1.8 (1834)	B18B1/①	0.039-0.043	16B	16B	40-47 ②	700-800	700-800	0.003-0.005	0.006-0.008
	1.8 (1797)	B18C1/③	0.051	16B	16B	48-55 ②	700-800	—	0.006-0.007	0.007-0.008
	3.0 (2977)	C30A1/NA1	0.043-0.047	15B	15B	47-53 ②	750-850	730-830	0.006-0.007	0.007-0.008
	3.2 (3206)	C32B1/NA2	0.043-0.047	—	15B	47-53 ②	—	750-850	0.006-0.007	0.007-0.008
	3.5 (3474)	C35A1/KA9	0.039-0.043	—	15B	43-50 ②	—	600-700	HYD	HYD
	3.2 (3210)	J32A1/UA5	0.039-0.043	—	10B	41-48 ②	—	630-730	0.008-0.009	0.011-0.013
2001	1.8 (1834)	B18B1/①	0.039-0.043	16B	16B	40-47 ②	700-800	700-800	0.003-0.005	0.006-0.008
	1.8 (1797)	B18C1/③	0.051	16B	16B	48-55 ②	700-800	—	0.006-0.007	0.007-0.008
	3.0 (2977)	C30A1/NA1	0.043-0.047	15B	15B	47-53 ②	750-850	730-830	0.006-0.007	0.007-0.008
	3.2 (3206)	C32B1/NA2	0.043-0.047	—	15B	47-53 ②	—	750-850	0.006-0.007	0.007-0.008
	3.5 (3474)	C35A1/KA9	0.039-0.043	—	15B	43-50 ②	—	600-700	HYD	HYD
	3.2 (3206)	J32A1/UA5	0.039-0.043	—	10B	41-48 ②	—	630-730	0.008-0.009	0.011-0.013
	3.2 (3210)	J32A2/YA4	0.039-0.043	—	10B	41-48 ②	—	700-800	0.008-0.009	0.011-0.013
2002	2.0 (1999)	K20A3	0.039-0.043	8B	8B	47-52 ②	600-700	600-700	0.003-0.005	0.006-0.008
	2.0 (1999)	K20A2	0.039-0.043	8B	8B	47-52 ②	650-750	650-750	0.006-0.007	0.007-0.008
	3.0 (2977)	C30A1/NA1	0.043-0.047	15B	15B	47-53 ②	750-850	730-830	0.006-0.007	0.007-0.008
	3.2 (3206)	C32B1/NA2	0.043-0.047	—	15B	47-53 ②	—	750-850	0.006-0.007	0.007-0.008
	3.5 (3474)	C35A1/KA9	0.039-0.043	—	15B	43-50 ②	—	600-700	HYD	HYD
	3.2 (3206)	J32A1/UA5	0.039-0.043	—	10B	41-48 ②	—	630-730	0.008-0.009	0.011-0.013
	3.2 (3210)	J32A2/YA4	0.039-0.043	—	10B	41-48 ②	—	700-800	0.008-0.009	0.011-0.013

42356-INTE-C03

ENGINE TUNE-UP SPECIFICATIONS

Year	Engine Displacement Liters (cc)	Engine ID/VIN	Spark Plug Gap (in.)	Ignition Timing (deg.)		Fuel Pump (psi)	Idle Speed (rpm)		Valve Clearance	
				MT	AT		MT	AT	In.	Ex.
2003-04	2.0 (1999)	K20A3	0.039-0.043	8B	8B	47-52 ②	600-700	600-700	0.003-0.005	0.006-0.008
	2.0 (1999)	K20A2	0.039-0.043	8B	8B	47-52 ②	650-750	650-750	0.006-0.007	0.007-0.008
	3.0 (2977)	C30A1/NA1	0.043-0.047	15B	15B	47-53 ②	750-850	730-830	0.006-0.007	0.007-0.008
	3.2 (3206)	C32B1/NA2	0.043-0.047	—	15B	47-53 ②	—	750-850	0.006-0.007	0.007-0.008
	3.5 (3474)	C35A1/KA9	0.039-0.043	—	15B	43-50 ②	—	600-700	HYD	HYD
	3.2 (3206)	J32A1/UA5	0.039-0.043	—	10B	41-48 ②	—	630-730	0.008-0.009	0.011-0.013
	3.2 (3210)	J32A2/YA4	0.039-0.043	—	10B	41-48 ②	—	700-800	0.008-0.009	0.011-0.013

NOTE: The Vehicle Emission Control Information label reflects specification changes during production and must be used if they differ from this chart.

B: Before Top Dead Center

HYD: Hydraulic

① DB7: 4 door
   DC4: 3 door

② At idle, pressure regulator vacuum hose disconnected

③ DB8: 4 door (Except Type R)
   DC2: 3 door

42356-INTE-C04

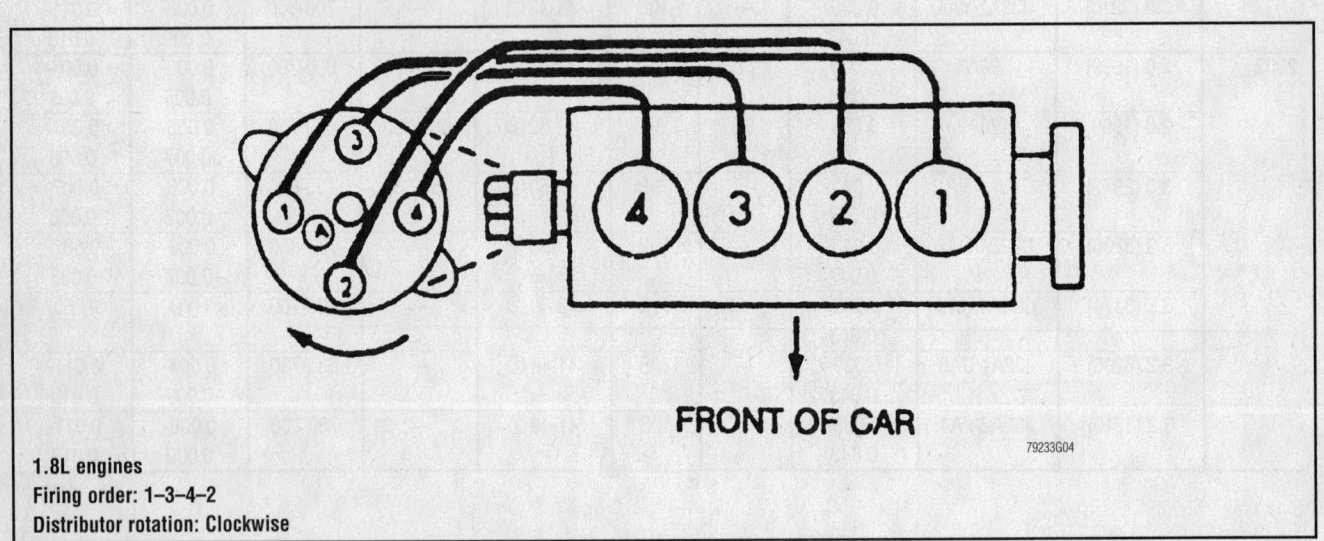

FRONT OF CAR

79233G04

1.8L engines
Firing order: 1–3–4–2
Distributor rotation: Clockwise

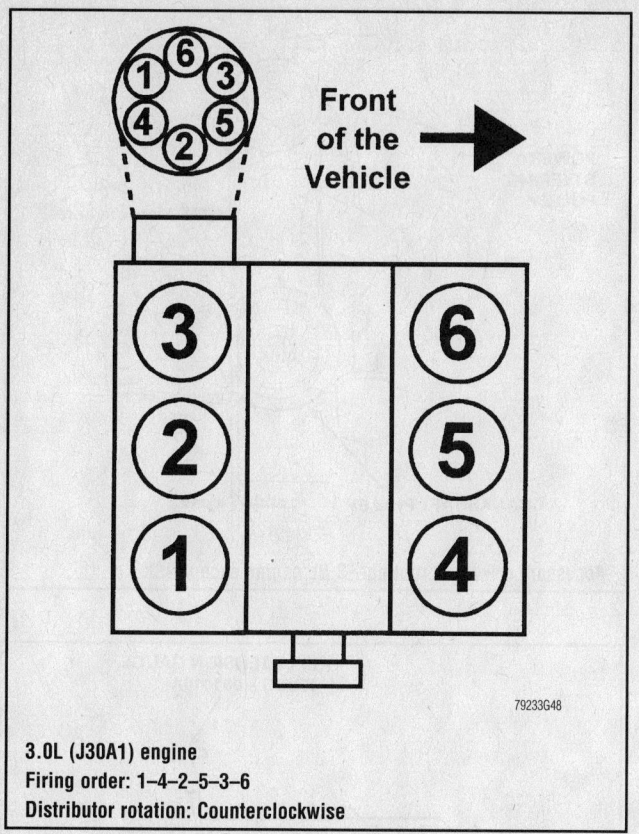

3.0L (J30A1) engine
Firing order: 1–4–2–5–3–6
Distributor rotation: Counterclockwise

79233G48

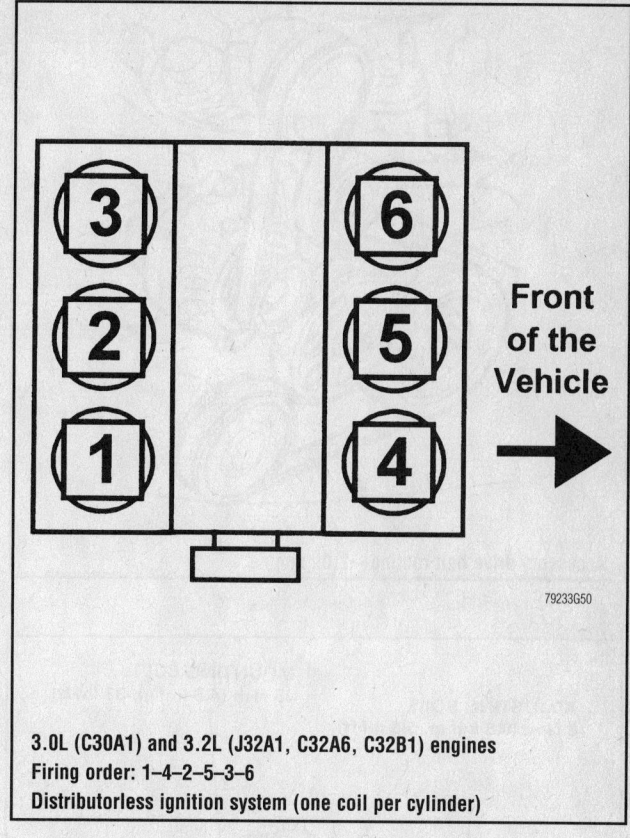

3.0L (C30A1) and 3.2L (J32A1, C32A6, C32B1) engines
Firing order: 1–4–2–5–3–6
Distributorless ignition system (one coil per cylinder)

79233G50

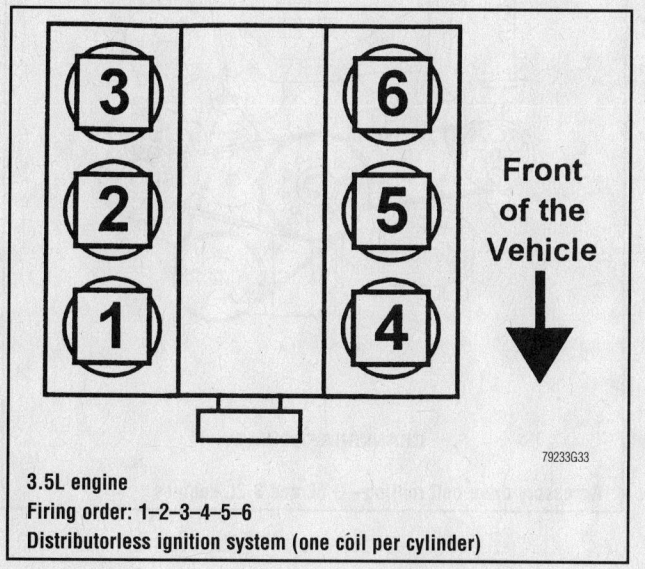

3.5L engine
Firing order: 1–2–3–4–5–6
Distributorless ignition system (one coil per cylinder)

79233G33

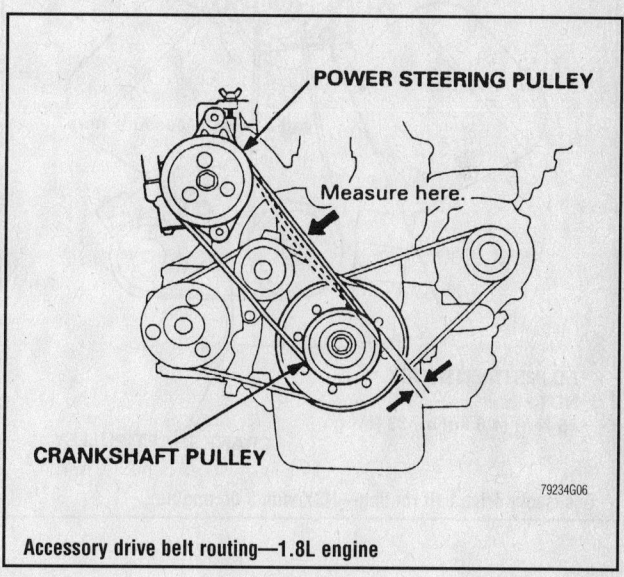

POWER STEERING PULLEY

Measure here.

CRANKSHAFT PULLEY

Accessory drive belt routing—1.8L engine

79234G06

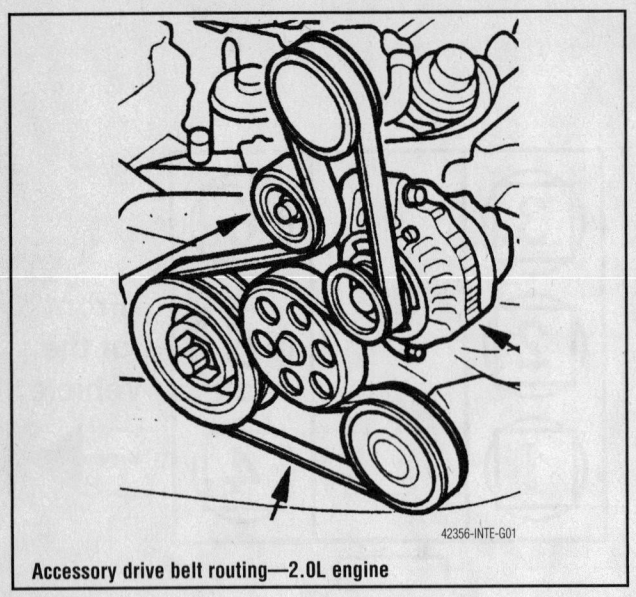

Accessory drive belt routing—2.0L engine

42356-INTE-G01

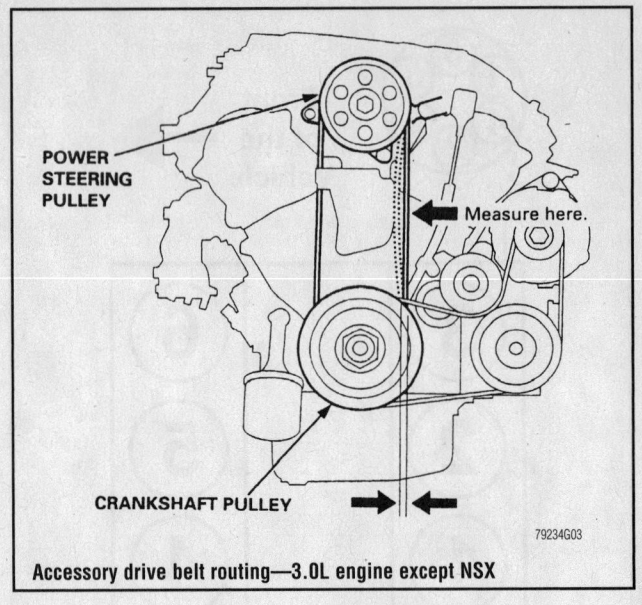

POWER STEERING PULLEY

Measure here.

CRANKSHAFT PULLEY

Accessory drive belt routing—3.0L engine except NSX

79234G03

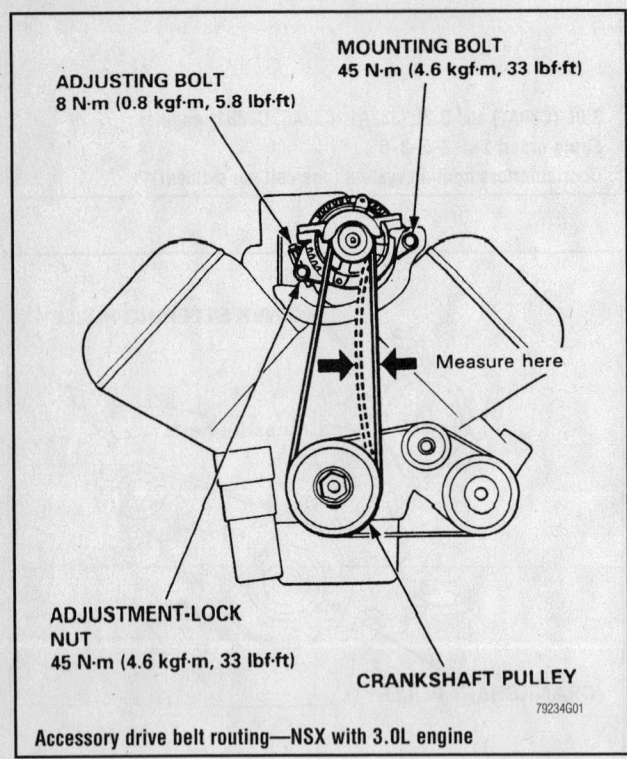

ADJUSTING BOLT
8 N·m (0.8 kgf·m, 5.8 lbf·ft)

MOUNTING BOLT
45 N·m (4.6 kgf·m, 33 lbf·ft)

Measure here

ADJUSTMENT-LOCK NUT
45 N·m (4.6 kgf·m, 33 lbf·ft)

CRANKSHAFT PULLEY

Accessory drive belt routing—NSX with 3.0L engine

79234G01

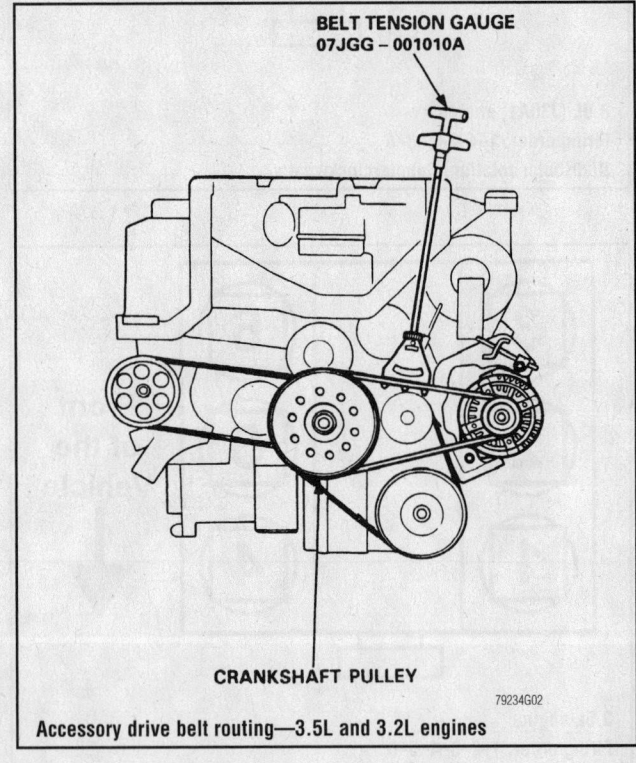

BELT TENSION GAUGE
07JGG – 001010A

CRANKSHAFT PULLEY

Accessory drive belt routing—3.5L and 3.2L engines

79234G02

CAPACITIES

Year	Model	Engine Displacement Liters (cc)	Engine ID/VIN	Engine Oil with Filter (qts.)	Transmission (pts.) 5-Spd	6-Spd	Auto.	Transfer Case (pts.)	Drive Axle Front (pts.)	Rear (pts.)	Fuel Tank (gal.)	Cooling System (qts.)
2000	Integra	1.8 (1834)	B18B1/①	4.0	4.6	—	5.8	—	—	—	13.2	②
	Integra GSR	1.8 (1797)	B18C1/③	4.2	4.6	—	—	—	—	—	13.2	5.0
	NSX	3.0 (2977)	C30A1/NA1	5.3	—	—	6.2	—	—	—	18.5	12.7
	NSX	3.2 (3206)	C32B1/NA2	5.3	—	5.6	—	—	—	—	18.5	12.7
	3.2TL	3.2 (3210)	J32A1/UA5	4.6	—	—	6.2	—	—	—	17.1	5.9
	3.5RL	3.5 (3474)	C35A1/KA9	4.9	—	—	6.4	—	2.2	—	18.0	6.4
2001	Integra	1.8 (1834)	B18B1/①	4.0	4.6	—	5.8	—	—	—	13.2	②
	Integra GSR	1.8 (1797)	B18C1/③	4.2	4.6	—	—	—	—	—	13.2	5.0
	NSX	3.0 (2977)	C30A1/NA1	5.3	—	—	6.2	—	—	—	18.5	12.7
	NSX	3.2 (3206)	C32B1/NA2	5.3	—	5.6	—	—	—	—	18.5	12.7
	3.2CL	3.2 (3210)	J32A2/YA4	5	—	—	7.6	—	—	—	17.1	7.9
	3.2TL	3.2 (3210)	J32A1/UA5	4.6	—	—	6.2	—	—	—	17.1	5.9
	3.5RL	3.5 (3474)	C35A1/KA9	4.9	—	—	6.4	—	2.2	—	18.0	6.4
2002	RSX	2.0 (1999)	K20A3	4.4	3.2	—	5.8	—	—	—	13.2	⑥
	RSX ⑤	2.0 (1999)	K20A2	5	—	3.2	—	—	—	—	13.2	5.4
	NSX	3.0 (2977)	C30A1/NA1	5.3	—	—	6.2	—	—	—	18.5	12.7
	NSX	3.2 (3206)	C32B1/NA2	5.3	—	5.6	—	—	—	—	18.5	12.7
	3.2CL	3.2 (3210)	J32A2/YA4	5	—	—	7.6	—	—	—	17.1	7.9
	3.2TL	3.2 (3210)	J32A1/UA5	4.6	—	—	6.2	—	—	—	17.1	5.9
	3.5RL	3.5 (3474)	C35A1/KA9	4.9	—	—	6.4	—	2.2	—	18.0	6.4
2003-04	RSX	2.0 (1999)	K20A3	4.4	3.2	—	5.8	—	—	—	13.2	⑥
	RSX ⑤	2.0 (1999)	K20A2	5	—	3.2	—	—	—	—	13.2	5.4
	NSX	3.0 (2977)	C30A1/NA1	5.3	—	—	6.2	—	—	—	18.5	12.7
	NSX	3.2 (3206)	C32B1/NA2	5.3	—	5.6	—	—	—	—	18.5	12.7
	3.2CL	3.2 (3210)	J32A2/YA4	5	—	—	7.6	—	—	—	17.1	7.9
	3.2TL	3.2 (3210)	J32A1/UA5	4.6	—	—	6.2	—	—	—	17.1	5.9
	3.5RL	3.5 (3474)	C35A1/KA9	4.9	—	—	6.4	—	2.2	—	18.0	6.4

NOTE: All capacities are approximate. Add fluid gradually and ensure a proper fluid level is obtained.

NOTE: Capacities given are service, not overhaul capacities

① DB7: 4 door
   DC4: 3 door
③ DB8: 4 door (Except Type R)
   DC2: 3 door
② Automatic transmission: 5.0
   Manual transmission: 4.6
⑤ Type-S
⑥ Automatic transmission: 5.3
   Manual Transmission: 5.4

42356-INTE-C05

VALVE SPECIFICATIONS

Year	Engine Displacement Liters (cc)	Engine ID/VIN	Seat Angle (deg.)	Face Angle (deg.)	Spring Test Pressure (lbs. @ in.)	Spring Installed Height (in.)	Stem-to-Guide Clearance (in.)		Stem Diameter (in.)	
							Intake	Exhaust	Intake	Exhaust
2000	1.8 (1834)	B18B1/①	45	45	NA	NA	0.0010-0.0020	0.0020-0.0030	0.2591-0.2594	0.2579-0.2583
	1.8 (1797)	B18C1/②	45	45	NA	NA	0.0010-0.0022	0.0020-0.0031	0.2156-0.2159	0.2146-0.2150
	3.0 (2977)	C30A1/NA1	45	45	NA	NA	0.0010-0.0020	0.0020-0.0030	0.2156-0.2159	0.2146-0.2150
	3.2 (3206)	C32B1/NA2	45	45	NA	NA	0.0010-0.0029	0.0020-0.0030	0.2156-0.2159	0.2146-0.2150
	3.5 (3474)	C35A1/KA9	45	45	NA	NA	0.0010-0.0020	0.0020-0.0030	0.2157-0.2161	0.2146-0.2150
	3.2 (3210)	J32A1/UA5	45	45	NA	NA	0.0008-0.0018	0.0022-0.0031	0.2159-0.2163	0.2146-0.2150
2001	1.8 (1834)	B18B1/①	45	45	NA	NA	0.0010-0.0020	0.0020-0.0030	0.2591-0.2594	0.2579-0.2583
	1.8 (1797)	B18C1/②	45	45	NA	NA	0.0010-0.0022	0.0020-0.0031	0.2156-0.2159	0.2146-0.2150
	3.0 (2977)	C30A1/NA1	45	45	NA	NA	0.0010-0.0020	0.0020-0.0030	0.2156-0.2159	0.2146-0.2150
	3.2 (3206)	C32B1/NA2	45	45	NA	NA	0.0010-0.0029	0.0020-0.0030	0.2156-0.2159	0.2146-0.2150
	3.5 (3474)	C35A1/KA9	45	45	NA	NA	0.0010-0.0020	0.0020-0.0030	0.2157-0.2161	0.2146-0.2150
	3.2 (3210)	J32A2/YA4	45	45	NA	NA	0.0008-0.0018	0.0022-0.0031	0.2159-0.2163	0.2146-0.2150
	3.2 (3210)	J32A1/UA5	45	45	NA	NA	0.0008-0.0018	0.0022-0.0031	0.2159-0.2163	0.2146-0.2150
2002	2.0 (1999)	K20A3	45	45	NA	NA	0.0012-0.0022	0.0022-0.0031	0.2156-0.2159	0.2146-0.2150
	2.0 (1999)	K20A4	45	45	NA	NA	0.0012-0.0022	0.0022-0.0031	0.2156-0.2159	0.2146-0.2150
	3.0 (2977)	C30A1/NA1	45	45	NA	NA	0.0010-0.0020	0.0020-0.0030	0.2156-0.2159	0.2146-0.2150
	3.2 (3206)	C32B1/NA2	45	45	NA	NA	0.0010-0.0029	0.0020-0.0030	0.2156-0.2159	0.2146-0.2150
	3.5 (3474)	C35A1/KA9	45	45	NA	NA	0.0010-0.0020	0.0020-0.0030	0.2157-0.2161	0.2146-0.2150
	3.2 (3210)	J32A2/YA4	45	45	NA	NA	0.0008-0.0018	0.0022-0.0031	0.2159-0.2163	0.2146-0.2150
	3.2 (3210)	J32A1/UA5	45	45	NA	NA	0.0008-0.0018	0.0022-0.0031	0.2159-0.2163	0.2146-0.2150

42356-INTE-C06

VALVE SPECIFICATIONS

Year	Engine Displacement Liters (cc)	Engine ID/VIN	Seat Angle (deg.)	Face Angle (deg.)	Spring Test Pressure (lbs. @ in.)	Spring Installed Height (in.)	Stem-to-Guide Clearance (in.)		Stem Diameter (in.)	
							Intake	Exhaust	Intake	Exhaust
2003-04	2.0 (1999)	K20A3	45	45	NA	NA	0.0012-0.0022	0.0022-0.0031	0.2156-0.2159	0.2146-0.2150
	2.0 (1999)	K20A4	45	45	NA	NA	0.0012-0.0022	0.0022-0.0031	0.2156-0.2159	0.2146-0.2150
	3.0 (2977)	C30A1/NA1	45	45	NA	NA	0.0010-0.0020	0.0020-0.0030	0.2156-0.2159	0.2146-0.2150
	3.2 (3206)	C32B1/NA2	45	45	NA	NA	0.0010-0.0029	0.0020-0.0030	0.2156-0.2159	0.2146-0.2150
	3.5 (3474)	C35A1/KA9	45	45	NA	NA	0.0010-0.0020	0.0020-0.0030	0.2157-0.2161	0.2146-0.2150
	3.2 (3210)	J32A2/YA4	45	45	NA	NA	0.0008-0.0018	0.0022-0.0031	0.2159-0.2163	0.2146-0.2150
	3.2 (3210)	J32A1/UA5	45	45	NA	NA	0.0008-0.0018	0.0022-0.0031	0.2159-0.2163	0.2146-0.2150

NA: Not Available

① DB7: 4 door
   DC4: 3 door

② DB8: 4 door (Except Type R)
   DC2: 3 door

42356-INTE-C07

TORQUE SPECIFICATIONS
All readings in ft. lbs.

Year	Engine Displacement Liters (cc)	Engine ID/VIN	Cylinder Head Bolts	Main Bearing Bolts	Rod Bearing Bolts	Crankshaft Damper Bolts	Flywheel Bolts	Manifold Intake	Manifold Exhaust	Spark Plugs	Lug Nut
2000	1.8 (1834)	B18B1/①	②	56	③	130	③	17	23	13	80
	1.8 (1797)	B18C1/④	②	⑤	⑥	130	76	17	23	13	80
	3.0 (2977)	C30A1/NA1	56	⑦	⑧	181	⑨	16	25	13	80
	3.2 (3206)	C32B1/NA2	56	⑩	29	181	76	16	22	13	80
	3.5 (3474)	C35A1/KA9	56	⑪	33	181	54	16	22	13	80
	3.2 (3210)	J32A1/UA5	⑫	⑬	⑭	181	54	16	23	13	80
2001	1.8 (1834)	B18B1/①	②	56	③	130	⑨	17	23	13	80
	1.8 (1797)	B18C1/④	②	⑤	⑥	130	76	17	23	13	80
	3.0 (2977)	C30A1/NA1	56	⑦	⑧	181	⑨	16	25	13	80
	3.2 (3206)	C32B1/NA2	56	⑩	29	181	76	16	22	13	80
	3.5 (3474)	C35A1/KA9	56	⑪	33	181	54	16	22	13	80
	3.2 (3210)	J32A2/YA4	⑫	⑬	⑭	181	54	16	23	13	80
	3.2 (3210)	J32A1/UA5	⑫	⑬	⑭	181	54	16	23	13	80
2002	2.0 (1999)	K20A3	⑮	⑯	14	181	⑰	16	33	13	80
	2.0 (1999)	K20A2	⑮	⑯	22	181	90	16	33	⑱	80
	3.0 (2977)	C30A1/NA1	56	⑦	⑧	181	⑨	16	25	13	80
	3.2 (3206)	C32B1/NA2	56	⑩	29	181	76	16	22	13	80
	3.5 (3474)	C35A1/KA9	56	⑪	33	181	54	16	22	13	80
	3.2 (3210)	J32A2/YA4	⑫	⑬	⑭	181	54	16	23	13	80
	3.2 (3210)	J32A1/UA5	⑫	⑬	⑭	181	54	16	23	13	80
2003-04	2.0 (1999)	K20A3	⑮	⑯	14	181	⑰	16	33	13	80
	2.0 (1999)	K20A2	⑮	⑯	22	181	90	16	33	⑱	80
	3.0 (2977)	C30A1/NA1	56	⑦	⑧	181	⑨	16	25	13	80
	3.2 (3206)	C32B1/NA2	56	⑩	29	181	76	16	22	13	80
	3.5 (3474)	C35A1/KA9	56	⑪	33	181	54	16	22	13	80
	3.2 (3210)	J32A2/YA4	⑫	⑬	⑭	181	54	16	23	13	80
	3.2 (3210)	J32A1/UA5	⑫	⑬	⑭	181	54	16	23	13	80

① DB7: 4 door
   DC4: 3 door

② Step 1: 22 ft. lbs.
   Step 2: 63 ft. lbs.

③ Step 1: 14 ft. lbs.
   Step 2: 23 ft. lbs.

④ DB8: 4 door (Except Type R)
   DC2: 3 door

⑤ Step 1: 22 ft. lbs.
   Step 2: Cap Nos. 1, 5: 56 ft. lbs.
       Cap Nos. 2, 3 and 4: 49 ft. lbs.

⑥ Step 1: 14 ft. lbs.
   Step 2: 33 ft. lbs.

⑦ Cap bolts: 29 ft. lbs.
   Cap bridge bolts: 48 ft. lbs.

⑧ 14 ft. lbs. plus 116 degrees

⑨ Manual transmission: 76 ft. lbs.
   Automatic transmission: 54 ft. lbs.

⑩ Step 1: Cap bolts 48 ft. lbs.
   Step 2: Side bolts 36 ft. lbs.

⑪ Step 1: Outer (9mm) 29 ft. lbs.
   Step 2: Inner (11mm) 56 ft. lbs.
   Step 3: Side (10mm) 36 ft. lbs.

⑫ Step 1: 29 ft. lbs.
   Step 2: 51 ft. lbs.
   Step 3: 72.3 ft. lbs.

⑬ Step 1: Cap bolts 56 ft. lbs.
   Step 2: Side bolts 36 ft. lbs.

⑭ Step 1: 14 ft. lbs.
   Step 2: Rotate 90 degrees

⑮ Step 1: 29 ft. lbs.
   Step 2: Rotate 90 degrees
   Step 3: Rotate an additional 90 degrees
   Step 4 (new bolts only): additional 90 degrees

⑯ Step 1: 22 ft. lbs.
   Step 2: Rotate 56 degrees

⑰ Manual trans.: 54 ft. lbs.
   Auto trans.: 76 ft. lbs.

⑱ NGK type IFRG-11KS & Denso type SK22PRM11S: 18 ft. lbs.
   All others: 13 ft. lbs.

42356-INTE-C08

CRANKSHAFT AND CONNECTING ROD SPECIFICATIONS

All measurements are given in inches.

Year	Engine Displacement Liters (cc)	Engine ID/VIN	Crankshaft				Connecting Rod		
			Main Brg. Journal Dia.	Main Brg. Oil Clearance	Shaft End-play	Thrust on No.	Journal Diameter	Oil Clearance	Side Clearance
2000	1.8 (1834)	B18B1/①	②	③	0.0040-0.0180	4	1.7707-1.7717	0.0008-0.0015	0.0060-0.0160
	1.8 (1797)	B18C1/④	⑤	⑥	0.0040-0.0180	4	1.7707-1.7717	0.0013-0.0020	0.0060-0.0160
	3.0 (2977)	C30A1/NA1	2.5187-2.5197	0.0009-0.0019	0.0040-0.0110	3	2.0463-2.0472	0.0016-0.0025	0.0060-0.0120
	3.2 (3206)	C32B1/NA2	2.5187-2.5197	0.0009-0.0019	0.0040-0.0110	3	2.0463-2.0472	0.0016-0.0025	0.0060-0.0120
	3.2 (3210)	J32A1/UA5	2.8337-2.8346	0.0008-0.0017	0.0040-0.0140	3	2.1644-2.1654	0.0008-0.0017	0.0060-0.0140
	3.5 (3474)	C35A1/KA9	2.6762-2.6772	0.0008-0.0017	0.0040-0.0110	3	2.1248-2.1257	0.0009-0.0018	0.0060-0.0120
2001	1.8 (1834)	B18B1/①	②	③	0.0040-0.0180	4	1.7707-1.7717	0.0008-0.0015	0.0060-0.0160
	1.8 (1797)	B18C1/④	⑤	⑥	0.0040-0.0180	4	1.7707-1.7717	0.0013-0.0020	0.0060-0.0160
	3.0 (2977)	C30A1/NA1	2.5187-2.5197	0.0009-0.0019	0.0040-0.0110	3	2.0463-2.0472	0.0016-0.0025	0.0060-0.0120
	3.2 (3206)	C32B1/NA2	2.5187-2.5197	0.0009-0.0019	0.0040-0.0110	3	2.0463-2.0472	0.0016-0.0025	0.0060-0.0120
	3.2 (3210)	J32A1/UA5	2.8337-2.8346	0.0008-0.0017	0.0040-0.0140	3	2.1644-2.1654	0.0008-0.0017	0.0060-0.0140
	3.2 (3210)	J32A2/YA4	2.8337-2.8346	0.0008-0.0017	0.0040-0.0140	3	2.1644-2.1654	0.0008-0.0017	0.0060-0.0140
	3.5 (3474)	C35A1/KA9	2.6762-2.6772	0.0008-0.0017	0.0040-0.0110	3	2.1248-2.1257	0.0009-0.0018	0.0060-0.0120
2002	2.0 (1999)	K20A3	2.1648-2.1657	⑦	0.0040-0.0140	4	1.7707-1.7717	0.0008-0.0019	0.0060-0.0160
	2.0 (1999)	K20A2	2.1646-2.1655	⑦	0.0040-0.0140	4	1.8888-1.8898	0.0013-0.0024	0.0060-0.0160
	3.0 (2977)	C30A1/NA1	2.5187-2.5197	0.0009-0.0019	0.0040-0.0110	3	2.0463-2.0472	0.0016-0.0025	0.0060-0.0120
	3.2 (3206)	C32B1/NA2	2.5187-2.5197	0.0009-0.0019	0.0040-0.0110	3	2.0463-2.0472	0.0016-0.0025	0.0060-0.0120
	3.2 (3210)	J32A1/UA5	2.8337-2.8346	0.0008-0.0017	0.0040-0.0140	3	2.1644-2.1654	0.0008-0.0017	0.0060-0.0140
	3.2 (3210)	J32A2/YA4	2.8337-2.8346	0.0008-0.0017	0.0040-0.0140	3	2.1644-2.1654	0.0008-0.0017	0.0060-0.0140
	3.5 (3474)	C35A1/KA9	2.6762-2.6772	0.0008-0.0017	0.0040-0.0110	3	2.1248-2.1257	0.0009-0.0018	0.0060-0.0120
2003-04	2.0 (1999)	K20A3	2.1648-2.1657	⑦	0.0040-0.0140	4	1.7707-1.7717	0.0008-0.0019	0.0060-0.0160
	2.0 (1999)	K20A2	2.1646-2.1655	⑦	0.0040-0.0140	4	1.8888-1.8898	0.0013-0.0024	0.0060-0.0160
	3.0 (2977)	C30A1/NA1	2.5187-2.5197	0.0009-0.0019	0.0040-0.0110	3	2.0463-2.0472	0.0016-0.0025	0.0060-0.0120
	3.2 (3206)	C32B1/NA2	2.5187-2.5197	0.0009-0.0019	0.0040-0.0110	3	2.0463-2.0472	0.0016-0.0025	0.0060-0.0120
	3.2 (3210)	J32A1/UA5	2.8337-2.8346	0.0008-0.0017	0.0040-0.0140	3	2.1644-2.1654	0.0008-0.0017	0.0060-0.0140
	3.2 (3210)	J32A2/YA4	2.8337-2.8346	0.0008-0.0017	0.0040-0.0140	3	2.1644-2.1654	0.0008-0.0017	0.0060-0.0140
	3.5 (3474)	C35A1/KA9	2.6762-2.6772	0.0008-0.0017	0.0040-0.0110	3	2.1248-2.1257	0.0009-0.0018	0.0060-0.0120

① DB7: 4 door
   DC4: 3 door

② Nos. 1, 2, 4 and 5: 2.1644-2.1654
   No. 3: 2.1642-2.1651

③ Nos. 1,2,4, and 5: 0.0009-0.0017
   No. 3: 0.0012-0.0019

④ DB8: 4 door (Except Type R)
   DC2: 3 door

⑤ Nos. 1, 2, 4 and 5: 2.1644-2.1654
   No. 3: 2.1643-2.1653

⑥ Nos. 1, 2, 4 and 5: 0.0009-0.0017
   No. 3: 0.0012-0.0019

⑦ Nos. 1, 2, 4 and 5: 0.0007-0.0016
   No. 3: 0.0010-0.0019

42356-INTE-C09

PISTON AND RING SPECIFICATIONS
All measurements are given in inches

Year	Engine Displacement Liters (cc)	Engine ID/VIN	Piston Clearance	Ring Gap			Ring Side Clearance		
				Top Compression	Bottom Compression	Oil Control	Top Compression	Bottom Compression	Oil Control
1999	1.8 (1834)	B18B1/①	0.0004-0.0016	0.0080-0.0140	0.0160-0.0220	0.0080-0.0200	0.0018-0.0028	0.0018-0.0026	NA
	1.8 (1797)	B18C1/②	0.0004-0.0016	0.0080-0.0140	0.0160-0.0220	0.0080-0.0200	0.0018-0.0028	0.0016-0.0026	NA
	3.0 (2997)	J30A1/YA2	0.0006-0.00160	0.0080-0.0140	0.0160-0.0220	0.0080-0.0280	0.0014-0.0024	0.0012-0.0022	NA
	3.2 (3206)	C32B1/NA2	0.0002-0.0012	0.0080-0.0120	0.0140-0.0200	0.0080-0.0280	0.0014-0.0026	0.0012-0.0024	NA
	3.2 (3210)	J32A1/UA5	0.0006-0.00160	0.0080-0.0140	0.0160-0.0220	0.0080-0.0280	0.0014-0.0024	0.0012-0.0022	NA
	3.5 (3474)	C35A1/KA	0.0010-0.0020	0.0100-0.0160	0.0160-0.0220	③	0.0022-0.0031	0.0012-0.0022	NA
2000	1.8 (1834)	B18B1/①	0.0004-0.0016	0.0080-0.0140	0.0160-0.0220	0.0080-0.0200	0.0018-0.0028	0.0018-0.0026	NA
	1.8 (1797)	B18C1/②	0.0004-0.0016	0.0080-0.0140	0.0160-0.0220	0.0080-0.0200	0.0018-0.0028	0.0016-0.0026	NA
	3.0 (2977)	C30A1/NA1	0.0002-0.0014	0.0100-0.0160	0.0140-0.0200	0.0080-0.0280	0.0012-0.0022	0.0012-0.0022	NA
	3.0 (2997)	J30A1/YA2	0.0006-0.00160	0.0080-0.0140	0.0160-0.0220	0.0080-0.0280	0.0014-0.0024	0.0012-0.0022	NA
	3.2 (3206)	C32B1/NA2	0.0002-0.0012	0.0080-0.0120	0.0140-0.0200	0.0080-0.0280	0.0014-0.0026	0.0012-0.0024	NA
	3.2 (3210)	J32A1/UA5	0.0006-0.00160	0.0080-0.0140	0.0160-0.0220	0.0080-0.0280	0.0014-0.0024	0.0012-0.0022	NA
	3.5 (3474)	C35A1/KA	0.0010-0.0020	0.0100-0.0160	0.0160-0.0220	③	0.0022-0.0031	0.0012-0.0022	NA
2001	1.8 (1834)	B18B1/①	0.0004-0.0016	0.0080-0.0140	0.0160-0.0220	0.0080-0.0200	0.0018-0.0028	0.0018-0.0026	NA
	1.8 (1797)	B18C1/②	0.0004-0.0016	0.0080-0.0140	0.0160-0.0220	0.0080-0.0200	0.0018-0.0028	0.0016-0.0026	NA
	3.0 (2977)	C30A1/NA1	0.0002-0.0014	0.0100-0.0160	0.0140-0.0200	0.0080-0.0280	0.0012-0.0022	0.0012-0.0022	NA
	3.0 (2997)	J30A1/YA2	0.0006-0.00160	0.0080-0.0140	0.0160-0.0220	0.0080-0.0280	0.0014-0.0024	0.0012-0.0022	NA
	3.2 (3206)	C32B1/NA2	0.0002-0.0012	0.0080-0.0120	0.0140-0.0200	0.0080-0.0280	0.0014-0.0026	0.0012-0.0024	NA
	3.2 (3210)	J32A1/UA5	0.0006-0.00160	0.0080-0.0140	0.0160-0.0220	0.0080-0.0280	0.0014-0.0024	0.0012-0.0022	NA
	3.2 (3210)	J32A2/YA4	0.0006-0.00160	0.0080-0.0140	0.0160-0.0220	0.0080-0.0280	0.0014-0.0024	0.0012-0.0022	NA
	3.5 (3474)	C35A1/KA	0.0010-0.0020	0.0100-0.0160	0.0160-0.0220	③	0.0022-0.0031	0.0012-0.0022	NA
2002	2.0 (1999)	K20A3	0.0008-0.0016	0.0080-0.0014	0.0160-0.0220	0.0100-0.0260	0.0014-0.0024	0.0012-0.0022	NA
	2.0 (1999)	K20A2	0.0008-0.0016	0.0080-0.0140	0.0200-0.0260	0.0080-0.0280	0.0016-0.0026	0.0018-0.0028	NA
	3.0 (2977)	C30A1/NA1	0.0002-0.0014	0.0100-0.0160	0.0140-0.0200	0.0080-0.0280	0.0012-0.0022	0.0012-0.0022	NA
	3.0 (2997)	J30A1/YA2	0.0006-0.00160	0.0080-0.0140	0.0160-0.0220	0.0080-0.0280	0.0014-0.0024	0.0012-0.0022	NA
	3.2 (3206)	C32B1/NA2	0.0002-0.0012	0.0080-0.0120	0.0140-0.0200	0.0080-0.0280	0.0014-0.0026	0.0012-0.0024	NA
	3.2 (3210)	J32A1/UA5	0.0006-0.00160	0.0080-0.0140	0.0160-0.0220	0.0080-0.0280	0.0014-0.0024	0.0012-0.0022	NA
	3.2 (3210)	J32A2/YA4	0.0006-0.00160	0.0080-0.0140	0.0160-0.0220	0.0080-0.0280	0.0014-0.0024	0.0012-0.0022	NA
	3.5 (3474)	C35A1/KA	0.0010-0.0020	0.0100-0.0160	0.0160-0.0220	③	0.0022-0.0031	0.0012-0.0022	NA
2003-04	2.0 (1999)	K20A3	0.0008-0.0016	0.0080-0.0014	0.0160-0.0220	0.0100-0.0260	0.0014-0.0024	0.0012-0.0022	NA
	2.0 (1999)	K20A2	0.0008-0.0016	0.0080-0.0140	0.0200-0.0260	0.0080-0.0280	0.0016-0.0026	0.0018-0.0028	NA
	3.0 (2977)	C30A1/NA1	0.0002-0.0014	0.0100-0.0160	0.0140-0.0200	0.0080-0.0280	0.0012-0.0022	0.0012-0.0022	NA
	3.0 (2997)	J30A1/YA2	0.0006-0.00160	0.0080-0.0140	0.0160-0.0220	0.0080-0.0280	0.0014-0.0024	0.0012-0.0022	NA
	3.2 (3206)	C32B1/NA2	0.0002-0.0012	0.0080-0.0120	0.0140-0.0200	0.0080-0.0280	0.0014-0.0026	0.0012-0.0024	NA
	3.2 (3210)	J32A1/UA5	0.0006-0.00160	0.0080-0.0140	0.0160-0.0220	0.0080-0.0280	0.0014-0.0024	0.0012-0.0022	NA
	3.2 (3210)	J32A2/YA4	0.0006-0.00160	0.0080-0.0140	0.0160-0.0220	0.0080-0.0280	0.0014-0.0024	0.0012-0.0022	NA
	3.5 (3474)	C35A1/KA	0.0010-0.0020	0.0100-0.0160	0.0160-0.0220	③	0.0022-0.0031	0.0012-0.0022	NA

NA; Not Applicable

① DB7: 4 door
DC4: 3 door

② DB8: 4 door
DC2: 3 door (Except Type R)

③ RIKEN: 0.0080-0.0280 inches
TEIKOKU: 0.0080-0.0200 inches

42356-INTE-C10

WHEEL ALIGNMENT

Year	Model		Caster Range (+/-Deg.)	Caster Preferred Setting (Deg.)	Camber Range (+/-Deg.)	Camber Preferred Setting (Deg.)	Toe-in (in.)	Steering Axis Inclination (Deg.)
2000	3.2 TL	F	1.00	+2.80	1.00	0	0 +/- 0.06	—
		R	—	—	0.50	-0.50	0.06 +/- 0.06	—
	3.5 RL	F	1.00	+2.81	1.00	0	0 +/- 0.06	—
		R	—	—	1.00	-0.50	0.06 +/- 0.06	—
	Integra ①	F	1.00	+1.16	1.00	+0.16	0 +/- 0.06	—
		R	—	—	1.00	+0.75	0.16 +/- 0.16	—
	Integra ②	F	1.00	+1.16	1.00	+0.16	0 +/- 0.06	—
		R	—	—	1.00	+0.50	0.06 +/- 0.06	—
	NSX ③	F	0.25	+8.00	0.17	-0.33	0.14 +/- 0.04	—
		R	—	—	0.50	-1.50	0.18 +/- 0.06	—
	NSX ④	F	0.25	+8.23	0.17	+0.50	0.14 +/- 0.04	—
		R	—	—	0.50	-2.00	0.18 +/- 0.06	—
2001	3.2 TL	F	1.00	+2.80	1.00	0	0 +/- 0.06	—
		R	—	—	0.50	-0.50	0.06 +/- 0.06	—
	3.2 CL	F	1.00	+2.80	1.00	0	0 +/- 0.06	—
		R	—	—	1.00	-0.50	0.06 +/- 0.06	—
	3.5 RL	F	1.00	+2.81	1.00	0	0 +/- 0.06	—
		R	—	—	1.00	-0.50	0.06 +/- 0.06	—
	Integra ①	F	1.00	+1.16	1.00	+0.16	0 +/- 0.16	—
		R	—	—	1.00	+0.75	0.16 +/- 0.16	—
	Integra ②	F	1.00	+1.16	1.00	+0.16	0 +/- 0.06	—
		R	—	—	1.00	+0.50	0.06 +/- 0.06	—
	NSX ③	F	0.25	+8.00	0.17	-0.33	0.14 +/- 0.04	—
		R	—	—	0.50	-1.50	0.18 +/- 0.06	—
	NSX ④	F	0.25	+8.23	0.17	+0.50	0.14 +/- 0.04	—
		R	—	—	0.50	-2.00	0.18 +/- 0.06	—
2002	3.2 TL	F	1.00	+2.80	1.00	0	0 +/- 0.06	—
		R	—	—	0.50	-0.50	0.06 +/- 0.06	—
	3.2 CL	F	1.00	+2.80	1.00	0	0 +/- 0.06	—
		R	—	—	1.00	-0.50	0.06 +/- 0.06	—
	3.5 RL	F	1.00	+2.81	1.00	0	0 +/- 0.06	—
		R	—	—	1.00	-0.50	0.06 +/- 0.06	—
	RSX ③	F	1.00	+1.16	1.00	+0.16	0 +/- 0.16	—
		R	—	—	1.00	+0.75	0.16 +/- 0.16	—
	RSX ④	F	1.00	+1.16	1.00	+0.16	0 +/- 0.06	—
		R	—	—	1.00	+0.50	0.06 +/- 0.06	—
	NSX ③	F	0.25	+8.00	0.17	-0.33	0.14 +/- 0.04	—
		R	—	—	0.50	-1.50	0.18 +/- 0.06	—
	NSX ④	F	0.25	+8.23	0.17	+0.50	0.14 +/- 0.04	—
		R	—	—	0.50	-2.00	0.18 +/- 0.06	—

42356-INTE-C11

WHEEL ALIGNMENT

Year	Model		Caster Range (+/-Deg.)	Caster Preferred Setting (Deg.)	Camber Range (+/-Deg.)	Camber Preferred Setting (Deg.)	Toe-in (in.)	Steering Axis Inclination (Deg.)
2003-04	3.2 TL	F	1.00	+2.80	1.00	0	0 +/- 0.06	—
		R	—	—	0.50	-0.50	0.06 +/- 0.06	—
	3.2 CL	F	1.00	+2.80	1.00	0	0 +/- 0.06	—
		R	—	—	1.00	-0.50	0.06 +/- 0.06	—
	3.5 RL	F	1.00	+2.81	1.00	0	0 +/- 0.06	—
		R	—	—	1.00	-0.50	0.06 +/- 0.06	—
	RSX ③	F	1.00	+1.16	1.00	+0.16	0 +/- 0.16	—
		R	—	—	1.00	+0.75	0.16 +/- 0.16	—
	RSX ④	F	1.00	+1.16	1.00	+0.16	0 +/- 0.06	—
		R	—	—	1.00	+0.50	0.06 +/- 0.06	—
	NSX ③	F	0.25	+8.00	0.17	-0.33	0.14 +/- 0.04	—
		R	—	—	0.50	-1.50	0.18 +/- 0.06	—
	NSX ④	F	0.25	+8.23	0.17	+0.50	0.14 +/- 0.04	—
		R	—	—	0.50	-2.00	0.18 +/- 0.06	—

① Except Type R
② Type R
③ Except Type S
④ Type S

42356-INTE-C12

TIRE, WHEEL AND BALL JOINT SPECIFICATIONS

Year	Model	OEM Tires Standard	OEM Tires Optional	Tire Pressures (psi) Front	Tire Pressures (psi) Rear	Wheel Size	Ball Joint Inspection
2000	Integra	P195/55VR15	None	32	30	6-JJ	NS
	3.2TL	P205/60VR16	None	29	29	6-JJ	NS
	3.5RL	P225/55/R16	None	32	29	6-JJ	NS
	NSX	①	None	32	29	6-JJ	NS
2001	Integra	P195/55VR15	None	32	30	6-JJ	NS
	3.2TL	P205/60VR16	None	29	29	6-JJ	NS
	3.2CL	P205/60VR16	None	29	29	6-JJ	NS
	3.5RL	P225/55/R16	None	32	29	6-JJ	NS
	NSX	①	None	32	29	6-JJ	NS
2002	RSX	P205/55R16	None	32	30	6-JJ	NS
	3.2TL	P205/60VR16	None	29	29	6-JJ	NS
	3.2CL	P205/60VR16	None	29	29	6-JJ	NS
	3.5RL	P225/55/R16	None	32	29	6-JJ	NS
	NSX	①	None	32	29	6-JJ	NS
2003-04	RSX	P205/55R16	None	32	30	6-JJ	NS
	3.2TL	P205/60VR16	None	29	29	6-JJ	NS
	3.2CL	P205/60VR16	None	29	29	6-JJ	NS
	3.5RL	P225/55/R16	None	32	29	6-JJ	NS
	NSX	①	None	32	29	6-JJ	NS

OEM: Original Equipment Manufacturer

PSI: Pounds Per Square Inch

STD: Standard

OPT: Optional

NS: Not Specified by manufacturer

① Front tires: P215/40R17

Rear tires: P255/40R17

42356-INTE-C13

BRAKE SPECIFICATIONS
All measurements in inches unless noted

Year	Model		Brake Disc Original Thickness	Brake Disc Minimum Thickness	Brake Disc Maximum Runout	Brake Drum Diameter Original Inside Diameter	Brake Drum Diameter Max. Wear Limit	Brake Drum Diameter Maximum Machine Diameter	Minimum Lining Thickness Front	Minimum Lining Thickness Rear	Brake Caliper Bracket Bolts (ft. lbs.)	Brake Caliper Mounting Bolts (ft. lbs.)
2000	3.5RL	F	0.910	0.830	0.004	—	—	—	0.06	—	80	36
		R	0.350	0.300	0.004	—	—	—	—	0.06	28	17
	3.2TL	F	1.100	1.020	0.004	—	—	—	0.06	—	80	36
		R	0.350	0.310	0.004	①	6.693 ②	②	—	③	41	17
	Integra	F	0.830	0.750	0.004	—	—	—	0.06	—	—	24
		R	0.350	0.310	0.004	—	—	—	—	0.06	—	24
	NSX	F	1.100	1.020	0.004	—	—	—	0.06	—	80	36
		R	0.830	0.750	0.004	—	—	—	—	0.06	80	36
2001	3.5RL	F	0.910	0.830	0.004	—	—	—	0.06	—	80	36
		R	0.350	0.300	0.004	—	—	—	—	0.06	28	17
	3.2TL	F	1.100	1.020	0.004	—	—	—	0.06	—	80	36
		R	0.350	0.310	0.004	①	6.693 ②	②	—	③	41	17
	3.2CL	F	1.100	1.020	0.004	—	—	—	0.06	—	80	36
		R	0.350	0.310	0.004	①	6.693 ②	②	—	③	41	17
	Integra	F	0.830	0.750	0.004	—	—	—	0.06	—	—	24
		R	0.350	0.310	0.004	—	—	—	—	0.06	—	24
	NSX	F	1.100	1.020	0.004	—	—	—	0.06	—	80	36
		R	0.830	0.750	0.004	—	—	—	—	0.06	80	36
2002	3.5RL	F	0.910	0.830	0.004	—	—	—	0.06	—	80	36
		R	0.350	0.300	0.004	—	—	—	—	0.06	28	17
	3.2TL	F	1.100	1.020	0.004	—	—	—	0.06	—	80	36
		R	0.350	0.310	0.004	①	6.693 ②	②	—	③	41	17
	3.2CL	F	1.100	1.020	0.004	—	—	—	0.06	—	80	36
		R	0.350	0.310	0.004	①	6.693 ②	②	—	③	41	17
	RSX	F	0.830	0.750	0.004	—	—	—	0.06	—	—	24
		R	0.350	0.310	0.004	—	—	—	—	0.06	—	24
	RSX Type-S	F	0.980	0.910	0.004	—	—	—	0.06	—	—	24
		R	0.350	0.310	0.004	—	—	—	—	0.06	—	24
	NSX	F	1.100	1.020	0.004	—	—	—	0.06	—	80	36
		R	0.830	0.750	0.004	—	—	—	—	0.06	80	36
2003-04	3.5RL	F	0.910	0.830	0.004	—	—	—	0.06	—	80	36
		R	0.350	0.300	0.004	—	—	—	—	0.06	28	17
	3.2TL	F	1.100	1.020	0.004	—	—	—	0.06	—	80	36
		R	0.350	0.310	0.004	①	6.693 ②	②	—	③	41	17
	3.2CL	F	1.100	1.020	0.004	—	—	—	0.06	—	80	36
		R	0.350	0.310	0.004	①	6.693 ②	②	—	③	41	17
	RSX	F	0.830	0.750	0.004	—	—	—	0.06	—	—	24
		R	0.350	0.310	0.004	—	—	—	—	0.06	—	24
	RSX Type-S	F	0.980	0.910	0.004	—	—	—	0.06	—	—	24
		R	0.350	0.310	0.004	—	—	—	—	0.06	—	24
	NSX	F	1.100	1.020	0.004	—	—	—	0.06	—	80	36
		R	0.830	0.750	0.004	—	—	—	—	0.06	80	36

NA: Not Available

F: Front

R: Rear

① Rear parking brake drum: 6.693 inches

② Rear parking brake drum maximum diameter: 6.732 inches

③ Rear pad: 0.06 inches

Rear parking brake shoes: 0.04 inches

SCHEDULED MAINTENANCE INTERVALS
ACURA—3.2CL, 3.2TL, 3.5RL, INTEGRA, RSX & NSX

TO BE SERVICED	TYPE OF SERVICE	VEHICLE MILEAGE INTERVAL (x1000)												
		7.5	15	22.5	30	37.5	45	52.5	60	67.5	75	82.5	90	97.5
Engine oil	R	✓	✓	✓	✓	✓	✓	✓	✓	✓	✓	✓	✓	✓
Rear brake discs, calipers & pads	S/I		✓		✓		✓		✓		✓		✓	
Rotate tires	S/I	✓	✓	✓	✓	✓	✓	✓	✓	✓	✓	✓	✓	✓
A/C filter	R		✓		✓		✓		✓		✓		✓	
A/C filter (3.5RL)	R				✓				✓				✓	
Brake hoses & lines (including ABS)	S/I		✓		✓		✓		✓		✓		✓	
Cooling system hoses & connections	S/I		✓		✓		✓		✓		✓		✓	
Driveshaft boots	S/I		✓		✓		✓		✓		✓		✓	
Exhaust system	S/I		✓		✓		✓		✓		✓		✓	
Front brake discs & calipers	S/I		✓		✓		✓		✓		✓		✓	
Fuel pipes, hoses & connections	S/I		✓		✓		✓		✓		✓		✓	
Suspension components	S/I		✓		✓		✓		✓		✓		✓	
Suspension mounting bolts	S/I		✓		✓		✓		✓		✓		✓	
Tie rods, steering gear box & boots	S/I		✓		✓		✓		✓		✓		✓	
Steering operation, tie rod ends, steering gearbox & boots	S/I		✓		✓				✓				✓	
Valve clearance (NSX)	S/I				✓				✓				✓	
Parking brake	S/I		✓		✓		✓		✓		✓		✓	
Air cleaner element	R				✓				✓				✓	
Automatic transmission fluid	R				✓				✓				✓	
Brake fluid (including ABS) (Integra & RSX)	R				✓				✓				✓	
Brake fluid (including ABS) (3.5RL)	R								✓				✓	
Brake fluid (including ABS) (3.2TL & NSX)	R						✓				✓			
Front differential fluid (3.2TL & 3.5RL)	R								✓				✓	
Manual transmission fluid	R				✓				✓				✓	
ABS operation	S/I				✓				✓				✓	
Drive belt(s)	S/I				✓				✓				✓	
Spark plugs (Integra exc. GSR & RSX exc. Type S)	R				✓				✓				✓	
Spark plugs (3.2TL, Integra GSR & NSX)	R								✓				✓	
Spark plugs (3.5.RL)	R								✓					
Engine coolant	R						✓				✓			

42356-INTE-C15

SCHEDULED MAINTENANCE INTERVALS
ACURA—3.2CL, 3.2TL, 3.5RL, INTEGRA, RSX & NSX

TO BE SERVICED	TYPE OF SERVICE	VEHICLE MILEAGE INTERVAL (x1000)												
		7.5	15	22.5	30	37.5	45	52.5	60	67.5	75	82.5	90	97.5
ABS high pressure hose (NSX)	R								✓					
Fuel filter	R								✓					
PCV valve	S/I								✓					
Timing belt (except as noted below)	R												✓	
Timing belt & timing balancer belt (3.5RL) ①	R													
Transmission fluid	R												✓	
Distributor, ignition cap & rotor (Integra)	S/I								✓					
Idle speed (3.2TL, Integra & NSX)	S/I								✓					
Idle speed (3.5RL) ②	S/I		✓		✓		✓		✓		✓		✓	
Ignition wires	S/I		✓		✓				✓				✓	
TWC converter heat shield	S/I		✓		✓		✓		✓		✓		✓	
Water pump	S/I				✓				✓				✓	
Water pump (3.5RL) ②	S/I		✓		✓		✓		✓		✓		✓	

R: Replace S/I: Service or Inspect

① Replace at 105,000 miles.

② Service or inspect at 105,000 miles.

FREQUENT OPERATION MAINTENANCE (SEVERE SERVICE)

If a vehicle is operated under any of the following conditions it is considered severe service:

- Extremely dusty areas.

-50% or more of the vehicle operation is in 32°C (90°F) or higher temperatures, or constant operation in temperatures below 0°C (32°F).

-Prolonged idling (vehicle operation in stop and go traffic).

-Frequent short running periods (engine does not warm to normal operating temperatures).

-Police, taxi, delivery usage or trailer towing usage.

Oil & oil filter: change every 3750 miles.

Brake hoses & lines (including ABS) (3.5RL): service or inspect every 7500 miles.

Cooling system hoses & connections (3.5RL): service or inspect every 7500 miles.

Driveshaft boots (3.2TL & 3.5RL): check every 7500 miles.

Exhaust system (3.5RL): check every 7500 miles.

Brake discs, calipers & pads: service or inspect every 7500 miles.

Fuel pipes, hoses & connections (3.5RL): check every 7500 miles.

Power steering system: service or inspect every 7500 miles.

Suspension components: service or inspect every 7500 miles.

Tie rod ends, steering gear box & boots (3.2TL & 3.5RL): service or inspect every 7500 miles.

Air cleaner element (NSX): service or inspect every 7500 miles.

Air cleaner element (except NSX): service or inspect every 15,000 miles.

Front differential fluid (3.2TL, 3.5RL): replace every 15,000 miles.

Transmission fluid (NSX): replace every 15,000 miles.

Transmission fluid (3.2TL & 3.5RL): replace every 30,000 miles

Timing belt (3.2TL): replace every 60,000 miles.

Water pump (3.2TL): service or inspect every 60,000 miles.

42356-INTE-C16

PRECAUTIONS

Before servicing any vehicle, please be sure to read all of the following precautions, which deal with personal safety, prevention of component damage and important points to take into consideration when servicing a motor vehicle:

• Never open, service or drain the radiator or cooling system when the engine is hot; serious burns can occur from the steam and hot coolant.

• Observe all applicable safety precautions when working around fuel. Whenever servicing the fuel system, always work in a well-ventilated area. Do not allow fuel spray or vapors to come in contact with a spark, open flame, or excessive heat (a hot drop light, for example). Keep a dry chemical fire extinguisher near the work area. Always keep fuel in a container specifically designed for fuel storage; also, always properly seal fuel containers to avoid the possibility of fire or explosion. Refer to the additional fuel system precautions later in this section.

• Fuel injection systems often remain pressurized, even after the engine has been turned **OFF**. The fuel system pressure must be relieved before disconnecting any fuel lines. Failure to do so may result in fire and/or personal injury.

• Brake fluid often contains polyglycol ethers and polyglycols. Avoid contact with the eyes and wash your hands thoroughly after handling brake fluid. If you do get brake fluid in your eyes, flush your eyes with clean, running water for 15 minutes. If eye irritation persists, or if you have taken brake fluid internally, seek medical assistance IMMEDIATELY.

• The EPA warns that prolonged contact with used engine oil may cause a number of skin disorders, including cancer! You should make every effort to minimize your exposure to used engine oil. Protective gloves should be worn when changing oil. Wash your hands and any other exposed skin areas as soon as possible after exposure to used engine oil. Soap and water, or waterless hand cleaner should be used.

• All new vehicles are now equipped with an air bag system. The system must be disabled before performing service on or around system components, steering column, instrument panel components, wiring and sensors. Failure to follow safety and disabling procedures could result in accidental air bag deployment, possible personal injury and unnecessary system repairs.

• Always wear safety goggles when working with, or around, the air bag system. When carrying a non–deployed air bag, be sure the bag and trim cover are pointed away from your body. When placing a non–deployed air bag on a work surface, always face the bag and trim cover upward, away from the surface. This will reduce the motion of the module if it is accidentally deployed. Refer to the additional air bag system precautions later in this section.

• Clean, high quality brake fluid from a sealed container is essential to the safe and proper operation of the brake system. You should always buy the correct type of brake fluid for your vehicle. If the brake fluid becomes contaminated, completely flush the system with new fluid. Never reuse any brake fluid. Any brake fluid that is removed from the system should be discarded. Also, do not allow any brake fluid to come in contact with a painted surface; it will damage the paint.

• Never operate the engine without the proper amount and type of engine oil; doing so WILL result in severe engine damage.

• Timing belt maintenance is extremely important! Many models utilize an interference type, non–freewheeling engine. If the timing belt breaks, the valves in the cylinder head may strike the pistons, causing potentially serious (also time consuming and expensive) engine damage. Refer to the maintenance interval charts in the front of this section for the recommended replacement interval for the timing belt and to the timing belt procedure for belt replacement and inspection.

• Disconnecting the negative battery cable on some vehicles may interfere with the functions of the on–board computer system(s) and may require the computer to undergo a relearning process once the negative battery cable is reconnected.

• When servicing drum brakes, only disassemble and assemble one side at a time, leaving the remaining side intact for reference.

• Only an MVAC–trained, EPA–certified automotive technician should service the air conditioning system or its components.

• The radio may contain a coded theft protection circuit. Always obtain the code number before disconnecting the battery.

ENGINE REPAIR

➡**Disconnecting the negative battery cable on some vehicles may interfere with the functions of the on board computer system. The computer may undergo a relearning process once the negative battery cable is reconnected.**

Distributor

REMOVAL & INSTALLATION

1.8L Engines

1. Before servicing the vehicle, refer to the precautions in the beginning of this section.
2. Remove or disconnect the following:
   • Negative battery cable

• Engine wiring harness and connectors from the distributor
• Spark plug wires from the distributor cap

3. If removing the ignition coil, remove the distributor cap, rotor, and cap seal, then remove the leak cover.
4. Remove or disconnect the following:
   • 2 screws to disconnect the wires from the coil
   • Ignition coil screws and slide the ignition coil out of the distributor housing
   • Distributor hold–down bolts
   • Distributor from the cylinder head

To install:
5. Use a new O–ring coated with engine oil, on the distributor housing.
6. Slip the distributor into position.

➡**Lugs on the end of the distributor and the matching grooves in the camshaft end are offset to eliminate any possibility of installing the distributor 180 degrees out of time.**

7. Install the hold–down bolts and hand tighten them.
8. Slide the ignition coil into the distributor housing and install the 2 mounting screws.
9. Install or connect the following:
   • 2 wires to the coil and install the 2 screws.
   • Distributor leak cover, rotor, cap seal, and cap
   • Engine wiring harness and connector to distributor
   • Spark plug wires
   • Negative battery cable

10. Set the timing, then tighten the hold down bolts to 16 ft. lbs. (22 Nm).

3.0L Engines

1. Before servicing the vehicle, refer to the precautions in the beginning of this section.
2. Remove or disconnect the following:
   - The negative battery cable
   - The spark plug and coil wires from the distributor cap
   - Harness connector(s) from the distributor
   - Distributor mounting bolts
   - Distributor from the cylinder head

To install:
3. Install or connect the following:
   - New O–ring coated with engine oil, on the distributor housing
   - Distributor
   - Mounting bolts and tighten them to 13 ft. lbs. (18 Nm)
   - Spark plug and coil wires
   - Negative battery cable
4. Check the ignition timing with a timing light. The timing marks are located on the crankshaft pulley and lower timing cover. If the timing is not within specification replace the PCM module.

Alternator

REMOVAL & INSTALLATION

1.8L and 3.5L Engines

1. Before servicing the vehicle, refer to the precautions in the beginning of this section.
2. Remove or disconnect the following:
   - Both battery cables
   - 4 prong connector and the black wire from the rear of the alternator
   - Alternator adjusting bolt(s)
   - Mounting bolt(s)
   - Alternator belt
   - Alternator assembly

To install:
3. Install or connect the following:
   - Alternator assembly
   - Mounting bolts and tighten them to 33 ft. lbs. (44 Nm)
   - Alternator adjusting bolt, hand tight
   - Alternator belt
4. Adjust alternator belt to tension.
5. Tighten the 8 x 1.25mm locknut/lock bolt(s) to 16 ft. lbs. (22 Nm).
6. Install or connect the following:
   - 4 prong connector and the black wire to the rear of the alternator
   - Both battery cables

✳✳ WARNING

Be sure to adjust the alternator belt to the proper tension or alternator bearing failure may occur.

➡The Powertrain Control Module (PCM) idle memory must be reset after reconnecting the battery. Start the engine and hold it at 3000 rpm until the cooling fan comes on. Then allow the engine to idle for about 5 minutes with all accessories OFF and with the transmission in Park or Neutral.

2.0L Engines

1. Before servicing the vehicle, refer to the precautions in the beginning of this section.
2. Remove or disconnect the following:
   - Both battery cables
   - Drive belt
   - Auto tensioner
   - Connector and the black wire from the rear of the alternator
   - Mounting bolt(s)
   - Alternator assembly

To install:
3. Install or connect the following:
   - Alternator assembly
   - Mounting bolts and tighten them to 16 ft. lbs. (22 Nm)
   - Connector and the black wire to the rear of the alternator
   - Auto tensioner
   - Drive belt
   - Both battery cables

✳✳ WARNING

Be sure to adjust the alternator belt to the proper tension or alternator bearing failure may occur.

➡The Powertrain Control Module (PCM) idle memory must be reset after reconnecting the battery. Start the engine and hold it at 3000 rpm until the cooling fan comes on. Then allow the engine to idle for about 5 minutes with all accessories OFF and with the transmission in Park or Neutral.

3.0L Engine

1. Before servicing the vehicle, refer to the precautions in the beginning of this section.
2. Remove or disconnect the following:
   - Battery cover
   - Both battery cables
   - Engine cover
   - Accessory drive belt
   - Fan motor connector and A/C compressor clutch wiring connector from the fan shroud
   - A/C condenser fan shroud assembly
   - Ground cable
   - 4 prong connector
   - Alternator and bracket assembly

To install:
3. Install or connect the following:
   - Alternator and bracket assembly and tighten the 10 x 1.25 mounting bolts to 33 ft. lbs. (44 Nm) and the 8 x 1.25mm locknut/lock bolt(s) to 16 ft. lbs. (22 Nm)
   - 4 prong connector
   - Ground cable
   - A/C condenser fan shroud assembly
   - Accessory drive belt
   - Front engine cover

➡There is no belt tension adjustment due to the use of an automatic tensioner.

   - Both battery cables
   - Battery cover
4. The Powertrain Control Module (PCM) idle memory must be reset after reconnecting the battery. Start the engine and hold it at 3000 rpm until the cooling fan comes on. Then allow the engine to idle for about 5 minutes with all accessories OFF and with the transmission in Park or Neutral.

3.2L Engine

1. Before servicing the vehicle, refer to the precautions in the beginning of this section.
2. Remove or disconnect the following:
   - Both battery cables
   - Adjusting bolts
   - Accessory drive belt
   - Mounting bolts
3. Turn the alternator 90° in a counter–clockwise direction.
   - Alternator
   - 4 prong connector
   - Harness clip and bracket assembly
   - Black wire from the terminal
   - Alternator from the vehicle

To install:
4. Install or connect the following:
   - Alternator
   - Black wire
   - Harness clip and bracket assembly and t tighten the bolt to 8.7 ft. lbs. (12 Nm)
   - 4 prong connector
   - Mounting bolts and tighten the 10 x 1.25mm bolts to 33 ft. lbs. (44 Nm), the 8 x 1.25mm locknut/lock bolt(s) to 16 ft. lbs. (22 Nm) and the 6 x 1.0mm 104 inch lbs. (12 Nm)
   - Accessory drive belt

➥There is no belt tension adjustment due to the use of an automatic tensioner.

5. The Powertrain Control Module (PCM) idle memory must be reset after reconnecting the battery. Start the engine and hold it at 3000 rpm until the cooling fan comes on. Then allow the engine to idle for about 5 minutes with all accessories OFF and with the transmission in Park or Neutral.

6. Connect the positive, then the negative battery cable.

Ignition Timing

ADJUSTMENT

1.8L Engines

1. Before servicing the vehicle, refer to the precautions in the beginning of this section.

2. Start the engine and hold the engine speed at 3000 rpm, until the radiator fan comes on. The engine should be at idle speed and at normal operating temperature. Be sure all electrical devices (radio, air conditioning, lights, etc.) are turned OFF.

3. Locate the Service Check (SCS) connector:
  • 1.8L engines: behind the right kick panel

4. Connect the SCS service connector (part number 07PAZ–0010100) to the service check connector.

5. Connect a timing light to No. 1 ignition wire and point the light toward the pointer on the timing belt cover.

6. Check the idle speed and adjust if necessary.

7. The red mark on the crankshaft pulley should be aligned with the pointer on the timing belt cover.

➥The white mark on the crank pulley is Top Dead Center (TDC).

8. Adjust the ignition timing by loosening the distributor mounting bolts and rotating the distributor housing to adjust the timing. Set as follows:
  • 1.8L (Except Type R): 16 degrees Before Top Dead Center (BTDC) at 700–800 rpm
  • 1.8L (Type R): 16 degrees BTDC at 750–850 rpm

9. Tighten the distributor bolts to 17 ft. lbs. (24 Nm) and recheck the timing.

10. Remove the SCS service connector from the service check connector.

2.0L and 3.0L Engines

➥The ignition timing is controlled by the Powertrain Control Module (PCM) and can be checked for diagnostic purposes. If the timing is out of specification, all mechanical and electrical systems should be checked for proper operation before replacing the PCM.

1. Before servicing the vehicle, refer to the precautions in the beginning of this section.

2. To check the ignition timing, start the engine and allow it to fast idle at 3000 rpm with all electrical accessories off and the transmission in **N** or **P**. Allow the engine to warm up and reach normal operating temperature. The engine cooling fan should cycle at least 1 time.

3. Locate the Service Check (SCS) connector under the glove box. Connect the service connector tool part number 07PAZ–0010100 to the SCS terminals.

4. Check the idle speed and adjust if necessary.

5. Connect a timing light to the No. 1

plug wire. While engine idles, point the light toward the pointer on the timing belt cover.

6. Inspect the ignition timing at idle. The specification is 6–10 degrees Before Top Dead Center (BTDC) at idle for the 2.0L engines, or 8–12 degrees BTDC at 700–800 for 3.0L engines

➥All mechanical and electrical systems should checked for proper operation before replacing the PCM.

7. If the ignition timing is incorrect, replace the PCM.

8. Remove the service connector.

3.2L and 3.5L Engines

➥The ignition timing is controlled by the Powertrain Control Module (PCM) and can be checked for diagnostic purposes. If the timing is out of specification, all mechanical and electrical systems should checked for proper operation before replacing the PCM.

1. Before servicing the vehicle, refer to the precautions in the beginning of this section.

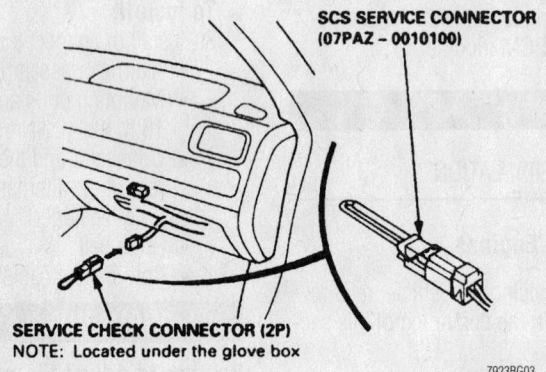

SCS SERVICE CONNECTOR
(07PAZ – 0010100)

SERVICE CHECK CONNECTOR (2P)
NOTE: Located under the glove box

7923BG03

Service check connector—3.2TL and 3.5RL

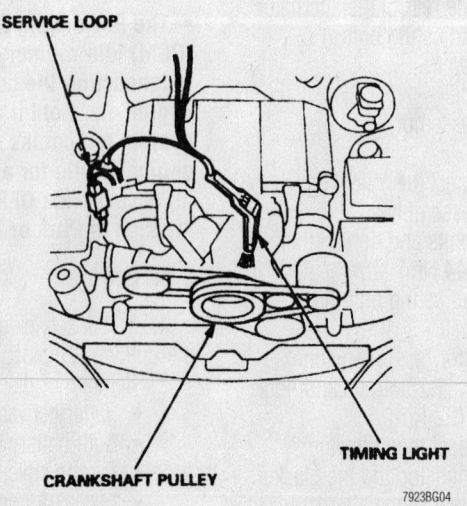

SERVICE LOOP

CRANKSHAFT PULLEY

TIMING LIGHT

7923BG04

Timing light attachment—3.2TL

2. To check the ignition timing, start the engine and allow it to fast idle at 3000 rpm with all electrical accessories off and the transmission in **N** or **P**. Allow the engine to warm up and reach normal operating temperature. The engine cooling fan should cycle at least 1 time.

3. Locate the Service Check (SCS) connector under the glove box and connect the service connector tool part number 07PAZ–0010100 to it.

4. Check the idle speed and adjust if necessary.

5. Connect a timing light to the No. 1 plug wire. With the engine idling at normal operating temperature point the timing light toward the pointer on the timing belt cover.

6. Inspect the ignition timing. The specifications are as follows:
- 3.2L: 13–17 degrees Before Top Dead Center (BTDC) at 590–690 rpm
- 3.5L: 13–17 degrees BTDC at 700–800 rpm

➡**All mechanical and electrical systems should checked for proper operation before replacing the PCM.**

7. If the ignition timing is incorrect, replace the PCM.

Only replace the PCM as a last resort.

8. Remove the timing light.

9. Disconnect the special tool (SCS service connector) from the service check connector.

Engine Assembly

REMOVAL & INSTALLATION

1.8L Engines

1. Before servicing the vehicle, refer to the precautions in the beginning of this section.

2. Relieve the fuel system pressure.

3. Remove or disconnect the following:
- Both battery cables
- Hood
- Strut brace (if equipped)
- Battery cables from the under–hood fuse/relay box and under–hood Antilock Brake (ABS) System fuse/relay box
- Air cleaner assembly and mounting bracket
- Evaporative emission (EVAP) control canister hose and vacuum hose from the intake manifold
- Brake booster vacuum hose
- Fuel return hose

- Engine wiring harness connectors on the right side of the engine compartment
- Fuel feed hose
- Throttle cable

➡**Be careful not to bend the cable when removing it. Replace the cable if it gets kinked.**

- Engine wiring harness connectors on the left side of the engine compartment
- Cruise control actuator
- Engine ground cable
- Power steering belt
- Power steering pump without disconnecting the hoses
- Air conditioning compressor belt
- Clutch slave cylinder and pipe/hose assembly (if equipped). Do not disconnect the pipe/hose assembly.
- Transmission ground cable and hose clamp
- Front wheels
- Lower splash shield

4. Drain the engine coolant, engine oil, and transmission fluid into sealable containers. Reinstall the drain plugs using new washers. Be careful not to overtighten the drain plugs.

5. Remove or disconnect the following:
- Upper and lower radiator hoses and the heater hoses from the engine
- Transmission oil cooler hoses (if equipped)
- Radiator assembly
- Air conditioning compressor without disconnecting the hoses
- Heated Oxygen (HO_2S) sensor connector
- Front exhaust pipe from the exhaust manifold
- Shift rod and extension rod from the transaxle (if equipped)
- Shift cable cover, then disconnect the shift cable from the transaxle (if equipped)
- Damper fork
- Lower ball joints
- Halfshafts from the transaxle

6. Attach a hoist to the engine.
- Left and right front engine mounts and brackets
- Rear engine mount and bracket
- Side engine mount
- Transmission mount

7. Check that the engine is completely clear of all vacuum, fuel and coolant hoses, and electrical wiring.

8. Raise the engine and transaxle assembly all the way and remove it from the vehicle.

9. Separate the engine and transaxle.

To install:

10. Installation is the reverse of the removal procedure, while using the following torque values:
- Transaxle bolts: 47 ft. lbs. (64 Nm) (if equipped with a manual transaxle), or 43 ft. lbs. (59 Nm) (if equipped with an automatic transaxle)
- Rear mounting bracket: 87 ft. lbs. (118 Nm)
- Upper mounting bolts: 54 ft. lbs. (74 Nm)
- Torque converter bolts: 104 inch lbs. (12 Nm) (if equipped)
- Transmission mount bolts: 47 ft. lbs. (64 Nm)
- Engine side mount bolt: 47 ft. lbs. (64 Nm)
- Rear mount bracket bolts: 40 ft. lbs. (54 Nm)
- Right front mount/bracket bolts: 47 ft. lbs. (64 Nm)
- Damper fork bolts: 47 ft. lbs. (64 Nm)
- Lower ball joints nuts: 40 ft. lbs. (54 Nm)
- Shift cable bolt: 10 ft. lbs. (14 Nm)
- Shift rod and extension rod bolt: 16 ft. lbs. (22 Nm)
- Front exhaust pipe nuts: 40 ft. lbs. (54 Nm)
- A/C compressor bolts: 17 ft. lbs. (24 Nm)
- Power steering pump bolts: 17 ft. lbs. (24 Nm)
- Cruise control actuator bolts: 17 ft. lbs. (24 Nm)

2.0L Engines

1. Before servicing the vehicle, refer to the precautions in the beginning of this section.

2. Obtain the anti–theft code for the radio.

3. Drain the engine oil, coolant, and transmission oil (or fluid) into sealable containers and carefully reinstall the drain plugs using new sealing washers.

4. Properly relieve the fuel system pressure.

5. Remove or disconnect the following:
- Negative and positive battery cables
- Battery
- Intake manifold cover

- Intake Air Temperature (IAT) sensor connector
- Breather hose
- Air cleaner housing
- Intake air duct
- Battery cable from the underhood fuse/relay box, then the harness clamps and ground cable
- Throttle cover, if equipped

6. Fully open the throttle link and cruise control link by hand, then remove the cables from the links. Loosen the locknuts and remove the cables from the bracket.

- Engine Control Module (ECM)/Powertrain Control Module (PCM) connectors and main wire harness connector
- Harness clamps and grommet, then pull the engine wire harness through the bulkhead
- Fuel feed hose
- Evaporative Emission (EVAP) canister hose and brake booster vacuum hose
- Clutch slave cylinder and clutch line bracket mounting bolt, if equipped with a manual transaxle
- Shift cable and select cable, on manual transaxles
- Drive belt
- Power steering pump and position aside without disconnecting the hoses
- Bolt holding the power steering hose bracket
- Radiator cap
- Front wheels and tires
- Splash shield
- Air Fuel (A/F) ratio sensor connector
- Secondary Heated Oxygen (HO$_2$) sensor connector
- Three way catalytic converter assembly
- Lower ball joints and stabilizer links
- Halfshafts. Coat all machined surfaces with clean engine oil, then tie plastic bags over the halfshaft ends to protect them.
- Shift cable holder and the shift cable cover. To avoid damaging the control lever joint, be sure to remove the bolts holding the shift cable cover, on automatic transaxles.
- Spring clip and control pin, then the shift cable from the control lever, on automatic transaxles
- Lower hose
- Automatic Transaxle Fluid (ATF) filter mounting bolt, if equipped

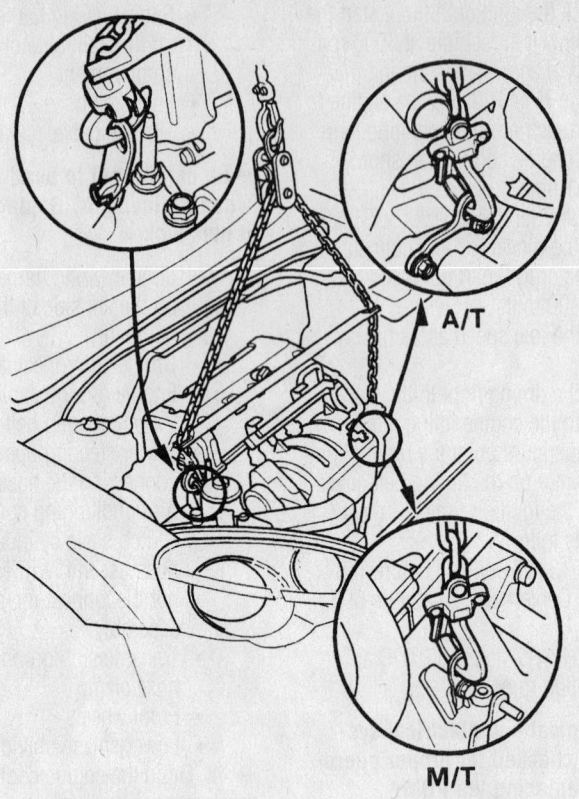

A/T

M/T

42356-INTE-G02

Attach a suitable engine tilt hanger to the engine

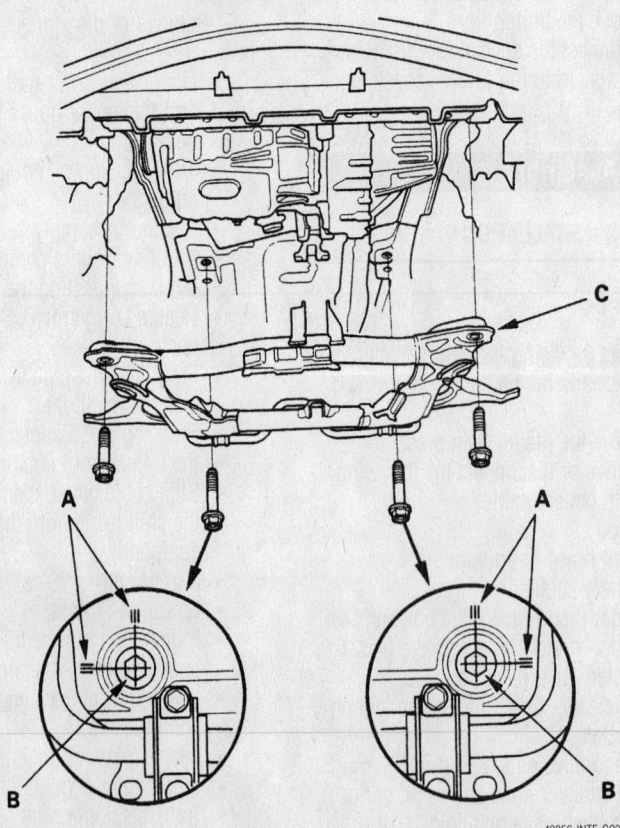

C

A

A

B

B

42356-INTE-G03

Make alignment marks on the reference lines (A) that align with the centers of the rear subframe mounting bolts (B), then remove the front subframe (C)

- ATF cooler hoses, then plug the hoses and lines
- Upper hose and heater hoses
- Radiator

7. Attach a suitable engine tilt hanger to the engine.

- Transaxle mount bracket support bolts/nuts
- Upper bracket mounting bolt and nut

8. Make sure the hoist brackets are positioned properly, then raise the hoist to its full height.

- Rear mount mounting bolts
- Front mount bracket mounting bolt

9. Make alignment marks on the reference lines that align with the centers of the rear subframe mounting bolts, then remove the subframe.

- Compressor clutch connector and position the compressor aside without disconnect the refrigerant lines
- Any remaining electrical connectors, vacuum, fuel or coolant hoses

10. Slowly lower the engine about 6 in. (15 cm). Check that all hoses and wires are disconnected from the engine/transaxle.

11. Lower the engine all the way, then remove the chain hoist from the engine.

12. Remove the engine from under the vehicle.

To install:

13. Installation is the reverse of the removal procedure, while using the following torque values:

- Engine mount bracket: 33 ft. lbs. (44 Nm)
- A/C compressor bracket: 33 ft. lbs. (44 Nm)
- A/C compressor bolts: 16 ft. lbs. (22 Nm)
- New subframe bolts: 76 ft. lbs. (103 Nm)
- Rear mount mounting bolts: 43 ft. lbs. (59 Nm)
- Upper bracket mounting bolt: 40 ft. lbs. (54 Nm)
- Support bolts/nuts: 40 ft. lbs. (54 Nm)
- Front engine mount bracket mounting bolts: 47 ft. lbs. (64 Nm)
- A/T shift control cable: 7.2 ft. lbs. (9.8 Nm)
- New catalytic converter self-locking nuts: 25 ft. lbs. (33 Nm)
- Catalytic converter bolt: 16 ft. lbs. (22 Nm)
- ATF filter mounting bolt: 8.7 ft. lbs. (12 Nm)
- Power steering pump bolts: 16 ft. lbs. (22 Nm)

- Power steering hose bracket bolt: 7.2 ft. lbs. (9.8 Nm)
- M/T cable bolts: 7.2 ft. lbs. (9.8 Nm)
- Clutch slave cylinder mounting bolts: 17 ft. lbs. (24 Nm)
- Clutch line bracket mounting bolt: 7.2 ft. lbs. (9.8 Nm)
- Battery cable harness clamp bolt: 8.7 ft. lbs. (12 Nm)
- Air cleaner housing bolts: 8.7 ft. lbs. (12 Nm)
- Intake manifold cover bolts: 8.7 ft. lbs. (12 Nm)

3.0L Engines

1. Before servicing the vehicle, refer to the precautions in the beginning of this section.

2. Obtain the anti-theft code for the radio.

3. Drain the engine oil, coolant, and transmission oil (or fluid) into sealable containers and carefully reinstall the drain plugs using new sealing washers.

4. Properly relieve the fuel system pressure.

5. Remove or disconnect the following:

- Air cleaner assembly
- Hood support struts and support the hood in a vertical position
- Negative battery cable and then the positive cable
- Strut brace
- Battery, battery tray and the engine ground cable
- Accelerator and cruise control cables from the throttle body and bracket
- Battery cables from the underhood Antilock Brake System (ABS) fuse/relay box and underhood fuse/relay box assemblies
- Engine wiring harness on the right side of the engine
- Evaporative emission (EVAP) control canister hose
- Brake booster vacuum hose
- Fuel feed and return hoses
- Engine wiring harness on the left side of the engine
- Accessory drive belt(s)
- Power steering pump leaving the hoses attached
- Shift cable (if equipped with a manual transmission)
- Clutch slave cylinder leaving (if equipped) the hydraulic line attached
- Reverse light switch connector (if equipped with a manual transmission)

- Cruise control vacuum tank (if equipped)
- Front wheels
- Lower splash shield
- Center support beam
- Heated Oxygen (HO$_2$S) sensor connector
- Front exhaust pipe from the manifold
- Shift selector cable and cover (if equipped with an automatic transmission)
- Damper fork
- Lower ball joints
- Driveshafts from the transaxle
- Crankshaft pulley (if equipped with 3.0L engine)
- VTEC/oil filter housing (if equipped with 3.0L engine)
- Upper and lower radiator hoses
- Transmission oil cooler hoses (if equipped)
- Radiator
- Air conditioning compressor leaving the hoses attached
- Heater hoses

6. Attach a suitable engine lifting hoist to the engine lifting hooks and secure the engine.

- Front engine mount
- Rear engine mount
- Side engine mount
- Transmission mount

7. Lift the engine slightly and check that all hoses, cables and wires have been properly disconnected.

8. Carefully raise the engine/transmission from the vehicle.

To install:

9. Installation is the reverse of the removal procedure, while using the following torque values:

- Transmission mount bolts: 47 ft. lbs. (64 Nm)
- Front engine mount bolts: 47 ft. lbs. (64 Nm)
- Side engine mount bolts: 47 ft. lbs. (64 Nm)
- Rear engine mount bolts: 47 ft. lbs. (64 Nm)
- A/C compressor bolts: 16 ft. lbs. (22 Nm)
- VTEC/oil filter housing bolts (if equipped with 3.0L engine): 16 ft. lbs. (22 Nm)
- Crankshaft pulley bolt (if equipped with 3.0L engine): 181 ft. lbs. (245 Nm)
- Lower ball joint nuts: 40 ft. lbs. (54 Nm)
- Damper fork bolts: 47 ft. lbs. (64 Nm)

- Front exhaust pipe nuts: 40 ft. lbs. (54 Nm)
- Center support beam bolts: 37 ft. lbs. (50 Nm)
- Splash shield bolts: 84 inch lbs. (9.5 Nm)
- Power steering pump bolts: 16 ft. lbs. (22 Nm)
- Strut brace bolts (if equipped): 16 ft. lbs. (22 Nm)

3.2L Engine

1. Before servicing the vehicle, refer to the precautions in the beginning of this section.

2. Do not remove the hood. Disconnect the hood support strut and reconnect it to hold the hood in a vertical position.

3. Properly relieve the fuel pressure.

4. Drain the engine oil, coolant, transmission fluid and the differential fluid.

5. Remove or disconnect the following:

- Negative battery cable, then the positive battery cable
- Battery and battery tray
- Air cleaner assembly
- Water bypass hoses (on 3.2CL)
- Traction Control System (TCS) control valve actuator connector (on 3.2CL)
- TCS control valve angel sensor connector (on 3.2CL)
- Throttle cable and cruise control cable from the throttle and bracket
- Left side engine wiring harness connectors
- Fuel feed and return hoses
- Brake booster vacuum hose
- Battery cables from the under–hood fuse/relay box
- Under–hood fuse/relay box
- Powertrain Control Module (PCM) electrical connectors
- Accessory drive belts
- Power steering pump without disconnecting the hoses
- Vehicle Speed sensor (VSS) connector, then remove the VSS/power steering sensor leaving the fluid hoses attached
- Front wheels
- Lower splash shield
- Heated Oxygen (HO$_2$S) sensor connector
- Front exhaust pipe from the manifold
- Front damper forks
- Lower ball joints from the steering knuckles
- Halfshafts from the differential and the intermediate shaft

- Shift cable cover
- Shift cable with the control lever from the transaxle
- Power steering hose clamps and the engine mount control vacuum hose
- Upper and lower radiator hoses
- Heater hoses
- Transmission fluid cooler hoses
- Ground cable
- Power steering hose clamp from the rear beam assembly

6. Attach a suitable chain hoist to the engine lifting hooks and support the engine.

➡**The engine and transmission assembly is removed by lowering it from the vehicle. Be sure the vehicle is in a position that will allow the engine and transmission assembly enough clearance to be moved from the vehicle once it is lowered away from the vehicle.**

7. Remove or disconnect the following:

- Side, rear and front engine mount support fasteners
- Front suspension radius rod bolts

8. Make alignment marks on the front beam and remove the front beam.

9. Remove or disconnect the following:

- Air conditioning compressor leaving the hoses attached
- Rear mounts from the engine and transmission

10. Check that all hoses, cables and wires have been properly disconnected and slowly lower the engine about 6 inches (150 mm). Recheck that all hoses, cables and wires have been properly disconnected.

11. Carefully lower the engine/transmission assembly from the vehicle.

To install:

12. Installation is the reverse of the removal procedure, while using the following torque values:

- Transmission rear mount bolts: 28 ft. lbs. (38 Nm)
- Air conditioning compressor bolts: 16 ft. lbs. (22 Nm)
- Front beam nuts: 28 ft. lbs. (38 Nm)
- Front beam bolts: 76 ft. lbs. (103 Nm)
- Rear bolts: 28 ft. lbs. (38 Nm)
- Front suspension radius rod bolts: 119 ft. lbs. (162 Nm)
- Front engine mount nut: 40 ft. lbs. (54 Nm)
- Rear engine mount nut: 40 ft. lbs. (54 Nm)
- Rear engine mount bolt: 28 ft. lbs. (38 Nm)

- Side engine mount bracket bolts: 33 ft. lbs. (44 Nm)
- Side engine mount through bolt: 40 ft. lbs. (54 Nm)
- Lower ball joint nuts: 40 ft. lbs. (54 Nm)
- Damper fork bolts: 47 ft. lbs. (64 Nm)
- Front exhaust pipe nuts: 40 ft. lbs. (54 Nm)
- Shift cable and control lever bolts: 10 ft. lbs. (14 Nm)
- Power steering pump bolts: 16 ft. lbs. (22 Nm)

3.5L Engine

1. Before servicing the vehicle, refer to the precautions in the beginning of this section.

2. Move the front passenger's seat forward.

3. Relieve the fuel system pressure.

4. Drain the engine oil, coolant, transmission fluid and differential fluid.

5. Remove or disconnect the following:

- Hood
- Negative battery cable, then the positive battery cable
- Strut brace
- Engine cover
- Air cleaner assembly and intake duct
- Throttle cover
- Throttle cable and cruise control cable from the throttle and bracket
- Battery
- Battery tray
- Relay box
- Ground cable and wiring harness clips from the firewall
- Alternator and battery cables from the under–hood fuse/relay box
- Underhood fuse/relay box
- Left side engine wiring harness connectors
- Fuel feed and return hoses
- Brake booster vacuum hose
- Evaporative emissions (EVAP) canister hose
- Transmission sub–harness connector
- Control box
- Right side engine wiring harness connectors
- Spark plug voltage detection module
- Engine ground cables
- Accessory drive belts
- Power Steering Pressure (PSP) switch connector

- Power steering pump leaving the hoses attached

6. Pull the carpet back under the front passengers seat and detach the secondary Heated Oxygen (HO₂S) sensor connector.

7. Remove or disconnect the following:

- Front wheels
- Splash shield
- Front suspension damper forks
- Lower ball joints from the steering knuckles
- Halfshafts from the transmission

- Air conditioning compressor without disconnecting the hoses
- Vehicle Speed Sensor (VSS) leaving the hoses attached
- Transmission stop collars
- Front exhaust pipe from the vehicle
- Wire harness cover and grommet
- Three way catalytic converter
- Converter heat shield
- Transmission fluid cooler lines
- Shift cable cover
- Shift cable from the transmission
- Control lever from the control shaft

- Upper and lower radiator hoses
- Radiator
- Heater hoses

8. Loosen the locknut on the fuel pressure regulator and rotate it 180 degrees.

9. Attach a chain hoist to the engine lifting eyelets.

10. Raise and safely support the vehicle.

11. Remove or disconnect the following:

- Shift cable guide
- Transmission beam
- Transmission mount and bracket

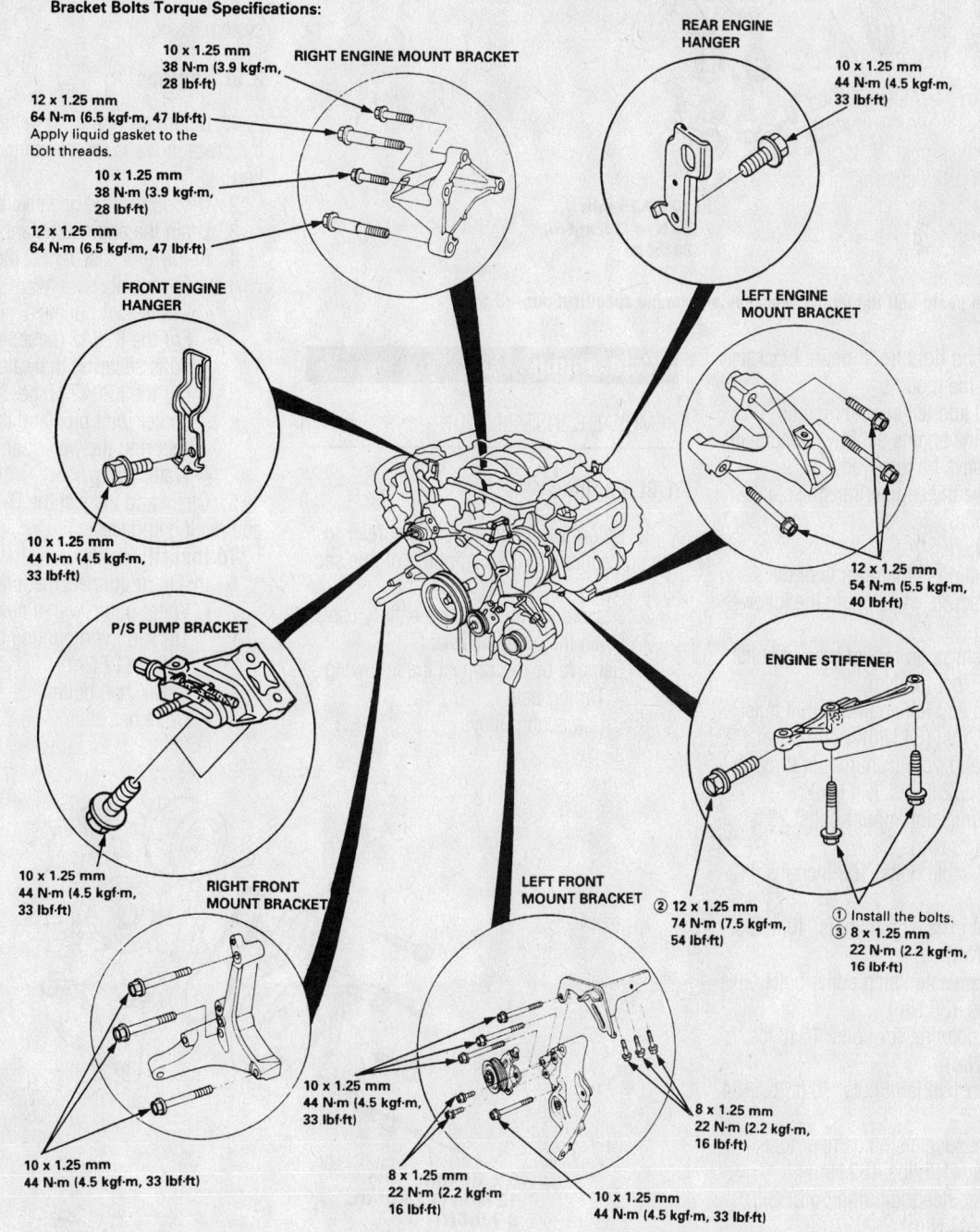

View of the engine mounting bracket showing torque specifications—3.5RL

7923BG77

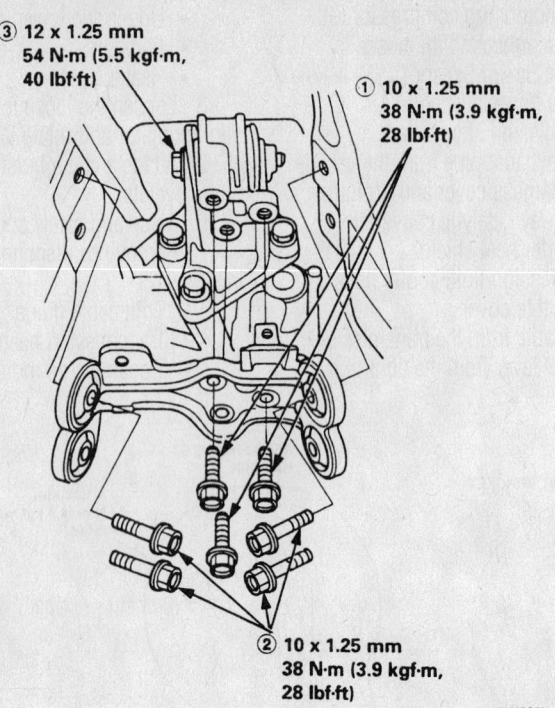

③ 12 x 1.25 mm
54 N·m (5.5 kgf·m, 40 lbf·ft)

① 10 x 1.25 mm
38 N·m (3.9 kgf·m, 28 lbf·ft)

② 10 x 1.25 mm
38 N·m (3.9 kgf·m, 28 lbf·ft)

7923BG78

Transmission beam bolt tightening sequence and torque specifications—3.5RL

- Left and right front mount brackets from the mounts
- Right and left engine mounts

12. Raise the engine slightly, be sure all connections have be removed.

13. Remove the engine/transmission from the vehicle.

To install:

14. Installation is the reverse of the removal procedure, while using the following torque values:

- Transmission mount bracket bolts: 28 ft. lbs. (38 Nm)
- Right and left engine mount nuts: 47 ft. lbs. (64 Nm)
- Left and right front mount through bolts: 52 ft. lbs. (74 Nm)
- Transmission mount bolts: 28 ft. lbs. (38 Nm)
- Shift cable bolts: 108 inch lbs. (12 Nm)
- Front exhaust pipe nuts: 40 ft. lbs. (54 Nm)
- Transmission stop collar bolts: 28 ft. lbs. (38 Nm)
- A/C compressor bolts: 16 ft. lbs. (22 Nm)
- Lower ball joint nuts: 40 ft. lbs. (54 Nm)
- Front suspension damper fork bolts: 41 ft. lbs. (69 Nm)
- Power steering pump bolt: 33 ft. lbs. (44 Nm)
- Power steering nut: 16 ft. lbs. (22 Nm)

Water Pump

REMOVAL & INSTALLATION

1.8L Engines

1. Before servicing the vehicle, refer to the precautions in the beginning of this section.
2. Disconnect the negative battery cable.
3. Drain the engine coolant.
4. Remove or disconnect the following:
   - Timing belt
   - Camshaft pulleys
   - Rear timing belt cover
   - 5 water pump mounting bolts and remove the water pump

To install:

5. Install or connect the following:
   - Water pump and tighten the bolts to 108 inch lbs. (12 Nm)
   - Rear timing belt cover
   - Camshaft pulleys
   - Timing belt

6. Fill the engine with coolant and bleed the air from the cooling system.

7. Connect the negative battery cable and enter the radio security code.

8. Run the engine and check for cooling system leaks.

2.0L Engines

1. Before servicing the vehicle, refer to the precautions in the beginning of this section.
2. Disconnect the negative battery cable.
3. Drain the engine coolant.
4. Remove or disconnect the following:
   - Drive belt
   - Crankshaft pulley
   - For the K20A3 (base) engine, 6 bolts securing the water pump
   - For the K20A2 (Type–S) engine, oil cooler joint pipe and the 7 bolts securing the water pump
   - Water pump

5. Clean and inspect the O–ring groove and mating surfaces.

To install:

6. Install or connect the following:
   - Water pump with a new O–ring. Tighten the mounting bolts to 8.7 ft. lbs. (12 Nm).
   - Crankshaft pulley
   - Drive belt

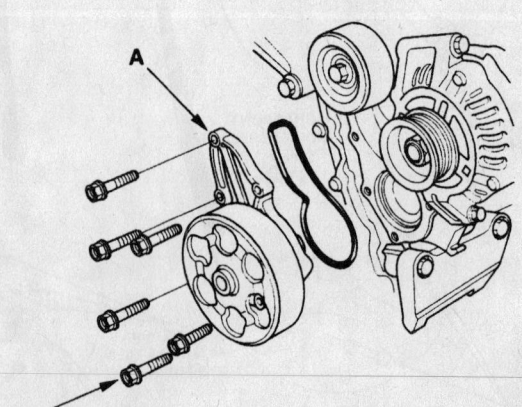

A

6 x 1.0 mm
12 N·m (1.2 kgf·m, 8.7 lbf·ft)

42356-INTE-G04

Water pump mounting—2.0L (K20A3) engine

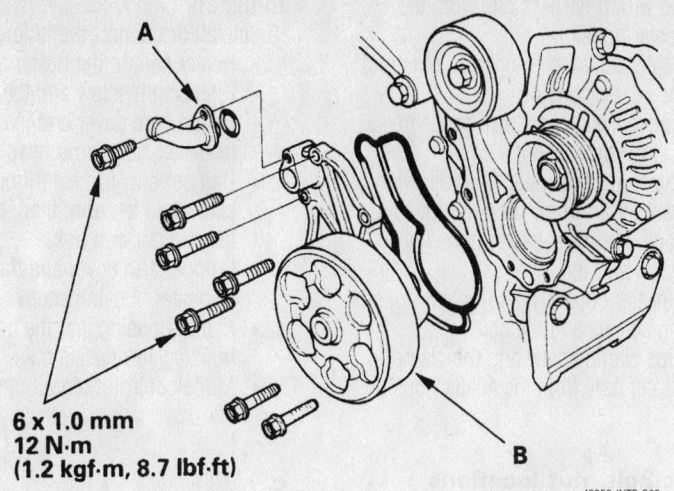

6 x 1.0 mm
12 N·m
(1.2 kgf·m, 8.7 lbf·ft)

42356-INTE-G05

Water pump mounting—2.0L (K20A2) engine

7. Fill the engine with coolant and bleed the air from the cooling system.

8. Connect the negative battery cable and enter the radio security code.

9. Run the engine and check for cooling system leaks.

3.0L, 3.2L and 3.5L Engines

➡Perform this service operation with the engine cold.

1. Before servicing the vehicle, refer to the precautions in the beginning of this section.

2. Disconnect the negative battery cable.

3. Remove the front splash panel.

4. Drain the cooling system.

5. Remove the timing belt. Inspect the timing belt for any signs of damage or oil and coolant contamination. Replace the tim-

ing belt if there is any doubt about its condition.

6. On 3.5 RL models, remove the left camshaft pulley and back cover.

7. On 2000–03 3.2 TL models and 2001–03 3.2 CL models, remove the timing belt tensioner.

8. Remove the water pump bolts. Then, remove the water pump and sprocket assembly from the engine block.

To install:

9. Install the water pump with a new O-ring. Use new bolts and tighten the 6mm mounting bolts evenly to 104 inch lbs. (12 Nm) and the 8mm bolts to 16 ft. lbs. (22 Nm).

10. If removed, install the timing belt rear cover and camshaft pulley.

11. If removed, install the timing belt tensioner.

12. Install or connect the following:
   • Timing belt and timing belt covers
   • Accessory drive belts

13. Close the cooling system drain plug. Refill and bleed the cooling system.

14. Connect the negative battery cable.

15. Start the engine, allow it to reach normal operating temperature, check for leaks, and top off as necessary.

Heater Core

REMOVAL & INSTALLATION

Integra and RSX

➡**Be sure to acquire the anti–theft code for the radio; then, write down the frequencies for the preset buttons.**

1. Before servicing the vehicle, refer to the precautions in the beginning of this section.

2. Disconnect the negative battery cable.

✱✱ CAUTION

After disconnecting the negative battery cable, wait for at least 3 minutes for the SRS module to deplete its energy.

3. Drain the cooling system into a clean container for reuse.

4. Remove or disconnect the following:
   • Heater hoses from the heater core
   • Heater housing-to-cowl nut, located in the engine compartment

5. Remove the instrument panel by removing or disconnecting the following:
   • Front seats
   • Front and rear consoles

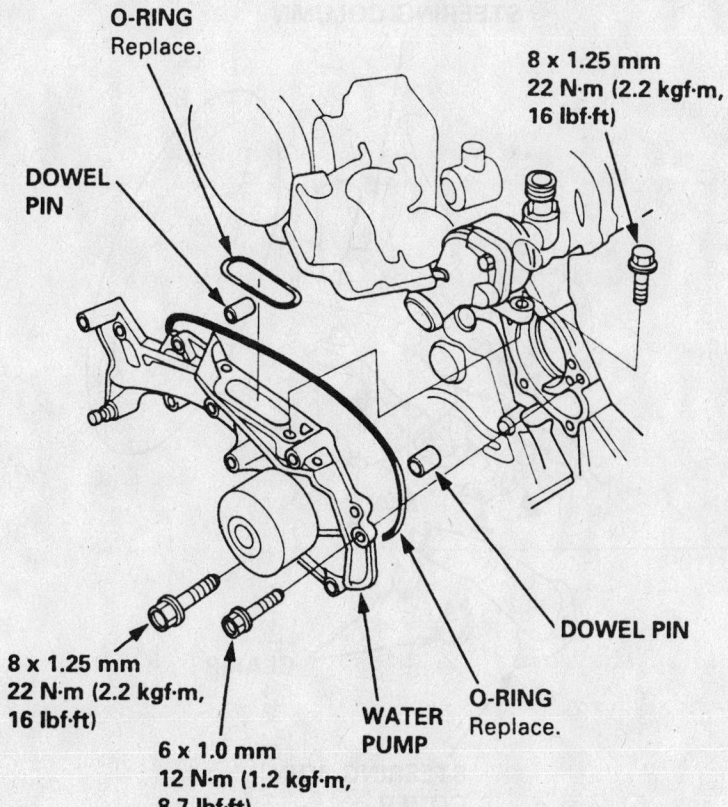

O-RING
Replace.

DOWEL
PIN

8 x 1.25 mm
22 N·m (2.2 kgf·m, 16 lbf·ft)

DOWEL PIN

8 x 1.25 mm
22 N·m (2.2 kgf·m, 16 lbf·ft)

6 x 1.0 mm
12 N·m (1.2 kgf·m, 8.7 lbf·ft)

WATER
PUMP

O-RING
Replace.

7923BG79

Water pump mounting and bolt torque specifications—3.5L engine

- Lower dashboard-to-instrument panel screws and the lower dashboard, located on the driver's side
- Knee bolster
- Glove box
- 4 glove box frame-to-instrument panel bolts and the frame
- Clock
- Moon roof switch
- Stereo radio/cassette
- Lower steering column cover clamps and the cover
- Wrap a shop towel around the steering column to prevent damage to the column.
- Steering column-to-instrument panel nuts/bolts and lower the steering column
- SRS module-to-instrument panel nuts (located on the driver's side); then, carefully, remove the SRS module and disconnect the electrical connector
- Air mix control cable and electrical connectors, located at the center of the instrument panel
- Antenna lead
- Electrical connectors, located at the under–dash fuse/relay box
- Pry out the access panels, from both sides of the instrument panel
- Instrument panel-to-chassis bolts and remove the instrument panel
- Wiring harness clips; then, remove the 4 heater housing-to-blower motor duct screws and the duct, (models not equipped with air conditioning only)

6. If equipped with air conditioning, remove the evaporator housing removing or disconnecting the following:
- Discharge and recover the air conditioning system refrigerant
- Refrigerant lines from the evaporator core, discard the O–rings and plug the openings to prevent contamination
- Cut the insulation pad (at the firewall) at the lower evaporator housing-to-chassis location
- Thermostat electrical connector, located at the evaporator housing
- Wiring harness clips from the evaporator housing
- Evaporator housing-to-chassis screws, the nut and bolt
- Drain hose
- Evaporator housing, carefully
- 2 SRS support beam bolts, nut and the SRS beam, located at the passenger's side
- Mode control motor connector and

the wiring harness clip from the heater housing
- Heater housing-to-chassis nuts and the heater housing

7. Remove the damper arm cover-to-heater housing screw and the cover.
8. Remove or disconnect the following:
- Damper arm link; then, remove the damper arm-to-heater housing screw and the arm
- 2 heater core cover-to-heater housing screws and the cover
- Pipe clamp screw and the clamp
- Heater core from the heater housing

To install:
9. Install or connect the following:
- Heater core to the heater housing
- Pipe clamp screw and the clamp
- Heater core cover and the 2 cover-to-heater housing screws
- Damper arm-to-heater housing screw and the arm; then, connect the damper arm link
- Damper arm cover and the cover-to-heater housing screw
- Heater housing and the heater housing-to-chassis nuts
- Mode control motor connector and

▲ : Bolt, nut locations

A▲ : Bolt, 2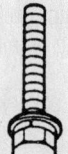

8 x 1.25 mm
22 N·m (2.2 kgf·m, 16 lbf·ft)

B▲ : Nut, 2

8 x 1.25 mm
13 N·m (1.3 kgf·m, 9 lbf·ft)
Replace.

STEERING COLUMN

CLAMP

STEERING JOINT COVER

93112GP5

View of the steering column and related components—Integra

the wiring harness clip to the heater housing
• SRS support beam, nut and the 2 SRS beam bolts on the passengers side

▲ : Nut locations, 4

6 x 1.0 mm
9.8 N·m
(1.0 kgf·m,
7.2 lbf·ft)

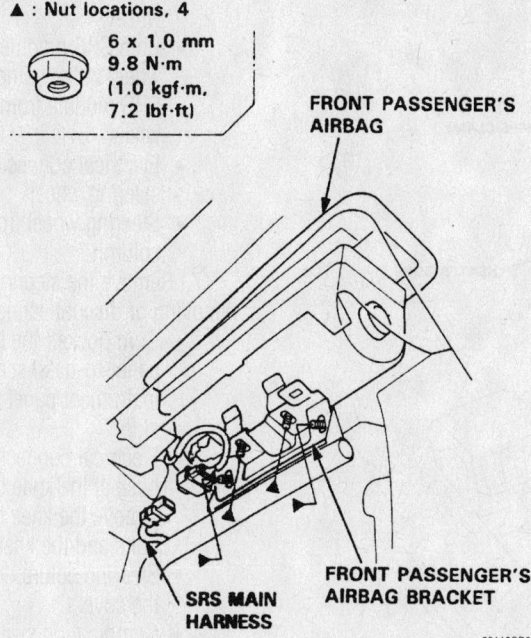

View of the passenger's side SRS module and related components—Integra

▼ : Bolt locations, 6

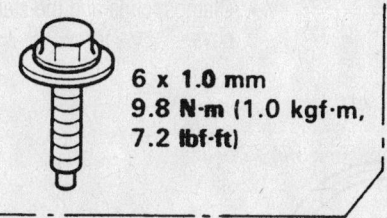

6 x 1.0 mm
9.8 N·m (1.0 kgf·m,
7.2 lbf·ft)

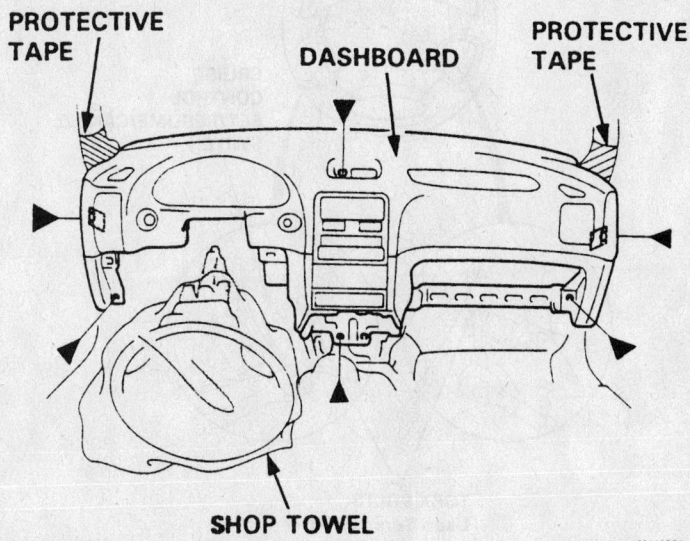

View of the instrument panel fastener locations—Integra

10. If equipped with air conditioning, install the evaporator housing by installing or connecting the following:
• Evaporator housing
• Drain hose

• 4 evaporator housing-to-chassis screws, the nut and bolt
• Wiring harness clips to the evaporator housing
• Thermostat electrical connector at the evaporator housing
• Refrigerant lines to the evaporator core making sure to use new O–rings
• Heater housing-to-blower motor duct and the 4 duct screws; then, connect the wiring harness clips (models not equipped with air conditioning only)

11. Install the instrument panel by installing or connecting the following:
• Instrument panel and the instrument panel-to-chassis bolts
• Access panels on both sides of the instrument panel
• Electrical connectors at the under-dash fuse/relay box
• Antenna lead
• Air mix control cable and electrical connectors at the center of the instrument panel
• SRS module electrical connector (located on the passenger's side) and torque the SRS module-to-instrument panel nuts to 84 inch lbs. (9.8 Nm)
• Steering column and torque the steering column-to-instrument panel nuts to 108 ft. lbs. (13 Nm) and the bolts to 16 ft. lbs. (22 Nm)
• Lower steering column cover and the cover clamps
• Stereo radio/cassette
• Moon roof switch
• Clock
• Glove box frame and the 4 frame-to-instrument panel bolts
• Glove box
• Knee bolster
• Lower dashboard and the lower dashboard-to-instrument panel screws on the drivers side
• Front and rear consoles
• Front seats
• Heater housing-to-cowl nut (located in the engine compartment) and torque to 16 ft. lbs. (22 Nm)
• Heater hoses to the heater core
12. Refill the cooling system.
13. Connect the negative battery cable.
14. Evacuate, charge and leak test the air conditioning system refrigerant.
15. Operate the engine to normal operating temperatures; then, check the climate control operation and check for leaks.

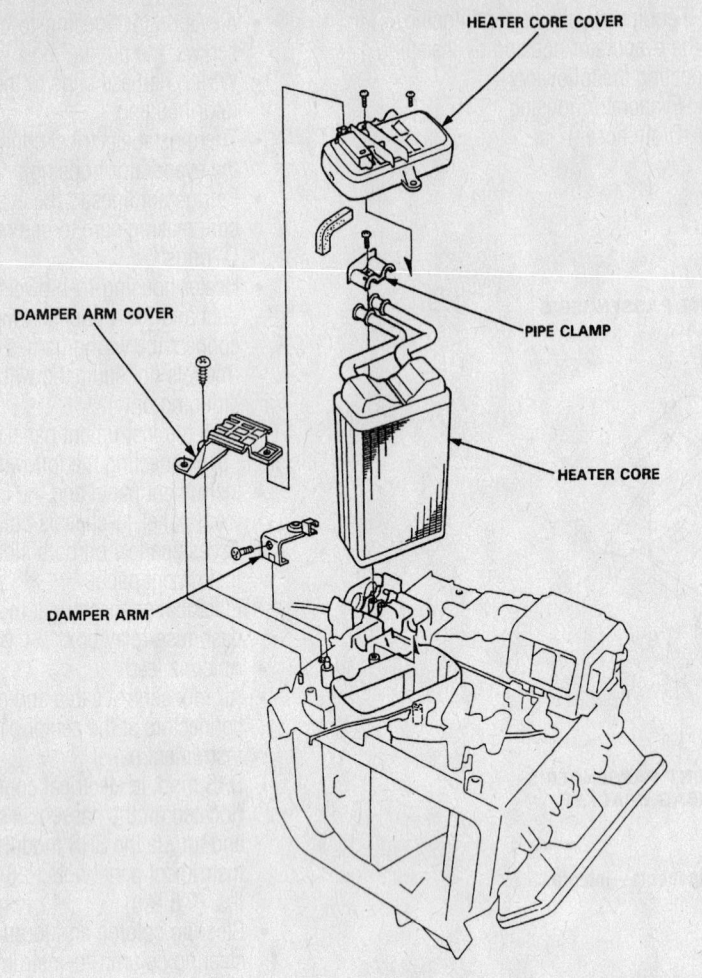

HEATER CORE COVER

DAMPER ARM COVER

PIPE CLAMP

HEATER CORE

DAMPER ARM

93112GP8

Exploded view of the heater core, heater housing and related components—Integra

3.2CL

➡ **Be sure to acquire the anti-theft code for the radio; then, write down the frequencies for the preset buttons.**

1. Before servicing the vehicle, refer to the precautions in the beginning of this section.

2. Disconnect the negative battery cable.

☀☀ CAUTION

After disconnecting the negative battery cable, wait for at least 3 minutes for the SRS module to deplete its energy.

3. Drain the cooling system into a clean container for reuse.

4. Remove or disconnect the following:
- Heater hoses from the heater core
- 2 heater housing-to-cowl nuts, located in the engine compartment

5. Place the front wheel in the straight-ahead position.

6. Remove the SRS module and the

steering wheel by removing or disconnecting the following:
- Access panel-to-steering wheel screws and the panel
- SRS module electrical connector
- SRS module-to-steering wheel covers from both sides of the steering wheel
- Both SRS module-to-steering wheel bolts, using a T30 Torx®bit
- SRS module from the steering wheel
- Electrical connectors from the steering wheel
- Steering wheel from the steering column

7. Remove the steering column by removing or disconnecting the following:
- Coin pocket, the lower instrument panel-to-dash screw and the lower instrument panel from the driver's side
- Electrical connector and the air hose at the knee bolster; then, remove the knee bolster-to-dash bolts and the knee bolster
- Steering column cover screws and the covers
- Combination switch-to-steering column screws, disconnect the electrical connectors and remove the combination switch
- Ignition switch connectors
- Clamps, clips and the steering joint cover

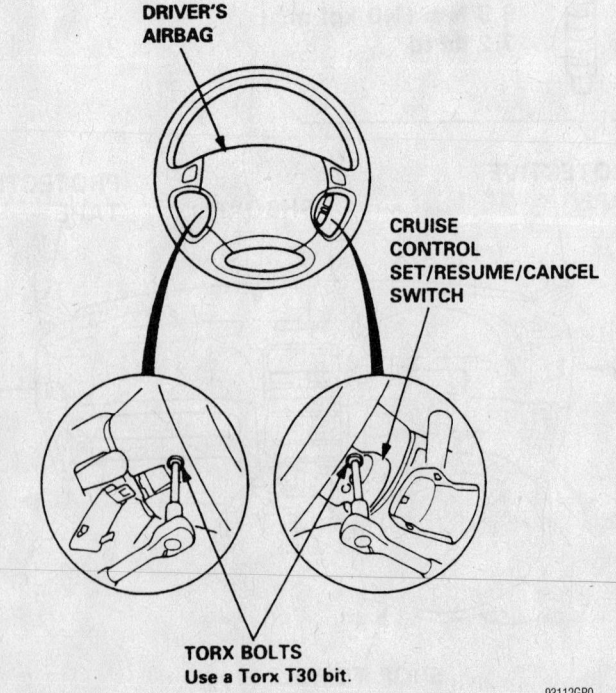

DRIVER'S AIRBAG

CRUISE CONTROL SET/RESUME/CANCEL SWITCH

TORX BOLTS
Use a Torx T30 bit.

93112GP0

Removing the driver's side SRS module-to-steering wheel bolts—3.2CL

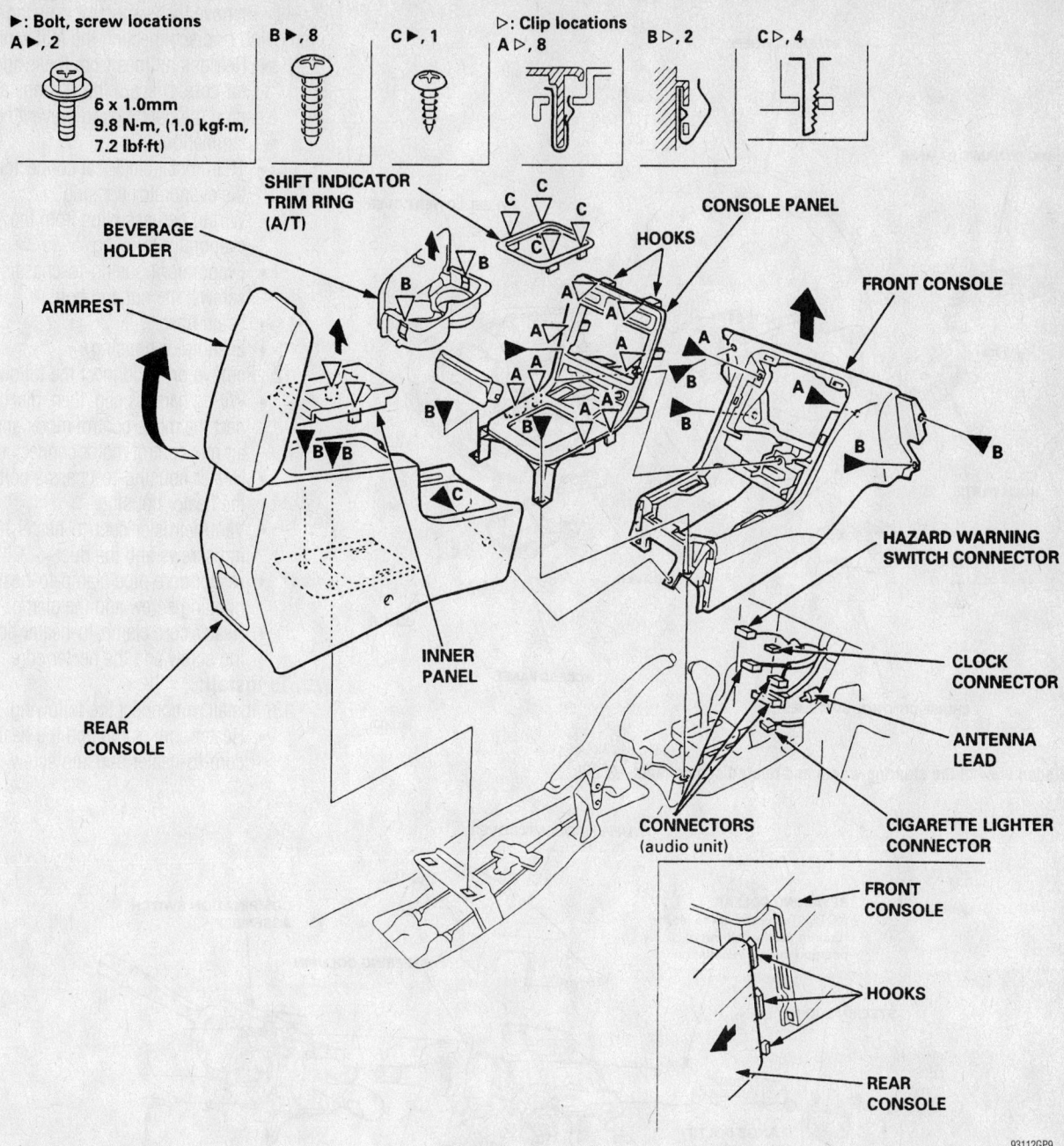

▶: Bolt, screw locations
A ▶, 2

6 x 1.0mm
9.8 N·m, (1.0 kgf·m,
7.2 lbf·ft)

B ▶, 8

C ▶, 1

▷: Clip locations
A ▷, 8

B ▷, 2

C ▷, 4

SHIFT INDICATOR
TRIM RING
(A/T)

BEVERAGE
HOLDER

ARMREST

CONSOLE PANEL

HOOKS

FRONT CONSOLE

INNER
PANEL

REAR
CONSOLE

HAZARD WARNING
SWITCH CONNECTOR

CLOCK
CONNECTOR

ANTENNA
LEAD

CONNECTORS
(audio unit)

CIGARETTE LIGHTER
CONNECTOR

FRONT
CONSOLE

HOOKS

REAR
CONSOLE

93112GP9

View of the front and rear consoles and related components—3.2CL

- Steering joint from the steering column shaft
- Steering column-to-instrument panel nuts/bolts and the steering column
8. Remove the passenger's side SRS module by removing or disconnecting the following:
    - SRS module electrical connector
    - 5 SRS module-to-instrument panel nuts and carefully remove the SRS module
9. Remove the instrument panel by removing or disconnecting the following:
    - Front and rear consoles

- Cruise control master switch and the panel brightness controller; then, disconnect the electrical connectors
- Instrument cluster-to-instrument panel screws and pry out the instrument cluster and disconnect the electrical connectors
- Glove box-to-damper screw
- 2 glove box-to-dash screws and the glove box
- Audio unit-to-instrument panel fasteners and the audio unit
- 3 passenger's dashboard panel screws (located on the passengers

side); then, carefully pry out the dashboard panel and the climate control unit as an assembly
- Side defogger trim from both sides of the instrument panel
- Instrument panel electrical connectors, the recirculation control motor connector and the resistor connector
- Instrument panel-to-chassis bolts and the instrument panel
- Steering hanger beam-to-chassis nuts, bolts and the beam
10. Discharge and recover the air conditioning system refrigerant.

11. Remove the evaporator housing by removing or disconnecting the following:
- Refrigerant lines from the evaporator core. Discard the O–rings and plug the openings to prevent contamination.
- Thermostat electrical connector at the evaporator housing
- Wiring harness clips from the evaporator housing
- Evaporator housing-to-chassis screws, the nut and bolt
- Drain hose
- Evaporator housing

12. Remove or disconnect the following:
- Wiring harness clip; then, disconnect the mode control motor and the air mix control motor connectors
- Heater housing-to-chassis bolt and the heater housing
- Vent/defroster duct-to-heater housing screws and the duct
- Heater core pipe clamp-to-heater housing screw and the clamp
- Heater core clamp-to-heater housing screw and the heater core

To install:

13. Install or connect the following:
- Heater core clamp and the heater core-to-heater housing screw

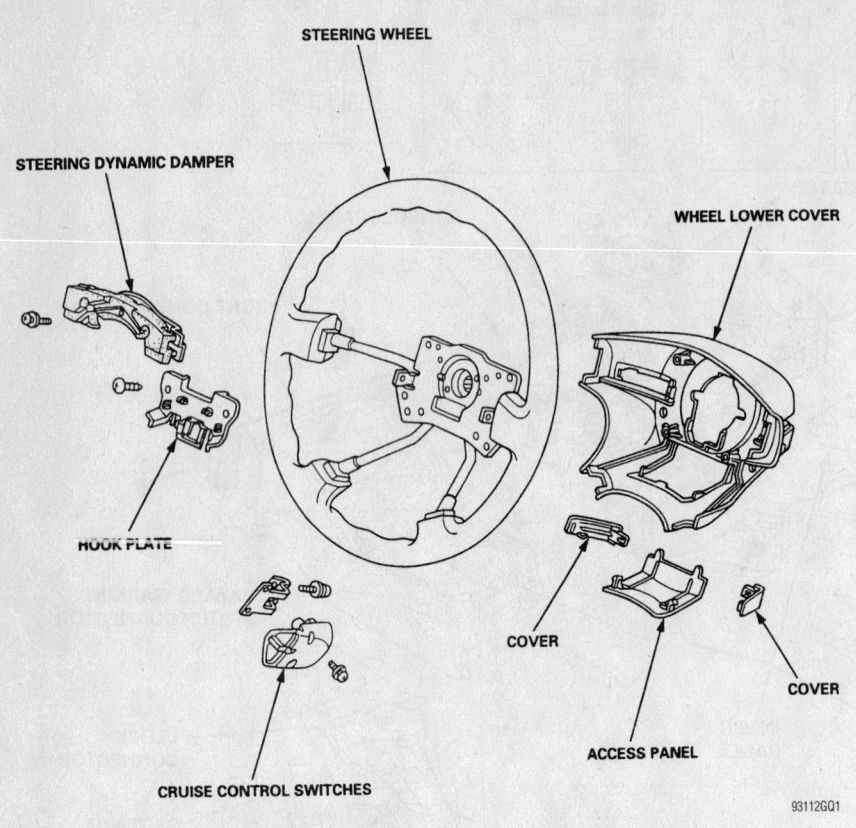

Exploded view of the steering wheel and related components—3.2CL

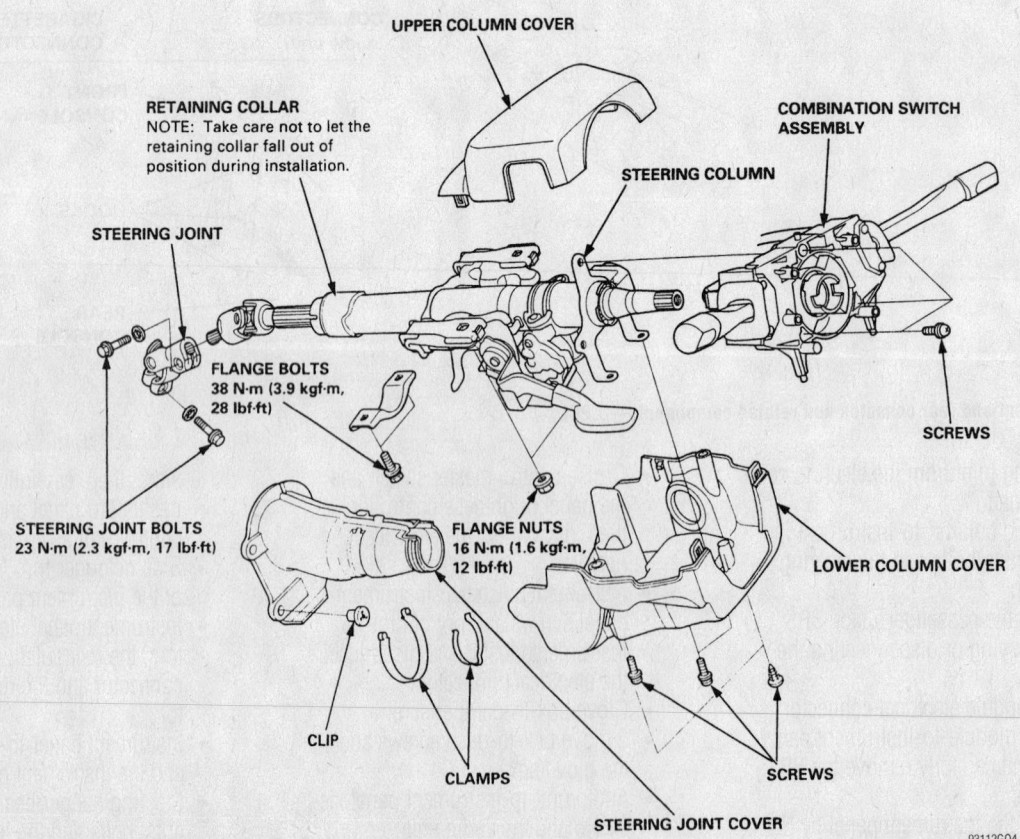

Exploded view of the steering wheel column and related components—3.2CL

- Heater core pipe clamp and the clamp-to-heater housing screw
- Vent/defroster duct and the duct-to-heater housing screws
- Heater housing and the heater housing-to-chassis bolt
- Wiring harness clip; then, connect the mode control motor and the air mix control motor connectors

14. Install the evaporator housing by installing or connecting the following:
   - Evaporator housing
   - Drain hose
   - Evaporator housing-to-chassis screws, the nut and bolt
   - Wiring harness clips to the evaporator housing
   - Thermostat electrical connector, at the evaporator housing
   - Refrigerant lines to the evaporator core, make sure to use new O–rings
   - Steering hanger beam and the beam-to-chassis nuts and bolts

15. Install the instrument panel by installing or connecting the following:
   - Instrument panel and the instrument panel-to-chassis bolts
   - Instrument panel electrical connec-

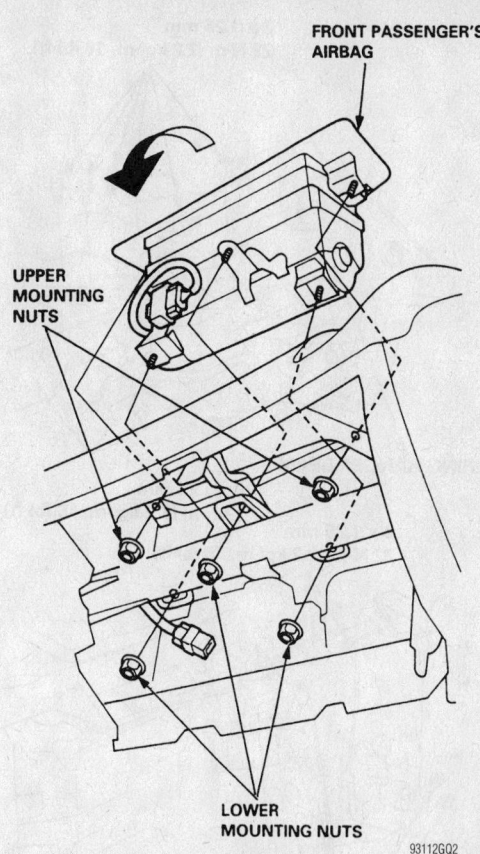

Exploded view of the passenger's side SRS module—3.2CL

▶: Bolt locations, 6

6 x 1.0 mm
9.8 N·m (1.0 kgf·m, 7.2 lbf·ft)

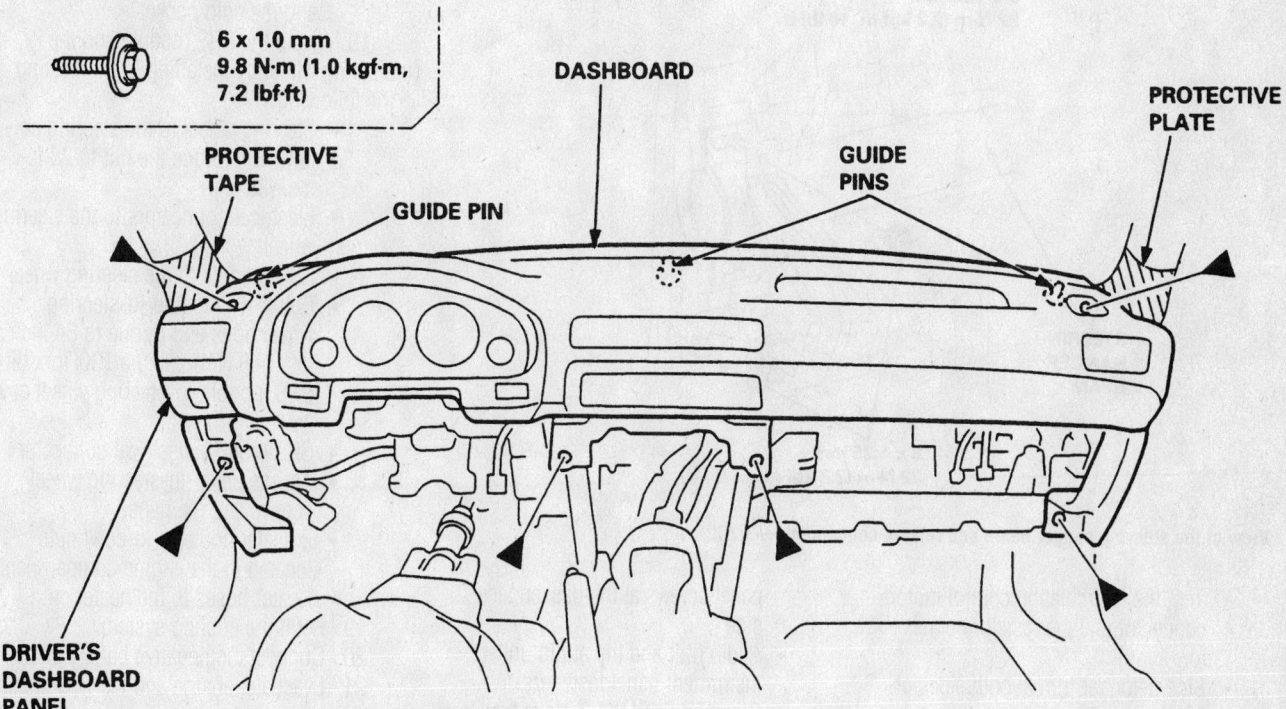

View of the dashboard bolt locations—3.2CL

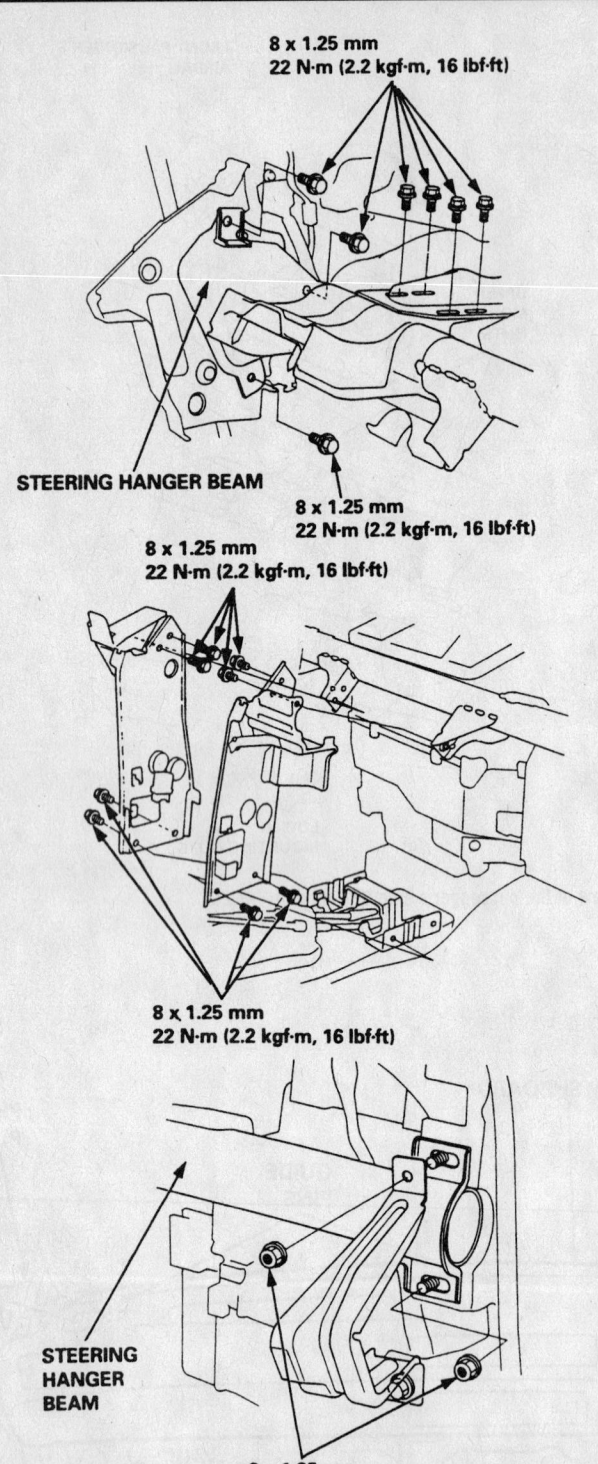

8 x 1.25 mm
22 N·m (2.2 kgf·m, 16 lbf·ft)

STEERING HANGER BEAM

8 x 1.25 mm
22 N·m (2.2 kgf·m, 16 lbf·ft)

8 x 1.25 mm
22 N·m (2.2 kgf·m, 16 lbf·ft)

8 x 1.25 mm
22 N·m (2.2 kgf·m, 16 lbf·ft)

STEERING HANGER BEAM

8 x 1.25 mm
22 N·m (2.2 kgf·m, 16 lbf·ft)

93112GQ5

View of the steering hanger beam and related components—3.2CL

tors, the recirculation control motor connector and the resistor connector
- Side defogger trim at both sides of the instrument panel
- Dashboard panel and the climate control unit as an assembly; then, install the 3 passenger's dashboard panel screws at the passenger's side
- Audio unit and the audio unit-to-instrument panel fasteners
- Glove box and the 2 glove box-to-dash screws
- Glove box-to-damper screw
- Instrument cluster and the instrument cluster-to-instrument panel screws; then, connect the electrical connectors
- Cruise control master switch and the panel brightness controller; then, connect the electrical connectors
- Front and rear consoles

16. Install the passenger's side SRS module by installing or connecting the following:
- SRS module and the 5 SRS module-to-instrument panel nuts and torque to 84 inch lbs. (9.8 Nm)
- SRS module electrical connector

17. Install the steering column by installing or connecting the following:
- Steering column and the steering column-to-instrument panel nuts/bolts
- Steering joint to the steering column shaft
- Clamps, clips and the steering joint cover
- Ignition switch connectors
- Combination switch, connect the electrical connectors and install the combination switch-to-steering column screws
- Steering column covers and the cover screws
- Knee bolster and the knee bolster-to-dash bolts; then, connect the electrical connector and the air hose.
- Lower instrument panel, the lower instrument panel-to-dash screw and the coin pocket

18. Install the SRS module and the steering wheel by installing or connecting the following:
- Steering wheel to the steering column and torque the nut to 28 ft. lbs. (38 Nm)
- Electrical connectors to the steering wheel
- SRS module to the steering wheel
- Both SRS module-to-steering wheel bolts and torque to 84 inch lbs. (9.8 Nm) using a T30 Torx bit
- SRS module-to-steering wheel covers
- SRS module electrical connector
- Access panel-to-steering wheel screws and the panel.
- 2 heater housing-to-cowl nuts located in the engine compartment
- Heater hoses to the heater core

19. Refill the cooling system.
20. Connect the negative battery cable.
21. Evacuate, charge and leak test the air conditioning system.
22. Operate the engine to normal operating temperatures; then, check the climate control operation and check for leaks.

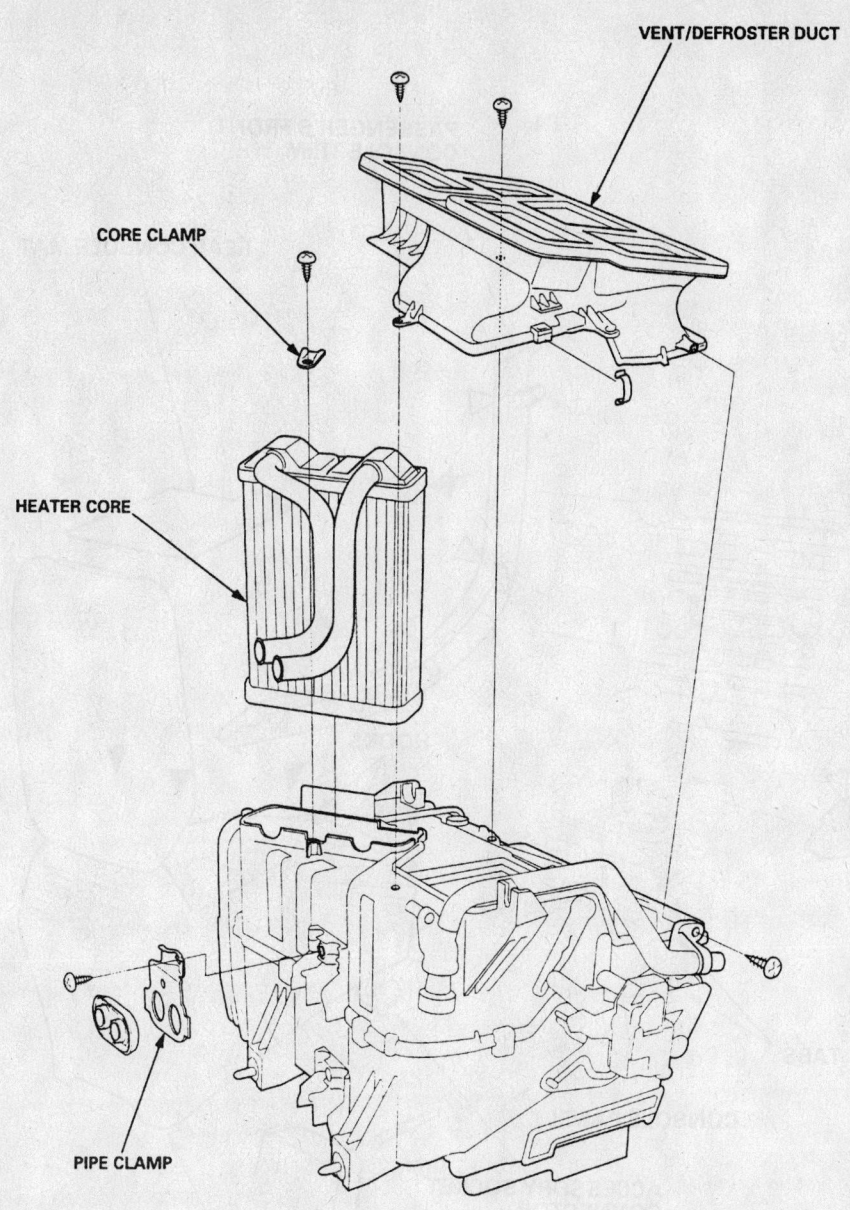

VENT/DEFROSTER DUCT

CORE CLAMP

HEATER CORE

PIPE CLAMP

93112GQ6

Exploded view of the heater core, heater housing and related components—3.2CL

3.2TL

➡**Be sure to acquire the anti–theft code for the radio; then, write down the frequencies for the preset buttons.**

1. Before servicing the vehicle, refer to the precautions in the beginning of this section.

2. Disconnect the negative battery cable.

✳✳ CAUTION

After disconnecting the negative battery cable, wait for at least 3 minutes for the SRS module to deplete its energy.

3. Drain the cooling system into a clean container for reuse.

4. Disconnect the heater hoses from the heater core.

5. In the engine compartment, remove the heater housing-to-cowl nut.

6. Remove the instrument panel by removing or disconnecting the following:
- Clips and remove the dashboard side cover, located at the driver's side
- Lower dashboard cover screw, detach the clips and remove the lower dashboard cover
- Console panel screws and the console panel
- Glove box-to-instrument panel

screw; then, remove the damper from the glove box
- Glove box stop located at each side while holding the glove box
- Both glove box-to-instrument panel screws and the glove box
- Clips and remove the dashboard side cover located at the passenger's side
- Glove box cover-to-instrument panel screws, disconnect the electrical connectors and remove the glove box cover
- Rear console and the front console cover
- 4 audio unit-to-instrument panel screws, disconnect the electrical connectors and remove the audio unit
- Pry out the front pillar trim at both sides
- Bolts at the rear vent duct, disconnect the clips and the OBD–II connector; then, detach the tabs and remove the rear vent duct
- Steering column cover screws and the covers
- Steering column electrical connectors
- Steering joint cover clamp, screws and the cover
- Steering column-to-instrument panel nuts/bolts and lower the steering column
- Instrument panel's electrical connectors and clips
- Instrument panel-to-chassis bolts and the instrument panel

7. Discharge and recover the air conditioning system refrigerant.

8. Remove the evaporator housing by removing or disconnecting the following:
- Refrigerant lines from the evaporator core. Discard the O–rings and plug the openings to prevent contamination.
- Electrical connectors from the evaporator housing
- Evaporator housing-to-chassis screws, nut and bolts; then, remove the evaporator housing

9. Remove or disconnect the following:
- Heater housing-to-chassis bolts and the heater housing
- Mode control motor-to-heater housing screws, the linkage and the motor
- Bracket-to-heater housing screws and the brackets
- Upper-to-lower heater housing screws and separate the housings
- Heater core

Fastener Locations

▶ : Screw, 4   ▷ : Clip, 10

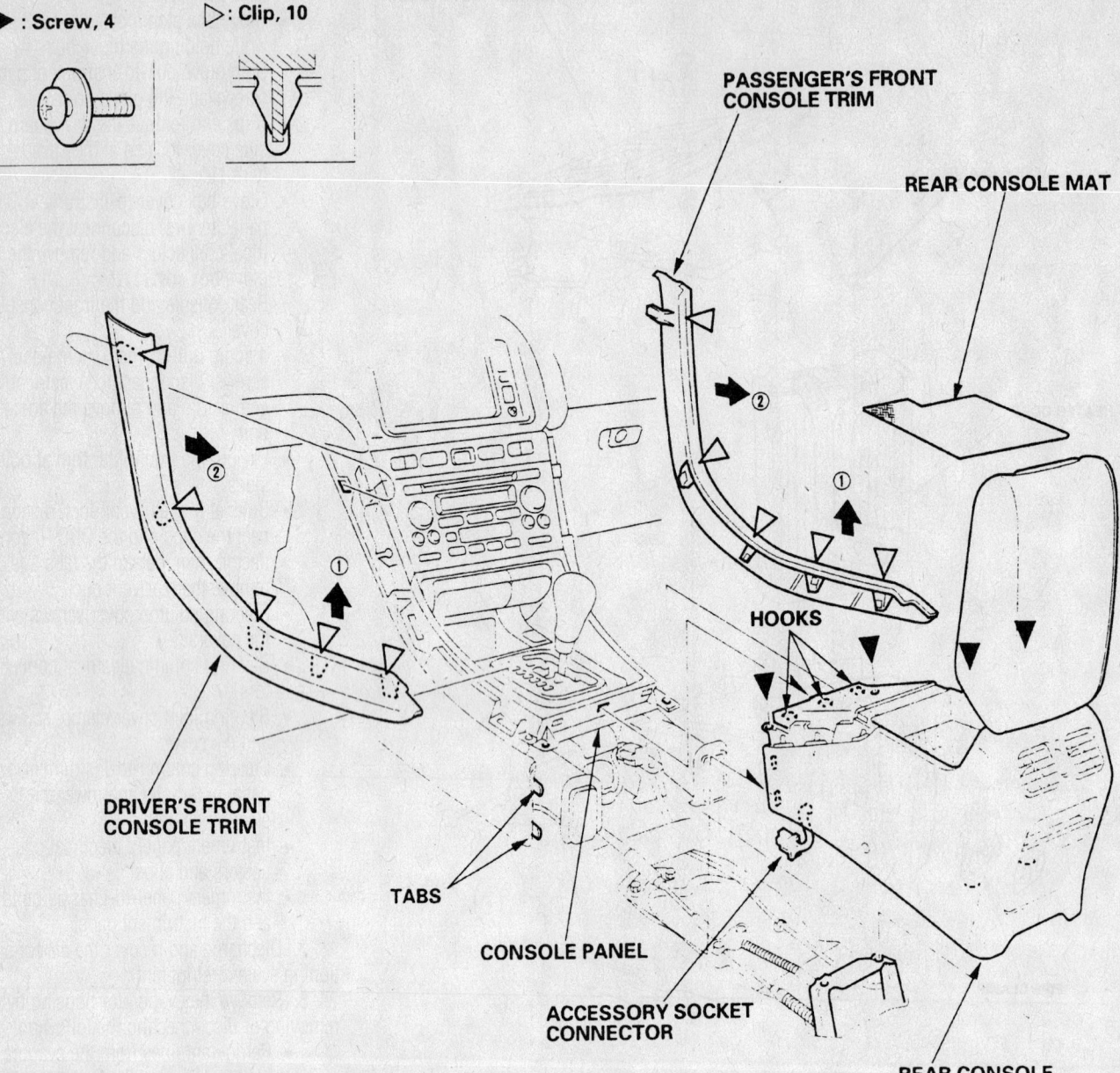

Exploded view of the rear console and related components—3.2TL

To install:

10. Install or connect the following:
  - Heater core
  - Upper-to-lower heater housing and install the heater housing screws
  - Brackets and the bracket-to-heater housing screws
  - Mode control motor, the linkage and the motor-to-heater housing screws
  - Heater housing and the heater housing-to-chassis bolts

11. Install the evaporator housing by installing or connecting the following:
  - Evaporator housing and the evaporator housing-to-chassis screws, nut and bolts
  - Electrical connectors to the evaporator housing
  - Refrigerant lines to the evaporator core, make sure to use new O–rings

12. Install the instrument panel by installing or connecting the following:
  - Instrument panel and the instrument panel-to-chassis bolts
  - Instrument panel's electrical connectors and clips
  - Steering column and torque the steering column-to-instrument panel nuts to 12 ft. lbs. (16 Nm) and bolts to 17 ft. lbs. (23 Nm)
  - Steering joint cover and the cover clamp and screws
  - Steering column electrical connectors
  - Steering column covers and the cover screws
  - Rear vent duct, connect the clips and the OBD–II connector, attach the tabs and install the bolts
  - Front pillar trim. On both sides.
  - Audio unit, connect the electrical

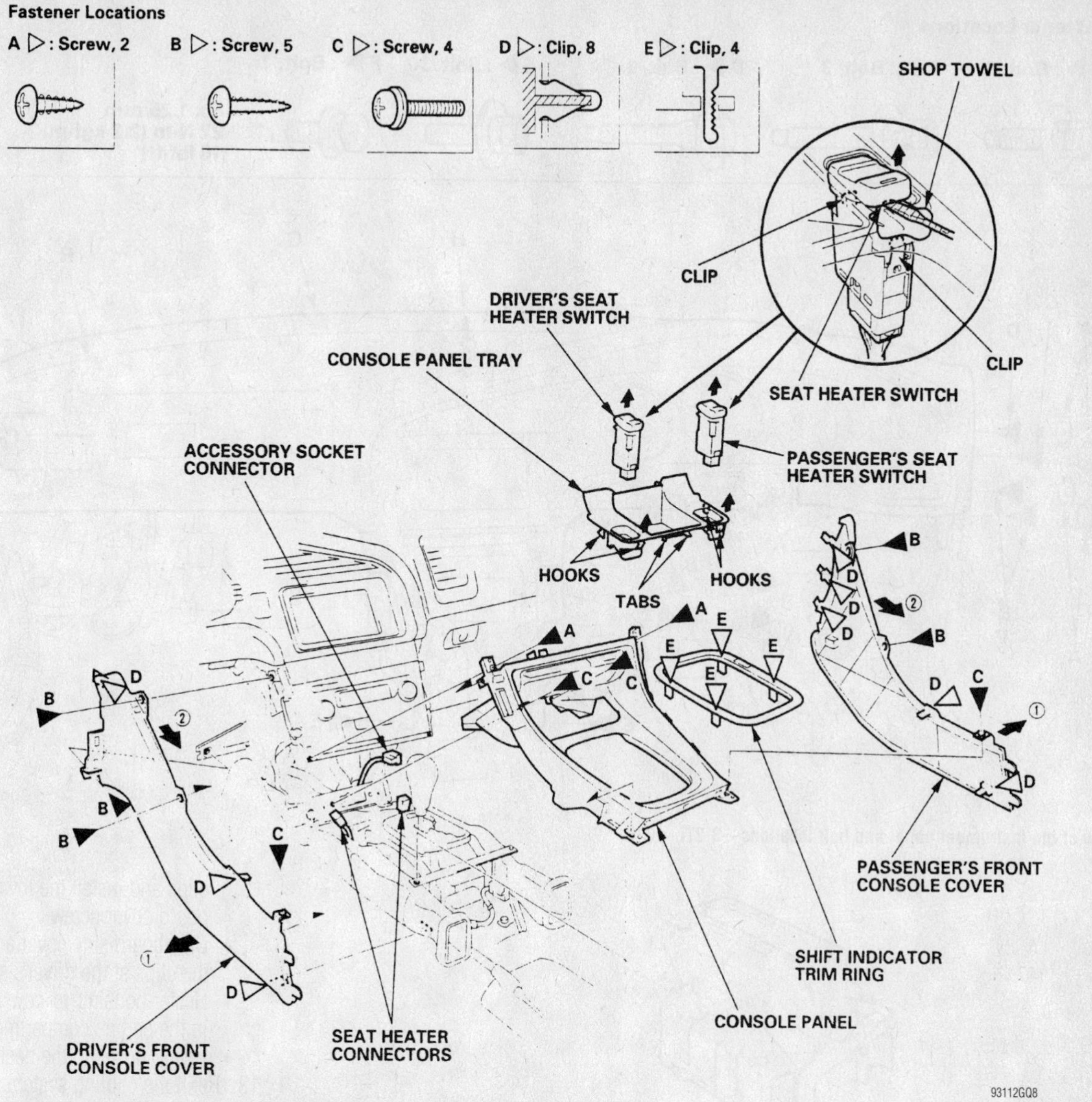

Fastener Locations

A ▷ : Screw, 2   B ▷ : Screw, 5   C ▷ : Screw, 4   D ▷ : Clip, 8   E ▷ : Clip, 4

SHOP TOWEL

CLIP

CLIP

DRIVER'S SEAT
HEATER SWITCH

CONSOLE PANEL TRAY

SEAT HEATER SWITCH

ACCESSORY SOCKET
CONNECTOR

PASSENGER'S SEAT
HEATER SWITCH

HOOKS   HOOKS

TABS

PASSENGER'S FRONT
CONSOLE COVER

SHIFT INDICATOR
TRIM RING

CONSOLE PANEL

DRIVER'S FRONT
CONSOLE COVER

SEAT HEATER
CONNECTORS

93112GQ8

Exploded view of the console panel and related components—3.2TL

connectors and install the 4 audio
unit-to-instrument panel screws
- Rear console and the front console
cover
- Glove box cover, connect the electrical connectors and install the
glove box cover-to-instrument
panel screws
- Dashboard side cover and attach
the clips at the passenger's side
- Both glove box and the glove box-
to-instrument panel screws
- Glove box stop (located at each
side), while holding the glove box
- Damper to the glove box and the
glove box-to-instrument panel
screw
- Console panel and the console
panel screws
- Lower dashboard cover, attach the

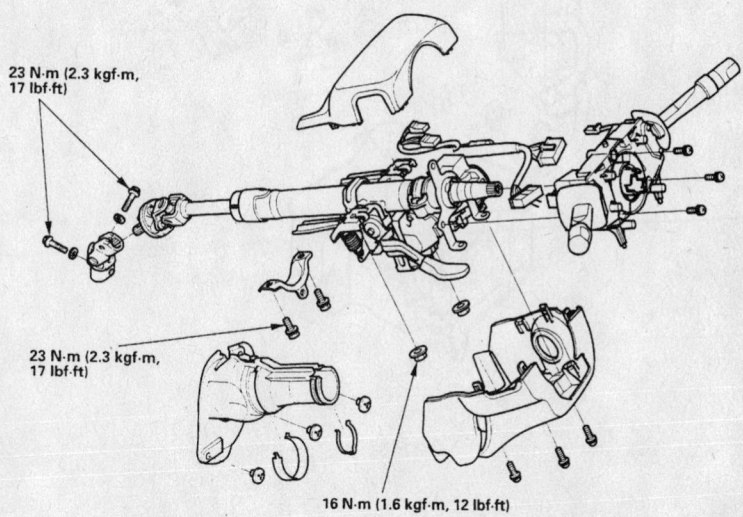

23 N·m (2.3 kgf·m,
17 lbf·ft)

23 N·m (2.3 kgf·m,
17 lbf·ft)

16 N·m (1.6 kgf·m, 12 lbf·ft)

93112GQ9

Exploded view of the steering column and related components—3.2TL

Fastener Locations

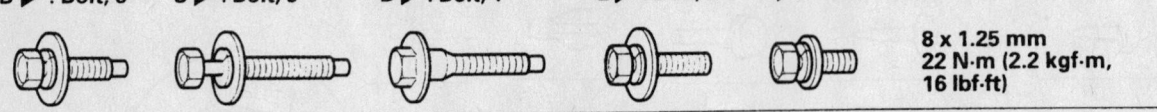

B ▶ : Bolt, 6   C ▶ : Bolt, 3   D ▶ : Bolt, 1   E ▶ : Bolt, 3   F ▶ : Bolt, 1

8 x 1.25 mm
22 N·m (2.2 kgf·m, 16 lbf·ft)

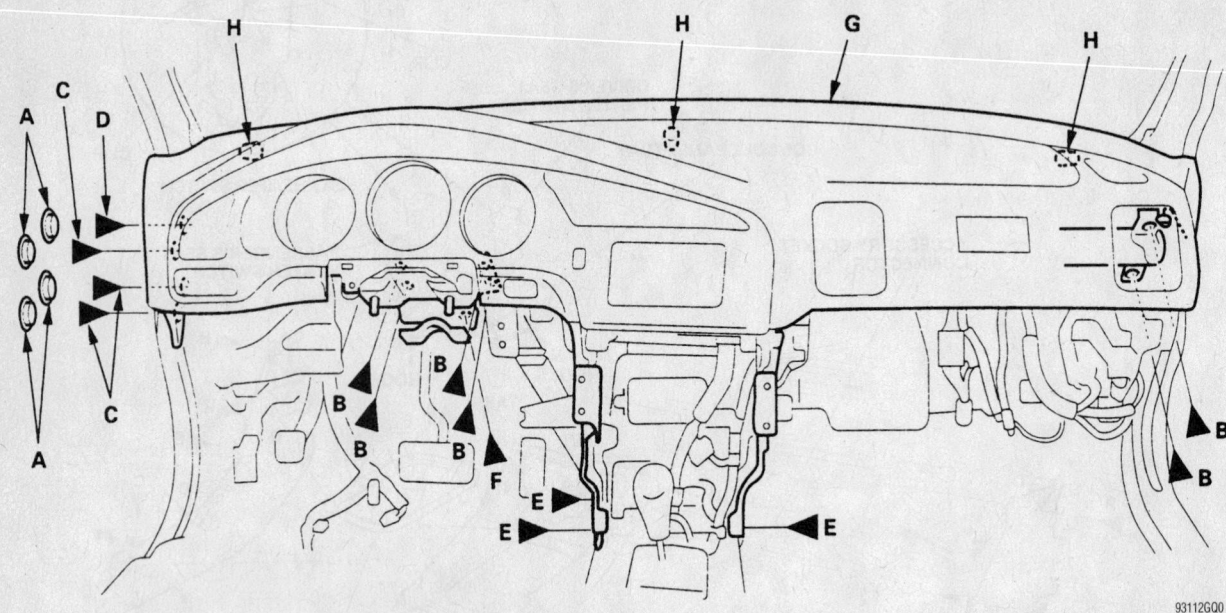

View of the instrument panel and bolt locations—3.2TL

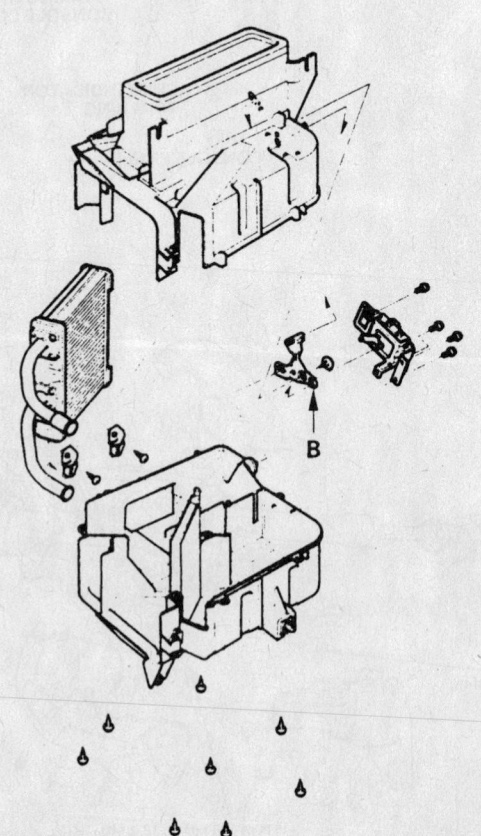

Exploded view of the heater core, heater housing and related components—3.2TL

---

clips and install the lower dashboard cover screw
• Dashboard side cover and attach the clips at the driver's side
• Heater housing-to-cowl nut located in the engine compartment
• Heater hoses to the heater core
13. Refill the cooling system.
14. Connect the negative battery cable.
15. Evacuate, charge and leak test the air conditioning system.
16. Operate the engine to normal operating temperatures; then, check the climate control operation and check for leaks.

3.5RL

➡Be sure to acquire the anti–theft code for the radio; then, write down the frequencies for the preset buttons.

1. Before servicing the vehicle, refer to the precautions in the beginning of this section.
2. Disconnect the negative battery cable.

❋❋ CAUTION

After disconnecting the negative battery cable, wait for at least 3 minutes for the SRS module to deplete its energy.

3. Drain the cooling system into a clean container for reuse.

4. Disconnect the heater hoses from the heater core.

5. Place the front wheel in the straight–ahead position.

6. Remove the SRS module and the steering wheel by removing or disconnecting the following:

- Access panel-to-steering wheel screws and the panel
- RS module electrical connector
- SRS module-to-steering wheel covers, located at both sides of the steering wheel

- Both SRS module-to-steering wheel bolts using a T30 Torx bit
- SRS module from the steering wheel
- Electrical connectors from the steering wheel
- Steering wheel from the steering column

7. Remove the passenger's side SRS module by removing or disconnecting the following:

- SRS module electrical connector
- 2 SRS module-to-instrument panel nuts and carefully remove the SRS module

✵✵ CAUTION

Store the SRS module in a safe place with the front facing upward.

8. Remove the instrument panel by removing or disconnecting the following:

- Console panel and the rear console
- Center air vent using a suitable prytool
- 4 climate control unit/audio assembly-to-instrument panel bolts; then, disconnect the electrical connectors and remove the climate control unit/audio assembly

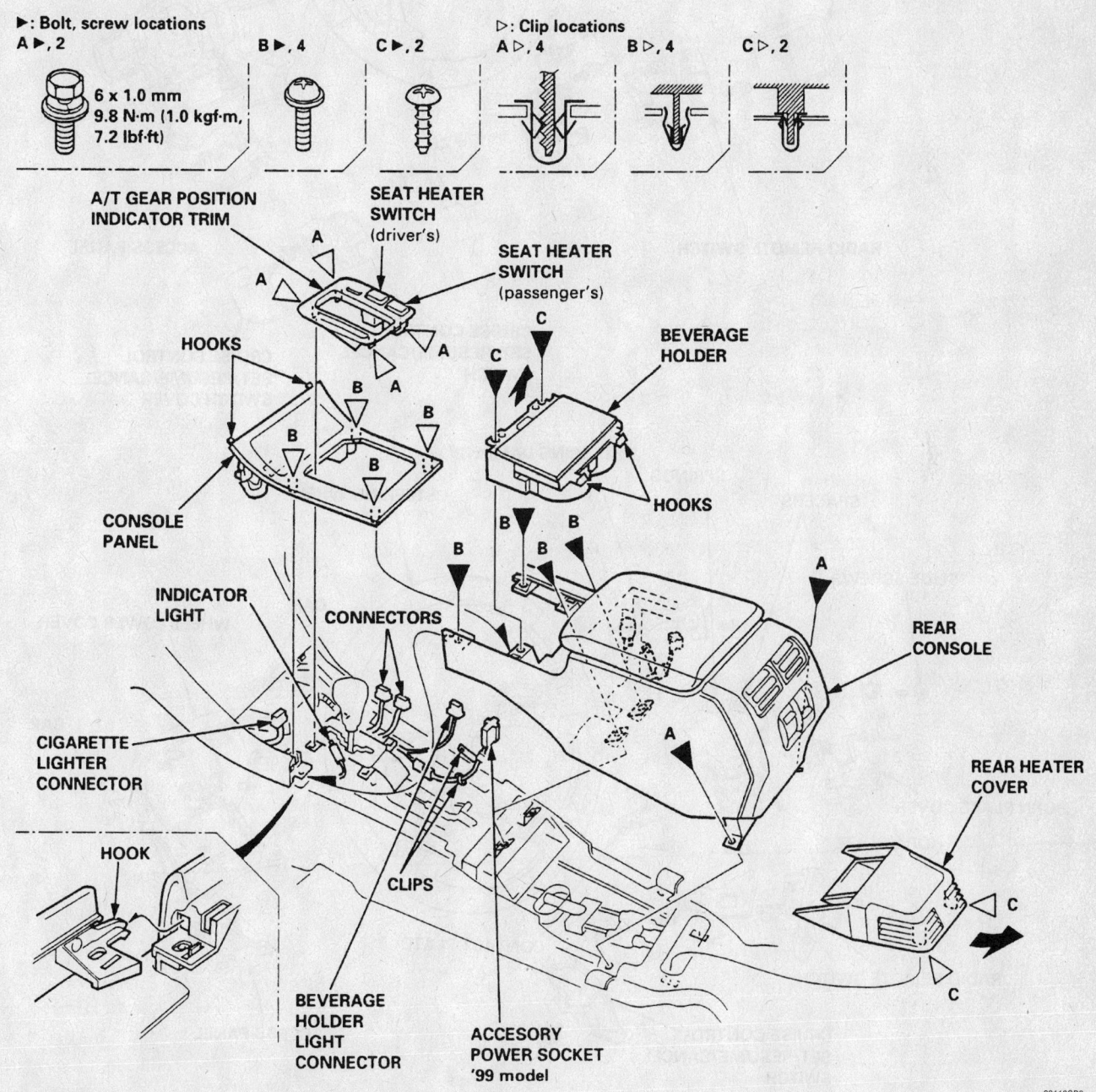

▶: Bolt, screw locations
A ▶, 2   B ▶, 4   C ▷, 2

6 x 1.0 mm
9.8 N·m (1.0 kgf·m, 7.2 lbf·ft)

▷: Clip locations
A ▷, 4   B ▷, 4   C ▷, 2

A/T GEAR POSITION INDICATOR TRIM

SEAT HEATER SWITCH (driver's)

SEAT HEATER SWITCH (passenger's)

HOOKS

BEVERAGE HOLDER

CONSOLE PANEL

HOOKS

INDICATOR LIGHT

CONNECTORS

REAR CONSOLE

CIGARETTE LIGHTER CONNECTOR

REAR HEATER COVER

HOOK

CLIPS

BEVERAGE HOLDER LIGHT CONNECTOR

ACCESORY POWER SOCKET '99 model

93112GR2

Exploded view of the console panel, rear console and related components—3.5RL

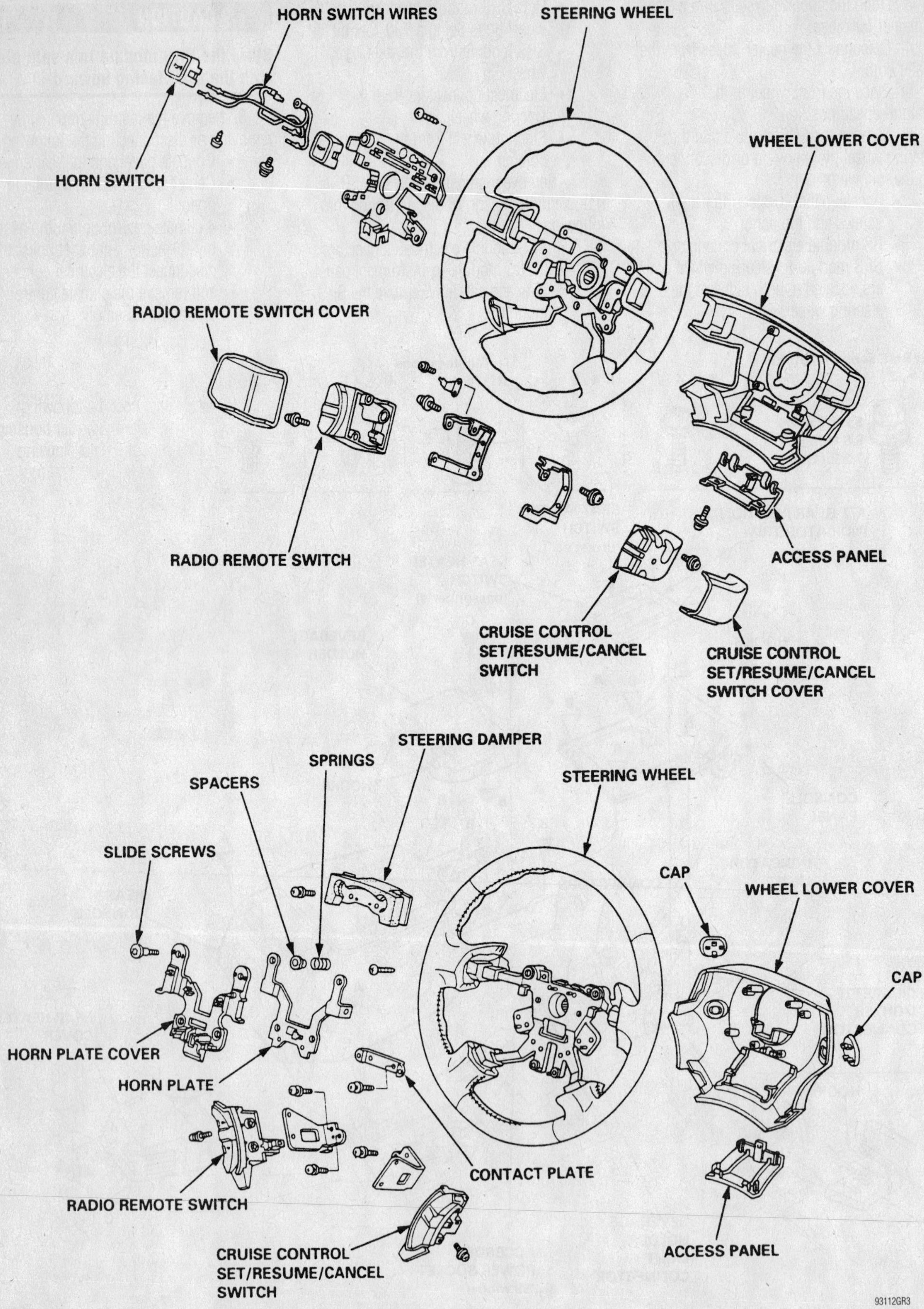

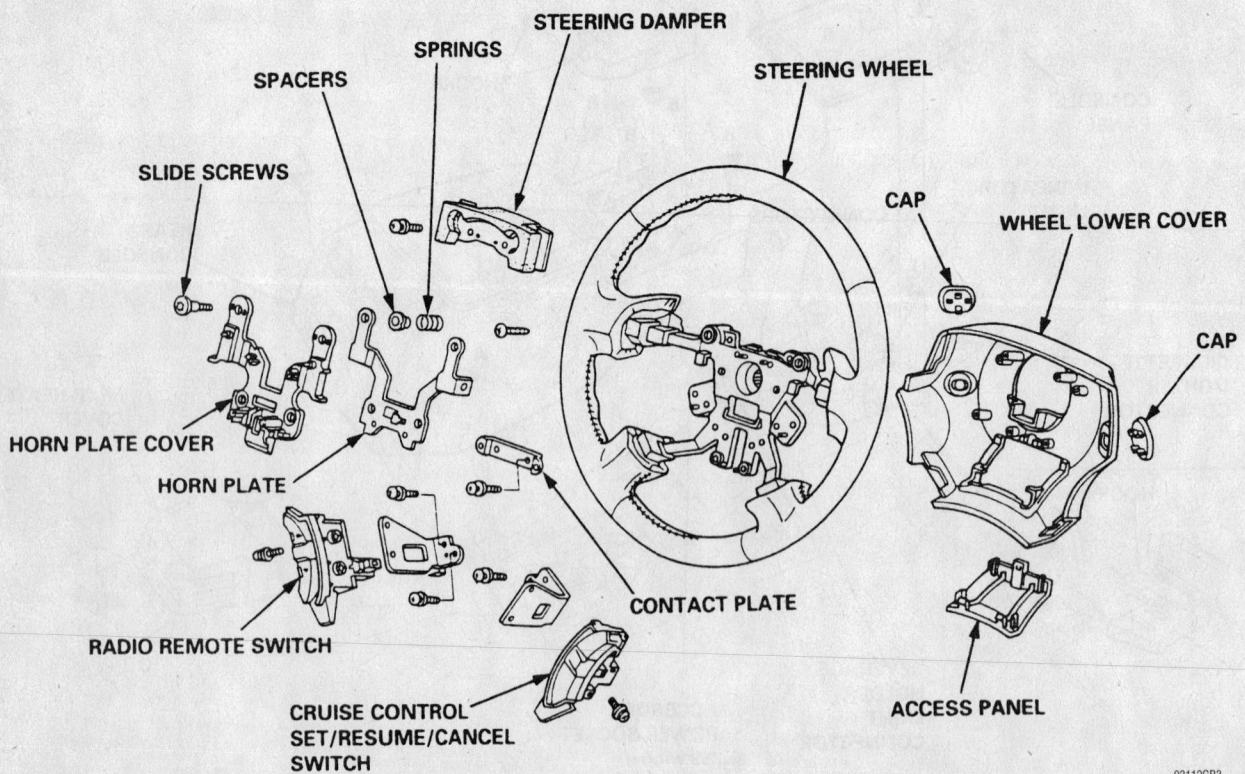

HORN SWITCH WIRES

STEERING WHEEL

WHEEL LOWER COVER

HORN SWITCH

RADIO REMOTE SWITCH COVER

RADIO REMOTE SWITCH

ACCESS PANEL

CRUISE CONTROL
SET/RESUME/CANCEL
SWITCH

CRUISE CONTROL
SET/RESUME/CANCEL
SWITCH COVER

STEERING DAMPER

SPACERS

SPRINGS

STEERING WHEEL

SLIDE SCREWS

CAP

WHEEL LOWER COVER

CAP

HORN PLATE COVER

HORN PLATE

RADIO REMOTE SWITCH

CONTACT PLATE

CRUISE CONTROL
SET/RESUME/CANCEL
SWITCH

ACCESS PANEL

93112GR3

Exploded view of the SRS module, the steering wheel and related components—3.5RL

- Lower instrument panel cover at the passenger's side
- The stop located at each side of the glove box
- Damper clip and the electrical connector while holding the glove box
- Glove box-to-instrument panel bolts and the glove box
- 3 glove box back cover screws, disconnect the clips and remove the cover
- Lower carpet at both sides of the console
- Lower instrument panel cover and the kick panel at the driver's side
- Steering column cover screws and the covers
- Combination switch-to-steering column screws, the combination switch and disconnect the electrical connectors
- Steering joint cover clamp, screws and the cover
- Steering column-to-instrument panel nuts and bolts; then, lower the steering column

- Instrument panel wiring harness connectors, the clip and the air hose
- Instrument panel-to-chassis bolts, the screws and the instrument panel
- Steering beam-to-chassis bolts at the left side
- Steering beam-to-chassis nuts, the bolt and the steering hanger beam at the right side

9. Discharge and recover the air conditioning system refrigerant.

10. Remove the evaporator/blower housing by removing or disconnecting the following:

- Refrigerant lines from the evaporator core
- Electrical connectors from the evaporator/blower housing
- Evaporator/blower housing-to-chassis fasteners and the evaporator/blower housing
- Clips and the floor heater duct

11. At the left side of the heater housing, remove the cruise control unit -to-heater housing bolt and the cruise control unit.

12. Remove or disconnect the following:

- Mode control motor and the air mix control motor connectors
- Wiring harness clips and the wiring harness
- Heater housing-to-chassis nuts, the bolt and the heater housing
- Vent/defroster duct-to-heater housing screws and the duct
- Heater core-to-heater housing pipe clamp screws and the clamps
- Heater core-to-heater housing clamp screws and the clamp
- Heater core from the heater housing

To install:

13. Install or connect the following:

- Heater core to the heater housing
- Heater core-to-heater housing clamp and the clamp screws
- Heater core-to-heater housing pipe clamps and the clamp screws
- Vent/defroster duct and the duct-to-heater housing screws

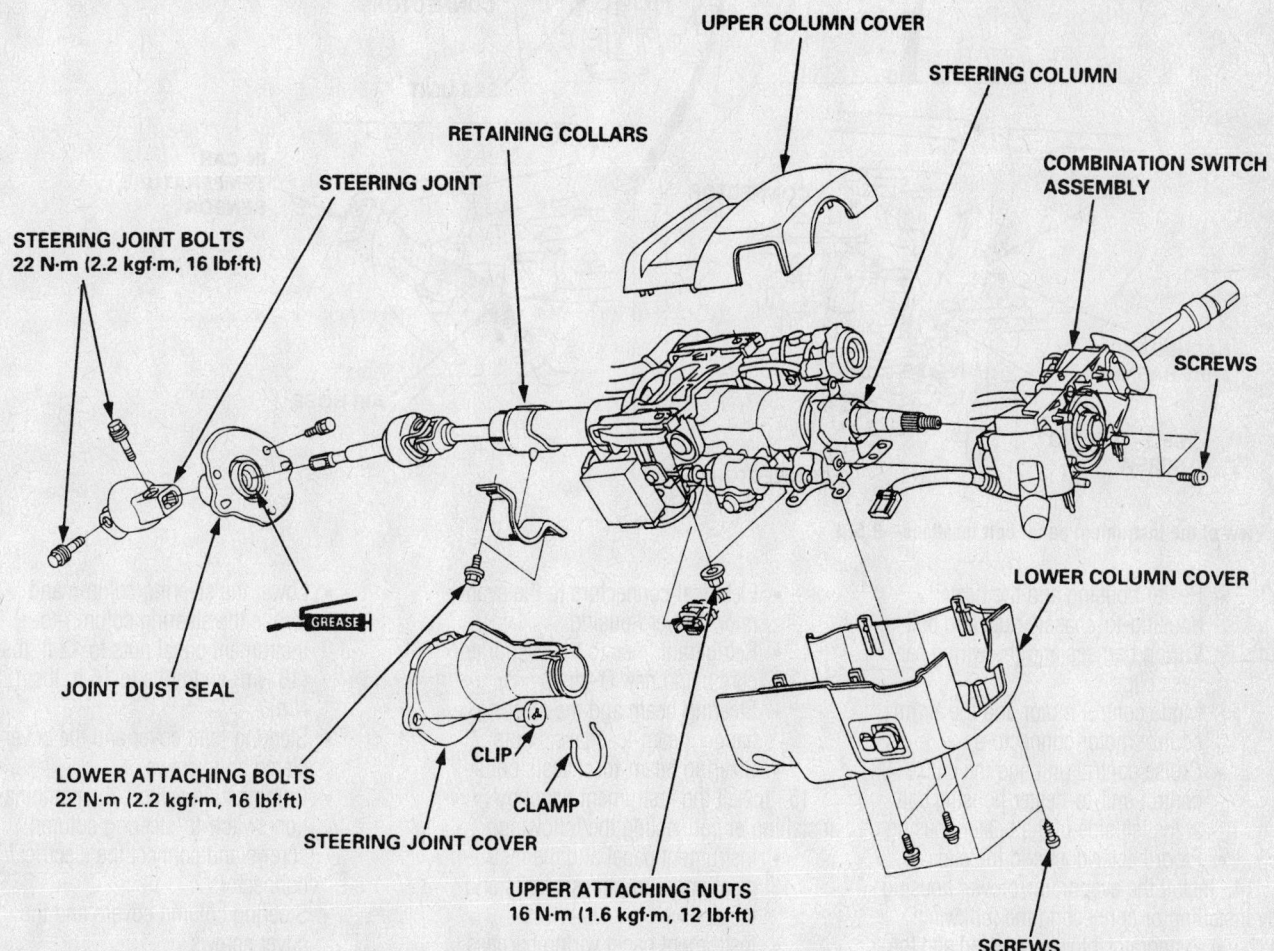

Exploded view of the steering column and related components—3.5RL

93112GR4

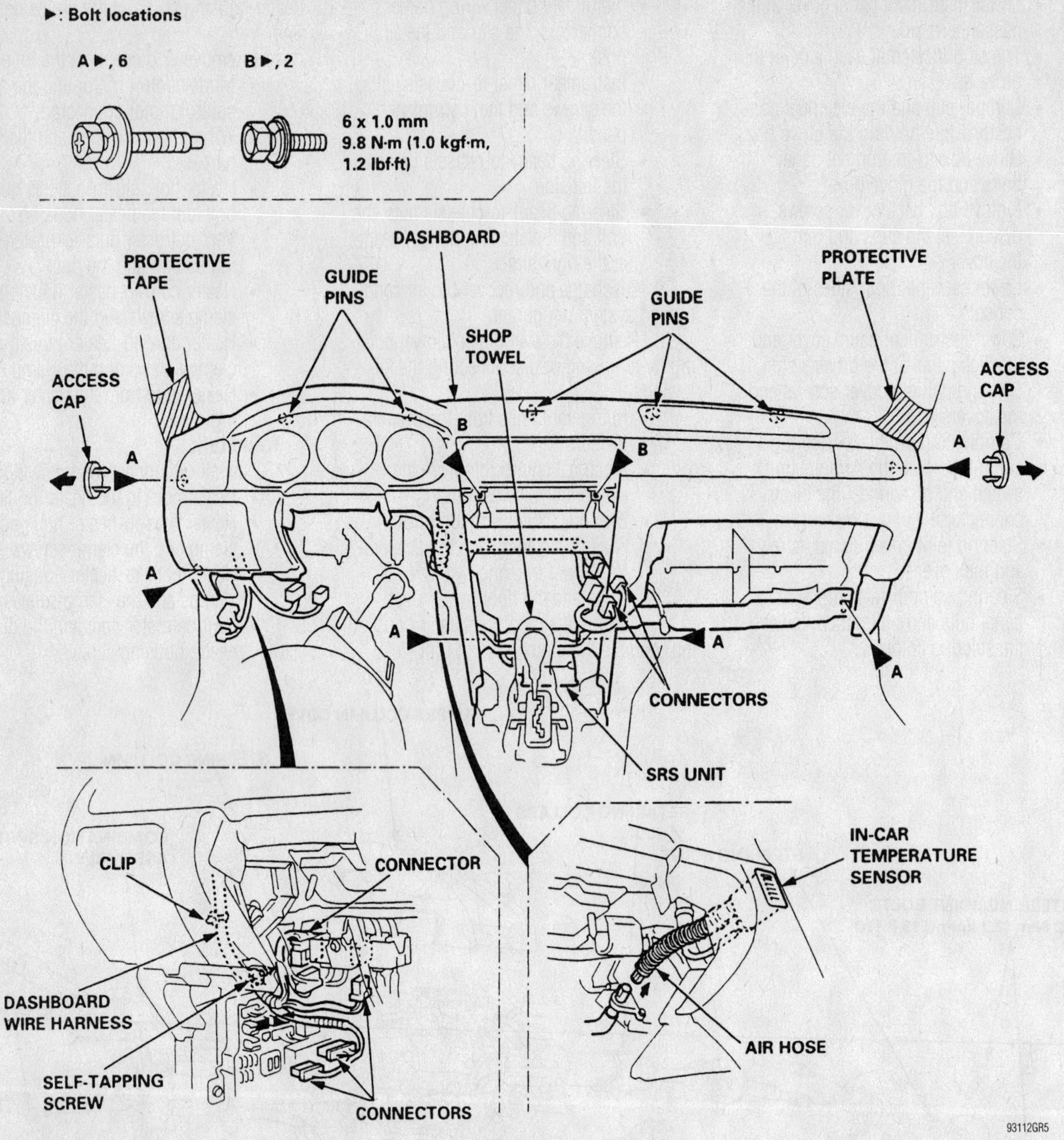

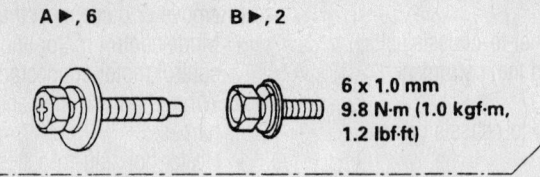

►: Bolt locations

A ►, 6   B ►, 2

6 x 1.0 mm
9.8 N·m (1.0 kgf·m,
1.2 lbf·ft)

PROTECTIVE TAPE

GUIDE PINS

DASHBOARD

SHOP TOWEL

GUIDE PINS

PROTECTIVE PLATE

ACCESS CAP

ACCESS CAP

B   B

A   A

A   A   A

A

CONNECTORS

SRS UNIT

CLIP   CONNECTOR

IN-CAR TEMPERATURE SENSOR

DASHBOARD WIRE HARNESS

SELF-TAPPING SCREW

CONNECTORS

AIR HOSE

93112GR5

View of the instrument panel bolt locations—3.5RL

- Heater housing and the heater housing-to-chassis nuts and bolt
- Wiring harness and the wiring harness clips
- Mode control motor and the air mix control motor connectors
- Cruise control unit and the cruise control unit-to-heater housing bolt at the left side of the heater housing
- Floor heater duct and the clips

14. Install the evaporator/blower housing by installing or connecting the following:
- Evaporator/blower housing and the evaporator/blower housing-to-chassis fasteners

- Electrical connectors to the evaporator/blower housing
- Refrigerant lines to the evaporator core using new O–rings
- Steering beam and the steering hanger beam-to-chassis nuts
- Steering beam-to-chassis bolts

15. Install the instrument panel by installing or connecting the following:
- Instrument panel and the instrument panel-to-chassis bolts and the screws
- Instrument panel wiring harness connectors, the clip and the air hose

- Lower the steering column and torque the steering column-to-instrument panel nuts to 12 ft. lbs. (16 Nm) and bolts to 16 ft. lbs. (22 Nm)
- Steering joint cover and the cover clamp and screws
- Combination switch, the combination switch-to-steering column screws and connect the electrical connectors
- Steering column covers and the cover screws
- Lower instrument panel cover and the kick panel at the driver's side

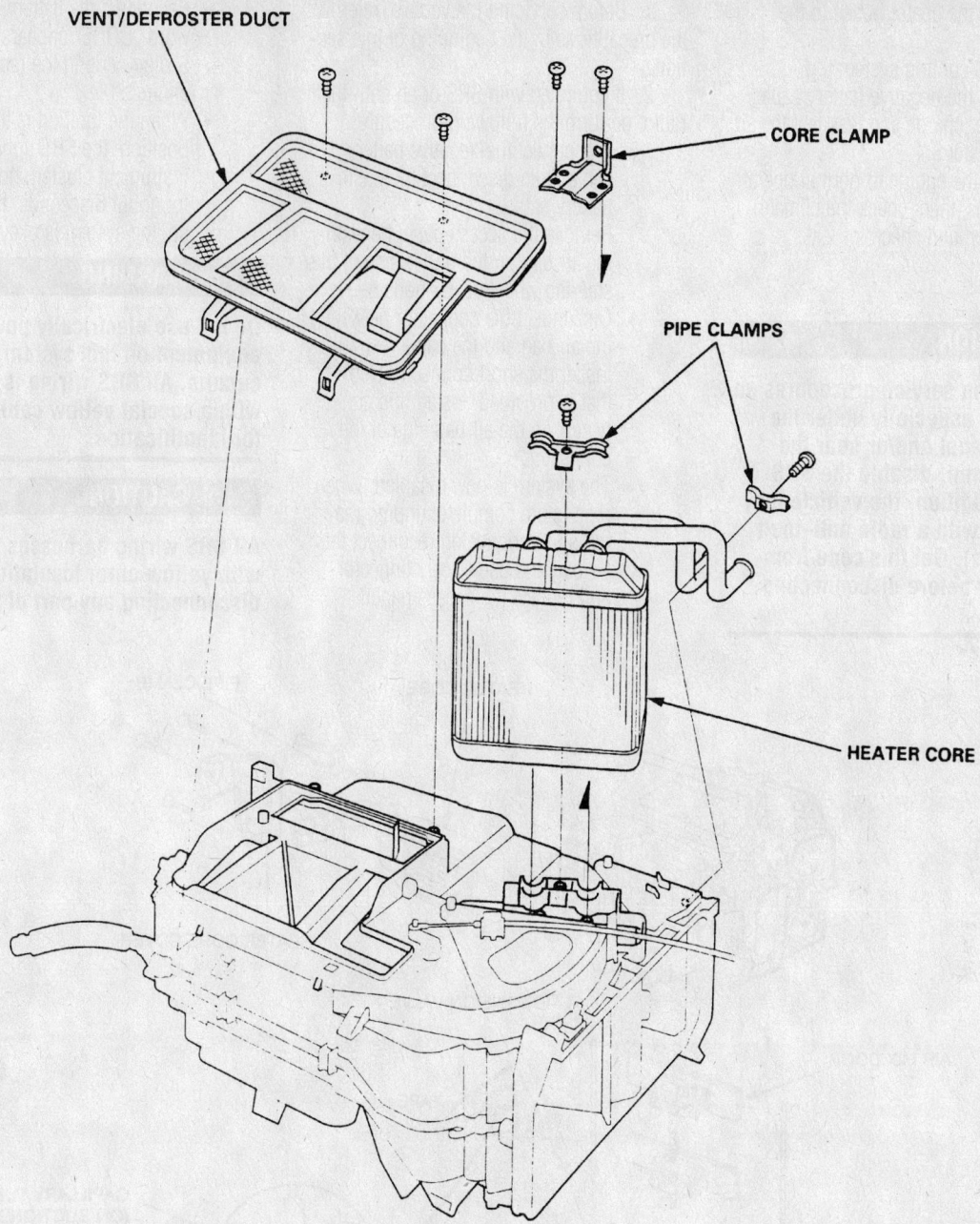

VENT/DEFROSTER DUCT

CORE CLAMP

PIPE CLAMPS

HEATER CORE

93112GR6

Exploded view of the heater core, heater housing and related components—3.5RL

- Lower carpet at both sides of the console
- Glove box back cover, connect the clips and install the 3 cover screws
- Glove box and the glove box-to-instrument panel bolts
- Damper clip and the electrical connector while holding the glove box
- The stop at each side of the glove box
- Lower instrument panel cover
- Climate control unit/audio assembly; then, connect the electrical connectors and install the 4 climate

control unit/audio assembly-to-instrument panel bolts
- Center air vent.
- Console panel and the rear console

16. Install the passenger's side SRS module by installing or connecting the following:

- SRS module and the 2 SRS module-to-instrument panel nuts; then, torque the nuts to 84 inch lbs. (9.8 Nm)
- SRS module electrical connector

17. Install the SRS module and the steering wheel by installing or connecting the following:

- Steering wheel to the steering column and torque the nut to 36 ft. lbs. (49 Nm)
- Electrical connectors to the steering wheel
- SRS module to the steering wheel
- Both SRS module-to-steering wheel bolts and torque to 84 inch lbs. (9.8 Nm) using a T30 Torx bit
- SRS module-to-steering wheel covers
- SRS module electrical connector
- Access panel-to-steering wheel screws and the panel

18. Connect the heater hoses to the heater core.
19. Refill the cooling system.
20. Connect the negative battery cable.
21. Evacuate, charge and leak test the air conditioning system.
22. Operate the engine to normal operating temperatures; then, check the climate control operation and check for leaks.

NSX

⁂ CAUTION

Before starting service procedures on components, especially under the instrument panel and/or near the steering column, disable the SRS system. In addition, the vehicle may be equipped with a radio anti-theft code (5 digits). Get this code from the customer before disconnecting the battery.

1. Before servicing the vehicle, refer to the precautions in the beginning of this section.

2. If equipped with SRS or an anti-theft radio, perform the following procedures:
- Disconnect the negative battery cable, then disconnect the positive battery cable.
- Remove the access panel beneath the air bag on the under-side of the steering wheel center housing.
- Disconnect the connector between the air bag and the cable reel and install the short connector (red) that is provided. Install this connector on the air bag side of the connector.
- The system is now disabled. When repairs are complete, unplug the red short connector, reconnect the air bag and cable reel connector and replace the access panel.

- Reconnect the battery. When the word "CODE" appears, re-enter the 5-digit code in the radio, if equipped.
- When the ignition is turned to the **II** position, the SRS light on the instrument cluster should come ON for about 6 seconds, then go out. If so, the system is okay.

⁂ WARNING

Do not use electrically powered test equipment on this system or related circuits. All SRS wiring is covered with a special yellow cable housing for identification.

⁂ CAUTION

All SRS wiring harnesses are covered with yellow outer insulation. Before disconnecting any part of the SRS

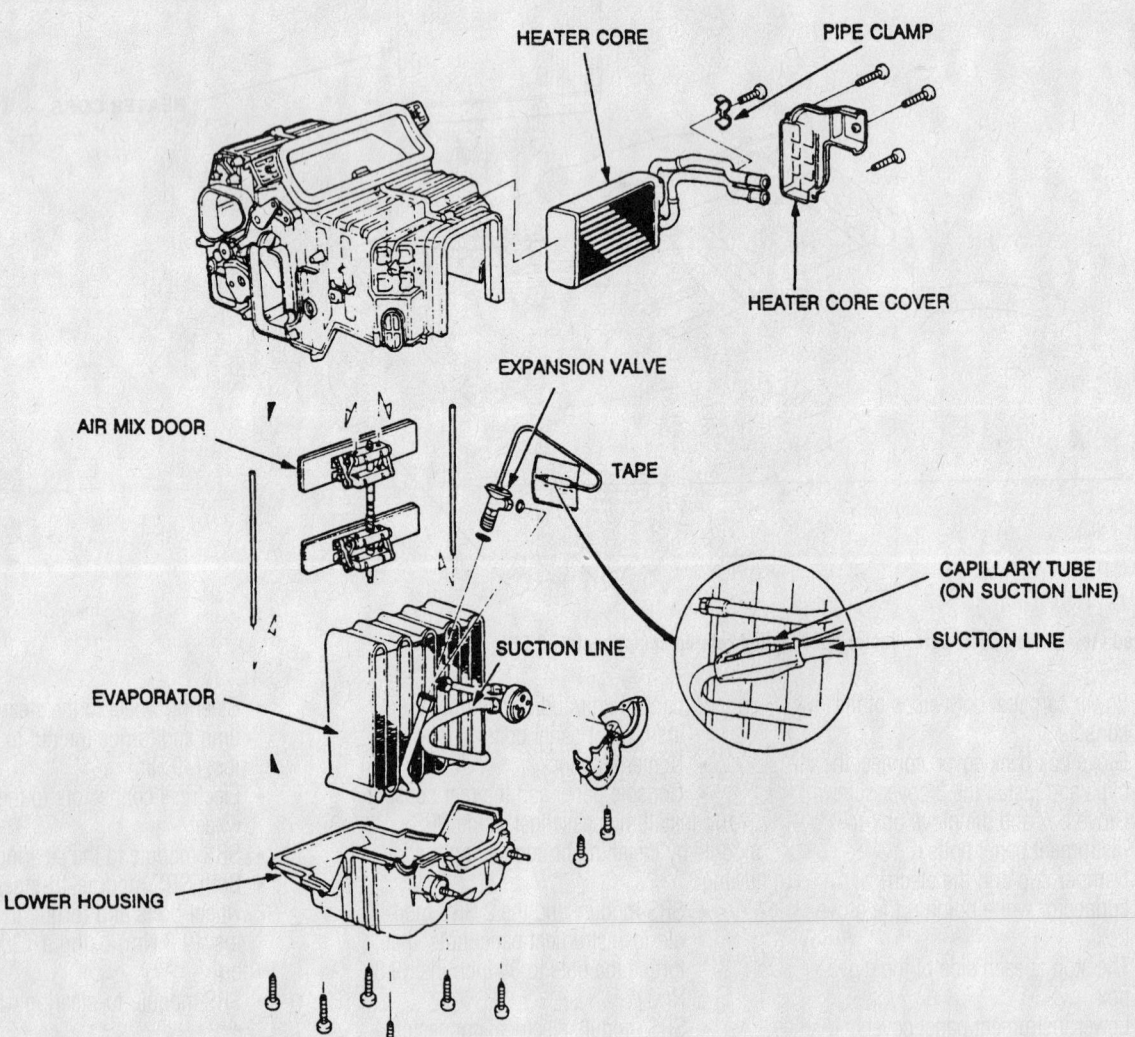

HEATER CORE   PIPE CLAMP

HEATER CORE COVER

EXPANSION VALVE

AIR MIX DOOR

TAPE

CAPILLARY TUBE
(ON SUCTION LINE)

SUCTION LINE

SUCTION LINE

EVAPORATOR

LOWER HOUSING

93112G07

Exploded view of the heater core/evaporator assembly—NSX

wire harness, install short connection to prevent accidental discharge of the air bag system. If the SRS harness assembly should have an open circuit or damaged wiring, replace the entire affected harness.

3. Disconnect the negative battery cable.

4. Properly drain the cooling system. Properly discharge the air conditioning system using an approved refrigerant recover/recycling system.

5. Remove or disconnect the following:
- Blower assembly
- Refrigerant lines from their connection at the firewall. Cap all opening immediately to keep dirt and moisture from contaminating the system.

6. Remove the dashboard by removing or disconnecting the following:

- Front seats
- Knee bolster, pad, dashboard stay and center armrest
- Clock, center air vent and center console panel
- Heater/air conditioning control panel and the radio assembly
- Right dashboard lower panel. Remove the glove box
- Lower the steering column
- Instrument cluster bezel and the instrument cluster
- Air vents from both ends of the dashboard
- U–plate at the foot of the center console, rearward of the parking brake
- Defrost outlet grille, remove the mounting bolts and lift out the dashboard. Note the position of the center guide pin for reinstallation.

7. Remove or disconnect the following:

- Heater duct
- Connectors from the control unit and from the evaporator temperature sensor, then remove the control unit and bracket
- Sound system speaker
- Connectors from the control unit and from the evaporator temperature sensor, then remove the control unit and bracket
- Sound system speaker
- Actuator connectors and sensor connectors from the heater-evaporator unit
- 2 under-dash mounting bolts and remove the heater-evaporator unit through the passenger's door

8. The unit can now be disassembled and the heater core removed.

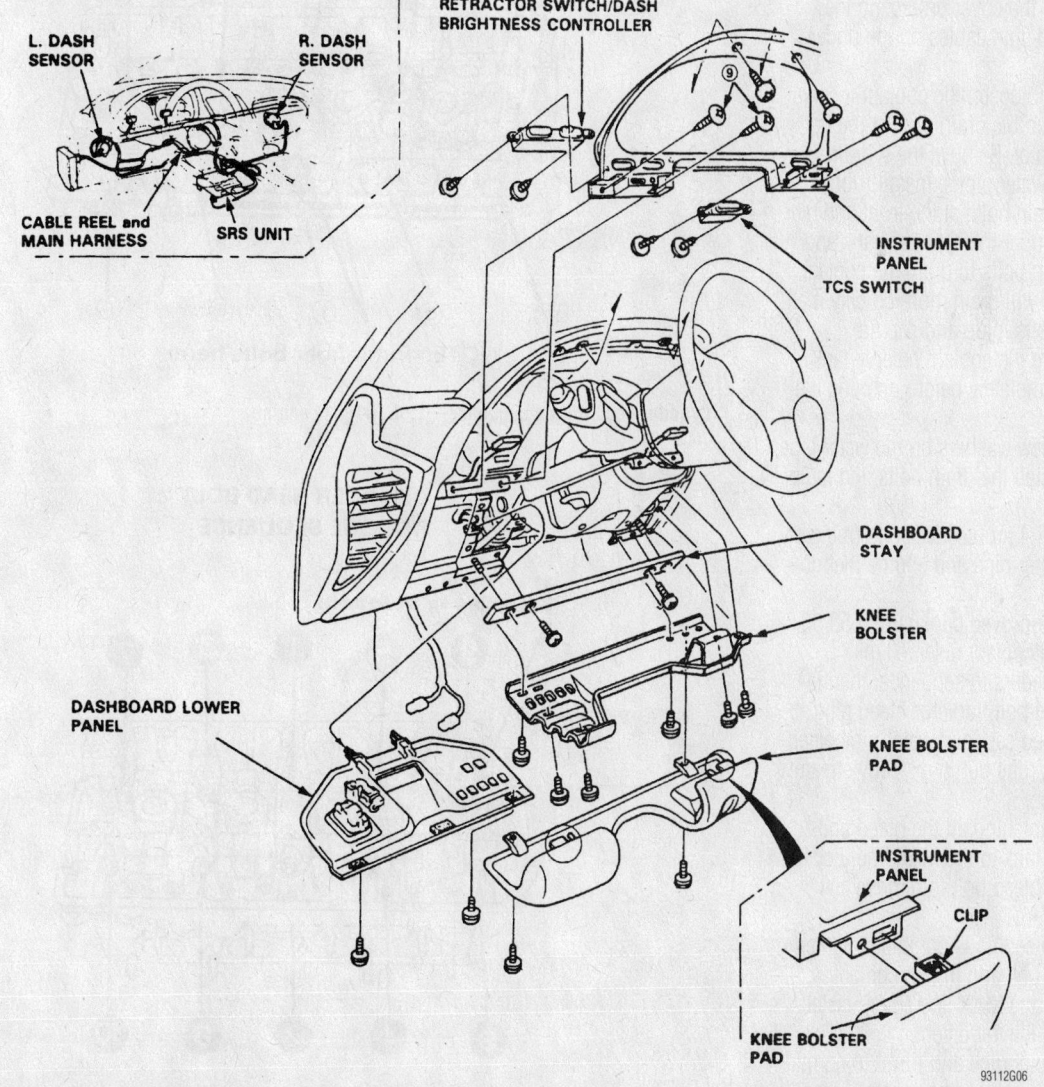

Exploded view of the dashboard and related components—NSX

To install:

9. Installation is the reverse of the removal procedures.

10. Add 0.3 oz. of extra refrigerant oil if the evaporator core was replaced.

11. Adjust the heater valve cable, if needed.

12. Refill the cooling system and bleed the system by performing the following procedure:

✳✳ WARNING

Failure to properly bleed the air from the cooling system could cause engine damage.

13. Turn the ignition switch **ON** and slowly set the climate control temperature knob to **90**. This will allow the coolant in the heater to drain out with the rest of the system.

14. Open the hood, rear hatch and the engine cover.

15. Remove the cover protecting the water pipes and shift cables on the underside of the car.

16. Carefully loosen the coolant reserve tank cap. Loosen the drain plug at the bottom of the radiator. Remove the 2 drain bolts from the water pipes. Install rubber hoses to the drain bolts at the front and rear of the engine under the cylinder bank and loosen the drain bolts to drain the coolant.

17. Coolant will drain more quickly if all the air bleed bolts, plug and cap are opened. Be sure the coolant reserve tank has drained completely before opening the air bleed bolts.

18. Using new washers on the water pipe drain bolts, install the drain bolts and radiator drain plug.

19. Open all 4 air bleeder bolts (radiator, heater pipe, water pipe and engine thermostat cover).

20. Using approved coolant in a 50/50 mixture, fill the coolant reserve tank. Tighten the bleeders in sequence; thermostat cover bleed bolt, radiator bleed plug, heater pipe bleed cap, and water pipe bleed bolt as coolant runs out in a steady stream with no bubbles.

21. After tightening all the bleed bolts, fill the coolant tank to the MAX line. Loosen the thermostat bleed bolt to remove any remaining air.

22. When bleeding is complete, tighten the thermostat bolt and fill the coolant reserve tank to the MAX line again. Install the tank cap to the 1st detent.

23. Start the engine and run it to normal operating temperatures (thermostat opens and radiator cooling fan runs).

24. Turn the engine **OFF**, check and adjust the coolant to the MAX line if needed.

25. Install the coolant tank cap securely and install the car's undercover.

26. Evacuate, charge and leak test the air conditioning system.

27. Connect the negative battery cable.

Cylinder Head

REMOVAL & INSTALLATION

1.8L Engines

1. Before servicing the vehicle, refer to the precautions in the beginning of this section.

2. Before removing the cylinder head, be sure the engine temperature is below 100°F (38°C); a fully cooled engine is best.

3. Disconnect the negative battery cable.

4. Be sure the crankshaft is at TDC on No. 1 cylinder by aligning the white mark on the crankshaft pulley with the pointer on the lower timing belt cover.

5. Drain the engine coolant.

6. Properly relieve the fuel system pressure.

7. Remove or disconnect the following:
   • Cylinder head cover
   • Crankshaft pulley
   • Middle and lower timing belt covers
   • Timing belt

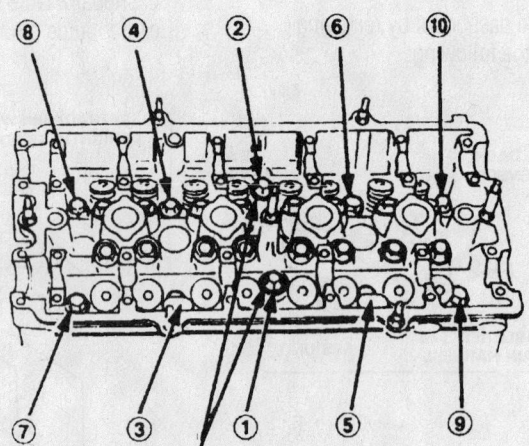

NOTE: Put longer bolts here.

7923BG07

Cylinder head torque sequence—1.8L (B18B1) engine

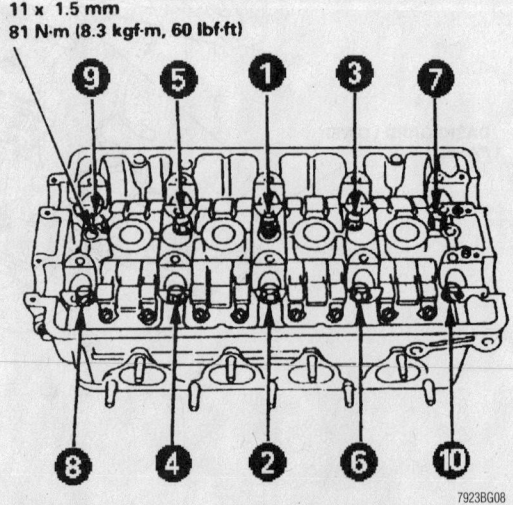

**CYLINDER HEAD BOLT
TORQUE SEQUENCE**

11 x 1.5 mm
81 N·m (8.3 kgf·m, 60 lbf·ft)

7923BG08

Cylinder head torque sequence—1.8L (B18C1, B18C5) engines

- Camshaft pulleys
- Back cover
- Exhaust manifold cover
- Exhaust manifold and bracket

8. Loosen the locknuts and adjusting screws, then remove the camshaft holder bolts. Remove the camshaft holders, camshafts and rocker arms.

9. Remove the cylinder head bolts, then remove the cylinder head. To prevent warpage, unscrew the bolts in the reverse of the torque sequence, ⅓ turn at a time. Repeat the sequence until all bolts are loosened.

To install:

10. Install the cylinder head with a new gasket. Be sure to pay attention to the following points:

- Be sure the No. 1 cylinder is at top dead center and the camshaft pulley **UP** mark is on the top before positioning the head in place.
- The cylinder head dowel pins and oil control orifice must be cleaned and aligned.
- Replace the washer if damaged or deteriorated.
- Apply engine oil to the cylinder head bolts and the washers.
- Use the longer cylinder head bolts at the No. 1 and No. 2 positions.

11. Tighten the cylinder head bolts in 2 steps. In the first step tighten all bolts in sequence to 22 ft. lbs. (29 Nm), then in the second step tighten all bolts in the same sequence to 63 ft. lbs. (85 Nm).

12. Install or connect the following:

- Intake manifold with a new gasket and torque the bolts in a criss–cross pattern to 17 ft. lbs. (24 Nm)
- Intake manifold bracket and torque the bolts to 17 ft. lbs. (24 Nm)
- Exhaust manifold and tighten the new self–locking nuts in a criss-cross pattern in to 23 ft. lbs. (31 Nm)
- Exhaust pipe with a new gasket and tighten the new nuts to 40 ft. lbs. (54 Nm)

13. Be sure that the keyways on the camshafts are facing up and that the rocker arms are in their original position. The valve locknuts should be loosened and the adjusting screw backed off before installation.

14. Place the rocker arms on the pivot bolts and the valve stems.

15. Install the camshafts, then install the camshaft seals with the open side facing in. Install the rubber cap with liquid gasket applied. If the rubber cap has 2 horizontal marks, align the marks with the cylinder head upper surface.

16. Apply liquid gasket to the cylinder head mating surfaces of the No. 1 and No. 6 intake and exhaust camshaft holders and install them, along with No. 2, 3, 4 and 5. Be sure to pay attention to the following points:

17. Do not apply oil to the holder mating surface of camshaft seals.

➡•I or E marks are stamped on the camshaft holders.

18. The arrows marked on the camshaft holders should point to the timing belt.

19. Tighten the camshaft holders temporarily. Be sure that the rocker arms are properly positioned on the valve stems.

20. Tighten each bolt in 2 steps to ensure that the rockers do not bind on the valves. Tighten the 6mm bolts to 86 inch lbs. (9.8 Nm) and the 8mm bolts to 20 ft. lbs. (27 Nm) working from the middle outward.

21. Install the keys into the camshaft grooves. To set the No. 1 piston at TDC, align the holes on the camshaft with the holes in the No. 1 camshaft holders and insert 5.0mm pin punches into the holes.

22. Install or connect the following:

- Rear timing belt cover and torque the bolts to 84 inch lbs. (9.5 Nm)
- Camshaft pulleys and tighten the retaining bolts to 27 ft. lbs. (37 Nm) (B18B1 engine) or 41 ft. lbs. (56 Nm) (B18C1 and B18C5 engines)
- Timing belt and adjust the tension
- Timing belt cover(s) and torque the bolts to 84 inch lbs. (9.5 Nm)

23. Adjust the valve clearance.

- Cylinder head cover and torque the nuts to 86 inch lbs. (9.8 Nm)
- Engine side mount and torque the mounting bolts to 38 ft. lbs. (52 Nm) and the through bolt to 54 ft. lbs. (74 Nm)
- Crankshaft pulley and torque the bolt to 130 ft. lbs. (177 Nm)

24. Connect the negative battery cable and enter the radio security code.

25. After installation, check to see that all hoses and wires are installed correctly.

26. Fill and bleed the air from the cooling system.

27. Attach the negative battery cable.

28. Enter the radio security code.

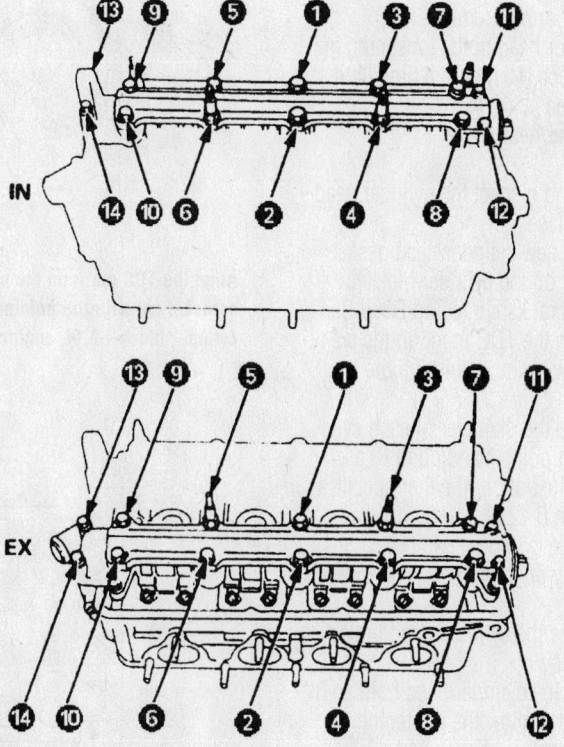

❶ – ❿: 8 x 1.25 mm 27 N·m (2.8 kgf·m, 20 lbf·ft)
⓫ – ⓮: 6 x 1.0 mm 9.8 N·m (1.0 kgf·m, 7.2 lbf·ft)

7923BG09

Camshaft plate torque sequence—1.8L engines

2.0L Engines

1. Before servicing the vehicle, refer to the precautions in the beginning of this section.

2. Disconnect the negative battery cable.

3. Drain the coolant.

4. Relieve the fuel system pressure.

5. Remove or disconnect the following:
   - Fuel feed hose, K20A3 engine only
   - Drive belt
   - Intake manifold
   - Water bypass hose
   - Exhaust manifold
   - Timing chain

6. Disconnect the following engine wire harness connectors and wire harness clamps from the cylinder head:
   - Fuel injector connectors, K20A3 engine
   - Engine Coolant Temperature (ECT) sensor connector
   - Exhaust and intake Camshaft Position (CMP) sensor connectors

7. Remove or disconnect the following:
   - Upper radiator hose and heater hose
   - Harness holder from the bracket
   - Connecting pipe mounting bolt and water bypass line mounting bolts
   - Water bypass hose
   - Rocker arm assembly
   - Cylinder head bolts, loosening in sequence, ⅓ turn at a time until removed
   - Cylinder head

To install:

8. Clean and inspect the cylinder head and mating surfaces.

9. Install a new cylinder head gasket and dowel pins on the cylinder block

10. Set the crankshaft to Top Dead Center (TDC). Align the TDC mark on the crankshaft sprocket with the pointer on the cylinder block.

11. Measure the diameter of each cylinder head bolt at point A and point B, as shown in the illustration. If either specification is less than 0.42 in. (10.6mm), you must replace the cylinder head bolt.

12. Carefully position the cylinder heads on the engine.

13. Lubricate the cylinder head bolts with clean engine oil.

14. Tighten the cylinder head bolts in 3 steps. Be sure to follow the tightening torque sequence:
   a. Step 1: 29 ft. lbs. (39 Nm).
   b. Step 2: 90 degrees.
   c. Step 3: 90 degrees.
   d. Step 4 (new bolts only): 90 degrees.

15. Install or connect the following:
   - Rocker arm assembly
   - Water bypass hose
   - Connecting pipe mounting bolt and water bypass line mounting bolts
   - Harness holder on the bracket
   - Upper radiator hose and heater hose
   - Water bypass hose
   - Intake manifold
   - Exhaust manifold
   - Timing chain
   - Fuel feed hose, K20A3 engine only

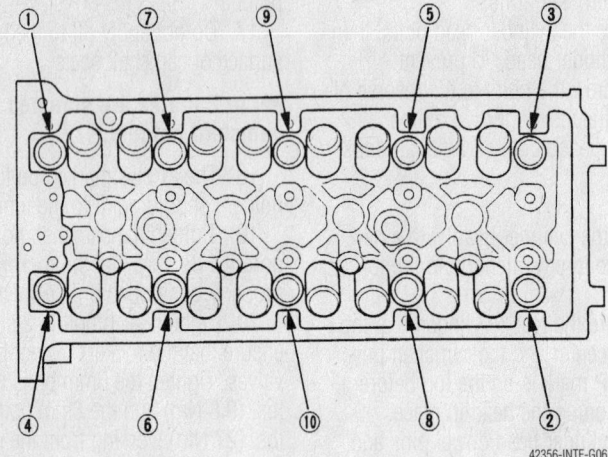

Cylinder head bolt loosening sequence—2.0L engines

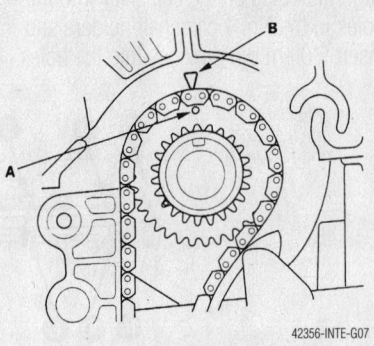

Align the TDC mark on the crankshaft sprocket (A) with the pointer (B) on the cylinder block—2.0L engines

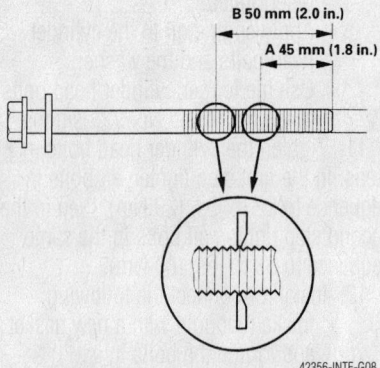

Measure the diameter of the head bolts to determine if they are reusable—2.0L engine

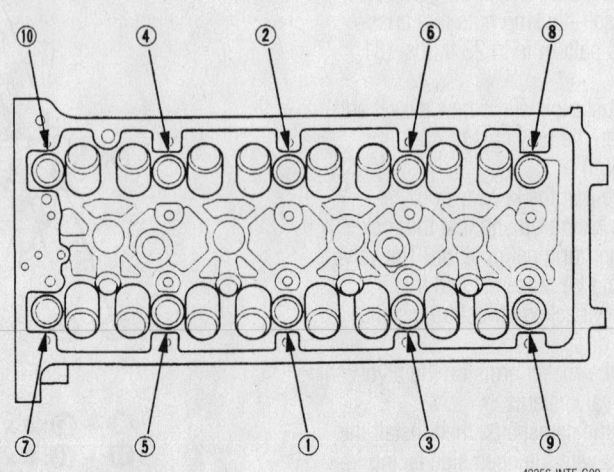

Cylinder head bolt tightening sequence—2.0L engines

16. Adjust the valve clearance.
  • Drive belt
  • Negative battery cable and enter the radio security code.
17. After installation, check to see that all hoses and wires are installed correctly.
18. Fill and bleed the air from the cooling system.
19. Attach the negative battery cable.
20. Enter the radio security code.

3.0L, 3.2L and 3.5L Engines

1. Before servicing the vehicle, refer to the precautions in the beginning of this section.
2. Disconnect the negative battery cable.
3. Drain the coolant.
4. Relieve the fuel system pressure.
5. Remove or disconnect the following:
  • Engine covers
  • Strut brace
  • Water bypass hose
  • Traction Control System (TCS) control valve from the throttle body (if equipped)
  • Evaporative Emissions (EVAP) canister hose from the throttle body
  • Intake air duct
  • Upper engine covers
  • Accelerator and cruise control cables from the throttle body
  • Fuel feed and return hoses
  • Brake booster vacuum hose
  • PCV hose
  • Intake Manifold Runner Control (IMRC) actuator
  • Wire harness holder
  • Side engine mount bracket
  • Accessory drive belts
  • Power steering pump without disconnecting the lines
  • Ground cable from the engine
  • Alternator
  • Spark plug wires
  • Distributor
  • Intake Air Temperature (IAT) sensor connector
  • Idle Air Control (IAC) valve connector
  • Throttle Position (TPS) sensor connector
  • Manifold Absolute Pressure (MAP) sensor connector
  • Engine Coolant Temperature (ECT) sensor connector
  • Radiator fan switch connectors
  • Crankshaft Position (CKP) sensor connector
  • Top Dead Center (TDC) sensor connector

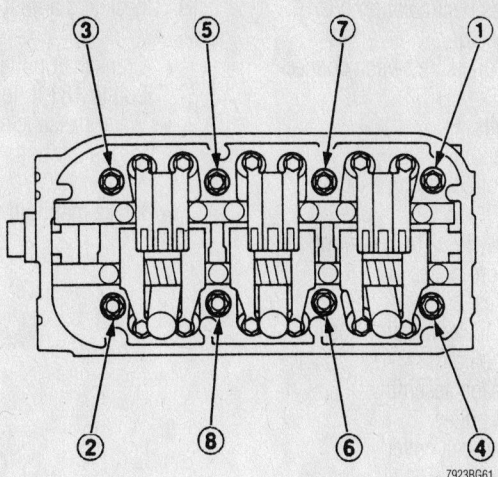
Loosen the cylinder head bolts in the sequence shown to prevent damage to the head—3.0L engine

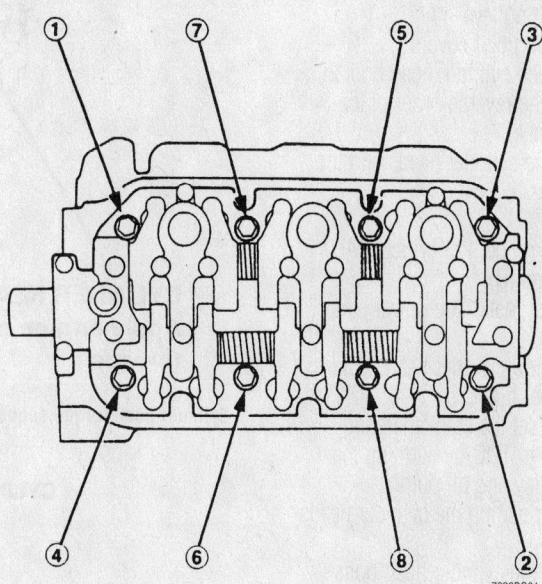
Loosen the cylinder head bolts in the sequence shown—3.5L engine

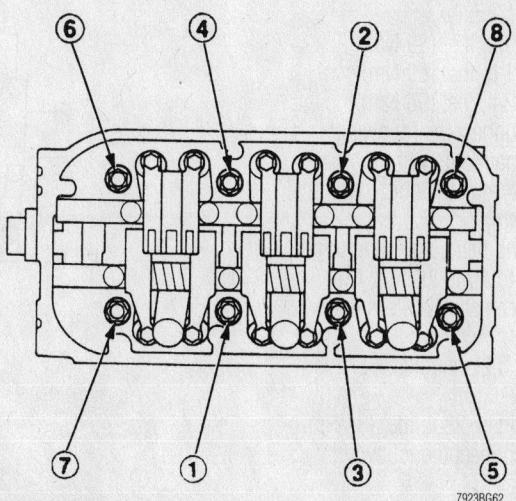
Cylinder head torque sequence—3.0L engine

- Exhaust Gas Recirculation (EGR) valve connector
- Engine oil pressure switch connector
- Ignition coils
- Intake manifold
- Fuel injector connectors
- Fuel rails
- Intake Air Bypass (IAB) control valve vacuum hoses
- Heater hoses
- Upper and lower radiator hoses
- Exhaust manifolds
- Water passage assembly
- Crankshaft pulley
- Front timing belt cover

6. Set the engine to TDC by aligning the marks on the crankshaft and camshaft pulleys.

- Timing belt
- Camshaft pulleys
- Rear timing belt covers

7. Loosen each cylinder head bolt ⅓ turn at a time in the reverse order of the tightening sequence.

8. Remove the cylinder heads and the oil control orifices.

To install:

9. Install the oil control orifices (A) using new O-rings (B).

10. If removed, install the dowel pins (C).

11. Position new cylinder head gaskets (D) on the cylinder block.

12. If moved, set the crankshaft and camshaft pulleys to TDC by aligning the marks on the pulley and oil pump.

13. Carefully position the cylinder heads on the engine.

14. Lubricate the cylinder head bolts with clean engine oil.

15. Tighten the cylinder head bolts in 3 steps. Be sure to follow the tightening torque sequence:

    a. Step 1: 29 ft. lbs. (39 Nm).
    b. Step 2: 51 ft. lbs. (69 Nm).
    c. Step 3: 72 ft. lbs. (98 Nm).

16. Install or connect the following:

- Exhaust manifolds with new gaskets and torque the nuts to 23 ft. lbs. (31 Nm)
- Rear timing belt covers and torque the bolts to 16 ft. lbs. (22 Nm)

17. Install the camshaft pulleys and torque to:

    a. 3.0L and 3.2L engines: 67 ft. lbs. (90 Nm).
    b. 3.5L engines: 23 ft. lbs. (31 Nm.

18. Install or connect the following:

- Timing belt
- Front timing belt cover and torque the bolts to 108 inch lbs. (12 Nm)

19. Check and adjust the valve clearance.

- Crankshaft pulley and torque the bolt to 181 ft. lbs. (245 Nm)
- Water passage assembly and torque the bolts to 16 ft. lbs. (22 Nm)
- Intake manifold with new gaskets

and O-rings and tighten the bolts to 16 ft. lbs. (22 Nm)

- Cylinder head cover and tighten the bolts to 108 inch lbs. (12 Nm)
- Ignition coils and torque the bolts to 108 inch lbs. (12 Nm)
- Exhaust manifolds and torque the bolts to 23 ft. lbs. (31 Nm)

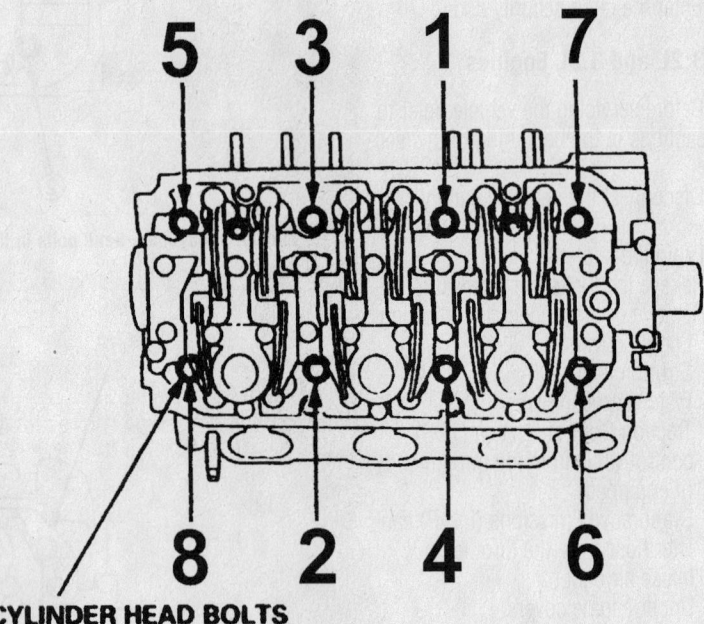

CYLINDER HEAD BOLTS
Apply engine oil to the bolt threads.

7923BG12

Cylinder head torque sequence—3.2L engine

CYLINDER HEAD BOLTS TORQUE SEQUENCE

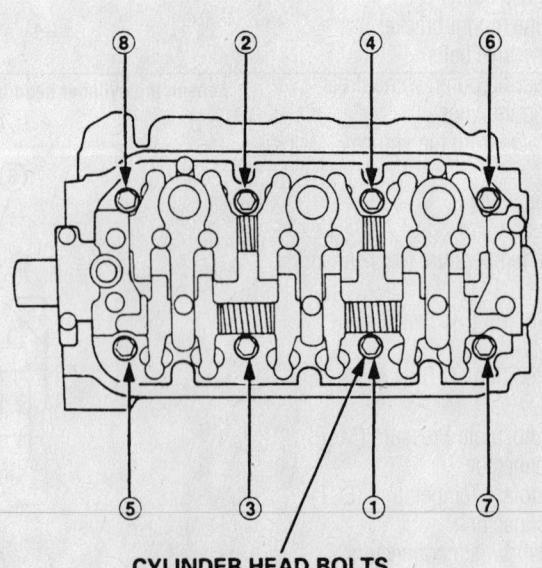

CYLINDER HEAD BOLTS
Apply engine oil to the bolt threads.

7923BG82

Cylinder head torque sequence—3.5L engine

- Upper and lower radiator hoses
- Heater hoses
- IAB vacuum hoses
- Fuel rails and torque the bolts to 84 inch lbs. (9.5 Nm)
- Fuel injector connectors
- Intake manifold and torque the bolts to 16 ft. lbs. (22 Nm)
- Engine oil pressure switch connector
- EGR valve connector
- TDC sensor connector
- CKP sensor connector
- Radiator fan switch connectors
- MAP sensor connector
- ECT sensor connector
- TPS sensor connector
- IAC sensor connector
- IAT sensor connector
- Spark plug wires
- Distributor
- Alternator and torque the upper bolt to 16 ft. lbs. (22 Nm) and the lower bolt to 33 ft. lbs. (44 Nm)
- Ground cable

- Power steering pump and torque the bolts to 17 ft. lbs. (24 Nm)
- Accessory drive belts
- Side engine mount and torque the mounting bolts to 33 ft. lbs. (44 Nm) and the through bolt to 40 ft. lbs. (54 Nm)
- Wire harness holder
- IMRC actuator
- PCV hose
- Brake booster vacuum hose
- Fuel feed and return lines and torque the fitting to 16 ft. lbs. (22 Nm)
- Accelerator and cruise control cables
- Engine covers
- Intake air duct
- Evaporative Emissions (EVAP) canister hose from the throttle body
- TCS control valve
- Strut brace and torque the bolts to 16 ft. lbs. (22 Nm)
- Engine covers and torque the bolts to 108 inch lbs. (12 Nm)

20. Change the engine oil and filter.
21. Fill and bleed the cooling system.
22. Connect the negative battery cable.
23. Bring the engine to operating temperature and inspect for any fluid leaks. Top off all fluid levels as necessary.
24. Enter the security code for the radio.

➥The PCM idle memory must be reset after reconnecting the battery. Start the engine and hold it at 3000 rpm until the cooling fan comes on. Then allow the engine to idle for about 5 minutes with all accessories OFF and with the transmission in Park or Neutral.

Rocker Arms/Shafts

REMOVAL & INSTALLATION

1.8L (B18B1) Engine

1. Before servicing the vehicle, refer to the precautions in the beginning of this section.

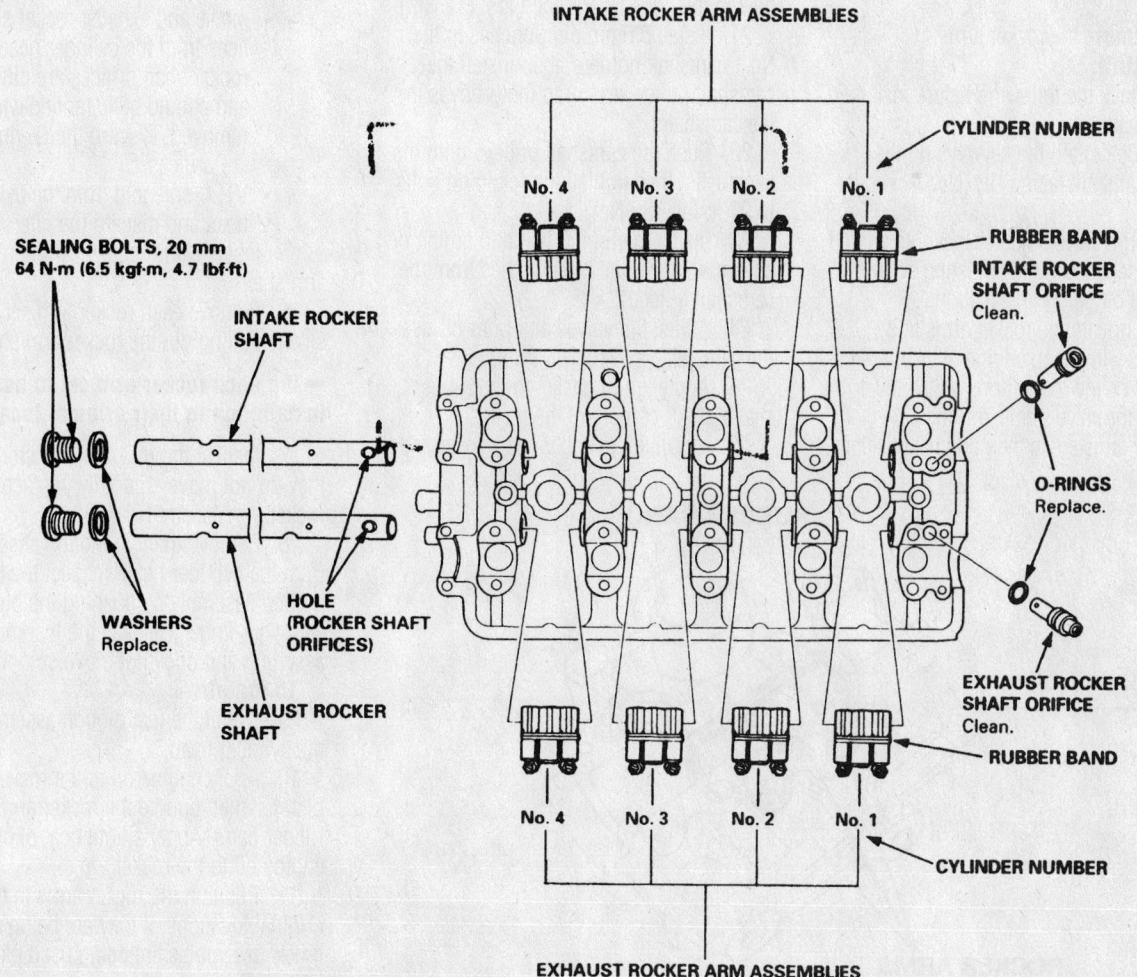

Rocker arms and shafts—1.8L (B18B1, B18C1, and B18C5) engines

7923BG13

2. Remove or disconnect the following:
- Negative battery cable
- Spark plug wires
- Cylinder head cover
- Timing belt cover
- Timing belt
- Distributor

3. Install 5.0mm pin punches to the No.1 camshaft holders, then remove the camshaft sprockets.

4. Loosen the valve adjusters to remove as much spring tension as possible.

5. Remove the pin punches from the camshaft holders.

6. To remove the camshaft bearing caps, loosen each bolt 2 turns at a time in a crisscross pattern to avoid damage to the valves or rockers. Mark the caps so they can be replaced in their original position.

7. Lift the camshafts from the cylinder head, wipe them clean and inspect the lift ramps. Replace the camshafts and rockers if the lobes are pitted, scored or excessively worn.

8. Label the rocker arms before removing so they can be installed in their original locations.

9. Remove the rocker arms.

To install:

10. Check the following before installing the camshafts:

  a. Be certain the keyways on the camshafts are facing UP (No. 1 cylinder at TDC).

  b. The valve adjuster locknuts should be loosened and the adjusting screws backed off before installation.

11. Lubricate the rocker arms and camshafts with clean engine oil.

12. Place the rocker arms on the pivot bolts and the valve stems, making sure that the rocker arms are in their original positions.

13. Install the camshaft seals with the open side (spring) facing in. Lubricate the lip of the seal.

14. Be sure the keyways on the camshafts are facing up and install the camshafts to the cylinder head.

15. Apply liquid gasket to the head mating surfaces of the No. 1 and No. 6 camshaft holders, then install them along with Nos. 2, 3, 4 and 5 camshaft holders. The arrows stamped on the holders should point toward the timing belt. Do not apply oil to the holder mating surface where the camshaft seals are housed.

16. Tighten the camshaft holders temporarily and be sure that the rocker arms are properly positioned.

17. Press the oil seals into the No.1 camshaft holders with a seal driver.

18. Tighten the bolts in a crisscross pattern to 84 inch lbs. (10 Nm). Check that the rockers do not bind on the valves.

19. Install the cylinder head plug to the end of the cylinder head. If the plug has alignment marks, align the marks.

20. Install the rear timing belt cover and tighten the bolts to 108 inch lbs. (12 Nm).

21. Install 5.0mm pin punches to the No.1 camshaft holders, then install the camshaft pulley keys onto the grooves in the camshafts.

22. Push the camshaft pulleys onto the camshafts, then tighten the retaining bolts to 27 ft. lbs. (38 Nm).

23. Install the timing belt and timing belt covers. Remove the pin punches from the camshaft holders.

24. Adjust the valves and pour oil over the camshafts and rocker arms.

25. Apply liquid gasket to the rubber seal at the 8 corners of the recesses.

26. Install the cylinder head cover and engine ground cable. Be sure the contact surfaces are clean and do not touch surfaces where liquid gasket has been applied.

27. Tighten the cylinder head cover nuts in 2–3 steps. In the final step, tighten all nuts in sequence, to 84 inch lbs. (10 Nm).

28. Install the distributor to the cylinder head and reconnect the spark plug wires to the spark plugs.

29. Connect the negative battery cable and enter the radio security code.

30. Change the engine oil. Wait at least 20 minutes for the sealant to cure before filling the engine with oil.

1.8L (B18C1, B18C5) Engine and NSX

1. Before servicing the vehicle, refer to the precautions in the beginning of this section.

2. Remove or disconnect the following:
- Negative battery cable
- Cylinder head from the vehicle

3. Hold each rocker arm assembly together with a rubber band to prevent them from separating.

4. Remove or disconnect the following:
- Intake and exhaust rocker shaft orifices from the cylinder head—The rocker shaft orifices are different and should be identified when removed. Discard the O–rings on the orifices.
- VTEC solenoid from the cylinder head and discard the filter

5. Insert 12mm bolts into the rocker arm shafts.

6. Remove each rocker arm set while slowly pulling out the rocker arm shaft.

➡Tag each rocker arm set to assure installation in their original locations.

7. Inspect the rocker arm pistons. If they do not move smoothly, replace the rocker arm assembly.

8. Remove the lost motion assembly from the cylinder head. Inspect the lost motion assembly by pushing the plunger with your finger. Replace the lost motion assembly if it does not move smoothly.

To install:

9. Install the lost motion assembly to the cylinder head.

10. Apply engine oil to the rocker arm pistons, then bundle the rocker arms with a rubber band. Apply a light coat of clean engine oil to the rocker arms.

11. Position the rocker arms in their original locations, if they are being reused. If new assembles are being used place them in the cylinder head.

12. Lightly coat the rocker arm shafts

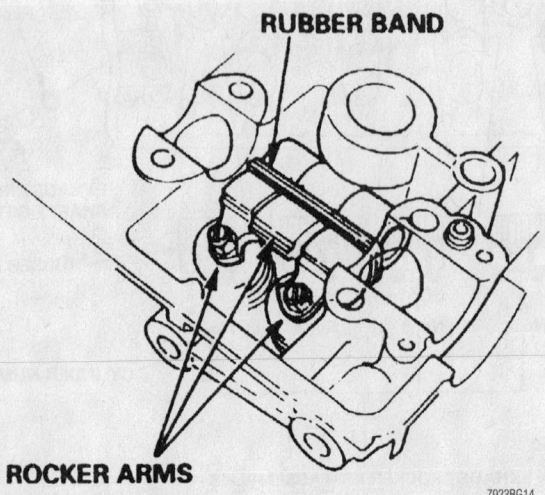

RUBBER BAND

ROCKER ARMS

7923BG14

Rocker arms with rubber band installed—1.8L (B18B1, B18C1 and B18C5) engines

with clean engine oil, then install the rocker arm shafts into the cylinder head. A 12mm bolt can be installed into the end of the rocker arm shafts to aid in their installation. Be sure to install the shafts in the proper positions. Remove the 12mm bolts from the rocker arm shafts.

13. Clean and install the rocker arm shaft orifices with new O−rings. If the holes in the rocker arm shafts are not aligned screw a 12mm bolt into the end of the shaft to position the it.

14. Install the sealing bolts with new washers and tighten them to 47 ft. lbs. (64 Nm).

15. Install the cylinder head into the vehicle.

2.0L Engines

1. Before servicing the vehicle, refer to the precautions in the beginning of this section.

2. Remove or disconnect the following:
   - Negative battery cable
   - Cylinder head from the vehicle
   - Loosen the rocker arm adjusting screws
   - Camshaft holder bolts, 2 turns at a time in a criss−cross pattern
   - Timing chain guide B, camshaft holders and camshafts

3. Insert the bolts into the rocker shaft holder, then remove the rocker arm assembly.

4. Disassemble the rocker shafts as necessary.

To install:

5. Clean and dry the No. 5 rocker shaft holder mating surface.

6. Apply liquid gasket part No. 08718−0009 evenly to the cylinder head mating surface of the No. 5 rocker shaft holder and install within 5 minutes.

7. Reassemble the rocker arm assembly.

8. Insert the bolts into the rocker shaft holder, then install the rocker arm assembly on the cylinder head. Remove the bolts from the rocker shaft holder.

9. Make sure the punch marks on the VTC actuator and exhaust camshaft sprocket are facing up, then set the camshafts in the holder.

10. Set the camshaft holders and timing chain guide B in place.

11. Tighten the bolts, in sequence, to the following specifications:
   - 8mm bolts: 16 ft. lbs. (22 Nm)
   - 6mm bolts (nos. 21, 22 & 23): 8.7 ft. lbs. (12 Nm)

12. Install or connect the following:
   - Timing chain, then adjust the valve clearance

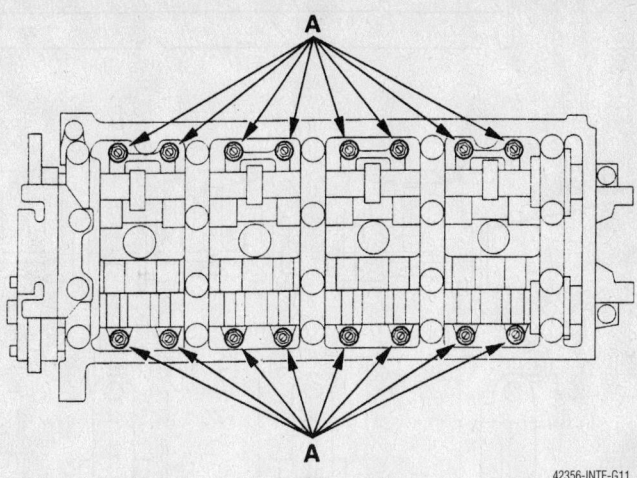

42356-INTE-G11

Loosen the rocker arm adjusting screws (A)—2.0L engines

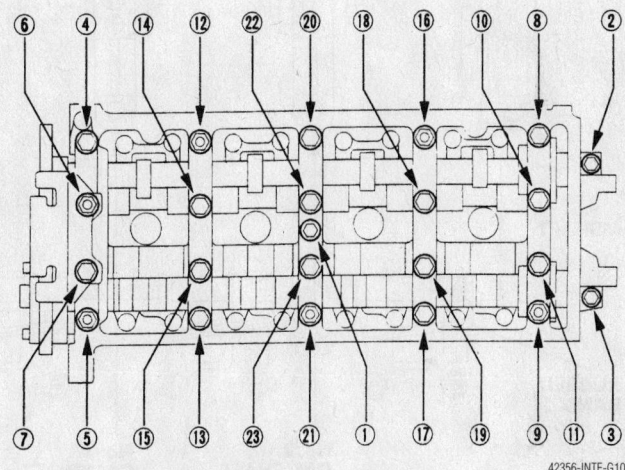

42356-INTE-G10

Camshaft holder bolt loosening sequence—2.0L engines

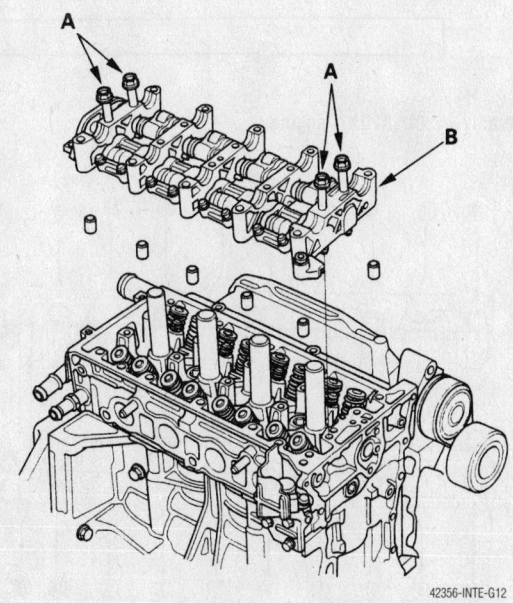

42356-INTE-G12

Insert the bolts (A) into the rocker shaft holder, then remove the rocker arm assembly (B)— 2.0L engines

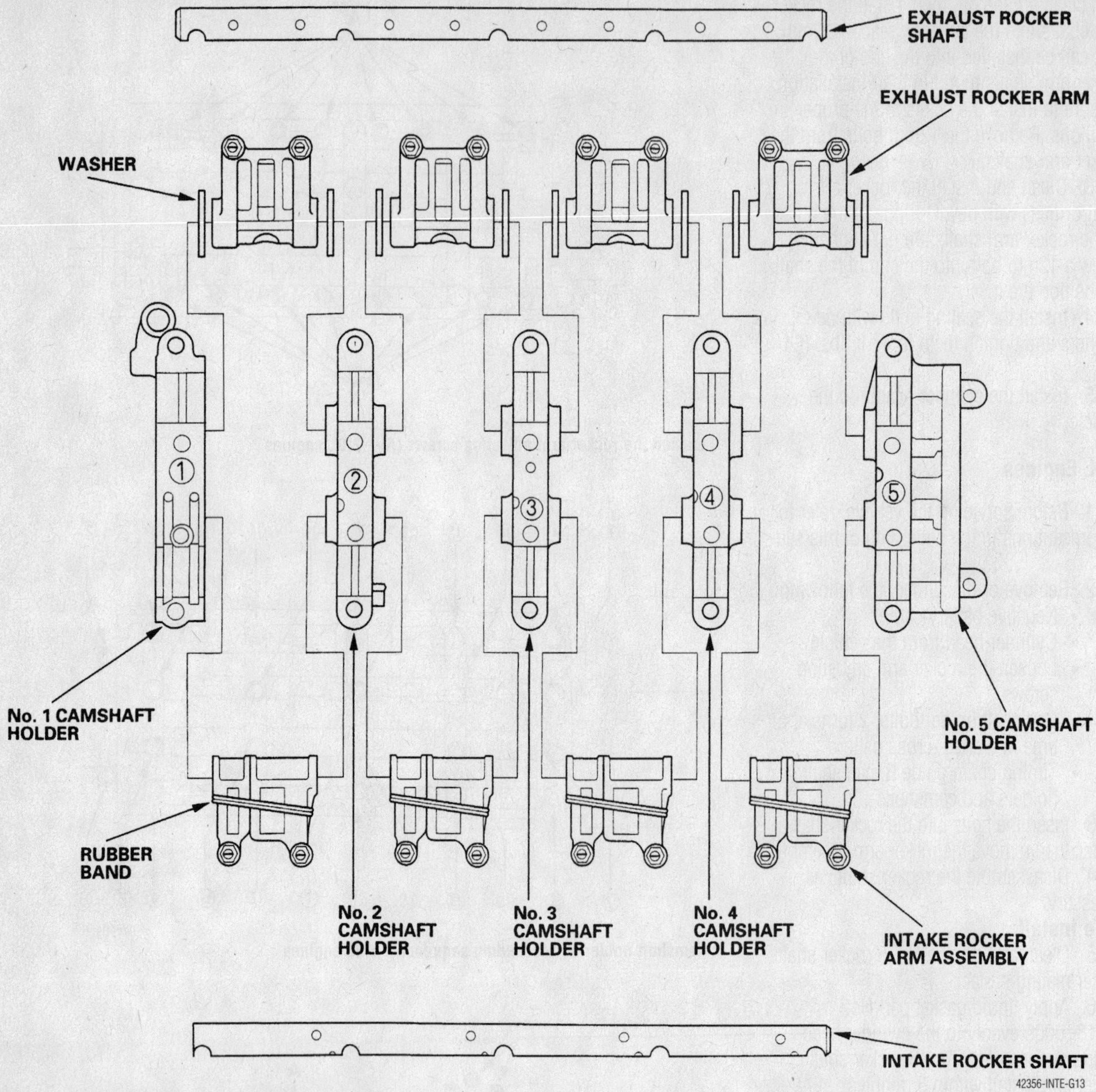

EXHAUST ROCKER SHAFT

EXHAUST ROCKER ARM

WASHER

No. 1 CAMSHAFT HOLDER

No. 5 CAMSHAFT HOLDER

RUBBER BAND

No. 2 CAMSHAFT HOLDER

No. 3 CAMSHAFT HOLDER

No. 4 CAMSHAFT HOLDER

INTAKE ROCKER ARM ASSEMBLY

INTAKE ROCKER SHAFT

42356-INTE-G13

Rocker arm disassembly—2.0L (K20A3) engine

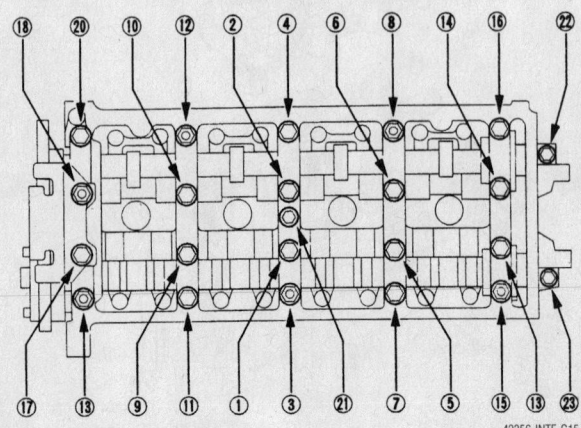

42356-INTE-G15

Rocker arm assembly tightening sequence—2.0L engines

- Cylinder head
- Negative battery cable

13. Check for proper engine and valve train operation.

3.0L, 3.2L and 3.5L Engines (except NSX)

1. Before servicing the vehicle, refer to the precautions in the beginning of this section.

2. Disconnect the negative battery cable.

3. Remove the cylinder head from the vehicle.

4. Loosen the rocker shaft holder bolts 1 turn at a time in the opposite of the instal-

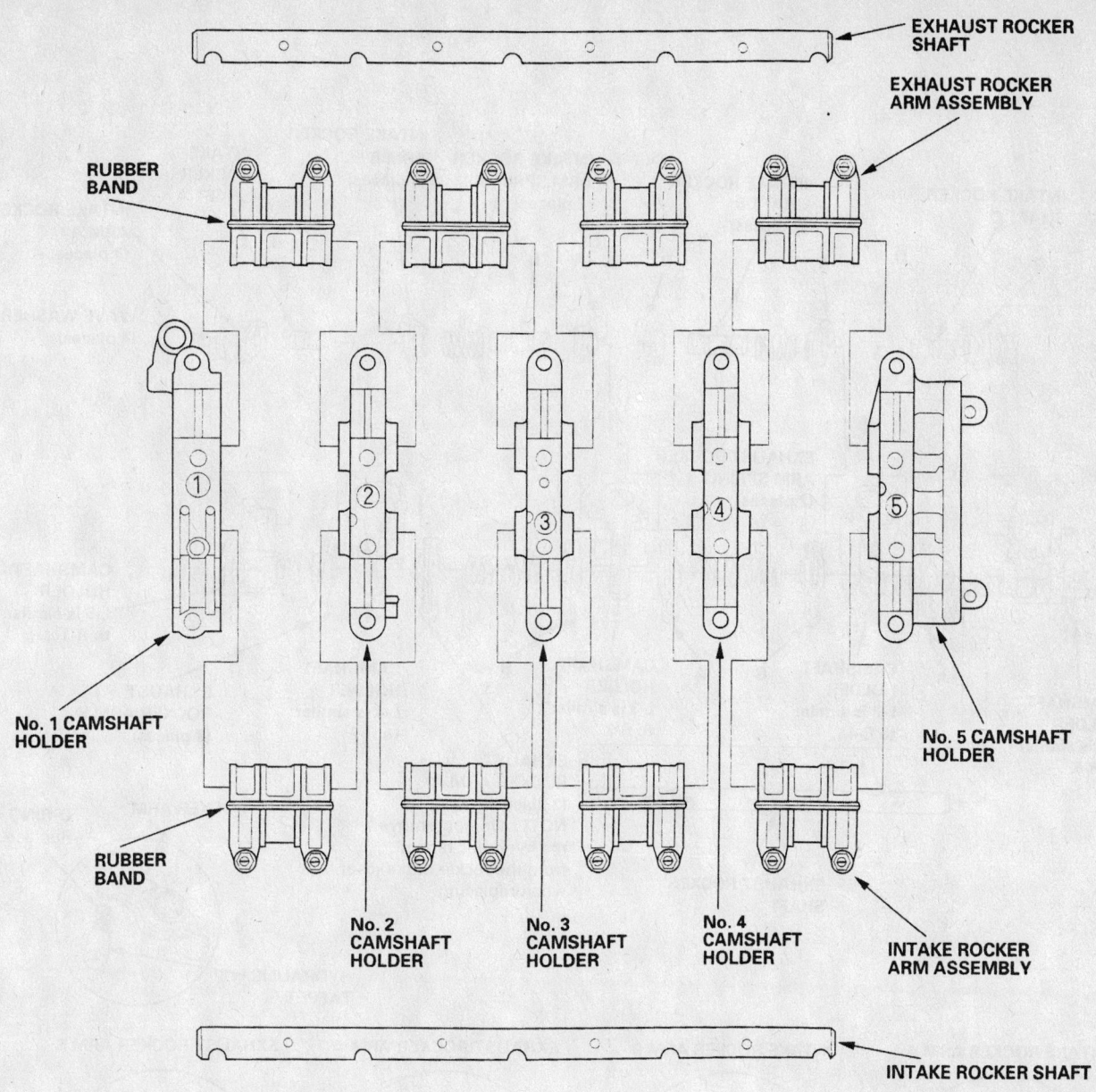

EXHAUST ROCKER SHAFT

EXHAUST ROCKER ARM ASSEMBLY

RUBBER BAND

No. 1 CAMSHAFT HOLDER

No. 5 CAMSHAFT HOLDER

RUBBER BAND

No. 2 CAMSHAFT HOLDER

No. 3 CAMSHAFT HOLDER

No. 4 CAMSHAFT HOLDER

INTAKE ROCKER ARM ASSEMBLY

INTAKE ROCKER SHAFT

42356-INTE-G14

Rocker arm disassembly—2.0L (K20A2) engine

lation sequence. Following this procedure will prevent the camshafts and rocker assemblies from warping.

5. After all bolts are loose, remove the rocker arm shafts as an assembly with the bolts still in the holders.

6. If the rocker shafts are to be disassembled, note that each rocker arm has a letter **A** or **B** stamped into the side. Before disassembling the rocker arms, make a note of the position of each letter so the arms can be reassembled the same way.

7. For 3.2L and 3.5L engines, do not remove the hydraulic tappets from the rocker arms unless they are to be replaced. Handle the rocker arms carefully so the oil does not drain out of the tappets.

8. Lift the camshafts from the cylinder head, wipe them clean and inspect the lift ramps. Replace the camshafts and rockers if the lobes are pitted, scored, or excessively worn.

To install:

9. Lubricate the camshaft and its journals with fresh engine oil.

10. Place a new camshaft seal on the end of the camshaft. The spring side of the seal must face in. Lubricate the journals and set the camshaft in place on the head.

11. Install the camshaft onto the cylinder head with the keyway pointed up.

12. Apply liquid gasket to the mounting surfaces of the camshaft end holders.

13. Set the rocker arm assemblies in

place and start all the cam holder bolts. Be sure the rocker arms are properly positioned and turn each bolt in sequence 2 turns at a time until the holders are seated on the head. Follow this procedure to avoid damaging the camshaft and rocker assemblies.

14. When all the camshaft and rocker holders are seated, tighten the bolts in the same sequence. Tighten the 8mm bolts to 16 ft. lbs. (22 Nm) and the 6mm bolts to 104 inch lbs. (12 Nm).

15. Install or connect the following:
   • Cylinder head
   • Negative battery cable

16. Check for proper engine and valve train operation.

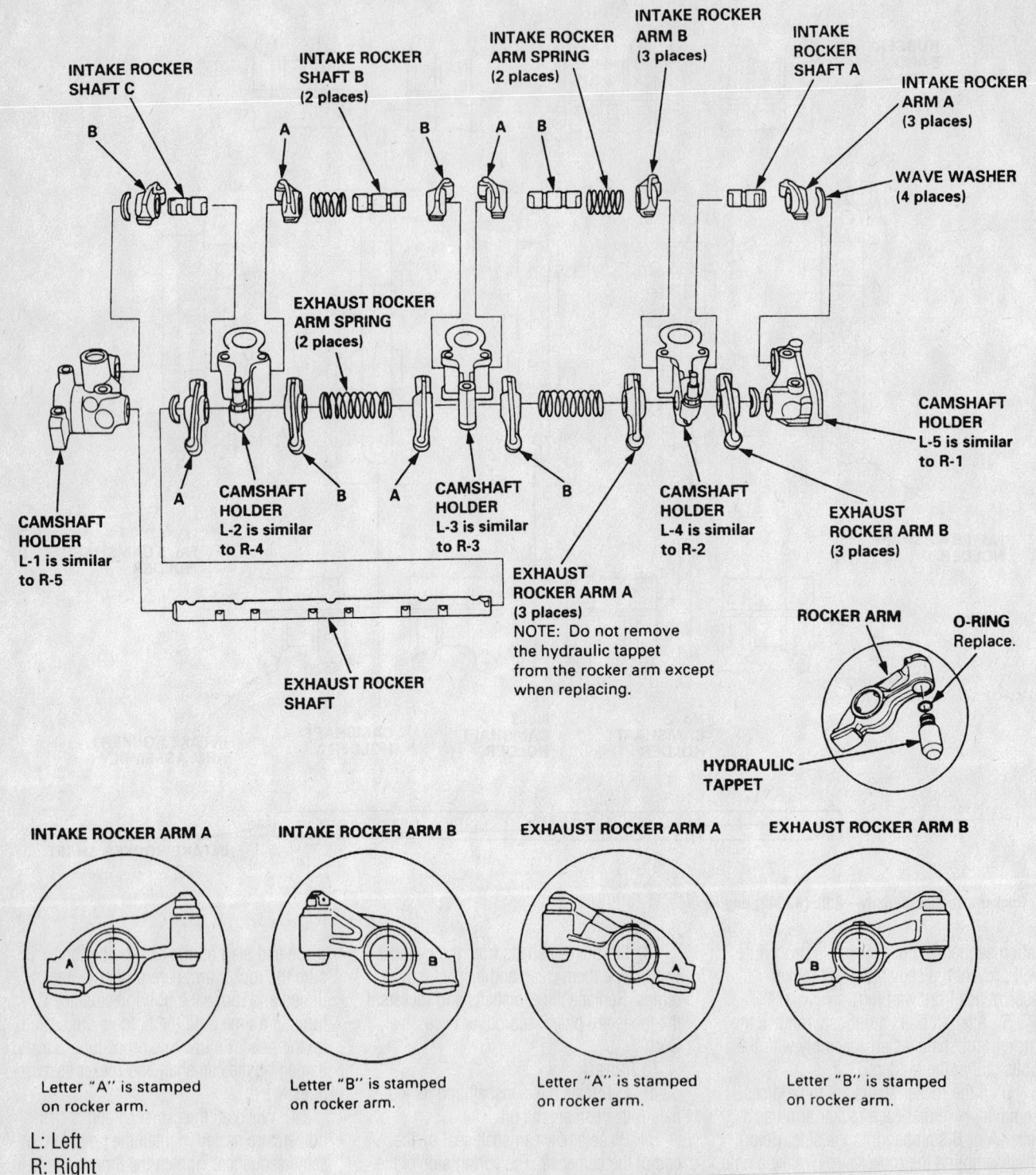

Exploded view of the rocker arms and related components—3.5L engine

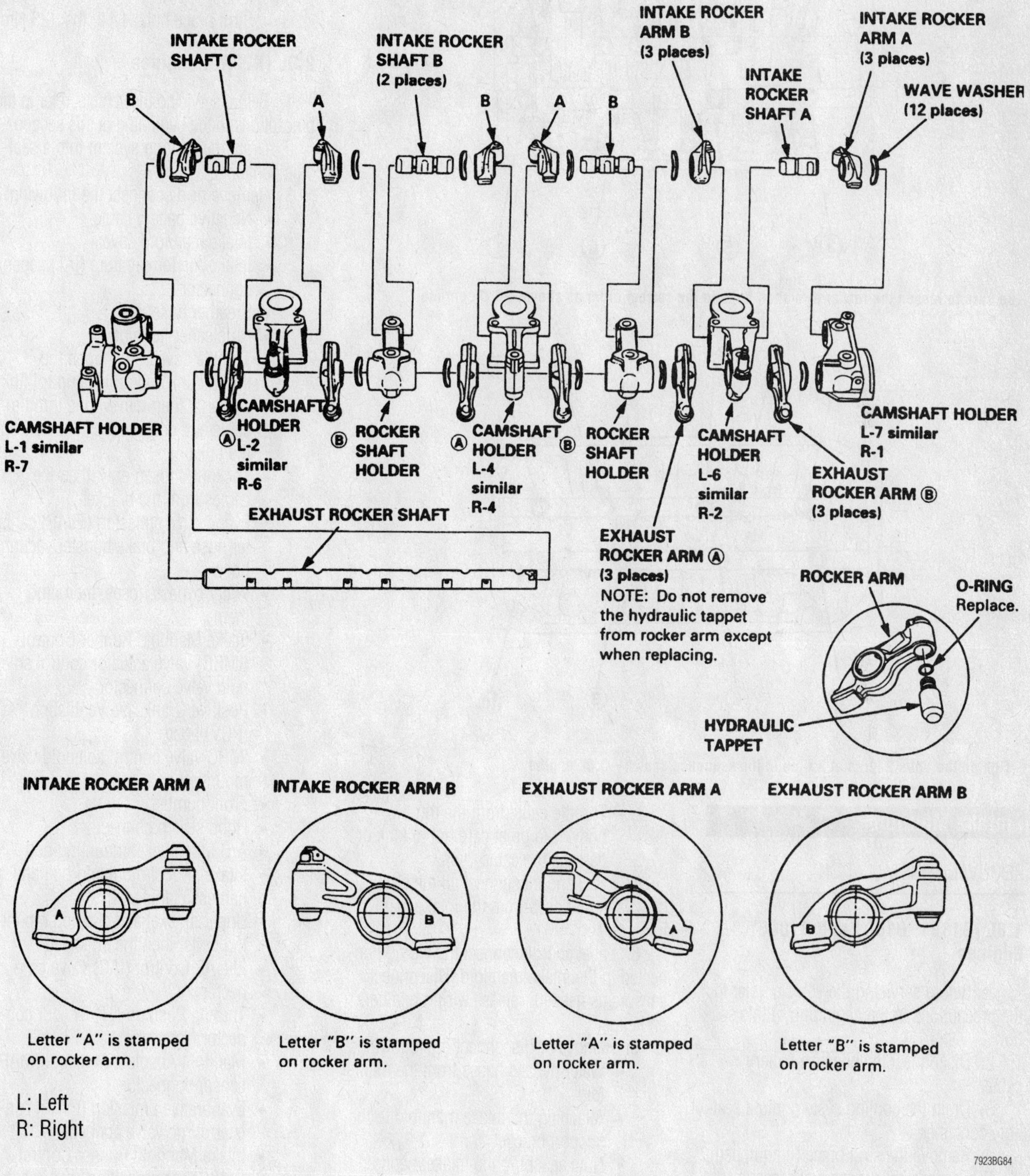

INTAKE ROCKER
SHAFT C

INTAKE ROCKER
SHAFT B
(2 places)

INTAKE ROCKER
ARM B
(3 places)

INTAKE
ROCKER
SHAFT A

INTAKE ROCKER
ARM A
(3 places)

WAVE WASHER
(12 places)

B   A   B   A   B

CAMSHAFT HOLDER
L-1 similar
R-7

CAMSHAFT
HOLDER
Ⓐ L-2
similar
R-6

Ⓑ

ROCKER
SHAFT
HOLDER

CAMSHAFT
HOLDER
Ⓐ L-4
similar
R-4

Ⓑ

ROCKER
SHAFT
HOLDER

CAMSHAFT
HOLDER
L-6
similar
R-2

CAMSHAFT HOLDER
L-7 similar
R-1

EXHAUST
ROCKER ARM Ⓑ
(3 places)

EXHAUST ROCKER SHAFT

EXHAUST
ROCKER ARM Ⓐ
(3 places)
NOTE: Do not remove
the hydraulic tappet
from rocker arm except
when replacing.

ROCKER ARM

O-RING
Replace.

HYDRAULIC
TAPPET

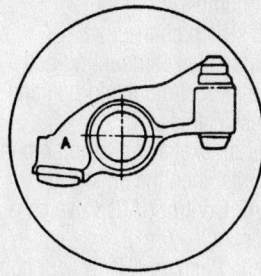

INTAKE ROCKER ARM A

Letter "A" is stamped
on rocker arm.

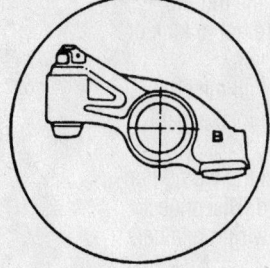

INTAKE ROCKER ARM B

Letter "B" is stamped
on rocker arm.

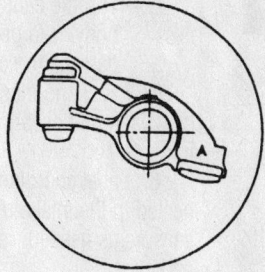

EXHAUST ROCKER ARM A

Letter "A" is stamped
on rocker arm.

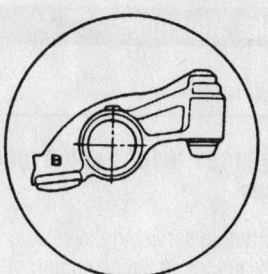

EXHAUST ROCKER ARM B

Letter "B" is stamped
on rocker arm.

L: Left
R: Right

7923BG84

Exploded view of the rocker arms and related components—3.2L engine

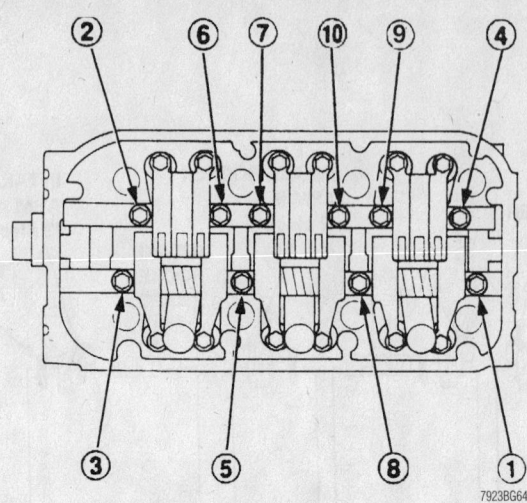

Be sure to loosen the rocker arm shaft bolts in the correct order as shown—3.0L engine

7923BG64

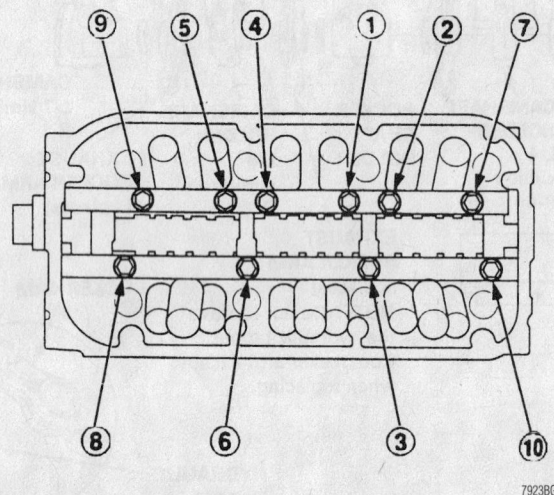

Tighten the bolts 2 turns at a time in the sequence shown—3.0L engine

7923BG65

Intake Manifold

REMOVAL & INSTALLATION

1.8L (B18B1, B18C1 and B18C5) Engines

1. Before servicing the vehicle, refer to the precautions in the beginning of this section.
2. Disconnect the negative battery cable.
3. Drain the cooling system into a sealable container.
4. Remove the strut brace (if equipped).
5. Properly relieve the fuel pressure.
6. Remove or disconnect the following:
   - Intake air duct
   - Fuel feed and return hoses
   - PCV hose
   - Brake booster vacuum hose
   - Throttle cable from the throttle body, take great care not to kink or damage the cable
7. Label and disconnect all the emission vacuum hoses from the intake manifold.
8. Label and disconnect the wiring connected to the intake manifold. Disconnect sensors as needed; release wiring retainers and clips.
9. Remove or disconnect the following:
   - Water bypass hoses from the manifold
   - Attaching the intake manifold to the support bracket
   - Nuts attaching the intake manifold to the cylinder head in a crisscross pattern, beginning from the center and moving out to both ends
   - Intake manifold

To install:

10. Installation is the reverse of the removal procedure, while using the following torque values:
   - Intake manifold nuts: 17 ft. lbs. (23 Nm)
   - Manifold support bracket bolts: 17 ft. lbs. (23 Nm)
   - Strut brace nuts: 17 ft. lbs. (23 Nm)

2.0L (K20A3) Engines

1. Before servicing the vehicle, refer to the precautions in the beginning of this section.
2. Drain the cooling system into a sealable container.
3. Remove or disconnect the following:
   - Negative battery cable
   - Intake manifold cover
   - Intake Air Temperature (IAT) sensor connector
   - Breather hose
   - Air cleaner housing
   - Throttle cover. Fully open the throttle link and cruise control link by hand, then remove the throttle cable and cruise control cable from the links
   - Locknut, loosen only, then the cables from the bracket
   - Evaporative emission (EVAP) canister hose and brake booster vacuum hose
   - Water bypass hoses, then plug them
   - Intake Manifold Runner Control (IMRC) valve actuator control solenoid valve connector
   - Positive Crankcase Ventilation (PCV) hose
   - IMRC valve control solenoid valve mounting bolt
   - Front bumper
   - Hood switch connector
   - A/C line bracket mounting bolt
   - Intake airduct mounting bolt and harness clamps
   - Upper bracket and cushion mounting bolts, then the bulkhead
   - Idle Air Control (IAC) valve connector
   - Throttle Position (TP) sensor connector
   - Manifold Absolute Pressure (MAP) sensor connector
   - Evaporative Emission (EVAP) canister purge valve connector
   - Intake Manifold Runner Control (IMRC) valve position sensor connector
   - Intake manifold and O–rings

To install:

4. Install or connect the following:
   - Intake manifold, with new O–rings.

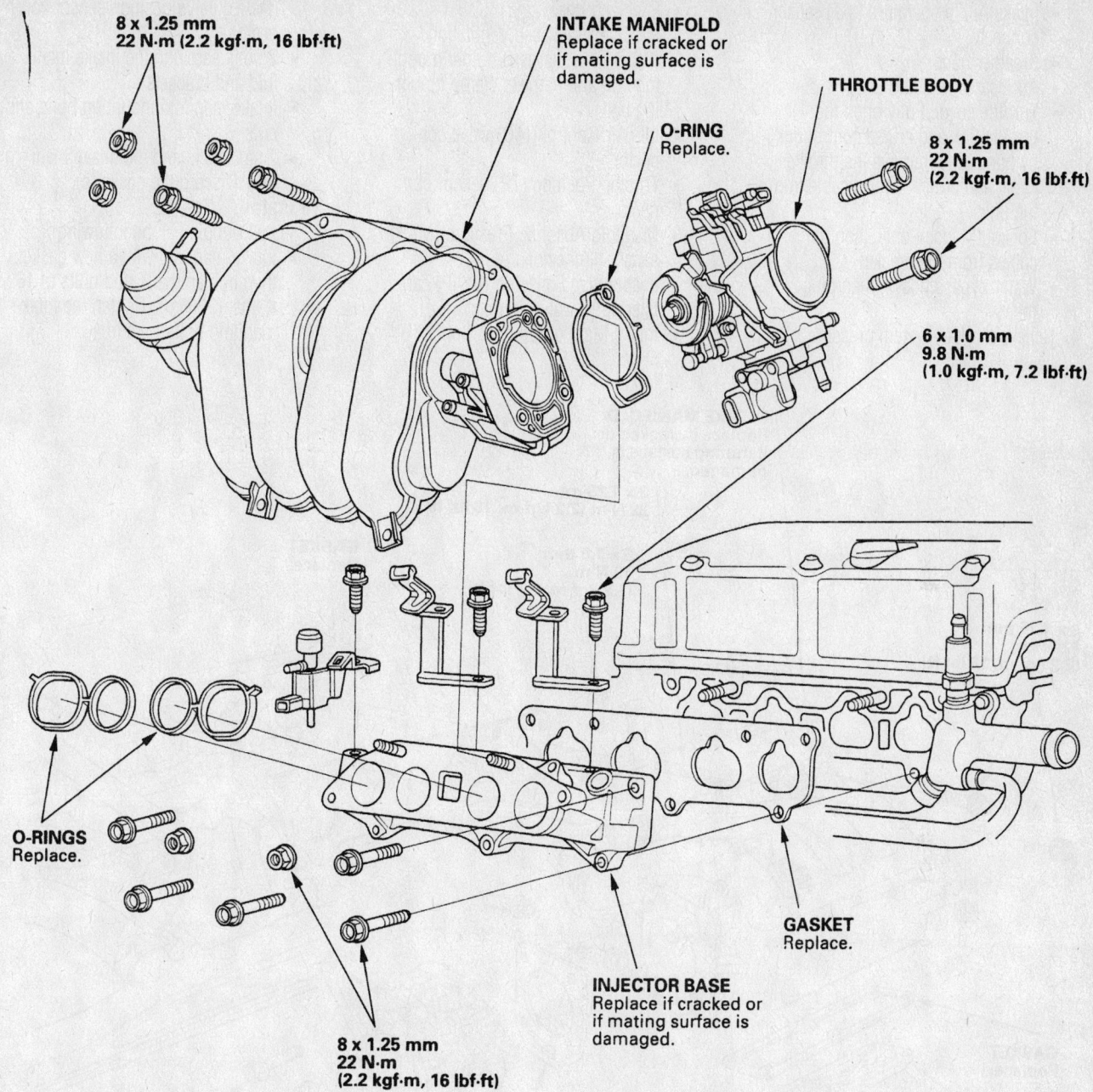

8 x 1.25 mm
22 N·m (2.2 kgf·m, 16 lbf·ft)

INTAKE MANIFOLD
Replace if cracked or
if mating surface is
damaged.

THROTTLE BODY

O-RING
Replace.

8 x 1.25 mm
22 N·m
(2.2 kgf·m, 16 lbf·ft)

6 x 1.0 mm
9.8 N·m
(1.0 kgf·m, 7.2 lbf·ft)

O-RINGS
Replace.

GASKET
Replace.

INJECTOR BASE
Replace if cracked or
if mating surface is
damaged.

8 x 1.25 mm
22 N·m
(2.2 kgf·m, 16 lbf·ft)

42356-INTE-G16

Exploded view of the intake manifold—2.0L (K20A3) engine

Tighten the bolts/nuts in a
criss–cross pattern in 2 or 3 steps
to 16 ft. lbs. (22 Nm).
- Bulkhead, and upper bracket and
cushion mounting bolts
- Hood switch connector
- A/C line bracket mounting bolt and
tighten to 8.7 ft. lbs. (10 Nm)
- Intake airduct mounting bolt and
tighten to 8.7 ft. lbs. (10 Nm)
- Harness clamps
- Front bumper
- IMRC valve actuator control sole-
noid valve connector

- PCV hose and IMRC valve actuator
control solenoid valve mounting
bolt
- Water bypass hoses
- EVAP canister hose and brake
booster vacuum hose
- Throttle and cruise control cables.
Adjust the cables.
- Air cleaner housing
- IAT sensor connector
- Breather hose
- Intake manifold cover and tighten
to 8.7 ft. lbs. (10 Nm)
- Negative battery cable

5. Fill the engine with coolant, then start
the engine and check for leaks.

2.0L (K20A2) Engine

1. Before servicing the vehicle, refer to the
precautions in the beginning of this section.
2. Drain the cooling system into a sealable container.
3. Remove or disconnect the following:
- Negative battery cable
- Intake manifold cover
- Evaporative emission (EVAP) canister hose and brake booster vacuum
hose

- Intake Air Temperature (IAT) sensor connector
- Breather hose
- Air cleaner housing
- Throttle cover. Fully open the throttle link and cruise control link by hand, then remove the throttle cable and cruise control cable from the links.
- Locknut, loosen only, then the cables from the bracket
- Water bypass hoses, then plug them

4. Relieve the fuel system pressure.

- Fuel feed hose
- Positive Crankcase Ventilation (PCV) hose, harness holder mounting bolt and harness clamp mounting bolt
- Idle Air Control (IAC) valve connector
- Throttle Position (TP) sensor connector
- Manifold Absolute Pressure (MAP) sensor connector
- Evaporative Emission (EVAP) canister purge valve connector
- Intake Manifold Runner Control

(IMRC) valve position sensor connector
- 2 bolts securing the intake manifold and brackets
- Intake manifold mounting bolts and nuts
- 2 stud bolts and the intake manifold. Discard the gasket(s).

To install:

5. Install or connect the following:
- Intake manifold with a new gasket, then tighten the 2 stud bolts to 16 ft. lbs. (22 Nm). Tighten the intake manifold bolts/nuts in a

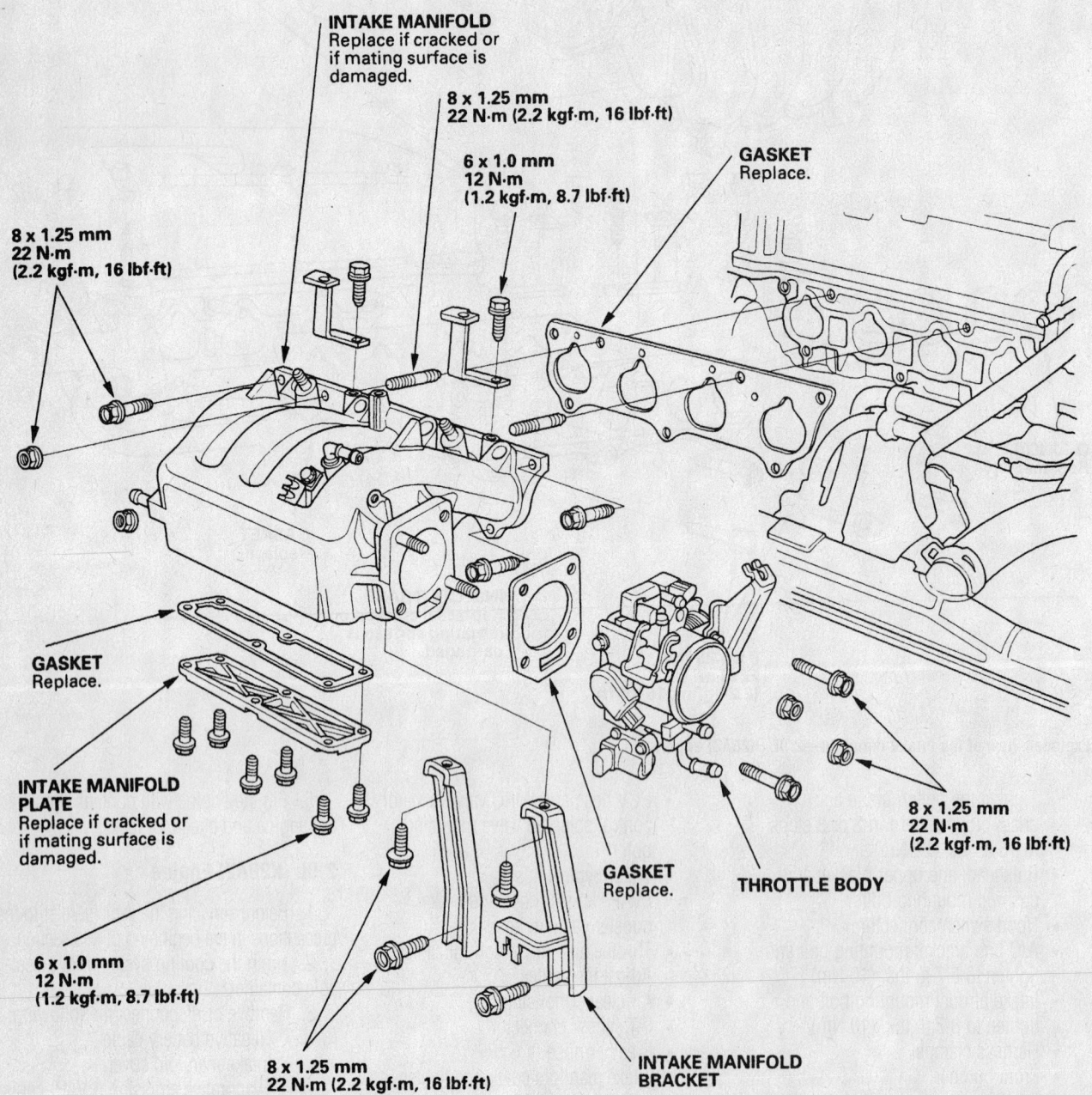

Exploded view of the intake manifold—2.0L (K20A3) engine

42356-INTE-G17

criss–cross pattern, beginning with the inner bolt to 16 ft. lbs. (22 Nm).
- 2 bolts securing the intake manifold brackets and tighten to 16 ft. lbs. (22 Nm)
- PCV hose, harness holder mounting bolt and harness clamp mounting bolt. Tighten the bolts to 8.7 ft. lbs. (10 Nm).
- Fuel feed hose
- Water bypass hoses
- Throttle and cruise control cables. Adjust the cables.
- Air cleaner housing
- IAT sensor connector
- Breather hose
- EVAP canister hose, brake booster vacuum hose and vacuum hoses
- Intake manifold cover and tighten to 8.7 ft. lbs. (10 Nm)
- Negative battery cable

6. Fill the engine with coolant, then start the engine and check for leaks.

3.0L Engines

1. Before servicing the vehicle, refer to the precautions in the beginning of this section.
2. Remove or disconnect the following:
   - Negative battery cable
   - Engine covers
   - Strut brace
   - Intake air duct
   - Throttle body and intake manifold covers
   - Throttle and cruise control cables
   - Fuel feed and return hoses
   - Brake booster vacuum hose
   - Electrical connectors from the manifold
   - Intake manifold

To install:

3. Installation is the reverse of the removal procedure, while using the following torque values:
   - Intake manifold bolts:16 ft. lbs. (22 Nm)
   - Strut brace bolts: 16 ft. lbs. (22 Nm)

3.2L and 3.5L Engines

1. Before servicing the vehicle, refer to the precautions in the beginning of this section.
2. Disconnect the negative battery cable.
3. Drain the cooling system.
4. Properly relieve the fuel system pressure.
5. Remove or disconnect the following:
   - Intake air duct
   - Strut brace (if equipped)
   - Intake manifold cover
   - Water bypass hose

- Traction Control System (TCS) actuator (if equipped)
- Evaporative Emission (EVAP) canister hose
- Throttle and cruise control cables
- Brake booster vacuum hose
- Upper intake manifold cover
- Intake manifold nuts and bolts in a crisscross pattern, beginning from the center and moving out

6. Verify that all vacuum lines are disconnected and remove the intake manifold and throttle body as a unit.

7. Inspect the manifold for cracks, flatness, or other damage; replace any damaged parts. If the intake manifold is to be replaced, transfer all the necessary components to the new manifold.

To install:
- Intake fasteners:16 ft. lbs. (22 Nm)
- Upper intake manifold cover fasteners: 108 inch lbs. (12 Nm)
- TCS actuator bolts: 16 ft. lbs. (22 Nm)
- Strut brace bolts: 16 ft. lbs. (22 Nm)

Exhaust Manifold

REMOVAL & INSTALLATION

1.8L Engines

1. Before servicing the vehicle, refer to the precautions in the beginning of this section.

2. Remove or disconnect the following:
   - Negative battery cable
   - Exhaust manifold cover
   - Front exhaust pipe from the manifold
   - Exhaust manifold support bracket
   - Exhaust manifold from the engine
   - Rear cover from the exhaust manifold (if necessary)

To install:

➡**Use new exhaust manifold nuts for installation.**

3. Install or connect the following:
   - Rear heat shield bolts: 17 ft. lbs. (24 Nm)
   - Exhaust manifold nuts: 23 ft. lbs. (31 Nm)
   - Bracket bolts: 33 ft. lbs. (44 Nm)
   - Front exhaust pipe nuts: 40 ft. lbs. (54 Nm)
   - Exhaust manifold heat shield bolts: 17 ft. lbs. (24 Nm)

2.0L Engine

1. Before servicing the vehicle, refer to the precautions in the beginning of this section.
2. Remove or disconnect the following:
   - VTEC solenoid valve
   - Intermediate shaft cover
   - Cover and exhaust manifold bracket
   - Mounting nuts and bolts and exhaust manifold. Discard the gasket.

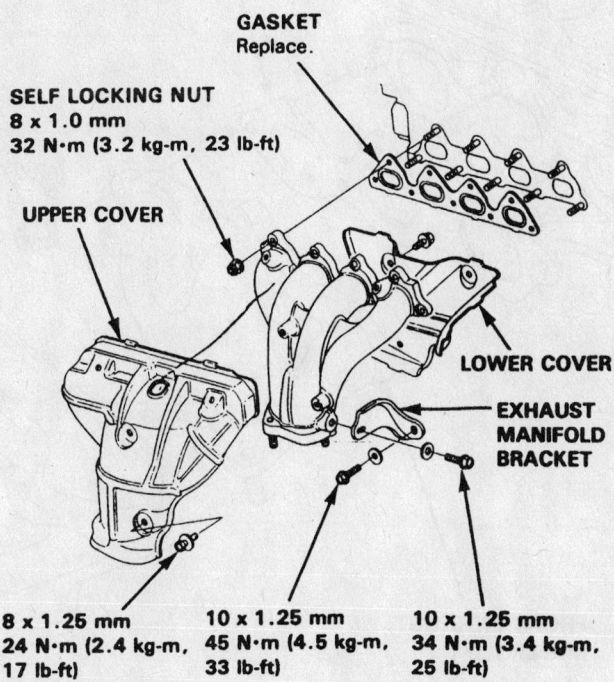

GASKET
Replace.

SELF LOCKING NUT
8 x 1.0 mm
32 N·m (3.2 kg-m, 23 lb-ft)

UPPER COVER

LOWER COVER

EXHAUST MANIFOLD BRACKET

8 x 1.25 mm
24 N·m (2.4 kg-m, 17 lb-ft)

10 x 1.25 mm
45 N·m (4.5 kg-m, 33 lb-ft)

10 x 1.25 mm
34 N·m (3.4 kg-m, 25 lb-ft)

Exhaust manifold—1.8L engines

7923BG18

To install:

3. Install or connect the following:
- Exhaust manifold with a new gasket. Tighten the retainers to 33 ft. lbs. (45 Nm), in a criss–cross pattern starting at the inner bolt.
- Cover and exhaust manifold bracket
- Intermediate shaft cover
- VTEC solenoid valve

3.0L Engine

1. Before servicing the vehicle, refer to the precautions in the beginning of this section.

2. Remove or disconnect the following:
- Manifold cover
- Exhaust pipe from the manifold to be removed
- Mounting nuts and the exhaust manifold

To install:

3. Install or connect the following:
- Exhaust manifold with a new gasket and tighten the nuts to 23 ft. lbs. (31 Nm)
- Exhaust pipe to the manifold using a new gasket and tighten the nuts to 40 ft. lbs. (54 Nm)
- Manifold cover and tighten the bolts to 16 ft. lbs. (22 Nm)

3.2L and 3.5L Engines

1. Before servicing the vehicle, refer to the precautions in the beginning of this section.

2. Remove or disconnect the following:
- Negative battery cable
- Exhaust manifold covers
- Small heat shields from the cylinder heads (if equipped)
- Exhaust pipe from the manifold
- Heated Oxygen (HO2S) sensors
- Exhaust manifold nuts in a criss-cross pattern starting from the center of the manifold
- Exhaust manifold

To install:

3. Install or connect the following:
- Exhaust manifold with a new gasket and new nuts and tighten the nuts in a crisscross pattern starting from the center to 22 ft. lbs. (30 Nm)
- Small heat shields and tighten the attaching bolts to 16 ft. lbs. (22 Nm) (if equipped)
- Exhaust pipe to the manifold with a new gasket and tighten the nuts to 40 ft. lbs. (55 Nm)
- HO2S sensor and tighten it to 33 ft. lbs. (45 Nm)
- Manifold covers and tighten the bolts to 16 ft. lbs. (22 Nm)

4. Verify that all vacuum lines and wiring are properly connected.
5. Reconnect the negative battery cable.
6. Start the engine and check for leaks.

Front Crankshaft Seal

REMOVAL & INSTALLATION

1. Before servicing the vehicle, refer to the precautions in the beginning of this section.

2. Disconnect negative cable at the battery.

3. Raise and safely support the vehicle. Drain the engine oil and properly dispose of it.

4. Be sure the crankshaft is at TDC on No. 1 cylinder by aligning the white mark on the crankshaft pulley with the pointer on the lower timing belt cover.

5. Remove or disconnect the following:
- Crankshaft pulley
- Cylinder head cover
- Timing belt cover
- Timing belt
- Crankshaft Speed Fluctuation (CKF) sensor (if equipped)
- Timing belt drive gear from the crankshaft

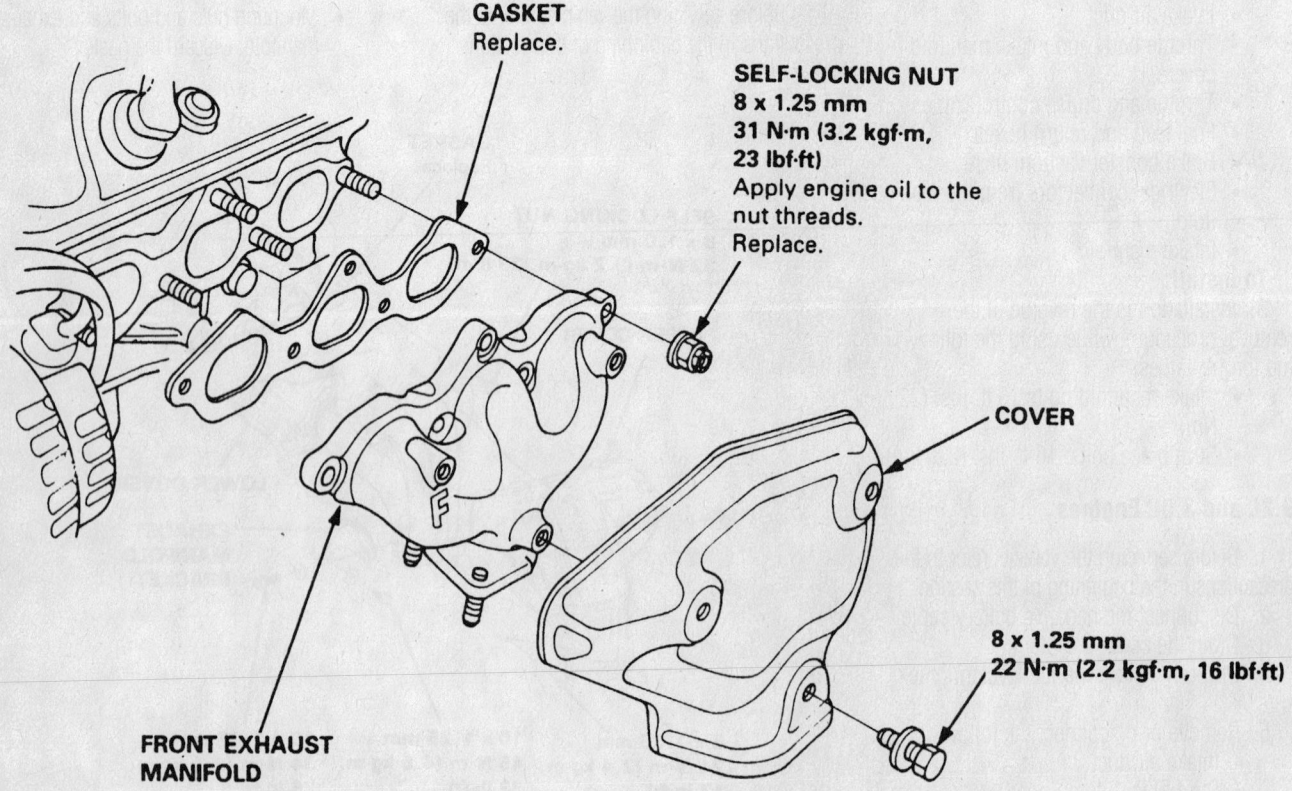

GASKET
Replace.

SELF-LOCKING NUT
8 x 1.25 mm
31 N·m (3.2 kgf·m,
23 lbf·ft)
Apply engine oil to the nut threads.
Replace.

COVER

8 x 1.25 mm
22 N·m (2.2 kgf·m, 16 lbf·ft)

FRONT EXHAUST MANIFOLD

7923BG68

Exploded view of the front exhaust manifold mounting—3.0L engine

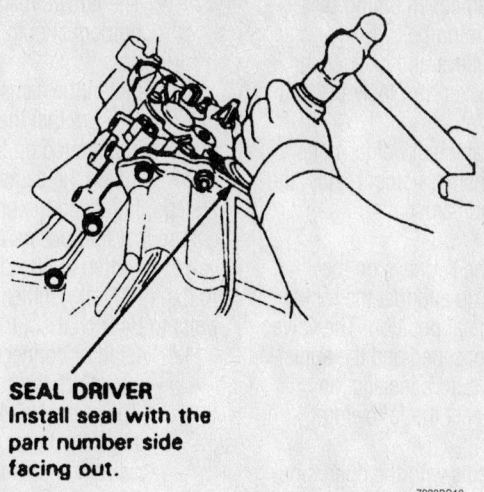

SEAL DRIVER
Install seal with the
part number side
facing out.

7923BG19

Installing the seal

6. Using a suitable prytool, carefully remove the seal.

To install:

7. Apply a light coat of oil to the seal lip.

8. Position the seal, then using a seal driver, install the seal into the housing.

9. Install or connect the following:
- Timing belt drive gear
- Timing belt
- Timing belt cover
- Cylinder head cover
- CKF sensor and tighten the attaching bolts to 96 inch lbs. (11 Nm) (if equipped)
- Crankshaft pulley

10. Lower the vehicle and check and fill the engine with oil as necessary.

11. Connect the negative battery cable and enter the radio security code.

12. Run the engine and check for leaks.

13. Turn off engine and check the oil level. Top off the oil level if necessary.

Camshaft and Valve Lifters

REMOVAL & INSTALLATION

➡**The radio may have a coded theft protection circuit. Obtain the code from the owner before disconnecting the battery, removing the radio fuse, or removing the radio.**

1.8L (B18B1) Engines

1. Before servicing the vehicle, refer to the precautions in the beginning of this section.

2. Remove or disconnect the following:
- Negative battery cable
- Spark plug wires

- Cylinder head cover and timing belt cover

3. Rotate the crankshaft to Top Dead center (TDC), compression of No. 1 piston and remove the timing belt.

4. Remove the distributor from the cylinder head.

5. Install 5.0mm pin punches to the No.1 camshaft holders, then remove the camshaft sprockets.

6. Loosen the valve adjusters to remove as much spring tension as possible.

7. Remove the pin punches from the camshaft holders.

To install:

8. Check the following before installing the camshafts:

   a. Be certain the keyways on the camshafts are facing UP (No. 1 cylinder at TDC).

   b. The valve adjuster locknuts should be loosened and the adjusting screws backed off before installation.

9. Lubricate the rocker arms and camshafts with clean oil.

10. Place the rocker arms on the pivot bolts and the valve stems, making sure that the rocker arms are in their original positions.

11. Install the camshaft seals with the open side (spring) facing in. Lubricate the lip of the seal.

12. Be sure the keyways on the camshafts are facing up and install the camshafts to the cylinder head.

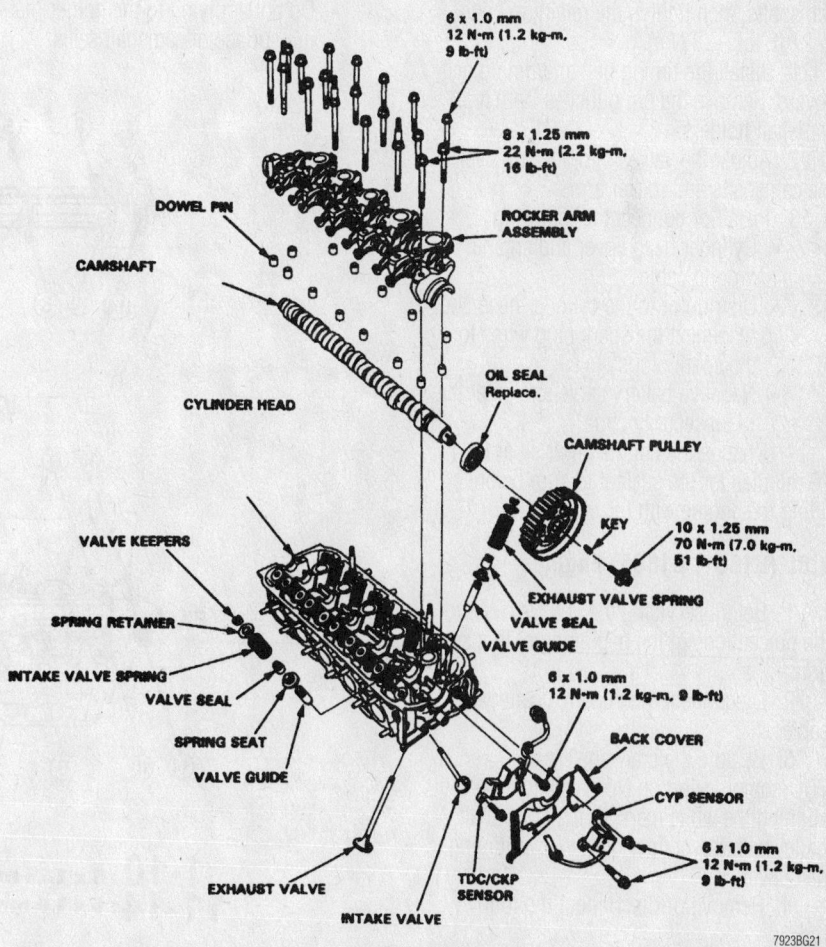

Exploded view of the cylinder head—1.8L (B18B1) engine

7923BG21

13. Apply liquid gasket to the head mating surfaces of the No. 1 and No. 6 camshaft holders, then install them along with No. 2, 3, 4 and 5 camshaft holders. The arrows stamped on the holders should point toward the timing belt. Do not apply oil to the holder mating surface where the camshaft seals are housed.

14. Tighten the camshaft holders temporarily and be sure that the rocker arms are properly positioned.

15. Press the oil seals into the No.1 camshaft holders with a seal driver.

16. Tighten the bolts in a crisscross pattern to 84 inch lbs. (10 Nm). Check that the rockers do not bind on the valves.

17. Install the cylinder head plug to the end of the cylinder head. If the plug has alignment marks, align the marks with the cylinder head upper surface.

18. If equipped with a timing belt back cover, install the cover and tighten the bolts to 84 inch lbs. (10 Nm).

19. Install 5.0mm pin punches to the No.1 camshaft holders, then install the camshaft pulley keys onto the grooves in the camshafts.

20. Push the camshaft pulleys onto the camshafts, then tighten the retaining bolts to 27 ft. lbs. (38 Nm).

21. Install the timing belt and timing belt covers. Remove the pin punches from the camshaft holders.

22. Adjust the valves and pour oil over the camshafts and rocker arms.

23. Install or connect the following:
- Cylinder head cover and engine ground cable
- Distributor to the cylinder head and reconnect the spark plug wires to the spark plugs
- Negative battery cable and enter the radio security code

24. Change the engine oil. Wait at least 20 minutes for the sealant to cure before filling the engine with oil.

1.8L (B18C1, B18C5) Engines

1. Before servicing the vehicle, refer to the precautions in the beginning of this section.

2. Disconnect the negative battery cable.

3. Be sure the crankshaft is at TDC/compression on No. 1 cylinder by aligning the white mark on the crankshaft pulley with the pointer on the lower timing belt cover.

4. Remove or disconnect the following:
- Strut brace

- Cylinder head cover, timing belt cover, and timing belt
- Camshaft pulleys and back cover

5. Loosen the rocker arm locknuts and adjusting screws.

6. Remove the camshaft holder bolts, then, remove the camshaft holder plates, the camshaft holders, and camshafts.

To install:

7. Be sure that the keyways on the camshafts are facing up and that the rocker arms are in their original position. The valve locknuts should be loosened and the adjusting screw backed off before installation

8. Install or connect the following:
- Camshafts
- Camshaft seals with the open side facing in
- Rubber cap with liquid gasket applied
- O–ring and the dowel pin to the oil passage of the No. 3 camshaft holder

9. Apply liquid gasket to the head of the mating surfaces of the No. 1 and No. 5 camshaft holders, then install them, along with No. 2, 3, and 4. Be sure to pay attention to the following points:
- Do not apply oil to the holder mating surface of camshaft seals

- The arrows marked on the camshaft holders should point to the timing belt

10. Tighten the camshaft holders temporarily. Be sure that the rocker arms are properly positioned on the valve stems.

11. Tighten the camshaft holder bolts in 2 steps, following the proper sequence, to ensure that the rockers do not bind on the valves. Tighten the 8 x 1.25mm bolts to 20 ft. lbs. (27 Nm). Tighten the 6 x 1.0mm bolts to 84 inch lbs. (10 Nm).

12. Install or connect the following:
- Keys into the camshaft grooves—To set the No. 1 piston at TDC, align the holes on the camshaft with the holes in the No. 1 camshaft holders and insert 5.0mm pin punches into the holes.
- Back cover and push the camshaft pulleys onto the camshafts, then tighten the retaining bolts to 27 ft. lbs. (37 Nm)
- Timing belt and adjust the tension, then install the timing belt covers

13. Adjust the valve clearance.

14. Install or connect the following:
- Cylinder head cover, be sure that the seal and groove are thoroughly clean first

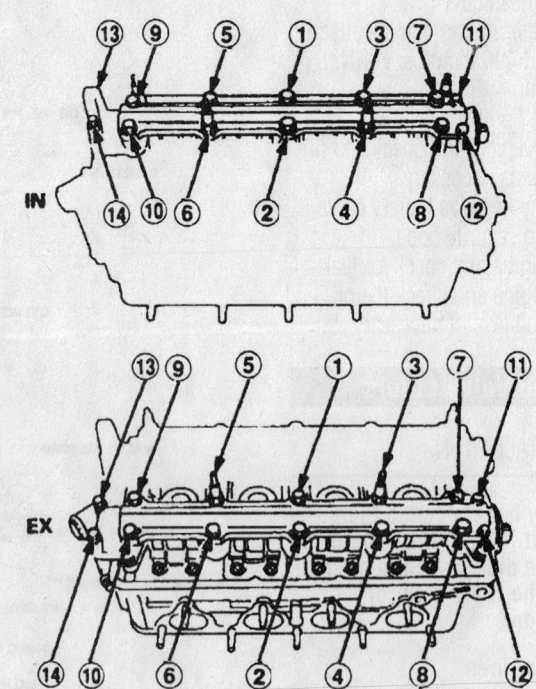

1-10 : 8 x 1.25 mm 27 N·m (2.8 kgf-m, 20 lbf-ft)
11-14 : 6 x 1.0 mm 9.8 N·m (1.0 kgf-m, 7.2 lbf-ft)

7923BG20

Camshaft holder plates torque sequence—1.8L (B18C1, B18C5) engines

- Engine side mount, tighten the 2 new nuts and new bolt to the engine to 38 ft. lbs. (52 Nm) and tighten the bolt attaching the mount to the vehicle to 54 ft. lbs. (74 Nm)
- Distributor to the cylinder head and reconnect the spark plug wires to the spark plugs
- Intake air duct
- Strut brace, tighten the nuts to 17 ft. lbs. (24 Nm)
- Negative battery cable and enter the radio security code

15. Drain the engine oil. Wait at least 20 minutes before filling the engine with oil; the time delay allows the sealant to cure.

2.0L Engine

1. Before servicing the vehicle, refer to the precautions in the beginning of this section.

2. Remove or disconnect the following:
- Negative battery cable
- Cylinder head from the vehicle
- Loosen the rocker arm adjusting screws
- Camshaft holder bolts, 2 turns at a time in a criss-cross pattern
- Timing chain guide B, camshaft holders and camshafts

To install:

3. Make sure the punch marks on the VTC actuator and exhaust camshaft sprocket are facing up, then set the camshafts in the holder.

4. Set the camshaft holders and timing chain guide B in place.

5. Tighten the bolts, in sequence, to the following specifications:
- 8mm bolts: 16 ft. lbs. (22 Nm)
- 6mm bolts (nos. 21, 22 & 23): 8.7 ft. lbs. (12 Nm)

6. Install or connect the following:
- Timing chain, then adjust the valve clearance
- Cylinder head
- Negative battery cable

7. Check for proper engine and valve train operation.

3.0L Engine

1. Before servicing the vehicle, refer to the precautions in the beginning of this section.

2. Remove or disconnect the following:
- Negative battery cable
- Timing belt
- Cylinder head
- Camshaft sprocket and rear cover
- Rocker arm/shaft assembly
- Camshaft thrust cover and O-ring

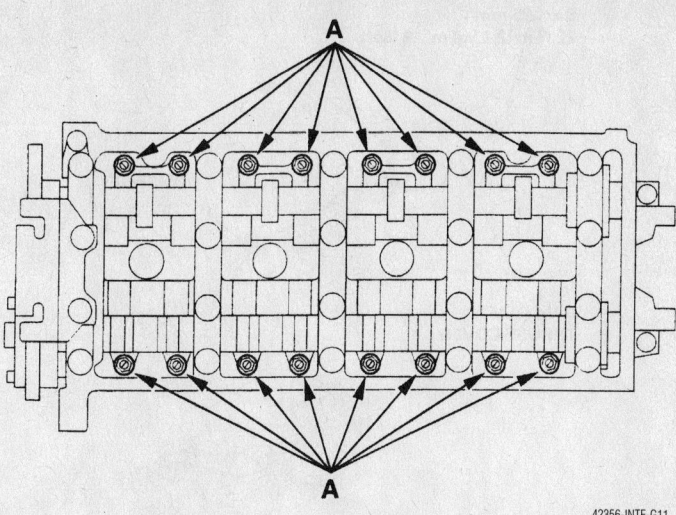

Loosen the rocker arm adjusting screws (A)—2.0L engines

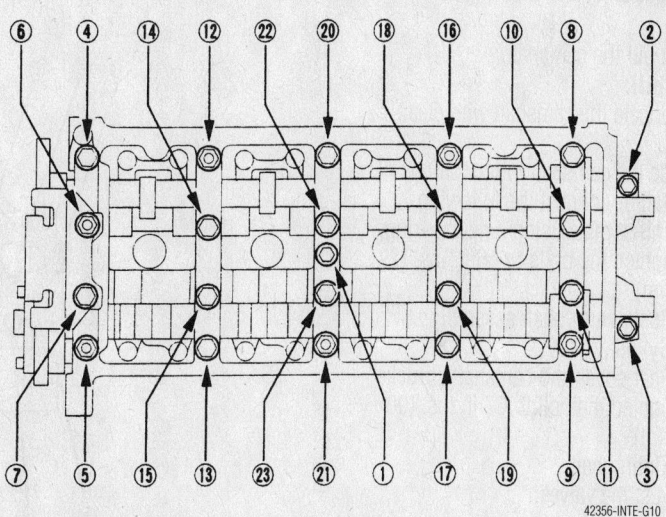

Camshaft holder bolt loosening sequence—2.0L engines

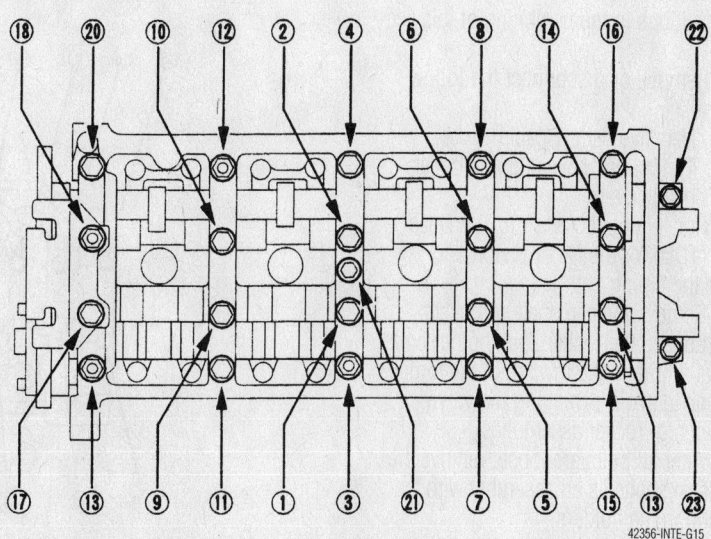

Rocker arm assembly tightening sequence—2.0L engines

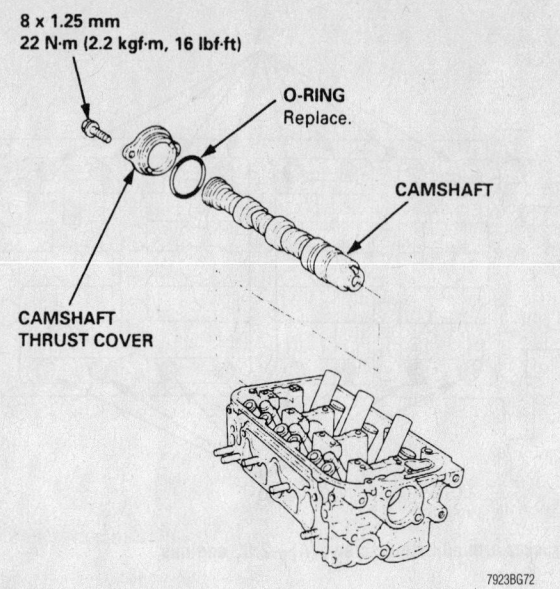

8 x 1.25 mm
22 N·m (2.2 kgf·m, 16 lbf·ft)

O-RING
Replace.

CAMSHAFT

CAMSHAFT
THRUST COVER

7923BG72

Camshaft installation—3.0L engine

3. Pull out the camshaft.

To install:

4. Lubricate the camshaft with clean engine oil.

5. Slide the camshaft into position.

6. Install or connect the following:
- Thrust plate using a new O–ring, tighten the bolts to 16 ft. lbs. (22 Nm)
- Rocker arm/shaft assembly
- Cylinder head
- Rear cover and camshaft sprocket, tighten the bolt to 67 ft. lbs. (90 Nm)
- Timing belt

7. Adjust the valves.

3.2L and 3.5L Engines

1. Before servicing the vehicle, refer to the precautions in the beginning of this section.

2. Remove or disconnect the following:
- Negative battery cable
- Timing belt covers and cylinder head covers

3. Rotate the crankshaft to Top Dead Center (TDC) for the No. 1 piston and remove the timing belt.

4. Remove the camshaft sprockets.

5. Loosen the rocker shaft holder bolts 1 turn at a time in the reverse of the torque sequence to avoid damaging the valves, camshafts, or rocker assemblies.

6. After all bolts are loose, remove the rocker arm shafts as an assembly with the bolts still in the holders.

7. If the rocker shafts are to be disassembled, note that each rocker arm has a

letter **A** or **B** stamped into the side. Before disassembling the rocker arms, make a note of the position of each letter so that the arms can be reassembled in the same position.

8. Do not remove the hydraulic tappets from the rocker arms unless they are to be replaced. Handle the rocker arms carefully so the oil does not drain out of the tappets.

9. Lift the camshafts from the cylinder head, wipe them clean and inspect the lift ramps. Replace the camshafts and rockers if the lobes are pitted, scored, or excessively worn.

To install:

10. Place a new seal on the end of the camshaft, lubricate the journals and set the camshaft in place on the head.

➡ **The pin hole in the front of the camshaft designates the top position.**

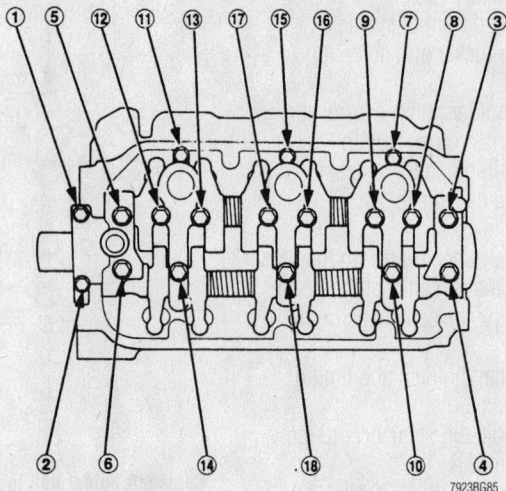

Loosen the camshaft holder bolts in the specified sequence—3.5L engine

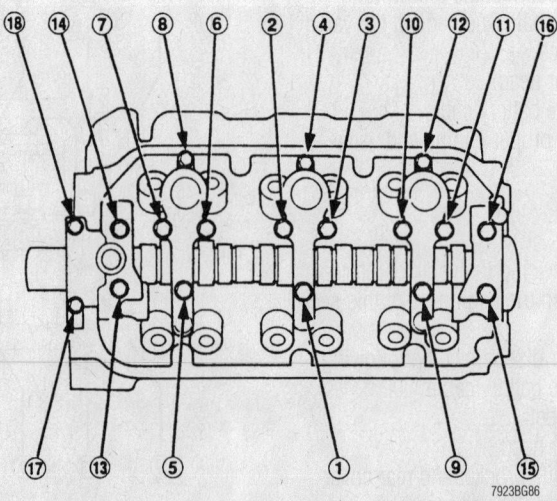

Tighten the camshaft holder bolts in the specified sequence—3.5L engine

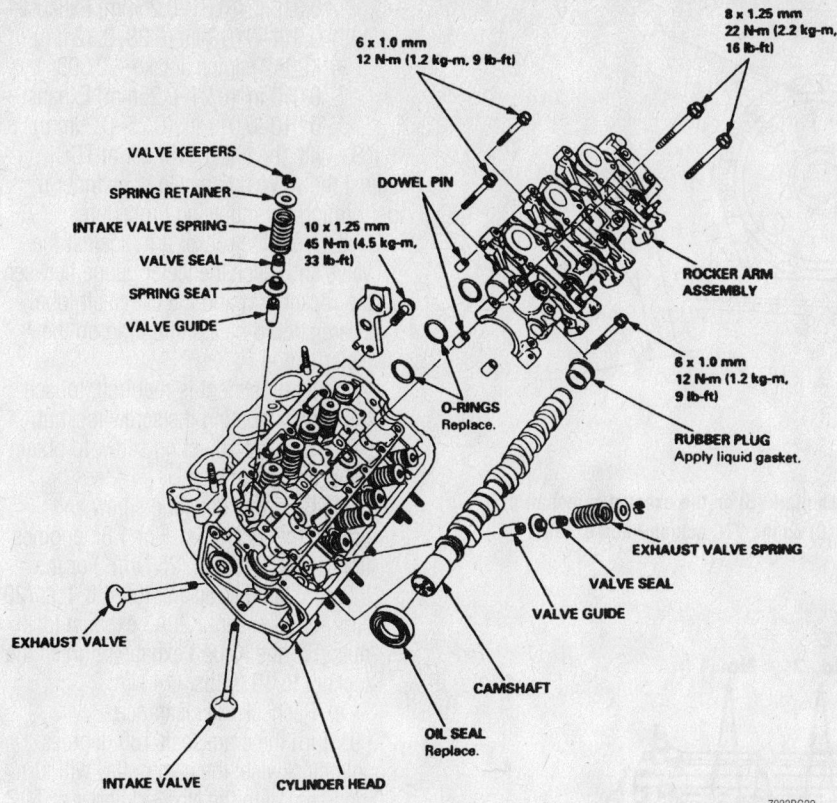

Camshaft and rocker arm assembly—3.2L engines

Specified torque:
8 mm bolts: 22 N·m (2.2 kg-m, 16 lb-ft)
6 mm bolts: 12 N·m (1.2 kg-m, 9 lb-ft)

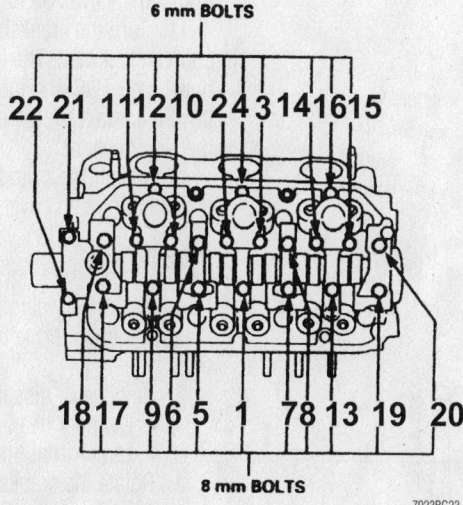

Camshaft holder bolt tightening sequence—3.2L engine

11. Apply liquid gasket to the mounting surfaces of the camshaft end holders.

12. Set the rocker arm assemblies in place and start all of the camshaft holder bolts. Be sure the rocker arms are properly positioned and turn each bolt in sequence 2 turns at a time until the holders are seated

on the head to avoid damaging the valves or rocker assemblies.

13. When all the camshaft and rocker holders are seated, tighten the bolts in the same sequence. Tighten the 8mm bolts to 16 ft. lbs. (22 Nm) and the 6mm bolts to 104 inch lbs. (12 Nm).

14. Install or connect the following:
 • Camshaft pulleys and tighten the bolts to 23 ft. lbs. (32 Nm)
 • Timing belt and pour oil over the camshafts
 • Cylinder head cover and reassemble accessory components

15. Verify that all electrical connections and vacuum lines are connected.

16. Reconnect the negative battery cable.

17. Run the engine and check for leaks and proper operation..

Valve Lash

ADJUSTMENT

1.8L and 2.0L Engines

➡ **While all valve adjustments must be as accurate as possible, it is better to have the valve adjustment slightly loose rather than too tight. Burned valves may result from overly tight adjustments. Perform the valve adjustment for each cylinder in the same sequence as the firing order: 1–3–4–2.**

1. Before servicing the vehicle, refer to the precautions in the beginning of this section.

2. Be sure the engine is cold; cylinder head temperature must be below 100° F (38° C). Overnight cold is best.

3. Remove the cylinder head cover.

4. Remove the upper timing belt cover, 1.8L engines only

5. Set the No. 1 cylinder to Top Dead Center (TDC).

 a. On 1.8L engines, the word **UP** should appear at the top and the TDC grooves on the pulley should align with the cylinder head surface or the mark on the rear belt cover.

 b. On 2.0L engines, the punch mark (arrow) on the Variable Valve Timing Control (VTC) actuator and the punch mark on the exhaust camshaft sprockets should be at the top. Align the TDC marks on the VTC actuator and exhaust camshaft sprocket.

6. For 1.8L engines, valve clearances are:
 • B18B1 engine: Intake—0.003–0.005 in. (0.08–0.12mm) Exhaust—0.006–0.008 in. (0.16–0.20mm)
 • B18C1 and B18C5 VTEC engine: Intake—0.006–0.007 in. (0.15–0.19mm) Exhaust—0.007–0.008 in. (0.17–0.20mm)

7. For 2.0L engines, valve clearances are:
 • K20A3 engine: Intake—0.008–

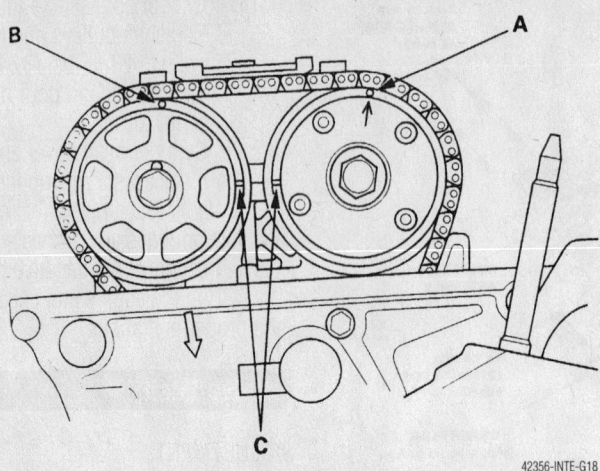

The punch mark (A) on the VTC actuator and the punch mark (B) on the exhaust camshaft sprockets should be at the top. Align the TDC marks (C) on the VTC actuator and exhaust camshaft sprocket.

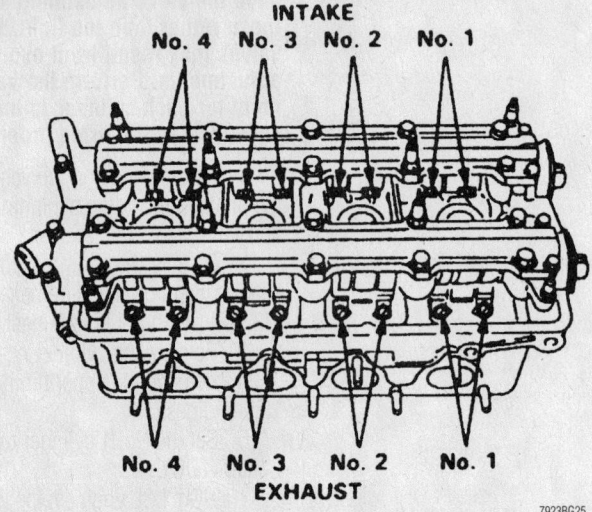

Valve arrangement—1.8L (B18B1, B18C1, and B18C5) engines

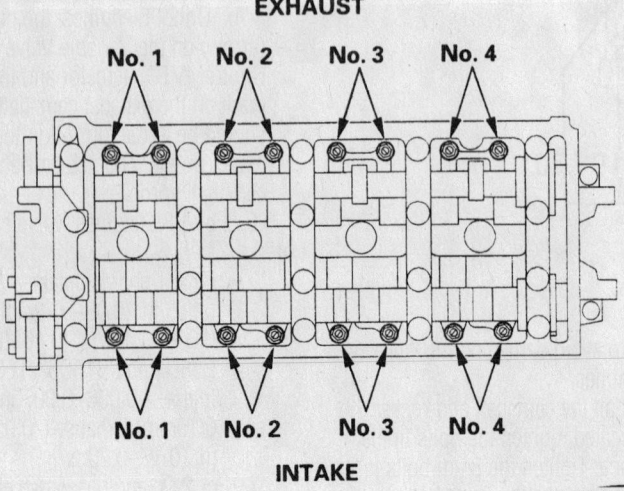

Valve arrangement—2.0L engines

0.010 in. (0.21–0.25mm) Exhaust—0.011–0.013 in. (0.28–0.32mm)
- K20A2 engine: Intake—0.008–0.010 in. (0.21–0.25mm) Exhaust—0.010–0.011 in. (0.25–0.29mm)

8. With the No. 1 cylinder at TDC, adjust the valves of the No. 1 cylinder by performing the following procedures:

   a. Hold the rocker arm against the valve and place the feeler gauge between the rocker arm and the camshaft lobe. There should be a slight drag on the feeler gauge.

   b. If adjustment is required, loosen the valve adjusting the screw locknut.

   c. Turn the adjusting screw to obtain the proper clearance.

   d. Hold the adjusting screw and tighten the locknut(s). For 1.8L engines tighten to 18 ft. lbs. (25 Nm). For the 2.0L tighten the locknut to 14 ft. lbs. (20 Nm) for all except K20A3 exhaust locknuts. For the K20A3 exhaust, tighten the locknut to 10 ft. lbs. (14 Nm).

   e. Recheck the clearance.

9. Turn the crankshaft 180 degrees counterclockwise; the cam pulley will turn 90 degrees. With the No. 3 cylinder at TDC, the **UP** marks should be at the exhaust side. Adjust the valves on the No. 3 cylinder.

10. Turn the crankshaft 180 degrees counterclockwise; the cam pulley will turn 90 degrees. With the No. 4 cylinder at TDC, both **UP** marks should be at the bottom. Adjust the valves on the No. 4 cylinder.

11. Turn the crankshaft 180 degrees counterclockwise. The No. 2 cylinder will now be on TDC and the **UP** marks should be at the intake side. Adjust the valves on the No. 2 cylinder.

12. Install the cylinder head cover and upper timing belt cover.

3.0L Engine

1. Before servicing the vehicle, refer to the precautions in the beginning of this section.

2. Remove or disconnect the following:
- Cylinder head cover
- Upper front timing belt cover

3. Rotate the crankshaft so the No. 1 piston is at Top Dead Center (TDC) on compression to adjust the valves for the No. 1 cylinder.

4. Loosen the locknuts and adjust the screws until a slight drag can be felt with the feeler gage when the gage is placed between the valve and rocker arm tip as shown. The specifications are as follows:
- Intake: 0.006–0.007 in. (0.15–0.18mm)

REAR:

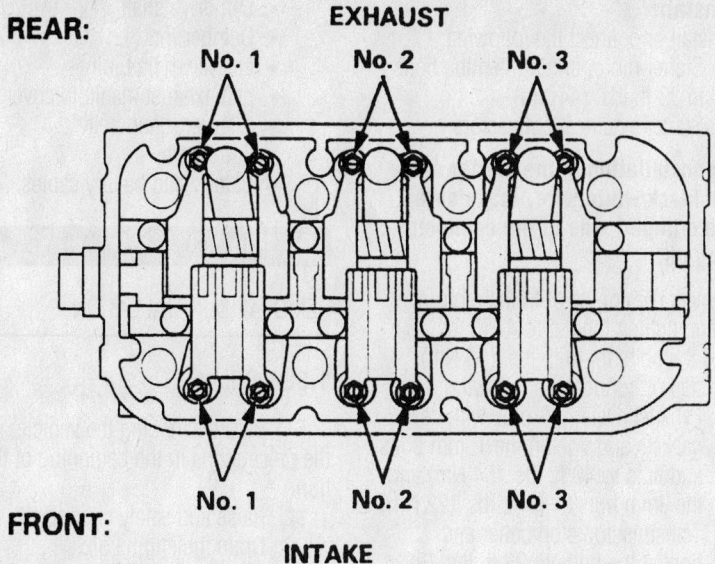

Adjusting screw locations for valve lash adjustment—3.0L engine

• Exhaust: 0.007–0.008 in. (0.18–0.20mm)

5. Rotate the crankshaft clockwise until the No. 4 on the camshaft sprocket is near the pointer on the rear cover. This is the No. 4 cylinder firing position.

6. Adjust the valves for the No. 4 cylinder while the sprocket is in this position. Tighten the locknuts to 14 ft. lbs. (20 Nm).

7. Continue to rotate the crankshaft and adjust the valves for cylinders 3 and 2.

8. Install the timing belt and cylinder head covers.

3.2L and 3.5L Engines

These engines are equipped with hydraulic valve lash adjusters on the rocker arms. No valve clearance adjustments are possible or necessary.

Starter Motor

REMOVAL & INSTALLATION

Except 2.0L and 3.5L Engines

1. Before servicing the vehicle, refer to the precautions in the beginning of this section.

2. Disconnect the negative battery cable.

3. On the 1.8L engine, remove the intake air duct.

4. On 3.0L engines, remove the automatic transmission cooler hose from the bracket on the starter motor.

5. Remove or disconnect the following:
   • Starter electrical connectors
   • 2 starter mounting bolts
   • Starter motor

To install:

6. Installation is the reverse of the

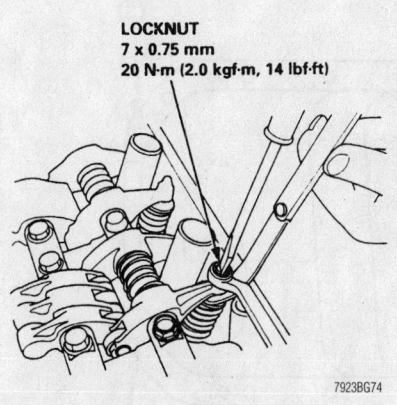

Slide the feeler gauge between the valve and rocker arm while turning the adjusting screw—3.0L engine

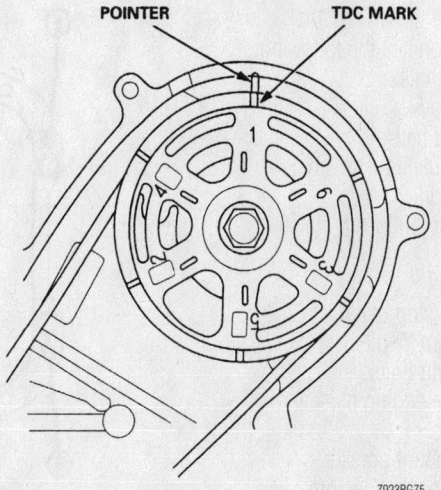

Camshaft sprocket position when No. 1 piston is at TDC—3.0L engine

removal procedure. Tighten the starter motor bolts to 33 ft. lbs. (44 Nm).

➡**When installing the starter cable, be sure to place the closed loop connector over the stud on the starter with the crimped side of the connector facing up. This is to ensure a proper fit against the stud.**

2.0L Engine

1. Before servicing the vehicle, refer to the precautions in the beginning of this section.
2. Remove or disconnect the following:
   • Negative, then the positive battery cables
   • Knock Sensor (KS) connector
   • Bolt securing the harness bracket, K20A3 engine
   • Bolt securing the harness bracket, then the intake manifold brackets, K20A2 engine
   • Starter cable from the B terminal on the solenoid
   • Black and white wire from the S terminal
   • 2 starter mounting bolts and starter

To install:

3. Installation is the reverse of the removal procedure, noting the following:
   • Make sure the crimped side of the B terminal connector ring terminal is facing out
   • Tighten the top starter mounting bolt to 33 ft. lbs. (44 Nm) and the bottom bolt to 47 ft. lbs. (64 Nm)

3.5L Engine

➡**This procedure requires the use of an engine hoist to lift the engine slightly.**

1. Before servicing the vehicle, refer to the precautions in the beginning of this section.
2. Remove or disconnect the following:
   • Both battery cables
   • Battery and tray
   • Alternator and belt
   • Left exhaust manifold cover
   • Left damper fork
   • Left lower ball joint from the suspension
   • Left drive shaft
   • Transmission stop collar
   • Exhaust system Y–pipe
   • Front mounting bolts
3. Attach a suitable engine hoist and slightly lift the engine.
   • Left engine mount bracket
   • Starter electrical connectors
   • Starter motor

To install:

4. Install or connect the following:
   • Starter motor and tighten the bolts to 33 ft. lbs. (44 Nm)
   • Starter electrical connectors

➡**Upon installation of the starter cable and the black/white wire, make sure that the crimped side of the connector is facing up.**

5. Lower the engine onto the motor mount and tighten the nut to 47 ft. lbs. (64 Nm) and the bolts to 40 ft. lbs. (54 Nm)
6. Install or connect the following:
   • Exhaust system Y–pipe with new gaskets and tighten the 10mm bolts and nuts to 40 ft. lbs. (54 Nm) and the 8mm nuts to 16 ft. lbs. (22 Nm)
   • Transmission stop collar and tighten the bolts to 28 ft. lbs. (38 Nm)

   • Left drive shaft
   • Damper fork
   • Left lower ball joint
   • Left exhaust manifold cover
   • Alternator and belt
   • Battery tray
   • Battery and battery cables

Oil Pan

REMOVAL & INSTALLATION

1.8L Engines

1. Before servicing the vehicle, refer to the precautions in the beginning of this section.
2. Raise and safely support the vehicle.
3. Drain the engine oil.
4. Remove or disconnect the following:
   • Negative battery cable

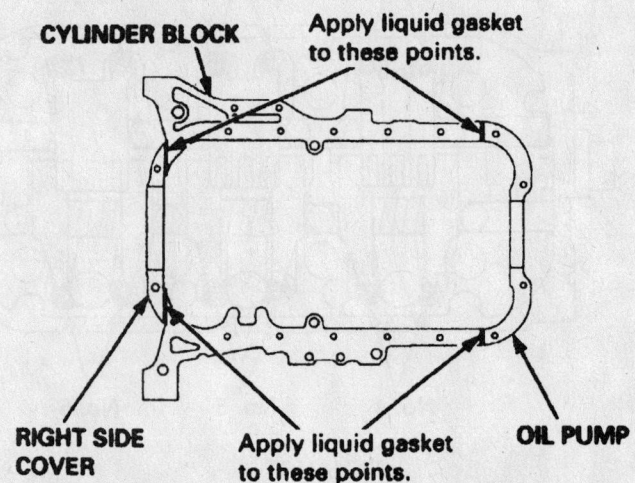

Apply liquid gasket to the oil pan as shown—1.8L (B18B1, B18C1 and B18C5) engines

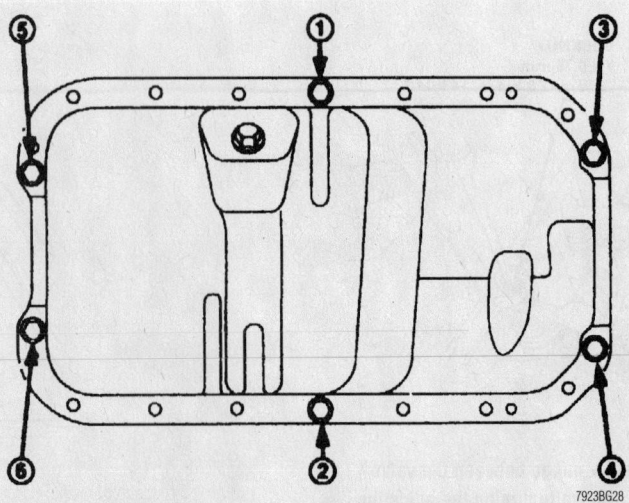

Oil pan bolt tightening sequence—1.8L (B18B1, B18C1 and B18C5) engines

- Splash shield
- Heated Oxygen (HO₂S) sensor connector
- Front exhaust pipe from the vehicle
- Center beam from the subframe

5. Loosen the oil pan bolts in a crisscross pattern. To remove the oil pan, lightly tap the corners of the oil pan with a rubber or plastic faced mallet.

To install:

6. Apply liquid gasket to the oil pan mating surface where the oil pump and the right side cover meet the engine block.

7. Install or connect the following:
- Oil pan with a new gasket
- Oil pan nuts and bolts and torque them in sequence to 10 ft. lbs. (14 Nm)

✱✱ CAUTION

Excessive tightening can cause distortion of the oil pan gasket and oil leakage.

- Oil drain plug with a new gasket and torque it to 33 ft. lbs. (44 Nm)
- Front exhaust pipe using new gaskets and locknuts and tighten the manifold nuts to 40 ft. lbs. (54 Nm) and the others to 16 ft. lbs. (22 Nm)
- HO₂S sensor connector
- Lower splash shield

8. Fill the engine with oil.

9. Connect the negative battery cable and enter the radio security code.

10. Run the engine and check for leaks.

11. Turn off engine and check the oil level. Top off the oil level if necessary.

2.0L Engine

K20A3 ENGINE

1. Before servicing the vehicle, refer to the precautions in the beginning of this section.

2. Raise and safely support the vehicle.

3. Drain the engine oil.

4. Attach a chain hoist to the engine.

5. Remove or disconnect the following:
- Lower ball joints
- Rear mount mounting bolts
- Front mount mounting bolts
- Automatic Transaxle Fluid (ATF) filter mounting bolt, if equipped with an automatic transaxle

6. Make alignment marks on the reference lines that align with the centers of the rear subframe mounting bolts.
- Front subframe
- Bolts/nuts securing the oil pan
- Oil pan by driving an oil pan seal

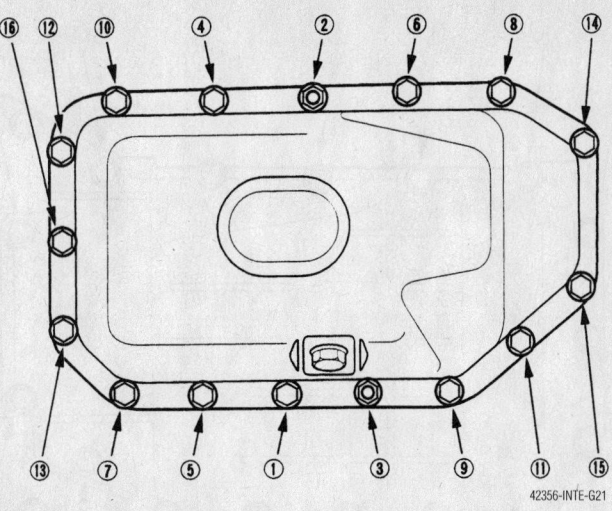

Oil pan tightening sequence—2.0L (K20A3) engine

cutter between the oil pan and cylinder block, then cutting the oil pan seal by striking the side of the cutter to slide the cutter along the oil pan

To install:

7. Thoroughly clean the mating surfaces, bolts and bolt holes. Apply liquid gasket part no. 08718–0009 evenly to the cylinder block mating surface of the oil pan and inner threads of the bolt holes.

➡**Make sure to install the oil pan within 5 minutes of applying the liquid gasket.**

8. Install or connect the following:
- Oil pan
- Oil pan mounting bolts. Tighten, in sequence, in 2 or 3 steps to 8.7 ft. lbs. (12 Nm).
- Subframe. Align the reference lines on the subframe with the bolt head center, tighten the bolts to 76 ft. lbs. (103 Nm).

- ATF filter mounting bolt
- Front mounting bolt
- Rear mounting bolts
- Lower ball joints
- Negative battery cable

9. Wait 30 minutes, then fill the engine with oil. Start the engine and check for leaks.

K20A2 ENGINE

1. Before servicing the vehicle, refer to the precautions in the beginning of this section.

2. Raise and safely support the vehicle.

3. Drain the engine oil.

4. Attach a chain hoist to the engine.

5. Remove or disconnect the following:
- Lower ball joints
- Rear mount mounting bolts
- Front mount mounting bolts

6. Make alignment marks on the reference lines that align with the centers of the rear subframe mounting bolts.

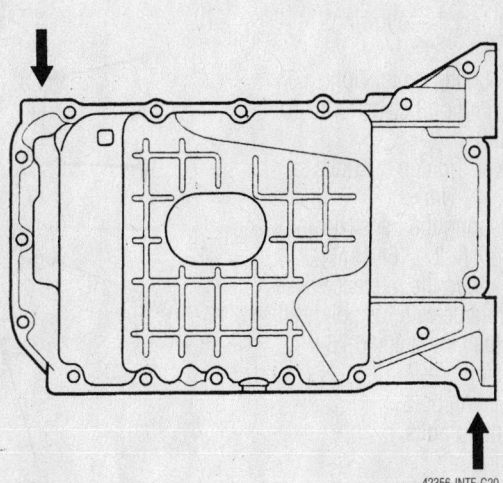

Insert a flat tip prytool where indicated by the arrows to carefully separate the oil pan from the block—2.0 (K20A2) engine

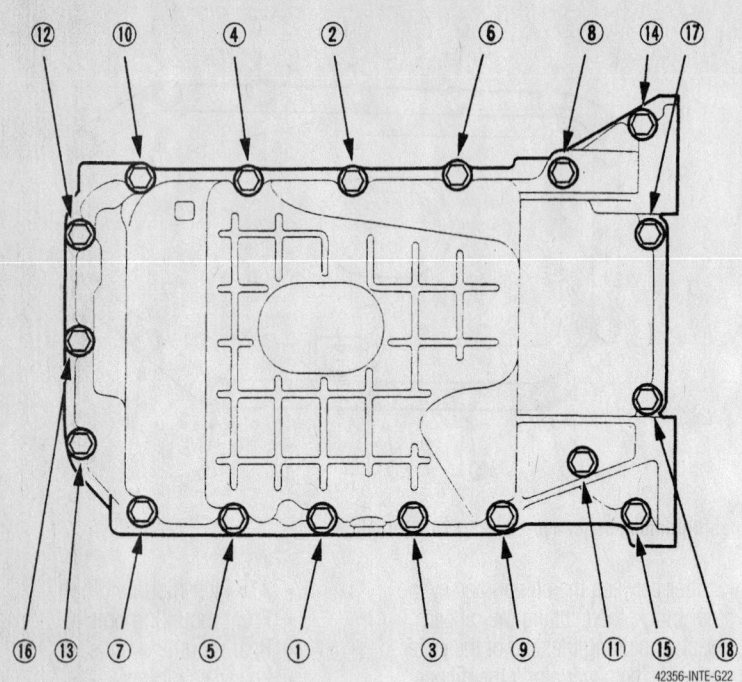

Oil pan tightening sequence—2.0L (K20A2) engine

- Front subframe
- Clutch cover and 2 bolts securing the transaxle
- Bolts/nuts securing the oil pan
- Oil pan by inserting a flat tip pry-toolr where shown in the illustration and separate the oil pan from the block

To install:

7. Thoroughly clean the mating surfaces, bolts and bolt holes. Apply liquid gasket part no. 08718–0009 evenly to the cylinder block mating surface of the oil pan and inner threads of the bolt holes.

➡ **Make sure to install the oil pan within 5 minutes of applying the liquid gasket.**

8. Install or connect the following:
- Oil pan
- Oil pan mounting bolts. Tighten, in sequence, in 2 or 3 steps to 8.7 ft. lbs. (12 Nm).
- Clutch cover. Tighten the bolts to 8.7 ft. lbs. (12 Nm).
- 2 bolts securing the transaxle and tighten to 47 ft. lbs. (64 Nm)
- Subframe. Align the reference lines on the subframe with the bolt head center, tighten the bolts to 76 ft. lbs. (103 Nm).
- Front mounting bolt
- Rear mounting bolts
- Lower ball joints
- Negative battery cable

9. Wait 30 minutes, then fill the engine with oil. Start the engine and check for leaks.

3.0L Engine

1. Before servicing the vehicle, refer to the precautions in the beginning of this section.
2. Raise and safely support the vehicle.
3. Drain the engine oil.
4. Remove or disconnect the following:
- Front exhaust pipe
- Oil pan mounting bolts
- Oil pan

To install:

5. Apply a bead of sealant to the oil pan flange and install the oil pan. Tighten the bolts in the sequence shown to 108 inch lbs. (12 Nm).
6. Install the front exhaust pipe with new gaskets and torque the manifold nuts to 40 ft. lbs. (54 Nm) and the catalytic converter nuts to 25 ft. lbs. (33 Nm).
7. Add the correct amount of engine oil to the crankcase.
8. Start the engine and check for leaks.

3.2L and 3.5L Engines

1. Before servicing the vehicle, refer to the precautions in the beginning of this section.
2. Drain the engine oil.
3. Drain the differential oil.
4. Remove or disconnect the following:
- Negative battery cable
- Accessory drive belts
- Power steering pump without disconnecting the lines
- Exhaust manifold covers
- Front wheels
- Splash shield
- Strut forks
- Lower ball joints
- Halfshafts from the differential
- Intermediate shaft
- Vehicle speed/power steering speed sensor without disconnecting the hoses
- Lower plate from the rear beam

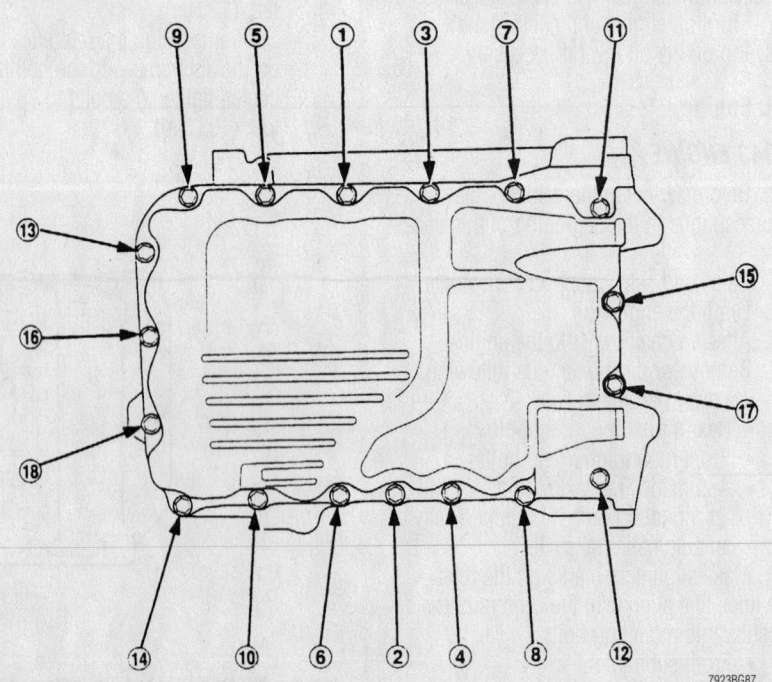

Oil pan bolt tightening sequence—3.0L engine

- A/C compressor without disconnecting the lines

5. Attach a chain hoist to the engine.
- Left engine mount bracket
- Right engine mount bracket
- Right engine mount
- 36mm sealing bolt on the transaxle. Ensure that the transaxle is in 1st gear (manual) or **P** (automatic).
- Extension shaft from the differential with an extension shaft puller
- Differential mounting bolts and the 26mm shim, then remove the differential
- Rear engine stiffener
- Flywheel cover or the torque converter covers
- Oil pan

➡**Do not lose the dowel pins from the oil pan**

To install:

6. Apply liquid gasket to the cylinder block. Be sure that the mating surfaces are clean and dry before installing the liquid gasket. Do not apply liquid gasket to the O–ring grooves.

7. Install or connect the following:
- Oil pan with new O–rings and torque the bolts to 16 ft. lbs. (22 Nm)

➡**Be sure the dowel pins are still in place.**

- Flywheel or torque converter cover and torque the bolts to 108 inch lbs. (12 Nm)
- Rear engine stiffener and tighten the bolt attaching the engine stiff-

ener to the transaxle first, to 47 ft. lbs. (64 Nm), then tighten the bolts to the engine block to 16 ft. lbs. (22 Nm)
- Differential to the engine and torque the bolts to 47 ft. lbs. (64 Nm)

➡**Be sure to install the shim in the original location**

- Air conditioning compressor to the engine block and tighten the mounting bolts to 16 ft. lbs. (22 Nm)
- New set ring to the extension shaft—Using an extension shaft installer tool, install the shaft to the differential.
- Extension shaft with a new set ring

8. Fill the secondary gear with super high temperature grease. Applying sealer to the threads of the 36mm sealing bolt, then install the bolt and tighten it to 58 ft. lbs. (78 Nm).
- Right engine mount and torque the bolts to 28 ft. lbs. (38 Nm)
- Right engine mount bracket and torque the nut and bolts to 47 ft. lbs. (64 Nm)
- Left engine mount bracket and torque the bolts to 40 ft. lbs. (54 Nm)
- A/C compressor and torque the bolts to 16 ft. lbs. (22 Nm)
- Lower plate and torque the bolts to 28 ft. lbs. (38 Nm)
- Vehicle speed/power steering speed sensor and torque the bolts to 108 inch lbs. (12 Nm)

- Intermediate shaft and the half-shafts
- Lower ball joints and tighten the nuts to 40 ft. lbs. (54 Nm)
- Strut forks and torque the bolts to 51 ft. lbs. (69 Nm)
- Engine splash shield and tighten the bolts to 84 inch lbs. (9.5 Nm)
- Front wheels
- Exhaust manifold covers and torque the bolts to 16 ft. lbs. (22 Nm)
- Power steering pump and torque the bolt to 33 ft. lbs. (44 Nm) and the nut to 16 ft. lbs. (22 Nm)
- Accessory drive belts
- Negative battery cable

9. Fill the engine with oil.
10. Fill the differential with oil.
11. Run the engine and check for leaks.
12. Check the front wheel alignment.

Oil Pump

REMOVAL & INSTALLATION

Except RSX and NSX

1. Before servicing the vehicle, refer to the precautions in the beginning of this section.

2. Drain the engine oil.

3. Be sure the crankshaft is at Top Dead Center (TDC) on the No. 1 cylinder.

4. Remove or disconnect the following:
- Negative battery cable
- Cylinder head cover
- Timing belt cover
- Timing belt
- Crankshaft Position (CKP) sensor (if necessary)
- Crankshaft timing belt gear
- Oil pan
- Pickup screen
- Oil pump from the front of the engine

➡**Any time the oil pump is removed, the front oil seal should be replaced.**

To install:

5. Install a new oil seal in the oil pump.

6. Apply liquid gasket to the mounting surface of the oil pump.

7. Install the oil pump, using new O–rings. For all engines, except the 1.8L (B18B1, B18C1, B18C5) engines, tighten the 6mm bolts to 108 inch lbs. (12 Nm) and the 8mm bolts to 16 ft. lbs. (22 Nm). For 1.8L (B18B1, B18C1, B18C5) engines, tighten the 8 x 1.25mm bolts to 17 ft. lbs. (24 Nm), tighten the 6 x 1.0mm bolts to 96 inch lbs. (11 Nm).

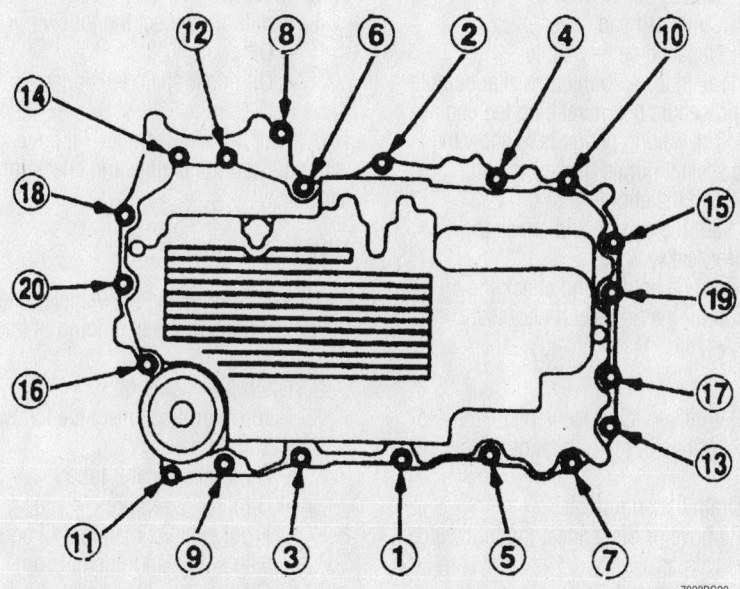

7923BG30

Be sure to tighten the oil pan bolts in the sequence shown—3.2L engines

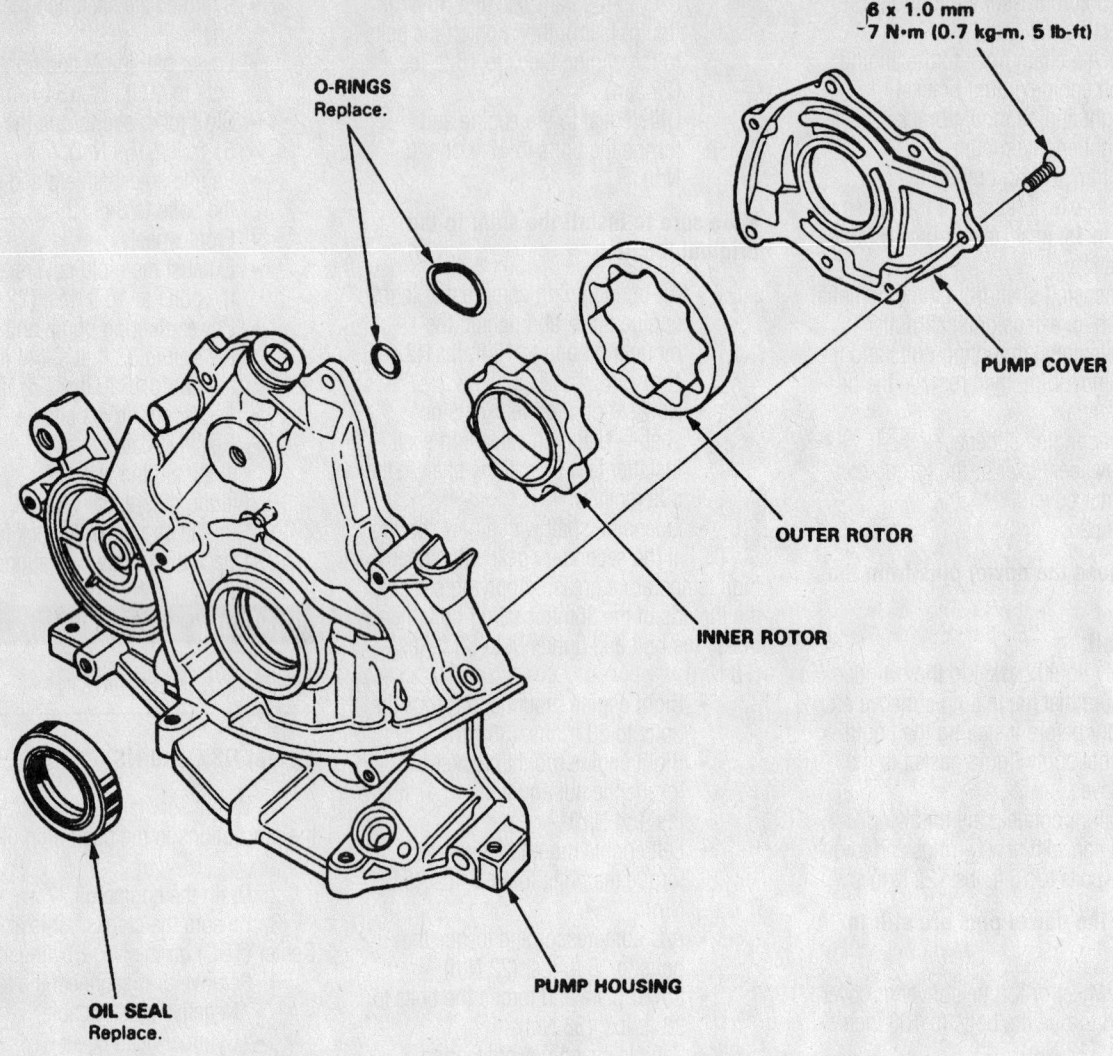

6 x 1.0 mm
7 N·m (0.7 kg-m, 5 lb-ft)

O-RINGS
Replace.

PUMP COVER

OUTER ROTOR

INNER ROTOR

PUMP HOUSING

OIL SEAL
Replace.

7923BG31

Exploded view of the oil pump—1.8L (B18B1) engines

✳✳ WARNING

The B18B1, B18C1 and B18C5 engines use different oil pumps, be sure that you have the correct oil pump. Match the crankshaft timing mark on the new oil pump with the timing mark on the old oil pump, because the timing marks are in different locations. If an oil pump is used with the timing mark in the wrong position the pistons may contact the valves.

8. Install or connect the following:
  • Oil pump pickup screen
  • Oil pan and tighten the bolts to 108 inch lbs. (12 Nm)
  • Crankshaft timing belt gear
  • Timing belt
  • Crankshaft Position (CKP) sensor and torque the bolt to 108 inch lbs. (12 Nm)

  • Timing belt cover
  • Cylinder head cover
  • Negative battery cable
9. Wait at least 30 minutes after completion of procedure before refilling the engine with oil. The waiting period is to allow the silicone sealant (liquid gasket) to cure.
10. Refill the engine with oil.
11. Start the engine and check the engine for leaks.
12. Turn off engine and check the oil level. Top off the oil level if necessary.

RSX

1. Before servicing the vehicle, refer to the precautions in the beginning of this section.
2. Drain the engine oil.
3. Remove or disconnect the following:
  • Oil pan
  • Oil pump chain tensioner
  • Oil pump

To install:
4. Install or connect the following:
  • Oil pump
  • Oil pump chain tensioner
  • Oil pan
5. Wait 30 minutes, then fill the engine with oil. Start the engine and check for leaks.

NSX

1. Before servicing the vehicle, refer to the precautions in the beginning of this section.
2. Drain the engine oil.
3. Remove or disconnect the following:
  • Timing belt
  • Oil level indicator tube
  • Oil filter assembly
  • Front exhaust manifold (if equipped with a manual transmission)
  • Oil pan
  • Oil screen

NOTE:
- Use new O-rings when reassembling.
- Apply oil to O-rings before installation.
- Use liquid gasket, Part No. 08718 – 0001 or 08718 – 0003.
- Clean the oil control orifice before installing.

CAUTION: Do not overtighten the drain bolt.

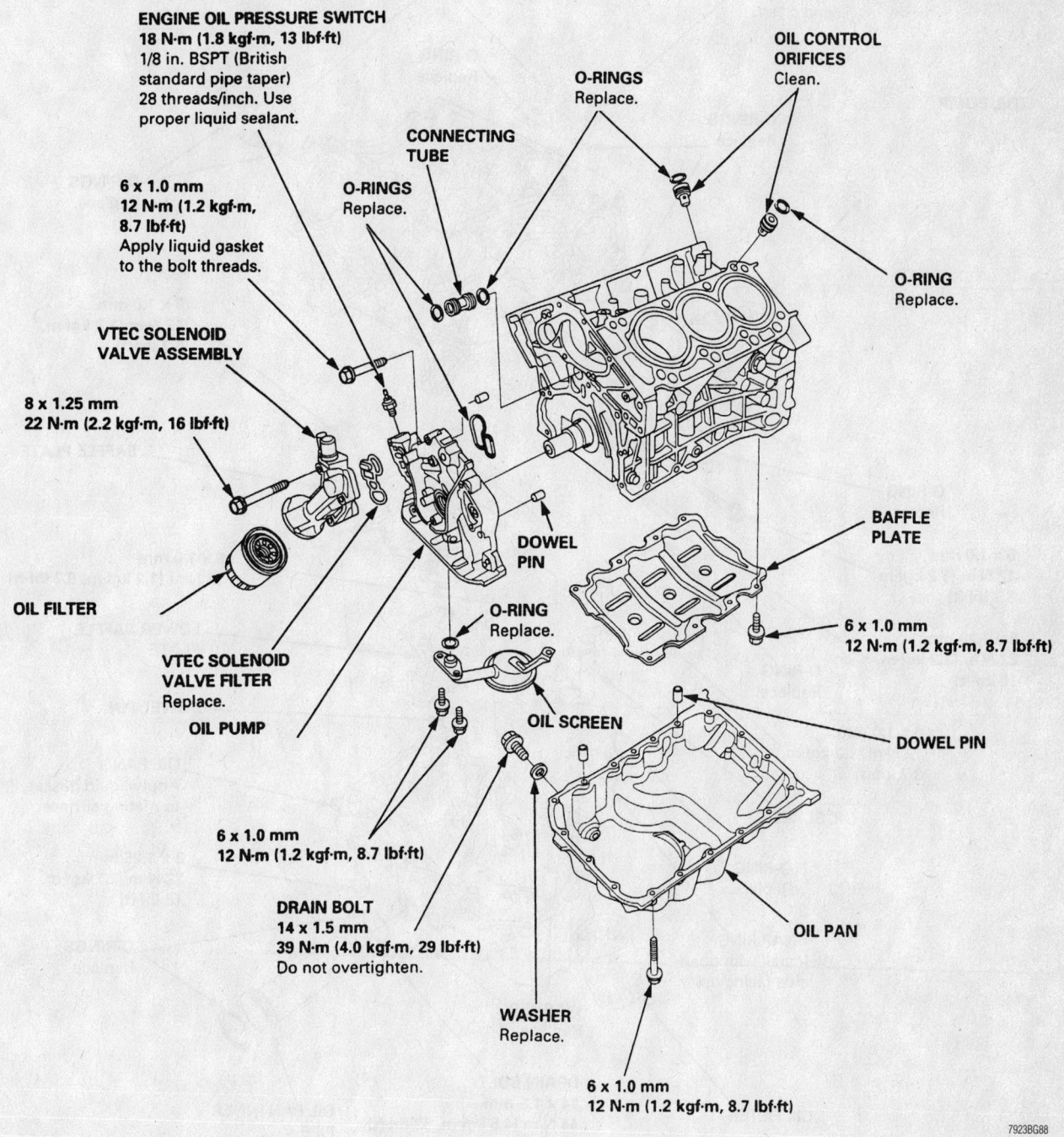

ENGINE OIL PRESSURE SWITCH
18 N·m (1.8 kgf·m, 13 lbf·ft)
1/8 in. BSPT (British standard pipe taper) 28 threads/inch. Use proper liquid sealant.

6 x 1.0 mm
12 N·m (1.2 kgf·m, 8.7 lbf·ft)
Apply liquid gasket to the bolt threads.

VTEC SOLENOID VALVE ASSEMBLY

8 x 1.25 mm
22 N·m (2.2 kgf·m, 16 lbf·ft)

OIL FILTER

VTEC SOLENOID VALVE FILTER
Replace.

OIL PUMP

6 x 1.0 mm
12 N·m (1.2 kgf·m, 8.7 lbf·ft)

DRAIN BOLT
14 x 1.5 mm
39 N·m (4.0 kgf·m, 29 lbf·ft)
Do not overtighten.

WASHER
Replace.

6 x 1.0 mm
12 N·m (1.2 kgf·m, 8.7 lbf·ft)

CONNECTING TUBE

O-RINGS Replace.

O-RINGS Replace.

OIL CONTROL ORIFICES Clean.

O-RING Replace.

DOWEL PIN

O-RING Replace.

OIL SCREEN

BAFFLE PLATE

6 x 1.0 mm
12 N·m (1.2 kgf·m, 8.7 lbf·ft)

DOWEL PIN

OIL PAN

7923BG88

Lubrication system—3.0L engine

NOTE:
- Use new O-rings when reassembling.
- Apply oil to O-rings before installation.
- Use liquid gasket, Part No. 08718 – 0001 or 08718 – 0003.
- Clean the oil control orifice before installing.
- Remove the balancer shaft

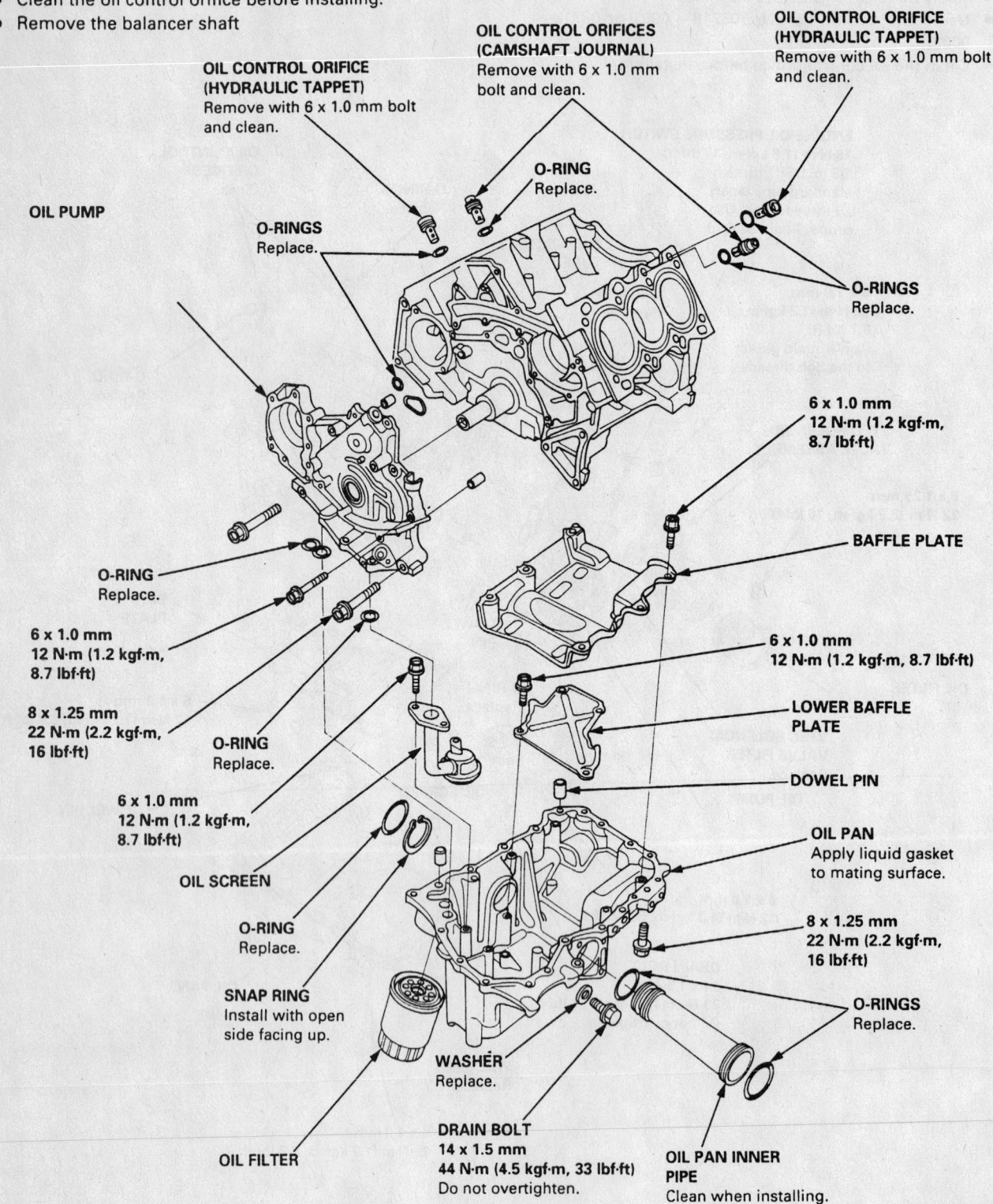

OIL CONTROL ORIFICES (CAMSHAFT JOURNAL)
Remove with 6 x 1.0 mm bolt and clean.

OIL CONTROL ORIFICE (HYDRAULIC TAPPET)
Remove with 6 x 1.0 mm bolt and clean.

OIL CONTROL ORIFICE (HYDRAULIC TAPPET)
Remove with 6 x 1.0 mm bolt and clean.

OIL PUMP

O-RING
Replace.

O-RINGS
Replace.

O-RINGS
Replace.

6 x 1.0 mm
12 N·m (1.2 kgf·m, 8.7 lbf·ft)

BAFFLE PLATE

O-RING
Replace.

6 x 1.0 mm
12 N·m (1.2 kgf·m, 8.7 lbf·ft)

6 x 1.0 mm
12 N·m (1.2 kgf·m, 8.7 lbf·ft)

8 x 1.25 mm
22 N·m (2.2 kgf·m, 16 lbf·ft)

LOWER BAFFLE PLATE

O-RING
Replace.

DOWEL PIN

6 x 1.0 mm
12 N·m (1.2 kgf·m, 8.7 lbf·ft)

OIL SCREEN

OIL PAN
Apply liquid gasket to mating surface.

O-RING
Replace.

8 x 1.25 mm
22 N·m (2.2 kgf·m, 16 lbf·ft)

SNAP RING
Install with open side facing up.

O-RINGS
Replace.

WASHER
Replace.

OIL FILTER

DRAIN BOLT
14 x 1.5 mm
44 N·m (4.5 kgf·m, 33 lbf·ft)
Do not overtighten.

OIL PAN INNER PIPE
Clean when installing.

7923BG89

Exploded view of the lubrication system—3.5L engine

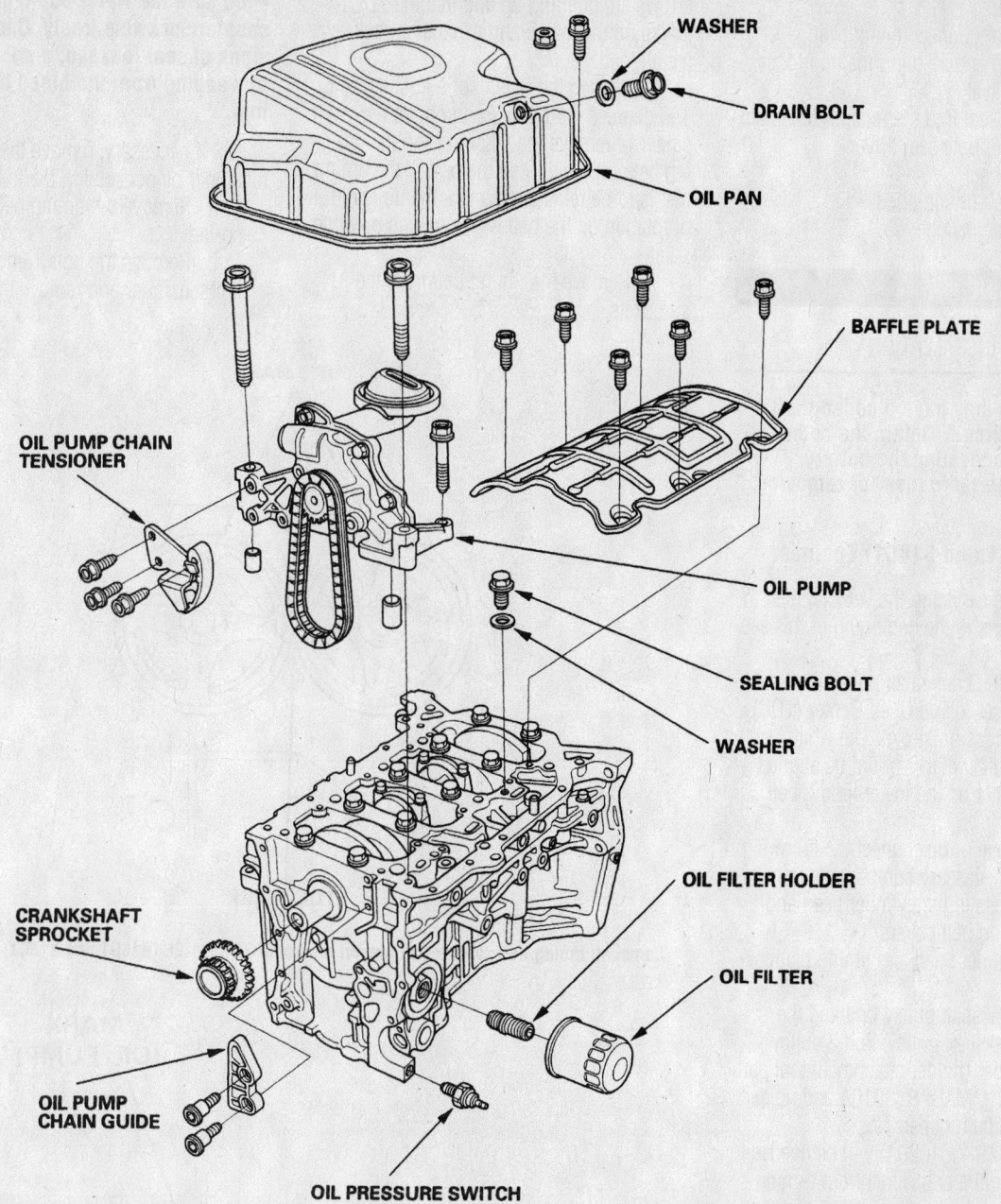

WASHER

DRAIN BOLT

OIL PAN

BAFFLE PLATE

OIL PUMP CHAIN TENSIONER

OIL PUMP

SEALING BOLT

WASHER

OIL FILTER HOLDER

OIL FILTER

CRANKSHAFT SPROCKET

OIL PUMP CHAIN GUIDE

OIL PRESSURE SWITCH

42356-INTE-G23

Exploded view of the lubrication system—2.0L (K20A3) engine shown

- Baffle plate
- Oil pass pipe and joint
- Oil pump
- Oil seal from the oil pump

To install:

4. Install a new oil seal in the oil pump.
5. Apply liquid gasket to the mounting surface of the oil pump.
6. Install or connect the following:
   - Oil pump and torque the bolts to 16 ft. lbs. (22 Nm)
   - Oil pass pipe and joint with new O–rings and torque the bolts to 108 inch lbs. (12 Nm)

- Baffle plate and torque the bolts to 108 inch lbs. (12 Nm)
- Oil screen and torque the bolts to 108 inch lbs. (12 Nm)
- Oil pan and torque the bolts to 16 ft. lbs. (22 Nm)
- Front exhaust manifold (if removed) and torque the nuts to 23 ft. lbs. (31 Nm)
- Oil filter assembly
- Oil level indicator tube
- Timing belt

7. Refill the engine oil.
8. Start the engine and check for leaks.

Rear Main Seal

REMOVAL & INSTALLATION

1. Before servicing the vehicle, refer to the precautions in the beginning of this section.

2. Remove or disconnect the following:
   - Transaxle
   - Clutch (if equipped)
   - Flexplate
   - Crankshaft seal by prying it out of the retainer

To install:

3. Install or connect the following:
   - Clean engine oil to the lip of the new seal
   - New seal into the retainer using an appropriate seal driver
   - Flywheel
   - Clutch (if equipped)
   - Transmission

Timing Belt

REMOVAL & INSTALLATION

➡ The radio may have a coded theft protection circuit. Obtain the code before disconnecting the battery, removing the radio fuse, or removing the radio.

1.8L (B18B1 and B18C1) Engines

1. Before servicing the vehicle, refer to the precautions in the beginning of this section.

2. Turn the crankshaft pulley until cylinder No. 1 is set to Top Dead Center (TDC) on the compression stroke. The white crankshaft pulley mark should be aligned with the pointer on the lower timing belt cover.

3. Remove or disconnect the following:
   - All necessary components to gain access to the cylinder head and timing belt covers
   - Cylinder head and timing belt covers
   - Crankshaft pulley bolt and the crankshaft pulley. Bolt a pulley holder (holder attachment tool part No. 07MAB–PY3010A and holder handle tool part No. 07JAB–001020A) will be needed to keep the crankshaft from turning.

✲✲ WARNING

Do not use the timing belt covers to store small parts. Grease or oil can transfer from the parts to the cover, then to the belt. Clean the covers thoroughly before installation.

4. Recheck that the No. 1 piston is at TDC on its compression stroke. Align the groove on the toothed side of the crankshaft timing belt drive sprocket to the arrow pointer on the oil pump.

5. To set the camshafts to top dead center for the No. 1 cylinder, align the hole in each camshaft with the holes in the No. 1 camshaft holders, then push 5.0mm pin punches into the holes. Be sure that the **UP** arrows are pointing up and that the TDC marks on the intake and exhaust sprockets are aligned.

6. Loosen the tensioner adjusting bolt 180 degrees (½ turn). Push on the tensioner to remove the tension from the timing belt, then retighten the bolt. If the timing belt is to be reinstalled, mark the direction of rotation on the belt with a crayon or white paint.

7. Remove the timing belt from the sprockets.

➡ Be sure the water pump pulley turns counterclockwise freely. Check for signs of seal leakage; a small amount of weeping from the bleed hole is normal.

8. If necessary, remove the timing belt tensioner by performing the following:
   a. Remove the timing belt tensioner spring.
   b. Remove the bolt from the timing belt tensioner and remove the tensioner.

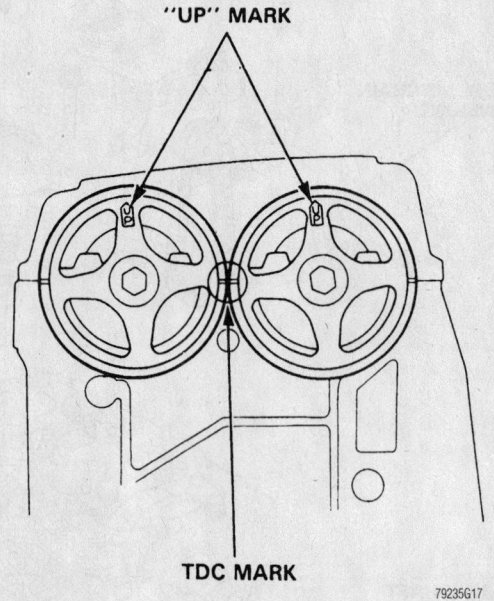

Camshaft timing belt sprocket alignment marks for TDC—1.8L (B18B1 and B18C1) engines

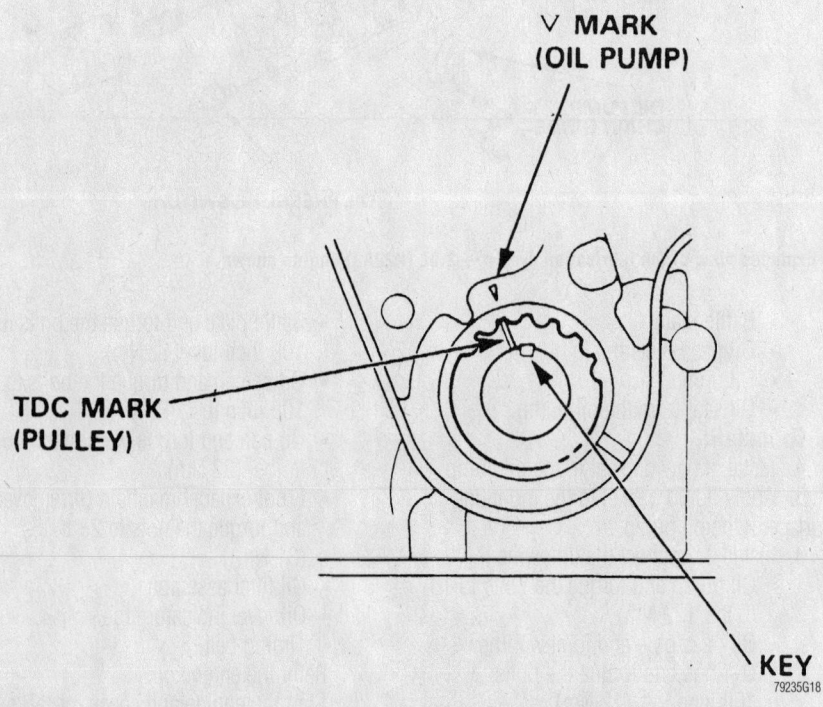

Crankshaft timing belt sprocket alignment marks for TDC—1.8L (B18B1 and B18C1) engines

To install:

9. If the timing belt tensioner was removed, perform the following:

 a. Position the timing belt tensioner on the engine and install the attaching bolt loosely.

 b. Install the timing belt tensioner spring.

 c. Push the tensioner down, then snug the tensioner bolt to hold this position.

➡**Before reinstallation, check every component for cleanliness. All covers, pulleys, shields, etc. must be completely free of grease and oil.**

➡**Install the timing belt in the correct sequence. Also, if installing the old belt, be sure it is turning the same direction.**

10. Install the timing belt first to the crankshaft pulley, then to the adjuster, then to the water pump pulley, the exhaust camshaft and finally to the intake camshaft pulley.

11. Install the lower belt cover. Install the crankshaft pulley, tightening the bolt to 130 ft. lbs. (177 Nm). Lubricate the threads and the flange of the bolt with engine oil before installation.

12. Loosen the adjusting bolt, allowing the adjuster to tension the belt. Retighten the bolt to 40 ft. lbs. (54 Nm).

13. Remove the pin punches from the camshafts.

14. Rotate the crankshaft 4–6 turns counterclockwise. This allows the belt to equalize tension across all of the pulleys.

15. Once again, set the engine to TDC compression for cylinder No. 1. Check that all timing marks for the cam and crankshaft are properly aligned. If any mark is out of alignment, remove the timing belt and reinstall it.

16. Loosen the adjusting bolt 180 degrees (½ turn). Rotate the crankshaft counterclockwise until the camshaft pulleys have moved 3 teeth. Retighten the adjusting bolt to 40 ft. lbs. (54 Nm).

17. Check the torque of the crankshaft pulley bolt.

18. Install or connect the following:
 • Other timing belt covers
 • Rubber seal in the groove of the cylinder head cover. Be sure that the seal and groove are thoroughly clean first.
 • Liquid gasket to the rubber seal at the eight corners of the recesses. Do not install the parts if 20 minutes or more have elapsed since applying the liquid gasket. Instead, reapply liquid gasket after removing the old residue.
 • Cylinder head cover and all other applicable components. Tighten the cylinder head cover nuts in 2 steps to 88 inch lbs. (10 Nm).

3.2L (C32A6) Engines

➡**Under normal driving conditions, the timing belt and timing balancer belt are to be replaced at 105,000 miles (168,000 km).**

1. Before servicing the vehicle, refer to the precautions in the beginning of this section.

2. Be sure to acquire the anti–theft code for the radio and the frequencies for the radio's preset buttons.

3. Remove or disconnect the following:
 • Negative battery cable
 • Left wheel well splash shield
 • Power steering pump belt by loosening the power steering pump's adjusting bolt, locknut and mounting
 • Alternator belt by loosening the alternator adjusting bolt, locknut and mounting nut
 • Alternator terminal and connector

4. To remove the engine-to-chassis center bracket, located at the front of the engine, perform the following procedure:

 a. Using a floor jack, position a cushion between the jack and the oil pan.

 b. Raise the engine slightly to take the weight off of the center bracket.

 c. Remove the center bracket-to-engine bolts, the center bracket-to-chassis bolts and the center bracket.

5. Remove or disconnect the following:
 • TCS upper and lower brackets
 • TCS throttle sensor and actuator connectors and remove the TCS control valve assembly
 • Oil pressure switch connector, the engine ground cable and the engine wire harness cover, from the front of the engine
 • Dipstick and the dipstick tube-to-engine bolt; then, pull the tube from its O–ring mount. Discard the O–ring.

6. Turn the engine to align the timing marks and set cylinder No. 1 to Top Dead Center (TDC) on the compression stroke. The white mark on the crankshaft pulley should align with the pointer on the timing belt cover. Remove the inspection caps on the upper timing belt covers to check the alignment of the timing marks. The pointers for the camshafts should align with the marks on the camshaft sprockets.

7. Using the Holder Handle and the 50mm Offset Holder Attachment tool 07MAB–PY3010A and a 19mm socket, secure the crankshaft pulley and remove the crankshaft pulley bolt.
 • Washer and pull the crankshaft pulley from the engine
 • Necessary components for access to the timing belt covers
 • Upper and lower timing belt covers. Clean any dirt, oil or grease from the covers. Do not use the covers for storing removed items.

8. Loosen the timing belt tensioner adjusting bolt 180 degrees (½ turn). Push on the tensioner to remove tension from the timing belt, then tighten the adjusting bolt.
 • Timing belt

9. If necessary, remove the timing belt tensioner by performing the following:
 a. Remove the spring from the tensioner.
 b. Remove the bolt mounting the tensioner, then remove the tensioner.

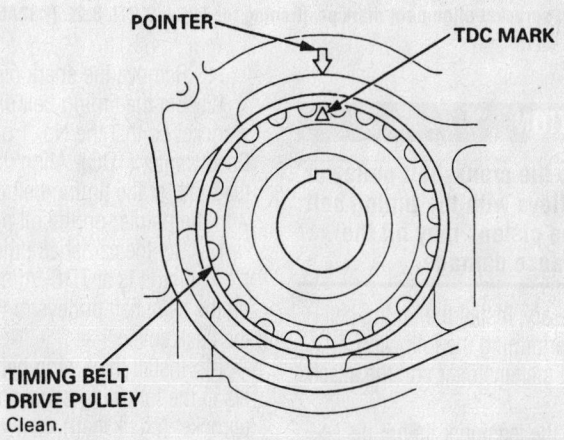

POINTER — TDC MARK

TIMING BELT DRIVE PULLEY Clean.

79235G04

Crankshaft timing belt sprocket alignment mark locations—3.2TL 3.2L (C32A6) engines

LEFT:

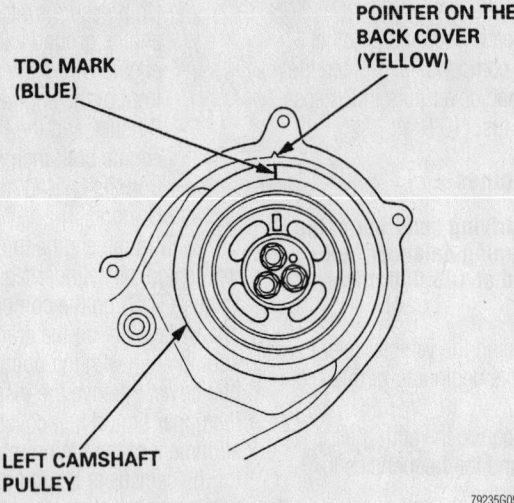

Left camshaft sprocket alignment mark positioning for TDC—3.2TL 3.2L (C32A6) engines

RIGHT:

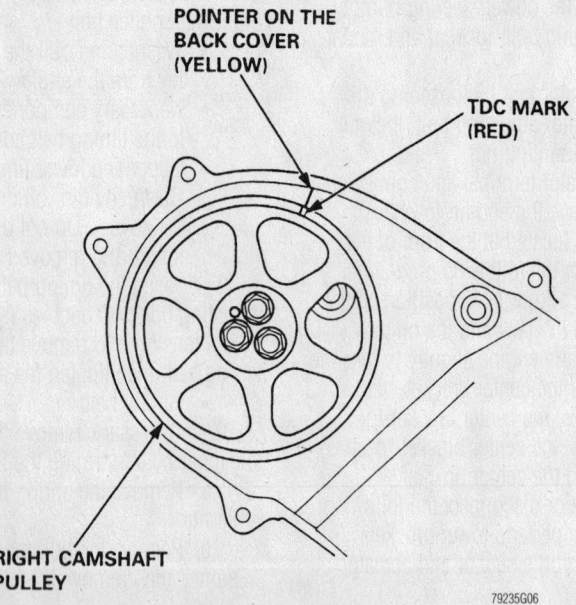

Right camshaft sprocket alignment mark positioning for TDC—3.2TL 3.2L (C32A6) engines

To install:

> **⁂ CAUTION**
>
> **Do not rotate the crankshaft pulley or camshaft pulleys with the timing belt removed. The pistons may hit the valves and cause damage.**

10. If necessary, install the timing belt tensioner by performing the following:

   a. Install the tensioner and the attaching bolt.

   b. Move the tensioner its full deflection to the left and tighten the bolt.

   c. Install the spring to the tensioner.

11. Remove the spark plugs.

12. Set the timing belt drive (crankshaft) sprocket so that the No. 1 piston is at Top Dead Center (TDC). Align the TDC mark on the tooth of the timing belt drive sprocket with the pointer on the oil pump.

13. Set the camshaft pulleys so that the No. 1 piston is at TDC. Align the TDC mark on the camshaft pulleys to the pointers on the back covers.

14. Install the timing belt on the sprockets in the following sequence: drive sprocket (crankshaft), tensioner pulley, left camshaft sprocket, water pump pulley, right camshaft sprocket.

15. Loosen, then retighten the timing belt adjuster bolt to tension the timing belt.

16. Install the lower timing belt cover.

17. Install the crankshaft pulley and finger-tighten the bolt and washer. Using the Holder Handle and the 50mm Offset Holder Attachment tool 07MAB–PY3010A and a 19mm socket with a torque wrench, tighten the crankshaft pulley bolt to181 ft. lbs. (245 Nm).

18. Rotate the crankshaft 5–6 turns clockwise so that the timing belt positions itself properly on the sprockets.

19. Set cylinder No. 1 to TDC by aligning the timing marks. If the timing marks do not align, remove the timing belt, then adjust the components and reinstall the timing belt.

20. Rotate the crankshaft clockwise enough to move the camshaft pulley nine teeth (the blue mark on the crankshaft pulley should align with the pointer on the lower cover).

21. Loosen the timing belt adjusting bolt 180 degrees (½ turn), then tighten the bolt to 31 ft. lbs. (42 Nm).

22. Install the upper timing belt covers, then install all applicable components. When installing the center bracket, tighten the bolts attaching the brackets to 40 ft. lbs. (54 Nm), then the mount through–bolt to 40 ft. lbs. (54 Nm).

23. To complete the installation, reverse the removal procedures. Adjust the tension of the drive belts.

3.0L (J30A1) and 3.2L (J32A1) Engines

1. Before servicing the vehicle, refer to the precautions in the beginning of this section.

2. Remove or disconnect the following:
   - Negative battery cable
   - Ignition coil cover
   - Front tire/wheel assemblies
   - Splash shield from under the vehicle
   - Drive belts

3. To remove the engine-to-chassis side mount, located at the front of the engine, perform the following procedure:

   a. Using a floor jack, position a cushion between the jack and the oil pan.

   b. Raise the engine slightly to take the weight off of the side mount.

   c. Remove the side mount-to-engine bracket bolt, the side mount-to-chassis bolts and the side mount.

4. Remove the dipstick and the dipstick tube-to-engine bolt; then, pull the tube from its O–ring mount. Discard the O–ring.

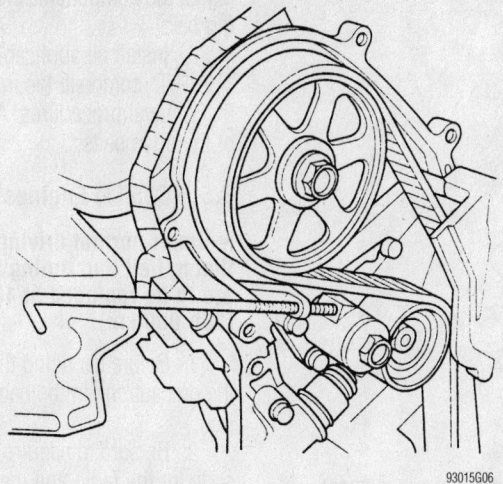

Using battery clamp bolt to hold timing belt adjuster in position—3.0L (J30A1) and 3.2L (J32A1) engines

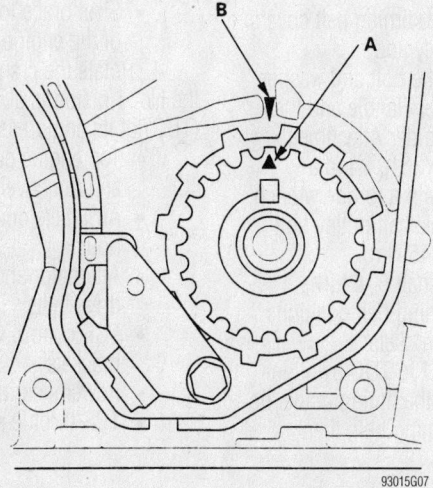

Crankshaft sprocket alignment mark positioning for TDC—3.0L (J30A1) and 3.2L (J32A1) engines

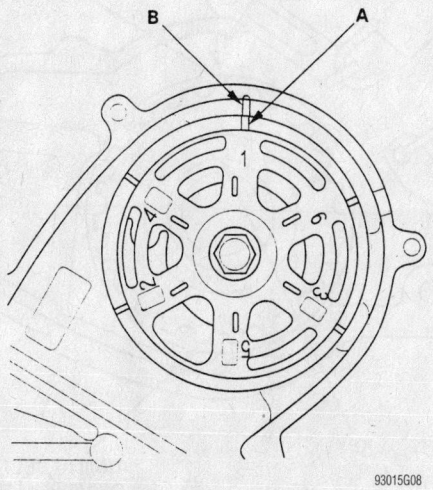

Left camshaft sprocket alignment mark positioning for TDC—3.0L (J30A1) and 3.2L (J32A1) engines

5. Turn the engine to align the timing marks and set cylinder No. 1 to Top Dead Center (TDC) on the compression stroke. The white mark on the crankshaft pulley should align with the pointer on the timing belt cover. Remove the inspection caps on the upper timing belt covers to check the alignment of the timing marks. The pointers for the camshaft pulleys should align with the marks on rear upper cover mark.

6. Using the Holder Handle and the 50mm Offset Holder Attachment tool 07MAB–PY3010A and a 19mm socket, secure the crankshaft pulley and remove the crankshaft pulley bolt.

7. Remove or disconnect the following:

- Washer and pull the crankshaft pulley from the engine
- All necessary components for access to the timing belt covers
- Upper and lower timing belt covers. Clean any dirt, oil or grease from the covers. Do not use the covers for storing removed items.
- One of the battery clamp bolts and grind a 45° bevel on the threaded end. Screw the battery clamp bolt into hole provided at the base of the right cylinder head to hold the timing belt adjuster in it's current position; tighten the bolt by hand, Do NOT use a wrench.
- Engine mount bracket

8. At the base of the left cylinder head, loosen the idler pulley bolt about 5–6 turns; then, remove the timing belt.

To install:

⚹⚹ **CAUTION**

Do not rotate the crankshaft pulley or camshaft pulleys with the timing belt removed. The pistons may hit the valves and cause damage.

9. Remove the spark plugs.

10. Set the timing belt drive (crankshaft) sprocket so that the No. 1 piston is at Top Dead Center (TDC). Align the TDC mark on the tooth of the timing belt drive sprocket with the pointer on the oil pump.

11. Set the camshaft pulleys so that the No. 1 piston is at TDC. Align the TDC mark on the camshaft pulleys to the pointers on the back covers.

12. Remove the battery clamp bolt from the back cover. Remove the auto-tensioner.

13. Service the auto-tensioner by performing the following procedure:

  a. Position the auto-tensioner in a soft jawed vise with the maintenance bolt

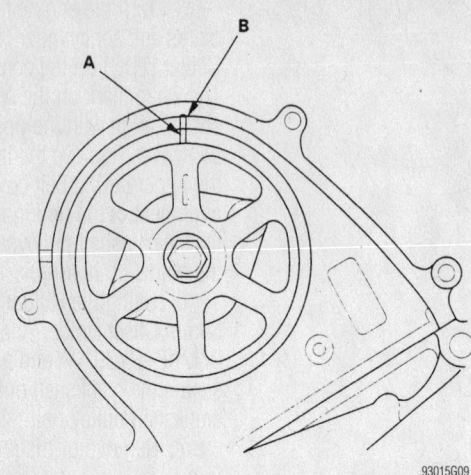

Right camshaft sprocket alignment mark positioning for TDC—3.0L (J30A1) and 3.2L (J32A1) engines

facing upward. DO NOT grip the body of the auto-tensioner.

b. Remove the maintenance bolt.

c. Be careful not to spill the oil from inside the tensioner. If oil is spilled, replenish it; the total capacity is 0.22 fl. oz. (6.5 ml).

d. Using Stopper tool 14540–P8A–A01, position it on the auto-tensioner while turning the internal screw.

e. Insert a flat–blade screwdriver into the maintenance hole and turn it clockwise to compress the bottom.

✷✷ WARNING

Be careful not to damage the threads or the gasket contact surface with the screwdriver.

f. Using a new gasket, reinstall the maintenance bolt and torque it to 6 ft. lbs. (8 Nm).

g. Make sure that no oil is leaking around the maintenance bolt and install the auto-tensioner; torque the bolts 33 ft. lbs. (44 Nm).

➡**Make sure that the Stopper tool 14540–P8A–A01 stays in place.**

14. Install the timing belt on the sprockets in the following sequence: drive sprocket (crankshaft), idler pulley, left camshaft sprocket, water pump pulley, right camshaft sprocket and adjusting pulley.

15. Torque the idler pulley bolt to 33 ft. lbs. (44 Nm).

16. Remove the Stopper tool from the auto-tensioner.

17. Install or connect the following:
- Engine mount-to-engine and torque the No. 10 bolts to 33 ft. lbs. (44 Nm) and the No. 6 bolt to 8.7 ft. lbs. (12 Nm)
- Lower and upper timing belt covers
- Crankshaft pulley and finger–tighten the bolt and washer. Using the Holder Handle and the 50mm Offset Holder Attachment tool 07MAB–PY3010A and a 19mm socket with a torque wrench, tighten the crankshaft pulley bolt to181 ft. lbs. (245 Nm).

18. Rotate the crankshaft 5–6 turns clockwise so that the timing belt positions itself properly on the sprockets.

19. Set cylinder No. 1 to TDC by aligning the timing marks. If the timing marks do not align, remove the timing belt, then adjust the components and reinstall the timing belt.

20. Install all applicable components.

21. To complete the installation, reverse the removal procedures. Adjust the tension of the drive belts.

3.5L (C35A1) Engines

➡**Under normal driving conditions, the timing belt and timing balancer belt are to be replaced at 105,000 miles (168,000 km).**

1. Before servicing the vehicle, refer to the precautions in the beginning of this section.

2. Be sure to acquire the anti–theft code for the radio and the frequencies for the radio's preset buttons.

3. Remove or disconnect the following:
- Negative battery cable
- Strut brace located at the top rear of the engine, if necessary

4. Rotate the crankshaft pulley so that the No. 1 piston is at Top Dead Center (TDC) of its compression stroke.
- Top engine cover-to-engine bolts and the cover
- Air intake duct and air cleaner housing
- Alternator and A/C compressor drive belts
- TCS control valve upper and lower brackets
- Power steering belt
- TCS throttle sensor connector, the

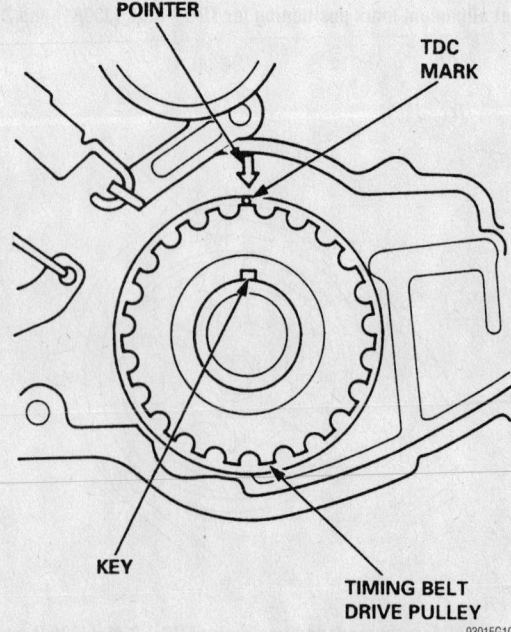

Crankshaft timing belt pulley alignment mark locations—3.5L (C35A1) engine

TCS throttle actuator connector and the Throttle Position (TP) sensor
- TCS control valve assembly.
- Vehicle Speed Sensor (VSS) harness connector and remove the wire harness holder
- Breather and vacuum hoses
- Ignition Control Module (ICM) bracket from the right timing belt cover
- Idler pulley bracket from the left side of the crankshaft pulley
- Dipstick and the dipstick tube-to-engine bolt; then, pull the tube

from its O–ring mount. Discard the O–ring.

5. Using the Holder Handle and the 50mm Offset Holder Attachment tool 07MAB–PY3010A and a 19mm socket, secure the crankshaft pulley and remove the crankshaft pulley bolt.
- Washer and pull the crankshaft pulley from the engine
- Upper and lower timing belt cover

6. Loosen the balancer belt adjusting nut 180° (½) turn. Push the tensioner to relieve the tension from the balancer belt;

then, retighten the adjusting bolt. Remove the balancer belt.

7. Loosen the timing belt adjusting nut 180° (½) turn. Push the tensioner to relieve the tension from the timing belt; then, retighten the adjusting bolt. Remove the timing belt.

To install:

8. Remove the spark plugs.

9. Remove the balancer belt drive pulley and the timing belt guide plate from the crankshaft.

10. Clean the upper and lower timing belt covers.

11. Position the timing belt pulley so the No. 1 piston is at TDC of its compression stroke. Align the mark on the pulley (near keyway) with the pointer mark on the oil pump.

12. Adjust the camshaft pulley so that the No. 1 piston is the TDC of the compression stroke. Align the TDC marks on the pulley with the upper surface of the back cover; the arrow mark on the left camshaft pulley and the "1" on the right camshaft pulley should be facing the back cover pointer.

13. Install the timing belt in the following sequence:
 a. Crankshaft timing belt pulley sprocket.
 b. Adjusting pulley.
 c. Left camshaft pulley.
 d. Water pump pulley.
 e. Right camshaft pulley.

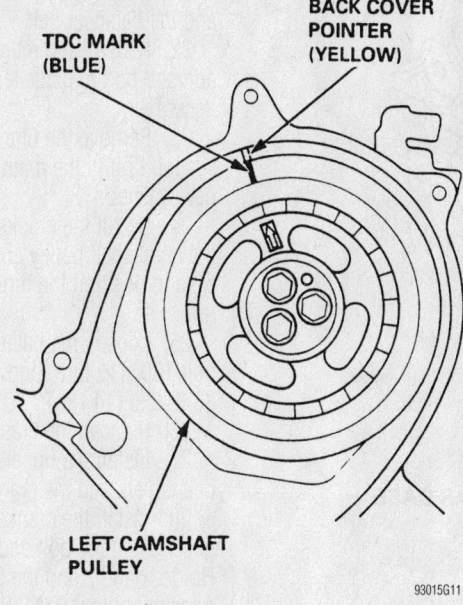

Left camshaft timing belt pulley alignment mark locations—3.5L (C35A1) engine

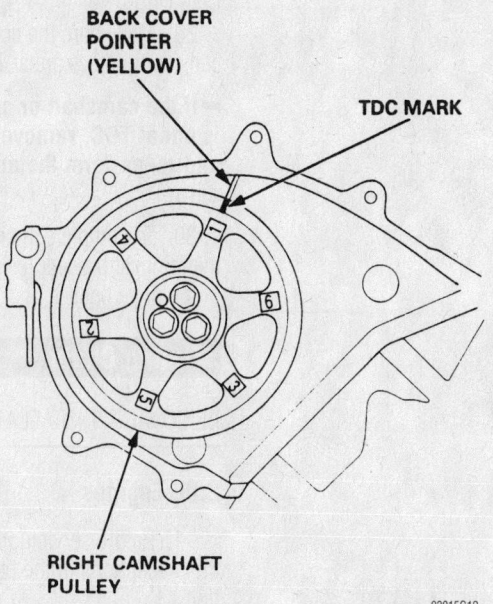

Right camshaft timing belt pulley alignment mark locations—3.5L (C35A1) engine

❊❊ WARNING

Make sure that the camshaft and crankshaft pulleys are at TDC.

➡**For easier installation, turn the right camshaft pulley about ½ tooth from TDC.**

14. Loosen and retighten the timing belt adjusting bolt to tension the timing belt.

15. Install the lower cover and the crankshaft pulley.

16. Rotate the crankshaft pulley about 5–6 turns clockwise to position the timing belt on the pulleys.

17. To adjust the timing belt tension, perform the following procedure:
 a. Make sure that the No. 1 piston is at TDC of its compression stroke.
 b. Rotate the crankshaft clockwise ten teeth on the camshaft pulley; the blue mark on the crankshaft pulley should align with the lower cover pointer.
 c. Loosen the adjusting nut 180° (½ turn).
 d. Tighten the adjusting nut to 31 ft. lbs. (42 Nm).

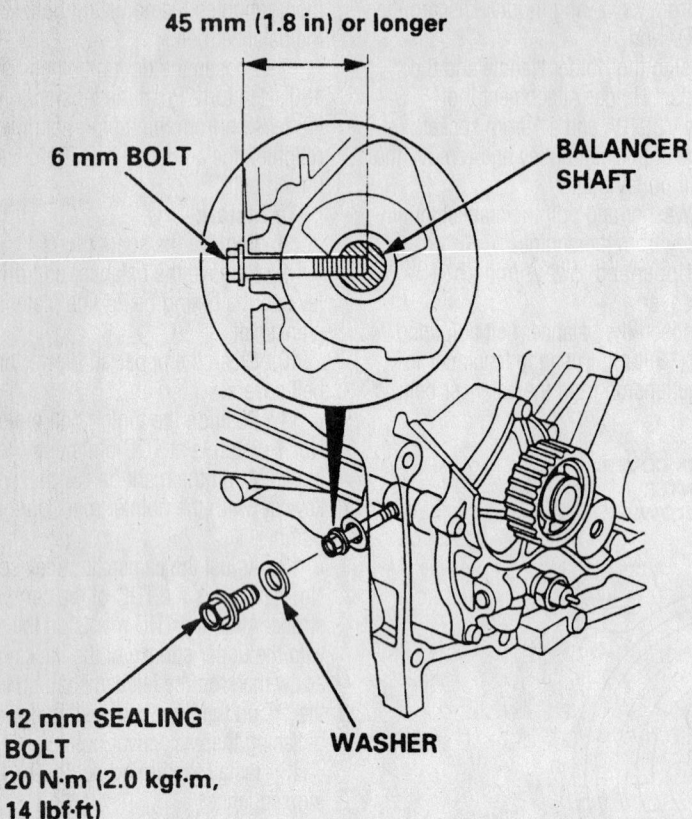

45 mm (1.8 in) or longer

6 mm BOLT

BALANCER SHAFT

12 mm SEALING BOLT
20 N·m (2.0 kgf·m, 14 lbf·ft)

WASHER

93015G13

Securing the balancer shaft—3.5L (C35A1) engine

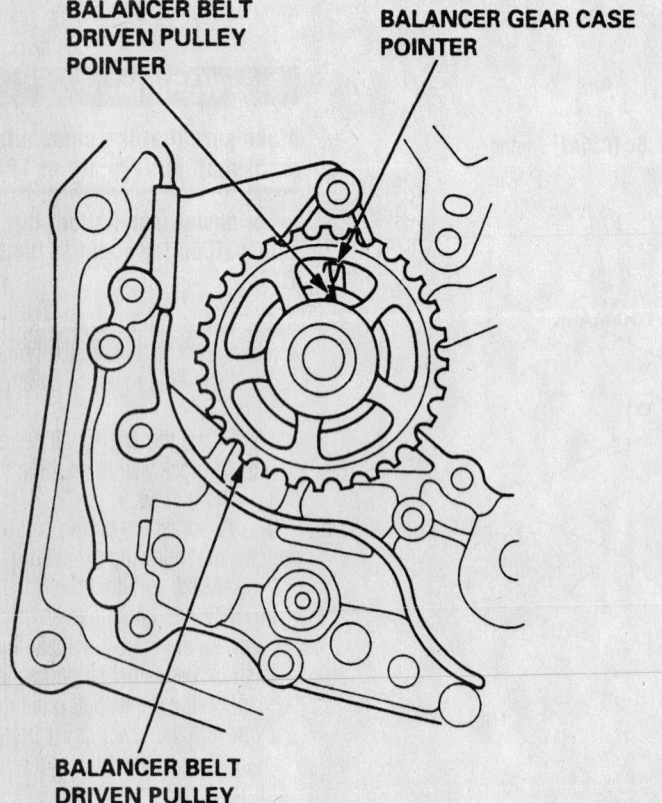

BALANCER BELT DRIVEN PULLEY POINTER

BALANCER GEAR CASE POINTER

BALANCER BELT DRIVEN PULLEY

93015G14

Balancer shaft alignment mark locations—3.5L (C35A1) engine

18. Remove the crankshaft pulley and the lower cover; then, install the timing belt guide plate and the balancer belt drive pulley.

19. Align the balancer shaft pulley by performing the following procedure:

　a. Using a 6 x 45mm bolt, insert it into the maintenance hole and the balancer shaft.

　b. Align the pointer on the balancer belt pulley with the pointer on the balancer gear case.

20. Adjust the timing belt drive pulley so that the No. 1 piston is at TDC of the compression stroke.

21. Install the balancer belt drive pulley and the balancer belt.

22. Loosen and retighten the balancer adjuster bolt to place tension on the balancer belt.

23. Remove the 6mm bolt and install the sealing bolt in the maintenance hole using a new washer.

24. Install the crankshaft pulley. Rotate the crankshaft pulley about 5–6 turns clockwise to position the timing belt on the pulleys.

25. Loosen the balancer belt adjuster bolt 180° (½ turn) and retighten the bolt to 33 ft. lbs. (44 Nm).

26. Remove the crankshaft pulley.

27. Install the upper and lower timing belt covers and the crankshaft pulley.

28. Install the crankshaft pulley and finger–tighten the bolt and washer. Using the Holder Handle and the 50mm Offset Holder Attachment tool 07MAB–PY3010A and a 19mm socket with a torque wrench, tighten the crankshaft pulley bolt to 181 ft. lbs. (245 Nm).

29. Make sure the crankshaft and camshaft pulleys are at TDC.

➡**If the camshaft or crankshaft pulley is not at TDC, remove the timing belt and re–perform the adjustment procedure.**

30. To complete the installation, reverse the removal procedures. Adjust the tension of the drive belts.

Timing Chain

REMOVAL & INSTALLATION

2.0L Engines

1. Before servicing the vehicle, refer to the precautions in the beginning of this section.

2. Rotate the crankshaft pulley so that

the No. 1 piston is at Top Dead Center (TDC) of its compression stroke.

3. Remove or disconnect the following:

- Front wheels and tires
- Splash shield
- Drive belt
- Cylinder head cover

4. Hold the crankshaft pulley with holder handle 07JAB–001020A and holder attachment 07NAB–001040A, then remove the pulley bolt with a 19mm socket and breaker bar.

- Oil cooler hose joint pipe from the water pump, K20A2 engines
- Crankshaft Position (CKP) sensor connector and Variable Valve Timing Control (VTC) oil control solenoid valve connector
- VTC oil control solenoid valve

5. Support the engine under the oil pan with a jack and block of wood.

- Ground cable and upper bracket
- Side engine mounting bracket
- Retaining bolts and timing chain case

6. Loosely install the crankshaft pulley.

7. Turn the crankshaft counterclockwise to compress the auto-tensioner.

8. Align the holes on the lock and auto-tensioner, then insert a 0.06 in. (1.5mm) pin into the hole. Turn the crankshaft clockwise to hold the pin.

9. Auto-tensioner

- Timing chain guide B
- Timing chain guide A and tensioner arm
- Timing chain

➡**Do not let the timing chain near any magnetic fields.**

To install:

10. Set the crankshaft to TDC. Align the TDC mark on the crankshaft sprocket with the pointer on the cylinder block.

11. Set the camshafts to TDC. The punch mark (arrow) on the VTC actuator and punch mark on the exhaust camshaft sprocket should be at the top. Align the TDC marks on the VTC actuator and exhaust camshaft sprocket.

12. Install or connect the following:

- Timing chain on the crankshaft sprocket with the colored piece aligned with the punch mark on the crankshaft sprocket
- Timing chain on the VTC actuator and exhaust camshaft sprocket with the punch marks aligned with the 2 colored pieces
- Timing chain guide A and tighten the bolts to 8.7 ft. lbs. (12 Nm)

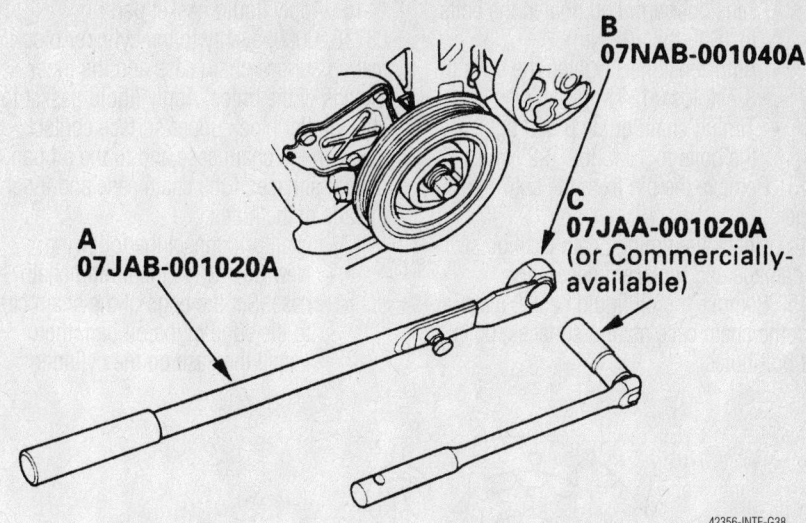

B
07NAB-001040A

C
07JAA-001020A
(or Commercially-available)

A
07JAB-001020A

42356-INTE-G38

Hold the crankshaft pulley with holder handle (A) and holder attachment (B)—2.0L engines

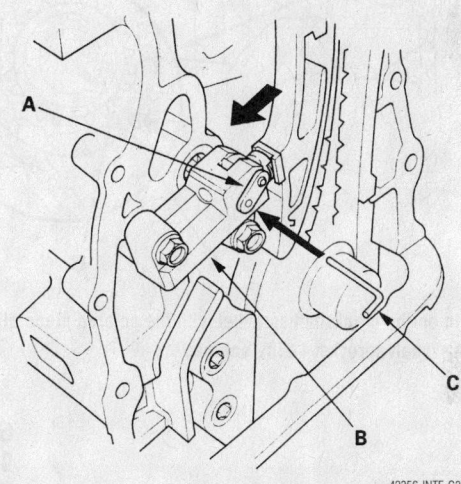

42356-INTE-G37

Align the holes on the lock (A) and auto-tensioner (B), then insert a 0.06 in. (1.5mm) pin (C) into the hole–2.0L engine

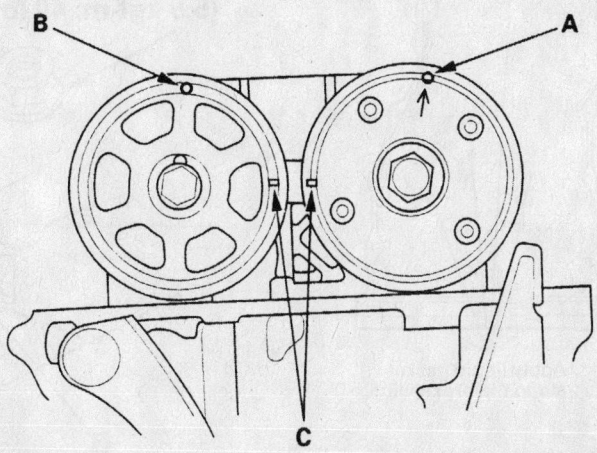

42356-INTE-G24

The punch mark (arrow) (A) on the VTC actuator and punch mark (B) on the exhaust camshaft sprocket should be at the top. Align the TDC marks (C)on the VTC actuator and exhaust camshaft sprocket

- Tensioner arm and tighten the bolts to 16 ft. lbs. (22 Nm)
- Auto-tensioner. Tighten the bolts to 8.7 ft. lbs. (12 Nm).
- Timing chain guide B and tighten the bolts to 16 ft. lbs. (22 Nm)

13. Remove the pin from the auto-tensioner.

14. Check the timing chain case oil seal for damage and replace if necessary.

15. Remove the oil liquid gasket material from the chain case mating surfaces, bolts and bolt holes.

16. Apply liquid gasket part no. 08718-0009, evenly to the cylinder block mating surface chain case and the inner threads of the holes. Apply liquid gasket to the cylinder block upper surface contact areas on the chain case and to the oil pan mating surface of the chain case and inner threads of the holes.

17. Install or connect the following:
- New O-ring on the timing chain case. Set the edge of the chain case to the edge of the oil pan, then install the case on the cylinder block. Tighten the case bolts to 8.7 ft. lbs. (12 Nm).

> ❋❋ **WARNING**
>
> **When installing the chain case, do not slide the bottom surface on the oil pan mounting surface.**

- Side engine mount bracket and tighten to 33 ft. lbs. (44 m)
- Upper bracket, then tighten the bolts/nuts in sequence as shown in the illustration
- Ground cable
- VTC coil control solenoid valve
- CKP sensor connector and VTC oil control solenoid valve connector
- Oil cooler hose using a new O-ring

18. Clean the crankshaft pulley and pulley bolt, then apply lubrication to the pulley bolt and washer.
- Crankshaft pulley and hold the pulley with the holder handle and holder attachment. Using a 19mm socket and torque wrench, tighten the pulley bolt to 181 ft. lbs. (245 Nm).
- Cylinder head cover
- Drive belt
- Splash shield
- Front wheels and tires

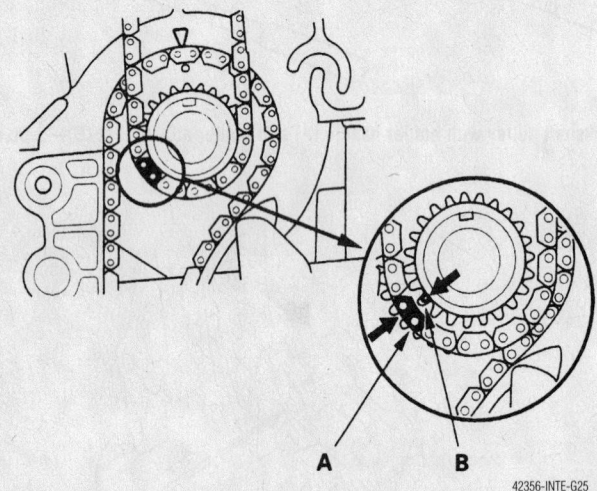

42356-INTE-G25

Install the timing chain on the crankshaft sprocket with the colored piece aligned with the punch mark on the crankshaft sprocket—2.0L engines

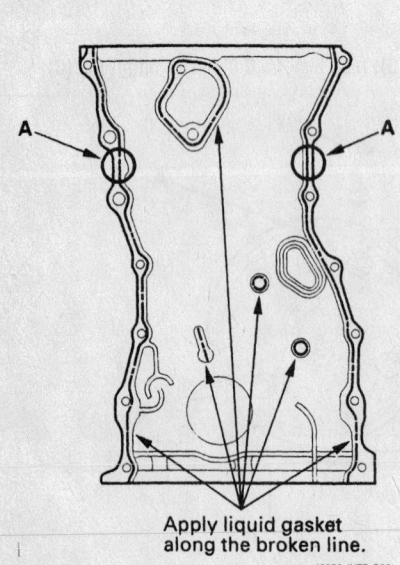

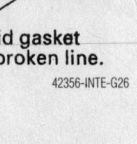

Apply liquid gasket along the broken line.

42356-INTE-G26

Apply liquid gasket to the points shown—2.0L engines

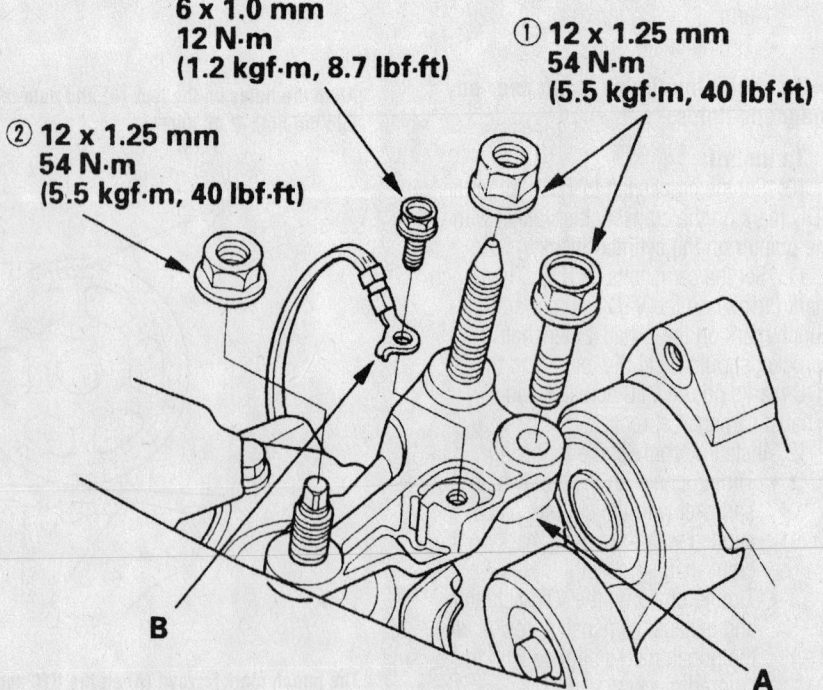

6 x 1.0 mm
12 N·m
(1.2 kgf·m, 8.7 lbf·ft)

② 12 x 1.25 mm
54 N·m
(5.5 kgf·m, 40 lbf·ft)

① 12 x 1.25 mm
54 N·m
(5.5 kgf·m, 40 lbf·ft)

42356-INTE-G27

Upper bracket tightening sequence and specifications—2.0L engines

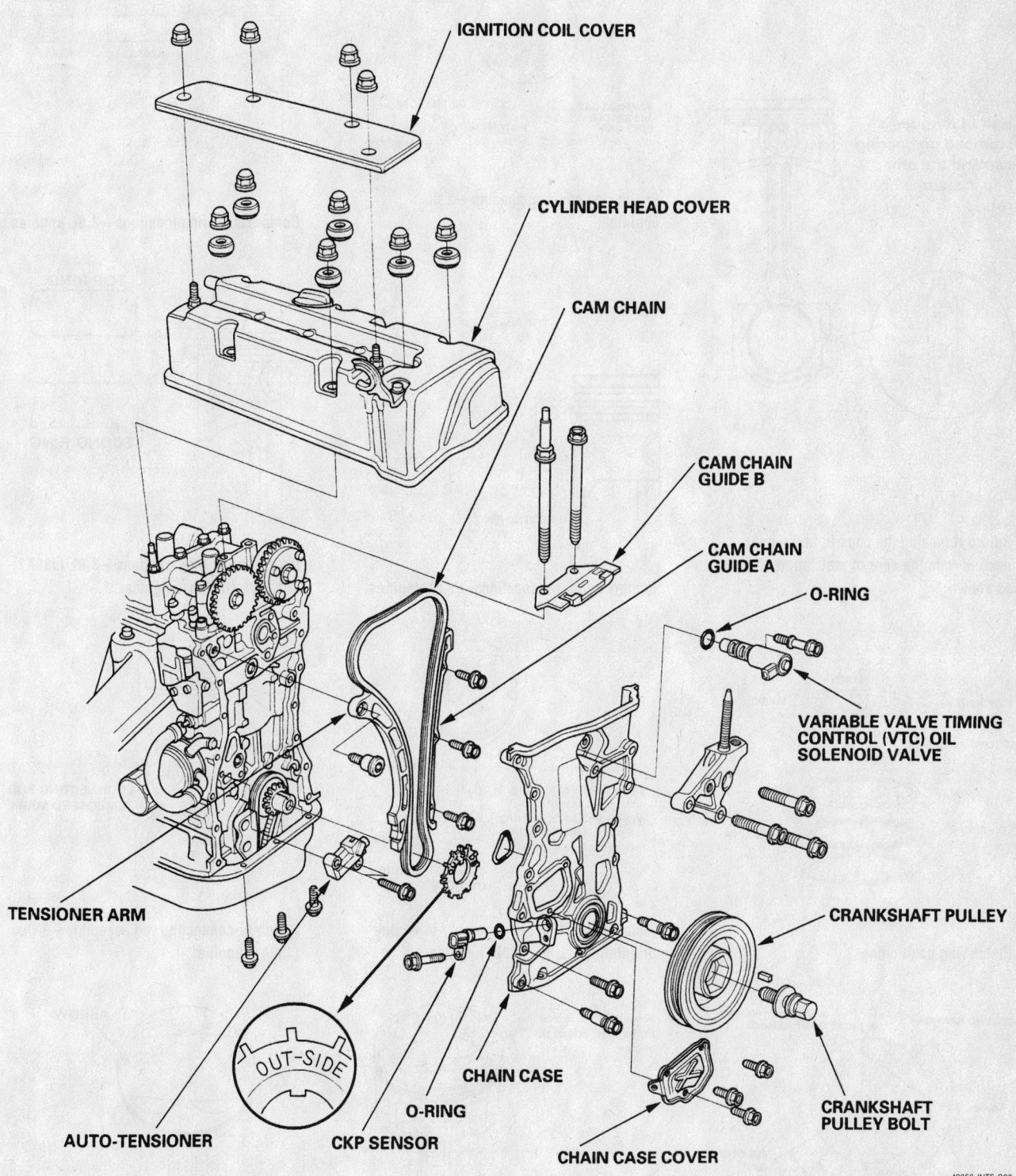

IGNITION COIL COVER

CYLINDER HEAD COVER

CAM CHAIN

CAM CHAIN GUIDE B

CAM CHAIN GUIDE A

O-RING

VARIABLE VALVE TIMING CONTROL (VTC) OIL SOLENOID VALVE

CRANKSHAFT PULLEY

CRANKSHAFT PULLEY BOLT

CHAIN CASE COVER

CHAIN CASE

O-RING

CKP SENSOR

AUTO-TENSIONER

OUT-SIDE

TENSIONER ARM

42356-INTE-G28

Exploded view of the timing (cam) chain and related components—2.0L engines

Piston and Ring

POSITIONING

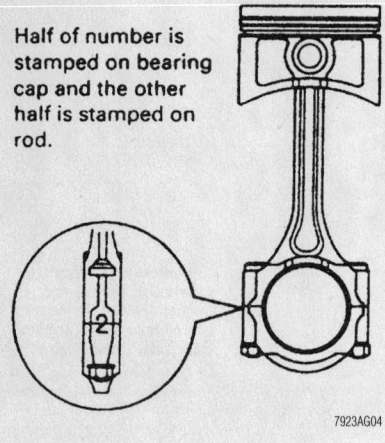

Half of number is stamped on bearing cap and the other half is stamped on rod.

Before removing the caps from the connecting rods, be sure to matchmark them as shown

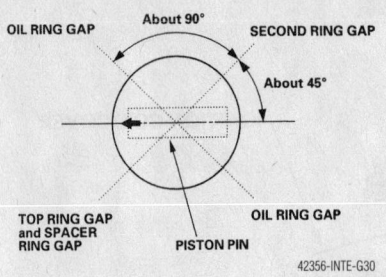

Piston ring end-gap spacing—2.0L engines

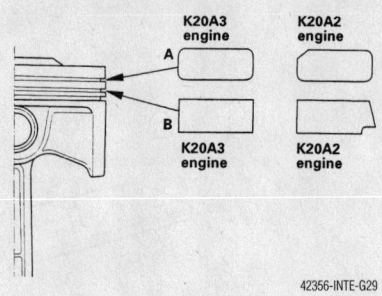

Compression ring locations—2.0L engines

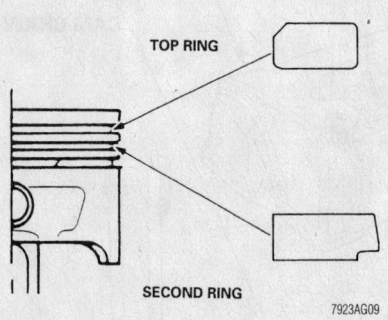

Compression ring locations—1.8L engines

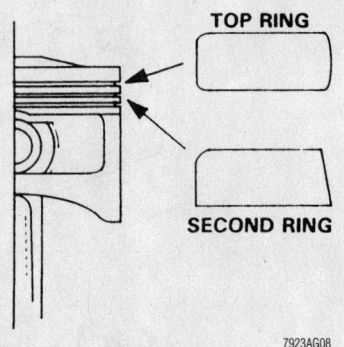

Compression ring locations—3.0L (J30A1) and 3.2L (C32A1) engines

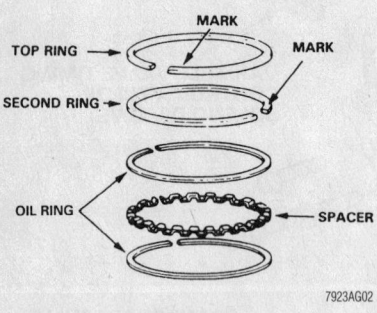

Piston ring positioning

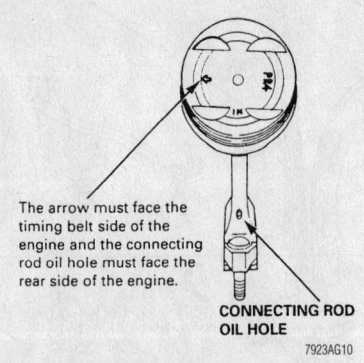

Piston/connecting rod assembly-to-engine orientation—1.8L (B18B1) engine

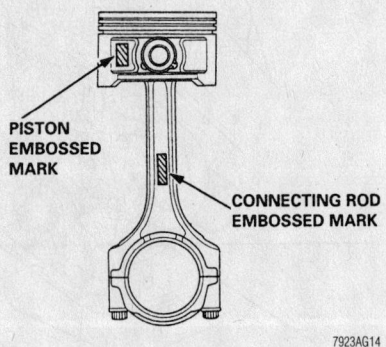

Piston-to-connecting rod assembly—3.0L (J30A1) engine

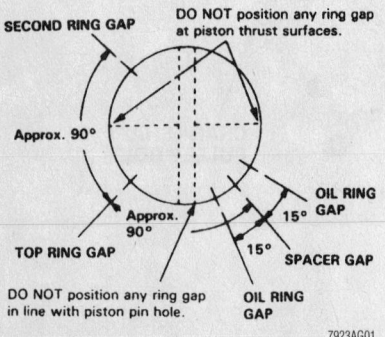

Piston ring end-gap spacing—except 2.0L engines

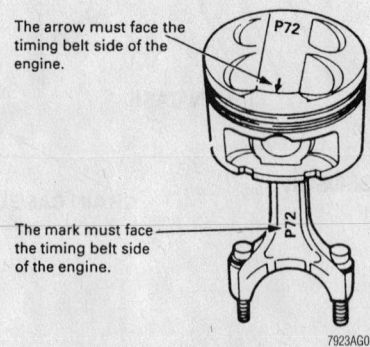

Piston/connecting rod assembly-to-engine orientation—1.8L (B18C1) engine

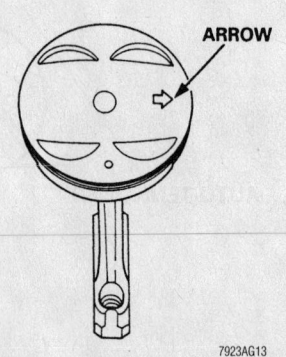

Piston directional arrow location—3.0L (J30A1) engine

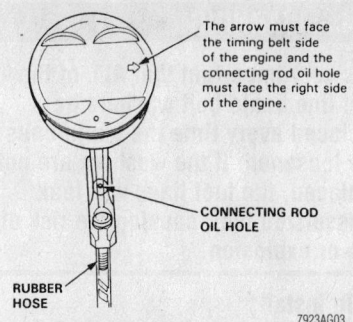

Piston/connecting rod assembly-to-engine orientation—3.2L (C32A6) and (J32A1) engines

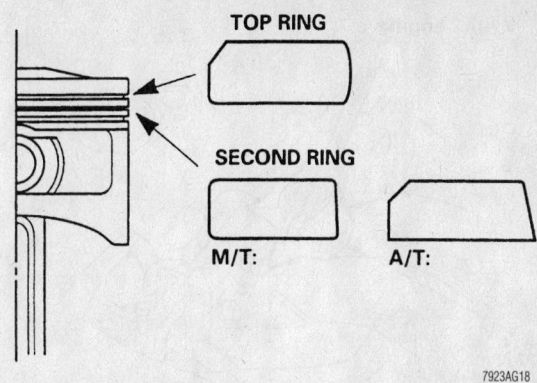

Compression ring locations—3.0L (C30A1), (C32A6), (J32A1) and 3.2L (C32B1) engines

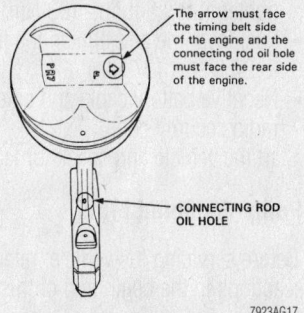

Piston/connecting rod assembly-to-engine orientation—3.0L (C30A1) and 3.2L (C32B1) engines

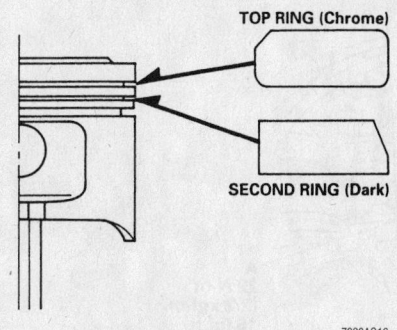

Compression ring locations—3.5L engine

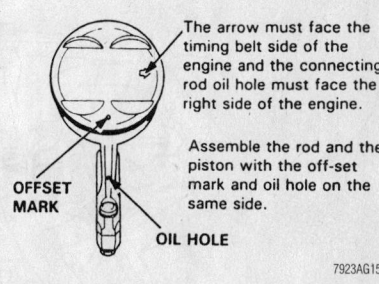

Piston/connecting rod assembly-to-engine orientation—3.5L engine

FUEL SYSTEM

Fuel System Service Precautions

Safety is the most important factor when performing not only fuel system maintenance but also any type of maintenance. Failure to conduct maintenance and repairs in a safe manner may result in serious personal injury or death. Maintenance and testing of the vehicle's fuel system components can be accomplished safely and effectively by adhering to the following rules and guidelines.

• To avoid the possibility of fire and personal injury, always disconnect the negative battery cable unless the repair or test procedure requires that battery voltage be applied.

• Always relieve the fuel system pressure prior to disconnecting any fuel system component (injector, fuel rail, pressure regulator, etc.), fitting or fuel line connection. Exercise extreme caution whenever relieving fuel system pressure, to avoid exposing skin, face and eyes to fuel spray. Please be advised that fuel under pressure may penetrate the skin or any part of the body that it contacts.

• Always place a shop towel or cloth around the fitting or connection prior to loosening to absorb any excess fuel due to spillage. Ensure that all fuel spillage (should it occur) is quickly removed from engine surfaces. Ensure that all fuel soaked cloths or towels are deposited into a suitable waste container.

• Always keep a dry chemical (Class B) fire extinguisher near the work area.

• Do not allow fuel spray or fuel vapors to come into contact with a spark or open flame.

• Always use a back—up wrench when loosening and tightening fuel line connection fittings. This will prevent unnecessary stress and torsion to fuel line piping. Always follow the proper torque specifications.

• Always replace worn fuel fitting O—rings with new. Do not substitute fuel hose or equivalent, where fuel pipe is installed.

Fuel System Pressure

RELIEVING

1. Before servicing the vehicle, refer to the precautions in the beginning of this section.

2. Disconnect the negative battery cable.

3. Remove the fuel fill cap.

4. Use a box wrench on the 6mm service bolt or fuel pulsation damper, as applicable, on the fuel rail while holding the special banjo bolt with another wrench.

5. Place a rag or shop towel over the bolt/pulsation damper.

6. Slowly loosen the bolt/damper 1 complete turn.

❋❋ CAUTION

Do not allow fuel spray or fuel vapors to come in contact with a spark or open flame. Keep a dry chemical fire extinguisher nearby. Never store fuel in an open container due to risk of fire or explosion.

➡A fuel pressure gauge may be attached at the 6mm service bolt/pulsation damper location. Always replace the washer between the service bolt/pulsation damper and the banjo bolt whenever it is loosened.

7. Properly dispose of the rag or shop towel.

8. Remove the service bolt/damper and install a new washer. Tighten the 6mm service bolt to 104 inch lbs. (12 Nm), if

K20A2 engine

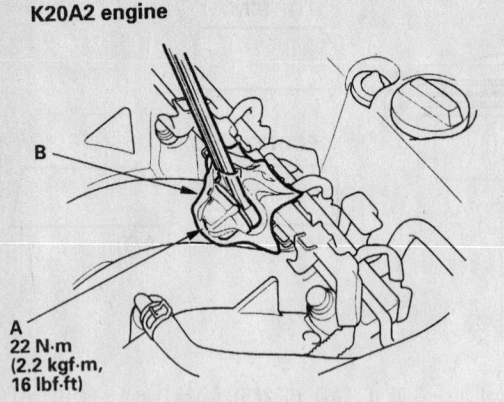

B

A
22 N·m
(2.2 kgf·m,
16 lbf·ft)

K20A3 engine

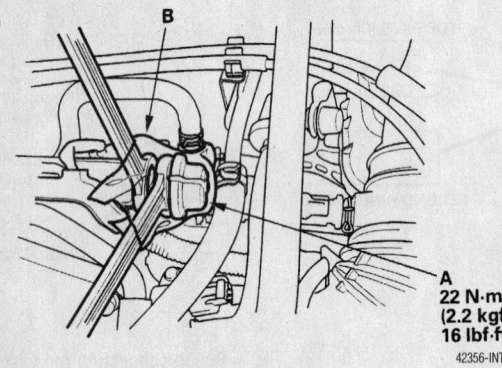

B

A
22 N·m
(2.2 kgf·m,
16 lbf·ft)

42356-INTE-G31

View of the fuel pulsation damper (A)—2.0L engines

equipped. If equipped with a fuel pulsation damper, tighten to 16 ft. lbs. (22 Nm).

9. Clean up any fuel spilled on the engine and intake manifold.

10. Install the fuel fill cap.

11. Reconnect the negative battery cable.

12. After servicing the vehicle, turn the ignition **ON**, but don't start the engine. Repeat this process 2 or 3 times to pressurize the fuel system. Check for fuel leaks.

13. Enter the radio security code.

Fuel Filter

REMOVAL & INSTALLATION

Except Fuel Pump–Mounted Filters

1. Before servicing the vehicle, refer to the precautions in the beginning of this section.

2. Disconnect the negative battery cable.

3. Properly relieve the fuel pressure.

4. Wrap a shop towel around the filter fittings. Use a properly sized wrench to slowly loosen the fuel line fittings.

5. Remove or disconnect the following:
- Fuel pipes from the fuel filter
- Fuel filter clamp
- Filter from the vehicle

To install:

6. Install or connect the following:
- New filter in position and tighten the clamp mounting bolt to 89 inch lbs. (10 Nm)
- Banjo bolt with new washers and tighten it to 25 ft. lbs. (33 Nm)
- Fuel feed line and tighten the fitting to 27 ft. lbs. (37 Nm)
- Negative battery cable and enter the radio security code

7. Start the vehicle and check for leaks.

Fuel Pump–Mounted Filter

1. Before servicing the vehicle, refer to the precautions in the beginning of this section.

2. Properly relieve the fuel pressure.

3. Remove or disconnect the following:
- Negative battery cable

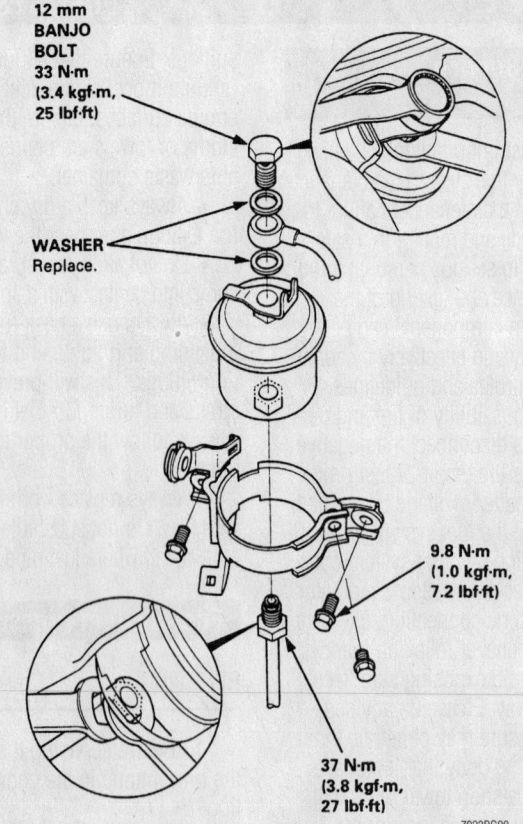

12 mm
BANJO
BOLT
33 N·m
(3.4 kgf·m,
25 lbf·ft)

WASHER
Replace.

9.8 N·m
(1.0 kgf·m,
7.2 lbf·ft)

37 N·m
(3.8 kgf·m,
27 lbf·ft)

7923BG90

Fuel filter assembly

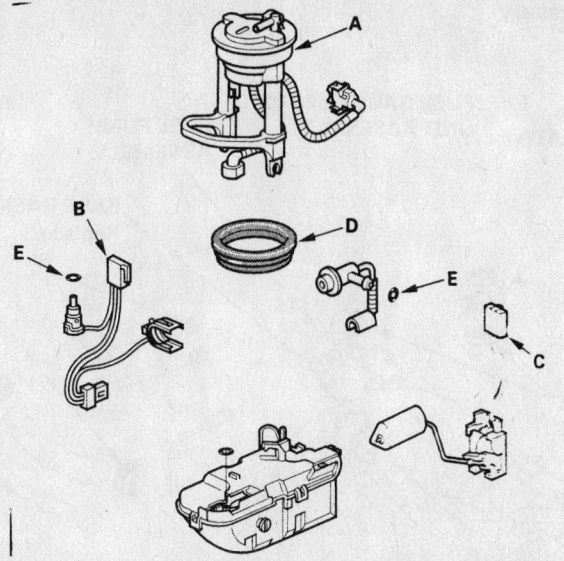

42356-INTE-G32

View of the fuel pump–mounted fuel filter (A)—2.0L engine shown, others similar

- Fuel pump
- Fuel filter set

To install:

4. Install or connect the following:
- Fuel filter set, using a new base gasket and new O–rings
- Fuel pump. When installing the fuel gauge sending unit, make sure the is secure and the connector is locked into place.
- Negative battery cable

Fuel Pump

REMOVAL & INSTALLATION

1.8L and 2.0L Engines

1. Before servicing the vehicle, refer to the precautions in the beginning of this section.
2. Disconnect the negative battery cable.
3. Properly relieve the fuel system pressure.
4. Remove the rear seat to gain access to the fuel pump access panel.
5. Remove the maintenance access cover.
6. Disconnect the electrical connector from the fuel pump.
7. If equipped with quick–connect fittings, hold the fuel line connector with one hand and press down the retainer tabs with the other hand, and then pull the connector off. Check the contact area of the pipe for dirt or damage, clean or replace the pipe or pump as required. Remove the old retainer from the pipe and discard. Cover the connector and pipe with plastic bags to prevent damage and keep foreign material out.

8. Remove the fuel pump mounting nuts, then remove the fuel pump from the fuel tank.

To install:

9. Installation is the reverse of the removal procedure. Tighten the fuel pump mounting nuts to 53 inch lbs. (6 Nm).

3.0L Engines

1. Before servicing the vehicle, refer to the precautions in the beginning of this section.
2. Properly relieve the fuel system pressure.
3. Lower the fuel tank and detach the electrical connector and fuel lines from the pump assembly.
4. Remove the nuts and the fuel pump from the tank.

To install:

5. Installation is the reverse of the removal procedure. Tighten the fuel pump mounting nuts to 53 inch lbs. (6 Nm).

3.2L Engines

1. Before servicing the vehicle, refer to the precautions in the beginning of this section.
2. Remove or disconnect the following:
- Negative battery cable
- Left rear wheel
- Tank drain bolt and drain the fuel into an approved container
- Pump and float wiring connectors located under the trunk floor
- Fuel hose and pipe covers from the inside of the quarter panel
3. Support the tank with a transmission jack, remove the straps and lower the tank out of the vehicle. If it sticks on the under-

coating, carefully pry it free using a blunt or wooden instrument as a lever.

4. Disconnect the fuel line by removing the banjo bolt or uncoupling the quick–connect fittings.
5. Remove the fuel pump mounting nuts. Remove the fuel pump from the fuel tank.

To install:

6. Installation is the reverse of the removal procedure, while using the following torque values:
- Fuel pump mounting nuts: 48 inch lbs. (6 Nm)
- Fuel tank strap bolts: 28 ft. lbs. (38 Nm)
- Fuel tank drain bolt: 36 ft. lbs. (49–50 Nm)

3.5L Engine

1. Before servicing the vehicle, refer to the precautions in the beginning of this section.
2. Properly relieve the fuel system pressure.
3. Remove or disconnect the following:
- Rear seat cushion
- Access panel from the floor
- Fuel line and wiring from the fuel pump assembly
- Mounting nuts and the fuel pump from the fuel tank

To install:

4. Installation is the reverse of the removal procedure. Tighten the fuel pump mounting nuts to 53 inch lbs. (6 Nm).

Fuel Injector

REMOVAL & INSTALLATION

✳✳ CAUTION

Fuel injection systems remain under pressure, even after the engine has been turned OFF. The fuel system pressure must be relieved before disconnecting any fuel lines. Failure to do so may result in fire and/or personal injury. Observe all applicable safety precautions when working around fuel. Whenever servicing the fuel system, always work in a well-ventilated area. Do not allow fuel spray or vapors to come in contact with a spark or open flame. Keep a dry chemical fire extinguisher near the work area. Always keep fuel in a container specifically designed for fuel storage; also, always properly seal fuel containers to avoid the possibility of fire or explosion.

NOTE: Check all hose clamps and retighten if necessary.

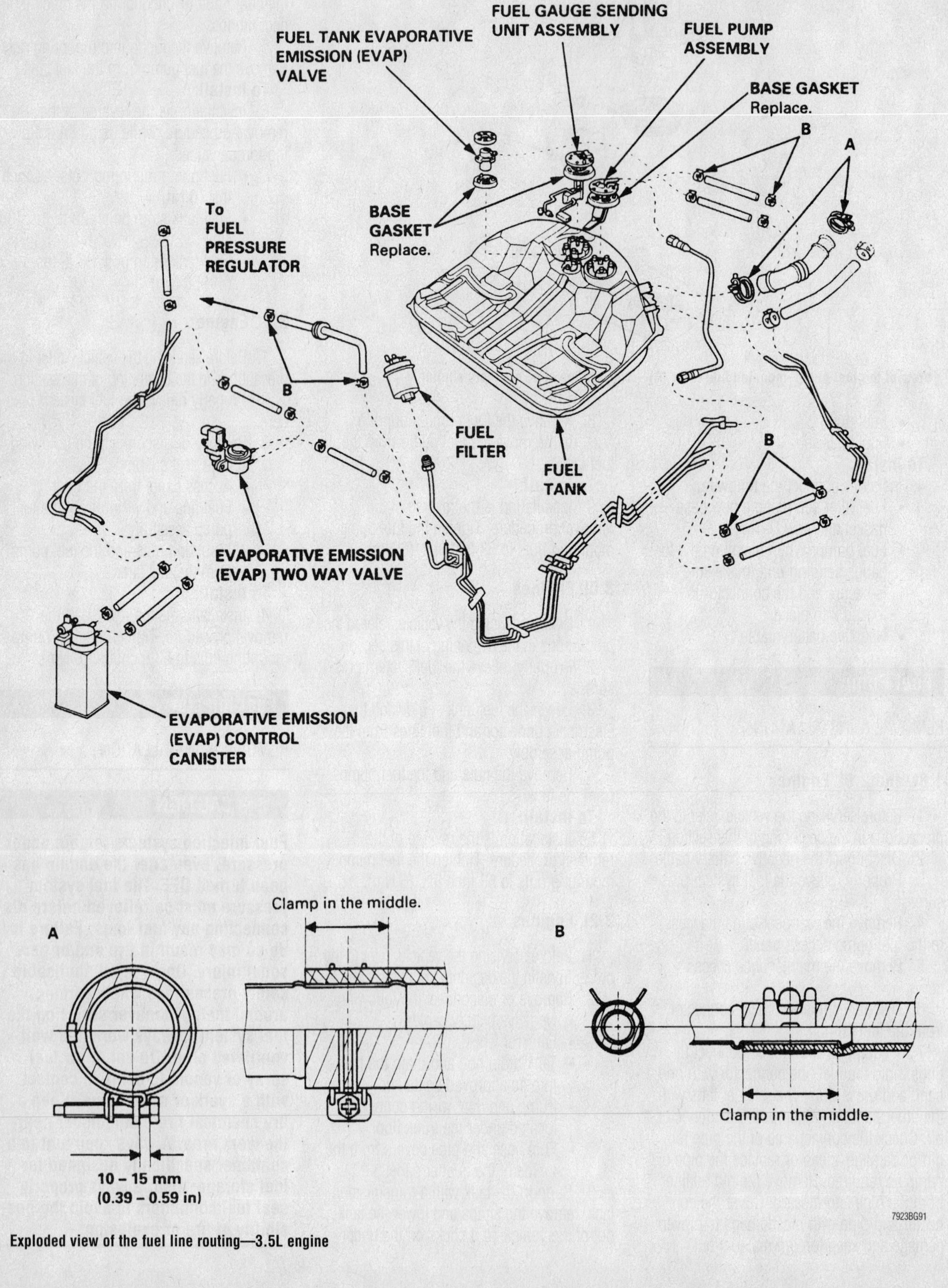

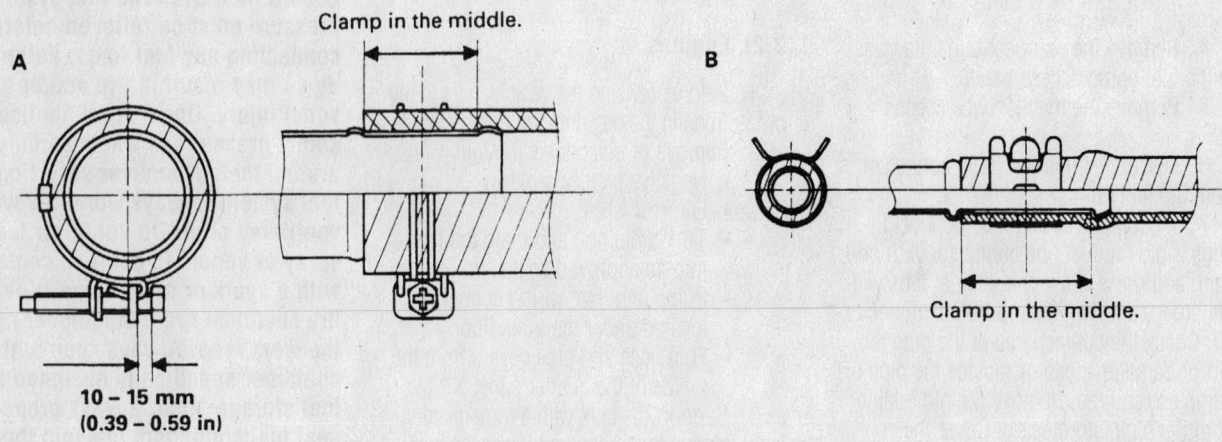

FUEL TANK EVAPORATIVE
EMISSION (EVAP)
VALVE

FUEL GAUGE SENDING
UNIT ASSEMBLY

FUEL PUMP
ASSEMBLY

BASE GASKET
Replace.

B

A

To
FUEL
PRESSURE
REGULATOR

BASE
GASKET
Replace.

B

B

FUEL
FILTER

FUEL
TANK

B

EVAPORATIVE EMISSION
(EVAP) TWO WAY VALVE

EVAPORATIVE EMISSION
(EVAP) CONTROL
CANISTER

Clamp in the middle.

A

B

10 – 15 mm
(0.39 – 0.59 in)

Clamp in the middle.

7923BG91

Exploded view of the fuel line routing—3.5L engine

1. Disconnect the negative battery cable.
2. Relieve the fuel system pressure.
3. Remove or disconnect the following:
   - Fuel injector electrical connectors
   - Fuel feed line from the fuel rail
   - Vacuum hose and fuel return line from the fuel pressure regulator
   - Fuel rail
4. Grasp the fuel injectors body and pull up while gently rocking the fuel injector from side to side.

5. Once removed, inspect the fuel injector cap and body for signs of deterioration. Replace as required.
6. Remove the O–rings and discard.

To install:

7. Install or connect the following:
   - New O–rings onto each injector and apply a small amount of clean engine oil to the O–rings
   - Injectors using a slight twisting downward motion

- Injector retaining clips
- Fuel rail
- Fuel feed line
- Vacuum hose and fuel return line to the fuel pressure regulator
- Fuel injector electrical connectors
- Negative battery cable

8. Run the engine at idle for 2 minutes, then turn the engine **OFF** and check for fuel leaks and proper operation.

DRIVE TRAIN

Transaxle Assembly

REMOVAL & INSTALLATION

Manual

INTEGRA

1. Before servicing the vehicle, refer to the precautions in the beginning of this section.
2. Disconnect the negative battery cable, then the positive battery cable.
3. Drain the transaxle oil. Install the drain plug with a new washer.
4. Remove or disconnect the following:
   - Air cleaner housing and the intake air tube
   - Backup light switch connector
   - Transaxle ground wire
   - Lower radiator hose clamp from the transaxle hanger
   - Wiring harness clips
   - Starter motor electrical connectors
   - Vehicle Speed (VSS) sensor electrical connector
   - Clutch pipe bracket and slave cylinder

✳✳ CAUTION

Do not operate the clutch pedal once the slave cylinder has been removed.

- 3 upper transaxle mounting bolts and the lower starter mounting bolt
- Engine splash shield
- Heated Oxygen (HO$_2$S) sensor connector
- Front exhaust pipe from the vehicle
- Ball joints from the lower control arms
- Right strut fork
- Both halfshafts
- Intermediate shaft
- Extension rod
- Shift rod
- Front and rear engine stiffeners
- Clutch cover
- Right front mount/bracket

5. Place a transmission jack under the transaxle and a jackstand under the engine.
   - Transaxle mount
   - Rear mounting bracket
   - Transaxle mounting bolts
   - Transaxle from the vehicle

To install:

6. Installation is the reverse of the removal procedure, while using the following torque values:
   - Transaxle mounting bolts: 47 ft. lbs. (64 Nm)
   - Rear mounting bracket bolts: 87 ft. lbs. (118 Nm)
   - Transaxle mount fasteners: 47 ft. lbs. (64 Nm)
   - Transaxle mount through bolt: 54 ft. lbs. (74 Nm)
   - Starter motor bolts: 33 ft. lbs. (44 Nm)
   - Right front mount/bracket self–locking bolt: 61 ft. lbs. (83 Nm)
   - Right front bracket and mount bolts, except self–locking bolt: 33 ft. lbs. (44 Nm)
   - Clutch cover 6mm bolts: 108 inch lbs. (12 Nm)
   - Clutch cover 8mm bolt: 17 ft. lbs. (24 Nm)
   - Clutch cover 12mm bolt: 42 ft. lbs. (57 Nm)
   - Front and rear engine stiffener transaxle bolts: 42 ft. lbs. (57 Nm)
   - Front and rear engine stiffener engine bolts: 17 ft. lbs. (24 Nm)
   - Extension rod bolt: 16 ft. lbs. (22 Nm)
   - Intermediate shaft mounting bolts: 29 ft. lbs. (39 Nm)
   - Halfshafts
   - Ball joint nuts: 36–43 ft. lbs. (49–59 Nm)
   - Right strut fork pinch bolt: 32 ft. lbs. (43 Nm)
   - Right strut fork lower nut and bolt: 47 ft. lbs. (64 Nm)

- Exhaust pipe nuts: 40 ft. lbs. (54 Nm)
- Catalytic converter nuts: 25 ft. lbs. (33 Nm)
- Slave cylinder bolts: 16 ft. lbs. (22 Nm)

RSX

1. Before servicing the vehicle, refer to the precautions in the beginning of this section.
2. Disconnect the negative battery cable, then the positive battery cable.
3. Drain the transaxle oil. Install the drain plug with a new washer.
4. Remove or disconnect the following:
   - Intake manifold cover
   - Air cleaner housing
   - Intake air duct
   - Battery tray
   - Transmission ground cable
   - Back–up light switch connector
   - Vehicle Speed Sensor (VSS) connector
   - Reverse lockout solenoid connector
   - Cable bracket and cable from the top of the transaxle. Remove the cables and bracket together to avoid bending the cables.
   - Harness clips
   - Slave cylinder, being careful not to bend the clutch line. Do not depress the clutch pedal after the slave cylinder has been removed.
   - Engine wire harness cover by lifting up on the locktab, then sliding the harness forward off the air cleaner housing mounting bracket
   - Water pipe mounting bolt and lower the pipe slightly
   - Loosen the air cleaner housing bracket mounting bolt, then the mounting bolt
   - Brake booster and Evaporative emission (EVAP) line brake mounting bolts and attach special tool EQS00BRSX0 to the threaded hole in the cylinder head

5. Install the engine support hanger to the vehicle and attach the hook to the special tool.
- 2 upper transmission mounting bolts
- Transmission mount bracket and transmission mounting bolt
- Air cleaner bracket
- Splash shield
- Three–way catalytic converter
- Halfshafts
- Intermediate shaft
- Front engine mount bracket mounting bolt
- 3 bolts securing the transaxle rear mount

6. Support the subframe with the subframe adapter and a jack.

7. Make reference marks of the installed position on the front suspension subframe and mounting bolts, then remove the subframe.

8. Remove or disconnect the following:
- Clutch cover. The cover will have 2 bolts on 5–speed transaxles and 3 bolts on 6–speed transaxles.
- Front engine mount
- Transaxle mounting bolts, after placing a jack under the transaxle. 5–speed models will have 4 lower bolts and 6–speed models will have 2 rear and 2 lower bolts.
- Harness bracket and intake manifold bracket, 6–speed transaxles only
- 2 front transaxle mounting bolts, 5–speed transaxles only

9. Pull the transaxle away from the engine until the mainshaft clears the clutch pressure plate, then lower the transmission on the jack.
- Transaxle rear mount and rear mount bracket
- Boot, release fork and release bearing from the transaxle

10. Installation is the reverse of the removal procedure, while using the following torque values:
- Transaxle rear mount bracket bolts: 47 ft. lbs. (64 Nm)
- 2 front transaxle mounting bolts (5–speed only): 47 ft. lbs. (64 Nm)
- 4 lower transaxle mounting bolts (5–speed only): 47 ft. lbs. (64 Nm)
- 2 rear transaxle mounting bolts (6–speed only): 47 ft. lbs. (64 Nm)
- 2 lower transaxle mounting bolts (6–speed only): 33 ft. lbs. (44 Nm)
- Front engine mount bolts: 47 ft. lbs. (64 Nm)
- Clutch cover bolts: 9 ft. lbs. (12 Nm)

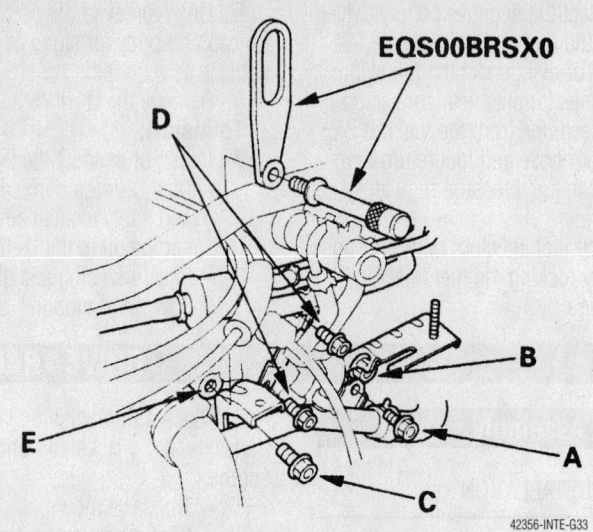

Attach the special tool to the threaded hole (E) in the cylinder head —RSX

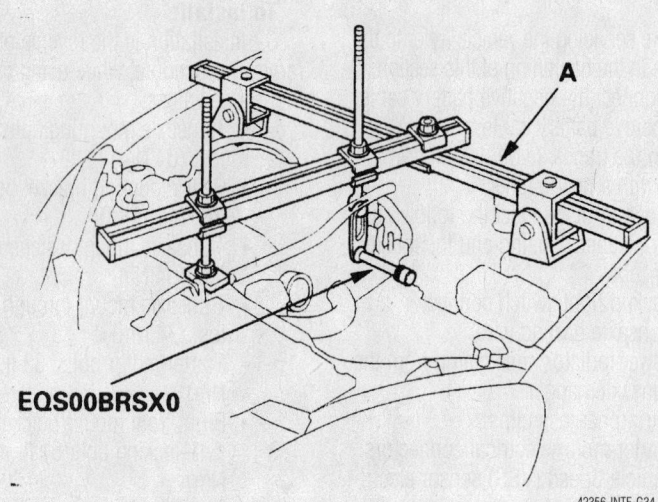

Install the engine support hanger (A) to the vehicle and attach the hook to the special tool—RSX

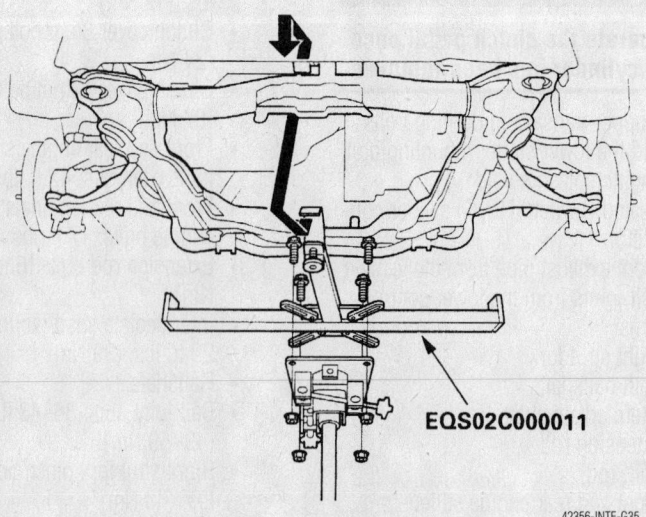

Support the subframe with the subframe adapter and a jack—RSX

- Subframe mounting bolts: 76 ft. lbs. (98 Nm)
- 3 transaxle rear mount mounting bolts: 43 ft. lbs. (59 Nm)
- Transaxle mount bracket and transmission mounting bolt: 40 ft. lbs. (54 Nm)
- Air cleaner bracket bolts: 9 ft. lbs. (12 Nm)
- 2 upper transaxle mounting bolts: 47 ft. lbs. (64 Nm)
- 2 Slave cylinder mounting bolts: 16 ft. lbs. (22 Nm)
- Transmission ground cable: 16 ft. lbs. (22 Nm)

NSX

1. Before servicing the vehicle, refer to the precautions in the beginning of this section.
2. Drain the transaxle fluid.
3. Remove or disconnect the following:
   - Negative battery cable, then the positive battery cable
   - Strut bar
   - Air cleaner assembly
   - Control box
   - Transmission ground cable
   - Back up light switch electrical connector
   - Neutral position switch electrical connector
   - Differential speed sensor electrical connector
   - Reverse lockout solenoid electrical connector
   - Vehicle Speed (VSS) sensor electrical connector
   - Starter motor
   - Transmission mount
   - Parking brake cable holders from the rear beam rod
   - Rear beam rod
   - Parking brake cable holder from the rear sub frame
   - Wheel sensor wire clamps from the lower control arms
   - Toe control arms from the side beams
   - Strut forks
   - Lower control arm from the side beam
   - Half shafts from the differential
   - Intermediate shaft from the differential
   - Lower cover
   - Change wire bracket
   - Upper cover
   - Shift and select cables
   - Slave cylinder
   - Release fork from the clutch release hanger

4. Attach a chain hoist to the transmission hangers
   - Front engine mounting bolts on the transmission side
   - Transmission mounting bolts and stiffener
   - Transmission housing mounting bolts
   - Transmission from the vehicle

To install:

5. Installation is the reverse of the removal procedure, while using the following torque values:
   - Transmission and engine stiffener bolts: 47 ft. lbs. (64 Nm)
   - Front engine mounting bolts: 43 ft. lbs. (60 Nm)
   - Slave cylinder bolts: 16 ft. lbs. (22 Nm)
   - Upper cover bolts: 108 inch lbs. (12 Nm)
   - Change wire bracket bolts: 19 ft. lbs. (25 Nm)
   - Lower cover bolts: 108 inch lbs. (12 Nm)
   - Lower control arm bolts: 90 ft. lbs. (123 Nm)
   - Strut fork bolts: 69 ft. lbs. (93 Nm)
   - Toe control arms and torque the bolts to 69 ft. lbs. (93 Nm)
   - Parking brake cable holder bolts: 16 ft. lbs. (22 Nm)
   - Rear beam rod bolts: 43 ft. lbs. (59 Nm)
   - Transmission mount bolts: 43 ft. lbs. (59 Nm)
   - Starter motor bolts: 54 ft. lbs. (74 Nm)
   - Control box bolt: 16 ft. lbs. (22 Nm)
   - Strut bar bolts: 16 ft. lbs. (22 Nm)

Automatic

INTEGRA & RSX

1. Before servicing the vehicle, refer to the precautions in the beginning of this section.
2. Drain the transaxle fluid.
3. Remove or disconnect the following:
   - Negative battery cable, then the positive battery cable
   - Air cleaner housing assembly with intake air duct
   - Starter motor cables and holder
   - Transaxle ground cable from the transaxle hanger
   - Lockup control solenoid valve connector
   - Shift control solenoid valve connector
   - Harness clamp on the lockup control solenoid harness from the harness stay
   - Vehicle Speed (VSS) sensor electrical connector
   - Main shaft speed sensor electrical connector
   - Counter shaft speed sensor electrical connector
   - Upper transaxle mounting bolts
   - Splash shield
   - Front wheels
   - Ball joints from the lower control arms
   - Right strut fork
   - Both halfshafts from the vehicle
   - Heated Oxygen (HO$_2$S) sensor connector
   - Front exhaust pipe from the vehicle
   - Intermediate shaft
   - Shift cable cover
   - Shift cable by removing the control lever

✷✷ WARNING

Do not bend the shift control cable when removing it.

   - Right front mount/bracket
   - End of the throttle control cable from the throttle control drum
   - Transmission oil cooler hoses from the joint pipes
   - Engine stiffener
   - Torque converter cover
   - 8 drive plate bolts
4. Support the transaxle.
   - Transaxle mounting bolts and rear engine mounting bolts
   - Transaxle from the engine
   - Starter motor from the transaxle (if necessary)

To install:

5. Installation is the reverse of the removal procedure, while using the following torque values:
   - Starter motor bolts: 33 ft. lbs. (45 Nm)
   - Transaxle mounting bolts: 47 ft. lbs. (64 Nm)
   - Rear engine mount bolts: 87 ft. lbs. (118 Nm)
   - Transmission mount bolt: 54 ft. lbs. (74 Nm)
   - Transmission mount nuts: 47 ft. lbs. (64 Nm)
   - Drive plate bolts: 108 inch lbs. (12 Nm)
   - Torque converter cover 6mm bolts: 108 inch lbs. (12 Nm)
   - Torque converter cover 10mm bolt: 33 ft. lbs. (44 Nm)

- Engine stiffener transaxle bolt: 32 ft. lbs. (43 Nm)
- Engine stiffener engine bolts: 17 ft. lbs. (24 Nm)
- Right front mount/bracket 12mm bolts: 47 ft. lbs. (64 Nm)
- Right front mount/bracket 10mm bolts: 33 ft. lbs. (44 Nm)
- Control lever bolt: 10 ft. lbs. (14 Nm)
- Intermediate shaft mounting bolts: 29 ft. lbs. (39 Nm)
- Front exhaust pipe manifold nuts: 40 ft. lbs. (54 Nm)
- Catalytic converter nuts: 16 ft. lbs. (22 Nm)
- Strut fork pinch bolt: 32 ft. lbs. (43 Nm)
- Strut fork lower bolt: 47 ft. lbs. (64 Nm)
- Splash shield bolts: 108 inch lbs. (12 Nm)

3.2CL AND 3.2TL

1. Before servicing the vehicle, refer to the precautions in the beginning of this section.
2. Shift the transmission into **P**.
3. Drain the transmission fluid.
4. Remove or disconnect the following:
    - Both battery cables
    - Battery and the tray
    - Intake air duct
    - Transmission oil cooler hoses
    - Starter motor
    - Transmission ground cable
    - Shift control solenoid valve connectors
    - Clutch pressure switch electrical connector
    - Mainshaft speed sensor electrical connector
    - Clutch pressure control solenoid valve electrical connector
    - Lock–up control solenoid valve electrical connector
    - Countershaft speed sensor electrical connector
    - Gear position switch connector
    - Vehicle Speed (VSS) sensor without disconnecting the hoses
    - Front mount nut
    - Engine cover
    - Splash shield
    - Strut forks
    - Ball joints from the control arms
    - Radius rods from the control arms
    - Both halfshafts
    - Front beam
5. Raise the transmission with a jack to take the pressure off of the mounts.
    - Lower rear mount
    - Intermediate shaft

- Shift cable holder
- Shift cable cover
- Shift cable with the control lever
- Torque converter cover
- Drive plate bolts
- Engine stiffener
- Front mount bracket
- Transmission-to-engine bolts
- Transmission from the vehicle

To install:

6. Installation is the reverse of the removal procedure, while using the following torque values:
    - Transmission mounting bolts: 47 ft. lbs. (64 Nm)
    - Front mount bracket bolts: 28 ft. lbs. (38 Nm)
    - Engine stiffener bolts: 28 ft. lbs. (38 Nm)
    - Drive plate bolts: 20 ft. lbs. (26 Nm)
    - Torque converter cover bolts: 108 inch lbs. (12 Nm)
    - Control lever bolt: 10 ft. lbs. (14 Nm)
    - Cable cover bolts: 16 ft. lbs. (22 Nm)
    - Shift cable holder bolts: 84 inch lbs. (9.5 Nm)
    - Intermediate shaft bolts: 30 ft. lbs. (39 Nm)
    - Lower transmission mount bolts: 28 ft. lbs. (38 Nm)
    - Front beam 10mm bolts: 28 ft. lbs. (38 Nm)
    - Front beam 12mm bolts: 47 ft. lbs. (64 Nm)
    - Front beam 14mm bolts: 76 ft. lbs. (103 Nm)
    - Lower transmission mount nuts: 28 ft. lbs. (38 Nm)
    - Strut fork pinch bolts: 32 ft. lbs. (43 Nm)
    - Strut fork through bolts to 47 ft. lbs. (64 Nm)
    - Ball joint nuts: 36–43 ft. lbs. (49–59 Nm)
    - Radius rod bolts: 132 ft. lbs. (179 Nm)
    - Front mount nut: 40 ft. lbs. (54 Nm)
    - VSS sensor bolts: 20 ft. lbs. (26 Nm)
    - Transmission ground cable bolt: 20 ft. lbs. (26 Nm)
    - Starter motor bolts: 33 ft. lbs. (44 Nm)

3.5RL

1. Before servicing the vehicle, refer to the precautions in the beginning of this section.

2. Disconnect the negative, then the positive battery cables.
3. Shift the transaxle into **P**.
4. Drain the transmission.
5. Remove or disconnect the following:
    - Strut brace
    - Control box
    - Transaxle sub–harness connectors, and remove the sub–harness clamp
    - 3 bolts securing the transaxle dipstick pipe bracket
    - Upper transaxle mounting bolts
6. Pull the carpet back under the passenger seat to expose the secondary Heated Oxygen (HO2S) sensor connector. Detach the connector and push it out from the inside of the vehicle.
7. Remove or disconnect the following:
    - Transmission stop collars
    - Front exhaust pipe from the vehicle
    - HO2S sensor wiring harness cover and grommet
    - Catalytic converter
    - Exhaust heat shield from the floor of the vehicle
    - Transaxle oil cooler hoses
    - Shift solenoid valve electrical connector
    - Shift cable cover from the transaxle
    - Shift cable holder from the holder base
    - Control lever from the control shaft
    - Transaxle dipstick pipe from the torque converter housing
    - Range switch connector
    - Lower plate from under the steering gear, then re–install the 2 steering gear mounting bolts
    - Shift cable guide bracket from the transaxle beam
8. Raise the transmission slightly to take the weight off of the mounts.
    - Transaxle beam
    - Rear transaxle mount bracket and the mount
    - Exhaust pipe bracket
    - Sealing bolt from the differential
    - Extension shaft from the differential
    - Transmission-to-engine bolts
    - Engine stiffener
    - Torque converter covers
    - Drive plate bolts
    - Transmission from the vehicle

To install:

9. Installation is the reverse of the removal procedure, while using the following torque values:
    - Drive plate bolts: Step 1: 108 inch lbs. (12 Nm). Step 2: 20 ft. lbs. (26 Nm)
    - Torque converter cover bolts: 108 inch lbs. (12 Nm)

- Transaxle housing 8mm bolts: 16 ft. lbs. (22 Nm)
- Transaxle housing 12mm bolts: 47 ft. lbs. (64 Nm)
- Engine stiffener bolts: 16 ft. lbs. (22 Nm)
- Transaxle beam bolts: 28 ft. lbs. (38 Nm)
- Rear transaxle mount bracket bolts: 28 ft. lbs. (38 Nm)
- Rear transaxle mount bolts: 40 ft. lbs. (54 Nm)
- Shift cable guide bolt: 84 inch lbs. (9.5 Nm)
- Differential sealing bolt: 58 ft. lbs. (78 Nm)
- Lower plate bolts: 28 ft. lbs. (38 Nm)
- Steering gear bolts: 43 ft. lbs. (59 Nm)
- Shift control lever nut: 12 ft. lbs. (16 Nm)
- Shift cable holder bolts: 108 inch lbs. (12 Nm)
- Shift cable cover bolts: 108 inch lbs. (12 Nm)
- Exhaust heat shield bolts: 84 inch lbs. (9.5 Nm)
- Catalytic converter nuts: 25 ft. lbs. (33 Nm)
- Front exhaust pipe manifold nuts: 40 ft. lbs. (54 Nm)
- Transmission stop collar bolts: 28 ft. lbs. (38 Nm)
- Control box mounting bolts: 108 inch lbs. (12 Nm)
- Strut brace bolts: 16 ft. lbs. (22 Nm)

NSX

1. Before servicing the vehicle, refer to the precautions in the beginning of this section.
2. Drain the transmission.
3. Remove or disconnect the following:
   - Both battery cables
   - Strut brace
   - Air cleaner assembly
   - Control box
   - Vehicle Speed (VSS) sensor electrical connector
   - Transmission ground cable
   - Starter cables
   - Lock–up solenoid valve electrical connector
   - Shift control solenoid valve electrical connector
   - Transaxle oil cooler
   - Starter motor
   - Upper transmission housing and mounting bolts
   - Parking brake cable holders from the rear beam
   - Rear beam

- Front exhaust pipe
- Toe control arms from the side beams
- Damper forks
- Lower control arms from the side beams
- Both halfshafts and the intermediate shaft
- Shift cable cover and holder
- Shift cable from the control lever
- Torque converter cover
- Drive plate bolts

4. Place a jack under the transmission and raise it just enough to take the weight off of the mounts.
   - Front engine mount bolts on the transaxle side
   - Rear transaxle mount bolts
   - Transaxle-to-engine bolts
   - Transmission from the vehicle

To install:

5. Installation is the reverse of the removal procedure, while using the following torque values:
   - Transaxle mounting bolts: 54 ft. lbs. (74 Nm)
   - Rear transaxle mount bolts: 43 ft. lbs. (59 Nm)
   - Front engine mount bolts: 43 ft. lbs. (59 Nm)
   - Drive plate bolts: 108 inch lbs. (12 Nm)
   - Torque converter cover bolts: 108 inch lbs. (12 Nm)
   - Shift cable cover bolts: 6 ft. lbs. (8 Nm)

- Intermediate shaft bolts: 16 ft. lbs. (22 Nm)
- Intermediate shaft heat shield bolts: 84 inch lbs. (9.5 Nm)
- Lower control arm bolts: 90 ft. lbs. (123 Nm)
- Damper fork bolts: 69 ft. lbs. (93 Nm)
- Toe control arm bolts: 69 ft. lbs. (93 Nm)
- Front exhaust pipe manifold nuts: 40 ft. lbs. (54 Nm)
- Catalytic converter nuts: 25 ft. lbs. (33 Nm)
- Rear beam 10mm bolts: 43 ft. lbs. (59 Nm)
- Rear beam nuts and 12mm bolts: 69 ft. lbs. (93 Nm)
- Parking brake cable holder bolts: 16 ft. lbs. (22 Nm)
- Starter motor bolts: 33 ft. lbs. (44 Nm)
- Transaxle oil cooler bolts: 13 ft. lbs. (18 Nm)
- Strut bar bolts: 28 ft. lbs. (38 Nm)

Clutch

REMOVAL & INSTALLATION

1. Before servicing the vehicle, refer to the precautions in the beginning of this section.
2. Disconnect the negative battery cable.

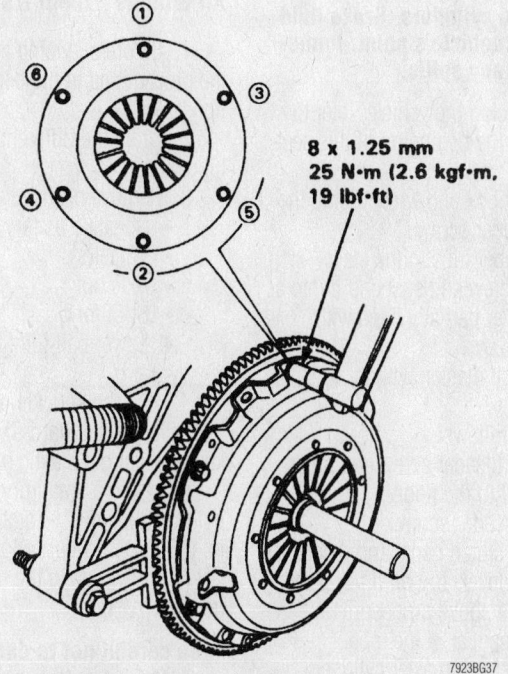

8 x 1.25 mm
25 N·m (2.6 kgf·m, 19 lbf·ft)

7923BG37

Pressure plate bolt torque sequence—Integra and RSX

3. Remove the transaxle assembly from the vehicle.

4. Insert a clutch alignment tool. Use a feeler gauge and measure the clearance between the pressure plate spring fingers and the clutch alignment disc. There should be a maximum of 0.02 in. (0.6mm) of clearance for a new pressure plate with 0.03 in. (0.8mm) limit for a used pressure plate. If the height is more than the service limit, replace the pressure plate.

5. Remove the clutch alignment disc.

6. Install a flywheel holder to aid in the removal of the pressure plate and clutch disc.

7. Matchmark the flywheel and pressure plate for easy reassembly. Remove the pressure plate bolts in a crisscross pattern 2 turns at a time to prevent warping the plate.

8. Remove the pressure plate, then the clutch disc with the alignment shaft.

To install:

9. Installation is the reverse of the removal procedure, while using the following torque values:

- Flywheel mounting bolts: 76 ft. lbs. (103 Nm) for 5–speed transaxles or 90 ft. lbs. (122 Nm) for 6–speed transaxles
- Pressure plate bolts: 19 ft. lbs. (25 Nm)

Hydraulic Clutch System

BLEEDING

➡**Use DOT 3 brake fluid in the clutch master and slave cylinders. Brake fluid will damage the vehicle's paint. Immediately clean up any spills.**

1. Before servicing the vehicle, refer to the precautions in the beginning of this section.

2. Fit a flare or box end wrench onto the slave cylinder bleeder screw.

3. Attach a rubber tube to the slave cylinder bleeder screw and suspend it into a clear drain container partially filled with brake fluid.

4. Fill the clutch master cylinder with brake fluid.

5. Proceed as follows:
   a. Open the bleeder screw and press the clutch pedal to the floor.

6. Close the bleeder screw.

7. Release the clutch pedal and recheck the reservoir fluid level. Top off if necessary.

8. Continue the above procedure until no more bubbles appear in the tube.

9. Top off the clutch master cylinder reservoir with brake fluid.

Halfshaft

REMOVAL & INSTALLATION

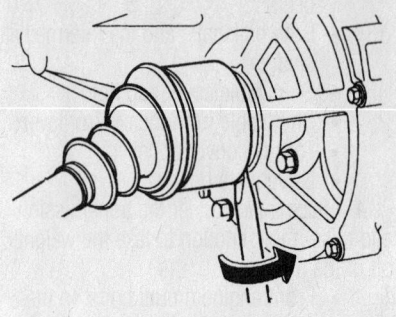

Carefully pry the inboard joint from the transaxle

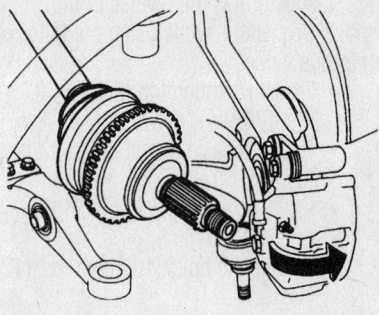

Pull the hub assembly from the outboard joint

All Models Except NSX

1. Before servicing the vehicle, refer to the precautions in the beginning of this section.

2. Drain the differential or transmission lubricant.

3. Remove or disconnect the following:
   - Negative battery cable
   - Wheel(s)
   - Axle nut
   - Strut fork
   - Lower ball joint from the control arm

4. Pull the knuckle outward and remove the halfshaft outboard CV–joint from the knuckle using a plastic mallet.

5. Using a small pry bar carefully pry out the inboard CV–joint approximately ½ in. (13mm) in order to force the spring clip out of the groove in the differential side gears.

➡**Be careful not to damage the oil seal. Do not pull on the inboard CV–joint, it may come apart.**

6. Pull the halfshaft out of the differential or the intermediate shaft.

7. Remove the halfshaft from the wheel hub.

To install:

➡**Always use a new set ring whenever the driveshaft is being installed. Be sure the driveshaft locks in the differential side gear groove and that the CV–joint stub–axle bottoms in the differential or the intermediate shaft.**

8. Install or connect the following:
   - Outboard joint to the wheel hub
   - Inboard joint with a new set ring into the differential or intermediate shaft until the set ring locks in the groove
   - Ball joint to the control arm and torque the nut to 36–43 ft. lbs. (49–59 Nm)
   - Strut fork and torque the pinch bolt to 32 ft. lbs. (43 Nm) and the lower bolt to 47 ft. lbs. (64 Nm)
   - Axle nut and torque it to 181 ft. lbs. (245 Nm)
   - Wheel(s)

9. Refill the transmission or differential with the correct amount and type of fluid.

10. Reconnect the battery cable and enter the radio security code.

11. Measure and adjust the wheel alignment.

NSX

1. Before servicing the vehicle, refer to the precautions in the beginning of this section.

2. Drain the differential or transmission lubricant.

3. Remove or disconnect the following:
   - Negative battery cable
   - Rear wheel(s)
   - Axle nut
   - Brake hose from the caliper and from the knuckle
   - Wheel sensor from the knuckle and lower control arm
   - Parking brake cable from the control arm
   - Toe control arm from the knuckle
   - Damper mounting nut
   - Stabilizer link
   - Lower control arm from the sub–frame
   - Outboard joint from the knuckle
   - Inboard joint from the differential or intermediate shaft using a suitable prytool

To install:

4. Install or connect the following:
   - Inboard joint with a new set ring to the differential or intermediate shaft

- Outboard joint to the knuckle
- Lower control arm to the sub–frame and torque the bolts to 90 ft. lbs. (123 Nm)
- Stabilizer link and torque the nut to 61 ft. lbs. (83 Nm)
- Damper mounting nut and torque it to 69 ft. lbs. (93 Nm)
- Toe control arm to the knuckle and torque the bolt to 69 ft. lbs. (93 Nm)
- Parking brake cable and torque the bolt to 16 ft. lbs. (22 Nm)
- Wheel sensor and torque the bolts to 16 ft. lbs. (22 Nm)
- Brake hose and torque the bracket bolts to 84 inch lbs. (10 Nm) and the banjo bolt to 25 ft. lbs. (34 Nm)
- Axle nut and torque it to 242 ft. lbs. (329 Nm)
- Rear wheel(s)
- Negative battery cable

5. Fill the transaxle with the proper amount and type of fluid.

6. Fill and bleed the brake system.

7. Measure and adjust the wheel alignment.

CV–Joints

OVERHAUL

Integra, RSX, 3.2TL and 3.5RL

➡**The outer joint is not serviceable, if wear or excessive play is found the joint must be replace.**

1. Before servicing the vehicle, refer to the precautions in the beginning of this section.

2. Remove or disconnect the following:

- Negative battery cable
- Halfshaft from the vehicle
- Set ring from the inner joint

- Inner joint boot bands
- Inboard joint after marking relationship of joint-to-rollers for later installation
- Rollers from the spider
- Circlip from the shaft
- Spider using a bearing remover
- Stop ring
- Inboard boot
- Dynamic damper band and damper (if equipped)
- Outer joint boot bands and boot (if necessary)

3. Wrap the splines with vinyl tape to prevent damage to the boots.

4. Install or connect the following:

- Outer joint boot
- Dynamic damper
- Inner joint boot then remove the tape

5. Pack the outer joint boot with grease included with the kit (approx. 4.7 oz.)

- Stop ring into the halfshaft groove
- Spider on the halfshaft
- Circlip into the halfshaft groove
- Rollers onto the spider

6. Pack the inner joint with the grease included with the kit (approx. 4.7 oz.)

- Inner joint on the halfshaft matching the marks made earlier

7. Adjust the length of the halfshaft to the specification shown in the illustration.

8. Adjust the boots to halfway between full compression and full extension.

9. Position the dynamic damper (if equipped) as shown in the illustration.

10. Install a new dynamic damper band and bend down both locking tabs.

11. Install new inner and outer joint boot bands.

12. Install a new set ring in the halfshaft groove.

13. Install the halfshaft in the vehicle.

NSX and 3.2CL

1. Before servicing the vehicle, refer to the precautions in the beginning of this section.

2. Remove or disconnect the following:

- Negative battery cable
- Halfshaft from the vehicle
- Boot bands
- Circlip from the shaft (outer joint only)
- Joint from the shaft after match-marking the rollers and the joint for re–installation

➡**There is a spring located under the outer joint.**

- Rollers from the spider
- Set ring
- Spider using a bearing puller
- Stop ring

To install:

3. Wrap the splines of the halfshaft with vinyl tape to protect the boots from damage.

4. Install or connect the following:

- CV boot, then remove the tape
- Stop ring
- Spider
- Set ring
- Rollers onto the spider
- Pack the outer joint with the grease provided (approx. 6.2 oz. for outer joint or 4.4 oz. for the inner)
- Spring and cap (outer joint only)
- CV joint onto the shaft, matching the marks made earlier
- Circlip into the outer joint inner groove

5. Adjust the length of the halfshaft as shown in the illustration.

6. Adjust the boots to halfway between full compression and full extension.

7. Install new boot bands on the boots and bend both sets of locking tabs down.

8. Install a new set ring in the halfshaft groove.

9. Install the halfshaft in the vehicle.

STEERING AND SUSPENSION

Air Bag

✳✳ CAUTION

Some vehicles are equipped with an air bag system, also known as the Supplemental Restraint System (SRS). The system must be disabled before performing service on or around system components, steering column, instrument panel components, wiring and sensors. Failure to follow safety and disabling proce-dures could result in accidental air bag deployment, possible personal injury and unnecessary system repairs.

PRECAUTIONS

Several precautions must be observed when handling the inflator module to avoid accidental deployment and possible personal injury.

- Never carry the inflator module by the wires or connector on the underside of the module.
- When carrying a live inflator module, hold securely with both hands, and ensure that the bag and trim cover are pointed away.
- Place the inflator module on a bench or other surface with the bag and trim cover facing up.
- With the inflator module on the bench, never place anything on or close to the module which may be thrown in the event of an accidental deployment.

DISARMING

Integra and NSX

❄❄ CAUTION

The Supplemental Restraint System (SRS, air bag system) must be disarmed before any of its components are disconnected or removed. Failing to disable the SRS before servicing its components may cause accidental deployment of the air bag, resulting in unnecessary repairs and possible personal injury.

1. Before servicing the vehicle, refer to the precautions in the beginning of this section.

2. Turn the ignition switch **OFF**.

3. Wait 3 minutes to let the capacitor in the backup circuit discharge.

4. Disconnect the negative battery cable, then disconnect the positive battery cable.

5. For the driver air bag:

a. Remove the access panel lid below the air bag assembly on the steering wheel and remove the red shorting connector.

b. Disconnect the connector between the air bag and cable reel.

c. Connect the red shorting connector to the air bag side of the connector.

6. For the passenger air bag:

a. If necessary, remove the glove box, then remove the red shorting connector from its holder.

b. Disconnect the 3 pin connector between the passenger air bag and the main harness.

c. Connect the shorting connector to the air bag side of the connector.

7. After installing the shorting connectors on the air bags, connect shorting connector 07–MAZ–SP0020A, or equivalent, on the cable reel connector and another on the main harness connector of the passenger's side air bag. This will prevent static electricity from setting off the seat belt pre-tensioners before you disconnect them.

8. For the seat belt pre-tensioners, disarm them one side at a time:

a. Remove the B–pillar trim panels.

b. Remove the red shorting connector from the short connector holder.

c. Disconnect the pre-tensioner 3–pin connector, then install the shorting connector to the pre-tensioner side of the connector.

To enable:

9. Enable the seat belt pre-tensioners:

a. Disconnect the shorting connector from the 3–pin connector. Then, reconnect the 3–pin connector.

b. Fit the shorting connector into its holder and reinstall the B–pillar trim panels.

10. Enable the passenger air bag:

a. Disconnect the shorting connectors from the air bag and main harness connectors.

b. Reconnect the air bag connector to the main harness.

c. Fit the short connector into its holder.

d. If removed, install the glove box.

11. Disconnect the shorting connector from the cable reel connection.

12. Enable the driver's air bag:

a. Disconnect the shorting connector from the air bag connector.

b. Reconnect the air bag and cable reel connectors.

c. Fit the shorting connector back into its holder.

d. Install the steering wheel access cover.

13. Reconnect the positive and negative battery cables.

14. Turn the ignition switch to the **ON** position, but don't start the engine. The SRS indicator light should turn on for 6 seconds, then turn off. If the SRS indicator light doesn't come on, or stays on longer than 6 seconds, the system fault must be diagnosed.

15. Enter the radio security code.

RSX

❄❄ CAUTION

The Supplemental Restraint System (SRS, air bag system) must be disarmed before any of its components are disconnected or removed. Failing to disable the SRS before servicing its components may cause accidental deployment of the air bag, resulting in unnecessary repairs and possible personal injury.

1. Before servicing the vehicle, refer to the precautions in the beginning of this section.

2. Turn the ignition switch **OFF**.

3. Disconnect the negative battery cable, then wait 3 minutes.

4. For the driver air bag:

a. Remove the access panel from the steering wheel and disconnect the driver's airbag 4P connector from the cable reel.

5. For the passenger air bag:

a. Remove the glove box.

b. Disconnect the front passenger's

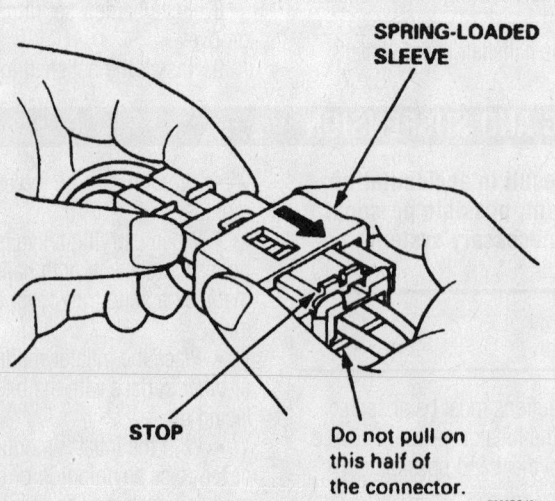

SPRING-LOADED SLEEVE

STOP

Do not pull on this half of the connector.

7923BG42

Spring-sleeve connectors

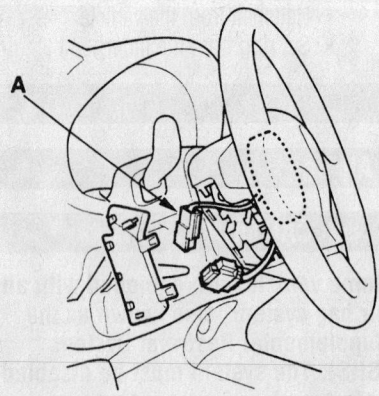

42356-INTE-G36

Remove the access panel from the steering wheel and disconnect the driver's airbag 4P connector from the cable reel— RSX

4P connector from the dashboard wire harness.

6. For the side airbag:

a. Disconnect both side airbag 2P connectors from the floor wire harness.

7. For the seat belt tensioners:

a. Remove the B–pillar trim panels.

b. Disconnect both seat belt tensioner 2P connectors from the floor wire harness.

8. For the seat belt buckle tensioners:

a. Disconnect both seat belt buckle tensioner 4P connectors.

To enable:

9. For the seat belt buckle tensioners:

a. Connect both seat belt buckle tensioner 4P connectors.

10. For the seat belt tensioners:

a. Connect both seat belt tensioner 2P connectors to the floor wire harness.

b. Install the B–pillar trim panels.

11. For the side airbag:

a. Connect both side airbag 2P connectors to the floor wire harness.

12. For the passenger air bag:

a. Connect the front passenger's 4P connector to the dashboard wire harness.

b. Install the glove box.

13. For the driver air bag:

a. Connect the driver's airbag 4P connector to the cable reel. Install the access panel to the steering wheel.

14. Reconnect the negative battery cable.

15. Turn the ignition switch to the **ON** position, but don't start the engine. The SRS indicator light should turn on for 6 seconds, then turn off. If the SRS indicator light doesn't come on, or stays on longer than 6 seconds, the system fault must be diagnosed.

16. Enter the radio security code.

3.2CL and 3.2TL

✳✳ CAUTION

The SRS must be disarmed before any of its components are disconnected or removed. Failing to disable the SRS before servicing its components may cause accidental deployment of the air bag, resulting in unnecessary repairs and possible personal injury.

1. Before servicing the vehicle, refer to the precautions in the beginning of this section.

2. Turn the ignition switch to the **LOCK** position. Remove the key.

3. Disconnect the negative and positive battery cables.

4. Always wait at least 3 minutes after disconnecting the battery before working around the air bags.

5. Remove or disconnect the following:

• Steering wheel lower access cover
• Clip securing the air bag module/cable reel connection to the steering column
• Air bag and cable reel connection—Immediately install the red shorting connector onto the air bag module connector.

➡**The driver's side air bag connection contains a spring–contact self–disabling device. A shorting connector doesn't need to be installed on the driver's air bag connector.**

6. After servicing has been completed, couple the air bag and cable reel connectors.

7. Install or connect the following:

• Clip securing the air bag/cable reel connection to the steering column
• Access cover

To enable:

8. Reconnect the positive and negative battery cables.

9. Turn the ignition switch to the **ON** position, but don't start the engine. The SRS indicator light should turn on for 6 seconds, then turn off. If the SRS indicator light doesn't come on, or stays on longer than 6 seconds, the system fault must be diagnosed.

10. Enter the radio security code.

3.5RL

1. Before servicing the vehicle, refer to the precautions in the beginning of this section.

2. Disconnect the negative battery cable, then the positive cable.

3. Wait 3 minutes for the air bag reserve power to discharge before preceding with work.

To enable:

4. Reconnect the positive and negative battery cables.

5. Turn the ignition switch to the **ON** position, but don't start the engine. The SRS indicator light should turn on for 6 seconds, then turn off. If the SRS indicator light doesn't come on, or stays on longer than 6 seconds, the system fault must be diagnosed.

6. Enter the radio security code.

Rack and Pinion Steering Gear

REMOVAL & INSTALLATION

Power

3.2CL, INTEGRA AND RSX

1. Before servicing the vehicle, refer to the precautions in the beginning of this section.

2. Drain the power steering fluid.

3. Remove or disconnect the following:

• Both battery cables
• Wheels

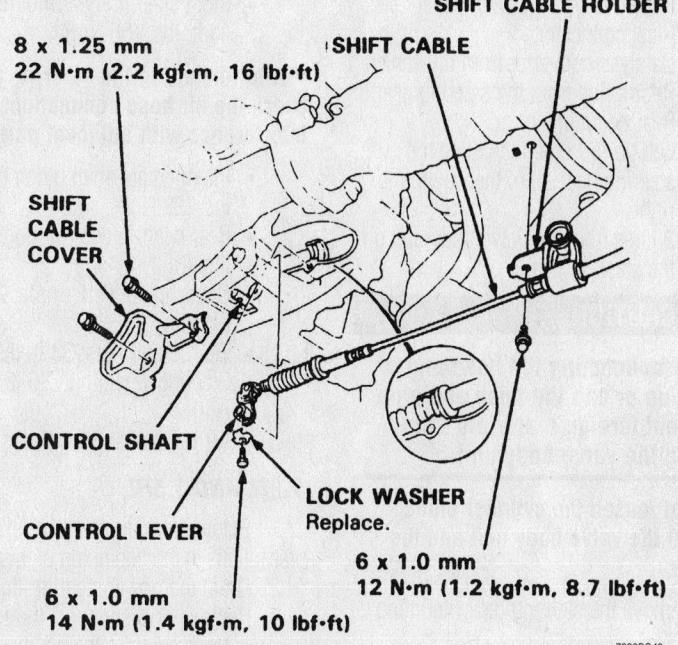

Automatic transaxle shift cable attachment—Integra

7923BG43

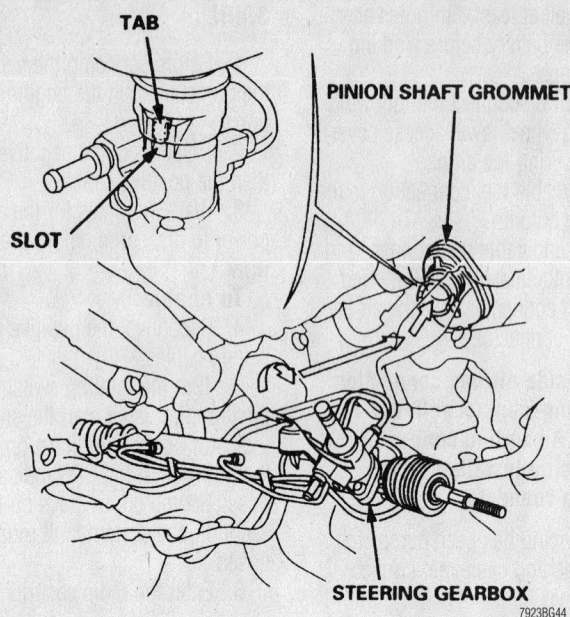

Installing the steering gear—Integra

- Steering wheel lower access panel
- Supplemental Inflatable Restraint (SIR) electrical connector
- Steering wheel side panels
- Air bag
- Horn electrical connector
- Cruise control electrical connector
- Steering wheel
- Steering joint cover
- Steering joint lower bolt and pull the joint toward the column
- Tie rods from the steering knuckles
- Shift linkage (if equipped with a manual transmission)
- Heated Oxygen (HO$_2$S) sensor electrical connector
- Catalytic converter from the vehicle
- Return line from the steering gear
- Rear beam brace rod
- Left tie rod end, then slide the steering gear all of the way to the right
- 2 lines from the valve body unit on the steering gear

✳✳ CAUTION

After disconnecting the hose and pipe, plug or cap the hose and pipe to prevent foreign materials from entering the valve body unit.

➡**Do not loosen the cylinder pipes between the valve body unit and the cylinder.**

4. Remove the steering gear mounting bolts.
5. Pull the steering gear all the way down to clear the pinion shaft from the bulkhead, and remove the pinion shaft grommet.
6. Slide the rack all of the way to the right, then place the left rack end below the rear beam.
7. Move the steering gear assembly to the left, and tilt the left side down to remove it from the vehicle.

To install:

8. Installation is the reverse of the removal procedure, while using the following torque values:
- Left steering gear mounting bolts: 28 ft. lbs. (38 Nm)
- Right steering gear mounting bolts: 43 ft. lbs. (58 Nm)

➡**After installing the steering gear, check the air hose connections for interference with adjacent parts.**

- Intermediate shaft pinch bolt: 16 ft. lbs. (22 Nm)
- Rear beam brace rod bolts: 28 ft. lbs. (38 Nm)
- Catalytic converter nuts: 25 ft. lbs. (33 Nm)
- Tie rod end nuts: 33 ft. lbs. (44 Nm)
- Steering wheel nut: 36 ft. lbs. (49 Nm)
- Air bag bolts: 84 inch lbs. (9.5 Nm)

3.2TL AND 3.5RL

1. Before servicing the vehicle, refer to the precautions in the beginning of this section.
2. Drain the power steering fluid.
3. Remove or disconnect the following:
- Negative then the positive battery cables

- Steering wheel lower access panel
- Supplemental Inflatable Restraint (SIR) electrical connector
- Steering wheel side panels
- Air bag
- Horn electrical connector
- Cruise control electrical connector
- Steering wheel
- Steering joint bolts then move the joint toward the column
- Front wheels
- Tie rods from the steering knuckles
- Splash guard
- Feed line from the valve body
- Line mounting clamps
- Feed line from the line mounting cushions
- Sensor inlet line and both return lines from the hoses
- Steering gear mounting brackets
- Steering gear from the vehicle

To install:

4. Installation is the reverse of the removal procedure, while using the following torque values:
- Line mounting clamps bolts: 84 inch lbs. (9.5 Nm)
- Feed line bolts: 96 inch lbs. (11 Nm)
- Mounting bracket bolts: 28 ft. lbs. (38 Nm)
- Splash shield short bolts: 28 ft. lbs. (38 Nm)
- Splash shield long bolts: 43 ft. lbs. (59 Nm)
- Steering joint pinch bolt: 16 ft. lbs. (22 Nm)
- Steering wheel nut: 36 ft. lbs. (49 Nm)
- Air bag bolts: 84 inch lbs. (9.5 Nm)

NSX

1. Before servicing the vehicle, refer to the precautions in the beginning of this section.
2. Drain the power steering fluid.
3. Remove or disconnect the following:
- Negative then the positive battery cables
- Battery
- Steering joint cover
- Steering joint bolts then move the joint toward the column to separate it
- Wheels
- Tie rods from the steering knuckles
- Spare tire
- Spare tire holder plate
- Spare tire holder
- Floor under cover
- Terminal guard
- Ground cable

- Wires from the steering gear terminals
- Radiator pipe bracket at the front compartment bulkhead
- Radiator pipe bracket at the floor

4. Support the steering gear and the front crossbeam
- Steering gear and front crossbeam bolts and nuts
- Steering gear and crossbeam from the vehicle

To install:

5. Installation is the reverse of the removal procedure, while using the following torque values:

- Steering gear and crossbeam fasteners: 43 ft. lbs. (59 Nm)
- Radiator pipe bracket bolts: 16 ft. lbs. (22 Nm)
- Ground cable bolt: 84 inch lbs. (9.5 Nm)
- Terminal guard bolts: 84 inch lbs. (9.5 Nm)
- Floor under cover bolts: 84 inch lbs. (9.5 Nm)
- Spare tire holder bolt: 18 ft. lbs. (25 Nm)
- Steering joint bolts: 16 ft. lbs. (22 Nm)

Strut

REMOVAL & INSTALLATION

Front

EXCEPT NSX

1. Before servicing the vehicle, refer to the precautions in the beginning of this section.
2. Support the lower suspension arm with a jack.
3. Remove or disconnect the following:
- Wheel(s)
- Brake hose from the strut

CAUTION:
- Replace the self-locking nuts after removal.
- The vehicle should be on the ground before any bolts or nuts connected to rubber mounts or bushings are tightened.
- Torque the castle nut to the lower torque specification, then tighten it only far enough to align the slot with the pin hole. Do not align the nut by loosening.

NOTE: Wipe off the grease before tightening the nut at the ball joint.

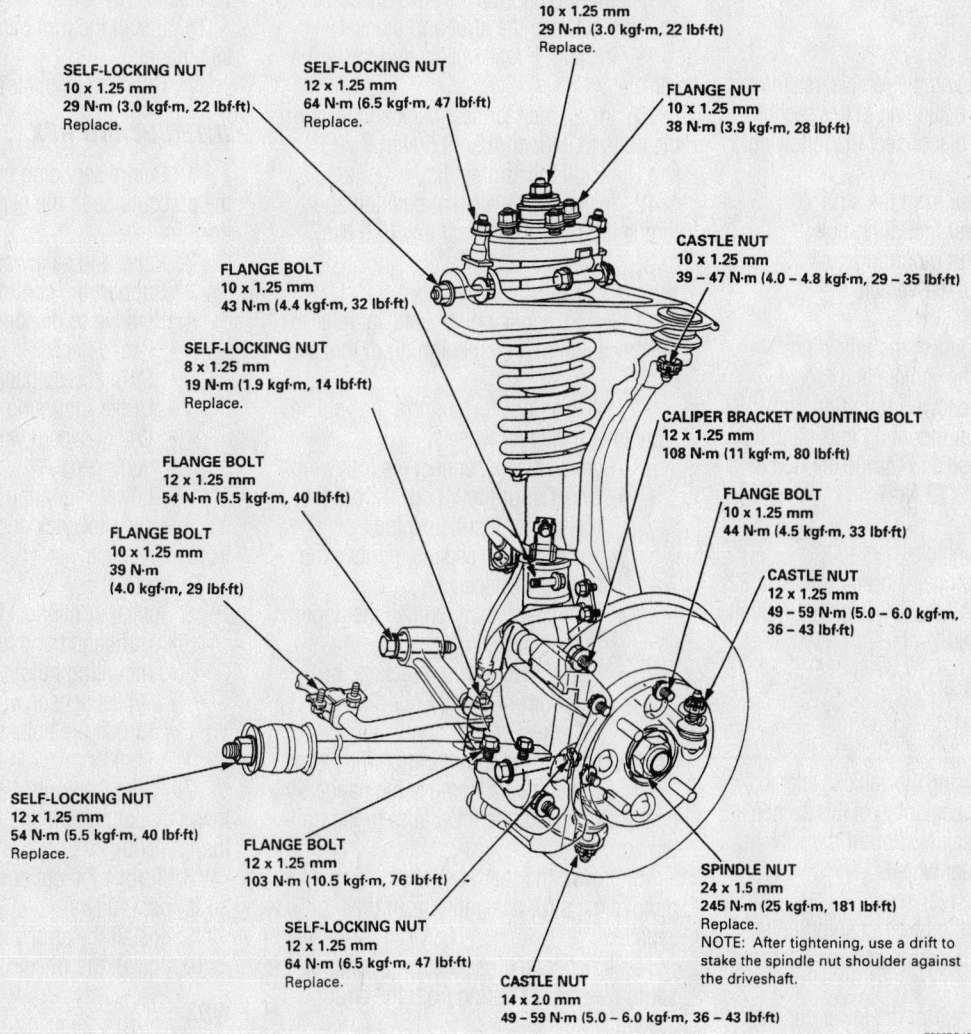

Front suspension showing the torque specifications

- Strut fork
- Upper strut mounting nuts
- Strut from the vehicle

To install:

4. Install the strut and loosely install the mounting nuts

5. Install the strut fork, do not tighten the bolts at this time

➡ **All suspension nuts and bolts should be tightened with the vehicle on the ground, or with a floor jack supporting the vehicle's weight.**

6. Tighten the pinch bolt to 32 ft. lbs. (43 Nm) (except Integra 16 ft. lbs. [22 Nm]).

7. Tighten the lower strut fork bolt to 47 ft. lbs. (64 Nm).

8. Install the brake hose to the strut fork and tighten the bolt to 16 ft. lbs. (22 Nm).

9. Install the wheels.

10. Tighten the upper strut nuts to 28 ft. lbs. (38 Nm) (except Integra and 3.2 models 37 ft. lbs. [50 Nm])

11. Measure and adjust the wheel alignment.

NSX

1. Before servicing the vehicle, refer to the precautions in the beginning of this section.

2. Remove or disconnect the following:
   - Wheel(s)
   - Brake hose from the strut
   - Lower strut mounting bolt
   - Upper strut mounting nuts
   - Strut from the vehicle

To install:

3. Install or connect the following:
   - Strut to the vehicle and loosely install the lower mounting bolt
   - New upper mounting nuts hand tight
   - Brake hose and torque the bolt to 16 ft. lbs. (22 Nm)
   - Wheel(s)

4. Lower the vehicle.

5. Torque the upper mounting nuts to 32 ft. lbs. (43 Nm) and the lower mounting bolt to 69 ft. lbs. (93 Nm).

Rear

3.2TL

1. Before servicing the vehicle, refer to the precautions in the beginning of this section.

2. Raise and safely support the vehicle and remove the rear wheels.

3. Remove the rear seat:
   a. Remove the lower cushion bolts located under the armrest and near the floor.
   b. Pull the rear of the lower cushion up and lift it forward to release it from the clips.

c. Pull down the trunk bulkhead trim and release the armrest lid clips.

d. Remove the bolts from the top and bottom of the back cushion, then lift it up and forward to disengage the securing hooks.

4. Place a floor jack under the lower arm and slightly compress the spring.

5. Remove the upper mounting nuts and the lower flange bolt.

6. Lower the jack to remove the strut. Be sure and mark the right and left struts so they can be reinstalled on the proper sides.

To install:

7. Install the strut into the vehicle. Loosely install the mounting nuts and mounting bolt, but do not tighten them until the weight of the vehicle is on the suspension.

8. Raise the rear suspension with a floor jack until the weight of the vehicle is on the strut. Tighten the upper mounting nuts to 28 ft. lbs. (39 Nm), then tighten the lower mounting bolts to 40 ft. lbs. (55 Nm). Be careful not to pinch the ABS speed sensor wire between the strut and bracket.

9. Install the rear wheels and lower the vehicle.

10. Install the rear seat back and torque the bolts to 84 inch lbs. (9.5 Nm).

11. Install the armrest lid.

12. Install the lower seat cushion and torque the bolts to 84 inch lbs. (9.5 Nm).

3.5RL

1. Before servicing the vehicle, refer to the precautions in the beginning of this section.

2. Raise and safely support the vehicle and remove the rear wheels.

3. Remove or disconnect the following:
   - Upper strut mount cover from the rear panel, just below the speaker—On sedans, remove the trunk side panel.
   - Trim cover, then remove the upper mount nuts
   - Wheel sensor wire brackets, on cars with ABS—do not disconnect the wheel sensor connector
   - Lower strut mounting bolt

4. On the Integra, remove the flange bolt that connects the lower arm to the trailing arm.

5. Lower the rear suspension and remove the strut assembly from the vehicle.

6. If necessary, use a spring compressor to remove the spring from the strut assembly.

To install:

7. Reassemble the strut and coil spring

assembly. Tighten the strut self–locking nut to 22 ft. lbs. (30 Nm).

8. Lower the rear suspension and position the strut assembly in the vehicle. The nut welded to the lower strut mounting should face the front of the vehicle.

9. Loosely install the upper mounting nuts.

10. On the Integra, raise the rear suspension and install the bolt connecting the lower arm to the trailing arm.

11. Install the strut lower mounting bolt.

12. Raise the vehicle until the vehicle just lifts off the safety stand and tighten the lower strut bolt and lower control arm bolt. On the integra, tighten the lower strut bolt and the control arm bolt to 40 ft. lbs. (54 Nm). In the 3.5RL, tighten the lower strut mounting bolt to 76 ft. lbs. (103 Nm).

13. Install the wheel sensor wire bracket on cars with ABS.

14. Tighten the upper mounting nuts to 36 ft. lbs. (49 Nm).

15. Install the rear wheels, then lower the vehicle.

16. Install the trim panel, or the trunk side panel.

17. Check the vehicle's alignment.

INTEGRA AND RSX

1. Before servicing the vehicle, refer to the precautions in the beginning of this section.

2. Raise and safely support the vehicle.

3. Support the control arm with a jack.

4. Remove or disconnect the following:
   - Rear wheels
   - Strut access panel
   - Upper mounting nuts
   - Wheel sensor wire brackets (if equipped)
   - Lower mounting bolt

5. Lower the jack and remove the strut from the vehicle.

To install:

6. Install or connect the following:
   - Strut and hand tighten the upper mounting nuts
   - Wheel sensor wire bracket and torque the bolts to 84 inch lbs. (9.5 Nm)

7. Raise the control arm and install the lower mounting bolt and torque it to 40 ft. lbs. (54 Nm).

8. Torque the upper mounting nuts to 36 ft. lbs. (49 Nm).

9. Install the strut access panel.

10. Install the wheels.

NSX

1. Before servicing the vehicle, refer to the precautions in the beginning of this section.

2. Remove or disconnect the following:
   - Rear wheels
   - Lower rear hatch glass trim
   - Upper strut mounting nuts
   - Rear strut brace (NSX–T model only)
   - Lower strut mounting nut
   - Stabilizer link
   - Strut from the vehicle

To install:

3. Install or connect the following:

   - Strut
   - Stabilizer link and torque the nut (at the stabilizer bar) to 61 ft. lbs. (83 Nm)
   - Lower strut mounting bolt and torque it to 69 ft. lbs. (93 Nm)
   - New upper strut mounting nuts and torque them to 39 ft. lbs. (53 Nm)
   - Strut brace (if equipped) and torque the bolts to 16 ft. lbs. (22 Nm)
   - Rear hatch glass trim

   - Rear wheels

4. Measure and adjust the wheel alignment.

Coil Spring

REMOVAL & INSTALLATION

1. Before servicing the vehicle, refer to the precautions in the beginning of this section.

CAUTION:
- **Replace the self-locking nuts after removal.**
- **The vehicle should be on the ground before any bolts or nuts connected to rubber mounts or bushings are tightened.**
- **Torque the castle nut to the lower torque specification, then tighten it only far enough to align the slot with the pin hole. Do not align the nut by loosening.**

NOTE: Wipe off the grease before tightening the nut at the ball joint.

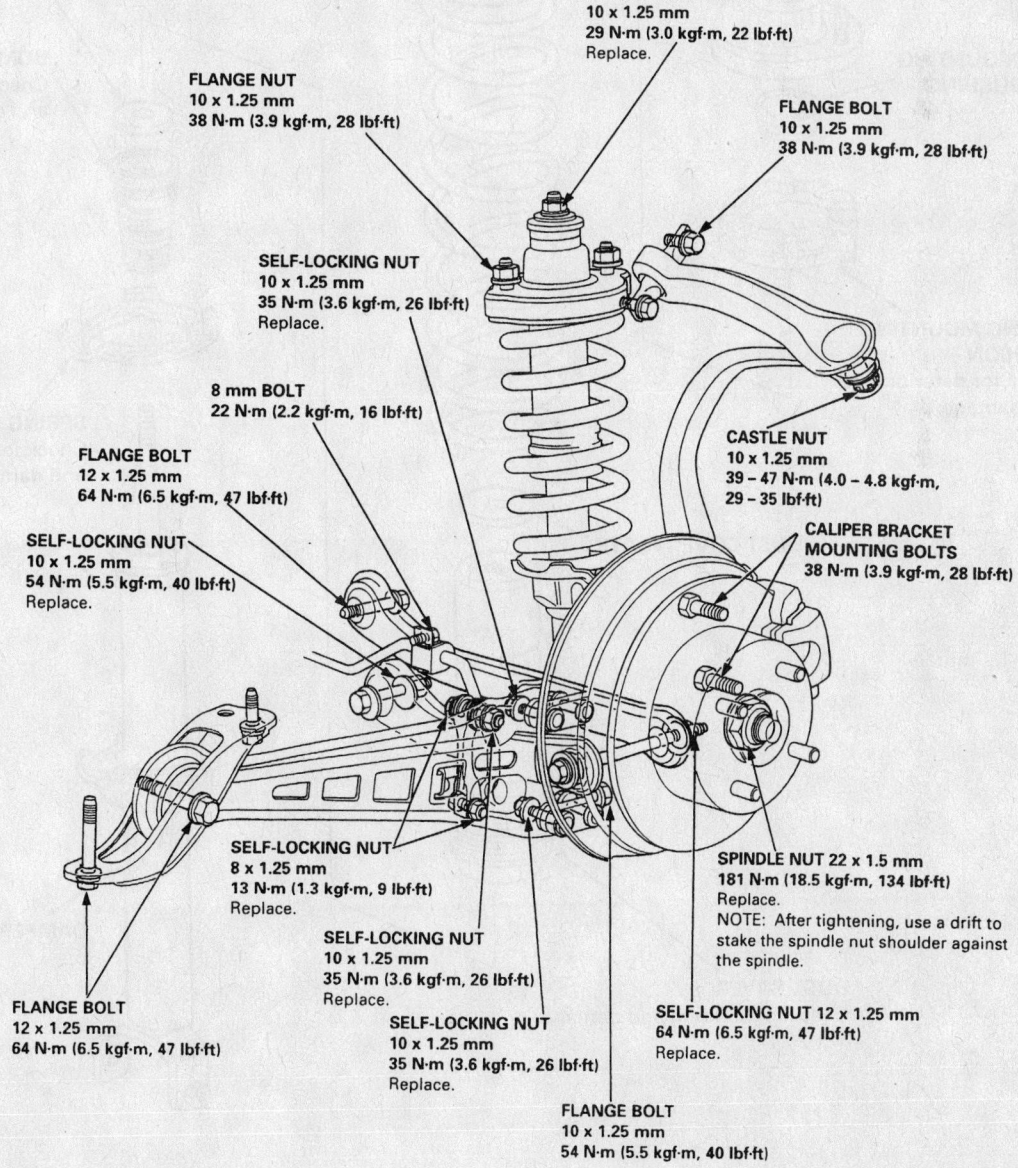

SELF-LOCKING NUT
10 x 1.25 mm
29 N·m (3.0 kgf·m, 22 lbf·ft)
Replace.

FLANGE NUT
10 x 1.25 mm
38 N·m (3.9 kgf·m, 28 lbf·ft)

FLANGE BOLT
10 x 1.25 mm
38 N·m (3.9 kgf·m, 28 lbf·ft)

SELF-LOCKING NUT
10 x 1.25 mm
35 N·m (3.6 kgf·m, 26 lbf·ft)
Replace.

8 mm BOLT
22 N·m (2.2 kgf·m, 16 lbf·ft)

CASTLE NUT
10 x 1.25 mm
39 – 47 N·m (4.0 – 4.8 kgf·m, 29 – 35 lbf·ft)

FLANGE BOLT
12 x 1.25 mm
64 N·m (6.5 kgf·m, 47 lbf·ft)

CALIPER BRACKET MOUNTING BOLTS
38 N·m (3.9 kgf·m, 28 lbf·ft)

SELF-LOCKING NUT
10 x 1.25 mm
54 N·m (5.5 kgf·m, 40 lbf·ft)
Replace.

SELF-LOCKING NUT
8 x 1.25 mm
13 N·m (1.3 kgf·m, 9 lbf·ft)
Replace.

SPINDLE NUT 22 x 1.5 mm
181 N·m (18.5 kgf·m, 134 lbf·ft)
Replace.
NOTE: After tightening, use a drift to stake the spindle nut shoulder against the spindle.

SELF-LOCKING NUT
10 x 1.25 mm
35 N·m (3.6 kgf·m, 26 lbf·ft)
Replace.

SELF-LOCKING NUT
10 x 1.25 mm
35 N·m (3.6 kgf·m, 26 lbf·ft)
Replace.

SELF-LOCKING NUT 12 x 1.25 mm
64 N·m (6.5 kgf·m, 47 lbf·ft)
Replace.

FLANGE BOLT
12 x 1.25 mm
64 N·m (6.5 kgf·m, 47 lbf·ft)

FLANGE BOLT
10 x 1.25 mm
54 N·m (5.5 kgf·m, 40 lbf·ft)

7923BG92

Rear suspension showing the torque specifications

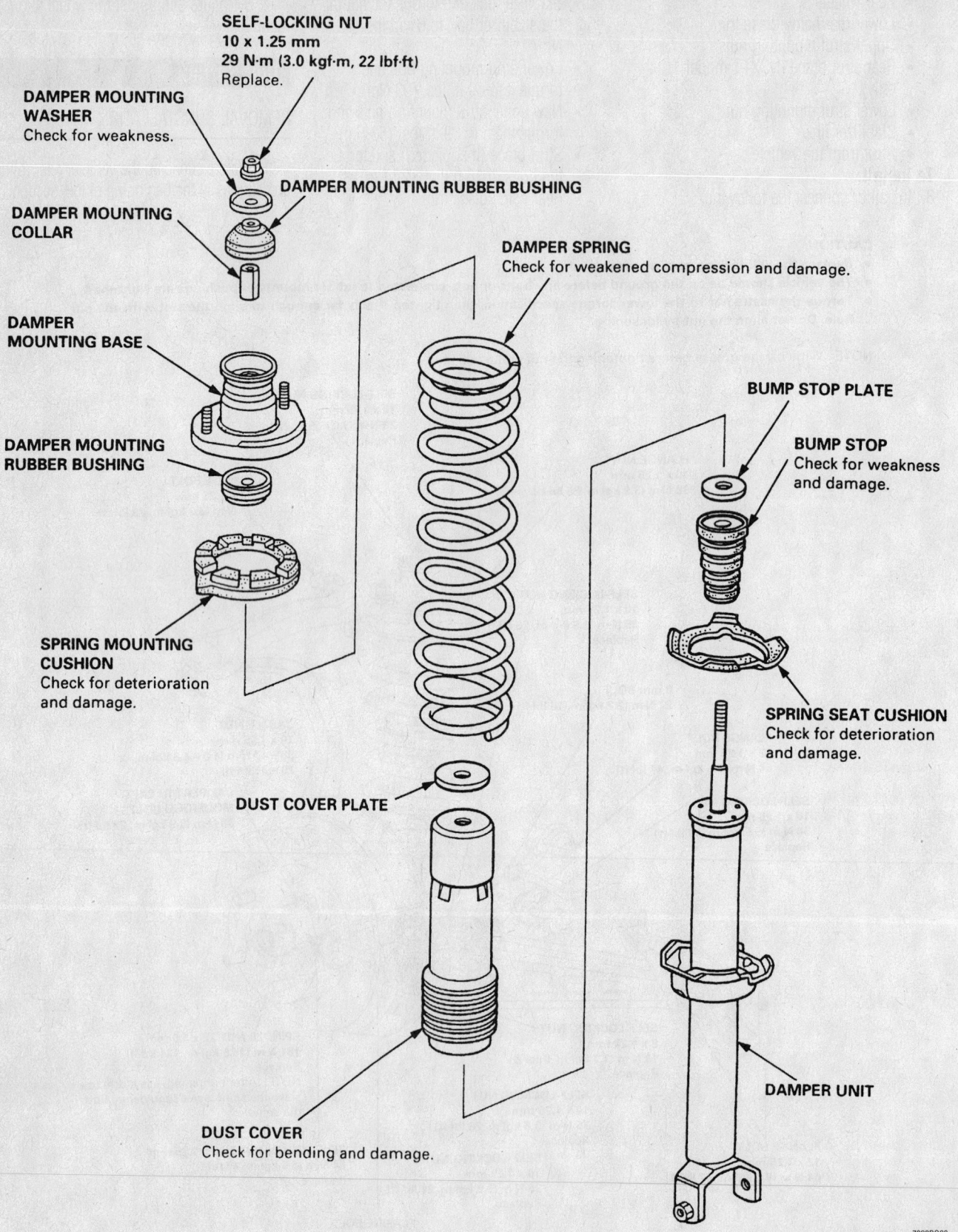

SELF-LOCKING NUT
10 x 1.25 mm
29 N·m (3.0 kgf·m, 22 lbf·ft)
Replace.

DAMPER MOUNTING WASHER
Check for weakness.

DAMPER MOUNTING RUBBER BUSHING

DAMPER MOUNTING COLLAR

DAMPER SPRING
Check for weakened compression and damage.

DAMPER MOUNTING BASE

BUMP STOP PLATE

BUMP STOP
Check for weakness and damage.

DAMPER MOUNTING RUBBER BUSHING

SPRING MOUNTING CUSHION
Check for deterioration and damage.

SPRING SEAT CUSHION
Check for deterioration and damage.

DUST COVER PLATE

DAMPER UNIT

DUST COVER
Check for bending and damage.

7923BG93

Exploded view of the rear strut (damper)

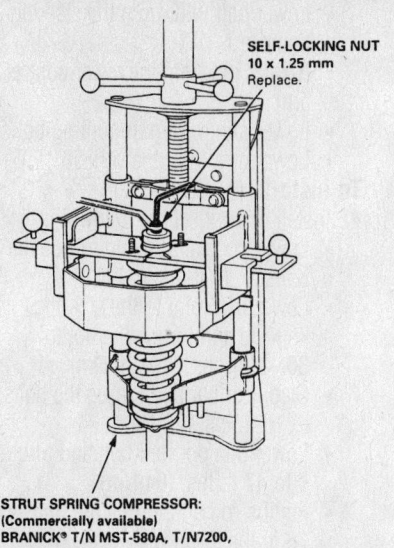

SELF-LOCKING NUT
10 x 1.25 mm
Replace.

STRUT SPRING COMPRESSOR:
(Commercially available)
BRANICK® T/N MST-580A, T/N7200,
or equivalent

7923BG94

Compress the coil spring until the spring moves away from the seat and use a hex wrench to hold the piston rod while removing the nut

2. Raise and support the vehicle and remove the front wheels.

3. Remove the strut (damper).

4. Place the strut assembly in a coil spring compressor.

5. Compress the coil spring and remove the locking nut from the top of the strut.

6. Release the pressure from the spring compressor.

7. Remove the coil spring and related pieces from the strut.

To install:

➡Use new self–locking nuts and bolts when assembling the strut.

8. Install the strut, coil spring and related components on the spring compressor.

9. Compress the spring.

10. Install the mounting washer, and loosely install a new self–locking nut.

11. Hold the strut piston rod with a hex wrench and tighten the self–locking nut to 22 ft. lbs. (30 Nm).

12. Install the strut in the vehicle.

13. Check and adjust the vehicle's front wheel alignment.

Upper Ball Joint

REMOVAL & INSTALLATION

➡The upper ball joint cannot be removed from the control arm. If the ball joint is damaged, the upper arm assembly must be replaced.

Lower Ball Joint

REMOVAL & INSTALLATION

Integra and RSX

1. Before servicing the vehicle, refer to the precautions in the beginning of this section.

2. Raise and support the vehicle safely.

3. Remove or disconnect the following:
   • Front wheels
   • Axle nut
   • Brake hose from the knuckle
   • Brake caliper
   • Brake rotor
   • Wheel sensor wire bracket
   • Wheel sensor from the knuckle
   • Tie rod from the knuckle
   • Lower ball joint from the knuckle
   • Upper ball joint from the knuckle

   • Steering knuckle
   • Boot by prying off the snapring

4. Press the ball joint from the knuckle.

To install:

5. Place the ball joint in position by hand. Install the ball joint into the tool and press in the new ball joint.

✳✳ WARNING

After installing the boot, check the ball joint pin tapered section for grease contamination and wipe it if necessary.

6. Install or connect the following:
   • Ball joint boot and snapring
   • Steering knuckle to the vehicle
   • Lower ball joint and torque the nut to 40 ft. lbs. (54 Nm)
   • Tie rod and torque the nut to 32 ft. lbs. (43 Nm)

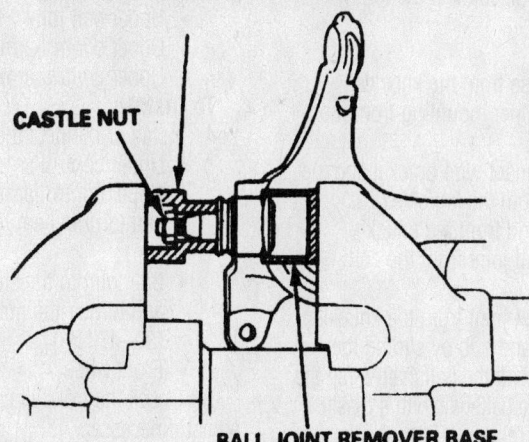

BALL JOINT REMOVER/INSTALLER

CASTLE NUT

BALL JOINT REMOVER BASE

7923BG50

Use a vise or press to remove the ball joint from the steering knuckle

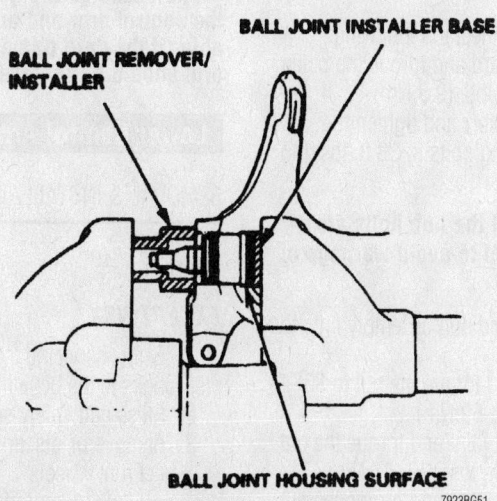

BALL JOINT INSTALLER BASE

BALL JOINT REMOVER/
INSTALLER

BALL JOINT HOUSING SURFACE

7923BG51

Press the new ball joint into the steering knuckle

- Upper ball joint and torque the nut to 32 ft. lbs. (43 Nm)
- Wheel sensor and torque the bolts to 84 inch lbs. (9.5 Nm)
- Brake rotor and torque the bolts to 84 inch lbs. (9.5 Nm)
- Brake caliper and torque the bolts to 80 ft. lbs. (108 Nm)
- Brake hose and torque the bolts to 84 inch lbs. (9.5 Nm)
- New axle nut and torque it to 134 ft. lbs. (181 Nm)
- Front wheels

7. Check the front wheel alignment and adjust if necessary.

3.2TL, 3.2CL, 3.5RL and NSX

➡The lower ball joint cannot be removed from the steering knuckle.

1. Before servicing the vehicle, refer to the precautions in the beginning of this section.
2. Raise and safely support the vehicle
3. Remove or disconnect the following:
   - Wheels
   - Axle nut
   - Brake hose from the knuckle
   - Brake caliper mounting from the knuckle
   - Wheel sensor wire bracket and the sensor from the knuckle
   - Tie rod end from the knuckle
   - Lower ball joint from the control arm
   - Upper ball joint from the knuckle
   - Knuckle and hub by sliding the assembly off the halfshaft—Tap the end of the halfshaft with a plastic mallet to release it from the knuckle.
   - Hub and rotor assembly from the knuckle
   - Splash guard from the knuckle

To install:
4. Install or connect the following:
   - Splash guard and torque the bolts to 84 inch lbs. (9.5 Nm)
   - Hub assembly and tighten the self–locking bolts to 33 ft. lbs. (45 Nm)

➡Be sure that all the hub bolts are properly tightened to avoid warpage of the brake disc.

   - Knuckle and hub assembly onto the halfshaft
   - Tie rod and torque the nut to 36–43 ft. lbs. (49–59 Nm)
   - Upper ball joint and torque the nut to 29–35 ft. lbs. (39–47 Nm)
   - Lower ball joint and torque the nut to 36–43 ft. lbs. (49–59 Nm)

- Wheel sensor and torque the bolts to 16 ft. lbs. (22 Nm)
- Wheel sensor wire and torque the bolts to 84 inch lbs. (9.5 Nm)
- Brake caliper and torque the bolts to 80 ft. lbs. (108 Nm)
- Brake hoses and torque the bolts to 84 inch lbs. (9.5 Nm)
- New axle nut and torque it to 181 ft. lbs. (245 Nm)
- Wheel

5. Measure and adjust the wheel alignment.

Upper Control Arm

REMOVAL & INSTALLATION

1. Before servicing the vehicle, refer to the precautions in the beginning of this section.
2. Raise and safely support the vehicle.
3. Remove or disconnect the following:
   - Front wheel
   - Upper ball joint from the knuckle
   - Upper control arm nuts
   - Upper control arm from the vehicle

To install:
4. Install or connect the following:
   - Upper control arm
   - Upper control arm-to-chassis nuts and torque them to 47 ft. lbs. (64 Nm)
   - Ball joint to the steering knuckle and torque the nut to 29–35 ft. lbs. (39–47 Nm)
   - Front wheel

5. Check the front wheel alignment and adjust if necessary.

CONTROL ARM BUSHING REPLACEMENT

➡The bushings are an integral part of the control arm and are not replaceable. If they are damaged the control arm should be replaced.

Lower Control Arm

REMOVAL & INSTALLATION

Front

EXCEPT NSX

1. Before servicing the vehicle, refer to the precautions in the beginning of this section.
2. Raise and safely support the vehicle.
3. Remove or disconnect the following:
   - Front wheels
   - Lower damper fork bolt
   - Stabilizer bar from the arm

- Lower ball joint from the steering knuckle
- Radius rod from the lower control arm
- Lower control arm mounting bolts
- Lower arm from the vehicle

To install:
4. Install or connect the following:
   - Lower control arm and torque the bolts to 40 ft lbs. (54 Nm)
   - Lower ball joint to the steering knuckle and torque the nut to 36–43 ft. lbs. (49–59 Nm)
   - Stabilizer bar and torque the bolts to 16 ft. lbs. (22 Nm)
   - Lower damper fork bolt and torque it to 47 ft. lbs. (64 Nm)
   - Radius rod and torque the bolts to 76 ft. lbs. (103 Nm)
   - Front wheels

5. Measure and adjust the wheel alignment.

NSX

1. Before servicing the vehicle, refer to the precautions in the beginning of this section.
2. Raise and safely support the vehicle.
3. Remove or disconnect the following:
   - Front wheels
   - Steering knuckle from the control arm
   - Lower strut mounting bolt
   - Stabilizer link from the control arm
   - Ball joint from the compliance pivot assembly
   - Control arm adjusting bolt
   - Control arm from the vehicle

To install:
4. Install or connect the following:
   - Lower control arm to the vehicle and torque the adjusting bolt to 90 ft. lbs. (123 Nm)
   - Ball joint to the compliance pivot assembly and torque the nut to 40–47 ft. lbs. (54–64 Nm)
   - Stabilizer link and torque the nut to 61 ft. lbs. (83 Nm)
   - Lower strut mounting bolt and torque it to 69 ft. lbs. (93 Nm)
   - Steering knuckle to the control arm and torque the nut to 40–47 ft. lbs. (54–64 Nm)
   - Front wheels

5. Measure and adjust the wheel alignment.

Rear

INTEGRA AND RSX

1. Before servicing the vehicle, refer to the precautions in the beginning of this section.

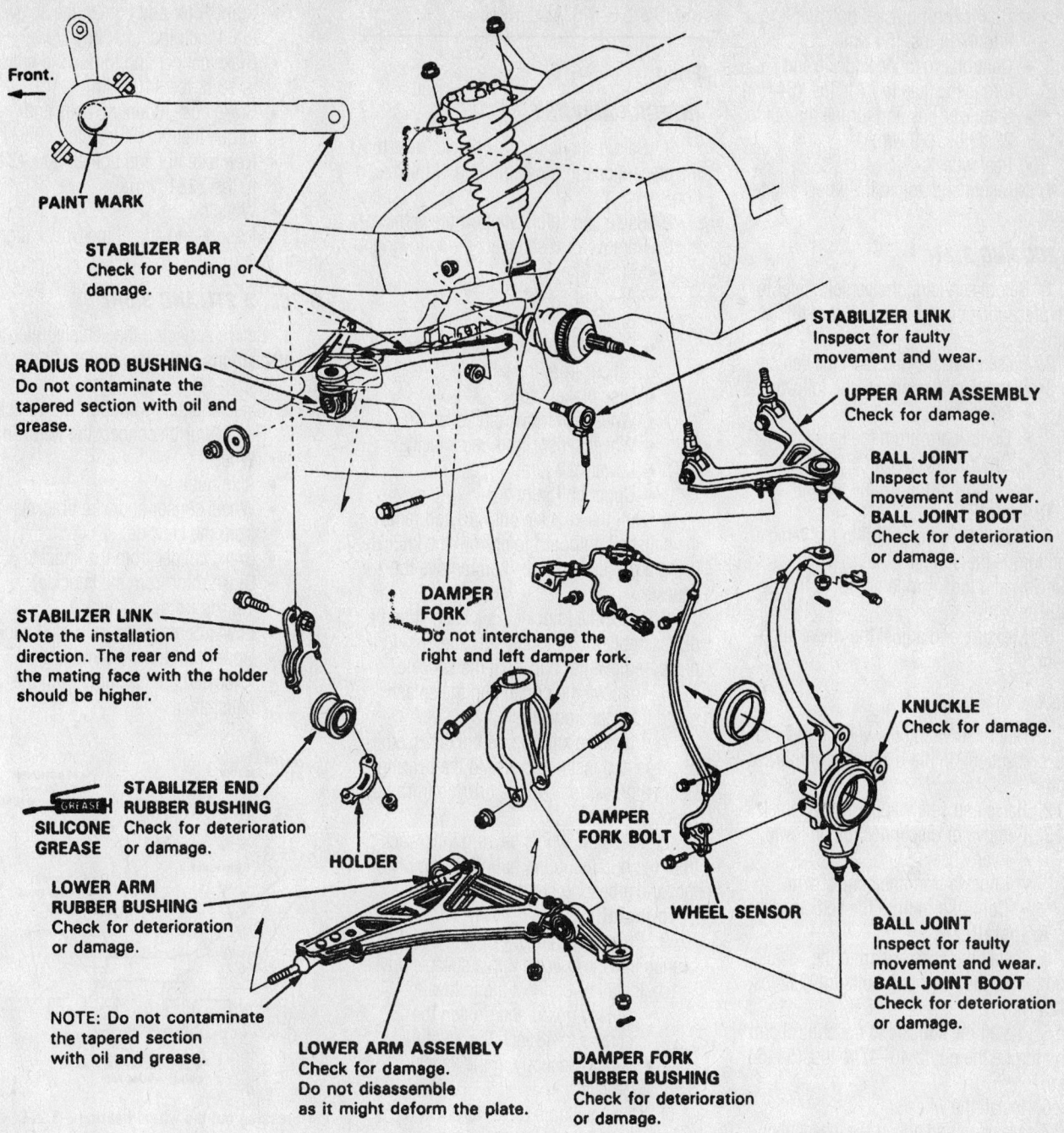

Front.

PAINT MARK

STABILIZER BAR
Check for bending or damage.

RADIUS ROD BUSHING
Do not contaminate the tapered section with oil and grease.

STABILIZER LINK
Note the installation direction. The rear end of the mating face with the holder should be higher.

GREASE

SILICONE GREASE

STABILIZER END RUBBER BUSHING
Check for deterioration or damage.

LOWER ARM RUBBER BUSHING
Check for deterioration or damage.

NOTE: Do not contaminate the tapered section with oil and grease.

HOLDER

DAMPER FORK
Do not interchange the right and left damper fork.

DAMPER FORK BOLT

LOWER ARM ASSEMBLY
Check for damage.
Do not disassemble as it might deform the plate.

DAMPER FORK RUBBER BUSHING
Check for deterioration or damage.

WHEEL SENSOR

STABILIZER LINK
Inspect for faulty movement and wear.

UPPER ARM ASSEMBLY
Check for damage.

BALL JOINT
Inspect for faulty movement and wear.
BALL JOINT BOOT
Check for deterioration or damage.

KNUCKLE
Check for damage.

BALL JOINT
Inspect for faulty movement and wear.
BALL JOINT BOOT
Check for deterioration or damage.

7923BG49

A common upper control arm and ball joint assembly

2. Raise and safely support the vehicle.
3. Remove or disconnect the following:
   - Rear wheels
   - Hub/bearing assembly
   - Splash guard
   - Lower strut mounting bolt
   - Upper arm from the control arm
   - Lower arm from the control arm
   - Compensator arm
   - Control arm mounting bolts
   - Control arm from the vehicle

To install:
4. Install or connect the following:

- Control arm and torque the bolts to 47 ft. lbs. (64 Nm)
- Compensator arm and torque the bolt to 47 ft. lbs. (64 Nm)
- Lower arm and torque the bolt to 40 ft. lbs. (54 Nm)
- Upper arm and torque the bolt to 40 ft. lbs. (54 Nm)
- Lower strut mounting bolt and torque it to 40 ft. lbs. (54 Nm)
- Hub/bearing assembly and torque the nut to 134 ft. lbs. (181 Nm)

3.5RL

1. Before servicing the vehicle, refer to the precautions in the beginning of this section.
2. Raise and safely support the vehicle.
3. Remove or disconnect the following:
   - Rear wheels
   - Stabilizer link from the control arm
   - Control arm from the knuckle
   - Control arm from the bracket
   - Control arm from the vehicle

To install:
- Control arm to the vehicle

- Control arm bracket bolt and torque it to 47 ft. lbs. (64 Nm)
- Control arm to the knuckle and torque the nuts to 47 ft. lbs. (64 Nm)
- Stabilizer link and torque the nut to 22 ft. lbs. (29 Nm)
- Rear wheels

4. Measure and adjust the wheel alignment.

3.2CL AND 3.2TL

1. Before servicing the vehicle, refer to the precautions in the beginning of this section.
2. Raise and safely support the vehicle.
3. Remove or disconnect the following:
   - Rear wheels
   - Control arm from the knuckle
   - Control arm from the subframe
   - Control arm from the vehicle

To install:

4. Install the control arm to the vehicle and torque the subframe nut to 40 ft. lbs. (54 Nm) and the knuckle nut to 43 ft. lbs. (59 Nm).
5. Measure and adjust the wheel alignment.

NSX

1. Before servicing the vehicle, refer to the precautions in the beginning of this section.
2. Raise and safely support the vehicle.
3. Remove or disconnect the following:
   - Rear wheels
   - Knuckle from the control arm
   - Control arm from the vehicle

To install:

4. Install the control arm to the vehicle and torque the mounting bolts to 90 ft. lbs. (123 Nm).
5. Install the knuckle to the control arm and torque the nut to 40–47 ft. lbs. (54–64 Nm).
6. Install the wheels.
7. Measure and adjust the wheel alignment.

CONTROL ARM BUSHING REPLACEMENT

➡The bushings are an integral part of the control arm and are not replaceable. If they are damaged the control arm should be replaced.

Wheel Bearings

ADJUSTMENT

The front and rear wheel bearings are not adjustable or repairable and should be replaced if found defective.

REMOVAL & INSTALLATION

Front

INTEGRA AND RSX

1. Before servicing the vehicle, refer to the precautions in the beginning of this section.
2. Raise and safely support the vehicle.
3. Remove or disconnect the following:
   - Front wheels
   - Axle nut
   - Brake hose mounting bolts
   - Brake caliper bolts and remove the caliper from the knuckle
   - Disc brake rotor
   - Wheel sensor wire bracket
   - Wheel sensor from the knuckle
   - Lower ball joint
   - Upper ball joint

4. Pull the knuckle outward and remove the halfshaft outboard joint from the knuckle using a plastic hammer, then remove the knuckle.
5. Place the knuckle on a base. Insert a disassembly tool into the hub, then using a press, remove the hub from the knuckle.
6. Remove the circlip and the splash guard from the knuckle.
7. Place the knuckle on the disassembly base and install a driver to the bearing. Using a press, remove the bearing from the knuckle.
8. Press the wheel bearing inner race from the hub using the hub disassembly tool and a bearing separator.

To install:

9. Press a new inner race and wheel bearing into the knuckle with a suitable driver.
10. Install or connect the following:
    - Splash guard and tighten the screws to 43 inch lbs. (5 Nm)
    - Circlip securely in the knuckle groove
11. Use a suitable driver to press the knuckle onto the hub.
12. Install or connect the following:
    - Knuckle/hub assembly onto the halfshaft
    - Lower ball joint to the knuckle and torque the nut to 36–43 ft. lbs. (49–59 Nm)
    - Tie rod and tighten the nut to 29–35 ft. lbs. (39–47 Nm)
    - Upper ball joint and tighten the nut to 29–35 ft. lbs. (39–47 Nm)
    - Wheel sensor and torque the bolts to 84 inch lbs. (9.5 Nm)
    - Wheel sensor wire bracket and torque the bolts to 84 inch lbs. (9.5 Nm)
    - Brake rotor and tighten the screws to 84 inch lbs. (10 Nm)
    - Brake caliper and torque the bolts to 80 ft. lbs. (108 Nm)
    - Brake hose mounting bolts and torque them to 84 inch lbs. (9.5 Nm)
    - New axle nut and tighten it to 134 ft. lbs. (181 Nm)
    - Wheels
13. Measure and adjust the wheel alignment.

3.2CL, 3.2TL AND 3.5RL

1. Before servicing the vehicle, refer to the precautions in the beginning of this section.
2. Raise and safely support the vehicle.
3. Remove or disconnect the following:
   - Wheel
   - Axle nut
   - Wheel sensor and wire brackets from the knuckle
   - Brake caliper from the knuckle
   - Brake rotor from the knuckle
   - Tie rod from the knuckle
   - Lower control arm from the knuckle
   - Upper arm from the knuckle
   - Knuckle/hub assembly from the halfshaft

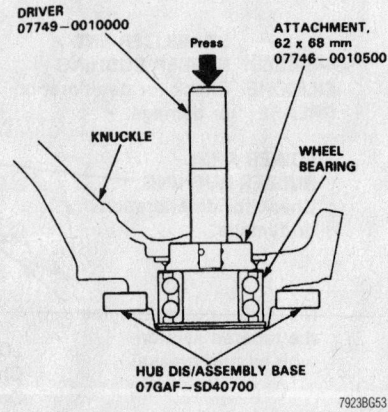

Pressing out the wheel bearing—3.2TL

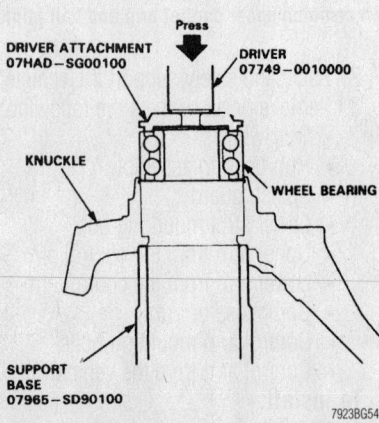

Installing the wheel bearing—3.2TL

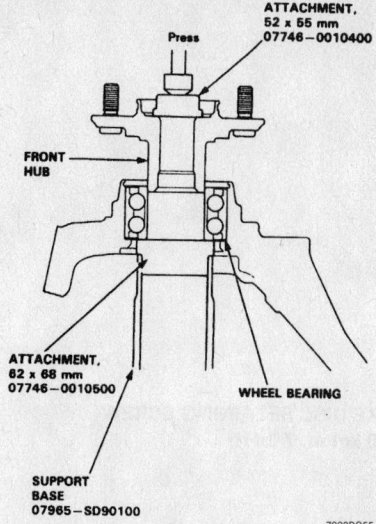

Hub installation—3.2TL

4. Clamp the knuckle in a vise and secure a slide hammer to the wheel studs to separate the hub from the knuckle.

5. Remove the splash guard.

6. Remove the snapring from the knuckle.

7. Support the knuckle and press the bearing out towards the wheel side.

8. If the inner bearing race stayed on the hub, use a puller to remove it.

To install:

9. Press a new inner race on the hub.

10. Press a new bearing into the knuckle.

11. Install the outer snapring.

12. Install the splash guard and torque the bolts to 4 ft. lbs. (4.9 Nm).

13. Properly support the knuckle and press the hub into the bearing.

✳✳ CAUTION

Do not press on the wheel studs or they will press out of the hub.

14. Install or connect the following:
  - Knuckle/hub assembly onto the halfshaft
  - Lower ball joint and torque the nut to 51–58 ft. lbs. (69–78 Nm)
  - Upper ball joint and torque the nut to 29–35 ft. lbs. (40–48 Nm)
  - Tie rod end and torque the nut to 36–43 ft. lbs. (50–60 Nm)
  - Brake rotor and torque the bolts to 84 inch lbs. (9.5 Nm)
  - Brake caliper and torque the bolts to 80 ft. lbs. (108 Nm)
  - Brake hose brackets and torque the bolts to 16 ft. lbs. (22 Nm)
  - Wheel sensor and torque the bolts to 84 inch lbs. (9.5 Nm)
  - Wheel sensor wire brackets and

torque the bolts to 84 inch lbs. (9.5 Nm)
  - New axle nut and torque it to 181 ft. lbs. (245 Nm)
  - Wheel

15. Measure and adjust the wheel alignment

NSX

1. Before servicing the vehicle, refer to the precautions in the beginning of this section.

2. Raise and safely support the vehicle.

3. Remove or disconnect the following:
  - Wheel
  - Wheel sensor and wire brackets from the knuckle
  - Brake caliper and hose brackets from the knuckle
  - Brake rotor from the knuckle
  - Hub assembly from the knuckle
  - Spindle nut
  - Pulser from the hub using a puller

4. Press the bearing from the hub.

5. Press the inner race from the hub.

To install:

6. Press a new inner race on the hub

7. Press a new bearing into the hub.

8. Install or connect the following:
  - Pulser
  - New spindle nut and torque it to 242 ft. lbs. (329 Nm)
  - Hub assembly and torque the nuts to 47 ft. lbs. (64 Nm)
  - Brake rotor and torque the bolts to 84 inch lbs. (9.5 Nm)
  - Brake caliper and torque the bolts to 80 ft. lbs.(108 Nm)
  - Brake hose brackets and torque the bolts to 16 ft. lbs. (22 Nm)
  - Wheel sensor and torque the bolts to 16 ft. lbs. (22 Nm)
  - Wheel sensor wire brackets and torque the bolts to 84 inch lbs. (9.5 Nm)
  - Wheel

9. Measure and adjust the wheel alignment.

Rear

EXCEPT NSX

➡**The rear wheel bearings on these vehicles is part of the wheel hub and is not serviceable. If the bearing is bad the wheel hub must be replaced.**

1. Before servicing the vehicle, refer to the precautions in the beginning of this section.

2. Be sure the emergency brake is disengaged.

3. Raise and safely support the vehicle.

4. Remove or disconnect the following:

  - Wheel
  - Spindle hub cap
  - Spindle nut
  - Caliper shield, if equipped
  - Brake hose mounting bolts from the knuckle
  - Brake caliper
  - Brake rotor
  - Hub assembly from the knuckle

To install:

5. Install or connect the following:
  - Hub/bearing assembly
  - Brake disc and tighten the bolts to 84 inch lbs. (9.5 Nm)
  - Brake caliper and torque the bolts to 41 ft. lbs. (56 Nm)
  - Brake hose clamps and torque the bolts to 16 ft. lbs. (22 Nm)
  - Brake caliper shield and torque the mounting bolts to 84 inch lbs. (9.5 Nm),, if equipped
  - New spindle nut and torque it to 134 ft. lbs. (181 Nm)
  - Spindle hub cap
  - Rear wheels

NSX

1. Before servicing the vehicle, refer to the precautions in the beginning of this section.

2. Raise and safely support the vehicle.

3. Remove or disconnect the following:
  - Wheel
  - Spindle nut
  - Wheel sensor and wire brackets
  - Brake hose from the caliper
  - Brake caliper
  - Brake rotor
  - Hub/bearing from the knuckle

4. Press the bearing from the hub.

5. Press the inner race from the hub.

To install:

6. Press a new inner race onto the hub.

7. Press a new wheel bearing onto the hub.

8. Install or connect the following:
  - Hub/bearing and torque the bolts to 47 ft. lbs. (64 Nm)
  - Brake rotor and torque the bolts to 84 inch lbs. (9.5 Nm)
  - Brake caliper and torque the bolts to 80 ft. lbs. (108 Nm)
  - Brake caliper hose and torque the bolt to 25 ft. lbs. (34 Nm)
  - Wheel sensor and torque the bolts to 16 ft. lbs. (22 Nm)
  - Wheel sensor wire brackets and torque the bolts to 84 inch lbs. (9.5 Nm)
  - New spindle nut and torque it to 242 ft. lbs. (329 Nm)
  - Wheel

9. Fill and bleed the brake system.

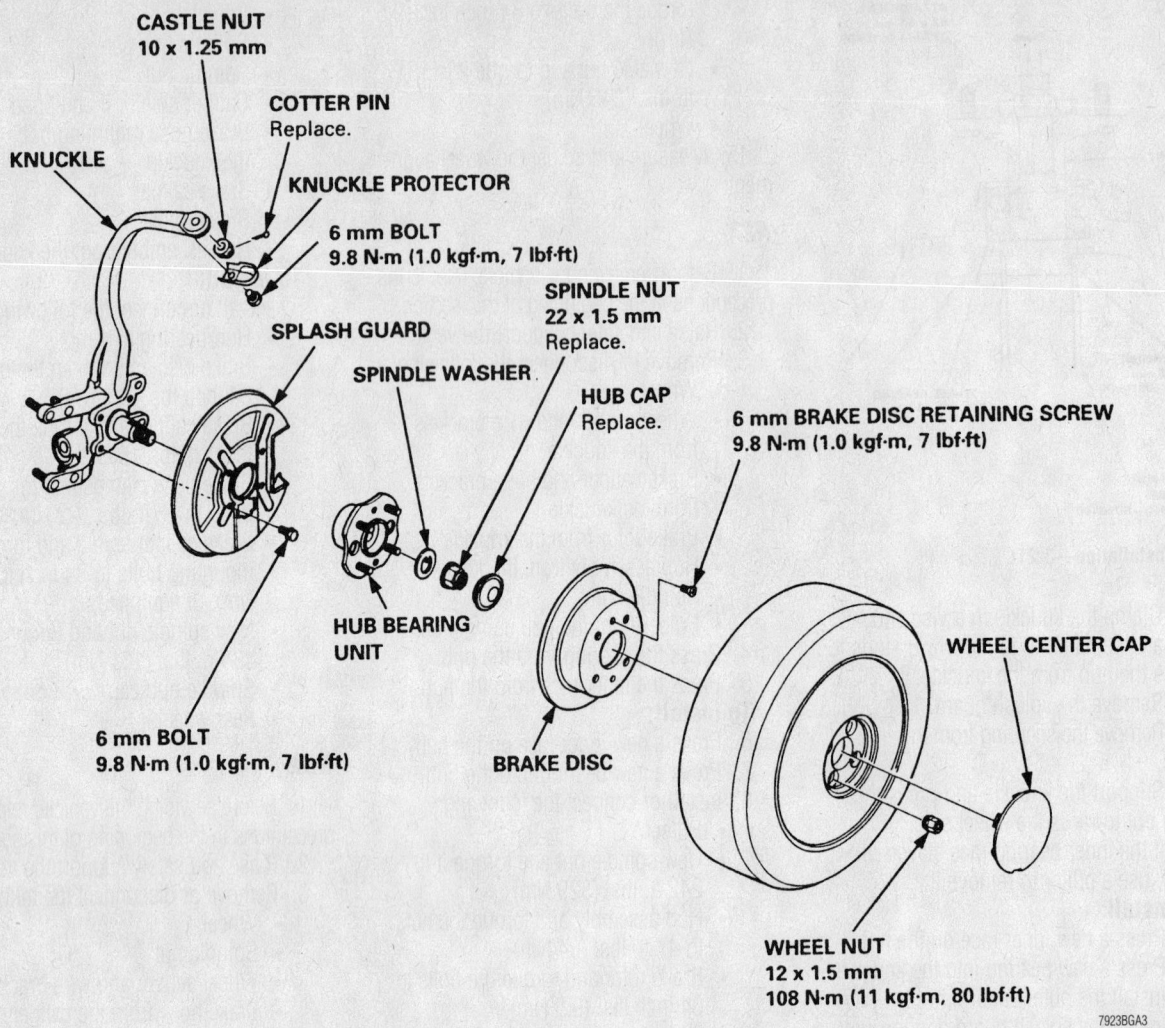

CASTLE NUT
10 x 1.25 mm

COTTER PIN
Replace.

KNUCKLE

KNUCKLE PROTECTOR

6 mm BOLT
9.8 N·m (1.0 kgf·m, 7 lbf·ft)

SPINDLE NUT
22 x 1.5 mm
Replace.

SPLASH GUARD

SPINDLE WASHER

HUB CAP
Replace.

6 mm BRAKE DISC RETAINING SCREW
9.8 N·m (1.0 kgf·m, 7 lbf·ft)

HUB BEARING UNIT

WHEEL CENTER CAP

6 mm BOLT
9.8 N·m (1.0 kgf·m, 7 lbf·ft)

BRAKE DISC

WHEEL NUT
12 x 1.5 mm
108 N·m (11 kgf·m, 80 lbf·ft)

7923BGA3

Exploded view of the rear bearing and related components

BRAKES

Brake Caliper

REMOVAL & INSTALLATION

Front

1. Before servicing the vehicle, refer to the precautions in the beginning of this section.
2. Remove or disconnect the following:
   - Wheels
   - Banjo bolt and disconnect the brake line from the caliper
   - Caliper mounting bolts
   - Caliper
   - Brake pads and shims
   - Pad spring from the caliper body, if equipped
   - Caliper bracket mounting bolts and bracket

To install:

3. Install or connect the following:
   - Bracket and torque the bolts to 80 ft. lbs. (109 Nm)
   - Pad spring, brake pads, shims, caliper and slide mounting bolts
4. Caliper slide mounting bolt torque:
   - Integra and RSX: 24 ft. lbs. (33 Nm)
   - TL and RL: 36 ft. lbs. (49 Nm)
   - CL: 54 ft. lbs. (74 Nm)
5. Install or connect the following:
   - Brake line and the banjo bolt. Replace the crush washers and torque the banjo bolt to 25 ft. lbs. (35 Nm).
6. Bleed the brake system. Torque the bleed screws to 84 inch lbs. (9 Nm).
   - Front wheels and tighten the wheel nuts to 80 ft. lbs. (109 Nm)

Rear

1. Before servicing the vehicle, refer to the precautions in the beginning of this section.
2. Remove or disconnect the following:
   - Wheels
   - Caliper dust shield
   - Parking brake cable from the caliper arm, if equipped
   - Brake line from the caliper
   - Caliper mounting bolts and pull the caliper off the bracket
   - Pads, shim, and pad retainer spring.
   - Caliper bracket mounting bolts
   - Bracket from the rotor

To install:

3. Install or connect the following:
   - Caliper bracket. Torque the mounting bolts to 28 ft. lbs. (39 Nm).

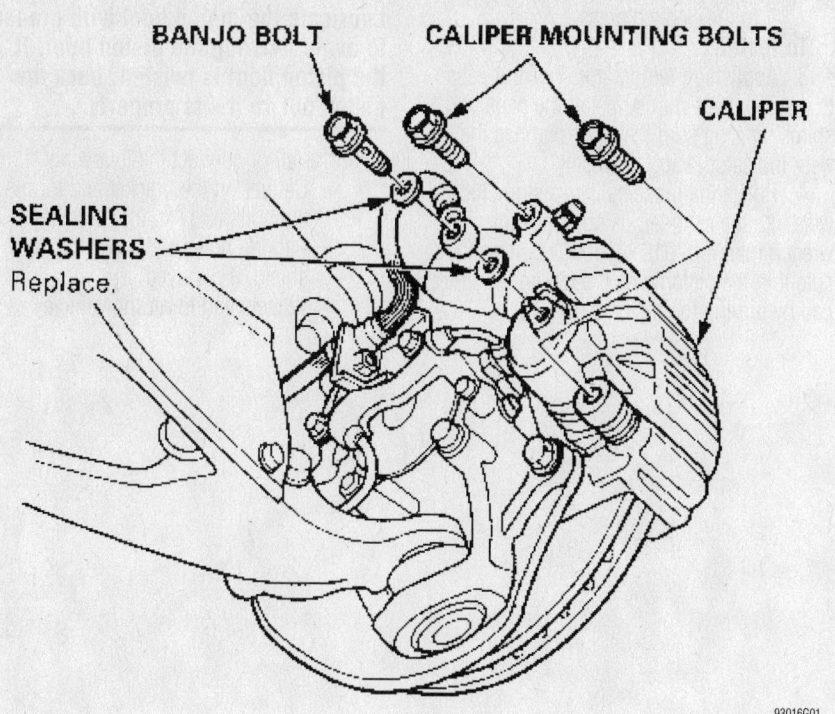

Front caliper mounting

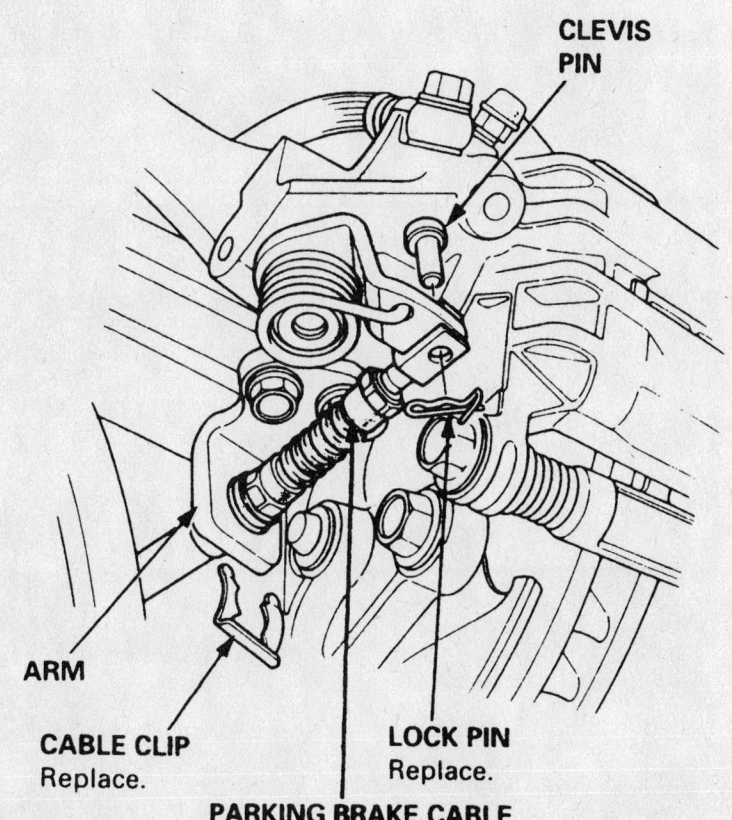

Rear caliper mounting and parking brake cable attachment

- Pads, shims, and pad retainer springs
- Caliper. Torque the mounting bolts to 17 ft. lbs. (23 Nm).
- Brake hose with new crush washers and torque the banjo bolt to 25 ft. lbs. (35 Nm).
- Parking brake cable, if equipped
4. Bleed the brake system.
- Caliper dust shield and tighten the bolts to 84 inch lbs. (10 Nm)
- Rear wheels and torque the wheel nuts to 80 ft. lbs. (109 Nm)

Disc Brake Pads

REMOVAL & INSTALLATION

Front

1. Before servicing the vehicle, refer to the precautions in the beginning of this section.
2. Remove or disconnect the following:
- Wheels
- Lower caliper bolt and pivot the caliper up and away from the rotor
- Pads, shims, and pad retainer springs

To install:

3. Clean all points where the pads and shims touch the caliper and mount. Apply a thin film of silicone grease to the cleaned areas.
4. Install or connect the following:
- Pad retainers in position on the caliper bracket
- High temperature brake grease to the back side of the pads and both sides of shims and wipe off the excess.
- Pads and shims
- Inner brake pad with the wear indicator facing upward

5. Loosen the bleed screw slightly and push in the caliper piston to allow mounting of the caliper over the rotor. Torque the bleed screw to 84 inch lbs. (9 Nm).

6. Pivot the caliper down over the rotor and install the caliper bolts. Torque the bolts to the following:
- Integra and RSX: 24 ft. lbs. (33 Nm)
- TL and RL: 36 ft. lbs. (50 Nm)
- CL: 54 ft. lbs. (74 Nm)

7. If disconnected, install the brake pad wear indicator connector. Install the wheels.

Rear

1. Before servicing the vehicle, refer to the precautions in the beginning of this section.
2. Remove or disconnect the following:
   - Wheels
   - Caliper dust shield, if equipped
   - Both caliper mounting bolts and pull the caliper off the bracket. Be sure to hang the caliper with a piece of wire so no tension is on the brake line.
   - Pads, shim and pad retainer spring. Clean all points where the pads and shims touch the caliper and mount. Apply a thin film of silicone grease to the cleaned areas.

To install:

3. Apply high temperature brake grease to the pads and shims. Install the pads and shims, making sure the inner pad has the wear indicator facing down.
4. Rotate the brake caliper piston clockwise into the cylinder, using a locknut wrench (part # 07916–6390001). Align the cutout in the piston with the tab on the inner pad by turning the piston back.

✳✳ WARNING

Lubricate the piston boot with grease to avoid twisting the piston boot. If the piston boot is twisted, back the piston out so it sits properly.

5. Install or connect the following:
   - Caliper on the bracket and torque the bolts to 17 ft. lbs. (23 Nm)
   - Parking brake cable and the dust shield, if removed
   - Wheels and lower the vehicle

SPECIFICATION CHARTS

ENGINE AND VEHICLE IDENTIFICATION CHART

Engine Code							Model Year	
Code	Liters (cc)	Cu. In.	Cyl.	Fuel Sys.	Engine Type	Eng. Mfg.	Code ①	Year
J35A3	3.5 (3471)	212	6	SMFI	SOHC	Honda	1	2001
SOHC: Single Overhead Cam							2	2002
SMFI: Sequential Multi-port Fuel Injection							3	2003
① 10th position of VIN							4	2004

42356-AMDX-C01

GENERAL ENGINE SPECIFICATIONS

Year	Model	Engine Displacement Liters (cc)	Engine ID/VIN	Fuel System Type	Net Horsepower @ rpm	Net Torque @ rpm (ft. lbs.)	Bore x Stroke (in.)	Compression Ratio	Oil Pressure @ rpm
2001	MDX	3.5 (3471)	J35A3	SMFI	210@5200	229@4300	3.50x3.66	9.4:1	71@3000
2002	MDX	3.5 (3471)	J35A3	SMFI	210@5200	229@4300	3.50x3.66	9.4:1	71@3000
2003	MDX	3.5 (3471)	J35A3	SMFI	210@5200	229@4300	3.50x3.66	9.4:1	71@3000

SMFI: Sequential Multi-port Fuel Injection

42356-AMDX-C02

ENGINE TUNE-UP SPECIFICATIONS

Year	Engine Displacement Liters (cc)	Engine ID/VIN	Spark Plug Gap (in.)	Ignition Timing (deg.) MT	Ignition Timing (deg.) AT	Fuel Pump (psi)	Idle Speed (rpm) MT	Idle Speed (rpm) AT	Valve Clearance (in.) In.	Valve Clearance (in.) Ex.
2001	3.5 (3471)	J35A1	0.039-0.043	—	8-12 ①	32-40	—	680-780	0.008-0.009	0.011-0.013
2002	3.5 (3471)	J35A1	0.039-0.043	—	8-12 ①	32-40	—	680-780	0.008-0.009	0.011-0.013
2003	3.5 (3471)	J35A1	0.039-0.043	—	8-12 ①	32-40	—	680-780	0.008-0.009	0.011-0.013

NOTE: The Vehicle Emission Control Information label often reflects changes made during production and must be used if they differ from this chart.

NOTE: The fuel pressure readings are given with the vacuum hose connected to the regulator and the engine running

① Before top dead center

42356-AMDX-C03

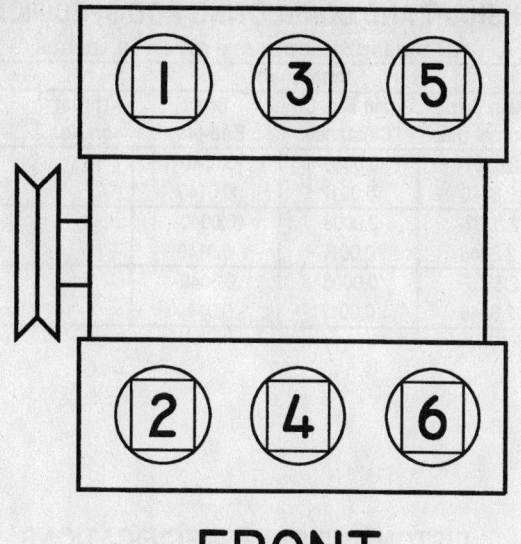

FRONT

9308MG32

3.5L Engine
Firing Order: 1–2–3–4–5–6
Distributorless ignition system (One coil per cylinder)

CAPACITIES

Year	Model	Engine Displacement Liters (cc)	Engine ID/VIN	Engine Oil with Filter (qts.)	Transmission (pts.) 5-Spd	Transmission (pts.) Auto.	Transfer Case (pts.)	Drive Axle Front (pts.)	Drive Axle Rear (pts.)	Fuel Tank (gal.)	Cooling System (qts.)
2001	MDX	3.5 (3471)	J35A3	5.0	—	9.0	—	—	—	19.2	8.0
2002	MDX	3.5 (3471)	J35A3	5.0	—	9.0	—	—	—	19.2	8.0
2003	MDX	3.5 (3471)	J35A3	5.0	—	9.0	—	—	—	19.2	8.0

NOTE: All capacities are approximate. Add fluid gradually and check to be sure a proper fluid level is obtained.

42356-AMDX-C04

VALVE SPECIFICATIONS

Year	Engine Displacement Liters (cc)	Engine ID/VIN	Seat Angle (deg.)	Face Angle (deg.)	Spring Test Pressure (lbs. @ in.)	Spring Installed Height (in.)	Stem-to-Guide Clearance (in.) Intake	Stem-to-Guide Clearance (in.) Exhaust	Stem Diameter (in.) Intake	Stem Diameter (in.) Exhaust
2001	3.5 (3471)	J35A3	45	45	NA	①	0.0008-0.0018	0.0022-0.0031	0.2159-0.2163	0.2146-0.2150
2002	3.5 (3471)	J35A3	45	45	NA	①	0.0008-0.0018	0.0022-0.0031	0.2159-0.2163	0.2146-0.2150
2003	3.5 (3471)	J35A3	45	45	NA	①	0.0008-0.0018	0.0022-0.0031	0.2159-0.2163	0.2146-0.2150

NA: Not Available

① Valve spring free length:
Intake: 1.9713 in.
Exhaust: 2.1060 in.

42356-AMDX-C05

CRANKSHAFT AND CONNECTING ROD SPECIFICATIONS
All measurements are given in inches

Year	Engine Displacement Liters (cc)	Engine ID/VIN	Crankshaft				Connecting Rod		
			Main Brg. Journal Dia.	Main Brg. Oil Clearance	Shaft End-play	Thrust on No.	Journal Diameter	Oil Clearance	Side Clearance
2001	3.5 (3471)	J35A3	2.8337-2.8346	0.0008-0.0017	0.0040-0.0140	3	2.1644-2.1654	0.0008-0.0017	0.0060-0.0140
2002	3.5 (3471)	J35A3	2.8337-2.8346	0.0008-0.0017	0.0040-0.0140	3	2.1644-2.1654	0.0008-0.0017	0.0060-0.0140
2003	3.5 (3471)	J35A3	2.8337-2.8346	0.0008-0.0017	0.0040-0.0140	3	2.1644-2.1654	0.0008-0.0017	0.0060-0.0140

42356-AMDX-C06

PISTON AND RING SPECIFICATIONS
All measurements are given in inches

Year	Engine Displacement Liters (cc)	Engine ID/VIN	Piston Clearance	Ring Gap			Ring Side Clearance		
				Top Compression	Bottom Compression	Oil Control	Top Compression	Bottom Compression	Oil Control
2001	3.5 (3471)	J35A1	0.0006-0.0016	0.0080-0.0140	0.0160-0.0220	0.0080-0.0280	0.0014-0.0024	0.0012-0.0022	NA
2002	3.5 (3471)	J35A1	0.0006-0.0016	0.0080-0.0140	0.0160-0.0220	0.0080-0.0280	0.0014-0.0024	0.0012-0.0022	NA
2003	3.5 (3471)	J35A1	0.0006-0.0016	0.0080-0.0140	0.0160-0.0220	0.0080-0.0280	0.0014-0.0024	0.0012-0.0022	NA

NA: Not Available

42356-AMDX-C07

TORQUE SPECIFICATIONS
All readings in ft. lbs.

Year	Engine Displacement Liters (cc)	Engine ID/VIN	Cylinder Head Bolts	Main Bearing Bolts	Rod Bearing Bolts	Crankshaft Damper Bolts	Flywheel Bolts	Manifold		Spark Plugs	Lug Nut
								Intake	Exhaust		
2001	3.5 (3471)	J35A1	①	②	③	181	54	16	23	13	80
2002	3.5 (3471)	J35A1	①	②	③	181	54	16	23	13	80
2003	3.5 (3471)	J35A1	①	②	③	181	54	16	23	13	80

NOTE: Dip main bearing bolts and crankshaft damper bolt in clean engine oil prior to tightening.

① Step 1: 29 ft. lbs.
   Step 2: 51 ft. lbs.
   Step 3: 72 ft. lbs.

② 11mm bolt 56 ft. lbs.
   10mm bolt 36 ft. lbs.

③ Step 1: 14 ft. lbs.
   Step 2: 90 degrees

42356-AMDX-C08

WHEEL ALIGNMENT

Year	Model		Caster Range (+/-Deg.)	Caster Preferred Setting (Deg.)	Camber Range (+/-Deg.)	Camber Preferred Setting (Deg.)	Toe-in (in.)	Steering Axis Inclination (Deg.)
2001	MDX	F	1.00	1.00	1.00	30	0+/-1/16	—
		R	—	—	0	30	0+/-1/16	—
2002	MDX	F	1.00	1.00	1.00	30	0+/-1/16	—
		R	—	—	0	30	0+/-1/16	—
2003	MDX	F	1.00	1.00	1.00	30	0+/-1/16	—
		R	—	—	0	30	0+/-1/16	—

42356-AMDX-C09

TIRE, WHEEL AND BALL JOINT SPECIFICATIONS

Year	Model	OEM Tires Standard	OEM Tires Optional	Tire Pressures (psi) Front	Tire Pressures (psi) Rear	Wheel Size	Ball Joint Inspection
2001	MDX	P235/65R17	None	32	32	R17	NS
2002	MDX	P235/65R17	None	32	32	R17	NS
2003	MDX	P235/65R17	None	32	32	R17	NS

OEM: Original Equipment Manufacturer

PSI: Pounds Per Square Inch

NS: Not specified by manufacturer

42356-AMDX-C10

BRAKE SPECIFICATIONS
Acura MDX
All measurements in inches unless noted

Year	Model		Brake Disc Original Thickness	Brake Disc Minimum Thickness	Brake Disc Maximum Runout	Brake Drum Diameter Original Inside Diameter	Brake Drum Diameter Max. Wear Limit	Brake Drum Diameter Maximum Machine Diameter	Minimum Lining Thickness Front	Minimum Lining Thickness Rear	Brake Caliper Bracket Bolts (ft. lbs.)	Brake Caliper Mounting Bolts (ft. lbs.)
2001	MDX	F	1.100	1.020	0.004	—	—	—	0.060	—	80	27
		R	0.430	0.350	0.004	—	—	—	—	0.41	41	27
2002	MDX	F	1.100	1.020	0.004	—	—	—	0.060	—	80	27
		R	0.430	0.350	0.004	—	—	—	—	0.41	41	27
2003	MDX	F	1.100	1.020	0.004	—	—	—	0.060	—	80	27
		R	0.430	0.350	0.004	—	—	—	—	0.41	41	27

F: Front

R: Rear

42356-AMDX-C11

SCHEDULED MAINTENANCE INTERVALS
ACURA—MDX

TO BE SERVICED	TYPE OF SERVICE	7.5	15	22.5	30	37.5	45	52.5	60	67.5	75	82.5	90	97.5	105	112.5	120
Accessory drive belts	I & A				✓				✓				✓				✓
Air cleaner element	R				✓				✓				✓				✓
Brake fluid	R	Every 3 years															
Brake hoses & lines (including ABS)	I		✓		✓		✓		✓		✓		✓		✓		✓
Cooling system hoses & connections	I		✓		✓		✓		✓		✓		✓		✓		✓
Engine coolant ①	R							✓					✓				
Engine oil	R	✓	✓	✓	✓	✓	✓	✓	✓	✓	✓	✓	✓	✓	✓	✓	✓
Engine oil and coolant levels	I	Inspect at each fuel stop															
Engine oil filter	R		✓		✓		✓		✓		✓		✓		✓		✓
Exhaust system	I		✓		✓		✓		✓		✓		✓		✓		✓
Fluid levels and condition	I		✓		✓		✓		✓		✓		✓		✓		✓
Front and rear brakes	I		✓		✓		✓		✓		✓		✓		✓		✓
Fuel lines & connection	I		✓		✓		✓		✓		✓		✓		✓		✓
Halfshaft boots	I		✓		✓		✓		✓		✓		✓		✓		✓
Idle speed	I & A														✓		
Parking brake system	I & A		✓		✓		✓		✓		✓		✓				✓
Rear differential fluid	R	✓			✓		✓						✓				✓
Rotate and inspect tires	I	✓	✓	✓	✓	✓	✓	✓	✓	✓	✓	✓	✓	✓	✓	✓	✓
Spark plugs	R														✓		
Supplemental Restrain system (SRS)	I	Inspect the SRS 10 years after production															
Suspension components	I		✓		✓		✓		✓		✓		✓		✓		✓
Tie rod ends, steering gear box & boots	I		✓		✓		✓		✓		✓		✓		✓		✓
Timing belt	R														✓		
Transmission fluid	R							✓			✓				✓		
Valve clearance	I	Adjust if valves are noisy															
Water pump	S/I														✓		

R: Replace   I: Inspect   A: Adjust

① Every 12,000 miles or 10 years, then every 60,000 miles or 5 years

FREQUENT OPERATION MAINTENANCE (SEVERE SERVICE)

If a vehicle is operated under any of the following conditions it is considered severe service:

- Towing a trailer or using a camper or car-top carrier.
- Repeated short trips of less than 5 miles in temperatures below freezing, or trips of less than 10 miles in any temperature.
- Extensive idling or low-speed driving for long distances as in heavy commercial use, such as delivery, taxi or police cars.
- Operating on rough, muddy or salt-covered roads.
- Operating on unpaved or dusty roads.
- Driving in extremely hot (over 90°) conditions.

Air cleaner element: replace every 15,000 miles

Engine oil and filter: replace every 3750 miles or 6 months, whichever occurs first.

Timing belt: replace every 60,000 miles if the vehicle is regularly driven in temperatures above 110°F or below -20°F, or if frequently towing a trailer.

Transmission fluid: replace every 30,000 miles.

Rear differential fluid: replace every 60,000 miles.

Front and rear brakes: inspect every 7500 miles or 6 months, whichever occurs first.

Locks and hinges: lubricate every 15,000 miles.

Tie rods, steering gear box, boots: inspect every 7500 miles or 6 months, whichever occurs first.

Suspension components: inspect every 7500 miles or 6 months, whichever occurs first.

Halfshaft boots: inspect every 7500 miles or 6 months, whichever occurs first.

42356-AMDX-C12

PRECAUTIONS

Before servicing any vehicle, please be sure to read all of the following precautions, which deal with personal safety, prevention of component damage, and important points to take into consideration when servicing a motor vehicle:

• Never open, service or drain the radiator or cooling system when the engine is hot; serious burns can occur from the steam and hot coolant.

• Observe all applicable safety precautions when working around fuel. Whenever servicing the fuel system, always work in a well-ventilated area. Do not allow fuel spray or vapors to come in contact with a spark, open flame or excessive heat (a hot drop light, for example). Keep a dry chemical fire extinguisher near the work area. Always keep fuel in a container specifically designed for fuel storage; also, always properly seal fuel containers to avoid the possibility of fire or explosion. Refer to the additional fuel system precautions later in this section.

• Fuel injection systems often remain pressurized, even after the engine has been turned **OFF**. The fuel system pressure must be relieved before disconnecting any fuel lines. Failure to do so may result in fire and/or personal injury.

• Brake fluid often contains polyglycol ethers and polyglycols. Avoid contact with the eyes and wash your hands thoroughly after handling brake fluid. If you do get brake fluid in your eyes, flush your eyes with clean, running water for 15 minutes. If eye irritation persists, or if you have taken

brake fluid internally, IMMEDIATELY seek medical assistance.

• The EPA warns that prolonged contact with used engine oil may cause a number of skin disorders, including cancer! You should make every effort to minimize your exposure to used engine oil. Protective gloves should be worn when changing oil. Wash your hands and any other exposed skin areas as soon as possible after exposure to used engine oil. Soap and water, or waterless hand cleaner should be used.

• All new vehicles are now equipped with an air bag system. The system must be disabled before performing service on or around system components, steering column, instrument panel components, wiring and sensors. Failure to follow safety and disabling procedures could result in accidental air bag deployment, possible personal injury and unnecessary system repairs.

• Always wear safety goggles when working with, or around, the air bag system. When carrying a non-deployed air bag, be sure the bag and trim cover are pointed away from your body. When placing a non-deployed air bag on a work surface, always face the bag and trim cover upward, away from the surface. This will reduce the motion of the module if it is accidentally deployed. Refer to the additional air bag system precautions later in this section.

• Clean, high quality brake fluid from a sealed container is essential to the safe and proper operation of the brake system. You

should always buy the correct type of brake fluid for your vehicle. If the brake fluid becomes contaminated, completely flush the system with new fluid. Never reuse any brake fluid. Any brake fluid that is removed from the system should be discarded. Also, do not allow any brake fluid to come in contact with a painted surface; it will damage the paint.

• Never operate the engine without the proper amount and type of engine oil; doing so WILL result in severe engine damage.

• Timing belt maintenance is extremely important! Many models utilize an interference-type, non-freewheeling engine. If the timing belt breaks, the valves in the cylinder head may strike the pistons, causing potentially serious (also time-consuming and expensive) engine damage. Refer to the maintenance interval charts in the front of this section for the recommended replacement interval for the timing belt, and to the timing belt procedure in this section for belt replacement and inspection.

• Disconnecting the negative battery cable on some vehicles may interfere with the functions of the on-board computer system(s) and may require the computer to undergo a relearning process once the negative battery cable is reconnected.

• When servicing drum brakes, only disassemble and assemble one side at a time, leaving the remaining side intact for reference.

• Only an MVAC-trained, EPA-certified automotive technician should service the air conditioning system or its components.

ENGINE REPAIR

➡**Disconnecting the negative battery cable on some vehicles may interfere with the functions of the on board computer system. The computer may undergo a relearning process once the negative battery cable is reconnected.**

Distributor

The MDX is equipped with a Distributorless Ignition System (DIS).

Alternator

REMOVAL

1. Before servicing the vehicle, refer to the precautions in the beginning of this section.

2. Remove or disconnect the following:
   • Negative battery cable
   • Intake manifold and ignition coil covers
   • Accessory drive belt
   • Alternator wiring harness connectors
   • Alternator mounting bolts
   • Wiring harness clamp
   • Alternator

INSTALLATION

1. Install or connect the following:
   • Alternator
   • Wiring harness clamp. Tighten the bolt to 105 inch lbs. (12 Nm).
   • Alternator mounting bolts. Tighten the 10mm bolt to 33 ft. lbs. (44

Nm) and the 8mm bolt to 16 ft. lbs. (22 Nm).
   • Alternator wiring harness connectors. Tighten the battery terminal nut to 105 inch lbs. (12 Nm).
   • Accessory drive belt
   • Intake manifold and ignition coil covers
   • Negative battery cable

Ignition Timing

ADJUSTMENT

The MDX is equipped with a Distributorless Ignition System (DIS). The ignition timing is controlled by the Powertrain Control module (PCM). No adjustment is necessary.

Engine Assembly

REMOVAL & INSTALLATION

→ **The engine and transaxle are removed from the vehicle as a unit.**

1. Before servicing the vehicle, refer to the precautions in the beginning of this section.
2. Drain the cooling system.
3. Drain the power steering system.
4. Drain the transaxle fluid.
5. Drain the engine oil.
6. Relieve fuel system pressure.
7. Remove or disconnect the following:

- Negative battery cable
- Battery and tray
- Intake and ignition coil covers
- Air intake duct
- Left engine wire harness connectors
- Relay bracket
- Starter cable and harness clamp
- Accelerator cable
- Cruise control cable
- Fuel lines
- EVAP canister hose

8. Remove the drivers side center console lower panel and pull back the cover to access steering joint cover.

- Steering joint bolt
- Powertrain Control Module (PCM) connectors
- Heated Oxygen (HO$_2$S) sensor connector and grommet. Pull the PCM harness through the firewall.
- Brake booster vacuum line
- Power steering hose's clamps and clips
- Fuse/Relay box battery cable
- Accessory drive belts
- Front wheels
- Splash shield
- Front sub-frame stiffener
- Exhaust front pipe
- Propeller shaft
- Shift control cable
- Transfer assembly
- Ball joints
- Stabilizer bar links
- Power steering hose and pressure switch connector
- Transaxle lower front mount
- Transaxle lower rear mount
- A/C compressor
- Heater hoses
- Radiator hoses
- Ground cable
- Transaxle oil cooler lines
- Radiator

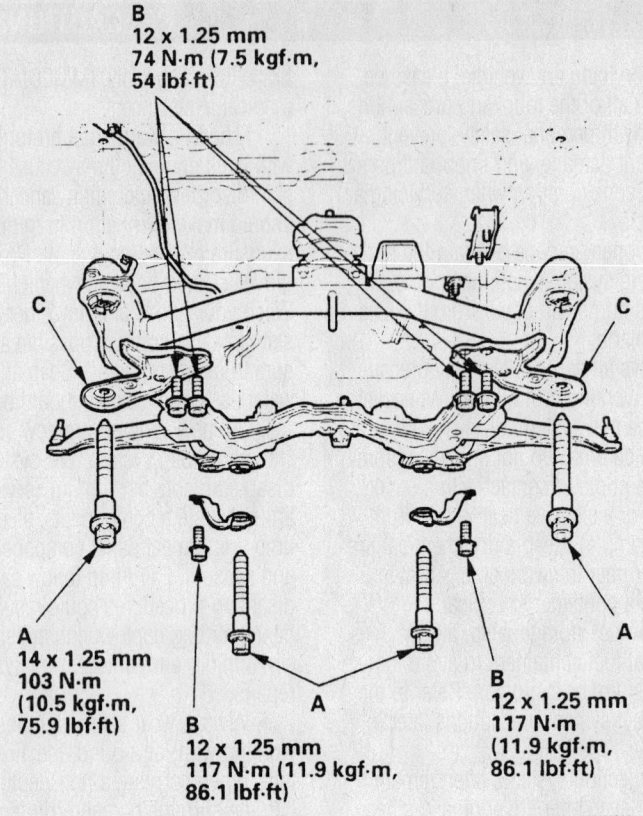

B
12 x 1.25 mm
74 N·m (7.5 kgf·m,
54 lbf·ft)

A
14 x 1.25 mm
103 N·m
(10.5 kgf·m,
75.9 lbf·ft)

B
12 x 1.25 mm
117 N·m (11.9 kgf·m,
86.1 lbf·ft)

B
12 x 1.25 mm
117 N·m
(11.9 kgf·m,
86.1 lbf·ft)

9302MG69

Sub-frame fastener locations and tightening torque—MDX

9. Attach a hoist to the engine lifting eyes and support the powertrain weight.
10. Remove or disconnect the following:

- Side engine mount bracket
- Front mount bracket support nut

11. Matchmark the front subframe to the mounting points.
12. Remove or disconnect the following:

- Front subframe
- All remaining hoses and electrical connections

13. Lower the powertrain away from the vehicle.

To install:

14. Raise the powertrain into position.
15. Installation is the reverse of removal but please note the following steps:

- A/C compressor bolts to 16 ft. lbs. (22 Nm)
- Front subframe. Use new bolts and tighten the 14mm bolts to 76 ft. lbs. (103 Nm). Tighten the front brace bolts to 54 ft. lbs. (74 Nm) and the rear brace bolts to 86 ft. lbs. (117 Nm).
- Transaxle lower front mount nuts to 28 ft. lbs. (38 Nm)
- Transaxle lower rear mount bolts to 28 ft. lbs. (38 Nm)
- Front mount bracket support nut to 40 ft. lbs. (54 Nm)
- Side engine mount bracket bolts to 33 ft. lbs. (44 Nm) and the through bolt to 40 ft. lbs. (54 Nm)

16. Fill the engine crankcase to the correct level.
17. Fill the transaxle to the correct level.
18. Fill the cooling system.
19. Fill the power steering system.
20. Start the engine and check for leaks.
21. Check the wheel alignment and adjust as necessary.

Water Pump

REMOVAL & INSTALLATION

1. Before servicing the vehicle, refer to the precautions in the beginning of this section.
2. Drain the cooling system.
3. Remove or disconnect the following:

- Negative battery cable
- Accessory drive belts
- Front cover
- Timing belt
- Timing belt tensioner
- Water pump

To install:

4. Install or connect the following:

- Water pump. Use a new O-ring seal

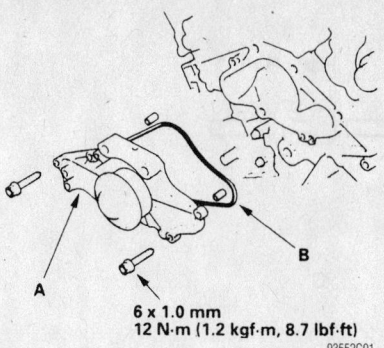

6 x 1.0 mm
12 N·m (1.2 kgf·m, 8.7 lbf·ft)
93552G01

Exploded view of the water pump mounting

and tighten the bolts to 105 inch lbs. (12 Nm).
- Timing belt tensioner
- Timing belt
- Front cover
- Accessory drive belts
- Negative battery cable

5. Fill the cooling system.
6. Start the engine and check for leaks.

Heater Core

REMOVAL & INSTALLATION

1. Before servicing the vehicle, refer to the precautions in the beginning of this section.
2. Drain the cooling system.
3. Remove or disconnect the following:
- Negative battery cable

4. Recover the refrigerant using approved equipment.
- Heater valve cable from the valve arm. Turn the valve arm to the fully opened position.
- Heater hoses from the heater unit
- Mounting nut from the heater unit. Be careful not to bend or damage fuel or brake lines.

5. Remove the dashboard as follows:
a. Remove the center console by unlatching the clips.
b. Remove the dashboard lower cover screw, gently pull down on the cover to disengage the clips and disconnect the electrical connections.
c. Remove the dashboard side cover by gently pulling and turning to unfasten the clips.
d. While holding the glove box, remove the box stop from each side, then disconnect the lock from the damper.
e. Remove the glove ox bolts and the glove box.
f. Remove the shift lever assembly.
g. Remove the front door trim, lick panel and A-pillar trim from both sides.

h. Remove the cap from the front pillar corner trim. Unfasten the screw, slide the trim upward along the pillar and remove it. Remove the remaining clips from the body.
i. On the drivers side, remove the fuel/relay box nut and pull out the box.
j. Remove the steering column
k. On the passenger side remove the fuse/relay bolt and pull out the box.
l. Disconnect all electrical connections from the dashboard.
m. If equipped with a navigation system, pull back the carpet, remove the harness cushions and then pull out the GPS harness.
n. Remove all harness and connector clips.
o. Remove all the bolts and lift up on the dashboard to release the dashboard and steering hanger beam from the guide pins.
p. Remove the dashboard through the door.

6. Remove the evaporator as follows:
a. Disconnect the receiver and suction lines from the evaporator.
b. Remove the mounting nuts and plug the lines to avoid system contamination.
c. Remove the plastic brace and glove box frame.
d. Disconnect the wire harness and evaporator temperature sensor connector.
e. Remove the self-tapping screws, the nuts and the evaporator.

7. Remove or disconnect the following:
- Mounting bolts and the heater unit
- Self-tapping screws and the clamp, then pull the heater core from the case being careful not to bend the pipes

To install:
8. Install or connect the following:
- Heater core in the case
- Clamp and the screws
- Heater unit and tighten the bolts to 7 ft. lbs. (10 Nm)
- Evaporator in the reverse order of removal. Tighten all the retainers to 7 ft. lbs. (10 Nm) .

9. Install the dashboard in the reverse order of removal keeping in mind the following points:
a. Make sure the dashboard is seated properly and that the wiring harness and steering hanger beam wire harness are not pinched.
b. Referring to the accompanying illustration, tighten bolts **(A)** to 7 ft. lbs. (10 Nm). Tighten all the other bolts to 16 ft. lbs. (22 Nm). Apply thread lock to the **B** bolts before installation.

c. Ensure that all electrical connectors are properly connected.
10. Install or connect the following:
- Mounting nut to the heater unit and tighten to 9 ft. lbs. (13 Nm)
- Heater hoses
11. Connect the heater valve cable and adjust as follows:
a. In the engine compartment, open the cable clamp (A), then disconnect the heater valve cable (B) from the valve arm (C).
b. Under the dashboard, disconnect the valve cable housing from the clamp (A) and the cable (B) from the mix control linkage (C).
c. Set the temperature control button to the MAX COOL position with the ignition switch in the on position.
d. Attach the valve cable (B) to the mix control linkage (C) as shown in the illustration, hold the end of the cable housing against the stop, then snap the cable housing into the clamp.
Sin the engine compartment, turn the

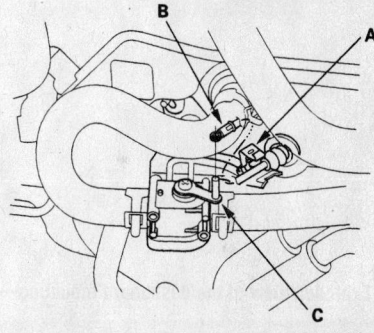

42356-AMDX-G01

In the engine compartment, open the cable clamp (A), then disconnect the heater valve cable (B) from the valve arm (C)

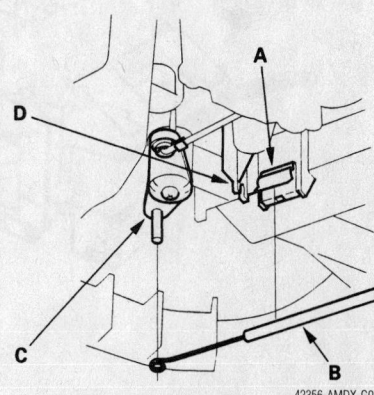

42356-AMDX-G02

Under the dashboard, disconnect the valve cable housing from the clamp (A) and the cable (B) from the mix control linkage (C)

Fastener Locations

A ▶ : Bolt, 2   B ▶ : Bolt, 5   C ▶ : Bolt, 3   D ▶ : Bolt, 2   E ▶ : Bolt, 2   F ▶ : Bolt, 1

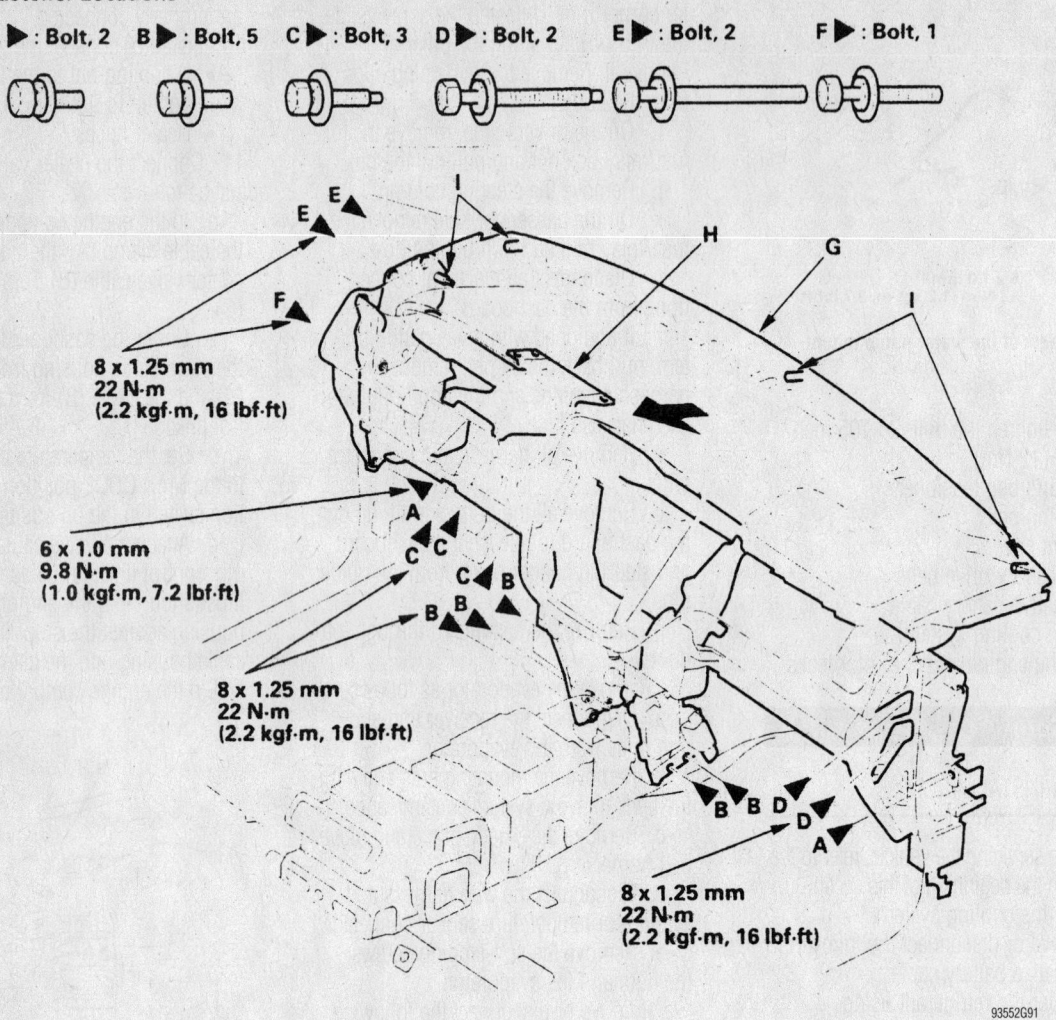

8 x 1.25 mm
22 N·m
(2.2 kgf·m, 16 lbf·ft)

6 x 1.0 mm
9.8 N·m
(1.0 kgf·m, 7.2 lbf·ft)

8 x 1.25 mm
22 N·m
(2.2 kgf·m, 16 lbf·ft)

8 x 1.25 mm
22 N·m
(2.2 kgf·m, 16 lbf·ft)

93552G91

Exploded view of the dashboard mounting—Acura MDX

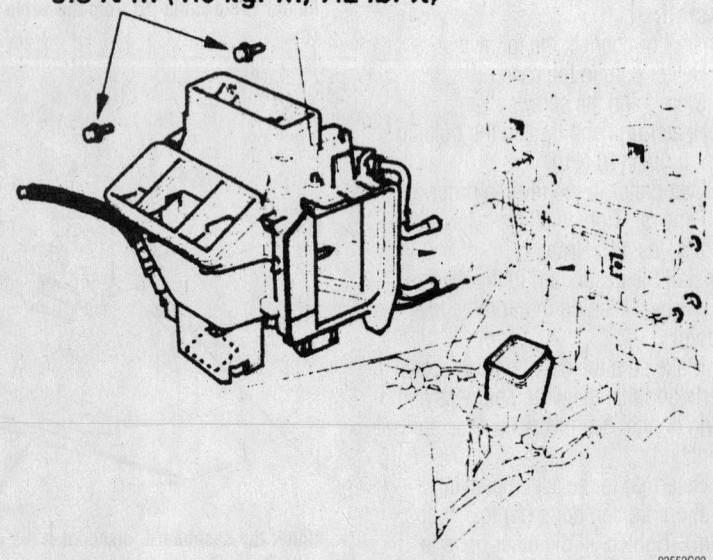

6 x 1.0 mm
9.8 N·m (1.0 kgf·m, 7.2 lbf·ft)

93552G92

Exploded view of the evaporator mounting—Acura MDX

valve arm (C) to the fully closed position as shown in the accompanying illustration and hold it there. Attach the cable (B) to the vale arm and pull gently on the cable housing to take up the slack, then install the cable housing into the clamp (A).

12. Fill the cooling system
13. Connect the battery cable.

Cylinder Head

REMOVAL & INSTALLATION

1. Before servicing the vehicle, refer to the precautions in the beginning of this section.

2. Drain the cooling system.

3. Relieve the fuel system pressure.

4. Remove or disconnect the following:

- Negative battery cable
- Ignition coil covers

- Intake manifold cover
- Air intake tube
- Accelerator cable
- Cruise control cable
- EVAP canister, breather, water bypass hoses and the bypass pipe bolt
- Fuel lines
- Brake booster vacuum line
- Intake manifold vacuum line
- Positive Crankcase Ventilation (PCV) valve and hose
- Power steering hose clamp
- Intake Manifold Runner Control (IMRC) actuator
- Wiring harness holder and joint connector

5. Support the engine with a jack and a block of wood.
- Side engine mount bracket
- Accessory drive belts
- Power steering pump
- Alternator
- Intake Air Temperature (IAT) sensor connector
- Idle Air Control (IAC) valve connector
- Throttle Position (TP) sensor connector
- Manifold Absolute Pressure (MAP) sensor connector
- Evaporative Emission (EVAP) canister purge valve connector
- Engine Coolant Temperature (ECT) sensor connector
- Radiator fan switch connectors
- ECT gauge sending unit connector
- Crankshaft Position (CKP) sensor connector
- Top Dead Center (TDC) sensor connector
- Exhaust Gas Recirculation (EGR) connector
- Variable Valve Timing and Valve Lift Electronic Control (VTEC) solenoid valve connector
- VTEC oil pressure switch connector
- Oil pressure switch connector
- Ignition coils
- Intake manifold
- Fuel injector connectors
- Fuel supply manifold
- Fuel injection air control valve vacuum lines
- Front cover
- Timing belt
- Radiator hoses
- Heater hoses
- Front and rear exhaust manifolds
- Coolant cross-over pipe

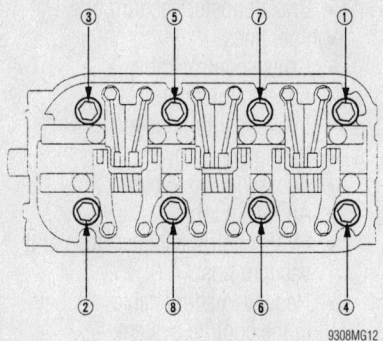

9308MG12

Cylinder head bolt loosening sequence— MDX

- Valve covers

6. Loosen the cylinder head bolts in sequence and ⅓ turns until all bolts are loose.

7. Remove the cylinder head.

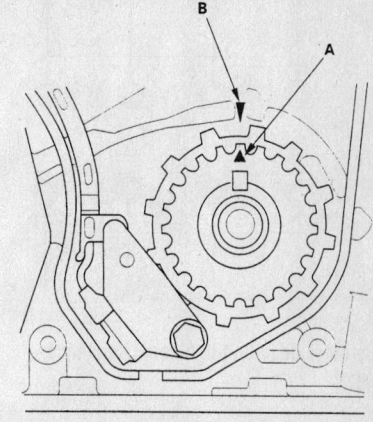

9302MG74

Crankshaft timing belt sprocket TDC marks. Align sprocket mark (A) with pointer (B)—MDX

FRONT:

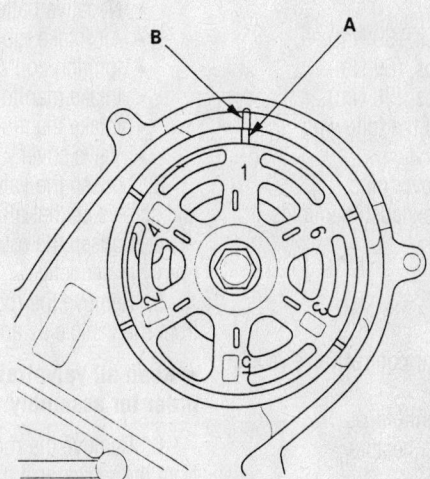

REAR:

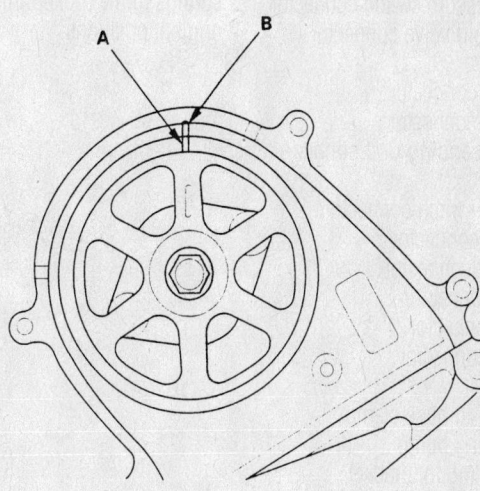

9302MG85

Camshaft TDC marks. Align sprocket mark (A) with the back cover pointer (B)—MDX

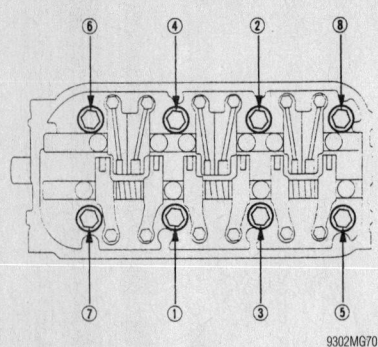

Cylinder head bolt tightening sequence—MDX

9302MG70

To install:

8. Align the crankshaft and camshaft sprocket TDC marks as shown.

9. Install the cylinder heads with new gaskets.

10. Apply clean engine oil to the cylinder head bolt threads and flanges.

11. Tighten the cylinder head bolts in sequence as follows:

   a. Step 1: 29 ft. lbs. (39 Nm).
   b. Step 2: 51 ft. lbs. (69 Nm).
   c. Step 3: 72 ft. lbs. (98 Nm).

12. Install or connect the following:
- Valve covers
- Coolant cross-over pipe
- Front and rear exhaust manifolds
- Heater hoses
- Radiator hoses
- Timing belt
- Front cover
- Fuel injection air control valve vacuum lines
- Fuel supply manifold
- Fuel injector connectors
- Intake manifold
- Ignition coils
- Oil pressure switch connector
- VTEC oil pressure switch connector
- VTEC solenoid valve connector
- EGR connector
- TDC sensor connector
- CKP sensor connector
- ECT gauge sending unit connector
- Radiator fan switch connectors
- ECT sensor connector
- MAP sensor connector
- TP sensor connector
- IAC valve connector
- IAT sensor connector
- Alternator
- Power steering hose clamp
- Power steering pump
- Side engine mount bracket
- PCV valve and hose
- Intake manifold vacuum line

- Brake booster vacuum line
- Fuel lines
- Cruise control cable
- Accelerator cable
- Intake manifold cover
- Ignition coil covers
- Accessory drive belts
- Air intake tube
- EVAP control canister hose and vacuum hose
- Negative battery cable

13. Fill the cooling system.

14. Start the engine and check for leaks.

Rocker Arms/Shafts

REMOVAL & INSTALLATION

2001–02 Models

1. Before servicing the vehicle, refer to the precautions in the beginning of this section.

2. Remove or disconnect the following:
- Negative battery cable
- Air intake tube
- Ignition coil covers
- Intake manifold cover
- Intake manifold
- Valve cover

3. Loosen the valve adjuster locknuts and screws so that all valves are closed.

4. Loosen the rocker arm shaft bolts evenly in sequence.

5. Remove the rocker arms and shafts from the vehicle as an assembly.

➡ **Keep all valvetrain components in order for assembly.**

6. Remove the rocker arms and springs from the rocker arm shafts.

To install:

7. Assemble the rocker arms and springs to the rocker arm shafts in their original positions.

8. Install the rocker arm assemblies. Tighten the bolts in sequence and in multiple passes to 17 ft. lbs. (24 Nm).

9. Adjust the valve clearance.

10. Install or connect the following:
- Valve covers
- Intake manifold
- Intake manifold cover
- Ignition coil covers
- Air intake tube
- Negative battery cable

11. Start the engine and check for leaks.

2003–04 Models

1. Before servicing the vehicle, refer to the precautions in the beginning of this section.

2. Remove or disconnect the following:
- Negative battery cable
- Intake manifold cover
- Ignition coil covers
- Ignition coils
- Dipstick
- 2 bolts attaching the harness holder
- Positive Crankcase ventilation (PCV) hose
- Injector electrical connections
- Power steering hose bracket bolt and the harness holder bolts
- Harness clamps and the breather hose
- Valve cover
- Rocker arm adjusting screws. refer to the illustration for location.

3. Remove the rocker arm assembly as follows:

   a. Unscrew the rocker shaft bolts 2 turns at a time in a criss-cross pattern to avoid damaging the vales or rocker assembly.

   b. Do not remove the rocker shaft bolts. These bolts keep the springs and rocker arms on the shafts.

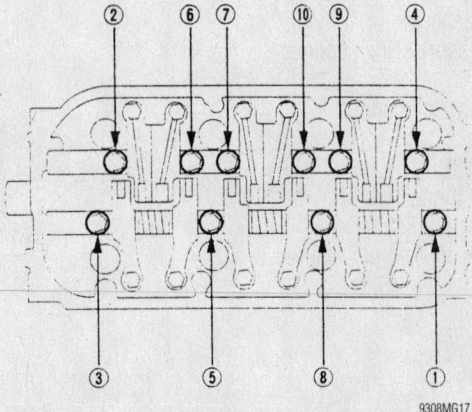

9308MG17

Rocker arm shaft loosening sequence—2001–02 models

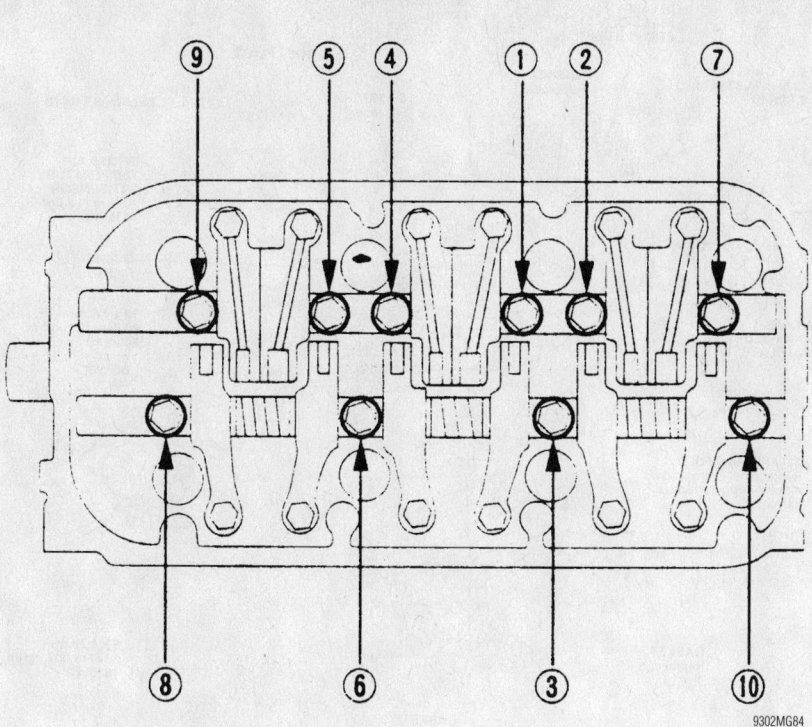

Rocker shaft tightening sequence—2001–02 models

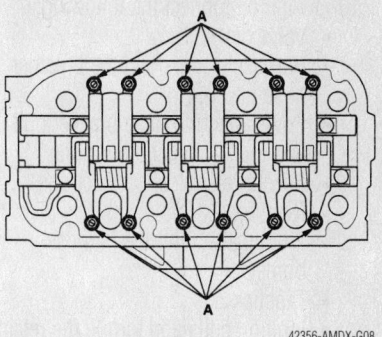

42356-AMDX-G08

Rocker arm shaft adjusting screw locations—2003–04 models

4. Loosen the valve adjuster locknuts and screws so that all valves are closed.

5. Remove the rocker arms and shafts from the vehicle as an assembly.

➡ **Keep all valvetrain components in order for assembly.**

6. Remove the rocker arms and springs from the rocker arm shafts.

To install:

7. Assemble the rocker arms and springs to the rocker arm shafts in their original positions.

8. Install the rocker arm assemblies. Tighten the bolts in sequence and in multiple passes to 17 ft. lbs. (24 Nm).

9. Adjust the valve clearance.

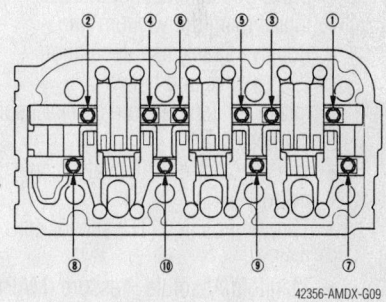

42356-AMDX-G09

Rocker arm shaft loosening sequence—2003–04 models

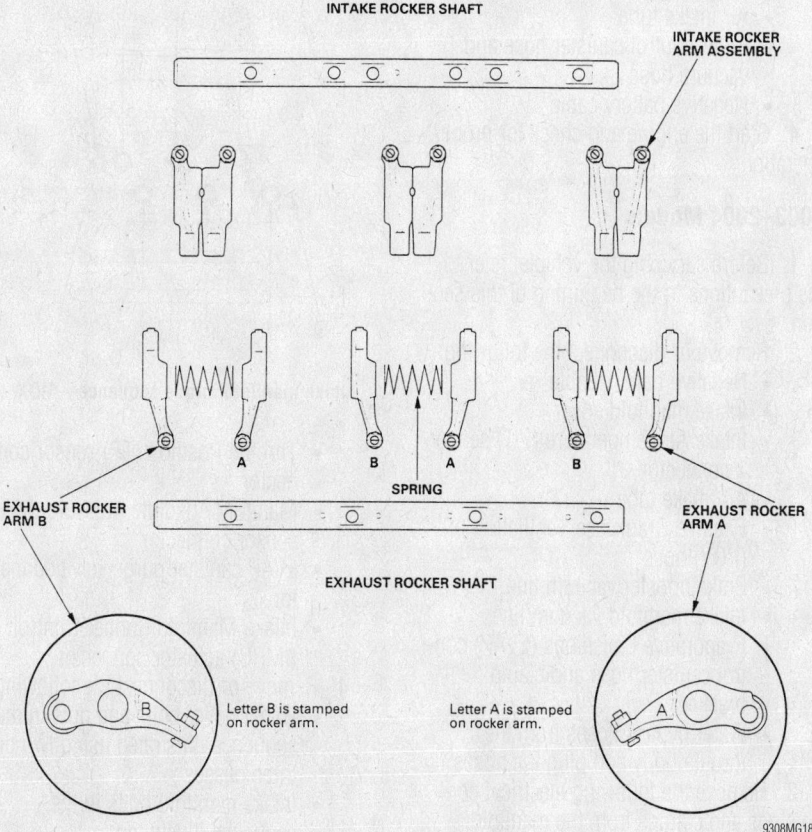

INTAKE ROCKER SHAFT

INTAKE ROCKER ARM ASSEMBLY

A   B   A   B

SPRING

EXHAUST ROCKER ARM B

EXHAUST ROCKER SHAFT

EXHAUST ROCKER ARM A

Letter B is stamped on rocker arm.

Letter A is stamped on rocker arm.

9308MG18

Exploded view of the rocker arms and shafts—MDX

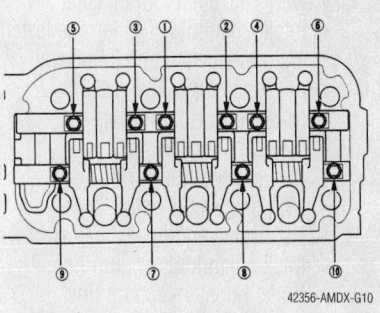

42356-AMDX-G10

Rocker shaft tightening sequence—2003–04 models

10. Install or connect the following:
- Valve covers
- Harness clamps and the breather hose
- Power steering hose bracket bolt and the harness holder bolts
- Injector electrical connections
- PCV hose
- 2 bolts attaching the harness holder
- Dipstick
- Ignition coils and torque the retainers to 9 ft. lbs. (12 Nm)
- Ignition coil covers
- Intake manifold cover
- Negative battery cable

Intake Manifold

REMOVAL & INSTALLATION

2001–2002 Models

1. Before servicing the vehicle, refer to the precautions in the beginning of this section.
2. Remove or disconnect the following:
- Negative battery cable
- Evaporative Emissions (EVAP) control canister hose and vacuum hose
- Air intake tube
- Intake manifold cover
- Accelerator cable
- Cruise control cable
- Brake booster vacuum line
- Intake manifold vacuum line
- Positive Crankcase Ventilation (PCV) valve and hose
- Intake Air Temperature (IAT) sensor connector
- Idle Air Control (IAC) valve connector
- Throttle Position (TP) sensor connector
- Manifold Absolute Pressure (MAP) sensor connector
- Intake manifold

To install:
3. Install or connect the following:
- New intake manifold gasket
- Intake manifold. Tighten the fasteners in sequence and in several passes to 16 ft. lbs. (22 Nm).
- MAP sensor connector
- TP sensor connector
- IAC valve connector
- IAT sensor connector
- PCV valve and hose
- Intake manifold vacuum line
- Brake booster vacuum line
- Cruise control cable
- Accelerator cable

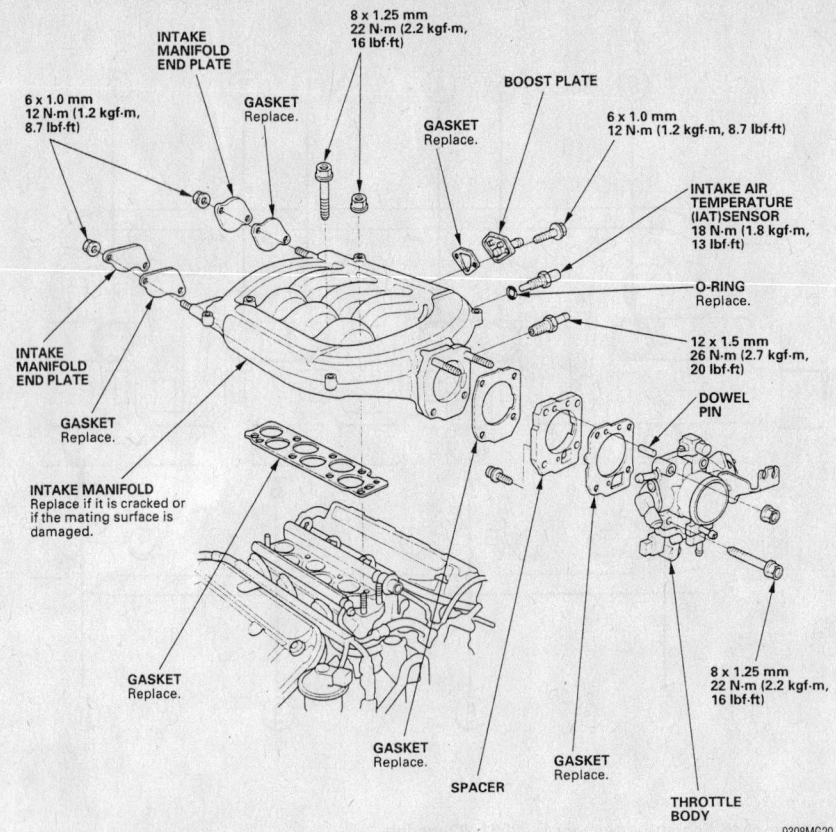

Exploded view of the intake manifold—2001–02 models

- Intake manifold cover
- Air intake tube
- EVAP control canister hose and vacuum hose
- Negative battery cable
4. Start the engine and check for proper operation.

2003–2004 Models

1. Before servicing the vehicle, refer to the precautions in the beginning of this section.
2. Remove or disconnect the following:
- Negative battery cable
- Intake manifold cover
- Intake Air Temperature (IAT) sensor 2 connector
- Air intake tube
- Positive Crankcase Ventilation (PCV) hose
- Brake booster vacuum line
- Intake manifold vacuum line
- Evaporative Emissions (EVAP) control canister hose and clamp bracket
- Water bypass hoses from the throttle body and plug the hoses
3. Remove the following electrical connections and clamps from the manifold:
- IAT sensor 1 connector

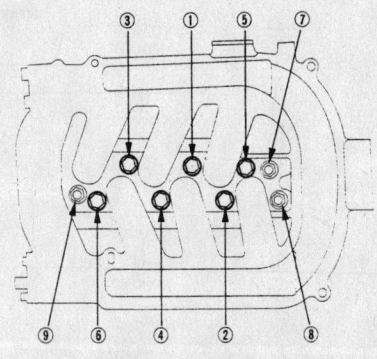

Intake manifold torque sequence—MDX

- Throttle Position (TP) sensor connector
- Manifold Absolute Pressure (MAP) sensor connector
- EVAP canister purge valve connector
- Intake Manifold Runner Control (IMRC) actuator connector
4. Remove or disconnect the following:
- Upper cover bolts and nuts in the sequence illustrated using two or three passes
- Intake manifold bolts in the sequence illustrated
- Intake manifold and spacer

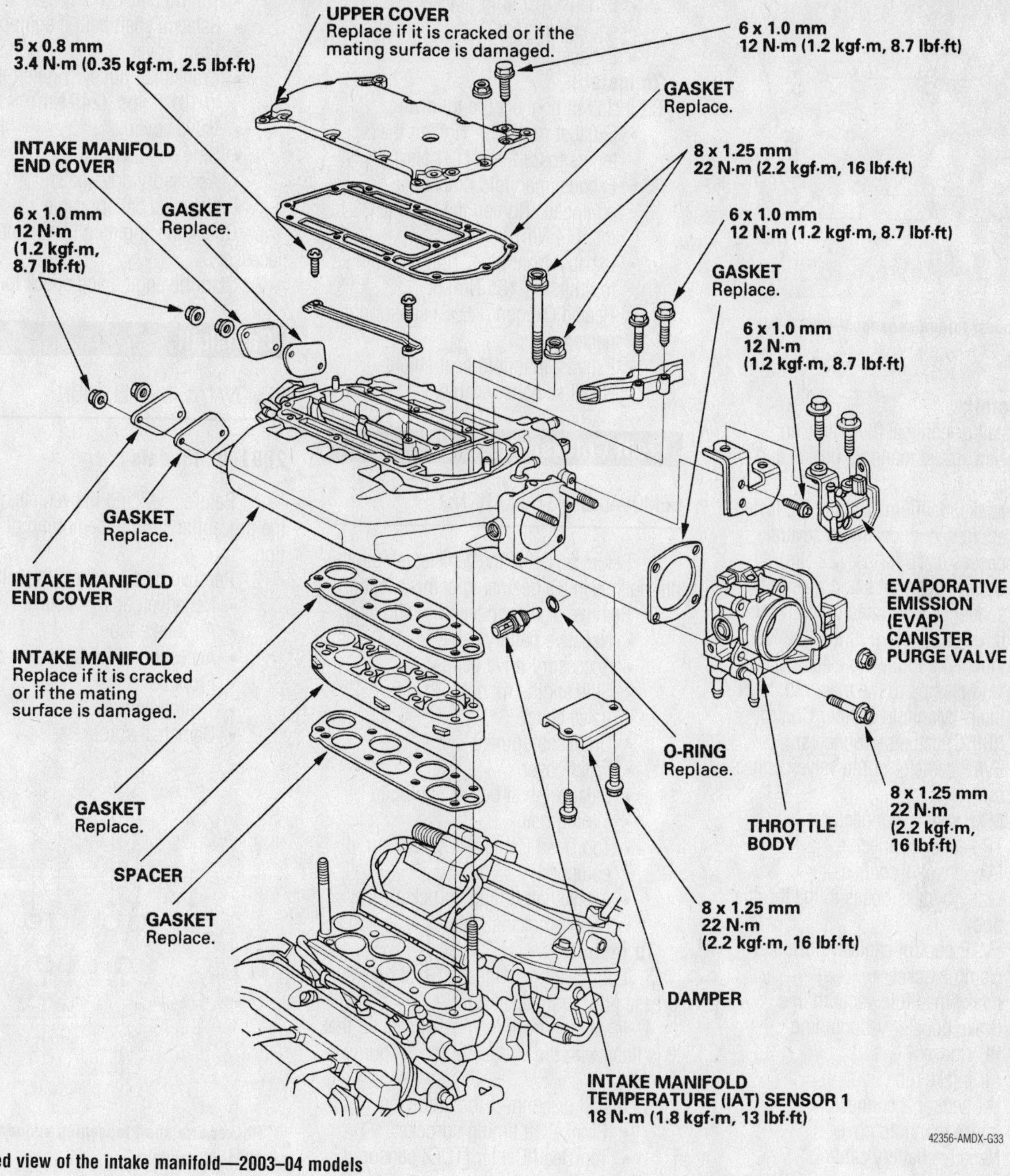

UPPER COVER
Replace if it is cracked or if the mating surface is damaged.

5 x 0.8 mm
3.4 N·m (0.35 kgf·m, 2.5 lbf·ft)

6 x 1.0 mm
12 N·m (1.2 kgf·m, 8.7 lbf·ft)

GASKET
Replace.

**INTAKE MANIFOLD
END COVER**

8 x 1.25 mm
22 N·m (2.2 kgf·m, 16 lbf·ft)

6 x 1.0 mm
12 N·m
(1.2 kgf·m,
8.7 lbf·ft)

GASKET
Replace.

6 x 1.0 mm
12 N·m (1.2 kgf·m, 8.7 lbf·ft)

GASKET
Replace.

6 x 1.0 mm
12 N·m
(1.2 kgf·m, 8.7 lbf·ft)

GASKET
Replace.

**INTAKE MANIFOLD
END COVER**

**EVAPORATIVE
EMISSION
(EVAP)
CANISTER
PURGE VALVE**

INTAKE MANIFOLD
Replace if it is cracked or if the mating surface is damaged.

GASKET
Replace.

O-RING
Replace.

8 x 1.25 mm
22 N·m
(2.2 kgf·m,
16 lbf·ft)

GASKET
Replace.

SPACER

**THROTTLE
BODY**

GASKET
Replace.

8 x 1.25 mm
22 N·m
(2.2 kgf·m, 16 lbf·ft)

DAMPER

**INTAKE MANIFOLD
TEMPERATURE (IAT) SENSOR 1**
18 N·m (1.8 kgf·m, 13 lbf·ft)

42356-AMDX-G33

Exploded view of the intake manifold—2003–04 models

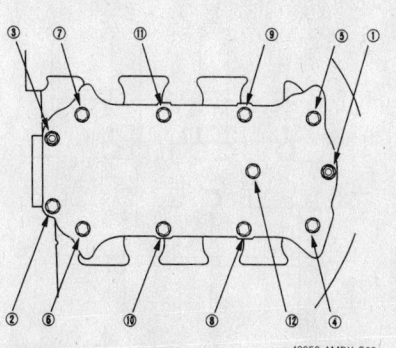

42356-AMDX-G03

Upper cover loosening sequence—
2003–04 models

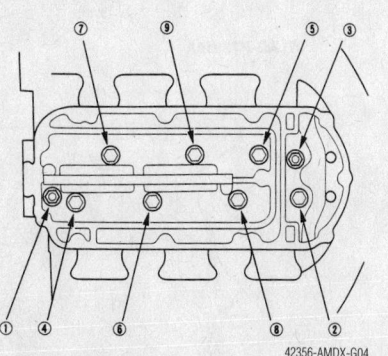

42356-AMDX-G04

Intake manifold loosening sequence—
2003–04 models

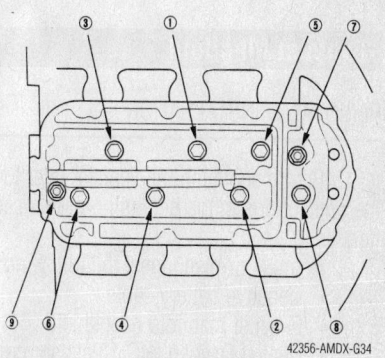

42356-AMDX-G34

Intake manifold torque sequence—
2003–04 models

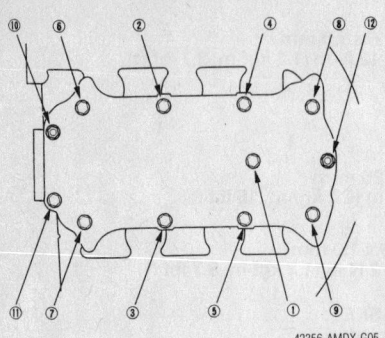

Upper cover torque sequence—2003–04 models

42356-AMDX-G05

To install:

5. Install or connect the following:
- New intake manifold gasket and spacer
- Intake manifold. Tighten the fasteners in sequence and in several passes to 16 ft. lbs. (22 Nm).
- Upper cover bolts and nuts in the sequence illustrated using two or three passes to 9 ft. lbs. (12 Nm)

6. Connect the following electrical connections and clamps to the manifold:
- Intake Manifold Runner Control (IMRC) actuator connector
- EVAP canister purge valve connector
- MAP sensor connector
- TP sensor connector
- IAT sensor 1 connector
- Water bypass hoses to the throttle body
- EVAP control canister hose and clamp bracket
- Intake manifold vacuum line
- Brake booster vacuum line
- PCV hose
- Air intake tube
- IAT sensor 2 connector
- Intake manifold cover
- Negative battery cable

7. Start the engine and check for proper operation.

Exhaust Manifold

REMOVAL & INSTALLATION

1. Before servicing the vehicle, refer to the precautions in the beginning of this section.

2. Remove or disconnect the following:
- Negative battery cable
- Exhaust manifold heat shield
- Heated Oxygen (HO$_2$S) sensor connector
- Exhaust front pipe

- Exhaust manifold bracket, if equipped
- Exhaust manifold

To install:

3. Install or connect the following:
- Exhaust manifold. Tighten the fasteners to 23 ft. lbs. (31 Nm).
- Exhaust manifold bracket, if equipped. Tighten the bolts to 33 ft. lbs. (44 Nm).
- Exhaust front pipe. Tighten the nuts to 40 ft. lbs. (55 Nm).
- Heated Oxygen (HO$_2$S) sensor connector
- Exhaust manifold heat shield
- Negative battery cable

Front Crankshaft Seal

REMOVAL & INSTALLATION

1. Before servicing the vehicle, refer to the precautions in the beginning of this section.

2. Remove or disconnect the following:
- Negative battery cable
- Accessory drive belts
- Side engine mount
- Valve cover
- Crankshaft pulley
- Front cover
- Balance shaft belt, if equipped
- Timing belt
- Top Dead Center (TDC) sensor, if equipped
- Crankshaft timing sprocket
- Front crankshaft seal

To install:

3. Lubricate the crankshaft seal lip with grease prior to installation.

4. Install the front crankshaft seal so that it is flush with the surface of the oil pump housing.

5. Install or connect the following:
- Crankshaft timing sprocket
- Top Dead Center (TDC) sensor, if equipped

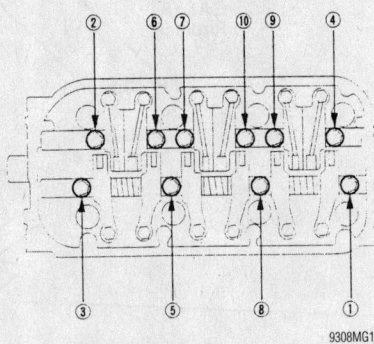

07LAD-PT3010A

93552G02

Front crankshaft seal installation

- Timing belt
- Balance shaft belt, if equipped
- Front cover
- Crankshaft pulley. Tighten the bolt to 181 ft. lbs. (245 Nm).
- Valve cover
- Side engine mount
- Accessory drive belts
- Negative battery cable

6. Check the engine oil level and add if necessary.

7. Start the engine and check for leaks.

Camshaft

REMOVAL & INSTALLATION

2001–02 Models

1. Before servicing the vehicle, refer to the precautions in the beginning of this section.

2. Remove or disconnect the following:
- Negative battery cable
- Air intake tube
- Accessory drive belts
- Front cover
- Timing belt
- Camshaft sprockets

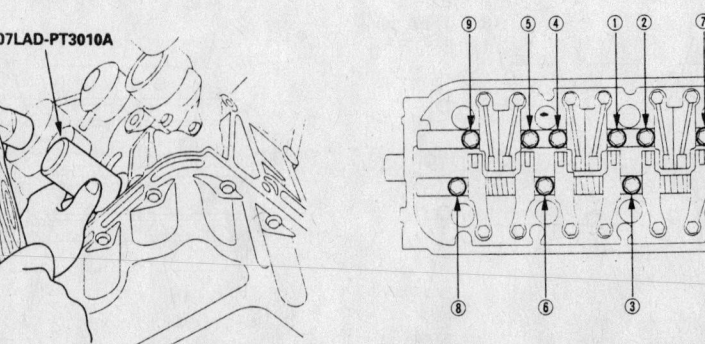

9308MG17

Rocker arm shaft loosening sequence—2001–02 models

9302MG84

Rocker shaft tightening sequence—2001–02 models

- Timing belt rear covers
- Ignition coil covers
- Intake manifold cover
- Intake manifold
- Valve cover
- Rocker arms and shaft assembly
- Camshaft thrust cover
- Camshaft

To install:

➡**Use new O-rings, seals and gaskets when installing the camshaft.**

3. Install or connect the following:
   - Camshaft
   - Camshaft thrust cover. Tighten the bolts to 16 ft. lbs. (22 Nm).
   - Rocker arms and shaft assembly
   - Valve cover
   - Intake manifold
   - Intake manifold cover
   - Ignition coil covers
   - Timing belt rear covers
   - Camshaft sprockets. Tighten the bolts to 67 ft. lbs. (90 Nm).
   - Timing belt
   - Front cover
   - Accessory drive belts
   - Air intake tube
   - Negative battery cable
4. Start the engine and check for leaks.

2003–04 Models

FRONT

1. Before servicing the vehicle, refer to the precautions in the beginning of this section.
2. Remove or disconnect the following:
   - Negative and positive battery cables
   - Battery
3. Drain the coolant.
   - Exhaust Gas Recirculation (EGR) vale
   - Timing belt
   - Rocker arm assembly
   - Front camshaft pulley
   - Thrust plate and camshaft

To install:

4. Install or connect the following:
   - Camshaft using a new O-ring. Tighten the thrust plate to 16 ft. lbs. (22 Nm).
   - Front camshaft pulley
   - Rocker arm assembly
   - Timing belt
   - Exhaust Gas Recirculation (EGR) vale
   - Battery
   - Positive, then negative battery cables

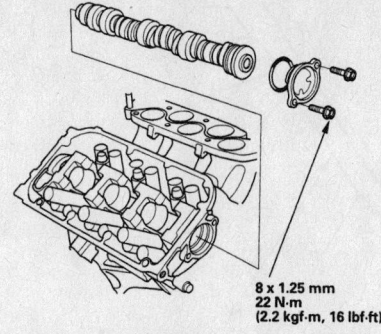

8 x 1.25 mm
22 N·m
(2.2 kgf·m, 16 lbf·ft)

42356-AMDX-G06

Front camshaft assembly—2003–04 models

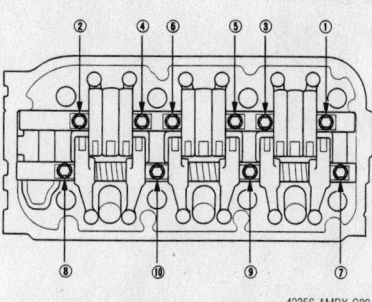

42356-AMDX-G09

Rocker arm shaft loosening sequence— 2003–04 models

5. Fill the cooling system.
   - Camshaft
6. Start the engine and check for leaks.

REAR

1. Before servicing the vehicle, refer to the precautions in the beginning of this section.
2. Drain the cooling system.
3. Relieve the fuel system pressure.
4. Remove or disconnect the following:
   - Negative battery cable
   - Under-hood fuse box
   - Fuel feed hose
   - Nuts securing the fuel line
   - Brake lines from the master cylinder
   - Timing belt
   - Rocker arm assembly
   - Rear camshaft pulley
   - Thrust plate and camshaft

To install:

5. Install or connect the following:
   - Camshaft using a new O-ring. Tighten the thrust plate to 16 ft. lbs. (22 Nm).
   - Rear camshaft pulley
   - Rocker arm assembly
   - Timing belt
   - Brake lines to the master cylinder

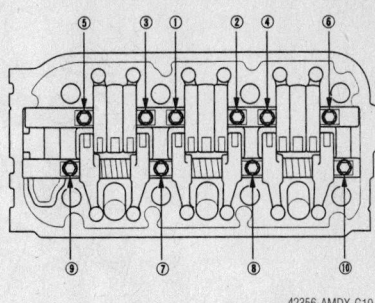

42356-AMDX-G10

Rocker shaft tightening sequence— 2003–04 models

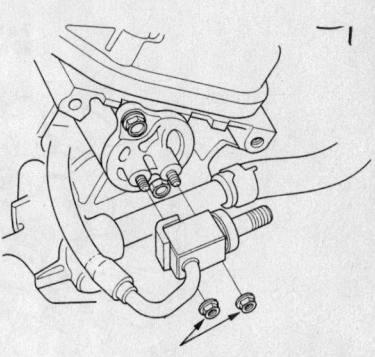

6 x 1.0 mm
12 N·m (1.2 kgf·m, 8.7 lbf·ft)

42356-AMDX-G07

Remove the nuts attaching the fuel line when removing the rear camshaft— 2003–04 models

- Nuts securing the fuel line
- Fuel feed hose
- Under-hood fuse box
- Negative battery cable

Valve Lash

ADJUSTMENT

Adjust the valves only when the cylinder head temperature is less than 100°F (38°C).

1. Before servicing the vehicle, refer to the precautions in the beginning of this section.
2. Remove or disconnect the following:
   - Negative battery cable
   - Air intake tube
   - Intake manifold
   - Valve cover
3. Rotate the crankshaft so that the valves to be adjusted are closed and the rocker arm is contacting the camshaft lobe base circle.
4. Measure the valve clearance. If adjustment is necessary, loosen the locknut and turn the adjusting screw as necessary to achieve the correct valve clearance.

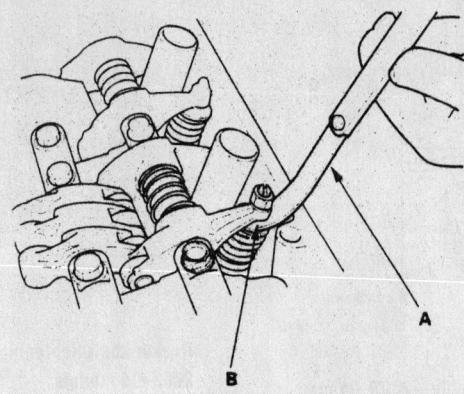

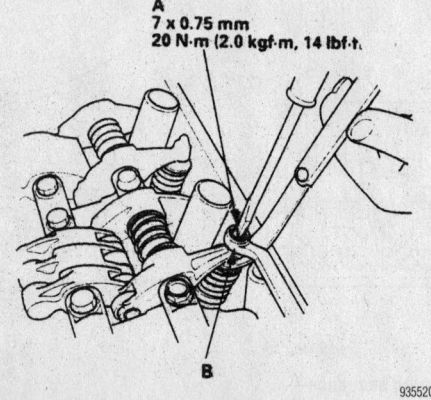

A
7 x 0.75 mm
20 N·m (2.0 kgf·m, 14 lbf·ft)

93552G04

Inspect the valve clearance, adjust to specification and tighten the retainer to specification

Adjusting screw locations:

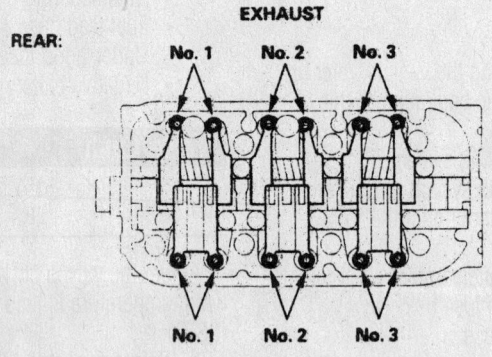

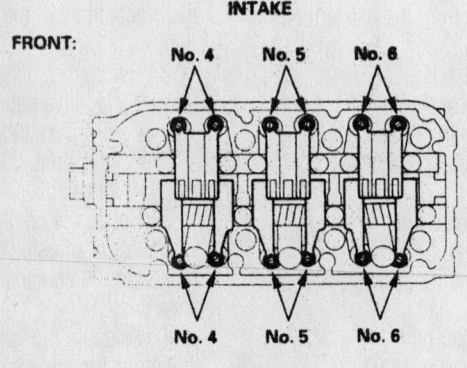

93552G03

Valve adjusting retainer locations

5. The correct valve clearance is:
 • Intake valves: 0.008–0.009 inches (0.20–0.24mm)
 • Exhaust valves: 0.011–0.013 inches (0.28–0.32mm)
6. After adjustment, tighten the locknuts to 14 ft. lbs. (20 Nm).
7. Install or connect the following:
 • Valve cover
 • Intake manifold
 • Air intake tube
 • Negative battery cable
8. Start the engine and check for proper operation.

Starter Motor

REMOVAL & INSTALLATION

2001–02 Models

1. Before servicing the vehicle, refer to the precautions in the beginning of this section.
2. Remove or disconnect the following:

 • Negative battery cable and wait at least 3 minutes
 • Transmission fluid cooler line clamp
 • Starter wiring harness connectors
 • Starter motor

To install:
3. Install or connect the following:
 • Starter motor. Tighten the 10mm bolt to 33 ft. lbs. (44 Nm) and the 12mm bolt to 47 ft. lbs. (64 Nm).
 • Starter wiring harness connectors. Tighten the battery cable nut to 79 inch lbs. (9 Nm).
 • Transmission fluid cooler line clamp
 • Negative battery cable

2003–04 Models

1. Before servicing the vehicle, refer to the precautions in the beginning of this section.
2. Remove or disconnect the following:

 • Negative battery cable and wait at least 3 minutes
 • Battery and battery tray
 • Harness clamps
 • Starter wiring harness connectors
 • Starter motor

To install:
3. Install or connect the following:
 • Starter motor. Tighten the bolts to 54 ft. lbs. (74 Nm).

- Starter wiring harness connectors. Tighten the battery cable nut to 79 inch lbs. (9 Nm).
- Harness clamps
- Battery tray
- Negative battery cable

Oil Pan

REMOVAL & INSTALLATION

1. Before servicing the vehicle, refer to the precautions in the beginning of this section.
2. Drain the engine oil.
3. Remove or disconnect the following:

- Negative battery cable
- Front splash shield
- Heated Oxygen (HO$_2$S) sensor connector
- Exhaust front pipe
- Torque converter cover, if equipped with an automatic transaxle
- Oil pan

To install:

4. Install the oil pan. Apply liquid gasket as shown.
5. Tighten the bolts in sequence to 105 inch lbs. (12 Nm) using several passes.
6. Install or connect the following:

- Torque converter cover, if removed
- Exhaust front pipe
- Subframe center beam, if removed
- HO$_2$S sensor connector
- Front splash shield
- Negative battery cable

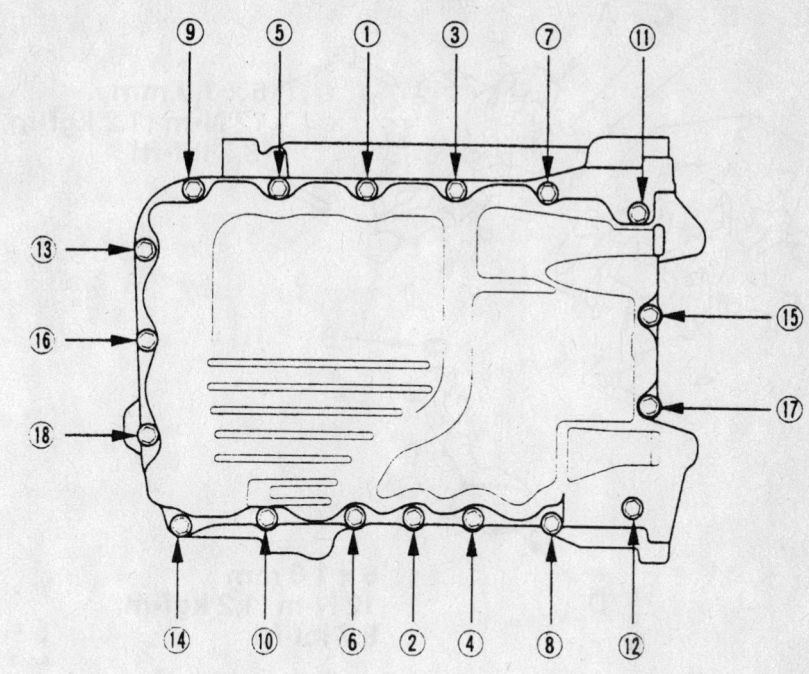

9302MG76

Oil pan tightening sequence—MDX

Oil Pump

REMOVAL & INSTALLATION

1. Before servicing the vehicle, refer to the precautions in the beginning of this section.
2. Drain the engine oil.
3. Remove or disconnect the following:

- Negative battery cable

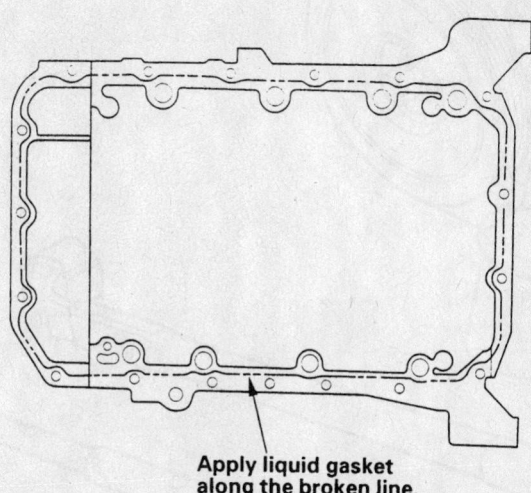

Apply liquid gasket along the broken line.

9302MG75

Apply liquid gasket to the inner threads of the bolt holes and the engine block along the area indicated by the broken line—MDX

- Accessory drive belts
- Front cover
- Timing belt
- Timing belt idler pulley
- Crankshaft Position (CKP) sensor
- Crankshaft timing sprocket
- Variable Valve Timing and Valve Lift Electronic Control (VTEC) solenoid valve connector
- Oil filter adapter
- Oil pan
- Oil pump pickup tube
- Oil pump

To install:

➡**Use new gaskets and O-ring seals for assembly.**

4. Apply liquid gasket to the oil pump and to the bolt hole threads.
5. Install or connect the following:

- Oil pump. Tighten the bolts to 105 inch lbs. (12 Nm).
- Oil pump pickup tube. Tighten the bolts to 105 inch lbs. (12 Nm).
- Oil pan
- Oil filter adapter
- VTEC solenoid valve connector
- Crankshaft timing sprocket
- CKP sensor
- Timing belt idler pulley
- Timing belt
- Front cover
- Accessory drive belts

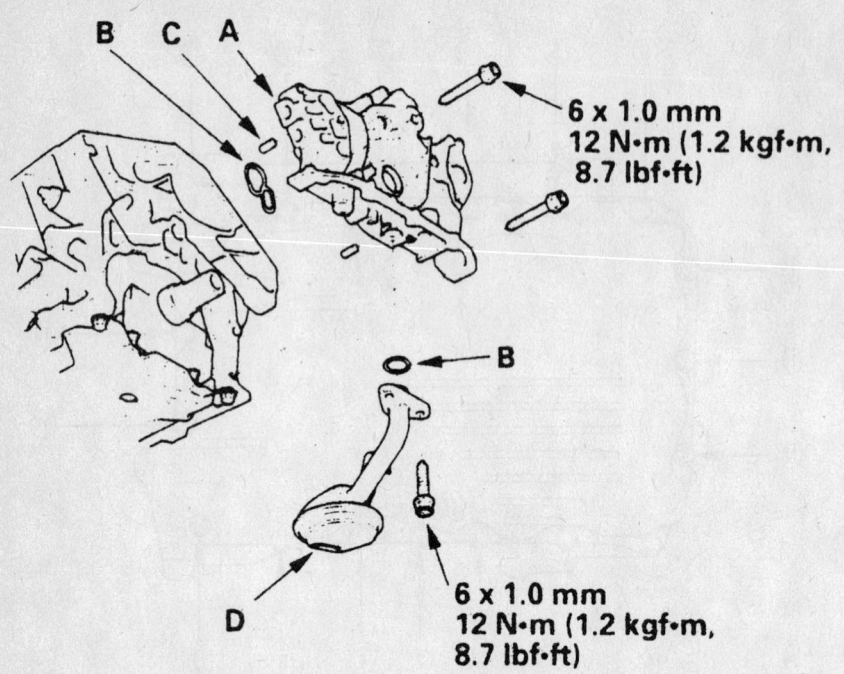

B C A

**6 x 1.0 mm
12 N•m (1.2 kgf•m,
8.7 lbf•ft)**

B

D

**6 x 1.0 mm
12 N•m (1.2 kgf•m,
8.7 lbf•ft)**

93552G05

Exploded view of the oil pump assembly

• Negative battery cable
6. Fill the crankcase to the correct level.
7. Start the engine and check for leaks.

Rear Main Seal

REMOVAL & INSTALLATION

1. Before servicing the vehicle, refer to the precautions in the beginning of this section.
2. Remove or disconnect the following:

• Transaxle
• Clutch pressure plate and disc, if equipped
• Flywheel
• Oil seal

To install:
3. Install or connect the following:

• Oil seal. Drive the seal square into the seal case.
• Flywheel. Tighten the bolts in a crossing pattern to 54 ft. lbs. (73 Nm).
• Clutch pressure plate and disc, if equipped
• Transaxle

4. Check the fluid levels.
5. Start the engine and check for leaks.

Timing Belt

REMOVAL & INSTALLATION

1. Before servicing the vehicle, refer to the precautions in the beginning of this section.

2. Turn the crankshaft so the white mark aligns with the pointer.
3. Make sure the number 1 piston is at Top Dead center (TDC).
4. Remove or disconnect the following:

• Negative battery cable
• Wheels and splash shield
• Drive belt

5. Support the engine with a block of wood and a jack under the oil pan.

• Upper side engine mount

6. Remove the crankshaft pulley using holder tool shown in the accompanying illustration and a breaker and socket, loosen the 19mm bolt and remove the pulley.

• Front upper cover, rear upper cover and the lower cover
• One of the battery clamp bolts and grind the end as illustrated

7. Screw the battery clamp bolt as illustrated to hold the belt adjuster in position. Do not use a wrench, hand tighten only.

• Lower side engine mount
• Idler pulley bolt and the pulley
• Timing belt

To install:
8. If installing a new belt, perform the following steps:

a. Clean the pulleys, belt guide plate and the upper and lower covers.

b. Set the timing belt drive pulley to TDC by aligning the TDC mark on the tooth of the belt drive pulley with the pointer on the oil pump.

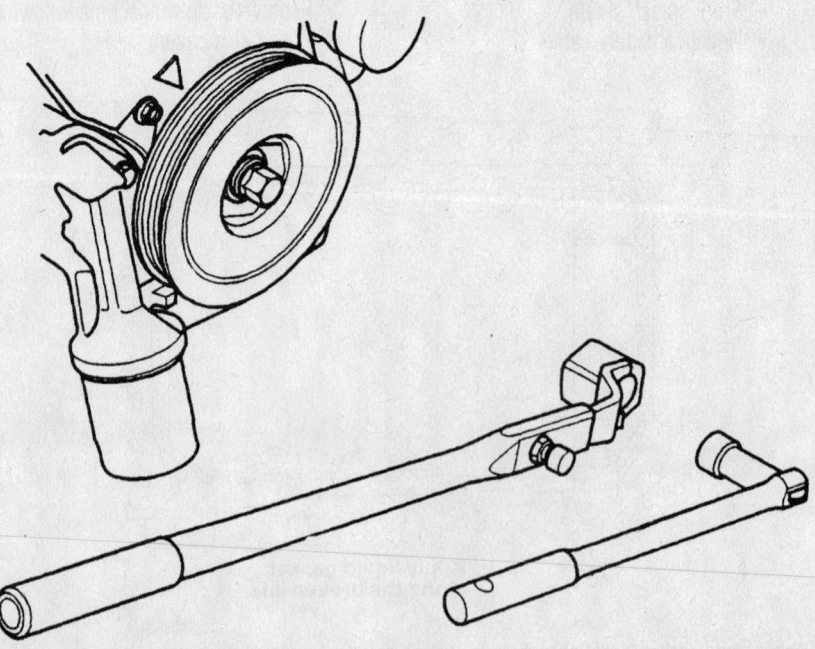

42356-AMDX-G15

Remove the crankshaft pulley using holder tool shown

c. Set the camshaft pulleys to TDC by aligning the TDC marks on the camshaft pulleys with the pointers on the back covers.

d. Remove the battery clamp bolt.

e. Remove the belt tensioner.

f. Align the holes on the rod and housing of the tensioner.

g. Using a press or other suitable device, slowly compress the tensioner and insert a 0.08 inch (2mm) pin through the housing and rod.

h. Install the tensioner making sure the pin is still installed.

i. Apply thread locker to idler pulley bolt then hand tighten the bolt.

j. Install the belt over the pulleys in this sequence; drive pulley, idler pulley, front camshaft pulley, water pump pulley, rear camshaft pulley and adjusting pulley.

k. Tighten the idler pulley bolt to 33 ft. lbs. (44 Nm).

l. Remove the pin from the tensioner.

9. Install or connect the following:

- Lower half of the side mount and tighten the 3 long bolts to 33 ft. lbs. (44 Nm) and the one short bolt to 9 ft. lbs. (12 Nm)
- Timing belt guide plate as illustrated
- Lower timing cover and tighten the bolts to 9 ft. lbs. (12 Nm)
- Front and rear upper timing covers and tighten the bolts to 9 ft. lbs. (12 Nm)
- Crankshaft pulley and tighten the bolts to 181 ft. lbs. (245 Nm), using the holding tool to prevent the unit from turning

10. Rotate the crankshaft pulley about 5 or 6 degrees clockwise so the belt positions itself on the pulleys.

11. Turn the crankshaft pulley so the white mark aligns with the pointer.

12. Check the camshaft pulley marks are aligned. If the marks are aligned, proceed to the next step. If the marks are not aligned, remove the timing belt and reinstall using the steps outlined before this step.

13. Remove or disconnect the following:

- Drive belt
- Upper side mount and tighten the bolts in the sequence illustrated to the specifications in the illustration

14. Using a suitable scan tool, perform the Powertrain Control Module (PCM) reset and the Crankshaft position (CKP) pattern clear/learn procedures, following the scan tool manufactures instructions.

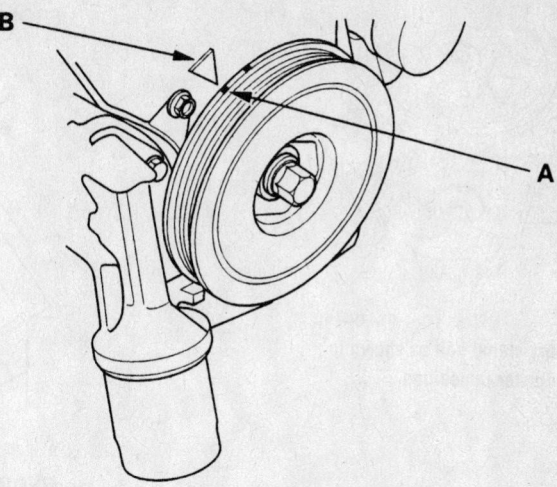

42356-AMDX-G11

Turn the crankshaft so the white mark (A) aligns with the pointer (B)

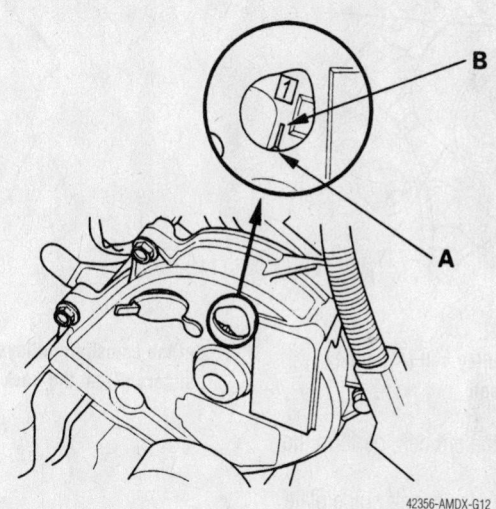

42356-AMDX-G12

Make sure the number 1 piston is at top dead center (A) on the front camshaft pulley and pointer (B)

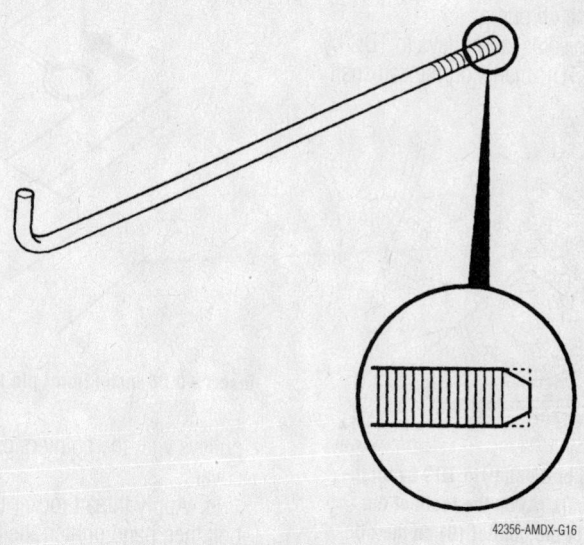

42356-AMDX-G16

Remove a battery clamp bolt and grind the end as shown

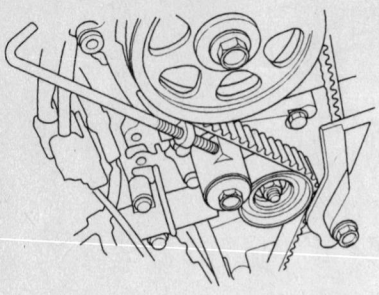

42356-AMDX-G17

Install the battery clamp bolt as shown to hold the belt adjuster in position

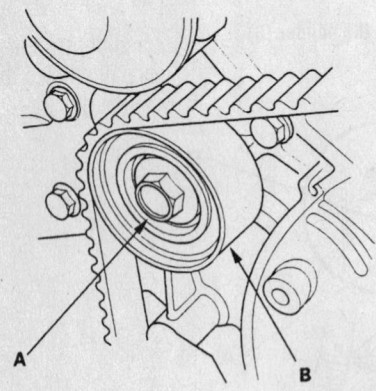

42356-AMDX-G18

Remove the idler pulley bolt (A), pulley (B) and the timing belt

15. If installing the old belt, perform the following steps:

a. Clean the pulleys, belt guide plate and the upper and lower covers.

b. Set the timing belt drive pulley to TDC by aligning the TDC mark on the tooth of the belt drive pulley with the pointer on the oil pump.

c. Set the camshaft pulleys to TDC by aligning the TDC marks on the camshaft

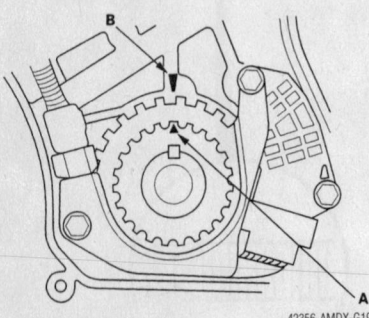

42356-AMDX-G19

Set the timing belt pulley to TDC by aligning the TDC mark (A) on the tooth of the belt pulley with the pointer (B) on the oil pump

FRONT:

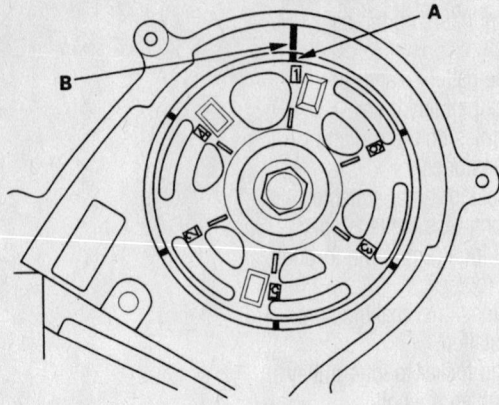

REAR:

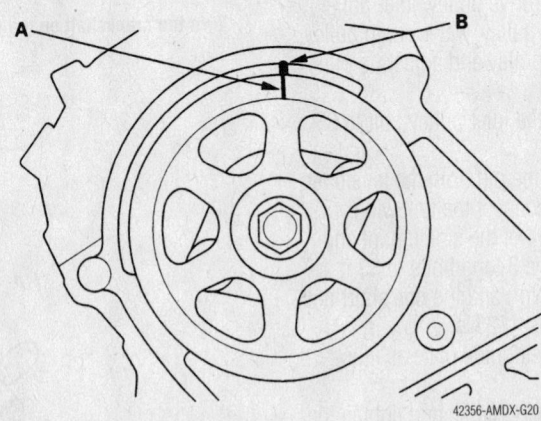

42356-AMDX-G20

Set the camshaft pulleys to TDC by aligning the TDC marks (A) on the camshaft pulleys with the pointers (B) on the back covers

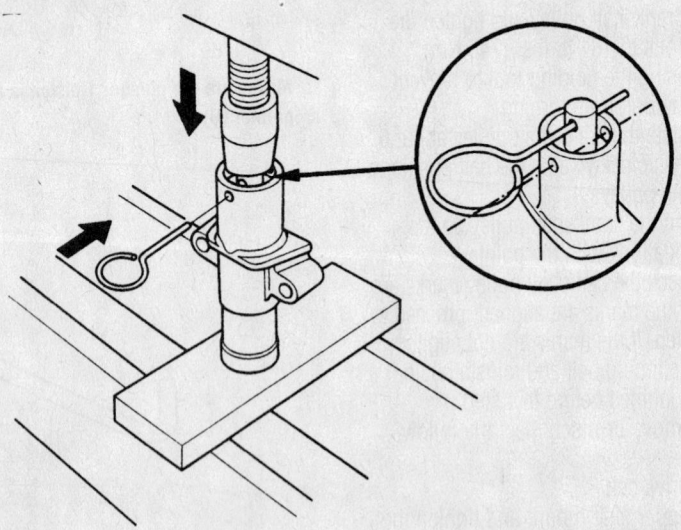

42356-AMDX-G21

Insert a 0.08 inch (2mm) pin through the tensioner housing and rod

pulleys with the pointers on the back covers.

d. Apply thread locker to idler pulley bolt then hand tighten the bolt.

16. If the tensioner was extended and the

belt cannot be installed, perform the steps above for the new belt installation.

a. Install the belt over the pulleys in this sequence; drive pulley, idler pulley, front camshaft pulley, water pump pulley,

-1 Drive pulley (A).
-2 Idler pulley (B).
-3 Front camshaft pulley (C).
-4 Water pump pulley (D).
-5 Rear camshaft pulley (E).
-6 Adjusting pulley (F).

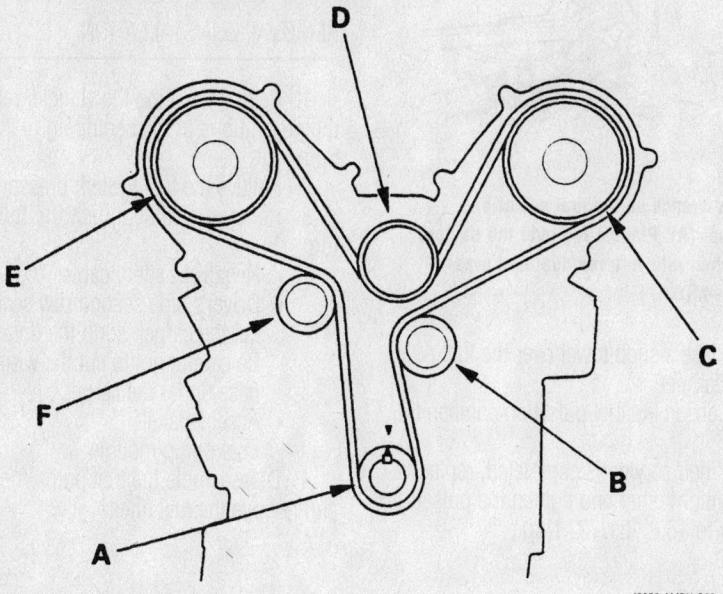

42356-AMDX-G22

Route the belt as shown in the sequence listed

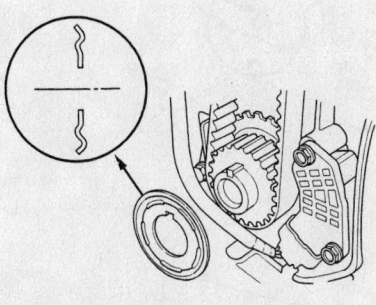

42356-AMDX-G23

Install the timing belt guide plate as shown

rear camshaft pulley and adjusting pulley.

    b. Tighten the idler pulley bolt to 33 ft. lbs. (44 Nm).

    c. Remove the battery clamp bolt.

17. Install or connect the following:
- Lower half of the side mount and tighten the 3 long bolts to 33 ft. lbs. (44 Nm) and the one short bolt to 9 ft. lbs. (12 Nm)
- Timing belt guide plate as illustrated
- Lower timing cover and tighten the bolts to 9 ft. lbs. (12 Nm)
- Front and rear upper timing covers

FRONT CAMSHAFT PULLEY:

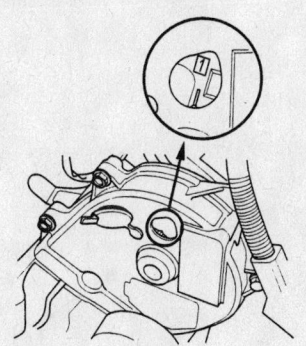

REAR CAMSHAFT PULLEY:

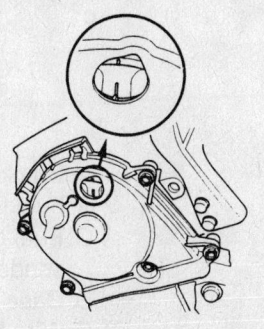

42356-AMDX-G32

Check that the camshaft pulley marks are aligned as shown

and tighten the bolts to 9 ft. lbs. (12 Nm)
- Crankshaft pulley and tighten the bolts to 181 ft. lbs. (245 Nm), using the holding tool to prevent the

18. Rotate the crankshaft pulley about 5 or 6 degrees clockwise so the belt positions itself on the pulleys.

19. Turn the crankshaft pulley so the white mark aligns with the pointer.

20. Check the camshaft pulley marks are aligned. If the marks are aligned, proceed to the next step. If the marks are not aligned, remove the timing belt and reinstall using the steps outlined before this step.

21. Remove or disconnect the following:
- Drive belt
- Upper side mount and tighten the bolts in the sequence illustrated to the specifications in the illustration

22. Using a suitable scan tool, perform the Powertrain Control Module (PCM) reset and the Crankshaft position (CKP) pattern clear/learn procedures, following the scan tool manufactures instructions.

Piston and Ring

POSITIONING

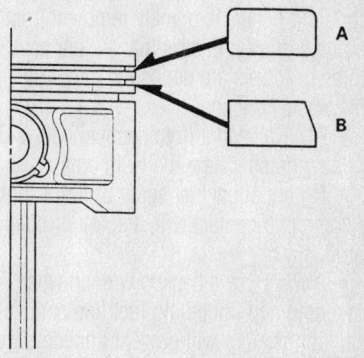

9302AG06

Compression ring identification—3.5L engine

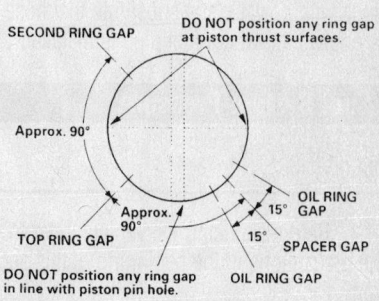

93552G06

Ring end gap positioning

FUEL SYSTEM

Fuel System Service Precautions

Safety is the most important factor when performing not only fuel system maintenance, but any type of maintenance. Failure to conduct maintenance and repairs in a safe manner may result in serious personal injury or death. Maintenance and testing of the vehicle's fuel system components can be accomplished safely and effectively by adhering to the following rules and guidelines:

• To avoid the possibility of fire and personal injury, always disconnect the negative battery cable unless the repair or test procedure requires that battery voltage be applied.

• Always relieve the fuel system pressure prior to disconnecting any fuel system component (injector, fuel rail, pressure regulator, etc.), fitting or fuel line connection. Exercise extreme caution whenever relieving fuel system pressure to avoid exposing skin, face and eyes to fuel spray. Please be advised that fuel under pressure may penetrate the skin or any part of the body that it contacts.

• Always place a shop towel or cloth around the fitting or connection prior to loosening to absorb any excess fuel due to spillage. Ensure that all fuel spillage (should it occur) is quickly removed from engine surfaces. Ensure that all fuel soaked cloths or towels are deposited into a suitable waste container.

• Always keep a dry chemical (Class B) fire extinguisher near the work area.

• Do not allow fuel spray or fuel vapors to come into contact with a spark or open flame.

• Always use a backup wrench when loosening and tightening fuel line connection fittings. This will prevent unnecessary stress and torsion to fuel line piping. Always follow the proper torque specifications.

• Always replace worn fuel fitting O-rings with new. Do not substitute fuel hose or equivalent, where fuel pipe is installed.

Fuel System Pressure

RELIEVING

1. Before servicing the vehicle, refer to the precautions in the beginning of this section.
2. Disconnect the negative battery cable.
3. Remove the fuel filler cap.

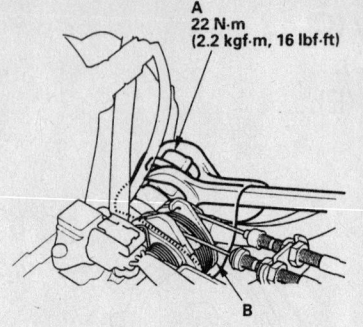

A
22 N·m
(2.2 kgf·m, 16 lbf·ft)

B

9302MG77

Use a wrench on the fuel pulsation damper (A). Place a rag over the damper (B) when relieving residual fuel pressure—MDX

4. Place a shop towel over the fuel pulsation damper.
5. Loosen the fuel pulsation damper 1 turn.
6. When service is completed, replace the sealing washer and tighten the pulsation damper to 16 ft. lbs. (22 Nm).

7. Replace the fuel filler cap.
8. Connect the negative battery cable.
9. Start the engine and check for leaks.

Fuel Filter

REMOVAL & INSTALLATION

1. Before servicing the vehicle, refer to the precautions in the beginning of this section.
2. Relieve the fuel system pressure.
3. Remove or disconnect the following:

• Negative battery cable
• Driver's side second row seat and cut the carpet along the dotted line. Be careful not to cut the wiring harness under the carpet.
• Access panel
• Fuel pump module

4. Disassemble the fuel pump module and remove the fuel filter.

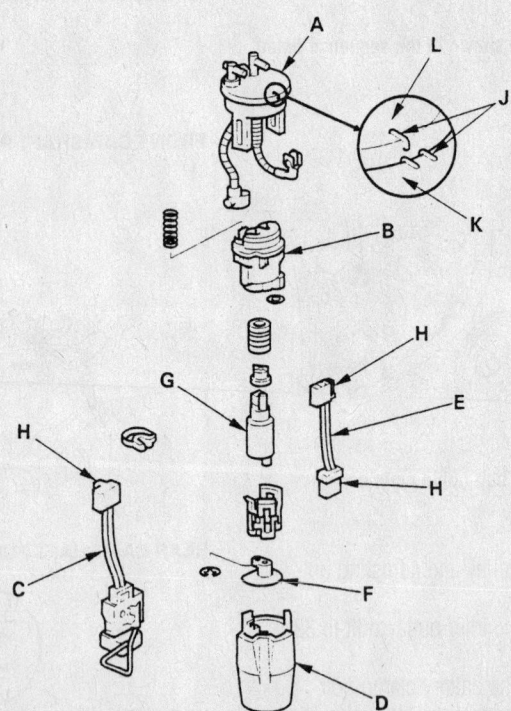

A. Bracket
B. Fuel filter
C. Fuel gauge sender
D. Case
E. Wire harness
F. Suction filter
G. Fuel pump
H. Connectors
J. Alignment marks
K. Fuel tank
L. Fuel pump module

9308MG26

Exploded view of the fuel pump module—MDX

To install:

5. Install the fuel filter and assemble the fuel pump module.

6. Install or connect the following:
- Fuel pump module
- Access panel
- Carpet and seat
- Negative battery cable

7. Start the engine and check for leaks.

Fuel Pump

REMOVAL & INSTALLATION

1. Before servicing the vehicle, refer to the precautions in the beginning of this section.

2. Relieve the fuel system pressure.

3. Remove or disconnect the following:
- Negative battery cable
- Driver's side second row seat and cut the carpet along the dotted line. Be careful not to cut the wiring harness under the carpet.
- Access panel
- Fuel pump module wiring connector

- Fuel supply and return lines
- Fuel pump locknut
- Fuel pump module

To install:

4. Install or connect the following:
- Fuel pump module. Use a new seal and align the matchmarks.
- Fuel pump locknut
- Fuel supply and return lines
- Fuel pump module wiring connector
- Access panel
- Carpet and seat
- Negative battery cable

5. Start the engine and check for leaks.

Fuel Injector

REMOVAL & INSTALLATION

1. Before servicing the vehicle, refer to the precautions in the beginning of this section.

2. Relieve the fuel system pressure.

3. Remove or disconnect the following:

- Negative battery cable
- Intake manifold
- Fuel lines
- Fuel injector connectors
- Fuel pressure regulator vacuum line
- Fuel supply manifold

4. Separate the fuel injectors from the fuel supply manifold.

To install:

5. Install the fuel injectors to the fuel supply manifold with new cushion rings and O-rings.

6. Install new seal rings to the intake manifold.

7. Install or connect the following:
- Fuel supply manifold and injector assembly. Tighten the bolts to 86 inch lbs. (10 Nm).
- Fuel pressure regulator vacuum line
- Fuel injector connectors
- Fuel lines
- Intake manifold
- Negative battery cable

8. Start the engine and check for leaks.

DRIVE TRAIN

Transaxle Assembly

REMOVAL & INSTALLATION

1. Before servicing the vehicle, refer to the precautions in the beginning of this section.

2. Drain the transaxle.

3. Drain the power steering system.

4. Remove the engine appearance covers.

5. Remove the drivers side center console lower panel and pull back the cover to access steering joint cover.

6. Remove or disconnect the following:
- Steering joint bolt
- Steering joint from the steering gearbox pinion shaft
- Air intake assembly
- Battery
- Battery tray
- Power steering pump hose and the clamp bolt
- Transmission breather tube
- Cooler hose from the clamp on the starter
- Transaxle oil cooler lines
- Starter motor
- Shift control solenoid valve connectors
- Transaxle ground cable
- 8P connector from the bracket and the connector

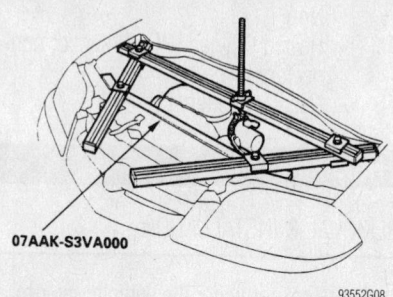

07AAK-S3VA000

93552G08

Support the engine while removing the transaxle—MDX

- Clutch pressure switch connectors
- Joint connector and transmission range switch connector from the brackets
- Countershaft speed sensor connector
- Heated Oxygen (HO$_2$S) sensor connectors
- Transmission housing mounting bolts
- Nut from the front mount and the ground cable from the engine
- Bulkhead cover, windshield wiper arms, cowl cover sealing and cover
- Install a support fixture to the engine lifting eyes.
- Splash shield
- Front sub-frame stiffener

- Exhaust front pipe
- Lower control arms from the knuckle
- Stabilizer bar links
- Tie rod ends from the knuckle
- Left driveshaft from the differential
- Right driveshaft from the intermediate shaft
- Propeller shaft from the companion flange
- Shift cable cover and holder
- Shift control cable and lever

7. Install a 6 x 1 x 14mm bolt and nut on the cable cover, then reinstall the cable cover to the torque converter housing. If this is not done, the bolt head of the cable cover may prevent torque converter removal.
- Transfer assembly
- Engine-to-torque converter bolts
- Power steering pressure switch connection
- Power steering hose clamp, then the hose from the pipe at the sub-frame
- Transmission lower mount nuts

8. Matchmark the front subframe to the vehicle body.
- Rear mount bracket bolts

9. Support the sub-frame with a 4 x 4 x 50 inch piece of wood and a jack.
- Sub-frame
- Transaxle lower mounts

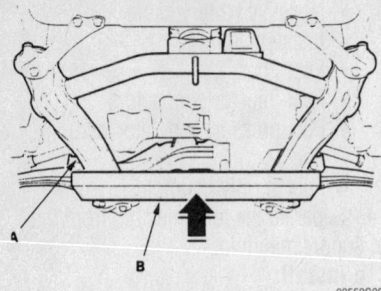

Support the sub-frame with a 4 x 4 x 50 inch piece of wood and a jack

93552G09

- Driveshafts from the differential and intermediate shaft
- Intermediate shaft
- Transmission front mount bracket
- Transmission flange bolts
- Transmission

To install:

➡**Use new circlips, split pins and self-locking nuts for assembly.**

10. Installation is the reverse of removal. Please note the following specifications:
- Transmission housing bolts and harness clamp bolts to 47 ft. lbs. (64 Nm)
- Transmission housing bolts to 40 ft. lbs. (54 Nm)
- Front mount bracket bolts to 28 ft. lbs. (38 Nm)
- Intermediate shaft bolts to 29 ft. lbs. (39 Nm)

11. Raise the subframe into position and align the matchmarks. Tighten the subframe bolts to 76 ft. lbs. (103 Nm). Tighten the front subframe bracket bolts to 54 ft. lbs. (74 Nm) and the rear bracket bolts to 86 ft. lbs. (117 Nm).
- Rear engine mount bolts to 28 ft. lbs. (38 Nm)
- Engine-to-torque converter bolts. Tighten the 6 x 1 mm bolts to 105 inch lbs. (12 Nm), 10 x 1.25mm bolt to 28 ft. lbs. (38 Nm).
- Front motor mount nut to 40 ft. lbs. (54 Nm)

12. Fill the transaxle to the correct level.
13. Start the engine and check for leaks.
14. Check the wheel alignment and adjust as necessary.

Transfer Assembly

REMOVAL & INSTALLATION

1. Before servicing the vehicle, refer to the precautions in the beginning of this section.

2. Drain the transmission fluid.
3. Remove or disconnect the following:
- Negative battery cable
- Heated Oxygen (HO$_2$S) sensor connectors
- Front sub-frame stiffener
- Exhaust front pipe
- Breather tube bracket bolt, then the tube from the breather pipe
- Propeller shaft from the transfer assembly
- Transfer assembly bolts and the assembly

To install:

4. Install or connect the following:
- New O-ring on the transfer cover
- Dowel pin on the assembly
- Transfer assembly and tighten the bolts to 33 ft. lbs. (44 Nm) on 2001–02 models and 38 ft. lbs. (51 Nm) on 2003–04 models
- Propeller shaft
- Breather tube bracket, attach the tube with the dot facing outwards and tighten the bolt to 9 ft. lbs. (12 Nm)
- Exhaust front pipe
- Front sub-frame stiffener and tighten the bolts to 40 ft. lbs. (54 Nm)
- Heated Oxygen (HO$_2$S) sensor connectors
- Negative battery cable

Halfshaft

REMOVAL & INSTALLATION

1. Before servicing the vehicle, refer to the precautions in the beginning of this section.

2. Drain the transaxle if removing the left halfshaft. If is not necessary to drain the fluid if removing the right halfshaft.

3. Remove or disconnect the following:
- Negative battery cable
- Front wheels
- Spindle nut
- Stabilizer bar link
- Lower ball joint

4. Pry the inboard joint from the transaxle or intermediate shaft.

5. Remove the outer CV-joint stub shaft from the hub by tapping the stub shaft with a plastic hammer.

To install:

➡**Use new circlips, split pins and self-locking nuts for assembly.**

6. Install the outer CV-joint stub shaft into the hub.

7. Install the inner CV-joint to the transaxle or intermediate shaft until the circlip locks in the retaining groove.

8. Install or connect the following:
- Lower ball joint. Tighten the nut to 43–51 ft. lbs. (59–69 Nm).
- Stabilizer bar link. Tighten the nut to 58 ft. lbs. (78 Nm).
- Spindle nut. Tighten the nut to 181 ft. lbs. (245 Nm) on 2001–02 models, or 210 ft. lbs. (285 Nm) on 2003–04 models.
- Front wheels
- Negative battery cable

9. Fill the transaxle to the correct level and check for leaks.

CV-Joint

OVERHAUL

Front

OUTBOARD JOINT

1. Before servicing the vehicle, refer to the precautions in the beginning of this section.

2. Remove or disconnect the following:
- Axle halfshaft from the vehicle and place it in a vise
- Outboard joint boot clamps and push the boot back
- Outboard joint by driving it off the axle shaft with a brass drift and hammer
- Outboard joint boot

To install:

➡**Use new circlips and boot clamps for assembly.**

3. Install the outboard joint boot and clamps to the axle shaft.

4. Fill the outboard joint with grease. Install the outboard joint to the axle shaft. Tap the stub shaft with a brass hammer to seat the circlip.

5. Fill the outboard joint boot with grease and install the boot clamps.

6. Install the axle halfshaft to the vehicle.

INBOARD JOINT

1. Before servicing the vehicle, refer to the precautions in the beginning of this section.

2. Remove or disconnect the following:
- Axle halfshaft from the vehicle.

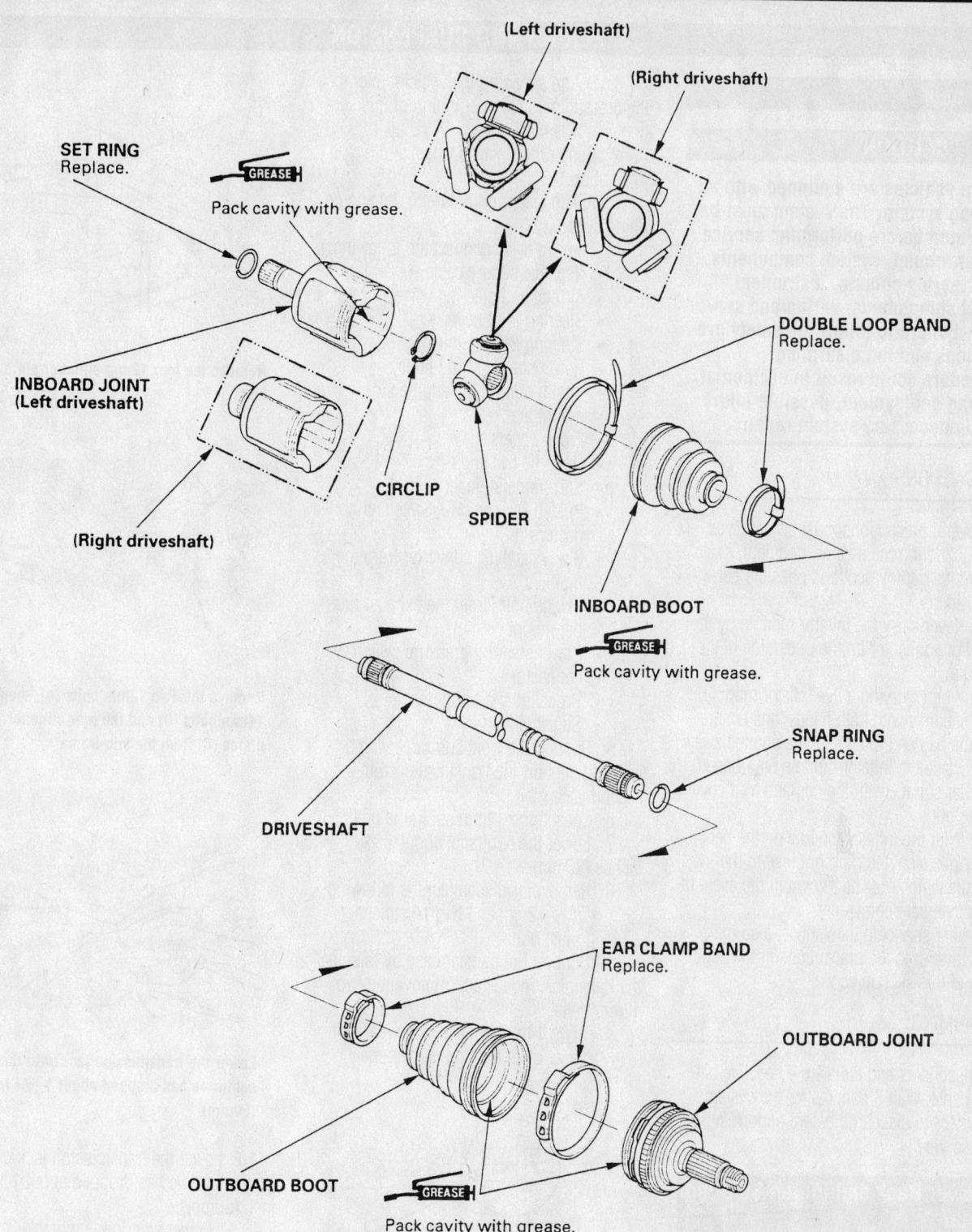

(Left driveshaft)

(Right driveshaft)

SET RING
Replace.

GREASE

Pack cavity with grease.

INBOARD JOINT
(Left driveshaft)

(Right driveshaft)

CIRCLIP

SPIDER

DOUBLE LOOP BAND
Replace.

INBOARD BOOT

GREASE

Pack cavity with grease.

SNAP RING
Replace.

DRIVESHAFT

EAR CLAMP BAND
Replace.

OUTBOARD JOINT

OUTBOARD BOOT

GREASE

Pack cavity with grease.

9308MG29

Front axle exploded view—MDX

- Inboard joint boot clamps and push the boot back
- Inboard joint housing from the axle
- Rollers from the spider
- Snapring and the spider from the axle shaft
- Inboard joint boot

To install:

→ **Use new circlips and boot clamps for assembly.**

3. Install or connect the following:
   - Inboard joint boot and clamps to the axle shaft

   - Spider with a new snapring
   - Rollers to the spider
4. Fill the joint housing with grease and install it.
5. Fill the inboard joint boot with grease and install the boot clamps.
6. Install the axle halfshaft to the vehicle.

STEERING AND SUSPENSION

Air Bag

✳✳ CAUTION

Some vehicles are equipped with an air bag system. The system must be disarmed before performing service on, or around, system components, the steering column, instrument panel components, wiring and sensors. Failure to follow the safety precautions and the disarming procedure could result in accidental air bag deployment, possible injury and unnecessary system repairs.

PRECAUTIONS

Several precautions must be observed when handling the inflator module to avoid accidental deployment and possible personal injury.

- Never carry the inflator module by the wires or connector on the underside of the module.
- When carrying a live inflator module, hold securely with both hands, and ensure that the bag and trim cover are pointed away.
- Place the inflator module on a bench or other surface with the bag and trim cover facing up.
- With the inflator module on the bench, never place anything on or close to the module which may be thrown in the event of an accidental deployment.

Before servicing the vehicle, also make sure to refer to the precautions in the beginning of this section as well.

DISARMING

Disconnect and isolate the negative battery cable. Wait 3 minutes for the system capacitor to discharge before performing any service.

Power Rack and Pinion Steering Gear

REMOVAL & INSTALLATION

✳✳ WARNING

Do not permit the steering wheel to turn whenever the steering gear is disconnected from the steering column. Damage to the air bag wiring can result.

1. Before servicing the vehicle, refer to the precautions in the beginning of this section.
2. Center the steering wheel and lock it in position.
3. Attach a support fixture to the engine lifting eyes.
4. Remove or disconnect the following:
   - Negative battery cable
   - Air bag and steering wheel
   - Steering joint cover
   - Steering flexible joint
   - Power steering fluid lines
   - 10mm bolt on the engine side mount bracket
   - Front wheels
   - Outer tie rod ends
   - Sub-frame stiffener
   - Heated Oxygen (HO_2S) sensor connectors
   - 3 way catalytic converter from the mufflers
   - Flange bolts from the exhaust rubber mount
   - Power steering pressure switch connector
   - Propeller shaft protector
   - Splash shield
5. Support the front subframe with a jack and support the transmission with a second jack.
6. Loosen the 14mm subframe bolts.
7. Lower the subframe about 1 3/16 inches (30mm).
8. Remove or disconnect the following:
   - Two 12mm and two 14 stiffener plate bolts
9. Support the transfer case by raising the transmission jack and remove the two 12mm bolts.
   - Two 14mm bolts and the rear stiffener plats from the sub-frame

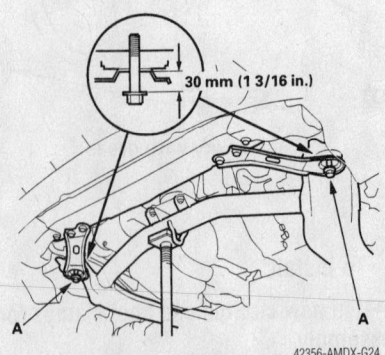

Loosen the 14mm subframe bolts and lower the subframe about 1 3/16 inches (30mm)

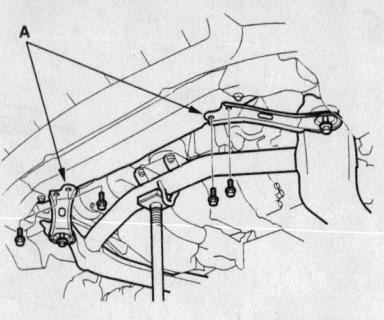

Remove the four 12mm stiffener plate bolts

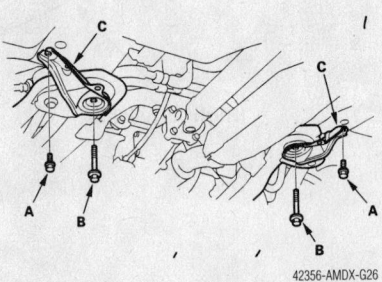

Remove the Two 12mm bolts (A), then the 14mm bolts (B) and the rear stiffener plates (C) from the sub-frame

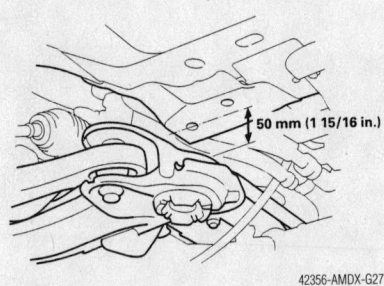

Lower the transmission jack until the front subframe has dropped about 1 15/16 inch (50mm)

10. Lower the transmission jack until the front subframe has dropped about 1 15/16 inch (50mm).
    - Power steering line brackets
    - Feed line
    - Return hose
    - Two 10mm bolts from the right side gearbox
    - Mounting bracket and cushion
    - Two 10mm bolts from the left side gearbox
11. Lower the transmission jack until the front subframe has dropped about 3 15/16 inch (100mm).

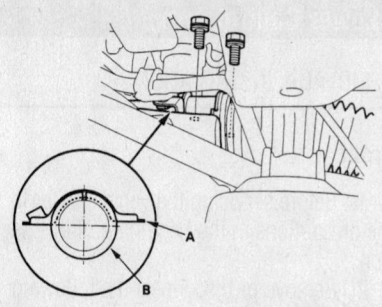

42356-AMDX-G28

Remove the two 10mm bolts from the right side gearbox and the mounting bracket and cushion

- Gearbox stiffener bracket

12. Slide the gearbox between the body and front sub-frame towards the left and from the vehicle.

To install:

13. Position the steering gear in the vehicle.

14. Install or connect the following:
- Left steering gear mounting bolts. Tighten the bolts to 43 ft. lbs. (58 Nm).
- Right steering gear mounting bracket. Tighten the bolts to 29 ft. lbs. (39 Nm).
- Return hose
- Feed line
- Power steering line mounting brackets and tighten the bolts to 7 ft. lbs. (10 Nm)

15. Raise the subframe into position. Tighten the 14mm bolts to 76 ft. lbs. (103 Nm) and the 12mm bolts to 86 ft. lbs. (117 Nm).

16. Install or connect the following:
- Front stiffener plates. Tighten the 14mm bolts to 76 ft. lbs. (103 Nm) and the 12mm bolts to 54 ft. lbs. (74 Nm).
- Splash shield
- Propeller shaft protector
- Power steering pressure switch
- 3 way catalytic converter and mufflers. Tighten the nuts to 25 ft. lbs. (33 Nm)
- Rubber exhaust mount and tighten the bolts to 28 ft. lbs. (38 Nm)
- HO_2S sensor connectors
- Sub-frame stiffener plate
- 10mm flange bolts on the engine side mount bracket to 33 ft. lbs. (44 Nm)
- Power steering hoses
- Outer tie rod ends
- Front wheels. Position the wheels straight-ahead.
- Steering flexible joint. Tighten the pinch bolts to 16 ft. lbs. (22 Nm).

- Steering joint cover
- Negative battery cable

17. Fill the power steering system.

18. Check the wheel alignment and adjust as necessary.

Strut

REMOVAL & INSTALLATION

Front

1. Before servicing the vehicle, refer to the precautions in the beginning of this section.

2. Remove or disconnect the following:
- Front wheel
- Wheel speed sensor wiring bracket
- Brake hose bracket
- Stabilizer bar link
- Strut pinch bolts
- Upper mount nuts
- Strut

To install:

3. Install or connect the following:
- Strut. Tighten the upper mount nuts to 43 ft. lbs. (59 Nm).

- Strut pinch bolts. Tighten the nuts to 116 ft. lbs. (157 Nm).
- Stabilizer bar link. Tighten the nut to 58 ft. lbs. (78 Nm).
- Brake hose bracket
- Wheel speed sensor wiring bracket
- Front wheel

4. Check the wheel alignment and adjust as necessary.

Shock Absorber

REMOVAL & INSTALLATION

Rear

1. Before servicing the vehicle, refer to the precautions in the beginning of this section.

2. Support the vehicle under the lower control arm.

3. Remove or disconnect the following:
- Rear wheel
- Upper shock absorber flange bolt
- Lower shock absorber nut
- Shock absorber

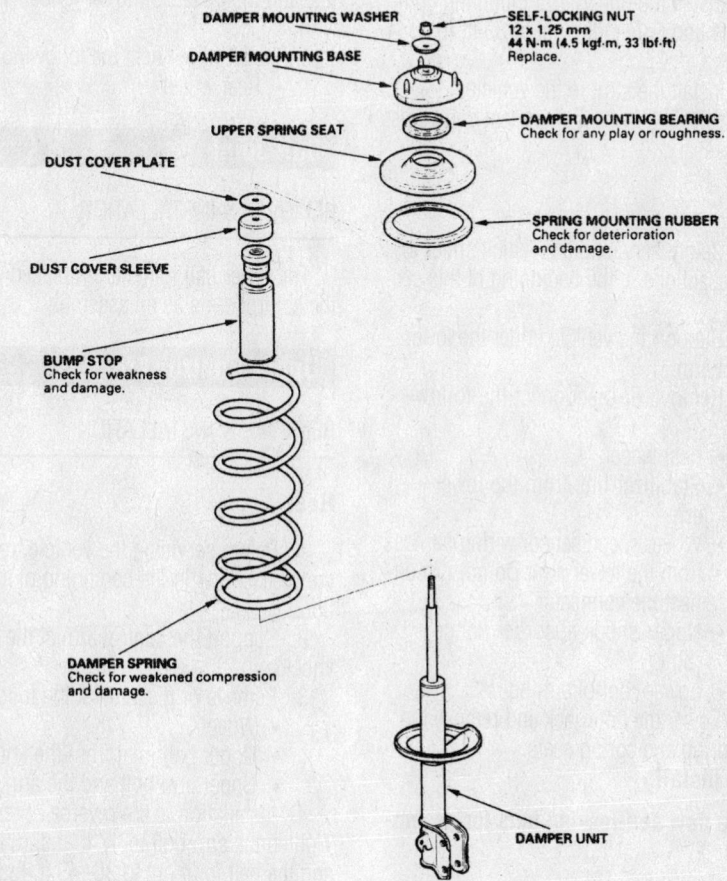

DAMPER MOUNTING WASHER

DAMPER MOUNTING BASE

UPPER SPRING SEAT

DUST COVER PLATE

DUST COVER SLEEVE

BUMP STOP
Check for weakness and damage.

DAMPER SPRING
Check for weakened compression and damage.

SELF-LOCKING NUT
12 x 1.25 mm
44 N·m (4.5 kgf·m, 33 lbf·ft)
Replace.

DAMPER MOUNTING BEARING
Check for any play or roughness.

SPRING MOUNTING RUBBER
Check for deterioration and damage.

DAMPER UNIT

93552G11

Exploded view of the front strut assembly

To install:

4. Install or connect the following:
- Shock absorber. Tighten the fasteners to 47 ft. lbs. (64 Nm).
- Rear wheel

Coil Spring

REMOVAL & INSTALLATION

Front

1. Before servicing the vehicle, refer to the precautions in the beginning of this section.
2. Remove the strut from the vehicle and install in a strut spring compressor. Compress the spring until the end of the spring comes away from the spring seat.
3. Remove the upper strut mount, spring seat and related components.
4. Remove the coil spring from the strut spring compressor.

To install:

➡**Use a new self-locking nut.**

5. Compress the spring and position the strut so that the end of the spring aligns with the notch in the spring seat.
6. Install the upper strut mounting components and tighten the nut to 33 ft. lbs. (44 Nm).
7. Install the strut to the vehicle.
8. Check the wheel alignment and adjust as necessary.

Rear

1. Before servicing the vehicle, refer to the precautions in the beginning of this section.
2. Support the vehicle under the lower control arm.
3. Remove or disconnect the following:
- Rear wheel
- Stabilizer link from the lower arm
- Wheel speed sensor wiring harness from the lower arm. Do not disconnect the connector.
- Upper shock absorber flange bolt
- Lower control arm bolts
4. Lower the floor jack and remove the coil spring and spring seats.

To install:

➡**Use new self-locking nuts for assembly.**

5. Place the coil spring and spring seats on the lower control arm and raise into

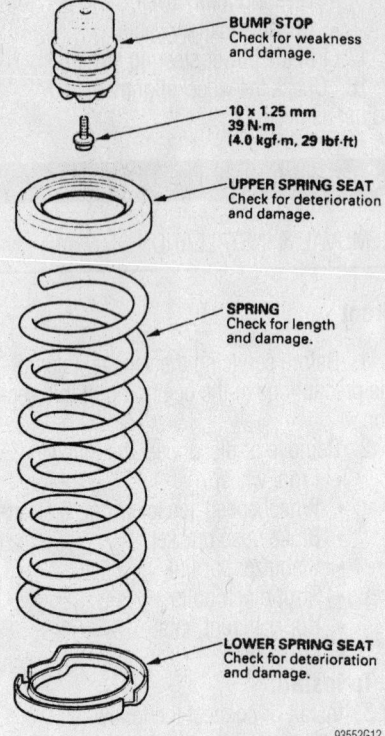

BUMP STOP
Check for weakness and damage.

10 x 1.25 mm
39 N·m
(4.0 kgf·m, 29 lbf·ft)

UPPER SPRING SEAT
Check for deterioration and damage.

SPRING
Check for length and damage.

LOWER SPRING SEAT
Check for deterioration and damage.

93552G12

Exploded view of the rear spring assembly

position. Tighten the inboard bolt to 61 ft. lbs. and the outer bolt to 54 ft. lbs. (74 Nm).
6. Install or connect the following:
- Rear wheel

Ball Joint

REMOVAL & INSTALLATION

The lower ball joints are replaced with the control arms as an assembly.

Upper Control Arm

REMOVAL & INSTALLATION

Rear

1. Before servicing the vehicle, refer to the precautions in the beginning of this section.
2. Support the control arm at the knuckle.
3. Remove or disconnect the following:
- Wheel
- Upper ball joint from the knuckle
- Upper arm bolt and the arm
4. Installation is the reverse of removal. Tighten the arm bolt to 47 ft. lbs. (64 Nm) and the ball joint nut to 36–43 ft. lbs. (49–59 Nm).

Lower Control Arm

REMOVAL & INSTALLATION

Front

1. Before servicing the vehicle, refer to the precautions in the beginning of this section.
2. Remove or disconnect the following:
- Front wheel
- Lower ball joint
- Front inner flange bolt
- Rear inner flange bolt
- Lower control arm

To install:

➡**Use a new split pin for assembly.**

3. Install or connect the following:
- Lower control arm. Tighten the inner flange bolts to 69 ft. lbs. (93 Nm).
- Lower ball joint. Tighten the nut to 43–51 ft. lbs. (59–69 Nm).
- Front wheel
4. Check the wheel alignment and adjust as necessary.

Rear

LOWER ARM (A)

1. Before servicing the vehicle, refer to the precautions in the beginning of this section.
2. Remove or disconnect the following:
- Lower arm mounting bolt and nut
- Lower arm
3. Installation is the reverse of removal. Tighten the bolt to 105 ft. lbs. (142 Nm) and the nut to 47 ft. lbs. (64 Nm).

LOWER ARM (B)

1. Before servicing the vehicle, refer to the precautions in the beginning of this section.
2. Support the control arm with a jack.
3. Remove or disconnect the following:
- Wheel

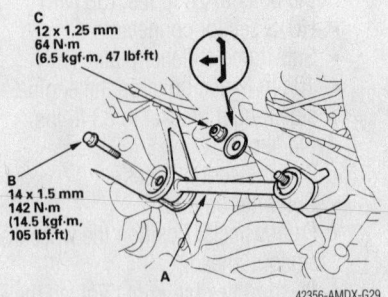

C
12 x 1.25 mm
64 N·m
(6.5 kgf·m, 47 lbf·ft)

B
14 x 1.5 mm
142 N·m
(14.5 kgf·m, 105 lbf·ft)

A

42356-AMDX-G29

Rear lower arm (A) mounting

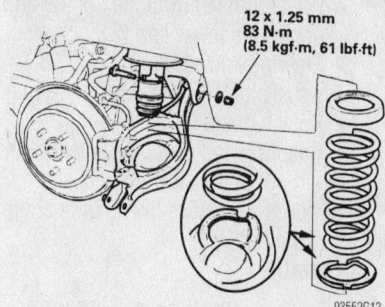

Rear lower arm (B) mounting

- Stabilizer link from the lower arm
- Wheel speed sensor wiring harness from the lower arm. Do not disconnect the connector.
- Flange bolts that attaches the lower arm to the knuckle
4. Spring assembly
  - Inner nuts and bolts and the arm
5. Install or connect the following:
  - Arm, inner bolt and loosely install the nut
  - Spring assembly
6. Raise the arm into position and install the flange bolt.
7. Raise the rear suspension with a floor jack to load the vehicle weight.
  - Tighten the flange bolt to 54 ft. lbs. (74 Nm) and the inner nut and bolt to 61 ft. lbs. (83 Nm).
  - Wheel speed sensor harness
  - Wheel
8. Check the vehicle alignment.

CONTROL ARM BUSHING REPLACEMENT

The control arm bushings are serviced with the control arms as an assembly.

Wheel Bearings

ADJUSTMENT

The wheel bearings are sealed units and are not adjustable.

REMOVAL & INSTALLATION

Front

1. Before servicing the vehicle, refer to the precautions in the beginning of this section.
2. Remove or disconnect the following:
  - Front wheel
  - Brake hose mounting bolt
  - Brake caliper
  - Wheel speed sensor
  - Spindle nut

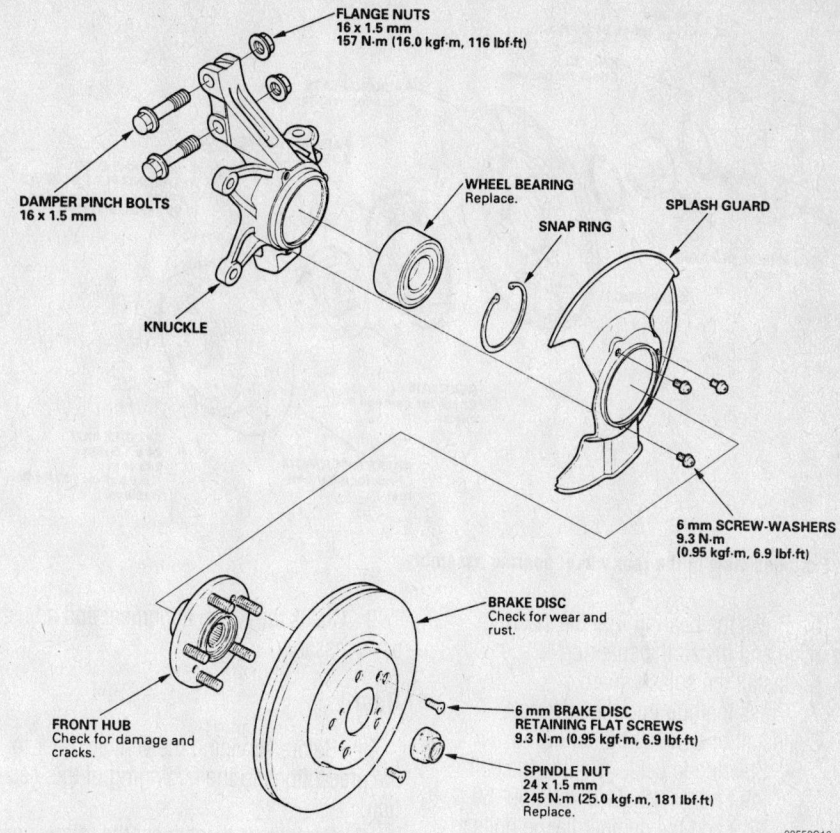

Front wheel bearing assembly, spindle nut torque shown is for 2001–02, on 2003–04 torque is 210 ft. lbs. (285 Nm)

- Brake rotor
- Outer tie rod end
- Lower ball joint
- Steering knuckle
3. Press the hub out of the wheel bearing.

- Splash guard
- Snapring and press the wheel bearing out of the steering knuckle
4. If necessary, press the inner bearing race off of the hub.

To install:

➡ **Use a new ball joint nut, split pin, snapring and spindle nut for assembly.**

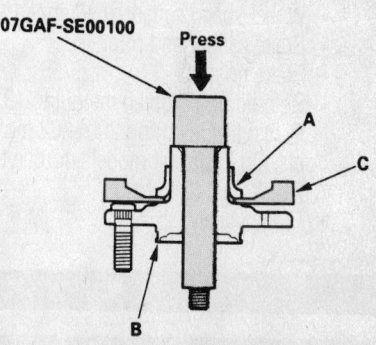

Press the wheel bearing inner race from the hub

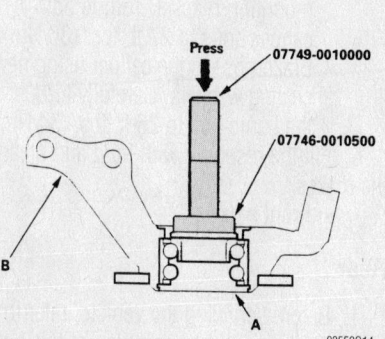

Press the wheel bearing out of the knuckle

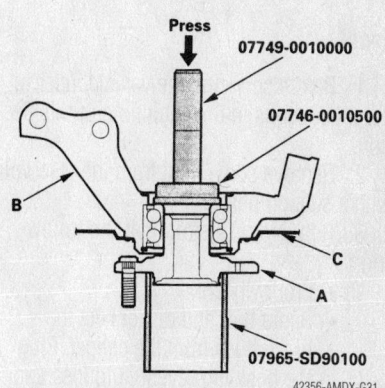

Press the wheel bearing into the knuckle

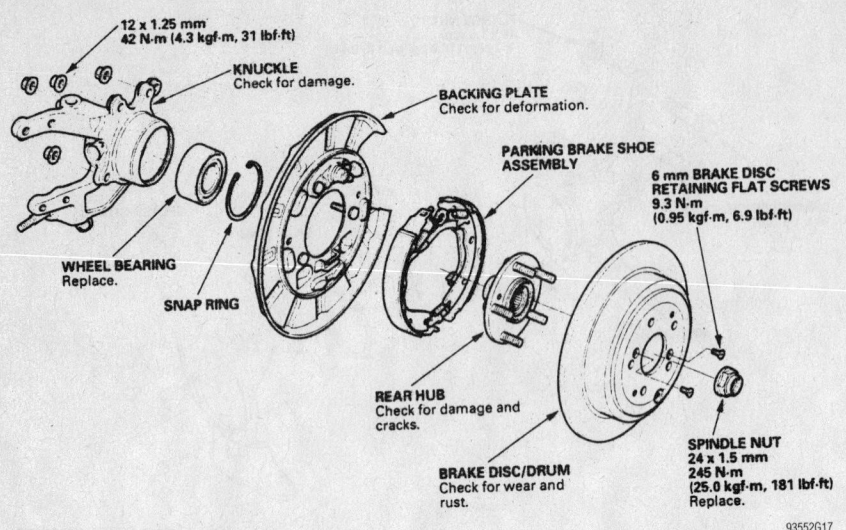

12 x 1.25 mm
42 N·m (4.3 kgf·m, 31 lbf·ft)

KNUCKLE
Check for damage.

BACKING PLATE
Check for deformation.

**PARKING BRAKE SHOE
ASSEMBLY**

6 mm BRAKE DISC
RETAINING FLAT SCREWS
9.3 N·m
(0.95 kgf·m, 6.9 lbf·ft)

WHEEL BEARING
Replace.

SNAP RING

REAR HUB
Check for damage and
cracks.

BRAKE DISC/DRUM
Check for wear and
rust.

SPINDLE NUT
24 x 1.5 mm
245 N·m
(25.0 kgf·m, 181 lbf·ft)
Replace.

93552G17

Exploded view of the rear wheel bearing assembly

5. Press the bearing into the steering knuckle and install the snapring.
6. Install the splash guard.
7. Press the hub into the bearing.
8. Install or connect the following:
   - Steering knuckle. Tighten the ball joint nut to 43–51 ft. lbs. (59–69 Nm) and the damper flange bolts to 116 ft. lbs. (157 Nm).
   - Outer tie rod end. Tighten the nut to 40 ft. lbs. (54 Nm).
   - Wheel speed sensor, if equipped
   - Brake caliper and rotor
   - Brake hose
   - Spindle nut. Tighten the nut to 181 ft. lbs. (245 Nm) on 2001–02 models. On 2003–04 models, torque to 210 ft. lbs. (285 Nm).
   - Front wheel

9. Check the wheel alignment and adjust as necessary.

Rear

1. Before servicing the vehicle, refer to the precautions in the beginning of this section.
2. Remove or disconnect the following:
   - Rear wheel
   - Brake hose bracket mounting bolts from the trailing arm and the knuckle
   - Brake caliper
   - Wheel speed sensor
   - Spindle nut
   - Brake rotor
   - Upper ball joint
   - Lower arm (A)

   - Lower arm (B) from the trailing arm
3. Support the lower arm (B)
   - Steering knuckle
4. Press the hub out of the wheel bearing.
   - Splash guard
   - Snapring and press the wheel bearing out of the steering knuckle
5. If necessary, press the inner bearing race off of the hub.

To install:

➡**Use a new ball joint nut, split pin, snapring and spindle nut for assembly.**

6. Press the bearing into the steering knuckle and install the snapring.
7. Install the splash guard.
8. Press the hub into the bearing.
9. Install or connect the following:
   - Steering knuckle. Tighten the flange bolt to 54 ft. lbs. (74 Nm) and the lower shock nut to 47 ft. lbs. (64 Nm)
   - Lower arm (B) to the trailing arm and tighten the bolts to 47 ft. lbs. (64 Nm)
   - Lower arm (A)
   - Upper ball joint and tighten the nut to 40 ft. lbs. (54 Nm)
   - Brake rotor and tighten the screws to 7 ft. lbs. (9 Nm)
   - Spindle nut and tighten to 181 ft. lbs. (245 Nm)
   - Wheel speed sensor
   - Brake caliper and tighten the bolts to 41 ft. lbs. (55 Nm)
   - Brake hose bracket mounting bolts to the knuckle and trailing arm
   - Rear wheel
10. Check the wheel alignment and adjust as necessary.

BRAKES

Brake Caliper

REMOVAL & INSTALLATION

Front

1. Before servicing the vehicle, refer to the precautions in the beginning of this section.
2. Remove some fluid from the reservoir with a suction pump.
3. Remove or disconnect the following:
   - Front wheels
   - Banjo bolt and disconnect the brake hose from the caliper. Plug the hose to prevent fluid loss and contamination.

   - Mounting bolts and the caliper from its mounting bracket

To Install:
4. Install or connect the following:
   - Caliper over the pads and onto its mounting bracket. Torque both caliper bolts to 27 ft. lbs. (36 Nm).
   - Brake hose to the caliper using new sealing washers. Carefully torque the banjo bolt to 25 ft. lbs. (34 Nm).
5. Fill the reservoir with fluid and bleed the brakes.
   - Front wheels

Rear

1. Before servicing the vehicle, refer to the precautions in the beginning of this section.

2. Remove some fluid from the reservoir with a suction pump.
3. Remove or disconnect the following:
   - Rear wheels
   - Banjo bolt and disconnect the brake hose from the caliper. Plug the hose to prevent fluid loss and contamination.
   - 2 caliper mounting bolts and the caliper from its mounting bracket

To Install:
4. Install or connect the following:
   - Caliper over the pads and onto its mounting bracket. Tighten the caliper bolts to 27 ft. lbs. (37 Nm).
   - Brake hose with new sealing washers. Tighten the banjo bolt to 25 ft. lbs. (34 Nm).

Front Brake Caliper Overhaul

⚠CAUTION

Frequent inhalation of brake pad dust, regardless of material composition, could be hazardous to your health.
* Avoid breathing dust particles.
* Never use an air hose or brush to clean brake assemblies. Use an OSHA-approved vacuum cleaner.

Remove, disassemble, inspect, reassemble, and install the caliper, and note these items:

* Do not spill brake fluid on the vehicle; it may damage the paint; if brake fluid gets on the paint, wash it off immediately with water.
* To prevent dripping brake fluid, cover disconnected hose joints with rags or shop towels.
* Clean all parts in brake fluid and air dry; blow out all passages with compressed air.
* Before reassembling, check that all parts are free of dirt and other foreign particles.
* Replace parts with new ones as specified in the illustration.
* Make sure no dirt or other foreign matter gets in the brake fluid.
* Make sure no grease or oil gets on the brake discs or pads.
* When reusing pads, always reinstall them in their original positions to prevent loss of braking efficiency.
* Do not reuse drained brake fluid. Use only clean Genuine Honda DOT 3 Brake Fluid. Non-Honda brake fluid can cause corrosion and shorten the life of the system.
* Coat the piston, piston seal groove, and caliper bore with clean brake fluid.
* Replace all rubber parts with new ones.
* After installing the caliper, check the brake hose and line for leaks, interference, and twisting.

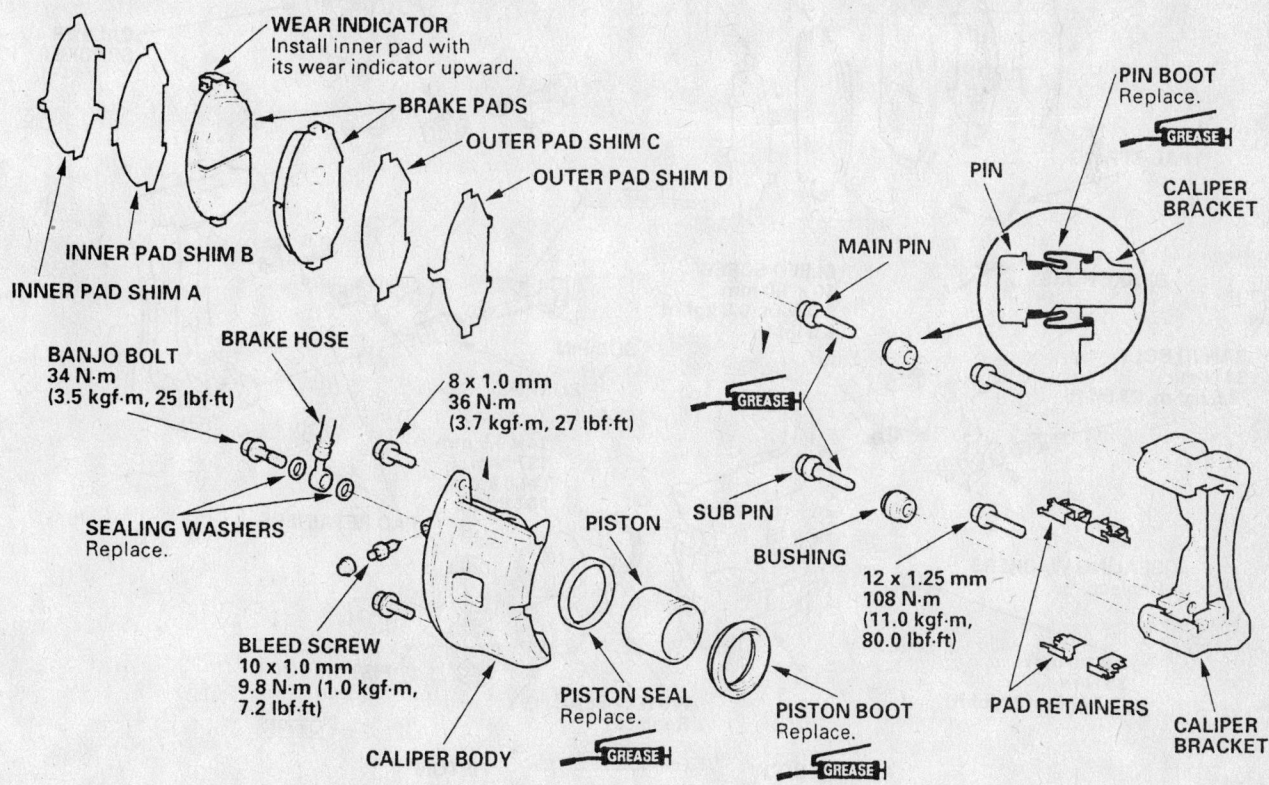

Exploded view of the front caliper components—2001–02 models

⚠CAUTION

Frequent inhalation of brake pad dust, regardless of material composition, could be hazardous to your health.
• Avoid breathing dust particles.
• Never use an air hose or brush to clean brake assemblies. Use an OSHA-approved vacuum cleaner.

Remove, disassemble, inspect, reassemble, and install the caliper, and note these items:

• Do not spill brake fluid on the vehicle; it may damage the paint; if brake fluid gets on the paint, wash it off immediately with water.
• To prevent dripping brake fluid, cover disconnected hose joints with rags or shop towels.
• Clean all parts in brake fluid and air dry; blow out all passages with compressed air.
• Before reassembling, check that all parts are free of dirt and other foreign particles.
• Replace parts with new ones as specified in the illustration.
• Make sure no dirt or other foreign matter gets in the brake fluid.
• Make sure no grease or oil gets on the brake discs or pads.
• When reusing pads, always reinstall them in their original positions to prevent loss of braking efficiency.
• Do not reuse drained brake fluid. Use only clean Honda DOT 3 Brake Fluid. Non-Honda brake fluid can cause corrosion and shorten the life of the system.
• Coat the piston, piston seal groove, and caliper bore with clean brake fluid.
• Replace all rubber parts with new ones.
• After installing the caliper, check the brake hose and line for leaks, interference, and twisting.

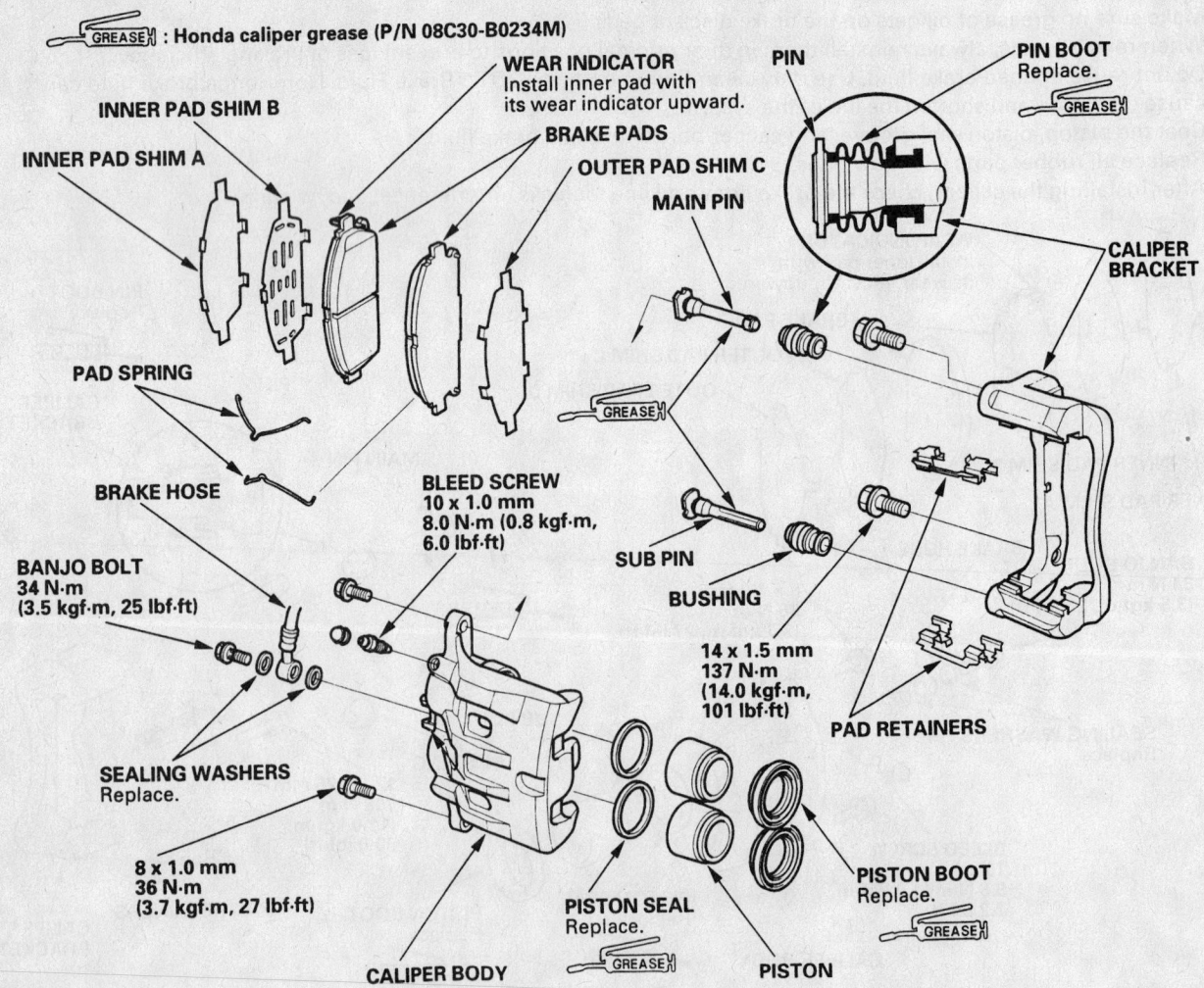

Exploded view of the front caliper components—2003–04 models

32562G30

Rear Brake Caliper Overhaul

⚠CAUTION

Frequent inhalation of brake pad dust, regardless of material composition, could be hazardous to your health.
- Avoid breathing dust particles.
- Never use an air hose or brush to clean brake assemblies. Use an OSHA-approved vacuum cleaner.

Remove, disassemble, inspect, reassemble, and install the caliper, and note these items:

- Do not spill brake fluid on the vehicle; It may damage the paint; If brake fluid gets on the paint, wash it off immediately with water.
- To prevent dripping brake fluid, cover disconnected hose joints with rags or shop towels.
- Clean all parts in brake fluid and air dry; blow out all passages with compressed air.
- Before reassembling, check that all parts are free of dirt and other foreign particles.
- Replace parts with new ones as specified in the illustration.
- Make sure no dirt or other foreign matter gets into the brake fluid.
- Make sure no grease or oil gets on the brake discs or pads.
- When reusing pads, always reinstall them in their original positions to prevent loss of braking efficiency.
- Do not reuse drained brake fluid. Use only clean Genuine Honda DOT 3 Brake Fluid. Non-Honda brake fluid can cause corrosion and shorten the life of the system.
- Coat the piston, piston seal groove, and caliper bore with clean brake fluid.
- Replace all rubber parts with new ones.
- After installing the caliper, check the brake hose and line for leaks, interference, and twisting.

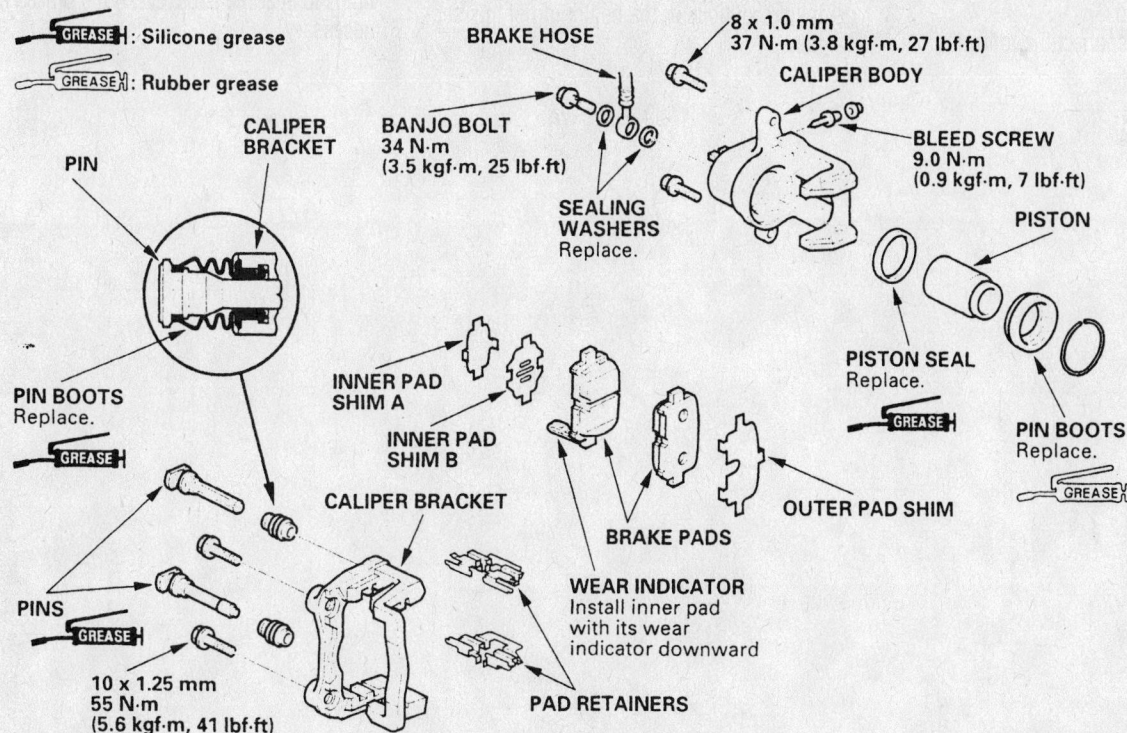

93352GZB

Exploded view of the rear caliper components—MDX

5. Fill the reservoir with fluid and bleed the brake system. Adjust the parking brake if necessary.

- Rear wheels

Disc Brake Pads

REMOVAL & INSTALLATION

Front

1. Before servicing the vehicle, refer to the precautions in the beginning of this section.
2. Remove or disconnect the following:
   - Front wheels
3. Remove a small amount of brake fluid from the reservoir using a suction pump.
   - Brake hose clamp from the knuckle by unfastening the retaining bolts
   - Lower caliper retaining bolt and pivot the caliper upward, off of the pads
   - Pad shim and pad retainers
   - Disc brake pads from the caliper

To install:

4. Clean the caliper thoroughly; remove any rust from the lip of the disc or rotor. Check the brake rotor for grooves or cracks. If any heavy scoring is present, the rotor must be replaced.
5. Install or connect the following:

- Pad retainers. Apply molybdenum brake grease to both surfaces of the shims and the back of the disc brake pads.
- Pads and shims. The pad with the wear indicator goes in the inboard position.

6. Push in the caliper piston so the caliper will fit over the pads. This is most easily accomplished with a pad spreader or large C-clamp.

- Caliper down into position and tighten the mounting bolt to 27 ft. lbs. (37 Nm)
- Brake hose to the knuckle, if removed
- Wheels

7. Add brake fluid to the master cylinder reservoir and install the cap.
8. Depress the brake pedal several times and make sure that the movement feels normal. The first brake pedal application may result in a very long pedal action due to the pistons being retracted. Always make several brake applications before starting the vehicle. Bleed the system if necessary.

Rear

1. Before servicing the vehicle, refer to the precautions in the beginning of this section.

2. Remove a small amount of brake fluid from the reservoir using a suction pump.
3. Remove or disconnect the following:
   - Rear wheels
   - 2 caliper mounting bolts and the caliper from the bracket
   - pads, shims, and pad retainers

To install:

4. Clean the caliper thoroughly; remove any dirt or dust. Check the brake rotor for grooves or cracks and machine or replace, as necessary.
5. Install or connect the following:
   - Pad retainers. Apply molybdenum brake grease to both surfaces of the shims and the back of the disc brake pads.
   - Pads and shims. The wear retainer on the inboard pad faces down.

6. Use a suitable tool to push caliper piston into its bore and enable the caliper to fit over the pads. Lubricate the piston boot with silicon grease. Avoid twisting the boot.
   - Brake caliper and tighten the mounting bolts to 27 ft. lbs. (37 Nm)
   - Rear wheels

7. Add brake fluid to the master cylinder reservoir. Depress the brake pedal several times to seat the pads. Bleed the brakes if necessary.

HONDA

Accord • Civic • Prelude • S2000

3

SPECIFICATION CHARTS

ENGINE AND VEHICLE IDENTIFICATION

Engine						Model Year	
Code	Liters	Cu. In. (cc)	Cyl.	Fuel Sys.	Eng. Mfg.	Code	Year
B16A2	1.6	97 (1595)	4	PGM-FI	Honda	Y	2000
D16Y5	1.6	97 (1590)	4	PGM-FI	Honda	1	2001
D16Y7	1.6	97 (1590)	4	PGM-FI	Honda	2	2002
D16Y8	1.6	97 (1590)	4	PGM-FI	Honda	3	2003
D17A1	1.7	101.7 (1668)	4	PGM-FI	Honda	4	2004
D17A2	1.7	101.7 (1668)	4	PGM-FI	Honda		
F20C1	2.0	121.9 (1997)	4	PGM-FI	Honda		
F23A1	2.3	137 (2254)	4	PGM-FI	Honda		
F23A4	2.3	137 (2254)	4	PGM-FI	Honda		
F23A5	2.3	137 (2254)	4	PGM-FI	Honda		
H22A4	2.2	132 (2157)	4	PGM-FI	Honda		
K24A4	2.4	144 (2354)	4	PGM-FI	Honda		
J30A1	3.0	183 (2997)	6	PGM-FI	Honda		
J30A4	3.0	183 (2997)	6	PGM-FI	Honda		

PGM-FI: Programmed Fuel Injection

43256-ACCO-C01

GENERAL ENGINE SPECIFICATIONS

Year	Model	Engine Displacement Liters (cc)	Engine ID/VIN	Fuel System Type	Net Horsepower @ rpm	Net Torque @ rpm (ft. lbs.)	Bore X Stroke (in.)	Compression Ratio	Oil Pressure @ rpm
2000	Civic	1.6 (1590)	D16Y5	PGM-FI	115@6300	104@5400	2.95X3.54	9.6:1	50@3000
	Civic	1.6 (1590)	D16Y7	PGM-FI	106@6200	103@4600	2.95X3.54	9.4:1	50@3000
	Civic	1.6 (1590)	D16Y8	PGM-FI	127@6600	107@5500	2.95X3.54	9.4:1	50@3000
	Civic	1.6 (1595)	B16A2	PGM-FI	160@7600	111@7000	3.19X3.05	10.2:1	50@3000
	Prelude	2.2 (2157)	H22A4	PGM-FI	①	158@5500	3.43X3.57	10.0:1	50@3000
	Prelude SH	2.2 (2157)	H22A4	PGM-FI	①	158@5500	3.43X3.57	10.0:1	50@3000
	Accord Coupe (EX, LX)	2.3 (2254)	F23A1	PGM-FI	150@5700	152@4900	3.39X3.82	9.3:1	50@3000
	Accord Coupe (EX, LX)	2.3 (2254)	F23A4	PGM-FI	150@5700	152@4900	3.39X3.82	9.3:1	50@3000
	Accord Sedan (DX)	2.3 (2254)	F23A5	PGM-FI	150@5700	152@4900	3.39X3.82	9.3:1	50@3000
	Accord Sedan (EX, LX)	2.3 (2254)	F23A1	PGM-FI	150@5700	152@4900	3.39X3.82	9.3:1	50@3000
	Accord Sedan (EX, LX)	2.3 (2254)	F23A4	PGM-FI	150@5700	152@4900	3.39X3.82	9.3:1	50@3000
	Accord Coupe (EX, LX)	3.0 (2997)	J30A1	PGM-FI	200@5500	195@4700	3.39X3.39	9.4:1	50@3000
	Accord Sedan (EX, LX)	3.0 (2997)	J30A1	PGM-FI	200@5500	195@4700	3.39X3.39	9.4:1	50@3000
	S2000	2.0 (1997)	F20C1	PGM-FI	240@8300	153@7500	3.43X3.31	11:01	85@3000
2001	Civic DX	1.7 (1668)	D17A1	PGM-FI	115@6100	110X4500	2.95X3.72	9.5:1	50@3000
	Civic GX	1.7 (1668)	D17A1	PGM-FI	115@6100	110X4500	2.95X3.72	9.5:1	50@3000
	Civic LX	1.7 (1668)	D17A1	PGM-FI	115@6100	110X4500	2.95X3.72	9.5:1	50@3000
	Civic EX	1.7 (1668)	D17A2	PGM-FI	127@6100	117X4500	2.95X3.72	9.9:1	50@3000
	Prelude	2.2 (2157)	H22A4	PGM-FI	①	158@5500	3.43X3.57	10.0:1	50@3000
	Prelude SH	2.2 (2157)	H22A4	PGM-FI	①	158@5500	3.43X3.57	10.0:1	50@3000
	Accord Coupe (EX, LX)	2.3 (2254)	F23A1	PGM-FI	150@5700	152@4900	3.39X3.82	9.3:1	50@3000
	Accord Coupe (EX, LX)	2.3 (2254)	F23A4	PGM-FI	150@5700	152@4900	3.39X3.82	9.3:1	50@3000
	Accord Sedan (DX)	2.3 (2254)	F23A5	PGM-FI	150@5700	152@4900	3.39X3.82	9.3:1	50@3000
	Accord Sedan (EX, LX)	2.3 (2254)	F23A1	PGM-FI	150@5700	152@4900	3.39X3.82	9.3:1	50@3000
	Accord Sedan (EX, LX)	2.3 (2254)	F23A4	PGM-FI	150@5700	152@4900	3.39X3.82	9.3:1	50@3000
	Accord Coupe (EX, LX)	3.0 (2997)	J30A1	PGM-FI	200@5500	195@4700	3.39X3.39	9.4:1	50@3000
	Accord Sedan (EX, LX)	3.0 (2997)	J30A1	PGM-FI	200@5500	195@4700	3.39X3.39	9.4:1	50@3000
	S2000	2.0 (1997)	F20C1	PGM-FI	240@8300	153@7500	3.43X3.31	11:01	85@3000
2002	Civic DX	1.7 (1668)	D17A1	PGM-FI	115@6100	110X4500	2.95X3.72	9.5:1	50@3000
	Civic GX	1.7 (1668)	D17A1	PGM-FI	115@6100	110X4500	2.95X3.72	9.5:1	50@3000
	Civic LX	1.7 (1668)	D17A1	PGM-FI	115@6100	110X4500	2.95X3.72	9.5:1	50@3000
	Civic EX	1.7 (1668)	D17A2	PGM-FI	127@6100	117X4500	2.95X3.72	9.9:1	50@3000
	Accord Coupe (EX, LX)	2.3 (2254)	F23A1	PGM-FI	150@5700	152@4900	3.39X3.82	9.3:1	50@3000
	Accord Coupe (EX, LX)	2.3 (2254)	F23A4	PGM-FI	150@5700	152@4900	3.39X3.82	9.3:1	50@3000
	Accord Sedan (DX)	2.3 (2254)	F23A5	PGM-FI	150@5700	152@4900	3.39X3.82	9.3:1	50@3000
	Accord Sedan (EX, LX)	2.3 (2254)	F23A1	PGM-FI	150@5700	152@4900	3.39X3.82	9.3:1	50@3000
	Accord Sedan (EX, LX)	2.3 (2254)	F23A4	PGM-FI	150@5700	152@4900	3.39X3.82	9.3:1	50@3000
	Accord Coupe (EX, LX)	3.0 (2997)	J30A1	PGM-FI	200@5500	195@4700	3.39X3.39	9.4:1	50@3000
	Accord Sedan (EX, LX)	3.0 (2997)	J30A1	PGM-FI	200@5500	195@4700	3.39X3.39	9.4:1	50@3000
	S2000	2.0 (1997)	F20C1	PGM-FI	240@8300	153@7500	3.43X3.31	11:01	85@3000

43256-ACCO-C02

GENERAL ENGINE SPECIFICATIONS

Year	Model	Engine Displacement Liters (cc)	Engine ID/VIN	Fuel System Type	Net Horsepower @ rpm	Net Torque @ rpm (ft. lbs.)	Bore X Stroke (in.)	Compression Ratio	Oil Pressure @ rpm
2003-04	Civic DX	1.7 (1668)	D17A1	PGM-FI	115@6100	110X4500	2.95X3.72	9.5:1	50@3000
	Civic GX	1.7 (1668)	D17A1	PGM-FI	115@6100	110X4500	2.95X3.72	9.5:1	50@3000
	Civic LX	1.7 (1668)	D17A1	PGM-FI	115@6100	110X4500	2.95X3.72	9.5:1	50@3000
	Civic EX	1.7 (1668)	D17A2	PGM-FI	127@6100	117X4500	2.95X3.72	9.9:1	50@3000
	Accord Coupe (EX, LX)	2.4 (2354)	K24A4	PGM-FI	160@5500	161@4500	3.42X3.89	9.7:1	50@3000
	Accord Sedan (DX)	2.4 (2354)	K24A4	PGM-FI	160@5500	161@4500	3.42X3.89	9.7:1	50@3000
	Accord Sedan (EX, LX)	2.4 (2354)	K24A4	PGM-FI	160@5500	161@4500	3.42X3.89	9.7:1	50@3000
	Accord Coupe (EX, LX)	3.0 (2997)	J30A4	PGM-FI	240@6250	212@5000	3.38X3.38	10.0:1	50@3000
	Accord Sedan (EX, LX)	3.0 (2997)	J30A4	PGM-FI	240@6250	212@5000	3.39X3.39	10.0:1	50@3000
	S2000	2.0 (1997)	F20C1	PGM-FI	240@8300	153@7500	3.43X3.31	11:01	85@3000

PGM-FI: Programmed Fuel Injection

① Manual transaxle: 195@7000
    Automatic transaxle: 190@6600

43256-ACCO-C03

ENGINE TUNE-UP SPECIFICATIONS

Year	Engine Displacement Liters (cc)	Engine ID/VIN	Spark Plugs Gap (in.)	Ignition Timing (deg.) MT	AT	Fuel Pump (psi)	Idle Speed (rpm) MT	AT	Valve Clearance In.	Ex.
2000	1.6 (1595)	B16A2	0.047-0.051	16B	—	31-38	650-750	—	0.006-0.007	0.007-0.008
	1.6 (1590)	D16Y5	0.039-0.043	12B	12B	28-36	620-720	650-750	0.007-0.009	0.009-0.011
	1.6 (1590)	D16Y7	0.039-0.043	12B	12B	28-36	620-720	650-750	0.007-0.009	0.009-0.011
	1.6 (1590)	D16Y8	0.039-0.043	12B	12B	28-36	620-720	650-750	0.007-0.009	0.009-0.011
	2.0 (1997)	F20C1	0.039-0.043	5B	5B	47-54	750-850	—	0.008-0.010	0.010-0.011
	2.3 (2254)	F23A1	0.039-0.043	12B	12B	40-47	650-750	650-750	0.009-0.011	0.011-0.013
	2.3 (2254)	F23A4	0.039-0.043	12B	12B	40-47	650-750	650-750	0.009-0.011	0.011-0.013
	2.3 (2254)	F23A5	0.039-0.043	12B	12B	40-47	650-750	650-750	0.009-0.011	0.011-0.013
	2.2 (2157)	H22A4	0.039-0.043	15B	15B	47-54	650-750	650-750	0.006-0.007	0.007-0.008
	3.0 (2997)	J30A1	0.039-0.043	—	10B	41-48	—	630-730	0.008-0.009	0.011-0.013
2001	1.7 (1668)	D17A1	0.039-0.043	8B	8B	40-47	650-750	650-750	0.007-0.009	0.009-0.011
	1.7 (1668)	D17A2	0.039-0.043	8B	8B	40-47	650-750	650-750	0.007-0.009	0.009-0.011
	2.0 (1997)	F20C1	0.039-0.043	5B	5B	47-54	750-850	—	0.008-0.010	0.010-0.011
	2.3 (2254)	F23A1	0.039-0.043	12B	12B	40-47	650-750	650-750	0.009-0.011	0.011-0.013
	2.3 (2254)	F23A4	0.039-0.043	12B	12B	40-47	650-750	650-750	0.009-0.011	0.011-0.013
	2.3 (2254)	F23A5	0.039-0.043	12B	12B	40-47	650-750	650-750	0.009-0.011	0.011-0.013
	2.2 (2157)	H22A4	0.039-0.043	15B	15B	47-54	650-750	650-750	0.006-0.007	0.007-0.008
	3.0 (2997)	J30A1	0.039-0.043	—	10B	41-48	—	630-730	0.008-0.009	0.011-0.013
2002	1.7 (1668)	D17A1	0.039-0.043	8B	8B	40-47	650-750	650-750	0.007-0.009	0.009-0.011
	1.7 (1668)	D17A2	0.039-0.043	8B	8B	40-47	650-750	650-750	0.007-0.009	0.009-0.011
	2.0 (1997)	F20C1	0.039-0.043	5B	5B	47-54	750-850	—	0.008-0.010	0.010-0.011
	2.3 (2254)	F23A1	0.039-0.043	12B	12B	40-47	650-750	650-750	0.009-0.011	0.011-0.013
	2.3 (2254)	F23A4	0.039-0.043	12B	12B	40-47	650-750	650-750	0.009-0.011	0.011-0.013
	2.3 (2254)	F23A5	0.039-0.043	12B	12B	40-47	650-750	650-750	0.009-0.011	0.011-0.013
	3.0 (2997)	J30A1	0.039-0.043	—	10B	41-48	—	630-730	0.008-0.009	0.011-0.013

43256-ACCO-C04

ENGINE TUNE-UP SPECIFICATIONS

Year	Engine Displacement Liters (cc)	Engine ID/VIN	Spark Plugs Gap (in.)	Ignition Timing (deg.)		Fuel Pump (psi)	Idle Speed (rpm)		Valve Clearance	
				MT	AT		MT	AT	In.	Ex.
2003-04	1.7 (1668)	D17A1	0.039-0.043	8B	8B	40-47	650-750	650-750	0.007-0.009	0.009-0.011
	1.7 (1668)	D17A2	0.039-0.043	8B	8B	40-47	650-750	650-750	0.007-0.009	0.009-0.011
	2.0 (1997)	F20C1	0.039-0.043	5B	5B	47-54	750-850	—	0.008-0.010	0.010-0.011
	2.4 (2354)	K24A4	0.039-0.043	8B	8B	①	650-750	750-850	0.008-0.010	0.011-0.013
	3.0 (2997)	J30A4	0.039-0.043	—	10B	41-48	—	630-730	0.008-0.009	0.011-0.013

NOTE: The Vehicle Emission Control Information label often reflects specification changes made during production. The label figures must be used if they differ from those in this chart.

B: Before Top Dead Center

① Except SULEV: 48-55 psi
   SULEV: 47-54

43256-ACCO-C05

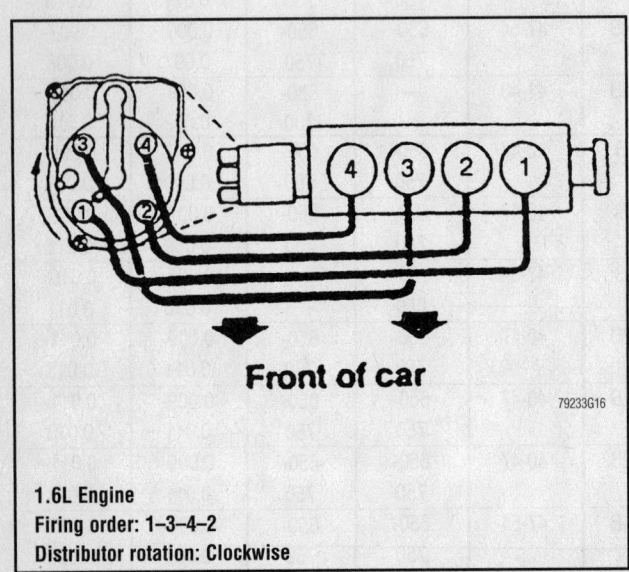

1.6L Engine
Firing order: 1–3–4–2
Distributor rotation: Clockwise

79233G16

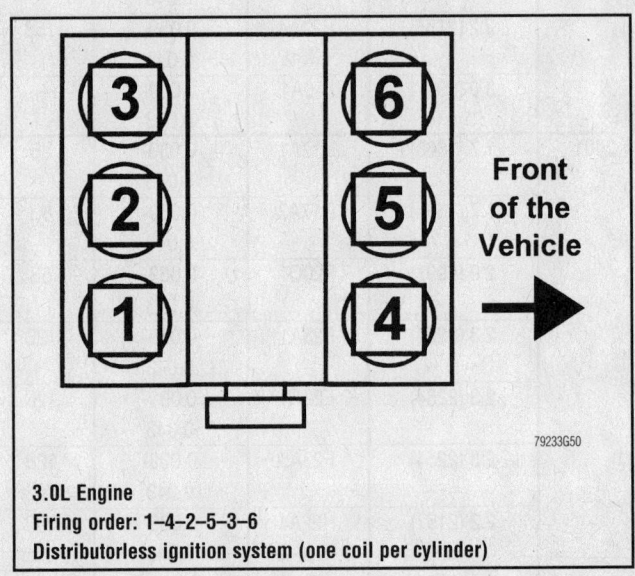

Front of the Vehicle

3.0L Engine
Firing order: 1–4–2–5–3–6
Distributorless ignition system (one coil per cylinder)

79233G50

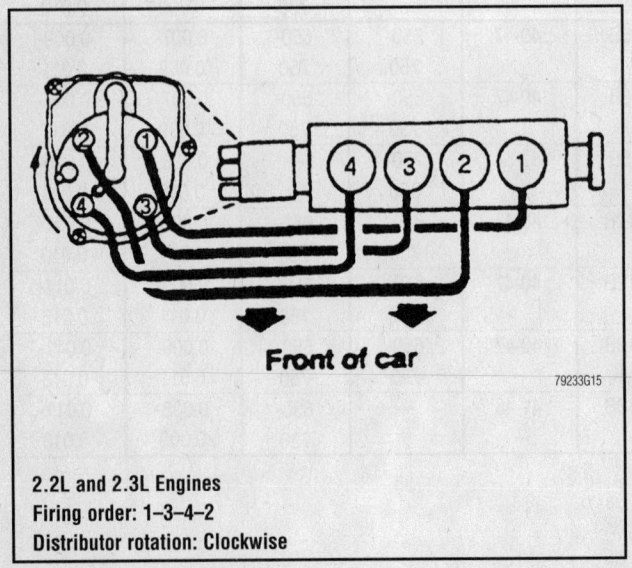

2.2L and 2.3L Engines
Firing order: 1–3–4–2
Distributor rotation: Clockwise

79233G15

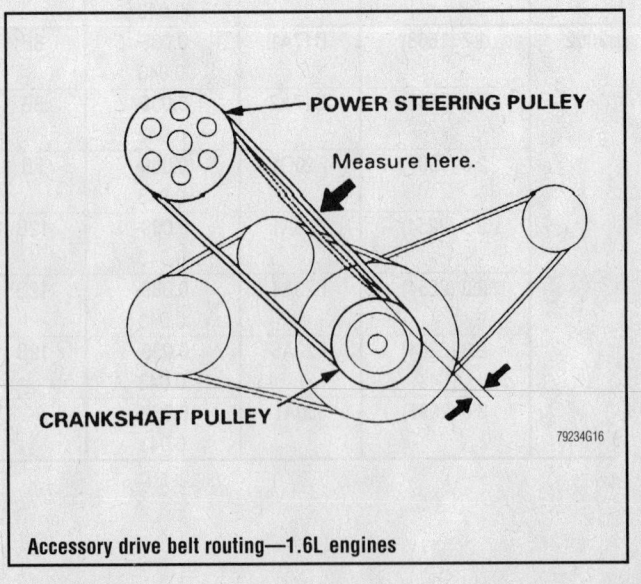

POWER STEERING PULLEY

Measure here.

CRANKSHAFT PULLEY

Accessory drive belt routing—1.6L engines

79234G16

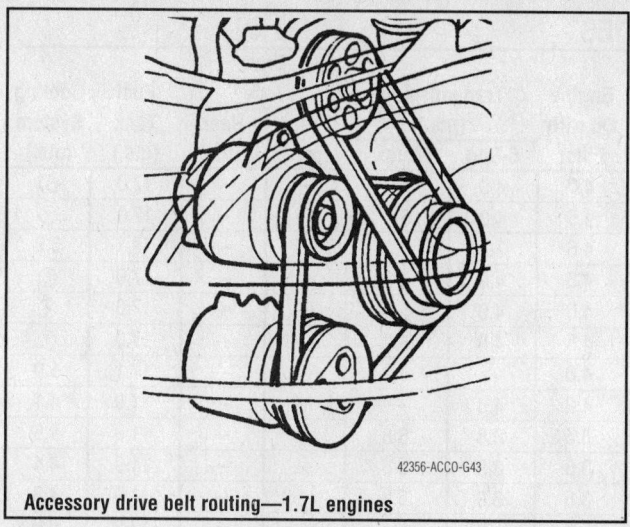

Accessory drive belt routing—1.7L engines

42356-ACCO-G43

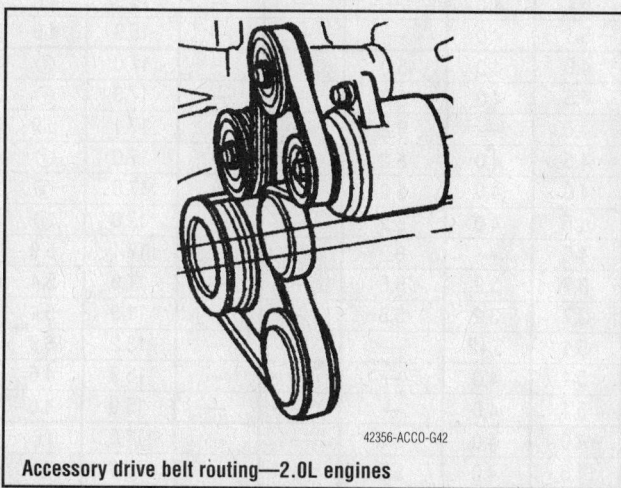

Accessory drive belt routing—2.0L engines

42356-ACCO-G42

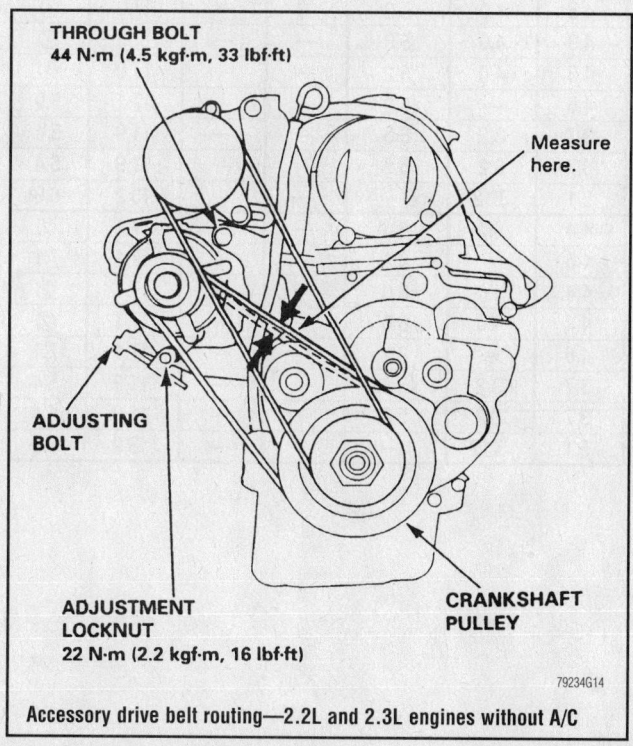

THROUGH BOLT
44 N·m (4.5 kgf·m, 33 lbf·ft)

Measure here.

ADJUSTING
BOLT

ADJUSTMENT
LOCKNUT
22 N·m (2.2 kgf·m, 16 lbf·ft)

CRANKSHAFT
PULLEY

Accessory drive belt routing—2.2L and 2.3L engines without A/C

79234G14

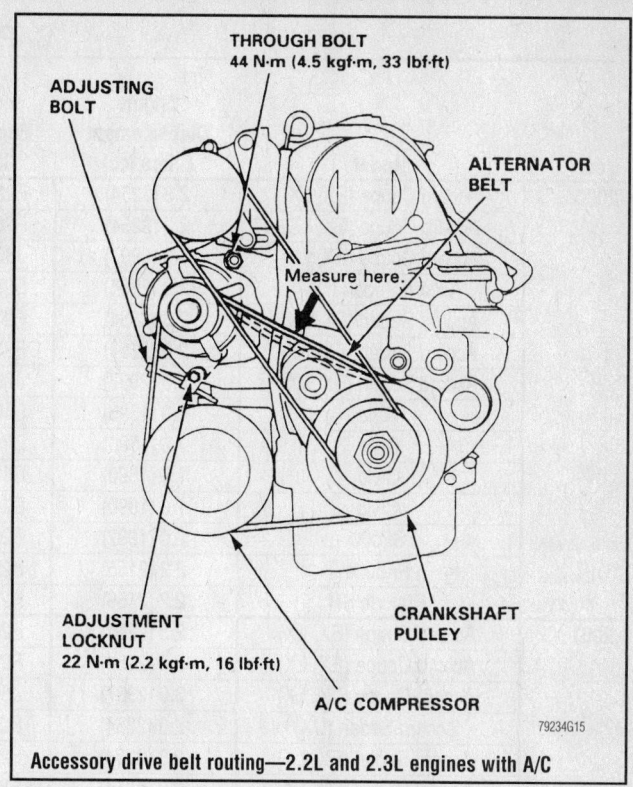

THROUGH BOLT
44 N·m (4.5 kgf·m, 33 lbf·ft)

ADJUSTING
BOLT

ALTERNATOR
BELT

Measure here.

ADJUSTMENT
LOCKNUT
22 N·m (2.2 kgf·m, 16 lbf·ft)

A/C COMPRESSOR

CRANKSHAFT
PULLEY

Accessory drive belt routing—2.2L and 2.3L engines with A/C

79234G15

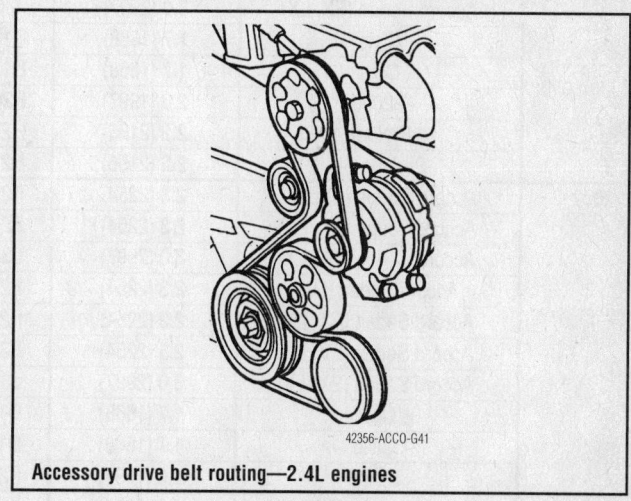

Accessory drive belt routing—2.4L engines

42356-ACCO-G41

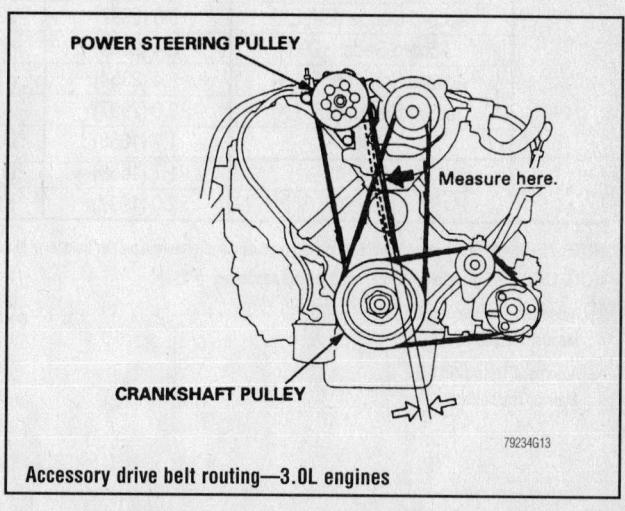

POWER STEERING PULLEY

Measure here.

CRANKSHAFT PULLEY

Accessory drive belt routing—3.0L engines

79234G13

CAPACITIES

Year	Model	Engine Displacement Liters (cc)	Engine ID	Engine Oil with Filter	Transmission (pts.) 5-Spd	Transmission (pts.) Auto.	Drive Axle Front (pts.)	Drive Axle Rear (pts.)	Fuel Tank (gal.)	Cooling System (qts.)
2000	Accord Coupe (EX, LX)	2.3 (2254)	F23A1	4.0	4.0	5.0	—	—	17.0	①
	Accord Coupe (EX, LX)	2.3 (2254)	F23A4	4.5	4.0	5.0	—	—	17.0	①
	Accord Coupe (EX, LX)	3.0 (2997)	J30A1	4.6	—	6.2	—	—	17.1	5.9
	Accord Sedan (DX)	2.3 (2254)	F23A5	4.5	4.0	5.2	—	—	17.0	①
	Accord Sedan (EX, LX)	2.3 (2254)	F23A1	4.0	4.0	5.0	—	—	17.0	①
	Accord Sedan (EX, LX)	2.3 (2254)	F23A4	4.5	4.0	5.2	—	—	17.0	①
	Accord Sedan (EX, LX)	3.0 (2997)	J30A1	4.6	—	6.2	—	—	17.1	5.9
	Civic	1.6 (1595)	B16A2	4.2	4.8	—	—	—	11.9	4.1
	Civic	1.6 (1590)	D16Y5	3.5	3.8	5.8	—	—	11.9	4.5
	Civic	1.6 (1590)	D16Y7	3.5	3.8	5.8	—	—	11.9	4.4
	Civic	1.6 (1590)	D16Y8	3.5	3.8	5.8	—	—	11.9	4.3
	S2000	2.0 (1997)	F20C1	5.1	3.12	—	—	—	13.2	6.9
	Prelude	2.2 (2156)	H22A4	5.1	4.0	—	—	—	15.9	4.6
	Prelude SH	2.2 (2156)	H22A4	5.1	4.0	—	—	—	15.9	4.6
2001	Accord Coupe (EX, LX)	2.3 (2254)	F23A1	4.0	4.0	5.0	—	—	17.0	①
	Accord Coupe (EX, LX)	2.3 (2254)	F23A4	4.5	4.0	5.0	—	—	17.0	①
	Accord Coupe (EX, LX)	3.0 (2997)	J30A1	4.6	—	6.2	—	—	17.1	5.9
	Accord Sedan (DX)	2.3 (2254)	F23A5	4.5	4.0	5.2	—	—	17.0	①
	Accord Sedan (EX, LX)	2.3 (2254)	F23A1	4.0	4.0	5.0	—	—	17.0	①
	Accord Sedan (EX, LX)	2.3 (2254)	F23A4	4.5	4.0	5.2	—	—	17.0	①
	Accord Sedan (EX, LX)	3.0 (2997)	J30A1	4.6	—	6.2	—	—	17.1	5.9
	Civic	1.7 (1668)	D17A1	3.7	3.2	5.8	—	—	11.9	5.4
	Civic	1.7 (1668)	D17A2	3.7	3.2	5.8	—	—	11.9	5.4
	S2000	2.0 (1997)	F20C1	5.1	3.12	—	—	—	13.2	6.9
	Prelude	2.2 (2156)	H22A4	5.1	4.0	—	—	—	15.9	4.6
	Prelude SH	2.2 (2156)	H22A4	5.1	4.0	—	—	—	15.9	4.6
2002	Accord Coupe (EX, LX)	2.3 (2254)	F23A1	4.0	4.0	5.0	—	—	17.0	①
	Accord Coupe (EX, LX)	2.3 (2254)	F23A4	4.5	4.0	5.0	—	—	17.0	①
	Accord Coupe (EX, LX)	3.0 (2997)	J30A1	4.6	—	6.2	—	—	17.1	5.9
	Accord Sedan (DX)	2.3 (2254)	F23A5	4.5	4.0	5.2	—	—	17.0	①
	Accord Sedan (EX, LX)	2.3 (2254)	F23A1	4.0	4.0	5.0	—	—	17.0	①
	Accord Sedan (EX, LX)	2.3 (2254)	F23A4	4.5	4.0	5.2	—	—	17.0	①
	Accord Sedan (EX, LX)	3.0 (2997)	J30A1	4.6	—	6.2	—	—	17.1	5.9
	Civic	1.7 (1668)	D17A1	3.7	3.2	5.8	—	—	11.9	5.4
	Civic	1.7 (1668)	D17A2	3.7	3.2	5.8	—	—	11.9	5.4
	S2000	2.0 (1997)	F20C1	5.1	3.12	—	—	—	13.2	6.9
2003-04	Accord Coupe (EX, LX)	2.4 (2354)	K24A4	4.4	4.0	6.0	—	—	17.1	②
	Accord Coupe (EX, LX)	3.0 (2997)	J30A4	4.6	—	6.2	—	—	17.1	7.1
	Accord Sedan (DX)	2.4 (2354)	K24A4	4.4	4.0	6.0	—	—	17.1	②
	Accord Sedan (EX, LX)	2.4 (2354)	K24A4	4.4	4.0	6.0	—	—	17.1	②
	Accord Sedan (EX, LX)	3.0 (2997)	J30A4	4.6	—	6.2	—	—	17.1	7.1
	Civic	1.7 (1668)	D17A1	3.7	3.2	5.8	—	—	11.9	5.4
	Civic	1.7 (1668)	D17A2	3.7	3.2	5.8	—	—	11.9	5.4
	S2000	2.0 (1997)	F20C1	5.1	3.12	—	—	—	13.2	6.9

NOTE: All capacities are approximate. Add fluid gradually and ensure a proper fluid level is obtained.

NOTE: Capacities given are service, not overhaul capacities.

① Automatic Transaxle: 5.7
Manual Transaxle: 5.8

② Automatic Transaxle: 5.3
Manual Transaxle: 5.4

43256-ACCO-C10

CRANKSHAFT AND CONNECTING ROD SPECIFICATIONS

All measurements are given in inches.

Year	Engine Displacement Liters (cc)	Engine ID/VIN	Crankshaft				Connecting Rod		
			Main Brg. Journal Dia.	Main Brg. Oil Clearance	Shaft End-play	Thrust on No.	Journal Diameter	Oil Clearance	Side Clearance
2000	1.6 (1595)	B16A2	2.1644- ① 2.1354	0.0009- ② 0.0020	0.004-0.018	3	1.7707-1.7717	0.0013-0.0022	0.0006-0.0016
	1.6 (1590)	D16Y5	2.1644-2.1654	0.0007- ③ 0.0016	0.004-0.018	3	1.7707-1.7717	0.0008-0.0017	0.0006-0.0016
	1.6 (1590)	D16Y7	2.1644-2.1654	0.0007- ③ 0.0016	0.004-0.018	3	1.7707-1.7717	0.0008-0.0017	0.0006-0.0016
	1.6 (1590)	D16Y8	2.1644-2.1654	0.0007- ③ 0.0016	0.004-0.018	3	1.7707-1.7717	0.0008-0.0017	0.0006-0.0016
	2.0 (1997)	F20C1	2.1644-2.1654	0.0007-0.0016	0.004-0.014	4	1.8888-1.8898	0.0012-0.0020	0.0006-0.0016
	2.3 (2254)	F23A1	④	⑤	0.004-0.018	4	1.7707-1.7717	0.0008-0.0024	0.0006-0.0016
	2.3 (2254)	F23A4	④	⑤	0.004-0.018	4	1.7707-1.7717	0.0008-0.0024	0.006-0.018
	2.3 (2254)	F23A5	④	⑤	0.004-0.018	4	1.7707-1.7717	0.0008-0.0024	0.0006-0.0016
	2.2 (2157)	H22A4	⑥	⑦	0.004-0.018	4	1.8888-1.8898	0.0011-0.0020	0.0006-0.0016
	3.0 (2997)	J30A1	2.8337-2.8346	0.0008-0.0020	0.004-0.018	3	2.0857-2.0866	0.0008-0.0020	0.006-0.018
2001	1.7 (1668)	D17A1	2.1644-2.1654	0.0007- ③ 0.0014	0.004-0.014	4	1.7707-1.7717	0.0009-0.0017	0.0006-0.0016
	1.7 (1668)	D17A2	2.1644-2.1654	0.0007- ③ 0.0014	0.004-0.014	4	1.7707-1.7717	0.0009-0.0017	0.0006-0.0016
	2.0 (1997)	F20C1	2.1644-2.1654	0.0007-0.0016	0.004-0.014	4	1.8888-1.8898	0.0012-0.0020	0.0006-0.0016
	2.3 (2254)	F23A1	④	⑤	0.004-0.018	4	1.7707-1.7717	0.0008-0.0024	0.0006-0.0016
	2.3 (2254)	F23A4	④	⑤	0.004-0.018	4	1.7707-1.7717	0.0008-0.0024	0.006-0.018
	2.3 (2254)	F23A5	④	⑤	0.004-0.018	4	1.7707-1.7717	0.0008-0.0024	0.0006-0.0016
	2.2 (2157)	H22A4	⑥	⑦	0.004-0.018	4	1.8888-1.8898	0.0011-0.0020	0.0006-0.0016
	3.0 (2997)	J30A1	2.8337-2.8346	0.0008-0.0020	0.004-0.018	3	2.0857-2.0866	0.0008-0.0020	0.006-0.018
2002	1.7 (1668)	D17A1	2.1644-2.1654	0.0007- ③ 0.0014	0.004-0.014	4	1.7707-1.7717	0.0009-0.0017	0.0006-0.0016
	1.7 (1668)	D17A2	2.1644-2.1654	0.0007- ③ 0.0014	0.004-0.014	4	1.7707-1.7717	0.0009-0.0017	0.0006-0.0016
	2.0 (1997)	F20C1	2.1644-2.1654	0.0007-0.0016	0.004-0.014	4	1.8888-1.8898	0.0012-0.0020	0.0006-0.0016
	2.3 (2254)	F23A1	④	⑤	0.004-0.018	4	1.7707-1.7717	0.0008-0.0024	0.0006-0.0016
	2.3 (2254)	F23A4	④	⑤	0.004-0.018	4	1.7707-1.7717	0.0008-0.0024	0.006-0.018

43256-ACCO-C06

CRANKSHAFT AND CONNECTING ROD SPECIFICATIONS
All measurements are given in inches.

Year	Engine Displacement Liters (cc)	Engine ID/VIN	Crankshaft				Connecting Rod		
			Main Brg. Journal Dia.	Main Brg. Oil Clearance	Shaft End-play	Thrust on No.	Journal Diameter	Oil Clearance	Side Clearance
2002 (cont.)	2.3 (2254)	F23A5	④	⑤	0.004-0.018	4	1.7707-1.7717	0.0008-0.0024	0.0006-0.0016
	3.0 (2997)	J30A1	2.8337-2.8346	0.0008-0.0020	0.004-0.018	3	2.0857-2.0866	0.0008-0.0020	0.006-0.018
2003-04	1.7 (1668)	D17A1	2.1644-2.1654	0.0007-③ 0.0014	0.004-0.014	4	1.7707-1.7717	0.0009-0.0017	0.0006-0.0016
	1.7 (1668)	D17A2	2.1644-2.1654	0.0007-③ 0.0014	0.004-0.014	4	1.7707-1.7717	0.0009-0.0017	0.0006-0.0016
	2.0 (1997)	F20C1	2.1644-2.1654	0.0007-0.0016	0.004-0.014	4	1.8888-1.8898	0.0012-0.0020	0.0006-0.0016
	2.4 (2254)	K24A4	⑧	⑨	0.004-0.014	4	1.8888-1.8898	0.0002-0.0006	0.0006-0.0016
	3.0 (2997)	J30A4	2.8337-2.8346	0.0008-0.0020	0.004-0.018	3	2.0857-2.0866	0.0008-0.0020	0.006-0.018

① Journals 1, 2, 4 and 5
  Journal 3: 2.1642 - 2.1651

② Journals 1, 2, 4 and 5
  Journal 3: 0.0012 - 0.0021

③ Journals 1 and 5
  Journals 2, 3 and 4: 0.0009 - 0.0019

④ Journals 1, 2 and 4: 2.1646 - 2.1655
  Journal 3: 2.1644 - 2.1654
  Journal 5: 2.1650 - 2.1660

⑤ Journals 1, 2 and 4: 0.0008 - 0.0020
  Journal 3: 0.0010 - 0.0022
  Journal 5: 0.0004 - 0.0016

⑥ Journals 1 and 2: 1.9676 - 1.9685
  Journal 3: 1.9674 - 1.9683
  Journal 4: 1.9679 - 1.9688
  Journal 5: 1.9680 - 1.9690

⑦ Journals 1 and 2: 0.0008 - 0.0020
  Journal 3: 0.0010 - 0.0022
  Journal 4: 0.0005 - 0.0020
  Journal 5: 0.0004 - 0.0016

⑧ Journals 1, 2 4 and 5: 2.1648 - 2.1657
  Journal 3: 2.1644 - 2.1654

⑨ Journals 1, 2 4 and 5: 0.0007-0.0016
  Journal 3: 0.0010-0.0019

43256-ACCO-C07

PISTON AND RING SPECIFICATIONS

All measurements are given in inches.

Year	Engine Displacement Liters (cc)	Engine ID/VIN	Piston Clearance	Ring Gap			Ring Side Clearance		
				Top Compression	Bottom Compression	Oil Control	Top Compression	Bottom Compression	Oil Control
2000	1.6 (1595)	B16A2	0.0004-0.0020	0.008-0.024	0.016-0.028	0.008-0.028	0.0018-0.0050	0.0016-0.0050	N/A
	1.6 (1590)	D16Y5	0.0004-0.0020	0.006-0.024	0.012-0.028	0.008-0.031	0.0014-0.0050	0.0012-0.0050	N/A
	1.6 (1590)	D16Y7	0.0004-0.0020	0.006-0.024	0.012-0.028	0.008-0.031	0.0014-0.0050	0.0012-0.0050	N/A
	1.6 (1590)	D16Y8	0.0004-0.0020	0.006-0.024	0.012-0.028	0.008-0.031	0.0014-0.0050	0.0012-0.0050	N/A
	2.0 (1997)	F20C1	0.0002-0.0011	0.010-0.024	0.024-0.035	0.008-0.031	0.0018-0.0035	0.0016-0.0028	N/A
	2.3 (2254)	F23A1	0.0008-0.0020	0.008-0.024	0.016-0.028	0.008-0.031	0.0014-0.0050	0.0012-0.0050	N/A
	2.3 (2254)	F23A4	0.0008-0.0020	0.008-0.024	0.016-0.028	0.008-0.031	0.0014-0.0050	0.0012-0.0050	N/A
	2.3 (2254)	F23A5	0.0008-0.0020	0.008-0.024	0.016-0.028	0.008-0.031	0.0014-0.0050	0.0012-0.0050	N/A
	2.2 (2157)	H22A4	0.0002-0.0020	0.010-0.024	0.024-0.035	0.008-0.031	0.0022-0.0050	0.0016-0.0050	N/A
	3.0 (2997)	J30A1	0.0006-0.0030	0.008-0.024	0.016-0.028	0.008-0.031	0.0014-0.0050	0.0012-0.0050	N/A
2001	1.7 (1668)	D17A1	0.0004-0.0016	0.006-0.024	0.012-0.024	0.008-0.031	0.0014-0.0024	0.0012-0.0022	N/A
	1.7 (1668)	D17A2	0.0004-0.0016	0.006-0.024	0.012-0.024	0.008-0.031	0.0014-0.0024	0.0012-0.0022	N/A
	2.0 (1997)	F20C1	0.0002-0.0011	0.010-0.024	0.024-0.035	0.008-0.031	0.0018-0.0035	0.0016-0.0028	N/A
	2.3 (2254)	F23A1	0.0008-0.0020	0.008-0.024	0.016-0.028	0.008-0.031	0.0014-0.0050	0.0012-0.0050	N/A
	2.3 (2254)	F23A4	0.0008-0.0020	0.008-0.024	0.016-0.028	0.008-0.031	0.0014-0.0050	0.0012-0.0050	N/A
	2.3 (2254)	F23A5	0.0008-0.0020	0.008-0.024	0.016-0.028	0.008-0.031	0.0014-0.0050	0.0012-0.0050	N/A
	2.2 (2157)	H22A4	0.0002-0.0020	0.010-0.024	0.024-0.035	0.008-0.031	0.0022-0.0050	0.0016-0.0050	N/A
	3.0 (2997)	J30A1	0.0006-0.0030	0.008-0.024	0.016-0.028	0.008-0.031	0.0014-0.0050	0.0012-0.0050	N/A
2002	1.7 (1668)	D17A1	0.0004-0.0016	0.006-0.024	0.012-0.024	0.008-0.031	0.0014-0.0024	0.0012-0.0022	N/A
	1.7 (1668)	D17A2	0.0004-0.0016	0.006-0.024	0.012-0.024	0.008-0.031	0.0014-0.0024	0.0012-0.0022	N/A
	2.0 (1997)	F20C1	0.0002-0.0011	0.010-0.024	0.024-0.035	0.008-0.031	0.0018-0.0035	0.0016-0.0028	N/A
	2.3 (2254)	F23A1	0.0008-0.0020	0.008-0.024	0.016-0.028	0.008-0.031	0.0014-0.0050	0.0012-0.0050	N/A
	2.3 (2254)	F23A4	0.0008-0.0020	0.008-0.024	0.016-0.028	0.008-0.031	0.0014-0.0050	0.0012-0.0050	N/A

43256-ACCO-C08

PISTON AND RING SPECIFICATIONS

All measurements are given in inches.

Year	Engine Displacement Liters (cc)	Engine ID/VIN	Piston Clearance	Ring Gap			Ring Side Clearance		
				Top Compression	Bottom Compression	Oil Control	Top Compression	Bottom Compression	Oil Control
2002 (cont.)	2.3 (2254)	F23A5	0.0008-0.0020	0.008-0.024	0.016-0.028	0.008-0.031	0.0014-0.0050	0.0012-0.0050	N/A
	3.0 (2997)	J30A1	0.0006-0.0030	0.008-0.024	0.016-0.028	0.008-0.031	0.0014-0.0050	0.0012-0.0050	N/A
2003-04	1.7 (1668)	D17A1	0.0004-0.0016	0.006-0.024	0.012-0.024	0.008-0.031	0.0014-0.0024	0.0012-0.0022	N/A
	1.7 (1668)	D17A2	0.0004-0.0016	0.006-0.024	0.012-0.024	0.008-0.031	0.0014-0.0024	0.0012-0.0022	N/A
	2.0 (1997)	F20C1	0.0002-0.0011	0.010-0.024	0.024-0.035	0.008-0.031	0.0018-0.0035	0.0016-0.0028	N/A
	2.4 (2354)	K24A4	0.0008-0.0016	0.008-0.0014	0.016-0.022	0.008-0.028	0.0018-0.0028	0.0020-0.0030	N/A
	3.0 (2997)	J30A4	0.0006-0.0030	0.008-0.024	0.016-0.028	0.008-0.031	0.0014-0.0050	0.0012-0.0050	N/A

NA: Not Available

43256-ACCO-C09

VALVE SPECIFICATIONS

Year	Engine Displacement Liters (cc)	Engine ID/VIN	Seat Angle (deg.)	Face Angle (deg.)	Spring Test Pressure (lbs. @ in.)	Spring Installed Height (in.)	Stem-to-Guide Clearance (in.)		Stem Diameter (in.)	
							Intake	Exhaust	Intake	Exhaust
2000	1.6 (1595)	B16A2	45	45	NA	NA	0.0010-0.0022	0.0020-0.0031	0.2156-0.2159	0.2146-0.2150
	1.6 (1590)	D16Y5	45	45	NA	NA	0.0010-0.0020	0.0020-0.0030	0.2157-0.2161	0.2146-0.2150
	1.6 (1590)	D16Y7	45	45	NA	NA	0.0010-0.0020	0.0020-0.0030	0.2157-0.2161	0.2146-0.2150
	1.6 (1590)	D16Y8	45	45	NA	NA	0.0010-0.0020	0.0020-0.0030	0.2157-0.2161	0.2146-0.2150
	2.0 (1997)	F20C1	45	45	NA	NA	0.0010-0.0020	0.0020-0.0030	0.2157-0.2162	0.2146-0.2150
	2.3 (2254)	F23A1	45	45	NA	NA	0.0008-0.0018	0.0022-0.0031	0.2159-0.2163	0.2146-0.2150
	2.3 (2254)	F23A4	45	45	NA	NA	0.0008-0.0018	0.0022-0.0031	0.2159-0.2163	0.2146-0.2150
	2.3 (2254)	F23A5	45	45	NA	NA	0.0008-0.0018	0.0022-0.0031	0.2159-0.2163	0.2146-0.2150
	2.2 (2157)	H22A4	45	45	NA	NA	0.0010-0.0022	0.0020-0.0031	0.2156-0.2159	0.2156-0.2159
	3.0 (2997)	J30A1	45	45	NA	NA	0.0008-0.0018	0.0022-0.0031	0.2159-0.2163	0.2146-0.2150
2001	1.7 (1668)	D17A1	45	45	NA	NA	0.0008-0.0020	0.0020-0.0031	0.2157-0.2161	0.2146-0.2150
	1.7 (1668)	D17A2	45	45	NA	NA	0.0008-0.0020	0.0020-0.0031	0.2157-0.2161	0.2146-0.2150
	2.0 (1997)	F20C1	45	45	NA	NA	0.0010-0.0020	0.0020-0.0030	0.2157-0.2162	0.2146-0.2150
	2.3 (2254)	F23A1	45	45	NA	NA	0.0008-0.0018	0.0022-0.0031	0.2159-0.2163	0.2146-0.2150
	2.3 (2254)	F23A4	45	45	NA	NA	0.0008-0.0018	0.0022-0.0031	0.2159-0.2163	0.2146-0.2150
	2.3 (2254)	F23A5	45	45	NA	NA	0.0008-0.0018	0.0022-0.0031	0.2159-0.2163	0.2146-0.2150
	2.2 (2157)	H22A4	45	45	NA	NA	0.0010-0.0022	0.0020-0.0031	0.2156-0.2159	0.2156-0.2159
	3.0 (2997)	J30A1	45	45	NA	NA	0.0008-0.0018	0.0022-0.0031	0.2159-0.2163	0.2146-0.2150
2002	1.7 (1668)	D17A1	45	45	NA	NA	0.0008-0.0020	0.0020-0.0031	0.2157-0.2161	0.2146-0.2150
	1.7 (1668)	D17A2	45	45	NA	NA	0.0008-0.0020	0.0020-0.0031	0.2157-0.2161	0.2146-0.2150
	2.0 (1997)	F20C1	45	45	NA	NA	0.0010-0.0020	0.0020-0.0030	0.2157-0.2162	0.2146-0.2150
	2.3 (2254)	F23A1	45	45	NA	NA	0.0008-0.0018	0.0022-0.0031	0.2159-0.2163	0.2146-0.2150
	2.3 (2254)	F23A4	45	45	NA	NA	0.0008-0.0018	0.0022-0.0031	0.2159-0.2163	0.2146-0.2150

43256-ACCO-C11

VALVE SPECIFICATIONS

Year	Engine Displacement Liters (cc)	Engine ID/VIN	Seat Angle (deg.)	Face Angle (deg.)	Spring Test Pressure (lbs. @ in.)	Spring Installed Height (in.)	Stem-to-Guide Clearance (in.)		Stem Diameter (in.)	
							Intake	Exhaust	Intake	Exhaust
2002 (cont.)	2.3 (2254)	F23A5	45	45	NA	NA	0.0008-0.0018	0.0022-0.0031	0.2159-0.2163	0.2146-0.2150
	3.0 (2997)	J30A1	45	45	NA	NA	0.0008-0.0018	0.0022-0.0031	0.2159-0.2163	0.2146-0.2150
2003-04	1.7 (1668)	D17A1	45	45	NA	NA	0.0008-0.0020	0.0020-0.0031	0.2157-0.2161	0.2146-0.2150
	1.7 (1668)	D17A2	45	45	NA	NA	0.0008-0.0020	0.0020-0.0031	0.2157-0.2161	0.2146-0.2150
	2.0 (1997)	F20C1	45	45	NA	NA	0.0010-0.0020	0.0020-0.0030	0.2157-0.2162	0.2146-0.2150
	2.4 (2354)	K24A4	45	45	NA	NA	0.0012-0.0022	0.0022-0.0031	0.2156-0.2159	0.2146-0.2150
	3.0 (2997)	J30A4	45	45	NA	NA	0.0008-0.0018	0.0022-0.0031	0.2159-0.2163	0.2146-0.2150

NA: Not Available

43256-ACCO-C12

TORQUE SPECIFICATIONS
All readings in ft. lbs.

Year	Engine Displacement Liters (cc)	Engine ID/VIN	Cylinder Head Bolts	Main Bearing Bolts	Rod Bearing Bolts	Crankshaft Damper Bolts	Flywheel Bolts	Manifold Intake	Manifold Exhaust	Spark Plugs	Lug Nut
2000	1.6 (1585)	B16A2	①	②	30	130	76	17	23	13	80
	1.6 (1590)	D16Y5	③	④	23	134	87	17	23	13	80
	1.6 (1590)	D16Y7	③	④	23	134	87	17	23	13	80
	1.6 (1590)	D16Y8	③	④	23	134	87	17	23	13	80
	2.0 (1997)	F20C1	⑤	⑥	⑦	181	94	16	23	13	80
	2.2 (2157)	H22A4	⑧	54	34	181	⑨	16	23	13	80
	2.3 (2254)	F23A1	⑤	⑩	⑪	181	⑨	16	23	13	80
	2.3 (2254)	F23A4	⑤	⑩	⑪	181	⑨	16	23	13	80
	2.3 (2254)	F23A5	⑤	⑩	⑪	181	⑨	16	23	13	80
	3.0 (2997)	J30A1	⑧	⑫	⑪	181	⑨	16	23	13	80
2001	1.7 (1668)	D17A1	③	④	24	⑨	87	16	23	13	80
	1.7 (1668)	D17A2	③	④	24	⑨	87	16	23	13	80
	2.0 (1997)	F20C1	⑤	⑥	⑦	181	94	16	23	13	80
	2.2 (2157)	H22A4	⑧	54	34	181	⑨	16	23	13	80
	2.3 (2254)	F23A1	⑤	⑩	⑪	181	⑨	16	23	13	80
	2.3 (2254)	F23A4	⑤	⑩	⑪	181	⑨	16	23	13	80
	2.3 (2254)	F23A5	⑤	⑩	⑪	181	⑨	16	23	13	80
	3.0 (2997)	J30A1	⑧	⑫	⑪	181	⑨	16	23	13	80
2002	1.7 (1668)	D17A1	③	④	24	⑪	87	16	23	13	80
	1.7 (1668)	D17A2	③	④	24	⑪	87	16	23	13	80
	2.0 (1997)	F20C1	⑤	⑥	⑦	181	94	16	23	13	80
	2.3 (2254)	F23A1	⑤	⑩	⑪	181	⑨	16	23	13	80
	2.3 (2254)	F23A4	⑤	⑩	⑪	181	⑨	16	23	13	80
	2.3 (2254)	F23A5	⑤	⑩	⑪	181	⑨	16	23	13	80
	3.0 (2997)	J30A1	⑧	⑫	⑪	181	⑨	16	23	13	80
2003-04	1.7 (1668)	D17A1	③	④	24	⑪	87	16	23	13	80
	1.7 (1668)	D17A2	③	④	24	⑪	87	16	23	13	80
	2.0 (1997)	F20C1	⑤	⑥	⑦	181	94	16	23	13	80
	2.4 (2354)	K24A4	⑬	⑭	16	⑮	⑨	16	33	13	80
	3.0 (2997)	J30A4	⑧	⑫	⑪	181	⑨	16	23	13	80

① Step 1: 22 ft. lbs.
Step 2: 61 ft. lbs.

② Step 1: 18 ft. lbs.
Step 2: 54 ft. lbs.

③ Step 1: 14 ft. lbs.
Step 2: 36 ft. lbs.
Step 3: 49 ft. lbs.
Step 4: Bolts 1-2, retorque to 49 ft. lbs.

④ Step 1: 18 ft. lbs.
Step 2: 38 ft. lbs.

⑤ Step 1: 22 ft. lbs.
Step 2: Rotate 90 degrees
Step 3: Rotate 90 degrees
Step 4: If new bolt rotate additional 90 degrees

⑥ Step 1: Bearing cap bolts to 22 ft. lbs.
Step 2: Bearing cap bolts plus 60 degrees
Step 3: 8mm bolts to 16 ft. lbs.

⑦ Step 1: 18 ft. lbs.
Step 2: Plus 90 degrees

⑧ Step 1: 29 ft. lbs.
Step 2: 51 ft. lbs.
Step 3: 72.3 ft. lbs.

⑨ Automatic transaxle: 54 ft. lbs.
Manual transaxle: 76 ft. lbs.

⑩ Step 1: 11mm bolts, 29 ft. lbs.
Step 2: 11mm bolts, 58 ft. lbs.
Step 3: 6mm bolts, 8.7 ft. lbs.

⑪ Step 1: 14 ft. lbs.
Step 2: Rotate 90 degrees

⑫ Step 1: Cap bolts, 56 ft. lbs.
Step 2: Side bolts, 36 ft. lbs.

⑬ Step 1: 29 ft. lbs.
Step 2: Rotate 90 degrees
Step 3: Rotate 90 degrees
Step 4: If new bolt rotate additional 90 degrees

⑭ Step 1: 22 ft. lbs.
Step 2: Rotate 56 degrees

⑮ Step 1: 36 ft. lbs.
Step 2: Rotate 90 degrees

43256-ACCO-C18

WHEEL ALIGNMENT

Year	Model		Caster Range (+/-Deg.)	Caster Preferred Setting (Deg.)	Camber Range (+/-Deg.)	Camber Preferred Setting (Deg.)	Toe-in (in.)	Steering Axis Inclination (Deg.)
2000	Accord	F	1.00	+2.80	1.00	+0.06	0 +/- 0.03	—
		R	—	—	0.50	-0.50	0.03 +/- 0.03	—
	Civic	F	—	+1.66	—	0	0.03	—
		R	—	—	—	-1.00	0.03	—
	Prelude	F	1.00	+2.66	1.00	0	0 +/- 0.03	—
		R	—	—	1.00	-0.45	0.03 +/- 0.03	—
	S2000	F	0.75	+6.00	0.50	-0.50	0 +/- 0.03	—
		R	—	—	0.50	-1.50	0.12 +/- 0.03	—
2001	Accord	F	1.00	+2.80	1.00	+0.06	0 +/- 0.03	—
		R	—	—	0.50	-0.50	0.03 +/- 0.03	—
	Civic	F	—	+1.66	—	0	0.03	—
		R	—	—	—	-1.00	0.03	—
	Prelude	F	1.00	+2.66	1.00	0	0 +/- 0.03	—
		R	—	—	1.00	-0.45	0.03 +/- 0.03	—
	S2000	F	0.75	+6.00	0.50	-0.50	0 +/- 0.03	—
		R	—	—	0.50	-1.50	0.12 +/- 0.03	—
2002	Accord	F	1.00	+2.80	1.00	+0.06	0 +/- 0.03	—
		R	—	—	0.50	-0.50	0.03 +/- 0.03	—
	Civic	F	—	+1.66	—	0	0.03	—
		R	—	—	—	-1.00	0.03	—
	S2000	F	0.75	+6.00	0.50	-0.50	0 +/- 0.03	—
		R	—	—	0.50	-1.50	0.12 +/- 0.03	—
2003-04	Accord	F	1.00	+2.80	1.00	+0.06	0 +/- 0.03	—
		R	—	—	0.50	-0.50	0.03 +/- 0.03	—
	Civic	F	—	+1.66	—	0	0.03	—
		R	—	—	—	-1.00	0.03	—
	S2000	F	0.75	+6.00	0.50	-0.50	0 +/- 0.03	—
		R	—	—	0.50	-1.50	0.12 +/- 0.03	—

43256-ACCO-C13

WHEEL ALIGNMENT

Year	Model		Caster Range (+/-Deg.)	Caster Preferred Setting (Deg.)	Camber Range (+/-Deg.)	Camber Preferred Setting (Deg.)	Toe-in (in.)	Steering Axis Inclination (Deg.)
2000	Accord	F	1.00	+2.80	1.00	+0.06	0 +/- 0.03	—
		R	—	—	0.50	-0.50	0.03 +/- 0.03	—
	Civic	F	—	+1.66	—	0	0.03	—
		R	—	—	—	-1.00	0.03	—
	Prelude	F	1.00	+2.66	1.00	0	0 +/- 0.03	—
		R	—	—	1.00	-0.45	0.03 +/- 0.03	—
	S2000	F	0.75	+6.00	0.50	-0.50	0 +/- 0.03	—
		R	—	—	0.50	-1.50	0.12 +/- 0.03	—
2001	Accord	F	1.00	+2.80	1.00	+0.06	0 +/- 0.03	—
		R	—	—	0.50	-0.50	0.03 +/- 0.03	—
	Civic	F	—	+1.66	—	0	0.03	—
		R	—	—	—	-1.00	0.03	—
	Prelude	F	1.00	+2.66	1.00	0	0 +/- 0.03	—
		R	—	—	1.00	-0.45	0.03 +/- 0.03	—
	S2000	F	0.75	+6.00	0.50	-0.50	0 +/- 0.03	—
		R	—	—	0.50	-1.50	0.12 +/- 0.03	—
2002	Accord	F	1.00	+2.80	1.00	+0.06	0 +/- 0.03	—
		R	—	—	0.50	-0.50	0.03 +/- 0.03	—
	Civic	F	—	+1.66	—	0	0.03	—
		R	—	—	—	-1.00	0.03	—
	S2000	F	0.75	+6.00	0.50	-0.50	0 +/- 0.03	—
		R	—	—	0.50	-1.50	0.12 +/- 0.03	—
2003-04	Accord	F	1.00	+2.80	1.00	+0.06	0 +/- 0.03	—
		R	—	—	0.50	-0.50	0.03 +/- 0.03	—
	Civic	F	—	+1.66	—	0	0.03	—
		R	—	—	—	-1.00	0.03	—
	S2000	F	0.75	+6.00	0.50	-0.50	0 +/- 0.03	—
		R	—	—	0.50	-1.50	0.12 +/- 0.03	—

43256-ACCO-C14

BRAKE SPECIFICATIONS
All measurements in inches unless noted

Year	Model		Brake Disc			Brake Drum Diameter			Minimum Lining Thickness		Brake Caliper	
			Original Thickness	Minimum Thickness	Maximum Runout	Original Inside Diameter	Max. Wear Limit	Maximum Machine Diameter	Front	Rear	Bracket Bolts (ft. lbs.)	Mounting Bolts (ft. lbs.)
2000	Accord	F	0.910	0.830	0.004	—	—	—	0.060	—	—	①
		R	0.358	0.310	0.004	8.66	8.70	8.70	—	0.080 ②	—	17
	Civic	F	0.840	0.750	0.004	—	—	—	0.060	—	—	①
		R	—	—	—	7.87	7.91	7.91	—	0.080	—	—
	Prelude	F	0.910	0.830	0.004	—	—	—	0.060	—	83	①
		R	0.390	0.320	0.004	—	—	—	—	0.060	—	17
	S2000	F	0.990	0.910	0.004	—	—	—	0.060	—	80	24
		R	0.476	0.390	0.004	—	—	—	—	0.060	41	17
2001	Accord	F	0.910	0.830	0.004	—	—	—	0.060	—	—	①
		R	0.358	0.310	0.004	8.66	8.70	8.70	—	0.080 ②	—	17
	Civic	F	0.840	0.750	0.004	—	—	—	0.060	—	—	①
		R	—	—	—	7.87	7.91	7.91	—	0.080	—	—
	Prelude	F	0.910	0.830	0.004	—	—	—	0.060	—	83	①
		R	0.390	0.320	0.004	—	—	—	—	0.060	—	17
	S2000	F	0.990	0.910	0.004	—	—	—	0.060	—	80	24
		R	0.476	0.390	0.004	—	—	—	—	0.060	41	17
2002	Accord	F	0.910	0.830	0.004	—	—	—	0.060	—	—	①
		R	0.358	0.310	0.004	8.66	8.70	8.70	—	0.080 ②	—	17
	Civic	F	0.840	0.750	0.004	—	—	—	0.060	—	—	①
		R	—	—	—	7.87	7.91	7.91	—	0.080	—	—
	S2000	F	0.990	0.910	0.004	—	—	—	0.060	—	80	24
		R	0.476	0.390	0.004	—	—	—	—	0.060	41	17
2003-04	Accord	F	0.910	0.830	0.004	—	—	—	0.060	—	—	①
		R	0.400	0.310	0.004	8.66	8.70	8.70	—	0.080 ②	—	17
	Civic	F	0.840	0.750	0.004	—	—	—	0.060	—	—	①
		R	—	—	—	7.87	7.91	7.91	—	0.080	—	—
	S2000	F	0.990	0.910	0.004	—	—	—	0.060	—	80	24
		R	0.476	0.390	0.004	—	—	—	—	0.060	41	17

NA: Not Available

F: Front

R: Rear

① Calipers with long pins beyond bolt threads, 54 ft. lbs.
Calipers with no pin beyond threads, 20 ft. lbs.

② With rear disc: 0.060

43256-ACCO-C15

SCHEDULED MAINTENANCE INTERVALS
Honda—Civic, Accord, Prelude & S2000

TO BE SERVICED	TYPE OF SERVICE	VEHICLE MILEAGE INTERVAL (x1000)												
		7.5	15	22.5	30	37.5	45	52.5	60	67.5	75	82.5	90	97.5
Engine oil & filter	R	✓	✓	✓	✓	✓	✓	✓	✓	✓	✓	✓	✓	✓
Front brake pads	S/I	✓	✓	✓	✓	✓	✓	✓	✓	✓	✓	✓	✓	✓
Rotate tires	S/I	✓	✓	✓	✓	✓	✓	✓	✓	✓	✓	✓	✓	✓
Cooling system, hoses & connections	S/I		✓		✓		✓		✓		✓		✓	
Driveshaft boots	S/I		✓		✓		✓		✓		✓		✓	
Exhaust system	S/I		✓		✓		✓		✓		✓		✓	
Front brake discs & calipers	S/I		✓		✓		✓		✓		✓		✓	
Front wheel alignment	S/I		✓		✓		✓		✓		✓		✓	
Front & rear wheel alignment (Prelude w/4WS)	S/I		✓		✓		✓		✓		✓		✓	
Fuel pipes, hoses & connections	S/I		✓		✓		✓		✓		✓		✓	
Parking brake adjustment	S/I		✓		✓		✓		✓		✓		✓	
Power steering system	S/I		✓		✓		✓		✓		✓		✓	
Rear brake discs, calipers & pads	S/I		✓		✓		✓		✓		✓		✓	
Suspension components	S/I		✓		✓		✓		✓		✓		✓	
Suspension mounting bolts	S/I		✓		✓		✓		✓		✓		✓	
Tie rods, steering gear box & boots	S/I		✓		✓		✓		✓		✓		✓	
Valve clearance (Prelude VTEC) ①	S/I		✓		✓		✓		✓		✓		✓	
Valve clearance (Accord L4, Civic & Prelude non-VTEC)	S/I				✓				✓				✓	
Parking brake	S/I		✓		✓				✓				✓	
Air cleaner element	R				✓				✓				✓	
Transmission fluid (Civic CVT)	R				✓		✓		✓		✓		✓	
Transmission fluid (A/T or M/T) (except as noted below)	R				✓				✓				✓	
Transmission fluid (Prelude L4)	R												✓	
Brake fluid (including ABS) (Accord V6)	R				✓				✓				✓	
Brake fluid (including ABS) (Accord L4, Civic, & Prelude)	R						✓						✓	
Spark plugs (non-VTEC)	R				✓				✓				✓	
Spark plugs (VTEC) ①	R								✓				✓	
ABS operation	S/I				✓				✓					
Alternator drive belt	S/I				✓				✓					
Power steering pump belt	S/I				✓				✓				✓	
Rear brake drums, wheel cylinders & linings (except Prelude)	S/I				✓				✓				✓	
Engine coolant	R						✓				✓			

43256-ACCO-C16

SCHEDULED MAINTENANCE INTERVALS
Honda—Civic, Accord, Prelude & S2000

TO BE SERVICED	TYPE OF SERVICE	VEHICLE MILEAGE INTERVAL (x1000)												
		7.5	15	22.5	30	37.5	45	52.5	60	67.5	75	82.5	90	97.5
ABS high pressure hose	R								✓					
Fuel filter	R								✓					
Timing belt	R												✓	
Timing balancer belt	R												✓	
Distributor, ignition cap & rotor	S/I								✓					
Idle speed	S/I								✓					
Ignition wires	S/I								✓					
PCV valve	S/I								✓					
TWC converter heat shield	S/I								✓					
Water pump	S/I												✓	

R: Replace   S/I: Service or Inspect
① S2000: 105,000 miles

FREQUENT OPERATION MAINTENANCE (SEVERE SERVICE)

If a vehicle is operated under any of the following conditions it is considered severe service:

- Extremely dusty areas.

- 50% or more of the vehicle operation is in 32°C (90°F) or higher temperatures, or constant operation in temperatures below 0°C (32°F).

- Prolonged idling (vehicle operation in stop and go traffic).

- Frequent short running periods (engine does not warm to normal operating temperatures).

- Police, taxi, delivery usage or trailer towing usage.

Oil & oil filter: change every 3750 miles.

Driveshaft boots: service or inspect every 7500 miles.

Front brake discs & calipers, & rear brake discs, calipers & pads: service or inspect every 7500 miles.

Power steering system: service or inspect every 7500 miles.

Suspension components: service or inspect every 7500 miles.

Tie rods, steering gear box & boots: service or inspect every 7500 miles.

Air cleaner element: service or inspect every 15,000 miles.

Transmission fluid (Accord V6 & Civic CVT): replace every 15,000 miles.

Transmission fluid (Accord L4, Civic, & Prelude): replace every 30,000 miles.

Timing balancer belt: replace every 60,000 miles.

Timing belt: replace every 60,000 miles.

Water pump: service or inspect every 60,000 miles.

43256-ACCO-C17

PRECAUTIONS

Before servicing any vehicle, please be sure to read all of the following precautions, which deal with personal safety, prevention of component damage, and important points to take into consideration when servicing a motor vehicle:

• Never open, service or drain the radiator or cooling system when the engine is hot; serious burns can occur from the steam and hot coolant.

• Observe all applicable safety precautions when working around fuel. Whenever servicing the fuel system, always work in a well-ventilated area. Do not allow fuel spray or vapors to come in contact with a spark, open flame, or excessive heat (a hot drop light, for example). Keep a dry chemical fire extinguisher near the work area. Always keep fuel in a container specifically designed for fuel storage; also, always properly seal fuel containers to avoid the possibility of fire or explosion. Refer to the additional fuel system precautions later in this section.

• Fuel injection systems often remain pressurized, even after the engine has been turned **OFF**. The fuel system pressure must be relieved before disconnecting any fuel lines. Failure to do so may result in fire and/or personal injury.

• Brake fluid often contains polyglycol ethers and polyglycols. Avoid contact with the eyes and wash your hands thoroughly after handling brake fluid. If you do get brake fluid in your eyes, flush your eyes with clean, running water for 15 minutes. If

eye irritation persists, or if you have taken brake fluid internally, IMMEDIATELY seek medical assistance.

• The EPA warns that prolonged contact with used engine oil may cause a number of skin disorders, including cancer! You should make every effort to minimize your exposure to used engine oil. Protective gloves should be worn when changing oil. Wash your hands and any other exposed skin areas as soon as possible after exposure to used engine oil. Soap and water, or waterless hand cleaner should be used.

• All new vehicles are now equipped with an air bag system. The system must be disabled before performing service on or around system components, steering column, instrument panel components, wiring and sensors. Failure to follow safety and disabling procedures could result in accidental air bag deployment, possible personal injury, and unnecessary system repairs.

• Always wear safety goggles when working with, or around, the air bag system. When carrying a non-deployed air bag, be sure the bag and trim cover are pointed away from your body. When placing a non-deployed air bag on a work surface, always face the bag and trim cover upward, away from the surface. This will reduce the motion of the module if it is accidentally deployed. Refer to the additional air bag system precautions later in this section.

• Clean, high quality brake fluid from a sealed container is essential to the safe and

proper operation of the brake system. You should always buy the correct type of brake fluid for your vehicle. If the brake fluid becomes contaminated, completely flush the system with new fluid. Never reuse any brake fluid. Any brake fluid that is removed from the system should be discarded. Also, do not allow any brake fluid to come in contact with a painted surface; it will damage the paint.

• Never operate the engine without the proper amount and type of engine oil; doing so WILL result in severe engine damage.

• Timing belt maintenance is extremely important! Many models utilize an interference-type, non-freewheeling engine. If the timing belt breaks, the valves in the cylinder head may strike the pistons, causing potentially serious (also time-consuming and expensive) engine damage. Refer to the maintenance interval charts in the front of this section for the recommended replacement interval for the timing belt, and to the timing belt procedure in this section for belt replacement and inspection.

• Disconnecting the negative battery cable on some vehicles may interfere with the functions of the on-board computer system(s) and may require the computer to undergo a relearning process once the negative battery cable is reconnected.

• When servicing drum brakes, only disassemble and assemble one side at a time, leaving the remaining side intact for reference.

ENGINE REPAIR

➡**Disconnecting the negative battery cable on some vehicles may interfere with the functions of the on board computer systems and may require the computer to undergo a relearning process, once the negative battery cable is reconnected.**

Distributor

REMOVAL

➡**The radio may contain a coded theft protection circuit. Always obtain the code number before disconnecting the battery. If the vehicle is equipped with 4WS, the steering control unit is shut down when the battery is disconnected. After connecting the battery, turn the steering wheel lock-to-lock to reset the steering control unit.**

1. Before servicing the vehicle, refer to the precautions in the beginning of this section.
2. Disconnect the negative battery cable.
3. Rotate the crankshaft to bring No. 1 cylinder to TDC, and align the white mark on the crankshaft pulley with the pointer on the timing belt cover.
4. Remove or disconnect the following:
   • Distributor cap with the ignition wires attached
   • Electrical connectors from the distributor
5. Mark the direction the ignition rotor is pointing on the distributor housing to aid in installation.
6. Matchmark the distributor housing with the cylinder head to aid in installation.
   • Distributor mounting bolts and remove the distributor
   • O-ring from the distributor housing, then discard the O-ring

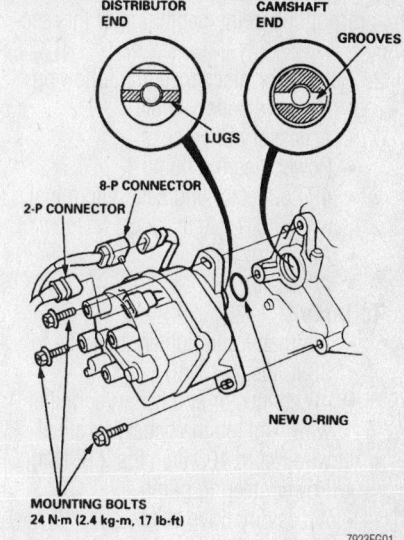

7923FG01

Distributor components—1.6L engines

INSTALLATION

1. Coat a new O-ring with clean engine oil and install it to the distributor housing.

2. Align the ignition rotor with the mark made on the distributor housing. The drive lugs are offset so the distributor cannot be installed incorrectly. Fit the distributor into place and turn the rotor until the drive lugs engage and the distributor seats in the cylinder head.

➡ **The lugs on the end of the distributor and their mating grooves in the camshaft end, are offset to eliminate the possibility of installing the distributor 180° out of time.**

3. Align the matchmark on the distributor housing and the cylinder head and install the mounting bolts snugly.

4. Install or connect the following:
- Distributor cap with the ignition wires
- Distributor electrical connectors
- Negative battery cable and enter the radio security code.

5. If equipped with 4-Wheel Steering (4WS), start the engine and turn the steering wheel lock-to-lock to reset the 4WS control unit.

6. Adjust the ignition timing.

7. Tighten the distributor mounting bolts to 16 ft. lbs. (22 Nm)

Alternator

REMOVAL & INSTALLATION

1.6L and 1.7L Engines

1. Before servicing the vehicle, refer to the precautions in the beginning of this section.

2. Remove or disconnect the following:
- Negative battery cable
- Accessory drive belts
- Power steering pump
- 4P connector and battery terminal wire
- Alternator bolts
- Alternator

To install:
- Alternator and tighten the bolts to 33 ft. lbs. (44 Nm)
- 4P connector and battery terminal wire. Tighten the battery terminal wire nut to 108 inch lbs. (12 Nm).
- Power steering pump
- Accessory drive belts
- Negative battery cable

2.0L Engine

1. Before servicing the vehicle, refer to the precautions in the beginning of this section.

2. Remove or disconnect the following:
- Negative battery cable
- Accessory drive belt
- 4P connector and battery terminal wire
- Alternator bolts
- Alternator

To install:
- Alternator and tighten the bolts to 33 ft. lbs. (44 Nm)
- 4P connector and battery terminal wire. Tighten the battery terminal wire nut to 108 inch lbs. (12 Nm).
- Accessory drive belt
- Negative battery cable

2.2L Engine

1. Before servicing the vehicle, refer to the precautions in the beginning of this section.

2. Remove or disconnect the following:
- Negative battery cable, then the positive
- Power steering pump

- Cruise control actuator but do not remove the cable

3. Loosen the through bolt and then loosen adjusting bolt.
- Alternator belt
- Adjusting bolt
- Through bolt and then the alternator

To install:

4. Installation is the reverse of removal.

5. Adjust the alternator belt tension.

2.3L Engines

1. Before servicing the vehicle, refer to the precautions in the beginning of this section.

2. Note the radio security code and the radio presets.

3. Remove or disconnect the following:
- Negative battery cable, then the positive
- 4P connector and battery terminal wire from the alternator
- Adjusting bolt, locknut and the mounting bolt
- Alternator belt
- Alternator from the bracket

To install:

4. Installation is the reverse of removal.

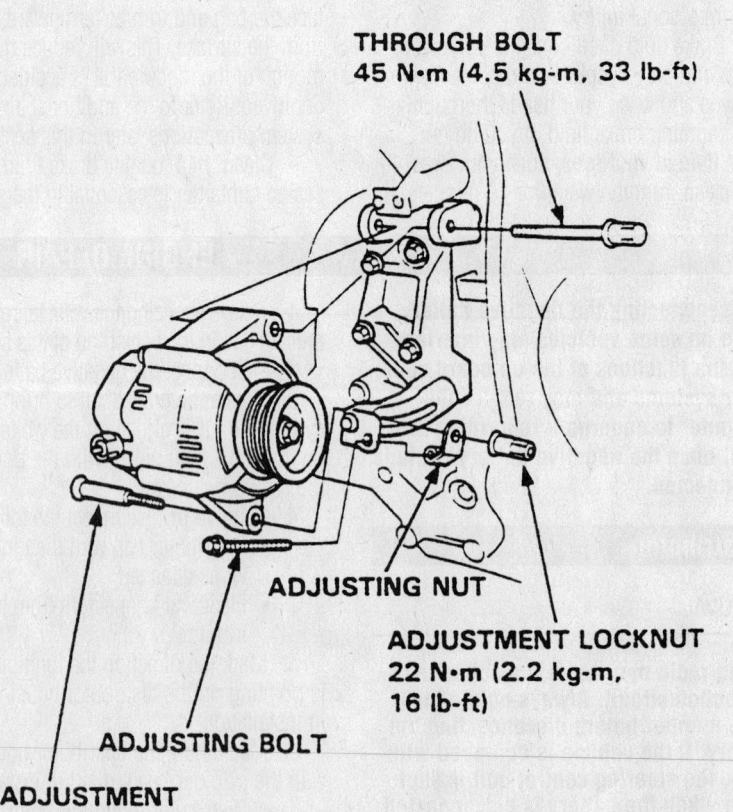

THROUGH BOLT
45 N·m (4.5 kg-m, 33 lb-ft)

ADJUSTING NUT

ADJUSTMENT LOCKNUT
22 N·m (2.2 kg-m, 16 lb-ft)

ADJUSTING BOLT

ADJUSTMENT BOLT

91182G24

Alternator mounting bolt locations

5. Adjust the alternator belt tension.
6. Enter the anti-theft code for the radio.

2.4L Engine

1. Before servicing the vehicle, refer to the precautions in the beginning of this section.
2. Note the radio security code and the radio presets.
3. Remove or disconnect the following:
   - Negative battery cable, then the positive
   - Drive belt
   - Auto-tensioner
   - Connectors from the alternator
   - 3 alternator mounting bolts and the alternator

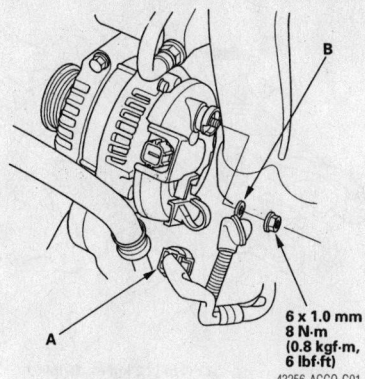

6 x 1.0 mm
8 N·m
(0.8 kgf·m,
6 lbf·ft)
43256-ACCO-G01

Alternator mounting—2.4L engine

To install:
- Alternator and 3 mounting bolts. Torque the bolts to 16 ft. lbs. (22 Nm).
- Electrical connectors
- Auto-tensioner
- Drive belt
- Positive, then negative battery cables

4. Enter the security code and radio presets

3.0L Engine

1. Before servicing the vehicle, refer to the precautions in the beginning of this section.
2. Note the radio security code and the radio presets.
3. Remove or disconnect the following:
   - Negative battery cable, then the positive
   - Alternator belt tension by pulling back on the adjuster and then remove the belt
   - Condenser fan motor connector from the shroud
   - Condenser fan assembly
   - Four prong connector from the rear of the alternator

- Alternator mounting bolts
- Wiring harness clamp
- Alternator assembly

To install:
4. Alternator installation is the reverse of the removal procedure.
5. Connect the positive battery cable, then the negative battery cable. Enter the radio security code and station presets.

⁑ WARNING

Be sure to adjust the alternator belt to the proper tension or alternator bearing failure may occur.

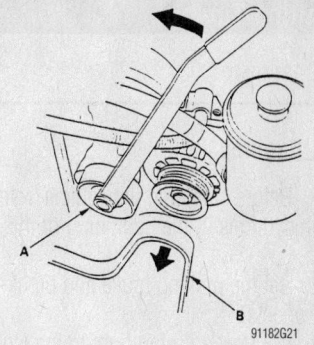

91182G21

Relieve the belt tension by pulling back on the tensioner—3.0L engine

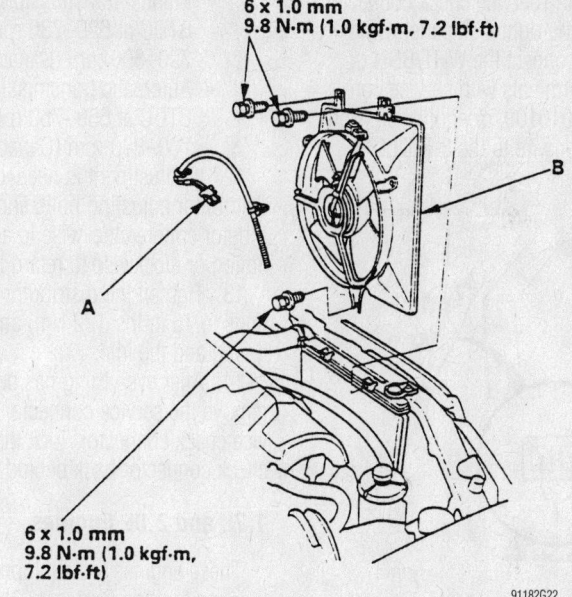

6 x 1.0 mm
9.8 N·m (1.0 kgf·m, 7.2 lbf·ft)

6 x 1.0 mm
9.8 N·m (1.0 kgf·m, 7.2 lbf·ft)

91182G22

Remove the condenser fan—3.0L engine

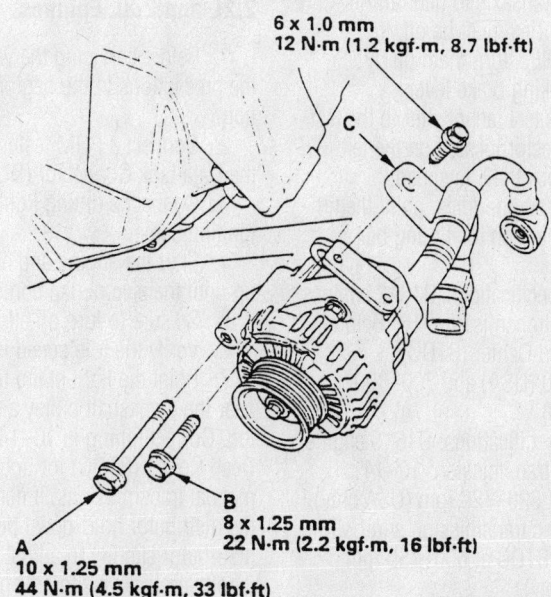

6 x 1.0 mm
12 N·m (1.2 kgf·m, 8.7 lbf·ft)

8 x 1.25 mm
22 N·m (2.2 kgf·m, 16 lbf·ft)

10 x 1.25 mm
44 N·m (4.5 kgf·m, 33 lbf·ft)

91182G23

Torque the alternator bolts to the specs shown—3.0L engine

Ignition Timing

ADJUSTMENT

1.6L Engines

1. Before servicing the vehicle, refer to the precautions in the beginning of this section.

2. Set the parking brake and block the front wheels.

3. Connect a timing light to the No. 1 spark plug wire.

4. Start the engine and allow it to warm up.

5. Pull out the service check connector located behind the right kick panel. On the 2-P connector, connect the WHT/BGN or BRN and BLK terminals with service connector 07PAZ-0010100, or equivalent. Don't connect a jumper wire to the 3-P DLC.

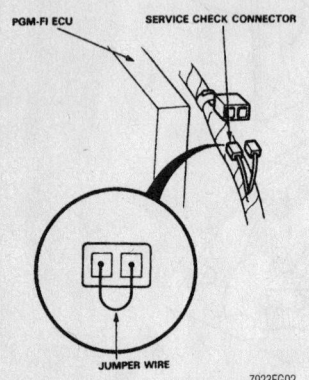

Service check connector—1.6L engines

6. Shift the transaxle to neutral. All electrical accessories must be off. If equipped with DRL's, turn them off by engaging the parking brake lever.

7. Connect a test tachometer to the test tachometer connector located on the left shock tower. Check the idle speed.

8. While the engine idles, point the timing light at the mark on the timing belt cover.

9. Timing specifications: B16A2 engines
- Manual transmission: 16° Before Top Dead Center (BTDC) at 650–750 (USA) and 700–800 (Canada)

10. Timing specifications: D16Y5 engine
- Manual transmission: 10–14° BTDC at 620–720 rpm (USA only)
- Automatic transmission and CVT: 10–14° BTDC at 650–750 rpm (USA only)

11. Timing specifications: D16Y7 and D16Y8 engines

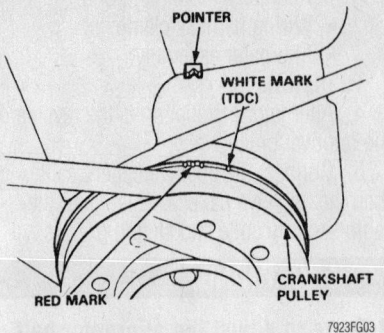

Typical crankshaft pulley timing mark location

- Manual transmission: 10–14° BTDC at 620–730 rpm (USA) or 700–800 rpm (Canada)
- Automatic transmission: 10–14° BTDC at 650–750 rpm (USA) or 700–800 rpm (Canada)

12. If adjustment is needed, loosen the distributor adjusting bolts and turn the distributor counterclockwise to advance the timing or clockwise to retard the timing.

13. Tighten the distributor adjusting bolts to 16 ft. lbs. (22 Nm) and recheck the timing and the idle.

14. After everything has been rechecked, remove the service connector from the service check connector. Tuck the service check connector back behind the kick panel.

1.7L and 2.0L Engines

These engines are equipped with Distributorless Ignition Systems (DIS). No adjustment is necessary.

2.2L and 2.3L Engines

1. Before servicing the vehicle, refer to the precautions in the beginning of this section.

2. Connect a PGM tester (scan tool) to the Data Link Connector (DLC).

3. Connect a timing light to the No. 1 ignition cable.

4. Start the engine and allow it to warm up until the electric fan comes on.

5. Be sure to turn off all accessories.

6. Verify the idle speed is 650–750 rpm.

7. Point the light at the timing belt cover near the crankshaft pulley and read the timing. Correct timing is 10–14° Before Top Dead Center (BTDC) for both automatic and manual transmissions. If necessary, loosen the distributor hold-down bolt and rotate the distributor slightly to adjust the timing. Turn it counterclockwise to advance and clockwise to retard the timing.

8. Tighten the hold-down bolt to 16 ft.

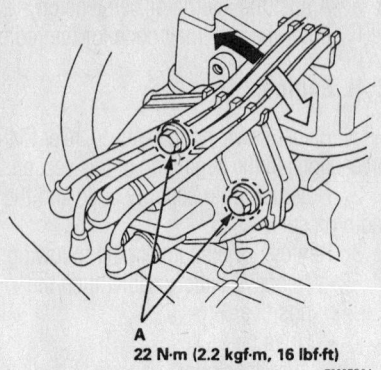

22 N·m (2.2 kgf·m, 16 lbf·ft)

Distributor hold-down bolt locations—2.3L (F23A1 and F23A4) engine

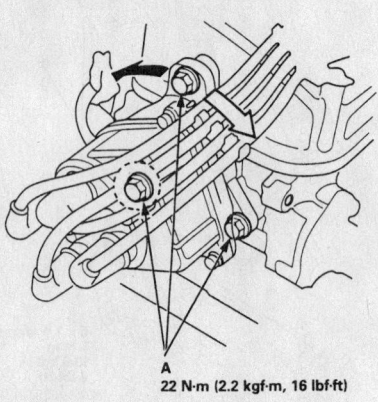

22 N·m (2.2 kgf·m, 16 lbf·ft)

Distributor hold-down bolt locations—2.3L (F23A5) engine

lbs. (22 Nm). Recheck the timing after the bolt is tight to confirm the correct timing.

9. Disconnect the PGM tester.

2.4L and 3.0L Engines

The ignition timing is only adjustable by the Powertrain Control Module (PCM), but the ignition base timing can be checked by performing the following:

1. Before servicing the vehicle, refer to the precautions in the beginning of this section.

2. Connect a PGM tester (scan tool) to the data link connector.

3. Connect a timing light to the No. 1 ignition cable.

4. Start the engine and allow it to warm up until the electric fan comes on.

5. Be sure to turn off all accessories.

6. Verify that the idle speed is 630–730 rpm.

7. Point the light at the timing belt cover near the crankshaft pulley and read the timing. Correct timing is 8–12° Before Top Dead Center (BTDC). If the ignition timing is different from the specification, replace the PCM.

Engine Assembly

REMOVAL & INSTALLATION

➡**The original radio contains a coded anti-theft circuit. Obtain the security code number before disconnecting the battery cables.**

Civic

1. Before servicing the vehicle, refer to the precautions in the beginning of this section.

2. Disconnect the negative and positive battery cables. Wait at least 3 minutes before working around the air bags.

➡**The engine and transaxle are removed from the vehicle as 1 unit.**

3. Support the hood as far open as possible. If the hood is to be removed, first matchmark the hinge plates with a felt-tipped marker.

4. Remove the battery from the vehicle. Unbolt and remove the battery tray.

5. Disconnect the battery and alternator cables from the underhood fuse and relay box on the right shock tower.

6. Remove the lower right kick panel to expose the Powertrain Control Module (PCM).

7. Label and disconnect the 5 wiring harness connections from the PCM.

8. Unbolt the main wiring harness retainer from the rear of the fuse and relay box on the right side of the bulkhead. Carefully pull the grommet out of its bulkhead opening. Next, pull the PCM harness and connectors through the opening. Be careful not to damage the wiring, insulation, or connectors.

9. Relieve the fuel pressure:
   a. Loosen the fuel filler cap.
   b. Use a box-end wrench and a flare nut wrench to hold the fuel filter banjo fitting.
   c. Place a shop towel over the fuel filter to catch the fuel spray.
   d. Slowly loosen the fuel filter service bolt 1 full turn.
   e. Clean up any spilled fuel.

10. Remove or disconnect the following:
   • Intake air duct and air cleaner
   • Intake Air Temperature (IAT) sensor connector from the air cleaner case, if equipped
   • Fuel feed hose from the fuel filter
   • Fuel return hose from the fuel rail
   • Intake manifold/throttle body vacuum hoses
   • Brake booster vacuum hose
   • Evaporative emissions (EVAP) canister vacuum hose
   • Power Steering Pressure (PSP) switch and detach its clamp from the bracket below the brake booster
   • Transaxle ground cable
   • Radiator hose bracket

11. Loosen the throttle cable locknut, then disconnect the cable from the throttle body linkage. Move the cable aside without kinking it.

12. Loosen the power steering pump mounting bolts. Slip power steering belt off its pulleys. Unbolt the steering pump and move it out of the work area. Don't disconnect the hydraulic hoses.

➡**Label the connectors before detaching them.**

13. Remove or disconnect the following:
   • Engine wiring harness connectors at the left side of the engine compartment

14. Drain the coolant from the radiator and engine block.

15. Remove or disconnect the following:
   • Upper and lower radiator hoses
   • Heater hoses from the cylinder head

16. If equipped with a CVT transaxle, loosen the shift cable locknut. Remove the spring clip and washers and disconnect the shift cable from its linkage. Be careful not to kink the cable or damage its boot.

17. Remove or disconnect the following:
   • Hydraulic line brackets from the top of the transaxle case, if equipped with a manual transaxle

18. Attach a chain hoist to the engine lifting brackets. Don't raise the hoist to lift the engine yet.

19. Raise the vehicle and support it safely. Remove the front wheels.

20. Remove or disconnect the following:
   • Engine splash shield

21. Drain the engine oil.

22. Drain the fluid from the transaxle.

23. Remove or disconnect the following:
   • Left front engine mount bracket from the shock tower, if equipped with air conditioning

24. Loosen the compressor idler pulley and adjusting bolt. Slip the belt around the engine mount stud to remove it.

25. Remove or disconnect the following:
   • Compressor mounting bolts to separate the compressor from its mounting plate. Move the compressor out of the work area. Do not disconnect the air conditioning refrigerant lines.

26. Remove or disconnect the following:
   • ATF cooler lines, if equipped. Plug the cooler lines to prevent fluid leakage and contamination.
   • Slave cylinder from the transaxle case without disconnecting its hydraulic line, if equipped with manual transaxle
   • Front exhaust pipe from the exhaust manifold and catalytic converter. Unbolt its hanger bracket and remove the exhaust pipe.
   • Shift cable from the transaxle control shaft, if equipped with automatic transaxle
   • Shift rod and extension rod from the transaxle, if equipped with manual transaxle
   • Strut damper fork
   • Steering knuckle ball joint from the lower control arm using a ball joint separator

27. Pry the inboard CV-joints from the transaxle. Then, move the halfshafts away from the transaxle and wire them to the undercarriage of the vehicle. Tie plastic bags over the inboard CV-joints to prevent damage to the boots and splined shafts.

28. Raise the hoist slightly to take up the weight of the engine and transaxle assembly.

29. Disconnect the engine mounts in the following order:
   a. Unbolt and remove the left front engine mount.
   b. Unbolt and remove the right front engine mount and bracket assembly.
   c. Remove the rear engine mount through-bolt. Then, unbolt the rear mount bracket from the engine block.

30. If necessary, lower the vehicle slightly to gain access to the side engine and transaxle mounts. Do not release the tension on the chain hoist. The engine must be securely supported.

31. Unbolt the side engine mount bracket from the engine block bracket and mount damper.

32. Unbolt the transaxle mount bracket from the transaxle case. Then, unbolt the mount from the shock tower.

33. Raise the chain hoist to lift the engine a few inches off of its mounts.

34. Verify that all electrical, vacuum, and fuel lines have been disconnected.

35. Raise the engine and transaxle assembly and remove it from the vehicle.

To install:

➡**Use new self-locking nuts and gaskets when installing the front exhaust pipe and when assembling the front suspension. Use new set rings on the inboard CV-joint splined shafts.**

36. Lower the engine and transaxle assembly into the vehicle.

37. Install and connect the engine and transaxle mounts and brackets. Use new self-locking nuts and color-coded bolts. At this point, only tighten the mounting nuts and bolts by hand.

38. Before installing the left front engine mount, fit the air conditioning compressor back into place and install the compressor belt. Tighten the compressor bolts to 17 ft. lbs. (24 Nm).

➡**Failure to tighten the bolts in the proper sequence can cause excessive noise and vibration and reduce bushing life. Be sure to check that the bushings are not twisted or offset.**

39. The engine and transaxle mount and bracket fasteners must be tightened in the proper sequence with the weight of the engine resting upon them. This step is important for engine mount pre-loading. Tighten the engine mount bolts in the following sequence:

   a. Transaxle mount bolts: 47 ft. lbs. (64 Nm); or 28 ft. lbs. (38 Nm) for CVT-equipped vehicles.

   b. Side engine mount bracket nuts: 54 ft. lbs. (74 Nm).

   c. Rear mount bracket bolts: 61 ft. lbs. (83 Nm); or 43 ft. lbs. (59 Nm) for CVT-equipped vehicles.

   d. Rear mount through-bolt: 43 ft. lbs. (59 Nm).

   e. Transaxle mount bracket nuts or bolts: 47 ft. lbs. (64 Nm).

   f. Transaxle mount through-bolt: 54 ft. lbs. (74 Nm).

   g. Right front mount bracket bolts: 33 ft. lbs. (44 Nm).

   h. Right front mount carrier bolts: 33 ft. lbs. (44 Nm).

   i. Left front mount: stud: 61 ft. lbs. (85 Nm); carrier bolts: 33 ft. lbs. (44 Nm); nut: 43 ft. lbs. (59 Nm).

40. Remove the chain hoist from the engine lifting hooks.

41. Install or connect the following:
- New set rings on the inboard splined shafts of each halfshaft. Check that the set ring on each inboard CV-joint clicks into place when the halfshafts are installed into the transaxle.
- Damper fork and reconnect the lower ball joint. When the weight of the vehicle is resting on its suspension, tighten the pinch bolt to 32 ft. lbs. (44 Nm) and the fork bolt to 47 ft. lbs. (65 Nm). Tighten the ball joint castle nut to 36–43 ft. lbs. (50–60

Nm). Next, tighten the castle nut only enough to install a new cotter pin.
- Slave cylinder, if equipped. Tighten the slave cylinder mounting bolts to 16 ft. lbs. (22 Nm). If the clutch hydraulic line was disconnected, air must be bled from the system.
- Transaxle shift and extension rods to the linkage at the transaxle case, if equipped. Install a new 8mm spring pin into the shift rod linkage. Then, install the retainer clip and boot. Tighten the extension rod bolt to 16 ft. lbs. (22 Nm).
- Shift cable to the control shaft, if equipped with an automatic transaxle. Use a new lockwasher and tighten the lockbolt to 10 ft. lbs. (14 Nm). Tighten the shift cable cover bolts to 16 ft. lbs. (22 Nm). Install the shift cable cover and tighten its bolts to 16 ft. lbs. (22 Nm).

42. Install the front exhaust pipe using new self-locking nuts:

   a. If equipped with the D16Y8 engine, tighten the converter flange nuts to 16 ft. lbs. (22 Nm)

   b. Tighten the exhaust manifold nuts to 40 ft. lbs. (55 Nm).

   c. If equipped with the D16Y5 or D16Y7 engine, tighten the converter flange nuts to 25 ft. lbs. (33 Nm).

43. Install or connect the following:
- Tighten the exhaust flange bolts to 16 ft. lbs. (22 Nm).
- ATF cooler lines. If the rubber cooler lines are cracked or stressed, they must be replaced.
- Engine splash shield

44. Refill the engine with fresh oil.

45. Refill the transaxle with the proper fluid.

46. Lower the vehicle.

47. If equipped, fit the clutch hydraulic line brackets back into place. Tighten the 8mm bolts to 17 ft. lbs. (24 Nm). Tighten the 6mm bolts to 96 inch lbs. (11 Nm).

48. If equipped with a CVT transaxle, reconnect the shift cable to the linkage. Use new plastic washers and a new spring clip. Tighten the locknut to 22 ft. lbs. (29 Nm).

49. Adjust the alternator and air conditioning compressor belt tensions.

50. Install or connect the following:
- Upper and lower radiator hoses and the heater hoses
- Power steering pump into its mounts. Adjust the pump belt tension, then tighten the mounting bolts to 17 ft. lbs. (24 Nm).
- PSP switch connector and attach its harness clamp

- Intake manifold/throttle body vacuum hoses
- Brake booster vacuum hose
- EVAP canister vacuum hose
- Fuel line fittings to the fuel filter and fuel rail. Use new sealing washers. Tighten the banjo fittings to 25 ft. lbs. (33 Nm), and the service bolts to 11 ft. lbs. (15 Nm). Don't overtighten the fittings
- Throttle cable and adjust its deflection to 10–12mm (0.39–0.47 in.).

51. Feed the PCM harness through the hole in the bulkhead. Apply sealant to the grommet, then install the retainer.

52. Install or connect the following:
- Engine wiring harness and ground cables that were disconnected during removal. Be sure the grounds are free of corrosion to ensure good contact.
- Fuse and relay box back into position
- Battery and alternator cables
- Air cleaner case and air intake duct
- IAT connector
- 5 PCM connectors and kick panel
- Battery tray and the battery

53. Verify that all wiring harnesses and grounds, vacuum lines, fuel lines have been reconnected.

54. Refill the radiator with fresh coolant.

55. If it was removed, install the hood. Reconnect the windshield washer tubing. After installation, check to be sure that the hood, fender, and grille panel gaps are equal.

56. Reconnect the positive and negative battery cables.

57. Turn the ignition switch to the **ON** position, but don't start the engine. Then, turn the ignition **OFF**. Repeat this procedure 2 or 3 times and check for fuel leaks.

58. Start the engine and allow it to warm up to its normal operating temperature.

59. Bleed the air from the cooling system with the heater valve open.

60. Check the throttle cable deflection and operation.

61. Check and adjust the ignition timing.

62. Shut the engine off and check the drive belt adjustments.

63. Check all fluid levels and top up as necessary.

64. Check and adjust the front wheel alignment.

65. Road test the vehicle.

Prelude

1. Before servicing the vehicle, refer to the precautions in the beginning of this section.

2. Secure the hood as far open as possible.

3. Remove or disconnect the following:
- Negative battery cable, then the positive battery cable
- Radiator cap

4. Raise and safely support the vehicle. Remove the front wheels and the engine splash shield.

5. Drain the engine coolant into a sealable container.

6. Drain the transaxle fluid into a sealable container. Install the drain plug with a new gasket.

7. Lower the vehicle to a working level.

8. Remove or disconnect the following:
- Air intake duct and the air cleaner case
- PAIR vacuum tank and bracket
- Battery and the battery base
- Battery cable and starter cable harnesses from the body

9. Relieve the pressure from the fuel system.

✳✳ CAUTION

The fuel injection system remains under pressure after the engine has been turned OFF. Properly relieve fuel pressure before disconnecting any fuel lines. Failure to do so may result in fire or personal injury.

10. Remove or disconnect the following:
- Fuel feed hose from the fuel rail and the fuel return line from the fuel pressure regulator
- Injector resistor connector on the left side of the engine compartment
- Throttle cable by loosening the locknut, then slip the cable end out of the throttle linkage. Take care not to bend the cable when removing it. Always replace any kinked cable with a new one.
- Engine wiring harness connectors, terminal, and clamps on the right side of the engine
- Power cable from the under-hood fuse/relay box
- Brake booster vacuum hose and emissions control vacuum tubes from the intake manifold
- Cruise control actuator electrical connector and vacuum tube, then the actuator
- Engine ground cable from the body side
- Power steering pump drive belt, then the pump
- Air conditioning condenser fan,

then install a protector plate on the radiator
- Alternator mounting bolt, nut, and adjusting nut, then the alternator drive belt
- Air conditioning compressor electrical connector
- Compressor without disconnecting the air conditioning hoses. Support the compressor with a strong wire out of the way.
- Upper and lower radiator hoses, and the heater hoses from the engine
- Transaxle ground cable
- Cooler hoses, if equipped with an automatic transaxle

11. If equipped with a manual transaxle perform the following:

a. Disconnect the shift cable and the select cable from the transaxle. Do not bend the cables when removing them. Replace any kinked cable with a new one.

b. Remove the clutch slave cylinder and the pipe/hose assembly. Do not depress the clutch pedal once the slave cylinder has been removed.

c. Remove the clutch damper assembly.

12. Remove or disconnect the following:
- Vehicle Speed/Power Steering Speed (VSS/PSS) sensor assembly. Do not disconnect the hoses
- Nuts attaching the exhaust pipe to the exhaust manifold and the catalytic converter
- Bolts from the exhaust pipe hanger, then the exhaust pipe and discard the gaskets

13. If equipped with an automatic transaxle, remove the shift cable cover, then disconnect the shift cable. Do not bend the cable and replace the cable if it becomes kinked.

14. Remove or disconnect the following:
- Left and the right side damper forks
- Lower ball joints from the lower control arms

15. Pry the halfshafts from the transaxle. Cover the inner CV-joints with plastic bags to protect them.

16. Swing the halfshafts under the fender out of the way.

17. Attach an engine hoist to the engine lifting points and raise the hoist to remove all slack from the chain.

18. Remove or disconnect the following:
- Rear engine mount bracket
- Front engine mount bracket
- Left side engine mount
- Transaxle mount and the mount bracket

19. Check that the engine is completely free of vacuum hoses, fuel and coolant hoses, and electrical wiring.

20. Slowly raise the engine approximately 6 in. (150mm). Check once again that all hoses and wires have been disconnected from the engine.

21. Raise the engine all the way and remove it from the vehicle.

22. Remove the transaxle.

23. If equipped with a manual transaxle, remove the clutch cover (pressure plate) and clutch disc.

24. Mount the engine on an engine stand, making sure the mounting bolts are tight. If an engine stand is not available, support the engine in an upright position with blocks. Never leave an engine hanging from a lift or hoist.

To install:

25. Install or connect the following:
- Clutch disc and pressure plate to the flywheel for manual transaxle vehicles
- Transaxle
- Engine into position and lower it into the car, aligning the mounts and bushings.

➡**When installing the engine mounts and vibration dampers in the following steps, they must be tightened to the correct tension in the correct order if they are to damp vibration properly.**

- Side engine mount and the through-bolt. Do not tighten the through-bolt at this time
- Nut and bolt attaching the side mount to the engine: 40 ft. lbs. (55 Nm)
- Transaxle mount and through-bolt. Do not tighten the through-bolt at this time
- Rear engine mount and new bolts attaching the mount to the engine: 40 ft. lbs. (55 Nm)
- New rear engine mount through-bolt: 47 ft. lbs. (65 Nm)
- Front mount and the 3 bolts attaching the mount to the engine assembly, only snug the bolts in place
- New through-bolt to the front mount: 47 ft. lbs. (65 Nm)
- Nuts to the transaxle mount: 28 ft. lbs. (39 Nm)

26. Tighten the side engine mount through-bolt to 47 ft. lbs. (65 Nm).

27. Tighten the transaxle mount through-bolt to 47 ft. lbs. (65 Nm).

28. Tighten the 3 bolts attaching the front mount to the engine to 28 ft. lbs. (39 Nm).

29. Remove the hoist equipment from the engine.

30. Install or connect the following:
- New spring clips to the inner CV-joints
- Halfshafts into the transaxle. Be sure that the inner joint spring clips click into place
- Lower ball joints to the lower control arms. Tighten the nuts to 36–43 ft. lbs. (50–60 Nm). Install a new cotter pin to the ball joint stud.

31. Using new self-locking bolts, attach the damper forks to the struts and tighten to 32 ft. lbs. (44 Nm). Tighten the new nut and bolt attaching the damper fork to the lower control arm to 47 ft. lbs. (65 Nm).

32. If equipped with an automatic transaxle, connect the shift cable to the transaxle. Install a new lockwasher and tighten the attaching bolt to 84 inch lbs. (10 Nm). Install the shift cable cover. Tighten the shift cable cover attaching bolts to 13 ft. lbs. (18 Nm).

33. Install or connect the following:
- Exhaust pipe with new gaskets
- New nuts attaching the exhaust pipe to the exhaust manifold: 40 ft. lbs. (55 Nm)
- New nuts attaching the exhaust pipe to the catalytic converter: 25 ft. lbs. (34 Nm)
- New attaching bolts to the exhaust pipe hanger: 13 ft. lbs. (18 Nm)
- VSS/PSS, connect the electrical connector, and tighten the mounting bolt to 13 ft. lbs. (18 Nm)

34. If equipped with a manual transaxle perform the following:
  a. Install the clutch damper assembly and tighten the attaching bolts to 16 ft. lbs. (22 Nm).
  b. Install the clutch slave cylinder and the pipe/hose assembly and tighten the slave cylinder mounting bolts to 16 ft. lbs. (22 Nm).
  c. Connect the shift cable and the select cable to the transaxle. Adjust the shift cable and select cable.

35. Install or connect the following:
- Cooler hoses, if equipped with an automatic transaxle
- Transaxle ground cable
- Upper and lower radiator hoses and heater hoses to the engine
- Air conditioning compressor, connect the electrical connector. Tighten the mounting bolts to 16 ft. lbs. (22 Nm).
- Alternator drive belt, then adjust

36. Remove the protector plate from the radiator and install the air conditioning condenser fan.

37. Install or connect the following:
- Power steering pump and drive belt. Adjust the drive belt tension, then tighten the attaching nuts and bolts to 16 ft. lbs. (22 Nm).
- Engine ground cable to the body
- Cruise control actuator, electrical connector and vacuum tube. Tighten the mounting bolts to 84 inch lbs. (10 Nm)
- Brake booster vacuum hose and the emissions control vacuum tubes to the intake manifold
- Engine wiring harness connectors, terminal, and clamps
- Throttle cable, then adjust
- Injector resistor connector on the left side of the engine compartment
- Fuel return hose to the regulator
- Fuel feed hose to the fuel rail with new washers. Tighten the cap nut to 16 ft. lbs. (22 Nm)
- Battery and starter cables to the body
- Battery base and the battery. Tighten the battery base attaching bolts to 16 ft. lbs. (22 Nm)
- PAIR vacuum tank and bracket. Tighten the mounting bolts to 96 inch lbs. (10 Nm)
- Air cleaner duct and housing
- Engine splash shield and the front wheels

38. Lower the vehicle.

39. Fill the engine with oil and the transaxle with fluid.

40. Fill and bleed the air from the cooling system.

41. Connect the positive, then the negative battery cable and enter the radio security code.

42. Switch the ignition **ON** but do not engage the starter. The fuel pump should run for approximately 2 seconds, building pressure within the lines. Switch the ignition **OFF**, then **ON** 2 or 3 more times to build full system pressure. Check for fuel leaks.

43. Start the engine, allowing it to idle. Check the hoses and lines carefully for any sign of leakage.

44. Check the timing and idle speed.

45. After the engine has warmed up fully and the fan(s) have come on at least once, recheck the engine for fluid leaks. Switch the engine OFF.

46. Adjust the belts and throttle cable as necessary.

47. If equipped with 4WS, start the engine and turn the steering wheel lock-to-lock to reset the 4WS control unit.

48. Road test the vehicle, then loosen

and retighten the 3 bolts attaching the front engine mount to the engine. Tighten the bolts to 28 ft. lbs. (39 Nm).

Accord

1. Before servicing the vehicle, refer to the precautions in the beginning of this section.

2. Obtain the anti-theft code for the radio, then disconnect the battery cables. Be sure to disconnect the negative cable first.

3. Remove the air intake duct.

4. Secure the hood in the open position with a long prop rod such as P/N 74145-S84-A00.

5. Remove or disconnect the following:
- Both battery cables and the connector from the underhood relay box
- Battery and tray, on the 3.0L engine
- Bolt securing the relay box to the body
- Accelerator and cruise control cables from the throttle body and bracket

6. Properly relieve the fuel system pressure.
- Fuel hoses from the fuel rail
- Brake booster vacuum and evaporative emissions (EVAP) canister hoses
- Vacuum hose from the canister
- Hose securing the power steering hose on the engine
- Power steering pump belt, then remove the pump and position it out of the way. Use wire if necessary.
- Powertrain Control Module (PCM) connectors from the control module. Remove the grommet and pull the connectors through.

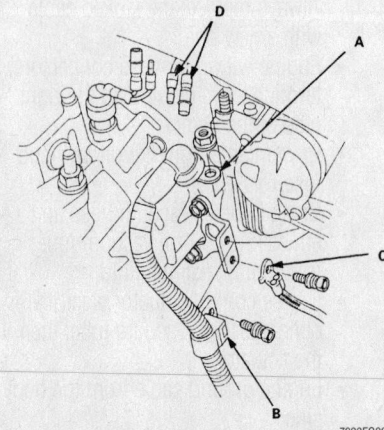

Starter cable, clamp, ground cable and back-up light switch connector locations— Accord with 2.3L engine

- Wiring harness connectors at the right side of the engine compartment for 2.3L and 2.4L engines and on the left side for the 3.0L engine.

7. On the 2.3L and 2.4L engine, remove the starter cable (A) and clamp (B). Remove the ground cable (C) and back-up light switch connectors (D). On the 3.0L engine, remove the starter wiring from the engine compartment attaching points.

8. On vehicles with a manual transaxle, disconnect the shift and select cables from the transaxle. Remove the slave cylinder mounting bolts and position the cylinder out of the way. Be sure not to bend the line.

9. Remove or disconnect the following:
- Rear engine mount through-bolt and stiffener
- Front engine mount bracket mounting bolts and loosen the through-bolt
- Radiator cap

10. Raise and safely support the vehicle.

11. Remove or disconnect the following:
- Front tires
- Engine under cover

12. Loosen the radiator drain plug and drain the coolant.

13. Drain the transaxle oil or fluid, then reinstall the plug using a new washer.

14. Drain the engine oil, then reinstall the plug using a new washer.

15. Lower the vehicle and remove the upper and lower radiator hoses and heater hoses from the engine.

16. On vehicles with an automatic transaxle, disconnect the ATF fluid cooler lines.

17. Remove the air conditioning compressor from the engine and position it to the side without disconnecting the hoses.

18. Raise the vehicle and remove the front exhaust pipe.

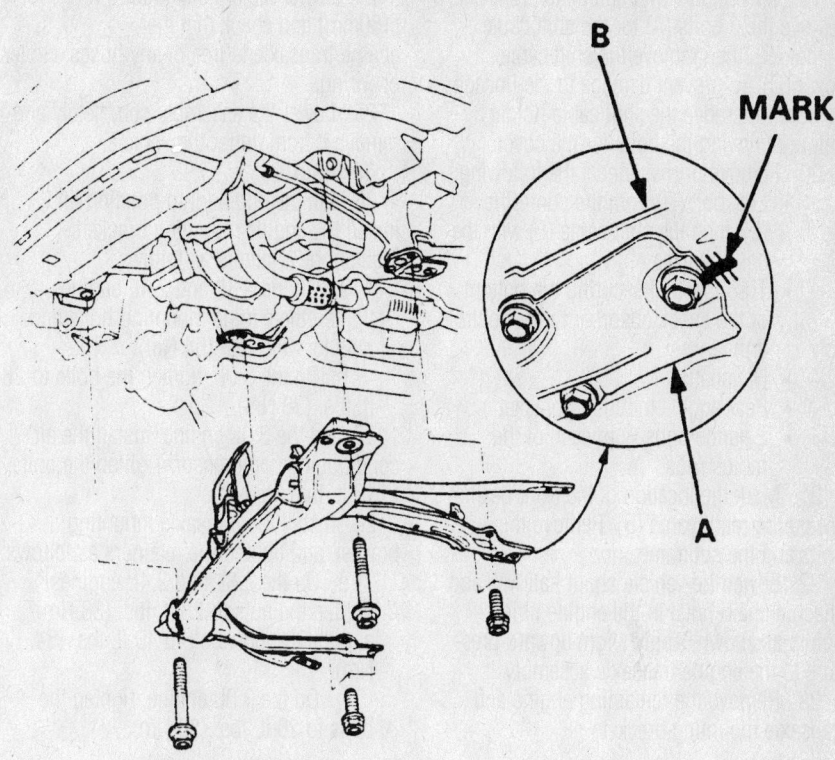

Be sure to mark the location of the front beams (A) on the rear beams (B) before removing the subframe—2.3L Accord

7923FG08

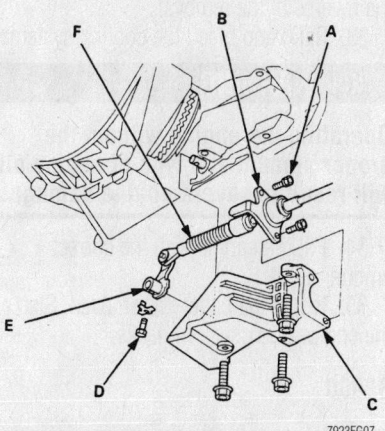

Automatic transaxle linkage components—2.3L Accord

7923FG07

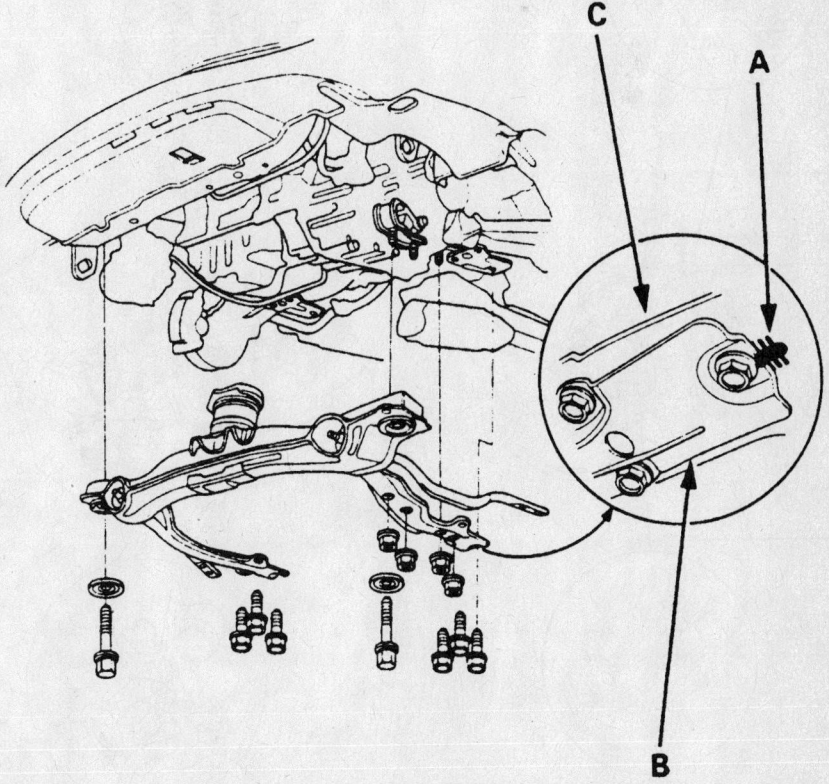

Mark the location of the front beams (A) on the rear beams (B) before removing the subframe—3.0L Accord

7923FG10

19. On vehicles with automatic transaxle, remove the 2 bolts (A) for the shift cable holder (B), then remove the shift cable cover (C). To prevent damage to the linkage, be sure to remove the shift cable holder before removing the bolts for the cover.

20. Remove or disconnect the following:
- Lockbolt (D) from the control lever (E), then the shift cable (F) with the control lever
- Through-bolt securing the bottom of the shock absorber to the control arm
- Halfshafts
- Rear engine mounting bracket
- 2 flange bolts from each of the radius rods

21. Mark the location of the front beams (A) on the rear beams (B). Remove the 4 bolts and the subframe.

22. Lower the vehicle about half way and attach a chain hoist to the engine lifting points as shown. Apply slight upward pressure to the engine/transaxle assembly.

23. Remove the remaining engine and transaxle mounting brackets.

24. Lower the engine about 6 in. (150mm) and check that the engine/transaxle is free of any hoses, cables or wiring.

25. Lower the assembly completely and remove it from under the vehicle.

To install:

26. Lift the engine into position and install the engine mounting brackets. Tighten the retainers as follows:

a. On the 2.3L and 2.4L engines, tighten the engine mounting bolts and nuts to 40 ft. lbs. (54 Nm).

b. On the 3.0L, tighten the bolts to 28 ft. lbs. (38 Nm).

27. On the 3.0L engine, install the air conditioning compressor. Tighten the bolts to 16 ft. lbs. (22 Nm).

28. Install the transaxle mounting bracket, and tighten the retainers as follows:

a. On the 2.3L and 2.4L engines, tighten the nuts to 28 ft. lbs. (38 Nm) and the through-bolt to 40 ft. lbs. (54 Nm).

b. On the 3.0L engine, tighten the bolts to 28 ft. lbs. (38 Nm).

29. Install the sub-frame in its original position, and tighten the retainers as follows:

a. On the 2.3L and 2.4L engines, tighten the rear bolts to 47 ft. lbs. (64 Nm) and the front bolts to 76 ft. lbs. (103 Nm).

b. On the 3.0L engine, tighten the rear bolts to 40 ft. lbs. (54 Nm), front bolts to 76 ft. lbs. (103 Nm) and the nuts to 28 ft. lbs. (38 Nm).

30. On the 2.3L and 2.4L engines, install or connect the following:
- Radius rod bolts: 119 ft. lbs. (162 Nm)
- Rear mount bracket: 40 ft. lbs. (54 Nm)
- Stiffener. Tighten the through-bolt to 47 ft. lbs. (64 Nm) for manual transaxles or the nut and bolt to 28 ft. lbs. (38 Nm) for automatic transaxles.
- 3 front mounting bracket bolts: 28 ft. lbs. (38 Nm). Then, tighten the through-bolt to 47 ft. lbs. (64 Nm).
- Air conditioning compressor: 16 ft. lbs. (22 Nm)

31. On the 3.0L engine, install or connect the following:
- Radius rod bolts: 119 ft. lbs. (162 Nm)
- Front mounting bracket support nut: 40 ft. lbs. (54 Nm)
- Rear mounting bracket nut and bolt. Tighten the nut to 40 ft. lbs. (54 Nm) and the bolt to 28 ft. lbs. (38 Nm).
- Side mounting bracket. Tighten the bolts to 40 ft. lbs. (54 Nm) and the through-bolt to 40 ft. lbs. (54 Nm).
- Exhaust system
- Shift linkage, if equipped with an automatic transaxle

32. The remainder of the installation is the reverse of the removal.

33. Refill and bleed the cooling system.

❈❈ WARNING

Operating the engine without the proper amount and type of engine oil will result in severe engine damage.

34. Fill the engine with the correct amount of oil.

35. Install the battery if removed. Start the engine and check for leaks.

S2000

1. Before servicing the vehicle, refer to the precautions in the beginning of this section.

7923FG09

Engine lifting points—2.3L Accord shown, 2.4L similar

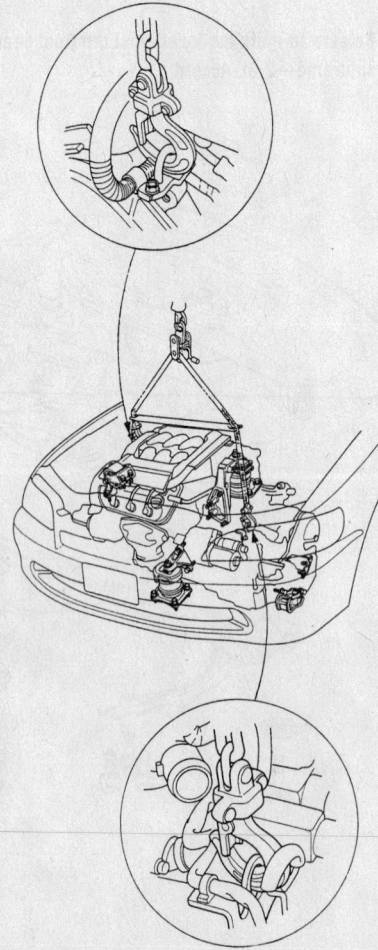

7923FG11

Engine lifting points—3.0L Accord

2. Drain the cooling system.
3. Relieve the fuel system pressure.
4. Remove or disconnect the following:
   - Battery
   - Front wheels
   - Transmission
   - Engine Control Module (ECM) connectors and the main wire harness connector. Pass the connectors through the cowl panel.
   - Vacuum tank
   - Throttle cable
   - Electrical Power Steering (EPS) control unit
   - Battery cable at the main underhood fuse/relay box
   - Battery cable at the auxiliary fuse box
   - Ground cable and harness clamps
   - Fuel lines
   - Brake booster vacuum line
   - Evaporative Emissions (EVAP) canister hose
   - Front motor mount and bracket
   - Heater hoses
   - Radiator hoses
   - Left motor mount
   - Right motor mount bracket
5. Carefully raise the engine out of the vehicle.

To install:

6. Installation is the reverse of the removal procedure, while using the following torque values:
   - Right motor mount bracket bolts: 28 ft. lbs. (38 Nm)
   - Motor mount nuts: 40 ft. lbs. (54 Nm)
   - Front motor mount bolts: 16 ft. lbs. (22 Nm)

Water Pump

REMOVAL & INSTALLATION

1.6L, 2.2L and 2.3L Engines

➡The original radio contains a coded anti-theft circuit. Obtain the security code number before disconnecting the battery cables.

1. Before servicing the vehicle, refer to the precautions in the beginning of this section.
2. Remove or disconnect the following:
   - Negative battery cable
3. Drain the cooling system.
4. Remove or disconnect the following:
   - Accessory drive belts, the valve cover, and the upper timing belt cover

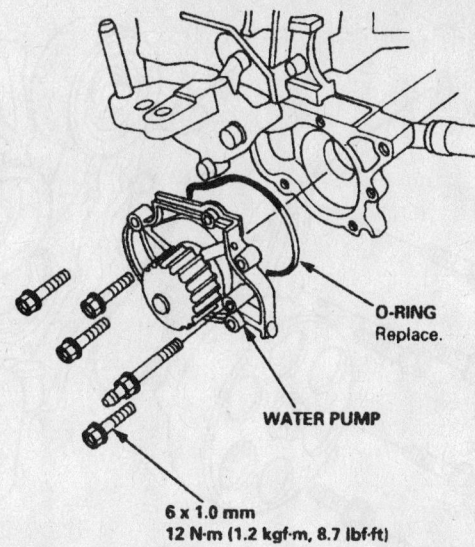

O-RING
Replace.

WATER PUMP

6 x 1.0 mm
12 N·m (1.2 kgf·m, 8.7 lbf·ft)

7923FG12

Water pump—2.2L and 2.3L engines

5. Set the timing at Top Dead Center (TDC)/compression for No. 1 piston.
6. Remove or disconnect the following:
   - Crankshaft pulley and lower timing belt cover
   - Timing belt. Replace the timing belt if it is contaminated with oil or coolant or shows any signs of wear and damage.
   - Crankshaft Speed Fluctuation (CKF) sensor bracket and move the sensor out of the way, if equipped. Cover the sensor with a shop towel to keep coolant off of it.
   - Water pump from the engine block. On 1.6L engines, the top right water pump mounting bolt also secures the alternator adjusting bracket. Leave the bracket attached to the alternator.

To install:

7. Clean the water pump and O-ring mating surfaces before installation.
8. Install or connect the following:
   - Water pump with a new O-ring. Coat only the bolt threads with liquid gasket and tighten them to 108 inch lbs. (12 Nm). On 1.6L engines, tighten the bracket bolt to 33 ft. lbs. (44 Nm).
   - Timing belt. Be sure it is fitted and adjusted properly.
   - CKF sensor, if equipped, and tighten the bracket bolts to 108 inch lbs. (12 Nm).
   - Lower belt cover and crankshaft pulley
   - Upper timing belt cover, the valve cover, and the accessory drive belts
9. Be sure the cooling system drain

plug is closed. Refill and bleed the cooling system.
10. Connect the negative battery cable and enter the radio security code.
11. Start the engine, allow it to reach normal operating temperature, and check for coolant leaks.
12. If equipped with 4WS, and turn the steering wheel lock-to-lock to reset the 4WS control unit.

2.0L Engine

1. Before servicing the vehicle, refer to the precautions in the beginning of this section.
2. Drain the cooling system.
3. Remove or disconnect the following:
   - Negative battery cable
   - Accessory drive belt
   - Water pump pulley
   - Water pump

To install:

4. Install or connect the following:
   - Water pump with a new O ring seal. Tighten the 8mm bolts to 16 ft. lbs. (22 Nm) and the 10mm bolt to 33 ft. lbs. (44 Nm).
   - Water pump pulley and tighten the bolts to 10 ft. lbs. (14 Nm).
   - Accessory drive belt
   - Negative battery cable
5. Fill the cooling system.
6. Start the engine and check for leaks.

2.4L Engine

1. Before servicing the vehicle, refer to the precautions in the beginning of this section.
2. Drain the cooling system.

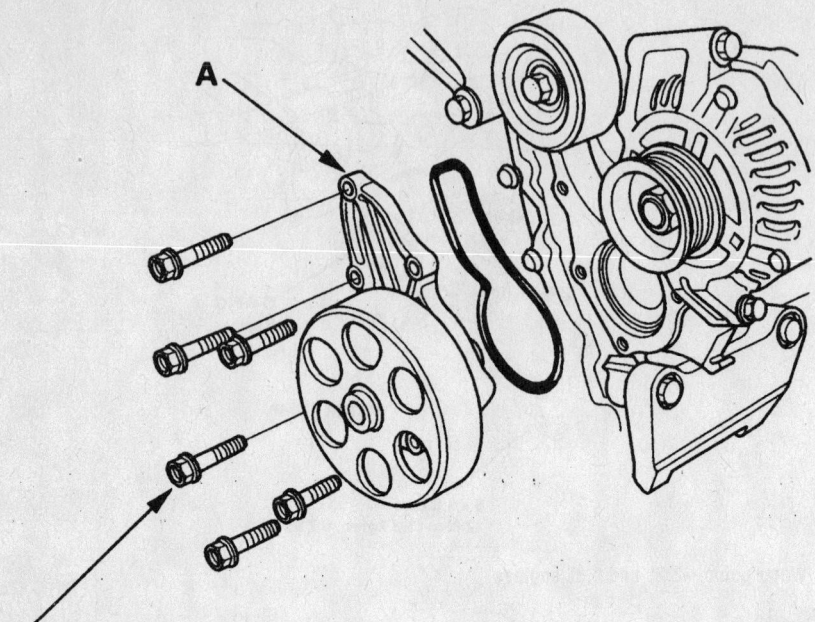

6 x 1.0 mm
12 N·m (1.2 kgf·m, 8.7 lbf·ft)

Water pump mounting—2.4L engine

43256-ACCO-G02

3. Remove or disconnect the following:
  • Crankshaft pulley
  • 6 water pump mounting bolts, then the pump and O-ring seal
4. Clean the seal groove and mating surfaces.
5. Install or connect the following:
  • Water pump, with a new seal. Tighten the bolts to 104 inch lbs. (12 Nm).
  • Crankshaft pulley
6. Fill the cooling system.
7. Start the engine and check for leaks.

3.0L Engine

1. Before servicing the vehicle, refer to the precautions in the beginning of this section.
2. Remove or disconnect the following:
  • Timing belt.
  • Timing belt tensioner
  • 5 water pump mounting bolts, then remove the pump and seal
To install:
3. Clean the seal groove and mating surfaces.
4. Install or connect the following:
  • Water pump, with a new seal. Tighten the bolts to 104 inch lbs. (12 Nm).

  • Timing belt tensioner
  • Timing belt.
5. Refill the cooling system.
6. Start the engine and check for leaks.
7. Top off the cooling system if necessary after the engine has cooled.

Heater Core

REMOVAL & INSTALLATION

Accord

➡Make sure to acquire the anti-theft code form the radio and write down the frequencies for the radio's preset buttons.

1. Disconnect the negative battery cable.

❋❋ CAUTION

After disconnecting the negative battery cable, wait for at least 3 minutes for the air bag module to deplete its energy before working the on the instrument panel or steering wheel.

2. Drain the cooling system into a clean container for reuse.
3. In the engine compartment, open the heater valve cable clamp and disconnect the cable from the heater valve arm. Then, turn the heater valve to the fully opened position.
4. Remove or disconnect the following:
  • Heater hoses from the heater core
  • Heater housing-to-chassis nut

➡**When removing the heater housing nut, be careful not to damage or bend the fuel lines, the brake lines or etc.**

  • Center console

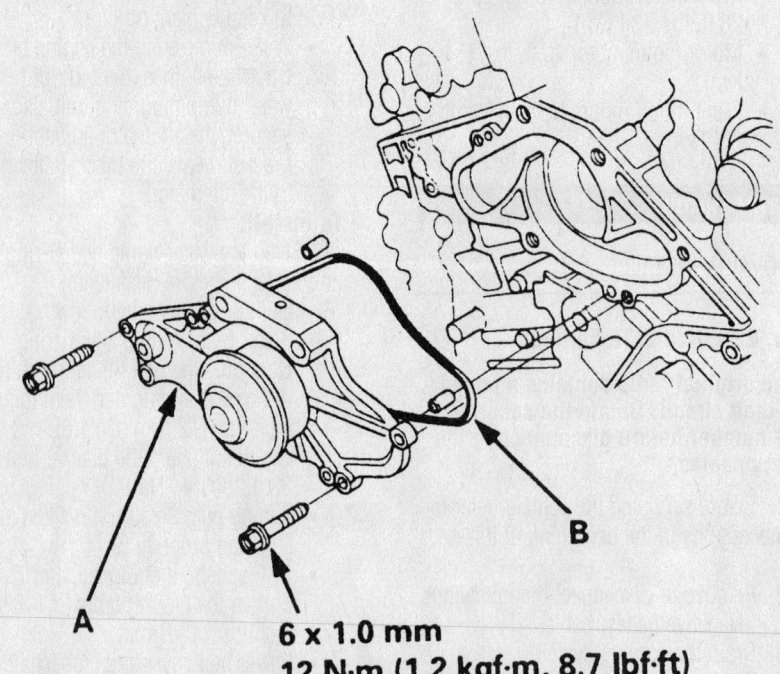

6 x 1.0 mm
12 N·m (1.2 kgf·m, 8.7 lbf·ft)

7923FG14

Exploded view of the water pump mounting—3.0L engine

5. Remove the instrument panel by removing or disconnecting the following:
- Center lower cover
- Passenger's side lower cover
- Lower glove box screws
- Hooks at the top inner side of the glove box by placing a flat tipped screwdriver in the cap notch and pry out the cap
- 4 glove box screws, release the hooks and remove the glove box
- Front door opening trim and the front door pillar trim at both sides
- Combination switch, the ignition switch and the air bag connectors

- Steering column-to-instrument panel nuts and lower the steering column
- Side wiring harness connectors, cabin wire harness and door wire harness from the fuse box at the driver's side
- Instrument panel wiring harness connector
- Wiring harness, the steering hanger beam wiring harness and the brake switch connectors under the dash
- Clutch switch connectors, if equipped with a manual transmission

- Pull back the carpet at the passenger's side
- SRS wiring harness, the ECM/PCM, the air mix control motor, the evaporator temperature sensor and the antenna lead
- ECM/PCM and the antenna lead harness clips
- Ground bolt using a Torx®bit T30
- Parking brake switch positive (+) terminal and harness clips at the driver's side
- Parking pin shift and the shift lock solenoid connectors, if equipped with an automatic transmission

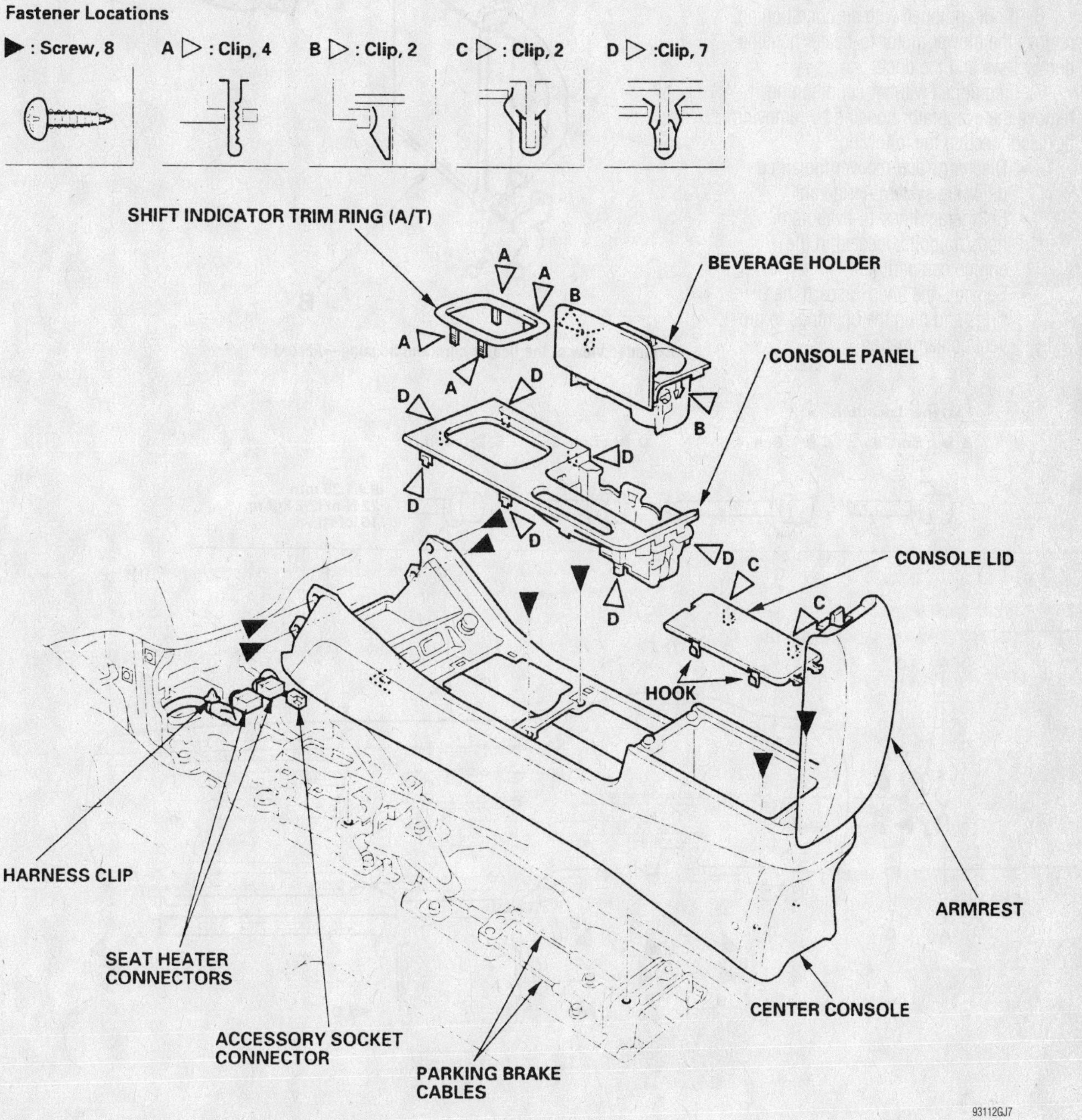

Fastener Locations

▶ : Screw, 8   A ▷ : Clip, 4   B ▷ : Clip, 2   C ▷ : Clip, 2   D ▷ :Clip, 7

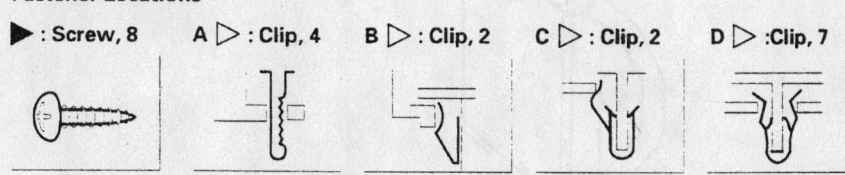

SHIFT INDICATOR TRIM RING (A/T)

BEVERAGE HOLDER

CONSOLE PANEL

CONSOLE LID

HOOK

ARMREST

CENTER CONSOLE

HARNESS CLIP

SEAT HEATER CONNECTORS

ACCESSORY SOCKET CONNECTOR

PARKING BRAKE CABLES

93112GJ7

Exploded view of the center console and related components—Accord

- Side wiring harness, the cabin wiring harness, the roof wiring harness and the door wiring harness from the passenger's fuse box at the passenger's side
- Instrument panel wiring harness
- Instrument pane harness, the blower motor and the recirculation control motor connectors under the passenger's side
- Harness, the harness holder and the connector clips
- Instrument panel-to-chassis cap and bolts
- Instrument panel from the guide pins and remove it

6. If not equipped with air conditioning, remove the blower motor-to-heater housing duct screws and the duct.

7. If equipped with air conditioning, remove the evaporator housing by removing or disconnecting the following:

- Discharge and recover the air conditioning system refrigerant
- Refrigerant lines-to-evaporator housing bolts, located in the engine compartment
- Separate the lines, discard the O-rings and plug the openings to prevent contamination

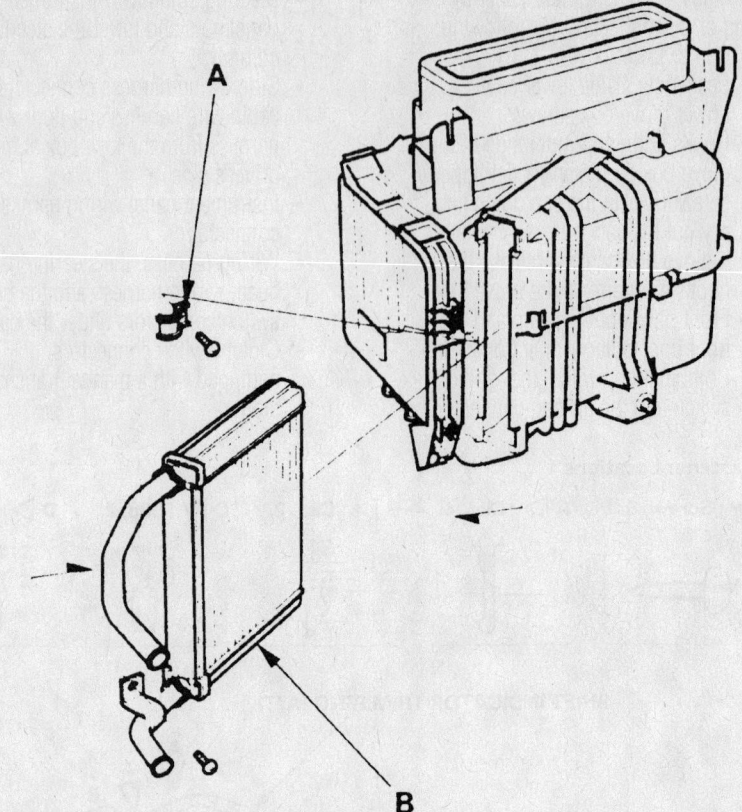

93112GJ9

Exploded view of the heater core and housing—Accord

Fastener Locations

B ▶ : Bolt, 6	C ▶ : Bolt, 4	D ▶ : Bolt, 3	E ▶ : Bolt, 1

8 x 1.25 mm
22 N·m (2.2 kgf·m, 16 lbf·ft)

View of the instrument panel bolt locations—Accord

93112GJ8

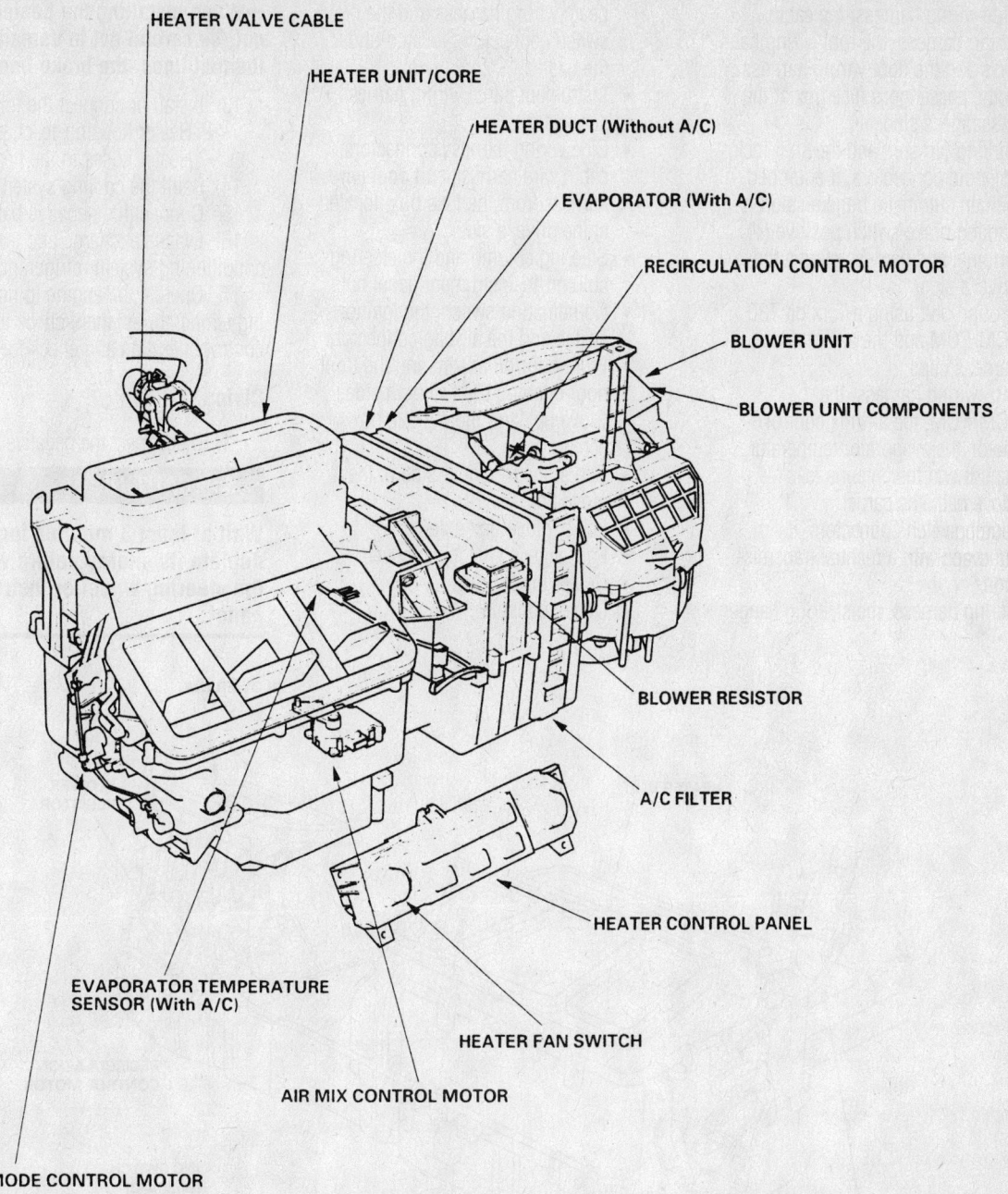

HEATER VALVE CABLE

HEATER UNIT/CORE

HEATER DUCT (Without A/C)

EVAPORATOR (With A/C)

RECIRCULATION CONTROL MOTOR

BLOWER UNIT

BLOWER UNIT COMPONENTS

BLOWER RESISTOR

A/C FILTER

HEATER CONTROL PANEL

HEATER FAN SWITCH

AIR MIX CONTROL MOTOR

EVAPORATOR TEMPERATURE
SENSOR (With A/C)

MODE CONTROL MOTOR

93112GJ0

View of the heater housing, evaporator housing and related components—Accord

- Evaporator housing's temperature sensor connector
- Evaporator housing-to-chassis nut/bolts and remove the evaporator housing

8. Remove or disconnect the following:
- Heater housing-to-chassis bolts and the heater housing
- Heater core-to-heater housing bracket screws and the brackets
- Heater core from the heater housing

To install:
9. Install or connect the following:
- Heater core to the heater housing

- Heater core-to-heater housing bracket and the bracket screws
- Heater housing and the heater housing-to-chassis bolts

10. If not equipped with air conditioning, install the blower motor-to-heater housing duct and the duct screws.

11. If equipped with air conditioning, install the evaporator housing by installing or connecting the following:
- Evaporator housing and the evaporator housing-to-chassis nut/bolts
- Evaporator housing's temperature sensor connector
- Refrigerant lines using new O-rings

- Refrigerant lines-to-evaporator housing bolts, located in the engine compartment

12. Install the instrument panel by installing or connecting the following:
- Instrument panel to the guide pins
- Instrument panel-to-chassis cap and bolts
- Harness holder and the connector clips
- Instrument pane harness, the blower motor and the recirculation control motor connectors under the passenger's side
- Instrument panel wiring harness

- Side wiring harness, the cabin wiring harness, the roof wiring harness and the door wiring harness to the passenger's fuse box at the passenger's side
- Parking pin shift and the shift lock solenoid connectors, if equipped with an automatic transmission
- Parking brake switch positive (+) terminal and harness clips a the driver's side
- Ground bolt using a Torx bit T30
- ECM/PCM and the antenna lead harness clips
- SRS wiring harness, the ECM/PCM, the air mix control motor, the evaporator temperature sensor and the antenna lead
- Move back the carpet.
- Clutch switch connectors, if equipped with a manual transmission
- Wiring harness, the steering hanger

beam wiring harness and the brake switch connectors, located under the dash
- Instrument panel wiring harness connector
- Side wiring harness connectors, cabin wire harness and door wire harness from the fuse box, located at the driver's side
- Steering column and the steering column-to-instrument panel nuts
- Combination switch, the ignition switch and the air bag connectors
- Front door pillar trim and the front door opening trim on both sides
- Glove box and the 4 glove box screws
- Cap at the top inner side of the glove box
- Lower glove box screws
- Passenger's side lower cover
- Center lower cover
- Center console

➡**When installing the heater housing nut, be careful not to damage or bend the fuel lines, the brake lines or etc.**

13. Install or connect the following:
    - Heater housing-to-chassis nut
    - Heater hoses to the heater core
14. Refill the cooling system.
15. Connect the negative battery cable.
16. Evacuate, charge and leak test the air conditioning system refrigerant.
17. Operate the engine to normal operating temperatures; then, check the climate control operation and check for leaks.

Civic

1. Disconnect the negative battery cable.

❊❊ **CAUTION**

Wait at least 3 minutes for the SRS to deplete its energy before working on the steering wheel or instrument panel.

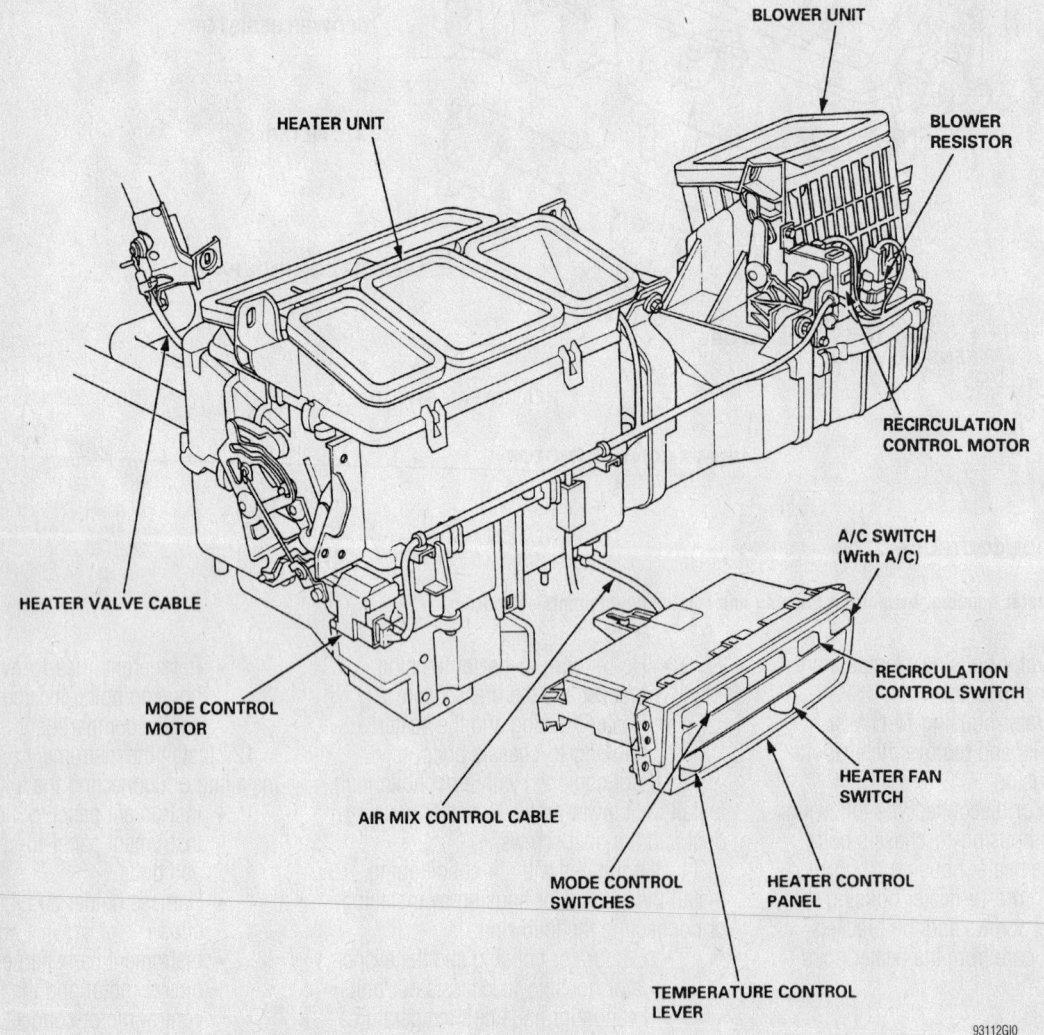

View of the heater housing, evaporator housing and related components—Civic

93112GI0

2. In the engine compartment, remove the heater valve cable clamp; then, disconnect the heater valve cable and rotate the heater valve to the fully open position.

3. Drain the engine coolant into a clean container for reuse.

4. Remove or disconnect the following:
- Heater hoses from the heater unit
- Heater housing-to-chassis nut
- Instrument panel.

5. If not equipped with air conditioning, remove the wiring harness from the heater duct; then, remove the 2 screws and the heater duct.

6. If equipped with air conditioning, remove or disconnect the following:
- Discharge and recover the air conditioning system refrigerant
- Refrigerant lines-to-evaporator bolts. Disconnect the lines. Discard the O-rings. Plug the openings to prevent contamination
- Thermostat electrical connector and

the wiring harness from the evaporator
- 4 evaporator housing screws, the bolt and nut
- Drain hose and remove the evaporator housing

7. On 2000–01 models, disconnect the electrical connectors from the blower motor, the power transistor, the blower motor high relay and the recirculation control motor.

8. Remove or disconnect the following:
- Electrical connectors from the mode control motor and the air mix control motor (for 2000–01); then, remove the electrical harness clips and wiring harness from the heater housing
- Heater duct clip
- 2 heater housing-to-chassis nuts and the heater housing
- Heater core cover-to-heater hous-

ing screws, the cover, the clamp and the heater core

To install:

9. Install or connect the following:
- Heater core, the clamp, the cover and the heater core cover-to-heater housing screws
- Heater housing and the 2 heater housing-to-chassis nuts
- Heater duct clip
- Electrical connectors to the mode control motor and the air mix control motor (for 2000–01); then, install the electrical harness clips and wiring harness to the heater housing

10. On 2000–01 models, connect the electrical connectors to the blower motor, the power transistor, the blower motor high relay and the recirculation control motor.

11. If not equipped with air conditioning, install the heater duct, the 2 screws and the wiring harness to the heater duct.

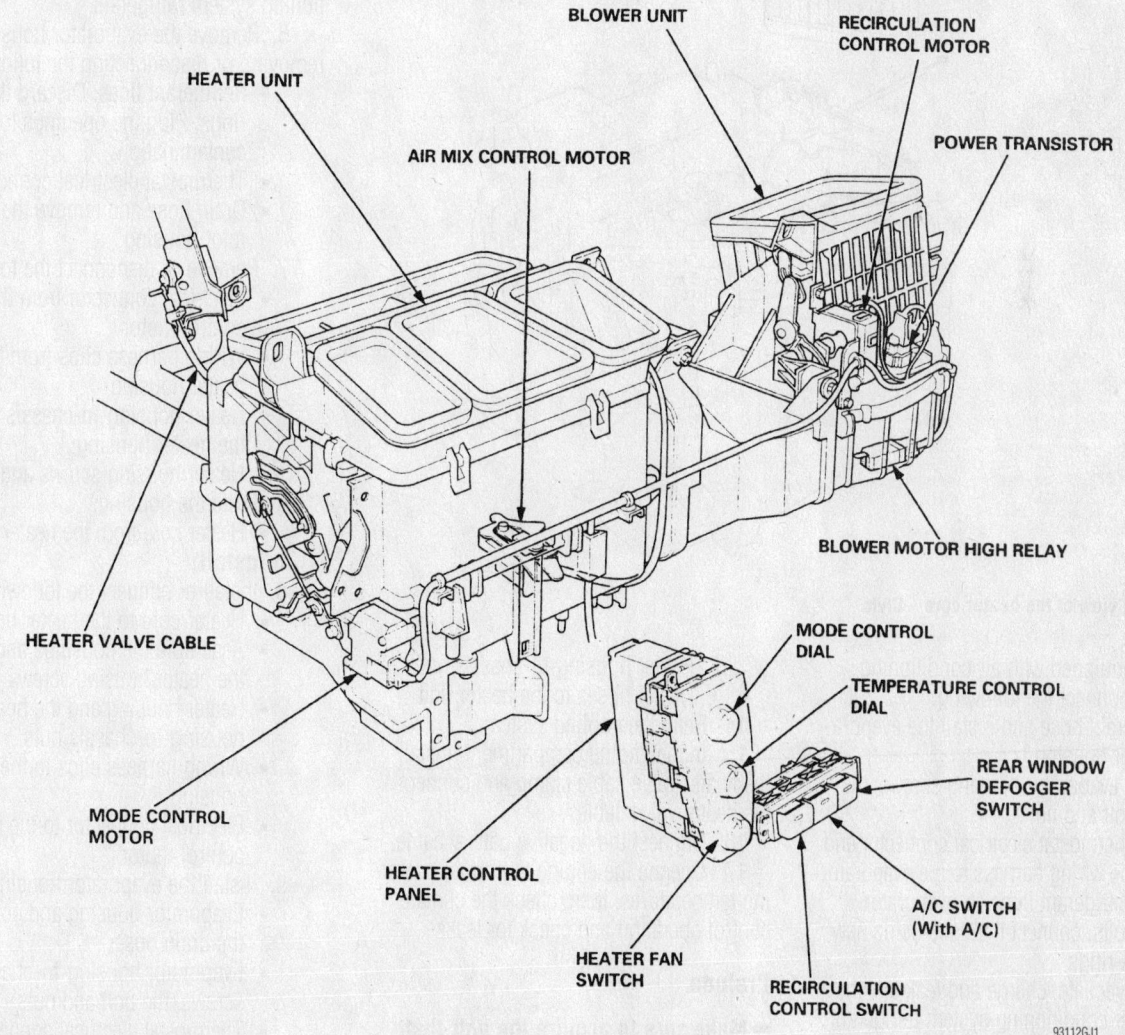

93112GJ1

View of the heater housing, evaporator housing and related components—Civic

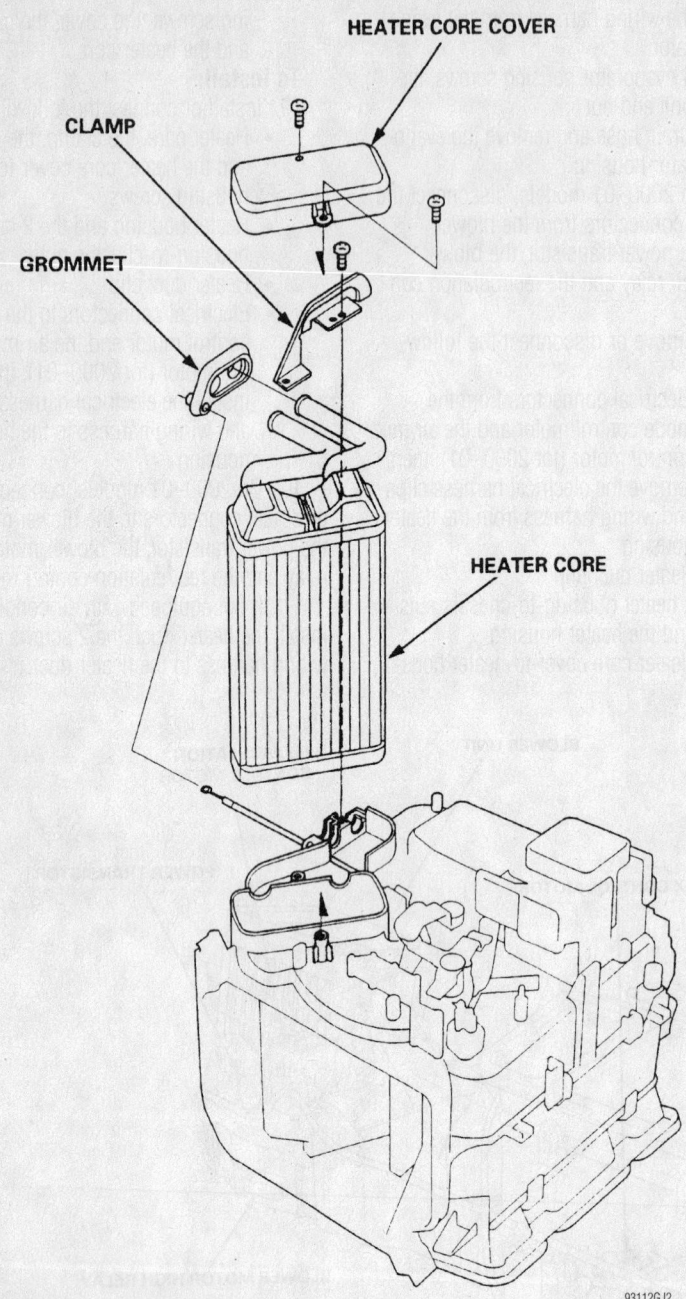

HEATER CORE COVER

CLAMP

GROMMET

HEATER CORE

93112GJ2

Exploded view of the heater core—Civic

12. If equipped with air conditioning, install or connect the following:
- Drain hose and install the evaporator housing
- 4 evaporator housing screws, the bolt and nut
- Thermostat electrical connector and the wiring harness to the evaporator
- Refrigerant lines-to-evaporator bolts, connect the lines using new O-rings
- Evacuate, charge and leak test the air conditioning system refrigerant.

13. Install or connect the following:
- Instrument panel
- Heater housing-to-chassis nut
- Heater hoses to the heater unit

14. Refill the cooling system.

15. In the engine compartment, install the heater valve cable clamp and connect the heater valve cable.

16. Connect the negative battery cable.

17. Operate the engine to normal operating temperatures; then, check the climate control operation and check for leaks.

Prelude

➡**Make sure to acquire the anti-theft code form the radio and write down the**

frequencies for the radio's preset buttons.

1. Disconnect the negative battery cable.

✳✳ CAUTION

Wait at least 3 minutes for the SRS to deplete its energy before working on the steering wheel or instrument panel.

2. In the engine compartment, remove the heater valve cable clamp; then, disconnect the heater valve cable and rotate the heater valve to the fully open position.

3. Drain the engine coolant into a clean container for reuse.

4. Remove or disconnect the following:
- Heater hoses from the heater unit
- Heater housing-to-chassis nuts
- Instrument panel
- Steering hanger beam mounting bolts and the steering hanger beam

5. Discharge and recover the air conditioning system refrigerant

6. Remove the evaporator housing by removing or disconnecting the following:
- Refrigerant lines. Discard the O-rings. Plug the openings to prevent contamination.
- Thermostat electrical connector
- Drain hose and remove the evaporator housing.

7. Remove or disconnect the following:
- Electrical connector from the mode control motor
- Wiring harness clips from the heater housing
- Heater housing-to-chassis nuts and the heater housing
- Heater housing screws and separate the housings
- Heater core from the heater housing

To install:

8. Install or connect the following:
- Heater core to the heater housing
- Assemble the housings and install the heater housing screws
- Heater housing and the heater housing-to-chassis nuts
- Wiring harness clips to the heater housing
- Electrical connector to the mode control motor

9. Install the evaporator housing:
- Evaporator housing and connect the drain hose
- Evaporator housing-to-chassis screws, the bolt and nuts
- Thermostat electrical connector and the wiring harness to the evaporator

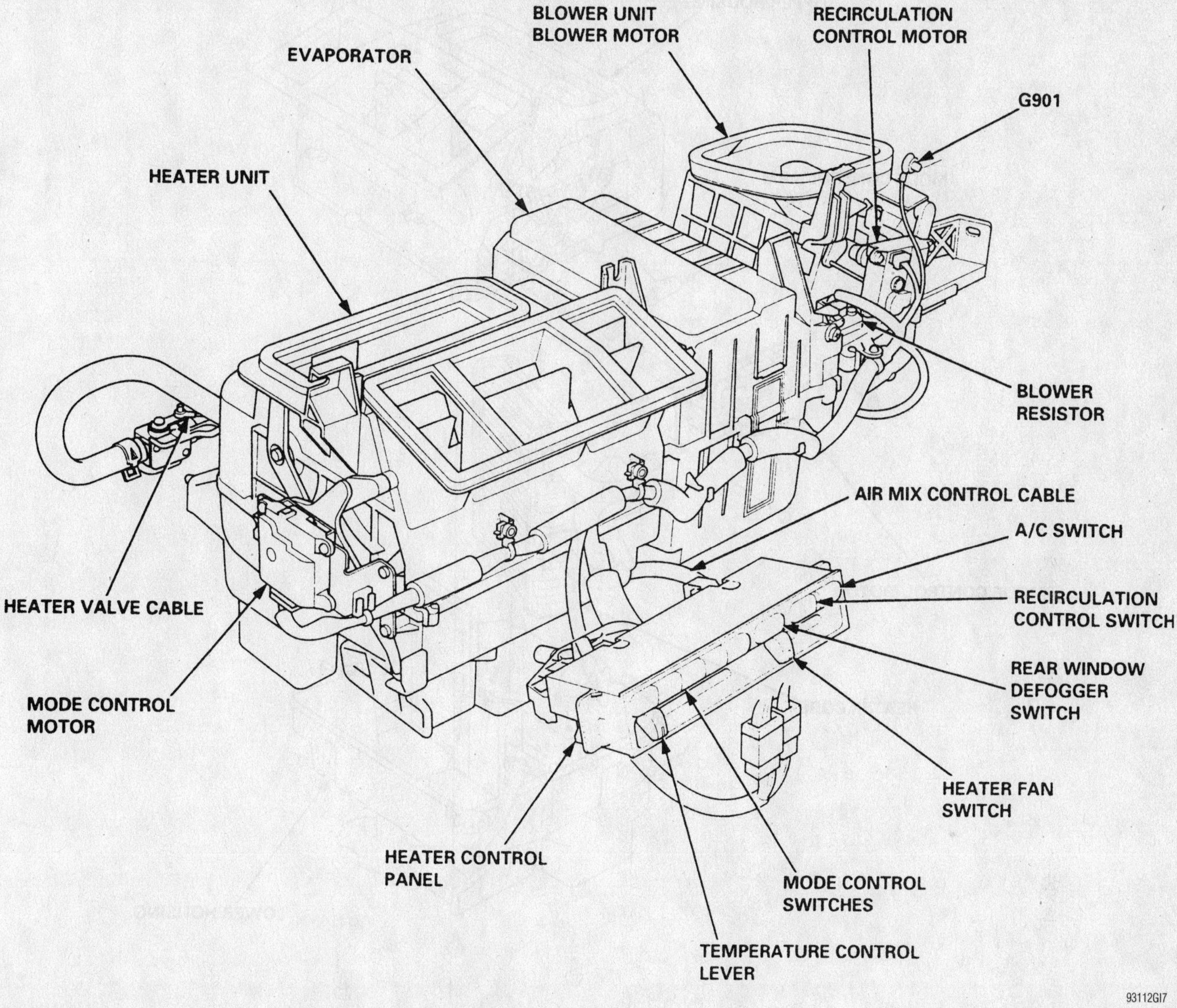

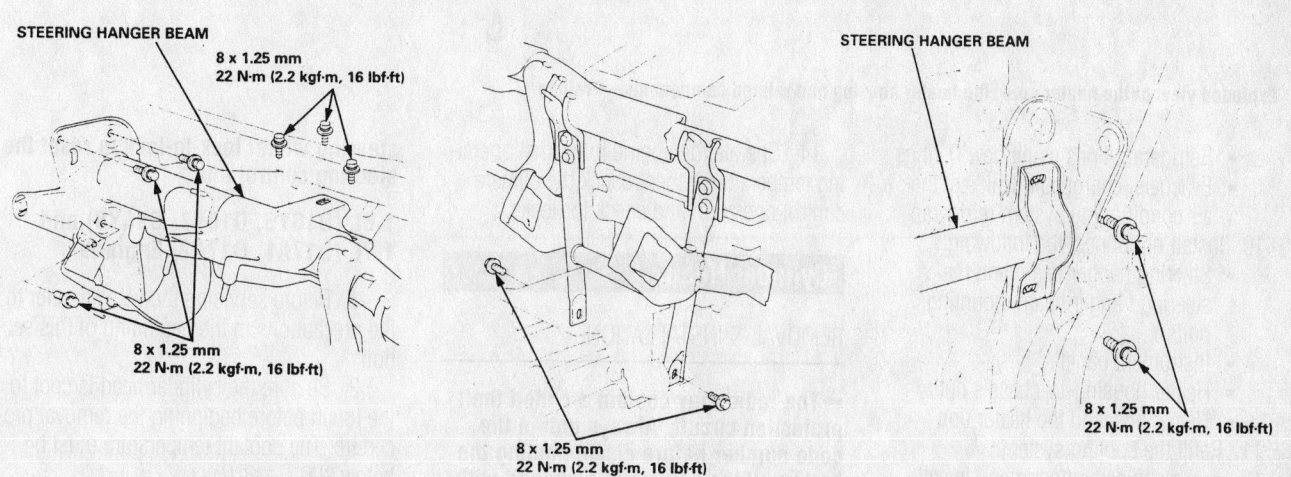

EVAPORATOR

HEATER UNIT

BLOWER UNIT
BLOWER MOTOR

RECIRCULATION
CONTROL MOTOR

G901

BLOWER
RESISTOR

AIR MIX CONTROL CABLE

A/C SWITCH

RECIRCULATION
CONTROL SWITCH

REAR WINDOW
DEFOGGER
SWITCH

HEATER FAN
SWITCH

MODE CONTROL
SWITCHES

TEMPERATURE CONTROL
LEVER

HEATER CONTROL
PANEL

HEATER VALVE CABLE

MODE CONTROL
MOTOR

93112GI7

View of the heater housing, evaporator housing and related components—Prelude

STEERING HANGER BEAM

8 x 1.25 mm
22 N·m (2.2 kgf·m, 16 lbf·ft)

8 x 1.25 mm
22 N·m (2.2 kgf·m, 16 lbf·ft)

8 x 1.25 mm
22 N·m (2.2 kgf·m, 16 lbf·ft)

STEERING HANGER BEAM

8 x 1.25 mm
22 N·m (2.2 kgf·m, 16 lbf·ft)

93112GI8

View of the steering hanger beam and related components—Prelude

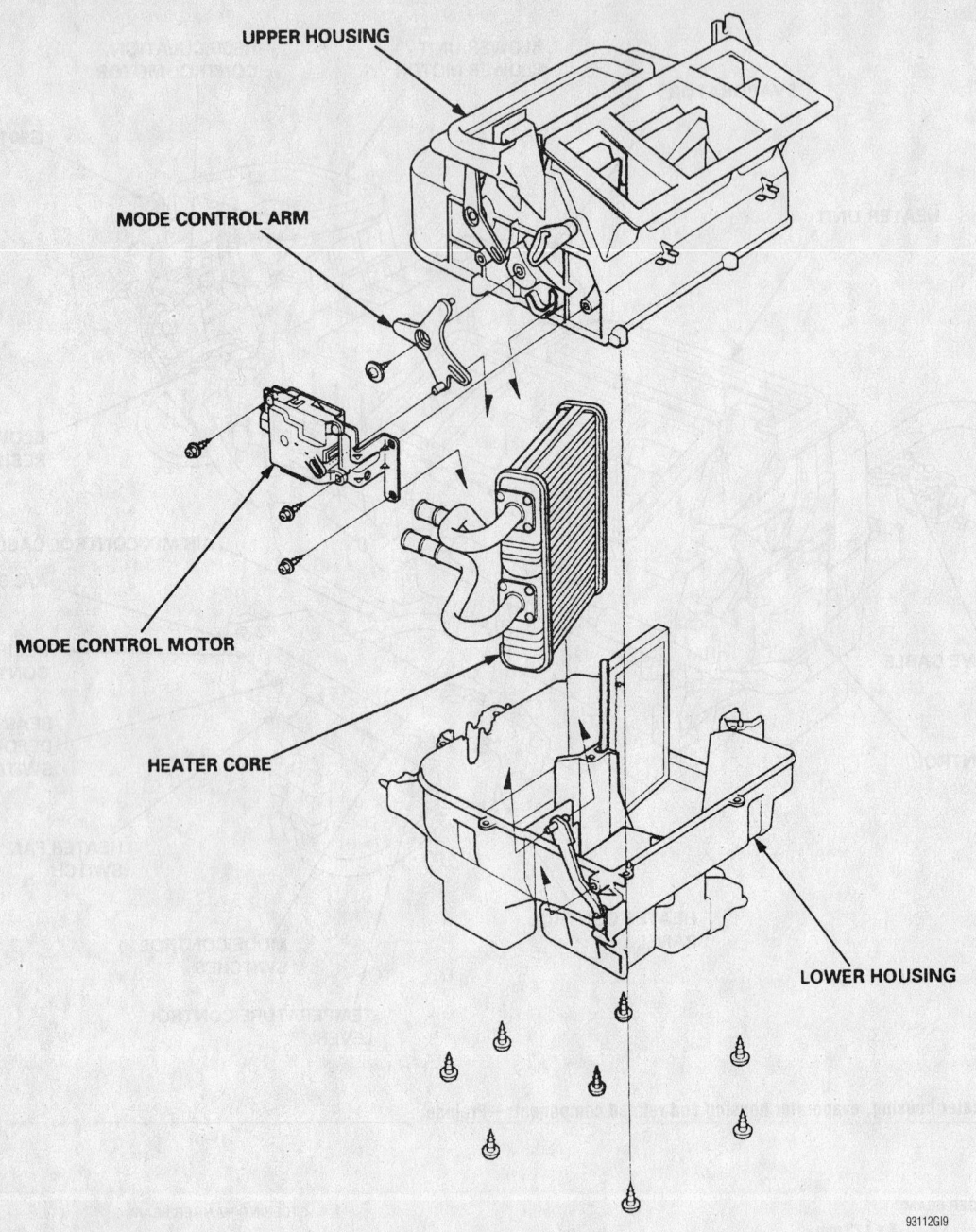

Exploded view of the heater core, the heater housing and related components—Prelude

- Refrigerant lines using new O-rings
- Evacuate, charge and leak test the air conditioning system refrigerant
10. Install or connect the following:
   - Steering hanger beam and the steering hanger beam mounting bolts
   - Instrument panel
   - Heater housing-to-chassis nuts
   - Heater hoses to the heater unit
11. Refill the cooling system.
12. In the engine compartment, Install the heater valve cable clamp; then, connect the heater valve cable.
13. Connect the negative battery cable.

14. Operate the engine to normal operating temperatures; then, check the climate control operation and check for leaks.

Cylinder Head

REMOVAL & INSTALLATION

➡**The radio may contain a coded theft protection circuit. Always obtain the code number before disconnecting the battery. If the vehicle is equipped with 4WS, the steering control unit is shut down when the battery is disconnected. After connecting the battery, turn the**

steering wheel lock-to-lock to reset the steering control unit.

1.6L (D16Y5, D16Y7, D16Y8) and 1.7L (D17A1, D17A2) Engines

1. Before servicing the vehicle, refer to the precautions in the beginning of this section.
2. Be sure the cylinder head is cool to the touch before beginning the removal procedure. The coolant temperature must be below 100°F (38°C).
3. Remove or disconnect the following:
   - Negative battery cable
4. Drain the cooling system.

5. Remove or disconnect the following:

➡ **Label the wires before disconnecting them.**

- Ignition wires
- Air intake duct and the air cleaner assembly

6. Relieve the fuel pressure.

7. Clean up any fuel that may have spilled on the engine or intake manifold.

8. Remove or disconnect the following:

- Upper radiator hose from the coolant inlet
- Coolant bypass hoses and the heater hose from the intake manifold
- Power steering pump belt
- Power steering pump from its mounting bracket and lift the power steering reservoir from its mount. Move the pump and reservoir out of the work area and secure them. Don't disconnect the hydraulic lines.

9. Place a block of wood on the pad of a floor jack. Place the floor jack under the engine for support.

10. Remove or disconnect the following:

- Left-front engine mount bracket, if equipped with air conditioning

➡ **Slip the air conditioning compressor belt around the engine mount to remove it.**

- Air conditioning compressor belt
- Alternator belt

11. Be sure the engine is supported with the padded floor jack. Loosen the nuts from left side engine mount. Remove the engine mount bracket.

12. Remove or disconnect the following:

- Valve cover and the upper timing belt cover
- Crankshaft pulley and the lower timing belt cover
- Dipstick tube from its catches on the timing cover
- Timing belt.

13. With the timing belt removed, inspect the water pump and replace it if necessary.

14. Remove or disconnect the following:

- Distributor, if equipped
- Camshaft sprocket
- Fuel lines from the intake manifold fuel rail. Immediately plug the lines to prevent fuel leakage and contamination
- Throttle cable from the linkage by first loosening its locknut, then slipping it out of its holder.

➡ **Label the electrical connectors, before disconnecting them.**

- Fuel injector wiring harness connectors
- VTEC solenoid valve and pressure switch connectors, if equipped
- Idle Air Control (IAC) valve connector
- Throttle Position (TP) sensor connector
- Exhaust Gas Recirculation (EGR) valve lift sensor connectors, if equipped
- Engine Coolant Temperature (ECT) sensor, switch, and gauge sender connectors
- Manifold Absolute Pressure (MAP) sensor connector
- Primary and secondary Heated Oxygen Sensor (HO$_2$S) connectors
- Vacuum hoses and Positive Crankcase Ventilation (PCV) hose from the intake manifold and throttle body
- EVAP and breather hoses from the intake manifold
- Intake manifold together with the throttle body and plenum
- Exhaust manifold
- Power steering pump bracket

15. Loosen the cylinder head bolts in a 3-step crisscross pattern in the reverse order of the tightening sequence. Start with the outermost bolts and work toward the middle of the cylinder head. Loosen the bolts in the reverse order of installation.

16. Remove the cylinder head. If the head sticks to the engine block, tap it with a plastic or wooden mallet.

17. Inspect the cylinder head for warpage and cracking. Repair, machine, or replace as necessary. The warpage limit is 0.002 in. (0.05mm). Standard cylinder head height is 3.659–3.663 in. (92.95–93.05mm).

18. Remove the old cylinder head gasket and thoroughly clean the mating surfaces.

19. Cover the engine block with a sheet of plastic to keep out dust and foreign objects.

To install:

➡ **Use new O-ring, seals, and gaskets when installing the cylinder head and its components.**

20. Be sure the cylinder head and the engine block surfaces are clean, level, and straight.

21. Be sure the cylinder head dowel pins and control orifice are aligned. Clean the oil control orifice and reinstall it with a new O-ring.

22. Install or connect the following:

- New head gasket onto the engine block

- Camshaft, if removed, with the keyway facing up so that the engine will remain at Top Dead Center (TDC)/compression for the No. 1 cylinder
- New lubricated camshaft seal

➡ **Use new cylinder head bolts and washers. Used or previously-tightened bolts may be stretched, and therefore they have reduced clamping and sealing power under compression. Apply clean engine oil to the threads of each head bolt.**

- Cylinder head into position
- Cylinder head bolts and hand-tighten

23. Tighten the cylinder head bolts to their final torque specification in 4 steps. Use a crisscross sequence starting with the bolts at the middle of the head and working toward the outer bolts as follows:

a. Step 1: Tighten each bolt to 14 ft. lbs. (20 Nm).

b. Step 2: Tighten each bolt to 36 ft. lbs. (49 Nm).

c. Step 3: Tighten each bolt to 49 ft. lbs. (67 Nm).

d. Step 4: Retighten only the 2 center bolts to 49 ft. lbs. (67 Nm).

24. Apply oil to the camshaft sprocket bolt. Install the sprocket with the UP mark and the keyway pointing straight up. Tighten the sprocket bolt to 27 ft. lbs. (37 Nm).

25. Install or connect the following:

- Intake manifold with a new gasket, and tighten the nuts in a crisscross pattern in 2 or 3 steps to 17 ft. lbs. (24 Nm) starting with the inner nuts
- Bolts that secure the intake manifold to its bracket: 17 ft. lbs. (24 Nm)
- Power steering pump bracket: 33 ft. lbs. (44 Nm)
- Exhaust manifold with a new gasket. Apply anti-seize paste to the studs, and tighten the nuts to 23 ft. lbs. (31 Nm) in a crisscross sequence.
- Exhaust manifold to the front exhaust pipe. Tighten the self-locking nuts to 25 ft. lbs. (33 Nm). On vehicles with the D16Y8 engine, tighten the nuts to 40 ft. lbs. (55 Nm).

26. Verify that the engine is at Top Dead Center (TDC)/compression for the No. 1 cylinder.

27. Install or connect the following:

- Timing belt. After the timing belt has been properly tensioned, tighten the adjusting bolt to 33 ft. lbs. (44 Nm).

- Lower timing belt cover
- Crankshaft pulley; tighten its bolt to 134 ft. lbs. (181 Nm)
- Dipstick tube back into its catches

28. Adjust the valves. If equipped with a VTEC engine, also check the rocker arms for free and smooth motion..

29. If equipped with a VTEC engine, remove the VTEC solenoid valve and its filter. Install a new filter, then reinstall the VTEC solenoid valve and tighten its bolt to 108 inch lbs. (12 Nm).

30. Install or connect the following:
- Distributor, if equipped

31. Be sure all the spark plug tube sealing gaskets are fully seated.

32. Install or connect the following:
- New gasket onto to the valve cover. Apply liquid gasket to the corner recesses of the gasket. Don't let the sealant cure before installing the valve cover onto the cylinder head.
- Valve cover. Gently wiggle the valve cover to be sure it is fully seated. Tighten the valve cover bolts in a crisscross pattern to 84 inch lbs. (10 Nm).
- New spark plugs
- Ignition wires
- Upper radiator hose, heater hoses, and intake manifold coolant bypass hoses
- Intake manifold vacuum lines, PCV, EVAP canister, and breather hoses
- Fuel lines to the fuel rail. Use new sealing washers on the banjo fitting. Carefully tighten the banjo fitting to 21 ft. lbs. (28 Nm) for the D16Y5 engine, or to 16 ft. lbs. (22 Nm) for all other engines. Tighten the service bolt to 10–11 ft. lbs. (13–15 Nm).
- Throttle cable. Adjust its tension so the cable has a deflection of 10–12mm (0.39–0.47 in.).

33. Installation of the remaining components is the reverse of removal.

1.6L (B16A2) Engine

1. Before servicing the vehicle, refer to the precautions in the beginning of this section.

2. Before beginning the cylinder head removal procedure, be sure the engine temperature is below 100°F (38°C). To prevent warping, the cylinder head should be removed when the engine is cold.

3. Remove or disconnect the following:
- Negative battery cable

➡**Label the wires before disconnecting them.**

- Ignition wires

4. Drain the engine coolant. Remove the radiator cap to speed draining.

5. Remove or disconnect the following:
- Strut brace
- Intake air duct and the breather hose

6. Relieve the fuel pressure as follows:
   a. Loosen the fuel filler cap.
   b. Hold the fuel filter banjo bolt with a back-up wrench. Hold the fuel filter service bolt with a box end wrench.
   c. Place a shop rag over the fuel filter to absorb fuel spray.
   d. Slowly loosen the fuel filter service bolt 1 complete turn.

7. Clean up any fuel that may have spilled on the engine or intake manifold.

8. Remove or disconnect the following:
- Upper radiator hose from the coolant inlet
- Coolant bypass hoses and the heater hose from the intake manifold
- Power steering pump belt
- Power steering pump from its mounting bracket and lift the power steering reservoir from its mount. Move the pump and reservoir out of the work area and secure them. Don't disconnect the hydraulic lines.

9. Place a block of wood on the pad of a floor jack. Place the floor jack under the engine for support.

10. Remove or disconnect the following:
- Left-front engine mount bracket, if equipped with air conditioning

➡**Slip the air conditioning compressor belt around the engine mount to remove it.**

- Air conditioning compressor drive belt
- Alternator belt

➡**Be sure the engine is supported with the padded floor jack. Loosen the left side engine mount nuts. Remove the engine mount bracket.**

- Valve cover and the upper timing belt cover
- Crankshaft pulley and the lower timing belt cover
- Timing belt.

11. With the timing belt removed, inspect the water pump and replace it if necessary.

12. Remove or disconnect the following:
- Distributor from the cylinder head as an assembly
- Fuel lines from the intake manifold

fuel rail. Immediately plug the lines to prevent fuel leakage and contamination.
- Throttle cable from the linkage by first loosening its locknut, then slipping it out of its holder.

➡**Label all connectors and vacuum lines before disconnecting them.**

- Fuel injector wiring harness connectors
- VTEC solenoid valve and pressure switch connectors
- Idle Air Control (IAC) valve connector
- Throttle Position (TP) sensor connector
- Engine Coolant Temperature (ECT) sensor, switch, and gauge sender connectors
- Manifold Absolute Pressure (MAP) sensor connector
- Primary Heated Oxygen Sensor (HO$_2$S) connector
- Vacuum hoses and Positive Crankcase Ventilation (PCV) hose from the intake manifold and throttle body
- Evaporative emissions (EVAP) and breather hoses from the intake manifold

13. Loosen the intake manifold nuts in a crisscross sequence. Then, remove the intake manifold together with the throttle body and plenum.

14. Remove or disconnect the following:
- Exhaust manifold heat shield
- Exhaust manifold nuts in a crisscross sequence
- Exhaust manifold

➡**Be careful not to damage the oxygen sensors when removing the manifold. Cover the front exhaust pipe flange with a shop towel to keep dirt out**

- Power steering pump bracket
- Camshaft pulleys and back cover
- Camshaft holder plate bolts in a crisscross sequence working toward the middle of the cylinder head.

15. Loosen the valve adjusting screws.

16. Remove or disconnect the following:
- Camshaft holder plates and holders from the cylinder head. The holder bolts will keep the components together. Note the positions of each camshaft holder for reassembly.
- Camshafts from the cylinder head. Mark the exhaust and intake camshafts so that they will not be confused.

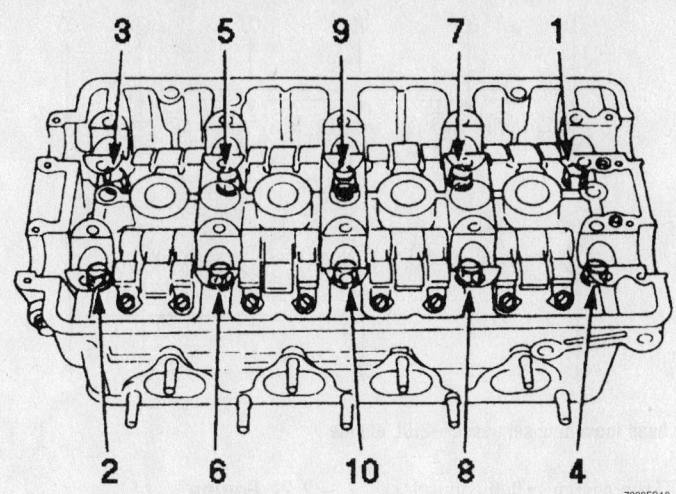

Cylinder head bolt loosening sequence—1.6L (B16A2) engines

- Cylinder head bolts in a 3-step crisscross pattern. Start with the outermost bolts and work toward the middle of the cylinder head.
- Cylinder head. If the head sticks to the engine block, tap it with a plastic-faced or wooden mallet.

17. Inspect the cylinder head for warpage and cracking. Repair, machine, or replace as necessary. The warpage limit is 0.002 in. (0.05mm). Standard cylinder head height is 5.589–5.593 in. (141.95–142.05mm).

To install:

➡**Use new O-ring, seals, and gaskets when installing the cylinder head and its components.**

18. Be sure the cylinder head and the engine block surfaces are clean, level, and straight.

19. Be sure the cylinder head dowel pins and oil control orifice are aligned. Clean the oil control orifice and reinstall it with a new O-ring.

20. Install or connect the following:
- New head gasket onto the engine block

➡**Use new cylinder head bolts and washers. Used or previously-tightened bolts may be stretched; and therefore, they have reduced clamping and sealing power under compression. Apply clean engine oil to the threads of each head bolt.**

- Cylinder head into position. Hand-tighten all the cylinder head bolts.

21. Tighten the cylinder head bolts to their final torque specification in 2 steps. Use a crisscross sequence starting with the bolts at the middle of the head and working toward the outer bolts:
 a. Step 1: Tighten each bolt to 22 ft. lbs. (30 Nm).
 b. Step 2: Tighten each bolt to 61 ft. lbs. (85 Nm).

22. Install or connect the following:
- Dowel pin in the No. 3 cylinder head camshaft holder with a new O-ring

23. Thoroughly clean the intake and exhaust camshaft oil control orifices. Reinstall them with new O-rings.

24. Install or connect the following:
- Camshafts
- Intake manifold with a new gasket, and tighten the nuts in a crisscross pattern in 2 or 3 steps to 17 ft. lbs. (24 Nm) starting with the inner nuts
- Bolts that secure the intake manifold to its bracket: 17 ft. lbs. (24 Nm)
- Power steering pump bracket: 33 ft. lbs. (44 Nm)
- Exhaust manifold with a new gasket. Apply anti-seize paste to the studs, and tighten the nuts to 23 ft. lbs. (31 Nm) in a crisscross sequence. Tighten the exhaust manifold bracket bolts to 17 ft. lbs. (24 Nm)
- Exhaust manifold to the front exhaust pipe. Tighten the self-locking nuts to 40 ft. lbs. (55 Nm).

25. Verify that the engine is at Top Dead Center (TDC)/compression for the No. 1 cylinder.
- Timing belt. After the timing belt has been properly tensioned, tighten the adjusting bolt to 40 ft. lbs. (55 Nm).
- Lower timing belt cover
- Crankshaft pulley and tighten its bolt to 130 ft. lbs. (180 Nm)

26. Adjust the valves.

27. Inspect the VTEC rocker arms for free and smooth motion.

28. Remove the VTEC solenoid valve and its filter. Install a new filter, then reinstall the VTEC solenoid valve and tighten its bolts to 108 inch lbs. (12 Nm).

29. Install or connect the following:
- Distributor

30. Be sure all the spark plug tube sealing gaskets are fully seated.
- New gasket onto to the valve cover. Apply liquid gasket to the corners of the gasket that meet the camshaft holders. Don't let the sealant cure

CYLINDER HEAD BOLTS
12 x 1.25 mm
100 N·m (10.0 kg-m, 72 lb-ft)
Apply clean engine oil to bolt threads and under bolt heads.

Cylinder head torque sequence—1.6L (B16A2) engine

before installing the valve cover onto the cylinder head.

- Valve cover. Gently wiggle the valve cover to be sure it is fully seated. Tighten the valve cover bolts in a crisscross pattern to 84 inch lbs. (10 Nm).
- New spark plugs
- Ignition wires
- Upper radiator hose, heater hoses, and intake manifold coolant bypass hoses
- Intake manifold vacuum lines, PCV, EVAP canister, and breather hoses
- Fuel lines to the fuel rail. Use new sealing washers on the banjo fitting. Carefully tighten the banjo fitting to 25 ft. lbs. (33 Nm). Tighten the service bolt to 11 ft. lbs. (15 Nm).
- Throttle cable. Adjust its tension so the cable has a deflection of 0.39–0.47 in. (10–12mm).

31. Installation of the remaining components is the reverse of removal.

32. After the installation procedure is complete, check that all tubes, hoses, and connectors are installed correctly.

2.0L Engine

1. Before servicing the vehicle, refer to the precautions in the beginning of this section.
2. Drain the cooling system.
3. Relieve the fuel system pressure.
4. Remove or disconnect the following:
   - Negative battery cable
   - Air cleaner housing
   - Accessory drive belt
   - Throttle cable
   - Fuel lines
   - Brake booster vacuum hose
   - Evaporative Emissions (EVAP) canister hose
   - Intake manifold bracket and air hose
   - Water outlet housing
   - Water bypass hose
   - Fuel injector connectors
   - Intake Air Temperature (IAT) sensor connector
   - Idle Air Control (IAC) valve connector
   - Throttle Position (TP) sensor connector
   - Manifold Absolute Pressure (MAP) sensor connector
   - Engine Coolant Temperature (ECT) sensor connector
   - Heated Oxygen (HO2S) sensor connector
   - VTEC solenoid valve connector

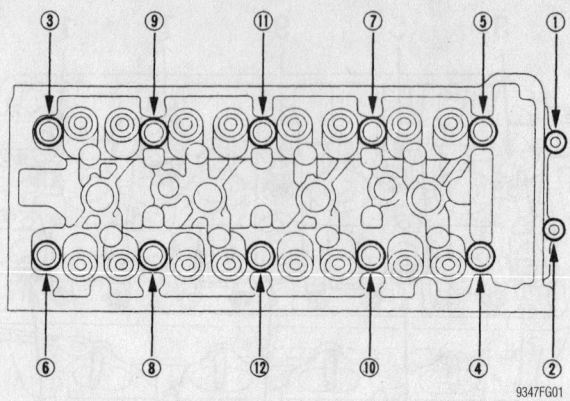

Cylinder head loosening sequence—2.0L engine

- VTEC pressure switch connector
- Crankshaft Position (CKP) sensor connector
- Exhaust manifold cover
- Exhaust manifold heat shield
- Exhaust manifold
- Oil level dipstick
- Positive Crankcase Ventilation (PCV) valve and hose
- Ignition coil cover
- Ignition coils
- Intake manifold
- Valve cover
- Timing chain auto-tensioner
- Camshafts
- Timing chain idler gear
- Cylinder head. Loosen the bolts in sequence and in 1/3 turns.

To install:

5. Install the cylinder head with a new gasket. Tighten the bolts in sequence as follows:
   a. Step 1: 22 ft. lbs. (29 Nm).
   b. Step 2: Plus 90 degrees.
   c. Step 3: Plus 90 degrees.
   d. Step 4: If using new cylinder head bolts, add an additional 90 degrees.

6. The remainder of the installation is the reverse of the removal procedure.

2.2L Engine

1. Before servicing the vehicle, refer to the precautions in the beginning of this section.
2. Disconnect the negative battery cable.
3. Bring the No. 1 cylinder to TDC.
4. Drain the engine coolant into a sealable container.
5. Relieve the fuel system pressure.
6. Remove or disconnect the following:
   - Vacuum hose, breather hose and air intake duct
   - Water bypass hose from the cylinder head
   - Fuel feed and return hose from the fuel rail
   - Evaporative emissions (EVAP) control canister hose from the intake manifold
   - Brake booster vacuum hose from the intake manifold
   - Vacuum hose mount, on automatic transaxle equipped vehicles
   - Throttle cable from the throttle body
   - Throttle control cable from the throttle body, on automatic transaxle equipped vehicles

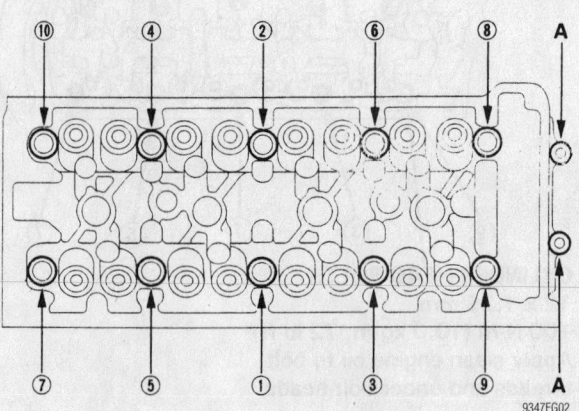

Cylinder head torque sequence—2.0L engine

➡**Be careful not to bend the cable when removing. Do not use pliers to remove the cable from the linkage. Always replace a kinked cable with a new one.**

- Ignition coil

➡**Label the connectors before disconnecting them.**

- Electrical connectors from the distributor and the spark plug wires from the spark plugs

➡**Matchmark the installed position of the distributor before removal.**

- Distributor
- Ignition coil wire from the distributor
- Connector and the terminal from the alternator, then the engine wiring harness from the valve cover
- Fuel injector connectors
- Intake Air Temperature (IAT) sensor connector, if equipped
- Idle Air Control (IAC) valve connector
- Throttle Position (TP) sensor connector
- Exhaust Gas Recirculation (EGR) valve lift sensor
- Ground cable terminals
- Engine Coolant Temperature (ECT) switch B connector, if equipped
- Heated Oxygen Sensor (HO2S) connector
- ECT sensor
- ECT gauge sending unit connector
- Crankshaft Position Sensor

(CKP)/Cylinder Position (CYP) sensor connector, if equipped
- Vehicle Speed Sensor (VSS) connector
- ECT switch **A** connector
- Upper radiator hose and the heater inlet hose from the cylinder head
- Lower radiator hose and heater outlet hose from the intake manifold
- Bypass hose from the thermostat housing and intake manifold
- Thermostat. Discard the O-rings.

➡**Tag all vacuum hoses before disconnecting them.**

- Emissions vacuum hoses from the intake manifold assembly
- Cruise control actuator electrical connector and the vacuum tube, then the actuator
- Engine ground cable from the body
- Mounting bolts and drive belt from the power steering pump. Pull the pump away from the mounting bracket without disconnecting the hoses. Support the pump out of the way.

7. Raise and safely support the vehicle.
8. Remove or disconnect the following:
- Front wheel and tire assemblies
- Splash shield
- Intake manifold bracket bolts
- Intake manifold
- Exhaust pipe from the exhaust manifold
- Exhaust manifold and the exhaust manifold heat insulator
- Power steering pump mounting bracket

- Positive Crankcase Ventilation (PCV) hose, then remove the cylinder head cover. Replace the rubber seals if damages or deteriorated.
- Timing belt.
- Cylinder head bolts in the reverse order of installation

➡**To prevent warpage, unscrew the bolts in sequence ⅓ turn at a time. Repeat the sequence until all bolts are loosened.**

9. Separate the cylinder head from the engine block with a suitable flat bladed pry-tool.

   To install:

10. Be sure all cylinder head and block gasket surfaces are clean. Check the cylinder head for warpage. If warpage is less than 0.002 in. (0.05mm), cylinder head resurfacing is not required. Maximum resurface limit is 0.008 in. (0.2mm) based on a cylinder head height of 3.94 in. (100mm).
11. Always use a new head gasket.
12. The **UP** mark on the camshaft pulley should be at the top.
13. Be sure the No. 1 cylinder is at Top Dead Center (TDC).
14. Clean the oil control orifice and install a new O-ring. Install and align the cylinder head dowel pins and oil control jet.
15. Install the bolts that secure the intake manifold to its bracket but do not tighten them.
16. Position the camshaft correctly.
17. Install the cylinder head, then tighten the cylinder head bolts sequentially in 3 steps:
   a. Step 1: 29 ft. lbs. (40 Nm).
   b. Step 2: 51 ft. lbs. (70 Nm).
   c. Step 3: 72 ft. lbs. (100 Nm).
18. Install or connect the following:
- Intake manifold and tighten the nuts in a crisscross pattern, in 2 or 3 steps, beginning with the inner nuts. Final torque should be 16 ft. lbs. (22 Nm). Always use a new intake manifold gasket.
- Intake manifold bracket to the intake manifold: 16 ft. lbs. (22 Nm)
- Heat insulator to the cylinder head and the block
- Power steering pump mounting bracket to the cylinder head. Tighten the 2 10mm bolts to 36 ft. lbs. (50 Nm). Torque the 8mm bolt to 16 ft. lbs. (22 Nm).
- Exhaust manifold and tighten the nuts in a crisscross pattern in 2 or 3 steps, beginning with the inner nut. Final torque should be 23 ft.

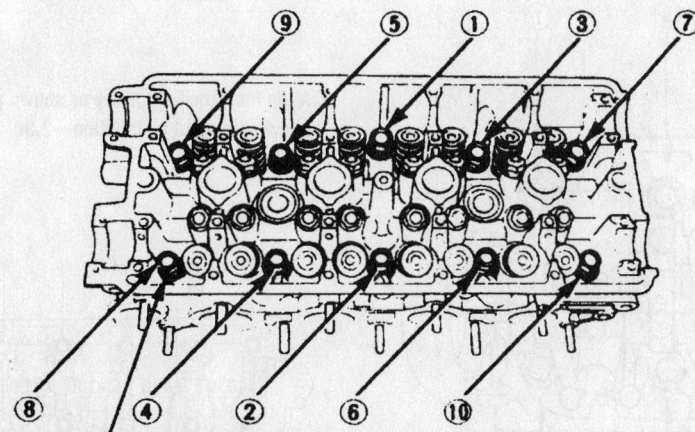

CYLINDER HEAD BOLTS
12 x 1.25 mm
100 N·m (10.0 kg-m, 72 lb-ft)
Apply clean engine oil bolt threads and under bolt heads.

7923FG19

Cylinder head torque sequence—2.2L engine

lbs. (32 Nm). Always use a new exhaust manifold gasket.

- Exhaust manifold bracket, then the exhaust pipe, bracket and upper shroud

➡**Be sure the camshaft sprocket and the crankshaft pulleys are aligned to TDC.**

- Timing belt.
- Splash shield and the front wheels

19. Lower the vehicle.

20. Check and adjust the valves, as necessary.

21. Tighten the crankshaft pulley bolt to 181 ft. lbs. (250 Nm).

22. Installation of the remaining components is the reverse of removal.

23. Fill the cooling system.

24. Connect the negative battery cable and enter the radio security code.

25. Start the engine and check carefully for any leaks.

26. Check the ignition timing and tighten the distributor bolts to 13 ft. lbs. (18 Nm).

27. If equipped with engine and turn the steering wheel lock-to-lock to reset the 4WS control unit.

2.3L Engines

1. Before servicing the vehicle, refer to the precautions in the beginning of this section.

2. Drain the cooling system.

3. Relieve the fuel system pressure.

4. Remove or disconnect the following:

- Negative battery cable

- Air intake duct
- Throttle and cruise control cables
- Positive Crankcase Ventilation (PCV) valve and hose
- Fuel lines
- Brake booster vacuum hose
- Evaporative Emissions (EVAP) canister hoses
- Water bypass hoses
- Accessory drive belts
- Power steering pump
- Alternator wiring harness
- Radiator hoses
- Heater hoses
- Fuel injector connectors
- Intake Air Temperature (IAT) sensor connector
- Idle Air Control (IAC) valve connector
- Throttle Position (TP) sensor connector
- Manifold Absolute Pressure (MAP) sensor connector
- Heated Oxygen (HO$_2$S) sensor connector (F23A1, F23A5 engines)
- Air/Fuel ratio sensor connector (F23A4 engine)
- Engine Coolant Temperature (ECT) sensor connector
- Radiator fan switch connector
- Coolant temperature gauge sender connector
- Exhaust Gas Recirculation (EGR) valve connector
- Crankshaft Position (CKP) sensor connector
- VTEC solenoid valve connector (F23A1, F23A4 engines)

- VTEC oil pressure switch connector (F23A1, F23A4 engines)
- Distributor
- Front motor mount bracket
- Valve cover
- Timing belt
- Camshaft pulley and back cover
- Intake manifold
- Exhaust manifold
- Cylinder head. Loosen the bolts in sequence and in 1/3 turns.

To install:

5. Set the crankshaft pulley to Top Dead Center (TDC).

6. Set the camshaft pulley to TDC.

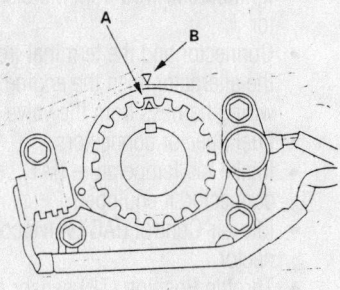

9347FG04

Set the crankshaft to TDC by aligning pointers A and B—2.3L engines

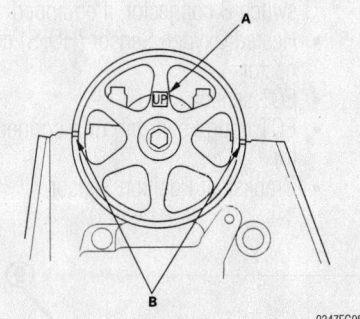

9347FG05

Align the camshaft pulley as shown prior to cylinder head installation—2.3L engines

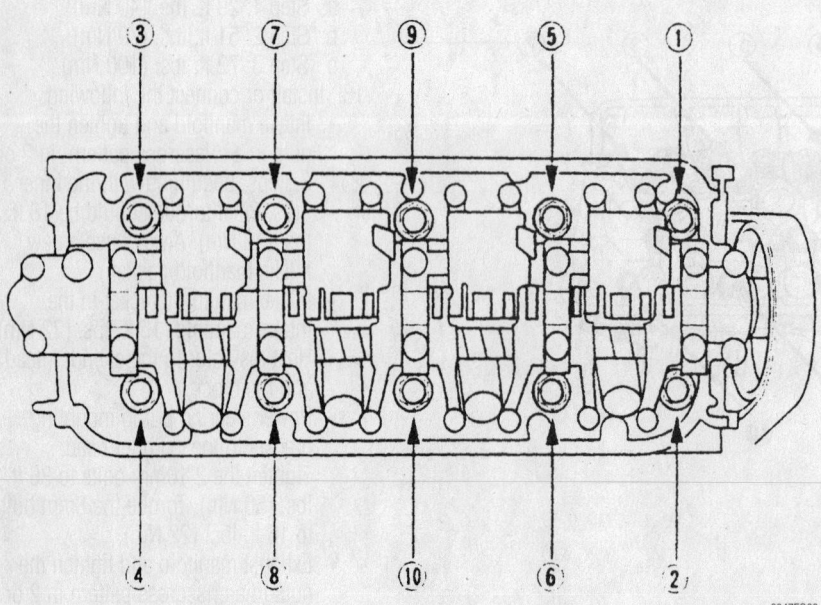

9347FG03

Cylinder head loosening sequence—2.3L engines

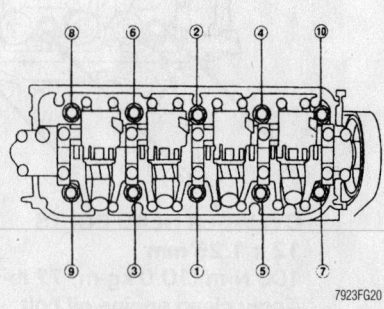

7923FG20

Cylinder head torque sequence—2.3L engines

7. Install the cylinder head with a new gasket. Tighten the bolts in sequence as follows:

    a. Step 1: 22 ft. lbs. (29 Nm).

    b. Step 2: Plus 90 degrees.

    c. Step 3: Plus 90 degrees.

    d. Step 4: If using new cylinder head bolts, add an additional 90 degrees.

8. The remainder of the installation is the reverse of the removal procedure.

2.4L Engine

1. Before servicing the vehicle, refer to the precautions in the beginning of this section.

2. Obtain the security code for the radio.

3. Disconnect the negative battery cable.

4. Drain the coolant.

5. Remove or disconnect the following:

- Intake manifold cover
- 4 ignition coils
- 2 bolts securing the vacuum line
- Bolt securing the power steering hose bracket
- Dipstick and breather hose
- Retainers and cylinder head cover
- Fuel line
- Drive belt
- Intake Air Temperature (IAT) sensor connector
- Vacuum hose (B) and breather pipe (C), then the intake air duct (D)
- Bolt securing the connecting pipe
- Evaporative emission (EVAP) canister hose and brake booster vacuum hose
- Intake manifold
- Exhaust manifold
- Positive Crankcase Ventilation (PCV) hose, vacuum hose and ground cable
- Upper radiator hose, heater hoses and water bypass hose
- Engine wire harness connectors and wire harness clamps from the cylinder head
- 4 injector connectors
- Engine Coolant Temperature (ECT) sensor connector
- Camshaft Position (CMP) sensor connectors
- Exhaust Gas Recirculation (EGR) valve connector
- VTEC solenoid valve connector
- Engine Oil Pressure (EOP) sensor connector
- 2 bolts securing the EVAP canister purge valve bracket and the bolt securing the harness bracket

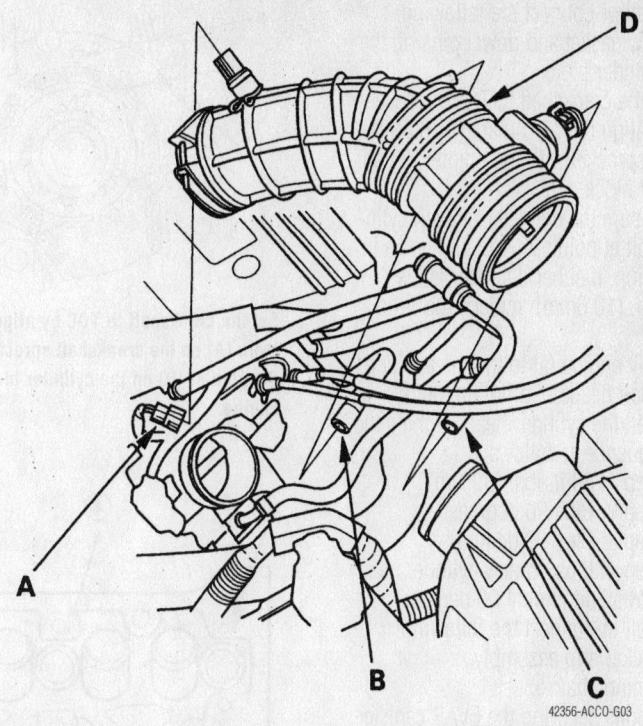

Remove the vacuum hose (B), breather pipe (C), then remove the intake air duct (D)—2.4L engine

- Timing chain
- Rocker arm assembly
- Cylinder head bolts, in sequence, ⅓ turn at a time until completely loosened
- Cylinder head. Discard the gasket.

To install:

6. Be sure all cylinder head and block gasket surfaces are clean. Check the cylinder head for warpage. If warpage is less than 0.002 in. (0.05mm), cylinder head resurfacing is not required. Maximum resurface limit is 0.008 in. (0.2mm) based on a cylinder head height of 3.94 in. (100mm).

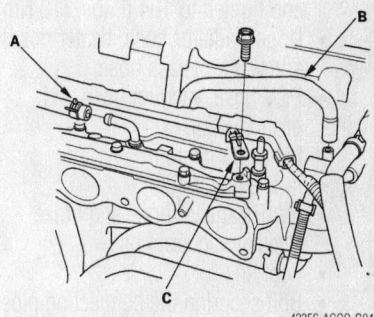

Disconnect the PCV hose (A), vacuum hose (B) and ground cable (C)—2.4L engine

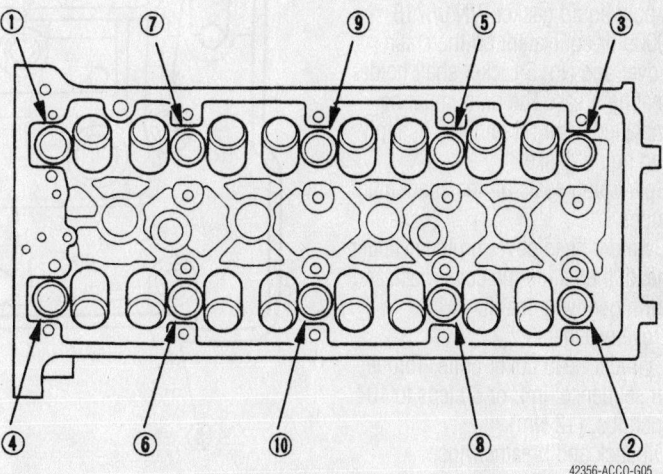

Cylinder head bolt loosening sequence—2.4L engine

7. Install or connect the following:
- New gasket and dowel pins on the cylinder block

8. Set the crankshaft to Top Dead Center (TDC). Align the TDC mark (A) on the crankshaft sprocket with the pointer (B) on the cylinder block.

9. Measure the diameter of each cylinder head bolt at points A & B, as shown in the illustration. If either diameter is less than 0.42 in. (10.6mm), replace the head bolt

10. Apply engine oil to the threads and under the bolt heads of all of the bolts.

11. Install the cylinder head. Tighten the bolts in sequence as follows:
   - a. Step 1: 29 ft. lbs. (39 Nm).
   - b. Step 2: Plus 90 degrees.
   - c. Step 3: Plus 90 degrees.
   - d. Step 4: If using new cylinder head bolts, add an additional 90 degrees.

12. Install or connect the following:
- Rocker arm assembly
- Timing chain
- 2 bolts securing the EVAP canister purge valve bracket and tighten to 16 ft. lbs. (22 Nm)
- Bolt securing the harness bracket and tighten to 104 ft. lbs. (12 Nm)
- Upper radiator hose, heater hoses and water bypass hose
- PCV hose, vacuum hose and ground cable
- Exhaust manifold
- Intake manifold
- EVAP canister hose and brake booster vacuum hose
- Fuel line
- Bolt securing the connecting pipe and tighten to 16 ft. lbs. (22 Nm)
- Intake air duct, IAT sensor connector, vacuum hose and breather pipe
- Cylinder head cover gasket in the groove of the cylinder head cover
- Apply liquid gasket P/N 08718-0009 or equivalent on the chain cover and No. 5 rocker shaft holder mating areas. The parts must be installed within 5 minutes of applying liquid gasket.
- Spark plug seals on the spark plug tubes
- Cylinder head cover on the cylinder head, then slide the cover back and forth gently to seat it
- Cover washers
- Cylinder head cover bolts. Torque, in sequence, in 2 or 3 steps to 104 inch lbs. (12 Nm).
- Dipstick and breather hose
- Bolt securing the power steering hose bracket

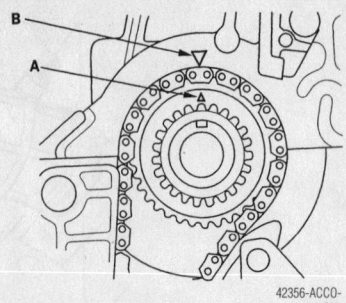

Set the crankshaft to TDC by aligning the mark (A) on the crankshaft sprocket with the pointer (B) on the cylinder block—2.4L engine

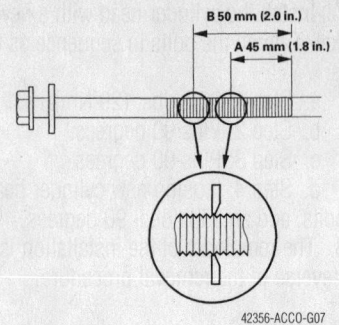

You must measure the cylinder head bolts to see if they can be reused or need to be replaced—2.4L engine

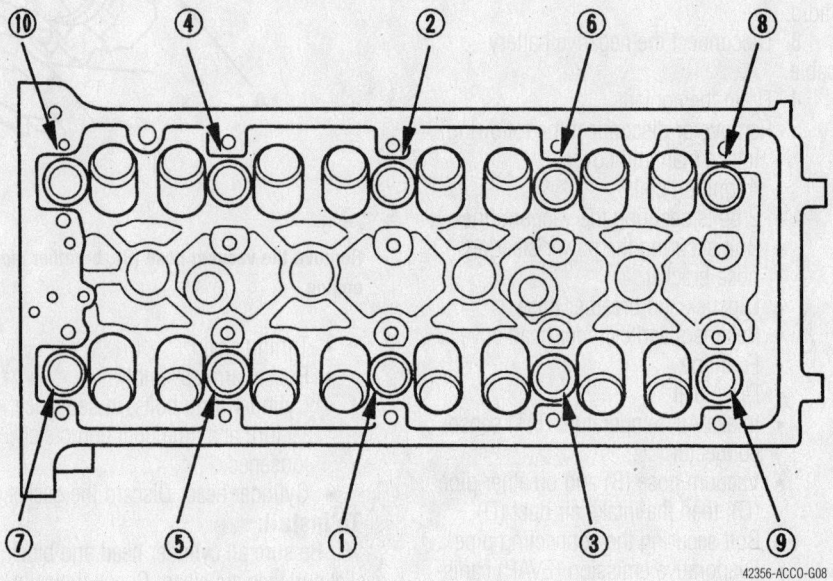

Cylinder head bolt tightening sequence—2.4L engine

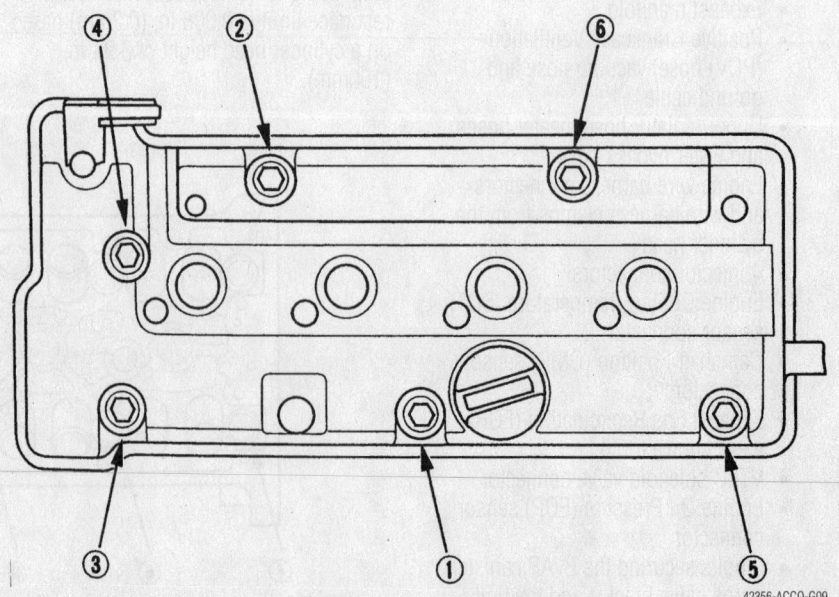

Cylinder head cover bolt tightening sequence—2.4L engine

- 2 bolts securing the vacuum line
- 4 ignition coils
- Intake manifold cover, and tighten the retainers to 104 inch lbs. (12 Nm)
- All of the remaining hoses, tubes, and connectors are installed correctly.

13. Fill the cooling system.

14. Connect the negative battery cable and enter the radio security code.

15. Start the engine and check carefully for any leaks.

3.0L Engine

1. Before servicing the vehicle, refer to the precautions in the beginning of this section.

2. Obtain the security code for the radio.

3. Disconnect the negative battery cable.

4. Drain the coolant.

5. Remove or disconnect the following:
- Evaporative emissions (EVAP) canister hose from the throttle body
- Air intake duct
- Upper engine covers
- Accelerator and cruise control cables from the throttle body
- Spark plug wire holder, cover, and intake manifold covers

6. Properly relieve the fuel system pressure.

7. Remove or disconnect the following:
- Fuel hoses from the supply rail
- Brake booster vacuum hose
- Positive Crankcase Ventilation (PCV) hose
- Breather hose
- Water bypass hose
- Vacuum hose from the throttle body
- Ground cable from the engine
- Alternator belt

8. Support the engine with a jack and a block of wood and remove the side engine mounting bracket.

9. Remove or disconnect the following:
- Power steering pump without disconnecting the hoses
- Alternator
- Wiring harness connectors from the components on the engine that may interfere with removing the cylinder head
- Distributor and spark plug wires
- Intake manifold
- Connectors from the fuel injectors
- Fuel supply rails
- Vacuum hoses from the fuel control valve

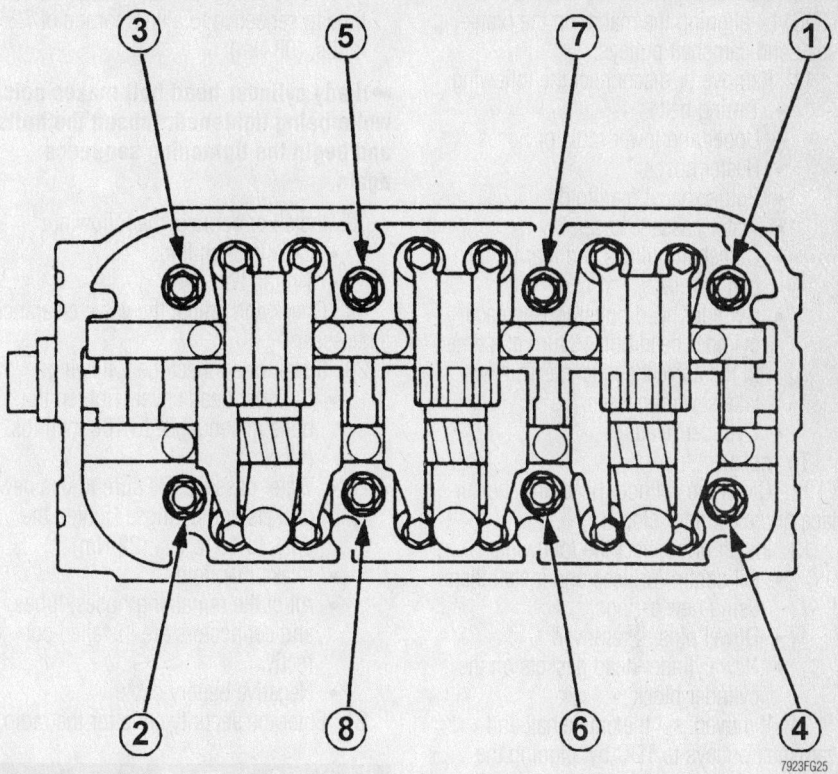

Loosen the cylinder head bolts in the sequence shown to prevent damage to the head—3.0L engine

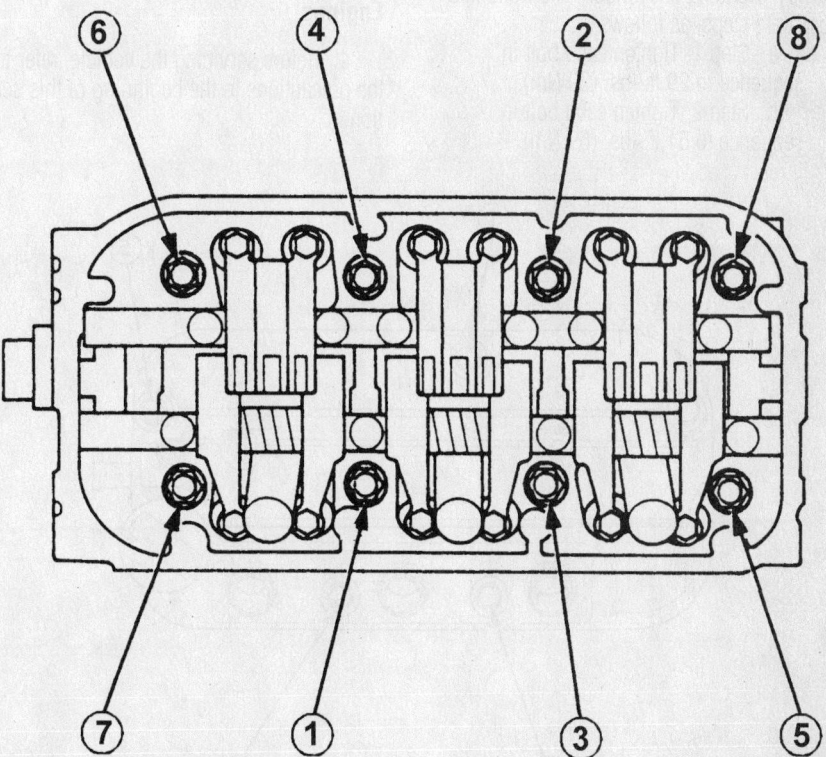

Cylinder head torque sequence—3.0L engine

10. Set the engine to Top Dead Center (TDC) by aligning the marks on the crankshaft and camshaft pulleys.

11. Remove or disconnect the following:
- Timing belt
- Upper and lower radiator hoses
- Heater hoses
- Both exhaust manifolds
- Water passage assembly
- Camshaft pulleys and rear timing belt covers
- Cylinder head bolts; loosen each cylinder head bolt ⅓ turn at a time in the correct sequence. This will take several passes.
- Cylinder heads

To install:

12. Clean the cylinder head and the surface of the cylinder block.

13. Install or connect the following:
- Oil control orifices and install them using new o-rings
- Dowel pins, if removed
- New cylinder head gaskets on the cylinder block

14. If moved, set the crankshaft and camshaft pulleys to TDC by aligning the marks on the pulley and oil pump.

15. Install or connect the following:
- Cylinder heads on the engine

16. Lubricate the cylinder head bolts with clean engine oil.

17. Tighten the cylinder head bolts in 3 separate steps, as follows:
   a. Step 1: Tighten each bolt in sequence to 29 ft. lbs. (39 Nm).
   b. Step 2: Tighten each bolt in sequence to 51 ft. lbs. (69 Nm).

   c. Step 3: Tighten each bolt a third time in sequence to a final torque of 72 ft. lbs. (98 Nm).

➡**If any cylinder head bolt makes noise while being tightened, loosen the bolts and begin the tightening sequence again.**

18. Install or connect the following:
- Exhaust manifolds
- Timing belt

19. Check and adjust the valve clearance if necessary.

20. Install or connect the following:
- Cylinder head cover. Tighten the bolts in sequence to 108 inch lbs. (12 Nm).
- Water passage. Be sure to use new gaskets and o-rings. Tighten the bolts to 16 ft. lbs. (22 Nm).
- Intake manifold
- All of the remaining hoses, tubes, and connectors are installed correctly.
- Negative battery cable

21. Enter the security code for the radio.

Rocker Arms/Shafts

REMOVAL & INSTALLATION

1.6L (D16Y5) and 1.7L (D17A2) Engines

1. Before servicing the vehicle, refer to the precautions in the beginning of this section.

2. Remove or disconnect the following:
- Negative battery cable

➡**Label the wires before disconnecting them.**

- Ignition wires
- Spark plugs and note their cylinder assignments.
- Valve cover

3. Rotate the crankshaft to set the No. 1 cylinder to Top Dead Center (TDC) for the compression stroke. The white TDC mark on the crankshaft pulley aligns with the pointers on the lower timing cover.

4. Remove or disconnect the following:
- Distributor, if equipped

5. Loosen the valve adjusting screws

6. Remove or disconnect the following:
- Variable Timing Electronic Control (VTEC) solenoid valve connector

7. Loosen the camshaft holder bolts 2 turns at a time in a crisscross pattern to prevent damaging the valves or rocker assembly.

8. Remove or disconnect the following:
- Rocker arm and shaft assemblies together with the camshaft holders. Do not remove the rocker shaft bolts yet. The bolts keep the bearing caps, springs, and rocker arms in place on the shafts.

➡**The rocker arms and shafts are an assembly. They must be removed from the engine as a unit. Always follow the torque sequence carefully when installing the rocker shaft assembly.**

- Camshaft holder bolts from the rocker arm and shaft assembly

9. Bundle the intake rocker arm assemblies with rubber bands so they don't separate when the intake rocker shaft is removed.

10. Disassemble the rocker arm and shaft assemblies. Label the parts as they are removed from the shafts to ensure reinstallation in the original location.

11. Disassemble the rocker arm assemblies taking care not to mix up any of the parts. Inspect the rocker arm synchronizing and timing pistons by pushing them with your fingers. If the pistons don't move smoothly in the rocker arm bores, replace the rocker arm assembly.

12. Apply oil to the synchronizing pistons, timing piston, and timing spring and reassemble the rocker arms. Bundle the rocker arm assemblies with rubber bands to prevent the parts from separating.

13. Inspect the timing plates and return springs which are located on the camshaft holders. Set each timing plate and return spring so that the C-shaped upper arm of

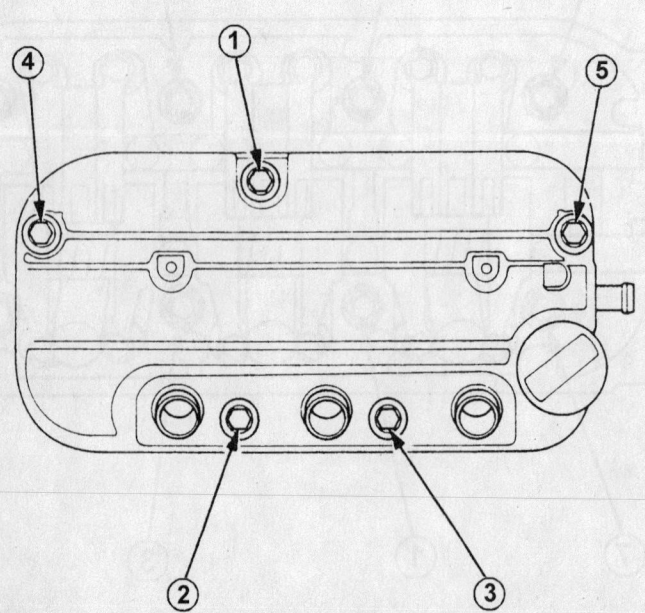

Tighten the cylinder head cover bolts in the sequence shown—3.0L engine

7923FG27

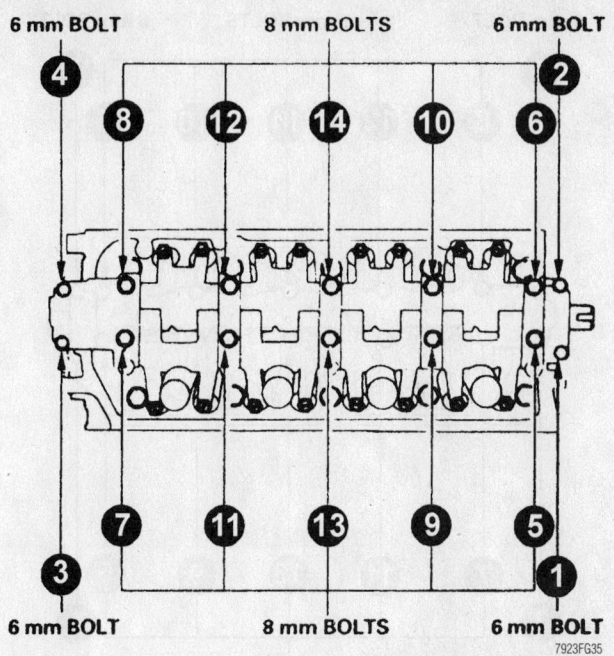

6 mm BOLT   8 mm BOLTS   6 mm BOLT

6 mm BOLT   8 mm BOLTS   6 mm BOLT

7923FG35

Rocker arm/shaft bolt loosening sequence—1.6L (D16Y5) and 1.7L (D17A2) engines

the plate is position parallel to the top of the camshaft holder.

To install:

14. Verify that the engine is set at Top Dead Center (TDC)/compression for the No. 1 cylinder.

15. Lubricate the camshaft journals and lobes. Coat the rocker shafts and camshaft holders with oil.

16. Remove the oil control orifice. Thoroughly clean it and reinstall it with a new O-ring.

17. Install or connect the following:
   • New camshaft seal if necessary

18. Assemble the rocker arm and shaft assemblies. Be sure the intake shaft collars and exhaust shaft springs are in the proper locations.

19. After the rocker arms and shafts are assembled, cut the rubber bands and remove them from the intake rockers. Be sure that no rubber band fragments are left in the engine.

20. Install or connect the following:
   • Fresh oil to the threads of camshaft holder bolts
   • Camshaft holder bolts
   • Liquid gasket to the cylinder head mating surfaces of the No. 1 and No. 5 camshaft holders. Do not allow the sealant to cure before installation.
   • Rocker arm and shaft assembly in place
   • Rocker and bolts, hand-tight

21. Tighten each bolt 2 turns at a time in the crisscross sequence so that the rockers are evenly tightened and don't bind on the valves. Tighten the 8mm rocker arm bolts to 14 ft. lbs. (20 Nm). Tighten the 6mm bolts to 108 inch lbs. (12 Nm).

22. Starting with the No. 1 cylinder at Top Dead Center (TDC)/compression, adjust the valve clearances. After the clearance has been reached, tighten the locknuts to 14 ft. lbs. (20 Nm). Set the No. 3, No. 4, and No. 2 cylinders at Top Dead Center (TDC)/compression, and adjust their valve clearances.

• Intake: 0.007–0.009 in. (0.18–0.22mm)
• Exhaust: 0.009–0.011 in. (0.23–0.27mm)

23. Remove the VTEC solenoid valve, then remove the filter. Install a new VTEC solenoid valve filter. Tighten the solenoid valve bolts to 108 inch lbs. (12 Nm) and reconnect the solenoid valve connector.

24. Rotate the crankshaft to set the No. 1 cylinder at Top Dead Center (TDC)/compression. Then, manually inspect the operation of each of the VTEC intake rocker arms:

   a. Move the No. 1 cylinder's secondary intake rocker arm up and down.

   b. Verify that the secondary intake rocker arm moves independently of the primary intake rocker arm.

   c. Repeat the rocker arm inspection for the other 3 cylinders with each cylinder set at Top Dead Center (TDC)/compression.

25. Rotate the crankshaft back to TDC/compression for the No. 1 cylinder. Install the distributor, if equipped, but do not tighten the mounting bolts yet.

26. Tighten the crankshaft pulley bolt to 134 ft. lbs. (181 Nm).

27. Install or connect the following:
   • Valve cover. Be sure the gasket is in good condition, and apply sealant to the corners where the gasket meets the camshaft holders.
   • Spark plugs and the ignition wires

28. Drain the engine oil and remove the

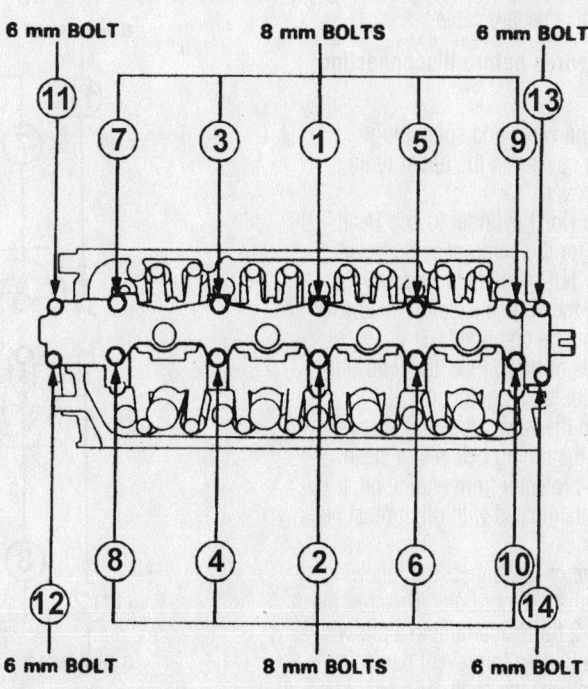

6 mm BOLT   8 mm BOLTS   6 mm BOLT

6 mm BOLT   8 mm BOLTS   6 mm BOLT

7923FG36

Rocker arm/shaft bolt torque sequence—1.6L (D16Y5) and 1.7L (D17A2) engine

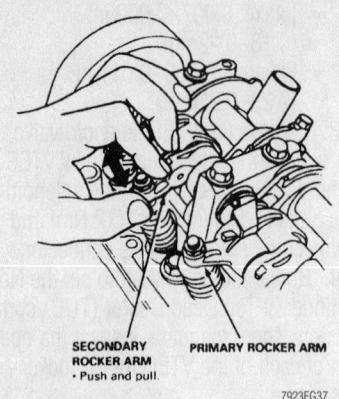

VTEC rocker arm inspection—1.6L
(D16Y5) and 1.7L (D17A2) engines

oil filter. Install a new oil filter and refill the engine with fresh oil.

29. Install or connect the following:
• Negative battery cable

30. Warm the engine up to normal operating temperature.

31. Check the ignition timing and adjust it if necessary. Then, tighten the distributor mounting bolts to 17 ft. lbs. (24 Nm).

32. Check all fluid levels. Test drive the vehicle and observe the engine RPM changes at various speeds.

1.6L (D16Y7) and 1.7L (D17A1) Engine

1. Before servicing the vehicle, refer to the precautions in the beginning of this section.

2. Remove or disconnect the following:
• Negative battery cable

➡Label the wires before disconnecting them.

• Ignition wires and spark plugs
• Valve cover and the upper timing belt cover

3. Set the No. 1 cylinder to Top Dead Center (TDC) for the compression stroke. Verify that the TDC marks are correctly aligned. Once the engine is set in this position, it must not be disturbed.

4. Remove or disconnect the following:
• Distributor, if equipped

5. Loosen the valve adjusting screws.

6. Cover the timing belt with a clean shop towel to protect it from engine oil. If the belt is contaminated with oil, it must be replaced.

7. Remove or disconnect the following:
• Camshaft holder bolts. Unscrew the bolts 2 turns at a time in a crisscross pattern to prevent damaging the valves, camshaft, or rocker arm assembly.

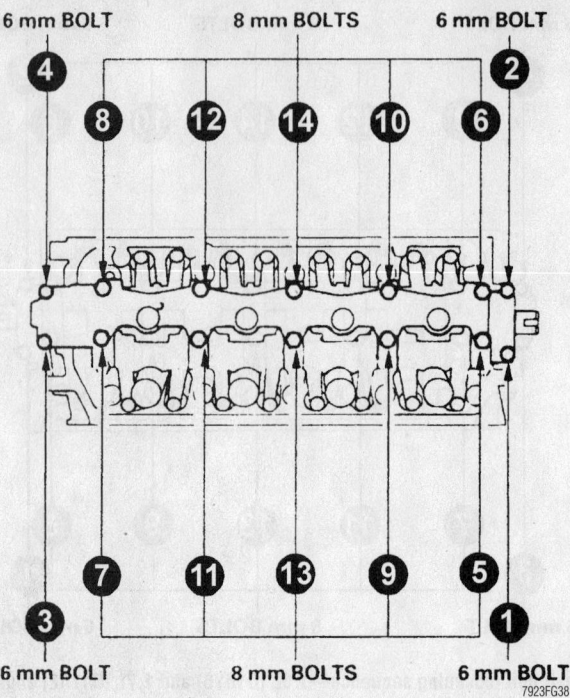

Rocker arm/shaft bolt loosening sequence—1.6L (D16Y7) and 1.7L (D17A1) engines

➡The rocker arms and shafts are an assembly; they must be removed from the engine as a unit. To prevent warpage, always follow the torque sequence carefully when removing or installing the rocker shaft assembly.

• Rocker arm and shaft assemblies. Do not remove the camshaft holder bolts. The bolts keep the camshaft

bearing caps, springs, and rocker arms in place on the shafts.

8. If the rocker arms or shafts are to be replaced, identify the parts as they are removed from the shafts to ensure reinstallation in the original location.

To install:

9. Verify that the engine is set to TDC/compression for the No. 1 cylinder.

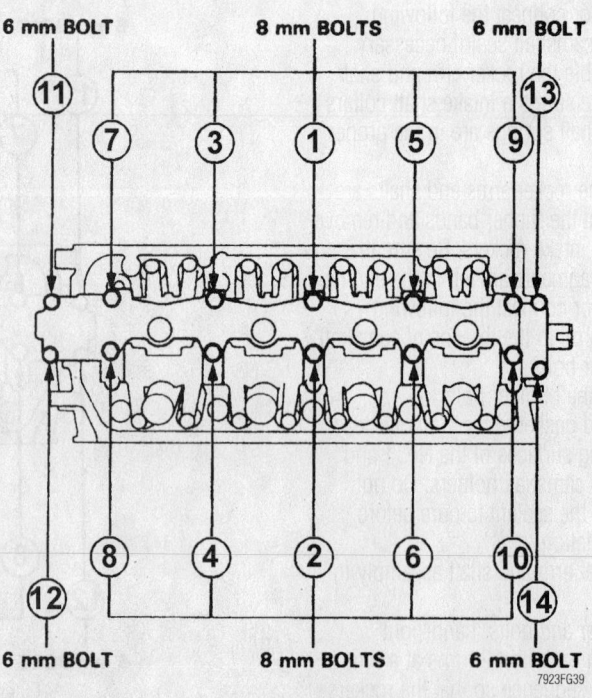

Rocker arm/shaft bolt tightening sequence—1.6L (D16Y7) and 1.7L (D17A1) engines

The camshaft keyway faces up when the engine is at Top Dead Center (TDC)/compression.

10. Lubricate the camshaft journals and lobes with clean engine oil. Install a new camshaft seal if necessary.

11. Remove the oil control orifice. Thoroughly clean it and install it with a new O-ring.

12. Assemble the rocker arms, shafts, and camshaft bearing caps.

13. Apply sealant to the mating surfaces of the No. 1 and No. 5 camshaft bearing caps. Do not allow the sealant to cure before the rocker arm assembly is installed.

14. Set the rocker arm assembly in place. Apply engine oil to the holder bolt threads, then loosely install the bolts. Tighten each bolt in a 2 step crisscross pattern to ensure that the rockers do not bind on the valves. Tighten the 8mm bolts to 14 ft. lbs. (20 Nm). Tighten the 6mm bolts to 104 inch lbs. (12 Nm).

15. Verify that the engine is at Top Dead Center (TDC)/compression for the No. 1 piston and install the distributor, if equipped.

16. Adjust the valves and tighten the locknuts to 14 ft. lbs. (20 Nm).

17. Install the valve cover and upper timing belt cover.

18. Reconnect the negative battery cable.

19. Check the ignition timing and adjust if necessary. Tighten the distributor mounting bolts to 17 ft. lbs. (24 Nm).

1.6L (D16Y8) Engine

1. Before servicing the vehicle, refer to the precautions in the beginning of this section.

2. Remove or disconnect the following:
   • Negative battery cable
   • Ignition wires and spark plugs
   • Valve cover

3. Rotate the crankshaft to set the No. 1 cylinder to Top Dead Center (TDC) for the compression stroke. The white TDC mark on the crankshaft pulley aligns with the TDC pointers on the lower timing belt cover.

4. Remove or disconnect the following:
   • Distributor from the cylinder head
   • Valve adjusting screws
   • Variable Timing Electronic Control (VTEC) solenoid valve connector
   • Camshaft holder bolts, by loosening them 2 turns at a time in a crisscross pattern to prevent damaging the valves or rocker assembly.
   • Rocker arm and shaft assemblies together with the camshaft holders and the lost motion assembly holder. Do not remove the rocker

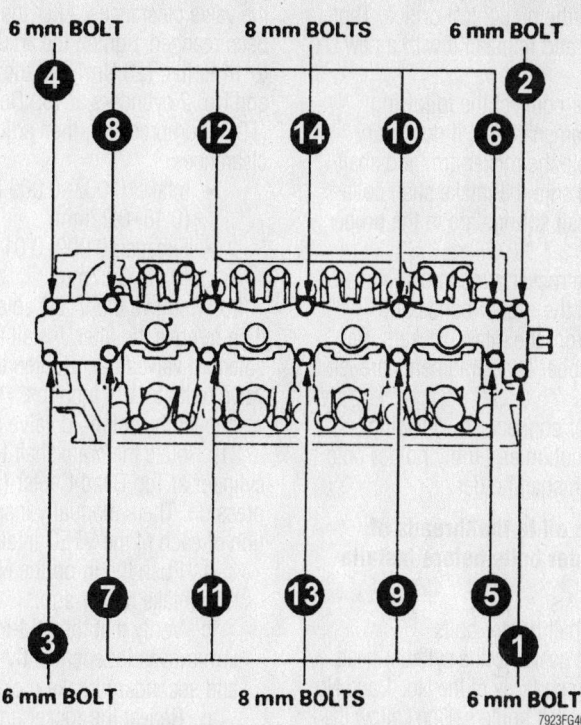

Rocker arm/shaft bolt loosening sequence—1.6L (D16Y8) engine

shaft bolts yet. The bolts keep the bearing caps, springs, and rocker arms in place on the shafts.

➡ **The rocker arms and shafts are an assembly; they must be removed from the engine as a unit. Always follow the torque sequence carefully when installing the rocker shaft assembly.**

   • Camshaft holder bolts from the rocker arm and shaft assembly
   • Lost motion assembly holder

5. Bundle the intake rocker arm assemblies with rubber bands so they don't separate when the intake rocker shaft is removed.

6. Disassemble the rocker arm and shaft assemblies. Label the parts as they are removed from the shafts to ensure reinstallation in the original location.

7. Disassemble the rocker arm assemblies taking care not to mix up any of the parts. Inspect the rocker arm synchronizing pistons by pushing them with your fingers. If the pistons don't move smoothly in their rocker arm bores, replace the rocker arm assembly.

8. Apply oil to the synchronizing pistons and reassemble the rocker arms. Bundle the rocker arm assemblies with rubber bands to prevent the parts from separating.

9. Remove or disconnect the following:
   • Lost motion assembly from its port in the lost motion assembly holder. Inspect each lost motion assembly

by pushing down on its piston. If the piston doesn't move smoothly, replace the lost motion assembly. Lost motion assemblies cannot be bled like hydraulic lash adjusters.

10. Install the lost motion assemblies back into the lost motion assembly holder.

To install:

11. Verify that the engine is set at Top Dead Center (TDC)/compression for the No. 1 cylinder.

12. Lubricate the camshaft journals and lobes. Coat the rocker shafts and camshaft holders with fresh oil.

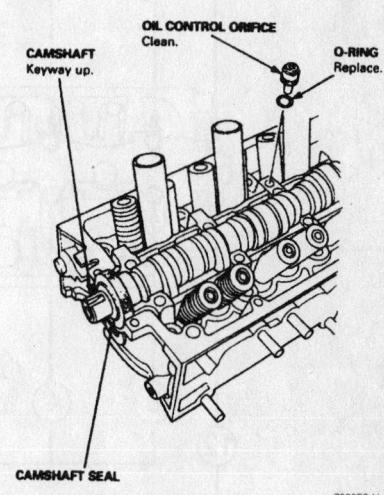

Camshaft seal and oil control orifice—1.6L (D16Y8) engine

13. Remove the oil control orifice. Thoroughly clean it and reinstall it with a new O-ring.

14. Install or connect the following:
- New camshaft seal if necessary

15. Assemble the rocker arm and shaft assemblies. Be sure the intake shaft collars and exhaust shaft springs are in the proper locations.

16. After the rocker arms and shafts are assembled, cut the rubber bands and remove them from the intake rockers. Be sure that no rubber band fragments are left in the engine.

17. Install or connect the following:
- Lost motion assembly holder onto the camshaft holder

➥Apply fresh oil to the threads of camshaft holder bolts before installation.

- Camshaft holder bolts
- Liquid gasket to the cylinder head mating surfaces of the No. 1 and No. 5 camshaft holders. Don't allow the sealant to cure before installation.

18. Set the rocker arm and shaft assembly in place. Install and hand-tighten the bolts. Tighten each bolt 2 turns at a time in the crisscross sequence to ensure that the rockers do not bind on the valves. Tighten the 8mm rocker arm bolts to 14 ft. lbs. (20 Nm). Tighten the 6mm bolts to 108 inch lbs. (12 Nm).

19. Starting with the No. 1 cylinder at Top Dead Center (TDC)/compression, adjust the valve clearances. After the clearance has been reached, tighten the adjuster locknuts to 14 ft. lbs. (20 Nm). Set the No. 3, No. 4, and No. 2 cylinders at Top Dead Center (TDC)/compression, then adjust their valve clearances.

- Intake: 0.007–0.009 in. (0.18–0.22mm)
- Exhaust: 0.009–0.011 in. (0.23–0.27mm)

20. Remove the VTEC solenoid valve, then remove the filter. Install a new VTEC solenoid valve filter. Tighten the solenoid valve bolts to 108 inch lbs. (12 Nm) and reconnect the solenoid valve connectors.

21. Rotate the crankshaft to set the No. 1 cylinder at Top Dead Center (TDC)/compression. Then, manually inspect the operation of each of the VTEC intake rocker arms:
  a. Push the in on the No. 1 cylinder's mid-intake rocker arm.
  b. Verify that the mid-intake rocker arm moves independently of the primary and secondary intake rocker arms.
  c. Repeat the rocker arm inspection for the other 3 cylinders with each cylinder set at Top Dead Center (TDC)/compression.

22. Rotate the crankshaft back to TDC/compression for the No. 1 cylinder. Install the distributor, but do not tighten the mounting bolts yet.

23. Tighten the crankshaft pulley to 134 ft. lbs. (181 Nm).

24. Install or connect the following:

- Valve cover. Be sure the gasket is in good condition, and apply sealant to the corners where the gasket meets the camshaft holders.
- Spark plugs and the ignition wires

25. Drain the engine oil and remove the oil filter. Install a new oil filter and refill the engine with fresh oil.

26. Connect the negative battery cable.

27. Warm the engine up to normal operating temperature.

28. Check the ignition timing and adjust it if necessary. Then, tighten the distributor mounting bolts to 17 ft. lbs. (24 Nm).

29. Check all fluid levels. Test drive the vehicle and observe the engine RPM changes at various speeds.

1.6L (B16A2) Engine

1. Before servicing the vehicle, refer to the precautions in the beginning of this section.

2. Remove or disconnect the following:
- Negative battery cable

➥Label the wires before disconnecting them.

- Ignition wires

3. Rotate the crankshaft to set the engine at Top Dead Center (TDC) for the compression stroke of the No. 1 cylinder. The white TDC mark on the crankshaft pulley should align with the pointer on the lower timing belt cover.

4. Remove or disconnect the following:
- Strut brace
- Intake air duct
- Drive belts

5. Use a floor jack padded with a block of wood to support the engine.

6. Remove or disconnect the following:
- Engine ground cable
- Engine side mount
- Valve cover and the upper timing belt cover

7. Verify that the engine is set at Top Dead Center (TDC)/compression. Loosen the timing belt tensioner bolt 180°. Then, remove the crankshaft pulley, the lower timing cover, and timing belt.

⚹⚹ WARNING

Inspect the timing belt for signs of cracked and broken teeth, as well as oil or coolant contamination. If the timing belt is damaged, or has been in contact with oil or coolant, it must be replaced to avoid potential failure.

8. Remove or disconnect the following:
- Distributor

6 mm BOLT   8 mm BOLTS   6 mm BOLT

(11) (13)
(7) (3) (1) (5) (9)

(8) (4) (2) (6) (10)
(12) (14)

6 mm BOLT   8 mm BOLTS   6 mm BOLT

7923FG42

Rocker arm/shaft bolt tightening sequence—1.6L (D16Y8) engine

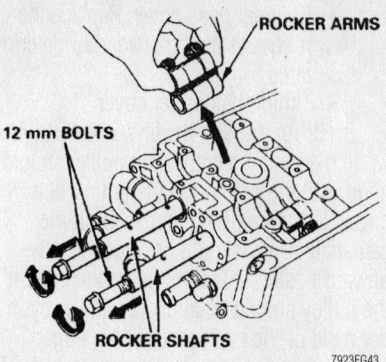

Removing the rocker arms—1.6L (B16A2) engines

- Variable Timing Electronic Control (VTEC) solenoid valve. Remove the solenoid filter and inspect it for clogging.
- Camshaft sprockets and back cover
- Camshaft holder plate bolts. Loosen in a crisscross sequence working toward the middle of the cylinder head.
- Valve adjusting screws
- Camshaft holder plates and holder from the cylinder head. The holder bolts will keep the components together. Note the positions of each camshaft holder for reassembly.
- Camshafts from the cylinder head. Mark the exhaust and intake camshafts so that they will not be confused.

9. Hold each rocker arm assembly together with a rubber band to prevent them from separating.

10. Remove or disconnect the following:
- Intake and exhaust rocker shaft orifices from the cylinder head. The rocker shaft orifices are different and should be identified when removed. Thoroughly clean the orifices and reinstall them with new O-rings.
- Rocker arm shaft sealing bolts and discard the washers

11. Insert 12mm bolts into the rocker arm shafts. Remove each rocker arm set while slowly pulling out the rocker arm shaft.

➡ **Tag each rocker arm set to assure installation in their original locations.**

12. Inspect the rocker arm pistons. If they do not move smoothly, replace the rocker arm assembly.

13. Remove or disconnect the following:
- 2 lost motion assemblies from the cylinder head. Inspect each lost motion assembly by pushing the plunger with your finger. Replace

the lost motion assembly if it does not move smoothly.

To install:

14. Install or connect the following:
- 2 lost motion assemblies to the cylinder head
- Engine oil to the rocker arm pistons, then bundle the rocker arms with a rubber band. Apply a light coat of clean engine oil to the rocker arms.
- Rocker arms in their original locations if they are being reused. If new assembles are being used, place them in the cylinder head.

15. Lightly coat the rocker arm shafts with clean engine oil, then install the rocker arm shafts into the cylinder head. A 12mm bolt can be installed into the end of the rocker arm shafts to aid in their installation. Be sure to install the shafts in the proper positions. Remove the 12mm bolts from the rocker arm shafts, if used.

16. Clean and install the rocker arm shaft orifices with new O-rings. If the holes in the rocker arm shafts are not aligned, screw a 12mm bolt into the end of the shaft to position the shaft.

17. Install or connect the following:
- Sealing bolts with new washers: 47 ft. lbs. (64 Nm)

18. Lubricate the camshaft lobes and journals with clean engine oil.

19. Install or connect the following:
- Camshafts into the cylinder head. Both the intake and exhaust camshafts should be installed with their keyways pointing straight up.
- New lubricated camshaft seals. Apply liquid gasket to a new camshaft end-plug and install it. If the end-plug is marked, the mark should be aligned with the cylinder head surface.

20. Apply liquid gasket to the cylinder head mating surfaces of the No. 1 and No. 5 camshaft holders, then install them, along with No. 2, 3, and 4 holders. Be sure to pay attention to the following points:
- Do not apply oil to the holder mating surface of camshaft seals.
- The arrows marked on the camshaft holders should point to the timing belt.

21. Install or connect the following:
- Camshaft holder plates

➡ **Lubricate the threads of the 10mm holder bolts before installation.**

- Camshaft holder bolts, but don't tighten them yet

22. Evenly hand-tighten the camshaft

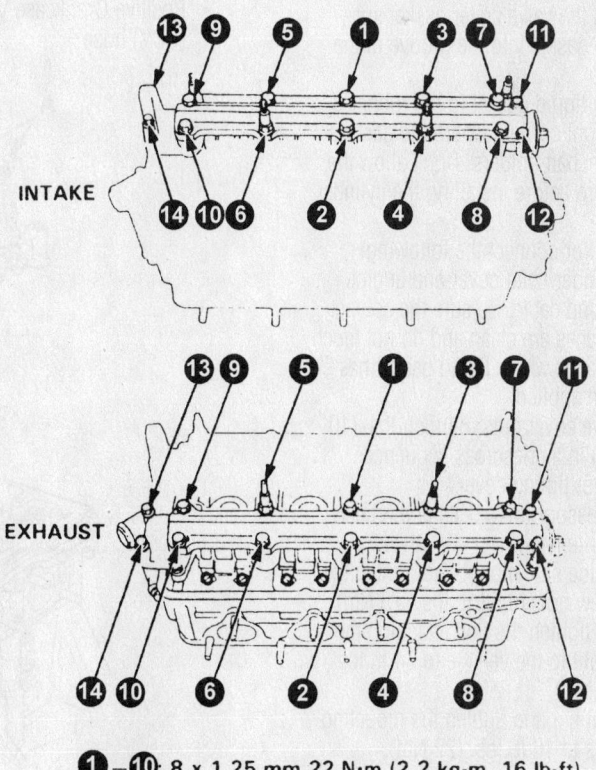

① — ⑩: 8 x 1.25 mm 22 N·m (2.2 kg-m, 16 lb-ft)
⑪ — ⑭: 6 x 1.0 mm 11 N·m (1.1 kg-m, 8 lb-ft)

Camshaft holder bolt torque sequences—1.6L (B16A2) engines

holders. Be sure that the rocker arms are properly positioned on the valve stems.

23. Use a 2-step crisscross pattern to tighten the camshaft holder bolts. Begin tightening with the bolts in the middle of the cylinder head, and work toward the outer edges. Final torque specifications are as follows:

    a. 8mm bolts 20 ft. lbs. (28 Nm).
    b. 6mm bolts to 84–96 inch lbs.
(10–11 Nm)

24. Verify that the camshaft keyways are pointing straight up and that the engine is at Top Dead Center (TDC)/compression for the No. 1 cylinder. Fit the camshaft sprocket keys into their keyways.

25. Install or connect the following:

- Back cover and push the camshaft pulleys onto the camshafts. Then, tighten the sprocket retaining bolts to 41 ft. lbs. (57 Nm).
- Timing belt, then tension the belt
- Lower timing cover and the crankshaft pulley. Tighten the pulley bolt to 130 ft. lbs. (180 Nm).

26. Adjust the valve clearance.

27. Inspect the VTEC rocker arms for smooth and independent movement.

28. Install or connect the following:

- VTEC solenoid valve with a new filter. Tighten the valve mounting bolts to 108 inch lbs. (12 Nm).
- Distributor

29. Clean the valve cover gasket surfaces. Fit the gasket into the groove of the valve cover.

30. Apply liquid gasket to the rubber seal at the eight corners where the gaskets meet the camshaft holders. Don't allow the sealant to cure before installing the cylinder head cover.

31. Install or connect the following:

- Cylinder head cover and engine ground cable. Be sure the contact surfaces are clean and do not touch surfaces where liquid gasket has been applied.
- Valve cover nuts: 84 inch lbs. (10 Nm) in a crisscross sequence
- Upper timing cover.
- Accessory drive belts and adjust their tensions
- Engine side mount and tighten the 2 new nuts to 54 ft. lbs. (75 Nm) and tighten the bolt attaching the mount to the vehicle to 54 ft. lbs. (74 Nm).
- Strut bar and tighten the mounting bolts to 16 ft. lbs. (22 Nm).
- Ignition wires

32. Drain the engine oil. Install a new oil filter and refill the engine with fresh oil.

33. Reconnect the negative battery cable.

34. Warm the engine up to its normal operating temperature. Then, check and adjust the ignition timing. Tighten the distributor mounting bolts to 17 ft. lbs. (24 Nm).

2.0L Engine

1. Before servicing the vehicle, refer to the precautions in the beginning of this section.

2. Remove or disconnect the following:

- Negative battery cable
- Valve cover
- Camshafts

3. Replace the front and rear camshaft holder bolts as shown, and remove the rocker arm assembly.

To install:

➡**Keep all valvetrain components in order. If reused, they must be installed in their original locations.**

4. Installation is the reverse of the removal procedure.

2.2L Engine

1. Before servicing the vehicle, refer to the precautions in the beginning of this section.

2. Remove or disconnect the following:

- Negative battery cable
- Air intake duct
- Positive Crankcase Ventilation (PCV) hose

- Cylinder head cover. Replace the rubber seals if damaged or deteriorated.
- Timing belt upper cover

3. Bring the No. 1 cylinder to TDC. The white mark on the crankshaft pulley should align with the pointer on the timing belt cover. The words **UP** embossed on the camshaft pulley should be aligned in the upward position. The marks on the edge of the pulley should be aligned with the cylinder head or the back cover upper edge. Once in this position, the engine must NOT be turned or disturbed.

4. Remove or disconnect the following:

- Electrical connectors from the distributor and the spark plug wires from the spark plugs.

➡**Matchmark the installed position distributor before removing it.**

- Distributor from the cylinder head
- Power steering pump drive belt

5. Mark the rotation of the timing belt if it is to be used again. Loosen the timing belt adjusting bolt ¾ to 1 turn, then release the tension on the timing belt. Push the tensioner to release tension from the belt, then tighten the adjusting bolt.

6. Remove or disconnect the following:

- Timing belt from the camshaft sprocket

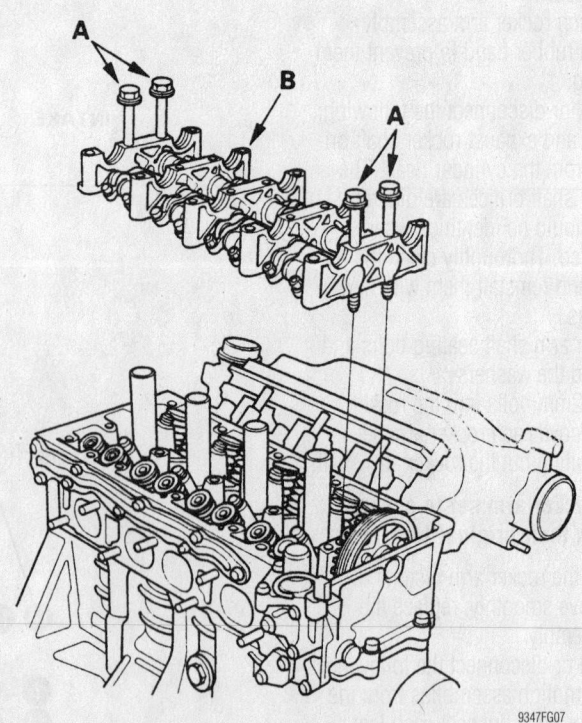

Use bolts (A) to hold the rocker arm assembly (B) together while removing or replacing the assembly—2.0L engine

NOTE:
- Identify parts as they are removed to ensure reinstallation in their original locations.
- Inspect rocker shafts and rocker arms
- Rocker arms must be installed in the same position if reused.
- Prior to reassembling, clean all the parts in solvent, dry them, and apply lubricant to any contact points.
- Bundle the rocker arms with rubber bands to keep them together as a set.

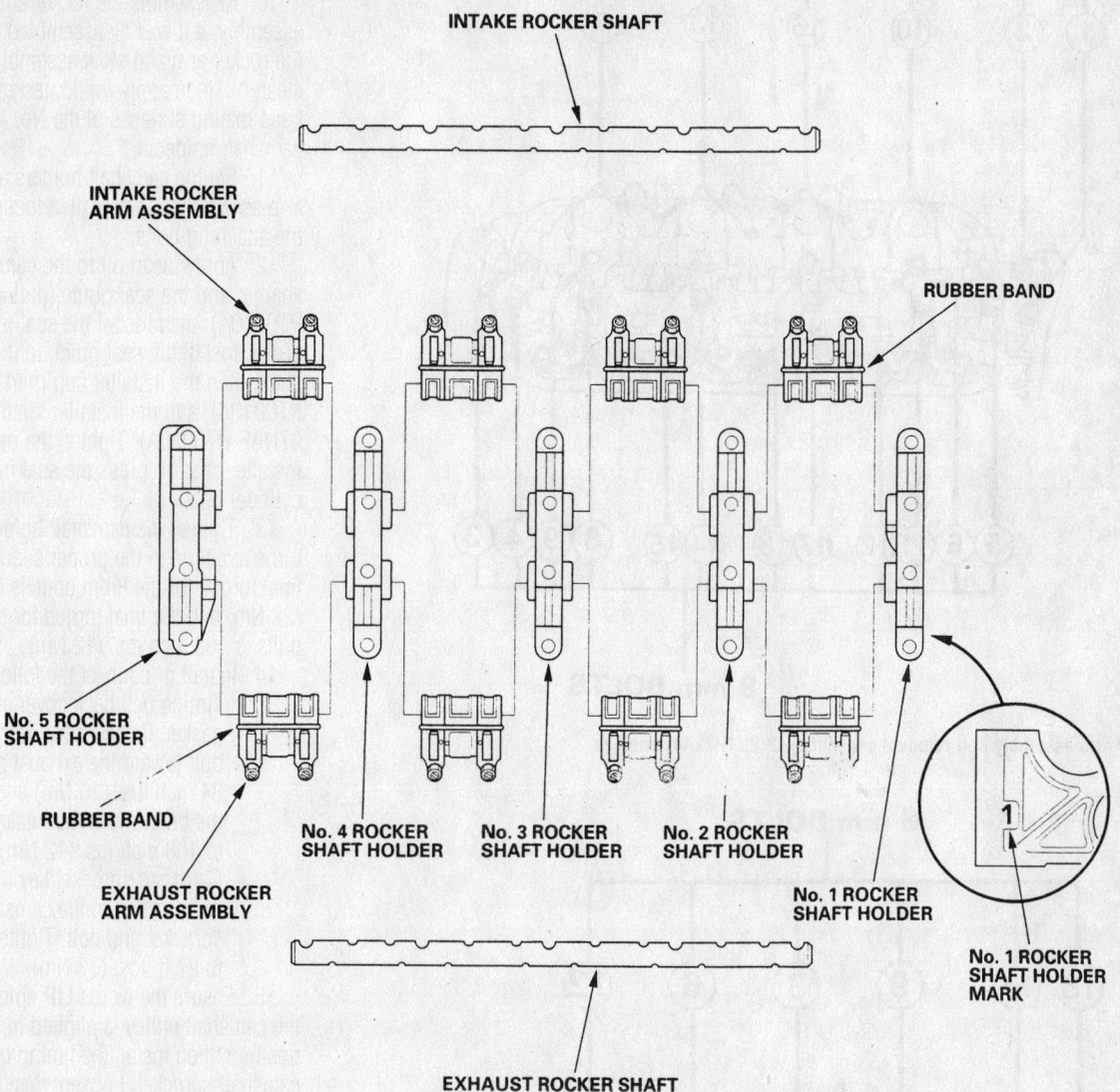

INTAKE ROCKER SHAFT

INTAKE ROCKER ARM ASSEMBLY

RUBBER BAND

No. 5 ROCKER SHAFT HOLDER

RUBBER BAND

EXHAUST ROCKER ARM ASSEMBLY

No. 4 ROCKER SHAFT HOLDER

No. 3 ROCKER SHAFT HOLDER

No. 2 ROCKER SHAFT HOLDER

No. 1 ROCKER SHAFT HOLDER

No. 1 ROCKER SHAFT HOLDER MARK

EXHAUST ROCKER SHAFT

9347FG08

Exploded view of the rocker arm assembly—2.0L engine

※ WARNING

Do not crimp or bend the timing belt more than 90°, or less than 1 inch (25mm) in diameter

➡ Ensure the words UP embossed on the camshaft pulley is aligned in the upward position.

- Camshaft sprocket bolt, sprocket and sprocket key
- Timing belt back cover
- Valve adjusting screws
- Camshaft holder attaching bolts.

Loosen the bolts 2 turns at a time in the proper sequence to prevent damaging the valves or rocker arm assemblies.

➡ When removing the rocker arm assembly, do not remove the camshaft holder bolts. The bolts will keep the camshaft holders, springs, and the rocker arms on the shafts.

- Camshaft holders and rocker arm assembly. If the rocker arm and shaft assembly needs to be disassembled for service, note the location of the components as

they are removed. Install a rubber band around the VTEC rocker arm assemblies to keep them from coming apart during disassembly of the rocker arm assembly. The rocker arms must be installed in the same position if reused.

- Camshaft from the cylinder head and discard the seal.
- Oil control orifice

To install:

7. Wipe the camshaft and the camshaft journals clean, then lubricate both surfaces, and install the camshaft.

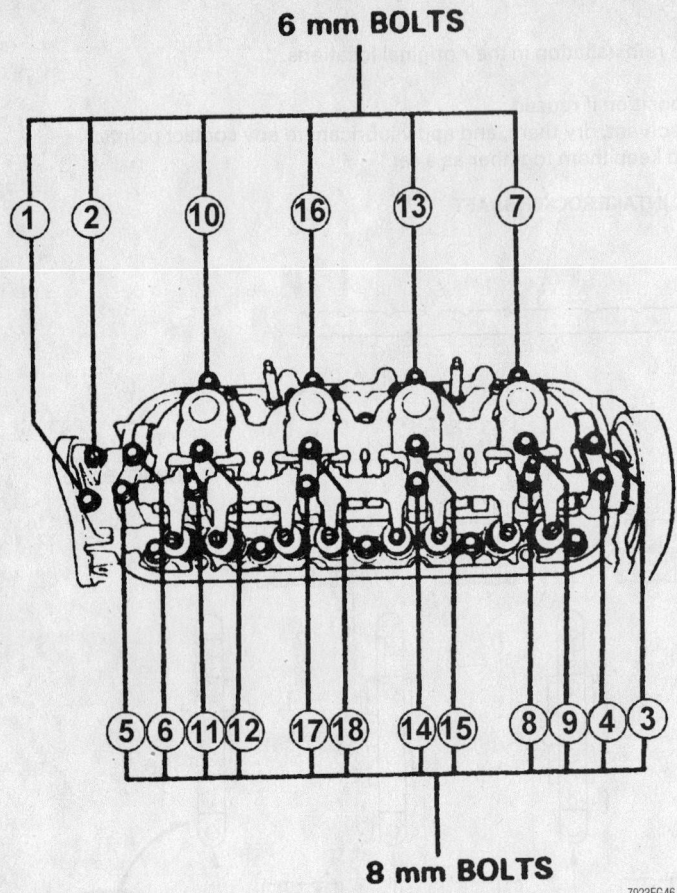

6 mm BOLTS

8 mm BOLTS

7923FG46

Rocker arm assembly bolt removal sequence—2.2L (H22A4) engines

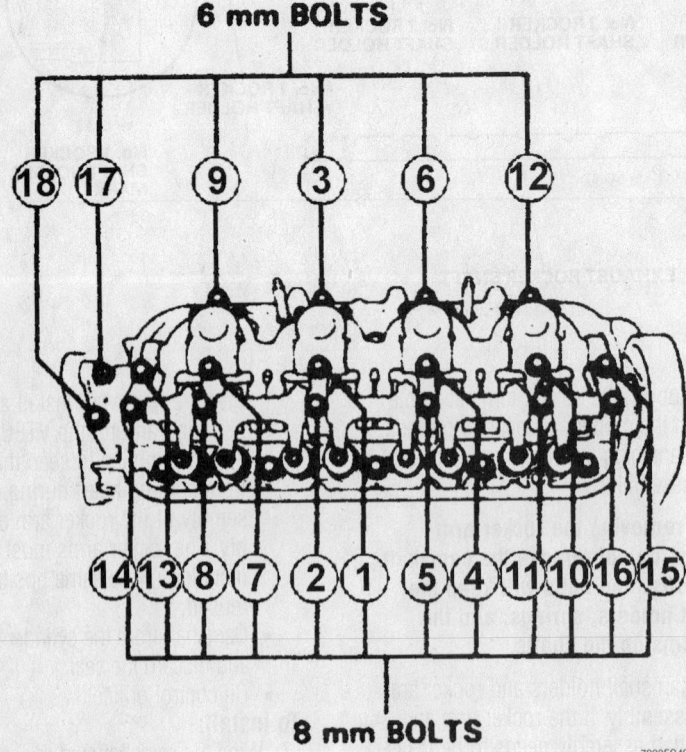

6 mm BOLTS

8 mm BOLTS

7923FG49

Rocker arm assembly torque sequence—2.2L (H22A4) engines

8. Turn the camshaft so that its keyway is facing up (No. 1 cylinder will be at Top Dead Center (TDC)).

9. Clean the oil control orifice and install a new O-ring, then install the oil control orifice.

10. Reassemble the rocker arm and shaft assembly, if it was disassembled. Lubricate the rocker arm and shaft assembly with clean oil, then apply liquid gasket to the head mating surfaces of the No. 1 and No. 6 camshaft holders.

11. Set the camshaft holders and rocker arm assembly in place, then loosely install the attaching bolts.

12. Apply clean oil to the camshaft oil seal lip and the seal guide (part # 07NAG-PT0010A), then install the seal to the seal guide. Install the seal guide to the camshaft, then the installer cup (part # 07NAF-PT0010A), and the installer shaft (part # 07NAF-PT0020A). Tighten the nut on the installer shaft to press the seal into the cylinder head.

13. Tighten the camshaft holder bolts 2 turns at a time in the proper sequence. The final torque for the 8mm bolts is 16 ft. lbs. (22 Nm) and the final torque for the 6mm bolts is 108 inch lbs. (12 Nm).

14. Install or connect the following:
- Timing belt back cover and a new gasket, if necessary. Tighten the bolt toward the exhaust manifold to 84 inch lbs. (10 Nm) and tighten the bolt toward the intake manifold to 108 inch lbs. (12 Nm).
- Camshaft sprocket key to the camshaft, then the camshaft sprocket and bolt. Tighten the bolt to 27 ft. lbs. (37 Nm).

15. Ensure the words **UP** embossed on the camshaft pulley is aligned in the upward position, then install the timing belt onto the camshaft sprocket. Loosen, then tighten the timing belt adjusting nut.

16. Rotate the crankshaft pulley 5 or 6 turns to position the timing belt on the pulleys.

17. Set the No. 1 cylinder to TDC and loosen the timing belt adjusting nut 1 turn. Turn the crankshaft counterclockwise until the cam pulley has moved 3 teeth; this creates tension on the timing belt. Loosen, then tighten the adjusting nut, and tighten it to 33 ft. lbs. (45 Nm).

18. Adjust the valves.

19. Tighten the crankshaft pulley bolt to 181 ft. lbs. (245 Nm).

20. Install or connect the following:
- Upper timing belt cover. Tighten the bolt toward the exhaust manifold to 84 inch lbs. (10 Nm) and tighten

the bolt toward the intake manifold to 108 inch lbs. (12 Nm).

- Cylinder head cover gasket cover to the groove of the cylinder head cover. Before installing the gasket, thoroughly clean the seal and the groove. Seat the recesses for the camshaft first, then work it into the groove around the outside edges. Be sure the gasket is seated securely in the corners of the recesses.

21. Apply liquid gasket to the 4 corners of the recesses of the cylinder head cover gasket. Do not install the parts if 5 minutes or more have elapsed since applying liquid gasket. After assembly, wait at least 20 minutes before filling the engine with oil.

22. Install or connect the following:
- Cylinder head (valve) cover. Tighten the bolts attaching the cylinder head cover in the proper sequence to 84 inch lbs. (10 Nm).
- PCV hose to the cylinder head cover
- Power steering belt, then adjust the belt.
- Distributor to the cylinder head. Snug the attaching bolts until the timing has been checked and adjusted.
- Spark plug wires and the distributor electrical connectors
- Air intake duct

23. Drain the oil from the engine into a sealable container. Install the drain plug and refill the engine with clean oil.

24. Connect the negative battery cable and enter the radio security code.

25. Start the engine and check carefully for any leaks.

26. Check and adjust the ignition timing as necessary, then tighten the distributor bolts to 13 ft. lbs. (18 Nm).

2.3L Engines

1. Before servicing the vehicle, refer to the precautions in the beginning of this section.

2. Disconnect the negative battery cable.

3. Turn the crankshaft so the No. 1 piston is at Top Dead Center (TDC).

➡**The No. 1 piston is at top dead center when the pointer on the block aligns with the white painted mark on the flywheel (manual transaxle) or driveplate (automatic transaxle).**

4. Remove or disconnect the following:
- Air intake duct
- Engine ground cable from the cylinder head cover

- Connector and the terminal from the alternator
- Engine wiring harness from the valve cover
- Ignition coil

➡**Label all electrical connectors before detaching them.**

- Electrical connectors from the distributor and the spark plug wires from the spark plugs.

➡**Matchmark the installed position if the distributor before removal.**

- Distributor from the cylinder head
- Ignition coil wire from the distributor
- Positive Crankcase Ventilation (PCV) hose
- Cylinder head cover. Replace the rubber seals if damaged or deteriorated.
- Timing belt middle cover

5. Ensure the words **UP** embossed on the camshaft pulleys are aligned in the upward position.

6. Mark the rotation of the timing belt if it is to be used again. Loosen the timing belt adjusting nut ½ turn, then release the

tension on the timing belt. Push the tensioner to release tension from the belt, then tighten the adjusting nut.

7. Remove the timing belt from the camshaft sprockets.

✶✶ WARNING

Do not crimp or bend the timing belt more than 90°, or less than 1 inch (25mm) in diameter

8. Insert a 5.0mm pin punch in each of the camshaft caps, nearest to the sprockets, through the holes provided. Remove the camshaft sprocket attaching bolts, then remove the sprockets. Do not lose the sprocket keys.

9. Remove or disconnect the following:
- Side engine mount bracket B, then the timing belt back cover from behind the camshaft sprockets.
- Rocker arm adjusting screws, then the pin punches from the camshaft caps

➡**Note the camshaft holders locations for ease of installation. Loosen the bolts in the reverse order of the holder bolts torque sequence.**

Specified torque:
Except Intake ⑤, ⑦, Exhaust ⑥, ⑧:
    10 N·m (1.0 kg-m, 7 lb-ft)
Intake ⑤, ⑦, Exhaust ⑥, ⑧:
    12 N·m (1.2 kg-m, 9 lb-ft)

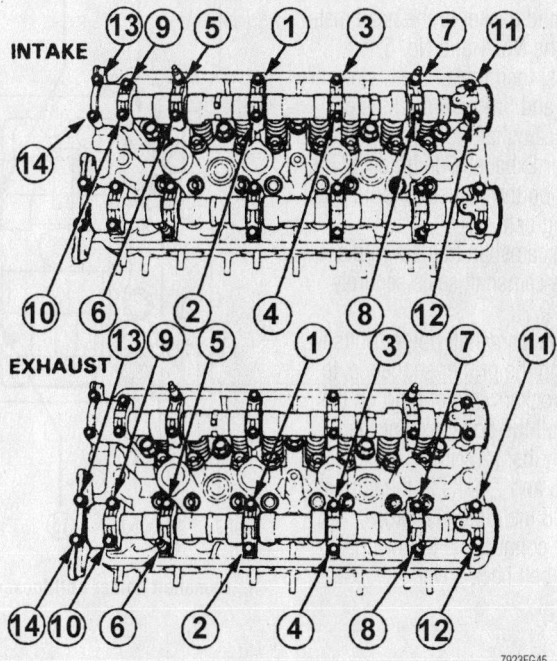

Camshaft holders torque sequence—2.3L engines

7923FG45

- Camshaft holders
- Camshafts from the cylinder head, then discard the camshaft seals.
- Rubber cap from the head, located at the end of the intake camshaft.
- Rocker arms from the cylinder head. Note the locations of the rocker arms.

➡ **The rocker arms have to be installed to their original locations if being reused.**

To install:

➡ **Lubricate the rocker arms with clean oil before installation.**

10. Install or connect the following:
- Rocker arms on the pivot bolts and the valve stems. If the rocker arms are being reused, install them to their original locations. The locknuts and adjustment screws should be loosened before installing the rocker arms.

11. Lubricate the camshafts with clean oil.

12. Install or connect the following:
- Camshaft seals to the end of the camshafts that the timing belt sprockets attach to. The open side (spring) should be facing into the cylinder head when installed.

➡ **Be sure the keyways on the camshafts are facing up and install the camshafts to the cylinder head.**

- Rubber plug to the cylinder head at the end of the intake camshaft

13. Apply liquid gasket to the head mating surfaces of the No. 1 and No. 6 camshaft holders, then install them along with No. 2, 3, 4 and 5. **I** or **E** marks are stamped on the camshaft holders to identify them as Intake or Exhaust side holders. The arrows stamped on the holders should point toward the timing belt.

14. Snug the camshaft holders in place.

15. Press the camshaft seals securely into place.

16. Tighten the camshaft holder bolts in 2 steps, following the proper sequence, to ensure that the rockers do not bind on the valves. Tighten all the bolts, except the 4 studs, to 84 inch lbs. (10 Nm). Tighten the studs (number 5 and 7 bolts in the correct sequence) to 108 inch lbs. (12 Nm).

17. Install or connect the following:
- Timing belt back cover.

- Side engine mount bracket B. Tighten the bolt attaching the bracket to the cylinder head to 33 ft. lbs. (45 Nm). Tighten the bolts attaching the bracket to the side engine mount to 16 ft. lbs. (22 Nm).

18. Insert a 5.0mm pin punch in each of the camshaft caps, nearest to the pulleys, through the holes provided. Install the keys into the camshaft grooves.

19. Push the camshaft sprockets onto the camshafts, then tighten the retaining bolts to 27 ft. lbs. (38 Nm).

20. Ensure the words **UP** embossed on the camshaft pulleys are aligned in the upward position. Install the timing belt to the camshaft sprockets, then remove the 2, 5.0mm pin punches from the camshaft bearing caps.

21. Loosen, then tighten the timing belt adjuster nut.

22. Turn the crankshaft counterclockwise until the cam pulley has moved 3 teeth; this creates tension on the timing belt. Loosen, then tighten the adjusting nut and tighten it to 33 ft. lbs. (45 Nm).

23. Adjust the valves.

24. Tighten the crankshaft pulley bolt to 181 ft. lbs. (250 Nm).

25. Install or connect the following:
- Middle timing belt cover and tighten the attaching bolts to 108 inch lbs. (12 Nm).
- Cylinder head cover and tighten the cap nuts to 84 inch lbs. (10 Nm).

- PCV hose to the cylinder head cover
- Distributor. Snug the attaching bolts until the timing has been checked and adjusted.
- Spark plug wires, then the distributor electrical connectors
- Ignition coil wire to the distributor.
- Ignition coil
- Alternator wiring harness to the cylinder head cover
- Terminal and connector to the alternator
- Engine ground cable to the cylinder head cover
- Air intake duct

26. Drain the oil from the engine into a sealable container. Install the drain plug and refill the engine with clean oil.

27. Connect the negative battery cable and enter the radio security code.

28. Start the engine and check carefully for any leaks.

29. Check and adjust the ignition timing. Tighten the distributor bolts to 13 ft. lbs. (18 Nm).

30. If equipped with 4WS, turn the steering wheel lock-to-lock to reset the 4WS control unit.

2.4L Engine

1. Before servicing the vehicle, refer to the precautions in the beginning of this section.

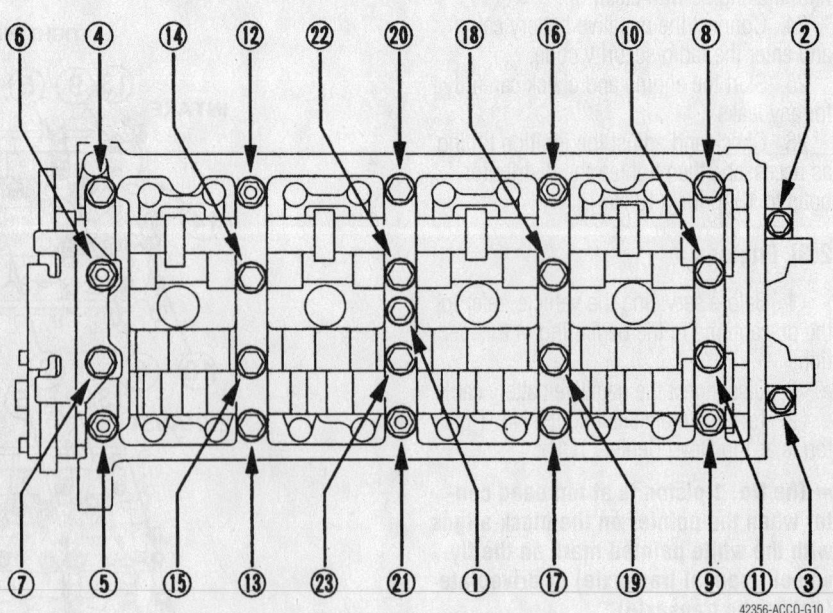

Camshaft holder bolt loosening sequence—2.4L engine

42356-ACCO-G10

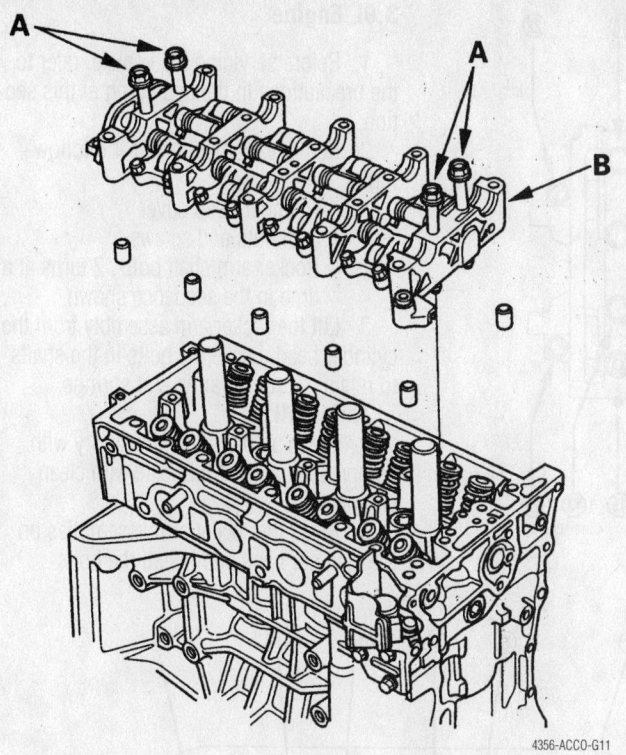

Insert the bolts (A) into the rocker shaft holder, then remove the rocker arm assembly (B)—2.4L engine

`4356-ACCO-G11`

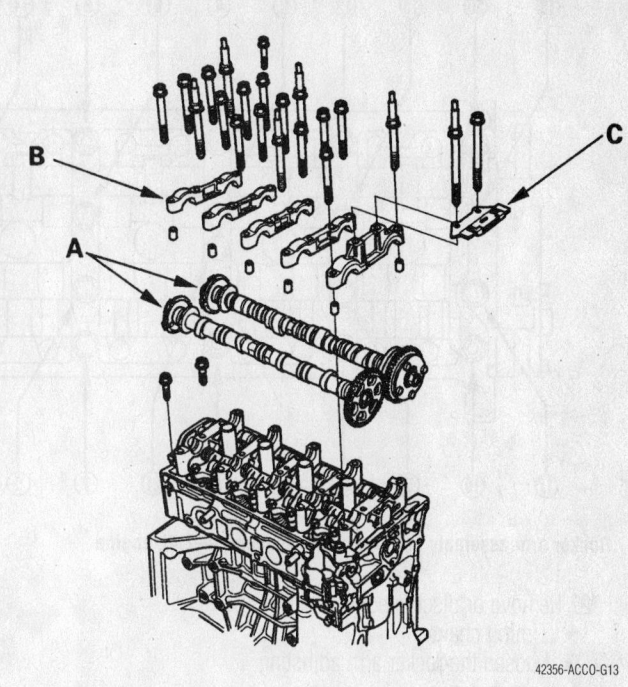

When installing the camshafts (A) make sure the punch marks on the VTC actuator and exhaust cam sprockets are facing up—2.4L engine

`42356-ACCO-G13`

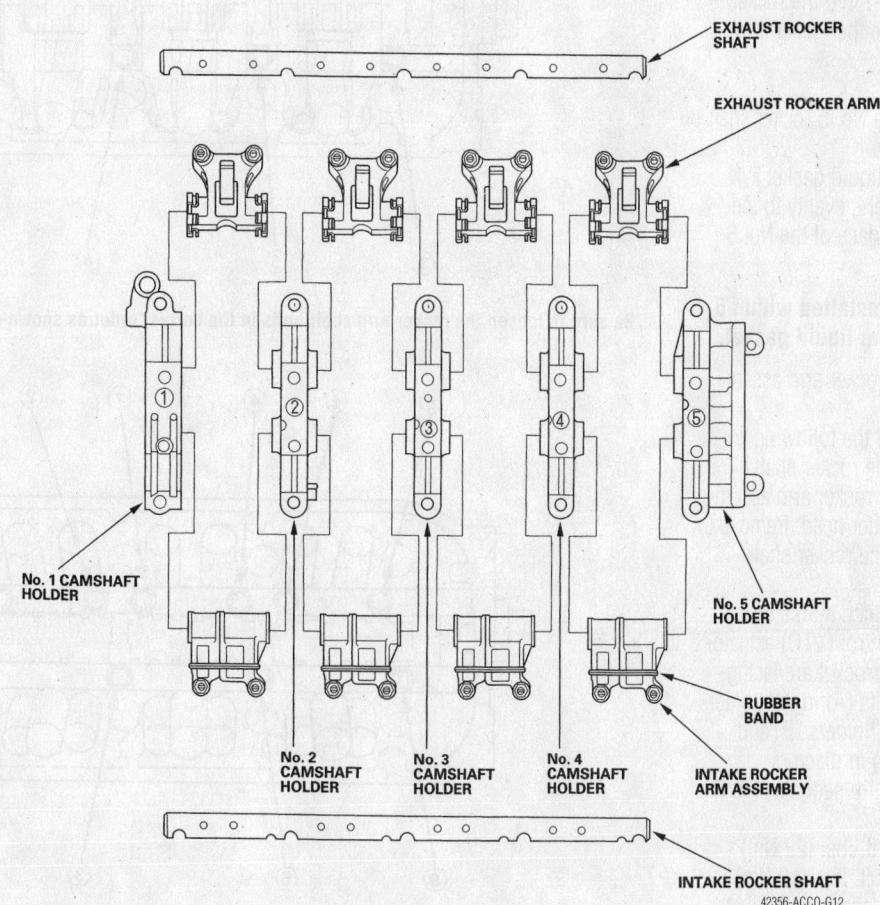

EXHAUST ROCKER SHAFT

EXHAUST ROCKER ARM

No. 1 CAMSHAFT HOLDER

No. 5 CAMSHAFT HOLDER

No. 2 CAMSHAFT HOLDER

No. 3 CAMSHAFT HOLDER

No. 4 CAMSHAFT HOLDER

RUBBER BAND

INTAKE ROCKER ARM ASSEMBLY

INTAKE ROCKER SHAFT

`42356-ACCO-G12`

Exploded view of the rocker arms and related components—2.4L engine

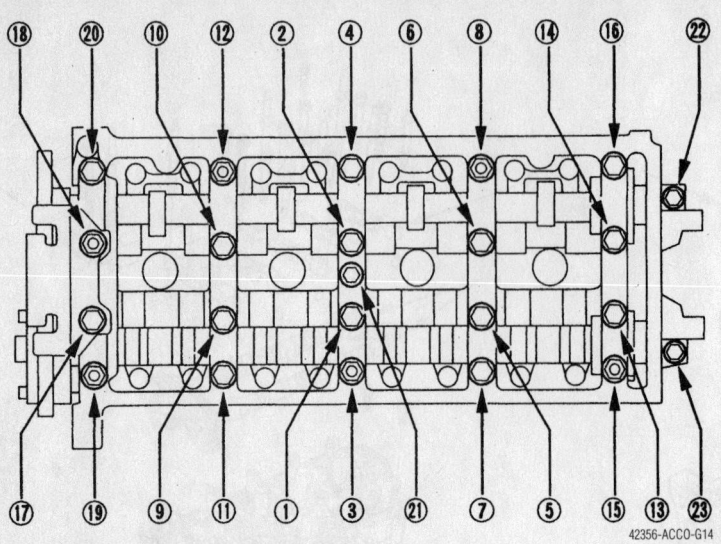

Rocker arm assembly bolt tightening sequence—2.4L engine

42356-ACCO-G14

3.0L Engine

1. Before servicing the vehicle, refer to the precautions in the beginning of this section.

2. Remove or disconnect the following:
   - Cylinder head cover
   - Jam nuts and screws
   - Rocker arm shaft bolts, 2 turns at a time in the sequence shown.

3. Lift the rocker arm assembly from the cylinder head. Leave the bolts in the shafts to retain the rocker arms and springs.

To install:

4. Clean all parts in solvent, dry with compressed air and lubricate with clean engine oil.

5. Place the rocker arm assemblies on the cylinder head and install the bolts

2. Remove or disconnect the following:
   - Timing chain
   - Loosen the rocker arm adjusting screws
   - Camshaft holder bolts, two turns at a time in sequence
   - Timing chain guide (B), camshaft and camshafts

3. Insert the bolts (A) into the rocker shaft holder, then remove the rocker arm assembly (B)

To install:

4. Clean and dry the No. 5 rocker shaft holding mating surface.

5. Apply a suitable liquid gasket P/N 08718-0009, or equivalent, evenly to the cylinder head mating surface of the No. 5 rocker shaft holder.

➡**The parts must be installed within 5 minutes of applying the liquid gasket.**

6. Reassemble the rocker arm assembly, as necessary.

7. Install or connect the following
   - Bolts (A) into the rocker shaft holder, then the rocker arm assembly on the cylinder head. Remove the bolts from the rocker shaft holder.

8. Make sure the punch marks on the variable valve timing control (VTC) actuator and exhaust camshaft sprocket are facing up, then set the camshafts (A) in the holder

9. Set the camshaft holders (B) and timing chain guide B (C) in place.

10. Tighten the bolts, in sequence, to the following specification:
    a. 8mm bolts: 16 ft. lbs. (22 Nm)
    b. 6mm bolts: 8.7 ft. lbs. (12 Nm)

11. Install the timing chain and adjust the valve lash.

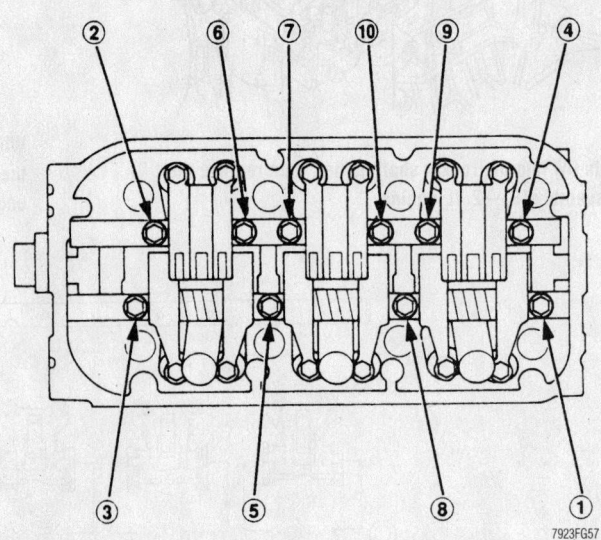

Be sure to loosen the rocker arm shaft bolts in the correct order as shown—3.0L engine

7923FG57

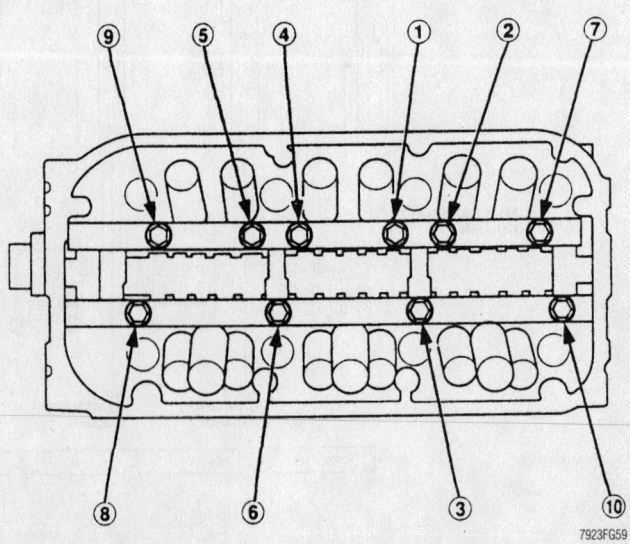

Tighten the bolts 2 turns at a time in the sequence shown—3.0L engine

7923FG59

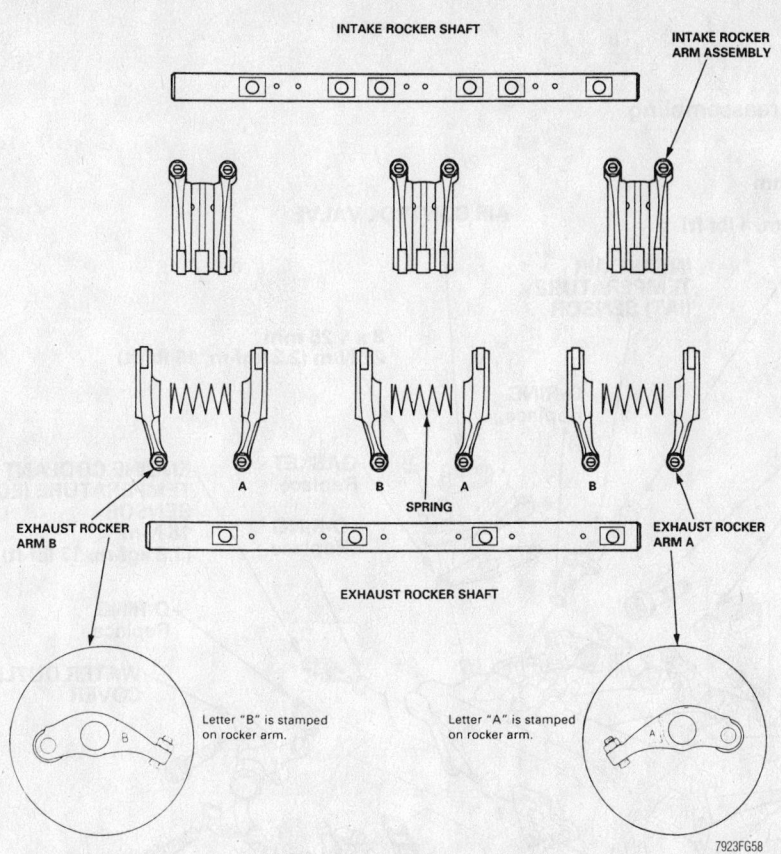

INTAKE ROCKER SHAFT

INTAKE ROCKER ARM ASSEMBLY

A   B   A   B

EXHAUST ROCKER ARM B

SPRING

EXHAUST ROCKER ARM A

EXHAUST ROCKER SHAFT

Letter "B" is stamped on rocker arm.

Letter "A" is stamped on rocker arm.

7923FG58

Exploded view of the rocker arms and related components—3.0L engine

loosely. Be sure that all rocker arms are in alignment with their valves.

6. Tighten each bolt 2 turns at a time in the correct sequence. Tighten the bolts to 17 ft. lbs. (24 Nm).

7. Adjust the valves and install the cylinder head covers.

Intake Manifold

REMOVAL & INSTALLATION

1.6L and 1.7L Engines

1. Before servicing the vehicle, refer to the precautions in the beginning of this section.

2. Remove or disconnect the following:
   • Negative battery cable

3. Drain the cooling system to a level below the upper radiator hose.

4. Relieve the fuel system pressure by loosening the fuel filter service bolt.

☀☀ CAUTION

The fuel injection system remains under pressure even after the engine has been turned off. The fuel system pressure must be relieved before dis-

connecting any fuel lines. Failure to do so may result in fire and personal injury.

5. Remove or disconnect the following:
   • Intake air duct
   • Air cleaner assembly, if equipped with the D16Y7 engine

➡**Cover the throttle body opening to keep dirt out.**

   • Fuel line from the fuel rail. Clean up any spilled fuel.
   • Fuel injector wiring harnesses
   • Fuel rail and injectors
   • Throttle cable from the linkage at the throttle body
   • Intake manifold cooling hoses. Use a drain pan to catch any spilled coolant. Also, be sure no coolant spills on electrical connections.

➡**Label all electrical connectors before detaching them.**

   • Engine wiring harness connectors from the intake manifold sensors
   • Idle Air Control (IAC) valve
   • Exhaust Gas Recirculation (EGR valve), if equipped

   • Throttle Position (TP) and Manifold Absolute Pressure (MAP) sensor
   • Manifold from its support bracket
   • Intake manifold nuts in a crisscross pattern.
   • Intake manifold assembly from the vehicle.

To install:

➡**Use new gaskets when installing the intake manifold. Use new O-rings when installing manifold sensors and components. Use new sealing washers when reconnecting the fuel lines.**

6. Clean all gasket mating surfaces.

7. Install or connect the following:
   • New intake manifold gaskets
   • Intake manifold

8. Tighten the intake manifold nuts in 2 or 3 steps in a crisscross pattern starting with the inside nuts. Tighten the nuts to 17 ft. lbs. (23 Nm).

9. Install or connect the following:
   • Support bracket bolts: 17 ft. lbs. (24 Nm)
   • Fuel rail and injectors
   • Fuel line using new washers
   • EGR valve and tighten its nuts to 15 ft. lbs. (21 Nm).
   • IAC valve. Tighten its mounting bolts to 16 ft. lbs. (22 Nm).
   • Fuel injector wiring harnesses
   • Intake manifold wiring harnesses
   • Intake manifold cooling hoses
   • Throttle cable
   • Intake air duct and air cleaner assembly

10. Refill and bleed the cooling system.

11. Connect the negative battery cable.

12. Verify that all sensors, valves, and vacuum lines are installed and connected properly. Be sure there are no loose electrical connections.

13. Turn the ignition on and off several times without starting the engine to pressurize the fuel system. Run the engine and check for proper operation. Check for vacuum leaks.

14. After the engine has warmed up, check the operation of the throttle cable and adjust it if necessary.

2.0L Engine

1. Before servicing the vehicle, refer to the precautions in the beginning of this section.

2. Drain the cooling system.

3. Relieve the fuel system pressure.

4. Remove or disconnect the following:
   • Negative battery cable
   • Cooling hoses from the intake manifold

NOTE: Use new O-rings and gaskets when reassembling.

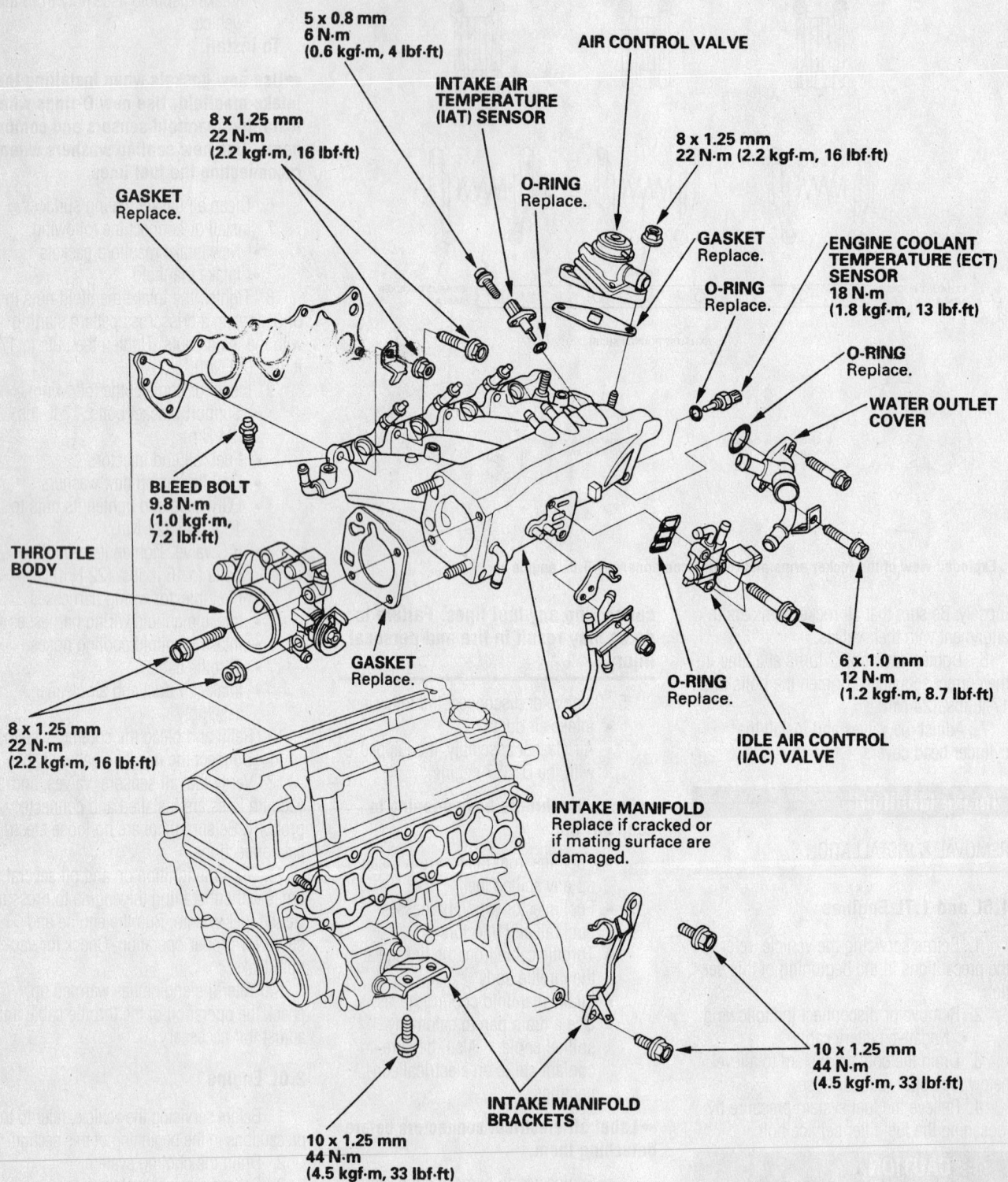

5 x 0.8 mm
6 N·m
(0.6 kgf·m, 4 lbf·ft)

**INTAKE AIR
TEMPERATURE
(IAT) SENSOR**

AIR CONTROL VALVE

8 x 1.25 mm
22 N·m
(2.2 kgf·m, 16 lbf·ft)

8 x 1.25 mm
22 N·m (2.2 kgf·m, 16 lbf·ft)

O-RING
Replace.

GASKET
Replace.

GASKET
Replace.

O-RING
Replace.

**ENGINE COOLANT
TEMPERATURE (ECT)
SENSOR**
18 N·m
(1.8 kgf·m, 13 lbf·ft)

O-RING
Replace.

**WATER OUTLET
COVER**

BLEED BOLT
9.8 N·m
(1.0 kgf·m,
7.2 lbf·ft)

**THROTTLE
BODY**

GASKET
Replace.

O-RING
Replace.

6 x 1.0 mm
12 N·m
(1.2 kgf·m, 8.7 lbf·ft)

**IDLE AIR CONTROL
(IAC) VALVE**

8 x 1.25 mm
22 N·m
(2.2 kgf·m, 16 lbf·ft)

INTAKE MANIFOLD
Replace if cracked or
if mating surface are
damaged.

10 x 1.25 mm
44 N·m
(4.5 kgf·m, 33 lbf·ft)

**INTAKE MANIFOLD
BRACKETS**

10 x 1.25 mm
44 N·m
(4.5 kgf·m, 33 lbf·ft)

9347FG09

Exploded view of the intake manifold—2.0L engine

- Vacuum hoses and electrical connectors from the manifold and throttle body
- Throttle cable from the throttle body
- Fuel rail and fuel injectors
- Intake manifold support brackets
- Intake manifold

To install:

5. Installation is the reverse of the removal procedure, while using the following torque values:

- Intake manifold fasteners: 16 ft. lbs. (22 Nm)
- Throttle body fasteners: 16 ft. lbs. (22 Nm)
- Intake manifold bracket bolts: 33 ft. lbs. (44 Nm)

2.2L and 2.3L Engines

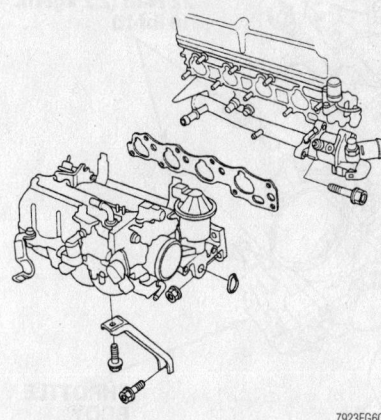

Intake manifold and related components— 2.3L engine

1. Before servicing the vehicle, refer to the precautions in the beginning of this section.
2. Disconnect the negative battery cable.
3. Drain the engine coolant into a sealable container.
4. Remove or disconnect the following:

- Cooling hoses from the intake manifold

➡**Label all vacuum hoses and electrical connectors before detaching them.**

- Vacuum hoses and electrical connectors from the manifold and throttle body
- Connector from the Exhaust Gas Recirculation (EGR) valve. Position the wiring harnesses out of the way.
- Throttle cable from the throttle body

5. Relieve the fuel pressure.
6. Remove or disconnect the following:

- Fuel rail and fuel injectors

- Thermostat housing from the intake manifold and the connecting pipe by pulling and twisting the housing. Discard the O-rings.

➡**It may be necessary to remove the upper intake manifold plenum and throttle body assembly in order to access the nuts securing the manifold to the head.**

- Intake manifold support bracket bolts and the bracket. If it is necessary to access it from under the vehicle; raise and support the vehicle safely.

7. While supporting the intake manifold, remove the nuts attaching the intake manifold to the cylinder head, then remove the manifold. Remove the old gasket from the cylinder head.
8. Clean any old gasket material from the cylinder head and the intake manifold. Check and clean the FIA chamber on the cylinder head.

To install:

9. Install or connect the following:

- New gasket
- Intake manifold, and support the manifold.
- Support bracket to the manifold. Tighten the retaining bolt to 16 ft. lbs. (22 Nm).

10. Starting with the inner or center nuts, tighten the nuts, in a crisscross pattern, to the correct torque. The tension must be even across the entire face of the manifold if leaks are to be prevented. Correct torque is 16 ft. lbs. (22 Nm).
11. Install or connect the following:

- New gasket
- Upper intake manifold and throttle body assembly, if removed as a separate unit. Tighten the nuts and bolts holding the chamber to 16 ft. lbs. (22 Nm).
- New O-ring to the coolant connecting pipe, and to the thermostat housing.
- Housing to the coolant pipe and the intake manifold. Tighten the mounting bolts to 16 ft. lbs. (22 Nm).
- Throttle cable and adjust.
- Fuel rail/injector assembly
- Fuel lines
- Wiring harnesses and the electrical connectors
- Vacuum hoses

12. Fill and bleed the air from the cooling system.
13. Connect the negative battery cable and enter the radio security code.
14. Start the engine and check carefully for any leaks of fuel, coolant or vacuum.

Check the manifold gasket areas carefully for any leakage of vacuum.

15. If equipped with 4WS, turn the steering wheel lock-to-lock to reset the 4WS control unit.

2.4L Engine

1. Before servicing the vehicle, refer to the precautions in the beginning of this section.
2. Disconnect the negative battery cable.
3. Drain the engine coolant into a sealable container.
4. Remove or disconnect the following:

- Intake Air Temperature (IAT) sensor electrical connector
- Vacuum hose and breather pipe and the air intake duct
- Intake manifold cover
- Throttle and cruise control cables by loosening the locknuts, then slipping the cable ends out of the accelerator linkage.

➡**Do not bend the cables during removal. Always replace any throttle or cruise control cables that get kinked during removal.**

- Evaporative emission (EVAP) canister hose and brake booster vacuum hose
- Idle Air Control (IAC) valve connectors
- Throttle Position (TP) sensor connector
- Manifold Absolute Pressure (MAP) sensor connector
- Necessary engine wire harness connectors and wire harness clamps from the intake manifold
- Bolt securing the harness holder and remove the harness clamps
- Water bypass hoses, then plug them
- Harness clamp and harness connector from the intake manifold bracket
- Intake manifold bracket
- A/T vacuum hose
- Retainer and intake manifold

To install:

5. Clean the mounting surfaces.
6. Install or connect the following:

- New gasket
- Intake manifold. Tighten the bolts, in a criss-cross pattern beginning with the inner bolt, to 16 ft. lbs. (22 Nm).
- A/T vacuum hose
- Intake manifold bracket
- Harness clamp and connector to the intake manifold bracket

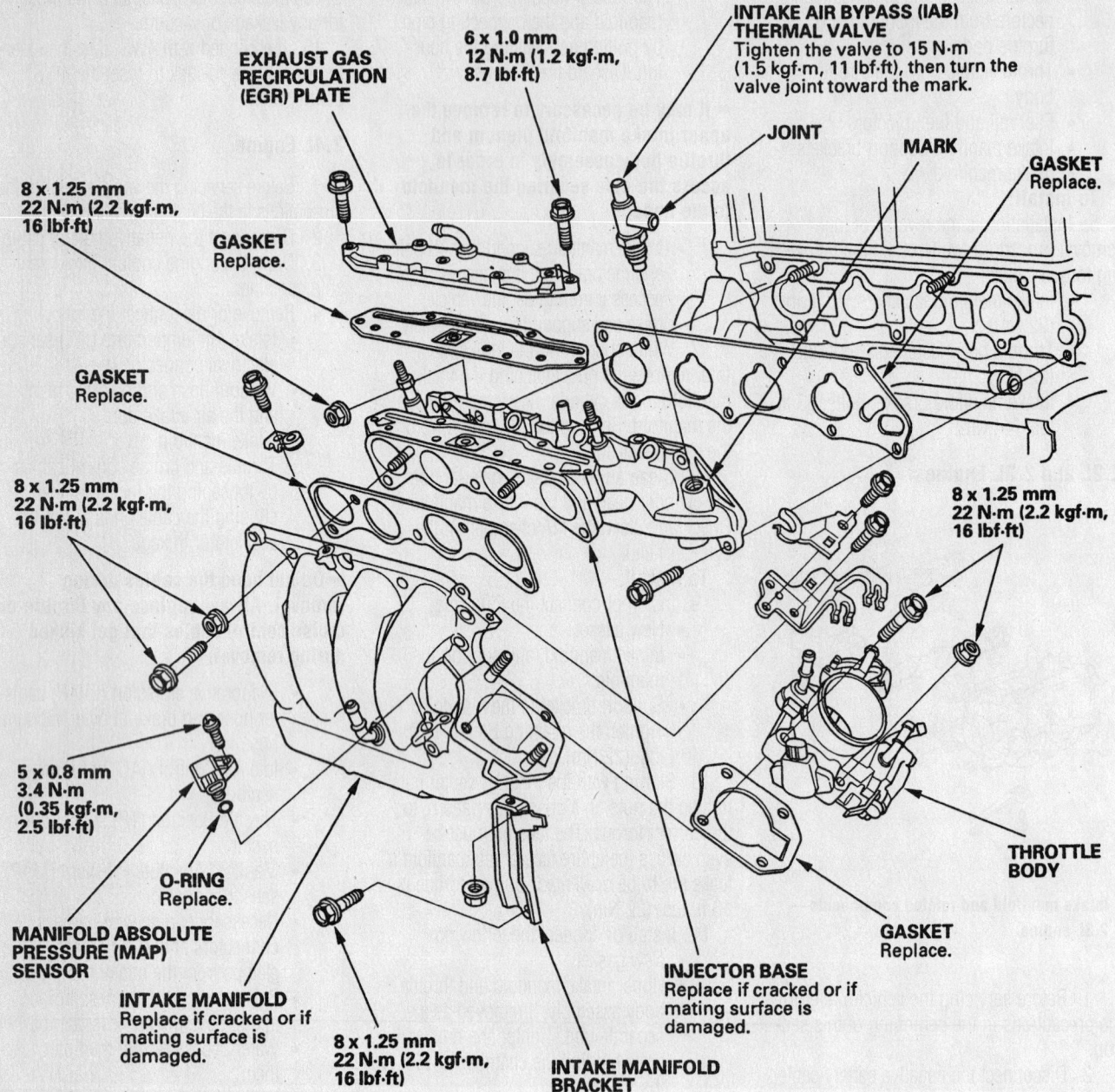

EXHAUST GAS RECIRCULATION (EGR) PLATE

6 x 1.0 mm 12 N·m (1.2 kgf·m, 8.7 lbf·ft)

INTAKE AIR BYPASS (IAB) THERMAL VALVE
Tighten the valve to 15 N·m (1.5 kgf·m, 11 lbf·ft), then turn the valve joint toward the mark.

JOINT

MARK

GASKET Replace.

8 x 1.25 mm 22 N·m (2.2 kgf·m, 16 lbf·ft)

GASKET Replace.

GASKET Replace.

8 x 1.25 mm 22 N·m (2.2 kgf·m, 16 lbf·ft)

8 x 1.25 mm 22 N·m (2.2 kgf·m, 16 lbf·ft)

5 x 0.8 mm 3.4 N·m (0.35 kgf·m, 2.5 lbf·ft)

THROTTLE BODY

O-RING Replace.

MANIFOLD ABSOLUTE PRESSURE (MAP) SENSOR

GASKET Replace.

INTAKE MANIFOLD Replace if cracked or if mating surface is damaged.

8 x 1.25 mm 22 N·m (2.2 kgf·m, 16 lbf·ft)

INTAKE MANIFOLD BRACKET

INJECTOR BASE Replace if cracked or if mating surface is damaged.

42356-ACCO-G15

Exploded view of the intake manifold and related components—2.4L engine

- Water bypass hoses
- Bolt securing the harness holder and tighten to 8.7 ft. lbs. (12 Nm)
- Harness clamps
- EVAP canister hose and brake booster vacuum hose
- Throttle and cruise control cables
- Intake manifold cover
- Intake air duct
- IAT sensor connector, vacuum hose and breather pipe

7. Refill the cooling system.
8. Connect the negative battery cable, start the engine, and check for leaks.

3.0L Engine

1. Before servicing the vehicle, refer to the precautions in the beginning of this section.
2. Obtain the security code for the radio.
3. Disconnect the negative battery cable.

4. Drain the coolant.
5. Remove or disconnect the following:
- Evaporative emissions (EVAP) canister hose from the throttle body.
- Air intake duct
- Upper engine covers
- Accelerator and cruise control cables from the throttle body.

➡**Ensure that all components have been removed from the intake manifold.**

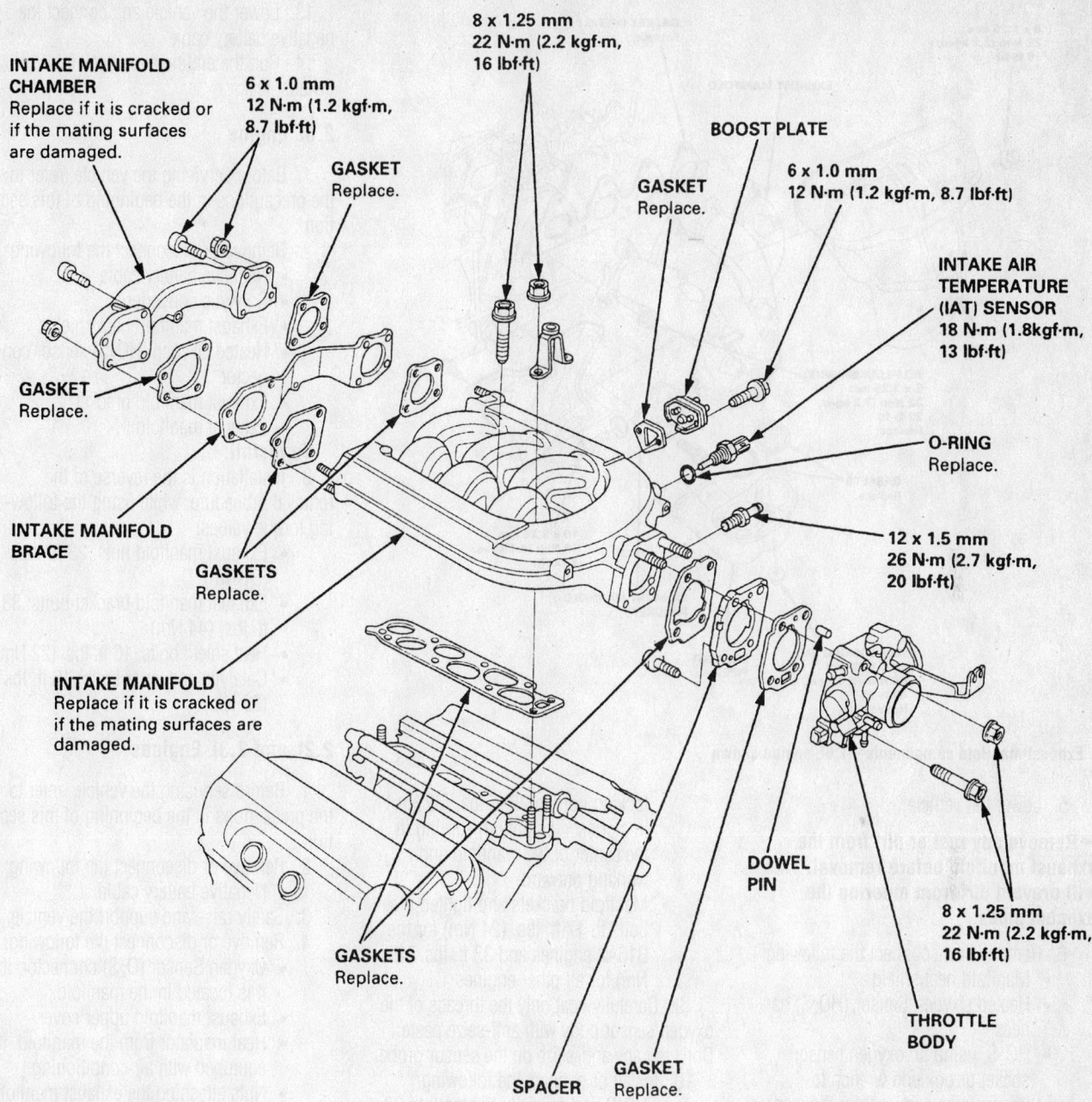

INTAKE MANIFOLD CHAMBER
Replace if it is cracked or if the mating surfaces are damaged.

6 x 1.0 mm
12 N·m (1.2 kgf·m, 8.7 lbf·ft)

8 x 1.25 mm
22 N·m (2.2 kgf·m, 16 lbf·ft)

BOOST PLATE

GASKET
Replace.

6 x 1.0 mm
12 N·m (1.2 kgf·m, 8.7 lbf·ft)

GASKET
Replace.

INTAKE AIR TEMPERATURE (IAT) SENSOR
18 N·m (1.8kgf·m, 13 lbf·ft)

GASKET
Replace.

O-RING
Replace.

INTAKE MANIFOLD BRACE

GASKETS
Replace.

12 x 1.5 mm
26 N·m (2.7 kgf·m, 20 lbf·ft)

INTAKE MANIFOLD
Replace if it is cracked or if the mating surfaces are damaged.

GASKETS
Replace.

DOWEL PIN

8 x 1.25 mm
22 N·m (2.2 kgf·m, 16 lbf·ft)

THROTTLE BODY

SPACER

GASKET
Replace.

7923FGC4

Exploded view of the intake manifold and related components—3.0L engine

- Intake manifold

To install:
6. Clean the mounting surfaces.
7. Install or connect the following:
   - New gasket
   - Intake manifold. Tighten the bolts to 16 ft. lbs. (22 Nm).
   - All removed hoses and wiring on the intake manifold and throttle body.
   - Engine covers
   - Intake air duct
8. Refill the cooling system.

9. Connect the negative battery cable, start the engine, and check for leaks.

Exhaust Manifold

REMOVAL & INSTALLATION

1.6L and 1.7L Engines

1. Before servicing the vehicle, refer to the precautions in the beginning of this section.

2. Disconnect the negative battery cable.

3. Raise and support the front of the vehicle and block the rear wheels.

4. Remove or disconnect the following:
   - Front exhaust pipe from the exhaust manifold/catalytic converter
   - Exhaust manifold support brackets if their bolts are accessible from this angle. The splash shield may be removed for better access.

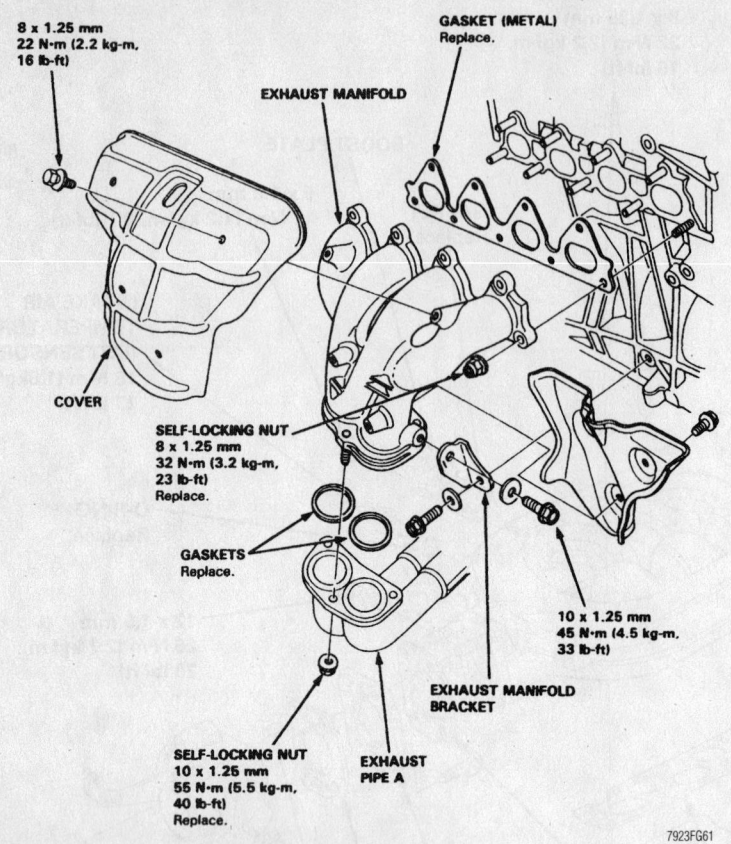

8 x 1.25 mm
22 N·m (2.2 kg-m, 16 lb-ft)

GASKET (METAL)
Replace.

EXHAUST MANIFOLD

COVER

SELF-LOCKING NUT
8 x 1.25 mm
32 N·m (3.2 kg-m, 23 lb-ft)
Replace.

GASKETS
Replace.

10 x 1.25 mm
45 N·m (4.5 kg-m, 33 lb-ft)

EXHAUST MANIFOLD
BRACKET

SELF-LOCKING NUT
10 x 1.25 mm
55 N·m (5.5 kg-m, 40 lb-ft)
Replace.

EXHAUST
PIPE A

7923FG61

Exhaust manifold components—1.6L engine shown

5. Lower the vehicle.

➡ **Remove any rust or dirt from the exhaust manifold before removal. This will prevent dirt from entering the exhaust pipes.**

6. Remove or disconnect the following:
- Manifold heat shield
- Heated Oxygen Sensor (HO2S) harness
- HO2S, using an oxygen sensor socket or box end wrench to unscrew the sensor from the manifold. Handle the sensor carefully.
- Exhaust manifold brackets
- Exhaust manifold and separate it from the cylinder head
- Exhaust manifold and gasket

To install:

➡ **Use new gaskets and self-locking nuts when installing the exhaust manifold.**

7. Clean the gasket mating surfaces of the manifold and cylinder head ports.
8. Install or connect the following:
- New gasket onto the cylinder head
- New gaskets onto the exhaust pipe flange
- Exhaust manifold. Apply anti-seize paste to the studs. Tighten the self-

locking nuts to 23 ft. lbs. (32 Nm) in a crisscross pattern starting in the center of the manifold and working outward.
- Manifold brackets and tighten their bolts to 17 ft. lbs. (24 Nm) for the B16A2 engines and 33 ft. lbs. (45 Nm) for all other engines

9. Carefully coat only the threads of the oxygen sensor body with anti-seize paste. Don't get any anti-seize on the sensor probe.
10. Install or connect the following:
- HO2S and carefully tighten it to 33 ft. lbs. (45 Nm)
- Heat shield and tighten the bolts to 16 ft. lbs. (22 Nm)
- HO2S connector

11. Raise and support the front of the vehicle and block the rear wheels.
12. Install or connect the following:
- Front exhaust pipe and the exhaust manifold/catalytic converter. Tighten the self-locking nuts to 40 ft. lbs. (55 Nm), if the converter is not attached to the manifold. If the converter is attached, tighten to 25 ft. lbs., (34 Nm). Install any manifold brackets and tighten them to 33 ft. lbs. (45 Nm).
- Splash shield if it was removed

13. Lower the vehicle and connect the negative battery cable.
14. Run the engine and check for exhaust leaks.

2.0L Engine

1. Before servicing the vehicle, refer to the precautions in the beginning of this section.
2. Remove or disconnect the following:
- Negative battery cable
- Catalytic converter
- Exhaust manifold heat shields
- Heated Oxygen (HO2S) sensor connector
- Exhaust manifold bracket
- Exhaust manifold

To install:

3. Installation is the reverse of the removal procedure, while using the following torque values:
- Exhaust manifold nuts: 23 ft. lbs. (31 Nm)
- Exhaust manifold bracket bolts: 33 ft. lbs. (44 Nm)
- Heat shield bolts: 16 ft. lbs. (22 Nm)
- Catalytic converter bolts: 16 ft. lbs. (22 Nm)

2.2L and 2.3L Engines

1. Before servicing the vehicle, refer to the precautions in the beginning of this section.
2. Remove or disconnect the following:
- Negative battery cable
3. Safely raise and support the vehicle.
4. Remove or disconnect the following:
- Oxygen Sensor (O2S) connector, if it is located in the manifold.
- Exhaust manifold upper cover
- Heat insulator from the manifold, if equipped with air conditioning.
- Nuts attaching the exhaust manifold to the front exhaust pipe.
- Pipe from the manifold and discard the gasket. Support the pipe with wire; do not allow it to hang by itself.
- Exhaust manifold bracket(s) bolts and bracket(s).
- Exhaust manifold attaching nuts, using a crisscross pattern (starting from the center).
- Manifold and discard the gasket. Clean the manifold and cylinder head mating surfaces.
- Lower manifold cover from the manifold, if equipped.

To install:

5. Install or connect the following:
- Lower manifold cover, if equipped,

NOTE: Use new gaskets and self-locking nuts when reassembling.

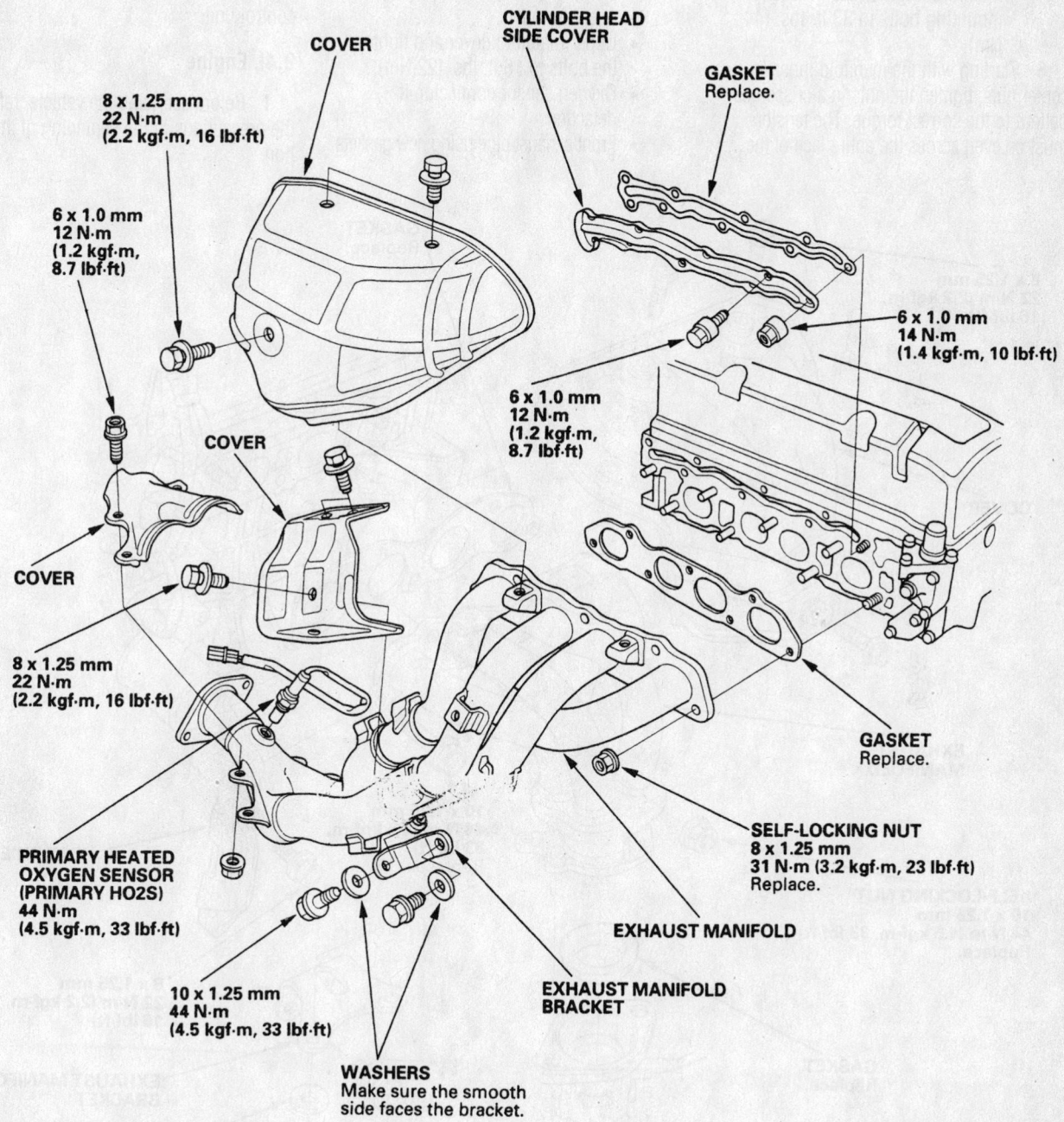

COVER

CYLINDER HEAD SIDE COVER

GASKET
Replace.

8 x 1.25 mm
22 N·m
(2.2 kgf·m, 16 lbf·ft)

6 x 1.0 mm
12 N·m
(1.2 kgf·m,
8.7 lbf·ft)

6 x 1.0 mm
14 N·m
(1.4 kgf·m, 10 lbf·ft)

6 x 1.0 mm
12 N·m
(1.2 kgf·m,
8.7 lbf·ft)

COVER

COVER

8 x 1.25 mm
22 N·m
(2.2 kgf·m, 16 lbf·ft)

GASKET
Replace.

SELF-LOCKING NUT
8 x 1.25 mm
31 N·m (3.2 kgf·m, 23 lbf·ft)
Replace.

**PRIMARY HEATED
OXYGEN SENSOR
(PRIMARY HO2S)**
44 N·m
(4.5 kgf·m, 33 lbf·ft)

EXHAUST MANIFOLD

**EXHAUST MANIFOLD
BRACKET**

10 x 1.25 mm
44 N·m
(4.5 kgf·m, 33 lbf·ft)

WASHERS
Make sure the smooth
side faces the bracket.

9347FG10

Exploded view of the exhaust manifold—2.0L engine

and tighten the attaching bolts to 16 ft. lbs. (22 Nm).
- New gasket
- Exhaust manifold into position and support it
- New nuts snug on the studs
- Support bracket(s) below the manifold. Tighten the bracket(s) mounting bolts to 33 ft. lbs. (44 Nm).

6. Starting with the manifold inner or center nuts, tighten the nuts in a crisscross pattern to the correct torque. The tension must be even across the entire face of the

manifold if leaks are to be prevented. Tighten the nuts to 23 ft. lbs. (31 Nm).

7. Install or connect the following:
- Heat insulator to the manifold, if equipped with air conditioning. Tighten the attaching bolts to 84 inch lbs. (10 Nm) on Prelude models and 108 inch lbs. (12 Nm) on Accord models.
- Upper manifold cover and tighten the bolts to 16 ft. lbs. (22 Nm).
- Oxygen sensor connector, if detached
- Front exhaust pipe using new gaskets

and nuts. Tighten the exhaust pipe attaching nuts to 40 ft. lbs. (55 Nm).
- Negative battery cable and enter the radio security code.

8. Start the engine and check for exhaust leaks.

9. If equipped with 4WS, turn the steering wheel lock-to-lock to reset the 4WS control unit.

2.4L Engine

1. Before servicing the vehicle, refer to the precautions in the beginning of this section.

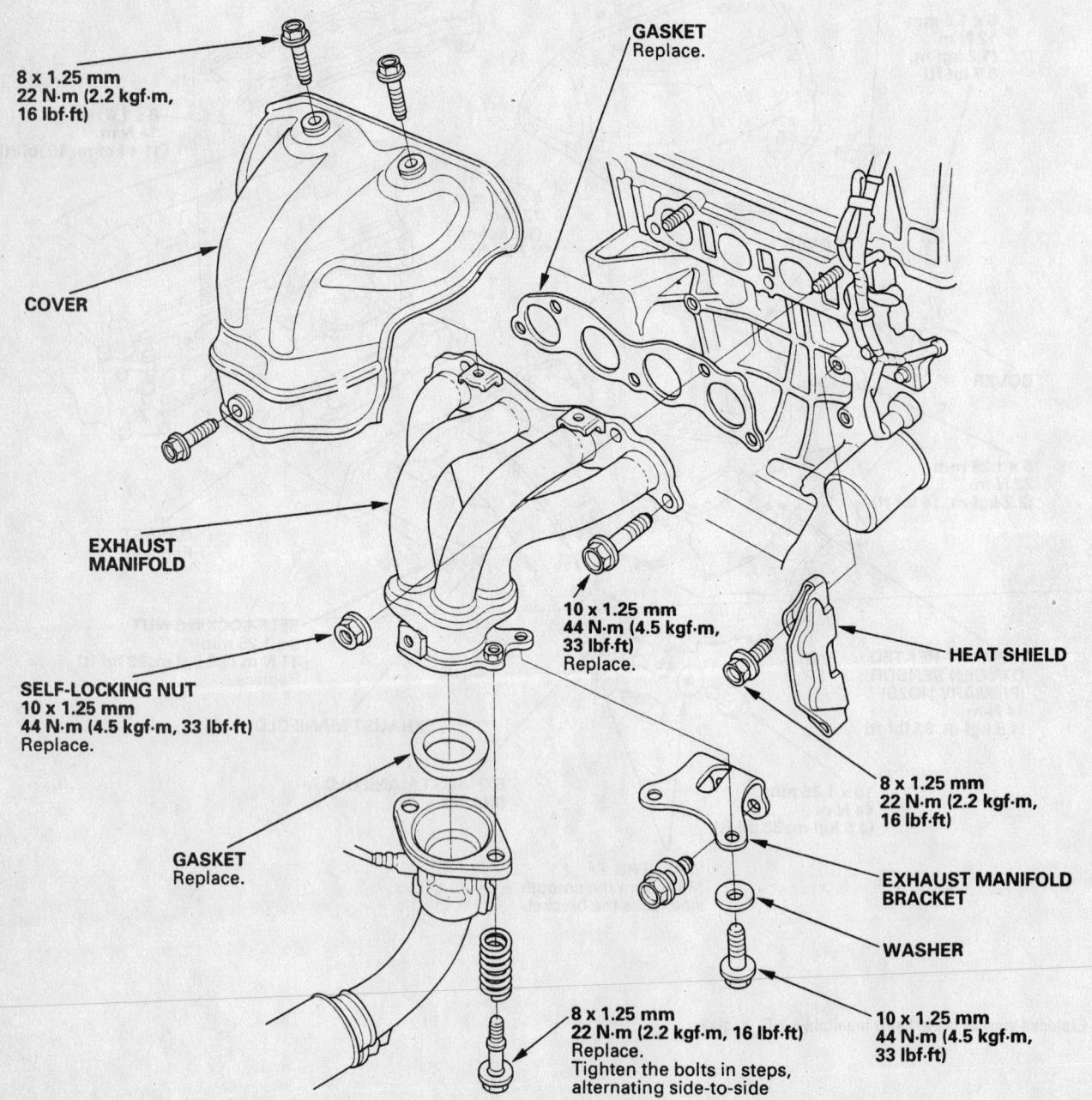

8 x 1.25 mm
22 N·m (2.2 kgf·m, 16 lbf·ft)

GASKET
Replace.

COVER

EXHAUST MANIFOLD

SELF-LOCKING NUT
10 x 1.25 mm
44 N·m (4.5 kgf·m, 33 lbf·ft)
Replace.

GASKET
Replace.

10 x 1.25 mm
44 N·m (4.5 kgf·m, 33 lbf·ft)
Replace.

HEAT SHIELD

8 x 1.25 mm
22 N·m (2.2 kgf·m, 16 lbf·ft)

EXHAUST MANIFOLD BRACKET

WASHER

8 x 1.25 mm
22 N·m (2.2 kgf·m, 16 lbf·ft)
Replace.
Tighten the bolts in steps, alternating side-to-side

10 x 1.25 mm
44 N·m (4.5 kgf·m, 33 lbf·ft)

42356-ACCO-G16

Exploded view of the exhaust manifold and related components—2.4L engine

2. Raise and safely support the vehicle.
3. Remove or disconnect the following:
   - VTEC solenoid valve
   - Driveshaft heat shield
   - Cover and exhaust manifold bracket
   - Exhaust manifold

To install:

4. Clean the mounting surfaces.
5. Install or connect the following:
   - New gasket on the cylinder head
   - Exhaust manifold. Tighten the nuts, in a criss-cross pattern starting with the inner nut, to 33 ft. lbs. (45 Nm).
   - Exhaust manifold bracket and cover
   - Driveshaft heat shield
   - VTEC solenoid valve

3.0L Engine

1. Before servicing the vehicle, refer to the precautions in the beginning of this section.
2. Raise and safely support the vehicle.
3. Remove or disconnect the following:
   - Engine undercover
   - Exhaust pipe from the manifold to be removed
4. Lower the vehicle.

5. Remove or disconnect the following:
   - Exhaust manifold heat shield
   - Mounting nuts and the exhaust manifold.

To install:

6. Clean the mounting surfaces.
7. Install or connect the following:
   - New gasket on the cylinder head
   - Exhaust manifold. Tighten the nuts to 23 ft. lbs. (31 Nm).
   - Heat shield. Tighten the bolts to 16 ft. lbs. (22 Nm).
8. Raise the vehicle and connect the exhaust pipe to the manifold using a new gasket. Tighten the nuts to 40 ft. lbs. (54 Nm).

Front Crankshaft Seal

REMOVAL & INSTALLATION

➡**The original radio may contain a coded anti-theft circuit. Obtain the security code number before disconnecting the battery cables.**

1. Before servicing the vehicle, refer to the precautions in the beginning of this section.
2. Disconnect the negative battery cable.

3. Safely raise and support the vehicle.
4. Remove or disconnect the following:
   - Splash shield
   - Engine accessory drive belts
5. Turn the engine to align the timing marks and set cylinder No.1 to TDC. The white mark on the crankshaft pulley should align with the pointer on the timing belt cover. Remove the inspection caps on the upper timing belt covers to check the alignment of the timing marks. The pointers for the camshafts should align with the green marks on the camshaft sprockets.
6. Remove or disconnect the following:
   - Upper timing belt covers and crankshaft pulley
   - Lower timing belt cover

➡**Mark the direction of the timing belt rotation if it is to be reinstalled.**

   - Timing belt
   - Crankshaft Position (CKP) sensor from the oil pump, if equipped
   - Stopper plate
   - Timing belt sprocket from the crankshaft. Do not lose the sprocket key.
   - Seal from the front of the engine, using a suitable seal removal tool

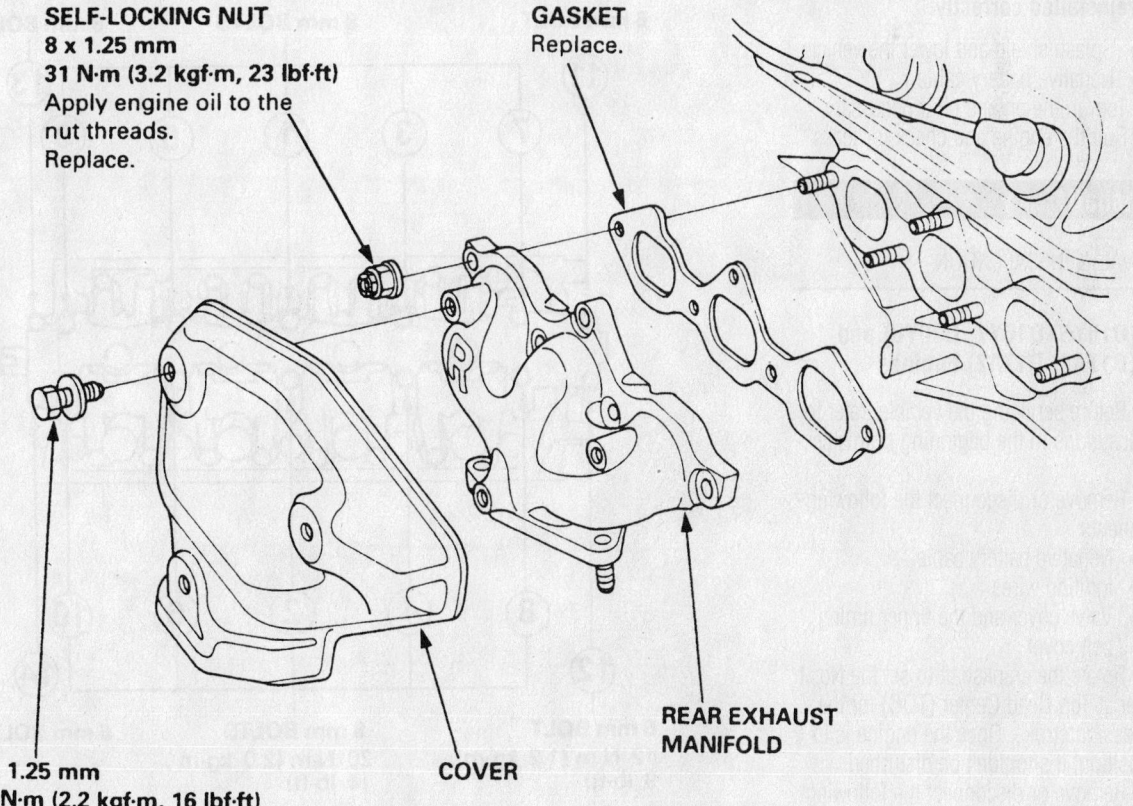

SELF-LOCKING NUT
8 x 1.25 mm
31 N·m (3.2 kgf·m, 23 lbf·ft)
Apply engine oil to the nut threads.
Replace.

GASKET
Replace.

8 x 1.25 mm
22 N·m (2.2 kgf·m, 16 lbf·ft)

COVER

REAR EXHAUST MANIFOLD

7923FG93

Exploded view of the rear exhaust manifold mounting—3.0L engine

To install:

7. Clean the seal mounting surfaces on the engine block.

8. Apply a thin coat of grease on the crankshaft and seal lips.

9. Install or connect the following:

- Seal with the part number facing out. Use a seal driver to seat the seal against the oil pump. Clean any excess grease off the crankshaft and be sure the seal lip is not distorted.
- Timing belt sprocket and key to the crankshaft
- Stopper plate and if equipped, the CKP sensor to the oil pump. Tighten the stopper plate and sensor mounting bolts to 108 inch lbs. (12 Nm).

➡ **Verify that the engine is at Top Dead Center (TDC) for the no. 1 cylinder on the compression stroke.**

- Timing belt
- Timing belt covers and crankshaft pulley. Tighten the crankshaft pulley bolt to 181 ft. lbs. (245 Nm), with the aid of a crank pulley holder.
- Accessory drive belts, then adjust.

➡ **Verify that all engine components that may have been removed have been reinstalled correctly.**

- Splash shield and lower the vehicle.
- Negative battery cable

10. Top up the engine oil if necessary.

11. Run the engine and check for leaks.

Camshaft

REMOVAL & INSTALLATION

1.6L (D16Y5, D16Y7, D16Y8) and 1.7L (D17A1, D17A2) Engines

1. Before servicing the vehicle, refer to the precautions in the beginning of this section.

2. Remove or disconnect the following components:

- Negative battery cable
- Ignition wires
- Valve cover and the upper timing belt cover.

3. Rotate the crankshaft to set the No. 1 cylinder at Top Dead Center (TDC) for the compression stroke. Once the engine is in this position, it shouldn't be disturbed.

4. Remove or disconnect the following components:

- Timing belt. If the timing belt is

contaminated with oil or coolant, it must be replaced. If the timing belt is to be reused, mark its direction of rotation.

- Distributor, if equipped
- Camshaft sprocket and its key. Remove the upper rear timing cover.
- Rocker arm locknuts and the valve adjusting screws.
- Camshaft holder bolts in a 2-step crisscross sequence, starting at the edges and working toward the center of the cylinder head.
- Rocker arm and shaft assembly. Leave the camshaft holder bolts in the camshaft holders to hold the rocker arm and shaft assembly together.

5. Wrap rubber bands around the VTEC rocker arm assemblies so that they do not separate.

6. Store the rocker arm and shaft assembly away from your work area. Cover the assembly with shop towels or a sheet of plastic to protect it from dust.

7. Lift the camshaft from the cylinder head. Remove the camshaft seal.

8. Inspect the camshaft journals and lobes for signs of scoring or other damage.

To install:

9. Remove the oil control orifice. Thoroughly clean it and reinstall it with a new O-ring.

10. Clean and inspect the camshaft bearing caps in the cylinder head.

11. Lubricate the lobes and journals of the camshaft prior to installation.

12. Install or connect the following:

- Camshaft with the keyway facing up so that the camshaft will be at Top Dead Center (TDC)/compression for the No. 1 cylinder.
- New, lightly lubricated, camshaft seal

13. Install the rocker arm and shaft assembly as follows:

a. Remove the rubber bands from the VTEC rocker arms.

b. Lubricate the rocker arm contact surfaces.

c. Apply liquid gasket to the head mating surfaces of the No. 1 and No. 5 camshaft holders. Don't allow the sealant to cure before installing the rocker arm assembly.

d. Set the rocker arm and shaft assembly in place. If equipped, install the lost motion assembly holder.

e. Coat the threads of the camshaft

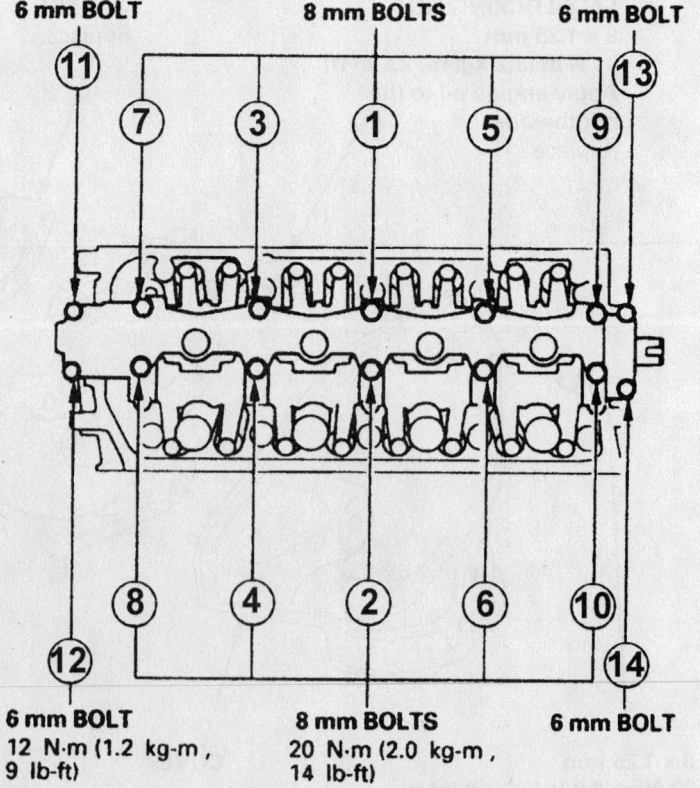

Camshaft holder bolt tightening sequence—1.6L (D16Y5, D16Y7, D16Y8) and 1.7L (D17A1, D17A2) engines

holder bolts with clean oil and loosely install them.

f. Tighten each bolt 2 turns at a time in the crisscross sequence to ensure that the rockers and camshaft holder do not bind on the camshaft journals.

g. Tighten the 8mm camshaft holder bolts to 14 ft. lbs. (20 Nm), and the 6mm camshaft holder bolts to 108 inch lbs. (12 Nm).

14. Install or connect the following:
- Camshaft sprocket and key. Tighten the retaining bolt to 27 ft. lbs. (38 Nm).

➡**Verify that the engine remains at Top Dead Center (TDC)/compression for the No. 1 cylinder.**

- Distributor, if equipped
- Timing belt. Tighten the tensioner bolt to 33 ft. lbs. (44 Nm) once the belt has been properly tensioned.
- Lower timing cover. Tighten the crankshaft pulley bolt to 134 ft. lbs. (181 Nm).

15. Adjust the valves.

16. Manually inspect the VTEC rocker arms for smooth motion.

17. Be sure all the spark plug tube sealing gaskets are fully seated.

18. Apply liquid gasket to the corner recesses of a new valve cover gasket.

19. Install or connect the following:
- Gasket to the valve cover. Don't allow the sealant to cure before installation.
- Valve cover. Gently wiggle the valve cover to be sure it is fully seated. Tighten the valve cover bolts in a crisscross pattern to 84 inch lbs. (10 Nm).
- Ignition wires

20. Refill the engine with fresh oil and install a new filter.

21. Reconnect the battery cable.

22. Warm the engine up to normal operating temperature. Check for oil leaks.

23. Check the ignition timing and adjust it if necessary. Then, tighten the distributor mounting bolts to 17 ft. lbs. (24 Nm).

1.6L (B16A2) Engine

1. Before servicing the vehicle, refer to the precautions in the beginning of this section.

2. Remove or disconnect the following:
- Negative battery cable
- Ignition wires

3. Rotate the crankshaft to set the engine at Top Dead Center (TDC) for the compression stroke of the No. 1 cylinder.

The white TDC mark on the crankshaft pulley should align with the pointer on the lower timing belt cover.

4. Remove or disconnect the following:
- Strut brace
- Intake air duct
- Accessory drive belts

5. Use a floor jack padded with a block of wood to support the engine.

6. Remove or disconnect the following:
- Engine ground cable
- Engine side mount
- Valve cover and the upper timing belt cover

7. Verify that the engine is set at Top Dead Center (TDC)/compression. Loosen the timing belt tensioner bolt 180°. Then, remove the crankshaft pulley, the lower timing cover, and timing belt.

✳✳ WARNING

Inspect the timing belt for signs of cracked and broken teeth, as well as oil or coolant contamination. If the timing belt is damaged, or has been in contact with oil or coolant, it must be replaced to avoid potential failure.

8. Remove or disconnect the following:
- Distributor
- Variable Timing Electronic Control (VTEC) solenoid valve.
- Solenoid valve's filter and inspect it for clogging.
- Camshaft sprockets and back cover.
- Camshaft holder plate bolts in a crisscross sequence working toward the middle of the cylinder head.
- Valve adjusting screws
- Camshaft holder plates and holders from the cylinder head. The holder bolts will keep the components together. Note the positions of each camshaft holder for reassembly.
- Camshafts from the cylinder head. Mark the exhaust and intake camshafts so that they will not be confused.
- Intake and exhaust oil control orifices. Thoroughly clean each, and reinstall them with new O-rings.

9. Inspect the camshaft lobes and journals for any signs of damage.

To install:

10. Install or connect the following:
- New O-ring and the dowel pin to the oil passage of the No. 3 camshaft holder.

➡**Lubricate the camshaft lobes and journals with clean engine oil.**

- Camshafts into the cylinder head. Both the intake and exhaust camshafts should be installed with their keyways pointing straight up.
- New camshaft seals. Apply liquid gasket to a new camshaft end-plug and install it. If the end-plug has a mark, the mark should be aligned with the cylinder head surface.

11. Apply liquid gasket to the cylinder head mating surfaces of the No. 1 and No. 5 camshaft holders, then install them, along with No. 2, 3, and 4 holders. Be sure to pay attention to the following points:
- Do not apply oil to the holder mating surface of camshaft seals.
- The arrows marked on the camshaft holders should point to the timing belt.

12. Install or connect the following:
- Camshaft holder plates
- Install all the lubricated camshaft holder bolts, but don't tighten them yet.

13. Evenly hand-tighten the camshaft holders. Be sure that the rocker arms are properly positioned on the valve stems.

14. Use a 2-step crisscross pattern to tighten the camshaft holder bolts. Begin tightening with the bolts in the middle of the cylinder head, and work toward the outer edges. Final torque specifications are as follows:

a. 8mm bolts: 20 ft. lbs. (28 Nm).
b. 6mm bolts: 84–96 inch lbs. (10–11 Nm)

15. Verify that the camshaft keyways are pointing straight up and that the engine is at Top Dead Center (TDC)/compression for the No. 1 cylinder. Fit the camshaft sprocket keys into their keyways.

16. Install or connect the following:
- Back cover and push the camshaft pulleys onto the camshafts. Then, tighten the sprocket retaining bolts to 41 ft. lbs. (57 Nm).
- Timing belt and tension.
- Lower timing cover and the crankshaft pulley. Tighten the pulley bolt to 130 ft. lbs. (180 Nm).

17. Adjust the valve clearance.

18. Inspect the VTEC rocker arms for smooth and independent movement.

19. Install or connect the following:
- VTEC solenoid valve with a new filter. Tighten the valve mounting bolts to 108 inch lbs. (12 Nm).
- Distributor

20. Clean the valve cover gasket surfaces. Fit the gasket into the groove of the valve cover.

21. Apply liquid gasket to the rubber

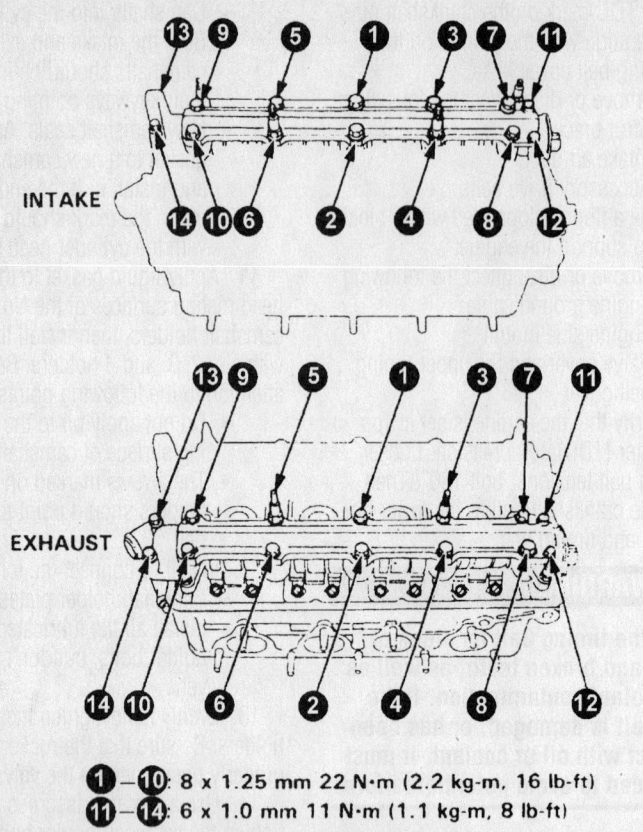

INTAKE

EXHAUST

1–**10**: 8 x 1.25 mm 22 N·m (2.2 kg-m, 16 lb-ft)
11–**14**: 6 x 1.0 mm 11 N·m (1.1 kg-m, 8 lb-ft)

7923FG64

Camshaft holder bolt tightening sequences—1.6L (B16A2) engines

seal at the 8 corners where the gasket meets the camshaft holders. Don't allow the sealant to cure before installing the cylinder head cover.

22. Install or connect the following:
- Cylinder head cover and engine ground cable. Be sure the contact surfaces are clean and do not touch surfaces where liquid gasket has been applied.
- Valve cover nuts: 84 inch lbs. (10 Nm) in a crisscross sequence.
- Upper timing cover
- Accessory drive belts and adjust their tensions.
- Engine side mount, tighten the 2 new nuts to 54 ft. lbs. (75 Nm) and tighten the bolt attaching the mount to the vehicle to 54 ft. lbs. (74 Nm).
- Strut bar and tighten the mounting bolts to 16 ft. lbs. (22 Nm).
- Ignition wires

23. Drain the engine oil. Install a new oil filter and refill the engine with fresh oil.

24. Reconnect the negative battery cable.

25. Warm the engine up to its normal

operating temperature. Then, check and adjust the ignition timing. Tighten the distributor mounting bolts to 17 ft. lbs. (24 Nm).

2.0L Engine

1. Before servicing the vehicle, refer to the precautions in the beginning of this section.

2. Loosen the valve adjustment screws so that all valves are closed and all rocker arms are loose.

3. Remove or disconnect the following:
- Negative battery cable
- Valve cover
- Camshaft bearing caps
- Camshafts

To install:

4. Set the engine to Top Dead Center (TDC) so that the timing chain sprocket timing marks are aligned with the cylinder head surface as shown.

5. Install or connect the following:
- Camshafts with the sprocket timing marks aligned as shown
- Camshaft bearing caps and tighten the bolts in sequence to 16 ft. lbs. (22 Nm). Adjust the valve clearance.
- Valve cover
- Negative battery cable

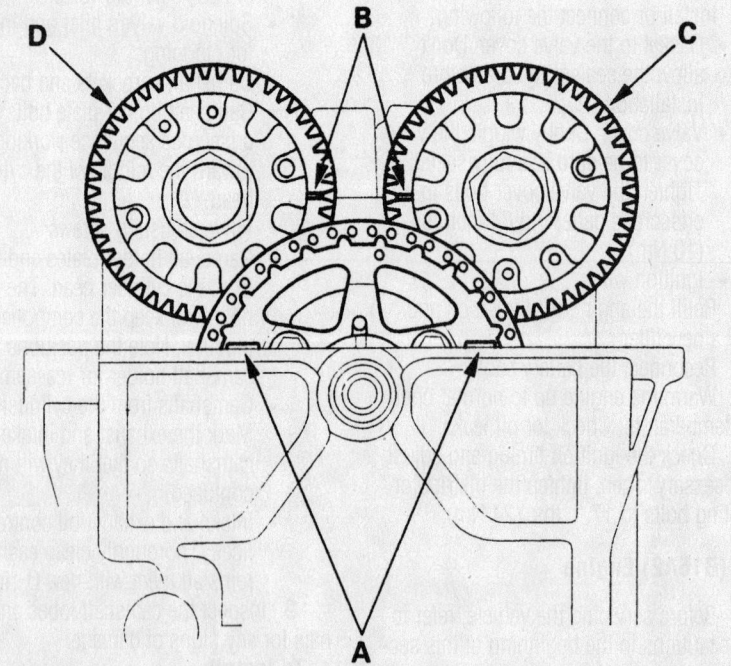

9347FG15

Timing chain sprocket alignment marks (A) and camshaft sprocket alignment marks (B)—2.0L engine

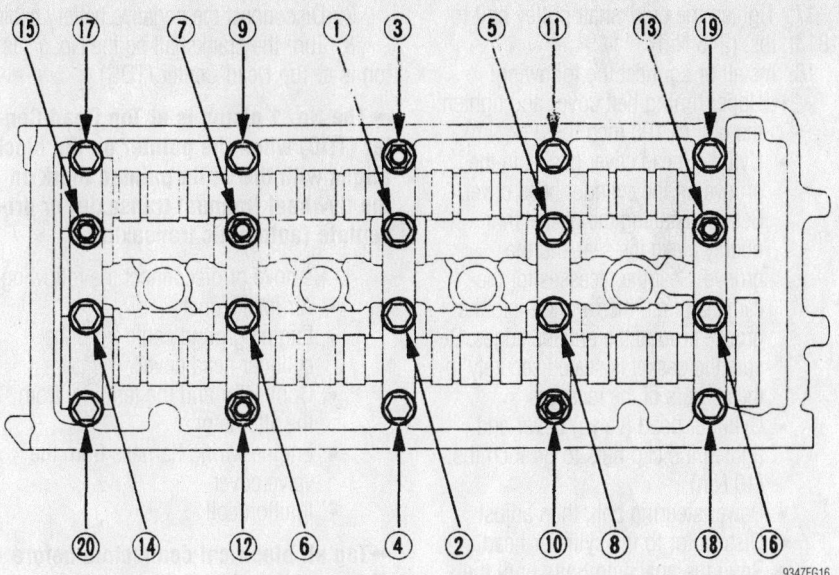

Camshaft bearing cap torque sequence—2.0L engine

9347FG16

2.2L Engine

1. Before servicing the vehicle, refer to the precautions in the beginning of this section.

2. Remove or disconnect the following:
- Negative battery cable
- Air intake duct
- Cylinder head cover and replace the rubber seals if damaged or deteriorated.
- Timing belt upper cover

3. Turn the engine to align the timing marks and set cylinder No.1 to TDC/compression. Once in this position, the engine must NOT be turned or disturbed.

4. Remove or disconnect the following:
- Electrical connectors from the distributor and the spark plug wires from the spark plugs.

➡ **Matchmark the installed position of the distributor before removing it.**

- Distributor from the cylinder head
- Power steering pump drive belt

5. Mark the rotation of the timing belt if it is to be used again. Loosen the timing belt adjusting bolt ¾ to 1 turn, then release the tension on the timing belt. Push the tensioner to release tension from the belt, then tighten the adjusting bolt.

6. Remove or disconnect the following:
- Timing belt from the camshaft sprocket

✳✳ WARNING

Do not crimp or bend the timing belt more than 90°, or less than 1 inch (25mm) in diameter

➡ **Ensure the words UP embossed on the camshaft pulley is aligned in the upward position before removing the sprocket bolt.**

To install:
- Camshaft sprocket bolt, sprocket and sprocket key
- Timing belt back cover
- Valve adjusting screws
- Camshaft holder attaching bolts 2 turns at a time, in the proper sequence to prevent damaging the valves or rocker arm assemblies

➡ **When removing the rocker arm assembly, do not remove the camshaft holder bolts. The bolts will keep the camshaft holders, springs, and the rocker arms on the shafts.**

- Camshaft holders and rocker arm assembly. If the rocker arm and shaft assembly needs to be disassembled for service, note the location of the components as they are removed. The rocker arms must be installed in the same position if reused
- Camshaft from the cylinder head and discard the seal

To install:

7. Wipe the camshaft and the camshaft journals clean, then lubricate both surfaces,

8. Install or connect the following:
- Camshaft

9. Turn the camshaft so that its keyway is facing up (No. 1 cylinder will be at Top Dead Center (TDC)).

10. Reassemble the rocker arm and shaft assembly if it was disassembled. Lubricate the rocker arm and shaft assembly with clean oil, then apply liquid gasket to the head mating surfaces of the No. 1 and No. 6 camshaft holders.

11. Install or connect the following:
- Camshaft holders and rocker arm assembly in place, then loosely install the attaching bolts
- Clean oil to the camshaft oil seal

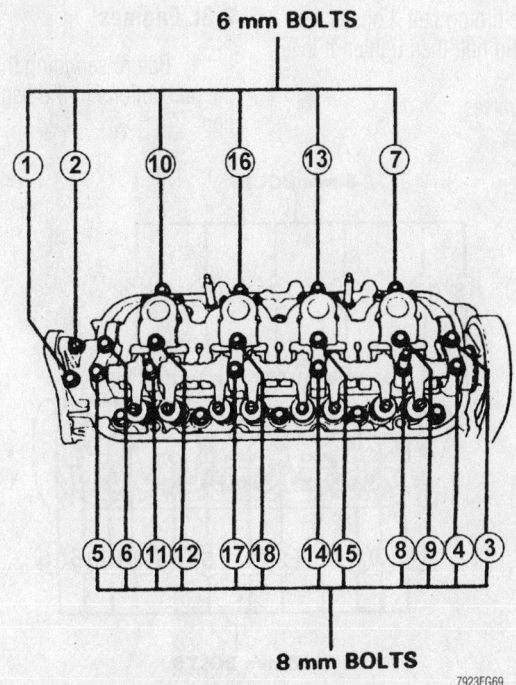

Rocker arm assembly bolt removal sequence—2.2L (H22A4) engines

7923FG69

lip and the seal guide (part # 07NAG-PT0010A)
- Seal to the seal guide
- Seal guide to the camshaft, then the installer cup (part # 07NAF-PT0010A) and the installer shaft (part # 07NAF-PT0020A). Tighten the nut on the installer shaft to press the seal into the cylinder head.

12. Tighten the camshaft holder bolts 2 turns at a time in the proper sequence. The final torque for the 8mm bolts is 16 ft. lbs. (22 Nm) and the final torque for the 6mm bolts is 108 inch lbs. (12 Nm).

13. Install or connect the following:
- Timing belt back cover and tighten the attaching bolt to 108 inch lbs. (12 Nm)
- Camshaft sprocket key to the camshaft, then the camshaft sprocket. Install the bolt and tighten it to 27 ft. lbs. (37 Nm)

➡ **Ensure the words UP embossed on the camshaft pulley is aligned in the upward position, before installing the timing belt.**

- Timing belt onto the camshaft sprocket. Loosen, then tighten the timing belt adjusting nut.

14. Rotate the crankshaft pulley 5 or 6 turns to position the timing belt on the pulleys.

15. Set the No. 1 cylinder to TDC and loosen the timing belt adjusting nut 1 turn. Turn the crankshaft counterclockwise until the cam pulley has moved 3 teeth; this creates tension on the timing belt. Loosen the timing belt adjusting nut, then tighten it to 33 ft. lbs. (45 Nm).

16. Adjust the valves.

17. Tighten the crankshaft pulley bolt to 181 ft. lbs. (245 Nm)

18. Install or connect the following:
- Upper timing belt cover and tighten the bolt to 108 inch lbs. (12 Nm)
- Cylinder head cover gasket in the groove on the cylinder head cover. Before installing the gasket thoroughly clean the seal and the groove. Seat the recesses for the camshaft first, then work it into the groove around the outside edges. Be sure the gasket is seated securely in the corners of the recesses.
- Cylinder head (valve) cover and tighten the cap nuts to 84 inch lbs. (10 Nm).
- Power steering belt, then adjust
- Distributor to the cylinder head. Snug the attaching bolts until the timing has been checked and adjusted.
- Spark plug wires and the distributor electrical connectors
- Ignition coil wire to the distributor
- Air intake duct

19. Drain the oil from the engine into a sealable container. Install the drain plug and refill the engine with clean oil.

20. Connect the negative battery cable and enter the radio security code.

21. Start the engine and check carefully for any leaks.

22. Check and adjust the ignition timing as necessary, then tighten the distributor bolts to 16 ft. lbs. (22 Nm).

2.3L Engines

1. Before servicing the vehicle, refer to the precautions in the beginning of this section.

2. Disconnect the negative battery cable.

3. Turn the crankshaft so the No. 1 piston is at Top Dead Center (TDC).

➡ **The No. 1 piston is at Top Dead Center (TDC) when the pointer on the block aligns with the white painted mark on the flywheel (manual transaxle) or driveplate (automatic transaxle).**

4. Remove or disconnect the following:
- Air intake duct
- Engine ground cable from the cylinder head cover
- Connector and the terminal from the alternator
- Engine wiring harness from the valve cover
- Ignition coil

➡ **Tag all electrical connectors before disconnecting them.**

- Electrical connectors from the distributor
- Spark plug wires from the spark plugs
- Distributor from the cylinder head
- Ignition coil wire from the distributor
- Positive Crankcase Ventilation (PCV) hose
- Cylinder head cover. Replace the rubber seals if damaged or deteriorated.
- Timing belt middle cover

5. Ensure the words **UP** embossed on the camshaft pulleys are aligned in the upward position.

6. Mark the rotation of the timing belt if it is to be used again. Loosen the timing belt adjusting nut ½ turn, then release the tension on the timing belt. Push the tensioner to release tension from the belt, then tighten the adjusting nut.

7. Remove or disconnect the following:
- Timing belt from the camshaft sprockets

✳✳ WARNING

Do not crimp or bend the timing belt more than 90°, or less than 1 inch (25mm) in diameter

8. Insert a 5.0mm pin punch in each of the camshaft caps nearest to the sprockets through the holes provided.

9. Remove or disconnect the following:
- Camshaft sprocket attaching bolts, then the sprockets. Do not lose the sprocket keys.
- Side engine mount bracket B, then the timing belt back cover from behind the camshaft sprockets
- Rocker arm adjusting screws, then

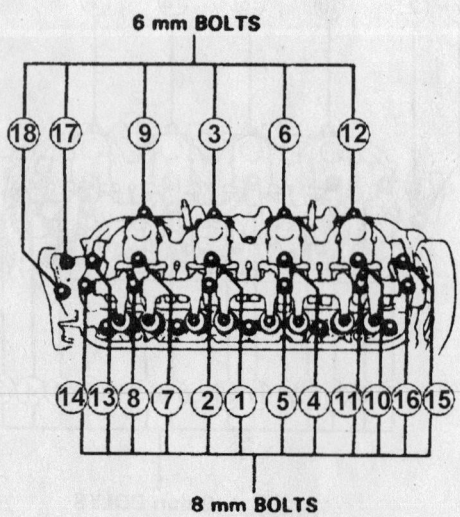

6 mm BOLTS

⑱ ⑰ ⑨ ③ ⑥ ⑫

⑭⑬⑧⑦②①⑤④⑪⑩⑯⑮

8 mm BOLTS

7923FG71

Rocker arm assembly torque sequence—2.2L (H22A4) engines

the pin punches from the camshaft caps
- Camshaft holders, note the holders locations for ease of installation. Loosen the bolts in the reverse order of the installation.
- Camshafts from the cylinder head, then discard the camshaft seals
- Rubber cap from the head, located at the end of the intake camshaft

10. Remove the rocker arms from the cylinder head. Note the locations of the rocker arms.

➡ **The rocker arms have to be installed to their original locations if being reused.**

To install:

11. Lubricate the rocker arms with clean oil.

12. Install or connect the following:
- Rocker arms on the pivot bolts and the valve stems. If the rocker arms are being reused, install them to their original locations. The locknuts and adjustment screws should be loosened before installing the rocker arms.

13. Lubricate the camshafts with clean oil.

14. Install or connect the following:
- Camshaft seals to the end of the camshafts that the timing belt sprockets attach to. The open side (spring) should be facing into the cylinder head when installed.

➡ **Be sure the keyways on the camshafts are facing up and install the camshafts to the cylinder head.**

- Rubber plug to the cylinder head at the end of the intake camshaft

15. Apply liquid gasket to the head mating surfaces of the No. 1 and No. 6 camshaft holders, then install them along with No. 2, 3, 4 and 5. I or E marks are stamped on the camshaft holders to identify them as Intake or Exhaust side holders. The arrows stamped on the holders should point toward the timing belt.

16. Snug the camshaft holders in place.

17. Press the camshaft seals securely into place.

18. Tighten the camshaft holder bolts in 2 steps, following the proper sequence, to ensure that the rockers do not bind on the valves. Tighten all the bolts, except the 4 studs, to 84 inch lbs. (10 Nm). Tighten the studs (number 5 and 7 bolts in the correct sequence) to 108 inch lbs. (12 Nm).

19. Install or connect the following:
- Timing belt back cover
- Side engine mount bracket B.

Tighten the bolt attaching the bracket to the cylinder head to 33 ft. lbs. (45 Nm). Tighten the bolts attaching the bracket to the side engine mount to 16 ft. lbs. (22 Nm).
- 5.0mm pin punch in each of the camshaft caps, nearest to the pulleys, through the holes provided.
- Keys into the camshaft grooves

20. Push the camshaft sprockets onto the camshafts, then tighten the retaining bolts to 27 ft. lbs. (38 Nm).

21. Ensure the words **UP** embossed on the camshaft pulleys are aligned in the upward position.

22. Install or connect the following:
- Timing belt to the camshaft sprockets, then remove the 2, 5.0mm pin punches from the camshaft bearing caps

23. Loosen, then tighten the timing belt adjuster nut.

24. Turn the crankshaft counterclockwise until the cam pulley has moved 3 teeth; this creates tension on the timing belt. Loosen the adjusting nut, then tighten it to 33 ft. lbs. (45 Nm).

25. Adjust the valves.

26. Tighten the crankshaft pulley bolt to 181 ft. lbs. (250 Nm).

Specified torque:
Except Intake ⑤, ⑦. Exhaust ⑥, ⑧:
    10 N·m (1.0 kg-m, 7 lb-ft)
Intake ⑤, ⑦. Exhaust ⑥, ⑧:
    12 N·m (1.2 kg-m, 9 lb-ft)

TIGHTENING SEQUENCE

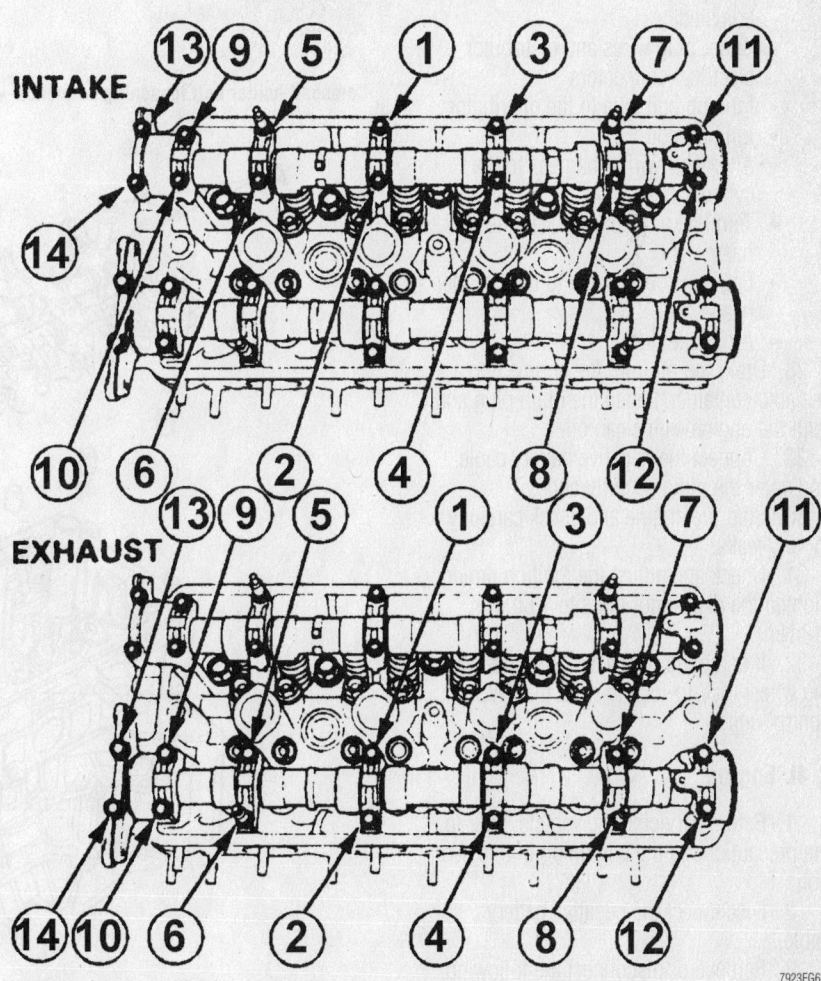

Camshaft holders torque sequence—2.3L engines

7923FG67

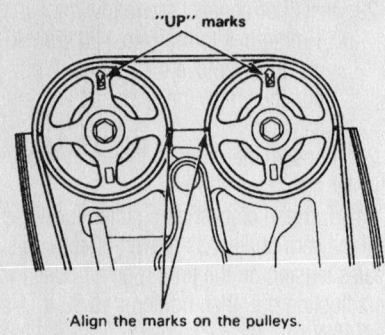

"UP" marks

Align the marks on the pulleys.

7923FG68

Camshaft sprockets alignment—2.3L engines

27. Install or connect the following:
   - Middle timing belt cover and tighten the attaching bolts to 108 inch lbs. (12 Nm)
   - Cylinder head cover and tighten the cap nuts to 84 inch lbs. (10 Nm)
   - PCV hose to the cylinder head cover
   - Distributor to the cylinder head, snug the attaching bolts until the timing has been checked and adjusted
   - Spark plug wires and distributor electrical connectors
   - Ignition coil wire to the distributor
   - Ignition coil
   - Alternator wiring harness to the cylinder head cover
   - Terminal and connector to the alternator
   - Engine ground cable to the cylinder head cover
   - Air intake duct

28. Drain the oil from the engine into a sealable container. Install the drain plug and refill the engine with clean oil.

29. Connect the negative battery cable and enter the radio security code.

30. Start the engine and check carefully for any leaks.

31. Check and adjust the ignition timing. Tighten the distributor bolts to 13 ft. lbs. (18 Nm).

32. If equipped with 4WS, turn the steering wheel lock-to-lock to reset the 4WS control unit.

2.4L Engine

1. Before servicing the vehicle, refer to the precautions in the beginning of this section.

2. Disconnect the negative battery cable.

3. Remove or disconnect the following:
   - Timing chain

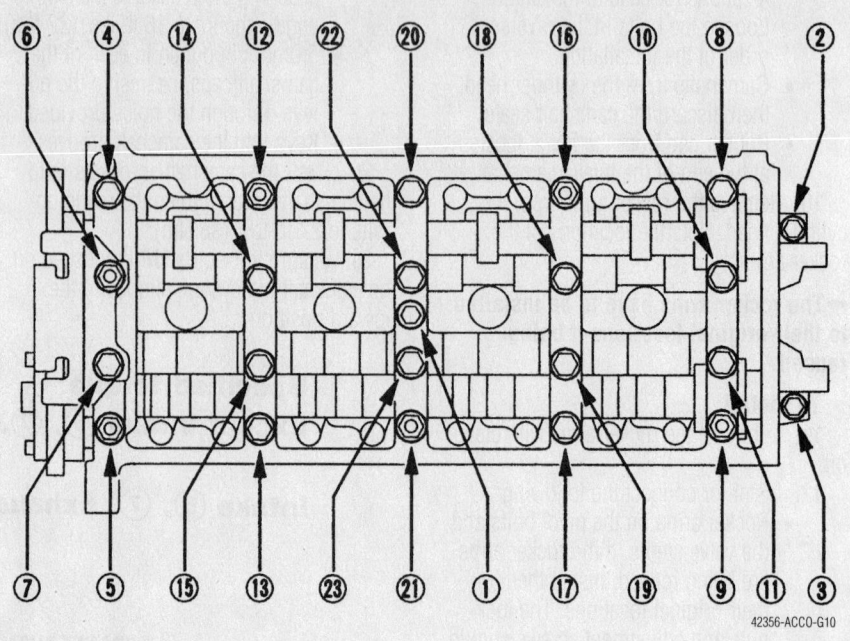

42356-ACCO-G10

Camshaft holder bolt loosening sequence—2.4L engine

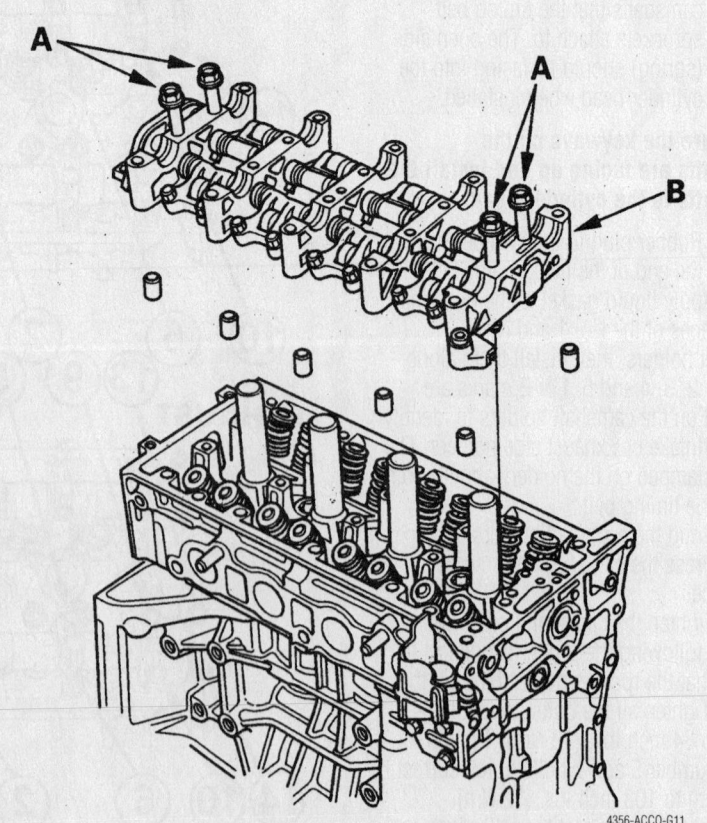

4356-ACCO-G11

Insert the bolts (A) into the rocker shaft holder, then remove the rocker arm assembly (B)—2.4L engine

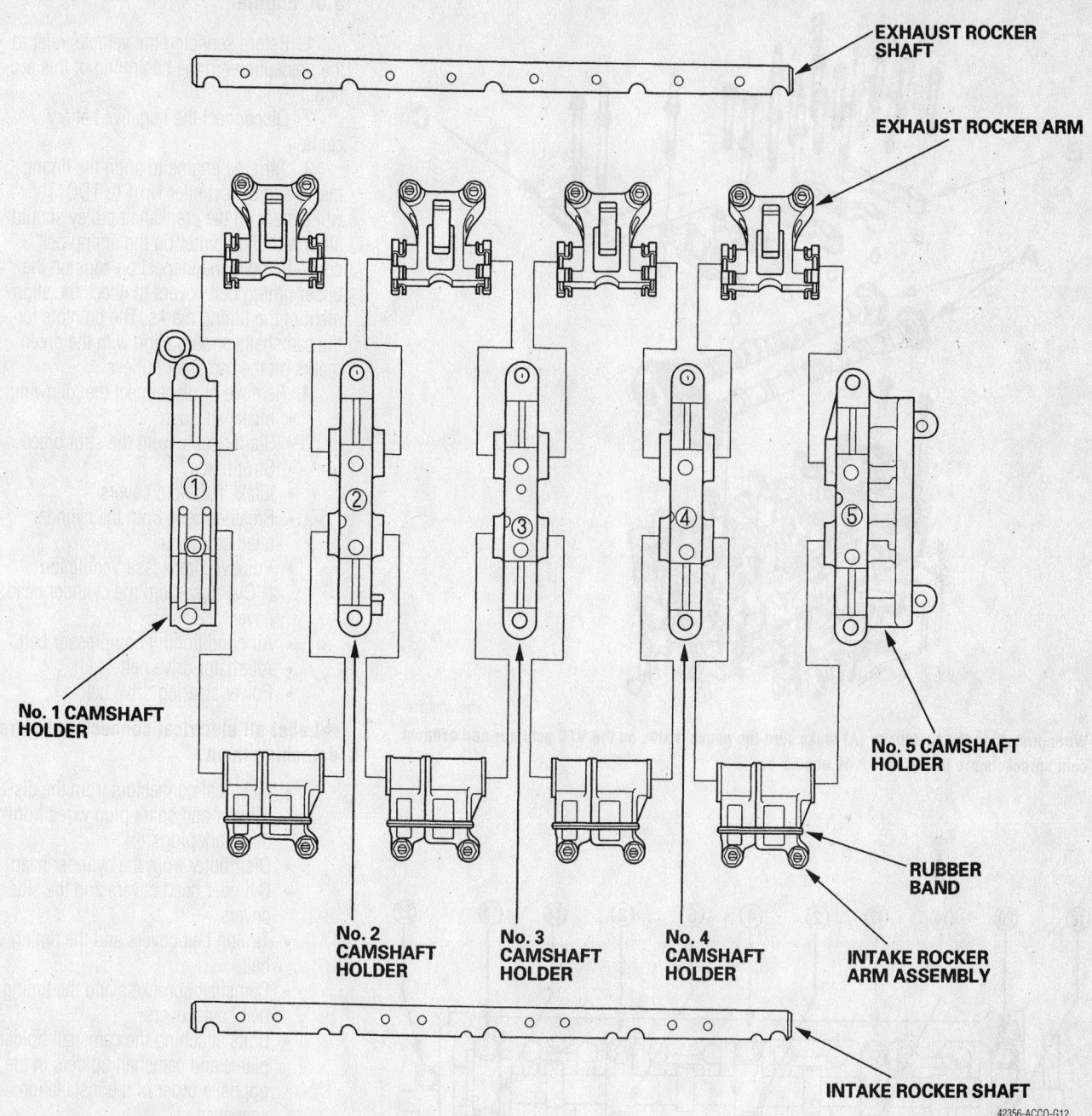

EXHAUST ROCKER SHAFT

EXHAUST ROCKER ARM

No. 1 CAMSHAFT HOLDER

No. 5 CAMSHAFT HOLDER

RUBBER BAND

No. 2 CAMSHAFT HOLDER

No. 3 CAMSHAFT HOLDER

No. 4 CAMSHAFT HOLDER

INTAKE ROCKER ARM ASSEMBLY

INTAKE ROCKER SHAFT

42356-ACCO-G12

Exploded view of the rocker arms and related components—2.4L engine

- Loosen the rocker arm adjusting screws
- Camshaft holder bolts, two turns at a time in sequence
- Timing chain guide (B), camshaft and camshafts

4. Insert the bolts (A) into the rocker shaft holder, then remove the rocker arm assembly (B)

- Camshafts by carefully lifting them out of the cylinder head

To install:

5. Clean and dry the No. 5 rocker shaft holding mating surface.

6. Apply a suitable liquid gasket P/N 08718-0009, or equivalent, evenly to the cylinder head mating surface of the No. 5 rocker shaft holder.

➡ **The parts must be installed within 5 minutes of applying the liquid gasket.**

7. Reassemble the rocker arm assembly, as necessary.

8. Install or connect the following

- Bolts (A) into the rocker shaft holder, then the rocker arm assembly on the cylinder head. Remove

the bolts from the rocker shaft holder.

9. Make sure the punch marks on the variable valve timing control (VTC) actuator and exhaust camshaft sprocket are facing up, then set the camshafts (A) in the holder

10. Set the camshaft holders (B) and timing chain guide B (C) in place.

11. Tighten the bolts, in sequence, to the following specification:

   a. 8mm bolts: 16 ft. lbs. (22 Nm)
   b. 6mm bolts: 8.7 ft. lbs. (12 Nm)

12. Install the timing chain and adjust the valve lash.

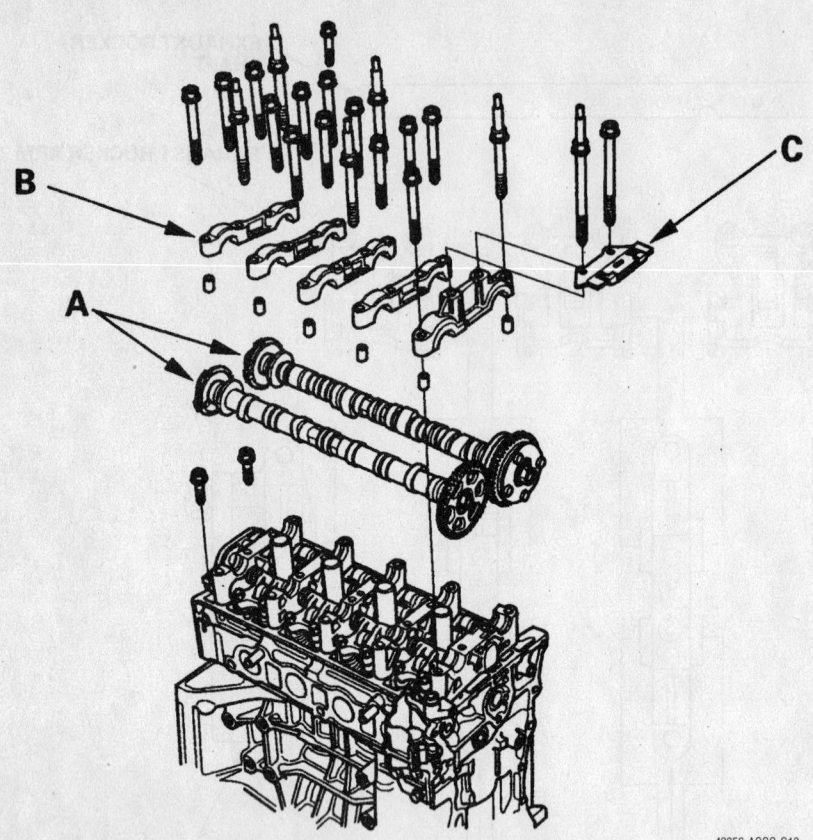

When installing the camshafts (A) make sure the punch marks on the VTC actuator and exhaust cam sprockets are facing up—2.4L engine

42356-ACCO-G13

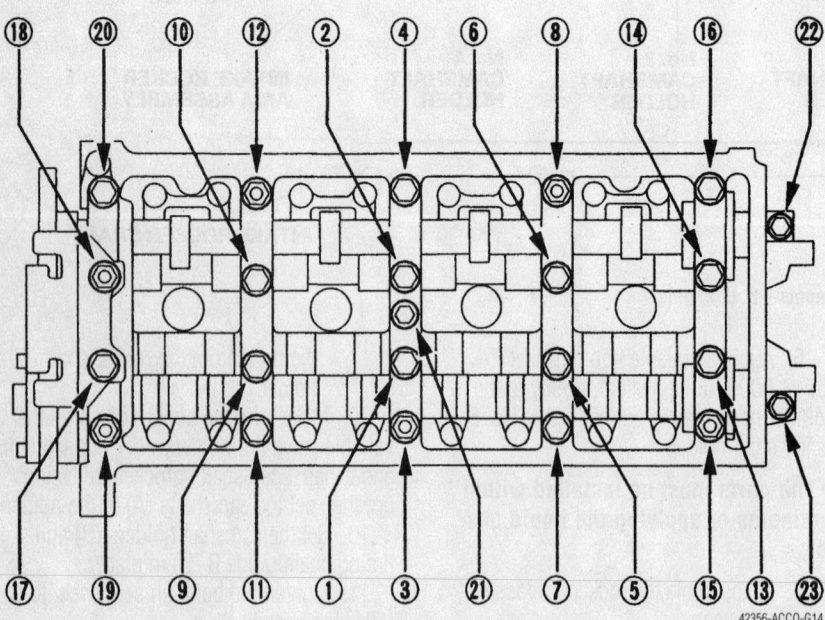

Rocker arm assembly bolt tightening sequence—2.4L engine

42356-ACCO-G14

3.0L Engine

1. Before servicing the vehicle, refer to the precautions in the beginning of this section.

2. Disconnect the negative battery cable.

3. Turn the engine to align the timing marks and set cylinder No.1 to TDC. The white mark on the crankshaft pulley should align with the pointer on the timing belt cover. Remove the inspection caps on the upper timing belt covers to check the alignment of the timing marks. The pointers for the camshafts should align with the green marks on the camshaft pulleys.

4. Remove or disconnect the following:

- Intake air duct
- Starter cable from the strut brace
- Strut brace
- Intake manifold covers
- Breather hose from the cylinder head cover
- Positive Crankcase Ventilation (PCV) hose from the cylinder head cover.
- Air conditioning compressor belt
- Alternator drive belt
- Power steering drive belt

➡ **Label all electrical connectors before detaching them.**

- Electrical connectors from the distributor and spark plug wires from the spark plugs
- Distributor from the cylinder head
- Cylinder head covers and the side covers
- Timing belt covers and the timing belt.
- Camshaft sprockets and the timing belt back covers.
- Bolts attaching the camshaft holder plates and camshaft holders in the opposite order of the installation sequence.
- Camshaft holder plates, camshaft holders, and the dowel pins. Discard the O-rings.
- Camshafts from the cylinder heads and the rubber cap from the rear cylinder head. Discard the camshaft seals.

To install:

5. Apply clean engine oil to the rocker arms and the camshafts.

6. Loosen the exhaust rocker arm adjusting screws and locknuts.

7. Be sure the rocker arms are properly positioned on the valve stems. Advance the crankshaft 30° from TDC to prevent interference between the pistons and valves, then

install the camshafts. Position the rear camshaft on the cylinder head so the cam is not pushing on any valves.

8. Apply liquid gasket around the rubber cap, then install it to the cylinder head.

9. Install or connect the following:
   - Camshaft seals to the camshafts with the open side (spring) facing in.
   - Liquid gasket the cylinder head and camshaft holder mating surfaces
   - Camshaft holders and the camshaft plates with the dowel pins
   - New O-rings to the camshaft holder plates
   - Clean oil to the camshaft holder bolts
   - Bolts and tighten them in the proper sequence to 17 ft. lbs. (24 Nm)
   - Timing belt back covers and tighten the attaching bolts to 108 inch lbs. (12 Nm).
   - Camshaft sprockets and tighten the attaching bolts to 23 ft. lbs. (31 Nm).

10. Set the camshaft sprockets so that the No. 1 piston is at Top Dead Center (TDC). Align the TDC marks (green mark) on the camshaft pulleys to the pointers on the back covers.

11. Turn the crankshaft counterclockwise to set it at Top Dead Center (TDC). Align the TDC mark on the tooth of the timing belt drive pulley with the pointer on the oil pump.

12. Install or connect the following:
   - Timing belt and timing belt covers.

13. Set No. 1 cylinder to TDC.

14. Tighten the valve adjusting screws for No. 1, No. 2 and No. 4 cylinders. Tighten the screw until it contacts the valve, then tighten the screw 1⅛ turns. Hold the screw in place and tighten the locknut to 14 ft. lbs. (20 Nm).

15. Rotate the crankshaft pulley 1 turn clockwise, then tighten the adjusting screws for No. 3, No. 5, and No. 6 cylinders. Tighten the screw until it contacts the valve, then tighten the screw 1⅛ turns. Hold the screw in place and tighten the locknut to 14 ft. lbs. (20 Nm).

16. Install or connect the following:
   - Cylinder head cover gasket into the groove of the cylinder head cover. Seat the recesses for the camshaft first, then work it into the groove around the outside edges.

➡ **Before installing the cylinder head cover gasket, thoroughly clean the seal groove.**

- Liquid gasket to the cylinder head cover gasket at the 4 corners of the recesses. Use a shop towel and wipe the cylinder heads where the cylinder head covers will come in contact.
- Cylinder head covers, hold the gasket in the groove by placing your fingers on the camshaft contacting surfaces. With the cylinder head cover on the cylinder heads, slide the covers slightly back and forth to seat the cylinder head cover gaskets. Replace the washers if damaged or deteriorated

17. Tighten the cylinder head cover bolts in 2 or 3 steps. In the final step, tighten all the bolts, in sequence, to 11 ft. lbs. (15 Nm).

18. Install or connect the following:
   - Cylinder head side covers with new O-rings and tighten the bolts to 108 inch lbs. (12 Nm).
   - Distributor to the cylinder head and tighten the mounting bolt to 16 ft. lbs. (22 Nm).
   - Spark plug wires to the correct spark plugs and the distributor electrical connectors.
   - Power steering belt, and adjust the tension.
   - Alternator belt and adjust the tension. Tighten the alternator mounting nut and bolt to 16 ft. lbs. (22 Nm).
   - Air conditioning belt, and adjust the belt tension. Tighten the idler center nut to 33 ft. lbs. (44 Nm).
   - PCV hose to the cylinder head cover
   - Breather hose to the cylinder head cover
   - Intake manifold cover and tighten the bolts to 108 inch lbs. (12 Nm).
   - Intake air duct
   - Strut brace and tighten the mounting bolts to 16 ft. lbs. (22 Nm).
   - Starter cable to the strut brace

19. Drain the engine oil into a sealable container, then refill the engine with clean oil.

20. Connect the negative battery cable and enter the radio security code.

21. Start the engine, allowing it to idle and check for any signs of leakage.

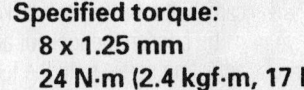

Specified torque:
8 x 1.25 mm
24 N·m (2.4 kgf·m, 17 lbf·ft)
Apply engine oil to the bolt threads and flange

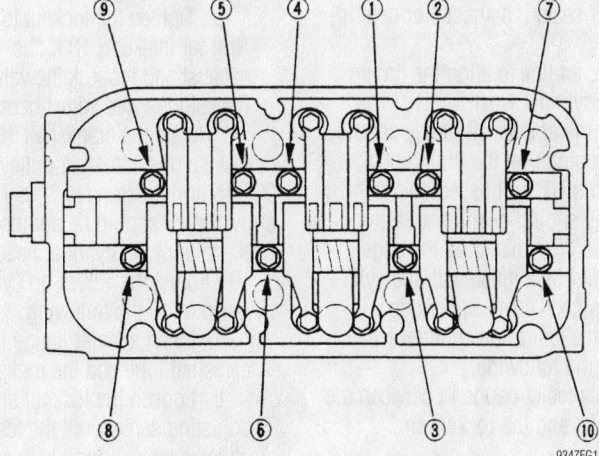

9347FG11

Camshaft holders torque sequence—3.0L engine

Valve Clearance

ADJUSTMENT

➡The radio may contain a coded theft protection circuit. Always obtain the code number before disconnecting the battery. If the vehicle is equipped with 4WS, the steering control unit is shut down when the battery is disconnected. After connecting the battery, turn the steering wheel lock-to-lock to reset the steering control unit.

Civic

1. Before servicing the vehicle, refer to the precautions in the beginning of this section.

2. Disconnect the negative battery cable.

3. Remove the cylinder head cover and the upper timing belt cover.

4. Rotate the crankshaft to align the white TDC mark on the crankshaft pulley with the pointer on the cover for the No. 1 cylinder compression stroke. Be sure the **UP** mark on the camshaft sprocket is up and the TDC marks align with the edge of the cylinder head.

5. Hold a No. 1 cylinder rocker arm against the camshaft and use a feeler gauge to check the clearance at the valve stem. Except on B16A2 engines, intake valve clearance should be 0.007–0.009 in. (0.18–0.26mm), exhaust valve clearance should be 0.009–0.011 in. (0.23–0.27mm). On B16A2 engines, the intake valve clearance should be 0.006–0.007 in. (0.15–0.19mm), exhaust valve clearance should be 0.007–0.008 in. (0.17–0.21mm). Loosen the locknut and turn the adjusting screw to adjust the clearance. Tighten the locknut to 14 ft. lbs. (20 Nm) and recheck the clearance. Don't overtighten the locknut, the aluminum rockers will strip easily.

6. The adjustment order is 1–3–4–2. Rotate the crankshaft counterclockwise 180° (the camshaft sprocket will rotate 90°) to bring each cylinder to TDC/compression. Adjust each set of valves.

   a. At Top Dead Center (TDC) for the No. 3 cylinder, the UP mark is pointed to the exhaust side of the cylinder head.

   b. At Top Dead Center (TDC) for the No. 4 cylinder, the UP mark is pointed down, and the TDC marks align with the edge of the cylinder head.

   c. At Top Dead Center (TDC) for the No. 2 cylinder, the UP mark is pointed to the intake side of the cylinder head.

7. After adjusting the valves of a VTEC engine, inspect the intake rocker arms for smooth and independent motion.

8. Apply sealant to the edges of the valve cover gasket where it meets the camshaft holders. Be sure the spark plug tube seals are properly seated.

9. Install the cylinder head and timing belt covers.

10. Tighten the crankshaft pulley bolt to 134 ft. lbs. (185 Nm).

11. Reconnect the negative battery cable. Enter the radio security code.

Prelude and Accord

➡The valve clearance should be adjusted when the engine is cold, the cylinder head temperature should be less than 100°F (38°C).

➡The radio may contain a coded theft protection circuit. Always obtain the code number before disconnecting the battery.

1. Before servicing the vehicle, refer to the precautions in the beginning of this section.

2. Remove or disconnect the following:

   • Negative battery cable

➡Label the wires before disconnecting them.

   • Spark plug wires from the spark plugs
   • Positive Crankcase Ventilation (PCV) hose
   • Cylinder head cover. Replace the rubber seals if damaged or deteriorated.

3. Turn the engine to align the timing marks and set cylinder No.1 to TDC. The white mark on the crankshaft pulley should align with the pointer on the timing belt cover. The words **UP** embossed on the camshaft pulley should be aligned in the upward position. The marks on the edge of the pulley should be aligned with the cylinder head or the back cover upper edge.

4. Adjust the valves on cylinder No. 1 by performing the following:

   a. Insert a feeler gauge in between the camshaft lobe and the rocker arm.

➡The intake valve clearance specification is 0.009–0.011 in (0.24–0.28mm) for 2000–02 vehicles or 0.008–0.010 in. (0.21–0.25mm) for 2003 vehicles. The exhaust valve clearance specification is 0.011–0.013 in. (0.27–0.32mm).

   b. Loosen the locknut and turn the adjusting screw until the feeler gauge slides back and forth with a slight amount of drag.

   c. Tighten the locknut to 14 ft. lbs. (20 Nm) for intake or 10 ft. lbs. (14 Nm) for exhaust and recheck the valve clearance. Repeat the valve adjustment if necessary.

5. Rotate the crankshaft 180° counterclockwise (the camshaft pulleys will turn 90°) The **UP** arrow marks should be pointing to the exhaust side of the cylinder head.

6. Adjust the valves on cylinder No. 3 by performing the following:

   a. Insert a feeler gauge in between the camshaft lobe and the rocker arm.

   b. Loosen the locknut and turn the adjusting screw until the feeler gauge slides back and forth with a slight amount of drag.

   c. Tighten the locknut to 14 ft. lbs. (20 Nm) for intake or 10 ft. lbs. (14 Nm) for exhaust and recheck the valve clearance. Repeat the valve adjustment if necessary.

7. Rotate the crankshaft 180° counterclockwise (the camshaft pulleys will turn 90°) to bring No. 4 piston to TDC. The **UP** arrow marks should be pointing down, toward the crankshaft.

8. Adjust the valves on cylinder No. 4 by performing the following:

   a. Insert a feeler gauge in between the camshaft lobe and the rocker arm.

   b. Loosen the locknut and turn the adjusting screw until the feeler gauge slides back and forth with a slight amount of drag.

   c. Tighten the locknut to 14 ft. lbs. (20 Nm) for intake or 10 ft. lbs. (14 Nm) for exhaust and recheck the valve clearance. Repeat the valve adjustment if necessary.

9. Rotate the crankshaft 180° counterclockwise (the camshaft pulleys will turn 90°) to bring piston No. 2 to TDC. The **UP** arrow marks should be pointing to the intake side of the cylinder head.

10. Adjust the valves on cylinder No. 2 by performing the following:

   a. Insert a feeler gauge in between the camshaft lobe and the rocker arm.

   b. Loosen the locknut and turn the adjusting screw until the feeler gauge slides back and forth with a slight amount of drag.

   c. Tighten the locknut to 14 ft. lbs. (20 Nm) and recheck the valve clearance. Repeat the valve adjustment if necessary.

11. Install the cylinder head cover gasket cover to the groove of the cylinder head cover. Before installing the gasket, thoroughly clean the seal and the groove. Seat the recesses for the camshaft first, then

EXHAUST

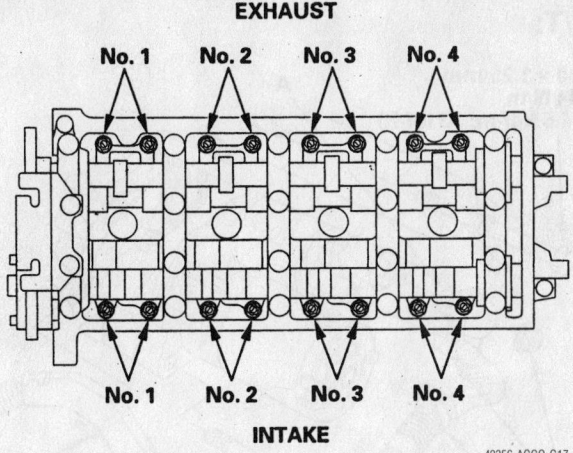

42356-ACCO-G17

Valve clearance adjusting screw locations

work it into the groove around the outside edges. Be sure the gasket is seated securely in the corners of the recesses.

12. Apply liquid gasket to the 4 corners of the recesses of the cylinder head cover gasket. Do not install the parts if 5 minutes or more have elapsed since applying liquid gasket. After assembly, wait at least 20 minutes before filling the engine with oil.

13. Install or connect the following:
- Cylinder head (valve) cover. Tighten the bolts attaching to 84 inch lbs. (10 Nm).
- Spark plug wires to the correct spark plugs.
- Positive, then the negative battery cable and enter the radio security code.

14. If equipped with 4WS, start the engine and turn the steering wheel lock-to-lock to reset the 4WS control unit.

S2000

➡The valve clearance is checked with the engine COLD.

1. Before servicing the vehicle, refer to the precautions at the beginning of this section.
2. Disconnect the negative battery cable.
3. Remove the valve covers
4. Rotate the crankshaft so that the camshaft lobe is directed away from the valve tappet to be measured.
5. Insert a feeler gauge under the camshaft lobe at a 90 degree angle to the camshaft. Clearance for the intake valves should be 0.008–0.010 inch (0.21–0.25mm). Clearance for the exhaust valves should be 0.010–0.011 inch (0.25–0.29mm).
6. If adjustment is necessary, loosen the locknut and turn the adjusting screw until the clearance is correct.

7. Tighten the locknut to 14 ft. lbs. (20 Nm) and recheck the valve clearance.
8. Repeat for each valve requiring adjustment.

Starter Motor

1.6L and 1.7L Engines

1. Before servicing the vehicle, refer to the precautions in the beginning of this section.

2. Remove or disconnect the following:
- Negative battery cable
- Air intake resonator
- Starter motor wiring harness
- Starter motor

To install:
3. Install or connect the following:
- Starter motor and tighten the bolts to 33 ft. lbs. (44 Nm)
- Starter motor wiring harness and tighten the battery cable terminal bolt to 84 inch lbs. (9 Nm)
- Negative battery cable

2.0L Engine

1. Before servicing the vehicle, refer to the precautions in the beginning of this section.
2. Remove or disconnect the following:
- Negative battery cable
- Accessory drive belt and tensioner
- Alternator
- Starter motor wiring harness
- Starter motor

To install:
3. Install or connect the following:
- Starter motor and tighten the bolts to 33 ft. lbs. (44 Nm)
- Starter motor wiring harness and tighten the battery cable terminal bolt to 84 inch lbs. (9 Nm)

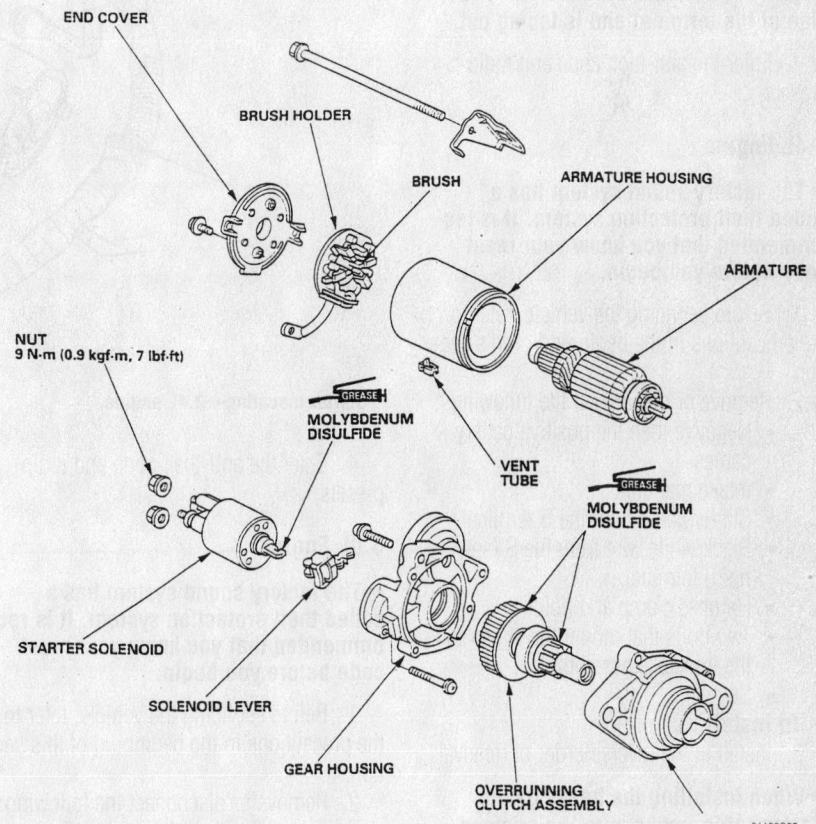

Exploded view of a typical Honda Starter

91182G25

- Alternator
- Accessory drive belt tensioner and tighten the bolts to 16 ft. lbs. (22 Nm)
- Accessory drive belt
- Negative battery cable

2.2L and 2.3L Engines

➡The factory sound system has a coded theft protection system. It is recommended that you know your reset code before you begin.

1. Before servicing the vehicle, refer to the precautions in the beginning of this section.
2. Remove or disconnect the following:
- Negative battery cable
- Wiring from harness
- Lower radiator hose from the bracket on the starter motor
- Starter cable from terminal B located on the back of the solenoid
- Black/white wire from the S (solenoid) terminal
- Two bolts that mount the starter to the transaxle assembly
- Starter

To install:
3. Install in the reverse order of removal.

➡When installing the heavy gauge starter cable, make sure the crimped side of the terminal end is facing out.

4. Enter the anti-theft code and radio presets.

2.4L Engine

➡The factory sound system has a coded theft protection system. It is recommended that you know your reset code before you begin.

1. Before servicing the vehicle, refer to the precautions in the beginning of this section.
2. Remove or disconnect the following:
- Negative then the positive battery cables
- Intake manifold
- Starter cable from the B terminal
- Black/white wire from the S (solenoid) terminal
- Harness clamp and holder
- Two bolts that mount the starter to the transaxle assembly
- Starter

To install:
3. Install in the reverse order of removal.

➡When installing the heavy gauge starter cable, make sure the crimped side of the terminal end is facing out.

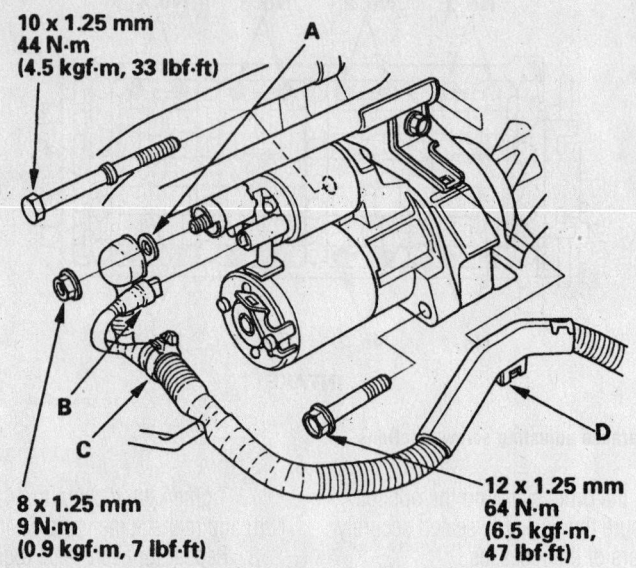

M/T:

10 x 1.25 mm
44 N·m
(4.5 kgf·m, 33 lbf·ft)

8 x 1.25 mm
9 N·m
(0.9 kgf·m, 7 lbf·ft)

12 x 1.25 mm
64 N·m
(6.5 kgf·m, 47 lbf·ft)

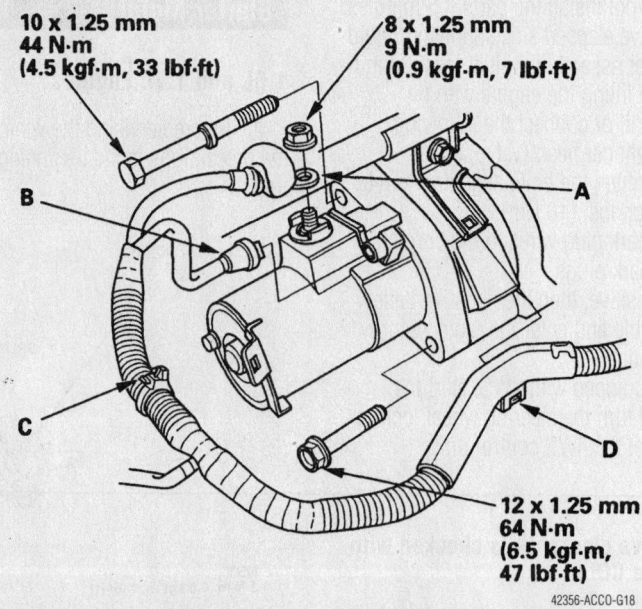

A/T:

10 x 1.25 mm
44 N·m
(4.5 kgf·m, 33 lbf·ft)

8 x 1.25 mm
9 N·m
(0.9 kgf·m, 7 lbf·ft)

12 x 1.25 mm
64 N·m
(6.5 kgf·m, 47 lbf·ft)

42356-ACCO-G18

Starter mounting—2.4L engine

4. Enter the anti-theft code and radio presets.

3.0L Engine

➡The factory sound system has a coded theft protection system. It is recommended that you know your reset code before you begin.

1. Before servicing the vehicle, refer to the precautions in the beginning of this section.
2. Remove or disconnect the following:
- Negative then the positive battery cables

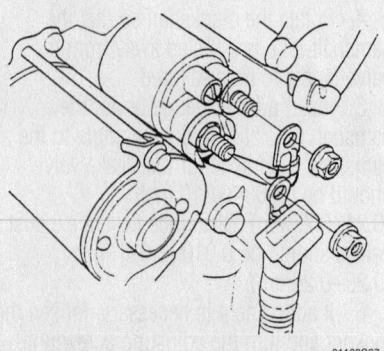

91182G27

Location of starter wiring—3.0L engine

- Automatic Transmission Fluid (ATF) cooler
- Starter cable from terminal B located on the back of the solenoid
- Black/white wire from the S (solenoid) terminal
- Two bolts that mount the starter to the transaxle assembly
- Starter

To install:

3. Install in the reverse order of removal.

➡**When installing the heavy gauge starter cable, make sure the crimped side of the terminal end is facing out.**

4. Enter the anti-theft code and radio presets.

Oil Pan

REMOVAL & INSTALLATION

➡**The radio may contain a coded theft protection circuit. Always obtain the code number before disconnecting the battery. If the vehicle is equipped with 4WS, the steering control unit is shut down when the battery is disconnected. After connecting the battery, turn the steering wheel lock-to-lock to reset the steering control unit.**

1.6L and 1.7L Engines

1. Before servicing the vehicle, refer to the precautions in the beginning of this section.
2. Disconnect the negative battery cable.
3. Raise and safely support the vehicle, then drain the oil.
4. Remove or disconnect the following:
- Lower splash panel
- Nuts and bolts connecting the exhaust pipe to the catalytic converter. Discard the gasket and the locknuts.
- Nuts attaching the exhaust pipe to the exhaust hanger
- Locknuts attaching the exhaust pipe to the exhaust manifold, then discard the nuts
- Remove the exhaust pipe from the vehicle. Discard the exhaust gaskets.
- Oil pan bolts in a crisscross pattern
- Oil pan. Lightly tap the corners of the oil pan with a rubber or plastic faced mallet. Clean off all the old gasket material.
5. Inspect the oil screen and pick-up tube for damaged and clogging. If the screen and tube are clogged with oil

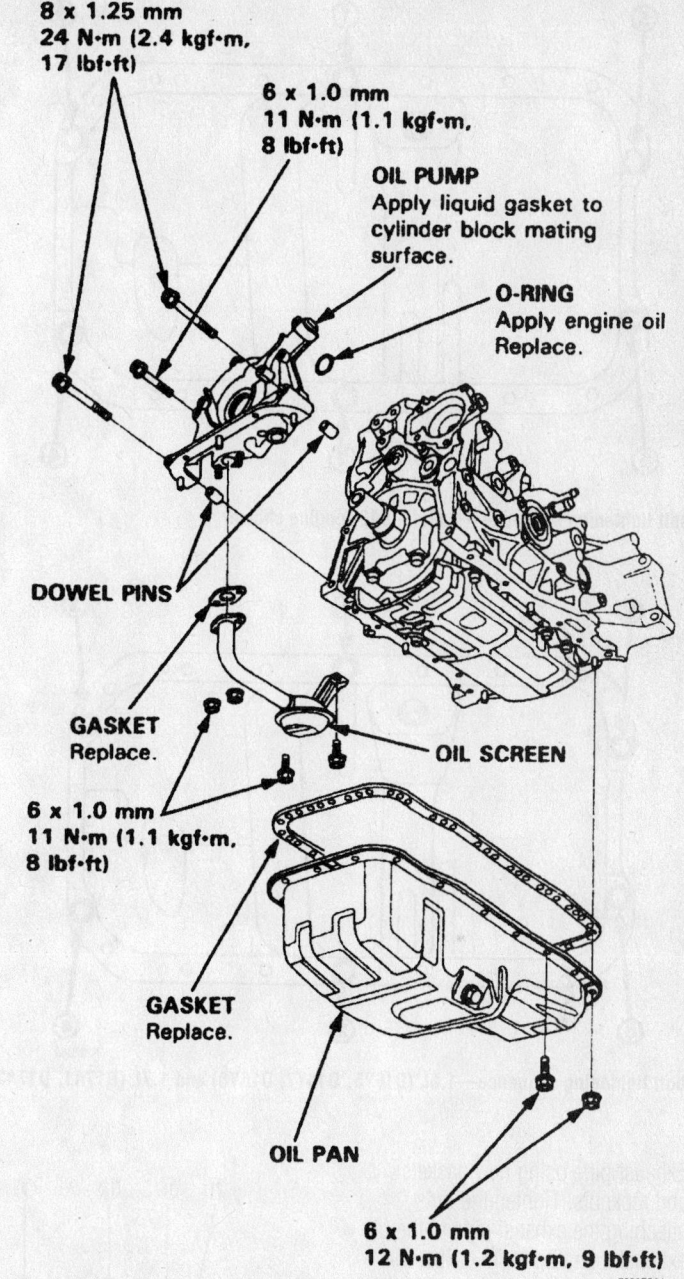

8 x 1.25 mm
24 N·m (2.4 kgf·m, 17 lbf·ft)

6 x 1.0 mm
11 N·m (1.1 kgf·m, 8 lbf·ft)

OIL PUMP
Apply liquid gasket to cylinder block mating surface.

O-RING
Apply engine oil Replace.

DOWEL PINS

GASKET
Replace.

OIL SCREEN

6 x 1.0 mm
11 N·m (1.1 kgf·m, 8 lbf·ft)

GASKET
Replace.

OIL PAN

6 x 1.0 mm
12 N·m (1.2 kgf·m, 9 lbf·ft)

7923FG81

Oil pan and oil screen—1.6L (B16A2) engine shown

residue, they should be thoroughly cleaned or replaced.

To install:

6. Install or connect the following:
- Oil screen and tube with a new gasket, if removed. Tighten the mounting nuts and bolts to 96 inch lbs. (11 Nm).
- Liquid gasket to the oil pan mating surface where the oil pump and the right side cover meet the engine block
- Oil pan gasket to the oil pan
- Oil pan
- Center and end mounting nuts and

bolts. Evenly hand-tighten the oil pan nuts and bolts.

7. Tighten the oil pan mounting nuts and bolts in a 3-step clockwise pattern starting with the center bolt next to the oil drain plug. The final torque value for the nuts and bolts is 10 ft. lbs. (14 Nm).

➡**Excessive tightening can cause distortion of the oil pan gasket and oil leakage.**

8. Install or connect the following:
- Oil drain plug with a new crush washer. Tighten the plug to 33 ft. lbs. (44 Nm).

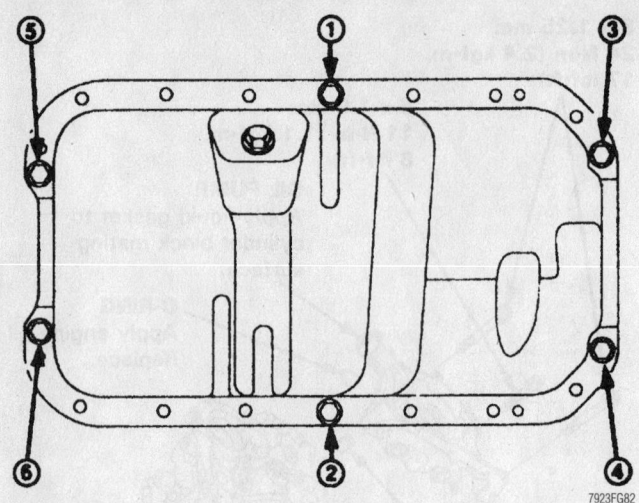

Oil pan bolt tightening sequence—1.6L (B16A2) engine shown

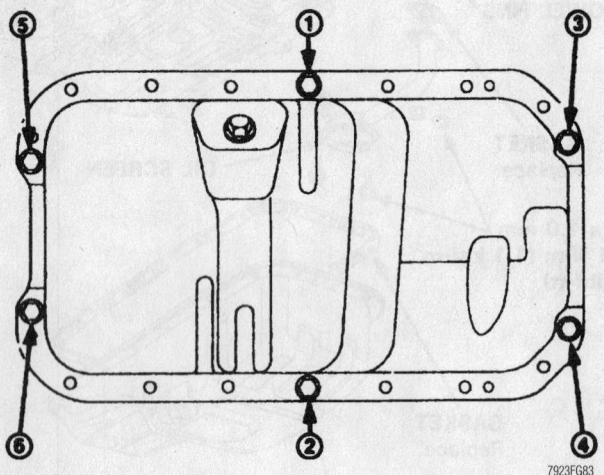

Oil pan bolt tightening sequence—1.6L (D16Y5, D16Y7, D16Y8) and 1.7L (D17A1, D17A2) engines

- Exhaust pipe using new gaskets and locknuts. Tighten the nuts attaching the exhaust pipe to the exhaust manifold to 40 ft. lbs. (54 Nm). Tighten the nuts attaching the exhaust pipe to the catalytic converter and the exhaust pipe hanger to 16 ft. lbs. (22 Nm).
- Lower splash panel
9. Lower the vehicle.
10. Refill the engine with clean oil.
11. Connect the negative battery cable and enter the radio security code.
12. Run the engine and check for leaks.
13. Turn off the engine and check the oil level. Top off the oil level if necessary.

2.0L Engine

1. Before servicing the vehicle, refer to the precautions in the beginning of this section.

2. Drain the engine oil.
3. Remove the oil pan bolts and the oil pan.
To install:
4. Remove old liquid gasket material from the oil pan mating surfaces and the bolt holes.
5. Clean and dry the oil pan mating surfaces.
6. Apply liquid gasket (PN 08718-0009) as shown.

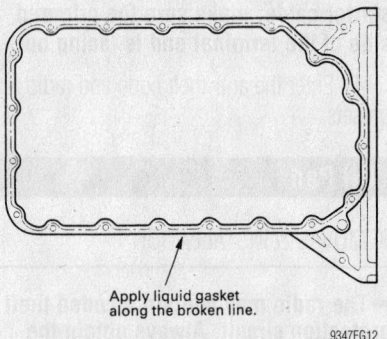

Apply liquid gasket along the broken line.

Apply liquid gasket along the broken line—2.0L engine

7. Install the oil pan. Tighten the bolts in sequence and in two or three steps to 104 inch lbs. (12 Nm).
8. Fill the crankcase to the correct level.
9. Start the engine and check for leaks.

2.2L and 2.3L Engines

1. Before servicing the vehicle, refer to the precautions in the beginning of this section.

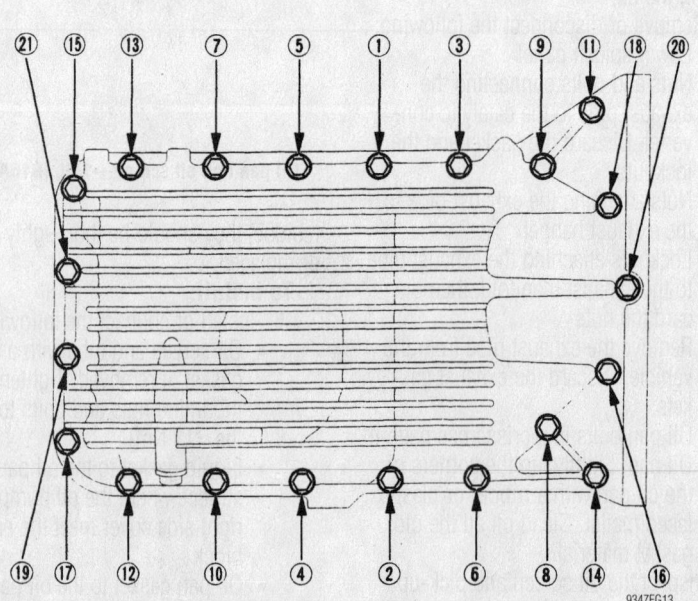

Oil pan torque sequence—2.0L engine

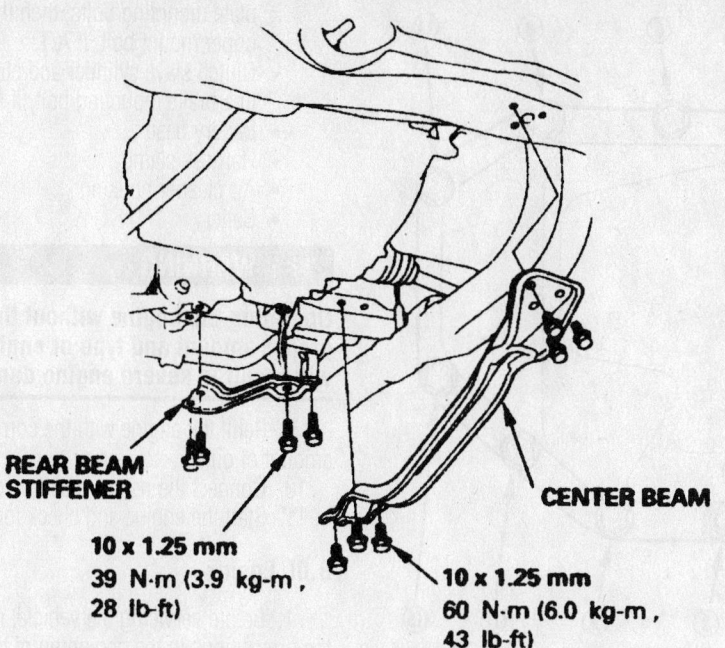

REAR BEAM
STIFFENER

10 x 1.25 mm
39 N·m (3.9 kg-m ,
28 lb-ft)

CENTER BEAM

10 x 1.25 mm
60 N·m (6.0 kg-m ,
43 lb-ft)

7923FG84

To gain access to the oil pan, remove the center beam—2.2L and 2.3L engines

2. Disconnect the negative battery cable.

3. Raise and safely support the vehicle.

4. Drain the engine oil into a sealable container. Install the drain bolt with a new gasket and tighten to 33 ft. lbs. (44 Nm).

5. Remove or disconnect the following:
- Front wheels and the splash shield
- Center beam
- Oxygen (O₂S) sensor electrical connector
- Bolts from the support bracket on the exhaust pipe
- Nuts attaching the exhaust pipe to the exhaust manifold and the catalytic converter
- Exhaust pipe and discard the gaskets
- Converter cover, if equipped with an automatic transaxle
- Clutch cover, if equipped with a manual transaxle
- Oil pan nuts and bolts (in a criss-cross pattern) and the oil pan; if necessary, use a mallet to tap the corners of the oil pan. DO NOT pry on the pan to get it loose

6. Clean the oil pan mounting surface of old gasket material and engine oil.

To install:

7. Install or connect the following:
- New oil pan gasket to the oil pan. Apply liquid gasket to the corners of the curved section of the gasket.
- Oil pan to the engine
- Oil pan nuts and bolts and tighten the nuts and bolts in sequence.

Tighten the nuts and bolts in 2 steps to 10 ft. lbs. (14 Nm).
- Torque converter cover or clutch cover, as applicable. Tighten the bolts to 108 inch lbs. (12 Nm).
- Exhaust pipe with new gaskets and new locknuts. Tighten the nuts attaching the exhaust pipe to the manifold to 40 ft. lbs. (54 Nm) and tighten the nuts attaching the exhaust pipe to the catalytic converter to 25 ft. lbs. (33 Nm).
- Bolts to the exhaust pipe support

bracket and tighten to 13 ft. lbs. (18 Nm)
- O₂S electrical connector
- Center beam and tighten the mounting bolts to 43 ft. lbs. (60 Nm)
- Splash shield and tighten the mounting bolts to 84 inch lbs. (10 Nm)
- Front wheels

8. Lower the vehicle and fill the engine with oil.

9. Connect the negative battery cable and enter the radio security code.

10. Start the engine and check for leaks.

11. If equipped with 4WS, turn the steering wheel lock-to-lock to reset the 4WS control unit.

2.4L Engine

1. Before servicing the vehicle, refer to the precautions in the beginning of this section.

2. Remove or disconnect the following:
- Negative battery cable
- Engine oil
- Battery
- Air cleaner housing
- Harness clamp
- Battery base
- Clutch slave cylinder and clutch line bracket mounting bolt, if manual transaxle
- Ground cable and the transaxle upper mount bracket assembly
- Front and rear mount stops and mount bolts
- Front tire and wheel assemblies
- Stabilizer links

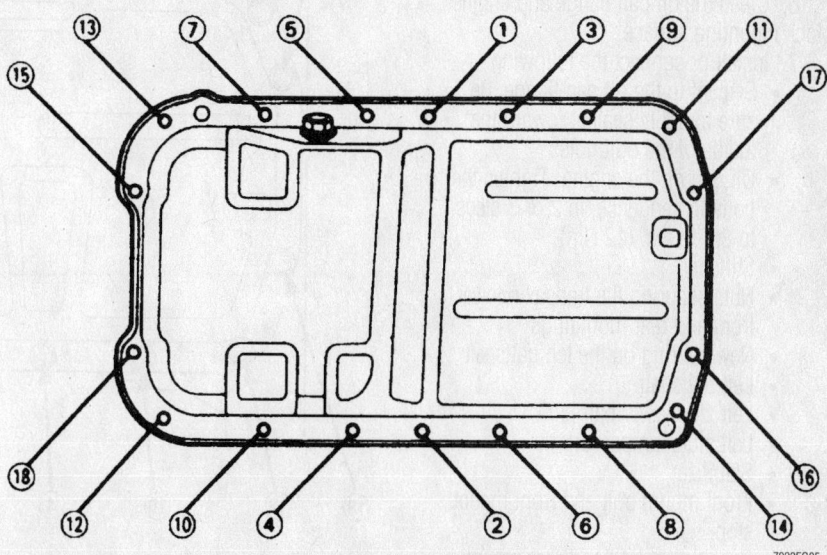

7923FG85

Oil pan mounting bolt tightening sequence—2.2L (H22A4) engines

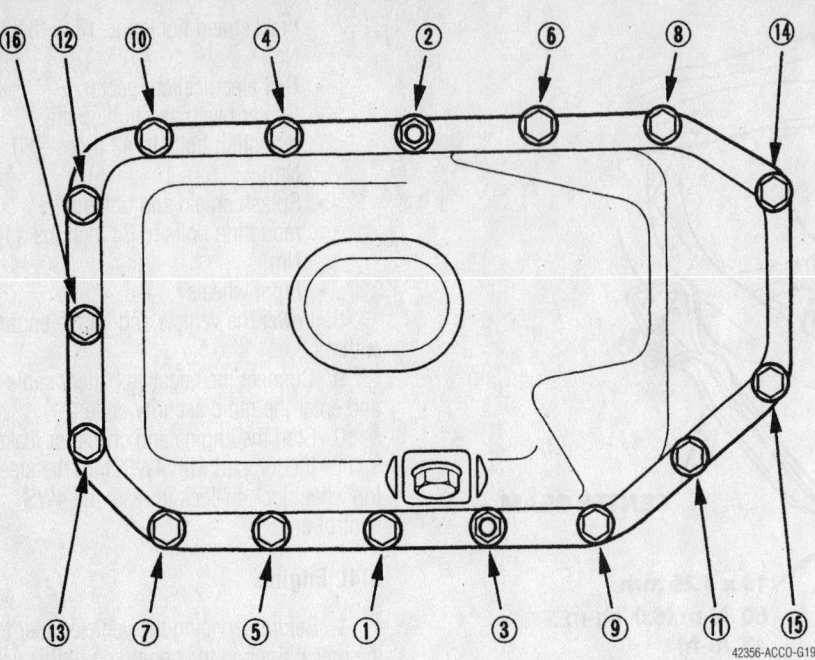

Oil pan bolt tightening sequence—2.4L engine

- Left side damper fork
- Left side lower ball joint
- Left side halfshaft. Coat all machined surfaces with clean engine oil and secure a plastic bag over the end of the halfshaft.
- Nuts securing the transaxle lower front and rear mounts

3. Use a suitable jack to lift the transaxle about 1.6—2.2 inches.
- Stiffener
- Oil pan bolts and nuts

4. Hammer a seal cutter between the engine block and oil pan to break the seal.

5. Remove the oil pan.

To install:

6. Clean the oil pan flange and engine block mounting surface.

7. Install or connect the following:
- Sealant to the oil pan flange. Be sure to apply sealant toward the inside of the bolt holes.
- Oil pan on the engine. Tighten the bolts in sequence, in 2 or 3 steps, to 8.7 ft. lbs. (12 Nm).
- Stiffener
- Nuts securing the transaxle lower front and rear mountings
- New set ring on the left halfshaft
- Left halfshaft
- Left side lower ball joint
- Left side damper fork
- Stabilizer
- Front mount bolt and front mount stop
- Rear mount bolt and rear mount stop

8. Loosen the mount bolt then install the transaxle upper mount/bracket assembly, clutch line clamp bracket (if M/T) and ground cable.
- Transaxle upper mount bolt if M/T
- Transaxle upper mount bracket

plate mounting bolts, then the upper mount bolt, if A/T
- Clutch slave cylinder and clutch line brake mounting bolt, if M/T
- Battery base
- Harness clamp
- Air cleaner housing
- Battery

✳✳ WARNING

Operating the engine without the proper amount and type of engine oil will result in severe engine damage.

9. Refill the engine with the correct amount of oil.

10. Connect the negative battery cable.

11. Start the engine and check for leaks.

3.0L Engine

1. Before servicing the vehicle, refer to the precautions in the beginning of this section.

2. Remove or disconnect the following:
- Negative battery cable

3. Raise and safely support the vehicle.

4. Remove or disconnect the following:
- Undercover

5. Drain the engine oil and replace the drain plug.

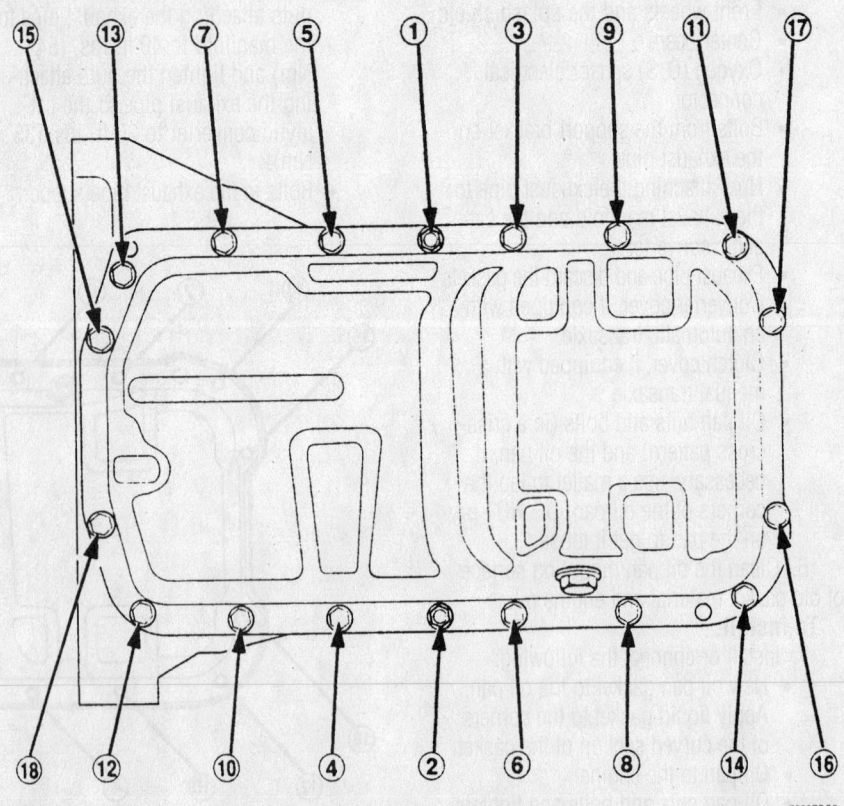

Oil pan mounting bolt tightening sequence—3.0L engine

6. Remove or disconnect the following:
   - Front exhaust pipe
   - Oil pan mounting bolts
7. Hammer a seal cutter between the engine block and oil pan to break the seal.
8. Remove the oil pan.

To install:

9. Clean the oil pan flange and engine block mounting surface.
10. Install or connect the following:
    - Sealant to the oil pan flange. Be sure to apply sealant toward the inside of the bolt holes.
    - Oil pan on the engine. Tighten the bolts in sequence to 10 ft. lbs. (14 Nm).
    - Exhaust pipe
    - Undercover
11. Lower the vehicle.

✳✳ WARNING

Operating the engine without the proper amount and type of engine oil will result in severe engine damage.

12. Refill the engine with the correct amount of oil.
13. Connect the negative battery cable.
14. Start the engine and check for leaks.

Oil Pump

REMOVAL & INSTALLATION

➡**The original radio may contain a coded anti-theft circuit. Always obtain the security code number before disconnecting the battery cables.**

1.6L (D16Y5, D16Y7, D16Y8) and 1.7l (D17A1, D17A2) Engines

1. Before servicing the vehicle, refer to the precautions in the beginning of this section.
2. Disconnect the negative battery cable.
3. Raise and safely support the vehicle.
4. Drain the engine oil.
5. Rotate the crankshaft to set the No. 1 cylinder to TDC for the compression stroke. The white TDC mark on the crankshaft pulley should align with the TDC pointers on the lower timing cover.

➡**Mark the direction of the timing belt rotation if it is to be reinstalled.**

6. Remove or disconnect the following:
   - Accessory drive belts and the crankshaft pulley
   - Valve cover and the upper and lower timing belt covers

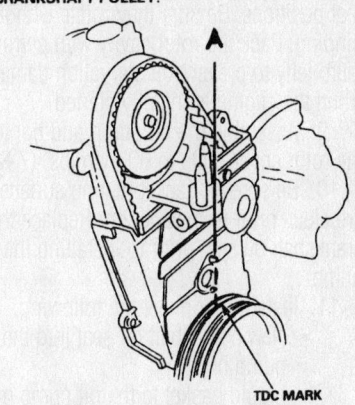

CRANKSHAFT PULLEY:

TDC MARK (WHITE)

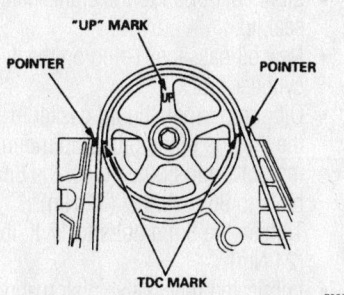

CAMSHAFT PULLEY:

"UP" MARK
POINTER   POINTER
TDC MARK

Crankshaft and camshaft TDC marks—1.6L (D16Y5, D16Y7, D16Y8) and 1.7L (D17A1, D17A2) engines

➡**Cover the rocker arm and shaft assemblies with a towel or sheet of plastic to keep out dirt and foreign objects.**

- Dipstick and its tube from the oil pump housing
- Timing belt
- Crankshaft Speed Fluctuation (CKF) sensor from the oil pump cover, and position it out of the way so that it will not come in contact with oil or become damaged.
- Crankshaft sprocket
- Oil pan
- Oil screen and pick-up tube from the oil pump housing and crankshaft buttress. If the screen and pick-up tube are blocked with oil residue, clean or replace them as necessary.
- Oil pump assembly

➡**If the rotors are to be reused, match-mark them with a felt-tipped marker for assembly.**

To install:

➡**Replace the rotors if they are worn or damaged. Use new O-rings when assembling and installing the oil pump.**

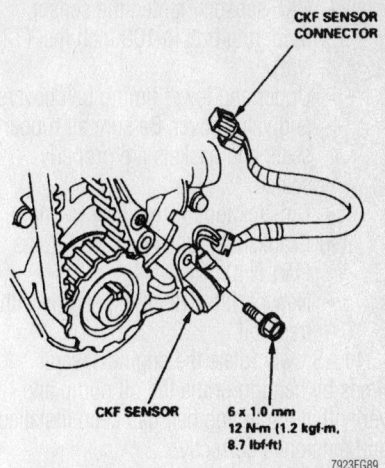

CKF SENSOR CONNECTOR

CKF SENSOR   6 x 1.0 mm
                12 N·m (1.2 kgf·m,
                8.7 lbf·ft)

CKF sensor location—1.6L (D16Y5, D16Y7, D16Y8) and 1.7L (D17A1, D17A2) engines

7. Install or connect the following:
   - Rotors back into their original positions. Be sure they move without binding. Pack the rotor cavity with petroleum jelly to prevent oil starvation damage when the engine is initially started.
8. Assemble the oil pump and tighten the rotor cover bolts to 60 inch lbs. (7 Nm).
9. Be sure all gasket mating surfaces are clean prior to installation.
10. Install or connect the following:
    - New crankshaft oil seal into the oil pump housing.
    - Liquid gasket to the cylinder block mating surface of the block.
    - Light coat of oil to the crankshaft seal lip.
    - New O-ring on the cylinder block
    - Oil pump
    - Liquid gasket to the threads of the oil pump mounting bolts and tighten them to 96 inch lbs. (11 Nm).
    - Lightly lubricated relief valve piston and spring. Tighten the sealing bolt (with a new crush washer) to 29 ft. lbs. (39 Nm).
    - Oil screen. Tighten the fastening nuts and bolts to 96 inch lbs. (11 Nm).
    - Oil pan. Tighten the oil pan nuts and bolts to 108 inch lbs. (12 Nm).
    - Crankshaft sprocket. The concave surface of the spacer must face the engine block.

➡**Verify that the engine is at Top Dead Center (TDC)/compression for the No. 1 cylinder.**

- Timing belt Tighten the tensioner adjusting bolt to 33 ft. lbs. (44 Nm).

- CKF sensor. Tighten the sensor mounting bolt to 108 inch lbs. (12 Nm).
- Upper and lower timing belt covers and valve cover. Be sure all rubber seals and gaskets are properly seated.
- Dipstick tube with a new O-ring.
- Crankshaft pulley bolt: 134 ft. lbs. (181 Nm).
- New oil filter. Refill the engine with fresh oil.

11. Slowly rotate the engine several times by hand to prime the oil pump and verify that the timing belt has been installed and tensioned correctly.

12. Install and adjust the accessory drive belts.

13. Connect the negative battery cable.

14. Run the engine and check for proper oil pressure.

15. Check for leaks. Top up the engine oil if necessary.

1.6L (B16A2) Engines

1. Before servicing the vehicle, refer to the precautions in the beginning of this section.

2. Disconnect the negative battery cable.

3. Raise and safely support the vehicle.

4. Drain the engine oil.

5. Remove or disconnect the following:
- Ignition wires

6. Set the No. 1 cylinder to TDC for the compression stroke. The mark on the crankshaft pulley should align with the index mark on the timing cover.

➡ **Mark the direction of the timing belt rotation if it is to be reinstalled.**

7. Remove or disconnect the following:
- Accessory drive belts
- Crankshaft pulley
- Valve cover and the upper and lower timing belt covers.

➡ **Cover the rocker arm and shaft assemblies with a towel or sheet of plastic to keep out dirt and foreign objects.**

- Timing belt
- Crankshaft Speed Fluctuation (CKF) sensor, if equipped
- Crankshaft sprocket
- Oil pan and oil screen
- Oil pump assembly

To install:

➡ **Replace the rotors if they are worn or damaged. Use new O-rings when assembling and installing the oil pump.**

8. Install the rotors back into their original positions. Be sure they rotate without binding. Pack the rotor cavity with petroleum jelly to prevent oil starvation damage when the engine is initially started.

9. Assemble the oil pump and tighten the rotor cover bolts to 60 inch lbs. (7 Nm).

10. Be sure all gasket mating surfaces are clean prior to installation. Replace the crankshaft oil seal prior to installing the oil pump.

11. Install or connect the following:
- New crankshaft oil seal into the oil pump housing
- Liquid gasket to the oil pump mating surface of the cylinder block.
- Light coat of oil to the crankshaft seal lip.
- New oil passage O-ring on the cylinder block
- Oil pump. Apply liquid gasket to the threads of the oil pump mounting bolts and tighten them the 6mm bolts to 96 inch lbs. (11 Nm). Tighten the 8mm bolts to 17 ft. lbs. (24 Nm).
- Lubricated relief valve piston and spring
- Sealing bolt (with a new crush washer) and tighten it to 29 ft. lbs. (39 Nm).
- Oil screen
- Oil pan. Wait for the sealant to cure before refilling the engine with oil.
- Crankshaft sprocket. The concave surface of the spacer faces out.

- Timing belt Tighten the tensioner adjusting bolt to 40 ft. lbs. (55 Nm).
- CKF sensor. Tighten the sensor mounting bolt to 108 inch lbs. (12 Nm).
- Upper and lower timing belt covers and valve cover. Be sure all rubber seals and gaskets are properly seated.
- Crankshaft pulley bolt: 130 ft. lbs. (180 Nm).
- New oil filter. Refill the engine with fresh oil.
- Accessory drive belts, and adjust the belt tension.

12. Connect the negative battery cable.

13. Run the engine and check for proper oil pressure.

14. Check for leaks. Top up the engine oil if necessary.

2.0L Engine

1. Before servicing the vehicle, refer to the precautions in the beginning of this section.

2. Drain the engine oil.

3. Remove or disconnect the following:
- Negative battery cable
- Oil pan
- Timing chain
- Oil pump chain guide and tensioner
- Baffle plate
- Oil pump, chain, and crankshaft sprocket

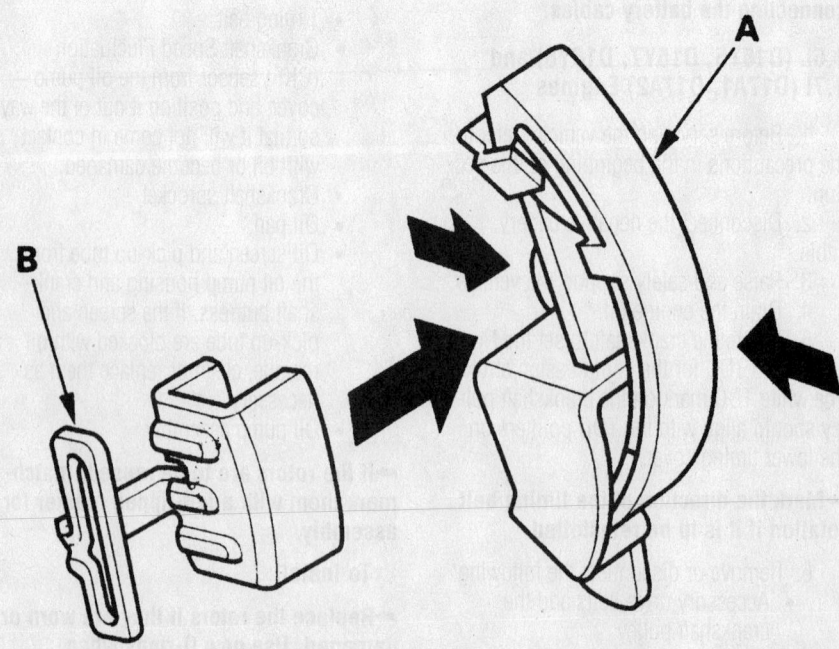

Compress the oil pump chain tensioner (A) and install the retaining clip (B)—2.0L engine

9347FG14

To install:

➡ **Use a new oil pump chain tensioner for assembly.**

4. Compress the oil pump chain tensioner and install the supplied retaining clip.

5. Install or connect the following:
- Oil pump, chain, and crankshaft sprocket. Tighten the 8mm bolts to 16 ft. lbs. (22 Nm) and the 6mm bolt to 104 inch lbs. (12 Nm).
- Baffle plate and tighten the bolts to 104 inch lbs. (12 Nm)
- Oil pump chain guide and tensioner. Tighten the 8mm bolt to 16 ft. lbs. (22 Nm) and the 6mm bolts to 104 inch lbs. (12 Nm).
- Timing chain
- Oil pan
- Negative battery cable

6. Fill the crankcase to the correct level.

7. Start the engine and check for leaks.

2.2L and 2.3L Engines

1. Before servicing the vehicle, refer to the precautions in the beginning of this section.

2. Disconnect the negative battery cable.

3. Drain the engine oil into a sealable container.

4. Turn the engine to align the timing marks and set cylinder No.1 to TDC. The white mark on the crankshaft pulley should align with the pointer on the timing belt cover.

5. Remove or disconnect the following:
- Valve cover and upper timing belt cover
- Power steering pump belt and the alternator belt, also the air conditioning belt if so equipped
- Crankshaft pulley and the lower timing belt cover
- Balancer belt and the timing belt. Be sure to mark the rotation of the timing belt if it is going to be reused.
- Timing belt and balancer belt tensioners
- Crankshaft Position (CKP) sensor, if equipped
- Timing belt drive pulley and key from the crankshaft
- Balancer driven pulley, by inserting a suitable tool into the maintenance hole in the front balancer shaft.

6. Align the rear timing balancer pulley using a 6 x 100mm bolt or rod. Mark the bolt or rod at a point 2.9 in. (74mm) from

the end. Remove the bolt from the maintenance hole on the side of the block; insert the bolt/rod into the hole. Align the 74mm mark with the face of the hole. This pin will hold the shaft in place.

7. Remove or disconnect the following:
- Balancer gear case and the dowel pins. Discard the O-ring.
- Balancer driven gear attaching bolt and the balancer driven gear
- Oil pan and the oil screen. Discard the screen gasket.
- Oil pump mounting bolts and oil pump assembly
- Dowel pins from the engine and clean the oil pump mating surfaces of old gasket material and oil. Discard the O-rings.

To install:

8. Install the 2 dowel pins and new O-rings to the cylinder block.

9. Be sure that the mating surfaces are clean and dry. Apply a liquid gasket evenly in a narrow bead, centered on the mating surface. Once the sealant is applied, do not wait longer than 20 minutes to install the parts; the sealant will become ineffective. After final assembly, wait at least 30 minutes before adding oil to the engine, giving the sealant time to set. To prevent leakage of oil, apply a suitable thread sealer to the inner threads of the bolt holes.

10. Install or connect the following:
- Oil pump to the engine block. Tighten the mounting bolts to 108 inch lbs. (12 Nm).
- Oil screen. Tighten the screen mounting bolts and nuts to 108 inch lbs. (12 Nm).
- Oil pan
- Balancer driven pulley to the front balancer belt, hold the balancer shaft in place with a suitable tool. Tighten the attaching bolt to 22 ft. lbs. (29 Nm).
- Balancer driven gear to the rear balancer shaft. Tighten the bolt to 18 ft. lbs. (25 Nm).

➡ **Before installing the balancer driven gear and the gear case, apply molybdenum disulfide (lithium grease) to the thrust surfaces of the balancer gears.**

11. Align the groove on the pulley edge to the pointer on the balancer gear case.

12. Install or connect the following:
- Balancer gear case to the engine and the mounting bolts and nut. The rear balancer shaft is being held in place with a 6 x 100mm

bolt. Tighten the mounting bolts and nut to 18 ft. lbs. (25 Nm).

13. Check the alignment of the pointer on the balancer pulley to the pointer on the oil pump.

14. Install or connect the following:
- Drive pulley to the crankshaft
- CKP sensor. Tighten the mounting bolts to 108 inch lbs. (12 Nm).
- Timing belt tensioners
- Timing belt and the balancer belt
- Crankshaft pulley and the lower timing belt cover
- Drive belts for the alternator, power steering, and air conditioning compressor. Tension the belts properly.
- Valve cover and upper timing belt cover

15. Refill the engine with clean, fresh oil.

16. Connect the negative battery cable and enter the radio security code.

2.4L Engine

1. Before servicing the vehicle, refer to the precautions in the beginning of this section.

2. Raise and safely support the vehicle.

3. Drain the engine oil.

4. Turn the crankshaft to position the No. 1 piston at Top Dead Center (TDC) on the compression stroke.

5. Remove or disconnect the following:
- Oil pan
- Oil pump chain tensioner, and discard

6. Secure the rear balancer shaft by inserting a 6mm pin into the maintenance hole in the lower balancer shaft holder and through the rear balancer shaft
- Oil pump sprocket mounting bolt
- Oil pump sprocket and oil pump

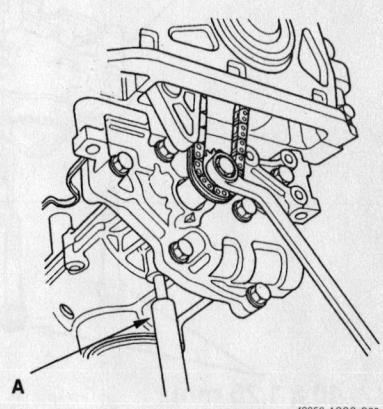

A

42356-ACC0-G20

Insert a 6mm pin into the maintenance hole in the lower balancer shaft holder and through the rear balancer shaft

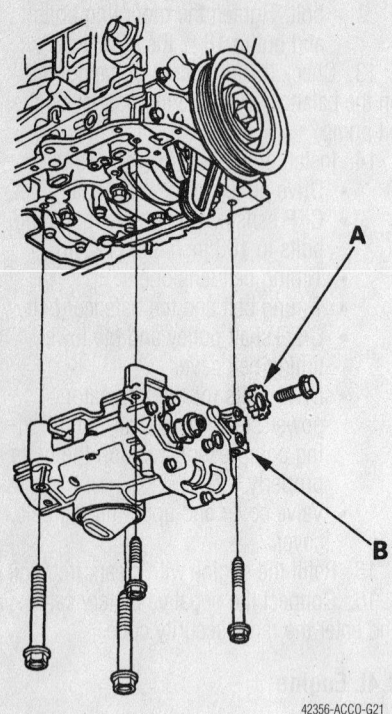

Exploded view of the oil pump sprocket (A) and oil pump (B)—2.4L engine

42356-ACCO-G21

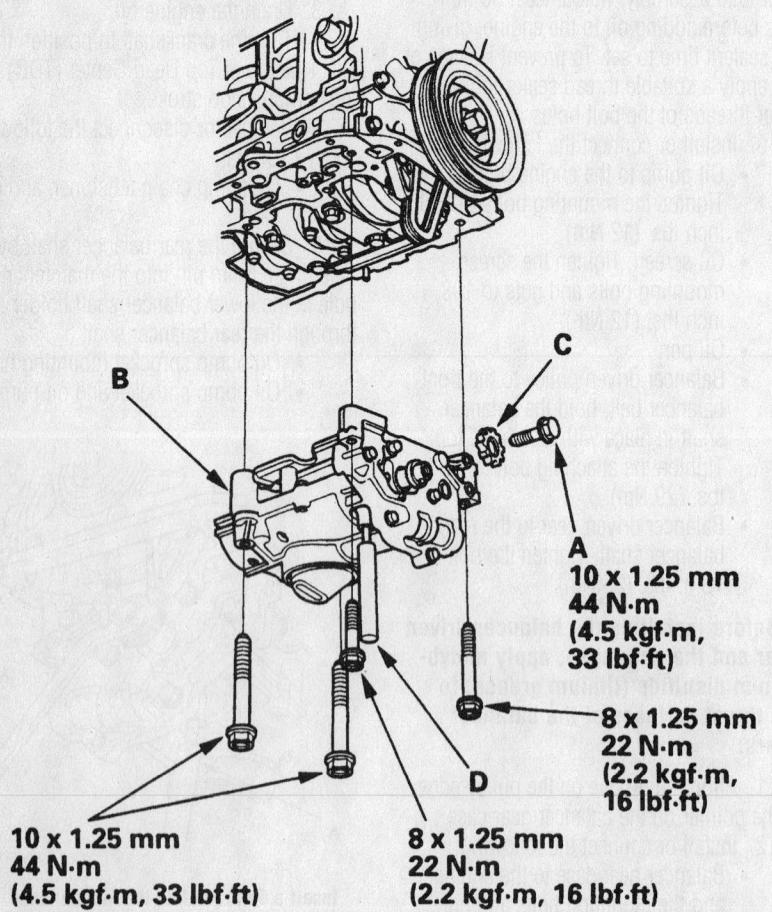

**10 x 1.25 mm
44 N·m
(4.5 kgf·m, 33 lbf·ft)**

**8 x 1.25 mm
22 N·m
(2.2 kgf·m, 16 lbf·ft)**

**A
10 x 1.25 mm
44 N·m
(4.5 kgf·m,
33 lbf·ft)**

**8 x 1.25 mm
22 N·m
(2.2 kgf·m,
16 lbf·ft)**

42356-ACCO-G22

Oil pump tightening specifications—2.4L engine

To install:

7. Make sure the No. 1 piston is still at TDC.

8. Align the dowel pin on the rear balancer shaft wit the mark on the oil pump

9. Secure the rear balancer shaft by inserting a 6mm pin into the maintenance hole in the lower balancer shaft holder and through the rear balancer shaft

10. Install or connect the following:
- Engine oil to the threads of the oil pump sprocket mounting bolt

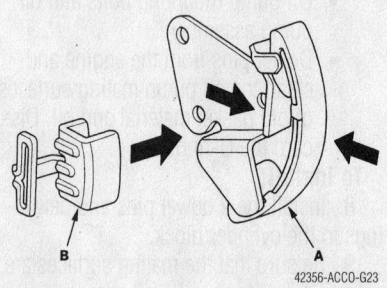

Squeeze a new oil pump chain tensioner (A) then install the set clip (B) on it—2.4L engine

42356-ACCO-G23

- Oil pump and pump sprocket, loosely. Remove the 6mm pin.
- Oil pump mounting bolts and tighten as shown in the illustration

11. Squeeze a new oil pump chain tensioner then install the set clip on it. The set clip is provided with the oil pump chain tensioner
- New oil pump chain tensioner and tighten the bolts to 8.7 ft. lbs. (12 Nm). Remove the set clip from the tensioner.
- Oil pan

12. Fill the crankcase with the proper amount of new engine oil.

3.0L Engine

1. Before servicing the vehicle, refer to the precautions in the beginning of this section.

2. Raise and safely support the vehicle.

3. Drain the engine oil.

4. Turn the crankshaft to position the No. 1 piston at Top Dead Center (TDC) on the compression stroke.

5. Remove or disconnect the following:
- Timing belt
- Idler pulley
- Crankshaft Position (CKP) sensor
- Variable Timing Electronic Control (VTEC) solenoid valve
- Oil filter
- Oil pan and pick-up
- Oil pump assembly

To install:

6. Install or connect the following:
- New crankshaft seal in the oil pump
- Sealant to the oil pump mounting surface and bolt holes on the engine block
- Grease to the lip of the new seal and engine oil to the o-ring
- Dowel pin and oil pump while aligning the inner rotor with the crankshaft. Tighten the bolts to 108 inch lbs. (12 Nm).
- Oil pump pick-up. Tighten the mounting bolts to 108 inch lbs. (12 Nm).
- Oil pan, VTEC solenoid, oil filter, CKP, and idler pulley
- Timing belt

✹✹ WARNING

Operating the engine without the proper amount and type of engine oil will result in severe engine damage.

7. Fill the crankcase with the proper amount of new engine oil.

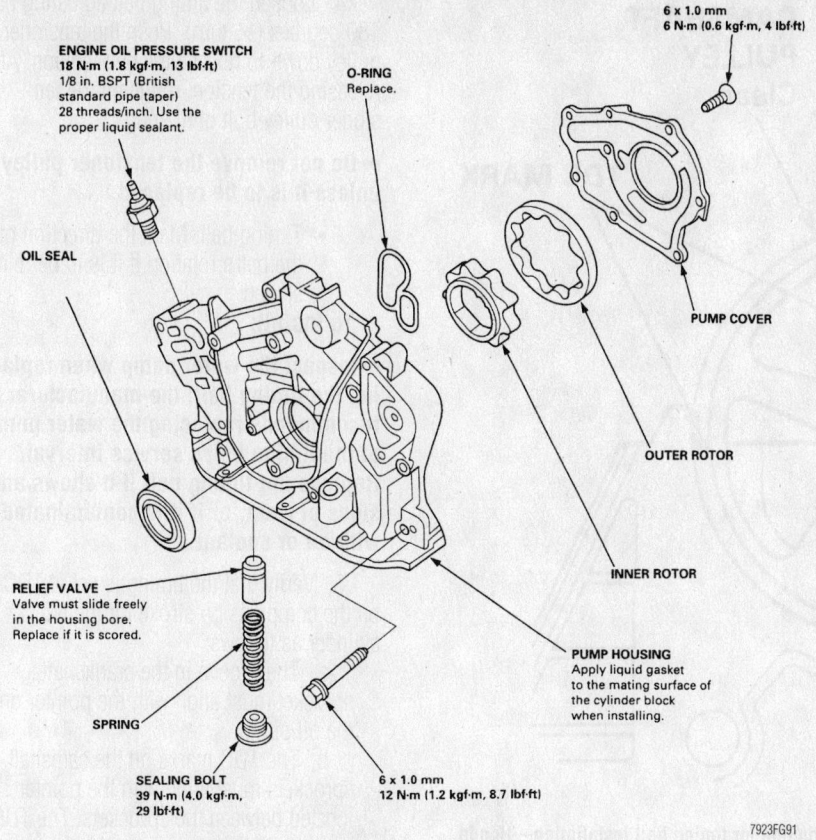

Exploded view of the oil pump—3.0L engine

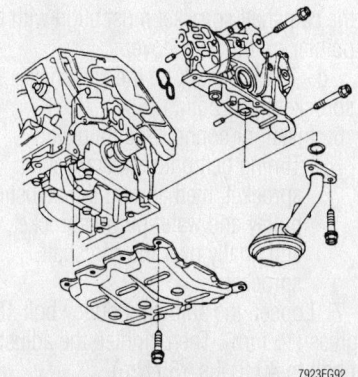

Oil pump mounting—3.0L engine

Rear Main Seal

REMOVAL & INSTALLATION

1. Before servicing the vehicle, refer to the precautions in the beginning of this section.
2. Remove the transmission.
3. Remove the driveplate from the crankshaft.
4. Carefully pry the crankshaft seal out of the retainer.

To install:

5. Apply clean engine oil to the lip of the new seal.

6. Install the seal onto the crankshaft and into the retainer using the appropriate seal driver.
7. Install the driveplate and the transmission.

Timing Belt, Sprockets, Front Cover and Seal

REMOVAL & INSTALLATION

1.6L and 1.7L Engines

1. Before servicing the vehicle, refer to the precautions in the beginning of this section.
2. Rotate the crankshaft to set the engine at Top Dead Center (TDC) on the compression stroke for the No. 1 piston. The white mark on the crankshaft pulley should align with the pointers on the timing cover. Once the engine is in this position, it must not be disturbed.
3. Remove or disconnect the following:
   - Necessary components for access to the cylinder head and timing belt covers. Cover the rocker arm and shaft assemblies with a towel or sheet of plastic to keep out dust and foreign objects.
   - Timing belt covers

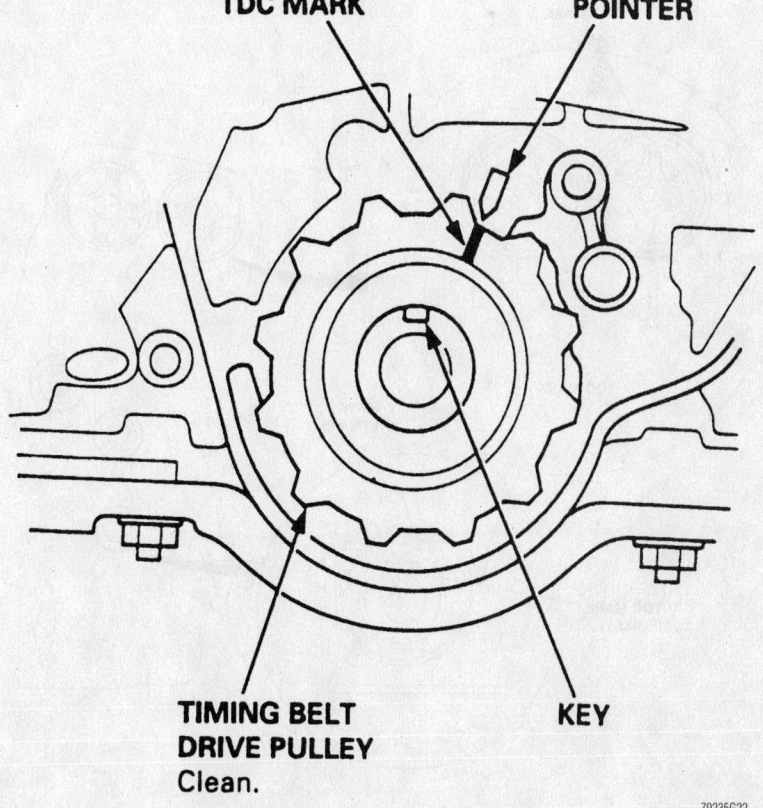

TDC alignment mark locations for the crankshaft sprocket—Honda 1.6L and 1.7L SOHC engines

CAMSHAFT PULLEY Clean.

"UP" MARK

TDC MARK

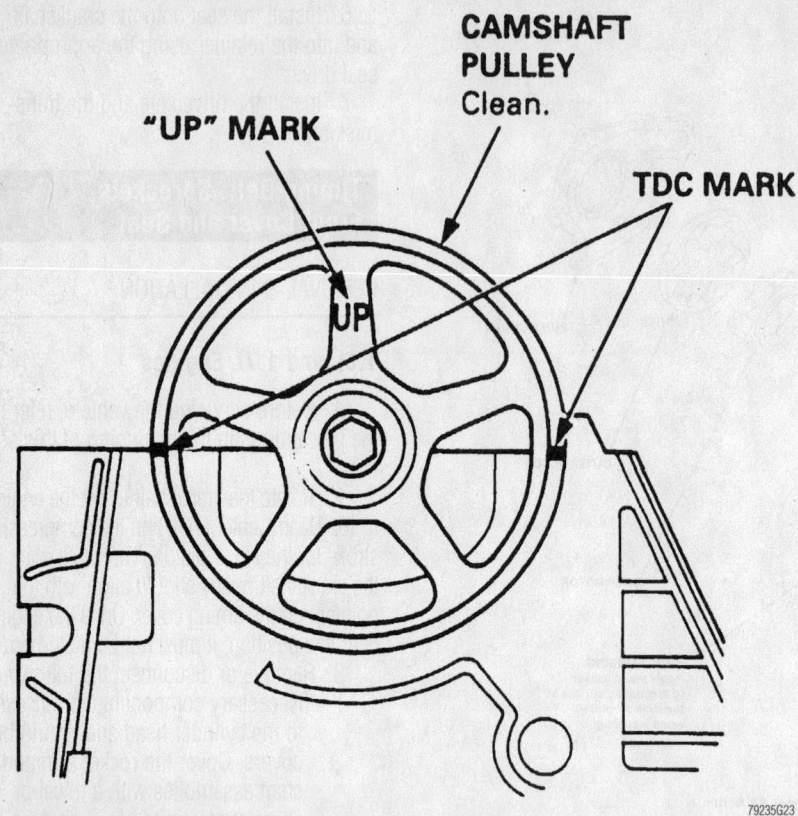

79235G23

Single camshaft timing belt sprocket TDC mark positioning for timing belt installation—Honda 1.6L and 1.7L SOHC engines

"UP" MARK

TDC MARK

▽ MARK (OIL PUMP)

TDC MARK (PULLEY)

KEY

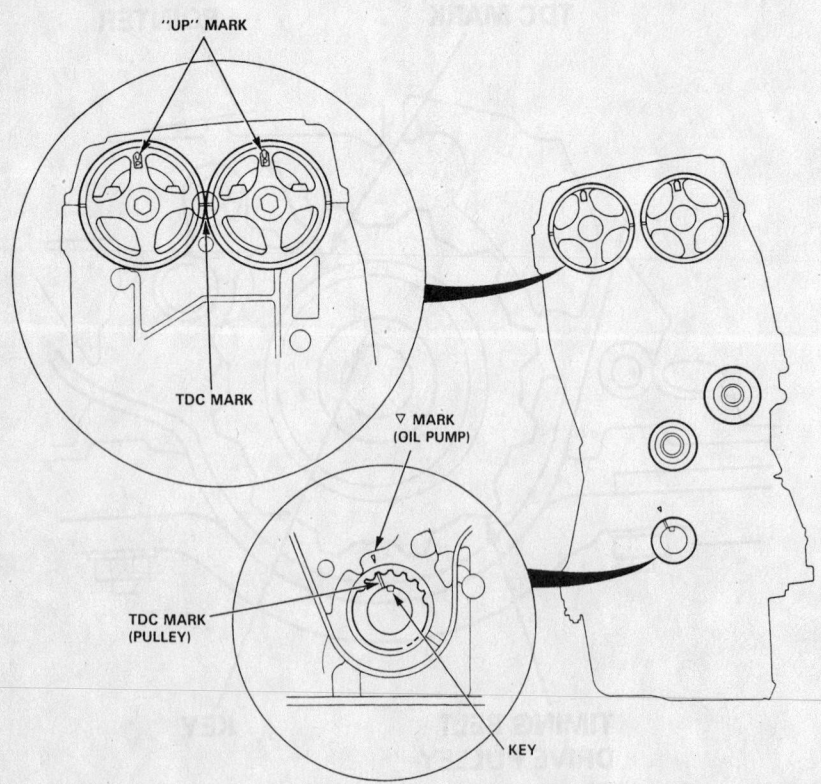

79235G32

Twin camshaft timing belt alignment marks—Honda 1.6L DOHC engines

4. Loosen the timing belt adjusting bolt 180 degrees (½ turn). Push the tensioner pulley down to release the belt tension. After releasing the tension, retighten the tensioner pulley bolt until snug.

➥**Do not remove the tensioner pulley unless it is to be replaced.**

• Timing belt. Mark the direction of the belt's rotation if it is to be reinstalled.

To install:

➥**Inspect the water pump when replacing the timing belt; the manufacturer recommends replacing the water pump at the timing belt's service interval. Replace the timing belt if it shows any signs of wear, or if it is contaminated with oil or coolant.**

5. Verify that the timing is set at TDC on the compression stroke for the No. 1 cylinder as follows:

a. The groove in the crankshaft sprocket must align with the pointer on the oil pump.

b. The TDC marks on the camshaft sprockets must align with the pointer located between the sprockets. The TDC marks will also be in line with the upper surface of the head.

c. On other engines, the TDC mark on the camshaft sprocket must align with the pointer on the back cover.

d. The **UP** mark on the camshaft sprocket must point up.

6. Install or connect the following:

• Timing belt onto the crankshaft sprocket, then around the adjusting pulley and water pump sprocket, and finally over the camshaft sprocket.

7. Loosen the adjusting pulley bolt 180 degrees (½ turn). Then, tighten the adjusting bolt to 40 ft. lbs. (55 Nm).

• Lower timing belt cover and the crankshaft pulley. Apply a light coat of fresh oil to the pulley bolt threads, then tighten it to 134 ft. lbs. (181 Nm).

8. Rotate the crankshaft 5–6 turns counterclockwise to position the belt on the sprockets.

9. Adjust the timing belt tension, as follows:

a. Set the No. 1 piston at TDC on the compression stroke for the No. 1 cylinder.

b. Loosen the adjusting pulley bolt 180 degrees (½ turn).

c. Rotate the crankshaft counterclockwise so that the camshaft sprocket moves 3 teeth from the TDC/compression position.

d. Tighten the adjusting bolt to 33 ft. lbs. (45 Nm).

e. Tighten the crankshaft pulley to 134 ft. lbs. (181 Nm).

10. Verify that the crankshaft and camshaft sprockets will align properly at the TDC/compression position. If the camshaft pulley is not at TDC/compression, remove the timing belt, adjust the sprocket positions and reinstall the belt.

11. Install the upper timing and cylinder head covers, and all other applicable components. When reattaching the side engine mount, tighten the support nuts to 54 ft. lbs. (75 Nm).

2.2L and 2.3L Engines

1. Before servicing the vehicle, refer to the precautions in the beginning of this section.

2. Remove the cylinder head (valve) and upper timing belt covers.

3. Turn the engine to align the timing marks and set cylinder No. 1 to Top Dead Center (TDC). The white mark on the crankshaft sprocket should align with the pointer on the timing belt cover. The words **UP** embossed on the camshaft sprocket should be aligned in the upward position. The marks on the edge of the sprocket should be aligned with the cylinder head or the back cover upper edge. Once in this position, the engine must NOT be turned or disturbed.

4. Remove all necessary components for access to the lower timing belt cover, then remove the cover.

5. There are two belts in this system; the one running to the camshaft sprocket is the timing belt. The other, shorter one drives the balance shaft and is referred to as the balancer shaft belt or timing balancer belt. Lock the timing belt adjuster in position, by installing one of the lower timing belt cover bolts to the adjuster arm.

6. Loosen the timing belt and balancer shafts tensioner adjuster nut, do not loosen the nut more than one turn. Push the tensioner for the balancer belt away from the belt to relieve the tension. Hold the tensioner and tighten the adjusting nut to hold the tensioner in place.

7. Carefully remove the balancer belt. Do not crimp or bend the belt; protect it from contact with oil or coolant.

8. Remove the balancer belt sprocket from the crankshaft.

9. Loosen the lockbolt installed to the timing belt adjuster and loosen the adjusting nut. Push the timing belt adjuster to remove the tension on the timing belt, then tighten the adjuster nut.

10. Remove the timing belt by sliding it off the sprockets. Do not crimp or bend the belt; protect it from contact with oil or coolant.

11. If defective, remove the belt tensioners by performing the following:

a. Remove the springs from the balancer belt and the timing belt tensioners.

b. Remove the adjusting nut from the belt tensioners.

c. Remove the bolt from the balancer belt adjuster lever, then remove the lever and the tensioner pulley.

d. Remove the lockbolt from the timing belt tensioner lever, then remove the tensioner pulley and lever from the engine.

12. This is an excellent time to check or replace the water pump. Even if the timing belt is only being replaced as part of a good

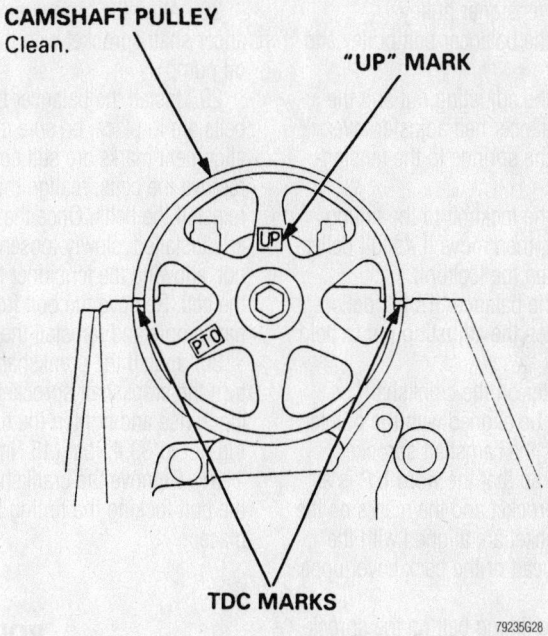

Position the camshaft sprocket as indicated for timing belt installation—Honda 2.2L and 2.3L engines

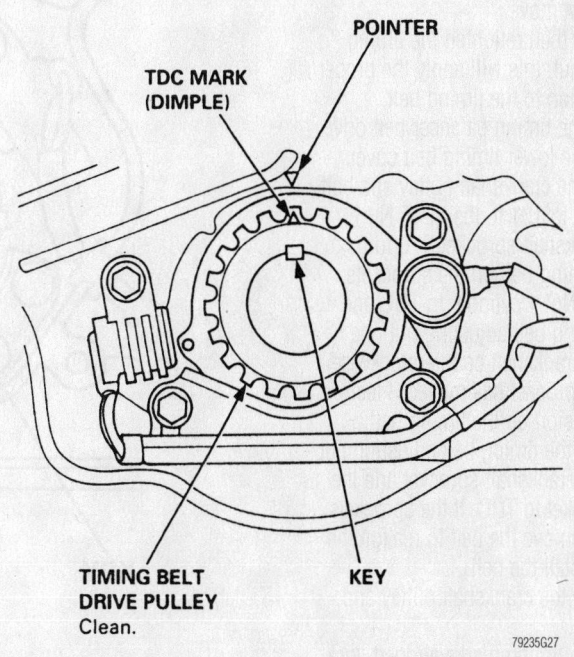

Before installing the timing belt, ensure the crankshaft sprocket marks are properly aligned—Honda 2.2L and 2.3L engines

maintenance schedule, consider replacing the pump at the same time.

To install:

13. If the water pump is to be replaced, install a new O-ring and make certain it is properly seated. Install the water pump and tighten the mounting bolts to 106 inch lbs. (12 Nm).

14. If the tensioners were removed, perform the following procedures:

a. Install the timing belt tensioner lever and the tensioner pulley.

b. Install the balancer belt pulley and adjuster lever.

c. Install the adjusting nut and the bolt to the balancer belt adjuster lever.

d. Install the springs to the tensioners.

e. Install the lockbolt to the timing belt tensioner, then move it its full deflection and tighten the lockbolt.

f. Move the balancer it's full deflection and tighten the adjusting nut to hold its position.

15. The pointer on the crankshaft sprocket should be aligned with the pointer on the oil pump; the camshaft sprocket must be aligned so that the word **UP** is at the top of the sprocket and the marks on the edge of the sprocket are aligned with the surfaces of the head or the back cover upper edge.

16. Install the timing belt on the sprockets in the following sequence: crankshaft sprocket, tensioner sprocket, water pump sprocket and camshaft sprocket.

17. Check the timing marks to be sure that they did not move.

18. Loosen, then retighten the timing belt adjusting nut; this will apply the proper amount of tension to the timing belt.

19. Install the timing balancer belt drive sprocket and the lower timing belt cover.

20. Install the crankshaft pulley and bolt, tighten the bolt to 181 ft. lbs. (245 Nm). Rotate the crankshaft sprocket 5–6 turns to position the timing belt on the sprockets.

21. Set the No. 1 cylinder to TDC and loosen the timing belt adjusting nut one turn. Turn the crankshaft counterclockwise until the cam sprocket has moved 3 teeth; this creates tension on the timing belt.

22. Tighten the timing belt adjusting nut.

23. Set the crankshaft sprocket and the camshaft sprocket to TDC. If the sprockets do not align, remove the belt to realign the marks, then install the belt.

24. Remove the crankshaft pulley and the lower cover.

25. With the timing marks aligned, lock the timing belt adjuster in place with one of the lower cover mounting bolts.

26. Loosen the adjusting nut and ensure the timing balancer belt adjuster moves freely.

27. Align the rear timing balancer sprocket using a 6 x 100mm bolt or rod. Mark the bolt or rod at a point 2.9 in. (74mm) from the end. Remove the bolt from the maintenance hole on the side of the block; insert the bolt/rod into the hole and align the 2.9 in. (74mm) mark with the face of the hole. This will hold the shaft in place during installation.

28. Align the groove on the front balancer shaft sprocket with the pointer on the oil pump.

29. Install the balancer belt. Once the belts are in place, be sure that all the engine alignment marks are still correct. If not, remove the belts, realign the engine and reinstall the belts. Once the belts are properly installed, slowly loosen the adjusting nut, allowing the tensioner to move against the belt. Remove the bolt from the maintenance hole and reinstall the bolt and washer.

30. Install the crankshaft pulley, then turn the crankshaft sprocket 1 turn counterclockwise and tighten the timing belt adjusting nut to 33 ft. lbs. (45 Nm).

31. Remove the crankshaft pulley and the bolt locking the timing belt adjuster in place.

32. Install the lower and upper timing belt covers, and all applicable components. When installing the crankshaft pulley, coat the threads and seating face of the pulley bolt with engine oil, then install and tighten the bolt to 181 ft. lbs. (250 Nm).

33. Install the cylinder head cover gasket cover to the groove of the cylinder head cover. Before installing the gasket thoroughly clean the seal and the groove. Seat the recesses for the camshaft first, then work it into the groove around the outside edges. Be sure the gasket is seated securely in the corners of the recesses.

34. Apply liquid gasket to the four corners of the recesses of the cylinder head cover gasket. Do not install the parts if 5 minutes or more have elapsed since applying liquid gasket. After assembly, wait at least 20 minutes before filling the engine with oil.

35. Install the cylinder head (valve) cover and all other applicable components.

3.0L Engines

1. Before servicing the vehicle, refer to the precautions in the beginning of this section.

2. Turn the engine to align the timing marks and set cylinder No. 1 to Top Dead

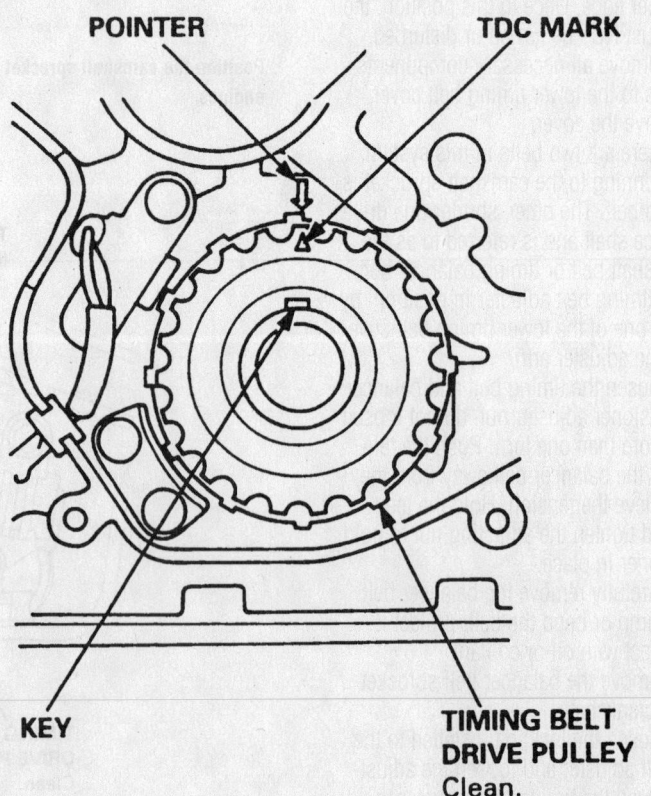

POINTER **TDC MARK**

KEY **TIMING BELT DRIVE PULLEY**
Clean.

Crankshaft timing belt sprocket alignment mark locations—Honda 3.0L engine

FRONT:

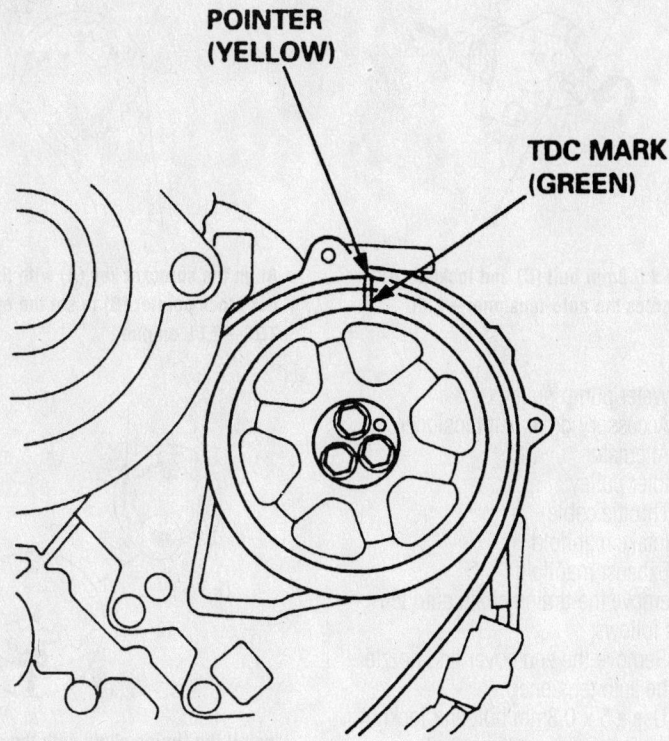

Left camshaft timing belt sprocket alignment mark location—Honda 3.0L engine

REAR:

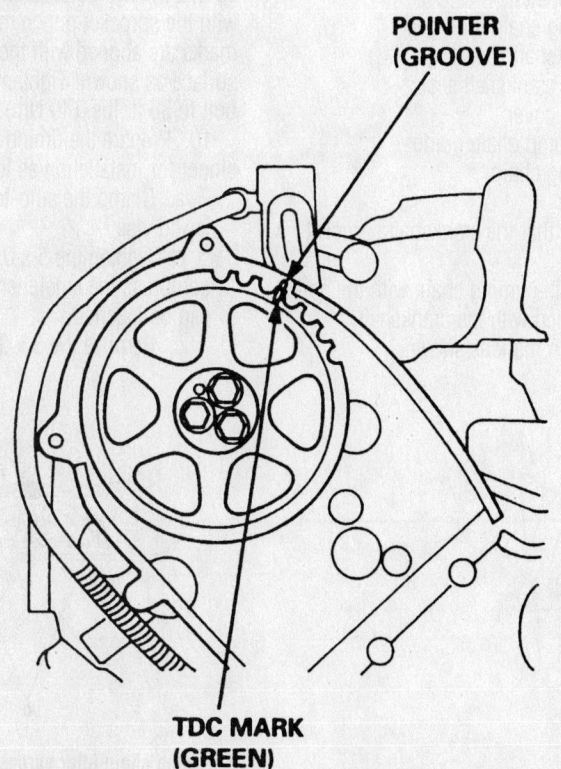

Rear camshaft timing belt sprocket alignment mark location—Honda 3.0L engine

Center (TDC). The white mark on the crankshaft pulley should align with the pointer on the timing belt cover. Remove the inspection caps on the upper timing belt covers to check the alignment of the timing marks. The pointers for the camshafts should align with the green marks on the camshaft sprockets.

3. Remove all necessary components for access to the timing belt covers, then remove the covers.

➡ **Do not use the covers to store removed items.**

4. Loosen the timing belt adjuster bolt 180 degrees (½ turn). Push the tensioner to remove the tension from the timing belt, then retighten the adjusting bolt.

5. Remove the timing belt. Do not crimp or bend the belt; protect it from contact with oil or coolant. Slide the belt off the sprockets.

6. Remove the bolts attaching the camshaft sprockets to the camshafts, then remove the sprockets.

7. If the timing belt tensioner is defective, remove the spring from the timing belt tensioner. Remove the tensioner pulley adjusting bolt and the adjuster assembly from the engine.

➡ **This is an excellent time to check or replace the water pump. Even if the timing belt is only being replaced as part of a good maintenance schedule, consider replacing the pump at the same time.**

To install:

8. If the water pump is to be replaced, install a new O-ring and make certain it is properly seated. Install the water pump and retaining bolts. Tighten the mounting bolts to 16 ft. lbs. (22 Nm).

9. If removed, install the tensioner pulley and the adjusting bolt, be sure the tensioner is properly positioned on its pivot pin. Install the spring to the tensioner, then push the tensioner to its full deflection and tighten the adjusting bolt.

10. Set the timing belt drive sprocket so that the No. 1 piston is at TDC. Align the TDC mark on the tooth of the timing belt drive sprocket with the pointer on the oil pump.

11. Set the camshaft sprockets so that the No. 1 piston is at TDC. Align the TDC marks (green mark) on the camshaft sprockets to the pointers on the back covers.

12. Install the timing belt onto the sprockets in the following sequence: crankshaft sprocket, tensioner pulley, front camshaft sprocket, water pump pulley and rear camshaft sprocket.

13. Loosen, then retighten the timing

belt adjuster bolt to tension the timing belt.

14. Install the lower timing belt cover.

15. Install the crankshaft sprocket and the crankshaft pulley bolt. Tighten the bolt to 181 ft. lbs. (245 Nm) with the aid of the crank pulley holder.

16. Rotate the crankshaft five or six turns clockwise so that the timing belt positions on the sprockets.

17. Set cylinder No. 1 to TDC by aligning the timing marks. If the timing marks do not align, remove the timing belt, then adjust the components and reinstall the timing belt.

18. Loosen the timing belt adjusting bolt 180 degrees (½ turn) and retighten the adjusting bolt. Tighten the adjusting bolt to 31 ft. lbs. (42 Nm).

19. Install the upper timing belt cover and all other applicable components. When installing the side engine mount to the engine, use 3 new attaching bolts. Tighten the new bolts to 40 ft. lbs. (54 Nm).

Timing Chain, Sprockets, Front Cover and Seal

REMOVAL & INSTALLATION

2.0L Engine

1. Before servicing the vehicle, refer to the precautions in the beginning of this section.

2. Set the engine to Top Dead Center (TDC).

3. Drain the cooling system.

4. Relieve the fuel system pressure.

5. Remove or disconnect the following:
- Negative battery cable
- Air cleaner housing
- Vacuum tank
- Accessory drive belt
- Water bypass hose and tube

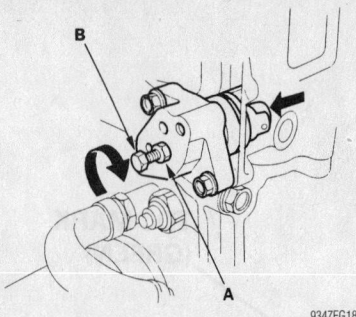

9347FG18

Use a 5 x 0.8mm bolt (B) and locknut (A) to compress the auto-tensioner—2.0L engine

- Water pump pulley
- Accessory drive belt tensioner
- Alternator
- Idler pulley
- Throttle cable
- Intake manifold
- Exhaust manifold

6. Remove the timing chain auto tensioner as follows:

a. Remove the end cover and nozzle from the auto tensioner.

b. Use a 5 x 0.8mm bolt and locknut as shown to compress the auto-tensioner.

c. Remove the timing chain auto-tensioner.

- Valve cover
- Camshafts
- Timing chain idler gear
- Crankshaft sprocket
- Front crankshaft seal
- Front cover
- Oil pump chain guide
- Timing chain

To install:

7. Ensure that the crankshaft sprocket is set to TDC.

8. Install the timing chain with the colored link aligned with the crankshaft sprocket punch mark as shown.

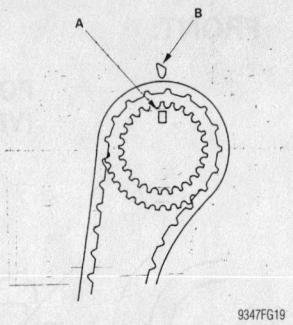

9347FG19

Align the sprocket key (A) with the cylinder block pointer (B) to set the engine to TDC—2.0L engine

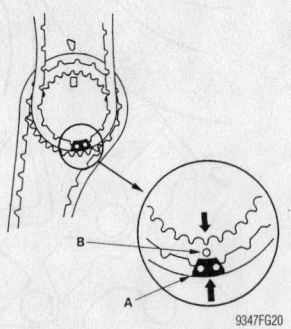

9347FG20

Install the timing chain with the colored link (A) aligned with the crankshaft sprocket punch mark (B)—2.0L engine

9. Install the timing chain idler sprocket so that the two colored links are aligned with the sprocket punch mark, and the TDC marks are aligned with the cylinder head surface as shown. Tighten the idler sprocket bolt to 36 ft. lbs. (49 Nm).

10. Prepare the timing chain auto-tensioner for installation as follows:

a. Clamp the auto-tensioner in a soft-jawed vise.

b. Tighten the 5 x 0.8mm bolt to compress the tensioner until a set pin can be inserted.

c. Remove the 5 x 0.8mm bolt and

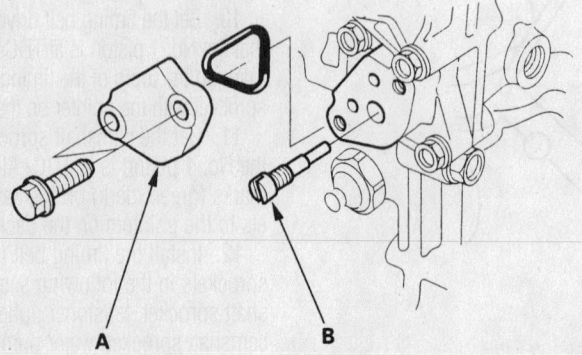

9347FG17

Timing chain auto-tensioner cover (A) and nozzle (B)—2.0L engine

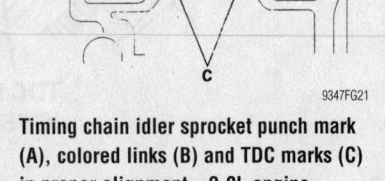

9347FG21

Timing chain idler sprocket punch mark (A), colored links (B) and TDC marks (C) in proper alignment—2.0L engine

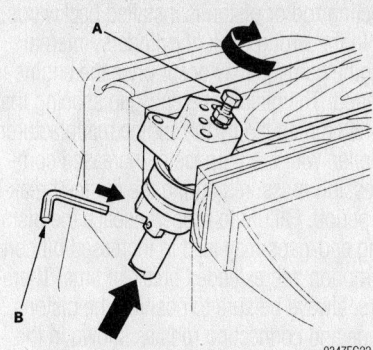

9347FG22

Tighten the 5 x 0.8mm bolt (A) and insert the set pin (B)—2.0L engine

install the nozzle and cover. Tighten the nozzle to 48 inch lbs. (5 Nm) and the cover bolts to 104 inch lbs. (12 Nm).

d. Install the auto-tensioner with new O ring seals and tighten the bolts to 104 inch lbs. (12 Nm).

e. Remove the service bolt from the cylinder head and remove the set pin.

f. Replace the service bolt and tighten it to 22 ft. lbs. (29 Nm).

11. The remainder of the installation is the reverse of the removal procedure, while using the following torque values:

- Oil pump chain guide bolts: 104 inch lbs. (12 Nm)
- Front cover: 10mm bolts to 33 ft. lbs. (44 Nm) and the 6mm bolts to 104 inch lbs. (12 Nm)
- Crankshaft sprocket bolt: 181 ft. lbs. (245 Nm)

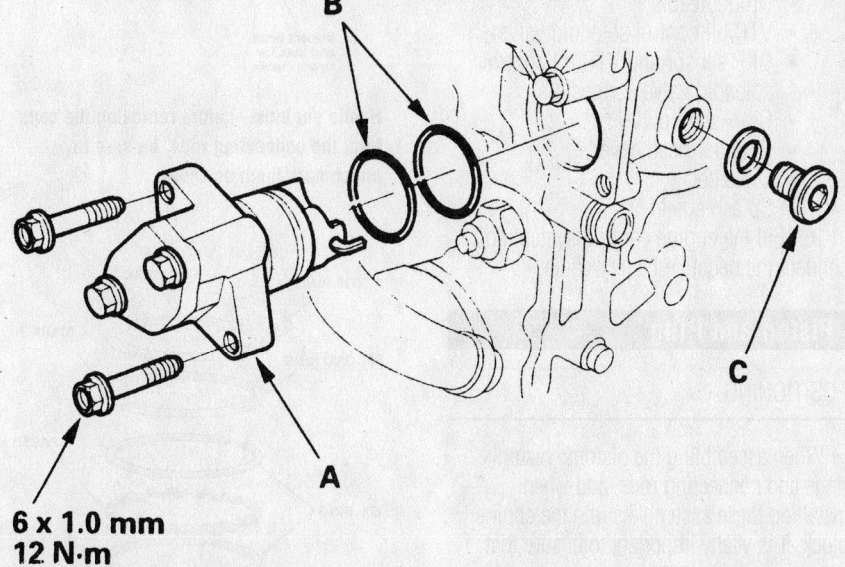

6 x 1.0 mm
12 N·m
(1.2 kgf·m, 8.7 lbf·ft)

9347FG23

Install the auto-tensioner (A) with new O ring seals (B), then remove the service bolt (C) and the set pin—2.0L engine

- Idler pulley: 10mm bolts to 33. Ft. lbs. (44 Nm) and the 6mm bolt to 104 inch lbs. (12 Nm)
- Water pump pulley bolts: 10 ft. lbs. (14 Nm)
- Bypass tube bolts: 104 inch lbs. (12 Nm)

2.4L Engine

1. Before servicing the vehicle, refer to the precautions in the beginning of this section.

2. Set the engine to Top Dead Center (TDC).

3. Drain the cooling system.

4. Relieve the fuel system pressure.

5. Remove or disconnect the following:

- Negative battery cable
- Front tires and wheels
- Splash shield
- Drive belt
- Cylinder head cover
- Crankshaft pulley
- Crankshaft Position (CKP) sensor connector
- Variable Valve Timing Control (CTV) oil control solenoid valve connector
- VTC oil control solenoid valve

6. Support the engine with a suitable jack with a wooden block under the oil pan.

- Ground cable and upper bracket
- Side engine mount bracket
- Chain cover

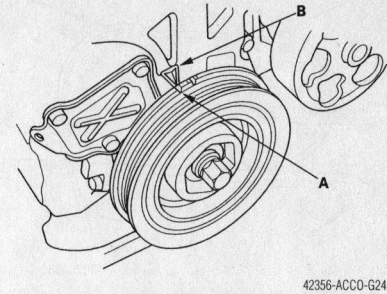

42356-ACCO-G24

Turn the crankshaft pulley so the TDC mark (A) is aligned with the pointer (B)— 2.4L engine

7. Loosely install the crankshaft pulley. Turn the crankshaft counterclockwise to compress the auto-tensioner.

8. Align the holes on the lock (A) and the auto-tensioner (B), then place a 1.5mm pin into the holes. Turn the crankshaft clockwise to secure the pin.

9. Remove or disconnect the following:

- Auto-tensioner
- Timing chain guide B (top guide)
- Timing chain guide A and tensioner arm
- Timing chain

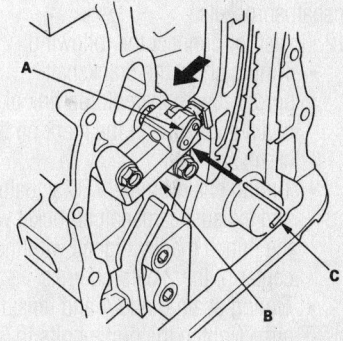

42356-ACCO-G25

Align the holes on the lock (A) and the auto-tensioner (B), then place a 1.5mm pin into the holes. Turn the crankshaft clockwise to secure the pin

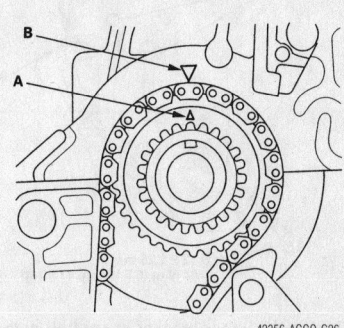

42356-ACCO-G26

Set the crankshaft to TDC. Align the TDC mark (A) on the crankshaft sprocket with the pointer (B) on the cylinder block

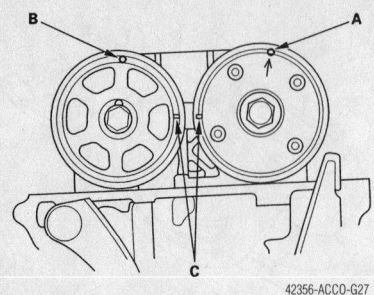

The mark (A) on the VTC actuator and the mark (B) on the exhaust cam (C) should be at the top. Align the TDC marks (C) on the VTC actuator and exhaust cam sprockets

✳✳ WARNING

Do not let the timing chain near any magnetic fields.

To install:

10. Set the crankshaft to TDC. Align the TDC mark (A) on the crankshaft sprocket with the pointer (B) on the cylinder block.

11. Set the camshafts to TDC. The punch mark (A) on the VTC actuator and the punch mark (B) on the exhaust camshaft (C) should be at the top. Align the TDC marks (C) on the VTC actuator and exhaust camshaft sprockets.

12. Install or connect the following:
- Timing chain the crankshaft sprocket with the colored link of the chain aligned with the mark on the crank sprocket
- Timing chain on the VTC actuator and exhaust camshaft sprocket with the punch marks aligned with the center of the 2 colored links
- Timing chain guide A and tensioner arm. Tighten the guide bolts to 8.7 ft. lbs. (12 Nm) and the tensioner arm retainer to 16 ft. lbs. (22 Nm).

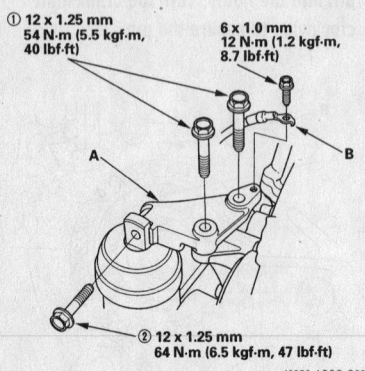

① 12 x 1.25 mm
54 N·m (5.5 kgf·m, 40 lbf·ft)

6 x 1.0 mm
12 N·m (1.2 kgf·m, 8.7 lbf·ft)

② 12 x 1.25 mm
64 N·m (6.5 kgf·m, 47 lbf·ft)

Tighten the upper bracket upper bolt/nuts in the proper order to the correct specification—2.4L engine

- Auto-tensioner and tighten the bolts to 8.7 ft. lbs. (12 Nm)
- Timing chain guide B and tighten the retainers to 16 ft. lbs. (22 Nm)

13. Remove the pin from the auto-tensioner.

14. Inspect the chain cover seal for damage and replace if necessary. Clean and dry the chain cover mating surfaces.

15. Install or connect the following:
- Liquid gasket, P/N 08718-0009 evenly to the cylinder block mating surface of the timing chain cover and the inner threads of the holes
- Liquid gasket to the cylinder block upper surface contact areas on the chain cover and the oil pan mating surface of the chain cover in the inner threads of the holes

➡ **Make sure to install the components within 5 minutes of applying the sealer.**

- New O-ring the timing chain cover. Set the edge of the cover to the edge of the oil pan, then install the cover on the engine block. Tighten the retainers to 8.7 ft. lbs. (12 Nm).

➡ **When installing the chain case, do not slide the bottom surface on the oil pan mounting surface.**

- Side engine mounting bracket and tighten the retainers to 33 ft. lbs. (44 Nm)
- Upper bracket, then tighten the bolts/nuts as shown in the illustration
- Ground cable
- VTC oil control solenoid valve
- CKP sensor and VTC oil control solenoid valve connectors
- Crankshaft pulley
- Cylinder head cover
- Drive belt
- Splash shield

16. Fill the engine cooling system and connect the negative battery cable.

Piston and Ring

POSITIONING

When assembling the pistons, piston rings and connecting rods, and when installing these assemblies into the engine block, it is vitally important to ensure that these three components are properly positioned with respect to each other. Often times the engine block is designed so that if a con-

necting rod or piston is installed backwards, or in the wrong bank of cylinders, internal engine damage may occur once the engine is started. The piston ring end-gap spacing that is recommended by the engine manufacturer is often with the purpose of increased compression pressures during the engine break-in period. Failure to properly space the piston ring end-gaps may lead to increased oil consumption and extended break-in time. Therefore, always be sure to position the pistons, rings and connecting rods as shown in the accompanying illustrations.

✳✳ WARNING

Always be sure to matchmark the connecting rods and caps prior to disassembly so that they may be reassembled with their original counterparts. If the caps are not installed on their original connecting rods, the assemblies will most likely need machining to avoid bearing, connecting rod and/or crankshaft damage.

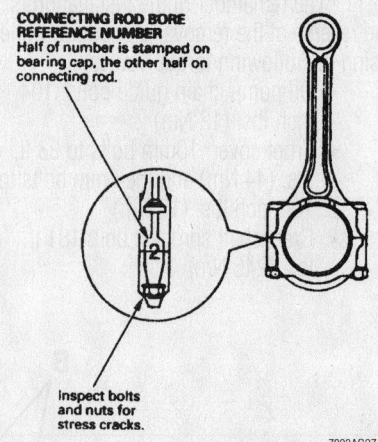

CONNECTING ROD BORE REFERENCE NUMBER
Half of number is stamped on bearing cap, the other half on connecting rod.

Inspect bolts and nuts for stress cracks.

Honda engines—before removing the caps from the connecting rods, be sure to matchmark them as shown

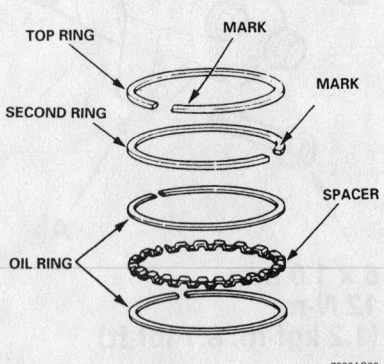

TOP RING
SECOND RING
OIL RING
MARK
MARK
SPACER

Honda engines—piston ring positioning

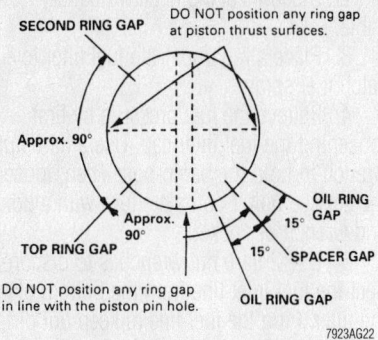

Honda engines—piston ring end-gap spacing

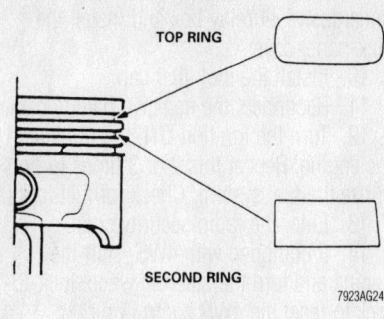

Honda 2.2L (H22A4) engines—compression ring locations

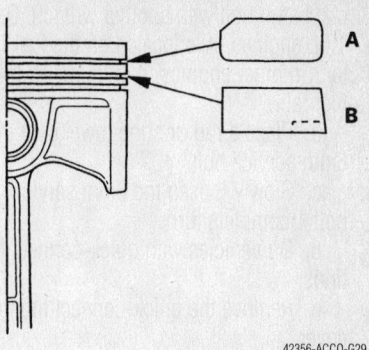

Honda 2.4 engine—compression ring locations

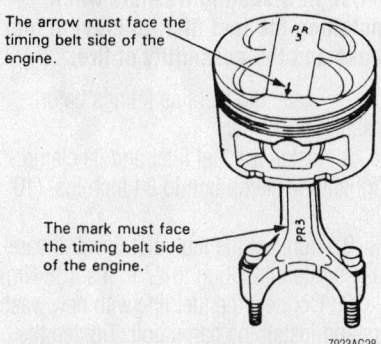

Honda 1.6L engines—piston/connecting rod assembly-to-engine orientation

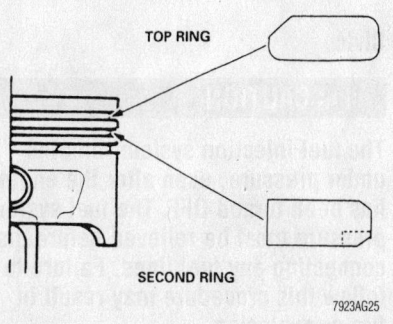

Honda 2.2L and 2.3L engines—compression ring locations

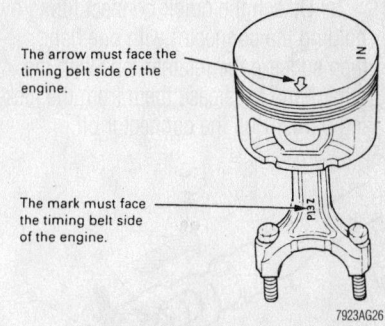

Honda 2.2L (H22A4) engines—piston/connecting rod assembly-to-engine orientation

FUEL SYSTEM

Fuel System Service Precautions

Safety is the most important factor when performing not only fuel system maintenance but any type of maintenance. Failure to conduct maintenance and repairs in a safe manner may result in serious personal injury or death. Maintenance and testing of the vehicle's fuel system components can be accomplished safely and effectively by adhering to the following rules and guidelines.

• To avoid the possibility of fire and personal injury, always disconnect the negative battery cable unless the repair or test procedure requires that battery voltage be applied.

• Always relieve the fuel system pressure prior to disconnecting any fuel system component (injector, fuel rail, pressure regulator, etc.), fitting or fuel line connection. Exercise extreme caution whenever relieving fuel system pressure, to avoid exposing skin, face and eyes to fuel spray. Please be advised that fuel under pressure may penetrate the skin or any part of the body that it contacts.

• Always place a shop towel or cloth around the fitting or connection prior to loosening to absorb any excess fuel due to spillage. Ensure that all fuel spillage (should it occur) is quickly removed from engine surfaces. Ensure that all fuel soaked cloths or towels are deposited into a suitable waste container.

• Always keep a dry chemical (Class B) fire extinguisher near the work area.

• Do not allow fuel spray or fuel vapors to come into contact with a spark or open flame.

• Always use a back-up wrench when loosening and tightening fuel line connection fittings. This will prevent unnecessary stress and torsion to fuel line piping. Always follow the proper torque specifications.

• Always replace worn fuel fitting O-rings with new. Do not substitute fuel hose or equivalent, where fuel pipe is installed.

Fuel System Pressure

RELIEVING

✳✳ CAUTION

The fuel injection system remains under pressure after the engine has been turned OFF. Properly relieve fuel pressure before disconnecting any fuel lines. Failure to do so may result in fire or personal injury.

➡**The radio may contain a coded theft protection circuit. Always obtain the code number before disconnecting the battery. If the vehicle is equipped with 4WS, the steering control unit is shut down when the battery is disconnected. After connecting the battery, turn the steering wheel lock-to-lock to reset the steering control unit.**

1. Before servicing the vehicle, refer to the precautions in the beginning of this section.

2. Disconnect the negative battery cable.

3. Remove the kick panel, then remove the PGM-FI main relay (FUEL PUMP) from the under-dash fuse/relay box. Start the engine and let it run until it stalls.

4. Remove the fuel filler cap.

5. On vehicles without quick connect fittings:

   a. Use a box wrench to loosen the 6mm service bolt while holding the special banjo bolt with another wrench. On 1.6L engines, it is located on the fuel filter. On other engines, it is found on the fuel rail.

b. Place a rag or shop towel over the 6mm service bolt.

c. Slowly loosen the 6mm service bolt 1 complete turn.

d. On vehicles with quick-connect fittings:

e. Remove the quick-connect fitting cover.

f. Clean any dirt from the quick-connect fitting.

g. Place a rag or shop towel over quick-connect fitting.

h. Detach the quick-connect fitting by holding the connector with one hand, then squeeze the retainer tabs with the other hand to release them from the locking pawls. Pull the connector off.

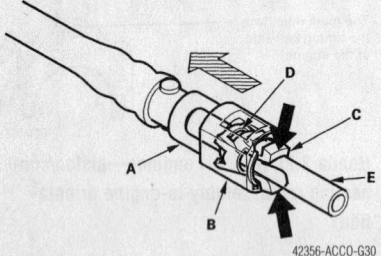

42356-ACCO-G30

Hold the quick-connect (A) connector (B) with one hand, then squeeze the retainer tabs (C) with the other hand to release them from the locking pawls (D)

✳✳ CAUTION

Do not allow fuel spray or fuel vapors to come in contact with a spark or open flame. Keep a dry chemical fire extinguisher nearby. Never store fuel in an open container due to risk of fire or explosion.

➥A fuel pressure gauge may be attached at the 6mm service bolt or quick-connect location. Always replace the washer between the service bolt and the banjo bolt whenever the service bolt is loosened.

6. If equipped, remove the service bolt and install a new washer. Tighten the 6mm service bolt to 108 inch lbs. (12 Nm). Don't overtighten the service bolts, their threads may strip and cause leaks.

7. If equipped with a quick-connect fitting, connect the fitting, making sure the locking pawls are properly engaged,

8. Clean up any fuel spilled on the engine and intake manifold.

9. Install the fuel pump relay to the underdash fuel/relay box and install the kick-panel cover.

10. Install the fuel filler cap.

11. Reconnect the negative battery cable.

12. Turn the ignition **ON**, but don't start the engine. Repeat this 2 or 3 times to pressurize the fuel system. Check for fuel leaks.

13. Enter the radio security code.

14. If equipped with 4WS, start the engine and turn the steering wheel lock-to-lock to reset the 4WS control unit.

Fuel Filter

REMOVAL & INSTALLATION

Civic

✳✳ CAUTION

The fuel injection system remains under pressure, even after the engine has been turned OFF. The fuel system pressure must be relieved before disconnecting any fuel lines. Failure to follow this procedure may result in fire or explosion.

➥The original radio contains a coded anti-theft circuit. Obtain the security code number before disconnecting the battery.

1. Before servicing the vehicle, refer to the precautions in the beginning of this section.

2. Disconnect the negative battery cable.

3. Place a rag under the fuel filter to catch fuel spray.

4. Relieve the fuel pressure by first loosening the fuel filler cap. Use a flare nut wrench to hold the banjo bolt. Then, loosen the service bolt 1 complete turn with a box-end wrench or socket.

5. Use 2 flare nut wrenches to disconnect the fuel inlet line from the bottom of the filter. Plug the fuel line to keep out dirt.

6. Unbolt and remove the fuel filter clamp. Remove the filter from its bracket.

To install:

➥**Use new sealing washers when installing the fuel filter to prevent fuel leaks and the possibility of fire.**

7. Clean the fuel line fittings before installing the filter.

8. Install the fuel filter and its clamp. Tighten the clamp bolt to 84 inch lbs. (10 Nm).

9. Connect the fuel inlet line and carefully tighten its fitting to 27 ft. lbs. (38 Nm).

10. Connect the fuel line with new washers and install the banjo bolt. Tighten the banjo bolt to 16 ft. lbs. (22 Nm). Install the service bolt and tighten it to 108 inch lbs. (12 Nm).

11. Connect the battery cable. Tighten the fuel filler cap.

12. Turn the ignition on and off several times to pressurize the fuel system. Start and run the engine and check for fuel leaks.

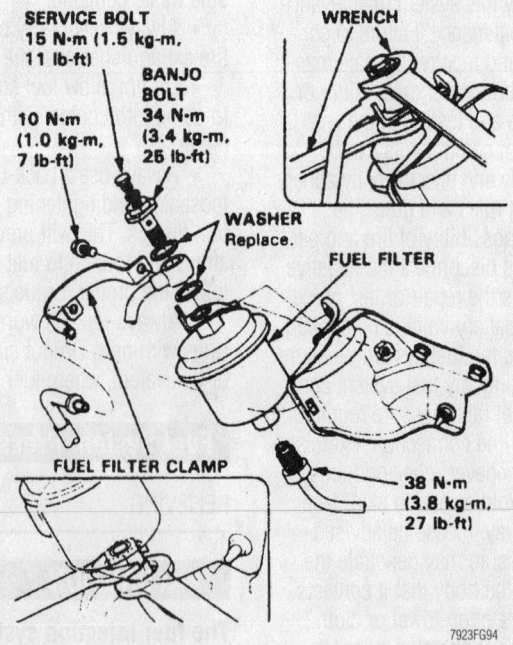

SERVICE BOLT 15 N·m (1.5 kg-m, 11 lb-ft)

BANJO BOLT 34 N·m (3.4 kg-m, 25 lb-ft)

10 N·m (1.0 kg-m, 7 lb-ft)

WRENCH

WASHER Replace.

FUEL FILTER

38 N·m (3.8 kg-m, 27 lb-ft)

FUEL FILTER CLAMP

7923FG94

Fuel filter components—1.6L engines

Prelude

➡The radio may contain a coded theft protection circuit. Always obtain the code number before disconnecting the battery. If the vehicle is equipped with 4WS, the steering control unit is shut down when the battery is disconnected. After connecting the battery, turn the steering wheel lock-to-lock to reset the steering control unit.

1. Before servicing the vehicle, refer to the precautions in the beginning of this section.
2. Disconnect the negative battery cable.
3. Place a shop towel under and around the fuel rail, then relieve the fuel pressure.

❄❄ CAUTION

Do not allow fuel spray or fuel vapors to come in contact with a spark or open flame. Keep a dry chemical fire extinguisher nearby. Never store fuel in an open container due to risk of fire or explosion.

4. Remove or disconnect the following:
   • 12mm banjo bolt and the fuel feed pipe from the fuel filter. Discard the washers.
   • Fuel filter clamp and the fuel filter.

To install:
5. Install or connect the following:
   • Fuel filter on the bracket and the filter clamp. Tighten the clamp bolts to 84 inch lbs. (10 Nm).

➡Clean the fuel fittings thoroughly before reconnecting them.

   • Fuel feed pipe to the filter and tighten the fitting to 28 ft. lbs. (38 Nm).
   • Fuel outlet pipe to the filter using new gaskets around the fitting. Tighten the banjo bolt to 20 ft. lbs. (28 Nm).
   • Negative battery cable and enter the radio security code.
6. Turn the ignition **ON** and check for fuel leaks.
7. If equipped with 4WS, start the engine and turn the steering wheel lock-to-lock to reset the 4WS control unit.

Accord

1. Before servicing the vehicle, refer to the precautions in the beginning of this section.
2. Relieve the fuel system pressure.
3. Remove or disconnect the following:
   • Fuel pump from the tank
   • Fuel filter from the pump module

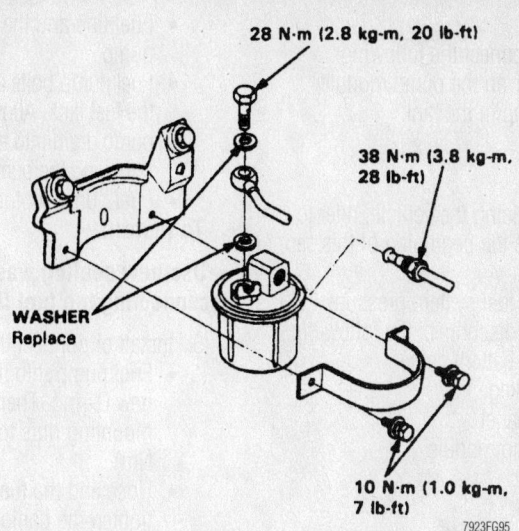

28 N·m (2.8 kg-m, 20 lb-ft)
38 N·m (3.8 kg-m, 28 lb-ft)
10 N·m (1.0 kg-m, 7 lb-ft)
WASHER Replace
7923FG95

Fuel filter mounting—Prelude

Except SULEV model:

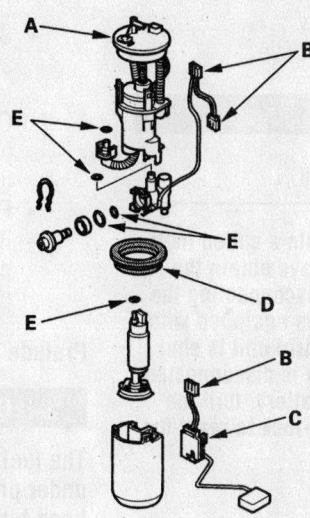

SULEV model:

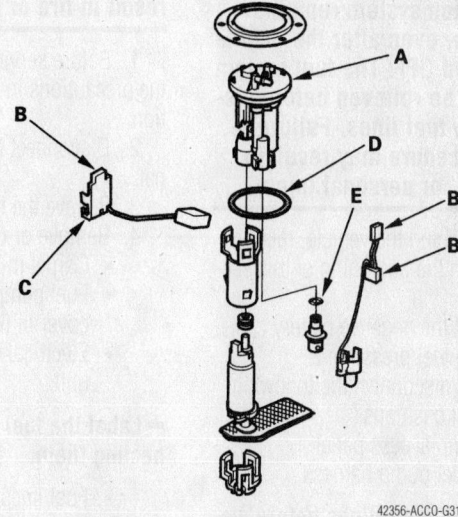

42356-ACCO-G31

Fuel filter (A) and related components—2003 Accord shown

To install:

4. Install or connect the following:
   - New filter on the pump module
   - Fuel pump in the tank

S2000

1. Before servicing the vehicle, refer to the precautions in the beginning of this section.
2. Relieve the fuel system pressure.
3. Remove or disconnect the following:
   - Negative battery cable
   - Rear package tray
   - Access panel
   - Fuel pump module
   - Fuel filter

To install:

4. Install or connect the following:
   - Fuel filter
   - Fuel pump module and tighten the bolts to 36 inch lbs. (4 Nm)
   - Access panel
   - Rear package tray
   - Negative battery cable

Fuel Pump

REMOVAL & INSTALLATION

➡ **The radio may contain a coded theft protection circuit. Always obtain the code number before disconnecting the battery. If the vehicle is equipped with 4WS, the steering control unit is shut down when the battery is disconnected. After connecting the battery, turn the steering wheel lock-to-lock to reset the steering control unit.**

Civic

❄ CAUTION

The fuel injection system remains under pressure, even after the engine has been turned OFF. The fuel system pressure must be relieved before disconnecting any fuel lines. Failure to follow this procedure may result in fire, explosion, or personal injury.

1. Before servicing the vehicle, refer to the precautions in the beginning of this section.
2. Disconnect the negative battery cable.
3. Relieve the fuel pressure.
4. Remove or disconnect the following:
   - Rear seat cushions
   - Fuel pump access panel
   - 2-wire fuel pump harness

➡ **Clean the fuel line fittings before disconnecting them.**

- Fuel line and the hose from the fuel pump.
- Fuel pump bolts and fuel pump from the fuel tank. Allow the fuel in the pump drain into the tank before removing the pump from the vehicle.
- Fuel pump motor from its bracket.

To install:

➡ **Use new sealing washers when reconnecting the fuel line banjo bolt.**

5. Install or connect the following:
   - Fuel pump into the fuel tank with a new O-ring. Then, tighten the mounting nuts to 48 inch lbs. (6 Nm).
   - Hose and the fuel line. Carefully tighten the banjo bolt to 20 ft. lbs. (28 Nm).
   - Fuel pump harness
   - Fuel filler cap
   - Fuel filter service bolt: 11 ft. lbs. (15 Nm)
   - Battery cable and turn the ignition switch **ON** and **OFF** several times to pressurize the fuel system. Check the connections at the fuel pump for any leaks. Check the fuel filter service bolt for leaks.
   - Fuel pump access cover
   - Rear seat cushions or rear compartment trim. Be sure the clips are properly seated.

Prelude

❄ CAUTION

The fuel injection system remains under pressure after the engine has been turned OFF. Properly relieve fuel pressure before disconnecting any fuel lines. Failure to do so may result in fire or personal injury.

1. Before servicing the vehicle, refer to the precautions in the beginning of this section.
2. Disconnect the negative battery terminal.
3. Relieve the fuel pressure.
4. Remove or disconnect the following:
   - Carpet in the luggage area.
   - Fuel pump maintenance access cover in the floor.
   - Electrical connector at the pump unit.

➡ **Label the fuel lines before disconnecting them.**

- Fuel lines. Discard the washers from the fuel feed connection.

- Retaining nuts holding the pump
- Pump up and out of the tank.

➡ **The pump sits on an angle and may require some manipulation to remove. If the pump still won't come out, loosen the fuel tank mounting nuts under the car, slide the tank downward a bit to give more clearance at the top.**

To install:

5. Install or connect the following:
   - New sealing ring
   - Fuel pump, making certain it is correctly seated and not wedged or jammed.
   - Retaining nuts and tighten them evenly and alternately to 48 inch lbs. (6 Nm).
   - Fuel lines. Make certain the clamp is secure; use new ones if necessary.
   - New washers to the fuel feed connection before installing the attaching bolt. Tighten the fuel feed attaching bolt to 20 ft. lbs. (28 Nm).
   - Fuel pump connector
   - Negative battery cable and enter the radio security code.
6. Switch the ignition **ON** but do not engage the starter. The fuel pump should run for approximately 2 seconds, building pressure within the lines. Switch the ignition **OFF**, then **ON** 2 or 3 more times to build full system pressure. Check for fuel leaks.
7. If equipped with 4WS, start the engine and turn the steering wheel lock-to-lock to reset the 4WS control unit.
8. Install the maintenance access cover and seal or gasket, if used.
9. Reposition the carpeting in the luggage compartment.

Accord

1. Before servicing the vehicle, refer to the precautions in the beginning of this section.
2. Relieve the fuel pressure.
3. Remove or disconnect the following:
   - Negative battery cable
   - Spare tire cover, if equipped
   - Trunk floor
   - Access panel from the floor
   - 5-pin connector from the pump assembly.
   - Quick-connect connections from the pump assembly.
   - Mounting bolts and the fuel tank unit
   - Strainer case, fuel gauge sending unit and wire harness, if necessary

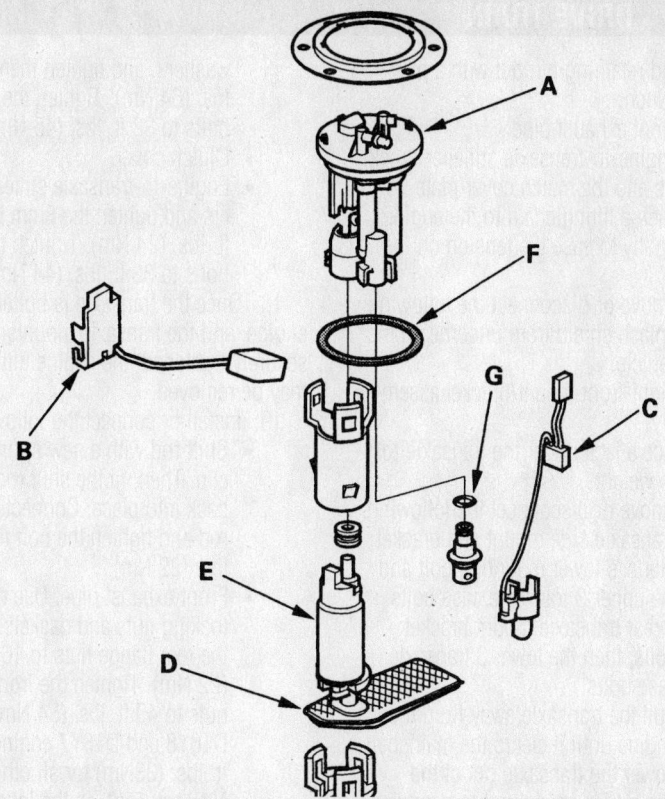

Exploded view of the strainer case (A), fuel gauge sending unit (B), wire harness (C), suction filter (D) and fuel pump (E)—Accord

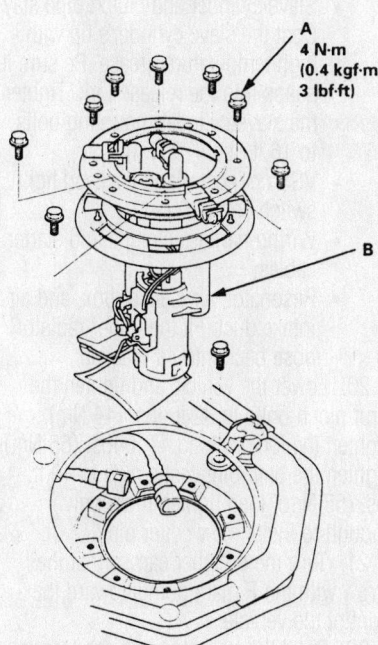

Fuel pump assembly mounting—Accord

To install:

4. Assembly the strainer case, fuel gauge sending unit and wire harness, if necessary

5. Install or connect the following:
   - Fuel pump, using a new gasket
   - Quick-connect fuel lines to the pump assembly
   - Fuel cap
   - 5-pin connector to the pump
   - Negative battery cable and enter the radio security code

6. Switch the ignition **ON** but do not engage the starter. The fuel pump should run for approximately 2 seconds, building pressure within the lines. Switch the ignition **OFF**, then **ON** 2 or 3 more times to build full system pressure. Check for fuel leaks.

7. If there are no leaks, install the access panel cover and the tire cover.

S2000

1. Before servicing the vehicle, refer to the precautions in the beginning of this section.

2. Relieve the fuel system pressure.

3. Remove or disconnect the following:
   - Negative battery cable
   - Rear package tray
   - Access panel
   - Fuel pump module
   - Fuel pump

To install:

4. Install or connect the following:
   - Fuel pump
   - Fuel pump module and tighten the bolts to 36 inch lbs. (4 Nm)
   - Access panel
   - Rear package tray
   - Negative battery cable

Fuel Injector

REMOVAL & INSTALLATION

1. Disconnect the negative battery cable.

2. Relieve the fuel system pressure.

3. Remove the fuel rail assembly

4. Carefully pull the injectors from the intake manifold.

5. Discard the seal rings, cushion rings and O-rings.

To install:

6. Slide new cushion rings onto the injectors.

7. Coat new O-rings with clean engine oil and put them on the injectors.

8. Insert the injectors into the fuel rail. Be sure to align the center line on the injector with the mark on the fuel rail.

9. Coat new seal rings with clean engine oil and insert them into the intake manifold.

10. Install the fuel rail assembly.

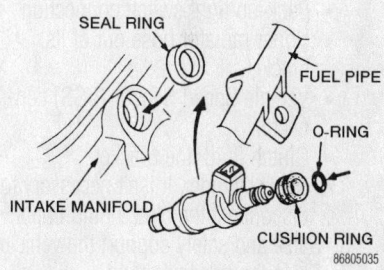

Always use new cushion rings, seal rings and O-rings

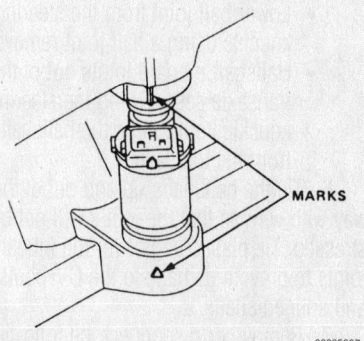

Be sure to align the center line on the injector with the mark on the fuel rail

DRIVE TRAIN

Transaxle Assembly

REMOVAL & INSTALLATION

Manual

CIVIC

❋❋ WARNING

Use only genuine Honda manual transaxle fluid (MTF)-it is specially formulated for use in Honda transaxles. If Honda MTF is not available, API SG/SJ 10W-30 or 10W-40 motor oil may be used as a temporary lubricant. However, motor oil will cause increased transaxle wear and shifting effort. Refill the transaxle with Honda MTF as soon as possible.

1. Before servicing the vehicle, refer to the precautions in the beginning of this section.
2. Remove or disconnect the following:
   • Negative and positive battery cables
   • Resonator, the air cleaner box, and the air intake duct.
   • Starter cables and the transaxle ground cable.
   • Back-up light switch connection
   • Upper radiator hose out of its bracket.
   • Vehicle Speed Sensor (VSS) connector.
   • Clutch fluid line bracket
   • Slave cylinder. It isn't necessary to disconnect the clutch fluid line.
3. Raise and safely support the vehicle.
4. Drain the transaxle fluid.
5. Remove or disconnect the following:
   • Front wheels
   • Strut pinch bolt and fork bolt
   • Lower ball joint from the steering knuckle using a ball joint remover.
   • Halfshaft inboard joints out of the transaxle case. Swing the steering knuckles out to free the halfshafts from the transaxle.
6. Tie the halfshafts up and out of the way with wire so that the joints will not be stressed. Tie plastic bags over the inboard joints to prevent damage to the CV-boots and splined shafts.
7. Remove or disconnect the following:
   • Shift rod and extension rod from the transaxle case. Drive the shift

rod retaining pin out with a pin punch.
   • Front exhaust pipe
   • Engine-to-transaxle stiffener brackets and the clutch cover plate
8. Attach a lifting chain to the engine and lift slightly to ease the tension on the mounts.
9. Remove or disconnect the following:
   • Splash shield from underneath the vehicle.
   • Right-front mount/bracket assembly
10. Place a jack under the transaxle to support its weight.
11. Remove or disconnect the following:
   • Transaxle side mount and bracket
   • Starter's lower mounting bolt and the upper 3 transaxle case bolts.
   • 3 rear transaxle mount bracket bolts, then the lower 3 transaxle case bolts.
   • Pull the transaxle away from the engine until it clears the mainshaft. Lower the transaxle out of the vehicle Be careful not to bend the clutch hydraulic line.

To install:

➡**Use new self-locking nuts and color-coded bolts when installing the transaxle and suspension components.**

12. Apply high temperature grease to the mainshaft splines, release fork contact points, and throw-out bearing. The manufacturer recommends part No. 08798-9002, Honda Super High temp Urea Grease.
13. Place the transaxle on a transaxle jack and raise it to the level of the engine.
14. Align the transaxle and engine. Be sure the transaxle case dowel pins are securely seated, and fit the transaxle onto the engine. Install the upper and lower transaxle case bolts and the 14mm rear mount bolts and washers. Only hand-tighten them at this time.
15. Raise the transaxle and install the side mount. Tighten the upper and lower transaxle case bolts to 47 ft. lbs. (64 Nm). Tighten the 14mm rear mount bracket bolts to 61 ft. lbs. (84 Nm).
16. First, tighten the transaxle side mount bracket nuts and bolt to 47 ft. lbs. (64 Nm) each. Next, tighten the mount bushing bolts to 47 ft. lbs. (64 Nm). Finally, tighten the through-bolt to 54 ft. lbs. (74 Nm).
17. Install or connect the following:
   • Right-front mount/bracket assembly. Use 3 new 12mm bolt and

washers, and tighten them to 47 ft. lbs. (64 Nm). Tighten the 2 10mm bolts to 33 ft. lbs. (45 Nm).
   • Clutch cover
   • Engine-to-transaxle stiffener brackets and tighten the 8mm bolts to 17 ft. lbs. (24 Nm). Tighten the 10mm bolts to 33 ft. lbs. (44 Nm).
18. Once the transaxle is bolted to the engine, and the transaxle mounts are securely tightened, the engine lifting chain may be removed.
19. Install or connect the following:
   • Shift rod with a new spring pin and clip. Then, fit the shift rod boot back into place. Connect the torque rod and tighten the bolt to 16 ft. lbs. (22 Nm).
   • Front exhaust pipe. Use new self-locking nuts and gaskets. Tighten the rear flange nuts to 16 ft. lbs. (22 Nm). Tighten the front flange nuts to 40 ft. lbs. (54 Nm) for D16Y8 and D16Y7 engines, or 25 ft. lbs. (33Nm) for all others.
   • New set rings on the inboard CV-joint splines
   • Halfshafts into the transaxle case and intermediate shaft. The inboard joints must snap into place.
   • Lower ball joint and damper fork
   • Front wheels
   • Slave cylinder and clutch pipe stay. Coat the slave cylinder's tip with high temperature grease. Be sure it snaps into the release fork. Tighten the slave cylinder mounting bolts to 16 ft. lbs. (22 Nm).
   • VSS connector and back-up light switch connectors
   • Wiring harness clamps and starter cables.
   • Resonator, air cleaner box, and air intake duct. Fit the upper radiator hose back into its bracket.
20. Lower the vehicle and tighten the strut pinch bolts to 32 ft. lbs. (44 Nm). Tighten the fork bolts to 47 ft. lbs. (65 Nm). Tighten the ball joint castle nuts to 40 ft. lbs. (55 Nm), then tighten them only enough to install new cotter pins.
21. Turn the breather cap so that the arrow with the **F** mark points toward the front of the vehicle.
22. Refill the transaxle with the Honda MTF fluid.
23. Reconnect the positive and negative battery cables.
24. Bleed the clutch hydraulic system.

25. Check the clutch and transaxle for smooth operation.
26. Check and adjust the front wheel alignment.

PRELUDE

1. Before servicing the vehicle, refer to the precautions in the beginning of this section.
2. Shift the transaxle to **R**.
3. Remove or disconnect the following:
   - Negative and positive battery cables
   - Battery
   - Intake duct, air cleaner case and battery base
   - Vacuum tank and bracket. Do not disconnect the hoses.
   - Starter wires and the starter.
4. Loosen, but do not remove the 2 upper transaxle mounting bolts.
5. Remove or disconnect the following:
   - Transaxle ground cable and the back-up light switch wire.
   - Engine harness clamp
   - Shift cables from the transaxle case, leaving them attached to their bracket, and wire them safely out of the work area.
   - Vehicle Speed Sensor (VSS) connector. Leave the sensor hoses connected and remove the sensor from the transaxle case.
   - Slave cylinder mounting bolts.
   - Slave cylinder from the release fork and move it out of the work area, leaving the hydraulic line connected to the slave cylinder

➡**Do not depress the clutch pedal once the slave cylinder has been removed. Be careful not to kink the metal hydraulic line.**

6. Raise and safely support the vehicle. Drain the transaxle fluid.
7. Remove or disconnect the following:
   - Clutch damper mounting bolts and raise the clutch damper.
   - Remove the rear engine mount bracket stay, if equipped.
   - Front wheels
   - Cotter pins and lower arm ball joint nuts
   - Ball joints and lower arms using a press-type ball joint tool.
   - Damper fork bolt and the radius rod on the right side of the vehicle only.
8. Use a suitable tool to pry the right and left halfshafts out of the differential and the intermediate shaft. Pull on the inboard joint and remove the right and left halfshafts.

9. Remove or disconnect the following:
   - Intermediate shaft from the differential. Tie plastic bags over the halfshaft inboard joints to prevent damage to the boots and splines. Wire the halfshafts to the underbody of the vehicle so that their weight doesn't hang on their outboard joints.
   - Center beam and remove the clutch cover
   - Rear beam stiffener and the intake manifold stay.
   - 3 rear engine mount bracket bolts.
10. Place a transaxle jack under the transaxle and raise the transaxle just enough to take its weight off the mounts.
11. Remove or disconnect the following:
    - Transaxle mount and mount bracket.
    - 2 upper transaxle housing mounting bolts and the 3 lower transaxle housing bolts.
12. Pull the transaxle away from the engine to clear the mainshaft.
13. Lower the transaxle from the vehicle.

To install:

➡**Use new self-locking nuts and set rings when assembling the front suspension components and halfshafts. Use new self-locking bolts when installing the center beam and rear engine mount bracket. These fasteners can be purchased from a Honda dealer.**

14. Be sure the dowel pins are installed into the transaxle case.
15. Apply heavy duty high temperature grease (use Honda part number 08798–9002) to the mainshaft splines, release fork bolt and paws, and the throwout bearing. Install the bearing and release fork. Be sure the release fork snaps into place.
16. Raise the transaxle into position with a transaxle jack.
17. Install or connect the following:
    - 3 lower and 2 upper transaxle mounting bolts and evenly tighten them to 47 ft. lbs. (65 Nm).
    - Transaxle mount and mount bracket
    - Through-bolt and tighten temporarily. Be sure the engine is level. First tighten the 3 bracket-to-mount nuts and 2 bolts to 28 ft. lbs. (39 Nm). Then, tighten the through-bolt to 47 ft. lbs. (65 Nm).
    - 3 new rear engine mount bracket bolts on the engine side and tighten them to 40 ft. lbs. (55 Nm).
    - Rear beam stiffener and tighten the bolts to 28 ft. lbs. (39 Nm).

    - Intake manifold stay and tighten the bolts to 16 ft. lbs. (22 Nm).
    - Clutch cover and tighten the bolts to 108 inch lbs. (12 Nm).
    - Center beam and tighten the bolts to 43 ft. lbs. (60 Nm).
    - Intermediate shaft. Tighten its mounting bolts to 28 ft. lbs. (39 Nm).
    - New set rings onto the halfshaft inboard joint splines.
    - Halfshafts, making sure that they lock into place.
    - Radius rod and damper fork. Only hand-tighten their fasteners at this time.
    - Ball joint to the lower arm. Tighten the castle nut to 36–43 ft. lbs. (50–60 Nm). Then, only tighten the nut enough to install a new cotter pin.
    - Rear engine mount bracket stay, if equipped. Tighten the nut to 15 ft. lbs. (21 Nm) and the bolt to 28 ft. lbs. (39 Nm).
    - Clutch damper and tighten the bolts to 16 ft. lbs. (22 Nm).
    - Front wheels
18. Lower the vehicle.
19. Use a floor jack placed under the right front control arm to raise the vehicle enough so that its weight is supported by the jack. Tighten the radius rod mounting bolts to 76 ft. lbs. (105 Nm) and the radius rod nut to 32 ft. lbs. (44 Nm). Tighten the damper pinch bolt to 32 ft. lbs. (44 Nm). Tighten the damper fork bolt to 47 ft. lbs. (65 Nm). After pre-loading the suspension, lower the vehicle and remove the floor jack.
20. Coat the tip of the slave cylinder with heavy duty high temperature grease.
21. Install or connect the following:
    - Clutch hose pipe and clutch slave cylinder to the transaxle housing. Be sure the slave cylinder snaps into the release fork. Tighten the slave cylinder mounting bolts to 16 ft. lbs. (22 Nm).
    - Speed sensor. Tighten the mounting bolt to 14 ft. lbs. (19 Nm).
    - Shift cable and select cable to the shift arm lever.
    - Shift cable assembly onto the transaxle case. Tighten the cable bracket mounting bolts to 16 ft. lbs. (22 Nm).
    - New cotter pins
    - Back-up light switch coupler and the transaxle ground cable.
    - Harness clamp
    - Starter. Tighten the 10mm bolt to

32 ft. lbs. (45 Nm) and the 12mm bolt to 54 ft. lbs. (75 Nm).
- Starter wires

22. Loosen the 3 front engine mount bracket bolts. Tighten them to 28 ft. lbs. (39 Nm).

23. Install or connect the following:
- Vacuum tank and its bracket
- Air cleaner case and intake duct

24. Fill the transaxle with the proper type and quantity of oil.

25. Install or connect the following:
- Battery base stay and the battery base. Tighten the battery base bolts to 16 ft. lbs. (22 Nm).

26. Install or connect the following:
- Battery and connect the battery cables.

27. Check the clutch pedal free-play.

28. Start the vehicle and check the transaxle and clutch for smooth operation.

29. On Preludes equipped with 4WS, and turn the steering wheel lock-to-lock to reset the steering control unit.

30. Check and adjust the front wheel alignment.

31. Enter the radio security code.

ACCORD

1. Before servicing the vehicle, refer to the precautions in the beginning of this section.

2. Shift the transaxle into **R**.

3. Remove or disconnect the following:
- Negative and positive battery cables and the battery.
- Idle Air Control (IAC) solenoid connector
- Intake duct, resonator, air cleaner case, and battery base.
- Starter wires and starter
- Transaxle ground cable and the back-up light switch wire.
- Cable stay, then the cables from the top housing of the transaxle
- Both cables and the stay together
- Vehicle Speed Sensor (VSS). Leave the speed sensor hoses connected.
- Shift cable bracket
- Shift and select cables from the top of the transaxle case. Leave the cables and bracket together, and wire them out of the work area.
- Mounting bolts and clutch slave cylinder with the clutch pipe and pushrod.
- Mounting bolt and clutch hose joint with the clutch pipe and clutch hose.

➡Do not depress the clutch pedal once the slave cylinder has been removed.

Be careful not to kink the metal hydraulic lines.

- 2 upper transaxle case bolts
4. Raise and safely support the vehicle.
5. Remove or disconnect the following:
- Front wheels
- Engine splash shield
6. Drain the transaxle fluid.
7. Remove or disconnect the following:
- Clutch damper bracket and raise it out of the way.
- Subframe center beam
- Cotter pins and lower arm ball joint nuts
- Ball joints and lower arms using a press type ball joint tool.
- Right damper fork bolt
- Right damper pinch bolt, then separate the damper fork and damper.
- Radius rod bolts and nut, then the right radius rod.
- Right and left halfshafts from the differential and the intermediate shaft, using a suitable prytool.
- Left halfshaft
- Intermediate shaft from the differential by removing its 3 bearing shaft mounting bolts.

8. Swing the right halfshaft out and wire it up inside the right fender well. Tie plastic bags over the inboard CV-joints to protect the boots and splines from damage.

9. Remove or disconnect the following:
- Engine stiffener and the clutch cover
- Intake manifold bracket
- Rear engine mount bracket
- 3 rear engine mount bracket mounting bolts, then discard.

10. Place a transaxle jack under the transaxle. Raise the transaxle just enough to take the weight off the its mounts.

➡A chain hoist may be attached the transaxle lifting hooks to steady it and aid in lowering it from the vehicle.

11. Remove or disconnect the following:
- Transaxle housing mounting bolt on the engine side
- Transaxle mount bolt and loosen the mount bracket nuts.
- 3 transaxle housing mounting bolts, then the transaxle from the vehicle.

To install:

➡Use new self-locking nuts when assembling the front suspension. Install new set rings onto the inboard CV-joints. Use new self-locking bolts when installing transaxle rear mount bracket (the bolts are color coded by

type). New fasteners are available from a Honda dealer.

12. Be sure the 2 dowel pins are installed into the transaxle case.

13. Apply heavy duty high temperature grease (use Honda part No. 08798–9002) to the release bearing, mainshaft splines, and the release fork pawls. Install the release fork and release bearing.

14. Raise the transaxle into position.

15. Install or connect the following:
- 3 lower transaxle case bolts and tighten to 47 ft. lbs. (65 Nm).
- Transaxle mount and mount bracket
- Through-bolt and tighten temporarily. Be sure the engine is level and tighten the 3 mount bracket nuts to 40 ft. lbs. (55 Nm). Tighten the through-bolt to 47 ft. lbs. (65 Nm).
- Upper transaxle case bolts on the engine side and tighten to 47 ft. lbs. (65 Nm).
- 3 new rear engine bracket mounting bolts and tighten to 40 ft. lbs. (55 Nm).
- Intake manifold bracket and tighten the bolts to 16 ft. lbs. (22 Nm).
- Clutch cover and tighten the bolts to 108 inch lbs. (12 Nm).
- Subframe center beam with new self-locking bolts. Evenly tighten the bolts to 37 ft. lbs. (50 Nm).
- Engine stiffener plate, if equipped. Loosely install the mounting bolts. Tighten the stiffener-to-transaxle case mounting bolt to 28 ft. lbs. (39 Nm), then tighten the 2 stiffener-to-engine block mounting bolts to 28 ft. lbs. (39 Nm) beginning with the bolt closest to the transaxle.
- Radius rod and the damper fork. Hand-tighten all the fasteners.
- Intermediate shaft and tighten its mounting bolts to 28 ft. lbs. (39 Nm).
- New set ring on the end of each halfshaft.
- Right and left halfshafts. Turn the right and left steering knuckle fully outward and slide the axle into the differential, until the set ring is felt engaging the differential side gear.
- Lower control arm ball joints. Tighten the castle nuts to 40 ft. lbs. (50 Nm). Then, tighten them only enough to install a new cotter pin.
- Clutch damper and tighten its mounting bolts to 16 ft. lbs. (22 Nm).

- Front wheels. Lower the vehicle.

16. Place a floor jack under the right front knuckle, and raise the jack until it is supporting the vehicle's weight.

17. Tighten the radius rod mounting bolts to 76 ft. lbs. (105 Nm) and the radius rod nut to 32 ft. lbs. (44 Nm). Tighten the damper fork nut while holding the damper fork bolt to 40 ft. lbs. (55 Nm). Tighten the damper pinch bolt to 32 ft. lbs. (44 Nm).

18. Coat the tip of the slave cylinder with high temperature grease. Install the clutch hose joint and clutch slave cylinder to the transaxle housing. Be sure the slave cylinders tip snaps into the release fork. Tighten the slave cylinder mounting bolts to 16 ft. lbs. (22 Nm).

19. Install or connect the following:
- Speed sensor. Tighten the mounting bolt to 13 ft. lbs. (18 Nm).
- Shift cable and select cable to the shift arm lever. Tighten the cable bracket mounting bolts to 20 ft. lbs. (27 Nm).
- New cotter pins
- Back-up light switch connector
- Starter. Tighten the 10mm bolt to 32 ft. lbs. (45 Nm) and the 12mm bolt to 54 ft. lbs. (75 Nm).
- Starter wires
- Transaxle ground cable

20. Fill the transaxle with the proper type and quantity of oil.

21. Install or connect the following:
- Air cleaner case and the resonator, then the intake duct.
- Battery tray bracket and battery tray and tighten the bolts to 16 ft. lbs. (22 Nm).
- Battery and the battery cables.

22. Check the clutch pedal free-play.

23. Check and adjust the front wheel alignment.

24. Road test the vehicle and check the transaxle for smooth operation.

25. Loosen the 3 front engine mount bracket mounting bolts, then retighten them to 28 ft. lbs. (38 Nm).

26. Enter the radio security code.

S2000

1. Before servicing the vehicle, refer to the precautions in the beginning of this section.

2. Remove or disconnect the following:
- Battery cables and the battery
- Shift lever knob
- Center console
- Shift lever boot
- Shift lever
- Air cleaner housing

- Steering shaft from the steering gear box
- Alternator
- A/C compressor
- Exhaust manifold heat shields
- Upper starter mounting bolt
- Upper intake manifold bracket mounting bolt
- Suction valve hose
- Camshaft Position (CMP) sensor connectors
- Splash shield
- Steering gear box electrical connector
- Torque sensor connector
- Intake manifold bracket
- Heated Oxygen (HO2S) sensor connectors
- Catalytic converter
- Exhaust manifold
- Driveshaft
- Shifter boot holder bolts
- Clutch slave cylinder
- Clutch release fork
- Lower transmission flange bolts

3. Support the front subframe with a floor jack and remove the two center mounting bolts.

4. Loosen the four outer mounting bolts 3 inches (75 mm).

5. Lower the front subframe until it is supported by the loosened bolts.

6. Support the transmission with the floor jack.

7. Remove or disconnect the following:
- Rear transmission mount
- Speed sensor connector and wiring harness
- Upper transmission flange bolts
- Transmission

To install:

➡**Use new subframe bolts for assembly.**

8. Installation is the reverse of the removal procedure, while using the following torque values:
- Transmission flange bolts: 47 ft. lbs. (64 Nm)
- Rear transmission mount bolts: 28 ft. lbs. (38 Nm)
- Subframe mounting bolts: 14mm bolts to 85 ft. lbs. (116 Nm) and 12mm bolts to 43 ft. lbs. (59 Nm)
- Clutch slave cylinder bolts: 16 ft. lbs. (22 Nm)
- A/C compressor bolts: 33 ft. lbs. (44 Nm)
- Steering shaft pinch bolt: 16 ft. lbs. (22 Nm)
- Shift lever bolts: 86 inch lbs. (10 Nm)

Automatic

CIVIC

1. Before servicing the vehicle, refer to the precautions in the beginning of this section.

2. Remove or disconnect the following:
- Negative and positive battery cables
- Resonator, the air cleaner box, and the air intake duct.
- Starter cables and the transaxle ground cable
- Engine wiring harness clip

➡**Label all electrical connectors before removing them.**

- Lock-up control solenoid connector
- Vehicle Speed Sensor and countershaft speed sensor connectors
- Upper transaxle case bolts and the rear engine mounting bolt

3. Raise and safely support the vehicle.

4. Remove or disconnect the following:
- Front wheels

5. Drain the automatic transaxle fluid. Then, install the drain plug with a new crush washer. Note the color, consistency, and odor of the drained fluid.

6. Remove or disconnect the following:
- Front splash shield
- Shift control and linear solenoid connectors
- Mainshaft speed sensor connector
- Strut pinch bolt and fork bolt
- Lower ball joint using a ball joint remover.

7. Pry the halfshaft inboard joints out of the transaxle case and intermediate shaft. Swing the steering knuckles out to free the halfshafts from the transaxle.

8. Tie the halfshafts up and out of the way with wire. Tie plastic bags over the inboard joints to prevent damage to the CV-boots and splined shafts.

9. Remove or disconnect the following:
- Front exhaust pipe
- Shift cable cover
- Shift cable from the transaxle control shaft. Move the shift cable out of the way, and tie it up with wire.
- Automatic Transaxle Fluid (ATF) cooler hoses from the cooler lines. Cap the lines to prevent fluid lose and contamination.
- Right-front mount and bracket assembly
- Engine stiffener and the torque converter cover plate.
- 8 torque converter-to-driveplate bolts 1 at a time by rotating the crankshaft pulley.

➡**There are no gear teeth on the drive-plate; the starter motor engages a ring gear on the inner edge of the torque converter.**

10. After unbolting the torque converter from the driveplate, rotate the crankshaft to set the engine at Top Dead Center (TDC)/compression for the No. 1 cylinder.

11. Remove or disconnect the following:
- Ignition wires
- Distributor, if equipped

12. Attach a lifting chain to the engine and lift slightly to ease the tension on the mounts.

13. Place a transaxle jack under the transaxle and remove the transaxle side mount and bracket.

14. With the transaxle supported, remove the transaxle rear mount bracket bolts and transaxle case bolts.

15. Pull the transaxle away from the engine until it clears the locating dowel pins. Carefully lower the transaxle from the vehicle with the torque converter angled upward so it doesn't drop out of the transaxle.

16. Remove the torque converter from the transaxle. Inspect the ring gear teeth for breakage and inspect the converter's hub for burrs and scoring. Check the condition of the converter fluid. Replace the torque converter if necessary.

17. Inspect the transaxles front oil pump bearing and seal for signs of leakage and scoring. Inspect the mainshaft for burrs, scoring, and roughness.

18. With the transaxle removed, carefully inspect the driveplate for stress cracks, enlarged bolt holes, and other defects. Replace it if necessary.

To install:

➡**Use new self-locking nuts and color-coded bolts when installing the transaxle and suspension components.**

19. Flush the transaxle cooler lines to remove any contaminated fluid and residual clutch material:

a. Use a pressurized flusher (Honda J38405-A or equivalent). Use only Honda flushing fluid (Honda J35944–20); other fluids may damage the system.

b. Fill the flusher with 21 ounces of fluid. Pressurize the flusher to 80–120 psi, following the procedure on the fluid container and flusher.

c. Clamp the discharge hose of the flusher to the cooler return line. Clamp the drain hose to the cooler inlet line and route it into a bucket or drain tank.

d. Connect the flusher to air and water lines. The air line use a water trap to keep excess moisture out.

e. Open the flusher water valve and flush the cooler for 10 seconds.

f. Depress the flusher trigger to mix flushing fluid with the water. Flush for 2 minutes, turning the air valve on and off for 5 seconds every 15–20 seconds to create a surging action.

g. After finishing 1 flushing cycle, reverse the hose and flush in the opposite direction.

h. Dry the cooler lines with compressed air for 2 minutes or longer to remove all excess moisture from the system.

20. Install or connect the following:
- Starter motor onto the transaxle case and tighten its mounting bolts to 33 ft. lbs. (45 Nm).
- Torque converter with a new hub O-ring.

21. Place the transaxle on a transaxle jack and raise it to the level of the engine.

22. Align the transaxle and engine. Install the transaxle case bolts. Install new 14mm rear mount bolts and washers.

23. Raise the transaxle and install the side mount. Tighten the case bolts to 47 ft. lbs. (64 Nm). Tighten all of the 14mm rear mount bolts to 61 ft. lbs. (85 Nm).

24. Install or connect the following:
- Transaxle side mount and bracket. Tighten the bracket nuts to 47 ft. lbs. (64 Nm). Tighten the mount through-bolt to 54 ft. lbs. (74 Nm).

25. Remove the transaxle jack.

26. Rotate the crankshaft and install the torque converter-to-driveplate bolts. Tighten the bolts to 108 inch lbs. (12 Nm) in a crisscross pattern. Tighten the bolts to the specification in 2 steps.

27. Rotate the crankshaft to reset the engine at Top Dead Center (TDC)/compression for the No. 1 cylinder. After the engine is set at Top Dead Center (TDC), it must not be disturbed until the distributor, if equipped, has been reinstalled.

28. Install or connect the following:
- Torque converter cover and tighten the bolts to 108 inch lbs. (12 Nm).
- Engine stiffener and tighten the 8mm bolts to 17 ft. lbs. (24 Nm). Tighten the 10mm bolts to 33 ft. lbs. (45 Nm).
- Right-front mount and bracket assembly. Tighten the 10mm bolt to 33 ft. lbs. (44 Nm), and the 12mm bolts to 47 ft. lbs. (64 Nm).

29. Remove the lifting chain and chain hooks.

30. Verify that the engine is at Top Dead Center (TDC)/compression for the No. 1 cylinder. Align the tabs on the distributor drive with the grooves on the end of the camshaft. Install the distributor and hand-tighten the mounting bolts. Reconnect the ignition wires.

31. Tighten the crankshaft pulley to 134 ft. lbs. (181 Nm).

32. Install or connect the following:
- Transaxle cooler lines
- New set rings on the inboard CV-joint splines
- Halfshafts into the transaxle case and intermediate shaft. The inboard joints must snap into place.
- Lower ball joint and damper fork
- Front wheels
- Shift cable linkage to the transaxle control shaft.
- New lockwasher and tighten the linkage bolt to 10 ft. lbs. (14 Nm).
- Shift cable cover and tighten its bolt to 16 ft. lbs. (22 Nm).
- Front exhaust pipe. Use new self-locking nuts and gaskets. Tighten the rear flange nuts to 16 ft. lbs. (22 Nm), and the front flange nuts to 47 ft. lbs. (64 Nm).
- VSS and countershaft speed sensor connectors
- Lock-up control solenoid connector.
- Shift control and linear solenoid connectors
- Mainshaft speed sensor connector
- Wiring harness clamps and starter cables
- Resonator, air cleaner box, and air intake duct
- Front splash shield

33. Lower the vehicle and tighten the strut pinch bolts to 32 ft. lbs. (44 Nm). Tighten the fork bolts to 47 ft. lbs. (65 Nm). Tighten the ball joint castle nuts to 40 ft. lbs. (55 Nm), then tighten them only enough to install new cotter pins.

34. Refill the transaxle with fresh ATF. Use only Honda Premium ATF or DEXRON®II or III ATF. Reconnect the positive and negative battery cables.

a. Leave the flusher drain hose attached to the cooler return line.

b. With the transaxle in park, run the engine for 30 seconds, or until approximately 1 quart of fluid is discharged. Immediately shut the engine off. This completes the cooler flushing process.

c. Remove the drain hose and reconnect the cooler return line.

d. Refill the transaxle to the proper level.

35. Check shift cable and throttle cable adjustments.

36. Check the ignition timing. Rotate the distributor counterclockwise to advance the timing, or clockwise to retard the timing. When the timing has been set, tighten the distributor mounting bolts to 13 ft. lbs. (18 Nm).

37. Start the engine and shift through all the gears 3 times.

38. Let the engine warm up to operating temperature and check the fluid level with the transaxle in the **P** or **N** position.

39. Check and adjust the front wheel alignment.

40. Road test the vehicle. Recheck the transaxle fluid level.

PRELUDE

1. Before servicing the vehicle, refer to the precautions in the beginning of this section.

2. Disconnect both cables from the battery.

3. Shift the transaxle into **N**.

4. Remove or disconnect the following:
   • Battery hold-down and the battery

5. Drain the transaxle fluid and reinstall the drain plug with a new crush washer.

6. Remove or disconnect the following:
   • Air intake duct, air cleaner case, and resonator
   • Connector from the vacuum tank and the vacuum tank and tank bracket. Do not remove the vacuum tube from the vacuum tank.
   • Transaxle-to-body ground cable
   • Battery base with the ground cable and the battery base stay.
   • Lock-up control solenoid valve and shift control solenoid valve connectors.
   • Throttle control cable from the throttle control lever.
   • Countershaft speed sensor connector
   • Vehicle Speed Sensor (VSS) connector
   • Rear stiffener, then remove the VSS and Power Steering Sensor (PSS).

➡**Do not disconnect the power steering pressure hoses from the VSS and PSS.**

   • Automatic Transaxle Fluid (ATF) cooler hoses at the joint pipes. Turn the ends of the cooler hoses upward to prevent fluid loss. Plug the joint pipes.
   • Starter motor
   • Upper transaxle housing mounting bolts

   • Front engine mount bracket bolts
   • Transaxle mount

7. Raise and support the vehicle safely.

8. Remove or disconnect the following:
   • Front wheels
   • Splash shield, subframe center beam and rear beam stiffener.
   • Cotter pins and castle nuts from the lower ball joints. Use a press-type ball joint tool to separate the ball joints from the lower arm.
   • Damper fork bolts, then separate the damper fork and the damper.

9. Use a suitable prytool to separate the right and left halfshafts from the differential.

10. Pull on the inboard joint and remove the right and left halfshafts. Tie plastic bags over the halfshaft ends to protect the boots and splined shafts from damage.

11. Remove or disconnect the following:
   • Right damper pinch bolt, then separate the right damper fork from the strut.
   • Right radius rod bolts and nut, then the radius rod
   • Torque converter cover and the shift cable cover.
   • Control lever lockbolt and the shift cable with the lever. Do not bend the shift control cable during removal. Wire the cable to the underbody of the vehicle our of the work area.
   • Driveplate bolts while rotating the crankshaft.

12. Place a transaxle jack below the transaxle and raise it enough to take the weight off the mounts.

13. Remove or disconnect the following:
   • Intake manifold bracket
   • Lower transaxle housing mounting bolts and lower rear engine mounting bolts.

14. Pull the transaxle away from the engine until it clears the dowel pins. Lower the transaxle out of the vehicle.

To install:

➡**Use new self-locking nuts when assembling the front suspension components. Use new set rings on the halfshaft inboard joints. Use new self-locking bolts for the subframe beams. These fasteners are available from a Honda dealer.**

15. Flush the transaxle cooling lines before installing the transaxle. Use a pressurized flushing canister, such as Honda tool No. J38405-A, or its equivalent. Use only biodegradable flushing fluid, Honda part No. J35944-20.
   a. Fill the flusher with 21 ounces of

fluid. Pressurize the flusher to 80–120 psi, following the procedure on the fluid container and flusher.
   b. Clamp the discharge hose of the flusher to the cooler return line. Clamp the drain hose to the cooler inlet line and route it into a bucket or drain tank.
   c. Connect the flusher to air and water lines. Open the flusher water valve and flush the cooler for 10 seconds.
   d. Depress the flusher trigger to mix flushing fluid with the water. Flush for 2 minutes, turning the air valve on and off for 5 seconds every 15–20 seconds.
   e. After finishing 1 flushing cycle, reverse the hose and flush in the opposite direction.
   f. Dry the cooler lines with compressed air so that no moisture remains in the cooler lines.

16. Install the starter motor onto the transaxle case. Install the torque converter with a new hub O-ring. Tighten the starter bolts to 33 ft. lbs. (45 Nm).

17. Place the transaxle on a transaxle jack and raise it to the level of the engine.

18. Align the transaxle to the engine and install the transaxle housing mounting bolts and lower rear engine mounting bolts. Tighten the rear engine mounting bolts to 40 ft. lbs. (55 Nm) and the transaxle mounting bolts to 47 ft. lbs. (65 Nm). Install the intake manifold bracket and tighten the bolts to 16 ft. lbs. (22 Nm).

19. Tighten the front engine mount bracket bolts to 28 ft. lbs. (39 Nm).

20. Install or connect the following:
   • Transaxle mount. Tighten the bolt to 47 ft. lbs. (65 Nm) and the nuts to 28 ft. lbs. (39 Nm).

21. Remove the transaxle jack.

22. Install or connect the following:
   • Torque converter to the driveplate and the mounting bolts. Turn the crankshaft to rotate the driveplate. Tighten the bolts in 2 steps, first to 50 inch lbs. (6 Nm) in a crisscross pattern and finally to 108 inch lbs. (12 Nm). Check for free rotation after tightening the last bolt.
   • Shift cable onto the control shaft and tighten the lockbolt to 10 ft. lbs. (14 Nm).
   • Torque converter cover and the shift cable cover.
   • New set ring onto the inboard joint of each halfshaft.
   • Damper fork bolts and ball joint nuts to the lower arms. Tighten the ball joint nut to 47 ft. lbs. (65 Nm)
   • New cotter pin
   • Radius rod and the damper fork.

Only hand-tighten the radius rod and damper fork fasteners at this point.

23. Turn the right steering knuckle fully outward and slide the axle into the differential until the spring clip is felt engaging the differential side gear. Repeat the procedure on the left side.

24. Install or connect the following:
   • Subframe rear beam stiffener and the center beam. Tighten the stiffener bolts to 28 ft. lbs. (39 Nm). Tighten the subframe center beam bolts to 43 ft. lbs. (60 Nm).
   • Front wheels and lower the vehicle.

25. Use a floor jack to place the weight of the vehicle onto the right front knuckle. Tighten the radius rod bolts to 76 ft. lbs. (105 Nm) and the nut to 40 ft. lbs. (55 Nm). Tighten the damper pinch bolt to 32 ft. lbs. (44 Nm). Tighten the nut to 47 ft. lbs. (65 Nm) while holding the damper fork bolt. Remove the floor jack.

26. Install or connect the following:
   • Speedometer sensor. Tighten the sensor bolt to 108 inch lbs. (12 Nm).
   • ATF cooler hoses to the joint pipes
   • Lock-up control solenoid and shift control solenoid valve connectors.
   • VSS and PSS sensor connectors
   • Starter motor cables and battery base and base stay.
   • Ground cables on the body and on the transaxle.
   • Vacuum tank, tank bracket, and tank connector.
   • Resonator, air cleaner case, and air intake duct.

27. Refill the transaxle with ATF. Use only Honda Premium ATF or DEXRON®II ATF. Connect the negative and positive battery cables.
   a. Leave the flusher drain hose attached to the cooler return line.
   b. With the transaxle in park, run the engine for 30 seconds, or until approximately 1 quart of fluid is discharged. This completes the cooler flushing process.
   c. Remove the drain hose and reconnect the cooler return line.
   d. Refill the transaxle to the proper level with ATF.

28. Start the engine, set the parking brake, and shift the transaxle through all gears 3 times. Check for proper control cable adjustment.

29. On Preludes equipped with 4WS, and turn the steering wheel lock-to-lock to reset the steering control unit.

30. Check and adjust the front wheel alignment.

31. Let the engine reach operating temperature with the transaxle in **N** or **P**, then turn the engine OFF and check the fluid level

32. After road testing the vehicle, loosen the front engine mount bolts, and tighten them to 28 ft. lbs. (39 Nm).

33. Enter the radio security code.

ACCORD—2.3L AND 2.4L ENGINES

1. Before servicing the vehicle, refer to the precautions in the beginning of this section.

2. Remove or disconnect the following:
   • Negative, then the positive battery cables
   • Battery

3. Shift the transaxle into **N**.

4. Remove or disconnect the following:
   • Air intake hose, air cleaner housing, and the resonator assembly.
   • Battery base and the base stay
   • Throttle cable from the throttle control lever.
   • Transaxle ground cable and the speed sensor connectors.
   • Solenoid valve connectors
   • Lock-up control solenoid valve and shift control solenoid valve connectors.
   • Transaxle cooler hoses from the joint pipes and plug the hoses.
   • Starter cables and starter
   • Countershaft Speed Sensor (CSS) and Vehicle Speed Sensor (VSS) connectors

5. Install a hoist to the engine.

6. Remove or disconnect the following:
   • 4 upper bolts attaching the transaxle to the engine block.
   • 3 bolts attaching the front engine mount bracket to the engine.
   • Transaxle mount

7. Raise and safely support the vehicle. Remove the front wheels.

8. Drain the transaxle fluid and reinstall the drain plug with a new washer.

9. Remove or disconnect the following:
   • Splash shield
   • Subframe center beam
   • Cotter pins and lower arm ball joint nuts, then separate the ball joints from the lower arms using a suitable tool.
   • Right damper pinch bolt, then separate the damper fork and damper.
   • Bolts and nut, then the right radius rod.

10. Using a small prying device, carefully pry the right and left halfshafts out of the differential. Remove the right and left halfshafts. Tie plastic bags over the halfshaft

ends to prevent damage to the CV boots and splines.

11. Remove or disconnect the following:
   • Bolts mounting the intermediate shaft, then the intermediate shaft from the differential.
   • Torque converter cover and shift cable cover.
   • Shift control cable by removing the lockbolt.
   • Shift cable lever from the control shaft. Don't disconnect the control lever from the shift cable. Wire the shift cable out of the work area and be careful not to kink it.
   • 8 drive plate bolts, one at a time while rotating the crankshaft pulley.

12. Place a suitable jack under the transaxle and raise the jack just enough to take weight off of the mounts.

13. Remove or disconnect the following:
   • Intake manifold bracket.
   • Transaxle housing mounting bolts
   • Mounting bolts from the rear engine mount bracket.
   • 4 transaxle housing mounting bolts and 3 mount bracket nuts.

14. Pull the transaxle away from the engine until it clears the 14mm dowel pins, then lower it using the jack.

To install:

➡ **Use new self-locking nuts when assembling the front suspension components. Install new set rings onto the halfshaft inboard joint splines. Replace any color-coded self-locking bolts.**

15. Flush the transaxle cooler lines before installing the transaxle. Use a pressurized flushing unit such as Honda J38405-A or equivalent. Use only Honda biodegradable flushing fluid, Honda J35944–20. Other fluids will damage the automatic transmission cooling system.
   a. Fill the flusher with 21 ounces of fluid. Pressurize the flusher to 80–120 psi, following the procedure on the fluid container and flusher.
   b. Clamp the discharge hose of the flusher to the cooler return line. Clamp the drain hose to the cooler inlet line and route it into a bucket or drain tank.
   c. Connect the flusher to air and water lines. Open the flusher water valve and flush the cooler for 10 seconds. The air line should be equipped with a water trap to keep the system dry.
   d. Depress the flusher trigger to mix flushing fluid with the water. Flush for 2 minutes, turning the air valve on and off for 5 seconds every 15–20 seconds to create a surging action.

e. After finishing 1 flushing cycle, reverse the hose and flush in the opposite direction following the same steps.

f. Dry the cooler lines with compressed air so that no moisture is left in the cooler system.

16. Be sure the 2, 14mm dowel pins are installed into the torque converter housing.

17. Install or connect the following:
- Torque converter onto the transaxle mainshaft with a new hub O-ring.
- Starter motor onto the transaxle case and tighten the mounting bolts to 33 ft. lbs. (44 Nm).
- Transaxle and transaxle housing mounting bolts: 47 ft. lbs. (65 Nm)
- Rear engine mounting bolts: 40 ft. lbs. (54 Nm)
- Intake manifold bracket and tighten the bolts to 16 ft. lbs. (22 Nm).
- Upper bolts attaching the transaxle to the engine: 47 ft. lbs. (64 Nm)
- Front engine mount bracket bolts: 28 ft. lbs. (38 Nm)
- Transaxle mount and the nuts and bolt that attach the mount. Tighten the nuts first to 28 ft. lbs. (38 Nm), then tighten the bolt to 47 ft. lbs. (64 Nm).

18. Remove the jack from the transaxle.

19. Install or connect the following:
- Torque converter to the drive plate with the 8 bolts. Tighten the bolts in 2 steps in a crisscross pattern: first to 54 inch lbs. (6 Nm), and finally to 108 inch lbs. (12 Nm). Check for free rotation after tightening the last bolt.
- Shift control cable and control cable holder. Tighten the shift cable lockbolt to 10 ft. lbs. (14 Nm). Tighten the shift cable cover bolts to 13 ft. lbs. (18 Nm).
- Torque converter cover and tighten the bolts to 108 inch lbs. (12 Nm).

20. Remove the engine hoist.

21. Install or connect the following:
- Radius rod and damper fork
- Intermediate shaft into the differential and tighten the mounting bolts to 28 ft. lbs. (38 Nm).
- New set ring on the end of each halfshaft.

22. Turn the right steering knuckle fully outward and slide the axle into the differential until the set ring snaps into the differential side gear. Repeat the procedure on the left side.

23. Install or connect the following:
- Damper fork bolts and ball joint

nuts to the lower arms: 40 ft. lbs. (55 Nm) with a new cotter pin.
- Subframe center beam and tighten the center beam bolts to 28 ft. lbs. (39 Nm).
- Splash shield
- Front wheels and lower the vehicle.
- Speed sensor connector

24. Support the right front knuckle with a floor jack, until the weight of the vehicle is held by the jack. Tighten the damper fork pinch bolt to 32 ft. lbs. (44 Nm). Tighten the radius rod bolts to 76 ft. lbs. (105 Nm), and the radius rod nut to 32 ft. lbs. (44 Nm). Hold the damper fork bolt with a wrench, and tighten the nut to 40 ft. lbs. (55 Nm). Remove the floor jack.

25. Install or connect the following:
- Cables to the starter
- Throttle control cable
- Lock-up control solenoid valve and shift control solenoid valve connectors
- Speed sensor connectors and the transaxle ground cable
- Transaxle cooler inlet hose to the joint pipe. Attach a drain hose to the return line.
- Battery base stay and the battery base

26. Install the resonator assembly, the air cleaner assembly, and the air intake hose.
- Battery, positive then the negative battery cables

27. Refill the transaxle with ATF. Use only Honda Premium ATF or DEXRON®II ATF.

a. With the flusher drain hose attached to the cooler return line.

b. Place the transaxle in **P**, run the engine for 30 seconds, or until approximately 1 quart of fluid is discharged. Immediately shut off the engine. This completes the cooler flushing process.

c. Remove the drain hose and reconnect the cooler return line.

d. Refill the transaxle to the proper level with ATF.

28. Start the engine, set the parking brake, and shift the transaxle through all gears 3 times. Check for proper shift cable adjustment.

29. Let the engine reach operating temperature with the transaxle in **P** or **N**. Then, shut off the engine and check the fluid level.

30. Road test the vehicle.

31. After road testing the vehicle, loosen the front engine mount bracket bolts, then retighten them to 28 ft. lbs. (39 Nm).

32. Check and adjust the vehicle's front end alignment.

33. Enter the radio security code.

ACCORD—3.0L ENGINES

1. Before servicing the vehicle, refer to the precautions in the beginning of this section.

2. Remove or disconnect the following:
- Negative, then the positive battery cables
- Battery and tray
- Clamps securing the battery cables to the base.
- Intake air duct and the air cleaner assembly

3. Raise the vehicle and drain the transaxle fluid. Replace the drain plug with a new washer.

4. Remove or disconnect the following:
- Starter wiring and harness clamps, then the breather and radiator hoses from the retainer.
- Wiring connectors from the transaxle assembly
- Cooler lines; point them up to prevent fluid drainage.
- Bolt and nut, then the rear stiffener.
- Bolts securing the transaxle to the engine.
- Front mounting bracket bolts
- Engine under cover
- Lower shock absorber mounting and the lower ball joints from the control arms.
- Bolts securing the radius rods to the lower arms.
- Halfshafts. Keep the splined ends of the shafts clean.

➡**Matchmark the installed position of the sub-frame on the main-frame before removing it.**

- Sub-frame from the main-frame
- Engine brace from the rear of the engine
- Shift cable cover, bracket and cable
- 8 bolts securing the drive plate to the torque converter

5. Attach a chain hoist to the engine and raise it slightly.

6. Place a jack under the transaxle.

7. Remove or disconnect the following:
- Transaxle mount bracket
- Intake manifold support bracket
- Rear mount bracket

8. Pull the transaxle back slightly until it comes off the dowels and lower it from the vehicle. Do not let the torque converter fall out of the transaxle.

To install:

9. Install or connect the following:
- Torque converter using a new O-ring, if removed.

- Dowel pins in the torque converter housing.
- Transaxle to the engine and the rear mount bracket. Tighten the 8mm bolt to 16 ft. lbs. (22 Nm) and the 12mm bolts to 40 ft. lbs. (54 Nm).
- Transaxle-to-engine bolts: 47 ft. lbs. (64 Nm)
- Breather tube with the dot facing up
- Transaxle mount bracket. Tighten the nuts to 28 ft. lbs. (38 Nm) and the through-bolt to 40 ft. lbs. (54 Nm).
- Driveplate-to-torque converter bolts: 108 inch lbs. (12 Nm) in a crisscross pattern.
- Shift cable, bracket and cover
- Engine brace on the rear of the engine.
- Halfshafts
- Sub-frame after aligning the matchmarks. Tighten the rear bolts to 47 ft. lbs. (64 Nm) and the front bolts to 76 ft. lbs. (103 Nm).
- Front mount. Tighten the bolts to 28 ft. lbs. (38 Nm).
- Shock absorbers and the radius rods to the lower control arms.
- Engine under cover
- All the wiring connectors
- Starter wiring and harness clamps
- Battery
- Air cleaner assembly and intake duct

10. Refill the transaxle with Genuine Honda® premium automatic transmission fluid.

Clutch

REMOVAL & INSTALLATION

1. Before servicing the vehicle, refer to the precautions in the beginning of this section.
2. Remove or disconnect the following:
- Negative battery cable
3. Raise and safely support the vehicle.
4. Remove or disconnect the following:
- Transmission from the vehicle. Matchmark the flywheel and clutch for reassembly.
5. Use a flywheel ring-gear holder to lock the flywheel in position.
6. Remove or disconnect the following:
- Pressure plate bolts, 2 turns at a time working in a crisscross pattern to prevent warping the pressure plate.
- Pressure plate and clutch disc
7. Inspect the flywheel, disc, and pres-

sure plate for wear, cracks, and warpage. Light scoring of the flywheel may be polished out; gouges, warpage, burn marks, cracks, or chipped teeth require replacement of the flywheel.

➡ **If the flywheel is to be removed, but is going to be reused, matchmark it to the engine block prior to removal. Aligning the matchmarks upon reassembly will preserve driveline balance.**

8. Inspect the flywheel's ball bearing: turn the inner race of the bearing with your finger, and be sure it turns smoothly and quietly. If the bearing is loose or noisy, or exhibits rough motion, replace it.
9. Remove or disconnect the following:
- Release fork boot. Squeeze the release fork retaining spring to disengage the fork from its pivot.
- Release fork from the clutch housing.
- Release bearing. Spin the bearing by hand to check its degree of play. Replace the release bearing if it has excessive play or is leaking grease.
10. Inspect the rear main bearing oil seal for signs of leakage. If necessary, replace

the seal to prevent oil leakage onto the clutch's friction surfaces.

To install:
11. If necessary, drive out the flywheel bearing, then use a suitably-sized bearing driver to install a new one. Use a crisscross pattern to tighten the flywheel mounting bolts in several steps to 87 ft. lbs. (118 Nm) for vehicles with SOHC engines. If equipped with the B16A2 engine, tighten the flywheel bolts to 76 ft. lbs. (105 Nm). For 2.0L engines, tighten the flywheel bolts to 94 ft. lbs. (127 Nm).
12. Install or connect the following:
- Clutch disc and pressure plate by aligning the dowels on the flywheel with the dowel holes in the pressure plate. If a new pressure plate is not being installed, align the matchmarks that were made during removal.
- Pressure plate bolts, hand-tight
13. Insert a suitable clutch disc alignment tool into the splined hole in the clutch disc. Align the clutch and pressure plate.
14. Tighten the pressure plate bolts in a crisscross pattern 2 turns at a time to prevent warping the pressure plate. The final torque is 19 ft. lbs. (26 Nm).

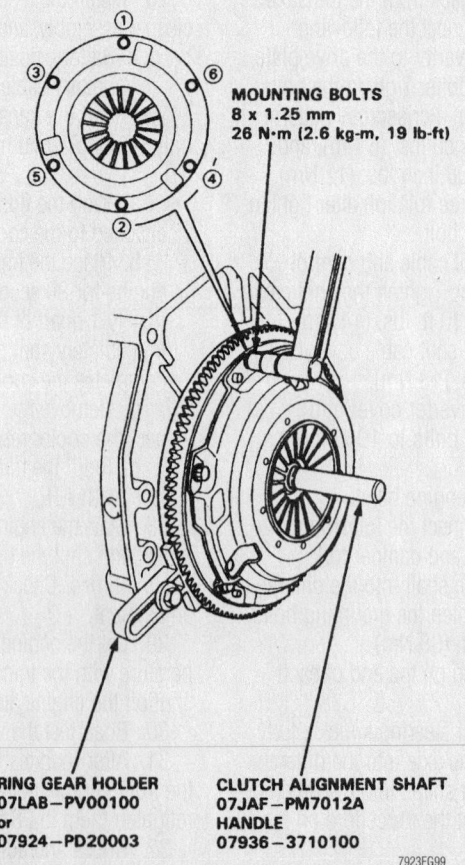

MOUNTING BOLTS
8 x 1.25 mm
26 N·m (2.6 kg-m, 19 lb-ft)

RING GEAR HOLDER
07LAB—PV00100
or
07924—PD20003

CLUTCH ALIGNMENT SHAFT
07JAF—PM7012A
HANDLE
07936—3710100

7923FG99

Clutch alignment tools and pressure plate torque sequence

15. Remove the alignment tool and ring gear holder.

16. Coat the mainshaft with heavy-duty high-temperature grease. The manufacturer recommends part No. 08798–9002, Honda super high-temp urea grease.

17. Coat the release fork pawls and the inner race of the release bearing with high temperature grease and install them into the clutch housing. Be sure the release fork retainer spring snaps into place on the pivot. The bearing and fork must fit together properly and slide back and forth smoothly.

18. Coat the tip of the slave cylinder with grease. Install the release fork boot.

19. Install or connect the following:
- Transmission, making sure the mainshaft is properly aligned with the clutch disc splines, and the transmission case dowels are properly aligned with the engine block.
- Transmission case bolts: 47 ft. lbs. (65 Nm), sequentially

20. Bleed the clutch hydraulic system.

21. Adjust the clutch pedal free-play.

22. Verify that all engine and transaxle components are installed and connected properly.

23. Reconnect the negative battery cable.

24. Road test the vehicle.

Hydraulic Clutch System

BLEEDING

1. Before servicing the vehicle, refer to the precautions in the beginning of this section.

2. Fill the clutch master cylinder reservoir with clean DOT 3 or 4 brake fluid.

3. Attach a rubber tube to the clutch slave cylinder bleed screw. Route the tube into a container of clean brake fluid.

4. Loosen the bleed screw.

5. Slowly pump the clutch pedal until the fluid draining from the slave cylinder is free of air bubbles.

6. Tighten the bleed screw to 72–84 inch lbs. (8–10 Nm).

7. Refill the clutch master cylinder reservoir with brake fluid.

Halfshaft

REMOVAL & INSTALLATION

Except S2000

1. Before servicing the vehicle, refer to the precautions in the beginning of this section.

2. Loosen the front spindle nut.

3. Raise and safely support the vehicle.

4. Remove or disconnect the following:
- Front wheels and the spindle nut

5. Drain the transaxle fluid and install the drain plug with a new washer. If the halfshaft to be removed is installed into the intermediate shaft, the transaxle fluid does not need to be drained.
- Flange nut, while holding the stabilizer ball joint pin with a hex wrench
- Front stabilizer link from the lower arm
- Damper fork nut and damper pinch bolt(s)
- Damper fork
- Cotter pin and castle nut from the lower arm ball joint

6. Install a hex nut flush onto the ball joint stud to prevent the ball joint tool from damaging the stud threads.

7. Using a ball joint tool, separate the lower arm from the knuckle.

8. Pull the knuckle outward.

9. Remove or disconnect the following:
- Heat shield, if equipped
- Halfshaft outboard joint from the hub by tapping it with a plastic hammer.
- Inner CV-joint away from the transaxle case to force the halfshaft set ring out of the groove.
- Halfshaft from the differential case or intermediate shaft by pulling on the inboard CV-joint

➡ Do not pull on the halfshaft as the CV-joint may come apart. Use care when prying out the assembly and pull it straight to avoid damaging the differential oil seal or intermediate shaft oil or dust seals.

To install:

10. Replace the differential oil seal or intermediate shaft seal if either were damaged during removal.

11. Install or connect the following:
- New set rings on the ends of the halfshafts

12. For the 2003 Accord, apply 0.02–0.04 oz. of grease to the whole splined surface of the right side halfshaft. Then, remove the grease from the splined grooves at intervals of 2–3 splines and from the set ring groove to let air bleed from the intermediate shaft.
- Halfshafts and be sure the set ring locks in the differential gear groove and the halfshaft bottoms in the differential or intermediate shaft.
- Outboard joint into the hub. Be sure

the splines mesh together and the joint is fully seated into the hub.
- Heat shield, if equipped
- Ball joint stud into the lower control arm.

13. Torque the new castle nut to the lower of the following specifications, then tighten only enough to install a new cotter pin. Never loosen the nut to install the cotter pin.

    a. 2000–02 Civic: 36–43 ft. lbs. (49–59 Nm)

    b. 2003 Civic: 43–51 ft. lbs. (59–69 Nm)

    c. Prelude: 36–43 ft. lbs. (49–59 Nm)

    d. 2000–02 Accord: 36–43 ft. lbs. (49–59 Nm)

    e. 2003 Accord: 65–72 ft. lbs. (88–98 Nm)

14. Install or connect the following:
- Damper fork into position. Make sure the aligning tab is lined up with the slot in the damper fork. Tighten the retainers loosely
- Front stabilizer link to the lower arm. Hold the stabilizer link ball joint pin with a hex wrench and tighten the new flange nut to 22 ft. lbs. (30 Nm).
- Tighten the upper damper pinch bolt to 32 ft. lbs. (44 Nm) and the fork nut to 47 ft. lbs. (65 Nm).
- Front wheels
- New spindle nut; don't tighten it yet.

15. Lower the vehicle.

16. Tighten the spindle nut to 134 ft. lbs. (181 Nm) for the Civic, and 4-cylinder Accord with A/T. For all other models, tighten the spindle nut to 181 ft. lbs. (245 Nm) and stake its tab. Tighten the wheel nuts to 80 ft. lbs. (110 Nm).

17. Fill the transaxle with the proper type and quantity of fluid.

18. Warm the engine up, check the transaxle fluid level, and road test the vehicle.

S2000

1. Before servicing the vehicle, refer to the precautions in the beginning of this section.

2. Remove or disconnect the following:
- Rear wheel
- Spindle nut
- Lower ball joint
- Wheel speed sensor harness
- Inboard joint mounting bolts

3. Pull the knuckle outward to separate the inboard joint from the differential.

4. Remove the outboard joint from the wheel hub by tapping it with a plastic-faced hammer.

To install:

5. Installation is the reverse of the

removal procedure, while using the following torque values:

- Inboard joint mounting bolts: 61 ft. lbs. (83 Nm)
- Lower ball joint nut: 43–51 ft. lbs. (69–78 Nm)
- Wheel speed sensor harness bolts: 88 inch lbs. (10 Nm)
- Spindle nut: 181 ft. lbs. (245 Nm)
- Wheel lug nuts: 80 ft. lbs. (108 Nm)

CV-Joints

OVERHAUL

1. Before servicing the vehicle, refer to the precautions in the beginning of this section.

2. Remove the halfshaft.

3. Remove the large retaining band from the inboard boot. Remove the smaller band from the inboard boot and slide the boot off the joint.

4. Carefully remove the stub end of the inboard joint. Check the splines for cracks, wear or damage. Check the inside bore for any sign of wear.

5. Remove and discard the snapring from the end of the halfshaft. This will allow removal of the spider assembly.

6. Mark the rollers, spider and the stub end of the axle so that all parts may be reassembled in the same position. Remove the rollers from the spider.

7. Remove the second snapring from the shaft. Remove the joint boot. If equipped, remove the dynamic damper from the shaft.

8. If the outer joint's boot is to be replaced, remove the boot clamps and slide the boot off the joint, then off the shaft. Hold the outer joint and swivel the end. If the joint is noisy, it must be replaced. The replacement joint will come with a new shaft; the inner joint must be assembled onto the shaft.

9. Clean and inspect all disassembled parts. Any sign of wear requires replacement.

To install:

10. Thoroughly pack the inboard and outboard joints with moly grease. Use only moly grease; other lubricants will not last. Wrap the splines of the shaft in vinyl or electrical tape to protect the boots as they are installed.

11. Slide the boot for the outer joint over the shaft and onto the joint. Do not install the bands yet.

Left driveshaft:
Japan-produced models:
  M/T: 788—793 mm (31.0—31.2 in.)
  A/T: 792—797 mm (31.2—31.4 in.)
U.S.-and Canada-produced models:
  M/T: 788.2—793.2 mm (31.0—31.2 in.)
  A/T: 791.7—796.7 mm (31.2—31.4 in.)

Right driveshaft:
Japan-produced models:
  '01-02 models: 502—507 mm (19.8—20.0 in.)
  '03 model M/T: 497—502 mm (19.6—19.8 in.)
  '03 model A/T: 502—507 mm (19.8—20.0 in.)
U.S.-and Canada-produced models:
      501.2—506.2 mm (19.7—19.9 in.)

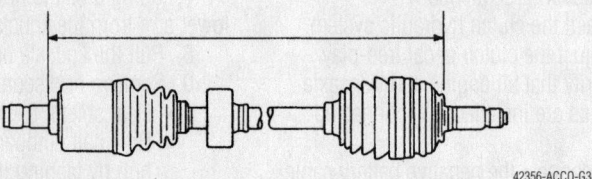

42356-ACCO-G34

Halfshaft length specifications—Civic

LEFT DRIVESHAFT

RIGHT DRIVESHAFT

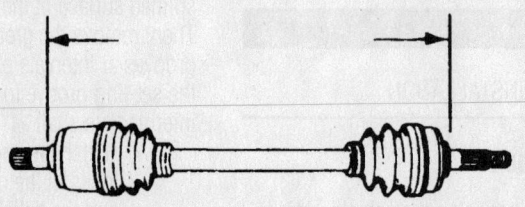

86807027

Halfshafts must be set to the correct length before installing boot bands

12. Slide the inner boot onto the shaft. Install the dynamic damper if it was removed.

13. Install the inboard snapring on the shaft. Install the rollers and bearing races on the spider shafts. Hold the shaft upright, then slide the spider assembly into the inboard shaft joint. Install the outer snapring.

14. Slide the boots over both joints. Position the small end of the boot so that the band will be centered between the locating humps on the shaft. Install the band; bend both sets of locking tabs. Once the band is in place, expand and compress the boots once or twice; allow the boots to return to their normal size and length.

15. Adjust the length of the halfshaft to the specifications listed below, then adjust the boots to halfway between full compression and full extension. Make sure the ends of the boots seat in the groove of the driveshaft and joint. Correct shaft lengths are:

- 2000–02 Accord with M/T—Left

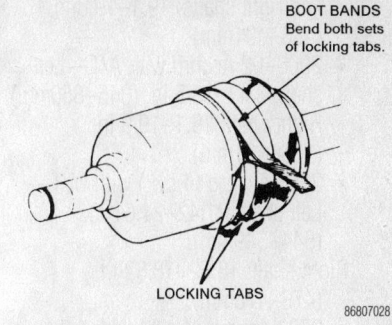

BOOT BANDS
Bend both sets of locking tabs.

LOCKING TABS

86807028

Always use new boot bands

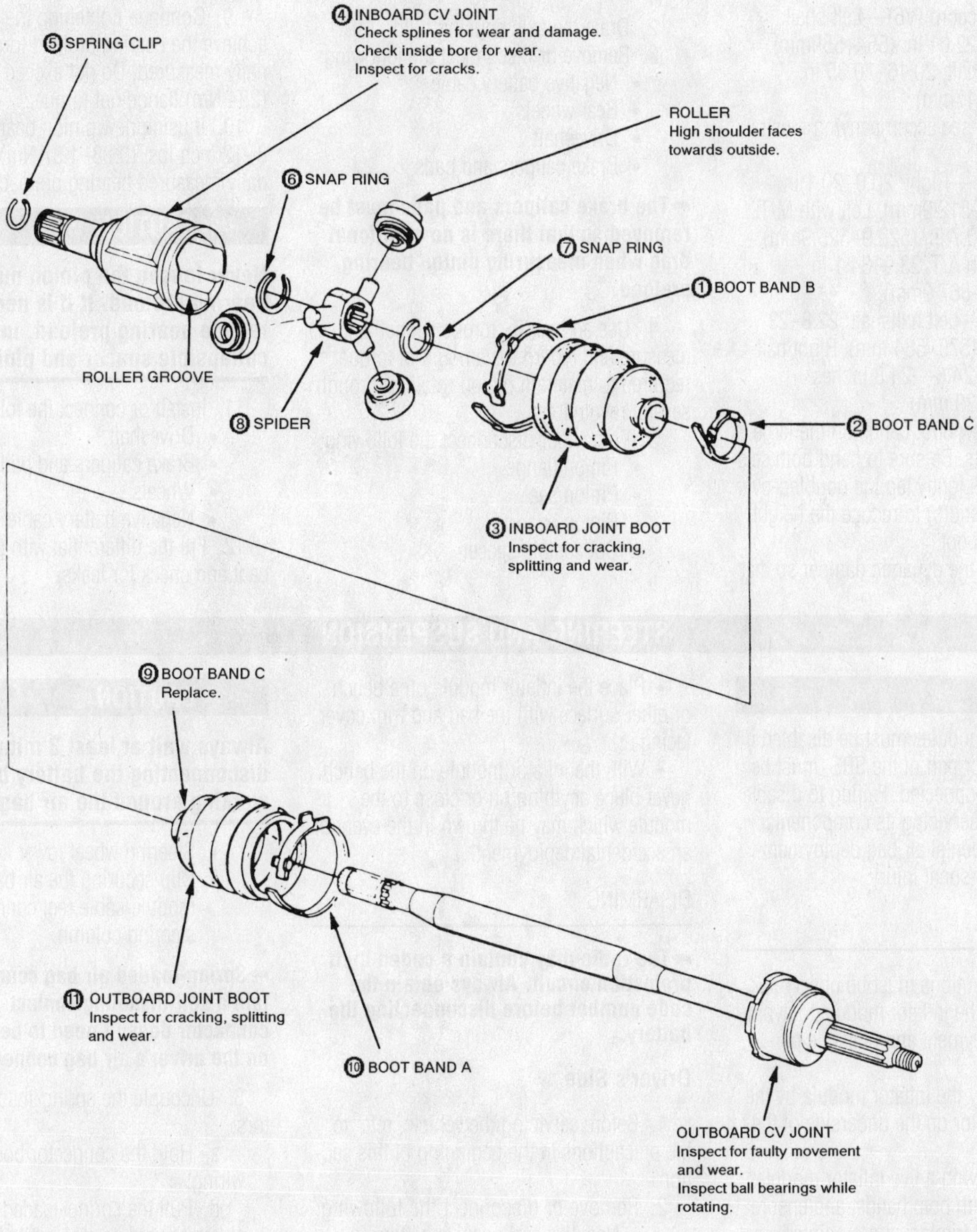

⑤ SPRING CLIP

④ INBOARD CV JOINT
Check splines for wear and damage.
Check inside bore for wear.
Inspect for cracks.

ROLLER
High shoulder faces towards outside.

⑥ SNAP RING

⑦ SNAP RING

① BOOT BAND B

ROLLER GROOVE

⑧ SPIDER

② BOOT BAND C

③ INBOARD JOINT BOOT
Inspect for cracking, splitting and wear.

⑨ BOOT BAND C
Replace.

⑪ OUTBOARD JOINT BOOT
Inspect for cracking, splitting and wear.

⑩ BOOT BAND A

OUTBOARD CV JOINT
Inspect for faulty movement and wear.
Inspect ball bearings while rotating.

86807026

Exploded view of the halfshaft

and right shafts: 19.1–19.3 in. (486–491mm).
- 2000–02 Accord with A/T—Left shaft: 33.3–33.5 in. (845–850mm). Right shaft: 19.1–19.9 in. (486–491mm).
- 2003 Accord (4 cyl.) with M/T— Left shaft: 21.42–21.61 in. (544–549mm). Right shaft: 18.62–18.82 in. (473–478mm).
- 2003 Accord (4 cyl.) with A/T— Left shaft: 21.61–21.81 in. (549–554mm). Right shaft: 33.43–33.62 in. (849–854mm).
- 2003 Accord (V6)—Left shaft: 21.81–22.01 in. (554–559mm). Right shaft: 20.16–20.35 in. (512–517mm)
- Civic—see accompanying illustration.
- Prelude—Right: 20.0–20.1 in. (507.9–512.9mm). Left with M/T: 20.5–20.7 in. (520.9–525.9mm). Left with A/T: 33.9–34.1 in. (862.9–867.9mm).
- S2000—Left halfshaft: 22.8–23 inches (579–584 mm). Right halfshaft—24.6—24.8 inches (624–629 mm).

16. Install new boot bands on the large ends of the boots. Be sure to bend both sets of locking tabs. Lightly tap the doubled-over portion of the band(s) to reduce the height. Do NOT hit the boot.

17. Position the dynamic damper so that it is 0.1–1.2 in. (3–7mm) from the CV-boot. Install a new retaining band in the same fashion as the boot bands.

18. Install a new snapring on the inboard end of the joint, then install the halfshaft.

Pinion Seal

REMOVAL & INSTALLATION

S2000

1. Before servicing the vehicle, refer to the precautions in the beginning of this section.
2. Drain the axle housing fluid.
3. Remove or disconnect the following:
   - Negative battery cable
   - Rear wheels
   - Driveshaft
   - Brake calipers and pads

➡ **The brake calipers and pads must be removed so that there is no additional drag when measuring pinion bearing preload.**

4. Use an inch lb. torque wrench and measure and record the amount of torque required to maintain pinion rotation through several revolutions.
5. Remove or disconnect the following:
   - Pinion flange
   - Pinion seal
   - Pinion bearing
   - Collapsible spacer

To install:

➡ **Use a new collapsible spacer and flange nut for assembly.**

6. Install or connect the following:
   - Collapsible spacer
   - Pinion bearing
   - Pinion seal
   - Pinion flange
7. Rotate the pinion flange occasionally while tightening the flange nut to make sure the pinion bearings seat correctly.
8. Tighten the flange nut to 94 ft. lbs. (127 Nm) and then measure bearing preload torque.
9. Continue tightening the flange nut to achieve the bearing preload torque originally measured. Do not exceed 210 ft. lbs. (284 Nm) flange nut torque.
10. If using new pinion bearings, add 8–12 inch lbs. (0.88–1.37 Nm) to the originally measured bearing preload.

❋❋ CAUTION

Never loosen the pinion nut to reduce bearing preload. If it is necessary to reduce bearing preload, install a new collapsible spacer and pinion nut.

11. Install or connect the following:
   - Driveshaft
   - Brake calipers and pads
   - Wheels
   - Negative battery cable
12. Fill the differential with gear lubricant and check for leaks.

STEERING AND SUSPENSION

Air Bag

The air bag modules must be disabled if they, or any other part of the SRS, must be serviced or disconnected. Failing to disable the SRS before servicing its components may cause accidental air bag deployment and possible personal injury.

PRECAUTIONS

Several precautions must be observed when handling the inflator module to avoid accidental deployment and possible personal injury.
- Never carry the inflator module by the wires or connector on the underside of the module.
- When carrying a live inflator module, hold securely with both hands, and ensure that the bag and trim cover are pointed away.

- Place the inflator module on a bench or other surface with the bag and trim cover facing up.
- With the inflator module on the bench, never place anything on or close to the module which may be thrown in the event of an accidental deployment.

DISARMING

➡ **The radio may contain a coded theft protection circuit. Always obtain the code number before disconnecting the battery.**

Driver's Side

1. Before servicing the vehicle, refer to the precautions in the beginning of this section.
2. Remove or disconnect the following:
   - Negative and positive battery cables

❋❋ CAUTION

Always wait at least 3 minutes after disconnecting the battery before working around the air bag.

- Steering wheel lower access cover
- Clip securing the air bag module/cable reel connection to the steering column.

➡ **Spring-loaded air bag connectors contain a self-disabling contact. A shorting connector doesn't need to be installed on the driver's air bag connector.**

3. Uncouple the spring-loaded connectors:
   a. Hold the connector body, not the wiring.
   b. Pull the spring-loaded locking sleeve toward its stop while holding the opposite half of the connector.

c. After releasing the locking sleeve, uncouple the connectors.

Passenger's Side

1. Before servicing the vehicle, refer to the precautions in the beginning of this section.
2. Remove or disconnect the following:
   • Negative and positive battery cables

❄❄ CAUTION

Always wait at least 3 minutes after disconnecting the battery before working around the air bag.

   • Glove box door and frame
   • Lower mounting brackets that may cover the air bag connection, if equipped.
   • Passenger's air bag connector. Pull the spring-loaded sleeve toward the stop while holding the opposite half of the connector and pull the connector apart.

REARMING

Driver's Side

1. After servicing has been completed, couple the air bag and cable reel connectors. Press the sleeve side of the connector into the pawl side until the sleeve locks the connectors together.
2. Install or connect the following:
   • Clip securing the air bag/cable reel connection to the steering column.
   • Access cover
   • Positive and negative battery cables
3. Turn the ignition switch to the **ON** position, but don't start the engine. The air bag indicator light should turn on for 6 seconds, then turn off. If the air bag indicator light doesn't come on, or stays on longer than 6 seconds, the system fault must be diagnosed.
4. Enter the radio security code.

Passenger's Side

1. After servicing has been completed, immediately couple the air bag and cable reel connectors.
2. Install or connect the following:
   • Any lower mounting brackets that may have been removed.
   • Glove box frame and glove box door
   • Positive and negative battery cables
3. Turn the ignition switch to the **ON** position, but don't start the engine. The air

bag indicator light should turn on for 6 seconds, then turn off. If the air bag indicator light doesn't come on, or stays on longer than 6 seconds, the system fault must be diagnosed.
4. Enter the radio security code.

Rack and Pinion Steering Gear

REMOVAL & INSTALLATION

Manual

CIVIC

❄❄ CAUTION

The air bag must be disabled before removing the steering wheel to center the cable reel. Failure to disarm the air bag system may cause accidental air bag deployment, resulting in unnecessary air bag system repairs and the risk of personal injury.

1. Before servicing the vehicle, refer to the precautions in the beginning of this section.
2. Position the front wheels straight ahead. Lock the steering column and remove the ignition key.
3. Remove or disconnect the following:
   • Negative, then the positive battery cables.
4. Disable the air bag.
5. Remove or disconnect the following:
   • Steering joint cover
   • Upper and lower steering joint bolts
6. Raise and support the vehicle safely.
7. Remove or disconnect the following:
   • Front wheels
   • Tie-rod end cotter pins and castle nuts
   • Tie-rod ends from the steering knuckles, using a ball joint tool.
   • Left tie-rod end and slide the rack all the way to the right.
   • Self-locking nuts, then separate the catalytic converter or front exhaust pipe from the rear exhaust pipes.
   • Catalytic converter or front exhaust pipe
8. If equipped with a manual transaxle, remove or disconnect the following:
   • Shift lever extension rod from the clutch housing
   • Pin retainer out of the way, then drive out the spring pin
   • Shift rod
9. If equipped with an automatic

transaxle, remove or disconnect the following:
   • Shift cable bracket and holder
   • Shift cable from the control shaft. Suspend the cable from the underbody with a piece of wire.
10. Remove or disconnect the following:
   • Steering rack stiffener plate
   • Steering rack mounting bracket
   • Steering rack from the pinion shaft, by pulling the rack down.
11. Drop the steering rack far enough to permit the end of the pinion shaft and the grommet to come out of the hole in the bulkhead.
12. Slide the gearbox to the right until the left tie rod clears the subframe, then drop it down and out of the vehicle to the left.

To install:

➡**Use new self-locking nuts and gaskets when installing the catalytic converter.**

13. Install or connect the following:
   • Steering rack into position
   • Pinion shaft grommet, insert the pinion through the hole in the bulkhead.
   • Steering rack mounting cushion, bracket, and bolts. The arrow on the bracket faces the front of the vehicle. Tighten the bracket bolts to 28 ft. lbs. (39 Nm).
   • Steering rack stiffener plate. Tighten the steering rack mounting bolts to 43 ft. lbs. (59 Nm). Tighten the stiffener plate bolts to 28 ft. lbs. (39 Nm).
14. Center the rack ends within their steering strokes.
15. Install or connect the following:
   • Tie rod ends onto the rack ends
   • Tie rod ends to the steering knuckles, then the castle nuts.
   • Front wheels
   • Catalytic converter using new gaskets and self-locking nuts. Tighten the front nuts to 16 ft. lbs. (22 Nm), and the rear nuts to 25 ft. lbs. (34 Nm).
16. If equipped with a manual transaxle, install or connect the following:
   • Shift linkage with a new spring pin and clip.
   • Extension rod and tighten its bolt to 16 ft. lbs. (22 Nm).
17. If equipped with an automatic transaxle, install or connect the following:
   • Shift cable and brackets. Tighten the bracket bolts to 108 inch lbs. (12 Nm). Tighten the cable lockbolt to 10

ft. lbs. (14 Nm). Tighten the cable holder bolts to 16 ft. lbs. (22 Nm).

18. Verify that the rack is centered within its strokes. Lower the vehicle.

19. Center the air bag cable reel as follows:

    a. Remove the steering wheel.

    b. Turn the cable reel clockwise until it stops.

    c. Turn the steering wheel counterclockwise, approximately 2 turns, until the arrow on the label points straight up.

    d. Install the steering wheel.

20. During steering wheel installation, verify that the slot on the steering wheel shaft engages with the tabs on the turn signal canceling sleeve. The pins on the cable reel fit into the holes on the steering wheel body. Install a new steering wheel nut and tighten it to 36 ft. lbs. (50 Nm).

21. Line up the bolt hole in the steering joint with the groove in the pinion shaft. Slip the joint onto the pinion shaft. Pull the joint up and down to be sure the splines are fully seated. Tighten the joint bolts to 16 ft. lbs. (22 Nm).

➡**Connect the steering joint and pinion shaft with the cable reel and steering rack centered. Verify that the lower joint bolt is securely seated in the pinion shaft groove. If the steering wheel and rack are not centered, reposition the serrations at the lower end of the steering joint.**

22. Install or connect the following:
- Steering joint cover

23. Tighten the ball joint castle nuts to 29–35 ft. lbs. (40–48 Nm). Then, tighten them only enough to install new cotter pins.

24. Enable the air bag.

25. Install or connect the following:
- Steering wheel's lower access cover
- Negative and positive battery cables

26. Turn the ignition switch to the **ON** position. The air bag indicator light should come on for 6 seconds, then turn off. This light sequence indicates that the air bag system is enabled and functioning normally. If the air bag light stays on longer, or doesn't turn on, the system must be diagnosed.

27. Check the front wheel alignment and steering wheel spoke angle. Make adjustments by turning the left and right tie-rod ends equally.

28. Road test the vehicle.

Power

➡**The radio may contain a coded theft protection circuit. Always obtain the** code number before disconnecting the battery. If the vehicle is equipped with 4WS, the steering control unit is shut down when the battery is disconnected. After connecting the battery, turn the steering wheel lock-to-lock to reset the steering control unit.

CIVIC

⁂ CAUTION

The air bag must be disabled before removing the steering wheel to center the cable reel. Failure to disarm the air bag system may cause accidental air bag deployment, resulting in unnecessary air bag system repairs and the risk of personal injury.

1. Before servicing the vehicle, refer to the precautions in the beginning of this section.

2. Remove or disconnect the following:
- Power steering reservoir off of its mount
- Inlet hose

3. Insert a length of tubing into the inlet hose and route the tubing into a drain container.

4. With the engine running at idle, turn the steering wheel lock-to-lock several times until fluid stops running out of the hose.

5. Position the front wheels straight ahead. Shut off the engine and lock the steering column and remove the ignition key. Reconnect the reservoir inlet hose.

6. Remove or disconnect the following:
- Negative and positive battery cables. Wait 3 minutes before working around the air bags.
- Steering wheel's lower access cover

7. Uncouple the air bag connector from the cable reel connector as follows:

    a. Hold the cable reel connector. With your other hand, slide the spring-loaded sleeve toward the stop tab on the air bag connector.

    b. Separate the 2 connectors. There is no need to install a shorting connector, as the connectors are automatically grounded when they are uncoupled.

8. Remove or disconnect the following:
- Steering joint cover, then the upper and lower steering joint bolts.

9. Raise and support the vehicle safely.

10. Remove or disconnect the following:
- Front wheels
- Tie rod end cotter pins and castle nuts
- Tie rod ends from the steering knuckles, using a ball joint tool

11. If equipped with a manual transaxle, remove or disconnect the following:
- Shift lever extension rod from the clutch housing
- Pin retainer out of the way, then drive out the spring pin
- Shift rod

12. If equipped with an automatic transaxle, remove or disconnect the following:
- Shift cable bracket and holder
- Shift cable from the control shaft. Suspend the cable from the underbody with a piece of wire.

13. Remove or disconnect the following:
- Self-locking nuts and separate the catalytic converter from the exhaust pipes.
- Catalytic converter
- Hydraulic line and hose from the rack valve body using a flare nut wrench.
- Left tie rod end and slide the rack all the way to the right.
- Steering rack mounting bolts
- Steering rack from the pinion shaft by pulling the rack downward

14. Drop the gearbox far enough to permit the end of the pinion shaft to come out of the hole in the frame channel.

15. Slide the gearbox to the right until the left tie rod clears the subframe, then drop it down and out of the vehicle to the left.

To install:

➡**Use new self-locking nuts when installing the catalytic converter.**

⁂ WARNING

Use only genuine Honda power steering fluid. Any other type or brand of fluid will damage the power steering pump.

16. Install or connect the following:
- Steering rack into position
- Pinion shaft grommet, insert the pinion through the hole in the bulkhead.
- Rack mounting bolts. Tighten the bracket bolts to 28 ft. lbs. (39 Nm). Tighten the mounting bolt under the valve body to 43 ft. lbs. (59 Nm).
- 2 hydraulic lines to the rack valve body. Carefully tighten the hydraulic line fitting to 28 ft. lbs. (38 Nm). Securely tighten the return hose clamp.

17. Center the rack ends within their steering strokes.

18. Install or connect the following:
- Tie rod ends onto the rack ends
- Tie rod ends to the steering knuckles, then the castle nuts.
- Front wheels
- Catalytic converter using new gaskets and self-locking nuts. Tighten the front nuts to 16 ft. lbs. (22 Nm), and the rear nuts to 25 ft. lbs. (34 Nm).

19. If equipped with a manual transaxles, install or connect the following:
- Shift linkage with a new spring pin and clip.
- Extension rod and tighten its bolt to 16 ft. lbs. (22 Nm).

20. If equipped with an automatic transaxles, install or connect the following:
- Shift cable and brackets. Tighten the bracket bolts to 108 inch lbs. (12 Nm). Tighten the cable lockbolt to 10 ft. lbs. (14 Nm). Tighten the cable holder bolts to 16 ft. lbs. (22 Nm).

21. Verify that the rack is centered within its strokes. Lower the vehicle.

22. Center the air bag cable reel as follows:
 a. Remove the steering wheel.
 b. Turn the cable reel clockwise until it stops.
 c. Turn the steering wheel counterclockwise (approximately 2 turns) until the arrow on the label points straight up.
 d. Install the steering wheel.

23. During steering wheel installation, verify that the slot on the steering wheel shaft engages with the tabs on the turn signal canceling sleeve. The pins on the cable reel fit into the holes on the steering wheel body. Install a new steering wheel nut and tighten it to 36 ft. lbs. (50 Nm).

24. Line up the bolt hole in the steering joint with the groove in the pinion shaft. Slip the joint onto the pinion shaft. Pull the joint up and down to be sure the splines are fully seated. Tighten the joint bolts to 16 ft. lbs. (22 Nm).

➡ **Connect the steering joint and pinion shaft with the cable reel and steering rack centered. Verify that the lower joint bolt is securely seated in the pinion shaft groove. If the steering wheel and rack are not centered, reposition the serrations at the lower end of the steering joint.**

25. Install or connect the following:
- Steering joint cover

26. Tighten the ball joint castle nuts to 29–35 ft. lbs. (40–48 Nm). Then, tighten them only enough to install new cotter pins.

27. Install or connect the following:
- Air bag and cable reel connectors: Be sure the connectors fit squarely together. Then, press the connectors to couple them. The spring-loaded sleeve will lock into place as the 2 connectors are coupled.
- Steering wheel lower access cover
- Negative and positive battery cables

28. Turn the ignition switch to the **ON** position. The air bag indicator light should come on for 6 seconds, then turn off. This light sequence indicates that the air bag system is enabled and functioning normally. If the air bag light stays on longer, or doesn't turn on, the system must be diagnosed.

29. Be sure the reservoir inlet line has been reconnected. Fill the reservoir to the upper line with Honda power steering fluid. Run the engine at idle and turn the steering wheel lock-to-lock several times to bleed air from the system and fill the rack valve body. Recheck the fluid level and add more if necessary.

30. Check the power steering system for leaks.

31. Check the front wheel alignment and steering wheel spoke angle. Make adjustments by turning the left and right tie rod ends equally.

32. Road test the vehicle.

PRELUDE

➡ **The electronic neutral check must be performed on 4WS equipped Preludes any time the steering rack, steering wheel, or steering column is removed, and before the wheels are aligned.**

1. Before servicing the vehicle, refer to the precautions in the beginning of this section.

2. Remove or disconnect the following:
- Power steering reservoir off of its mount
- Inlet hose

3. Insert a length of tubing into the inlet hose and route the tubing into a drain container.

4. With the engine running at idle, turn the steering wheel lock-to-lock several times until fluid stops running out of the hose. Shut off the engine.

5. Position the front wheels straight ahead. Lock the steering column with the ignition key. Reconnect the reservoir inlet hose.

6. Remove or disconnect the following:
- Negative battery cable
- Steering joint cover, then the upper and lower steering joint bolts.

7. Raise and support the vehicle safely.

8. Remove or disconnect the following:
- Front wheels
- Tie rod end cotter pins and castle nuts. Install a 12mm nut onto the end of the ball joint stud to protect the threads from damage.
- Tie rod ends from the steering knuckles, using a ball joint tool
- Heated Oxygen Sensor (HO2S) sensor connector
- Self-locking nuts, then separate the catalytic converter from the exhaust pipe.
- Exhaust pipe from the intake manifold
- Exhaust pipe from the vehicle

9. If equipped with an automatic transaxle, remove or disconnect the following:
- Remove the shift cable cover
- Shift cable, and wire it up and out of the way.

➡ **Clean any oil or dirt off of the valve body with solvent.**

- Center beam from the subframe
- Valve body shield
- 4 hydraulic lines from the rack valve body, using a flare nut wrench. Plug the lines to keep dirt and moisture out.

10. On models with 4-Wheel Steering (4WS) remove or disconnect the following:
- Carefully cut the wire tie securing the cover to the front sub-steering angle sensor
- Cover
- Sensor wiring harness from the 2 securing clamps
- Sensor connector from the 4WS steering main wiring harness.

11. Remove or disconnect the following:
- Steering joint bolt, then slide the pinion shaft out of the joint.
- Left mounting bracket, then the right mounting brackets.
- Left tie rod end and slide the rack all the way to the right.

12. Pull the steering rack down to release it from the pinion shaft.

13. Slide the steering rack to the right until the left tie rod clears the subframe, then drop it down and out of the vehicle to the left.

To install:

➡ **Use new gaskets and self-locking nuts when installing the exhaust pipe.**

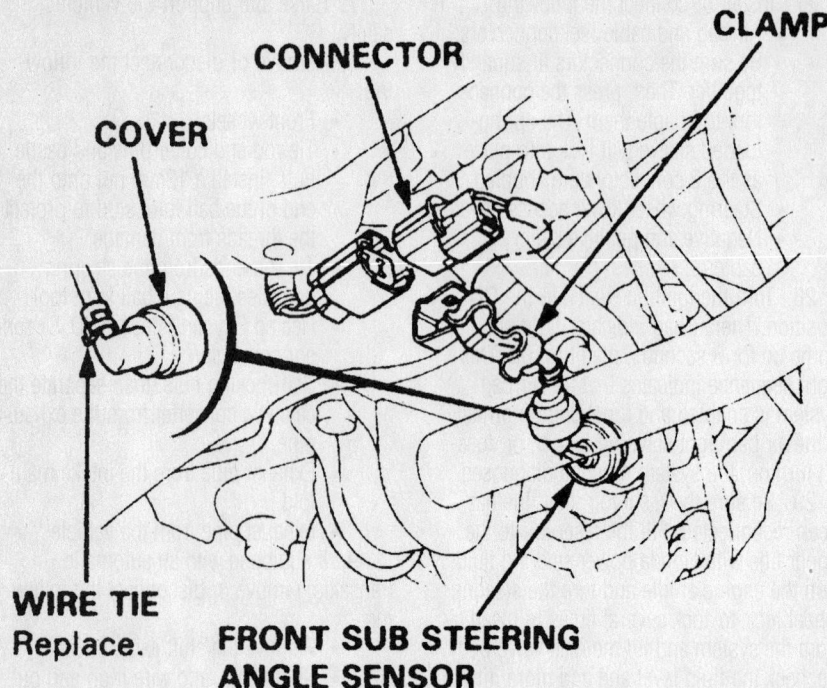

4WS front sub-steering angle sensor—Prelude

7923FGA2

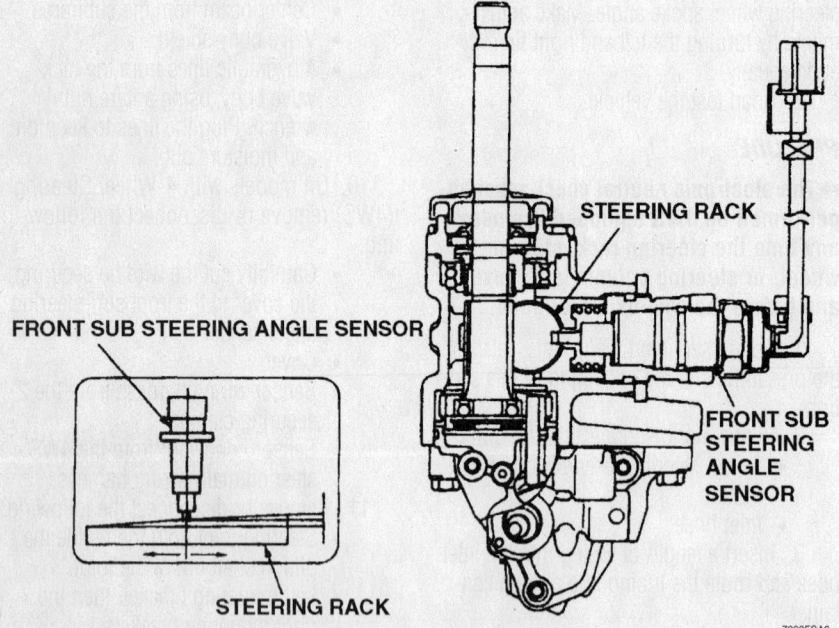

Front sub-steering angle sensor, harness, and steering rack—Prelude

7923FGA3

✳✳ WARNING

Use only genuine Honda power steering fluid. Any other type or brand of fluid will damage the power steering pump.

14. Install or connect the following:
  • Steering rack into position
  • Pinion shaft grommet and insert

the pinion through the hole in the firewall.
  • Right and left mounting brackets. Tighten the short bolts to 28 ft. lbs. (39 Nm), and the long bolts to 32 ft. lbs. (44 Nm).
15. Center the rack ends within their steering strokes.
16. Center the air bag cable reel as follows:

  a. Turn the steering wheel clockwise until it stops.
  b. Turn the steering wheel counter-clockwise until the yellow gear tooth lines up with the alignment mark on the lower column cover.
17. Line up the bolt hole in the steering joint with the groove in the pinion shaft. Slip the joint onto the pinion shaft. Pull the joint up and down to be sure the splines are fully seated. Tighten the joint bolts to 16 ft. lbs. (22 Nm).

➡**Connect the steering joint and pinion shaft with the cable reel and steering rack centered. Verify that the lower joint bolt is securely seated in the pinion shaft groove. If the steering wheel and rack are not centered, reposition the serrations at the lower end of the steering joint.**

18. Install or connect the following:
  • 4 hydraulic lines to the rack valve body. Carefully tighten the 12mm fittings to 108 inch lbs. (13 Nm), the 14mm inlet fitting to 28 ft. lbs. (37 Nm), and the 17mm oil cooler fitting to 21 ft. lbs. (29 Nm).
  • Front sub-steering angle sensor to the 4WS harness.
  • Wire back into its clamps, making sure that it doesn't interfere with the stabilizer bar.
  • Sensor cover with a new wire tie
  • Valve body shield
  • Center beam. Use new self-locking bolts and tighten them to 43 ft. lbs. (60 Nm).
19. If equipped with an automatic transaxle, install or connect the following:
  • Shift cable and tighten the locknut to 10 ft. lbs. (14 Nm).
  • Cable holder and tighten its bolts to 13 ft. lbs. (18 Nm).
20. Install or connect the following:
  • Catalytic converter using new gaskets and self-locking nuts. Tighten the exhaust manifold nuts to 40 ft. lbs. (55 Nm), and the rear nuts to 25 ft. lbs. (34 Nm).
  • HO2S connector
  • Tie rod ends onto the rack ends
  • Tie rod ends to the steering knuckles, then the castle nuts.
  • Front wheels
21. Verify that the rack is centered within its strokes. Lower the vehicle.
22. Install the steering joint cover.
23. Tighten the ball joint castle nuts to 36–43 ft. lbs. (50–60 Nm). Then, tighten them only enough to install new cotter pins.
24. Reconnect the negative battery cable.

25. Be sure the reservoir inlet line has been reconnected. Fill the reservoir to the upper line with Honda power steering fluid. Run the engine at idle and turn the steering wheel lock-to-lock several times to bleed air from the system and fill the rack valve body. Recheck the fluid level and add more if necessary.

26. Check the power steering system for leaks.

27. On Preludes without 4WS, check and adjust the front wheel alignment. On Preludes with 4WS, the electronic neutral check must be performed on the 4WS system.

ACCORD

1. Before servicing the vehicle, refer to the precautions in the beginning of this section.

2. Lift the power steering reservoir off of its mount and disconnect the inlet hose.

3. Insert a length of tubing into the inlet hose and route the tubing into a drain container.

4. With the engine running at idle, turn the steering wheel lock-to-lock several times until fluid stops running out of the hose. Immediately shut off the engine.

5. Position the front wheels straight ahead. Lock the steering column with the ignition key. Reconnect the reservoir inlet hose.

6. Remove or disconnect the following:
- Negative battery cable
- Steering joint cover and the upper and lower steering joint bolts.

7. Raise and support the vehicle safely.

8. Remove or disconnect the following:
- Front wheels
- Tie rod end cotter pins and castle nuts
- Tie rod ends from the steering knuckles, using a ball joint tool.
- Left tie rod end and slide the rack all the way to the right.
- Heated Oxygen Sensor (HO$_2$S) connector
- Self-locking nuts, then separate the catalytic converter from the exhaust pipe.
- Catalytic converter
- Shift linkage from the transaxle case, if equipped with a manual transaxle.
- Shift cable cover and cable (wire it up and out of the way), if equipped with an automatic transaxle.
- 2 hydraulic lines from the rack valve body, using a flare nut wrench. Plug the lines to keep dirt and moisture out. Carefully move the disconnected lines to the rear of

the rack assembly so that they are not damaged when the rack is removed.
- Rack stiffener plate, then the steering rack mounting bolts.

9. Pull the steering rack down to release it from the pinion shaft.

10. Drop the steering rack far enough to permit the end of the pinion shaft to come out of the hole in the frame channel.

11. Slide the steering rack to the right until the left tie rod clears the subframe, then drop it down and out of the vehicle to the left.

To install:

➡Use new gaskets and self-locking nuts when installing the catalytic converter.

Use only genuine Honda power steering fluid. Any other type or brand of fluid will damage the power steering pump.

12. Before installing the rack & pinion, slide the ends all the way to the right.

13. Install or connect the following:
- Pinion shaft grommet. The lug on the pinion shaft grommet aligns with the slot on the valve body.
- Steering rack into position
- Pinion shaft grommet and insert the pinion through the hole in the bulkhead.
- Rack mounting bolts. Tighten the

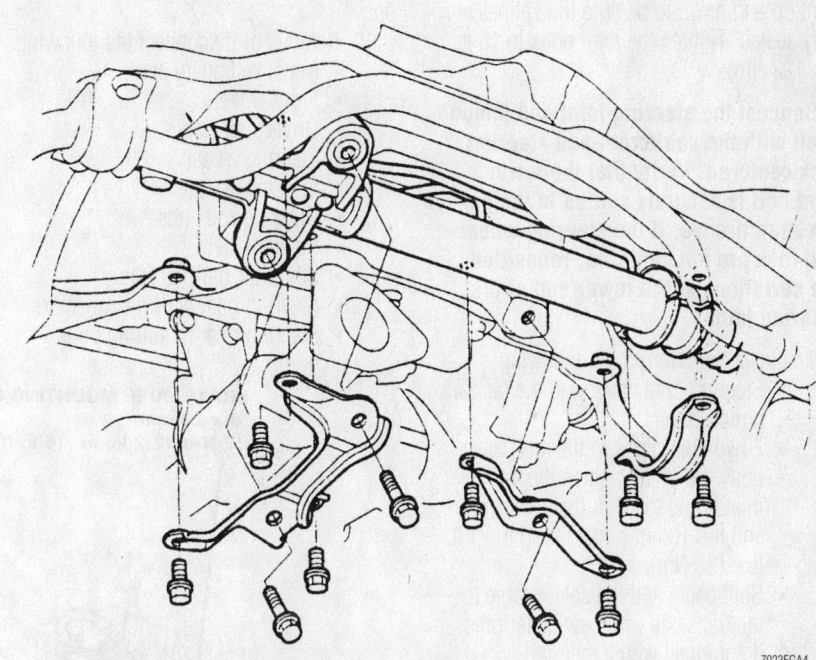

Power rack and pinion steering gear mounting—Accord

7923FGA4

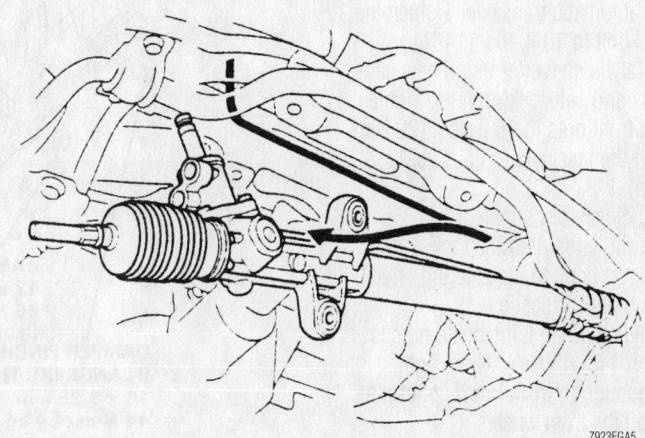

Move the steering rack to the right, then down and out of the vehicle—Accord

7923FGA5

bracket bolts to 28 ft. lbs. (39 Nm). Tighten the stiffener plate mounting bolts to 32 ft. lbs. (43 Nm).

14. Center the rack ends within their steering strokes.

15. Center the air bag cable reel, as follows:

a. Turn the steering wheel left approximately 150°, to check the cable reel position with the indicator.

b. If the cable reel is centered, the yellow gear tooth lines up with the alignment mark on the cover.

c. Return the steering wheel right approximately 150° to position the steering wheel in the straight-ahead position.

16. Line up the bolt hole in the steering joint with the groove in the pinion shaft. Slip the joint onto the pinion shaft. Pull the joint up and down to be sure the splines are fully seated. Tighten the joint bolts to 16 ft. lbs. (22 Nm).

➡ **Connect the steering joint and pinion shaft with the cable reel and steering rack centered. Verify that the lower joint bolt is securely seated in the pinion shaft groove. If the steering wheel and rack are not centered, reposition the serrations at the lower end of the steering joint.**

17. Install or connect the following:
- Steering joint cover and the rack & pinion cover
- 2 hydraulic lines to the rack valve body. Carefully tighten the 14mm inlet fitting to 27 ft. lbs. (37 Nm) and the 16mm outlet fitting to 21 ft. lbs. (28 Nm).
- Shift cable and the select cable to the transaxle with new cotter pins, if equipped with a manual transaxle.
- Shift cable to the transaxle using a new lockwasher, if equipped with an automatic transaxle. Tighten the lockbolt to 10 ft. lbs. (14 Nm).
- Catalytic converter using new gaskets and self-locking nuts. Tighten the front nuts to 16 ft. lbs. (22 Nm), and the rear nuts to 25 ft. lbs. (34 Nm).
- HO$_2$S sensor connector
- Tie rod ends onto the rack ends
- Tie rod ends to the steering knuckles, then the castle nuts.

18. Tighten the ball joint castle nuts to 29–35 ft. lbs. (40–48 Nm). Then, tighten them only enough to install new cotter pins.

19. Install the front wheels.

20. Lower the vehicle.

21. Reconnect the negative battery cable.

22. Be sure the reservoir inlet line has been reconnected. Fill the reservoir to the upper line with Honda power steering fluid. Run the engine at idle and turn the steering wheel lock-to-lock several times to bleed air from the system and fill the rack valve body. Recheck the fluid level and add more if necessary.

23. Check the power steering system for leaks.

24. Check the front wheel alignment and steering wheel spoke angle. Make adjustments by turning the left and right tie rod ends equally.

25. Road test the vehicle.

S2000

1. Before servicing the vehicle, refer to the precautions in the beginning of this section.

2. Remove or disconnect the following:
- Negative battery cable
- Front wheels
- Driver's air bag
- Steering wheel
- Steering coupler
- Outer tie rod ends
- Splash shield
- Stabilizer bar brackets
- Steering gear wiring connectors
- Steering gear mounting bolts

3. Move the steering gear forward and to the right to remove the steering gear.

To install:

4. Installation is the reverse of the removal procedure, while using the following torque values:
- Steering gear mounting bolts: 33 ft. lbs. (44 Nm)
- Steering gear ground cable bolt: 88 inch lbs. (10 Nm)
- Stabilizer bar bracket bolts: 61 ft. lbs. (83 Nm)
- Splash shield bolts: 88 inch lbs. (10 Nm)
- Outer tie rod end nuts: 40 ft. lbs. (54 Nm)
- Steering coupler pinch bolts: 16 ft. lbs. (22 Nm)

Strut

REMOVAL & INSTALLATION

Front

CIVIC

1. Before servicing the vehicle, refer to the precautions in the beginning of this section.

2. Raise and safely support the vehicle.

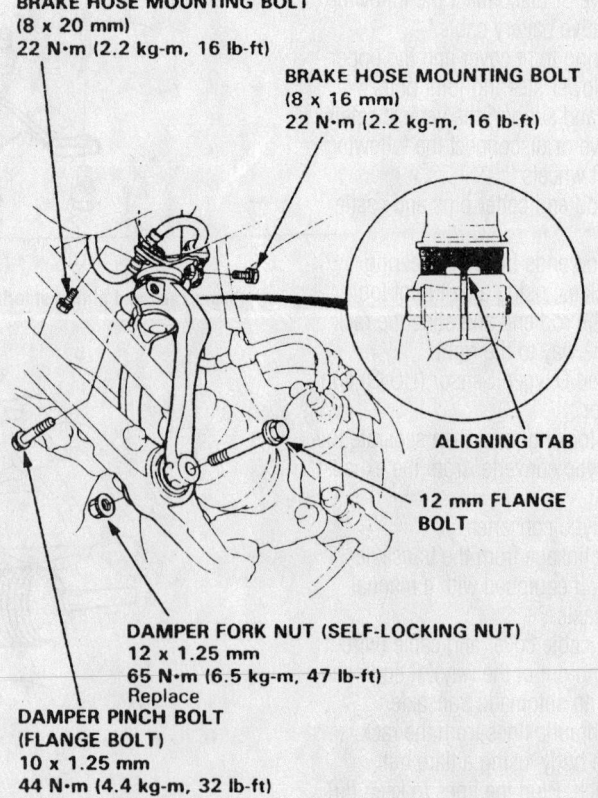

BRAKE HOSE MOUNTING BOLT
(8 x 20 mm)
22 N·m (2.2 kg-m, 16 lb-ft)

BRAKE HOSE MOUNTING BOLT
(8 x 16 mm)
22 N·m (2.2 kg-m, 16 lb-ft)

ALIGNING TAB

12 mm FLANGE BOLT

DAMPER FORK NUT (SELF-LOCKING NUT)
12 x 1.25 mm
65 N·m (6.5 kg-m, 47 lb-ft)
Replace

DAMPER PINCH BOLT
(FLANGE BOLT)
10 x 1.25 mm
44 N·m (4.4 kg-m, 32 lb-ft)

7923FGA6

Damper fork components—Civic

3. Remove or disconnect the following:
- Front wheels
- Brake hose brackets from the bottom of the strut tube. Do not disconnect the brake hoses.

➡ **Some Civic models may not have brake hose brackets on their struts. In these cases, there is no need to unbolt the brackets.**

- Damper pinch bolt
- Damper fork nut and bolt
- Damper fork
- 2 strut mounting bolts from the shock tower
- Strut from the vehicle

To install:

➡ **Use new self-locking nuts when installing the strut.**

4. Install or connect the following:
- Strut into the vehicle. Hand-tighten the strut mounting bolts. The alignment mark on the strut tube faces away from the wheel.
- Damper fork onto the strut and lower control arm
- Pinch and fork bolts
- Brake hose brackets to the strut tube and tighten them to 16 ft. lbs. (22 Nm).
- Front wheels and lower the vehicle.

5. Tighten the strut mount bolts to 36 ft. lbs. (50 Nm).

6. Tighten the pinch bolt to 32 ft. lbs. (44 Nm). Tighten the damper fork nut to 47 ft. lbs. (65 Nm).

7. Tighten the wheel nuts to 80 ft. lbs. (110 Nm).

8. Check the vehicle's front end alignment and adjust it if necessary.

PRELUDE AND ACCORD

1. Before servicing the vehicle, refer to the precautions in the beginning of this section.

2. Raise and safely support the vehicle.

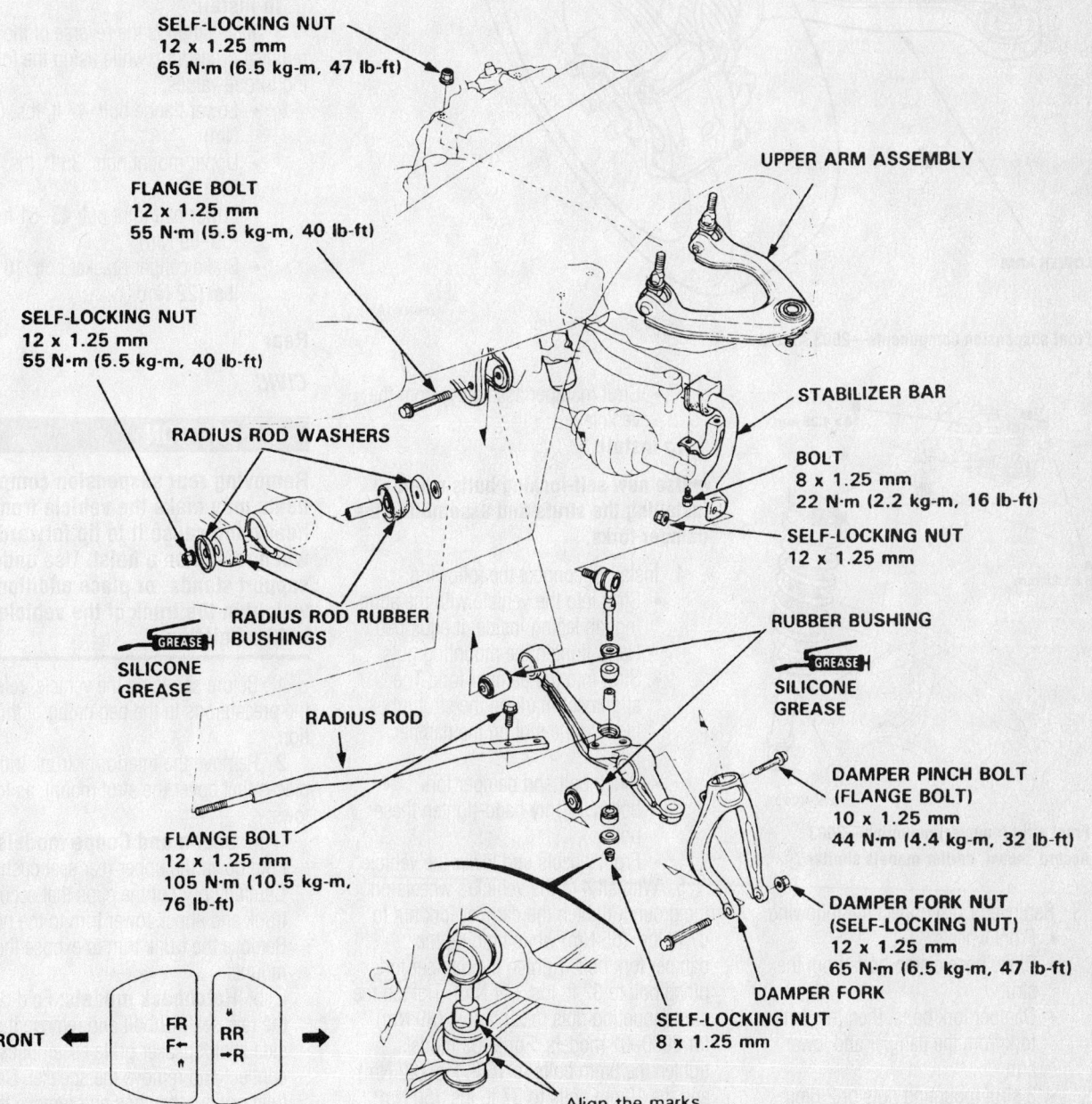

SELF-LOCKING NUT
12 x 1.25 mm
65 N·m (6.5 kg-m, 47 lb-ft)

FLANGE BOLT
12 x 1.25 mm
55 N·m (5.5 kg-m, 40 lb-ft)

SELF-LOCKING NUT
12 x 1.25 mm
55 N·m (5.5 kg-m, 40 lb-ft)

RADIUS ROD WASHERS

RADIUS ROD RUBBER BUSHINGS

GREASE
SILICONE GREASE

RADIUS ROD

FLANGE BOLT
12 x 1.25 mm
105 N·m (10.5 kg-m, 76 lb-ft)

FRONT ◀ FR F← RR →R ▶

UPPER ARM ASSEMBLY

STABILIZER BAR

BOLT
8 x 1.25 mm
22 N·m (2.2 kg-m, 16 lb-ft)

SELF-LOCKING NUT
12 x 1.25 mm

RUBBER BUSHING

GREASE
SILICONE GREASE

DAMPER PINCH BOLT
(FLANGE BOLT)
10 x 1.25 mm
44 N·m (4.4 kg-m, 32 lb-ft)

DAMPER FORK NUT
(SELF-LOCKING NUT)
12 x 1.25 mm
65 N·m (6.5 kg-m, 47 lb-ft)

DAMPER FORK

SELF-LOCKING NUT
8 x 1.25 mm

Align the marks.

Front suspension components—Prelude and 2000–02 Accord

7923FGA7

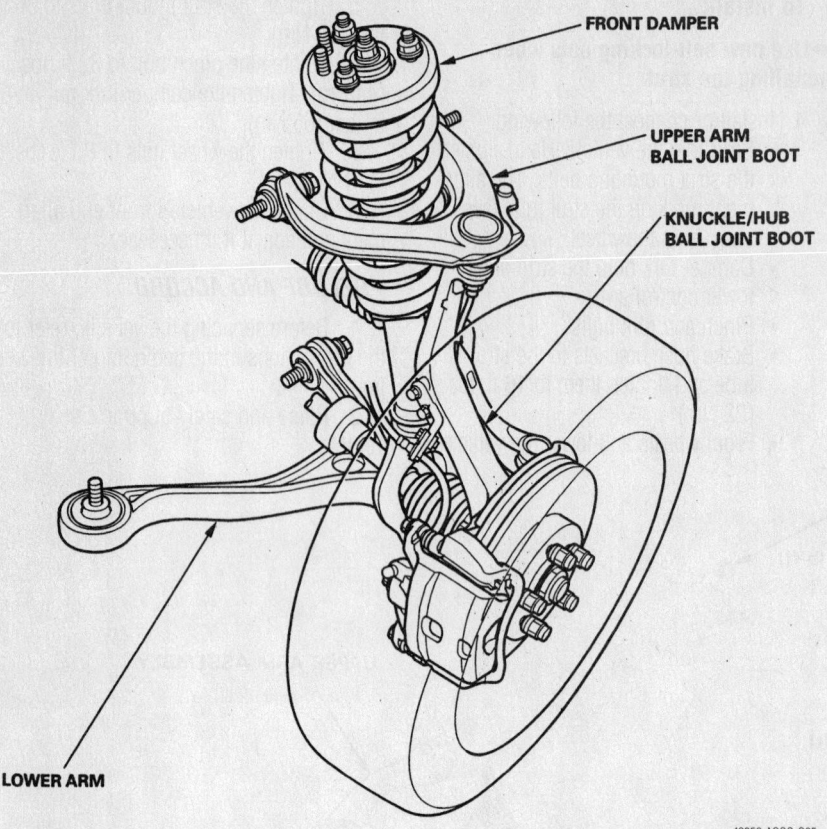

Front suspension components—2003 Accord

42356-ACCO-G35

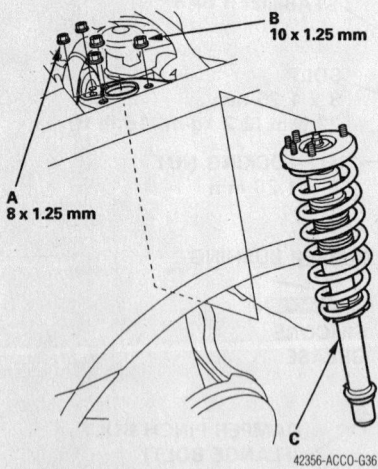

B
10 x 1.25 mm

A
8 x 1.25 mm

C

42356-ACCO-G36

Front strut (damper) mounting—2003 Accord shown, earlier models similar

3. Remove or disconnect the following:
- Front wheels
- Brake hose clamp bolts from the strut
- Damper fork bolts, then the damper fork from the damper and lower arm
- 3 strut mounting nuts or 2 8mm flange nuts and 3 10mm flange nuts, as applicable

- Strut (damper assembly) from the vehicle

To install:

➡Use new self-locking bolts when installing the struts and assembling the damper forks.

4. Install or connect the following:
- Strut into the vehicle with the aligning tab facing inside, if equipped. Hand-tighten the mounting nuts.
- Strut into the damper fork. The alignment mark on the strut tube fits into the slot on the damper fork.
- Pinch bolt and damper fork bolt/nut. Only hand-tighten these bolts.
- Front wheels and lower the vehicle.

5. With all 4 of the vehicle's wheels on the ground, tighten the damper fork nut to 47 ft. lbs. (65 Nm) while holding the damper fork bolt. Tighten the damper fork pinch bolt to 32 ft. lbs. (44 Nm). Tighten the strut mounting nuts to 28 ft. lbs. (39 Nm) for 2000–02 models. For 2003 models, tighten the 8mm bolts to 16 ft. lbs. (22 Nm) and the 10mm bolts to 37 ft. lbs. (50 Nm).

6. Tighten the wheel nuts to 80 ft. lbs. (110 Nm).

7. Check and adjust the vehicle's front end alignment. On Preludes equipped with 4WS, the electronic neutral check must be performed before aligning all 4 wheels.

S2000

1. Before servicing the vehicle, refer to the precautions in the beginning of this section.

2. Remove or disconnect the following:
- Front wheel
- Lower ball joint
- Brake caliper bracket bolt
- Upper strut mount nuts
- Lower flange bolt
- Strut

To install:

3. Installation is the reverse of the removal procedure, while using the following torque values:
- Lower flange bolt: 47 ft. lbs. (64 Nm)
- Upper mount nuts: 36 ft. lbs. (49 Nm)
- Lower ball joint nut: 43–51 ft. lbs. (59–69 Nm)
- Brake caliper bracket bolt: 16 ft. lbs. (22 Nm)

Rear

CIVIC

✳✳ CAUTION

Removing rear suspension components may make the vehicle front-heavy and cause it to tip forward when raised on a hoist. Use under-lift support stands, or place additional weight in the trunk of the vehicle before hoisting it.

1. Before servicing the vehicle, refer to the precautions in the beginning of this section.

2. Remove the interior or trunk trim pieces that cover the strut mount, as follows:

   a. **Sedan and Coupe models:** Fold down the upper rear seat cushion. Carefully pry out the clips that secure the trunk and shock tower trim to the body. Remove the trunk trim to expose the strut mounts.

   b. **Hatchback models:** Fold down the rear seat. Unbolt and remove the rear side shelf/speaker grille assemblies. Disconnect and remove the speaker. Carefully pry out the clips and remove the screws to remove shock tower trim panel.

3. Raise and support the vehicle.

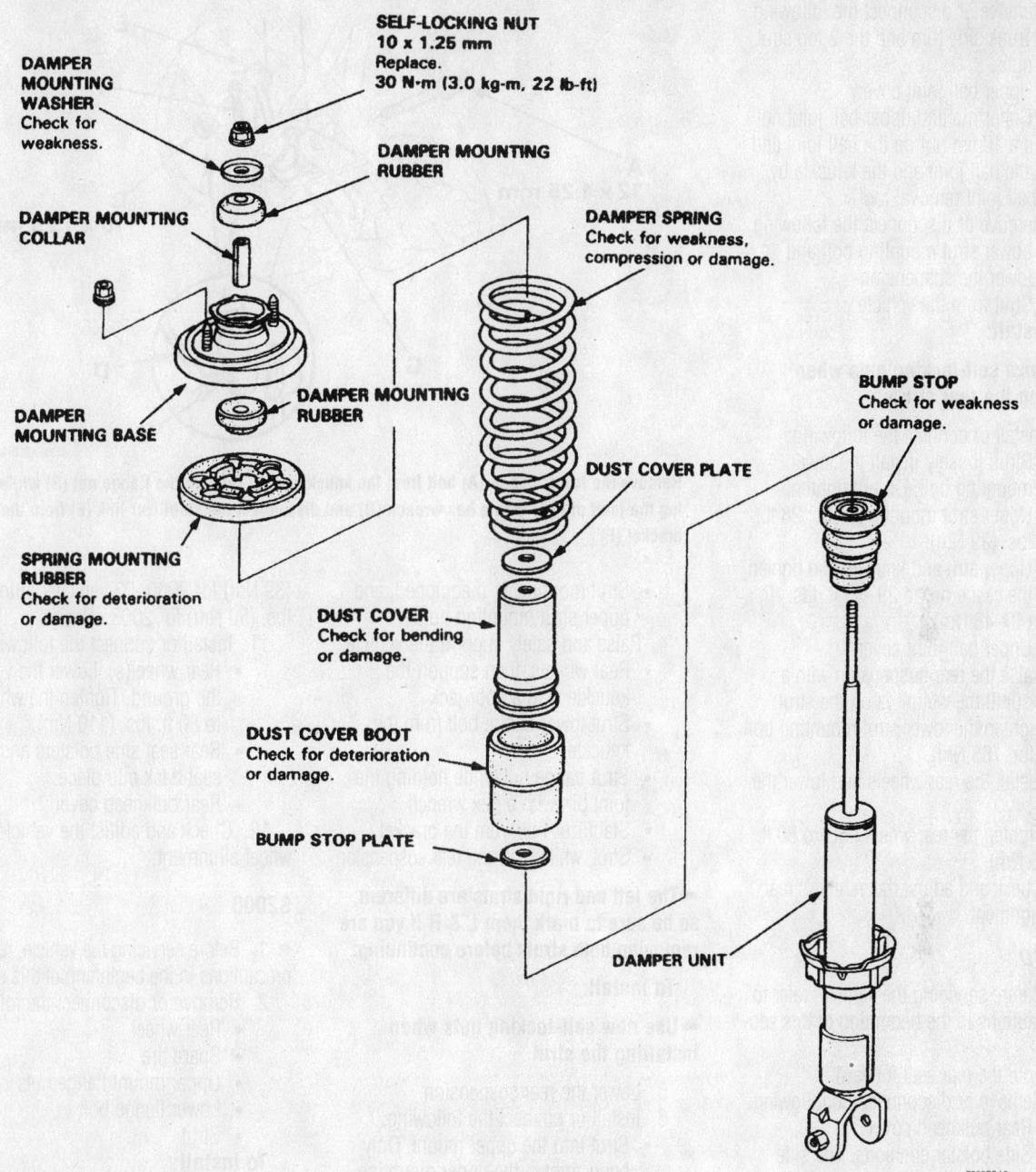

SELF-LOCKING NUT
10 x 1.25 mm
Replace.
30 N·m (3.0 kg-m, 22 lb-ft)

DAMPER MOUNTING WASHER
Check for weakness.

DAMPER MOUNTING RUBBER

DAMPER MOUNTING COLLAR

DAMPER SPRING
Check for weakness, compression or damage.

DAMPER MOUNTING RUBBER

DAMPER MOUNTING BASE

BUMP STOP
Check for weakness or damage.

DUST COVER PLATE

SPRING MOUNTING RUBBER
Check for deterioration or damage.

DUST COVER
Check for bending or damage.

DUST COVER BOOT
Check for deterioration or damage.

BUMP STOP PLATE

DAMPER UNIT

7923FGA9

Exploded view of the rear suspension strut—Civic

4. Remove or disconnect the following:
   - Rear wheels
   - 2 upper mounting nuts
   - Wheel sensor bracket from the lower control arm.
   - Lower strut bolt and the knuckle flange bolt.
   - Strut from the vehicle

To install:

➡All suspension nuts and bolts should be tightened with the vehicle on the ground. Alternatively, raise the lower control arm with a floor jack until the jack is supporting the weight of the

vehicle. This method pre-loads the suspension and allows room to work.

5. Install or connect the following:
   - Strut into the vehicle with the lock-nut facing the front of the vehicle. Hand-tighten the upper mounting nuts.
   - Wheel sensor bracket onto the lower control arm. Tighten the bolts to 84 inch lbs. (10 Nm).
   - Knuckle flange bolt and the lower strut bolt. Hand-tighten the bolts.
   - Wheels and lower the vehicle.
   - Upper mounting nuts to 36 ft. lbs.

(50 Nm). Tighten the knuckle flange bolt and strut bolts to 40 ft. lbs. (55 Nm). Tighten the wheel nuts to 80 ft. lbs. (110 Nm).
   - Trunk side trim panels
6. Check and adjust the vehicle's rear wheel alignment.

PRELUDE

1. Before servicing the vehicle, refer to the precautions in the beginning of this section.
2. Raise and safely support the vehicle.

3. Remove or disconnect the following:
- Trunk side trim and the 2 top strut nuts.
- Upper ball joint cover
- Cotter pin and upper ball joint nut

4. Fit a 10mm nut on the ball joint and separate the ball joint and the knuckle by using a ball joint removal tool.

5. Remove or disconnect the following:
- Lower strut mounting bolt and lower the suspension.
- Strut from the vehicle

To install:

➡**Use new self-locking nuts when installing the rear struts.**

6. Install or connect the following:
- Strut; loosely install the lower mounting bolt. Do not tighten.
- Upper strut mounting bolts: 28 ft. lbs. (39 Nm).
- Upper arm and knuckle and tighten the castle nut to 29–35 ft. lbs. (40–48 Nm).
- Upper ball joint cover

7. Raise the rear suspension with a floor jack until the weight is on the strut.

8. Tighten the lower strut mounting bolt to 47 ft. lbs. (65 Nm).

9. Install the rear wheels and lower the vehicle.

10. Tighten the rear wheel nuts to 80 ft. lbs. (110 Nm).

11. Check and adjust the vehicle's rear wheel alignment.

ACCORD

1. Before servicing the vehicle, refer to the precautions in the beginning of this section.

2. Fold the rear seat forward.

3. Remove or disconnect the following:
- Rear bulkhead cover
- Side bolster cushions. The side bolster cushions are secured by 1 screw at the bottom, and 2 clips at the top.

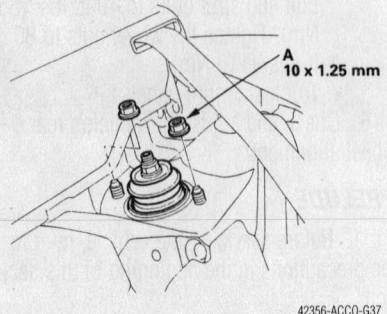

Rear strut upper mounting nut locations— Accord

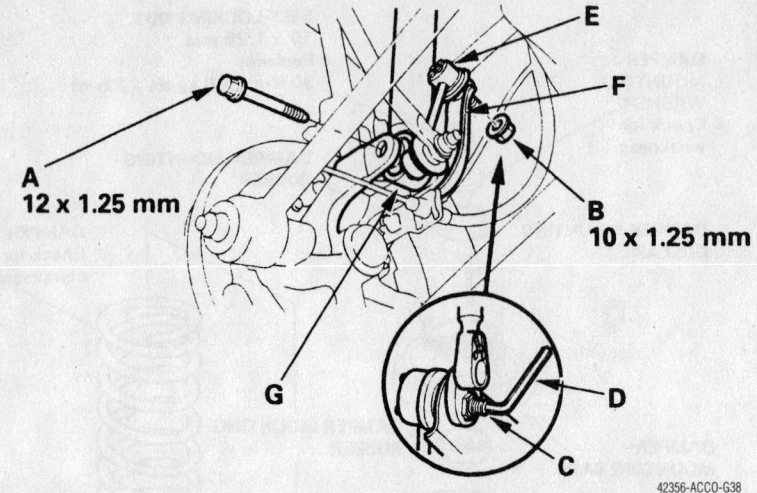

Remove the lower flange (A) bolt from the knuckle, then remove the flange nut (B) while holding the joint pin (C) with a hex wrench (D) and disconnect the stabilizer link (E) from the bracket (F)

- Strut mount cap, if equipped, and upper strut mounting nuts

4. Raise and safely support the vehicle.
- Rear wheels, then support the knuckle with a floor jack
- Strut lower flange bolt from the knuckle
- Strut flange nut while holding the joint pin with a hex wrench
- Stabilizer link from the bracket
- Strut, while lowering rear suspension

➡**The left and right struts are different, so be sure to mark them L & R if you are removing both struts before continuing.**

To install:

➡**Use new self-locking nuts when installing the strut.**

5. Lower the rear suspension.

6. Install or connect the following:
- Strut into the upper mount. Only hand-tighten the upper mounting nuts.
- Strut into position on the knuckle, then loosely install the flange bolt on the bottom of the strut
- Stabilizer link on the bracket and loosely install the flange nut

7. Place a jack under the lower strut mount. Raise the jack until the weight of the vehicle is on the jack.

8. With the suspension under load, tighten the lower mount bolt to 40 ft. lbs. (55 Nm) for 2000–02 vehicles or to 43 ft. lbs. (59 Nm) for 2003 vehicles.

9. While holding the joint pin with a cotter pin, tighten the flange nut to 29 ft. lbs. (39 Nm).

10. Tighten the upper nuts to 28 ft. lbs.

(39 Nm) for 2000–02 vehicles or to 37 ft. lbs. (50 Nm) for 2003 vehicles.

11. Install or connect the following:
- Rear wheel(s). Lower the vehicle to the ground. Tighten the wheel nuts to 80 ft. lbs. (110 Nm).
- Rear seat side bolsters and fold the seat back into place.
- Rear bulkhead cover

12. Check and adjust the vehicle's rear wheel alignment.

S2000

1. Before servicing the vehicle, refer to the precautions in the beginning of this section.

2. Remove or disconnect the following:
- Rear wheel
- Spare tire
- Upper mount flange nuts
- Lower flange bolt
- Strut

To install:

3. Installation is the reverse of the removal procedure, while using the following torque values:
- Lower flange bolt: 47 ft. lbs. (64 Nm)
- Upper mount flange nuts: 36 ft. lbs. (49 Nm)

Coil Spring

REMOVAL & INSTALLATION

Civic

FRONT

1. Before servicing the vehicle, refer to the precautions in the beginning of this section.

2. Raise and safely support the vehicle.

3. Remove or disconnect the following:
- Front wheels
- Brake hose brackets from the bottom of the strut tube. Do not disconnect the brake hoses.

➡**Some Civic models may not have brake hose brackets on their struts.**

- Damper fork pinch bolt and flange bolt, then the damper fork.
- Strut's upper mounting nuts
- Strut assembly from the vehicle.

4. Install a spring compressor onto the strut assembly and tighten the compressor according to the manufacturer's instructions.

5. Remove the locking nut from the top of the shock absorber piston. Disassemble the strut and remove the coil spring.

To install:

➡**Use new self-locking nuts when assembling the strut.**

6. Install or connect the following:
- Spring compressor onto the coil spring

7. Assemble the lower strut mounts, dust covers, coil spring, and upper strut mount onto the shock absorber. Position the strut bearing mounting studs so that they will line up with the mounting holes in the shock tower.

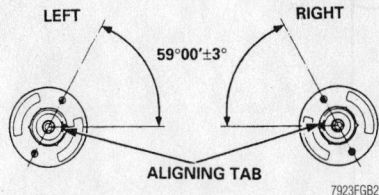

Strut bearing installation direction—Civic

8. Install or connect the following:
- Mounting washer, and a new self-locking nut (loosely).

9. Hold the shock absorber piston with a hex wrench and tighten the self-locking nut. Tighten the self-locking nut to 22 ft. lbs. (30 Nm).

➡**All suspension nuts and bolts should be tightened with the vehicle on the ground.**

10. Install or connect the following:
- Strut assembly into the vehicle. Tighten the upper mounting nuts to 36 ft. lbs. (50 Nm).
- Damper fork. Tighten the pinch bolt to 32 ft. lbs. (44 Nm), and the fork bolt to 47 ft. lbs. (64 Nm).

- Brake hose clamps. Tighten them to 16 ft. lbs. (22 Nm).
- Wheel, and tighten the wheel nuts to 80 ft. lbs. (110 Nm).

11. Check and adjust the vehicle's front wheel alignment.

REAR

✳✳ CAUTION

Removing rear suspension components may make the vehicle front-heavy and cause it to tip forward when raised on a hoist. Use under-lift support stands, or place additional weight in the trunk of the vehicle before hoisting it.

1. Before servicing the vehicle, refer to the precautions in the beginning of this section.

2. Remove the interior or trunk trim pieces that cover the strut mount:
   a. **Sedan and Coupe models:** Fold down the upper rear seat cushion. Carefully pry out the clips that secure the trunk and shock tower trim to the body. Remove the trunk trim to expose the strut mounts.
   b. **Hatchback models:** Fold down the rear seat. Unbolt and remove the rear side shelf/speaker grille assemblies. Disconnect and remove the speaker. Carefully pry out the clips and remove the screws to remove shock tower trim panel.

3. Raise and safely support the vehicle.

4. Remove or disconnect the following:
- 2 strut mounting bolts
- Wheel sensor brackets from the lower control arm. Do not disconnect the sensor.

5. Support the lower control arm with a floor jack.

6. Remove or disconnect the following:
- Strut mounting flange bolt and the knuckle flange bolt, then lower the floor jack
- Strut from the vehicle.

7. Install a spring compressor onto the strut assembly and tighten the compressor according to the manufacturer's instructions.

8. Remove the locking nut from the top of the shock absorber. Disassemble the strut and remove the coil spring.

To install:

➡**Use new self-locking nuts when assembling the strut.**

9. Install or connect the following:
- Spring compressor onto the coil spring.

- Upper and lower strut mounts, dust covers, and coil spring onto the shock absorber.
- Mounting washer, and a new self-locking nut (loosely)

10. Hold the shock absorber piston with a hex wrench and tighten the self-locking nut. Tighten the self-locking nut to 22 ft. lbs. (30 Nm).

➡**All suspension nuts and bolts should be tightened with the vehicle on the ground. Alternatively, raise the lower control arm with a floor jack until the jack is supporting the weight of the vehicle. This method pre-loads the suspension and allows room to work.**

11. Install or connect the following:
- Strut assembly into the vehicle. Tighten the upper mounting nuts to 36 ft. lbs. (50 Nm).
- Shock mounting bolt at the knuckle and tighten to 40 ft. lbs. (55 Nm).
- Knuckle flange bolt and tighten it to 40 ft. lbs. (55 Nm).
- Wheel sensor brackets
- Wheel, and tighten the wheel nuts to 80 ft. lbs. (110 Nm).
- Trunk side trim

12. Check and adjust the rear wheel alignment.

Accord and Prelude

FRONT

1. Before servicing the vehicle, refer to the precautions in the beginning of this section.

2. Raise and safely support the vehicle.

3. Remove or disconnect the following:
- Strut from the vehicle

4. Place the strut in vice and install a spring compressor onto the coil spring. Follow the spring compressor manufacturer's instructions.

5. Compress the spring and remove the self-locking nut from the top of the strut. Disassemble the strut mounts and remove the coil spring.

6. Inspect the strut mounts for wear and damage. Replace any damaged or worn parts.

To install:

➡**Use new self-locking nuts when assembling and installing the struts.**

7. Install or connect the following:
- Spring compressor onto the coil spring. Set the spring onto the strut cartridge. The flat part of the coil spring is its top.

8. Assemble the strut mount and

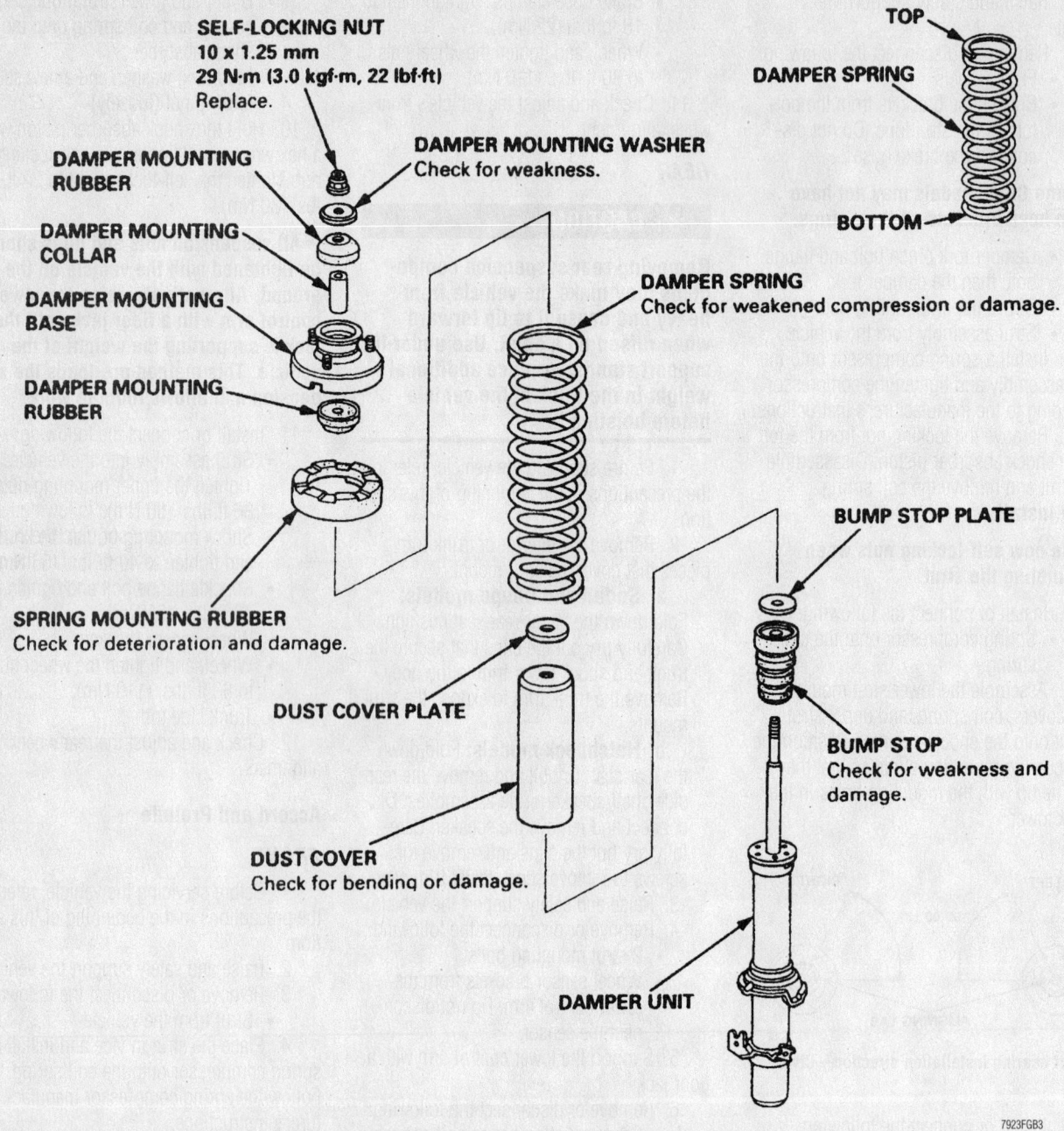

SELF-LOCKING NUT
10 x 1.25 mm
29 N·m (3.0 kgf·m, 22 lbf·ft)
Replace.

DAMPER MOUNTING RUBBER

DAMPER MOUNTING WASHER
Check for weakness.

DAMPER MOUNTING COLLAR

DAMPER MOUNTING BASE

DAMPER MOUNTING RUBBER

SPRING MOUNTING RUBBER
Check for deterioration and damage.

DUST COVER PLATE

DUST COVER
Check for bending or damage.

TOP

DAMPER SPRING

BOTTOM

DAMPER SPRING
Check for weakened compression or damage.

BUMP STOP PLATE

BUMP STOP
Check for weakness and damage.

DAMPER UNIT

7923FGB3

Coil spring, strut cartridge, and strut mount components—2000–02 Accord and Prelude shown, 2003 similar

washer onto the strut. Tighten the self-locking nut to 22 ft. lbs. (29 Nm). Remove the spring compressor.

9. Install or connect the following:
• Strut

10. Check and adjust the vehicle's front wheel alignment. On Preludes equipped with 4WS, the electronic neutral check must be performed before all 4 wheels are aligned.

REAR—ACCORD

1. Before servicing the vehicle, refer to the precautions in the beginning of this section.

2. Remove or disconnect the following:
• Strut

3. Place the strut in a vice and install a spring compressor onto the coil spring. Follow the spring compressor manufacturer's instructions.

4. Compress the spring and remove the self-locking nut from the strut. Disassemble the strut mounts and remove the coil spring.

5. Inspect the strut mounts for wear and damage. Replace any damaged or worn parts.

To install:

➡Use new self-locking nuts when assembling and installing the struts.

6. Install or connect, the following:
• Spring compressor onto the coil spring. Set the spring onto the strut cartridge. The flat part of the coil spring is its top.
• Strut mount and washer onto the strut. Tighten the self-locking nut to 22 ft. lbs. (29 Nm). Remove the spring compressor.
• Strut into the vehicle. Hand-tighten the mounting nuts.
• Strut into position on the knuckle.
• Mounting bolt

7. Place a jack under the lower strut

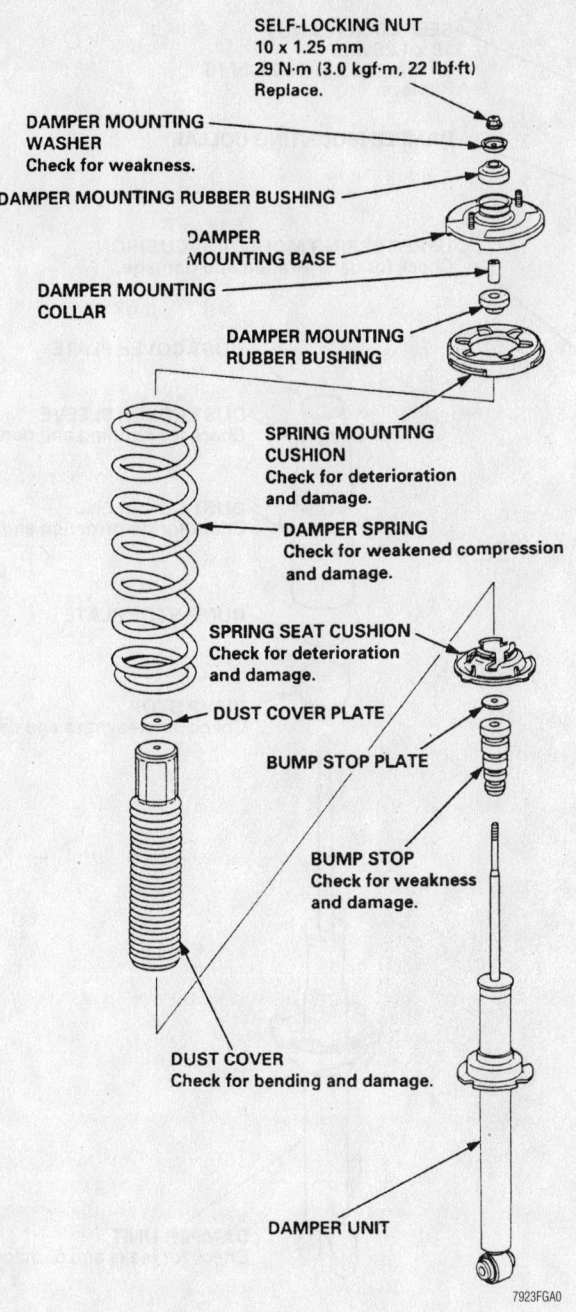

SELF-LOCKING NUT
10 x 1.25 mm
29 N·m (3.0 kgf·m, 22 lbf·ft)
Replace.

DAMPER MOUNTING
WASHER
Check for weakness.

DAMPER MOUNTING RUBBER BUSHING

DAMPER
MOUNTING BASE

DAMPER MOUNTING
COLLAR

DAMPER MOUNTING
RUBBER BUSHING

SPRING MOUNTING
CUSHION
Check for deterioration
and damage.

DAMPER SPRING
Check for weakened compression
and damage.

SPRING SEAT CUSHION
Check for deterioration
and damage.

DUST COVER PLATE

BUMP STOP PLATE

BUMP STOP
Check for weakness
and damage.

DUST COVER
Check for bending and damage.

DAMPER UNIT

7923FGA0

Exploded view of the rear suspension strut assembly—Accord

mount. Raise the jack until the weight of the vehicle is on the jack.

8. With the suspension under load, tighten the lower mount bolt to 40 ft. lbs. (55 Nm). Tighten the upper nuts to 28 ft. lbs. (39 Nm).

9. Install or connect, the following:
- Rear wheel. Lower the vehicle to the ground.
- Wheel nuts to 80 ft. lbs. (110 Nm).
- Rear seat side bolsters and fold the seat back into place.

10. Check and adjust the vehicle's rear wheel alignment.

REAR—PRELUDE

1. Before servicing the vehicle, refer to the precautions in the beginning of this section.

2. Raise and safely support the vehicle.

3. Remove or disconnect the following:
- Trunk side trim and remove the 2 strut mounting nuts.
- Upper ball joint cover
- Cotter pin and upper ball joint nut

4. Fit a 10mm nut on the ball joint and separate the ball joint and the knuckle by using a ball joint removal tool.

5. Remove or disconnect the following:
- Lower strut mounting bolt and lower the suspension.
- Strut from the vehicle.

6. Place the strut in vice and install a spring compressor onto the coil spring. Follow the spring compressor manufacturer's instructions.

7. Compress the spring and remove the self-locking nut from the strut. Disassemble the strut mounts and remove the coil spring.

8. Inspect the strut mounts for wear and damage. Replace any damaged or worn parts.

To install:

➡ **Use new self-locking nuts when installing the rear struts.**

9. Install or connect the following:
- Spring compressor onto the coil spring. Set the spring onto the strut cartridge. The flat part of the coil spring is its top.
- Strut mount and washer onto the strut. Tighten the self-locking nut to 22 ft. lbs. (29 Nm). Remove the spring compressor.
- Strut to the vehicle and the lower mounting bolt (loosely). Do not tighten at this time.
- Upper strut mounting bolts. Tighten the bolts to 28 ft. lbs. (39 Nm).
- Upper arm and knuckle and tighten the castle nut to 29–35 ft. lbs. (40–48 Nm).
- Upper ball joint cover

10. Raise the rear suspension with a floor jack until the weight is on the strut.

11. Tighten the lower strut mounting bolt to 47 ft. lbs. (65 Nm).

12. Install or connect the following:
- Rear wheels and lower the vehicle.
- Rear wheel nuts to 80 ft. lbs. (110 Nm)
- Trunk trim

13. Check and adjust the vehicle's rear wheel alignment. On Preludes equipped with 4WS, the electronic neutral check must be performed before all 4 wheels are aligned.

S2000

FRONT AND REAR

1. Before servicing the vehicle, refer to the precautions at the beginning of this section.

2. Remove the strut from the vehicle.

3. Compress the coil spring using a

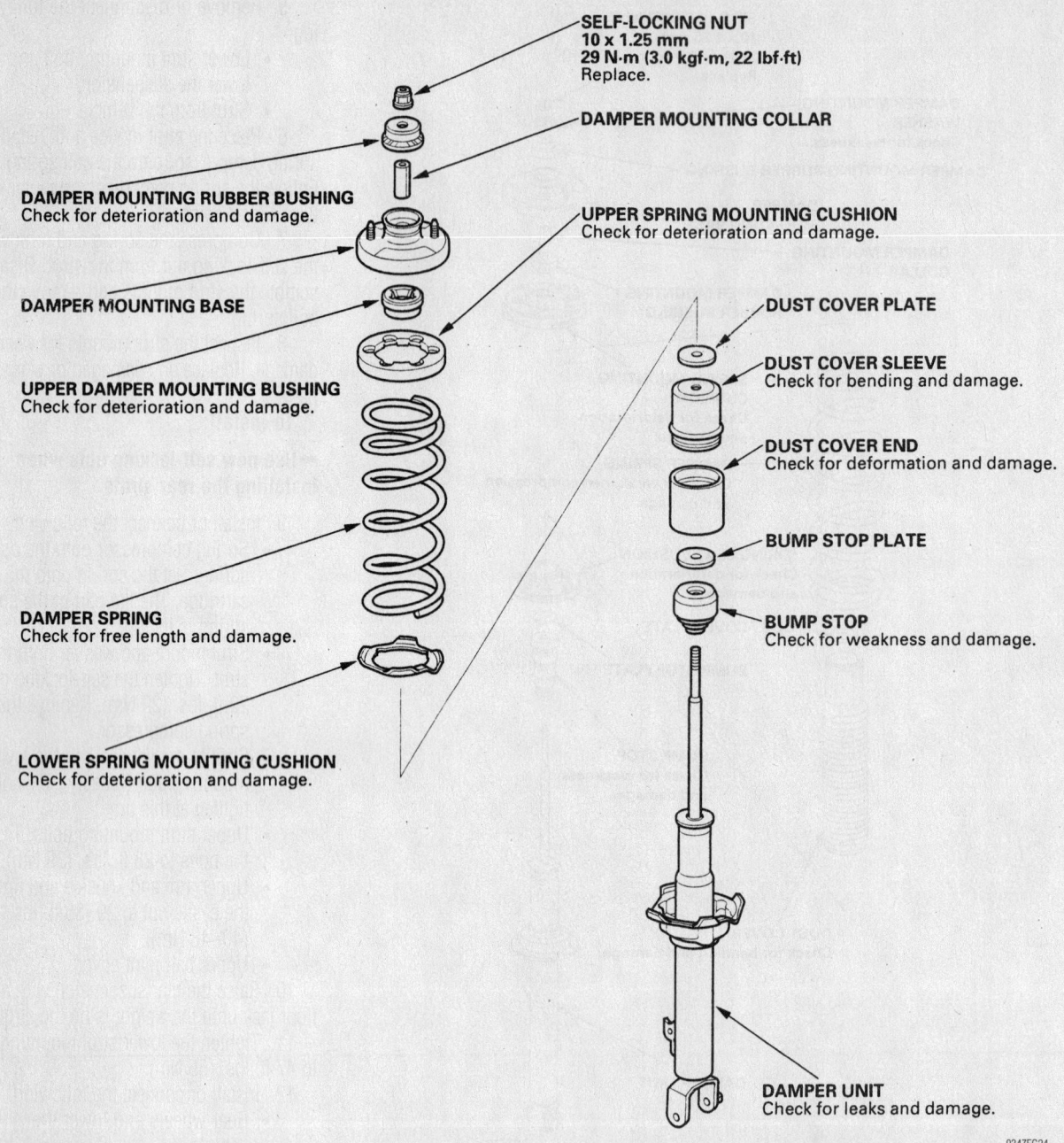

SELF-LOCKING NUT
10 x 1.25 mm
29 N·m (3.0 kgf·m, 22 lbf·ft)
Replace.

DAMPER MOUNTING COLLAR

DAMPER MOUNTING RUBBER BUSHING
Check for deterioration and damage.

UPPER SPRING MOUNTING CUSHION
Check for deterioration and damage.

DAMPER MOUNTING BASE

DUST COVER PLATE

DUST COVER SLEEVE
Check for bending and damage.

UPPER DAMPER MOUNTING BUSHING
Check for deterioration and damage.

DUST COVER END
Check for deformation and damage.

BUMP STOP PLATE

DAMPER SPRING
Check for free length and damage.

BUMP STOP
Check for weakness and damage.

LOWER SPRING MOUNTING CUSHION
Check for deterioration and damage.

DAMPER UNIT
Check for leaks and damage.

9347FG24

Exploded view of the strut and spring assembly—S2000—front shown

suitable spring compressor until the spring comes away from the seat.

4. Remove the center nut and slowly release the spring compressor.

To install:

5. Compress the spring and install it on the strut.

6. Install the lower washer and mounting bracket.

7. Install the upper washer and a new nut. Tighten the nut to 22 ft. lbs. (29 Nm).

8. Install the strut assembly in the vehicle.

Upper Ball Joint

REMOVAL & INSTALLATION

Front and Rear

ALL MODELS

The upper ball joint cannot be removed from the upper control arm. If the ball joint is faulty or worn, the entire control arm must be replaced. If the upper ball joint boot is damaged and the ball joint itself is still usable, the boot can be replaced.

Upper Control Arm

REMOVAL & INSTALLATION

Front

CIVIC

1. Before servicing the vehicle, refer to the precautions in the beginning of this section.

2. Raise and support the vehicle safely.

3. Remove or disconnect the following:
   • Front wheels

- Damper fork from the lower control arm
- Strut mounting nuts, then the strut from the vehicle
- Upper ball joint from the steering knuckle using a suitable ball joint remover.
- Self-locking nuts, then the upper arm from the vehicle.
- Upper arm bolts to separate the control arm from its anchor bolt assembly. Inspect the bushings for signs of deterioration and replace them if they are damaged.

4. Place the upper control arm anchor bolt assembly into a vice and drive out the upper arm bushings.

To install:

➡Use new self-locking nuts when assembling the anchor bolts and when installing the control arm into the vehicle.

5. Drive the new upper arm bushings into the upper arm anchor bolts. Center the bushing in the anchor bolt so that equal amounts of the bushing sleeve protrude on either side.

6. Install or connect the following:
- Anchor bolt assembly onto the

control arm. Align the marks on the arm and anchor assembly. Tighten the nuts to 22 ft. lbs. (30 Nm).
- Upper control arm assembly into the shock tower.
- Strut into the vehicle
- Damper fork bolt and nut
- Steering arm and upper ball joint
- Front wheels. Lower the vehicle to the ground.

7. Torque the strut mounting nuts to 36 ft. lbs. (50 Nm).

8. Torque the upper control arm mounting nuts to 47 ft. lbs. (65 Nm).

9. Torque the damper fork nut to 47 ft. lbs. (65 Nm).

10. Torque the upper ball joint castle nut to 29–35 ft. lbs. (40–48 Nm). Then, tighten the nut only enough to install a new cotter pin.

11. Tighten the wheel nuts to 80 ft. lbs. (108 Nm).

12. Check the vehicle's front end alignment and adjust it if necessary. Road test the vehicle.

ACCORD AND PRELUDE—2000–02 VEHICLES

➡Do not disassemble the upper arm. If the ball joint or bushings are faulty, or

the upper arm is damaged, the entire upper arm must be replaced.

1. Before servicing the vehicle, refer to the precautions in the beginning of this section.

2. Raise and support the vehicle safely.

3. Remove or disconnect the following:
- Front wheels. Support the lower control arm assembly with a floor jack.
- Upper ball joint from the steering knuckle using a ball joint separator tool.
- Self-locking nuts from the upper arm anchor bolts.
- Upper arm from the vehicle

➡Do not disassemble the upper arm. If the ball joint or bushings are faulty, or the upper arm is damaged, the entire upper arm must be replaced.

To install:

➡Use new self-locking nuts when installing the upper arm and strut.

4. Install or connect the following:
- Upper control arm assembly into the strut tower.
- Upper ball joint
- Front wheels and lower the vehicle.

5. With all 4 of the vehicle's wheels on the ground, torque the upper control arm nuts to 47 ft. lbs. (65 Nm). Torque the castle nut to 32 ft. lbs. (44 Nm); then, only tighten it only enough to install a new cotter pin.

6. Tighten the wheel nuts to 80 ft. lbs. (110 Nm).

7. Check and adjust the vehicle's front end alignment. On Preludes equipped with 4WS, the electronic neutral check must be performed before all 4 wheels are aligned.

ACCORD—2003 VEHICLES

1. Before servicing the vehicle, refer to the precautions in the beginning of this section.

2. Remove or disconnect the following:
- Front wheel
- Front strut (damper) assembly
- Wheel Speed Sensor (WSS) from the upper arm
- Cotter pin from the upper arm ball joint, then loosen the nut
- Upper ball joint from the knuckle
- Upper control arm mounting bolts
- Upper control arm

To install:

3. Install or connect the following:
- Upper control arm and loosely install the mounting bolts
- Upper ball joint to the knuckle.

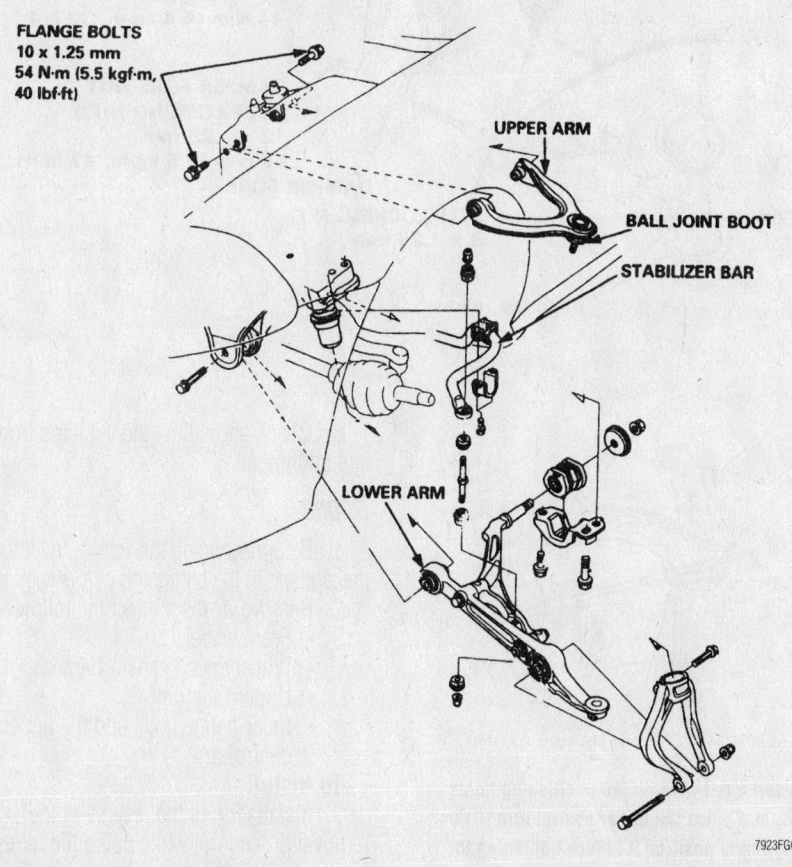

FLANGE BOLTS
10 x 1.25 mm
54 N·m (5.5 kgf·m, 40 lbf·ft)

UPPER ARM

BALL JOINT BOOT

STABILIZER BAR

LOWER ARM

7923FGC6

Front suspension components—Civic

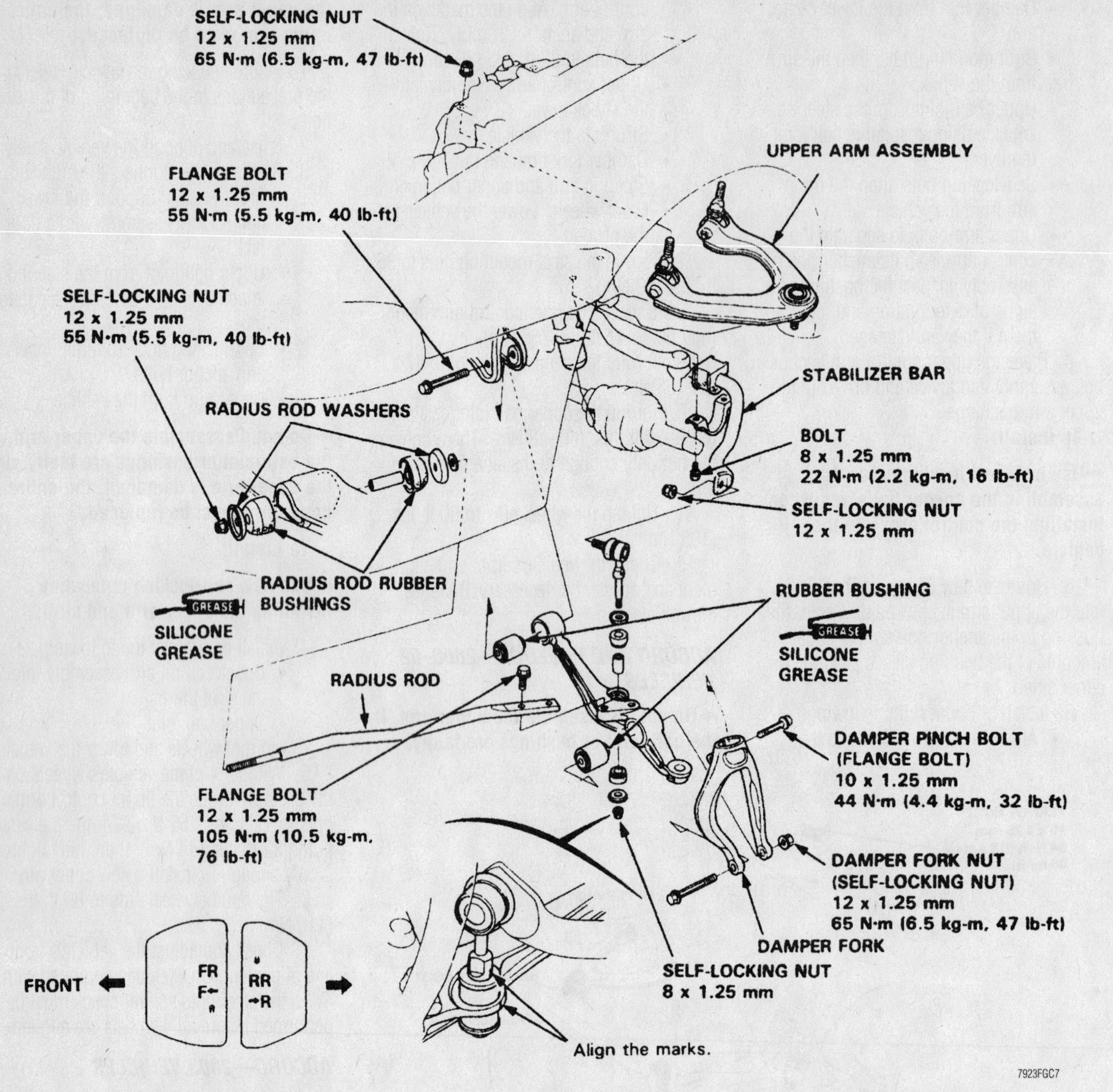

SELF-LOCKING NUT
12 x 1.25 mm
65 N·m (6.5 kg-m, 47 lb-ft)

FLANGE BOLT
12 x 1.25 mm
55 N·m (5.5 kg-m, 40 lb-ft)

SELF-LOCKING NUT
12 x 1.25 mm
55 N·m (5.5 kg-m, 40 lb-ft)

RADIUS ROD WASHERS

RADIUS ROD RUBBER BUSHINGS

SILICONE GREASE

RADIUS ROD

FLANGE BOLT
12 x 1.25 mm
105 N·m (10.5 kg-m, 76 lb-ft)

UPPER ARM ASSEMBLY

STABILIZER BAR

BOLT
8 x 1.25 mm
22 N·m (2.2 kg-m, 16 lb-ft)

SELF-LOCKING NUT
12 x 1.25 mm

RUBBER BUSHING

SILICONE GREASE

DAMPER PINCH BOLT (FLANGE BOLT)
10 x 1.25 mm
44 N·m (4.4 kg-m, 32 lb-ft)

DAMPER FORK NUT (SELF-LOCKING NUT)
12 x 1.25 mm
65 N·m (6.5 kg-m, 47 lb-ft)

DAMPER FORK

SELF-LOCKING NUT
8 x 1.25 mm

FRONT ←

FR
F←
↑

RR
→R

Align the marks.

7923FGC7

Front suspension components—Prelude and Accord

Tighten the nut to 29 ft. lbs. (39 Nm), then tighten it up to 35 ft. lbs. (47 Nm) to align the holes to install a new cotter pin. Never loosen the nut to install the cotter pin.

4. To tighten the upper control arm bolts, insert a rod about 6mm in diameter by 300mm long into the positioning holes and place the upper arm on the rock to position it. Tighten the bolts to 23 ft .lbs. (31 Nm).

- WSS to the upper arm
- Front strut (damper) assembly
- Front wheel

5. Tighten the wheel nuts to 80 ft. lbs. (110 Nm).

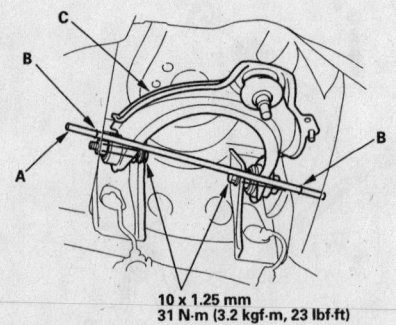

10 x 1.25 mm
31 N·m (3.2 kgf-m, 23 lbf·ft)

42356-ACCO-G39

Insert a rod (A) into the positioning holes (B) and place the upper control arm (C) on the rod to position it before tightening the bolts—2003 Accord

6. Check and adjust the vehicle's front end alignment.

S2000

1. Before servicing the vehicle, refer to the precautions in the beginning of this section.
2. Remove or disconnect the following:
- Front wheel
- Wheel speed sensor harness
- Upper ball joint
- Inner flange bolts and the upper control arm

To install:

3. Installation is the reverse of the removal procedure, while using the following torque values:

- Inner flange bolts: 76 ft. lbs. (103 Nm)
- Upper ball joint nut: 36–43 ft. lbs. (49–59 Nm)
- Wheel speed sensor harness bolts: 88 inch lbs. (10 Nm)

Rear

CIVIC

✳✳ CAUTION

Removing rear suspension components may make the vehicle front-heavy and cause it to tip forward when raised on a hoist. Use under-lift support stands, or place additional weight in the trunk of the vehicle before hoisting it.

1. Before servicing the vehicle, refer to the precautions in the beginning of this section.
2. Raise and safely support the vehicle.

3. Remove or disconnect the following:
   - Rear wheels
4. Support the lower control arm with a floor jack.
5. Remove or disconnect the following:
   - Upper control arm from the trailing arm
   - Upper control arm flange bar from its vehicle body mount
   - Upper control arm
6. Inspect the upper control arm and bushings for signs of wear and distortion. The bushings are replaced as follows:
   a. Press the bushings out of the upper control arm using suitably sized press fixtures.
   b. Matchmark the bolt flange bar to the body of the upper control arm.
   c. Lubricate the new bushings with silicon grease before installation.
   d. Press the new bushings into the control arm. Make sure the bolt flange bar matchmarks align. The leading edges of the control arm bushings must be flush with the edges of the control arm body.

To install:

➡**Use new self-locking nuts and color-coded bolts when assembling suspension components.**

7. Install or connect the following:
   - Control arm to its body mount. Hand-tighten the flange bolts.
   - Control arm to the trailing arm. Hand-tighten the flange bolt.
   - Rear wheel and lower the vehicle.
8. Torque the bolts with the vehicle on the ground. Tighten the control arm bolts-to-body to 29 ft. lbs. (40 Nm). Tighten the control arm-to-trailing arm bolt to 40 ft. lbs. (55 Nm).
9. Check and adjust the vehicle's rear wheel alignment.
10. Tighten the wheel nuts to 80 ft. lbs. (110 Nm).

PRELUDE

1. Before servicing the vehicle, refer to the precautions in the beginning of this section.

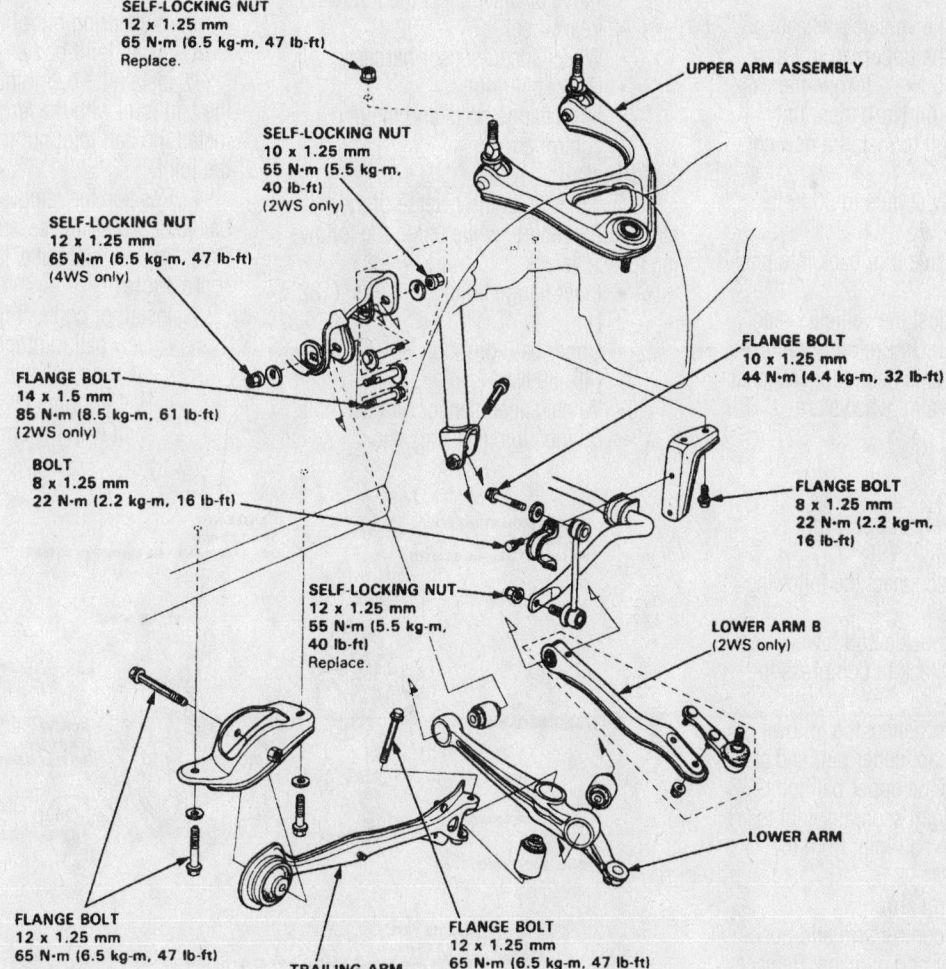

Rear suspension components—Prelude

7923FGC8

2. Raise and support the vehicle safely.

3. Remove or disconnect the following:
- Rear wheels. Support the knuckle and lower control arm assembly with a jack.
- Upper ball joint from the knuckle using a ball joint separator tool.
- Trunk side trim
- 2 strut mounting nuts
- Self-locking nuts from the upper arm anchor bolts
- Upper arm from the vehicle

➡️**Do not disassemble the upper arm. If the ball joint or bushings are faulty, or the upper arm is damaged, the entire upper arm must be replaced.**

To install:

➡️**Use new self-locking nuts when installing the upper arm and strut.**

4. Install or connect the following:
- Upper control arm assembly into the strut tower.
- Upper ball joint
- Rear wheels and lower the vehicle.

5. With all 4 of the vehicle's wheels on the ground, torque the upper control arm nuts to 47 ft. lbs. (65 Nm). Torque the castle nut to 32 ft. lbs. (44 Nm); then, only tighten it only enough to install a new cotter pin.

6. Tighten the wheel nuts to 80 ft. lbs. (110 Nm).

7. Put the trunk side trim back into position.

8. Check and adjust the vehicle's rear end wheel alignment. On Preludes equipped with 4WS, the electronic neutral check must be performed before all 4 wheels are aligned.

ACCORD

1. Raise and safely support the vehicle.

2. Remove or disconnect the following:
- Rear wheels

3. Support the knuckle and lower control arm with a floor jack to compress the strut.

4. Remove or disconnect the following:
- Castle nut cap, cotter pin, and castle nut from the upper ball joint. Use a ball joint separator tool to separate the ball joint from the knuckle.
- Upper control arm

5. Check upper control arm and bushing for signs of wear and damage. Replace the upper control arm if the ball joint is faulty.

To install:

➡️**Use new self-locking nuts when assembling suspension components.**

6. Install or connect the following:
- Upper arm into the vehicle
- Mounting bolts and only hand-tighten them.
- Upper arm to the knuckle.
- Castle nut at the ball joint to 32 ft. lbs. (44 Nm). Tighten the castle nut only enough to install a new cotter pin.
- Castle nut cap
- Rear wheels and lower the vehicle.

7. Tighten the upper mounting bolts to 28 ft. lbs. (39 Nm).

8. Tighten the wheel nuts to 80 ft. lbs. (110 Nm).

9. Check and adjust the vehicle's rear wheel alignment.

S2000

1. Before servicing the vehicle, refer to the precautions in the beginning of this section.

2. Remove or disconnect the following:
- Rear wheel
- Wheel speed sensor harness
- Upper ball joint
- Inner flange bolts and the upper control arm

To install:

3. Installation is the reverse of the removal procedure, while using the following torque values:
- Inner flange bolts: 98 ft. lbs. (132 Nm)
- Upper ball joint nut: 36–43 ft. lbs. (49–59 Nm)
- Wheel speed sensor harness bolts: 88 inch lbs. (10 Nm)

Lower Ball Joint

REMOVAL & INSTALLATION

Civic

➡️**The steering knuckle must be removed from the vehicle for the ball joint to be replaced. The following special tools or their equivalents are needed to press the ball joint in and out of the knuckle: ball joint installer base tool 07965-SB00200, ball joint installer/remover tool 07965-SB00100, and ball joint remover base tool 07965-SH20200. A large vise will be required to hold the knuckle and the press tools. A ball joint clip guide tool 07974-SA50700 or 07GAG-SD40700 is used to install the retaining clip on the joint boot.**

1. Before servicing the vehicle, refer to the precautions in the beginning of this section.

2. Remove or disconnect the following:
- Steering knuckle assembly from the vehicle.
- Ball joint boot snapring and the boot.
- Snapring out of the groove in the ball joint body.

3. Install the ball joint removal tool onto the ball joint with the large end facing out. Install the ball joint nut to attach the tool to the joint.

4. Position the removal base tool on the ball joint and set the assembly in a large vise. Press the ball joint out of the steering knuckle.

To install:

5. Install or connect the following:
- New ball joint into the hole of the steering knuckle.
- Ball joint installer tool over the ball joint with the small end facing out.

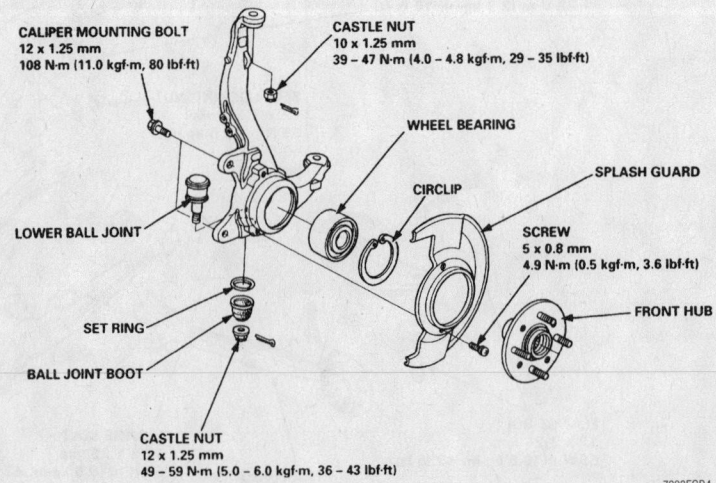

Knuckle components—Civic

7923FGB4

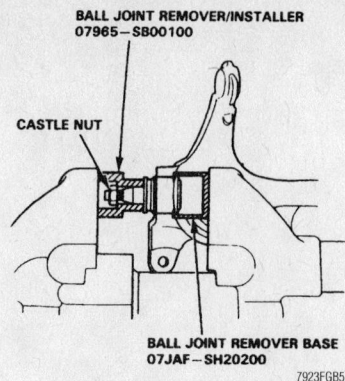

Ball joint removal tools—Civic

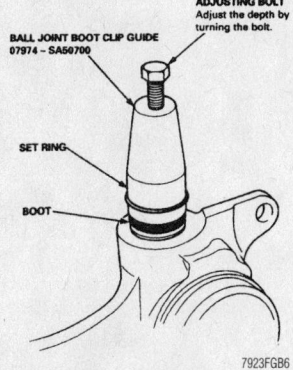

Ball joint boot clip guide—Civic

- Installation base tool on the ball joint and set the assembly in a large vise. Press the ball joint into the steering knuckle.
- Snapring in the groove of the ball joint.

6. Adjust the boot clip tool with the adjusting bolt until the end of the tool aligns with the groove on the boot. Slide the clip over the tool and into position.

7. Install the ball joint stud in the steering knuckle. Tighten the nut to 44 ft. lbs. (60 Nm).

Wheel Bearings

ADJUSTMENT

The wheel bearings are not adjustable or repairable and should be replaced if found defective.

REMOVAL & INSTALLATION

Front

CIVIC

➡**A hydraulic press and several bearing drivers and attachments are needed to remove and install the hub and bearing.**

1. Before servicing the vehicle, refer to the precautions in the beginning of this section.

2. Pry the spindle nut stake away from the spindle, then loosen the nut.

3. Raise and safely support the vehicle.

4. Remove or disconnect the following:
- Front wheel and the spindle nut
- Wheel sensor wire bracket from the knuckle, but don't disconnect it.
- Caliper mounting bolts and the caliper. Support the caliper out of the way with a length of wire. Do not let the caliper hang from the brake hose.
- 6mm brake disc retaining screws. Screw 2, 12mm bolts into the disc to push it away from the hub.
- Tie rod castle nut
- Tie rod ball joint using a suitable ball joint remover.
- Cotter pin and loosen the lower arm ball joint nut half the length of the joint threads.
- Ball joint and lower arm using a suitable puller with the pawls applied to the lower arm.

➡**Avoid damaging the ball joint boot. If necessary, apply penetrating type lubricant to loosen the ball joint.**

- Ball joint nut cover
- Cotter pin and the upper ball joint nut.
- Upper ball joint and knuckle using a ball joint remover.

5. Use a plastic mallet to free the halfshaft from the knuckle. Pull the knuckle out to remove it.

➡**A new wheel bearing must be used when the hub is removed.**

6. Place the knuckle in a press and use a base and pilot to press the hub assembly out of the wheel bearing.

7. Remove the knuckle ring seal and circlip. Remove the splash guard from the knuckle.

8. Press the wheel bearing out of the knuckle using a driving attachment.

To install:

9. Clean the knuckle and hub assembly and inspect them for damage.

10. Install or connect the following:
- New wheel bearing into the hub using a driving tool.
- Circlip in the outer groove of the knuckle.
- Splash guard
- Hub assembly into the steering knuckle using a base and a driving and guide tool.

- Knuckle ring seal
- Knuckle onto the spindle
- Knuckle onto the upper and lower ball joints and tighten the castle nuts.
- Tie rod ball joint onto the steering knuckle.

11. Tighten the upper ball joint nut and tie rod nut to 29–35 ft. lbs. (40–48 Nm) and the lower ball joint castle nut to 36–43 ft. lbs. (50–60 Nm).

12. Install or connect the following:
- Anti-lock Brake System (ABS) wheel sensor wire brackets onto the knuckle. Tighten the mounting bolts to 84 inch lbs. (10 Nm).
- Brake disc; use 2 lug nuts to evenly draw the disc onto the hub.
- Retainer screws: 84 inch lbs. (10 Nm)
- Spindle washer and nut. Don't tighten the nut until the vehicle is on the ground.
- Brake caliper and tighten the bolts to 80 ft. lbs. (110 Nm).
- Front wheels and lower the vehicle.

13. Tighten the spindle nut to 134 ft. lbs. (185 Nm), stake the nut, and install the grease cap.

14. Check and adjust the vehicle's front wheel alignment.

➡**Avoid damaging the ball joint boot. If necessary, apply penetrating-type lubricant to loosen the ball joint.**

PRELUDE AND ACCORD

➡**Once the hub has been removed, the wheel bearings must be replaced.**

A hydraulic press and bearing drivers must be used to remove and install the bearing.

1. Before servicing the vehicle, refer to the precautions in the beginning of this section.

2. Pry the spindle nut stake away from the spindle and loosen the nut. Do not tighten or loosen a spindle nut unless the vehicle is sitting on all 4 wheels. The torque required is high enough to cause the vehicle to fall off the stands even when properly supported.

3. Raise and safely support the vehicle.

4. Remove or disconnect the following:
- Wheel and the spindle nut
- Caliper mounting bolts and the caliper. Support the caliper out of the way with a length of wire. Do not let the caliper hang from the brake hose.

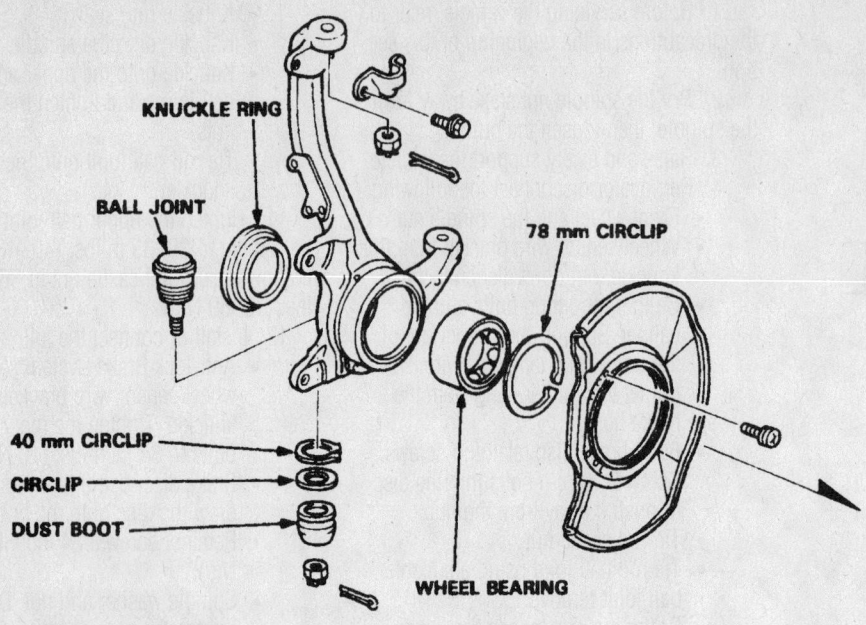

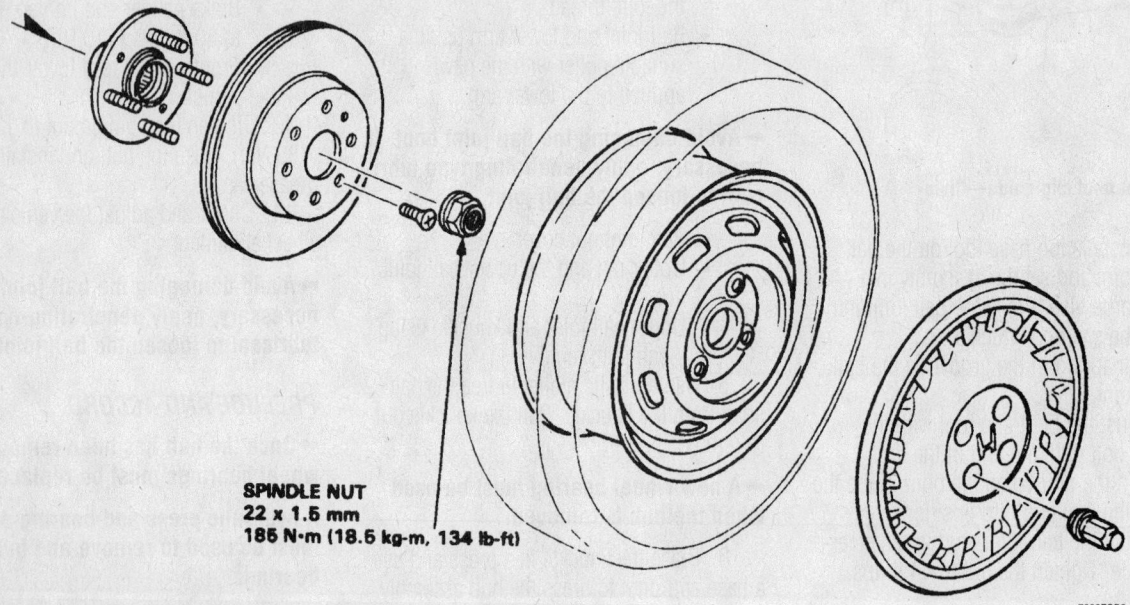

Wheel hub, bearing and steering knuckle components—Prelude and 2000–02 Accord

- 6mm brake disc retaining screws. Screw 2, 8 x 1.25mm bolts into the disc to push it away from the hub.

➡ **Turn each bolt 2 turns at a time to prevent cocking the brake disc.**

- Cotter pin from the tie rod castle nut, then the nut.
- Tie rod ball joint using a ball joint remover, then lift the tie rod out of the knuckle.
- Cotter pin, then loosen the lower arm ball joint nut half the length of the joint threads. The nut will retain the arm when the joint comes loose.

- Ball joint and lower arm using a puller with the pawls applied to the lower arm. Avoid damaging the ball joint boot. If necessary, apply penetrating lubricant to loosen the ball joint.
- Upper ball joint shield, if equipped.
- Cotter pin and the upper ball joint nut.
- Upper ball joint and knuckle
- Knuckle and hub by sliding them off the halfshaft.
- Splash guard screws from the knuckle.

5. Position the knuckle/hub assembly in a hydraulic press.

6. Remove or disconnect the following:

- Hub from the knuckle using a driver of the proper diameter while supporting the knuckle. The inner bearing race may stay on the hub.
- Splash guard and snapring from the knuckle.

7. Press the wheel bearing out of the knuckle while supporting the knuckle.

8. If necessary, remove the outboard bearing inner race from the hub using a bearing puller.

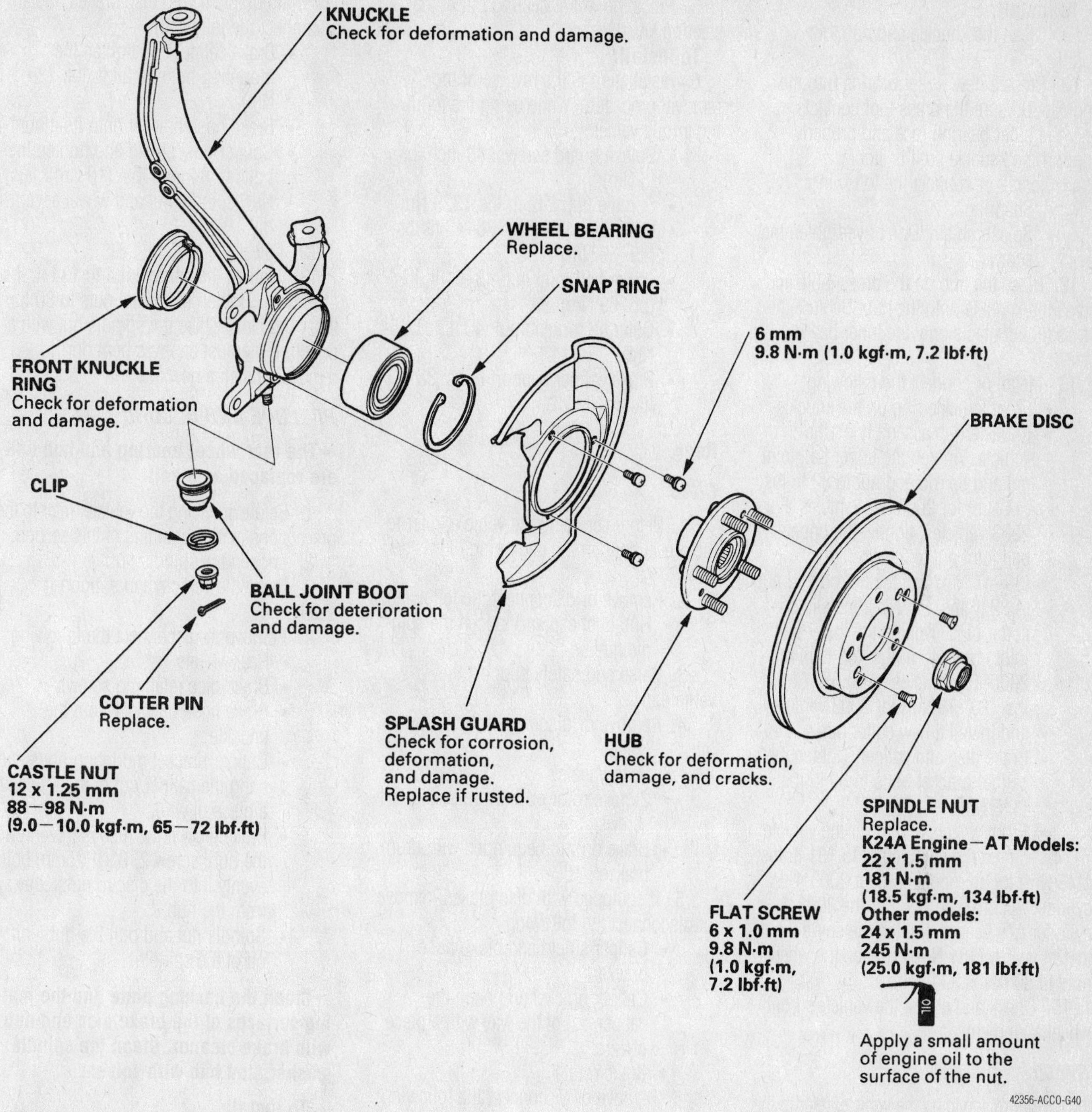

KNUCKLE
Check for deformation and damage.

WHEEL BEARING
Replace.

SNAP RING

6 mm
9.8 N·m (1.0 kgf·m, 7.2 lbf·ft)

BRAKE DISC

FRONT KNUCKLE RING
Check for deformation and damage.

CLIP

BALL JOINT BOOT
Check for deterioration and damage.

COTTER PIN
Replace.

CASTLE NUT
12 x 1.25 mm
88—98 N·m
(9.0—10.0 kgf·m, 65—72 lbf·ft)

SPLASH GUARD
Check for corrosion, deformation, and damage.
Replace if rusted.

HUB
Check for deformation, damage, and cracks.

SPINDLE NUT
Replace.
K24A Engine—AT Models:
22 x 1.5 mm
181 N·m
(18.5 kgf·m, 134 lbf·ft)
Other models:
24 x 1.5 mm
245 N·m
(25.0 kgf·m, 181 lbf·ft)

Apply a small amount of engine oil to the surface of the nut.

FLAT SCREW
6 x 1.0 mm
9.8 N·m
(1.0 kgf·m, 7.2 lbf·ft)

42356-ACCO-G40

Wheel hub, bearing and steering knuckle components—2003 Accord

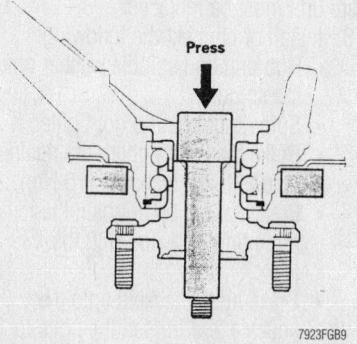

Press the hub out of the knuckle—Prelude and Accord

7923FGB9

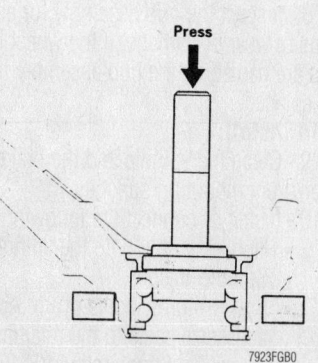

Press the bearing out of the knuckle—Prelude and Accord

7923FGB0

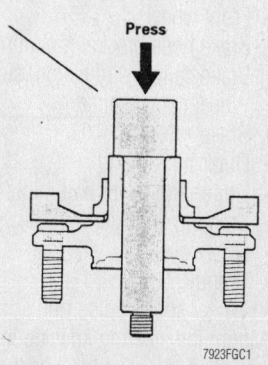

Use a press to remove the inner bearing race from the hub—Prelude and Accord

7923FGC1

To install:

9. Clean the knuckle and hub thoroughly.

10. Press a new wheel bearing into the knuckle. Be sure the press tool contacts only the outer bearing race and properly support the knuckle so it is stable.

11. Install or connect the following:
- Snapring
- Splash shield. Don't overtighten the screws.

12. Place the hub on the press table and press the knuckle onto the hub. Be sure the press tool contacts only the inner bearing race.

13. Install or connect the following:
- Front knuckle ring on the knuckle
- Knuckle/hub assembly on the vehicle. Tighten the upper ball joint nut and tie rod end nut to 32 ft. lbs. (44 Nm) for 2000–02 vehicles. For 2003 vehicles, tighten the upper ball joint nut to 29–35 ft. lbs. (39–47 Nm) and the tie rod end nut to 32 ft. lbs. (44 Nm). Install new cotter pins. Tighten the lower ball joint nut to 40 ft. lbs. (55 Nm) for 2000–02 vehicles, or to 65–72 ft. lbs. (88–98 Nm) for 2003 vehicles and install a new cotter pin.
- Brake disc and caliper. Tighten the caliper bracket bolts to 80 ft. lbs. (110 Nm).
- Front wheels and lower the vehicle.

14. Tighten the spindle nut to 181 ft. lbs. (245 Nm) for all models except 2003 4-cylinder Accord with A/T. For the 2003 4-cylinder A/T Accord, tighten the spindle nut to 134 ft. lbs. (181 Nm). Tighten the wheel nuts to 80 ft. lbs. (110 Nm).

15. Check and adjust the vehicle's front wheel alignment.

S2000

1. Before servicing the vehicle, refer to the precautions in the beginning of this section.

2. Remove or disconnect the following:
- Front wheel
- Brake hose bracket mounting bolts
- Brake caliper and caliper support
- Wheel speed sensor
- Brake rotor
- Outer tie rod end
- Upper and lower ball joints
- Steering knuckle from the vehicle
- Dust cover
- Spindle nut
- Wheel speed pulse ring

3. Mount the steering knuckle in a press and press the hub out of the wheel bearing.

4. Remove the splash guard and the wheel bearing snapring.

5. Press the wheel bearing out of the steering knuckle.

To install:

6. Installation is the reverse of the removal procedure, while using the following torque values:
- Splash guard screws: 48 inch lbs. (5 Nm)
- Spindle nut: 242 ft. lbs. (329 Nm)
- Upper ball joint nut: 36–43 ft. lbs. (49–59 Nm)
- Lower ball joint nut: 43–51 ft. lbs. (56–69 Nm)
- Outer tie rod end nut: 40 ft. lbs. (54 Nm)
- Brake caliper support bolts: 83 ft. lbs. (113 Nm)

Rear

CIVIC

1. Before servicing the vehicle, refer to the precautions in the beginning of this section.

2. Remove or disconnect the following:
- Hub dust cap and loosen the spindle nut.

3. Raise and safely support the vehicle.

4. Remove or disconnect the following:
- Rear wheels.
- 2 brake rotor or drum retaining screws
- Brake drum, if equipped with drum brakes.

5. If equipped with disc brakes, remove or disconnect the following:
- Caliper shield and brake hose bracket
- Caliper bracket and hang the caliper out of the way with a piece of wire.
- Brake rotor

6. Remove or disconnect the following:
- Hub assembly from the spindle

7. Clean the hub assembly in solvent.

8. Inspect the hub assembly for any signs of wear or damage. If the wheel bearings are damaged, the hub assembly must be replaced.

To install:

9. Clean the spindle and the brake rotor/drum mounting surfaces.

10. Install or connect the following:
- Hub assembly onto the spindle
- Spindle washer
- Brake rotor or brake drum. Apply anti-seize paste to the retaining screws and tighten them to 84 inch lbs. (10 Nm). Don't overtighten the retaining screws.

11. If equipped with disc brakes, install or connect the following:
- Brake caliper and tighten the mounting bolts to 28 ft. lbs. (39 Nm).
- Brake hose bracket onto its mount.
- Caliper dust shield and tighten the bolts to 84 inch lbs. (10 Nm).
- New spindle nut and wheel assembly.

12. Lower the vehicle.

13. Tighten the spindle nut to 134 ft. lbs. (185 Nm). Tighten the wheel nuts to 80 ft. lbs. (110 Nm). Stake the spindle nut with a punch. If the dust cap was bent during removal, install a new one.

PRELUDE AND ACCORD

➡ **The rear wheel bearing and hub unit are replaced as a unit.**

1. Before servicing the vehicle, refer to the precautions in the beginning of this section.

2. Loosen the spindle nut.

3. Raise the vehicle and support it safely.

4. Remove or disconnect the following:
- Rear wheels
- Brake disc retaining screws
- Brake hose brackets from the knuckle
- Caliper bracket mounting bolts and hang the caliper out of the way with a piece of wire.
- Brake disc. If the disc is frozen on the hub, screw 2, 8 x 1.25mm bolts evenly into the disc to push it away from the hub.
- Spindle nut and pull the hub unit off of the spindle.

➡ **Clean the backing plate and the mating surfaces of the brake disc and hub with brake cleaner. Clean the spindle, washer, and hub with solvent.**

To install:

5. Inspect the hub unit for signs of damage or wear. If the bearings are worn, the entire unit must be replaced.

6. Install or connect the following:
- Hub unit and spindle washer onto the spindle.
- Spindle nut but do not tighten it.
- Brake disc and tighten the retaining screws to 84 inch lbs. (10 Nm).
- Brake caliper and tighten the mounting bolts to 28 ft. lbs. (39 Nm).
- Brake hose brackets onto the knuckle and tighten the bolts to 16 ft. lbs. (22 Nm).
- Rear wheels and lower the vehicle.

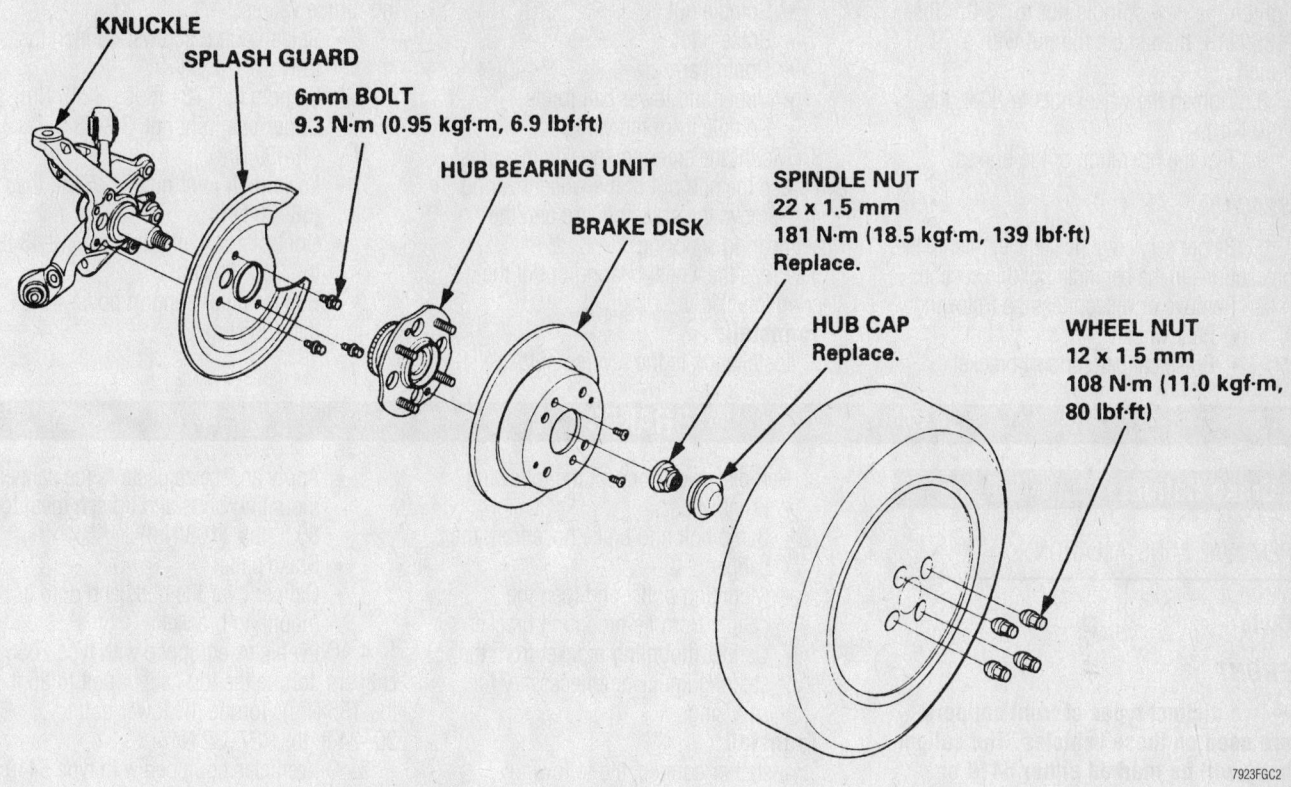

KNUCKLE

SPLASH GUARD

6mm BOLT
9.3 N·m (0.95 kgf·m, 6.9 lbf·ft)

HUB BEARING UNIT

BRAKE DISK

SPINDLE NUT
22 x 1.5 mm
181 N·m (18.5 kgf·m, 139 lbf·ft)
Replace.

HUB CAP
Replace.

WHEEL NUT
12 x 1.5 mm
108 N·m (11.0 kgf·m, 80 lbf·ft)

7923FGC2

Exploded view of the hub unit, drum brakes—Accord

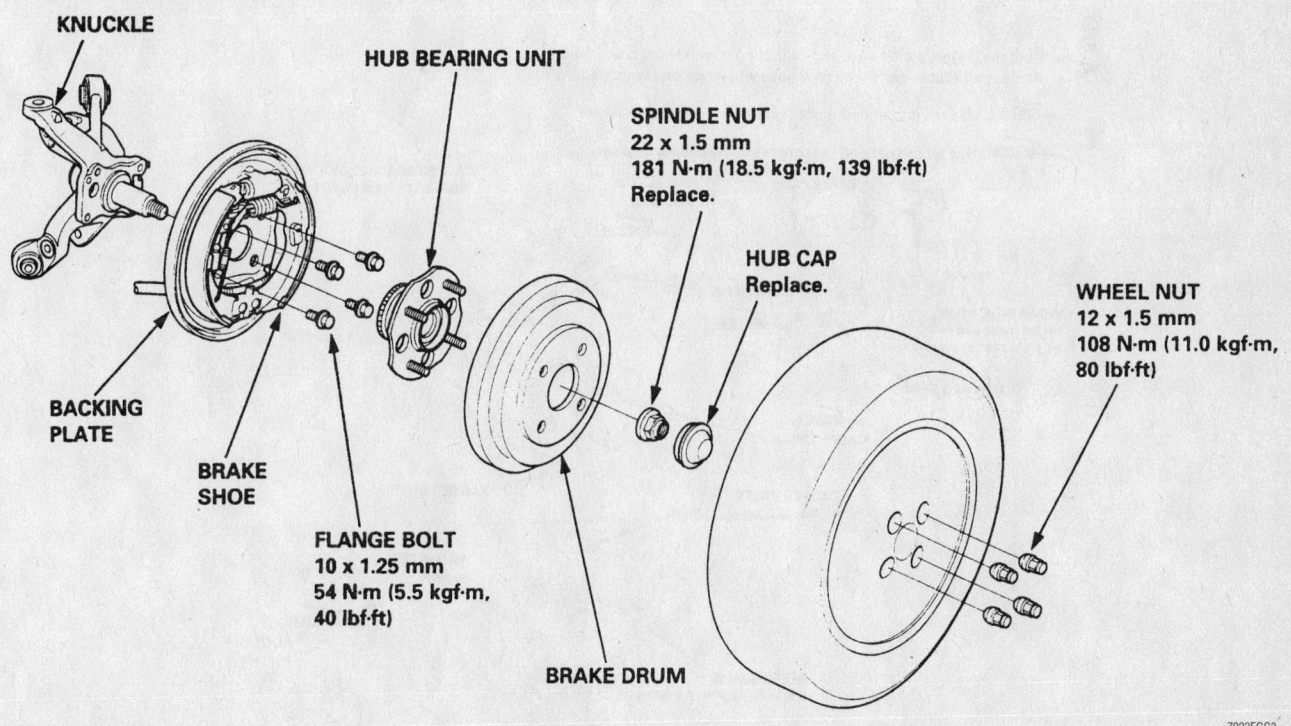

KNUCKLE

HUB BEARING UNIT

SPINDLE NUT
22 x 1.5 mm
181 N·m (18.5 kgf·m, 139 lbf·ft)
Replace.

HUB CAP
Replace.

WHEEL NUT
12 x 1.5 mm
108 N·m (11.0 kgf·m, 80 lbf·ft)

BACKING PLATE

BRAKE SHOE

FLANGE BOLT
10 x 1.25 mm
54 N·m (5.5 kgf·m, 40 lbf·ft)

BRAKE DRUM

7923FGC3

Hub unit, disc brakes—Accord and Prelude

7. With the vehicle on the ground, tighten the new spindle nut to 134 ft. lbs. (185 Nm), then stake the nut with a punch.

8. Tighten the wheel nuts to 80 ft. lbs. (110 Nm).

9. Test the operation of the brakes.

S2000

1. Before servicing the vehicle, refer to the precautions in the beginning of this section.

2. Remove or disconnect the following:
- Rear wheel
- Brake caliper support bracket
- Wheel speed sensor
- Spindle nut
- Brake rotor
- Control arm
- Upper and lower ball joints
- Spindle from the vehicle

3. Mount the steering knuckle in a press and press the hub out of the wheel bearing.

4. Remove the splash guard and the wheel bearing snapring.

5. Press the wheel bearing out of the steering knuckle.

To install:

6. Installation is the reverse of the removal procedure, while using the following torque values:
- Splash guard screws: 48 inch lbs. (5 Nm)
- Spindle nut: 181 ft. lbs. (245 Nm)
- Upper ball joint nut: 36–43 ft. lbs. (49–59 Nm)
- Lower ball joint nut: 51–58 ft. lbs. (68–78 Nm)
- Control arm ball joint nut: 36–43 ft. lbs. (49–59 Nm)
- Brake caliper support bolts: 41 ft. lbs. (55 Nm)

BRAKES

Brake Caliper

REMOVAL & INSTALLATION

Civic

FRONT

➡️ **Two distinct types of front calipers are used on these vehicles. The caliper body will be marked either 5410 or 2056 depending on type. Servicing procedures are similar, but the different calipers use different pads.**

1. Before servicing the vehicle, refer to the precautions in the beginning of this section.

2. Remove or disconnect the following:
- Front wheels
- Banjo bolt and brake hose from the caliper
- Mounting bolts, and then the caliper from its mounting bracket
- Caliper mounting bracket from the steering knuckle, if necessary for servicing

To install:

3. Install or connect the following:
- Caliper mounting bracket was removed, if removed
- Caliper pins with new pin boots, after applying brake seal grease to the pins and
- Apply anti-seize paste to the caliper mounting bolts and tighten them to 80 ft. lbs. (108 Nm)
- Brake pads
- Caliper over the pads and onto its mounting bracket

4. On vehicles equipped with type 2056 calipers, torque the top caliper bolt to 25 ft. lbs. (35 Nm). Torque the lower bolt to 20–24 ft. lbs. (27–32 Nm).

5. On vehicles equipped with type 5410 calipers, torque both caliper bolts to 24 ft. lbs. (33 Nm).

6. Install or connect the following:
- Brake hose to the caliper using new sealing washers. Carefully torque

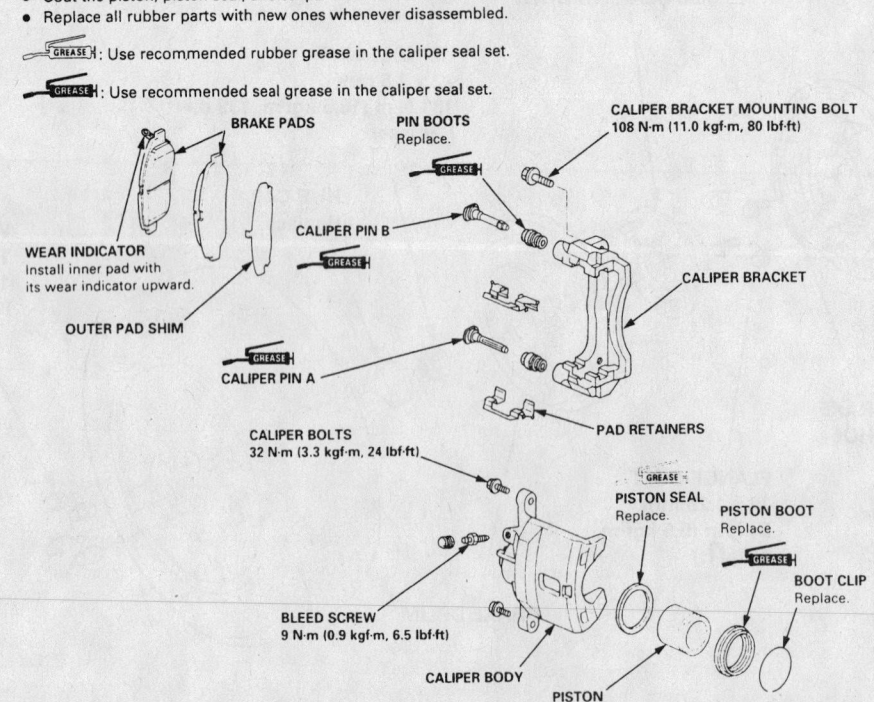

- Coat the piston, piston seal, and caliper bore with clean brake fluid.
- Replace all rubber parts with new ones whenever disassembled.

GREASE : Use recommended rubber grease in the caliper seal set.

GREASE : Use recommended seal grease in the caliper seal set.

BRAKE PADS

WEAR INDICATOR
Install inner pad with its wear indicator upward.

OUTER PAD SHIM

PIN BOOTS
Replace.

CALIPER PIN B

CALIPER BRACKET MOUNTING BOLT
108 N·m (11.0 kgf·m, 80 lbf·ft)

CALIPER BRACKET

CALIPER PIN A

PAD RETAINERS

CALIPER BOLTS
32 N·m (3.3 kgf·m, 24 lbf·ft)

PISTON SEAL
Replace.

PISTON BOOT
Replace.

BOOT CLIP
Replace.

BLEED SCREW
9 N·m (0.9 kgf·m, 6.5 lbf·ft)

CALIPER BODY

PISTON

93016G05

Exploded view of the front brakes—Honda 5410 Type shown

the banjo bolt to 25 ft. lbs. (35 Nm).

- Reservoir with fresh brake fluid and bleed the brake system
- Front wheels

REAR

1. Before servicing the vehicle, refer to the precautions in the beginning of this section.
2. Remove or disconnect the following:
   - Rear wheels
   - Caliper shield
   - Lock pin and clevis pin from the parking brake cable
   - Cable securing clip
   - Parking brake cable from the caliper
   - Banjo bolt and the brake hose from the caliper
   - 2 caliper mounting bolts
   - Caliper from its mounting bracket
   - Caliper mounting bracket from the trailing arm, if necessary for servicing

To install:

3. Install or connect the following:
   - Caliper mounting bracket, if removed.
   - Caliper pins with new pin boots, after applying brake seal grease to the pins
   - Anti-seize paste to the caliper bracket mounting bolts and tighten them to 28 ft. lbs. (39 Nm)

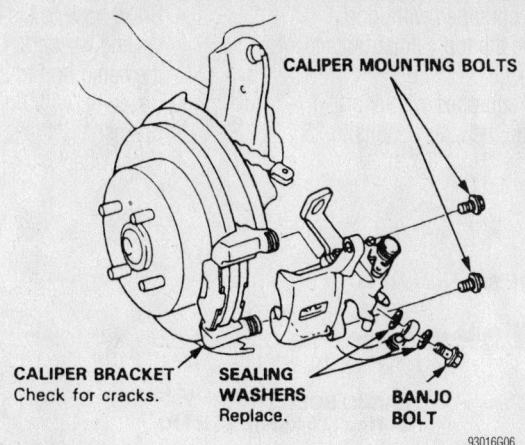

CALIPER MOUNTING BOLTS

CALIPER BRACKET Check for cracks.

SEALING WASHERS Replace.

BANJO BOLT

93016G06

Rear brake caliper mounting—Civic

- Brake pads

4. Rotate the caliper piston clockwise into place in the cylinder and then align the groove in the piston with the tab on inner pad.
   - Caliper over the pads and onto its mounting bracket. Tighten the caliper bolts to 17 ft. lbs. (23 Nm).
   - Parking brake cable, after greasing the parking brake linkage
   - Caliper shield
   - Reservoir with fresh brake fluid and bleed the brake system. Adjust the parking brake if necessary.
   - Rear wheels

Accord and Prelude

FRONT

1. Before servicing the vehicle, refer to the precautions in the beginning of this section.
2. Remove or disconnect the following:
   - Front wheels
   - Banjo bolt
   - Brake hose from the caliper
   - Mounting bolts and remove the caliper from its mounting bracket

To install:

3. Install or connect the following:
   - Caliper over the pads and onto its mounting bracket

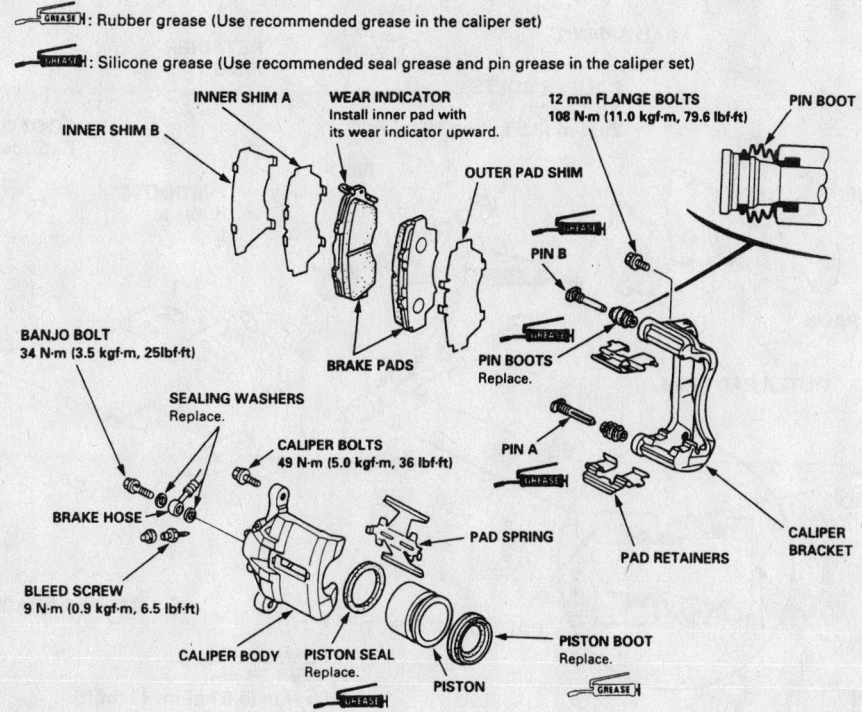

[GREASE]: Rubber grease (Use recommended grease in the caliper set)

[GREASE]: Silicone grease (Use recommended seal grease and pin grease in the caliper set)

INNER SHIM A

INNER SHIM B

WEAR INDICATOR Install inner pad with its wear indicator upward.

OUTER PAD SHIM

12 mm FLANGE BOLTS 108 N·m (11.0 kgf·m, 79.6 lbf·ft)

PIN BOOT

PIN B

BANJO BOLT 34 N·m (3.5 kgf·m, 25lbf·ft)

BRAKE PADS

PIN BOOTS Replace.

SEALING WASHERS Replace.

CALIPER BOLTS 49 N·m (5.0 kgf·m, 36 lbf·ft)

PIN A

BRAKE HOSE

PAD SPRING

PAD RETAINERS

CALIPER BRACKET

BLEED SCREW 9 N·m (0.9 kgf·m, 6.5 lbf·ft)

CALIPER BODY

PISTON SEAL Replace.

PISTON

PISTON BOOT Replace.

93016G07

Exploded view of the front brakes—Accord V6 shown

4. On vehicles equipped with long caliper pins, torque the top caliper bolts to 54 ft. lbs. (74 Nm).

5. On vehicles equipped with short caliper bolts, torque the caliper bolts to 36 ft. lbs. (50 Nm).

- Brake hose to the caliper using new sealing washers. Carefully torque the banjo bolt to 25 ft. lbs. (35 Nm).
- Reservoir with fluid and bleed the brakes
- Front wheels

REAR

1. Before servicing the vehicle, refer to the precautions in the beginning of this section.

2. Remove or disconnect the following:
   - Rear wheels

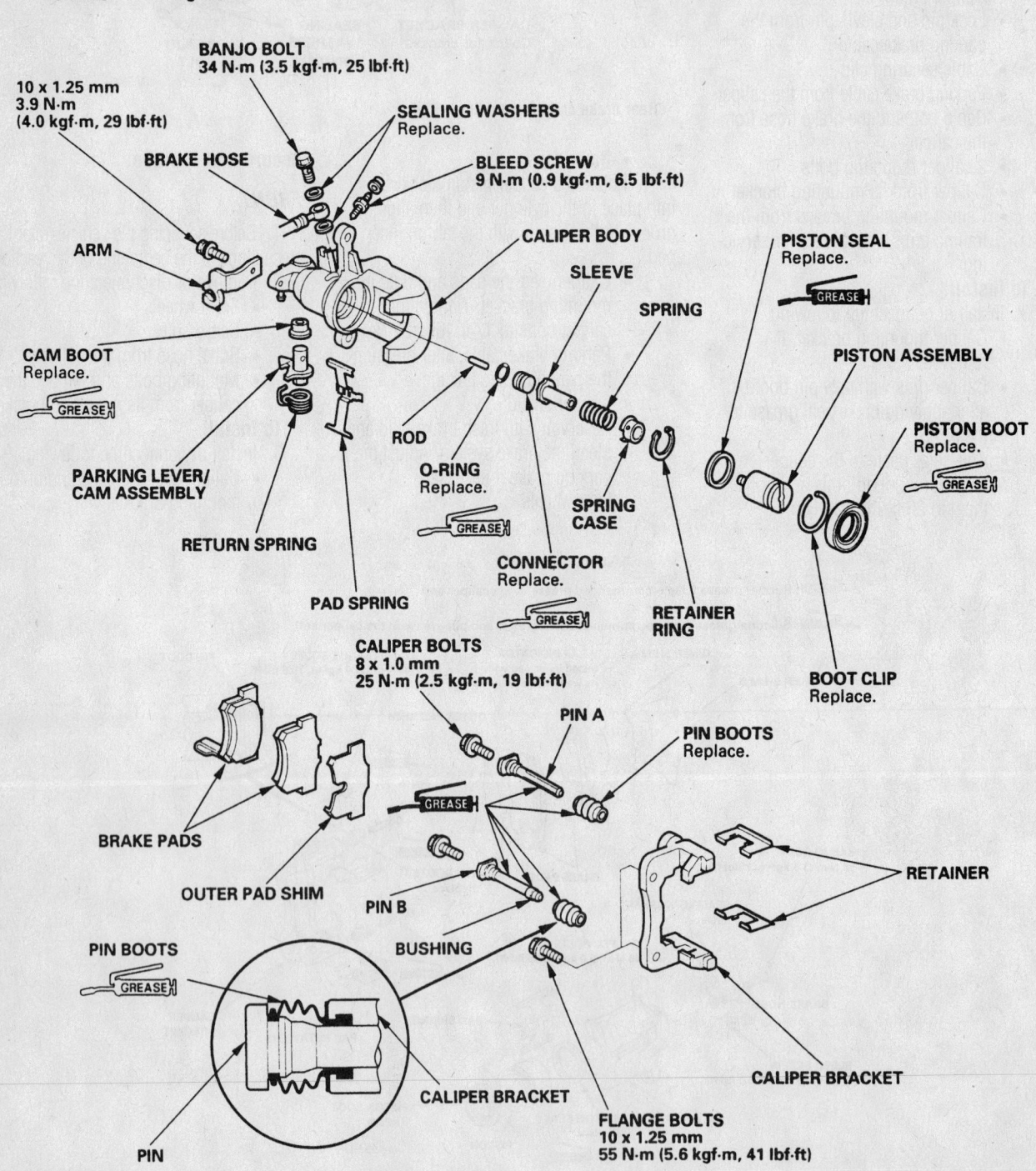

Sedan:

GREASE : Silicone grease

GREASE : Rubber grease

BANJO BOLT
34 N·m (3.5 kgf·m, 25 lbf·ft)

10 x 1.25 mm
3.9 N·m
(4.0 kgf·m, 29 lbf·ft)

SEALING WASHERS
Replace.

BRAKE HOSE

BLEED SCREW
9 N·m (0.9 kgf·m, 6.5 lbf·ft)

ARM

CALIPER BODY

SLEEVE

SPRING

PISTON SEAL
Replace.

GREASE

PISTON ASSEMBLY

CAM BOOT
Replace.

GREASE

PARKING LEVER/
CAM ASSEMBLY

RETURN SPRING

ROD

O-RING
Replace.

GREASE

SPRING
CASE

CONNECTOR
Replace.

GREASE

RETAINER
RING

PISTON BOOT
Replace.

GREASE

BOOT CLIP
Replace.

PAD SPRING

CALIPER BOLTS
8 x 1.0 mm
25 N·m (2.5 kgf·m, 19 lbf·ft)

PIN A

PIN BOOTS
Replace.

GREASE

BRAKE PADS

RETAINER

OUTER PAD SHIM

PIN B

BUSHING

PIN BOOTS

GREASE

CALIPER BRACKET

CALIPER BRACKET

PIN

FLANGE BOLTS
10 x 1.25 mm
55 N·m (5.6 kgf·m, 41 lbf·ft)

93016G08

Rear disc brakes—Accord Sedan shown

- Caliper shield
- Parking brake cable from the caliper
- Banjo bolt
- Brake hose from the caliper
- 2 caliper mounting bolts
- Caliper from the mounting bracket

To install:

3. Install or connect the following:
- Caliper over the pads and onto the mounting bracket. Rotate the piston clockwise into place in the cylinder and then align the groove in the piston with the tab on inner pad.
- Caliper bolts and tighten to 17 ft. lbs. (23 Nm)
- Brake hose to the caliper using new sealing washers. Torque the banjo bolt to 25 ft. lbs. (35 Nm).
- Parking brake cable
- Caliper shield
- Reservoir with fluid and bleed the brake system. Adjust the parking brake if necessary.
- Rear wheels

S2000

FRONT

1. Before servicing the vehicle, refer to the precautions in the beginning of this section.
2. Remove or disconnect the following:
- Front wheels
- Banjo bolt
- Brake hose from the caliper
- Mounting bolts and remove the caliper from its mounting bracket

To install:

3. Install or connect the following:
- Caliper over the pads and onto its mounting bracket

4. On vehicles equipped with long caliper pins, torque the top caliper bolts to 54 ft. lbs. (74 Nm).

5. On vehicles equipped with short caliper bolts, torque the caliper bolts to 36 ft. lbs. (50 Nm).
- Brake hose to the caliper using new sealing washers. Carefully torque the banjo bolt to 25 ft. lbs. (35 Nm).

- Reservoir with fluid and bleed the brakes
- Front wheels

REAR

1. Before servicing the vehicle, refer to the precautions in the beginning of this section.
2. Remove or disconnect the following:
- Rear wheels
- Caliper shield
- Brake hose mounting bracket from caliper
- Cable clip from the parking brake cable
- Parking brake cable from the parking brake arm
- Banjo bolt and brake hose
- Brake hose sealing washers and discard
- 2 caliper bolts, while holding the caliper pin with a wrench
- Caliper from the bracket

To install:

3. Install or connect the following:
- Caliper onto the bracket

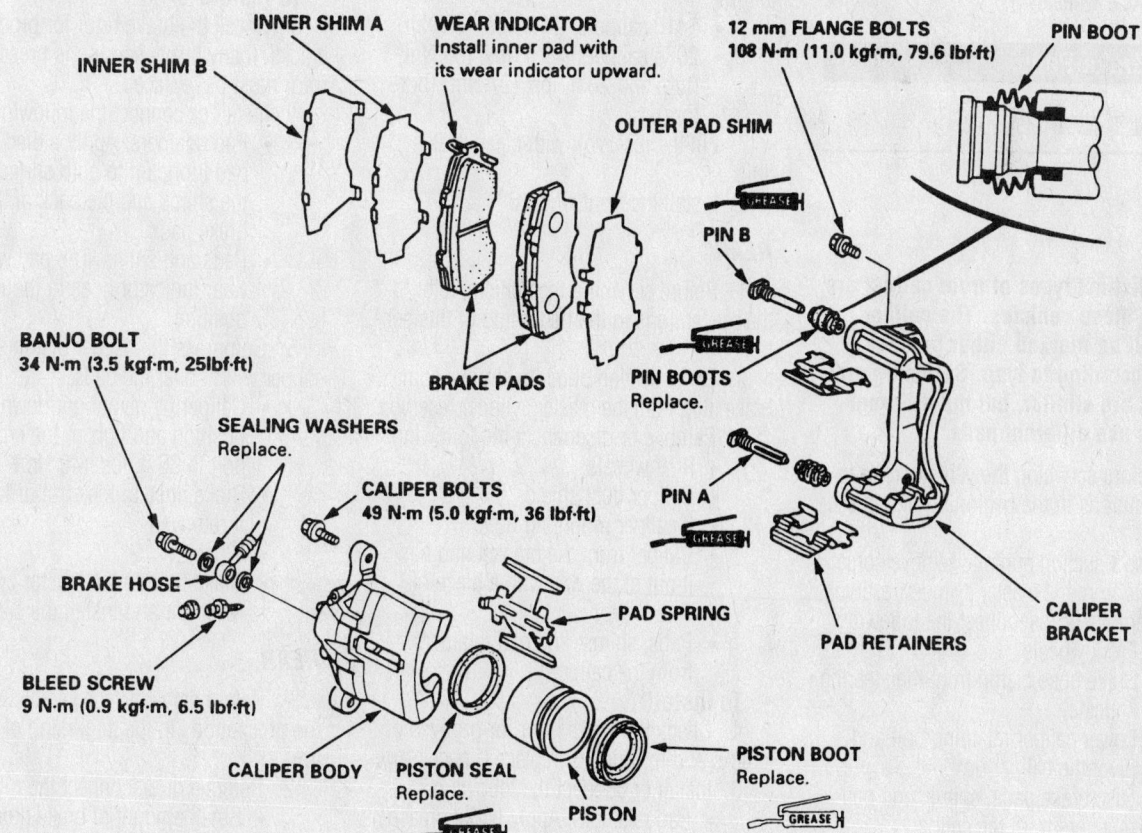

Exploded view of the front brakes—Accord V6 shown

93016G07

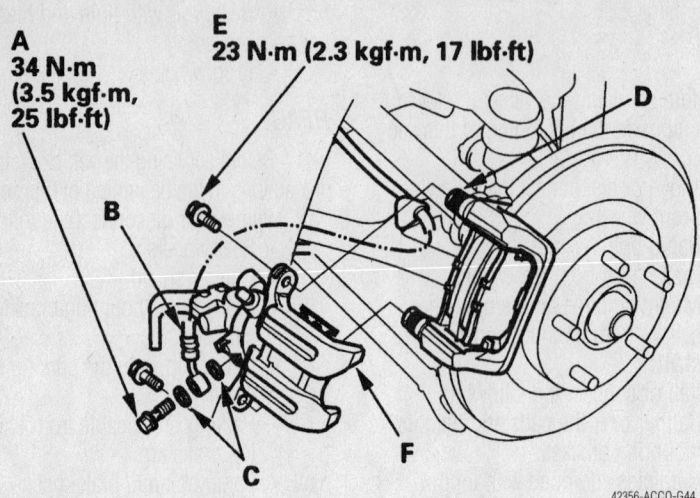

A
34 N·m
(3.5 kgf·m,
25 lbf·ft)

E
23 N·m (2.3 kgf·m, 17 lbf·ft)

D

B

C

F

42356-ACCO-G44

Rear caliper mounting—S2000

- Caliper bolts and tighten to 17 ft. lbs. (23 Nm)
- Brake hose, with new sealing washers
- Banjo bolt and tighten to 25 ft. lbs. (34 Nm)
- Parking brake cable
- Cable clip
- Brake hose mounting bracket
- Caliper shield
- Rear wheels

Disc Brake Pads

REMOVAL & INSTALLATION

Civic

FRONT

➡ Two distinct types of front caliper are used on these vehicles. The caliper body will be marked either 5410 or 2056, according to type. Servicing procedures are similar, but the different calipers use different pads.

1. Before servicing the vehicle, refer to the precautions in the beginning of this section.
2. Use a suction pump to remove some brake fluid from the master cylinder reservoir.
3. Remove or disconnect the following:
   - Front wheels
   - Brake hose clamp from the steering knuckle
   - Lower caliper retaining bolt and pivot the caliper up
   - Disc brake pads, shims, and pad retainers from the caliper

To install:

4. Install or connect the following:
   - Pad retainers. Apply brake grease

to the inner side of the shims and the back of the disc brake pads.
   - Pads, shims, and pad retainers. Make sure the wear indicator on the inner pad is facing up.
5. Compress the caliper piston with a suitable tool so that the caliper will fit over the pads.
6. Pivot the caliper down into position. Install the caliper bolts and tighten them as follows:
   - 5410 calipers: 24 ft. lbs. (33 Nm)
   - 2056 calipers: 25 ft. lbs. (35 Nm) (top) and 20 ft. lbs. (27 Nm) (bottom)
7. Fill the reservoir with clean brake fluid.
8. Install the front wheels.

REAR

1. Before servicing the vehicle, refer to the precautions in the beginning of this section.
2. Use a suction pump to remove some brake fluid from the master cylinder reservoir.
3. Remove or disconnect the following:
   - Rear wheels
   - Caliper dust shield
   - 2 caliper mounting bolts
   - Caliper from the bracket and hang it out of the way with a piece of wire
   - Pads, shims, and pad retainers from the caliper

To install:

4. Check the brake rotor for grooves or cracks and machine or replace if necessary.
5. Install or connect the following:
   - Pad retainers. Apply brake grease to the inner side of the shims and to the back of the disc brake pads.
   - Pads, shims, and pad retainers

6. Rotate the caliper piston clockwise into the caliper bore enough to allow the caliper to fit over the brake pads. Lubricate the piston boot with silicon grease. Avoid twisting the piston boot.
   - Brake caliper. Align the groove in the piston with the tab on the inner pad. Tighten the mounting bolts to 17 ft. lbs. (23 Nm).
   - Master cylinder reservoir with clean brake fluid
   - Rear wheels

Accord and Prelude

FRONT

1. Before servicing the vehicle, refer to the precautions in the beginning of this section.
2. Remove or disconnect the following:
   - Small amount of brake fluid from the reservoir using a suction pump
   - Wheels
   - Brake hose clamp from the strut or knuckle by removing the retaining bolts
   - Lower caliper retaining bolt and pivot the caliper upward
   - Pad shim and pad retainers
   - Disc brake pads from the caliper

To install:

3. Check the brake rotor for grooves or cracks. If any heavy scoring is present, the rotor must be replaced.
4. Install or connect the following:
   - Pad retainers. Apply a disc brake pad lubricant to both surfaces of the shims and the back of the disc brake pads.
   - Pads and shims. The pad with the wear indicator goes in the inboard position.
5. Compress the caliper piston so the caliper will fit over the pads.
   - Caliper by pivoting it down into position and tighten the mounting bolt to 36 ft. lbs. (49 Nm)
   - Brake hose to the strut or knuckle, if removed
   - Wheels
   - Brake fluid to the master cylinder reservoir and install the cap

REAR

1. Before servicing the vehicle, refer to the precautions in the beginning of this section.
2. Remove or disconnect the following:
   - Small amount of brake fluid from the reservoir using a suction pump
   - Rear wheels
   - Dust shield

- 2 caliper mounting bolts
- Caliper from the bracket
- Pads, shims and pad retainers

To install:

3. Clean the caliper thoroughly; remove any dirt or dust. Check the brake rotor for grooves or cracks and machine or replace, as necessary.

4. Install or connect the following:
- Pad retainers. Apply a disc brake pad lubricant to both surfaces of the shims and the back of the disc brake pads.
- Pads and shims. The wear retainer on the inboard pad faces down.

5. Use a suitable tool to rotate the caliper piston clockwise into the caliper bore, enough to enable the caliper to fit over the pads. Lubricate the piston boot with silicon grease, and avoid twisting the boot.
- Brake caliper, aligning the cutout in the piston with the tab on the inner pad. Tighten the mounting bolts to 17 ft. lbs. (23 Nm).
- Wheels
- Brake fluid to the master cylinder reservoir

S2000

FRONT

1. Before servicing the vehicle, refer to the precautions in the beginning of this section.

2. Remove or disconnect the following:
- Small amount of brake fluid from the reservoir using a suction pump
- Wheels
- Lower caliper bolt, while holding the pin with a wrench. Pivot the caliper up for access to the pads.
- Pad shim and pad retainers
- Disc brake pads from the caliper

To install:

3. Check the brake rotor for grooves or cracks. If any heavy scoring is present, the rotor must be replaced.

4. Install or connect the following:
- Pad retainers. Apply a disc brake pad lubricant to both surfaces of the shims and the back of the disc brake pads.
- Pads and shims. The pad with the wear indicator goes in the inboard position.

5. Compress the caliper piston so the caliper will fit over the pads.
- Caliper by pivoting it down into position and tighten the mounting bolt to 24 ft. lbs. (32 Nm)
- Wheels
- Brake fluid to the master cylinder reservoir and install the cap

REAR

1. Before servicing the vehicle, refer to the precautions in the beginning of this section.

2. Remove or disconnect the following:

- Small amount of brake fluid from the reservoir using a suction pump
- Rear wheels
- 2 caliper mounting bolts, while holding the caliper pin with a wrench
- Caliper from the bracket
- Pads, shims and pad retainers

To install:

3. Clean the caliper thoroughly; remove any dirt or dust. Check the brake rotor for grooves or cracks and machine or replace, as necessary.

4. Install or connect the following:
- Pad retainers. Apply a disc brake pad lubricant to both surfaces of the shims and the back of the disc brake pads.

- Pads and shims. The wear retainer on the inboard pad faces upward.

5. Use a suitable tool to rotate the caliper piston clockwise into the caliper bore, then align the cutout in the piston with the tab on the inner pad by turning the piston back. Lubricate the piston boot with silicon grease, and avoid twisting the boot.
- Brake caliper
- Caliper bolts, and tighten to 17 ft. lbs. (23 Nm), while holding the caliper pin with a wrench
- Wheels
- Brake fluid to the master cylinder reservoir

Brake Drums

REMOVAL & INSTALLATION

Accord and Civic

1. Before servicing the vehicle, refer to the precautions in the beginning of this section.

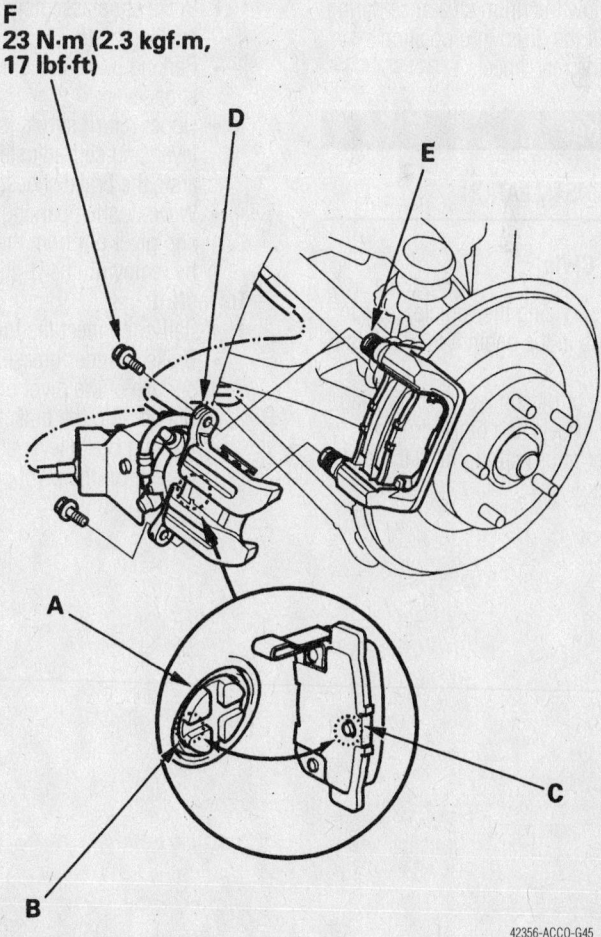

F
23 N·m (2.3 kgf·m, 17 lbf·ft)

42356-ACCO-G45

Rotate the caliper piston (A) clockwise into the caliper bore, then align the cutout (B) in the piston with the tab (C) on the inner pad by turning the piston back—S2000

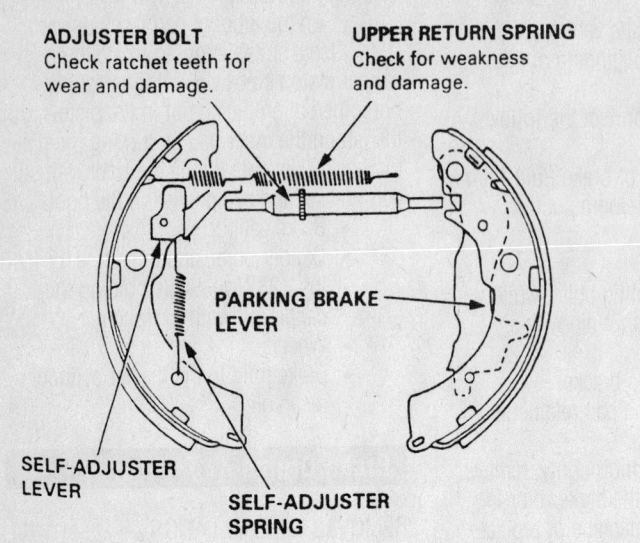

ADJUSTER BOLT
Check ratchet teeth for
wear and damage.

UPPER RETURN SPRING
Check for weakness
and damage.

PARKING BRAKE
LEVER

SELF-ADJUSTER
LEVER

SELF-ADJUSTER
SPRING

Rear drum brakes—Civic shown

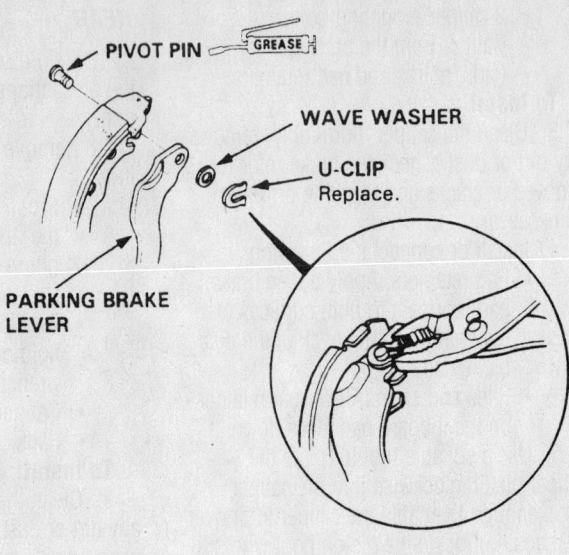

PIVOT PIN   GREASE

WAVE WASHER

U-CLIP
Replace.

PARKING BRAKE
LEVER

93016G09

2. Remove or disconnect the following:
- Rear wheels
- Rear brake drum

To install:

3. Make certain the brake shoes are adjusted to allow the drum clearance during installation. Fit the drum into position.

4. Install the rear wheels.

Brake Shoes

REMOVAL & INSTALLATION

Accord and Civic

1. Before servicing the vehicle, refer to the precautions in the beginning of this section.

2. Remove or disconnect the following:
- Rear wheels and brake drums
- Upper return spring from the brake shoes

3. Push the retainer springs and turn the tension pins to release the shoes from the backing plate.
- Lower the brake shoe assembly and remove the lower return spring
- Brake shoe assembly from the backing plate
- Parking brake cable from the brake shoe lever
- Upper return spring, self-adjuster lever, and self-adjuster spring. Separate the brake shoes.
- Wave washer, parking brake lever, and pivot pin from the brake shoe by removing the U-clip.

To install:

4. Install or connect the following:
- Brake cylinder grease to the sliding surface of the pivot pin and insert the pin into the brake shoe.
- Parking brake lever and wave washer on the pivot pin and pinch

the U-clip with a pair of pliers to secure the pivot pin to the shoe
- Parking brake cable to the parking brake lever
- Adjuster spring to the adjuster lever first, then to the brake shoe
- Adjuster bolt/clevis assembly and the upper return spring
- Brake shoes to the backing plate
- Lower return spring, the tension pins and retaining springs
- Connect the upper return spring

5. Turn the adjuster bolt to force the brake shoes out until the brake drum will not easily go on. Back off the adjuster bolt just enough that the brake drum will go on and turn easily.

6. Install the wheels.

7. Depress the brake pedal several times to set the self-adjusting brake. Adjust the parking brake.

SPECIFICATION CHARTS

ENGINE AND VEHICLE IDENTIFICATION CHART

		Engine Code					Model Year	
Code	Liters (cc)	Cu. In.	Cyl.	Fuel Sys.	Engine Type	Eng. Mfg.	Code ①	Year
B20Z2	2.0 (1973)	120	4	SMFI	DOHC	Honda	Y	2000
K24A1	2.4 (2354)	146	4	SMFI	DOHC	Honda	1	2001
							2	2002
							3	2003
							4	2004

DOHC: Double Overhead Cam

SMFI: Sequential Multi-port Fuel Injection

① 10th position of VIN

42356-HCRV-C01

GENERAL ENGINE SPECIFICATIONS

Year	Model	Engine Displacement Liters (cc)	Engine ID/VIN	Fuel System Type	Net Horsepower @ rpm	Net Torque @ rpm (ft. lbs.)	Bore x Stroke (in.)	Com- pression Ratio	Oil Pressure @ rpm
2000	CR-V	2.0 (1973)	B20Z2	SMFI	146@6200	133@4500	3.31x3.50	9.6:1	50@3000
2001	CR-V	2.0 (1973)	B20Z2	SMFI	146@6200	133@4500	3.31x3.50	9.6:1	50@3000
2002	CR-V	2.4 (2354)	K24A1	SMFI	160@6000	162@3600	3.43xNA	9.6:1	44@3000
2003	CR-V	2.4 (2354)	K24A1	SMFI	160@6000	162@3600	3.43xNA	9.6:1	44@3000

SMFI: Sequential Multi-port Fuel Injection

NA: Not available

42356-HCRV-C02

ENGINE TUNE-UP SPECIFICATIONS

Year	Engine Displacement Liters (cc)	Engine ID/VIN	Spark Plug Gap (in.)	Ignition Timing (deg.) MT	Ignition Timing (deg.) AT	Fuel Pump (psi)	Idle Speed (rpm) MT	Idle Speed (rpm) AT	Valve Clearance (in.) In.	Valve Clearance (in.) Ex.
2000	2.0 (1973)	B20Z2	0.039-0.043	14-18B	14-18B	38-46	700-800	700-800	0.003-0.005	0.006-0.008
2001	2.0 (1973)	B20Z2	0.039-0.043	14-18B	14-18B	38-46	700-800	700-800	0.003-0.005	0.006-0.008
2002	2.4 (2354)	K24A1	0.039-0.043	6-10B	6-10B	50	600-700	600-700	0.008-0.010	0.011-0.013
2003	2.4 (2354)	K24A1	0.039-0.043	6-10B	6-10B	50	600-700	600-700	0.008-0.010	0.011-0.013

NOTE: The Vehicle Emission Control Information label often reflects changes made during production and must be used if they differ from this chart.

NOTE: The fuel pressure readings are given with the vacuum hose connected to the regulator and the engine running

B: Before top dead center

HYD: Hydraulic

42356-HCRV-C03

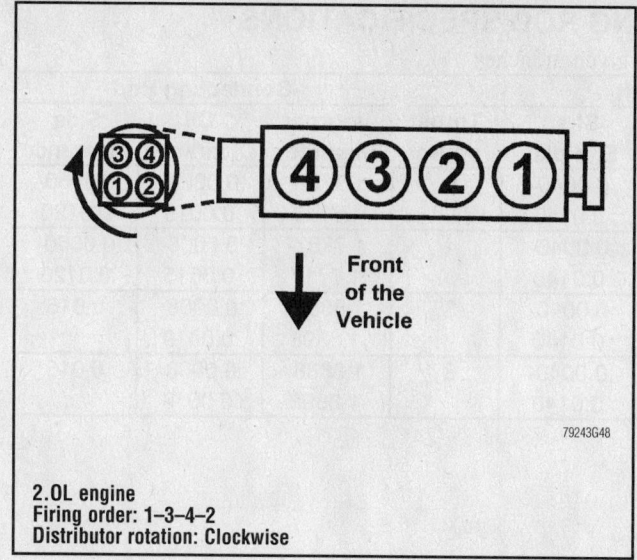

2.0L engine
Firing order: 1–3–4–2
Distributor rotation: Clockwise

79243G48

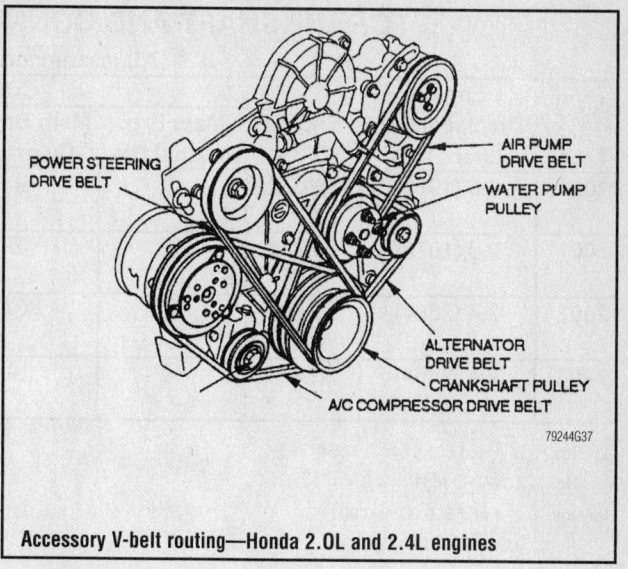

Accessory V-belt routing—Honda 2.0L and 2.4L engines

79244G37

CAPACITIES

Year	Model	Engine Displacement Liters (cc)	Engine ID/VIN	Engine Oil with Filter (qts.)	Transmission (pts.) 5-Spd	Transmission (pts.) Auto.	Transfer Case (pts.)	Drive Axle Front (pts.)	Drive Axle Rear (pts.)	Fuel Tank (gal.)	Cooling System (qts.)
2000	CR-V	2.0 (1973)	B20Z2	4.0	3.6	①	②	②	2.2	15.3	4.1
2001	CR-V	2.0 (1973)	B20Z2	4.0	3.6	①	②	②	2.2	15.3	4.1
2002	CR-V	2.4 (2354)	K24A1	4.4	4.0	③	②	②	2.2	15.3	5.8
2003	CR-V	2.4 (2354)	K24A1	4.4	4.0	③	②	②	2.2	15.3	5.8

NOTE: All capacities are approximate. Add fluid gradually and check to be sure a proper fluid level is obtained.

① 4WD: 6.2
   2WD: 5.8

② Included in transaxle refill figure

③ 4WD: 6.6 pts.
   2WD: 6.2 pts.

42356-HCRV-C04

CRANKSHAFT AND CONNECTING ROD SPECIFICATIONS

All measurements are given in inches

Year	Engine Displacement Liters (cc)	Engine ID/VIN	Crankshaft				Connecting Rod		
			Main Brg. Journal Dia.	Main Brg. Oil Clearance	Shaft End-play	Thrust on No.	Journal Diameter	Oil Clearance	Side Clearance
2000	2.0 (1973)	B20Z2	①	②	0.0040-0.0140	4	1.7707-1.7717	0.0008-0.0015	0.0060-0.0120
2001	2.0 (1973)	B20Z2	①	②	0.0040-0.0140	4	1.7707-1.7717	0.0008-0.0015	0.0060-0.0120
2002	2.4 (2354)	K24A1	③	④	0.0040-0.0140	3	1.8888-1.8898	0.0008-0.0019	0.016
2003	2.4 (2354)	K24A1	③	④	0.0040-0.0140	3	1.8888-1.8898	0.0008-0.0019	0.016

① Nos. 1, 2, 4 and 5: 2.1644-2.1654
  No. 3: 2.1642-2.1651
② Nos. 1, 2, 4 and 5: 0.0009-0.0017
  No. 3: 0.0012-0.0019
③ Except No. 3: 2.1648-2.1657
  No. 3: 2.1644-2.1654
④ Except No. 3: 0.0007-0.0016
  No. 3: 0.0010-0.0019

42356-HCRV-C05

VALVE SPECIFICATIONS

Year	Engine Displacement Liters (cc)	Engine ID/VIN	Seat Angle (deg.)	Face Angle (deg.)	Spring Test Pressure (lbs. @ in.)	Spring Installed Height (in.)	Stem-to-Guide Clearance (in.)		Stem Diameter (in.)	
							Intake	Exhaust	Intake	Exhaust
2000	2.0 (1973)	B20Z2	45	45	NA	①	0.0010-0.0020	0.0020-0.0030	0.2591-0.2594	0.2579-0.2583
2001	2.0 (1973)	B20Z2	45	45	NA	①	0.0010-0.0020	0.0020-0.0030	0.2591-0.2594	0.2579-0.2583
2002	2.4 (2354)	K24A1	NA	NA	NA	①	0.0012-0.0022	0.0022-0.0031	0.2156-0.2159	0.2146-0.2150
2003	2.4 (2354)	K24A1	NA	NA	NA	①	0.0012-0.0022	0.0022-0.0031	0.2156-0.2159	0.2146-0.2150

NA: Not Available

① Valve spring free length:
  Intake: 1.668 in.
  Exhaust: 1.745 in.

42356-HCRV-C06

PISTON AND RING SPECIFICATIONS
All measurements are given in inches

Year	Engine Displacement Liters (cc)	Engine ID/VIN	Piston Clearance	Ring Gap			Ring Side Clearance		
				Top Compression	Bottom Compression	Oil Control	Top Compression	Bottom Compression	Oil Control
2000	2.0 (1973)	B20Z2	0.0004-0.0016	0.0080-0.0120	0.0160-0.0220	0.0080-0.0200	0.0022-0.0031	0.0014-0.0024	NA
2001	2.0 (1973)	B20Z2	0.0004-0.0016	0.0080-0.0120	0.0160-0.0220	0.0080-0.0200	0.0022-0.0031	0.0014-0.0024	NA
2002	2.4 (2354)	K24A1	0.0008-0.0016	0.0080-0.0140	0.0160-0.0220	0.0080-0.0280	0.0018-0.0028	0.0020-0.0030	NA
2003	2.4 (2354)	K24A1	0.0008-0.0016	0.0080-0.0140	0.0160-0.0220	0.0080-0.0280	0.0018-0.0028	0.0020-0.0030	NA

NA: Not Applicable

42356-HCRV-C07

TORQUE SPECIFICATIONS
All readings in ft. lbs.

Year	Engine Displacement Liters (cc)	Engine ID/VIN	Cylinder Head Bolts	Main Bearing Bolts	Rod Bearing Bolts	Crankshaft Damper Bolts	Flywheel Bolts	Manifold		Spark Plugs	Lug Nut
								Intake	Exhaust		
2000	2.0 (1973)	B20Z2	①	②	23	130	54	17	23	13	80
2001	2.0 (1973)	B20Z2	①	②	23	130	54	17	23	13	80
2002	2.4 (2354)	K24A1	③	④	⑤	181	NA	16	33	13	80
2003	2.4 (2354)	K24A1	③	④	⑤	181	NA	16	33	13	80

NOTE: Dip main bearing bolts and crankshaft damper bolt in clean engine oil prior to tightening.

① Step 1: 22 ft. lbs.
  Step 2: 63 ft. lbs.

② Step 1: 18 ft. lbs.
  Step 2: 56 ft. lbs.

③ Step 1: 29 ft. lbs.
  Step 2: +90 degrees
  Step 3: +90 degrees
  Step 4: NEW BOLT ONLY +90 degrees

④ 22 ft. lbs. +56 degrees

⑤ 14 ft. lbs. +90 degrees

42356-HCRV-C08

TIRE, WHEEL AND BALL JOINT SPECIFICATIONS

Year	Model	OEM Tires Standard	OEM Tires Optional	Tire Pressures (psi) Front	Tire Pressures (psi) Rear	Wheel Size	Ball Joint Inspection
2000	CR-V	P205/70R15	None	26	26	6JJ	NS
2001	CR-V	P205/70R15	None	26	26	6JJ	NS
2002	CR-V	P205/70R15	None	26	26	6JJ	NS
2003	CR-V	P205/70R15	None	26	26	6JJ	NS

OEM: Original Equipment Manufacturer

PSI: Pounds Per Square Inch

NS: Not specified by manufacturer

42356-HCRV-C09

BRAKE SPECIFICATIONS
All measurements in inches unless noted

Year	Model		Brake Disc Original Thickness	Brake Disc Minimum Thickness	Brake Disc Maximum Runout	Brake Drum Diameter Original Inside Diameter	Brake Drum Diameter Max. Wear Limit	Brake Drum Diameter Maximum Machine Diameter	Minimum Lining Thickness Front	Minimum Lining Thickness Rear	Brake Caliper Bracket Bolts (ft. lbs.)	Brake Caliper Mounting Bolts (ft. lbs.)
2000	CR-V	F	0.929	0.830	0.004	—	—	—	0.060	—	80	36
		R	—	—	—	8.66	8.70	8.70	—	0.080	—	—
2001	CR-V	F	0.929	0.830	0.004	—	—	—	0.060	—	80	36
		R	—	—	—	8.66	8.70	8.70	—	0.080	—	—
2002	CR-V	F	0.910	0.830	0.004	—	—	—	0.060	—	80	25
		R	0.350	0.280	0.004	—	—	—	0.040	—	41	16
2003	CR-V	F	0.910	0.830	0.004	—	—	—	0.060	—	80	25
		R	0.350	0.280	0.004	—	—	—	0.040	—	41	16

F: Front

R: Rear

42356-HCRV-C10

WHEEL ALIGNMENT

Year	Model		Caster Range (+/-Deg.)	Caster Preferred Setting (Deg.)	Camber Range (+/-Deg.)	Camber Preferred Setting (Deg.)	Toe-in (in.)	Steering Axis Inclination (Deg.)
2000	CR-V	F	1.00	+2.10	1.00	0	0+/-0.12	—
		R	—	—	1.00	-1.00	0.08+/-0.08	—
2001	CR-V	F	1.00	+2.10	1.00	0	0+/-0.12	—
		R	—	—	1.00	-1.00	0.08+/-0.08	—
2002	CR-V	F	1.00	+1.75	0.75	0	0+/-0.08	—
		R	—	—	0.75	-1.00	0.08+/-0.08	—
2003	CR-V	F	1.00	+1.75	0.75	0	0+/-0.08	—
		R	—	—	0.75	-1.00	0.08+/-0.08	—

42356-HCRV-C11

SCHEDULED MAINTENANCE INTERVALS

2000-01 HONDA—CRV

TO BE SERVICED	SERVICE	VEHICLE MILEAGE INTERVAL (x1000)															
		8	15	22.5	30	37.5	45	52.5	60	67.5	75	82.5	90	97.5	105	112.5	120
Accessory drive belts	I & A				✓				✓				✓				✓
Air cleaner element	R				✓				✓				✓				✓
Air conditioning filter	R				✓				✓				✓				✓
Brake fluid	R						✓						✓				
Brake hoses & lines	I		✓		✓		✓		✓		✓		✓		✓		✓
Cooling system	I		✓		✓		✓		✓		✓		✓		✓		✓
Engine coolant	R						✓						✓				
Engine oil	R	✓	✓	✓	✓	✓	✓	✓	✓	✓	✓	✓	✓	✓	✓	✓	✓
Engine oil and coolant levels	I	Inspect at each fuel stop															
Engine oil filter	R		✓		✓		✓		✓		✓		✓		✓		✓
Exhaust system	I		✓		✓		✓		✓		✓		✓		✓		✓
Fluid levels and condition	I		✓		✓		✓		✓		✓		✓		✓		✓
Front and rear brakes	I		✓		✓		✓		✓		✓		✓		✓		✓
Fuel lines & connection	I		✓		✓		✓		✓		✓		✓		✓		✓
Halfshaft boots	I		✓		✓		✓		✓		✓		✓		✓		✓
Idle speed	I & A												✓				
Parking brake system	I & A		✓		✓		✓		✓		✓		✓		✓		✓
Rear differential fluid	R												✓				
Rotate and inspect tires	I	✓	✓	✓	✓	✓	✓	✓	✓	✓	✓	✓	✓	✓	✓	✓	✓
Spark plugs	R				✓				✓				✓				✓
Supplemental Restrain system	I	Inspect the SRS 10 years after production															
Suspension components	I		✓		✓		✓		✓		✓		✓		✓		✓
Tie rod ends, steering gear box & boots	I		✓		✓		✓		✓		✓		✓		✓		✓
Timing balancer belt ①	R														✓		
Timing belt	R														✓		
Transmission fluid	R						✓				✓				✓		
Valve clearance	I			✓					✓				✓				✓
Water pump	S/I														✓		

R: Replace   I: Inspect   A: Adjust

FREQUENT OPERATION MAINTENANCE (SEVERE SERVICE)

If a vehicle is operated under any of the following conditions it is considered severe service:

- Towing a trailer or using a camper or car-top carrier.
- Repeated short trips of less than 5 miles in temperatures below freezing, or trips of less than 10 miles in any temperature.
- Extensive idling or low-speed driving for long distances as in heavy commercial use, such as delivery, taxi or police cars.
- Operating on rough, muddy or salt-covered roads.
- Operating on unpaved or dusty roads.
- Driving in extremely hot (over 90°) conditions.

Air cleaner element: replace every 15,000 miles.

Engine oil and filter: replace every 3750 miles or 6 months, whichever occurs first.

Timing belt: replace every 60,000 miles if the vehicle is regularly driven in temperatures above 110°F or below -20°F.

Transmission fluid: replace every 30,000 miles.

Rear differential fluid: replace every 60,000 miles.

Front and rear brakes: inspect every 7500 miles or 6 months, whichever occurs first.

Locks and hinges: lubricate every 15,000 miles.

Tie rods, steering gear box, boots: inspect every 7500 miles or 6 months, whichever occurs first.

Suspension components: inspect every 7500 miles or 6 months, whichever occurs first.

Halfshaft boots: inspect every 7500 miles or 6 months, whichever occurs first.

SCHEDULED MAINTENANCE INTERVALS

2002-03 HONDA—CRV

TO BE SERVICED	TYPE OF SERVICE	VEHICLE MILEAGE INTERVAL (x1000)											
		10	20	30	40	50	60	70	80	90	100	110	120
Accessory drive belts	I & A			✓			✓			✓			✓
Air cleaner element	R			✓			✓			✓			✓
Air conditioning filter	R			✓			✓			✓			✓
Brake fluid	R											✓	
Brake hoses & lines (including ABS)	I		✓		✓		✓		✓		✓		
Cooling system hoses & connections	I		✓		✓		✓		✓		✓		
Engine coolant	R												✓
Engine oil	R	✓	✓	✓	✓	✓	✓	✓	✓	✓	✓	✓	✓
Engine oil and coolant levels	I	Inspect at each fuel stop											
Engine oil filter	R		✓		✓		✓		✓		✓		
Exhaust system	I		✓		✓		✓		✓		✓		
Fluid levels and condition	I		✓		✓		✓		✓		✓		
Front and rear brakes	I		✓		✓		✓		✓		✓		
Fuel lines & connection	I		✓		✓		✓		✓		✓		
Halfshaft boots	I		✓		✓		✓		✓		✓		
Idle speed	I & A											✓	
Parking brake system	I & A		✓		✓		✓		✓		✓		
Rear differential fluid	R										✓		
Rotate and inspect tires	I	✓	✓	✓	✓	✓	✓	✓	✓	✓	✓	✓	✓
Spark plugs	R											✓	
Suspension components	I		✓		✓		✓		✓		✓		
Tie rod ends, steering gear box & boots	I		✓		✓		✓		✓		✓		
Transmission fluid	R												✓
Valve clearance	I											✓	

R: Replace   I: Inspect   A: Adjust

FREQUENT OPERATION MAINTENANCE (SEVERE SERVICE)

If a vehicle is operated under any of the following conditions it is considered severe service:

- Towing a trailer or using a camper or car-top carrier.
- Repeated short trips of less than 5 miles in temperatures below freezing, or trips of less than 10 miles in any temperature.
- Extensive idling or low-speed driving for long distances as in heavy commercial use, such as delivery, taxi or police cars.
- Operating on rough, muddy or salt-covered roads.
- Operating on unpaved or dusty roads.
- Driving in extremely hot (over 90°) conditions.

Air cleaner element: replace every 15,000 miles

Engine oil and filter: replace every 3750 miles or 6 months, whichever occurs first.

Timing belt: replace every 60,000 miles if the vehicle is regularly driven in temperatures above 110°F or below -20°F.

Transmission fluid: replace every 30,000 miles.

Rear differential fluid: replace every 60,000 miles.

Front and rear brakes: inspect every 7500 miles or 6 months, whichever occurs first.

Locks and hinges: lubricate every 15,000 miles.

Tie rods, steering gear box, boots: inspect every 7500 miles or 6 months, whichever occurs first.

Suspension components: inspect every 7500 miles or 6 months, whichever occurs first.

Halfshaft boots: inspect every 7500 miles or 6 months, whichever occurs first.

42356-HCRV-C13

PRECAUTIONS

Before servicing any vehicle, please be sure to read all of the following precautions, which deal with personal safety, prevention of component damage, and important points to take into consideration when servicing a motor vehicle:

• Never open, service or drain the radiator or cooling system when the engine is hot; serious burns can occur from the steam and hot coolant.

• Observe all applicable safety precautions when working around fuel. Whenever servicing the fuel system, always work in a well-ventilated area. Do not allow fuel spray or vapors to come in contact with a spark, open flame, or excessive heat (a hot drop light, for example). Keep a dry chemical fire extinguisher near the work area. Always keep fuel in a container specifically designed for fuel storage; also, always properly seal fuel containers to avoid the possibility of fire or explosion. Refer to the additional fuel system precautions later in this section.

• Fuel injection systems often remain pressurized, even after the engine has been turned **OFF**. The fuel system pressure must be relieved before disconnecting any fuel lines. Failure to do so may result in fire and/or personal injury.

• Brake fluid often contains polyglycol ethers and polyglycols. Avoid contact with the eyes and wash your hands thoroughly after handling brake fluid. If you do get brake fluid in your eyes, flush your eyes with clean, running water for 15 minutes. If eye irritation persists, or if you have taken

brake fluid internally, seek medical assistance IMMEDIATELY.

• The EPA warns that prolonged contact with used engine oil may cause a number of skin disorders, including cancer. You should make every effort to minimize your exposure to used engine oil. Protective gloves should be worn when changing oil. Wash your hands and any other exposed skin areas as soon as possible after exposure to used engine oil. Soap and water, or waterless hand cleaner should be used.

• All new vehicles are now equipped with an air bag system. The system must be disabled before performing service on or around system components, steering column, instrument panel components, wiring and sensors. Failure to follow safety and disabling procedures could result in accidental air bag deployment, possible personal injury and unnecessary system repairs.

• Always wear safety goggles when working with, or around, the air bag system. When carrying a non-deployed air bag, be sure the bag and trim cover are pointed away from your body. When placing a non-deployed air bag on a work surface, always face the bag and trim cover upward, away from the surface. This will reduce the motion of the module if it is accidentally deployed. Refer to the additional air bag system precautions later in this section.

• Clean, high quality brake fluid from a sealed container is essential to the safe and proper operation of the brake system. You

should always buy the correct type of brake fluid for your vehicle. If the brake fluid becomes contaminated, completely flush the system with new fluid. Never reuse any brake fluid. Any brake fluid that is removed from the system should be discarded. Also, do not allow any brake fluid to come in contact with a painted surface; it will damage the paint.

• Never operate the engine without the proper amount and type of engine oil; doing so WILL result in severe engine damage.

• Timing belt maintenance is extremely important. Many models utilize an interference-type, non-freewheeling engine. If the timing belt breaks, the valves in the cylinder head may strike the pistons, causing potentially serious (also time-consuming and expensive) engine damage. Refer to the maintenance interval charts in the front of this section for the recommended replacement interval for the timing belt, and to the timing belt procedure in this section for belt replacement and inspection.

• Disconnecting the negative battery cable on some vehicles may interfere with the functions of the on-board computer system(s) and may require the computer to undergo a relearning process once the negative battery cable is reconnected.

• When servicing drum brakes, only disassemble and assemble one side at a time, leaving the remaining side intact for reference.

• Only an MVAC-trained, EPA-certified automotive technician should service the air conditioning system or its components.

ENGINE REPAIR

➡**Disconnecting the negative battery cable on some vehicles may interfere with the functions of the on board computer system. The computer may undergo a relearning process once the negative battery cable is reconnected.**

Distributor

The 2.4L engine does not have a distrubutor.

REMOVAL & INSTALLATION

1. Before servicing the vehicle, refer to the precautions in the beginning of this section.

2. Remove or disconnect the following:
   • Negative battery cable

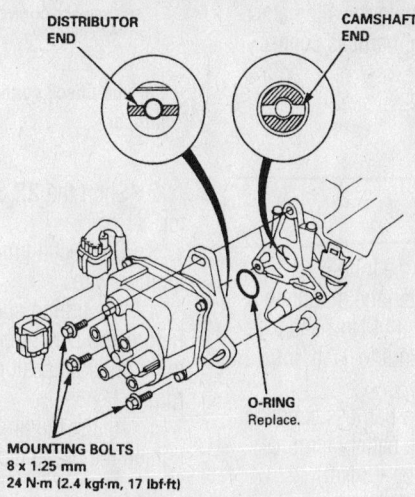

DISTRIBUTOR END   CAMSHAFT END

O-RING
Replace.

MOUNTING BOLTS
8 x 1.25 mm
24 N·m (2.4 kgf·m, 17 lbf·ft)

7924MG02

Exploded view of the distributor mounting—CR-V

- Cruise control cable
- Air intake duct
- Distributor harness connector
- Spark plug wires
- Distributor

To install:

3. Install or connect the following:
- Distributor. Use a new O-ring seal.
- Spark plug wires
- Distributor harness connector
- Air intake duct
- Cruise control cable
- Negative battery cable

4. Set the ignition timing and tighten the mounting bolts to 13 ft. lbs. (18 Nm).

Alternator

REMOVAL

2000–01

1. Before servicing the vehicle, refer to the precautions in the beginning of this section.
2. Remove or disconnect the following:

- Negative battery cable
- Accessory drive belts
- Alternator wiring harness connectors
- Alternator

2002–03

1. Before servicing the vehicle, refer to the precautions in the beginning of this section.
2. Remove or disconnect the following:

- Negative battery cable
- Front cover
- Accessory drive belt
- The 3 bolts holding the alternator
- Alternator wiring harness connectors
- Alternator

INSTALLATION

2000–01

1. Install or connect the following:
- Alternator. Tighten the mounting nut to 33 ft. lbs. (44 Nm) and the adjustment locknut to 17 ft. lbs. (24 Nm).
- Alternator wiring harness connectors. Tighten the battery terminal nut to 70 inch lbs. (8 Nm).
- Accessory drive belts
- Negative battery cable

2002–03

1. Install or connect the following:
- Alternator. Tighten the bolts to 16 ft. lbs. (22 Nm).
- Alternator wiring harness connectors. Tighten the battery terminal nut to 70 inch lbs. (8 Nm).
- Accessory drive belts
- Front cover
- Negative battery cable

Ignition Timing

ADJUSTMENT

Adjustment is not possible on the 2.4L engine.

➡**Timing adjustments are made with the engine at operating temperature.**

1. Before servicing the vehicle, refer to the precautions in the beginning of this section.

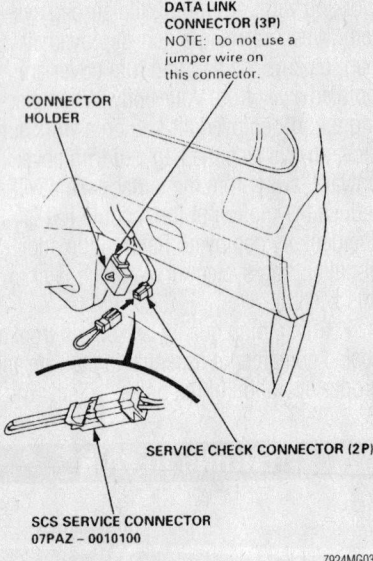

DATA LINK CONNECTOR (3P)
NOTE: Do not use a jumper wire on this connector.

CONNECTOR HOLDER

SERVICE CHECK CONNECTOR (2P)

SCS SERVICE CONNECTOR
07PAZ – 0010100

7924MG03

Service Check connector and shorting jumper

2. Short the **2P** Service Check connector.

3. Connect a timing light to the No. 1 ignition wire.

4. The timing should be 13–17 degrees Before Top Dead Center (BTDC) (red timing mark on crankshaft pulley) at 650–750 rpm.

5. Adjust the timing as necessary and tighten the distributor bolts to 13 ft. lbs. (18 Nm).

6. Remove the **2P** connector jumper.

Engine Assembly

REMOVAL & INSTALLATION

2000–01

➡**The engine and transaxle are removed from the vehicle as a unit.**

1. Before servicing the vehicle, refer to the precautions in the beginning of this section.
2. Drain the cooling system.
3. Drain the transaxle fluid.
4. Drain the engine oil.
5. Relieve fuel system pressure.
6. Remove or disconnect the following:

- Negative battery cable
- Fuse/Relay box battery cables
- Battery and tray
- Air intake assembly
- Powertrain Control Module (PCM) connectors and grommet. Pull the PCM harness through the firewall.
- Left engine wire harness connectors
- Cruise control actuator
- Power steering pump belt and pump
- A/C compressor drive belt
- Fuel lines
- Brake booster vacuum line
- Accelerator cable
- Power Steering Pressure (PSP) switch
- Splash shield
- Radiator hoses
- Heater hoses
- Heated Oxygen Sensor (HO_2S) connector

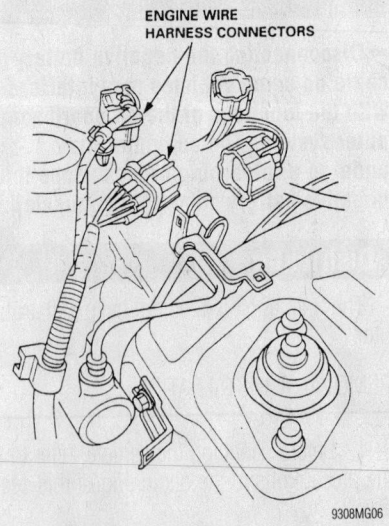

ENGINE WIRE HARNESS CONNECTORS

9308MG06

Left engine wire harness connectors— CR-V

- Exhaust front pipe
- Right damper fork
- Lower ball joints

7. Separate the inner CV-joints from the transaxle and support the axle halfshafts out of the work area with safety wire.

8. If equipped with a manual transaxle, remove or disconnect the following:
- Clutch slave cylinder
- Clutch hose bracket
- Transaxle ground cable
- Shift cables

9. If equipped with an automatic transaxle, remove or disconnect the following:
- Shift cable cover
- Shift cable
- Transaxle ground cable and hose clamp
- Transaxle fluid cooler lines

10. For all vehicles, remove or disconnect the following:
- A/C hose clamp
- Radiator
- A/C compressor
- Rear driveshaft, if equipped

11. Attach a hoist to the engine lifting eyes and support the powertrain weight.

12. Remove or disconnect the following:
- Left front mount and bracket

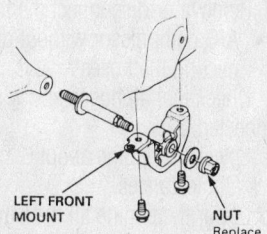

LEFT FRONT MOUNT
NUT
Replace.

M/T:

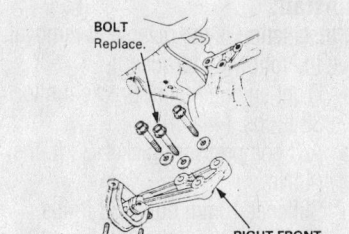

BOLT
Replace.

RIGHT FRONT MOUNT/BRACKET

A/T:

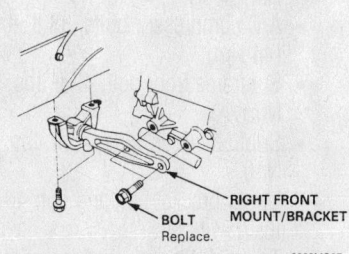

BOLT
Replace.

RIGHT FRONT MOUNT/BRACKET

9308MG07

Front mounts—CR-V

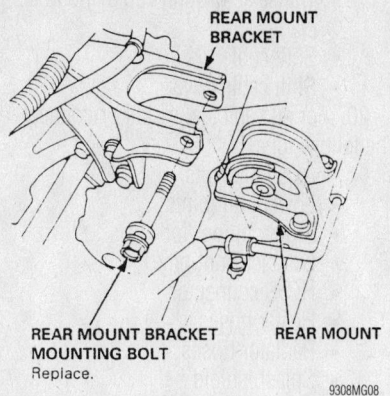

REAR MOUNT BRACKET

REAR MOUNT BRACKET MOUNTING BOLT
Replace.

REAR MOUNT

9308MG08

Rear mount—CR-V

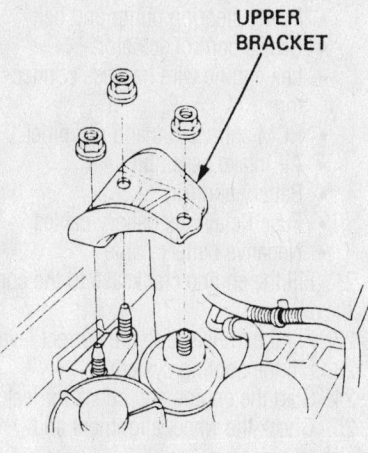

UPPER BRACKET

9308MG09

Upper bracket—CR-V

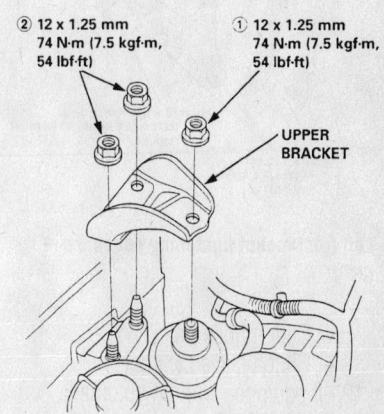

TRANSMISSION MOUNT

TRANSMISSION MOUNT BRACKET

9308MG10

Transaxle mount—CR-V

- Right front mount and bracket
- Rear mount bracket through bolt
- Upper bracket
- Transaxle mount and bracket

13. Lift the powertrain away from the vehicle.

To install:

→Use new self-locking nuts and color-coded self-locking bolts when installing the engine mounts and suspension components.

→Do not tighten the engine or transaxle mount fasteners until instructed to do so.

14. Lower the powertrain into position.
15. Install or connect the following:
- Transaxle mount and bracket. Tighten the frame mounting bolts to 47 ft. lbs. (64 Nm).
- Upper bracket. Tighten the nuts in sequence to 54 ft. lbs. (74 Nm).
- Rear mount bracket through bolt
- Right front mount and bracket
- Left front mount and bracket

16. Tighten the remaining mount fasteners as follows:

a. Transaxle mount fasteners to 47 ft. lbs. (64 Nm) and the through bolt to 54 ft. lbs. (74 Nm).

b. Rear mount bracket through bolt to 43 ft. lbs. (59 Nm).

c. Right front mount 12mm bolts to 47 ft. lbs. (64 Nm) and the 10mm bolts to 33 ft. lbs. (44 Nm).

d. Left front mount 12mm stud bolt to 61 ft. lbs. (83 Nm), 10mm bolts to 33 ft. lbs. (44 Nm), and 12mm nut to 43 ft. lbs. (59 Nm).

e. Right front mount 12mm nut to 43 ft. lbs. (59 Nm).

17. Install or connect the following:
- Rear driveshaft, if equipped
- A/C compressor
- Radiator
- A/C hose clamp

18. If equipped with a manual transaxle, install or connect the following:
- Shift cables
- Transaxle ground cable

② 12 x 1.25 mm
74 N·m (7.5 kgf·m, 54 lbf·ft)

① 12 x 1.25 mm
74 N·m (7.5 kgf·m, 54 lbf·ft)

UPPER BRACKET

7924MG05

Upper bracket tightening sequence—CR-V

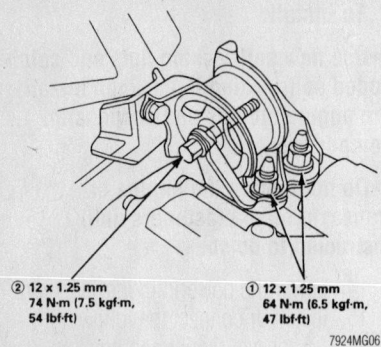

② 12 x 1.25 mm
74 N·m (7.5 kgf·m, 54 lbf·ft)

① 12 x 1.25 mm
64 N·m (6.5 kgf·m, 47 lbf·ft)

7924MG06

Transaxle mount fastener tightening sequence—CR-V

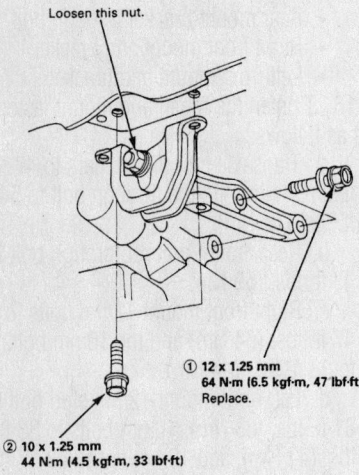

Loosen this nut.

① 12 x 1.25 mm
64 N·m (6.5 kgf·m, 47 lbf·ft)
Replace.

② 10 x 1.25 mm
44 N·m (4.5 kgf·m, 33 lbf·ft)

7924MG08

Right front mount tightening sequence—CR-V

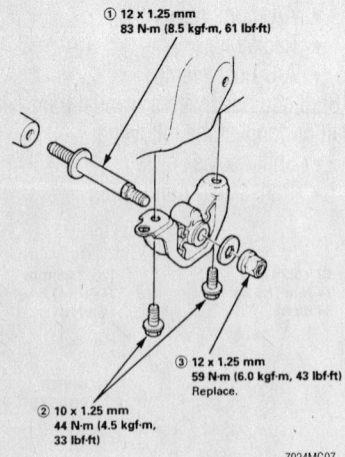

① 12 x 1.25 mm
83 N·m (8.5 kgf·m, 61 lbf·ft)

① 12 x 1.25 mm
59 N·m (6.0 kgf·m, 43 lbf·ft)
Replace.

② 10 x 1.25 mm
44 N·m (4.5 kgf·m, 33 lbf·ft)

7924MG07

Left front mount tightening sequence—CR-V

- Clutch hose bracket
- Clutch slave cylinder

19. If equipped with an automatic transaxle, install or connect the following:
- Transaxle fluid cooler lines
- Transaxle ground cable and hose clamp
- Shift cable
- Shift cable cover

20. For all vehicles, install or connect the following:
- Axle halfshafts
- Lower ball joints
- Right damper fork
- Exhaust front pipe
- HO2S connector
- Heater hoses
- Radiator hoses
- Splash shield
- PSP switch
- Accelerator cable
- Brake booster vacuum line
- Fuel lines
- A/C compressor drive belt
- Power steering pump and belt
- Cruise control actuator
- Left engine wire harness connectors
- PCM connectors and grommet
- Air intake assembly
- Battery and tray
- Fuse/Relay box battery cables
- Negative battery cable

21. Fill the engine crankcase to the correct level.

22. Fill the transaxle to the correct level.

23. Fill the cooling system.

24. Start the engine and check for leaks.

25. Check the wheel alignment and adjust as necessary.

2002–03

➡**The engine and transaxle are removed from the vehicle as a unit.**

1. Before servicing the vehicle, refer to the precautions in the beginning of this section.

2. Drain the cooling system.

3. Drain the transaxle fluid.

4. Drain the engine oil.

5. Relieve fuel system pressure.

6. Remove or disconnect the following:
- Negative battery cable
- Fuse/Relay box battery cables
- Battery and tray
- Intake manifold cover
- IAT sensor connector
- Breather hose
- Intake duct
- Cables from the power distribution center
- Throttle and cruise cables
- Powertrain Control Module (PCM) connectors and grommet. Pull the PCM harness through the firewall.
- Fuel lines
- EVAP canister
- Brake booster vacuum line
- Clutch slave cylinder
- Clutch hose bracket
- Shift cables
- Drive belt
- Power steering pump, leaving the hoses connected

7. Attach a hoist to the engine lifting eyes and support the powertrain weight.

8. Remove or disconnect the following:
- Splash shield
- Wheels
- Catalytic converter
- Rear driveshaft
- Stabilizer links
- Right damper fork
- Halfshafts
- Shift cable
- Radiator
- Upper bracket
- Transaxle mount and bracket
- Front mount bolt
- Rear mount bracket bolts. Match-mark the sub-frame mounting bolt centers.

There is a special tool necessary for sub-frame removal. The Honda tool number is EQS02C000011. Attach the tool as explained in the tool instructions, attach a floor jack with adapter, remove the 4 sub-frame bolts and lower the sub-frame.

9. Remove or disconnect the following:
- A/C compressor without disconnecting the hoses

10. Check that all hoses and wires are disconnected.

11. Lower the engine about 6 inches and recheck all clearances.

12. Lower the engine all the way.

13. Remove the chain hoist.

To install:

14. Installation is the reverse of removal. Observe the following torques:
- Front engine mount bracket bolts: 33 ft. lbs. (44Nm)
- A/C compressor bracket: 33 ft. lbs. (44Nm)
- Stiffener 10mm bolts: 33 ft. lbs. (44Nm); 6mm bolts 9 ft. lbs. (12 Nm)
- A/C compressor bolts: 33 ft. lbs. (44 Nm)
- Subframe front bolt: 47 ft. lbs. (64 Nm)
- Subframe rear bolts: 43 ft. lbs. (59 Nm)
- Upper bracket bolt and nut: 40 ft. lbs. (54 Nm)
- Transmission mount bracket support bolts/nuts: 40 ft. lbs. (54 Nm)
- PS pump bolts: 16 ft. lbs. (22 Nm)

- Intake manifold cover: 9 ft. lbs. (12 Nm)

➡️**Use new self-locking nuts and color-coded self-locking bolts when installing the engine mounts and suspension components.**

➡️**Do not tighten the engine or transaxle mount fasteners until instructed to do so.**

15. Lower the powertrain into position.
16. Install or connect the following:
   - Transaxle mount and bracket. Tighten the frame mounting bolts to 47 ft. lbs. (64 Nm).
   - Upper bracket. Tighten the nuts in sequence to 54 ft. lbs. (74 Nm).
   - Rear mount bracket through bolt
   - Right front mount and bracket
   - Left front mount and bracket
17. Tighten the remaining mount fasteners as follows:
   a. Transaxle mount fasteners to 47 ft. lbs. (64 Nm) and the through bolt to 54 ft. lbs. (74 Nm).
   b. Rear mount bracket through bolt to 43 ft. lbs. (59 Nm).
   c. Right front mount 12mm bolts to 47 ft. lbs. (64 Nm) and the 10mm bolts to 33 ft. lbs. (44 Nm).
   d. Left front mount 12mm stud bolt to 61 ft. lbs. (83 Nm), 10mm bolts to 33 ft. lbs. (44 Nm), and 12mm nut to 43 ft. lbs. (59 Nm).
   e. Right front mount 12mm nut to 43 ft. lbs. (59 Nm).
18. Install or connect the following:
   - Rear driveshaft, if equipped
   - A/C compressor
   - Radiator
   - A/C hose clamp
19. If equipped with a manual transaxle, install or connect the following:
   - Shift cables
   - Transaxle ground cable
   - Clutch hose bracket
   - Clutch slave cylinder
20. If equipped with an automatic transaxle, install or connect the following:
   - Transaxle fluid cooler lines
   - Transaxle ground cable and hose clamp
   - Shift cable
   - Shift cable cover
21. For all vehicles, install or connect the following:
   - Axle halfshafts
   - Lower ball joints
   - Right damper fork
   - Exhaust front pipe
   - HO2S connector

- Heater hoses
- Radiator hoses
- Splash shield
- PSP switch
- Accelerator cable
- Brake booster vacuum line
- Fuel lines
- A/C compressor drive belt
- Power steering pump and belt
- Cruise control actuator
- Left engine wire harness connectors
- PCM connectors and grommet
- Air intake assembly
- Battery and tray
- Fuse/Relay box battery cables
- Negative battery cable

22. Fill the engine crankcase to the correct level.
23. Fill the transaxle to the correct level.
24. Fill the cooling system.
25. Start the engine and check for leaks.
26. Check the wheel alignment and adjust as necessary.

Water Pump

REMOVAL & INSTALLATION

2000–01

1. Before servicing the vehicle, refer to the precautions in the beginning of this section.
2. Drain the cooling system.
3. Remove or disconnect the following:
   - Negative battery cable
   - Accessory drive belts
   - Front cover
   - Timing belt
   - Water pump

To install:

4. Install or connect the following:
   - Water pump. Use a new O-ring seal and tighten the bolts to 105 inch lbs. (12 Nm).

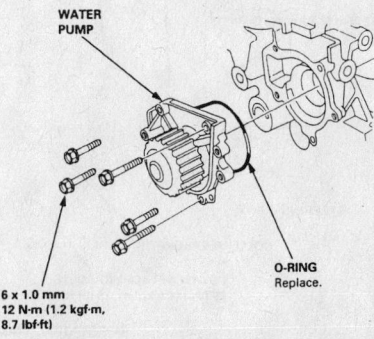

WATER PUMP

6 x 1.0 mm
12 N·m (1.2 kgf·m,
8.7 lbf·ft)

O-RING
Replace.

7924MG10

Exploded view of the water pump mounting

- Timing belt
- Front cover
- Accessory drive belts
- Negative battery cable
5. Fill the cooling system.
6. Start the engine and check for leaks.

2002–03

1. Before servicing the vehicle, refer to the precautions in the beginning of this section.
2. Drain the cooling system.
3. Remove or disconnect the following:
   - Negative battery cable
   - Accessory drive belt
   - Crankshaft pulley
   - Water pump (6 bolts)

Installation is the reverse of removal. Use new O-rings. Torque the bolts to 104 inch lbs. (12 Nm).

Heater Core

REMOVAL & INSTALLATION

1. Disconnect the negative battery cable.
2. Drain the cooling system into a clean container for reuse.
3. In the engine compartment, open the heater valve cable clamp and disconnect the cable from the heater valve arm. Then, turn the heater valve to the fully opened position.
4. Disconnect the heater hoses from the heater core.
5. Remove the heater housing-to-chassis nut.

➡️**When removing the heater housing nut, be careful not to damage or bend the fuel lines, the brake lines, etc.**

6. Remove the instrument panel by performing the following procedure:
   a. Remove the driver's side lower instrument panel cover screws, disengage the clips and remove the lower cover.
   b. Remove the knee bolster bolts and the knee bolster.
   c. Remove the glove box stops from each side of the glove box.
   d. Remove the glove box-to-instrument panel bolts and the glove box.
   e. Remove the lower console cover by disengaging the 4 clips and removing the cover.
   f. Remove the 6 center pocket-to-instrument panel screws; then, insert a flat tipped screwdriver at the upper right side corner of the center pocket, push down on the top of the hook and remove

the center pocket/beverage holder assembly.

g. Remove the center instrument panel lower cover screws and disengage the clips on the upper left side; then, disconnect the electrical connectors and remove the cover.

h. Gently, push the power window switch from the instrument panel's lower cover opening by hand. Disconnect the electrical connectors and remove the power window switch.

i. Close the driver's side air vent; then, gently, push out the clips and pull out the vent. Disconnect the electrical connectors and remove the vent.

j. Gently, push out the driver's side defogger trim; then, disconnect the electrical connector and remove the side defogger trim.

k. At the base of the steering wheel, remove the access panel and disconnect the air bag electrical connector.

l. Remove the steering column covers screws and the covers.

m. Remove the steering column-to-instrument panel nuts/bolts and lower the steering column.

n. Remove the instrument panel side covers.

o. Disconnect the wiring harness connector and remove the nuts.

p. Move the under-dash fuse/relay box.

q. Disconnect the antenna connector and the harness clips.

r. Remove the connector holder from the instrument panel frame.

s. Remove the control unit/relay bracket from behind the center of the instrument panel.

t. Remove the passenger's side lower instrument panel cover.

u. Disconnect the connectors and the harness clips.

v. Remove the instrument panel-to-chassis bolts.

w. Using an assistant, remove the instrument panel.

7. Remove the evaporator housing by performing the following procedure:

a. Discharge and recover the air conditioning system refrigerant.

b. In the engine compartment, remove the refrigerant lines-to-evaporator housing bolts.

c. Separate the lines, discard the grommets and plug the openings to prevent contamination.

d. Disconnect the evaporator housing's temperature sensor connector.

e. Remove the evaporator housing-to-

chassis screws/nut and remove the evaporator housing.

8. Disconnect the mode control motor and the air mix control motor electrical connectors and remove the wiring harness clips and the wiring harness from the heater housing.

9. Remove the heater duct clip, the heater housing-to-chassis nuts and the heater housing.

10. Remove the heater core cover screws and the cover.

11. Remove the heater core pipe clamp screws and the clamp.

12. Remove the heater core from the heater housing.

To install:

13. Install the heater core in the heater housing.

14. Install the heater core pipe clamp and the clamp screws.

15. Install the heater core cover and the cover screws.

16. Install the heater housing, the heater housing-to-chassis nuts and the heater duct clip.

17. Install the wiring harness clips and the wiring harness to the heater housing and connect the mode control motor and the air mix control motor electrical connectors.

18. Install the evaporator housing by performing the following procedure:

a. Install the evaporator housing and the evaporator housing-to-chassis screws/nut.

b. Connect the evaporator housing's temperature sensor connector.

c. Using new grommets, connect the refrigerant lines.

d. In the engine compartment, install the refrigerant lines-to-evaporator housing bolts.

19. Install the instrument panel by performing the following procedure:

a. Using an assistant, install the instrument panel.

b. Install the instrument panel-to-chassis bolts.

c. Connect the connectors and the harness clips.

d. Install the passenger's side lower instrument panel cover.

e. Install the control unit/relay bracket to the center of the instrument panel.

f. Install the connector holder to the instrument panel frame.

g. Connect the antenna connector and the harness clips.

h. Install the under-dash fuse/relay box.

i. Connect the wiring harness connector and install the nuts.

j. Install the instrument panel side covers.

k. Install the steering column and the column-to-instrument panel nuts/bolts. Torque the nuts to 12 ft. lbs. (16 Nm) and the bolts to 29 ft. lbs. (39 Nm).

l. Install the steering column covers and the cover screws.

m. At the base of the steering wheel, connect the air bag electrical connector and install the access panel.

n. Connect the electrical connector and install the driver's side defogger trim.

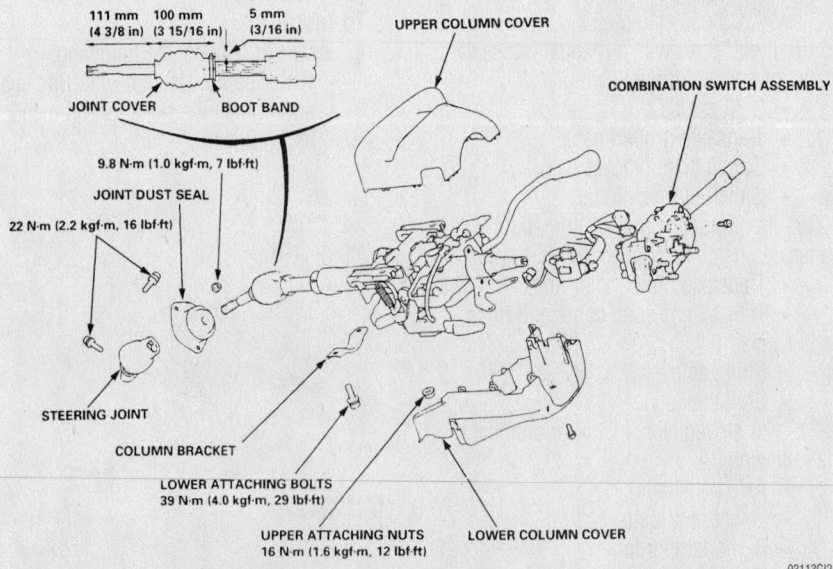

111 mm (4 3/8 in) 100 mm (3 15/16 in) 5 mm (3/16 in)

JOINT COVER   BOOT BAND

UPPER COLUMN COVER

COMBINATION SWITCH ASSEMBLY

9.8 N·m (1.0 kgf·m, 7 lbf·ft)

JOINT DUST SEAL

22 N·m (2.2 kgf·m, 16 lbf·ft)

STEERING JOINT

COLUMN BRACKET

LOWER ATTACHING BOLTS
39 N·m (4.0 kgf·m, 29 lbf·ft)

UPPER ATTACHING NUTS
16 N·m (1.6 kgf·m, 12 lbf·ft)

LOWER COLUMN COVER

93113GI2

Exploded view of the steering column and related components

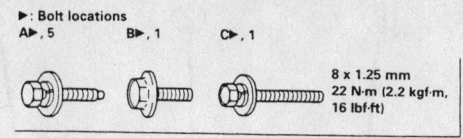

▶ : Bolt locations
A▶, 5   B▶, 1   C▶, 1

8 x 1.25 mm
22 N·m (2.2 kgf·m,
16 lbf·ft)

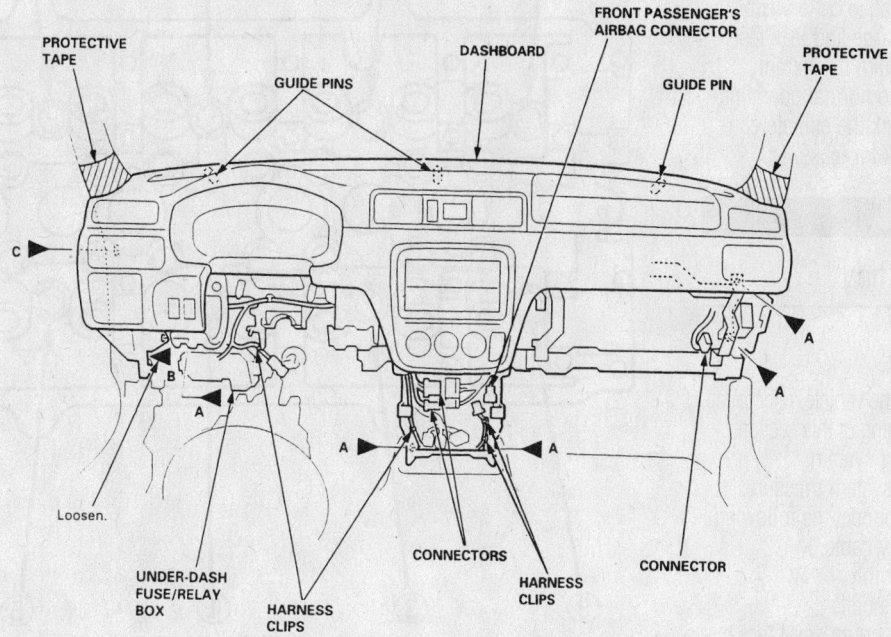

PROTECTIVE TAPE
GUIDE PINS
DASHBOARD
FRONT PASSENGER'S AIRBAG CONNECTOR
GUIDE PIN
PROTECTIVE TAPE
C
A
A
A
A
B
Loosen.
UNDER-DASH FUSE/RELAY BOX
HARNESS CLIPS
CONNECTORS
HARNESS CLIPS
CONNECTOR

93113GI3

Exploded view of the instrument panel and related components

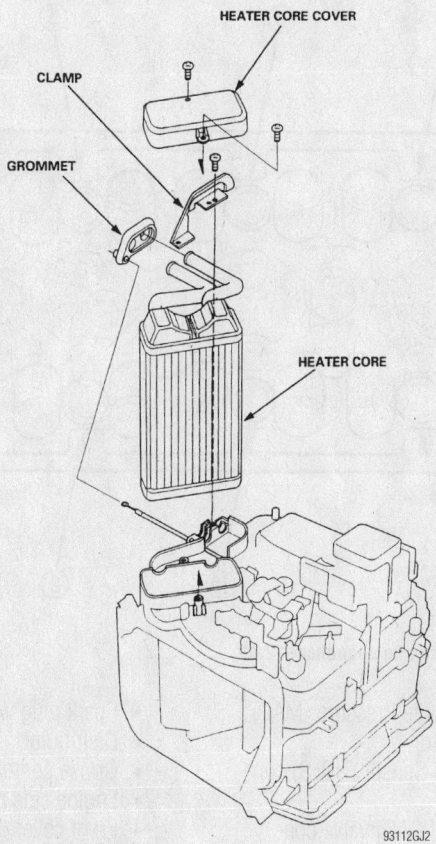

HEATER CORE COVER
CLAMP
GROMMET
HEATER CORE

93112GJ2

Exploded view of the heater core and housing

o. Connect the electrical connectors and install driver's side air vent.

p. Connect the electrical connectors and install the power window switch to the instrument panel's lower cover opening.

q. Install the center instrument panel lower cover and engage the clips on the upper left side; then, connect the electrical connectors and Install the cover screws.

r. Install the center pocket/beverage holder assembly and the 6 center pocket-to-instrument panel screws.

s. Install the lower console cover by engaging the 4 clips.

t. Install the glove box and the glove box-to-instrument panel bolts.

u. Install the glove box stops to each side of the glove box.

v. Install the knee bolster and the knee bolster bolts.

w. Install the driver's side lower instrument panel cover, engage the clips and install the lower cover screws.

➡ **When installing the heater housing nut, be careful not to damage or bend the fuel lines, the brake lines or etc.**

20. Install the heater housing-to-chassis nut.

21. Connect the heater hoses to the heater core.

22. In the engine compartment, connect the cable to the heater valve arm and close the heater valve cable clamp.

23. Refill the cooling system.

24. Connect the negative battery cable.

25. Evacuate and charge and leak test the air conditioning system refrigerant.

26. Run the engine to normal operating temperatures; then, check the climate control operation and check for leaks.

Cylinder Head

REMOVAL & INSTALLATION

2000–01

1. Before servicing the vehicle, refer to the precautions in the beginning of this section.

2. Drain the cooling system.

3. Relieve the fuel system pressure.

4. Remove or disconnect the following:
- Negative battery cable
- Air intake assembly
- Accessory drive belts
- Power steering pump and bracket
- Accelerator cable
- Fuel lines
- Evaporative Emissions (EVAP) control canister hose and vacuum hose
- Brake booster vacuum line
- Intake manifold vacuum line
- Positive Crankcase Ventilation (PCV) valve and hose
- Upper radiator hose
- Heater hose
- Bypass hoses
- Fuel injector connectors
- Engine Coolant Temperature (ECT) sensor connector
- Radiator fan switch connector
- ECT gauge sending unit connector
- Throttle Position (TP) sensor connector

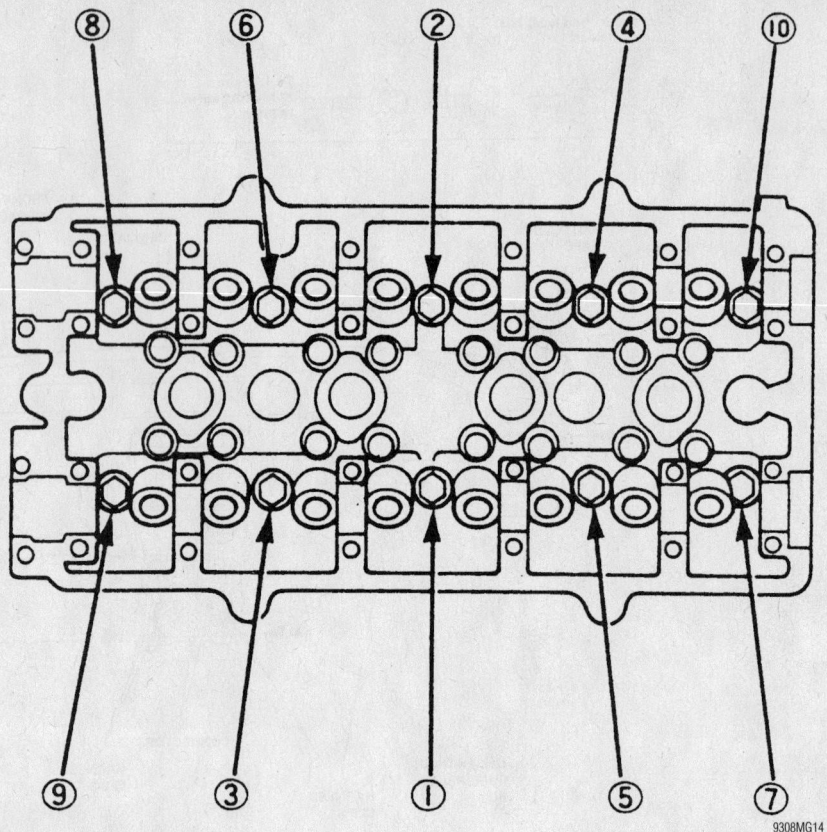

Cylinder head bolt tightening sequence—2.0L

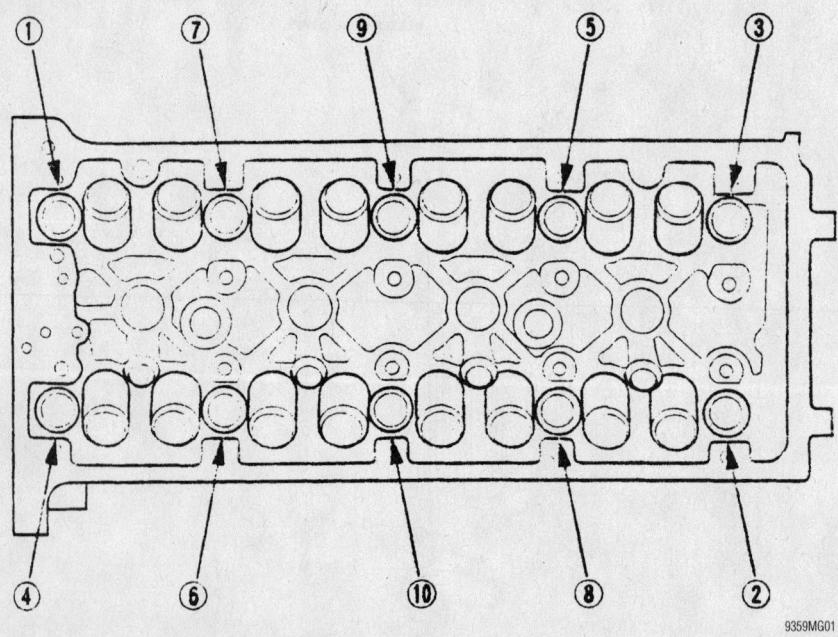

Cylinder head bolt loosening sequence—2.4L

- Manifold Absolute Pressure (MAP) sensor connector
- Heated Oxygen Sensor (HO$_2$S) connector
- Idle Air Control (IAC) valve connector
- Spark plug wires
- Distributor
- Cruise control actuator
- Engine side mount bracket
- Front cover
- Timing belt

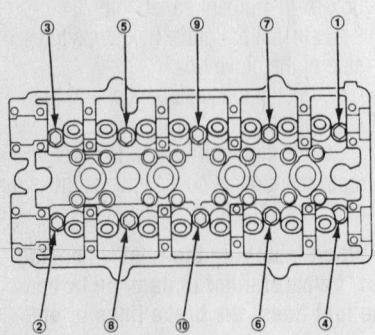

Cylinder head bolt loosening sequence—2.0L

- Camshaft sprockets
- Rear timing belt cover
- Valve cover

5. Loosen the valve adjuster locknuts and screws so that all valves are closed.

➡**Keep all valvetrain components in order for assembly.**

6. Remove or disconnect the following:
- Camshafts
- Rocker arms
- Exhaust front pipe
- Exhaust manifold bracket
- Intake manifold

7. Loosen the cylinder head bolts in sequence and ⅓ turns until all bolts are loose.

8. Remove the cylinder head.

To install:

9. Install the cylinder head with a new gasket.

10. Apply clean engine oil to the bolt threads and under the bolt heads.

11. Tighten the cylinder head bolts in sequence as follows:
   a. Step 1: 22 ft. lbs. (29 Nm)
   b. Step 2: 63 ft. lbs. (85 Nm)

12. Install or connect the following:
- Intake manifold. Tighten the bolts to 17 ft. lbs. (24 Nm).
- Exhaust manifold bracket
- Exhaust front pipe
- Rocker arms in their original positions
- Camshafts
- Rear timing belt cover
- Camshaft sprockets. Tighten the bolts to 27 ft. lbs. (37 Nm).
- Timing belt

13. Adjust the valve clearance.

14. Install or connect the following:
- Valve cover
- Front cover
- Engine side mount bracket
- Cruise control actuator
- Distributor
- Spark plug wires
- IAC valve connector
- HO$_2$S connector
- MAP sensor connector
- TP sensor connector
- ECT gauge sending unit connector
- Radiator fan switch connector
- ECT sensor connector
- Fuel injector connectors
- Bypass hoses
- Heater hose
- Upper radiator hose
- PCV valve and hose
- Intake manifold vacuum line
- Brake booster vacuum line
- EVAP control canister hose and vacuum hose

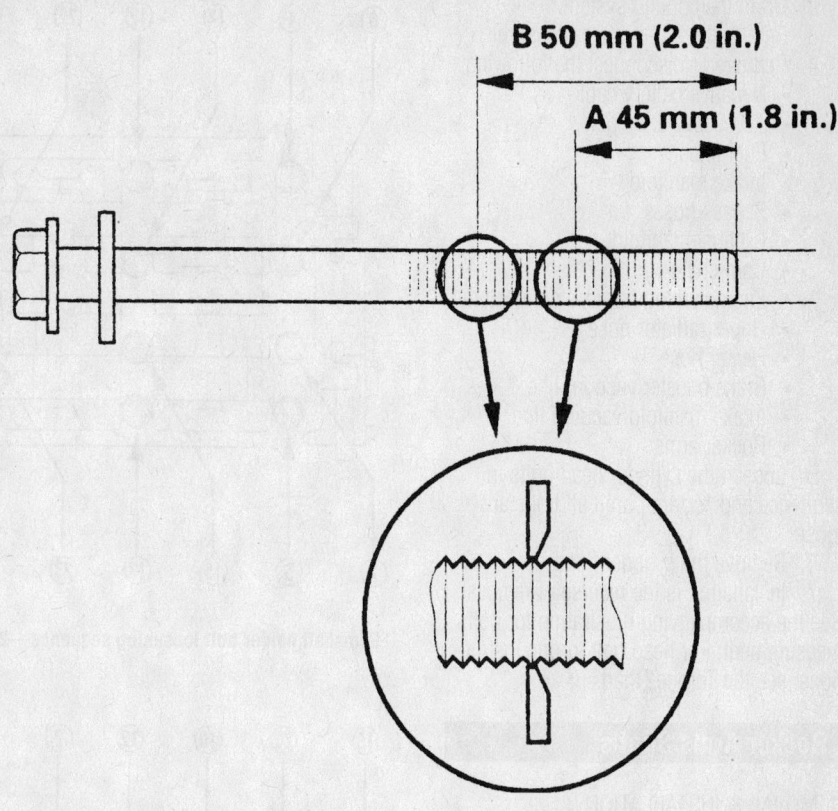

B 50 mm (2.0 in.)
A 45 mm (1.8 in.)

Cylinder head bolt inspection—2.4L

9359MG02

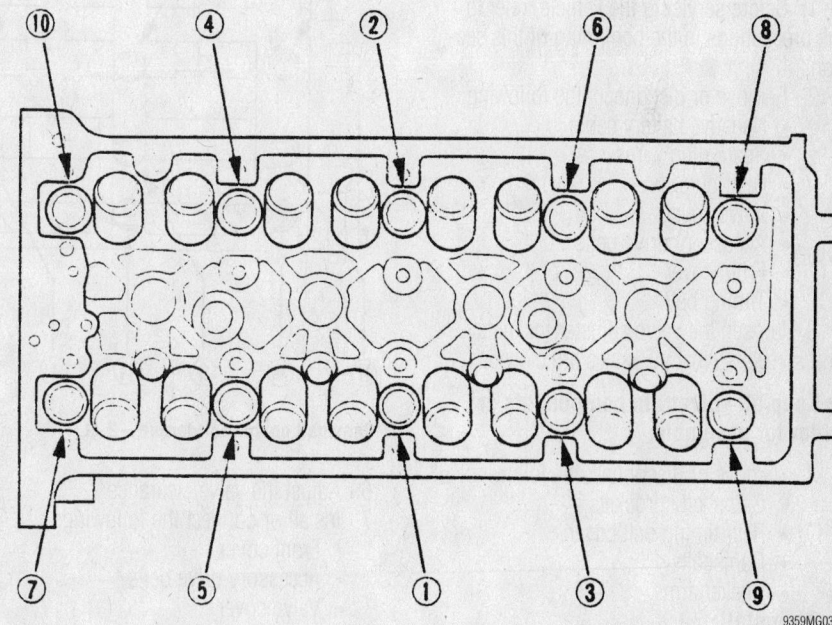

Cylinder head bolt torque sequence—2.4L

9359MG03

- Fuel lines
- Accelerator cable
- Power steering pump and bracket
- Accessory drive belts
- Air intake assembly
- Negative battery cable

15. Fill the cooling system.
16. Start the engine and check for leaks.

2002–03

1. Before servicing the vehicle, refer to the precautions in the beginning of this section.

2. Drain the cooling system.
3. Relieve the fuel system pressure.
4. Remove or disconnect the following:
  • Negative battery cable
  • Accessory drive belt
  • Fuel lines
  • Intake manifold
  • Bypass hoses
  • Exhaust manifold
  • Cam chain
  • Engine wiring harness connectors
  • Upper radiator hose
  • Heater hose
  • Brake booster vacuum line
  • Intake manifold vacuum line
  • Rocker arms

5. Loosen the cylinder head bolts in sequence and ⅓ turns until all bolts are loose.
6. Remove the cylinder head.
7. Installation is the reverse of removal. See the accompanying illustration for bolt measurement. For head bolt torque instructions, see the Torque Chart.

Rocker Arms/Shafts

REMOVAL & INSTALLATION

2.0L

1. Before servicing the vehicle, refer to the precautions in the beginning of this section.
2. Remove or disconnect the following:
  • Negative battery cable
  • Spark plug wires
  • Distributor
  • Valve cover
  • Accessory drive belts
  • Front cover
  • Timing belt.
3. Loosen the valve adjuster locknuts and screws so that all valves are closed.

➡ **Keep all valvetrain components in order for assembly.**

4. Remove or disconnect the following:
  • Camshaft sprockets
  • Rear timing belt cover
  • Camshafts
  • Rocker arms

To install:

5. Install or connect the following:
  • Rocker arms in their original positions
  • Camshafts
  • Rear timing belt cover
  • Camshaft sprockets. Tighten the bolts to 27 ft. lbs. (37 Nm).
  • Timing belt

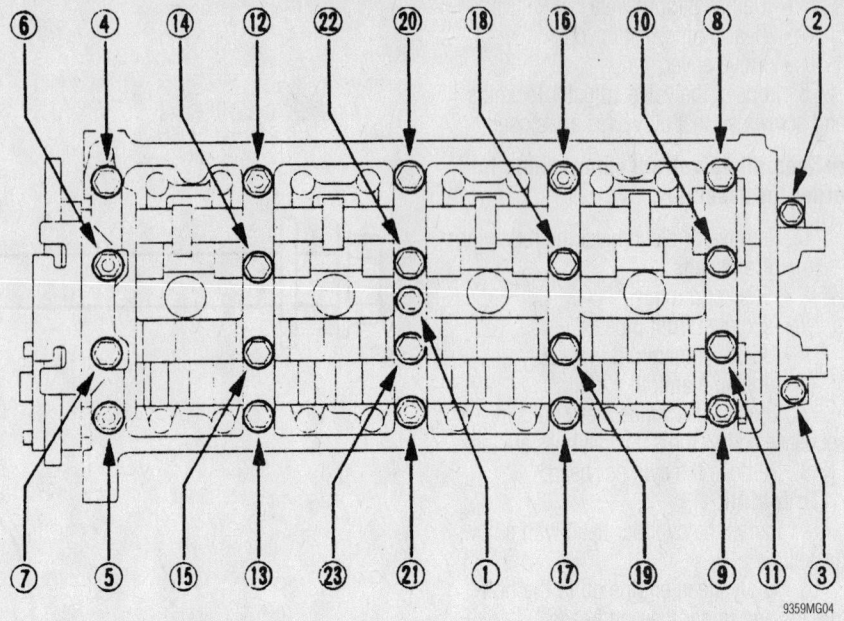

Camshaft holder bolt loosening sequence—2.4L

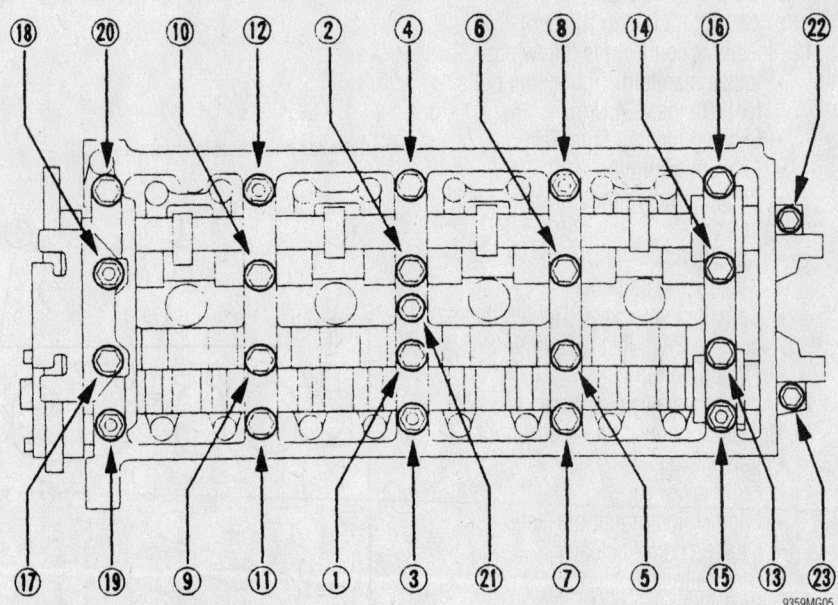

Camshaft holder bolt torque—2.4L

6. Adjust the valve clearance.
7. Install or connect the following:
  • Front cover
  • Accessory drive belts
  • Valve cover
  • Distributor
  • Spark plug wires
  • Negative battery cable
8. Start the engine and check for leaks.

2.4L

1. Remove the camshaft chain
2. Loosen the rocker adjusting screws

3. Remove the camshaft holder bolts 2 turns at a time in the sequence shown.
4. Remove the camshaft chain guide, camshaft holders and camshafts.
5. Remove the rocker arm assembly.
6. Installation is the reverse of removal. Prior to installation, clean the No.5 rocker shaft holder mating surface and apply RTV gasket sealer to the mounting point on the head. See the illustration for the torque sequence. Torque the 8mm bolts to 16 ft. lbs. (22 Nm) and the 6mm bolts to 9 ft. lbs. (12 Nm).

Intake Manifold

REMOVAL & INSTALLATION

2.0L

1. Before servicing the vehicle, refer to the precautions in the beginning of this section.
2. Drain the cooling system.
3. Relieve the fuel system pressure.
4. Remove or disconnect the following:
   • Negative battery cable
   • Air intake assembly
   • Intake manifold resonator chamber and bracket
   • Accelerator cable
   • Fuel lines
   • Evaporative Emissions (EVAP) control canister hose and vacuum hose
   • Brake booster vacuum line
   • Intake manifold vacuum line
   • Positive Crankcase Ventilation (PCV) valve and hose
   • Bypass hoses
   • Fuel injector connectors
   • Throttle Position (TP) sensor connector
   • Manifold Absolute Pressure (MAP) sensor connector
   • Idle Air Control (IAC) valve connector
   • Cruise control actuator
   • Intake manifold brackets
   • Intake manifold

To install:

5. Install or connect the following:
   • New intake manifold gasket
   • Intake manifold. Tighten the fasteners to 17 ft. lbs. (23 Nm).
   • Intake manifold brackets
   • Cruise control actuator

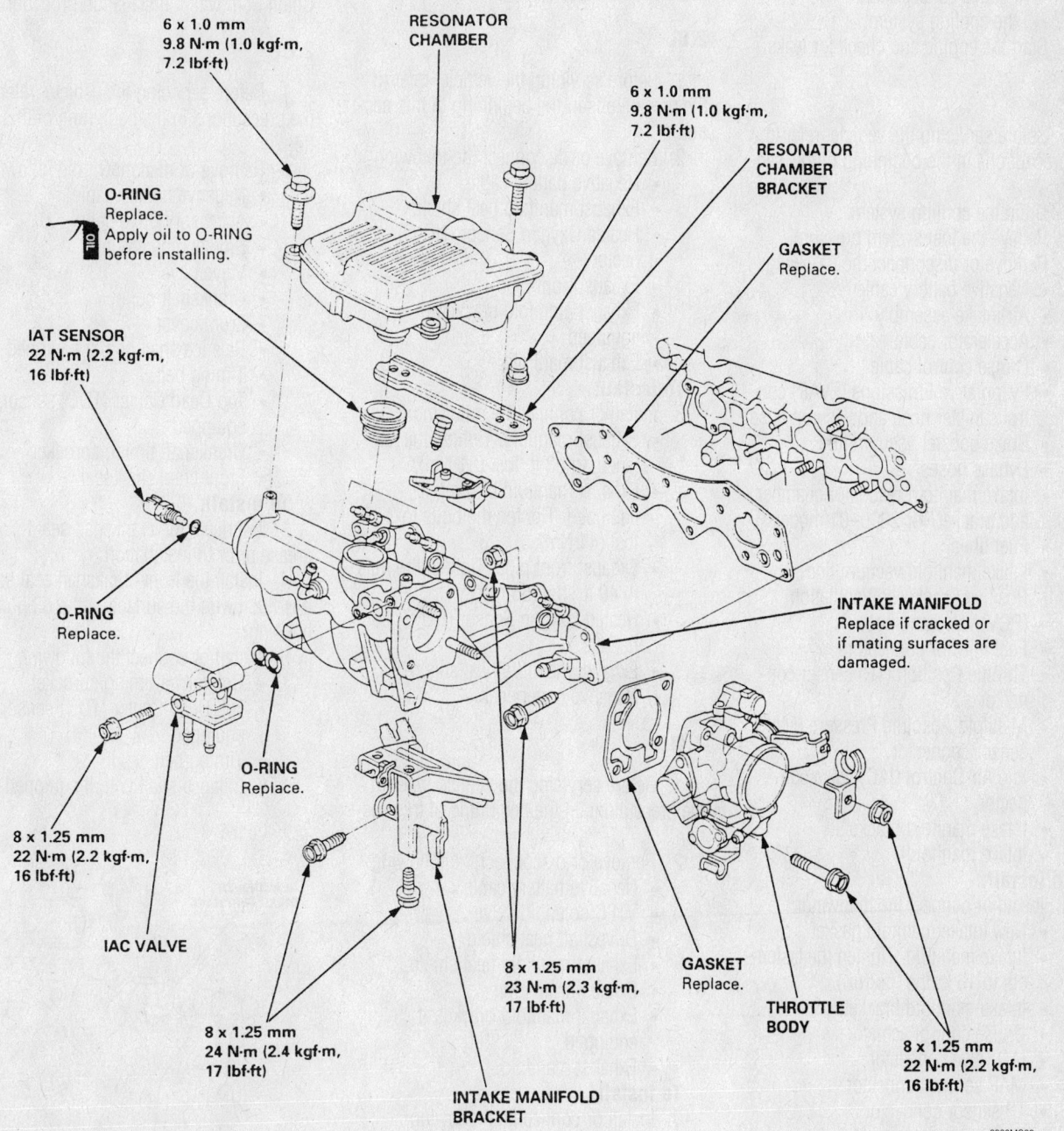

**6 x 1.0 mm
9.8 N·m (1.0 kgf·m, 7.2 lbf·ft)**

RESONATOR CHAMBER

**6 x 1.0 mm
9.8 N·m (1.0 kgf·m, 7.2 lbf·ft)**

RESONATOR CHAMBER BRACKET

O-RING
Replace.
Apply oil to O-RING before installing.

GASKET
Replace.

**IAT SENSOR
22 N·m (2.2 kgf·m, 16 lbf·ft)**

O-RING
Replace.

INTAKE MANIFOLD
Replace if cracked or if mating surfaces are damaged.

**8 x 1.25 mm
22 N·m (2.2 kgf·m, 16 lbf·ft)**

IAC VALVE

O-RING
Replace.

**8 x 1.25 mm
24 N·m (2.4 kgf·m, 17 lbf·ft)**

INTAKE MANIFOLD BRACKET

**8 x 1.25 mm
23 N·m (2.3 kgf·m, 17 lbf·ft)**

GASKET
Replace.

THROTTLE BODY

**8 x 1.25 mm
22 N·m (2.2 kgf·m, 16 lbf·ft)**

9308MG22

Intake manifold exploded view—2000–01 CR-V

- IAC valve connector
- MAP sensor connector
- TP sensor connector
- Fuel injector connectors
- Bypass hoses
- PCV valve and hose
- Intake manifold vacuum line
- Brake booster vacuum line
- EVAP control canister hose and vacuum hose
- Fuel lines
- Accelerator cable
- Intake manifold resonator chamber and bracket
- Air intake assembly
- Negative battery cable

6. Fill the cooling system.
7. Start the engine and check for leaks.

2.4L

1. Before servicing the vehicle, refer to the precautions in the beginning of this section.
2. Drain the cooling system.
3. Relieve the fuel system pressure.
4. Remove or disconnect the following:
   - Negative battery cable
   - Air intake assembly
   - Accelerator cable
   - Cruise control cable
   - Evaporative Emissions (EVAP) control canister hose and vacuum hose
   - Brake booster vacuum line
   - Bypass hoses
   - Intake manifold resonator chamber and bracket, for 2000–01 models
   - Fuel lines
   - Intake manifold vacuum line
   - Positive Crankcase Ventilation (PCV) valve and hose
   - Fuel injector connectors
   - Throttle Position (TP) sensor connector
   - Manifold Absolute Pressure (MAP) sensor connector
   - Idle Air Control (IAC) valve connector
   - Intake manifold brackets
   - Intake manifold

To install:
5. Install or connect the following:
   - New intake manifold gasket
   - Intake manifold. Tighten the fasteners to 16 ft. lbs. (22Nm).
   - Intake manifold brackets
   - Cruise control actuator
   - IAC valve connector
   - MAP sensor connector
   - TP sensor connector
   - Fuel injector connectors
   - Bypass hoses

- PCV valve and hose
- Intake manifold vacuum line
- Brake booster vacuum line
- EVAP control canister hose and vacuum hose
- Fuel lines
- Accelerator cable
- Air intake assembly
- Negative battery cable

6. Fill the cooling system.
7. Start the engine and check for leaks.

Exhaust Manifold

REMOVAL & INSTALLATION

2.0L

1. Before servicing the vehicle, refer to the precautions in the beginning of this section.
2. Remove or disconnect the following:
   - Negative battery cable
   - Exhaust manifold heat shield
   - Heated Oxygen Sensor (HO2S) connector
   - Exhaust front pipe
   - Exhaust manifold bracket, if equipped
   - Exhaust manifold

To install:
3. Install or connect the following:
   - Exhaust manifold. Tighten the fasteners to 23 ft. lbs. (31 Nm).
   - Exhaust manifold bracket, if equipped. Tighten the bolts to 33 ft. lbs. (44 Nm).
   - Exhaust front pipe. Tighten the nuts to 40 ft. lbs. (55 Nm).
   - Heated Oxygen Sensor (HO2S) connector
   - Exhaust manifold heat shield
   - Negative battery cable

2.4L

1. Before servicing the vehicle, refer to the precautions in the beginning of this section.
2. Remove or disconnect the following:
   - Negative battery cable
   - VTEC solenoid valve
   - Driveshaft heat shield
   - Exhaust manifold heat shield
   - Exhaust front pipe
   - Exhaust manifold bracket, if equipped
   - Exhaust manifold

To install:
3. Install or connect the following:
   - Exhaust manifold. Tighten the fasteners to 3 ft. lbs. (44m).

- Exhaust manifold bracket, if equipped. Tighten the bolts to 33 ft. lbs. (44 Nm).
- Exhaust front pipe. Tighten the nuts to 16t. lbs. (22m).
- Exhaust manifold heat shield
- Driveshaft heat shield
- VTEC solenoid valve
- Negative battery cable

Front Crankshaft Seal

REMOVAL & INSTALLATION

For the 2.4L engine, see the Timing Chain Removal & Installation procedure.

2.0L

1. Before servicing the vehicle, refer to the precautions in the beginning of this section.
2. Remove or disconnect the following:
   - Negative battery cable
   - Accessory drive belts
   - Side engine mount
   - Valve cover
   - Crankshaft pulley
   - Front cover
   - Balance shaft belt, if equipped
   - Timing belt.
   - Top Dead Center (TDC) sensor, if equipped
   - Crankshaft timing sprocket
   - Front crankshaft seal

To install:
3. Lubricate the crankshaft seal lip with grease prior to installation.
4. Install the front crankshaft seal so that it is flush with the surface of the oil pump housing.
5. Install or connect the following:
   - Crankshaft timing sprocket
   - Top Dead Center (TDC) sensor, if equipped
   - Timing belt.
   - Balance shaft belt, if equipped

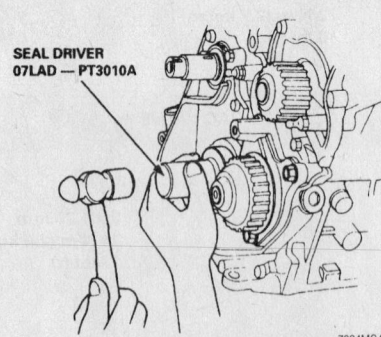

SEAL DRIVER
07LAD — PT3010A

7924MG48

Front crankshaft seal installation–2.0L

- Front cover
- Crankshaft pulley. Tighten the bolt to 130 ft. lbs. (177 Nm).
- Valve cover
- Side engine mount
- Accessory drive belts
- Negative battery cable

6. Check the engine oil level and add if necessary.

7. Start the engine and check for leaks.

Camshaft

REMOVAL & INSTALLATION

2.0L

1. Before servicing the vehicle, refer to the precautions in the beginning of this section.

2. Remove or disconnect the following:
   - Negative battery cable
   - Spark plug wires
   - Distributor
   - Valve cover
   - Accessory drive belts

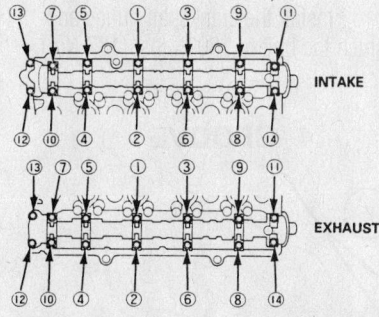

Camshaft bearing tightening sequence—2.0L

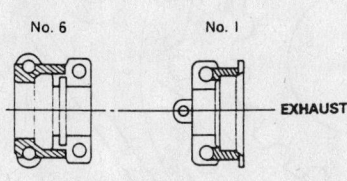
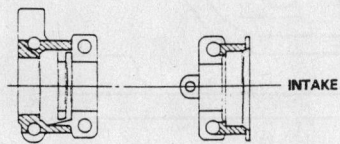

Apply liquid gasket to the shaded areas.

9302MG72

Apply liquid gasket to the shaded areas of the camshaft journals—2.0L

- Front cover
- Timing belt.

3. Loosen the valve adjuster locknuts and screws so that all valves are closed.

➡ **Keep all valvetrain components in order for assembly.**

4. Remove or disconnect the following:
   - Camshaft sprockets
   - Rear timing belt cover
   - Camshafts

To install:

➡ **Use new O-rings, seals and gaskets when installing the camshaft.**

5. Install or connect the following:
   - Camshafts. Tighten the bearing cap bolts in sequence to 86 inch lbs. (10 Nm).
   - Rear timing belt cover
   - Camshaft sprockets. Tighten the bolts to 27 ft. lbs. (37 Nm).
   - Timing belt

6. Adjust the valve clearance.

7. Install or connect the following:
   - Front cover
   - Accessory drive belts
   - Valve cover
   - Distributor
   - Spark plug wires
   - Negative battery cable

8. Start the engine and check for leaks.

2.4L

See the Rocker Arm Shaft Removal & Installation procedure.

Valve Lash

ADJUSTMENT

Adjust the valves only when the cylinder head temperature is less than 100°F (38°C).

2.0L

1. Before servicing the vehicle, refer to the precautions in the beginning of this section.

2. Remove or disconnect the following:
   - Negative battery cable
   - Air intake tube
   - Valve cover

3. Rotate the crankshaft so that the valves to be adjusted are closed and the rocker arm is contacting the camshaft lobe base circle.

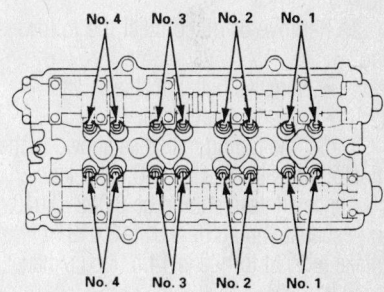

Intake and exhaust valve identification—2.0L

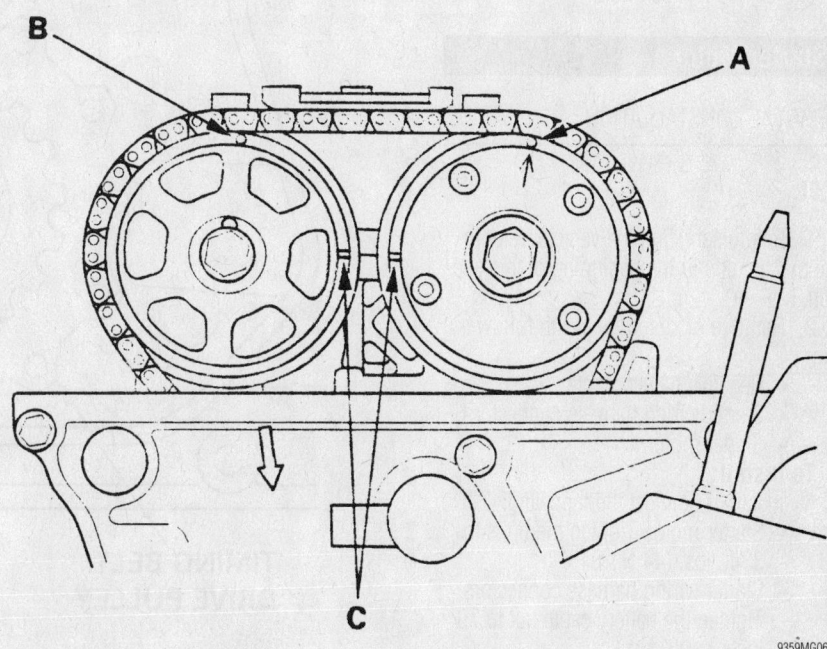

Align the timing marks—2.4L

4. Measure the valve clearance. If adjustment is necessary, loosen the locknut and turn the adjusting screw as necessary to achieve the correct valve clearance.

5. The correct valve clearance is:
- Intake valves: 0.003–0.005 inches (0.08–0.12mm)
- Exhaust valves: 0.006–0.008 inches (0.16–0.20mm)

6. After adjustment, tighten the locknuts to 18 ft. lbs. (24 Nm).

7. Install or connect the following:
- Valve cover
- Air intake tube
- Negative battery cable

8. Start the engine and check for proper operation.

2.4L

1. Before servicing the vehicle, refer to the precautions in the beginning of this section.

2. Remove or disconnect the following:
- Negative battery cable
- cylinder head cover

3. Set the timing marks as shown in the illustration with NO.1 at TDC. Check all clearances. Intake should be 0.008–0.010 in.; exhaust should be 0.011–0.013 in. Intake locknut torque is 14 ft. lbs.; exhaust is 10 ft. lbs.

4. Rotate the crankshaft 180 degrees clockwise and recheck No.3.

5. Rotate the crankshaft 180 degrees clockwise and recheck No.4.

6. Rotate the crankshaft 180 degrees clockwise and recheck No.2.

Starter Motor

REMOVAL & INSTALLATION

2.0L

1. Before servicing the vehicle, refer to the precautions in the beginning of this section.

2. Remove or disconnect the following:
- Negative battery cable
- Starter wiring harness connectors
- Starter motor

To install:

3. Install or connect the following:
- Starter motor. Tighten the bolts to 33 ft. lbs. (44 Nm).
- Starter wiring harness connectors. Tighten the battery cable nut to 79 inch lbs. (9 Nm).
- Negative battery cable

2.4L

1. Before servicing the vehicle, refer to the precautions in the beginning of this section.

2. Remove or disconnect the following:
- Negative battery cable
- Knock sensor connector
- Starter wiring harness connectors
- Starter motor

To install:

3. Install or connect the following:
- Starter motor. Tighten the upper bolt to 33 ft. lbs. (44 Nm); the lower bolt to 47 ft. lbs. (64 Nm).
- Starter wiring harness connectors. Tighten the battery cable nut to 84 inch lbs. (9 Nm).
- Knock sensor connector
- Negative battery cable

Timing Belt

REMOVAL & INSTALLATION

2.0L

1. Disconnect the negative battery cable.

2. Position crankshaft so that No. 1 piston is at Top Dead Center (TDC).

3. Remove the splash guard.

4. Remove the accessory drive belts.

5. If equipped, remove the cruise control actuator.

6. Place a piece of wood between the oil pan and the jack, support the engine with a jack.

7. Remove upper engine bracket.

8. Remove the valve cover.

9. Remove the timing belt covers.

10. Loosen the adjusting bolt 180 degrees. Release the tension from the belt by pushing on the tensioner, then retighten the adjusting bolt.

11. Remove the timing belt.

To install:

12. Be sure the timing marks are properly aligned.

13. Install the timing belt on the pulleys following this sequence:
   a. Crankshaft pulley.
   b. Adjusting pulley.
   c. Water pump pulley.
   d. Exhaust camshaft pulley.
   e. Intake camshaft pulley.

14. Loosen and retighten the adjusting bolt to allow tension to be applied to the belt.

15. Install the lower and middle timing covers.

16. Install the crankshaft pulley and tighten the bolt to 130 ft. lbs. (177 Nm).

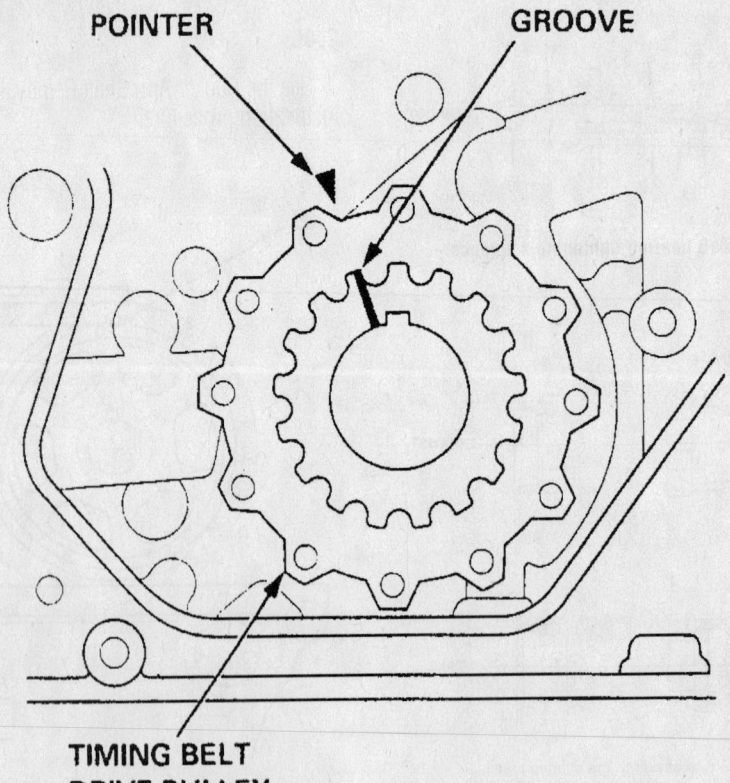

POINTER   GROOVE

TIMING BELT
DRIVE PULLEY

79245G44

Crankshaft timing mark will be easier to verify when clean—2.0L (B20B4) engine

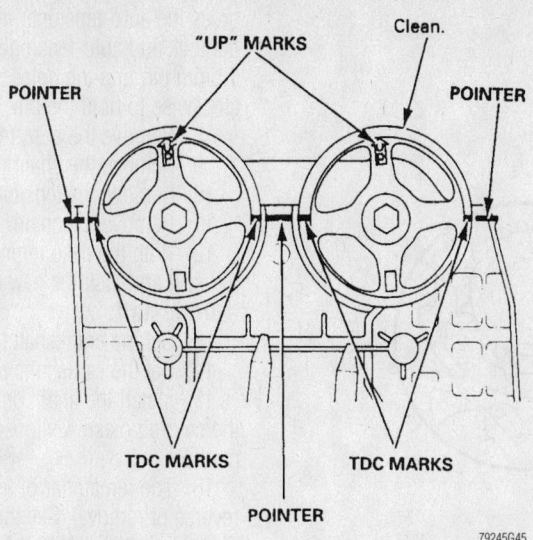

Intake and exhaust camshaft timing marks properly aligned at TDC—2.0L (B20B4) engine

✳✳ WARNING

If any binding is felt when adjusting the timing belt tension by turning the crankshaft, STOP turning the engine, because the pistons may be hitting the valves.

17. Rotate the crankshaft about 5–6 times counterclockwise to seat the timing belt.

18. Position the No. 1 piston to TDC.

19. Loosen the adjusting bolt ½ turn.

20. Rotate the crankshaft counterclockwise 3 teeth on the camshaft pulley.

21. Tighten the adjusting bolt to 40 ft. lbs. (54Nm).

22. Retighten the crankshaft pulley bolt to 130 ft. lbs. (177 Nm).

23. Install the valve cover.

24. Install the engine mounting bracket, then remove the jack.

25. If removed, install the cruise control actuator.

26. Install the accessory drive belts.

27. Install the splash guard.

28. Connect the negative battery cable.

29. Check the engine operation and road test.

Timing Chain and Front Seal

REMOVAL & INSTALLATION

2.4L

1. Before servicing the vehicle, refer to the precautions in the beginning of this section.

2. Drain the engine oil.

3. Align the timing marks at TDC No.1.

4. Remove or disconnect the following:
- Negative battery cable
- Front splash shield
- Drive belt
- Cylinder head cover
- Crankshaft pulley
- CKP sensor
- VTC oil control connector
- VTC oil control solenoid valve

5. Support the engine with a block of wood and jack.

6. Remove or disconnect the following:
- Ground cable
- Upper support bracket
- Side engine mount
- Chain case

7. Loosely install the crank pulley. Turn the crankshaft counterclockwise to com-

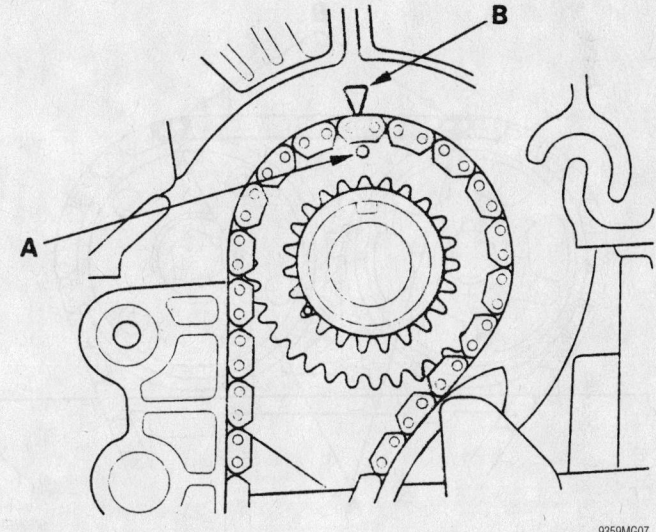

Align the crankshaft timing marks—2.4L

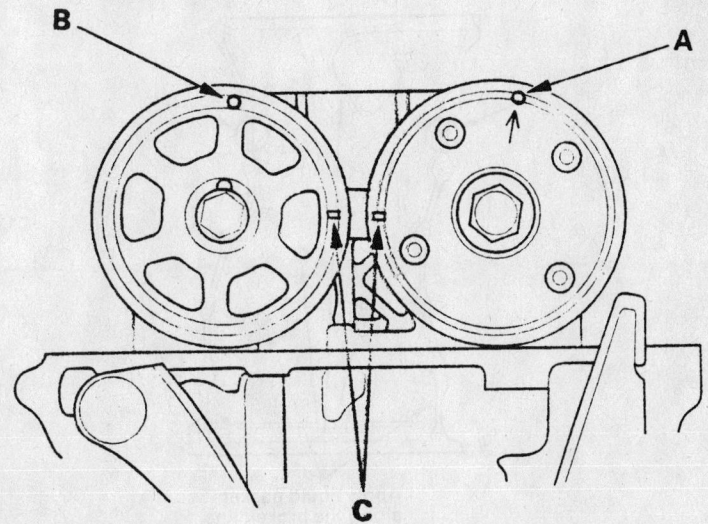

Align the camshaft timing marks—2.4L

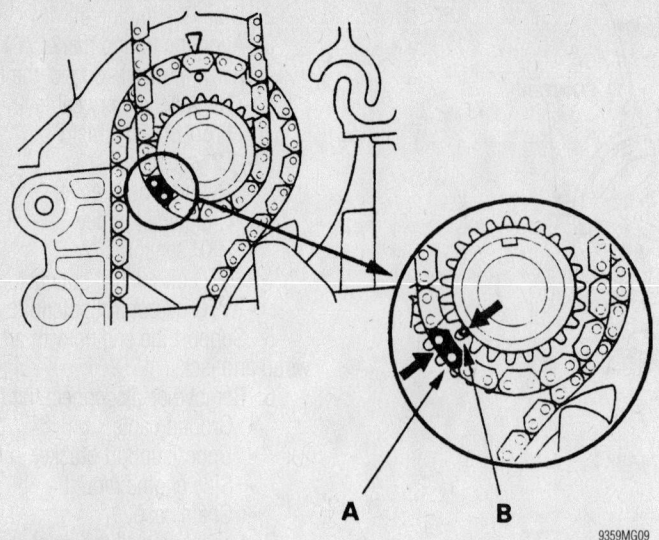

Chain installed on crankshaft—2.4L

9359MG09

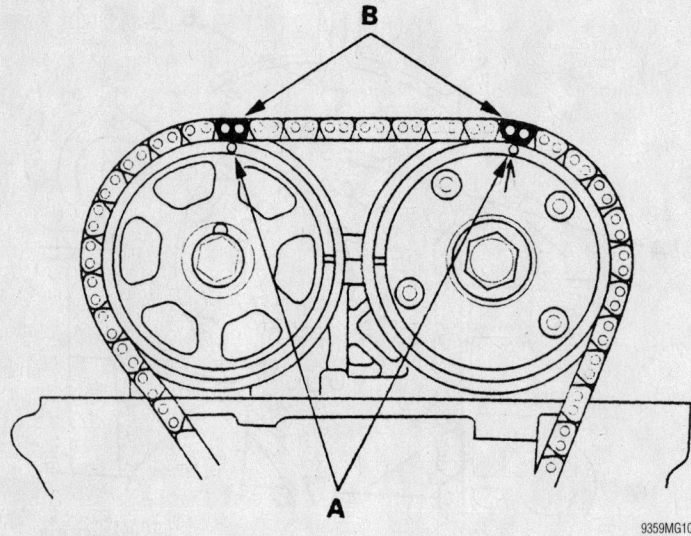

Chain installed on camshafts—2.4L

9359MG10

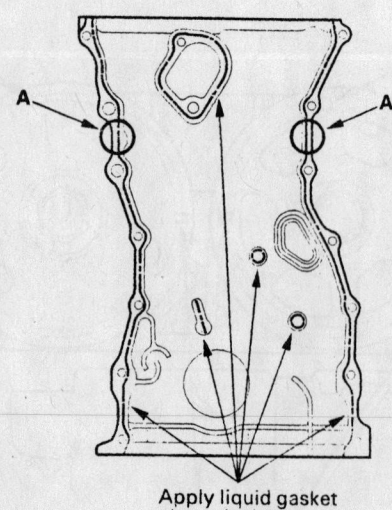

Apply liquid gasket
along the broken line.

9359MG11

Chain case sealer application—2.4L

press the auto-tensioner. Align the holes on the lock and auto-tensioner and insert a 1.5mm pin into the holes. Turn the crank clockwise to hold the pin.

8. Remove the auto-tensioner.
9. Remove the chain guides.
10. Remove the tensioner arm.
11. Remove the chain.
12. With the case removed, drive out the old seal and install a new one.

To install:

13. Set the crankshaft to TDC.
14. Set the camshafts to TDC.
15. Install the chain on the sprocket with the colored piece A aligned with the punch mark B.
16. The remainder of installation is the reverse of removal. See the accompanying illustration for sealer application. Observe the following torques:

- Chain guide: 9 ft. lbs. (12 Nm)
- Tensioner arm: 16 ft. lbs. (22 Nm)
- Auto-tensioner: 9 ft. lbs. (12 Nm)
- Upper chain guide: 16 ft. lbs. (22 Nm)
- Case: 9 ft. lbs. (12 Nm)
- Side mount: 33 ft. lbs. (44 Nm)
- Upper bracket: 40 ft. lbs. (54 Nm)

Oil Pan

REMOVAL & INSTALLATION

2.0L

1. Before servicing the vehicle, refer to the precautions in the beginning of this section.
2. Drain the engine oil.
3. Remove or disconnect the following:
- Negative battery cable
- Front splash shield
- Heated Oxygen Sensor (HO2S) connector
- Exhaust front pipe
- Torque converter cover, if equipped with an automatic transaxle
- Oil pan

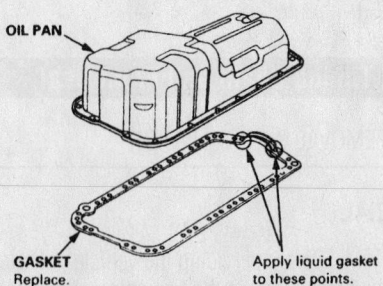

OIL PAN

GASKET
Replace.

Apply liquid gasket
to these points.

7924MG27

Oil pan gasket installation—2.0L

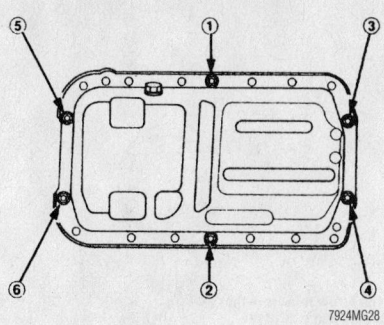

Oil pan fastener tightening sequence—2.0L

7924MG28

To install:

4. Install the oil pan.

5. Tighten the bolts in sequence to 105 inch lbs. (12 Nm).

6. Install or connect the following:
- Torque converter cover, if removed
- Exhaust front pipe
- Subframe center beam, if removed
- HO2S connector
- Front splash shield
- Negative battery cable

2.4L

1. Before servicing the vehicle, refer to the precautions in the beginning of this section.

2. Drain the engine oil.

3. Remove or disconnect the following:
- Subframe. See engine Removal and Installation.
- With MT, the stiffener
- Oil pan bolts
- Oil pan. A gasket cutter will be needed.

4. Installation is the reverse of removal. Torque the bolts, in sequence, in 2 or 3 steps, to 9 ft. lbs. (12 Nm).

Oil Pump

REMOVAL & INSTALLATION

2.0L

1. Before servicing the vehicle, refer to the precautions in the beginning of this section.

2. Drain the engine oil.

3. Remove or disconnect the following:
- Negative battery cable
- Accessory drive belts
- Front cover
- Timing belt.
- Crankshaft timing sprocket
- Oil pan
- Oil pump pickup tube
- Oil pump

To install:

➡**Use new gaskets and O-ring seals for assembly.**

4. Apply liquid gasket to the oil pump and to the bolt hole threads.

5. Install or connect the following:
- Oil pump. Tighten the 8mm bolts to 17 ft. lbs. (24 Nm) and the 6mm bolts to 86 inch lbs. (10 Nm).
- Oil pump pickup tube. Tighten the fasteners to 86 inch lbs. (10 Nm).
- Oil pan
- Crankshaft timing sprocket

- Timing belt.
- Front cover
- Accessory drive belts
- Negative battery cable

6. Fill the crankcase to the correct level.

7. Start the engine and check for leaks.

2.4L

1. Before servicing the vehicle, refer to the precautions in the beginning of this section.

2. Drain the engine oil.

3. Remove or disconnect the following:
- Negative battery cable
- Oil pan
- Pump chain
- Pump sprocket
- Pump

To install:

4. Make sure that No.1 piston is at TDC.

5. Align the dowel pin on the rear balance shaft with the mark on the pump.

6. Install the pump and sprocket loosely.

7. Remove the balance shaft holding pin.

8. Torque the 10mm mounting bolts to 33 ft. lbs. (44 Nm); the 8mm bolts to 16 ft. lbs. (22 Nm).

9. Torque the pulley bolt to 33 ft. lbs. (44 Nm).

10. Torque the tensioner bolts to 9 ft. lbs. (12 nm).

Rear Main Seal

REMOVAL & INSTALLATION

1. Before servicing the vehicle, refer to the precautions in the beginning of this section.

2. Remove or disconnect the following:
- Transaxle
- Clutch pressure plate and disc, if equipped
- Flywheel
- Oil seal

To install:

3. Install or connect the following:
- Oil seal. Drive the seal square into the seal case.
- Flywheel. Tighten the bolts in a crossing pattern to 54 ft. lbs. (73 Nm).
- Clutch pressure plate and disc, if equipped
- Transaxle

4. Check the fluid levels.

5. Start the engine and check for leaks.

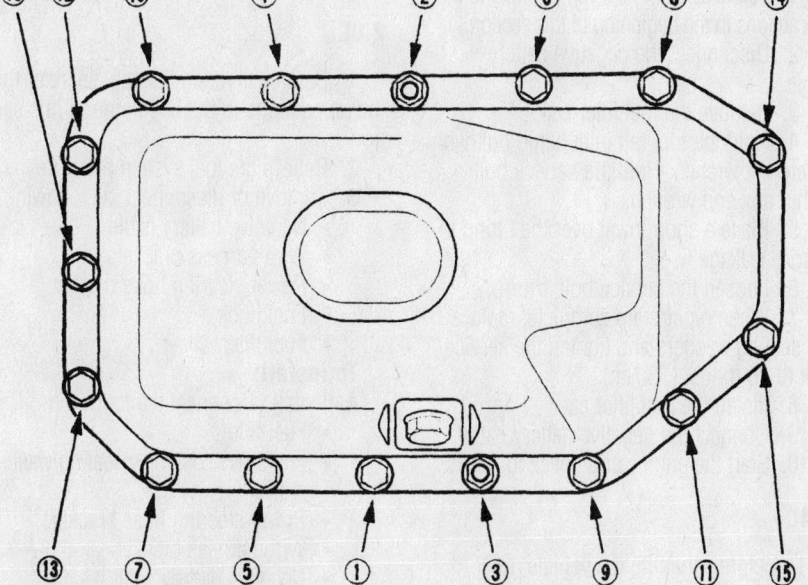

Oil pan fastener tightening sequence—2.4L

9359MG12

Piston and Ring

POSITIONING

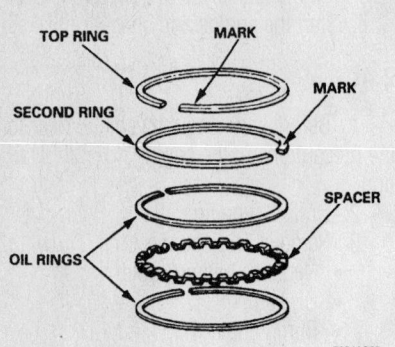

Piston ring positioning and top mark location

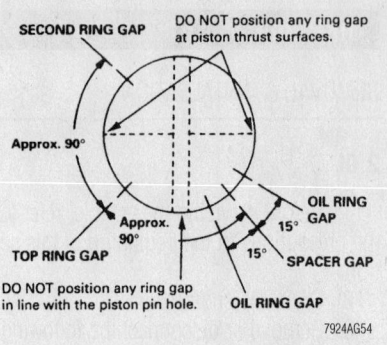

Piston ring end-gap spacing

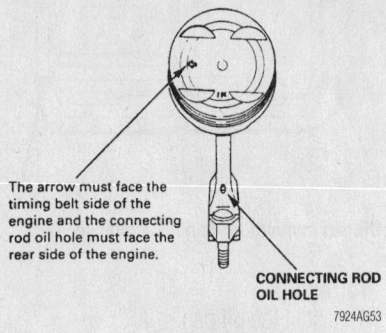

Piston and connecting rod assembly

FUEL SYSTEM

Fuel System Service Precautions

Safety is the most important factor when performing not only fuel system maintenance, but any type of maintenance. Failure to conduct maintenance and repairs in a safe manner may result in serious personal injury or death. Maintenance and testing of the vehicle's fuel system components can be accomplished safely and effectively by adhering to the following rules and guidelines:

• To avoid the possibility of fire and personal injury, always disconnect the negative battery cable unless the repair or test procedure requires that battery voltage be applied.

• Always relieve the fuel system pressure prior to disconnecting any fuel system component (injector, fuel rail, pressure regulator, etc.), fitting or fuel line connection. Exercise extreme caution whenever relieving fuel system pressure to avoid exposing skin, face and eyes to fuel spray. Please be advised that fuel under pressure may penetrate the skin or any part of the body that it contacts.

• Always place a shop towel or cloth around the fitting or connection prior to loosening to absorb any excess fuel due to spillage. Ensure that all fuel spillage (should it occur) is quickly removed from engine surfaces. Ensure that all fuel soaked cloths or towels are deposited into a suitable waste container.

• Always keep a dry chemical (Class B) fire extinguisher near the work area.

• Do not allow fuel spray or fuel vapors to come into contact with a spark or open flame.

• Always use a backup wrench when loosening and tightening fuel line connection fittings. This will prevent unnecessary stress and torsion to fuel line piping. Always follow the proper torque specifications.

• Always replace worn fuel fitting O-rings with new. Do not substitute fuel hose or equivalent, where fuel pipe is installed.

Fuel System Pressure

RELIEVING

2.0L

1. Before servicing the vehicle, refer to the precautions in the beginning of this section.
2. Disconnect the negative battery cable.
3. Remove the fuel filler cap.
4. Hold the fuel rail inlet banjo bolt with a flare nut wrench. Hold the service bolt with a box end wrench.
5. Place a shop towel over the fitting to absorb leakage.
6. Loosen the service bolt 1 turn.
7. When repairs are complete, replace the sealing washers and tighten the service bolt to 25 ft. lbs. (33 Nm).
8. Install the fuel filler cap.
9. Connect the negative battery cable.
10. Start the engine and check for leaks.

2.4L

1. Before servicing the vehicle, refer to the precautions in the beginning of this section.

2. Disconnect the negative battery cable.
3. Remove the fuel filler cap.
4. Remove the engine cover.
5. Using a back up wrench and shop towel, turn the fuel pulsation damper one complete turn, slowly.

➡Replace all washers whenever the pulsation damper is loosened or removed.

6. Tighten the damper to 16 ft. lbs. (22 Nm).

Fuel Filter

REMOVAL & INSTALLATION

2.0L

1. Before servicing the vehicle, refer to the precautions in the beginning of this section.
2. Relieve the fuel system pressure.
3. Remove or disconnect the following:
   • Negative battery cable
   • Wire harness bracket
   • Power steering hose bracket
   • Fuel lines
   • Fuel filter

To install:
4. Install or connect the following:
   • Fuel filter
   • Fuel lines. Use new sealing washers.
   • Power steering hose bracket
   • Wire harness bracket
   • Negative battery cable
5. Start the engine and check for leaks.

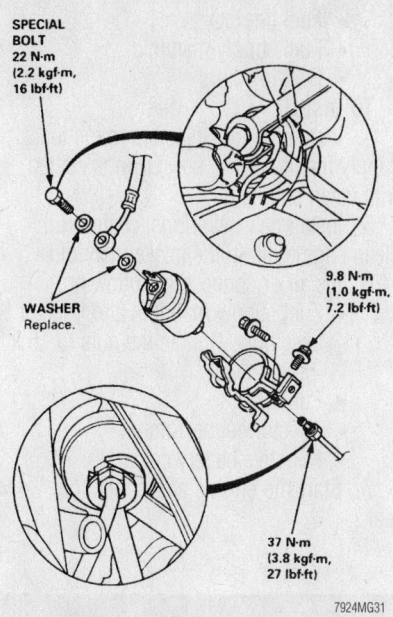

SPECIAL
BOLT
22 N·m
(2.2 kgf·m,
16 lbf·ft)

WASHER
Replace.

9.8 N·m
(1.0 kgf·m,
7.2 lbf·ft)

37 N·m
(3.8 kgf·m,
27 lbf·ft)

7924MG31

Exploded view of the fuel filter mounting—2.0L

2.4L

➡️**The fuel filter should be replaced whenever the fuel pressure drops below 48 psi, after making sure that the fuel pump and fuel pressure regulator are okay.**

1. Before servicing the vehicle, refer to the precautions in the beginning of this section.
2. Relieve the fuel system pressure.
3. Remove or disconnect the following:
   - Negative battery cable
   - Fuel pump
   - Fuel filter carrier (A)
   - Fuel filter

To install:

4. Install or connect the following:
   - Fuel filter
   - Fuel lines
   - New gasket (B)
   - New o-rings (E)
   - Connectors (C)
   - Sending unit (D)
5. Start the engine and check for leaks.

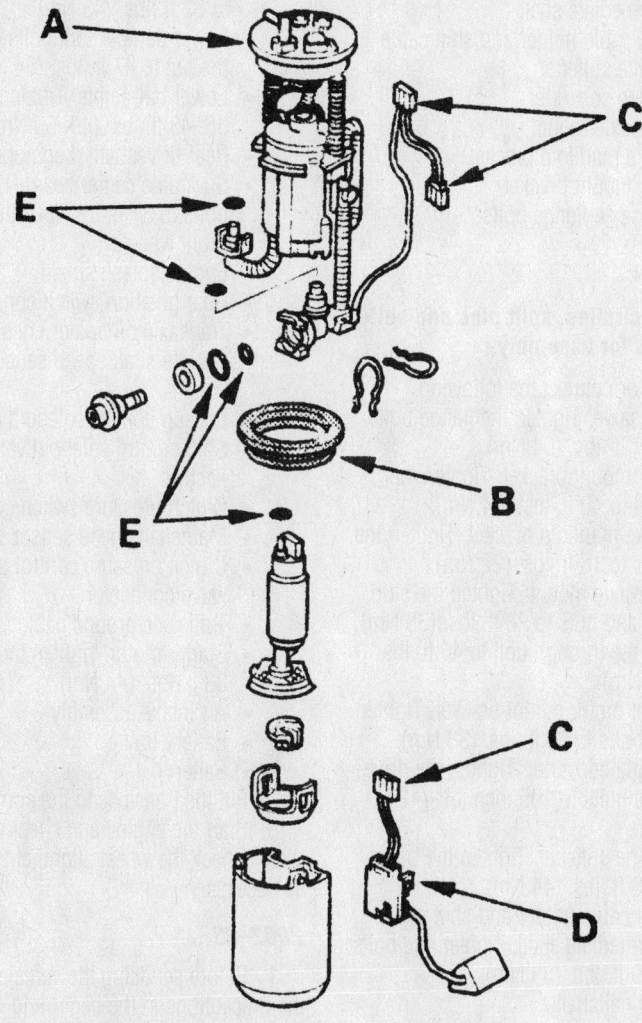

9359MG13

Exploded view of the fuel filter mounting—2.4L

Fuel Pump

REMOVAL & INSTALLATION

2.4L

1. Before servicing the vehicle, refer to the precautions in the beginning of this section.
2. Relieve the fuel system pressure.
3. Remove or disconnect the following:
   - Negative battery cable
   - Left rear seat cushion
   - Base frame cover
   - Access panel
   - Fuel pump module wiring connector
   - Fuel supply and return lines
   - Fuel pump module

To install:

4. Install or connect the following:
   - Fuel pump module. Use a new seal and tighten the nuts to 52 inch lbs. (6 Nm).
   - Fuel supply and return lines
   - Fuel pump module wiring connector
   - Access panel
   - Base frame cover
   - Left rear seat cushion
   - Negative battery cable
5. Start the engine and check for leaks.

2.4L

1. Before servicing the vehicle, refer to the precautions in the beginning of this section.
2. Relieve the fuel system pressure.
3. Remove or disconnect the following:
   - Negative battery cable
   - Fuel filler cap
   - Access panel, under the rear seats
   - Fuel pump connector
   - Fuel supply and return lines
   - Fuel pump locknut
   - Fuel pump module
4. Installation is the reverse of removal.

Fuel Injector

REMOVAL & INSTALLATION

2.0L

1. Before servicing the vehicle, refer to the precautions in the beginning of this section.
2. Relieve the fuel system pressure.
3. Remove or disconnect the following:
   - Negative battery cable

- Air intake resonator
- Injector connectors
- Positive Crankcase Ventilation (PCV) valve and hose
- Fuel pressure regulator vacuum line
- Fuel lines
- Fuel supply manifold
- Fuel injectors

To install:

4. Install the fuel injectors to the fuel supply manifold with new cushion rings and O-rings.

5. Install new seal rings to the intake manifold.

6. Install or connect the following:
  - Fuel supply manifold and injector assembly. Tighten the nuts to 105 inch lbs. (12 Nm).

- Fuel lines
- Fuel pressure regulator vacuum line
- PCV valve and hose
- Injector connectors
- Air intake resonator
- Negative battery cable

7. Start the engine and check for leaks.

2.4L

1. Before servicing the vehicle, refer to the precautions in the beginning of this section.

2. Relieve the fuel system pressure.

3. Remove or disconnect the following:
  - Negative battery cable
  - Injector connectors, ground cable and harness holder

- Fuel line
- Fuel supply manifold
- Fuel injectors

To install:

4. Install the fuel injectors to the fuel supply manifold with new O-rings coated with clean engine oil.

5. Install new seal rings, coated with clean engine oil, in the intake manifold.

6. Install or connect the following:
  - Fuel supply manifold and injector assembly. Tighten the nuts to 16 ft. lbs. (22 Nm).
  - Fuel lines
  - Injector connectors
  - Negative battery cable

7. Start the engine and check for leaks.

DRIVE TRAIN

Transaxle Assembly

REMOVAL & INSTALLATION

Automatic Transaxle

2000–01

1. Before servicing the vehicle, refer to the precautions in the beginning of this section.

2. Drain the transaxle.

3. Remove or disconnect the following:
  - Battery
  - Battery tray
  - Air intake assembly
  - Starter motor
  - Transaxle ground cable
  - Clutch pressure control solenoid valve connector
  - Mainshaft speed sensor connector
  - Clutch pressure switch connectors
  - Shift control solenoid valve connectors
  - Lockup control solenoid connector
  - Countershaft speed sensor connector
  - Transaxle oil cooler lines
  - Gear position switch connector
  - Engine splash shield
  - Front wheels
  - Subframe center beam
  - Rear driveshaft, if equipped
  - Front motor mount bracket
  - Lower ball joints
  - Lower damper fork bolts

4. Separate the inner CV-joints from the transaxle and intermediate shaft and support the axle halfshafts out of the work area with safety wire.

5. Remove or disconnect the following:
  - Right damper fork
  - Right radius rod
  - Intermediate shaft
  - Shift cable holder and shift cable
  - Engine stiffener
  - Torque converter
  - Transaxle mount
  - Intake manifold bracket
  - Rear mount bracket
  - Transaxle flange bolts
  - Transaxle

To install:

➡**Use new circlips, split pins and self-locking nuts for assembly.**

6. Install or connect the following:
  - Transaxle. Tighten the flange bolts to 47 ft. lbs. (64 Nm).
  - Rear mount bracket. Tighten the bolts to 40 ft. lbs. (54 Nm).
  - Intake manifold bracket. Tighten the bolts to 16 ft. lbs. (22 Nm).
  - Transaxle mount. Tighten the stud bolt and nuts to 28 ft. lbs. (38 Nm) and the through bolt to 40 ft. lbs. (54 Nm).
  - Front motor mount bracket. Tighten the bolts to 28 ft. lbs. (38 Nm).
  - Torque converter. Tighten the drive-plate bolts to 105 inch lbs. (12 Nm).
  - Engine stiffener. Tighten the bolts to 33 ft. lbs. (44 Nm).
  - Shift cable holder and shift cable
  - Intermediate shaft. Tighten the bolts to 29 ft. lbs. (39 Nm).
  - Axle halfshafts
  - Right damper fork. Tighten the

pinch bolt to 32 ft. lbs. (43 Nm).
  - Right radius rod. Tighten the bolts to 76 ft. lbs. (103 Nm) and the nut to 32 ft. lbs. (43 Nm).
  - Lower damper fork bolts. Tighten the nut to 47 ft. lbs. (64 Nm).
  - Lower ball joints. Tighten the nut to 36–43 ft. lbs. (49–59 Nm).
  - Rear driveshaft, if equipped
  - Subframe center beam. Tighten the bolts to 37 ft. lbs. (50 Nm).
  - Front wheels
  - Engine splash shield
  - Gear position switch connector
  - Transaxle oil cooler lines
  - Countershaft speed sensor connector
  - Lockup control solenoid connector
  - Shift control solenoid valve connectors
  - Clutch pressure switch connectors
  - Mainshaft speed sensor connector
  - Clutch pressure control solenoid valve connector
  - Transaxle ground cable
  - Starter motor. Tighten the bolts to 33 ft. lbs. (44 Nm).
  - Air intake assembly
  - Battery tray
  - Battery

7. Fill the transaxle to the correct level.

8. Start the engine and check for leaks.

9. Check the wheel alignment and adjust as necessary.

2002–03

1. Before servicing the vehicle, refer to the precautions in the beginning of this section.

2. Drain the transaxle.
3. Remove or disconnect the following:
- Battery
- Battery tray
- Air intake assembly
- Splash shield
- Transaxle ground cable
- Starter motor
- Clutch pressure control solenoid valve connector
- Mainshaft speed sensor connector
- Clutch pressure switch connectors
- Shift control solenoid valve connectors
- Lockup control solenoid connector
- Countershaft speed sensor connector
- Transaxle oil cooler lines
- Engine wiring harness from the air cleaner bracket
- Water pipe mounting bolt
- Brake booster and EVAP line mounting bolts. Attach an engine support hanger to the head to support the weight of the engine.
- Stabilizer link from the lower arm
- Lower arms from the knuckles
- Torque converter nuts
- Shift cable
- Front mount bolt and nut
- Rear mount bolts
- Subframe (see the Engine Removal and Installation procedure)
- Rear driveshaft, if equipped

4. Separate the inner CV-joints from the transaxle and intermediate shaft and support the axle halfshafts out of the work area with safety wire.
- Intermediate shaft
- Engine stiffener
- Transaxle mount bolts and nuts
- Transaxle

5. Installation is the reverse of removal. Observe the following torques:
- Air cleaner housing bracket bolt: 16 ft. lbs. (22 Nm)
- Front mount bolts: 47 ft. lbs. (64 Nm)
- Rear mount bracket bolts: 40 ft. lb. (54 Nm)
- Transmission-to-engine bolts: 47 ft. lbs. (64 Nm)
- Upper transmission mount bolt: 40 ft. lbs. (54 Nm)
- Upper transmission mount nuts: 40 ft. lbs. (54 Nm)
- Rear driveshaft bolts: 24 ft. lbs. (32 Nm)
- Subframe bolts: 76 ft. lbs. (103 Nm)

Manual Transaxle

2000–01

1. Before servicing the vehicle, refer to the precautions in the beginning of this section.
2. Remove or disconnect the following:
- Negative battery cable
- Air intake assembly
- Clutch slave cylinder and hose bracket
- Starter motor
- Transaxle ground cable
- Reverse lamp switch connector
- Wire harness bracket
- Shift cables and bracket
- Vehicle Speed Sensor (VSS) connector
- Splash shield
- Heated Oxygen Sensor (HO2S) connector
- Exhaust front pipe
- Rear driveshaft
- Lower ball joints
- Right damper fork

3. Separate the inner CV-joints from the transaxle and intermediate shaft and support the axle halfshafts out of the work area with safety wire.
4. Remove or disconnect the following:
- Intermediate shaft
- Rear engine stiffener
- Clutch housing cover
- Right front mount and bracket
- Transaxle mount and bracket
- Rear engine mounting bolts
- Transaxle flange bolts
- Transaxle

To install:

➡Use new circlips, split pins and self-locking nuts for assembly.

5. Install or connect the following:
- Transaxle. Tighten the flange bolts to 47 ft. lbs. (64 Nm) and the rear engine mounting bolts to 61 ft. lbs. (83 Nm).
- Transaxle mount and bracket. Tighten the bracket bolts to 47 ft. lbs. (64 Nm) and the through bolt to 54 ft. lbs. (74 Nm).
- Right front mount and bracket. Tighten the mount bolts to 33 ft. lbs. (44 Nm) and the bracket bolts to 47 ft. lbs. (64 Nm).
- Clutch housing cover. Tighten the 12mm bolts to 22 ft. lbs. (29 Nm) and the 6mm bolts to 105 inch lbs. (12 Nm).
- Rear engine stiffener. Tighten the 8mm bolts to 18 ft. lbs. (24 Nm)

and the 10mm bolt to 33 ft. lbs. (44 Nm).
- Intermediate shaft. Tighten the bolts to 29 ft. lbs. (39 Nm).
- Axle halfshafts
- Right damper fork. Tighten the pinch bolt to 32 ft. lbs. (43 Nm) and the nut to 47 ft. lbs. (64 Nm).
- Lower ball joints. Tighten the nut to 36–43 ft. lbs. (49–59 Nm).
- Rear driveshaft. Tighten the bolts to 24 ft. lbs. (32 Nm).
- Exhaust front pipe
- HO2S connector
- Splash shield
- VSS connector
- Shift cables and bracket. Tighten the bracket bolts to 20 ft. lbs. (27 Nm).
- Wire harness bracket
- Reverse lamp switch connector
- Transaxle ground cable
- Starter motor. Tighten the bolts to 32 ft. lbs. (44 Nm).
- Clutch slave cylinder and hose bracket
- Air intake assembly
- Negative battery cable

6. Fill the transaxle to the correct level.
7. Start the engine and check for leaks.
8. Check the wheel alignment and adjust as necessary.

Manual Transaxle

2002–03

1. Before servicing the vehicle, refer to the precautions in the beginning of this section.
2. Remove or disconnect the following:
- Negative battery cable
- Air intake assembly
- Transaxle ground cable
- Vehicle Speed Sensor (VSS) connector
- Splash shield
- Shift cables and bracket
- Clutch slave cylinder and hose bracket
- Wire harness bracket
- Water pipe mounting bolt
- Brake booster and EVAP line mounting bolts. Attach an engine support hanger to the head to support the weight of the engine.
- Upper transmission mounting bolts
- Transaxle mount and bracket

3. Separate the inner CV-joints from the transaxle and intermediate shaft and support the axle halfshafts out of the work area with safety wire.

4. Remove or disconnect the following:
- Intermediate shaft
- Right front mount and bracket
- Rear engine mounting bolts
- Rear driveshaft
- Subframe (see the Engine Removal and Installation procedure)
- Clutch housing cover
- Transaxle

5. Installation is the reverse of removal. Observe the following torques:
- Transaxle rear mount and bracket. Tighten the bracket bolts to 40 ft. lbs. (54 Nm) and the through bolt to 47 ft. lbs. (64 Nm)
- Transaxle. Tighten the flange bolts to 47 ft. lbs. (64 Nm)
- Front mount and bracket. Tighten the bolts to 47 ft. lbs. (64 Nm).
- Clutch housing cover. Tighten the bolts to 29 ft. lbs. (39 Nm)
- Subframe: 76 ft. lbs. (98 Nm)

Clutch

ADJUSTMENTS

The CR-V is equipped with a hydraulic clutch system. No adjustment is necessary.

REMOVAL & INSTALLATION

1. Before servicing the vehicle, refer to the precautions in the beginning of this section.
2. Remove or disconnect the following:
- Negative battery cable
- Transaxle
- Pressure plate. Loosen the bolts evenly in a crossing pattern.
- Clutch disc

To install:

3. Install the clutch disc and pressure plate. Tighten the pressure plate bolts in a crossing pattern and in several steps to 19 ft. lbs. (25 Nm).
4. Install or connect the following:
- Transaxle
- Negative battery cable

Hydraulic Clutch System

BLEEDING

1. Before servicing the vehicle, refer to the precautions in the beginning of this section.
2. Attach a hose to the bleeder screw and suspend the other end in a container of clean brake fluid.
3. Open the bleeder screw.

4. Slowly pump the clutch pedal until no more air bubbles appear at the bleeder hose.
5. Tighten the bleeder screw to 70 inch lbs. (8 Nm).
6. Refill the clutch master cylinder as necessary.
7. Check for leaks and proper clutch operation.

Transfer Assembly

REMOVAL & INSTALLATION

2000-01

1. Before servicing the vehicle, refer to the precautions in the beginning of this section.
2. Drain the transaxle fluid.
3. Remove or disconnect the following:
- Negative battery cable
- Heated Oxygen Sensor (HO2S) connector on 2000-01 models
- Exhaust front pipe on 2000-01 models
- Rear driveshaft
- Transfer assembly and bracket

To install:

4. Install the transfer assembly and bracket with a new O-ring seal. Tighten the 10mm bolts to 33 ft. lbs. (44 Nm) and the 8mm bolts to 17 ft. lbs. (24 Nm).
5. Install or connect the following:
- Rear driveshaft. Tighten the bolts to 24 ft. lbs. (32 Nm).
- Exhaust front pipe
- HO2S connector
- Negative battery cable
6. Fill the transaxle to the correct level and check for leaks.

Halfshaft

REMOVAL & INSTALLATION

Front

2000-01

1. Before servicing the vehicle, refer to the precautions in the beginning of this section.
2. Drain the transaxle.
3. Remove or disconnect the following:
- Negative battery cable
- Front wheels
- Damper fork
- Lower ball joint
- Spindle nut

4. Pry the inboard joint from the transaxle or intermediate shaft.
5. Remove the outer CV-joint stub shaft from the hub by tapping the stub shaft with a plastic hammer.

To install:

➡ **Use new circlips, split pins and self-locking nuts for assembly.**

6. Install the outer CV-joint stub shaft into the hub.
7. Install the inner CV-joint to the transaxle or intermediate shaft until the circlip locks in the retaining groove.
8. Install or connect the following:
- Lower ball joint. Tighten the nut to 36-43 ft. lbs. (49-59 Nm).
- Damper fork. Tighten the pinch bolt to 32 ft. lbs. (43 Nm) and the nut to 47 ft. lbs. (64 Nm).
- Spindle nut. Tighten the nut to 181 ft. lbs. (245 Nm).
- Front wheels
- Negative battery cable
9. Fill the transaxle to the correct level and check for leaks.

2002-03

1. Before servicing the vehicle, refer to the precautions in the beginning of this section.
2. Drain the transaxle.
3. Remove or disconnect the following:
- Negative battery cable
- Front wheels
- Stabilizer bar
- Lower ball joint
- Spindle nut
4. On the left side, pry the inboard joint from the case with a prybar.
5. On the right side, drive the inboard shaft off the intermediate shaft with a drift and hammer.
6. Installation is the reverse of removal. Observe the following torques:
- Ball stud nuts: 40 ft. lbs. (54 Nm)
- Stabilizer link nuts: 29 ft. lbs. (39 Nm)
- Spindle nut: 181 ft. lbs. (245 Nm)

Rear

2000-03

1. Before servicing the vehicle, refer to the precautions in the beginning of this section.
2. Drain the differential.
3. Remove or disconnect the following:
- Negative battery cable
- Rear wheels
- Spindle nut
4. Pry the inboard joint from the differential.

5. Remove the outer CV-joint stub shaft from the hub by tapping the stub shaft with a plastic hammer.

To install:

➡**Use new circlips and self-locking nuts for assembly.**

6. Install the outer CV-joint stub shaft into the hub.

7. Install the inner CV-joint to the differential until the circlip locks in the retaining groove.

8. Install or connect the following:
  • Spindle nut. Tighten the nut to 134 ft. lbs. (181 Nm).

• Rear wheels
• Negative battery cable

9. Fill the differential to the correct level and check for leaks.

CV-Joint

OVERHAUL

Front

OUTBOARD JOINT

1. Before servicing the vehicle, refer to the precautions in the beginning of this section.

2. Remove or disconnect the following:
  • Axle halfshaft from the vehicle and place it in a vise
  • Outboard joint boot clamps and push the boot back
  • Outboard joint by driving it off the axle shaft with a brass drift and hammer
  • Outboard joint boot

To install:

➡**Use new circlips and boot clamps for assembly.**

3. Install the outboard joint boot and clamps to the axle shaft.

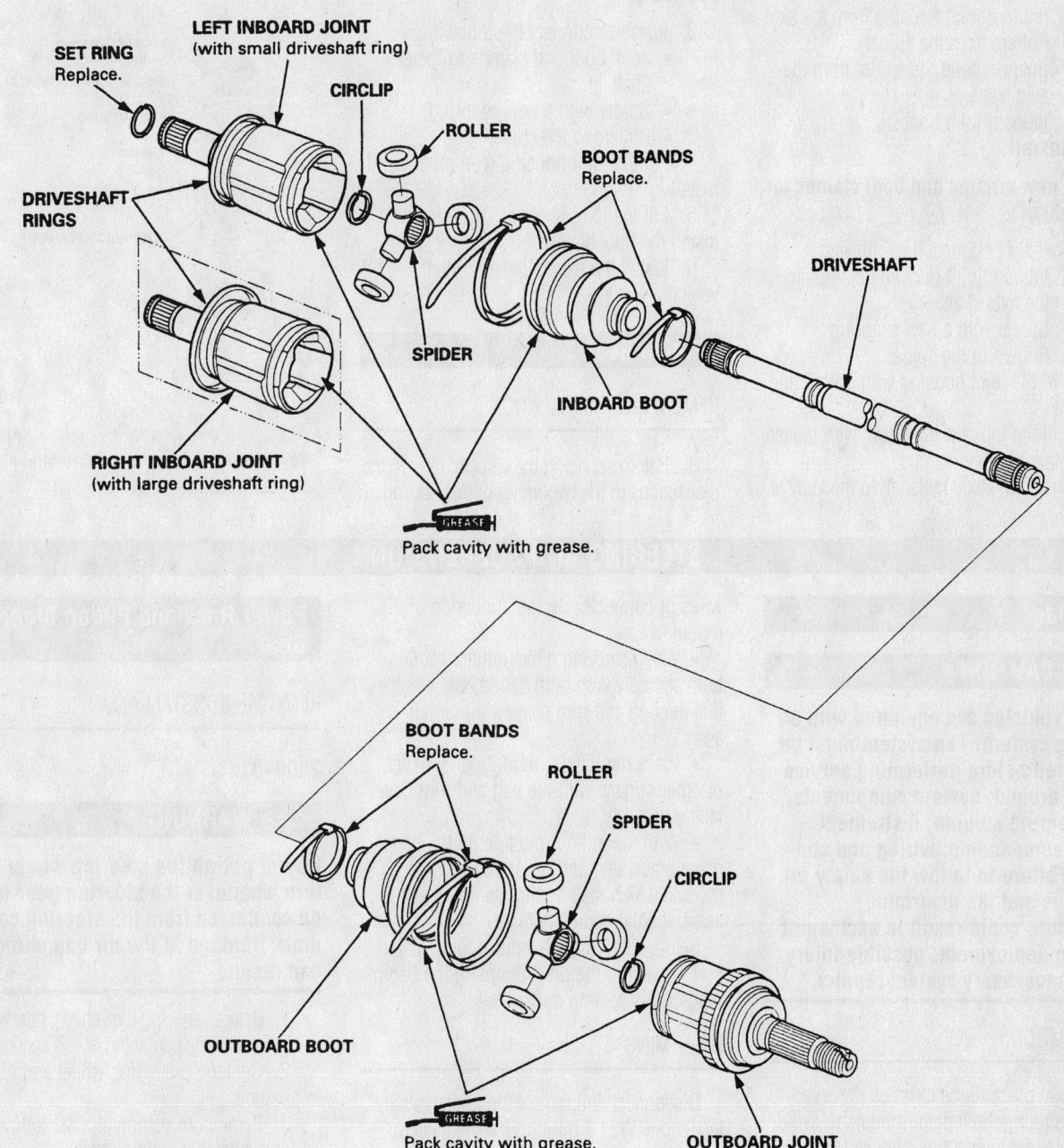

Exploded view of the rear axle—CR-V

9308MG30

4. Fill the outboard joint with grease. Install the outboard joint to the axle shaft. Tap the stub shaft with a brass hammer to seat the circlip.

5. Fill the outboard joint boot with grease and install the boot clamps.

6. Install the axle halfshaft to the vehicle.

INBOARD JOINT

1. Before servicing the vehicle, refer to the precautions in the beginning of this section.

2. Remove or disconnect the following:
- Axle halfshaft from the vehicle.
- Inboard joint boot clamps and push the boot back
- Inboard joint housing from the axle
- Rollers from the spider
- Snapring and the spider from the axle shaft
- Inboard joint boot

To install:

➡ Use new circlips and boot clamps for assembly.

3. Install or connect the following:
- Inboard joint boot and clamps to the axle shaft
- Spider with a new snapring
- Rollers to the spider

4. Fill the joint housing with grease and install it.

5. Fill the inboard joint boot with grease and install the boot clamps.

6. Install the axle halfshaft to the vehicle.

Rear

1. Before servicing the vehicle, refer to the precautions in the beginning of this section.

2. Remove or disconnect the following:
- Axle halfshaft from the vehicle
- Joint boot clamps and push the boot back
- Joint housing from the axle
- Rollers from the spider
- Snapring and the spider from the axle shaft
- Joint boot

To install:

➡ Use new circlips and boot clamps for assembly.

3. Install or connect the following:
- Joint boot and clamps to the axle shaft
- Spider with a new snapring
- Rollers to the spider

4. Fill the joint housing with grease and install it.

5. Fill the joint boot with grease and install the boot clamps.

6. Install the axle halfshaft to the vehicle.

Pinion Seal

REMOVAL & INSTALLATION

1. Before servicing the vehicle, refer to the precautions in the beginning of this section.

2. Remove or disconnect the following:
- Driveshaft
- Companion flange
- Pinion seal

To install:

➡ Use a new locknut and O-ring for assembly.

3. Install or connect the following:
- Pinion seal. Drive the seal square into the bore.
- Companion flange. Tighten the locknut to 87 ft. lbs. (118 Nm).
- Driveshaft. Tighten the flange bolts to 24 ft. lbs. (32 Nm).

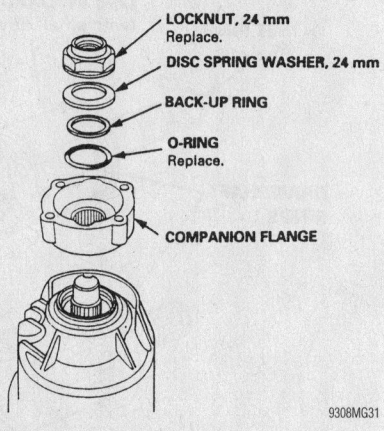

Exploded view of the rear differential pinion components—CR-V

STEERING AND SUSPENSION

Air Bag

✳✳ CAUTION

Some vehicles are equipped with an air bag system. The system must be disarmed before performing service on, or around, system components, the steering column, instrument panel components, wiring and sensors. Failure to follow the safety precautions and the disarming procedure could result in accidental air bag deployment, possible injury and unnecessary system repairs.

PRECAUTIONS

Several precautions must be observed when handling the inflator module to avoid accidental deployment and possible personal injury.
- Never carry the inflator module by the wires or connector on the underside of the module.
- When carrying a live inflator module, hold securely with both hands, and ensure that the bag and trim cover are pointed away.
- Place the inflator module on a bench or other surface with the bag and trim cover facing up.
- With the inflator module on the bench, never place anything on or close to the module which may be thrown in the event of an accidental deployment.

Before servicing the vehicle, also make sure to refer to the precautions in the beginning of this section as well.

DISARMING

Disconnect and isolate the negative battery cable. Wait 3 minutes for the system capacitor to discharge before performing any service.

Power Rack and Pinion Steering Gear

REMOVAL & INSTALLATION

2000–01

✳✳ WARNING

Do not permit the steering wheel to turn whenever the steering gear is disconnected from the steering column. Damage to the air bag wiring can result.

1. Before servicing the vehicle, refer to the precautions in the beginning of this section.

2. Center the steering wheel and lock it in position.

3. Remove or disconnect the following:
- Negative battery cable
- Air bag and steering wheel
- Steering flexible joint

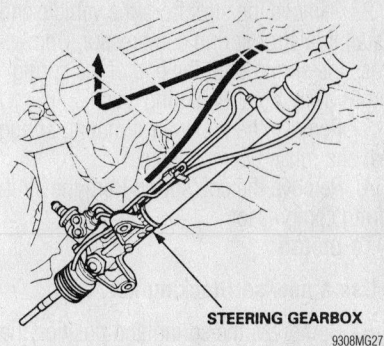

STEERING GEARBOX

9308MG27

Rack and pinion steering gear installation—2000–01

- Front wheels
- Outer tie rod ends
- Heated Oxygen Sensor (HO2S) connector
- Exhaust front pipe
- Transmission shift cable
- Power steering fluid lines

4. For 4 wheel drive vehicles, perform the following:

   a. Remove the rear driveshaft.

   b. Remove the right and left front mounts.

   c. Remove the rear mount and bracket.

   d. Tilt the engine back to lower the transfer assembly output flange about 1.57 inches (40mm).

5. For all vehicles, remove or disconnect the following:

- Stiffener plate
- Steering gear mounting brackets

6. Slide the steering gear to the right until the left end clears the subframe, then slide the gear to the left and out of the vehicle.

To install:

7. Position the steering gear in the vehicle, right end first.

8. Install or connect the following:

- Steering gear mounting brackets. Tighten the bolts to 29 ft. lbs. (39 Nm).
- Stiffener plate. Tighten the plate bolts to 28 ft. lbs. (38 Nm) and the steering bear bolt to 32 ft. lbs. (43 Nm).

➡**Use new self-locking nuts and bolts for engine mount installation.**

9. For 4 wheel drive vehicles, install or connect the following:

- Rear mount and bracket. Tighten the upper bracket bolt to 43 ft. lbs. (59 Nm), the lower bracket bolt to 61 ft. lbs. (83 Nm), the mount attaching bolts to 47 ft. lbs. (64 Nm), and the mount through bolt to 43 ft. lbs. (59 Nm).
- Left and right front mounts. Tighten

the bolts to 33 ft. lbs. (44 Nm) and the nuts to 43 ft. lbs. (59 Nm).

- Rear driveshaft. Tighten the bolts to 24 ft. lbs. (32 Nm).

10. For all vehicles, install or connect the following:

- Power steering fluid lines
- Transmission shift cable
- Exhaust front pipe
- HO2S connector
- Outer tie rod ends
- Front wheels. Position the wheels straight ahead.
- Steering flexible joint. Tighten the pinch bolts to 16 ft. lbs. (22 Nm).
- Steering wheel and air bag
- Negative battery cable

11. Fill the power steering system.

12. Check the wheel alignment and adjust as necessary.

2002

1. Before servicing the vehicle, refer to the precautions in the beginning of this section.

2. Center the steering wheel and lock it in position.

3. Remove or disconnect the following:

- Negative battery cable

- Air bag and steering wheel
- Front wheels
- Driver's side dashboard lower cover and undercover
- Air cleaner housing
- Steering joint bolts
- Tie rod ends
- Steering hoses
- Left side flange bolts
- Mounting brackets

4. Lower the unit so the pinion shaft points outward. Remove the pinion shaft grommet. The steering gear is removed through the driver's side.

5. Installation is the reverse of removal. Observe the following torques:

- Mounting bracket and side flange bolts: 46 ft. lbs. (62 Nm)
- Supply line flare nut: 27 ft. lbs. (37 Nm)
- Tie rod ball stud nuts: 32 ft. lbs. (43 Nm)
- Steering joint bolts: 21 ft. lbs. (28 Nm)

Strut

REMOVAL & INSTALLATION

Front

2000–01

1. Before servicing the vehicle, refer to the precautions in the beginning of this section.

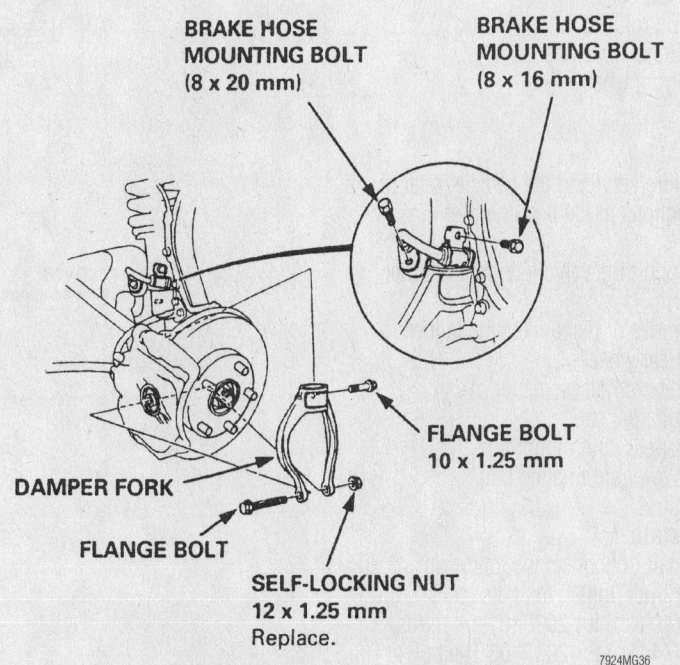

BRAKE HOSE MOUNTING BOLT (8 x 20 mm)

BRAKE HOSE MOUNTING BOLT (8 x 16 mm)

FLANGE BOLT 10 x 1.25 mm

DAMPER FORK

FLANGE BOLT

SELF-LOCKING NUT 12 x 1.25 mm Replace.

7924MG36

Identification of some of the front suspension components

2. Remove or disconnect the following:
- Front wheel
- Brake hose retainer
- Damper fork
- Strut

To install:

➡**Use new self-locking fasteners for assembly.**

3. Install or connect the following:
- Strut. Tighten the mounting nuts to 43 ft. lbs. (59 Nm).
- Damper fork. Tighten the pinch bolt to 32 ft. lbs. (43 Nm) and the lower bolt to 47 ft. lbs. (64 Nm).
- Brake hose retainer
- Front wheel

2002–03

1. Before servicing the vehicle, refer to the precautions in the beginning of this section.
2. Remove or disconnect the following:
- Front wheel
- Tie rod end
- Brake hose retainer
- ABS sensor
- Strut

To install:

➡**Use new self-locking fasteners for assembly.**

3. Install or connect the following:
- Strut. Tighten the upper mounting nuts to 33 ft. lbs. (44 Nm).
- Tighten the pinch bolts to 116 ft. lbs. (157 Nm)
- ABS sensor
- Tie rod end
- Brake hose retainer
- Front wheel

Rear

1. Before servicing the vehicle, refer to the precautions in the beginning of this section.
2. Support the vehicle under the lower control arm.
3. Remove or disconnect the following:
- Rear wheel
- Interior access panel
- Damper cap
- Upper strut mount nuts
- Lower strut flange bolt
- Strut

To install:

4. Install or connect the following:
- Strut. Tighten the nuts to 36 ft. lbs. (49 Nm) for 2000–01; 54 ft. lbs. (74 Nm) for 2002–03 The bolt to 40 ft. lbs. (54 Nm) for 2000–01; 69

ft. lbs. (93 Nm).
- Damper cap
- Interior access panel
- Rear wheel

Coil Spring

REMOVAL & INSTALLATION

Front

1. Before servicing the vehicle, refer to the precautions in the beginning of this section.

2. Remove the strut from the vehicle and install in a strut spring compressor. Compress the spring until the end of the spring comes away from the spring seat.
3. Remove the upper strut mount, spring seat and related components.
4. Remove the coil spring from the strut spring compressor.

To install:

➡**Use a new self-locking nut.**

5. Compress the spring and position the strut so that the end of the spring aligns with the notch in the spring seat.

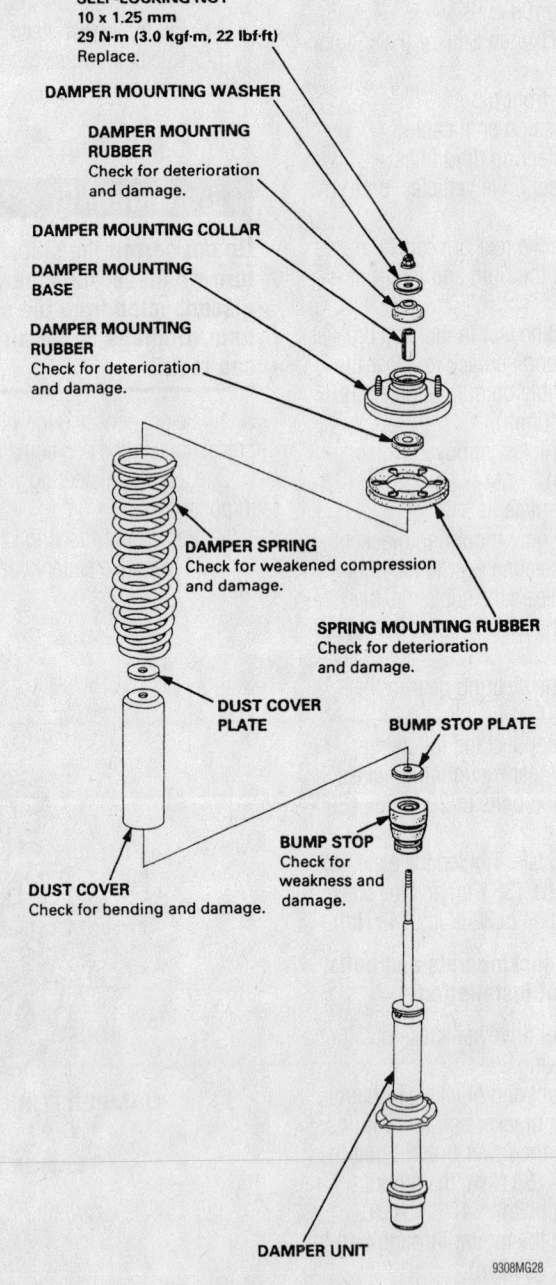

SELF-LOCKING NUT
10 x 1.25 mm
29 N·m (3.0 kgf·m, 22 lbf·ft)
Replace.

DAMPER MOUNTING WASHER

DAMPER MOUNTING RUBBER
Check for deterioration and damage.

DAMPER MOUNTING COLLAR

DAMPER MOUNTING BASE

DAMPER MOUNTING RUBBER
Check for deterioration and damage.

DAMPER SPRING
Check for weakened compression and damage.

SPRING MOUNTING RUBBER
Check for deterioration and damage.

DUST COVER PLATE

BUMP STOP PLATE

DUST COVER
Check for bending and damage.

BUMP STOP
Check for weakness and damage.

DAMPER UNIT

9308MG28

Front strut and spring exploded view—2000–01 shown

6. Install the upper strut mounting components and tighten the nut to 22 ft. lbs. (29 Nm) for 2000–01; 33 ft. lbs. (44 Nm) for 2002–03.

7. Install the strut to the vehicle.

8. Check the wheel alignment and adjust as necessary.

Rear

1. Before servicing the vehicle, refer to the precautions in the beginning of this section.

2. Remove the strut from the vehicle and install in a strut spring compressor. Compress the spring until the end of the spring comes away from the spring seat.

3. Remove or disconnect the following:
- Upper strut mount, spring seat and related components
- Coil spring from the strut spring compressor

To install:

➡**Use a new self-locking nut.**

4. Compress the spring and position the strut so that the end of the spring aligns with the notch in the spring seat.

5. Install or connect the following:
- Upper strut mounting components and tighten the nut to 22 ft. lbs. (29 Nm).
- Strut to the vehicle

6. Check the wheel alignment and adjust as necessary.

Upper Ball Joint

REMOVAL & INSTALLATION

The upper ball joints are replaced with the upper control arms as an assembly.

Lower Ball Joint

REMOVAL & INSTALLATION

2000–01

1. Before servicing the vehicle, refer to the precautions in the beginning of this section.

2. Remove or disconnect the following:
- Front wheel
- Spindle nut
- Brake hose bracket
- Brake caliper and rotor
- Wheel speed sensor, if equipped
- Outer tie rod end
- Upper and lower ball joints

- Steering knuckle
- Lower ball joint boot and set ring

3. Press the lower ball joint out of the steering knuckle.

To install:

➡**Use new ball joint nuts, split pins, and a new spindle nut for assembly.**

4. Press the lower ball joint into the steering knuckle.

5. Install or connect the following:
- Lower ball joint boot and set ring
- Steering knuckle. Tighten the upper ball joint nut to 29–35 ft. lbs. (39–47 Nm) and the lower ball joint nut to 36–43 ft. lbs. (49–59 Nm).
- Outer tie rod end. Tighten the nut to 32 ft. lbs. (43 Nm).
- Wheel speed sensor, if equipped
- Brake caliper and rotor. Tighten the caliper bracket bolts to 80 ft. lbs. (108 Nm).
- Brake hose bracket
- Spindle nut. Tighten the nut to 181 ft. lbs. (245 Nm).
- Front wheel

6. Check the wheel alignment and adjust as necessary.

2002–03

The ball joint is not replaceable.

Upper Control Arm

REMOVAL & INSTALLATION

1. Before servicing the vehicle, refer to the precautions in the beginning of this section.

2. Support the lower control arm assembly with a floor jack.

3. Remove or disconnect the following:
- Upper ball joint
- Inner control arm flange bolts.
- Upper control arm

To install:

➡**Use new self-locking nuts for assembly.**

4. Install the upper control arm. Tighten the ball joint nut to 29–35 ft. lbs. (39–47 Nm) and the inner flange bolts to 40 ft. lbs. (54 Nm).

CONTROL ARM BUSHING REPLACEMENT

The upper control arm bushings are serviced with the upper control arm as an assembly.

Lower Control Arm

REMOVAL & INSTALLATION

2000–01

1. Before servicing the vehicle, refer to the precautions in the beginning of this section.

2. Remove or disconnect the following:
- Front wheel
- Lower ball joint
- Damper fork lower bolt
- Stabilizer bar link
- Front inner flange bolt
- Rear bushing bracket bolts
- Lower control arm

To install:

➡**Use new self-locking nuts and split pins for assembly.**

3. Install or connect the following:
- Lower control arm. Tighten the front flange bolt to 76 ft. lbs. (103 Nm).
- Rear bushing bracket. Tighten the bolts to 66 ft. lbs. (89 Nm).
- Stabilizer bar link. Tighten the nut to 22 ft. lbs. (29 Nm).
- Damper fork lower bolt. Tighten the nut to 47 ft. lbs. (64 Nm).
- Lower ball joint. Tighten the nut to 36–43 ft. lbs. (49–59 Nm).
- Front wheel

4. Check the wheel alignment and adjust as necessary.

2002–03

1. Before servicing the vehicle, refer to the precautions in the beginning of this section.

2. Remove or disconnect the following:
- Front wheel
- Stabilizer link
- Lower arm from the knuckle
- Lower arm

3. Installation is the reverse of removal. Observe the following torques:
- Lower arm bolts: 61 ft. lbs. (83 nm)
- Ball stud nut: 51 ft. lbs. (69 Nm)
- Stabilizer link: 29 ft. lbs. (39 Nm)

CONTROL ARM BUSHING REPLACEMENT

The lower control arm front inner bushing and the damper fork bushing are serviced with the control arm as an assembly.

REAR INNER BUSHING

1. Before servicing the vehicle, refer to the precautions in the beginning of this section.

2. Remove or disconnect the following:
- Front wheel
- Rear bushing bracket
- Rear bushing

To install:

➡**Use a new self-locking nut for assembly.**

3. Install or connect the following:
- Rear bushing. Tighten the nut to 61 ft. lbs. (83 Nm).
- Rear bushing bracket. Tighten the bolts to 66 ft. lbs. (89 Nm).
- Front wheel

4. Check the wheel alignment and adjust as necessary.

Wheel Bearings

ADJUSTMENT

The wheel bearings are sealed units and are not adjustable.

REMOVAL & INSTALLATION

Front

2000–01

1. Before servicing the vehicle, refer to the precautions in the beginning of this section.
2. Remove or disconnect the following:
- Front wheel
- Spindle nut
- Brake hose bracket
- Brake caliper and rotor
- Wheel speed sensor, if equipped
- Outer tie rod end
- Upper and lower ball joints
- Steering knuckle

3. Press the hub out of the wheel bearing.
4. Remove the splash guard.
5. Remove the snapring and press the wheel bearing out of the steering knuckle.
6. If necessary, press the inner bearing race off of the hub.

To install:

➡**Use new ball joint nuts, split pins, snapring and a new spindle nut for assembly.**

7. Press the bearing into the steering knuckle and install the snapring.
8. Install the splash guard.
9. Press the hub into the bearing.
10. Install or connect the following:
- Steering knuckle. Tighten the upper ball joint nut to 29–35 ft. lbs. (39–47 Nm) and the lower ball

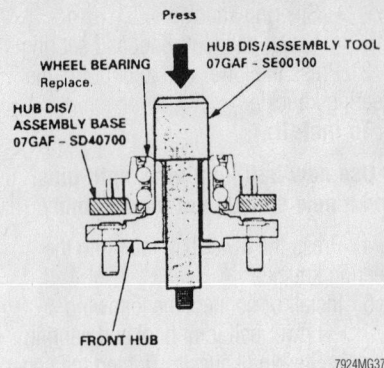

Removing the hub from the wheel bearing using the disassembly tools

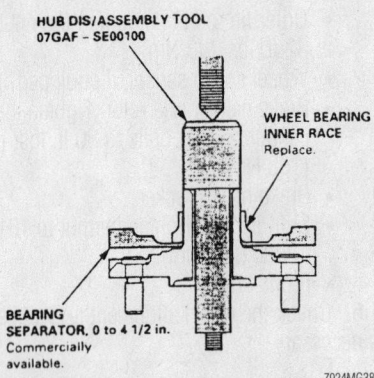

Pressing out the wheel bearing inner race

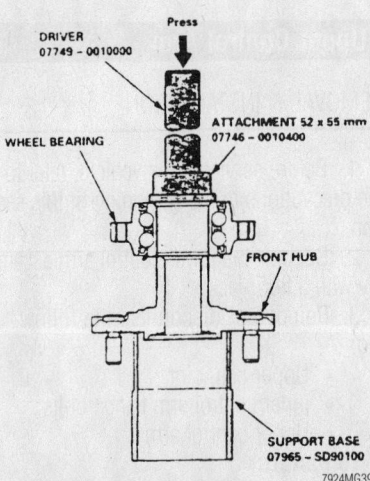

Utilizing the hub support base and driving attachment tools to install the new wheel bearing

joint nut to 36–43 ft. lbs. (49–59 Nm).
- Outer tie rod end. Tighten the nut to 32 ft. lbs. (43 Nm).
- Wheel speed sensor, if equipped
- Brake caliper and rotor. Tighten the caliper bracket bolts to 80 ft. lbs. (108 Nm).

- Brake hose bracket
- Spindle nut. Tighten the nut to 181 ft. lbs. (245 Nm).
- Front wheel

11. Check the wheel alignment and adjust as necessary.

2002–03

1. Before servicing the vehicle, refer to the precautions in the beginning of this section.
2. Remove or disconnect the following:
- Front wheel
- Spindle nut
- Brake caliper and rotor. Forcing screws are needed to remove the rotor.
- Brake hose bracket
- ABS sensor
- Stabilizer link
- Lower arm from the knuckle
- Strut-to-knuckle bolts
- Steering hub/knuckle assembly

3. Press the hub from the knuckle. The bearings and races can now be pressed out and replaced.

➡**With ABS, install the bearing with the magnetic encoder (brown color) toward the inside of the knuckle.**

4. Observe the following torques:
- Strut bolts: 116 ft. lbs. (157 Nm)
- Ball stud nuts: 51 ft. lbs. (69 Nm)
- Stabilizer bar link: 29 ft. lbs. (39 Nm)

Rear

2000–01

1. Before servicing the vehicle, refer to the precautions in the beginning of this section.
2. Remove or disconnect the following:
- Rear wheel
- Brake drum
- Spindle nut
- Brake shoes and parking brake cable
- Brake fluid line
- Wheel sensor, if equipped
- Wheel bearing flange bolts
- Hub, backing plate and bearing assembly

3. Press the hub out of the wheel bearing.
4. If necessary, press the inner bearing race off of the hub.
5. Remove the backing plate from the bearing assembly.

To install:

➡**Use a new spindle nut for assembly.**

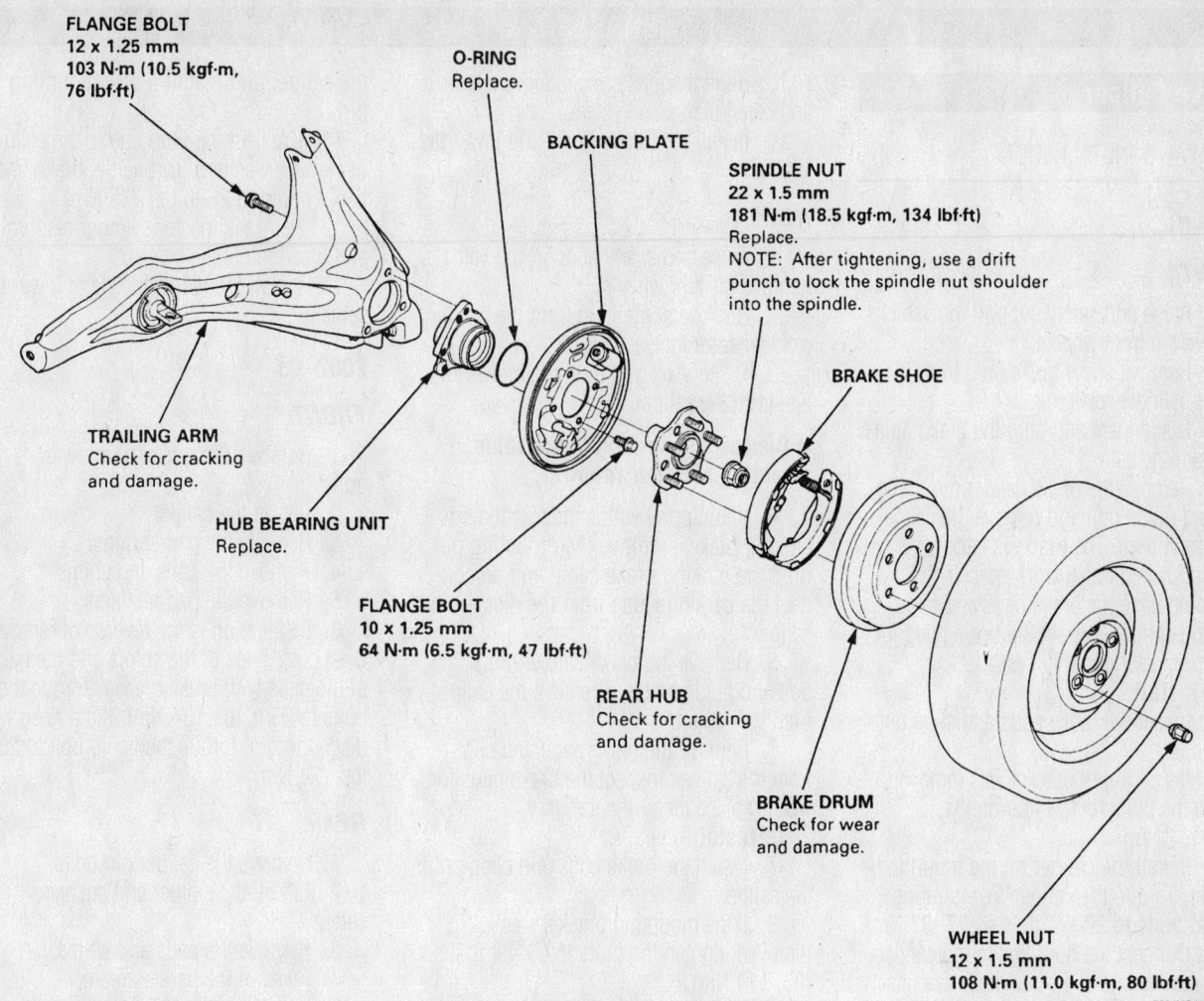

FLANGE BOLT
12 x 1.25 mm
103 N·m (10.5 kgf·m, 76 lbf·ft)

O-RING
Replace.

BACKING PLATE

SPINDLE NUT
22 x 1.5 mm
181 N·m (18.5 kgf·m, 134 lbf·ft)
Replace.
NOTE: After tightening, use a drift punch to lock the spindle nut shoulder into the spindle.

BRAKE SHOE

TRAILING ARM
Check for cracking and damage.

HUB BEARING UNIT
Replace.

FLANGE BOLT
10 x 1.25 mm
64 N·m (6.5 kgf·m, 47 lbf·ft)

REAR HUB
Check for cracking and damage.

BRAKE DRUM
Check for wear and damage.

WHEEL NUT
12 x 1.5 mm
108 N·m (11.0 kgf·m, 80 lbf·ft)

7924MG43

Exploded view of the rear hub and wheel bearing components—CR-V

6. Use a new O-ring and install the backing plate to the bearing assembly. Tighten the bolts to 47 ft. lbs. (64 Nm).

7. Press the hub into the bearing assembly.

8. Install or connect the following:
- Hub, backing plate and bearing assembly. Tighten the flange bolts to 76 ft. lbs. (103 Nm).
- Brake fluid line
- Parking brake cable and brake shoes
- Wheel sensor, if equipped
- Spindle nut. Tighten the nut to 134 ft. lbs. (181 Nm).
- Brake drum
- Rear wheel

9. Bleed the brake system.

10. Before servicing the vehicle, refer to the precautions in the beginning of this section.

11. Remove or disconnect the following:
- Rear wheel
- Brake caliper
- Rotor
- Spindle nut
- Axle shaft (2wd)
- Parking brake shoes
- Parking brake cable
- Wheel sensor, if equipped

12. Support the trailing arm.

13. Remove or disconnect the following:
- Upper arm from the knuckle

14. Matchmark the trailing arm cam

adjusting bolt and cam. Remove the bolt. Discard the nut.

15. Remove the flange bolt.

16. Remove the knuckle assembly.

17. Press the hub from the knuckle. The bearings and races can now be pressed out and replaced.

➡ **With ABS, install the bearing with the magnetic encoder (brown color) toward the inside of the knuckle.**

18. Observe the following torques:
- Flange bolt: 69 ft. lbs. (93 Nm)
- Cam bolts: 43 ft. lbs. (59 Nm)
- Spindle nut: 134 ft. lbs. (181 Nm)
- Caliper mounting bolts: 41 ft. lbs. (55 Nm)

BRAKES

Brake Caliper

REMOVAL & INSTALLATION

2000–01

FRONT

1. Raise and safely support the vehicle. Remove the front wheels.
2. Remove some brake fluid from the master cylinder reservoir.
3. Disconnect and plug the brake fluid line from the caliper.
4. Remove the brake caliper mounting bolt and guide bolt and remove the caliper from the mount. The brackets can be remove for additional work space.
5. Remove the brake pads and clips from the caliper. Inspect the brake pads for wear; replace them, if necessary.

To install:

6. Install the brake pads and clips onto the caliper.
7. If the caliper bracket was removed, tighten the bolts to 103–126 ft. lbs. (139–171 Nm).
8. Install the caliper on the mounting bracket. Torque the caliper-to-mounting bracket bolts to 20–27 ft. lbs. (27–37 Nm).
9. Connect the fluid line to the caliper using new washers. Torque the brake line banjo fitting to 26 ft. lbs. (35 Nm).

✳✳ WARNING

Be sure the hook end of the flexible brake line is positioned in the anti-rotation cavity.

10. Refill the master cylinder reservoir and bleed the brake system.
11. Install the front wheels and lower the vehicle.

REAR

1. Raise and safely support the vehicle. Remove the rear wheels.
2. Remove some fluid from the master cylinder reservoir.
3. Disconnect and plug the brake fluid line from the caliper.

➡ **Discard the parking brake cable mounting pin after removal.**

4. If equipped with caliper actuated parking brakes; remove the mounting pin from the parking brake cable and disconnect the parking cable from the disc caliper.
5. Remove the brake caliper mounting bolt and guide bolt and remove the caliper from the mount.
6. Remove the brake pads and clips from the caliper. Inspect the brake pads for wear; replace them, if necessary.

To install:

7. Install the brake pads and clips onto the caliper.
8. If the mounting bracket was removed, tighten the bolts to 69–84 ft. lbs. (93–114 Nm).
9. Install the caliper on the mounting bracket. Torque the caliper-to-mounting bracket bolts to 12–17 ft. lbs. (16–24 Nm), or 32 ft. lbs. (43 Nm) on vehicles with shoe-type parking brakes.
10. Connect the parking brake cable to

the caliper and install a new mounting pin.
11. Connect the fluid line to the caliper using new washers. Torque the brake line banjo fitting to 26 ft. lbs. (35 Nm).
12. Refill the master cylinder reservoir and bleed the brake system.
13. Install the rear wheels and lower the vehicle.

2002–03

FRONT

1. Remove the upper and lower bolts.
2. Lift off the caliper.
3. Remove the pad springs.
4. Remove the pads and shims.
5. Remove the pad retainers.
6. Installation is the reverse of removal. Coat both sides of the shims and the backs of the pads with brake grease. Torque the bolts to 25 ft. lbs. (34 Nm). If the hose was disconnected, torque the banjo bolt to 25 ft. lbs. (34 Nm).

REAR

1. Remove the caliper pin bolts.
2. Lift off the caliper and suspend it safely.
3. Remove the pads and shims.
4. Remove the pad retainers.
5. Installation is the reverse of removal. Coat both sides of the shims and the backs of the pads with brake grease. Torque the bolts to 16 ft. lbs. (22 Nm). If the hose was disconnected, torque the banjo bolt to 16 ft. lbs. (22 Nm).

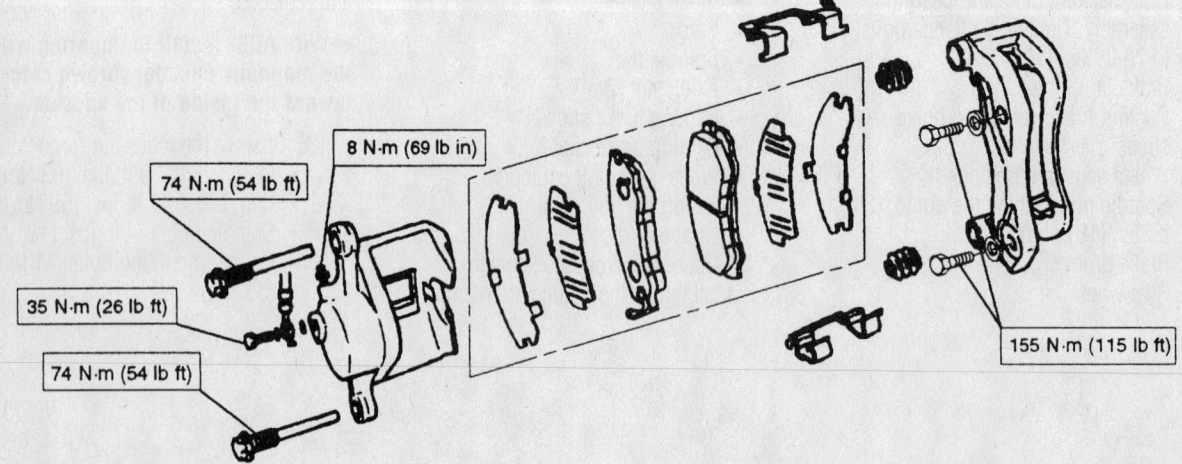

8 N·m (69 lb in)

74 N·m (54 lb ft)

35 N·m (26 lb ft)

74 N·m (54 lb ft)

155 N·m (115 lb ft)

93026G54

Exploded view of the front caliper components

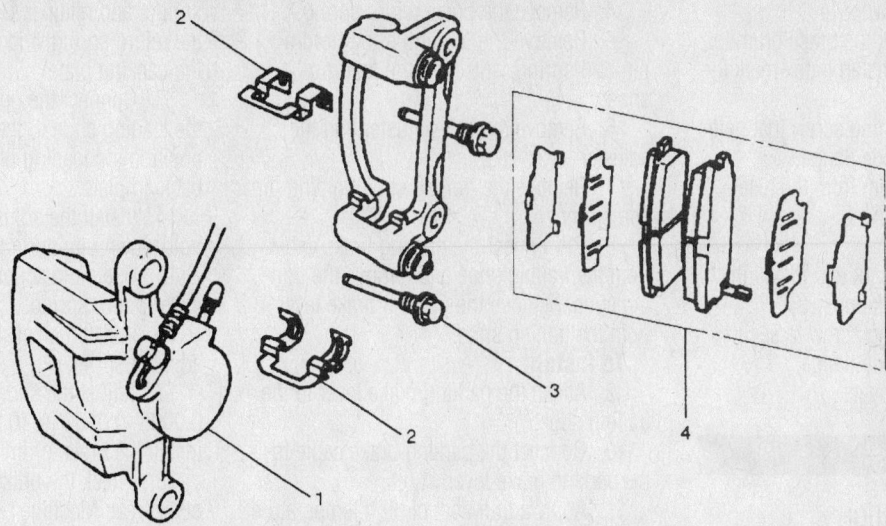

(1) Caliper Assembly
(2) Clip
(3) Lock Bolt
(4) Pad Assembly

93026G55

Exploded view of the rear caliper components

Disc Brake Pads

REMOVAL & INSTALLATION

2000–01

FRONT

Most disc brake pads are equipped with wear indicators. If a squealing noise occurs from the brakes while driving, check the pad wear indicator plate. If there is evidence of the indicator plate contacting the brake disc, the brake pad should be replaced.

1. Remove ½ of the volume of brake fluid from the master cylinder to prevent overflow when the caliper piston is compressed.
2. Raise and safely support the vehicle.
3. Remove the wheel and tire assemblies.
4. Remove the brake caliper without disconnecting the brake line. Support the caliper with a length of wire. Do not let the caliper hang from the brake hose.

➡On some disc brake systems it is not necessary to remove the caliper when installing new brake pads. Remove the lower slide bolt and rotate the caliper upward to remove the pads.

5. Remove the brake pads and shims. Inspect the brake rotor and machine or replace as necessary. Check the minimum thickness (specification is cast into the rotor) before machining.

To install:

6. Use a suitable tool to push the caliper piston into its bore.

7. Apply a thin coat of grease to the rear face of the brake pad and install the shim. Install the brake pads.
8. Install the calipers. Lubricate the caliper bolts and boots. Tighten the caliper mounting bolts to 24 ft. lbs. (33 Nm).
9. Install the wheel and tire assemblies and lower the vehicle.
10. Apply the brakes several times to seat the pads before moving the vehicle. Check the fluid in the master cylinder and add as necessary.

REAR

1. Raise and safely support the vehicle. Remove the rear wheels.
2. Remove the brake caliper mounting bolts and remove the caliper without disconnecting the brake fluid line. Support the caliper so it does not hang on the brake line.
3. Remove the brake pads and retaining clips from the caliper.
4. If equipped with caliper-activated parking brakes; use tool J-37617 or equivalent, to rotate the piston clockwise until it retracts into the bore. Align the notches of the piston face so the centerline of the notches is perpendicular to the centerline of the mounting bosses.

To install:

5. Install the new brake pads and clips in the caliper and install the caliper in the mounting bracket.
6. Tighten the caliper mounting bolts to 12–17 ft. lbs. (16–24 Nm), or 32 ft. lbs. (43 Nm) on vehicles with shoe-type parking brakes.

7. Install the rear wheels. Check the brake fluid level.
8. Pump the brake pedal until pressure is felt before moving the vehicle.

2002–03

FRONT

1. Remove the lower bolt.
2. Pivot the caliper up and hold the pads.
3. Remove the pad springs.
4. Remove the pads and shims.
5. Remove the pad retainers.
6. Installation is the reverse of removal. Coat both sides of the shims and the backs of the pads with brake grease. Torque the lower bolt to 25 ft. lbs. (34 Nm).

REAR

1. Remove the caliper pin bolts.
2. Lift off the caliper and suspend it safely.
3. Remove the pads and shims.
4. Remove the pad retainers.
5. Installation is the reverse of removal. Coat both sides of the shims and the backs of the pads with brake grease. Torque the bolts to 16 ft. lbs. (22 Nm).

Brake Drums

REMOVAL & INSTALLATION

1. Raise and safely support the vehicle. Release the parking brake.

2. Remove the rear wheels.

3. Use chalk to mark the brake drum to one of the wheel studs as an index mark for reinstallation.

4. Remove the retaining screw that holds the brake drum to the axle flange.

5. Pull the brake drum from the axle flange.

To install:

6. Align the index mark and install the brake drum to the axle flange.

7. Install the retaining screw to secure the brake drum to the axle flange.

8. Install the rear wheels.

Rear Brake Shoes

REMOVAL & INSTALLATION

1. Raise and safely support the vehicle.
2. Remove the rear wheels.
3. Remove the brake drums.

4. Remove the brake return springs.

5. Remove the leading shoe holding pin and spring, and then the leading shoe.

6. Remove the self-adjuster and the adjuster lever.

7. Remove the trailing shoe holding pin and spring.

8. Disconnect the parking brake cable from the trailing shoe and remove the trailing shoe. Remove the parking brake lever from the trailing shoe.

To install:

9. Attach the parking brake lever to the trailing shoe.

10. Connect the parking brake cable to the parking brake lever.

11. Apply a thin coat of high temperature grease to the shoe contact points on the brake backing plate contact surface (B), and self-adjuster (D).

12. Position the trailing shoe on the backing plate and install the hold-down pin,

spring, and retainer. Be careful not to stretch the return spring when fitting the shoes onto the backing plate.

13. Connect the upper return spring and the leading shoe to the trailing shoe and position the leading brake shoe on the backing plate.

14. Install the adjuster assembly and the hold-down pin, spring, and retainer.

15. Use a brake spring tool to install the lower return spring.

16. Install the self-adjuster lever and adjuster spring.

17. Adjust the shoe-to-drum clearance to 0.0098–0.0157 in. (0.25–0.40mm) and install the brake drum.

18. Check the brake drum for scoring or other wear. Machine or replace as necessary. Check the maximum brake drum diameter specification when machining.

19. Install the rear wheels. Lower the vehicle.

20. Road-test the vehicle.

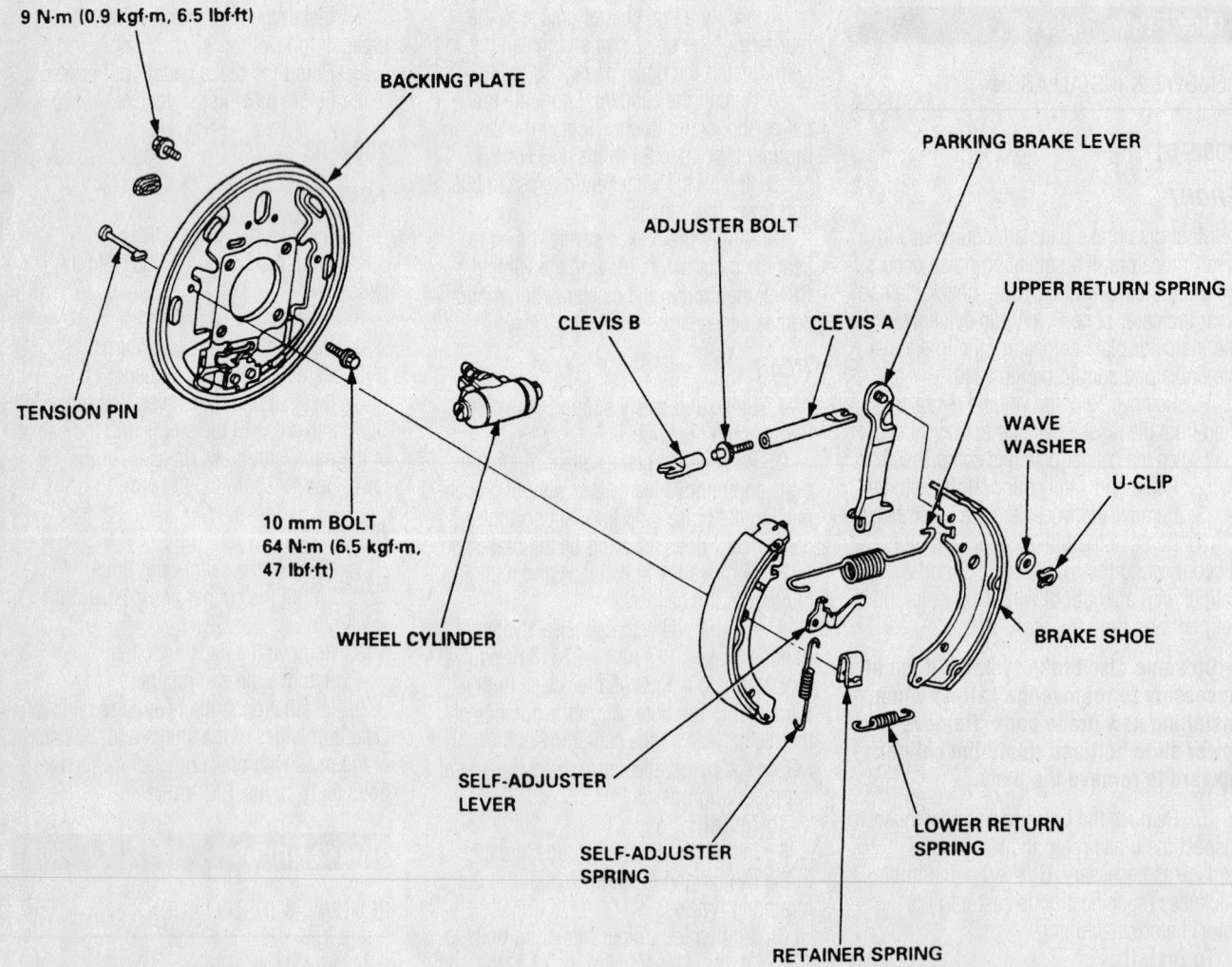

9 N·m (0.9 kgf·m, 6.5 lbf·ft)

BACKING PLATE

PARKING BRAKE LEVER

ADJUSTER BOLT

UPPER RETURN SPRING

CLEVIS B   CLEVIS A

TENSION PIN

WAVE WASHER

U-CLIP

10 mm BOLT
64 N·m (6.5 kgf·m, 47 lbf·ft)

WHEEL CYLINDER

BRAKE SHOE

SELF-ADJUSTER LEVER

SELF-ADJUSTER SPRING

LOWER RETURN SPRING

RETAINER SPRING

93026G59

Exploded view of the rear drum brakes

HONDA

Element

SPECIFICATION CHARTS

ENGINE AND VEHICLE IDENTIFICATION CHART

		Engine Code					Model Year	
Code	Liters (cc)	Cu. In.	Cyl.	Fuel Sys.	Engine Type	Eng. Mfg.	Code ①	Year
K24A1	2.4 (2354)	144	4	SMFI	DOHC	Honda	3	2003
							4	2004

DOHC: Double Overhead Cam

SMFI: Sequential Multi-port Fuel Injection

① 10th position of VIN

NA: Not Avalaible

42356-ELEM-C01

GENERAL ENGINE SPECIFICATIONS

Year	Model	Engine Displacement Liters (cc)	Engine ID/VIN	Fuel System Type	Net Horsepower @ rpm	Net Torque @ rpm (ft. lbs.)	Bore x Stroke (in.)	Compression Ratio	Oil Pressure @ rpm
2003-04	Element	2.4 (2354)	K24A1	SMFI	160@5500	161@4500	3.42x3.90	9.7:1	44@3000

NA: Not Avalaible

SMFI: Sequential Multi-port Fuel Injection

42356-ELEM-C02

ENGINE TUNE-UP SPECIFICATIONS

Year	Engine Displacement Liters (cc)	Engine ID/VIN	Spark Plug Gap (in.)	Ignition Timing (deg.) MT	Ignition Timing (deg.) AT	Fuel Pump (psi)	Idle Speed (rpm) MT	Idle Speed (rpm) AT	Valve Clearance (in.) In.	Valve Clearance (in.) Ex.
2003-04	2.4 (2354)	K24A1	0.039-0.043	6-10B	6-10B	48-55	650-750	650-750	NA	NA

NOTE: The Vehicle Emission Control Information label often reflects changes made during production and must be used if they differ from this chart.

NOTE: The fuel pressure readings are given with the vacuum hose connected to the regulator and the engine running

B: Before top dead center

HYD: Hydraulic

42356-ELEM-C03

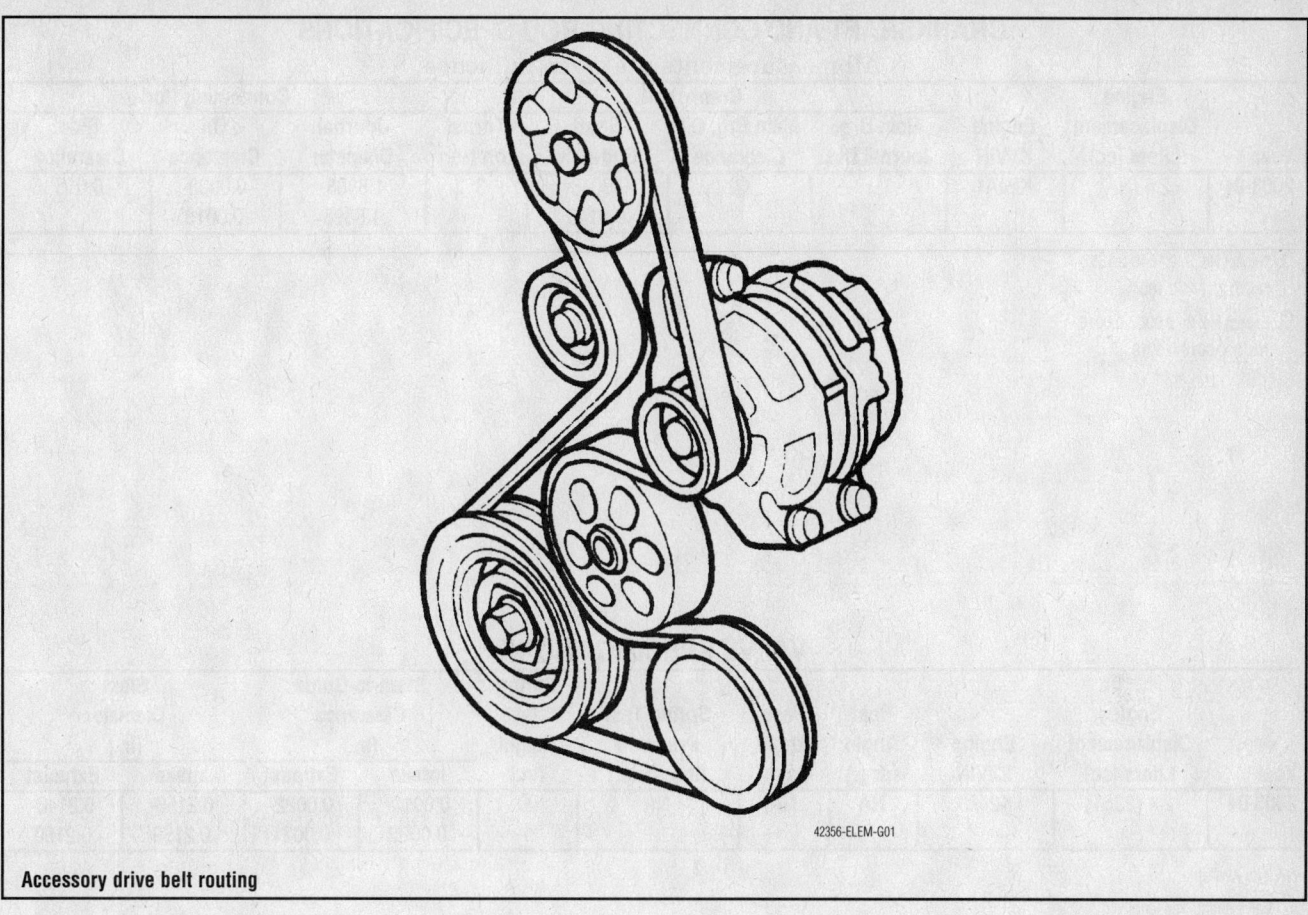

42356-ELEM-G01

Accessory drive belt routing

CAPACITIES

Year	Model	Engine Displacement Liters (cc)	Engine ID/VIN	Engine Oil with Filter (qts.)	Transmission (pts.) 5-Spd	Auto.	Transfer Case (pts.)	Drive Axle Front (pts.)	Rear (pts.)	Fuel Tank (gal.)	Cooling System (qts.)
2003-04	Element	2.4 (2354)	K24A1	4.4	4.0	①	②	②	2.2	15.9	③

NOTE: All capacities are approximate. Add fluid gradually and check to be sure a proper fluid level is obtained.

① 2WD: 6.6 pts. for fluid change, 15.2 for overhaul
  4WD: 6.3 pts. for fluid change, 13.8 for overhaul

② Included in transaxle refill figure

③ Manual trans: 7.6 qts
  Auto trans: 7.5 qts.

42356-ELEM-C04

CRANKSHAFT AND CONNECTING ROD SPECIFICATIONS
All measurements are given in inches

Year	Engine Displacement Liters (cc)	Engine ID/VIN	Crankshaft				Connecting Rod		
			Main Brg. Journal Dia.	Main Brg. Oil Clearance	Shaft End-play	Thrust on No.	Journal Diameter	Oil Clearance	Side Clearance
2003-04	2.4 (NA)	K24A1	①	②	0.0040-0.0140	3	1.8888-1.8898	0.0008-0.0019	0.016

① Except No. 3: 2.1648-2.1657
  No. 3: 2.1644-2.1654

② Except No. 3: 0.0007-0.0016
  No. 3: 0.0010-0.0019

42356-ELEM-C05

VALVE SPECIFICATIONS

Year	Engine Displacement Liters (cc)	Engine ID/VIN	Seat Angle (deg.)	Face Angle (deg.)	Spring Test Pressure (lbs. @ in.)	Spring Installed Height (in.)	Stem-to-Guide Clearance (in.)		Stem Diameter (in.)	
							Intake	Exhaust	Intake	Exhaust
2003-04	2.4 (2354)	K24A1	NA	NA	NA	①	0.0012-0.0022	0.0022-0.0031	0.2156-0.2159	0.2146-0.2150

NA: Not Available

① Valve spring free length:
  Intake: 1.668 in.
  Exhaust: 1.745 in.

42356-ELEM-C06

PISTON AND RING SPECIFICATIONS
All measurements are given in inches

Year	Engine Displacement Liters (cc)	Engine ID/VIN	Piston Clearance	Ring Gap			Ring Side Clearance		
				Top Compression	Bottom Compression	Oil Control	Top Compression	Bottom Compression	Oil Control
2003-04	2.4 (2354)	K24A1	0.0008-0.0016	0.0080-0.0140	0.0160-0.0220	0.0080-0.0280	0.0018-0.0028	0.0020-0.0030	NA

NA: Not Applicable

42356-ELEM-C07

TORQUE SPECIFICATIONS
All readings in ft. lbs.

| Year | Engine Displacement Liters (cc) | Engine ID/VIN | Cylinder Head Bolts | Main Bearing Bolts | Rod Bearing Bolts | Crankshaft Damper Bolts | Flywheel Bolts | Manifold | | Spark Plugs | Lug Nut |
								Intake	Exhaust		
2003-04	2.4 (NA)	K24A1	①	②	③	181	76	16	33	13	80

NOTE: Dip main bearing bolts and crankshaft damper bolt in clean engine oil prior to tightening.

① Step 1: 29 ft. lbs.
　Step 2: +90 degrees
　Step 3: +90 degrees
　Step 4: NEW BOLT ONLY +90 degrees
② 22 ft. lbs. +56 degrees
③ 14 ft. lbs. +90 degrees

42356-ELEM-C08

TIRE, WHEEL AND BALL JOINT SPECIFICATIONS

| Year | Model | OEM Tires | | Tire Pressures (psi) | | Wheel Size | Ball Joint Inspection |
		Standard	Optional	Front	Rear		
2003-04	Element	P215/70R16	None	26	26	6JJ	NS

OEM: Original Equipment Manufacturer
PSI: Pounds Per Square Inch
NS: Not specified by manufacturer

42356-ELEM-C09

BRAKE SPECIFICATIONS
HONDA ELEMENT
All measurements in inches unless noted

| Year | Model | | Brake Disc | | | Brake Drum Diameter | | | Minimum Lining Thickness | | Brake Caliper | |
			Original Thickness	Minimum Thickness	Maximum Runout	Original Inside Diameter	Max. Wear Limit	Maximum Machine Diameter	Front	Rear	Bracket Bolts (ft. lbs.)	Mounting Bolts (ft. lbs.)
2003-04	Element	F	0.910	0.830	0.004	—	—	—	0.060	—	80	25
		R	0.350	0.280	0.004	—	—	—	0.040	—	41	16

F: Front
R: Rear

42356-ELEM-C10

WHEEL ALIGNMENT

Year	Model		Caster Range (+/-Deg.)	Caster Preferred Setting (Deg.)	Camber Range (+/-Deg.)	Camber Preferred Setting (Deg.)	Toe-in (in.)	Steering Axis Inclination (Deg.)
2003-04	Element	F	1.00	+1.75	0.75	0	0+/-0.08	—
		R	—	—	0.75	-1.00	0.08+/-0.08	—

42356-ELEM-C11

SCHEDULED MAINTENANCE INTERVALS
HONDA—ELEMENT

TO BE SERVICED	TYPE OF SERVICE	VEHICLE MILEAGE INTERVAL (x1000)											
		10	20	30	40	50	60	70	80	90	100	110	120
Accessory drive belts	I & A			✓			✓			✓			✓
Air cleaner element	R			✓			✓			✓			✓
Air conditioning filter	R			✓			✓			✓			✓
Brake fluid	R											✓	
Brake hoses & lines (including ABS)	I		✓		✓		✓		✓		✓		
Cooling system hoses & connections	I		✓		✓		✓		✓		✓		
Engine coolant	R												✓
Engine oil	R	✓	✓	✓	✓	✓	✓	✓	✓	✓	✓	✓	✓
Engine oil and coolant levels	I	Inspect at each fuel stop											
Engine oil filter	R		✓		✓		✓		✓		✓		
Exhaust system	I		✓		✓		✓		✓		✓		
Fluid levels and condition	I		✓		✓		✓		✓		✓		
Front and rear brakes	I		✓		✓		✓		✓		✓		
Fuel lines & connection	I		✓		✓		✓		✓		✓		
Halfshaft boots	I		✓		✓		✓		✓		✓		
Idle speed	I & A											✓	
Parking brake system	I & A		✓		✓		✓		✓		✓		
Rear differential fluid	R										✓		
Rotate and inspect tires	I	✓	✓	✓	✓	✓	✓	✓	✓	✓	✓	✓	✓
Spark plugs	R											✓	
Suspension components	I		✓		✓		✓		✓		✓		
Tie rod ends, steering gear box & boots	I		✓		✓		✓		✓		✓		
Transmission fluid	R												✓
Valve clearance	I											✓	

R: Replace   I: Inspect   A: Adjust

FREQUENT OPERATION MAINTENANCE (SEVERE SERVICE)

If a vehicle is operated under any of the following conditions it is considered severe service:
- Towing a trailer or using a camper or car-top carrier.
- Repeated short trips of less than 5 miles in temperatures below freezing, or trips of less than 10 miles in any temperature.
- Extensive idling or low-speed driving for long distances as in heavy commercial use, such as delivery, taxi or police cars.
- Operating on rough, muddy or salt-covered roads.
- Operating on unpaved or dusty roads.
- Driving in extremely hot (over 90°) conditions.

Air cleaner element: replace every 15,000 miles

Engine oil and filter: replace every 3750 miles or 6 months, whichever occurs first.

Timing belt: replace every 60,000 miles if the vehicle is regularly driven in temperatures above 110°F or below -20°F.

Transmission fluid: replace every 30,000 miles.

Rear differential fluid: replace every 60,000 miles.

Front and rear brakes: inspect every 7500 miles or 6 months, whichever occurs first.

Locks and hinges: lubricate every 15,000 miles.

Tie rods, steering gear box, boots: inspect every 7500 miles or 6 months, whichever occurs first.

Suspension components: inspect every 7500 miles or 6 months, whichever occurs first.

Halfshaft boots: inspect every 7500 miles or 6 months, whichever occurs first.

42356-ELEM-C12

PRECAUTIONS

Before servicing any vehicle, please be sure to read all of the following precautions, which deal with personal safety, prevention of component damage, and important points to take into consideration when servicing a motor vehicle:

• Never open, service or drain the radiator or cooling system when the engine is hot; serious burns can occur from the steam and hot coolant.

• Observe all applicable safety precautions when working around fuel. Whenever servicing the fuel system, always work in a well-ventilated area. Do not allow fuel spray or vapors to come in contact with a spark, open flame, or excessive heat (a hot drop light, for example). Keep a dry chemical fire extinguisher near the work area. Always keep fuel in a container specifically designed for fuel storage; also, always properly seal fuel containers to avoid the possibility of fire or explosion. Refer to the additional fuel system precautions later in this section.

• Fuel injection systems often remain pressurized, even after the engine has been turned **OFF**. The fuel system pressure must be relieved before disconnecting any fuel lines. Failure to do so may result in fire and/or personal injury.

• Brake fluid often contains polyglycol ethers and polyglycols. Avoid contact with the eyes and wash your hands thoroughly after handling brake fluid. If you do get brake fluid in your eyes, flush your eyes with clean, running water for 15 minutes. If eye irritation persists, or if you have taken brake fluid internally, seek medical assistance IMMEDIATELY.

• The EPA warns that prolonged contact with used engine oil may cause a number of skin disorders, including cancer. You should make every effort to minimize your exposure to used engine oil. Protective gloves should be worn when changing oil. Wash your hands and any other exposed skin areas as soon as possible after exposure to used engine oil. Soap and water, or waterless hand cleaner should be used.

• All new vehicles are now equipped with an air bag system. The system must be disabled before performing service on or around system components, steering column, instrument panel components, wiring and sensors. Failure to follow safety and disabling procedures could result in accidental air bag deployment, possible personal injury and unnecessary system repairs.

• Always wear safety goggles when working with, or around, the air bag system. When carrying a non-deployed air bag, be sure the bag and trim cover are pointed away from your body. When placing a non-deployed air bag on a work surface, always face the bag and trim cover upward, away from the surface. This will reduce the motion of the module if it is accidentally deployed. Refer to the additional air bag system precautions later in this section.

• Clean, high quality brake fluid from a sealed container is essential to the safe and proper operation of the brake system. You should always buy the correct type of brake fluid for your vehicle. If the brake fluid becomes contaminated, completely flush the system with new fluid. Never reuse any brake fluid. Any brake fluid that is removed from the system should be discarded. Also, do not allow any brake fluid to come in contact with a painted surface; it will damage the paint.

• Never operate the engine without the proper amount and type of engine oil; doing so WILL result in severe engine damage.

• Timing belt maintenance is extremely important. Many models utilize an interference-type, non-freewheeling engine. If the timing belt breaks, the valves in the cylinder head may strike the pistons, causing potentially serious (also time-consuming and expensive) engine damage.

• Disconnecting the negative battery cable on some vehicles may interfere with the functions of the on-board computer system(s) and may require the computer to undergo a relearning process once the negative battery cable is reconnected.

• When servicing drum brakes, only disassemble and assemble one side at a time, leaving the remaining side intact for reference.

• Only an MVAC-trained, EPA-certified automotive technician should service the air conditioning system or its components.

ENGINE REPAIR

➡**Disconnecting the negative battery cable on some vehicles may interfere with the functions of the on board computer system. The computer may undergo a relearning process once the negative battery cable is reconnected.**

Distributor

The 2.4L engine does not have a distributor.

Alternator

REMOVAL

1. Before servicing the vehicle, refer to the precautions in the beginning of this section.
2. Remove or disconnect the following:
   • Negative, then the positive battery cables

• Accessory drive belt
• Auto-tensioner
• Alternator wiring harness connectors and harness clamp
• Positive Crankcase Ventilation (PCV) valve
• 3 bolts holding the alternator
• Alternator

INSTALLATION

1. Install or connect the following:
   • Alternator. Tighten the bolts to 16 ft. lbs. (22 Nm).
   • PCV valve
   • Alternator wiring harness connectors and harness clamp
   • Auto tensioner
   • Accessory drive belt
   • Negative battery cable

Ignition Timing

ADJUSTMENT

Adjustment is not possible on the 2.4L engine.

Engine Assembly

REMOVAL & INSTALLATION

➡**The engine and transaxle are removed from the vehicle as a unit.**

1. Before servicing the vehicle, refer to the precautions in the beginning of this section.
2. Drain the cooling system.
3. Drain the transaxle fluid.
4. Drain the engine oil.

5. Relieve fuel system pressure.
6. Remove or disconnect the following:
   - Negative battery cable
   - Fuse/Relay box battery cables
   - Battery and tray
   - Intake manifold cover
   - IAT sensor connector
   - Breather hose
   - Intake duct
   - Cables from the power distribution center
   - Throttle and cruise cables
   - Powertrain Control Module (PCM) connectors and grommet. Pull the PCM harness through the firewall.
   - Fuel lines
   - EVAP canister
   - Brake booster vacuum line
   - Clutch slave cylinder
   - Clutch hose bracket
   - Shift cables
   - Drive belt
   - Power steering pump, leaving the hoses connected

7. Attach a hoist to the engine lifting eyes and support the powertrain weight.
8. Remove or disconnect the following:
   - Splash shield
   - Wheels
   - Catalytic converter
   - Rear driveshaft
   - Stabilizer links
   - Right damper fork
   - Halfshafts
   - Shift cable
   - Radiator
   - Upper bracket
   - Transaxle mount and bracket
   - Front mount bolt
   - Rear mount bracket bolts. Match-mark the sub-frame mounting bolt centers.

There is a special tool necessary for sub-frame removal. The Honda tool number is EQS02C000011. Attach the tool as explained in the tool instructions, attach a floor jack with adapter, remove the 4 sub-frame bolts and lower the sub-frame.

9. Remove or disconnect the following:
   - A/C compressor without disconnecting the hoses
10. Check that all hoses and wires are disconnected.
11. Lower the engine about 6 inches and recheck all clearances.
12. Lower the engine all the way.
13. Remove the chain hoist.
To install:
14. Installation is the reverse of removal. Observe the following torques:
   - Front engine mount bracket bolts: 33 ft. lbs. (44Nm)

   - A/C compressor bracket: 33 ft. lbs. (44Nm)
   - Stiffener 10mm bolts: 33 ft. lbs. (44Nm); 6mm bolts 9 ft. lbs. (12 Nm)
   - A/C compressor bolts: 33 ft. lbs. (44 Nm)
   - Subframe front bolt: 47 ft. lbs. (64 Nm)
   - Subframe rear bolts: 43 ft. lbs. (59 Nm)
   - Upper bracket bolt and nut: 40 ft. lbs. (54 Nm)
   - Transmission mount bracket support bolts/nuts: 40 ft. lbs. (54 Nm)
   - PS pump bolts: 16 ft. lbs. (22 Nm)
   - Intake manifold cover: 9 ft. lbs. (12 Nm)

➡**Use new self-locking nuts and color-coded self-locking bolts when installing the engine mounts and suspension components.**

➡**Do not tighten the engine or transaxle mount fasteners until instructed to do so.**

15. Lower the powertrain into position.
16. Install or connect the following:
   - Transaxle mount and bracket. Tighten the frame mounting bolts to 47 ft. lbs. (64 Nm).
   - Upper bracket. Tighten the nuts in sequence to 54 ft. lbs. (74 Nm).
   - Rear mount bracket through bolt
   - Right front mount and bracket
   - Left front mount and bracket
17. Tighten the remaining mount fasteners as follows:
   a. Transaxle mount fasteners to 47 ft. lbs. (64 Nm) and the through bolt to 54 ft. lbs. (74 Nm).
   b. Rear mount bracket through bolt to 43 ft. lbs. (59 Nm).
   c. Right front mount 12mm bolts to 47 ft. lbs. (64 Nm) and the 10mm bolts to 33 ft. lbs. (44 Nm).
   d. Left front mount 12mm stud bolt to 61 ft. lbs. (83 Nm), 10mm bolts to 33 ft. lbs. (44 Nm), and 12mm nut to 43 ft. lbs. (59 Nm).
   e. Right front mount 12mm nut to 43 ft. lbs. (59 Nm).
18. Install or connect the following:
   - Rear driveshaft, if equipped
   - A/C compressor
   - Radiator
   - A/C hose clamp
19. If equipped with a manual transaxle, install or connect the following:
   - Shift cables
   - Transaxle ground cable

   - Clutch hose bracket
   - Clutch slave cylinder
20. If equipped with an automatic transaxle, install or connect the following:
   - Transaxle fluid cooler lines
   - Transaxle ground cable and hose clamp
   - Shift cable
   - Shift cable cover
21. For all vehicles, install or connect the following:
   - Axle halfshafts
   - Lower ball joints
   - Right damper fork
   - Exhaust front pipe
   - HO$_2$S connector
   - Heater hoses
   - Radiator hoses
   - Splash shield
   - PSP switch
   - Accelerator cable
   - Brake booster vacuum line
   - Fuel lines
   - A/C compressor drive belt
   - Power steering pump and belt
   - Cruise control actuator
   - Left engine wire harness connectors
   - PCM connectors and grommet
   - Air intake assembly
   - Battery and tray
   - Fuse/Relay box battery cables
   - Negative battery cable
22. Fill the engine crankcase to the correct level.
23. Fill the transaxle to the correct level.
24. Fill the cooling system.
25. Start the engine and check for leaks.
26. Check the wheel alignment and adjust as necessary.

Water Pump

REMOVAL & INSTALLATION

1. Before servicing the vehicle, refer to the precautions in the beginning of this section.
2. Drain the cooling system.
3. Remove or disconnect the following:
   - Negative battery cable
   - Accessory drive belt
   - Crankshaft pulley
   - Water pump (6 bolts)
To install:
4. Clean the water pump mating surfaces.
5. Install or connect the following:
   - Water pump with a new O-ring. Torque the bolts to 7.8 ft. lbs. (12 Nm).

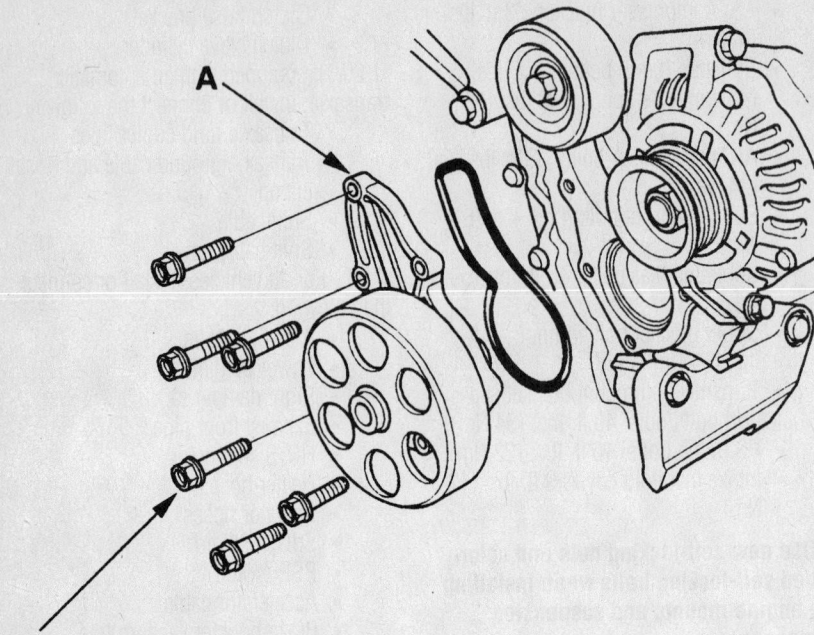

A

**6 x 1.0 mm
12 N·m (1.2 kgf·m, 8.7 lbf·ft)**

42356-ELEM-G02

Exploded view of the water pump mounting

- Crankshaft pulley
- Accessory drive belt
- Negative battery cable

6. Refill the engine cooling system.

Heater Core

REMOVAL & INSTALLATION

1. Before servicing the vehicle, refer to the precautions in the beginning of this section.

2. Disconnect the negative battery cable.

3. Drain the cooling system into a clean container for reuse.

4. In the engine compartment, open the heater valve cable clamp and disconnect the cable from the heater valve arm. Then, turn the heater valve to the fully opened position.

5. Remove or disconnect the following:
- Heater hoses from the heater core
- Heater housing-to-chassis nut

➡**When removing the heater housing nut, be careful not to damage or bend the fuel lines, the brake lines, etc.**

6. Remove the instrument panel by performing the following procedure:

a. Remove the driver's side lower instrument panel cover screws, disengage the clips and remove the lower cover.

b. Remove the knee bolster bolts and the knee bolster.

c. Remove the glove box stops from each side of the glove box.

d. Remove the glove box-to-instrument panel bolts and the glove box.

e. Remove the lower console cover by disengaging the 4 clips and removing the cover.

f. Remove the 6 center pocket-to-instrument panel screws; then, insert a flat tipped screwdriver at the upper right side corner of the center pocket, push down on the top of the hook and remove the center pocket/beverage holder assembly.

g. Remove the center instrument panel lower cover screws and disengage the clips on the upper left side; then, disconnect the electrical connectors and remove the cover.

h. Gently, push the power window switch from the instrument panel's lower cover opening by hand. Disconnect the electrical connectors and remove the power window switch.

i. Close the driver's side air vent; then, gently, push out the clips and pull out the vent. Disconnect the electrical connectors and remove the vent.

j. Gently, push out the driver's side defogger trim; then, disconnect the electrical connector and remove the side defogger trim.

k. At the base of the steering wheel, remove the access panel and disconnect the air bag electrical connector.

l. Remove the steering column covers screws and the covers.

m. Remove the steering column-to-instrument panel nuts/bolts and lower the steering column.

n. Remove the instrument panel side covers.

o. Disconnect the wiring harness connector and remove the nuts.

p. Move the under-dash fuse/relay box.

q. Disconnect the antenna connector and the harness clips.

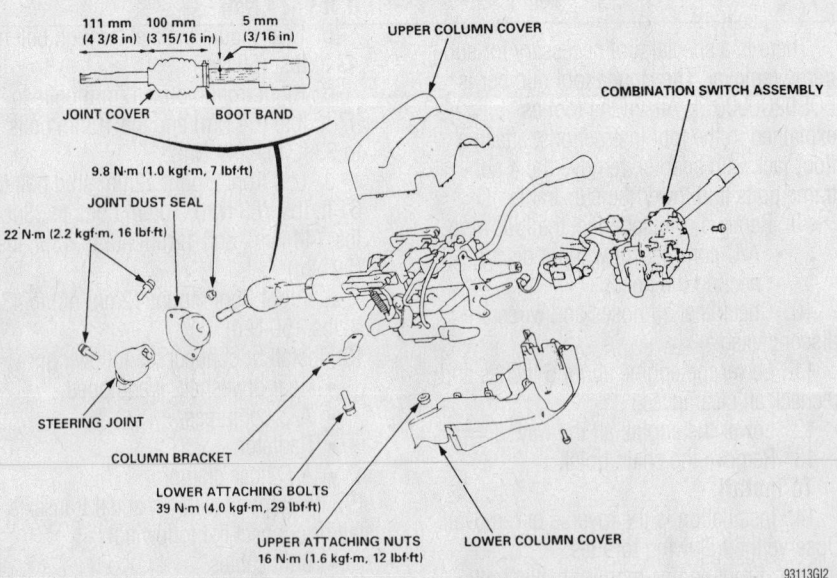

111 mm | 100 mm | 5 mm
(4 3/8 in) | (3 15/16 in) | (3/16 in)

JOINT COVER | BOOT BAND

UPPER COLUMN COVER

COMBINATION SWITCH ASSEMBLY

9.8 N·m (1.0 kgf·m, 7 lbf·ft)

JOINT DUST SEAL

22 N·m (2.2 kgf·m, 16 lbf·ft)

STEERING JOINT

COLUMN BRACKET

LOWER ATTACHING BOLTS
39 N·m (4.0 kgf·m, 29 lbf·ft)

UPPER ATTACHING NUTS
16 N·m (1.6 kgf·m, 12 lbf·ft)

LOWER COLUMN COVER

93113GI2

Exploded view of the steering column and related components

r. Remove the connector holder from the instrument panel frame.

s. Remove the control unit/relay bracket from behind the center of the instrument panel.

t. Remove the passenger's side lower instrument panel cover.

u. Disconnect the connectors and the harness clips.

v. Remove the instrument panel-to-chassis bolts.

w. Using an assistant, remove the instrument panel.

7. Remove the evaporator housing by performing the following procedure:

a. Discharge and recover the air conditioning system refrigerant.

b. In the engine compartment, remove the refrigerant lines-to-evaporator housing bolts.

c. Separate the lines, discard the grommets and plug the openings to prevent contamination.

d. Disconnect the evaporator housing's temperature sensor connector.

e. Remove the evaporator housing-to-chassis screws/nut and remove the evaporator housing.

8. Disconnect the mode control motor and the air mix control motor electrical connectors and remove the wiring harness clips and the wiring harness from the heater housing.

9. Remove the heater duct clip, the heater housing-to-chassis nuts and the heater housing.

10. Remove the heater core cover screws and the cover.

11. Remove the heater core pipe clamp screws and the clamp.

12. Remove the heater core from the heater housing.

To install:

13. Install the heater core in the heater housing.

14. Install the heater core pipe clamp and the clamp screws.

15. Install the heater core cover and the cover screws.

16. Install the heater housing, the heater housing-to-chassis nuts and the heater duct clip.

17. Install the wiring harness clips and the wiring harness to the heater housing

▶: Bolt locations

A▶, 5   B▶, 1   C▶, 1

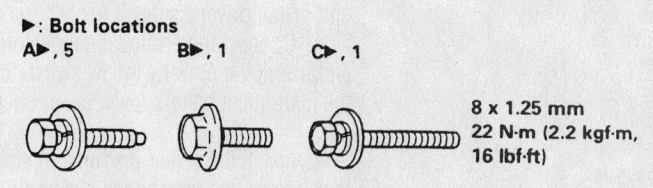

8 x 1.25 mm
22 N·m (2.2 kgf·m, 16 lbf·ft)

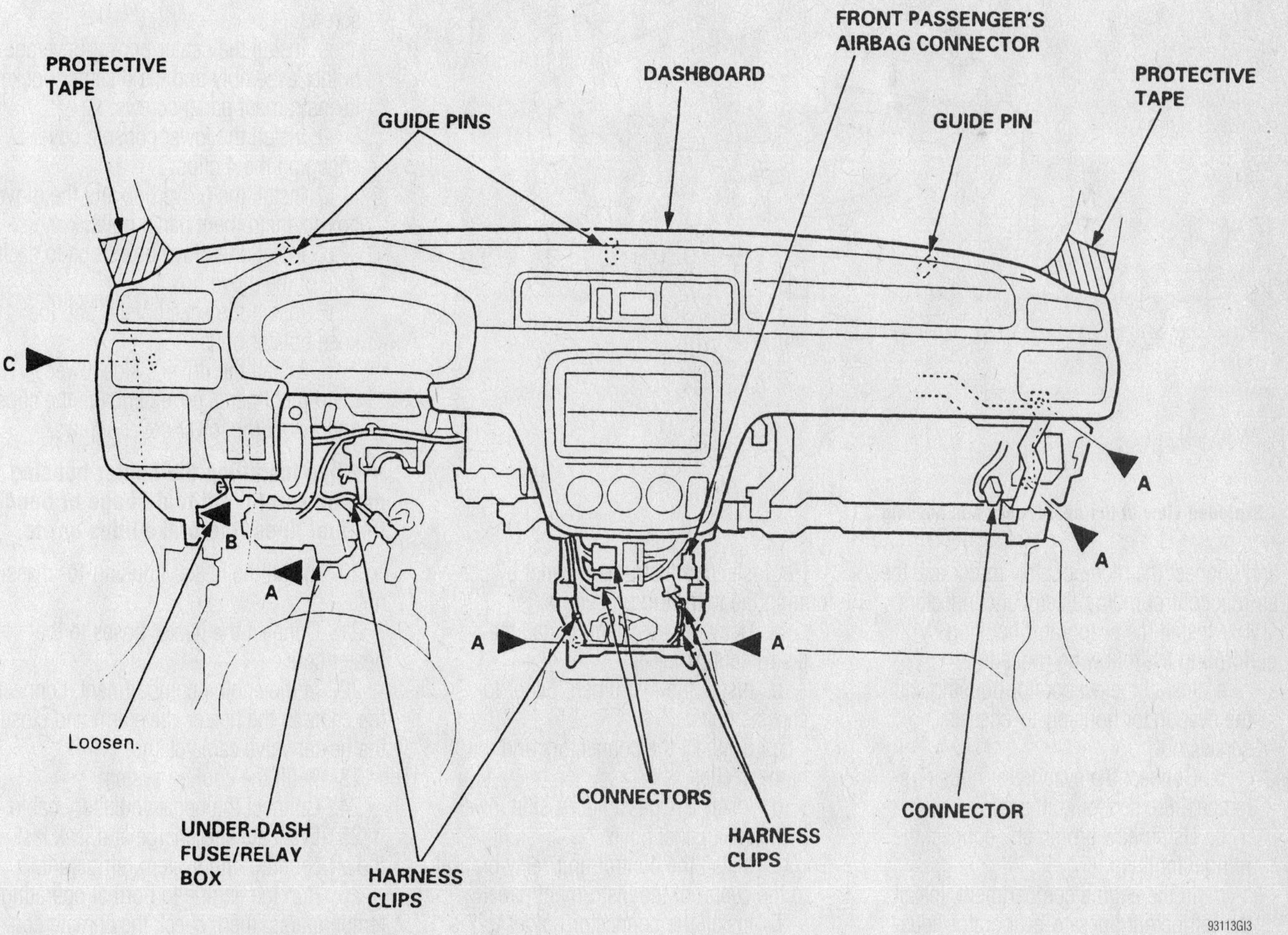

PROTECTIVE TAPE

GUIDE PINS

DASHBOARD

FRONT PASSENGER'S AIRBAG CONNECTOR

GUIDE PIN

PROTECTIVE TAPE

C ▶

A ▶

A ▶

B ▶

A ▶

Loosen.

A ▶   A ▶

UNDER-DASH FUSE/RELAY BOX

HARNESS CLIPS

CONNECTORS

HARNESS CLIPS

CONNECTOR

Exploded view of the instrument panel and related components

93113GI3

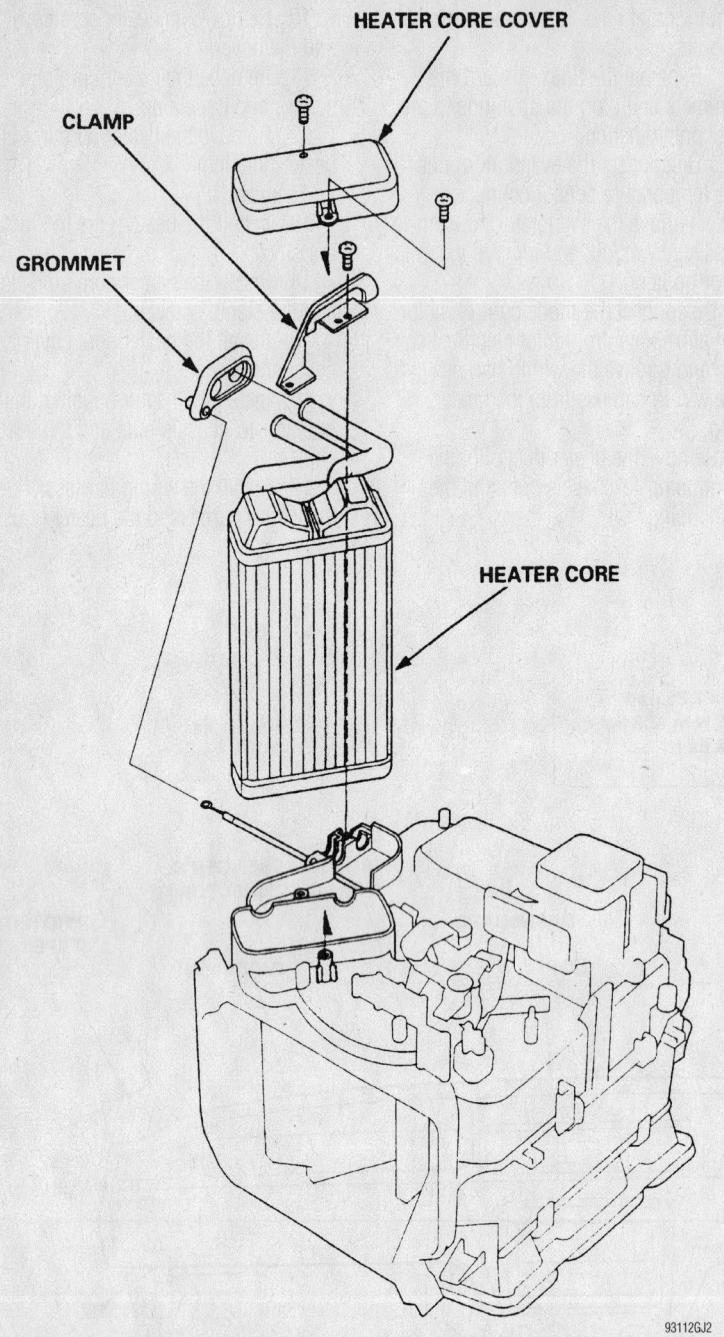

CLAMP

HEATER CORE COVER

GROMMET

HEATER CORE

93112GJ2

Exploded view of the heater core and housing

and connect the mode control motor and the air mix control motor electrical connectors.

18. Install the evaporator housing by performing the following procedure:

  a. Install the evaporator housing and the evaporator housing-to-chassis screws/nut.

  b. Connect the evaporator housing's temperature sensor connector.

  c. Using new grommets, connect the refrigerant lines.

  d. In the engine compartment, install the refrigerant lines-to-evaporator housing bolts.

19. Install the instrument panel by performing the following procedure:

  a. Using an assistant, install the instrument panel.

  b. Install the instrument panel-to-chassis bolts.

  c. Connect the connectors and the harness clips.

  d. Install the passenger's side lower instrument panel cover.

  e. Install the control unit/relay bracket to the center of the instrument panel.

  f. Install the connector holder to the instrument panel frame.

  g. Connect the antenna connector and the harness clips.

  h. Install the under-dash fuse/relay box.

  i. Connect the wiring harness connector and install the nuts.

  j. Install the instrument panel side covers.

  k. Install the steering column and the column-to-instrument panel nuts/bolts. Torque the nuts to 12 ft. lbs. (16 Nm) and the bolts to 29 ft. lbs. (39 Nm).

  l. Install the steering column covers and the cover screws.

  m. At the base of the steering wheel, connect the air bag electrical connector and install the access panel.

  n. Connect the electrical connector and install the driver's side defogger trim.

  o. Connect the electrical connectors and install driver's side air vent.

  p. Connect the electrical connectors and install the power window switch to the instrument panel's lower cover opening.

  q. Install the center instrument panel lower cover and engage the clips on the upper left side; then, connect the electrical connectors and Install the cover screws.

  r. Install the center pocket/beverage holder assembly and the 6 center pocket-to-instrument panel screws.

  s. Install the lower console cover by engaging the 4 clips.

  t. Install the glove box and the glove box-to-instrument panel bolts.

  u. Install the glove box stops to each side of the glove box.

  v. Install the knee bolster and the knee bolster bolts.

  w. Install the driver's side lower instrument panel cover, engage the clips and install the lower cover screws.

➡**When installing the heater housing nut, be careful not to damage or bend the fuel lines, the brake lines or etc.**

20. Install the heater housing-to-chassis nut.

21. Connect the heater hoses to the heater core.

22. In the engine compartment, connect the cable to the heater valve arm and close the heater valve cable clamp.

23. Refill the cooling system.

24. Connect the negative battery cable.

25. Evacuate and charge and leak test the air conditioning system refrigerant.

26. Run the engine to normal operating temperatures; then, check the climate control operation and check for leaks.

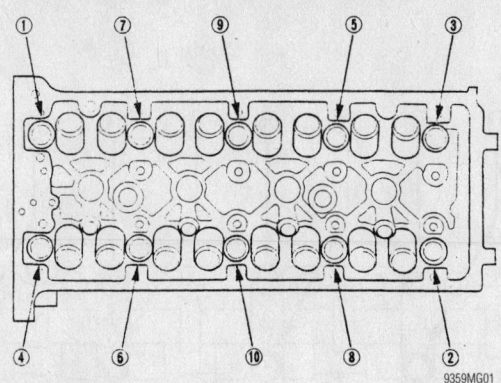

Cylinder head bolt loosening sequence

Cylinder Head

REMOVAL & INSTALLATION

1. Before servicing the vehicle, refer to the precautions in the beginning of this section.
2. Drain the cooling system.
3. Relieve the fuel system pressure.
4. Remove or disconnect the following:
- Negative battery cable
- Accessory drive belt
- Intake Air Temperature (IAT) sensor connector
- Vacuum hoses and breather pipe and air intake duct
- Fuel feed hose
- Bolt securing the connecting pipe support bracket to the engine block
- Evaporative emission (EVAP) canister hose and brake booster vacuum hose
- Intake manifold
- Exhaust manifold
- Cam chain
- Positive Crankcase Ventilation (PCV) hose and ground cable
- Upper radiator hose, heater hoses and water bypass hose
5. Remove the following engine wire

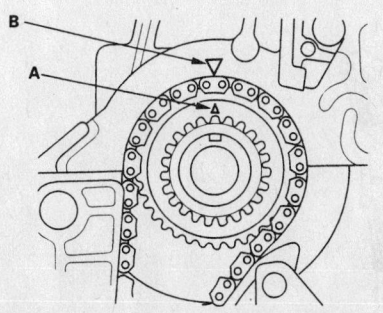

Set the crankshaft to TDC by aligning the mark (A) on the crankshaft sprocket with the pointer (B) on the cylinder block

harness connectors and wire harness clamps from the cylinder head:
- Four injector connector
- Engine Coolant Temperature (ECT) sensor connector
- Camshaft Position (CMP) sensor A & B (intake & exhaust) connectors
- VTEC solenoid valve connector
- Engine Oil Pressure (EOP) sensor connector
6. Remove or disconnect the following:

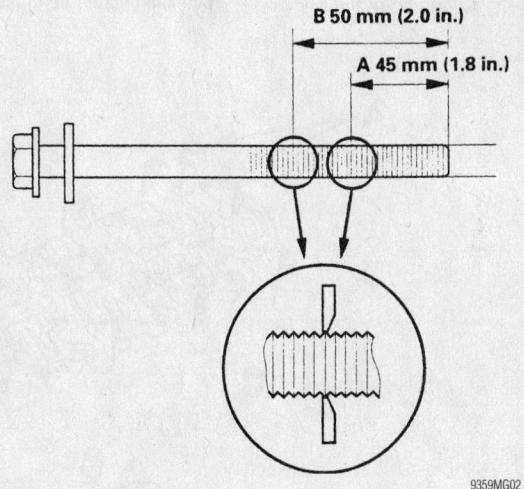

Cylinder head bolt inspection

- 3 bolts holding the EVAP canister purge valve bracket and remove the two bolts (B) securing the harness bracket
- Timing (cam) chain
- Rocker arm assembly
7. Loosen the cylinder head bolts in sequence and ⅓ turns until all bolts are loose.
8. Remove the cylinder head.
To install:
9. Be sure all cylinder head and block gasket surfaces are clean. Check the cylinder head for warpage. If warpage is less than 0.002 in. (0.05mm), cylinder head resurfacing is not required. Maximum resurface limit is 0.008 in. (0.2mm) based on a cylinder head height of 3.94 in. (100mm).
10. Install or connect the following:
- New gasket and dowel pins on the cylinder block
11. Set the crankshaft to Top Dead Center (TDC). Align the TDC mark (A) on the crankshaft sprocket with the pointer (B) on the cylinder block.

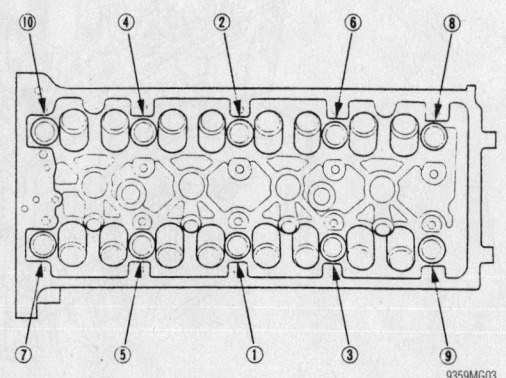

Cylinder head bolt torque sequence

12. Measure the diameter of each cylinder head bolt at points A & B, as shown in the illustration. If either diameter is less than 0.42 in. (10.6mm), replace the head bolt

13. Apply engine oil to the threads and under the bolt heads of all of the bolts.

14. Install the cylinder head. Tighten the bolts in sequence as follows:

    a. Step 1: 29 ft. lbs. (39 Nm).

    b. Step 2: Plus 90 degrees.

    c. Step 3: Plus 90 degrees.

    d. Step 4: If using new cylinder head bolts, add an additional 90 degrees.

15. The remainder of installation is the reverse of removal.

16. Fill the cooling system.

17. Connect the negative battery cable and enter the radio security code.

18. Start the engine and check carefully for any leaks.

Rocker Arms/Shafts

REMOVAL & INSTALLATION

1. Before servicing the vehicle, refer to the precautions in the beginning of this section.

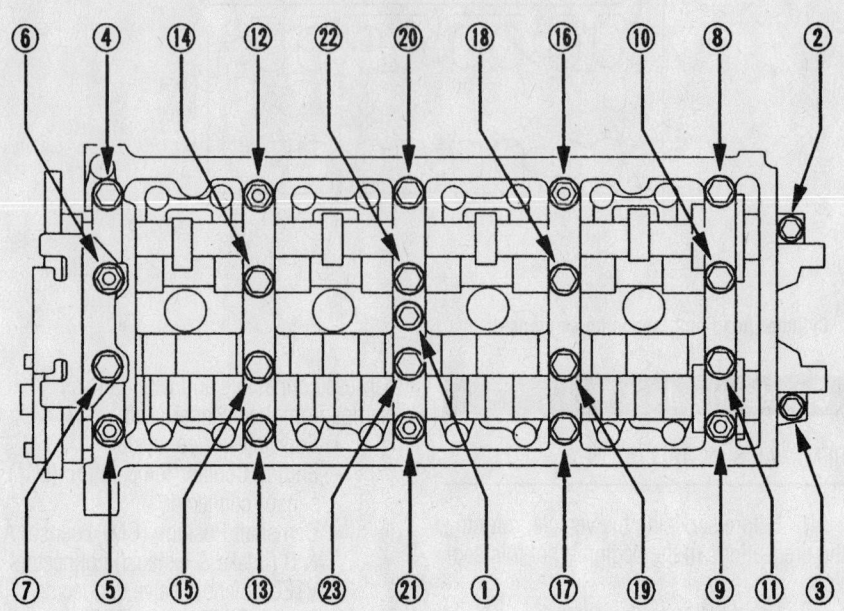

42356-ELEM-G16

Camshaft holder bolt loosening sequence

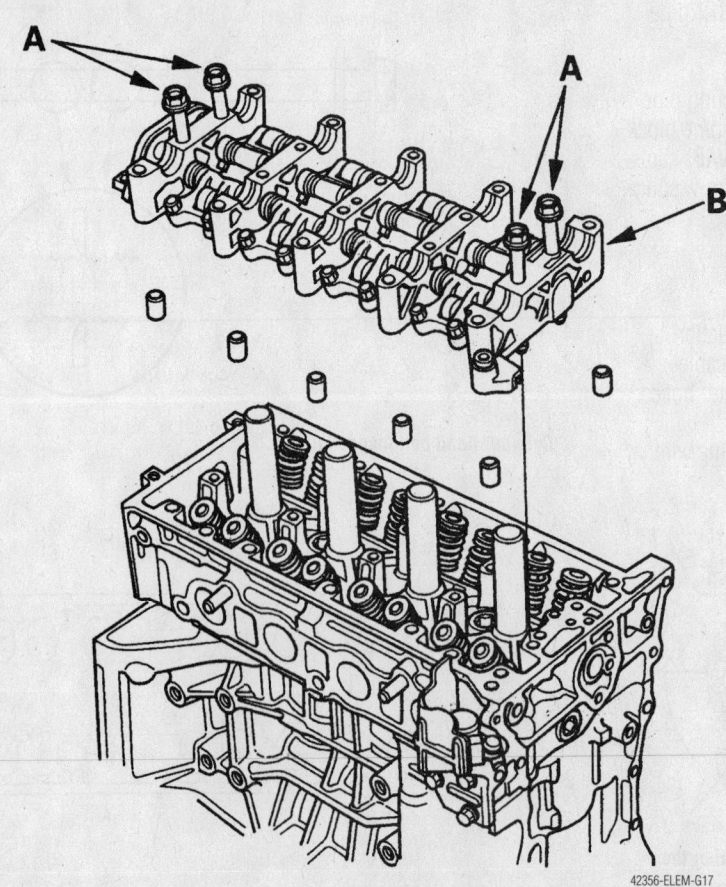

42356-ELEM-G17

Insert the bolts (A) into the rocker shaft holder, then remove the rocker arm assembly (B)

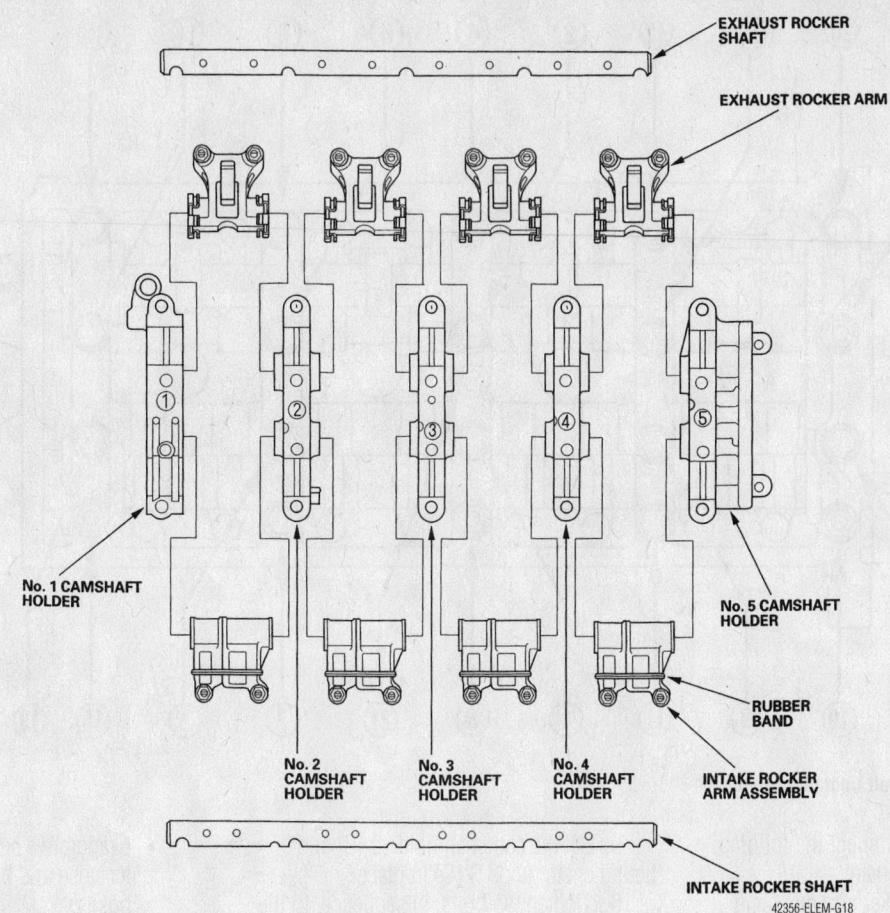

Exploded view of the rocker arms and related components

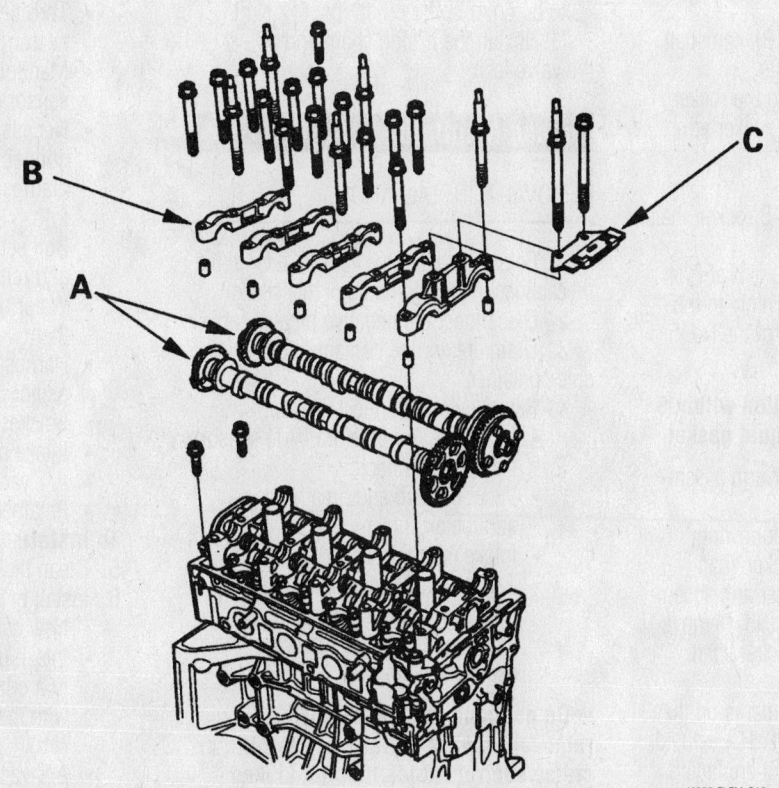

When installing the camshafts (A) make sure the punch marks on the VTC actuator and exhaust cam sprockets are facing up

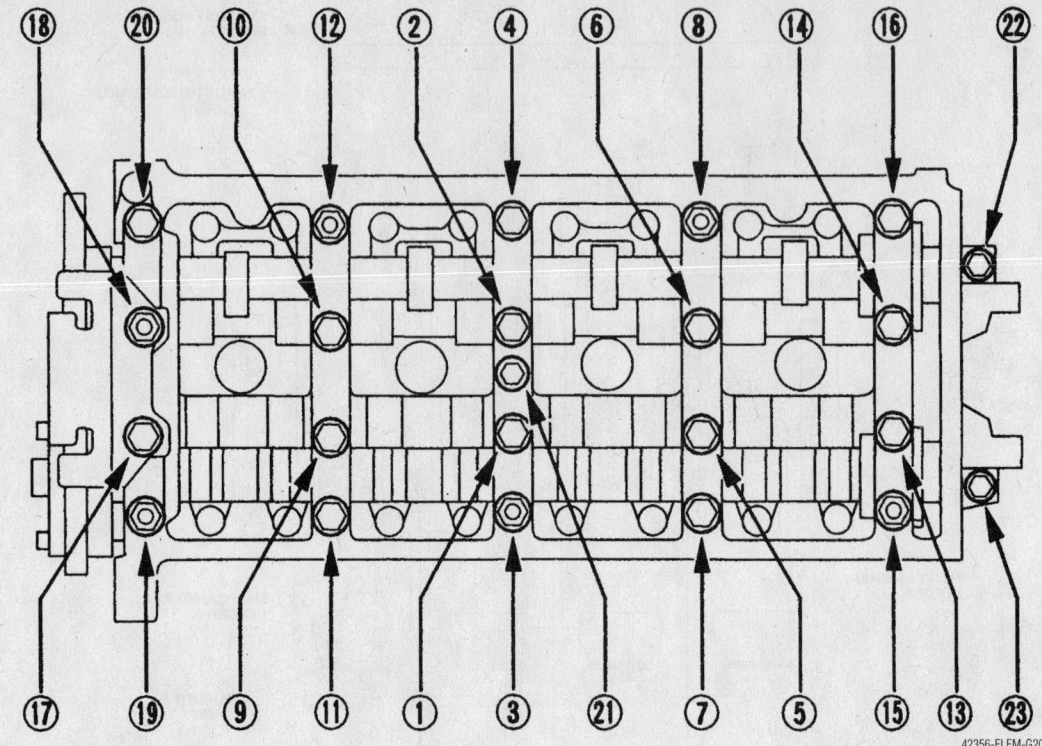

Rocker arm assembly bolt tightening sequence

2. Remove or disconnect the folloing:
- Timing (cam) chain
- Loosen the rocker arm adjusting screws
- Camshaft holder bolts, two turns at a time in sequence
- Timing chain guide (B), camshaft holders and camshafts

3. Insert the bolts (A) into the rocker shaft holder, then remove the rocker arm assembly (B)

To install:

4. Clean and dry the No. 5 rocker shaft holding mating surface.

5. Apply a suitable liquid gasket P/N 08718-0009, or equivalent, evenly to the cylinder head mating surface of the No. 5 rocker shaft holder.

➡ **The parts must be installed within 5 minutes of applying the liquid gasket.**

6. Reassemble the rocker arm assembly, as necessary.

7. Install or connect the following
- Bolts (A) into the rocker shaft holder, then the rocker arm assembly on the cylinder head. Remove the bolts from the rocker shaft holder.

8. Make sure the punch marks on the variable valve timing control (VTC) actuator and exhaust camshaft sprocket are facing up, then set the camshafts (A) in the holder

9. Set the camshaft holders (B) and timing chain guide B (C) in place.

10. Tighten the bolts, in sequence, to the following specification:
   a. 8mm bolts: 16 ft. lbs. (22 Nm)
   b. 6mm bolts: 8.7 ft. lbs. (12 Nm)

11. Install the timing chain and adjust the valve lash.

Intake Manifold

REMOVAL & INSTALLATION

1. Before servicing the vehicle, refer to the precautions in the beginning of this section.

2. Disconnect the negative battery cable.

3. Drain the engine coolant into a sealable container.

4. Remove or disconnect the following:
- Intake Air Temperature (IAT) sensor electrical connector
- Vacuum hose and breather pipe and the air intake duct
- Intake manifold cover
- Throttle and cruise control cables by loosening the locknuts, then slipping the cable ends out of the accelerator linkage.

➡ **Do not bend the cables during removal. Always replace any throttle or cruise control cables that get kinked during removal.**

- Evaporative emission (EVAP) canister hose and brake booster vacuum hose
- Idle Air Control (IAC) valve connectors
- Throttle Position (TP) sensor connector
- Manifold Absolute Pressure (MAP) sensor connector
- Necessary engine wire harness connectors and wire harness clamps from the intake manifold
- Bolt securing the harness holder and remove the harness clamps
- Water bypass hoses, then plug them
- Harness clamp and harness connector from the intake manifold bracket
- Intake manifold bracket
- A/T vacuum hose
- Retainer and intake manifold

To install:
5. Clean the mounting surfaces.
6. Install or connect the following:
- New gasket
- Intake manifold. Tighten the bolts, in a criss-cross pattern beginning with the inner bolt, to 16 ft. lbs. (22 Nm).
- A/T vacuum hose
- Intake manifold bracket

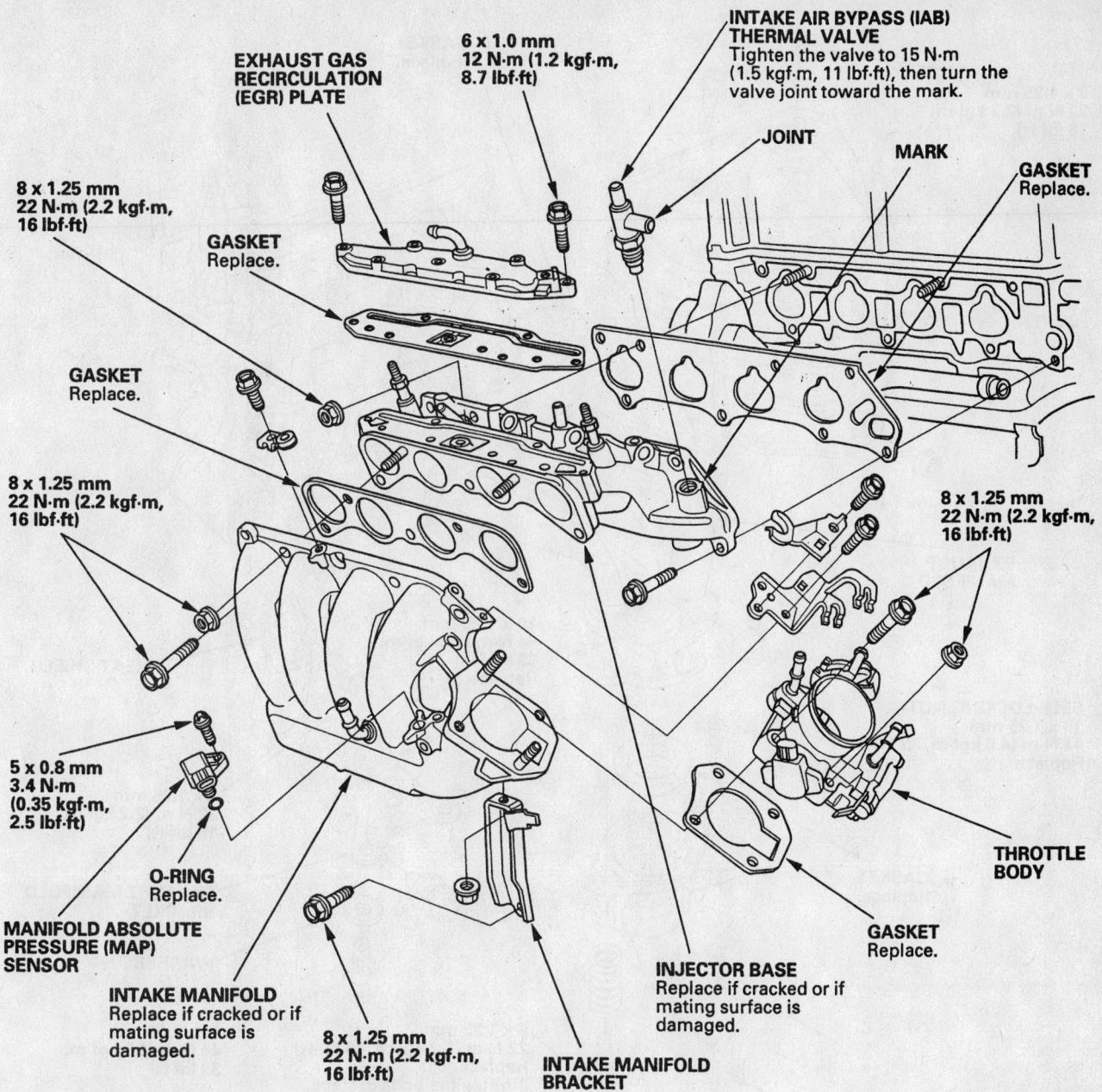

EXHAUST GAS RECIRCULATION (EGR) PLATE

6 x 1.0 mm
12 N·m (1.2 kgf·m, 8.7 lbf·ft)

INTAKE AIR BYPASS (IAB) THERMAL VALVE
Tighten the valve to 15 N·m (1.5 kgf·m, 11 lbf·ft), then turn the valve joint toward the mark.

JOINT

MARK

GASKET Replace.

8 x 1.25 mm
22 N·m (2.2 kgf·m, 16 lbf·ft)

GASKET Replace.

GASKET Replace.

8 x 1.25 mm
22 N·m (2.2 kgf·m, 16 lbf·ft)

8 x 1.25 mm
22 N·m (2.2 kgf·m, 16 lbf·ft)

5 x 0.8 mm
3.4 N·m (0.35 kgf·m, 2.5 lbf·ft)

O-RING Replace.

MANIFOLD ABSOLUTE PRESSURE (MAP) SENSOR

INTAKE MANIFOLD Replace if cracked or if mating surface is damaged.

8 x 1.25 mm
22 N·m (2.2 kgf·m, 16 lbf·ft)

INTAKE MANIFOLD BRACKET

INJECTOR BASE Replace if cracked or if mating surface is damaged.

THROTTLE BODY

GASKET Replace.

42356-ELEM-G21

Exploded view of the intake manifold and related components

- Harness clamp and connector to the intake manifold bracket
- Water bypass hoses
- Bolt securing the harness holder and tighten to 8.7 ft. lbs. (12 Nm)
- Harness clamps
- EVAP canister hose and brake booster vacuum hose
- Throttle and cruise control cables
- Intake manifold cover
- Intake air duct
- IAT sensor connector, vacuum hose and breather pipe
7. Refill the cooling system.

8. Connect the negative battery cable, start the engine, and check for leaks.

Exhaust Manifold

REMOVAL & INSTALLATION

1. Before servicing the vehicle, refer to the precautions in the beginning of this section.
2. Raise and safely support the vehicle.
3. Remove or disconnect the following:
- VTEC solenoid valve

- Intermediate shaft heat cover
- Cover and exhaust manifold bracket
- Exhaust manifold
To install:
4. Clean the mounting surfaces.
5. Install or connect the following:
- New gasket on the cylinder head
- Exhaust manifold. Tighten the nuts, in a criss-cross pattern starting with the inner nut, to 33 ft. lbs. (45 Nm).
- Exhaust manifold bracket and cover
- Intermediate shaft heat cover
- VTEC solenoid valve

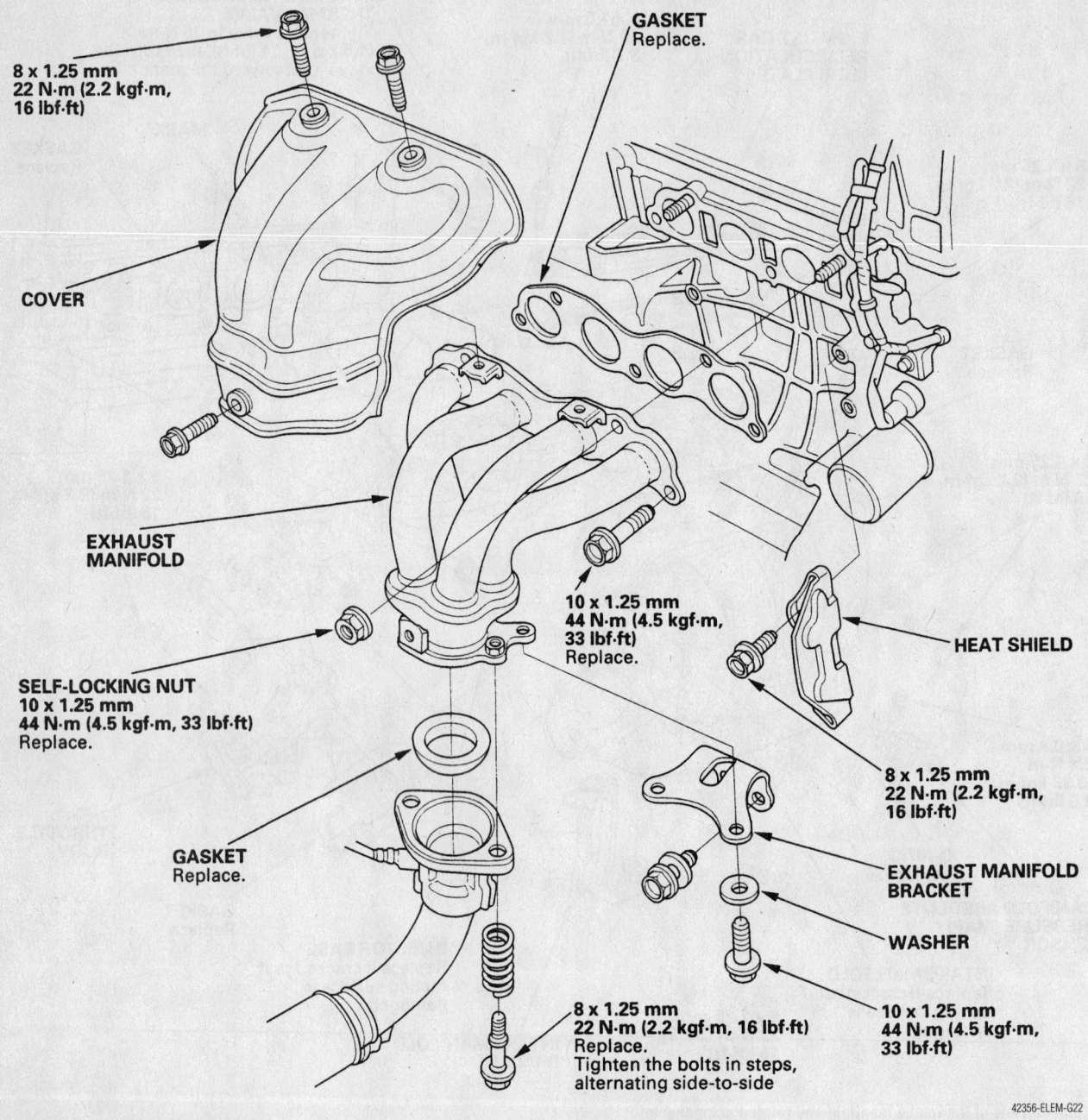

8 x 1.25 mm
22 N·m (2.2 kgf·m, 16 lbf·ft)

GASKET
Replace.

COVER

EXHAUST MANIFOLD

SELF-LOCKING NUT
10 x 1.25 mm
44 N·m (4.5 kgf·m, 33 lbf·ft)
Replace.

10 x 1.25 mm
44 N·m (4.5 kgf·m, 33 lbf·ft)
Replace.

HEAT SHIELD

8 x 1.25 mm
22 N·m (2.2 kgf·m, 16 lbf·ft)

EXHAUST MANIFOLD BRACKET

WASHER

GASKET
Replace.

8 x 1.25 mm
22 N·m (2.2 kgf·m, 16 lbf·ft)
Replace.
Tighten the bolts in steps, alternating side-to-side

10 x 1.25 mm
44 N·m (4.5 kgf·m, 33 lbf·ft)

42356-ELEM-G22

Exploded view of the exhaust manifold and related components

Front Crankshaft Seal

REMOVAL & INSTALLATION

For the 2.4L engine, see the Timing Chain Removal & Installation procedure.

Camshaft

REMOVAL & INSTALLATION

See the Rocker Arm Shaft Removal & Installation procedure.

Valve Lash

ADJUSTMENT

Adjust the valves only when the cylinder head temperature is less than 100°F (38°C).

1. Before servicing the vehicle, refer to the precautions in the beginning of this section.
2. Remove or disconnect the following:

- Negative battery cable
- Cylinder head cover

3. Set the timing marks as shown in the illustration with N0.1 at TDC. Check all clearances. Intake should be 0.008–0.010 in.; exhaust should be 0.011–0.013 in. Intake locknut torque is 14 ft. lbs. (19 Nm); exhaust is 10 ft. lbs. (14 Nm).
4. Rotate the crankshaft 180 degrees clockwise and recheck No.3.
5. Rotate the crankshaft 180 degrees clockwise and recheck No.4.
6. Rotate the crankshaft 180 degrees clockwise and recheck No.2.

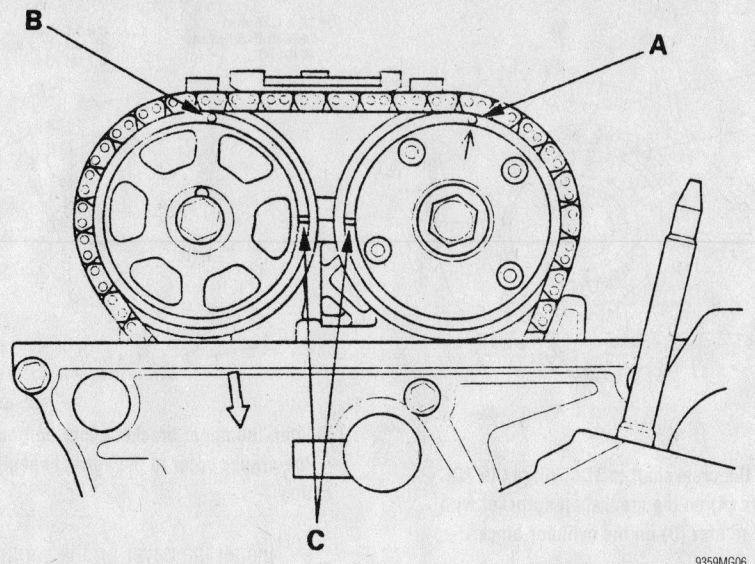

Align the timing marks

9359MG06

Starter Motor

REMOVAL & INSTALLATION

➡ **The factory sound system has a coded theft protection system. It is recommended that you know your reset code before you begin.**

1. Before servicing the vehicle, refer to the precautions in the beginning of this section.
2. Remove or disconnect the following:

- Negative then the positive battery cables
- Intake manifold
- Starter cable from the B terminal

M/T:

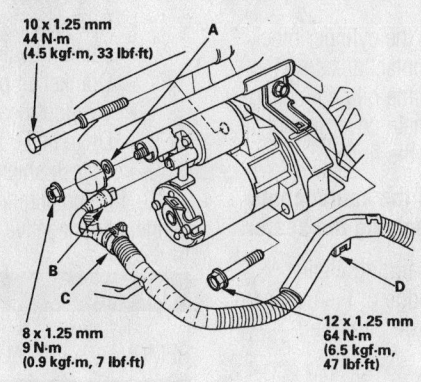

A/T:

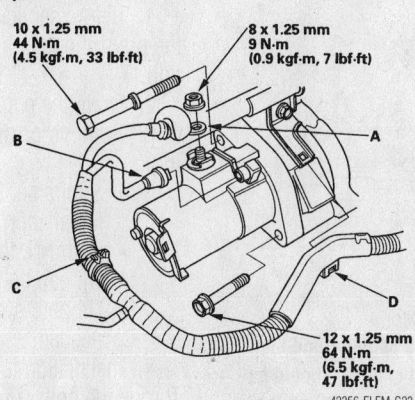

42356-ELEM-G23

Starter mounting

- Black/white wire from the S (solenoid) terminal
- Harness clamp and holder
- Two bolts that mount the starter to the transaxle assembly
- Starter

To install:

3. Install in the reverse order of removal.

➡ **When installing the heavy gauge starter cable, make sure the crimped side of the terminal end is facing out.**

4. Enter the anti-theft code and radio presets.

Timing Chain and Front Seal

REMOVAL & INSTALLATION

1. Before servicing the vehicle, refer to the precautions in the beginning of this section.
2. Set the engine to Top Dead Center (TDC).
3. Drain the cooling system.
4. Relieve the fuel system pressure.
5. Remove or disconnect the following:

- Negative battery cable
- Front tires and wheels
- Splash shield
- Drive belt
- Cylinder head cover. Check that the No. 1 piston TDC marks on the Variable Valve Timing Control (VTC) actuator and exhaust camshaft sprocket are aligned
- Crankshaft pulley
- Crankshaft Position (CKP) sensor connector
- VTC oil control solenoid valve connector
- VTC oil control solenoid valve

6. Support the engine with a suitable jack with a wooden block under the oil pan.

- Ground cable and upper bracket
- Side engine mount bracket
- Chain (case) cover

7. Loosely install the crankshaft pulley.

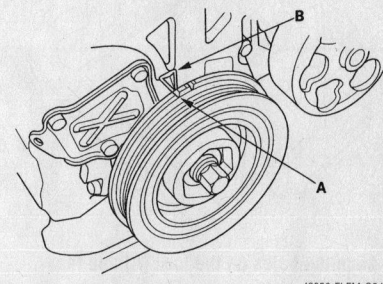

42356-ELEM-G24

Turn the crankshaft pulley so the TDC mark (A) is aligned with the pointer (B)

Turn the crankshaft counterclockwise to compress the auto-tensioner.

8. Align the holes on the lock (A) and the auto-tensioner (B), then place a 1.5mm pin into the holes. Turn the crankshaft clockwise to secure the pin.

9. Remove or disconnect the following:
- Auto-tensioner
- Timing chain guide B (top guide)
- Timing chain guide A and tensioner arm
- Timing chain

✳✳ WARNING

Do not let the timing chain near any magnetic fields.

To install:

10. Set the crankshaft to TDC. Align the TDC mark (A) on the crankshaft sprocket with the pointer (B) on the cylinder block.

11. Set the camshafts to TDC. The punch mark (A) on the VTC actuator and the punch mark (B) on the exhaust camshaft (C) should be at the top. Align the TDC marks (C) on the VTC actuator and exhaust camshaft sprockets.

12. Install or connect the following:
- Timing chain the crankshaft sprocket with the colored link of the chain aligned with the mark on the crank sprocket
- Timing chain on the VTC actuator and exhaust camshaft sprocket with the punch marks aligned with the center of the 2 colored links
- Timing chain guide A and tensioner arm. Tighten the guide bolts to 8.7 ft. lbs. (12 Nm) and the tensioner arm retainer to 16 ft. lbs. (22 Nm).
- Auto-tensioner and tighten the bolts to 8.7 ft. lbs. (12 Nm)

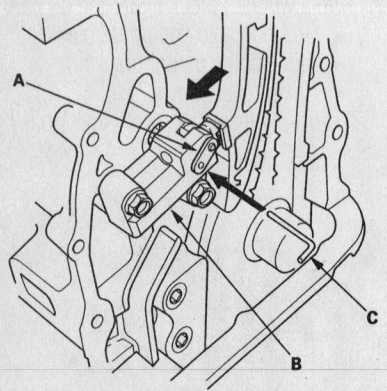

Align the holes on the lock (A) and the auto-tensioner (B), then place a 1.5mm pin into the holes. Turn the crankshaft clockwise to secure the pin

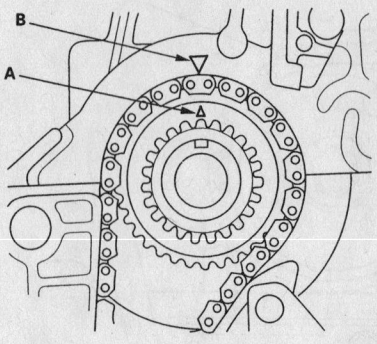

42356-ELEM-G26

Set the crankshaft to TDC. Align the TDC mark (A) on the crankshaft sprocket with the pointer (B) on the cylinder block

- Timing chain guide B and tighten the retainers to 16 ft. lbs. (22 Nm)

13. Remove the pin from the auto-tensioner.

14. Inspect the chain cover seal for damage and replace if necessary. Clean and dry the chain cover mating surfaces.

15. Install or connect the following:
- Liquid gasket, P/N 08718-0009 evenly to the cylinder block mating surface of the timing chain cover and the inner threads of the holes
- Liquid gasket to the cylinder block upper surface contact areas on the chain cover and the oil pan mating surface of the chain cover in the inner threads of the holes

➡**Make sure to install the components within 5 minutes of applying the sealer.**

- New O-ring the timing chain cover. Set the edge of the cover to the edge of the oil pan, then

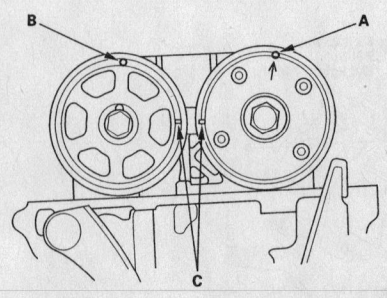

42356-ELEM-G27

The mark (A) on the VTC actuator and the mark (B) on the exhaust cam (C) should be at the top. Align the TDC marks (C) on the VTC actuator and exhaust cam sprockets

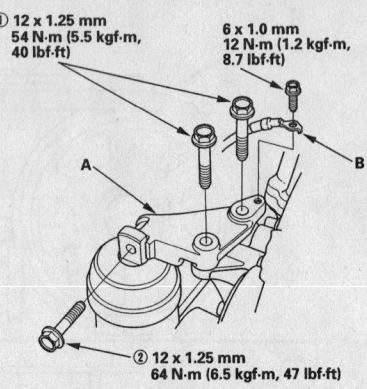

① 12 x 1.25 mm
54 N·m (5.5 kgf·m, 40 lbf·ft)

6 x 1.0 mm
12 N·m (1.2 kgf·m, 8.7 lbf·ft)

② 12 x 1.25 mm
64 N·m (6.5 kgf·m, 47 lbf·ft)

42356-ELEM-G28

Tighten the upper bracket upper bolt/nuts in the proper order to the correct specification

install the cover on the engine block. Tighten the retainers to 8.7 ft. lbs. (12 Nm).

➡**When installing the chain case, do not slide the bottom surface on the oil pan mounting surface.**

- Side engine mounting bracket and tighten the retainers to 33 ft. lbs. (44 Nm)
- Upper bracket, then tighten the bolts/nuts as shown in the illustration
- Ground cable
- VTC oil control solenoid valve
- CKP sensor and VTC oil control solenoid valve connectors
- Crankshaft pulley
- Cylinder head cover
- Drive belt
- Splash shield

16. Fill the engine cooling system and connect the negative battery cable.

Oil Pan

REMOVAL & INSTALLATION

1. Before servicing the vehicle, refer to the precautions in the beginning of this section.

2. Drain the engine oil.

3. Remove or disconnect the following:
- Subframe. See engine Removal and Installation.
- With MT, the stiffener
- Oil pan bolts
- Oil pan. A gasket cutter will be needed.

4. Installation is the reverse of removal. Torque the bolts, in sequence, in 2 or 3 steps, to 9 ft. lbs. (12 Nm).

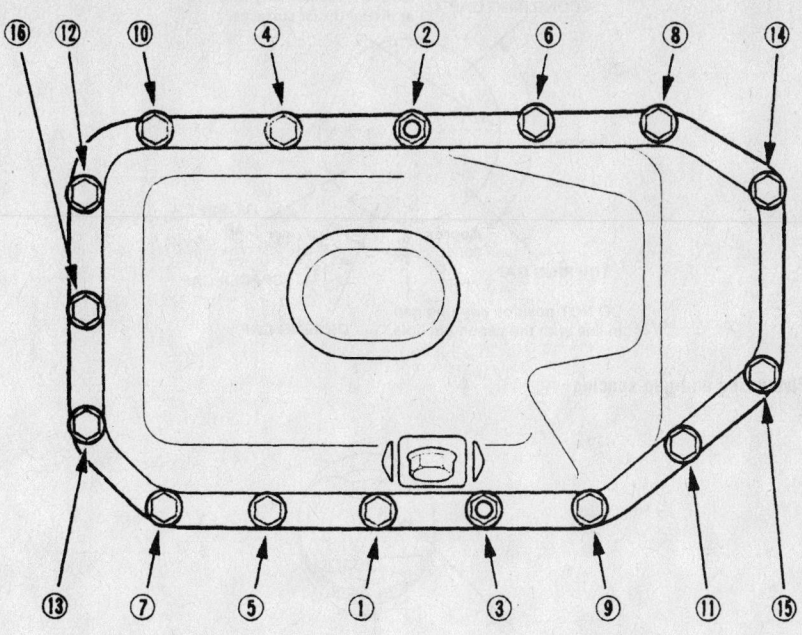

9359MG12

Oil pan fastener tightening sequence

Oil Pump

REMOVAL & INSTALLATION

1. Before servicing the vehicle, refer to the precautions in the beginning of this section.

2. Drain the engine oil.

3. Set the No. 1 piston to Top Dead Center (TDC).

4. Remove or disconnect the following:
   - Negative battery cable
   - Oil pan

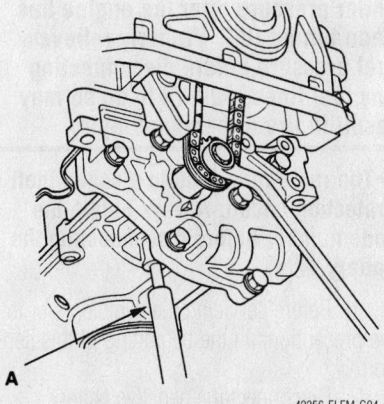

42356-ELEM-G04

Insert a 6mm pin into the maintenance hole in the lower balance shaft holder, through the rear balancer shaft to hold the shaft, then loosen the sprocket mounting bolt

   - Oil pump chain tensioner and discard

5. Insert a 6mm pin driver into the maintenance hole in the lower balance shaft holder and through the rear balancer shaft to hold the rear balancer shaft.

6. Loosen the oil pump sprocket mounting bolt.
   - Oil pump sprocket
   - Oil pump

To install:

7. Make sure that No.1 piston is at TDC.

8. Align the dowel pin on the rear balance shaft with the mark on the pump.

9. Insert a 6mm pin into the maintenance hole in the lower balance shaft holder, through the rear balancer shaft to hold the shaft.

10. Install or connect the following:
    - Engine oil to the threads of the oil pump sprocket mounting bolt
    - Oil pump and sprocket loosely

11. Remove the balance shaft holding pin.

12. Torque the 10mm mounting bolts to 33 ft. lbs. (44 Nm); the 8mm bolts to 16 ft. lbs. (22 Nm).

13. Torque the pulley bolt to 33 ft. lbs. (44 Nm).

14. Squeeze the new oil pump chain tensioner then install the set clip on it as shown in the illustration.

15. Install or connect the following:
    - New oil pump chain tensioner and torque the bolts to 9 ft. lbs. (12 Nm). Remove the set clip from the tensioner.
    - Oil pan

16. Fill the engine with oil.

Rear Main Seal

REMOVAL & INSTALLATION

1. Before servicing the vehicle, refer to the precautions in the beginning of this section.

2. Remove or disconnect the following:
   - Transaxle
   - Clutch pressure plate and disc, if equipped
   - Flywheel
   - Oil seal

To install:

3. Install or connect the following:

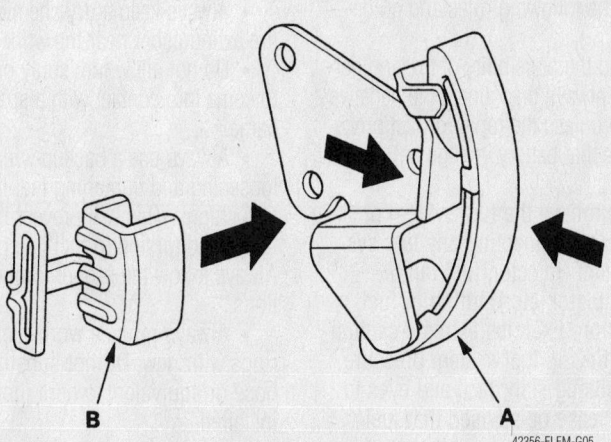

42356-ELEM-G05

Squeeze the new oil pump chain tensioner (A) then install the set clip (A) on it as shown. The clip is supplied with the new tensioner

- Oil seal. Drive the seal square into the seal case.
- Flywheel. Tighten the bolts in a crossing pattern to 76 ft. lbs. (103 Nm).
- Clutch pressure plate and disc, if equipped
- Transaxle

4. Check the fluid levels.

5. Start the engine and check for leaks.

Piston and Ring

POSITIONING

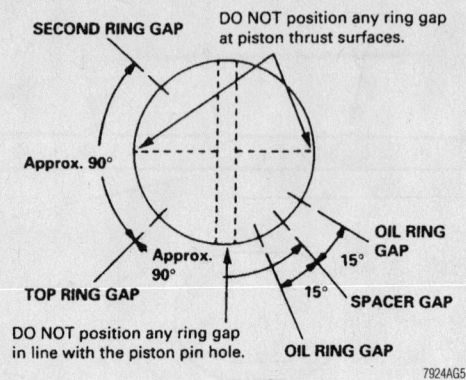

Piston ring end-gap spacing

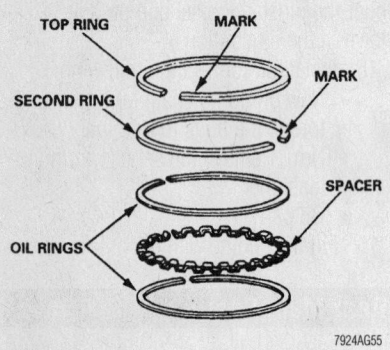

Piston ring positioning and top mark location

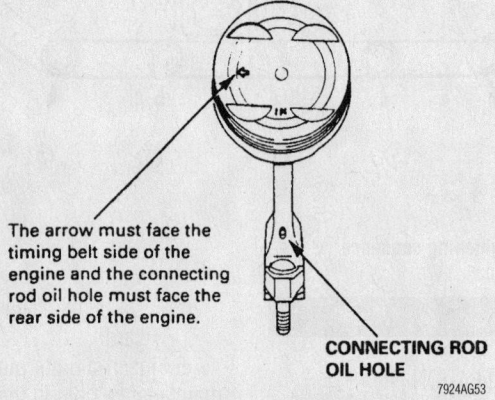

The arrow must face the timing belt side of the engine and the connecting rod oil hole must face the rear side of the engine.

CONNECTING ROD OIL HOLE

Piston and connecting rod assembly

FUEL SYSTEM

Fuel System Service Precautions

Safety is the most important factor when performing not only fuel system maintenance, but any type of maintenance. Failure to conduct maintenance and repairs in a safe manner may result in serious personal injury or death. Maintenance and testing of the vehicle's fuel system components can be accomplished safely and effectively by adhering to the following rules and guidelines:

- To avoid the possibility of fire and personal injury, always disconnect the negative battery cable unless the repair or test procedure requires that battery voltage be applied.
- Always relieve the fuel system pressure prior to disconnecting any fuel system component (injector, fuel rail, pressure regulator, etc.), fitting or fuel line connection. Exercise extreme caution whenever relieving fuel system pressure to avoid exposing skin, face and eyes to fuel spray. Please be advised that fuel

under pressure may penetrate the skin or any part of the body that it contacts.

- Always place a shop towel or cloth around the fitting or connection prior to loosening to absorb any excess fuel due to spillage. Ensure that all fuel spillage (should it occur) is quickly removed from engine surfaces. Ensure that all fuel soaked cloths or towels are deposited into a suitable waste container.
- Always keep a dry chemical (Class B) fire extinguisher near the work area.
- Do not allow fuel spray or fuel vapors to come into contact with a spark or open flame.
- Always use a backup wrench when loosening and tightening fuel line connection fittings. This will prevent unnecessary stress and torsion to fuel line piping. Always follow the proper torque specifications.
- Always replace worn fuel fitting O-rings with new. Do not substitute fuel hose or equivalent, where fuel pipe is installed.

Fuel System Pressure

RELIEVING

✳✳ CAUTION

The fuel injection system remains under pressure after the engine has been turned OFF. Properly relieve fuel pressure before disconnecting any fuel lines. Failure to do so may result in fire or personal injury.

➤ **The radio may contain a coded theft protection circuit. Always obtain the code number before disconnecting the battery.**

1. Before servicing the vehicle, refer to the precautions in the beginning of this section.

2. Disconnect the negative battery cable.

3. Remove the glove box, then remove the PGM-FI main relay (FUEL PUMP) from

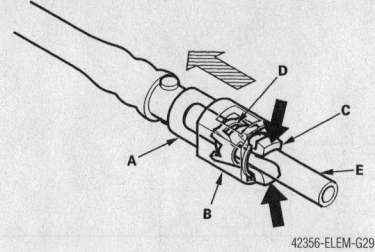

42356-ELEM-G29

Hold the quick-connect (A) connector (B) with one hand, then squeeze the retainer tabs (C) with the other hand to release them from the locking pawls (D)

the fuse/relay box. Start the engine and let it run until it stalls.

4. Turn the engine OFF.
5. Remove the fuel filler cap.
6. Remove the quick-connect fitting cover.
7. Clean any dirt from the quick-connect fitting.
8. Place a rag or shop towel over quick-connect fitting.
9. Detach the quick-connect fitting by holding the connector with one hand, then squeeze the retainer tabs with the other hand to release them from the locking pawls. Pull the connector off.

✳✳ CAUTION

Do not allow fuel spray or fuel vapors to come in contact with a spark or open flame. Keep a dry chemical fire extinguisher nearby. Never store fuel in an open container due to risk of fire or explosion.

➡**A fuel pressure gauge may be attached at the quick-connect location.**

10. Connect the quick-connect fitting, making sure the locking pawls are properly engaged.
11. Clean up any fuel spilled on the engine and intake manifold.
12. Install the fuel pump relay to the under dash fuel/relay box and install the glove box.
13. Install the fuel filler cap.
14. Reconnect the negative battery cable.
15. Turn the ignition **ON**, but don't start the engine. Repeat this 2 or 3 times to pressurize the fuel system. Check for fuel leaks.
16. Enter the radio security code.

Fuel Filter

REMOVAL & INSTALLATION

➡**The fuel filter should be replaced whenever the fuel pressure drops below 48 psi, after making sure that**

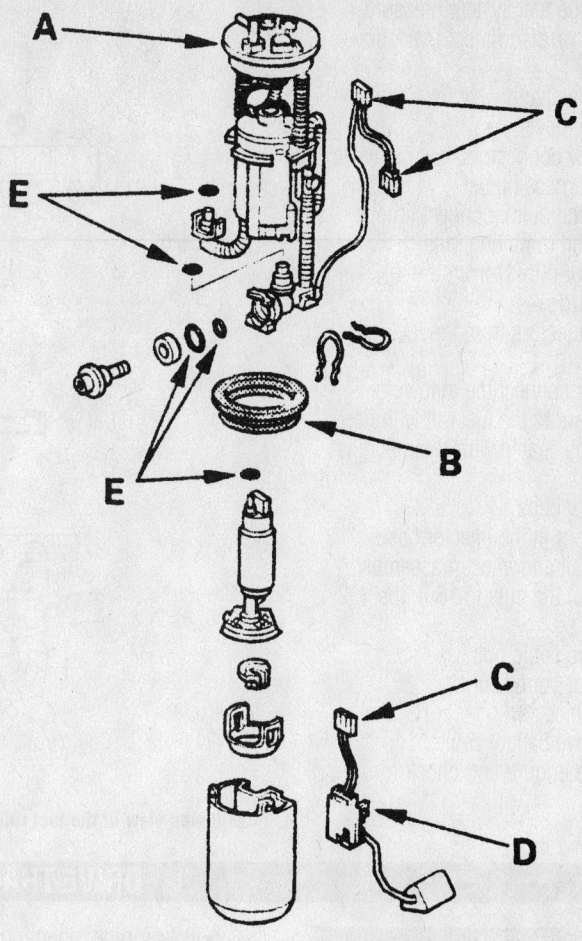

9359MG13

Exploded view of the fuel filter mounting

the fuel pump and fuel pressure regulator are okay.

1. Before servicing the vehicle, refer to the precautions in the beginning of this section.
2. Relieve the fuel system pressure.
3. Remove or disconnect the following:
   • Negative battery cable
   • Fuel pump
   • Fuel filter carrier (A)
   • Fuel filter

To install:
4. Install or connect the following:
   • Fuel filter
   • Fuel lines
   • New gasket (B)
   • New o-rings (E)
   • Connectors (C)
   • Sending unit (D)
5. Start the engine and check for leaks.

Fuel Pump

REMOVAL & INSTALLATION

1. Before servicing the vehicle, refer to the precautions in the beginning of this section.

2. Relieve the fuel system pressure.
3. Remove or disconnect the following:
   • Negative battery cable
   • Fuel filler cap
   • Center console, then both track floor covers and sill trims.
4. Fold back the floor covering until you can get to the access panel
   • Access panel from the floor
   • Fuel pump connector
   • Fuel supply and return line quick-connect fittings
   • Fuel pump locknut, using special tool No. 07XAA-001010A
   • Fuel pump sending assembly
5. Installation is the reverse of removal.

Fuel Injector

REMOVAL & INSTALLATION

1. Before servicing the vehicle, refer to the precautions in the beginning of this section.

2. Relieve the fuel system pressure.

3. Remove or disconnect the following:

- Negative battery cable
- Engine cover
- Injector connectors, ground cable and harness holder
- Fuel line quick-connect fittings
- Fuel rail mounting nuts
- Injector clip(s) from the injector(s)
- Fuel injectors from the fuel rail

To install:

4. Install or connect the following:

- Injectors to the fuel rail with new O-rings coated with clean engine oil.
- Injector clips
- Injectors in the injector base
- Fuel rail and injector assembly. Tighten the nuts to 16 ft. lbs. (22 Nm).
- Ground cable bolt
- Injector connectors
- Fuel lines
- Negative battery cable

5. Start the engine and check for leaks.

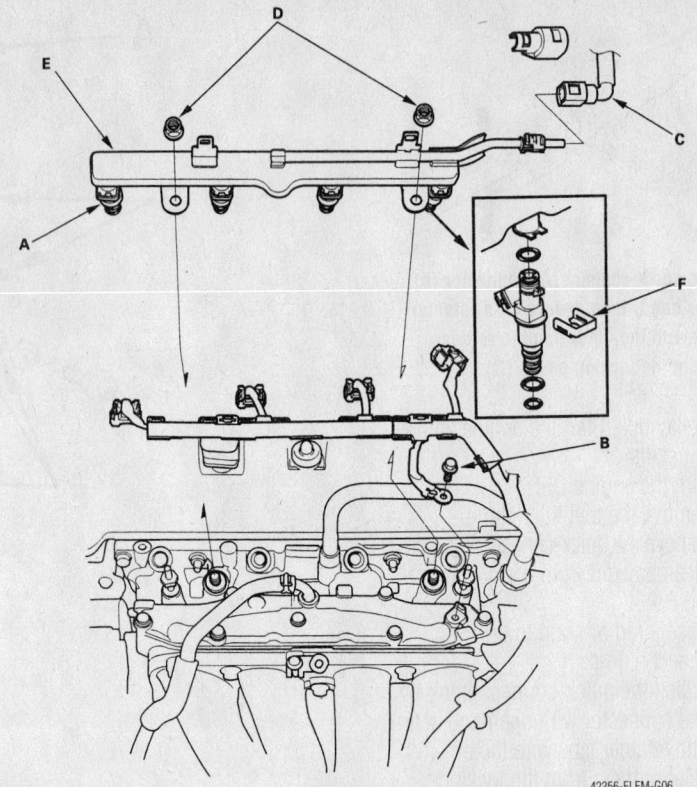

Exploded view of the fuel rail (E), injectors (A) and related components

---

DRIVE TRAIN

Transaxle Assembly

REMOVAL & INSTALLATION

Automatic Transaxle

1. Before servicing the vehicle, refer to the precautions in the beginning of this section.

2. Drain the transaxle.

3. Remove or disconnect the following:

- Battery
- Battery tray
- Air intake assembly
- Splash shield
- Transaxle ground cable
- Starter motor
- Clutch pressure control solenoid valve connector
- Mainshaft speed sensor connector
- Clutch pressure switch connectors
- Shift control solenoid valve connectors
- Lockup control solenoid connector
- Countershaft speed sensor connector
- Transaxle oil cooler lines
- Engine wiring harness from the air cleaner bracket

- Water pipe mounting bolt
- Brake booster and EVAP line mounting bolts. Attach an engine support hanger to the head to support the weight of the engine.
- Stabilizer link from the lower arm
- Lower arms from the knuckles
- Torque converter nuts
- Shift cable
- Front mount bolt and nut
- Rear mount bolts
- Subframe (see the Engine Removal and Installation procedure)
- Rear driveshaft, if equipped

4. Separate the inner CV-joints from the transaxle and intermediate shaft and support the axle halfshafts out of the work area with safety wire.

- Intermediate shaft
- Engine stiffener
- Transaxle mount bolts and nuts
- Transaxle

5. Installation is the reverse of removal. Observe the following torques:

- Air cleaner housing bracket bolt: 16 ft. lbs. (22 Nm)
- Front mount bolts: 47 ft. lbs. (64 Nm)
- Rear mount bracket bolts: 40 ft. lb. (54 Nm)

- Transmission-to-engine bolts: 47 ft. lbs. (64 Nm)
- Upper transmission mount bolt: 40 ft. lbs. (54 Nm)
- Upper transmission mount nuts: 40 ft. lbs. (54 Nm)
- Rear driveshaft bolts: 24 ft. lbs. (32 Nm)
- Subframe bolts: 76 ft. lbs. (103 Nm)

Manual Transaxle

1. Before servicing the vehicle, refer to the precautions in the beginning of this section.

2. Remove or disconnect the following:

- Negative battery cable
- Air intake assembly
- Transaxle ground cable
- Vehicle Speed Sensor (VSS) connector
- Splash shield
- Shift cables and bracket
- Clutch slave cylinder and hose bracket
- Wire harness bracket
- Water pipe mounting bolt
- Brake booster and EVAP line mounting bolts. Attach an engine

support hanger to the head to support the weight of the engine.
- Upper transmission mounting bolts
- Transaxle mount and bracket

3. Separate the inner CV-joints from the transaxle and intermediate shaft and support the axle halfshafts out of the work area with safety wire.

4. Remove or disconnect the following:
- Intermediate shaft
- Right front mount and bracket
- Rear engine mounting bolts
- Rear driveshaft
- Subframe (see the Engine Removal and Installation procedure)

- Clutch housing cover
- Transaxle

5. Installation is the reverse of removal. Observe the following torques:
- Transaxle rear mount and bracket. Tighten the bracket bolts to 40 ft. lbs. (54 Nm) and the through bolt to 47 ft. lbs. (64 Nm)
- Transaxle. Tighten the flange bolts to 47 ft. lbs. (64 Nm)
- Front mount and bracket. Tighten the bolts to 47 ft. lbs. (64 Nm).
- Clutch housing cover. Tighten the bolts to 29 ft. lbs. (39 Nm)
- Subframe: 76 ft. lbs. (98 Nm)

Clutch

ADJUSTMENTS

The Element is equipped with a hydraulic clutch system. No adjustment is necessary.

REMOVAL & INSTALLATION

1. Before servicing the vehicle, refer to the precautions in the beginning of this section.
2. Remove or disconnect the following:
- Negative battery cable

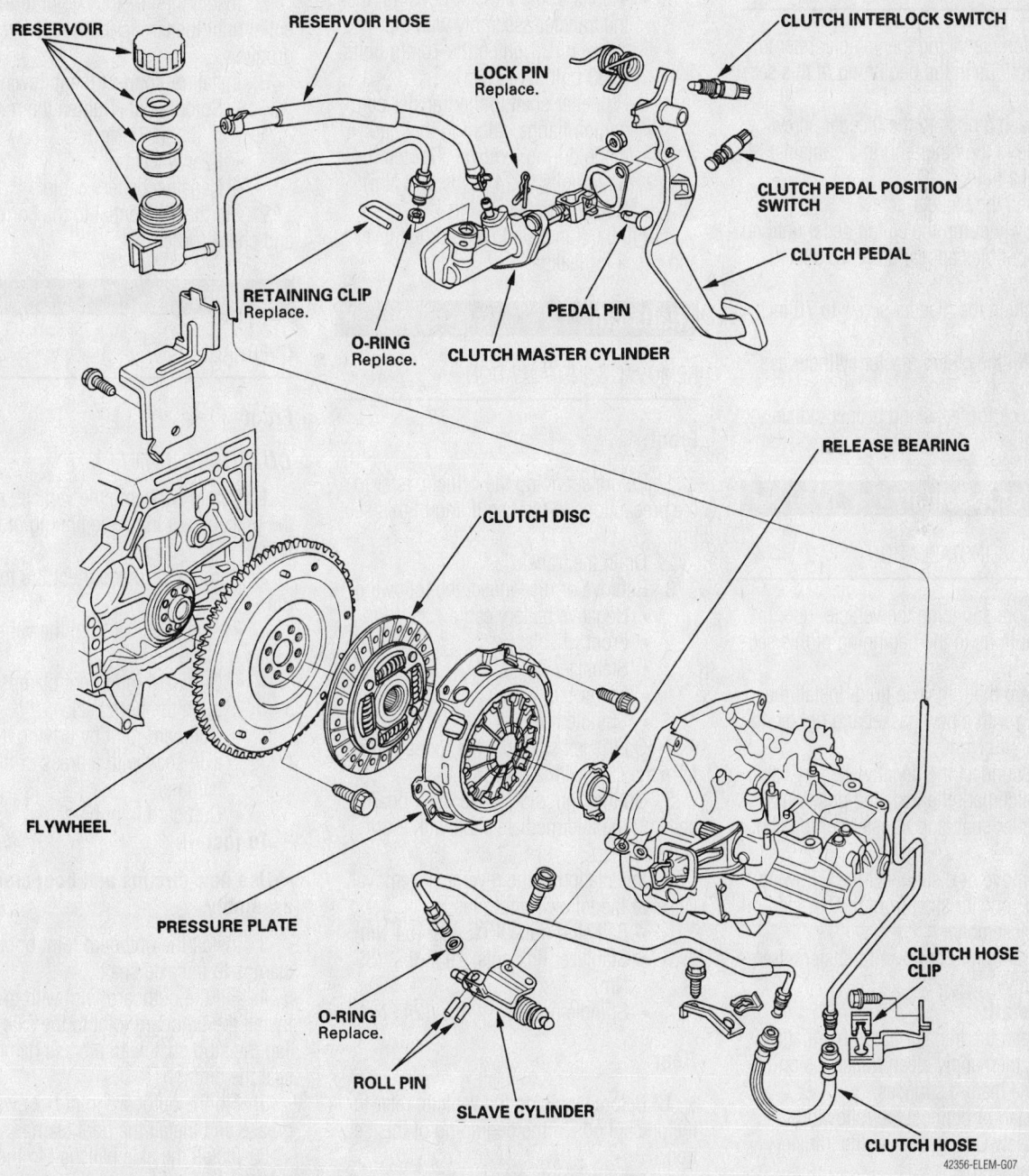

Exploded view of the clutch system components

42356-ELEM-G07

- Transaxle
- Pressure plate. Loosen the bolts evenly in a crossing pattern.
- Clutch disc

To install:

3. Install the clutch disc and pressure plate. Tighten the pressure plate bolts in a crisscross pattern, in several steps to 19 ft. lbs. (26 Nm).

4. Install or connect the following:
- Transaxle
- Negative battery cable

Hydraulic Clutch System

BLEEDING

1. Before servicing the vehicle, refer to the precautions in the beginning of this section.

2. Attach a hose to the bleeder screw and suspend the other end in a container of clean brake fluid.

3. Open the bleeder screw.

4. Slowly pump the clutch pedal until no more air bubbles appear at the bleeder hose.

5. Tighten the bleeder screw to 70 inch lbs. (8 Nm).

6. Refill the clutch master cylinder as necessary.

7. Check for leaks and proper clutch operation.

Transfer Assembly

REMOVAL & INSTALLATION

1. Before servicing the vehicle, refer to the precautions in the beginning of this section.

2. Drain the transaxle fluid. Install the drain plug with a new gasket and tighten to 36 ft. lbs. (49 Nm).

3. Disconnect the negative battery cable.

4. Matchmark the installed position of the propeller shaft and transfer companion flange.

5. Remove or disconnect the following:
- Propeller shaft from the transfer assembly
- Mounting bolts and transfer assembly

To install:

6. Clean the transfer assembly mating surfaces, then apply clean transmission fluid to the mating surfaces.

7. Install or connect the following:
- New O-ring seal on the transfer assembly

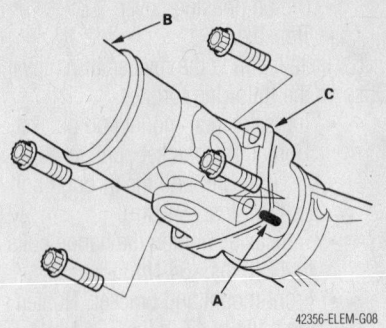

42356-ELEM-G08

Matchmark (A) the installed position of the propeller shaft (B) and transfer companion flange (C)

- 4 bolts in the transfer housing, then the transfer assembly with the dowel pin. Tighten the 10mm bolts to 33 ft. lbs. (44 Nm).
- Propeller shaft to the transfer companion flange, aligning the mark made during removal. Tighten the 8mm bolts to 24 ft. lbs. (33 Nm).
- Negative battery cable

8. Fill the transaxle to the correct level and check for leaks.

Halfshaft

REMOVAL & INSTALLATION

Front

1. Before servicing the vehicle, refer to the precautions in the beginning of this section.

2. Drain the transaxle.

3. Remove or disconnect the following:
- Negative battery cable
- Front wheels
- Stabilizer bar
- Lower ball joint
- Spindle nut

4. On the left side, pry the inboard joint from the case with a prybar.

5. On the right side, drive the inboard shaft off the intermediate shaft with a drift and hammer.

6. Installation is the reverse of removal. Observe the following torques:
- Ball stud nuts: 40 ft. lbs. (54 Nm)
- Stabilizer link nuts: 29 ft. lbs. (39 Nm)
- Spindle nut: 181 ft. lbs. (245 Nm)

Rear

1. Before servicing the vehicle, refer to the precautions in the beginning of this section.

2. Drain the differential.

3. Remove or disconnect the following:
- Negative battery cable
- Rear wheels
- Spindle nut

4. Pry the inboard joint from the differential.

5. Remove the outer CV-joint stub shaft from the hub by tapping the stub shaft with a plastic hammer.

To install:

➡ **Use new circlips and self-locking nuts for assembly.**

6. Install the outer CV-joint stub shaft into the hub.

7. Install the inner CV-joint to the differential until the circlip locks in the retaining groove.

8. Install or connect the following:
- Spindle nut. Tighten the nut to 134 ft. lbs. (181 Nm).
- Rear wheels
- Negative battery cable

9. Fill the differential to the correct level and check for leaks.

CV-Joint

OVERHAUL

Front

OUTBOARD JOINT

1. Before servicing the vehicle, refer to the precautions in the beginning of this section.

2. Remove or disconnect the following:
- Axle halfshaft from the vehicle and place it in a vise
- Outboard joint boot clamps and push the boot back
- Outboard joint by driving it off the axle shaft with a brass drift and hammer
- Outboard joint boot

To install:

➡ **Use new circlips and boot clamps for assembly.**

3. Install the outboard joint boot and clamps to the axle shaft.

4. Fill the outboard joint with grease. Install the outboard joint to the axle shaft. Tap the stub shaft with a brass hammer to seat the circlip.

5. Fill the outboard joint boot with grease and install the boot clamps.

6. Install the axle halfshaft to the vehicle.

INBOARD JOINT

1. Before servicing the vehicle, refer to the precautions in the beginning of this section.

2. Remove or disconnect the following:
   - Axle halfshaft from the vehicle.
   - Inboard joint boot clamps and push the boot back
   - Inboard joint housing from the axle
   - Rollers from the spider
   - Snapring and the spider from the axle shaft
   - Inboard joint boot

To install:

➡**Use new circlips and boot clamps for assembly.**

3. Install or connect the following:
   - Inboard joint boot and clamps to the axle shaft
   - Spider with a new snapring
   - Rollers to the spider

4. Fill the joint housing with grease and install it.

5. Fill the inboard joint boot with grease and install the boot clamps.

6. Install the axle halfshaft to the vehicle.

Rear

1. Before servicing the vehicle, refer to the precautions in the beginning of this section.

2. Remove or disconnect the following:
   - Axle halfshaft from the vehicle

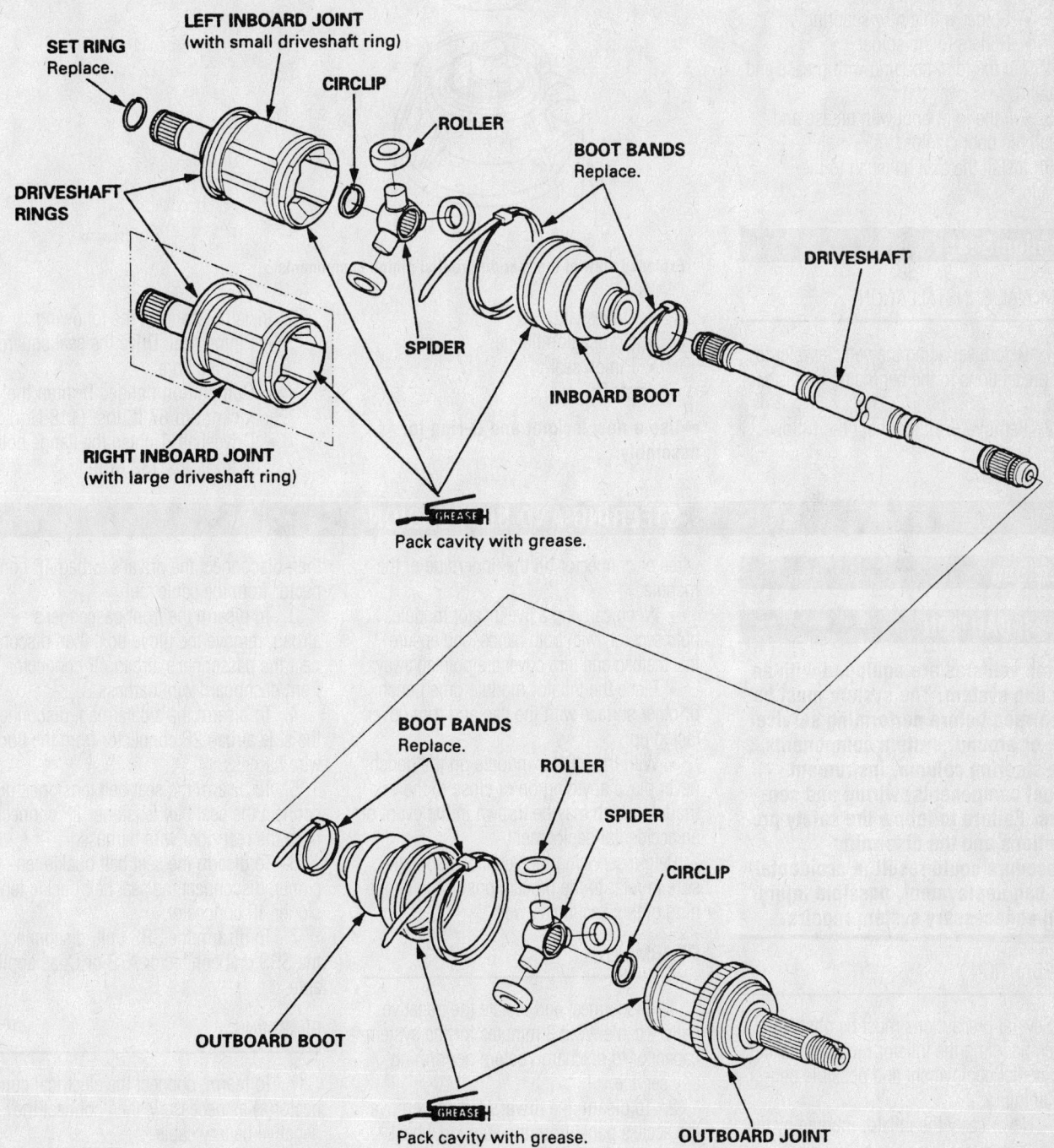

SET RING
Replace.

LEFT INBOARD JOINT
(with small driveshaft ring)

CIRCLIP

ROLLER

BOOT BANDS
Replace.

DRIVESHAFT
RINGS

DRIVESHAFT

SPIDER

INBOARD BOOT

RIGHT INBOARD JOINT
(with large driveshaft ring)

GREASE
Pack cavity with grease.

BOOT BANDS
Replace.

ROLLER

SPIDER

CIRCLIP

OUTBOARD BOOT

GREASE
Pack cavity with grease.

OUTBOARD JOINT

9308MG30

Exploded view of the rear axle

- Joint boot clamps and push the boot back
- Joint housing from the axle
- Rollers from the spider
- Snapring and the spider from the axle shaft
- Joint boot

To install:

➡ **Use new circlips and boot clamps for assembly.**

3. Install or connect the following:
- Joint boot and clamps to the axle shaft
- Spider with a new snapring
- Rollers to the spider

4. Fill the joint housing with grease and install it.

5. Fill the joint boot with grease and install the boot clamps.

6. Install the axle halfshaft to the vehicle.

Pinion Seal

REMOVAL & INSTALLATION

1. Before servicing the vehicle, refer to the precautions in the beginning of this section.

2. Remove or disconnect the following:

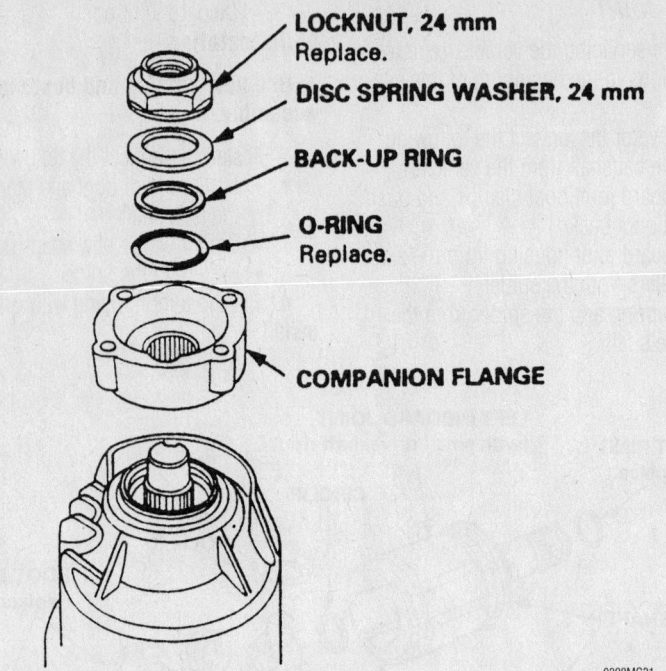

- LOCKNUT, 24 mm
  Replace.
- DISC SPRING WASHER, 24 mm
- BACK-UP RING
- O-RING
  Replace.
- COMPANION FLANGE

9308MG31

Exploded view of the rear differential pinion components

- Driveshaft
- Companion flange
- Pinion seal

To install:

➡ **Use a new locknut and O-ring for assembly.**

3. Install or connect the following:
- Pinion seal. Drive the seal square into the bore.
- Companion flange. Tighten the locknut to 87 ft. lbs. (118 Nm).
- Driveshaft. Tighten the flange bolts to 24 ft. lbs. (32 Nm).

STEERING AND SUSPENSION

Air Bag

✳✳ CAUTION

Some vehicles are equipped with an air bag system. The system must be disarmed before performing service on, or around, system components, the steering column, instrument panel components, wiring and sensors. Failure to follow the safety precautions and the disarming procedure could result in accidental air bag deployment, possible injury and unnecessary system repairs.

PRECAUTIONS

Several precautions must be observed when handling the inflator module to avoid accidental deployment and possible personal injury.

- Never carry the inflator module by the wires or connector on the underside of the module.
- When carrying a live inflator module, hold securely with both hands, and ensure that the bag and trim cover are pointed away.
- Place the inflator module on a bench or other surface with the bag and trim cover facing up.
- With the inflator module on the bench, never place anything on or close to the module which may be thrown in the event of an accidental deployment.

Before servicing the vehicle, also make sure to refer to the precautions in the beginning of this section as well.

DISARMING

1. Disconnect and isolate the negative battery cable. Wait 3 minutes for the system capacitor to discharge before performing any service.

2. To disarm the driver's airbag, remove the access panel from the steering wheel, then disconnect the driver's airbag 4P connector from the cable reel.

3. To disarm the front passenger's airbag, remove the glove box, then disconnect the passenger's airbag 4P connector from dashboard wire harness B.

4. To disarm the side airbag, disconnect the side airbag 2P connector from the floor wire harness.

5. To disarm the seat belt tensioner, disconnect the seat belt tensioner 2P connector from the rear door wire harness.

6. To disarm the seat belt buckle tensioner, disconnect the seat belt buckle tensioner 4P connector.

7. To disarm the SRS unit, disconnect the SRS unit connector A, B or C, as applicable.

REARMING

1. To rearm, connect the electrical connector(s) as necessary, then connect the negative battery cable.

Power Rack and Pinion Steering Gear

REMOVAL & INSTALLATION

✳✳ WARNING

Do not permit the steering wheel to turn whenever the steering gear is disconnected from the steering column. Damage to the air bag wiring can result.

1. Before servicing the vehicle, refer to the precautions in the beginning of this section.

2. Center the steering wheel and lock it in position.

3. Remove or disconnect the following:
   - Negative battery cable
   - Air bag and steering wheel
   - Front wheels
   - Driver's side dashboard lower cover and undercover
   - Air cleaner housing
   - Steering joint bolts
   - Tie rod ends
   - Steering hoses
   - Left side flange bolts
   - Mounting brackets

4. Lower the unit so the pinion shaft points outward. Remove the pinion shaft grommet. The steering gear is removed through the driver's side.

5. Installation is the reverse of removal. Observe the following torques:
   - Mounting bracket and side flange bolts: 46 ft. lbs. (62 Nm)
   - Supply line flare nut: 27 ft. lbs. (37 Nm)
   - Tie rod ball stud nuts: 32 ft. lbs. (43 Nm)
   - Steering joint bolts: 21 ft. lbs. (28 Nm)

Strut (Damper)

REMOVAL & INSTALLATION

Front

1. Before servicing the vehicle, refer to the precautions in the beginning of this section.

2. Remove or disconnect the following:
   - Front wheel
   - Tie rod end
   - Brake hose retainer
   - ABS sensor harness bracket and brake hose bracket. Do not disconnect the wheel sensor connector.
   - Pinch bolts from the damper, while holding the nuts
   - Flange nuts from the top of the damper
   - Strut (damper), after lowering the lower control arm

To install:

➡**Use new self-locking fasteners for assembly.**

3. Install or connect the following:
   - Strut (damper). Tighten the upper mounting nuts to 33 ft. lbs. (44 Nm).
   - Tighten the pinch bolts to 116 ft. lbs. (157 Nm)
   - ABS sensor
   - Tie rod end
   - Brake hose retainer
   - Front wheel

Rear

1. Before servicing the vehicle, refer to the precautions in the beginning of this section.

2. Support the vehicle under the lower control arm.

3. Remove or disconnect the following:
   - Rear wheel

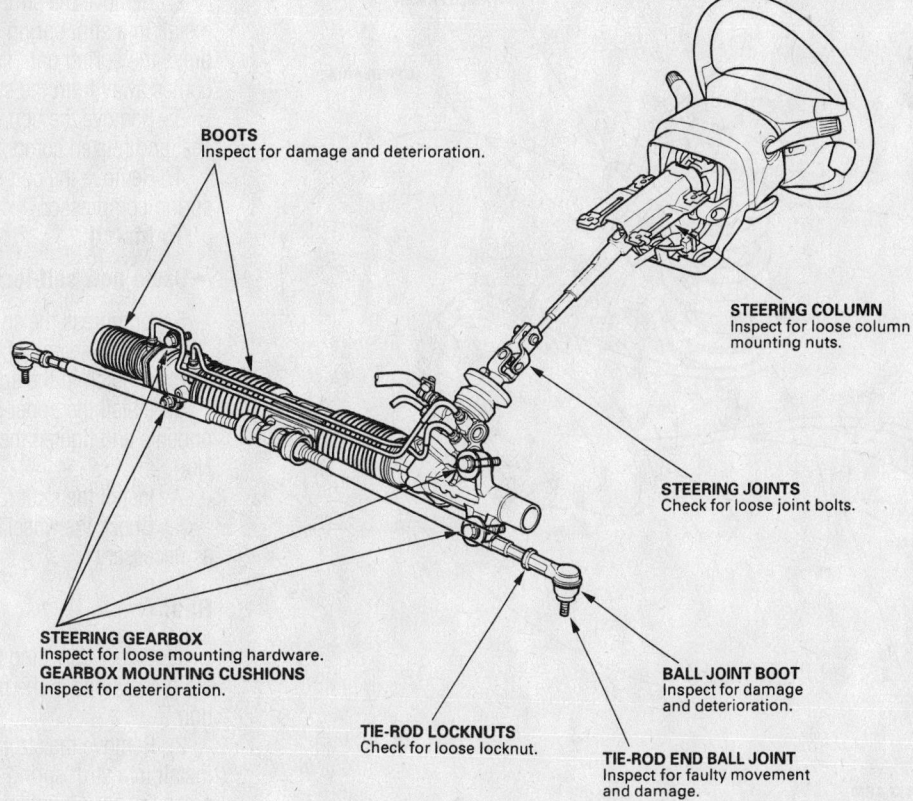

BOOTS
Inspect for damage and deterioration.

STEERING COLUMN
Inspect for loose column mounting nuts.

STEERING JOINTS
Check for loose joint bolts.

STEERING GEARBOX
Inspect for loose mounting hardware.
GEARBOX MOUNTING CUSHIONS
Inspect for deterioration.

TIE-ROD LOCKNUTS
Check for loose locknut.

BALL JOINT BOOT
Inspect for damage and deterioration.

TIE-ROD END BALL JOINT
Inspect for faulty movement and damage.

42356-ELEM-G09

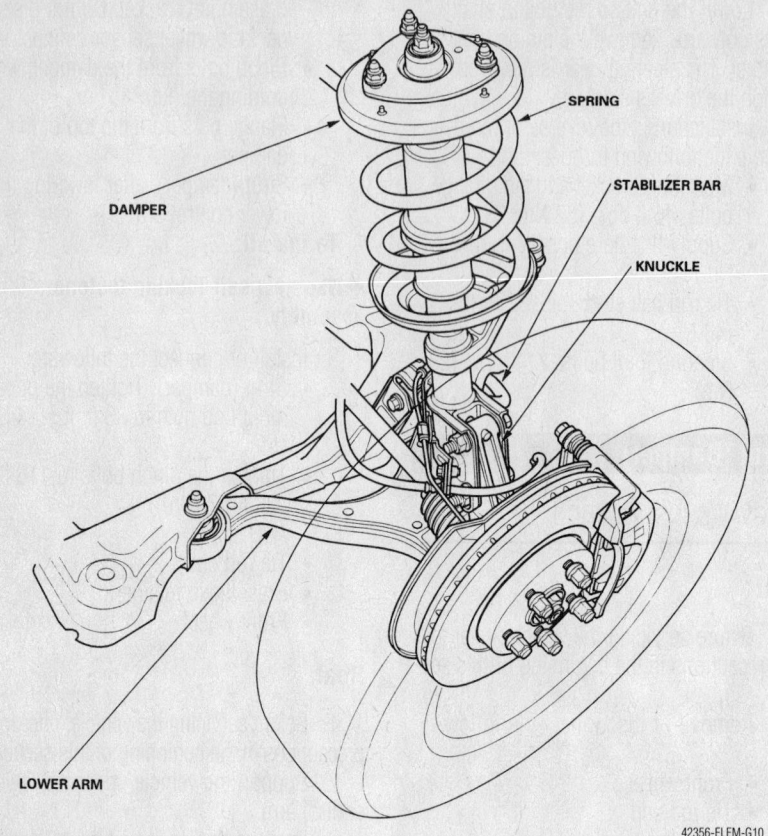

DAMPER

SPRING

STABILIZER BAR

KNUCKLE

LOWER ARM

Front suspension components

42356-ELEM-G10

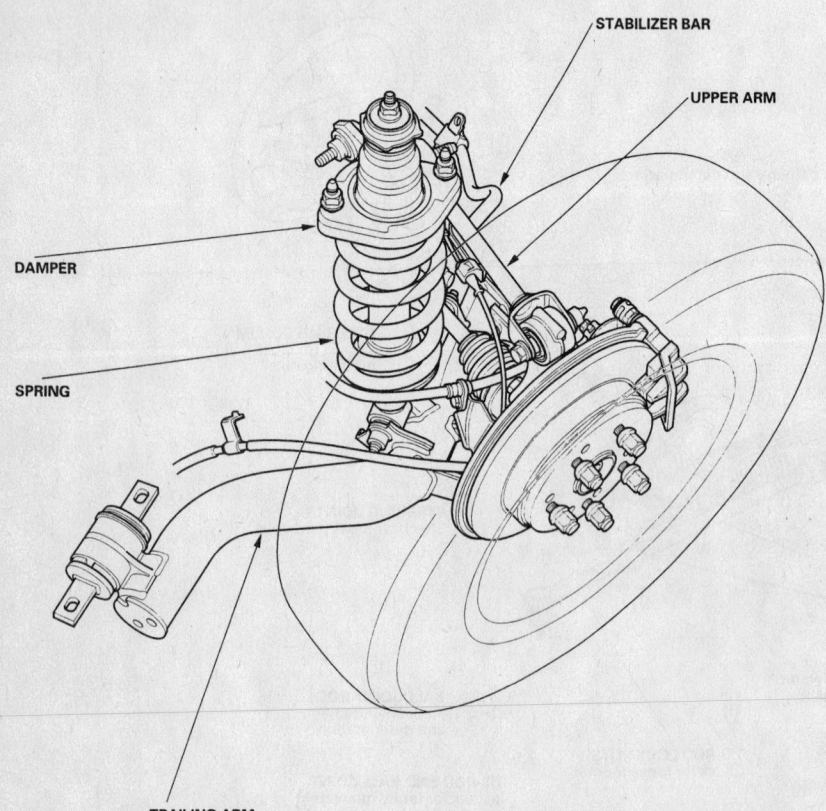

STABILIZER BAR

UPPER ARM

DAMPER

SPRING

TRAILING ARM

Rear suspension components

42356-ELEM-G12

- Flange bolt from the bottom of the damper (strut)
- Evaporative emission (EVAP) canister bolts, and loosen the EVAP canister mounting (left side only)
- Interior access panel, if necessary
- Flange nuts from the top of the damper in the cargo area
- Strut

To install:

4. Install or connect the following:
   - Strut. Position the damper mounting base so the indent mark is toward the inside of the vehicle.
   - Upper flange nuts, hand-tight only
   - Bottom flange bolt, hand-tight only

5. With the suspension raised with a jack to load it with the vehicles weight, tighten the bottom bolt to 69 ft. lbs. and the top nuts to 54 ft. lbs. (74 Nm).
   - Interior access panel, if necessary
   - EVAP canister mounting bolts
   - Rear wheel

Coil Spring

REMOVAL & INSTALLATION

Front

1. Before servicing the vehicle, refer to the precautions in the beginning of this section.

2. Remove the strut from the vehicle and install in a strut spring compressor. Compress the spring until the end of the spring comes away from the spring seat.

3. Remove the upper strut mount, spring seat and related components.

4. Remove the coil spring from the strut spring compressor.

To install:

➡**Use a new self-locking nut.**

5. Compress the spring and position the strut so that the end of the spring aligns with the notch in the spring seat.

6. Install the upper strut mounting components and tighten the nut to 33 ft. lbs. (44 Nm).

7. Install the strut to the vehicle.

8. Check the wheel alignment and adjust as necessary.

Rear

1. Before servicing the vehicle, refer to the precautions in the beginning of this section.

2. Remove the strut from the vehicle and install in a strut spring compressor. Compress the spring until the end of the spring comes away from the spring seat.

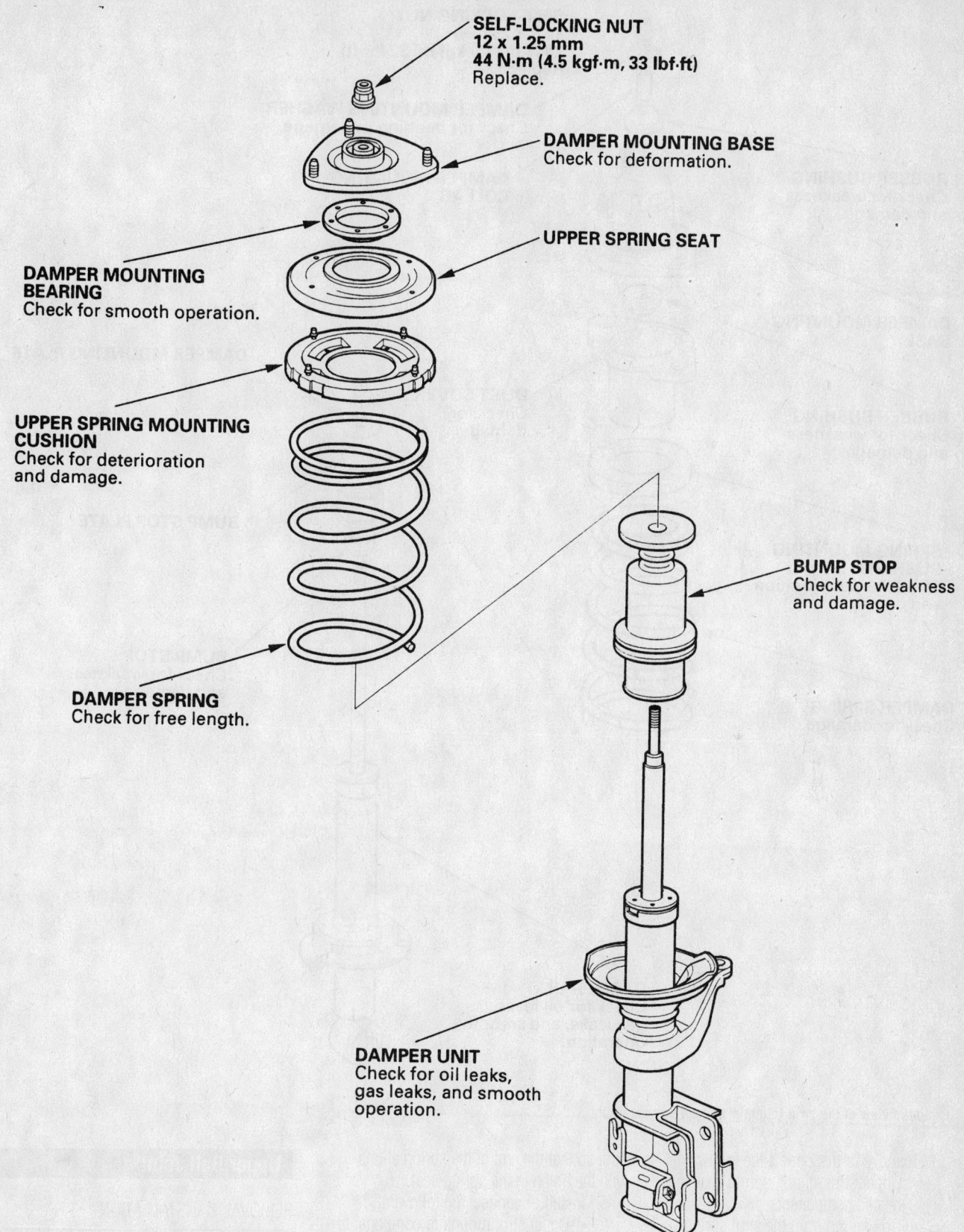

SELF-LOCKING NUT
12 x 1.25 mm
44 N·m (4.5 kgf·m, 33 lbf·ft)
Replace.

DAMPER MOUNTING BASE
Check for deformation.

UPPER SPRING SEAT

**DAMPER MOUNTING
BEARING**
Check for smooth operation.

**UPPER SPRING MOUNTING
CUSHION**
Check for deterioration
and damage.

BUMP STOP
Check for weakness
and damage.

DAMPER SPRING
Check for free length.

DAMPER UNIT
Check for oil leaks,
gas leaks, and smooth
operation.

42356-ELEM-G11

Exploded view of the front strut (damper and spring) assembly

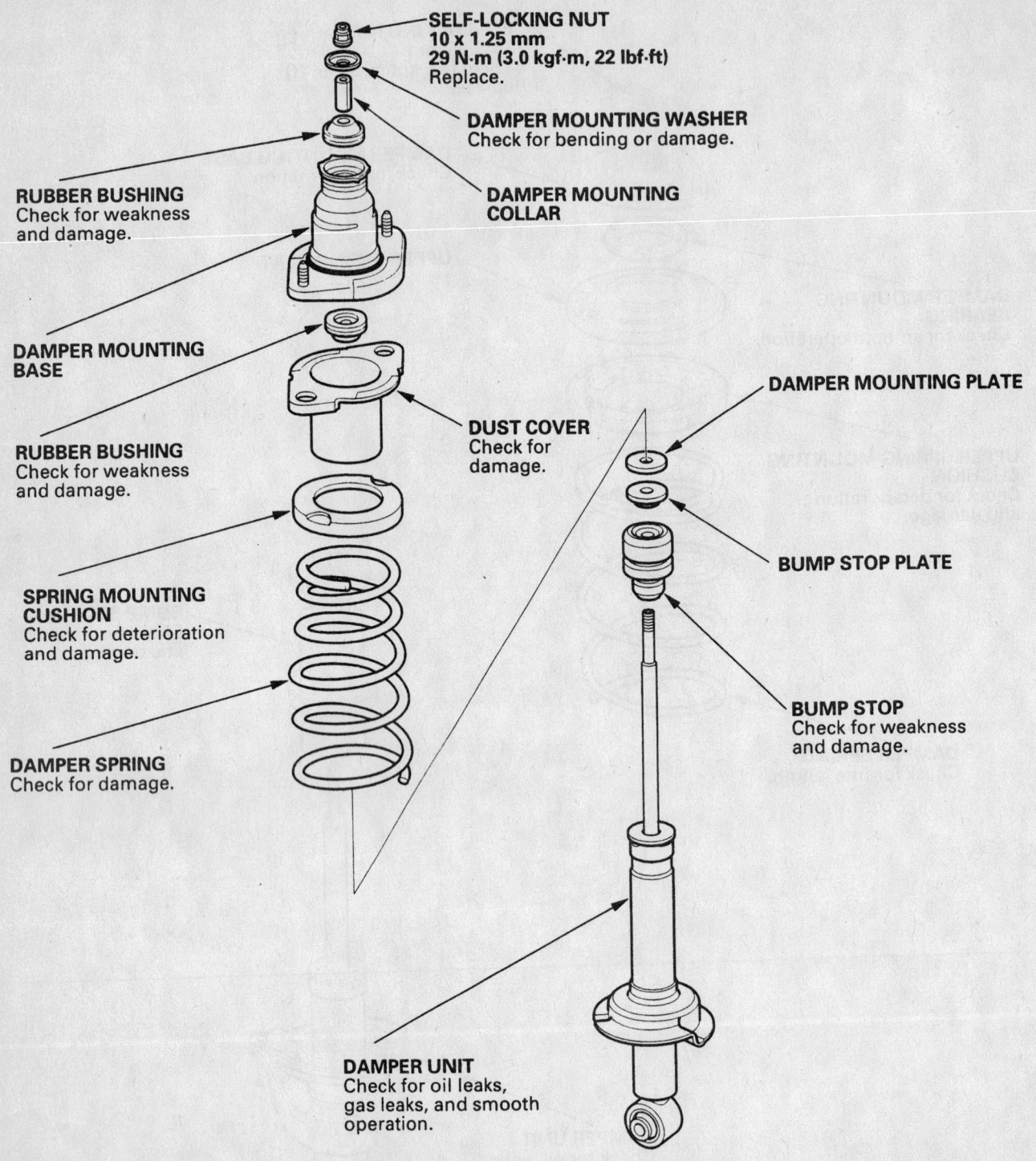

SELF-LOCKING NUT
10 x 1.25 mm
29 N·m (3.0 kgf·m, 22 lbf·ft)
Replace.

DAMPER MOUNTING WASHER
Check for bending or damage.

RUBBER BUSHING
Check for weakness
and damage.

**DAMPER MOUNTING
COLLAR**

**DAMPER MOUNTING
BASE**

RUBBER BUSHING
Check for weakness
and damage.

DUST COVER
Check for
damage.

**SPRING MOUNTING
CUSHION**
Check for deterioration
and damage.

DAMPER SPRING
Check for damage.

DAMPER MOUNTING PLATE

BUMP STOP PLATE

BUMP STOP
Check for weakness
and damage.

DAMPER UNIT
Check for oil leaks,
gas leaks, and smooth
operation.

42356-ELEM-G13

Exploded view of the strut (damper and spring) assembly

3. Remove or disconnect the following:
- Upper strut mount, spring seat and related components
- Coil spring from the strut spring compressor

To install:

➡**Use a new self-locking nut.**

4. Compress the spring and position the strut so that the end of the spring aligns with the notch in the spring seat.
5. Install or connect the following:
- Upper strut mounting components and tighten the nut to 22 ft. lbs. (29 Nm).
- Strut to the vehicle
6. Check the wheel alignment and adjust as necessary.

Upper Ball Joint

REMOVAL & INSTALLATION

The upper ball joints are replaced with the upper control arms as an assembly.

Lower Ball Joint

REMOVAL & INSTALLATION

The ball joint is not replaceable.

Upper Control Arm

REMOVAL & INSTALLATION

1. Before servicing the vehicle, refer to the precautions in the beginning of this section.
2. Support the lower control arm assembly with a floor jack.
3. Remove or disconnect the following:

   - Upper ball joint
   - Inner control arm flange bolts.
   - Upper control arm

To install:

➡**Use new self-locking nuts for assembly.**

4. Install the upper control arm. Tighten the ball joint nut to 29–35 ft. lbs. (39–47 Nm) and the inner flange bolts to 40 ft. lbs. (54 Nm).

CONTROL ARM BUSHING REPLACEMENT

The upper control arm bushings are serviced with the upper control arm as an assembly.

Lower Control Arm

REMOVAL & INSTALLATION

1. Before servicing the vehicle, refer to the precautions in the beginning of this section.
2. Remove or disconnect the following:

   - Front wheel
   - Stabilizer link
   - Lower arm from the knuckle
   - Lower arm

3. Installation is the reverse of removal. Observe the following torques:

   - Lower arm bolts: 61 ft. lbs. (83 Nm)
   - Ball stud nut: 51 ft. lbs. (69 Nm)
   - Stabilizer link: 29 ft. lbs. (39 Nm)

CONTROL ARM BUSHING REPLACEMENT

The lower control arm front inner bushing and the damper fork bushing are serviced with the control arm as an assembly.

REAR INNER BUSHING

1. Before servicing the vehicle, refer to the precautions in the beginning of this section.
2. Remove or disconnect the following:

   - Front wheel
   - Rear bushing bracket
   - Rear bushing

To install:

➡**Use a new self-locking nut for assembly.**

3. Install or connect the following:

   - Rear bushing. Tighten the nut to 61 ft. lbs. (83 Nm).
   - Rear bushing bracket. Tighten the bolts to 66 ft. lbs. (89 Nm).
   - Front wheel

4. Check the wheel alignment and adjust as necessary.

Wheel Bearings

ADJUSTMENT

The wheel bearings are sealed units and are not adjustable.

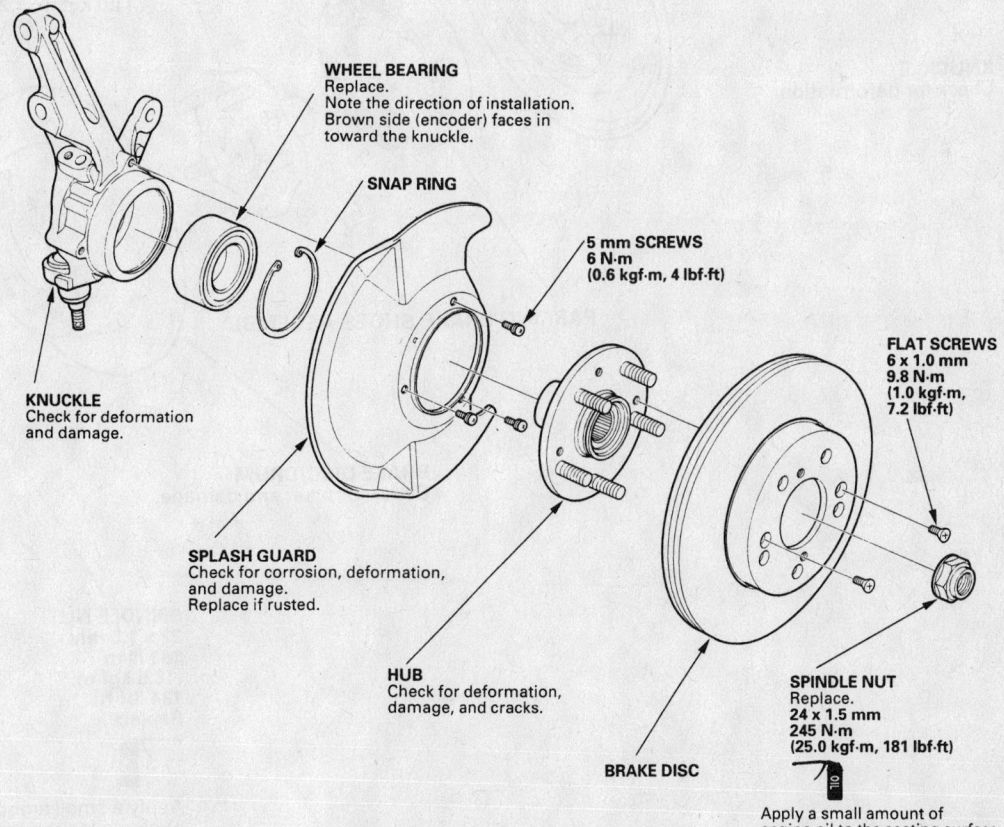

WHEEL BEARING
Replace.
Note the direction of installation.
Brown side (encoder) faces in toward the knuckle.

SNAP RING

5 mm SCREWS
6 N·m
(0.6 kgf·m, 4 lbf·ft)

FLAT SCREWS
6 x 1.0 mm
9.8 N·m
(1.0 kgf·m, 7.2 lbf·ft)

KNUCKLE
Check for deformation and damage.

SPLASH GUARD
Check for corrosion, deformation, and damage.
Replace if rusted.

HUB
Check for deformation, damage, and cracks.

BRAKE DISC

SPINDLE NUT
Replace.
24 x 1.5 mm
245 N·m
(25.0 kgf·m, 181 lbf·ft)

Apply a small amount of engine oil to the seating surface.

42356-ELEM-G14

Exploded view of the front hub, wheel bearing and related components

REMOVAL & INSTALLATION

Front

1. Before servicing the vehicle, refer to the precautions in the beginning of this section.

2. Remove or disconnect the following:
   - Front wheel
   - Spindle nut
   - Brake caliper and rotor. Forcing screws are needed to remove the rotor.
   - Brake hose bracket
   - ABS sensor
   - Stabilizer link
   - Lower arm from the knuckle
   - Strut-to-knuckle bolts
   - Steering hub/knuckle assembly

3. Press the hub from the knuckle. The bearings and races can now be pressed out and replaced.

➡ **With ABS, install the bearing with the magnetic encoder (brown color) toward the inside of the knuckle.**

4. Observe the following torques:
   - Strut bolts: 116 ft. lbs. (157 Nm)
   - Ball stud nuts: 51 ft. lbs. (69 Nm)
   - Stabilizer bar link: 29 ft. lbs. (39 Nm)

Rear

1. Before servicing the vehicle, refer to the precautions in the beginning of this section.

2. Remove or disconnect the following:
   - Rear wheel
   - Brake caliper
   - Rotor
   - Spindle nut

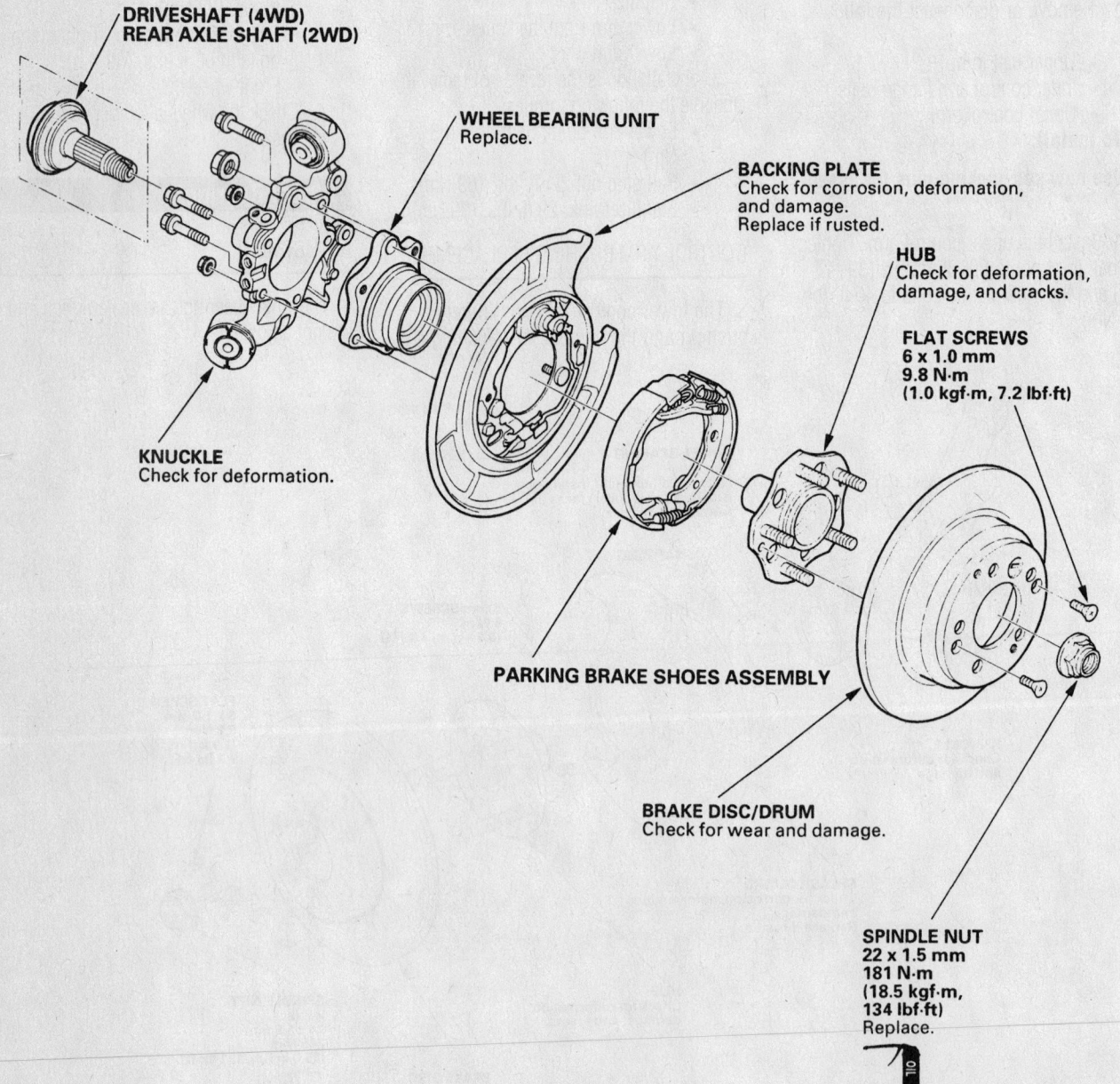

**DRIVESHAFT (4WD)
REAR AXLE SHAFT (2WD)**

WHEEL BEARING UNIT
Replace.

BACKING PLATE
Check for corrosion, deformation, and damage.
Replace if rusted.

HUB
Check for deformation, damage, and cracks.

FLAT SCREWS
6 x 1.0 mm
9.8 N·m
(1.0 kgf·m, 7.2 lbf·ft)

KNUCKLE
Check for deformation.

PARKING BRAKE SHOES ASSEMBLY

BRAKE DISC/DRUM
Check for wear and damage.

SPINDLE NUT
22 x 1.5 mm
181 N·m
(18.5 kgf·m, 134 lbf·ft)
Replace.

Apply a small amount of engine oil to the seating surface.

42356-ELEM-G15

Exploded view of the rear hub, wheel bearing and related components

- Axle shaft (2wd)
- Parking brake shoes
- Parking brake cable
- Wheel sensor, if equipped
3. Support the trailing arm.
4. Remove or disconnect the following:
  - Upper arm from the knuckle
5. Matchmark the trailing arm cam

adjusting bolt and cam. Remove the bolt. Discard the nut.
6. Remove the flange bolt.
7. Remove the knuckle assembly.
8. Press the hub from the knuckle. The bearings and races can now be pressed out and replaced.

➡ **With ABS, install the bearing with the magnetic encoder (brown**

color) **toward the inside of the knuckle.**

9. Observe the following torques:
  - Flange bolt: 69 ft. lbs. (93 Nm)
  - Cam bolts: 43 ft. lbs. (59 Nm)
  - Spindle nut: 134 ft. lbs. (181 Nm)
  - Caliper mounting bolts: 41 ft. lbs. (55 Nm)

BRAKES

Brake Caliper

REMOVAL & INSTALLATION

Front

1. Remove the upper and lower bolts.
2. Lift off the caliper.
3. Remove the pad springs.
4. Remove the pads and shims.
5. Remove the pad retainers.
6. Installation is the reverse of removal. Coat both sides of the shims and the backs of the pads with brake grease. Torque the bolts to 25 ft. lbs. (34 Nm). If the hose was disconnected, torque the banjo bolt to 25 ft. lbs. (34 Nm).

Rear

1. Remove the caliper pin bolts.
2. Lift off the caliper and suspend it safely.
3. Remove the pads and shims.
4. Remove the pad retainers.
5. Installation is the reverse of removal. Coat both sides of the shims and the backs of the pads with brake grease. Torque the bolts to 16 ft. lbs. (22 Nm). If the hose was disconnected, torque the banjo bolt to 16 ft. lbs. (22 Nm).

Disc Brake Pads

REMOVAL & INSTALLATION

Front

1. Remove the lower bolt.
2. Pivot the caliper up and hold the pads.
3. Remove the pad springs.
4. Remove the pads and shims.
5. Remove the pad retainers.
6. Installation is the reverse of removal. Coat both sides of the shims and the backs of the pads with brake grease. Torque the lower bolt to 25 ft. lbs. (34 Nm).

Rear

1. Remove the caliper pin bolts.
2. Lift off the caliper and suspend it safely.
3. Remove the pads and shims.
4. Remove the pad retainers.
5. Installation is the reverse of removal. Coat both sides of the shims and the backs of the pads with brake grease. Torque the bolts to 16 ft. lbs. (22 Nm).

Brake Drums

REMOVAL & INSTALLATION

1. Raise and safely support the vehicle. Release the parking brake.
2. Remove the rear wheels.
3. Use chalk to mark the brake drum to

one of the wheel studs as an index mark for reinstallation.
4. Remove the retaining screw that holds the brake drum to the axle flange.
5. Pull the brake drum from the axle flange.
To install:
6. Align the index mark and install the brake drum to the axle flange.
7. Install the retaining screw to secure the brake drum to the axle flange.
8. Install the rear wheels.

Rear Brake Shoes

REMOVAL & INSTALLATION

1. Raise and safely support the vehicle.
2. Remove the rear wheels.
3. Remove the brake drums.
4. Remove the brake return springs.

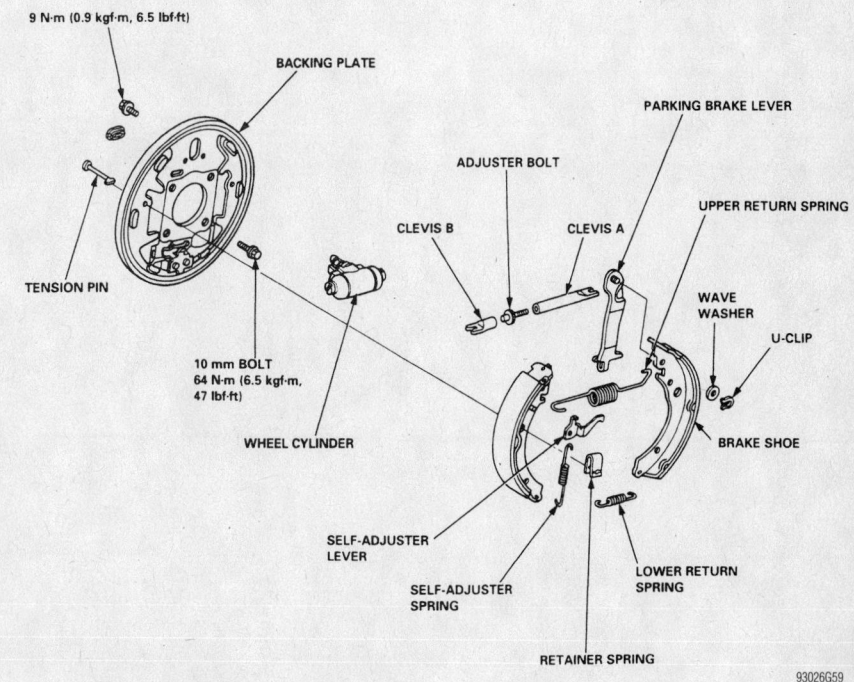

Exploded view of the rear drum brakes

5. Remove the leading shoe holding pin and spring, and then the leading shoe.

6. Remove the self-adjuster and the adjuster lever.

7. Remove the trailing shoe holding pin and spring.

8. Disconnect the parking brake cable from the trailing shoe and remove the trailing shoe. Remove the parking brake lever from the trailing shoe.

To install:

9. Attach the parking brake lever to the trailing shoe.

10. Connect the parking brake cable to the parking brake lever.

11. Apply a thin coat of high temperature grease to the shoe contact points on the brake backing plate contact surface (B), and self-adjuster (D).

12. Position the trailing shoe on the backing plate and install the hold-down pin, spring, and retainer. Be careful not to stretch the return spring when fitting the shoes onto the backing plate.

13. Connect the upper return spring and the leading shoe to the trailing shoe and position the leading brake shoe on the backing plate.

14. Install the adjuster assembly and the hold-down pin, spring, and retainer.

15. Use a brake spring tool to install the lower return spring.

16. Install the self-adjuster lever and adjuster spring.

17. Adjust the shoe-to-drum clearance to 0.0098–0.0157 in. (0.25–0.40mm) and install the brake drum.

18. Check the brake drum for scoring or other wear. Machine or replace as necessary. Check the maximum brake drum diameter specification when machining.

19. Install the rear wheels. Lower the vehicle.

20. Road-test the vehicle.

HONDA

Odyssey

6

SPECIFICATION CHARTS

ENGINE AND VEHICLE IDENTIFICATION CHART

		Engine Code						Model Year	
Code	Liters (cc)	Cu. In.	Cyl.	Fuel Sys.	Engine Type	Eng. Mfg.		Code ①	Year
J35A1	3.5 (3471)	212	6	SMFI	SOHC	Honda		Y	2000
J35A4	3.5 (3471)	212	6	SMFI	SOHC	Honda		1	2001
SOHC: Single Overhead Cam								2	2002
SMFI: Sequential Multi-port Fuel Injection								3	2003
① 10th position of VIN								4	2004

42356-ODYS-C01

GENERAL ENGINE SPECIFICATIONS

Year	Model	Engine Displacement Liters (cc)	Engine ID/VIN	Fuel System Type	Net Horsepower @ rpm	Net Torque @ rpm (ft. lbs.)	Bore x Stroke (in.)	Com-pression Ratio	Oil Pressure @ rpm
2000	Odyssey	3.5 (3471)	J35A1	SMFI	240@5200	242@3500	3.50x3.66	9.4:1	71@3000
2001	Odyssey	3.5 (3471)	J35A1	SMFI	240@5200	242@3500	3.50x3.66	9.4:1	71@3000
2002	Odyssey	3.5 (3471)	J35A4	SMFI	240@5200	242@3500	3.50x3.66	10.0:1	71@3000
2003	Odyssey	3.5 (3471)	J35A4	SMFI	250@5200	242@4500	3.50x3.66	10.0:1	71@3000

SMFI: Sequential Multi-port Fuel Injection

42356-ODYS-C02

ENGINE TUNE-UP SPECIFICATIONS

Year	Engine Displacement Liters (cc)	Engine ID/VIN	Spark Plug Gap (in.)	Ignition Timing (deg.) MT	Ignition Timing (deg.) AT	Fuel Pump (psi)	Idle Speed (rpm) MT	Idle Speed (rpm) AT	Valve Clearance (in.) In.	Valve Clearance (in.) Ex.
2000	3.5 (3471)	J35A1	0.039-0.043	—	10B	43-50	—	680-780	0.008-0.009	0.011-0.013
2001	3.5 (3471)	J35A1	0.039-0.043	—	10B	43-50	—	680-780	0.008-0.009	0.011-0.013
2002	3.5 (3471)	J35A4	0.039-0.043	—	10B	48-54	—	680-780	0.008-0.009	0.011-0.013
2003	3.5 (3471)	J35A4	0.039-0.040	—	10B	41-48	—	680-780	0.008-0.009	0.011-0.013

NOTE: The Vehicle Emission Control Information label often reflects changes made during production and must be used if they differ from this chart.

NOTE: The fuel pressure readings are given with the vacuum hose connected to the regulator and the engine running

B: Before top dead center

42356-ODYS-C03

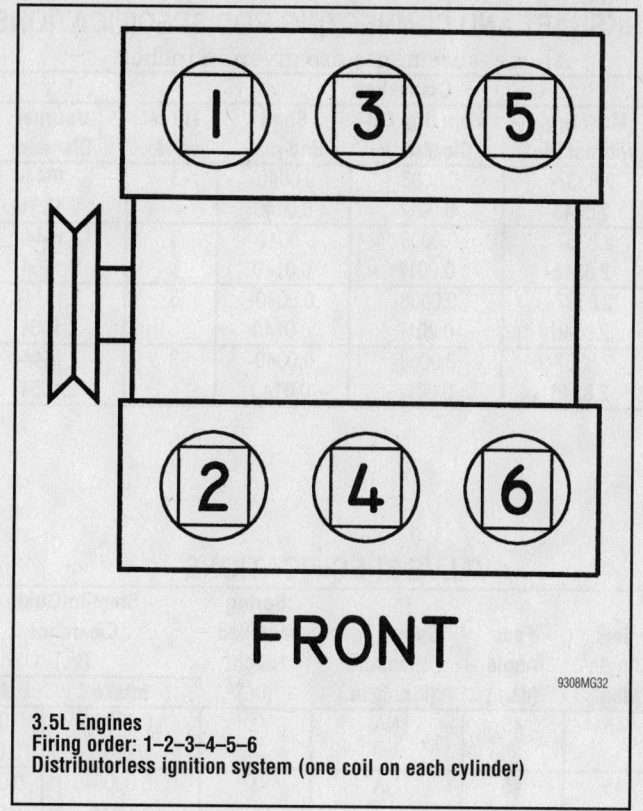

FRONT

9308MG32

3.5L Engines
Firing order: 1–2–3–4–5–6
Distributorless ignition system (one coil on each cylinder)

CAPACITIES

Year	Model	Engine Displacement Liters (cc)	Engine ID/VIN	Engine Oil with Filter (qts.)	Transmission (qts.)		Transfer Case (pts.)	Drive Axle		Fuel Tank (gal.)	Cooling System (qts.)
					5-Spd	Auto.		Front (pts.)	Rear (pts.)		
2000	Odyssey	3.5 (3471)	J35A1	4.6	—	6.2	—	—	—	20.0	7.0
2001	Odyssey	3.5 (3471)	J35A1	4.6	—	6.2	—	—	—	20.0	7.0
2002	Odyssey	3.5 (3471)	J35A4	4.6	—	8.3	—	—	—	20.0	7.0
2003	Odyssey	3.5 (3471)	J35A4	4.6	—	8.3	—	—	—	20.0	8.0

NOTE: All capacities are approximate. Add fluid gradually and check to be sure a proper fluid level is obtained.

42356-ODYS-C04

CRANKSHAFT AND CONNECTING ROD SPECIFICATIONS

All measurements are given in inches

Year	Engine Displacement Liters (cc)	Engine ID/VIN	Crankshaft				Connecting Rod		
			Main Brg. Journal Dia.	Main Brg. Oil Clearance	Shaft End-play	Thrust on No.	Journal Diameter	Oil Clearance	Side Clearance
2000	3.5 (3471)	J35A1	2.8337-2.8346	0.0008-0.0017	0.0040-0.0140	3	2.1644-2.1654	0.0008-0.0017	0.0060-0.0140
2001	3.5 (3471)	J35A1	2.8337-2.8346	0.0008-0.0017	0.0040-0.0140	3	2.1644-2.1654	0.0008-0.0017	0.0060-0.0140
2002	3.5 (3471)	J35A4	2.8337-2.8346	0.0008-0.0017	0.0040-0.0140	3	2.1644-2.1654	0.0008-0.0017	0.0060-0.0140
2003	3.5 (3471)	J35A4	2.8337-2.8346	0.0008-0.0017	0.0040-0.0140	3	2.1644-2.1654	0.0008-0.0017	0.0060-0.0140

42356-ODYS-C05

VALVE SPECIFICATIONS

Year	Engine Displacement Liters (cc)	Engine ID/VIN	Seat Angle (deg.)	Face Angle (deg.)	Spring Test Pressure (lbs. @ in.)	Spring Installed Height (in.)	Stem-to-Guide Clearance (in.)		Stem Diameter (in.)	
							Intake	Exhaust	Intake	Exhaust
2000	3.5 (3471)	J35A1	45	45	NA	①	0.0008-0.0018	0.0022-0.0031	0.2159-0.2163	0.2146-0.2150
2001	3.5 (3471)	J35A1	45	45	NA	①	0.0008-0.0018	0.0022-0.0031	0.2159-0.2163	0.2146-0.2150
2002	3.5 (3471)	J35A4	45	45	NA	②	0.0008-0.0018	0.0022-0.0031	0.2159-0.2163	0.2146-0.2150
2003	3.5 (3471)	J35A4	45	45	NA	②	0.0008-0.0018	0.0022-0.0031	0.2159-0.2163	0.2146-0.2150

NA: Not Available

① Valve spring free length:
Intake: 1.9713 in.
Exhaust: 2.1060 in.

② Valve spring free length:
Intake: 2.0290 in.
Exhaust: 2.1000 in.

42356-ODYS-C06

PISTON AND RING SPECIFICATIONS

All measurements are given in inches

Year	Engine Displacement Liters (cc)	Engine ID/VIN	Piston Clearance	Ring Gap			Ring Side Clearance		
				Top Compression	Bottom Compression	Oil Control	Top Compression	Bottom Compression	Oil Control
2000	3.5 (3471)	J35A1	0.0006-0.0016	0.0080-0.0140	0.0160-0.0220	0.0080-0.0280	0.0014-0.0024	0.0012-0.0022	NA
2001	3.5 (3471)	J35A1	0.0006-0.0016	0.0080-0.0140	0.0160-0.0220	0.0080-0.0280	0.0014-0.0024	0.0012-0.0022	NA
2002	3.5 (3471)	J35A4	0.0006-0.0016	0.0080-0.0140	0.0160-0.0220	0.0080-0.0280	0.0022-0.0031	0.0012-0.0022	NA
2003	3.5 (3471)	J35A4	0.0006-0.0016	0.0080-0.0140	0.0160-0.0220	0.0080-0.0280	0.0022-0.0031	0.0012-0.0022	NA

NA: Not Applicable

42356-ODYS-C07

TORQUE SPECIFICATIONS
All readings in ft. lbs.

Year	Engine Displacement Liters (cc)	Engine ID/VIN	Cylinder Head Bolts	Main Bearing Bolts	Rod Bearing Bolts	Crankshaft Damper Bolts	Flywheel Bolts	Manifold		Spark Plugs	Lug Nut
								Intake	Exhaust		
2000	3.5 (3471)	J35A1	①	②	③	181	54	16	23	13	80
2001	3.5 (3471)	J35A1	①	②	③	181	54	16	23	13	80
2002	3.5 (3471)	J35A4	①	②	③	181	54	16	23	13	80
2003	3.5 (3471)	J35A4	①	②	③	181	54	16	23	13	80

NOTE: Dip main bearing bolts and crankshaft damper bolt in clean engine oil prior to tightening.

① Step 1: 29 ft. lbs.
  Step 2: 51 ft. lbs.
  Step 3: 72 ft. lbs.

② 11mm bolt 56 ft. lbs.
  10mm bolt 36 ft. lbs.

③ Step 1: 14 ft. lbs.
  Step 2: 90 degrees

42356-ODYS-C08

WHEEL ALIGNMENT

Year	Model		Caster		Camber		Toe-in (in.)	Steering Axis Inclination (Deg.)
			Range (+/-Deg.)	Preferred Setting (Deg.)	Range (+/-Deg.)	Preferred Setting (Deg.)		
2000	Odyssey	F	1.00	+2.07	1.00	0	0+/-0.08	—
		R	—	—	0.75	-0.50	0+/-0.08	—
2001	Odyssey	F	1.00	+2.07	1.00	0	0+/-0.08	—
		R	—	—	0.75	-0.50	0+/-0.08	—
2002	Odyssey	F	1.00	+2.07	1.00	0	0+/-0.08	—
		R	—	—	0.75	-0.50	0+/-0.08	—
2003	Odyssey	F	1.00	+2.07	1.00	0	0+/-0.08	—
		R	—	—	0.75	-0.50	0+/-0.08	—

42356-ODYS-C09

BRAKE SPECIFICATIONS
All measurements in inches unless noted

Year	Model		Brake Disc Original Thickness	Brake Disc Minimum Thickness	Maximum Runout	Brake Drum Diameter Original Inside Diameter	Brake Drum Diameter Max. Wear Limit	Brake Drum Diameter Maximum Machine Diameter	Minimum Lining Thickness Front	Minimum Lining Thickness Rear	Brake Caliper Bracket Bolts (ft. lbs.)	Brake Caliper Mounting Bolts (ft. lbs.)
2000	Odyssey	F	1.100	1.020	0.004	—	—	—	0.060	—	80	27
		R	—	—	—	9.996	10.04	—	—	0.080	—	—
2001	Odyssey	F	1.100	1.020	0.004	—	—	—	0.060	—	80	27
		R	—	—	—	9.996	10.04	—	—	0.080	—	—
2002	Odyssey	F	1.100	1.020	0.004	—	—	—	0.060	—	80	27
		R	0.440	0.350	0.004	8.268 ①	8.272	—	0.060	0.040	41	27
2003	Odyssey	F	1.100	1.020	0.004	—	—	—	0.060	—	80	27
		R	0.440	0.350	0.004	8.268 ①	8.272	—	0.060	0.040	41	27

F: Front

R: Rear

① Parking brake drum

42356-ODYS-C010

TIRE, WHEEL AND BALL JOINT SPECIFICATIONS

Year	Model	OEM Tires Standard	OEM Tires Optional	Tire Pressures (psi) Front	Tire Pressures (psi) Rear	Wheel Size	Ball Joint Inspection
2000	Odyssey	P215/65R16	None	32	32	6JJ	NS
2001	Odyssey	P215/65R16	None	32	32	6JJ	NS
2002	Odyssey	P215/65R16	None	32	32	6JJ	NS
2003	Odyssey	P225/60R16	None	32	32	6.5JJ	NS

OEM: Original Equipment Manufacturer

PSI: Pounds Per Square Inch

STD: Standard

OPT: Optional

NS: Not specified by manufacturer

42356-ODYS-C011

SCHEDULED MAINTENANCE INTERVALS
Honda Odyssey

TO BE SERVICED	TYPE OF SERVICE	VEHICLE MILEAGE INTERVAL (x1000)															
		7.5	15	22.5	30	37.5	45	52.5	60	67.5	75	82.5	90	97.5	105	112	120
Accessory drive belts	I & A				✓				✓				✓				✓
Air cleaner element	R				✓				✓				✓				✓
Air conditioning filter	R				✓				✓				✓				✓
Brake fluid	R					✓							✓				
Brake hoses & lines	I		✓		✓		✓		✓		✓		✓		✓		✓
Cooling system hoses	I		✓		✓		✓		✓		✓		✓		✓		✓
Engine coolant	R					✓							✓				
Engine oil	R	✓	✓	✓	✓	✓	✓	✓	✓	✓	✓	✓	✓	✓	✓	✓	✓
Engine oil and coolant levels	I	Inspect at each fuel stop															
Engine oil filter	R		✓		✓		✓		✓		✓		✓		✓		✓
Exhaust system	I		✓		✓		✓		✓		✓		✓		✓		✓
Fluid levels and condition	I		✓		✓		✓		✓		✓		✓		✓		✓
Front and rear brakes	I		✓		✓		✓		✓		✓		✓		✓		✓
Fuel lines & connection	I		✓		✓		✓		✓		✓		✓		✓		✓
Halfshaft boots	I		✓		✓		✓		✓		✓		✓		✓		✓
Idle speed	I & A														✓		
Parking brake system	I & A		✓		✓		✓		✓		✓		✓		✓		
Rear differential fluid	R												✓				
Rotate and inspect tires	I	✓	✓	✓	✓	✓	✓	✓	✓	✓	✓	✓	✓	✓	✓	✓	✓
Spark plugs	R				✓								✓				✓
Supplemental Restraint system	I	Inspect the SRS 10 years after production															
Suspension components	I		✓		✓		✓		✓		✓		✓		✓		✓
Steering components	I		✓		✓		✓		✓		✓		✓		✓		✓
Timing balancer belt	R														✓		
Timing belt	R														✓		
Transmission fluid	R					✓					✓				✓		
Valve clearance	I			✓					✓				✓				✓
Water pump	S/I														✓		

R: Replace   I: Inspect   A: Adjust

FREQUENT OPERATION MAINTENANCE (SEVERE SERVICE)

If a vehicle is operated under any of the following conditions it is considered severe service:

- Towing a trailer or using a camper or car-top carrier.
- Repeated short trips of less than 5 miles in temperatures below freezing, or trips of less than 10 miles in any temperature.
- Extensive idling or low-speed driving for long distances as in heavy commercial use, such as delivery, taxi or police cars.
- Operating on rough, muddy or salt-covered roads.
- Operating on unpaved or dusty roads.
- Driving in extremely hot (over 90°) conditions.

Air cleaner element: replace every 15,000 miles

Engine oil and filter: replace every 3750 miles or 6 months, whichever occurs first.

Timing belt: replace every 60,000 miles if the vehicle is regularly driven in temperatures above 110°F or below -20°F.

Transmission fluid: replace every 30,000 miles.

Rear differential fluid: replace every 60,000 miles.

Front and rear brakes: inspect every 7500 miles or 6 months, whichever occurs first.

Locks and hinges: lubricate every 15,000 miles.

42356-ODYS-C012

PRECAUTIONS

Before servicing any vehicle, please be sure to read all of the following precautions, which deal with personal safety, prevention of component damage, and important points to take into consideration when servicing a motor vehicle:

• Never open, service or drain the radiator or cooling system when the engine is hot; serious burns can occur from the steam and hot coolant.

• Observe all applicable safety precautions when working around fuel. Whenever servicing the fuel system, always work in a well-ventilated area. Do not allow fuel spray or vapors to come in contact with a spark, open flame, or excessive heat (a hot drop light, for example). Keep a dry chemical fire extinguisher near the work area. Always keep fuel in a container specifically designed for fuel storage; also, always properly seal fuel containers to avoid the possibility of fire or explosion. Refer to the additional fuel system precautions later in this section.

• Fuel injection systems often remain pressurized, even after the engine has been turned **OFF**. The fuel system pressure must be relieved before disconnecting any fuel lines. Failure to do so may result in fire and/or personal injury.

• Brake fluid often contains polyglycol ethers and polyglycols. Avoid contact with the eyes and wash your hands thoroughly after handling brake fluid. If you do get brake fluid in your eyes, flush your eyes with clean, running water for 15 minutes. If eye irritation persists, or if you have taken brake fluid internally, seek medical assistance IMMEDIATELY.

• The EPA warns that prolonged contact with used engine oil may cause a number of skin disorders, including cancer. You should make every effort to minimize your exposure to used engine oil. Protective gloves should be worn when changing oil. Wash your hands and any other exposed skin areas as soon as possible after exposure to used engine oil. Soap and water, or waterless hand cleaner should be used.

• All new vehicles are now equipped with an air bag system. The system must be disabled before performing service on or around system components, steering column, instrument panel components, wiring and sensors. Failure to follow safety and disabling procedures could result in accidental air bag deployment, possible personal injury and unnecessary system repairs.

• Always wear safety goggles when working with, or around, the air bag system. When carrying a non-deployed air bag, be sure the bag and trim cover are pointed away from your body. When placing a non-deployed air bag on a work surface, always face the bag and trim cover upward, away from the surface. This will reduce the motion of the module if it is accidentally deployed. Refer to the additional air bag system precautions later in this section.

• Clean, high quality brake fluid from a sealed container is essential to the safe and proper operation of the brake system. You should always buy the correct type of brake fluid for your vehicle. If the brake fluid

becomes contaminated, completely flush the system with new fluid. Never reuse any brake fluid. Any brake fluid that is removed from the system should be discarded. Also, do not allow any brake fluid to come in contact with a painted surface; it will damage the paint.

• Never operate the engine without the proper amount and type of engine oil; doing so WILL result in severe engine damage.

• Timing belt maintenance is extremely important. Many models utilize an interference-type, non-freewheeling engine. If the timing belt breaks, the valves in the cylinder head may strike the pistons, causing potentially serious (also time-consuming and expensive) engine damage. Refer to the maintenance interval charts in the front of this section for the recommended replacement interval for the timing belt, and to the timing belt procedure in this section for belt replacement and inspection.

• Disconnecting the negative battery cable on some vehicles may interfere with the functions of the on-board computer system(s) and may require the computer to undergo a relearning process once the negative battery cable is reconnected.

• When servicing drum brakes, only disassemble and assemble one side at a time, leaving the remaining side intact for reference.

• Only an MVAC-trained, EPA-certified automotive technician should service the air conditioning system or its components.

ENGINE REPAIR

→Disconnecting the negative battery cable on some vehicles may interfere with the functions of the on board computer system. The computer may undergo a relearning process once the negative battery cable is reconnected.

Distributor

The 3.5L engine is equipped with a Distributorless Ignition System (DIS).

REMOVAL & INSTALLATION

Alternator

REMOVAL

1. Before servicing the vehicle, refer to the precautions in the beginning of this section.
2. Remove or disconnect the following:
   • Negative battery cable
   • Accessory drive belt
   • Alternator wiring harness connectors
   • Alternator mounting bolts

• Wiring harness clamp
• Alternator

INSTALLATION

1. Install or connect the following:
   • Alternator
   • Wiring harness clamp. Tighten the bolt to 105 inch lbs. (12 Nm).
   • Alternator mounting bolts. Tighten the 10mm bolt to 33 ft. lbs. (44 Nm) and the 8mm bolt to 16 ft. lbs. (22 Nm).

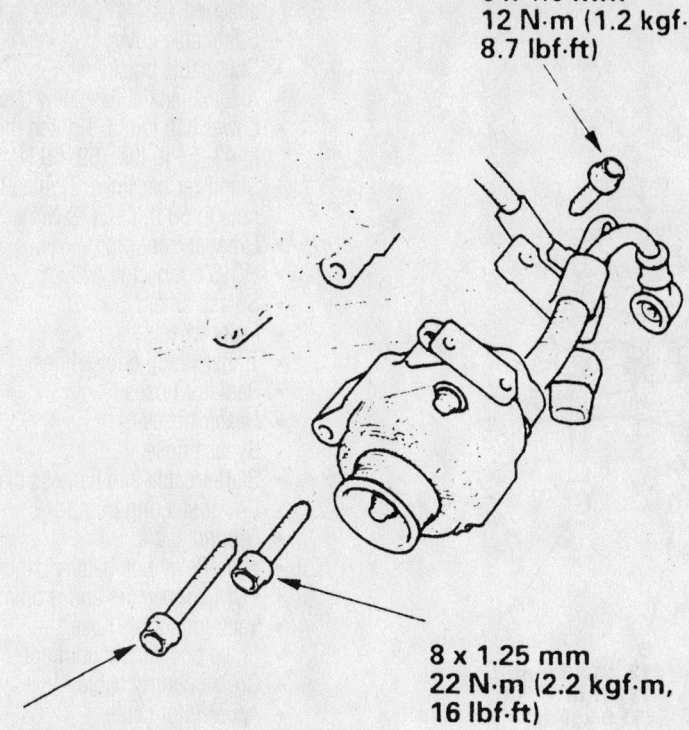

6 x 1.0 mm
12 N·m (1.2 kgf·m,
8.7 lbf·ft)

8 x 1.25 mm
22 N·m (2.2 kgf·m,
16 lbf·ft)

9358MG01

10 x 1.25 mm
44 N·m (4.5 kgf·m, 33 lbf·ft)

Exploded view of the alternator mounting

- Alternator wiring harness connectors. Tighten the battery terminal nut to 105 inch lbs. (12 Nm).
- Accessory drive belt
- Negative battery cable

Ignition Timing

ADJUSTMENT

The 3.5L engine is equipped with a Distributorless Ignition System (DIS). The ignition timing is controlled by the Powertrain Control module (PCM). No adjustment is necessary.

Engine Assembly

REMOVAL & INSTALLATION

➡**The engine and transaxle are removed from the vehicle as a unit.**

1. Before servicing the vehicle, refer to the precautions in the beginning of this section.
2. Drain the cooling system.
3. Drain the transaxle fluid.
4. Drain the engine oil.
5. Relieve fuel system pressure.

6. Remove or disconnect the following:
- Negative battery cable
- Intake manifold cover and ignition coil cover on 2002–03 models
- Evaporative Emissions (EVAP) control canister or vacuum hose, whichever applies
- Air intake duct
- Battery
- Left engine wire harness connectors
- Relay bracket
- Battery tray
- Accessory drive belts
- Accelerator cable
- Cruise control cable
- Fuel lines
- Brake booster vacuum line
- Vacuum supply hose
- Powertrain Control Module (PCM) connectors and grommet. Pull the PCM harness through the firewall.
- Fuse/Relay box battery cable
- Ground cable
- Power steering pump
- Starter cable and harness clamp
- Radiator hoses
- Heater hoses
- Bypass hose
- Transaxle oil cooler lines

- Front wheels
- Splash shield
- Heated Oxygen Sensor (HO$_2$S) connector
- Exhaust front pipe
- Stabilizer bar links
- Lower ball joints

7. Separate the inner CV-joints from the transaxle and support the axle halfshafts out of the work area with safety wire.
8. Remove or disconnect the following:
- Shift cable bracket
- Shift cable cover
- Shift control lever with cable attached
- Power steering hose clamp and clips
- Transaxle lower front mount
- Transaxle lower rear mount
- Steering rack and pinion gear. Support the steering gear with safety wire.

9. Attach a hoist to the engine lifting eyes and support the powertrain weight.
10. Remove or disconnect the following:
- Side engine mount bracket
- Front mount bracket support nut

11. Matchmark the front subframe to the mounting points.
12. Remove or disconnect the following:
- Front subframe
- A/C compressor

13. Lower the powertrain away from the vehicle.

To install:
14. Raise the powertrain into position.
15. Install or connect the following:
- A/C compressor. Tighten the bolts to 16 ft. lbs. (22 Nm).
- Front subframe. Use new bolts and tighten the 14mm bolts to 76 ft. lbs. (103 Nm). Tighten the front brace bolts to 54 ft. lbs. (74 Nm) and the rear brace bolts to 86 ft. lbs. (117 Nm).
- Transaxle lower front mount. Tighten the nuts to 28 ft. lbs. (38 Nm).
- Transaxle lower rear mount. Tighten the bolts to 28 ft. lbs. (38 Nm).
- Front mount bracket support nut. Tighten the nut to 40 ft. lbs. (54 Nm).
- Side engine mount bracket. Tighten the bracket bolts to 33 ft. lbs. (44 Nm) and the through bolt to 40 ft. lbs. (54 Nm).
- Steering rack and pinion gear. Tighten the bolts to 29 ft. lbs. (39 Nm).
- Power steering hose clamp and clips

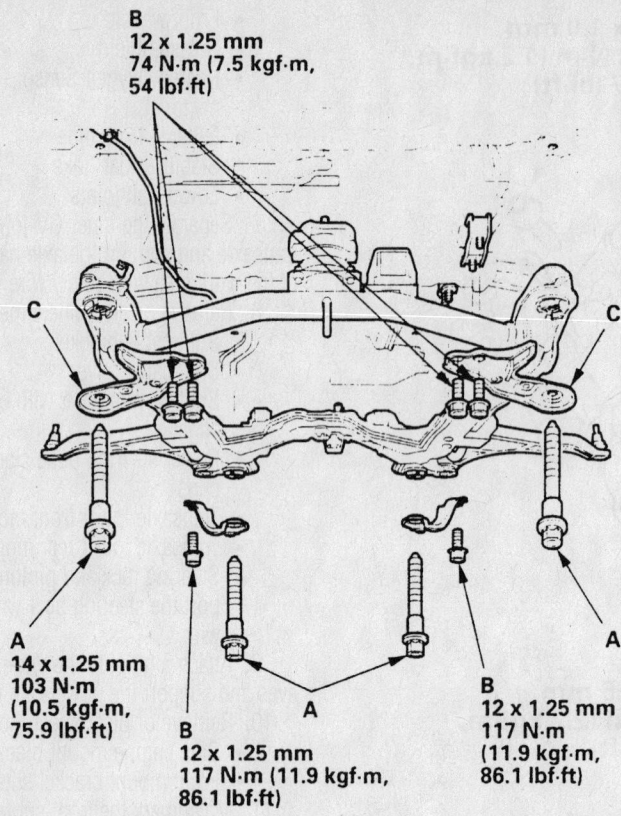

Sub-frame fastener locations and tightening torque—2000–01 models

B
12 x 1.25 mm
74 N·m (7.5 kgf·m, 54 lbf·ft)

A
14 x 1.25 mm
103 N·m
(10.5 kgf·m,
75.9 lbf·ft)

B
12 x 1.25 mm
117 N·m (11.9 kgf·m,
86.1 lbf·ft)

B
12 x 1.25 mm
117 N·m
(11.9 kgf·m,
86.1 lbf·ft)

9302MG69

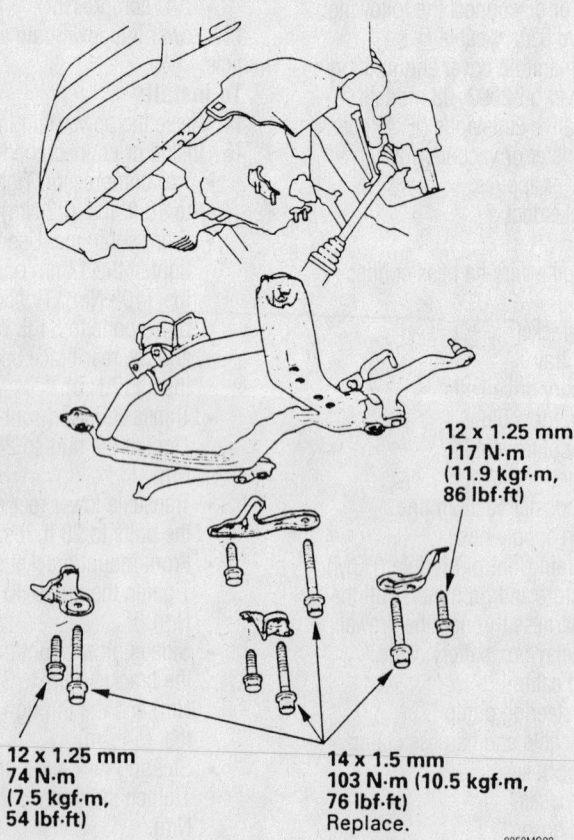

Sub-frame fastener locations and tightening torque—2002–03 models

12 x 1.25 mm
117 N·m
(11.9 kgf·m,
86 lbf·ft)

12 x 1.25 mm
74 N·m
(7.5 kgf·m,
54 lbf·ft)

14 x 1.5 mm
103 N·m (10.5 kgf·m,
76 lbf·ft)
Replace.

9358MG03

- Shift control lever with cable attached
- Shift cable cover
- Shift cable bracket
- Axle halfshafts. Use new circlips.
- Lower ball joints. Tighten the nuts to 43–51 ft. lbs. (59–69 Nm).
- Stabilizer bar links. Tighten the nuts to 58 ft. lbs. (78 Nm).
- Exhaust front pipe
- HO2S connector
- Splash shield
- Front wheels
- Transaxle oil cooler lines
- Radiator hoses
- Heater hoses
- Bypass hose
- Starter cable and harness clamp
- Power steering pump
- Ground cable
- Fuse/Relay box battery cable
- PCM connectors and grommet
- Vacuum supply hose
- Brake booster vacuum line
- Cruise control cable
- Accelerator cable
- Accessory drive belts
- Fuel lines
- Battery tray
- Relay bracket
- Left engine wire harness connectors
- Battery
- Air intake duct
- EVAP control canister hose
- Negative battery cable

16. Fill the engine crankcase to the correct level.

17. Fill the transaxle to the correct level.

18. Fill the cooling system.

19. Start the engine and check for leaks.

20. Check the wheel alignment and adjust as necessary.

Heater Core

REMOVAL & INSTALLATION

➡ **Make sure to acquire the anti-theft code for the radio and write down the frequencies for the radio's preset buttons.**

1. Disconnect the negative battery cable.

✳✳ CAUTION

Wait at least 3 minutes for the air bag to deplete its energy before working on the steering wheel or instrument panel.

2. In the engine compartment, remove the heater valve cable clamp; then, disconnect the heater valve cable and rotate the heater valve to the fully open position.

3. Drain the engine coolant into a clean container for reuse.

4. Disconnect the heater hoses from the heater unit.

5. Remove the heater housing-to-chassis nuts.

6. Remove the center console.

7. Remove the instrument panel.

8. Remove the steering hanger beam mounting bolts and the steering hanger beam.

9. Remove the evaporator housing by performing the following procedure:

   a. Discharge and recover the air conditioning system refrigerant.

   b. Remove the refrigerant lines. Discard the O-rings. Plug the openings to prevent contamination.

   c. Disconnect the thermostat electrical connector and the wiring harness from the evaporator.

   d. Remove the evaporator housing-to-chassis screws, the bolt and nuts.

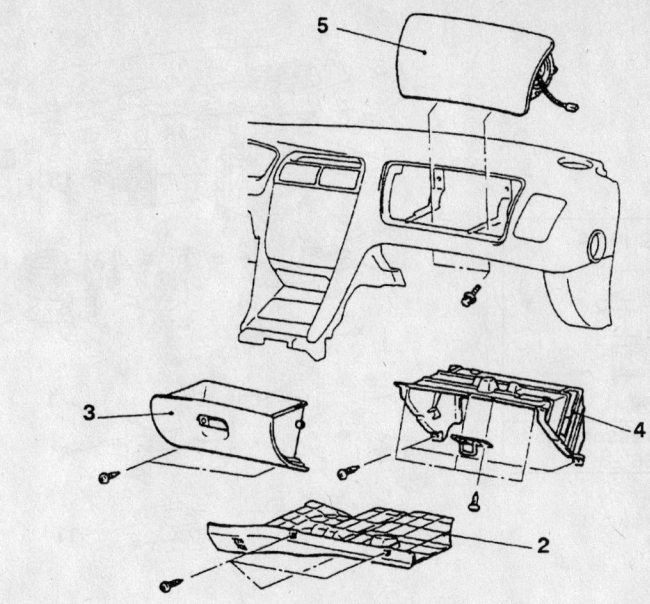

2. UNDERCOVER
3. GLOVE BOX ASSEMBLY
4. GLOVE BOX CASE
5. AIR BAG MODULE

93112GG1

View of the heater housing, evaporator housing and related components

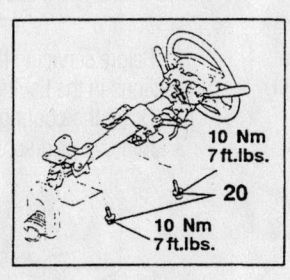

1. COLUMN COVER
2. HOOD LOCK RELEASE HANDLE
3. PARKING BRAKE RELEASE HANDLE
4. INSTRUMENT PANEL LOWER COVER ASSEMBLY (LH)
5. KEY CYLINDER PANEL
6. INSTRUMENT PANEL ECU
7. METER BEZEL
8. COMBINATION METER
9. CENTER AIR OUTLET ASSEMBLY
10. ASHTRAY
11. AIR CONTROL PANEL ASSEMBLY & AUDIO UNIT
12. UNDERCOVER ASSEMBLY
13. GLOVEBOX ASSEMBLY
14. GLOVEBOX OUTER CASE
15. PASSENGER SIDE AIRBAG MODULE
16. CONSOLE SIDE COVER ASSEMBLY
17. FLOOR CARPET REAR REINFORCEMENT
18. HARNESS CONNECTOR
19. PLUG
20. STEERING COLUMN MOUNTIN BOLT
21. INSTRUMENT PANEL

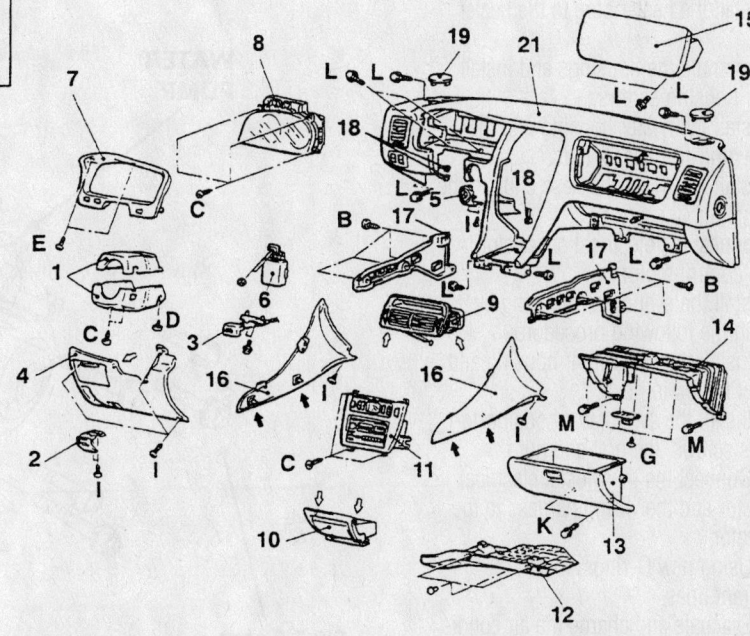

NOTE
(1) ⇐ : metal clip position
(2) ← : plastic clip position

93112GG2

View of the steering hanger beam and related components

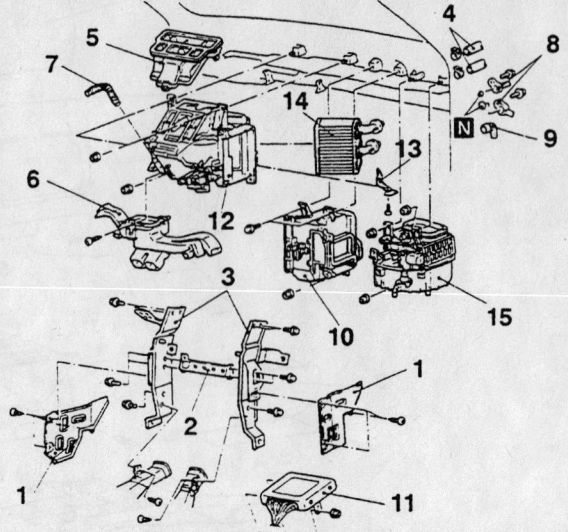

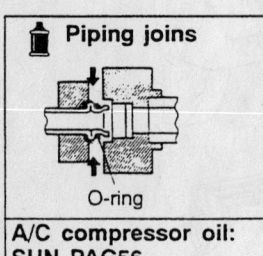

Piping joins

O-ring

A/C compressor oil: SUN PAG56

1. FLOOR CARPET FRONT REINFORCEMENT
3. ECU BRACKET
4. CENTER STAY ASSEMBLY
5. HEATER HOSE CONNECTION
6. CENTER DUCT ASSEMBLY
7. FOOT DISTRIBUTION DUCT
8. BREATHER HOSE
9. SUCTION PIPE, LIQUID PIPE B AND COOLING UNIT CONNECTION
10. DRAIN HOSE
11. EVAPORATOR
12. ENGINE CONTROL MODULE
13. HEATER UNIT
14. HEATER CORE SUPPORT
15. HEATER CORE

93112GG3

Exploded view of the heater core, the heater housing and related components

e. Disconnect the drain hose and remove the evaporator housing.

10. Disconnect the electrical connector from the mode control motor.

11. Remove the wiring harness clips from the heater housing.

12. Remove the heater housing-to-chassis nuts and the heater housing.

13. Remove the heater housing screws and separate the housings.

14. Remove the heater core from the heater housing.

To install:

15. Install the heater core to the heater housing.

16. Assemble the housings and install the heater housing screws.

17. Install the heater housing and the heater housing-to-chassis nuts.

18. Install the wiring harness clips to the heater housing.

19. Connect the electrical connector to the mode control motor.

20. Install the evaporator housing by performing the following procedure:

a. Install the evaporator housing and connect the drain hose.

b. Install the evaporator housing-to-chassis screws, the bolt and nuts.

c. Connect the thermostat electrical connector and the wiring harness to the evaporator.

d. Using new O-rings, install the refrigerant lines.

e. Evacuate and charge the air conditioning system refrigerant.

21. Install the steering hanger beam and the steering hanger beam mounting bolts.

22. Install the instrument panel.

➡**When installing the nuts, be careful not to damage or bend the fuel lines, the brake lines or etc.**

23. Install the heater housing-to-chassis nuts.

24. Connect the heater hoses to the heater unit.

25. Refill the cooling system.

26. In the engine compartment, Install the heater valve cable clamp; then, connect the heater valve cable.

27. Connect the negative battery cable.

28. Run the engine to normal operating temperatures; then, check the climate control operation and check for leaks.

Water Pump

REMOVAL & INSTALLATION

1. Before servicing the vehicle, refer to the precautions in the beginning of this section.

2. Drain the cooling system.

3. Remove or disconnect the following:

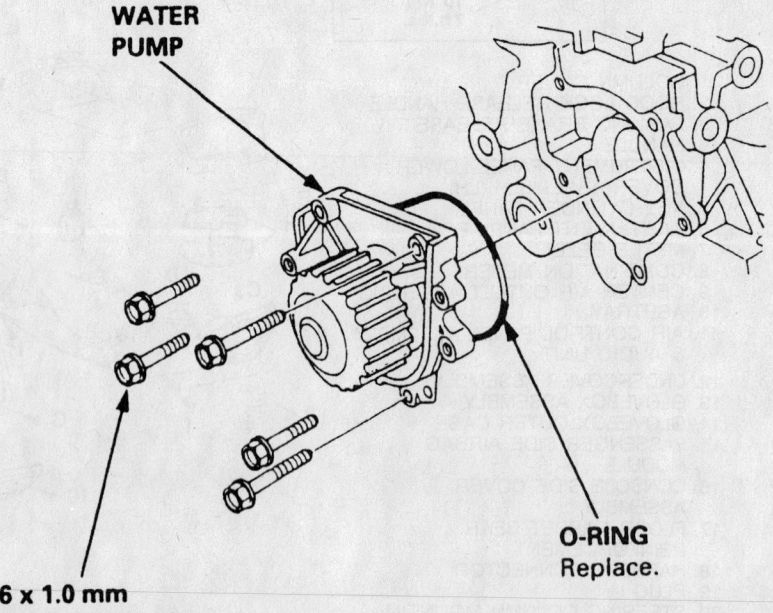

WATER PUMP

O-RING
Replace.

6 x 1.0 mm
12 N·m (1.2 kgf·m, 8.7 lbf·ft)

7924MG10

Exploded view of the water pump mounting

- Negative battery cable
- Accessory drive belts
- Front cover
- Timing belt. Refer to the Timing Belt unit repair section.
- Timing belt tensioner
- Water pump

To install:

4. Install or connect the following:
- Water pump. Use a new O-ring seal

and tighten the bolts to 105 inch lbs. (12 Nm).
- Timing belt tensioner
- Timing belt
- Front cover
- Accessory drive belts
- Negative battery cable

5. Fill the cooling system.
6. Start the engine and check for leaks.

Cylinder Head

REMOVAL & INSTALLATION

1. Before servicing the vehicle, refer to the precautions in the beginning of this section.
2. Drain the cooling system.
3. Relieve the fuel system pressure.

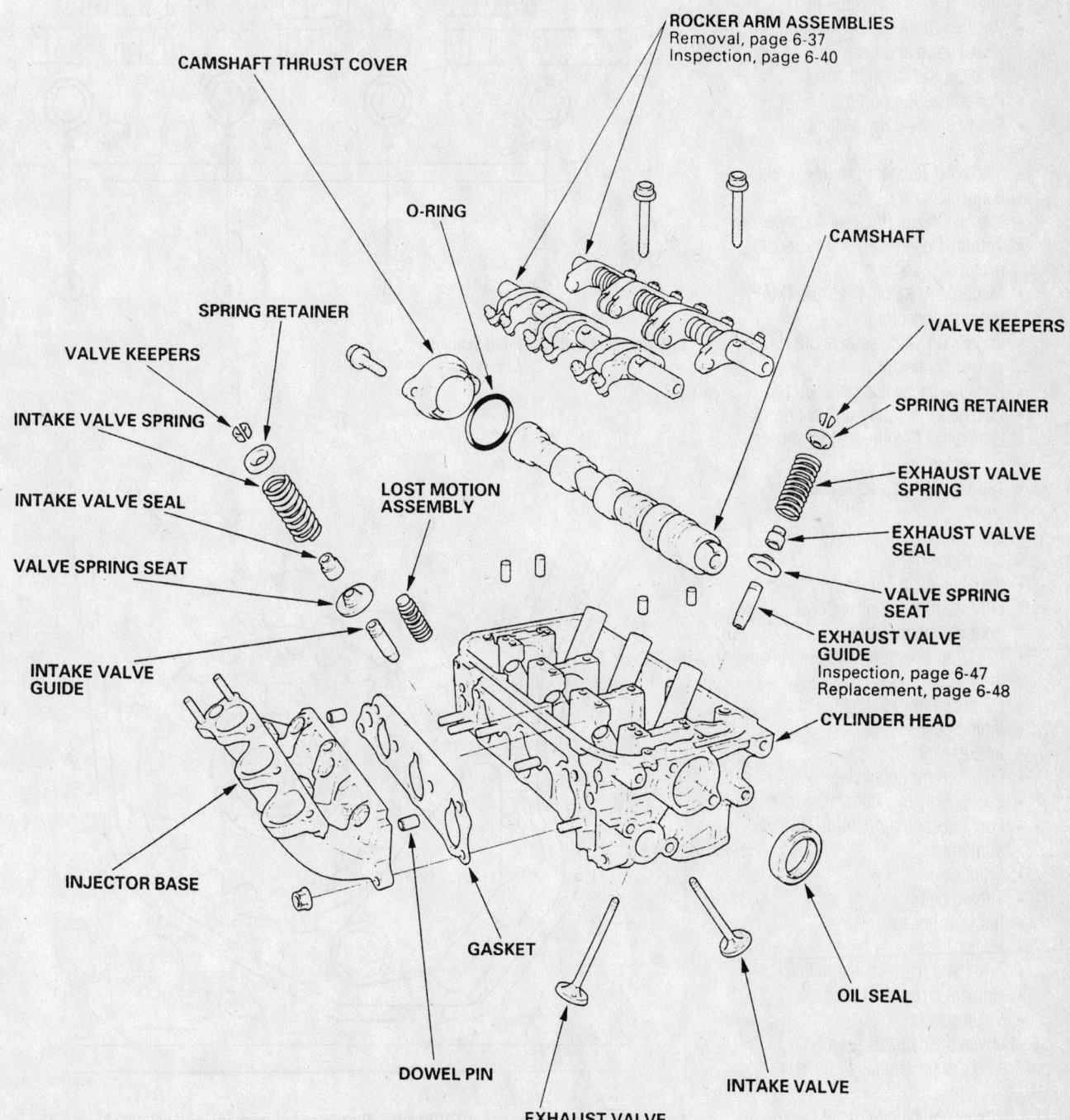

CAMSHAFT THRUST COVER

O-RING

ROCKER ARM ASSEMBLIES
Removal, page 6-37
Inspection, page 6-40

CAMSHAFT

VALVE KEEPERS

SPRING RETAINER

VALVE KEEPERS

SPRING RETAINER

EXHAUST VALVE SPRING

INTAKE VALVE SPRING

INTAKE VALVE SEAL

LOST MOTION ASSEMBLY

EXHAUST VALVE SEAL

VALVE SPRING SEAT

VALVE SPRING SEAT

EXHAUST VALVE GUIDE
Inspection, page 6-47
Replacement, page 6-48

INTAKE VALVE GUIDE

CYLINDER HEAD

INJECTOR BASE

OIL SEAL

GASKET

DOWEL PIN

EXHAUST VALVE

INTAKE VALVE

9358MG04

Exploded view of the cylinder head assembly and related components—2002–03 models

4. Remove or disconnect the following:
- Negative battery cable
- Accessory drive belts
- Evaporative Emissions (EVAP) control canister hose and vacuum hose
- Air intake tube
- Ignition coil covers
- Intake manifold cover
- Accelerator cable
- Cruise control cable
- Fuel lines
- Brake booster vacuum line
- Intake manifold vacuum line
- Positive Crankcase Ventilation (PCV) valve and hose
- Side engine mount bracket
- Power steering pump
- Power steering hose clamp
- Alternator
- Intake Air Temperature (IAT) sensor connector
- Idle Air Control (IAC) valve connector
- Throttle Position (TP) sensor connector
- Manifold Absolute Pressure (MAP) sensor connector
- Engine Coolant Temperature (ECT) sensor connector
- Radiator fan switch connectors
- ECT gauge sending unit connector
- Crankshaft Position (CKP) sensor connector
- Top Dead Center (TDC) sensor connector
- Exhaust Gas Recirculation (EGR) connector
- Variable Valve Timing and Valve Lift Electronic Control (VTEC) solenoid valve connector
- VTEC oil pressure switch connector
- Oil pressure switch connector
- Ignition coils
- Intake manifold
- Fuel injector connectors
- Fuel supply manifold
- Fuel injection air control valve vacuum lines
- Front cover
- Timing belt
- Radiator hoses
- Heater hoses
- Front and rear exhaust manifolds
- Coolant cross-over pipe
- Valve covers

5. Loosen the cylinder head bolts in sequence and ⅓ turns until all bolts are loose.

6. Remove the cylinder head.

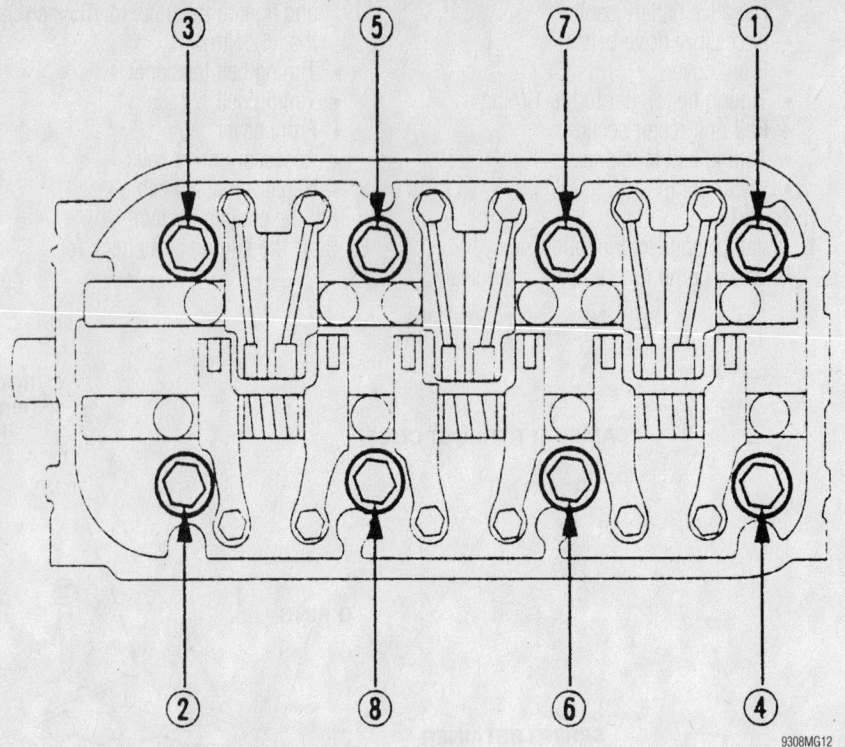

Cylinder head bolt loosening sequence

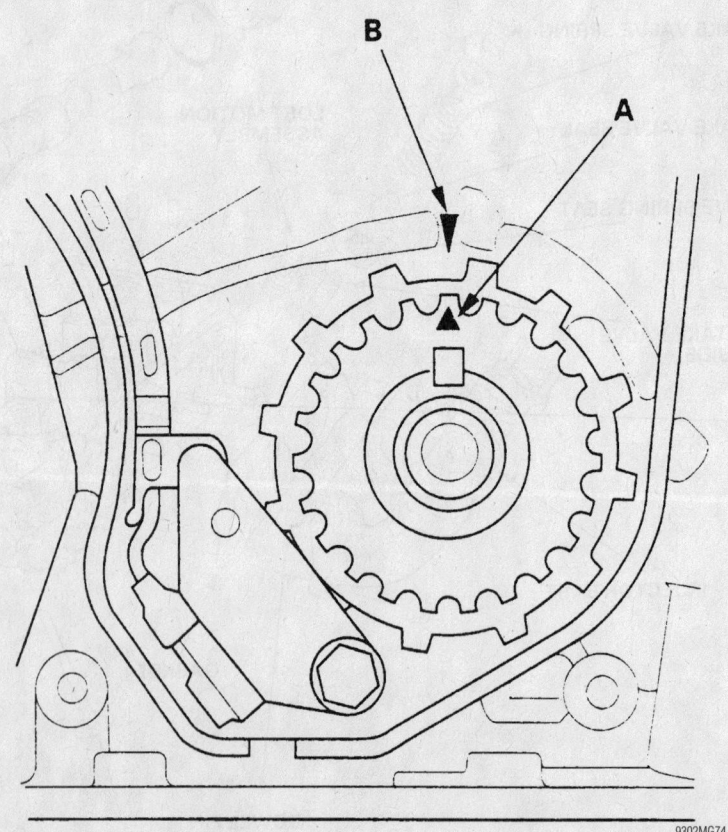

Crankshaft timing belt sprocket TDC marks. Align sprocket mark (A) with pointer (B)

FRONT:

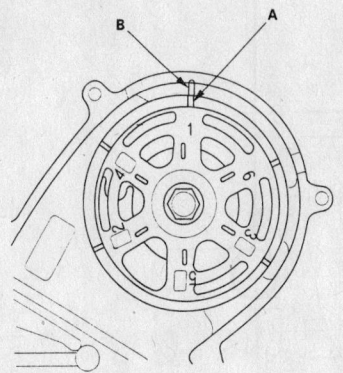

REAR:

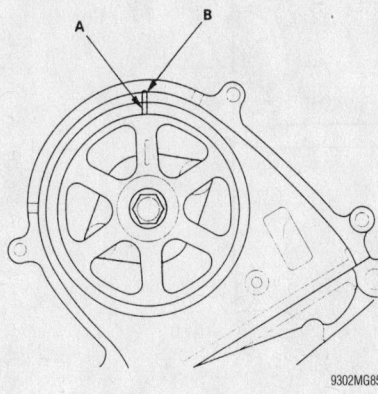

9302MG85

Camshaft TDC marks. Align sprocket mark (A) with the back cover pointer (B)

To install:

7. Align the crankshaft and camshaft sprocket TDC marks as shown.

8. Install the cylinder heads with new gaskets.

9. Apply clean engine oil to the cylinder head bolt threads and flanges.

10. Tighten the cylinder head bolts in sequence as follows:
   a. Step 1: 29 ft. lbs. (39 Nm)
   b. Step 2: 51 ft. lbs. (69 Nm)
   c. Step 3: 72 ft. lbs. (98 Nm)

11. Install or connect the following:
   • Valve covers
   • Coolant cross-over pipe
   • Front and rear exhaust manifolds
   • Heater hoses
   • Radiator hoses
   • Timing belt
   • Front cover
   • Fuel injection air control valve vacuum lines
   • Fuel supply manifold
   • Fuel injector connectors
   • Intake manifold
   • Ignition coils
   • Oil pressure switch connector
   • VTEC oil pressure switch connector
   • VTEC solenoid valve connector
   • EGR connector
   • TDC sensor connector
   • CKP sensor connector
   • ECT gauge sending unit connector
   • Radiator fan switch connectors
   • ECT sensor connector
   • MAP sensor connector
   • TP sensor connector
   • IAC valve connector
   • IAT sensor connector
   • Alternator
   • Power steering hose clamp
   • Power steering pump
   • Side engine mount bracket
   • PCV valve and hose
   • Intake manifold vacuum line
   • Brake booster vacuum line
   • Fuel lines
   • Cruise control cable
   • Accelerator cable
   • Intake manifold cover
   • Ignition coil covers
   • Accessory drive belts
   • Air intake tube
   • EVAP control canister hose and vacuum hose
   • Negative battery cable

12. Fill the cooling system.

13. Start the engine and check for leaks.

Rocker Arms/Shafts

REMOVAL & INSTALLATION

1. Before servicing the vehicle, refer to the precautions in the beginning of this section.

2. Remove or disconnect the following:
   • Negative battery cable
   • Air intake tube
   • Ignition coil covers
   • Intake manifold cover
   • Intake manifold
   • Valve cover

3. Loosen the valve adjuster locknuts and screws so that all valves are closed.

4. Loosen the rocker arm shaft bolts evenly in sequence.

5. Remove the rocker arms and shafts from the vehicle as an assembly.

➡ **Keep all valvetrain components in order for assembly.**

6. Remove the rocker arms and springs from the rocker arm shafts.

To install:

7. Assemble the rocker arms and springs to the rocker arm shafts in their original positions.

8. Install the rocker arm assemblies. Tighten the bolts in sequence and in multiple passes to 17 ft. lbs. (24 Nm).

9. Adjust the valve clearance.

10. Install or connect the following:

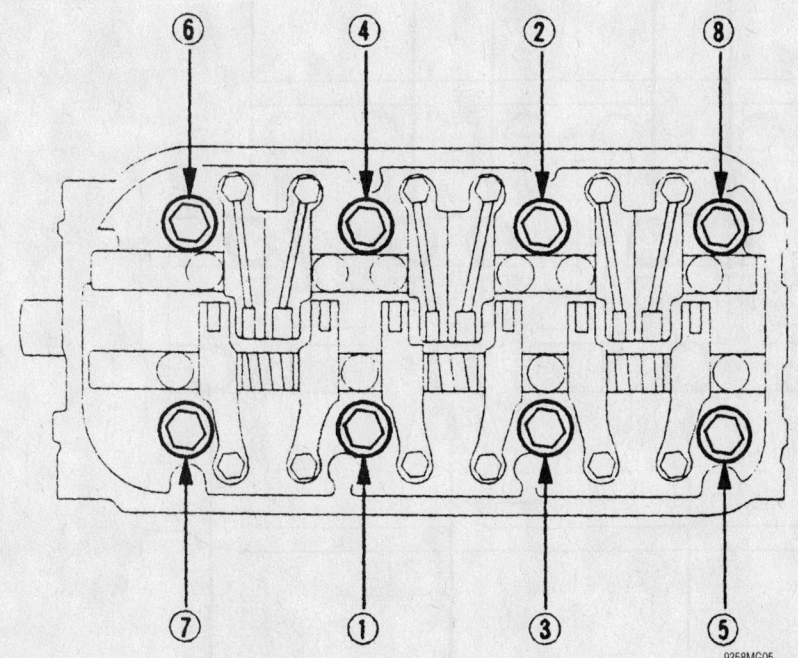

9358MG05

Cylinder head bolt torque sequence

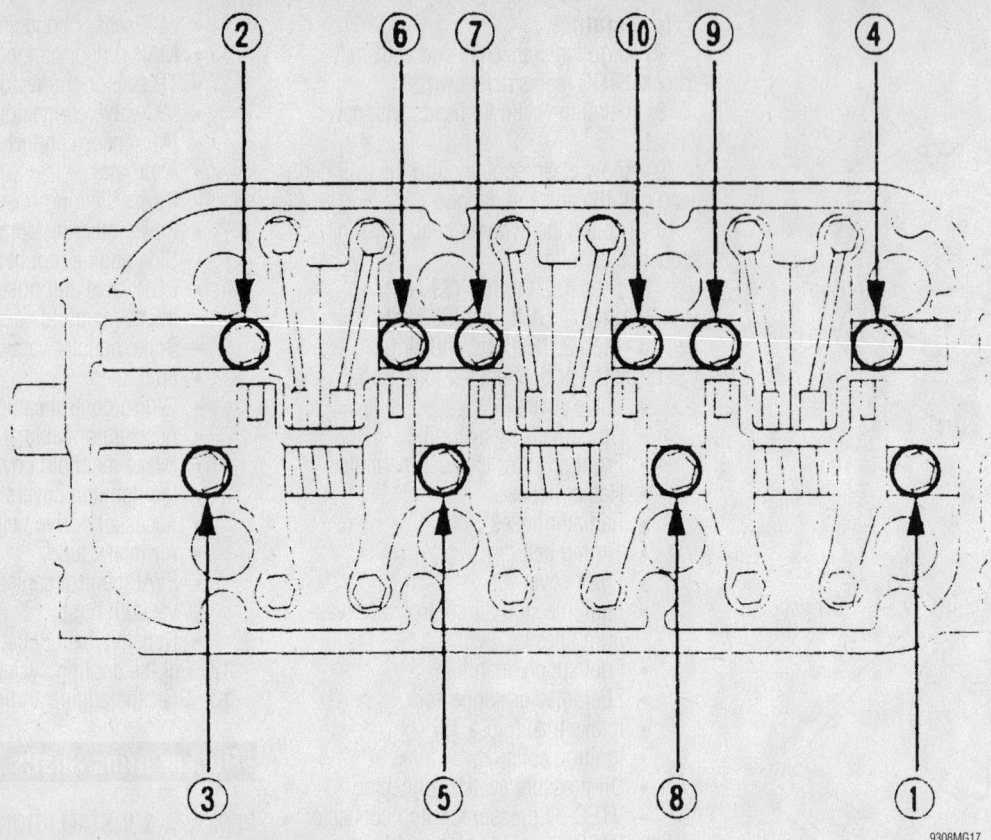

Rocker arm shaft loosening sequence—2000–01 models

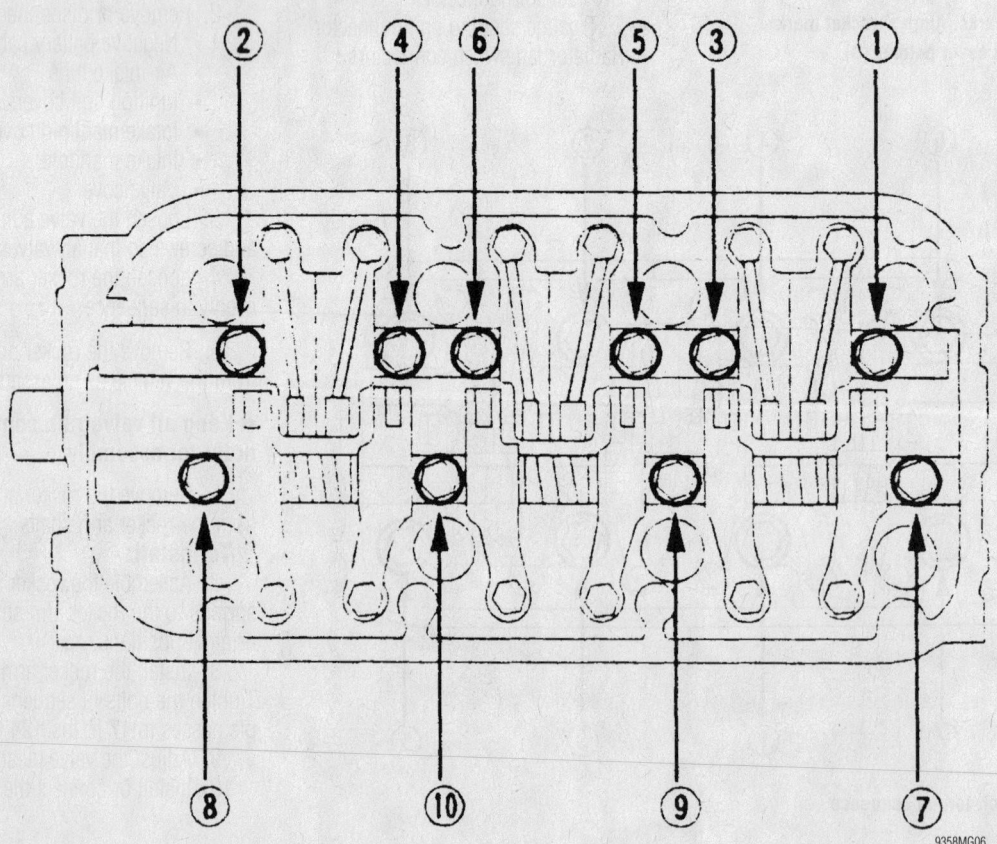

Rocker arm shaft loosening sequence—2002–03 models

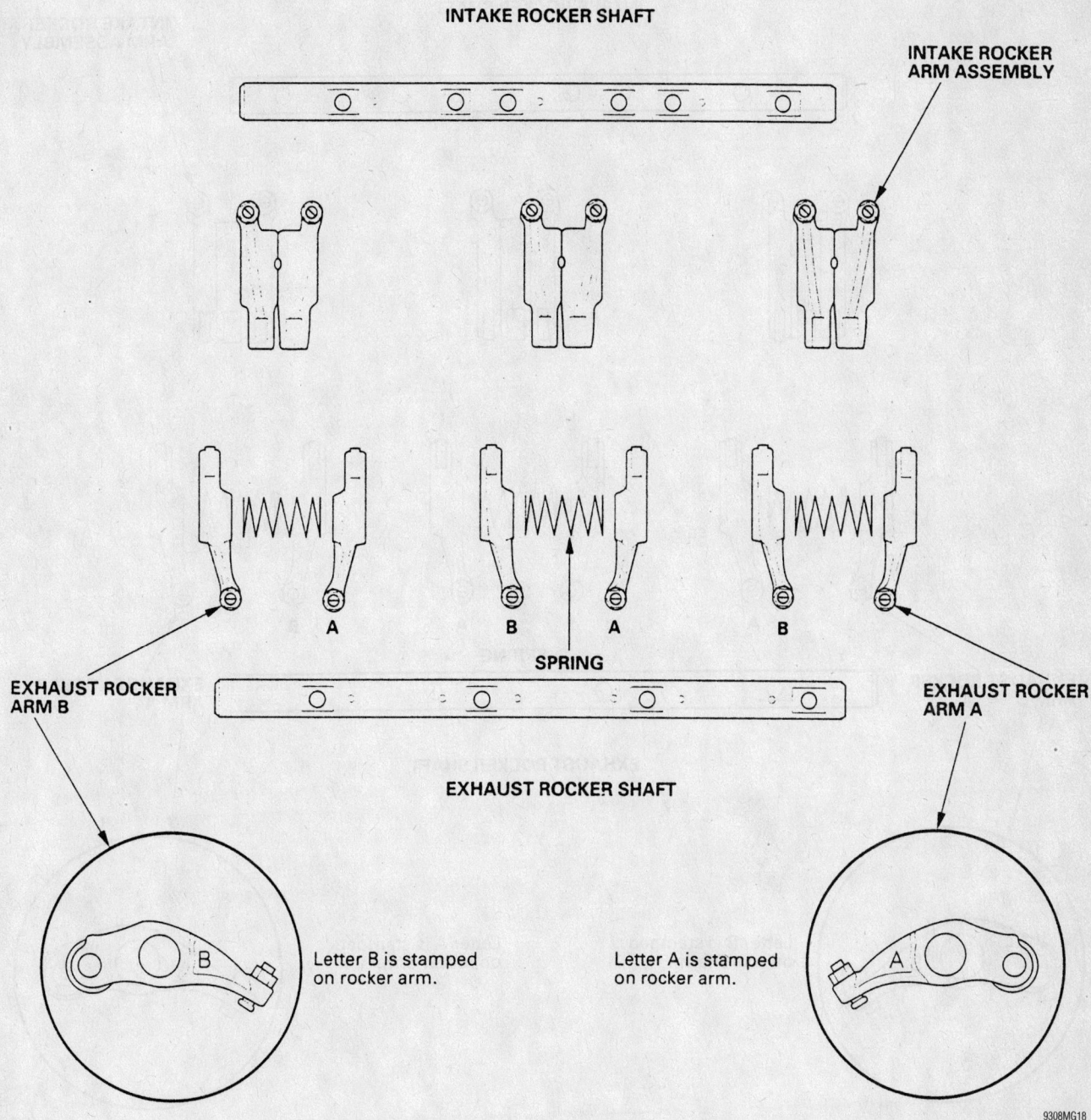

INTAKE ROCKER SHAFT

INTAKE ROCKER
ARM ASSEMBLY

A

B A

B A

B

SPRING

EXHAUST ROCKER
ARM B

EXHAUST ROCKER
ARM A

EXHAUST ROCKER SHAFT

B

Letter B is stamped
on rocker arm.

Letter A is stamped
on rocker arm.

A

9308MG18

Exploded view of the rocker arms and shafts—2000–01 models

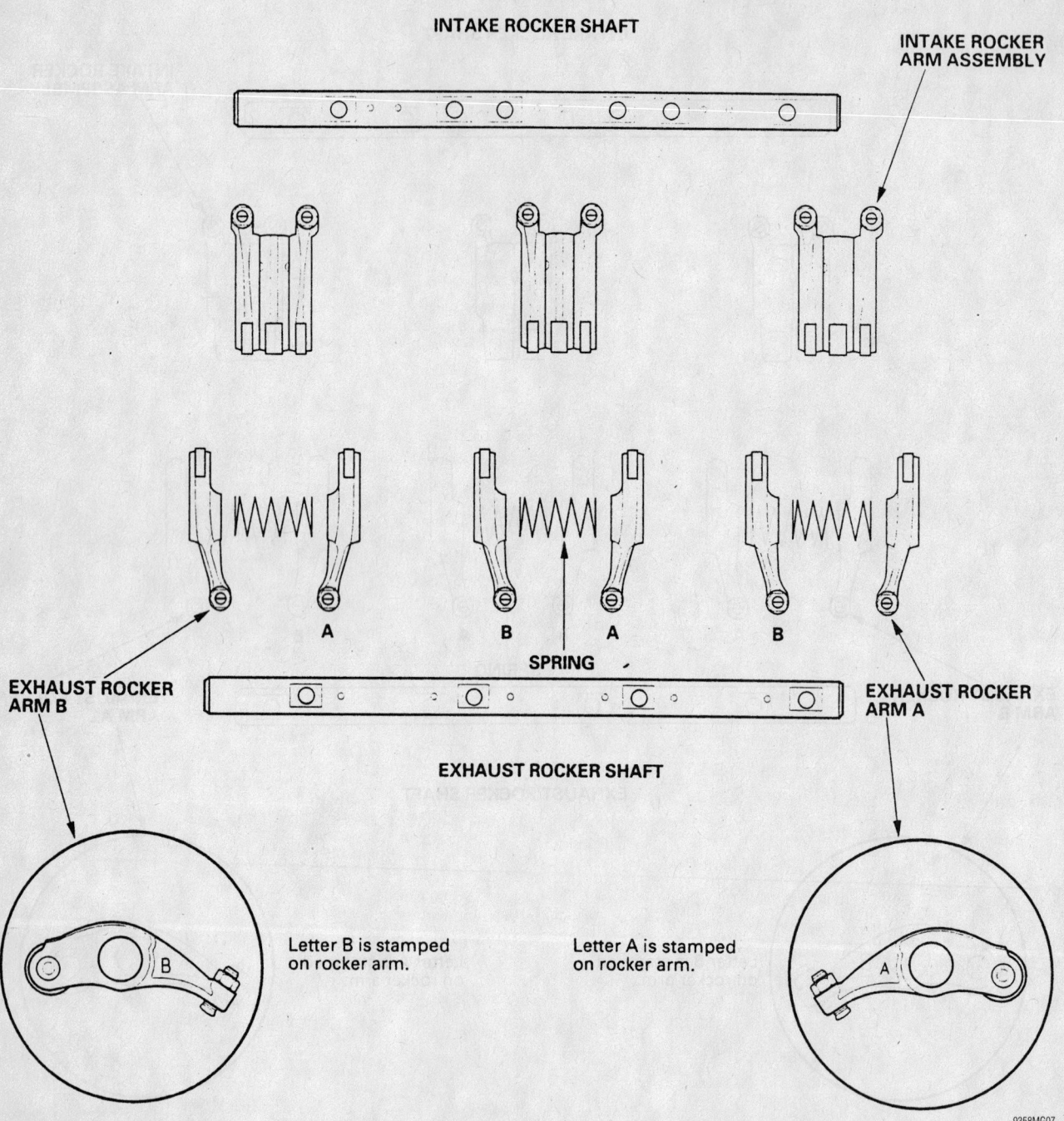

INTAKE ROCKER SHAFT

INTAKE ROCKER
ARM ASSEMBLY

A   B   A   B

SPRING

EXHAUST ROCKER
ARM B

EXHAUST ROCKER
ARM A

EXHAUST ROCKER SHAFT

Letter B is stamped
on rocker arm.

Letter A is stamped
on rocker arm.

9358MG07

Exploded view of the rocker arms and shafts—2002–03 models

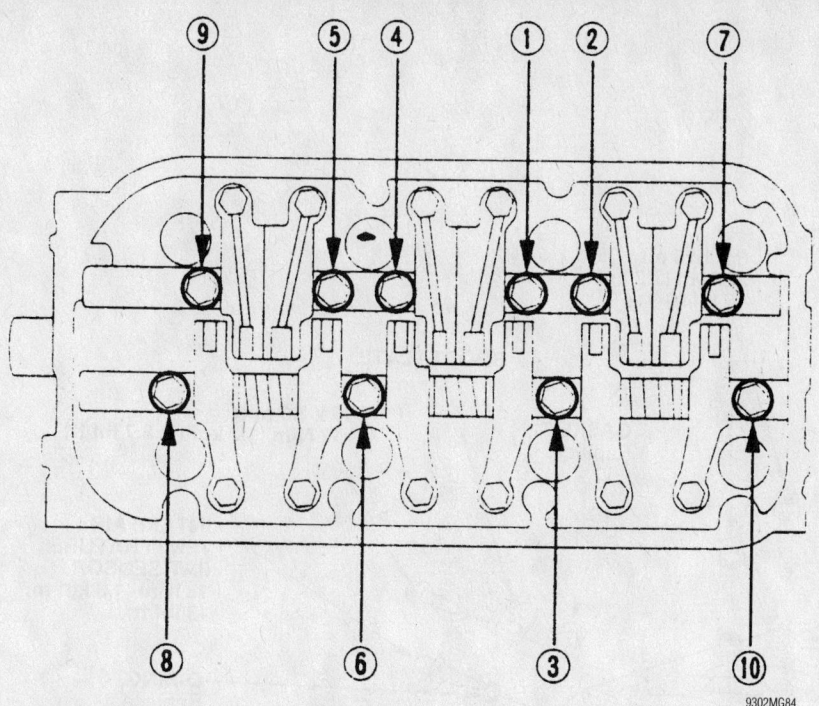

Rocker shaft tightening sequence—2000–01 models

9302MG84

- Valve covers
- Intake manifold
- Intake manifold cover
- Ignition coil covers
- Air intake tube
- Negative battery cable
11. Start the engine and check for leaks.

Intake Manifold

REMOVAL & INSTALLATION

1. Before servicing the vehicle, refer to the precautions in the beginning of this section.
2. Remove or disconnect the following:
   - Negative battery cable
   - Intake manifold cover
   - Evaporative Emissions (EVAP) control canister hose or vacuum hose
   - Air intake tube
   - Accelerator cable
   - Cruise control cable
   - Brake booster vacuum line
   - Intake manifold vacuum line
   - Positive Crankcase Ventilation (PCV) valve and hose

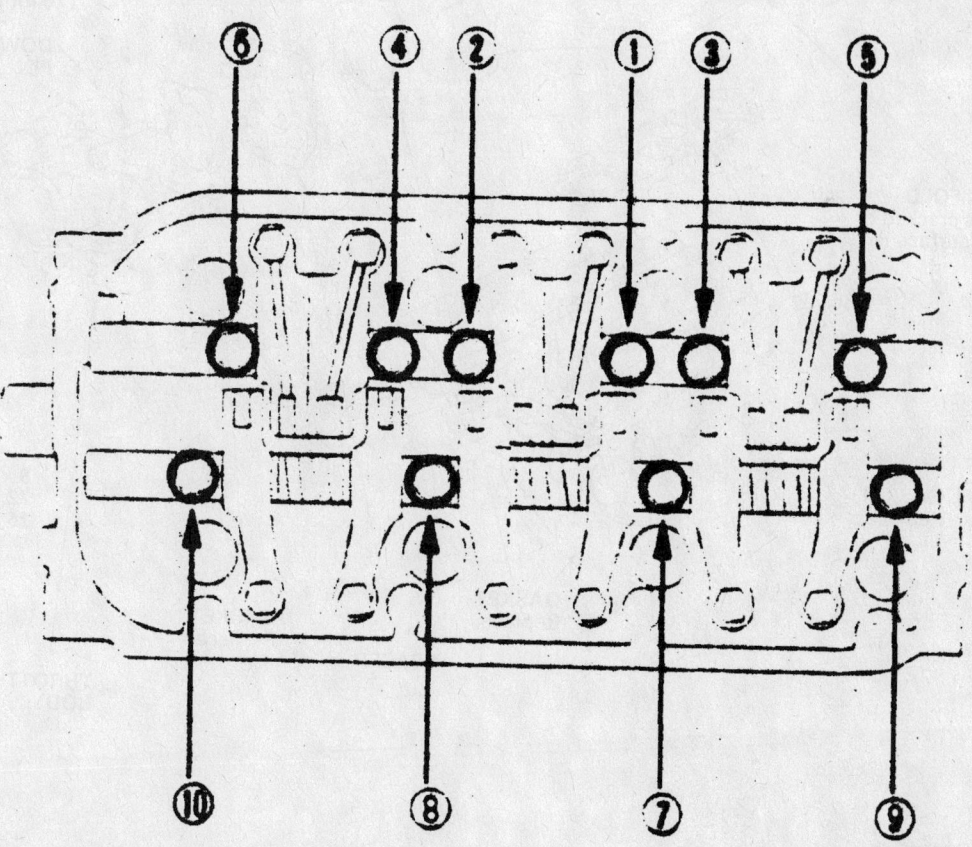

Rocker shaft tightening sequence—2002–03 models

9358MG08

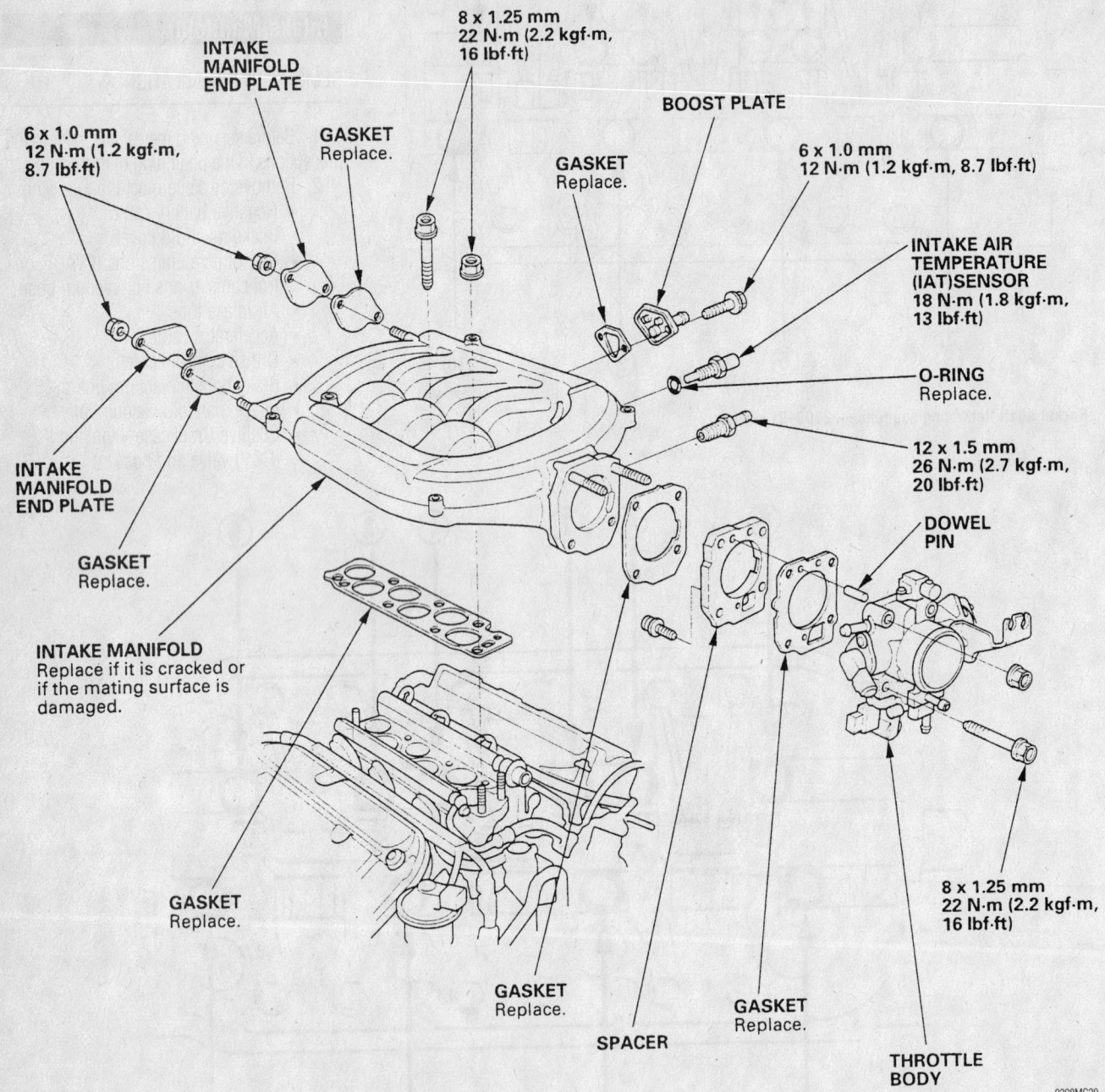

INTAKE
MANIFOLD
END PLATE

8 x 1.25 mm
22 N·m (2.2 kgf·m,
16 lbf·ft)

BOOST PLATE

6 x 1.0 mm
12 N·m (1.2 kgf·m,
8.7 lbf·ft)

GASKET
Replace.

GASKET
Replace.

6 x 1.0 mm
12 N·m (1.2 kgf·m, 8.7 lbf·ft)

INTAKE AIR
TEMPERATURE
(IAT)SENSOR
18 N·m (1.8 kgf·m,
13 lbf·ft)

O-RING
Replace.

INTAKE
MANIFOLD
END PLATE

GASKET
Replace.

12 x 1.5 mm
26 N·m (2.7 kgf·m,
20 lbf·ft)

DOWEL
PIN

INTAKE MANIFOLD
Replace if it is cracked or
if the mating surface is
damaged.

GASKET
Replace.

GASKET
Replace.

GASKET
Replace.

SPACER

THROTTLE
BODY

8 x 1.25 mm
22 N·m (2.2 kgf·m,
16 lbf·ft)

9308MG20

Exploded view of the intake manifold—2000–01 models

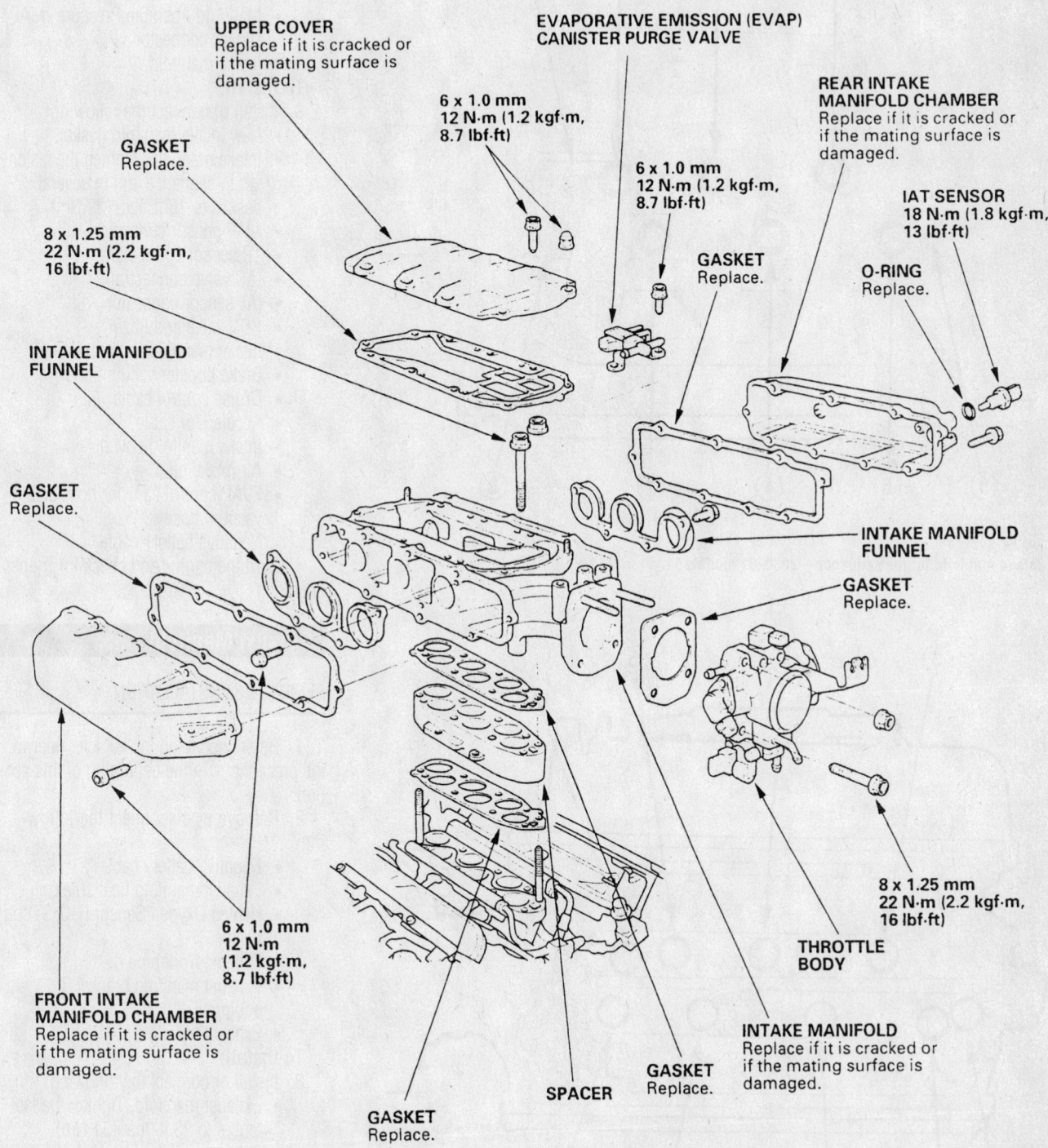

UPPER COVER
Replace if it is cracked or
if the mating surface is
damaged.

EVAPORATIVE EMISSION (EVAP)
CANISTER PURGE VALVE

REAR INTAKE
MANIFOLD CHAMBER
Replace if it is cracked or
if the mating surface is
damaged.

GASKET
Replace.

6 x 1.0 mm
12 N·m (1.2 kgf·m,
8.7 lbf·ft)

IAT SENSOR
18 N·m (1.8 kgf·m,
13 lbf·ft)

8 x 1.25 mm
22 N·m (2.2 kgf·m,
16 lbf·ft)

6 x 1.0 mm
12 N·m (1.2 kgf·m,
8.7 lbf·ft)

GASKET
Replace.

O-RING
Replace.

INTAKE MANIFOLD
FUNNEL

INTAKE MANIFOLD
FUNNEL

GASKET
Replace.

GASKET
Replace.

8 x 1.25 mm
22 N·m (2.2 kgf·m,
16 lbf·ft)

THROTTLE
BODY

6 x 1.0 mm
12 N·m
(1.2 kgf·m,
8.7 lbf·ft)

FRONT INTAKE
MANIFOLD CHAMBER
Replace if it is cracked or
if the mating surface is
damaged.

GASKET
Replace.

SPACER

GASKET
Replace.

INTAKE MANIFOLD
Replace if it is cracked or
if the mating surface is
damaged.

9358MG09

Exploded view of the intake manifold—2002–03 models

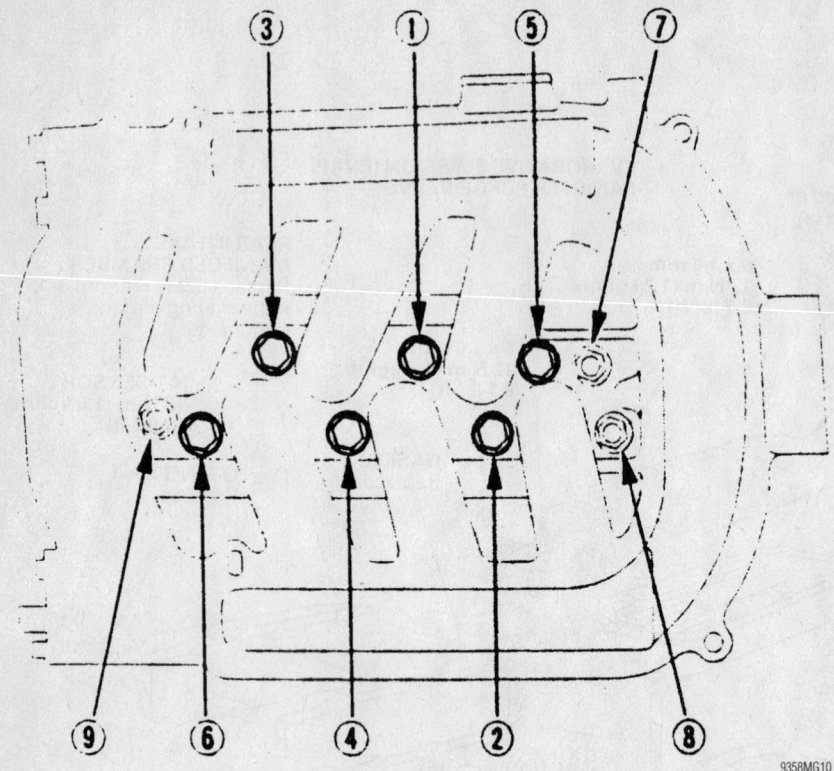

Intake manifold torque sequence—2000–01 models

9358MG10

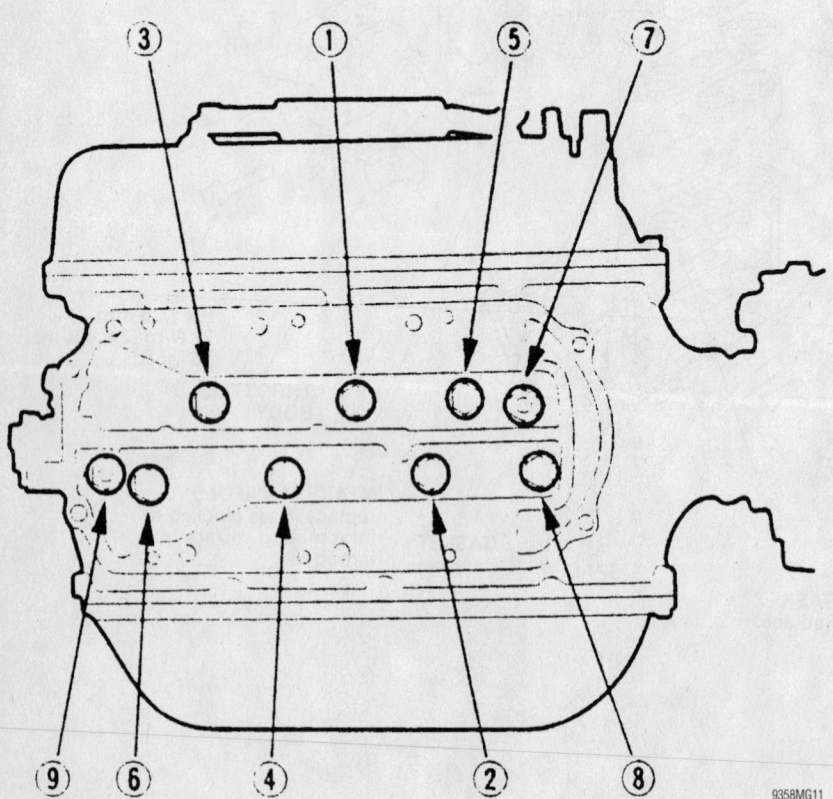

Intake manifold torque sequence—2002–03 models

9358MG11

- Intake Air Temperature (IAT) sensor connector
- Idle Air Control (IAC) valve connector
- Throttle Position (TP) sensor connector
- Manifold Absolute Pressure (MAP) sensor connector
- Intake manifold

To install:

3. Install or connect the following:
- New intake manifold gasket
- Intake manifold. Tighten the fasteners in sequence and in several passes to 16 ft. lbs. (22 Nm).
- MAP sensor connector
- TP sensor connector
- IAC valve connector
- IAT sensor connector
- PCV valve and hose
- Intake manifold vacuum line
- Brake booster vacuum line
- Cruise control cable
- Accelerator cable
- Intake manifold cover
- Air intake tube
- EVAP control canister hose and vacuum hose
- Negative battery cable

4. Start the engine and check for proper operation.

Exhaust Manifold

REMOVAL & INSTALLATION

1. Before servicing the vehicle, refer to the precautions in the beginning of this section.

2. Remove or disconnect the following:

- Negative battery cable
- Exhaust manifold heat shield
- Heated Oxygen Sensor (HO$_2$S) connector
- Exhaust front pipe
- Exhaust manifold bracket, if equipped
- Exhaust manifold

To install:

3. Install or connect the following:

- Exhaust manifold. Tighten the fasteners to 23 ft. lbs. (31 Nm).
- Exhaust manifold bracket, if equipped. Tighten the bolts to 33 ft. lbs. (44 Nm).
- Exhaust front pipe. Tighten the nuts to 40 ft. lbs. (55 Nm).
- HO$_2$S connector
- Exhaust manifold heat shield
- Negative battery cable

FRONT:

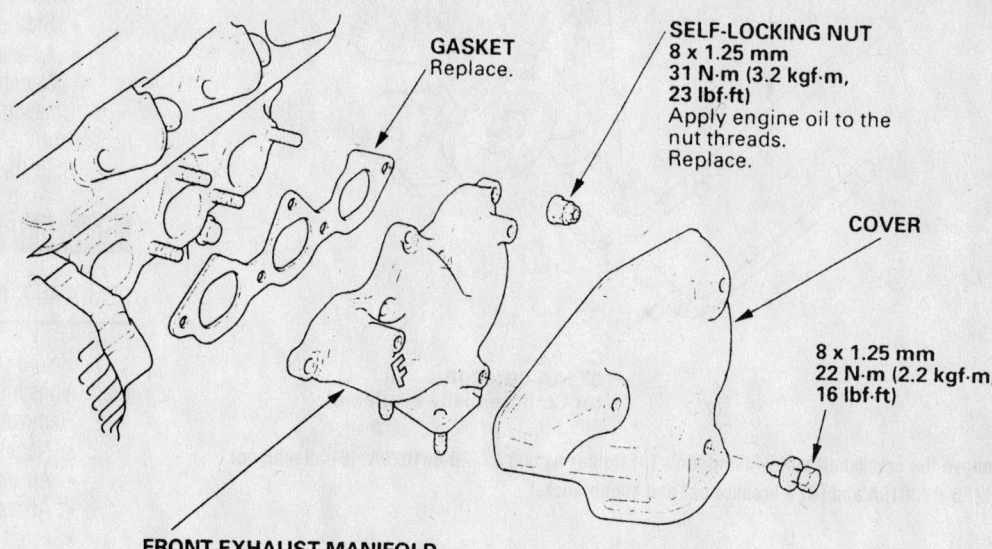

GASKET
Replace.

SELF-LOCKING NUT
8 x 1.25 mm
31 N·m (3.2 kgf·m, 23 lbf·ft)
Apply engine oil to the nut threads.
Replace.

COVER

8 x 1.25 mm
22 N·m (2.2 kgf·m, 16 lbf·ft)

FRONT EXHAUST MANIFOLD

REAR:

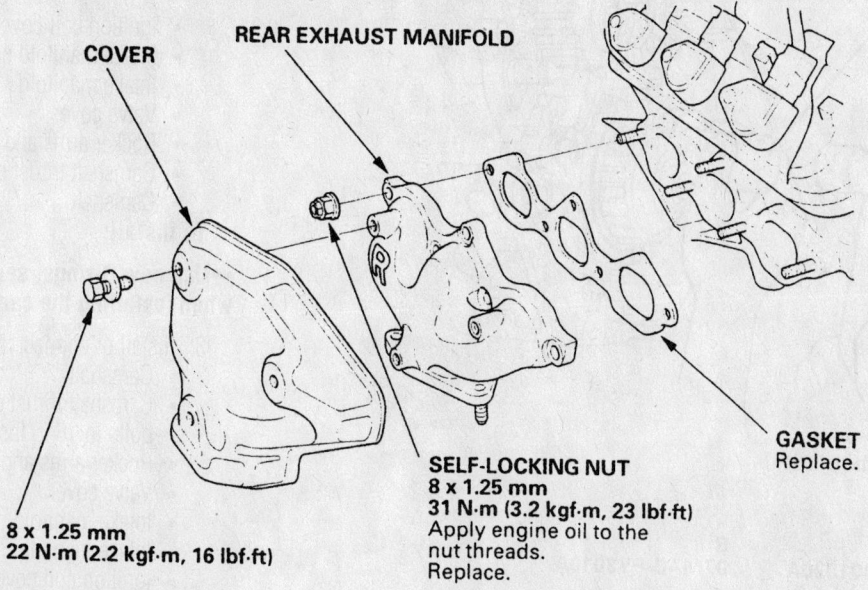

COVER

REAR EXHAUST MANIFOLD

SELF-LOCKING NUT
8 x 1.25 mm
31 N·m (3.2 kgf·m, 23 lbf·ft)
Apply engine oil to the nut threads.
Replace.

GASKET
Replace.

8 x 1.25 mm
22 N·m (2.2 kgf·m, 16 lbf·ft)

Exploded view of the exhaust manifolds

9358MG12

Front Crankshaft Seal

REMOVAL & INSTALLATION

1. Before servicing the vehicle, refer to the precautions in the beginning of this section.

2. Remove or disconnect the following:

- Negative battery cable
- Accessory drive belts
- Side engine mount
- Valve cover

- Crankshaft pulley using the tools in the accompanying illustration
- Front cover
- Balance shaft belt, if equipped
- Timing belt. Refer to the Timing Belt unit repair section.
- Top Dead Center (TDC) sensor, if equipped
- Crankshaft timing sprocket
- Front crankshaft seal

To install:

3. Lubricate the crankshaft seal lip with grease prior to installation.

4. Install the front crankshaft seal so that it is flush with the surface of the oil pump housing.

5. Install or connect the following:

- Crankshaft timing sprocket
- Top Dead Center (TDC) sensor, if equipped
- Timing belt. Refer to the Timing Belt unit repair section.
- Balance shaft belt, if equipped
- Front cover
- Crankshaft pulley. Tighten the bolt to 181 ft. lbs. (245 Nm) using the

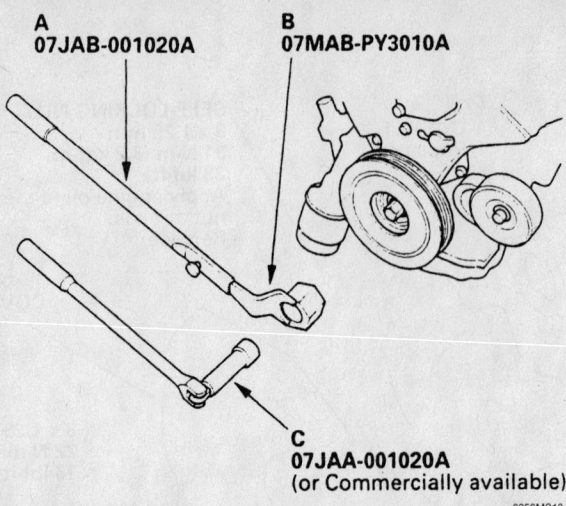

Remove the crankshaft pulley using tools (A) Holder handle 07JAB-001020A, (B) attachment 07MAB-PY3010A and (C) a breaker bar and 19mm socket

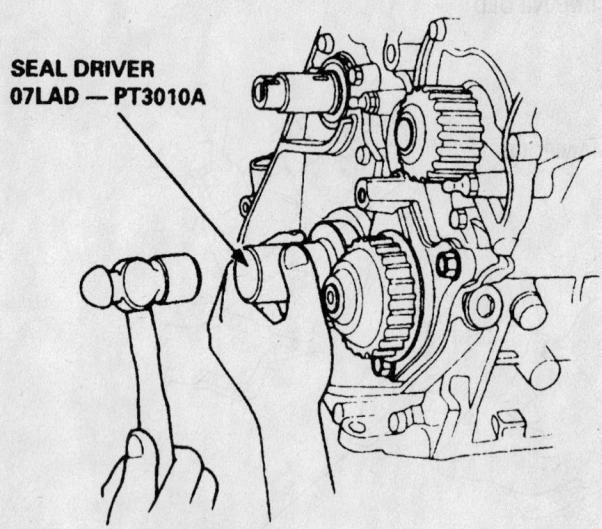

Typical front crankshaft seal installation

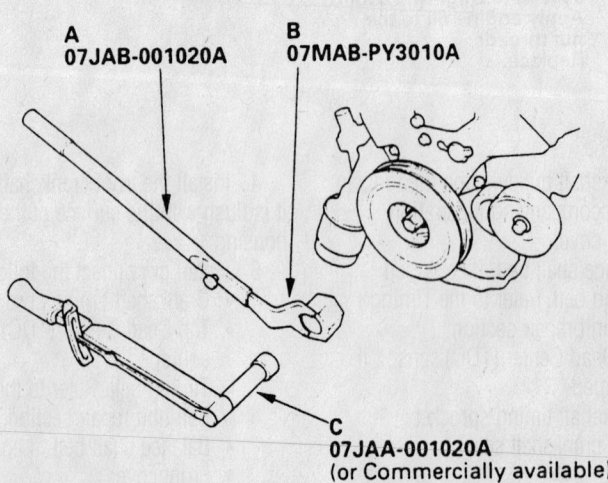

Install the crankshaft pulley using tools (A) Holder handle 07JAB-001020A, (B) attachment 07MAB-PY3010A and (C) a torque wrench and 19mm socket

tools in the accompanying illustration.
- Valve cover
- Side engine mount
- Accessory drive belts
- Negative battery cable

6. Check the engine oil level and add if necessary.

7. Start the engine and check for leaks.

Camshaft

REMOVAL & INSTALLATION

1. Before servicing the vehicle, refer to the precautions in the beginning of this section.

2. Remove or disconnect the following:
- Negative battery cable
- Air intake tube
- Accessory drive belts
- Front cover
- Timing belt. Refer to the Timing Belt unit repair section.
- Camshaft sprockets
- Timing belt rear covers
- Ignition coil covers
- Intake manifold cover
- Intake manifold
- Valve cover
- Rocker arms and shaft assembly
- Camshaft thrust cover
- Camshaft

To install:

→ **Use new O-rings, seals and gaskets when installing the camshaft.**

3. Install or connect the following:
- Camshaft
- Camshaft thrust cover. Tighten the bolts to 16 ft. lbs. (22 Nm).
- Rocker arms and shaft assembly
- Valve cover
- Intake manifold
- Intake manifold cover
- Ignition coil covers
- Timing belt rear covers
- Camshaft sprockets. Tighten the bolts to 67 ft. lbs. (90 Nm).
- Timing belt
- Front cover
- Accessory drive belts
- Air intake tube
- Negative battery cable

4. Start the engine and check for leaks.

Valve Lash

ADJUSTMENT

Adjust the valves only when the cylinder head temperature is less than 100°F (38°C).

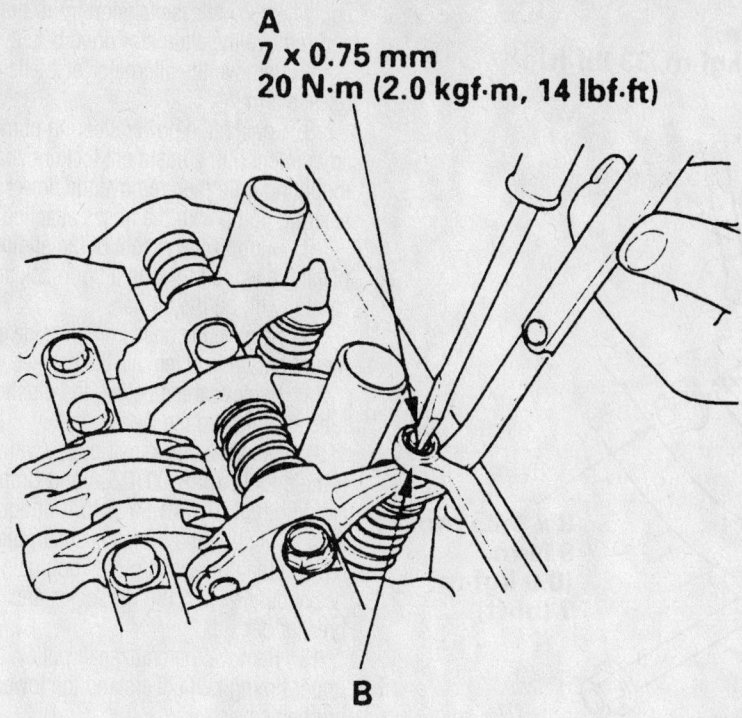

A
7 x 0.75 mm
20 N·m (2.0 kgf·m, 14 lbf·ft)

B

9358MG15

After adjustment tighten the locknut to specification

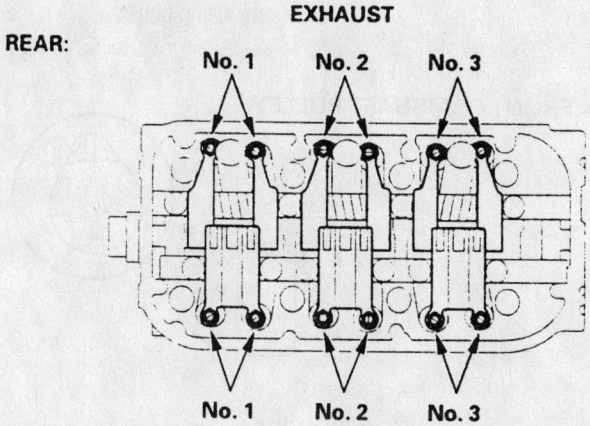

REAR:

EXHAUST

No. 1   No. 2   No. 3

No. 1   No. 2   No. 3

INTAKE

FRONT:

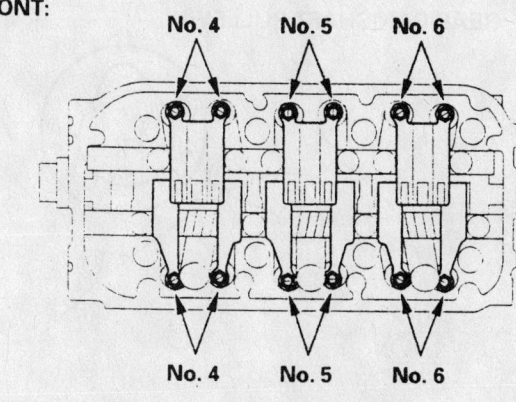

No. 4   No. 5   No. 6

No. 4   No. 5   No. 6

EXHAUST

9358MG16

Adjusting screw locations

1. Before servicing the vehicle, refer to the precautions in the beginning of this section.

2. Remove or disconnect the following:

- Negative battery cable
- Air intake tube
- Intake manifold
- Valve cover

3. Rotate the crankshaft so that the valves to be adjusted are closed and the rocker arm is contacting the camshaft lobe base circle.

4. Measure the valve clearance. If adjustment is necessary, loosen the locknut and turn the adjusting screw as necessary to achieve the correct valve clearance.

5. The correct valve clearance is:

- Intake valves: 0.008–0.009 inches (0.20–0.24mm)
- Exhaust valves: 0.011–0.013 inches (0.28–0.32mm)

6. After adjustment, tighten the locknuts to 14 ft. lbs. (20 Nm).

7. Install or connect the following:

- Valve cover
- Intake manifold
- Air intake tube
- Negative battery cable

8. Start the engine and check for proper operation.

Starter Motor

REMOVAL & INSTALLATION

1. Before servicing the vehicle, refer to the precautions in the beginning of this section.

2. Remove or disconnect the following:

- Negative battery cable
- Transmission fluid cooler line clamp
- Starter wiring harness connectors
- Starter motor

To install:

3. Install or connect the following:

- Starter motor. Tighten the bolts to 33 ft. lbs. (44 Nm).
- Starter motor. Tighten the upper bolt to 33 ft. lbs. (44 Nm) and the lower bolt to 47 ft. lbs. (64 Nm).
- Starter wiring harness connectors. Tighten the battery cable nut to 79 inch lbs. (9 Nm).
- Transmission fluid cooler line clamp
- Negative battery cable

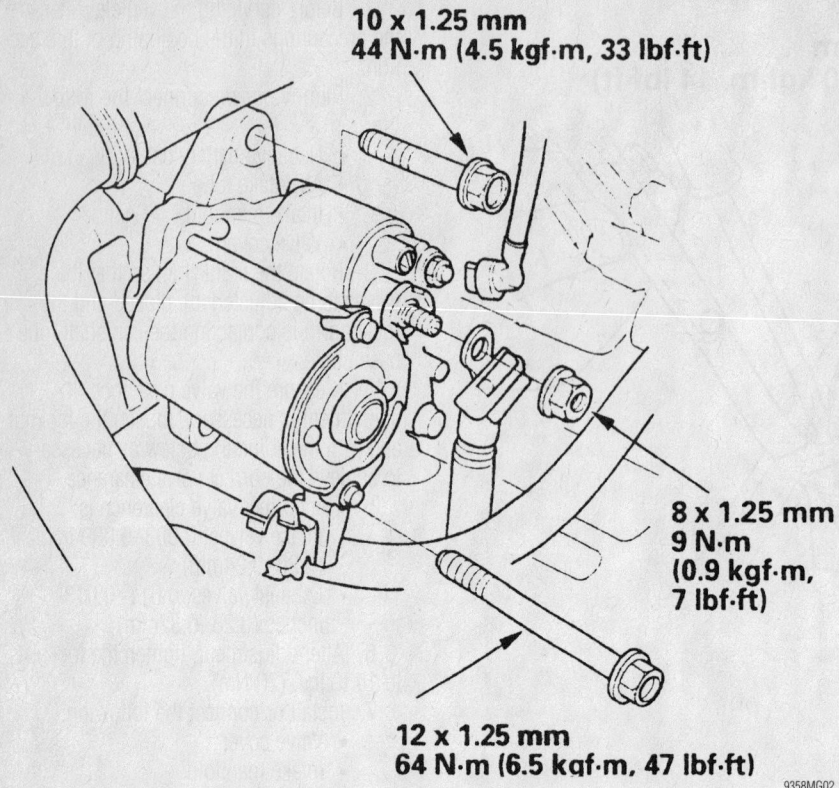

**10 x 1.25 mm
44 N·m (4.5 kgf·m, 33 lbf·ft)**

**8 x 1.25 mm
9 N·m
(0.9 kgf·m,
7 lbf·ft)**

**12 x 1.25 mm
64 N·m (6.5 kgf·m, 47 lbf·ft)**

9358MG02

Exploded view of the starter mounting

Timing Belt and Front Cover

REMOVAL & INSTALLATION

➡**The radio may contain a coded theft protection circuit. Always make note the code number before disconnecting the battery.**

1. Disconnect the negative battery terminal.

2. Turn the crankshaft so the white mark on the crankshaft pulley aligns with the pointer on the oil pump housing cover.

3. Open the inspection plugs on the upper timing belt covers and check that the camshaft sprocket marks align with the upper cover marks.

✳✳ WARNING

Align the camshaft and crankshaft sprockets with their alignment marks before removing the timing belt. Failure to align the timing marks correctly may result in valve damage.

4. Raise and safely support the vehicle and remove both front tires/wheels.

5. Remove the front lower splash shield.

6. Move the alternator tensioner with a Belt Tensioner Release Arm tool YA9317, or

equivalent, to release tension from the belt and remove the alternator drive belt.

7. Remove the alternator belt tensioner release arm.

8. Loosen the power steering pump adjustment nut, adjustment locknut and mounting bolt, then remove the power steering pump with the hoses attached.

9. Support the weight of the engine by placing a wood block on a floor jack and carefully lift on the oil pan.

10. Remove the bolts from the side engine mount bracket and remove the bracket.

11. Remove the dipstick, the dipstick tube and discard the O-ring.

12. Hold the crankshaft pulley with the Handle tool 07JAB-001020A and Crankshaft Holding tool 07MAB-PY3010A, or equivalent. While holding the crankshaft pulley, remove the crankshaft pulley bolt using a heavy duty ¾ in. (19mm) socket and breaker bar.

13. Remove the crankshaft pulley, the upper timing belt covers and the lower timing belt cover.

14. Remove one of the battery clamp fasteners from the battery tray and grind a 45 degree bevel on the threaded end of the battery clamp bolt.

FRONT CAMSHAFT PULLEY:

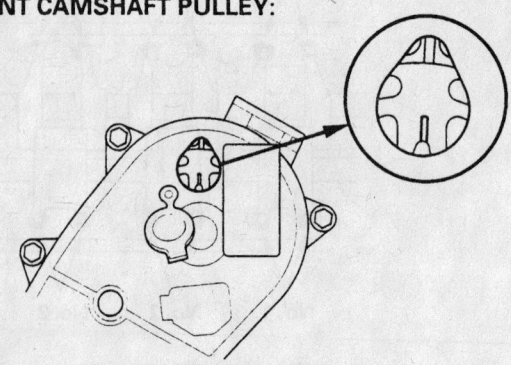

REAR CAMSHAFT PULLEY:

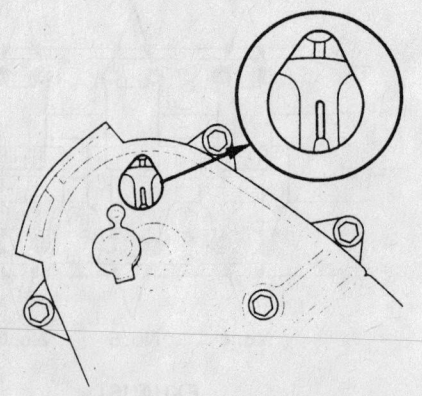

93025G04

Crankshaft and camshaft timing marks at Top Dead Center (TDC)

15. Screw in the battery hold-down bolt into the threaded bracket just above the auto-tensioner (automatic timing belt adjuster) and tighten the bolt hand-tight to hold the auto-tensioner adjuster in its current position.

16. Remove the engine mount bracket bolts and the bracket.

17. Loosen the timing belt idler pulley bolt (located on the right side across from the auto-tensioner pulley) about 5–6 revolutions and remove the timing belt.

To install:

18. Clean the timing belt sprockets and the timing belt covers.

✳✳ WARNING

Align the camshaft and crankshaft sprockets with their alignment marks before installing the timing belt. Failure to align the timing marks correctly may result in valve damage.

19. Align the timing mark on the crankshaft sprocket with the oil pump pointer.

20. Align the camshaft sprocket TDC timing marks with the pointers on the rear cover.

21. If installing a new belt or if the auto-tensioner has extended or if the timing belt cannot be reinstalled easily, the auto-tensioner must be collapsed before installation of the timing belt, perform the following procedures:

a. Remove the battery hold-down bolt from the auto-tensioner bracket.

b. Remove the timing belt auto-tensioner bolts and the auto-tensioner.

c. Secure the auto-tensioner in a soft jawed vise, clamping onto the flat surface of one of the mounting bolt holes with the maintenance bolt facing upward.

d. Remove the maintenance bolt and use caution not to spill oil from the tensioner assembly.

e. Should oil spill from the tensioner, be sure the tensioner is filled with 0.22 ounces (6.5 ml) of fresh engine oil.

f. Using care not to damage the threads or the gasket sealing surface, insert a flat-blade screwdriver through the tensioner maintenance hole and turn the screwdriver clockwise to compress the auto-tensioner bottom while the Tensioner Holder tool 14540-P8A-A01, or equivalent, is installed on the auto-tensioner assembly.

g. Install the auto-tensioner maintenance bolt with a new gasket and tighten to a torque 72 inch lbs. (8 Nm).

h. Install the auto-tensioner on the

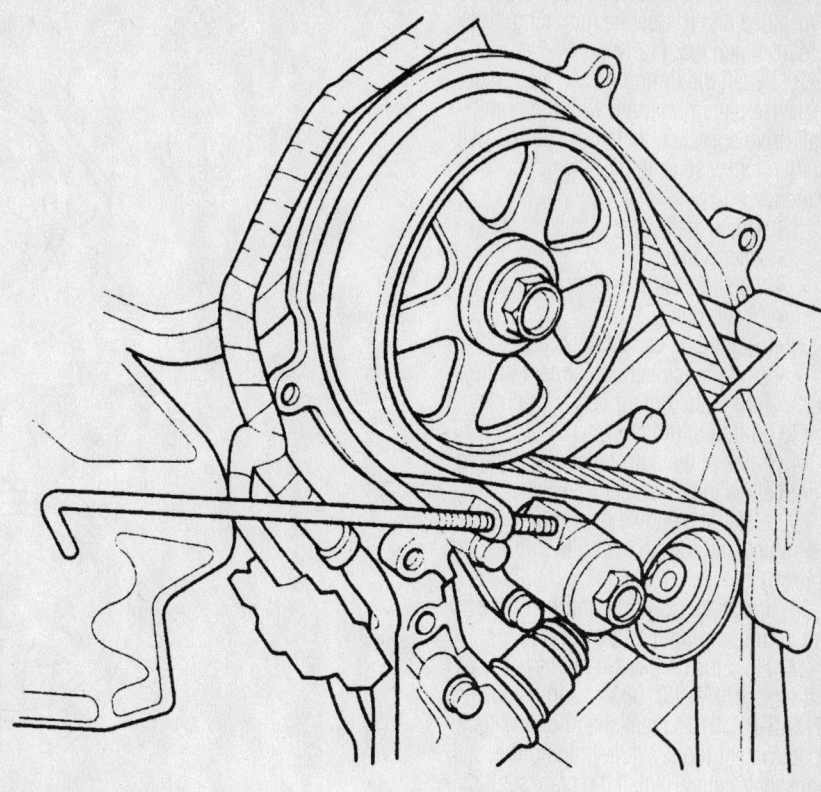

Battery hold-down bolt installed to hold auto-tensioner

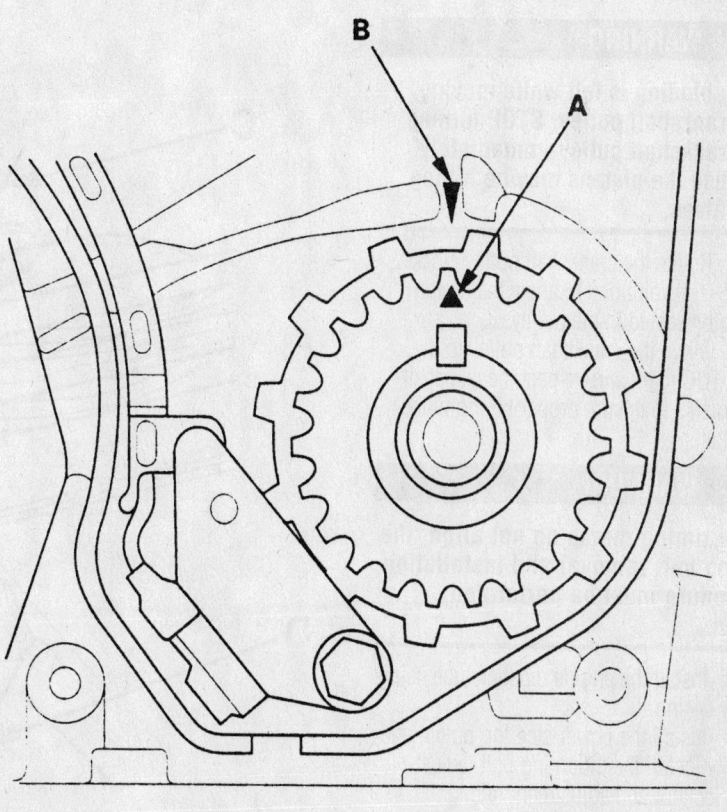

Crankshaft sprocket Top Dead Center (TDC) mark

engine with the tensioner holder tool installed and torque the mounting bolts to 104 inch lbs. (12 Nm).

22. Install the timing belt in a counterclockwise pattern starting with the crankshaft drive sprocket. Install the timing belt counterclockwise in the following sequence:

- Crankshaft drive sprocket.
- Idler pulley.
- Left side camshaft sprocket.
- Water pump.
- Right side camshaft sprocket.
- Auto-tensioner adjustment pulley.

23. Torque the timing belt idler pulley bolt to 33 ft. lbs. (44 Nm).

24. Remove the auto-tensioner holding tool to allow the tensioner to extend.

25. Install the engine mount bracket to the engine and torque the bolts to 33 ft. lbs. (44 Nm).

26. Install the lower timing belt cover and both upper timing belt covers.

27. Hold the crankshaft pulley with special tools 07JAB-001020A handle and 07MAB-PY3010A crankshaft holding tool, or equivalent tools. While holding the crankshaft pulley, install the crankshaft pulley bolt using a heavy duty ¾ in. (19mm) socket and a commercially available torque wrench and torque the bolt to 181 ft. lbs. (245 Nm).

❄❄ WARNING

If any binding is felt while moving the crankshaft pulley, STOP turning the crankshaft pulley immediately because the pistons may be hitting the valves.

28. Rotate the crankshaft pulley clockwise 5–6 revolutions to allow the timing belt to be seated in the pulleys.

29. Move the crankshaft pulley to the white TDC mark and inspect the camshaft TDC marks to ensure proper timing of the camshafts.

❄❄ WARNING

If the timing marks do not align, the timing belt removal and installation procedure must be performed again.

30. Install the engine dipstick tube using a new O-ring.

31. Install the power steering pump, and loosely install the mounting bolt, adjustment locknut and adjustment nut.

32. Adjust the power steering belt to a tension such that a 22 lb. (98 N) pull

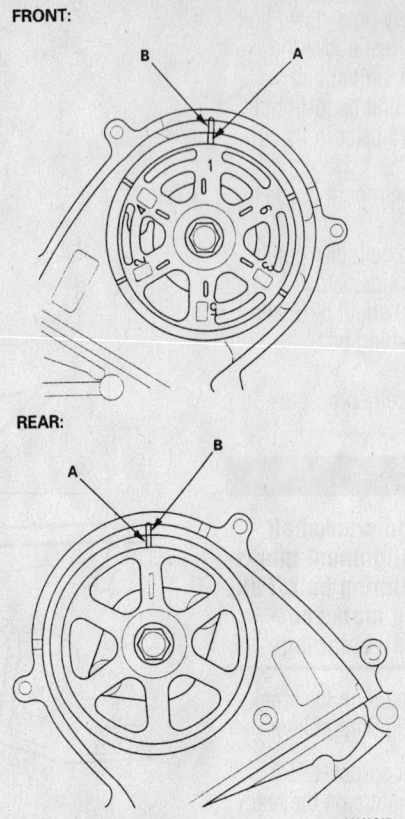

FRONT:

REAR:

Camshaft sprocket Top Dead Center (TDC) mark

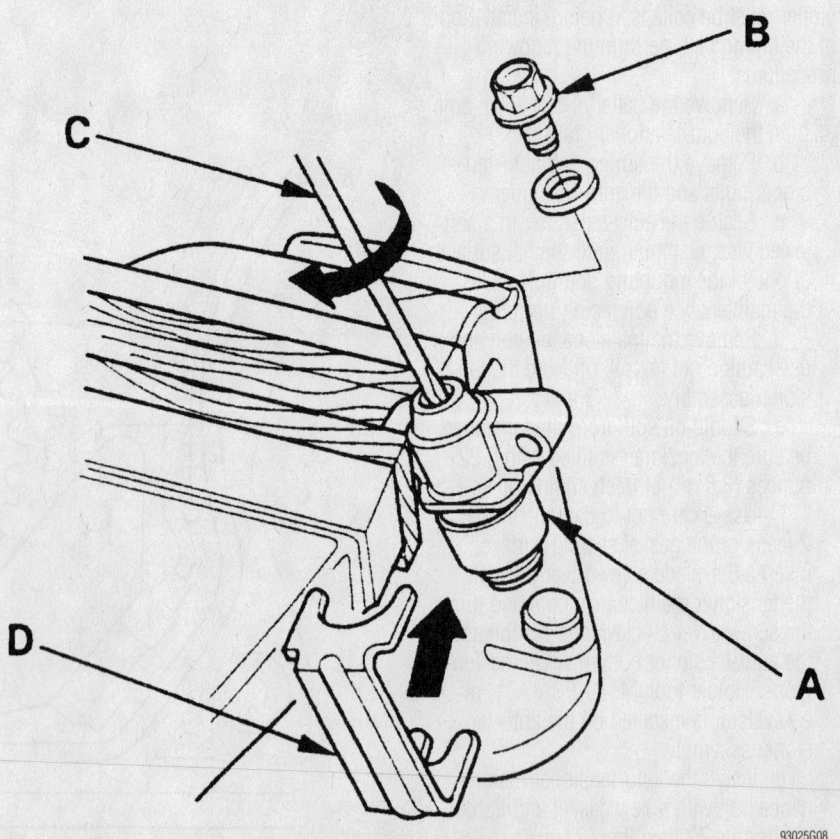

Adjusting the auto-tensioner

halfway between the 2 drive pulleys will allow the belt to move 0.51–0.65 in. (13.0–16.5mm).

33. Tighten the power steering pump mounting bolt and adjustment locknut.

➡**If a new belt is used, set the deflection to 0.33–0.43 in. (8.5–11.0mm) and after engine has run for 5 minutes, readjust the new belt to the used belt specification.**

34. Install the alternator belt tensioner arm.

35. Move the alternator tensioner with a Belt Tensioner Release Arm tool YA9317, or equivalent, to release tension from the belt and install the alternator drive belt.

36. Install both engine mount bracket bolts and torque to 33 ft. lbs. (44 Nm).

37. Install the bushing through bolt and tighten to 40 ft. lbs. (54 Nm).

38. Release and carefully remove the floor jack.

39. Install the front lower splash shield.

40. Install both front tires/wheels.

41. Carefully lower the vehicle.

42. Install the battery hold-down bolt in the battery tray.

43. Install the negative battery cable.

44. Enter the radio security code.

Oil Pan

REMOVAL & INSTALLATION

1. Before servicing the vehicle, refer to the precautions in the beginning of this section.

2. Drain the engine oil.

3. Set the engine at Top Dead Center (TDC).

4. Remove or disconnect the following:

- Negative battery cable
- Timing belt
- Idler pulley
- VTEC solenoid valve and oil filter
- Oil pan bolts and the pan.

To install:

5. Apply liquid gasket to the inner threads of the bolt holes and the engine block along the area indicated by the broken line in the accompanying illustration.

6. Install the oil and tighten the bolts to 105 inch lbs. (12 Nm).

7. Install the remaining components in the reverse order of removal.

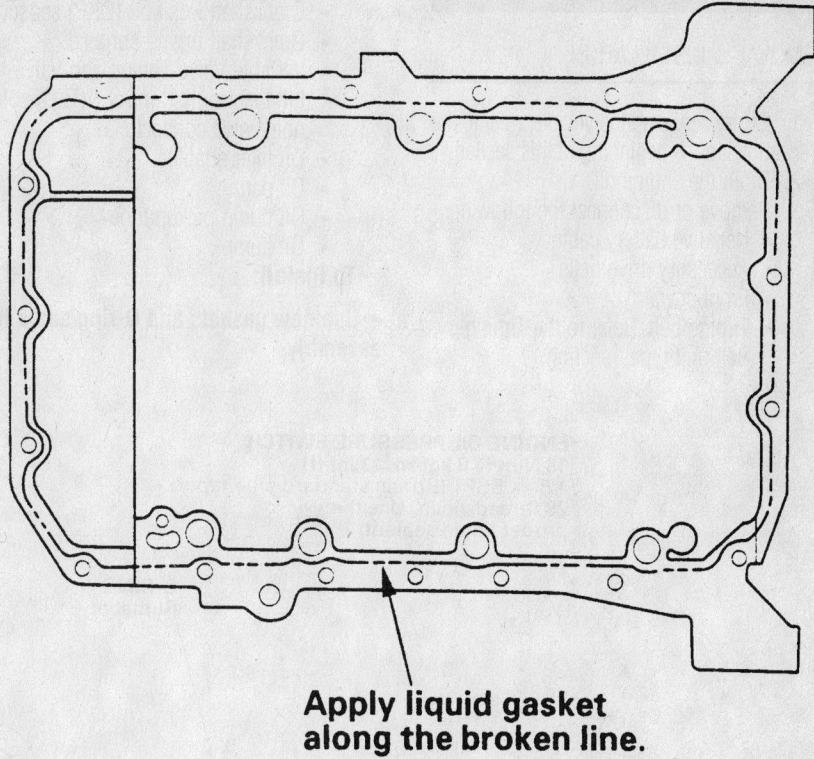

Apply liquid gasket along the broken line.

Apply liquid gasket to the inner threads of the bolt holes and the engine block along the area indicated by the broken line

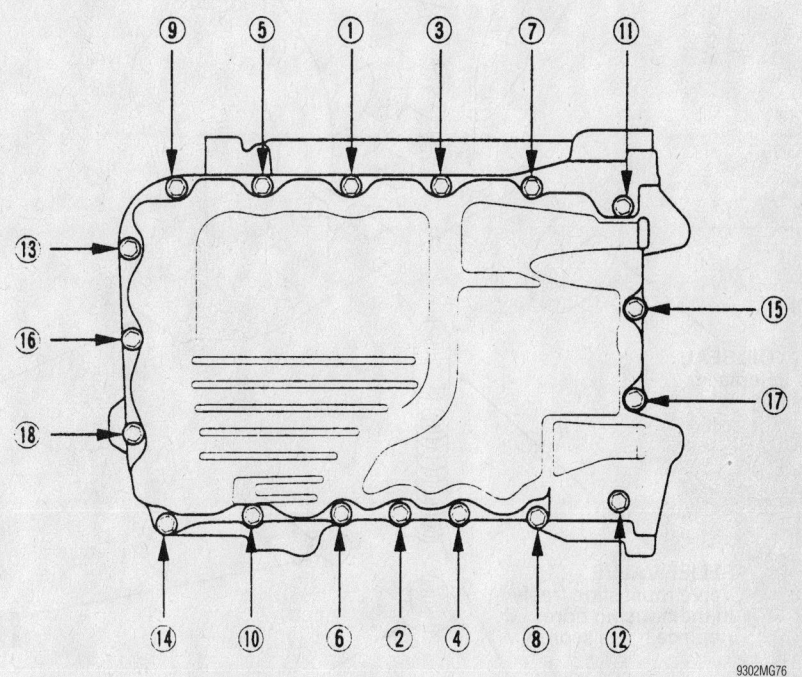

Oil pan tightening sequence

Oil Pump

REMOVAL & INSTALLATION

1. Before servicing the vehicle, refer to the precautions in the beginning of this section.
2. Drain the engine oil.
3. Remove or disconnect the following:
   - Negative battery cable
   - Accessory drive belts
   - Front cover
   - Timing belt. Refer to the Timing Belt unit repair section.
   - Timing belt idler pulley
   - Crankshaft Position (CKP) sensor
   - Crankshaft timing sprocket
   - Variable Valve Timing and Valve Lift Electronic Control (VTEC) solenoid valve connector
   - Oil filter adapter
   - Oil pan
   - Oil pump pickup tube
   - Oil pump

To install:

➡**Use new gaskets and O-ring seals for assembly.**

4. Apply liquid gasket to the oil pump and to the bolt hole threads.
5. Install or connect the following:
   - Oil pump. Tighten the bolts to 105 inch lbs. (12 Nm).
   - Oil pump pickup tube. Tighten the bolts to 105 inch lbs. (12 Nm).
   - Oil pan
   - Oil filter adapter
   - VTEC solenoid valve connector
   - Crankshaft timing sprocket
   - CKP sensor
   - Timing belt idler pulley

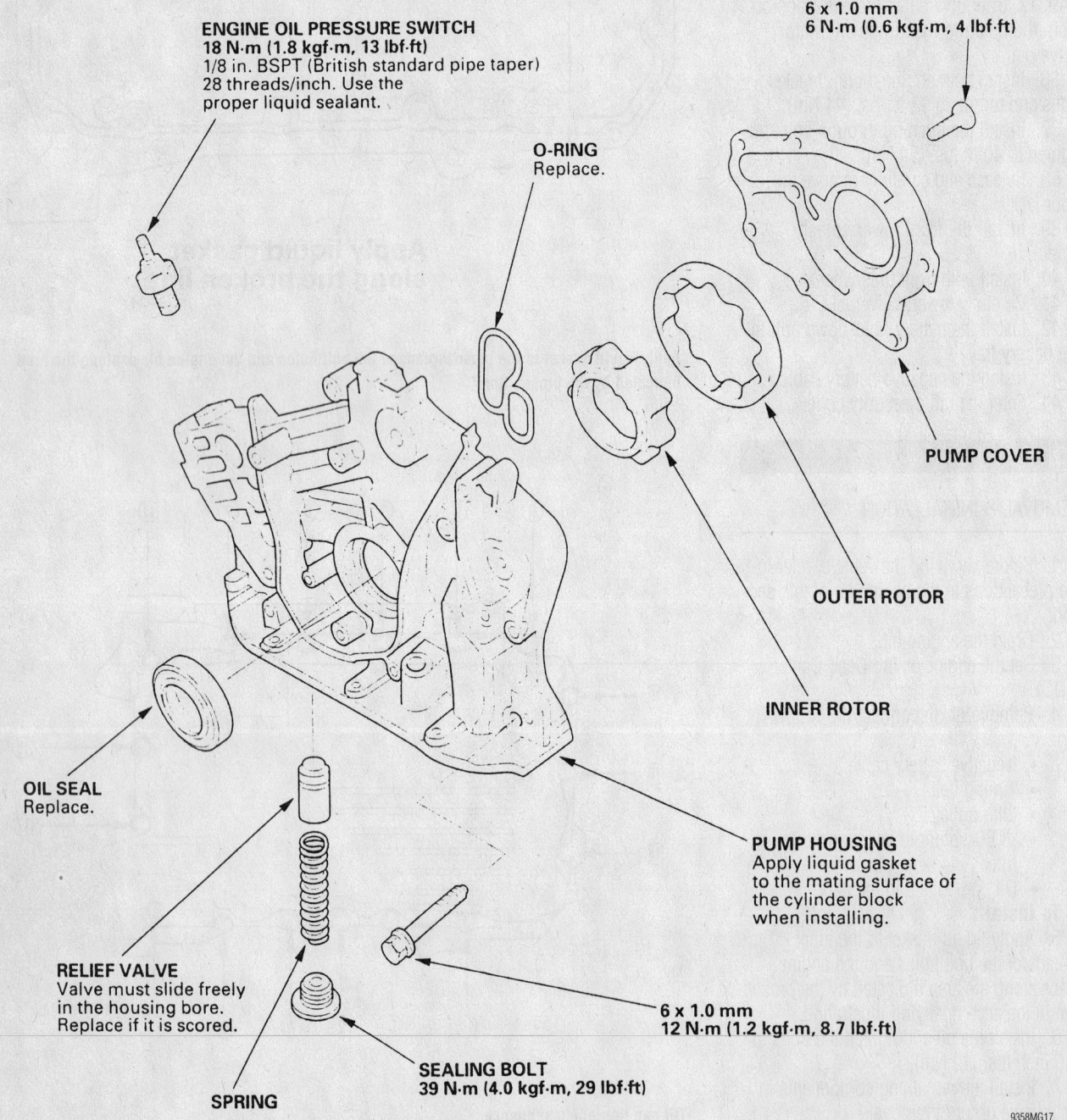

ENGINE OIL PRESSURE SWITCH
18 N·m (1.8 kgf·m, 13 lbf·ft)
1/8 in. BSPT (British standard pipe taper)
28 threads/inch. Use the
proper liquid sealant.

O-RING
Replace.

6 x 1.0 mm
6 N·m (0.6 kgf·m, 4 lbf·ft)

PUMP COVER

OUTER ROTOR

INNER ROTOR

PUMP HOUSING
Apply liquid gasket
to the mating surface of
the cylinder block
when installing.

OIL SEAL
Replace.

RELIEF VALVE
Valve must slide freely
in the housing bore.
Replace if it is scored.

6 x 1.0 mm
12 N·m (1.2 kgf·m, 8.7 lbf·ft)

SEALING BOLT
39 N·m (4.0 kgf·m, 29 lbf·ft)

SPRING

9358MG17

Exploded view of the oil pump assembly

**6 x 1.0 mm
12 N·m (1.2 kgf·m, 8.7 lbf·ft)**

**6 x 1.0 mm
12 N·m
(1.2 kgf·m, 8.7 lbf·ft)**

9358MG18

Oil pump assembly mounting and seal locations

- Timing belt
- Front cover
- Accessory drive belts
- Negative battery cable
6. Fill the crankcase to the correct level.
7. Start the engine and check for leaks.

Rear Main Seal

REMOVAL & INSTALLATION

1. Before servicing the vehicle, refer to the precautions in the beginning of this section.

2. Remove or disconnect the following:
- Transaxle
- Clutch pressure plate and disc, if equipped
- Flywheel
- Oil seal
To install:
3. Install or connect the following:
- Oil seal. Drive the seal square into the seal case.
- Flywheel. Tighten the bolts in a crossing pattern to 54 ft. lbs. (73 Nm).
- Clutch pressure plate and disc, if equipped

- Transaxle
4. Check the fluid levels.
5. Start the engine and check for leaks.

Piston and Ring

POSITIONING

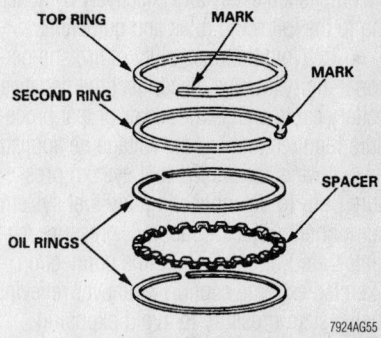

7924AG55

Piston ring positioning and top mark location

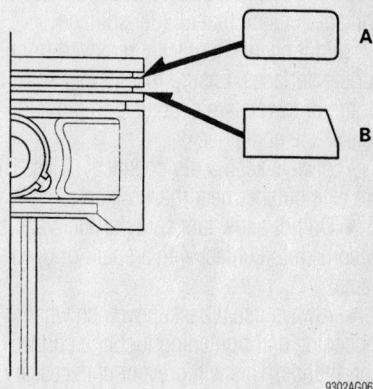

9302AG06

Compression ring identification

FUEL SYSTEM

Fuel System Service Precautions

Safety is the most important factor when performing not only fuel system maintenance, but any type of maintenance. Failure to conduct maintenance and repairs in a safe manner may result in serious personal injury or death. Maintenance and testing of the vehicle's fuel system components can be accomplished safely and effectively by adhering to the following rules and guidelines:

• To avoid the possibility of fire and personal injury, always disconnect the negative battery cable unless the repair or test procedure requires that battery voltage be applied.

• Always relieve the fuel system pressure prior to disconnecting any fuel system component (injector, fuel rail, pressure regulator, etc.), fitting or fuel line connection. Exercise extreme caution whenever relieving fuel system pressure to avoid exposing skin, face and eyes to fuel spray. Please be advised that fuel under pressure may penetrate the skin or any part of the body that it contacts.

• Always place a shop towel or cloth around the fitting or connection prior to loosening to absorb any excess fuel due to spillage. Ensure that all fuel spillage (should it occur) is quickly removed from engine surfaces. Ensure that all fuel soaked cloths or towels are deposited into a suitable waste container.

• Always keep a dry chemical (Class B) fire extinguisher near the work area.

• Do not allow fuel spray or fuel vapors to come into contact with a spark or open flame.

• Always use a backup wrench when loosening and tightening fuel line connection fittings. This will prevent unnecessary stress and torsion to fuel line piping. Always follow the proper torque specifications.

• Always replace worn fuel fitting O-rings with new. Do not substitute fuel hose or equivalent, where fuel pipe is installed.

Fuel System Pressure

RELIEVING

1. Before servicing the vehicle, refer to the precautions in the beginning of this section.
2. Disconnect the negative battery cable.
3. Remove the fuel filler cap.
4. Place a shop towel over the fuel pulsation damper.

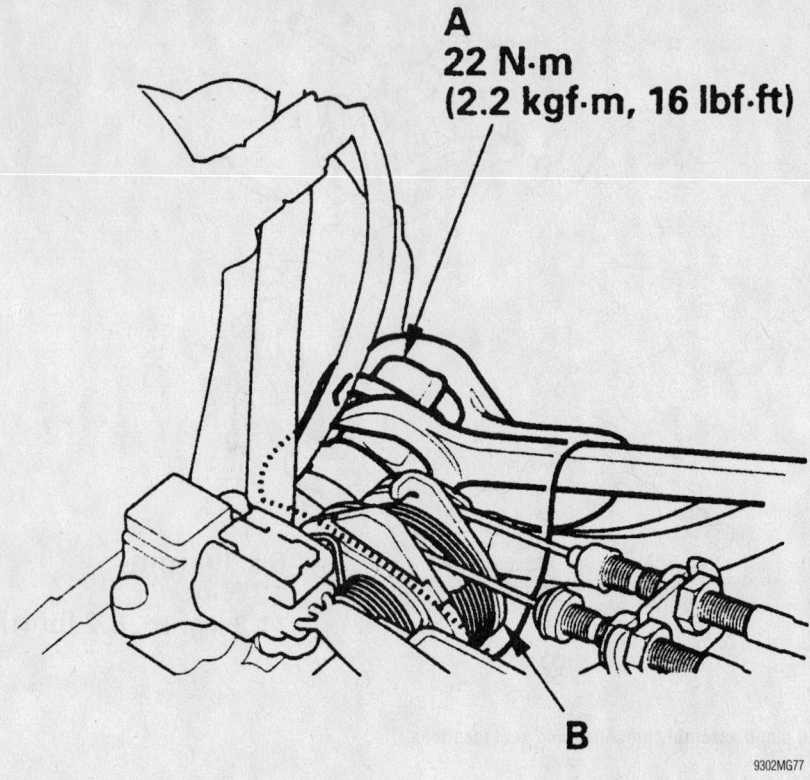

A 22 N·m (2.2 kgf·m, 16 lbf·ft)

9302MG77

Use a wrench on the fuel pulsation damper (A). Place a rag over the damper (B) when relieving residual fuel pressure

5. Loosen the fuel pulsation damper 1 turn.
6. When service is completed, replace the sealing washer and tighten the pulsation damper to 16 ft. lbs. (22 Nm).
7. Replace the fuel filler cap.
8. Connect the negative battery cable.
9. Start the engine and check for leaks.

Fuel Filter

REMOVAL & INSTALLATION

1. Before servicing the vehicle, refer to the precautions in the beginning of this section.
2. Relieve the fuel system pressure.
3. Remove or disconnect the following:
   • Negative battery cable
   • Rear seats and carpet
   • Access panel
   • Fuel lines
   • Fuel pump locknut using wrench 07XAA-001010A as shown in the accompanying illustration.
   • Fuel pump module

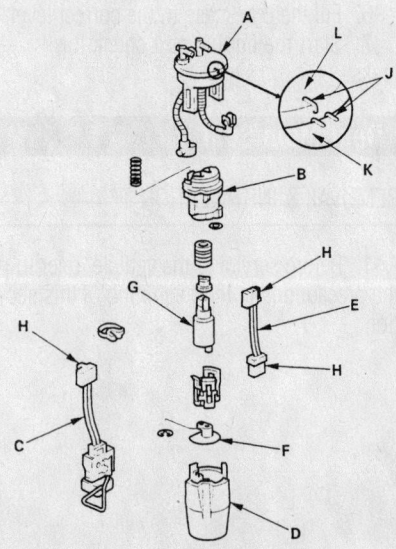

A. Bracket
B. Fuel filter
C. Fuel gauge sender
D. Case
E. Wire harness
F. Suction filter
G. Fuel pump
H. Connectors
J. Alignment marks
K. Fuel tank
L. Fuel pump module

9308MG26

Exploded view of the fuel pump module

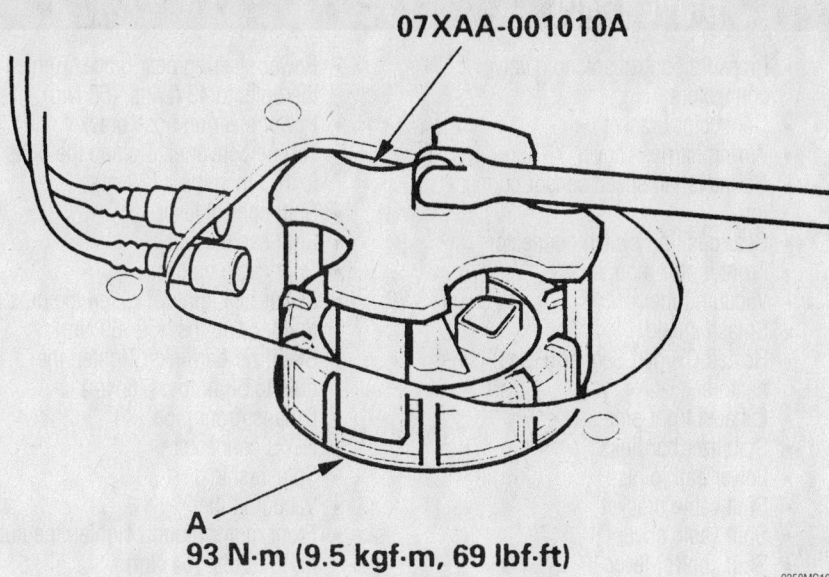

07XAA-001010A

A
93 N·m (9.5 kgf·m, 69 lbf·ft)

9358MG19

Use wrench 07XAA-001010A to remove and install the fuel pump locknut

4. Disassemble the fuel pump module and remove the fuel filter.
To install:
5. Install the fuel filter and assemble the fuel pump module.
6. Install or connect the following:
- Fuel pump module
- Fuel pump locknut and tighten to 69 ft. lbs. (93 Nm) using wrench 07XAA-001010A
- Fuel lines
- Access panel
- Rear seats and carpet
- Negative battery cable
7. Start the engine and check for leaks.

Fuel Pump

REMOVAL & INSTALLATION

1. Before servicing the vehicle, refer to the precautions in the beginning of this section.

2. Relieve the fuel system pressure.
3. Remove or disconnect the following:
- Negative battery cable
- Rear seats and carpet
- Access panel
- Fuel pump module wiring connector
- Fuel supply and return lines
- Fuel pump locknut using wrench 07XAA-001010A as shown in the accompanying illustration.
- Fuel pump module
To install:
4. Install or connect the following:
- Fuel pump module. Use a new seal and align the matchmarks.
- Fuel pump locknut and tighten to 69 ft. lbs. (93 Nm) using wrench 07XAA-001010A
- Fuel supply and return lines

- Fuel pump module wiring connector
- Access panel
- Rear seats and carpet
- Negative battery cable
5. Start the engine and check for leaks.

Fuel Injector

REMOVAL & INSTALLATION

1. Before servicing the vehicle, refer to the precautions in the beginning of this section.
2. Relieve the fuel system pressure.
3. Remove or disconnect the following:
- Negative battery cable
- Intake manifold
- Fuel lines
- Fuel injector connectors
- Fuel pressure regulator vacuum line
- Fuel supply manifold
4. Separate the fuel injectors from the fuel supply manifold.
To install:
5. Install the fuel injectors to the fuel supply manifold with new cushion rings and O-rings.
6. Install new seal rings to the intake manifold.
7. Install or connect the following:
- Fuel supply manifold and injector assembly. Tighten the bolts to 86 inch lbs. (10 Nm).
- Fuel pressure regulator vacuum line
- Fuel injector connectors
- Fuel lines
- Intake manifold
- Negative battery cable
8. Start the engine and check for leaks.

DRIVE TRAIN

Transaxle Assembly

REMOVAL & INSTALLATION

Automatic Transaxle

1. Before servicing the vehicle, refer to the precautions in the beginning of this section.
2. Drain the transaxle.
3. Remove the engine appearance covers and install a support fixture to the engine lifting eyes.
4. Remove or disconnect the following:
   - Air intake assembly
   - Battery
   - Battery tray
   - Transaxle oil cooler lines
   - Starter motor
   - Transaxle ground cable
   - Shift control solenoid valve connectors
   - Clutch pressure switch connectors
   - Mainshaft speed sensor connector

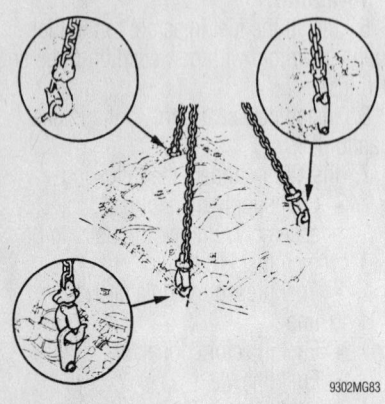

Support the engine while removing the transaxle

9302MG83

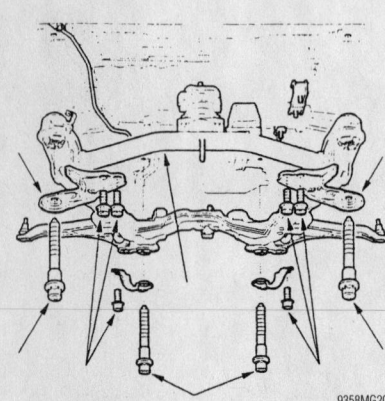

Remove the subframe bolts

9358MG20

- Pressure control solenoid valve connectors
- Connector bracket
- Wiring harness cover
- Countershaft speed sensor connector
- Gear position switch connector
- Front motor mount
- Vacuum tube
- Splash shield
- Heated Oxygen Sensor (HO2S) connectors
- Exhaust front pipe
- Stabilizer bar links
- Lower ball joints
- Shift cable bracket
- Shift cable cover
- Shift control lever
- Torque converter
- Power steering hose bracket
- Power steering gear and brace
- Rear engine mount
- Transaxle lower mounts

5. Matchmark the front subframe to the vehicle body.
6. Support the subframe with a jack.
7. Support the steering gear with safety wire and remove the subframe.
8. Separate the inner CV-joints from the transaxle and intermediate shaft and support the axle halfshafts out of the work area with safety wire.
9. Remove or disconnect the following:
   - Intermediate shaft
   - Transaxle flange bolts
   - Transaxle

To install:

➡**Use new circlips, split pins and self-locking nuts for assembly.**

10. Install or connect the following:
    - Transaxle. Tighten the flange bolts to 47 ft. lbs. (64 Nm).
    - Intermediate shaft. Tighten the bolts to 29 ft. lbs. (39 Nm).
    - Axle halfshafts
11. Raise the subframe into position and align the matchmarks. Tighten the subframe bolts to 76 ft. lbs. (103 Nm). Tighten the front subframe bracket bolts to 54 ft. lbs. (74 Nm) and the rear bracket bolts to 86 ft. lbs. (117 Nm).
12. Install or connect the following:
    - Transaxle lower mounts. Tighten the nuts to 28 ft. lbs. (38 Nm).
    - Rear engine mount. Tighten the bolts to 28 ft. lbs. (38 Nm).
    - Power steering gear. Tighten the bolts to 29 ft. lbs. (39 Nm).

- Power steering gear brace. Tighten the bolts to 43 ft. lbs. (58 Nm).
- Power steering hose bracket
- Torque converter. Tighten the bolts to 105 inch lbs. (12 Nm).
- Shift control lever
- Shift cable cover
- Shift cable bracket
- Lower ball joints. Tighten the nuts to 43–51 ft. lbs. (59–69 Nm).
- Stabilizer bar links. Tighten the nuts to 58 ft. lbs. (78 Nm).
- Exhaust front pipe
- HO2S connectors
- Splash shield
- Vacuum tube
- Front motor mount. Tighten the nut to 40 ft. lbs. (54 Nm).
- Gear position switch connector
- Countershaft speed sensor connector
- Wiring harness cover
- Connector bracket
- Pressure control solenoid valve connectors
- Mainshaft speed sensor connector
- Clutch pressure switch connectors
- Shift control solenoid valve connectors
- Transaxle ground cable
- Starter motor
- Transaxle oil cooler lines
- Battery tray
- Battery
- Air intake assembly

13. Fill the transaxle to the correct level.
14. Start the engine and check for leaks.
15. Check the wheel alignment and adjust as necessary.

Halfshaft

REMOVAL & INSTALLATION

1. Before servicing the vehicle, refer to the precautions in the beginning of this section.
2. Drain the transaxle.
3. Remove or disconnect the following:
   - Negative battery cable
   - Front wheels

➡**If removing the left halfshaft, drain the transmission fluid, it is not necessary to drain the fluid if removing the right halfshaft.**

- Stabilizer bar link
- Lower ball joint
- Spindle nut

4. Pry the inboard joint from the transaxle or intermediate shaft.

5. Remove the outer CV-joint stub shaft from the hub by tapping the stub shaft with a plastic hammer.

To install:

➡**Use new circlips, split pins and self-locking nuts for assembly.**

6. Install the outer CV-joint stub shaft into the hub.

7. Install the inner CV-joint to the transaxle or intermediate shaft until the circlip locks in the retaining groove.

8. Install or connect the following:
- Lower ball joint. Tighten the nut to 43–51 ft. lbs. (59–69 Nm).
- Stabilizer bar link. Tighten the nut to 58 ft. lbs. (78 Nm).
- Spindle nut. Tighten the nut to 181 ft. lbs. (245 Nm).
- Front wheels
- Negative battery cable

9. Fill the transaxle to the correct level and check for leaks.

CV-Joint

OVERHAUL

OUTBOARD JOINT

1. Before servicing the vehicle, refer to the precautions in the beginning of this section.

2. Remove or disconnect the following:
- Axle halfshaft from the vehicle and place it in a vise
- Outboard joint boot clamps and push the boot back
- Outboard joint by driving it off the axle shaft with a brass drift and hammer or tool 07XAC-001020A, if available
- Outboard joint boot

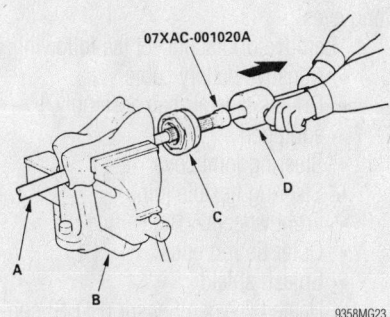

Removing the outboard joint using tool 07XAC-001020A

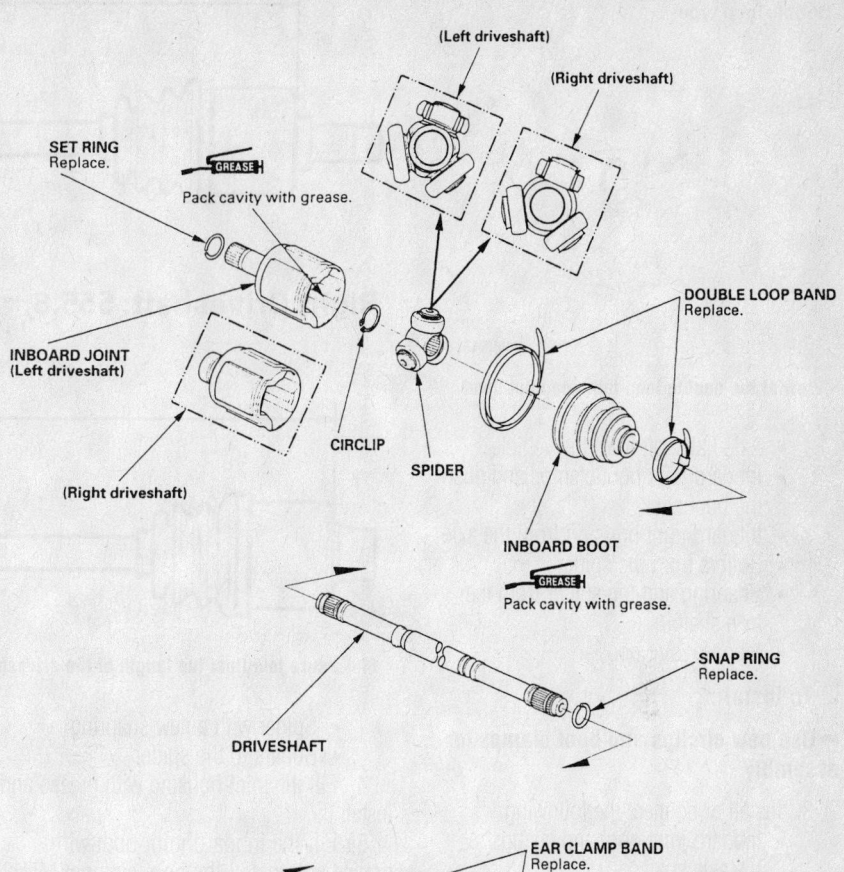

Front axle exploded view

To install:

➡**Use new circlips and boot clamps for assembly.**

3. Install the outboard joint boot and clamps to the axle shaft.

4. Fill the outboard joint with grease. Install the outboard joint to the axle shaft. Tap the stub shaft with a brass hammer to seat the circlip.

5. Fill the outboard joint boot with grease and install the boot clamps.

6. Install the axle halfshaft to the vehicle.

INBOARD JOINT

1. Before servicing the vehicle, refer to the precautions in the beginning of this section.

2. Remove or disconnect the following:

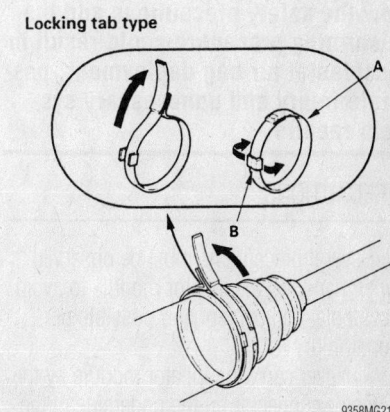

View of the locking tab type boot band

Double loop type

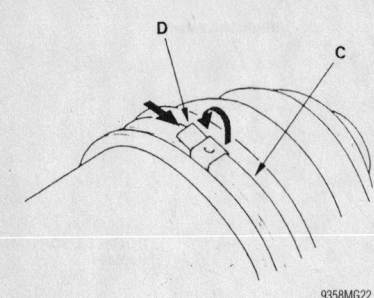

9358MG22

View of the double loop tab type boot band

- Axle halfshaft from the vehicle.
- Inboard joint boot clamps and push the boot back
- Inboard joint housing from the axle
- Rollers from the spider
- Snapring and the spider from the axle shaft
- Inboard joint boot

21

To install:

➡**Use new circlips and boot clamps for assembly.**

3. Install or connect the following:
- Inboard joint boot and clamps to the axle shaft

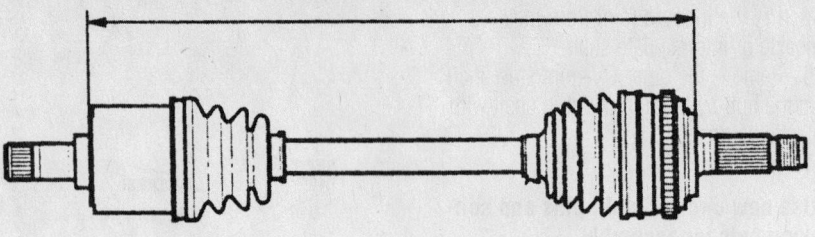

Right Driveshaft: 555.8—560.8 mm (21.9—22.1 in.)

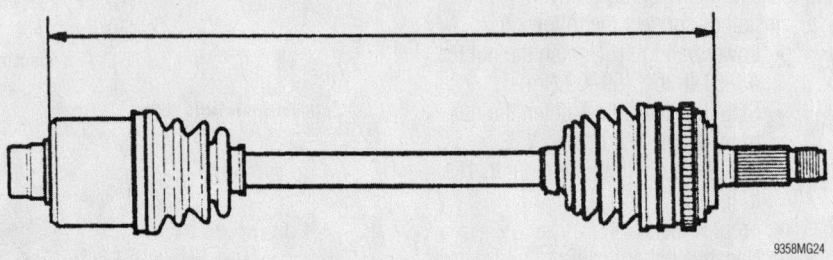

9358MG24

Make sure to adjust the length of the driveshafts as shown

- Spider with a new snapring
- Rollers to the spider
4. Fill the joint housing with grease and install it.
5. Fill the inboard joint boot with grease and install the boot clamps. Make

sure to adjust the length of the driveshafts as shown in the accompanying illustration.
6. Install the axle halfshaft to the vehicle.

STEERING AND SUSPENSION

Air Bag

✳✳ CAUTION

Some vehicles are equipped with an air bag system. The system must be disarmed before performing service on, or around, system components, the steering column, instrument panel components, wiring and sensors. Failure to follow the safety precautions and the disarming procedure could result in accidental air bag deployment, possible injury and unnecessary system repairs.

PRECAUTIONS

Several precautions must be observed when handling the inflator module to avoid accidental deployment and possible personal injury.
- Never carry the inflator module by the wires or connector on the underside of the module.

- When carrying a live inflator module, hold securely with both hands, and ensure that the bag and trim cover are pointed away.
- Place the inflator module on a bench or other surface with the bag and trim cover facing up.
- With the inflator module on the bench, never place anything on or close to the module that may be thrown in the event of an accidental deployment.

Before servicing the vehicle, also make sure to refer to the precautions in the beginning of this section as well.

DISARMING

Disconnect and isolate the negative battery cable. Wait 3 minutes for the system capacitor to discharge before performing any service.

Power Rack and Pinion Steering Gear

REMOVAL & INSTALLATION

✳✳ WARNING

Do not permit the steering wheel to turn whenever the steering gear is disconnected from the steering column. Damage to the air bag wiring can result.

1. Before servicing the vehicle, refer to the precautions in the beginning of this section.
2. Center the steering wheel and lock it in position.
3. Attach a support fixture to the engine lifting eyes.
4. Remove or disconnect the following:
- Negative battery cable
- Drivers side air bag assembly, if equipped
- Steering joint cover
- Steering flexible joint
- Front wheels
- Outer tie rod ends
- Splash shield
- Heated Oxygen Sensor (HO2S) connectors
- Exhaust front pipe

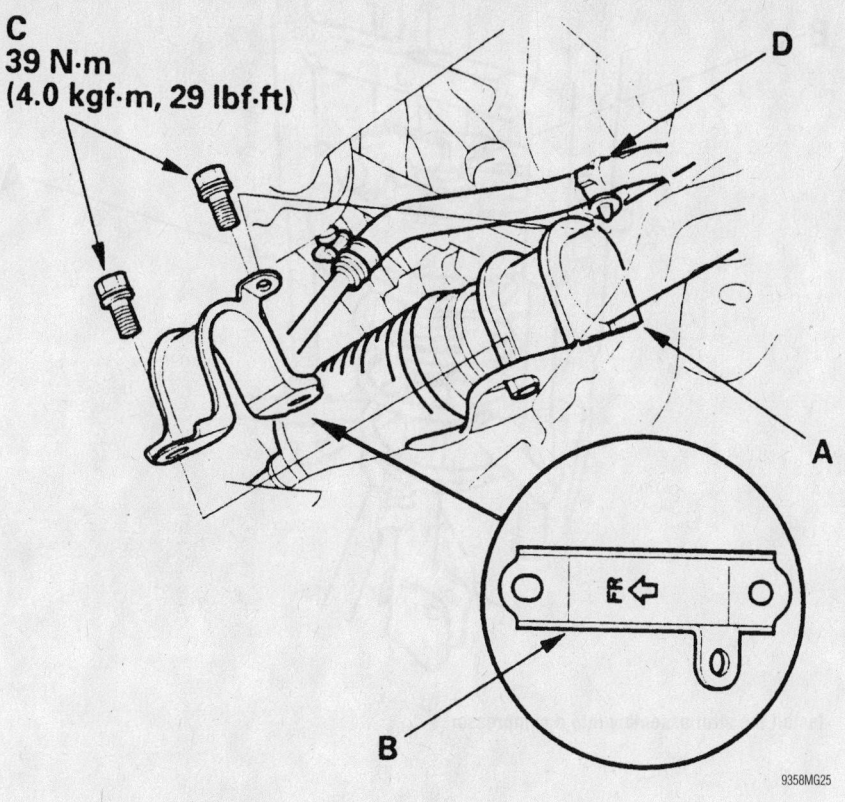

C
39 N·m
(4.0 kgf·m, 29 lbf·ft)

9358MG25

Make sure the right steering gear mounting bracket is oriented properly

- Fule feed line
- Power steering fluid lines
- Rear engine mount

5. Support the front subframe with a jack.

6. Loosen the 14mm subframe bolts and remove the 12mm stiffener plate bolts.

7. Lower the subframe about 1 3/16 inches (30mm).

8. Remove or disconnect the following:
- Right steering gear mounting bracket
- Left steering gear mounting bolts
- Steering gear

To install:

9. Position the steering gear in the vehicle.

10. Install or connect the following:
- Left steering gear mounting bolts. Tighten the bolts to 43 ft. lbs. (58 Nm).
- Right steering gear mounting bracket. Tighten the bolts to 29 ft. lbs. (39 Nm).

11. Raise the subframe into position. Tighten the 14mm bolts to 76 ft. lbs. (103 Nm) and the 12mm bolts to 54 ft. lbs. (74 Nm).

12. Install or connect the following:

- Rear engine mount
- Power steering fluid lines
- Exhaust front pipe
- HO$_2$S connectors
- Splash shield
- Outer tie rod ends
- Front wheels. Position the wheels straight-ahead.
- Steering flexible joint. Tighten the pinch bolts to 16 ft. lbs. (22 Nm).
- Steering joint cover
- Negative battery cable

13. Fill the power steering system.

14. Check the wheel alignment and adjust as necessary.

Strut

REMOVAL & INSTALLATION

Front

1. Before servicing the vehicle, refer to the precautions in the beginning of this section.

2. Remove or disconnect the following:
- Front wheel
- Wheel speed sensor wiring bracket
- Brake hose bracket
- Stabilizer bar link

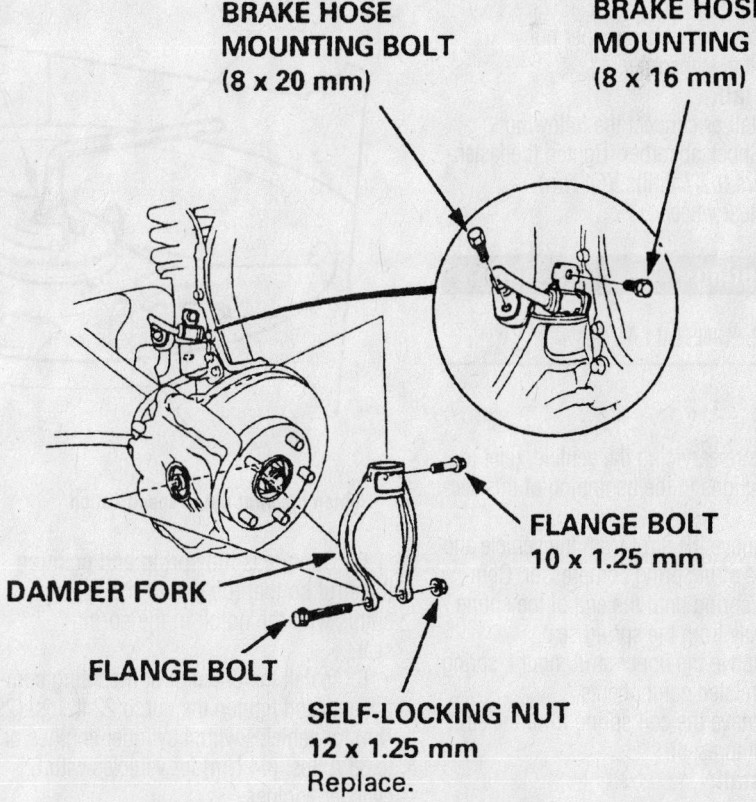

BRAKE HOSE MOUNTING BOLT (8 x 20 mm)

BRAKE HOSE MOUNTING BOLT (8 x 16 mm)

FLANGE BOLT 10 x 1.25 mm

DAMPER FORK

FLANGE BOLT

SELF-LOCKING NUT 12 x 1.25 mm
Replace.

7924MG36

Identification of some of the front suspension components

- Strut pinch bolts
- Upper mount nuts
- Strut

To install:

3. Install or connect the following:
   - Strut. Tighten the upper mount nuts to 43 ft. lbs. (59 Nm).
   - Strut pinch bolts. Tighten the nuts to 116 ft. lbs. (157 Nm).
   - Stabilizer bar link. Tighten the nut to 58 ft. lbs. (78 Nm).
   - Brake hose bracket
   - Wheel speed sensor wiring bracket
   - Front wheel

4. Check the wheel alignment and adjust as necessary.

Shock Absorber

REMOVAL & INSTALLATION

Rear

1. Before servicing the vehicle, refer to the precautions in the beginning of this section.

2. Support the vehicle under the lower control arm.

3. Remove or disconnect the following:
   - Rear wheel
   - Upper shock absorber flange bolt
   - Lower shock absorber nut
   - Shock absorber

To install:

4. Install or connect the following:
   - Shock absorber. Tighten the fasteners to 47 ft. lbs. (64 Nm).
   - Rear wheel

Coil Spring

REMOVAL & INSTALLATION

Front

1. Before servicing the vehicle, refer to the precautions in the beginning of this section.

2. Remove the strut from the vehicle and install in a strut spring compressor. Compress the spring until the end of the spring comes away from the spring seat.

3. Remove the upper strut mount, spring seat and related components.

4. Remove the coil spring from the strut spring compressor.

To install:

➡ **Use a new self-locking nut.**

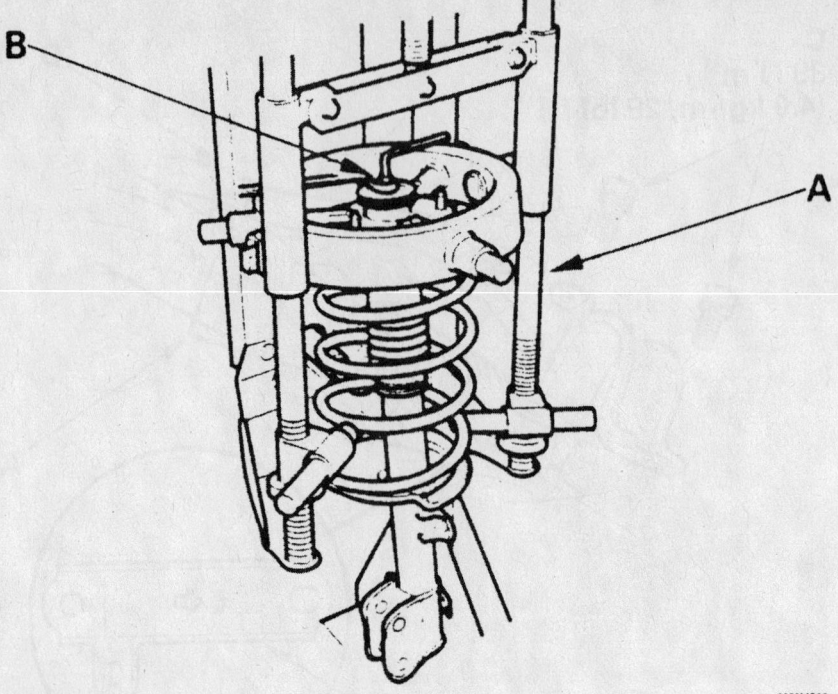

9358MG26

Install the strut assembly into a compressor

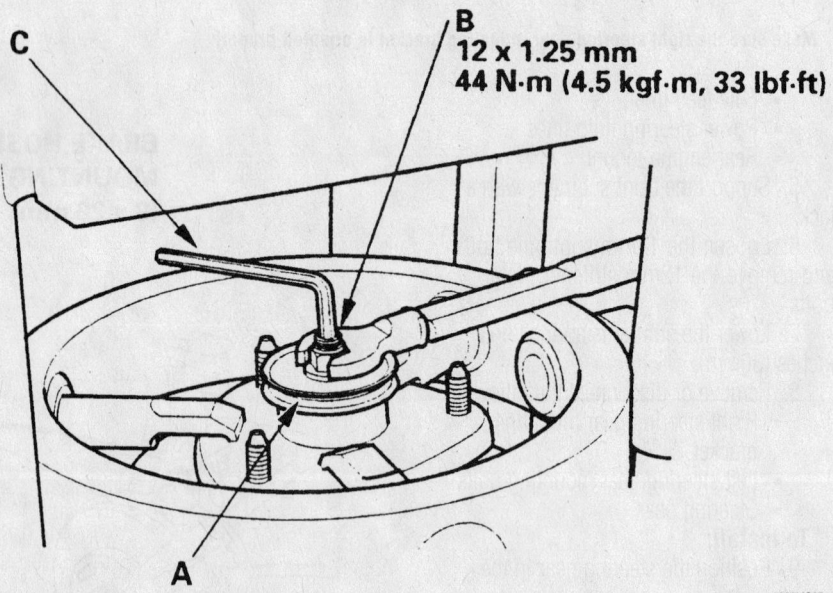

9358MG27

Tighten the strut nut to specification

5. Compress the spring and position the strut so that the end of the spring aligns with the notch in the spring seat.

6. Install the upper strut mounting components and tighten the nut to 22 ft. lbs. (29 Nm) for vehicles with 4 cylinder engines or to 33 ft. lbs. (44 Nm) for vehicles with 6 cylinder engines.

7. Install the strut to the vehicle.

8. Check the wheel alignment and adjust as necessary.

Rear

1. Before servicing the vehicle, refer to the precautions in the beginning of this section.

2. Support the vehicle under the lower control arm.

3. Remove or disconnect the following:

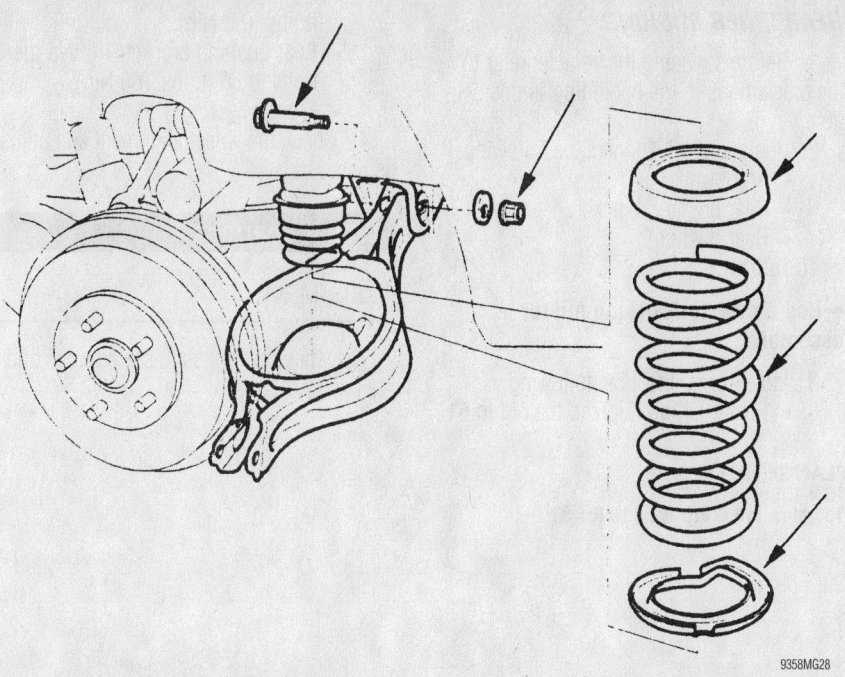

Exploded view of the rear spring assembly

- Rear wheel
- Upper shock absorber flange bolt
- Lower shock absorber nut
- Shock absorber
- Wheel speed sensor wiring harness
- Lower control arm bolts

4. Lower the floor jack and remove the coil spring and spring seats.

To install:

➡**Use new self-locking nuts for assembly.**

5. Place the coil spring and spring seats on the lower control arm and raise into position. Tighten the inboard bolt to 61 ft. lbs. and the outer bolt to 47 ft. lbs. (64 Nm).

6. Install or connect the following:
- Shock absorber. Tighten the fasteners to 47 ft. lbs. (64 Nm).
- Rear wheel

Upper Ball Joint

REMOVAL & INSTALLATION

The upper ball joints are replaced with the upper control arms as an assembly.

Lower Ball Joint

REMOVAL & INSTALLATION

The lower ball joints are replaced with the lower control arms as an assembly.

Upper Control Arm

REMOVAL & INSTALLATION

1. Before servicing the vehicle, refer to the precautions in the beginning of this section.

2. Support the lower control arm assembly with a floor jack.

3. Remove or disconnect the following:
- Upper ball joint
- Inner control arm flange bolts.
- Upper control arm

To install:

➡**Use new self-locking nuts for assembly.**

4. Install the upper control arm. Tighten the ball joint nut to 29–35 ft. lbs. (39–47 Nm) and the inner flange nuts to 47 ft. lbs. (64 Nm).

CONTROL ARM BUSHING REPLACEMENT

The upper control arm bushings are serviced with the upper control arm as an assembly.

Lower Control Arm

REMOVAL & INSTALLATION

1. Before servicing the vehicle, refer to the precautions in the beginning of this section.

2. Remove or disconnect the following:
- Front wheel
- Lower ball joint
- Front inner flange bolt
- Rear inner flange bolt
- Lower control arm

To install:

➡**Use a new split pin for assembly.**

3. Install or connect the following:
- Lower control arm. Tighten the inner flange bolt (A) to 69 ft. lbs. (93 Nm) and bolt (B) to 90 ft. lbs.

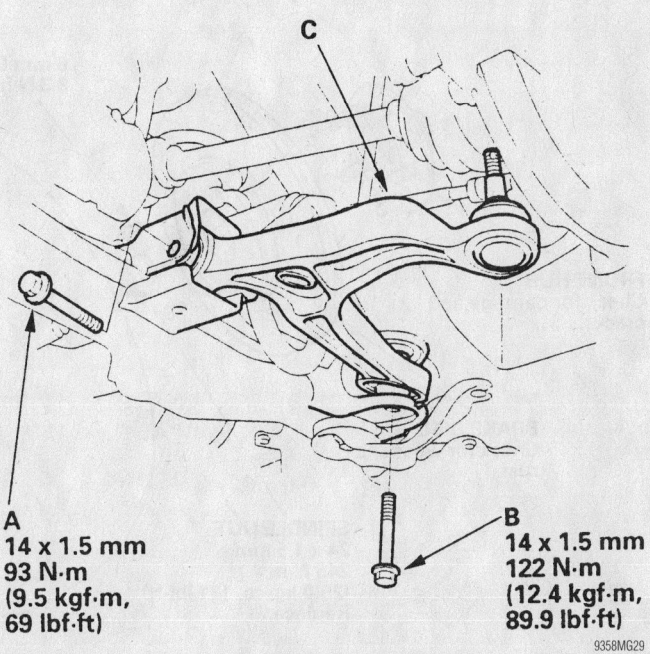

A
14 x 1.5 mm
93 N·m
(9.5 kgf·m,
69 lbf·ft)

B
14 x 1.5 mm
122 N·m
(12.4 kgf·m,
89.9 lbf·ft)

Tighten the lower control arm bolts to the specifications shown

(122 Nm). Refer to the accompanying illustration for the locations of bolts A and B.
- Lower ball joint. Tighten the nut to 43–51 ft. lbs. (59–69 Nm).
- Front wheel

4. Check the wheel alignment and adjust as necessary.

CONTROL ARM BUSHING REPLACEMENT

The lower control arm bushings are serviced with the lower control arm as an assembly.

REAR INNER BUSHING

1. Before servicing the vehicle, refer to the precautions in the beginning of this section.
2. Remove or disconnect the following:
   - Front wheel
   - Rear bushing bracket
   - Rear bushing

To install:

➡Use a new self-locking nut for assembly.

3. Install or connect the following:
   - Rear bushing. Tighten the nut to 61 ft. lbs. (83 Nm).
   - Rear bushing bracket. Tighten the bolts to 66 ft. lbs. (89 Nm).
   - Front wheel

4. Check the wheel alignment and adjust as necessary.

Wheel Bearings

ADJUSTMENT

The wheel bearings are sealed units and are not adjustable.

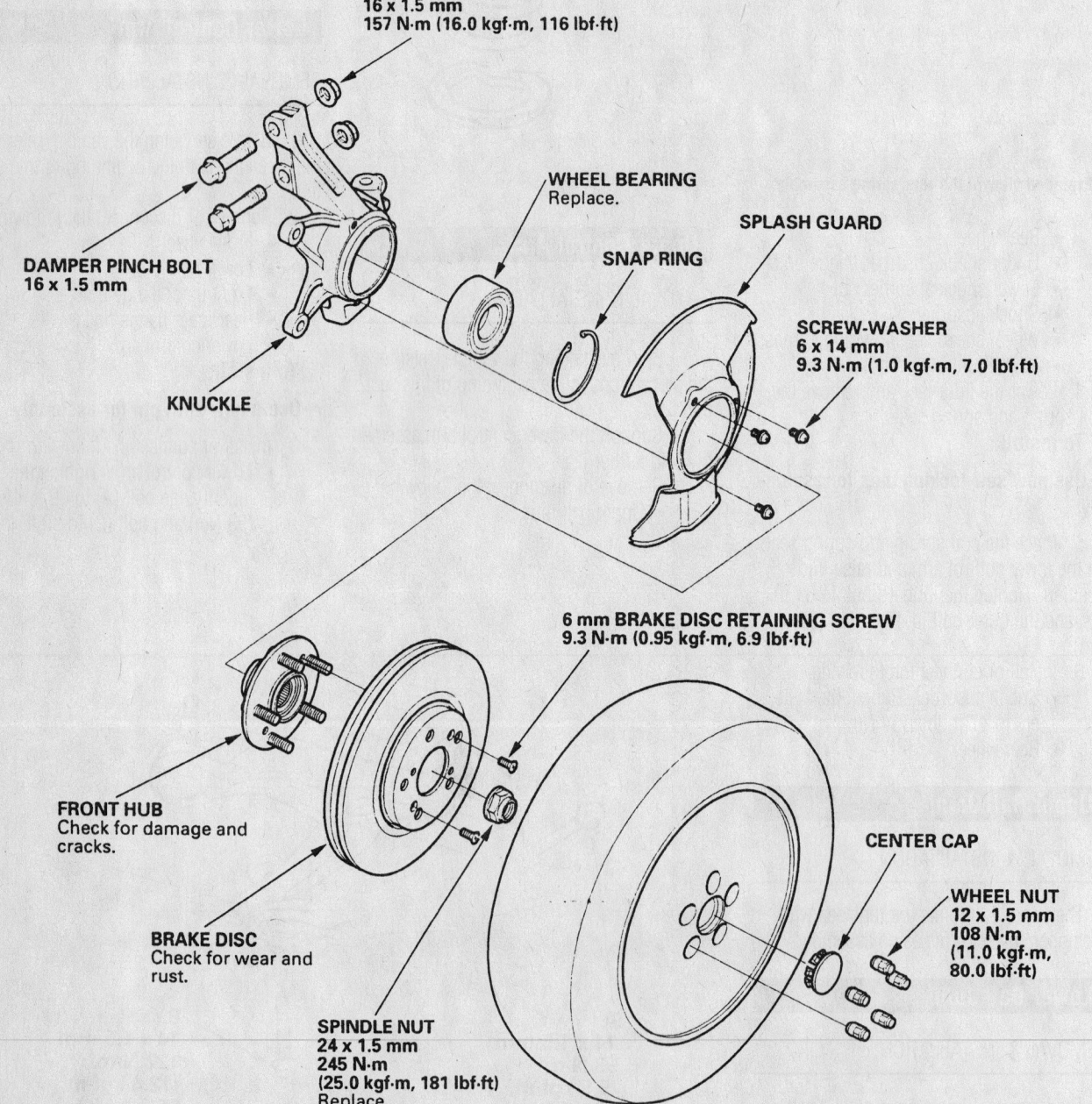

Exploded view of the front hub/knuckle assembly

REMOVAL & INSTALLATION

Front

1. Before servicing the vehicle, refer to the precautions in the beginning of this section.

2. Remove or disconnect the following:

- Front wheel
- Spindle nut
- Brake hose bracket
- Brake caliper and rotor
- Wheel speed sensor, if equipped
- Outer tie rod end
- Lower ball joint
- Steering knuckle

3. Press the hub out of knuckle.

4. Remove the splash guard.

5. Remove the snapring and press the wheel bearing out of the steering knuckle.

6. If necessary, press the inner bearing race off of the hub.

To install:

➡**Use a new ball joint nut, split pin, snapring and spindle nut for assembly.**

7. Press the bearing into the steering knuckle and install the snapring.

8. Install the splash guard.
9. Press the hub into the bearing.
10. Install or connect the following:

- Steering knuckle. Tighten the ball joint nut to 43–51 ft. lbs. (59–69 Nm) and the damper flange bolts to 116 ft. lbs. (157 Nm).
- Outer tie rod end. Tighten the nut to 32 ft. lbs. (43 Nm).
- Wheel speed sensor, if equipped
- Brake caliper and rotor. Tighten the caliper bracket bolts to 80 ft. lbs. (108 Nm).
- Brake hose bracket
- Spindle nut. Tighten the nut to 181 ft. lbs. (245 Nm).
- Front wheel

11. Check the wheel alignment and adjust as necessary.

Rear

WITH DRUM BRAKES

1. Before servicing the vehicle, refer to the precautions in the beginning of this section.

2. Remove or disconnect the following:

- Rear wheel
- Brake drum,
- Spindle cap and nut
- Hub and bearing assembly

To install:

➡**Use a new spindle nut for assembly.**

3. Install or connect the following:

- Hub and bearing assembly. Tighten the spindle nut to 181 ft. lbs. (245 Nm).
- Spindle cap
- Brake drum
- Rear wheel

WITH DISC BRAKES

1. Before servicing the vehicle, refer to the precautions in the beginning of this section.

2. Remove or disconnect the following:

- Rear wheel
- Brake caliper and rotor
- Spindle cap, nut and washer
- Hub and bearing assembly

To install:

➡**Use a new spindle nut for assembly.**

3. Install or connect the following:

- Hub and bearing assembly. Tighten the spindle nut to 181 ft. lbs. (245 Nm).
- Spindle cap
- Brake caliper and rotor. Tighten the caliper bracket bolts to 28 ft. lbs. (38 Nm).
- Rear wheel

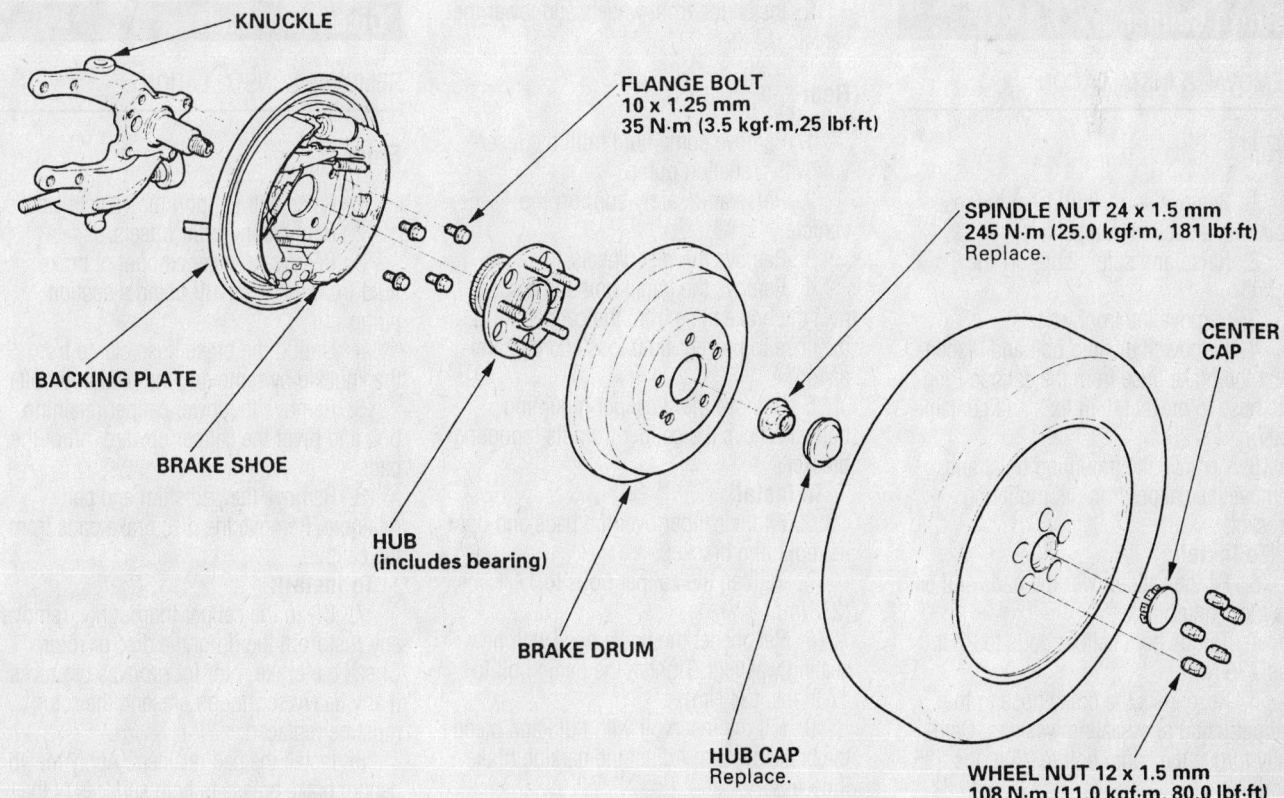

Exploded view of the rear hub/bearing assembly on models equipped with drum brakes

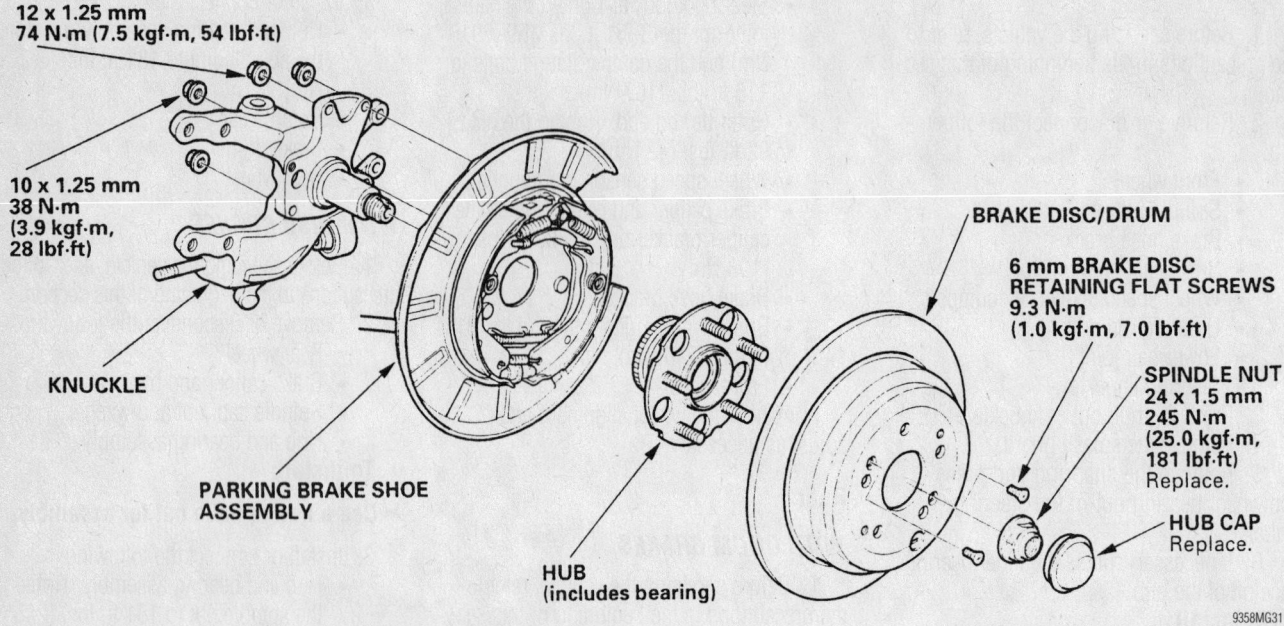

12 x 1.25 mm
74 N·m (7.5 kgf·m, 54 lbf·ft)

10 x 1.25 mm
38 N·m
(3.9 kgf·m,
28 lbf·ft)

KNUCKLE

PARKING BRAKE SHOE
ASSEMBLY

HUB
(includes bearing)

BRAKE DISC/DRUM

6 mm BRAKE DISC
RETAINING FLAT SCREWS
9.3 N·m
(1.0 kgf·m, 7.0 lbf·ft)

SPINDLE NUT
24 x 1.5 mm
245 N·m
(25.0 kgf·m,
181 lbf·ft)
Replace.

HUB CAP
Replace.

9358MG31

Exploded view of the rear hub/bearing assembly on models equipped with disc brakes

BRAKES

Brake Caliper

REMOVAL & INSTALLATION

Front

1. Remove some fluid from the reservoir with a suction pump.
2. Raise and safely support the vehicle.
3. Remove the front wheels.
4. Remove the banjo bolt and disconnect the brake hose from the caliper. Plug the hose to prevent fluid loss and contamination.
5. Remove the mounting bolts and remove the caliper from its mounting bracket.

To Install:

6. Fit the caliper over the pads and onto its mounting bracket.
7. Torque both caliper bolts to 27 ft. lbs. (36 Nm).
8. Reconnect the brake hose to the caliper using new sealing washers. Carefully torque the banjo bolt to 25 ft. lbs. (35 Nm).
9. Fill the reservoir with fluid and bleed the brakes.

10. Install the front wheels and lower the vehicle.

Rear

1. Remove some fluid from the reservoir with a suction pump.
2. Raise and safely support the vehicle.
3. Remove the rear wheels.
4. Remove the banjo bolt and disconnect the brake hose from the caliper. Plug the hose to prevent fluid loss and contamination.
5. Remove the 2 caliper mounting bolts. Remove the caliper from its mounting bracket.

To Install:

6. Fit the caliper over the pads and onto its mounting bracket.
7. Tighten the caliper bolts to 17 ft. lbs. (23 Nm).
8. Reconnect the brake hose with new sealing washers. Tighten the banjo bolt to 17 ft. lbs. (34 Nm).
9. Fill the reservoir with fluid and bleed the brake system. Adjust the parking brake if necessary.
10. Install the rear wheels and lower the vehicle.

Disc Brake Pads

REMOVAL & INSTALLATION

Front

1. Raise and support the vehicle safely.
2. Remove the front wheels.
3. Remove a small amount of brake fluid from the reservoir using a suction pump.
4. Unbolt the brake hose clamp from the knuckle by removing the retaining bolts.
5. Remove the lower caliper retaining bolt and pivot the caliper upward, off of the pads.
6. Remove the pad shim and pad retainers. Remove the disc brake pads from the caliper.

To install:

7. Clean the caliper thoroughly; remove any rust from the lip of the disc or rotor. Check the brake rotor for grooves or cracks. If any heavy scoring is present, the rotor must be replaced.
8. Install the pad retainers. Apply molybdenum brake grease to both surfaces of the shims and the back of the disc brake pads.
9. Install the pads and shims. The pad

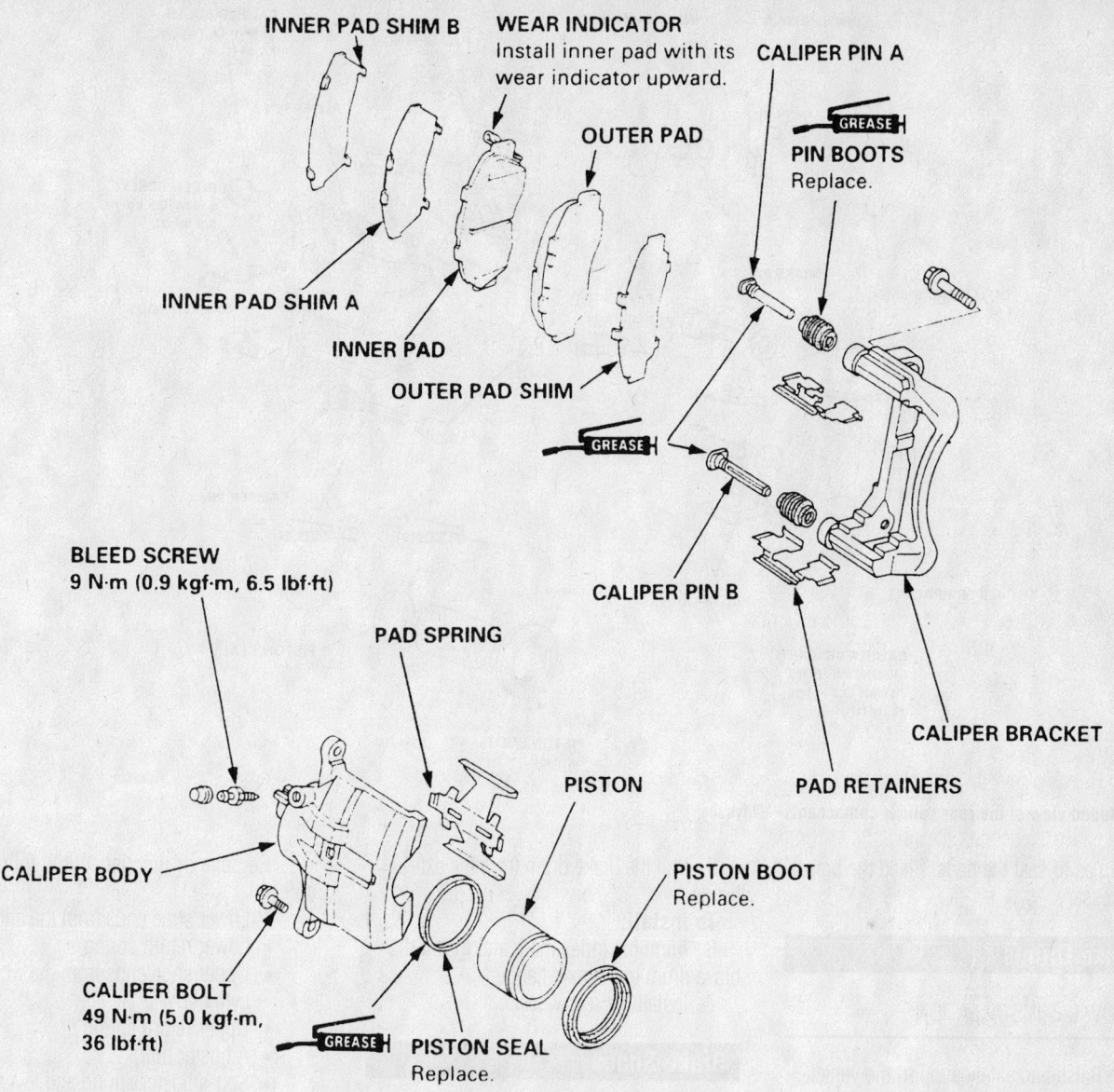

INNER PAD SHIM B

WEAR INDICATOR
Install inner pad with its wear indicator upward.

OUTER PAD

CALIPER PIN A
GREASE
PIN BOOTS
Replace.

INNER PAD SHIM A

INNER PAD

OUTER PAD SHIM

GREASE

CALIPER PIN B

BLEED SCREW
9 N·m (0.9 kgf·m, 6.5 lbf·ft)

PAD SPRING

CALIPER BRACKET

PAD RETAINERS

PISTON

PISTON BOOT
Replace.

CALIPER BODY

CALIPER BOLT
49 N·m (5.0 kgf·m, 36 lbf·ft)

GREASE

PISTON SEAL
Replace.

93026G57

Exploded view of the front caliper components—Odyssey

with the wear indicator goes in the inboard position.

10. Push in the caliper piston so the caliper will fit over the pads. This is most easily accomplished with a pad spreader or large C-clamp.

11. Pivot the caliper down into position and tighten the mounting bolt to 27 ft. lbs. (32 Nm).

12. Connect the brake hose to the knuckle, if removed.

13. Install the wheel and lower the vehicle to the ground.

14. Add brake fluid to the master cylinder reservoir and install the cap.

15. Depress the brake pedal several times and make sure that the movement feels normal. The first brake pedal application may result in a very long pedal action

due to the pistons being retracted. Always make several brake applications before starting the vehicle. Bleed the system if necessary.

Rear

1. Raise and safely support the vehicle.
2. Remove a small amount of brake fluid from the reservoir using a suction pump.
3. Remove the rear wheels.
4. Remove the 2 caliper mounting bolts and remove the caliper from the bracket.
5. Remove the pads, shims, and pad retainers.

To install:

6. Clean the caliper thoroughly; remove any dirt or dust. Check the brake rotor for

grooves or cracks and machine or replace, as necessary.

7. Install the pad retainers. Apply molybdenum brake grease to both surfaces of the shims and the back of the disc brake pads.

8. Install the pads and shims. The wear retainer on the inboard pad faces down.

9. Use a suitable tool to push caliper piston into its bore and enable the caliper to fit over the pads. Lubricate the piston boot with silicon grease. Avoid twisting the boot.

10. Install the brake caliper. Tighten the mounting bolts to 27 ft. lbs. (32 Nm).

11. Install the rear wheels. Lower the vehicle.

12. Add brake fluid to the master cylinder reservoir. Depress the brake pedal sev-

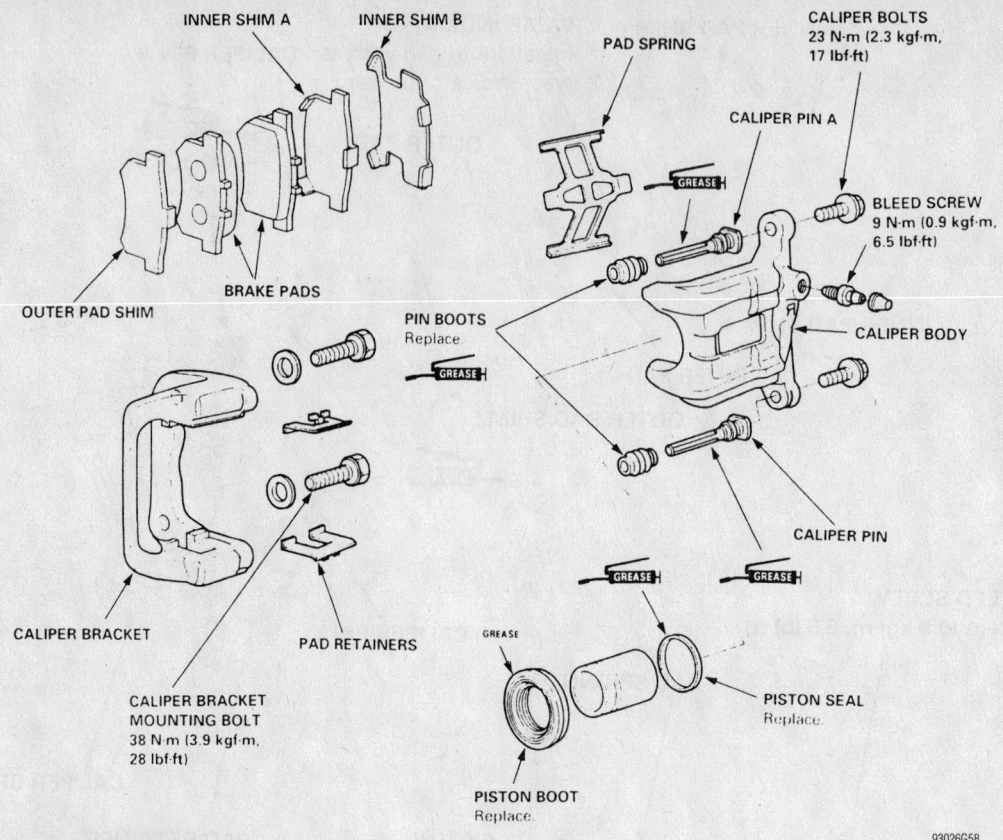

INNER SHIM A   INNER SHIM B   PAD SPRING   CALIPER BOLTS
23 N·m (2.3 kgf·m, 17 lbf·ft)

CALIPER PIN A

BLEED SCREW
9 N·m (0.9 kgf·m, 6.5 lbf·ft)

CALIPER BODY

OUTER PAD SHIM   BRAKE PADS

PIN BOOTS
Replace.
GREASE

GREASE

CALIPER PIN

CALIPER BRACKET   PAD RETAINERS

GREASE   GREASE

GREASE

CALIPER BRACKET
MOUNTING BOLT
38 N·m (3.9 kgf·m, 28 lbf·ft)

PISTON SEAL
Replace.

PISTON BOOT
Replace.

93026G58

Exploded view of the rear caliper components—Odyssey

eral times to seat the pads. Bleed the brakes if necessary.

Brake Drums

REMOVAL & INSTALLATION

1. Raise and safely support the vehicle. Release the parking brake.
2. Remove the rear wheels.
3. Use chalk to mark the brake drum to one of the wheel studs as an index mark for reinstallation.
4. Use forcing screws to remove the drum.

5. Pull the brake drum from the axle flange.

To install:

6. Align the index mark and install the brake drum to the axle flange.
7. Install the rear wheels.

Brake Shoes

REMOVAL & INSTALLATION

1. Remove the drum.
2. Compress, turn and remove the tension springs.

3. Remove or disconnect the following:

- Lower shoe ends from the anchor
- Lower return spring
- Upper shoe ends from the wheel cylinder
- Upper return spring
- Adjuster bolt
- Self-adjuster spring and lever
- Parking brake cable and lever

4. Installation is the reverse of removal. Clean and coat all threads and sliding surfaces with Molykote 44MA grease, or equivalent.

HONDA

Pilot

SPECIFICATION CHARTS

ENGINE AND VEHICLE IDENTIFICATION CHART

			Engine Code				Model Year	
Code	Liters (cc)	Cu. In.	Cyl.	Fuel Sys.	Engine Type	Eng. Mfg.	Code ①	Year
J35A3	3.5 (3471)	212	6	SMFI	SOHC	Honda	3	2003
							4	2004

SOHC: Single Overhead Cam

SMFI: Sequential Multi-port Fuel Injection

① 10th position of VIN

42356-HPIL-C01

GENERAL ENGINE SPECIFICATIONS

Year	Model	Engine Displacement Liters (cc)	Engine ID/VIN	Fuel System Type	Net Horsepower @ rpm	Net Torque @ rpm (ft. lbs.)	Bore x Stroke (in.)	Compression Ratio	Oil Pressure @ rpm
2003-04	Pilot	3.5 (3471)	J35A3	SMFI	210@5200	229@4300	3.50x3.66	9.4:1	71@3000

SMFI: Sequential Multi-port Fuel Injection

42356-HPIL-C02

ENGINE TUNE-UP SPECIFICATIONS

Year	Engine Displacement Liters (cc)	Engine ID/VIN	Spark Plug Gap (in.)	Ignition Timing (deg.) MT	Ignition Timing (deg.) AT	Fuel Pump (psi)	Idle Speed (rpm) MT	Idle Speed (rpm) AT	Valve Clearance (in.) In.	Valve Clearance (in.) Ex.
2003-04	3.5 (3471)	J35A1	0.039-0.043	—	8-12B	32-40	—	680-780	0.008-0.009	0.011-0.013

NOTE: The Vehicle Emission Control Information label often reflects changes made during production and must be used if they differ from this chart.

NOTE: The fuel pressure readings are given with the vacuum hose connected to the regulator and the engine running

B: Before top dead center

42356-HPIL-C03

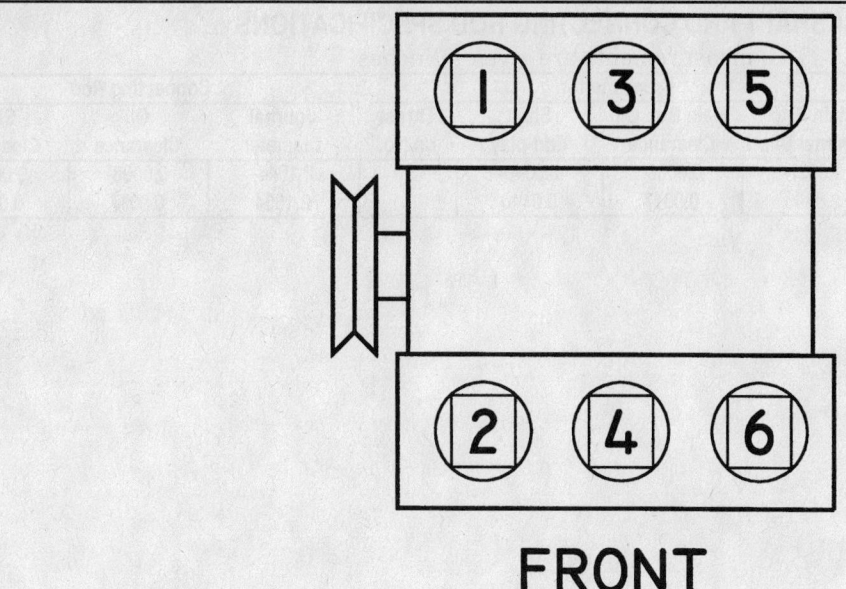

FRONT

9308MG32

3.5L Engine
Firing Order: 1–2–3–4–5–6
Distributorless ignition system (One coil per cylinder)

CAPACITIES

Year	Model	Engine Displacement Liters (cc)	Engine ID/VIN	Engine Oil with Filter (qts.)	Transmission (pts.)		Transfer Case (pts.)	Drive Axle		Fuel Tank (gal.)	Cooling System (qts.)
					5-Spd	Auto.		Front (pts.)	Rear (pts.)		
2003-04	Pilot	3.5 (3471)	J35A3	5.0	—	①	—	—	5.5	19.2	8.0

NOTE: All capacities are approximate. Add fluid gradually and check to be sure a proper fluid level is obtained.

① Without filter change: 6.4 pts;
   With filter Change: 16.4 pts.

42356-HPIL-C04

VALVE SPECIFICATIONS

Year	Engine Displacement Liters (cc)	Engine ID/VIN	Seat Angle (deg.)	Face Angle (deg.)	Spring Test Pressure (lbs. @ in.)	Spring Installed Height (in.)	Stem-to-Guide Clearance (in.)		Stem Diameter (in.)	
							Intake	Exhaust	Intake	Exhaust
2003-04	3.5 (3471)	J35A3	45	45	NA	①	0.0008-0.0018	0.0022-0.0031	0.2159-0.2163	0.2146-0.2150

NA: Not Available

① Valve spring free length:
   Intake: 1.9713 in.
   Exhaust: 2.1060 in.

42356-HPIL-C05

CRANKSHAFT AND CONNECTING ROD SPECIFICATIONS
All measurements are given in inches

Year	Engine Displacement Liters (cc)	Engine ID/VIN	Crankshaft				Connecting Rod		
			Main Brg. Journal Dia.	Main Brg. Oil Clearance	Shaft End-play	Thrust on No.	Journal Diameter	Oil Clearance	Side Clearance
2003-04	3.5 (3471)	J35A3	2.8337-2.8346	0.0008-0.0017	0.0040-0.0140	3	2.1644-2.1654	0.0008-0.0017	0.0060-0.0140

42356-HPIL-C06

PISTON AND RING SPECIFICATIONS
All measurements are given in inches

Year	Engine Displacement Liters (cc)	Engine ID/VIN	Piston Clearance	Ring Gap			Ring Side Clearance		
				Top Compression	Bottom Compression	Oil Control	Top Compression	Bottom Compression	Oil Control
2003-04	3.5 (3471)	J35A1	0.0006-0.0016	0.0080-0.0140	0.0160-0.0220	0.0080-0.0280	0.0014-0.0024	0.0012-0.0022	NA

NA: Not Available

42356-HPIL-C07

TORQUE SPECIFICATIONS
All readings in ft. lbs.

Year	Engine Displacement Liters (cc)	Engine ID/VIN	Cylinder Head Bolts	Main Bearing Bolts	Rod Bearing Bolts	Crankshaft Damper Bolts	Flywheel Bolts	Manifold		Spark Plugs	Lug Nut
								Intake	Exhaust		
2003-04	3.5 (3471)	J35A1	①	②	③	181	54	16	23	13	80

NOTE: Dip main bearing bolts and crankshaft damper bolt in clean engine oil prior to tightening.

① Step 1: 29 ft. lbs.
  Step 2: 51 ft. lbs.
  Step 3: 72 ft. lbs.

② 11mm bolt 56 ft. lbs.
  10mm bolt 36 ft. lbs.

③ Step 1: 14 ft. lbs.
  Step 2: 90 degrees

42356-HPIL-C08

WHEEL ALIGNMENT

Year	Model		Caster		Camber		Toe-in (in.)	Steering Axis Inclination (Deg.)
			Range (+/-Deg.)	Preferred Setting (Deg.)	Range (+/-Deg.)	Preferred Setting (Deg.)		
2003-04	Pilot	F	1.00	1.00	1.00	30	0+/-1/16	—
		R	—	—	0	30	0+/-1/16	—

42356-HPIL-C09

TIRE, WHEEL AND BALL JOINT SPECIFICATIONS

Year	Model	OEM Tires		Tire Pressures (psi)		Wheel Size	Ball Joint Inspection
		Standard	Optional	Front	Rear		
2003-04	Pilot	P235/65R17	None	32	32	R17	NS

OEM: Original Equipment Manufacturer

PSI: Pounds Per Square Inch

NS: Not specified by manufacturer

42356-HPIL-C10

BRAKE SPECIFICATIONS
Honda Pilot
All measurements in inches unless noted

Year	Model		Brake Disc			Brake Drum Diameter			Minimum Lining Thickness		Brake Caliper	
			Original Thickness	Minimum Thickness	Maximum Runout	Original Inside Diameter	Max. Wear Limit	Maximum Machine Diameter	Front	Rear	Bracket Bolts (ft. lbs.)	Mounting Bolts (ft. lbs.)
2003-04	Pilot	F	1.100	1.020	0.004	—	—	—	0.060	—	80	27
		R	0.430	0.350	0.004	—	—	—	—	0.41	41	27

F: Front

R: Rear

42356-HPIL-C11

SCHEDULED MAINTENANCE INTERVALS
HONDA—PILOT

TO BE SERVICED	TYPE OF SERVICE	VEHICLE MILEAGE INTERVAL (x1000)															
		7.5	15	22.5	30	37.5	45	52.5	60	67.5	75	82.5	90	97.5	105	112.5	120
Accessory drive belts	I & A				✓				✓				✓				✓
Air cleaner element	R				✓				✓				✓				✓
Brake fluid	R	Every 3 years															
Brake hoses & lines (including ABS)	I		✓		✓		✓		✓		✓		✓		✓		✓
Cooling system hoses & connections	I		✓		✓		✓		✓		✓		✓		✓		✓
Engine coolant ①	R							✓					✓				
Engine oil	R	✓	✓	✓	✓	✓	✓	✓	✓	✓	✓	✓	✓	✓	✓	✓	✓
Engine oil and coolant levels	I	Inspect at each fuel stop															
Engine oil filter	R		✓		✓		✓		✓		✓		✓		✓		✓
Exhaust system	I		✓		✓		✓		✓		✓		✓		✓		✓
Fluid levels and condition	I		✓		✓		✓		✓		✓		✓		✓		✓
Front and rear brakes	I		✓		✓		✓		✓		✓		✓		✓		✓
Fuel lines & connection	I		✓		✓		✓		✓		✓		✓		✓		✓
Halfshaft boots	I		✓		✓		✓		✓		✓		✓		✓		✓
Idle speed	I & A														✓		
Parking brake system	I & A		✓		✓		✓		✓		✓		✓		✓		✓
Rear differential fluid	R	✓			✓		✓		✓				✓				✓
Rotate and inspect tires	I	✓	✓	✓	✓	✓	✓	✓	✓	✓	✓	✓	✓	✓	✓	✓	✓
Spark plugs	R														✓		
Supplemental Restrain system (SRS)	I	Inspect the SRS 10 years after production															
Suspension components	I		✓		✓		✓		✓		✓		✓		✓		✓
Tie rod ends, steering gear box & boots	I		✓		✓		✓		✓		✓		✓		✓		✓
Timing belt	R														✓		
Transmission fluid	R							✓			✓				✓		
Valve clearance	I	Adjust if valves are noisy															
Water pump	S/I														✓		

R: Replace   I: Inspect   A: Adjust

① Every 12,000 miles or 10 years, then every 60,000 miles or 5 years

FREQUENT OPERATION MAINTENANCE (SEVERE SERVICE)

If a vehicle is operated under any of the following conditions it is considered severe service:

- Towing a trailer or using a camper or car-top carrier.
- Repeated short trips of less than 5 miles in temperatures below freezing, or trips of less than 10 miles in any temperature.
- Extensive idling or low-speed driving for long distances as in heavy commercial use, such as delivery, taxi or police cars.
- Operating on rough, muddy or salt-covered roads.
- Operating on unpaved or dusty roads.
- Driving in extremely hot (over 90°) conditions.

Air cleaner element: replace every 15,000 miles

Engine oil and filter: replace every 3750 miles or 6 months, whichever occurs first.

Timing belt: replace every 60,000 miles if the vehicle is regularly driven in temperatures above 110°F or below -20°F, or if frequently towing a trailer.

Transmission fluid: replace every 30,000 miles.

Rear differential fluid: replace every 60,000 miles.

Front and rear brakes: inspect every 7500 miles or 6 months, whichever occurs first.

Locks and hinges: lubricate every 15,000 miles.

Tie rods, steering gear box, boots: inspect every 7500 miles or 6 months, whichever occurs first.

Suspension components: inspect every 7500 miles or 6 months, whichever occurs first.

Halfshaft boots: inspect every 7500 miles or 6 months, whichever occurs first.

PRECAUTIONS

Before servicing any vehicle, please be sure to read all of the following precautions, which deal with personal safety, prevention of component damage, and important points to take into consideration when servicing a motor vehicle:

• Never open, service or drain the radiator or cooling system when the engine is hot; serious burns can occur from the steam and hot coolant.

• Observe all applicable safety precautions when working around fuel. Whenever servicing the fuel system, always work in a well-ventilated area. Do not allow fuel spray or vapors to come in contact with a spark, open flame or excessive heat (a hot drop light, for example). Keep a dry chemical fire extinguisher near the work area. Always keep fuel in a container specifically designed for fuel storage; also, always properly seal fuel containers to avoid the possibility of fire or explosion. Refer to the additional fuel system precautions later in this section.

• Fuel injection systems often remain pressurized, even after the engine has been turned **OFF**. The fuel system pressure must be relieved before disconnecting any fuel lines. Failure to do so may result in fire and/or personal injury.

• Brake fluid often contains polyglycol ethers and polyglycols. Avoid contact with the eyes and wash your hands thoroughly after handling brake fluid. If you do get brake fluid in your eyes, flush your eyes with clean, running water for 15 minutes. If eye irritation persists, or if you have taken

brake fluid internally, IMMEDIATELY seek medical assistance.

• The EPA warns that prolonged contact with used engine oil may cause a number of skin disorders, including cancer! You should make every effort to minimize your exposure to used engine oil. Protective gloves should be worn when changing oil. Wash your hands and any other exposed skin areas as soon as possible after exposure to used engine oil. Soap and water, or waterless hand cleaner should be used.

• All new vehicles are now equipped with an air bag system. The system must be disabled before performing service on or around system components, steering column, instrument panel components, wiring and sensors. Failure to follow safety and disabling procedures could result in accidental air bag deployment, possible personal injury and unnecessary system repairs.

• Always wear safety goggles when working with, or around, the air bag system. When carrying a non-deployed air bag, be sure the bag and trim cover are pointed away from your body. When placing a non-deployed air bag on a work surface, always face the bag and trim cover upward, away from the surface. This will reduce the motion of the module if it is accidentally deployed. Refer to the additional air bag system precautions later in this section.

• Clean, high quality brake fluid from a sealed container is essential to the safe and proper operation of the brake system. You

should always buy the correct type of brake fluid for your vehicle. If the brake fluid becomes contaminated, completely flush the system with new fluid. Never reuse any brake fluid. Any brake fluid that is removed from the system should be discarded. Also, do not allow any brake fluid to come in contact with a painted surface; it will damage the paint.

• Never operate the engine without the proper amount and type of engine oil; doing so WILL result in severe engine damage.

• Timing belt maintenance is extremely important! Many models utilize an interference-type, non-freewheeling engine. If the timing belt breaks, the valves in the cylinder head may strike the pistons, causing potentially serious (also time-consuming and expensive) engine damage. Refer to the maintenance interval charts in the front of this section for the recommended replacement interval for the timing belt, and to the timing belt procedure in this section for belt replacement and inspection.

• Disconnecting the negative battery cable on some vehicles may interfere with the functions of the on-board computer system(s) and may require the computer to undergo a relearning process once the negative battery cable is reconnected.

• When servicing drum brakes, only disassemble and assemble one side at a time, leaving the remaining side intact for reference.

• Only an MVAC-trained, EPA-certified automotive technician should service the air conditioning system or its components.

ENGINE REPAIR

➡**Disconnecting the negative battery cable on some vehicles may interfere with the functions of the on board computer system. The computer may undergo a relearning process once the negative battery cable is reconnected.**

Distributor

The Pilot is equipped with a Distributorless Ignition System (DIS).

Alternator

REMOVAL

1. Before servicing the vehicle, refer to the precautions in the beginning of this section.

2. Remove or disconnect the following:
• Negative battery cable
• Accessory drive belt
• Intake manifold and ignition coil covers
• Alternator wiring harness connectors
• Alternator mounting bolts
• Wiring harness clamp
• Alternator

INSTALLATION

1. Install or connect the following:
• Alternator
• Wiring harness clamp. Tighten the bolt to 105 inch lbs. (12 Nm).
• Alternator mounting bolts. Tighten the 10mm bolt to 33 ft. lbs. (44

Nm) and the 8mm bolt to 16 ft. lbs. (22 Nm).
• Alternator wiring harness connectors. Tighten the battery terminal nut to 105 inch lbs. (12 Nm).
• Accessory drive belt
• Intake manifold and ignition coil covers
• Negative battery cable

Ignition Timing

ADJUSTMENT

The Pilot is equipped with a Distributorless Ignition System (DIS). The ignition timing is controlled by the Powertrain Control module (PCM). No adjustment is necessary.

Engine Assembly

REMOVAL & INSTALLATION

➡ **The engine and transaxle are removed from the vehicle as a unit.**

1. Before servicing the vehicle, refer to the precautions in the beginning of this section.
2. Drain the cooling system.
3. Drain the power steering system.
4. Drain the transaxle fluid.
5. Drain the engine oil.
6. Relieve fuel system pressure.
7. Remove or disconnect the following:
   - Negative battery cable
   - Battery
   - Intake and ignition coil covers
   - Air intake duct
   - Left engine wire harness connectors
   - Relay bracket
   - Battery and tray
   - Starter cable and harness clamp
   - Accelerator cable
   - Cruise control cable
   - Fuel lines
   - EVAP canister hose
8. Remove the drivers side center console lower panel and pull back the cover to access steering joint cover.
   - Steering joint bolt
   - Powertrain Control Module (PCM) connectors
   - Heated Oxygen (HO2S) sensor connector and grommet. Pull the PCM harness through the firewall.
   - Brake booster vacuum line
   - Clamps and clips from power steering hoses
   - Fuse/relay box battery cable
   - Accessory drive belts
   - Front wheels
   - Splash shield
   - Front sub-frame stiffener
   - Exhaust front pipe
   - Propeller shaft
   - Shift control cable
   - Transfer assembly
   - Ball joints
   - Stabilizer bar links
   - Halfshafts
   - Power steering hose and pressure switch connector
   - Transaxle lower front mount
   - Transaxle lower rear mount
   - A/C compressor
   - Heater hoses
   - Radiator hoses
   - Ground cable
   - Transaxle oil cooler lines
   - Radiator

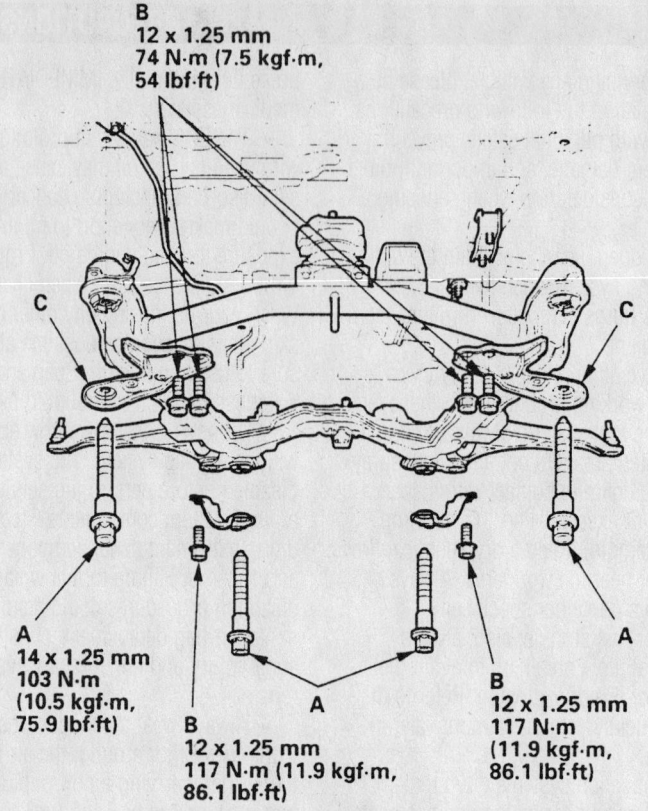

B
12 x 1.25 mm
74 N·m (7.5 kgf·m, 54 lbf·ft)

C

A
14 x 1.25 mm
103 N·m
(10.5 kgf·m,
75.9 lbf·ft)

B
12 x 1.25 mm
117 N·m (11.9 kgf·m,
86.1 lbf·ft)

A

B
12 x 1.25 mm
117 N·m
(11.9 kgf·m,
86.1 lbf·ft)

A

9302MG69

Sub-frame fastener locations and tightening torque—Pilot

9. Attach a hoist to the engine lifting eyes and support the powertrain weight.
10. Remove or disconnect the following:
   - Side engine mount bracket
   - Front mount bracket support nut
11. Matchmark the front subframe to the mounting points.
12. Remove or disconnect the following:
   - Front subframe
   - All remaining hoses and electrical connections
13. Lower the powertrain away from the vehicle.

To install:

14. Raise the powertrain into position.
15. Installation is the reverse of removal but please note the following steps:
   - A/C compressor bolts to 16 ft. lbs. (22 Nm)
   - Front subframe. Use new bolts and tighten the 14mm bolts to 76 ft. lbs. (103 Nm). Tighten the front brace bolts to 54 ft. lbs. (74 Nm) and the rear brace bolts to 86 ft. lbs. (117 Nm).
   - Transaxle lower front mount nuts to 28 ft. lbs. (38 Nm)
   - Transaxle lower rear mount bolts to 28 ft. lbs. (38 Nm)
   - Front mount bracket support nut to 40 ft. lbs. (54 Nm)
   - Side engine mount bracket bolts to 33 ft. lbs. (44 Nm) and the through bolt to 40 ft. lbs. (54 Nm)
16. Fill the engine crankcase to the correct level.
17. Fill the transaxle to the correct level.
18. Fill the cooling system.
19. Fill the power steering system.
20. Start the engine and check for leaks.
21. Check the wheel alignment and adjust as necessary.

Water Pump

REMOVAL & INSTALLATION

1. Before servicing the vehicle, refer to the precautions in the beginning of this section.
2. Drain the cooling system.
3. Remove or disconnect the following:
   - Negative battery cable
   - Accessory drive belts
   - Front cover
   - Timing belt
   - Timing belt tensioner
   - Water pump

To install:

4. Install or connect the following:
   - Water pump. Use a new O-ring seal

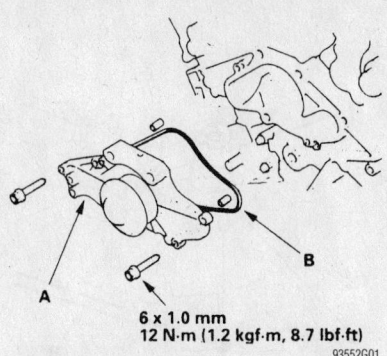

**6 x 1.0 mm
12 N·m (1.2 kgf·m, 8.7 lbf·ft)**
93552G01

Exploded view of the water pump mounting

and tighten the bolts to 105 inch lbs. (12 Nm).
- Timing belt tensioner
- Timing belt
- Front cover
- Accessory drive belts
- Negative battery cable
5. Fill the cooling system.
6. Start the engine and check for leaks.

Heater Core

REMOVAL & INSTALLATION

1. Before servicing the vehicle, refer to the precautions in the beginning of this section.
2. Drain the cooling system.
3. Remove or disconnect the following:
   - Negative battery cable
4. Recover the refrigerant using approved equipment.
   - Heater valve cable from the valve arm. Turn the valve arm to the fully opened position.
   - Heater hoses from the heater unit
   - Mounting nut from the heater unit. Be careful not to bend or damage fuel or brake lines.
5. Remove the dashboard as follows:
   a. Remove the center console by unlatching the clips.
   b. Remove the dashboard lower cover screw, gently pull down on the cover to disengage the clips and disconnect the electrical connections.
   c. Remove the dashboard side cover by gently pulling and turning to unfasten the clips.
   d. Remove the right kick panel.
   e. While holding the glove box, remove the box stop from each side, then disconnect the lock from the damper.
   f. Remove the glove box bolts and the glove box.
   g. Remove the front door trim, kick panels and A-pillar trim from both sides.

h. Remove the cap from the front pillar corner trim. Unfasten the screw, slide the trim upward along the pillar and remove it. Remove the remaining clips from the body.
i. On the drivers side, remove the fuel/relay box nut and pull out the box.
j. Remove the steering column
k. On the passenger side remove the fuse/relay bolt and pull out the box.
l. Disconnect all electrical connections from the dashboard.
m. If equipped with a navigation system, remove the passenger seat, pull back the carpet, remove the harness cushions and then pull out the GPS harness.
n. Remove all harness and connector clips.
o. Remove all the bolts and lift up on the dashboard to release the dashboard and steering hanger beam from the guide pins.
p. Remove the dashboard through the door.

6. Remove the evaporator as follows:
   a. Disconnect the receiver and suction lines from the evaporator.
   b. Remove the mounting nuts and plug the lines to avoid system contamination.
   c. Remove the plastic brace and glove box frame.
   d. Disconnect the wire harness and evaporator temperature sensor connector.
   e. Remove the self-tapping screws, the nuts and the evaporator.
7. Remove or disconnect the following:
   - Mounting bolts and the heater unit
   - Self-tapping screws and the clamp, then pull the heater core from the case being careful not to bend the pipes

To install:
8. Install or connect the following:
   - Heater core in the case
   - Clamp and the screws
   - Heater unit and tighten the bolts to 7 ft. lbs. (10 Nm)
   - Evaporator in the reverse order of

Fastener Locations

A ▶: Bolt, 2   B ▶: Bolt, 1   C ▶: Bolt, 3
D ▶: Bolt, 5   E ▶: Bolt, 2   F ▶: Bolt, 2

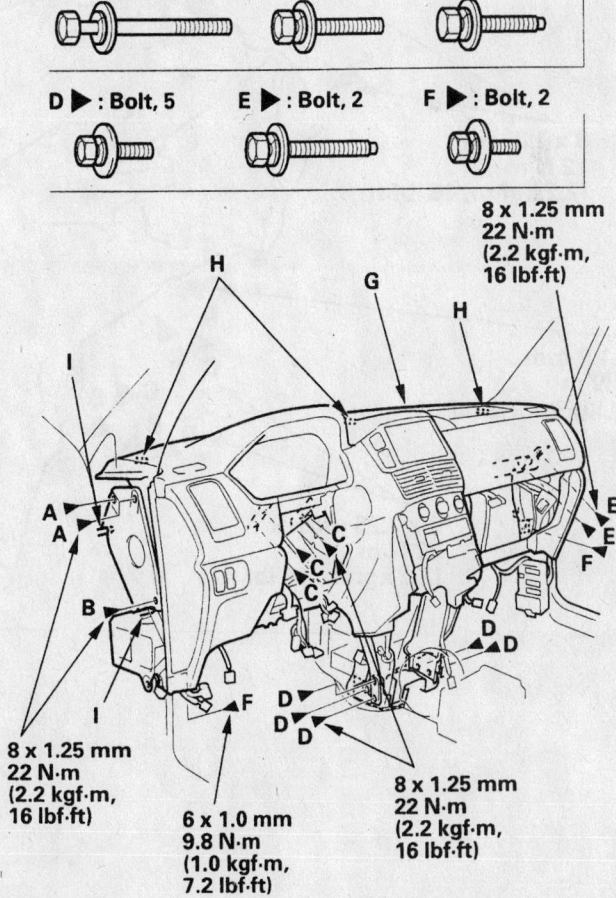

**8 x 1.25 mm
22 N·m
(2.2 kgf·m, 16 lbf·ft)**

**8 x 1.25 mm
22 N·m
(2.2 kgf·m, 16 lbf·ft)**

**6 x 1.0 mm
9.8 N·m
(1.0 kgf·m, 7.2 lbf·ft)**

**8 x 1.25 mm
22 N·m
(2.2 kgf·m, 16 lbf·ft)**

42356-HPIL-G03

Tighten the dashboard bolts as illustrated

removal. Tighten all the retainers to 7 ft. lbs. (10 Nm) .

9. Install the dashboard in the reverse order of removal keeping in mind the following points:

a. Make sure the dashboard is seated properly and that the wiring harness and steering hanger beam wire harness are not pinched.

b. Referring to the accompanying illustration, tighten bolts **(A, B, C, D and E)** to 16 ft. lbs. (22 Nm). Tighten bolts **F** to 7 ft. lbs. (10 Nm). Apply thread lock to the **B** bolts before installation.

c. Ensure that all electrical connectors are properly connected.

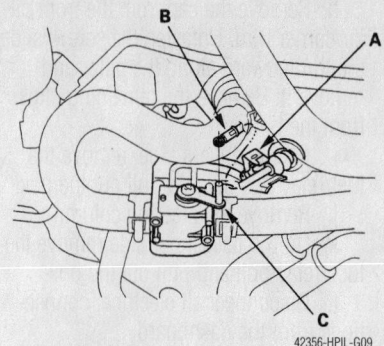

B
A
C
42356-HPIL-G09

In the engine compartment, open the cable clamp (A), then disconnect the heater valve cable (B) from the valve arm (C)

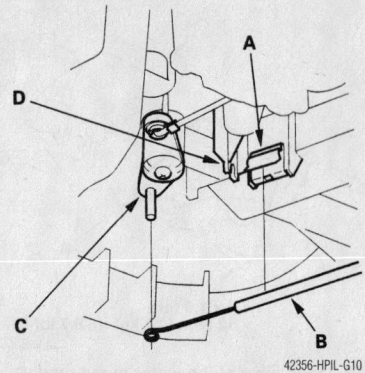

A
D
C
B
42356-HPIL-G10

Under the dashboard, disconnect the valve cable housing from the clamp (A) and the cable (B) from the mix control linkage (C)

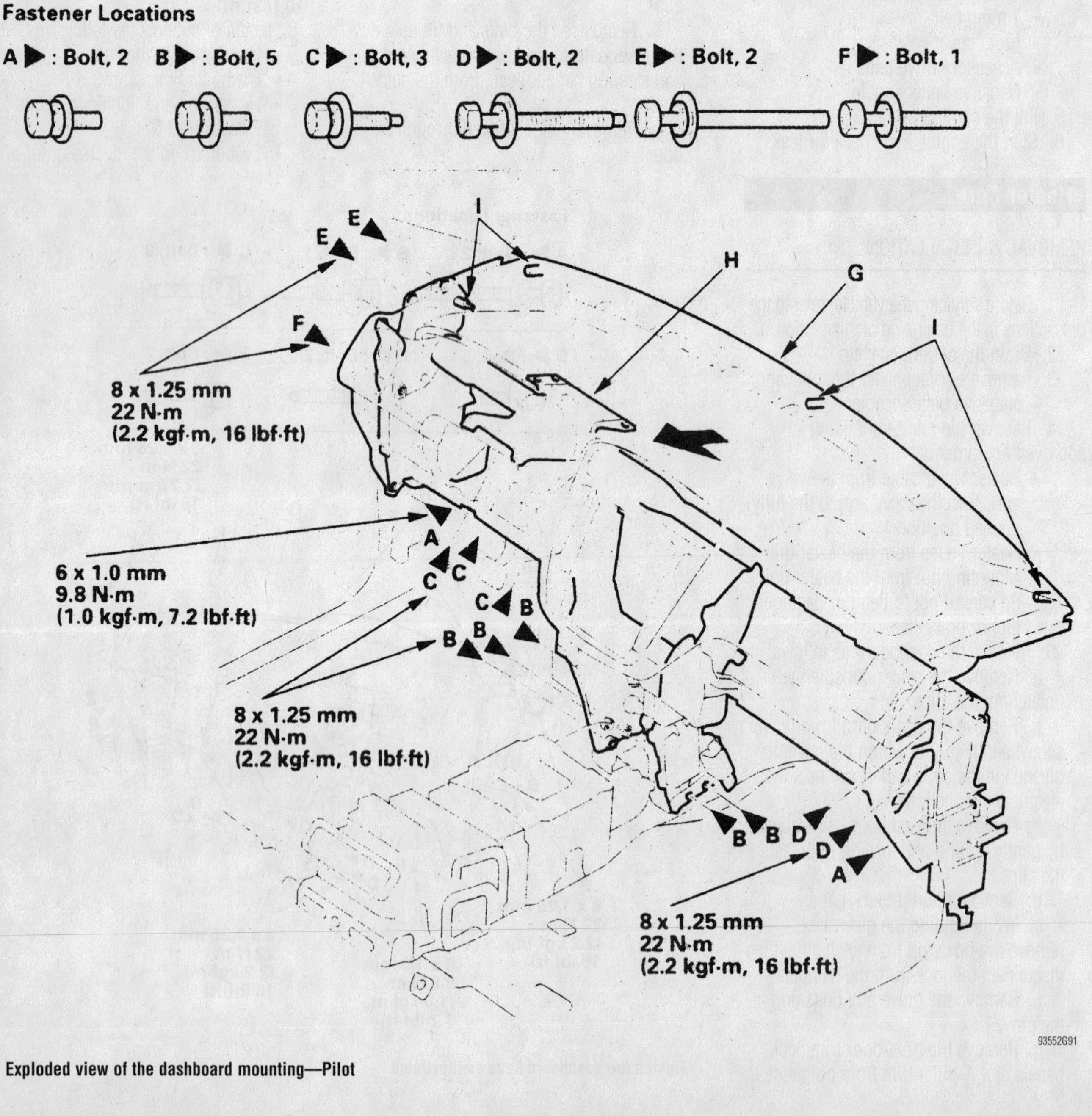

Fastener Locations
A : Bolt, 2
B : Bolt, 5
C : Bolt, 3
D : Bolt, 2
E : Bolt, 2
F : Bolt, 1
E
E
F
I
H
G
I
8 x 1.25 mm
22 N·m
(2.2 kgf·m, 16 lbf·ft)
A
C
C
C
B
B
B
6 x 1.0 mm
9.8 N·m
(1.0 kgf·m, 7.2 lbf·ft)
8 x 1.25 mm
22 N·m
(2.2 kgf·m, 16 lbf·ft)
B
B
D
D
A
8 x 1.25 mm
22 N·m
(2.2 kgf·m, 16 lbf·ft)
93552G91

Exploded view of the dashboard mounting—Pilot

6 x 1.0 mm
9.8 N·m (1.0 kgf·m, 7.2 lbf·ft)

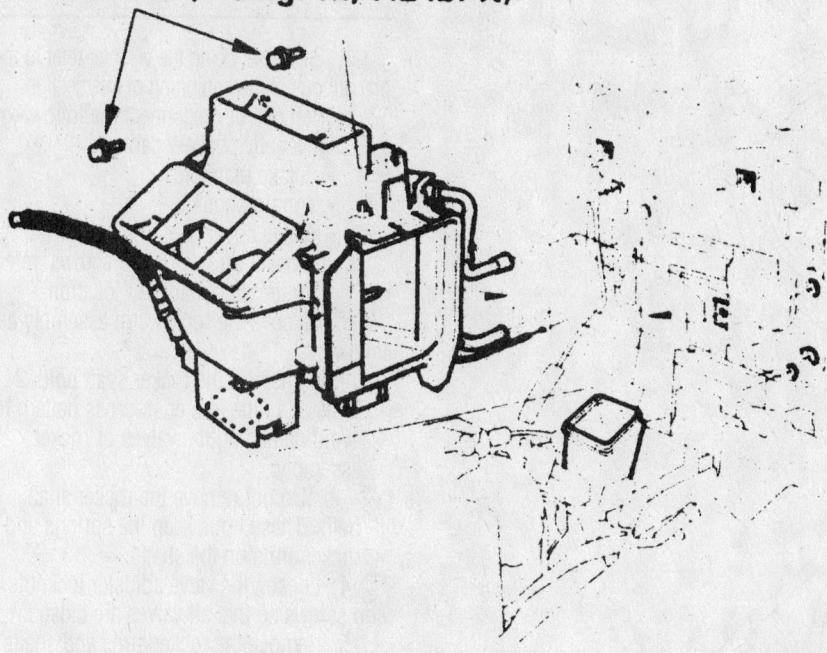

93552G92

Exploded view of the evaporator mounting

10. Install or connect the following:
   - Mounting nut to the heater unit and tighten to 7 ft. lbs. (10 Nm)
   - Heater hoses
11. Connect the heater valve cable and adjust as follows:

   a. In the engine compartment, open the cable clamp (A), then disconnect the heater valve cable (B) from the valve arm (C).

   b. Under the dashboard, disconnect the valve cable housing from the clamp (A) and the cable (B) from the mix control linkage (C).

   c. Set the temperature control button to the MAX COOL position with the ignition switch in the on position.

   d. Attach the valve cable (B) to the mix control linkage (C) as shown in the illustration, hold the end of the cable housing against the stop, then snap the cable housing into the clamp.

   e. In the engine compartment, turn the valve arm (C) to the fully closed position as shown in the accompanying illustration and hold it there. Attach the cable (B) to the valve arm and pull gently on the cable housing to take up the slack, then install the cable housing into the clamp (A).
12. Fill the cooling system
13. Connect the battery cable.

Cylinder Head

REMOVAL & INSTALLATION

1. Before servicing the vehicle, refer to the precautions in the beginning of this section.
2. Drain the cooling system.
3. Relieve the fuel system pressure.
4. Remove or disconnect the following:
   - Negative battery cable
   - Alternator belt
   - Intake manifold cover
   - Ignition coil covers
   - Power steering belt and pump
   - Power steering hose clamp
   - Alternator
   - Fuel feed and return lines
   - EVAP canister hose
   - Intake manifold
   - Ignition coils
   - Timing belt
   - Fuel injector connectors
   - Engine Coolant Temperature (ECT) sensor connector
   - Radiator fan switch connectors
   - Crankshaft Position (CKP) sensor connector
   - Camshaft Position (CMP) sensor connector
   - Exhaust Gas Recirculation (EGR) connector
   - Valve Lift Electronic Control (VTEC) solenoid valve connector and oil pressure switch connections
   - Oil pressure switch connector
   - Vacuum hoses from the intake air bypass control valve
   - Fuel rails
   - Heater hose
   - Radiator hoses
   - Ground cable
   - Exhaust manifolds
   - Water passage
   - Camshaft pulleys and back covers
   - Valve covers
5. Loosen the cylinder head bolts in sequence and ⅓ turns until all bolts are loose.
6. Remove the cylinder head.

To install:
7. Align the crankshaft and camshaft sprocket TDC marks as shown.
8. Install the cylinder heads with new gaskets.
9. Apply clean engine oil to the cylinder head bolt threads and flanges.

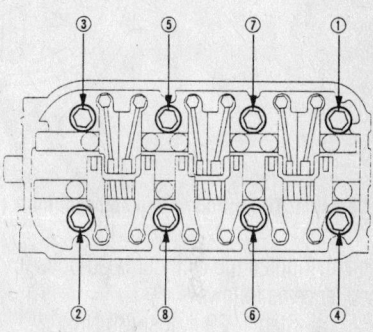

9308MG12

Cylinder head bolt loosening sequence—Pilot

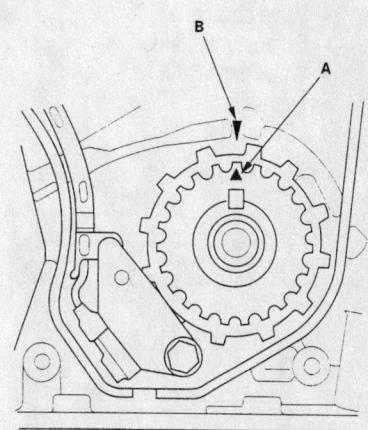

9302MG74

Crankshaft timing belt sprocket TDC marks. Align sprocket mark (A) with pointer (B)—Pilot

FRONT:

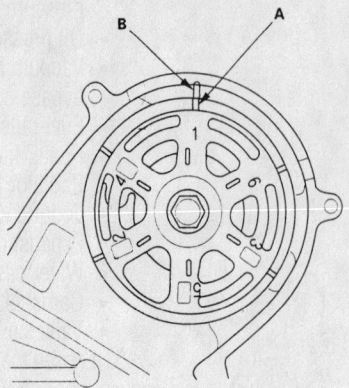

REAR:

9302MG85

Camshaft TDC marks. Align sprocket mark (A) with the back cover pointer (B)—Pilot

10. Tighten the cylinder head bolts in sequence as follows:
   a. Step 1: 29 ft. lbs. (39 Nm).
   b. Step 2: 51 ft. lbs. (69 Nm).
   c. Step 3: 72 ft. lbs. (98 Nm).
11. Install or connect the following:
   • Timing belt and adjust the valve clearance
   • Valve covers
   • Exhaust manifolds
   • Water passage
   • Fuel rails

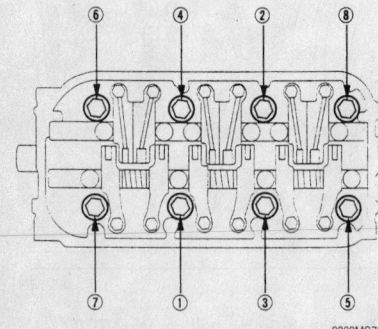

9302MG70

Cylinder head bolt tightening sequence—Pilot

   • Vacuum hoses to the intake air bypass control valve
   • Radiator hoses
   • Heater hose
   • Oil pressure switch connector
   • VTEC solenoid valve connector and oil pressure switch connections
   • EGR connector
   • CMP sensor connector
   • CKP sensor connector
   • Radiator fan switch connectors
   • ECT sensor connector
   • Fuel injector connectors
   • Ignition coils
   • Intake manifold
   • EVAP canister hose
   • Fuel feed and return lines
   • Alternator
   • Power steering hose clamp
   • Power steering pump and belt
   • Ground cable
   • Ignition coil covers
   • Intake manifold cover
   • Alternator belt
   • Negative battery cable
12. Fill the cooling system.
13. Start the engine and check for leaks.

Rocker Arms/Shafts

REMOVAL & INSTALLATION

1. Before servicing the vehicle, refer to the precautions in the beginning of this section.
2. Remove or disconnect the following:
   • Negative battery cable
   • Intake manifold
   • Ignition coils
   • Valve cover
   • Rocker arm adjusting screws. refer to the illustration for location.
3. Remove the rocker arm assembly as follows:
   a. Unscrew the rocker shaft bolts 2 turns at a time in a criss-cross pattern to avoid damaging the valves or rocker assembly.
   b. Do not remove the rocker shaft bolts. These bolts keep the springs and rocker arms on the shafts.
4. Loosen the valve adjuster locknuts and screws so that all valves are closed.
5. Remove the rocker arms and shafts from the vehicle as an assembly.

➡**Keep all valvetrain components in order for assembly.**

6. Remove the rocker arms and springs from the rocker arm shafts.
   To install:
7. Assemble the rocker arms and springs to the rocker arm shafts in their original positions.
8. Install the rocker arm assemblies. Tighten the bolts in sequence and in multiple passes to 17 ft. lbs. (24 Nm).
9. Adjust the valve clearance.
10. Install or connect the following:
   • Valve covers
   • Ignition coils and torque the retainers to 9 ft. lbs. (12 Nm)
   • Ignition coil covers
   • Intake manifold
   • Negative battery cable

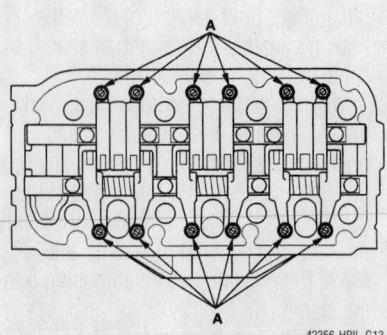

42356-HPIL-G13

Rocker arm shaft adjusting screw locations

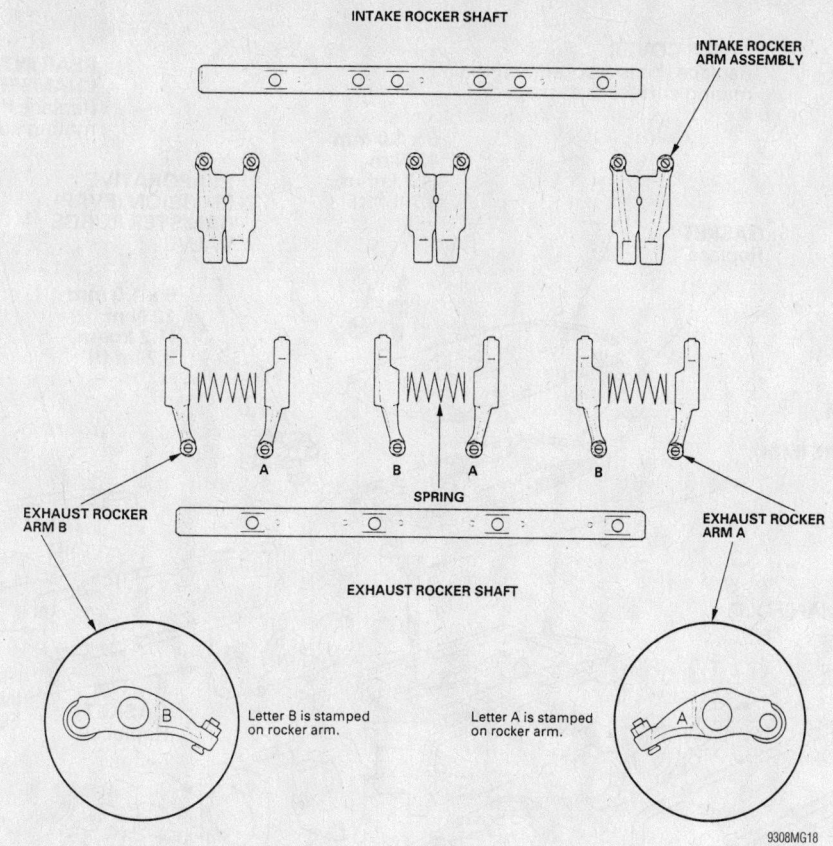

INTAKE ROCKER SHAFT

INTAKE ROCKER ARM ASSEMBLY

A B B A B

SPRING

EXHAUST ROCKER ARM B

EXHAUST ROCKER ARM A

EXHAUST ROCKER SHAFT

Letter B is stamped on rocker arm.

Letter A is stamped on rocker arm.

9308MG18

Exploded view of the rocker arms and shafts—Pilot

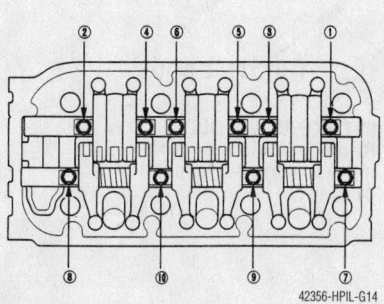

42356-HPIL-G14

Rocker arm shaft loosening sequence

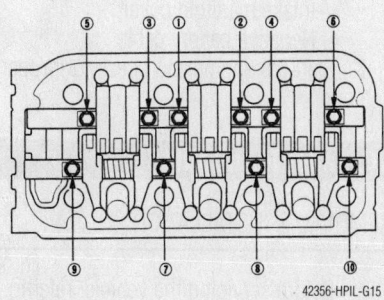

42356-HPIL-G15

Rocker shaft tightening sequence

Intake Manifold

REMOVAL & INSTALLATION

1. Before servicing the vehicle, refer to the precautions in the beginning of this section.
2. Remove or disconnect the following:
   - Negative battery cable
   - Intake manifold cover
   - Air intake tube
   - Throttle and cruise control cables
3. Remove the following electrical connections and clamps from the manifold:
   - Intake Air Temperature (IAT) sensor connector
   - Idle Air Control (IAC) valve connector
   - Throttle Position (TP) sensor connector
   - Manifold Absolute Pressure (MAP) sensor connector
   - Evaporative Emissions (EVAP) control canister purge valve connector
   - Brake booster vacuum line
   - Positive Crankcase Ventilation (PCV) hose
   - Water bypass hoses from the throttle body and plug the hoses
   - EVAP control canister hose

4. Remove or disconnect the following:
   - Upper cover bolts and nuts in the sequence illustrated using two or three passes
   - Intake manifold bolts using two or three passes
   - Intake manifold and spacer

To install:

5. Install or connect the following:
   - New intake manifold gasket and spacer
   - Intake manifold. Tighten the fasteners in sequence and in several passes to 16 ft. lbs. (22 Nm).
   - Upper cover bolts and nuts in the sequence illustrated using two or three passes to 9 ft. lbs. (12 Nm)
   - EVAP control canister hose
   - Water bypass hoses toe throttle body
   - Brake booster vacuum line
   - PCV hose

6. Connect the following electrical connections and clamps to the manifold:
   - EVAP control canister purge valve connector
   - MAP sensor connector
   - TP sensor connector
   - IAC valve connector
   - IAT sensor connector

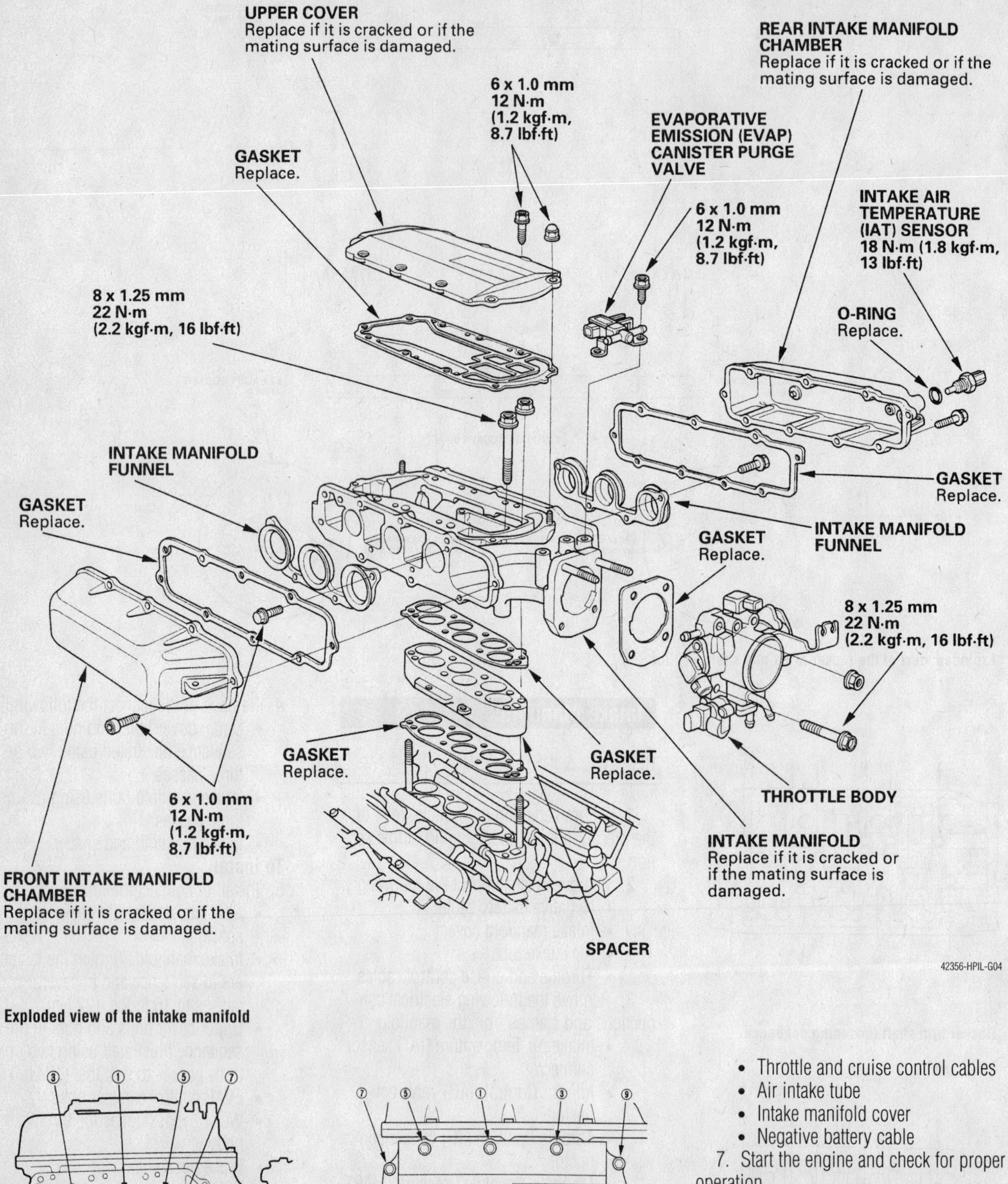

UPPER COVER
Replace if it is cracked or if the mating surface is damaged.

REAR INTAKE MANIFOLD CHAMBER
Replace if it is cracked or if the mating surface is damaged.

6 x 1.0 mm
12 N·m
(1.2 kgf·m, 8.7 lbf·ft)

EVAPORATIVE EMISSION (EVAP) CANISTER PURGE VALVE

GASKET
Replace.

6 x 1.0 mm
12 N·m
(1.2 kgf·m, 8.7 lbf·ft)

INTAKE AIR TEMPERATURE (IAT) SENSOR
18 N·m (1.8 kgf·m, 13 lbf·ft)

8 x 1.25 mm
22 N·m
(2.2 kgf·m, 16 lbf·ft)

O-RING
Replace.

INTAKE MANIFOLD FUNNEL

GASKET
Replace.

GASKET
Replace.

GASKET
Replace.

INTAKE MANIFOLD FUNNEL

GASKET
Replace.

8 x 1.25 mm
22 N·m
(2.2 kgf·m, 16 lbf·ft)

GASKET
Replace.

GASKET
Replace.

THROTTLE BODY

6 x 1.0 mm
12 N·m
(1.2 kgf·m, 8.7 lbf·ft)

FRONT INTAKE MANIFOLD CHAMBER
Replace if it is cracked or if the mating surface is damaged.

INTAKE MANIFOLD
Replace if it is cracked or if the mating surface is damaged.

SPACER

42356-HPIL-G04

Exploded view of the intake manifold

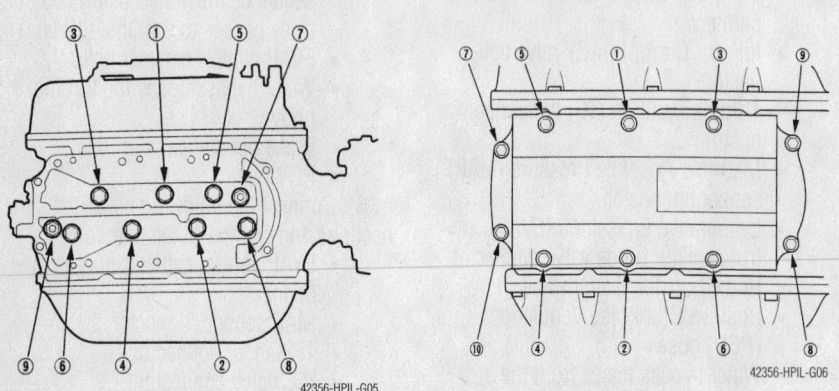

42356-HPIL-G05

Intake manifold torque sequence

42356-HPIL-G06

Upper cover torque sequence

- Throttle and cruise control cables
- Air intake tube
- Intake manifold cover
- Negative battery cable

7. Start the engine and check for proper operation.

Exhaust Manifold

REMOVAL & INSTALLATION

1. Before servicing the vehicle, refer to the precautions in the beginning of this section.

2. Remove or disconnect the following:
- Negative battery cable
- Exhaust manifold heat shield
- Heated Oxygen (HO2S) sensor connector
- Exhaust front pipe
- Exhaust manifold bracket, if equipped
- Exhaust manifold

To install:

3. Install or connect the following:
- Exhaust manifold. Tighten the fasteners to 23 ft. lbs. (31 Nm).
- Exhaust manifold bracket, if equipped. Tighten the bolts to 33 ft. lbs. (44 Nm).
- Exhaust front pipe. Tighten the nuts to 40 ft. lbs. (55 Nm).
- Heated Oxygen (HO2S) sensor connector
- Exhaust manifold heat shield
- Negative battery cable

Front Crankshaft Seal

REMOVAL & INSTALLATION

1. Before servicing the vehicle, refer to the precautions in the beginning of this section.

2. Remove or disconnect the following:
- Negative battery cable
- Accessory drive belts
- Side engine mount
- Valve cover
- Crankshaft pulley
- Front cover
- Balance shaft belt, if equipped
- Timing belt
- Top Dead Center (TDC) sensor, if equipped
- Crankshaft timing sprocket
- Front crankshaft seal

To install:

3. Lubricate the crankshaft seal lip with grease prior to installation.

4. Install the front crankshaft seal so that it is flush with the surface of the oil pump housing.

5. Install or connect the following:
- Crankshaft timing sprocket
- Top Dead Center (TDC) sensor, if equipped
- Timing belt
- Balance shaft belt, if equipped
- Front cover
- Crankshaft pulley. Tighten the bolt to 181 ft. lbs. (245 Nm).
- Valve cover
- Side engine mount
- Accessory drive belts
- Negative battery cable

6. Check the engine oil level and add if necessary.

7. Start the engine and check for leaks.

Camshaft

REMOVAL & INSTALLATION

Front

1. Before servicing the vehicle, refer to the precautions in the beginning of this section.

2. Remove or disconnect the following:
- Negative and positive battery cables
- Battery

3. Drain the coolant.
- Exhaust Gas Recirculation (EGR) valve
- Timing belt
- Rocker arm assembly
- Front camshaft pulley
- Thrust plate and camshaft

To install:

4. Install or connect the following:
- Camshaft using a new O-ring. Tighten the thrust plate to 16 ft. lbs. (22 Nm).
- Front camshaft pulley

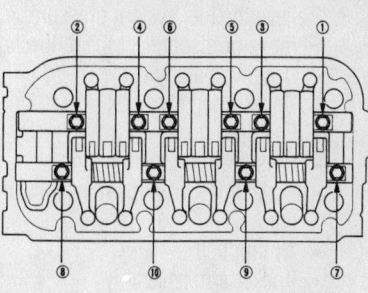

Rocker arm shaft loosening sequence

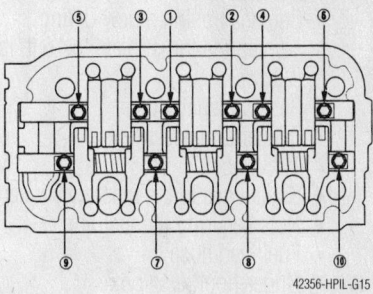

Rocker shaft tightening sequence

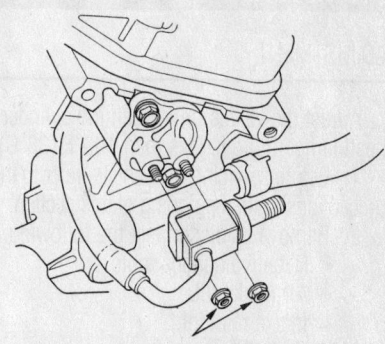

6 x 1.0 mm
12 N·m (1.2 kgf·m, 8.7 lbf·ft)

42356-HPIL-G12

Remove the nuts attaching the fuel line when removing the rear camshaft

- Rocker arm assembly
- Timing belt
- Exhaust Gas Recirculation (EGR) valve
- Battery
- Positive, then negative battery cables

5. Fill the cooling system.
- Camshaft

To install:

6. Start the engine and check for leaks.

Rear

1. Before servicing the vehicle, refer to the precautions in the beginning of this section.

2. Drain the cooling system.

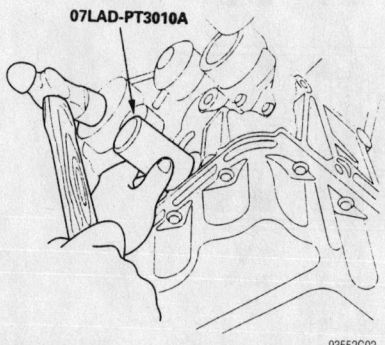

07LAD-PT3010A

Front crankshaft seal installation

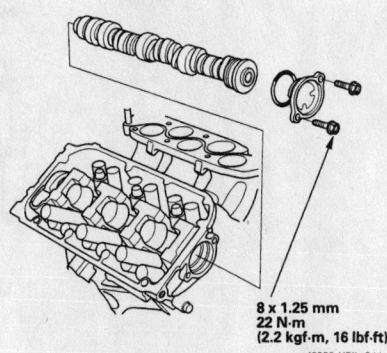

8 x 1.25 mm
22 N·m
(2.2 kgf·m, 16 lbf·ft)

42356-HPIL-G11

Front camshaft assembly

3. Relieve the fuel system pressure.
4. Remove or disconnect the following:
   - Negative battery cable
   - Under-hood fuse box
   - Fuel feed hose
   - Nuts securing the fuel line
   - Brake lines from the master cylinder
   - Timing belt
   - Rocker arm assembly
   - Rear camshaft pulley
   - Thrust plate and camshaft

To install:

5. Install or connect the following:
   - Camshaft using a new O-ring. Tighten the thrust plate to 16 ft. lbs. (22 Nm).
   - Rear camshaft pulley
   - Rocker arm assembly
   - Timing belt
   - Brake lines to the master cylinder
   - Nuts securing the fuel line
   - Fuel feed hose
   - Under-hood fuse box
   - Negative battery cable

Valve Lash

ADJUSTMENT

Adjust the valves only when the cylinder head temperature is less than 100°F (38°C).

1. Before servicing the vehicle, refer to the precautions in the beginning of this section.
2. Remove or disconnect the following:
   - Negative battery cable
   - Air intake tube
   - Intake manifold
   - Valve cover
3. Rotate the crankshaft so that the valves to be adjusted are closed and the rocker arm is contacting the camshaft lobe base circle.
4. Measure the valve clearance. If adjustment is necessary, loosen the locknut and turn the adjusting screw as necessary to achieve the correct valve clearance.
5. The correct valve clearance is:
   - Intake valves: 0.008–0.009 inches (0.20–0.24mm)
   - Exhaust valves: 0.011–0.013 inches (0.28–0.32mm)
6. After adjustment, tighten the locknuts to 14 ft. lbs. (20 Nm).
7. Install or connect the following:
   - Valve cover
   - Intake manifold
   - Air intake tube
   - Negative battery cable
8. Start the engine and check for proper operation.

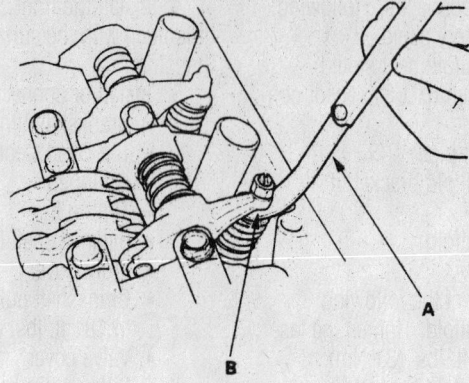

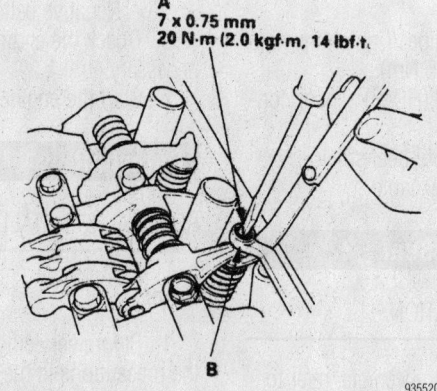

93552G04

Inspect the valve clearance, adjust to specification and tighten the retainer to specification

Adjusting screw locations:

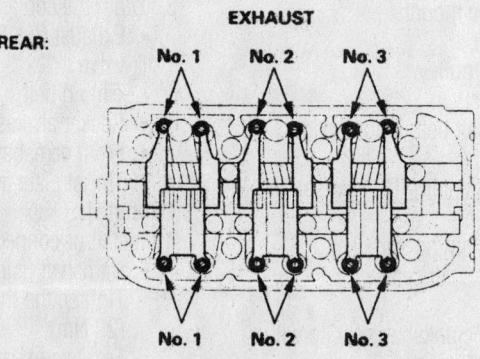

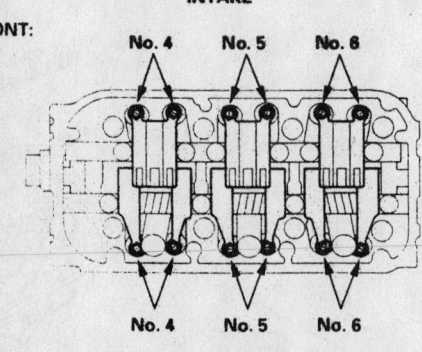

93552G03

Valve adjusting retainer locations

Starter Motor

REMOVAL & INSTALLATION

1. Before servicing the vehicle, refer to the precautions in the beginning of this section.
2. Remove or disconnect the following:
   - Negative battery cable and wait at least 3 minutes
   - Unlock the transmission fluid cooler hose clamp
   - Starter wiring harness connectors
   - Starter motor

To install:

3. Install or connect the following:
   - Starter motor. Tighten the 10mm bolt to 33 ft. lbs. (44 Nm) and the 12mm bolt to 47 ft. lbs. (64 Nm).
   - Starter wiring harness connectors. Tighten the battery cable nut to 79 inch lbs. (9 Nm).
   - Lock the transmission fluid cooler hose clamp
   - Negative battery cable

Oil Pan

REMOVAL & INSTALLATION

1. Before servicing the vehicle, refer to the precautions in the beginning of this section.
2. Drain the engine oil and power steering system.
3. Remove or disconnect the following:
   - Negative battery cable
   - Power steering pump outlet hose from the pump and the hose clamp
   - Steering joint cover, mark the steering joint-to-gearbox pinion shaft for reference
   - Steering joint from the pinion shaft
   - Splash shield
   - Transfer assembly
   - Tie rod ends from the knuckles
   - Lower arm ball joints from the knuckles
   - Power steering hose
   - Power steering pressure switch connector
   - Nuts attaching the transmission lower front and rear mount
   - Bolt attaching the rear mount
4. Support the engine with a hoist.
   - Nut attaching the front mount bracket

5. Make reference marks on the body across the marks on the edge of the front subframe.
   - Front subframe
   - Torque converter cover and the 2 bolts retaining the transmission
   - Oil pan

To install:

6. Install the oil pan. Apply liquid gasket as shown.
7. Tighten the bolts in sequence to 105 inch lbs. (12 Nm) using several passes.
8. Install or connect the following:
   - 2 bolts retaining the transmission and tighten to 28 ft. lbs. (38 Nm)
   - Torque converter cover and tighten the bolts to 9 ft. lbs. (12 Nm)
9. Align the reference marks on the body across the marks on the edge of the front subframe.
   - Front subframe. Use new bolts and tighten the 14mm bolts to 76 ft. lbs. (103 Nm). Tighten the front brace bolts to 54 ft. lbs. (74 Nm) and the rear brace bolts to 86 ft. lbs. (117 Nm).
   - Nut attaching the front mount bracket to 40 ft. lbs. (54 Nm)
   - Bolt attaching the rear mount to 28 ft. lbs. (38 Nm)
   - Nuts attaching the transmission lower front and rear mount to 28 ft. lbs. (38 Nm)

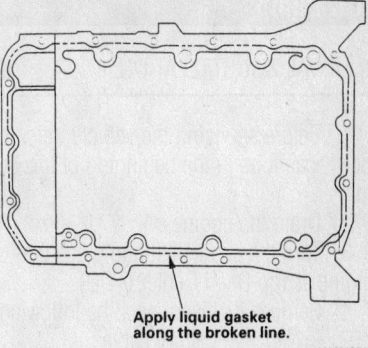

Apply liquid gasket along the broken line.

9302MG75

Apply liquid gasket to the inner threads of the bolt holes and the engine block along the area indicated by the broken line—Pilot

- Power steering pressure switch connector
- Power steering hose
- Lower arm ball joints to the knuckles
- Tie rod ends to the knuckles
- Transfer assembly
- Splash shield
- Steering joint to the pinion shaft
- Steering joint cover
- Power steering pump outlet hose clamp and hose
- Negative battery cable

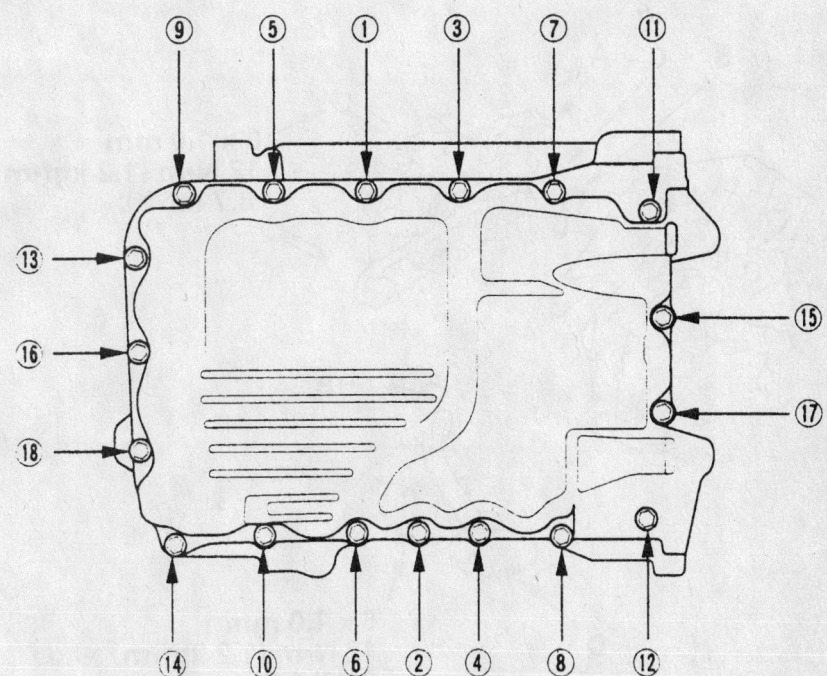

Oil pan tightening sequence—Pilot

9302MG76

Oil Pump

REMOVAL & INSTALLATION

1. Before servicing the vehicle, refer to the precautions in the beginning of this section.
2. Drain the engine oil.
3. Turn the crankshaft and place the engine at Top Dead Center (TDC).
4. Remove or disconnect the following:
   - Negative battery cable
   - Accessory drive belts
   - Front cover
   - Timing belt
   - Timing belt idler pulley
   - Crankshaft Position (CKP) sensor
   - Crankshaft timing sprocket
   - Variable Valve Timing and Valve Lift Electronic Control (VTEC) solenoid valve connector
   - Oil filter adapter
   - Oil pan
   - Oil pump pickup tube
   - Oil pump

To install:

➡Use new gaskets and O-ring seals for assembly.

5. Apply liquid gasket to the oil pump and to the bolt hole threads.
6. Install or connect the following:
   - Oil pump. Tighten the bolts to 9 ft. lbs. (12 Nm).

- Oil pump pickup tube. Tighten the bolts to 9 ft. lbs. (12 Nm).
- Oil pan
- Oil filter adapter
- VTEC solenoid valve connector
- Crankshaft timing sprocket
- CKP sensor
- Timing belt idler pulley
- Timing belt
- Front cover
- Accessory drive belts
- Negative battery cable

7. Fill the crankcase to the correct level.
8. Start the engine and check for leaks.

Rear Main Seal

REMOVAL & INSTALLATION

1. Before servicing the vehicle, refer to the precautions in the beginning of this section.
2. Remove or disconnect the following:
   - Transaxle
   - Flywheel
   - Oil seal

To install:

3. Install or connect the following:
   - Oil seal. Drive the seal square into the seal case.
   - Flywheel. Tighten the bolts in a crossing pattern to 54 ft. lbs. (73 Nm).
   - Transaxle

4. Check the fluid levels.
5. Start the engine and check for leaks.

Timing Belt

REMOVAL & INSTALLATION

1. Before servicing the vehicle, refer to the precautions in the beginning of this section.
2. Turn the crankshaft so the white mark aligns with the pointer.
3. Make sure the number 1 piston is at Top Dead center (TDC).
4. Remove or disconnect the following:
   - Negative battery cable
   - Wheels and splash shield
   - Drive belts
5. Support the engine with a block of wood and a jack under the oil pan.
   - Upper side engine mount
   - Dipstick tube
6. Remove the crankshaft pulley using holder tool shown in the accompanying illustration and a breaker and socket, loosen the 19mm bolt and remove the pulley.
   - Front upper cover, rear upper cover and the lower cover
   - One of the battery clamp bolts and grind the end as illustrated

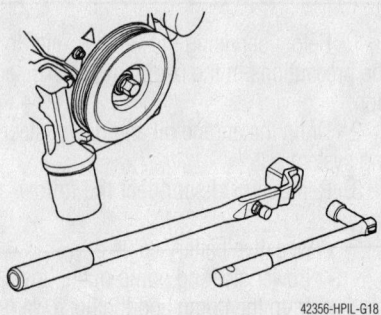

42356-HPIL-G18

Remove the crankshaft pulley using holder tool shown

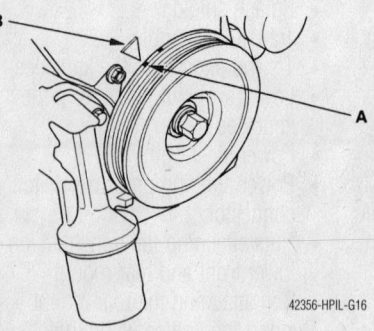

42356-HPIL-G16

Turn the crankshaft so the white mark (A) aligns with the pointer (B)

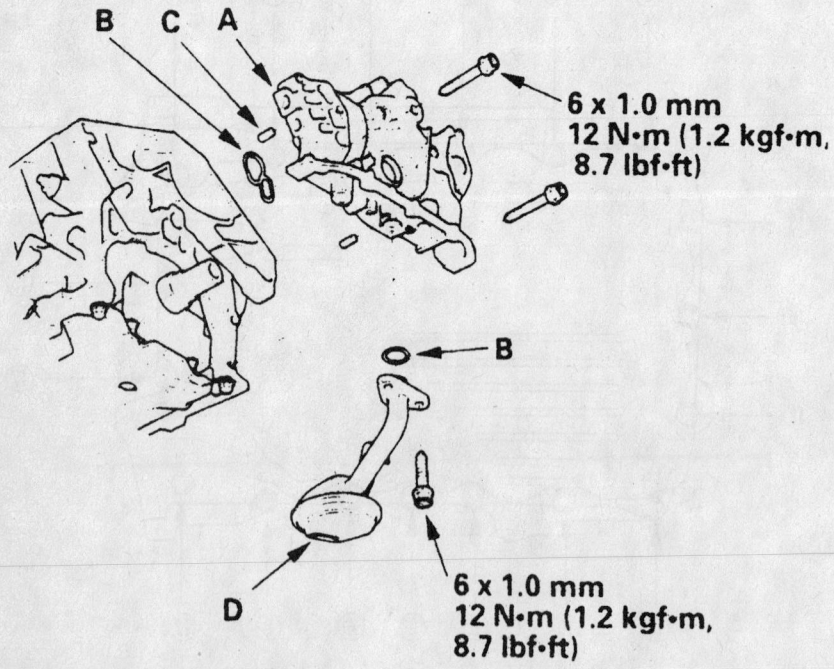

B   C   A

**6 x 1.0 mm
12 N•m (1.2 kgf•m, 8.7 lbf•ft)**

B

D

**6 x 1.0 mm
12 N•m (1.2 kgf•m, 8.7 lbf•ft)**

93552G05

Exploded view of the oil pump assembly

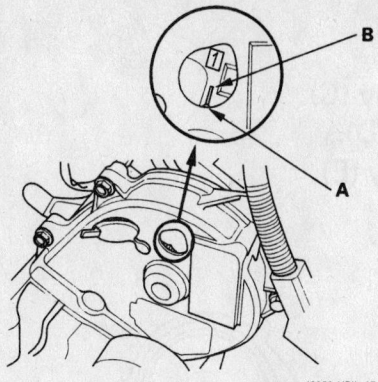

Make sure the number 1 piston is at top dead center (A) on the front camshaft pulley and pointer (B)

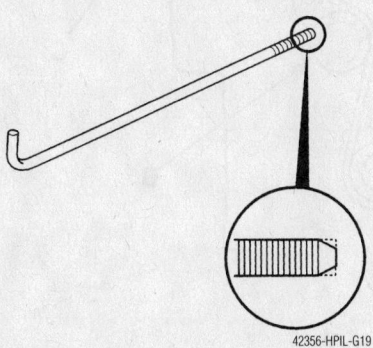

Remove a battery clamp bolt and grind the end as shown

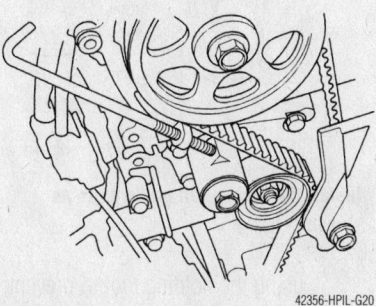

Install the battery clamp bolt as shown to hold the belt adjuster in position

7. Screw the battery clamp bolt as illustrated to hold the belt adjuster in position. Do not use a wrench, hand tighten only.
• Lower side engine mount
• Idler pulley bolt and the pulley
• Timing belt

To install:
8. If installing a new belt, perform the following steps:
a. Clean the pulleys, belt guide plate and the upper and lower covers.
b. Set the timing belt drive pulley to TDC by aligning the TDC mark on the

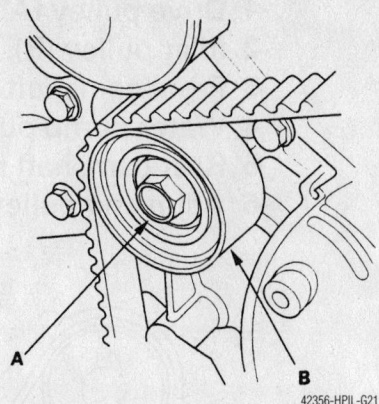

Remove the idler pulley bolt (A), pulley (B) and the timing belt

tooth of the belt drive pulley with the pointer on the oil pump.
c. Set the camshaft pulleys to TDC by aligning the TDC marks on the camshaft pulleys with the pointers on the back covers.
d. Remove the battery clamp bolt.
e. Remove the belt tensioner.
f. Align the holes on the rod and housing of the tensioner.
g. Using a press or other suitable

FRONT:

REAR:

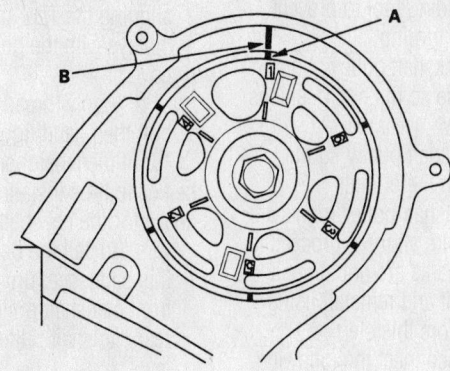
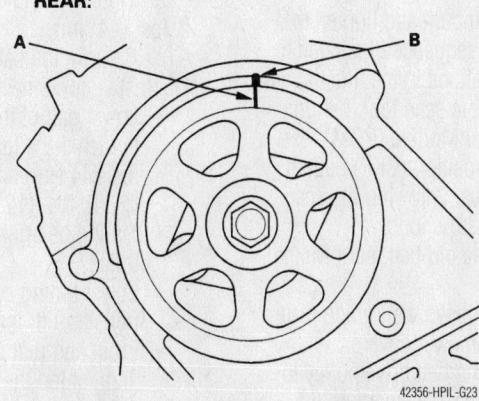

Set the camshaft pulleys to TDC by aligning the TDC marks (A) on the camshaft pulleys with the pointers (B) on the back covers

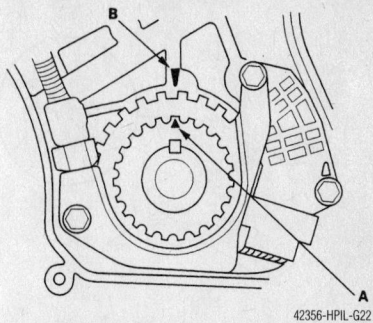

Set the timing belt pulley to TDC by aligning the TDC mark (A) on the tooth of the belt pulley with the pointer (B) on the oil pump

device, slowly compress the tensioner and insert a 0.08 inch (2mm) pin through the housing and rod.
h. Install the tensioner making sure the pin is still installed.
i. Apply thread locker to idler pulley bolt then hand tighten the bolt.
j. Install the belt over the pulleys in this sequence; drive pulley, idler pulley, front camshaft pulley, water pump pulley, rear camshaft pulley and adjusting pulley.

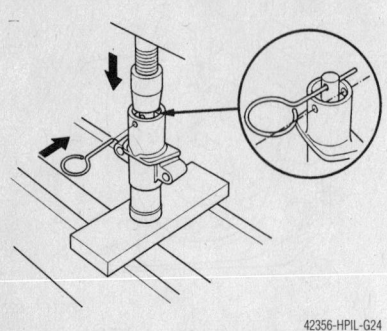

42356-HPIL-G24

Insert a 0.08 inch (2mm) pin through the tensioner housing and rod

    k. Tighten the idler pulley bolt to 33 ft. lbs. (44 Nm).

    l. Remove the pin from the tensioner.

9. Install or connect the following:

- Lower half of the side mount and tighten the 3 long bolts to 33 ft. lbs. (44 Nm) and the one short bolt to 9 ft. lbs. (12 Nm)
- Timing belt guide plate as illustrated
- Lower timing cover and tighten the bolts to 9 ft. lbs. (12 Nm)
- Front and rear upper timing covers and tighten the bolts to 9 ft. lbs. (12 Nm)
- Crankshaft pulley and tighten the bolts to 181 ft. lbs. (245 Nm), using the holding tool to prevent the unit from turning

10. Rotate the crankshaft pulley about 5 or 6 degrees clockwise so the belt positions itself on the pulleys.

11. Turn the crankshaft pulley so the white mark aligns with the pointer.

12. Check the camshaft pulley marks are aligned. If the marks are aligned, proceed to the next step. If the marks are not aligned, remove the timing belt and reinstall using the steps outlined before this step.

13. Remove or disconnect the following:

- Drive belt
- Upper side mount and tighten the bolts in the sequence illustrated to the specifications in the illustration

14. Using a suitable scan tool, perform the Powertain Control Module (PCM) reset and the Crankshaft position (CKP) pattern clear/learn procedures, following the scan tool manufactures instructions.

15. If installing the old belt, perform the following steps:

    a. Clean the pulleys, belt guide plate and the upper and lower covers.

    b. Set the timing belt drive pulley to TDC by aligning the TDC mark on the tooth of the belt drive pulley with the pointer on the oil pump.

-1 Drive pulley (A).
-2 Idler pulley (B).
-3 Front camshaft pulley (C).
-4 Water pump pulley (D).
-5 Rear camshaft pulley (E).
-6 Adjusting pulley (F).

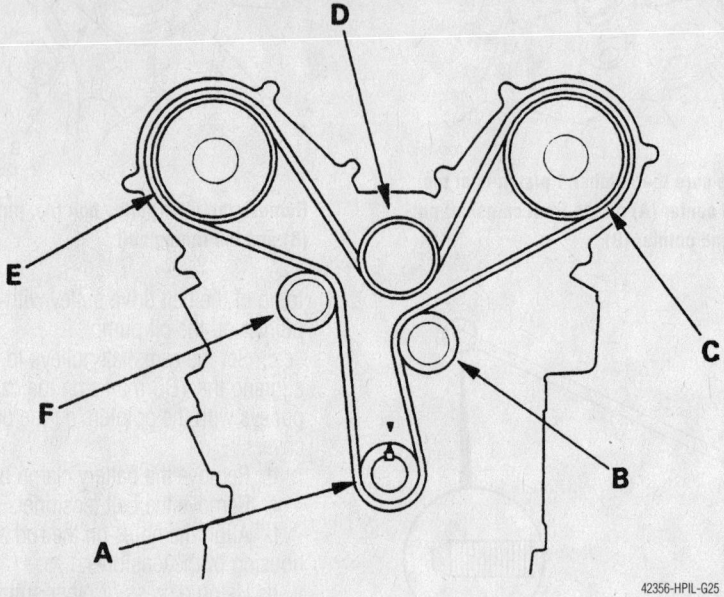

42356-HPIL-G25

Route the belt as shown in the sequence listed

    c. Set the camshaft pulleys to TDC by aligning the TDC marks on the camshaft pulleys with the pointers on the back covers.

    d. Apply thread locker to idler pulley bolt then hand tighten the bolt.

16. If the tensioner was extended and the belt cannot be installed, perform the steps above for the new belt installation.

    a. Install the belt over the pulleys in this sequence; drive pulley, idler pulley, front camshaft pulley, water pump pulley, rear camshaft pulley and adjusting pulley.

    b. Tighten the idler pulley bolt to 33 ft. lbs. (44 Nm).

    c. Remove the battery clamp bolt.

17. Install or connect the following:

- Lower half of the side mount and tighten the 3 long bolts to 33 ft. lbs. (44 Nm) and the one short bolt to 9 ft. lbs. (12 Nm)
- Timing belt guide plate as illustrated
- Lower timing cover and tighten the bolts to 9 ft. lbs. (12 Nm)
- Front and rear upper timing covers and tighten the bolts to 9 ft. lbs. (12 Nm)
- Crankshaft pulley and tighten the bolts to 181 ft. lbs. (245 Nm),

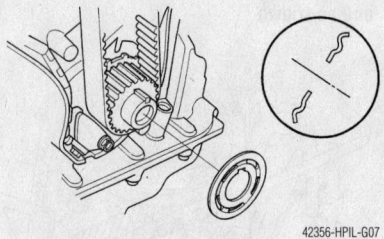

42356-HPIL-G07

Install the timing belt guide plate as shown

using the holding tool to prevent the

18. Rotate the crankshaft pulley about 5 or 6 degrees clockwise so the belt positions itself on the pulleys.

19. Turn the crankshaft pulley so the white mark aligns with the pointer.

20. Check the camshaft pulley marks are aligned. If the marks are aligned, proceed to the next step. If the marks are not aligned, remove the timing belt and reinstall using the steps outlined before this step.

21. Install or connect the following:

- Drive belt
- Upper side mount and tighten the bolts in the sequence illustrated to the specifications in the illustration
- Dipstick tube

FRONT CAMSHAFT PULLEY:

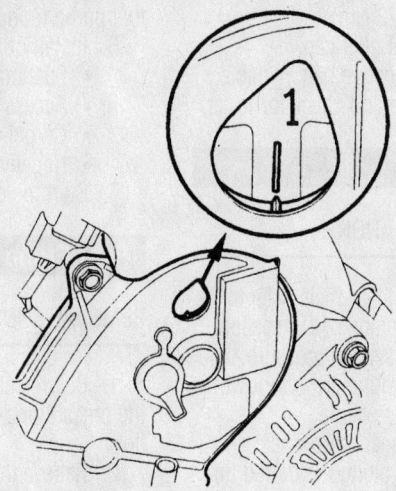

REAR CAMSHAFT PULLEY:

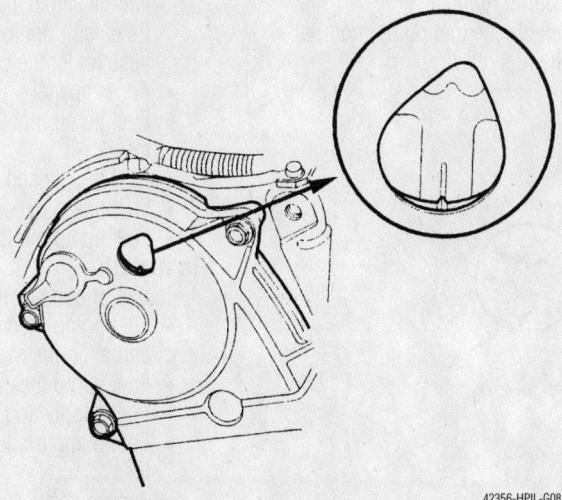

42356-HPIL-G08

Check that the camshaft pulley marks are aligned as shown

22. Using a suitable scan tool, perform the Powertain Control Module (PCM) reset and the Crankshaft position (CKP) pattern clear/learn procedures, following the scan tool manufactures instructions.

Piston and Ring

POSITIONING

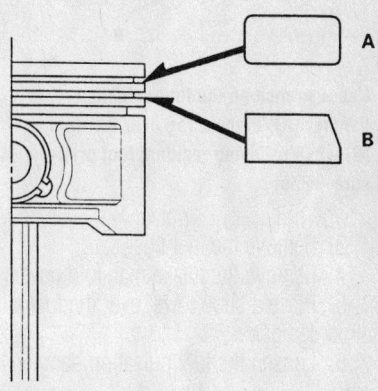

9302AG06

Compression ring identification—3.5L engine

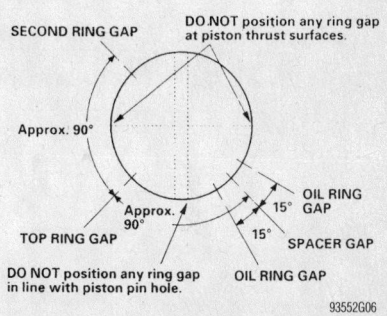

93552G06

Ring end gap positioning

FUEL SYSTEM

Fuel System Service Precautions

Safety is the most important factor when performing not only fuel system maintenance, but any type of maintenance. Failure to conduct maintenance and repairs in a safe manner may result in serious personal injury or death. Maintenance and testing of the vehicle's fuel system components can be accomplished safely and effectively by adhering to the following rules and guidelines:

• To avoid the possibility of fire and personal injury, always disconnect the negative battery cable unless the repair or test procedure requires that battery voltage be applied.

• Always relieve the fuel system pressure prior to disconnecting any fuel system component (injector, fuel rail, pressure regulator,

etc.), fitting or fuel line connection. Exercise extreme caution whenever relieving fuel system pressure to avoid exposing skin, face and eyes to fuel spray. Please be advised that fuel under pressure may penetrate the skin or any part of the body that it contacts.

• Always place a shop towel or cloth around the fitting or connection prior to loosening to absorb any excess fuel due to spillage. Ensure that all fuel spillage (should it occur) is quickly removed from engine surfaces. Ensure that all fuel soaked cloths or towels are deposited into a suitable waste container.

• Always keep a dry chemical (Class B) fire extinguisher near the work area.

• Do not allow fuel spray or fuel vapors to come into contact with a spark or open flame.

• Always use a backup wrench when loosening and tightening fuel line connection fittings. This will prevent unnecessary stress and torsion to fuel line piping. Always follow the proper torque specifications.

• Always replace worn fuel fitting O-rings with new. Do not substitute fuel hose or equivalent, where fuel pipe is installed.

Fuel System Pressure

RELIEVING

1. Before servicing the vehicle, refer to the precautions in the beginning of this section.
2. Disconnect the negative battery cable.

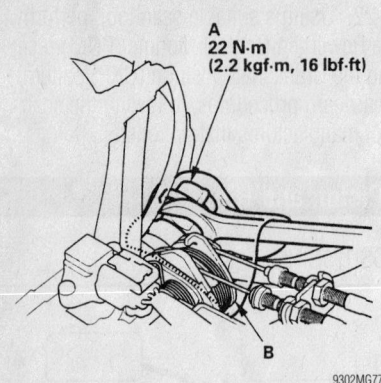

A
22 N·m
(2.2 kgf·m, 16 lbf·ft)

B

9302MG77

Use a wrench on the fuel pulsation damper (A). Place a rag over the damper (B) when relieving residual fuel pressure—Pilot

3. Remove the fuel filler cap.
4. Remove the intake manifold cover.
5. Place a shop towel over the fuel pulsation damper.
6. Loosen the fuel pulsation damper 1 turn.

7. When service is completed, replace the sealing washer and tighten the pulsation damper to 16 ft. lbs. (22 Nm).
8. Replace the fuel filler cap.
9. Connect the negative battery cable.
10. Start the engine and check for leaks.

Fuel Filter

REMOVAL & INSTALLATION

1. Before servicing the vehicle, refer to the precautions in the beginning of this section.
2. Relieve the fuel system pressure.
3. Remove or disconnect the following:
   - Negative battery cable
   - Driver's side second row seat and cut the carpet along the dotted line. Be careful not to cut the wiring harness under the carpet.
   - Access panel
   - Fuel pump module
4. Disassemble the fuel pump module and remove the fuel filter.

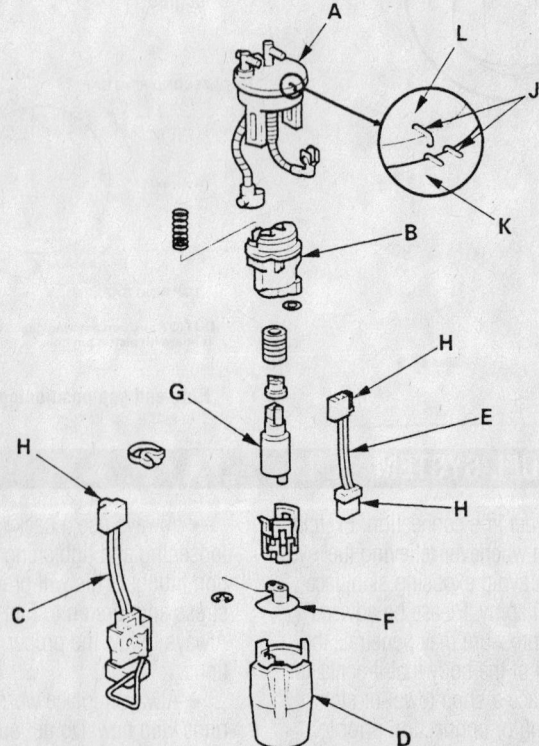

A. Bracket
B. Fuel filter
C. Fuel gauge sender
D. Case
E. Wire harness
F. Suction filter
G. Fuel pump
H. Connectors
J. Alignment marks
K. Fuel tank
L. Fuel pump module

9308MG26

Exploded view of the fuel pump module—Pilot

To install:
5. Install the fuel filter and assemble the fuel pump module.
6. Install or connect the following:
   - Fuel pump module
   - Access panel
   - Carpet and seat
   - Negative battery cable
7. Start the engine and check for leaks.

Fuel Pump

REMOVAL & INSTALLATION

1. Before servicing the vehicle, refer to the precautions in the beginning of this section.
2. Relieve the fuel system pressure.
3. Remove or disconnect the following:
   - Negative battery cable
   - Driver's side second row seat and cut the carpet along the dotted line. Be careful not to cut the wiring harness under the carpet.
   - Access panel
   - Fuel pump module wiring connector
   - Fuel supply and return lines
   - Fuel pump locknut
   - Fuel pump module

To install:
4. Install or connect the following:
   - Fuel pump module. Use a new seal and align the matchmarks.
   - Fuel pump locknut
   - Fuel supply and return lines
   - Fuel pump module wiring connector
   - Access panel
   - Carpet and seat
   - Negative battery cable
5. Start the engine and check for leaks.

Fuel Injector

REMOVAL & INSTALLATION

1. Before servicing the vehicle, refer to the precautions in the beginning of this section.
2. Relieve the fuel system pressure.
3. Remove or disconnect the following:
   - Negative battery cable
   - Intake manifold
   - Fuel lines
   - Fuel injector connectors
   - Fuel pressure regulator vacuum line
   - Fuel supply manifold
4. Separate the fuel injectors from the fuel supply manifold.

To install:

5. Install the fuel injectors to the fuel supply manifold with new cushion rings and O-rings.

6. Install new seal rings to the intake manifold.

7. Install or connect the following:
- Fuel supply manifold and injector assembly. Tighten the bolts to 86 inch lbs. (10 Nm).
- Fuel pressure regulator vacuum line

- Fuel injector connectors
- Fuel lines
- Intake manifold
- Negative battery cable

8. Start the engine and check for leaks.

DRIVE TRAIN

Transaxle Assembly

REMOVAL & INSTALLATION

1. Before servicing the vehicle, refer to the precautions in the beginning of this section.

2. Drain the transaxle.

3. Drain the power steering system.

4. Remove the engine appearance covers.

5. Remove the drivers side center console lower panel and pull back the cover to access steering joint cover.

6. Remove or disconnect the following:

- Steering joint bolt
- Steering joint from the steering gearbox pinion shaft
- Air intake assembly
- Battery
- Battery tray
- Power steering pump hose and the clamp bolt
- Transmission breather tube
- Cooler hose from the clamp on the starter
- Transaxle oil cooler lines
- Starter motor
- Shift control solenoid valve connectors
- Transaxle ground cable
- 8P connector from the bracket and the connector
- Clutch pressure switch connectors
- Joint connector and transmission range switch connector from the brackets

- Countershaft speed sensor connector
- Heated Oxygen (HO_2S) sensor connectors
- Transmission housing mounting bolts
- Nut from the front mount and the ground cable from the engine
- Bulkhead cover, windshield wiper arms, cowl cover sealing and cover
- Install a support fixture to the engine lifting eyes.
- Front sub-frame stiffener
- Primary HO_2S sensor clamp bracket from the transmission and harness from the clamp
- Exhaust front pipe
- Lower control arms from the knuckle
- Stabilizer bar links
- Tie rod ends from the knuckle
- Left driveshaft from the differential
- Right driveshaft from the intermediate shaft
- Propeller shaft from the companion flange
- Shift cable cover and holder

- Shift control cable and lever

7. Install a 6 x 1 x 14mm bolt and nut on the cable cover, then reinstall the cable cover to the torque converter housing. If this is not done, the bolt head of the cable cover may prevent torque converter removal.

- Transfer assembly
- Engine-to-torque converter bolts
- Power steering pressure switch connection
- Power steering hose clamp, then the hose from the pipe at the sub-frame
- Transmission lower mount nuts

8. Matchmark the front subframe to the vehicle body.

- Rear mount bracket bolts

9. Support the sub-frame with a 4 x 4 x 50 inch piece of wood and a jack.

- Sub-frame
- Transaxle lower mounts
- Driveshafts from the differential and intermediate shaft
- Intermediate shaft
- Transmission front mount bracket
- Transmission flange bolts
- Transmission

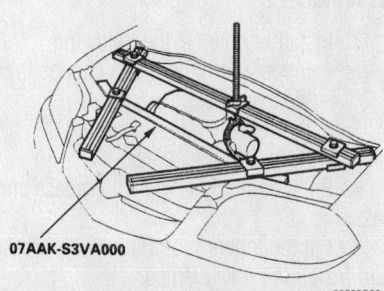

07AAK-S3VA000

93552G08

Support the engine while removing the transaxle—Pilot

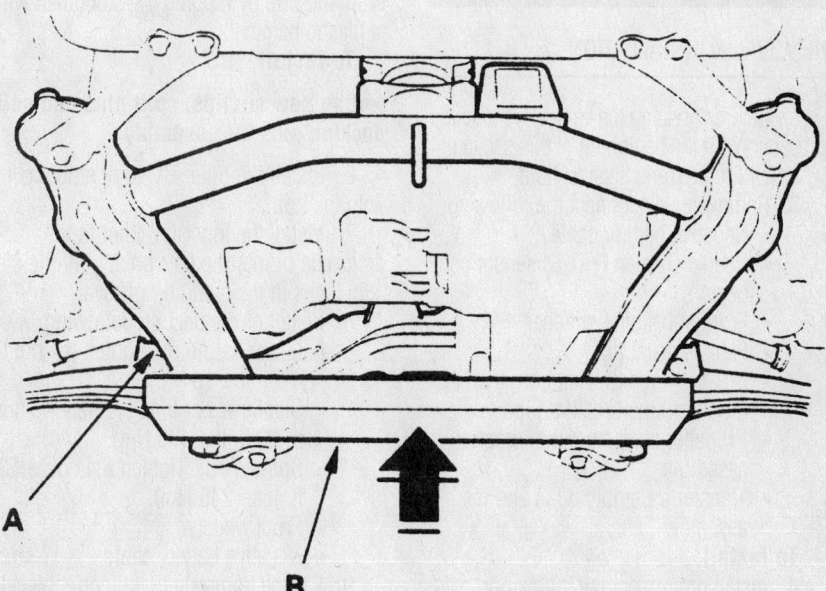

A

B

93552G09

Support the sub-frame with a 4 x 4 x 50 inch piece of wood and a jack

To install:

➡**Use new circlips, split pins and self-locking nuts for assembly.**

10. Installation is the reverse of removal. Please note the following specifications:

- Transmission housing bolts and harness clamp bolts to 47 ft. lbs. (64 Nm)
- Transmission housing bolts to 40 ft. lbs. (54 Nm)
- Front mount bracket bolts to 28 ft. lbs. (38 Nm)
- Intermediate shaft bolts to 29 ft. lbs. (39 Nm)
- Transfer assembly bolts to 33 ft. lbs. (44 Nm)

11. Raise the subframe into position and align the matchmarks. Tighten the subframe bolts to 76 ft. lbs. (103 Nm). Tighten the front subframe bracket bolts to 54 ft. lbs. (74 Nm) and the rear bracket bolts to 86 ft. lbs. (117 Nm).

- Rear engine mount bolts to 28 ft. lbs. (38 Nm)
- Engine-to-torque converter bolts. Tighten the 6 x 1 mm bolts to 105 inch lbs. (12 Nm), 10 x 1.25mm bolt to 28 ft. lbs. (38 Nm).
- Front motor mount nut to 40 ft. lbs. (54 Nm)

12. Fill the transaxle to the correct level.

13. Start the engine and check for leaks.

14. Check the wheel alignment and adjust as necessary.

Transfer Assembly

REMOVAL & INSTALLATION

1. Before servicing the vehicle, refer to the precautions in the beginning of this section.

2. Drain the transmission fluid.

3. Remove or disconnect the following:
- Negative battery cable
- Heated Oxygen (HO$_2$S) sensor connectors
- Front sub-frame stiffener
- Exhaust front pipe
- Breather tube bracket bolt, then the tube from the breather pipe
- Propeller shaft from the transfer assembly
- Transfer assembly bolts and the assembly

To install:

4. Install or connect the following:
- New O-ring on the transfer cover

- Dowel pin on the assembly
- Transfer assembly and tighten the bolts to 33 ft. lbs. (44 Nm) in a star pattern
- Propeller shaft
- Breather tube bracket, attach the tube with the dot facing outwards and tighten the bolt to 9 ft. lbs. (12 Nm)
- Exhaust front pipe
- Front sub-frame stiffener and tighten the bolts to 40 ft. lbs. (54 Nm)
- Heated Oxygen (HO$_2$S) sensor connectors
- Negative battery cable

Halfshaft

REMOVAL & INSTALLATION

1. Before servicing the vehicle, refer to the precautions in the beginning of this section.

2. Drain the transaxle if removing the left halfshaft. If is not necessary to drain the fluid if removing the right halfshaft.

3. Remove or disconnect the following:

- Negative battery cable
- Front wheels
- Spindle nut
- Stabilizer bar link
- Lower ball joint

4. Pry the inboard joint from the transaxle or intermediate shaft.

5. Remove the outer CV-joint stub shaft from the hub by tapping the stub shaft with a plastic hammer.

To install:

➡**Use new circlips, split pins and self-locking nuts for assembly.**

6. Install the outer CV-joint stub shaft into the hub.

7. Install the inner CV-joint to the transaxle or intermediate shaft until the circlip locks in the retaining groove.

8. Install or connect the following:
- Lower ball joint. Tighten the nut to 47 ft. lbs. (64 Nm).
- Stabilizer bar link. Tighten the nut to 58 ft. lbs. (78 Nm).
- Spindle nut. Tighten the nut to 181 ft. lbs. (245 Nm).
- Front wheels
- Negative battery cable

9. Fill the transaxle to the correct level and check for leaks.

CV-Joint

OVERHAUL

Front

OUTBOARD JOINT

1. Before servicing the vehicle, refer to the precautions in the beginning of this section.

2. Remove or disconnect the following:
- Axle halfshaft from the vehicle and place it in a vise
- Outboard joint boot clamps and push the boot back
- Outboard joint by driving it off the axle shaft with a brass drift and hammer
- Outboard joint boot

To install:

➡**Use new circlips and boot clamps for assembly.**

3. Install the outboard joint boot and clamps to the axle shaft.

4. Fill the outboard joint with grease. Install the outboard joint to the axle shaft. Tap the stub shaft with a brass hammer to seat the circlip.

5. Fill the outboard joint boot with grease and install the boot clamps.

6. Install the axle halfshaft to the vehicle.

INBOARD JOINT

1. Before servicing the vehicle, refer to the precautions in the beginning of this section.

2. Remove or disconnect the following:
- Axle halfshaft from the vehicle.
- Inboard joint boot clamps and push the boot back
- Inboard joint housing from the axle
- Rollers from the spider
- Snapring and the spider from the axle shaft
- Inboard joint boot

To install:

➡**Use new circlips and boot clamps for assembly.**

3. Install or connect the following:
- Inboard joint boot and clamps to the axle shaft
- Spider with a new snapring
- Rollers to the spider

4. Fill the joint housing with grease and install it.

5. Fill the inboard joint boot with grease and install the boot clamps.

6. Install the axle halfshaft to the vehicle.

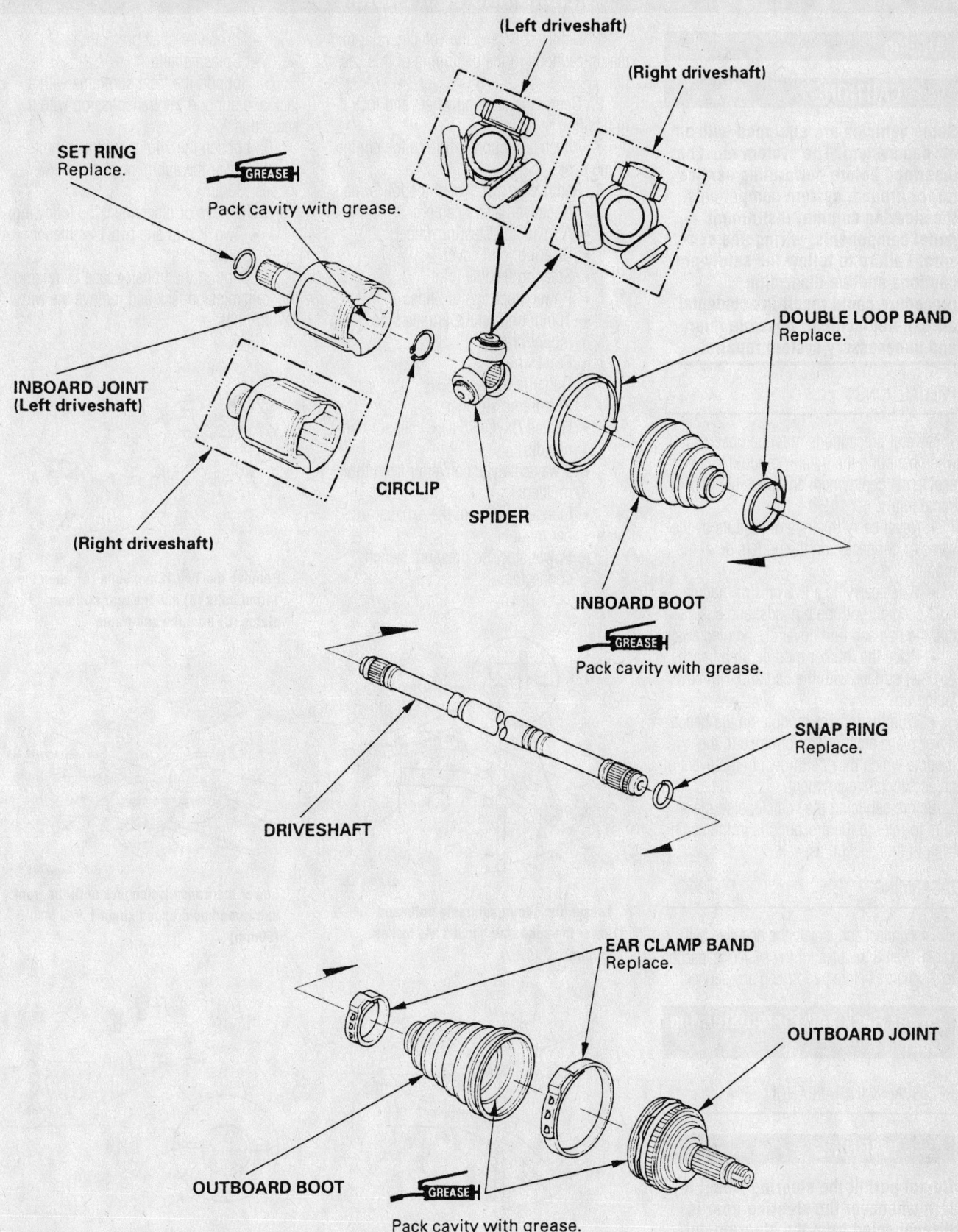

SET RING
Replace.

GREASE
Pack cavity with grease.

(Left driveshaft)

(Right driveshaft)

INBOARD JOINT
(Left driveshaft)

DOUBLE LOOP BAND
Replace.

(Right driveshaft)

CIRCLIP

SPIDER

INBOARD BOOT
GREASE
Pack cavity with grease.

SNAP RING
Replace.

DRIVESHAFT

EAR CLAMP BAND
Replace.

OUTBOARD JOINT

OUTBOARD BOOT

GREASE
Pack cavity with grease.

Front axle exploded view—Pilot

9308MG29

STEERING AND SUSPENSION

Air Bag

✳✳ CAUTION

Some vehicles are equipped with an air bag system. The system must be disarmed before performing service on, or around, system components, the steering column, instrument panel components, wiring and sensors. Failure to follow the safety precautions and the disarming procedure could result in accidental air bag deployment, possible injury and unnecessary system repairs.

PRECAUTIONS

Several precautions must be observed when handling the inflator module to avoid accidental deployment and possible personal injury.

• Never carry the inflator module by the wires or connector on the underside of the module.

• When carrying a live inflator module, hold securely with both hands, and ensure that the bag and trim cover are pointed away.

• Place the inflator module on a bench or other surface with the bag and trim cover facing up.

• With the inflator module on the bench, never place anything on or close to the module which may be thrown in the event of an accidental deployment.

Before servicing the vehicle, also make sure to refer to the precautions in the beginning of this section as well.

DISARMING

Disconnect and isolate the negative battery cable. Wait 3 minutes for the system capacitor to discharge before performing any service.

Power Rack and Pinion Steering Gear

REMOVAL & INSTALLATION

✳✳ WARNING

Do not permit the steering wheel to turn whenever the steering gear is disconnected from the steering column. Damage to the air bag wiring can result.

1. Before servicing the vehicle, refer to the precautions in the beginning of this section.

2. Center the steering wheel and lock it in position.

3. Attach a support fixture to the engine lifting eyes.

4. Remove or disconnect the following:
• Negative battery cable
• Air bag and steering wheel
• Steering joint cover
• Steering flexible joint
• Power steering fluid lines
• 10mm bolt on the engine side mount bracket
• Front wheels
• Outer tie rod ends
• Sub-frame stiffener
• Heated Oxygen (HO$_2$S) sensor connectors
• 3 way catalytic converter from the mufflers
• Flange bolts from the exhaust rubber mount
• Power steering pressure switch connector

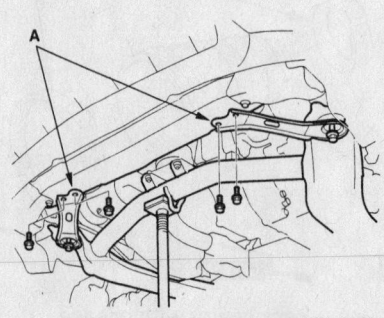

42356-HPIL-G26

Loosen the 14mm subframe bolts and lower the subframe about 1 ³⁄₁₆ inches (30mm)

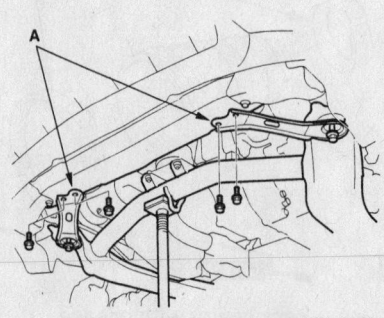

42356-HPIL-G27

Remove the Four 12mm stiffener plate bolts

• Propeller shaft protector
• Splash shield

5. Support the front subframe with a jack and support the transmission with a second jack.

6. Loosen the 14mm subframe bolts.

7. Lower the subframe about 1 ³⁄₁₆ inches (30mm).

8. Remove or disconnect the following:
• Two 12mm and two 14 stiffener plate bolts

9. Support the transfer case by raising the transmission jack and remove the two 12mm bolts.

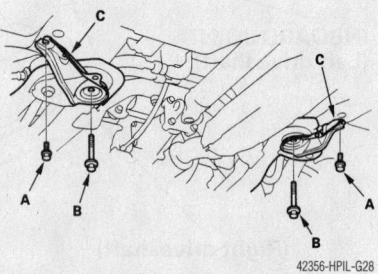

42356-HPIL-G28

Remove the Two 12mm bolts (A), then the 14mm bolts (B) and the rear stiffener plates (C) from the sub-frame

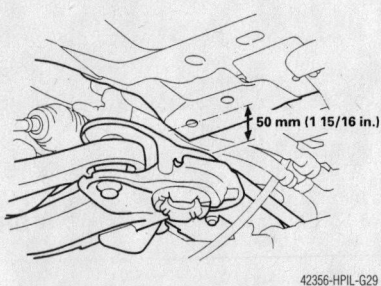

42356-HPIL-G29

Lower the transmission jack until the front subframe has dropped about 1 ¹⁵⁄₁₆ inch (50mm)

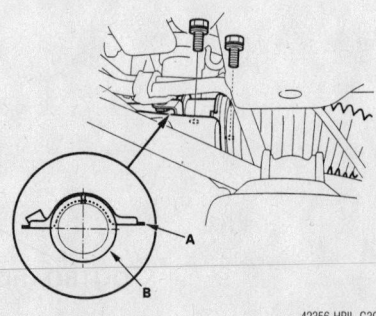

42356-HPIL-G30

Remove the two 10mm bolts from the right side gearbox and the mounting bracket and cushion

- Two 14mm bolts and the rear stiffener plats from the sub-frame

10. Lower the transmission jack until the front subframe has dropped about 1 ¹⁵⁄₁₆ inch (50mm).

- Power steering line brackets
- Feed line
- Return hose
- Two 10mm bolts from the right side gearbox
- Mounting bracket and cushion
- Two 10mm bolts from the left side gearbox

11. Lower the transmission jack until the front subframe has dropped about 3 ¹⁵⁄₁₆ inch (100mm).

- Gearbox stiffener bracket

12. Slide the gearbox between the body and front sub-frame towards the left and from the vehicle.

To install:

13. Position the steering gear in the vehicle.

14. Install or connect the following:

- Left steering gear mounting bolts. Tighten the bolts to 43 ft. lbs. (58 Nm).
- Right steering gear mounting bracket. Tighten the bolts to 29 ft. lbs. (39 Nm).
- Return hose
- Feed line
- Power steering line mounting brackets and tighten the bolts to 7 ft. lbs. (10 Nm)

15. Raise the subframe into position. Tighten the 14mm bolts to 76 ft. lbs. (103 Nm) and the 12mm bolts to 86 ft. lbs. (117 Nm).

16. Install or connect the following:

- Front stiffener plates. Tighten the 14mm bolts to 76 ft. lbs. (103 Nm) and the 12mm bolts to 54 ft. lbs. (74 Nm).
- Splash shield
- Propeller shaft protector
- Power steering pressure switch
- 3 way catalytic converter and mufflers. Tighten the nuts to 25 ft. lbs. (33 Nm)
- Rubber exhaust mount and tighten the bolts to 28 ft. lbs. (38 Nm)
- HO₂S sensor connectors
- Sub-frame stiffener plate
- 10mm flange bolts on the engine side mount bracket to 33 ft. lbs. (44 Nm)
- Power steering hoses
- Outer tie rod ends
- Front wheels. Position the wheels straight-ahead.
- Steering flexible joint. Tighten the pinch bolts to 16 ft. lbs. (22 Nm).

- Steering joint cover
- Negative battery cable

17. Fill the power steering system.

18. Check the wheel alignment and adjust as necessary.

Strut

REMOVAL & INSTALLATION

Front

1. Before servicing the vehicle, refer to the precautions in the beginning of this section.

2. Remove or disconnect the following:

- Front wheel
- Wheel speed sensor wiring bracket
- Brake hose bracket
- Stabilizer bar link
- Strut pinch bolts
- Upper mount nuts
- Strut

To install:

3. Install or connect the following:

- Strut. Tighten the upper mount nuts to 43 ft. lbs. (59 Nm).
- Strut pinch bolts. Tighten the nuts to 116 ft. lbs. (157 Nm).
- Stabilizer bar link. Tighten the nut to 58 ft. lbs. (78 Nm).
- Brake hose bracket
- Wheel speed sensor wiring bracket
- Front wheel

4. Check the wheel alignment and adjust as necessary.

Shock Absorber

REMOVAL & INSTALLATION

Rear

1. Before servicing the vehicle, refer to the precautions in the beginning of this section.

2. Support the vehicle under the lower control arm.

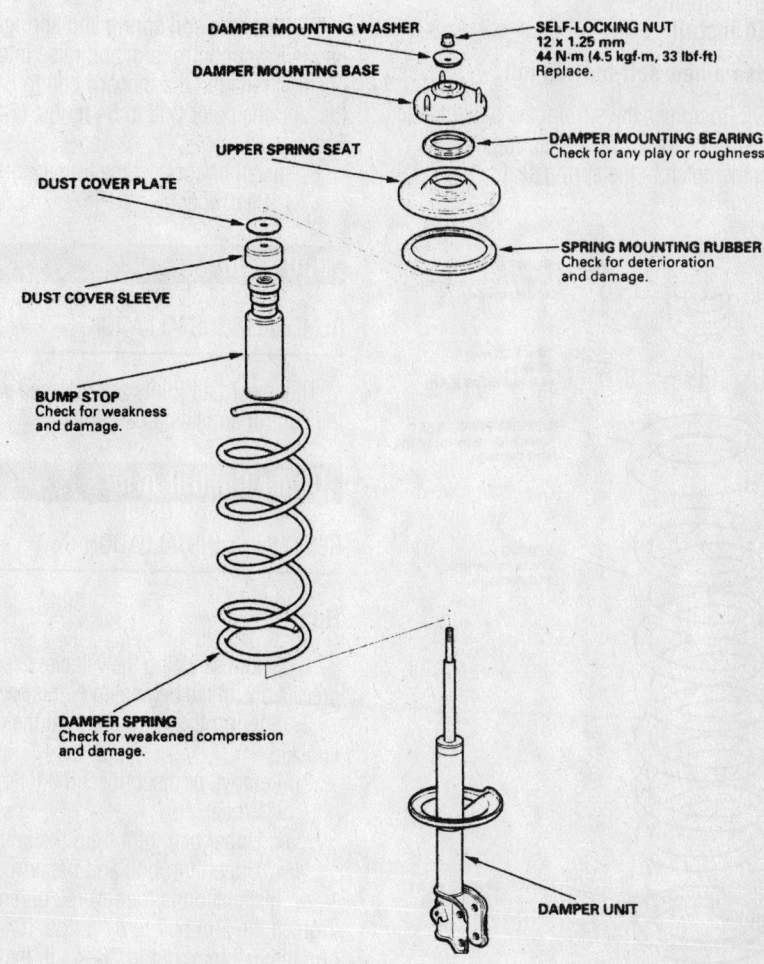

Exploded view of the front strut assembly

93552G11

3. Remove or disconnect the following:
- Rear wheel
- Upper shock absorber flange bolt
- Lower shock absorber nut
- Shock absorber

To install:

4. Install or connect the following:
- Shock absorber. Tighten the fasteners to 47 ft. lbs. (64 Nm).
- Rear wheel

Coil Spring

REMOVAL & INSTALLATION

Front

1. Before servicing the vehicle, refer to the precautions in the beginning of this section.

2. Remove the strut from the vehicle and install in a strut spring compressor. Compress the spring until the end of the spring comes away from the spring seat.

3. Remove the upper strut mount, spring seat and related components.

4. Remove the coil spring from the strut spring compressor.

To install:

➡**Use a new self-locking nut.**

5. Compress the spring and position the strut so that the end of the spring aligns with the notch in the spring seat.

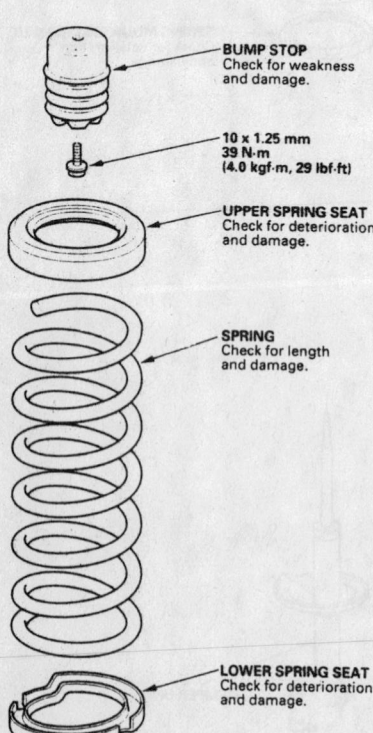

BUMP STOP
Check for weakness
and damage.

**10 x 1.25 mm
39 N·m
(4.0 kgf·m, 29 lbf·ft)**

UPPER SPRING SEAT
Check for deterioration
and damage.

SPRING
Check for length
and damage.

LOWER SPRING SEAT
Check for deterioration
and damage.

93552G12

Exploded view of the rear spring assembly

6. Install the upper strut mounting components and tighten the nut to 33 ft. lbs. (44 Nm).

7. Install the strut to the vehicle.

8. Check the wheel alignment and adjust as necessary.

Rear

1. Before servicing the vehicle, refer to the precautions in the beginning of this section.

2. Support the vehicle under the lower control arm.

3. Remove or disconnect the following:
- Rear wheel
- Stabilizer link from the lower arm
- Wheel speed sensor wiring harness from the lower arm. Do not disconnect the connector.
- Upper shock absorber flange bolt
- Lower control arm bolts

4. Lower the floor jack and remove the coil spring and spring seats.

To install:

➡**Use new self-locking nuts for assembly.**

5. Place the coil spring and spring seats on the lower control arm and raise into position. Tighten the inboard bolt to 61 ft. lbs. and the outer bolt to 54 ft. lbs. (74 Nm).

6. Install or connect the following:
- Rear wheel

Ball Joint

REMOVAL & INSTALLATION

The lower ball joints are replaced with the control arms as an assembly.

Upper Control Arm

REMOVAL & INSTALLATION

Rear

1. Before servicing the vehicle, refer to the precautions in the beginning of this section.

2. Support the control arm at the knuckle.

3. Remove or disconnect the following:
- Wheel
- Upper ball joint from the knuckle
- Upper arm bolt and the arm

4. Installation is the reverse of removal. Tighten the arm bolt to 47 ft. lbs. (64 Nm) and the ball joint nut to 36–43 ft. lbs. (49–59 Nm).

Lower Control Arm

REMOVAL & INSTALLATION

Front

1. Before servicing the vehicle, refer to the precautions in the beginning of this section.

2. Remove or disconnect the following:
- Front wheel
- Lower ball joint
- Front inner flange bolt
- Rear inner flange bolt
- Lower control arm

To install:

➡**Use a new split pin for assembly.**

3. Install or connect the following:
- Lower control arm. Tighten the inner flange bolts to 69 ft. lbs. (93 Nm).
- Lower ball joint. Tighten the nut to 43–51 ft. lbs. (59–69 Nm).
- Front wheel

4. Check the wheel alignment and adjust as necessary.

Rear

LOWER ARM (A)

1. Before servicing the vehicle, refer to the precautions in the beginning of this section.

2. Remove or disconnect the following:
- Lower arm mounting bolt and nut
- Lower arm

3. Installation is the reverse of removal. Tighten the bolt to 105 ft. lbs. (142 Nm) and the nut to 47 ft. lbs. (64 Nm).

LOWER ARM (B)

1. Before servicing the vehicle, refer to the precautions in the beginning of this section.

2. Support the control arm with a jack.

3. Remove or disconnect the following:
- Wheel
- Stabilizer link from the lower arm
- Wheel speed sensor wiring harness from the lower arm. Do not disconnect the connector.

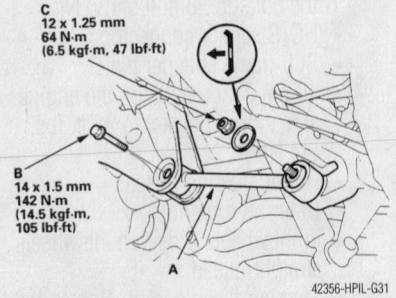

**C
12 x 1.25 mm
64 N·m
(6.5 kgf·m, 47 lbf·ft)**

**B
14 x 1.5 mm
142 N·m
(14.5 kgf·m,
105 lbf·ft)**

42356-HPIL-G31

Rear lower arm (A) mounting

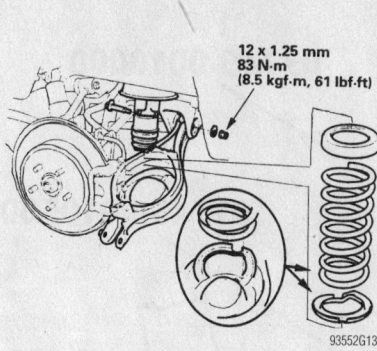

12 x 1.25 mm
83 N·m
(8.5 kgf·m, 61 lbf·ft)

93552G13

Rear lower arm (B) mounting

- Flange bolts that attaches the lower arm to the knuckle
4. Spring assembly
- Inner nuts and bolts and the arm
5. Install or connect the following:
- Arm, inner bolt and loosely install the nut
- Spring assembly
6. Raise the arm into position and install the flange bolt.
7. Raise the rear suspension with a floor jack to load the vehicle weight.
- Tighten the flange bolt to 54 ft. lbs. (74 Nm) and the inner nut and bolt to 61 ft. lbs. (83 Nm).

- Wheel speed sensor harness
- Wheel
8. Check the vehicle alignment.

CONTROL ARM BUSHING REPLACEMENT

The control arm bushings are serviced with the control arms as an assembly.

Wheel Bearings

ADJUSTMENT

The wheel bearings are sealed units and are not adjustable.

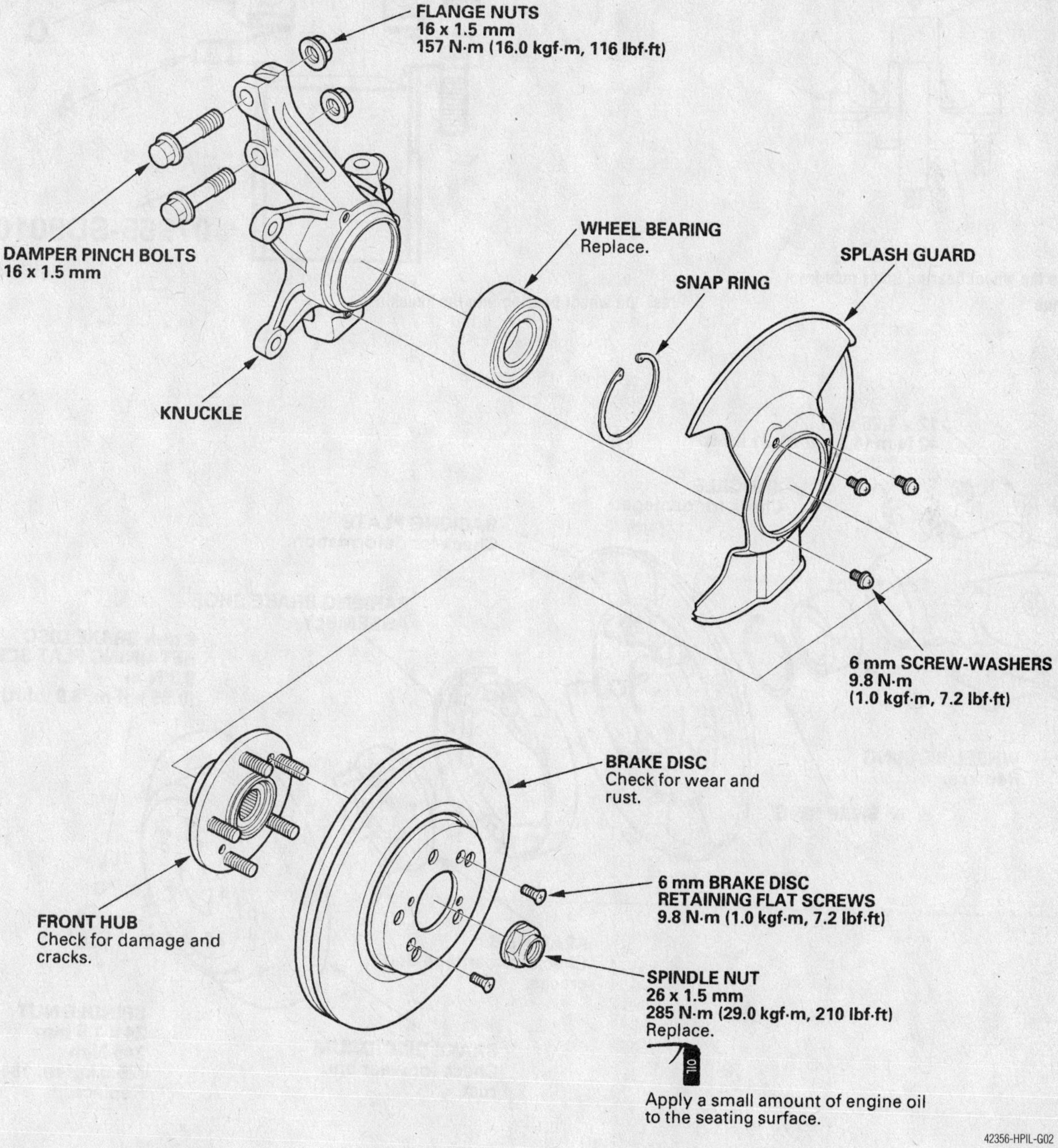

FLANGE NUTS
16 x 1.5 mm
157 N·m (16.0 kgf·m, 116 lbf·ft)

DAMPER PINCH BOLTS
16 x 1.5 mm

KNUCKLE

WHEEL BEARING
Replace.

SNAP RING

SPLASH GUARD

6 mm SCREW-WASHERS
9.8 N·m
(1.0 kgf·m, 7.2 lbf·ft)

FRONT HUB
Check for damage and cracks.

BRAKE DISC
Check for wear and rust.

6 mm BRAKE DISC RETAINING FLAT SCREWS
9.8 N·m (1.0 kgf·m, 7.2 lbf·ft)

SPINDLE NUT
26 x 1.5 mm
285 N·m (29.0 kgf·m, 210 lbf·ft)
Replace.

Apply a small amount of engine oil to the seating surface.

Front wheel bearing assembly

42356-HPIL-G02

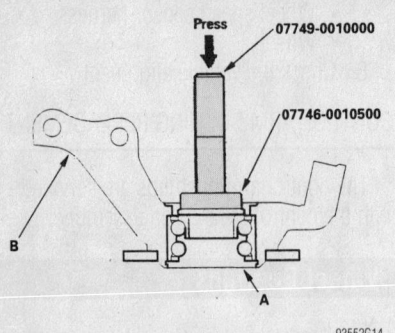

Press the wheel bearing out of the knuckle

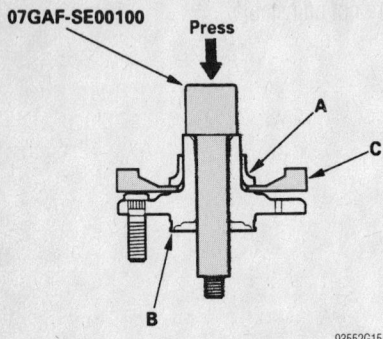

Press the wheel bearing inner race from the hub

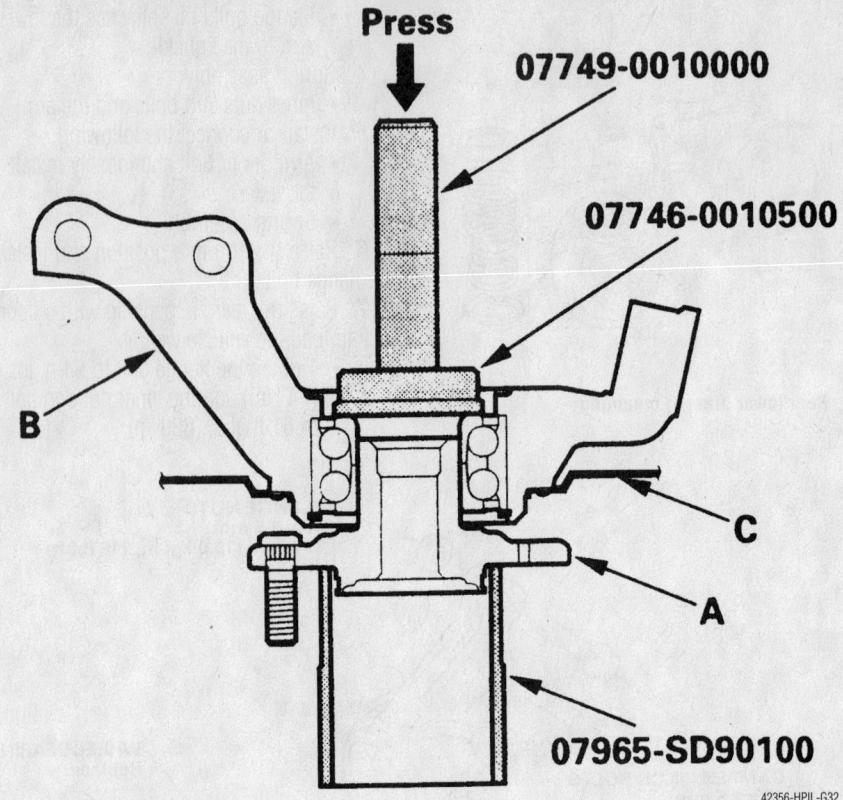

Press the wheel bearing into the knuckle

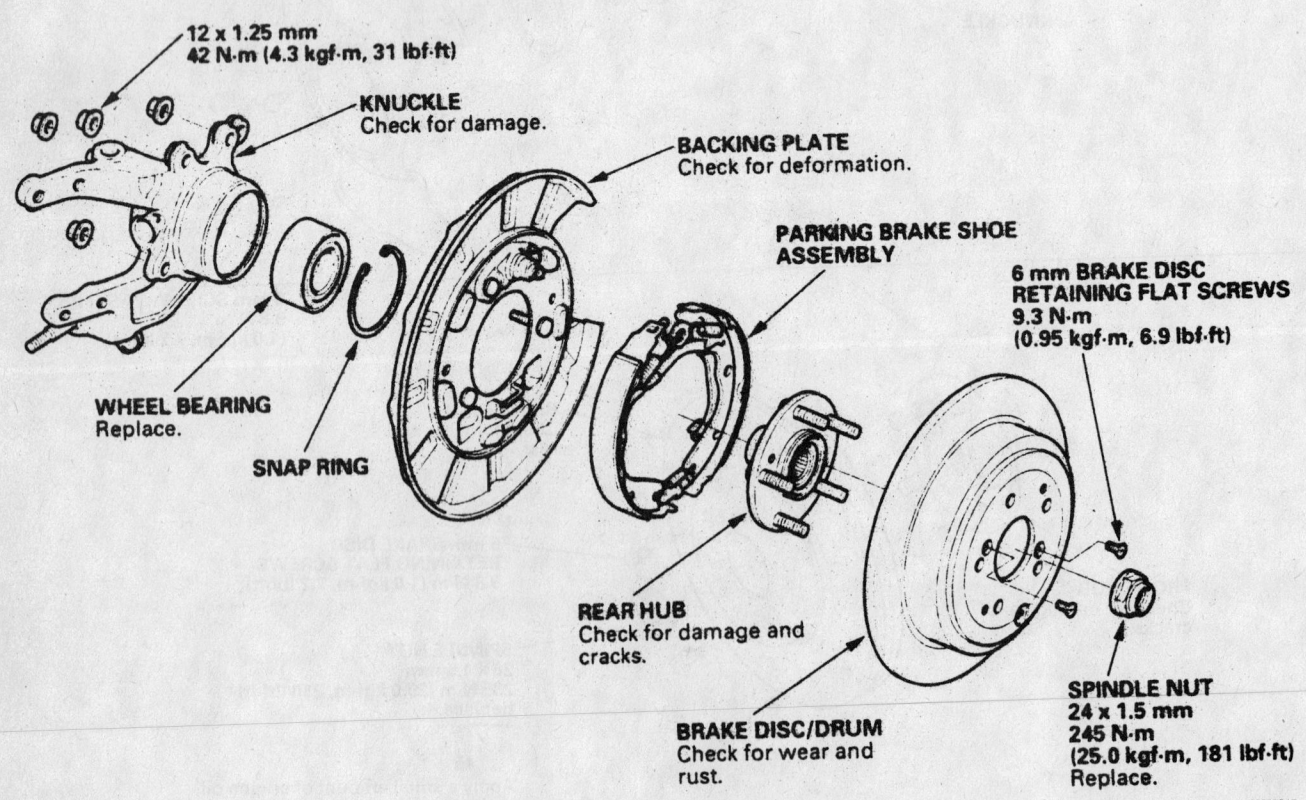

Exploded view of the rear wheel bearing assembly

REMOVAL & INSTALLATION

Front

1. Before servicing the vehicle, refer to the precautions in the beginning of this section.
2. Remove or disconnect the following:

- Front wheel
- Brake hose mounting bolt
- Brake caliper
- Wheel speed sensor
- Spindle nut
- Brake rotor
- Outer tie rod end
- Lower ball joint
- Steering knuckle

3. Press the hub out of the wheel bearing.

- Splash guard
- Snapring and press the wheel bearing out of the steering knuckle

4. If necessary, press the inner bearing race off of the hub.

To install:

➡**Use a new ball joint nut, split pin, snapring and spindle nut for assembly.**

5. Press the bearing into the steering knuckle and install the snapring.
6. Install the splash guard.
7. Press the hub into the bearing.
8. Install or connect the following:

- Steering knuckle. Tighten the ball

joint nut to 43–51 ft. lbs. (59–69 Nm) and the damper flange bolts to 116 ft. lbs. (157 Nm).

- Outer tie rod end. Tighten the nut to 40 ft. lbs. (54 Nm).
- Wheel speed sensor, if equipped
- Brake caliper and rotor
- Brake hose
- Spindle nut. Tighten the nut to 210 ft. lbs. (285 Nm).
- Front wheel

9. Check the wheel alignment and adjust as necessary.

Rear

1. Before servicing the vehicle, refer to the precautions in the beginning of this section.
2. Remove or disconnect the following:

- Rear wheel
- Brake hose bracket mounting bolts from the trailing arm and the knuckle
- Brake caliper
- Wheel speed sensor
- Spindle nut
- Brake rotor
- Upper ball joint
- Lower arm (A)
- Lower arm (B) from the trailing arm

3. Support the lower arm (B)

- Steering knuckle

4. Press the hub out of the wheel bearing.

- Splash guard

- Snapring and press the wheel bearing out of the steering knuckle

5. If necessary, press the inner bearing race off of the hub.

To install:

➡**Use a new ball joint nut, split pin, snapring and spindle nut for assembly.**

6. Press the bearing into the steering knuckle and install the snapring.
7. Install the splash guard.
8. Press the hub into the bearing.
9. Install or connect the following:

- Steering knuckle. Tighten the flange bolt to 54 ft. lbs. (74 Nm) and the lower shock nut to 47 ft. lbs. (64 Nm)
- Lower arm (B) to the trailing arm and tighten the bolts to 47 ft. lbs. (64 Nm)
- Lower arm (A)
- Upper ball joint and tighten the nut to 40 ft. lbs. (54 Nm)
- Brake rotor and tighten the screws to 7 ft. lbs. (9 Nm)
- Spindle nut and tighten to 181 ft. lbs. (245 Nm)
- Wheel speed sensor
- Brake caliper and tighten the bolts to 41 ft. lbs. (55 Nm)
- Brake hose bracket mounting bolts to the knuckle and trailing arm
- Rear wheel

10. Check the wheel alignment and adjust as necessary.

BRAKES

Brake Caliper

REMOVAL & INSTALLATION

Front

1. Before servicing the vehicle, refer to the precautions in the beginning of this section.
2. Remove some fluid from the reservoir with a suction pump.
3. Remove or disconnect the following:

- Front wheels
- Banjo bolt and disconnect the brake hose from the caliper. Plug the hose to prevent fluid loss and contamination.
- Mounting bolts and the caliper from its mounting bracket

To Install:
4. Install or connect the following:

- Caliper over the pads and onto its mounting bracket. Torque both caliper bolts to 27 ft. lbs. (36 Nm).
- Brake hose to the caliper using new sealing washers. Carefully torque the banjo bolt to 25 ft. lbs. (34 Nm).

5. Fill the reservoir with fluid and bleed the brakes.

- Front wheels

Rear

1. Before servicing the vehicle, refer to the precautions in the beginning of this section.
2. Remove some fluid from the reservoir with a suction pump.
3. Remove or disconnect the following:

- Rear wheels
- Banjo bolt and disconnect the brake hose from the caliper. Plug the hose to prevent fluid loss and contamination.
- 2 caliper mounting bolts and the caliper from its mounting bracket

To Install:
4. Install or connect the following:

- Caliper over the pads and onto its mounting bracket. Tighten the caliper bolts to 27 ft. lbs. (37 Nm).
- Brake hose with new sealing washers. Tighten the banjo bolt to 25 ft. lbs. (34 Nm).

5. Fill the reservoir with fluid and bleed the brake system. Adjust the parking brake if necessary.

- Rear wheels

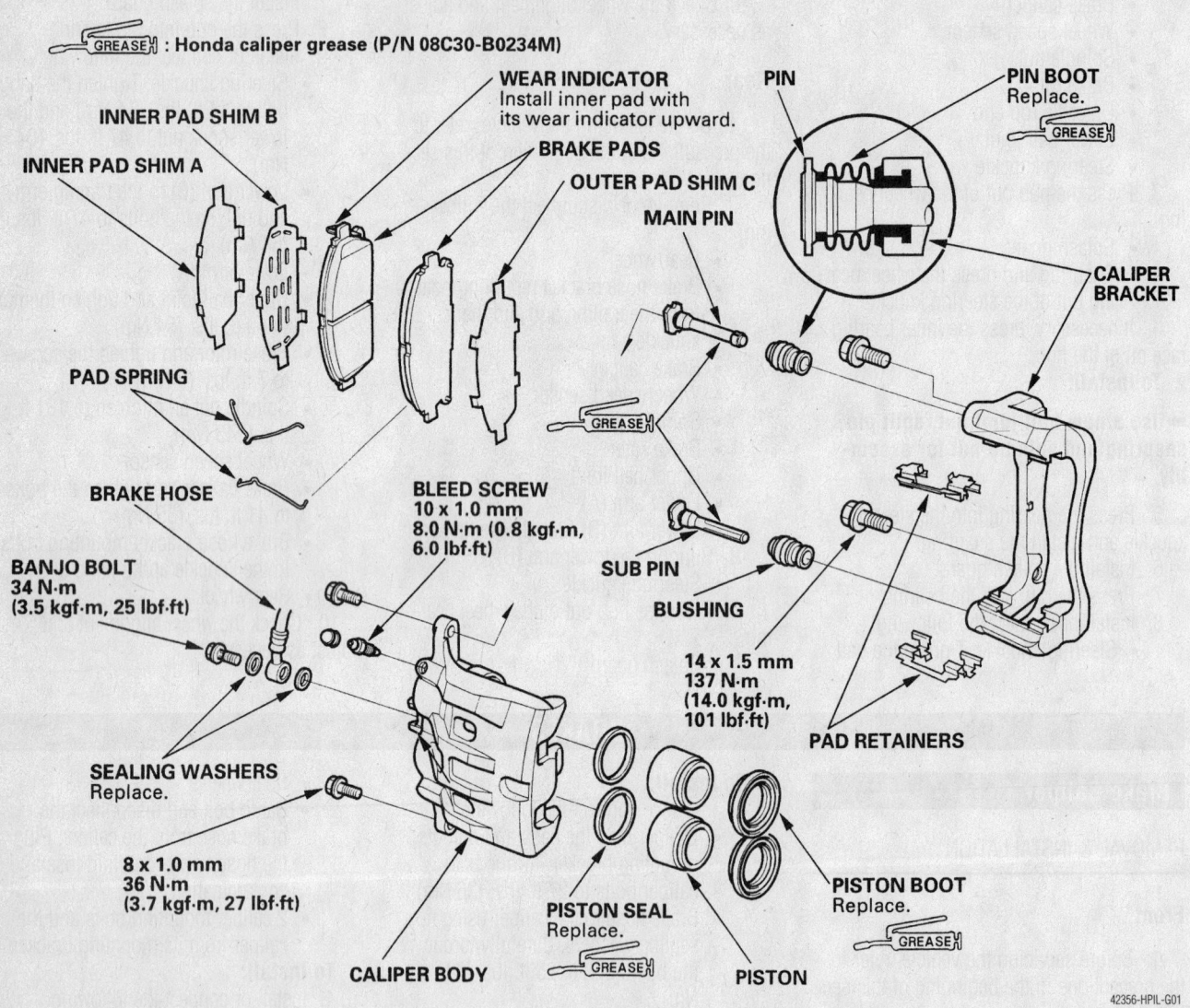

GREASE : Honda caliper grease (P/N 08C30-B0234M)

INNER PAD SHIM B

INNER PAD SHIM A

WEAR INDICATOR
Install inner pad with
its wear indicator upward.

BRAKE PADS

OUTER PAD SHIM C

MAIN PIN

PIN

PIN BOOT
Replace.

GREASE

CALIPER
BRACKET

PAD SPRING

BRAKE HOSE

BLEED SCREW
10 x 1.0 mm
8.0 N·m (0.8 kgf·m,
6.0 lbf·ft)

SUB PIN

BUSHING

14 x 1.5 mm
137 N·m
(14.0 kgf·m,
101 lbf·ft)

PAD RETAINERS

BANJO BOLT
34 N·m
(3.5 kgf·m, 25 lbf·ft)

SEALING WASHERS
Replace.

8 x 1.0 mm
36 N·m
(3.7 kgf·m, 27 lbf·ft)

CALIPER BODY

PISTON SEAL
Replace.

GREASE

PISTON

PISTON BOOT
Replace.

GREASE

42356-HPIL-G01

Exploded view of the front caliper components

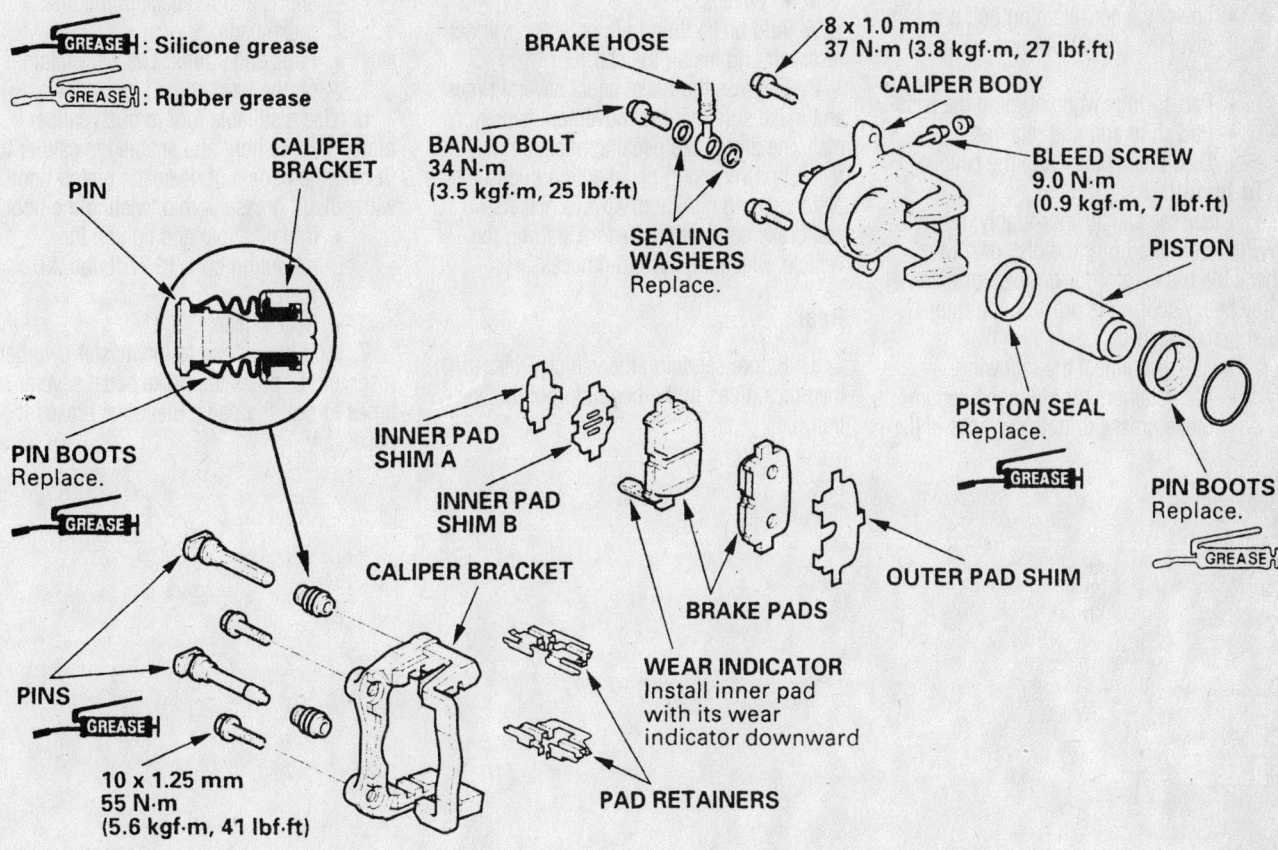

GREASE : Silicone grease

GREASE : Rubber grease

PIN

CALIPER
BRACKET

BRAKE HOSE

BANJO BOLT
34 N·m
(3.5 kgf·m, 25 lbf·ft)

8 x 1.0 mm
37 N·m (3.8 kgf·m, 27 lbf·ft)

CALIPER BODY

BLEED SCREW
9.0 N·m
(0.9 kgf·m, 7 lbf·ft)

SEALING
WASHERS
Replace.

PISTON

PIN BOOTS
Replace.

GREASE

INNER PAD
SHIM A

INNER PAD
SHIM B

CALIPER BRACKET

PISTON SEAL
Replace.

GREASE

PIN BOOTS
Replace.

GREASE

OUTER PAD SHIM

BRAKE PADS

PINS

GREASE

WEAR INDICATOR
Install inner pad
with its wear
indicator downward

10 x 1.25 mm
55 N·m
(5.6 kgf·m, 41 lbf·ft)

PAD RETAINERS

93552GZB

Exploded view of the rear caliper components—Pilot

Disc Brake Pads

REMOVAL & INSTALLATION

Front

1. Before servicing the vehicle, refer to the precautions in the beginning of this section.

2. Remove or disconnect the following:
   • Front wheels

3. Remove a small amount of brake fluid from the reservoir using a suction pump.
   • Brake hose clamp from the knuckle by unfastening the retaining bolts
   • Lower caliper retaining bolt and pivot the caliper upward, off of the pads
   • Pad springs while holding the pads
   • Pad shim and pad retainers
   • Disc brake pads from the caliper

To install:

4. Clean the caliper thoroughly; remove any rust from the lip of the disc or rotor. Check the brake rotor for grooves or cracks. If any heavy scoring is present, the rotor must be replaced.

5. Install or connect the following:
   • Pad retainers. Apply molybdenum brake grease to both surfaces of the shims and the back of the disc brake pads.
   • Pads and shims. The pad with the wear indicator goes in the inboard position.
   • Pad springs while holding the pads

6. Push in the caliper piston so the caliper will fit over the pads. This is most easily accomplished with a pad spreader or large C-clamp.
   • Caliper down into position and tighten the mounting bolt to 27 ft. lbs. (37 Nm)
   • Brake hose to the knuckle, if removed
   • Wheels

7. Add brake fluid to the master cylinder reservoir and install the cap.

8. Depress the brake pedal several times and make sure that the movement feels normal. The first brake pedal application may result in a very long pedal action due to the pistons being retracted. Always make several brake applications before starting the vehicle. Bleed the system if necessary.

Rear

1. Before servicing the vehicle, refer to the precautions in the beginning of this section.

2. Remove a small amount of brake fluid from the reservoir using a suction pump.

3. Remove or disconnect the following:
   • Rear wheels
   • 2 caliper mounting bolts and the caliper from the bracket
   • Pads, shims, and pad retainers

To install:

4. Clean the caliper thoroughly; remove any dirt or dust. Check the brake rotor for grooves or cracks and machine or replace, as necessary.

5. Install or connect the following:
   • Pad retainers. Apply molybdenum brake grease to both surfaces of the shims and the back of the disc brake pads.
   • Pads and shims. The wear retainer on the inboard pad faces down.

6. Use a suitable tool to push caliper piston into its bore and enable the caliper to fit over the pads. Lubricate the piston boot with silicon grease. Avoid twisting the boot.
   • Brake caliper and tighten the mounting bolts to 27 ft. lbs. (37 Nm)
   • Rear wheels

7. Add brake fluid to the master cylinder reservoir. Depress the brake pedal several times to seat the pads. Bleed the brakes if necessary.

HYUNDAI

Accent • Elantra • Sonata • Tiburon • XG 300

8

SPECIFICATION CHARTS

ENGINE AND VEHICLE IDENTIFICATION

		Engine							Model Year	
Code	Liters (cc)	Cu. In.	Cyl.	Fuel Sys.	Engine Type	Eng. Mfg.		Code	Year	
K	1.5 (1495)	91.17	4	MPFI	DOHC	Hyundai		Y	2000	
F ①	2.0 (1975)	120.52	4	MPFI	DOHC	Hyundai		1	2001	
F ②	2.0 (1997)	121.90	4	MPFI	DOHC	Hyundai		2	2002	
D	2.4 (2351)	143.46	4	MPFI	DOHC	Hyundai		3	2003	
E	2.5 (2493)	152.13	6	MPFI	DOHC	Hyundai		4	2004	
G	3.0 (2972)	181.40	6	MPFI	DOHC	Hyundai				

MPFI: Multi-Point Fuel Injection

SOHC: Single Overhead Camshaft

DOHC: Double Overhead Camshafts

① Elantra and Tiburon

② Sonata

42356-ELAN-C01

GENERAL ENGINE SPECIFICATIONS

Year	Engine Displacement Liters (cc)	Engine ID/VIN	Fuel System Type	Net Horsepower @ rpm	Net Torque @ rpm (ft. lbs.)	Bore x Stroke (in.)	Compression Ratio	Oil Pressure @ rpm
2000	1.5 (1495)	K	MFI	92@5500	97@4000	2.97 x 3.29	10.0:1	21@Idle
	2.0 (1975)	F	MFI	140@6000	133@4800	3.23 x 3.68	10.3:1	24@Idle
	2.4 (2351)	D	MFI	137@6000	129@4000	3.41 x 3.94	10.0:1	12@Idle
	2.5 (2493)	E	MFI	142@5000	168@2500	3.59 x 2.99	10.0:1	12@Idle
2001	1.5 (1495)	K	MFI	92@5500	97@4000	2.97 x 3.29	10.0:1	21@Idle
	2.0 (1975)	F	MFI	140@6000	133@4800	3.23 x 3.68	10.3:1	24@Idle
	2.4 (2351)	D	MFI	137@6000	129@4000	3.41 x 3.94	10.0:1	12@Idle
	2.5 (2493)	E	MFI	142@5000	168@2500	3.59 x 2.99	10.0:1	12@Idle
	3.0 (2972)	G	MFI	192@6000	178@4800	3.59 x 2.99	10.0:1	12@Idle
2002	1.5 (1495)	K	MFI	92@5500	97@4000	2.97 x 3.29	10.0:1	21@Idle
	2.0 (1975)	F	MFI	140@6000	133@4800	3.23 x 3.68	10.3:1	24@Idle
	2.4 (2351)	D	MFI	137@6000	129@4000	3.41 x 3.94	10.0:1	12@Idle
	2.5 (2493)	E	MFI	142@5000	168@2500	3.59 x 2.99	10.0:1	12@Idle
	3.0 (2972)	G	MFI	192@6000	178@4800	3.59 x 2.99	10.0:1	12@Idle
2003-04	1.5 (1495)	K	MFI	92@5500	97@4000	2.97 x 3.29	10.0:1	21@Idle
	2.0 (1975)	F	MFI	140@6000	133@4800	3.23 x 3.68	10.3:1	24@Idle
	2.4 (2351)	D	MFI	137@6000	129@4000	3.41 x 3.94	10.0:1	12@Idle
	2.5 (2493)	E	MFI	142@5000	168@2500	3.59 x 2.99	10.0:1	12@Idle
	3.0 (2972)	G	MFI	192@6000	178@4800	3.59 x 2.99	10.0:1	12@Idle

MFI : Multi-Port Fuel Injection

① Tiburon

② Sonata

42356-ELAN-C02

GASOLINE ENGINE TUNE-UP SPECIFICATIONS

Year	Engine Displacement Liters (cc)	Engine ID/VIN	Spark Plugs Gap (in.)	Ignition Timing (deg.) MT	AT	Fuel Pump (psi)	Idle Speed (rpm) MT	AT	Valve Clearance In.	Ex.
2000	1.5 (1495)	K	0.039-0.043	6-16B	6-16B	43	700-900	700-900	HYD	HYD
	2.0 (1975)	F	0.039-0.043	5-15B	5-15B	43	700-900	700-900	HYD	HYD
	2.4 (2351)	D	0.039-0.043	3-7B	3-7B	48	650-850	650-850	HYD	HYD
	2.5 (2493)	E	0.039-0.043	7-17B	7-17B	48	600-800	600-800	HYD	HYD
2001	1.5 (1495)	K	0.039-0.043	6-16B	6-16B	43	700-900	700-900	HYD	HYD
	2.0 (1975)	F	0.039-0.043	5-15B	5-15B	43	700-900	700-900	HYD	HYD
	2.4 (2351)	D	0.039-0.043	3-7B	3-7B	48	650-850	650-850	HYD	HYD
	2.5 (2493)	E	0.039-0.043	7-17B	7-17B	48	600-800	600-800	HYD	HYD
	3.0 (2972)	G	0.039-0.043	3-7B	3-7B	48	600-800	600-800	HYD	HYD
2002	1.5 (1495)	K	0.039-0.043	6-16B	6-16B	43	700-900	700-900	HYD	HYD
	2.0 (1975)	F	0.039-0.043	5-15B	5-15B	43	700-900	700-900	HYD	HYD
	2.4 (2351)	D	0.039-0.043	3-7B	3-7B	48	650-850	650-850	HYD	HYD
	2.5 (2493)	E	0.039-0.043	7-17B	7-17B	48	600-800	600-800	HYD	HYD
	3.0 (2972)	G	0.039-0.043	3-7B	3-7B	48	600-800	600-800	HYD	HYD
2003-04	1.5 (1495)	K	0.039-0.043	6-16B	6-16B	43	700-900	700-900	HYD	HYD
	2.0 (1975)	F	0.039-0.043	5-15B	5-15B	43	700-900	700-900	HYD	HYD
	2.4 (2351)	D	0.039-0.043	3-7B	3-7B	48	650-850	650-850	HYD	HYD
	2.5 (2493)	E	0.039-0.043	7-17B	7-17B	48	600-800	600-800	HYD	HYD
	3.0 (2972)	G	0.039-0.043	3-7B	3-7B	48	600-800	600-800	HYD	HYD

HYD: Hydraulic Valve Lifters

B: Before Top Dead Center

42356-ELAN-C03

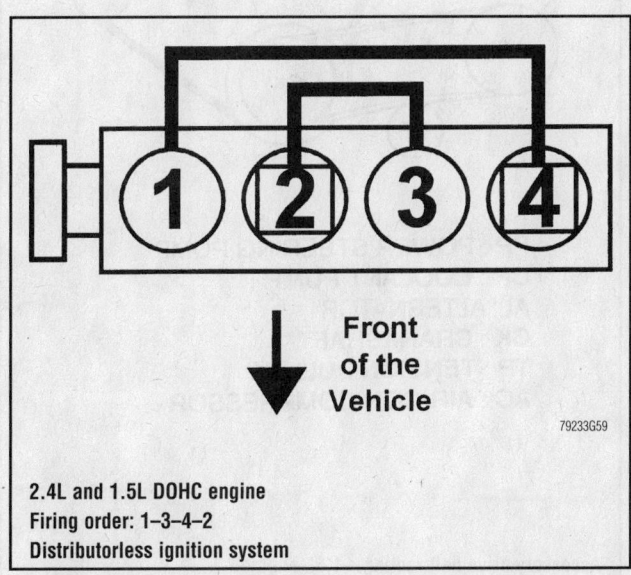

2.4L and 1.5L DOHC engine
Firing order: 1–3–4–2
Distributorless ignition system

79233G59

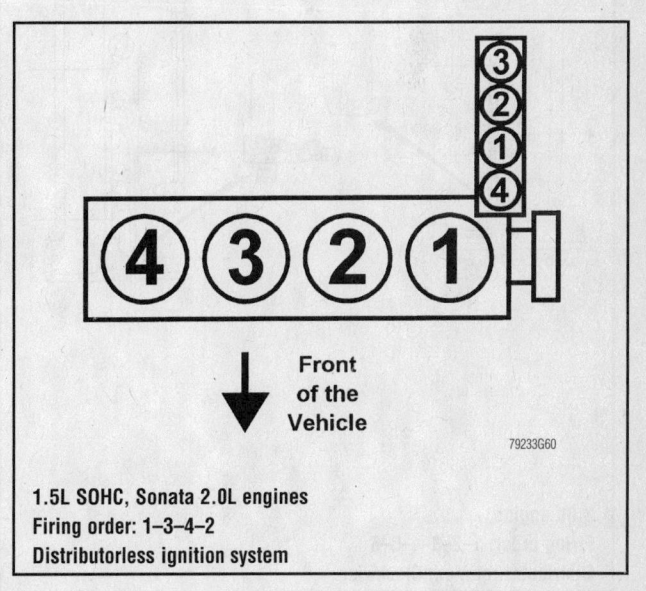

1.5L SOHC, Sonata 2.0L engines
Firing order: 1–3–4–2
Distributorless ignition system

79233G60

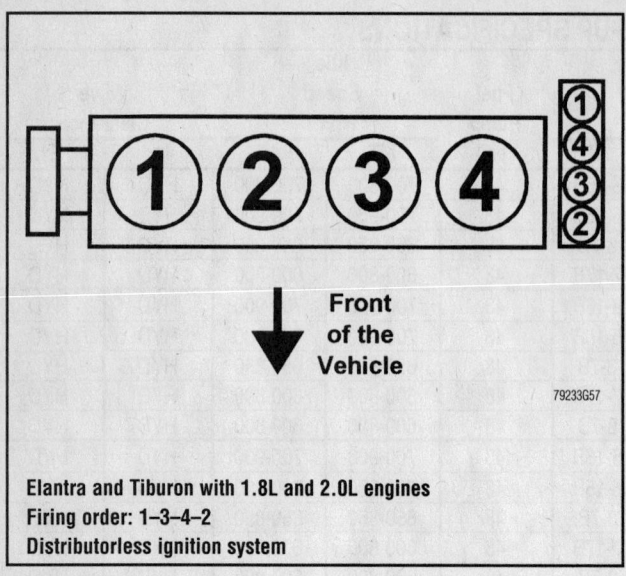

79233G57

Elantra and Tiburon with 1.8L and 2.0L engines
Firing order: 1–3–4–2
Distributorless ignition system

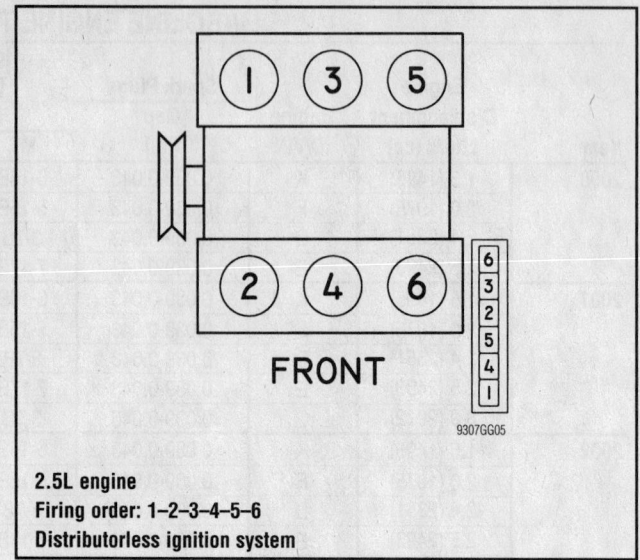

9307GG05

2.5L engine
Firing order: 1–2–3–4–5–6
Distributorless ignition system

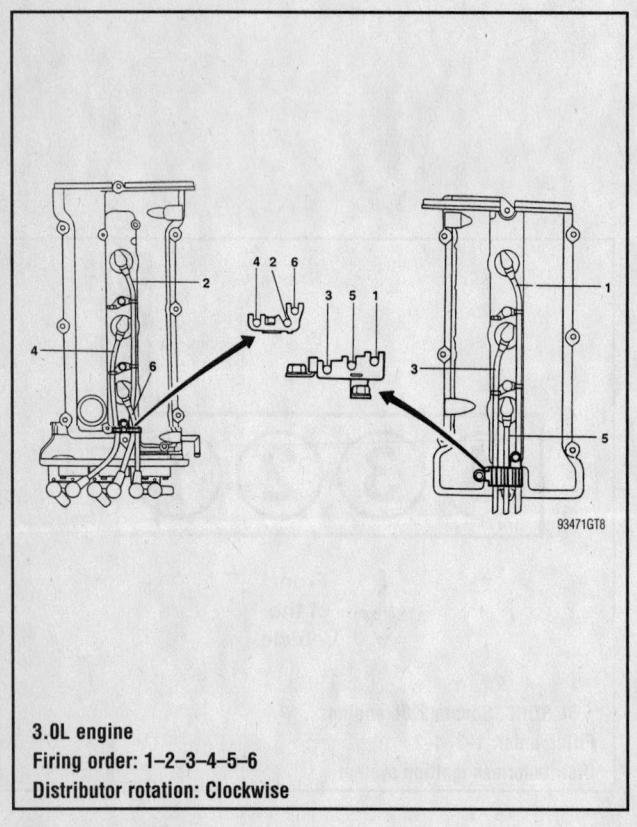

93471GT8

3.0L engine
Firing order: 1–2–3–4–5–6
Distributor rotation: Clockwise

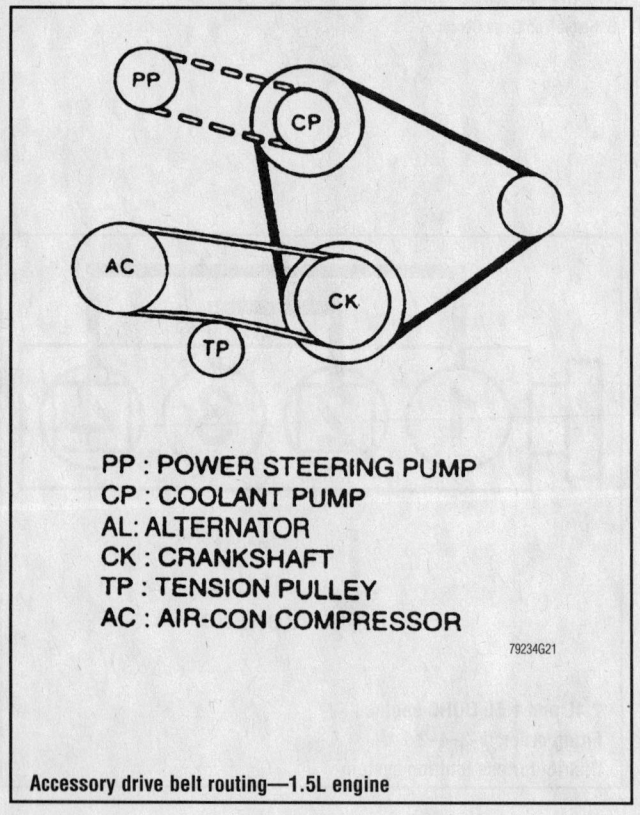

PP : POWER STEERING PUMP
CP : COOLANT PUMP
AL : ALTERNATOR
CK : CRANKSHAFT
TP : TENSION PULLEY
AC : AIR-CON COMPRESSOR

79234G21

Accessory drive belt routing—1.5L engine

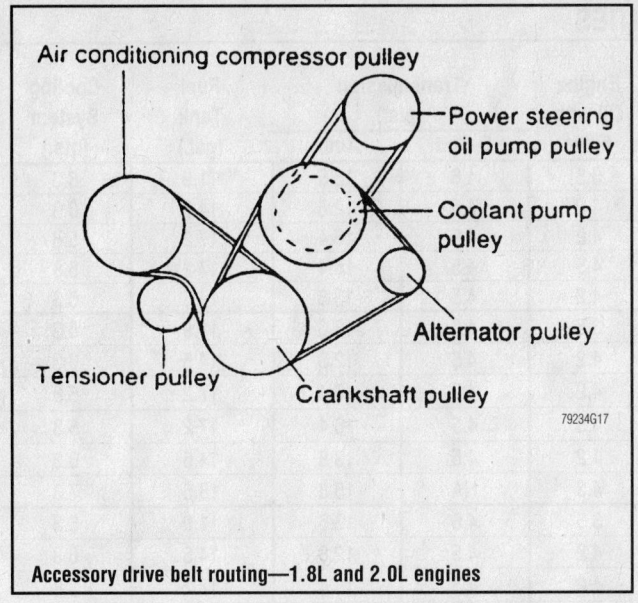

Accessory drive belt routing—1.8L and 2.0L engines

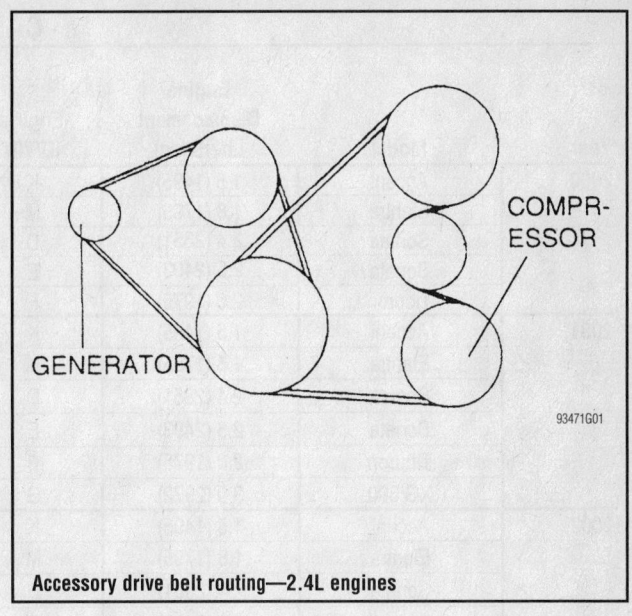

Accessory drive belt routing—2.4L engines

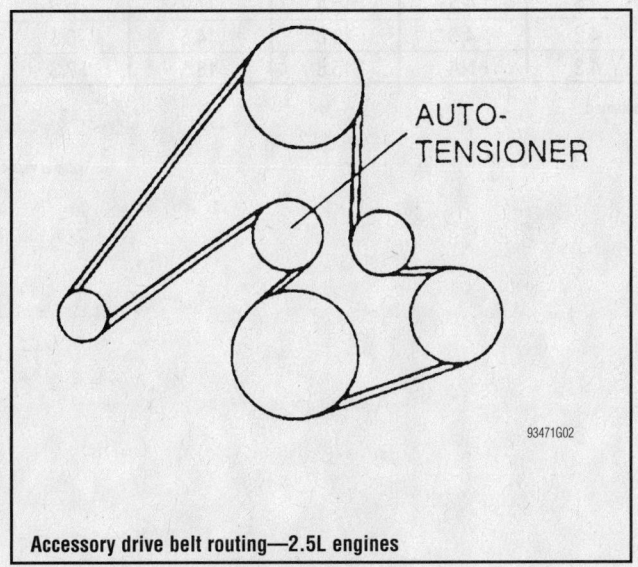

Accessory drive belt routing—2.5L engines

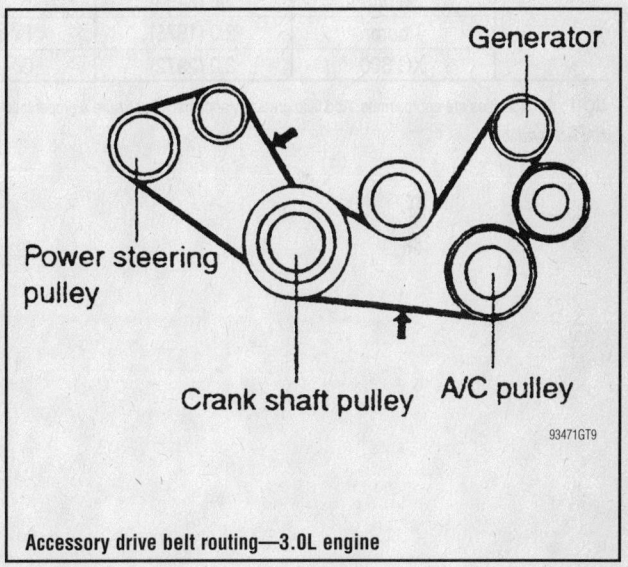

Accessory drive belt routing—3.0L engine

CAPACITIES

Year	Model	Engine Displacement Liters (cc)	Engine ID/VIN	Engine Oil with Filter	Transmission (pts.)		Fuel Tank (gal.)	Cooling System (qts.)
					5–Spd	Auto.		
2000	Accent	1.5 (1495)	K	3.5	4.6	13.6	11.9	6.3
	Elantra	1.8 (1795)	M	4.2	4.5	12.8	14.5	6.3
	Sonata	2.4 (2351)	D	4.2	4.5	16.4	17.2	5.8
	Sonata	2.5 (2493)	E	4.5	4.5	16.4	17.2	5.8
	Tiburon	2.0 (1975)	F	4.2	4.5	13.8	14.5	6.3
2001	Accent	1.5 (1495)	K	3.5	4.6	13.6	11.9	6.3
	Elantra	1.8 (1795)	M	4.2	4.5	12.8	14.5	6.3
	Sonata	2.4 (2351)	D	4.2	4.5	16.4	17.2	5.8
	Sonata	2.5 (2493)	E	4.5	4.5	16.4	17.2	5.8
	Tiburon	2.0 (1975)	F	4.2	4.5	13.8	14.5	6.3
	XG 300	3.0 (2972)	G	4.3	NA	15.8	18.5	7.3
2002	Accent	1.5 (1495)	K	3.5	4.6	13.6	11.9	6.3
	Elantra	1.8 (1795)	M	4.2	4.5	12.8	14.5	6.3
	Sonata	2.4 (2351)	D	4.2	4.5	16.4	17.2	5.8
	Sonata	2.5 (2493)	E	4.5	4.5	16.4	17.2	5.8
	Tiburon	2.0 (1975)	F	4.2	4.5	13.8	14.5	6.3
	XG 300	3.0 (2972)	G	4.3	NA	15.8	18.5	7.3
2003-04	Accent	1.5 (1495)	K	3.5	4.6	13.6	11.9	6.3
	Elantra	1.8 (1795)	M	4.2	4.5	12.8	14.5	6.3
	Sonata	2.4 (2351)	D	4.2	4.5	16.4	17.2	5.8
	Sonata	2.5 (2493)	E	4.5	4.5	16.4	17.2	5.8
	Tiburon	2.0 (1975)	F	4.2	4.5	13.8	14.5	6.3
	XG 300	3.0 (2972)	G	4.3	NA	15.8	18.5	7.3

NOTE: All capacities are approximate. Add fluid gradually and check to be sure a proper fluid level is obtained.

NA: Not Available

42356-ELAN-C04

CRANKSHAFT AND CONNECTING ROD SPECIFICATIONS
All measurements are given in inches.

Year	Engine Displacement Liters (cc)	Engine ID/VIN	Crankshaft Main Brg. Journal Dia.	Main Brg. Oil Clearance	Shaft End-play	Thrust on No.	Connecting Rod Journal Diameter	Oil Clearance	Side Clearance
2000	1.5 (1495)	K	2.2440	0.0011-0.0018	0.0019-0.0068	3	1.7700	0.0009-0.0016	0.0039-0.0098
	2.0 (1975)	F	2.2400	0.0011-0.0018	0.0023-0.0100	3	1.7700	0.0009-0.0016	0.0039-0.0098
	2.4 (2351)	D	2.2434-2.2442	0.0007-0.0014 ①	0.0020-0.0098	3	1.7709-1.7717	0.0008-0.0020	0.0040-0.0098
	2.5 (2493)	E	2.4402-2.4409	0.0002-0.0009	0.0028-0.0098	3	1.8891-1.8898	0.0007-0.0014	0.0039-0.0098
2001	1.5 (1495)	K	2.2440	0.0011-0.0018	0.0019-0.0068	3	1.7700	0.0009-0.0016	0.0039-0.0098
	2.0 (1975)	F	2.2400	0.0011-0.0018	0.0023-0.0100	3	1.7700	0.0009-0.0016	0.0039-0.0098
	2.4 (2351)	D	2.2434-2.2442	0.0007-0.0014 ①	0.0020-0.0098	3	1.7709-1.7717	0.0008-0.0020	0.0040-0.0098
	2.4 (2351)	D	2.2434-2.2442	0.0007-0.0014 ①	0.0020-0.0098	3	1.7709-1.7717	0.0008-0.0020	0.0040-0.0098
	3.0 (2972)	G	2.3617-2.3620	0.0007-0.0014	0.002-0.0098	3	1.9677-1.9685	0.0009-0.0020	0.0039-0.0098
2002	1.5 (1495)	K	2.2440	0.0011-0.0018	0.0019-0.0068	3	1.7700	0.0009-0.0016	0.0039-0.0098
	2.0 (1975)	F	2.2400	0.0011-0.0018	0.0023-0.0100	3	1.7700	0.0009-0.0016	0.0039-0.0098
	2.4 (2351)	D	2.2434-2.2442	0.0007-0.0014 ①	0.0020-0.0098	3	1.7709-1.7717	0.0008-0.0020	0.0040-0.0098
	2.4 (2351)	D	2.2434-2.2442	0.0007-0.0014 ①	0.0020-0.0098	3	1.7709-1.7717	0.0008-0.0020	0.0040-0.0098
	3.0 (2972)	G	2.3617-2.3620	0.0007-0.0014	0.002-0.0098	3	1.9677-1.9685	0.0009-0.0020	0.0039-0.0098
2003-04	1.5 (1495)	K	2.2440	0.0011-0.0018	0.0019-0.0068	3	1.7700	0.0009-0.0016	0.0039-0.0098
	2.0 (1975)	F	2.2400	0.0011-0.0018	0.0023-0.0100	3	1.7700	0.0009-0.0016	0.0039-0.0098
	2.4 (2351)	D	2.2434-2.2442	0.0007-0.0014 ①	0.0020-0.0098	3	1.7709-1.7717	0.0008-0.0020	0.0040-0.0098
	2.4 (2351)	D	2.2434-2.2442	0.0007-0.0014 ①	0.0020-0.0098	3	1.7709-1.7717	0.0008-0.0020	0.0040-0.0098
	3.0 (2972)	G	2.3617-2.3620	0.0007-0.0014	0.002-0.0098	3	1.9677-1.9685	0.0009-0.0020	0.0039-0.0098

① No. 3: 0.0009 - 0.0016

42356-ELAN-C05

VALVE SPECIFICATIONS

Year	Engine Displacement Liters (cc)	Engine ID/VIN	Seat Angle (deg.)	Face Angle (deg.)	Spring Test Pressure (lbs. @ in.)	Spring Installed Height (in.)	Stem-to-Guide Clearance (in.)		Stem Diameter (in.)	
							Intake	Exhaust	Intake	Exhaust
2000	1.5 (1495)	K	45	45	54@1.358	1.358	0.0012-0.0024	0.0014-0.0026	0.3920	0.3960
	2.0 (1975)	F	45	45	56@1.457	1.358	0.0008-0.0019	0.0019-0.0033	0.2348-0.2354	0.2334-0.2342
	2.4 (2351)	D	45-45.5	44-44.5	56@1.457	1.358	0.0008-0.0019	0.0020-0.0033	0.2585-0.2891	0.2571-0.2579
	2.5 (2493)	E	45	45	49@1.378	1.378	0.0009-0.0020	0.0014-0.0026	0.2348-0.2354	0.2343-0.2348
2001	1.5 (1495)	K	45	45	54@1.358	1.358	0.0012-0.0024	0.0014-0.0026	0.3920	0.3960
	2.0 (1975)	F	45	45	56@1.457	1.358	0.0008-0.0019	0.0019-0.0033	0.2348-0.2354	0.2334-0.2342
	2.4 (2351)	D	45-45.5	44-44.5	56@1.457	1.358	0.0008-0.0019	0.0020-0.0033	0.2585-0.2891	0.2571-0.2579
	2.5 (2493)	E	45	45	49@1.378	1.378	0.0009-0.0020	0.0014-0.0026	0.2348-0.2354	0.2343-0.2348
	3.0 (2972)	G	45-45.5	45-45.5	74@1.591	1.826	0.0009-0.0020	0.0020-0.0033	0.258-0.2594	0.257-0.2580
2002	1.5 (1495)	K	45	45	54@1.358	1.358	0.0012-0.0024	0.0014-0.0026	0.3920	0.3960
	2.0 (1975)	F	45	45	56@1.457	1.358	0.0008-0.0019	0.0019-0.0033	0.2348-0.2354	0.2334-0.2342
	2.4 (2351)	D	45-45.5	44-44.5	56@1.457	1.358	0.0008-0.0019	0.0020-0.0033	0.2585-0.2891	0.2571-0.2579
	2.5 (2493)	E	45	45	49@1.378	1.378	0.0009-0.0020	0.0014-0.0026	0.2348-0.2354	0.2343-0.2348
	3.0 (2972)	G	45-45.5	45-45.5	74@1.591	1.826	0.0009-0.0020	0.0020-0.0033	0.258-0.2594	0.257-0.2580
2003-04	1.5 (1495)	K	45	45	54@1.358	1.358	0.0012-0.0024	0.0014-0.0026	0.3920	0.3960
	2.0 (1975)	F	45	45	56@1.457	1.358	0.0008-0.0019	0.0019-0.0033	0.2348-0.2354	0.2334-0.2342
	2.4 (2351)	D	45-45.5	44-44.5	56@1.457	1.358	0.0008-0.0019	0.0020-0.0033	0.2585-0.2891	0.2571-0.2579
	2.5 (2493)	E	45	45	49@1.378	1.378	0.0009-0.0020	0.0014-0.0026	0.2348-0.2354	0.2343-0.2348
	3.0 (2972)	G	45-45.5	45-45.5	74@1.591	1.826	0.0009-0.0020	0.0020-0.0033	0.258-0.2594	0.257-0.2580

42356-ELAN-C06

PISTON AND RING SPECIFICATIONS

All measurements are given in inches.

Year	Engine Displacement Liters (cc)	Engine ID/VIN	Piston Clearance	Ring Gap			Ring Side Clearance		
				Top Compression	Bottom Compression	Oil Control	Top Compression	Bottom Compression	Oil Control
2000	1.5 (1495)	K	0.0008-0.0016	0.008-0.020	0.008-0.020	0.008-0.039	0.0016-0.0033	0.0016-0.0033	snug
	2.0 (1975)	F	0.0008-0.0016	0.009-0.015	0.013-0.019	0.008-0.024	0.0015-0.0031	0.0012-0.0027	snug
	2.4 (2351)	D	0.0008-0.0012	0.010-0.014	0.006-0.022	0.004-0.016	0.0008-0.0024	0.0008-0.0024	snug
	2.5 (2493)	E	0.0008-0.0016	0.008-0.014	0.015-0.020	0.008-0.028	0.0016-0.0031	0.0012-0.0028	snug
2001	1.5 (1495)	K	0.0008-0.0016	0.008-0.020	0.008-0.020	0.008-0.039	0.0016-0.0033	0.0016-0.0033	snug
	2.0 (1975)	F	0.0008-0.0016	0.009-0.015	0.013-0.019	0.008-0.024	0.0015-0.0031	0.0012-0.0027	snug
	2.4 (2351)	D	0.0008-0.0012	0.010-0.014	0.006-0.022	0.004-0.016	0.0008-0.0024	0.0008-0.0024	snug
	2.5 (2493)	E	0.0008-0.0012	0.008-0.014	0.015-0.020	0.008-0.028	0.0016-0.0031	0.0012-0.0028	snug
	3.0 (2972)	G	0.0008-0.0016	0.012-0.018	0.010-0.016	0.008-0.028	0.0012-0.0035	0.0008-0.0024	snug
2002	1.5 (1495)	K	0.0008-0.0016	0.008-0.020	0.008-0.020	0.008-0.039	0.0016-0.0033	0.0016-0.0033	snug
	2.0 (1975)	F	0.0008-0.0016	0.009-0.015	0.013-0.019	0.008-0.024	0.0015-0.0031	0.0012-0.0027	snug
	2.4 (2351)	D	0.0008-0.0012	0.010-0.014	0.006-0.022	0.004-0.016	0.0008-0.0024	0.0008-0.0024	snug
	2.5 (2493)	E	0.0008-0.0012	0.008-0.014	0.015-0.020	0.008-0.028	0.0016-0.0031	0.0012-0.0028	snug
	3.0 (2972)	G	0.0008-0.0016	0.012-0.018	0.010-0.016	0.008-0.028	0.0012-0.0035	0.0008-0.0024	snug
2003-04	1.5 (1495)	K	0.0008-0.0016	0.008-0.020	0.008-0.020	0.008-0.039	0.0016-0.0033	0.0016-0.0033	snug
	2.0 (1975)	F	0.0008-0.0016	0.009-0.015	0.013-0.019	0.008-0.024	0.0015-0.0031	0.0012-0.0027	snug
	2.4 (2351)	D	0.0008-0.0012	0.010-0.014	0.006-0.022	0.004-0.016	0.0008-0.0024	0.0008-0.0024	snug
	2.5 (2493)	E	0.0008-0.0012	0.008-0.014	0.015-0.020	0.008-0.028	0.0016-0.0031	0.0012-0.0028	snug
	3.0 (2972)	G	0.0008-0.0016	0.012-0.018	0.010-0.016	0.008-0.028	0.0012-0.0035	0.0008-0.0024	snug

42356-ELAN-C07

TORQUE SPECIFICATIONS
All readings in ft. lbs.

Year	Engine Displacement Liters (cc)	Engine ID/VIN	Cylinder Head Bolts	Main Bearing Bolts	Rod Bearing Bolts	Crankshaft Damper Bolts	Flywheel Bolts	Manifold Intake	Manifold Exhaust	Spark Plugs	Lug Nut
2000	1.5 (1495)	K	①	40-44	25-28	110-118	94-101	11-14	11-14	18	65-80
	2.0 (1975)	F	②	③	34-39	125-133	88-95	11-14	17-22	18	65-80
	2.4 (2351)	D	④	⑤	⑥	80-94	94-101	⑦	⑤	15-22	67-81
	2.5 (2493)	E	⑧	⑨	⑩	130-138	53-55	14-15	18-22	15-22	67-81
2001	1.5 (1495)	K	⑧	40-44	25-28	110-118	94-101	11-14	11-14	18	65-80
	2.0 (1975)	F	②	③	34-39	125-133	88-95	11-14	17-22	18	65-80
	2.4 (2351)	D	⑦	⑥	⑥	80-94	94-101	⑦	⑤	15-22	67-81
	2.4 (2351)	D	⑦	⑥	⑥	80-94	94-101	⑦	⑤	15-22	67-81
	3.0 (2972)	G	75-82	65-72	36-39	NA	NA	9-10	20-40	15-22	67-81
2002	1.5 (1495)	K	⑧	40-44	25-28	110-118	94-101	11-14	11-14	18	65-80
	2.0 (1975)	F	②	③	34-39	125-133	88-95	11-14	17-22	18	65-80
	2.4 (2351)	D	⑦	⑥	⑥	80-94	94-101	⑦	⑤	15-22	67-81
	2.4 (2351)	D	⑦	⑥	⑥	80-94	94-101	⑦	⑤	15-22	67-81
	3.0 (2972)	G	75-82	65-72	36-39	NA	NA	9-10	20-40	15-22	67-81
2003-04	1.5 (1495)	K	⑧	40-44	25-28	110-118	94-101	11-14	11-14	18	65-80
	2.0 (1975)	F	②	③	34-39	125-133	88-95	11-14	17-22	18	65-80
	2.4 (2351)	D	⑦	⑥	⑥	80-94	94-101	⑦	⑤	15-22	67-81
	2.4 (2351)	D	⑦	⑥	⑥	80-94	94-101	⑦	⑤	15-22	67-81
	3.0 (2972)	G	75-82	65-72	36-39	NA	NA	9-10	20-40	15-22	67-81

① Cold: 51-54 ft. lbs.; Warm: 58-61 ft. lbs.

② Step 1: M10 bolts to 22 ft. lbs. and M12 bolts to 26 ft. lbs.
  Step 2: Plus 60-65 degrees
  Step 3: Plus 60-65 degrees

③ Step 1: 20-24 ft. lbs.
  Step 2: Plus 60-65 degrees

④ Step 1: 14 ft. lbs.
  Step 2: Plus 90 degrees
  Step 3: Loosen to 0 ft. lbs.
  Step 4: 14 ft. lbs.
  Step 5: Plus 90 degrees
  Step 6: Plus 90 degrees

⑤ M8: 18-22 ft. lbs.
  M10: 25-40 ft. lbs.

⑥ 13-16 ft. lbs. plus 90-94 degrees

⑦ M8: 11-14 ft. lbs.
  M10: 13-18 ft. lbs.
  Nuts: 22-30 ft. lbs.

⑧ Step 1: 18 ft. lbs.
  Step 2: Plus 60-64 degrees
  Step 3: Plus 45-49 degrees

⑨ M10: 20-24 ft. lbs. plus 90-94 degrees
  M8: 10-13 ft. lbs. plus 90-94 degrees

⑩ 12-14 ft. lbs. plus 90-94 degrees

42356-ELAN-C08

WHEEL ALIGNMENT

Year	Model		Caster		Camber		Toe-in (in.)	Steering Axis Inclination (Deg.)
			Range (+/-Deg.)	Preferred Setting (Deg.)	Range (+/-Deg.)	Preferred Setting (Deg.)		
2000	Accent	F	0.50	+1.80	0.50	0	0 +/- 0.12	—
		R	—	—	0.50	-0.68	0.12 +/- 0.08	—
	Elantra	F	0.50	+2.35	2.00	+4.00	0 +/- 0.12	—
		R	—	—	0.50	-0.70	0.14 +/- 0.02	—
	Sonata	F	1.00	+3.25	0.50	0	0 +/- 0.12	—
		R	—	—	0.50	-0.50	0.08 +/- 0.08	—
	Tiburon	F	0.50	+2.45	2.00	+4.00	0 +/- 0.12	—
		R	—	—	0.50	-0.90	0.14 +/- 0.02	—
2001	Accent	F	0.50	+1.80	0.50	0	0 +/- 0.12	—
		R	—	—	0.50	-0.68	0.12 +/- 0.08	—
	Elantra	F	0.50	+2.35	2.00	+4.00	0 +/- 0.12	—
		R	—	—	0.50	-0.70	0.14 +/- 0.02	—
	Sonata	F	1.00	+3.25	0.50	0	0 +/- 0.12	—
		R	—	—	0.50	-0.50	0.08 +/- 0.08	—
	Tiburon	F	0.50	+2.45	2.00	+4.00	0 +/- 0.12	—
		R	—	—	0.50	-0.90	0.14 +/- 0.02	—
	XG 300	F	1.00	+3.15	0.50	0	0 +/- 0.12	—
		R	—	—	0.50	-0.50	0.08 +/- 0.08	—
2002	Accent	F	0.50	+1.80	0.50	0	0 +/- 0.12	—
		R	—	—	0.50	-0.68	0.12 +/- 0.08	—
	Elantra	F	0.50	+2.35	2.00	+4.00	0 +/- 0.12	—
		R	—	—	0.50	-0.70	0.14 +/- 0.02	—
	Sonata	F	1.00	+3.25	0.50	0	0 +/- 0.12	—
		R	—	—	0.50	-0.50	0.08 +/- 0.08	—
	Tiburon	F	0.50	+2.45	2.00	+4.00	0 +/- 0.12	—
		R	—	—	0.50	-0.90	0.14 +/- 0.02	—
	XG 300	F	1.00	+3.15	0.50	0	0 +/- 0.12	—
		R	—	—	0.50	-0.50	0.08 +/- 0.08	—
2003-04	Accent	F	0.50	+1.80	0.50	0	0 +/- 0.12	—
		R	—	—	0.50	-0.68	0.12 +/- 0.08	—
	Elantra	F	0.50	+2.35	2.00	+4.00	0 +/- 0.12	—
		R	—	—	0.50	-0.70	0.14 +/- 0.02	—
	Sonata	F	1.00	+3.25	0.50	0	0 +/- 0.12	—
		R	—	—	0.50	-0.50	0.08 +/- 0.08	—
	Tiburon	F	0.50	+2.45	2.00	+4.00	0 +/- 0.12	—
		R	—	—	0.50	-0.90	0.14 +/- 0.02	—
	XG 300	F	1.00	+3.15	0.50	0	0 +/- 0.12	—
		R	—	—	0.50	-0.50	0.08 +/- 0.08	—

42356-ELAN-C09

TIRE, WHEEL AND BALL JOINT SPECIFICATIONS

Year	Model	OEM Tires		Tire Pressures (psi)		Wheel Size	Ball Joint Inspection
		Standard	Optional	Front	Rear		
2000	Accent	P155/80R13	P175/70R13 P175/65R14	30	30	5-J	①
	Elantra	P195/60R14	None	30	30	5.5-JJ	①
	Sonata	P195/70R14	P205/60HR15	30	30	Std: 5.5-JJ Opt: 6-JJ	①
	Tiburon	P195/60R14	None	30	30	5.5-JJ	①
2001	Accent	P155/80R13	P175/70R13 P175/65R14	30	30	5-J	①
	Elantra	P195/60R14	None	30	30	5.5-JJ	①
	Sonata	P195/70R14	P205/60HR15	30	30	Std: 5.5-JJ Opt: 6-JJ	①
	Tiburon	P195/60R14	None	30	30	5.5JJ	①
	XG 300	P205/65R15	None	30		6.0J	①
2002	Accent	P155/80R13	P175/70R13 P175/65R14	30	30	5-J	①
	Elantra	P195/60R14	None	30	30	5.5-JJ	①
	Sonata	P195/70R14	P205/60HR15	30	30	Std: 5.5-JJ Opt: 6-JJ	①
	Tiburon	P195/60R14	None	30	30	5.5JJ	①
	XG 300	P205/65R15	None	30		6.0J	①
2003-04	Accent	P155/80R13	P175/70R13 P175/65R14	30	30	5-J	①
	Elantra	P195/60R14	None	30	30	5.5-JJ	①
	Sonata	P195/70R14	P205/60HR15	30	30	Std: 5.5-JJ Opt: 6-JJ	
	Tiburon	P195/60R14	None	30	30	5.5JJ	①
	XG 300	P205/65R15	None	30		6.0J	①

OEM: Original Equipment Manufacturer

PSI: Pounds Per Square Inch

STD: Standard

OPT: Optional

① Replace if any measurable movement is found.

42356-ELAN-C10

BRAKE SPECIFICATIONS
All measurements in inches unless noted

Year	Model		Brake Disc Original Thickness	Brake Disc Minimum Thickness	Brake Disc Maximum Run-out	Brake Drum Diameter Original Inside Diameter	Brake Drum Diameter Max. Wear Limit	Brake Drum Diameter Maximum Machine Diameter	Minimum Lining Thickness Front	Minimum Lining Thickness Rear	Brake Caliper Bracket Bolts (ft. lbs.)	Brake Caliper Mounting Bolts (ft. lbs.)
2000	Accent	F	0.750	0.670	0.002	—	—	—	0.039	—	50	①
		R	—	—	—	7.100	—	7.165	—	0.039	—	—
	Elantra	F	0.750	0.670	0.002	—	—	—	0.039	—	50	①
		R	—	—	—	8.000	—	8.079	—	0.039	—	—
	Elantra w/rear disc	F	0.750	0.670	0.002	—	—	—	0.039	—	50	①
		R	0.354	NA	NA	—	—	—	—	0.031	—	23
	Sonata	F	0.945	0.787	0.002	—	—	—	0.079	—	51-63	16-24
		R	—	—	—	9.000	—	8.936	—	0.059	—	—
	Sonata w/rear disc	F	0.945	0.880	0.003	—	—	—	0.079	—	51-63	16-24
		R	0.390	0.350	0.005	—	—	—	—	0.079	—	23
	Tiburon	F	0.866	0.787	0.002	—	—	—	0.079	—	55	①
		R	—	—	—	8.000	—	8.079	—	0.059	—	—
	Tiburon w/rear disc	F	0.866	0.787	0.002	—	—	—	0.079	—	55	①
		R	0.354	NA	NA	—	—	—	—	0.031	—	23
2001	Accent	F	0.750	0.670	0.002	—	—	—	0.039	—	50	①
		R	—	—	—	7.100	—	7.165	—	0.039	—	—
	Elantra	F	0.750	0.670	0.002	—	—	—	0.039	—	50	①
		R	—	—	—	8.000	—	8.079	—	0.039	—	—
	Elantra w/rear disc	F	0.750	0.670	0.002	—	—	—	0.039	—	50	①
		R	0.354	NA	NA	—	—	—	—	0.031	—	23
	Sonata	F	0.945	0.787	0.002	—	—	—	0.079	—	51-63	16-24
		R	—	—	—	9.000	—	8.936	—	0.059	—	—
	Sonata w/rear disc	F	0.945	0.880	0.003	—	—	—	0.079	—	51-63	16-24
		R	0.390	0.350	0.005	—	—	—	—	0.079	—	23
	Tiburon	F	0.866	0.787	0.002	—	—	—	0.079	—	55	①
		R	—	—	—	8.000	—	8.079	—	0.059	—	—
	Tiburon w/rear disc	F	0.866	0.787	0.002	—	—	—	0.079	—	55	①
		R	0.354	NA	NA	—	—	—	—	0.031	—	23
	XG 300	F	0.413	0.096	0.002	—	—	—	0.079	—	51-63	16-24
		R	0.390	0.080	NA	—	—	—	—	—	51-63	16-24
2002	Accent	F	0.750	0.670	0.002	—	—	—	0.039	—	50	①
		R	—	—	—	7.100	—	7.165	—	0.039	—	—
	Elantra	F	0.750	0.670	0.002	—	—	—	0.039	—	50	①
		R	—	—	—	8.000	—	8.079	—	0.039	—	—
	Elantra w/rear disc	F	0.750	0.670	0.002	—	—	—	0.039	—	50	①
		R	0.354	NA	NA	—	—	—	—	0.031	—	23
	Sonata	F	0.945	0.787	0.002	—	—	—	0.079	—	51-63	16-24
		R	—	—	—	9.000	—	8.936	—	0.059	—	—
	Sonata w/rear disc	F	0.945	0.880	0.003	—	—	—	0.079	—	51-63	16-24
		R	0.390	0.350	0.005	—	—	—	—	0.079	—	23
	Tiburon	F	0.866	0.787	0.002	—	—	—	0.079	—	55	①
		R	—	—	—	8.000	—	8.079	—	0.059	—	—
	Tiburon w/rear disc	F	0.866	0.787	0.002	—	—	—	0.079	—	55	①
		R	0.354	NA	NA	—	—	—	—	0.031	—	23
	XG 300	F	0.413	0.096	0.002	—	—	—	0.079	—	51-63	16-24
		R	0.390	0.080	NA	—	—	—	—	—	51-63	16-24

42356-ELAN-C11

BRAKE SPECIFICATIONS
All measurements in inches unless noted

Year	Model		Brake Disc Original Thickness	Brake Disc Minimum Thickness	Brake Disc Maximum Run-out	Brake Drum Diameter Original Inside Diameter	Brake Drum Diameter Max. Wear Limit	Brake Drum Diameter Maximum Machine Diameter	Minimum Lining Thickness Front	Minimum Lining Thickness Rear	Brake Caliper Bracket Bolts (ft. lbs.)	Brake Caliper Mounting Bolts (ft. lbs.)
2003-04	Accent	F	0.750	0.670	0.002	—	—	—	0.039	—	50	①
		R	—	—	—	7.100	—	7.165	—	0.039	—	—
	Elantra	F	0.750	0.670	0.002	—	—	—	0.039	—	50	①
		R	—	—	—	8.000	—	8.079	—	0.039	—	—
	Elantra w/rear disc	F	0.750	0.670	0.002	—	—	—	0.039	—	50	①
		R	0.354	NA	NA	—	—	—	—	0.031	—	23
	Sonata	F	0.945	0.787	0.002	—	—	—	0.079	—	51-63	16-24
		R	—	—	—	9.000	—	8.936	—	0.059	—	—
	Sonata w/rear disc	F	0.945	0.880	0.003	—	—	—	0.079	—	51-63	16-24
		R	0.390	0.350	0.005	—	—	—	—	0.079	—	23
	Tiburon	F	0.866	0.787	0.002	—	—	—	0.079	—	55	①
		R	—	—	—	8.000	—	8.079	—	0.059	—	—
	Tiburon w/rear disc	F	0.866	0.787	0.002	—	—	—	0.079	—	55	①
		R	0.354	NA	NA	—	—	—	—	0.031	—	23
	XG 300	F	0.413	0.096	0.002	—	—	—	0.079	—	51-63	16-24
		R	0.390	0.080	NA	—	—	—	—	—	51-63	16-24

NA: Not Available

F: Front

R: Rear

① Upper: 28 ft. lbs.
   Lower: 19 ft. lbs.

42356-ELAN-C12

SCHEDULED MAINTENANCE INTERVALS
HYUNDAI—ACCENT, ELANTRA, SONATA, TIBURON & XG300

TO BE SERVICED	TYPE OF SERVICE	VEHICLE MILEAGE INTERVAL (x1000)												
		7.5	15	22.5	30	37.5	45	52.5	60	67.5	75	82.5	90	97.5
Engine oil & filter	R	✓	✓	✓	✓	✓	✓	✓	✓	✓	✓	✓	✓	✓
Automatic transaxle fluid	S/I		✓		✓		✓		✓		✓		✓	
Brake pads, calipers & rotors	S/I		✓		✓		✓		✓		✓		✓	
Driveshafts & boots	S/I		✓		✓		✓		✓		✓			
Wheel bearing grease	S/I				✓				✓				✓	
Air cleaner filter	R				✓				✓				✓	
Automatic transaxle fluid & filter	R				✓				✓				✓	
Brake fluid	R				✓				✓				✓	
Engine coolant	R				✓				✓				✓	
Fuel hose, vapor hose & fuel filter cap	S/I							✓						
Spark plugs	R				✓				✓				✓	
Spark plugs (Sonata 3.0L V6)	R								✓					
Bolts & nuts on chassis & body (Accent)	S/I				✓				✓				✓	
Drive belts	S/I				✓				✓				✓	
Exhaust pipe connections, muffler & suspension bolts	S/I				✓				✓				✓	
Manual transaxle oil	S/I				✓				✓				✓	
Rear brake drums, linings & parking brake	S/I				✓				✓				✓	
Steering gear rack, linkage & boots	S/I				✓				✓				✓	
Suspension ball joints & dust covers (Accent)	S/I				✓				✓				✓	
Timing belt (Accent & Elantra)	S/I				✓				✓				✓	
Timing belt (except Accent & Elantra)	R								✓					
Fuel filter	R							✓						
Fuel lines & connections	S/I								✓					
Vacuum & crankcase ventilation hoses	S/I								✓					

R: Replace   S/I: Service or Inspect

FREQUENT OPERATION MAINTENANCE (SEVERE SERVICE)

If a vehicle is operated under any of the following conditions it is considered severe service

- Extremely dusty areas.

- 50% or more of the vehicle operation is in 32°C (90°F) or higher temperatures, or constant operation in temperatures below 0°C (32°F).

- Prolonged idling (vehicle operation in stop and go traffic).

- Frequent short running periods (engine does not warm to normal operating temperatures).

- Police, taxi, delivery usage or trailer towing usage.

Oil & oil filter: change every 3000 miles.

Brake pads, calipers & rotors: service or inspect every 7500 miles.

Driveshaft boots: service or inspect every 7500 miles

Steering gear rack, linkage & boots: service or inspect every 7500 miles.

Air cleaner filter: service or inspect every 15,000 miles.

Automatic transaxle fluid & filter: replace every 15,000 miles.

Rear brake drums & linings: service or inspect every 15,000 miles.

Spark plugs: replace every 24,000 miles.

42356-ELAN-C13

PRECAUTIONS

Before servicing any vehicle, please be sure to read all of the following precautions, which deal with personal safety, prevention of component damage, and important points to take into consideration when servicing a motor vehicle:

• Never open, service or drain the radiator or cooling system when the engine is hot; serious burns can occur from the steam and hot coolant.

• Observe all applicable safety precautions when working around fuel. Whenever servicing the fuel system, always work in a well-ventilated area. Do not allow fuel spray or vapors to come in contact with a spark, open flame, or excessive heat (a hot drop light, for example). Keep a dry chemical fire extinguisher near the work area. Always keep fuel in a container specifically designed for fuel storage; also, always properly seal fuel containers to avoid the possibility of fire or explosion. Refer to the additional fuel system precautions later in this section.

• Fuel injection systems often remain pressurized, even after the engine has been turned **OFF**. The fuel system pressure must be relieved before disconnecting any fuel lines. Failure to do so may result in fire and/or personal injury.

• Brake fluid often contains polyglycol ethers and polyglycols. Avoid contact with the eyes and wash your hands thoroughly after handling brake fluid. If you do get brake fluid in your eyes, flush your eyes with clean, running water for 15 minutes. If

eye irritation persists, or if you have taken brake fluid internally, IMMEDIATELY seek medical assistance.

• The EPA warns that prolonged contact with used engine oil may cause a number of skin disorders, including cancer. You should make every effort to minimize your exposure to used engine oil. Protective gloves should be worn when changing oil. Wash your hands and any other exposed skin areas as soon as possible after exposure to used engine oil. Soap and water, or waterless hand cleaner should be used.

• All new vehicles are now equipped with an air bag system. The system must be disabled before performing service on or around system components, steering column, instrument panel components, wiring and sensors. Failure to follow safety and disabling procedures could result in accidental air bag deployment, possible personal injury and unnecessary system repairs.

• Always wear safety goggles when working with, or around, the air bag system. When carrying a non-deployed air bag, be sure the bag and trim cover are pointed away from your body. When placing a non-deployed air bag on a work surface, always face the bag and trim cover upward, away from the surface. This will reduce the motion of the module if it is accidentally deployed. Refer to the additional air bag system precautions later in this section.

• Clean, high quality brake fluid from a

sealed container is essential to the safe and proper operation of the brake system. You should always buy the correct type of brake fluid for your vehicle. If the brake fluid becomes contaminated, completely flush the system with new fluid. Never reuse any brake fluid. Any brake fluid that is removed from the system should be discarded. Also, do not allow any brake fluid to come in contact with a painted surface; it will damage the paint.

• Never operate the engine without the proper amount and type of engine oil; doing so WILL result in severe engine damage.

• Timing belt maintenance is extremely important. Many models utilize an interference-type, non-freewheeling engine. If the timing belt breaks, the valves in the cylinder head may strike the pistons, causing potentially serious (also time-consuming and expensive) engine damage. Refer to the maintenance interval charts in the front of this section for the recommended replacement interval for the timing belt, and to the timing belt procedure in this section for belt replacement and inspection.

• Disconnecting the negative battery cable on some vehicles may interfere with the functions of the on-board computer system(s) and may require the computer to undergo a relearning process once the negative battery cable is reconnected.

• When servicing drum brakes, only disassemble and assemble one side at a time, leaving the remaining side intact for reference.

ENGINE REPAIR

➡**Disconnecting the negative battery cable on some vehicles may interfere with the functions of the on board computer system. The computer may undergo a relearning process once the negative battery cable is reconnected.**

Distributor

All engines except the 3.0L (VIN T) V6 are equipped with a Distributorless Ignition System (DIS).

REMOVAL

1. Before servicing the vehicle, refer to the precautions in the beginning of this section.
2. Remove or disconnect the following:
   • Distributor cap

• Distributor electrical connectors
3. Matchmark the rotor to the distributor housing.
4. Matchmark the distributor housing to the engine.
5. Unbolt and remove the distributor.

INSTALLATION

Timing Not Disturbed

1. Before servicing the vehicle, refer to the precautions in the beginning of this section.
2. Install or connect the following:
   • Distributor, with the rotor-to-housing and the housing-to-engine matchmarks aligned
   • Distributor electrical connectors
   • Distributor cap

3. Check the ignition timing and adjust as necessary.

Timing Disturbed

1. Before servicing the vehicle, refer to the precautions in the beginning of this section.
2. Set the crankshaft to Top Dead Center (TDC) of the compression stroke for the No. 1 cylinder.
3. Align the timing marks on the distributor gear and the distributor housing.
4. Install or connect the following:
   • Distributor, by aligning the groove of the installation flange with the center of the installation stud
   • Distributor electrical connectors
   • Distributor cap
5. Check the ignition timing and adjust as necessary.

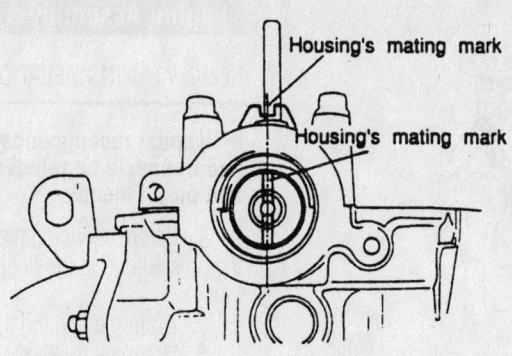

Aligning the distributor housing and gear mating marks—3.0L Sonata

Alternator

REMOVAL

1. Before servicing the vehicle, refer to the precautions in the beginning of this section.
2. Remove or disconnect the following:
   - Negative battery cable
   - Alternator drive belt
   - Alternator wiring harness connectors
   - Alternator. It may be necessary to raise the radiator

INSTALLATION

1. Before servicing the vehicle, refer to the precautions in the beginning of this section.
2. Install or connect the following:
   - Alternator and reposition the radiator, if raised
   - Alternator wiring harness connectors
   - Alternator drive belt
3. Adjust the alternator belt and torque the bolts to the following specifications:
   a. Accent: Pivot bolt to 14–18 ft. lbs. (19–25 Nm) and adjustment bolt to 14–20 ft. lbs. (19–28 Nm).
   b. 2.0L and 2.5L Sonata, Elantra and Tiburon: Pivot bolt to 14–18 ft. lbs. (19–25 Nm) and adjustment bolt to 105–132 inch lbs. (12–15 Nm).
   c. 2.4L Sonata: Pivot bolt to 26–41 ft. lbs. (34–54 Nm) and the adjustment bolt to 14–18 ft. lbs. (19–25 Nm).
   d. 3.0L Sonata and XG 300: Pivot bolt to 14–18 ft. lbs. (19–25 Nm) and adjustment bolt to 11–16 ft. lbs. (15–22 Nm).
4. Install the radiator mounting bolts, if removed.
5. Connect the negative battery cable.

Ignition Timing

ADJUSTMENT

Except 1.8L and 3.0L (VIN T) Engines

These engines are equipped with a Distributorless Ignition System (DIS). No adjustment is necessary.

1.8L and 3.0L (VIN T) Engines

➡ **Timing is adjusted with the engine at normal operating temperature and all electrical accessories OFF.**

1. Before servicing the vehicle, refer to the precautions in the beginning of this section.
2. Start the engine and allow it to reach operating temperature.
3. Turn the engine **OFF**.
4. Ground the ignition timing terminal.
5. Connect a timing light to the No. 1 spark plug wire.
6. Start the engine and check the base timing. Base timing should be 3–7 degrees Before Top Dead Center (BTDC).
7. Adjust the timing as necessary.
8. Turn the engine **OFF**.
9. Remove the ignition timing connector ground wire and the timing light.

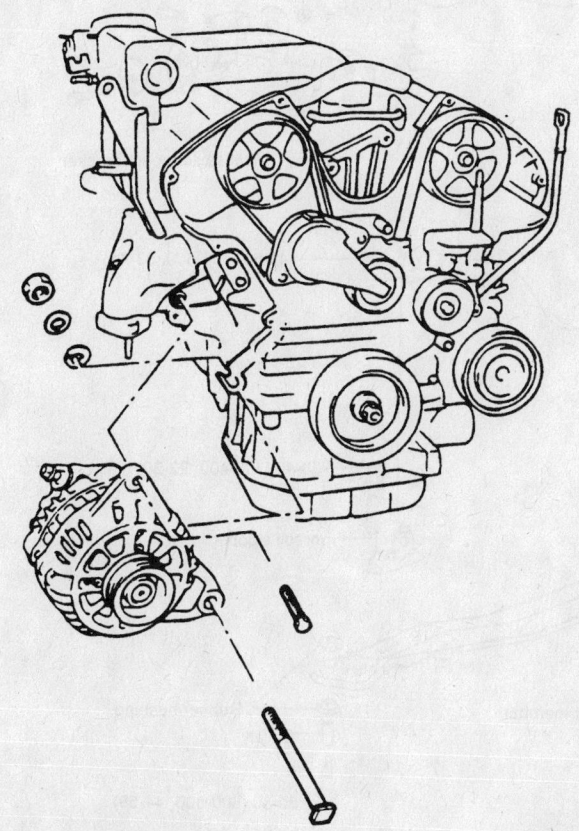

Exploded view of the alternator—XG 300

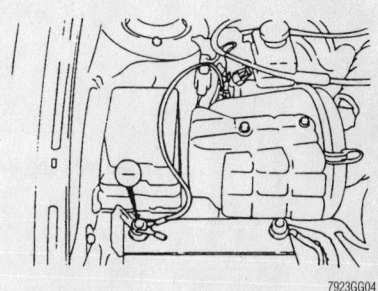

Connect the jumper wire from ground to the ignition timing terminal—3.0L engine

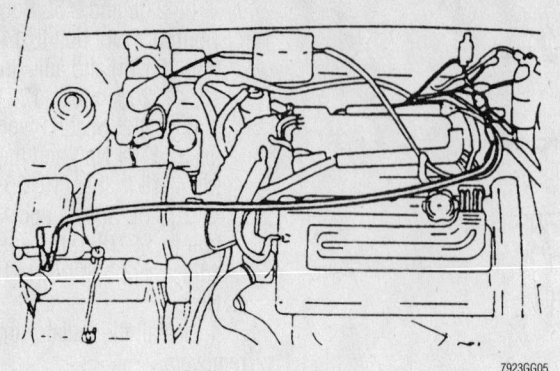

7923GG05

Connect the jumper wire from ground to the ignition timing terminal—1.8L engines

Engine Assembly

REMOVAL & INSTALLATION

➡**Hyundai recommends that the engine and transaxle be removed as a single unit on all models.**

1. Before servicing the vehicle, refer to the precautions in the beginning of this section.
2. Drain the cooling system.
3. Drain the transaxle.
4. Drain the engine oil.
5. Relieve fuel system pressure.

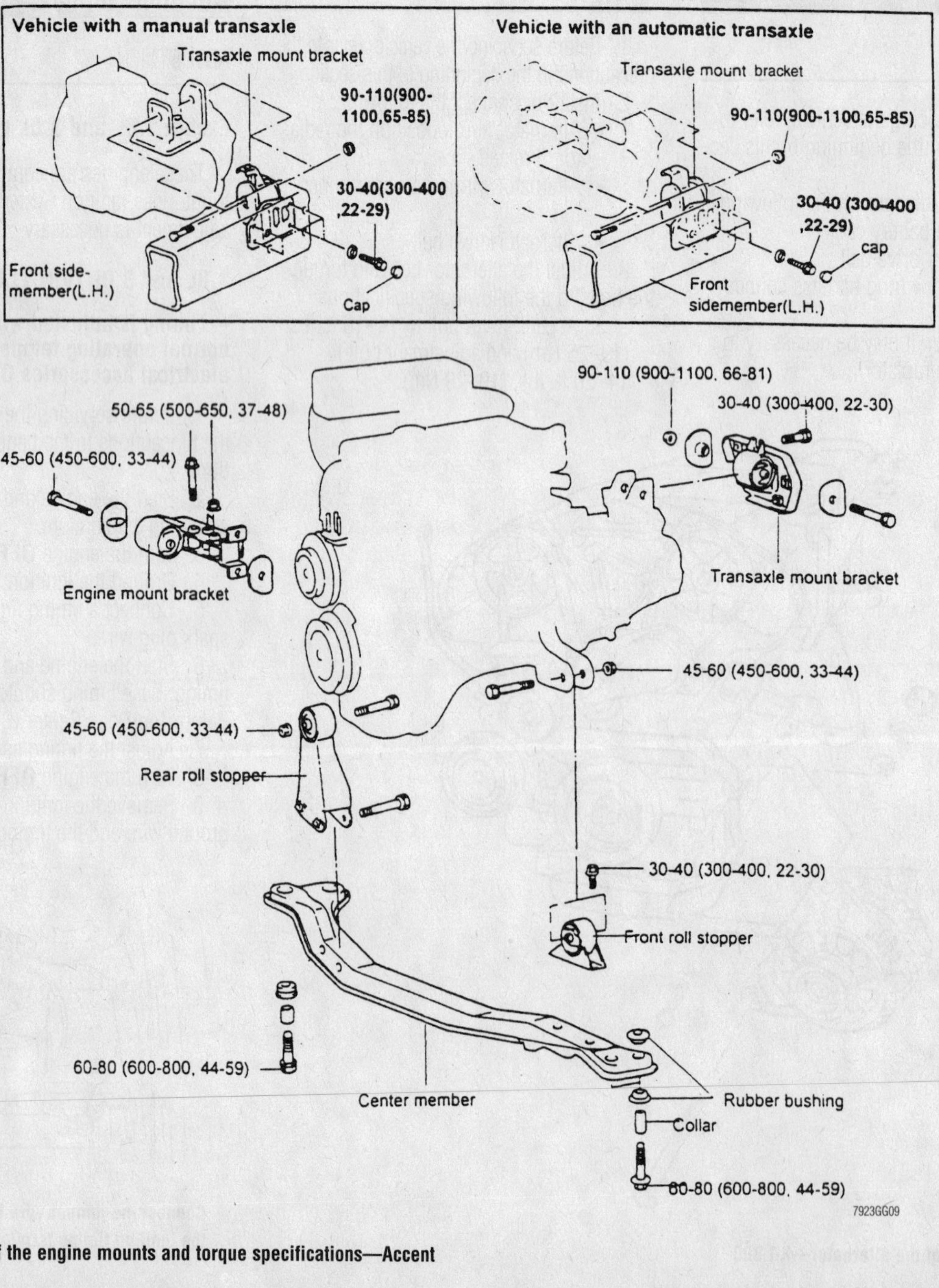

Vehicle with a manual transaxle

Transaxle mount bracket

90–110(900-1100,65-85)

30–40(300-400,22-29)

Front side-member(L.H.)

Cap

Vehicle with an automatic transaxle

Transaxle mount bracket

90–110(900-1100,65-85)

30–40 (300-400,22-29)

cap

Front sidemember(L.H.)

90-110 (900-1100, 66-81)

50-65 (500-650, 37-48)

30-40 (300-400, 22-30)

45-60 (450-600, 33-44)

Engine mount bracket

Transaxle mount bracket

45-60 (450-600, 33-44)

45-60 (450-600, 33-44)

Rear roll stopper

30-40 (300-400, 22-30)

Front roll stopper

60-80 (600-800, 44-59)

Center member

Rubber bushing

Collar

60-80 (600-800, 44-59)

7923GG09

Exploded view of the engine mounts and torque specifications—Accent

6. Remove or disconnect the following:
- Battery
- Hood
- Air intake assembly
- Accessory drive belts
- Engine wiring harness connectors
- Reverse lamp switch connector, if equipped
- Speedometer cable
- Alternator harness connectors
- Oil pressure gauge sender connector
- Radiator hoses
- Cooling fan
- Fuel lines
- Control cable, if equipped
- Brake booster vacuum line
- Intake manifold vacuum lines
- Heater hoses
- Accelerator cable
- Cruise control cable, if equipped
- Engine ground cable

7. If equipped with a manual transaxle, disconnect or remove the following:
- Clutch cable
- Select control valve connector
- Shift linkage rods

8. If equipped with an automatic transaxle, disconnect or remove the following:
- Transaxle oil cooler lines
- Shift cable
- Transaxle wiring connectors

9. For all vehicles, remove or disconnect the following:
- Radiator
- Power steering pump
- A/C compressor, if equipped
- Exhaust front pipe
- Lower ball joints
- Stabilizer bar links

10. Separate the inner CV-joints from the transaxle and suspend the halfshafts out of the work area with safety wire.

11. Attach a hoist to the engine lifting eyes.

12. Remove or disconnect the following:
- Front and rear roll stoppers
- Engine mount and bracket
- Transaxle mount and bracket

13. Lift the powertrain out of the vehicle.

To install:

14. Lower the powertrain into position.

15. Install the motor mount bracket and torque the fasteners as follows:

    a. V6 engines: 43–58 ft. lbs. (60–80 Nm).

    b. All others: 37–48 ft. lbs. (45–60 Nm).

16. Install the transaxle mount bracket and torque the fasteners as follows:

    a. Sonata: 29–36 ft. lbs. (40–50 Nm).

    b. Tiburon and Elantra: 33–43 ft. lbs. (45–60 Nm).

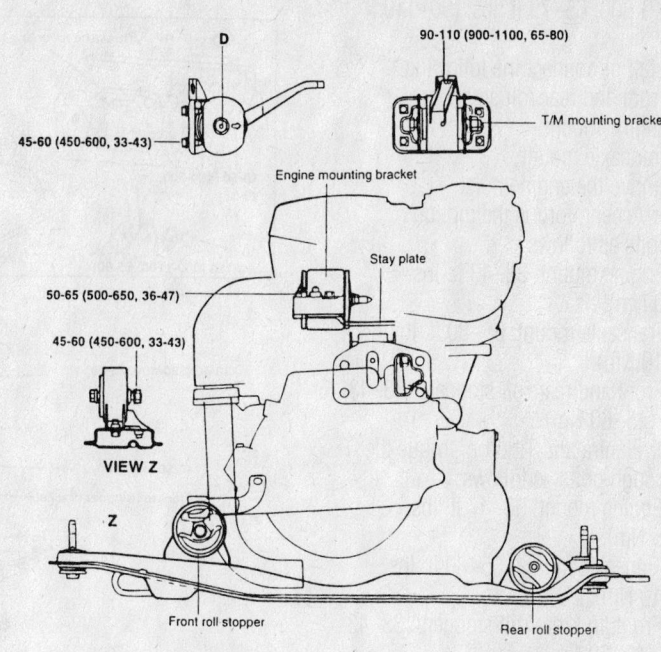

TORQUE: Nm (kg.cm, lb.ft)

7923GG13

Exploded view of the engine mounts and torque specifications—Tiburon and Elantra

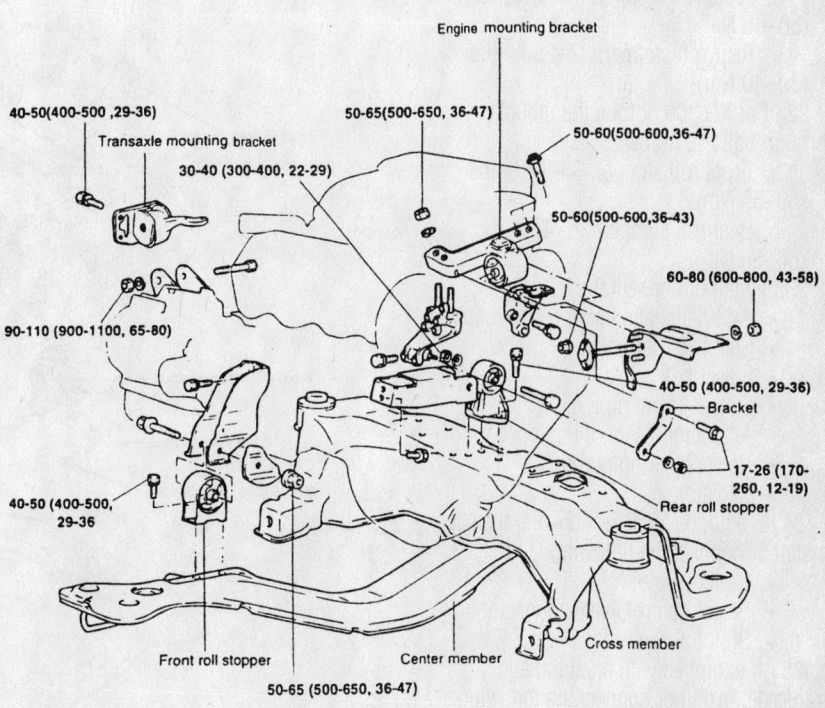

TORQUE : Nm (kg.cm, lb.ft)

7923GG15

Exploded view of the engine mounts and torque specifications—4 cylinder Sonata

c. Accent: 22–30 ft. lbs. (30–40 Nm).

d. XG 300: 65–79 ft. lbs. (90–110 Nm).

17. Install or connect the following:
- Front and rear roll stoppers
- Engine mount
- Transaxle mount

18. Remove the engine hoist.

19. For Accent, torque the mount through bolts as follows:

a. Engine mount: 33–43 ft. lbs. (45–60 Nm).

b. Transaxle mount: 65–80 ft. lbs. (90–110 Nm).

c. Front and rear roll stoppers: 33–43 ft. lbs. (45–60 Nm).

20. For Elantra and Tiburon, torque the mount through bolts as follows:

a. Engine mount: 36–47 ft. lbs. (50–65 Nm).

b. Transaxle mount: 65–80 ft. lbs. (90–110 Nm).

c. Front and rear roll stoppers: 33–43 ft. lbs. (45–60 Nm).

21. For Sonata, torque the mount through bolts as follows:

a. 4 cylinder engine mount: 43–58 ft. lbs. (60–80 Nm).

b. V6 engine mount: 65–80 ft. lbs. (90–110 Nm).

c. Transaxle mount: 65–80 ft. lbs. (90–110 Nm).

d. Front roll stopper: 36–47 ft. lbs. (50–65 Nm).

e. Rear roll stopper: 22–29 ft. lbs. (30–40 Nm).

22. For XG 300, torque the mount through bolts as follows:

a. Front roll stopper: 36–47 ft. lbs. (50–65 Nm).

b. Rear roll stopper: 36–47 ft. lbs. (50–65 Nm).

23. Install or connect the following:
- Axle halfshafts using new circlips
- Stabilizer bar links
- Lower ball joints
- Exhaust front pipe
- A/C compressor, if equipped
- Power steering pump
- Radiator

24. If equipped with a manual transaxle, install or connect the following:
- Clutch cable
- Select control valve connector
- Shift linkage rods

25. If equipped with an automatic transaxle, install or connect the following:
- Transaxle oil cooler lines
- Shift cable
- Transaxle wiring connectors

26. For all vehicles, install or connect the following:

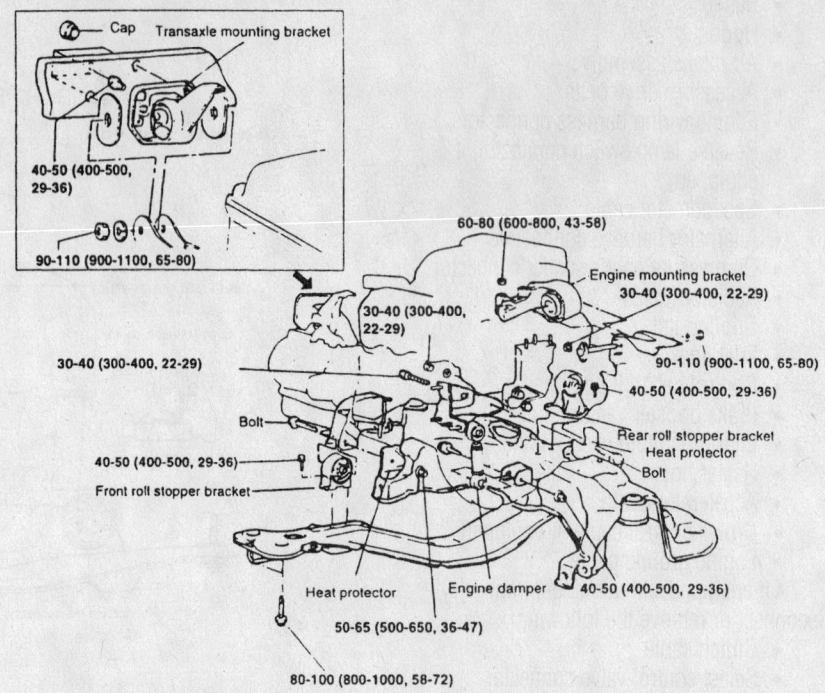

TORQUE : Nm (kg.cm, lb.ft)

Exploded view of the engine mounts and torque specifications—V6 Sonata

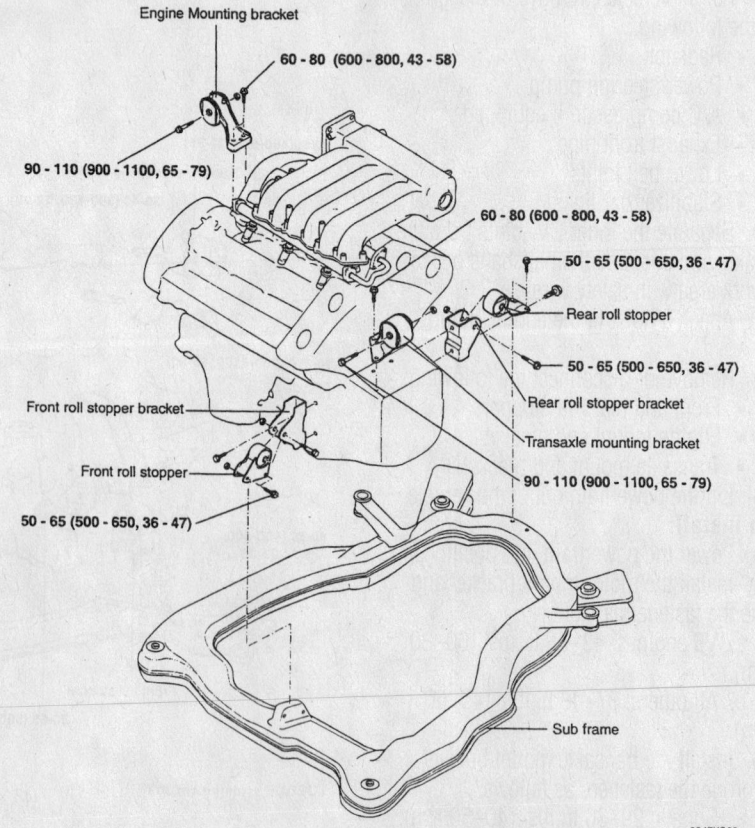

Exploded view of the engine mounts and torque specifications—XG 300

- Engine ground cable
- Cruise control cable, if equipped
- Accelerator cable
- Heater hoses
- Intake manifold vacuum lines
- Brake booster vacuum line
- Fuel lines
- Cooling fan
- Radiator hoses
- Oil pressure gauge sender connector
- Alternator harness connectors
- Speedometer cable
- Reverse lamp switch connector
- Engine wiring harness connectors
- Accessory drive belts
- Air intake assembly
- Hood
- Battery

27. Fill the engine with clean oil.
28. Fill the transaxle to the correct level.
29. Fill the cooling system to the proper level.
30. Start the engine and check for leaks.

Water Pump

REMOVAL & INSTALLATION

Except XG 300

1. Before servicing the vehicle, refer to the precautions in the beginning of this section.
2. Drain the cooling system.
3. Remove or disconnect the following:
   - Negative battery cable
   - Accessory drive belts
   - Radiator hose
   - Bypass hose, if equipped
   - Water pump pulley
   - Front cover
   - Timing belt
   - Alternator bracket
   - Water pump

➡ **The water pump bolts are different lengths. Note the bolt location for assembly.**

To install:

4. Install the water pump with new gaskets and O-rings. Tighten the bolts to the following specifications:
   a. 1.5L SOHC engine: 105–132 inch lbs. (12–15 Nm).
   b. 1.5L DOHC engine: 14–20 ft. lbs. (20–27 Nm).
   c. 1.8L, 2.0L, and 2.4L engines: 14–20 ft. lbs. (20–27 Nm).
   d. 2.5L engine: 11–16 ft. lbs. (15–22 Nm).
5. Install or connect the following:
   - Alternator bracket
   - Timing belt and front cover
   - Water pump pulley
   - Bypass hose, if equipped
   - Radiator hose
   - Accessory drive belts
   - Negative battery cable
6. Fill the cooling system to the proper level.
7. Start the engine and check for leaks.

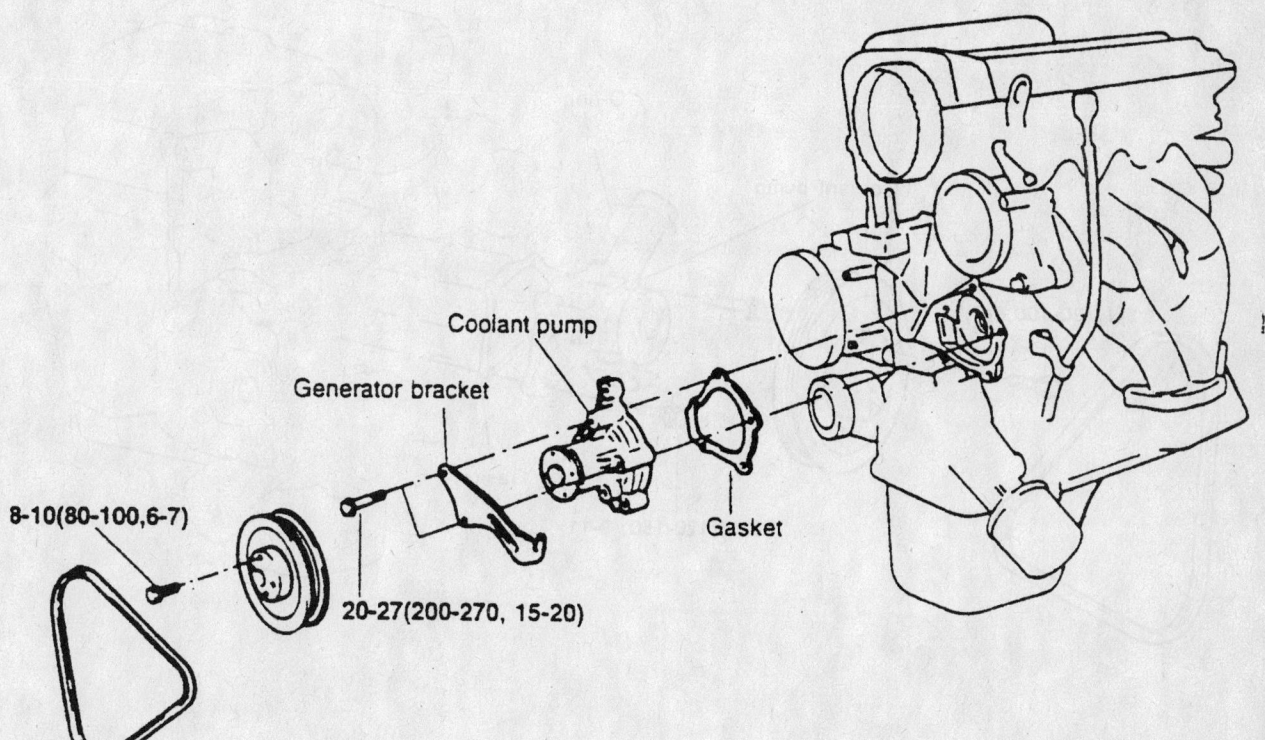

Coolant pump

Generator bracket

8-10(80-100,6-7)

20-27(200-270, 15-20)

Gasket

7923GG17

Exploded view of the water pump assembly—1.5L engines

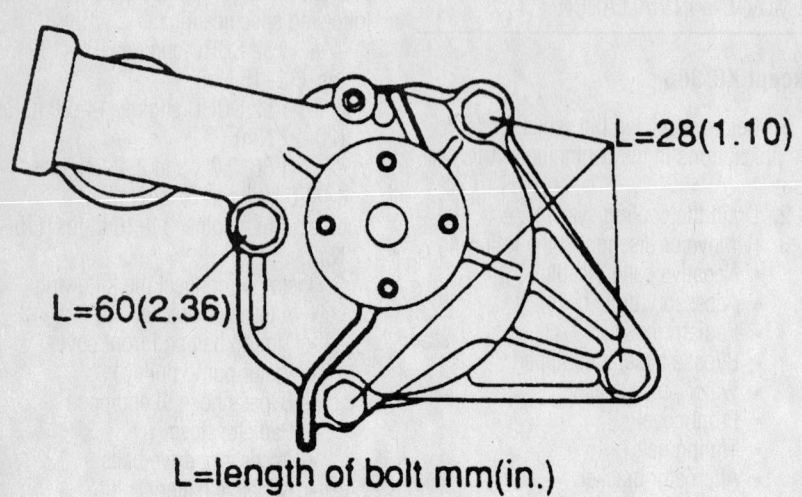

L=28(1.10)

L=60(2.36)

L=length of bolt mm(in.)

7923GG19

Water pump bolt lengths—1.5L engines

XG 300

1. Before servicing the vehicle, refer to the precautions in the beginning of this section.
2. Drain the cooling system.
3. Remove or disconnect the following:
   - Negative battery cable
   - Accessory drive belts
   - Water pump pulley
   - Timing belt, tensioner and idler pulley
   - Water pump

To install:

4. Clean all mating surfaces of any residual gasket material.
5. Install or connect the following:
   - Water pump with a new gasket and torque the bolts to 16 ft. lbs. (22 Nm)
   - Tensioner and timing belt and idler pulley
   - Water pump pulley and drive belts
   - Negative battery cable
6. Fill the cooling system to the proper level.

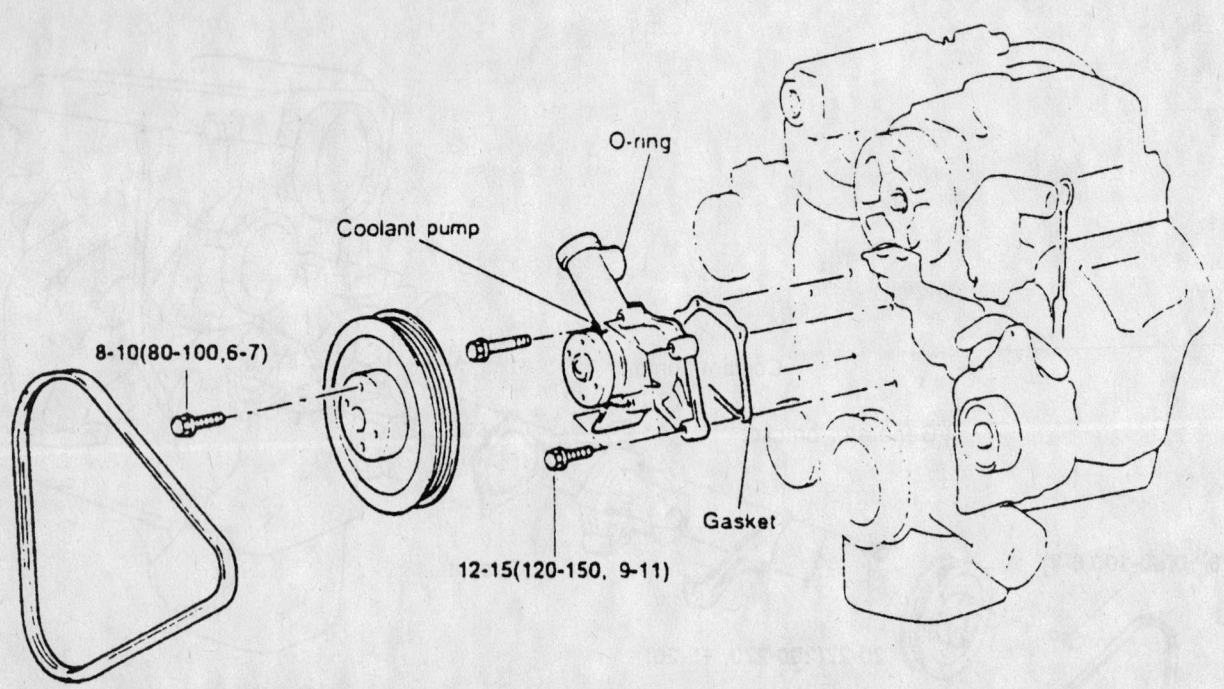

O-ring

Coolant pump

8-10(80-100,6-7)

12-15(120-150, 9-11)

Gasket

TORQUE : Nm (kg.cm, lb.ft)

7923GG20

Water pump assembly—1.8L and 2.0L engines

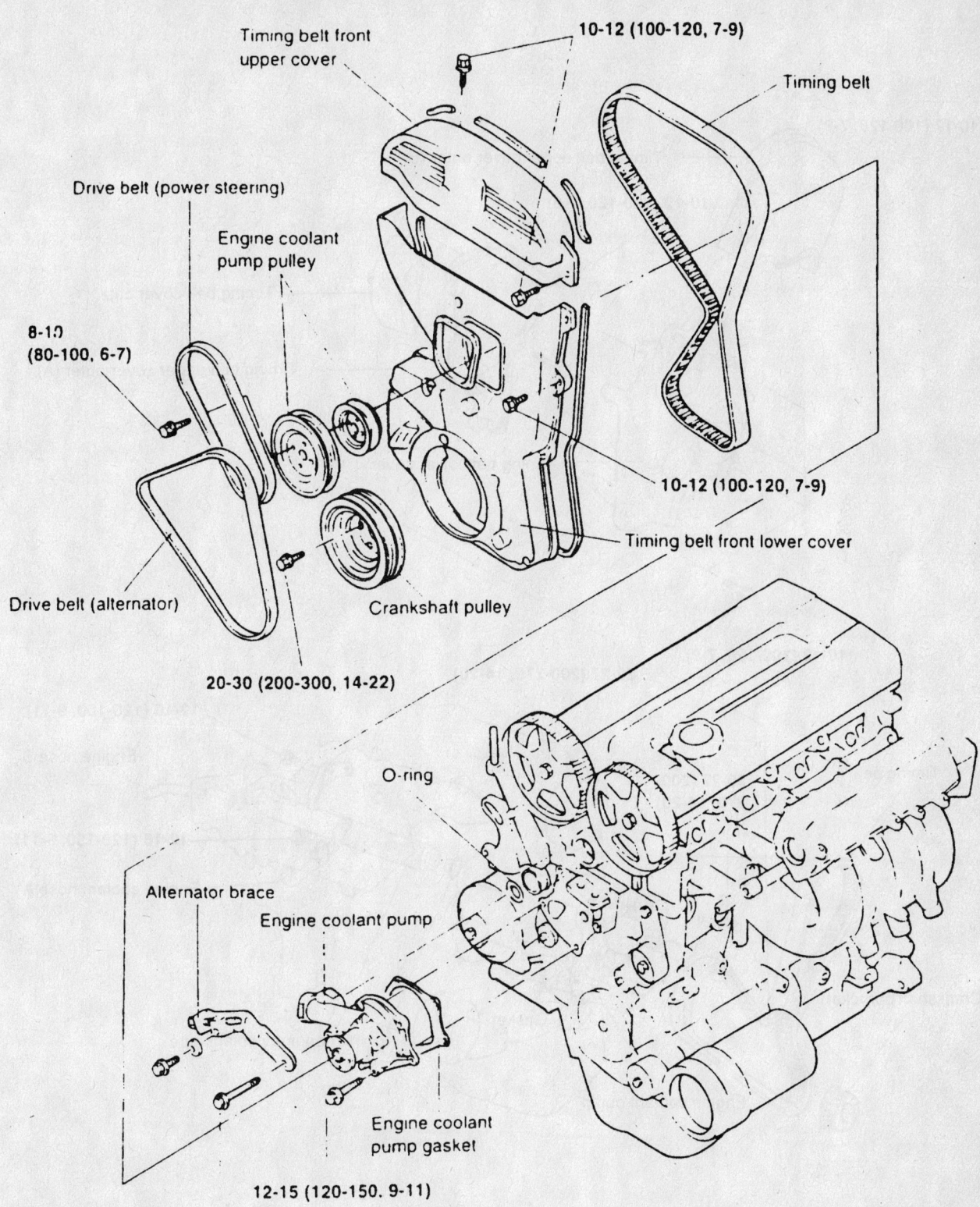

10-12 (100-120, 7-9)

Timing belt front upper cover

Timing belt

Drive belt (power steering)

Engine coolant pump pulley

8-10 (80-100, 6-7)

10-12 (100-120, 7-9)

Timing belt front lower cover

Drive belt (alternator)

Crankshaft pulley

20-30 (200-300, 14-22)

O-ring

Alternator brace

Engine coolant pump

Engine coolant pump gasket

12-15 (120-150. 9-11)

20-27 (200-270, 14-20)

TORQUE : Nm (kg.cm, lb.ft)

7923GG21

Exploded view of the water pump assembly and related components—2.0L engines

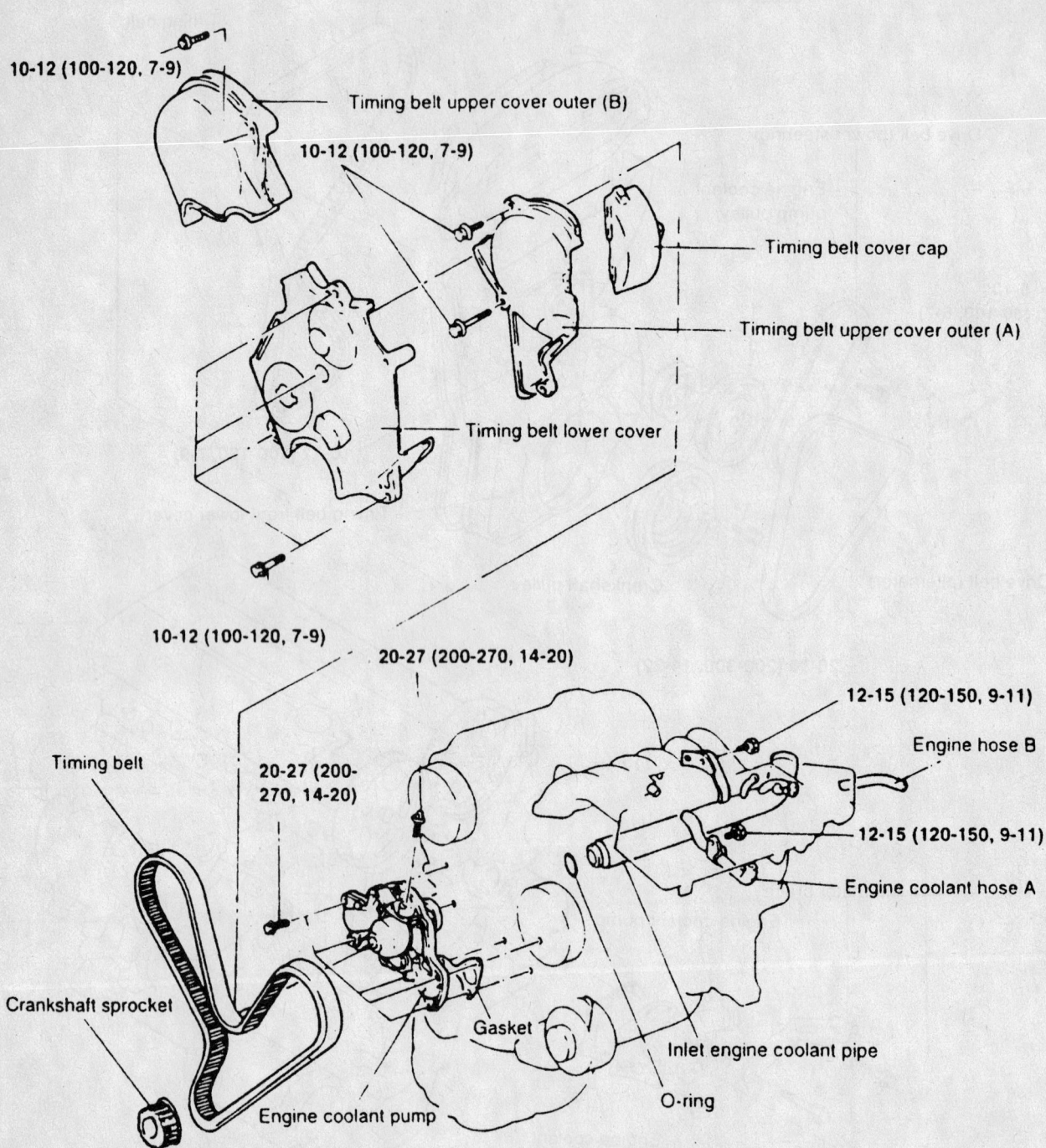

10-12 (100-120, 7-9)

Timing belt upper cover outer (B)

10-12 (100-120, 7-9)

Timing belt cover cap

Timing belt upper cover outer (A)

Timing belt lower cover

10-12 (100-120, 7-9)

20-27 (200-270, 14-20)

12-15 (120-150, 9-11)

Engine hose B

20-27 (200-270, 14-20)

12-15 (120-150, 9-11)

Engine coolant hose A

Timing belt

Crankshaft sprocket

Engine coolant pump

Gasket

Inlet engine coolant pipe

O-ring

TORQUE : Nm (kg.cm, lb.ft)

7923GG22

Water pump assembly—3.0L engine Sonata

V-6 ENGINE

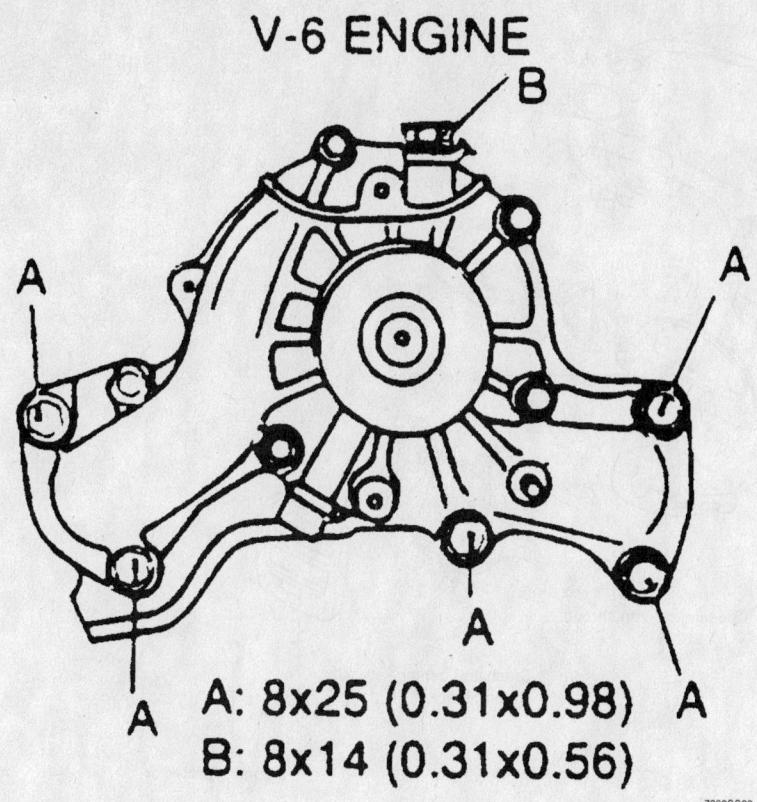

A: 8x25 (0.31x0.98)
B: 8x14 (0.31x0.56)

7923GG23

Water pump bolt lengths—3.0L Sonata

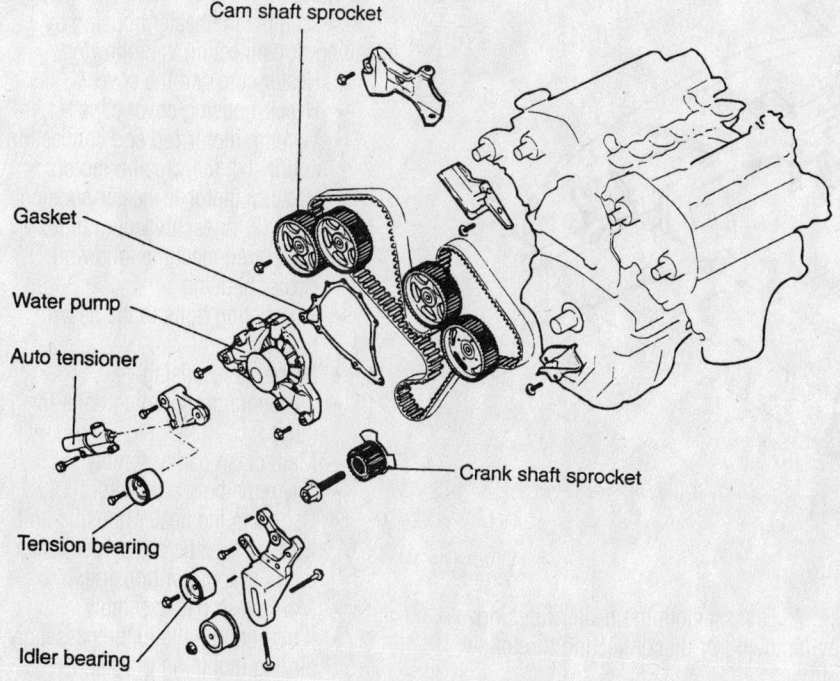

9347KG03

Water pump assembly—XG 300

7. Start the vehicle, check for leaks and repair if necessary.

Heater Core

REMOVAL & INSTALLATION

Accent

1. Before servicing the vehicle, refer to the precautions in the beginning of this section.

2. Disconnect the negative battery cable and wait 90 seconds for the SRS memory battery to drain.

※※ CAUTION

After disconnecting the negative battery cable, wait for at least 30 seconds for the SRS module to deplete its stored energy.

3. Drain the cooling system into a clean container for reuse.

4. Remove or disconnect the following:
- Heater hoses with the vacuum hose from the heater housing
- Discharge and recover the air conditioning system refrigerant
- Suction and discharge hoses from the evaporator assembly

5. Remove the SRS module and the steering wheel:
- Steering wheel-to-SRS module nuts
- SRS module from the steering wheel and disconnect the electrical connector
- Steering wheel-to-steering column nut
- Steering wheel from the steering column

6. Remove or disconnect the following:
- Multi-function switch assembly
- Front and rear console assemblies
- Lower left side crash pad
- Center fascia panel and disconnect the connectors and vacuum connector from the heater control assembly
- Heater control assembly and the audio system
- Glove box
- 4 mounting bolts from the passenger air bag mounting bracket, if equipped
- Main crash pad assembly
- Cables from the heater housing and the thermostatic switch connector from the evaporator housing
- Any remaining connectors

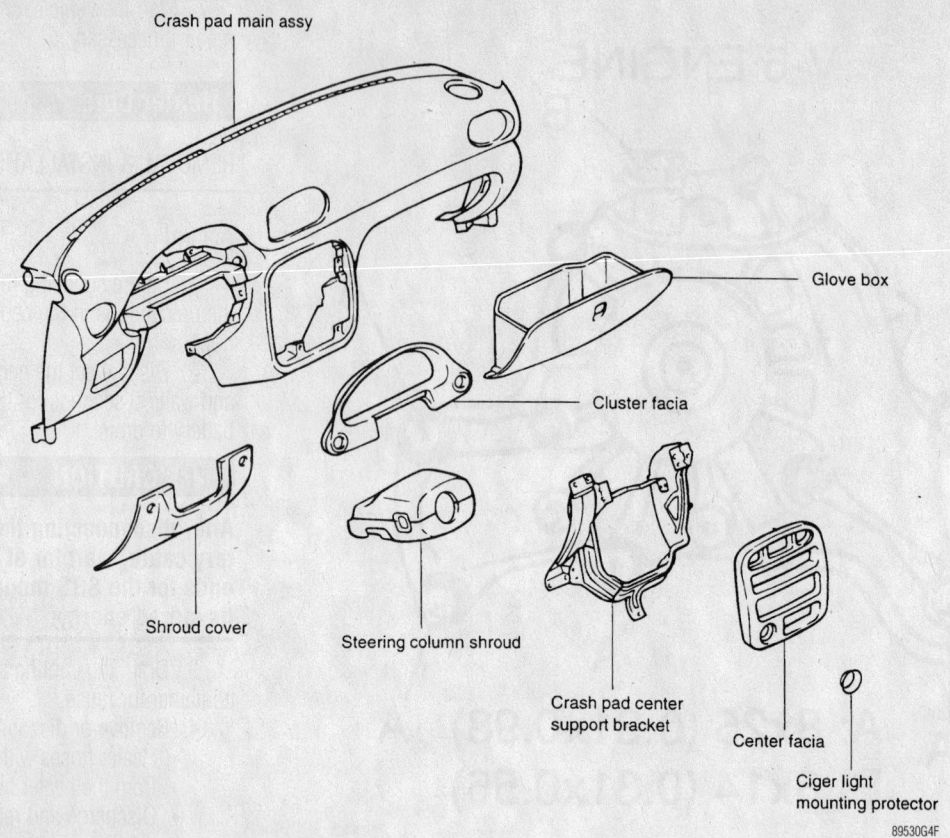

Crash pad main assy

Glove box

Cluster facia

Shroud cover

Steering column shroud

Crash pad center
support bracket

Center facia

Ciger light
mounting protector

89530G4F

Instrument panel assembly—Accent

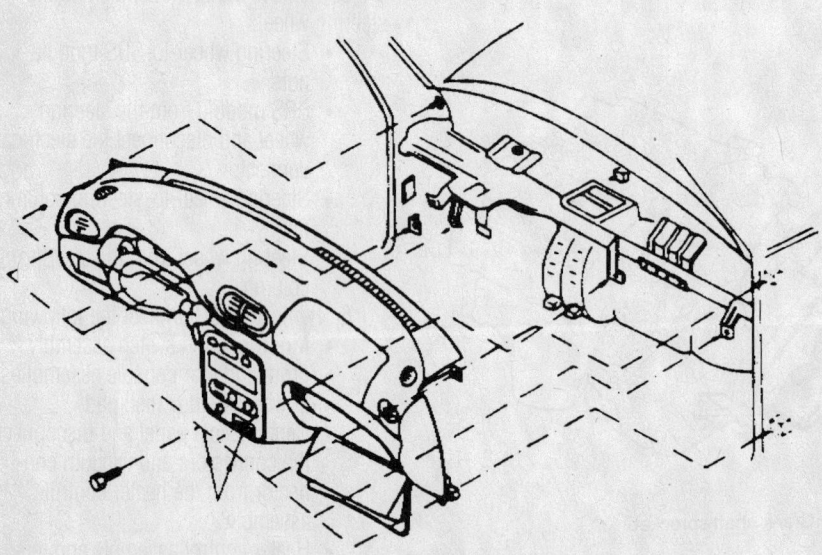

89530G4G

Instrument panel screw locations—Accent

- Main crash pad assembly
- 3 evaporator mounting bolts (or nuts)
- Evaporator housing
- 3 mounting bolts from the heater housing
- Heater housing

7. Disassemble the heater housing by removing or disconnecting the following:

- Vacuum motor-to-heater housing bolts (2 for each vacuum motor)
- Vacuum motor rod end connection and remove the vacuum motors

- Heater housing cover clips
- Cover and the heater core

To install:

8. Assemble the heater housing by installing or connecting the following:

- Heater core and the cover
- Heater housing cover clips
- Vacuum motor rod end connection and install the vacuum motors
- Vacuum motor-to-heater housing bolts (2 for each vacuum motor)

9. Install or connect the following:

- Heater housing
- 3 mounting bolts to the heater housing
- Evaporator housing
- 3 evaporator mounting bolts (or nuts)
- Main crash pad assembly
- Any remaining connectors
- Cables to the heater housing and the thermostatic switch connector to the evaporator housing
- Main crash pad assembly
- 4 mounting bolts to the passenger air bag mounting bracket, if equipped
- Glove box
- Heater control assembly and the audio system

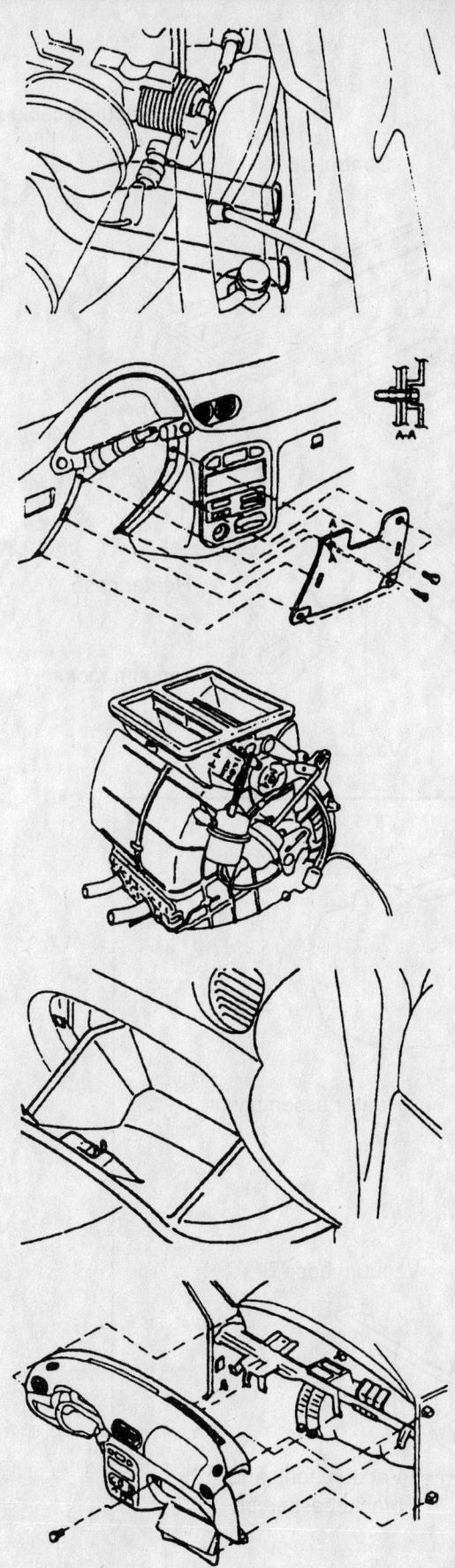

View of the heater housing assembly and related components—Accent

93112G30

- Connectors and vacuum connector to the heater control assembly. Install the center fascia panel.
- Lower left side crash pad
- Front and rear console assemblies

10. Install the SRS module and the steering wheel by installing or connecting the following:
- Steering wheel to the steering column
- Steering wheel-to-steering column nut and torque to 30–37 ft. lbs. (40–50 Nm)
- SRS module to the steering wheel and connect the electrical connector
- Steering wheel-to-SRS module nuts

11. Install or connect the following:
- Multi-function switch assembly
- Suction and discharge hoses to the evaporator assembly
- Heater hoses with the vacuum hose to the heater housing

12. Refill the cooling system.

13. Connect the negative battery cable.

14. Evacuate, charge and leak test the air conditioning system refrigerant.

15. Operate the engine to normal operating temperatures; then, check the climate control operation and check for leaks.

Elantra

1. Before servicing the vehicle, refer to the precautions in the beginning of this section.

2. Disconnect the negative battery cable.

✳✳ CAUTION

After disconnecting the negative battery cable, wait for at least 30 seconds for the SRS module to deplete its stored energy.

3. Discharge and recover the air conditioning system refrigerant.

4. Drain the engine coolant into a clean container for reuse.

5. Disconnect the heater hoses from the heater core. Plug the openings.

6. Disconnect the vacuum line from the heater housing vacuum nipple.

7. Remove the SRS module and the steering wheel by removing or disconnecting the following:
- Steering wheel-to-SRS module nuts
- SRS module from the steering wheel and disconnect the electrical connector

* Heater Assembly

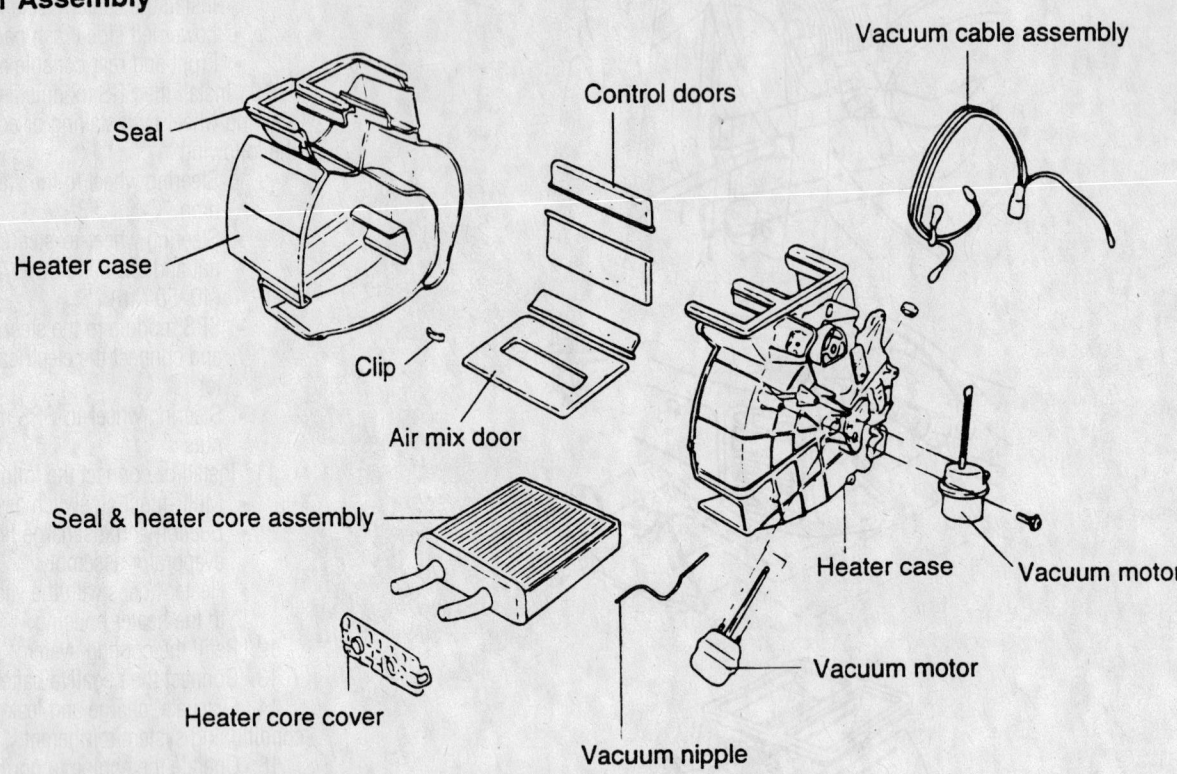

Seal

Control doors

Vacuum cable assembly

Heater case

Clip

Air mix door

Seal & heater core assembly

Heater core cover

Heater case

Vacuum motor

Vacuum motor

Vacuum nipple

* Vacuum Source Lines

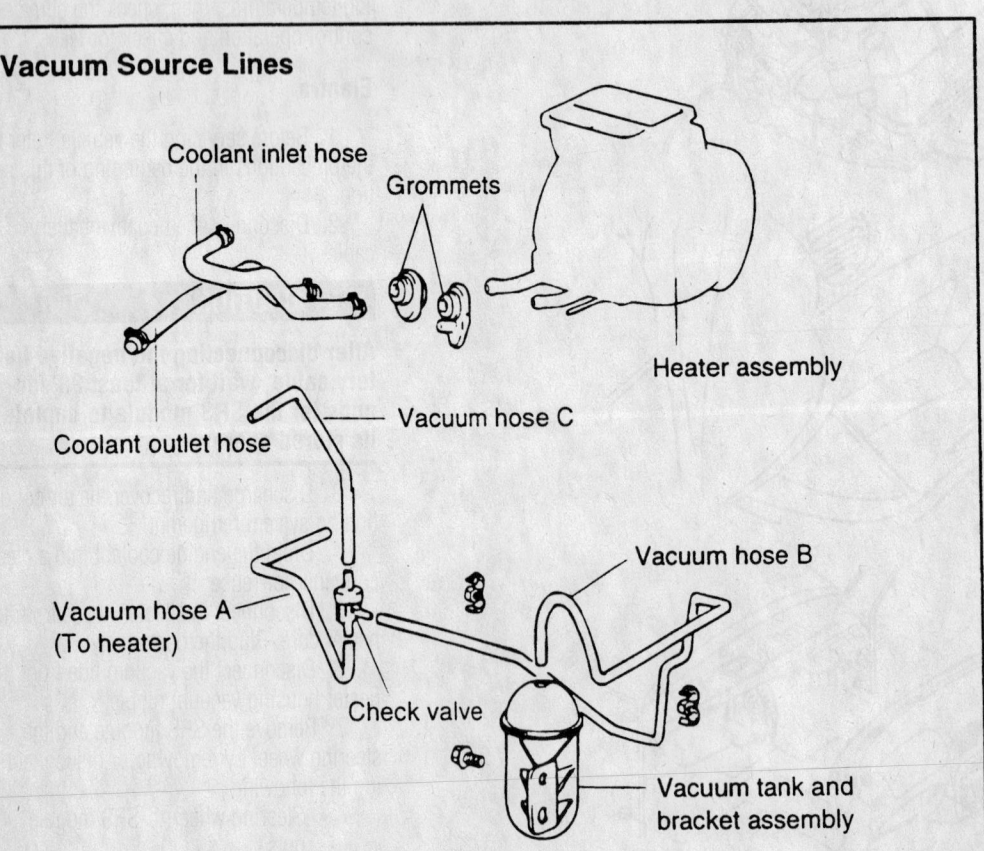

Coolant inlet hose

Grommets

Coolant outlet hose

Vacuum hose C

Heater assembly

Vacuum hose B

Vacuum hose A
(To heater)

Check valve

Vacuum tank and
bracket assembly

93112GP4

Exploded view of the heater core, heater housing and related components—Accent

- Steering wheel-to-steering column nut
- Steering wheel from the steering column

8. Remove the instrument panel by removing or disconnecting the following:
- Steering column cover screws and the covers
- Instrument panel lower cover at the driver's side
- Multi-function switch and disconnect the electrical connector at the steering column
- Instrument panel cluster fascia panel
- Instrument cluster-to-instrument panel screws, disconnect the electrical connectors and remove the instrument cluster
- Side fascia panel and disconnect the mirror control connector
- Hood release mounting screws
- Rheostat and the upper console cover
- Heater control cable
- Electrical connectors at the center of the instrument panel and remove the center fascia panel assembly
- Console

- Radio-to-chassis screws and the radio
- Glove box screws and the glove box
- Glove box striker screws and the upper glove box cover
- Defroster nozzle
- Loosen the speedometer drive gear sleeve and disconnect the speedometer cable from the instrument panel
- Passenger's side SRS module connector
- Instrument panel-to-chassis bolts
- Any remaining electrical connectors
- Ventilation ducts from the instrument panel
- Instrument panel

9. Remove or disconnect the following:
- Front right side heating duct from the heater housing
- Pull back the carpet and remove the right side console mounting bracket
- Front left side duct from the heater housing
- Pull back the carpet and remove the left side console mounting bracket
- Rear heating duct from the heater housing

- Control modules electrical connectors at the center fascia panel support bracket
- Center fascia panel support bracket screws, bolts and/or nuts; then, remove the center fascia panel support bracket
- Center support bars
- Glove box support bracket-to-instrument panel bolts and the bracket

10. Remove the evaporator housing by removing or disconnecting the following:
- Refrigerant lines from the evaporator housing and discard the O-rings
- Thermostatic switch connector
- Evaporator housing upper and lower bolts
- Evaporator housing

11. Remove or disconnect the following:
- Heater housing-to-chassis bolts and the housing
- Vacuum motor-to-heater housing bolts (2 for each vacuum motor)
- Vacuum motor rod end connection and remove the vacuum motors
- Heater housing cover clips
- Cover and the heater core

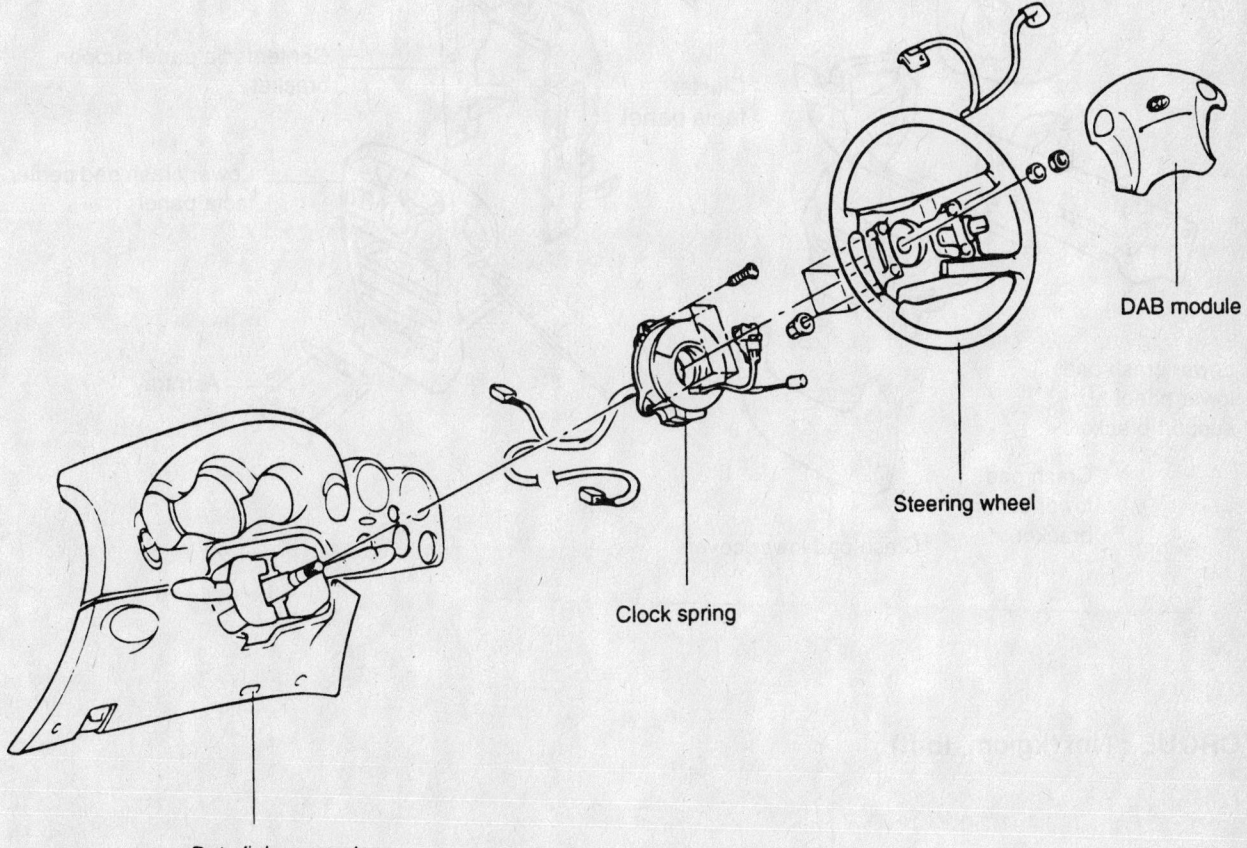

DAB module

Steering wheel

Clock spring

Data link connector

93112G07

Exploded view of the SRS module, steering wheel and related components—Elantra

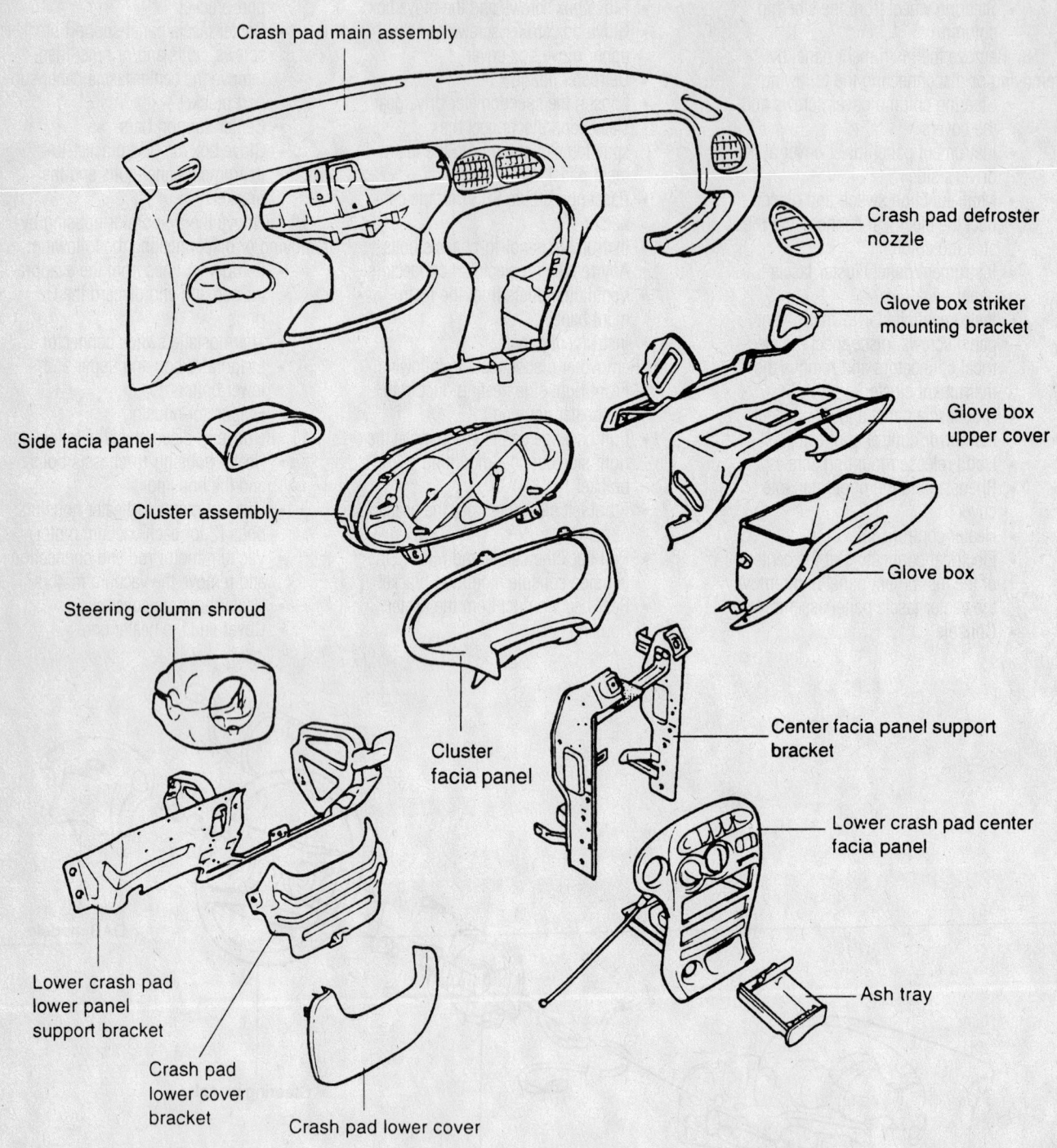

Crash pad main assembly

Crash pad defroster nozzle

Glove box striker mounting bracket

Glove box upper cover

Side facia panel

Cluster assembly

Glove box

Steering column shroud

Cluster facia panel

Center facia panel support bracket

Lower crash pad center facia panel

Lower crash pad lower panel support bracket

Crash pad lower cover bracket

Crash pad lower cover

Ash tray

TORQUE : Nm (kg·cm, lb·ft)

93112G08

Exploded view of the instrument panel and related components—Elantra

To install:

12. Assemble the heater housing by installing or connecting the following:
   - Heater core and the cover
   - Heater housing cover clips
   - Vacuum motor rod end connection and install the vacuum motors
   - Vacuum motor-to-heater housing bolts (2 for each vacuum motor)

13. Install or connect the following:
   - Heater housing and the housing-to-chassis bolts
   - Evaporator housing
   - Evaporator housing upper and lower bolts
   - Thermostatic switch connector
   - Connect the refrigerant lines to the evaporator housing

14. Install or connect the following:
   - Glove box support bracket and the bracket-to-instrument panel bolts
   - Center support bars
   - Center fascia panel support bracket; then, install the center fascia panel support bracket screws, bolts and/or nuts
   - Center fascia panel support bracket, connect the control modules electrical connectors
   - Rear heating duct to the heater housing
   - Left side console mounting bracket and install the carpet
   - Front left side duct to the heater housing
   - Right side console mounting bracket and install the carpet
   - Front right side heating duct to the heater housing

15. Install the instrument panel by installing or connecting the following:
   - Instrument panel
   - Ventilation ducts to the instrument panel

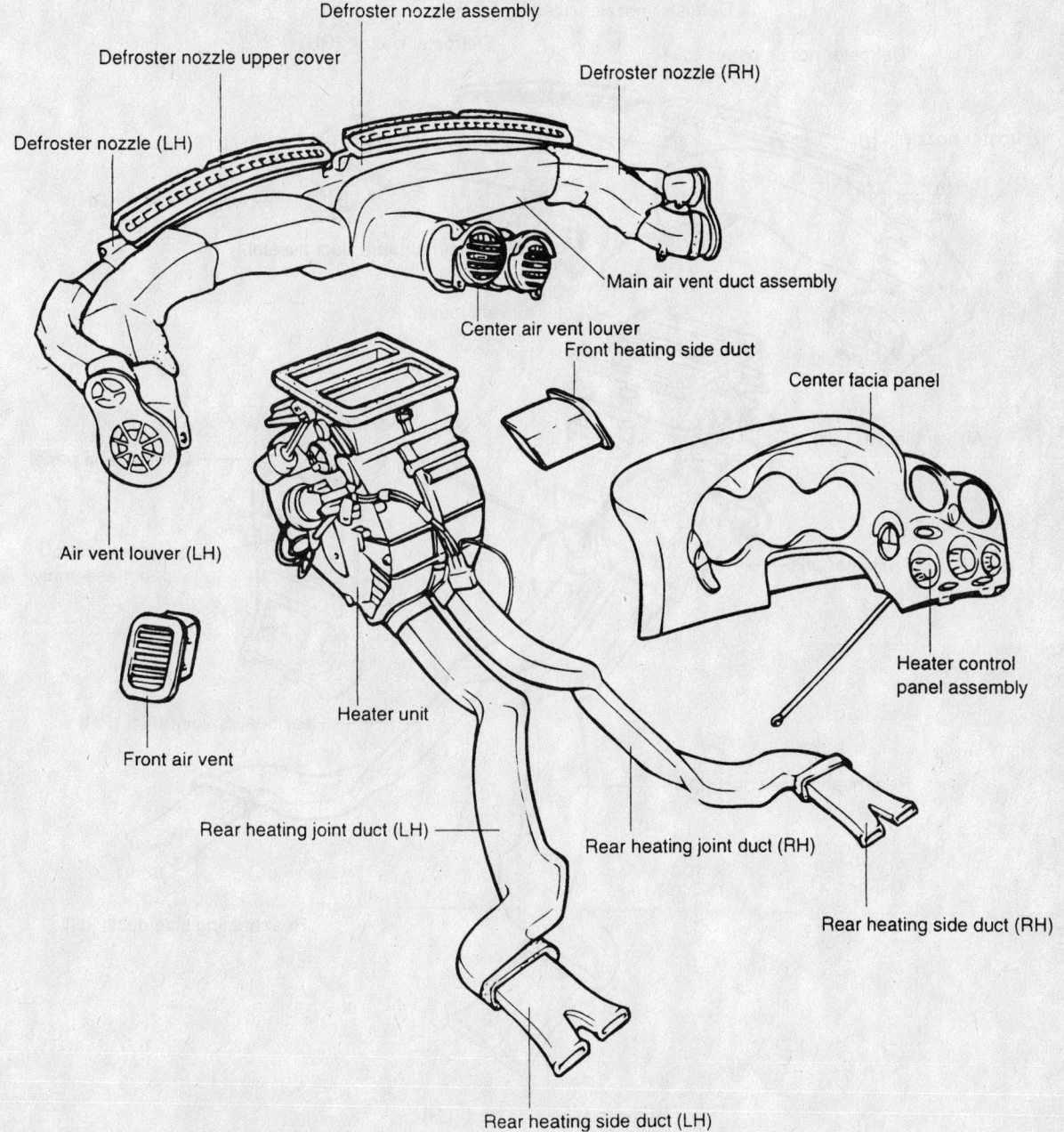

Exploded view of the heater housing, center fascia, distribution ducts and related components—Elantra Coupe

93112G00

- Instrument panel-to-chassis bolts
- Passenger's side SRS module connector
- Speedometer cable to the instrument panel and tighten the speedometer drive gear sleeve
- Defroster nozzle
- Upper glove box cover and the glove box striker screws
- Glove box and the glove box screws
- Radio and the radio-to-chassis screws
- Console
- Electrical connectors and install the center fascia panel assembly

- Heater control cable
- Rheostat and the upper console cover
- Hood release mounting screws
- Side fascia panel and connect the mirror control connector
- Instrument cluster, connect the electrical connectors and install the instrument cluster-to-instrument panel screws
- Instrument panel cluster fascia panel
- Multi-function switch and connect the electrical connector
- Instrument panel lower cover

- Steering column cover and the cover screws

16. Install the SRS module and the steering wheel by installing or connecting the following:

- Steering wheel to the steering column
- Steering wheel-to-steering column nut and torque to 30–37 ft. lbs. (40–50 Nm)
- SRS module to the steering wheel and connect the electrical connector
- Steering wheel-to-SRS module nuts

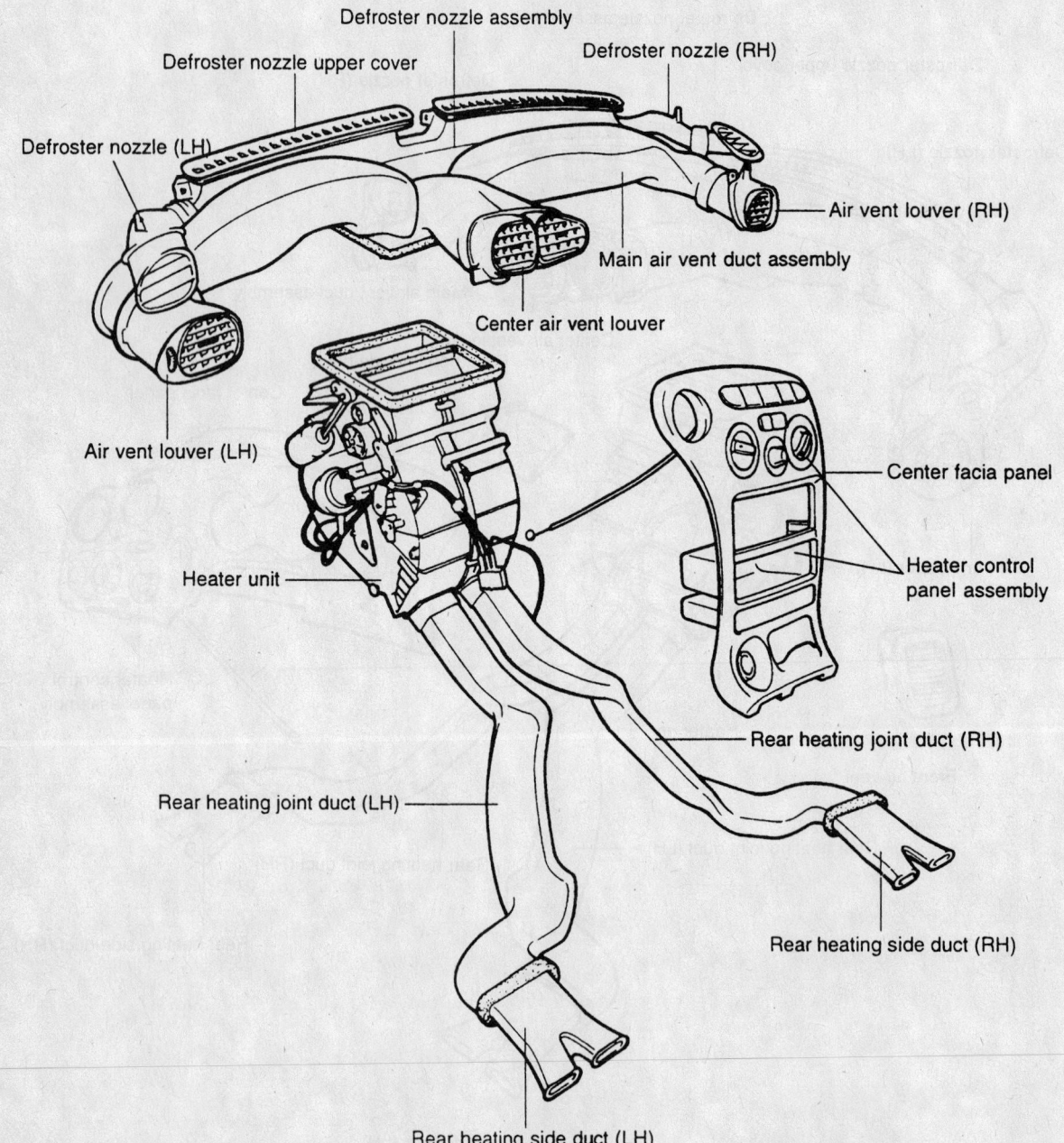

Exploded view of the heater housing, center fascia, distribution ducts and related components—Elantra Sedan and Wagon

93112GP1

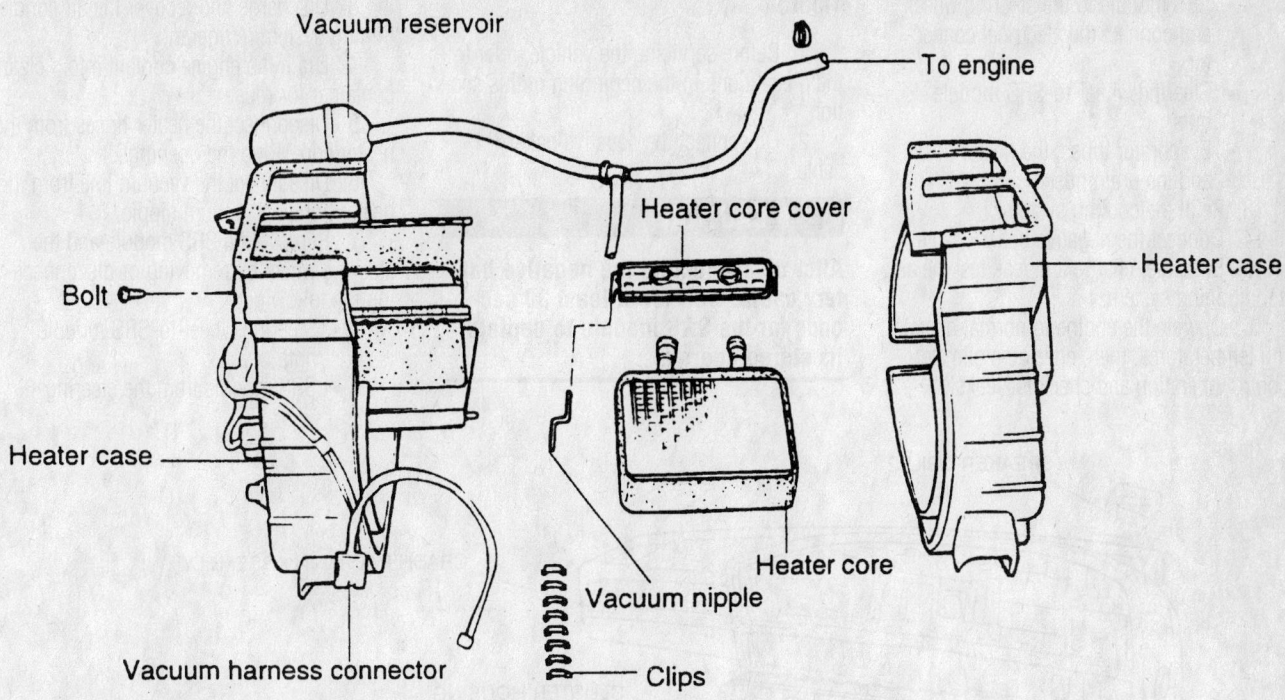

Exploded view of the heater core and heater housing and related components—Elantra

17. Connect the vacuum line to the heater housing vacuum nipple.

18. Connect the heater hoses to the heater core.

19. Refill the cooling system.

20. Connect the negative battery cable.

21. Evacuate, charge and leak test the air conditioning system.

22. Operate the engine to normal operating temperatures; then, check the climate control operation and check for leaks.

Sonata and XG 300

1. Before servicing the vehicle, refer to the precautions in the beginning of this section.

2. Disconnect the negative battery cable.

✳✳ CAUTION

After disconnecting the negative battery cable, wait for at least 30 seconds for the SRS module to deplete its stored energy.

3. Drain the cooling system into a clean container for reuse.

4. Remove the heater hoses from the heater housing.

5. Discharge and recover the air conditioning system refrigerant.

6. Remove the suction and discharge hoses from the evaporator assembly. Cap the hoses to minimize contamination.

7. Remove the evaporator drain hose.

8. Remove the SRS module and the steering wheel by removing or disconnecting the following:
   - Steering wheel-to-SRS module nuts
   - SRS module from the steering wheel and disconnect the electrical connector

✳✳ CAUTION

Store the SRS module in a safe place with the front facing upward.

   - Steering wheel-to-steering column nut
   - Steering wheel from the steering column

9. Remove or disconnect the following:
   - Front and rear console assembly and remove both side covers
   - Glove box, the center pad cover, the center crash pad and the cassette assembly
   - Lower crash pad. Remove the console mounting bracket and the center support bracket
   - Rear heater ducts from the heater housing
   - Control assembly
   - Blower speed control actuator connector and the blend door actuator connector, if equipped with semi-automatic temperature control

   - 4 retaining bolts and remove the heater assembly

10. Disassemble the heater housing by removing or disconnecting the following:
   - Vacuum motor-to-heater housing bolts (2 for each vacuum motor)
   - Vacuum motor rod end connection and remove the vacuum motors
   - Heater housing cover clips
   - Cover and the heater core

To install:

11. Install or connect the following:
   - Heater core and the cover
   - Heater housing cover clips
   - Vacuum motor rod end connection and install the vacuum motors
   - Vacuum motor-to-heater housing bolts (2 for each vacuum motor)
   - Heater assembly and attach it to the dash panel with the mounting bolts
   - Heater control assembly. Connect the ducts to the heater housing
   - Console mounting bracket and the center support bracket
   - Lower crash pad and both side covers
   - Front and rear console assembly

12. Install the SRS module and the steering wheel by installing or connecting the following:
   - Steering wheel to the steering column
   - Steering wheel-to-steering column nut and torque to 30–37 ft. lbs. (40–50 Nm)

- SRS module to the steering wheel and connect the electrical connector

- Steering wheel-to-SRS module nuts

- Evaporator tubes, the heater hoses and the drain hose

13. Refill the cooling system.

14. Connect the negative battery cable.

15. Evacuate, charge and leak test the air conditioning system.

16. Operate the engine to normal operating temperatures; then, check the climate control operation and check for leaks.

Tiburon

1. Before servicing the vehicle, refer to the precautions in the beginning of this section.

2. Disconnect the negative battery cable.

✳✳ CAUTION

After disconnecting the negative battery cable, wait for at least 30 seconds for the SRS module to deplete its stored energy.

3. Discharge and recover the air conditioning system refrigerant.

4. Drain the engine coolant into a clean container for reuse.

5. Disconnect the heater hoses from the heater core. Plug the openings.

6. Disconnect the vacuum line from the heater housing vacuum nipple.

7. Remove the SRS module and the steering wheel by removing or disconnecting the following:

- Steering wheel-to-SRS module nuts

- SRS module from the steering

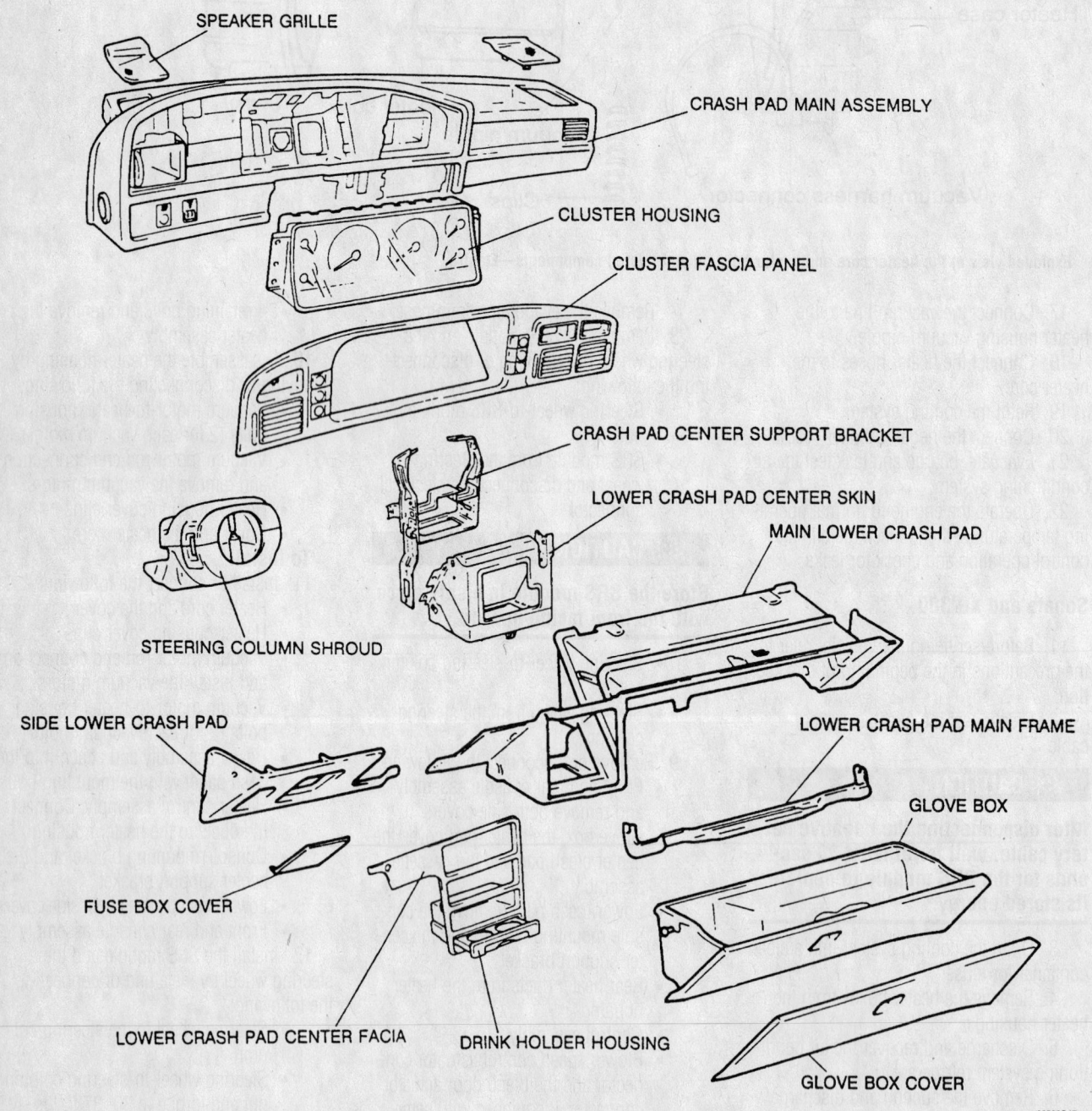

SPEAKER GRILLE

CRASH PAD MAIN ASSEMBLY

CLUSTER HOUSING

CLUSTER FASCIA PANEL

CRASH PAD CENTER SUPPORT BRACKET

LOWER CRASH PAD CENTER SKIN

MAIN LOWER CRASH PAD

STEERING COLUMN SHROUD

SIDE LOWER CRASH PAD

LOWER CRASH PAD MAIN FRAME

GLOVE BOX

FUSE BOX COVER

LOWER CRASH PAD CENTER FACIA

DRINK HOLDER HOUSING

GLOVE BOX COVER

89530G47

Instrument panel assembly—Sonata

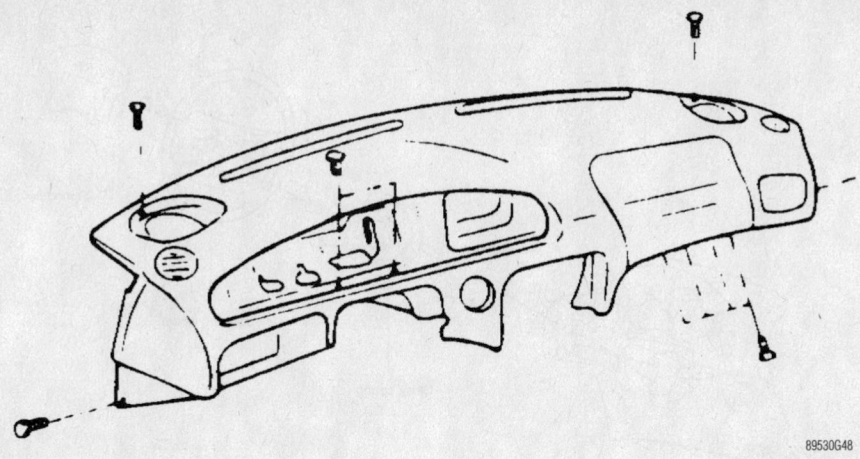

89530G48

Instrument panel screw locations—Sonata

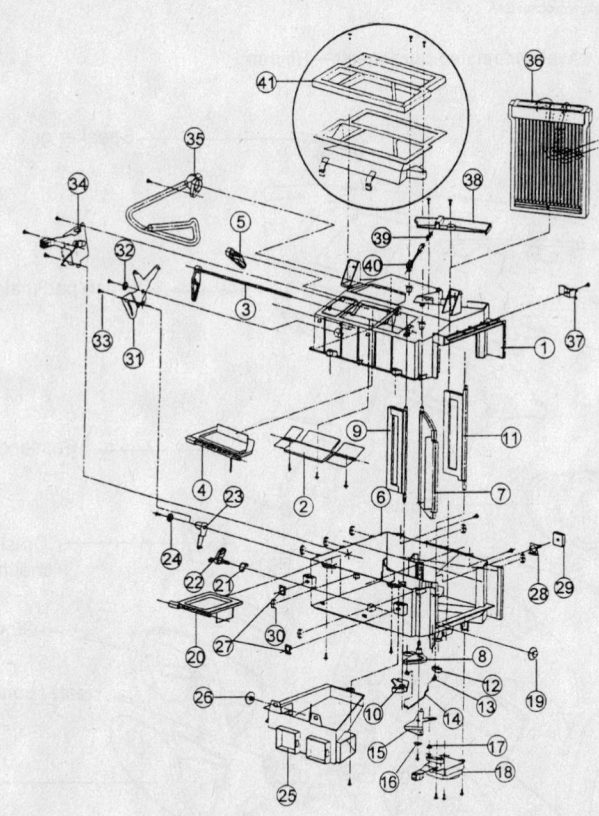

1 Case-heater upper	11 Door ass'y-by pass	21 Spring	31 Cam-mode
2 Door ass'y-vent	12 Arm-By pass door	22 Arm-floor door	32 Spring-washer
3 Shaft ass'y-vent door	13 Holder-rod link	23 Lever-floor door	33 Holder-rod link
4 Door ass'y-defrost	14 Link	24 Spring washer	34 Mode actuator
5 Arm defrost door	15 Lever-temp. door	25 Duct-floor	35 Aspirator & hose ass'y
6 Case-heater lower	16 Spring washer	26 Guide bush	36 Heater core
7 Door ass'y-temp.	17 Guide bush	27 U-nut	37 Clip
8 Arm-temp. door	18 Blend door actuator	28 Clip & Bolt ass'y	38 Cover-heater core
9 Door ass'y (A)-temp. door	(For AUTO A/C only)	29 Seal (A)-heater to D/panel	39 Stopper
10 Arm (A)-temp. door (A)	19 Guide bush	30 Clip	40 Sensor
	20 Door ass'y-floor		41 Plenum duct ass'y

93112GP3

Exploded view of the heater housing assembly—Sonata

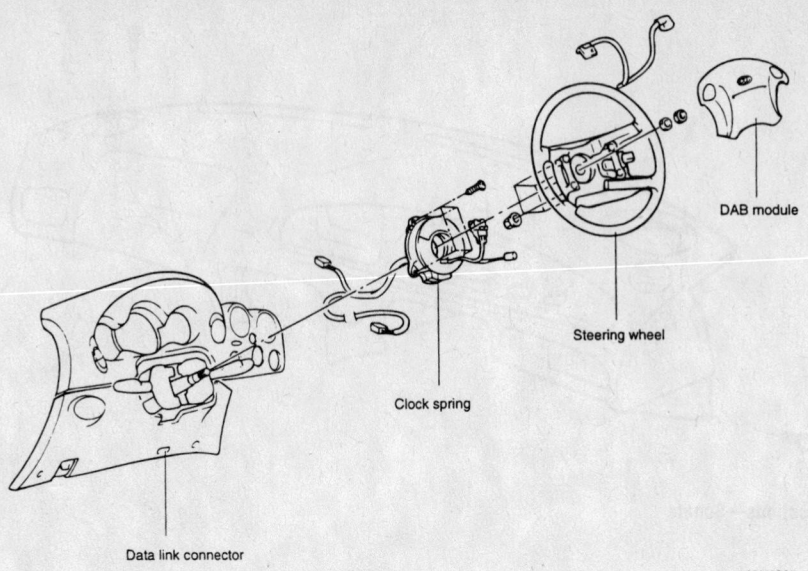

DAB module

Steering wheel

Clock spring

Data link connector

93112G07

Exploded view of the SRS module, steering wheel and related components—Tiburon

Speaker grill

Crash pad main assembly

Passenger airbag

Cluster

Crash pad lower mounting bracket

Glove box upper cover

Heater control assembly

Cluster facia panel assembly

Glove box housing

Steering column shroud

Center facia panel support bracket

Crash pad lower panel

Center facia panel

Rheostat

Ash tray

93112G09

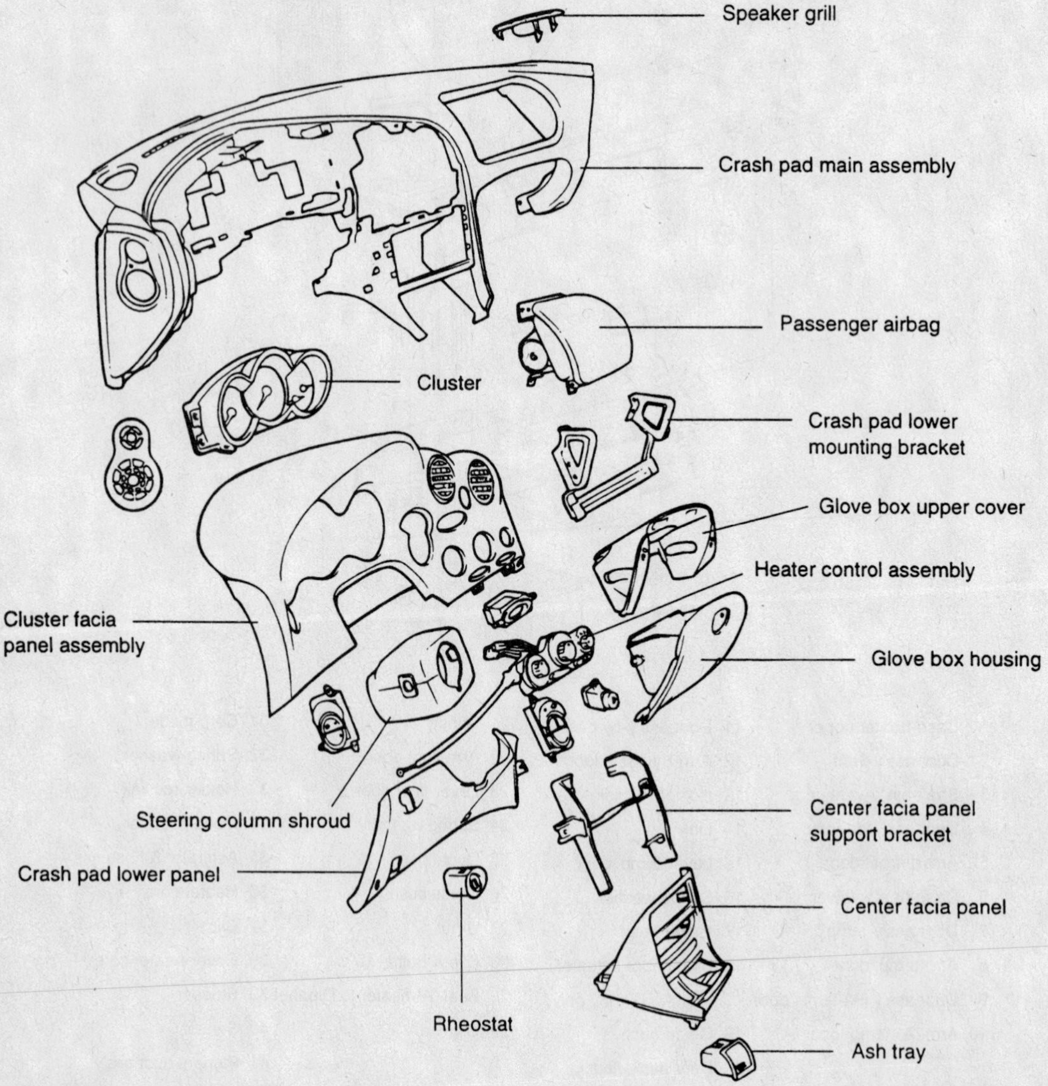

Exploded view of the instrument panel and related components—Tiburon

wheel and disconnect the electrical connector
- Steering wheel-to-steering column nut
- Steering wheel from the steering column

8. Remove the instrument panel by removing or disconnecting the following:
- Upper console cover
- Center fascia panel and disconnect the cigar lighter connector
- 3 lower instrument panel screws and the lower instrument panel
- Radio-to-bracket bolts and the radio

- Rheostat switch, the hood release handle and DLC from the lower instrument panel
- 5 cluster fascia panel-to-instrument panel screws; then, disconnect the heater control cable and the cluster electrical connectors and remove the cluster fascia panel
- 4 instrument cluster-to-instrument panel screws and the instrument cluster
- 2 glove box-to-instrument panel bolts and the glove box
- 4 upper glove box cover-to-instrument panel screws, the 2 glove box

striker screws and the upper glove box cover
- Upper instrument panel speaker grille
- 2 upper speaker-to-instrument panel screws
- Instrument panel-to-chassis bolts, disconnect the electrical connectors and remove the instrument panel

9. Remove or disconnect the following:
- Front right side heating duct from the heater housing
- Pull back the carpet and remove the right side console mounting bracket

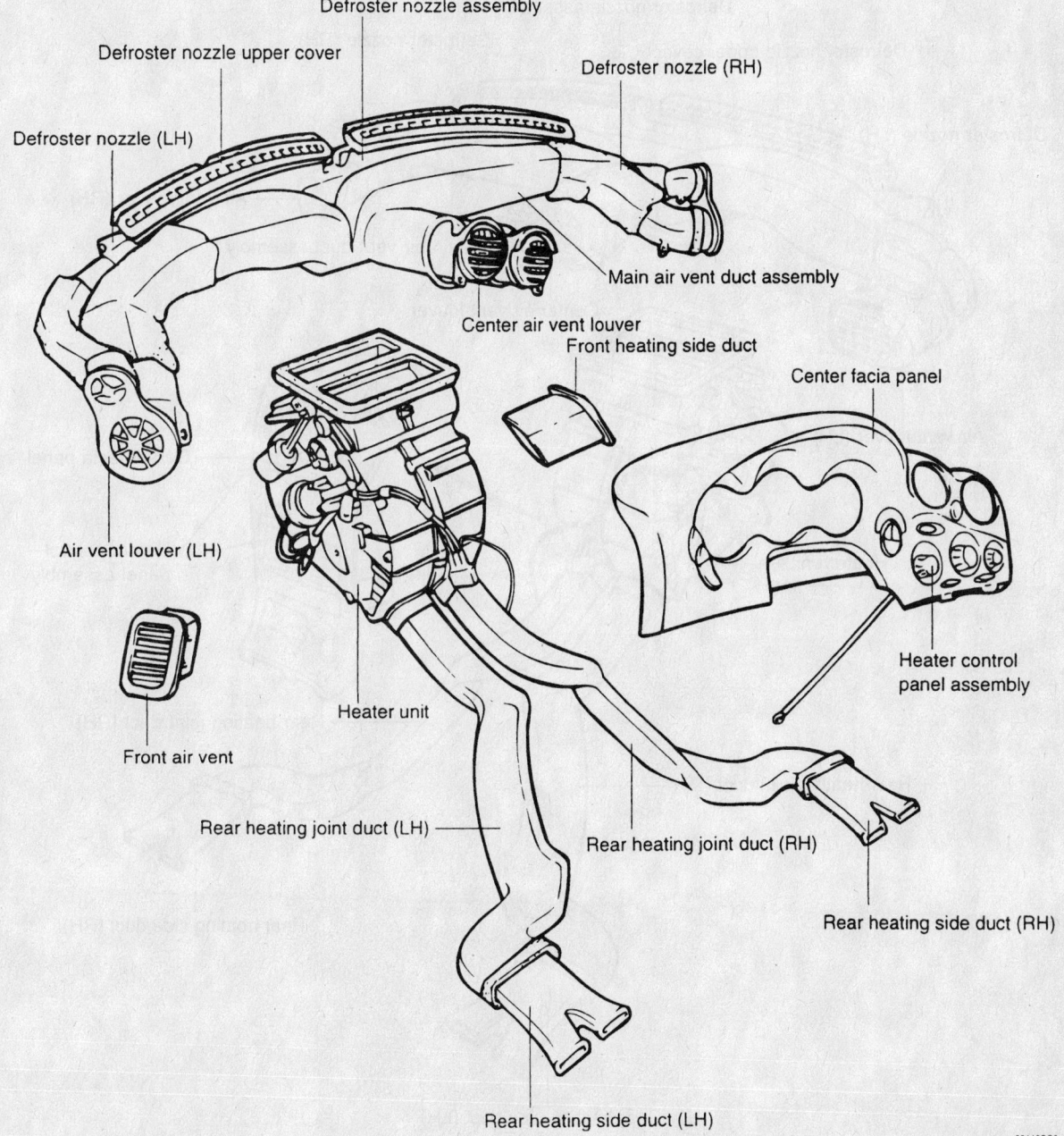

Exploded view of the heater housing, center fascia, distribution ducts and related components—Tiburon Coupe

93112G00

- Front left side duct from the heater housing
- Pull back the carpet and remove the left side console mounting bracket
- Rear heating duct from the heater housing
- Control modules electrical connectors at the center fascia panel support bracket
- Center fascia panel support bracket screws, bolts and/or nuts; then, remove the center fascia panel support bracket
- Center support bars
- Glove box support bracket-to-

instrument panel bolts and the bracket

10. Remove the evaporator housing by removing or disconnecting the following:

- Refrigerant lines (located in the engine compartment), from the evaporator housing and discard the O-rings
- Thermostatic switch connector
- Evaporator housing upper and lower bolts
- Evaporator housing
- Heater housing-to-chassis bolts and the housing

11. Disassemble the heater housing

by removing or disconnecting the following:

- Vacuum motor-to-heater housing bolts (2 for each vacuum motor)
- Vacuum motor rod end connection and remove the vacuum motors
- Heater housing cover clips
- Cover and the heater core

To install:

12. Assemble the heater housing by installing or connecting the following:

- Heater core and the cover
- Heater housing cover clips
- Vacuum motor rod end connection and install the vacuum motors

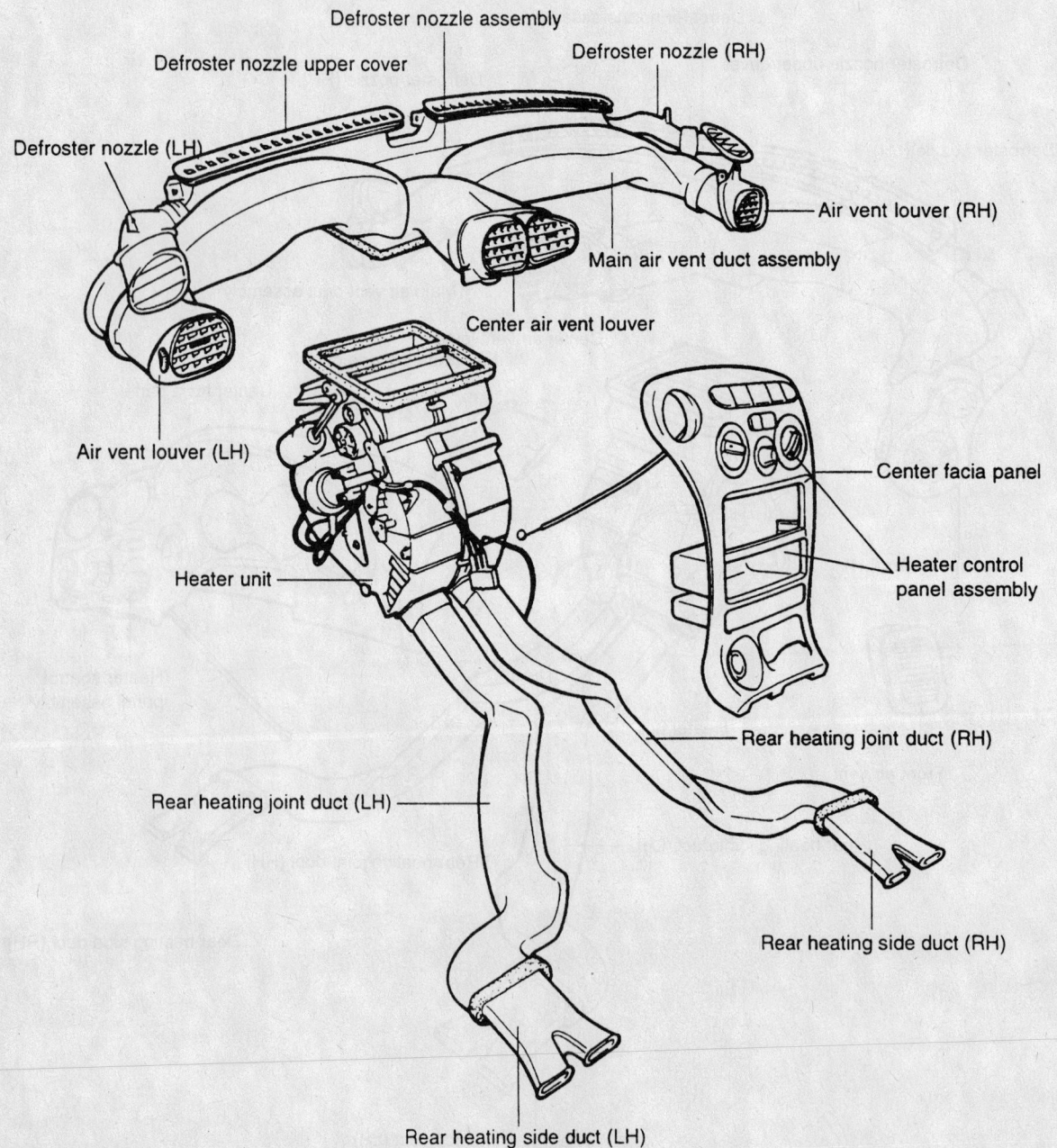

Exploded view of the heater housing, center fascia, distribution ducts and related components—Tiburon Sedan and Wagon

93112GP1

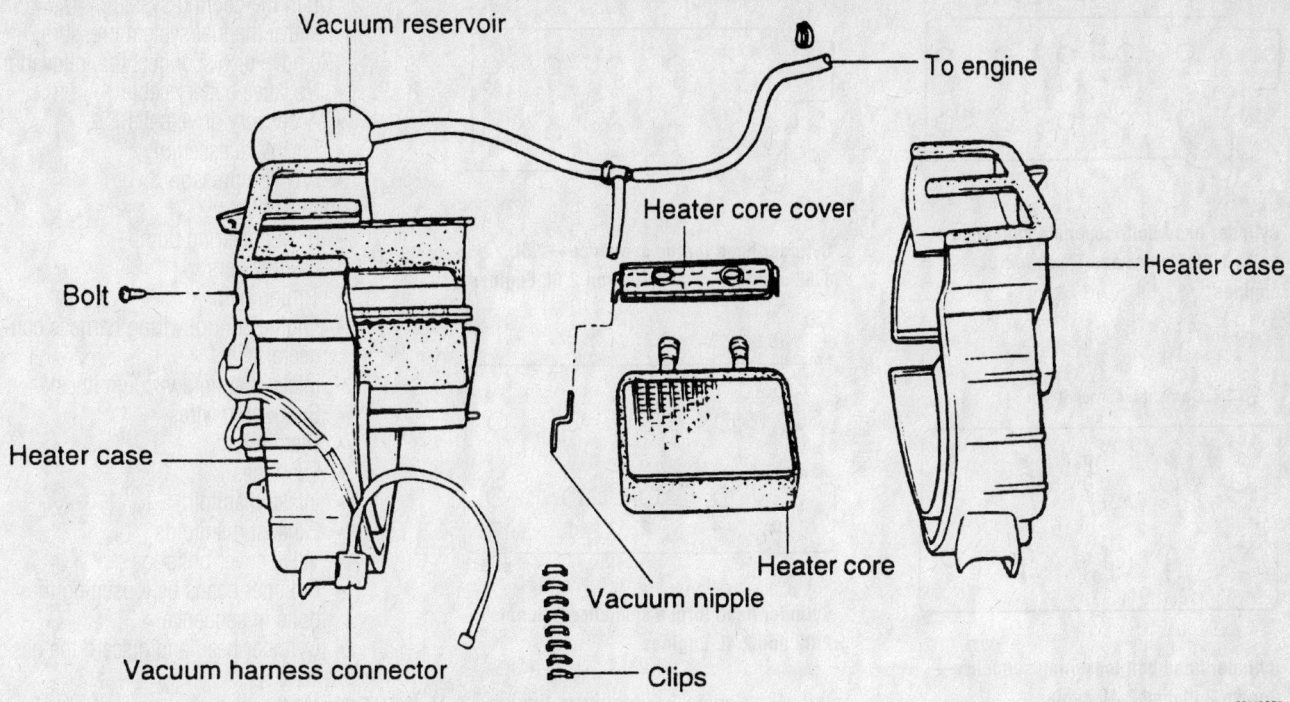

Exploded view of the heater core and heater housing and related components—Tiburon

- Vacuum motor-to-heater housing bolts (2 for each vacuum motor)
- Heater housing and the housing-to-chassis bolts

13. Install the evaporator housing by installing or connecting the following:
- Evaporator housing
- Evaporator housing upper and lower bolts
- Thermostatic switch connector
- Connect the refrigerant lines to the evaporator housing using new O-rings

14. Install or connect the following:
- Glove box support bracket and the bracket-to-instrument panel bolts
- Center support bars
- Center fascia panel support bracket; then, install the center fascia panel support bracket screws, bolts and/or nuts
- Control modules electrical connectors at the center fascia panel support bracket
- Rear heating duct to the heater housing
- Left side console mounting bracket and install the carpet
- Front left side duct to the heater housing
- Right side console mounting bracket and install the carpet
- Front right side heating duct to the heater housing

15. Install the instrument panel by installing or connecting the following:
- Instrument panel, connect the electrical connectors and install the instrument panel-to-chassis bolts
- 2 upper speaker-to-instrument panel screws
- Upper instrument panel speaker grille
- Upper glove box cover, the 2 glove box striker screws and the 4 upper glove box cover-to-instrument panel screws
- Glove box and the 2 glove box-to-instrument panel bolts
- Instrument cluster and the 4 instrument cluster-to-instrument panel screws
- Cluster fascia panel; then, connect the heater control cable and the cluster electrical connectors and install the 5 cluster fascia panel-to-instrument panel screws
- Rheostat switch, the hood release handle and DLC to the lower instrument panel
- Radio and the radio-to-bracket bolts
- Lower instrument panel and the 3 lower instrument panel screws
- Cigar lighter connector and install the center fascia panel
- Upper console cover

16. Install the SRS module and the steering wheel by installing or connecting the following:
- Steering wheel to the steering column
- Steering wheel-to-steering column nut and torque to 30–37 ft. lbs. (40–50 Nm)
- SRS module to the steering wheel and connect the electrical connector
- Steering wheel-to-SRS module nuts

17. Install or connect the following:
- Vacuum line to the heater housing vacuum nipple
- Heater hoses to the heater core

18. Refill the cooling system.
19. Connect the negative battery cable.
20. Evacuate, charge and leak test the air conditioning system.
21. Operate the engine to normal operating temperatures; then, check the climate control operation and check for leaks.

Cylinder Head

REMOVAL & INSTALLATION

4 Cylinder Engines

1. Before servicing the vehicle, refer to the precautions in the beginning of this section.
2. Drain the cooling system.
3. Relieve the fuel system pressure.

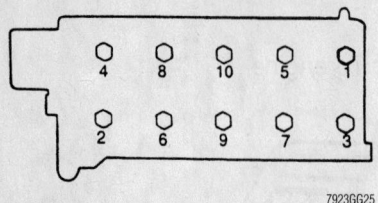

**Cylinder head bolt loosening sequence—
1.5L, 1.8L Elantra and Tiburon 2.0L
engines**

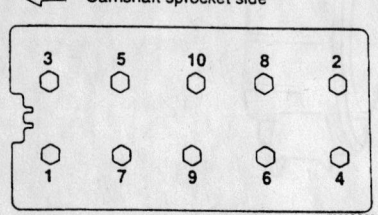

**Cylinder head bolt loosening sequence—
Sonata 2.0L and 2.4L engines**

4. Remove or disconnect the following:
  • Negative battery cable
  • Upper radiator hose
  • Heater hose
  • Air cleaner assembly
  • Intake manifold vacuum lines
  • Engine control wiring harness connectors
  • Spark plug wires
  • Distributor, if equipped
  • Ignition coil
  • Accessory drive belts
  • Power steering pump and bracket
  • Fuel lines
  • Intake manifold
  • Exhaust manifold
  • Front cover
  • Timing belt
  • Valve cover bolts
  • Cylinder head by loosening the bolts in sequence
  • Cylinder head and discard the gasket

To install:

5. Install the cylinder head with a new gasket.

6. For 1.5L engines, tighten the bolts in sequence to 51–54 ft. lbs. (71–75 Nm).

7. For 1.8L and 2.0L Elantra and Tiburon engines, tighten the bolts in sequence as follows:
  a. Step 1: M10 bolts to 22 ft. lbs. (30 Nm) and M12 bolts to 26 ft. lbs. (35 Nm).
  b. Step 2: Plus 60–65 degrees.
  c. Step 3: Plus 60–65 degrees.

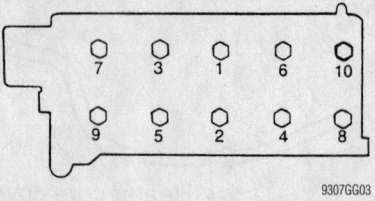

**Cylinder head torque sequence—1.5L,
1.8L and Elantra and Tiburon 2.0L Engines**

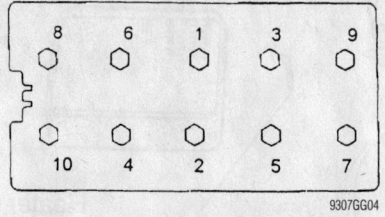

**Cylinder head torque sequence—Sonata
2.0L and 2.4L Engines**

8. For Sonata 2.0L engines, tighten the bolts in sequence to 65–72 ft. lbs. (90–100 Nm).

9. For 2.4L engines, tighten the bolts in sequence as follows:
  a. Step 1: 14 ft. lbs. (20 Nm).
  b. Step 2: Plus 90 degrees.
  c. Step 3: Loosen all bolts in reverse of tightening order.
  d. Step 4: 14 ft. lbs. (20 Nm).
  e. Step 5: Plus 90 degrees.
  f. Step 6: Plus 90 degrees.

10. Install or connect the following:
  • Valve cover
  • Timing belt and front cover
  • Exhaust manifold
  • Intake manifold
  • Fuel lines
  • Power steering pump and bracket
  • Accessory drive belts
  • Ignition coil
  • Distributor, if equipped
  • Spark plug wires
  • Engine control wiring harness connectors
  • Intake manifold vacuum lines
  • Air cleaner assembly
  • Heater hose
  • Upper radiator hose
  • Negative battery cable
11. Fill the cooling system.
12. Start the engine and check for leaks.

V6 Engines

2.5L ENGINE

1. Before servicing the vehicle, refer to the precautions in the beginning of this section.

2. Drain the cooling system.
3. Relieve the fuel system pressure.
4. Remove or disconnect the following:
  • Negative battery cable
  • Accessory drive belts
  • Air intake assembly
  • A/C compressor
  • Alternator
  • Power steering pump
  • Front covers
  • Timing belt
  • Engine control wiring harness connectors
  • Intake manifold vacuum lines
  • Spark plug wires
  • Distributor
  • Fuel lines
  • Intake manifold
  • Exhaust manifolds
  • Valve cover bolts
  • Cylinder heads by loosening the bolts in sequence
  • Cylinder head and discard the gasket

To install:

5. Install or connect the following:
  • New head gaskets
  • Cylinder heads

6. For 3.0L engines, tighten head bolts in sequence to 76–83 ft. lbs. (105–115 Nm).

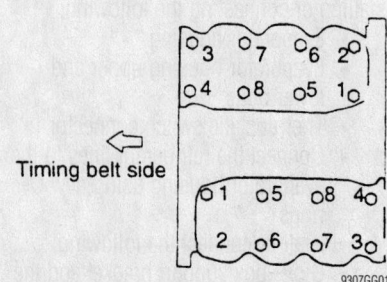

**Cylinder head bolt loosening sequence—
2.5L engine**

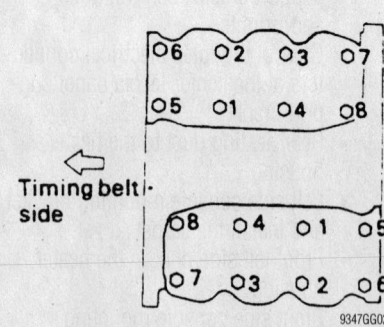

**Cylinder head torque sequence—2.5L and
3.0L Sonata**

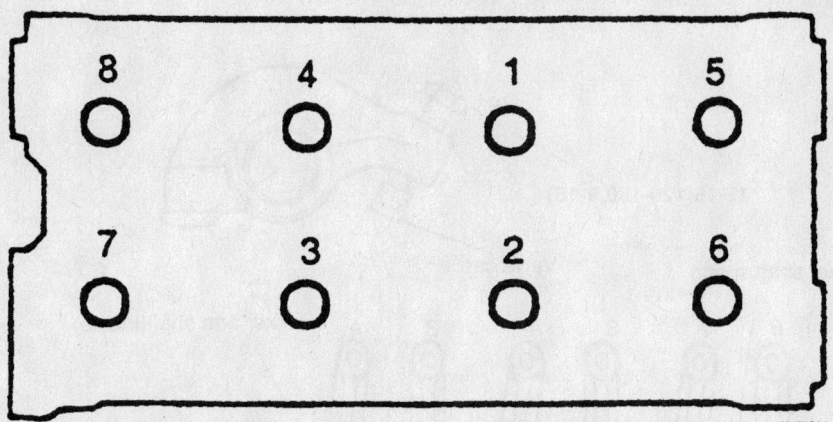

9347KG04

Cylinder head torque sequence—XG 300 with 3.0L engine

7. For 2.5L engines, tighten the bolts in sequence as follows:
   a. Step 1: 18 ft. lbs. (25 Nm).
   b. Step 2: Plus 60–64 degree turn.
   c. Step 3: Plus 45–49 degree turn.
8. Install or connect the following:
   • Valve covers
   • Exhaust manifolds
   • Intake manifold
   • Fuel lines
   • Distributor
   • Spark plug wires
   • Intake manifold vacuum lines
   • Engine control wiring harness connectors
   • Timing belt and front covers
   • Power steering pump
   • Alternator
   • A/C compressor
   • Air intake assembly
   • Accessory drive belts
   • Negative battery cable
9. Fill the cooling system.
10. Start the engine and check for leaks.

3.0L ENGINE

1. Before servicing the vehicle, refer to the precautions in the beginning of this section.
2. Drain the cooling system.
3. Relieve the fuel system pressure.
4. Remove or disconnect the following:
   • Negative battery cable
   • Upper radiator hose
   • Breather hose
   • Air intake hose
   • Vacuum hose
   • Fuel hoses
   • Intake manifold
   • Spark plug wires
   • Ignition coil

   • Upper and lower timing belt covers
   • Timing belt
   • Camshaft sprockets
   • Heat protector and exhaust manifold
   • Water pump
   • Rocker arm cover
   • Camshafts
   • Cylinder head and discard the gasket

5. Clean all mating surfaces of any residual gasket material.
To install:
6. Install or connect the following:
   • New gasket with the identification mark facing the cylinder head
   • Cylinder head and torque the bolts, in sequence, to 82 ft. lbs. (115 Nm)
   • Camshafts
   • Rocker arm cover and torque the bolts to 89 inch lbs. (10 Nm)
   • Water pump and torque the bolts to 16 ft. lbs. (22 Nm)
   • Heat protector and exhaust manifold and torque the bolts to 14 ft. lbs. (19 Nm)
   • Camshaft sprockets
   • Timing belt
   • Upper and lower timing belt covers
   • Ignition coil and spark plug wires
   • Intake manifold and torque the bolts to 10 ft. lbs. (15 Nm)
   • Fuel hoses
   • Vacuum hose
   • Air intake hose
   • Breather hose
   • Upper radiator hose
   • Negative battery cable

7. Fill the cooling system to the proper level.
8. Start the vehicle, check for leaks and repair if necessary.

REMOVAL & INSTALLATION

Except 1.5L SOHC and Sonata 2.0L Engines

These engines are not equipped with rocker arms. The camshaft acts directly on the valves through hydraulic lash adjusters.

1.5L SOHC Engine

1. Before servicing the vehicle, refer to the precautions in the beginning of this section.
2. Remove or disconnect the following:
   • Negative battery cable
   • Valve cover
   • Rocker arm shaft bolts by loosening them evenly in several steps
   • Rocker arm and shaft assemblies

➡**Keep all valvetrain components in order for assembly.**

   • Rocker arms and springs from the shafts
To install:
3. Install or connect the following:
   • Rocker arms and springs in their original positions
   • Rocker arm and shaft assemblies and torque the bolts evenly to 14–20 ft. lbs. (20–26 Nm)
   • Valve cover
   • Negative battery cable

Sonata 2.0L Engine

1. Before servicing the vehicle, refer to the precautions in the beginning of this section.
2. Remove or disconnect the following:
   • Negative battery cable
   • Valve cover
   • Accessory drive belts
   • Front cover
   • Timing belt
   • Camshaft
   • Rocker arms

➡**Keep all valvetrain components in order for assembly.**

To install:
3. Install or connect the following:
   • Rocker arms in their original positions
   • Camshaft
   • Timing belt
   • Front cover
   • Accessory drive belts
   • Valve cover
   • Negative battery cable

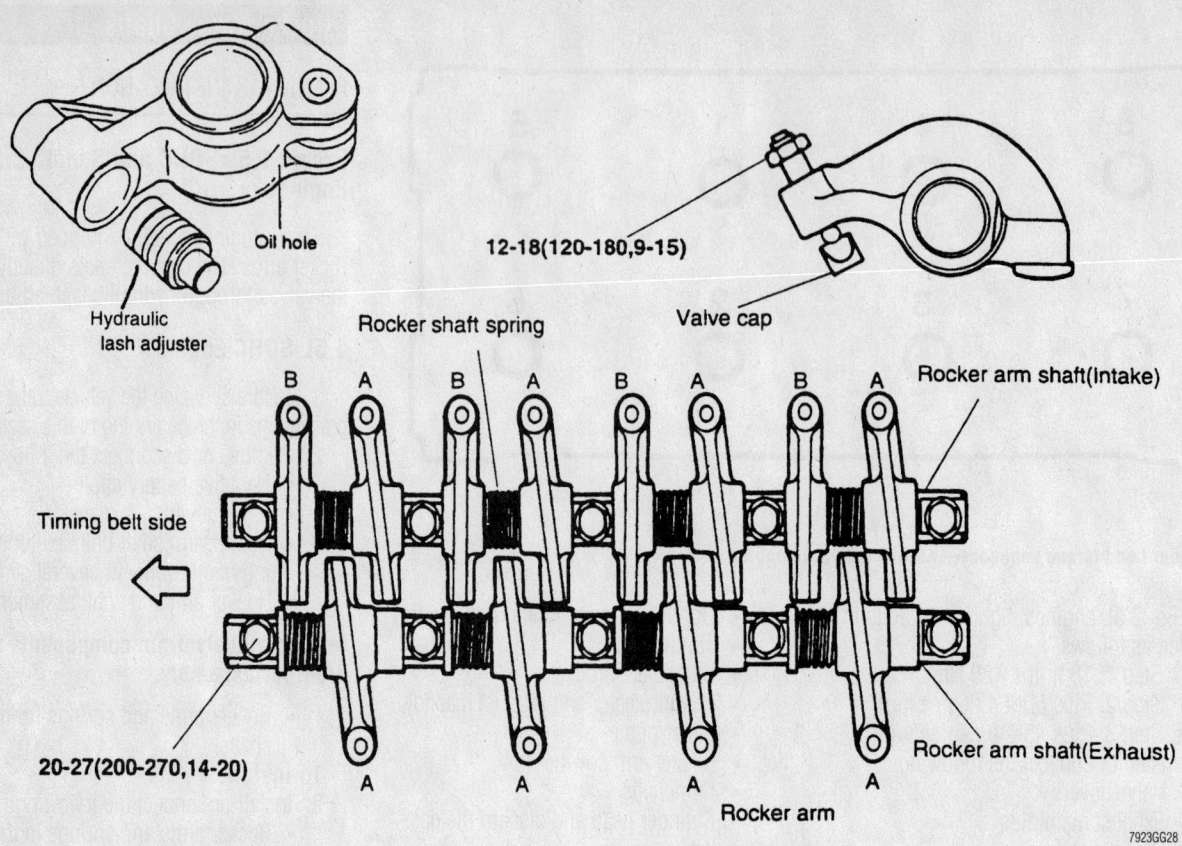

Oil hole

Hydraulic lash adjuster

12-18(120-180,9-15)

Valve cap

Rocker shaft spring

Rocker arm shaft(Intake)

B A B A B A B A

Timing belt side

20-27(200-270,14-20)

A A A A

Rocker arm shaft(Exhaust)

Rocker arm

7923GG28

Rocker assembly components and arrangement. Rockers marked "A" and "B" must be returned to their original positions—1.5L SOHC engines

Intake Manifold

REMOVAL & INSTALLATION

4 Cylinder Engines

1. Before servicing the vehicle, refer to the precautions in the beginning of this section.
2. Relieve the fuel system pressure.
3. Drain the cooling system.
4. Remove or disconnect the following:

- Negative battery cable
- Air intake hose
- Accelerator cable
- Upper radiator hose
- Engine control wiring harness connectors
- Throttle body
- Positive Crankcase Ventilation (PCV) valve and hose
- Brake booster vacuum line
- Intake manifold vacuum hoses
- Fuel lines
- Surge tank
- Fuel injector connectors
- Fuel supply manifold
- Heater hose

- Engine Coolant Temperature (ECT) sensor connector
- Spark plug wires
- Thermostat housing
- Distributor, if equipped
- Ignition coil
- Intake manifold bracket
- Intake manifold and discard the gasket

To install:

5. Install or connect the following:

- Intake manifold using a new gasket and torque the nuts to 11–14 ft. lbs. (15–20 Nm), starting from the center and working outwards
- Intake manifold bracket and torque the bolts to 13–18 ft. lbs. (18–25 Nm)
- Ignition coil
- Distributor, if equipped
- Thermostat housing and torque the bolts to 12–14 ft. lbs. (17–20 Nm)
- Spark plug wires
- ECT sensor connector
- Heater hose
- Fuel supply manifold
- Fuel injector connectors
- Surge tank using a new gasket and torque the bolts to 11–16 ft. lbs. (15–22 Nm)

- Fuel lines
- Intake manifold vacuum hoses
- Brake booster vacuum line
- PCV valve and hose
- Throttle body
- Engine control wiring harness connectors
- Upper radiator hose
- Accelerator cable
- Air intake hose
- Negative battery cable

6. Fill the cooling system.
7. Start the engine and check for leaks.

V6 Engines

1. Before servicing the vehicle, refer to the precautions in the beginning of this section.
2. Relieve the fuel system pressure.
3. Drain the cooling system.
4. Remove or disconnect the following:

- Negative battery cable
- Air intake hose
- Accelerator cable
- Upper radiator hose
- Engine control wiring harness
- Throttle body
- Positive Crankcase Ventilation (PCV) valve and hose

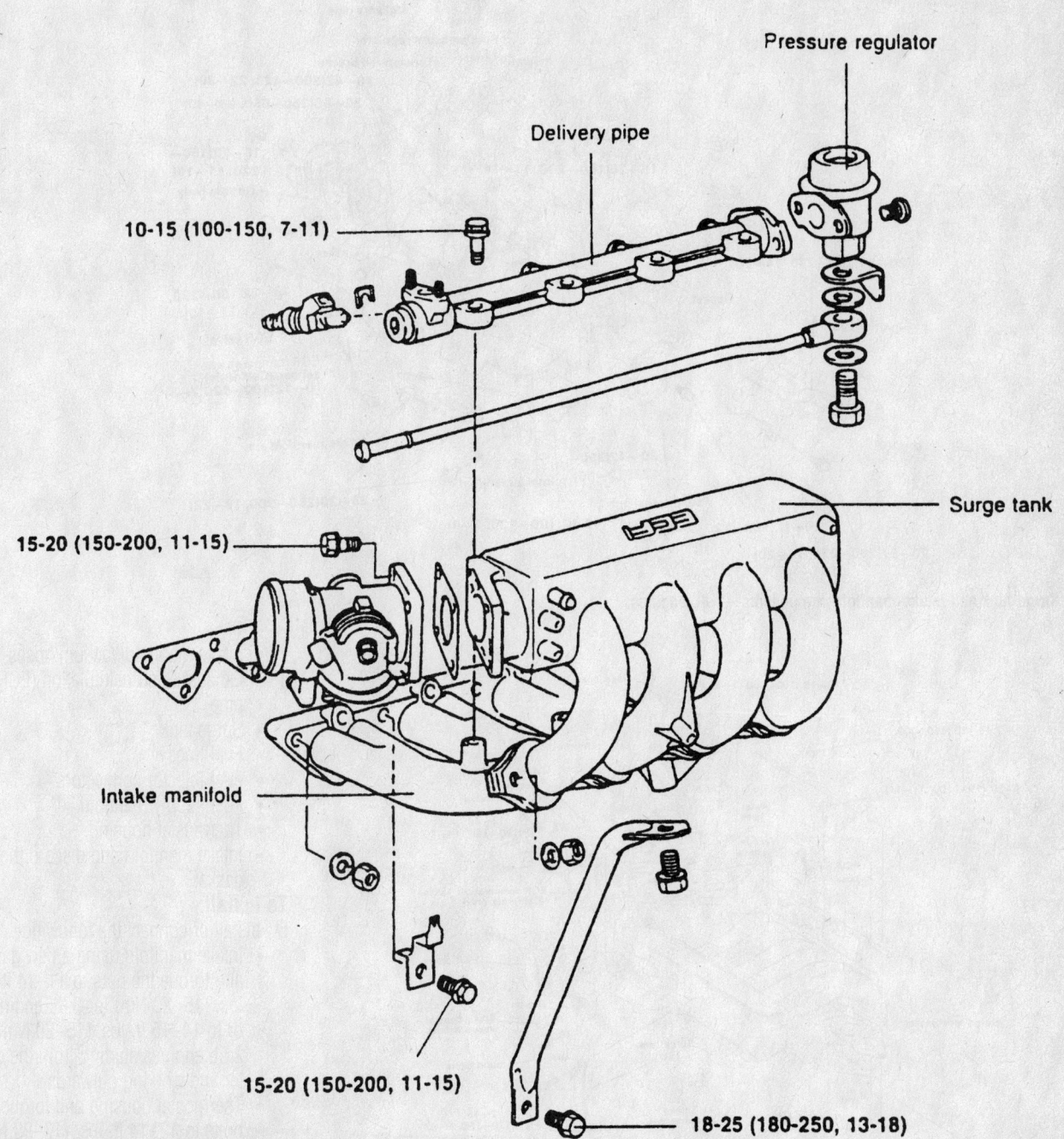

Pressure regulator

Delivery pipe

10-15 (100-150, 7-11)

Surge tank

15-20 (150-200, 11-15)

Intake manifold

15-20 (150-200, 11-15)

18-25 (180-250, 13-18)

TORQUE : Nm (kg.cm, lb.ft)

7923GG34

Surge tank and intake manifold components—1.5L engine

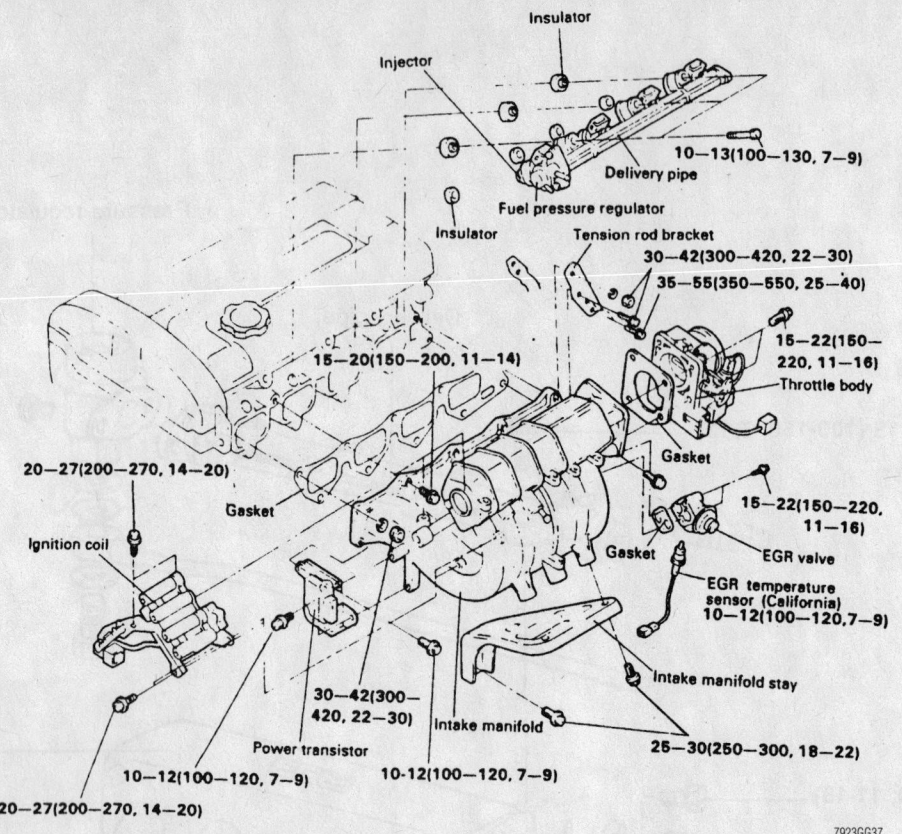

Insulator

Injector

10—13(100—130, 7—9)

Delivery pipe

Fuel pressure regulator

Insulator

Tension rod bracket

30—42(300—420, 22—30)

35—55(350—550, 25—40)

15—22(150—220, 11—16)

Throttle body

15—20(150—200, 11—14)

20—27(200—270, 14—20)

Gasket

Gasket

15—22(150—220, 11—16)

EGR valve

EGR temperature sensor (California)
10—12(100—120,7—9)

Ignition coil

Intake manifold stay

30—42(300—420, 22—30)

Intake manifold

Power transistor

10-12(100—120, 7—9)

25—30(250—300, 18—22)

20—27(200—270, 14—20)

7923GG37

Surge tank and intake manifold components—1.8L engines

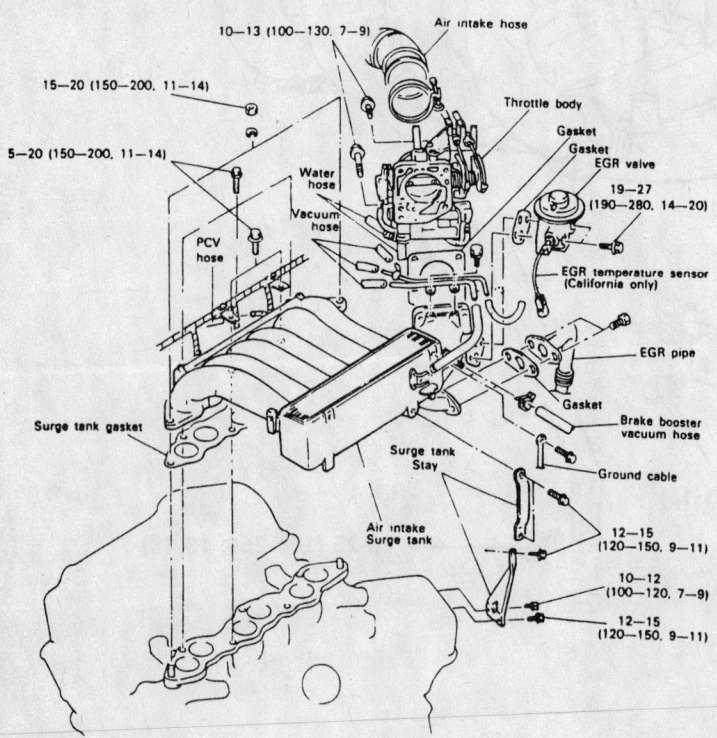

10—13 (100—130, 7—9)

Air intake hose

15—20 (150—200, 11—14)

Throttle body

5—20 (150—200, 11—14)

Gasket

Gasket

EGR valve

Water hose

19—27 (190—280, 14—20)

Vacuum hose

PCV hose

EGR temperature sensor (California only)

EGR pipe

Gasket

Brake booster vacuum hose

Surge tank gasket

Surge tank Stay

Ground cable

Air intake Surge tank

12—15 (120—150, 9—11)

10—12 (100—120, 7—9)

12—15 (120—150, 9—11)

TORQUE : Nm (kg.cm, lb.ft)

7923GG38

Surge tank and intake manifold components—Sonata 2.0L and 2.4L engines

- Intake manifold vacuum hoses
- Exhaust Gas Recirculation (EGR) pipe
- Surge tank
- Fuel lines
- Fuel injector connectors
- Fuel supply manifold
- Thermostat housing
- Intake manifold and discard the gasket

To install:

5. Install or connect the following:
   - Intake manifold using a new gasket and torque the nuts to 11–14 ft. lbs. (15–20 Nm) for 3.0L engines or to 14–15 ft. lbs. (15–20 Nm) for 2.5L engines starting from the center and working outwards
   - Thermostat housing and torque the bolts to 12–14 ft. lbs. (17–20 Nm)
   - Fuel supply manifold and torque the bolts to 84–108 inch lbs. (10–13 Nm)
   - Fuel injector connectors
   - Fuel lines
   - Surge tank and torque the bolts to 11–14 ft. lbs. (15–20 Nm)
   - EGR pipe
   - Intake manifold vacuum hoses
   - PCV valve and hose

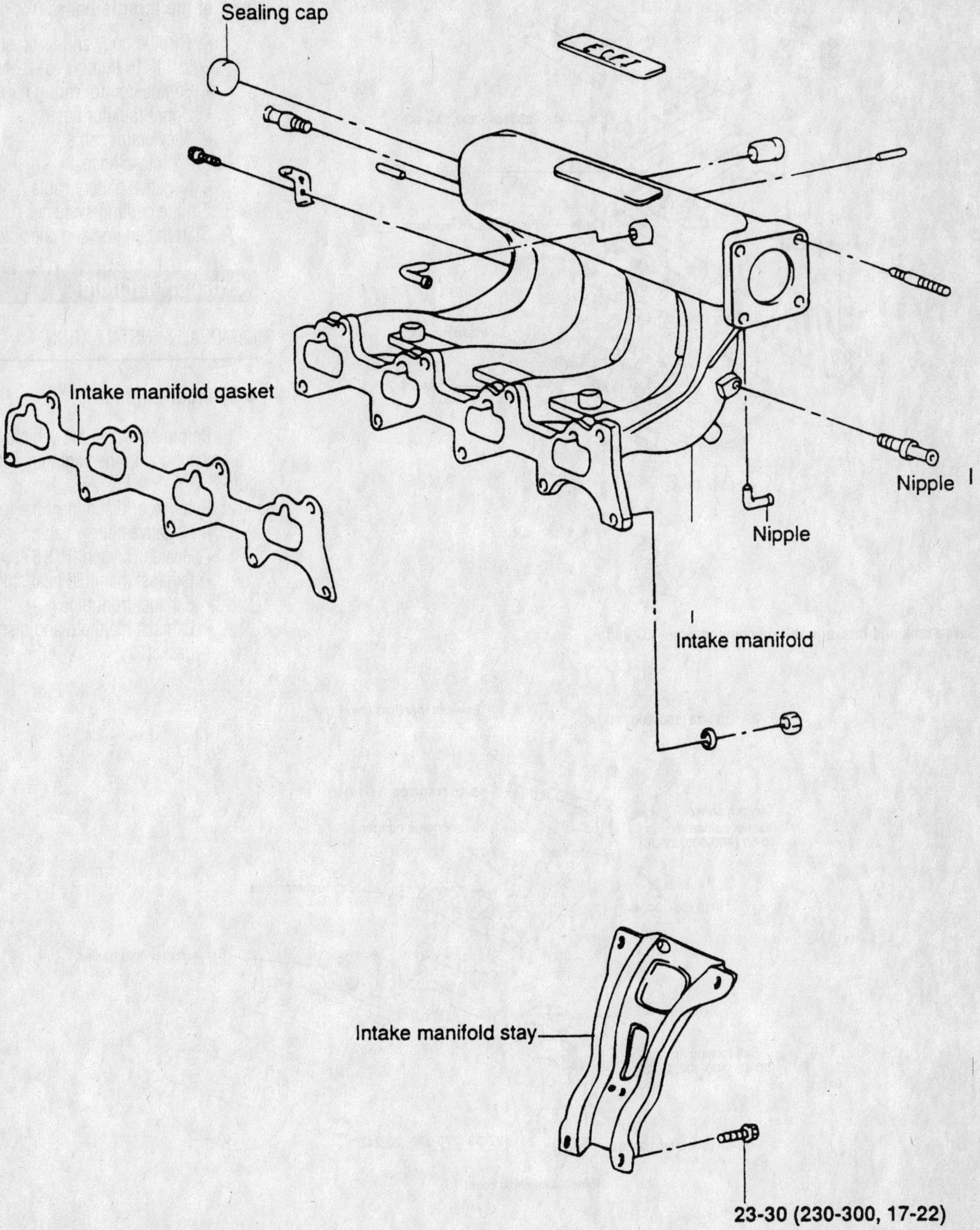

Sealing cap

Intake manifold gasket

Nipple

Nipple

Intake manifold

Intake manifold stay

23-30 (230-300, 17-22)

TORQUE : Nm (kg.cm, lb.ft)

7923GG39

Surge tank and intake manifold components—Elantra and Tiburon 2.0L engine

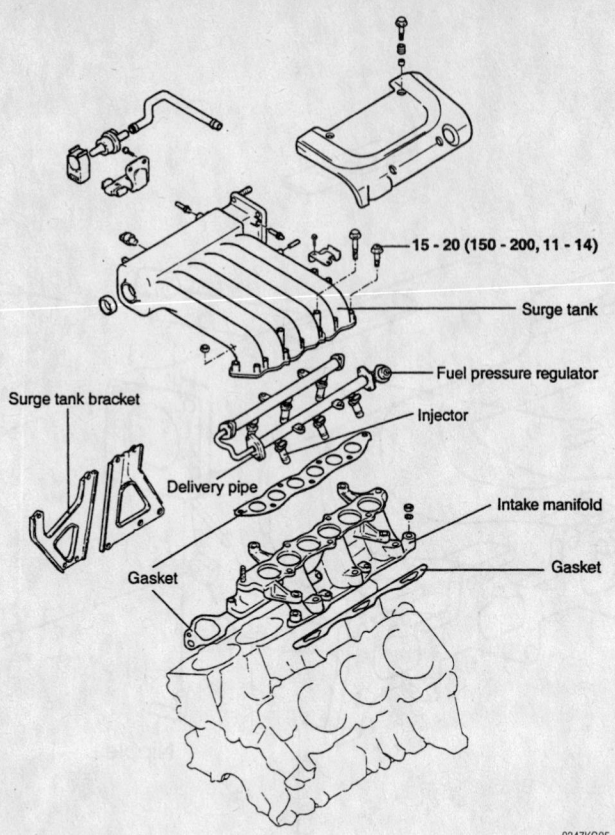

15 - 20 (150 - 200, 11 - 14)

Surge tank

Fuel pressure regulator

Injector

Surge tank bracket

Delivery pipe

Intake manifold

Gasket

Gasket

9347KG05

Surge tank and intake manifold components—XG 300

➡One throttle body bolt is shorter than the rest. This bolt is installed in the upper left hole when viewed from the front of the throttle body.

- Throttle body and torque the bolts to 11–16 ft. lbs. (15–22 Nm)
- Engine control wiring harness
- Upper radiator hose
- Accelerator cable
- Air intake hose
- Negative battery cable
6. Fill the cooling system.
7. Start the engine and check for leaks.

Exhaust Manifold

REMOVAL & INSTALLATION

4 Cylinder Engines

1. Before servicing the vehicle, refer to the precautions in the beginning of this section.
2. Remove or disconnect the following:
- Negative battery cable
- Heated Oxygen (HO$_2$S) sensor
- Exhaust manifold heat shield
- Exhaust front pipe
- Exhaust manifold and discard the gasket

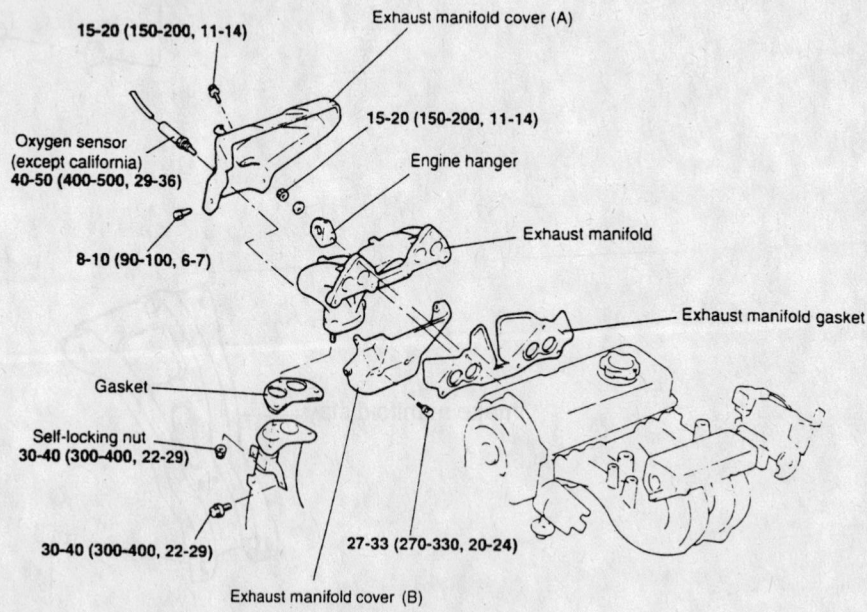

15-20 (150-200, 11-14)

Exhaust manifold cover (A)

15-20 (150-200, 11-14)

Oxygen sensor (except california) 40-50 (400-500, 29-36)

Engine hanger

8-10 (90-100, 6-7)

Exhaust manifold

Exhaust manifold gasket

Gasket

Self-locking nut 30-40 (300-400, 22-29)

30-40 (300-400, 22-29)

27-33 (270-330, 20-24)

Exhaust manifold cover (B)

TORQUE : Nm (kg.cm, lb.ft)

7923GG41

Exploded view of the exhaust manifold and related components—1.5L engines

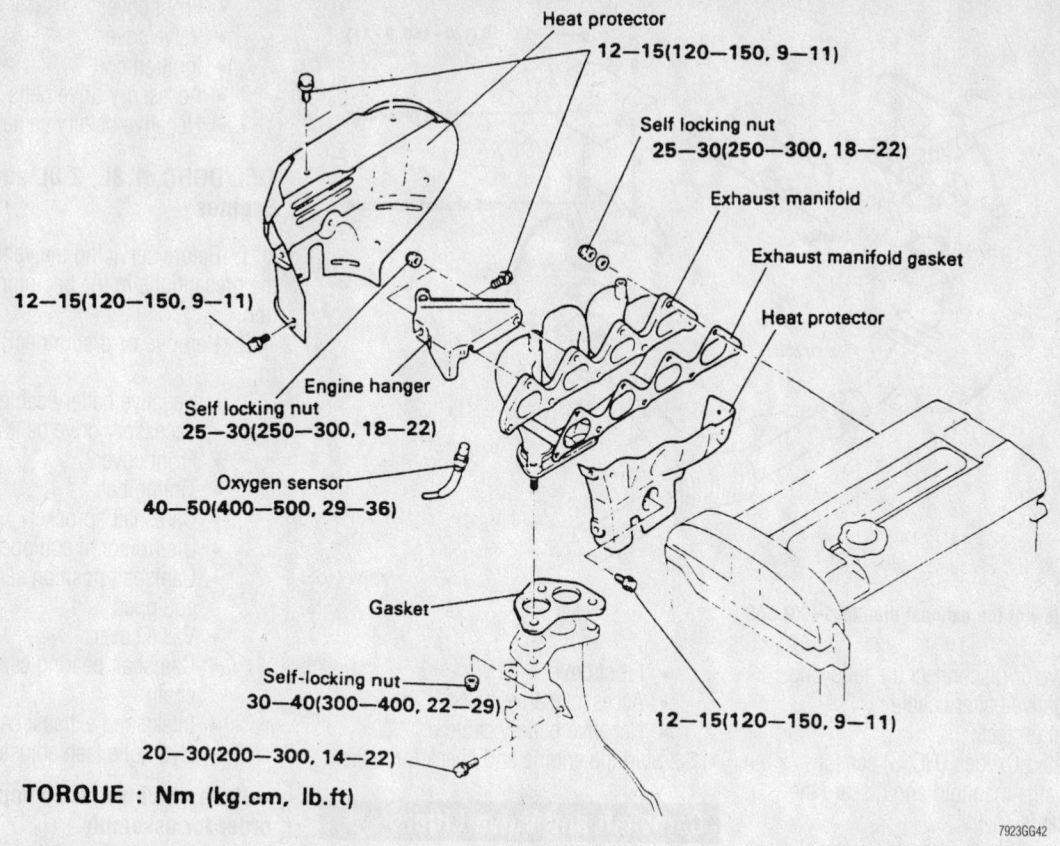

Heat protector
12—15(120—150, 9—11)

Self locking nut
25—30(250—300, 18—22)

Exhaust manifold

Exhaust manifold gasket

Heat protector

12—15(120—150, 9—11)

Engine hanger

Self locking nut
25—30(250—300, 18—22)

Oxygen sensor
40—50(400—500, 29—36)

Gasket

Self-locking nut
30—40(300—400, 22—29)

12—15(120—150, 9—11)

20—30(200—300, 14—22)

TORQUE : Nm (kg.cm, lb.ft)

7923GG42

Exploded view of the exhaust manifold components—1.8L and Sonata 2.0L engines

To install:

3. Install the exhaust manifold with a new gasket and torque the nuts to the following specifications:

    a. 1.5L engine: 11–15 ft. lbs. (15–20 Nm).

    b. 1.8L and Sonata 2.0L engines: 18–22 ft. lbs. (25–30 Nm).

    c. Elantra and Tiburon 2.0L engine: 32–41 ft. lbs. (43–50 Nm).

    d. 2.4L engine: M8 fasteners to 18–22 ft. lbs. (25–30 Nm) and M10 fasteners to 25–40 ft. lbs. (34–55 Nm).

4. Install or connect the following:
- Exhaust front pipe
- Exhaust manifold heat shield
- HO$_2$S sensor
- Negative battery cable

5. Start the engine and check for leaks.

6 Cylinder Engines

SONATA

1. Before servicing the vehicle, refer to the precautions in the beginning of this section.

2. Remove or disconnect the following:
- Negative battery cable
- Exhaust front pipe
- Exhaust Gas Recirculation (EGR) tube

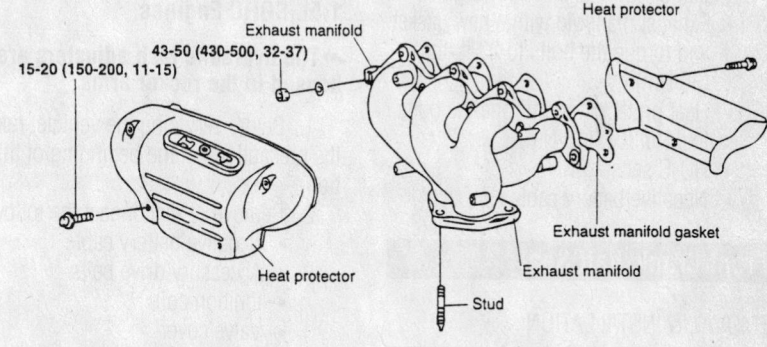

Exhaust manifold
43-50 (430-500, 32-37)

15-20 (150-200, 11-15)

Heat protector

Heat protector

Exhaust manifold gasket

Exhaust manifold

Stud

TORQUE: Nm (kg.cm, lb.ft)

7923GG43

Exploded view of the exhaust manifold components—Elantra and Tiburon with 2.0L engine

- Oil dipstick tube
- Exhaust manifold heat shields
- Exhaust manifolds and discard the gasket

To install:

3. Install or connect the following:
- Exhaust manifolds and torque the nuts to 11–16 ft. lbs. (15–22 Nm) for 3.0L engines or to 18–22 ft. lbs. (25–30 Nm) for 2.5L engines
- Exhaust manifold heat shields

- Oil dipstick tube
- EGR tube
- Exhaust front pipe and torque the nuts to 22–29 ft. lbs. (30–40 Nm)
- Negative battery cable

4. Start the engine and check for leaks.

XG 300

1. Before servicing the vehicle, refer to the precautions in the beginning of this section.

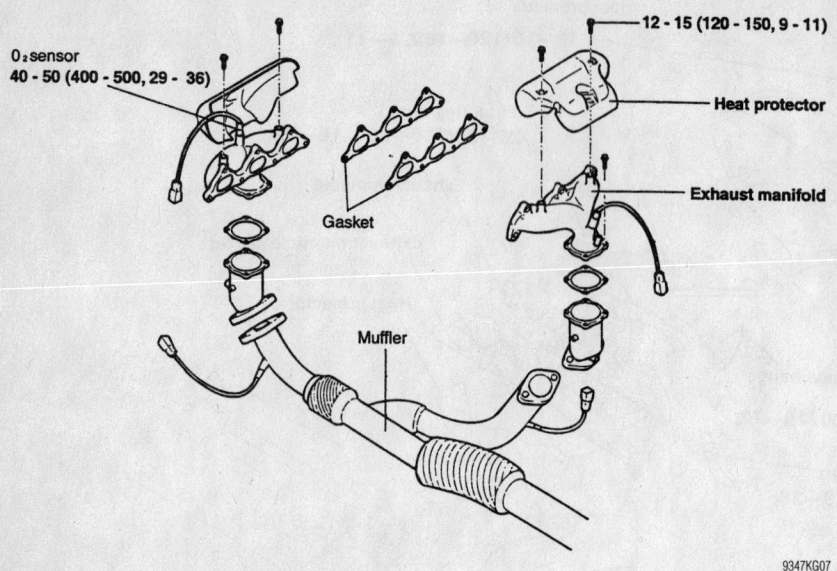

O₂ sensor
40 - 50 (400 - 500, 29 - 36)

12 - 15 (120 - 150, 9 - 11)

Heat protector

Exhaust manifold

Gasket

Muffler

9347KG07

Exploded view of the exhaust manifold—XG 300

2. Remove or disconnect the following:
 • Negative battery cable
 • Heat protector
 • Heated Oxygen (HO₂S) sensor
 • Exhaust manifold and discard the gaskets
3. Clean and residual gasket material from all mating surfaces.

To install:

4. Install or connect the following:
 • Exhaust manifold with a new gasket and torque the bolts to 22 ft. lbs. (30 Nm)
 • Heat protector and torque the bolts to 11 ft. lbs. (15 Nm)
 • HO₂S sensor
 • Negative battery cable

Front Crankshaft Seal

REMOVAL & INSTALLATION

1. Before servicing the vehicle, refer to the precautions in the beginning of this section.
2. Remove or disconnect the following:
 • Negative battery cable
 • Accessory drive belts
 • Front cover
 • Timing belt
 • Crankshaft timing sprocket
 • Front crankshaft seal

To install:

3. Install the front crankshaft seal so that it is flush with the oil pump housing.
4. Install or connect the following:
 • Crankshaft timing sprocket
 • Timing belt

 • Front cover
 • Accessory drive belts
 • Negative battery cable
5. Start the engine and check for leaks.

Camshaft and Valve Lifters

REMOVAL & INSTALLATION

1.5L SOHC Engines

➡The hydraulic lash adjusters are housed in the rocker arms.

1. Before servicing the vehicle, refer to the precautions in the beginning of this section.
2. Remove or disconnect the following:
 • Negative battery cable
 • Accessory drive belts
 • Ignition coil
 • Valve cover
 • Front cover
 • Timing belt
 • Camshaft timing belt sprocket
 • Rocker arm and shaft assembly
 • Camshaft bearing caps
 • Camshaft

To install:

3. Install or connect the following:
 • Camshaft
 • Camshaft bearing caps
 • Rocker arm and shaft assembly and torque the bolts evenly to 14–20 ft. lbs. (20–26 Nm)
 • Camshaft timing belt sprocket and torque the bolt to 58–72 ft. lbs. (80–100 Nm)
 • Timing belt

 • Front cover
 • Valve cover
 • Ignition coil
 • Accessory drive belts
 • Negative battery cable

1.5L DOHC, 1.8L, 2.0L and 2.4L Engines

1. Before servicing the vehicle, refer to the precautions in the beginning of this section.
2. Remove or disconnect the following:
 • Negative battery cable
 • Accessory drive belts
 • Front cover
 • Timing belt
 • Camshaft sprocket
 • Distributor, if equipped
 • Camshaft position sensor, if equipped
 • Valve cover
 • Camshaft bearing caps and timing chain
 • Intake and exhaust camshafts
 • Hydraulic lash adjusters

➡Keep all valvetrain components in order for assembly.

To install:

3. Install or connect the following:
 • Hydraulic lash adjusters in their original positions
 • Intake and exhaust camshafts with the secondary chain aligned as shown
 • Camshaft bearing caps and timing chain and torque the bolts to 15 ft. lbs. (21 Nm) for 2.0L (VIN P) engines or to 10 ft. lbs. (14 Nm) for all others
 • Valve cover
 • Distributor, if equipped
 • Camshaft position sensor, if equipped
 • Camshaft sprocket and torque the bolt to 60–74 ft. lbs. (80–100 Nm)
 • Timing belt
 • Front cover
 • Accessory drive belts
 • Negative battery cable

2.5L Engine

1. Before servicing the vehicle, refer to the precautions in the beginning of this section.
2. Remove or disconnect the following:
 • Negative battery cable
 • Accessory drive belts
 • Front cover

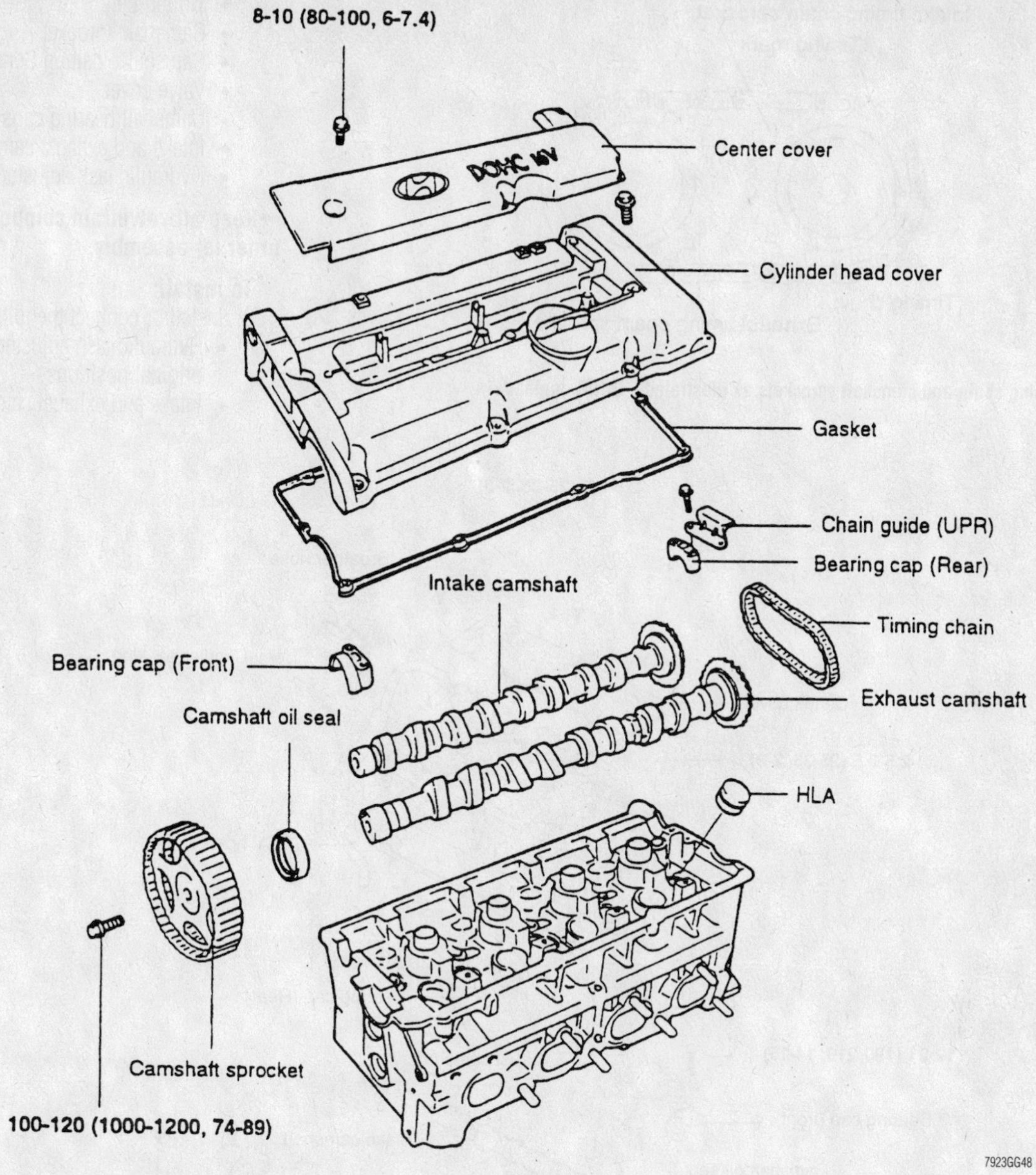

8-10 (80-100, 6-7.4)
Center cover
Cylinder head cover
Gasket
Chain guide (UPR)
Bearing cap (Rear)
Intake camshaft
Timing chain
Bearing cap (Front)
Exhaust camshaft
Camshaft oil seal
HLA
Camshaft sprocket
100-120 (1000-1200, 74-89)
7923GG48

Camshaft assembly components—1.5L DOHC, 1.8L and Elantra and Tiburon 2.0L engines

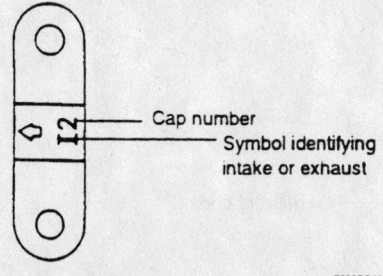

Cap number
Symbol identifying intake or exhaust
I2
7923GG49

The camshaft bearing caps are identified with a letter and number stamp. The letter indicates either intake or exhaust and the number is sequential from the cylinder head end opposite the timing chain—1.5L DOHC, 1.8L and Elantra and Tiburon 2.0L engines

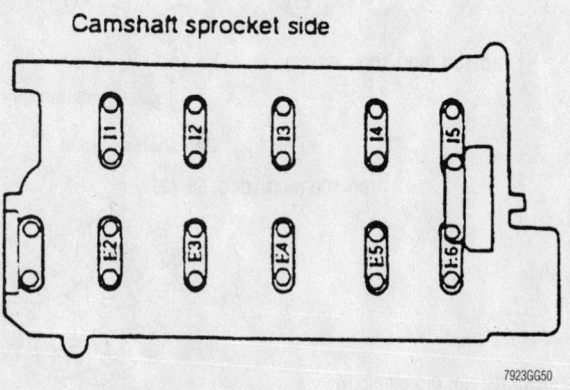

Camshaft sprocket side
I1
I2
I3
I4
I5
E2
E3
E4
E5
I-6
7923GG50

The camshaft bearing caps are arranged on the cylinder head as illustrated—1.5L DOHC, 1.8L and Elantra and Tiburon 2.0L engines

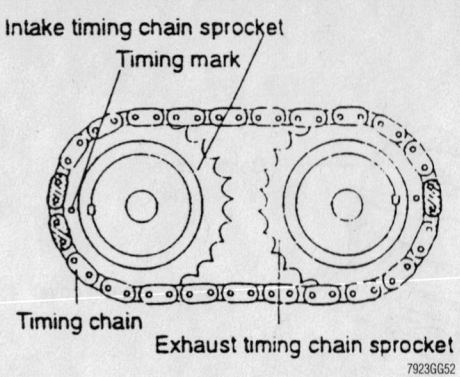

Align the timing chain and camshaft sprockets as illustrated—DOHC engines

- Timing belt
- Camshaft sprocket
- Camshaft Position Sensor (CMP)
- Valve cover
- Camshaft bearing caps
- Intake and exhaust camshafts
- Hydraulic lash adjusters

➡ **Keep all valvetrain components in order for assembly.**

To install:
3. Install or connect the following:
- Hydraulic lash adjusters in their original positions
- Intake and exhaust camshafts with

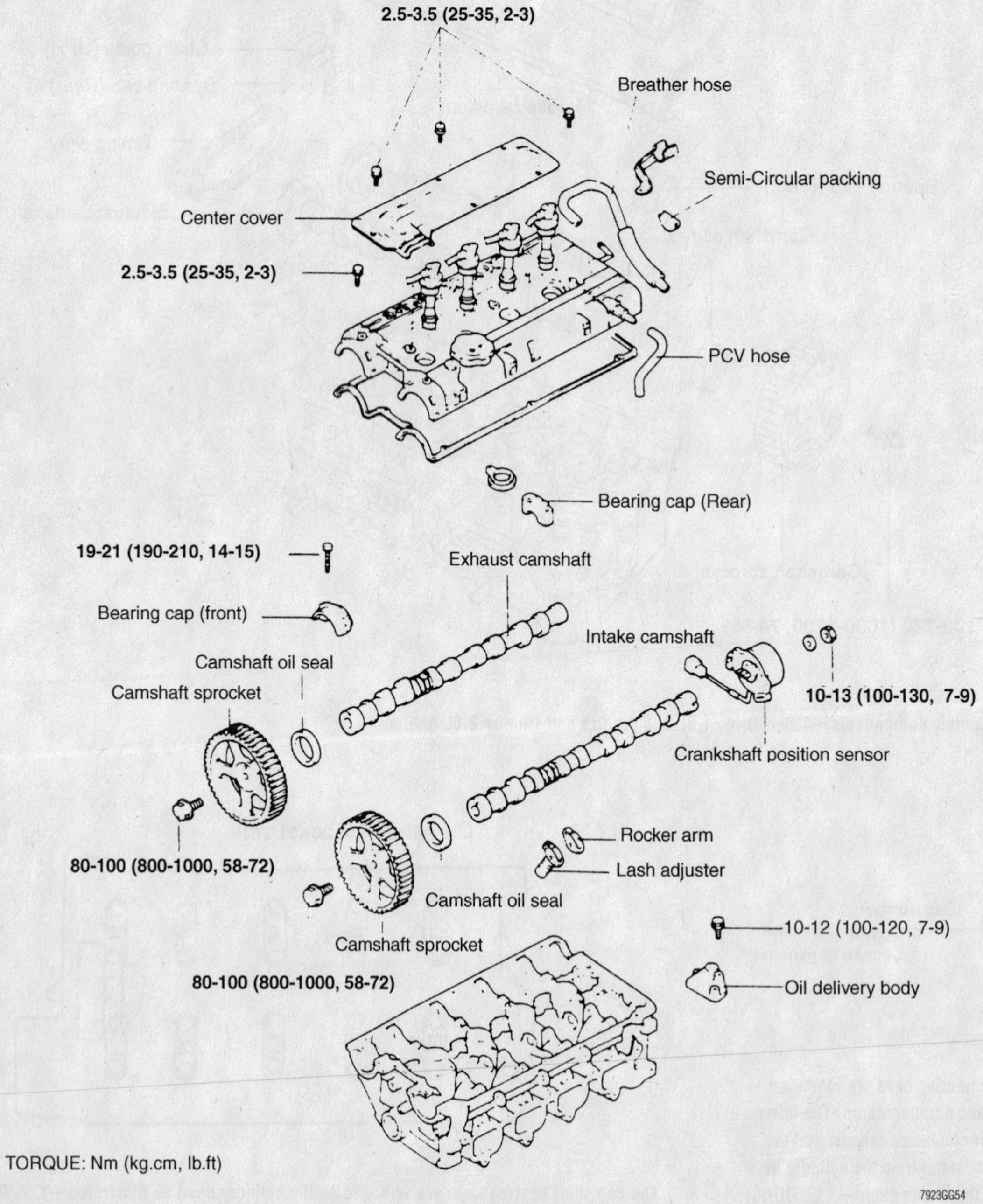

TORQUE: Nm (kg.cm, lb.ft)

Exploded view of the camshaft and rocker arm assembly components—Sonata with 2.0L engine

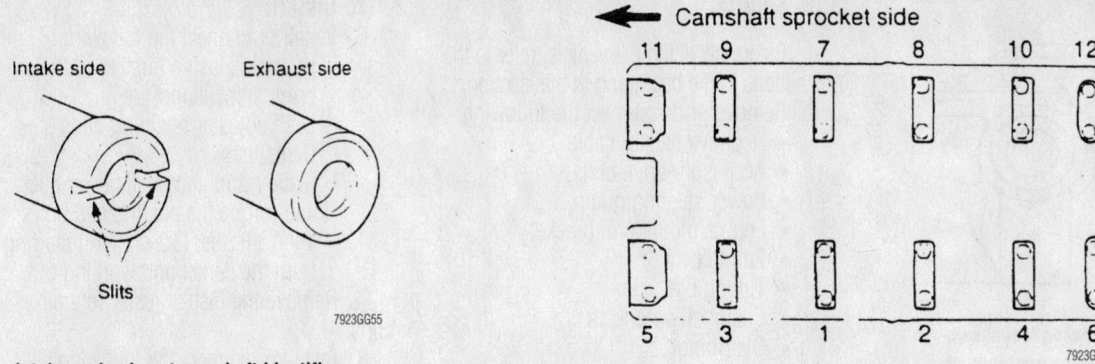

Intake side   Exhaust side

Slits

7923GG55

Intake and exhaust camshaft identification—Sonata with 2.0L engine

← Camshaft sprocket side

| 11 | 9 | 7 | 8 | 10 | 12 |

| 5 | 3 | 1 | 2 | 4 | 6 |

7923GG58

Bearing cap torque sequence—Sonata with 2.0L engine

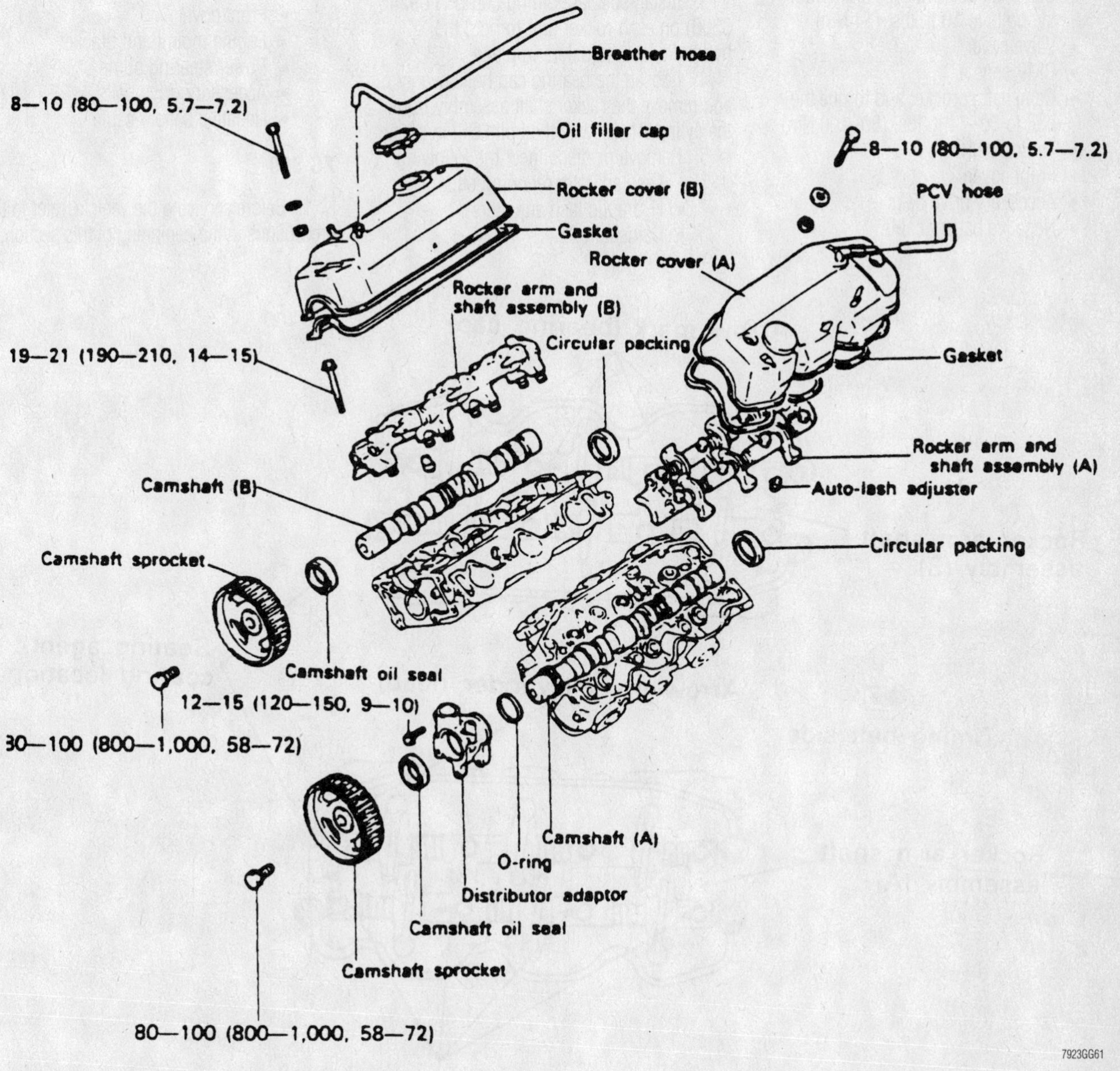

Breather hose

8—10 (80—100, 5.7—7.2)

Oil filler cap

8—10 (80—100, 5.7—7.2)

PCV hose

Rocker cover (B)

Gasket

Rocker cover (A)

Rocker arm and shaft assembly (B)

Circular packing

Gasket

19—21 (190—210, 14—15)

Rocker arm and shaft assembly (A)

Auto-lash adjuster

Camshaft (B)

Circular packing

Camshaft sprocket

Camshaft oil seal

12—15 (120—150, 9—10)

30—100 (800—1,000, 58—72)

Camshaft (A)

O-ring

Distributor adaptor

Camshaft oil seal

Camshaft sprocket

80—100 (800—1,000, 58—72)

7923GG61

Exploded view of the camshaft and rocker arm assembly components—3.0L engine

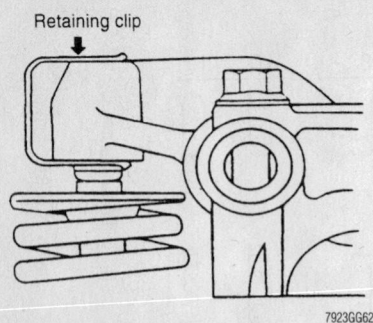

Before removing the rocker arm assemblies, install lash adjuster retaining clips (PN 09246-32000)—3.0L engine

the secondary chain aligned as shown
- Camshaft bearing caps and torque the bolts to 10 ft. lbs. (14 Nm)
- Valve cover
- CMP sensor
- Camshaft sprocket and torque the bolt to 60–74 ft. lbs. (80–100 Nm)
- Timing belt
- Front cover
- Accessory drive belts
- Negative battery cable

3.0L Sonata

1. Before servicing the vehicle, refer to the precautions in the beginning of this section.
2. Remove or disconnect the following:
   - Negative battery cable
   - Accessory drive belts
   - Power steering pump
   - Engine mount and bracket
   - Front cover
   - Timing belt
   - Camshaft sprockets
   - Distributor
   - Distributor adapter housing
   - Valve covers

➡ **Keep all valvetrain components in order for installation.**

3. Install special holding clip (PN 09246 3200) on each rocker arm to hold the hydraulic lash adjusters in place.
4. Loosen the bearing cap bolts evenly and remove the rocker shaft assembly from the cylinder head with the bolts still in place.
5. Remove or disconnect the following:
   - Lash adjuster retaining clips
   - Hydraulic lash adjusters
   - Camshafts

To install:

6. Install or connect the following:
   - Hydraulic lash adjusters in their original locations
   - Lash adjuster retaining clips
   - Camshafts
   - Rocker arm and shaft assemblies and torque the bearing cap bolts 14–15 ft. lbs. (19–21 Nm) starting from the center and working out
7. Remove the lash adjuster retaining clips.
8. Install or connect the following:
   - Valve covers
   - Distributor adapter housing
   - Distributor
   - Camshaft sprockets
   - Timing belt
   - Front cover
   - Engine mount and bracket
   - Power steering pump
   - Accessory drive belts
   - Negative battery cable

XG 300

1. Before servicing the vehicle, refer to the precautions in the beginning of this section.

Arrow mark (bearing cap)

Rocker arm shaft assembly (B)

Sealing agent coating location

Arrow mark (cylinder head)

Timing belt side

Rocker arm shaft assembly (A)

Arrow mark (bearing cap)

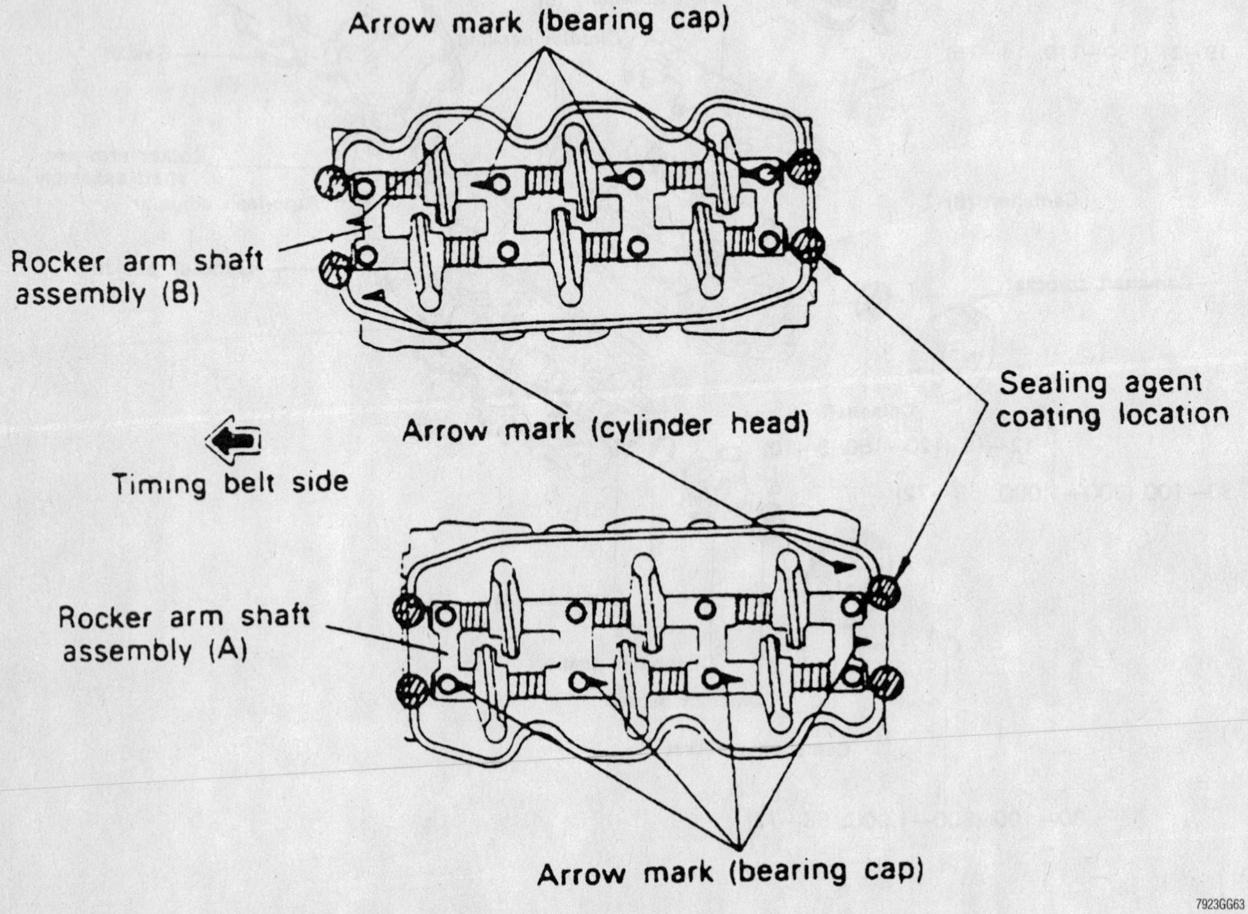

Install the rocker arm assemblies with the arrows as indicated. Place sealer at the corners—3.0L engine

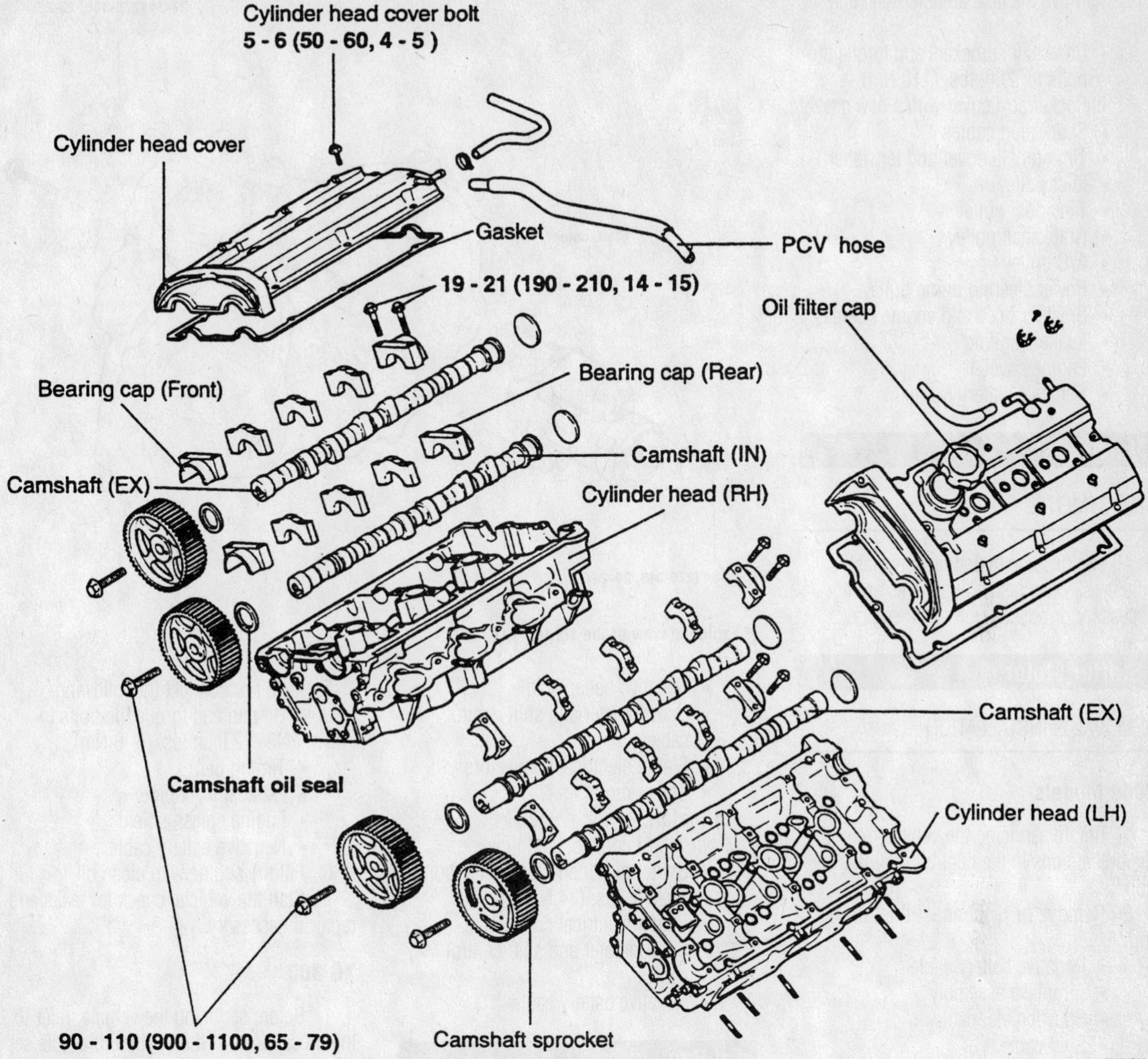

Cylinder head cover bolt
5 - 6 (50 - 60, 4 - 5)

Cylinder head cover

Gasket

PCV hose

19 - 21 (190 - 210, 14 - 15)

Oil filter cap

Bearing cap (Front)

Bearing cap (Rear)

Camshaft (IN)

Camshaft (EX)

Cylinder head (RH)

Camshaft oil seal

Camshaft (EX)

Cylinder head (LH)

90 - 110 (900 - 1100, 65 - 79)

Camshaft sprocket

9347KG08

Exploded view of the camshaft and related components—XG 300

2. Remove or disconnect the following:

- Negative battery cable
- Engine cover
- Intake manifold
- Breather hose and engine harness
- Power steering pulley
- A/C pulley
- Crankshaft pulley
- Idler pulley and tensioner pulley
- Timing belt cover and loosen the auto tensioner
- Timing belt
- Spark plug cables
- Rocker arm cover
- Camshaft sprockets
- Camshaft bearing caps
- Camshafts

To install:

3. Rotate the crankshaft so that the No. 1 cylinder is at the Top Dead Center (TDC) position.

4. Make certain that the rocker arm is installed properly on the lash adjuster and valve.

5. Install the camshaft dowel pin.

6. Install or connect the following:

- Hydraulic lash adjusters in their original locations
- Lash adjuster retaining clips
- Camshafts
- Rocker arm and shaft assemblies and torque the bearing cap bolts 14–15 ft. lbs. (19–21 Nm) starting from the center and working out

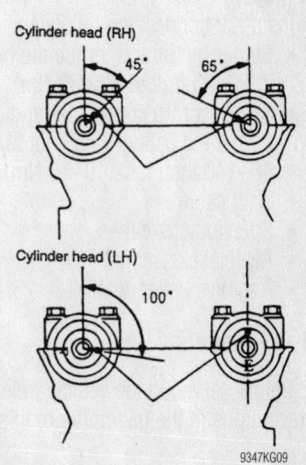

Cylinder head (RH)

45°   65°

Cylinder head (LH)

100°

9347KG09

Install the camshaft dowel pin as shown— XG 300

7. Remove the lash adjuster retaining clips.

- Camshaft sprockets and torque the bolts to 79 ft. lbs. (110 Nm)
- Rocker arm cover with a new gasket
- Spark plug cables
- Timing belt, cover and tensioner
- Idler pulley
- Tensioner pulley
- Crankshaft pulley
- A/C pulley
- Power steering pump pulley
- Breather hose and engine harness
- Intake manifold
- Engine cover
- Negative battery cable

Valve Lash

ADJUSTMENT

All engines use hydraulic valve lash adjusters. Valve lash adjustments are not necessary or possible on these engines.

Starter Motor

REMOVAL & INSTALLATION

2000 Models

1. Before servicing the vehicle, refer to the precautions in the beginning of this section.
2. Remove or disconnect the following:

- Negative battery cable
- Air intake assembly
- Speedometer cable
- Shift cable
- Starter wiring connectors
- Starter motor

To install:
3. Install or connect the following:

- Starter motor and torque the bolts to 20–25 ft. lbs. (26–33 Nm)
- Starter wiring connectors and torque the battery cable nut to 88–140 inch lbs. (10–16 Nm)
- Shift cable
- Speedometer cable
- Air intake assembly
- Negative battery cable

2001–03 Models

1. Before servicing the vehicle, refer to the precautions in the beginning of this section.
2. Remove or disconnect the follow-ing:

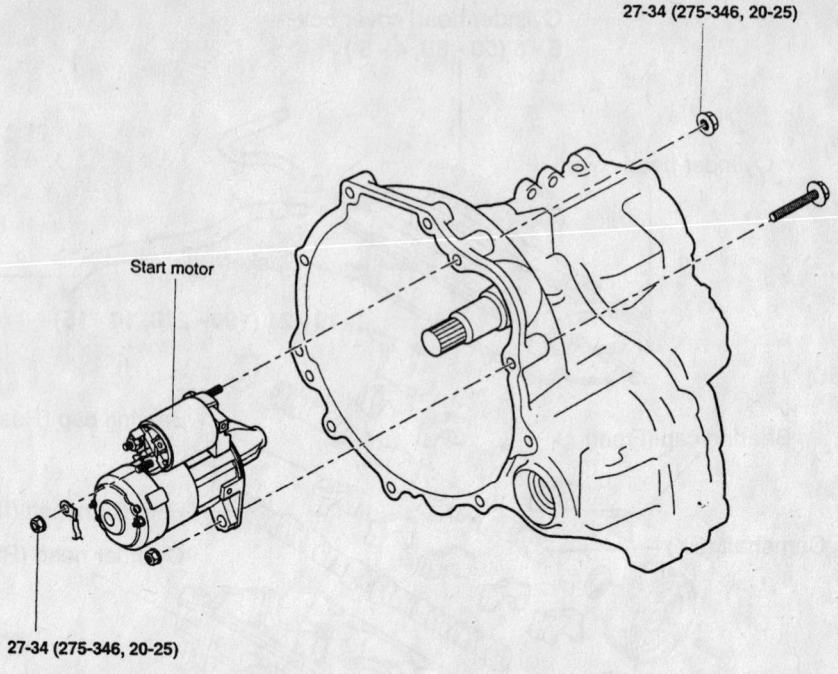

27-34 (275-346, 20-25)

Start motor

27-34 (275-346, 20-25)

9347KG10

Exploded view of the starter—XG 300

- Negative battery cable
- Speedometer and shift control cables
- Starter electrical connectors
- Starter motor

To install:
3. Install or connect the following:

- Starter motor and torque the bolts to 25 ft. lbs. (34 Nm)
- Starter electrical connectors
- Speedometer and shift control cables
- Negative battery cable

Oil Pan

REMOVAL & INSTALLATION

Except XG 300

1. Before servicing the vehicle, refer to the precautions in the beginning of this section.
2. Drain the engine oil.
3. Remove or disconnect the follow-ing:

- Negative battery cable
- Engine splash shield
- Exhaust front pipe
- Timing belt
- Oil pan

To install:
4. Apply a ⅛ in. (3mm) bead of RTV sealer along the groove in the oil pan.

5. Install or connect the following:

- Oil pan and torque the bolts to 48–72 inch lbs. (6–8 Nm)
- Timing belt
- Exhaust front pipe
- Engine splash shield
- Negative battery cable

6. Fill the engine with clean oil.
7. Start the vehicle, check for leaks and repair if necessary.

XG 300

1. Before servicing the vehicle, refer to the precautions in the beginning of this section.
2. Drain the engine oil.
3. Remove or disconnect the follow-ing:

- Negative battery cable
- Oil pressure switch
- Oil filter

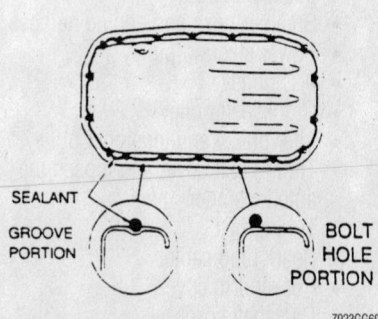

SEALANT

GROOVE PORTION

BOLT HOLE PORTION

7923GG69

Oil pan sealant applications points— except 3.0L engine

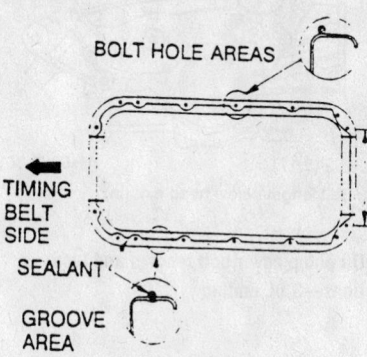

LIQUID-GASKET COATING AREA (TOP VIEW)

BOLT HOLE AREAS

TIMING BELT SIDE

SEALANT

GROOVE AREA

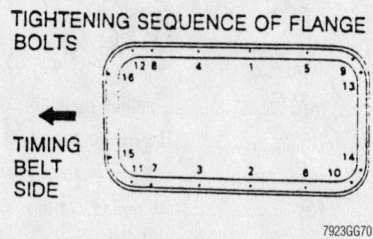

TIGHTENING SEQUENCE OF FLANGE BOLTS

TIMING BELT SIDE

7923GG70

Oil pan sealant applications points and tightening sequence—3.0L Sonata

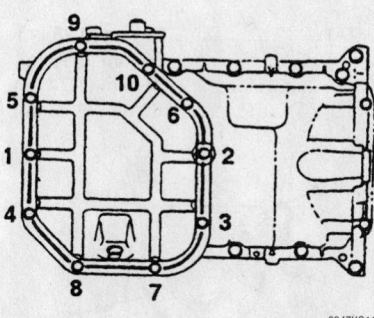

9347KG11

Tighten the oil pan bolts, in sequence, as shown

- Oil pan and discard the gasket
4. Clean all mating surfaces of any residual gasket material.

To install:

5. Apply a ⅛in. (3mm) bead of RTV sealer along the groove in the oil pan.
6. Install or connect the following:
- Oil pan and torque the bolts to 89 inch lbs. (10 Nm)
- Oil filter
- Oil pressure switch and torque the bolt to 89 inch lbs. (10 Nm)
- Negative battery cable
7. Fill the engine with clean oil.
8. Start the vehicle, check for leaks and repair if necessary.

Oil Pump

REMOVAL & INSTALLATION

Except Sonata 2.0L, 2.4L and XG 300 Models

1. Before servicing the vehicle, refer to the precautions in the beginning of this section.
2. Drain the engine oil.
3. Remove or disconnect the following:
- Negative battery cable
- Accessory drive belts
- Front cover
- Timing belt
- Crankshaft timing sprocket
- Oil pan
- Oil pickup tube
- Oil pump

To install:

4. Install or connect the following:
- Oil pump using a new gasket and torque the bolts to 11 ft. lbs. (15 Nm)
- Oil pickup tube and torque the bolts to 11 ft. lbs. (15 Nm)
- Oil pan
- Crankshaft timing sprocket
- Timing belt

- Front cover
- Accessory drive belts
- Negative battery cable
5. Fill the engine with clean oil.
6. Start the engine, check for leaks and repair if necessary.

Sonata 2.0L and 2.4L Engines

1. Before servicing the vehicle, refer to the precautions in the beginning of this section.
2. Drain the engine oil.
3. Remove or disconnect the following:
- Negative battery cable
- Accessory drive belts
- Front cover
- Timing belt
- Crankshaft timing sprocket
- Oil pan
- Oil pickup tube
- Pressure relief valve
- Oil pressure switch
- Oil filter adapter
- Oil pump plug cap
- Left cylinder block plug
4. Insert a prytool with a ⁵⁄₁₆in. (8mm) diameter shaft into the plug hole to hold the shaft while removing the balance shaft bolt.
5. Remove or disconnect the following:
- Left balance shaft retaining bolt

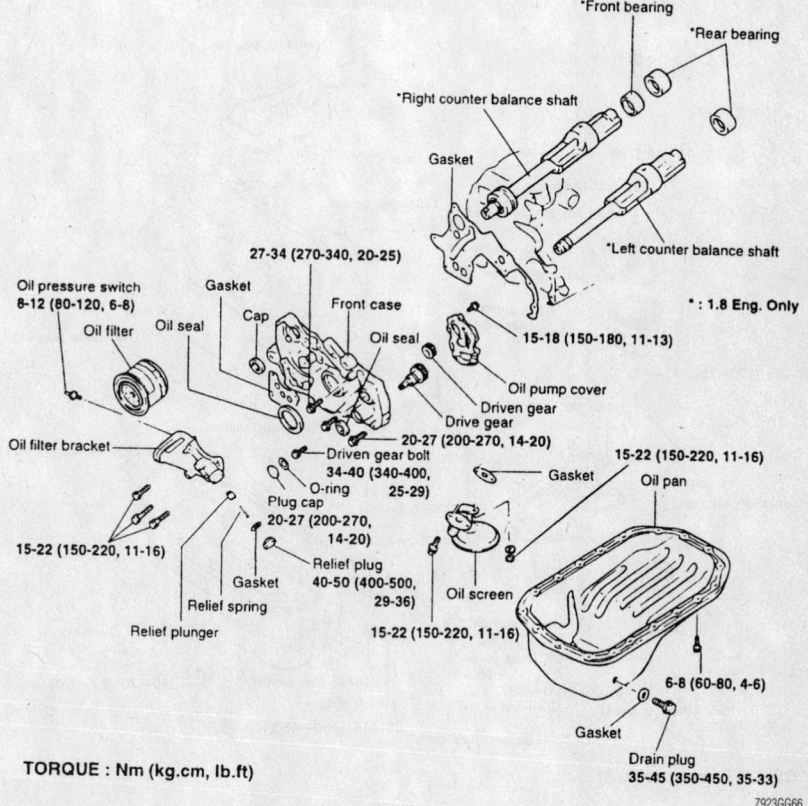

TORQUE : Nm (kg.cm, lb.ft)

Exploded view of the oil pump and pan—Sonata 2.0L engine

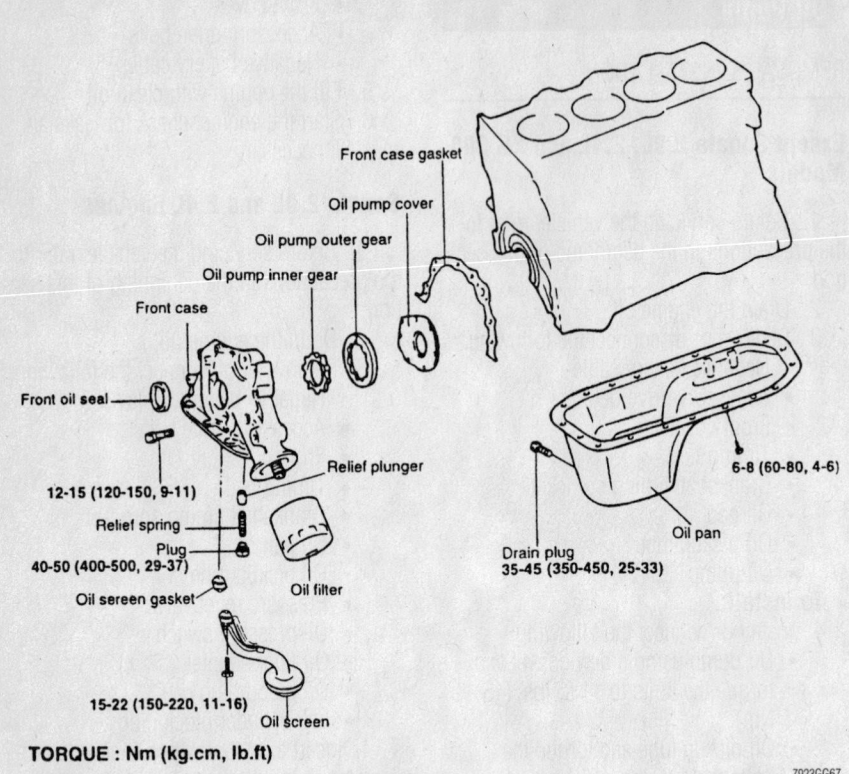

TORQUE : Nm (kg.cm, lb.ft)

Exploded view of the oil pump and pan—1.5L, 1.8L and Elantra and Tiburon with 2.0L engine

7923GG67

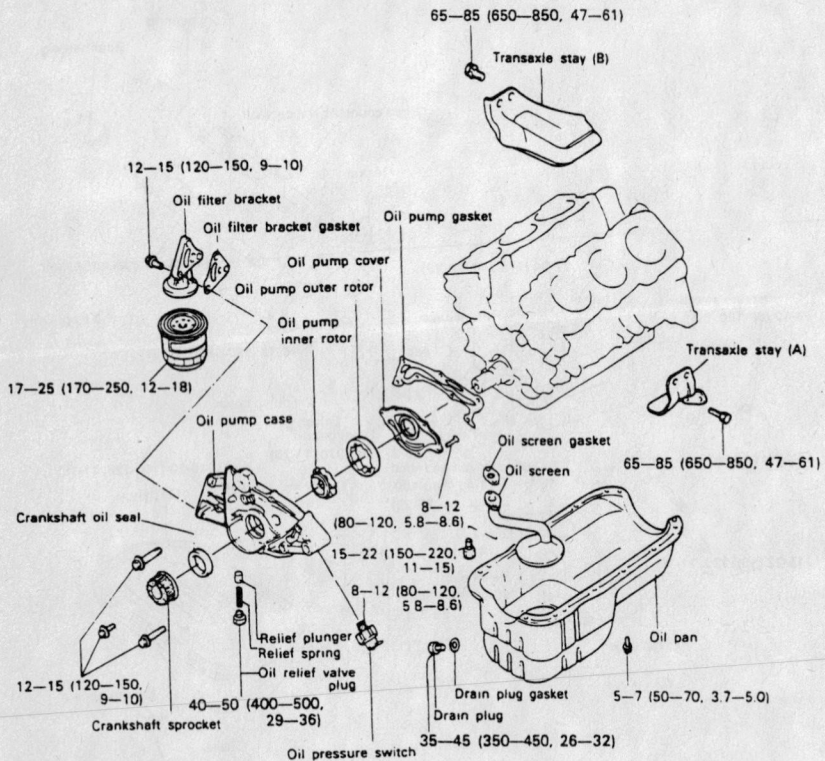

TORQUE : Nm (kg.cm, lb.ft)

7923GG68

Exploded view of the oil pump and pan—3.0L engine

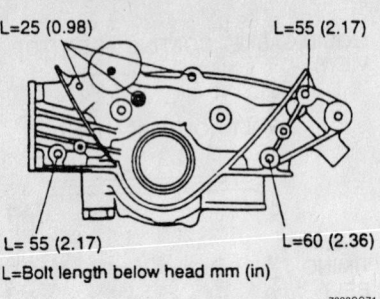

L=25 (0.98)   L=55 (2.17)
L= 55 (2.17)   L=60 (2.36)
L=Bolt length below head mm (in)

7923GG71

Oil pump cover bolt lengths and locations—3.0L engine

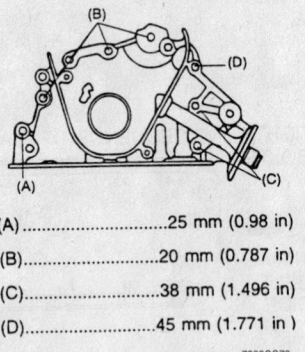

(A)............................25 mm (0.98 in)
(B)............................20 mm (0.787 in)
(C)............................38 mm (1.496 in)
(D)............................45 mm (1.771 in)

7923GG72

Oil pump cover bolt lengths and locations—1.8L and Elantra and Tiburon with 2.0L engines

(A) .... 25 mm (0.98 in.)
(B) .... 30 mm (1.18 in.)
(C) .... 45 mm (1.77 in.)
(D) .... 60 mm (2.36 in.)

7923GG73

Oil pump cover bolt lengths and locations—1.5L engines

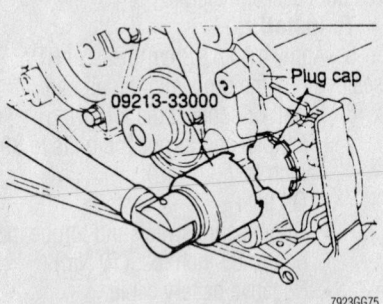

7923GG75

A special socket (PN 09213-33000) is available to remove the plug cap from the oil pump portion of the case—Sonata 2.0L and 2.4L engines

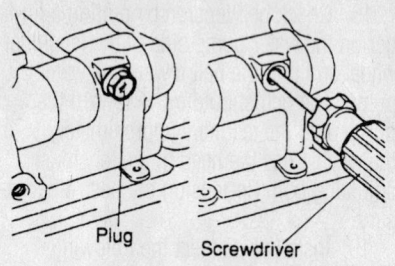

Remove the left side cylinder block plug and insert a screwdriver into the hole to hold the balance shaft from turning—Sonata 2.0L and 2.4L engines

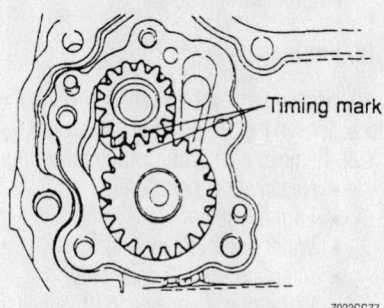

Align the timing marks on the gears during assembly—Sonata 2.0L and 2.4L engines

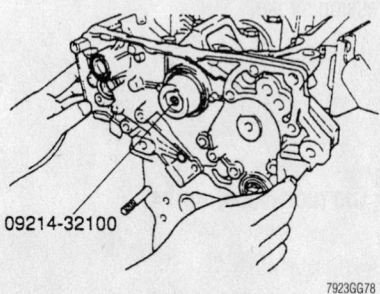

A special tool (PN 09214-32100) is used to center the front case hole on the crankshaft—Sonata 2.0L engines

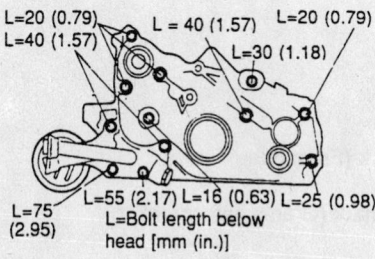

Oil pump cover bolt lengths and locations—Sonata 2.0L engines

• Front case assembly
• Left and right balance shafts
• Oil pump housing and gears

To install:

6. Install or connect the following:
  • Oil pump housing and gears
  • Left and right balance shafts

7. Lubricate and install Case Alignment Tool, PN 09214-32100 on the crankshaft.

8. Install or connect the following:
  • Front case assembly and torque the bolts to 20 ft. lbs. (27 Nm)
  • Left balance shaft retaining bolt and torque the bolt to 25–29 ft. lbs. (34–40 Nm)
  • Left cylinder block plug and torque the plug to 14–20 ft. lbs. (20–27 Nm)
  • Oil pump plug cap
  • Oil filter adapter
  • Oil pressure switch
  • Pressure relief valve
  • Oil pickup tube
  • Oil pan
  • Crankshaft timing sprocket
  • Timing belt
  • Front cover
  • Accessory drive belts
  • Negative battery cable

9. Fill the engine with clean oil.

10. Start the engine, check for leaks and repair if necessary.

XG 300

1. Before servicing the vehicle, refer to the precautions in the beginning of this section.

2. Drain the engine oil.

3. Remove or disconnect the following:
  • Negative battery cable
  • Oil pressure switch
  • Oil filter
  • Oil pan
  • Oil screen and gasket
  • Oil filter bracket and gasket
  • Oil relief plug
  • Oil pump case
  • Oil pump rotor and both covers

To install:

4. Install or connect the following:
  • Oil pump inner cover and torque the bolt to 11 ft. lbs. (15 Nm)
  • Oil pump outer cover and rotor
  • Oil pump case and torque the bolts to 11 ft. lbs. (15 Nm)
  • Oil relief plug and torque to 36 ft. lbs. (50 Nm)
  • Oil filter bracket with a new gasket
  • Oil screen and gasket and torque the bolt to 15 ft. lbs. (20 Nm)
  • Oil pan

• Oil filter
• Oil pressure switch and torque to 89 inch lbs. (10 Nm)
• Negative battery cable

5. Fill the engine with clean oil.

6. Start the vehicle, check for leaks and repair if necessary.

Rear Main Seal

REMOVAL & INSTALLATION

1. Before servicing the vehicle, refer to the precautions in the beginning of this section.

2. Remove or disconnect the following:
  • Transaxle
  • Flywheel
  • Oil seal case
  • Oil separator, if equipped
  • Oil seal

To install:

3. Install or connect the following:
  • Oil seal
  • Oil separator, if equipped
  • Oil seal case and torque the case bolts to 84–108 inch lbs. (8–10 Nm)
  • Flywheel
  • Transaxle

Timing Belt

REMOVAL & INSTALLATION

✳✳ CAUTION

Timing belt maintenance is extremely important. All Hyundai models use interference-type non-freewheeling engines. Should the timing belt break in these engines, the valves in the cylinder head will come in contact with the pistons, causing major engine damage. The recommended replacement interval for timing belts is 60,000 miles.

1.5L Engines

1. Before servicing the vehicle, refer to the precautions in the beginning of this section.

2. Remove or disconnect the following:
  • Negative battery cable
  • Engine coolant
  • Water pump pulley bolts
  • Alternator bolt, loosen only
  • Water pump pulley and drive belts
  • Crankshaft pulley
  • Timing belt cover

3. Rotate the crankshaft clockwise and align the timing marks so No. 1 piston will be at Top Dead Center (TDC) of the compression stroke.

4. Loosen the tensioning bolt and the pivot bolt on the timing belt tensioner. Move the tensioner as far as it will go toward the water pump. Tighten the adjusting bolt.

5. Mark the timing belt with an arrow showing direction of rotation.

6. Move the timing belt tensioner pulley toward the water pump and temporarily secure it.

7. Remove or disconnect the following:

- Timing belt
- Crankshaft sprocket bolts, sprocket and flange
- Timing belt tensioner, if defective

To install:

8. Install the flange and crankshaft sprocket. Tighten the crankshaft sprocket bolt to 103–111 ft. lbs. (140–150 Nm).

9. Align the timing marks of the camshaft sprocket and check that the crankshaft timing marks are still in alignment.

10. If removed, install the timing belt tensioner, spring and spacer with the bottom end of the spring free. Tighten the adjusting bolt slightly with the tensioner moved as far as possible away from water pump.

11. Install the free end of the spring into the locating tang on the front case.

12. Position the timing belt over the crankshaft sprocket, then over the camshaft sprocket. Slip the back of the belt over the tensioner wheel.

13. Turn the camshaft sprocket in the opposite of its normal direction of rotation until the straight side of the belt is tight and be sure the timing marks align.

➡ **If the timing marks are not properly aligned, shift the belt 1 tooth at a time in the appropriate direction until they are aligned.**

14. Loosen the tensioner mounting bolts so the tensioner works, without the interference of any friction, under spring pressure. Be sure the belt follows the curve of the camshaft pulley so the teeth are engaged all the way around. Correct the path of the belt, if necessary.

15. Tighten the tensioner adjusting bolt, then the tensioner pivot bolt to 15–18 ft. lbs. (20–26 Nm).

➡ **Bolts must be tightened in the stated order or tension won't be correct.**

16. Turn the crankshaft 1 turn clockwise until timing marks again align to seat the belt.

17. Loosen both tensioner attaching bolts and let the tensioner position itself under spring tension. Retighten the bolts.

18. Check belt tension by putting a finger on the water pump side of the tensioner wheel and pull the belt toward the water pump. The belt should move toward the pump until the teeth are approximately ½ of the way across the head of the tensioner adjusting bolt. Re-tension the belt, if necessary.

19. Install or connect the following:

- Timing belt cover
- Crankshaft pulley
- A/C compressor belt
- Water pump pulley
- V belt
- Negative battery cable
- Engine with coolant

2.0L Engines

1. Before servicing the vehicle, refer to the precautions in the beginning of this section.

2. Remove or disconnect the following:

- Negative battery cable
- Engine coolant
- Water pump pulley bolts
- Alternator bolt, loosen only
- Water pump pulley and drive belts
- Crankshaft pulley
- Timing belt cover(s)

3. Rotate the crankshaft clockwise and align the timing marks so No. 1 piston will be at Top Dead Center (TDC) of the compression stroke.

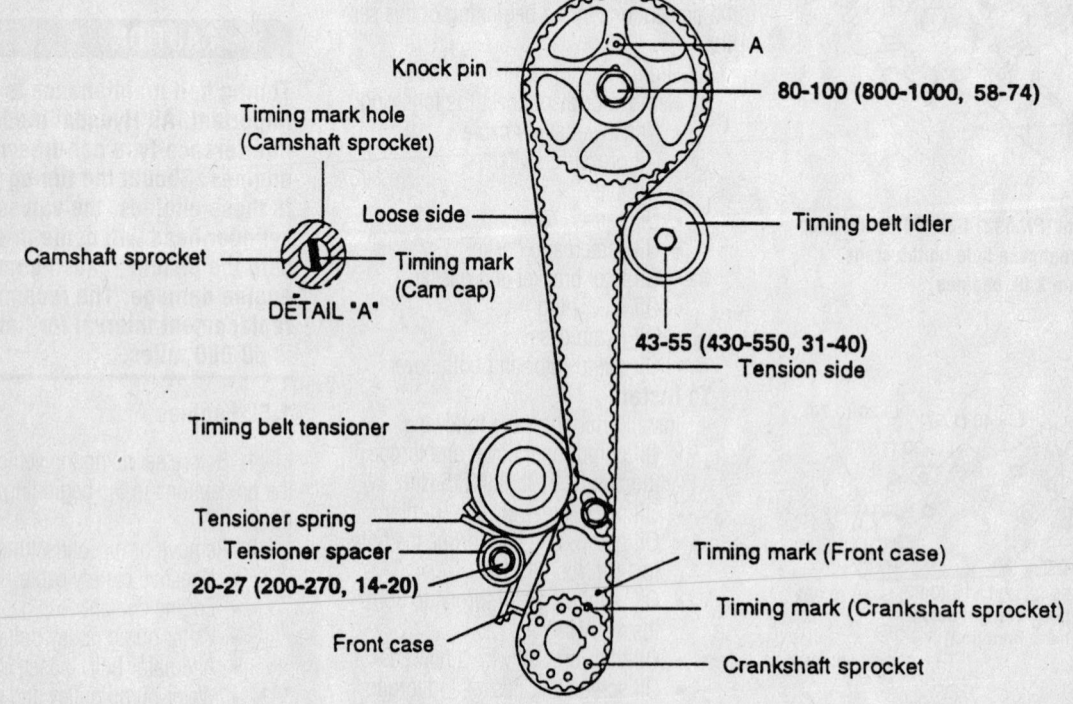

Proper pulley alignment for timing belt installation at TDC—Hyundai 1.5L DOHC engines

93015G25

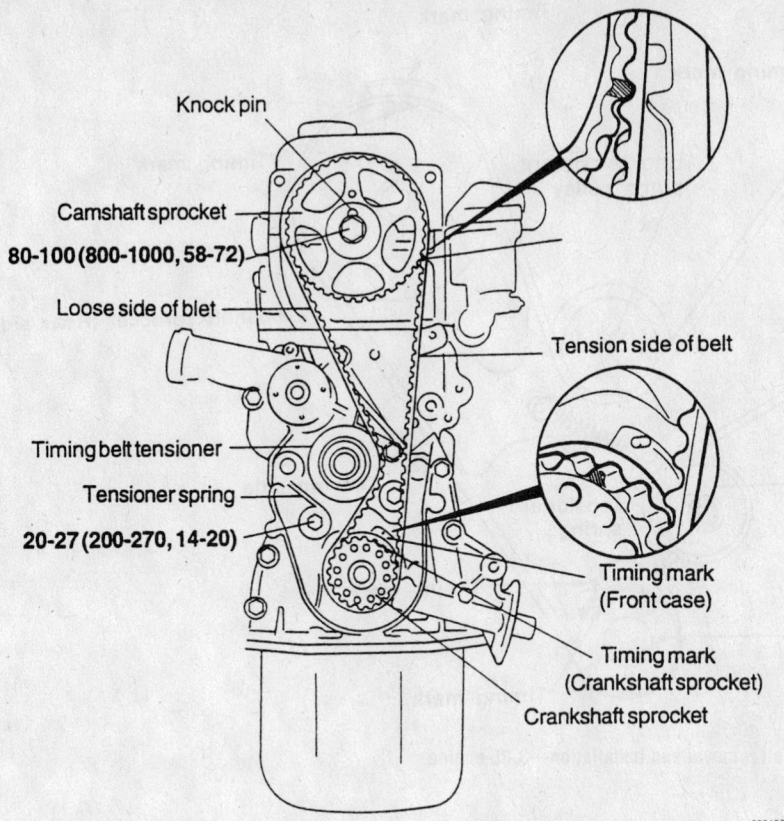

Proper pulley alignment for timing belt installation at TDC—Hyundai 1.5L SOHC engines

Knock pin

Camshaft sprocket
80-100 (800-1000, 58-72)

Loose side of belt

Timing belt tensioner

Tensioner spring
20-27 (200-270, 14-20)

Tension side of belt

Timing mark (Front case)

Timing mark (Crankshaft sprocket)

Crankshaft sprocket

93015G26

9. Install the idler pulley, if equipped. Tighten bolt to 32–41 ft. lbs. (43–55 Nm).

10. Position the timing belt over the camshaft sprocket, then over the crankshaft sprocket.

11. Tension the timing belt and tighten the tensioner pulley bolt to 32–41 ft. lbs. (43–55 Nm). When properly tensioned, the timing belt should deflect 0.16–0.24 in. (4–6mm) when a force of 5 lbs. (2.2kg) is placed on the longest span of the belt.

12. Turn the crankshaft sprocket one turn clockwise and realign the crankshaft sprocket timing mark.

13. Recheck the belt tension and adjust as necessary.

14. Install or connect the following:
- Timing belt cover(s)
- Crankshaft pulley
- A/C compressor belt
- Water pump pulley
- V belt
- Negative battery cable
- Engine with coolant

3.0L Engine

1. Before servicing the vehicle, refer to the precautions in the beginning of this section.

2. Remove or disconnect the following:
- Negative battery cable
- Engine coolant
- Water pump pulley bolts
- Alternator bolt, loosen only
- Water pump pulley and drive belts
- Crankshaft pulley
- Timing belt cover(s)

3. Turn the crankshaft until the timing marks on the camshaft sprocket and cylinder head are aligned.

4. Loosen the timing belt tensioner bolt and turn the tensioner counterclockwise as far as it will go. Tighten the adjusting bolt.

5. Mark the timing belt with an arrow showing direction of rotation.

6. Remove the timing belt.

7. If defective, remove the timing belt tensioner.

To install:

8. If necessary, install the timing belt tensioner.

9. Attach the top of the tensioner spring on the engine coolant pump pin. Ensure the hook on the pin is facing down and the hook on the tensioner is facing away from the engine

10. Rotate the timing belt tensioner to the extreme counterclockwise position. Temporarily lock the tensioner in place.

11. Align the timing marks of the camshaft and crankshaft sprockets.

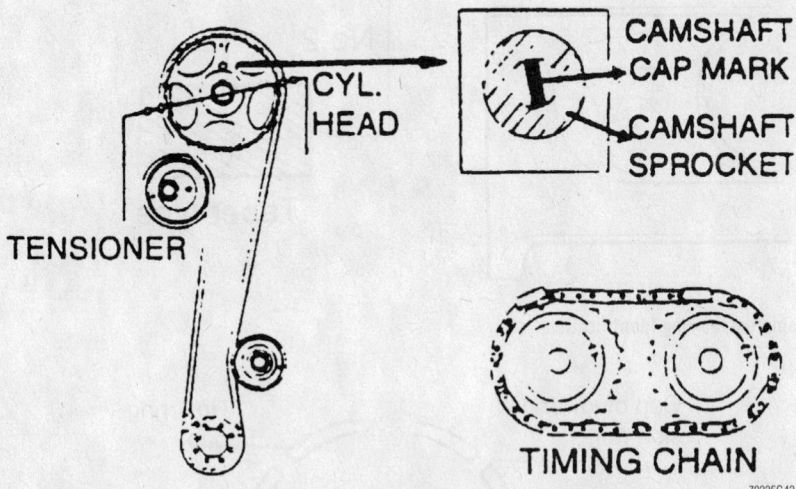

Proper alignment of the timing belt alignment marks for belt removal and installation— Hyundai 1.8L and 2.0L engines

CYL. HEAD

CAMSHAFT CAP MARK

CAMSHAFT SPROCKET

TENSIONER

TIMING CHAIN

79235G43

4. Remove the timing belt tensioner and idler pulley.

5. Mark the timing belt with an arrow showing direction of rotation.

6. Remove the timing belt.

To install:

7. Align the timing marks of the camshaft sprocket and check that the crankshaft timing marks are still in alignment.

8. Install the timing belt tensioner.

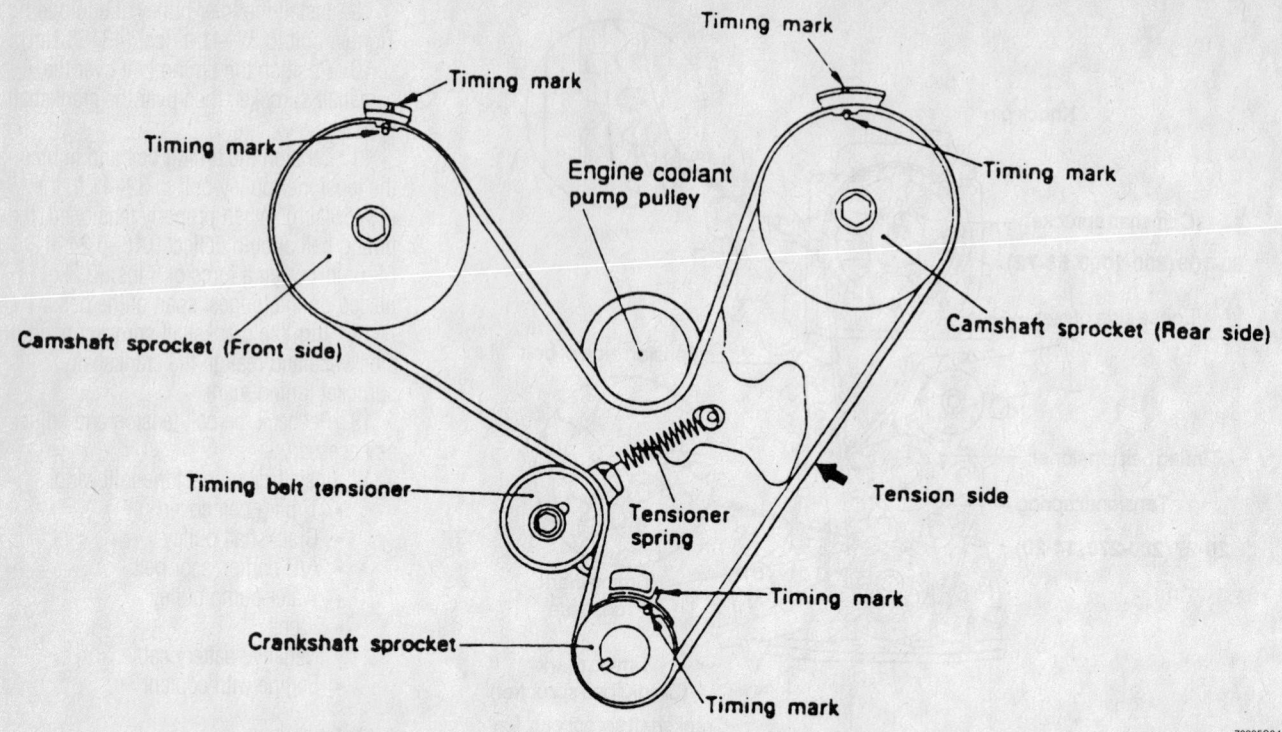

Timing belt sprocket alignment mark positioning for belt removal and installation—3.0L engine

12. Install the timing belt on the crankshaft sprocket, then onto the rear camshaft sprocket.

13. Route the belt to the coolant pump pulley, the front camshaft sprocket and the timing belt tensioner.

14. Apply force counterclockwise to the rear camshaft sprocket with tension on the tight side of the belt and check that timing marks are aligned.

15. Loosen the tensioner bolt 1–2 turns and tighten the timing belt to a tension of 57–84 lbs. (260–380 N).

16. Turn the crankshaft two turns clockwise.

17. Readjust the sprocket timing marks and tighten the tensioner bolts.

18. Install the timing belt covers.

19. Install the crankshaft pulley and tighten to 108–116 ft. lbs. (150–160 Nm).

20. Install or connect the following:
- Timing belt cover(s)
- Crankshaft pulley
- A/C compressor belt
- Water pump pulley
- V belt
- Negative battery cable
- Engine with coolant

Piston and Ring

POSITIONING

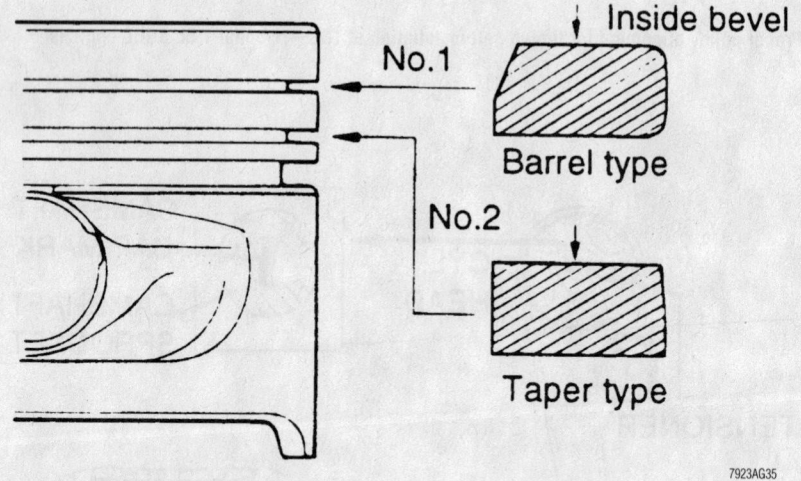

Compression ring identification

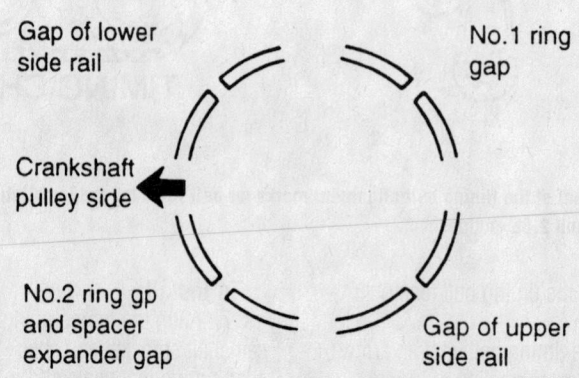

Piston ring end-gap spacing

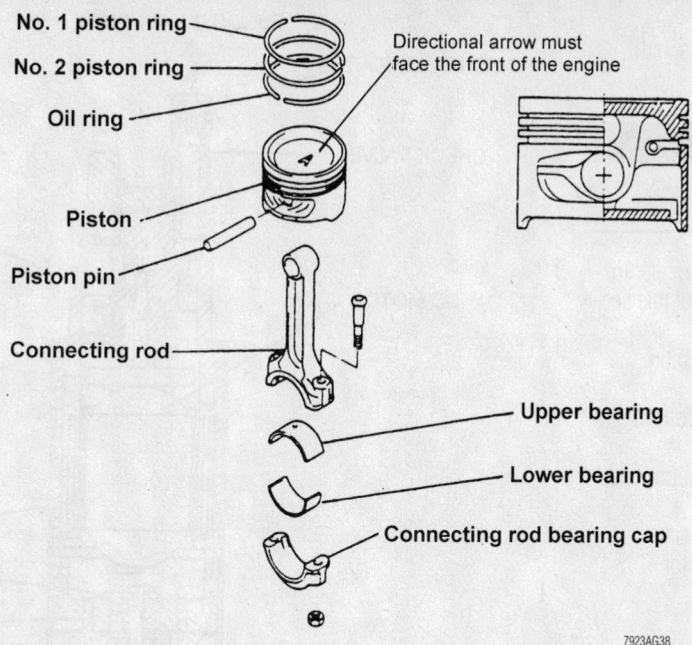

No. 1 piston ring
No. 2 piston ring
Oil ring
Piston
Piston pin
Connecting rod

Directional arrow must face the front of the engine

Upper bearing
Lower bearing
Connecting rod bearing cap

7923AG38

Piston and connecting rod assembly

FUEL SYSTEM

Fuel System Service Precautions

Safety is the most important factor when performing not only fuel system maintenance, but any type of maintenance. Failure to conduct maintenance and repairs in a safe manner may result in serious personal injury or death. Maintenance and testing of the vehicle's fuel system components can be accomplished safely and effectively by adhering to the following rules and guidelines.

• To avoid the possibility of fire and personal injury, always disconnect the negative battery cable unless the repair or test procedure requires that battery voltage be applied.

• Always relieve the fuel system pressure prior to disconnecting any fuel system component (injector, fuel rail, pressure regulator, etc.), fitting or fuel line connection. Exercise extreme caution whenever relieving the fuel system pressure, to avoid exposing skin, face and eyes to fuel spray. Please be advised that fuel under pressure may penetrate the skin or any part of the body that it contacts.

• Always place a shop towel or cloth around the fitting or connection prior to loosening to absorb any excess fuel due to spillage. Ensure that all fuel spillage (should it occur) is quickly removed from engine surfaces. Ensure that all fuel soaked cloths or towels are deposited into a suitable waste container.

• Always keep a dry chemical (Class B) fire extinguisher near the work area.

• Do not allow fuel spray or fuel vapors to come into contact with a spark or open flame.

• Always use a back-up wrench when loosening and tightening fuel line connection fittings. This will prevent unnecessary stress and torsion to fuel line piping. Always follow the proper torque specifications.

• Always replace worn fuel fitting O-rings with new. Do not substitute fuel hose where fuel pipe is installed.

Fuel System Pressure

RELIEVING

1. Before servicing the vehicle, refer to the precautions in the beginning of this section.
2. Remove or disconnect the following:

• Rear seat cushion
• Access panel
• Fuel pump module connector

3. Start the engine and allow it to run until it stalls.
4. Turn the ignition switch to the **OFF-** position.
5. Disconnect the negative battery cable.
6. Attach the fuel pump harness connector.

Fuel Filter

REMOVAL & INSTALLATION

➡**The fuel filter is located underneath the car, near the fuel tank.**

1. Before servicing the vehicle, refer to the precautions in the beginning of this section.
2. Relieve the fuel system pressure.
3. Remove or disconnect the following:

• Negative battery cable
• Fuel supply and pressure lines
• Fuel filter bracket, if equipped
• Fuel filter

To install:

4. Install or connect the following:

• Fuel filter and torque the mounting bolts to 18–25 ft. lbs. (25–35 Nm)
• Fuel filter bracket, if equipped
• Fuel supply and pressure lines

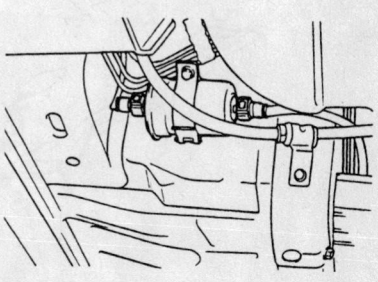

9347KG12

Remove the fuel filter

• Negative battery cable
5. Start the engine, check leaks and repair if necessary.

Fuel Pump

REMOVAL & INSTALLATION

1. Before servicing the vehicle, refer to the precautions in the beginning of this section.
2. Relieve the fuel system pressure.
3. Drain the fuel tank.
4. Remove or disconnect the following:
   • Negative battery cable
   • Fuel supply, return and vapor lines
   • Fuel fill and vent hoses
   • Fuel pump module connector
   • Fuel level sender connector
   • Fuel tank straps
   • Fuel tank
   • Fuel pump module

To install:
5. Install or connect the following:
   • Fuel pump module and torque the mounting bolts to 12–24 inch lbs. (1–3 Nm)
   • Fuel tank
   • Fuel tank straps
   • Fuel level sender connector

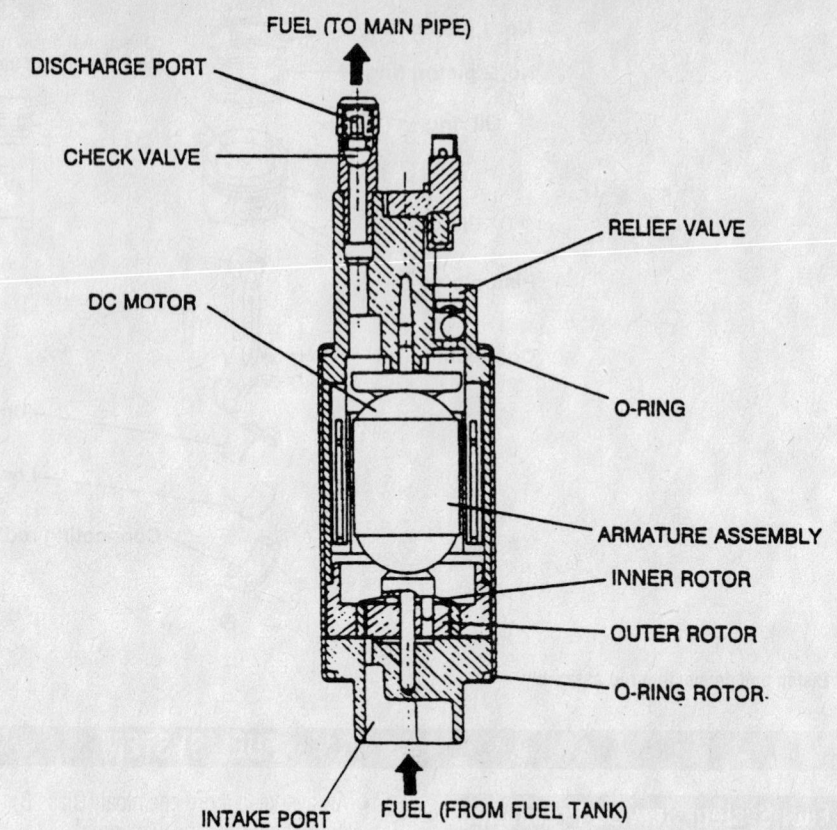

Cut away view of the electric fuel pump

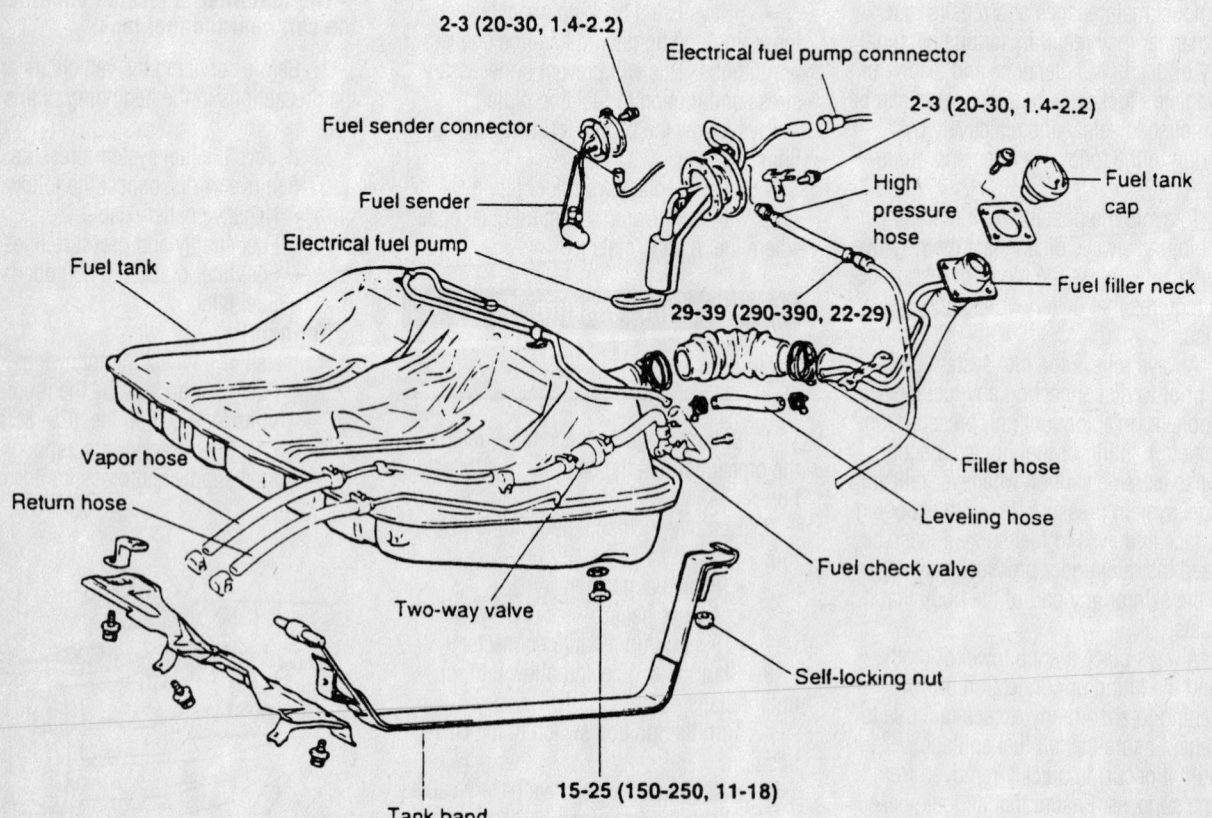

Exploded view of the fuel tank and fuel pump assembly—Sonata

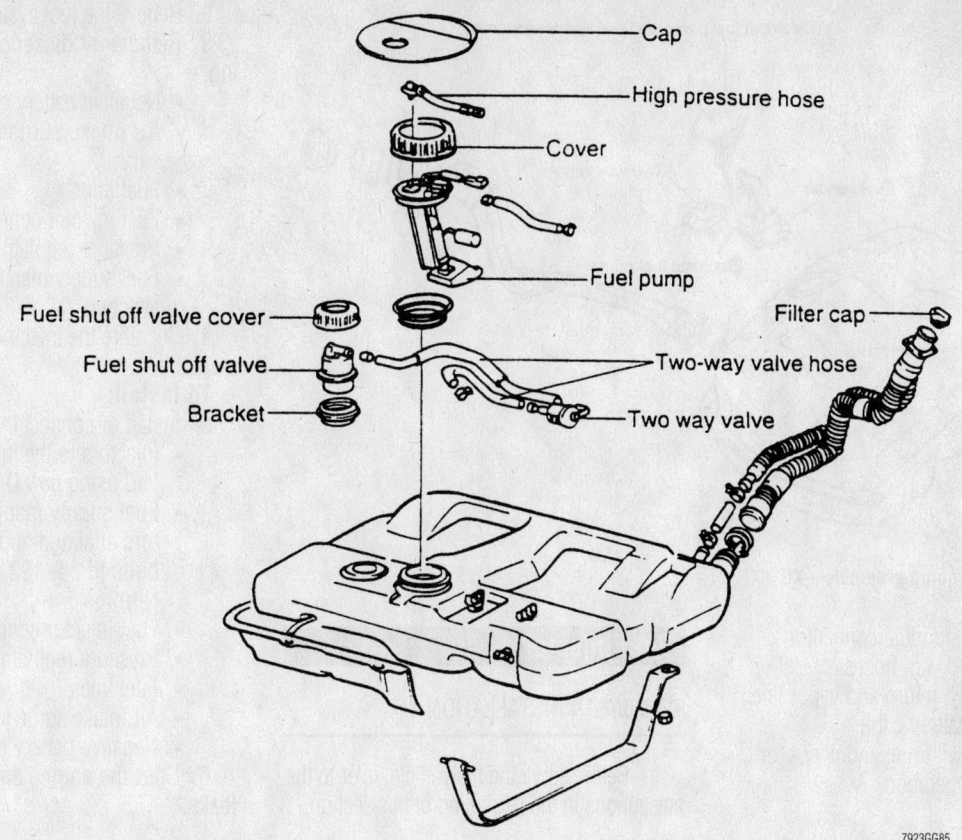

Exploded view of the fuel tank and fuel pump assembly—Accent

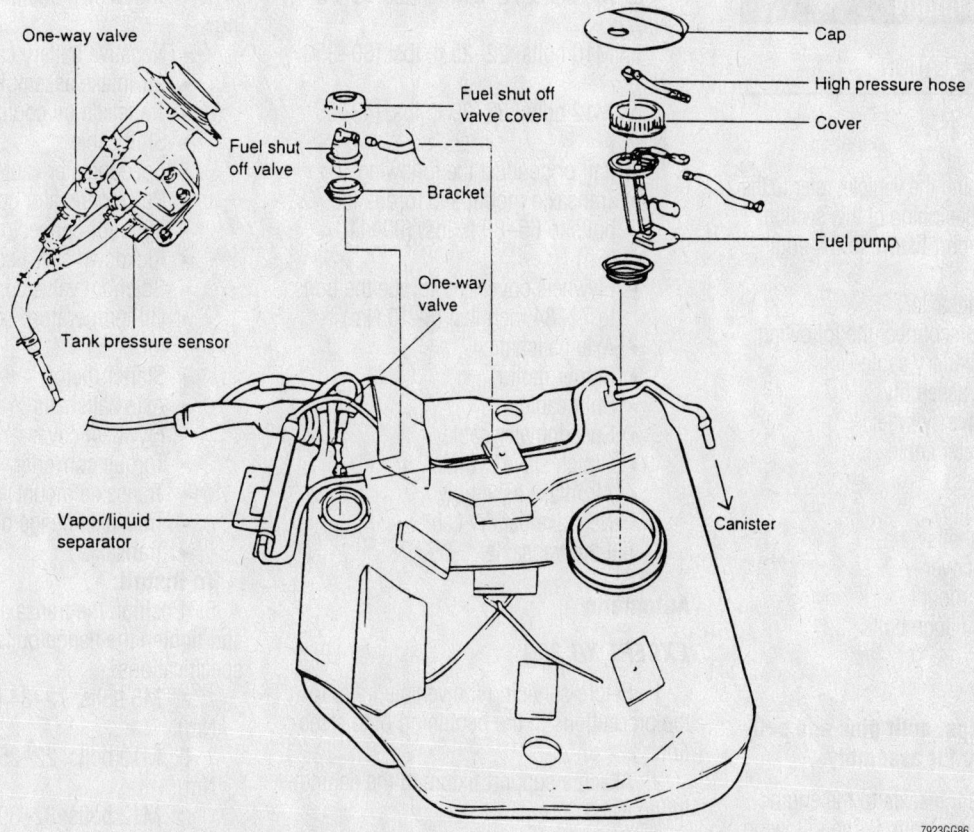

Fuel tank and fuel pump assembly—Tiburon and Elantra

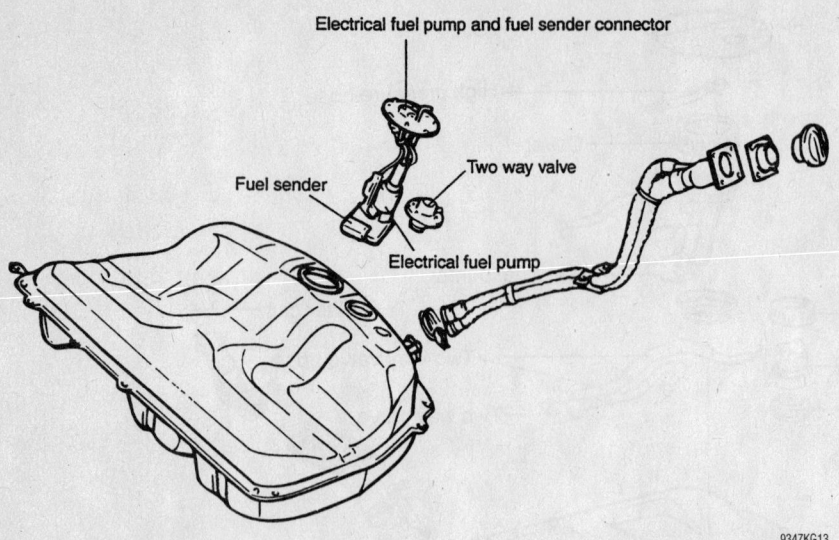

Fuel tank and fuel pump assembly—XG 300

- Fuel pump module connector
- Fuel fill and vent hoses
- Fuel supply, return and vapor lines
- Negative battery cable

6. Fill the tank with fuel and check for proper fuel pump operation.

Fuel Injector

REMOVAL & INSTALLATION

1. Before servicing the vehicle, refer to the precautions in the beginning of this section.

2. Relieve the fuel system pressure.
3. Remove or disconnect the following:

- Negative battery cable
- Air intake surge tank, if necessary
- Fuel lines
- Fuel injector connectors
- Pressure regulator vacuum line
- Fuel supply manifold with injectors attached

4. Separate the injectors from the supply manifold.

To install:

5. Install or connect the following:

- Injectors to the fuel supply manifold using new O-rings
- Fuel supply manifold with injectors attached and torque the bolts to 84–132 inch lbs. (10–15 Nm)
- Fuel injector connectors
- Pressure regulator vacuum line
- Fuel lines
- Air intake surge tank, if removed
- Negative battery cable

6. Start the engine and check for leaks.

DRIVE TRAIN

Transaxle Assembly

REMOVAL & INSTALLATION

Manual

1. Before servicing the vehicle, refer to the precautions in the beginning of this section.
2. Attach a support fixture to the engine lifting eyes.
3. Drain the transaxle.
4. Remove or disconnect the following:

- Negative battery cable
- Air intake assembly
- Clutch slave cylinder
- Speedometer cable
- Shift cables
- Starter motor
- Axle halfshafts
- Flywheel cover
- Transaxle mount
- Transaxle flange bolts
- Transaxle

To install:

➡️**Use new circlips, split pins and self-locking fasteners for assembly.**

5. Position the transaxle to the engine and tighten the flange bolts to the following specifications:

a. M8 bolts: 72–84 inch lbs. (8–10 Nm).
b. M10 bolts: 22–25 ft. lbs. (30–35 Nm).
c. M12 bolts: 32–39 ft. lbs. (43–55 Nm).

6. Install or connect the following:

- Transaxle mount and torque the bolts to 65–80 ft. lbs. (90–110 Nm)
- Flywheel cover and torque the bolts to 72–84 inch lbs. (8–10 Nm)
- Axle halfshafts
- Starter motor
- Shift cables
- Speedometer cable
- Clutch slave cylinder
- Air intake assembly
- Negative battery cable

7. Fill the transaxle.

Automatic

EXCEPT XG 300

1. Before servicing the vehicle, refer to the precautions in the beginning of this section.
2. Attach a support fixture to the engine lifting eyes.
3. Drain the transaxle.

4. Remove or disconnect the following:

- Negative battery cable
- Air intake assembly
- Transaxle oil cooler lines
- Shift cable
- Speedometer cable
- Pulse generator connector
- Inhibitor connector
- Kickdown servo connector
- Solenoid valve connector
- Oil temperature sensor connector
- Starter motor
- Axle halfshafts
- Flywheel cover
- Torque converter
- Transaxle mount
- Transaxle flange bolts
- Transaxle

To install:

5. Position the transaxle to the engine and tighten the flange bolts to the following specifications:

a. M8 bolts: 72–84 inch lbs. (8–10 Nm).
b. M10 bolts: 22–25 ft. lbs. (30–35 Nm).
c. M12 bolts: 32–39 ft. lbs. (43–55 Nm).

6. Install or connect the following:
- Transaxle mount and torque the bolts to 65–80 ft. lbs. (90–110 Nm)
- Torque converter and torque the bolts to 34–39 ft. lbs. (46–53 Nm)
- Flywheel cover and torque the bolts to 72–84 inch lbs. (8–10 Nm)
- Axle halfshafts
- Starter motor
- Oil temperature sensor connector
- Solenoid valve connector
- Kickdown servo connector
- Inhibitor connector
- Pulse generator connector
- Speedometer cable
- Shift cable
- Transaxle oil cooler lines
- Air intake assembly
- Negative battery cable

7. Fill the transaxle to the correct level.

XG 300

1. Before servicing the vehicle, refer to the precautions in the beginning of this section.

2. Attach a support fixture to the engine lifting eyes.

3. Drain the transaxle.

4. Remove or disconnect the following:
- Negative battery cable
- Air cleaner assembly
- Control cable
- Speedometer sensor connector
- Transaxle range switch connector
- Solenoid connector
- Oil temperature sensor connector
- Oil cooler lines
- Steering gear
- Sway bar
- Tie rod end
- Ball joints and drive shafts
- Steering U-joint and return tube mounting bolts
- Subframe
- Starter
- Engine-to-transaxle bolts
- Transaxle

To install:

5. Install or connect the following:
- Engine-to-transaxle and torque the bolts to 39 ft. lbs. (54 Nm)
- Starter
- Subframe and torque the subframe to transaxle bolts to 58 ft. lbs. (80 Nm) and the roll stopper bolts to 38 ft. lbs. (55 Nm)
- Steering U-joint and return tube mounting bolts
- Ball joints and driveshafts

- Tie rod end
- Sway bar
- Gear box
- Oil cooler lines
- Oil temperature sensor connector
- Solenoid connector
- Transaxle range switch connector
- Speedometer sensor connector and torque to19 ft. lbs. (26 Nm)
- Control cable and torque the bracket bolt to 18 ft. lbs. (25 Nm)
- Air cleaner assembly
- Negative battery cable

6. Fill the transaxle fluid to the proper level.

7. Start the vehicle, check for leaks and repair if necessary.

Clutch

ADJUSTMENTS

These vehicles are equipped with a hydraulic clutch system. No adjustment is necessary.

REMOVAL & INSTALLATION

1. Before servicing the vehicle, refer to the precautions in the beginning of this section.

2. Remove or disconnect the following:
- Transaxle
- Pressure plate bolts by loosening them evenly in a crossing pattern
- Pressure plate and clutch disc

To install:

3. Install or connect the following:
- Clutch disc on the flywheel
- Pressure plate and torque the bolts evenly in a crossing pattern to 11–15 ft. lbs. (15–21 Nm)
- Transaxle

Hydraulic Clutch System

BLEEDING

1. Connect a hose to the bleeder screw and place the other end of hose into a container of clean brake fluid. Open the bleeder screw.

2. Have an assistant pump the clutch pedal slowly until no air bubbles are present at the bleeder screw.

3. Close the bleeder screw.

4. Fill the clutch master cylinder.

5. Check the clutch operation.

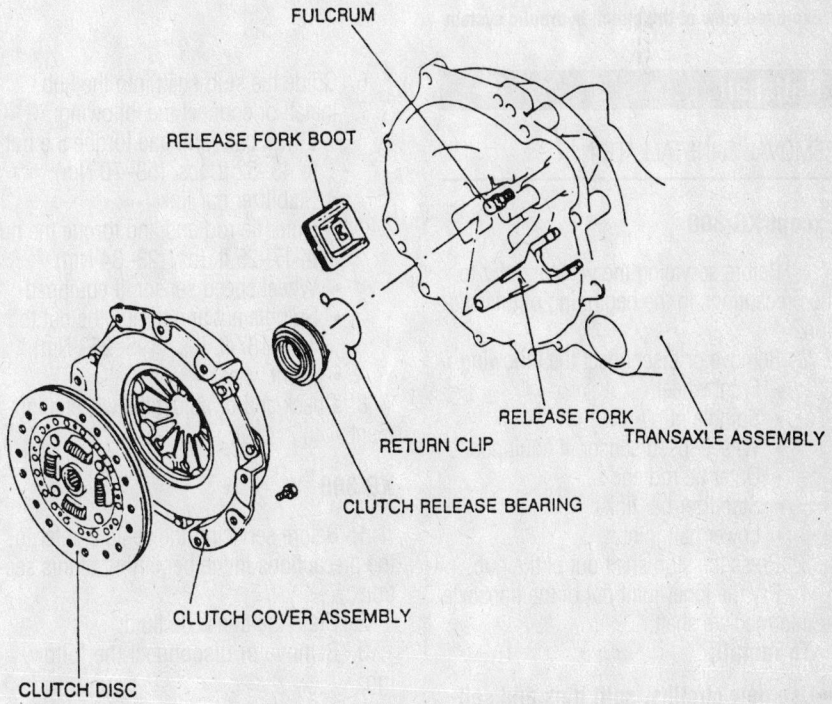

FULCRUM

RELEASE FORK BOOT

RETURN CLIP

RELEASE FORK

TRANSAXLE ASSEMBLY

CLUTCH RELEASE BEARING

CLUTCH COVER ASSEMBLY

CLUTCH DISC

7923GG93

Exploded view of the clutch disc and pressure plate components

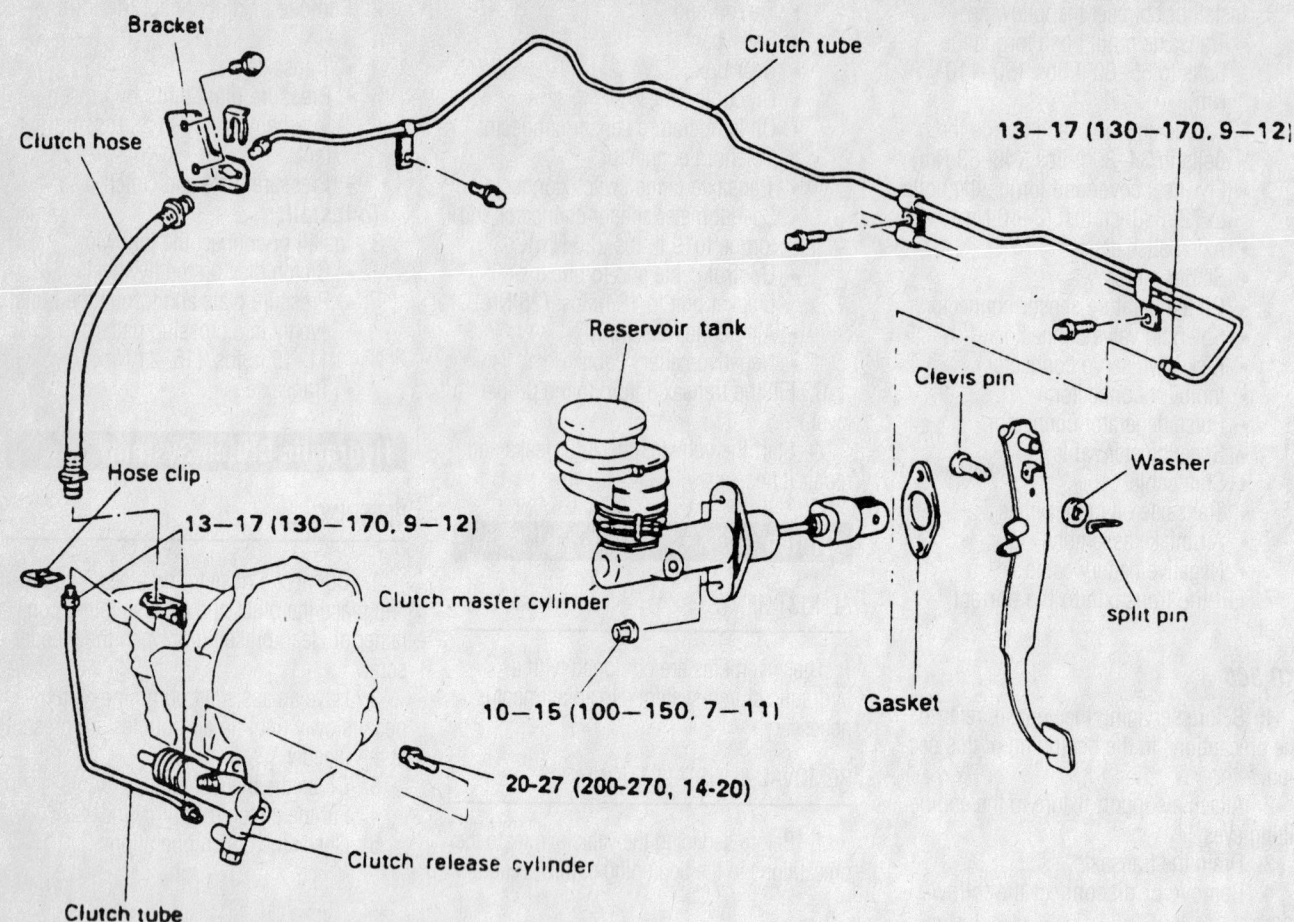

Bracket

Clutch tube

13—17 (130—170, 9—12)

Clutch hose

Reservoir tank

Clevis pin

Washer

Hose clip

13—17 (130—170, 9—12)

split pin

Clutch master cylinder

Gasket

10—15 (100—150, 7—11)

20-27 (200-270, 14-20)

Clutch release cylinder

Clutch tube

TORQUE : Nm (kg.cm, lb.ft)

7923GG90

Exploded view of the clutch hydraulic system

Halfshaft

REMOVAL & INSTALLATION

Except XG 300

1. Before servicing the vehicle, refer to the precautions in the beginning of this section.
2. Remove or disconnect the following:
   - Front wheel
   - Spindle nut
   - Wheel speed sensor, if equipped
   - Outer tie rod end
   - Stabilizer bar link
   - Lower ball joint
3. Press the stub shaft out of the hub.
4. Pry the inner joint out of the transaxle or intermediate shaft.

To install:

➡ Use new circlips, split pins and self-locking nuts for assembly.

5. Install the inner joint so that the circlip is felt to seat in the retaining groove.

6. Guide the stub shaft into the hub.
7. Install or connect the following:
   - Lower ball joint and torque the nut to 43–52 ft. lbs. (58–70 Nm)
   - Stabilizer bar link
   - Outer tie rod end and torque the nut to 17–25 ft. lbs. (23–34 Nm)
   - Wheel speed sensor, if equipped
   - Spindle nut and torque the nut to 144–187 ft. lbs. (195–253 Nm)
   - Front wheel
8. Check and/or adjust the wheel alignment.

XG 300

1. Before servicing the vehicle, refer to the precautions in the beginning of this section.
2. Drain the transaxle fluid.
3. Remove or disconnect the following:
   - Front wheel
   - Split pin and halfshaft nut
   - Ball joint from the steering knuckle
   - Halfshaft from the axle hub

4. Install a prybar between the center bearing bracket and the halfshaft and pry the halfshaft from the transaxle.
5. Remove the center bearing bracket bolts and install a prybar between bracket and the engine and disconnect the bracket from the engine.
6. Remove the inner shaft from the transaxle.

To install:

7. Install or connect the following:
   - Inner halfshaft to the transaxle
   - Center bearing bracket and torque the bolt to 36 ft. lbs. (50 Nm)
   - Halfshaft to the axle hub
   - Ball joint to the steering knuckle and torque 88 ft. lbs. (110 Nm)
   - Split pin and halfshaft nut and torque the nut to 206 ft. lbs. (280 Nm)
   - Front wheel
8. Fill the transaxle fluid to the proper level.
9. Check and/or adjust the wheel alignment.

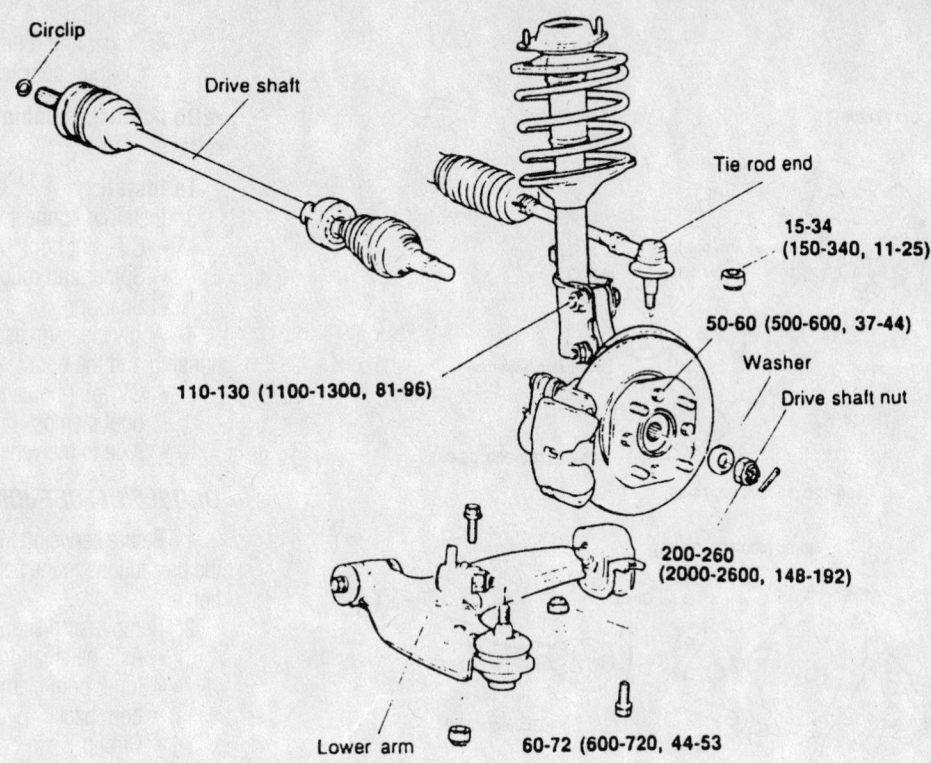

TORQUE : Nm (kg.cm, lb.ft)

7923GG94

Halfshaft components—except V6 Sonata

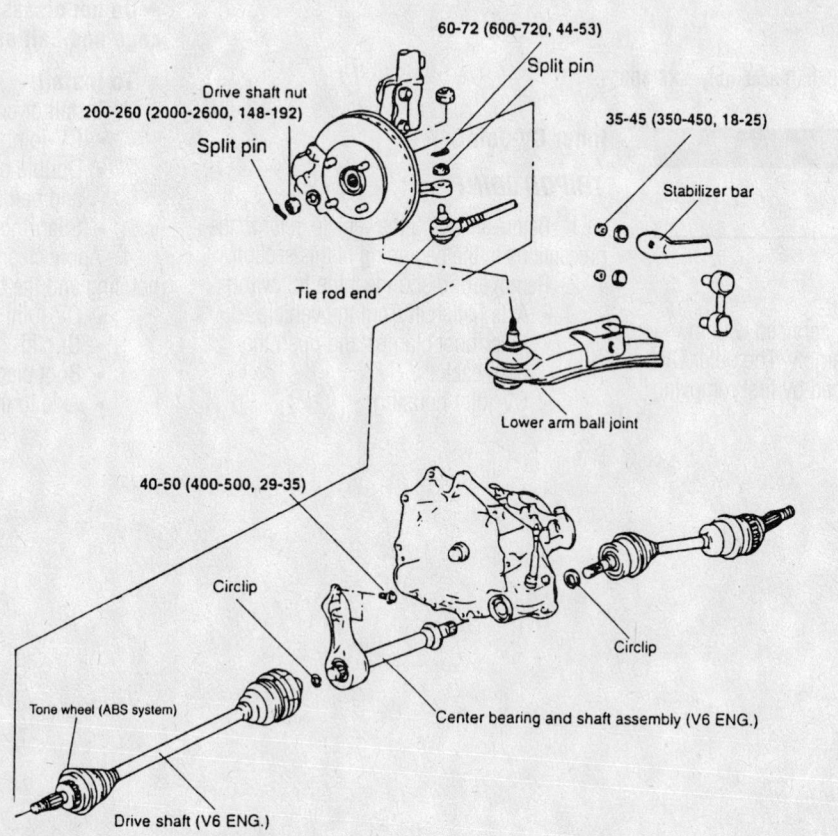

7923GG95

Halfshaft components—V6 Sonata

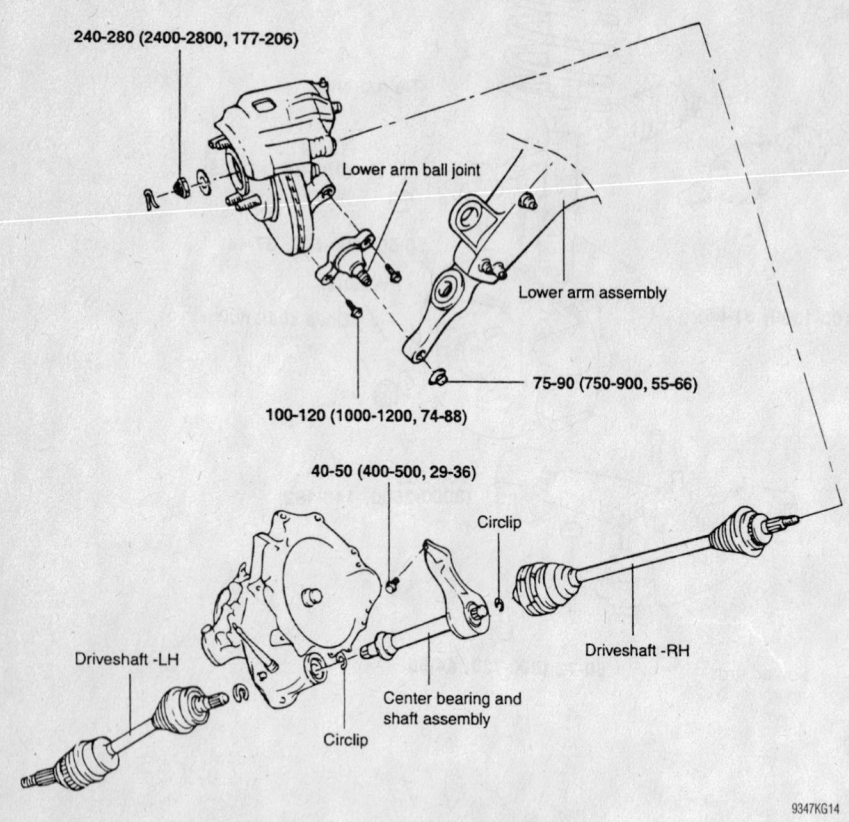

240-280 (2400-2800, 177-206)

Lower arm ball joint

Lower arm assembly

75-90 (750-900, 55-66)

100-120 (1000-1200, 74-88)

40-50 (400-500, 29-36)

Circlip

Driveshaft -LH

Driveshaft -RH

Center bearing and shaft assembly

Circlip

9347KG14

Exploded view of the halfshaft assembly—XG 300

CV-Joints

OVERHAUL

Outer CV-Joint

The outer CV-joint is serviced with the axle halfshaft as an assembly. The outer CV-joint boot may be replaced by first removing the inner joint.

Inner CV-Joint

TRIPOD JOINT

1. Before servicing the vehicle, refer to the precautions in the beginning of this section.
2. Remove or disconnect the following:
   • Axle halfshaft from the vehicle
   • Inner boot clamps and push the boot back
   • CV-joint housing
   • Snapring
   • Spider and rollers
   • CV-joint boot

➡ **Do not disassemble the spider and rollers.**

To install:
3. Install or connect the following:
   • CV-joint boot
   • Spider and rollers
   • Snapring
4. Apply clean grease to the CV-joint housing and the boot.
   • CV-joint housing and tighten the boot clamps
   • Axle to the vehicle

DOUBLE OFFSET JOINT

1. Before servicing the vehicle, refer to the precautions in the beginning of this section.
2. Remove or disconnect the following:
   • Axle halfshaft from the vehicle
   • Inner boot clamps and push the boot back
   • Circlip
   • CV-joint housing
   • Snapring
   • Double offset joint inner race, cage and ball assembly
   • CV-joint boot

➡ **Do not disassemble the inner race, cage and ball assembly.**

To install:
3. Install or connect the following:
   • CV-joint boot
   • Double offset joint inner race, cage and ball assembly
   • Snapring
4. Apply clean grease to the CV-joint housing and the boot.
   • CV-joint housing
   • Circlip
   • Boot clamps by tightening them
   • Axle to the vehicle

STEERING AND SUSPENSION

Air Bag

⚹⚹ CAUTION

Some vehicles are equipped with an air bag system. The system must be disarmed before performing service on, or around, system components, the steering column, instrument panel components, wiring and sensors. Failure to follow the safety precautions and the disarming procedure could result in accidental air bag deployment, possible injury and unnecessary system repairs.

PRECAUTIONS

Several precautions must be observed when handling the inflator module to avoid accidental deployment and possible personal injury.

• Never carry the inflator module by the wires or connector on the underside of the module.

• When carrying a live inflator module, hold securely with both hands, and ensure that the bag and trim cover are pointed away.

• Place the inflator module on a bench or other surface with the bag and trim cover facing up.

• With the inflator module on the bench, never place anything on or close to the module which may be thrown in the event of an accidental deployment.

Before servicing the vehicle, also make sure to refer to the precautions in the beginning of this section as well.

DISARMING

Disconnect and isolate the negative battery cable. Wait 3 minutes for the system capacitor to discharge before performing any service.

Rack and Pinion Steering Gear

REMOVAL & INSTALLATION

1. Before servicing the vehicle, refer to the precautions in the beginning of this section.
2. Remove or disconnect the following:
   • Negative battery cable
   • Front wheels

COMPONENTS

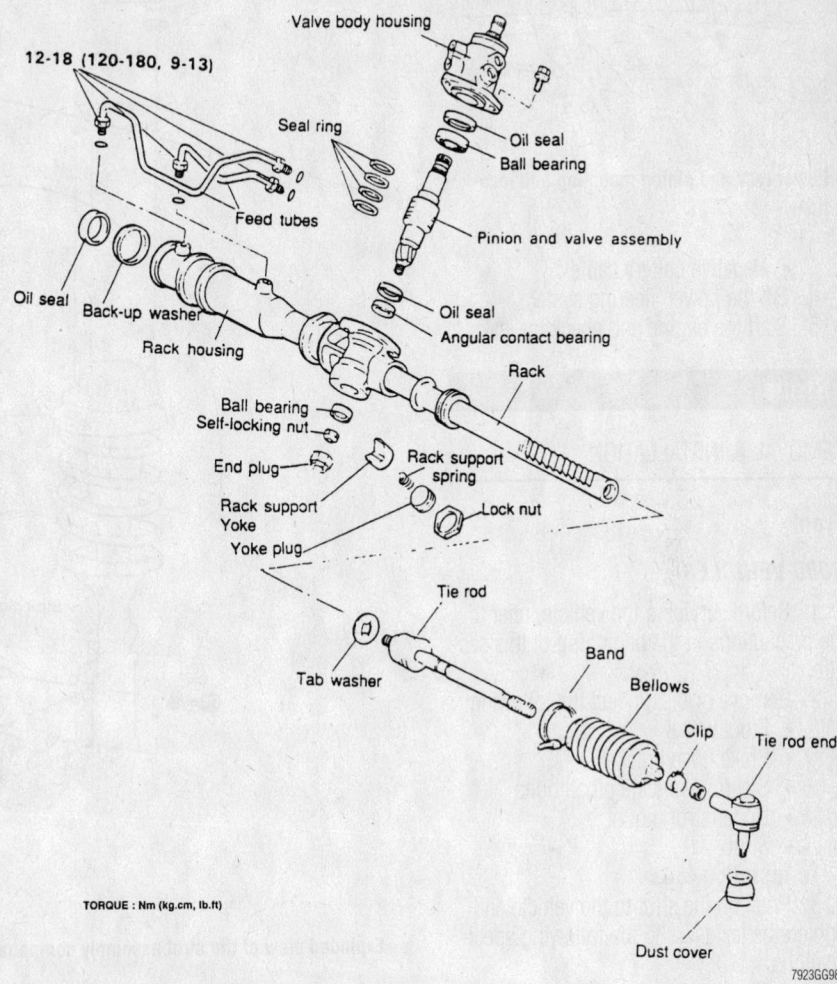

TORQUE : Nm (kg.cm, lb.ft)

Exploded view of the rack and pinion assembly

• Outer tie rod ends
• Steering column flexible coupler
• Power steering fluid hoses, if equipped with power steering
• Subframe center beam
• Exhaust front pipe
• Left lower control arm
• Stabilizer bar
• Steering gear

To install:

3. Install or connect the following:
   • Steering gear and torque the bolts to 44–59 ft. lbs. (60–80 Nm)
   • Stabilizer bar
   • Left lower control arm
   • Exhaust front pipe
   • Subframe center beam
   • Power steering fluid hoses, if equipped with power steering
   • Steering column flexible coupler

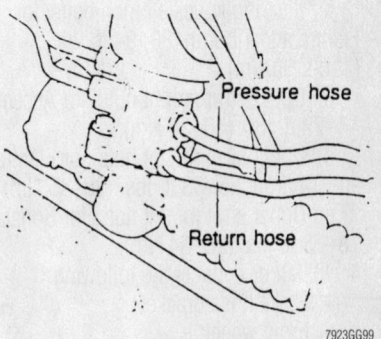

Pressure and return hose location on the rack

and torque the bolt to 11–14 ft. lbs. (15–19 Nm)
• Outer tie rod ends and torque the nuts to 17–25 ft. lbs. (23–34 Nm)
• Front wheels

Power rack and pinion mounting bolt locations

- Negative battery cable
4. Fill the power steering system.
5. Start the engine and check for leaks.

Struts

REMOVAL & INSTALLATION

Front

2000 VEHICLES

1. Before servicing the vehicle, refer to the precautions in the beginning of this section.

2. Remove or disconnect the following:
- Front wheel
- Brake hose bracket
- Steering knuckle pinch bolts
- Upper strut mount
- Strut

To install:

3. Position the strut to the vehicle and tighten the fasteners to the following specifications:

    a. Steering knuckle pinch bolts, for Accent and Sonata: 65–76 ft. lbs. (95–105 Nm).

    b. Steering knuckle pinch bolts, for Elantra and Tiburon: 80–94 ft. lbs. (110–130 Nm).

    c. Upper strut mount nuts, for Accent: 14–22 ft. lbs. (20–30 Nm).

    d. Upper strut mount nuts, for Elantra and Tiburon: 25–33 ft. lbs. (35–45 Nm).

    e. Upper strut mount nuts, for Sonata: 18–25 ft. lbs. (25–34 Nm).

4. Install or connect the following:
- Brake hose bracket
- Front wheel

5. Check and/or adjust the wheel alignment.

2001–03 ACCENT, ELANTRA AND TIBURON

1. Before servicing the vehicle, refer to the precautions in the beginning of this section.

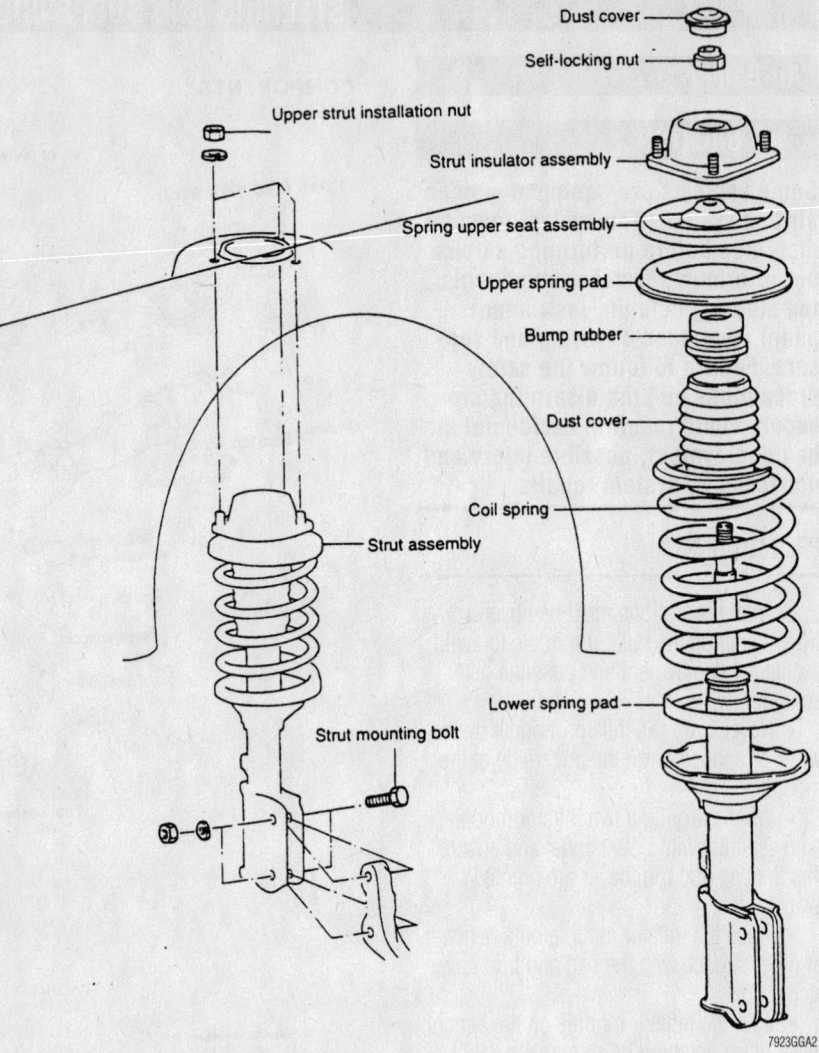

Exploded view of the strut assembly components

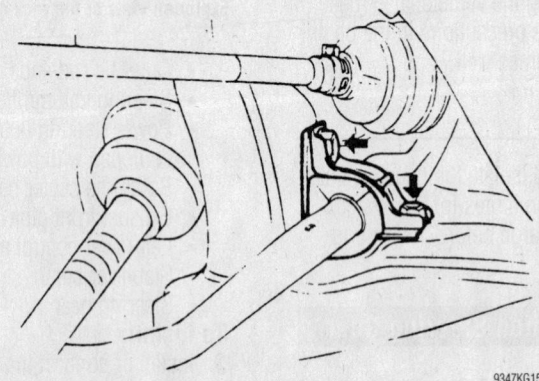

Exploded view of the strut assembly

2. Remove or disconnect the following:
- Front wheel
- Brake hose bracket
- Strut upper mounting bolts
- Lower mounting bolts
- Strut assembly

To install:

3. Install or connect the following:
- Strut assembly and torque the lower bolts to 66 ft. lbs. (90 Nm)
- Strut upper mounting bolts and torque to 22 ft. lbs. (30 Nm)
- Brake hose bracket

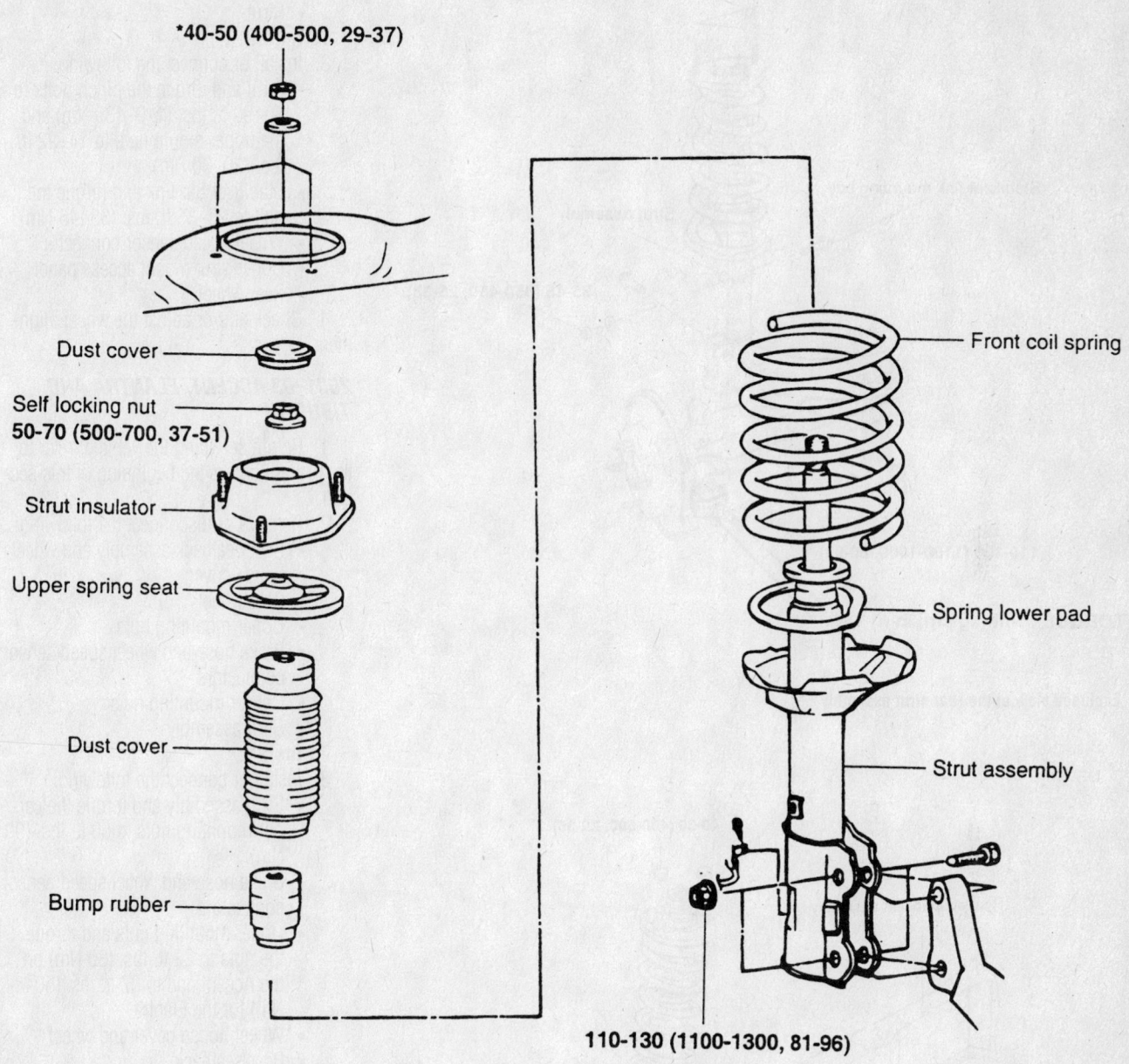

*40-50 (400-500, 29-37)

Dust cover

Self locking nut
50-70 (500-700, 37-51)

Strut insulator

Upper spring seat

Dust cover

Bump rubber

Front coil spring

Spring lower pad

Strut assembly

110-130 (1100-1300, 81-96)

9347KG16

Exploded view of the front strut assembly

- Front wheel
4. Check and/or adjust the wheel alignment.

2001–03 SONATA AND XG 300

1. Before servicing the vehicle, refer to the precautions in the beginning of this section.
2. Remove or disconnect the following:
- Front wheel
- Brake hose bracket from the mounting fork
- Mounting fork and lower arm connecting bolt

- Strut upper mounting nuts
- Strut assembly

To install:
3. Install or connect the following:
- Strut assembly
- Fork to the strut and torque the bolts to 59 ft. lbs. (80 Nm)
- Fork to the lower arm and torque the bolt to 88 ft. lbs. (120 Nm)
- Upper strut mounting nuts and torque to 36 ft. lbs. (50 Nm)
- Brake hose bracket
- Front wheel

4. Check and/or adjust the wheel alignment.

Rear

2000 MODELS

1. Before servicing the vehicle, refer to the precautions in the beginning of this section.
2. Remove or disconnect the following:
- Rear wheel
- Upper strut mount access panel
- Upper strut mount nuts
- Wheel speed sensor connector

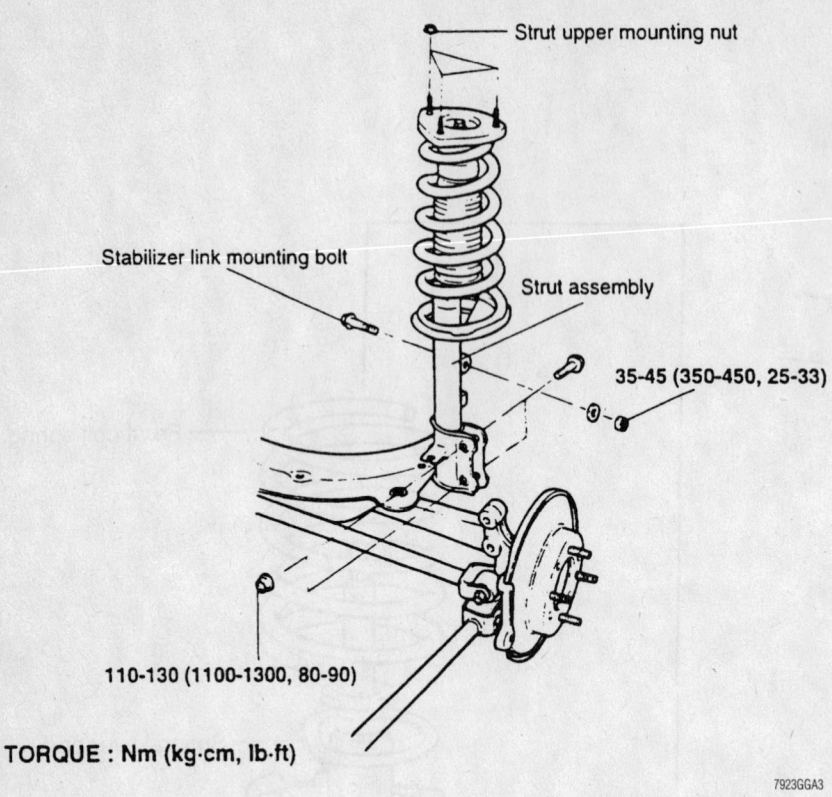

Strut upper mounting nut

Stabilizer link mounting bolt

Strut assembly

35-45 (350-450, 25-33)

110-130 (1100-1300, 80-90)

TORQUE : Nm (kg·cm, lb·ft)

7923GGA3

Exploded view of the rear strut assembly

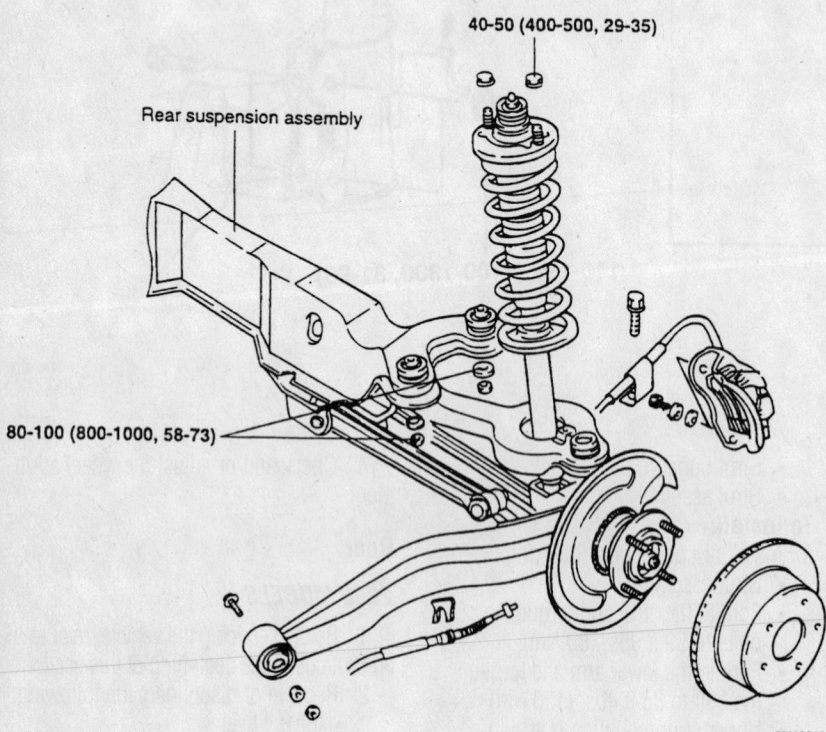

40-50 (400-500, 29-35)

Rear suspension assembly

80-100 (800-1000, 58-73)

7923GGA9

Rear suspension components—Sonata

- Stabilizer bar link
- Hub knuckle pinch bolts
- Strut

To install:

3. Install or connect the following:
- Strut and torque the pinch bolts to 80–90 ft. lbs. (110–130 Nm) and the upper mount nuts to 14–22 ft. lbs. (20–30 Nm)
- Stabilizer bar link and torque the bolt to 25–33 ft. lbs. (35–45 Nm)
- Wheel speed sensor connector
- Upper strut mount access panel
- Rear wheel

4. Check and/or adjust the wheel alignment.

2001–03 ACCENT, ELANTRA AND TIBURON

1. Before servicing the vehicle, refer to the precautions in the beginning of this section.

2. Remove or disconnect the following:
- Rear seatback assembly and wheel house cover
- Rear wheel
- Upper mounting nuts
- Brake hose and wheel speed sensor connectors
- Carrier mounting nuts
- Strut assembly

To install:

3. Install or connect the following:
- Strut assembly and torque the carrier mounting nuts to 66 ft. lbs. (90 Nm)
- Brake hose and wheel speed sensor connectors
- Upper mounting nuts and torque the nuts to 22 ft. lbs. (30 Nm) on the Accent and to 37 ft. lbs. (50 Nm) for the Elantra
- Wheel house cover and wheel
- Rear seatback

2001–03 SONATA AND XG 300

1. Before servicing the vehicle, refer to the precautions in the beginning of this section.

2. Remove or disconnect the following:
- Rear wheel
- Lower mounting bolt
- Upper arm and rear carrier bolt
- Strut mounting bracket
- Strut assembly

To install:

3. Install or connect the following:
- Strut assembly and mounting bracket and torque the bolt to 36 ft. lbs. (50 Nm)
- Upper arm and rear carrier bolt and torque the bolt to 88 ft. lbs. (120 Nm)

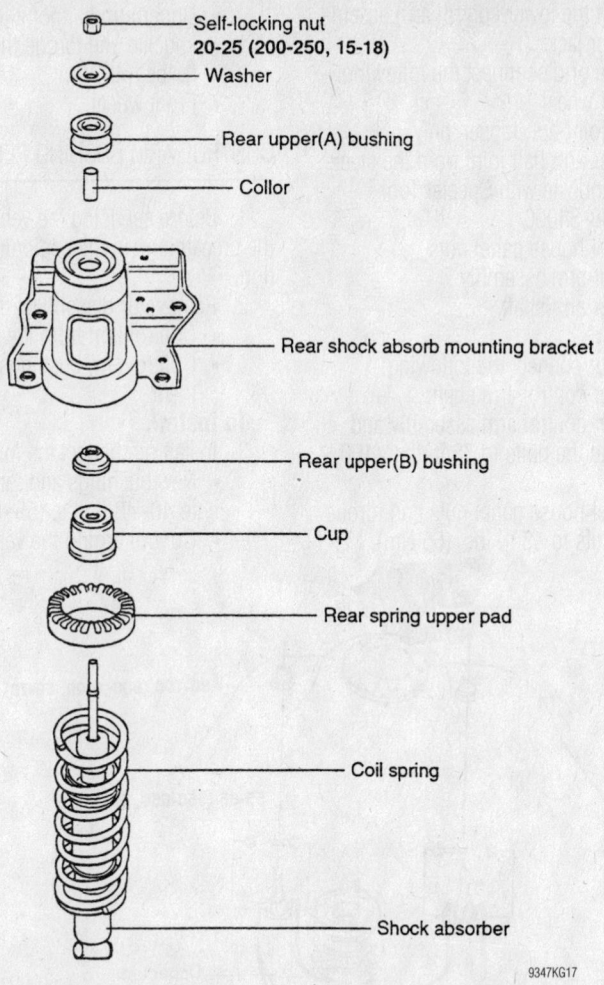

Exploded view of the rear strut assembly

- Lower mounting bolt
- Rear wheel

Coil Spring

REMOVAL & INSTALLATION

Front

ACCENT, ELANTRA AND TIBURON

1. Before servicing the vehicle, refer to the precautions in the beginning of this section.

2. Remove the strut from the vehicle and install a spring compressor.

3. Compress the coil spring so that the end of the spring comes away from the spring seat.

4. Remove or disconnect the following:
- Upper strut mount
- Upper spring seat
- Compressed spring from the strut
- Spring from the spring compressor

To install:

5. Compress the spring and install it on the strut.

6. Install or connect the following:
- Upper spring seat and the upper strut mount and torque the nut to 29–36 ft. lbs. (40–50 Nm)
- Strut to the vehicle

7. Check and/or adjust the wheel alignment.

SONATA AND XG 300

1. Before servicing the vehicle, refer to the precautions in the beginning of this section.

2. Remove the strut from the vehicle and install a Spring Compressor Tool, such as J38402.

3. Compress the coil spring so that the end of the spring comes away from the spring seat.

4. Remove or disconnect the following:
- Self locking nut

5. Install the Compressor Tool, J38402.

6. Remove the bracket, spring pad and coil spring.

To install:

7. Compress the coil spring with Compressor Tool J38402.

8. Install or connect the following:
- Coil spring to the strut
- Dust cover, upper spring pad, bushing and hand tighten the lock nut

9. Remove the compressor tool when the coil spring is properly aligned and torque the lock nut to 18 ft. lbs. (25 Nm).

10. Install the strut assembly.

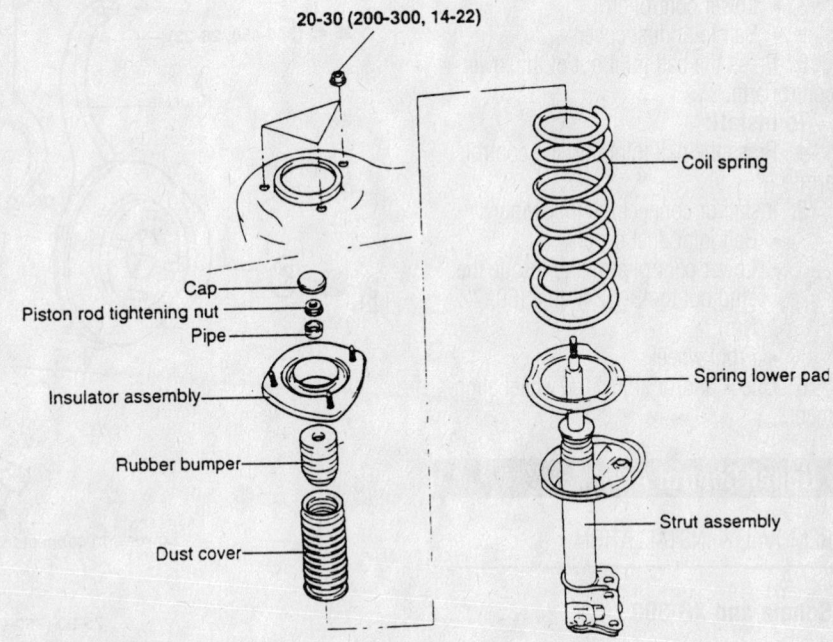

TORQUE : Nm (kg·cm, lb·ft)

Rear strut components

Upper Ball Joint

The upper ball joints are replaced with the upper control arms as an assembly.

Lower Ball Joint

REMOVAL & INSTALLATION

Bolt-On Type

1. Before servicing the vehicle, refer to the precautions in the beginning of this section.
2. Remove or disconnect the following:
   • Front wheel
   • Ball joint stud from the knuckle
   • Ball joint from the lower control arm

   To install:

➡ **Use a new split pin for assembly.**

3. Install or connect the following:
   • Ball joint and torque the mounting bolts to 69–87 ft. lbs. (95–120 Nm)
   • Stud nut and torque to 43–52 ft. lbs. (60–72 Nm)
   • Front wheel
4. Check and/or adjust the wheel alignment.

Press-In Type

1. Before servicing the vehicle, refer to the precautions in the beginning of this section.
2. Remove or disconnect the following:
   • Front wheel
   • Lower control arm
   • Ball joint dust cover
3. Press the ball joint out of the lower control arm.

   To install:

4. Press the ball joint into the control arm.
5. Install or connect the following:
   • Ball joint dust cover
   • Lower control arm and torque the stud nut to 43–52 ft. lbs. (60–72 Nm)
   • Front wheel
6. Check and/or adjust the wheel alignment.

Upper Control Arm

REMOVAL & INSTALLATION

Sonata and XG 300

1. Before servicing the vehicle, refer to the precautions in the beginning of this section.

---

2. Support the lower control arm assembly with a floor jack.
3. Remove or disconnect the following:
   • Front wheel
   • Ball joint nut, loosen only
   • Upper arm ball joint from the steering knuckle with Special Tool 09568-34000
   • Wheel house panel nuts
   • Upper arm assembly
   • Upper arm shaft

   To install:

4. Install or connect the following:
   • Upper control arm shaft
   • Upper control arm assembly and torque the bolts to 73 ft. lbs. (100 Nm)
   • Wheel house panel nuts and torque the nuts to 48 ft. lbs. (65 Nm)

   • Upper arm ball joint to the steering knuckle and torque the bolts to 33 ft. lbs. (45 Nm)
   • Front wheel

CONTROL ARM BUSHING REPLACEMENT

1. Before servicing the vehicle, refer to the precautions in the beginning of this section.
2. Remove or disconnect the following:
   • Control arm from the vehicle
   • Control arm bushings by unbolting them

   To install:

3. Install or connect the following:
   • New bushings and torque the bolts to 40–48 ft. lbs. (55–65 Nm)
   • Control arm to the vehicle

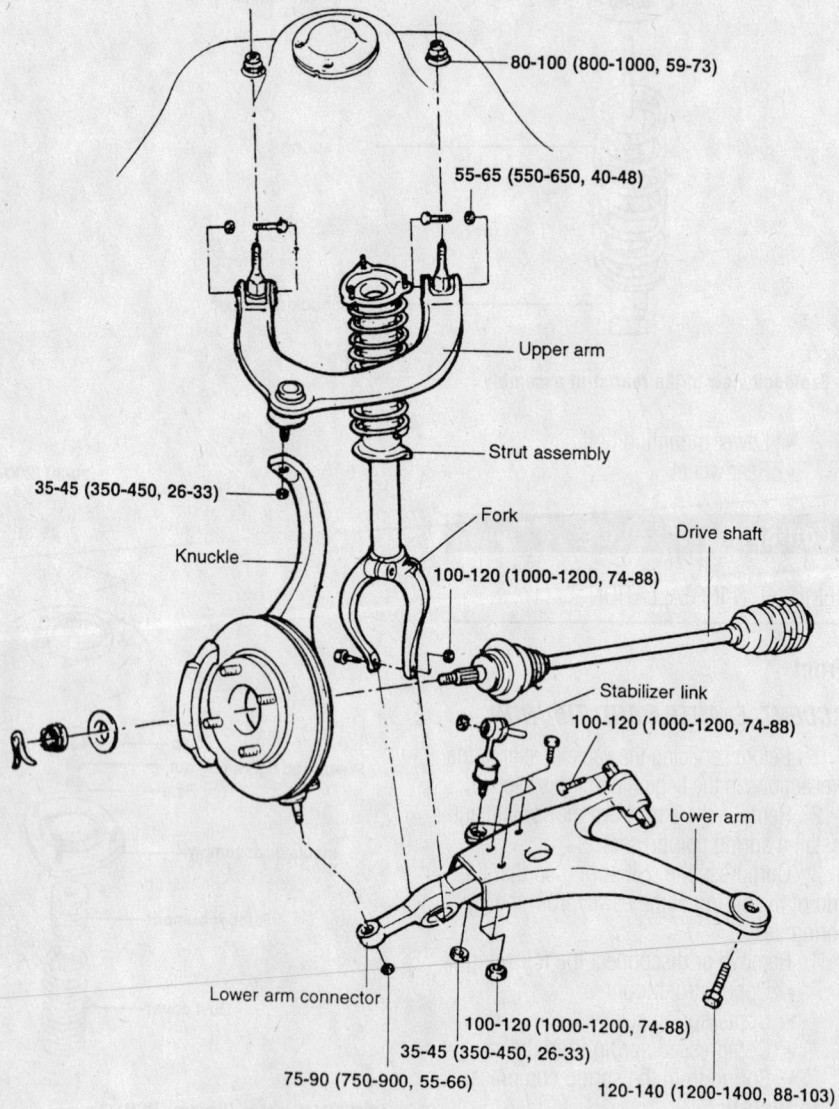

Exploded view of the upper control arm assembly—Sonata

80-100 (800-1000, 59-73)

55-65 (550-650, 40-48)

Upper arm

Strut assembly

35-45 (350-450, 26-33)

Fork

Drive shaft

Knuckle

100-120 (1000-1200, 74-88)

Stabilizer link

100-120 (1000-1200, 74-88)

Lower arm

Lower arm connector

100-120 (1000-1200, 74-88)

35-45 (350-450, 26-33)

75-90 (750-900, 55-66)

120-140 (1200-1400, 88-103)

9347KG18

Lower Control Arm

REMOVAL & INSTALLATION

Except Sonata and XG 300

1. Before servicing the vehicle, refer to the precautions in the beginning of this section.
2. Remove or disconnect the following:
   - Front wheel
   - Stabilizer bar link
   - Lower ball joint
   - Rear bushing bracket
   - Front bolt
   - Lower control arm

To install:

3. Install or connect the following:
   - Lower control arm and torque the front bolt to 72–87 ft. lbs. (100–120 Nm)
   - Rear bushing bracket. Tighten the bolts to 58–72 ft. lbs. (80–100 Nm)
   - Lower ball joint and torque the nut to 43–52 ft. lbs. (60–72 Nm)
   - Stabilizer bar link and torque the nut to 25–33 ft. lbs. (35–45 Nm)
   - Front wheel
4. Check and/or adjust the wheel alignment.

Sonata and XG 300

1. Before servicing the vehicle, refer to the precautions in the beginning of this section.
2. Remove or disconnect the following:
   - Front wheel
   - Lower ball joint nut, loosen only
   - Lower arm ball joint from the lower arm connector with Special Tool 09445-21000
   - Ball joint
   - Fork from the lower arm connector
   - Stabilizer bar link
   - Control arm inner bushing bolts
   - Lower control arm

To install:

3. Install or connect the following:
   - Lower control arm and torque the front bushing bolts to 74–88 ft. lbs. (100–120 Nm) and the rear bushing bolt to 88–103 ft. lbs. (120–140 Nm)
   - Stabilizer bar link and torque the nut to 26–33 ft. lbs. (35–45 Nm)
   - Damper fork lower bolt and torque the nut to 74–88 ft. lbs. (100–120 Nm)
   - Lower ball joint and torque the nut to 55–66 ft. lbs. (75–90 Nm)

   - Front wheel
4. Check and/or adjust the wheel alignment.

CONTROL ARM BUSHING REPLACEMENT

Except Sonata and XG 300

FRONT BUSHING

1. Before servicing the vehicle, refer to the precautions in the beginning of this section.
2. Remove the lower control arm from the vehicle.
3. Press the front bushing out of the control arm.

To install:

4. Lubricate the front bushing with soap and press into the control arm.
5. Install the control arm to the vehicle.
6. Check and/or adjust the wheel alignment.

REAR BUSHING

1. Before servicing the vehicle, refer to the precautions in the beginning of this section.
2. Remove or disconnect the following:
   - Front wheel
   - Rear bushing bracket
   - Rear bushing nut
   - Rear bushing

To install:

3. Install or connect the following:
   - Rear bushing and torque the nut to 25–33 ft. lbs. (35–45 Nm)
   - Rear bushing bracket and torque the bolts to 58–72 ft. lbs. (80–100 Nm)
   - Front wheel
4. Check and/or adjust the wheel alignment.

Sonata and XG 300

FRONT BUSHING

The front control arm bushing is serviced with the control arm as an assembly.

REAR BUSHING AND DAMPER FORK BUSHING

1. Before servicing the vehicle, refer to the precautions in the beginning of this section.
2. Remove the control arm from the vehicle.
3. Press the rear bushing and the damper fork bushing out of the control arm.

To install:

4. Press the rear bushing and the damper fork bushing into the control arm.
5. Install the control arm to the vehicle.

---

6. Check and/or adjust the wheel alignment.

Wheel Bearings

ADJUSTMENT

Front

The front wheel bearing is a sealed unit and is not adjustable.

Rear

WITH REAR DRUM BRAKES

1. Before servicing the vehicle, refer to the precautions in the beginning of this section.
2. Remove the rear wheels.
3. Loosen the spindle nut.
4. Torque the nut to 108–145 ft. lbs. (150–200 Nm). Check for correct bearing end-play by placing a dial indicator on the hub surface and moving the hub outward. Note the movement of the gauge and compare it to the desired reading of 0.008 in. (0.2mm) or less. If end-play exceeds the desired reading, retighten the rear hub bearing nut and recheck the end-play. If the reading is still excessive, replace the hub unit.
5. If end-play is correct, check the starting torque by attaching a spring balance to the hub lug bolts and pulling at a 90 degree angle while noting the required force to turn the hub. If the force required is above the desired reading of 5 lbs. (2.3 kg) or less, loosen the nut and again tighten to the desired torque. Recheck the starting torque. If the torque is still above the desired reading, replace the rear bearings.
6. Install the rear wheels.

WITH REAR DISC BRAKES

The rear wheel bearing is an integral part of the rear hub. No adjustment is possible.

REMOVAL & INSTALLATION

Front

1. Before servicing the vehicle, refer to the precautions in the beginning of this section.
2. Remove or disconnect the following:
   - Front wheel
   - Brake caliper
   - Lower ball joint
   - Spindle nut
   - Knuckle pinch bolts
   - Steering knuckle
3. Press the hub out of the wheel bearing.

4. Press the wheel bearings out of the steering knuckle.

5. If necessary, press the inner race off the hub.

To install:

6. Press the wheel bearings into the steering knuckle.

7. Install the outer grease seal and press the hub into the wheel bearings.

8. Install or connect the following:
- Inner grease seal
- Steering knuckle and torque the knuckle pinch bolts to 65–76 ft. lbs. (95–105 Nm)
- Lower ball joint and torque the stud nut to 43–52 ft. lbs. (60–72 Nm)
- Spindle nut and torque the nut to 144–187 ft. lbs. (195–253 Nm)
- Brake caliper and torque the bracket bolts to 50 ft. lbs. (68 Nm)
- Front wheel

9. Check and/or adjust the wheel alignment.

Rear

DRUM BRAKES

1. Before servicing the vehicle, refer to the precautions in the beginning of this section.

2. Remove or disconnect the following:
- Rear wheel
- Speed sensor, if equipped
- Grease cap
- Flange nut
- Outer bearing
- Brake drum
- Inner grease seal
- Inner bearing

3. Drive the bearing races out of the drum hub.

To install:

4. Install the inner and outer bearing races.

5. Apply grease to the bearings and to the cavity in the hub.

6. Install or connect the following:
- Inner bearing
- Inner grease seal
- Brake drum
- Outer bearing
- Flange nut and torque the nut to 159–192 ft. lbs. (200–260 Nm)
- Grease cap
- Wheel speed sensor, if equipped
- Rear wheel

DISC BRAKES

1. Before servicing the vehicle, refer to the precautions in the beginning of this section.

2. Release the parking brake.

3. Remove or disconnect the following:
- Rear wheel
- Wheel speed sensor, if equipped

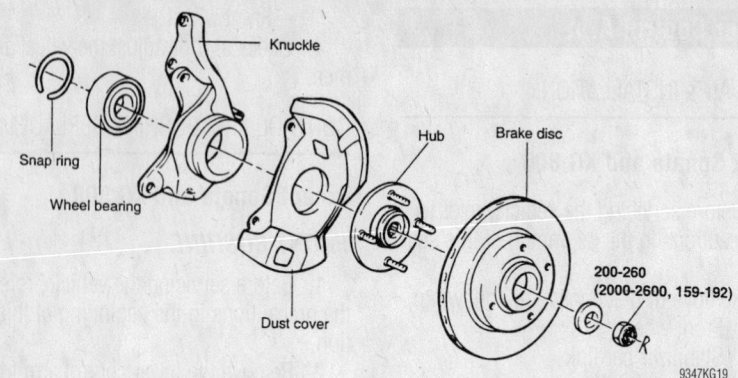

Exploded view of the front hub assembly

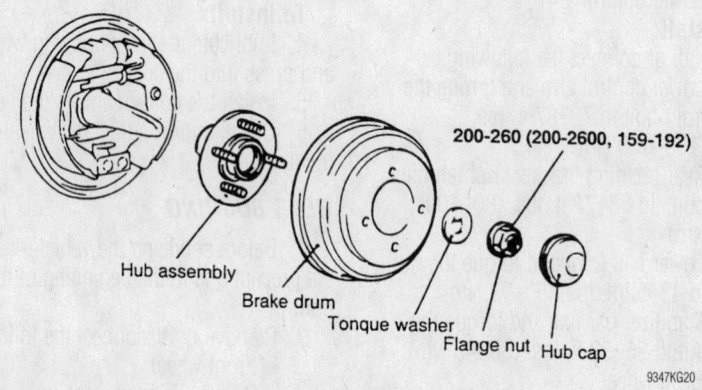

Exploded view of the rear hub assembly—with drum brakes

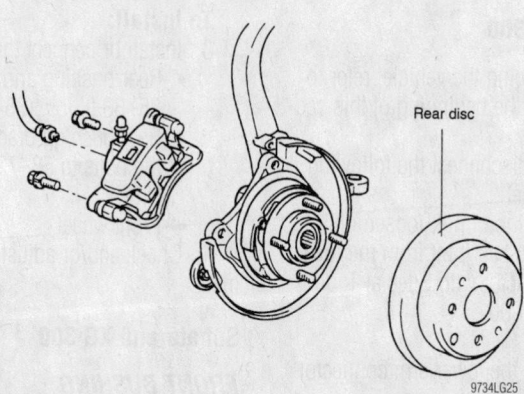

Exploded view of the rear wheel bearing assembly—with disc brakes

- Brake caliper and rotor
- Rear axle hub bolts
- Tone wheel with Tool 09445-21000
- Carrier assembly
- Nut after unstaking it

4. Press out the rear axle hub.

5. Remove the bearing inner race with Tool 09445-21000.

6. Remove the bushings from the carrier with Tools 09453-33000B and 09545-21100.

To install:

7. Press in the bushings to the carrier with Tools 09453-33000B and 09545-21100.

8. Press in the bearing to the hub with Tool 09221-21000.

9. Tighten the flange nut to meet the concave portion of the spindle.

10. Press in the tone wheel with Tool 09221-21000. Torque the nut to 191 ft. lbs. (260 Nm).

11. Install or connect the following:
- Hub and bearing assembly to the backing plate and torque the bolts to 88 ft. lbs. (120 Nm)
- Brake caliper and rotor
- Wheel speed sensor, if equipped
- Rear wheel

BRAKES

Brake Caliper

REMOVAL & INSTALLATION

Accent

1. Before servicing the vehicle, refer to the precautions in the beginning of this section.
2. Remove or disconnect the following:
   - Front wheels
   - Brake line at the caliper
   - Brake pads
   - Pin and sleeve boots
   - Lower caliper bolt and raise the caliper up and out to remove it

To install:
3. Install or connect the following:
   - Caliper onto its mounting and install the lower mounting bolt. Torque the bolt to 16–24 ft. lbs. (22–32 Nm).
   - Pin boots, sleeve boots and brake pads
   - Brake line to the caliper with 2 new metal gaskets. Torque the brake line union bolt to 18–22 ft. lbs. (25–30 Nm).
4. Bleed the system.
   - Front wheels

Elantra

FRONT

1. Before servicing the vehicle, refer to the precautions in the beginning of this section.
2. Remove or disconnect the following:
   - Front wheels
   - Brake line at the caliper
   - Brake pads
   - Pin and sleeve boots
   - Lower caliper bolt and raise the caliper up and out to remove it

To install:
3. Install or connect the following:
   - Caliper onto its mounting
   - Lower mounting bolt. Torque the bolt to 16–24 ft. lbs. (22–32 Nm).
   - Pin boots, sleeve boots and brake pads
   - Brake line to the caliper with 2 new metal gaskets. Torque the brake line union bolt to 18–22 ft. lbs. (24–30 Nm).
4. Bleed the system.
   - Front wheels

REAR

1. Before servicing the vehicle, refer to the precautions in the beginning of this section.

2. Remove or disconnect the following:
   - Center console and loosen the parking brake adjustment
   - Wheels
   - Brake hose
   - Caliper assembly mounting bolts
   - Caliper
   - Parking brake cable

To install:
3. Install or connect the following:
   - Parking brake cable
   - Caliper. Tighten the mounting bolts to 16–23 ft. lbs. (22–32 Nm).
   - Brake hose
   - Master cylinder with clean fluid and bleed the hydraulic system
   - Wheel. Adjust the parking brake.
   - Center console

Sonata and XG 300

FRONT

1. Before servicing the vehicle, refer to the precautions in the beginning of this section.
2. Remove or disconnect the following:
   - Front wheels
   - Brake tube from the brake hose
   - Brake hose clip
   - Brake hose from the strut
   - Brake line from the caliper
   - Small retaining pin from the lower part of the caliper
3. Swing the caliper up until it clears the rotor and pads.
4. Slide the caliper inboard until the locating pin disengages from its groove in the caliper. Pull the caliper from the locating pin.

To install:
5. Lubricate the locating pin bore with white silicone compound and mount the caliper onto the locating pin.
6. Lower the caliper until the small retaining pin holes are aligned. Install a new retaining pin into the lower part of the caliper. Tighten the pin.
7. Install or connect the following:
   - Brake line to the caliper and bleed the brakes
   - Brake hose to the strut
   - Brake hose clip
   - Brake tube to the brake hose
   - Front wheels

REAR

1. Before servicing the vehicle, refer to the precautions in the beginning of this section.

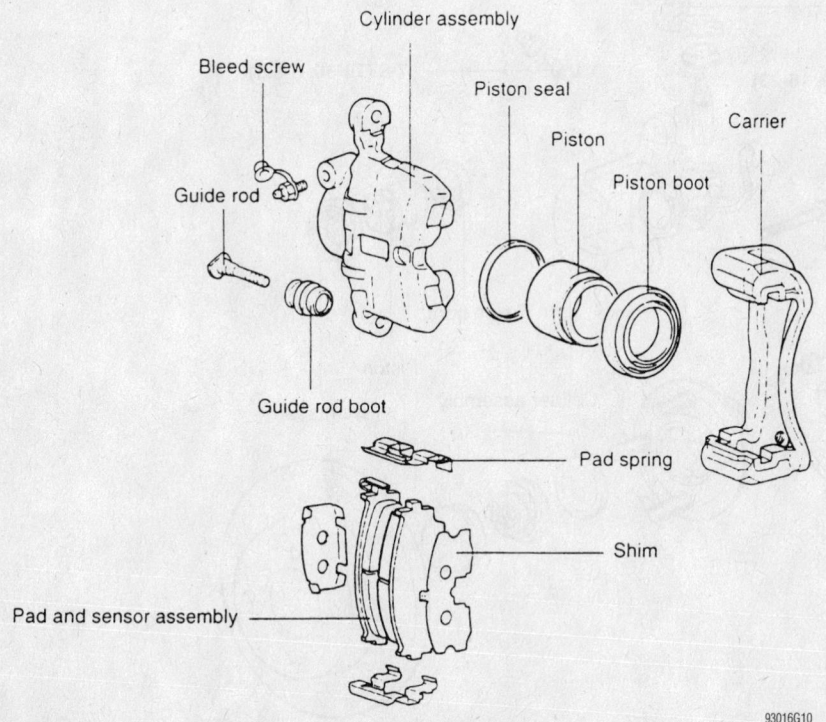

Bleed screw
Cylinder assembly
Guide rod
Piston seal
Piston
Carrier
Piston boot
Guide rod boot
Pad spring
Shim
Pad and sensor assembly

93016G10

Front caliper—Sonata shown

2. Remove or disconnect the following:
- Center console and loosen the parking brake adjustment
- Wheels
- Brake hose
- Caliper assembly mounting bolts
- Caliper
- Parking brake cable

To install:

3. Install or connect the following:
- Parking brake cable
- Caliper. Tighten the mounting bolts to 16–23 ft. lbs. (22–32 Nm).
- Brake hose
- Master cylinder with clean fluid and bleed the hydraulic system
- Wheel. Adjust the parking brake.
- Center console

Tiburon

FRONT

1. Before servicing the vehicle, refer to the precautions in the beginning of this section.

2. Remove or disconnect the following:
- Wheels
- Brake hose
- Caliper
- Caliper support

To install:

3. Install or connect the following:
- Caliper support and tighten the bolts to 44–63 ft. lbs. (69–85 Nm)
- Caliper. Tighten the guide rod bolts to 16–24 ft. lbs. (22–32 Nm).
- Brake hose and tighten to 18–22 ft. lbs. (25–30 Nm)
- Wheels

4. Fill the master cylinder with clean brake fluid and bleed the hydraulic system.

REAR

1. Before servicing the vehicle, refer to the precautions in the beginning of this section.

2. Remove or disconnect the following:
- Wheels
- Brake hose
- Caliper
- Parking brake cable

To install:

3. Install or connect the following:
- Parking brake cable
- Caliper. Tighten the mounting bolts to 16–24 ft. lbs. (22–32 Nm).
- Brake hose

4. Fill the master cylinder with clean brake fluid and bleed the hydraulic system.
- Wheels

Disc Brake Pads

REMOVAL & INSTALLATION

Accent

1. Before servicing the vehicle, refer to the precautions in the beginning of this section.

2. Remove or disconnect the following:
- Front wheels
- Lower caliper mounting bolt and rotate the caliper upward
- Pads from the caliper support
- Pad clips, if necessary

To install:

3. Install or connect the following:
- Pad clips
- Pads onto the pad clips

4. Compress the caliper piston using a C-clamp. Rotate the caliper downward and install the mounting bolt.
- Wheels

Elantra

FRONT

1. Before servicing the vehicle, refer to the precautions in the beginning of this section.

2. Remove or disconnect the following:

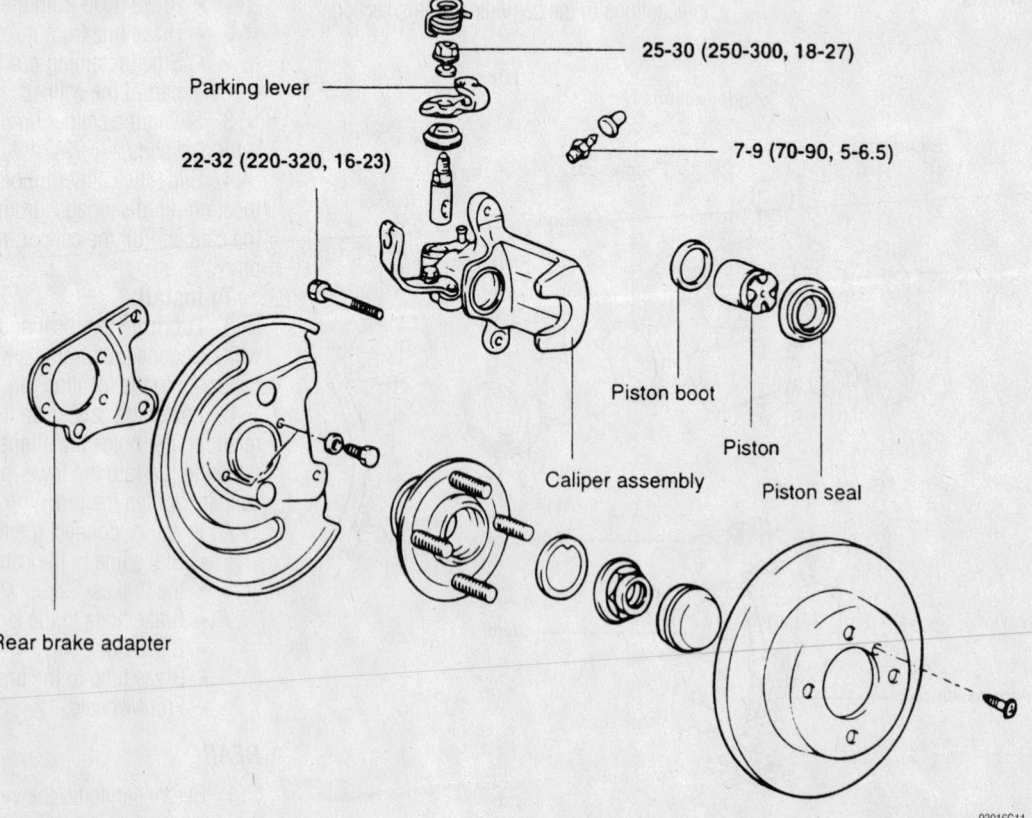

25-30 (250-300, 18-27)

Parking lever

22-32 (220-320, 16-23)

7-9 (70-90, 5-6.5)

Piston boot

Piston

Caliper assembly

Piston seal

Rear brake adapter

93016G11

Rear caliper—Elantra shown

- Front wheels
- Lower caliper mounting bolt and rotate the caliper upward
- Pads from the caliper support
- Pad clips

To install:

3. Install or connect the following:
   - Pad clips
   - Pads onto the pad clips

4. Compress the caliper piston using a C-clamp. Rotate the caliper downward and install the mounting bolt.
   - Wheels

REAR

1. Before servicing the vehicle, refer to the precautions in the beginning of this section.

2. Remove or disconnect the following:
   - ½ of the fluid from the brake master cylinder
   - Wheels
   - Caliper mounting bolts
   - Caliper
   - Brake pads and retaining clips

To install:

3. Install or connect the following:
   - New pads and retainers

4. Compress the caliper piston using special tool 09580-3400.
   - Caliper. Tighten the mounting bolts to 16–24 ft. lbs. (22–32 Nm).
   - Master cylinder with clean brake

fluid and bleed the hydraulic system
   - Wheels

Sonata, Tiburon and XG 300

FRONT

1. Before servicing the vehicle, refer to the precautions in the beginning of this section.

2. Remove or disconnect the following:
   - ½ of the fluid from the brake master cylinder
   - Front wheels
   - Small retaining pin from the lower part of the caliper

3. Swing the caliper up until it clears the rotor and pads.
   - Pads and anti-squeal spring from the caliper support

To install:

4. Install or connect the following:
   - Pads and anti-rattle spring

5. Compress the caliper piston using special tool 09580-3400.
   - Caliper. Tighten the mounting bolts to 16–24 ft. lbs. (22–32 Nm).
   - Master cylinder with clean brake fluid and bleed the hydraulic system
   - Wheels

REAR

1. Before servicing the vehicle, refer to the precautions in the beginning of this section.

2. Remove or disconnect the following:
   - ½ of the fluid from the brake master cylinder
   - Rear wheels
   - Caliper mounting bolts and remove the caliper
   - Brake pads and retaining clips

To install:

3. Install or connect the following:
   - New pads and retainers

4. Compress the caliper piston using special tool 09580-3400.
   - Caliper. Tighten the mounting bolts to 16–24 ft. lbs. (22–32 Nm).
   - Master cylinder with clean brake fluid and bleed the hydraulic system
   - Wheels

Brake Drums

REMOVAL & INSTALLATION

1. Before servicing the vehicle, refer to the precautions in the beginning of this section.

2. Remove or disconnect the following:

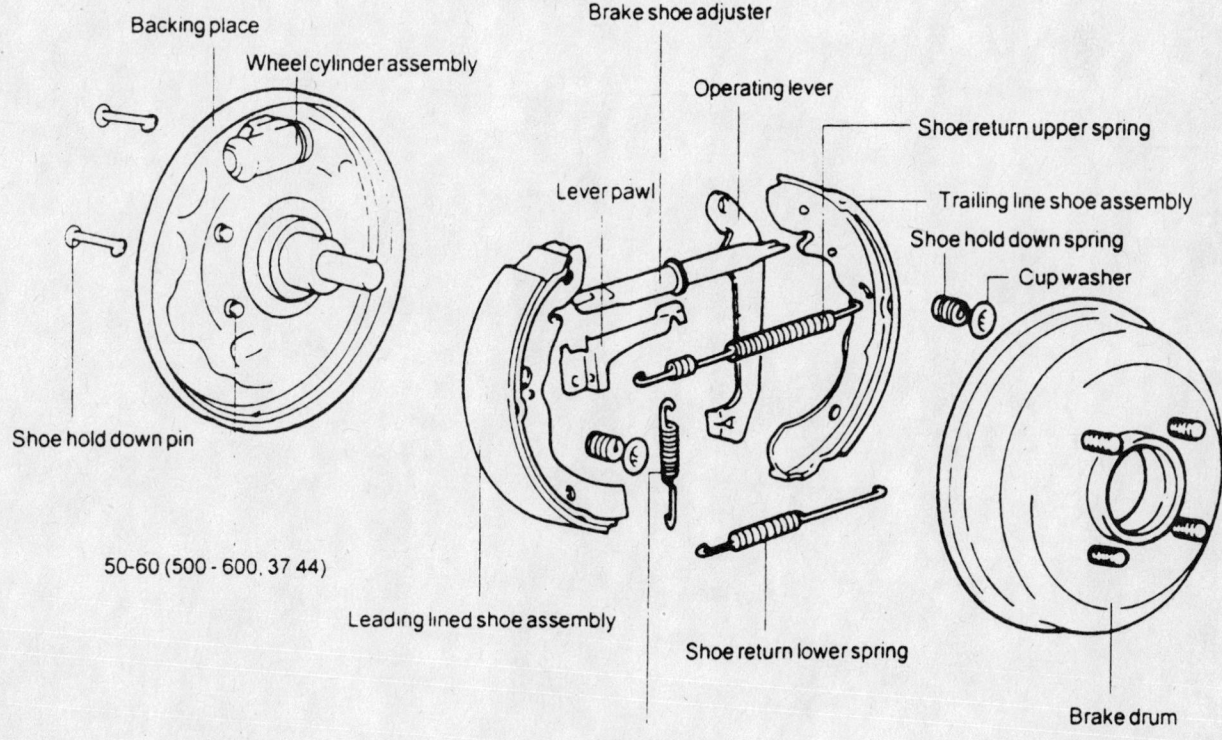

50-60 (500 - 600, 37 44)

Rear drum brakes—Accent shown

93016G12

- Wheels
- Dust cap, cotter pin, nut lock, wheel bearing nut and washer from the spindle
- Outer wheel bearing
- Drum with the inner wheel bearing from the spindle

To install:

3. Install or connect the following:
- Lubricated inner wheel bearing
- New grease seal
- Drum to the spindle
- Lubricated outer wheel bearing, washer and nut. Adjust the bearing preload as required.
- Nut lock and a new cotter pin
- Grease cap
- Wheels. Adjust the rear brakes as required.

Brake Shoes

REMOVAL & INSTALLATION

1. Before servicing the vehicle, refer to the precautions in the beginning of this section.
2. Remove or disconnect the following:
- Wheels
- Brake drum
- Self-adjuster spring and the adjuster lever
- Spread the shoes and remove the adjuster strut
- Shoe-to-shoe spring and the hold-down springs
- Primary brake shoe
- Horseshoe clip and the parking brake lever from the secondary brake shoe

To install:

3. Clean the backing plate with brake cleaning solvent.
4. Install or connect the following:
- Light coating of lithium grease to the friction points on the backing plate
- Primary shoe on the backing plate
- Hold-down spring and pin
- Parking brake lever to the secondary shoe
- Secondary shoe to the backing plate
- Adjuster strut assembly and the adjuster lever and spring
- Lower shoe-to-shoe spring
- Brake drum and the wheel
5. Adjust the brake shoes.

HYUNDAI

9

Santa Fe

SPECIFICATION CHARTS

ENGINE AND VEHICLE IDENTIFICATION CHART

	Engine Code							Model Year	
Code	Liters (cc)	Cu. In.	Cyl.	Fuel Sys.	Engine Type	Eng. Mfg.		Code ①	Year
B	2.4 (2351)	120	4	MFI	DOHC	Hyundai		1	2001
D	2.7 (2656)	120	6	MFI	DOHC	Hyundai		2	2002
DOHC: Double Overhead Cam								3	2003
MFI: Multi-port Fuel Injection								4	2004

① 8th position of VIN

42356-HSFE-C01

GENERAL ENGINE SPECIFICATIONS

Year	Model	Engine Displacement Liters (cc)	Engine ID/VIN	Fuel System Type	Net Horsepower @ rpm	Net Torque @ rpm (ft. lbs.)	Bore x Stroke (in.)	Com-pression Ratio	Oil Pressure @ rpm
2001	Santa Fe	2.4 (2351)	B	MFI	149@5500	156@3000	3.41x3.94	10:01	①
		2.7 (2656)	D	MFI	181@6000	177@4000	3.41x2.95	10:01	②
2002	Santa Fe	2.4 (2351)	B	MFI	149@5500	156@3000	3.41x3.94	10:01	①
		2.7 (2656)	D	MFI	181@6000	177@4000	3.41x2.95	10:01	②
2003-04	Santa Fe	2.4 (2351)	B	MFI	149@5500	156@3000	3.41x3.94	10:01	①
		2.7 (2656)	D	MFI	181@6000	177@4000	3.41x2.95	10:01	②

MFI: Multi-port Fuel Injection

① 11.6 Psi (80 kPa) @ idle.

② 7.3 Psi (50 kPa) or more @ idle.

42356-HSFE-C02

ENGINE TUNE-UP SPECIFICATIONS

Year	Engine Displacement Liters (cc)	Engine ID/VIN	Spark Plug Gap (in.)	Ignition Timing (deg.)		Fuel Pump (psi)	Idle Speed (rpm)		Valve Clearance (in.)	
				MT	AT		MT	AT	In.	Ex.
2001	2.4 (2351)	B	0.039-0.043	2-12B	2-12B	37	625-825	625-825	NA	NA
	2.7 (2656)	D	0.040-0.043	7-19B	7-19B	37	625-825	625-825	NA	NA
2002	2.4 (2351)	B	0.039-0.043	2-12B	2-12B	37	625-825	625-825	NA	NA
	2.7 (2656)	D	0.040-0.043	7-19B	7-19B	37	625-825	625-825	NA	NA
2003-04	2.4 (2351)	B	0.039-0.043	2-12B	2-12B	37	625-825	625-825	NA	NA
	2.7 (2656)	D	0.040-0.043	7-19B	7-19B	37	625-825	625-825	NA	NA

NOTE: The Vehicle Emission Control Information label often reflects changes made during production and must be used if they differ from this chart.

NOTE: The fuel pressure readings are given with the vacuum hose connected to the regulator and the engine running

B: Before top dead center

HYD: Hydraulic

NA;: Not Availible

42356-HSFE-C03

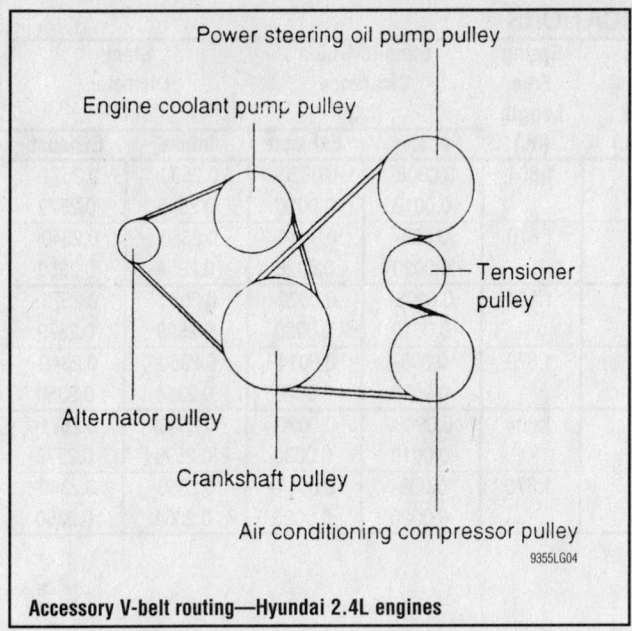

Accessory V-belt routing—Hyundai 2.4L engines

Power steering oil pump pulley

Engine coolant pump pulley

Tensioner pulley

Alternator pulley

Crankshaft pulley

Air conditioning compressor pulley

9355LG04

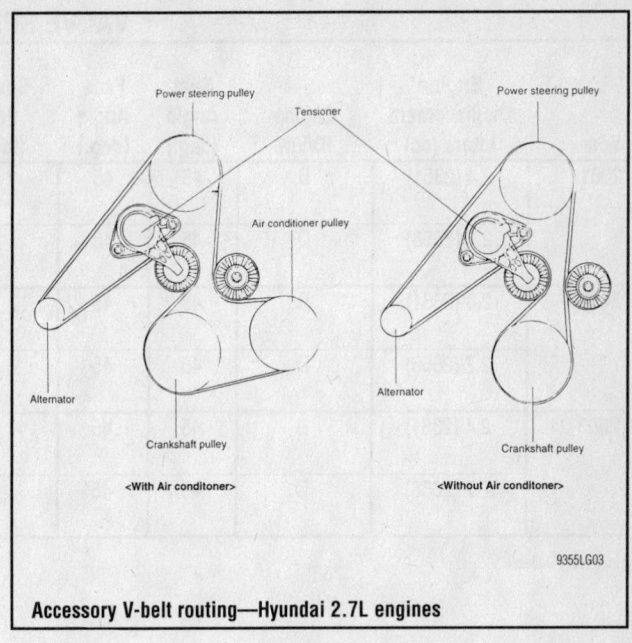

Accessory V-belt routing—Hyundai 2.7L engines

Power steering pulley

Tensioner

Air conditioner pulley

Alternator

Crankshaft pulley

<With Air conditoner>

Power steering pulley

Alternator

Crankshaft pulley

<Without Air conditoner>

9355LG03

CAPACITIES

Year	Model	Engine Displacement Liters (cc)	Engine ID/VIN	Engine Oil with Filter (qts.)	Transmission (pts.)		Transfer Case (pts.)	Drive Axle		Fuel Tank (gal.)	Cooling System (qts.)
					Man.	Auto.		Front (pts.)	Rear (pts.)		
2001	Santa Fe	2.4 (2351)	B	4.53	2.2	8.2	—	—	2.2	14.3	7.35
		2.7 (2656)	D	4.76	2.2	8.94	—	—	2.2	14.3	8.94
2002	Santa Fe	2.4 (2351)	B	4.53	2.2	8.2	—	—	2.2	14.3	7.35
		2.7 (2656)	D	4.76	2.2	8.94	—	—	2.2	14.3	8.94
2003-04	Santa Fe	2.4 (2351)	B	4.53	2.2	8.2	—	—	2.2	14.3	7.35
		2.7 (2656)	D	4.76	2.2	8.94	—	—	2.2	14.3	8.94

NOTE: All capacities are approximate. Add fluid gradually and check to be sure a proper fluid level is obtained.

42356-HSFE-C04

VALVE SPECIFICATIONS

Year	Engine Displacement Liters (cc)	Engine ID/VIN	Seat Angle (deg.)	Face Angle (deg.)	Spring Test Pressure (lbs. @ in.)	Spring Free Length (in.)	Stem-to-Guide Clearance (in.)		Stem Diameter (in.)	
							Intake	Exhaust	Intake	Exhaust
2001	2.4 (2351)	B	45	45	NA	1.804	0.0008-0.0019	0.0020-0.0030	0.2580-0.2590	0.2571-0.2579
	2.7 (2656)	D	45	45	NA	1.670	0.008-0.0020	0.0014-0.0026	0.2350-0.2354	0.2340-0.2350
2002	2.4 (2351)	B	45	45	NA	1.804	0.0008-0.0019	0.0020-0.0030	0.2580-0.2590	0.2571-0.2579
	2.7 (2656)	D	45	45	NA	1.670	0.008-0.0020	0.0014-0.0026	0.2350-0.2354	0.2340-0.2350
2003-04	2.4 (2351)	B	45	45	NA	1.804	0.0008-0.0019	0.0020-0.0030	0.2580-0.2590	0.2571-0.2579
	2.7 (2656)	D	45	45	NA	1.670	0.008-0.0020	0.0014-0.0026	0.2350-0.2354	0.2340-0.2350

NA: Not Available

42356-HSFE-C05

CRANKSHAFT AND CONNECTING ROD SPECIFICATIONS
All measurements are given in inches

Year	Engine Displacement Liters (cc)	Engine ID/VIN	Crankshaft				Connecting Rod		
			Main Brg. Journal Dia.	Main Brg. Oil Clearance	Shaft End-play	Thrust on No.	Journal Diameter	Oil Clearance	Side Clearance
2001	2.4 (2351)	B	NA	①	0.0020-0.0098	3	2.2434-2.2411	0.0007-0.0014	0.004-0.0098
	2.7 (2656)	D	NA	0.0002-0.0009	0.0024-0.0094	3	2.2434-2.2411	0.0007-0.0014	0.0039-0.0098
2002	2.4 (2351)	B	NA	①	0.0020-0.0098	3	2.2434-2.2411	0.0007-0.0014	0.004-0.0098
	2.7 (2656)	D	NA	0.0002-0.0009	0.0024-0.0094	3	2.2434-2.2411	0.0007-0.0014	0.0039-0.0098
2003-04	2.4 (2351)	B	NA	①	0.0020-0.0098	3	2.2434-2.2411	0.0007-0.0014	0.004-0.0098
	2.7 (2656)	D	NA	0.0002-0.0009	0.0024-0.0094	3	2.2434-2.2411	0.0007-0.0014	0.0039-0.0098

NA: Not Available

① Nos. 1, 2, 4 and 5: 0.0007-0.0014
No. 3: 0.0009-0.0016

42356-HSFE-C06

PISTON AND RING SPECIFICATIONS
All measurements are given in inches

Year	Engine Displacement Liters (cc)	Engine ID/VIN	Piston Clearance	Ring Gap			Ring Side Clearance		
				Top Compression	Bottom Compression	Oil Control	Top Compression	Bottom Compression	Oil Control
2001	2.4 (2351)	B	0.0008-0.0016	0.0098-0.0138	0.0157-0.0216	0.0039-0.0157	0.0012-0.0028	0.0008-0.0024	0.0024-0.0059
	2.7 (2656)	D	0.0004-0.0012	0.0079-0.0138	0.0146-0.0205	0.0079-0.0276	0.0016-0.0031	0.0012-0.0028	NA
2002	2.4 (2351)	B	0.0008-0.0016	0.0098-0.0138	0.0157-0.0216	0.0039-0.0157	0.0012-0.0028	0.0008-0.0024	0.0024-0.0059
	2.7 (2656)	D	0.0004-0.0012	0.0079-0.0138	0.0146-0.0205	0.0079-0.0276	0.0016-0.0031	0.0012-0.0028	NA
2003-04	2.4 (2351)	B	0.0008-0.0016	0.0098-0.0138	0.0157-0.0216	0.0039-0.0157	0.0012-0.0028	0.0008-0.0024	0.0024-0.0059
	2.7 (2656)	D	0.0004-0.0012	0.0079-0.0138	0.0146-0.0205	0.0079-0.0276	0.0016-0.0031	0.0012-0.0028	NA

NA: Not Applicable

42356-HSFE-C07

TORQUE SPECIFICATIONS
All readings in ft. lbs.

Year	Engine Displacement Liters (cc)	Engine ID/VIN	Cylinder Head Bolts	Main Bearing Bolts	Rod Bearing Bolts	Crankshaft Damper Bolts	Flywheel Bolts	Manifold		Spark Plugs	Lug Nut
								Intake	Exhaust		
2001	2.4 (2351)	B	①	②	③	14-22	94-101	④	⑤	15-22	66-83
	2.7 (2656)	D	⑥	⑦	⑧	—	53-56	14-15	18-22	15-22	66-83
2002	2.4 (2351)	B	①	②	③	14-22	94-101	④	⑤	15-22	66-83
	2.7 (2656)	D	⑥	⑦	⑧	—	53-56	14-15	18-22	15-22	66-83
2003-04	2.4 (2351)	B	①	②	③	14-22	94-101	④	⑤	15-22	66-83
	2.7 (2656)	D	⑥	⑦	⑧	—	53-56	14-15	18-22	15-22	66-83

NOTE: Dip main bearing bolts and crankshaft damper bolt in clean engine oil prior to tightening.

① If using used parts:
Step 1: 14 ft. lbs. (20 Nm).
Step 2: plus an additional 90 degrees.
Step 3: plus an additional 90 degrees.
If using new parts:
Step 1: 46 ft. lbs. (64 Nm)
Step 2: Release the bolts.
Step 3: 14 ft. lbs. (20 Nm)
Step 4: plus an additional 90 degrees.
Step 5: plus an additional 90 degrees.

② 15 ft. lbs. Plus 90 degrees

③ 14 ft. lbs. Plus 90 degrees

④ Bolt (M8): 11-14 ft. lbs.
Nut: 22-30 ft. lbs.

⑤ Bolt (M8): 18-2 ft. lbs.
Bolt (M10): 25-40

⑥ Step 1: 14 ft. lbs.
Step 2: plus an additional 90 degrees.
Step 3: plus an additional 90 degrees.

⑥ Step 1: 18 ft. lbs.
Step 2: plus an additional 58<en dash>62 degrees.
Step 3: plus an additional 43<en dash>47 degrees.

⑦ Bolt (M10): 10-12 ft. lbs.
Bolt (M7): 7-9

⑧ 12-15 ft. lbs. Plus 90-94 degrees

42356-HSFE-C08

WHEEL ALIGNMENT

Year	Model		Caster Range (+/-Deg.)	Caster Preferred Setting (Deg.)	Camber Range (+/-Deg.)	Camber Preferred Setting (Deg.)	Toe-in (in.)	Steering Axis Inclination (Deg.)
2001	Santa Fe	F	2 + or - 30'	2.00	0 + or - 30'	0	0.008	—
		R	—	—	0 + or - 30'	0	0.008	—
2002	Santa Fe	F	2 + or - 30'	2.00	0 + or - 30'	0	0.008	—
		R	—	—	0 + or - 30'	0	0.008	—
2003-04	Santa Fe	F	2 + or - 30'	2.00	0 + or - 30'	0	0.008	—
		R	—	—	0 + or - 30'	0	0.008	—

F: Front
R: Rear

42356-HSFE-C09

TIRE, WHEEL AND BALL JOINT SPECIFICATIONS

Year	Model	OEM Tires Standard	OEM Tires Optional	Tire Pressures (psi) Front	Tire Pressures (psi) Rear	Wheel Size	Ball Joint Inspection
2001	Santa Fe	P225/70R16	None	30	30	6.5Jx16	NS
2002	Santa Fe	P225/70R16	None	30	30	6.5Jx16	NS
2003-04	Santa Fe	P225/70R16	None	30	30	6.5Jx16	NS

OEM: Original Equipment Manufacturer
PSI: Pounds Per Square Inch
NS: Not specified by manufacturer

42356-HSFE-C10

BRAKE SPECIFICATIONS
Hyundai Santa Fe
All measurements in inches unless noted

Year	Model		Brake Disc Original Thickness	Brake Disc Minimum Thickness	Brake Disc Maximum Runout	Brake Drum Original Inside Diameter	Brake Drum Max. Wear Limit	Brake Drum Maximum Machine Diameter	Min Lining Thickness Pad	Min Lining Thickness Shoe	Bracket Bolts (ft. lbs.)	Mounting Bolts (ft. lbs.)
2001	Santa Fe	F	1.0200	0.960	0.002	—	—	—	0.079	—	58-73	16-24
		R	0.390	0.330	0.040	10.00	10.08	—	0.080	0.590	36-43	16-24
2002	Santa Fe	F	1.0200	0.960	0.002	—	—	—	0.079	—	58-73	16-24
		R	0.390	0.330	0.040	10.00	10.08	—	0.080	0.590	36-43	16-24
2003-04	Santa Fe	F	1.0200	0.960	0.002	—	—	—	0.079	—	58-73	16-24
		R	0.390	0.330	0.040	10.00	10.08	—	0.080	0.590	36-43	16-24

F: Front
R: Rear

42356-HSFE-C11

PRECAUTIONS

Before servicing any vehicle, please be sure to read all of the following precautions, which deal with personal safety, prevention of component damage, and important points to take into consideration when servicing a motor vehicle:

• Never open, service or drain the radiator or cooling system when the engine is hot; serious burns can occur from the steam and hot coolant.

• Observe all applicable safety precautions when working around fuel. Whenever servicing the fuel system, always work in a well-ventilated area. Do not allow fuel spray or vapors to come in contact with a spark, open flame or excessive heat (a hot drop light, for example). Keep a dry chemical fire extinguisher near the work area. Always keep fuel in a container specifically designed for fuel storage; also, always properly seal fuel containers to avoid the possibility of fire or explosion. Refer to the additional fuel system precautions later in this section.

• Fuel injection systems often remain pressurized, even after the engine has been turned **OFF**. The fuel system pressure must be relieved before disconnecting any fuel lines. Failure to do so may result in fire and/or personal injury.

• Brake fluid often contains polyglycol ethers and polyglycols. Avoid contact with the eyes and wash your hands thoroughly after handling brake fluid. If you do get brake fluid in your eyes, flush your eyes with clean, running water for 15 minutes. If

eye irritation persists, or if you have taken brake fluid internally, IMMEDIATELY seek medical assistance.

• The EPA warns that prolonged contact with used engine oil may cause a number of skin disorders, including cancer! You should make every effort to minimize your exposure to used engine oil. Protective gloves should be worn when changing oil. Wash your hands and any other exposed skin areas as soon as possible after exposure to used engine oil. Soap and water, or waterless hand cleaner should be used.

• All new vehicles are now equipped with an air bag system. The system must be disabled before performing service on or around system components, steering column, instrument panel components, wiring and sensors. Failure to follow safety and disabling procedures could result in accidental air bag deployment, possible personal injury and unnecessary system repairs.

• Always wear safety goggles when working with, or around, the air bag system. When carrying a non-deployed air bag, be sure the bag and trim cover are pointed away from your body. When placing a non-deployed air bag on a work surface, always face the bag and trim cover upward, away from the surface. This will reduce the motion of the module if it is accidentally deployed. Refer to the additional air bag system precautions later in this section.

• Clean, high quality brake fluid from a sealed container is essential to the safe and proper operation of the brake system. You should always buy the correct type of brake fluid for your vehicle. If the brake fluid becomes contaminated, completely flush the system with new fluid. Never reuse any brake fluid. Any brake fluid that is removed from the system should be discarded. Also, do not allow any brake fluid to come in contact with a painted surface; it will damage the paint.

• Never operate the engine without the proper amount and type of engine oil; doing so WILL result in severe engine damage.

• Timing belt maintenance is extremely important! Many models utilize an interference-type, non-freewheeling engine. If the timing belt breaks, the valves in the cylinder head may strike the pistons, causing potentially serious (also time-consuming and expensive) engine damage.

• Disconnecting the negative battery cable on some vehicles may interfere with the functions of the on-board computer system(s) and may require the computer to undergo a relearning process once the negative battery cable is reconnected.

• When servicing drum brakes, only disassemble and assemble one side at a time, leaving the remaining side intact for reference.

• Only an MVAC-trained, EPA-certified automotive technician should service the air conditioning system or its components.

ENGINE REPAIR

➡**Disconnecting the negative battery cable on some vehicles may interfere with the functions of the on board computer system. The computer may undergo a relearning process once the negative battery cable is reconnected.**

Distributor

The Santa Fe is equipped with a Distributorless Ignition System (DIS).

Alternator

REMOVAL

2.4L Engine

1. Before servicing the vehicle, refer to the precautions in the beginning of this section.

2. Remove or disconnect the following:

• Negative battery cable
• Drive belt
• Alternator electrical connections
• Alternator mounting bolts
• Alternator

2.7L Engine

1. Before servicing the vehicle, refer to the precautions in the beginning of this section.

2. Remove or disconnect the following:

• Negative battery cable
• Drive belt
• Alternator electrical connections
• Alternator mounting bolts
• Alternator

INSTALLATION

2.4L Engine

1. Install or connect the following:
• Alternator
• Alternator mounting bolts. Tighten the adjuster (upper) bolt to 15–18 ft. lbs. (20–25 Nm) and the lower bolt and nut to 26–41 ft. lbs. (34–54 Nm).
• Alternator electrical connections
• Drive belt
• Negative battery cable

2.7L Engine

1. Install or connect the following:
• Alternator
• Alternator mounting bolts. Tighten the adjuster (upper) bolt to 9–11 ft.

[2.4 I4]

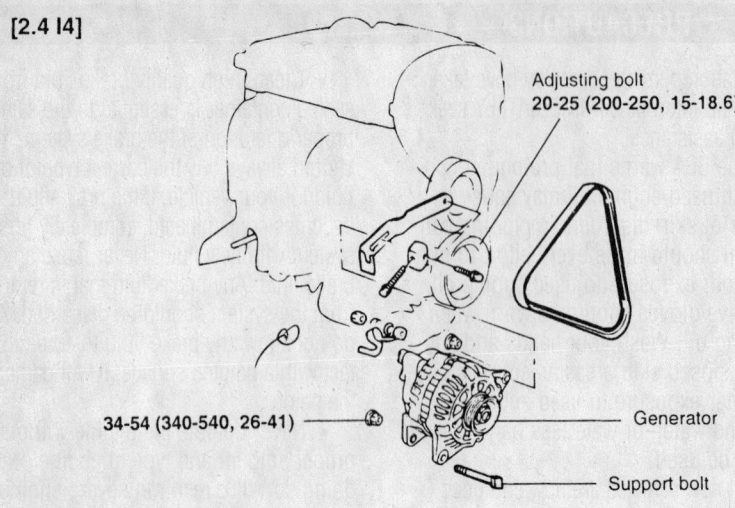

Adjusting bolt
20-25 (200-250, 15-18.6)

34-54 (340-540, 26-41) — Generator

Support bolt

[2.7 V6]

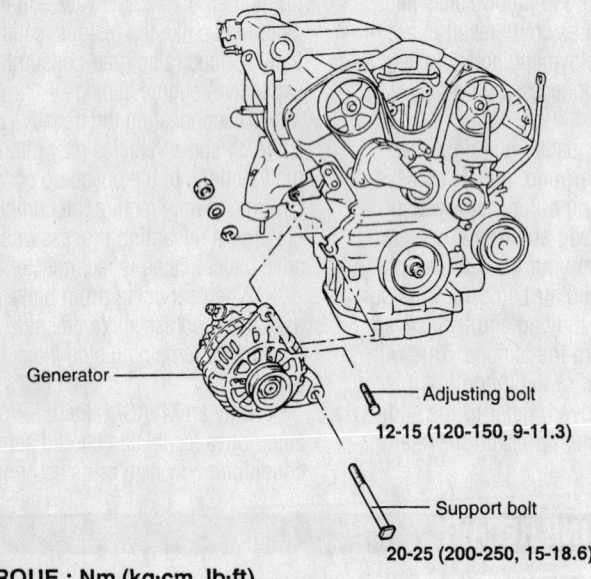

Generator

Adjusting bolt

12-15 (120-150, 9-11.3)

Support bolt

20-25 (200-250, 15-18.6)

9355LG01

TORQUE : Nm (kg·cm, lb·ft)

Exploded view of the alternator mounting for both engines used in the Santa Fe

lbs. (12–15 Nm) and the lower bolt and nut to 15–18 ft. lbs. (20–25 Nm).
- Alternator electrical connections
- Drive belt
- Negative battery cable

Ignition Timing

ADJUSTMENT

The Santa Fe is equipped with a Distributorless Ignition System (DIS). The ignition timing is controlled by the Powertrain Control module (PCM). No adjustment is necessary.

Engine Assembly

REMOVAL & INSTALLATION

1. Before servicing the vehicle, refer to the precautions in the beginning of this section.
2. Remove the battery and air cleaner assembly.
3. Drain the cooling system.
4. Drain the engine oil.
5. Drain the transaxle fluid.
6. Relieve fuel system pressure.
7. Disconnect the following electrical connections:
   - Starter

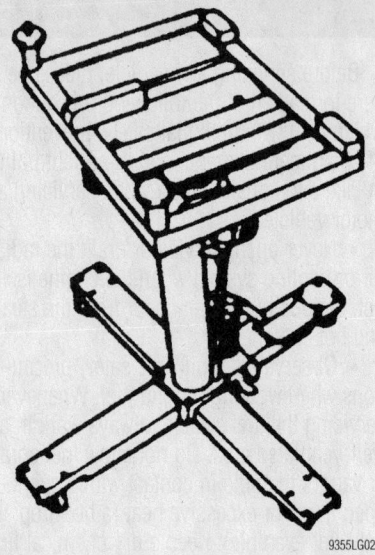

9355LG02

Attach the special tool to the transmission jack and support the transmission

- Alternator
- Throttle Position Sensor (TPS)
- Power steering switch connector
- Oil pressure gauge connector
- Back-up lamp switch connector
- A/T solenoid inhibitor switch connector
- Coolant Temperature Sensor (CTS)
- Ignition coil
- Idle Speed Control (ISC) valve connector
- Manifold Absolute Pressure (MAP) sensor
- Oxygen (O_2S) sensor connector

8. If equipped with an automatic transmission, disconnect the oil cooler lines.
9. Remove or disconnect the following:
   - Radiator hoses from the engine
   - Radiator
   - Engine ground
   - Brake vacuum hose
   - Heater hoses at the engine
   - Throttle cable at the engine
   - Cruise control cable at the engine, if equipped
   - Main fuel line at the supply/return pipe
   - Speedometer cable at the transaxle
   - Clutch or control cable from the transaxle
   - Power steering hoses from the pump
   - Steering dust cover in the engine compartment
   - Gear box universal joint bolt
   - Front wheel
   - Brake caliper and support with wire
   - Strut lower bolt
   - Front muffler bolts

- Transaxle control rod and extension rod, if equipped with a manual transmission

10. Support the transmission with a jack using the special attachment shown in the accompanying illustration.

11. Make sure all cable, harness connectors and hoses are disconnected from the engine and transmission.

- Engine and transaxle mounting brackets
- Sub frame bolts
- Drive shaft

12. Lower the engine and transaxle assembly enough so the front and rear roll stoppers can be removed.

13. Remove the engine assembly

14. Installation is the reverse of removal but please note the following steps:

- Tighten the roll stopper bolts to 36–47 ft. lbs. (50–65 Nm
- Tighten the transaxle mounting bracket bolts to 65–80 ft. lbs. (90–110 Nm)
- Tighten the engine mount bracket bolts to 43–58 ft. lbs. (60–80 Nm)

15. Fill the engine crankcase to the correct level.

16. Fill the transaxle to the correct level.

17. Fill the cooling system.

18. Fill the power steering system.

19. Start the engine and check for leaks.

20. Check the wheel alignment and adjust as necessary.

Water Pump

REMOVAL & INSTALLATION

2.4L Engine

1. Before servicing the vehicle, refer to the precautions in the beginning of this section.

2. Drain the cooling system.

3. Remove or disconnect the following:

- Negative battery cable
- Water pump inlet pipe
- Drive belt and water pump pulley
- Timing belt covers
- Timing belt tensioner
- Water pump bolts
- Alternator brace
- Water pump and gasket

4. Clean the gasket mating surfaces.

To install:

5. Install or connect the following:

- New O-ring onto the groove on the front end of the coolant pipe and wet the O-ring with water
- Water pump and gasket

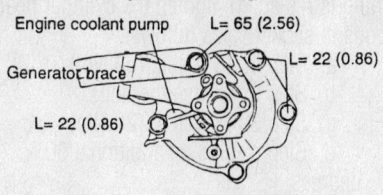

[DOHC ENGINE]

Engine coolant pump L= 65 (2.56)

Generator brace L= 22 (0.86)

L= 22 (0.86)

L= 22 (0.86)

L=Length of bolt mm (in.)

9355LG05

Make sure the water pump bolts are positioned in their original positions as the bolts are different lengths—2.4L engine

- Bolts and alternator bracket. Tighten the bolts to 14–20 ft. lbs. (20–27 Nm).
- Timing belt tensioner
- Timing belt covers
- Water pump pulley and drive belt
- Water pump inlet pipe
- Negative battery cable

6. Fill the cooling system.

7. Start the engine and check for leaks.

2.7L Engine

1. Before servicing the vehicle, refer to the precautions in the beginning of this section.

2. Drain the cooling system.

3. Remove or disconnect the following:

- Negative battery cable
- Drive belt and water pump pulley
- Timing belt covers
- Timing belt tensioner
- Idler pulley
- Water pump bolts
- Water pump and gasket

4. Clean the gasket mating surfaces.

To install:

5. Install or connect the following:

- Water pump and gasket
- Bolts and alternator bracket. Tighten the bolts to 11–16 ft. lbs. (15–22 Nm).
- Idler pulley
- Timing belt tensioner
- Timing belt covers
- Water pump pulley and drive belt
- Negative battery cable

6. Fill the cooling system.

7. Start the engine and check for leaks.

Cylinder Head

REMOVAL & INSTALLATION

2.4L Engine

1. Before servicing the vehicle, refer to the precautions in the beginning of this section.

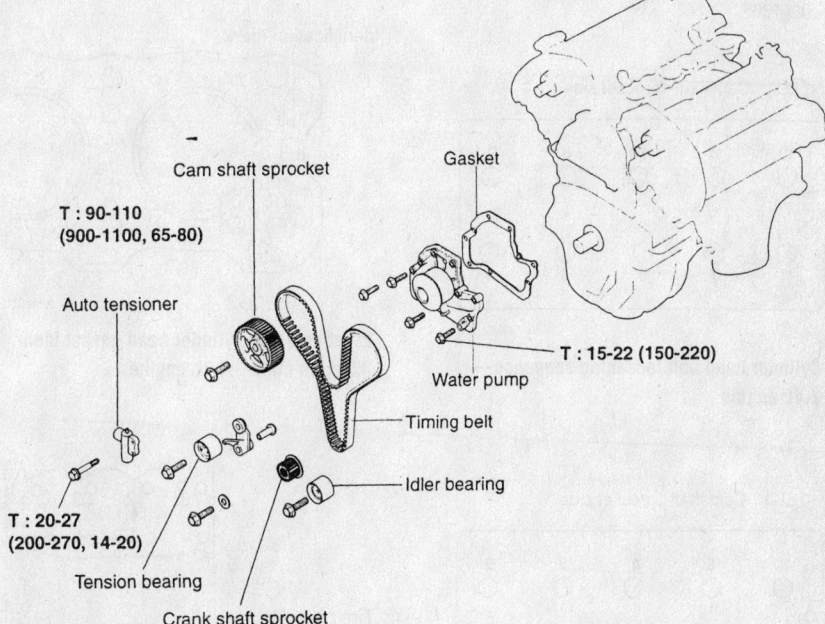

Cam shaft sprocket

Gasket

T : 90-110 (900-1100, 65-80)

Auto tensioner

T : 15-22 (150-220)

Water pump

Timing belt

T : 20-27 (200-270, 14-20)

Idler bearing

Tension bearing

Crank shaft sprocket

TORQUE : Nm (kg·cm, lb·ft)

9355LG06

Exploded view of the water pump mounting and related components—2.7L engine

2. Drain the cooling system.

3. Relieve the fuel system pressure.

4. Remove or disconnect the following:
- Negative battery cable
- All necessary electrical connections, hoses and cables
- Air cleaner
- Intake manifold
- Ignition coil
- Timing belt
- Exhaust manifold
- Rocker cover
- Camshafts

5. Loosen the cylinder head bolts in the sequence illustrated

6. Remove the cylinder head and gasket.

To install:

7. Clean the gasket mating surfaces

8. Install the cylinder head gasket so the surface with the identification mark faces towards the head.

9. Measure the head bolts. The bolt length should be 3.9 inch (99.4mm). If the bolts do not meet specification they must be replaced.

10. Install the cylinder head.

11. If using used parts (bolts, head or block), tighten the cylinder head bolts in sequence as follows:
   a. Step 1: 14 ft. lbs. (20 Nm).
   b. Step 2: plus an additional 90 degrees.
   c. Step 3: plus an additional 90 degrees.

12. If using new parts (even if only one thing is replaced), tighten the cylinder head bolts in sequence as follows:
   a. Step 1: 46 ft. lbs. (64 Nm).
   b. Step 2: release the bolts.
   c. Step 3: 14 ft. lbs. (20 Nm).
   d. Step 4: plus an additional 90 degrees.
   e. Step 5: plus an additional 90 degrees.

13. Install or connect the following:
- Camshafts
- Timing belt
- Rocker cover
- Ignition coil
- Exhaust manifold
- Intake manifold
- Air cleaner
- All necessary electrical connections, hoses and cables
- Negative battery cable

14. Fill the cooling system.

15. Start the engine and check for leaks.

2.7L Engine

1. Before servicing the vehicle, refer to the precautions in the beginning of this section.

2. Drain the cooling system.

3. Relieve the fuel system pressure.

4. Remove or disconnect the following:
- Negative battery cable

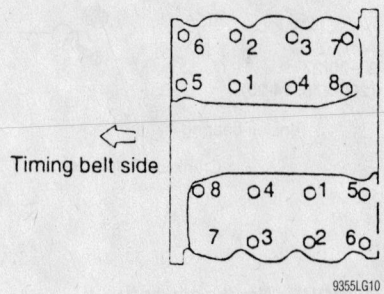

Identification mark

9355LG11

Location of the cylinder head gasket identification mark—2.7L engine

- All necessary electrical connections, hoses and cables
- Air cleaner
- Intake manifold
- Ignition coil
- Timing belt
- Exhaust manifold
- Rocker cover
- Camshafts

5. Loosen the cylinder head bolts in the reverse order of the tightening sequence.

6. Remove the cylinder head and gasket.

To install:

7. Clean the gasket mating surfaces

8. Install the cylinder head gasket so the surface with the identification mark faces towards the head.

9. Install the cylinder head.

10. Tighten the cylinder head bolts in sequence as follows:
   a. Step 1: 18 ft. lbs. (25 Nm).
   b. Step 2: plus an additional 58–62 degrees.
   c. Step 3: plus an additional 43–47 degrees.

11. Install or connect the following:
- Camshafts
- Timing belt
- Rocker cover
- Ignition coil
- Exhaust manifold
- Intake manifold
- Air cleaner
- All necessary electrical connections, hoses and cables
- Negative battery cable

12. Fill the cooling system.

13. Start the engine and check for leaks.

Rocker Arms/Shafts

REMOVAL & INSTALLATION

Refer to the camshaft removal and installation procedure.

Intake Manifold

REMOVAL & INSTALLATION

2.4L Engine

1. Before servicing the vehicle, refer to the precautions in the beginning of this section.

2. Remove or disconnect the following:
- Negative battery cable
- Air breather hose from the throttle body
- Throttle cable

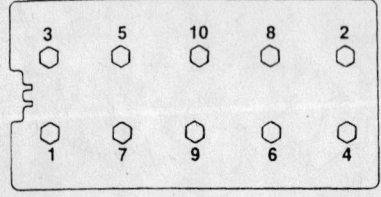

9355LG08

Cylinder head bolt loosening sequence—2.4L engine

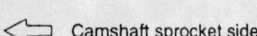
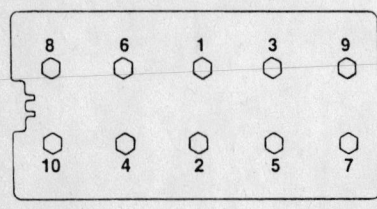

9355LG09

Cylinder head bolt tightening sequence—2.4L engine

9355LG10

Cylinder head bolt tightening sequence—2.7L engine

- Engine coolant hose and throttle body
- Positive Crankcase Ventilation (PCV) valve and brake booster vacuum hose
- Vacuum hose connector
- Injector cover
- High pressure fuel hose
- Fuel injector harness connector
- Delivery pipe with the injectors and the pressure regulator as an assembly
- Intake manifold stay
- Intake manifold

To install:

3. Install or connect the following:
- New intake manifold gasket
- Intake manifold and bolts and nuts. Tighten the bolts to 11–14 ft. lbs. (15–20 Nm) and the nuts to 22–30 ft. lbs. (30–42 Nm).
- Delivery pipe and injector assembly

- Intake manifold stay and tighten the bolts to 13–18 ft. lbs. (18–25 Nm)
- Fuel injector harness connector
- High pressure fuel hose
- Injector cover
- Vacuum hose connector
- PCV valve and brake booster vacuum hose
- Throttle body and engine coolant hose
- Throttle cable
- Air breather hose from the throttle body
- Negative battery cable

4. Start the engine and check for proper operation.

2.7L Engine

1. Before servicing the vehicle, refer to the precautions in the beginning of this section.

2. Remove or disconnect the following:

- Negative battery cable
- Air breather hose from the throttle body
- Throttle and cruise control cables
- Engine coolant hose and throttle body
- Positive Crankcase Ventilation (PCV) valve and brake booster vacuum hose
- Vacuum hose connector
- Surge tank stay
- High pressure fuel hose
- Surge tank and gasket
- Fuel injector harness connector
- Delivery pipe with the injectors and the pressure regulator as an assembly
- Coolant Temperature Sensor (CTS) electrical connector
- Intake manifold

To install:

3. Install or connect the following:
- New intake manifold gasket

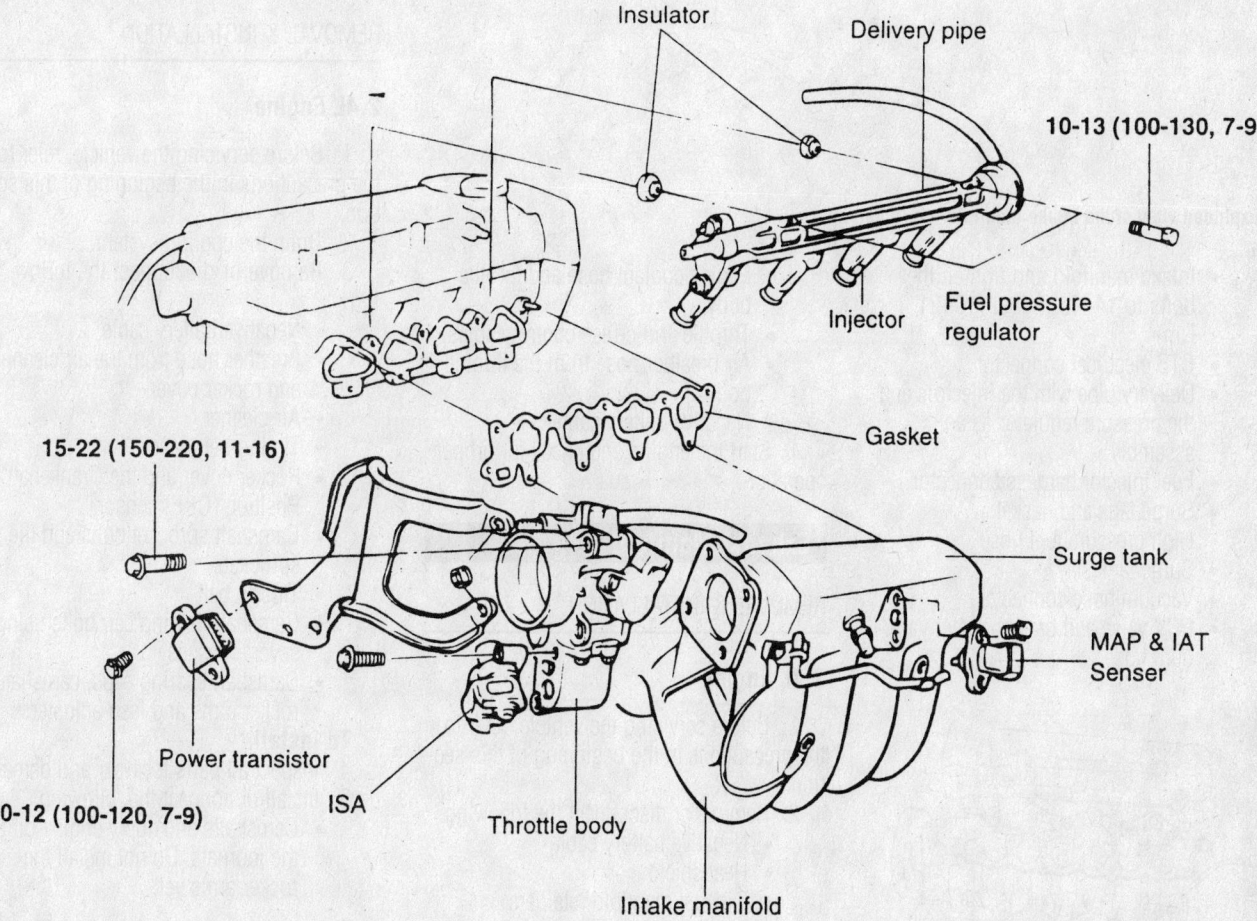

TORQUE : Nm (kg·cm, lb·ft)

9355LG12A

Exploded view of the intake manifold—2.4L engine

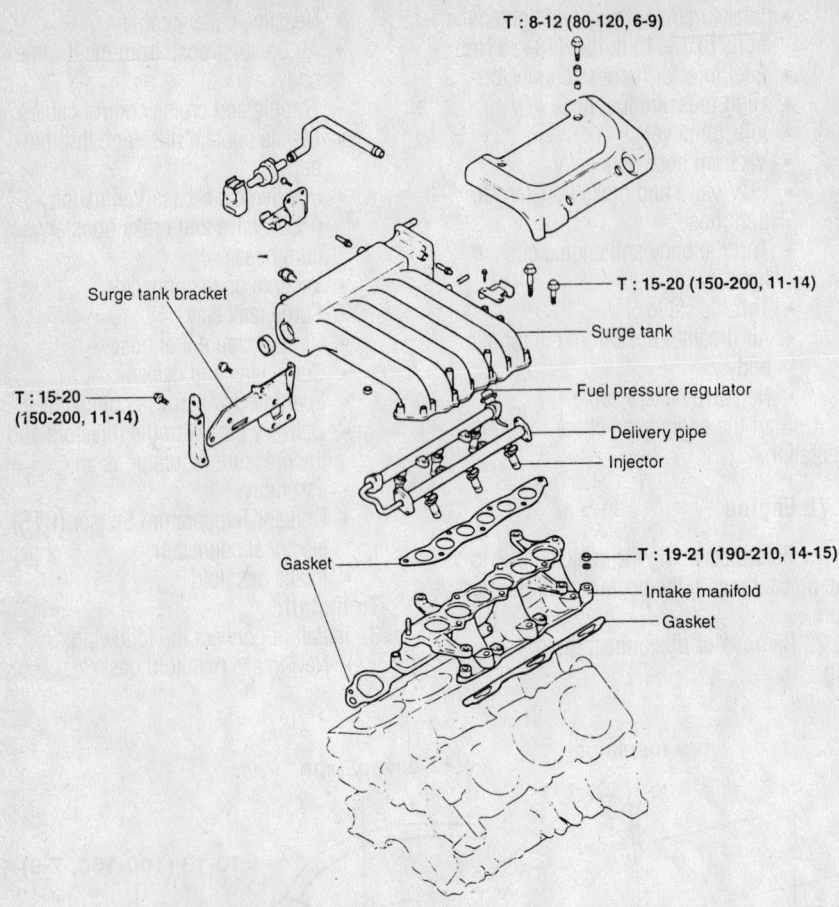

T : 8-12 (80-120, 6-9)

T : 15-20 (150-200, 11-14)
— Surge tank

Surge tank bracket

— Fuel pressure regulator

— Delivery pipe

T : 15-20
(150-200, 11-14)
— Injector

Gasket

T : 19-21 (190-210, 14-15)
— Intake manifold
— Gasket

9355LG12

Exploded view of the intake manifold—2.7L engine

- Intake manifold and tighten the bolts to 14–15 ft. lbs. (19–21 Nm)
- CTS electrical connector
- Delivery pipe with the injectors and the pressure regulator as an assembly
- Fuel injector harness connector
- Surge tank and gasket
- High pressure fuel hose
- Surge tank stay
- Vacuum hose connector
- PCV valve and brake booster vacuum hose

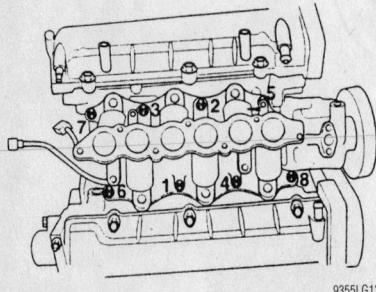

9355LG13

Intake manifold torque sequence—2.7L engine

- Engine coolant hose and throttle body
- Throttle and cruise control cables
- Air breather hose from the throttle body
- Negative battery cable
4. Start the engine and check for proper operation.

Exhaust Manifold

REMOVAL & INSTALLATION

2.4L Engine

1. Before servicing the vehicle, refer to the precautions in the beginning of this section.
2. Remove or disconnect the following:
    - Negative battery cable
    - Heat shield
    - Exhaust manifold retainers
    - Manifold and gasket
To install:
3. Install or connect the following:
    - New gasket and the manifold. Tighten the manifold M8 bolts to

18–22 ft. lbs. (25–30 Nm) and the M10 bolts to 25–40 ft. lbs. (35–55 Nm).
    - Heat shield
    - Negative battery cable

2.7L Engine

1. Before servicing the vehicle, refer to the precautions in the beginning of this section.
2. Remove or disconnect the following:
    - Negative battery cable
    - Heat shield
    - Exhaust manifold retainers
    - Manifold and gasket
To install:
3. Install or connect the following:
    - New gasket and the manifold. Tighten the manifold bolts to 18–22 ft. lbs. (25–30 Nm).
    - Heat shield
    - Negative battery cable

Camshaft

REMOVAL & INSTALLATION

2.4L Engine

1. Before servicing the vehicle, refer to the precautions in the beginning of this section.
2. Drain the cooling system.
3. Remove or disconnect the following:

- Negative battery cable
- Breather hose from the air cleaner and rocker cover
- Air cleaner
- Timing belt cover
- Rocker cover and the Crankshaft Position (CKP) sensor
- Camshaft sprocket bolts and the sprockets
- Timing belt
- Camshaft bearing cap bolts using several passes
- Camshaft bearing caps, camshafts, rocker arms and lash adjusters
To install:
4. Inspect all parts for wear and damage.
5. Install or connect the following:
    - Camshafts and apply engine oil to the journals. Do not install the rocker arms yet.

➡**The exhaust camshaft has a slit on the rear end for the CKP sensor.**

- Bearing caps. The caps are marked **I** for intake and **E** for exhaust, they also contain the cap number. For example **I2** would be intake cap number 2.

6. Make sure the camshafts can turn freely, then remove the caps and the camshafts.

7. Make sure the dowel pins on the ends of camshaft sprockets are facing up.

- Rocker arms
- Camshafts and bearing caps. Tighten the bearing cap bolts uniformly to 14–15 ft. lbs. (19–21 Nm).

8. Using special tools, camshaft oil seal installer and guide 09221-21000, 09221-

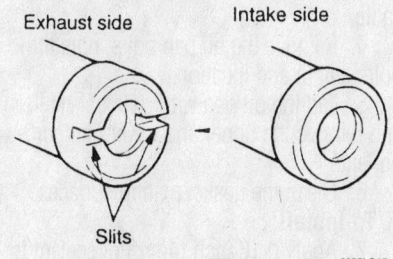

The exhaust camshaft has a slit on the rear end for the CKP sensor—2.4L engine

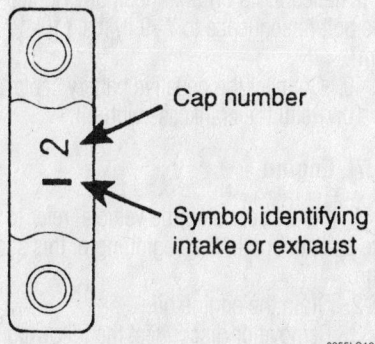

Cap number

Symbol identifying intake or exhaust

The caps are marked I for intake and E for exhaust, they also contain the cap number. For example I2 would be intake cap number 2—2.4L engine

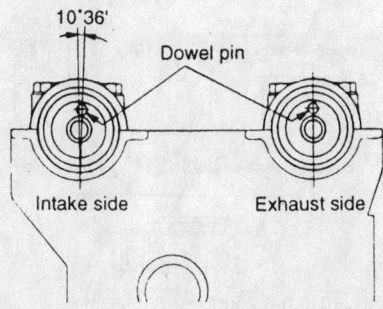

10˚36'

Dowel pin

Intake side
Exhaust side

Make sure the dowel pins on the ends of camshaft sprockets are facing up—2.4L engine

21100 install the oil seal. Coat the outside of the seal with oil prior to installation, then slide the seal along the front end of the camshaft and using the driver and a hammer install the seal until it is full seated.

- Camshaft sprockets and bolts. Tighten the bolts to 58–72 ft. lbs. (80–100 Nm).
- CKP sensor and rocker cover
- Breather hose to the air cleaner and rocker cover
- Air cleaner
- Negative battery cable

9. Start the engine and check for leaks.

2.7L Engine

1. Before servicing the vehicle, refer to the precautions in the beginning of this section.

2. Remove or disconnect the following:
- Negative battery cable
- Engine cover
- Intake manifold
- Breather hose and engine harness
- Power steering pulley
- A/C pulley
- Crankshaft pulley
- Idler pulley
- Tensioner pulley
- Timing belt cover
- Timing belt from the camshaft sprocket(s)
- Spark plug wires

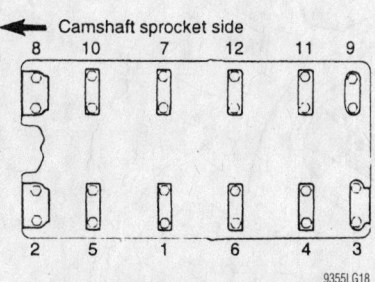

← Camshaft sprocket side

Camshaft bearing cap locations—2.4L engines

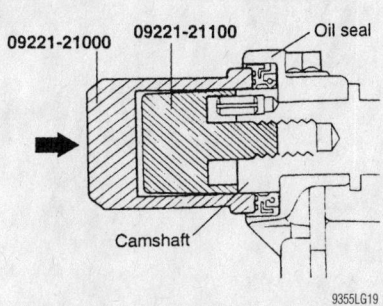

09221-21000   09221-21100   Oil seal

Camshaft

Installing the camshaft oil seal—2.4L engine

- Rocker arm cover
- Camshaft sprockets
- Camshaft bearing caps
- Camshafts

To install:

3. Align the camshaft timing chain with the intake timing chain sprocket and exhaust sprocket as shown in the accompanying illustration.

4. Lubricate the camshaft journals with oil and install them.

➡**To check the press fit, the camshaft (IN) and timing chain sprocket should be separable by a force greater than 1000kg minimum at room temperature.**

5. Install the bearing caps. The caps are marked **I** for intake and **E** for exhaust, they

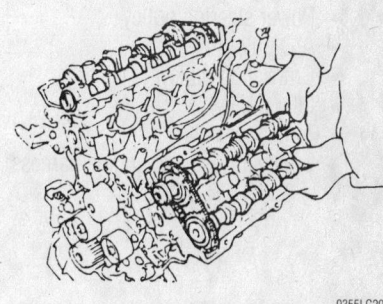

Remove the camshafts from the head—2.7L engine

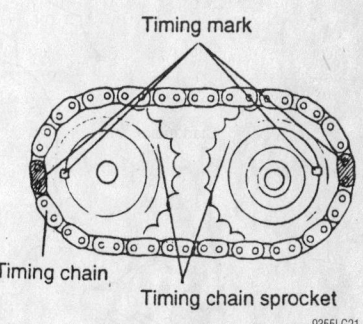

Timing mark

Timing chain

Timing chain sprocket

Align the camshaft timing chain with the intake timing chain sprocket and exhaust sprocket—2.7L engine

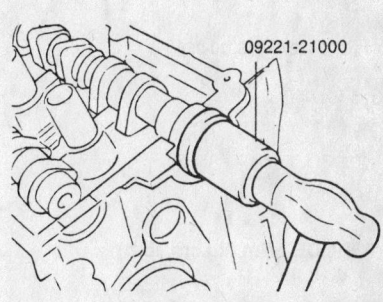

09221-21000

Install the camshaft oil seal—2.7L engine

also contain the cap number. For example **I2** would be intake cap number 2.

6. Tighten the bearing caps M10 bolts to 10–12 ft. lbs. (14–16 Nm) and the M7 bolts to 7–9 ft. lbs. (10–12 Nm) using several passes.

7. Using special tool, camshaft oil seal installer 09221-21000 install the oil seal. Coat the outside of the seal with oil prior to installation, then insert the seal along the front end of the camshaft and using the driver and a hammer install the seal until it is full seated.

8. Install or connect the following:
- Camshaft sprocket and tighten the bolt to 65–80 ft. lbs. (90–110 Nm)
- Rocker cover
- Spark plug wires
- Timing belt
- Timing belt cover
- Power steering pulley
- A/C pulley
- Crankshaft pulley
- Idler pulley
- Tensioner pulley
- Breather hose and engine harness

- Intake manifold
- Engine cover
- Negative battery cable

Valve Lash

ADJUSTMENT

The valve lash is controlled by hydraulic adjusters and no adjustment is possible.

Starter Motor

REMOVAL & INSTALLATION

1. Before servicing the vehicle, refer to the precautions in the beginning of this section.

2. Remove or disconnect the following:
- Negative battery cable and wait at least 3 minutes
- Speedometer cable and shift cable from the transaxle
- Starter motor wiring
- Starter motor bolts and the starter

27-34 (275-346, 20-25)

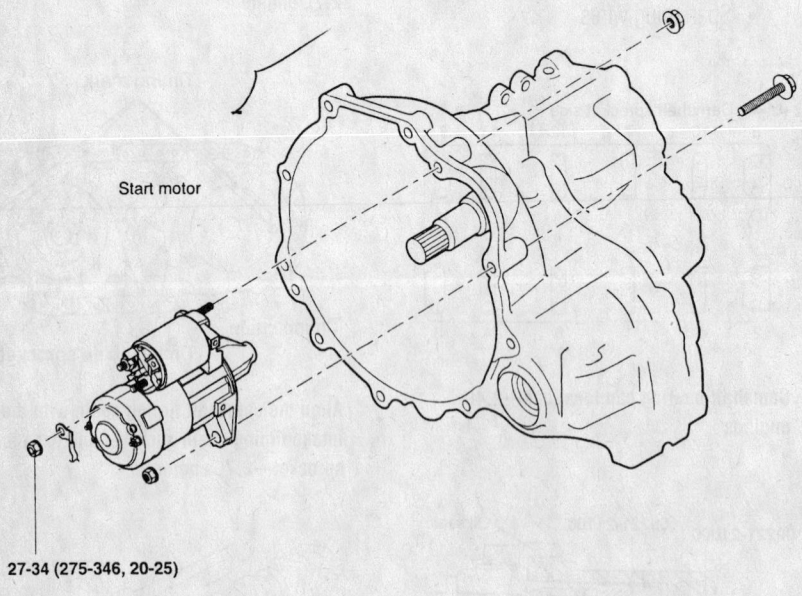

Start motor

127-34 (275-346, 20-25)

TORQUE : Nm (kg.cm, lb.ft)

9355LG23

Starter motor mounting

To install:

3. Installation is the reverse of removal. Tighten motor retainers to 20–25 ft. lbs. (27–34 Nm).

Oil Pan

REMOVAL & INSTALLATION

2.4L Engine

1. Before servicing the vehicle, refer to the precautions in the beginning of this section.

2. Drain the engine oil.

3. Disconnect the negative battery cable.

4. Remove the oil pan bolts, note the bolts length and location.

5. Tap the oil pan with a rubber mallet and remove the upper and lower pan components.

6. Clean the gasket mating surfaces.

To install:

7. Apply 0.16 inch (4mm) of sealant to the oil pan groove as illustrated. Install the within 15 minutes of sealant installation.

8. Install the upper and lower oil pans and the bolts making sure the proper length is installed in its original position. Tighten the bolt in sequence to 7–9 ft. lbs. (10–12 Nm).

9. Connect the negative battery cable.

10. Refill the crankcase with oil.

2.7L Engine

1. Before servicing the vehicle, refer to the precautions in the beginning of this section.

2. Drain the engine oil.

3. Remove or disconnect the following:
- Negative battery cable
- Lower oil pan bolts and the pan
- Upper oil pan bolts and the pan

4. Clean the gasket mating surfaces

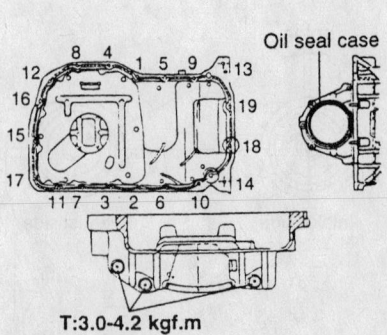

Oil seal case

T:3.0-4.2 kgf.m

9355LG31

Oil pan torque sequence and sealant application points—2.4L engine

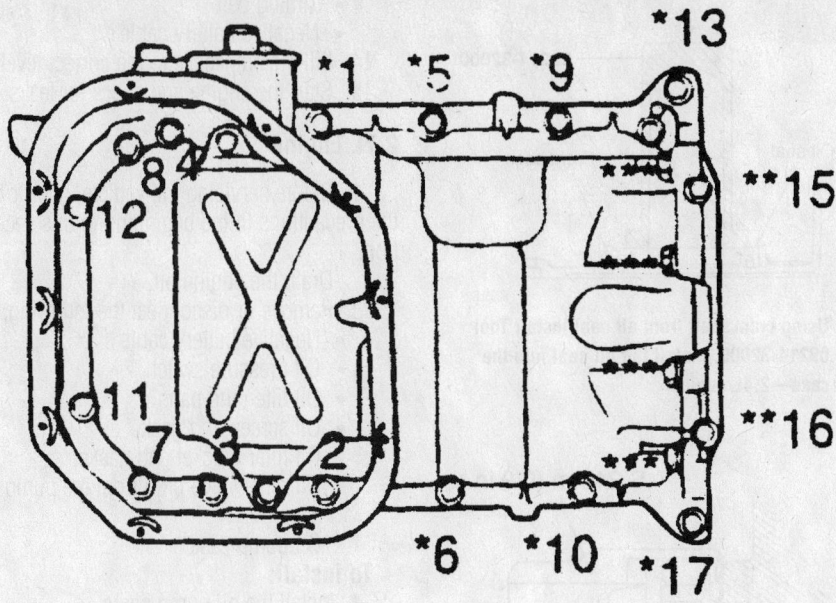

Upper oil pan torque sequence (tighten bolts indicated with * to 14–20 inch lbs. (19–28 Nm), bolts indicated with a ** to 4–5 ft. lbs. (5–7 Nm) and bolts indicated with a *** to 22–30 ft. lbs. (30–42 Nm)—2.7L engine

To install:

5. Apply 0.16 inch (4mm) of sealant to the lower oil pan groove. Install the within 15 minutes of sealant installation.

6. Install the upper oil pan and the tighten the bolts in sequence as follows:

  a. 0.937 x 1.4961 inch (10 x 38mm) bolt to 22–30 ft. lbs. (30–42 Nm).

  b. 0.3150 x 0.866 inch (8 x 22mm) bolt 14–20 inch (19–28 Nm).

  c. 6.7519 inch (171.5mm) bolt to 4–5 ft. lbs. (5–7 Nm).

  d. 6.7520 inch (152.5mm) bolt to 4–5 ft. lbs. (5–7 Nm).

7. Install the lower pan and tighten the bolts to 7–9 ft. lbs. (10–12 Nm).

8. Connect the negative battery cable.

9. Refill the crankcase with oil.

Oil Pump

REMOVAL & INSTALLATION

2.4L Engine

1. Before servicing the vehicle, refer to the precautions in the beginning of this section.

2. Drain the engine oil.

3. Remove or disconnect the following:
  - Negative battery cable
  - Timing belt
  - Oil pan
  - Oil screen and gasket
  - Oil pressure switch
  - Oil filter bracket and gasket

4. Using Tool 09213-33000, remove the plug cap from the oil pump portion of the case.

  - Plug from the left side of the block and insert an 0.32 inch (8mm) screwdriver into the plug hole. The screwdriver must be inserted at least 2.4 inch (60mm).
  - Pump driven gear the left counter balance shaft bolt
  - Front case bolts (noting the bolt length and location), the case and gasket.
  - Two counter balance shafts from the block
  - Oil pump cover from the case
  - Oil pump gears from the case
  - Screwdriver from the plug hole

To install:

5. Install the oil pump gears.

6. Inspect the tip clearance of the gears using a feeler gauge. The specifications are as follows:

  a. Standard value drive gear: 0.0063–0.0083 inch (0.16–0.21mm).

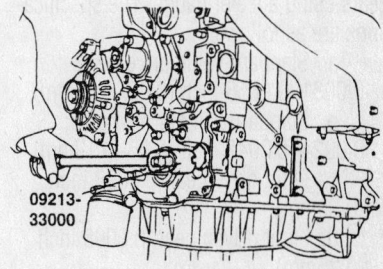

Using Tool 09213-33000, remove the plug cap from the oil pump portion of the case—2.4L engine

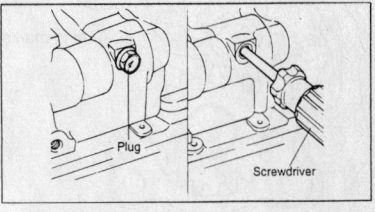

Insert an 0.32 inch (8mm) screwdriver at least 2.4 inch (60mm) into the plug hole—2.4L engine

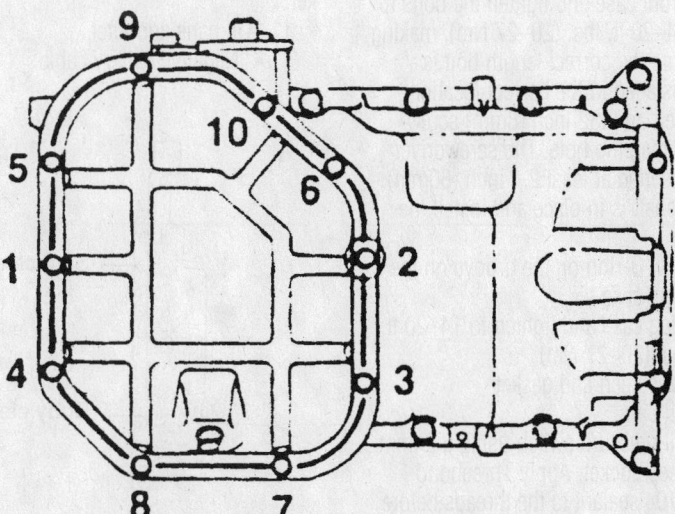

Lower oil pan bolt torque sequence—2.7L engine

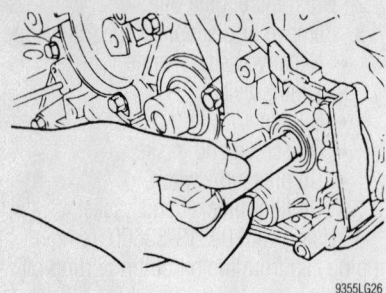

Remove the pump driven gear the left counter balance shaft bolt—2.4L engine

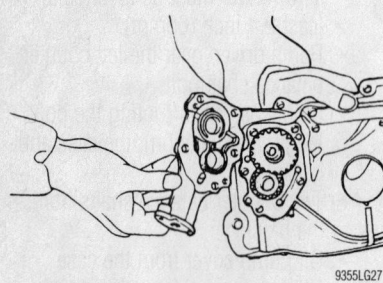

Remove the oil pump cover from the case—2.4L engine

    b. Standard value driven gear: 0.0071–0.0083 inch (0.18–0.21mm).
    c. Limit drive gear: 0.0098 inch (0.25mm).
    d. Limit driven gear: 0.0098 inch (0.25mm).
  7. Inspect the side clearance of the gears using a feeler gauge. The specifications are as follows:
    a. Standard value drive gear: 0.0031–0.0055 inch (0.08–0.14mm).
    b. Standard value driven gear: 0.0024–0.0047 inch (0.06–0.12mm).
    c. Limit drive gear: 0.0098 inch (0.25mm).
    d. Limit driven gear: 0.0098 inch (0.25mm).
  8. Apply engine oil to both gears and align the gear timing marks.
  9. Install the oil pump case.

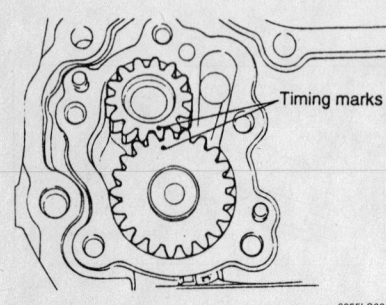

Apply engine oil to both gears and align the gear timing marks—2.4L engine

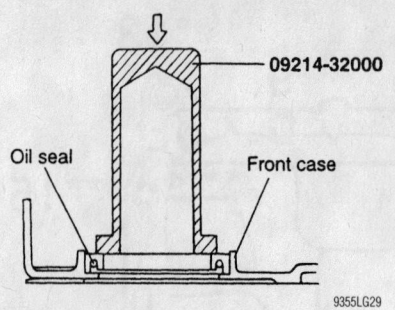

Using crankshaft front oil seal install Tool 09214-32000, install the oil seal into the case—2.4L engine

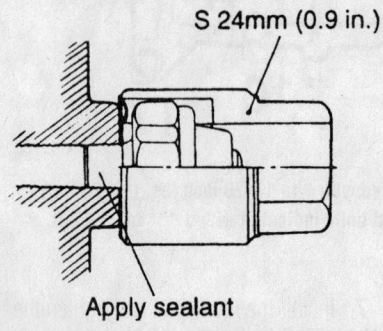

Front case bolt length and location—2.4L engine

  10. Using crankshaft front oil seal install Tool 09214-32000, install the oil seal into the case.
  11. Place special tool 09214-32100 on the front of the crankshaft and apply a coat of oil to the outside of the tool to aid in case installation.
  12. Install or connect the following:
- New front case gasket and temporarily tighten the flange bolts
- Front case and tighten the bolts to 14–20 ft. lbs. (20–27 Nm), making sure the correct length bolt is installed in the correct location.

  13. Insert an 0.32 inch (8mm) screwdriver into the plug hole. The screwdriver must be inserted at least 2.4 inch (60mm). Verify the shaft is in place and install the bolt.
- New O-ring on the groove on the front case
- Plug case and tighten to 14–20 ft. lbs. (20–27 Nm)
- Oil screen and gasket
- Oil pan
- Oil pressure switch using a 24mm deep socket. Apply Threebond 1104 sealant to the threads before installation and tighten to 6–9 ft. lbs. (8–12 Nm).

- Timing belt
- Negative battery cable

  14. Fill the crankcase to the correct level.
  15. Start the engine and check for leaks.

2.7L Engine

  1. Before servicing the vehicle, refer to the precautions in the beginning of this section.
  2. Drain the engine oil.
  3. Remove or disconnect the following:
- Negative battery cable
- Oil pressure switch
- Oil filter and pans
- Oil screen and gasket
- Oil filter bracket and gasket
- Oil relief valve plug from the pump case
- Oil pump case

To install:

  4. Install the oil pump gears.
  5. Inspect the side clearance of the gears using a feeler gauge. The specifications are as follows:
    a. Standard body clearance: 0.0039–0.0071 inch (0.100–0.181mm).
    b. Standard side clearance: 0.0016–0.0037 inch (0.040–0.095mm).
  6. Install the oil pump case with a new gasket. Tighten the bolt to 9–11 ft. lbs. (12–15 Nm) and the screw to 6–9 ft. lbs. (8–12 Nm).
  7. Install a new oil seal into the pump as tightly as possible.
  8. Using crankshaft front oil seal install Tool 09214-33000, install the oil seal into the case.
  9. Install the relief plunger and spring and tighten the valve plug to 29–36 ft. lbs. (40–50 Nm).
  10. Install the oil screen and a new gasket.
  11. Oil pans and filter.
- Negative battery cable

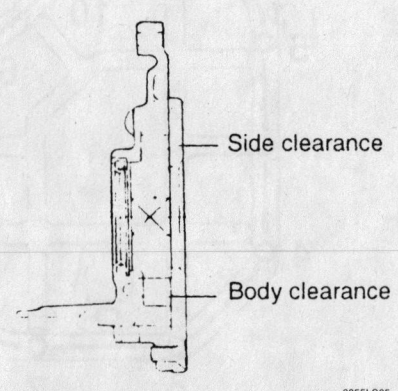

Check the oil pump gears side and body clearance—2.7L engine

12. Fill the crankcase to the correct level.
13. Start the engine and check for leaks.

Rear Main Seal

REMOVAL & INSTALLATION

2.4L Engine

1. Before servicing the vehicle, refer to the precautions in the beginning of this section.
2. Remove or disconnect the following:

- Transaxle
- Clutch pressure plate and disc, if equipped
- Flywheel
- Oil seal case bolts and the case
- Oil seal

To install:

3. Install or connect the following:

- Oil seal. Drive the seal square into the seal case.
- Oil seal case so that the oil hole in the separator may be directed downwards. Tighten the bolts to 7–9 ft. lbs. (10–12 Nm).
- Flywheel
- Clutch pressure plate and disc, if equipped
- Transaxle

4. Check the fluid levels.
5. Start the engine and check for leaks.

2.7L Engine

1. Before servicing the vehicle, refer to the precautions in the beginning of this section.
2. Remove or disconnect the following:

- Transaxle
- Clutch pressure plate and disc, if equipped
- Flywheel
- Drive plate and adapter plate

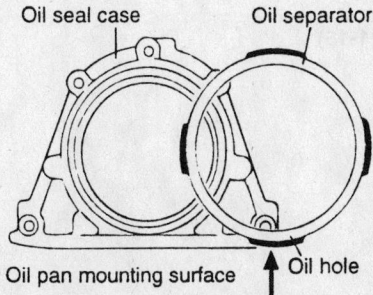

Position the oil seal case so that the oil hole in the separator may be directed downwards—2.4L engine

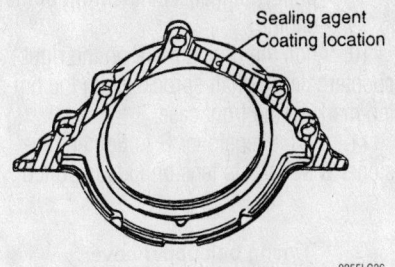

Apply sealant to the areas shown—2.7L engine

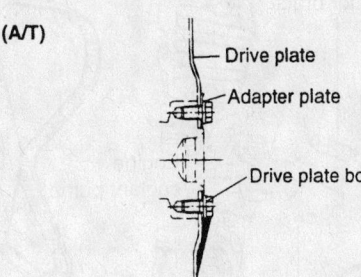

Install the drive plate and adapter plate—2.7L engine

- Oil seal case bolts and the case
- Oil seal

To install:

3. Install or connect the following:

- Oil seal. Drive the seal square into the seal case.
- Oil seal case so that the oil hole in the separator may be directed downwards. Tighten the bolts to 7–9 ft. lbs. (10–12 Nm).
- Drive plate and adapter plate. Tighten the bolt to 53–56 ft. lbs. (73–77 Nm).
- Flywheel
- Clutch pressure plate and disc, if equipped
- Transaxle

4. Check the fluid levels.
5. Start the engine and check for leaks.

Timing Belt

REMOVAL & INSTALLATION

2.4L Engine

1. Before servicing the vehicle, refer to the precautions in the beginning of this section.
2. Align the timing marks to set the No. 1 piston to Top Dead Center (TDC) by rotating the crankshaft clockwise. The timing

marks of the camshaft sprocket and the cylinder head cover should be aligned and the dowel pin of the camshaft sprocket should be at the upper side.

3. Remove or disconnect the following:

- Crankshaft pulley, water pump pulley and drive belt
- Timing belt cover
- Auto tensioner

4. Mark the timing belt is being reused, mark an arrow on the belt noting the direction of rotation or the front of the engine to make sure the belt is reinstalled in its original position.

- Timing belt

5. Hold the camshaft with a wrench and loosen the camshaft sprocket bolts.

- Sprockets

6. When removing the oil pump socket nut, first remove the plug at the side of the block and insert a 0.3 inch (8mm) diameter screwdriver to keep the left counterbalance shaft in position. Insert the screwdriver at least 2.36 inch (60 mm).

- Oil pump sprocket nut and the sprocket
- Loosen the right counterbalance shaft sprocket bolt until you can loosen it by hand
- Tensioner **B** and timing belt **B**. Refer to the accompanying illustration for tensioner and belt identification.

✳✳ CAUTION

Do not attempt to loosen bolts while holding the sprocket with pliers or any tool after removing timing belt B.

- Crankshaft sprocket **B** from the crankshaft

To install:

7. Install or connect the following:

- Crankshaft sprocket **B** to the crankshaft

✳✳ CAUTION

Pay attention to the direction of the flange. If it is installed in the wrong direction, the belt will break.

8. Apply engine oil to the outer surface of the spacer lightly and install the spacer to the right counterbalance shaft. Be sure to install spacer correctly.

- Counterbalance shaft sprocket onto the right counterbalance shaft and then tighten the flange bolt by hand until it is tight

9. Align the timing mark on each

sprocket with its corresponding timing mark on the front case.

- Timing belt **B** and make sure there is no slack
- Tensioner **B** so that the center of the pulley is located on the left side of the mounting bolt and the

pulley flange faces the front of the engine

10. Align the timing mark on the right counterbalance shaft sprocket with the timing mark on the front case.

11. Lift the tensioner **B** to tighten tensioner **B** so that its tension side is pulled

tight. Tighten the bolt on tensioner **B**. As the bolt is being tightened, make sure the shaft does not turn. If the shaft turns, the belt will be overtightened.

12. Make sure the timing marks are aligned.

13. Check the belt tension by depressing

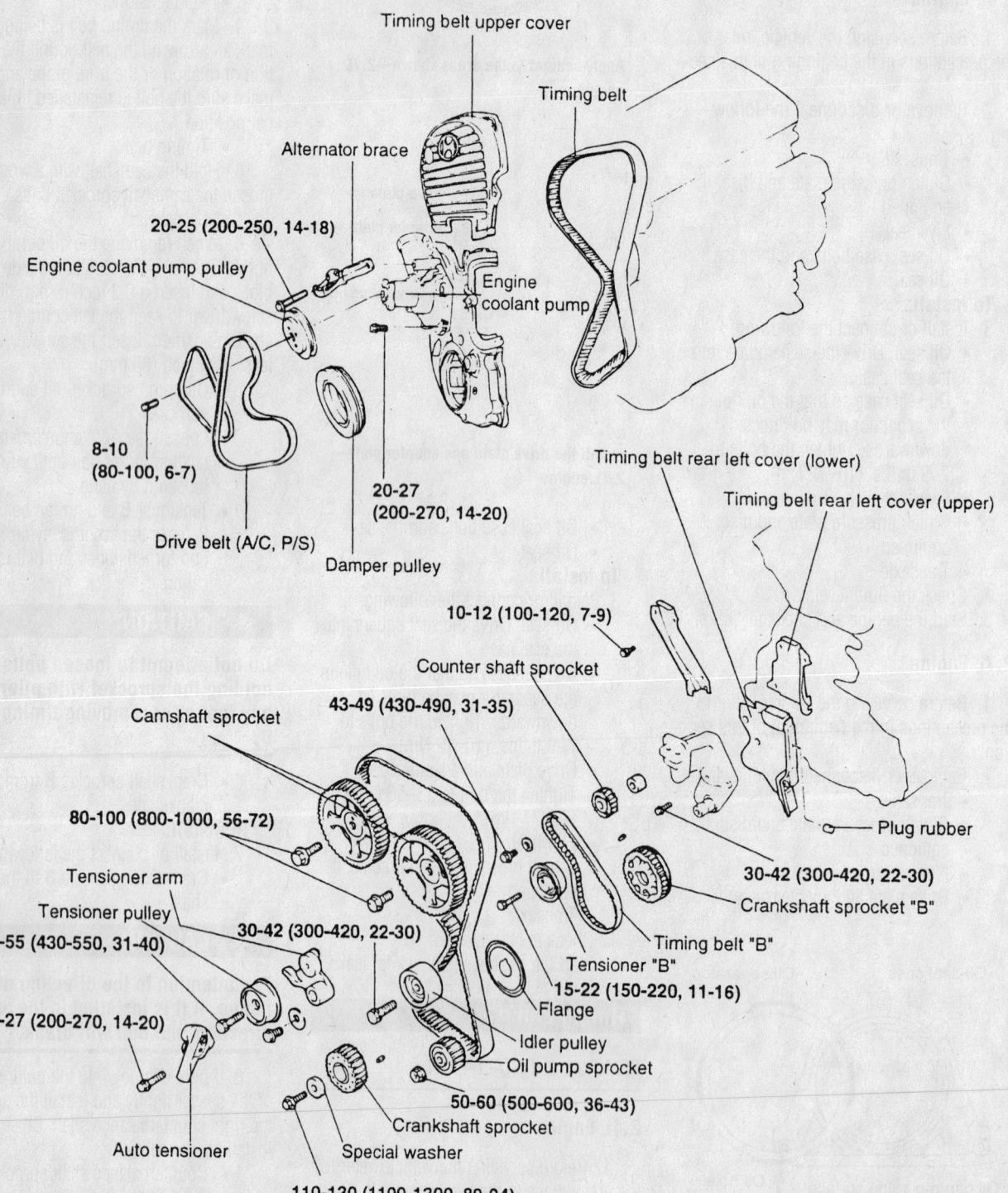

TORQUE : Nm (kg·cm, lb·ft)

9355LG73

Exploded view of the timing belt assembly and related components—2.4L engine

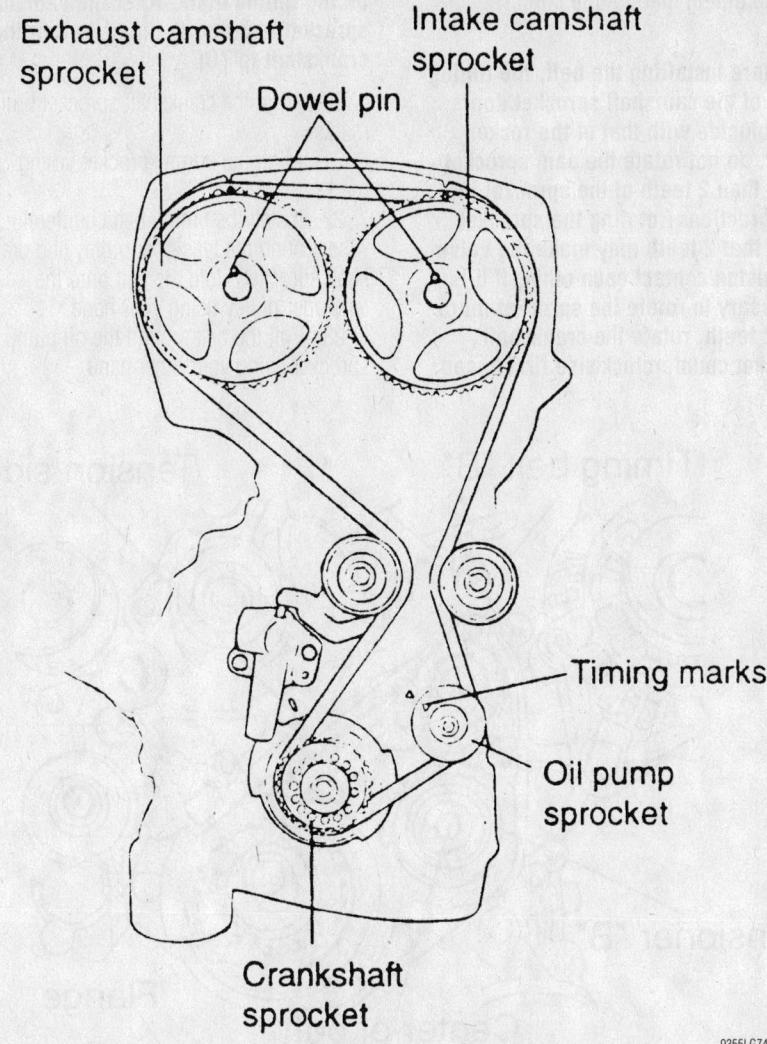

Correct sprocket alignment when the belt is installed—2.4L engine

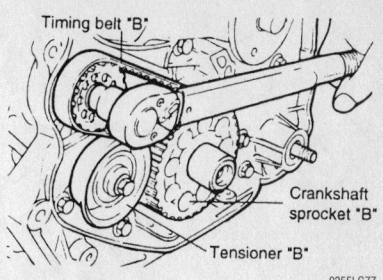

Remove tensioner B and timing belt B—2.4L engine

screwdriver through the plug hole on the left side of the block to keep the left counterbalance shaft in position. Insert the screwdriver at least 2.36 inch (60 mm).
- Oil pump sprocket and tighten the nut to 36–43 ft. lbs. (50–60 Nm)
- Camshaft sprockets and the bolts
16. Hold the camshaft with a wrench and tighten the camshaft sprocket bolts to 56–72 ft. lbs. (80–100 Nm).
17. Reset the auto tensioner as follows:

a. Place the tensioner in a soft jawed vice in a level position. If there is a plug at the bottom of the tensioner use a plain washer.

b. Compress the rod slowly using the vice until the set hole in the rod is aligned with the set hole on the cylinder.

c. Insert a set pin through the body and rod and leave the pin installed.
18. Install or connect the following:
- Tensioner and tighten the bolts to 14–20 ft. lbs. (20–27 Nm). Leave the set pin in place.
- Tensioner pulley and tighten the bolt to 31–40 ft. lbs. (43–55 Nm)
19. Rotate the camshaft sprockets so that the dowel pin is at the upper side.

the center of the belt span with an index finger. The deflection should be 0.20–0.28 inch (5–7mm).
14. Install or connect the following:
- Flange and crankshaft sprocket making sure it is installed properly. Installing the flange incorrectly will cause the belt to break.
- Crankshaft washer and bolt. Tighten the bolt to 80–94 ft. lbs. (110–130 Nm).
15. Insert a 0.3 inch (8mm) diameter

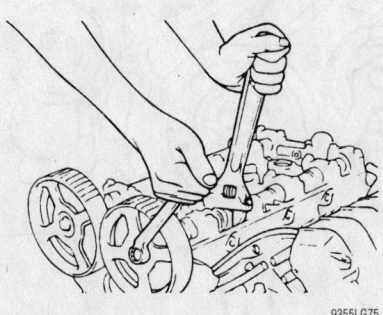

Hold the camshaft with a wrench and loosen the camshaft sprocket bolts—2.4L engine

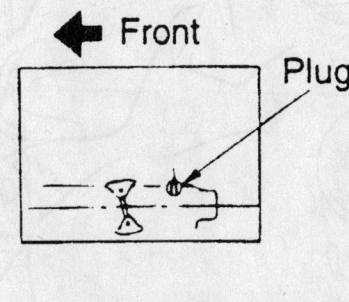

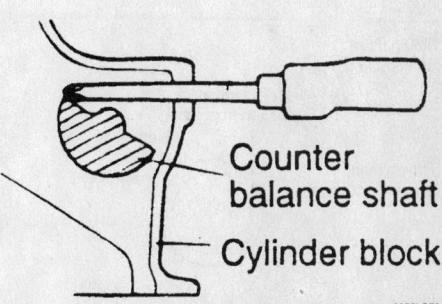

Insert a 0.3 inch (8mm) diameter screwdriver to keep the left counterbalance shaft in position—2.4L engine

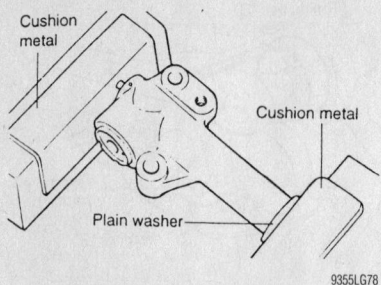

9355LG78

Place the tensioner in a soft jawed vice in a level position. If there is a plug at the bottom of the tensioner use a plain washer—2.4L engine

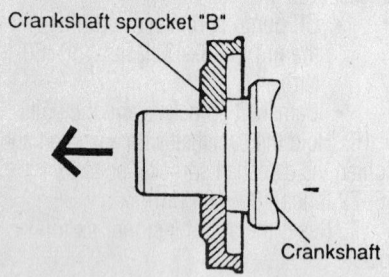

Crankshaft sprocket "B"

Crankshaft

9355LG79

Pay attention to the direction of the flange, if it is installed in the wrong direction, the belt will break—2.4L engine

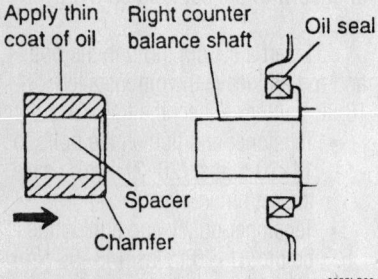

Apply thin coat of oil

Right counter balance shaft

Oil seal

Spacer

Chamfer

9355LG80

Apply engine oil to the outer surface of the spacer lightly and install the spacer to the right counterbalance shaft. Be sure to install spacer correctly—2.4L engine

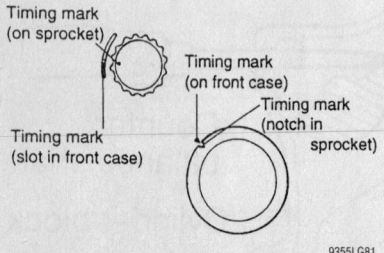

Timing mark (on sprocket)

Timing mark (on front case)

Timing mark (notch in sprocket)

Timing mark (slot in front case)

9355LG81

Align the timing mark on each sprocket with its corresponding timing mark on the front case—2.4L engine

Set the timing mark of the sprocket correctly.

➡ **Before installing the belt, the timing mark of the camshaft sprocket doers not coincide with that of the rocker cover, do not rotate the cam sprocket more than 2 teeth of the sprocket in any direction. Rotating the sprocket more that 2 teeth may make the valve and piston contact each other. If it is necessary to rotate the sprocket more that 2 teeth, rotate the crankshaft sprocket counterclockwise first based on the timing mark. After the camshaft sprocket is properly timed, return the crankshaft to TDC.**

20. Align the crankshaft sprocket timing marks.

21. Align the pump sprocket timing marks.

22. Install the timing belt counterclockwise around the tensioner pulley and crankshaft sprocket. Hold the belt onto the tensioner pulley using your hand.

23. Pull the belt around the oil pump sprocket using your other hand.

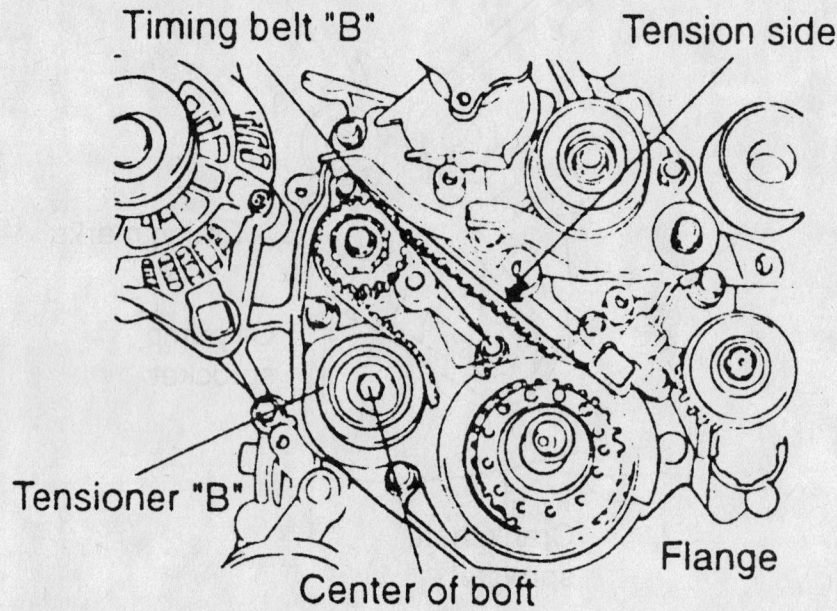

Timing belt "B" Tension side

Tensioner "B"

Center of boft

Flange

9355LG82

Align the timing mark on the right counterbalance shaft sprocket with the timing mark on the front case—2.4L engine

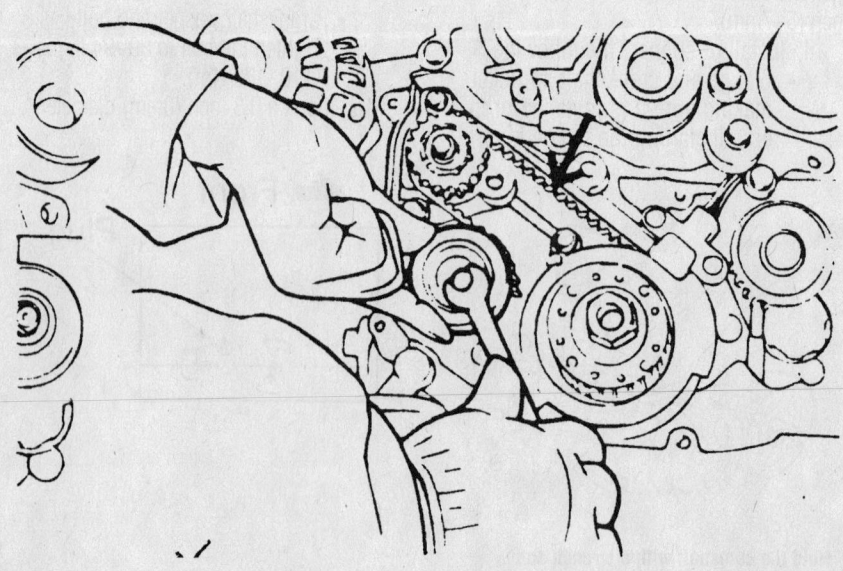

9355LG83

Lift the tensioner B to tighten tensioner B so that its tension side is pulled tight—2.4L engine

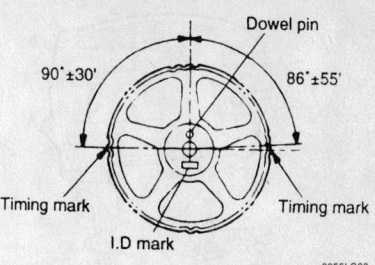

Camshaft sprocket alignment—2.4L

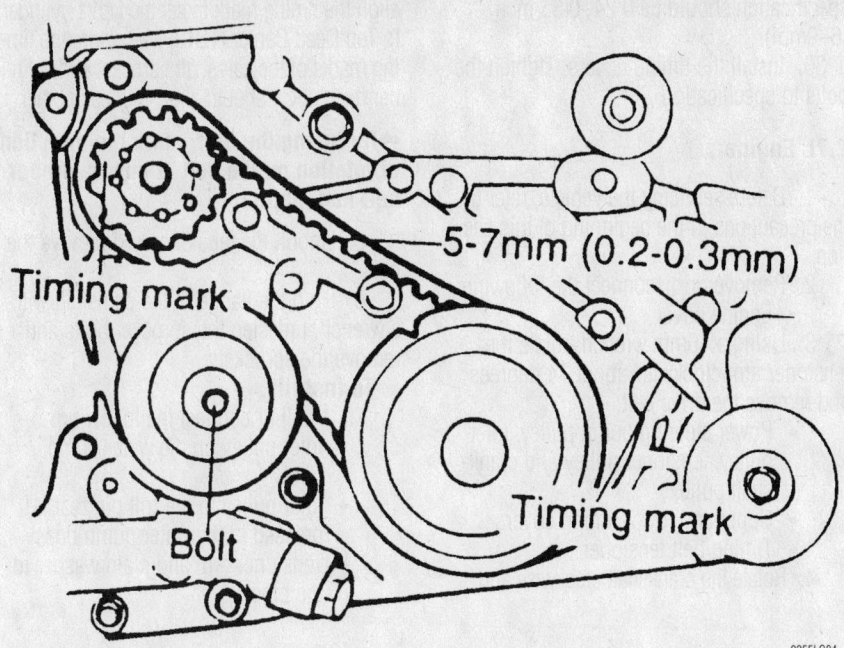

5-7mm (0.2-0.3mm)

Timing mark

Bolt

Timing mark

9355LG84

Make sure the timing belt B marks are aligned and the belt tensioner is correct—2.4L engine

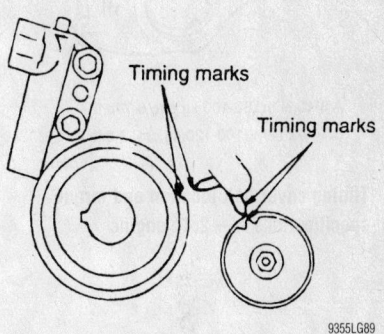

9355LG89

Align the oil pump sprocket timing marks—2.4L engine

24. Install the belt around the right-hand idler pulley, then the intake camshaft sprocket.

25. Turn the exhaust camshaft sprocket one tooth clockwise to align its timing mark with the cylinder top surface, then pull the belt around the exhaust camshaft sprocket.

26. Raise the tensioner pulley gently so that the belt does not sag and temporarily tighten the pulley center bolt.

27. Recheck that all timing marks are correct.

28. Remove the set pin from the auto tensioner.

29. Rotate the crankshaft two turns clockwise and let it sit for around 15 minutes. After 15 minutes, measure the auto tensioner protrusion **A** (the distance between the tensioner arm and tensioner) as shown in the accompanying illustration. The

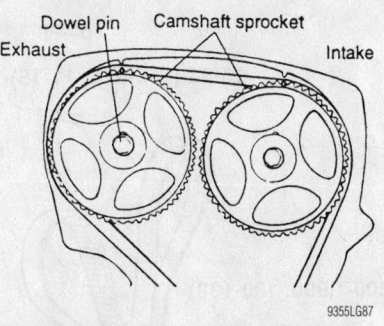

Dowel pin   Camshaft sprocket
Exhaust   Intake

9355LG87

Rotate the camshaft sprockets so that the dowel pin is at the upper side, set the timing mark of the sprocket correctly—2.4L engine

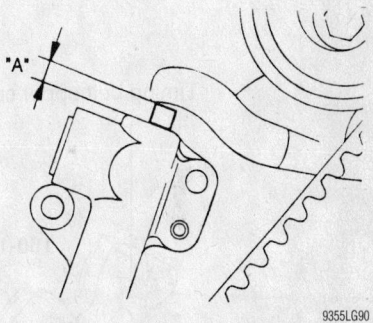

"A"

9355LG90

Measure the auto tensioner protrusion A —2.4L engine

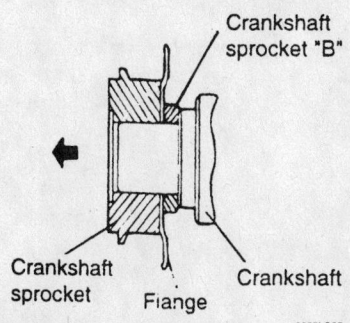

Crankshaft sprocket "B"

Crankshaft sprocket   Flange   Crankshaft

9355LG85

Install the flange and crankshaft sprocket making sure it is installed properly—2.4L engine

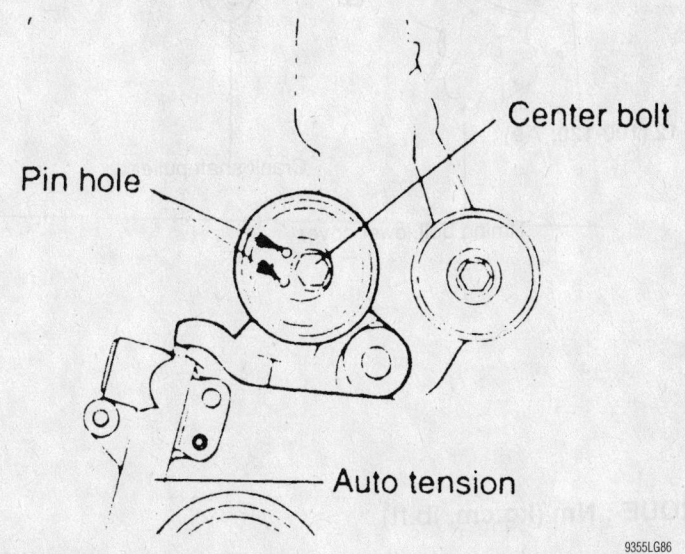

Center bolt

Pin hole

Auto tension

9355LG86

Install the tensioner pulley—2.4L engine

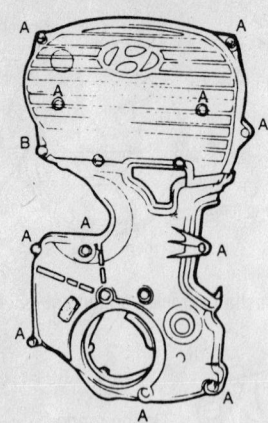

A:8-10 N.m (80-100 kg.cm, 6-7 lb.ft)
B:10-12 N.m (100-120 kg.cm, 7-9 lb.ft)

9355LG91

Timing cover bolt location and torque specifications A —2.4L engine

specification should be 0.24–0.35 inch (6–9mm).

30. Install the timing covers. Tighten the bolts to specification.

2.7L Engine

1. Before servicing the vehicle, refer to the precautions in the beginning of this section.
2. Remove or disconnect the following:
   • Engine cover
3. Using a 16mm wrench, rotate the tensioner arm clockwise about 14 degrees and remove the drive belt.
   • Power steering pump pulley, idler pulley, tensioner pulley and crankshaft pulley
   • Upper and lower timing covers
   • Timing belt tensioner
4. Rotate the crankshaft clockwise and

align the timing mark to set the No. 1 cylinder to Top Dead Center (TDC). Make sure the timing marks of the camshaft sprocket and cylinder head cover should align with each other.

➡**If reusing the belt, mark the direction of rotation on the belt to ensure proper belt installation.**

5. Unbolt the tensioner and remove the belt.
6. Hold the flange of the camshaft with a wrench, unfasten the sprocket bolts and remove the sprockets.
   To install:
7. Install or connect the following:
   • Idler pulley on the water pump boss
   • Idler pulley to the roll pin that is pressed in the water pump boss
   • Tensioner arm and plain washer to the block

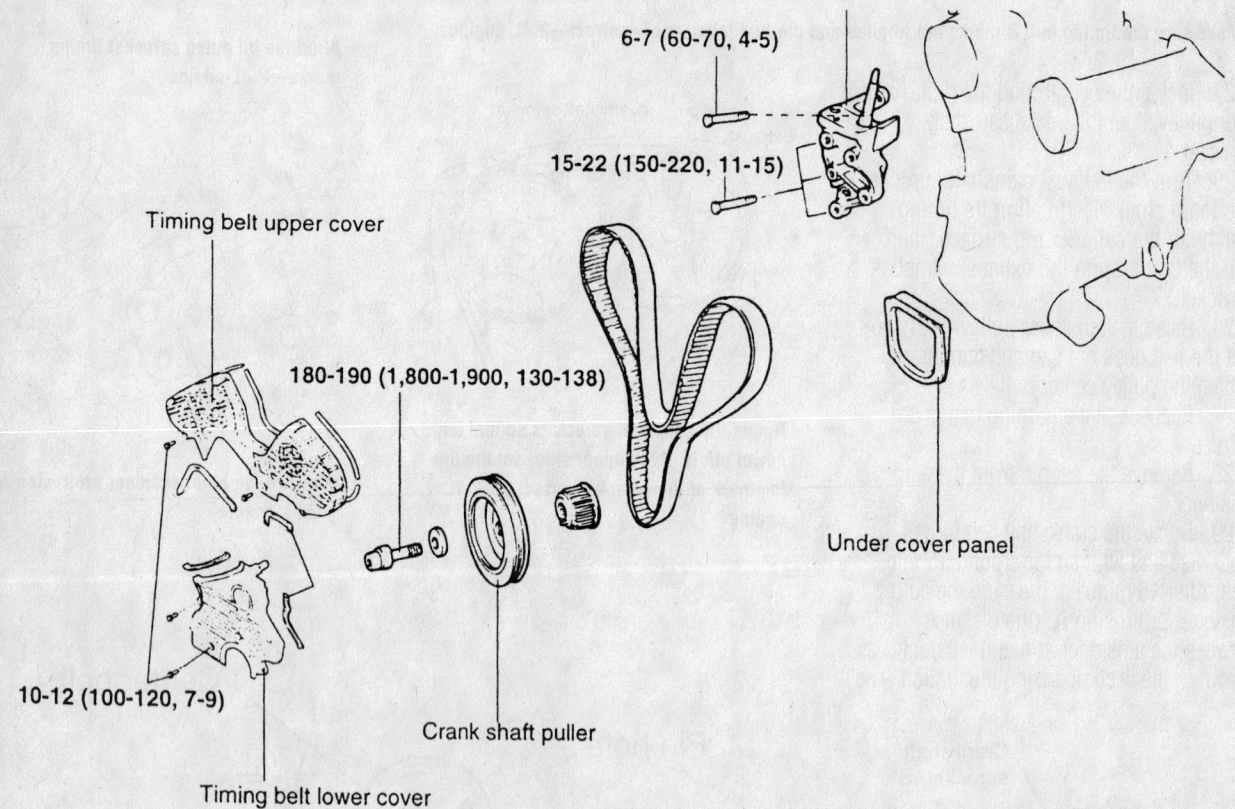

6-7 (60-70, 4-5)
15-22 (150-220, 11-15)
Timing belt upper cover
180-190 (1,800-1,900, 130-138)
Under cover panel
10-12 (100-120, 7-9)
Crank shaft puller
Timing belt lower cover

TORQUE : Nm (kg.cm, lb.ft)

9355LG92

Exploded view of the timing belt assembly—2.7L engine

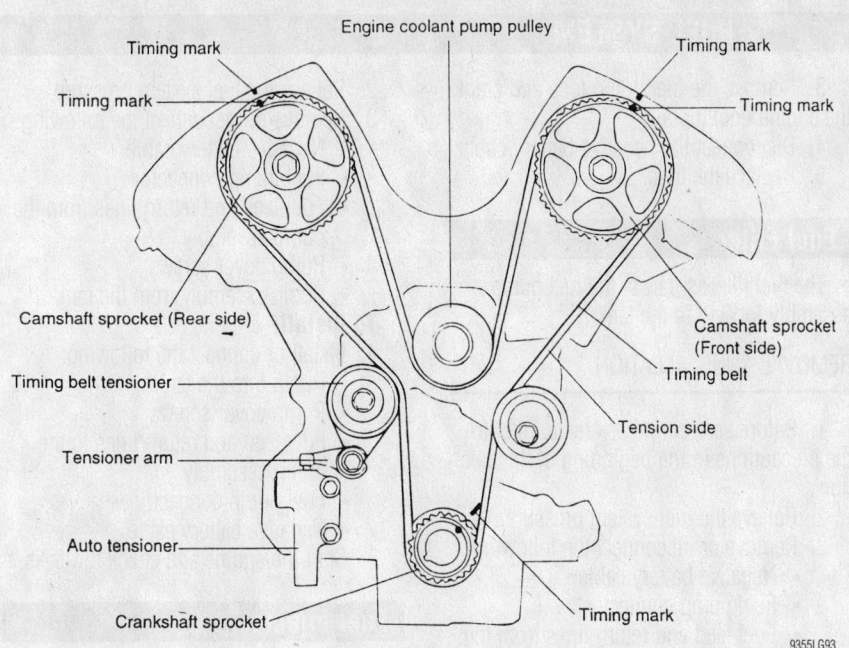

Correct timing marks alignment with the timing belt installed—2.7L engine

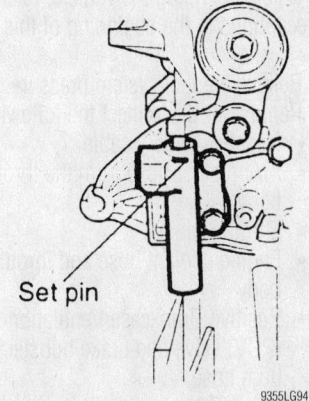

Set pin

9355LG94

Remove the set pin from the auto tensioner—2.7L engine

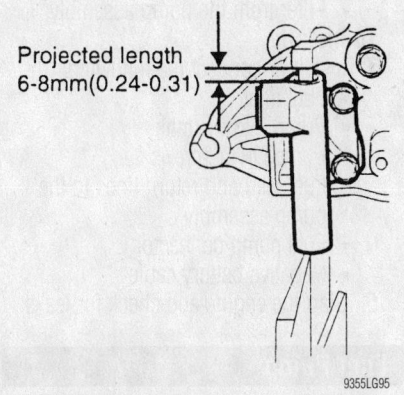

Projected length
6-8mm(0.24-0.31)

9355LG95

The projected length of the auto tensioner at Top Dead Center (TDC) should be 6-8mm—2.7L engine

- Tensioner pulley to the tensioner arm
- Camshaft sprockets and align the timing marks. Tighten the bolts to 65–80 ft. lbs. (90–110 Nm).
8. Compress the tensioner and install a set pin to keep the plunger in position.
- Tensioner and tighten the bolt to 14–20 ft. lbs. (20–27 Nm)
9. Align the sprocket timing marks and install the belt in the following order:
- Crankshaft sprocket
- Idler pulley
- Camshaft sprocket on the left hand side
- Water pump pulley
- Camshaft sprocket on the right hand side
- Tensioner pulley
10. Remove the tensioner set pin.
11. Adjust the timing belt tension as follows:

    a. Rotate the crankshaft 2 turns clockwise and measure the projected length of the auto tensioner at TDC after being installed for 5 minutes.

    b. The projected length should be 6–8mm.

    Make sure the timing marks are in their specified positions or remove and reinstall the belt.
12. Install or connect the following:
- Upper and lower timing belt covers
- Power steering pump pulley, idler pulley, tensioner pulley and crankshaft pulley
- Drive belt
- Engine cover

Piston and Ring

POSITIONING

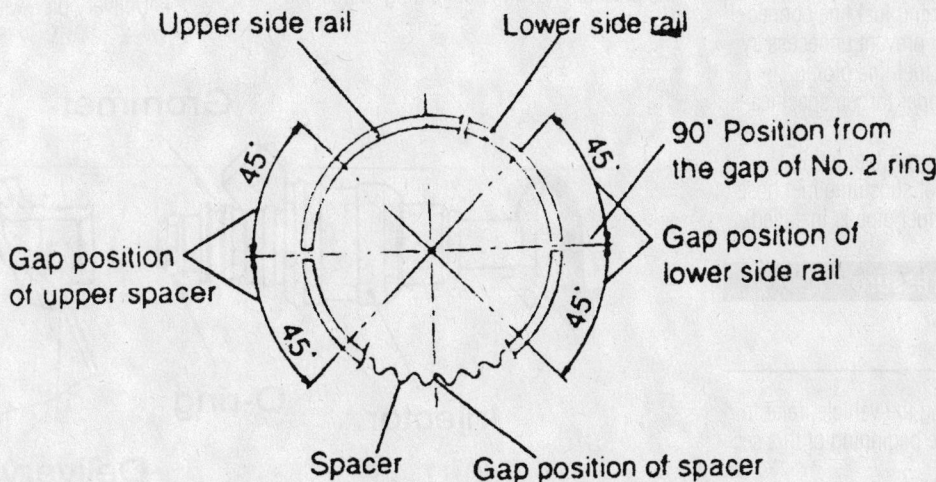

Ring end gap positioning

FUEL SYSTEM

Fuel System Service Precautions

Safety is the most important factor when performing not only fuel system maintenance, but any type of maintenance. Failure to conduct maintenance and repairs in a safe manner may result in serious personal injury or death. Maintenance and testing of the vehicle's fuel system components can be accomplished safely and effectively by adhering to the following rules and guidelines:

• To avoid the possibility of fire and personal injury, always disconnect the negative battery cable unless the repair or test procedure requires that battery voltage be applied.

• Always relieve the fuel system pressure prior to disconnecting any fuel system component (injector, fuel rail, pressure regulator, etc.), fitting or fuel line connection. Exercise extreme caution whenever relieving fuel system pressure to avoid exposing skin, face and eyes to fuel spray. Please be advised that fuel under pressure may penetrate the skin or any part of the body that it contacts.

• Always place a shop towel or cloth around the fitting or connection prior to loosening to absorb any excess fuel due to spillage. Ensure that all fuel spillage (should it occur) is quickly removed from engine surfaces. Ensure that all fuel soaked cloths or towels are deposited into a suitable waste container.

• Always keep a dry chemical (Class B) fire extinguisher near the work area.

• Do not allow fuel spray or fuel vapors to come into contact with a spark or open flame.

• Always use a backup wrench when loosening and tightening fuel line connection fittings. This will prevent unnecessary stress and torsion to fuel line piping. Always follow the proper torque specifications.

• Always replace worn fuel fitting O-rings with new. Do not substitute fuel hose or equivalent, where fuel pipe is installed.

Fuel System Pressure

RELIEVING

1. Before servicing the vehicle, refer to the precautions in the beginning of this section.
2. Remove the fuel filler cap.

3. Remove the fuel pump fuse and crank the engine until it stalls.
4. Disconnect the negative battery cable.
5. Replace the fuse.

Fuel Filter

The fuel filter is part of the fuel pump assembly located in the tank.

REMOVAL & INSTALLATION

1. Before servicing the vehicle, refer to the precautions in the beginning of this section.
2. Relieve the fuel system pressure.
3. Remove or disconnect the following:
   • Negative battery cable
   • Fuel pump connector
   • Fuel feed and return lines from the pump assembly
   • Pump cover screws
   • Pump assembly from the tank
   • Filter from the pump assembly

To install:
4. Install or connect the following:
   • Fuel filter
   • Pump into the tank
   • Pump cover screws
   • Fuel feed and return lines to the pump assembly
   • Fuel pump connector
   • Negative battery cable
5. Start the engine and check for leaks.

Fuel Pump

REMOVAL & INSTALLATION

1. Before servicing the vehicle, refer to the precautions in the beginning of this section.

2. Relieve the fuel system pressure.
3. Remove or disconnect the following:
   • Negative battery cable
   • Fuel pump connector
   • Fuel feed and return lines from the pump assembly
   • Pump cover screws
   • Pump assembly from the tank

To install:
4. Install or connect the following:
   • Pump into the tank
   • Pump cover screws
   • Fuel feed and return lines to the pump assembly
   • Fuel pump connector
   • Negative battery cable
5. Start the engine and check for leaks.

Fuel Injector

REMOVAL & INSTALLATION

1. Before servicing the vehicle, refer to the precautions in the beginning of this section.
2. Relieve the fuel system pressure.
3. Remove or disconnect the following:
   • Negative battery cable
   • Air breather hose from the throttle body
   • Throttle cable
   • Engine coolant hose and throttle body
   • Positive Crankcase Ventilation (PCV) valve and brake booster vacuum hose
   • Vacuum hose connector
   • Injector cover
   • High pressure fuel hose
   • Fuel injector harness connector
   • Delivery pipe with the injectors and

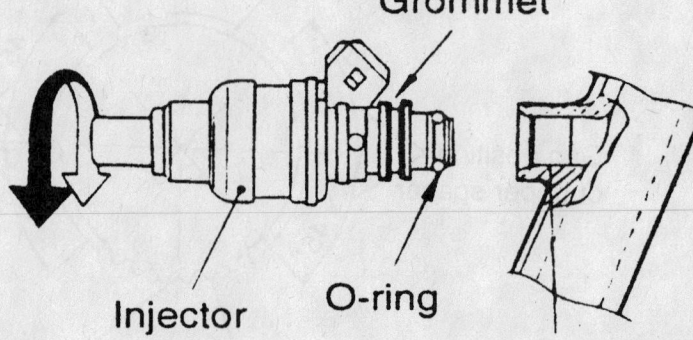

Install the injector using a twisting motion

9355LG39

the pressure regulator as an assembly
- Injector from the delivery pipe
- Injector O-ring, grommet and discard

To install:

4. Install a new grommet and O-ring
5. Apply a coating of spindle oil or gasoline to the injector O-ring
6. Install the injector into the delivery pipe while turning the injector left and right

making sure the injector turns smoothly. If the injector does not turn smoothly check for a jammed O-ring, remove the injector and reinsert it again.

7. Install or connect the following:
- Delivery pipe and injector assembly
- Intake manifold stay and tighten the bolts to 13–18 ft. lbs. (18–25 Nm)
- Fuel injector harness connector
- High pressure fuel hose
- Injector cover

- Vacuum hose connector
- PCV valve and brake booster vacuum hose
- Throttle body and engine coolant hose
- Throttle cable
- Air breather hose from the throttle body
- Negative battery cable

8. Start the engine and check for proper operation.

DRIVE TRAIN

Transaxle Assembly

REMOVAL & INSTALLATION

Manual

1. Before servicing the vehicle, refer to the precautions in the beginning of this section.
2. Drain the transaxle.
3. Remove or disconnect the following:
- Negative battery cable
- Air cleaner duct
- Air cleaner and air flow hose
- Back-up light connector
- Clutch line and clip
- Clutch release cylinder
- Speedometer cable
- Gear select or shift cables
- Starter motor mounting bolts
- Transaxle upper bolts

4. Attach an engine support fixture to the engine.
- Transaxle mounting bracket and insulator
- Front wheels
- Tie rod end, lower ball joint and drive shaft
- Gear box u-joint bolt and the return tube mounting bolts
- Front muffler
- Sub-frame mounting bolts and the frame
- Transaxle front and rear mounting brackets
- Transaxle side mounting bolts

5. Using a suitable jack, remove the transaxle from the vehicle.

To install:

6. Installation is the reverse of removal, keeping in mind the following torques:
- Transaxle case bolts to 15–20 ft. lbs. (20–27 Nm)
- Transaxle mounting sub-bracket nut 43–58 ft. lbs. (60–80 Nm)
- Transaxle mounting bracket bolts 29–40 ft. lbs. (40–55 Nm)

- Transaxle mounting insulator bolt 65–80 ft. lbs. (90–110 Nm)
- Clutch release cylinder retainers to 11–16 ft. lbs. (15–22 Nm)

7. Fill the transaxle to the correct level.
8. Start the engine and check for leaks.
9. Check the wheel alignment and adjust as necessary.

Automatic

1. Before servicing the vehicle, refer to the precautions in the beginning of this section.
2. Drain the transaxle.
3. Remove or disconnect the following:
- Negative battery cable
- Air cleaner assembly
- Control cable
- Speedometer sensor connector
- Transaxle range switch, solenoid, and oil temperature sensor connectors
- Oil cooler hose

4. Attach an engine support fixture to the engine.
- Gear box, stabilizer bar, tie rod end, lower ball joint and drive shaft
- Gear box u-joint bolt and the return tube mounting bolts
- Sub-frame mounting bolts and the frame
- Starter motor
- Transaxle mounting bolts

5. Using a suitable jack, remove the transaxle from the vehicle.

To install:

6. Installation is the reverse of removal, keeping in mind the following torques:
- Transaxle case bolts to 15–20 ft. lbs. (20–27 Nm)
- Transaxle mounting sub-bracket nut 43–58 ft. lbs. (60–80 Nm)
- Transaxle mounting bracket bolts 29–40 ft. lbs. (40–55 Nm)
- Transaxle mounting insulator bolt 65–80 ft. lbs. (90–110 Nm)

7. Adjust the control cable as follows:

a. Move the shift lever and the transaxle range switch to the **N** position and install the cable.

b. When attaching the control cable to the bracket, make sure the clip is installs so it contacts the cable.

c. Adjust the nut to remove any free-play in the cable and make sure the lever moves freely.

8. Fill the transaxle to the correct level.
9. Start the engine and check for leaks.
10. Check the wheel alignment and adjust as necessary.

Transfer Case

REMOVAL & INSTALLATION

1. Before servicing the vehicle, refer to the precautions in the beginning of this section.
2. Attach an engine support fixture to the engine.
3. Drain the transfer case fluid.
4. Remove or disconnect the following:
- Negative battery cable
- Transfer case and sub-frame mounting brackets
- Wheels
- Front muffler
- Steering tube from the sub-frame
- Sub-frame assembly

5. Support the transfer case with a suitable jack.
- Driveshafts
- Transfer case bolts and the case assembly

6. Installation is the reverse of removal

Clutch

ADJUSTMENTS

1. Before servicing the vehicle, refer to the precautions in the beginning of this section.

2. Measure the clutch pedal height from the face of the pedal pad to the floorboard. The proper measurement 218.9mm.

3. Measure the clutch pedal clevis pin play. The measurement should be 0.04–0.11 inch (1–3mm).

4. Adjust the height and clevis pin if necessary as follows:

   a. Turn and adjust the bolt, then secure it by tightening the lock nut.

➡ **After adjustment, tighten the bolt until it reaches the pedal stopper and tighten the lock nut.**

   b. Turn the push rod until the proper specification is reached and tighten the lock nut.

➡ **When adjusting the clevis pin play or the pedal height, make sure not to push the push rod towards the master cylinder.**

   c. After the adjustments are made, check that the clutch pedal free play is 0.2–0.5 inch (6–13mm). The free play is measured at the face of the pedal pad

5. If the adjustments do not bring the pedal into specifications, check the hydraulic system for air or a faulty component.

REMOVAL & INSTALLATION

1. Before servicing the vehicle, refer to the precautions in the beginning of this section.

2. Disconnect the negative battery cable.

3. Remove the transmission assembly.

4. If the pressure plate is attached to the flywheel, remove the release bearing using snap-ring pliers as follows:

   a. Insert the snap-ring pliers under the wave washer and place it in the center of the snap-ring

   b. Spread the snap-ring by pushing down on the bearing assembly.

   c. Once the snap-ring is fully expanded, remove the release bearing

5. Install a clutch disc alignment tool.

6. Remove pressure plate retaining bolts using a star pattern in several passes.

7. Remove the pressure plate with clutch disc.

To install:

8. Apply multi-purpose grease to clutch splines.

9. Install or connect the following:
   - Clutch plate to the flywheel
   - Pressure plate and cover. Do not torque the bolts at this time.

10. Align the bearing to the release fork and install it until it is fully engaged.

11. Torque the pressure plate bolts using a star pattern torque sequence to 11–16 ft. lbs. (15–22 Nm).

12. Install transmission.

13. Connect the negative battery cable.

Hydraulic Clutch System

BLEEDING

Bleeding air from the hydraulic clutch system is necessary whenever any part of the system has been disconnected or the fluid level (in the reservoir) has been allowed to fall so low that air has been drawn into the master cylinder.

❄ WARNING

NEVER use fluid that has been bled from a clutch system to fill the master cylinder reservoir, as it may be aerated, contain excessive moisture and/or be contaminated in some other way.

1. Before servicing the vehicle, refer to the precautions in the beginning of this section.

2. Fill the clutch master cylinder reservoir with new hydraulic clutch fluid.

3. Attach a hose to the bleeder on the clutch actuator and submerge the other end of the hose in a container of hydraulic clutch fluid.

4. Have an assistant slowly depress and hold the clutch pedal.

5. Loosen the bleeder to purge air.

6. Tighten the bleeder.

7. Repeat the above 3 steps until all air is completely purged from the system.

8. Refill the clutch master cylinder reservoir.

Halfshaft

REMOVAL & INSTALLATION

Front

1. Before servicing the vehicle, refer to the precautions in the beginning of this section.

2. Drain the transaxle.

3. Remove or disconnect the following:
   - Negative battery cable
   - Aluminum wheel cover
   - Split pin and halfshaft nut
   - Wheel Speed Sensor (WSS) from the bracket, if equipped

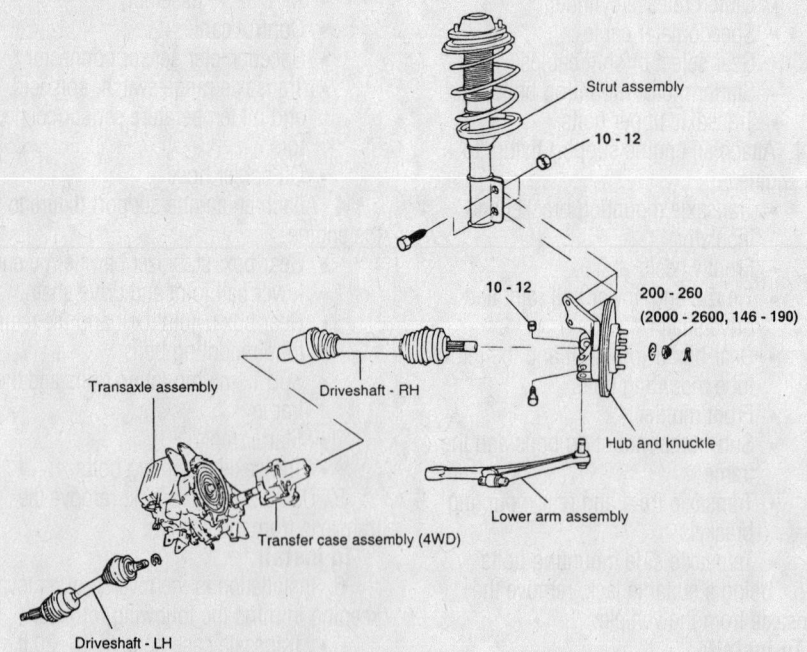

TORQUE : Nm (kg·cm, lb·ft)

9355LG40

Exploded view of the halfshaft mounting and related components

- Brake hose from the bracket
- Knuckle from the strut by removing the flange bolts
- Halfshaft from the hub by tapping the end with a plastic mallet
- Halfshaft from the transaxle using a pry bar.

4. Insert a plug in the transaxle opening.

To install:

➡**Use new circlips, split pins and self-locking nuts for assembly.**

5. Remove the plug from the transaxle opening.

6. Coat the halfshaft splines and sliding surfaces with gear oil.

7. Make sure the gap of the circlip is facing downwards.

8. Install the inner CV-joint to the transaxle until the circlip locks in the retaining groove. Pull on the shaft by hand to make sure it is properly engaged.

9. Install or connect the following:
- Halfshaft into the hub
- Knuckle to the strut
- Flange bolts and tighten to 74–88 ft. lbs. (100–120 Nm)
- Halfshaft nut and tighten to 146–190 ft. lbs. (200–260 Nm)
- New split pin
- Brake hose to the bracket
- WSS to the bracket, if equipped
- Aluminum wheel cover
- Negative battery cable

10. Fill the transaxle to the correct level and check for leaks.

Rear

1. Before servicing the vehicle, refer to the precautions in the beginning of this section.

2. Remove or disconnect the following:
- Rear wheel
- Split pin and nut
- Spare tire and support the hanger of the main muffler to avoid interfering with the carrier during right hand shaft removal

3. Matchmark the propeller rubber coupling and differential flange, then remove the bolts and nuts.

4. Support the differential carrier with a jack and remove the differential carrier mounting nuts and bolts.
- Shaft from the carrier by inserting a prybar between the carrier and the shaft
- Differential carrier to the rear after lowering the jack
- Shaft from the axle hub using a plastic mallet

- Shaft from the vehicle

To install:

5. Install or connect the following:
- Shaft into the axle hub
- Differential carrier to the rear using a jack
- Shaft to the carrier
- Differential carrier mounting nuts and bolts. Tighten the carrier nuts and bolts to 51–58 ft. lbs. (70–80 Nm) and the carrier rear bracket bolt to 58–73 ft. lbs. (80–100 Nm).
- Propeller shaft
- Spare tire and remove the support from the main muffler hanger
- New shaft nut and tighten to 146–253 ft. lbs. (200–260 Nm)
- New split pin
- Rear wheel

CV-Joint

OVERHAUL

DOJ-BJ Type

The DOJ-BJ type joint is used on both the front and rear of the vehicle and the following overhaul procedure is used.

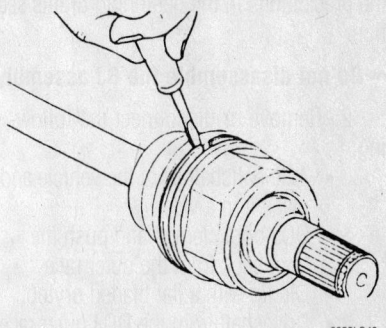

Remove the boot clamps from the DOJ

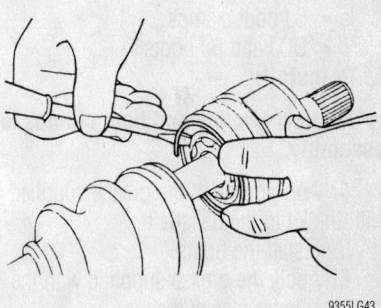

Remove the circlip with a suitable tool

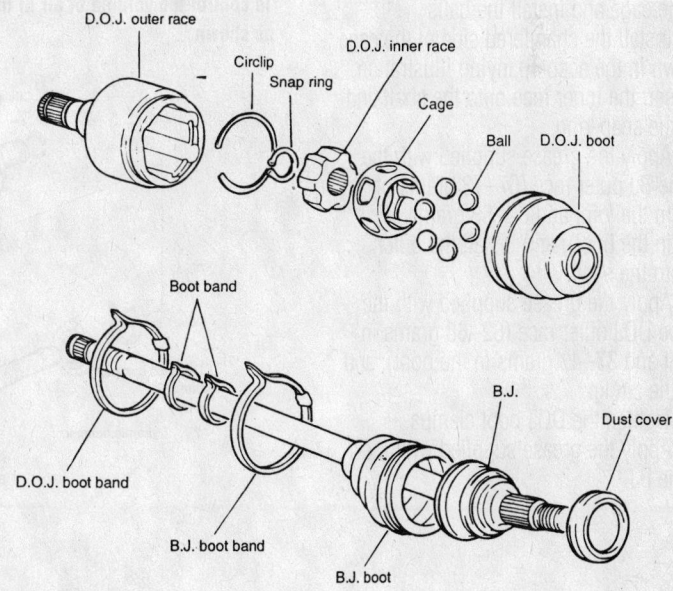

D.O.J. outer race — Circlip — Snap ring — D.O.J. inner race — Cage — Ball — D.O.J. boot

Boot band — B.J. — Dust cover

D.O.J. boot band — B.J. boot band — B.J. boot

TORQUE : Nm (kg·cm, lb·ft)

Exploded view of the DOJ-BJ type CV-joint assembly

9355LG41

1. Before servicing the vehicle, refer to the precautions in the beginning of this section.

➡ **Do not disassemble the BJ assembly.**

2. Remove or disconnect the following:

- Axle halfshaft from the vehicle and place it in a vise
- DOJ boot clamps and push the boot back from the outer race
- Circlip with a flat bladed prytool
- Driveshaft from the DOJ outer race
- Snap-ring and take out the inner race, cage and balls as an assembly

3. Clean the outer race, cage and balls without disassembling them.

- BJ boot clamps
- DOJ and BJ boots

To install:

➡ **Use new circlips and boot clamps for assembly.**

4. Apply some of the grease supplied with the kit to the halfshaft..

5. Install the boots.

6. Apply the grease supplied with the kit to the inner race and cage.

7. Install the cage so that it is offset on the race.

8. Apply the grease supplied with the kit to the cage and install the balls

9. Install the chamfered side of the cage as shown in the accompanying illustration, then insert the inner race onto the shaft and install the snap-ring.

10. Apply the grease supplied with the kit to the BJ outer race (67–73 grams of grease in the joint and 62–68 grams of grease in the boot); and install the outer race onto the shaft.

11. Apply the grease supplied with the kit to the DOJ outer race (62–68 grams in the joint and 37–43 grams in the boot); and install the circlip.

12. Tighten the DOJ boot clamps.

13. Apply the grease supplied with the kit to the BJ.

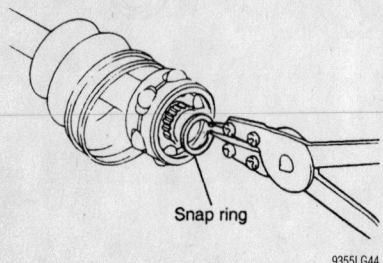

Snap ring

9355LG44

Remove snap-ring and take out the inner race, cage and balls as an assembly

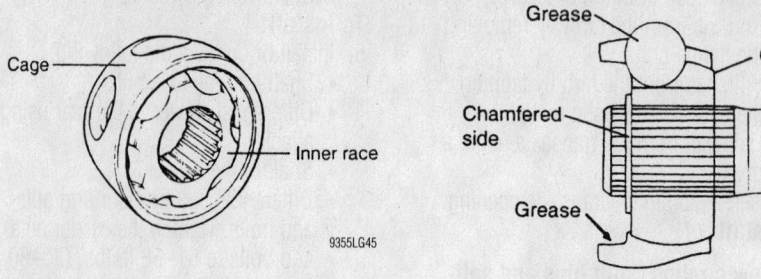

Cage

Inner race

9355LG45

Install the cage so that it is offset on the race

Grease

Grease

Chamfered side

Grease

9355LG46

Install the chamfered side of the cage as shown

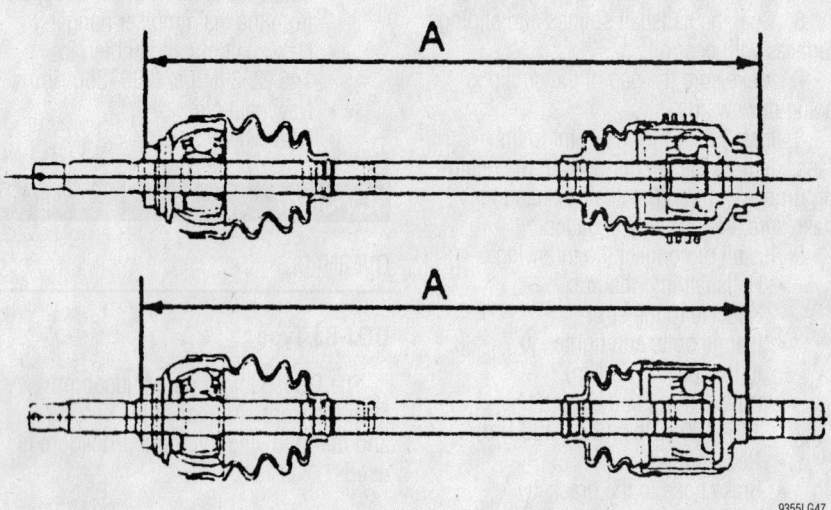

A

A

9355LG47

To control the volume of air in the DOJ boot, make sure the distance between the boot bands is as shown

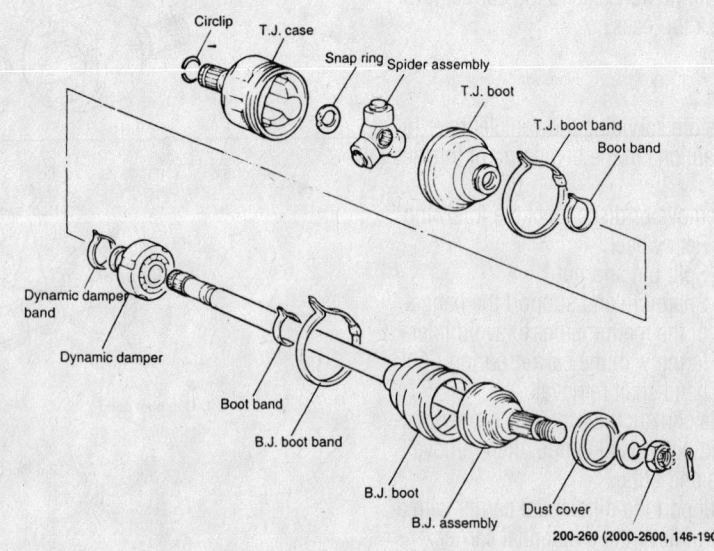

Circlip

T.J. case

Snap ring

Spider assembly

T.J. boot

T.J. boot band

Boot band

Dynamic damper band

Dynamic damper

Boot band

B.J. boot band

B.J. boot

B.J. assembly

Dust cover

200–260 (2000–2600, 146–190)

TORQUE : Nm (kg·cm, lb·ft)

9355LG48

Exploded view of the TJ-BJ type CV assembly

14. Install the boots.
15. Tighten the BJ boot clamps

➡**To control the volume of air in the DOJ boot, make sure the distance between the boot bands is as shown in the accompanying illustration.**

16. Install the axle halfshaft to the vehicle.

TJ-BJ Type

1. Before servicing the vehicle, refer to the precautions in the beginning of this section.

2. Remove or disconnect the following:
• Axle halfshaft from the vehicle.
• TJ boot clamps
• TJ boot from the TJ case
• Snap-ring and spider assembly
• BJ boot clamps, then pull out the TJ boot and the BJ boot

3. Clean all the components properly

To install:

➡**Use new circlips and boot clamps for assembly.**

4. Apply the grease supplied with the kit to the shaft and install the boots.

5. If installing the dynamic damper, keep the BJ shaft in a straight line and attach the damper in the direction shown in the accompanying illustration, then install the boot clamp.

6. Apply the grease supplied with the

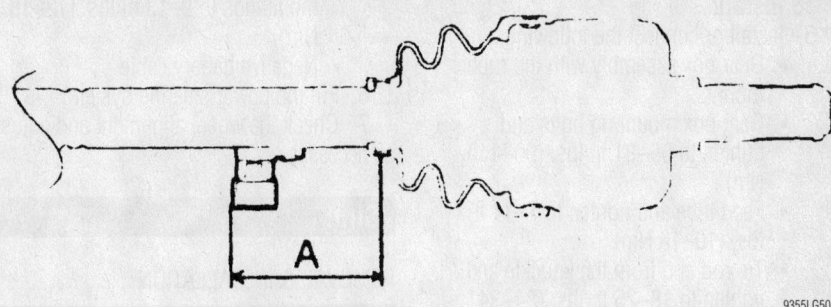

9355LG50

If installing the dynamic damper, keep the BJ shaft in a straight line and attach the damper in the direction shown

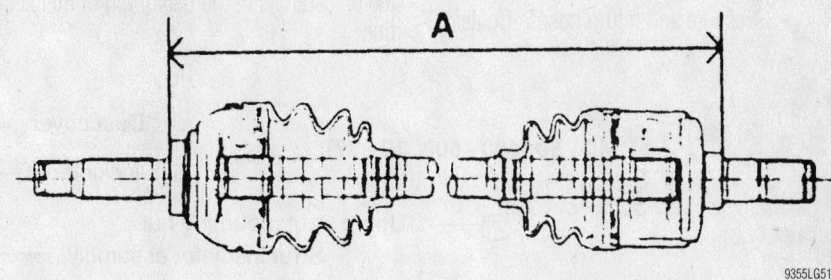

9355LG51

To control the volume of air in the TJ boot, make sure the distance between the boot bands is as shown

kit to the TJ boot (97–103 grams in the joint and 42–48 grams in the boot), then install the boot.

7. Tighten the TJ boot clamps.

8. Add the grease supplied with the kit to the BJ as much as was wiped out during cleaning and inspection.

9. Install the boots.
10. Tighten the BJ boot clamps.

➡**To control the volume of air in the TJ boot, make sure the distance between the boot bands is as shown in the accompanying illustration**

STEERING AND SUSPENSION

Air Bag

✳✳ CAUTION

Some vehicles are equipped with an air bag system. The system must be disarmed before performing service on, or around, system components, the steering column, instrument panel components, wiring and sensors. Failure to follow the safety precautions and the disarming procedure could result in accidental air bag deployment, possible injury and unnecessary system repairs.

PRECAUTIONS

Several precautions must be observed when handling the inflator module to avoid accidental deployment and possible personal injury.
• Never carry the inflator module by the

wires or connector on the underside of the module.
• When carrying a live inflator module, hold securely with both hands, and ensure that the bag and trim cover are pointed away.
• Place the inflator module on a bench or other surface with the bag and trim cover facing up.
• With the inflator module on the bench, never place anything on or close to the module which may be thrown in the event of an accidental deployment.
Before servicing the vehicle, also make sure to refer to the precautions in the beginning of this section as well.

DISARMING

1. Turn the wheel to the straight ahead position.
2. Disconnect the negative battery cable and wait at least 30 seconds for the air bag energy to deplete.

3. After work has been completed, connect the battery cable.

Power Rack and Pinion Steering Gear

REMOVAL & INSTALLATION

1. Before servicing the vehicle, refer to the precautions in the beginning of this section.
2. Center the steering wheel and lock it in position.
3. Drain the power steering system.
4. Remove or disconnect the following:
• Negative battery cable
• Pressure and return hoses
• Joint assembly connecting bolt
• Tie rod end from the knuckle
• Feed tube
• Gear box mounting bolts
• Gear box assembly with the rubber mounts

To install:

5. Install or connect the following:
- Gear box assembly with the rubber mounts
- Gear box mounting bolts and tighten to 66–81 ft. lbs. (90–110 Nm)
- Feed tube and tighten to 7–11 ft. lbs. (10–16 Nm)
- Tie rod end from the knuckle and tighten to 18–25 ft. lbs. (24–34 Nm)
- Joint assembly connecting bolt and tighten to 11–14 ft. lbs. (15–20 Nm)
- Pressure and return hoses. Tighten the fittings to 9–13 ft. lbs. (12–18 Nm).
- Negative battery cable

6. Fill the power steering system.

7. Check the wheel alignment and adjust as necessary.

Strut

REMOVAL & INSTALLATION

Front

1. Before servicing the vehicle, refer to the precautions in the beginning of this section.

2. Remove or disconnect the following:
- Front wheel
- Brake hose clip from the strut mounting bracket
- Wheel Speed Sensor (WSS) from the knuckle
- Stabilizer bar link
- Strut-to-knuckle bolts
- 3 upper mounting nuts
- Strut

To install:

3. Install or connect the following:
- Strut. Tighten the upper mount nuts to 30–37 ft. lbs. (40–50 Nm).
- Strut-to-knuckle bolts and tighten to 74–88 ft. lbs. (100–120 Nm)

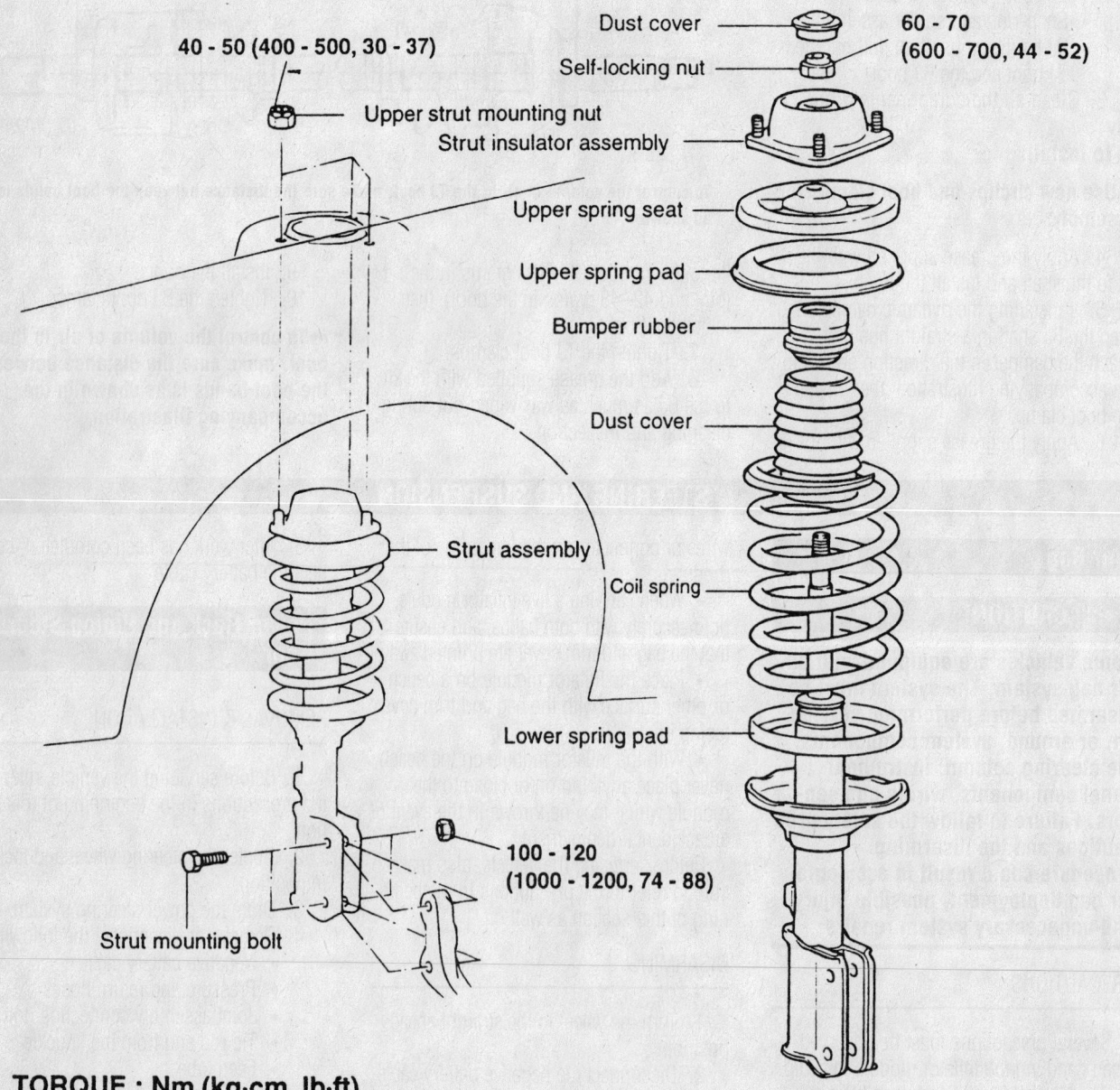

40 - 50 (400 - 500, 30 - 37)

Upper strut mounting nut

Dust cover   60 - 70 (600 - 700, 44 - 52)

Self-locking nut

Strut insulator assembly

Upper spring seat

Upper spring pad

Bumper rubber

Dust cover

Strut assembly

Coil spring

Lower spring pad

Strut mounting bolt   100 - 120 (1000 - 1200, 74 - 88)

TORQUE : Nm (kg·cm, lb·ft)

Exploded view of the front strut assembly and mounting

9355LG52

- Stabilizer bar link and tighten to 30–37 ft. lbs. (30–50 Nm)
- WSS from the knuckle
- Brake hose clip to the strut mounting bracket
- Front wheel

4. Check the wheel alignment and adjust as necessary.

Shock Absorber

REMOVAL & INSTALLATION

Rear

1. Before servicing the vehicle, refer to the precautions in the beginning of this section.
2. Support the vehicle under the lower control arm.
3. Remove or disconnect the following:
   - Rear wheel
   - Upper shock absorber upper nut
   - Lower shock absorber nut
   - Shock absorber

To install:

4. Install or connect the following:
   - Shock absorber. Tighten the upper

nut to 15–22 ft. lbs. (20–30 Nm) and the lower nut to 104–118 ft. lbs. (140–160 Nm).
   - Rear wheel

Coil Spring

REMOVAL & INSTALLATION

Front

1. Before servicing the vehicle, refer to the precautions in the beginning of this section.
2. Remove the strut from the vehicle and install in a strut spring compressor. Compress the spring until the end of the spring comes away from the spring seat.
3. Remove the upper strut mount, spring seat and related components.
4. Remove the coil spring from the strut spring compressor.

To install:

➡**Use a new self-locking nut.**

5. Compress the spring and position the strut so that the end of the spring aligns with the notch in the spring seat.

6. Install the upper strut mounting components and tighten the nut to 44–52 ft. lbs. (60–70 Nm).
7. Install the strut to the vehicle.
8. Check the wheel alignment and adjust as necessary.

Rear

1. Before servicing the vehicle, refer to the precautions in the beginning of this section.
2. Support the vehicle under the lower control arm.
3. Remove or disconnect the following:
   - Rear wheel
   - Flange nut and brake caliper assembly
   - Parking brake assembly
   - Wheel Speed Sensor (WSS) and the parking brake cable
   - Rear shock assembly

4. Lower the jack assembly and remove the spring.

To install:

5. Install or connect the following:
   - Spring and raise the jack into position
   - Rear shock assembly
   - Parking brake cable and WSS
   - Parking brake assembly
   - Flange nut and brake caliper assembly
   - Rear wheel

Ball Joint

REMOVAL & INSTALLATION

1. Before servicing the vehicle, refer to the precautions in the beginning of this section.
2. Remove the control arm.
3. Using tools 09551-3100 and 09216-21100, remove the bushing.

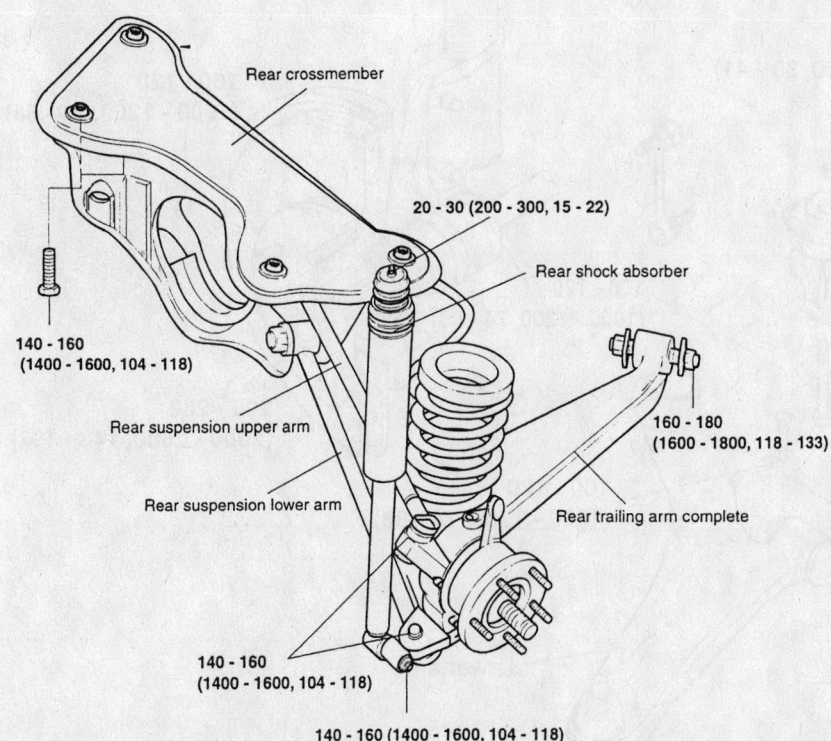

Rear crossmember

20 - 30 (200 - 300, 15 - 22)

Rear shock absorber

140 - 160 (1400 - 1600, 104 - 118)

Rear suspension upper arm

160 - 180 (1600 - 1800, 118 - 133)

Rear suspension lower arm

Rear trailing arm complete

140 - 160 (1400 - 1600, 104 - 118)

140 - 160 (1400 - 1600, 104 - 118)

TORQUE : Nm (kg·cm, lb·ft)

9355LG53

Exploded view of the rear suspension assembly

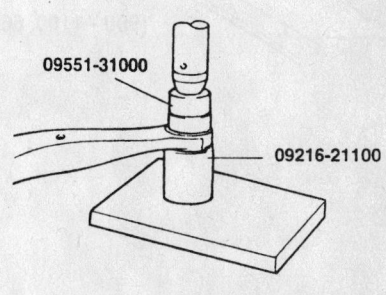

09551-31000

09216-21100

9355LG55

Removing the bushing from the control arm using the appropriate removal and installation tools

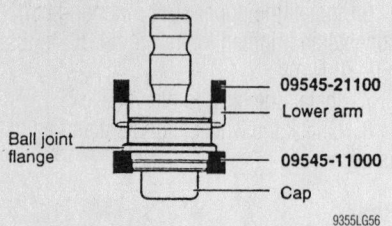

09545-21100
Lower arm

Ball joint flange

09545-11000
Cap

9355LG56

Installing the ball joint dust cover

4. Using a suitable prytool and remove the dust cover from the ball joint.

5. Remove the snap-ring.

6. Remove the ball joint from the arm using a plastic hammer.

To install:

7. Install the ball joint.

8. Install the bushings into the arm using the appropriate tools. Make sure that the ball joint flange is supported while pressing down on the bushing until the flange touches the arm surface.

9. Install the ball joint snap-ring. Be careful to keep the snap-ring expansion as small as possible during installation.

10. Apply multi-purpose grease to the dust cover lip and inside of the cover.

11. Using tool 09545-11000, install the dust cover until it is completely seated on the snap ring.

12. Install the control arm.

Lower Control Arm

REMOVAL & INSTALLATION

Front

1. Before servicing the vehicle, refer to the precautions in the beginning of this section.

2. Remove or disconnect the following:
   - Front wheel
   - Ball joint-to-knuckle bolt
   - Sub frame bolts and the frame
   - Lower arm bolts and the arm

3. Installation is the reverse of removal. Tighten the fasteners as follows:

   a. Tighten the arm bolt **A** to 74–88 ft. lbs. (100–120 Nm) and bolt **B** to 66–81 ft. lbs. (90–110 Nm).

   b. Tighten the sub frame bolts to 118–148 ft. lbs. (160–200 Nm).

   c. Tighten the ball joint-to-knuckle bolt to 74–88 ft. lbs. (100–120 Nm)

4. Check the wheel alignment and adjust as necessary.

Trailing Arm

REMOVAL & INSTALLATION

Rear

1. Before servicing the vehicle, refer to the precautions in the beginning of this section.

2. Support the vehicle under the lower control arm.

3. Remove or disconnect the following:
   - Rear wheel
   - Flange nut and brake caliper assembly
   - Parking brake assembly
   - Wheel Speed Sensor (WSS) and the parking brake cable
   - Rear shock assembly
   - Spring
   - Rear driveshaft from the rear axle
   - Upper and lower arm using tool 09517-43001.
   - Trailing arm bolt and the arm

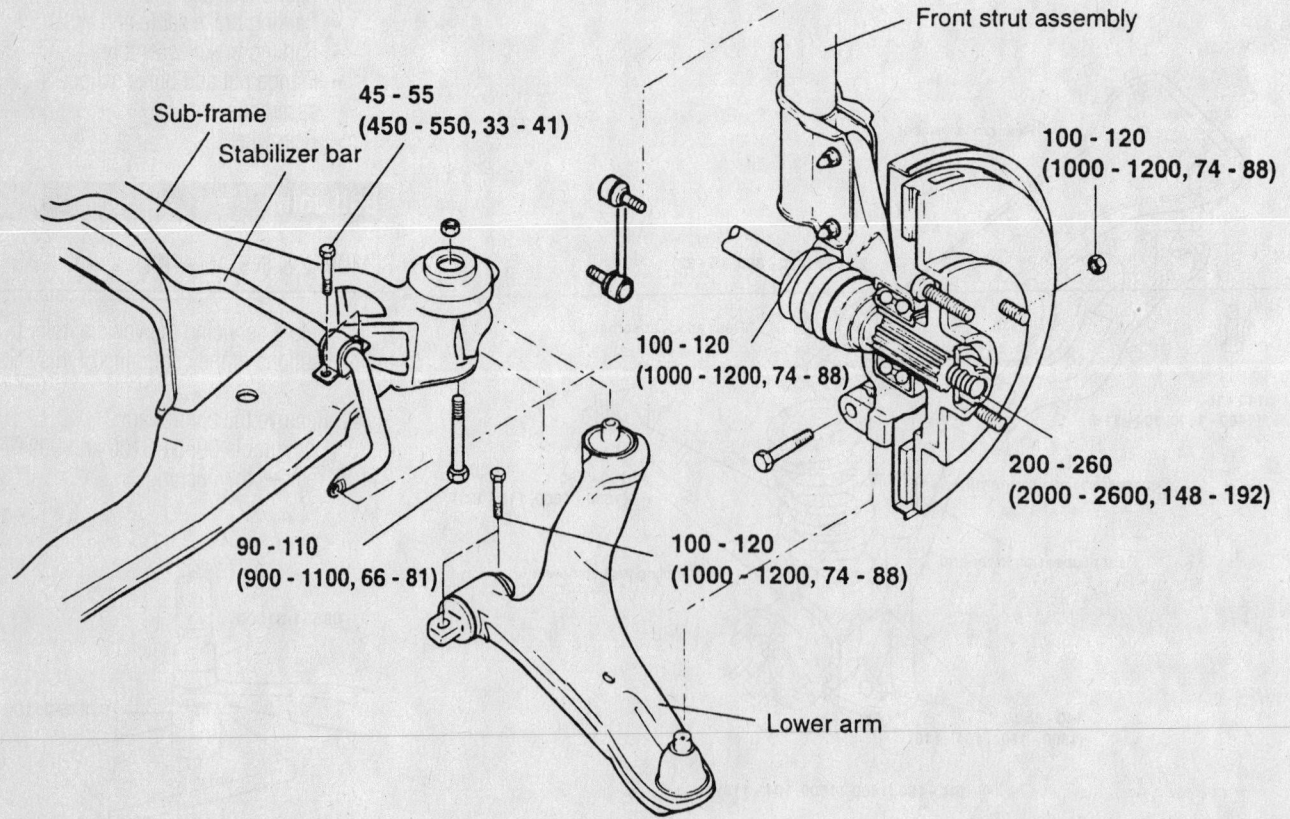

Front strut assembly

Sub-frame

Stabilizer bar

45 - 55
(450 - 550, 33 - 41)

100 - 120
(1000 - 1200, 74 - 88)

100 - 120
(1000 - 1200, 74 - 88)

200 - 260
(2000 - 2600, 148 - 192)

90 - 110
(900 - 1100, 66 - 81)

100 - 120
(1000 - 1200, 74 - 88)

Lower arm

TORQUE : Nm (kg·cm, lb·ft)

Lower control arm assembly

9355LG54

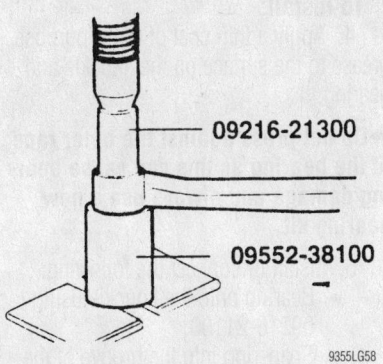

9355LG58

Using tools 09216-21300 and 09552-38100, press fit the trailing arm bushing

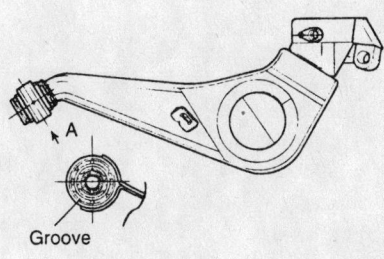

Groove

9355LG59

Position the groove in the arm bushing so that it is aligned as shown, before pressing the bushing into position

To install:
4. Install or connect the following:
- Trailing arm and tighten the trailing arm bolt to 118–133 (160–180 Nm)
- Upper and lower arm. Tighten the ball joint nut to 104–11 ft. lbs. (140–118 Nm).
- Spring and raise the jack into position
- Rear shock assembly
- Parking brake cable and WSS
- Parking brake assembly
- Flange nut and brake caliper assembly
- Rear wheel

TRAILING ARM BUSHING REPLACEMENT

1. Before servicing the vehicle, refer to the precautions in the beginning of this section.
2. Remove the trailing arm.
3. Press the bushing from the trailing arm.

➡**Position the groove in the arm bushing so that it is aligned as shown in the accompanying illustration, then press fit the bushing.**

4. Using tools 09216-21300 and 09552-38100, press fit the bushing.

Wheel Bearings

ADJUSTMENT

The wheel bearings are sealed units and are not adjustable.

REMOVAL & INSTALLATION

Front

1. Before servicing the vehicle, refer to the precautions in the beginning of this section.
2. Remove or disconnect the following:
- Front wheel
- Wheel Speed Sensor (WSS) from the knuckle
- Brake caliper and suspend it aside using wire
- Split pin and nut from the axle

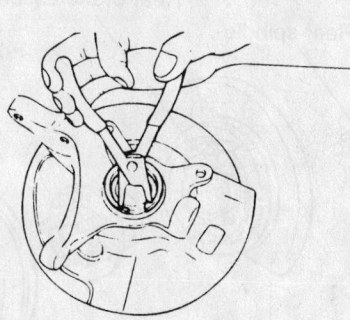

9355LG60

Remove the snap-ring from the hub

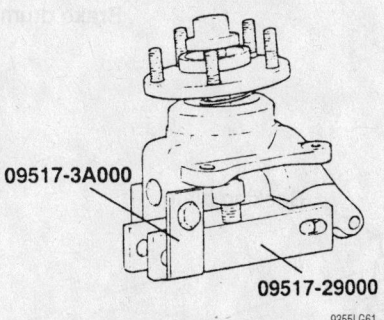

09517-3A000

09517-29000

9355LG61

Remove the hub from the knuckle

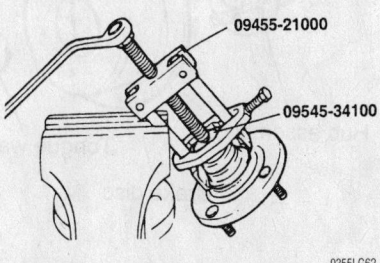

09455-21000

09545-34100

9355LG62

Remove the wheel bearing inner race from the hub

- Strut from the knuckle
- Tie rod end from the knuckle
- Lower ball joint bolt
- Axle shaft from the knuckle using a plastic hammer

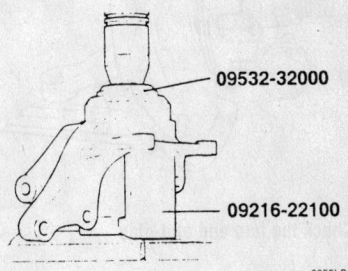

09532-32000

09216-22100

9355LG63

Remove the wheel bearing outer race from the knuckle

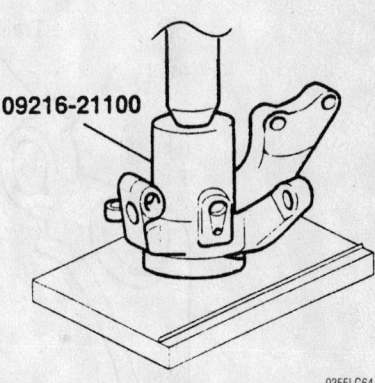

09216-21100

9355LG64

Install the bearing onto the knuckle

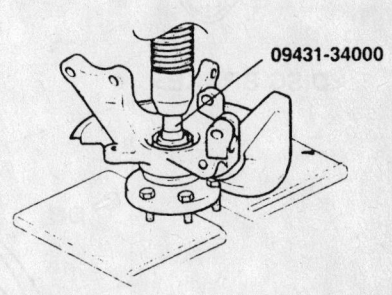

09431-34000

9355LG65

Press the hub onto the knuckle

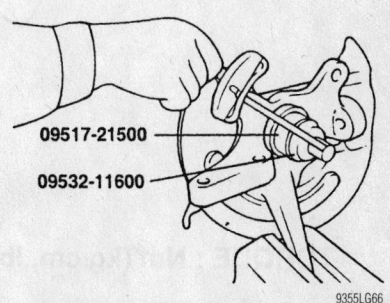

09517-21500

09532-11600

9355LG66

Check the wheel bearing starting torque

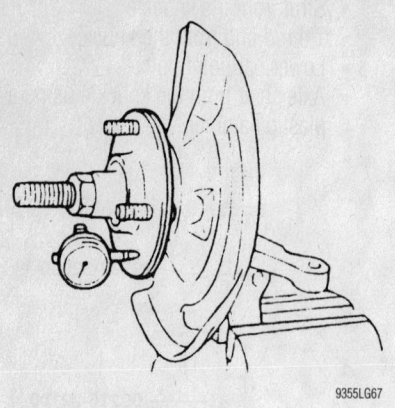

Check the hub end end-play

9355LG67

- Brake disc
- Knuckle assembly
- Snap-ring from the hub
- Hub from the knuckle by installing tools 09517-3A00 and 09517-2900, then tighten the nut of the tool to separate the tool from the knuckle
- Wheel bearing inner race from the hub using tools 09455-2100 and 09545-34100
- Wheel bearing outer race from the knuckle using tools 09532-3200 and 09216-22100

3. Check all components for wear or damage and replace as necessary.

To install:

4. Apply a thin coat of multi-purpose grease to the surface on the knuckle and bearing.

➡ **Do not press against the outer race of the bearing as this can cause bearing damage and always use a new bearing kit.**

5. Install or connect the following:
- Bearing onto the knuckle using tool 09216-21100.
- Snap-ring into the groove of the knuckle
- Backing plate onto the knuckle

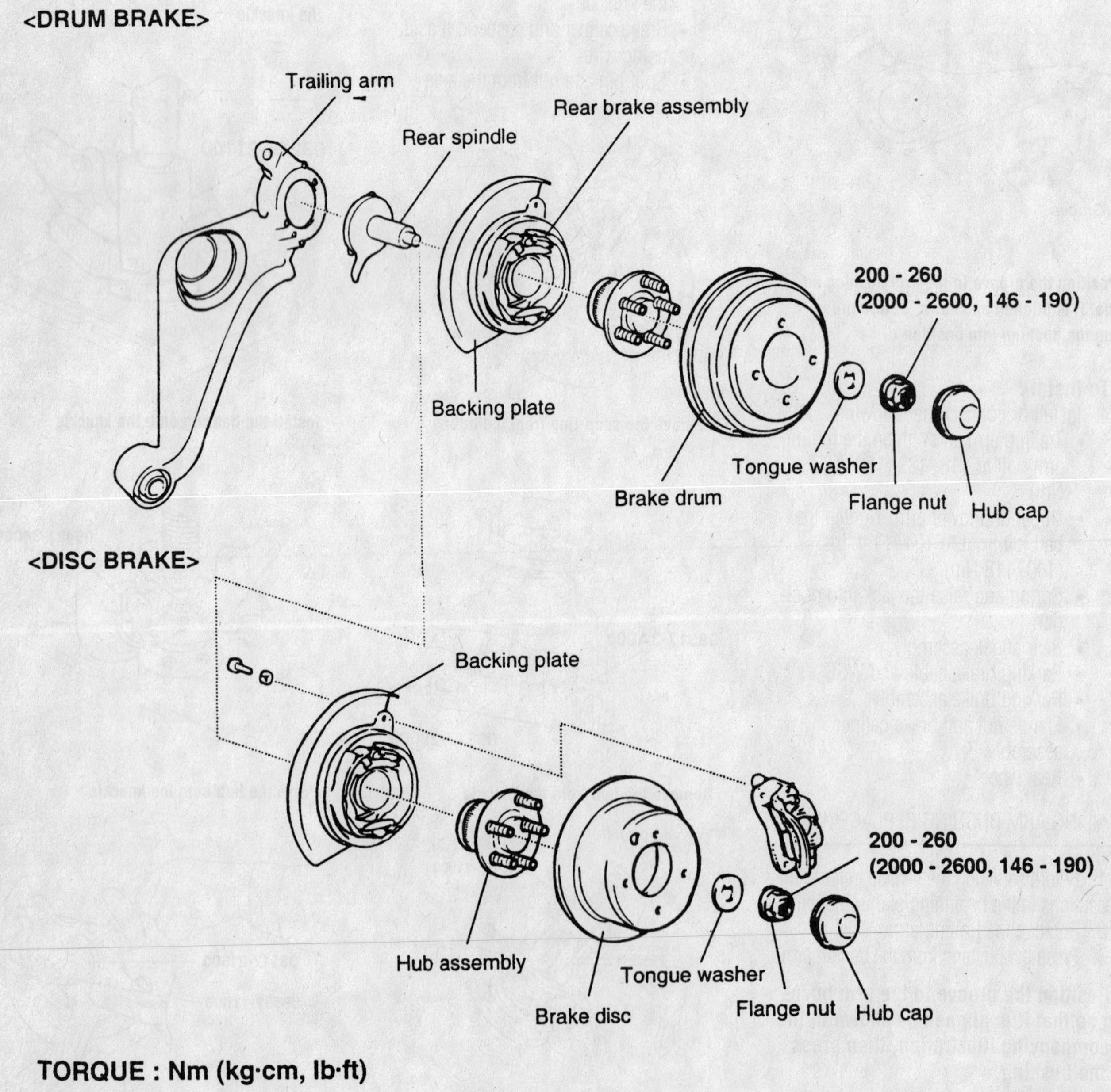

<DRUM BRAKE>

Trailing arm

Rear spindle

Rear brake assembly

Backing plate

200 - 260
(2000 - 2600, 146 - 190)

Brake drum

Tongue washer

Flange nut

Hub cap

<DISC BRAKE>

Backing plate

200 - 260
(2000 - 2600, 146 - 190)

Hub assembly

Brake disc

Tongue washer

Flange nut

Hub cap

TORQUE : Nm (kg·cm, lb·ft)

Exploded view of the rear hub assembly

9355LG68

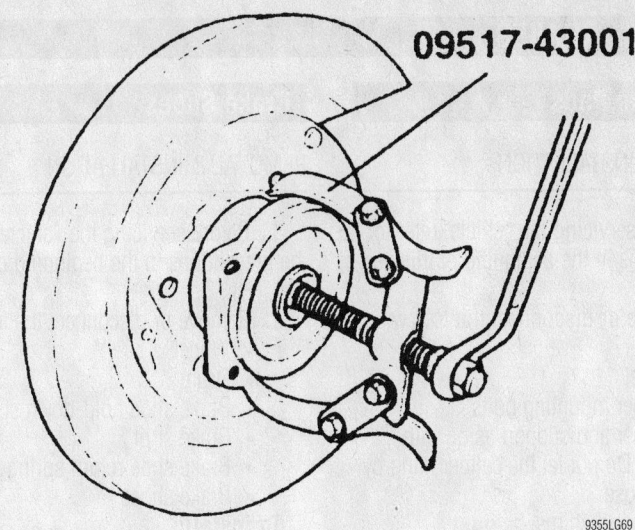

09517-43001

Removing the rear hub from housing

9355LG69

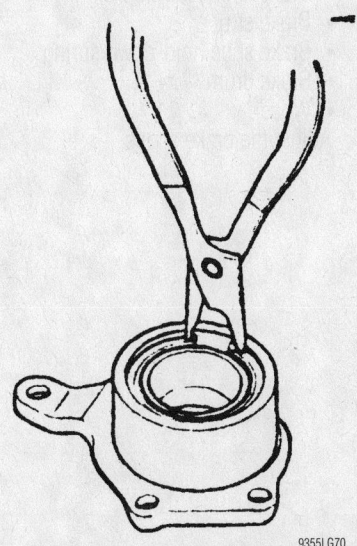

9355LG70

Removing the wheel bearing snap-ring

➡️**Do not press against the outer race of the bearing as this can cause bearing damage and always use a new bearing kit.**

• Hub onto the knuckle by pressing it into position using tool 09431-3400

6. Rotate the bearing several times to seat the bearing.

7. Measure the wheel bearing torque using an inch lb. torque wrench. The measurement is 16.64 inch lbs. (1.88 Nm).

8. Measure the end-play of the hub using a dial gauge. The specification is 0.003-0.008mm.

9. Install the remaining components in the reverse order of removal.

10. Check the wheel alignment and adjust as necessary.

Rear

1. Before servicing the vehicle, refer to the precautions in the beginning of this section.

2. Remove or disconnect the following:

- Rear wheel
- Flange nut and washer
- Drum or rotor
- Brake line
- Parking brake assembly
- Parking brake cable
- Spindle bolts and the spindle
- Rear hub from the housing using tool 09517-43001
- Wheel bearing snap-ring
- Wheel bearing inner race from the housing using tools 09500-2100, 09527-33000 and 09216-22100

3. Inspect the components for damage and replace as necessary.

To install:

4. Apply a thin coat of multi-purpose grease to the surface on the housing and bearing.

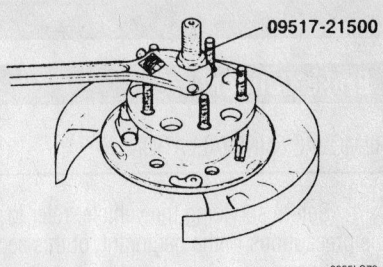

09517-21500

9355LG72

Press the hub onto the housing

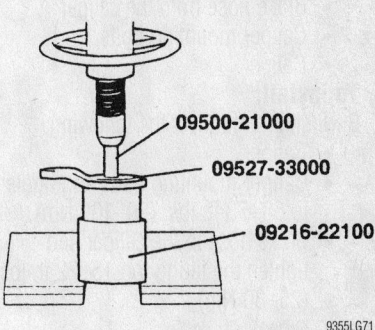

09500-21000

09527-33000

09216-22100

9355LG71

Removing the wheel bearing inner race from housing

➡️**Do not press against the outer race of the bearing as this can cause bearing damage and always use a new bearing kit.**

5. Install or connect the following:

- Bearing onto the spindle using tools 09216-21100 and 09532-3200
- Snap-ring
- Backing plate, then press the hub onto the housing using tool 09517-21500

6. Rotate the bearing several times to seat the bearing.

7. Measure the wheel bearing torque using an inch lb. torque wrench. The measurement is 16.64 inch lbs. (1.88 Nm).

8. Measure the end-play of the hub using a dial gauge. The specification is 0.003-0.008mm.

9. Install the remaining components in the reverse order of removal.

10. Check the wheel alignment and adjust as necessary.

BRAKES

Brake Caliper

REMOVAL & INSTALLATION

1. Before servicing the vehicle, refer to the precautions in the beginning of this section.
2. Remove or disconnect the following:
   - Wheel
   - Brake hose from the caliper
   - Caliper mounting bolts
   - Caliper

To install:

3. Install or connect the following:
   - Caliper
   - Caliper mounting bolts and tighten to 58–73 ft. lbs. (80–100 Nm)
   - Brake hose to the caliper and tighten the fitting too 18–22 ft. lbs. (25–30 Nm)
   - Wheel
4. Bleed the brake system.

Disc Brake Pads

REMOVAL & INSTALLATION

1. Before servicing the vehicle, refer to the precautions in the beginning of this section.
2. Remove or disconnect the following:
   - Wheel
   - Caliper mounting bolts
   - Caliper and support aside with wire. Do not let the caliper hang by the hose.
   - Pads and shims

To install:

3. Install or connect the following:
   - Pads, clips and shims.
4. Bottom the caliper piston using tool 09581-11000 or a C-clamp
   - Caliper mounting bolts and tighten to 16–24 ft. lbs. (22–32Nm)
   - Wheel

Brake Shoes

REMOVAL & INSTALLATION

1. Before servicing the vehicle, refer to the precautions in the beginning of this section.
2. Remove or disconnect the following:
   - Wheel
   - Drum
   - Brake shoe hold-down spring
   - Brake strut
   - Brake shoe return spring
   - Brake shoes

To install:

3. Install or connect the following:
   - Brake shoes
   - Brake shoe return spring
   - Brake strut
   - Brake shoe hold-down spring
   - Brake drum
   - Wheel
4. Adjust the brake shoes.

SPECIFICATION CHARTS

ENGINE AND VEHICLE IDENTIFICATION

| Engine | | | | | | | Model Year | |
Code ①	Liters (cc)	Cu. In.	Cyl.	Fuel Sys.	Engine Type	Eng. Mfg.	Code ②	Year
SR20DE	2.0 (1998)	122	4	MFI	DOHC	Nissan	Y	2000
VQ30DE	3.0 (2988)	182	6	MFI	DOHC	Nissan	1	2001
VQ35DE	3.5 (3498)	213	6	MFI	DOHC	Nissan	2	2002
VH41DE	4.1 (4130)	252	8	MFI	DOHC	Nissan	3	2003
VK45DE	4.5 (4494)	274	8	MFI	DOHC	Nissan	4	2004

MFI: Multi-port Fuel Injection

DOHC: Double Overhead Camshaft

① 4th digit of the Vehicle Identification Number (VIN)

② 10th digit of the Vehicle Identification Number (VIN)

42356-INFC-C01

GENERAL ENGINE SPECIFICATIONS

Year	Model	Engine Displacement Liters (cc)	Engine ID/VIN	Fuel System Type	Net Horsepower @ rpm	Net Torque @ rpm (ft. lbs.)	Bore x Stroke (in.)	Compression Ratio	Oil Pressure @ rpm
2000	G20	2.0 (1998)	SR20DE	MFI	140@6400	132@4800	3.39x3.39	9.5:1	46-57@3200
	I30	3.0 (2988)	VQ30DE	MFI	190@5600	205@4000	3.66x2.89	10.1:1	63-80@3000
	Q45	4.1 (4130)	VH41DE	MFI	266@5600	278@4000	3.66x2.99	10.2:1	67-81@3000
2001	G20	2.0 (1998)	SR20DE	MFI	140@6400	132@4800	3.39x3.39	9.5:1	46-57@3200
	I30	3.0 (2988)	VQ30DE	MFI	190@5600	205@4000	3.66x2.89	10.1:1	63-80@3000
	Q45	4.1 (4130)	VH41DE	MFI	266@5600	278@4000	3.66x2.99	10.2:1	67-81@3000
2002	G20	2.0 (1998)	SR20DE	MFI	140@6400	132@4800	3.39x3.39	9.5:1	46-57@3200
	G35	3.5 (3498)	VQ35DE	MFI	260@6000	260@4800	3.76X3.20	10.3:1	43@2000
	I35	3.5 (3498)	VQ35DE	MFI	255@5800	246@4400	3.76X3.20	10.3:1	43@2000
	Q45	4.5 (4494)	VK45DE	MFI	340@6400	333@4000	3.66x3.25	10.5:1	43@3000
2003	G20	2.0 (1998)	SR20DE	MFI	140@6400	132@4800	3.39x3.39	9.5:1	46-57@3200
	G35	3.5 (3498)	VQ35DE	MFI	260@6000	260@4800	3.76X3.20	10.3:1	43@2000
	I35	3.5 (3498)	VQ35DE	MFI	255@5800	246@4400	3.76X3.20	10.3:1	43@2000
	Q45	4.5 (4494)	VK45DE	MFI	340@6400	333@4000	3.66x3.25	10.5:1	43@3000

MFI: Multi-port Fuel Injection

42356-INFC-C02

GASOLINE ENGINE TUNE-UP SPECIFICATIONS

Year	Engine Displacement Liters (cc)	Engine ID/VIN	Spark Plug Gap (in.)	Ignition Timing (deg.)		Fuel Pump (psi) ①	Idle Speed (rpm)		Valve Clearance	
				MT	AT		MT	AT	Intake	Exhaust
2000	2.0 (1998)	SR20DE	0.031-0.035	15B	15B	34	800	800	HYD	HYD
	3.0 (2988)	VQ30DE	0.039-0.043	15B	15B	34	625	700	HYD	HYD
	4.1 (4130)	VH41DE	0.039-0.041	—	15B	34	—	650	HYD	HYD
2001	2.0 (1998)	SR20DE	0.031-0.035	15B	15B	34	800	800	HYD	HYD
	3.0 (2988)	VQ30DE	0.039-0.043	15B	15B	34	625	700	HYD	HYD
	4.1 (4130)	VH41DE	0.039-0.041	—	15B	34	—	650	HYD	HYD
2002	2.0 (1998)	SR20DE	0.031-0.035	15B	15B	34	800	800	HYD	HYD
	3.5 (3498)	VQ35DE	0.043	—	15B	34	—	600-700	HYD	HYD
	4.5 (4494)	VK45DE	0.039-0.041	—	15B	34	—	650	HYD	HYD
2003	2.0 (1998)	SR20DE	0.031-0.035	15B	15B	34	800	800	HYD	HYD
	3.5 (3498)	VQ35DE	0.043	—	15B	34	—	600-700	HYD	HYD
	4.5 (4494)	VK45DE	0.039-0.041	—	15B	34	—	650	HYD	HYD

NOTE: The Vehicle Emission Control Information label often reflects specification changes made during production. The label figures must be used if they differ from those in this chart.

B: Before top dead center

42356-INFC-C03

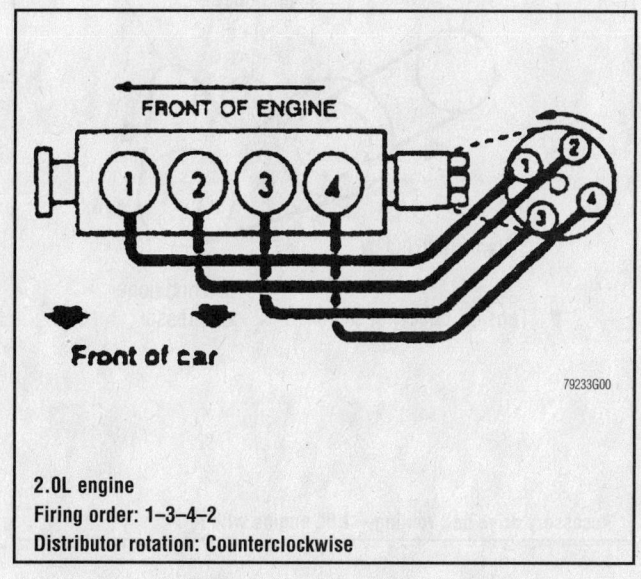

2.0L engine
Firing order: 1–3–4–2
Distributor rotation: Counterclockwise

79233G00

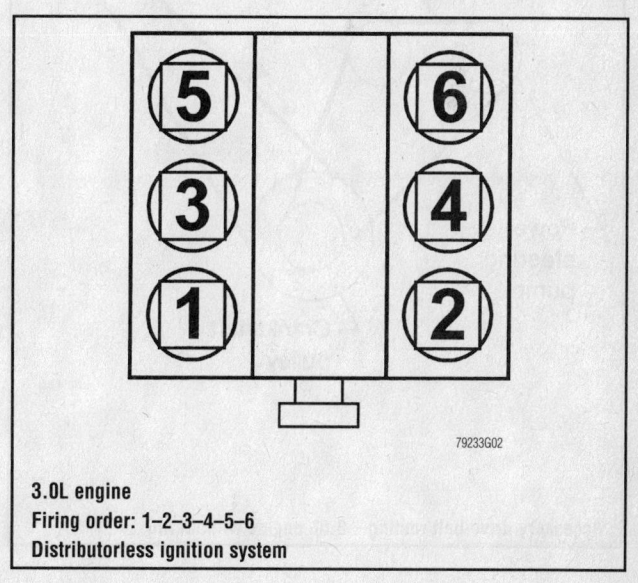

3.0L engine
Firing order: 1–2–3–4–5–6
Distributorless ignition system

79233G02

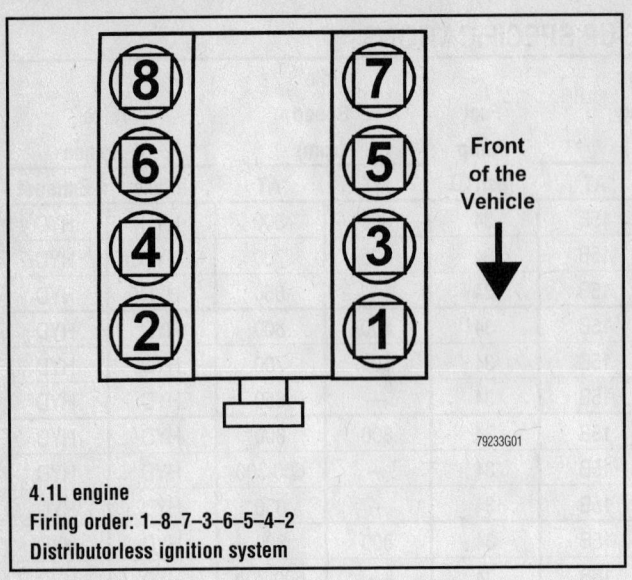

4.1L engine
Firing order: 1–8–7–3–6–5–4–2
Distributorless ignition system

79233G01

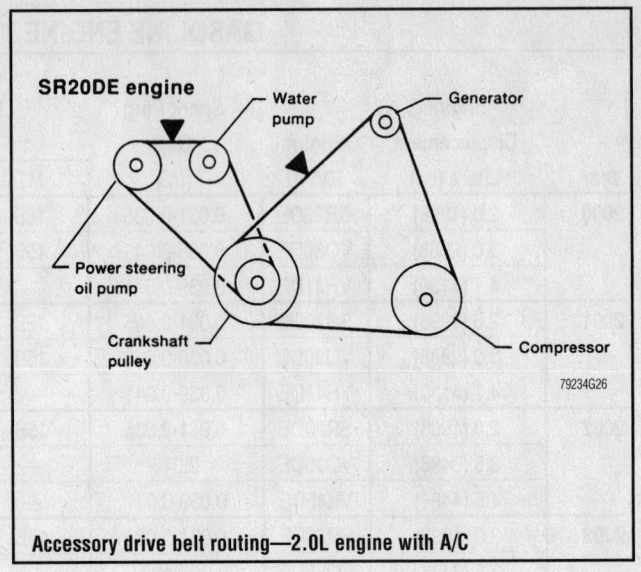

Accessory drive belt routing—2.0L engine with A/C

79234G26

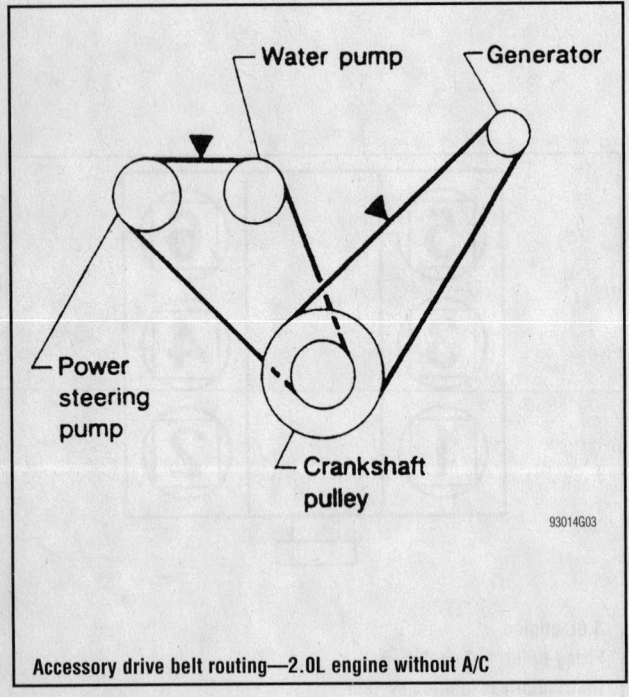

Accessory drive belt routing—2.0L engine without A/C

93014G03

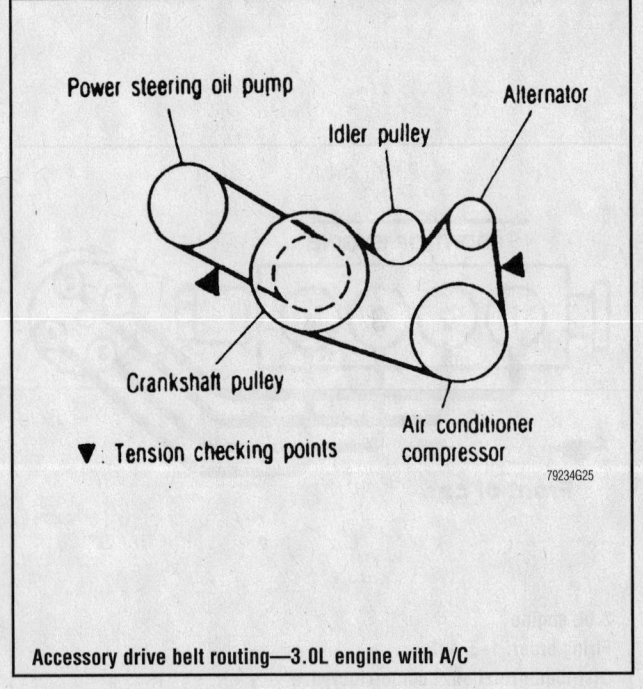

▼: Tension checking points

Accessory drive belt routing—3.0L engine with A/C

79234G25

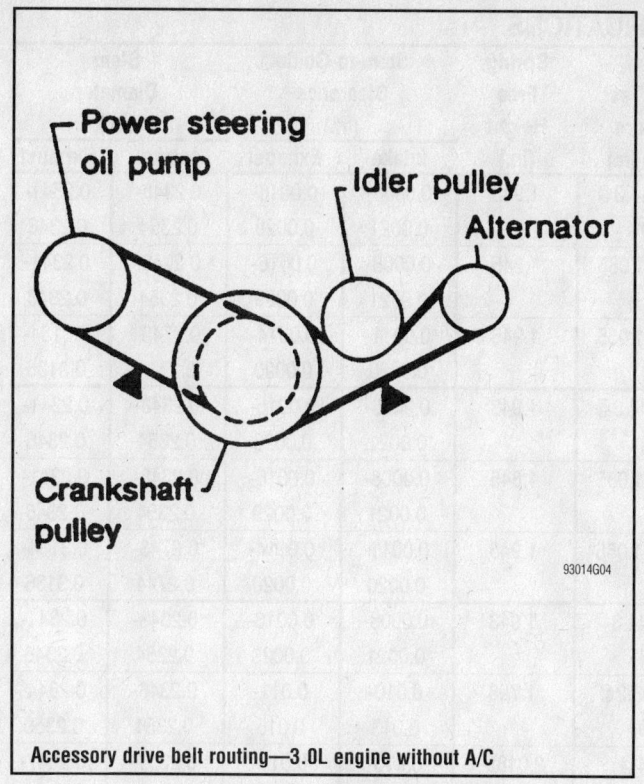

Accessory drive belt routing—3.0L engine without A/C

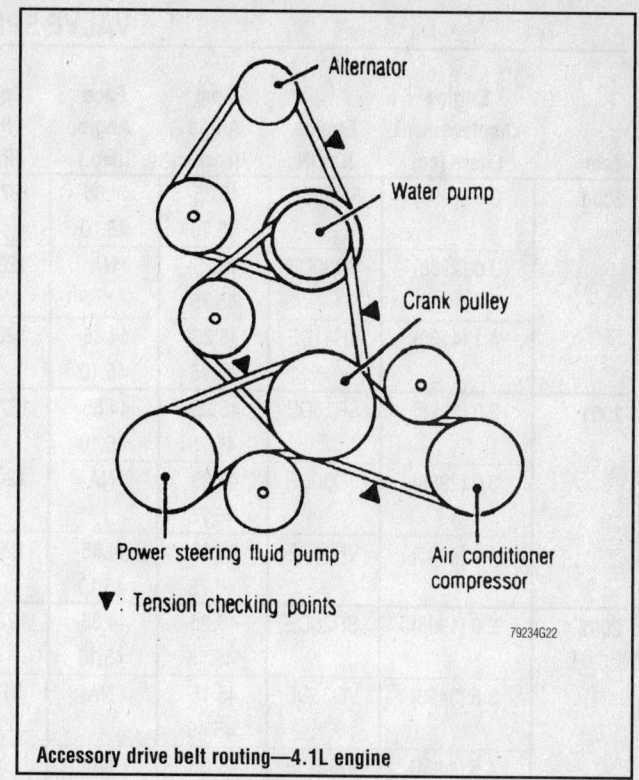

▼: Tension checking points

Accessory drive belt routing—4.1L engine

CAPACITIES

Year	Model	Engine Displacement Liters (cc)	Engine ID/VIN	Engine Oil with Filter	Transmission (pts.)		Drive Axle (pts.)	Fuel Tank (gal.)	Cooling System (qts.)
					5-Spd	Auto.			
2000	G20	2.0 (1998)	SR20DE	3.62	①	14.8	—	15.9	6.50
	I30	3.0 (2988)	VQ30DE	4.25	②	19.8	—	18.5	9.88
	Q45	4.1 (4130)	VH41DE	5.63	—	22.2	2.75	21.1	13.00
2001	G20	2.0 (1998)	SR20DE	3.62	①	14.8	—	15.9	6.50
	I30	3.0 (2988)	VQ30DE	4.25	②	19.8	—	18.5	9.88
	Q45	4.1 (4130)	VH41DE	5.63	—	22.2	2.75	21.1	13.00
2002	G20	2.0 (1998)	SR20DE	3.62	②	14.8	—	15.9	6.50
	G35	3.5 (3498)	VQ35DE	5.00	—	21.8	—	18.5	9.00
	I35	3.5 (3498)	VQ35DE	5.00	—	21.8	—	18.5	9.00
	Q45	4.5 (4494)	VK45DE	5.63	—	21.3	2.75	21.4	10.38
2003	G20	2.0 (1998)	SR20DE	3.62	②	14.8	—	15.9	6.50
	G35	3.5 (3498)	VQ35DE	5.00	—	21.8	—	18.5	9.00
	I35	3.5 (3498)	VQ35DE	5.00	—	21.8	—	18.5	9.00
	Q45	4.5 (4494)	VK45DE	5.63	—	21.3	2.75	21.4	10.38

NOTE: All capacities are approximate. Add fluid gradually and check to be sure a proper fluid level is obtained.

① RS5F32A: 7.62 - 8.00
   RS5F32V: 7.86 - 8.25

② RSF50V: 9.13-9.50
   RSF50A: 9.50-10.13

VALVE SPECIFICATIONS

Year	Engine Displacement Liters (cc)	Engine ID/VIN	Seat Angle (deg.)	Face Angle (deg.)	Spring Test Pressure (lbs. @ in.)	Spring Free Height (in.)	Stem-to-Guide Clearance (in.)		Stem Diameter (in.)	
							Intake	Exhaust	Intake	Exhaust
2000	2.0 (1998)	SR20DE	45.25-45.75	44.85-45.10	127.9-144.3@ 1.181	1.943	0.0008-0.0021	0.0016-0.0029	0.2348-0.2354	0.2341-0.2346
	3.0 (2988)	VQ30DE	45.25-45.75	NA	120.1@1.085	1.845	0.0008-0.0021	0.0016-0.0029	0.2348-0.2354	0.2341-0.2346
	4.1 (4130)	VH41DE	45.25-45.75	44.85-45.10	120.4@1.055	1.946	0.0011-0.0020	0.0014-0.0020	0.2743-0.2744	0.3134-0.3136
2001	2.0 (1998)	SR20DE	45.25-45.75	44.85-45.10	127.9-144.3@ 1.181	1.943	0.0008-0.0021	0.0016-0.0029	0.2348-0.2354	0.2341-0.2346
	3.0 (2988)	VQ30DE	45.25-45.75	NA	120.1@1.085	1.845	0.0008-0.0021	0.0016-0.0029	0.2348-0.2354	0.2341-0.2346
	4.1 (4130)	VH41DE	45.25-45.75	44.85-45.10	120.4@1.055	1.946	0.0011-0.0020	0.0014-0.0020	0.2743-0.2744	0.3134-0.3136
2002	2.0 (1998)	SR20DE	45.25-45.75	44.85-45.10	127.9-144.3@ 1.181	1.943	0.0008-0.0021	0.0016-0.0029	0.2348-0.2354	0.2341-0.2346
	3.5 (3498)	VQ35DE	45.15-45.45	NA	91.5-103.2@ 1.094	1.796	0.010-0.013	0.011-0.015	0.2348-0.2354	0.2344-0.2350
	4.5 (4494)	VK45DE	45.25-45.75	NA	65-74@ 1.331	2.0189-2.0268	0.010-0.013	0.011-0.015	0.2351-0.2354	0.2347-0.2350
2003	2.0 (1998)	SR20DE	45.25-45.75	44.85-45.10	127.9-144.3@ 1.181	1.943	0.0008-0.0021	0.0016-0.0029	0.2348-0.2354	0.2341-0.2346
	3.5 (3498)	VQ35DE	45.15-45.45	NA	91.5-103.2@ 1.094	1.796	0.010-0.013	0.011-0.015	0.2348-0.2354	0.2344-0.2350
	4.5 (4494)	VK45DE	45.25-45.75	NA	65-74@ 1.331	2.0189-2.0268	0.010-0.013	0.011-0.015	0.2351-0.2354	0.2347-0.2350

NA: Not Available

42356-INFC-C05

TORQUE SPECIFICATIONS
All readings in ft. lbs.

Year	Engine Displacement Liters (cc)	Engine ID/VIN	Cylinder Head Bolts	Main Bearing Bolts	Rod Bearing Bolts	Crankshaft Damper Bolts	Flywheel Bolts	Manifold		Spark Plugs	Lug Nuts
								Intake	Exhaust		
2000	2.0 (1998)	SR20DE	①	②	③	105-112	61-69	13-15	27-35	14-22	72-87
	3.0 (2988)	VQ30DE	④	⑤	⑥	⑦	61-69	20-23	21-24	14-22	72-87
	4.1 (4130)	VH41DE	⑧	⑨	⑩	260-275	61-69	12-15	20-23	14-22	72-87
2001	2.0 (1998)	SR20DE	①	②	③	105-112	61-69	13-15	27-35	14-22	72-87
	3.0 (2988)	VQ30DE	④	⑤	⑥	⑦	61-69	20-23	21-24	14-22	72-87
	4.1 (4130)	VH41DE	⑧	⑨	⑩	260-275	61-69	12-15	20-23	14-22	72-87
2002	2.0 (1998)	SR20DE	①	②	③	105-112	61-69	13-15	27-35	14-22	72-87
	3.5 (3498)	VQ35DE	④	⑤	⑥	⑦	61-69	⑪	21-23	15-21	72-87
	4.5 (4494)	VK45DE	⑫	⑬	③	⑭	62-68	18-23	19-22	15-21	72-87
2003	2.0 (1998)	SR20DE	①	②	③	105-112	61-69	13-15	27-35	14-22	72-87
	3.5 (3498)	VQ35DE	④	⑤	⑥	⑦	61-69	⑪	21-23	15-21	72-87
	4.5 (4494)	VK45DE	⑫	⑬	③	⑭	62-68	18-23	19-22	15-21	72-87

① Step 1: 29 ft. lbs.
Step 2: 58 ft. lbs.
Step 3: Loosen bolts completely
Step 4: 25-33 ft. lbs.
Step 5: Tighten an additional 90-100 degrees
Step 6: Repeat Step 5.

② Step 1: 24-28 ft. lbs.
Step 2: Tighten an additional 45-50 degrees or 54-61 ft. lbs.

③ Step 1: 10-12 ft. lbs.
Step 2: Tighten an additional 60-65 degrees or 28-33 ft. lbs.

④ Step 1: 72 ft. lbs.
Step 2: Loosen bolts completely
Step 3: 25-33 ft. lbs.
Step 4: Tighten an additional 90-95 degrees
Step 5: Repeat Step 4

⑤ Step 1: Shift crankshaft to align the bearing beam
Step 2: Tighten all bolts to 24-28 ft. lbs.
Step 3: Tighten an additional 90-95 degrees

⑥ Step 1: Tighten to 15 ft. lbs.
Step 2: Tighten an additional 90-95 degrees

⑦ Step 1: 29-36 ft. lbs.
Step 2: Tighten an additional 60-66 degrees

⑧ Step 1: 22 ft. lbs.
Step 2: 69 ft. lbs.
Step 3: Loosen bolts completely
Step 4: 18-25 ft. lbs.
Step 5: Tighten an additional 90-95 degrees or 69-72 ft. lbs.

⑨ Step 1: Shift crankshaft back and forth to seat bearing caps
Step 2: Tighten inner cap bolts to 27-31 ft. lbs.
Step 3: Tighten outer cap bolts to 20-24 ft. lbs.
Step 4: Tighten No. 1-3, 5 inner cap bolts an additional 60 degrees
Step 5: Tighten No. 4 inner cap bolt and additional 35 degrees
Step 6: Tighten outer cap bolts an additional 35 degrees
Step 7: Tighten bearing cap side bolts to 34-38 ft. lbs.

⑩ Step 1: 10-12 ft. lbs.
Step 2: Tighten an additional 60-65 degrees or 43-48 ft. lbs.

⑪ Step 1: Tighten to 4-7 ft. lbs.
Step 2: Tighten to 20-23 ft. lbs.
Step 3: Tighten, again, to 20-23 ft. lbs.

⑫ Step 1: 72 ft. lbs.
Step 2: Loosen bolts completely
Step 3: 29-36 ft. lbs.
Step 4: Tighten an additional 60-65 degrees
Step 5: Repeat step 4

⑬ Step 1: M12 bolts: 27-31 ft. lbs., plus an additional 40-45 degrees
Step 2: M9 bolts: 20-23 ft. lbs., plus an additional 30-35 degrees
Step 3: M10 side bolts: 10-12 ft. lbs.

⑭ Step 1: 65-72 ft. lbs.
Step 2: Tighten an additional 90-96 degrees

42356-INFC-C06

PISTON AND RING SPECIFICATIONS

All measurements are given in inches.

Year	Engine Displacement Liters (cc)	Engine ID/VIN	Piston Clearance	Ring Gap			Ring Side Clearance		
				Top Compression	Bottom Compression	Oil Control	Top Compression	Bottom Compression	Oil Control
2000	2.0 (1998)	SR20DE	0.0004-0.0012	0.0079-0.0118	0.0138-0.0197	0.0079-0.0236	0.0018-0.0031	0.0012-0.0026	SNUG
	3.0 (2988)	VQ30DE	0.0004-0.0012	0.0087-0.0161	0.0197-0.0291	0.0079-0.0272	0.0016-0.0031	0.0012-0.0028	SNUG
	4.1 (4130)	VH41DE	0.0004-0.0012	0.0106-0.0181	0.0154-0.0248	0.0079-0.0272	0.0016-0.0031	0.0012-0.0028	SNUG
2001	2.0 (1998)	SR20DE	0.0004-0.0012	0.0079-0.0118	0.0138-0.0197	0.0079-0.0236	0.0018-0.0031	0.0012-0.0026	SNUG
	3.0 (2988)	VQ30DE	0.0004-0.0012	0.0087-0.0161	0.0197-0.0291	0.0079-0.0272	0.0016-0.0031	0.0012-0.0028	SNUG
	4.1 (4130)	VH41DE	0.0004-0.0012	0.0106-0.0181	0.0154-0.0248	0.0079-0.0272	0.0016-0.0031	0.0012-0.0028	SNUG
2002	2.0 (1998)	SR20DE	0.0004-0.0012	0.0079-0.0118	0.0138-0.0197	0.0079-0.0236	0.0018-0.0031	0.0012-0.0026	SNUG
	3.5 (3498)	VQ35DE	0.0004-0.0012	0.0091-0.0130	0.0130-0.0189	0.0079-0.0197	0.0018-0.0031	0.0012-0.0028	0.0026-0.0053
	4.5 (4494)	VK45DE	0.0002-0.0007	0.0087-0.0126	0.0126-0.0185	0.0079-0.0236	0.0018-0.0039	0.0012-0.0028	0.0026-0.0053
2003	2.0 (1998)	SR20DE	0.0004-0.0012	0.0079-0.0118	0.0138-0.0197	0.0079-0.0236	0.0018-0.0031	0.0012-0.0026	SNUG
	3.5 (3498)	VQ35DE	0.0004-0.0012	0.0091-0.0130	0.0130-0.0189	0.0079-0.0197	0.0018-0.0031	0.0012-0.0028	0.0026-0.0053
	4.5 (4494)	VK45DE	0.0002-0.0007	0.0087-0.0126	0.0126-0.0185	0.0079-0.0236	0.0018-0.0039	0.0012-0.0028	0.0026-0.0053

42356-INFC-C07

CRANKSHAFT AND CONNECTING ROD SPECIFICATIONS
All measurements are given in inches.

Year	Engine Displacement Liters (cc)	Engine ID/VIN	Crankshaft				Connecting Rod		
			Main Brg. Journal Dia.	Main Brg. Oil Clearance	Shaft End-play	Thrust on No.	Journal Diameter	Oil Clearance	Side Clearance
2000	2.0 (1998)	SR20DE	2.1643-2.1646	0.0002-0.0009	0.0039-0.0102	3	1.8885-1.8887	0.0008-0.0018	0.0079-0.0138
	3.0 (2988)	VQ30DE	2.3610-2.3612	0.0014-0.0021	0.0039-0.0098	3	1.7704-1.7706	0.0013-0.0023	0.0079-0.0138
	4.1 (4130)	VH41DE	2.5181-2.5183	0.0002-0.0008	0.0039-0.0102	3	2.0460-2.0462	0.0008-0.0018	0.0079-0.0138
2001	2.0 (1998)	SR20DE	2.1643-2.1646	0.0002-0.0009	0.0039-0.0102	3	1.8885-1.8887	0.0008-0.0018	0.0079-0.0138
	3.0 (2988)	VQ30DE	2.3610-2.3612	0.0014-0.0021	0.0039-0.0098	3	1.7704-1.7706	0.0013-0.0023	0.0079-0.0138
	4.1 (4130)	VH41DE	2.5181-2.5183	0.0002-0.0008	0.0039-0.0102	3	2.0460-2.0462	0.0008-0.0018	0.0079-0.0138
2002	2.0 (1998)	SR20DE	2.1643-2.1646	0.0002-0.0009	0.0039-0.0102	3	1.8885-1.8887	0.0008-0.0018	0.0079-0.0138
	3.5 (3498)	VQ35DE	2.3603-2.3612	0.0014-0.0021	0.0039-0.0098	3	1.7704-1.7706	0.0013-0.0023	0.0079-0.0138
	4.5 (4494)	VK45DE	2.5173-2.5183	①	0.0039-0.0102	3	2.1654-2.1659	0.0008-0.0018	0.0079-0.0138
2003	2.0 (1998)	SR20DE	2.1643-2.1646	0.0002-0.0009	0.0039-0.0102	3	1.8885-1.8887	0.0008-0.0018	0.0079-0.0138
	3.5 (3498)	VQ35DE	2.3603-2.3612	0.0014-0.0021	0.0039-0.0098	3	1.7704-1.7706	0.0013-0.0023	0.0079-0.0138
	4.5 (4494)	VK45DE	2.5173-2.5183	①	0.0039-0.0102	3	2.1654-2.1659	0.0008-0.0018	0.0079-0.0138

① Nos. 1 and 5: 0.0004-0.0004 in.

Nos. 2, 3 and 4: 0.0003-0.0007

42356-INFC-C08

WHEEL ALIGNMENT

Year	Model		Caster Range (+/-Deg.)	Caster Preferred Setting (Deg.)	Camber Range (+/-Deg.)	Camber Preferred Setting (Deg.)	Toe-in (in.)	Steering Axis Inclination (Deg.)
2000	G20	F	0.75	+1.92	0.75	0	0.04 +/- 0.04	—
		R	—	—	0.75	-1.03	0.16 +/- 0.16	—
	I30	F	0.75	+2.75	0.75	-0.25	0.04 +/- 0.04	14.25
		R	—	—	0.75	-1.00	0.04 +/- 0.15	—
	Q45	F	0.75	+6.42	0.75	-0.66	0.08 +/- 0.04	—
		R	—	—	0.50	-0.75	0.09 +/- 0.10	—
2001	G20	F	0.75	+1.92	0.75	0	0.04 +/- 0.04	—
		R	—	—	0.75	-1.03	0.16 +/- 0.16	—
	I30	F	0.75	+2.75	0.75	-0.25	0.04 +/- 0.04	14.25
		R	—	—	0.75	-1.00	0.04 +/- 0.15	—
	Q45	F	0.75	+6.42	0.75	-0.66	0.08 +/- 0.04	—
		R	—	—	0.50	-0.75	0.09 +/- 0.10	—
2002	G20	F	0.75	+1.92	0.75	0	0.04 +/- 0.04	—
		R	—	—	0.75	-1.03	0.16 +/- 0.16	—
	G35	F	0.75	+7.75	0.75	-0.08	0.04 +/- 0.04	14.25
		R	—	—	0.75	-1.00	0.04 +/- 0.15	—
	I35	F	0.75	+7.75	0.75	-0.08	0.04 +/- 0.04	14.25
		R	—	—	0.75	-1.00	0.04 +/- 0.15	—
	Q45	F	0.75	+6.42	0.75	-0.66	0.08 +/- 0.04	—
		R	—	—	0.50	-0.75	0.09 +/- 0.10	—
2003	G20	F	0.75	+1.92	0.75	0	0.04 +/- 0.04	—
		R	—	—	0.75	-1.03	0.16 +/- 0.16	—
	G35	F	0.75	+7.75	0.75	-0.08	0.04 +/- 0.04	14.25
		R	—	—	0.75	-1.00	0.04 +/- 0.15	—
	I35	F	0.75	+7.75	0.75	-0.08	0.04 +/- 0.04	14.25
		R	—	—	0.75	-1.00	0.04 +/- 0.15	—
	Q45	F	0.75	+6.42	0.75	-0.66	0.08 +/- 0.04	—
		R	—	—	0.50	-0.75	0.09 +/- 0.10	—

42356-INFC-C09

TIRE, WHEEL AND BALL JOINT SPECIFICATIONS

Year	Model	OEM Tires		Tire Pressures (psi)		Wheel Size	Ball Joint Inspection
		Standard	Optional	Front	Rear		
2000	G20	P195/65HR15	None	35	35	6-JJ	①
	Q45	P215/60VR16	P225/50VR17	35	35	Std: 7-JJ Opt: 7.5-JJ	①
	I30	P205/65R15	P215/60HR15	35	35	Std: 6-JJ Opt: 6.5-JJ	①
2001	G20	P195/65HR15	None	35	35	6-JJ	①
	Q45	P215/60VR16	P225/50VR17	35	35	Std: 7-JJ Opt: 7.5-JJ	①
	I30	P205/65R15	P215/60HR15	35	35	Std: 6-JJ Opt: 6.5-JJ	①
2002	G20	P195/65HR15	None	35	35	6-JJ	①
	G35	P205/65R16	P215/55R17	30	30	Std: 6.5-JJ Opt: 7-JJ	①
	I35	P205/65R16	P215/55R17	30	30	Std: 6.5-JJ Opt: 7-JJ	①
	Q45	P225/55VR17	P245/45VR18	33	33	Std: 7.5-JJ Opt: 7.5-JJ	①
2003	G20	P195/65HR15	None	35	35	6-JJ	①
	G35	P205/65R16	P215/55R17	30	30	Std: 6.5-JJ Opt: 7-JJ	①
	I35	P205/65R16	P215/55R17	30	30	Std: 6.5-JJ Opt: 7-JJ	①
	Q45	P225/55VR17	P245/45VR18	33	33	Std: 7.5-JJ Opt: 7.5-JJ	①

OEM: Original Equipment Manufacturer

PSI: Pounds Per Square Inch

STD: Standard

OPT: Optional

① Replace if any measurable movement is found.

42356-INFC-C10

BRAKE SPECIFICATIONS
All measurements in inches unless noted

| Year | Model | Front Brake Disc | | | Rear Brake Disc | | | Minimum Lining Thickness | | Brake Caliper | |
		Original Thickness	Minimum Thickness	Maximum Run-out	Original Thickness	Minimum Thickness	Maximum Run-out	Front	Rear	Bracket Bolts (ft. lbs.)	Mounting Bolts (ft. lbs.)
2000	G20	0.870	0.787	0.003	0.350	0.310	0.003	0.079	0.059	①	16-23
	I30	0.870	0.787	0.003	0.350	0.315	0.003	0.079	0.059	①	16-23
	Q45	1.100	1.024	0.003	0.350	0.551	0.003	0.079	0.079	②	24-31
2001	G20	0.870	0.787	0.003	0.350	0.310	0.003	0.079	0.059	①	16-23
	I30	0.870	0.787	0.003	0.350	0.315	0.003	0.079	0.059	①	16-23
	Q45	1.100	1.024	0.003	0.350	0.551	0.003	0.079	0.079	②	24-31
2002	G20	0.870	0.787	0.003	0.350	0.310	0.003	0.079	0.059	①	16-23
	G35	0.945	0.866	0.003	0.630	0.551	0.004	0.079	0.079	①	16-23
	I35	0.945	0.866	0.003	0.630	0.551	0.004	0.079	0.079	①	16-23
	Q45	1.100	1.020	0.003	0.630	0.550	0.004	0.079	0.079	②	24-31
2003	G20	0.870	0.787	0.003	0.350	0.310	0.003	0.079	0.059	①	16-23
	G35	0.945	0.866	0.003	0.630	0.551	0.004	0.079	0.079	①	16-23
	I35	0.945	0.866	0.003	0.630	0.551	0.004	0.079	0.079	①	16-23
	Q45	1.100	1.020	0.003	0.630	0.550	0.004	0.079	0.079	②	24-31

① Front: 53-72
   Rear: 28-38

② Front: 103-118
   Rear: 28-38

42356-INFC-C11

SCHEDULED MAINTENANCE INTERVALS
Infinity—G20, G35 I30, I35 & Q45

TO BE SERVICED	TYPE OF SERVICE	VEHICLE MILEAGE INTERVAL (x1000)												
		7.5	15	22.5	30	37.5	45	52.5	60	67.5	75	82.5	90	97.5
Engine oil & filter	R	✓	✓	✓	✓	✓	✓	✓	✓	✓	✓	✓	✓	✓
Automatic transaxle fluid	S/I		✓		✓		✓		✓		✓		✓	
Brake lines & cables	S/I		✓		✓		✓		✓		✓		✓	
Brake pads & discs	S/I		✓		✓		✓		✓		✓		✓	
Differential gear oil (Q45)	S/I		✓		✓		✓		✓		✓		✓	
Driveshaft boots (I30 & G20)	S/I		✓		✓		✓		✓		✓		✓	
Full-active suspension fluid (Q45) ①	S/I		✓		✓		✓		✓		✓		✓	
Manual transaxle oil (G20 & I30)	S/I		✓		✓		✓		✓		✓		✓	
Air cleaner filter	R				✓				✓				✓	
Exhaust system	S/I				✓				✓				✓	
Fuel lines	S/I				✓				✓				✓	
Steering gear linkage axle & suspension parts	S/I				✓				✓				✓	
SUPER HICAS linkage (J30 & Q45)	S/I				✓				✓				✓	
Vapor lines	S/I				✓				✓				✓	
Engine coolant	R								✓				✓	
Spark plugs	R								✓					
Timing belt	R								✓					
Drive belts	S/I								✓					

R: Replace   S/I: Service or Inspect

① Replace at 60,000 miles (if not previously replaced).

FREQUENT OPERATION MAINTENANCE (SEVERE SERVICE)

If a vehicle is operated under any of the following conditions it is considered severe service

- Extremely dusty areas.

- 50% or more of the vehicle operation is in 32°C (90°F) or higher temperatures, or constant operation in temperatures below 0°C (32°F).

- Prolonged idling (vehicle operation in stop and go traffic).

- Frequent short running periods (engine does not warm to normal operating temperatures).

- Police, taxi, delivery usage or trailer towing usage.

Oil & oil filter: change every 3750 miles.

Brake pads & discs: service or inspect every 7500 miles.

Driveshaft boots (G20 & I30): service or inspect every 7500 miles

Exhaust system: service or inspect every 7500 miles.

Steering gear, linkage, axle & suspension ball joints: service or inspect every 7500 miles.

Steering linkage, ball joints & front suspension ball joints: service or inspect every 7500 miles.

SUPER HICAS linkage (Q45): service or inspect every 7500 miles.

42356-INFC-C12

PRECAUTIONS

PRECAUTIONS

Before servicing any vehicle, please be sure to read all of the following precautions, which deal with personal safety, prevention of component damage, and important points to take into consideration when servicing a motor vehicle:

• Never open, service or drain the radiator or cooling system when the engine is hot; serious burns can occur from the steam and hot coolant.

• Observe all applicable safety precautions when working around fuel. Whenever servicing the fuel system, always work in a well-ventilated area. Do not allow fuel spray or vapors to come in contact with a spark, open flame, or excessive heat (a hot drop light, for example). Keep a dry chemical fire extinguisher near the work area. Always keep fuel in a container specifically designed for fuel storage; also, always properly seal fuel containers to avoid the possibility of fire or explosion. Refer to the additional fuel system precautions later in this section.

• Fuel injection systems often remain pressurized, even after the engine has been turned **OFF**. The fuel system pressure must be relieved before disconnecting any fuel lines. Failure to do so may result in fire and/or personal injury.

• Brake fluid often contains polyglycol ethers and polyglycols. Avoid contact with the eyes and wash your hands thoroughly after handling brake fluid. If you do get

brake fluid in your eyes, flush your eyes with clean, running water for 15 minutes. If eye irritation persists, or if you have taken brake fluid internally, IMMEDIATELY seek medical assistance.

• The EPA warns that prolonged contact with used engine oil may cause a number of skin disorders, including cancer! You should make every effort to minimize your exposure to used engine oil. Protective gloves should be worn when changing oil. Wash your hands and any other exposed skin areas as soon as possible after exposure to used engine oil. Soap and water, or waterless hand cleaner should be used.

• All new vehicles are now equipped with an air bag system. The system must be disabled before performing service on or around system components, steering column, instrument panel components, wiring and sensors. Failure to follow safety and disabling procedures could result in accidental air bag deployment, possible personal injury and unnecessary system repairs.

• Always wear safety goggles when working with, or around, the air bag system. When carrying a non-deployed air bag, be sure the bag and trim cover are pointed away from your body. When placing a non-deployed air bag on a work surface, always face the bag and trim cover upward, away from the surface. This will reduce the motion of the module if it is accidentally deployed. Refer to the additional air bag system precautions later in this section.

• Clean, high quality brake fluid from a

sealed container is essential to the safe and proper operation of the brake system. You should always buy the correct type of brake fluid for your vehicle. If the brake fluid becomes contaminated, completely flush the system with new fluid. Never reuse any brake fluid. Any brake fluid that is removed from the system should be discarded. Also, do not allow any brake fluid to come in contact with a painted surface; it will damage the paint.

• Never operate the engine without the proper amount and type of engine oil; doing so WILL result in severe engine damage.

• Timing belt maintenance is extremely important! Many models utilize an interference-type, non-freewheeling engine. If the timing belt breaks, the valves in the cylinder head may strike the pistons, causing potentially serious (also time-consuming and expensive) engine damage. Refer to the maintenance interval charts in the front of this section for the recommended replacement interval for the timing belt, and to the timing belt procedure in this section for belt replacement and inspection.

• Disconnecting the negative battery cable on some vehicles may interfere with the functions of the on-board computer system(s) and may require the computer to undergo a relearning process once the negative battery cable is reconnected.

• When servicing drum brakes, only disassemble and assemble one side at a time, leaving the remaining side intact for reference.

ENGINE REPAIR

Distributor

REMOVAL

2.0L Engine

1. Before servicing the vehicle, refer to the precautions in the beginning of this section.
2. Remove or disconnect the following:
   • Negative battery cable
   • Splash shield, if equipped
   • Distributor connections but leave the ignition wires in place
   • Distributor cap hold-down screws and lift off the distributor cap with all ignition wires still connected
3. Matchmark the rotor to the distributor housing, and the distributor housing to the engine.

➡Do not crank the engine during this procedure. If the engine is cranked, the matchmark must be disregarded.

   • Hold-down bolt
   • Distributor from the engine

3.0L, 3.5L and 4.1L Engines

These engines are equipped with a distributorless ignition.

INSTALLATION

2.0L Engine

ENGINE NOT DISTURBED

1. If the engine was not disturbed, install or connect the following:
   • New distributor housing O-ring
   • Distributor in the engine so the

rotor is aligned with the matchmark on the housing and the housing is aligned with the matchmark on the engine. Be sure the distributor is fully seated and the distributor gear is fully engaged.
   • Snug the hold-down bolt
   • Distributor pick-up lead wires
   • Distributor cap and tighten the screws
   • Splash shield
   • Negative battery cable
2. Check and/or adjust the ignition timing and tighten the hold-down bolt to 10–12 ft. lbs. (14–16 Nm).

ENGINE DISTURBED

1. If the engine was disturbed (cranked or turned over with the distributor removed), install or connect the following:

- New distributor housing O-ring

2. Position the engine so the No. 1 piston is at TDC of its compression stroke and the mark on the vibration damper is aligned with **0** on the timing indicator.

- Distributor in the engine so the rotor is aligned with the position of the No. 1 ignition wire on the distributor cap. Be sure the distributor is fully seated and that the distributor shaft is fully engaged.

➡**There are distributor cap runners inside the cap on 2.0L engine. Be sure the rotor is pointing to where the No. 1 runner originates inside the cap.**

- Snug the hold-down bolt
- Distributor pick-up lead wires
- Distributor cap and tighten the screws
- Splash shield, if equipped
- Negative battery cable

3. Check and/or adjust the ignition timing and tighten the hold-down bolt to 10–12 ft. lbs. (14–16 Nm).

Alternator

REMOVAL

2.0L Engine

1. Before servicing the vehicle, refer to the precautions in the beginning of this section.
2. Remove or disconnect the following:

- Negative battery cable
- Drive belt
- Alternator harness connector
- Alternator bracket, if necessary
- Alternator retainers
- Alternator

3.0L Engine

1. Before servicing the vehicle, refer to the precautions in the beginning of this section.
2. Remove or disconnect the following:

- Negative battery cable
- Engine right-hand undercover
- Right-hand side inspection cover
3. Loosen the belt idler pulley.
- Drive belt
- 4 A/C compressor mounting bolts
- Cooling fan and fan shroud
4. Slide the A/C compressor forward.
- Alternator harness connector
- Upper and lower alternator bolts
- Alternator

3.5L Engines

I35 MODELS

1. Before servicing the vehicle, refer to the precautions in the beginning of this section.
2. Remove or disconnect the following:

- Negative battery cable
- Right side engine undercover and side inspection cover
- Radiator
- Drive belt
- Alternator and A/C compressor harness connectors
- Upper and lower alternator bolts
- Alternator

G35 MODELS

1. Before servicing the vehicle, refer to the precautions in the beginning of this section.
2. Remove or disconnect the following:

- Negative battery cable
- Engine undercover
- Stabilizer bar clamps, then slide stabilizer downward
- Loosen drive belts
- Alternator electrical connector
- Oil pressure switch harness connector
- B terminal mounting nut
- Upper and lower alternator mounting bolt
- Both alternator bracket bolts
- Alternator from the vehicle

4.1L Engine

1. Before servicing the vehicle, refer to the precautions in the beginning of this section.
2. Remove or disconnect the following:

- Negative battery cable
- Engine upper cover
- Engine drive belt
- Alternator electrical connector
- Alternator

INSTALLATION

2.0L Engine

1. Install or connect the following:

- Alternator
- Alternator bracket, if removed
- Alternator retainers as shown in the accompanying illustration

3.0L Engine

1. Install or connect the following:

- Alternator
- Upper and lower alternator bolts. Torque the upper bolt to 11–15 ft. lbs. (15–20 Nm) and the lower bolt to 32–38 ft. lbs. (44–52 Nm)
- Alternator harness connector
2. Slide the A/C compressor rearward.
- Fan shroud and cooling fan
- 4 A/C compressor mounting bolts
- Drive belt
3. Tighten the belt idler pulley.
- Right-hand side inspection cover
- Engine right-hand undercover
- Negative battery cable

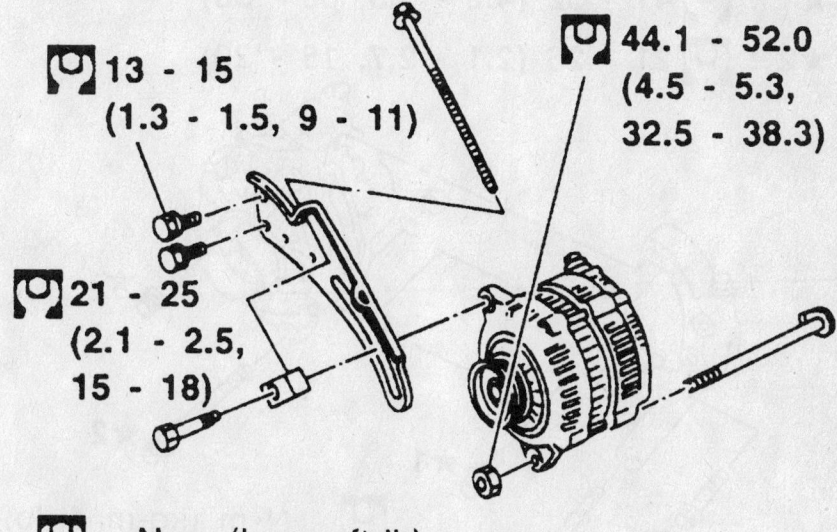

13 - 15
(1.3 - 1.5, 9 - 11)

21 - 25
(2.1 - 2.5, 15 - 18)

44.1 - 52.0
(4.5 - 5.3, 32.5 - 38.3)

🔧 : N•m (kg-m, ft-lb)

9307HG20

Alternator and bracket retainer locations and torque specifications—G20 models

3.5L Engine

I35 MODELS

1. Install or connect the following:
   - Alternator
   - Upper and lower alternator bolts. Tighten the upper bolt to 12–15 ft. lbs. (16 –20 Nm) and the lower bolt to 32–38 ft. lbs. (44–52 Nm).
   - Alternator and A/C compressor harness connectors
   - Drive belt
   - Radiator
   - Right side engine undercover and side inspection cover
   - Negative battery cable

G35 MODELS

1. Install or connect the following:
   - Alternator into the vehicle
   - Alternator bracket bolts and torque to 18–23 ft. lbs. (24–31 Nm)
   - Upper and lower alternator mounting bolts and tighten to 45–51 ft. lbs. (60–70 Nm)
   - B terminal mounting nut. Torque the nut to 83–95 inch lbs. (9–11 Nm).
   - Oil pressure switch harness connector
   - Alternator electrical connector
   - Drive belts
   - Stabilizer into position and secure with the clamps
   - Engine undercover
   - Negative battery cable

4.1L Engine

1. Install or connect the following:
   - Alternator
   - Alternator bolts. Torque the alternator bolts marked **1**, (shown in the accompanying illustration) to 30–38 ft. lbs. (41–52 Nm) and the bolt marked **2**, (shown in the same illustration) to 15–20 ft. lbs. (21–26 Nm).
   - Alternator electrical connector
   - Engine drive belt
   - Engine upper cover
   - Negative battery cable

Ignition Timing

ADJUSTMENT

2.0L Engine

➡**The engine should be in good mechanical condition and all electrical connectors and vacuum hoses connected before making this adjustment.**

1. Before servicing the vehicle, refer to the precautions in the beginning of this section.
2. Start the engine and let it warm up to normal operating temperature.
3. Open the hood and run the engine under no load at about 2,000 rpm for about 2 minutes.
4. Perform Diagnostic Test Mode II and repair any causes of trouble codes as needed.

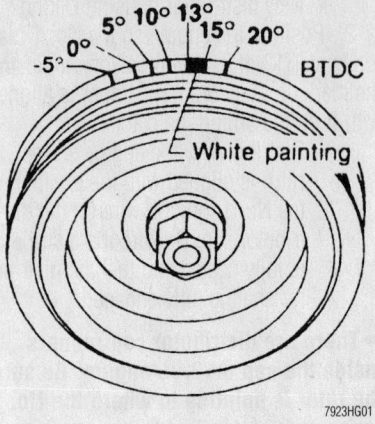

Crankshaft pulley and timing marks—2.0L engine

5. Run the engine under no load at 2,000 rpm for about 2 minutes. Rev the engine 2 or 3 times and let it idle for 1 minute.
6. Turn **OFF** the engine and disconnect the Throttle Position (TP) sensor connector. Connect a timing light to the No. 1 spark plug wire. Start the engine.
7. Adjust the timing to 13–17° BTDC by loosening the distributor mounting bolts and turning the distributor. When the timing is correct, tighten the mounting bolts and turn the engine **OFF**.
8. Reconnect the TP sensor connector. Start the engine and check the ignition timing again.

3.0L, 3.5L and 4.1 Engines

➡**The engine should be in good mechanical condition and all electrical connectors and vacuum hoses attached before making this adjustment.**

1. Before servicing the vehicle, refer to the precautions in the beginning of this section.
2. Start the engine and let it warm up to normal operating temperature.
3. Open the hood and run the engine under no load at about 2,000 rpm for about 2 minutes.
4. Perform Diagnostic Test Mode II and repair any causes of trouble codes as needed.
5. Run the engine under no load at 2,000 rpm for about 2 minutes. Rev the engine 2 or 3 times and let it idle for 1 minute.
6. Turn **OFF** the engine and disconnect the Throttle Position sensor connector. Remove the No. 1 ignition coil. Connect the coil to the spark plug using a spare piece of high-tension wire so you have a place to connect your timing light. Start the engine.

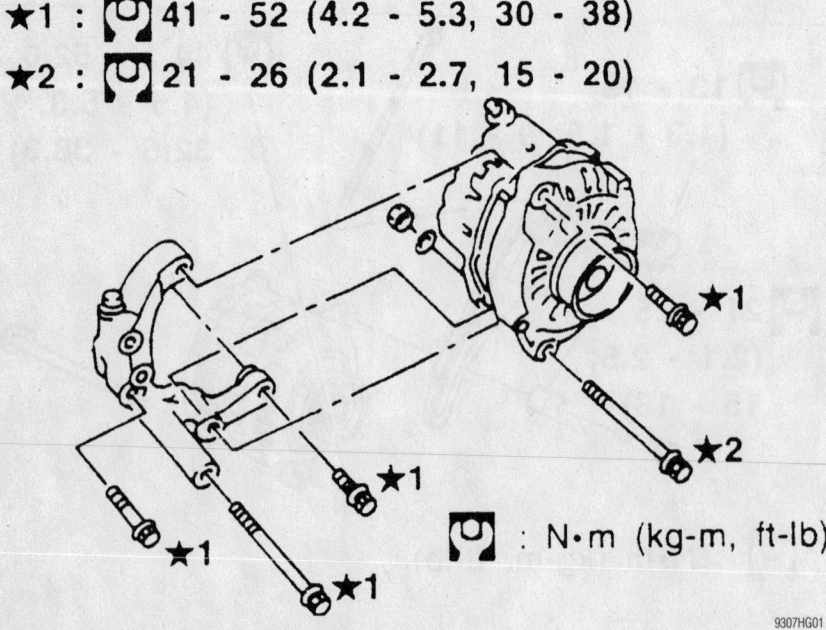

★1 : 41 - 52 (4.2 - 5.3, 30 - 38)

★2 : 21 - 26 (2.1 - 2.7, 15 - 20)

: N•m (kg-m, ft-lb)

Alternator bolt locations—Q45 models

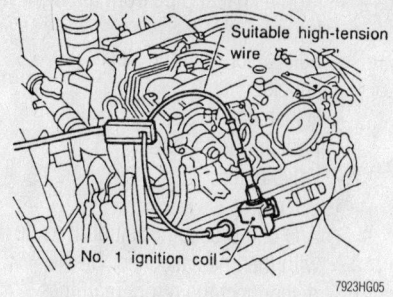

Connect the No. 1 ignition coil to the spark plug with the spare piece of high-tension wire—4.1L engine shown

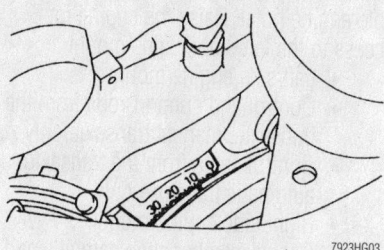

Location of timing marks—3.0L and 3.5L engines

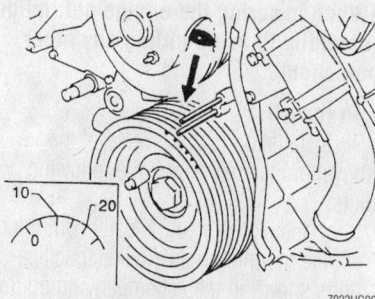

Location of timing marks—4.1L engine

7. Run the engine under no load at 2,000 rpm for about 2 minutes. Rev the engine 2 or 3 times and let it idle.

8. Check the ignition timing and adjust if needed.

9. The correct ignition timing is as follows:

　a. Correct ignition timing for 3.0L and 3.5L engines is 8–12° BTDC.

　b. Correct ignition timing for the 4.1L engine is 13–17° BTDC.

- Adjustment is made by loosening the screws and turning the Camshaft Position (CMP) sensor until the mark on the crankshaft pulley is pointing at 10° BTDC. Tighten the mounting screws and confirm ignition timing has not changed.
- Turn the engine **OFF** and connect the TP sensor connector.

Engine Assembly

REMOVAL & INSTALLATION

2.0L Engine

1. Before servicing the vehicle, refer to the precautions in the beginning of this section.

2. Drain the coolant system.

3. Drain the engine oil.

4. Drain the transaxle fluid.

5. Release fuel system pressure and remove fuel line.

6. Remove or disconnect the following:
- Hood and hinges
- Negative battery cable
- Both front wheels
- Engine undercover
- Air cleaner assembly and duct
- Battery and battery tray
- All vacuum lines and wiring harness connectors
- Heater hoses
- Oil cooler lines, if equipped
- Power steering hoses
- Fuel lines
- Throttle cable
- Cruise control cable, if equipped
- A/T control cable, if equipped
- Cooling fans, radiator and recovery tank
- Drive shafts
- Front exhaust pipe
- Starter and intake manifold support
- Drive belts
- Alternator, A/C compressor, and the power steering pump from the engine and lay them aside. Do not disconnect the compressor or power steering pump lines.

7. Support the engine with a hoist and the transaxle with a suitable jack. Raise the engine and transaxle slightly and remove the center member.

- Engine mounting bolts from both sides and slowly lower the hoist and transaxle jack
- Engine and transaxle from beneath the vehicle

To install:

8. Install or connect the following:
- Center member bracket (manual transmission) on the engine, if removed. Ensure that all insulators are correctly positioned on the brackets. Torque the insulator through-bolts to 32–41 ft. lbs. (43–55 Nm).

9. If equipped with manual transaxle, ensure that the distance between the center

of the insulator through-bolt and the center member is 2.28–2.36 in. (58–60mm). Torque the through-bolt to 46–58 ft. lbs. (62–78 Nm).

- Engine. Torque the center member-to-frame bolts to 57–72 ft. lbs. (77–98 Nm).
- Alternator, air conditioning compressor, and power steering pump
- Drive belts
- Starter and intake manifold support
- Front exhaust pipe
- Drive shafts
- Cooling fans, radiator and recovery tank
- A/T control cable, if equipped
- Cruise control cable, if equipped
- Throttle cable
- Fuel lines
- Power steering hoses
- Oil cooler lines, if equipped
- Heater hoses
- All vacuum lines and wiring harness connectors
- Engine undercover
- Front wheels
- Battery and battery tray
- Air cleaner assembly and duct
- Negative battery cable
- Hood and hinges

10. Fill and bleed the cooling system.

11. Fill the engine with clean oil.

12. Fill the transaxle to the proper level.

13. Start the vehicle, check for leaks and repair if necessary.

3.0L Engine

It is recommended the engine and transaxle be removed as a single unit. If necessary, the units may be separated after removal.

➡**The engine and transaxle assembly must be removed from the under side of the vehicle.**

1. Drain the cooling system.

2. Drain the engine oil.

3. Drain the transaxle fluid.

4. Properly relieve the fuel system pressure.

5. Before servicing the vehicle, refer to the precautions in the beginning of this section.

6. Remove or disconnect the following:
- Negative battery cable
- Hood
- Both front wheels
- Engine undercover
- All necessary vacuum hoses, fuel hoses and electrical connections that would interfere with engine removal

- Front exhaust tubes
- Ball joints
- Driveshafts
- Radiator and fans
- Drive belts
- Alternator, compressor and power steering oil pump from the engine compartment

7. Set a suitable transmission jack under the transaxle. Hoist the engine with a suitable engine slinger.

- Rear engine mounting
- Control cable
- Front engine mounting
- Center member and slowly lower the transmission jack

8. Remove the engine and the transaxle assembly from the engine as shown in the accompanying illustration.

To install:

9. Install or connect the following:

- Center member bracket (manual transmission) on the engine, if removed. Ensure that all insulators are correctly positioned on the brackets. Torque the insulator through-bolts to 72 ft. lbs. (98 Nm).
- Engine. Torque the center member-to-frame bolts to 57–72 ft. lbs. (77–98 Nm).
- Front engine mount. Torque the bolts to 72 ft. lbs. (98 Nm).
- Control cable
- Rear engine mount. Torque the bolts to 72 ft. lbs. (98 Nm) and remove the transaxle jack
- Alternator, air conditioning compressor, and power steering pump
- Radiator and fans
- Drive belts
- Driveshafts

- Ball joints
- Front exhaust tubes
- Vacuum hoses, electrical connectors, fuel hoses and wiring which was removed
- Engine undercover
- Front wheels
- Battery and battery tray
- Negative battery cable
- Hood and hinges

10. Fill and bleed the cooling system.
11. Fill the engine with clean oil.
12. Fill the transaxle to the proper level.
13. Start the vehicle, check for leaks and repair if necessary.

3.5L Engines

I35 MODELS

It is recommended the engine and transaxle be removed as a single unit. If need be, the units may be separated after removal.

➡**The engine and transaxle assembly must be removed from the underside of the vehicle.**

1. Before servicing the vehicle, refer to the precautions in the beginning of this section.
2. Release the fuel system pressure.
3. Drain the cooling system.
4. Drain the engine oil.
5. Drain the automatic transaxle, if equipped.
6. Remove or disconnect the following:

- Negative battery cable
- Hood
- Engine undercover
- All vacuum hoses, fuel lines, wires and connectors; tag before disconnecting

- Front exhaust pipe from the manifold
- Ball joints from the steering knuckle
- Halfshafts
- Radiator and fans
- Drive belts
- Alternator
- A/C compressor. Position it aside with the lines attached. Do NOT disconnect the refrigerant lines.
- Power steering pump and position aside with the lines attached. Do NOT disconnect the fluid lines.

7. Place a suitable jack under the transaxle. Install engine slingers and a suitable engine hoist. Raise the engine for access to the left side engine mount.

- Left side engine mount
- Control and support rods from the transaxle, manual transaxle only
- Control cable from the transaxle, automatic transaxle only
- Right side engine mount
- Center member, then carefully and slowly lower the transmission jack

8. Lower the engine/transaxle assembly onto an engine stand.

➡**When lowering the engine out, guide it carefully to avoid hitting any other components.**

To install:

9. Installation is the reverse of the removal procedure, noting the following points:

   a. Install the electronically controlled engine mount harness to the specifications shown in the accompanying figure.

   b. Make sure to connect all vacuum hoses, lines, and electrical connectors as tagged during removal.

   c. Fill the cooling system.

   d. Fill the engine with clean oil.

   e. Start the vehicle, check for leaks and repair if necessary.

G35 MODELS

1. Before servicing the vehicle, refer to the precautions in the beginning of this section.
2. Evacuate the A/C system.
3. Release the fuel system pressure.
4. Drain the engine oil.
5. Drain the cooling system.
6. Drain the transaxle fluid.
7. Remove or disconnect the following:

- Negative battery cable
- Hood
- Engine cover
- Battery cover
- Engine undercover

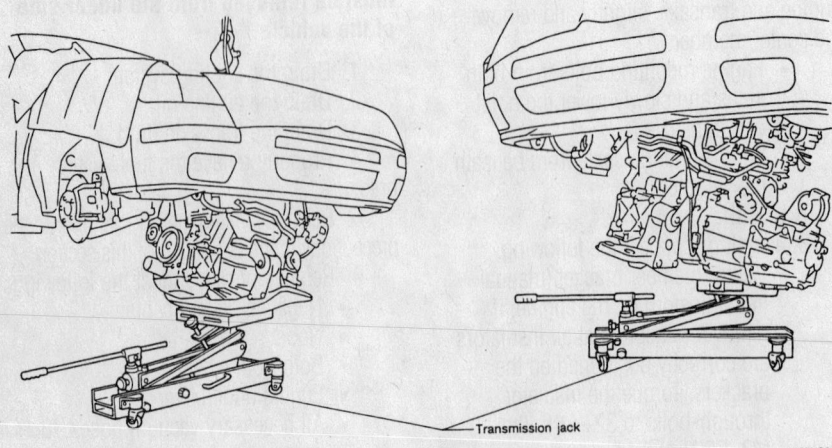

Transmission jack

9307HG02

Remove the engine and the transaxle as an assembly from beneath the vehicle using a transaxle jack and a slinger

Vehicle front

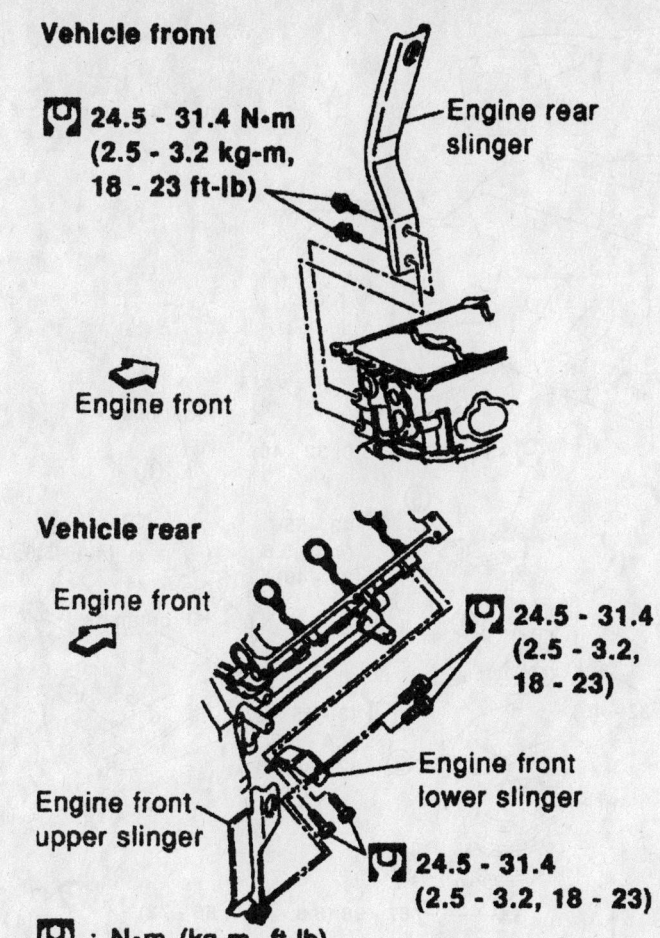

24.5 - 31.4 N•m
(2.5 - 3.2 kg-m,
18 - 23 ft-lb)

Engine rear slinger

Engine front

Vehicle rear

Engine front

24.5 - 31.4
(2.5 - 3.2,
18 - 23)

Engine front lower slinger

Engine front upper slinger

24.5 - 31.4
(2.5 - 3.2, 18 - 23)

: N•m (kg-m, ft-lb)

9357RG01

Installation of engine slingers to lift the engine

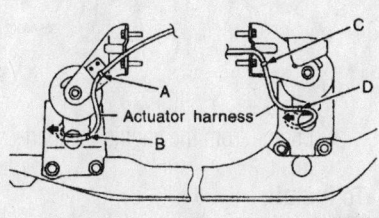

C

A
Actuator harness
B

D

9357MG01

For the electronically controlled engine mount harness, the proper length from A to B is 6.69 in. (170mm) and from C to D is 5.12 in. (130mm)

- Front wheels
- Wiper arm and cowl top cover
- Air cleaner assembly, including air ducts
- Fan, radiator shroud, radiator, reservoir tank and hoses
- Heater hose from the engine and plug to avoid leaks
- Ground wire from left hand cylinder head
- Positive battery cable from the vehicle and temporarily fasten it on the engine

- Battery
- Engine harness connector; tag before disconnecting
- A/C lines from the A/C compressor
- Body ground cables
- Brake booster vacuum hose
- Fuel feed and Evaporative Emission (EVAP) hoses. Plug the fuel lines to prevent fuel from leaking out.
- Power steering pump reservoir tank and pipes

8. Disconnect the connectors from the passenger compartment as follows:

a. Remove the passenger side kick plate, dashboard side trim and glove box.

b. From the engine compartment, detach the connectors from the Transmission Control Module (TCM), and Engine Control Module (ECM).

c. Unfasten the wire harnesses, pull the harnesses out into the engine compartment, then temporarily secure them to the engine. Cover all connectors with plastic or similar material to protect them.

9. Remove or disconnect the following:

- Front exhaust pipe from the manifold
- Steering lower joint and release the steering shaft
- Propeller shaft from the transmission
- Shift control linkage from the gear selector. Secure it temporarily to the transmission, so it doesn't drag or catch on any other components.
- Rear plate cover from upper oil pan
- Bolts securing the drive plate to the torque converter
- Bolts securing the transmission to the lower rear side of the oil pan
- Front stabilizer shaft
- Left and right tie rod ends from the steering knuckle
- Disconnect the lower strut from the lower control arms on the left and right sides
- Left and right control arms from the suspension crossmember

10. Position a suitable engine table or other suitable tool under the engine. Securely support the bottom of the suspension member and transmission.

- Rear member mounting bolt
- Suspension member mounting bolt and nut

11. Carefully lower the jack, or raise the lift to remove the engine, transmission and suspension member assembly. Make sure that all lines, hoses, connectors etc. have been disconnected. If necessary, support the rear of the vehicle at the rear jacking point, as it's center of gravity has changed with the engine removed.

12. Remove and components and separate the engine and transmission as necessary.

To install:

13. Installation is the reverse of the removal procedure, noting the following points:

- Tighten all engine mounts and related components to the specification shown in the accompanying figure
- Tighten the rear member mounting bolts in sequence.
- Make sure to connect all vacuum hoses, lines, and electrical connectors as tagged during removal.
- Fill the cooling system.
- Fill the engine with clean oil.
- Fill the transmission to the proper level.
- Recharge the A/C system.

14. Start the vehicle, check for leaks and repair if necessary.

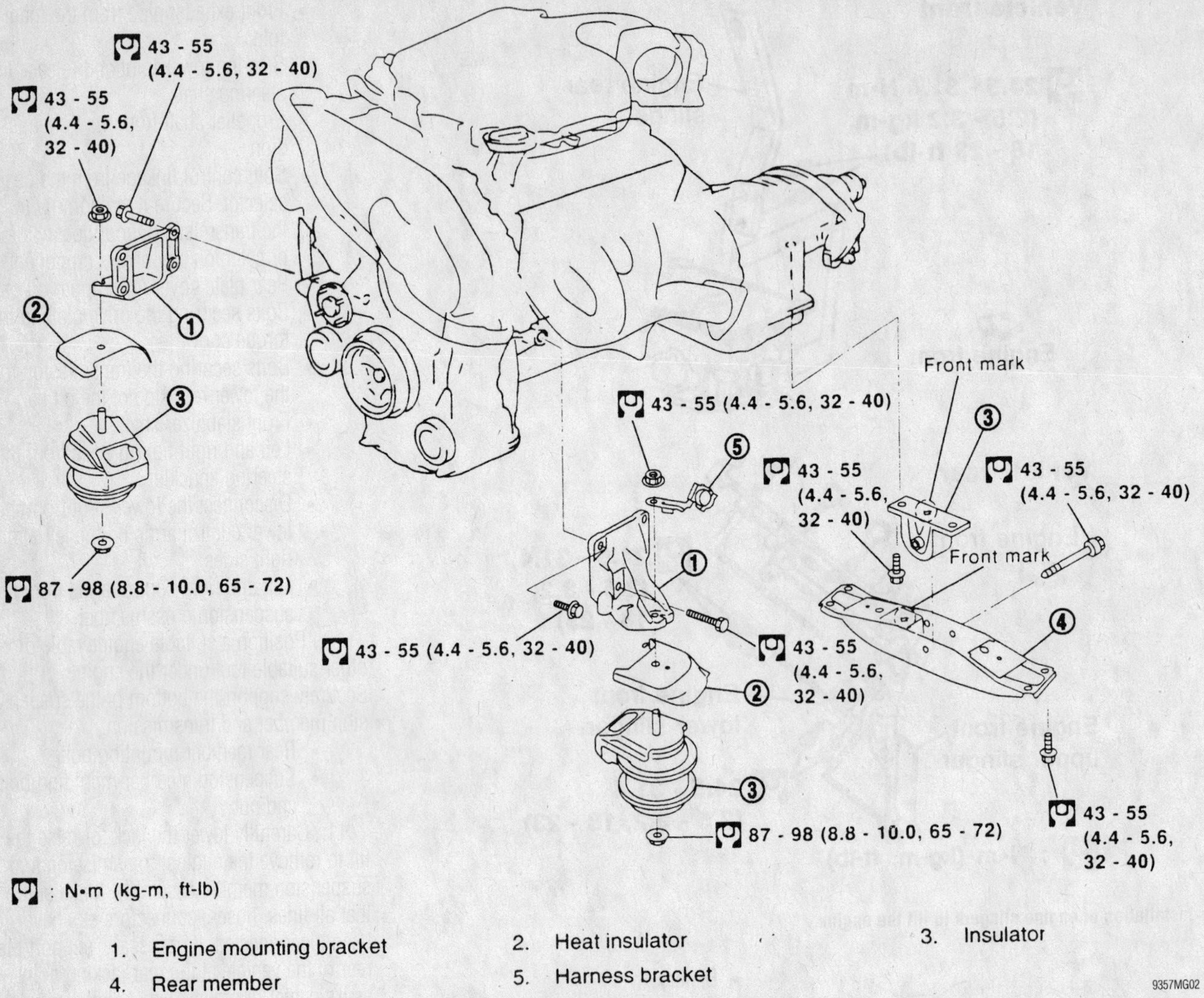

43 - 55
(4.4 - 5.6, 32 - 40)

43 - 55
(4.4 - 5.6,
32 - 40)

Front mark

43 - 55 (4.4 - 5.6, 32 - 40)

43 - 55
(4.4 - 5.6,
32 - 40)

43 - 55
(4.4 - 5.6, 32 - 40)

Front mark

87 - 98 (8.8 - 10.0, 65 - 72)

43 - 55 (4.4 - 5.6, 32 - 40)

43 - 55
(4.4 - 5.6,
32 - 40)

43 - 55
(4.4 - 5.6,
32 - 40)

87 - 98 (8.8 - 10.0, 65 - 72)

: N•m (kg-m, ft-lb)

1. Engine mounting bracket
2. Heat insulator
3. Insulator
4. Rear member
5. Harness bracket

9357MG02

Exploded view of the engine mounts and related components—G35 models

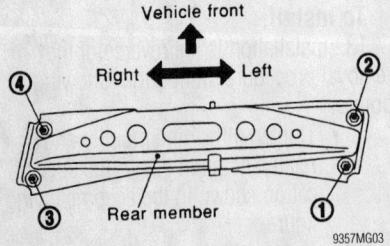

Vehicle front

Right | Left

Rear member

9357MG03

Rear member bolt torque sequence—G35 models

4.1L Engines

1. Before servicing the vehicle, refer to the precautions in the beginning of this section.
2. Evacuate the A/C system.
3. Release the fuel system pressure.
4. Drain the engine oil.
5. Drain the cooling system.

6. Drain the transmission fluid.
7. Remove or disconnect the following:
- Negative battery cable
- Hood
- Engine undercover
- Transmission
- All necessary vacuum hoses, fuel hoses and electrical connections that would interfere with engine removal
- Front exhaust tubes
- Radiator and shroud
- Drive belts
- Power steering oil pump from the engine compartment
8. Attach engine slingers to the cylinder head and attach a suitable hoist to the slinger
- Engine mounting bolts from both sides and then slowly raise the engine

- Engine from the engine compartment

To install:
9. Install or connect the following:
- Engine. Torque the front mounting bolts to 41 ft. lbs. (55 Nm) and the rear mounting bolts to 21 ft. lbs. (28 Nm). Remove the engine supports
- Power steering pump
- Radiator and shroud
- Drive belts
- Front exhaust tubes
- Vacuum hoses, electrical connectors, fuel hoses and wiring
- Transmission
- Engine undercover
- Negative battery cable
- Hood
10. Fill the transmission to the proper level.
11. Fill and bleed the cooling system.

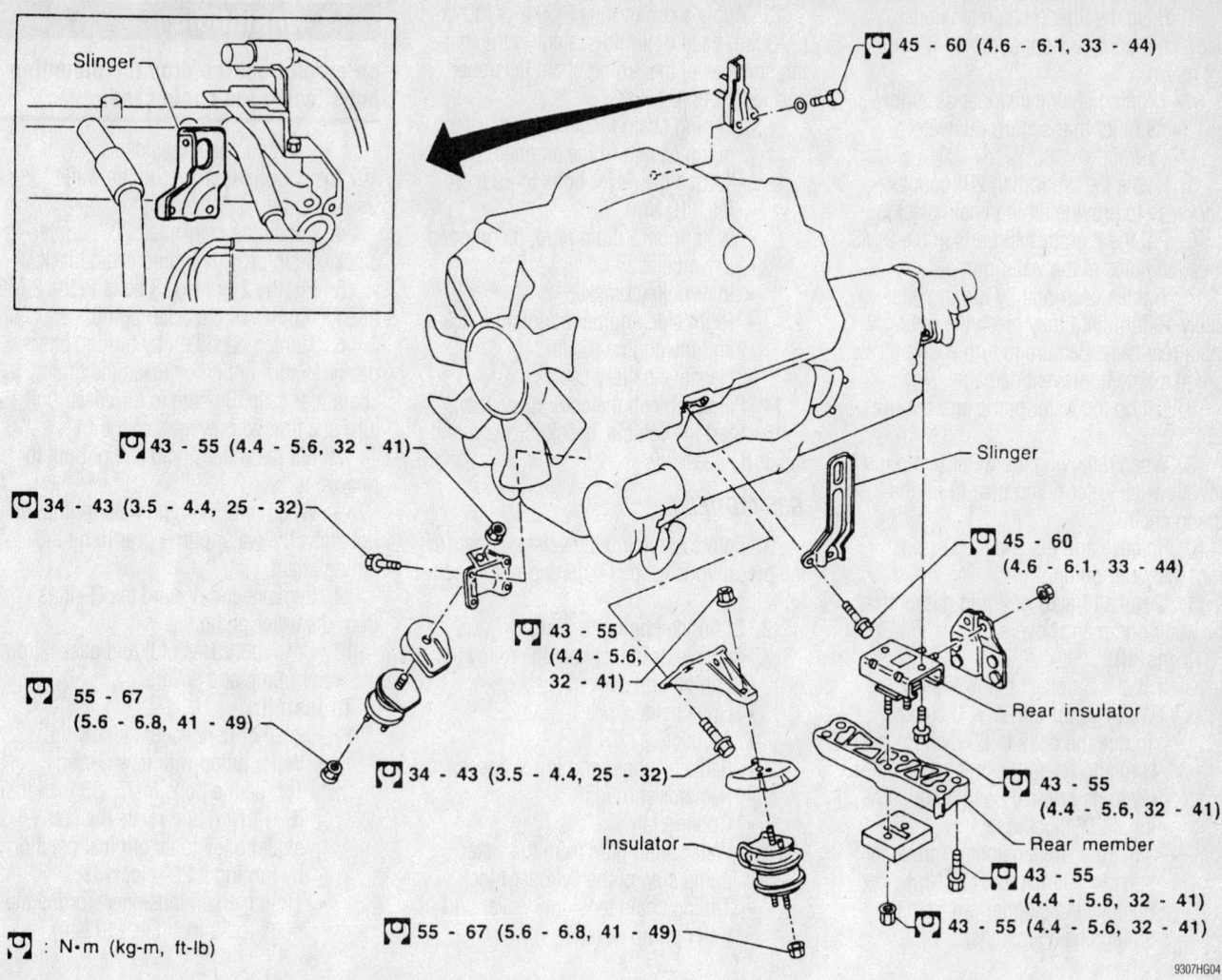

: N•m (kg-m, ft-lb)

45 - 60 (4.6 - 6.1, 33 - 44)

43 - 55 (4.4 - 5.6, 32 - 41)

34 - 43 (3.5 - 4.4, 25 - 32)

55 - 67 (5.6 - 6.8, 41 - 49)

43 - 55 (4.4 - 5.6, 32 - 41)

34 - 43 (3.5 - 4.4, 25 - 32)

Insulator

55 - 67 (5.6 - 6.8, 41 - 49)

Slinger

45 - 60 (4.6 - 6.1, 33 - 44)

Rear insulator

43 - 55 (4.4 - 5.6, 32 - 41)

Rear member

43 - 55 (4.4 - 5.6, 32 - 41)

43 - 55 (4.4 - 5.6, 32 - 41)

9307HG04

Engine mounting and torque specifications—4.1L engines

12. Fill the engine with clean oil.
13. Recharge the A/C system.
14. Start the vehicle, check for leaks and repair if necessary.

Water Pump

REMOVAL & INSTALLATION

2.0L Engine

1. Before servicing the vehicle, refer to the precautions in the beginning of this section.
2. Drain the coolant from the radiator and engine block. The drain plug in the engine block is located at the left front of the cylinder block.
3. Remove or disconnect the following:
   • Negative battery cable
   • Right front wheel
   • Engine side cover
   • Drive belts
   • Right front engine mount
   • Water pump

To install:
4. Clean all mating surfaces and place a 2–3mm bead of liquid gasket on the water pump mating surface.
5. Install or connect the following:
   • Water pump. Torque the bolts to 15 ft. lbs. (21 Nm).
   • Drive belts
   • Front engine mount

Liquid gasket

16 - 21 N•m
(1.6 - 2.1 kg-m,
12 - 15 ft-lb)

Water pump

7923HG08

Exploded view of the water pump mounting—2.0L engine

   • Engine side cover
   • Right front wheel
   • Negative battery cable
6. Fill and bleed the cooling system.
7. Start the vehicle, check for leaks and repair if necessary.

3.0L and 3.5L Engines

EXCEPT G35 MODELS

1. Before servicing the vehicle, refer to the precautions in the beginning of this section.
2. Drain the cooling system.
3. Remove or disconnect the following:
   • Negative battery cable
   • Right side engine mount and bracket
   • Drive belts
   • Idler pulley bracket
   • Water pump drain plug, if equipped
   • Timing chain tensioner cover
   • Water pump cover

4. Push the timing chain tensioner sleeve and apply a stopper pin so it does not return.
- Timing chain tensioner assembly
- 3 bolts that secure the water pump

5. Rotate the crankshaft 20° counter-clockwise to provide timing chain slack.

6. Put the 2 grade M8 bolts in the 2 M8 threaded holes of the water pump.

7. Tighten each bolt by turning alternately ½ turn until they reach the timing chain rear case. Be sure to turn each bolt ½ turn at a time to prevent damage.

8. Lift up the water pump and remove it.

9. When removing the water pump, do not allow the water pump gear to hit the timing chain.

10. Remove and discard the O-rings from the water pump.

11. Clean all traces of liquid gasket from the water pump and covers.

To install:

12. Install or connect the following:
- Water pump with new O-rings. Torque the bolts to 89 inch lbs. (10 Nm) and rotate the crankshaft pulley to its original position by turning it 20° clockwise.
- Timing chain tensioner. Torque the bolts to 89 inch lbs. (10 Nm). Remove the stopper pin from the timing chain tensioner.

13. Apply a continuous 0.091–0.130 in. (2–3mm) bead of liquid sealant to the mating surfaces of the timing chain tensioner and water pump covers.
- Timing chain tensioner and water pump covers to the engine block. Torque the cover bolts to 89 inch lbs. (10 Nm).
- Water pump drain plug, if equipped
- Drive belts
- Idler pulley bracket
- Right side engine mounting bracket and the engine mount
- Negative battery cable

14. Fill and bleed the cooling system.

15. Start the vehicle, check for leaks and repair if necessary.

G35 MODELS

1. Before servicing the vehicle, refer to the precautions in the beginning of this section.

2. Drain the cooling system.

3. Remove or disconnect the following:
- Engine undercover
- Drive belts
- Air duct
- Radiator upper and lower hoses
- Radiator shrouds
- Cooling fan
- Water drain plug from the water pump side of the cylinder block
- Timing chain tensioner cover and water pump cover

✳✳ WARNING

Be careful not the drop the mounting bolts inside the chain case.

- Timing chain tensioner
- 3 bolts that secure the water pump

4. Rotate the crankshaft 20° counter-clockwise to provide timing chain slack.

5. Put the 2 grade M8 bolts in the 2 M8 threaded holes of the water pump.

6. Tighten each bolt by turning alternately ½ turn until they reach the timing chain rear case. Be sure to turn each bolt ½ turn at a time to prevent damage.

7. Lift the water pump straight out to remove it.

8. When removing the water pump, do not allow the water pump gear to hit the timing chain.

9. Remove and discard the O-rings from the water pump.

10. Clean all traces of liquid gasket from the water pump and covers.

To install:

11. Install or connect the following:
- Water pump with new O-rings. Torque the bolts to 75–95 inch lbs. (8–11 Nm) and rotate the crankshaft pulley to its original position by turning it 20° clockwise.
- Timing chain tensioner. Torque the bolts to 89 inch lbs. (10 Nm).

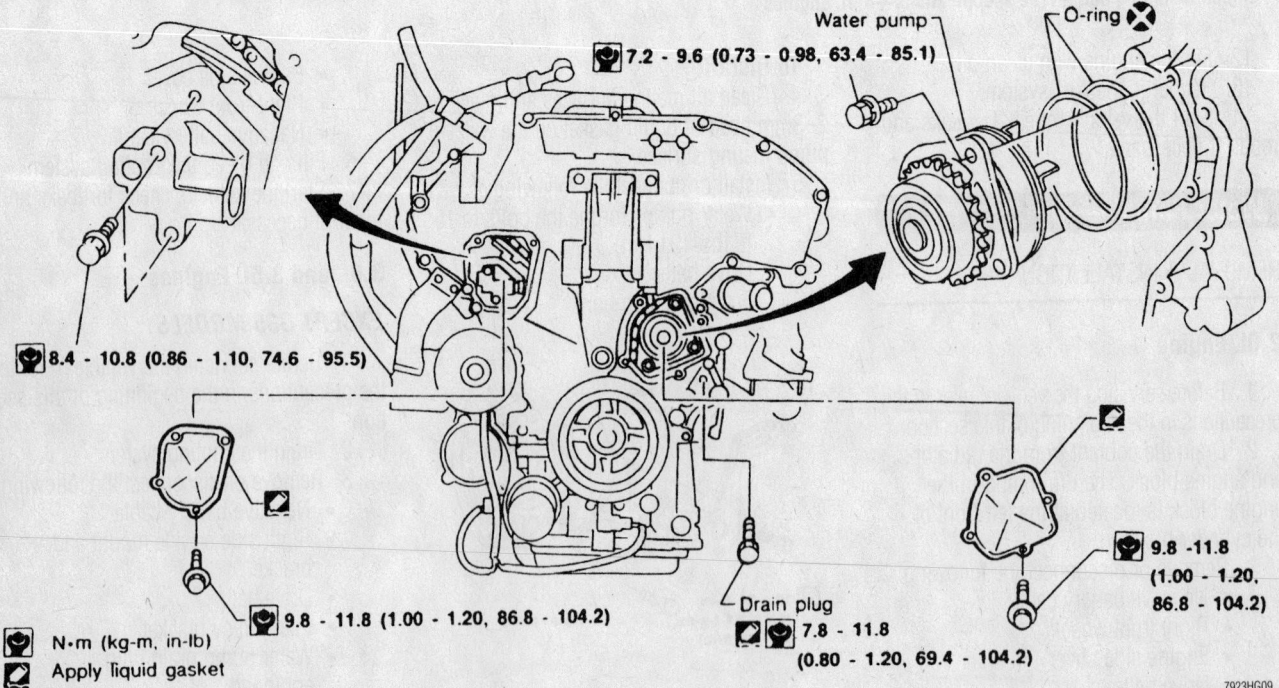

Water pump and timing cover assembly—3.0L and 3.5L engines, except G35

7.2 - 9.6 (0.73 - 0.98, 63.4 - 85.1)

8.4 - 10.8 (0.86 - 1.10, 74.6 - 95.5)

9.8 - 11.8 (1.00 - 1.20, 86.8 - 104.2)

Drain plug
7.8 - 11.8 (0.80 - 1.20, 69.4 - 104.2)

9.8 - 11.8 (1.00 - 1.20, 86.8 - 104.2)

N·m (kg-m, in-lb)
Apply liquid gasket

7923HG09

SEC. 130•135•210

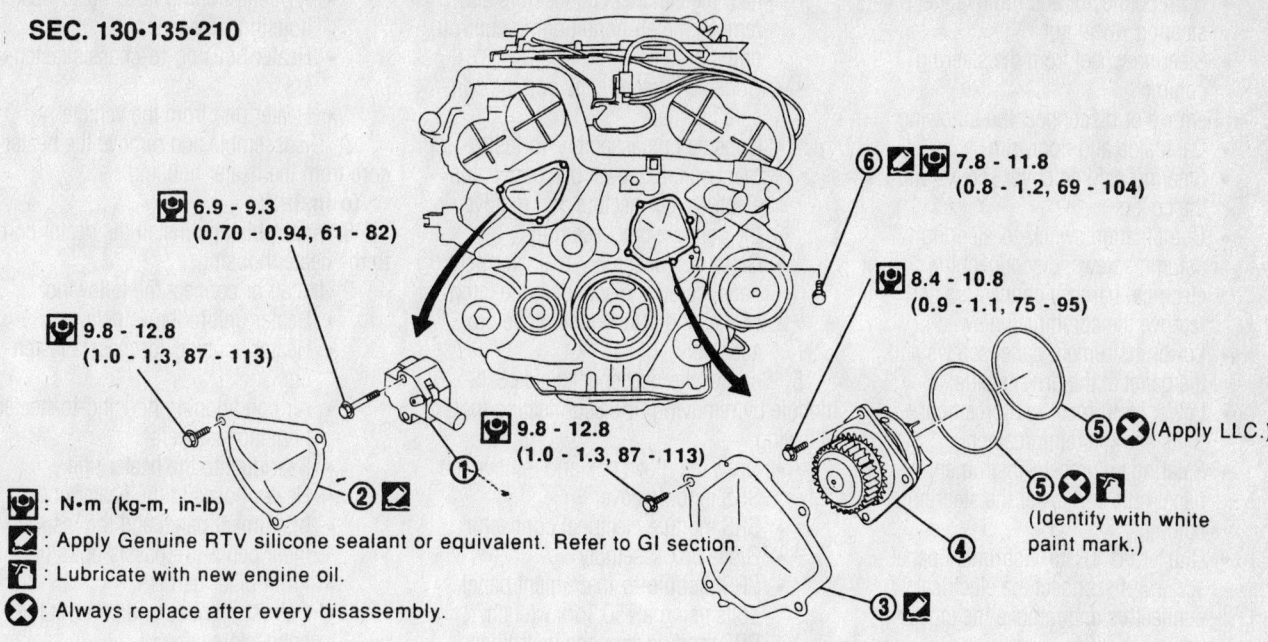

6.9 - 9.3
(0.70 - 0.94, 61 - 82)

7.8 - 11.8
(0.8 - 1.2, 69 - 104)

8.4 - 10.8
(0.9 - 1.1, 75 - 95)

9.8 - 12.8
(1.0 - 1.3, 87 - 113)

9.8 - 12.8
(1.0 - 1.3, 87 - 113)

(Apply LLC.)

(Identify with white paint mark.)

: N•m (kg-m, in-lb)

: Apply Genuine RTV silicone sealant or equivalent. Refer to GI section.

: Lubricate with new engine oil.

: Always replace after every disassembly.

1. Chain tensioner
4. Water pump
2. Chain tensioner cover
5. O - rings
3. Water pump cover
6. Drain plug

9357MG04

Exploded view of the water pump mounting—G35 models

Remove the stopper pin from the timing chain tensioner.

12. Apply a continuous 0.091–0.130 in. (2–3mm) bead of liquid sealant to the mating surfaces of the timing chain tensioner and water pump covers.

- Timing chain tensioner and water pump covers to the engine block. Torque the cover bolts to 89 inch lbs. (10 Nm).
- Water drain plug to the water pump side of the cylinder block
- Cooling fan
- Radiator shrouds
- Radiator upper and lower hoses
- Air duct
- Drive belts
- Engine undercover
- Negative battery cable

13. Fill and bleed the cooling system.

14. Start the vehicle, check for leaks and repair if necessary.

4.1L Engine

1. Before servicing the vehicle, refer to the precautions in the beginning of this section.

2. Drain the cooling system.

3. Remove or disconnect the following:

- Negative battery cable
- Loosen the drive belts
- Fan coupling and fan assembly

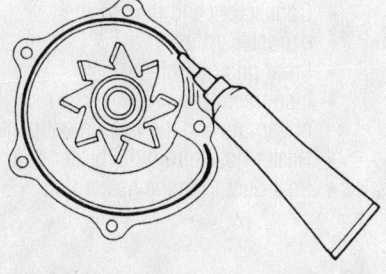

7923HG11

Apply a continuous bead of RTV sealant to the mounting surface of the water pump assembly

- Idler pulley bracket and drive belt
- Water pump

To install:

4. Thoroughly clean and dry the mating surfaces, bolts and bolt holes.

5. Apply liquid gasket to the water pump.

6. Install or connect the following:

- Water pump. Torque the bolts to 13 ft. lbs. (18 Nm).
- Idler pulley bracket. Torque the bolts to 89 inch lbs. (10 Nm).
- Drive belts
- Fan and coupling assembly
- Negative battery cable

7. Fill and bleed the cooling system.

8. Start the vehicle, check for leaks and repair if necessary.

Heater Core

REMOVAL & INSTALLATION

G20

1. Disconnect both battery cables, the negative (() cable first.

✸✸ CAUTION

After disconnecting the battery, wait for a least 3 minutes for the SRS module to deplete its energy before working on the steering column or instrument panel.

2. Drain the cooling system into a clean container for reuse.

3. Remove the driver's side SRS module and the steering wheel by removing or disconnecting the following:

- Lower cover at the base of the steering wheel
- SRS module electrical connector
- Side covers at both sides of the steering wheel
- SRS module-to-steering wheel bolts using a T50 Torx wrench
- SRS module from the steering wheel
- Place the front wheel in the straight-ahead position

- Horn connector and remove the steering wheel nut
- Steering wheel from the steering column

4. Remove or disconnect the following:
- Dash side and floor trim
- Steering column cover screws and the covers
- Combination switch-to-steering column screws, disconnect the electrical harness connectors and remove the combination switch
- Lower instrument panel screws and the panel at the driver's side
- Lower instrument reinforcement bolts and the reinforcement
- Steering column-to-instrument panel nuts and lower the steering column
- Cluster lid "C"-to-instrument panel screws, disconnect the electrical connectors and remove the cluster lid
- Cluster lid "A"-to-instrument panel screws, disconnect the electrical connectors and remove the cluster lid
- Combination meter-to-instrument panel screws, disconnect the electrical connectors and remove the combination meter
- Audio center-to-instrument panel bolts, disconnect the electrical connectors and remove the audio center
- Air conditioning control unit-to-instrument panel screws, discon-

nect the electrical connectors and remove the air conditioning control unit
- Console finisher clips and the console finisher
- Console box assembly-to-instrument panel screws, disconnect the electrical connectors and remove the console box assembly
- Glove box assembly-to-instrument panel screws, disconnect the lamp socket and remove the glove box assembly

5. Remove the passenger's side SRS module by removing or disconnecting the following:
- Open the glove box and remove the SRS module cover
- SRS module electrical connector
- Glove box assembly
- SRS module-to-instrument panel bolts using a T50 Torx wrench
- SRS module from the instrument panel

6. Remove or disconnect the following:
- Lower instrument cover clip and the cover
- Lower instrument panel center-to-instrument panel screws and the panel
- Connectors and remove the defroster grilles
- Front pillar garnish
- Instrument panel-to-chassis bolts/nuts and the instrument panel
- Heater hoses from the heater core
- Rear duct from the heater unit

- Air conditioning housing-to-heater housing fasteners
- Heater housing-to-chassis fasteners
- Heater unit from the vehicle

7. Disassemble and remove the heater core from the heater housing.

To install:

8. Assemble and install the heater core to the heater housing.

9. Install or connect the following:
- Heater unit to the vehicle
- Heater housing-to-chassis fasteners
- Air conditioning housing-to-heater housing fasteners
- Rear duct to the heater unit
- Heater hoses to the heater core
- Instrument panel and the instrument panel-to-chassis bolts/nuts
- Front pillar garnish
- Defroster grilles and connect the connectors
- Lower instrument panel center and the panel-to-instrument panel screws
- Lower instrument cover and the cover clip

10. Install the passenger's side SRS module by installing or connecting the following:
- SRS module to the instrument panel
- SRS module-to-instrument panel bolts and torque the bolts using a T50 Torx wrench to 11–18 ft. lbs. (15–25 Nm)

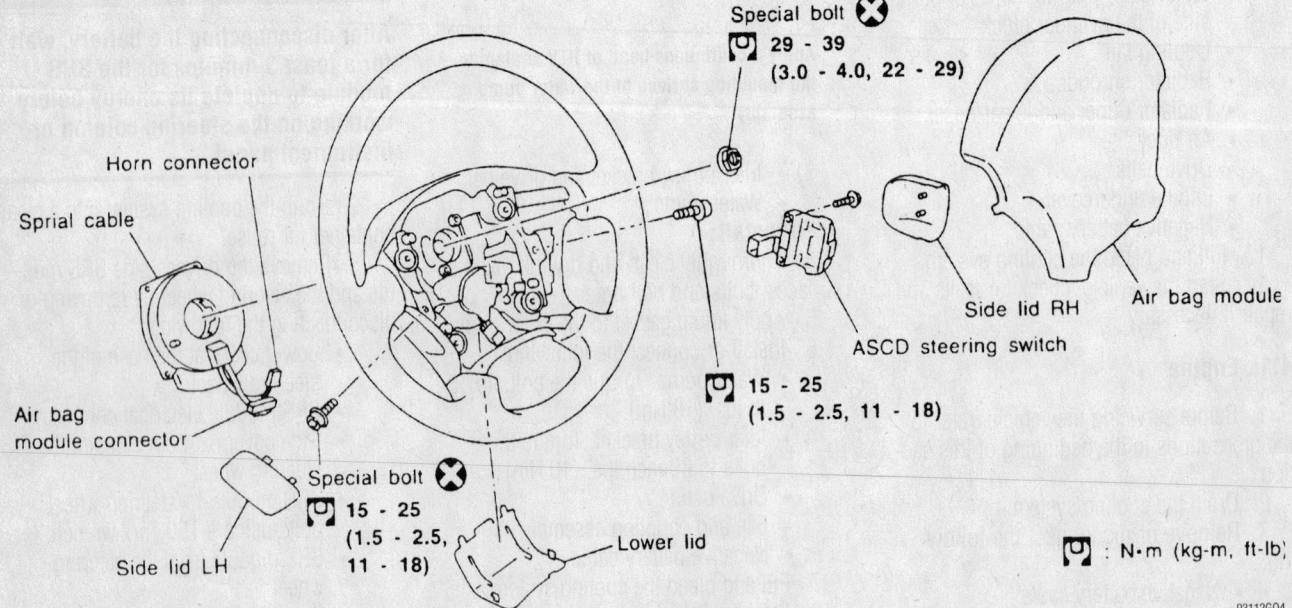

Special bolt ⊗
29 - 39
(3.0 - 4.0, 22 - 29)

Horn connector

Sprial cable

Air bag module connector

Air bag module

Side lid RH

ASCD steering switch

15 - 25
(1.5 - 2.5, 11 - 18)

Special bolt ⊗
15 - 25
(1.5 - 2.5, 11 - 18)

Side lid LH

Lower lid

: N•m (kg-m, ft-lb)

93112G04

Exploded view of the air bag module, the steering wheel and related components—G20

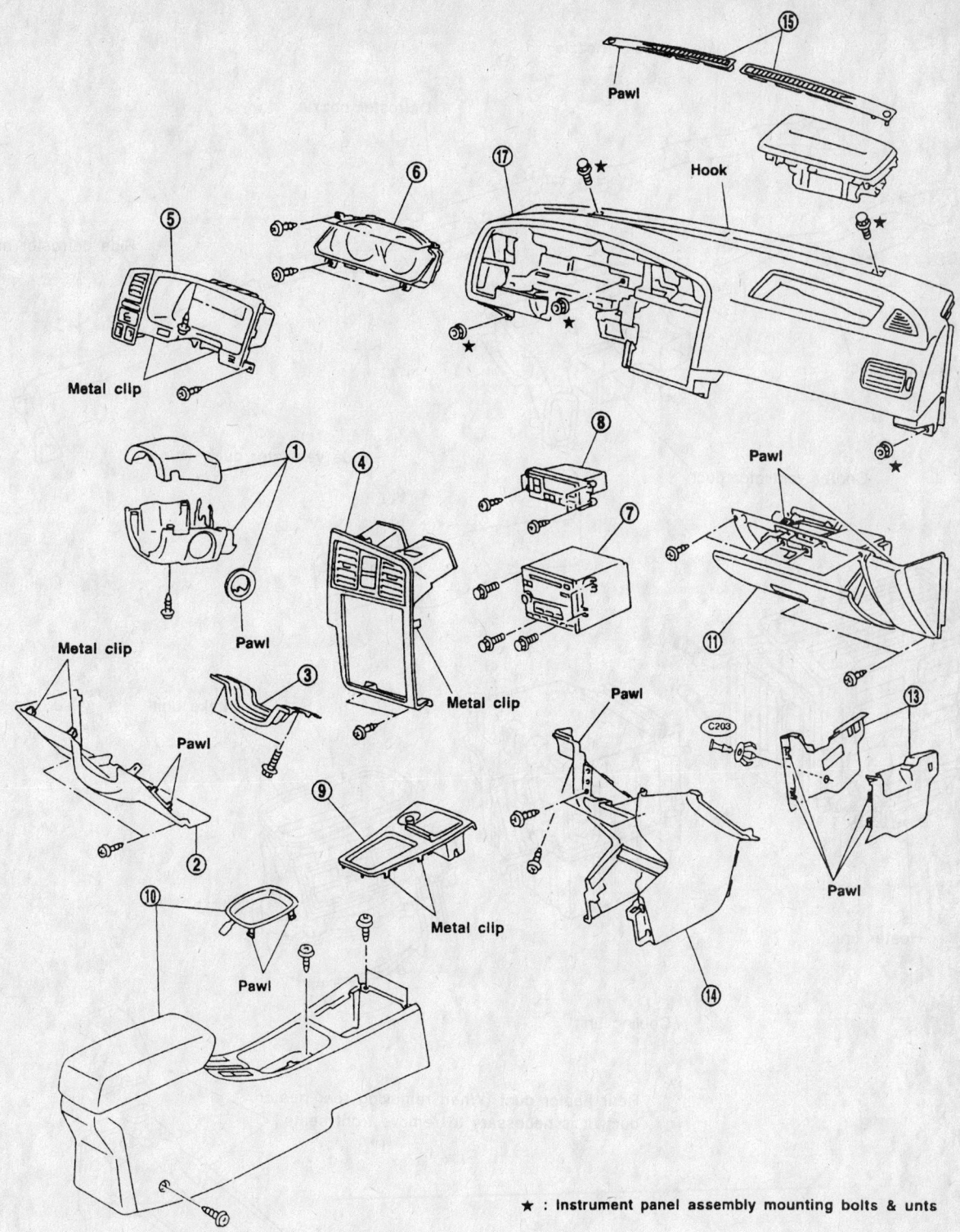

★ : Instrument panel assembly mounting bolts & unts

1. Steering column cover
   and combination switch
2. Instrument lower panel
   on driver side
3. Instrument lower reinforcement
4. Cluster lid C
5. Cluster lid A

6. Combination meter
7. Audio
8. A/C control unit
9. Console M/T or A/T finisher
10. Console box assembly
11. Glove box assembly

12. Passenger air bag module
13. Lower instrument cover
14. Lower instrument panel center
15. Defroster grille
16. Front pillar garnish
17. Instrument panel and pads

Exploded view of the instrument panel—G20

93112G05

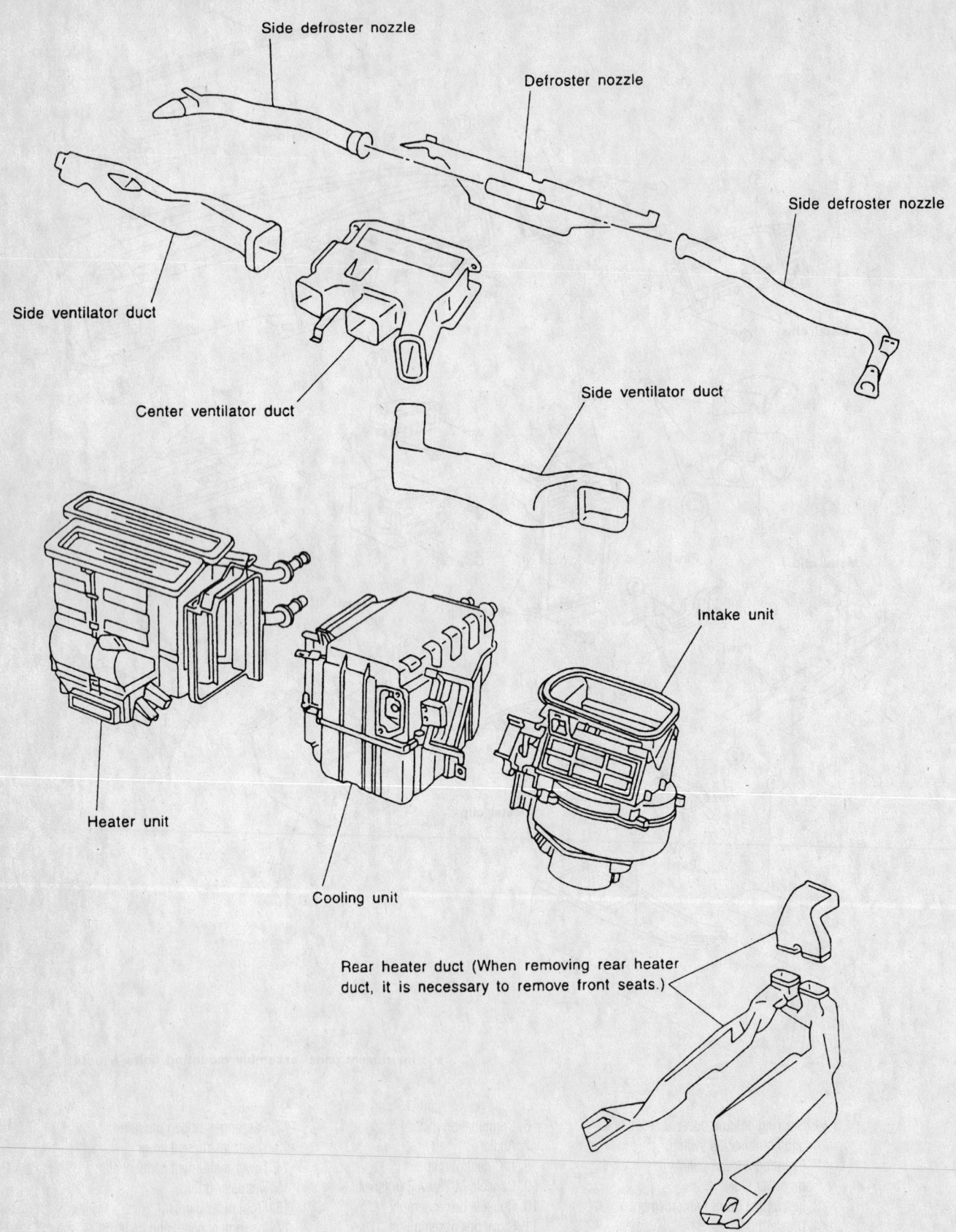

Side defroster nozzle

Defroster nozzle

Side defroster nozzle

Side ventilator duct

Center ventilator duct

Side ventilator duct

Intake unit

Heater unit

Cooling unit

Rear heater duct (When removing rear heater duct, it is necessary to remove front seats.)

93112G06

Exploded view of the heater housing, air conditioning housing, blower motor and ventilation ducts—G20

- Glove box assembly
- SRS module electrical connector
- SRS module cover
11. Install or connect the following:
   - Glove box assembly, connect the lamp socket and install the glove box assembly-to-instrument panel screws
   - Console box assembly, connect the electrical connectors and install the console box assembly-to-instrument panel screws
   - Console finisher and engage the clips
- Air conditioning control unit, connect the electrical connectors and Install the air conditioning control unit-to-instrument panel screws
- Audio center, connect the electrical connectors and Install the audio center-to-instrument panel bolts
- Combination meter, connect the electrical connectors and Install the combination meter-to-instrument panel screws
- Cluster lid "A", connect the electrical connectors and install the cluster lid-to-instrument panel screws
- Cluster lid "C", connect the electrical connectors and install the cluster lid-to-instrument panel screws
- Steering column and torque the steering column-to-instrument panel nuts to 11–14 ft. lbs. (15–19 Nm)
- Lower instrument reinforcement and the reinforcement bolts
- Lower instrument panel and the panel screws at the driver's side
- Combination switch and the combination switch-to-steering column screws and connect the electrical harness connectors

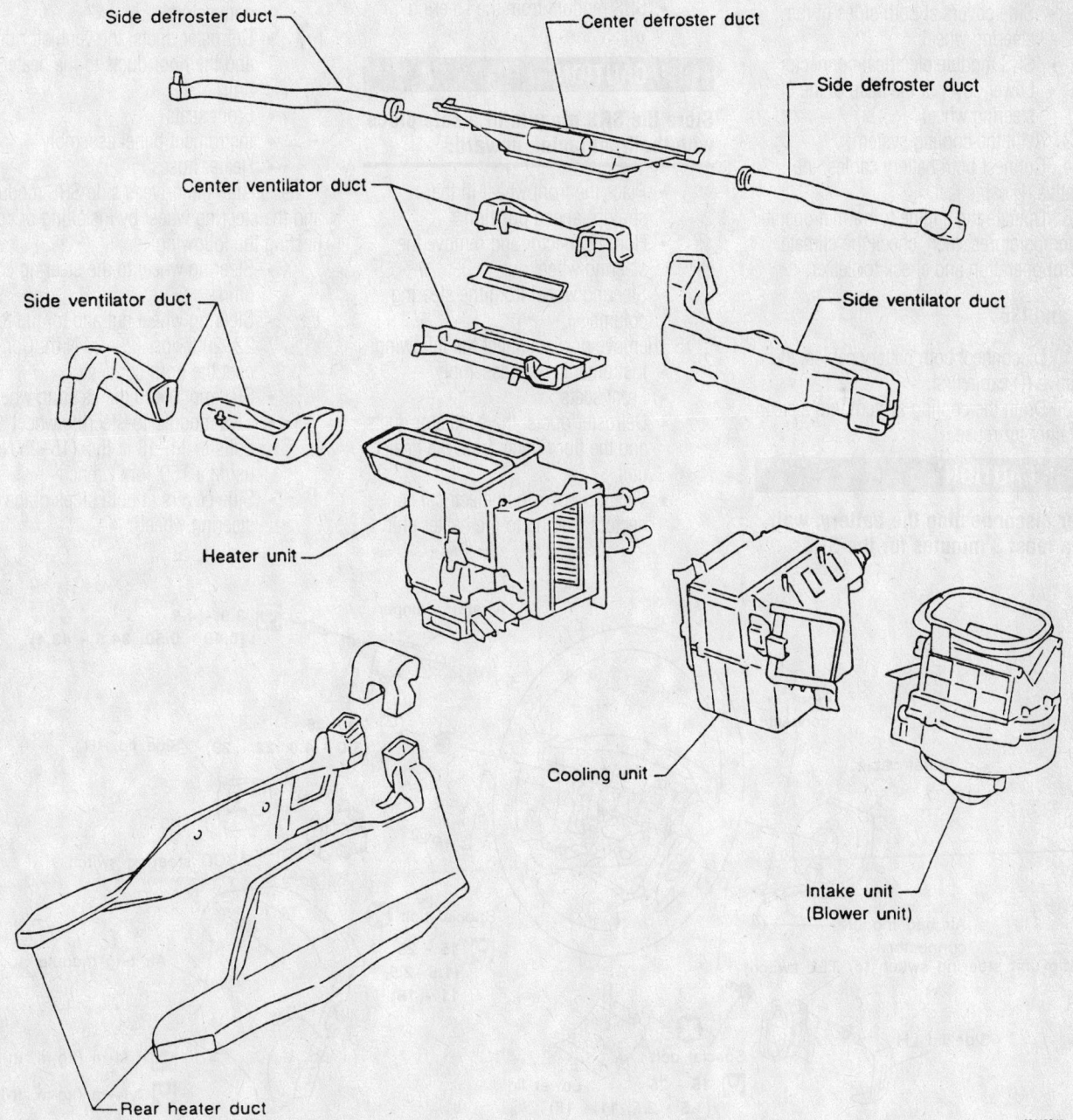

Side defroster duct
Center defroster duct
Side defroster duct
Center ventilator duct
Side ventilator duct
Side ventilator duct
Heater unit
Cooling unit
Intake unit (Blower unit)
Rear heater duct

93112G45

Exploded view of the heating-air conditioning system assemblies and related components—I30

- Steering column covers and the cover screws
- Dash side and floor trim

12. Install the driver's side SRS module and the steering wheel by installing or connecting the following:

- Steering wheel to the steering column
- Steering wheel nut and torque to 22–29 ft. lbs. (29–39 Nm). Connect the horn connector
- SRS module to the steering wheel
- Torque the SRS module-to-steering wheel bolts to 11–18 ft. lbs. (15–25 Nm) using a T50 Torx wrench
- Side covers at both sides of the steering wheel
- SRS module electrical connector
- Lower cover at the base of the steering wheel

13. Refill the cooling system.

14. Connect both battery cables, the negative (()) cable last.

15. Operate the engine to normal operating temperatures; then, check the climate control operation and check for leaks.

I30 and I35

1. Disconnect both battery cables, the negative (()) cable first.

2. Drain the cooling system into a clean container for reuse.

✳✳ CAUTION

After disconnecting the battery, wait for a least 3 minutes for the SRS

module to deplete its energy before working on the steering column or instrument panel.

3. Drain the cooling system into a clean container for reuse.

4. Remove the driver's side SRS module and the steering wheel by removing or disconnecting the following:

- Lower cover at the base of the steering wheel
- SRS module electrical connector
- Side covers at both sides of the steering wheel
- SRS module-to-steering wheel bolts using a T50 Torx wrench
- SRS module from the steering wheel

✳✳ CAUTION

Store the SRS module in a safe place with the front facing upward.

- Place the front wheel in the straight-ahead position
- Horn connector and remove the steering wheel nut
- Steering wheel from the steering column

5. Remove or disconnect the following:
- Instrument panel assembly
- Front seats
- Defroster ducts, the ventilator ducts and the floor ducts from the heater unit
- Vacuum hoses and electrical connectors leading to the heater unit

- Heater unit from the cooling unit. Take care not to damage the air conditioning tubes
- Heater unit attaching bolts. Remove the heater unit from the passenger compartment

6. Disassemble the heater unit and remove the heater core.

To install:

7. Install the heater core and assemble the heater unit.

8. Install or connect the following:
- Heater unit in the passenger compartment and tighten the attaching bolts securely
- All vacuum hoses and electrical connectors
- Defroster ducts, the ventilator ducts and the floor ducts to the heater unit
- Front seats
- Instrument panel assembly
- Heater hoses

9. Install the driver's side SRS module and the steering wheel by installing or connecting the following:

- Steering wheel to the steering column
- Steering wheel nut and torque to 22–29 ft. lbs. (29–39 Nm). Connect the horn connector
- SRS module to the steering wheel
- SRS module-to-steering wheel bolts to 11–18 ft. lbs. (15–25 Nm) using a T50 Torx wrench
- Side covers at both sides of the steering wheel

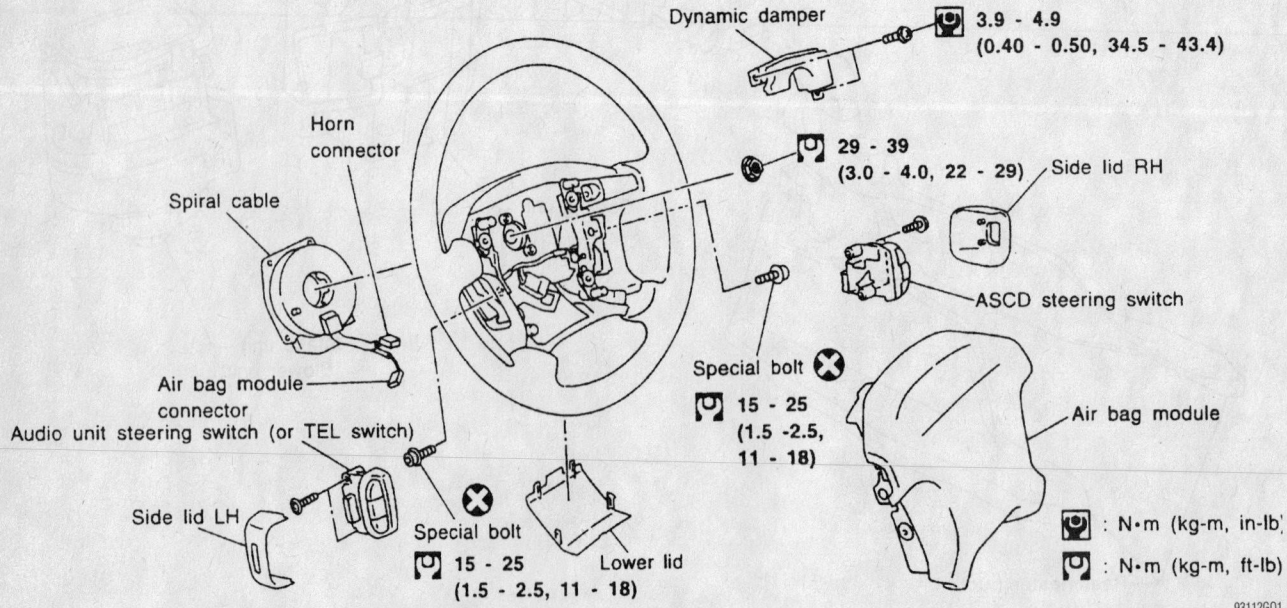

Exploded view of the air bag module, the steering wheel and related components—Q45

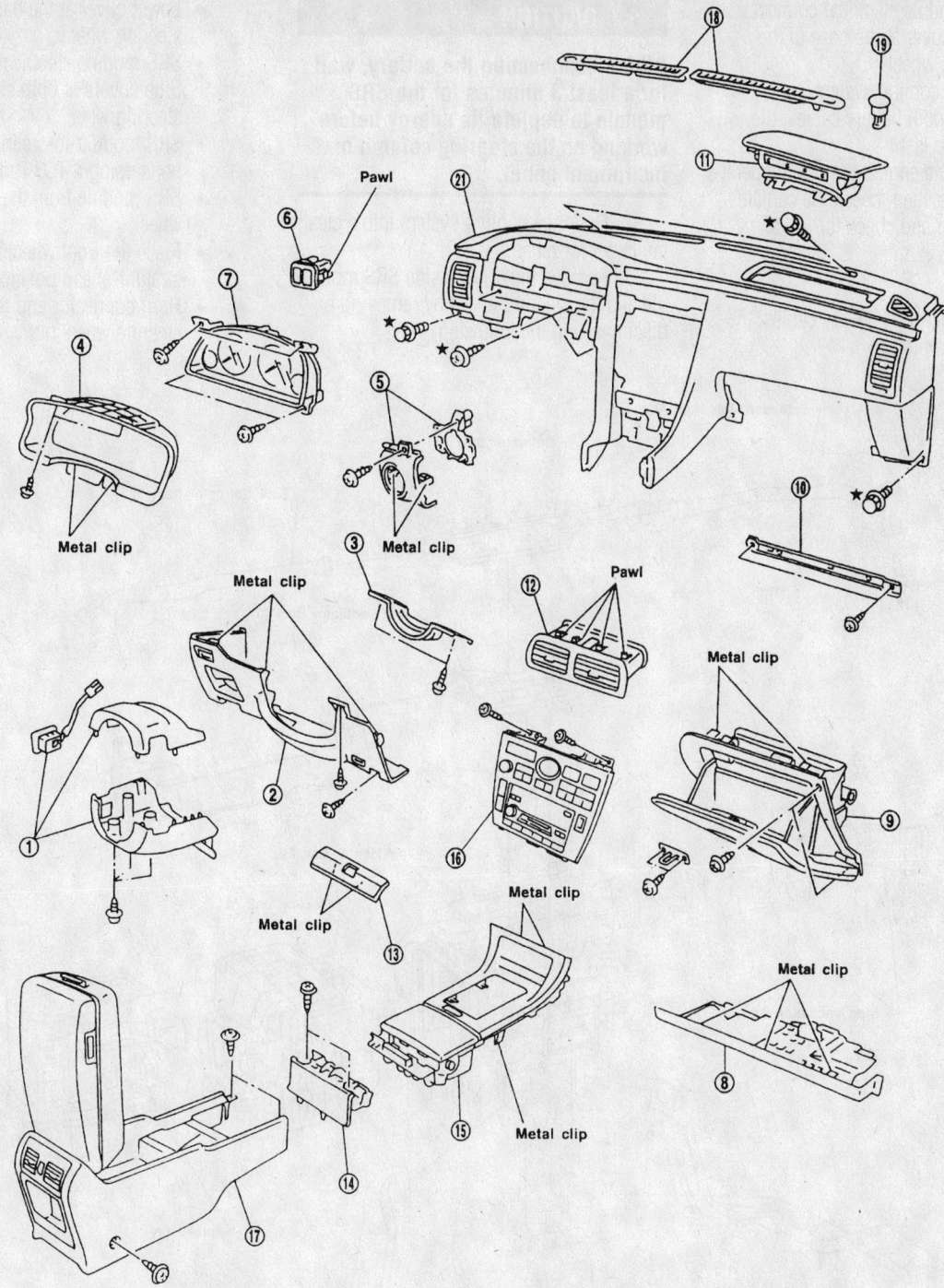

★ : Instrument panel assembly mounting bolts and screws

1. Steering column cover and combination switch
2. Instrument lower panel on driver side
3. Instrument lower reinforcement
4. Cluster lid A
5. Steering lock escutcheon
6. Cluster lid D
7. Combination meter
8. Instrument lower cover on passenger side
9. Glove box assembly
10. Instrument panel reinforcement
11. Passenger air bag module

12. Center ventilation grille
13. Lock console
14. Card pocket assembly
15. Console A/T finisher
16. Audio, cluster lid C and A/C control unit
17. Console box assembly
18. Defroster grille
19. Sunload sensor
20. Front pillar garnish
21. Instrument panel and pads

93112G02

Exploded view of the instrument panel—Q45

- SRS module electrical connector
- Lower cover at the base of the steering wheel
10. Refill the cooling system.
11. Connect both battery cables, the negative (() cable last.
12. Operate the engine to normal operating temperatures; then, check the climate control operation and check for leaks.

Q45

1. Disconnect both battery cables, the negative (() cable first.

❈❈ CAUTION

After disconnecting the battery, wait for a least 3 minutes for the SRS module to deplete its energy before working on the steering column or instrument panel.

2. Drain the cooling system into a clean container for reuse.
3. Remove the driver's side SRS module and the steering wheel by removing or disconnecting the following:

- Lower cover at the base of the steering wheel
- SRS module electrical connector
- Side covers at both sides of the steering wheel
- SRS module-to-steering wheel bolts using a T50 Torx wrench
- SRS module from the steering wheel
- Place the front wheel in the straight-ahead position
- Horn connector and remove the steering wheel nut

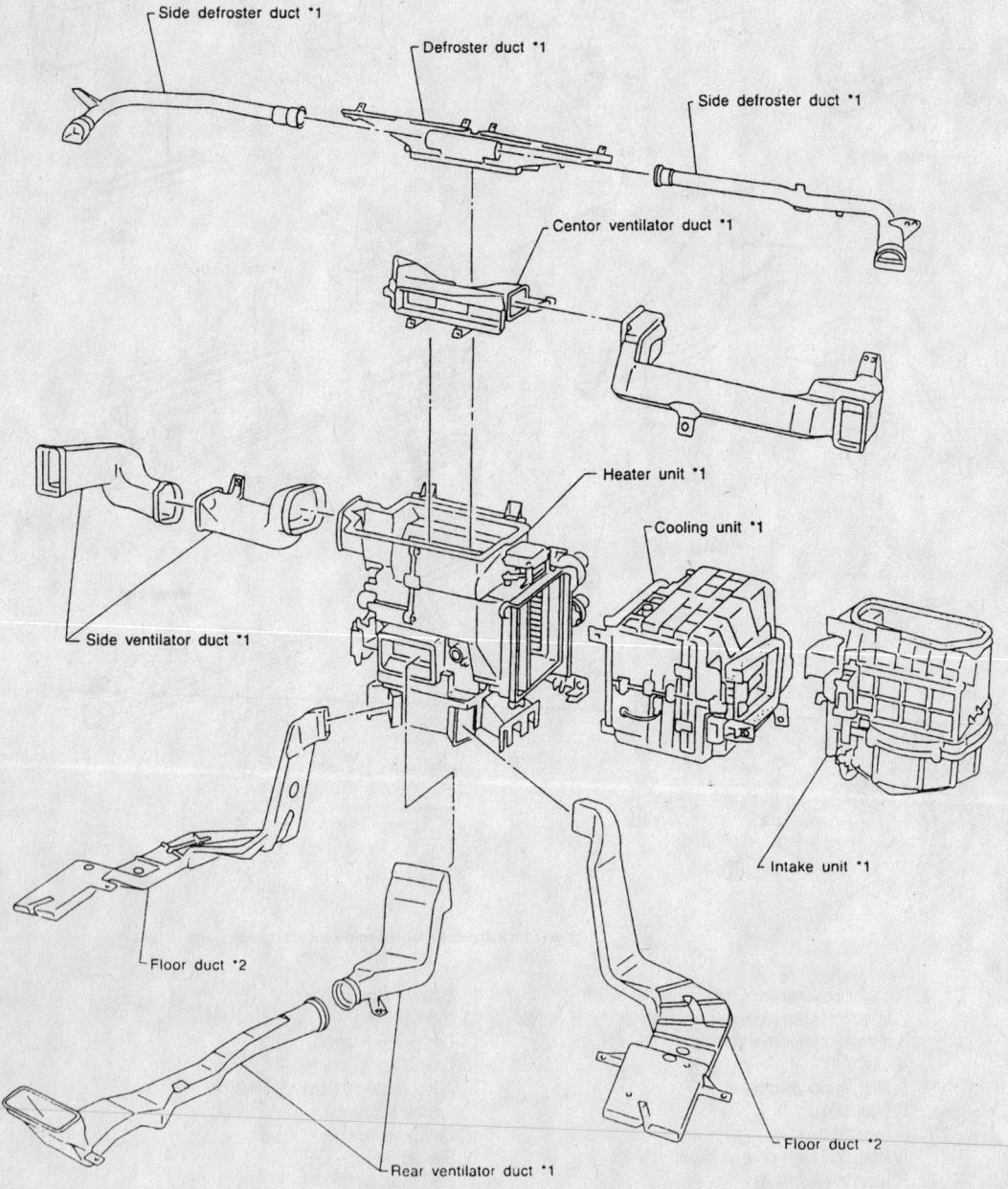

*1 : For removal, it is necessary to remove instrument assembly.

*2 : For removal, it is necessary to remove front seat.

93112G03

Exploded view of the heater housing, air conditioning housing, blower motor and ventilation ducts—Q45

- Steering wheel from the steering column
4. Remove or disconnect the following:
  - Dash side lower finishers
  - Steering column cover screws and the covers
  - Combination switch-to-steering column screws, disconnect the electrical harness connectors and remove the combination switch
  - Lower instrument panel screws/bolts and the panel. Disconnect the electrical harness connectors and the in-vehicle sensor at the driver's side
  - Lower instrument reinforcement bolts and the reinforcement
  - Steering column-to-instrument panel nuts and lower the steering column
  - Cluster lid "A"-to-instrument panel screws and remove the cluster lid
  - Steering lock escutcheon screws and the escutcheon
  - Cluster lid "D"
  - Combination meter-to-instrument panel screws, disconnect the electrical connectors and remove the combination meter
  - Glove box assembly-to-instrument panel screws, disconnect the lamp socket and remove the glove box assembly
5. Remove the passenger's side SRS module by removing or disconnecting the following:
  - Lower instrument panel cover
  - SRS module electrical connector
  - SRS module-to-instrument panel bolts using a T50 Torx wrench
  - SRS module from the instrument panel

✳✳ CAUTION

Store the SRS module in a safe place with the front facing upward.

6. Remove or disconnect the following:
  - Center ventilation grille using a suitable prytool
  - Lock console
  - Card pocket assembly screws and the assembly
  - Console finisher clips and the console finisher
  - Audio center, the cluster lid "C" and the air conditioning control unit-to-instrument panel screws, disconnect the electrical connectors and remove the panel
  - Console box assembly-to-instru-

ment panel screws, disconnect the electrical connectors and remove the console box assembly
  - Connectors and remove the defroster grilles
  - Connectors and remove the sunload sensor
  - Front pillar garnish
  - Instrument panel-to-chassis bolts/nuts and the instrument panel
  - Heater hoses from the heater core
  - Rear duct from the heater unit
  - Air conditioning housing-to-heater housing fasteners
  - Heater unit from the vehicle
7. Disassemble and remove the heater core from the heater housing.

To install:

8. Assemble and install the heater core to the heater housing.
9. Install or connect the following:
  - Heater unit to the vehicle
  - Heater housing-to-chassis fasteners
  - Air conditioning housing-to-heater housing fasteners
  - Rear duct to the heater unit
  - Heater hoses to the heater core
  - Instrument panel and the instrument panel-to-chassis bolts/nuts
  - Front pillar garnish
  - Defroster grilles and connect the connectors
  - Lower instrument panel center and the panel-to-instrument panel screws
  - Lower instrument cover and the cover clip
  - Center ventilation grille
  - Lock console
  - Card pocket assembly and the assembly screws
  - Console finisher and engage the clips
  - Audio center, the cluster lid "C" and the air conditioning control unit panel, connect the electrical connectors and install the panel-to-instrument panel screws
  - Console box assembly, connect the electrical connectors and install the console box assembly-to-instrument panel screws
  - Defroster grilles and connect the connector
  - Sunload sensor and connectors
10. Install the passenger's side SRS module by installing or connecting the following:
  - SRS module to the instrument panel

- SRS module-to-instrument panel bolts and torque the bolts to 11–18 ft. lbs. (15–25 Nm) using a T50 Torx wrench
  - SRS module electrical connector
  - Lower instrument panel cover
  - Glove box assembly, connect the lamp socket and install the glove box assembly-to-instrument panel screws
11. Install or connect the following:
  - Combination meter, connect the electrical connectors and install the combination meter-to-instrument panel screws
  - Cluster lid "D"
  - Steering lock escutcheon and the escutcheon screws
  - Cluster lid "A" and the cluster lid-to-instrument panel screws
  - Steering column and torque the steering column-to-instrument panel nuts to 11–14 ft. lbs. (15–19 Nm)
  - Lower instrument reinforcement and the reinforcement bolts.
  - Electrical harness connectors and the in-vehicle sensor; then, install the lower instrument panel and the panel screws/bolts at the driver's side
  - Combination switch and the combination switch-to-steering column screws and connect the electrical harness connectors
  - Steering column covers and the cover screws
  - Dash side lower finishers
12. Install the driver's side SRS module and the steering wheel by installing or connecting the following:
  - Steering wheel to the steering column
  - Steering wheel nut and torque to 22–29 ft. lbs. (29–39 Nm). Connect the horn connector
  - SRS module to the steering wheel
  - Torque the SRS module-to-steering wheel bolts to 11–18 ft. lbs. (15–25 Nm) using a T50 Torx wrench
  - Side covers at both sides of the steering wheel
  - SRS module electrical connector
  - Lower cover at the base of the steering wheel
13. Refill the cooling system.
14. Connect both battery cables, the negative (()) cable last.
15. Operate the engine to normal operating temperatures; then, check the climate control operation and check for leaks.

Cylinder Head

REMOVAL & INSTALLATION

2.0L Engine

1. Before servicing the vehicle, refer to the precautions in the beginning of this section.
2. Relieve the fuel system pressure.
3. Drain the cooling system.
4. Remove or disconnect the following:
   - Negative battery cable
   - Right front wheel
   - Engine side cover
   - Radiator
   - Air duct to intake manifold
   - ASCD actuator
   - Vacuum and fuel hoses
   - Electrical connectors and wiring
   - Spark plugs
   - Rocker cover bolts in sequence
   - Rocker cover
   - Steering pump
   - Intake manifold supports
   - Water pipe assembly and set the No. 1 piston at Top Dead Center (TDC) of its compression stroke

➡ **Rotate the crankshaft until the mating mark on the camshaft sprocket is properly set.**

   - Timing chain tensioner
   - Distributor. Do not turn the rotor with the distributor removed.
   - Camshaft sprockets and brackets
   - Starter
   - Heater hoses

   - Cylinder head bolts in the proper sequence

To install:

5. Apply liquid gasket to the top of the chain cover where it meets the cylinder block before installing the head gasket.
6. Install the gasket and cylinder head on the block.

➡ **Cylinder head bolts may be reused providing the dimension from the bottom of the head to the end of the bolt does not exceed 6.228 in. (158.2mm). If the dimension exceeds the specification, replace the cylinder head bolts.**

7. Tighten the cylinder head bolts as follows:
   a. Tighten all bolts to 29 ft. lbs. (39 Nm) using the proper sequence.
   b. Tighten all bolts to 58 ft. lbs. (78 Nm) using the proper sequence.
   c. Loosen all bolts completely.
   d. Tighten all bolts to 25–33 ft. lbs. (34–44 Nm) using the proper sequence.
   e. Tighten all bolts 90–95°.
   f. Tighten all bolts an additional 90–95°.
8. Install or connect the following:
   - Heater hoses
   - Camshafts and brackets. Ensure that the camshaft keys are at 12 o'clock.
9. The procedure for tightening camshaft bolts must be followed exactly to prevent camshaft damage. Tighten the bolts as follows:
   a. Tighten the right camshaft bolts Nos. 9 and 10 to 18 inch lbs. (2 Nm).

Tighten bolts 1 through 8 to the same amount.
   b. Tighten the left camshaft bolts No. 11 and No. 12 to 18 inch lbs. (2 Nm). Tighten bolts 1 through 10 to the same amount.
   c. Tighten all bolts in sequence to 54 inch lbs. (6 Nm).
   d. Tighten all bolts in sequence again. Tighten type A, B, and D bolts to 78–102 inch lbs. (9–12 Nm) and type C bolts to 13–19 ft. lbs. (18–25 Nm).
10. Line up the mating marks on the timing chain and camshaft sprockets and install the sprockets. Tighten the sprocket bolts to 101–116 ft. lbs. (137–157 Nm).
11. Install or connect the following:
   - Timing chain guide, distributor, chain tensioner, oil filter bracket and power steering oil pump bracket
   - Intake manifold supports
   - Rocker cover. Torque the bolts, in sequence, to 89 inch lbs. (10 Nm).
   - Spark plugs, power steering pump, alternator, water pump pulley and drive belts, air duct to the intake manifold and the radiator
   - Vacuum and fuel hoses and reconnect all electrical connections
   - Engine undercover
   - Right front wheel
   - Negative battery cable
12. Fill and bleed the cooling system.
13. Start the vehicle, check for leaks and repair if necessary.

3.0L Engine

1. Before servicing the vehicle, refer to the precautions in the beginning of this section.
2. Relieve the fuel system pressure.
3. Drain the engine oil.
4. Drain the cooling system.

➡ **Before disconnecting any hoses or connectors, note the locations for reassembly.**

5. Remove or disconnect the following:
   - Negative battery cable
   - Right front wheel
   - Left side ornament cover
   - Air duct to intake manifold hose, collector hose, blow-by hose, and vacuum hoses
   - Fuel hoses and the harness connections
   - Evaporative emissions (EVAP) canister purge hose
   - Ignition coils from the spark plugs
   - Exhaust Gas Recirculation (EGR) tube

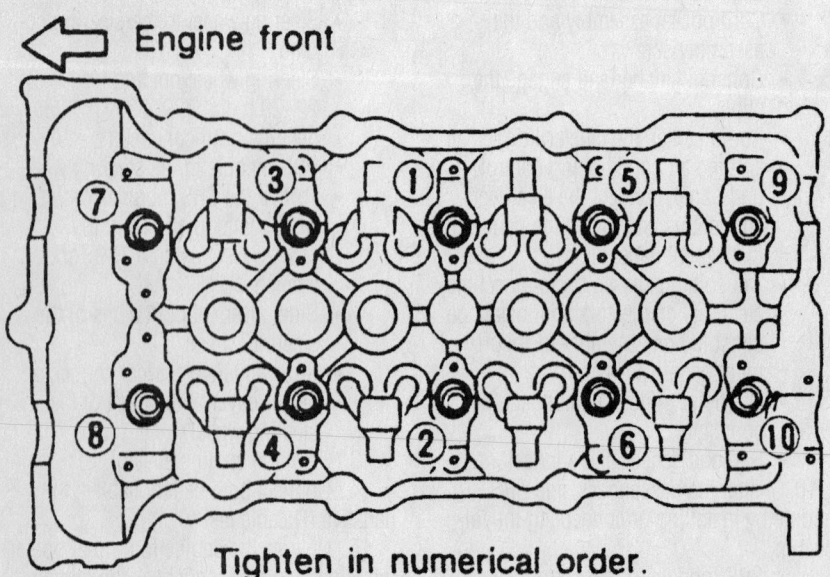

⬅ **Engine front**

Tighten in numerical order.

7923HG12

Cylinder head torque sequence—2.0L engine

- Right side intake manifold collector supports and remove the collector. Remove the manifold from the cylinder head.
- Fuel tube assembly
- Rocker arm covers
- Engine undercover
- Engine side cover
- Power steering oil pump and belt
- Camshaft Position (CMP) sensor (PHASE) and Crankshaft Position (CKP) sensors (REF)/(POS).

6. Set the No. 1 piston to Top Dead Center (TDC) of compression stroke by rotating the crankshaft.

- Crankshaft pulley
- Air compressor and bracket
- Timing chain tensioner and slack side chain guide
- Engine oil pan

7. Remove the camshaft sprockets first. Be sure to hold the flats of the camshafts while removing the sprocket bolts.

8. Loosen the camshaft bearing caps in several steps. The bearing caps MUST be loosened in sequence.

➡**Keep all bearing caps and camshafts in proper order for reinstallation.**

9. Loosen the cylinder head bolts in sequence.

To install:

10. Turn the crankshaft until the No. 1 piston is set 240° Before Top Dead Center (BTDC) on compression stroke.

11. Using new head gaskets, install the cylinder heads.

➡**If possible, replacement of the head bolts is suggested.**

12. If replacement of the head bolts is not possible, perform the following bolt measurement:

 a. Measure the diameter of the head bolt (11mm) from the bottom of the bolt.

 b. Measure the diameter of the head bolt (48mm) from the bottom of the bolt.

 c. Whenever the size difference between the 2 measurements exceeds 0.0043 in. (0.11mm) the head bolts must be replaced.

13. Lubricate the head bolt threads and the bolt seat surfaces with new engine oil.

14. Tighten the cylinder head bolts in sequence using the following steps:

 a. Step 1: All bolts in sequence to 72 ft. lbs. (98 Nm).

 b. Step 2: Completely loosen all bolts.

 c. Step 3: Tighten all bolts in sequence to 25–33 ft. lbs. (34–44 Nm).

 d. Step 4: Tighten all bolts in sequence an additional 90° clockwise.

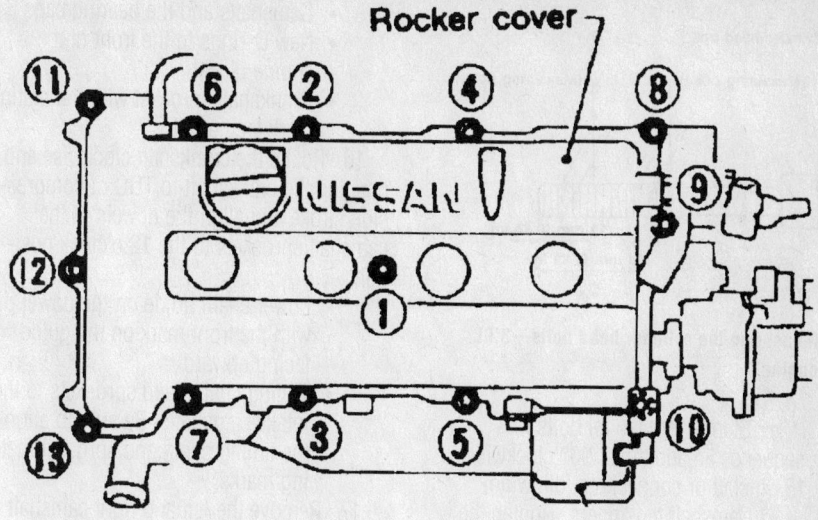

Tighten the rocker cover bolts according to the sequence shown—2.0L engine

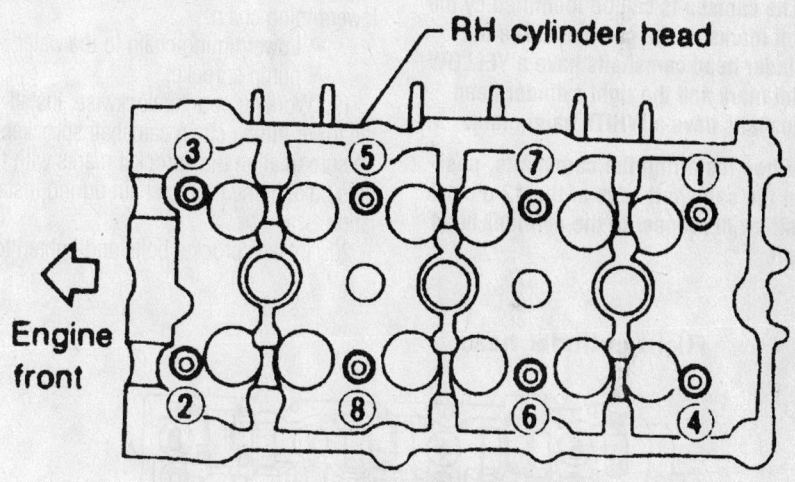

Right cylinder head bolt loosening sequence—3.0L engine

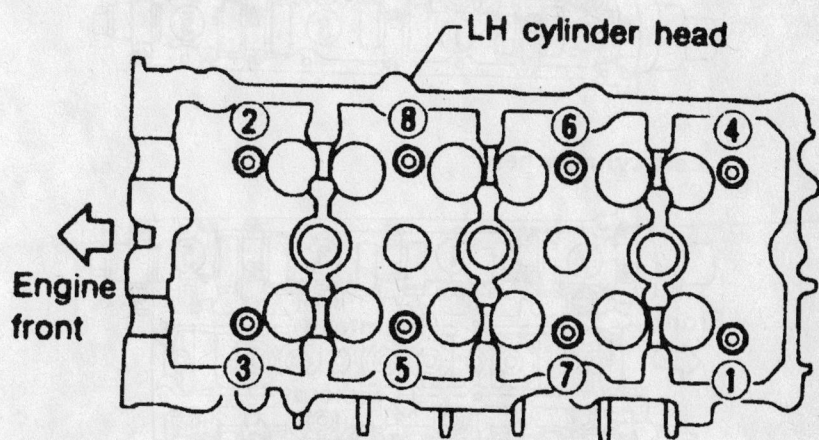

Left cylinder head bolt loosening sequence—3.0L engine

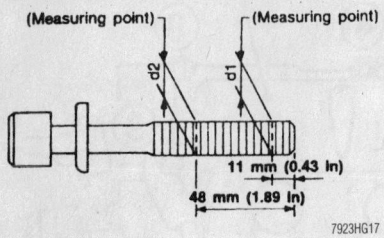

Cylinder head bolt

(Measuring point)   (Measuring point)

11 mm (0.43 in)
48 mm (1.89 in)

7923HG17

Measuring the cylinder head bolts—3.0L engine

e. Step 5: Tighten all bolts in sequence an additional 90° clockwise.

15. Install or connect the following:
- Camshaft tensioners. Tighten the tensioner mounting bolts to 75–96 inch lbs. (8.4–10.8 Nm).

➡ The camshafts can be identified by the paint marks on the camshaft. The left cylinder head camshafts have a YELLOW paint mark and the right cylinder head camshafts have a WHITE paint mark.

➡ When installing the camshafts, position the camshaft keys at the 12 o'clock position in respect to the cylinder head angle.

- Camshafts and the bearing caps
- New O-rings to the front of the engine block
- Crankshaft sprocket with the mating mark facing out

16. Rotate the crankshaft clockwise and position the crankshaft to TDC of compression stroke and align the dowels of the camshaft sprockets to the 12 o'clock position.
- Lower chain guide on the dowel pin with the front mark on the guide facing upward
- Timing chains and sprockets to the intake camshafts. Be sure to align the timing chain and sprocket mating marks.

17. Remove the left and right camshaft tensioner stopper pins.

18. Align the mating mark on the crankshaft with the matchmark (gold link) on the lower timing chain.
- Lower timing chain to the water pump sprocket

19. Working counterclockwise, install the lower timing chain camshaft sprockets. Be sure to align the sprocket marks with the blue links of the timing chain during installation.

20. Intake sprocket bolts and tighten to

88–95 ft. lbs. (119–128 Nm). Be sure to secure the camshafts while tightening the bolts

21. Timing chain guide, upper timing chain guide, lower timing chain tensioner and slack side timing chain guide

22. Timing cover evenly and gently. Be sure to align the dowel pin holes. Tighten the mounting bolts in sequence
- Front exhaust pipe and its support
- A/C compressor and bracket
- Crankshaft pulley to the crankshaft and install the mounting bolt

23. Torque the mounting bolt to 14–22 ft. lbs. (20–29 Nm). Torque the crankshaft bolt an additional 60–66° clockwise. This is approximately the angle from 1 hexagon bolt head corner to another.
- Ring gear cover plate
- CKP (PHASE) and CMP sensors (REF/POS)
- Power steering pump assembly
- Drive belts and the idler pulley
- Right front inner wheel cover and the right front wheel
- Engine undercovers
- Intake manifold, using new gaskets. Torque the nuts and bolts in sequence.
- Intake manifold collector gasket with the arrow facing forward
- Intake manifold collector assembly and torque the mounting bolts to 16–18 ft. lbs. (22–25 Nm)
- Intake manifold collector support brackets
- EGR tube, using new gaskets and torque the mounting bolts to 15–20 ft. lbs. (21–26 Nm) in 2 progressive steps
- Spark plugs
- Ignition coils and torque the mounting bolts to 27–33 inch lbs. (3–4 Nm)
- Cylinder head cover ornament on the left side
- Water hoses to the cylinder head and intake manifold
- EVAP canister purge hoses
- Fuel hoses and wiring harness connections to the fuel rail
- Air duct-to-intake manifold hose, collector hose, blow-by hose, and vacuum hoses
- Negative battery cable

24. Fill the engine with clean oil.
25. Fill and bleed the cooling system.
26. Connect the negative battery cable.
27. Start the engine and run at 3000 rpm under no load to purge the air from the high pressure chamber. The engine may produce a rattling noise. This indicates that air still

Right cylinder head

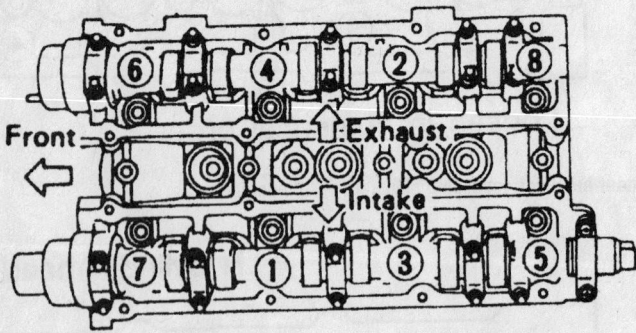

Left cylinder head

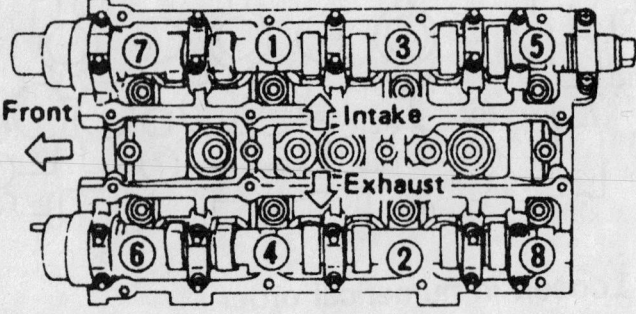

7923HG14

Cylinder head torque sequence—3.0L engine

remains in the chamber and is not a matter of concern.

28. Inspect the vehicle for leaks and repair if necessary.

3.5L Engine

I35 MODELS

➡**For this procedure, you must remove the engine from the vehicle in order to remove the cylinder head.**

1. Before servicing the vehicle, refer to the precautions in the beginning of this section.
2. Relieve the fuel system pressure.
3. Drain the engine oil.
4. Drain the cooling system.

➡**Before detaching any hoses or connectors, note the locations for reassembly.**

5. Remove or disconnect the following:
   - Negative battery cable
   - Engine assembly
   - Exhaust manifold
6. Place the engine on a suitable workstand.
   - Oil pan
   - Timing chain
   - Intake manifold
   - Water outlet
   - Rear timing chain case bolts, in the reverse of the sequence shown
   - Rear timing chain case
   - O-rings from the cylinder head and block
   - Intake valve timing control solenoid valves

➡**For installation purposes, matchmark the camshaft brackets before removing them.**

   - Intake and exhaust camshafts and brackets. Loosen the bracket bolts in several steps, in the sequence shown.
   - Right and left side cam chain tensioner from the cylinder head
   - Cylinder head bolts. Loosen in several steps, in the sequence shown.

➡**A warped or cracked cylinder head could result from removing the bolts in incorrect order.**

   - Cylinder heads from the vehicle
   - Discard the head gaskets
7. Remove all traces of liquid gasket from the timing chain case and from the water pump covers.
8. Remove all traces of liquid gasket from the engine block.
9. Inspect the timing chain for exces-

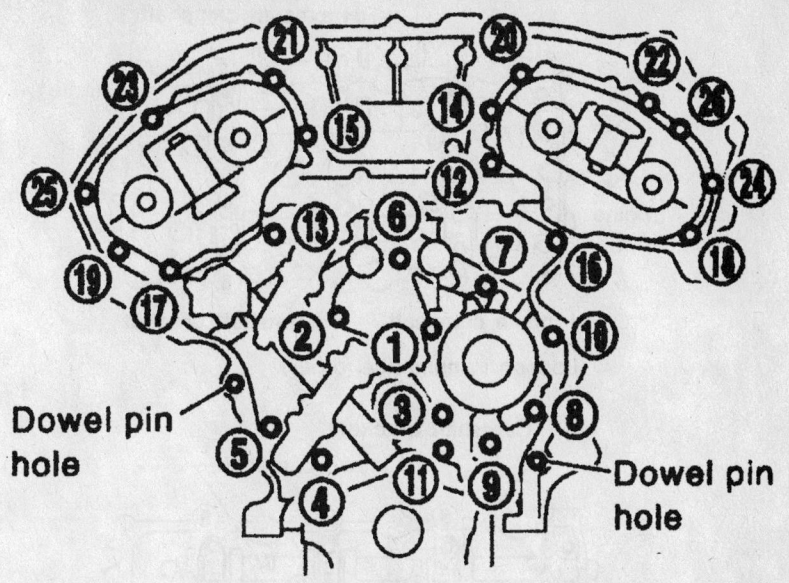

Loosen the rear timing chain case bolts in the reverse of this sequence—I35 with 3.5L engine

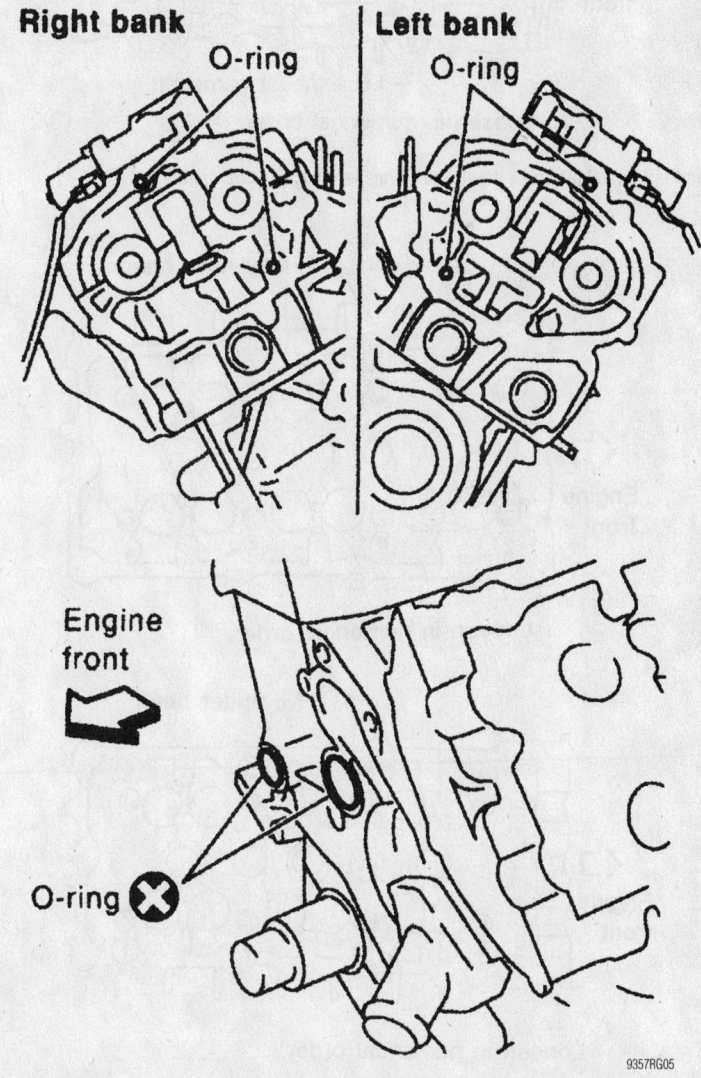

Remove the O-rings from the cylinder head and block— I35 with 3.5L engine

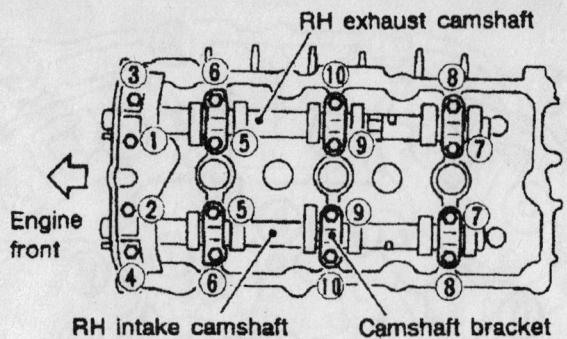

Loosen in numerical order.

Right and left camshaft bracket bolt loosening sequence—3.5L engine

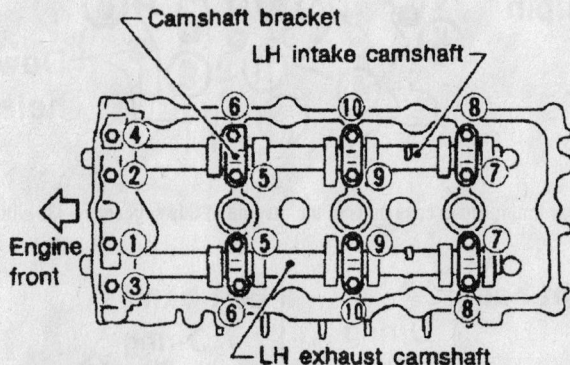

Loosen in numerical order.

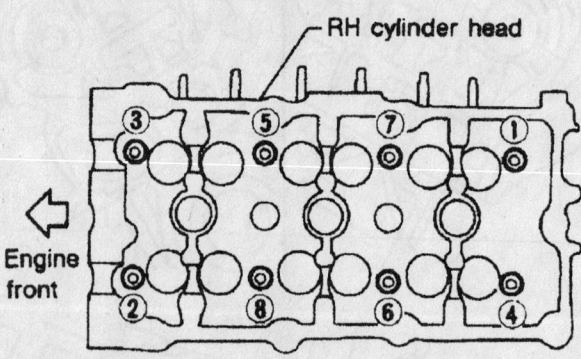

Loosen in numerical order.

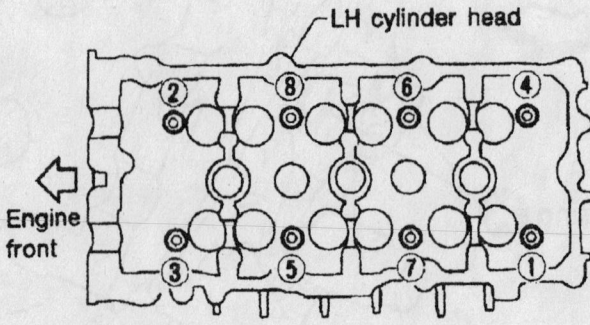

Loosen in numerical order.

Cylinder head bolt loosening sequence— I35 and G35 with 3.5L engine

sive wear or damage and replace as necessary.

To install:

10. Turn the crankshaft until the No. 1 piston is a Top Dead Center (TDC) on compression stroke. The crankshaft key should face toward the right bank.

11. Using new head gaskets, install the cylinder heads.

➡**If possible, replacement of the head bolts is suggested.**

12. If replacement of the head bolts is not possible, perform the following bolt measurement:

a. Measure the diameter of the head bolt 0.43 in. (11mm) from the bottom of the bolt.

b. Measure the diameter of the head bolt 1.89 in. (48mm) from the bottom of the bolt.

c. Whenever the size difference between the 2 measurements exceeds 0.0043 in. (0.11mm) the head bolts must be replaced.

13. Install the cylinder head bolts and torque in sequence as follows:

a. Step 1: 72 ft. lbs. (98 Nm).

b. Step 2: Completely loosen all bolts.

c. Step 3: 26–32 ft. lbs. (34–44 Nm).

d. Step 4: plus 90–95 degrees clockwise.

e. Step 5: plus 90–95 degrees clockwise.

14. Install or connect the following:

• Camshafts and related components
• Intake valve timing control solenoid valves
• New O-rings to the front of the engine block and cylinder head

15. Apply sealant to the hatched portion of the of the rear timing chain case.

16. Align the rear timing chain case with the dowel pins and install onto the cylinder heads and engine block.

17. Torque the rear timing chain case mounting bolts in sequence to 105–121 inch lbs. (11.8–13.7 Nm).

18. Install or connect the following:

• Water outlet
• Intake manifold
• Timing chain
• Oil pan
• Exhaust manifold
• Engine assembly into the vehicle
• Negative battery cable

19. Fill the cooling system.

20. Fill the engine with clean oil.

21. Start the vehicle, check for leaks and repair if necessary.

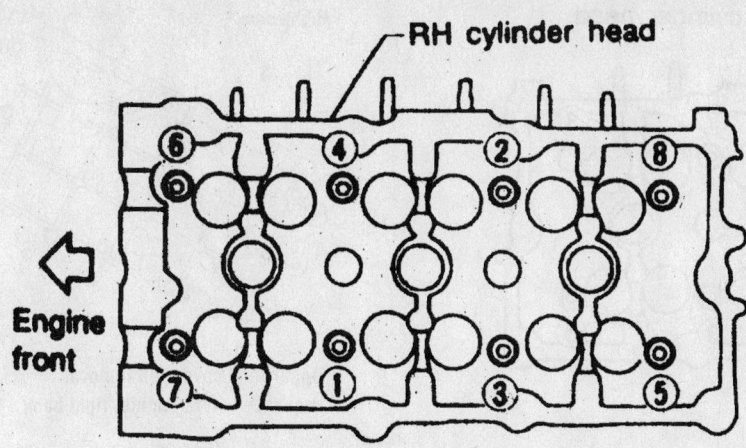

Tighten in numerical order.

79230G11

Right cylinder head bolt torque sequence— I35 with 3.5L engine

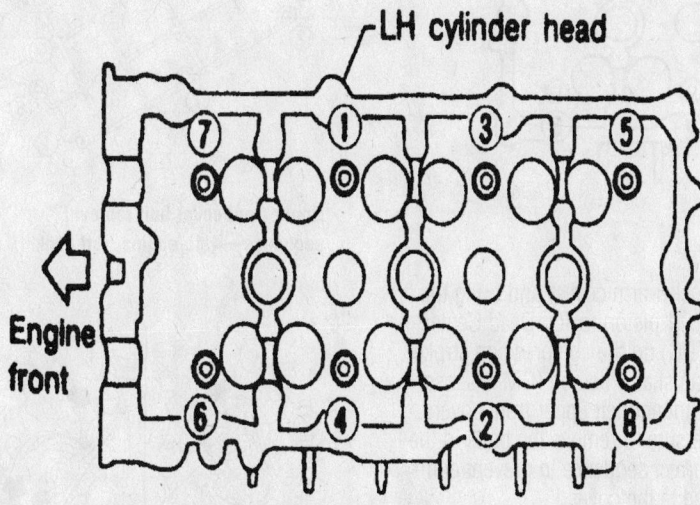

Tighten in numerical order.

79230G12

Left cylinder head bolt torque sequence— I35 with 3.5L engine

G35 MODELS

→For this procedure, you must remove the engine from the vehicle in order to remove the cylinder head.

1. Before servicing the vehicle, refer to the precautions in the beginning of this section.

2. Properly relieve the fuel system pressure.

3. Drain the cooling system.

4. Drain the engine oil.

5. Remove or disconnect the following:

- Engine and place on a suitable stand
- Intake manifold collector
- Fuel rail and injector assembly
- Intake and exhaust manifolds
- Ignition coil
- Rocker arm (valve) cover
- Water inlet and thermostat housing
- Water outlet and hoses
- Upper and lower oil pans and strainer
- Front timing chain case, timing chain and rear timing chain case
- Camshaft
- Cylinder head bolts. Loosen in several steps, in the sequence shown.

→A warped or cracked cylinder head could result from removing the bolts in incorrect order.

- Cylinder heads from the vehicle
- Discard the head gaskets

6. Remove all traces of liquid gasket from the timing chain case and from the water pump covers.

7. Remove all traces of liquid gasket from the engine block.

8. Inspect the timing chain for excessive wear or damage and replace as necessary.

To install:

9. Turn the crankshaft until the No. 1 piston is a Top Dead Center (TDC) on compression stroke. The crankshaft key should face toward the right bank.

10. Using new head gaskets, install the cylinder heads.

→If possible, replacement of the head bolts is suggested.

11. If replacement of the head bolts is not possible, perform the following bolt measurement:

a. Measure the diameter of the head bolt 0.43 in. (11mm) from the bottom of the bolt.

b. Measure the diameter of the head bolt 1.89 in. (48mm) from the bottom of the bolt.

c. Whenever the size difference between the 2 measurements exceeds 0.0043 in. (0.11mm) the head bolts must be replaced.

12. Install the cylinder head bolts and torque in sequence as follows:

a. Step 1: 72 ft. lbs. (98 Nm).

b. Step 2: Completely loosen all bolts, in the reverse of the tightening sequence.

c. Step 3: 26–32 ft. lbs. (34–44 Nm).

d. Step 4: plus 90–95 degrees clockwise.

e. Step 5: plus 90–95 degrees clockwise.

13. After installing the cylinder head, measure the distance between the front end faces of the cylinder block and cylinder head. If the specification does not fall within 0.555–0.587 in. (14.1–14.9mm), you must reinstall the cylinder head.

14. Install or connect the following:

- Camshaft
- Front timing chain case, timing chain and rear timing chain case
- Oil strainer and upper and lower oil pans
- Water outlet and hoses
- Thermostat housing and water inlet
- Rocker arm (valve) cover
- Ignition coil
- Intake and exhaust manifolds
- Fuel rail and injector assembly

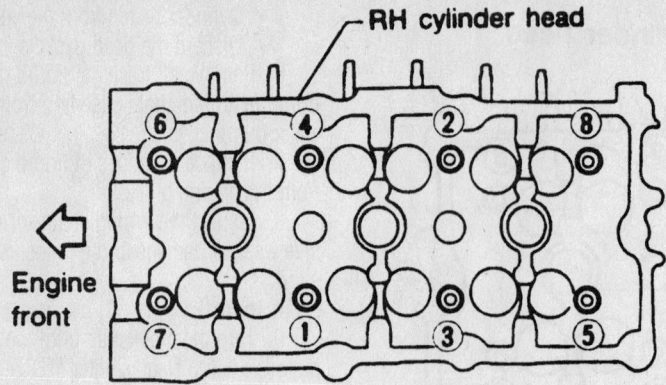

Cylinder head bolt torque sequence— G35 with 3.5L engine

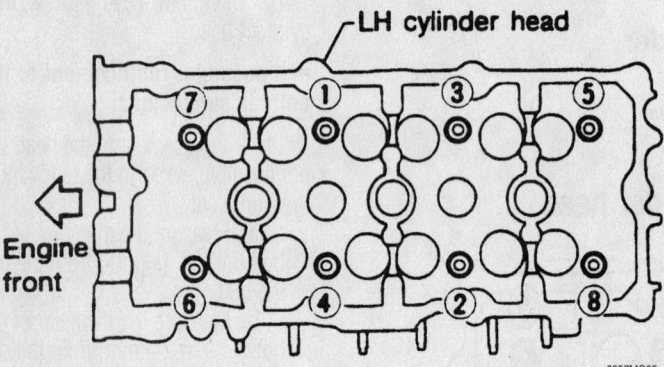

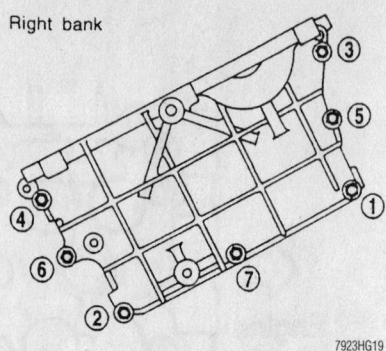

Upper front cover bolt removal sequence—4.1L engine, right bank

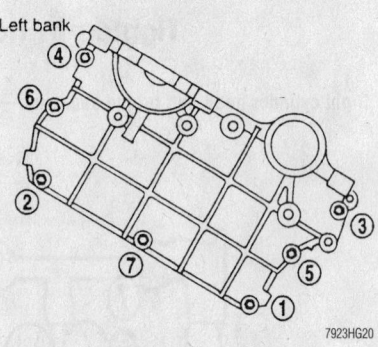

Upper front cover bolt removal sequence—4.1L engine, left bank

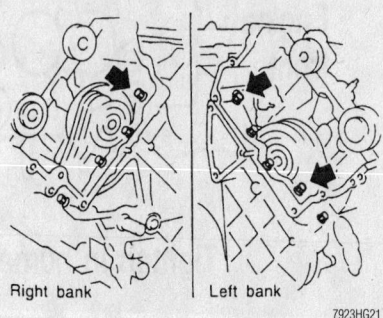

Be sure to install the long bolts in the positions indicated by the arrows—4.1L

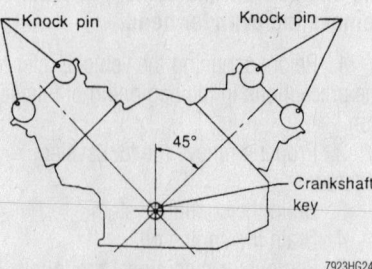

Be sure to position the camshaft knock pins as shown before installing the cylinder heads—4.1L engine

- Intake manifold collector
- Engine into the vehicle
- Negative battery cable
15. Fill the cooling system.
16. Fill the engine with clean oil.
17. Start the vehicle, check for leaks and repair if necessary.

4.1L Engine

1. Before servicing the vehicle, refer to the precautions in the beginning of this section.

2. Properly relieve the fuel system pressure.

3. Drain the cooling system.

4. Drain the engine oil.

5. Remove or disconnect the following:
- Both battery cables
- Engine assembly from the vehicle and place it on a workstand
- Exhaust manifold
- Drain plugs on the sides of the engine and drain the coolant
- Intake manifold collector
- Ignition coil sub-harness
- Ignition coils
- Spark plugs
- Fuel rail with injectors. Do not disassemble the fuel hose.
- Intake manifold

- Rocker arm covers and bring the No. 1 piston to Top Dead Center (TDC) on the compression stroke
- Camshaft Position (CMP) sensor
- Right and left upper front covers. Be sure to remove the bolts in the correct sequence to prevent damage to the cover.
- Upper chain tensioners

6. Apply paint marks on the upper timing chains, camshaft and idler sprockets so they can be installed in their original positions.

- Camshaft sprocket
- Idler sprocket bolt
- Cylinder head sub-bolts. The bolts are different lengths, note their differences so they can be installed in their original positions.

7. Loosen the cylinder head bolts gradually in the proper removal sequence, then remove the cylinder head.

To install:

8. Be sure all mating surfaces are clean before installation.

9. Check the cylinder head surface for warpage using a feeler gauge and a suitable straightedge. If the cylinder head is warped more than 0.004 in. (0.1mm), it must be resurfaced or replaced. The total amount machined from the head or head and block

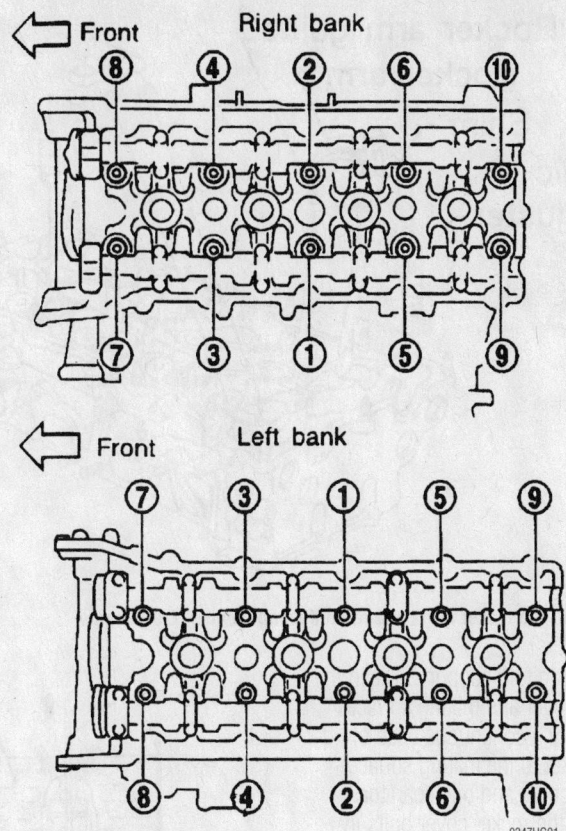

Cylinder head torque sequence—4.1L engine

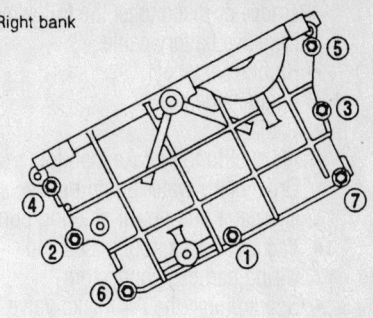

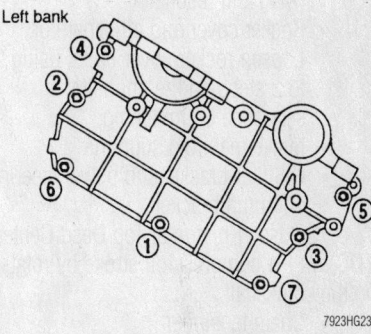

7923HG23

Torque the upper front cover bolts in the sequence shown—4.1L engine

combined, cannot total more than 0.008 in. (0.2mm).

10. Place new gaskets on the cylinder block.

11. Be sure the knock pins on the camshafts are in the positions shown. Carefully place the cylinder heads on the engine. Do not damage the head gasket.

12. Lubricate the cylinder head bolt threats and seat surfaces with engine oil. Torque the bolts in sequence using the following sub-steps:

   a. Step 1: 22 ft. lbs. (29 Nm).
   b. Step 2: 69 ft. lbs. (93 Nm).
   c. Step 3: loosen the bolts completely.
   d. Step 4: 18–25 ft. lbs. (25–35 Nm).
   e. Step 5: Plus 90–95°.

13. Install or connect the following:
- Cylinder head sub-bolts. Torque the bolts to 56–74 inch lbs. (6–8 Nm). Be sure the long bolts are returned to the positions shown.
- Idler and camshaft sprockets
- Chain tensioners

14. Using new gaskets, install the cylinder head covers. Be sure to apply RTV silicone sealant to the gasket arch and rubber plugs.

15. Torque the cylinder head cover

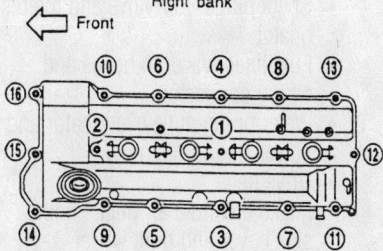

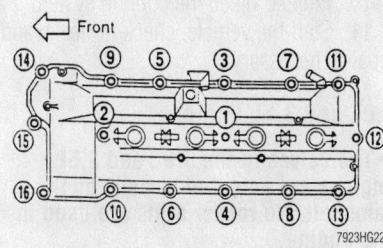

7923HG22

Cylinder head cover bolt tightening sequence—4.1L engine

bolts in sequence using the following sub-steps:

   a. Step 1: Nos. 1 through 16: 35–52 inch lbs. (4–6 Nm).
   b. Step 2: Nos. 1 through 16: 61–78 inch lbs. (7–9 Nm).

   c. Step 3: Nos. 1 and 2: 61–78 inch lbs. (7–9 Nm).

16. Install or connect the following:
- Intake valve timing control solenoid with a new O-ring. Torque the solenoid to 18–25 ft. lbs. (25–34 Nm).

17. Apply a bead of RTV silicone sealant to the upper front covers, then install them. Torque the bolts to 56–74 inch lbs. (6–8 Nm).

- CMP sensor
- Rocker arm covers
- Intake manifold with new gaskets
- Fuel rail and injectors
- Spark plugs and ignition coils
- Intake manifold collector
- Exhaust manifold with new gaskets
- Engine assembly to the vehicle
- Battery and cables

18. Fill and bleed the cooling system.
19. Fill the engine with clean oil.
20. Start the vehicle, check for leaks and repair if necessary.

Rocker Arms/Shafts

REMOVAL & INSTALLATION

2.0L Engine

1. Before servicing the vehicle, refer to the precautions in the beginning of this section.

2. Relieve the fuel system pressure.
3. Drain the cooling system.

4. Remove or disconnect the following:
- Negative battery cable
- Right front wheel
- Engine side cover
- Radiator
- Air duct to the intake manifold
- Drive belts, water pump pulley, alternator and power steering pump
- Vacuum hoses, fuel hoses and wiring harness connectors
- Spark plugs, the Air Intake Valve (AIV) and resonator
- Rocker cover and oil separator. Loosen rocker cover bolts, using 2 to 3 steps, in the opposite sequence of tightening.
- Intake manifold supports
- Oil filter bracket and power steering oil pump bracket

5. Set No. 1 piston at Top Dead Center (TDC) on the compression stroke by rotating the crankshaft.
- Chain tensioner
- Distributor. Do not turn the rotor with the distributor removed.
- Timing chain guide, camshaft sprockets, camshafts, brackets, oil tubes and baffle plate. The camshaft bracket bolts must be loosened in sequence to prevent damage to the camshafts or the head.
- Rocker arm assembly

To install:

6. Check the hydraulic lash adjusters to ensure they did not bleed down during disassembly by trying to compress them. If the lash adjuster can be compressed 0.04 in. (1mm), air has entered and it must be bleed.

➡ **Air cannot be bled from the lash adjusters by running the engine.**

7. Clean the camshaft end bracket and coat with liquid gasket. Install the camshafts, camshaft brackets, oil tubes and baffle plate. Ensure the left camshaft key is at 12 o'clock and the right camshaft key is at 10 o'clock.

➡ **The procedure for tightening camshaft bracket bolts must be followed exactly to prevent camshaft damage.**

8. Line up the mating marks on the timing chain and camshaft sprockets and install the sprockets. Torque the sprocket bolts to 101–116 ft. lbs. (137–157 Nm).

9. Install or connect the following:
- Timing chain guide, distributor (ensure that rotor head is at 5 o'clock position) and chain tensioner

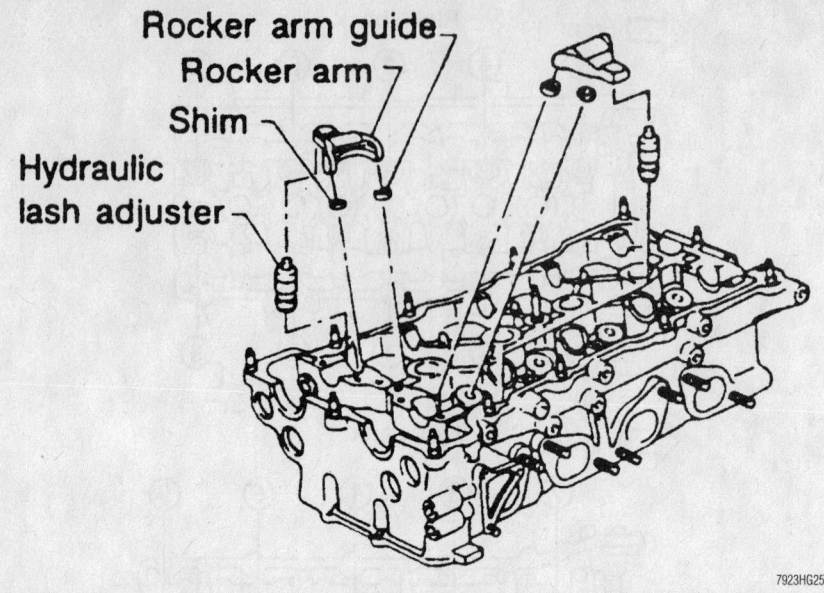

Exploded view of the rocker arms and related components—2.0L engine

7923HG25

- Intake manifold supports. Clean the rocker cover and mating surfaces and apply a continuous bead of liquid gasket to the mating surface.
- Rocker cover and oil separator. Tighten the rocker cover bolts in sequence.
- Oil filter bracket and the power steering pump bracket
- Spark plugs, AIV valve and the resonator
- Fuel lines, vacuum hoses and wiring connectors
- Water pump pulley, alternator and power steering pump
- Drive belts
- Intake manifold air duct, engine side cover and right wheel
- Radiator
- Negative battery cable

10. Fill and bleed the cooling system.

11. Start the vehicle, check for leaks and repair if necessary.

3.0L and 3.5L Engines

➡ **The valves in the 3.0L and 3.5L engines are actuated directly by the camshaft. No rocker arms are used in this engine.**

4.1 Engines

1. Before servicing the vehicle, refer to the precautions in the beginning of this section.

2. Remove or disconnect the following:

- Negative battery cable
- Engine and transmission assembly from the vehicle

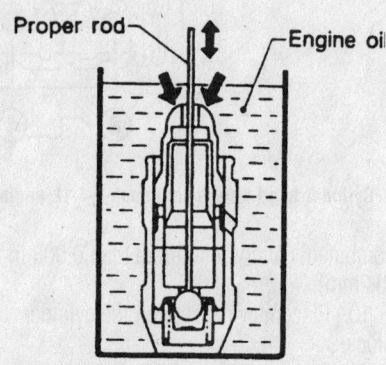

7923HG26

Submerge the lash adjuster in engine oil, lightly unseat the check ball with a thin rod and push on the plunger to release the

- Suspension member and engine mounts from the engine
- Air compressor bracket
- Cooling fan with coupling and the engine gusset

3. Separate the engine from the transmission and mount the engine on a suitable workstand.
- Oil pan
- Crank angle sensor and the Valve Timing Control (VTC) solenoid
- Chain tensioners and the upper front covers
- Front timing chain cover

➡ **The timing chain will not be disengaged or dislocated from the crankshaft sprocket unless the front cover is removed. The cast portion of the front cover is located on the lower side of the crankshaft sprocket so the timing chain is not disengaged from the sprocket.**

- VTC assembly and the camshaft sprocket
- Oil pump chain and the timing chains

→Do not attempt to disassemble the VTC assembly since they are difficult to reassemble accurately in the field. If it should be disassembled, the VTC assembly must be replaced with a new one.

- Camshaft brackets and the camshafts. Mark the parts so they can be reinstalled in their original positions.
- Rocker arms. Be sure to identify each rocker arm so they can be reinstalled in their original positions.

To install:

4. Be sure all mating surfaces are clean before installation.

5. Install or connect the following:

- Rocker arms, camshafts and camshaft brackets on the right bank. Properly lubricate the rocker arms and camshafts prior to installation.
- VTC assembly and the exhaust cam sprocket on the right bank

6. Be sure the camshafts are still correctly positioned and the piston in the No. 1 cylinder is still at Top Dead Center (TDC).

- Timing chain on the right bank, aligning the mating marks on the chain with those on the crankshaft and camshaft sprockets
- Chain tensioner on the right bank

7. Turn the crankshaft approximately 120° clockwise from the point where the No. 1 piston is at TDC on the compression stroke. At this point, the valves on the left bank still remain closed.

8. Correctly position the camshafts and rocker arms for the left cylinder head. Properly lubricate the rocker arms and camshafts prior to installation. Install the VTC assembly and the exhaust cam sprocket.

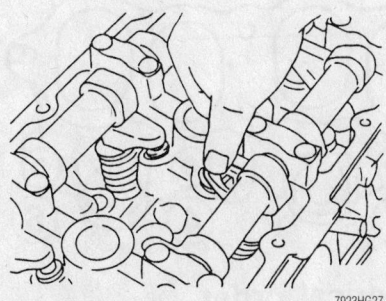

Press down on the lash adjuster to check for bleed-down—4.1L engines

9. Install the timing chain on the left bank, aligning the mating marks on the chain with those on the crankshaft and camshaft sprockets.

10. Install or connect the following:

- Oil pump chain and sprockets
- Oil pump chain guides. Place a 0.04 in. (1.0mm) feeler gauge between the upper chain guide and chain before assembling the chain guides. The force applied to the chain is equivalent to the upper chain guide weight.

11. Apply suitable sealer and install the front covers.

- Chain tensioner for the left bank

12. Apply suitable sealer to the rubber plugs and install them on the cylinder head.

- Crank angle sensor, VTC solenoid, rocker cover and crank pulley
- Transmission on the engine and install the engine assembly in the vehicle

Intake Manifold

REMOVAL & INSTALLATION

2.0L Engine

1. Before servicing the vehicle, refer to the precautions in the beginning of this section.

2. Properly relieve the fuel system pressure.

3. Drain the cooling system.

4. Remove or disconnect the following:

- Negative battery cable
- Fuel lines, vacuum hoses and electrical connectors
- Throttle linkage
- Intake manifold supports from the front and rear
- Intake manifold collector. Loosen the bolts in the sequence illustrated.

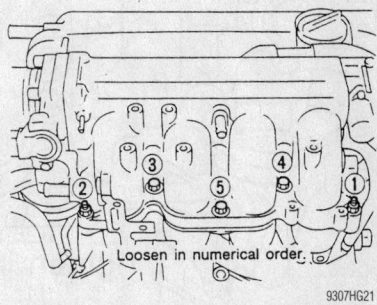

Intake manifold collector bolt loosening sequence—2.0L engine

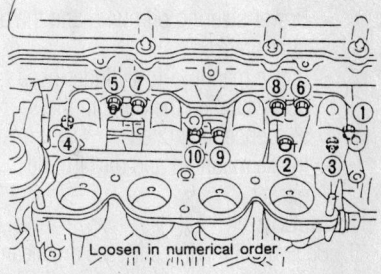

Intake manifold bolt loosening sequence—2.0L engine

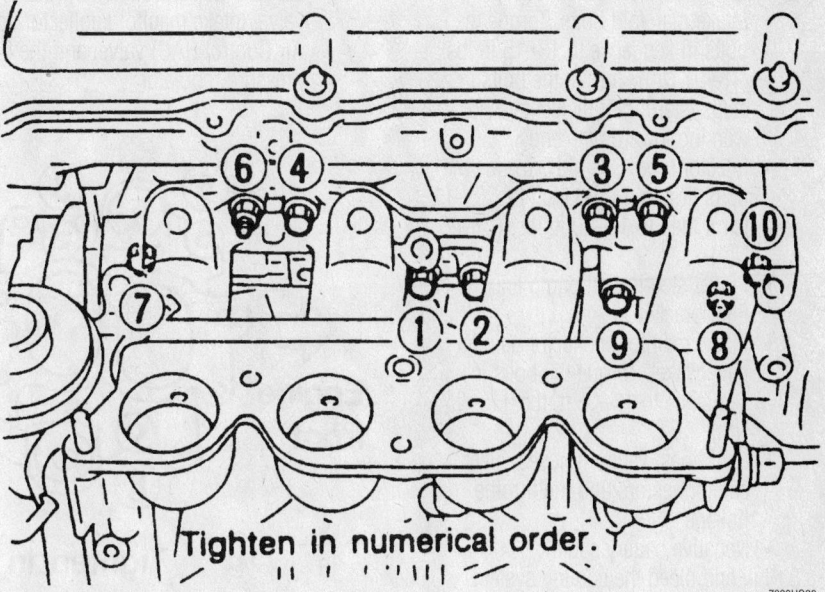

Lower intake manifold torque sequence—2.0L engine

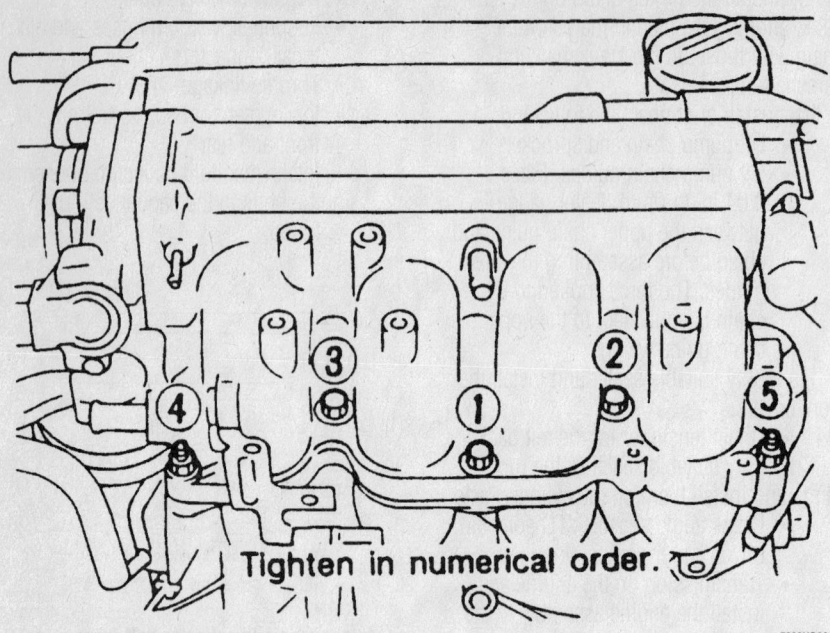

Upper intake manifold torque sequence—2.0L engine

- Injector tube assembly
- Power steering oil pump and the oil filter bracket
- Intake manifold-to-cylinder head bolts. Loosen the bolts in the sequence illustrated.
- Intake manifold and discard the gasket

To install:

5. Be sure all mating surfaces are clean prior to installation.

6. Fit a new gasket and the manifold into place. Start the support bolts to hold the manifold in place.

7. Install or connect the following:
- Intake manifold bolts. Torque the bolts in sequence to 13–15 ft. lbs. (18–21 Nm). Torque the bolts in 2 steps, starting at the center and working towards the ends.
- Injector tube assembly. Torque the bolts first to 84–96 inch lbs. (9–11 Nm), then to 15–20 ft. lbs. (21–26 Nm).
- Power steering oil pump and the oil filter bracket
- Intake manifold collector using a new gasket. Torque the bolts in sequence to 13–15 ft. lbs. (18–21 Nm).
- Fuel lines, vacuum hoses, electrical connectors and the throttle linkage
- Negative battery cable

8. Fill and bleed the cooling system.

9. Start the vehicle, check for leaks and repair if necessary.

3.0L and 3.5L Engines

EXCEPT G35 MODELS

1. Before servicing the vehicle, refer to the precautions in the beginning of this section.

2. Release the fuel system pressure.

3. Drain the cooling system.

4. Remove or disconnect the following:
- Negative battery cable
- Throttle body coolant hoses
- Electrical connectors from the Throttle Position (TP) sensor
- Hoses from the throttle body, the Exhaust Gas Recirculation (EGR) valve, intake manifold collector, Idle Air Control (IAC) valve, and the fuel pressure regulator

- Evaporative emissions (EVAP) canister purge hose and blow-by hose
- EGR guide tube
- Accelerator cable from the throttle body
- Intake manifold collector support brackets
- Right side electrical connectors from the ignition coils
- Electrical connector from the crank angle sensor and the power transistor, if necessary
- Intake manifold collector-to-intake manifold bolts/nuts and remove the intake manifold collector

5. Fuel injector assembly by removing or disconnecting the following:
- Electrical connectors from the fuel injectors
- Fuel lines from the fuel injector assembly
- Fuel rail-to-cylinder head bolts
- Fuel rail assembly from the engine
- Intake manifold bolts/nuts in the reverse sequence of the torque procedure

6. Remove the intake manifold from the engine and discard the gaskets

7. Clean all gasket mounting surfaces.

To install:

8. Using new gaskets, install the intake manifold to the engine.

9. Tighten the bolts/nuts in sequence as follows:

a. Step 1: Tighten nuts and bolts to 44–86 inch lbs. (5–10 Nm).

b. Step 2: Tighten nuts and bolts to 20–23 ft. lbs. (26–31 Nm).

c. Step 3: Repeat step 2 at least 5 times until all nuts and bolts have a final torque of 20–23 ft. lbs. (26–31 Nm).

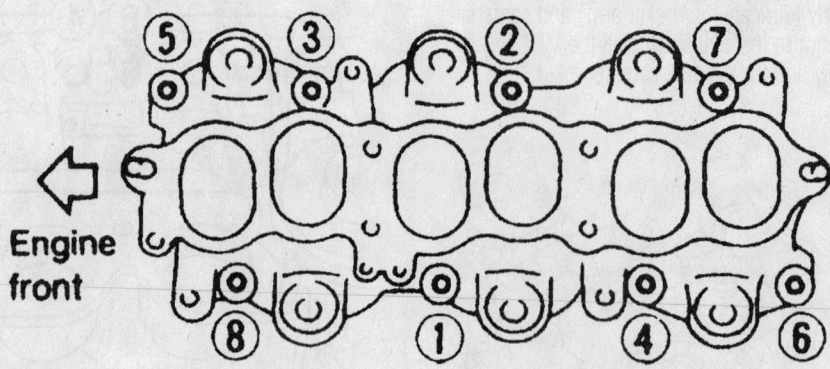

Engine front

Tighten in numerical order.

Lower intake manifold torque sequence—3.0L and 3.5L engine, except G35

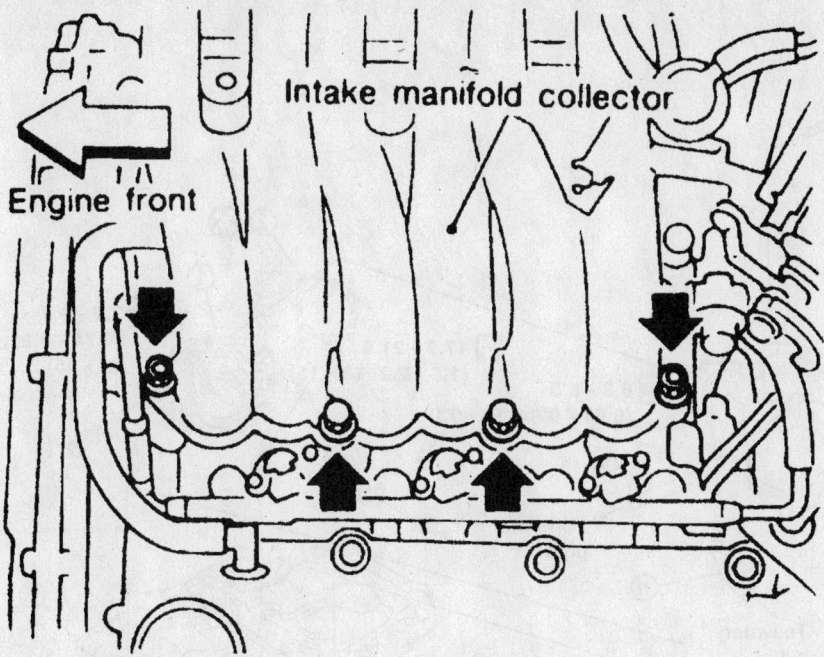

Upper intake manifold torque sequence—3.0L and 3.5L engine, except G35

7923HG30

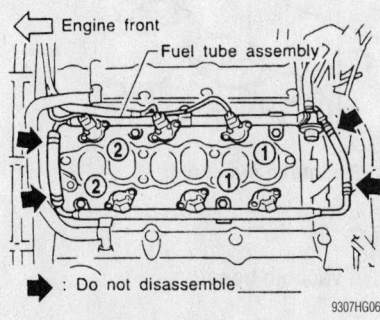

▶ : Do not disassemble

9307HG06

Tighten the fuel rail (tube) bolts in the sequence illustrated—3.0L and 3.5L engines, except G35

10. Install the fuel injector assembly by installing or connecting the following:
  • Fuel rail assembly to the engine
11. Install the fuel rail-to-cylinder head bolts and torque the bolts to 15–20 ft. lbs. (21–26 Nm) in the following sequence:
  a. Step 1: Tighten bolts in sequence to 84–96 inch lbs. (9–10 Nm).
  b. Step 2: Tighten the bolts in sequence to 15–20 ft. lbs. (21–26 Nm).
  • Fuel lines to the fuel injector assembly
  • Electrical connectors to the fuel injectors
  • Intake manifold collector using a new gasket, and torque the intake manifold collector-to-intake manifold bolts/nuts to 13–16 ft. lbs. (18–22 Nm) in the sequence illustrated
  • Intake manifold collector supports.

Torque the bolts to 14–18 ft. lbs. (20–25 Nm).
  • Electrical connector to the crank angle sensor and the power transistor, if disconnected
  • Electrical connectors to the ignition coils and torque the mounting bolts to 27–33 inch lbs. (3–4 Nm)
  • Accelerator cable to the throttle body
  • EGR guide tube. Torque the bolts to 15–20 ft. lbs. (21–26 Nm) in 2 progressive steps
  • EVAP canister purge hose and blow-by hose
  • Hoses to the throttle body, EGR valve, intake manifold collector, IAC valve, and the fuel pressure regulator
  • Electrical connectors to the TP sensor
  • Throttle body coolant hoses
  • Negative battery cable
12. Fill and bleed the cooling system.
13. Start the vehicle, check for leaks and repair if necessary.

G35 MODELS

1. Before servicing the vehicle, refer to the precautions in the beginning of this section.
2. Release the fuel system pressure.
3. Drain the cooling system.
4. Remove or disconnect the following:
  • Negative battery cable
  • Engine cover

  • Air cleaner and duct
  • Electric throttle control actuator, loosening the bolts in a criss-cross pattern
  • Fuel sub-tube mounting bolt to disconnect it from the rear of the lower intake manifold (collector)
  • Vacuum hose and water hose from the intake manifold collector
  • Evaporative emissions (EVAP) canister purge volume control solenoid valve bracket mounting bolt from the intake manifold collector
  • Intake manifold collector (upper) bolts in the sequence shown, then remove the collector
  • Positive Crankcase Ventilation (PCV) hose between the intake manifold collector and right side rocker arm cover
  • Intake manifold collector (lower) bolts in the reverse of the sequence shown
  • Intake manifold collector (lower) cover, gaskets and lower manifold collector
  • Fuel rail and injector assembly
  • Intake manifold bolts and nuts, in the reverse of the installation sequence
5. Discard all gaskets and clean all gasket mounting surfaces.

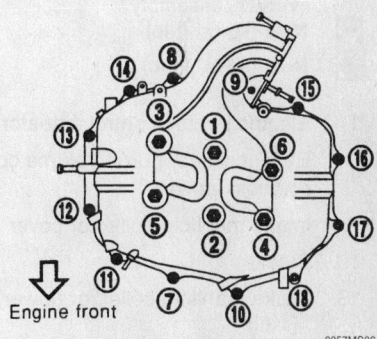

9357MG06

Intake manifold collector (upper) bolt loosening sequence—G35 models

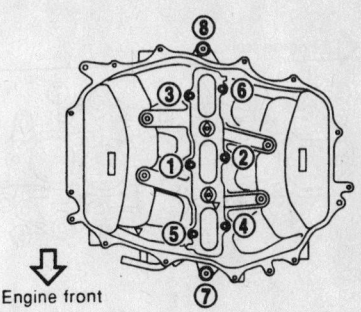

9357MG08

Intake manifold collector (lower) bolt tightening sequence. Use the reverse sequence for bolt removal—G35 models

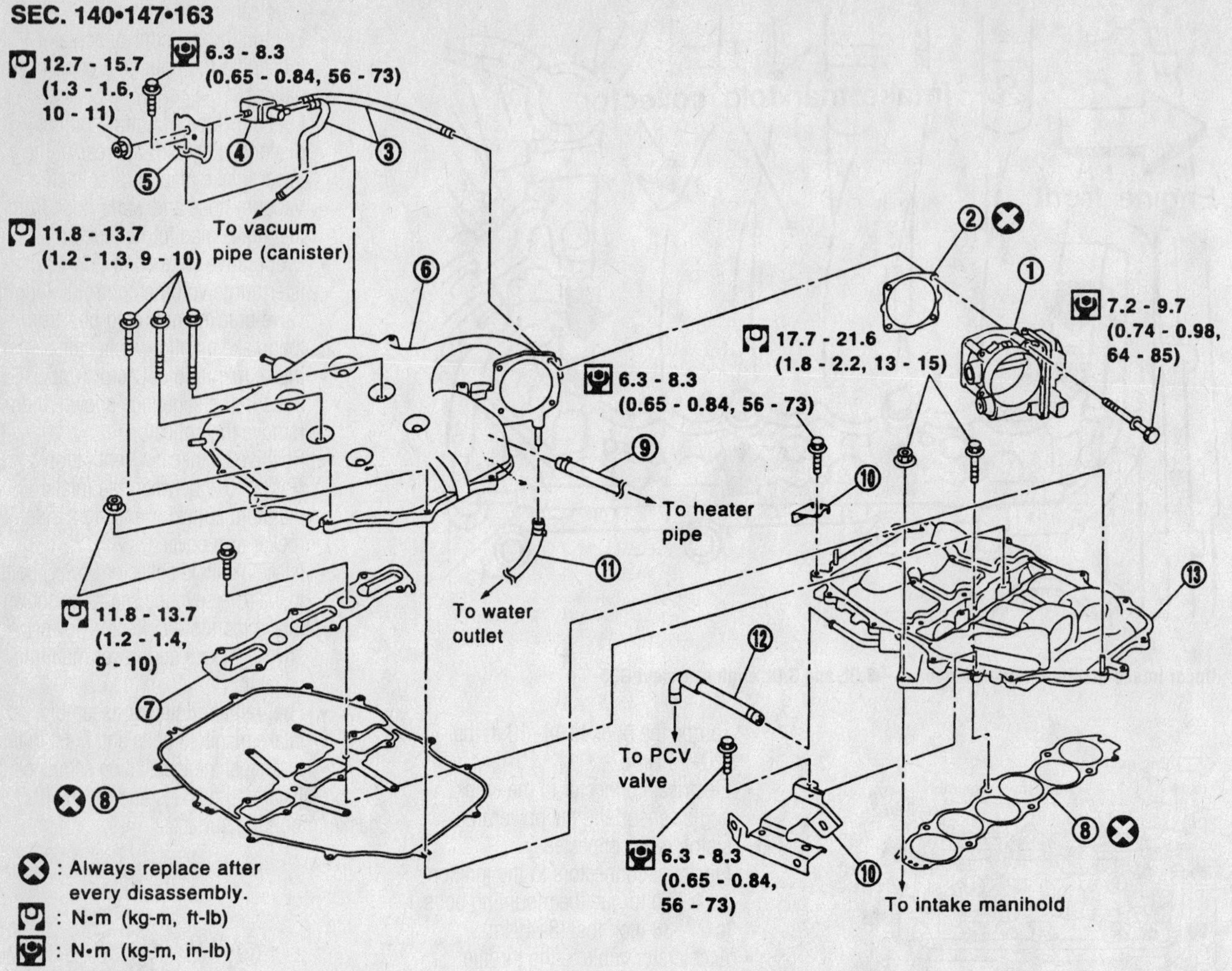

SEC. 140•147•163

12.7 - 15.7 (1.3 - 1.6, 10 - 11)

6.3 - 8.3 (0.65 - 0.84, 56 - 73)

11.8 - 13.7 (1.2 - 1.3, 9 - 10)

To vacuum pipe (canister)

17.7 - 21.6 (1.8 - 2.2, 13 - 15)

6.3 - 8.3 (0.65 - 0.84, 56 - 73)

7.2 - 9.7 (0.74 - 0.98, 64 - 85)

11.8 - 13.7 (1.2 - 1.4, 9 - 10)

To heater pipe

To water outlet

To PCV valve

6.3 - 8.3 (0.65 - 0.84, 56 - 73)

To intake manihold

⊗ : Always replace after every disassembly.

▣ : N•m (kg-m, ft-lb)

▣ : N•m (kg-m, in-lb)

1. Electric throttle control actuator
2. Gasket
3. Vacuum hose
4. EVAP canister purge volume control solenoid valve
5. Bracket
6. Intake manifold collector (upper)
7. Intake manifold collector cover
8. Gasket
9. Water hose
10. Bracket
11. Water hose
12. PCV hose
13. Intake manifold collector (lower)

9357MG07

Exploded view of the upper and lower intake manifold collectors and tightening specifications—G35 models

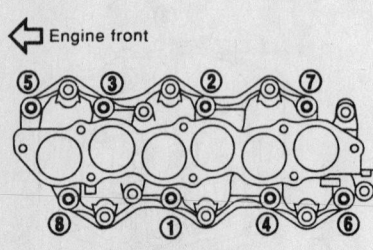

Engine front

9357MG11

Intake manifold bolt tightening sequence. Use the reverse sequence when removing the bolts—G35 models

To install:

6. Install the intake manifold, with new gaskets. Install the intake manifold nut and bolts, in sequence, and tighten as follows:

a. Stud bolts, if removed: 7.3–8.7 ft. lbs. (9.8–11.8 Nm).

b. Step 1: 4–7 ft. lbs. (5–10 Nm).

c. Step 2: 20–23 ft. lbs. (26–31 Nm).

d. Step 3: 20–23 ft. lbs. (26–31 Nm).

7. Install or connect the following:

• Fuel rail and injector assembly
• New gaskets, lower manifold collector, and collector cover, as shown. Tighten the bolts, in

sequence to 13–15 ft. lbs. (17.7–21.6 Nm).

• Intake manifold collector (upper). Torque the bolts and nuts, in sequence, to the specifications shown in the illustration.

8. The remainder of installation is the reverse of the removal procedure.

9. Fill and bleed the cooling system.

10. Start the vehicle, check for leaks and repair if necessary.

4.1L Engine

1. Before servicing the vehicle, refer to the precautions in the beginning of this section.

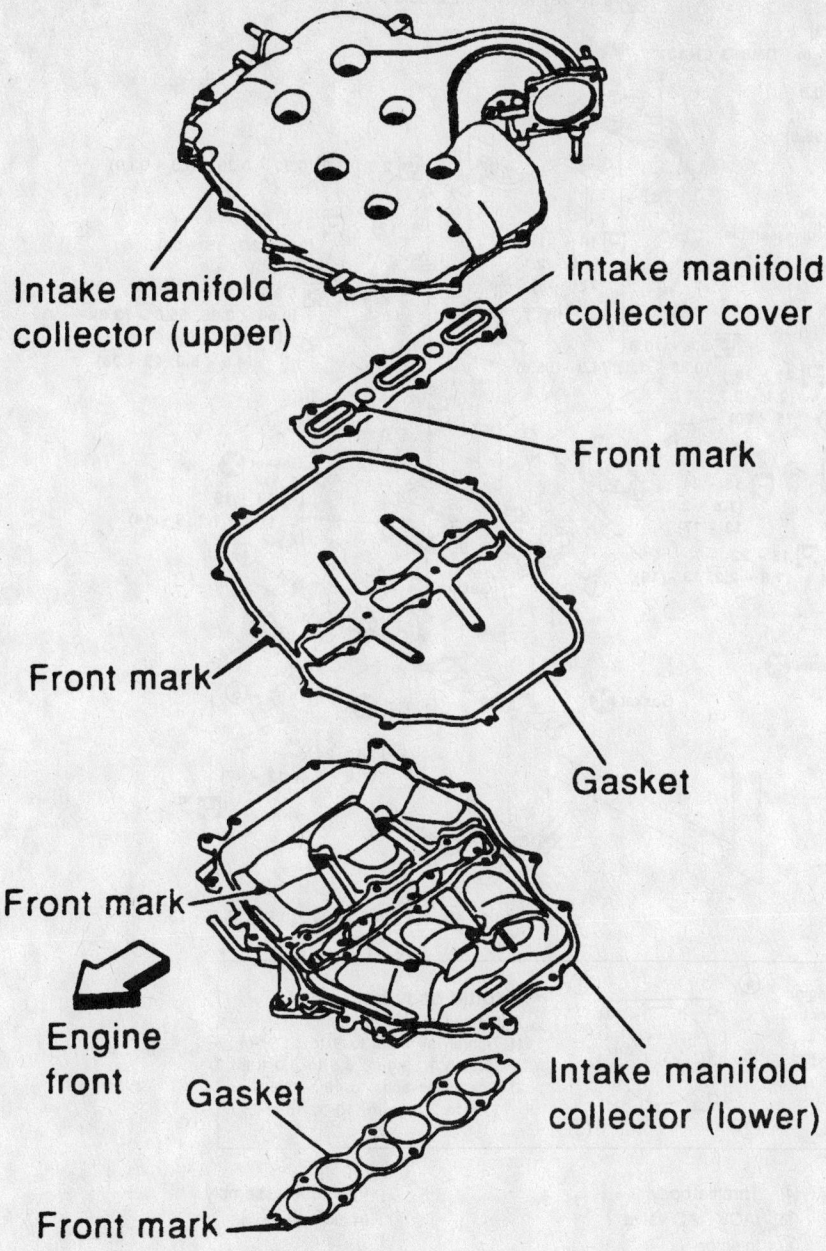

Intake manifold collector (upper)

Intake manifold collector cover

Front mark

Front mark

Gasket

Front mark

Engine front

Gasket

Intake manifold collector (lower)

Front mark

9357MG09

Exploded view of the intake manifold collector installation—G35 models

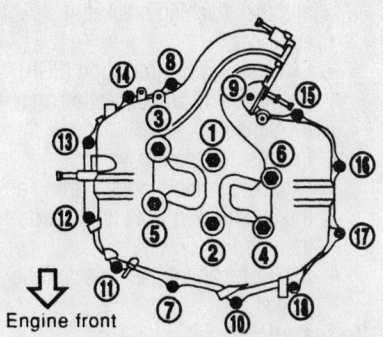

Engine front

9357MG10

Intake manifold collector (upper) bolt tightening sequence—G35 models

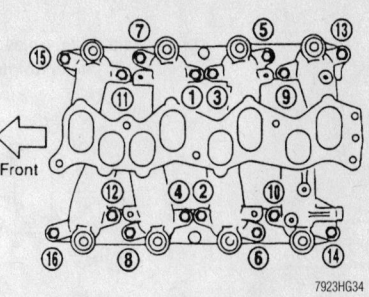

Front

7923HG34

Lower intake manifold torque sequence— 4.1L engine

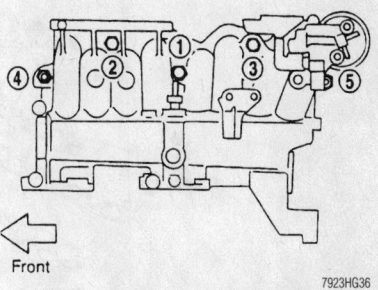

Front

7923HG36

Upper intake manifold torque sequence— 4.1L engine

2. Drain the cooling system.
3. Relieve the fuel system pressure.
4. Remove or disconnect the following:

- Negative battery cable
- Intake manifold collector
- Exhaust Gas Recirculation (EGR) valve
- EGR temperature sensor
- Throttle body
- Fuel injectors and lift up the fuel rail assembly with the injectors. Do not disconnect the fuel hose.

- Intake manifold and discard the gaskets

To install:
5. Clean the intake manifold and intake manifold collector mounting surfaces
6. Install or connect the following:
- Intake manifold using new gaskets. Torque the bolts in sequence to 13–17 ft. lbs. (18–21 Nm).
- Fuel tube assembly. Torque the bolts in sequence first to 84–96 inch lbs. (9–11 Nm), then to 15–20 ft. lbs. (21–26 Nm).
- Throttle body
7. Torque the throttle body bolts in the following steps:
   a. Step 1: bolts in sequence to 80–96 inch lbs. (9–11 Nm).
   b. Step 2: Bolts in sequence to 13–16 ft. lbs. (18–22 Nm).
- EGR temperature sensor
- EGR valve
- Intake manifold collector using a new gasket. Torque the bolts in sequence to 13–16 ft. lbs. (18–22 Nm).
- Negative battery cable
8. Fill and bleed the cooling system.
9. Start the vehicle, check for leaks and repair if necessary.

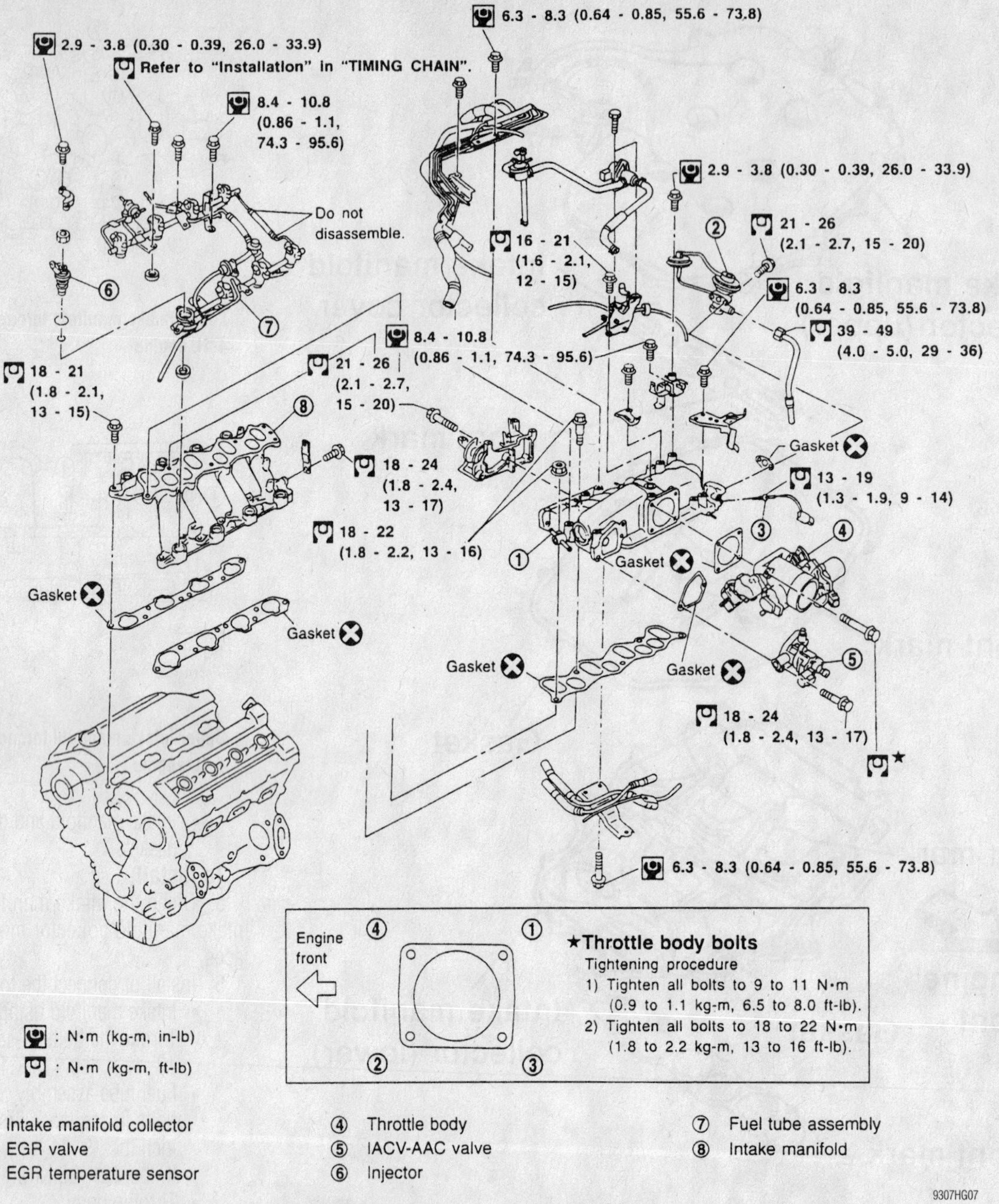

2.9 - 3.8 (0.30 - 0.39, 26.0 - 33.9)
Refer to "Installation" in "TIMING CHAIN".

6.3 - 8.3 (0.64 - 0.85, 55.6 - 73.8)

8.4 - 10.8 (0.86 - 1.1, 74.3 - 95.6)

2.9 - 3.8 (0.30 - 0.39, 26.0 - 33.9)

Do not disassemble.

21 - 26 (2.1 - 2.7, 15 - 20)

16 - 21 (1.6 - 2.1, 12 - 15)

6.3 - 8.3 (0.64 - 0.85, 55.6 - 73.8)

39 - 49 (4.0 - 5.0, 29 - 36)

18 - 21 (1.8 - 2.1, 13 - 15)

21 - 26 (2.1 - 2.7, 15 - 20)

8.4 - 10.8 (0.86 - 1.1, 74.3 - 95.6)

Gasket

13 - 19 (1.3 - 1.9, 9 - 14)

18 - 24 (1.8 - 2.4, 13 - 17)

18 - 22 (1.8 - 2.2, 13 - 16)

Gasket

Gasket

18 - 24 (1.8 - 2.4, 13 - 17)

Gasket

Gasket

Gasket

6.3 - 8.3 (0.64 - 0.85, 55.6 - 73.8)

Engine front

★ Throttle body bolts
Tightening procedure
1) Tighten all bolts to 9 to 11 N·m (0.9 to 1.1 kg-m, 6.5 to 8.0 ft-lb)
2) Tighten all bolts to 18 to 22 N·m (1.8 to 2.2 kg-m, 13 to 16 ft-lb)

: N·m (kg-m, in-lb)

: N·m (kg-m, ft-lb)

① Intake manifold collector
② EGR valve
③ EGR temperature sensor
④ Throttle body
⑤ IACV-AAC valve
⑥ Injector
⑦ Fuel tube assembly
⑧ Intake manifold

9307HG07

Exploded view of the intake manifold assembly, related components and the throttle body torque sequence—4.1L engines

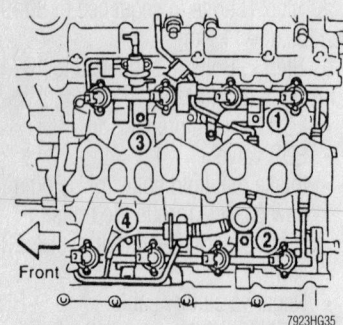

7923HG35

To prevent possible tube breakage, be sure to tighten the fuel tube mounting bolts according to the sequence shown

Exhaust Manifold

REMOVAL & INSTALLATION

2.0L Engine

1. Before servicing the vehicle, refer to the precautions in the beginning of this section.
2. Remove or disconnect the following:
   • Negative battery cable
   • Undercover and dust covers, if equipped
   • Exhaust pipe at the manifold flange
   • Air Injection Valve (AIV), AIV tube,

and the attaching bracket, if equipped
   • Exhaust Gas Recirculation (EGR) sensor electrical connection and the sensor
   • Exhaust manifold cover
   • Exhaust manifold nuts, starting at the outside and working towards the middle
   • Exhaust manifold and discard the gasket

To install:

3. Clean the gasket mating surface and install a new exhaust manifold gasket.

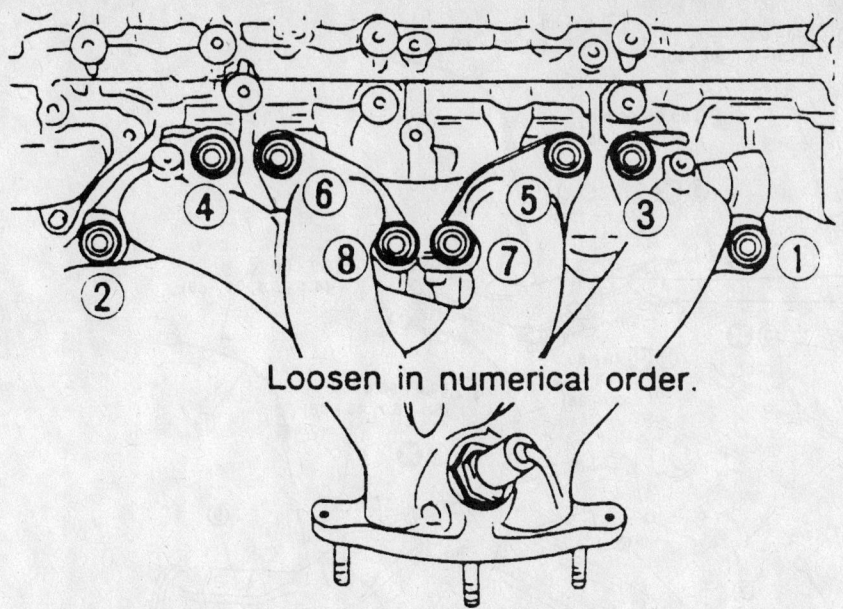

Loosen in numerical order.

9307HG23

Be sure to tighten the exhaust manifold nuts in the proper sequence—2.0L engines

4. Install or connect the following:
   - Exhaust manifold. Torque the nuts in sequence, in 2 steps, to 27–35 ft. lbs. (37–48 Nm).
   - Exhaust manifold cover and EGR sensor
   - Exhaust gas sensor electrical connection
   - AIV, AIV tube, and the attaching bracket, if equipped
   - Exhaust pipe to the manifold

flange. Torque the nuts to 30–35 ft. lbs. (41–48 Nm).
   - Negative battery cable
5. Start the engine, check for leaks and repair if necessary.

3.0L and 3.5L Engines

1. Before servicing the vehicle, refer to the precautions in the beginning of this section.
2. Remove or disconnect the following:
   - Negative battery cable

Right bank

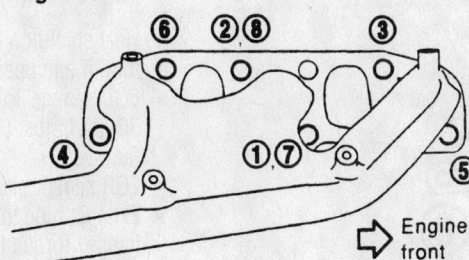

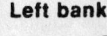

Left bank

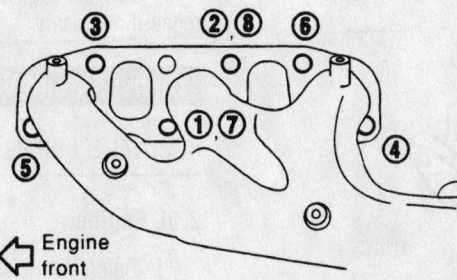

9357MG12

Exhaust manifold tightening sequence. Use the reverse of the sequence for removal—G35 shown

→ **If necessary, soak the exhaust pipe retaining nuts with penetrating oil to loosen them.**

- Engine cover, G35 only
- Air cleaner assembly and duct, G35 only
- Front exhaust pipe from the exhaust manifolds
- Heated Oxygen Sensors (HO$_2$S) from the manifold, as necessary
- Protective covers from the manifolds
- Exhaust manifold-to-engine mounting nuts
- Manifold from the engine and discard the gaskets

To install:
3. Clean all gasket mounting surfaces.
4. Install or connect the following:
   - Exhaust manifold with new gaskets. Torque the bolts, in sequence, in 2 steps to 21–23 ft. lbs. (28–32 Nm).
   - Protective shields. Torque the bolts in 2 steps to 46–57 inch lbs. (5–6 Nm).
   - HO$_2$S to the manifold, as necessary
   - Exhaust manifolds to the exhaust pipes. Torque the bolts/nuts to 32–37 ft. lbs. (43–50 Nm) for all models except G35. For the G35, torque the nuts to 45–48 ft. lbs. (60–66 Nm).
   - Air cleaner assembly and engine cover, G35 only
   - Negative battery cable
5. Start the engine, check for exhaust leaks and repair if necessary.

4.1L Engines

1. Before servicing the vehicle, refer to the precautions in the beginning of this section.
2. Remove or disconnect the following:
   - Negative battery cable
   - Engine undercovers
   - Exhaust pipe at the manifold flange
   - Heat shield from the exhaust manifold, if equipped
   - Exhaust Gas Recirculation (EGR) sensor electrical connection and if necessary, remove the sensor
   - Exhaust manifold nuts, starting at the ends and working towards the center
   - Exhaust manifold and discard the gasket

To install:
3. Clean the gasket mating surfaces.
4. Install or connect the following:
   - Exhaust manifold with new gaskets. Torque the nuts to 16–21 ft. lbs. (22–28 Nm).

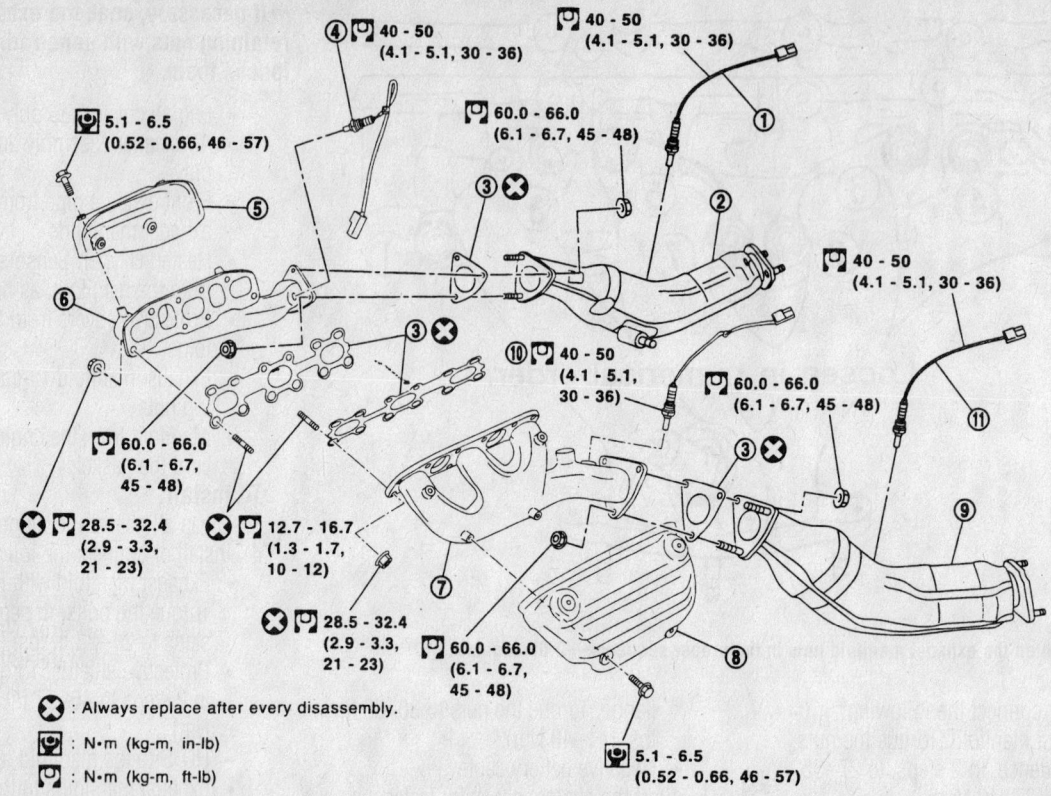

④ 🔧 40 - 50
(4.1 - 5.1, 30 - 36)

🔧 40 - 50
(4.1 - 5.1, 30 - 36) — ①

🔧 5.1 - 6.5
(0.52 - 0.66, 46 - 57)

🔧 60.0 - 66.0
(6.1 - 6.7, 45 - 48)

③❌

⑤

② ⑥

🔧 40 - 50
(4.1 - 5.1, 30 - 36)

③❌

⑩ 🔧 40 - 50
(4.1 - 5.1, 30 - 36)

🔧 60.0 - 66.0
(6.1 - 6.7, 45 - 48)

⑪

🔧 60.0 - 66.0
(6.1 - 6.7, 45 - 48)

❌ 🔧 28.5 - 32.4
(2.9 - 3.3, 21 - 23)

❌ 🔧 12.7 - 16.7
(1.3 - 1.7, 10 - 12)

③❌

⑨

⑦

❌ 🔧 28.5 - 32.4
(2.9 - 3.3, 21 - 23)

🔧 60.0 - 66.0
(6.1 - 6.7, 45 - 48)

⑧

🔧 5.1 - 6.5
(0.52 - 0.66, 46 - 57)

❌ : Always replace after every disassembly.

🔧 : N•m (kg-m, in-lb)

🔧 : N•m (kg-m, ft-lb)

1. Heated oxygen sensor 2 (rear) (bank1)
2. Three way catalyst (RH bank)
3. Gasket
4. Heated oxygen sensor 1 (front) (bank1)
5. Exhaust manifold cover (RH bank)
6. Exhaust manifold (RH bank)
7. Exhaust manifold (LH bank)
8. Exhaust manifold cover (LH bank)
9. Three way catalyst (LH bank)
10. Heated oxygen sensor 1 (front) (bank2)
11. Heated oxygen sensor 2 (rear) (bank2)

9357MG13

Exploded view of the exhaust manifold and related components—G35 shown

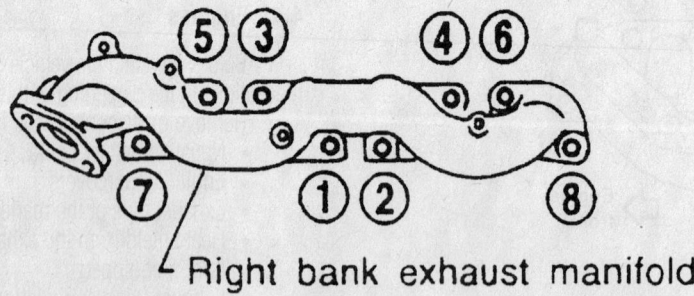

Right bank exhaust manifold

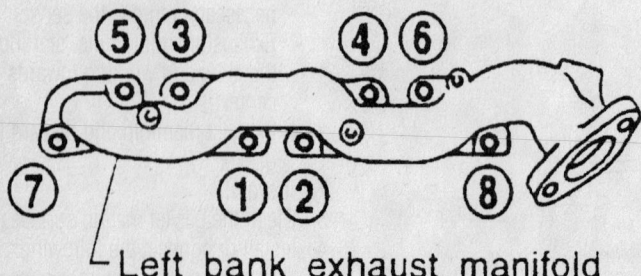

Left bank exhaust manifold

7923HG39

To avoid leaks, tighten the exhaust manifold nuts in the sequence shown—4.1L engine

- Heat shield on the exhaust manifold, if equipped
- EGR sensor. Torque the fastener to 30–37 ft. lbs. (40–50 Nm), if removed.
- EGR sensor electrical connection
- Exhaust pipe to the manifold flange. Torque the nuts to 33–44 ft. lbs. (45–60 Nm).
- Negative battery cable

5. Start the engine, check for leaks and repair if necessary.

Camshaft and Valve Lifters

REMOVAL & INSTALLATION

2.0L Engine

1. Before servicing the vehicle, refer to the precautions in the beginning of this section.
2. Relieve the fuel system pressure.

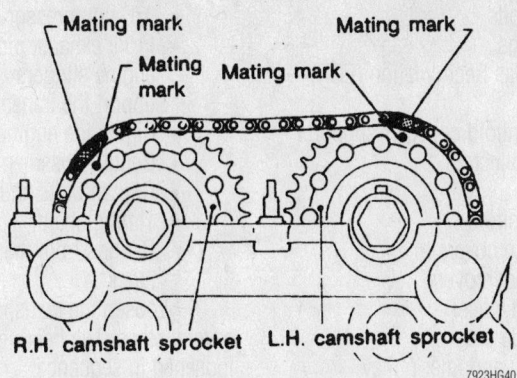

Be sure to align the marks on the sprockets with the marks on the chain—2.0L engine

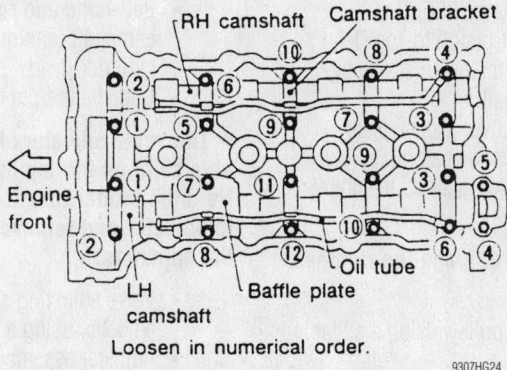

Camshaft bearing cap bolt loosening sequence—2.0L engine

3. Remove or disconnect the following:
- Negative battery cable
- Rocker arm cover
- Oil separator

4. Rotate the crankshaft until the No. 1 piston is at Top Dead Center (TDC) on the compression stroke and the mating marks on the camshaft sprockets line up with the mating marks on the timing chain.
- Timing chain tensioner
- Distributor
- Timing chain guide
- Camshaft sprockets. Use a wrench to hold the camshaft while loosening the sprocket bolt.

5. Loosen the camshaft bearing cap bolts in sequence.
- Camshaft from the cylinder head

6. When removing the rocker arm, be careful not to drop the valve shims into the cylinder head. After removing the adjuster, set them upright or lay them down in a pan of clean engine oil. Do not lay them down on the bench or the oil will drain out and the adjuster will become air bound. Keep all of these parts in order so they can be installed in the same locations.

To install:

7. Install the adjusters, shims and rockers into their original locations.

8. Clean the left-hand camshaft end bearing cap and coat the mating surface with liquid gasket. Install the camshafts, bearing caps, oil tubes and baffle plate. Ensure the left camshaft key is at 12 o'clock and the right camshaft key is at 10 o'clock.

9. The procedure for tightening bearing cap bolts must be followed exactly to prevent camshaft damage. Torque bolts as follows:
   a. Torque the right camshaft bolts 9 and 10 (in that order) to 17 inch lbs. (2 Nm), then bolts 1 through 8 (in that order) to the same specification.
   b. Torque the left camshaft bolts 11 and 12 (in that order) to 17 inch lbs. (2 Nm), then bolts 1 through 10 (in that order) to the same specification.
   c. Torque all bolts in sequence to 52 inch lbs. (6 Nm).
   d. Torque all bolts again in sequence to 78–102 inch lbs. (9–12 Nm), then bolts 8 and 9 on the left camshaft to 13–19 ft. lbs. (18–25 Nm).

10. Line up the mating marks on the timing chain and camshaft sprockets and install the sprockets. Torque sprocket bolts to 101–116 ft. lbs. (137–157 Nm).

11. Install or connect the following:
- Timing chain guide and chain tensioner
- Distributor making certain that the rotor head is at the 5 o'clock position

12. Clean the rocker cover and mating surfaces and apply a continuous bead of liquid gasket to the mating surface.

13. Install the rocker cover and oil separator. Tighten the rocker cover bolts as follows:
   a. Torque the nuts 1, 10, 11 and 8, in that order to 36 inch lbs. (4 Nm).
   b. Torque the nuts 1 through 13 as

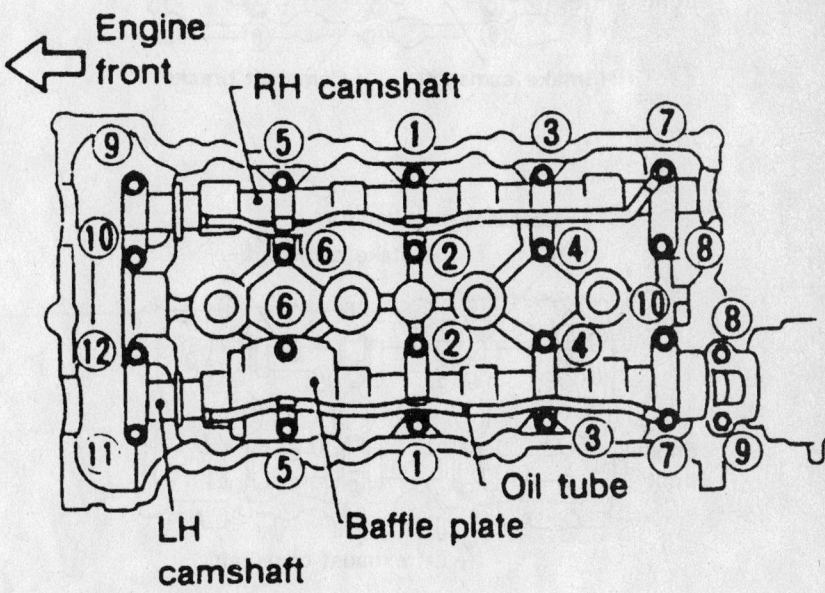

Camshaft bearing cap bolt torque sequence—2.0L engine

indicated in the figure to 72–84 inch lbs. (8–10 Nm).

14. Connect the negative battery cable.

3.0L and 3.5L Engines

EXCEPT G35 MODELS

1. Before servicing the vehicle, refer to the precautions in the beginning of this section.
2. Relieve the fuel system pressure.
3. Drain the engine oil.
4. Drain the cooling system.
5. Remove or disconnect the following:
   - Negative battery cable
   - Left side rocker cover ornament

➡ **Before disconnecting any hoses or connectors, note the locations for reassembly.**

- Air duct to intake manifold hose, collector hose, blow-by hose, and vacuum hoses
- Fuel hoses and disconnect the harness connection
- Evaporative emissions (EVAP) canister purge hoses
- Water hoses from the cylinder head and intake manifold

- Ignition coils
- Spark plugs
- Exhaust Gas Recirculation (EGR) tube
- Intake manifold collector supports and the collector
- Fuel tube
- Intake manifold
- Rocker arm covers
- Engine undercovers
- Right front wheel
- Engine side covers
- Drive belts and idler pulley
- Power steering oil pump and belt
- Camshaft Position (CMP) sensor (PHASE) and Crankshaft Position (CKP) sensors (REF)/(POS)

6. Set the No. 1 piston to Top Dead center (TDC) of compression stroke by rotating the crankshaft.
   - Ring gear cover access plate
7. Loosen the crankshaft pulley bolt while securing the ring gear so the crankshaft cannot rotate.

➡ **Use care not to damage the ring gear teeth.**

- Crankshaft pulley, using a suitable puller

- A/C compressor and bracket
- Front exhaust pipe and install engine slingers

8. Support the transaxle with jack.
   - Right side engine mounting bracket
   - Center crossmember assembly
   - Oil pan bolts and oil pans
   - Timing chain
   - O-rings from the front of the engine block

9. Loosen the camshaft bearing caps in several steps. The bearing caps MUST be loosened in sequence.

➡ **Keep all bearing caps and camshafts in proper order for reinstallation.**

- Left-hand and right-hand camshaft tensioners from the cylinder head
- Camshafts from the cylinder heads

➡ **The valve adjusters have a replaceable shim on the top of the adjuster. Note the proper locations of each shim to adjuster and remove the shims from the adjusters.**

- Valve adjusting shim from the adjuster, using a magnet
- Adjuster assembly from the bore. Be sure to note the locations from where each adjuster came.

10. Check the diameter of the valve adjuster and the valve adjuster guide bore.
11. The diameter of the adjuster should be 1.3764–1.3770 in. (34.960–34.975mm) and the diameter of the bore should be 1.3780–1.3788 in. (35.000–35.021mm).
   - All traces of liquid gasket from the timing chain case and from the water pump covers
   - All traces of liquid gasket from the engine block

12. Inspect the camshafts for excessive wear or damage and replace as necessary.

To install:

13. Lubricate the valve adjusters with clean engine oil and install the adjusters into the bore from which they were removed.
14. Lubricate the valve adjuster shims with clean engine oil and install the shims into the adjuster from which they were removed.
15. Turn the crankshaft clockwise until the No. 1 piston is set 240° before TDC on compression stroke.
16. Install or connect the following:
   - Camshaft tensioners on both sides of the cylinder heads. Torque the tensioner mounting bolts to 75–96 inch lbs. (8–11 Nm).

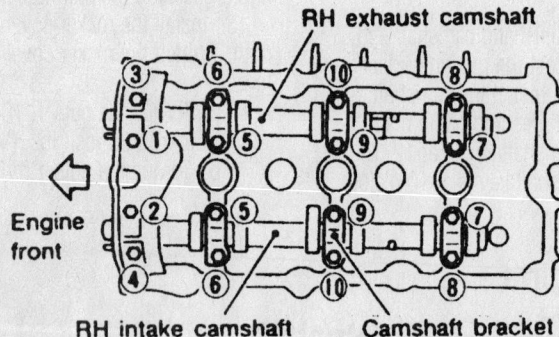

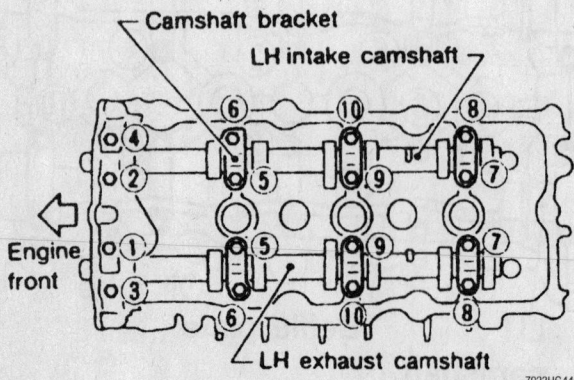

7923HG44

To avoid camshaft damage, loosen the bearing cap bolts in the sequence shown—3.0L and 3.5L engines, except G35

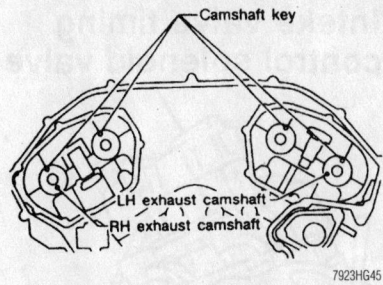

Positioning of the camshaft keys during installation—3.0L and 3.5L engines, except G35

➡The camshafts can be identified by the paint marks on the camshaft. The left cylinder head camshafts have a YELLOW paint mark and the right cylinder head camshafts have a WHITE paint mark. When installing the camshafts, position the camshaft keys at the 12 o'clock position in respect to the cylinder head angle.

- Exhaust and intake camshafts and install the bearing caps. Before installing the No. 1 bearing cap, apply liquid gasket to the corners of the cap.

17. Torque the camshaft bearing caps as follows:

a. Nos. 7 through 10 then, Nos. 1 through 6 to 17 inch lbs. (2 Nm).

b. All bolts in order to 52 inch lbs. (6 Nm).

c. Nos. 1 though 6 to 80–104 inch lbs. (9–11 Nm).

d. Nos. 7 though 10 to 74–91 inch lbs. (9–11 Nm).

- New O-rings to the front of the engine block

18. Apply sealant to the hatched portion of the of the rear timing chain case.

19. Align the rear timing chain case with the dowel pins and install onto the cylinder heads and engine block. Torque the rear timing chain case mounting bolts in sequence to 105–121 inch lbs. (11.8–13.7 Nm).

- Crankshaft sprocket with the mating mark facing out

20. Rotate the crankshaft clockwise and position the crankshaft to TDC of compression stroke and align the dowels of the camshaft sprockets to the 12 o'clock position in respect to the cylinder head

- Lower chain guide on the dowel pin with the front mark on the guide facing upward

21. On a workbench, align the marks on the intake and exhaust camshaft sprockets with the marks of the chain.

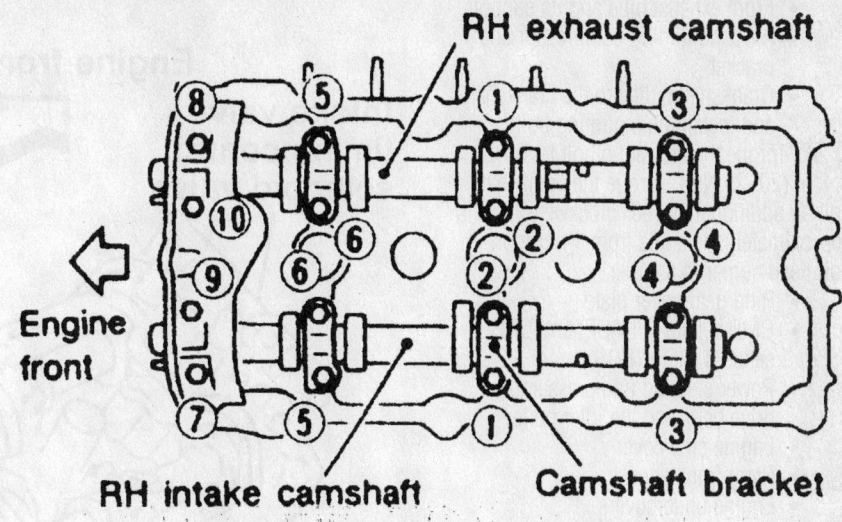

Tighten in numerical order.

Be sure to tighten the camshaft bearing cap bolts in the correct sequence—3.0L and 3.5L engines (except G35), right cylinder head

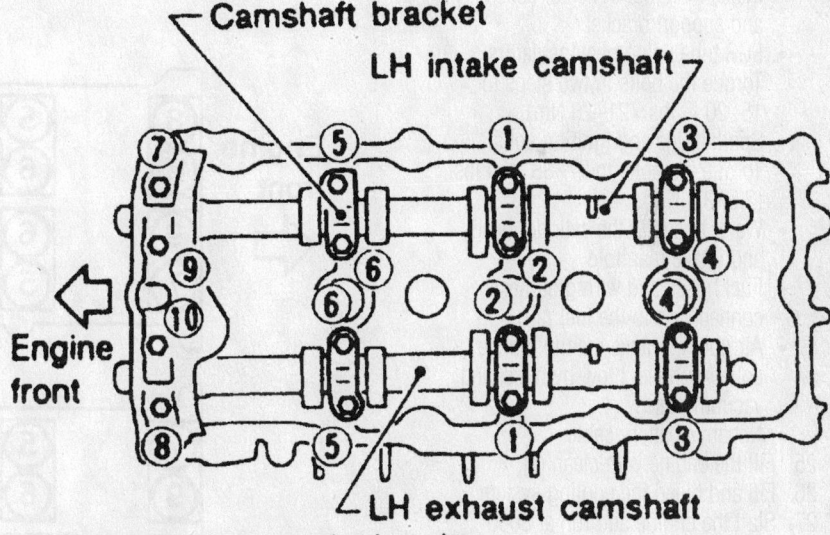

Tighten in numerical order.

Tighten the camshaft bearing cap bolts in the correct sequence—3.0L and 3.5L engines (except G35), left cylinder head

- Exhaust camshaft sprockets onto the dowel pin. Torque the bolts to 88–95 ft. lbs. (120–129 Nm). Be sure to secure the camshafts while tightening the bolts.
- Align and install the timing chains and sprockets to the camshafts
- Timing cover evenly and gently. Be sure to align the dowel pin holes

➡Leave the bolts unattended for 30 minutes or more after tightening.

22. Apply a 0.091–0.130 in. (2.3–3.3mm) continuous bead of liquid gasket to the water pump cover and install the cover. Tighten the bolts to 84–108 inch lbs. (10–13 Nm).

- Rocker covers

23. Apply sealant to the front and rear seal of the oil pan and install the oil pan.

- Center crossmember assembly
- Right side engine mounting bracket and mount assembly
- Engine slinger assembly

- Front exhaust pipe and its support
- Air conditioning compressor and bracket
- Crankshaft pulley to the crankshaft and install the mounting bolt

24. Torque the mounting bolt to 14–22 ft. lbs. (20–29 Nm). Torque the crankshaft bolt an additional 60–66° clockwise. This is approximately the angle from 1 hexagon bolt head corner to another.

- Ring gear cover plate
- CMP sensor (PHASE) and CKP sensors (REF)/(POS)
- Power steering pump assembly, drive belts and the idler pulley
- Engine side cover
- Right front wheel
- Engine undercovers
- Intake manifold, using new gaskets
- Fuel tube assembly using new insulators. Torque the bolts in several steps to 15–20 ft. lbs. (21–26 Nm).
- Intake manifold collector gasket with the arrow facing forward
- Intake manifold collector assembly and support bracket
- EGR tube using new insulators. Torque the bolts in two steps to 15–20 ft. lbs. (21–26 Nm).
- Spark plugs and ignition coils. Torque the bolts to 27–33 inch lbs. (2.9–3.8 Nm).
- Water hoses to the cylinder head and intake manifold
- Fuel hoses and wiring harness connections to the fuel rail
- Air duct to intake manifold hose, collector hose, blow-by hose, and vacuum hoses
- Negative battery cable

25. Fill the engine with clean oil.
26. Fill and bleed the cooling system.
27. Start the engine and run at 3000 RPM under no load to purge the air from the high pressure chamber. The engine may produce a rattling noise. This indicates that air still remains in the chamber and is not a matter of concern.

G35 MODELS

1. Before servicing the vehicle, refer to the precautions in the beginning of this section.
2. Drain the engine oil and cooling system.
3. Relieve the fuel system pressure.
4. Remove or disconnect the following:
   - Negative battery cable
   - Front timing chain case
   - Camshaft sprocket
   - Timing chain

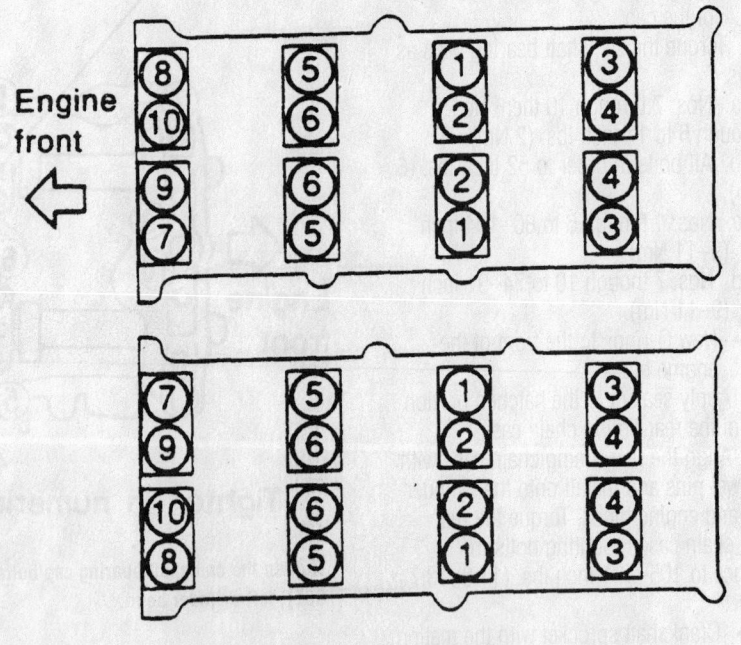

Location of the intake valve timing control solenoid valves—G35 models

Intake and exhaust camshaft bracket bolts tightening sequence. Use the reverse sequence for removal—G35 models

- Rear timing chain case
- Camshaft Position (CMP) sensors (PHASE) from the back side of the cylinder heads
- Intake valve timing control solenoid valves from the No. 1 camshaft bracket

→Before removal, matchmark the position of the camshafts, brackets and bolts, so they are reinstalled in their original locations.

- Intake and exhaust camshaft brackets. Loosen the bolts in several

stages, in reverse of the sequence shown.
- Camshafts
- Valve lifters if necessary, noting their installed positions

To install:

5. Install or connect the following:
- Valve lifters, in their original positions
- Camshafts with the dowel pin attached to its front end face on the exhaust side
- Camshaft brackets, as shown in the illustration

6. Torque the camshaft bracket bolts, in sequence, as follows:
   a. Step 1: Nos. 7–10, then Nos. 1–6 to 17 inch lbs. (2 Nm).
   b. Step 2: Nos. 1–10 to 52 inch lbs. (5.9 Nm).
   c. Step 3: Nos. 1–6 to 80–104 inch lbs. (9–12 Nm).
   d. Step 4: Nos. 7–10 to 74–91 inch lbs.

7. Measure the difference in levels between the front end faces of the No. 1 camshaft bracket and the cylinder head. If the measurement falls out of the range of -0.0055–0.0055 in. (-0.14–0.14mm), you must reinstall the camshaft and brackets.

8. Check and adjust the valve clearance.
- Intake valve timing control solenoid valves
- CMP sensors (PHASE)
- Rear timing chain case, timing chain, camshaft sprocket and front timing chain case
- Negative battery cable

9. Fill the engine with oil.
10. Fill and bleed the cooling system.
11. Start the vehicle, check for leaks and repair if necessary.

4.1L Engine

1. Before servicing the vehicle, refer to the precautions in the beginning of this section.
2. Drain the cooling system.
3. Relieve the fuel system pressure.
4. Remove or disconnect the following:
- Negative battery cable
- Ornament cover
- Undercover
- Radiator and cooling fan
- Inlet and outlet hoses
- Alternator belt and idler bracket
- Air duct and intake manifold collector
- Intake valve timing control solenoid
- Rocker arm covers

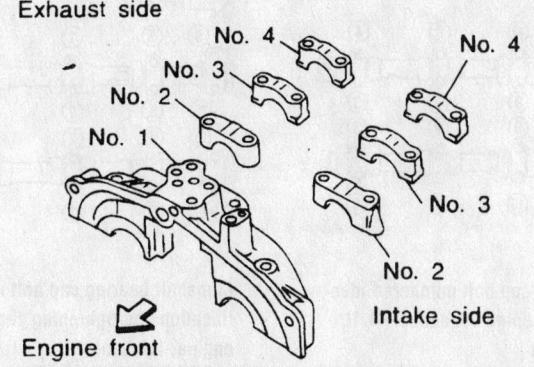

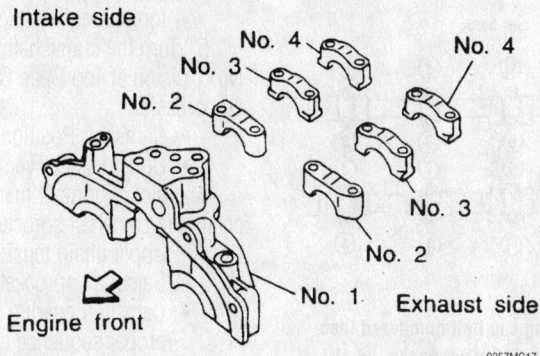

Camshaft identification and installation—G35 models

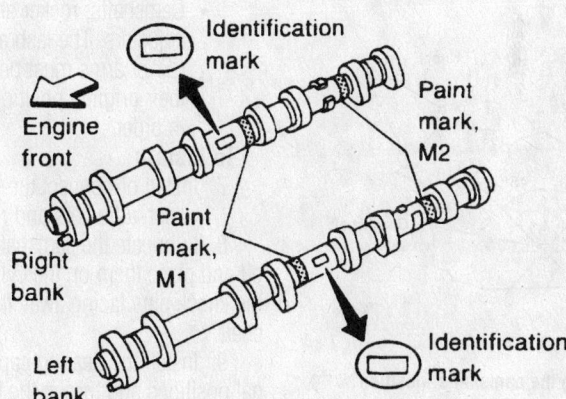

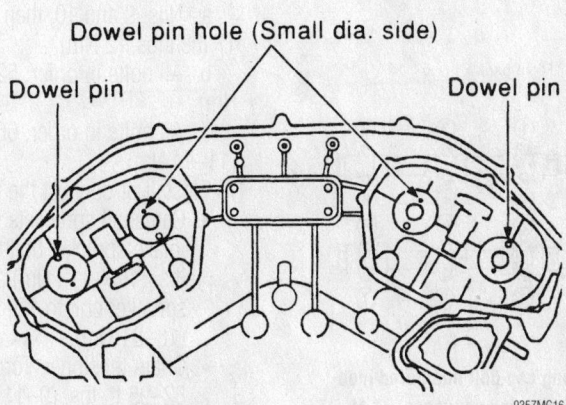

Camshaft bracket installation—G35 models

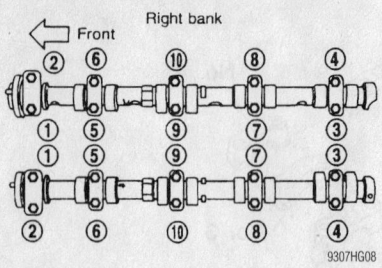

Camshaft bearing cap bolt numbered identification and loosening sequence—4.1L engine, right bank

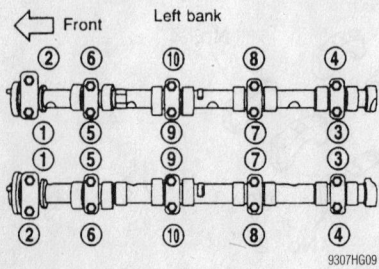

Camshaft bearing cap bolt numbered identification and loosening sequence—4.1L engine, left bank

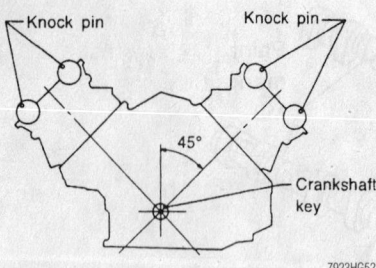

When installing the camshafts, position the knock pins as shown—4.1L engine

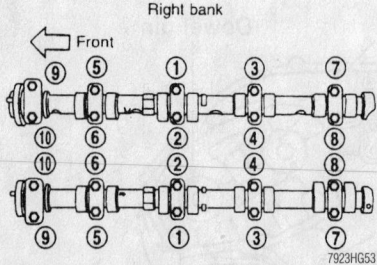

Camshaft bearing cap bolt numbered identification and tightening sequence—4.1L engine, right bank

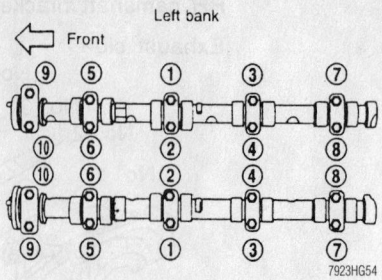

Camshaft bearing cap bolt numbered identification and tightening sequence—4.1L engine, left bank

- Vacuum pipe
- Ignition coils and spark plugs

5. Turn the crankshaft to position the No. 1 piston at Top Dead Center TDC on compression.
- Camshaft Position (CMP) sensor
- Upper front covers

6. Paint alignment marks on the timing chain and camshaft sprockets.
- Upper chain tensioners
- Camshaft sprockets
- Camshaft bearing cap bolts in the proper sequence to prevent damage to the camshaft. Keep the caps in order so they can be installed in the correct locations.
- Camshafts, rocker arms and lash adjusters. The lash adjusters and rocker arms must be installed in their original positions. Keep them in order.

To install:

7. Install or connect the following:
- Lash adjusters and rocker arms

8. Lubricate the camshafts with engine oil and place them on the cylinder head with the knock pins facing away from the crankshaft.

9. Install the bearing caps in their original positions and torque the bolts as follows:

   a. Nos. 9 and 10, then 1 through 8: 17 inch lbs. (2 Nm).

   b. All bolts in order: 52 inch lbs. (6 Nm).

   c. All bolts in order: 96–121 inch lbs. (11–14 Nm).

10. Install or connect the following:
- Camshaft sprockets. Torque the intake sprocket bolt to 76–83 ft. lbs. (103–113 Nm) and the exhaust sprocket bolt to 12–15 ft. lbs. (16–21 Nm).
- Chain tensioner. Torque the bolts to 82–95 ft. lbs. (9–11 Nm).
- Upper chain covers

- Rocker covers using new gaskets
- Valve timing control solenoid valve using a new O-ring. Torque the solenoid to 18–25 ft. lbs. (25–34 Nm).
- Air duct and intake manifold collector
- Alternator belt and idler bracket
- Inlet and outlet hoses
- Spark plugs and ignition coils
- Rocker arm covers
- Vacuum pipe
- Radiator and cooling fan
- Engine undercover
- Ornament cover
- Negative battery cable

11. Fill and bleed the cooling system.

12. Start the vehicle, check for leaks and repair if necessary.

Valve Lash

ADJUSTMENT

2.0L Engine

➡**A special gauge plate and collar will be needed to complete this procedure.**

1. Before servicing the vehicle, refer to the precautions in the beginning of this section.

2. Remove the camshafts.

3. Install the J38957–1 gauge plate to the cylinder head. Use the bolts supplied in the kit to secure the plate to the cam bearing journals.

4. Install the collar J38957–2 on the dial indicator. Be sure the dished side of the collar is toward the gauge and tighten the setscrew.

5. Place the gauge on the No. 1 intake valve (shim side). Be sure the shim has been removed. Place the tip of the dial gauge on the top of the valve stem and the collar on the gauge plate. Zero the dial gauge.

6. Move the dial gauge to the other intake valve (rocker guide side). Place the tip of the dial gauge on the rocker guide and the collar of the gauge plate. Record the measurement.

7. Select the correct size shim using the chart. Shims are available in 17 different sizes ranging from 0.1102 in. (2.800mm) to 0.1260 in. (3.200mm) in increments of 0.001 in. (0.025mm).

3.0L, 3.5L and 4.1L Engines

➡**Check and adjust the valve clearances while the engine is cold and not running.**

Available shim

Thickness mm (in)	Identification mark
2.800 (0.1102)	28 00
2.825 (0.1112)	28 25
2.850 (0.1122)	28 50
2.875 (0.1132)	28 75
2.900 (0.1142)	29 00
2.925 (0.1152)	29 25
2.950 (0.1161)	29 50
2.975 (0.1171)	29 75
3.000 (0.1181)	30 00
3.025 (0.1191)	30 25
3.050 (0.1201)	30 50
3.075 (0.1211)	30 75
3.100 (0.1220)	31 00
3.125 (0.1230)	31 25
3.150 (0.1240)	31 50
3.175 (0.1250)	31 75
3.200 (0.1260)	32 00

7923HG56

Select the correct valve lash adjusting shim using the chart—2.0L engine

1. Before servicing the vehicle, refer to the precautions in the beginning of this section.

2. Remove the intake manifold collector.

3. Remove the left and right rocker covers.

4. Remove the spark plugs.

5. Set the No. 1 cylinder at Top Dead Center (TDC) on its compression stroke. Align the pointer with the TDC mark on the crankshaft pulley. Check that the valve adjusters on the No. 1 cylinder are loose and valve adjusters on the No. 4 cylinder are tight. If not, turn the crankshaft 1 revolution (360°) and align the pointer with the TDC mark on the crankshaft pulley.

6. Check the following valves:
- Both No. 1 intake valves
- Both No. 2 exhaust valves
- Both No. 3 exhaust valves
- Both No. 6 intake valves

7. Using a feeler gauge, measure the clearance between the valve adjuster and the camshaft. Record any valve clearance measurements that are out of specification. Intake valve clearance (cold) is 0.010–0.013 in. (0.26–0.34mm) and exhaust valve clearance (cold) is 0.011–0.015 in. 0.29–0.37mm).

8. Turn the crankshaft 240° and set the No. 3 cylinder to TDC of its compression stroke.

9. Check the following valves:
- Both No. 2 intake valves
- Both No. 3 intake valves
- Both No. 4 exhaust valves
- Both No. 5 exhaust valves

10. Using a feeler gauge, measure the clearance between the valve adjuster and the camshaft. Record any valve clearance measurements that are out of specification. Intake valve clearance (cold) is 0.010–0.013 in. (0.26–0.34mm) and exhaust valve clearance (cold) is 0.011–0.015 in. (0.29–0.37mm).

11. Turn the crankshaft 240° and set the No. 5 cylinder to TDC of its compression stroke.

12. Check the following valves:
- Both No. 1 exhaust valves
- Both No. 4 intake valves
- Both No. 5 intake valves
- Both No. 6 exhaust valves

13. Using a feeler gauge, measure the clearance between the valve adjuster and the camshaft. Record any valve clearance measurements that are out of specification. Intake valve clearance (cold) is 0.010–0.013 in. (0.26–0.34mm) and exhaust valve clearance (cold) is 0.011–0.015 inches (0.29–0.37mm).

14. If all the valve clearances are within specification, install the cylinder head cover, spark plugs, and the intake manifold collector.

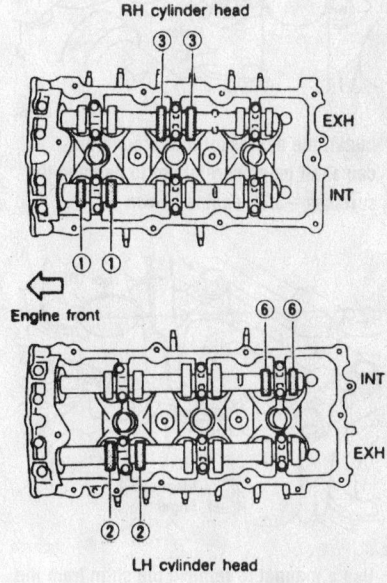

7923HG57

Valve lash checking sequence at TDC of cylinder No. 1—3.0L engine

15. If an adjustment is necessary, adjust the valve clearance while engine is cold by removing the adjusting shim. The adjusting shim can be removed by using the following procedures:

    a. Turn the crankshaft so the camshaft lobe of the valve to be adjusted is pointed straight up.

    b. Turn the adjuster so the notch is pointed towards the center of the cylinder head; this will facilitate the shim removal process.

    c. Using a depressor tool No. KV10115110 push down on the adjuster and insert a keeper tool on the edge of the adjuster to keep the adjuster in the depressed position.

    d. Remove the depressor tool and remove the shim with a magnet.

➡ **Compressed air can be blown into the hole of the adjuster to separate the adjusting shim from the adjuster.**

16. Determine the replacement adjusting shim size by using the following procedures and formula:

    a. Using a micrometer determine thickness of the removed shim.

    b. Calculate the thickness of a new adjusting shim so valve clearance is within the specified values.

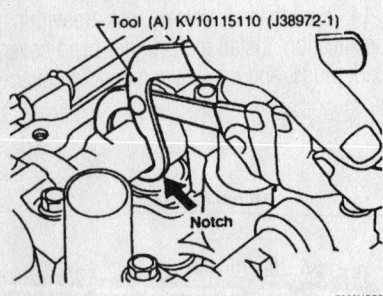

Install the depressor tool around the camshaft being careful not to damage the surfaces—3.0L engine shown

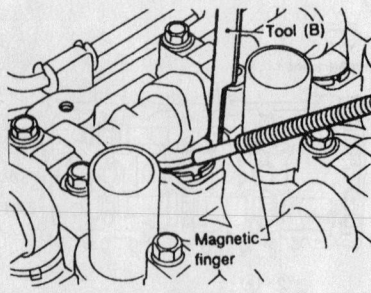

Use a magnet to remove the shim from the adjuster. Sometimes a shot of compressed air can help lift the shim up—3.0L engine shown

- R= thickness of the removed shim
- N= thickness of the new shim
- M= measured valve clearance
- Calculate the Intake Shim as follows: N = R + M—0.0118 in. (0.30mm)
- Calculate the Exhaust Shim as follows: N = R + M—0.0130 in. (0.33mm)

17. Shims are available in 64 sizes from 0.0913–0.1161 in. (2.32–2.95mm) in steps of 0.004 in. (0.01mm). The thickness is stamped on the shim; this side is always installed facing down. Select new shims with thickness as close as possible to calculated valve and install it in the adjuster.

18. Install the new shim onto the adjuster.

19. Depress the adjuster and remove the keeper tool. Remove the depressor tool and recheck the valve clearance. Repeat this procedure for any other valves requiring adjustment.

20. When all valve adjustments are finished, install the cylinder head cover, spark plugs, and the intake manifold collector.

Starter Motor

REMOVAL & INSTALLATION

2.0L Engine

1. Before servicing the vehicle, refer to the precautions in the beginning of this section.

2. Remove or disconnect the following:
- Negative battery cable
- Starter insulator
- Starter harness connector and cable
- Starter mounting bolt and nut
- Starter

To install:

3. Install or connect the following:
- Starter. Torque the bolts to 27 ft. lbs. (36 Nm).
- Starter harness connector and cable
- Starter insulator
- Negative battery cable

3.0L and 3.5L Engines

EXCEPT G35 MODELS

1. Before servicing the vehicle, refer to the precautions in the beginning of this section.

2. Remove or disconnect the following:
- Negative battery cable
- Air duct assembly
- Harness protector
- Starter harness
- Both starter bolts
- Starter

To install:

3. Install or connect the following:
- Starter
- Both starter bolts. Tighten the long bolt to 57–72 ft. lbs. (77–98 Nm) and the short bolt to 22–30 ft. lbs. (30–41 Nm).
- Starter harness
- Harness protector
- Air duct assembly
- Negative battery cable

G35 MODELS

1. Before servicing the vehicle, refer to the precautions in the beginning of this section.

2. Remove or disconnect the following:
- Negative battery cable
- Engine undercover
- S and B terminals from the starter
- Starter mounting bolts and harness bracket
- Starter

To install:

3. Install or connect the following:
- Starter motor. Tighten the mounting bolts to 37–45 ft. lbs. (49–62 Nm).
- Terminals to the starter. Tighten the B terminal nut to 83–95 inch lbs. (9–10.8 Nm).
- Engine undercover
- Negative battery cable

4.1L Engine

1. Before servicing the vehicle, refer to the precautions in the beginning of this section.

2. Remove or disconnect the following:
- Negative battery cable
- Steering gear and linkage assembly
- Harness connector
- Starter retainers
- Starter

To install:

3. Install or connect the following:
- Starter
- Starter retainers. Tighten the starter retainers to 30–37 ft. lbs. (40–50 Nm).
- Harness connector
- Steering gear and linkage assembly
- Negative battery cable

Oil Pan

REMOVAL & INSTALLATION

2.0L Engine

1. Before servicing the vehicle, refer to the precautions in the beginning of this section.

2. Raise and support the vehicle safely.

3. Drain the engine oil.

4. Remove or disconnect the following:

- Negative battery cable
- Engine undercover
- Steel oil pan bolts in the proper sequence
- Steel oil pan. Insert tool KV10111100 between steel oil pan and aluminum oil pan to break the seal.
- Front exhaust tube and support the transaxle with a suitable jack and raise the engine with an engine hoist
- Center crossmember
- Transaxle shift control cable, if equipped with an automatic transaxle
- A/C compressor bracket gussets and the rear cover plate
- Aluminum oil pan bolts in sequence
- Baffle plate
- 2 engine-to-transaxle bolts and install them into vacant bolt holes on the oil pan. Tighten the bolts to release the oil pan from the cylinder block. Use tool KV10111100 to break the remaining seal.

To install:

5. Remove the 2 bolts previously installed in the oil pan.

6. Clean the oil pan rail of all liquid gasket and apply a new bead of ⅛ inch thickness to the oil pan rail.

7. Install or connect the following:

- Aluminum oil pan.

8. Torque the bolts in the proper sequence as follows:

   a. Bolts 1 through 16 to 12–14 ft. lbs. (16–19 Nm).

   b. Bolts 17 and 18 to 56–66 inch lbs. (6.5–7.5 Nm).

- 2 engine-to-transaxle bolts, rear cover plate, compressor bracket gussets, automatic transmission shift control cable (if equipped), center member, front exhaust tube and baffle plate

9. Clean the oil pan rail of all liquid gasket and apply a new bead of ⅛ inch thickness to the oil pan rail.

- Steel oil pan. Torque the bolts in numbered sequence to 56–66 inch lbs. (6–8 Nm). Wait 30 minutes before refilling engine case with oil.
- Negative battery cable

10. Fill the engine with clean oil.

- Start the vehicle, check for leaks and repair if necessary.

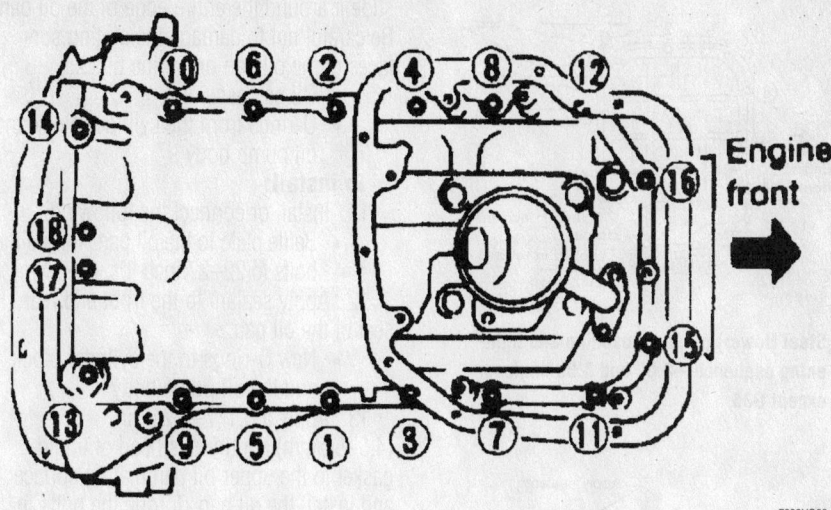

Tighten the aluminum oil pan mounting bolts in the sequence shown—2.0L engine

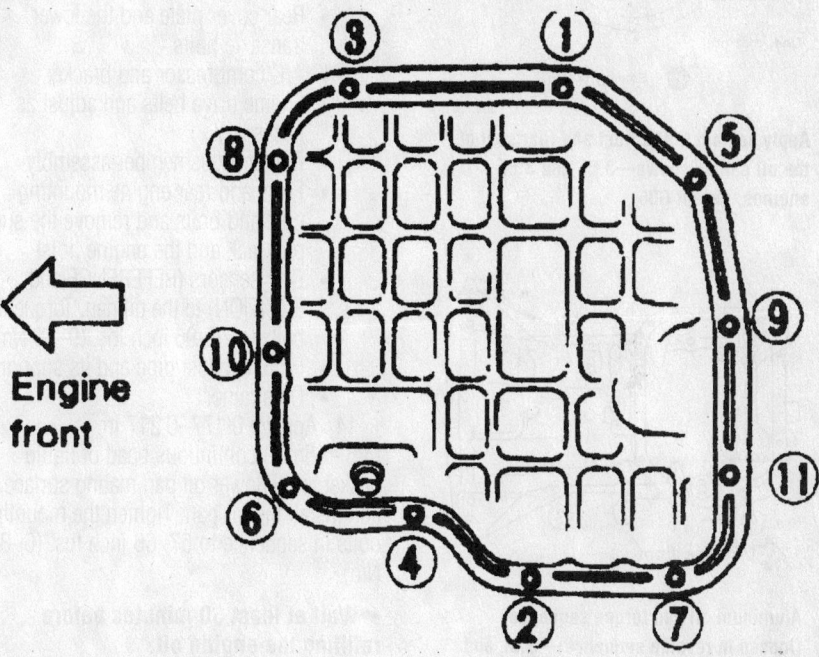

Be sure to tighten the steel oil pan mounting bolts in the proper order to prevent leakage—2.0L engine

3.0L and 3.5L Engines

EXCEPT G35 MODELS

1. Before servicing the vehicle, refer to the precautions in the beginning of this section.

2. Drain the engine oil.

3. Remove or disconnect the following:

- Negative battery cable
- Engine undercover(s)
- Steel (lower) oil pan bolts in the reverse sequence of the torque sequence

4. Insert a seal cutter between the steel and aluminum oil pan.

5. Tapping the cutter with a hammer, slide it around the entire edge of the oil pan. Be careful not to damage the aluminum mating surface of the upper oil pan.

- Steel oil pan and the oil strainer
- Front exhaust pipe and its support

6. Hang the engine at the right and left side engine slingers with a suitable hoist.

7. Position a suitable jack under the transaxle.

8. Remove or disconnect the following:

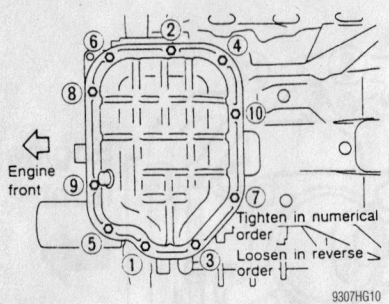

Steel (lower) oil pan loosening and tightening sequence—3.0L and 3.5L engines, except G35

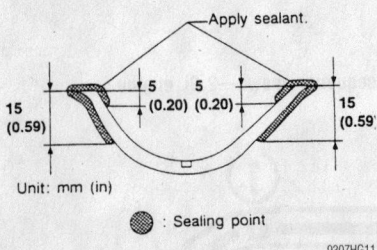

Apply sealant to the front and rear seal of the oil pan as shown—3.0L and 3.5L engines, except G35

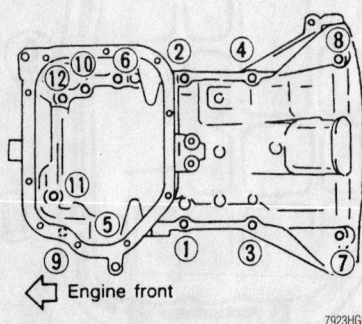

Aluminum oil pan torque sequence (loosen in reverse sequence)—3.0L and 3.5L engines, except G35

- Crankshaft Position (CKP) sensors (REFERENCE and POSITION) from the oil pan
- Front and rear engine mounting nuts and bolts
- Center crossmember assembly
- Engine drive belts
- A/C compressor and bracket
- Rear cover plate
- Aluminum (upper) oil pan bolts in the reverse sequence of the torque sequence
- 4 engine-to-transaxle bolts

9. Insert a seal cutter between the aluminum oil pan and the engine block.
10. Tapping the cutter with a hammer, slide it around the entire edge of the oil pan. Be careful not to damage the mating surfaces of the oil pan or engine block.

- Oil pan assembly
- O-rings from the cylinder block and oil pump body

To install:

11. Install or connect the following:
- Baffle plate to the oil pan. Torque the bolts to 22–27 inch lbs. (2–3 Nm).

12. Apply sealant to the front and rear seal of the oil pan.
- New O-rings to the cylinder block and the oil pump body

13. Apply a 0.177–0.217 in. (4.5–5.5mm) continuous bead of liquid gasket to the upper oil pan mating surface and install the oil pan. Torque the bolts in sequence to 12–14 ft. lbs. (16–19 Nm).
- Oil pan strainer. Torque the bolts to 12–14 ft. lbs. (16–19 Nm).
- Rear cover plate and the lower transaxle bolts
- A/C compressor and bracket
- Engine drive belts and adjust as necessary
- Center crossmember assembly
- Front and rear engine mounting nuts and bolts and remove the support jack and the engine hoist
- CKP sensors (REFERENCE and POSITION) to the oil pan. Torque the bolts to 75–96 inch lbs. (9–10 Nm).
- Front exhaust pipe and its support
- Oil strainer

14. Apply a 0.177–0.217 in. (4.5–5.5mm) continuous bead of liquid gasket to the lower oil pan mating surface and install the oil pan. Tighten the mounting bolts in sequence to 57–66 inch lbs. (6–8 Nm).

➡**Wait at least 30 minutes before refilling the engine oil.**

- Engine undercover(s)
- Negative battery cable

15. Fill the engine with clean oil.
16. Start the engine, check for leaks and repair if necessary.

G35 MODELS

1. Before servicing the vehicle, refer to the precautions in the beginning of this section.
2. Drain the engine oil and coolant.
3. Remove the hood.
4. Install a suitable engine slinger to secure the engine for crossmember removal.
5. Remove or disconnect the following:
- Front suspension crossmember
- Drive belts

- Alternator and starter
- Idler pulley and bracket
- Crankshaft Position (CKP) sensor
- Oil filter and oil cooler, as necessary
- Lower oil pan bolts in the reverse sequence of the torque sequence

6. Insert a seal cutter between the upper and lower oil pan. Tapping the cutter with a hammer, slide it around the entire edge of the oil pan. Be careful not to damage the aluminum mating surface of the upper oil pan.
- Transmission joint bolts which pierce upper oil pan
- Rear cover plate
- Upper oil pan bolts in the reverse of the torque sequence

7. Insert a seal cutter between the steel and aluminum oil pan. Tapping the cutter with a hammer, slide it around the entire edge of the oil pan. Be careful not to damage the mating surfaces.
- Oil strainer
- O-rings and discard

8. Clean all gasket mating surfaces.

To install:

9. Install or connect the following:
- Oil strainer to the oil pump
- New O-rings to the cylinder block and oil pump side
- Oil pan gasket, applying RTV sealant as shown. Align the protrusion of the oil pan gasket with the notches of the front timing chain case and rear oil seal retainer.

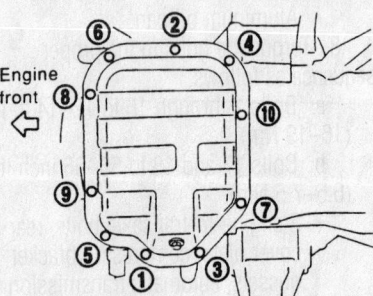

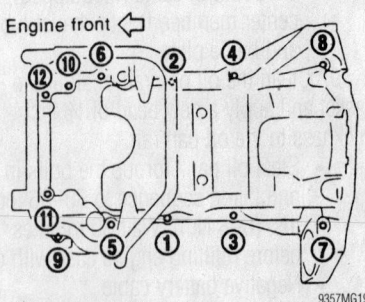

Lower and upper oil pan tightening sequence (use reverse for loosening)—G35 models

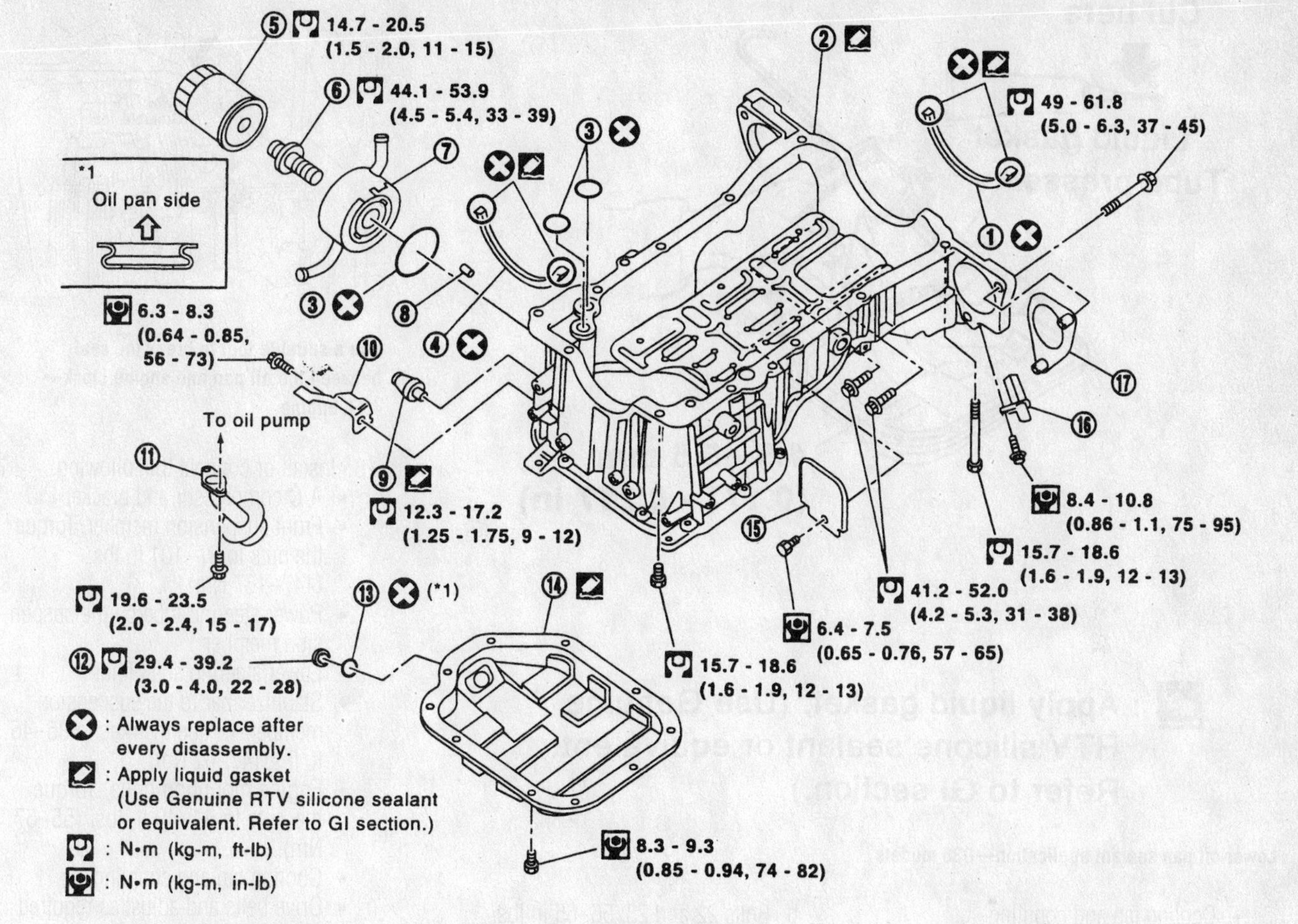

⑤ 🔧 14.7 - 20.5
(1.5 - 2.0, 11 - 15)

⑥ 🔧 44.1 - 53.9
(4.5 - 5.4, 33 - 39)

②

✕ 🖊

🔧 49 - 61.8
(5.0 - 6.3, 37 - 45)

③ ✕

*1
Oil pan side ⬆

🔧 6.3 - 8.3
(0.64 - 0.85,
56 - 73)

⑦

✕ 🖊

① ✕

⑰

To oil pump

③ ✕
⑧
④ ✕

⑩

🔧 8.4 - 10.8
(0.86 - 1.1, 75 - 95)

⑯

🔧 15.7 - 18.6
(1.6 - 1.9, 12 - 13)

⑨ 🖊
🔧 12.3 - 17.2
(1.25 - 1.75, 9 - 12)

⑮

🔧 41.2 - 52.0
(4.2 - 5.3, 31 - 38)

⑪

🔧 19.6 - 23.5
(2.0 - 2.4, 15 - 17)

⑬ ✕ (*1)

⑭ 🖊

🔧 6.4 - 7.5
(0.65 - 0.76, 57 - 65)

🔧 15.7 - 18.6
(1.6 - 1.9, 12 - 13)

⑫ 🔧 29.4 - 39.2
(3.0 - 4.0, 22 - 28)

✕ : Always replace after
every disassembly.

🖊 : Apply liquid gasket
(Use Genuine RTV silicone sealant
or equivalent. Refer to GI section.)

🔧 : N•m (kg-m, ft-lb)

🔧 : N•m (kg-m, in-lb)

🔧 8.3 - 9.3
(0.85 - 0.94, 74 - 82)

1.	Oil pan gasket	2.	Oil pan (upper)	3.	O-ring
4.	Oil pan gasket	5.	Oil filter	6.	Connector bolt
7.	Oil cooler	8.	Relief valve	9.	Oil pressure switch
10.	Bracket	11.	Oil strainer	12.	Drain plug
13.	Drain plug washer	14.	Oil pan (lower)	15.	Rear plate
16.	Crankshaft position sensor (POS)	17.	Rear cover plate		

9357MG18

Exploded view of the upper and lower oil pans and related components—G35 models

10. Apply sealant as shown in the illustration. Install the upper oil pan and tighten the bolts, in sequence, to 12–13 ft. lbs. (15.7–18.6 Nm).
 • Transmission joint bolts
11. For the lower oil pan, apply sealant as shown in the illustrations, then install and tighten the bolts, in sequence to 74–82 inch lbs. (8.3–9.3 Nm).
12. The remainder of installation is the reverse of the removal procedure.

➡**Wait at least 30 minutes before refilling the engine oil.**

13. Connect the negative battery cable
14. Fill the engine with clean oil and coolant.
15. Start the engine, check for leaks and repair if necessary.

4.1L Engine

1. Before servicing the vehicle, refer to the precautions in the beginning of this section.
2. Drain the engine oil.
3. Attach an engine support fixture to the engine so the right and left engine mounts can be removed.
4. Remove or disconnect the following:
 • Negative battery cable
 • Drive belts

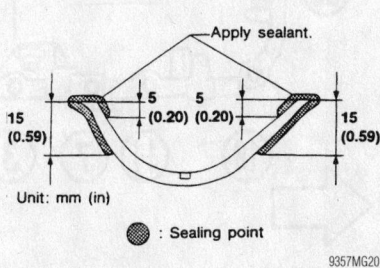

Proper sealant application for upper oil pan gasket—G35 models

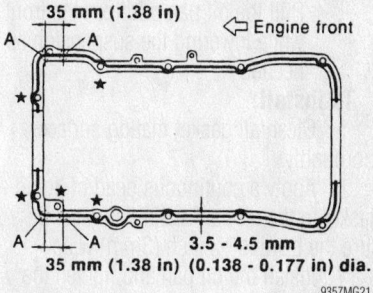

Upper oil pan sealant application—G35 models

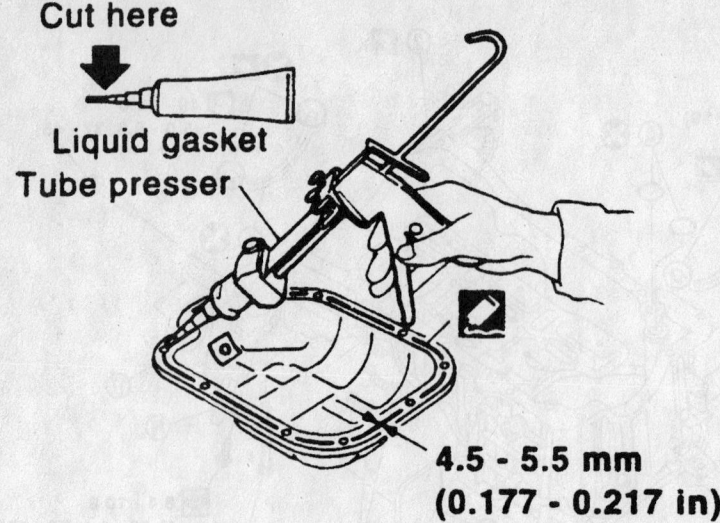

Cut here

Liquid gasket
Tube presser

4.5 - 5.5 mm
(0.177 - 0.217 in)

Use a suitable tool to break the seal between the oil pan and engine block—4.1L engine

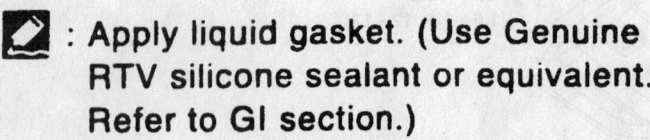

: Apply liquid gasket. (Use Genuine RTV silicone sealant or equivalent. Refer to GI section.)

Lower oil pan sealant application—G35 models

- Cooling fan and coupling
- Power steering oil pump
- Front stabilizer bar brackets from the side members
- Right and left engine mounting bolts
- Steering shaft lower joint
- Power steering tube bracket and support the front suspension member
- Lower the front suspension member
- A/C compressor and bracket
- Oil pan mounting bolts, then insert a tool into the notch on the oil pan and break the seal between the pan and engine block. Be careful not to damage the sealing surface.
- Pull the oil pan out from the front while lowering the suspension as needed

To install:

5. Clean all gasket mating surfaces thoroughly.

6. Apply a continuous bead of liquid gasket to the oil pan mating surface. Be sure the bead is ⅛ inch (3mm) wide.

7. Install the oil pan and tighten the retainers as follows:

a. Bolts 1 through 21 in sequence: 12–14 ft. lbs. (16–19 Nm).

b. Bolts 22 and 23: 56–65 in lbs. (6–7 Nm).

➡ Wait at least 30 minutes for the sealant to cure before filling the engine with oil.

8. Install or connect the following:
- A/C compressor and bracket
- Front suspension member. Torque the nuts to 87–101 ft. lbs. (147–167 Nm).
- Power steering tube on the suspension member
- Lower steering shaft joint
- Stabilizer bar to the suspension member. Torque the nuts to 35–46 ft. lbs. (47–62 Nm).
- Engine mounting bolts. Torque the nuts to 41–49 ft. lbs. (55–67 Nm).
- Cooling fan and coupling
- Drive belts and adjust as required
- Negative battery cable

9. Fill the engine with clean oil.

10. Start the vehicle, check for leaks and repair if necessary.

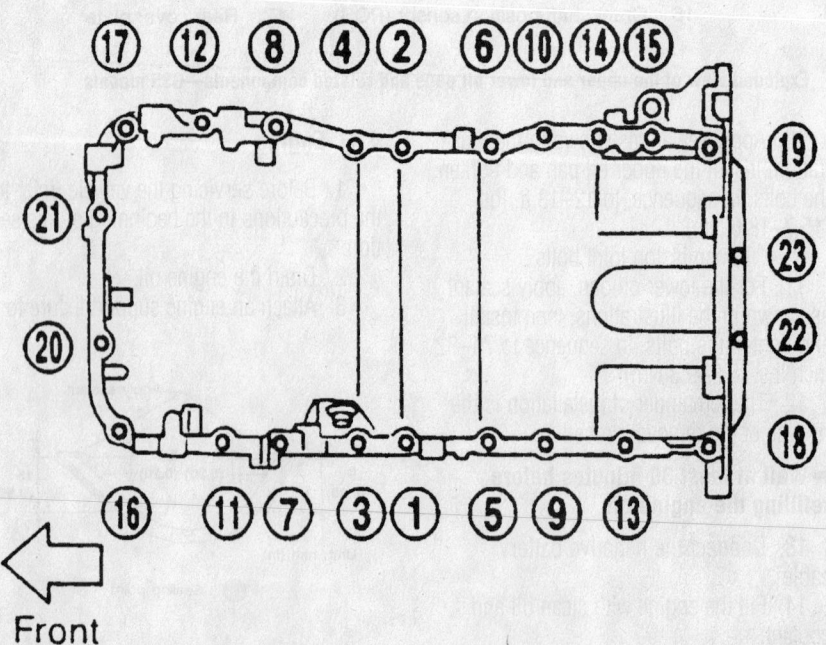

Tighten the oil pan bolts in the sequence shown—4.1L engine

Oil Pump

REMOVAL & INSTALLATION

2.0L Engine

1. Before servicing the vehicle, refer to the precautions in the beginning of this section.
2. Relieve the fuel system pressure.
3. Drain the engine oil.
4. Remove or disconnect the following:
   • Negative battery cable
   • Drive belts
   • Cylinder head with the intake and exhaust manifolds attached
   • Oil pans
   • Oil strainer and baffle plate
   • Crankshaft pulley and the front cover assembly
   • Oil pump from the inside of the front cover

To install:

5. Coat the oil pump gears with oil and fit the pump to the cover, using a new oil seal and O-ring.
6. Clean the mating surfaces of liquid gasket and apply a fresh bead of 1/8 inch (3mm) sealer to the sealing surface of the front cover.

7. Install or connect the following:
   • Front cover assembly
   • Crankshaft pulley
   • Oil strainer, baffle plate, oil pans, cylinder head and drive belts
   • Negative battery cable
8. Fill the engine with clean oil.
9. Start the vehicle, check for leaks and repair if necessary.

3.0L and 3.5L Engines

➡ **The oil pump bolts to the front of the engine block and is driven by the crankshaft. Removal of the timing cover and chains are necessary for oil pump service.**

1. Before servicing the vehicle, refer to the precautions in the beginning of this section.
2. Drain the engine oil.
3. Rotate the engine and position it to Top Dead Center (TDC) compression stroke of cylinder No. 1.
4. Remove or disconnect the following:
   • Negative battery cable
   • Drive belts
   • Camshaft Position (CMP) sensor (PHASE) and the Crankshaft Position (CKP) sensor (REF/POS)
   • Right front wheel and inner fender cover

   • Engine undercovers
   • Crankshaft pulley
   • Front exhaust pipe and its support and support the engine at the left and right side slingers with a suitable hoist
   • Engine right side mounting insulator and bracket nuts and bolts
   • Center crossmember assembly
   • A/C compressor and mounting bracket
   • Lower and upper oil pans
   • Oil strainer from the oil pump
   • Water pump cover and the front cover assembly
   • Lower timing chain assembly
   • Oil pump

To install:

➡ **When installing the oil pump, be sure to apply engine oil to the gears.**

5. Install or connect the following:
   • Oil pump. Torque the bolts to 57–66 inch lbs. (6.4–7.5 Nm) and the mounting screws to 53–69 inch lbs. (5.9–7.8 Nm).
   • Lower timing chain assembly
   • Front timing cover and water pump covers
   • Oil strainer using a new gasket.

SEC. 150

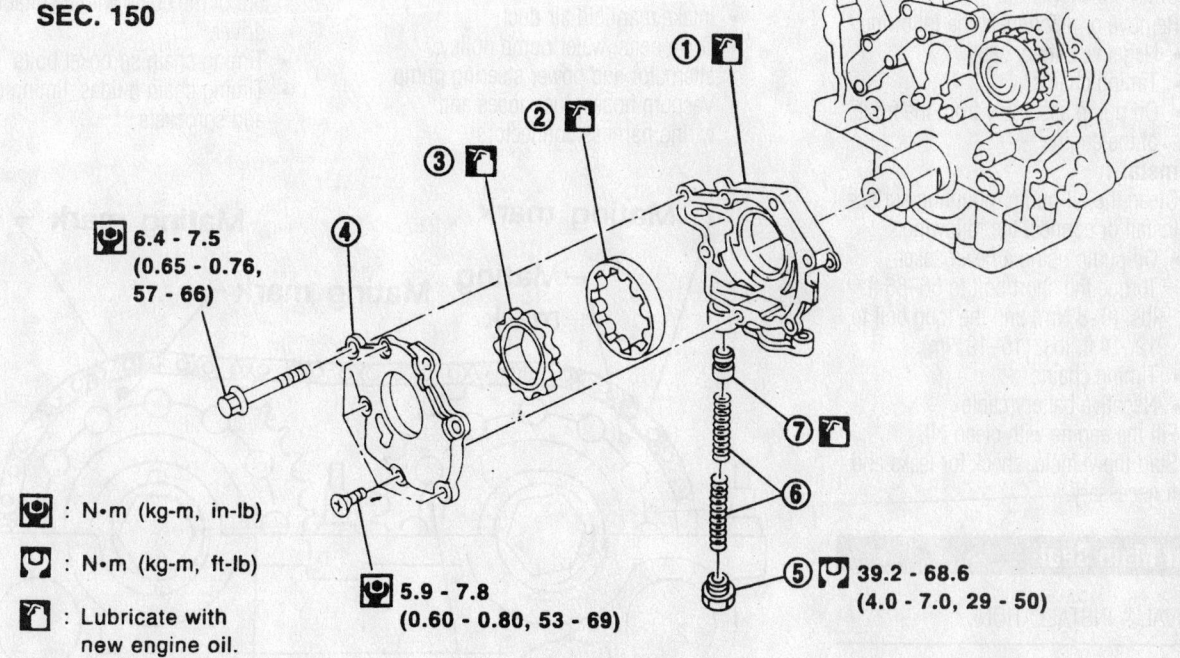

6.4 - 7.5
(0.65 - 0.76, 57 - 66)

5.9 - 7.8
(0.60 - 0.80, 53 - 69)

39.2 - 68.6
(4.0 - 7.0, 29 - 50)

🔧 : N•m (kg-m, in-lb)
🔧 : N•m (kg-m, ft-lb)
🛢 : Lubricate with new engine oil.

1. Oil pump body
2. Outer rotor
3. Inner rotor
4. Oil pump cover
5. Regulator valve
6. Spring
7. Regulator valve

9357MG23

Exploded view of the oil pump—G35 shown, other models similar

Torque the bolts to 12–14 ft. lbs. (16–19 Nm).
- Upper and lower oil pans. Be sure to use new O-rings at the oil pump to upper oil pan mating surface.
- A/C compressor and mounting bracket
- Center crossmember assembly
- Engine right side mounting insulator and bracket and remove the engine support hoist
- Front exhaust pipe and its support
- Crankshaft pulley
- Engine undercovers and the right side inner fender cover
- Right front wheel
- CMP sensor (PHASE) and the CKP sensor (REF/POS)
- Engine drive belts and adjust as necessary
- Negative battery cable

6. Fill the engine with clean oil.
7. Start the engine, check the oil pressure, and check for oil leaks.

4.1L Engine

➡The oil pump is mounted in the cylinder block below the left bank and behind the left timing chain.

1. Before servicing the vehicle, refer to the precautions in the beginning of this section.
2. Drain the engine oil.
3. Remove or disconnect the following:
   - Negative battery cable
   - Timing chains
   - Oil pump assembly from the front of the engine

To install:
4. Clean the oil pump mounting surface.
5. Install or connect the following:
   - Oil pump using a new gasket. Torque the short bolt to 56–66 ft. lbs. (6–8 Nm) and the long bolt to 12–14 ft. lbs. (16–19 Nm).
   - Timing chains
   - Negative battery cable

6. Fill the engine with clean oil.
7. Start the vehicle, check for leaks and repair if necessary.

Rear Main Seal

REMOVAL & INSTALLATION

1. Before servicing the vehicle, refer to the precautions in the beginning of this section.
2. Remove or disconnect the following:
   - Transmission or transaxle
   - Drive plate from the crankshaft

3. Carefully pry the seal out of the retainer without damaging the crankshaft or the seal retainer.

To install:
4. Lubricate the seal with clean engine oil.
5. Install or connect the following:
   - Seal into the retainer using the appropriate seal driver
   - Driveplate and transmission or transaxle

Timing Chain, Sprockets, Front Cover and Seal

REMOVAL & INSTALLATION

2.0L Engine

1. Before servicing the vehicle, refer to the precautions in the beginning of this section.
2. Relieve the fuel system pressure.
3. Raise and support the vehicle safely.
4. Drain the cooling system.
5. Remove or disconnect the following:
   - Negative battery cable
   - Engine undercovers
   - Right front wheel
   - Engine side cover and lower the vehicle
   - Radiator
   - Intake manifold air duct
   - Drive belts, water pump pulley, alternator and power steering pump
   - Vacuum hoses, fuel hoses and wiring harness connectors

- Spark plugs
- Cylinder head cover and oil separator
- Intake manifold supports
- Oil filter bracket and the power steering oil pump bracket

6. Place the No. 1 piston at Top Dead Center (TDC) on the compression stroke.
   - Chain tensioner
   - Distributor. Do not turn the rotor while the distributor is removed.
   - Timing chain guide
   - Camshaft sprockets
   - Camshafts, camshaft brackets, oil tubes and baffle plate
   - Starter
   - Heater hoses and the water hoses from the cylinder head
   - Knock sensor harness connector
   - Cylinder head outside bolts
   - Cylinder head with the intake and exhaust manifolds and raise and support the vehicle safely
   - Oil pans
   - Oil strainer and baffle plate
   - Crankshaft pulley and place a transmission jack under the main bearing beam and raise the engine slightly to take the weight off of the front engine mount
   - Front engine mount
   - Timing chain cover. Tap the seal out of the cover with a suitable seal driver.
   - Timing chain sprocket bolts
   - Timing chain guides, timing chain and sprockets

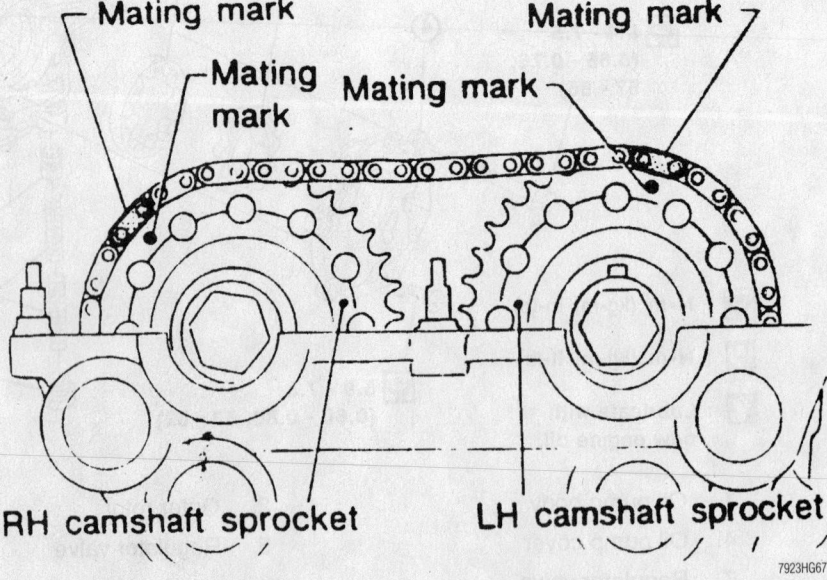

During disassembly, be sure to align the timing chain and camshaft sprocket mating marks—2.0L engine

7923HG67

To install:

7. Be sure all sealing surfaces are clean and prepared for assembly.

8. Install or connect the following:

- Crankshaft sprocket. Position the crankshaft so No. 1 piston is set at TDC (keyway at 12 o'clock, mating mark at 4 o'clock).

9. Fit the timing chain to crankshaft sprocket with the gold mating mark on the chain aligned with the mark on the sprocket. (The mating marks for the camshaft sprockets are silver).

- Timing chain guides and hang the chain off the left (front) guide. If necessary, secure the chain so it does not disengage from the crankshaft sprocket during assembly.
- New seal in the front cover and apply engine oil to the lip of the seal and apply a bead of liquid gasket to the front cover
- Oil pump drive spacer and front cover. Torque the bolts evenly to 60 inch lbs. (6.7 Nm) and wipe away any excess liquid gasket.
- Front engine mount
- Crankshaft pulley and temporarily tighten the bolt to hold the sprocket in place. The timing mark should align with the TDC mark.
- Oil strainer, baffle plate and oil pan
- Cylinder head, camshafts, oil tubes and baffles. Position the left camshaft key at 12 o'clock and the right camshaft key at 10 o'clock.
- Camshaft sprockets by lining up the mating marks on the timing chain with the mating marks on the camshaft sprockets. Torque the camshaft bolts to 101–116 ft. lbs. (137–157 Nm) and the crankshaft pulley bolt to 105–112 ft. lbs. (142–152 Nm).
- Upper timing chain guide and distributor. Ensure that the rotor is at the 5 o'clock position.

10. Before installing the chain tensioner, press the cam stopper down and the push in the sleeve until the hook can be engaged on the pin. When tensioner is bolted in position, the hook will release automatically. Ensure the arrow on the outside faces the front of the engine.

- Oil filter bracket and the power steering pump bracket
- Intake manifold supports
- Oil separator and the cylinder head cover
- Spark plugs
- Vacuum hoses, fuel hoses, and wiring harness connectors

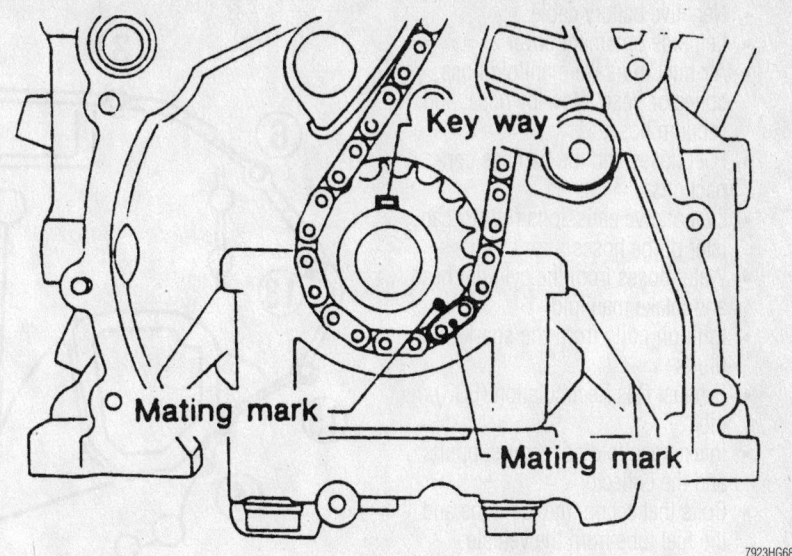

Crankshaft sprocket and timing chain alignment marks—2.0L engine

7923HG68

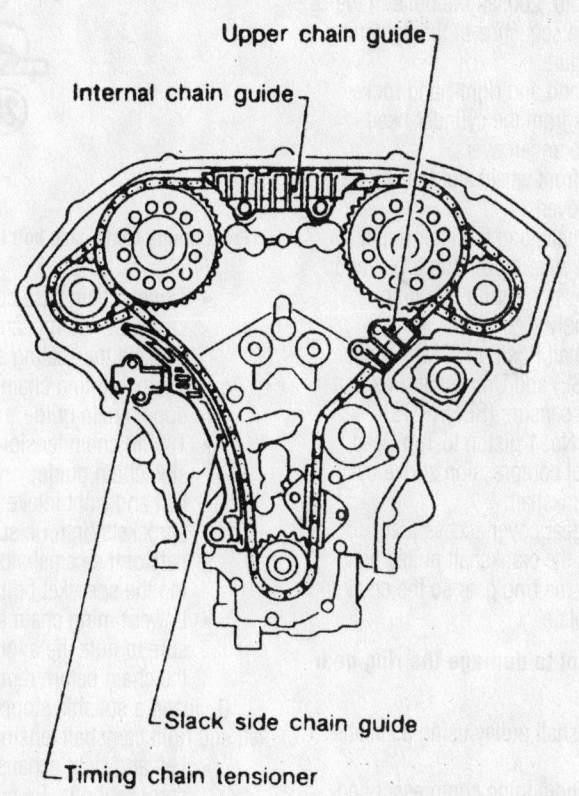

Timing chain tensioner and guide locations—3.0L engine

7923HG69

- Alternator and power steering pump
- Water pump pulley
- Drive belts
- Radiator
- Engine undercovers
- Right front wheel
- Negative battery cable

11. Fill and bleed the cooling system.

12. Start the vehicle, check for leaks and repair if necessary.

3.0L Engine

1. Before servicing the vehicle, refer to the precautions in the beginning of this section.

2. Drain the engine oil.

3. Drain the cooling system.

4. Relieve the fuel system pressure.

5. Remove or disconnect the following:

- Negative battery cable
- Left side ornament cover
- Air duct to intake manifold hose, collector hose, blow-by hose, and vacuum hoses
- Fuel hoses and the harness connections
- Evaporative emissions (EVAP) canister purge hoses
- Water hoses from the cylinder head and intake manifold
- Ignition coils from the spark plugs
- Exhaust Gas Recirculation (EGR) tube
- Intake manifold collector supports and the collector
- Bolts that secure the fuel tube and the fuel tube from the vehicle
- Bolts that secure the intake manifold to the engine block and the manifold. Loosen the bolts in the reverse sequence of the tightening procedure.
- Left-hand and right-hand rocker covers from the cylinder head
- Engine undercovers
- Right front wheel and the engine side covers
- Drive belts and the idler pulley
- Power steering oil pump belt and the power steering oil pump assembly
- Camshaft Position (CMP) sensor (PHASE) and Crankshaft Position (CKP) sensors (REF)/(POS)

6. Set the No. 1 piston to Top Dead Center (TDC) of compression stroke by rotating the crankshaft.
- Ring gear cover access plate

7. Loosen the crankshaft pulley bolt while securing the ring gear so the crankshaft cannot rotate.

➡ **Use care not to damage the ring gear teeth.**

- Crankshaft pulley using a suitable puller
- Air conditioning compressor and bracket
- Front exhaust pipe and its support

8. Hang the engine at the right and left side engine slingers with a suitable hoist.

9. Support the transaxle with jack.
- Right side engine mount and bracket
- Center crossmember assembly
- Steel oil pan and the oil strainer
- Aluminum oil pan
- Water pump cover
- Front timing chain case bolts in the proper sequence

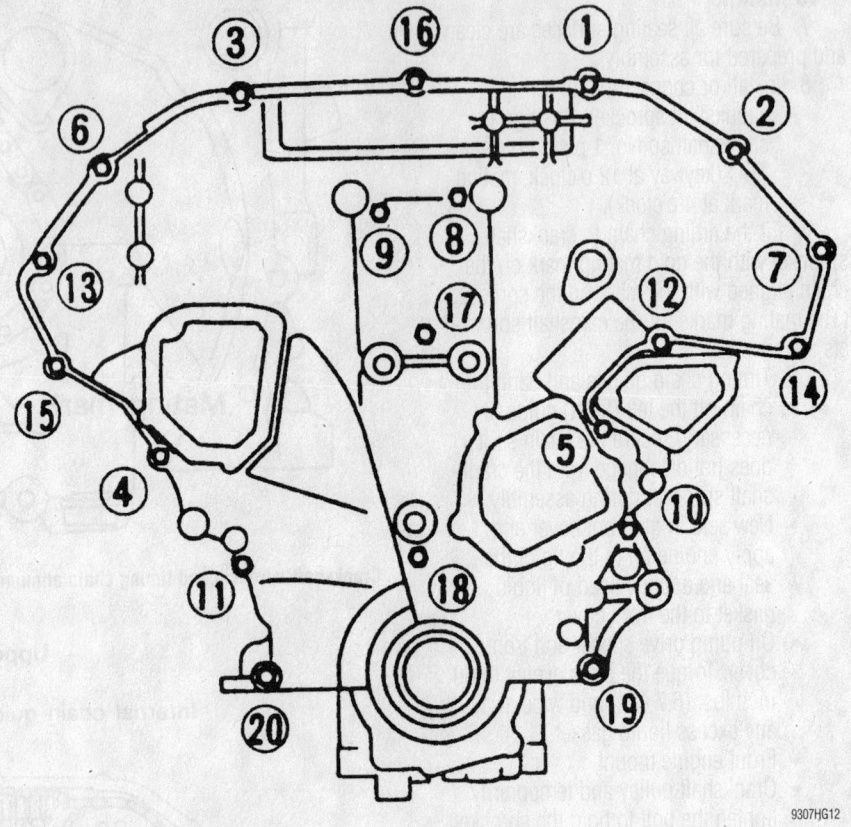

Front timing chain case bolt loosening sequence—3.0L engine

- Timing chain case cover using the seal cutter being careful not to damage the sealing surfaces
- Internal timing chain guide and the upper chain guide
- Timing chain tensioner and slack side chain guide
- Left and right intake camshaft sprockets first. Be sure to hold the flats of the camshafts while removing the sprocket bolts.
- Lower timing chain assembly. Be sure to note the aligning marks of the chain before removal.

10. Insert a suitable stopper pin for the left and right camshaft tensioners.
- Left and right exhaust camshaft sprocket bolts. Be sure to hold the flats of the camshafts while removing the sprocket bolts.
- Upper timing chain assembly. Be sure to note the aligning marks of the chain before removal.
- Lower timing chain guide
- Crankshaft sprocket
- All traces of liquid gasket from the front timing chain case and from the water pump

11. Inspect the timing chain for excessive wear or damage and replace if necessary.

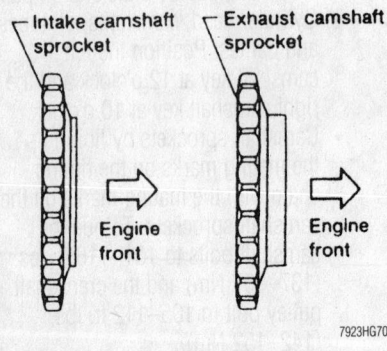

Identification of the intake and exhaust camshaft sprockets—3.0L engine

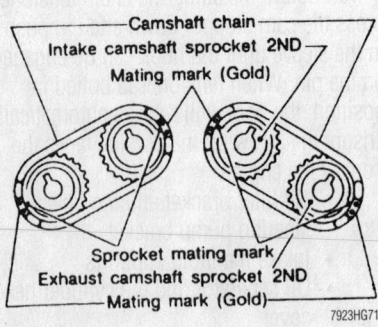

Upper timing chain alignment marks—3.0L engine

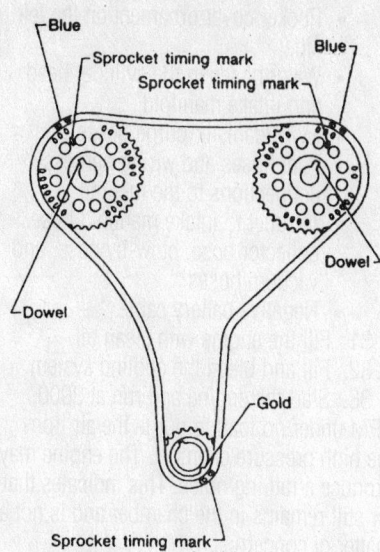

Lower timing chain alignment marks—
3.0L engine

To install:

12. Install or connect the following:
   • Crankshaft sprocket with the mating mark facing out
13. Position the crankshaft to TDC of compression stroke and align the dowels of the camshaft sprockets to the 12 o'clock position in respect to the cylinder head
   • Lower timing chain guide. The front mark on the guide should face upwards.
14. On a workbench, align the marks on the intake and exhaust camshaft sprockets with the marks of the chain.
15. Put the exhaust camshaft sprockets onto the dowel pin and torque the mounting bolts to 88–95 ft. lbs. (119–128 Nm). Be sure to secure the camshafts while tightening the bolts.
   • Timing chains and sprockets to the intake camshafts. Be sure to align the timing chain and sprocket mating marks.
16. Remove the left and right camshaft tensioner stopper pins.
17. Align the mating mark on the crankshaft with the matchmark (gold link) on the lower timing chain.
   • Lower timing chain to the water pump sprocket
18. Working counterclockwise, install the lower timing chain camshaft sprockets. Be sure to align the sprocket marks with the blue links of the timing chain during installation.
   • Intake sprocket bolts. Torque the bolts to 88–95 ft. lbs. (119–128 Nm). Be sure to secure the

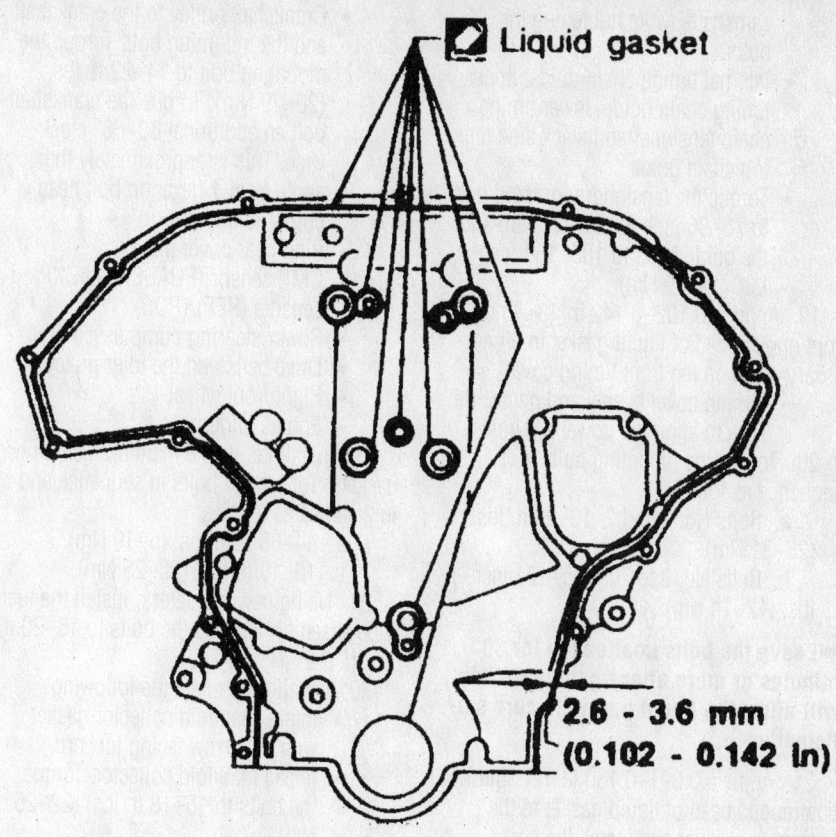

Application of liquid gasket to the front timing case—3.0L engine

① - ② 8 mm dia. bolts
25.5 - 31.4 N·m
(2.6 - 3.2 kg-m, 18.8 - 23.1 ft-lb)
③ - ⑳ 6 mm dia. bolts
11.8 - 13.7 N·m
(1.2 - 1.4 kg-m, 8.7 - 10.1 ft-lb)

Front timing chain case bolt tightening sequence—3.0L engine

camshafts while tightening the bolts.

- Internal timing chain guide, upper timing chain guide, lower timing chain tensioner and slack side timing chain guide
- Torque the tensioner mounting bolt to 75–96 inch lbs. (8–11 Nm) and the guide bolts to 108–168 inch lbs. (13–19 Nm)

19. Apply a 0.102–0.142 in. (3–4mm) continuous bead of liquid gasket to all necessary areas on the front timing cover.

- Timing cover evenly and gently. Be sure to align the dowel pin holes.

20. Torque the mounting bolts in sequence as follows:

    a. Bolts No. 1 and 2: 19–23 ft. lbs. (26–31 Nm).

    b. Bolts No. 3 to 20: 105–121 inch lbs. (12–14 Nm).

➡ **Leave the bolts unattended for 30 minutes or more after tightening. This will allow the liquid gasket to cure sufficiently.**

21. Apply a 0.091–0.130 in. (2–3mm) continuous bead of liquid gasket to the water pump cover and install the cover. Torque the bolts to 84–108 inch lbs. (10–13 Nm).

22. Apply a 0.12 in. (3mm) continuous bead of liquid gasket to the rocker covers and install the covers. Torque the mounting bolts in sequence as follows:

    a. Bolts No. 1 to 10: 9–26 inch lbs. (1–3 Nm).

    b. Bolts No. 1 to 10: 52–69 inch lbs. (6–8 Nm).

23. Apply sealant to the front and rear seal of the oil pan.

24. Apply a 0.177–0.217 in. (4.5–5.5mm) continuous bead of liquid gasket to the upper oil pan mating surface and install the oil pan. Torque the bolts in sequence to 12–14 ft. lbs. (16–19 Nm).

25. Install or connect the following:

- Transaxle bolts that secure the oil pan
- Oil pan strainer. Torque the bolts to 12–14 ft. lbs. (16–19 Nm).

26. Apply a 0.177–0.217 in. (5–6mm) continuous bead of liquid gasket to the lower oil pan mating surface and install the oil pan. Torque the bolts in sequence to 57–66 inch lbs. (6–8 Nm)

- Center crossmember assembly
- Right side engine mounting bracket and mount assembly

27. Remove the engine slinger assembly.

- Front exhaust pipe and its support
- A/C compressor and bracket

- Crankshaft pulley to the crankshaft and the mounting bolt. Torque the mounting bolt to 14–22 ft. lbs. (20–29 Nm). Torque the crankshaft bolt an additional 60–66° clockwise. This is approximately the angle from 1 hexagon bolt head corner to another.
- Ring gear cover plate
- CMP sensor (PHASE) and CKP sensors (REF)/(POS)
- Power steering pump assembly
- Drive belts and the idler pulley
- Right front wheel
- Engine undercovers

28. Install the intake manifold using new gaskets. Torque the bolts in sequence and in 2 stages as follows:

    a. 44–86 inch lbs. (5–10 Nm).

    b. 16–18 ft. lbs. (22–25 Nm).

29. Using new insulators, install the fuel tube assembly. Torque the bolts to 15–20 ft. lbs. (21–26 Nm).

30. Install or connect the following:

- Intake manifold collector gasket with the arrow facing forward
- Intake manifold collector. Torque the bolts to 16–18 ft. lbs. (22–25 Nm).
- Intake manifold collector support brackets
- EGR tube using new gaskets. Torque the bolts to 15–20 ft. lbs. (21–26 Nm) in 2 progressive steps.
- Ignition coils. Torque the bolts to 27–33 inch lbs. (3–4 Nm).

- Rocker cover ornament on the left side
- Water hoses to the cylinder head and intake manifold
- EVAP canister purge hoses
- Fuel hoses and wiring harness connections to the fuel rail
- Air duct to intake manifold hose, collector hose, blow-by hose, and vacuum hoses
- Negative battery cable

31. Fill the engine with clean oil.

32. Fill and bleed the cooling system.

33. Start the engine and run at 3000 RPM under no load to purge the air from the high pressure chamber. The engine may produce a rattling noise. This indicates that air still remains in the chamber and is not a matter of concern.

3.5L Engines

1. Before servicing the vehicle, refer to the precautions in the beginning of this section.

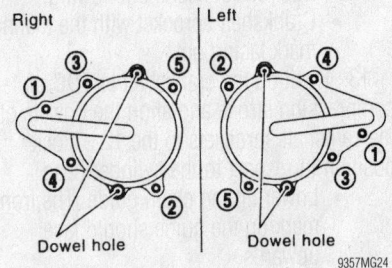

9357MG24

Intake valve timing control cover tightening sequence—3.5L engine

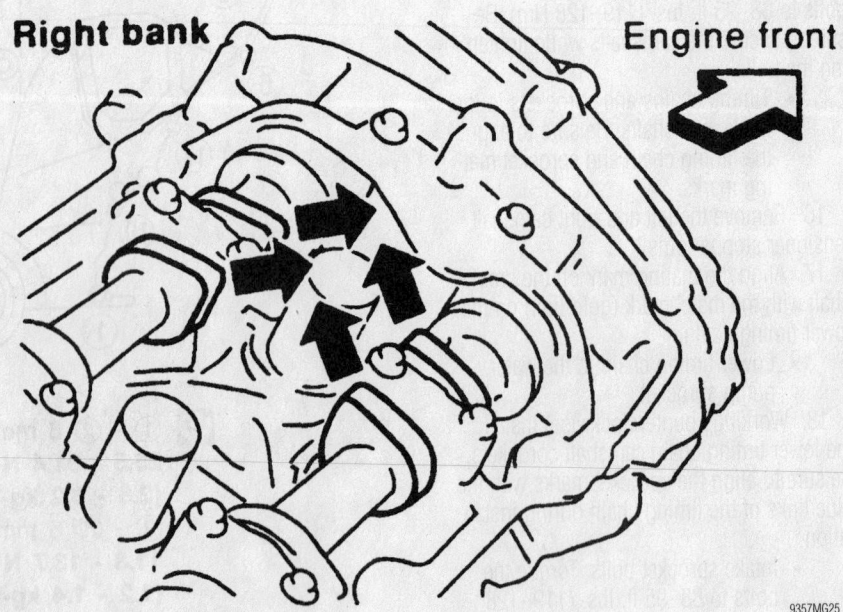

9357MG25

Proper orientation of the intake and exhaust cams—3.5L engine

2. Properly relieve the fuel system pressure.

3. Drain the engine oil and cooling system.

4. Remove or disconnect the following:
- Negative battery cable
- Right and left side rocker covers
- Cooling fan and radiator
- A/C compressor from bracket and position side with the lines attached
- A/C compressor bracket
- Power steering pump from its bracket and position aside with the lines attached
- Power steering pump bracket
- Water bypass hose and cooling fan bracket from the front timing chain case
- Lower and upper oil pans
- Right and left side intake valve timing control covers. Loosen the bolts in the reverse of the tightening sequence. Use a seal cutter to cut the gasket.

5. Set the No. 1 piston to Top Dead Center (TDC) on its compression stroke. Align the timing mark (grooved line) on the crankshaft pulley with the timing indicator on the front cover.

6. Make sure the intake and exhaust cam nose for the No. 1 cylinder is positioned as shown in the illustration. If not, turn the crankshaft on revolution (360°) and align.

7. Remove or disconnect the following:
- Crankshaft pulley using a suitable puller
- Front timing chain case. Loosen the bolts in the reverse of the torque sequence.

✳✳ WARNING

Do not use a screwdriver to pry the case off!

- Timing chain case. Insert the proper tool into the notch at the top of the case as shown. Pry the case off by levering the tool as shown. Use a seat cutter to cut the liquid gasket.
- Water pump cover and chain tensioner cover from the case
- Front oil seal from the case using a suitable prytool, being careful not the damage the case
- Internal chain guide, timing chain tensioner and slack guide. Remove the upper chain tensioner by pressing the tensioner in and inserting a 0.098 inch (2.5mm) diameter pin in

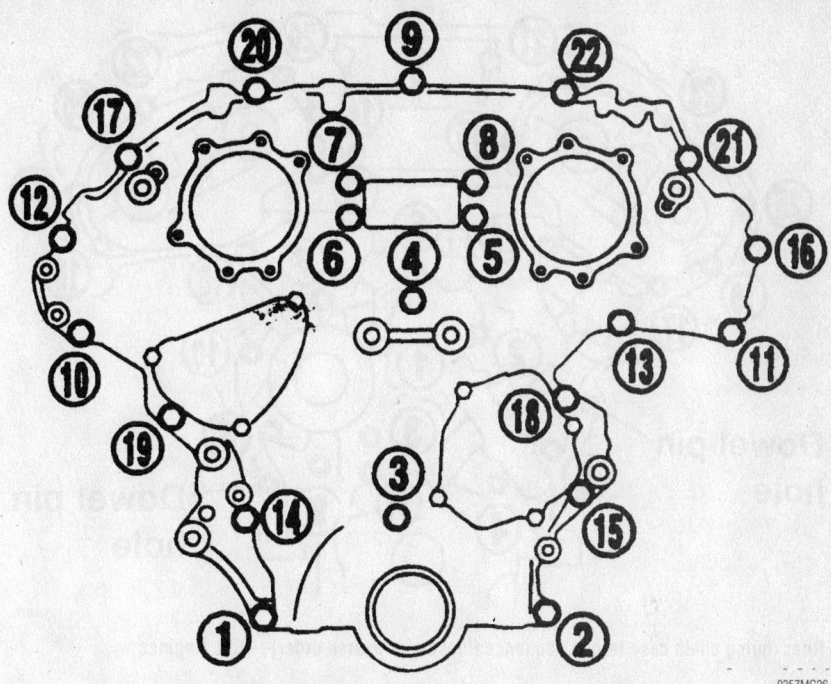

Front timing chain case torque sequence (loosen in reverse order)—3.5L engine

the pin hole. Once secured remove the bolts and the tensioner.

✳✳ WARNING

After the timing chain is removed, do NOT turn the crankshaft and camshaft separately, or the valve will strike the pistons.

8. Remove the timing chain and camshaft sprocket, as follows:

a. Attach a suitable stopper pin to the right and left camshaft chain tensioners (for secondary timing chains).

b. Hold the hex part of the camshaft secure with a wrench, then remove the camshaft sprocket mounting bolts.

c. Remove the primary and secondary timing chains with the camshaft sprockets.

9. Remove or disconnect the following:
- Tension guide and crankshaft sprocket
- Rear timing chain case. Use the reverse of the tightening sequence, then use a seal cutter to separate the gasket.

To install:

10. Install the rear timing chain case as follows:

a. Install new O-rings onto the cylinder block and head.

b. Apply suitable RTV sealant to the back side of the rear timing chain case as shown in the illustration.

c. Install the rear case and tighten the

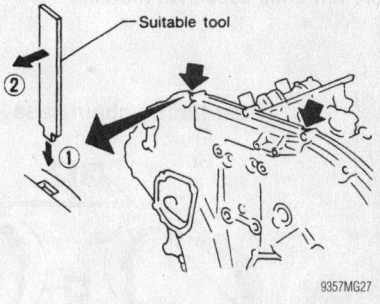

Front timing chain case removal, using a suitable tool—3.5L engine

bolts, in sequence, to 9–10 ft. lbs. (11.7–13.7 Nm). There are 2 bolts lengths: Bolts 1, 2, 3, 6–10 are 0.79 in. (20mm) long and the other bolts are 0.63 in. (16mm) long.

11. Set the No. 1 piston to Top Dead Center (TDC) on its compression stroke.

12. Install the crankshaft sprocket, making sure the mating marks on the sprocket face the front of the engine.

13. Push the plunger of the secondary chain tensioner and keep it pressed in with a stopper pin.

14. Install the secondary timing chains and camshaft sprockets, as follows:

a. Align the matchmarks on the secondary timing chain (gold link) with the stamped marks on the intake and exhaust sprockets, then install them.

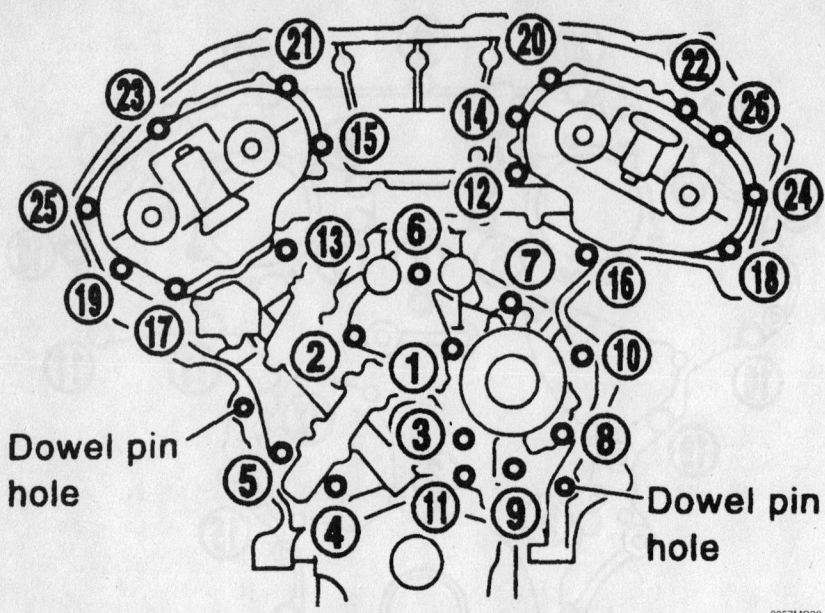

9357MG28

Rear timing chain case torque sequence (loosen in reverse order)—3.5L engines

➡**Matchmarks for the intake sprocket are on the back of the secondary sprocket. There are 2 kinds of marks, the right bank uses round marks and the left bank uses oval marks.**

b. Align the dowel pin and pin hole on the camshaft with the groove and dowel pin on the sprocket, then install them.

c. On the intake side, align the pin hold on the small diameter side of the

camshaft front end with the dowel pin on the on the back side of the camshaft sprocket, and install them.

d. On the exhaust side, align the dowel pin on the camshaft front end with the pin groove on the camshaft sprocket, then install them.

e. Tighten the camshaft sprocket bolts hand-tight to prevent the dowel pins from dislocating.

15. Install the primary timing chain, as follows:

a. Install the primary chain so the punched mating mark on the cam sprocket is aligned with the yellow link on the timing chain, while the notched mating mark on the crankshaft sprocket is aligned with the orange one on the timing chain.

b. Use a wrench on the hex portion of the camshaft to secure it in place, tighten the camshaft sprocket mounting bolts to 73–78 ft. lbs. (98–108 Nm).

c. Remove the stopper pins from secondary chain tensioners.

16. Install the internal chain guide, timing chain tensioner, tension guide and slack guide. Do not overtighten the slack guide mounting bolts. It's normal to have a gap under the bolt seats when the mounting

Rear timing chain case

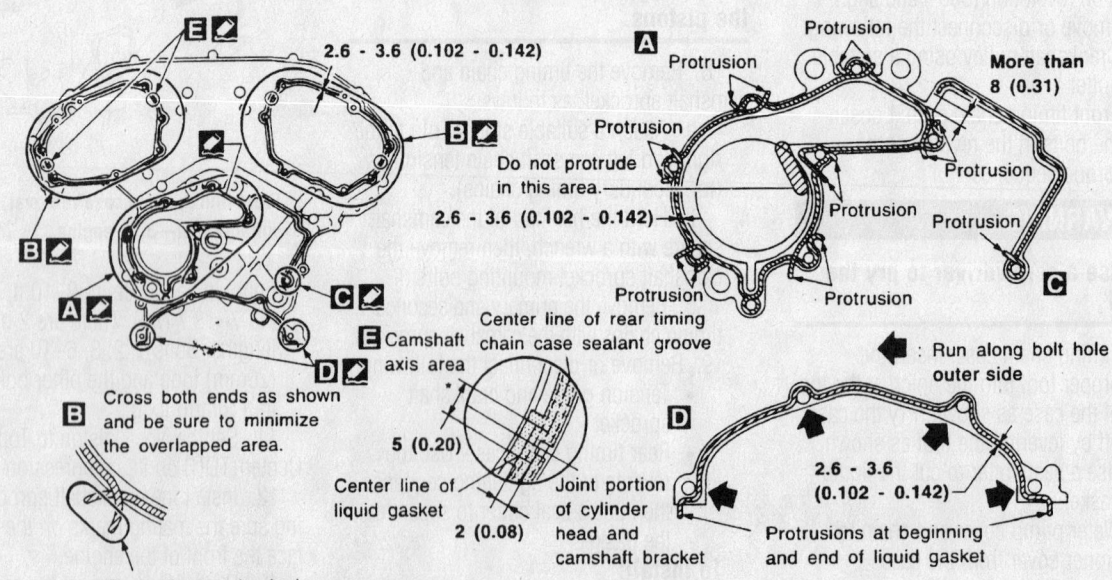

* : Apply liquid gasket to the chamfered surface between camshaft bracket and cylinder head.

✎ : Apply liquid gasket

Unit: mm (in)

9357MG40

Rear timing chain case sealant application—3.5L engines

Example: Right bank side (Rear view)

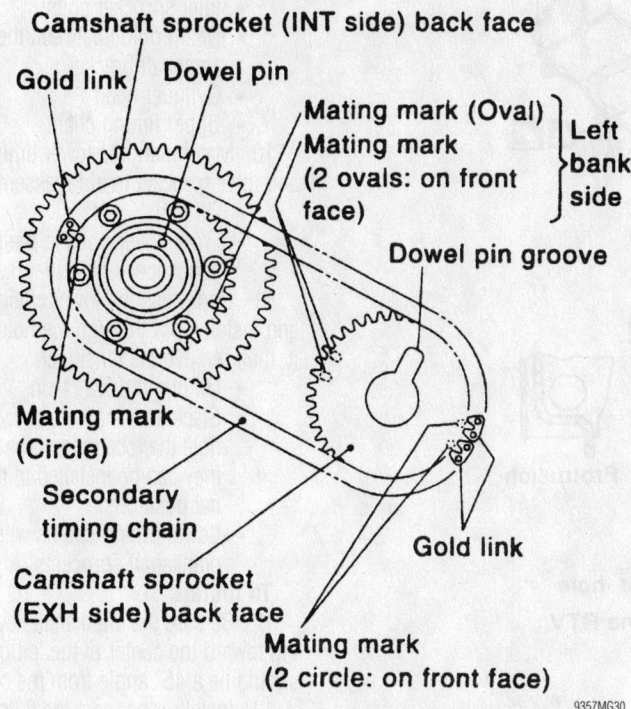

Camshaft sprocket alignment, right side bank shown—3.5L engine

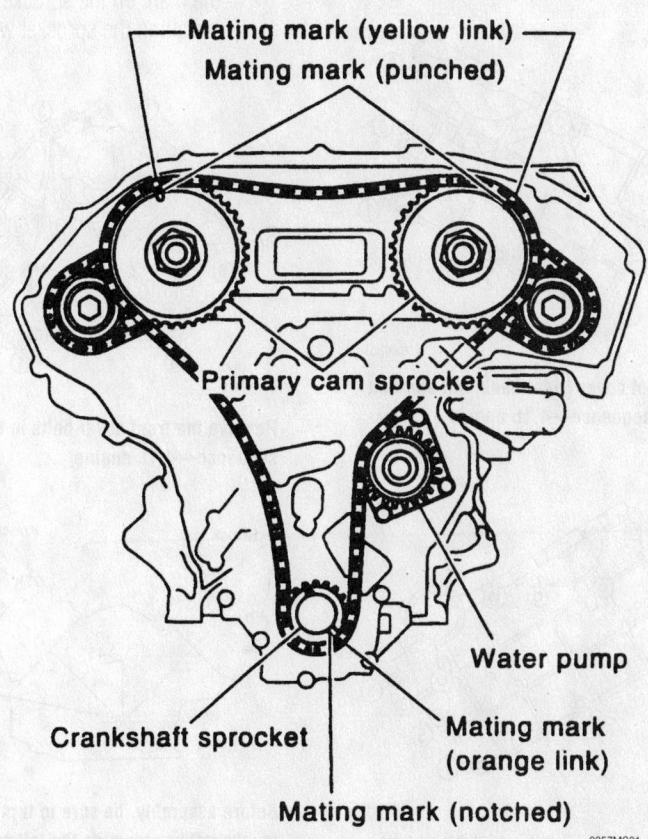

Proper alignment for the primary timing chain—3.5L engine

bolts are properly tightened to 10–13 ft. lbs. (14–18 Nm).

17. Recheck that all matchmarks are still aligned.

18. Install or connect the following:
- New O-rings on the rear timing chain case
- New front oil seal in to the timing chain cover
- Water pump and chain tensioner covers to the front cover
- Suitable liquid gasket to the back side of the timing chain case
- Dowel pin on the rear timing chain case into the dowel pin hold on the front chain case
- Front timing chain case bolts, in sequence. Tighten the M6 bolts to 9–10 ft. lbs. (11.7–13.7 Nm) and the M8 bolts to 19–23 ft. lbs. (25–31 Nm).
- Right and left intake valve timing control cover
- Crankshaft pulley to the crankshaft and the mounting bolt. Torque the mounting bolt to 29–36 ft. lbs. (39–49 Nm). Torque the crankshaft bolt an additional 60–66° clockwise. This is approximately the angle from 1 hexagon bolt head corner to another.

19. Installation of the remaining components is the reverse of the removal procedure.

20. Fill the engine with clean oil.

21. Fill and bleed the cooling system.

22. Start the engine and check for proper operation.

4.1L Engine

1. Before servicing the vehicle, refer to the precautions in the beginning of this section.

2. Properly relieve the fuel system pressure.

3. Remove or disconnect the following:
- Negative battery cable
- Engine from the vehicle
- Alternator
- A/C compressor
- Exhaust manifold and place the engine on a suitable stand
- Intake manifold collector
- Injector harness and the fuel tube assembly with the injector

➡ **Do not disassemble the fuel hose.**

- Intake manifold
- Valve cover

4. Set the No. 1 piston to Top Dead Center (TDC) on its compression stroke.

Front timing chain case

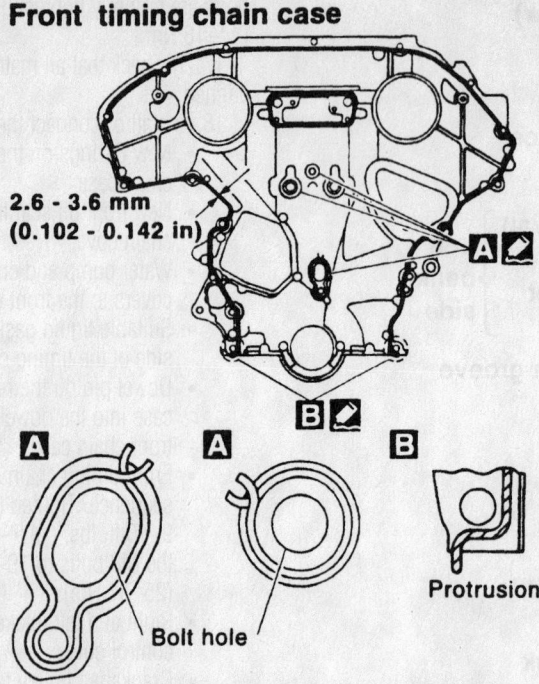

Apply proper sealant as shown on the back side of the front timing chain case—3.5L engine

2.6 - 3.6 mm
(0.102 - 0.142 in)

Sealant protrusion away from bolt hole

🔲 : Apply liquid gasket (Use Genuine RTV silicone sealant or equivalent. Refer to GI section.)

Align the timing mark (orange paint) on the crankshaft pulley with the timing indicator on the front cover.

5. Make sure the intake camshaft lobe for the No. 1 cylinder faces the intake port and the exhaust lobe faces the exhaust port.

➡️ **It is possible to check the camshaft positions by checking the notch positions on the camshaft when the No. 1 piston is at TDC of the compression stroke. At this position the cylinder head bolts can be removed.**

6. Remove or disconnect the following:
- Crankshaft pulley
- Crankshaft Position (CKP) sensor
- Upper front cover bolts and nuts in the sequence illustrated
- Front cover

7. Remove the upper chain tensioner by pressing the tensioner in and inserting a 0.04 inch (1mm) diameter pin in the pin hole. Once secured remove the bolts and the tensioner

8. Matchmark the upper timing chain and camshaft sprockets to aid reassembly.
- Upper camshaft sprocket bolt while holding the hexagonal part of the camshaft with a wrench
- Camshaft sprockets

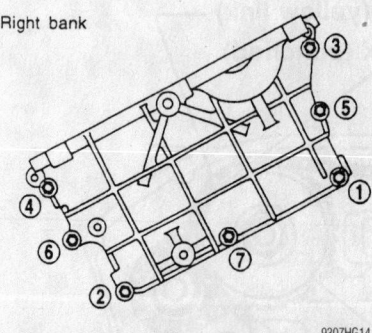

Upper front cover (right bank) nut and bolt removal sequence—4.1L engine

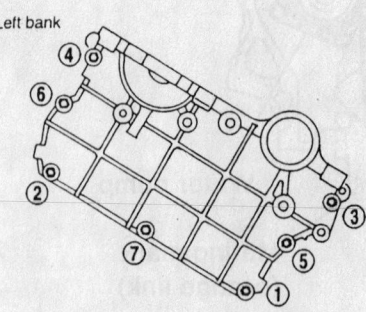

Upper front cover (left bank) nut and bolt removal sequence—4.1L engine

9. Remove the left and right tensioner covers from the front covers by pressing the tensioner in and inserting a 0.04 inch (1mm) diameter pin in the pin hole.
- Idler sprocket bolts
- Chain guide between the No. 1 camshaft bracket
- Cylinder head
- Upper timing chain

10. Matchmark the lower timing chain and idler sprocket to aid reassembly.
- Oil pan
- Front cover bolts in the proper sequence

11. Compress the lower chain tensioners and install a pin through the hole to secure it, then remove the tensioner.
- Oil pump drive chain
- Slack and chain guides. Be sure to note the locations of the bolts so they can be installed in their original positions.
- Lower timing chains with the crankshaft sprockets

To install:

12. Be sure the crankshaft key is pointing toward the center of the left bank. This should be a 45° angle from the center.

13. Install or connect the following:
- Lower right bank timing chain by aligning the mark on the chain with the mark on the sprocket and installing the sprocket with chain

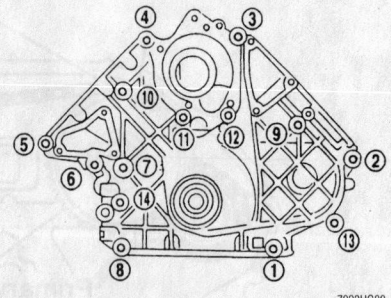

Remove the front cover bolts in the correct sequence—4.1L engine

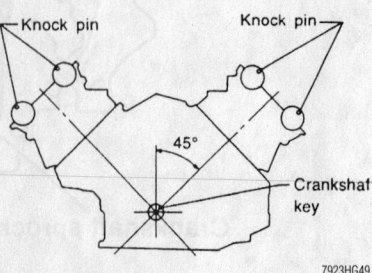

Before assembly, be sure to turn the crankshaft key towards the left cylinder head—4.1L engine

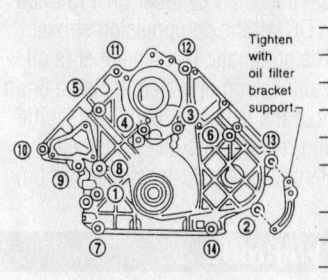

Position	Bolt dimensions	Tightening torque
①, ③, ⑤, ⑦	M6 x 45	
②	M6 x 47	
⑥, ⑩	M6 x 65	6.3 - 8.3 N·m (0.64 - 0.85 kg-m, 55.6 - 73.8 in-lb)
⑬	M6 x 67	
⑫	M6 x 84	
④, ⑧	M8 x 50	16 - 21 N·m (1.6 - 2.1 kg-m, 12 - 15 ft-lb)
⑨	M10 x 52	
⑭	M10 x 60	30 - 40 N·m (3.1 - 4.1 kg-m, 22 - 30 ft-lb)
⑪	M10 x 62	

Front cover bolt location and torque specifications—4.1L engine

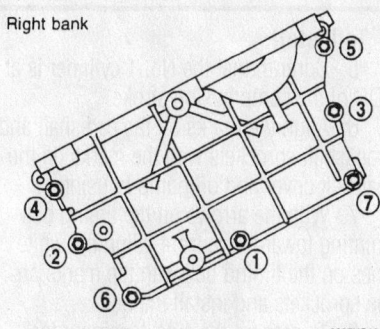

Upper front cover (right bank) nut and bolt tightening sequence—4.1L engine

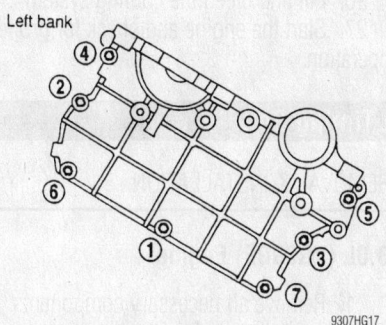

Upper front cover (left bank) nut and bolt tightening sequence—4.1L engine

on the crankshaft. Be sure the thick side of the sprocket faces the cylinder block to provide clearance between the block and chain.

- Slack and chain guides. Be sure to install the bolts in the correct locations. Torque the bolts to 10–14 ft. lbs. (13–19 Nm).
- Lower chain for the left bank in the same manner as the right one
- Left slack and chain guide. Torque the bolts to 10–14 ft. lbs. (13–19 Nm).
- Oil pump drive chain and sprockets. Torque the bolt on the driven gear to 22–30 ft. lbs. (30–40 Nm).
- Lower oil pump drive chain guide

14. Install the upper guide by installing the bolts loosely, then inserting a 0.04 in. (1mm) feeler gauge between the guide and chain. Press on the guide lightly with the same force as the weight of the guide and torque the bolts to 56–74 inch lbs. (6–8 Nm).

- Chain tensioner with the pins installed using new gaskets. Torque the bolts to 82–95 inch lbs. (9–11 Nm).

15. Confirm that the timing marks on the crankshaft sprockets and chains are still aligned.

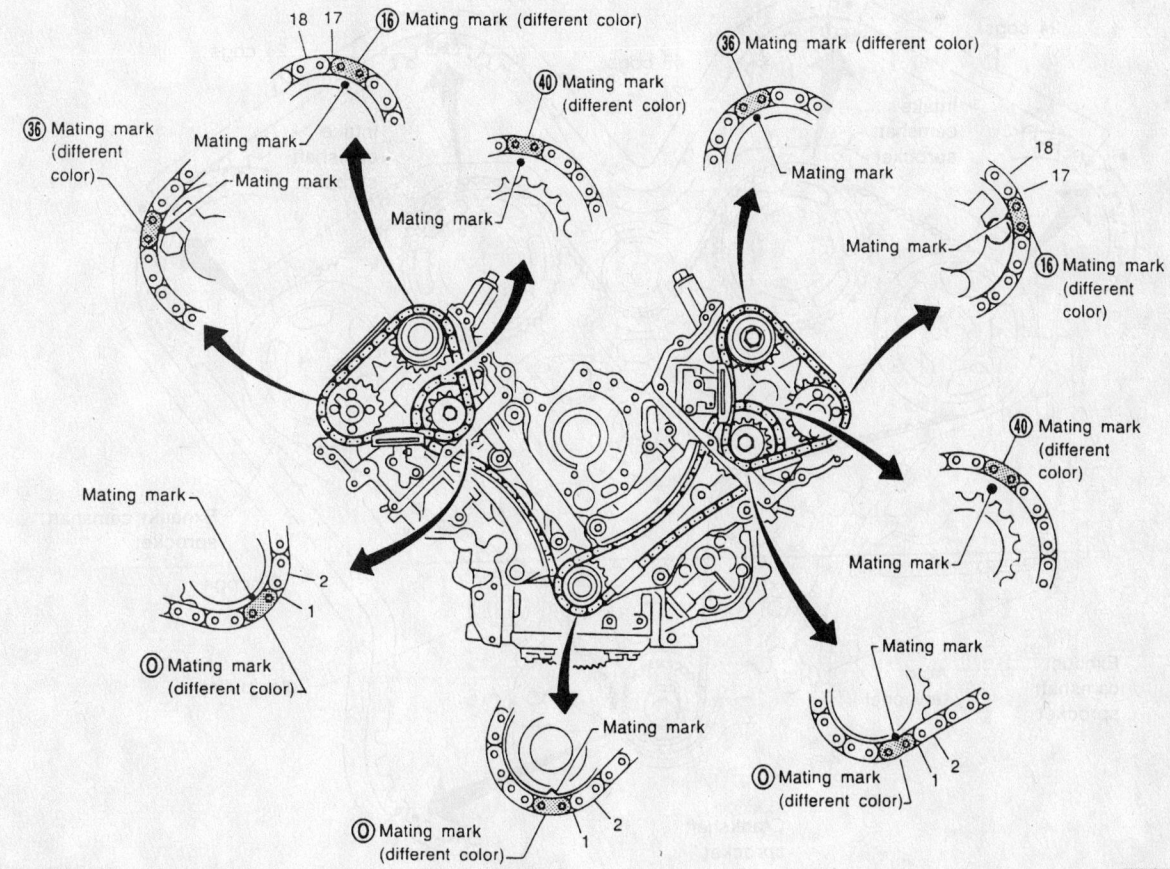

Timing chain and sprocket alignment marks—4.1L engine

- Front cover. Be sure to install the bolts in the correct locations.

16. Apply engine oil to the idler shaft and install it on the idler sprocket.

17. Align the mark on the chain with the mark on the idler sprocket and install the sprocket.

- Place the upper chains on the idler sprockets. It is not necessary to align the mating marks at this time. The marks can be aligned after the cylinder head is installed.

18. Install the cylinder heads

19. Align the marks on the upper chains with the marks on the sprockets, then install the sprockets on the camshafts while keeping the marks aligned.

- Idler shaft bolts. Torque the bolts to 32–43 ft. lbs. (43–58 Nm).

20. Remove the lower chain tensioner pins.

- Chain guide between the No. 1 camshaft bracket

21. Align the upper timing chain mating marks with the marks on the sprockets and install the sprockets. Torque the intake sprocket bolt to 76–83 ft. lbs. (103–113

Nm) and the exhaust sprocket bolt to 12–15 ft. lbs. (16–21 Nm)

22. Compress the upper chain tensioners and install a pin through it to secure it in position.

- Tensioners. Torque the bolts to 82–95 inch lbs. (9–11 Nm).

23. Lubricate the timing chains and related parts with clean engine oil and install the upper covers. Torque the cover bolts in sequence.

24. Installation of the remaining components is the reverse of the removal procedure.

25. Fill the engine with clean oil.

26. Fill and bleed the cooling system.

27. Start the engine and check for proper operation.

Timing Belt

REMOVAL & INSTALLATION

3.0L (VG30DE) Engine

1. Remove all necessary components for access to the front timing covers, then remove the covers.

2. Set the No. 1 cylinder on Top Dead Center (TDC) of the compression stroke.

3. The automatic belt tensioner is oil damped and spring operated. Install a 6mm bolt to hold the tensioner back against the spring and release tension on the belt.

4. Remove the auto-tensioner and timing belt.

✳✳ WARNING

Do not rotate the crankshaft or camshaft separately because the pistons will strike the valves causing engine damage.

To install:

5. Confirm that the No. 1 cylinder is at TDC of the compression stroke.

6. Align the marks on the camshaft and crankshaft sprockets with the marks on the rear belt cover and oil pump housing.

7. With the arrows on the timing belt pointing towards the front, align the white lines on the timing belt with the marks on the sprockets and install the belt.

8. To prepare the auto-tensioner for installation, perform the following:

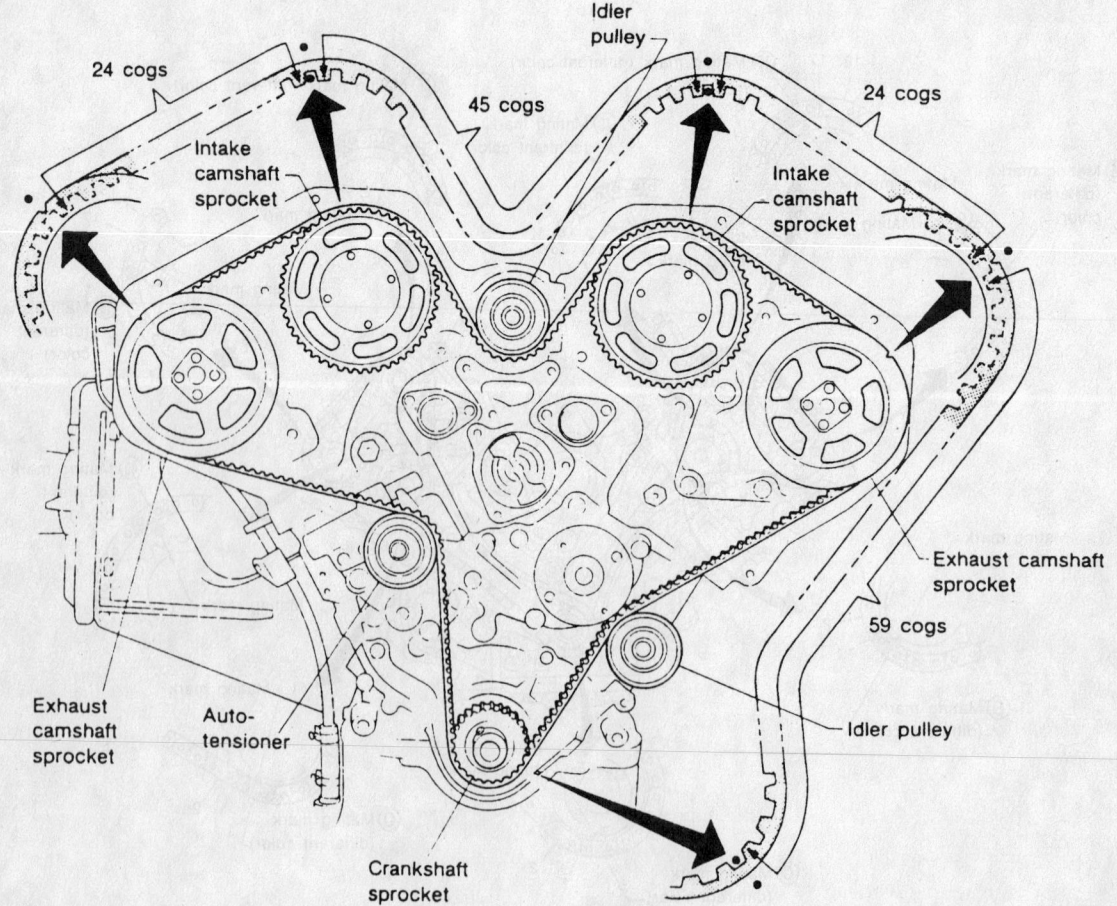

For proper timing belt positioning, ensure the number of teeth between each sprocket is as indicated—Infiniti 3.0L (VG30DE) engine

a. Remove the bolt holding the tensioner in position.

b. Use a vise to adjust the gap between the tensioner arm and pusher body to 0.160 in. (4mm).

c. Install the bolt again to hold the arm in this position. Do not try to use the bolt to adjust the gap or the threads will be damaged.

9. Install the auto-tensioner, push it towards the belt to just take up the slack, then tighten the bolts finger-tight.

10. Before adjusting the timing belt tension, the slack must be properly distributed:

a. Turn the crankshaft 10 degrees clockwise and tighten the tensioner bolts and nut to 12–15 ft. lbs. (16–21 Nm). Do not push the auto-tensioner hard or the belt will be adjusted too tight.

b. Turn the crankshaft 120 degrees (⅓ turn) counterclockwise.

c. Loosen the tensioner bolts and nut ½ turn and move the tensioner body away from the timing belt as far as it will move.

d. Turn the crankshaft clockwise to TDC again.

e. Push the tensioner against the belt with a force of 13 lbs. (59 N) using a spring scale or similar tool and tighten the bolts again to 12–15 ft. lbs. (16–21 Nm). The pressure specification is important and a special spring scale tool, J-38387, is available to measure the tensioner force.

11. To check the timing belt tension:

a. Turn the crankshaft 120 degrees (⅓ turn) clockwise, then turn counterclockwise and return the engine to TDC.

b. Prepare a steel plate that is approximately ⅜ in. (10mm) wide and longer than the width of the belt.

c. Set the plate on the timing belt between two camshaft sprockets and push against the plate with a force of 11 lbs. (49 N). Note the belt deflection.

d. Repeat the procedure between the other camshaft sprockets and between the exhaust sprockets and idler/tensioner pulleys. There will be a total of four measurements.

e. Add the deflection measurements and divide by four. The average deflection must be 0.240–0.280 in. (6–7mm). If belt tension is not correct, start the entire adjustment procedure again.

12. Confirm the auto-tensioner mounting nuts are tightened to 12–15 ft. lbs. (16–21 Nm) and remove the auto-tensioner stopper bolt.

13. After 5 minutes, measure the clearance between the tensioner arm and the pusher. It should be 0.138–0.205 in. (3.5–5.2mm).

14. Be sure all the sprocket timing marks are correctly aligned. Install the timing belt covers and tighten the bolts to 24–38 inch lbs. (3–5 Nm).

15. Install all applicable components.

Piston and Ring

POSITIONING

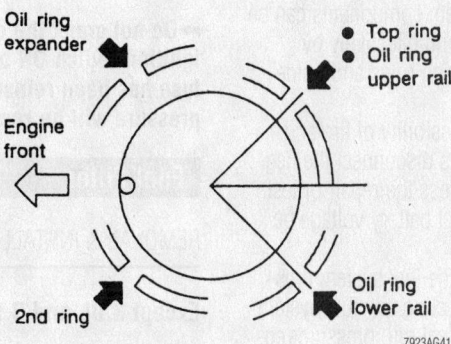

Piston ring end-gap spacing—Infiniti engines

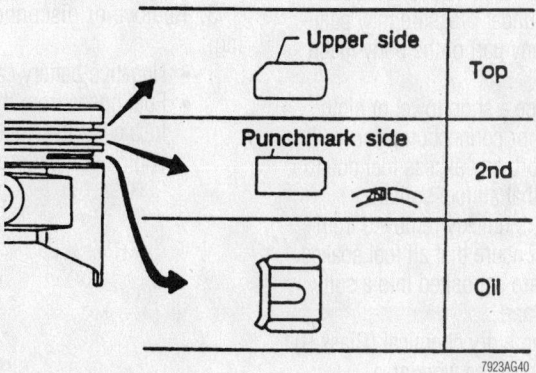

Piston ring positioning—Infiniti engines

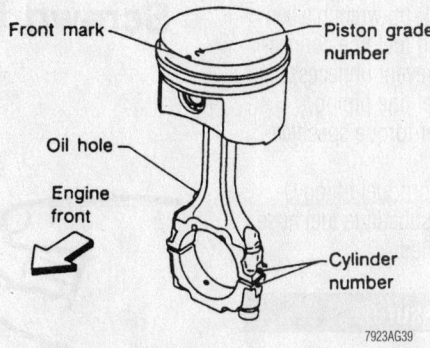

Piston/connecting rod assembly-to-engine orientation—Infiniti engines

FUEL SYSTEM

Fuel System Service Precautions

Safety is the most important factor when performing not only fuel system maintenance but any type of maintenance. Failure to conduct maintenance and repairs in a safe manner may result in serious personal injury or death. Maintenance and testing of the vehicle's fuel system components can be accomplished safely and effectively by adhering to the following rules and guidelines.

1. To avoid the possibility of fire and personal injury, always disconnect the negative battery cable unless the repair or test procedure requires that battery voltage be applied.

2. Always relieve the fuel system pressure prior to disconnecting any fuel system component (injector, fuel rail, pressure regulator, etc.), fitting or fuel line connection. Exercise extreme caution whenever relieving fuel system pressure, to avoid exposing skin, face and eyes to fuel spray. Please be advised that fuel under pressure may penetrate the skin or any part of the body that it contacts.

3. Always place a shop towel or cloth around the fitting or connection prior to loosening to absorb any excess fuel due to spillage. Ensure that all fuel spillage (should it occur) is quickly removed from engine surfaces. Ensure that all fuel soaked cloths or towels are deposited into a suitable waste container.

4. Always keep a dry chemical (Class B) fire extinguisher near the work area.

5. Do not allow fuel spray or fuel vapors to come into contact with a spark or open flame.

6. Always use a back-up wrench when loosening and tightening fuel line connection fittings. This will prevent unnecessary stress and torsion to fuel line piping. Always follow the proper torque specifications.

7. Always replace worn fuel fitting O-rings with new. Do not substitute fuel hose where fuel pipe is installed.

Fuel System Pressure

RELIEVING

1. Before servicing the vehicle, refer to the precautions in the beginning of this section.

2. Remove the fuel pump fuse.
3. Start the engine.

4. Allow the engine to run until it stalls.

5. After the engine stalls, crank the engine 2 or 3 times to release the remaining fuel pressure.

6. Turn the ignition switch **OFF**. Reinstall the fuel pump fuse into the fuse block.

➡**Do not crank the engine or turn the ignition switch ON after the fuel pump fuse has been reinstalled, or the fuel pressure will be reestablished.**

Fuel Filter

REMOVAL & INSTALLATION

Except 3.0L and 3.5L Engines

1. Before servicing the vehicle, refer to the precautions in the beginning of this section.

2. Relieve the fuel system pressure.
3. Remove or disconnect the following:

- Negative battery cable
- Fuel hoses from the fuel filter, located at the right side of the engine compartment

- Filter mounting screws
- Filter from the vehicle

To install:

4. Inspect all hoses and clamps for damage of any type. Replace parts, as required.

➡**The fuel filters are directional and should be installed with the arrow facing the direction of fuel flow.**

5. Install or connect the following:
- New filter in the bracket and install new hose clamp
- Negative battery cable

6. Start the vehicle, check for fuel leaks and repair if necessary.

➡**On some vehicles, a code will be set and/or the check engine light will remain on after starting the vehicle. This is because a code was set for an open fuel pump circuit when the fuel pressure was released. If you did not disconnect the negative battery cable during this procedure, do it now so the code will be erased. The negative battery cable should be disconnected for at least 1 minute. Also, remember to reset the clock and radio stations when finished.**

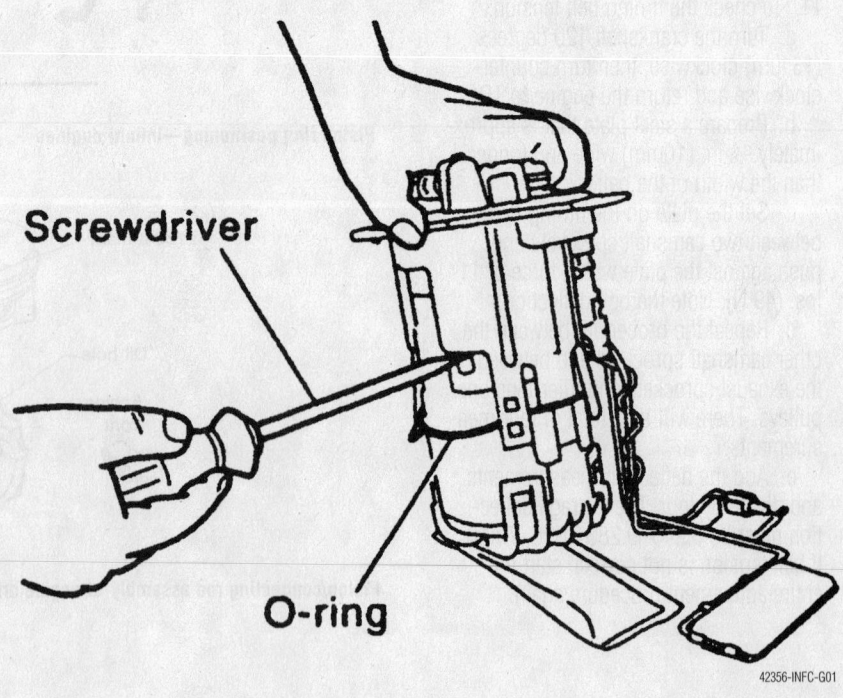

Remove the fuel filter from the fuel chamber–3.0L and 3.5L engines

42356-INFC-G01

3.0L and 3.5L Engines

1. Before servicing the vehicle, refer to the precautions in the beginning of this section.
2. Properly relieve fuel system pressure.
3. Remove or disconnect the following:
   - Negative battery cable
   - Rear seat bottom
   - Inspection hole cover
   - Electrical and quick connectors
   - Six screws
   - Fuel level sensor unit and fuel pump assembly
   - Flange and snap fit portion of the fuel pump
   - Fuel tank temperature sensor harness
   - Fuel level sensor flange
   - Fuel pump connector
   - Quick connectors from the fuel level sensor
   - Fuel level sensor from the chamber
   - Fuel filter from the chamber

To install:

4. Install or connect the following:
   - Fuel filter to the chamber
   - Fuel level sensor to the chamber
   - Quick connectors to the fuel level sensor
   - Fuel pump connector
   - Fuel level sensor flange
   - Fuel tank temperature sensor harness
   - Fuel pump assembly to the fuel tank
   - Screws and electrical connectors
   - Quick connectors
   - Negative battery cable
5. Start the vehicle, check for leaks and repair if necessary.
   - Inspection hole cover
   - Rear seat bottom

Fuel Pump

REMOVAL & INSTALLATION

2.0L Engine

1. Before servicing the vehicle, refer to the precautions in the beginning of this section.
2. Release the fuel system pressure.
3. Remove or disconnect the following:
   - Negative battery cable
   - Rear seat back and bottom
   - Inspection hole cover located beneath the rear seat
   - Connectors and fuel tubes
   - Six screws
   - Fuel pump/gauge assembly and disconnect the tubes and connector

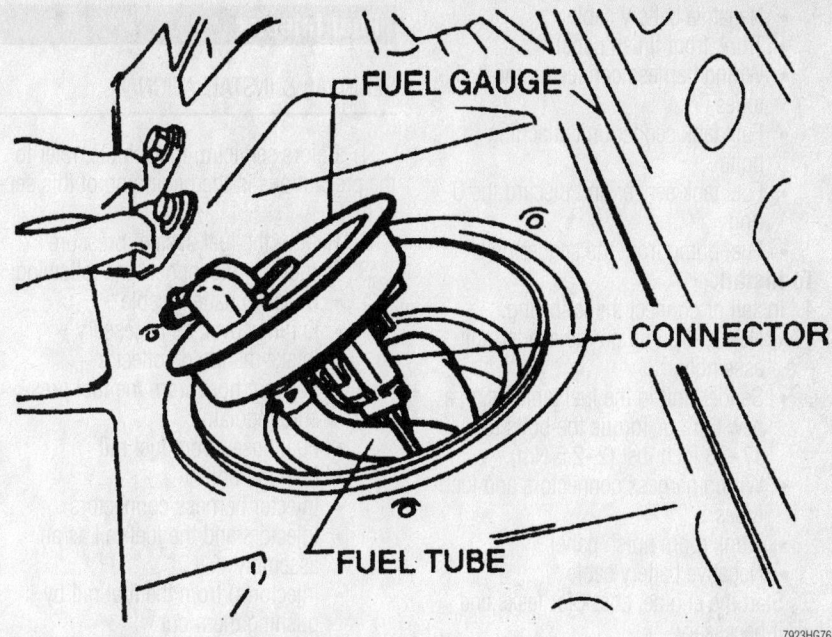

The fuel pump is located inside the tank—2.0L engine

   - Fuel pump by sliding it out on an angle

To install:

4. Install or connect the following:
   - Fuel pump/gauge assembly
   - All fuel lines and connectors
   - Six screws
   - Negative battery cable
5. Start the vehicle, check for leaks and repair if necessary.
6. Install the inspection cover.
7. Rear seat back and bottom.

3.0L and 3.5L Engines

1. Before servicing the vehicle, refer to the precautions in the beginning of this section.
2. Relieve the fuel system pressure
3. Remove or disconnect the following:
   - Negative battery cable
   - Rear seat bottom
   - Inspection hole cover from under the rear seat
   - Electrical and quick connectors
   - Six screws
   - Fuel level sensor/fuel pump assembly

➡**If replacement of the fuel filter is required, proceed with the following steps.**

   - Fuel level sensor/fuel pump assembly flange
   - Fuel tank temperature sensor harness
   - Fuel level sensor flange and raise the fuel level sensor

   - Fuel pump electrical connector
   - Quick connectors from the fuel level sensor
   - Fuel level sensor from the assembly
   - Snap fit connectors and remove the fuel filter from the fuel pump assembly

To install:

4. Install or connect the following:
   - Fuel filter to the fuel pump assembly
   - Fuel level sensor to the fuel pump assembly
   - Fuel level sensor quick connectors
   - Electrical connectors
   - Fuel tank temperature sensor harness
   - Fuel level sensor unit and fuel pump flanges. Make certain that they snap together.
   - Fuel pump assembly to the fuel tank
   - Six screws
   - Quick and electrical connectors
   - Negative battery cable
5. Start the vehicle, check for leaks and repair if necessary.
6. Install the inspection hole cover plate.
7. Install the rear seat bottom.

4.1L Engine

1. Before servicing the vehicle, refer to the precautions in the beginning of this section.
2. Relieve the fuel system pressure.
3. Remove or disconnect the following:

- Negative battery cable
- Trunk front finish panel
- Wiring harness connector and fuel tubes
- Fuel tank sender unit attaching bolts
- Fuel tank sender and discard the O-ring
- Fuel pump from the sender unit

To install:

4. Install or connect the following:
- New fuel pump on the sender unit assembly
- Sender unit in the fuel tank, using a new O-ring. Torque the bolts to 17–23 inch lbs. (2–2.5 Nm).
- Wiring harness connectors and fuel tubes
- Trunk room finish panel
- Negative battery cable

5. Start the engine, check for leaks and repair if necessary.

Fuel Injector

REMOVAL & INSTALLATION

1. Before servicing the vehicle, refer to the precautions in the beginning of this section.
2. Relieve the fuel system pressure
3. Remove or disconnect the following:
- Negative battery cable
- Engine cover, as necessary
- Intake manifold collector
- Vacuum hose from the fuel pressure regulator
- Fuel hoses from fuel rail
- Fuel rail bolts
- Injector harness connectors
- Injectors and the fuel rail as an assembly
- Injector(s) from the fuel rail by pushing them out

4. Remove and discard the fuel injector O-rings

To install:

5. Lubricate the new O-rings with clean engine oil and install the O-rings on the injector(s).
6. Install or connect the following:
- Fuel injectors to the fuel rail
- Fuel rail and injectors as an assembly to the intake manifold

7. Tighten the fuel rail bolts in the following sequence;
   a. Step 1: 7–8 ft. lbs. (9–10 Nm).
   b. Step 2: 15–20 ft. lbs. (21–26 Nm).
- Fuel hoses to the fuel rail
- Vacuum hose to the fuel pressure regulator
- Intake manifold collector
- Engine cover, as necessary
- Negative battery cable

8. Start the vehicle, check for leaks and repair if necessary.

DRIVE TRAIN

Transaxle Assembly

REMOVAL & INSTALLATION

Manual

G20 MODELS

1. Before servicing the vehicle, refer to the precautions in the beginning of this section.
2. Drain the transaxle fluid.
3. Remove or disconnect the following:
- Negative battery cable
- Air cleaner and air duct assembly
- Clutch operating cylinder from the transaxle
- Back-up light switch, neutral switch and ground harness connectors
- Speedometer sensor
- Starter
- Air bleeder hose
- Shift control and support rods
- Front exhaust tube
- Halfshafts and support the engine with a suitable jack under the oil pan
- Rear and left engine mount

4. Raise the jack and remove the lower transaxle housing bolts. Lower jack and remove the upper housing bolts. Keep the bolts in order as they are different lengths and must be returned to the same position.
5. Lower the transaxle.

To install:

6. Raise the transaxle into place and install the attaching bolts. Torque the shortest bolt to 22–30 ft. lbs. (30–40 Nm) and

the remaining bolts to 51–59 ft. lbs. (70–79 Nm).
7. Install or connect the following:
- Rear and left engine mounts
- Driveshafts
- Shift control rods, support rod, bleeder air hose and starter
- Air bleeder hose
- Starter
- Speedometer sensor
- Back-up light switch, neutral switch and ground harness connectors
- Clutch operating cylinder. Torque the bolts to 22–27 ft. lbs. (29–37 Nm).
- Negative battery cable

8. Fill the transaxle to the proper level.
9. Start the vehicle, check for leaks and repair if necessary.

I30 MODELS

1. Before servicing the vehicle, refer to the precautions in the beginning of this section.
2. Drain the transaxle fluid.
3. Remove or disconnect the following:
- Battery, battery bracket and tray
- Air cleaner assembly with the Mass Air Flow (MAF) sensor
- Clutch operating cylinder; do not disconnect the hydraulic line from the cylinder
- Clutch hose clamp
- Speedometer pinion and the neutral position switch connectors and the ground harness connectors

- Starter
- Back-up lamp switch and the neutral position switch
- Crankshaft Position (CKP) sensor (POS) from the transaxle front side
- Shifter control rod and the support rod bracket from the transaxle
- Both driveshafts from the transaxle assembly and support the transaxle

4. Support the engine of the transaxle by placing a jack under the oil pan. Be sure to use a block of wood between the oil pan and jack.
- Bolts that secure the center crossmember
- Left-hand engine mounts

➡ **The transaxle bolts are of different lengths, be sure to note the location of the bolts for reassembly.**

- Transaxle bolts
- Transaxle from the vehicle by sliding the transaxle input shaft out of the clutch, lowering the rear of the transaxle, then lowering the transaxle from of the vehicle

To install:

5. Install or connect the following:
- Transaxle assembly to the bell housing while aligning the output shaft of the transaxle with the clutch disc. Tighten the transaxle bolts to the specifications illustrated.
- Left-hand engine mount. Torque the through-bolt to 32–41 ft. lbs. (43–55 Nm).

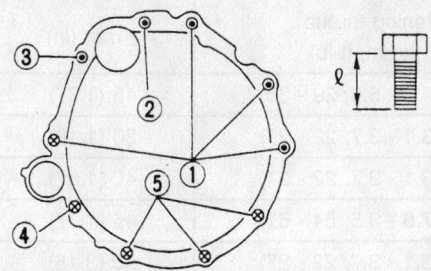

Bolt No.	Tightening torque N·m (kg-m, ft-lb)	"ℓ" mm (in)
①	70 - 79 (7.1 - 8.1, 51 - 59)	52 (2.05)
②	70 - 79 (7.1 - 8.1, 51 - 59)	65 (2.56)
③	70 - 79 (7.1 - 8.1, 51 - 59)	124 (4.88)
④	35.1 - 47.1 (3.58 - 4.80, 25.89 - 34.74)	40 (1.57)
⑤	35.1 - 47.1 (3.58 - 4.80, 25.89 - 34.74)	40 (1.57)

◉ M/T to engine
⊗ Engine to M/T

③ with starter
④ with support rod bracket

9307HG31

Manual transaxle-to-engine bolt torque sequence and specifications—I30 models

- Center crossmember assembly. Torque the bolts to 57–72 ft. lbs. (77–98 Nm).
- Both driveshafts to the transaxle assembly
- Shifter control rod and the support rod bracket to the transaxle
- CKP sensor to the transaxle front side
- Back-up lamp switch and the neutral position switch
- Starter motor assembly to the transaxle
- Speedometer pinion and the ground harness connectors
- Clutch hose clamp
- Clutch operating cylinder. Torque the bolts to 22–30 ft. lbs. (30–40 Nm).
- Air cleaner assembly with the MAF sensor
- Battery tray, bracket and battery
- Positive and the negative battery cables

6. Fill the transaxle with proper amount and type of fluid.

7. Start the vehicle, check for leaks and repair if necessary.

Automatic

G20 MODELS

1. Before servicing the vehicle, refer to the precautions in the beginning of this section.
2. Drain the transaxle fluid.
3. Remove or disconnect the following:
- Battery and bracket
- Air duct assembly
- Transaxle solenoid harness connector, Park Neutral Position (PNP) switch connector and revolution switch connector
- Crankshaft Position (CKP) sensor from the transaxle
- Control cable and transaxle coolant lines

- Halfshafts, the intake manifold support bracket and the starter
- Upper bolts attaching transaxle to the engine

4. Support the engine with a suitable stand and use a suitable jack to support the transaxle.

➡Bolts are of different lengths, note the locations that the bolts are removed from.

- Center member
- Rear cover plate and the bolts securing the torque converter to the driveplate. Rotate the crankshaft to gain access to the bolts.
- Transaxle mounts
- Lower transaxle mounting bolts and lower the transaxle

To install:

5. Place a straightedge across the bell housing of the transaxle and measure the distance to the mounting bosses on the torque converter. The distance should be 0.626 in. (16mm). If not, the torque converter is not installed correctly.

6. Check the driveplate run-out with a dial indicator. Maximum allowable run-out is 0.008 in. (0.2mm).

7. Raise the transaxle into position and install the transaxle mounting bolts. Tighten the 50, 55, and 65mm long bolts to 51–59 ft. lbs. (70–79 Nm). Torque the 35 and 45 mm long bolts to 12–15 ft. lbs. (16–21 Nm).

8. Install or connect the following:
- Torque converter bolts. Torque the bolts to 33–43 ft. lbs. (44–59 Nm). Rotate the crankshaft to gain access to the bolts.
- Rear cover and center member
- Transaxle mounts
- Halfshafts, intake manifold support bracket and the starter
- Control cable and transaxle coolant lines

- CKP sensor to the transaxle
- Transaxle solenoid harness connector, PNP switch connector and revolution switch connector
- Air duct assembly
- Battery and bracket

9. Fill the transaxle with fluid.

10. Start the vehicle, check for leaks and repair if necessary.

I30 AND I35 MODELS

➡The radio may contain a coded theft protection circuit. Always obtain the code number from the customer before disconnecting the battery.

1. Before servicing the vehicle, refer to the precautions in the beginning of this section.
2. Drain the transaxle fluid.
3. Remove or disconnect the following:
- Battery and bracket
- Air cleaner and resonator
- Terminal cord assembly harness connector and Park Neutral Position (PNP) switch harness connector
- Revolution and Vehicle Speed Sensor (VSS) electrical connections
- Crankshaft Position (CKP) sensor from the transaxle
- Left-hand mounting bracket from transaxle and body
- Control cable from the transaxle
- Driveshafts
- Oil cooler pipes and cap pipes to avoid contamination
- Starter motor and place a jack under the oil pan to support the engine. Do not place the jack under the oil pan drain plug.
- Crossmember
- Rear cover plate and bolts attaching the torque converter to the drive plate

4. Support the transaxle with a jack.

5. Remove the transaxle-to-engine bolts and lower the transaxle using the jack.

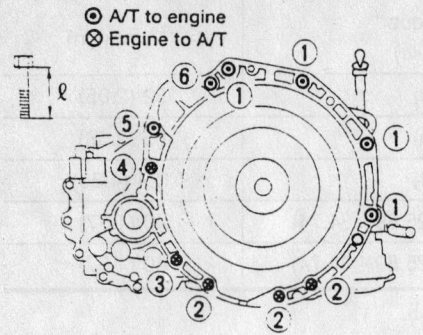

⊙ A/T to engine
⊗ Engine to A/T

Bolt No.	Tightening torque N·m (kg-m, ft-lb)	ℓ mm (in)
①	39 - 49 (4.0 - 5.0, 29 - 36)	45 (1.77)
②	30 - 36 (3.1 - 3.7, 22 - 27)	30 (1.18)
③	30 - 36 (3.1 - 3.7, 22 - 27)	40 (1.57)
④	74 - 83 (7.5 - 8.5, 54 - 61)	45 (1.77)
⑤	30 - 36 (3.1 - 3.7, 22 - 27)	80 (3.15)
⑥	30 - 36 (3.1 - 3.7, 22 - 27)	65 (2.56)

9307HG19

Transaxle bolts tightening sequence and torque specifications–I30 and I35 models with automatic transaxle

To install:

6. Install or connect the following:
- Torque converter in the transmission. Be sure the torque converter is fully seated in the front pump assembly. The distance from the front edge of the transmission to the bolt hole of the torque converter should be 0.75 in. (19mm).
- Position the transmission to the engine and install a few bolts to hold the transmission in place. Do not fully tighten the bolts at this time.
- Torque converter-to-flexplate bolts. Torque the bolts to 33–43 ft. lbs. (44–59 Nm).

7. Secure the transmission to the engine. Torque the bolts in the sequence illustrated.

8. Install all remaining components in the reverse order of removal.

9. Fill the transmission with new fluid. Use the same amount of fluid that was drained before removal.

10. Connect negative battery cable and start the engine. Allow the engine to reach normal operating temperature and check the transmission fluid level. Add fluid as needed.

Transmission Assembly

REMOVAL & INSTALLATION

Automatic

G35 MODELS

1. Before servicing the vehicle, refer to the precautions in the beginning of this section.

2. Remove or disconnect the following:
- Negative battery cable
- Engine cover
- Positive battery cable and battery

- Exhaust pipe
- Propeller shaft
- Automatic Transmission (A/T) control rod and solenoid valve harness connector
- Crankshaft Position (CKP) sensor from the transmission
- Oil cooler and transmission fluid pipes, then plug the lines
- Air breather hose
- Starter motor
- Dust cover from converter housing

➡When turning the crankshaft, turn it clockwise as viewed from the front of the engine.

3. Turn the crankshaft, then remove the 4 tightening bolts for the drive plate and torque converter.

4. Support the transmission assembly with a suitable jack. Be careful not the let the jack hit the drain plug.

5. Remove or disconnect the following:
- Rear member
- Engine-to-transmission bolts

✱✱ WARNING

Before removal, secure the transmission to the jack and secure the torque converter to prevent it from falling.

- Transmission from the vehicle by carefully lowering it with the jack

To install:
6. Installation is the reverse of the removal procedure, noting the following:
a. Tighten the transmission-to-engine bolts as shown in the illustration.
b. Align the positions of the tightening bolts for the drive plate with those of the torque converter and hand-tighten. Then, tighten to 33–42 ft. lbs. (44–58 Nm).
c. After the converter is installed, rotate the crankshaft a few times to be sure the transmission rotates without any binding.
d. Fill the transmission with fluid.

e. Start the vehicle, check for leaks and repair if necessary.

Q45 MODELS

1. Before servicing the vehicle, refer to the precautions in the beginning of this section.

2. Remove or disconnect the following:
- Negative battery cable
- Crankshaft Position (CKP) sensor
- Rear Heated Oxygen Sensor (HO2S) connector
- Exhaust tubes
- Fluid charging pipe
- Oil cooler lines. Plug fluid charging and oil cooler fittings after removing the lines.
- Control linkage from the selector lever
- Neutral safety switch and solenoid harness connectors
- Speed sensor connection
- Driveshaft (make matchmarks to ease in installation). Insert plug into rear seal opening to prevent loss of fluid.

3. Support the transmission safely.
- Bolts securing the torque converter to the flexplate
- Gussets securing the transmission to the engine
- Bolts attaching the transmission to the engine

➡The bolts securing the transmission to the engine are of different lengths. Note the length of the bolts as they are removed.

4. Support the engine safely. Avoid jacking directly under the oil pan drain plug.

5. Remove the transmission from the vehicle.

To install:
6. Install or connect the following:
- Transmission in the vehicle and install the torque converter-to-flex-

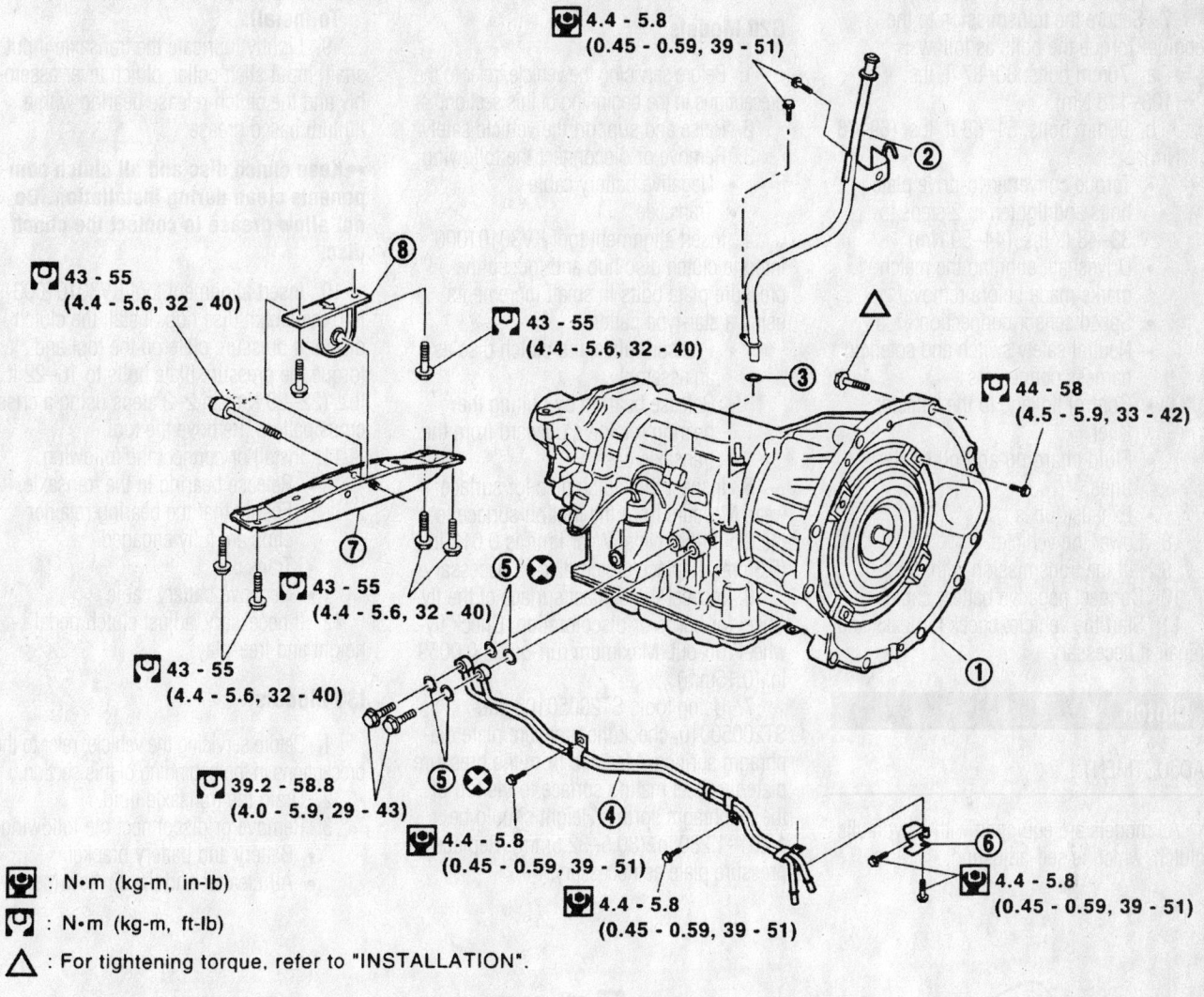

Exploded view of the automatic transmission mounting—G35 models

1. Transmission assembly
2. A/T fluid charging pipe
3. O-ring
4. Oil cooler tube
5. Copper washer
6. Bracket
7. Engine rear member
8. Insulator

9357MG33

Bolt No.	1	2	3
Number of bolts	1	5	2
Bolt length "ℓ"mm (in)	55 (2.17)	65 (2.56)	35 (1.38)
Tightening torque N·m (kg-m, ft-lb)	70 - 80 (7.2 - 8.1, 52 - 59)		41.2 - 52.0 (4.2 - 5.3, 31 - 38)

⊙ A/T to engine
⊗ Engine to A/T

View from vehicle front

9357MG34

Automatic transmission-to-engine bolt tightening specifications—G35 models

plate bolts. Torque the bolts to 33–43 ft. lbs. (44–59 Nm).

7. Secure the transmission to the engine. Torque the bolts as follows:

a. 70mm bolts: 80–87 ft. lbs. (108–118 Nm).

b. 90mm bolts: 51–58 ft. lbs. (69–78 Nm).

- Torque converter-to-drive plate bolts and tighten in 2 steps to 33–43 ft. lbs. (44–59 Nm)
- Driveshaft, aligning the match-marks made before removal
- Speed sensor connection
- Neutral safety switch and solenoid harness connectors
- Control linkage to the selector lever
- Fluid charging and oil cooler lines
- Exhaust tubes

8. Lower the vehicle.

9. Fill the transmission with fluid.

10. Connect negative battery cable.

11. Start the vehicle, check for leaks and repair if necessary.

Clutch

ADJUSTMENT

All models are equipped with a hydraulic clutch, which is self-adjusting.

REMOVAL & INSTALLATION

G20 Models

1. Before servicing the vehicle, refer to the precautions in the beginning of this section.

2. Raise and support the vehicle safely.

3. Remove or disconnect the following:
- Negative battery cable
- Transaxle

4. Insert alignment tool KV30101000 into the clutch disc hub and loosen the pressure plate bolts in small increments using a star-type pattern.
- Pressure plate and clutch disc as an assembly
- Release bearing by pulling the bearing retainers outward from the transaxle case

5. Inspect the clutch disc for surface wear. Measure from the friction surface to the top of the rivets. Wear limit is 0.012 in. (0.3mm). Replace clutch disc as necessary.

6. Inspect the contact surface of the flywheel for burns or discoloration. Check flywheel run-out. Maximum run-out is 0.0059 in. (0.15mm).

7. Using tools ST20050100 and ST20050010, check the pressure plate diaphragm springs. Measure from the pressure plate/flywheel mating surface to the top of the diaphragm spring. Height should be 1.201–1.280 in. (30.5–32.5mm). Replace pressure plate as necessary.

8. Inspect the release bearing for damage. Spin the bearing to see that it rolls freely.

To install:

9. Lightly lubricate the transaxle input shaft, input shaft collar, clutch lever assembly and the clutch release bearing with a lithium based grease.

➡**Keep clutch disc and all clutch components clean during installation. Do not allow grease to contact the clutch disc.**

10. Insert alignment tool KV30101000 into the clutch disc hub. Install the clutch disc and pressure plate on the tool and torque the pressure plate bolts to 16–22 ft. lbs. (22–29 Nm) in 2–3 steps using a criss-cross pattern. Remove the tool.

11. Install or connect the following:
- Release bearing in the transaxle. Ensure that the bearing retainer clips are fully engaged.
- Transaxle
- Negative battery cable

12. If necessary, adjust clutch pedal height and free-play.

I30 Models

1. Before servicing the vehicle, refer to the precautions in the beginning of this section.

2. Drain the transaxle fluid.

3. Remove or disconnect the following:
- Battery and battery bracket
- Air cleaner and the air flow meter

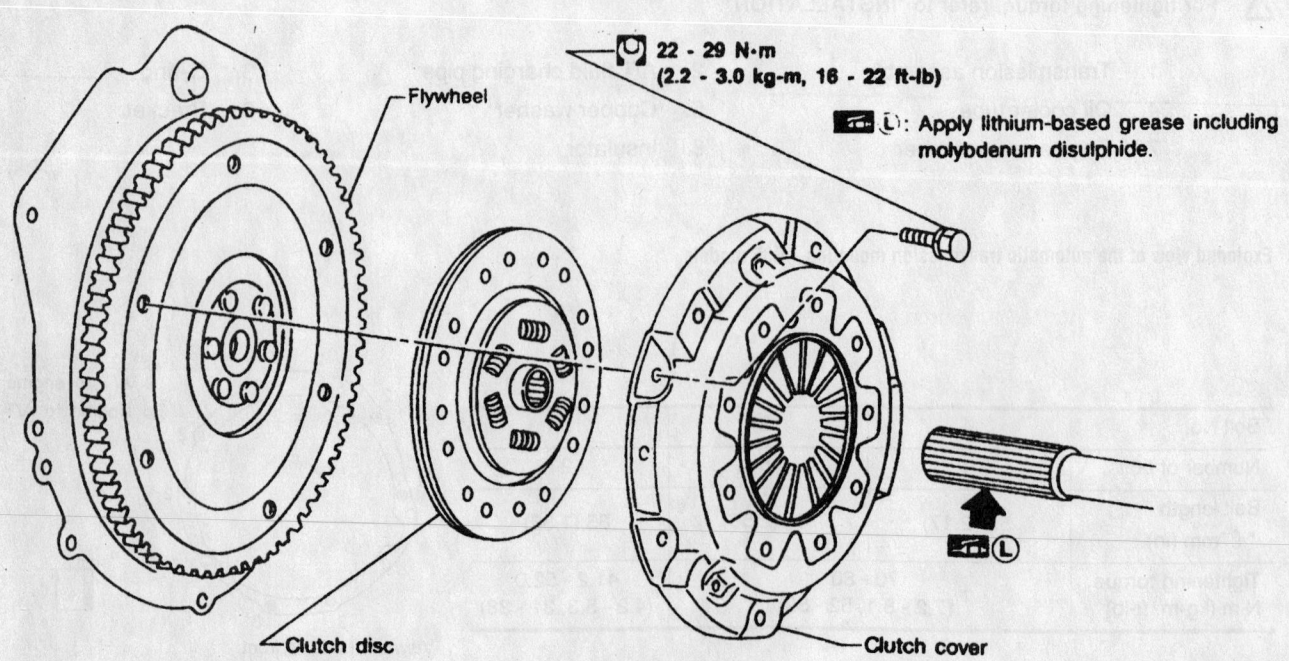

22 - 29 N·m
(2.2 - 3.0 kg-m, 16 - 22 ft-lb)

🔧 (L): Apply lithium-based grease including molybdenum disulphide.

Flywheel

Clutch disc

Clutch cover

7923HG79

Clutch disc and pressure plate—G20 models

4. Raise and safely support the vehicle so there is clearance to remove the transaxle from underneath. Securely support the engine via the oil pan using a cushioning wooden block and jack.

- Transaxle from the engine and lower to the floor

5. Insert a clutch aligning bar or similar tool all the way into the clutch disc hub. This must be done so as to support the weight of the clutch disc during removal. Mark the clutch assembly-to-flywheel relationship with paint or a center punch so the clutch assembly can be assembled in the same position from which it is removed.

- Bolts in reverse order of tightening sequence, a turn at a time
- Pressure plate and clutch disc
- Release mechanism from the transaxle housing

6. Inspect the pressure plate for wear, scoring, etc., and resurface or replace, as necessary.

7. Measure the thickness of the clutch plate lining to the rivet heads; if the it is worn to a minimum of 0.012 in. (0.3mm), replace the clutch plate.

8. Inspect the release bearing and replace as necessary.

9. Using a dial indicator, mount it to the engine and inspect the flywheel run-out; if the run-out exceeds 0.0059 in. (0.15mm), replace it.

To install:

10. Apply a small amount of grease to the transaxle input shaft splines.

11. Install the disc on the splines and slide back and forth a few times. Remove the disc and remove excess grease on hub. Be sure no grease contacts the disc or pressure plate.

12. Apply lithium based molybdenum disulfide grease to the bearing sleeve inside groove, the contact point of the withdrawal lever and bearing sleeve, the contact surface of the lever ball pin and lever.

13. Install or connect the following:

- Release mechanism and release bearing
- Pressure plate and clutch disc, aligning it with a splined dummy shaft tool KV301010000

14. Torque the pressure plate bolts in sequence as follows:

a. Step 1: torque in sequence to: 7–14 ft. lbs. (10–20 Nm).

b. Step 2: torque in sequence to: 25–33 ft. lbs. (34–44 Nm).

15. Remove the dummy shaft.

16. Install or connect the following:

- Transaxle in the correct position. Tighten the transaxle-to-engine bolts.
- Rear and left-hand mounts
- Speedometer cable
- Electrical harness connector
- Clutch release cylinder
- Starter assembly

17. Securely support the transaxle and install the driveshafts.

18. Fill the transaxle with the required amount of approved fluid.

19. Install or connect the following:

- Air flow meter and the air cleaner
- Battery and battery bracket

20. Road test the vehicle for proper shift operation.

Hydraulic Clutch System

BLEEDING

G20 and I30 Models

Bleeding is required to remove air trapped in the hydraulic system. The bleed screw is located on the clutch slave (release) cylinder.

1. Before servicing the vehicle, refer to the precautions in the beginning of this section.

2. Remove the bleed screw dust cap.

3. Attach a transparent vinyl tube to the bleed screw, immersing the free end in a clean container of clean brake fluid.

4. Fill the clutch master cylinder with the proper fluid.

5. Slowly depress the clutch pedal all the way several times and hold it down.

6. Have an assistant open the bleeder valve about ¾ turn to release the air. Then, close the bleeder valve while the pedal is still depressed.

7. Repeat the above procedure until no more air bubbles are seen in the fluid container.

8. Remove the bleed tube.

9. Replace the dust cap and refill the master cylinder.

10. Bleed the clutch damper, if equipped.

Halfshaft

REMOVAL & INSTALLATION

G20 Models

FRONT

1. Before servicing the vehicle, refer to the precautions in the beginning of this section.

2. Raise and support the vehicle safely.

3. Remove or disconnect the following:

- Front wheel
- Wheel bearing locknut
- Brake caliper assembly and rotor. Using a piece of wire, position the caliper so that it is not supported by the brake line.
- Tie-rod from the ball joint
- Kingpin from the knuckle
- Halfshaft from the wheel hub/knuckle by lightly tapping it

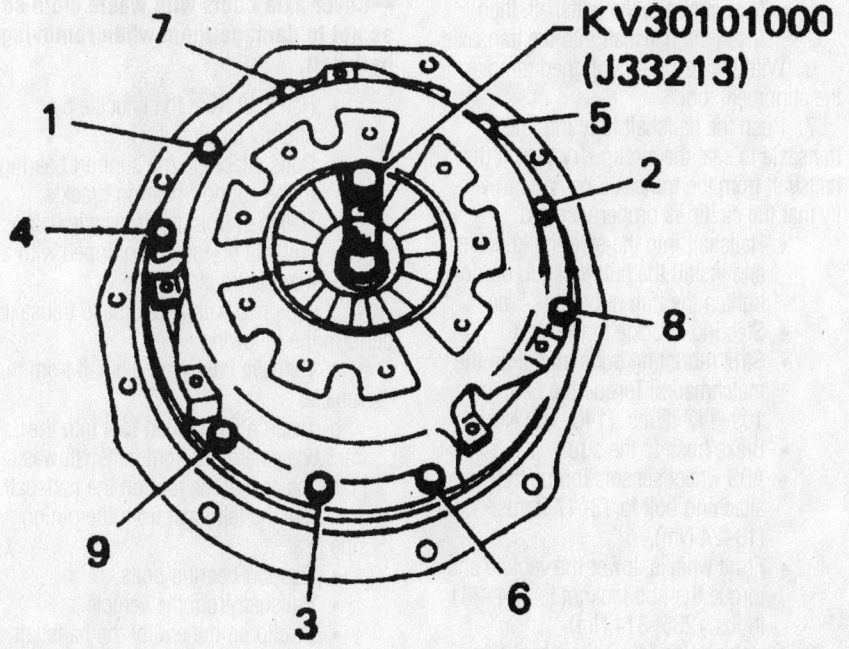

KV30101000
(J33213)

7923HG80

Tighten the pressure plate bolts according to the sequence shown—I30 models

with a wood drift. Take care not to damage the CV-boots.

- Halfshaft from the transaxle by prying outward with a suitable tool at the transaxle case

4. On automatic transaxle models, remove the left halfshaft by tapping it out with a drift from the right side of the transaxle case. Take care not to damage the pinion mate shaft and side gear.

To install:

5. Drive a new oil seal into the transaxle. For the right side use tool KV38106800 along the inner circumference of the oil seal. For the left side use tool KV38106700.

6. Install or connect the following:

- Halfshaft into the transaxle. Ensure that the serrations are aligned. Remove the tool.

7. Push the halfshaft inward and install the circular clip in the groove of the side gear. After inserting the clip, pull outward on the flange of the slide joint to ensure the clip is properly meshed with the side gear. If it pulls out, the clip was not installed properly.

- Halfshaft into the wheel hub/knuckle. Torque the upper knuckle nut to 72–87 ft. lbs. (98–118 Nm) and wheel bearing locknut to 174–231 ft. lbs. (235–314 Nm).

8. Using a dial indicator, check wheel bearing axial end-play. Specification calls for 0.0020 in. (0.05mm) or less.

- Rotor and brake caliper
- Wheel

I30 and I35 Models

FRONT—RIGHT SIDE

1. Before servicing the vehicle, refer to the precautions in the beginning of this section.

2. Raise and support the front of the vehicle safely.

3. Remove or disconnect the following:

- Front wheel
- Anti-lock Brake System (ABS) wheel sensor and move it out of the way
- Brake hose from the strut
- Wheel bearing locknut

4. Matchmark and remove the bolts attaching the steering knuckle to the strut

➡**Cover axle boots with waste cloth so as not to damage them when removing halfshaft.**

- Halfshaft from the knuckle by slightly tapping it
- Halfshaft from the transaxle using a suitable flat-bladed tool

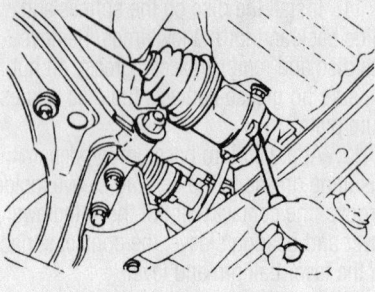

Separating the right halfshaft from the transaxle—I30 models

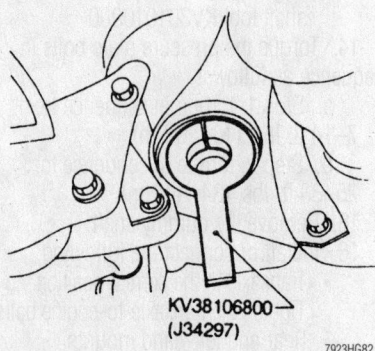

KV38106800
(J34297)

7923HG82

Right halfshaft alignment tool—I30 mod-

- Circlip on the end of the halfshaft and discard circlip
- Seal from the transaxle

To install:

5. Install or connect the following:

- New seal into the transaxle and install a halfshaft alignment tool KV38106800 into the transaxle seal
- New circlip to the halfshaft, then insert the halfshaft into the transaxle

6. With the serration's aligned remove the alignment tool.

7. Push the halfshaft fully into the transaxle to seat the circlip. Try to pull the halfshaft from the transaxle by hand to verify that the circlip is properly seated.

- Halfshaft into the steering knuckle and install the hub locknut, do not tighten the hub nut at this time
- Steering knuckle to the strut
- Strut mounting bolts and align the matchmarks. Torque the bolts to 103–117 ft. lbs. (140–159 Nm).
- Brake hose to the strut
- ABS wheel sensor. Torque the attaching bolt to 13–17 ft. lbs. (18–24 Nm).
- Front wheels, lower the vehicle and torque the hub locknut to 174–231 ft. lbs. (235–314 Nm)

8. Check and/or adjust the wheel alignment as necessary.

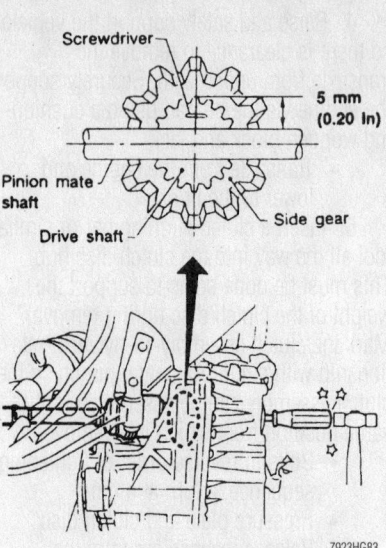

Separating the left halfshaft from an automatic transaxle—I30 models

FRONT—LEFT SIDE

1. Before servicing the vehicle, refer to the precautions in the beginning of this section.

2. Raise and support the front of the vehicle safely.

3. Remove or disconnect the following:

- Front wheel
- Anti-lock Brake System (ABS) wheel sensor and move it out of the way
- Brake hose from the strut
- Wheel bearing locknut
- Bolts attaching the steering knuckle to the strut. Matchmark the bolts prior to removal

➡**Cover axle boots with waste cloth so as not to damage them when removing halfshaft.**

- Halfshaft from the knuckle by slightly tapping it
- Bolts attaching the support bearing to the support bearing bracket
- Halfshaft from the transaxle using a suitable prytool, if equipped with a manual transaxle

4. If equipped with a automatic transaxle perform the following:

a. Remove the right halfshaft from the vehicle.

b. Insert a flat-bladed tool into the transaxle where the right halfshaft was, place the end of the tool on the halfshaft and drive the left shaft from the pinion side gear.

- Support bearing bolts
- Halfshaft from the vehicle
- Circlip on the end of the halfshaft and discard circlip
- Seal from the transaxle

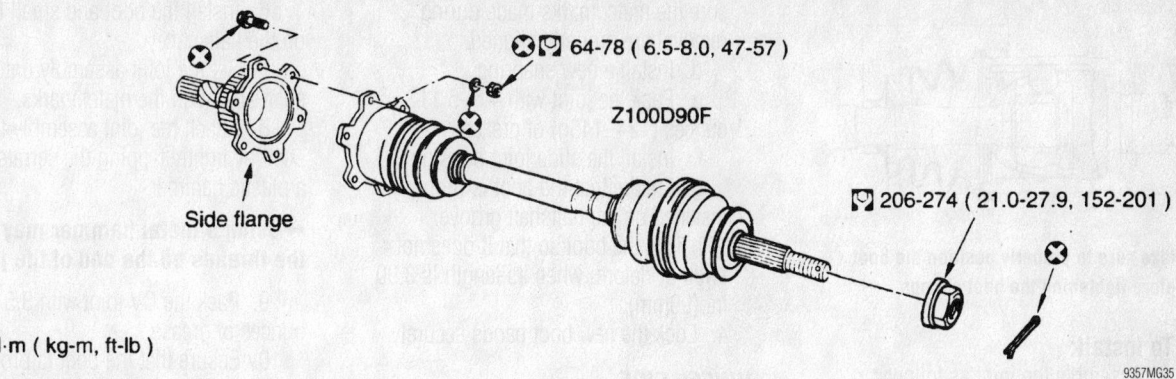

❌ ⏳ 64-78 (6.5-8.0, 47-57)

Z100D90F

⏳ 206-274 (21.0-27.9, 152-201)

Side flange

⏳ : N·m (kg-m, ft-lb)

9357MG35

Rear halfshaft removal and installation—G35 shown

To install:

5. Install or connect the following:
- New seal into the transaxle and install a halfshaft alignment tool KV38106700 into the transaxle seal
- New circlip to the halfshaft, then insert the halfshaft into the transaxle

6. With the serration's aligned remove the alignment tool.
- Halfshaft fully into the transaxle to seat the circlip. Try to pull the halfshaft from the transaxle by hand to verify that the circlip is properly seated.
- Support bearing bolts and torque the bolts to 10–14 ft. lbs. (13–19 Nm)
- Halfshaft into the steering knuckle and install the hub locknut, do not tighten the hub nut at this time
- Steering knuckle to the strut
- Strut mounting bolts and align the matchmarks. Torque the bolts to 103–117 ft. lbs. (140–159 Nm).
- Brake hose to the strut
- ABS wheel sensor. Torque the attaching bolt to 13–17 ft. lbs. (18–24 Nm).
- Front wheels, lower the vehicle and torque hub locknut to 174–231 ft. lbs. (235–314 Nm)

7. Check and/or adjust the wheel alignment as necessary.

G35 and Q45 Models

REAR

❋❋ CAUTION

The amount of force need to loosen the rear wheel bearing nut is high enough to cause the vehicle to fall off the jack. Loosen and tighten this nut with the vehicle on the ground.

1. Before servicing the vehicle, refer to the precautions in the beginning of this section.

2. Remove or disconnect the following:
- Rear wheel cotter pin, adjusting cap and insulator. Loosen the wheel bearing nut with the brakes applied and the vehicle sitting on the ground.

3. Raise the vehicle and support safely.
- Rear wheel
- Differential side flange bolts and nuts and separate shaft from the differential
- Wheel bearing locknut and washer from halfshaft
- Halfshaft by lightly tapping it with a copper hammer
- Halfshaft assembly from the vehicle

To install:

4. Install or connect the following:
- Halfshaft into wheel hub and install washer and wheel bearing locknut. Temporarily tighten the locknut.
- Halfshaft with the differential side flange. Install the nuts and bolts and tighten to 61–69 ft. lbs. (83–93 Nm) for the Q45 or to 47–57 ft. lbs. (64–78 Nm) for the G35.
- Wheels and lower the vehicle to the ground
- Torque the wheel bearing locknut with the brakes applied to 152–201 ft. lbs. (206–274 Nm)
- Insulator, adjusting cap and a new cotter pin

CV-Joints

OVERHAUL

G20 Models

TRANSAXLE SIDE—DS83 TYPE

1. Before servicing the vehicle, refer to the precautions in the beginning of this section.

2. Disassemble the joint as follows:
- a. Remove the boot bands.
- b. Matchmark the slide joint housing and inner race before separating the assembly.
- c. Using a suitable prytool, remove the stopper ring and pull out the slide joint.
- d. Matchmark the inner race and drive shaft.

3. Remove or disconnect the following:
- Snapring
- Ball cage, inner race and balls as a unit
- Boot

➡**Cover the halfshaft serration's with tape, so as not to damage the boot.**

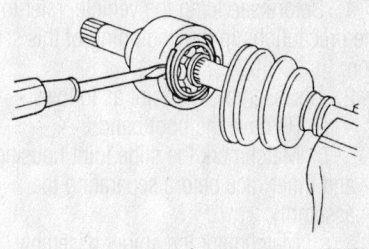

89617G07

The inner CV joint uses a large C-clip to retain the ball and cage assembly in the outer housing

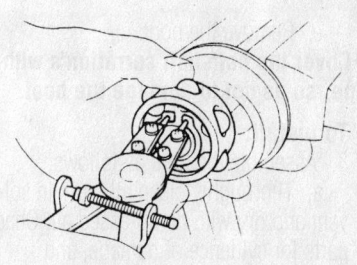

89617G08

After the outer housing is removed, the ball and cage assembly can slide from the shaft by removing the C-clip

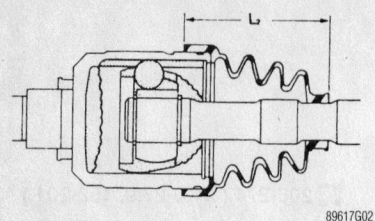

Make sure to properly position the boot before tightening the boot clamps

To install:

4. Assemble the joint as follows:

a. Thoroughly clean all parts in solvent and dry with compressed air. Check parts for evidence of damage, and replace as necessary.

b. Install the boot and new boot band on the halfshaft.

c. Install a new inner snapring.

d. Install the ball cage, inner race and balls as a unit. Confirm that the matchmarks are aligned.

e. Install a new outer snapring.

f. Pack the CV joint with 5.0–6.0 ounces of grease.

g. Ensure that the boot is properly installed on the halfshaft groove.

h. Set the boot so that it does not swell or deform when its length is 3.82–3.90 in. (97–99mm).

i. Lock the new boot bands securely.

TRANSAXLE SIDE—TS83 TYPE

1. Before servicing the vehicle, refer to the precautions in the beginning of this section.

2. Disassemble the joint as follows:

a. Remove the boot bands.

b. Matchmark the slide joint housing and inner race before separating the assembly.

c. Matchmark the spider assembly and driveshaft.

d. Remove the snapring and the spider assembly.

➡ **Do not disassemble the spider assembly.**

e. Remove the boot.

➡ **Cover the halfshaft serration's with tape, so as not to damage the boot.**

To install:

3. Assemble the joint as follows:

a. Thoroughly clean all parts in solvent and dry with compressed air. Check parts for evidence of damage, and replace as necessary.

b. Install the boot and new boot band on the halfshaft.

c. Install the spider assembly making

sure the matchmarks made during removal are properly aligned.

d. Install a new snapring.

e. Pack the joint with 4.5–5.11 ounces (124–145g) of grease.

f. Install the slide joint housing.

g. Ensure that the boot is properly installed on the halfshaft groove.

h. Set the boot so that it does not swell or deform when its length is 3.90 in. (99mm).

4. Lock the new boot bands securely.

WHEEL SIDE

➡ **The joint on the wheel side cannot be disassembled.**

1. Before servicing the vehicle, refer to the precautions in the beginning of this section.

2. Prior to separating the joint assembly, matchmark the halfshaft and joint assembly.

3. Separate the joint using a slide hammer.

4. Remove the boot bands.

To assemble:

5. Thoroughly clean all parts in solvent and dry with compressed air. Check parts for evidence of damage and replace as necessary.

➡ **Cover the halfshaft serration's with tape, so as not to damage the boot.**

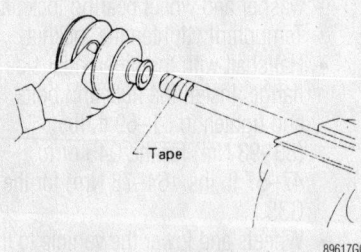

Use vinyl tape and wrap the end of the shaft to protect the boot during installation

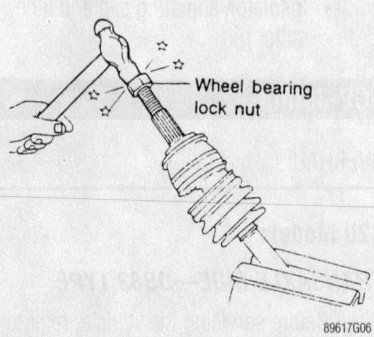

Use an old nut to protect the threads when tapping the outer CV joint onto the shaft

6. Install the boot and small boot band on the halfshaft.

7. Set the joint assembly onto the halfshaft and align the matchmarks.

8. Attach the joint assembly to the halfshaft by lightly tapping the serrated end with a plastic hammer.

➡ **Using a metal hammer may damage the threads on the end of the joint.**

9. Pack the CV joint with 3.5–4.0 ounces of grease.

10. Ensure that the boot is properly installed on the halfshaft groove.

11. Set the boot so that it does not swell or deform when its length is 3.327–3.406 in. (84.5–86.5mm).

12. Lock the new boot bands securely.

I30 and I35 Models

TRANSAXLE SIDE

1. Before servicing the vehicle, refer to the precautions in the beginning of this section.

2. Disassemble the joint as follows:

a. Remove the boot bands.

b. Matchmark the slide joint housing and inner race before separating the assembly.

c. Using a suitable prytool, remove the stopper ring and pull out the slide joint.

d. Matchmark the inner race and drive shaft.

e. Remove the snapring.

f. Remove the ball cage, inner race and balls as a unit.

g. Remove the boot.

➡ **Cover the halfshaft serration's with tape, so as not to damage the boot.**

To install:

3. Assemble the joint as follows:

a. Thoroughly clean all parts in solvent and dry with compressed air. Check parts for evidence of damage, and replace as necessary.

b. Install the boot and new boot band on the halfshaft.

4. Install a new inner snapring.

5. Install the ball cage, inner race and balls as a unit. Confirm that the matchmarks are aligned.

6. Install a new outer snapring.

7. Pack the CV joint with 5.8–6.17 ounces of grease.

8. Ensure that the boot is properly installed on the halfshaft groove.

9. Set the boot so that it does not swell or deform when its length is 3.82–3.90 in. (97–99mm).

10. Lock the new boot bands securely.

Circular clip:

Circular clips should be properly meshed with differential side gear (transaxle side) and with joint assembly (wheel side). Make sure they will not come out.

Be careful not to damage boots. Use suitable protector or cloth during removal and installation.

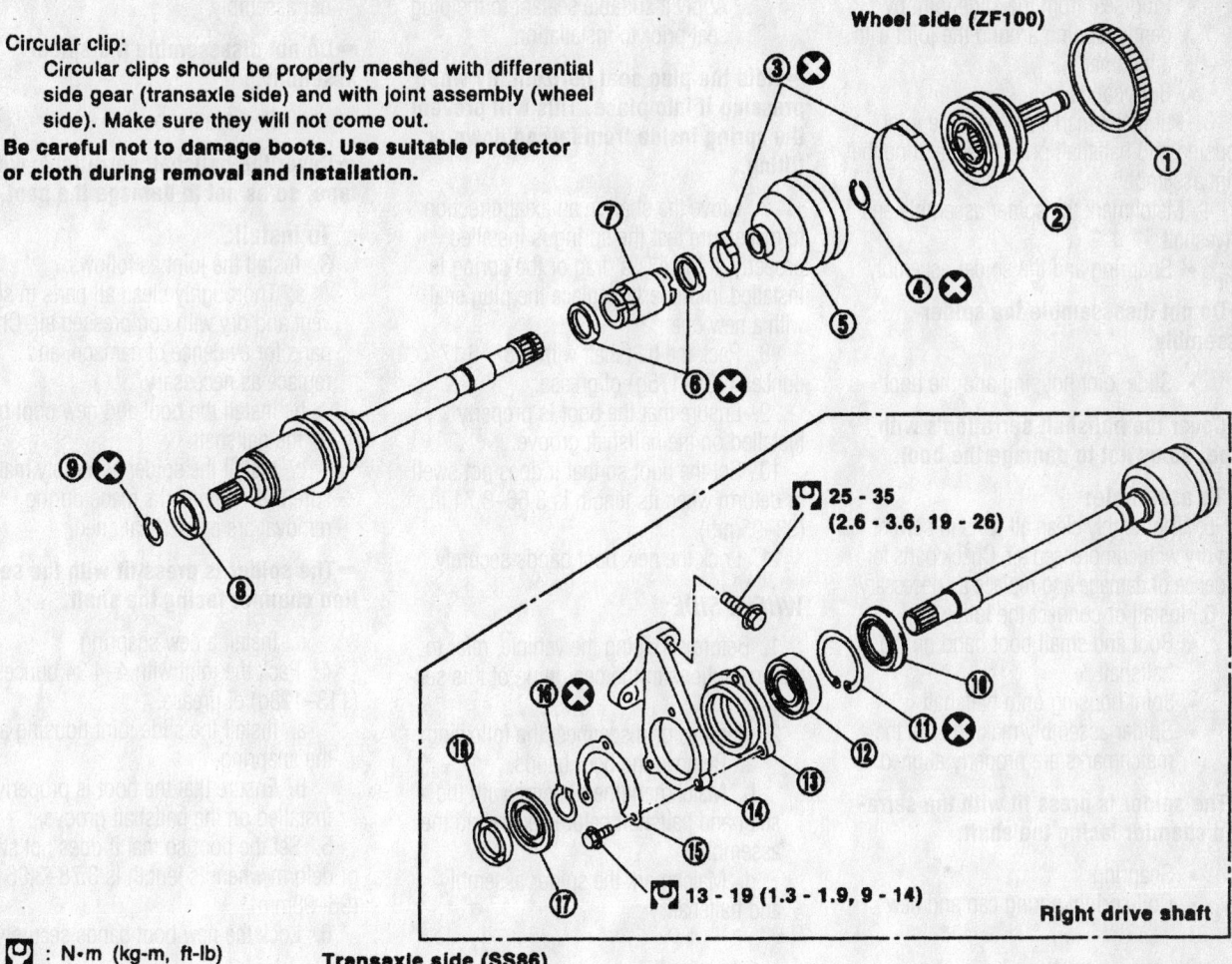

Wheel side (ZF100)

25 - 35 (2.6 - 3.6, 19 - 26)

13 - 19 (1.3 - 1.9, 9 - 14)

Right drive shaft

🔧 : N•m (kg-m, ft-lb)

Transaxle side (SS86)

1. ABS sensor rotor	7. Dynamic damper
2. Joint assembly	8. Dust shield
3. Boot band	9. Circular clip
4. Circular clip	10. Support bearing dust shield
5. Boot	11. Snap ring
6. Dynamic damper band	12. Support bearing

13. Support bearing retainer	
14. Support bearing bracket	
15. Shield heat plate	
16. Snap ring	
17. Support bearing dust shield	
18. Dust shield	

9357MG36

Exploded view of the halfshafts—I35 shown, I30 similar

WHEEL SIDE

➡**The joint on the wheel side cannot be disassembled.**

1. Before servicing the vehicle, refer to the precautions in the beginning of this section.

2. Prior to separating the joint assembly, matchmark the halfshaft and joint assembly.

3. Separate the joint using a slide hammer.

4. Remove the boot bands.

To assemble:

5. Thoroughly clean all parts in solvent and dry with compressed air. Check parts for evidence of damage and replace as necessary.

➡**Cover the halfshaft serration's with tape, so as not to damage the boot.**

6. Install the boot and small boot band on the halfshaft.

7. Set the joint assembly onto the halfshaft and align the matchmarks.

8. Attach the joint assembly to the halfshaft by lightly tapping the serrated end with a plastic hammer.

➡**Using a metal hammer may damage the threads on the end of the joint.**

9. Pack the CV joint with 4.7–5.11 ounces of grease.

10. Ensure that the boot is properly installed on the halfshaft groove.

11. Set the boot so that it does not swell or deform when its length is 3.78–3.86 in. (96–98mm).

12. Lock the new boot bands securely.

G35 and Q45 Models

TRANSMISSION SIDE

1. Before servicing the vehicle, refer to the precautions in the beginning of this section.

2. Remove or disconnect the following:
- Plug seal from the slide joint by gently tapping around the joint with a hammer
- Boot bands

3. Put matchmarks on the slide joint housing and halfshaft prior to separating the joint assembly.

4. Matchmark the spider assembly and driveshaft.
- Snapring and the spider assembly

➡️**Do not disassemble the spider assembly.**

- Slide joint housing and the boot

➡️**Cover the halfshaft serration's with tape, so as not to damage the boot.**

To assemble:

5. Thoroughly clean all parts in solvent and dry with compressed air. Check parts for evidence of damage and replace as necessary.

6. Install or connect the following:
- Boot and small boot band on the halfshaft
- Joint housing onto halfshaft
- Spider assembly making sure the matchmarks are properly aligned

➡️**The spider is press fit with the serration chamfer facing the shaft.**

- Snapring
- Coil spring, spring cap and new

plug seal to the slide joint housing. Apply a suitable sealant to the plug seal prior to installation

➡️**Hold the plug seal horizontally when pressing it into place. This will prevent the spring inside from falling down or tilting.**

7. Move the shaft in an axial direction to make sure that the spring is installed properly. If there is a drag or the spring is installed improperly, replace the plug seal with a new one.

8. Pack the halfshaft with 5.82–6.17 ounces (165–175g) of grease.

9. Ensure that the boot is properly installed on the halfshaft groove.

10. Set the boot so that it does not swell or deform when its length is 3.66–3.74 in. (93–95mm).

11. Lock the new boot bands securely.

WHEEL SIDE

1. Before servicing the vehicle, refer to the precautions in the beginning of this section.

2. Remove or disconnect the following:
   a. Remove the boot bands.
   b. Matchmark the housing with the shaft and halfshaft before separating the assembly.
   c. Matchmark the spider assembly and halfshaft.

d. Remove the snapring and the spider assembly.

➡️**Do not disassemble the spider assembly.**

   e. Remove the boot.

➡️**Cover the halfshaft serration's with tape, so as not to damage the boot.**

To install:

3. Install the joint as follows:
   a. Thoroughly clean all parts in solvent and dry with compressed air. Check parts for evidence of damage, and replace as necessary.
   b. Install the boot and new boot band on the halfshaft.
   c. Install the spider assembly making sure the matchmarks made during removal are properly aligned.

➡️**The spider is press fit with the serration chamfer facing the shaft.**

   d. Install a new snapring.

4. Pack the joint with 4–4.34 ounces (113–123g) of grease.
   a. Install the slide joint housing and the snapring.
   b. Ensure that the boot is properly installed on the halfshaft groove.

5. Set the boot so that it does not swell or deform when its length is 3.78–3.86 in. (96–98mm).

6. Lock the new boot bands securely.

STEERING AND SUSPENSION

Air Bag

PRECAUTIONS

Several precautions must be observed when handling the inflator module to avoid accidental deployment and possible personal injury.

1. Never carry the inflator module by the wires or connector on the underside of the module.

2. When carrying a live inflator module, hold securely with both hands, and ensure that the bag and trim cover are pointed away.

3. Place the inflator module on a bench or other surface with the bag and trim cover facing up.

4. With the inflator module on the bench, never place anything on or close to the module that may be thrown in the event of an accidental deployment.

DISARMING

➡️**All Air Bag electrical wiring harnesses and connectors are covered with YELLOW outer insulation. Do not use electrical test equipment on any circuit related to the Air Bag sensors. When installing Air Bag components, always install with the arrow marks facing the front of the vehicle.**

1. Before servicing the vehicle, refer to the precautions in the beginning of this section.

2. Turn the ignition switch to the **OFF** position.

3. Disconnect both battery cables starting with the negative cable first and wait at least 10 minutes after the cables are disconnected. Be sure to insulate the battery terminal ends.

REARMING

1. Before servicing the vehicle, refer to the precautions in the beginning of this section.

2. Turn the ignition switch to the **OFF** position.

3. Connect both battery cables starting with the positive cable first.

➡️**The Air Bag or Air Bag system is equipped with a self-diagnostic operation. After turning the ignition key to the ON or START position, the AIR BAG warning lamp will illuminate for 7 seconds. After 7 seconds, the AIR BAG lamp will extinguish if no malfunction is detected. If the AIR BAG lamp does not extinguish after 7 seconds, check the Air Bag self-diagnostic system for a malfunction.**

Power Rack and Pinion Steering Gear

REMOVAL & INSTALLATION

G20 Models

1. Before servicing the vehicle, refer to the precautions in the beginning of this section.

❋❋ CAUTION

The air bag system must be disarmed before removing the steering wheel. Failure to do so may cause accidental deployment, property damage or personal injury.

2. Point the front tires straight ahead and lock the steering in this position.

❋❋ WARNING

Do not turn the steering wheel or column with the lower joint removed from the steering column or the spiral cable may be damaged.

3. Remove the steering wheel.

➡ **The steering wheel must be removed before disconnecting the steering column lower joint to avoid damaging the Supplemental Restraint System (SRS) spiral cable.**

4. Raise and support the vehicle safely and remove the front wheels.

5. Remove or disconnect the following:
- Tie rod ends from the steering knuckles
- Carbon canister and properly support the engine
- Bolts attaching the engine mounts to the engine mounting center member
- Engine mounting center member
- Front stabilizer bar from the vehicle, if necessary
- Nuts attaching the hole cover to the bulkhead

6. Move the hole cover aside and disconnect the lower joint from the rack and pinion. Matchmark the pinion shaft and the pinion housing to record the steering neutral position.
- Power steering fluid pipes from the rack and pinion
- Bolts attaching the mounting brackets and the rack and pinion from the vehicle

To install:

7. Install or connect the following:
- Rack and pinion in the vehicle
- Mounting brackets and tighten the mounting nuts and bolts in the proper sequence
- New O-rings to the power steering fluid pipes and connect them to the rack and pinion. Torque the low

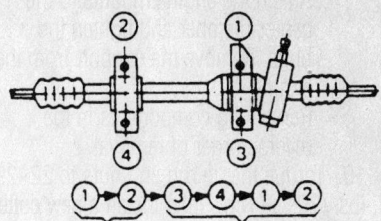

Temporary tightening Secure tightening

9307HG26

Exploded view of the steering gear assembly—G20 Models

pressure line 20–29 ft. lbs. (27–39 Nm) and the high pressure line to 11–18 ft. lbs. (15–25 Nm).

8. Align the lower steering joint to the pinion shaft and install the joint onto the pinion shaft. Torque the bolt to 17–22 ft. lbs. (24–29 Nm).

9. Properly position the hole cover. Torque the nuts to 2.9–3.6 ft. lbs. (4–5 Nm).

- Front stabilizer
- Engine mounting center member and tighten the attaching bolts.

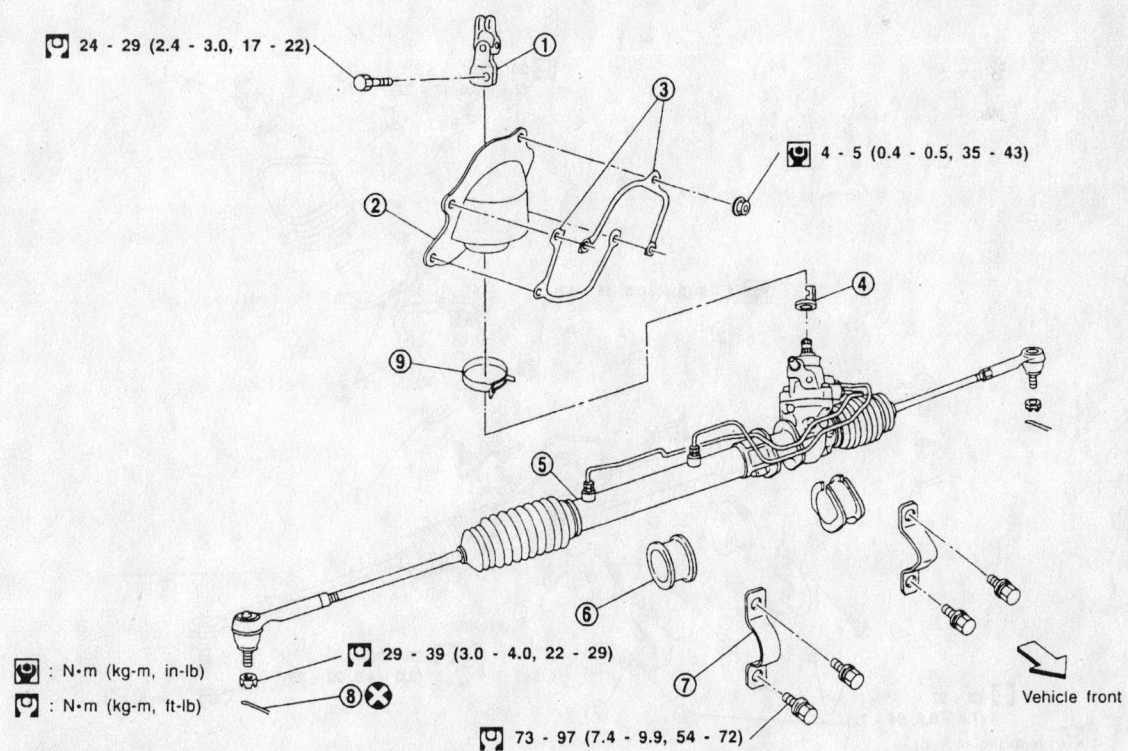

24 - 29 (2.4 - 3.0, 17 - 22)

4 - 5 (0.4 - 0.5, 35 - 43)

⬚ : N•m (kg-m, in-lb)
⬚ : N•m (kg-m, ft-lb)

29 - 39 (3.0 - 4.0, 22 - 29)

73 - 97 (7.4 - 9.9, 54 - 72)

Vehicle front

1. Lower joint
2. Hole cover
3. Insulator bracket
4. Rear cover cap
5. Gear and linkage assembly
6. Rack mounting insulator
7. Gear housing mounting bracket
8. Cotter pin
9. Clamp

9307HG25

Tighten the steering rack fasteners in this order—G20 Models

Attach the engine mounts to the center member and tighten the bolts. Remove the support from the engine.
- Remaining components in the reverse order of removal

10. Torque the tie rod end nuts to 22–29 ft. lbs. (29–39 Nm), then install a new cotter pin.

11. Fill the power steering reservoir with fluid and bleed the air from the power steering system.

12. Check the vehicle front end alignment and adjust as necessary.

I30 Models

1. Before servicing the vehicle, refer to the precautions in the beginning of this section.

2. Point the front tires straight ahead and lock the steering in this position.

3. Remove the steering wheel.

→The steering wheel must be removed before disconnecting the steering column lower joint to avoid damaging the Supplemental Restraint System (SRS) spiral cable.

4. Remove or disconnect the following:
- Both front wheels
- Tie rod ends from the steering knuckles
- Carbon canister from the vehicle and properly support the engine
- Bolts attaching the engine mounts to the engine mounting center member
- Engine mounting center member
- Front stabilizer bar from the vehicle, if necessary
- Nuts attaching the hole cover to the bulkhead

5. Move the hole cover aside and disconnect the lower joint from the rack and pinion. Matchmark the pinion shaft and the pinion housing to record the steering neutral position.
- Power steering fluid pipes from the rack and pinion
- Bolts attaching the mounting brackets and the rack and pinion from the vehicle

To install:

6. Install or connect the following:
- Rack and pinion in the vehicle
- Mounting brackets and tighten the mounting nuts and bolts in the proper sequence
- New O-rings to the power steering fluid pipes and connect them to the rack and pinion. Torque the low pressure line 20–29 ft. lbs. (27–39 Nm) and the high pressure line to 11–18 ft. lbs. (15–25 Nm).

7. Align the lower steering joint to the pinion shaft and install the joint onto the pinion shaft. Torque the bolt to 17–22 ft. lbs. (24–29 Nm).

8. Properly position the hole cover and install the attaching nuts. Torque the nuts to 2.9–3.6 ft. lbs. (4–5 Nm).
- Front stabilizer
- Engine mounting center member

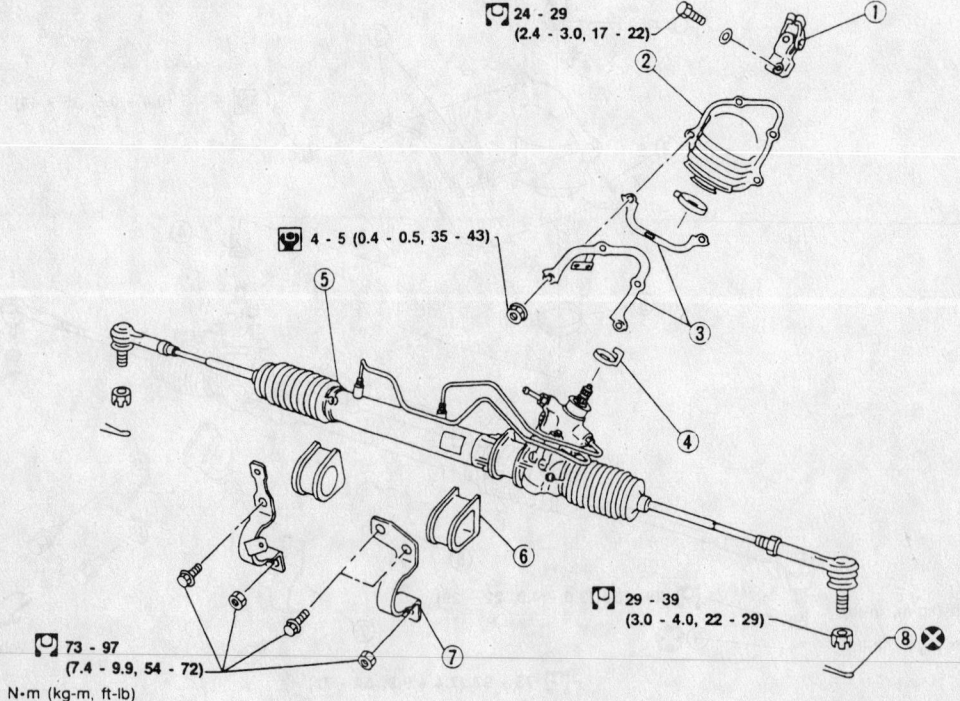

24 - 29 (2.4 - 3.0, 17 - 22)

4 - 5 (0.4 - 0.5, 35 - 43)

73 - 97 (7.4 - 9.9, 54 - 72)

29 - 39 (3.0 - 4.0, 22 - 29)

: N•m (kg-m, ft-lb)
: N•m (kg-m, in-lb)

① Lower joint
② Hole cover
③ Insulator bracket
④ Rear cover cap
⑤ Gear and linkage assembly
⑥ Rack mounting insulator
⑦ Gear housing mounting bracket
⑧ Cotter pin

9307HG32

Exploded view of the steering gear assembly—I30 models

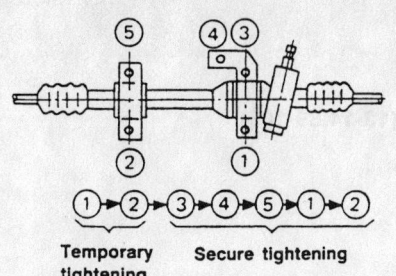

Tighten the steering rack fasteners in this order—I30 Models

and tighten the attaching bolts. Attach the engine mounts to the center member and tighten the bolts. Remove the support from the engine.

- Remaining components in the reverse order of removal

9. Torque the tie rod end nuts to 22–29 ft. lbs. (29–39 Nm), then install a new cotter pin.

10. Fill the power steering reservoir with fluid and bleed the air from the power steering system.

11. Check the vehicle front end alignment and adjust as necessary.

I35 Models

1. Before servicing the vehicle, refer to the precautions in the beginning of this section.

✳✳ CAUTION

The air bag system must be disarmed before removing the steering wheel. Failure to do so may cause accidental deployment, property damage or personal injury.

2. Point the front tires straight ahead and lock the steering in this position.

✳✳ WARNING

Do not turn the steering wheel or column with the lower joint removed from the steering column or the spiral cable may be damaged.

3. Remove the steering wheel.

➡The steering wheel must be removed before disconnecting the steering column lower joint to avoid damaging the Supplemental Restraint System (SRS) spiral cable.

4. Remove or disconnect the following:
- Front exhaust pipe

5. Place a suitable jack under the transaxle.
- Center member and rear engine mount
- Front stabilizer bar
- Separate the tie rod ends from the steering knuckle
- Power steering lines and plug them to prevent contaminants from entering
- Steering gear retaining bolts
- Steering gear assembly

To install:

6. Installation is the reverse of the removal procedure, noting the following:

a. Tighten all fasteners as shown in the illustration.

b. Fill the power steering reservoir with fluid and bleed the air from the power steering system.

c. Check the vehicle front end alignment and adjust as necessary.

G35 Models

1. Before servicing the vehicle, refer to the precautions in the beginning of this section.

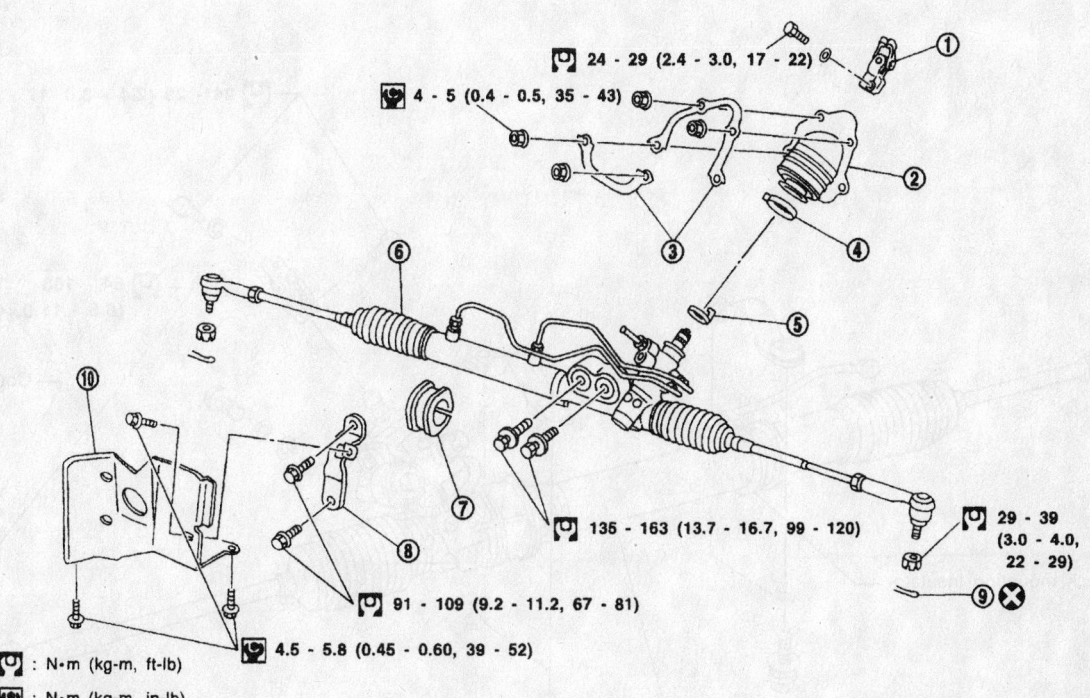

Key	Part	Key	Part	Key	Part
1.	Lower joint	5.	Rear cover cap	8.	Gear housing mounting bracket
2.	Hole cover	6.	Gear and linkage assembly	9.	Cotter pin
3.	Insulator bracket	7.	Rack mounting insulator	10.	Heat insulator
4.	Clamp				

Exploded view of the power steering gear assembly—I35 models

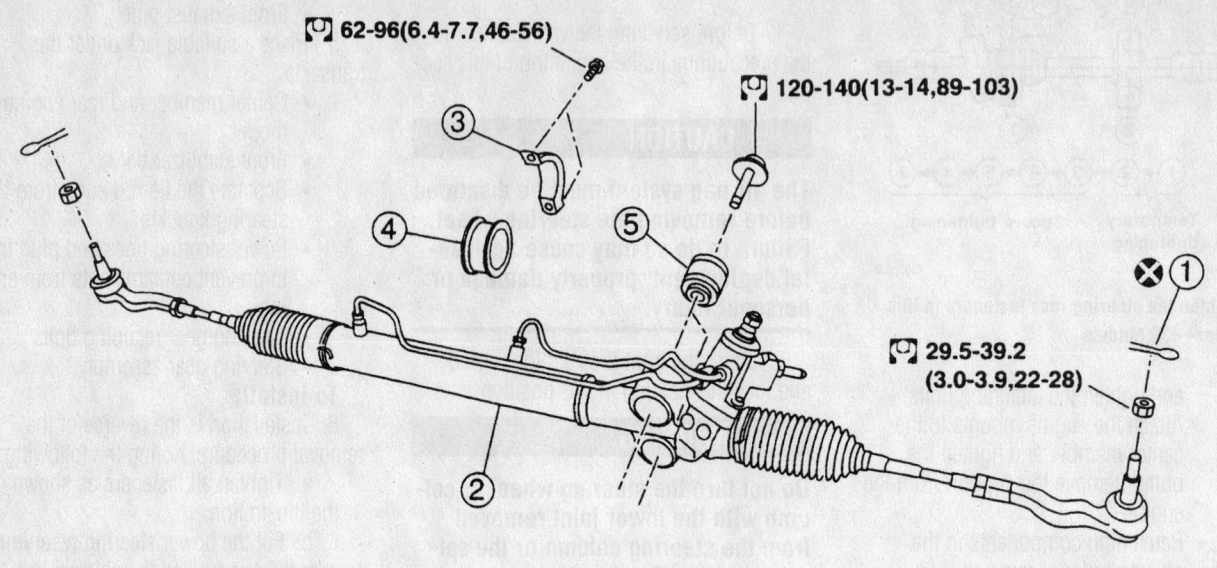

62-96(6.4-7.7,46-56)

120-140(13-14,89-103)

③

④

⑤

①

29.5-39.2
(3.0-3.9,22-28)

②

: N·m(kg-m,ft-lb)

1. Cotter pin
2. Gear and linkage assembly
3. Rack mounting bracket
4. Rack mounting insulator
5. Rack mount housing insulator

9357MG38

Exploded view of the power steering gear assembly—G35 models

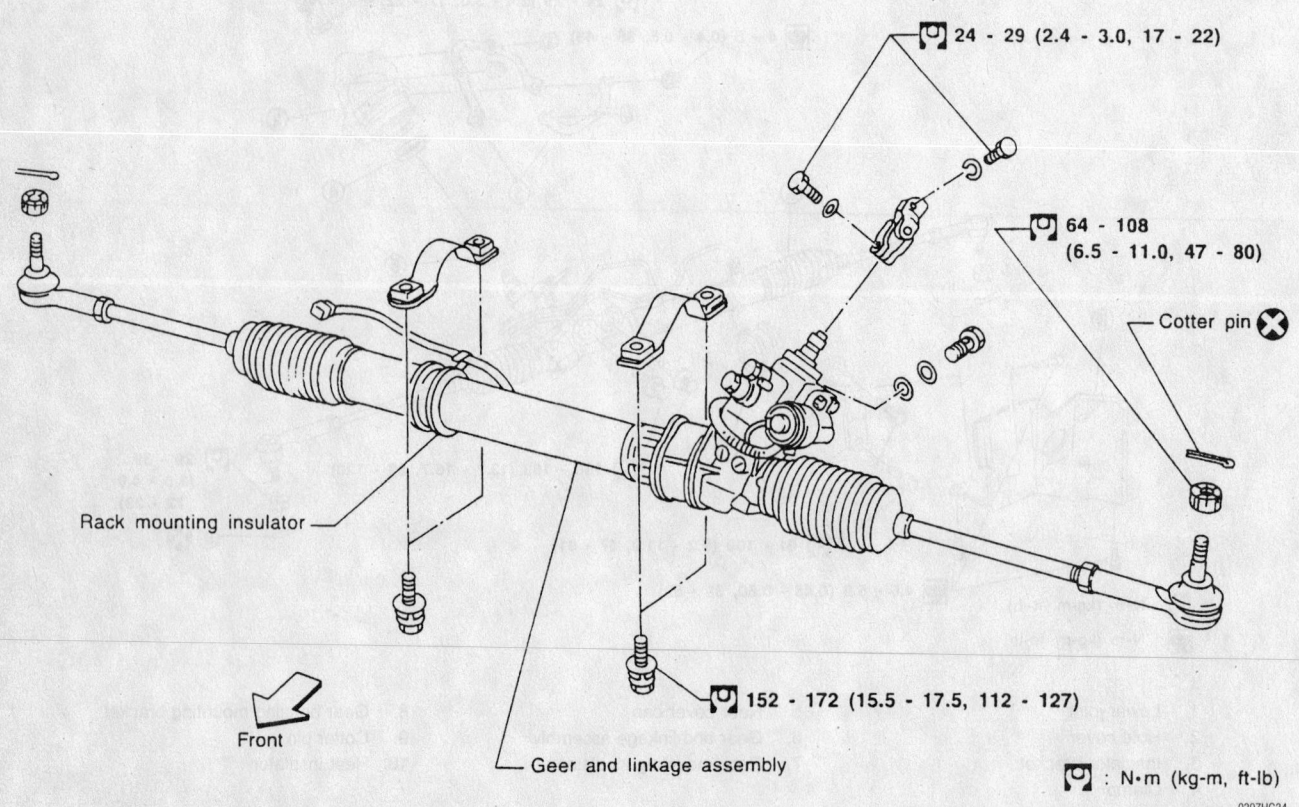

24 - 29 (2.4 - 3.0, 17 - 22)

64 - 108
(6.5 - 11.0, 47 - 80)

Cotter pin

Rack mounting insulator

Front

Gear and linkage assembly

152 - 172 (15.5 - 17.5, 112 - 127)

: N·m (kg-m, ft-lb)

9307HG34

Exploded view of the steering gear assembly—Q45 Models

⁕⁕ WARNING

Do not turn the steering wheel or column with the steering gear is removed.

2. Drain the power steering fluid.
3. Remove or disconnect the following:
   • Both front wheels
   • Engine undercover
   • Tie rod ends from the steering knuckles
   • Pinch bolts from upper and lower sides
   • Power steering fluid pipes from the steering gear assembly
   • Bolt from rack mounting bracket insulator
   • Bolts attaching the mounting brackets and the steering gear from the vehicle

To install:
4. Installation is the reverse of the removal procedure, noting the following:
   a. Tighten all fasteners as shown in the illustration.
   b. Fill the power steering reservoir with fluid and bleed the air from the power steering system.
   c. Check the vehicle front end alignment and adjust as necessary.

Q45 Models

1. Before servicing the vehicle, refer to the precautions in the beginning of this section.

⁕⁕ WARNING

Do not turn the steering wheel or column with the steering gear is removed.

2. Remove or disconnect the following:
   • Both front wheels
   • Tie rod ends from the steering knuckles
   • Carbon canister from the vehicle and properly support the engine
   • Bolts attaching the engine mounts to the engine mounting center member
   • Front stabilizer bar from the vehicle
   • Lower joint bolts
   • Power steering fluid pipes from the rack and pinion
   • Bolts attaching the mounting brackets and the rack and pinion from the vehicle

To install:
3. Install or connect the following:
   • Rack and pinion in the vehicle.
   • Mounting brackets. Torque the

bolts to 112–127 ft. lbs. (152–172 Nm).
   • New O-rings to the power steering fluid pipes and connect them to the rack and pinion. Torque the low pressure line 30–33 ft. lbs. (40–44 Nm) and the high pressure line to 11–18 ft. lbs. (15–25 Nm).
4. Align the lower steering joint to the pinion shaft and install the joint onto the pinion shaft. Install the bolt and torque to 17–22 ft. lbs. (24–29 Nm).
   • Front stabilizer, if removed
   • Engine mounting center member and tighten the attaching bolts. Attach the engine mounts to the center member and tighten the bolts if removed. Remove the support from the engine.
   • Remaining components in the reverse order of removal
5. Torque the tie rod end nuts to 47–80 ft. lbs. (64–108 Nm) and install a new cotter pin.
6. Fill the power steering reservoir with fluid and bleed the air from the power steering system.
7. Check the front end alignment and adjust as necessary.

Strut

REMOVAL & INSTALLATION

Front

G20 MODELS

1. Before servicing the vehicle, refer to the precautions in the beginning of this section.
2. Raise and support the vehicle safely.
3. Remove or disconnect the following:
   • Strut mounting bolt at the lower suspension member and the 3 nuts inside the engine compartment. Do not remove the piston rod locknut
   • Strut assembly and place in a suitable holding device
4. Using a prybar to hold the upper spring mount, loosen but do not remove the piston rod locknut.
5. Compress the spring with a spring compressor so the strut mounting insulator can be turned by hand.
   • Piston rod locknut
   • Coil spring from strut assembly

To install:
6. Inspect all components carefully for damage or wear. Replace as necessary.

7. Install or connect the following:
   • Compressed coil spring on the strut. Torque the locknut to 13–17 ft. lbs. (18–24 Nm).
   • Strut. Ensure the bend in the lower strut bracket faces rearward on the left side and forward on the right side of the vehicle.
   • Upper spring seat with the cutout facing the inside of the vehicle. Torque the upper mounting bolts to 31–40 ft. lbs. (42–54 Nm) and the lower through-bolt to 82–93 ft. lbs. (112–126 Nm). Final tightening must take place with the suspension loaded (vehicle at normal ride height).

I30, I35 AND G35 MODELS

1. Before servicing the vehicle, refer to the precautions in the beginning of this section.
2. Remove or disconnect the following:
   • Wheel. Matchmark the position of the strut-to-steering knuckle location
   • Brake hose from the strut
   • Anti-lock Brake System (ABS) wheel sensor and move it out of the way
   • Bolts attaching the steering knuckle transverse link to the strut. Matchmark the assembly prior to removing the bolts
   • Strut attaching nuts while holding the strut from inside the engine compartment

⁕⁕ CAUTION

Do not remove the center locknut from the strut assembly until the strut is safely compressed.

   • Strut from the vehicle
3. Place the strut assembly in a vise with the special holding tool ST35652000 or in a spring compressor.
4. Loosen the piston rod locknut.

⁕⁕ CAUTION

Do not remove the piston rod locknut, the spring is under tension and can cause serious personal injury.

5. Compress the spring with the spring compressor, then remove the piston rod locknut.

➡**Before removing the strut from the coil spring, note the positioning of the strut in relationship to the coil spring for reassembly.**

- Strut mounting insulator bracket, strut mounting bearing, upper spring seat, and the upper spring rubber seat
- Strut, leaving the coil spring compressed
- Piston boot and rebound bumper from the strut

To install:

6. Install or connect the following:
- Rebound bumper and the boot to the strut piston
- Strut into the coil spring, be sure the strut and spring are properly positioned
- Upper spring rubber seat, upper spring seat, strut mounting bearing, and the strut mounting insulator bracket. Be sure that the cutout on the upper spring seat is facing the outside of the vehicle.
- Piston rod locknut, then remove the spring compressor. Torque the piston rod locknut to 43–65 ft. lbs. (59–88 Nm) for I30 and I35 models, or to 40–47 ft. lbs. (54–65 Nm) for G35 models.
- Strut into the strut tower and install new attaching nuts. Torque the nuts to 29–40 ft. lbs. (39–54 Nm) for I30 models, to 32–38 ft. lbs. (43–51 Nm) for I35 models, or to 26–30 ft. lbs. (35–42 Nm) for G35 models.
- Bolts attaching the steering knuckle or transverse line to the strut and align the matchmarks. Torque the bolts to 103–117 ft. lbs. (140–159 Nm) for I30 models, to 130–139 ft. lbs. (176–189 Nm) for I35 models, or to 52–62 ft. lbs. (70–85 Nm) for G35 models.
- ABS wheel sensor. Torque the bolt to 13–17 ft. lbs. (18–24 Nm).
- Brake hose to the strut
- Front wheels and lower the vehicle

7. Check and/or adjust the wheel alignment as necessary.

Q45 MODELS WITH A STANDARD SUSPENSION

1. Before servicing the vehicle, refer to the precautions in the beginning of this section.

2. Remove or disconnect the following:
- Front wheel
- Brake caliper and rotor
- Tie rod ball joint and lower ball joint with tool ST29020001
- Stabilizer connecting rod upper nut and separate the strut from the connecting rod

- Upper mounting insulator bolts
- Strut assembly

3. Secure the strut in a suitable holding fixture.

4. Loosen the piston rod locknut. Do not remove the locknut.

5. Compress the spring with the proper tool so the strut assembly mounting insulator can be turned by hand.

6. Remove the piston rod locknut. Remove the spring assembly, dust cover and rubber seat. Remove the strut insert.

To install:

7. Inspect the rubber parts for deterioration. If the rubber is pulling away from the metal, the mounting insulator should be replaced.

8. Fit the spring into the lower seat, install the dust cover/bumper and upper seat and mounting insulator.

9. Install or connect the following:
- Piston rod locknut and torque to 13–17 ft. lbs. (17–23 Nm).
- Strut into place and torque the upper mounting nuts to 30–35 ft. lbs. (40–47 Nm).

10. Torque the lower mounting bolt to 80–94 ft. lbs. (108–128 Nm).
- Brake rotor and caliper
- Front wheel

Q45 MODELS WITH ACTIVE SUSPENSION

➡The Nissan Consult or a scan tool that can issue commands to the control unit is required for bleeding the hydraulics in the Full Active Suspension system.

1. Before servicing the vehicle, refer to the precautions in the beginning of this section.

2. Relieve the hydraulic pressure as follows:
   a. Raise all 4 wheels off the ground and wait at least 3 minutes for the system to stabilize.
   b. Remove both front inner fenders and the rear pressure control unit cover.
   c. Loosen the locknut and slowly open the bypass valve on each pressure control unit. Open the valves all the way and leave them open until the job is finished.

3. Remove the flange joint from the top of the actuator.

4. Install 2, 15mm bolts into the actuator in the flange joint mounting bolt holes.

5. Insert a bar between the bolts and loosen the joint adapter. Do not remove it yet.

6. Remove or disconnect the following:

- Upper mount insulator nuts
- Hydraulic lines, then cap the lines to keep the system clean
- Lower actuator mounting nut and remove the assembly

7. Secure the actuator/spring assembly in a suitable holding fixture. Scribe alignment marks on the spring, upper mount insulator and actuator unit.

8. Compress the spring with the proper tool so the joint adapter can be turned by hand. Remove the joint adapter and lift off the mount insulator, spring, and any other components necessary.

To install:

9. If the actuator is being replaced, the rubber bumper should also be replaced. Fit the bumper, dust cover and rubber seat onto the actuator.

10. Fit the spring into the lower seat with the matchmarks aligned. Install the upper seat/mounting insulator with the marks aligned and start the joint adapter. The joint adapter will be tightened after installing the actuator assembly.

11. Fit the strut into place and torque the upper mounting nuts to 30–41 ft. lbs. (40–55 Nm).

12. Torque the lower mounting bolt to 76–94 ft. lbs. (103–128 Nm).

13. Torque the joint adapter to 63–72 ft. lbs. (85–98 Nm).

14. Install the flange adapter and torque the bolts to 11–13 ft. lbs. (15–18 Nm).

15. Close the bypass valves on the pressure control units.

16. Bleed the system as follows:
   a. With all 4 wheels about 2 in. (50mm) off the ground, run the engine for about 2 minutes.
   b. Connect the Consult scan tool and enter "WORK SUPPORT" mode. Select "4. AIR BLEEDING".
   c. Check the fluid level in the reservoir. It should be slightly overfilled.
   d. Touch "START" on the scan tool. The display will show a regular rise and fall in system pressure. When the pressure stabilizes, stop the engine.
   e. Connect a clear tube to the air bleeder at the actuator and place the other end in a container.

➡Do not allow the fluid to contact the body or the paint will be damaged.

   f. Open the bleeder and watch the fluid move through the tube. If there are still air bubbles in the fluid when the flow stops, check the fluid level, pressurize the system again and repeat the process.

Rear

G20 MODELS

1. Before servicing the vehicle, refer to the precautions in the beginning of this section.

2. Remove or disconnect the following:
   • Wheels

✳✳ WARNING

Be sure to disconnect the Anti-lock Brake System (ABS) wheel sensor from the assembly. Failure to do so may result in damage to the sensor wire and the sensor becoming inoperative.

   • Brake calipers and suspend them with a piece of wire. Do not let them hang by the hose.

3. Using a transmission jack, raise the torsion beam slightly
   • Strut lower mounting bolt

4. Open the trunk, remove the trim and remove the two nuts attaching the strut to the vehicle.
   • Strut

✳✳ CAUTION

Do not remove the center locknut from the strut assembly until the strut is safely compressed.

5. Place the strut assembly in a vise with the special holding tool HT71780000 or in a spring compressor.

6. Loosen the piston rod locknut.

✳✳ CAUTION

Do not remove the piston rod locknut, the spring is under tension and can cause serious personal injury.

7. Compress the spring with the spring compressor then remove the piston rod locknut.

➡ **Before removing the strut from the coil spring, note the positioning of the**

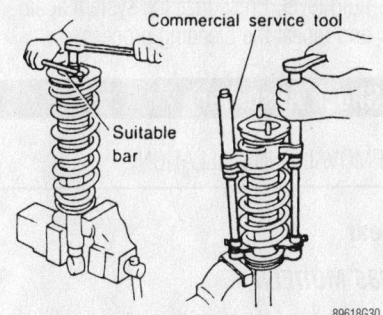

Be sure the spring is compressed before removing the piston locknut—G20 models

strut in relationship to the coil spring for reassembly.

8. Remove or disconnect the following:
   • Bushing, strut mounting bracket, and the upper spring seat rubber
   • Strut, leaving the coil spring compressed
   • Bushing, bound bumper cover and the bound bumper

To install:

9. Install or connect the following:
   • Bound bumper, bound bumper cover and the bushing
   • Strut into the coil spring, make sure the strut and spring are properly positioned
   • Upper spring seat rubber, strut mounting bracket, and the bushing. Make sure that the mounting bracket is properly positioned.
   • Piston rod locknut then remove the spring compressor. Torque the locknut to 13–17 ft. lbs. (18–24 Nm).
   • Strut with new attaching nuts. Torque the nuts to 14–16 ft. lbs. (19–22 Nm).
   • Strut on the rear torsion beam. Torque the bolt to 80–94 ft. lbs. (108–127 Nm).

10. Remove the support from the rear torsion beam.

11. Install the rear wheels and lower the vehicle.

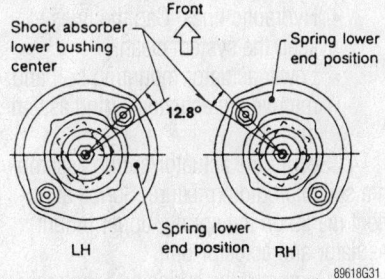

Align the spring seats as shown—G20 models

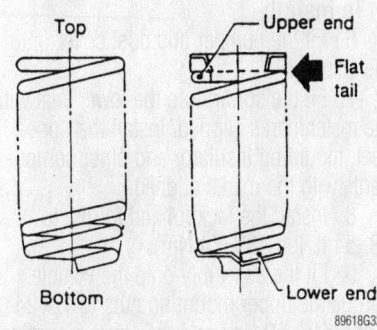

Make sure the springs are installed as shown—G20 models

12. Check the vehicle's alignment and adjust as necessary.

I30 AND I35 MODELS

1. Before servicing the vehicle, refer to the precautions in the beginning of this section.

2. Remove or disconnect the following:
   • Rear wheels

3. Support the rear torsion beam assembly with a jack.

✳✳ CAUTION

Do not remove the center locknut from the strut assembly until the strut is safely compressed.

   • 2 nuts attaching the strut to the vehicle located inside the trunk
   • Bolt attaching the strut to the rear torsion beam assembly and remove the strut

4. Place the strut assembly in a vise with the special holding tool HT71780000 or in a spring compressor.
   • Piston rod locknut

✳✳ CAUTION

Do not remove the piston rod locknut, the spring is under tension and can cause serious personal injury.

5. Compress the spring with the spring compressor, then remove the piston rod locknut.

➡ **Before removing the strut from the coil spring, note the positioning of the strut in relationship to the coil spring for reassembly.**

6. Remove or disconnect the following:
   • Bushing, strut mounting bracket, and the upper spring seat rubber
   • Strut, leaving the coil spring compressed
   • Bushing, bound bumper cover, and the bound bumper

To install:

7. Install or connect the following:
   • Bound bumper, bound bumper cover, and the bushing
   • Strut into the coil spring, be sure the strut and spring are properly positioned
   • Upper spring seat rubber, strut mounting bracket, and the bushing. Be sure that the mounting bracket is properly positioned.
   • Piston rod locknut, then remove the spring compressor. Torque the locknut to 13–17 ft. lbs. (18–24 Nm)
   • Strut with new attaching nuts.

Torque the nuts to 12–16 ft. lbs. (16–22 Nm).
- Strut on the rear torsion beam. Torque the bolt to 80–94 ft. lbs. (108–127Nm).
- Support from the rear torsion beam
- Rear wheels and lower the vehicle

8. Check the vehicle's alignment and adjust as necessary.

Q45 MODELS WITH A STANDARD SUSPENSION

✳ CAUTION

Do not remove piston rod locknut with the shock absorber on vehicle.

1. Before servicing the vehicle, refer to the precautions in the beginning of this section.
2. Remove the upper strut mounting nuts.
3. Raise and safely support the vehicle and remove the lower mounting bolt. Remove coil spring/strut absorber assembly.
4. Place the assembly into a suitable holding fixture and matchmark the spring, strut and upper seat. Loosen but do not remove the piston rod locknut.
5. Install a spring compressor and compress the spring until the upper spring seat can be turned by hand.
6. Remove the locknut, spring seat components, spring, bushings and bumper.
To install:
7. Fit the bumper, spring, upper seat and other components onto the strut with the matchmarks aligned. The top of the spring is flat.
8. Install the locknut and torque it to 13–17 ft. lbs. (18–24 Nm) and remove the spring compressor.
9. Install strut assembly. Torque the upper shock mounting nuts to 12–14 ft. lbs. (16–19 Nm) and the lower bolt to 57–72 ft. lbs. (77–98 Nm).

Q45 MODELS WITH A FULL ACTIVE SUSPENSION

➡The Nissan Consult or scan tool that can issue commands to the control unit is required for bleeding the hydraulics in the Full Active Suspension system.

1. Before servicing the vehicle, refer to the precautions in the beginning of this section.
2. Relieve the hydraulic pressure as follows:
    a. Raise and safely support the vehicle with all 4 wheels off the ground and wait at least 3 minutes for the system to stabilize.
    b. Remove both front inner fenders and the rear pressure control unit cover.

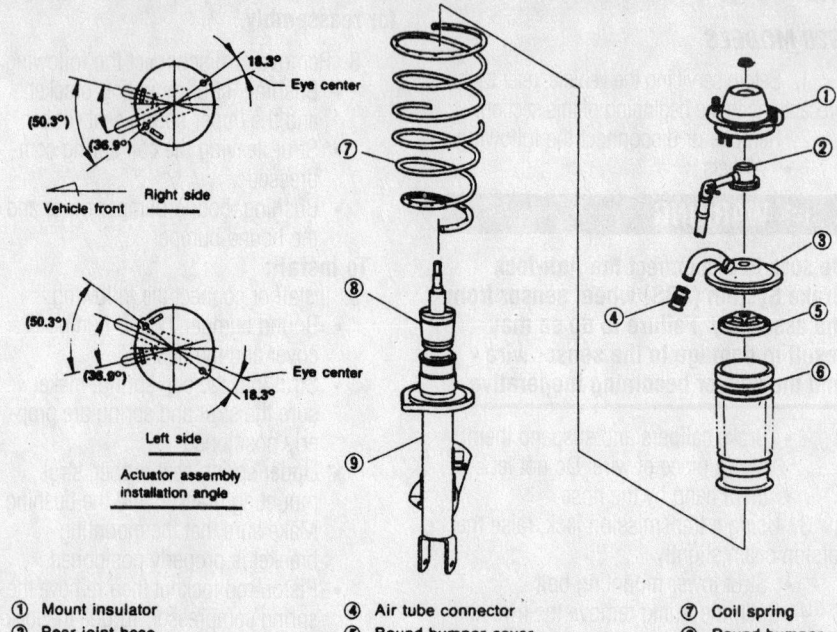

① Mount insulator
② Rear joint hose
③ Spring upper seat
④ Air tube connector
⑤ Bound bumper cover
⑥ Rear actuator dust cover
⑦ Coil spring
⑧ Bound bumper
⑨ Rear actuator

7923HG85

Rear actuator removal with Active Suspension—Q45 models

    c. Loosen the locknut and slowly open the bypass valve on each pressure control unit. Do not open the bleeder valves.
    d. Open the bypass valves all the way, and leave them open, until the job is finished.
3. Remove or disconnect the following:
    - Upper mount insulator nuts
    - Hydraulic lines. Cap the lines to keep the system clean
    - Lower actuator mounting bolt and remove the actuator/spring assembly
4. Secure the actuator/spring assembly in a suitable holding fixture. Scribe alignment marks on the spring, upper mount insulator and actuator unit.
5. Compress the spring with the proper tool. Remove the piston rod locknut lift off the mount insulator, hose joint adapter, spring, and any other components necessary.
To install:
6. Fit the bumper and dust cover onto the actuator.
7. Fit the spring into the lower seat with the matchmarks aligned. Install the upper seat, mounting insulator and other components with the marks aligned.
8. Install the locknut and torque to 43–54 ft. lbs. (59–74 Nm).
9. Fit the assembly onto the vehicle. Torque the upper mounting nuts to 12–24 ft. lbs. (16–19 Nm) and the lower mounting bolt to 58–72 ft. lbs. (78–98 Nm).
10. Bleed the system as follows:

    a. With all 4 wheels off the ground, run the engine for about 2 minutes.
    b. Connect the scan tool and enter "WORK SUPPORT" mode. Select "4. AIR BLEEDING".
    c. Check the fluid level in the reservoir and make it slightly overfilled.
    d. Touch "START" on the scan tool. The display will show a regular rise and fall in system pressure that may last for several minutes. When the pressure stabilizes, stop the engine.
    e. Connect a clear tube to the air bleeder at the actuator and place the other end in a container. Do not allow fluid to contact the body or the paint will be damaged.
    f. Open the bleeder and watch the fluid move through the tube. If there are still air bubbles in the fluid when the flow stops, close the bleeder and check the fluid level. Pressurize the system again and repeat the bleeding process.

Shock Absorber

REMOVAL & INSTALLATION

Rear

G35 MODELS

1. Before servicing the vehicle, refer to the precautions in the beginning of this section.

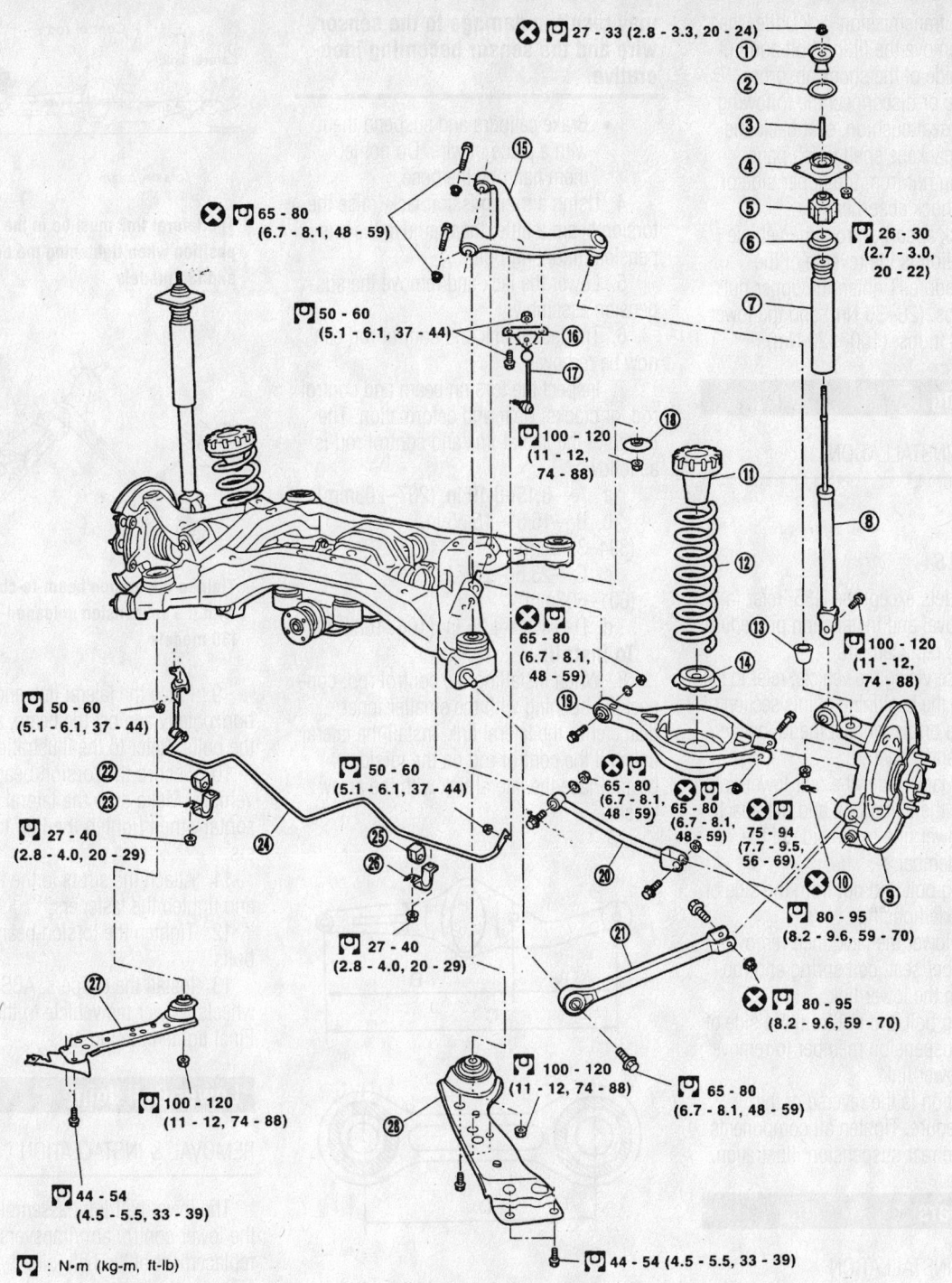

⊗ ⃞ 27 - 33 (2.8 - 3.3, 20 - 24)

⊗ ⃞ 65 - 80
(6.7 - 8.1, 48 - 59)

⃞ 26 - 30
(2.7 - 3.0, 20 - 22)

⃞ 50 - 60
(5.1 - 6.1, 37 - 44)

⃞ 100 - 120
(11 - 12, 74 - 88)

⃞ 100 - 120
(11 - 12, 74 - 88)

⊗ ⃞ 65 - 80
(6.7 - 8.1, 48 - 59)

⃞ 50 - 60
(5.1 - 6.1, 37 - 44)

⃞ 50 - 60
(5.1 - 6.1, 37 - 44)

⊗ ⃞ 65 - 80
(6.7 - 8.1, 48 - 59)

⊗ ⃞ 65 - 80
(6.7 - 8.1, 48 - 59)

⊗ ⃞ 75 - 94
(7.7 - 9.5, 56 - 69)

⃞ 27 - 40
(2.8 - 4.0, 20 - 29)

⃞ 80 - 95
(8.2 - 9.6, 59 - 70)

⃞ 27 - 40
(2.8 - 4.0, 20 - 29)

⊗ ⃞ 80 - 95
(8.2 - 9.6, 59 - 70)

⃞ 100 - 120
(11 - 12, 74 - 88)

⊗ ⃞ 65 - 80
(6.7 - 8.1, 48 - 59)

⃞ 100 - 120 (11 - 12, 74 - 88)

⃞ 44 - 54
(4.5 - 5.5, 33 - 39)

⃞ : N•m (kg-m, ft-lb)

⃞ 44 - 54 (4.5 - 5.5, 33 - 39)

1. Washer	2. Shock absorber mounting seal	3. Distance tube
4. Shock absorber mounting insulator	5. Bushing	6. Bound bumper cover
7. Bound bumper	8. Shock absorber	9. Axle assembly
10. Cotter pin	11. Upper seat	12. Coil spring
13. Ball seat	14. Rubber seat	15. Suspension arm
16. Connecting rod mounting bracket	17. Connecting rod	18. Mount stopper
19. Rear lower link	20. Front lower link	21. Radius rod
22. Bushing	23. Clamp	24. Stabilizer bar
25. Bushing	26. Clamp	27. Member stay
28. Member stay		

Exploded view of the rear suspension—G35 models

9357MG39

2. Place a transmission jack under the rear axle to remove the fitting bolt and nut in the lower side of the shock absorber.

3. Remove or disconnect the following:
- Rear seat cushion, seat back and rear package shelf finish panel
- Fitting nut from the upper side of the shock absorber
- Shock absorber from the vehicle

4. Installation is the reverse of the removal procedure. Tighten the upper nuts to 20–22 ft. lbs. (26–30 Nm) and the lower bolt to 74–88 ft. lbs. (100–120 Nm).

Coil Spring

REMOVAL & INSTALLATION

Rear

G35 MODELS

For all models except the G35, refer to the Strut removal and installation procedure for coil spring replacement.

1. Before servicing the vehicle, refer to the precautions in the beginning of this section.

2. Remove or disconnect the following:
- Tire and wheel

3. Place a jack under the rear lower link.

4. Loosen the fixing bolt and nut attaching the rear lower line to the side of the suspension member.
- Fixing bolt and nut from the side of the axle housing

5. Slowly lower the jack, then remove the upper rubber seat, coil spring and rubber sheet from the lower link.
- Fixing bolt and nut from the side of the suspension member to remove the lower link.

6. Installation is the reverse of the removal procedure. Tighten all components, as listed in the rear suspension illustration.

Torsion Bars

REMOVAL & INSTALLATION

G20 and I30 Models

1. Before servicing the vehicle, refer to precautions in the beginning of this section.

2. Loosen the lug nuts.

3. Remove or disconnect the following:
- Wheels

❋❋ WARNING

Be sure to disconnect the Anti-lock Brake System (ABS) wheel sensor from the assembly. Failure to do so

may result in damage to the sensor wire and the sensor becoming inoperative.

- Brake calipers and suspend them with a piece of wire. Do not let them hang by the hose

4. Using a transmission jack, raise the torsion beam a little, then remove the suspension mounting bolts.

5. Lower the jack and remove the suspension assembly.

6. The lateral link and control rod can now be removed.

7. Inspect the torsion beam and control rod for cracks, wear and deformation. The length of the lateral link and control rod is as follows:

    a. A—8.15–8.19 in. (207–208mm).

    b. B—15.51–15.55 in. (394–395mm).

    c. C—23.66–23.74 in. (601–603mm).

    d. D—4.17–4.25 in. (106–108mm).

To install:

8. When installing the control rod, connect the bushing with the smaller inner diameter to the lateral link. Install the lateral link and the control rod on the torsion beam. Place the lateral link with the arrow topside.

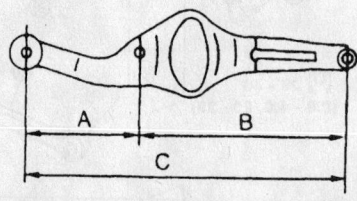

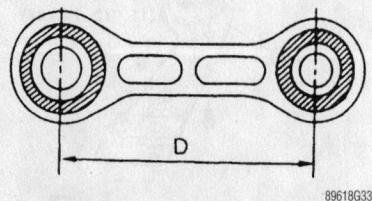

Measure the control rod and lateral links at these points—G20 and I30 models

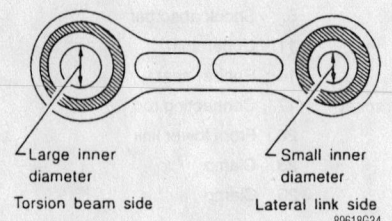

Be sure to install the control rod correctly—G20 and I30 models

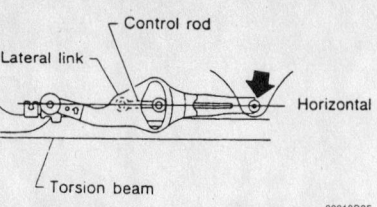

The lateral link must be in the horizontal position when tightening the bolts—G20 and I30 models

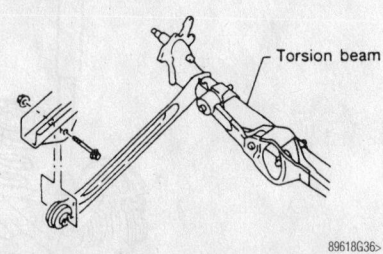

Tighten the torsion beam-to-chassis bolts with the suspension unloaded—G20 and I30 models

9. Place the lateral link and control rod horizontally against the beam, and tighten the bolts. Refer to the illustration.

10. Secure the torsion beam to the vehicle. Make sure the lateral link is horizontal , then tighten the link to the chassis.

11. Attach the struts to the torsion beam and tighten the fasteners.

12. Tighten the torsion beam-to-chassis bolts.

13. Install the calipers, ABS sensor and wheels. Lower the vehicle to the ground. Final tighten the lug nuts.

Lower Ball Joint

REMOVAL & INSTALLATION

The lower ball joint assembly is part of the lower control arm/transverse link. If replacement of the ball joint is required, the lower control arm needs to be replaced.

Lower Control Arms

REMOVAL & INSTALLATION

G20 Models

1. Before servicing the vehicle, refer to the precautions in the beginning of this section.

2. Remove or disconnect the following:
- Stabilizer bar

➡**Take note of paint mark and clamp position when removing stabilizer bar for correct reinstallation.**

3. Support the steering knuckle with a suitable jack and remove the lower ball joint nut. Separate the ball joint from the knuckle.
- Bolts attaching the lower control arm to the chassis
- Lower control arm

To install:

4. If the lower ball joint is worn or damaged, the lower control arm must be replaced. The ball joint is not serviceable separately.

5. Install or connect the following:
- Lower control arm to the chassis with the attaching bolts and nut
- Ball joint stud in the knuckle. Torque the nut to 52–64 ft. lbs. (71–86 Nm).
- Stabilizer bar and wheel

6. Lower the vehicle.

➡**Final tightening must be done with the vehicle at normal ride height, tires on the ground and the chassis loaded.**

7. Torque front control arm bolts to 87–108 ft. lbs. (118–147 Nm) and rear gusset nut to 69–87 ft. lbs. (93–118 Nm).

I30 and I35 Models

1. Before servicing the vehicle, refer to the precautions in the beginning of this section.

2. Remove or disconnect the following:
- Front wheels
- Anti-lock Brake System (ABS) wheel sensor and move it out of the way
- Wheel bearing locknut
- Tie rod from the steering knuckle
- Bolts attaching the strut to the steering knuckle. Matchmark prior to removal.
- Halfshaft from the steering knuckle by lightly tapping the end of the shaft
- Steering knuckle and the lower ball joint
- Stabilizer bar from the lower control arm
- Bolts attaching the link bushing pin to the chassis
- Nut attaching the link to the control arm and remove the link
- Bolts attaching the compression rod bushing clamp
- Lower control arm/traverse link

To install:

3. Install or connect the following:
- Lower control arm and the compression rod bushing clamp into the vehicle

- Link bushing pin, if removed from the control arm

4. Tighten all bolts and nuts until they are snug enough to support the weight of the vehicle but not fully tight. The bolts should be tightened to specification with the vehicle on the floor.
- Steering knuckle to the lower control arm and connect the ball joint. Torque the nut to 46–56 ft. lbs. (62–76 Nm).

➡**Always use a new nut when installing the ball joint to the control arm.**

- Steering knuckle to the strut and to the halfshaft
- Strut mounting bolts and align the matchmarks. Torque the bolts to 103–117 ft. lbs. (140–159 Nm).
- Tie rod ball joint and tighten the nut to 46–54 ft. lbs. (63–73 Nm)
- Wheel bearing locknut
- ABS wheel sensor. Torque the attaching bolt to 13–17 ft. lbs. (18–24 Nm).
- Front wheels

5. Lower the vehicle and torque the hub locknut to 174–231 ft. lbs. (235–314 Nm).

6. Torque the bolts attaching the compression rod bushing clamp and the link bushing pin, in the proper sequence to 87–108 ft. lbs. (118–147 Nm).

7. If the link bushing pin was removed from the control arm, torque the attaching nut to 87–108 ft. lbs. (118–147 Nm).

8. Tighten the sway bar attaching nut to 30–35 ft. lbs. (41–47 Nm).

9. Check the vehicle alignment.

G35 and Q45 Models

1. Before servicing the vehicle, refer to the precautions in the beginning of this section.

2. Remove or disconnect the following:
- Negative battery cable
- Front wheel
- Engine undercover, if necessary
- Nuts securing the tension rod to the transverse link (control arm)
- Nut and separate the ball joint stud from the knuckle
- Transverse link from the sub-frame

To install:

3. Install or connect the following:
- Transverse link on the sub-frame. Temporarily install the bolt and nut.
- Tension rod on the transverse link. Torque the nuts to 87–94 ft. lbs. (118–127 Nm).
- Nut on the ball joint stud. Torque the nut to 71–88 ft. lbs. (96–120

Nm) for Q45 models or to 59–69 ft. lbs. (75–94 Nm) for G35 models.
- Front wheel and lower the vehicle to the floor
- Transverse link mounting bolt. Torque the bolt to 72–87 ft. lbs. (98–118 Nm).
- Engine undercover, if necessary
- Negative battery cable

CONTROL ARM BUSHING REPLACEMENT

The bushing are part of the transverse assembly, if they are defective the whole assembly must be replaced.

Wheel Bearings

ADJUSTMENT

The front and rear wheel bearing assemblies on all models are pressed in and are not adjustable. If the bearing assembly does not turn smoothly or has more than 0.002 in. (0.05mm) of axial play, replace the bearing assembly.

REMOVAL & INSTALLATION

Front

G20 MODELS

1. Before servicing the vehicle, refer to the precautions in the beginning of this section.

2. The axle nut torque is very high and should be loosened and tightened with the vehicle on the ground. Remove the cotter pin, adjusting cap and insulator and loosen the front axle nut.

3. Remove or disconnect the following:
- Brake caliper, carrier, and the rotor. Hang the caliper from the body with wire; do not let it hang by the brake hose.
- Cotter pin and nut and use a ball joint press to disconnect the tie rod end
- Cap and the upper king pin mounting nut and separate the kingpin from the third link

4. Hold a block of wood against the axle stub and strike it with a hammer to release it from the hub. Withdraw the axle from the hub and fold the steering knuckle down on the ball joint.
- Cotter pin and nut and use a ball joint press to disconnect the ball joint
- Steering knuckle

➡**Wheel bearings must be replaced any time the hub is removed.**

5. Pry the grease seals out of the steering knuckle.

6. Support the steering knuckle and press the hub out of the bearing.

7. Remove the snaprings and press the bearing out towards the inside of the knuckle.

To install:

8. Be sure all parts are clean and dry. The hub and steering knuckle should be inspected for cracks using dye or a magnetic crack detection process.

9. Install or connect the following:

- Inner snapring and carefully press the new bearing into the steering knuckle. Be sure the press tool contacts only the outer bearing race or the bearing will be damaged.
- Outer snapring. Pack the new grease seals with clean grease and install them. If removed, install the splash guard.

10. Support the inner race on the press table and carefully press the hub into the bearing. Be sure the hub turns smoothly in both directions.

- Steering knuckle onto the lower ball joint and start the nut. Fit the

axle shaft through the hub and start the nut.

11. Pack the king pin bearing housing with grease and fit the third link into place. Torque the kingpin nut to 72–87 ft. lbs. (98–118 Nm) and install the dust cap.

12. Torque the lower ball joint nut to 52–64 ft. lbs. (71–86 Nm). Install a new cotter pin.

- Tie rod end. Torque the nut to 22–29 ft. lbs. (29–39 Nm). Torque as needed to install a new cotter pin but do not exceed 36 ft. lbs. (49 Nm).
- Brake caliper, carrier, rotor and the wheel

13. Lower the vehicle to the ground.

14. Torque the front axle nut to 174–231 ft. lbs. (235–314 Nm). Install the insulator, adjusting cap and cotter pin.

I30 AND I35 MODELS

➡**Whenever the hub or bearing assembly is removed, the wheel bearing assembly must be replaced. Never reuse the old bearing assembly.**

1. Before servicing the vehicle, refer to the precautions in the beginning of this section.

2. Remove the knuckle assembly from the vehicle by separating the ball joint and

tie rod end, then removing the retaining hardware securing the knuckle to the strut.

3. Using a shop press and a suitable tool, press the hub with the inner race from the steering knuckle.

4. Using a shop press and a suitable tool, press the bearing inner race from the hub and remove the outer grease seal.

5. Use snapring pliers to remove the snaprings from the steering knuckle.

6. Inspect the hub, steering knuckle and snaprings for cracks and/or wear; if necessary, replace the damaged part(s).

To install:

7. Install the inner snapring in the steering knuckle groove.

8. Using a shop press and a suitable tool, press the new wheel bearing assembly into the steering knuckle, until it seats, using a maximum pressure of 3 tons (2722kg).

9. Install the outer snapring.

10. Pack the new grease seal lips with multi-purpose grease.

11. Using a shop press and a suitable tool, press the new outer grease seal into the steering knuckle.

12. Using a shop press and a suitable tool, press the new inner grease seal into the steering knuckle.

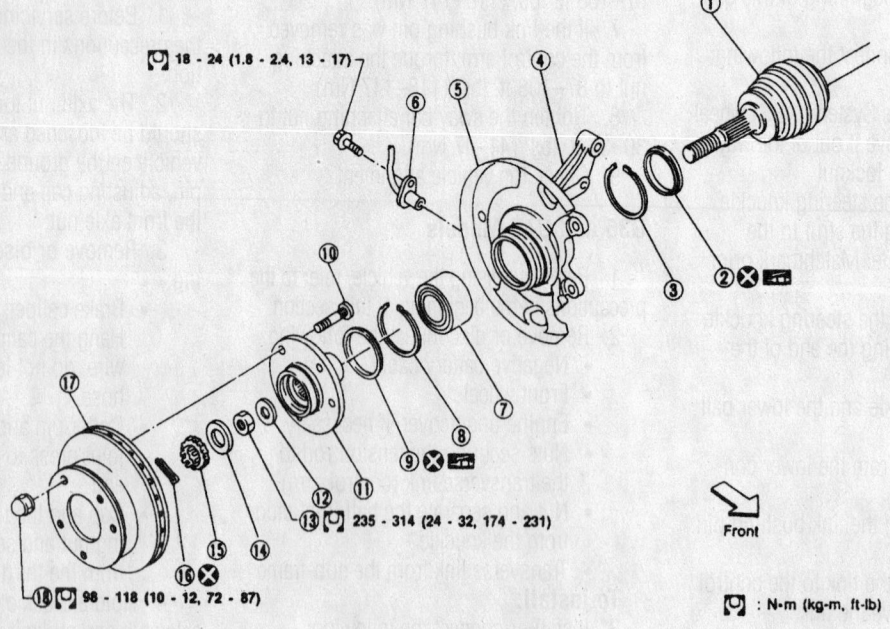

18 - 24 (1.8 - 2.4, 13 - 17)

235 - 314 (24 - 32, 174 - 231)

98 - 118 (10 - 12, 72 - 87)

Front

⊡ : N•m (kg-m, ft-lb)

① Drive shaft	⑦ Wheel bearing assembly	⑬ Wheel bearing lock nut
② Inner grease seal	⑧ Snap ring	⑭ Insulator
③ Snap ring	⑨ Outer grease seal	⑮ Adjusting cap
④ Knuckle	⑩ Hub bolt	⑯ Cotter pin
⑤ Baffle plate	⑪ Wheel hub	⑰ Disc rotor
⑥ ABS sensor	⑫ Plain washer	⑱ Wheel nut

Exploded view of the front knuckle assembly—I30 models

7923HG87

13. Using a shop press and a suitable tool, press the hub into the steering knuckle, until it seats, using a maximum pressure of 5.5 tons (4990kg); be careful not to damage the grease seal.

14. To check the bearing operation, perform the following procedures:

a. Increase the press pressure to 3.5–5.0 tons (3175–4536kg).

b. Spin the steering knuckle, several turns, in both directions.

c. Be sure the wheel bearings operate smoothly.

15. If the wheel bearings do not operate smoothly, replace the wheel bearing assembly.

16. Install the knuckle assembly.

17. Install the halfshaft into the hub. Torque the locknut to 174–231 ft. lbs. (235–314 Nm) for I30 models, or to 188–245 ft. lbs. (255–333 Nm) for I35 models.

18. Install the wheel assembly and lower the vehicle.

19. Road test the vehicle and verify proper operation.

G35 AND Q45 MODELS

1. Before servicing the vehicle, refer to the precautions in the beginning of this section.

2. Support the hub assembly with a suitable jack.

3. Remove or disconnect the following:

- Brake caliper, carrier and rotor. Hang the caliper from the body with wire, do not let it hang by the brake hose.
- Cotter pins and nuts and use a ball joint press to disconnect the lower ball joint and tie rod end
- Kingpin lower mounting nut to remove the steering knuckle assembly

➡Wheel bearings must be replaced any time the hub is removed.

4. Use a vise or a wheel to hold the hub and remove the hub cap and nut from the back of the hub. Remove the wheel speed sensor rotor.

5. Use a press or large drift pin to press the hub out of the steering knuckle.

6. Remove the snapring and press the bearings and grease seal out of the steering knuckle.

To install:

7. Be sure all parts are clean. Carefully press the new bearing into the steering knuckle. Be sure the press tool contacts only the outer bearing race or the bearing will be damaged.

8. Install a new grease seal and the snapring. If removed, install the splash guard.

9. Lightly lubricate the lips of the seal with clean grease. Be careful not to grease the bearing or hub mating surfaces.

10. Carefully press the hub into the bearing. Support the inner race on the press table or the bearing will be damaged. Do not exceed 3.9 tons (3538kg) pressure.

11. Install or connect the following:

- Speed sensor rotor and nut on the hub and torque to 152–210 ft. lbs. (206–284 Nm). Stake the nut into place.

12. Lightly tap the cap into place and install the bolts.

- Steering knuckle to the king pin. Torque the nut to 108–137 ft. lbs. (88–108 Nm).
- Lower ball joint. Torque the nut to 65–80 ft. lbs. (88–108 Nm). Install a new cotter pin.
- Tie rod end. Torque the nut to 22–29 ft. lbs. (29–39 Nm). Torque as needed to install a new cotter pin but do not exceed 36 ft. lbs. (49 Nm).
- Rotor and the brake caliper
- Wheel and tire assembly

Rear

G20 MODELS

1. Before servicing the vehicle, refer to the precautions in the beginning of this section.

2. Raise and support the vehicle safely.

3. Remove or disconnect the following:

- Rear caliper and rotor. Hang the caliper from the body with wire, do not let hang by the brake hose.
- Rear wheel hub cap, cotter pin and locknut
- Hub off the stub axle

➡The wheel bearing is integral with the hub and cannot be serviced separately.

To install:

4. Install or connect the following:

- New hub assembly onto the axle stub

5. Replace the washer and wheel bearing locknut. Torque the locknut to 137–174 ft lbs. (186–235 Nm). Install a new cotter pin.

- Brake rotor, caliper and wheel

6. Lower the vehicle to the ground.

I30 MODELS

➡If the vehicle is equipped with Anti-lock Brakes (ABS), the sensor must be removed to protect the sensor and its wiring.

1. Before servicing the vehicle, refer to the precautions in the beginning of this section.

2. Remove or disconnect the following:

- Both rear wheels
- Wheel speed sensor
- Brake caliper and hang it by a piece of wire
- Brake caliper support
- Disc brake pads
- Brake disc
- Grease cap
- Cotter pin, wheel bearing locknut, washer, and the wheel hub bearing assembly. A slide hammer may be needed to remove the hub bearing assembly.

➡The wheel hub bearing assembly is not repairable; it must be replaced when defective.

To install:

3. Install or connect the following:

- Wheel hub bearing assembly, the washer and the wheel bearing locknut. Torque the wheel bearing locknut to 137–188 ft. lbs. (186–255 Nm).

4. Verify that the wheel bearings operate smoothly.

- New cotter pin into the spindle to hold the wheel bearing locknut

5. Install a dial micrometer to the rear wheel hub bearing assembly and check the axial end-play; it should be less than 0.0020 in. (0.05mm).

- Grease cap
- ABS wheel sensor and its wiring
- Brake assembly and the wheels

G35 AND Q45 MODELS

1. Before servicing the vehicle, refer to the precautions in the beginning of this section.

2. Remove or disconnect the following:

- Cotter pin and adjusting cap and loosen the wheel bearing nut. Carefully tap the end of the axle shaft or use a puller to loosen the shaft from the hub.
- Brake caliper and rotor. Do not let the caliper hang by the brake hose, support it with wire.
- Parking brake assembly
- Nuts and through-bolts to remove the axle housing from the suspension. If equipped with rear wheel steering, use a ball joint press to separate the tie rod end.
- 4 bolts at the back and remove the bearing flange and hub from the bearing housing

3. Press the hub out of the bearing flange and use a puller to remove the bearing from the hub. If it is not damaged, the hub can be used again but the bearing and flange are supplied as a single unit.

To install:

→ **The wheel bearing and flange are supplied as an assembly.**

4. Place the hub on a press table and press the new bearing and flange onto the hub. Be sure the press tool contacts only the inner bearing race and take care not to damage the seal.

5. Install or connect the following:
- Bearing flange onto the axle housing. Torque the bolts to 58–72 ft. lbs. (78–98 Nm).
- Axle housing onto the lower ball joint, torque the nut to 58–69 ft. lbs. (78–93 Nm) and install a new cotter pin
- Torque the tie rod end nut to 33–44 ft. lbs. (45–60 Nm) and install a new cotter pin, if equipped with rear wheel steering
- Axle shaft into the hub and install the bolts through the suspension bushings. Tighten the bolts temporarily, they will be tightened with the vehicle resting on the wheels.
- Brake components and apply the brake to hold the hub from turning
- Wheel bearing locknut and torque it to 152–203 ft. lbs. (206–275 Nm)
- Insulator and adjusting cap and a new cotter pin
- Wheel and lower the vehicle to the ground. Torque the suspension bushing bolts to 57–72 ft. lbs. (77–98 Nm).

BRAKES

Brake Caliper

REMOVAL & INSTALLATION

G20 Models

FRONT

1. Before servicing the vehicle, refer to the precautions in the beginning of this section.
2. Remove the wheels.
3. Loosen the brake hose connecting bolt.
4. Remove the bolts connecting the caliper to the torque member.
5. Slide the caliper out from the rotor.
6. Remove the brake hose connecting bolt from the caliper.
7. Remove the caliper from the vehicle.

To install:

8. Fit the caliper onto the torque member and torque the bolts to 16–23 ft. lbs. (22–31 Nm).
9. Using new copper washers, connect the hydraulic hose to the caliper. Torque the union bolt to 12–14 ft. lbs. (17–19 Nm).
10. Bleed the air from the system.

REAR

1. Before servicing the vehicle, refer to the precautions in the beginning of this section.
2. Remove the rear wheels.
3. Remove the brake cable mounting bolt and lock spring.
4. Disconnect the parking brake cable from the caliper.
5. Disconnect the brake fluid hose.
6. Remove the torque member mounting bolts and remove the caliper assembly.

To install:

7. Fit the caliper onto the torque member and torque the bolts to 16–23 ft. lbs. (22–31 Nm).

8. Using new copper washers, connect the hydraulic hose to the caliper. Torque the union bolt to 12–14 ft. lbs. (17–19 Nm).
9. Connect the parking brake cable to the rear caliper.
10. Bleed the brake system.
11. Install the rear wheels.

130, 135 and G35 Models

FRONT

1. Before servicing the vehicle, refer to the precautions in the beginning of this section.
2. Remove the front wheels.
3. Remove both guide pin bolts securing the caliper to the steering knuckle.
4. Loosen and remove the brake hose connector from the caliper.
5. Remove the caliper assembly from the vehicle.

To install:

6. Using new copper washers, install the brake line to the brake caliper and torque the connecting bolt to 12–14 ft. lbs. (17–20 Nm).
7. Install the caliper to the steering knuckle using the guide pins bolts.
8. Install the wheels and tighten the lug nuts to the proper specification.
9. Bleed the brake system and top off the master cylinder as necessary.

REAR

1. Before servicing the vehicle, refer to the precautions in the beginning of this section.
2. Remove the rear wheels.
3. Remove the parking brake cable stay fixing bolt and the lock spring.
4. Remove the brake fluid hose from the caliper.
5. Remove the guide pin bolts and remove the caliper.

To install:

6. Install the caliper body into position and torque the caliper-to-torque member pin bolts to 16–23 ft. lbs. (22–31 Nm).
7. Reconnect the brake fluid hose and tighten the flare nut to 12–14 ft. lbs. (17–20 Nm).
8. Install the lock spring and the parking brake stay attaching bolt.
9. Bleed the brake system and top off the master cylinder as necessary.
10. Install the wheels.

Q45 Models

FRONT

1. Before servicing the vehicle, refer to the precautions in the beginning of this section.
2. Remove the wheels.
3. Remove the brake hose connecting bolt.
4. Remove the torque member mounting bolts and disconnect the brake fluid hose from the caliper.
5. Slide the caliper off of the rotor.
6. Remove the caliper from the vehicle.

To install:

7. Position the torque member on the knuckle assembly and install the mounting bolts. Torque the bolts to 118–137 ft. lbs. (160–186 Nm).
8. Using new copper washers, connect the hydraulic hose to the caliper. Torque the union bolt to 12–14 ft. lbs. (17–19 Nm).
9. Bleed the air from the system and fill the master cylinder with clean brake fluid.

REAR

1. Before servicing the vehicle, refer to the precautions in the beginning of this section.
2. Remove the rear wheels.
3. Disconnect the brake fluid hose.

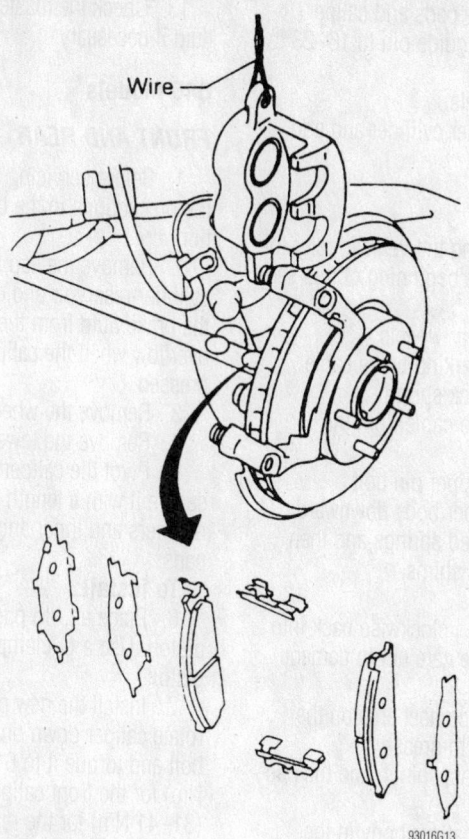

Front brake caliper—Q45 shown

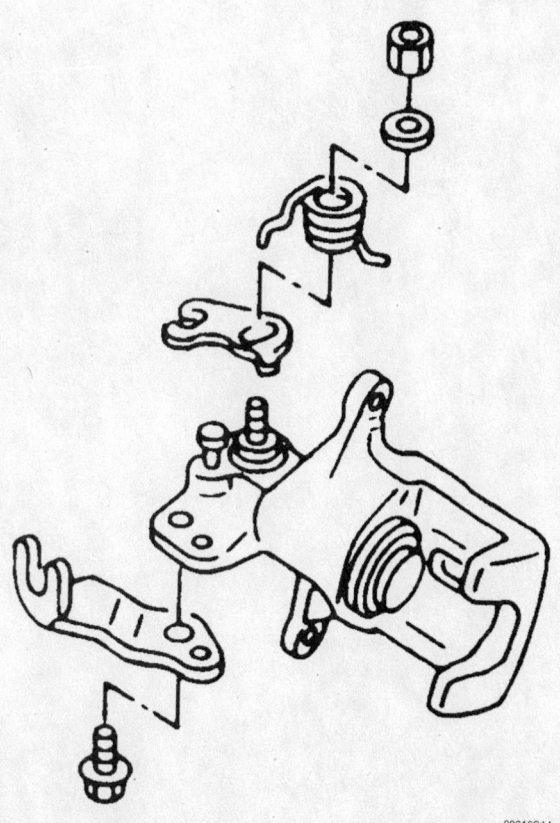

Rear caliper parking brake lever—G20 shown

93016G14

4. Remove the torque member mounting bolts and remove the caliper assembly.

To install:

5. Fit the caliper over the rotor and install the mounting bolts. Torque the bolts to 28–38 ft. lbs. (38–52 Nm).

6. Using new copper washers, connect the hydraulic hose to the caliper. Torque the union bolt to 12–14 ft. lbs. (17–19 Nm).

7. Bleed the brake system.

8. Install the rear wheels and lower the vehicle to the floor.

Disc Brake Pads

REMOVAL & INSTALLATION

G20 Models

FRONT

1. Before servicing the vehicle, refer to the precautions in the beginning of this section.

2. Remove the cap from the master cylinder reservoir and extract about ⅓ of the brake fluid from the reservoir to prevent overflow when the caliper piston is compressed.

3. Remove the wheels.

4. Remove the lower pin bolt.

5. Pivot the caliper body upward and secure it with a length of wire. Remove the retainers and inner and outer shims and pads.

To install:

6. Place an old pad over the caliper piston. Use a C-clamp to compress the piston.

7. Install the new pads and shims and rotate caliper down onto rotor. Install the pin bolt and torque it to 16–23 ft. lbs. (22–31 Nm).

8. Install the wheels and lower the vehicle to the floor.

9. Check and then refill the master cylinder if needed.

REAR

1. Before servicing the vehicle, refer to the precautions in the beginning of this section.

2. Remove the cap from the master cylinder reservoir and extract about ⅓ of the brake fluid from the reservoir to prevent overflow when the caliper piston is compressed.

3. Remove the wheels.

4. Remove the brake cable mounting bracket bolt and lock spring.

5. Disconnect the parking brake cable.

6. Remove the lower pin bolt.

7. Pivot the caliper body upward and

secure it with a length of wire. Remove the retainers and inner and outer shims and pads.

To install:

8. Push the piston into the cylinder body by turning the piston clockwise.

9. Install the new pads and shims and rotate the caliper down onto rotor. Install the pin bolt and torque it to 16–23 ft. lbs. (22–31 Nm).

10. Connect the parking brake cable and install the bracket.

11. Install the wheels.

12. Check and then refill the master cylinder if needed.

I30, I35 and G35 Models

FRONT

1. Before servicing the vehicle, refer to the precautions in the beginning of this section.

2. Remove the wheels.

3. Remove the bottom guide pin from the caliper and swing the caliper cylinder body upward. Support the caliper with a wire.

4. Remove the brake pad retainers and the pads.

To install:

5. Compress the piston of the disc brake caliper.

6. Install the brake pads and caliper assembly. Torque the guide pin to 16–23 ft. lbs. (22–31 Nm).

7. Install the wheels.

8. Check the master cylinder and add fluid if necessary.

REAR

1. Before servicing the vehicle, refer to the precautions in the beginning of this section.

2. Remove the rear wheels.

3. Remove the parking brake cable mounting bolt and lock spring.

4. Disconnect the cable from the caliper.

5. Remove the upper pin bolt.

6. Pivot the caliper body downward.

7. Pull out the pad springs and then remove the pads and shims.

To install:

8. Turn the piston clockwise back into the caliper body. Take care not to damage the piston boot.

9. Coat the pad contact area on the mounting support with grease.

10. Install the pads, shims, and the pad springs.

11. Position the caliper body in the mounting support and tighten the pin bolts to 16–23 ft. lbs. (22–31 Nm).

12. Install the wheels.

13. Check the master cylinder and add fluid if necessary.

Q45 Models

FRONT AND REAR

1. Before servicing the vehicle, refer to the precautions in the beginning of this section.

2. Remove the cap from the master cylinder reservoir and extract about ⅓ of the brake fluid from the reservoir to prevent overflow when the caliper piston is compressed.

3. Remove the wheels.

4. Remove the lower pin bolt.

5. Pivot the caliper body upward and secure it with a length of wire. Remove the retainers and inner and outer shims and pads.

To install:

6. Place an old pad over the caliper piston. Use a C-clamp to compress the piston.

7. Install the new pads and shims and rotate caliper down onto rotor. Install the pin bolt and torque it to 61–69 ft. lbs. (83–93 Nm) for the front caliper and 23–30 ft. lbs. (31–41 Nm) for the rear caliper.

8. Install the wheels.

9. Check and refill master cylinder if needed.

ISUZU

Hombre

11

SPECIFICATION CHARTS

ENGINE AND VEHICLE IDENTIFICATION

Engine							Model Year	
Code ①	Liters (cc)	Cu. In.	Cyl.	Fuel Sys.	Engine Type	Eng. Mfg.	Code ②	Year
4	2.2 (2189)	134	4	MFI	OHV	CPC	Y	2000
W	4.3 (4293)	263	6	MFI	OHV	CPC		

CPC: Chevrolet/Pontiac/Canada

MFI: Multi-port Fuel Injection

① 8th position of VIN

② 10th position of VIN

42356-HOMB-C01

GENERAL ENGINE SPECIFICATIONS

All measurements are given in inches.

Year	Model	Engine Displacement Liters (cc)	Engine Series (ID/VIN)	Fuel System	Net Horsepower @ rpm	Net Torque @ rpm (ft. lbs.)	Bore x Stroke (in.)	Compression Ratio	Oil Pressure @ rpm
2000	Hombre	2.2 (2189)	4	MFI	118@5200	130@2800	3.50x3.46	9.0:1	56@3000
		4.3 (4293)	W	MFI	205@5400	214@3000	4.00x3.48	9.1:1	60-80@3000

MFI: Multi-port Fuel Injection

42356-HOMB-C02

GASOLINE ENGINE TUNE-UP SPECIFICATIONS

Year	Engine Displacement Liters (cc)	Engine ID/VIN	Spark Plugs Gap (in.)	Ignition Timing (deg.)		Fuel Pump (psi)	Idle Speed (rpm)		Valve Clearance	
				MT	AT		MT	AT	In.	Ex.
2000	2.2 (2189)	4	0.060	①	①	41-47	②	②	HYD	HYD
	4.3 (4293)	W	0.060	①	①	58-64 ③	600	625	HYD	HYD

NOTE: The Vehicle Emission Control Information label often reflects specification changes made during production. The label figures must be used if they differ from those in this chart.

HYD: Hydraulic

① Ignition timing is preset and cannot be adjusted

② Idle speed is maintained by the PCM

③ With key ON and engine OFF

42356-HOMB-C03

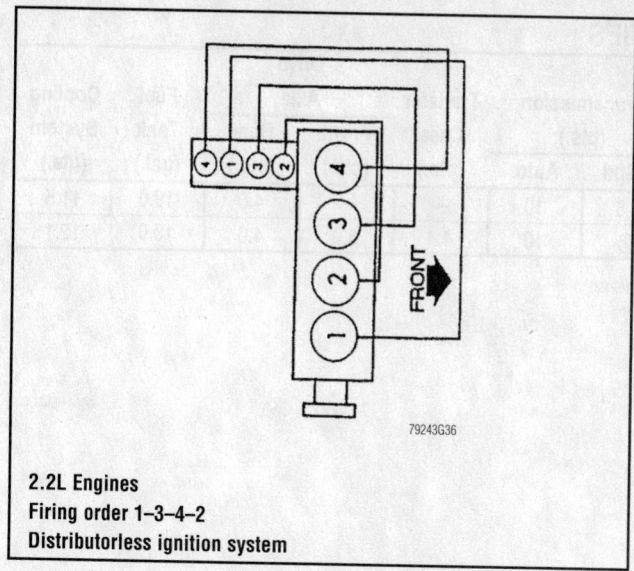

2.2L Engines
Firing order 1–3–4–2
Distributorless ignition system

79243G36

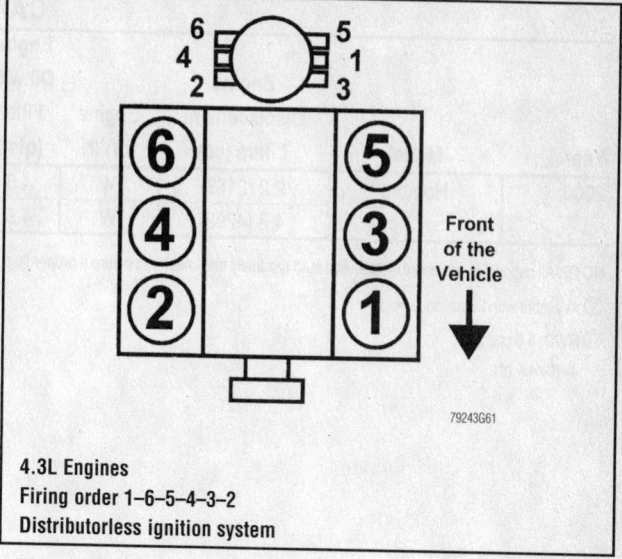

Front of the Vehicle

4.3L Engines
Firing order 1–6–5–4–3–2
Distributorless ignition system

79243G61

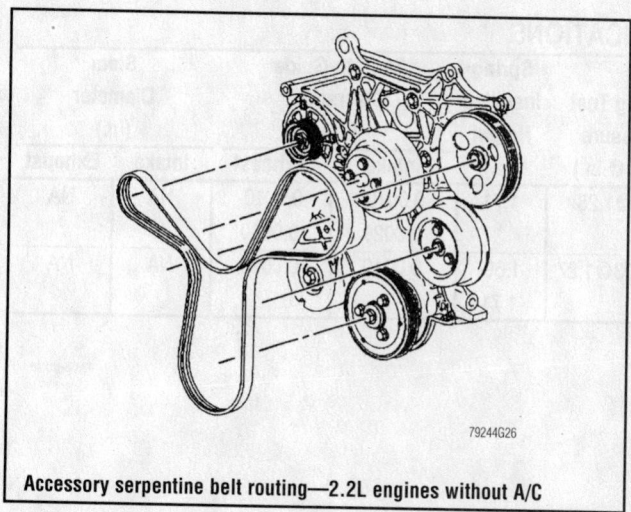

79244G26

Accessory serpentine belt routing—2.2L engines without A/C

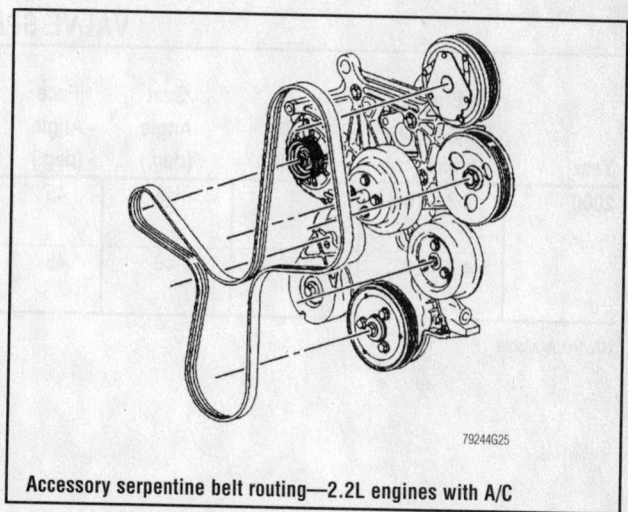

79244G25

Accessory serpentine belt routing—2.2L engines with A/C

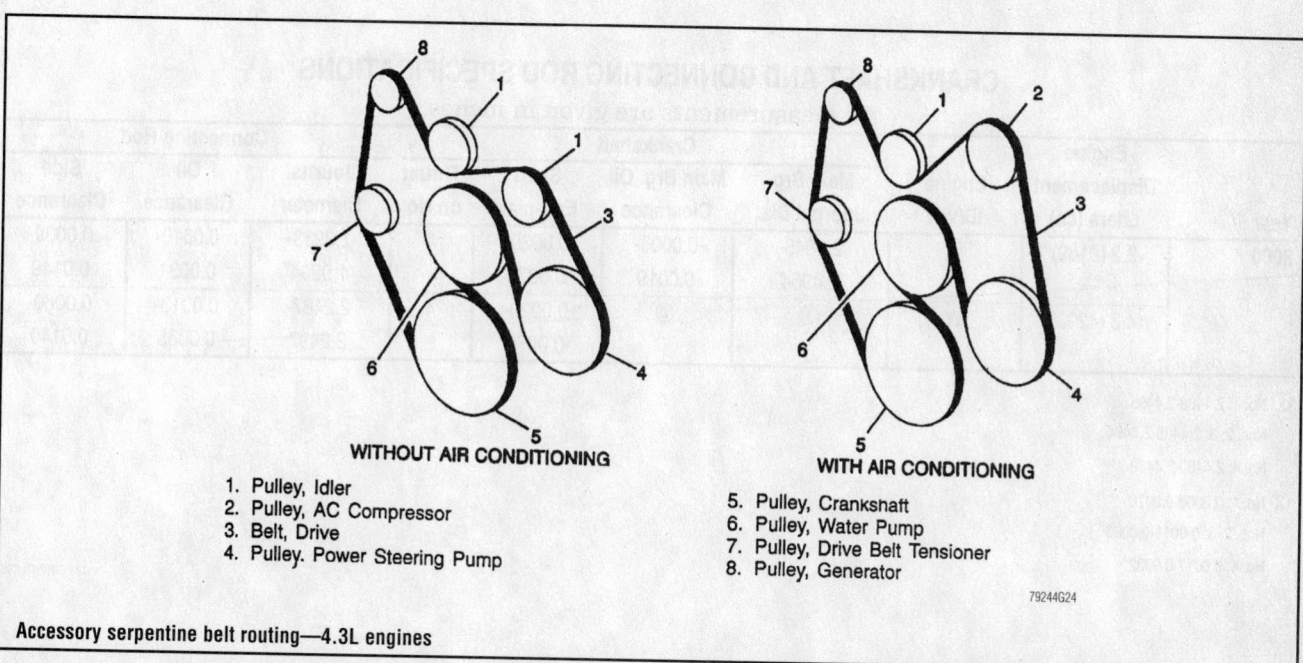

WITHOUT AIR CONDITIONING

WITH AIR CONDITIONING

1. Pulley, Idler
2. Pulley, AC Compressor
3. Belt, Drive
4. Pulley, Power Steering Pump

5. Pulley, Crankshaft
6. Pulley, Water Pump
7. Pulley, Drive Belt Tensioner
8. Pulley, Generator

79244G24

Accessory serpentine belt routing—4.3L engines

CAPACITIES

Year	Model	Engine Displacement Liters (cc)	Engine ID/VIN	Engine Oil with Filter (qts.)	Transmission (pts.) 5-Spd	Transmission (pts.) Auto.	Transfer Case (pts.)	Drive Axle Front (pts.)	Drive Axle Rear (pts.)	Fuel Tank (gal.)	Cooling System (qts.)
2000	Hombre	2.2 (2189)	4	4.5	5.8	10	—	—	4.0	19.0	11.5
		4.3 (4293)	W	4.5	②	10	4.4	3.0	4.0	19.0	12.1

NOTE: All capacities are approximate. Add fluid gradually and check to be sure a proper fluid level is obtained.

① Available with 20 gallon tank

② RWD: 5.0 pts.
   4WD: 4.4 pts.

42356-HOMB-C04

VALVE SPECIFICATIONS

Year	Engine Displacement Liters (cc)	Engine ID/VIN	Seat Angle (deg.)	Face Angle (deg.)	Spring Test Pressure (lbs. @ in.)	Spring Installed Height (in.)	Stem-to-Guide Clearance (in.) Intake	Stem-to-Guide Clearance (in.) Exhaust	Stem Diameter (in.) Intake	Stem Diameter (in.) Exhaust
2000	2.2 (2189)	4	46	45	228@1.28	1.71	0.0010-0.0020	0.0010-0.0030	NA	NA
	4.3 (4293)	W	46	45	187-203@1.27	1.69-1.71	0.0010	0.0020	NA	NA

NA: Not Available

42356-HOMB-C05

CRANKSHAFT AND CONNECTING ROD SPECIFICATIONS

All measurements are given in inches.

Year	Engine Displacement Liters (cc)	Engine ID/VIN	Crankshaft Main Brg. Journal Dia.	Crankshaft Main Brg. Oil Clearance	Crankshaft Shaft End-play	Thrust on No.	Connecting Rod Journal Diameter	Connecting Rod Oil Clearance	Connecting Rod Side Clearance
2000	2.2 (2189)	4	2.4945-2.4954	0.0006-0.0019	0.0020-0.0070	4	1.9983-1.9994	0.0010-0.0031	0.0039-0.0149
	4.3 (4293)	W	①	②	0.0020-0.0070	4	2.2487-2.2497	0.0013-0.0035	0.0060-0.0140

① No. 1: 2.4488-2.4495
   Nos. 2, 3: 2.4485-2.4494
   No. 4: 2.4480-2.4489

② No. 1: 0.0008-0.0020
   Nos. 2, 3: 0.0011-0.0023
   No. 4: 0.0017-0.0032

42356-HOMB-C06

PISTON AND RING SPECIFICATIONS
All measurements are given in inches.

Year	Engine Displacement Liters (cc)	Engine ID/VIN	Piston Clearance	Ring Gap			Ring Side Clearance		
				Top Compression	Bottom Compression	Oil Control	Top Compression	Bottom Compression	Oil Control
2000	2.2 (2189)	4	0.0007-0.0017	0.010-0.020	0.010-0.020	0.010-0.03	0.0019-0.0027	0.0019-0.0027	0.0019-0.0082
	4.3 (4293)	W	0.0007-0.0017	0.010-0.030	0.018-0.026	0.065 Max.	0.0042 Max.	0.0042 Max.	0.0020-0.0070

42356-HOMB-C07

TORQUE SPECIFICATIONS
All readings in ft. lbs.

Year	Engine Displacement Liters (cc)	Engine ID/VIN	Cylinder Head Bolts	Main Bearing Bolts	Rod Bearing Bolts	Crankshaft Damper Bolts	Flywheel Bolts	Manifold		Spark Plugs	Lug Nut
								Intake *	Exhaust		
2000	2.2 (2189)	4	①	70	38	77	55	②	10	11	100
	4.3 (4293)	W	③	77	④	74	74	⑤	⑥	11	100

* NOTE: Applies to Lower Manifold only.

① Short bolts: 43 ft. lbs. plus 90 degrees
Long bolts: 46 ft. lbs. plus 90 degrees

② Lower intake manifold nuts: 24 ft. lbs.
Lower intake manifold studs: 22 ft. lbs.
Upper intake manifold bolts: 22 ft. lbs.

③ 1st pass: 22 ft. lbs.
2nd pass:
Short bolt: Plus 55 degrees
Medium bolt: Plus 65 degrees
Long bolt: Plus 75 degrees

④ 20 ft. lbs. plus 70 degrees

⑤ Lower intake manifold:
1st pass: 27 inch lbs.
2nd pass: 106 inch lbs.
Final pass: 11 ft. lbs.
Upper manifold bolts:
1st pass: 44 inch lbs.
2nd pass: 88 inch lbs.

⑥ Tighten bolts to 12 ft. lbs.
Retorque to 22 ft. lbs.

42356-HOMB-C08

WHEEL ALIGNMENT

Year	Model	Caster		Camber		Toe-in (in.)	Steering Axis Inclination (Deg.)
		Range (+/-Deg.)	Preferred Setting (Deg.)	Range (+/-Deg.)	Preferred Setting (Deg.)		
2000	exc. ZR2/Z85 ZM6/G51	1.0	+3.0	1.0	0	0.10+/-0.10	—
	ZR2/Z85 ZM6/G51	1.0	+2.0	1.0	0	0.10+/-0.10	—

42356-HOMB-C09

TIRE, WHEEL AND BALL JOINT SPECIFICATIONS

Year	Model	OEM Tires		Tire Pressures (psi)		Wheel Size	Ball Joint Inspection
		Standard	Optional	Front	Rear		
2000	Hombre 2wd	P205/75R15	none	35	35	6-JJ	U: 4-28 ① L: 4-55
	Hombre 4wd	P235/75R15	31x10.5R15LT	35	35	6-JJ	U: 4-28 ① L: 4-55

OEM: Original Equipment Manufacturer

PSI: Pounds Per Square Inch

STD: Standard

OPT: Optional

L: Lower

U: Upper

① Do not lift truck. Inspect the boss into which the grease fitting is threaded. Replace if the boss is flush or receded below the surface of the ball joint

42356-HOMB-C10

BRAKE SPECIFICATIONS

All measurements in inches unless noted

Year	Model		Brake Disc			Brake Drum Diameter			Minimum Lining Thickness	Brake Caliper	
			Original Thickness	Minimum Thickness	Maximum Runout	Original Inside Diameter	Max. Wear Limit	Maximum Machine Diameter		Bracket Bolts (ft. lbs.)	Mounting Bolts (ft. lbs.)
2000	Hombre	F	1.027 ①	0.965 ②	0.004	—	—	—	0.030	—	38
		R	0.787	0.735	0.004	9.50	9.59	9.56	0.030	52	23

NA: Not Available

① Heavy duty models: 1.140 in.

② Heavy duty models: 1.080 in.

42356-HOMB-C11

SCHEDULED MAINTENANCE INTERVALS
ISUZU—HOMBRE

TO BE SERVICED	TYPE OF SERVICE	VEHICLE MILEAGE INTERVAL (x1000)															
		7.5	15	22.5	30	37.5	45	52.5	60	67.5	75	82.5	90	97.5	105	112.5	120
Accelerator linkage ①	L	✓	✓	✓	✓	✓	✓	✓	✓	✓	✓	✓	✓	✓	✓	✓	✓
Accessory drive belts ②	S/I				✓				✓				✓				✓
Air cleaner filter	R				✓				✓				✓				✓
Auto cruise control linkage & hose ③	S/I		✓		✓		✓		✓		✓		✓		✓		✓
Automatic transmission fluid level ③	S/I	✓		✓		✓		✓		✓		✓		✓		✓	
Battery fluid level ③	S/I	✓	✓	✓	✓	✓	✓	✓	✓	✓	✓	✓	✓	✓	✓	✓	✓
Body and chassis ①	L	✓	✓	✓	✓	✓	✓	✓	✓	✓	✓	✓	✓	✓	✓	✓	✓
Brake fluid level ③	S/I	✓	✓	✓	✓	✓	✓	✓	✓	✓	✓	✓	✓	✓	✓	✓	✓
Brake lines & hoses ③	S/I	✓	✓	✓	✓	✓	✓	✓	✓	✓	✓	✓	✓	✓	✓	✓	✓
Brake pedal play ③	S/I		✓		✓		✓		✓		✓		✓		✓		✓
Clutch fluid level ③	S/I	✓	✓	✓	✓	✓											
Clutch lines & hose ③	S/I				✓				✓				✓				✓
Clutch pedal free-play ③	S/I		✓		✓		✓		✓		✓		✓		✓		✓
Clutch pedal spring, bushing and clevis pin ①	S/I		✓		✓		✓		✓		✓		✓				✓
Cooling and heating system hoses ③	S/I		✓		✓		✓		✓		✓		✓		✓		✓
Driveshaft flange torque ③	S/I	✓		✓		✓		✓		✓		✓		✓		✓	
Drum and disc brakes ③	S/I		✓		✓		✓		✓		✓		✓		✓		✓
Engine coolant	R				✓				✓				✓				✓
Engine coolant level ③	S/I	✓	✓	✓	✓	✓	✓	✓	✓	✓	✓	✓	✓	✓	✓	✓	✓
Engine oil & filter ③	R	✓	✓	✓	✓	✓	✓	✓	✓	✓	✓	✓	✓	✓	✓	✓	✓
Exhaust system ③	S/I	✓	✓	✓	✓	✓	✓	✓	✓	✓	✓	✓	✓	✓	✓	✓	✓
Front and rear axle lubricant	R		✓		✓				✓				✓				✓
Front and rear driveshafts ①	S/I	✓	✓	✓	✓	✓	✓	✓	✓	✓	✓	✓	✓	✓	✓	✓	✓
Front wheel bearings	S/I & L				✓				✓				✓				✓
Fuel lines & tank cap ③	S/I								✓								✓
Inspect for fluid leaks ③	S/I	✓	✓	✓	✓	✓	✓	✓	✓	✓	✓	✓	✓	✓	✓	✓	✓
Key lock cylinder ③	L		✓		✓		✓		✓		✓		✓		✓		✓
Parking brake system ③	S/I		✓		✓		✓		✓		✓		✓		✓		✓
Power steering fluid	R				✓				✓				✓				✓
Radiator core and A/C condenser	S/I & C								✓								✓
Rotate tires	S/I	✓	✓	✓	✓	✓	✓	✓	✓	✓	✓	✓	✓	✓	✓	✓	✓
Shift-on-the-fly system gear fluid ③	S/I		✓		✓		✓		✓		✓		✓		✓		✓
Spark plugs	R	Every 100,000 miles															
Starter safety switch ③	S/I	✓	✓	✓	✓	✓	✓	✓	✓	✓	✓	✓	✓	✓	✓	✓	✓

SCHEDULED MAINTENANCE INTERVALS
ISUZU—HOMBRE

TO BE SERVICED	TYPE OF SERVICE	VEHICLE MILEAGE INTERVAL (x1000)															
		7.5	15	22.5	30	37.5	45	52.5	60	67.5	75	82.5	90	97.5	105	112.5	120
Steering operation ③	S/I	✓	✓	✓	✓	✓	✓	✓	✓	✓	✓	✓	✓	✓	✓	✓	✓
Suspension & steering ③	S/I	✓	✓	✓	✓	✓	✓	✓	✓	✓	✓	✓	✓	✓	✓	✓	✓
Throttle linkage ③	S/I		✓		✓		✓				✓		✓		✓		✓
Timing belt	R										✓						
Tires and wheels ③	S/I	✓	✓	✓	✓	✓	✓	✓	✓	✓	✓	✓	✓	✓	✓	✓	✓

R: Replace   S/I: Service or Inspect   L: Lubricate   A: Adjust   C: Clean

① Perform this at the mileage indicated or every 6 months, whichever occurs first.

② Perform this at the mileage indicated or every 24 months, whichever occurs first.

③ Perform this at the mileage indicated or every 12 months, whichever occurs first.

FREQUENT OPERATION MAINTENANCE (SEVERE SERVICE)

If a vehicle is operated under any of the following conditions it is considered severe service:

- Towing a trailer or using a camper or car-top carrier.

- Repeated short trips of less than 5 miles in temperatures below freezing.

- Extensive idling or low-speed driving for long distances as in heavy commercial use, such as delivery, taxi or police cars.

- Operating on rough, muddy or salt-covered roads.

- Operating on unpaved or dusty roads.

Air cleaner element: replace every 15,000 miles

Engine oil and filter: replace every 3000 miles or 3 months, whichever occurs first.

Automatic transmission fluid: replace every 20,000 miles.

Rear axle lubricant: replace every 15,000 miles.

42356-HOMB-C13

PRECAUTIONS

Before servicing any vehicle, please be sure to read all of the following precautions, which deal with personal safety, prevention of component damage, and important points to take into consideration when servicing a motor vehicle:

• Never open, service or drain the radiator or cooling system when the engine is hot; serious burns can occur from the steam and hot coolant.

• Observe all applicable safety precautions when working around fuel. Whenever servicing the fuel system, always work in a well-ventilated area. Do not allow fuel spray or vapors to come in contact with a spark, open flame or excessive heat (a hot drop light, for example). Keep a dry chemical fire extinguisher near the work area. Always keep fuel in a container specifically designed for fuel storage; also, always properly seal fuel containers to avoid the possibility of fire or explosion. Refer to the additional fuel system precautions later in this section.

• Fuel injection systems often remain pressurized, even after the engine has been turned **OFF**. The fuel system pressure must be relieved before disconnecting any fuel lines. Failure to do so may result in fire and/or personal injury.

• Brake fluid often contains polyglycol ethers and polyglycols. Avoid contact with the eyes and wash your hands thoroughly after handling brake fluid. If you do get brake fluid in your eyes, flush your eyes with clean, running water for 15 minutes. If eye irritation persists, or if you have taken brake fluid internally, IMMEDIATELY seek medical assistance.

• The EPA warns that prolonged contact with used engine oil may cause a number of skin disorders, including cancer! You should make every effort to minimize your exposure to used engine oil. Protective gloves should be worn when changing oil. Wash your hands and any other exposed skin areas as soon as possible after exposure to used engine oil. Soap and water, or waterless hand cleaner should be used.

• All new vehicles are now equipped with an air bag system. The system must be disabled before performing service on or around system components, steering column, instrument panel components, wiring and sensors. Failure to follow safety and disabling procedures could result in accidental air bag deployment, possible personal injury and unnecessary system repairs.

• Always wear safety goggles when working with, or around, the air bag system. When carrying a non-deployed air bag, be sure the bag and trim cover are pointed away from your body. When placing a non-deployed air bag on a work surface, always face the bag and trim cover upward, away from the surface. This will reduce the motion of the module if it is accidentally deployed. Refer to the additional air bag system precautions later in this section.

• Clean, high quality brake fluid from a sealed container is essential to the safe and proper operation of the brake system. You should always buy the correct type of brake fluid for your vehicle. If the brake fluid becomes contaminated, completely flush the system with new fluid. Never reuse any brake fluid. Any brake fluid that is removed from the system should be discarded. Also, do not allow any brake fluid to come in contact with a painted surface; it will damage the paint.

• Never operate the engine without the proper amount and type of engine oil; doing so WILL result in severe engine damage.

• Timing belt maintenance is extremely important! Many models utilize an interference-type, non-freewheeling engine. If the timing belt breaks, the valves in the cylinder head may strike the pistons, causing potentially serious (also time-consuming and expensive) engine damage.

• Disconnecting the negative battery cable on some vehicles may interfere with the functions of the on-board computer system(s) and may require the computer to undergo a relearning process once the negative battery cable is reconnected.

• When servicing drum brakes, only disassemble and assemble one side at a time, leaving the remaining side intact for reference.

• Only an MVAC-trained, EPA-certified automotive technician should service the air conditioning system or its components.

ENGINE REPAIR

Alternator

REMOVAL

2.2L Engine

1. Before servicing the vehicle, refer to the precautions in the beginning of this section.

2. Remove or disconnect the following:

• Negative battery cable
• Passenger side wheel assembly
• Alternator brace-to-block bolt, the brace-to-intake nut and the brace-to-engine stud
• Alternator wiring
• Accessory belt
• Mounting bolts
• Alternator

4.3L Engine

1. Before servicing the vehicle, refer to the precautions in the beginning of this section.

2. Remove or disconnect the following:

• Negative battery cable
• Air inlet duct, if necessary
• Accessory belt
• Heater hose brace
• Wires
• Mounting bolts
• Alternator

INSTALLATION

2.2L Engine

Install or connect the following:
• Alternator
• Mounting bolts. Torque the left bolt to 22 ft. lbs. (30 Nm) and the right bolt to 32 ft. lbs. (43 Nm).
• Wires. Torque the battery feed wire nut to 71 inch lbs. (8 Nm).
• Alternator brace. Torque the nuts and bolts to 22 ft. lbs. (30 Nm).
• Accessory belt
• Negative battery cable

4.3L Engine

Install or connect the following:
• Alternator and loosely install the mounting bolts
• Tighten the rear bolt to 37 ft. lbs. (50 Nm) and the front bolt to 18 ft. lbs. (25 Nm)
• Tighten the brace-to-alternator and brace-to-intake retainers to 18 ft. lbs. (25 Nm). Tighten the brace-to-engine stud nut to 37 ft. lbs. (50 Nm).
• Wires and the battery feed wire nut
• Heater hose bracket

- Accessory belt
- Negative battery cable

Ignition Timing

ADJUSTMENT

The ignition timing is preset and cannot be adjusted.

Engine Assembly

REMOVAL & INSTALLATION

2.2L Engine

➡In certain cases on some models the A/C system will have to be evacuated because the compressor may need to be removed from the vehicle to allow clearance for engine removal. On other models you maybe able to set the compressor and lines to one side and still have enough clearance to remove the engine. In this case the system does not have to be evacuated because the lines do not have to be disconnected from the compressor. To check if your system has to be evacuated, unplug the electrical connectors from the compressor, then unbolt the compressor assembly. Unfasten any brackets holding the refrigerant lines and try to set the components aside so that you will have enough clearance for engine removal. If there is not enough clearance for engine removal you must recover the refrigerant from the A/C system with an approved recovery station before attempting to remove the engine from your vehicle. DO NOT attempt this without the proper equipment. R-134a should NOT be mixed with R-12 refrigerant and, depending on your local laws, attempting to service this system could be illegal.

1. Disconnect the negative battery cable and properly relieve the fuel system pressure.
2. Drain the engine cooling system and the engine oil into separate drain pans.
3. Remove or disconnect the following:
   - Hood
   - Oxygen (O2S) sensor electrical connection
   - Exhaust pipe from the manifold

➡On some models it may also be necessary to disconnect the catalytic converter from the exhaust pipe.

- Braces from the engine and the transmission, if equipped
- Starter motor

- Transmission and separate it from the engine or, if necessary, remove it from the vehicle
- Alternator rear brace by unfastening the bolt and nuts
- Ground straps from the engine block
- Drive belt
- A/C compressor and bracket. If possible, set the compressor and bracket to one side without disconnecting the lines.
- Hoses and transmission coolant lines engaged to the radiator
- Radiator
- Power steering pump and cap the power steering lines to avoid contamination
- Heater hoses from the heater core
- 12 volt supply from the mega fuse, if necessary
- All electrical connections and wiring harnesses
- All vacuum lines
- Throttle cable, and if equipped the cruise control cable
- Exhaust Gas Recirculation (EGR) pipe and the EGR valve
- Fuel lines

4. Install a suitable lifting device to the engine.
5. Remove the engine mount bolts and carefully lift the engine from the vehicle. Pause several times while lifting the engine to make sure no wires or hoses have become snagged.

To install:

6. Carefully lower the engine into the vehicle and install the engine mount bolts. Remove the engine lifting device.
7. Install or connect the following:
   - Fuel lines
   - 12 volt supply to the mega fuse, if removed
   - All vacuum lines, electrical connections and wiring harnesses
   - EGR valve and pipe, if removed
   - Throttle and if equipped, the cruise control cable
   - Heater hoses to the heater core
   - Power steering pump and attach the lines
   - A/C compressor
   - Radiator, all hoses and fluid cooler lines
   - Water pump, if removed
   - Drive belt
   - Ground strap to the engine
   - Alternator rear brace and tighten the bolt and nuts, if removed
   - Transmission to the engine
   - Starter motor, if removed

- Braces to the engine and the transmission, if equipped
- Exhaust pipe to the manifold
- Catalytic converter to the exhaust pipe, if removed
- O2S sensor electrical connection
- Battery
- Hood

8. Check all powertrain fluid levels and add, as necessary. Be sure to properly fill the engine crankcase with clean engine oil.
9. Connect the battery cables and properly fill the engine cooling system.
10. Start and run the engine, then check for leaks.

4.3L Engines

1. Before servicing the vehicle, refer to the precautions in the beginning of this section.
2. Drain the engine cooling system
3. Drain the engine oil.
4. Remove or disconnect the following:
   - Negative battery cable
   - Fuel system pressure
   - Vacuum reservoir and/or the under-hood light from the hood, as equipped
   - Outer cowl vent grilles
   - Hood
   - Oxygen (O2S) sensor and/or wiring
   - Exhaust pipes at the manifolds and loosen the hanger at the catalytic converter. This is necessary to remove the rear catalytic converter cushion mounts for removal of the exhaust assembly.
   - Skid plate, if equipped
   - Engine-to-transmission pencil braces
   - Slave cylinder and position aside, if equipped
   - Line clamp at the bell housing
   - Wiring from the starter
   - Starter
   - Transfer case
   - Oil filter
   - Engine mount through bolts
   - Rear engine mount crossbar, nut and washer
   - Bell housing bolts, except the upper left.
   - Battery ground (negative) cable from the engine
   - Front drive axle bolts and roll the axle downward, on 4WD vehicles
   - Air cleaner assembly
   - Upper radiator shroud
   - Fan assembly
   - Drive belt assembly
   - Water pump pulley

- Upper radiator hose
- Air conditioning compressor, if equipped, and position aside with the lines intact
- Lower radiator hose
- Oil cooler and overflow lines from the radiator, plug the openings to prevent system contamination or excessive fluid loss
- Radiator and lower radiator shroud
- Power steering hoses from the steering gear, then cap the openings to prevent system contamination or excessive fluid loss.
- Heater hoses from the intake manifold and the water pump
- Wiring harness and vacuum lines from the engine
- Throttle cables
- Remaining bell housing bolt
- Fuel lines and the bracket
- Ground strap(s) from the rear of the cylinder head
- Front body mount bolts, on 4WD vehicles

5. Support the transmission.
6. Install a lifting device and lift the engine.

To install:

7. Install or connect the following:
- Engine into the vehicle
- Front body mount bolts, on 4WD vehicles
- Ground strap(s) to the rear of the cylinder head
- Fuel lines and the bracket
- Upper left bell-housing bolt
- Throttle cables
- Vacuum lines and wiring harness connectors
- Heater hoses
- Power steering hoses
- Lower shroud and radiator
- Oil cooler lines to the radiator and overflow hose
- Lower radiator hose
- Air conditioning compressor to the engine, if equipped
- Upper radiator hose
- Water pump pulley
- Drive belt assembly
- Fan assembly
- Upper radiator shroud
- Air cleaner assembly
- Front drive axle, for 4WD vehicles
- Battery ground strap to the engine block
- Remaining bell housing bolts
- Engine mount through-bolts. Torque them to 49 ft. lbs. (66 Nm).
- Rear engine mount crossbar nut

and washer. Tighten the nut to 33 ft. lbs. (45 Nm).
- Oil filter
- Starter motor
- Flywheel cover
- Clutch slave cylinder, if equipped
- Pencil brace and the skid plate, as equipped
- Catalytic converter Y-pipe assembly and hangers
- Hood
- Outer cowl vent grilles
- Vacuum reservoir and/or the underhood light to the hood, as equipped
- Negative battery cable

8. Check all powertrain fluid levels and add, as necessary.
9. Refill the engine crankcase.
10. Refill the engine cooling system.
11. Start and run the engine, then check for leaks.

Water Pump

REMOVAL & INSTALLATION

1. Before servicing the vehicle, refer to the precautions in the beginning of this section.
2. Disconnect the negative battery cable.
3. Drain the engine cooling system.
4. Relieve the belt tension and remove the accessory drive belts or the serpentine drive belt, as applicable.
5. Remove or disconnect the following:
- Upper fan shroud
- Fan or fan and clutch assembly, as applicable
- Water pump pulley
- Coolant hose(s) from the water pump

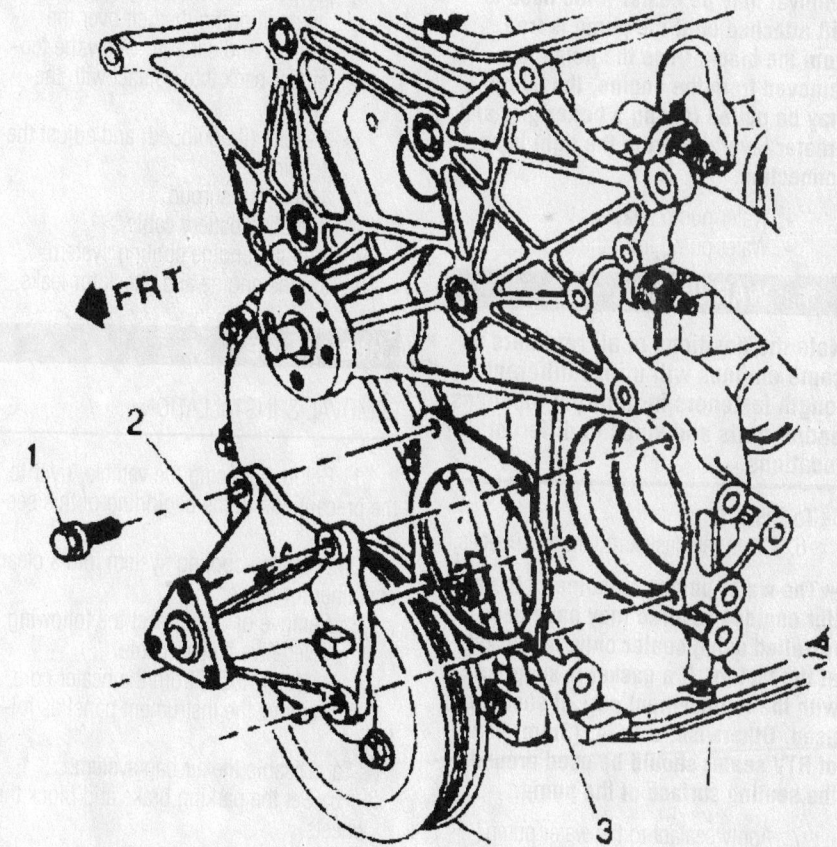

1. BOLT
2. PUMP, COOLANT
3. GASKET

Exploded view of the water pump mounting—2.2L engine

7924JG05

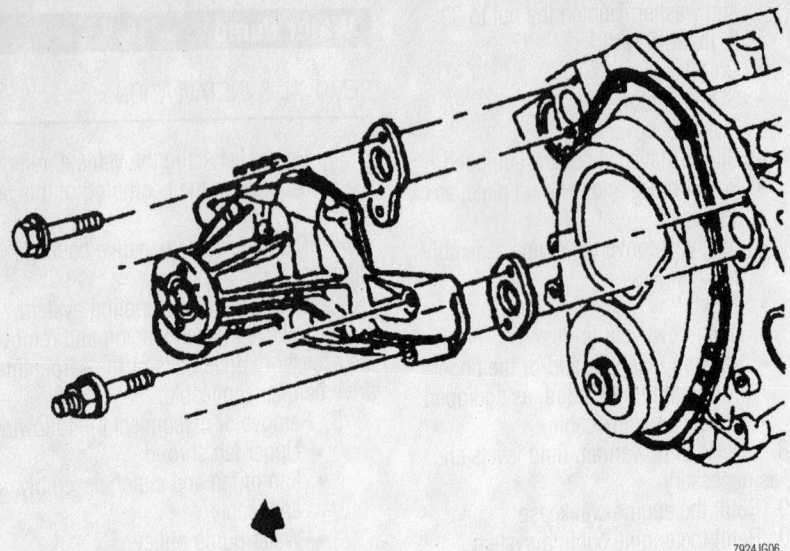

7924JG06

Exploded view of the water pump assembly mounting—4.3L engine

➡ **For the hoses on some engines, removal may be easier if the hose is left attached until the pump is free from the block. Once the pump is removed from the engine, the pump may be pulled (giving a better grip and greater leverage) from the tight hose connection.**

- Water pump retainers
- Water pump from the engine

✴✴ WARNING

Note the positions of all retainers as some engines will utilize different length fasteners in different locations and/or bolts and studs in different locations.

To install:

6. Clean the gasket mounting surfaces.

➡ **The water pumps on some of the earlier engines covered may have been installed using sealer only, no gasket, at the factory. If a gasket is supplied with the replacement part, it should be used. Otherwise, a ⅛ in. (3mm) bead of RTV sealer should be used around the sealing surface of the pump.**

7. Apply sealant to the water pump retainer threads.

8. Install or connect the following:
- Water pump using a new gasket. Tighten the water pump retainers to 18 ft. lbs. (25 Nm) for 2.2L engine or to 30 ft. lbs. (41 Nm) for 4.3L engine.
- Coolant hose(s)
- Water pump pulley

- Fan or fan and clutch assembly
- Serpentine drive belt (if equipped) by positioning the belt over the pulleys and carefully allow the tensioner back into contact with the belt
- V-belts (if equipped) and adjust the tension
- Upper fan shroud
- Negative battery cable

9. Refill the engine cooling system.
10. Run the engine and check for leaks.

Heater Core

REMOVAL & INSTALLATION

1. Before servicing the vehicle, refer to the precautions in the beginning of this section.

2. Drain the cooling system into a clean container for reuse.

3. Remove or disconnect the following:
- Negative battery cable
- Heater hoses from the heater core

4. Remove the instrument panel as follows:

a. Disable the air bag system.

b. Set the parking brake and block the wheels.

c. Disconnect the parking brake release cable from the parking brake lever.

d. Unfasten the screws that retain the DLC instrument panel left side sound insulator. Feed the DLC through the hole in the sound insulator.

e. Unfasten the right side sound insulator panel screws and remove the panel.

f. Unfasten the screws that attach the instrument panel left side sound insulator to the knee bolster and cowl panel.

g. Unfasten the nut that attaches the left side sound insulator to the accelerator pedal bracket.

h. Unplug the remote control door lock receiver module electrical connector.

i. Remove the door lock receiver module from the left side sound insulator. Remove the left side sound insulator.

j. Unfasten the screws that attach the instrument panel center sound insulator to the knee bolster, instrument panel, heater assembly and floor duct.

k. Remove the center sound insulator.

l. Unfasten the screws that attach the courtesy lamp to the knee bolster.

m. Unfasten the screws that attach the knee bolster to the instrument panel.

n. Disconnect the lap cooler duct from the knee bolster.

o. Unplug the lighter electrical connection and remove the knee bolster.

p. Unfasten the steering column-to-instrument panel nuts and lower the column.

q. Unfasten the screws that attach the instrument panel accessory trim plate to the instrument panel.

r. Remove the trim plate and unplug all necessary electrical connection.

s. Remove the heater and/or air conditioning control assembly.

t. Remove the radio and the storage compartment assembly (if equipped).

u. If necessary, remove the instrument cluster.

v. Unfasten the left and right instrument panel pivot bolts and the panel lower support bolt.

w. Unfasten the speaker grilles retaining screws and remove the speaker grilles.

x. Remove the windshield defroster grille using a flat-bladed prytool. Start at one end of the grille and work your way down the grille.

y. Unfasten the four instrument panel upper support screws.

z. Tag and unplug all necessary electrical connections.

aa. Remove the instrument panel from the vehicle.

5. Remove or disconnect the following:
- Air inlet assembly, if equipped
- Vacuum hoses
- Heater assembly studs, from inside the engine compartment
- Blower motor resistor
- Stud, from inside the heater case

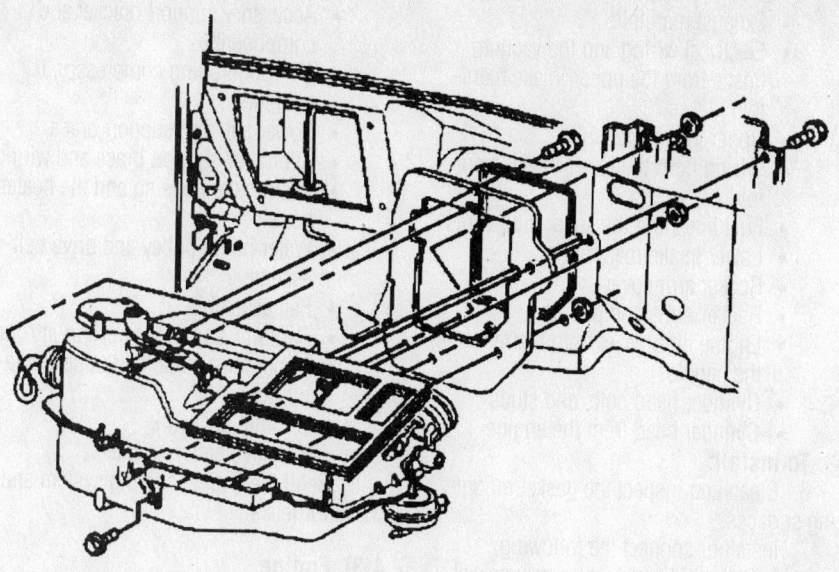

View of the heater case assembly—Isuzu Hombre

93113G77

assembly; the stud is located behind the blower motor resistor
• Heater assembly-to-chassis screws
• Heater assembly from the vehicle
• Access cover screws and cover from the heater assembly
• Heater core from the heater case assembly

To install:

6. Install or connect the following:

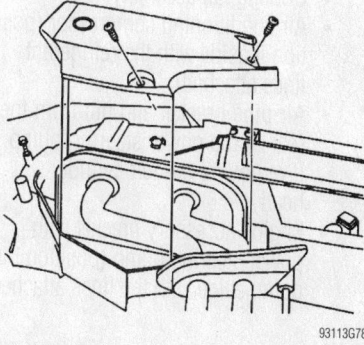

View of the heater case cover—Isuzu Hombre

93113G78

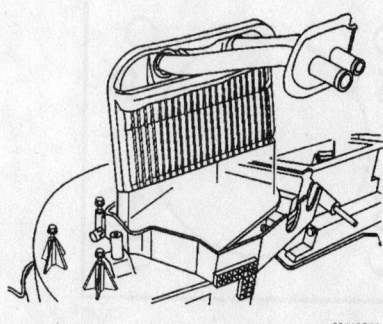

View of the heater core—Isuzu Hombre

93113G79

• Heater core to the heater case assembly
• Access cover to the heater assembly and the cover screws
• Heater assembly to the vehicle
• Heater assembly-to-chassis screws and torque them to 40 inch lbs. (4.5 Nm)
• Stud, located inside the heater case assembly; the stud is located behind the blower motor resistor
• Blower motor resistor.
• Heater assembly studs, located inside the engine compartment and torque them to 17 inch lbs. (1.9 Nm)
• Vacuum hoses
• Air inlet assembly, if equipped

7. Install the instrument panel as follows:

a. Rest the instrument panel on the lower pivot studs.

b. Attach the electrical connections.

c. Install but do not tighten the four upper instrument panel support screws.

d. Install the left and right panel pivot bolts. Tighten the bolts to 102 inch lbs. (11.5 Nm).

e. Install the panel lower support bolt. Tighten the bolt to 102 inch lbs. (11.5 Nm).

f. Tighten the upper support screws to 17 inch lbs. (1.9 Nm).

g. Install the windshield defroster grille and the speaker grilles.

h. Install the radio and storage compartment assembly (if equipped).

i. If removed, install the instrument cluster.

j. Install the heater and/or air conditioning control assembly.

k. Attach the electrical connections to the instrument panel accessory trim plate.

l. Place the trim plate in position and install its retaining screws. Tighten the screws to 17 inch lbs. (1.9 Nm).

m. Place the steering column into position and install its retaining nuts. Tighten the nuts to 22 ft. lbs. (30 Nm).

n. Attach the lighter electrical connection and the lap cooler duct to the knee bolster.

o. Place the knee bolster into position and install its retaining screws. Tighten the Torx® head screws to 80 inch lbs. (9 Nm) and the hex head screws to 17 inch lbs. (1.9 Nm).

p. Place the courtesy lamp in position and install its screws. Tighten the screws to 17 inch lbs. (1.9 Nm).

q. Place the instrument panel center sound insulator in position. Install the screws that attach the center sound insulator to the knee bolster, instrument panel and the floor duct. Tighten the screws to 17 inch lbs. (1.9 Nm).

r. Install the screw that attaches the center sound insulator to the heater assembly. Tighten the screw to 13 inch lbs. (1.5 Nm).

s. Install the remote control door lock receiver module to the instrument panel left side sound insulator.

t. Attach the door lock receiver electrical connection.

u. Install the nut that attaches the left side sound insulator to the accelerator pedal bracket. Tighten the nut to 35 inch lbs. (4 Nm).

v. Install the screw that attaches the left side sound insulator to cowl panel. Tighten the screw to 13 inch lbs. (1.5 Nm).

w. Install the screws that attach the left side sound insulator to knee bolster. Tighten the screw to 17 inch lbs. (1.9 Nm).

x. Feed the DLC through the hole in the sound insulator, place the DLC in position and install its retaining screws. Tighten the screws to 21 inch lbs. (2.4 Nm).

y. Install the right side sound insulator and tighten the screws

z. Connect the parking brake release cable to the lever.

aa. Enable the air bag system.

8. Install the heater hoses to the heater core.

9. Refill the cooling system.

10. Connect the negative battery cable.

11. Run the engine to normal operating temperatures; then, check the climate control operation and check for leaks.

Cylinder Head

REMOVAL & INSTALLATION

2.2L Engine

1. Before servicing the vehicle, refer to the precautions in the beginning of this section.

2. Relieve the fuel system pressure.

3. Disconnect the negative battery cable.

4. Drain the engine cooling system.

5. Remove or disconnect the following:

- Air duct from the air inlet
- Upper radiator hose and upper fan shroud
- Radiator assembly
- Lower fan shroud
- Fan assembly
- Drive belt assembly
- Water pump pulley
- Heater hose from the intake manifold and the thermostat housing
- Thermostat housing
- Alternator support brace and the alternator wiring
- Air conditioning compressor with brackets, if equipped, move it aside without disconnecting the lines
- Accessory bracket along with the alternator and power steering pump still attached. Be careful not to damage the steering pump lines.
- Throttle cable and cable support linkage
- Heater hose from the water pump
- Oil fill tube
- Exhaust pipe
- Oxygen (O2S) sensor

- Exhaust manifold
- Electrical wiring and the vacuum hoses from the upper intake manifold
- Upper intake manifold
- Wiring from the lower intake manifold
- Fuel lines and the spark plug wires
- Lower intake manifold
- Rocker arm cover
- Rocker arms and pushrods
- Engine lift bracket from the rear of the engine
- Cylinder head bolts and studs
- Cylinder head from the engine

To install:

6. Clean and inspect the gasket mounting surfaces.

7. Install or connect the following:

- Cylinder head using a new gasket
- Cylinder head bolt threads coated with sealer 1052080. Tighten the bolts within 15 minutes of sealer application, in sequence, to 46 ft. lbs. (63 Nm) for long bolts and to 43 ft. lbs. (58 Nm) for short bolts; then, tighten all bolts an additional 90 degree turn using a torque angle meter.
- Engine lift bracket
- Rocker arms and pushrods
- Rocker arm cover
- Lower intake manifold
- Spark plug wires and the fuel lines
- Lower intake manifold and wiring
- Upper intake manifold
- Vacuum hoses and electrical wiring to the upper intake
- Oil fill tube assembly
- Exhaust manifold
- Exhaust pipe and O2S sensor
- Heater hose to the water pump
- Throttle cable support and throttle cable

- Accessory support bracket and components
- Air conditioning compressor, if equipped
- Power steering support brace
- Alternator support brace and wiring
- Thermostat housing and the heater hose
- Water pump pulley and drive belt assembly
- Fan assembly
- Radiator and the lower fan shroud
- Upper fan shroud and upper radiator hose
- Air inlet ductwork
- Negative battery cable

8. Refill the engine cooling system and check for leaks.

4.3L Engine

1. Before servicing the vehicle, refer to the precautions in the beginning of this section.

2. Properly relieve the fuel system pressure, then disconnect the negative battery cable.

3. Drain the engine cooling system.

4. Remove or disconnect the following:

- Intake manifold

5. Remove the exhaust manifold.

- Alternator and bracket, if removing the right cylinder head
- Cooling fan assembly
- Air conditioning compressor (position it aside with the refrigerant lines attached)
- Air pipe bracket and nut from the rear of the power steering pump if removing the left cylinder head
- Engine accessory bracket with power steering pump (position the pump aside with the lines attached)

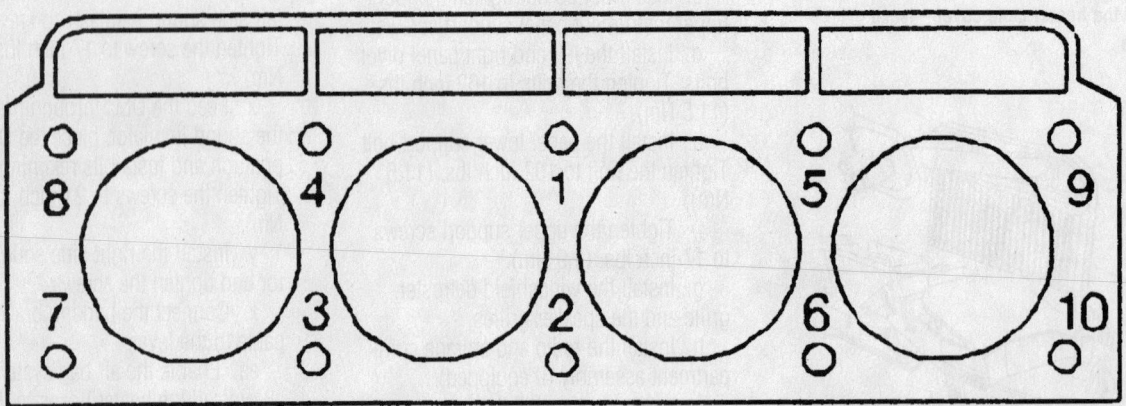

Cylinder head bolt torque sequence—2.2L engine

7924JG07

and brackets, if removing the left cylinder head
- Wiring harness and clip from the rear of the cylinder head
- Coolant sensor wire
- Wiring from the spark plugs
- Spark plugs, if necessary
- Ground wires and if necessary, the

fuel line bracket from the rear of the cylinder head
- Rocker arm cover

6. Loosen the rocker arms and remove the pushrods.

➡ **If valve train components, such as the rocker arms or pushrods, are to be**

reused, they must be tagged or arranged to insure installation in their original locations.

7. Unfasten the cylinder head bolts by loosening them in the reverse of the torque sequence, then carefully remove the cylinder head.

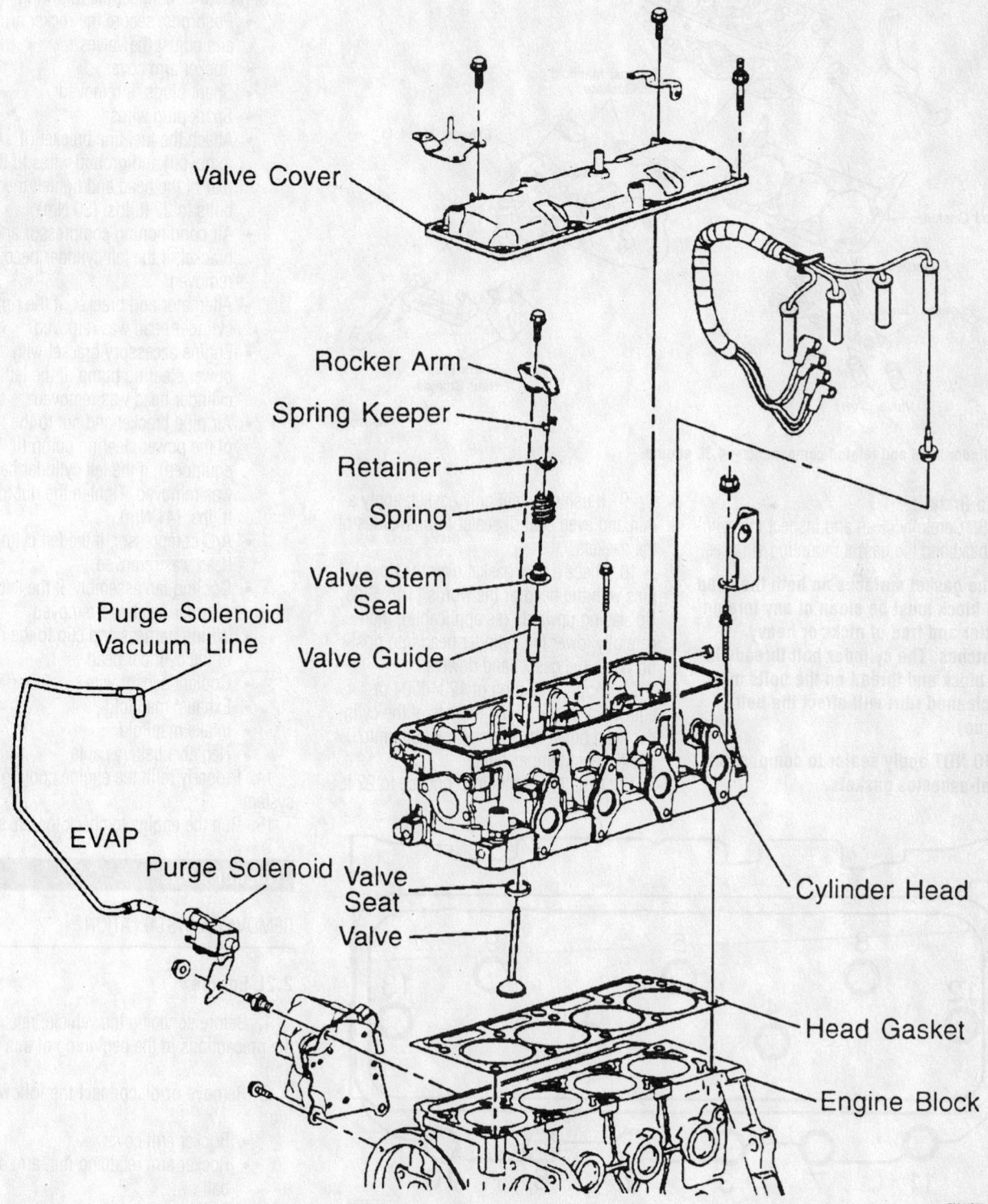

Valve Cover

Rocker Arm
Spring Keeper
Retainer
Spring
Valve Stem Seal
Valve Guide

Purge Solenoid Vacuum Line

EVAP
Purge Solenoid
Valve Seat
Valve

Cylinder Head

Head Gasket

Engine Block

Cylinder head and related components—2.2L engine

7924JG23

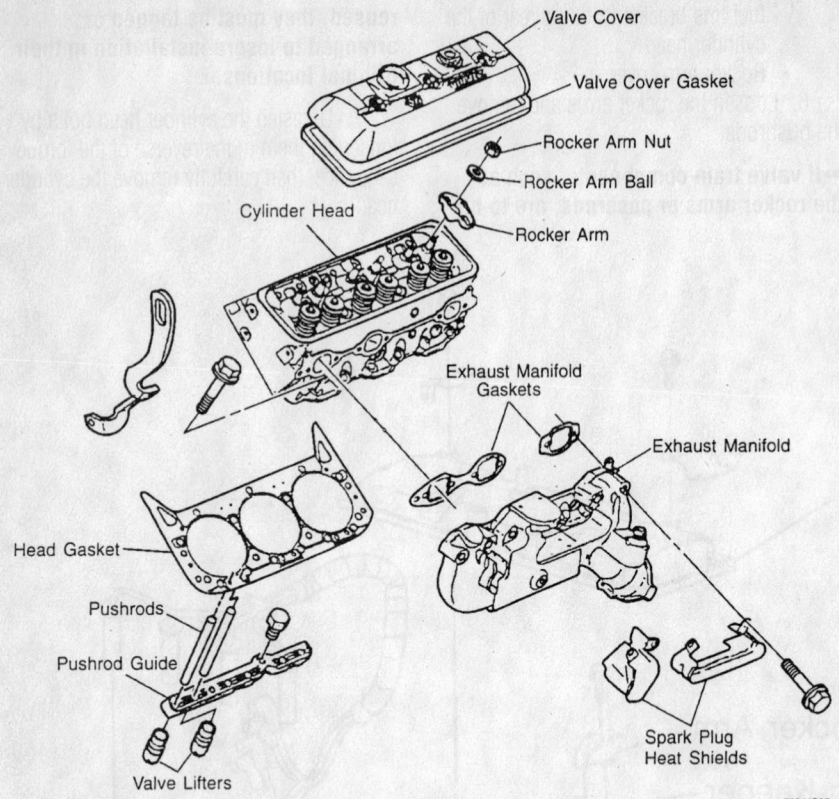

Cylinder head and related components—4.3L engine

To install:

8. Carefully clean and inspect the cylinder head and the gasket mounting surfaces.

➡**The gasket surfaces on both the head and block must be clean of any foreign matter and free of nicks or heavy scratches. The cylinder bolt threads in the block and thread on the bolts must be cleaned (dirt will affect the bolt torque).**

➡**DO NOT apply sealer to composition steel-asbestos gaskets.**

9. If using a steel only gasket, apply a thin and even coat of sealer to both sides of the gaskets.

10. Place a new gasket over the dowel pins with the bead or the words "This Side Up" facing upwards (as applicable), then carefully lower the cylinder head into position over the gasket and dowels.

11. Apply a coating of 12346004 or equivalent sealer to the threads of the cylinder head bolts, then thread the bolts into position until finger-tight.

12. Install the bolts in sequence to 22 ft.

lbs. (30 Nm). The bolts must then be tightened again in sequence in the following order:

    a. Short length bolts: (11, 7, 3, 2, 6, 10) 55 degrees.

    b. Medium length bolts: (12, 13) 65 degrees.

    c. Long length bolts: (1, 4, 8, 5, 9) 75 degrees.

13. Install or connect the following:

- Pushrods, secure the rocker arms and adjust the valves
- Rocker arm cover
- Spark plugs, if removed
- Spark plug wires
- Attach the fuel line bracket (if removed) and ground wires to the rear of the head and tighten the bolts to 22 ft. lbs. (30 Nm)
- Air conditioning compressor and bracket, if the left cylinder head was removed
- Alternator and bracket, if the right cylinder head was removed
- Engine accessory bracket with power steering pump. if the left cylinder head was removed
- Air pipe bracket and nut to the rear of the power steering pump (if equipped), if the left cylinder head was removed. Tighten the nut to 30 ft. lbs. (41 Nm).
- A/C compressor, if the left cylinder head was removed
- Cooling fan assembly, if the left cylinder head was removed
- Wiring harness and clip to the rear of the cylinder head
- Coolant sensor wire
- Exhaust manifold
- Intake manifold
- Negative battery cable

14. Properly refill the engine cooling system.

15. Run the engine to check for leaks.

Rocker Arms

REMOVAL & INSTALLATION

2.2L Engine

1. Before servicing the vehicle, refer to the precautions in the beginning of this section.

2. Remove or disconnect the following:

- Rocker arm cover
- Rocker arm retaining nut, arm and ball
- Pushrod, if necessary

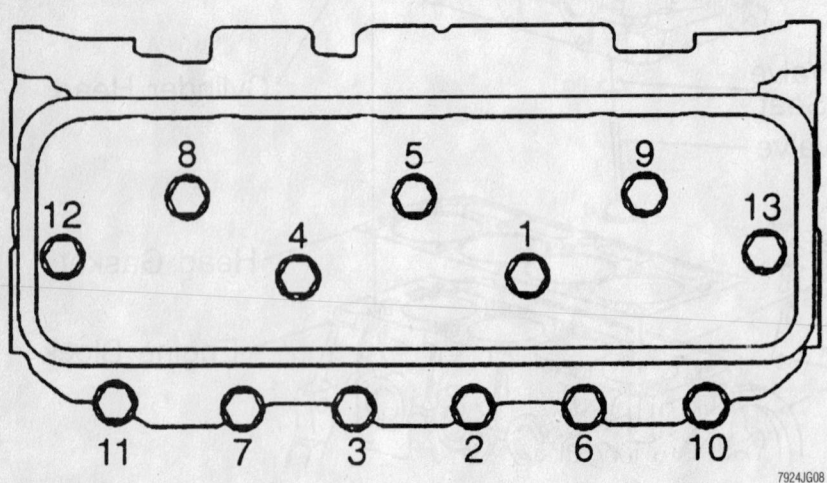

Cylinder head bolt torque sequence—4.3L engine

➡Valvetrain components, being reused, must be installed in their original positions. If removed, be sure to tag or arrange all rocker arms and pushrods to assure proper installation.

To install:

3. Inspect the rocker arms, balls and pushrods for damage or wear and replace as necessary.

4. Check the rocker arms, balls and their mating surfaces. Be sure the surfaces are smooth and free from scoring or other damage.

5. Check the rocker arm areas that contact the valve stems and the sockets that contact the pushrods, be sure these areas are smooth and free of both damage and wear.

6. Be sure the pushrods are not bent which can be determined by rolling them on a flat surface. Check the ends of the pushrods for scoring or roughness

7. Inspect the rocker arm bolts for thread damage. Check the rocker arm bolts in the shoulder area for contact damage with the rocker arm.

8. Install or connect the following:
- Pushrods making sure they are seated within the lifters, if removed
- New rocker arms and balls by coating the friction surfaces using Dri-Slide Molykote® or equivalent pre-lube

✳✳ WARNING

When tightening a rocker arm retainer, be sure the lifter for that valve is resting on the base circle of the camshaft and not on the lobe, otherwise the valve train can be damaged. Do not over-tighten the retainers.

- Rocker arms and ball. Tighten the nuts to 22 ft. lbs. (30 Nm).
- Rocker arm cover

9. Start and run the engine to check for leaks.

4.3L Engines

1. Before servicing the vehicle, refer to the precautions in the beginning of this section.

2. Remove or disconnect the following:
- Rocker arm cover(s)
- Rocker arm nut, rocker arm and ball washer

➡If only the pushrod is to be removed, loosen the rocker arm nut, swing the rocker arm to the side and remove the pushrod.

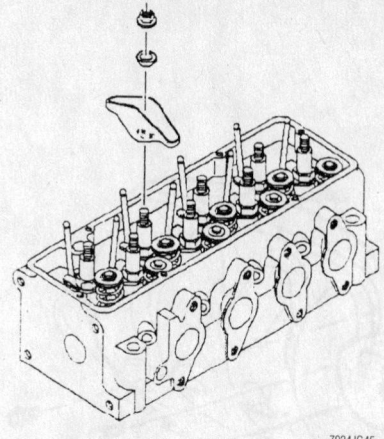

7924JG45

Exploded view of the rocker arm assembly—2.2L engine

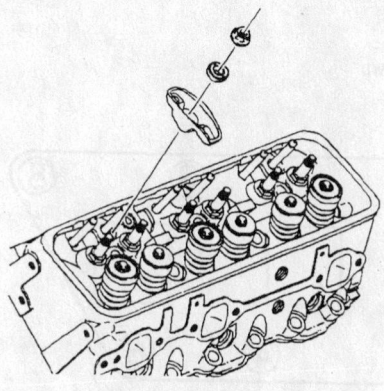

7924JG46

Exploded view of the rocker arm assembly—4.3L engine

- Pushrod(s)

To install:

3. Inspect and replace components if worn or damaged.

4. Coat the bearing surfaces of the rocker arms and the rocker arm ball washers with Molykote® or equivalent pre-lube.

5. Install or connect the following:
- Pushrods making sure they seat properly in the lifter
- Rocker arms, ball washers and the nuts

➡The 4.3L engines are equipped with screw-in rocker arm studs with positive stop shoulders.

- Rocker arm adjusting nuts. Tighten them against the stop shoulders to 18 ft. lbs. (24 Nm). No further adjustment is necessary or possible.

6. Install the rocker arm cover(s).

7. Start and run the engine, then check for leaks and for proper ignition timing adjustment.

Intake Manifold

REMOVAL & INSTALLATION

2.2L Engine

1. Disconnect the negative battery cable and remove the air cleaner resonator.

2. Tag and unplug the three vacuum hoses from the throttle body.

3. Remove the throttle cable support bracket and the throttle body assembly.

4. If necessary, remove the upper fan shroud and disconnect the vacuum brake booster hose.

5. If necessary, unfasten the EGR pipe-to-manifold bolts and the EGR pipe-to-EGR adapter bolt, then remove the EGR pipe.

6. If necessary, remove the EGR adapter.

7. Unplug the electrical connections from the following components:
   a. Idle Air Control (IAC) motor.
   b. Manifold Absolute Pressure (MAP) sensor.
   c. Throttle Position (TP) sensor.
   d. Fuel injector harness connector.

8. Remove the right fender wheelhouse extension.

9. Remove the retainers from the engine harness bracket, the transmission filler tube (if equipped) and the fuel system evaporator pipe.

10. Disconnect the fuel pipes from the fuel rail.

11. Disconnect the accelerator cable and if equipped, the cruise control cable.

12. Tag and disconnect the spark plug wires from the plugs.

13. Remove the spark plug wire harness retainer from the heater hose pipe and set aside the harness.

14. If necessary, remove the alternator rear brace by accessing the retaining nuts and bolts through the wheelhouse.

15. If equipped, remove the engine wiring harness bracket located at the rear of the cylinder head, by unfastening the bracket-to-valve cover and bracket-to-cylinder head retainers, then slide the bracket off the bolt at the rear of the cylinder head.

16. Unfasten the intake manifold bolts.

17. Remove the fuel rail bracket.

18. Remove the intake manifold and gasket.

To install:

19. Carefully remove all traces of gasket material from the mating surfaces. Check the EGR passage to be sure it is free of excessive carbon deposits and clean, as necessary.

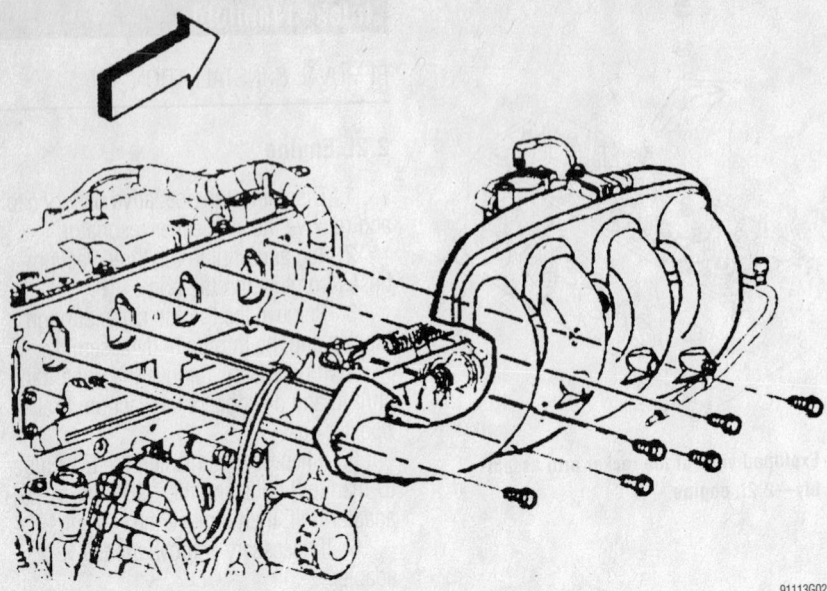

Typical intake manifold mounting—2.2L engine shown

91113G02

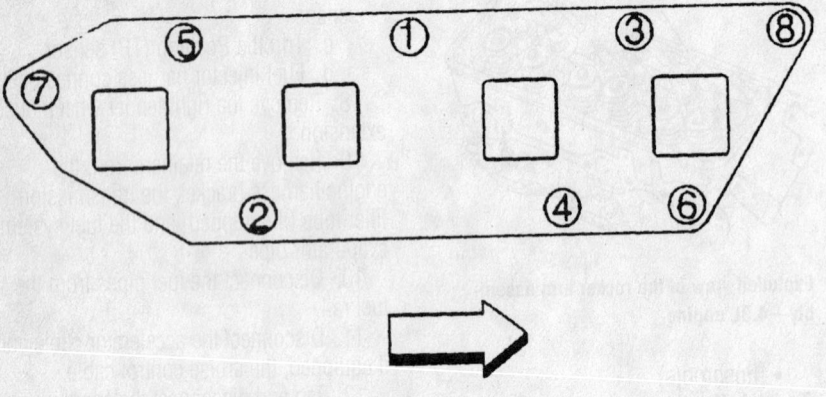

91113G03

Intake manifold bolt tightening sequence—2.2L engine

20. Install the lower intake manifold using a new gasket, then tighten the retaining bolts to 17 ft. lbs. (24 Nm) using the sequence illustrated.

21. If removed, install the engine wiring harness bracket. Tighten the bracket-to-valve cover bolts to 88 inch lbs. (10 Nm) and the bracket-to-cylinder head bolt to 18 ft. lbs. (25 Nm).

22. If removed, install the generator rear brace. Tighten the nuts and bolts to 18 ft. lbs. (25 Nm).

23. Install the spark plug wire harness and retainer and attach the spark plug wires to the plugs.

24. Install the throttle body assembly, if removed and the throttle cable support bracket.

25. Attach the three vacuum lines to the throttle body.

26. Attach the accelerator cable and if equipped, the cruise control cable.

27. Connect the fuel lines.

28. Install and tighten the retainers to the engine harness bracket, the transmission filler tube (if equipped) and the fuel system evaporator pipe.

29. Attach the following electrical connections:
  a. Idle Air Control (IAC) motor.
  b. Manifold Absolute Pressure (MAP) sensor.
  c. Throttle Position (TP) sensor.
  d. Fuel injector harness connector.

30. If removed, install the wheelhouse extension.

31. If removed, install the EGR adapter and tighten the retainers to 97 inch lbs. (11 Nm).

32. Install the EGR pipe to the EGR adapter and tighten the bolt to 18 ft. lbs. (25 Nm).

33. Install the EGR pipe-to-intake manifold bolts and tighten the bolts to 89 inch lbs. (10 Nm).

34. If removed, install the upper fan shroud and the brake booster hose.

35. Install the air cleaner resonator and connect the negative battery cable.

36. Start the engine and check for leaks.

4.3L Engines

➡If only the upper intake manifold is being removed, the fuel system pressure does not need to be released. ALWAYS release the pressure before disconnecting any fuel lines.

1. Before servicing the vehicle, refer to the precautions in the beginning of this section.

2. Remove the engine cover, if equipped

3. Properly relieve the fuel system pressure.

4. Disconnect the negative battery cable.

5. Drain the engine cooling system.

6. Remove or disconnect the following:
  • Air cleaner and air inlet duct
  • Wiring harness connectors and brackets
  • Throttle linkage from the upper intake manifold
  • Ignition coil
  • Fuel lines and bracket from the rear of the lower intake manifold
  • Brake booster vacuum hose at the upper intake manifold
  • Positive Crankcase Ventilation (PCV) hose at the rear of the upper intake manifold
  • Vacuum hoses from both the front and rear of the upper intake
  • Purge solenoid and bracket
  • Upper intake manifold
  • High Voltage Switch (HVS) assembly
  • Upper radiator hose at the thermostat housing
  • Heater hose at the lower intake manifold
  • Wiring harnesses and brackets.
  • Automatic transmission dipstick tube
  • Exhaust Gas Recirculation (EGR) tube, clamp and tube
  • Air conditioning compressor bracket-to-lower intake manifold pencil brace
  • Alternator bracket bolts near the thermostat housing

- Lower intake manifold

7. Insert clean rags into the openings in the cylinder head to prevent dirt and debris from entering the engine.

8. Clean the gasket mounting surfaces. Be sure to inspect the manifold for warpage and/or cracks. If necessary, replace it.

To install:

9. Remove the rags from the cylinder heads.

10. Position the gaskets on the cylinder head with the port blocking plates to the rear and the **this side up** stamps facing upward. Then apply a ³⁄₁₆ in. (5mm) bead of RTV sealant on the front and rear of the engine block at the block-to-manifold mating surface. Extend the bead ½ in. (13mm) up each cylinder head to seal and retain the gaskets.

11. Install the lower intake manifold. Tighten the bolts in sequence and in 3 steps, as follows:
   a. Step 1: 26 inch lbs. (3 Nm).
   b. Step 2: 106 inch lbs. (12 Nm).
   c. Step 3: 11 ft. lbs. (15 Nm).

12. Install or connect the following:
   - Alternator bracket bolt near the thermostat housing
   - EGR tube, clamp and bolt
   - Wiring harness to the lower manifold components, including the injector, EGR valve and ECT sensor
   - Air conditioning compressor bracket-to-the lower intake manifold pencil braces
   - Transmission oil dipstick tube, if necessary
   - Fuel supply and return lines to the rear of the lower intake

13. Temporarily reattach the negative battery cable, then pressurize the fuel system (by cycling the ignition without starting the engine) and check for leaks.

14. Disconnect the negative battery cable.

15. Install or connect the following:
   - Heater hose to the lower intake
   - Upper radiator hose to the thermostat housing
   - Vacuum hoses to the upper and lower intake manifold
   - New upper intake manifold gasket, making sure the green sealing lines are facing upward
   - Upper intake manifold being careful not to pinch the fuel injector wires between the manifolds
   - Manifold retainers. Tighten them to 88 inch lbs. (10 Nm) using two passes.
   - Purge solenoid and bracket

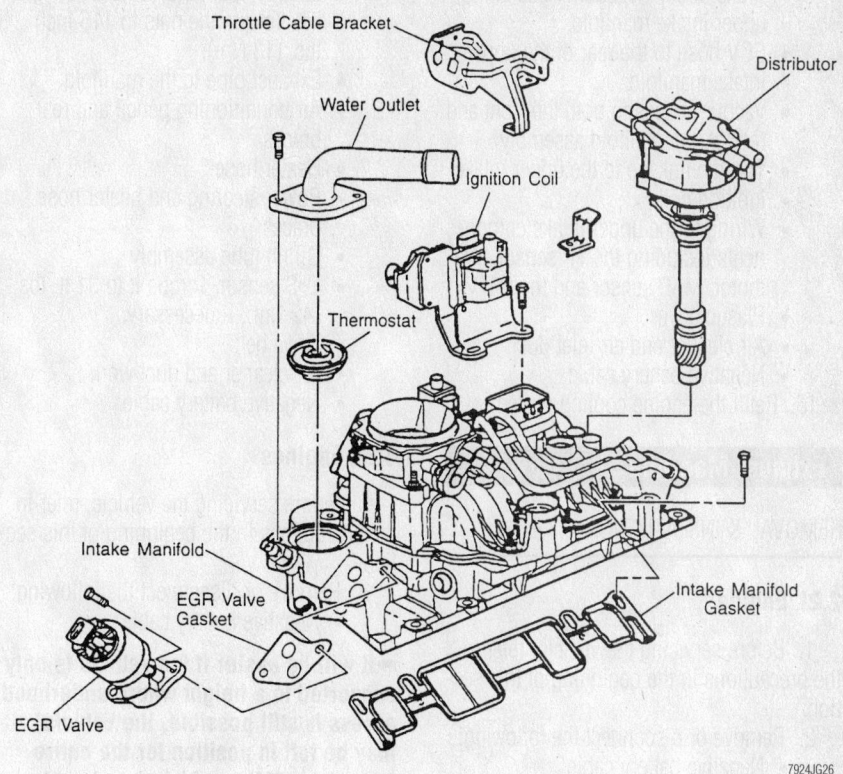

Intake manifold and related components—4.3L engine

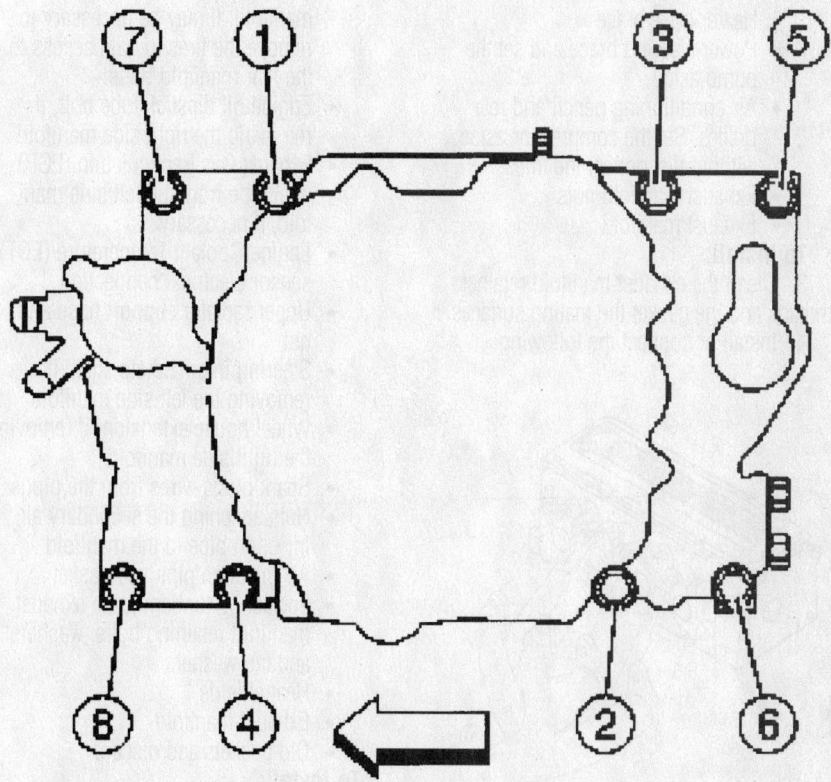

Lower intake manifold tightening sequence—4.3L engines

- Brake booster vacuum hose at the upper intake manifold.
- PCV hose to the rear of the upper intake manifold
- Vacuum hoses to both the front and rear of the manifold assembly
- Throttle linkage to the upper intake
- Ignition coil
- Wiring to the upper intake components including the TP sensor, IAC motor, MAP sensor and the IMTV.
- Plastic cover
- Air cleaner and air inlet duct
- Negative battery cable

16. Refill the engine cooling system.

Exhaust Manifold

REMOVAL & INSTALLATION

2.2L Engine

1. Before servicing the vehicle, refer to the precautions in the beginning of this section.
2. Remove or disconnect the following:
- Negative battery cable
- Air cleaner and duct work
- Oxygen (O_2S) sensor from the manifold, if replacing it
- Drive belt
- Oil fill tube assembly
- Heater hose brace
- Power steering brace and set the pump aside
- Air conditioning pencil and rear braces. Set the compressor aside without disconnect the lines.
- Exhaust manifold nuts
- Exhaust manifold

To install:

3. Clean the exhaust manifold retainer threads and the gasket the mating surfaces.
4. Install or connect the following:

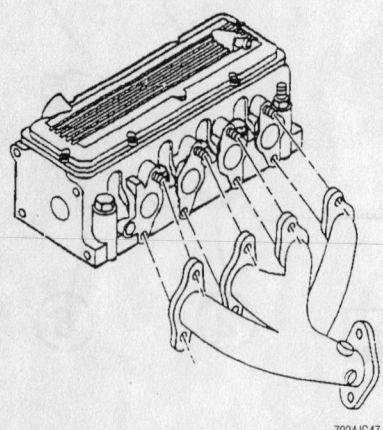

7924JG47

Exhaust manifold mounting—2.2L engine

- Exhaust manifold using a new gasket. Torque the nuts to 115 inch lbs. (13 Nm).
- Exhaust pipe to the manifold
- Air conditioning pencil and rear braces
- Heater hose
- Power steering and heater hose braces
- Oil fill tube assembly
- O_2S sensor. Torque it to 31 ft. lbs. (42 Nm), if necessary.
- Drive belt
- Air cleaner and duct work
- Negative battery cable

4.3L Engines

1. Before servicing the vehicle, refer to the precautions in the beginning of this section.
2. Remove or disconnect the following:
- Negative battery cable

➡**It will be easier if the vehicle is only supported to a height where underhood access is still possible, the vehicle may be left in position for the entire procedure. If the vehicle is raised too high for underhood access, it will have to lowered, raised and lowered again during the procedure.**

- Exhaust pipe from the exhaust manifold. It may be necessary to remove the tires to gain access to the rear manifold bolts.
- Engine oil dipstick tube bolt, if removing the right side manifold
- Exhaust Gas Recirculation (EGR) inlet pipe from the left side manifold, if necessary
- Engine Coolant Temperature (ECT) sensor electrical connection
- Upper radiator support hose and nut
- Steering intermediate shaft, if removing the left side manifold
- Wheel house extension, if removing the right side manifold
- Spark plugs wires from the plugs
- Nuts attaching the secondary air injection pipe to the manifold
- Air injection pipe and gasket
- Locktangs (unbend), the exhaust manifold retaining bolts, washers and tab washers
- Heat shields
- Exhaust manifold
- Old gaskets and discard

To install:

3. Using a putty knife, clean the gasket mounting surfaces. Inspect the exhaust

manifold for distortion, cracks or damage; replace if necessary.

4. Apply a threadlock such as GM 12345493 to the threads of the manifold retainers prior to installation.
5. Install or connect the following:
- Exhaust manifold to the cylinder using a new gasket, then tighten the center bolts to 11 ft. lbs. (15 Nm) and the front and rear manifold bolts to 22 ft. lbs. (30 Nm). Once the bolts are tightened, bend the tabs on the washers back over the heads of all bolts in order to lock them in position.
- Spark plug wires to the plugs
- Fender wheelhouse extension and the tire assembly, if removed
- Secondary air injection pipe with a NEW gasket to the manifold and tighten the nuts to 18 ft. lbs. (25 Nm), if removed
- EGR inlet pipe, if removed
- ECT sensor electrical connection, if removed
- Upper radiator hose support and nut, if removed
- Steering intermediate shaft, if removed (left side manifold only)
- Engine oil dipstick tube bolt to 106 inch lbs. (12 Nm), if removed
- Exhaust pipe to the manifold
- Negative battery cable

Camshaft and Valve Lifters

REMOVAL & INSTALLATION

2.2L Engine

1. Before servicing the vehicle, refer to the precautions in the beginning of this section.
2. Properly relieve the fuel system pressure.
3. Disconnect the negative battery cable.
4. Drain the engine cooling system and the engine oil.
5. Remove or disconnect the following:
- Radiator
- Rocker arm cover
- Cylinder head
- Anti-rotation bracket bolts and brackets
- Valve lifters
- Oil pump drive retaining bolt and the drive by lifting and twisting
- Camshaft Position (CMP) sensor, if equipped
- Crankshaft pulley and hub

- Drive belt idler pulley
- Timing cover from the engine
- Timing chain and camshaft sprocket
- Camshaft thrust plate
- Camshaft by pulling it straight out of the engine, while turning it slightly as it is withdrawn and taking care not to damage the bearings.

To install:

6. Inspect the camshaft, journals and lobes for wear and replace, if necessary.

7. If removed, use the camshaft bearing tool to install a new set of bearings.

8. Coat the camshaft lobes and journals with a high viscosity oil with zinc such as GM 12345501.

9. Install or connect the following:
- Camshaft by turning it slightly from side-to-side as it is inserted
- Thrust plate. Torque the bolts to 106 inch lbs. (12 Nm).
- Timing chain and camshaft sprocket
- Timing cover
- Serpentine drive belt idler pulley
- Crankshaft pulley and hub
- Oil pump drive by inserting while twisting. Torque the fasteners to 18 ft. lbs. (25 Nm).
- Valve lifters and the anti-rotation brackets
- Cylinder head. Torque the bolts to 46 ft. lbs. (62 Nm) plus an additional 90 degrees turn.
- Rocker arm cover
- Radiator
- Negative battery cable

10. Refill the engine cooling system.

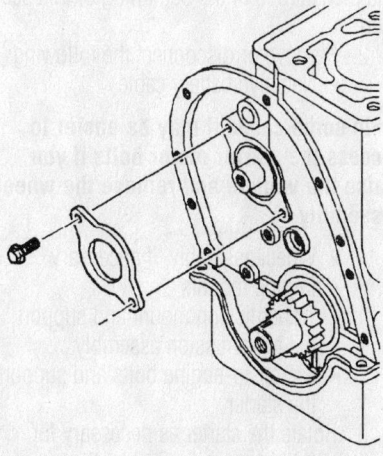

Remove the camshaft thrust plate and withdraw the camshaft from the engine— 2.2L engine

4.3L Engines

1. Before servicing the vehicle, refer to the precautions in the beginning of this section.

2. Properly relieve the fuel system pressure.

3. Disconnect the negative battery cable.

4. Drain the engine cooling system.

5. Discharge and recover the refrigerant from the air conditioning system.

6. Remove or disconnect the following:
- Radiator
- Air conditioning condenser
- Rocker arm covers
- Intake manifold assembly
- Rocker arms, pushrods and lifters
- Crankshaft pulley and hub
- Engine front (timing) cover

7. Align the timing marks on the crankshaft and camshaft sprockets.

8. Remove or disconnect the following:
- Camshaft sprocket and timing chain
- Balance shaft drive gear, if equipped
- Camshaft thrust plate
- Camshaft by installing the sprocket bolts or longer bolts the camshaft end to act as a handle; then, remove the camshaft while turning slightly from side to side, as necessary.

➡**Take care not to damage the camshaft bearings when removing the camshaft.**

To install:

9. Lubricate the camshaft journals with clean engine oil or a suitable pre-lube.

10. Install or connect the following:
- Camshaft being extremely careful not to contact the bearings with the cam lobes

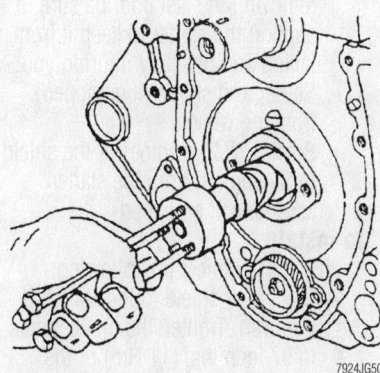

Thread 3 long bolts into the camshaft to use as a handle, then withdraw it from the engine

- Thrust plate. Torque the bolts to 106 inch lbs. (12 Nm).
- Balance shaft drive gear, if equipped
- Timing chain and camshaft sprocket
- Engine front (timing) cover
- Crankshaft pulley and hub
- Valve lifters, pushrods and rocker arms. Adjust the valve clearance.
- Intake manifold assembly
- Rocker arm covers
- Radiator
- Negative battery cable

11. Refill the engine cooling system.

Valve Lash

ADJUSTMENT

2.2L Engine

Because the rocker arm fasteners are secured and tightened, valve lash is not adjustable on the 2.2L engine. If a valvetrain problem is suspected, check that the rocker arm nuts are tightened to 22 ft. lbs. (30 Nm). Be very careful not to over-tighten the rocker arm nuts. ONLY tighten the nuts when the hydraulic lifter is resting on the base circle of the camshaft and not when it is held upward on the lobe. When valve lash falls out of specification (valve tap is heard), replace the rocker arm, pushrod and hydraulic lifter on the offending cylinder.

4.3L Engines

The 4.3L engines are equipped with screw-in rocker arm studs with positive stop shoulders. Because the shoulders that allow the rocker arms to be tightened into proper position, no adjustments are necessary or possible. If a valvetrain problem is suspected, check that the rocker arm nuts are tightened to 18 ft. lbs. (24 Nm). When valve lash falls out of specification (valve tap is heard), replace the rocker arm, pushrod and hydraulic lifter on the offending cylinder.

Starter Motor

REMOVAL & INSTALLATION

Two Wheel Drive

2.2L MODELS

1. Before servicing the vehicle, refer to the precautions in the beginning of this section.

2. Remove or disconnect the following:

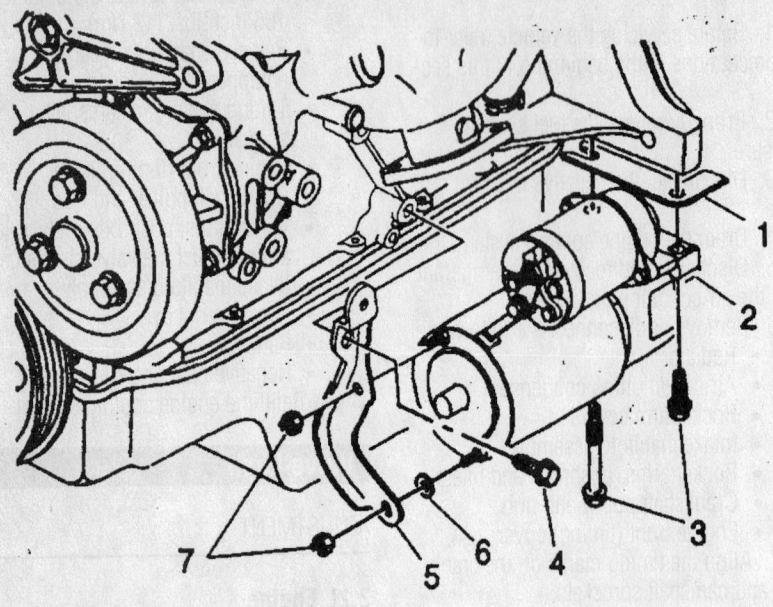

1. SHIM
2. STARTER ASSEMBLY
3. BOLT, 43 N·m (32 LBS. FT.)
4. BOLT, 43 N·m (32 LBS. FT.)
5. BRACKET, STARTER MOTOR
6. WASHER
7. NUT, 11 N·m (97 LBS. IN.)

88452G08

Starter motor and related components—2.2L engine

- Negative battery cable
- Front exhaust pipe, if necessary for access
- Starter heat shield, if equipped
- Brace rod from the front of the engine and the bell housing, on 2.2L engines

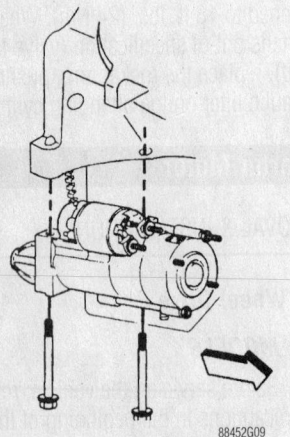

88452G09

The starter motor on later model 4.3L engines is retained by two long bolts

- Drivers side wheel to access the starter motor wires and the starter motor attaching bracket-to-engine bolt through the opening in the wheel well, on 2.2L engines
- Wires from the starter solenoid
- Attaching bracket-to-engine mount bolt, on 2.2L engines
- Starter-to-engine block bolts. When removing the last bolt, be sure to support the starter to keep it from falling and possibly injuring you.
- Starter and shims (if equipped) from the vehicle
- Bracket (2.2L engine) or the shield (4.3L engine) from the starter assembly, if equipped

To install:

3. Install or connect the following:
- Bracket or shield to the starter, if removed. Tighten the bracket nuts to 97 inch lbs. (11 Nm) or the shield nuts to 106 inch lbs. (12 Nm).
- Starter and shims (if equipped) into

position in the vehicle and thread one of the retaining bolts to hold it in position.
- Bracket-to-engine mount bolt (loosely), if equipped
- Starter mounting bolt, then tighten all mounting fasteners to 32 ft. lbs. (43 Nm)
- Wiring to the solenoid
- Brace rod and tighten the retainers, on 2.2L engines
- Front exhaust pipe and tighten the fasteners, if removed
- Starter heat shield, if equipped
- Driver's side wheel, if removed
- Negative battery cable

4.3L MODELS

1. Before servicing the vehicle, refer to the precautions in the beginning of this section.
2. Remove or disconnect the following:
- Negative battery cable
- Wires from the starter solenoid
- Starter motor mounting bolts
- Starter motor and if equipped, the shims

To install:

3. Install or connect the following:
- Starter motor into position
- Starter motor inboard bolt but do not tighten it at this time
- Starter motor shims, if equipped
- Outboard starter motor bolt. Tighten the bolts to 32 ft. lbs. (43 Nm).
- Wires to the solenoid
- Negative battery cable

Four Wheel Drive

1. Before servicing the vehicle, refer to the precautions in the beginning of this section.
2. Remove or disconnect the following:
- Negative battery cable

➡ In some cases it may be easier to access the starter motor bolts if you raise the vehicle and remove the wheel assembly.

- Wheel assembly, if necessary
- Engine mounts
- Transmission mount and support the transmission assembly
- Starter-to-engine bolts and support the starter

3. Rotate the starter as necessary for access, then tag and disconnect the solenoid wiring.
4. Carefully lower the starter and shims (if equipped) from the vehicle. Note the

location of any shims for installation purposes.

5. If necessary, remove the shield from the starter assembly.

To install:

6. Install or connect the following:
   - Starter into position in the vehicle along with any shims (making sure they are in their original positions), then tighten the mounting bolts to 32 ft. lbs. (43 Nm).
   - Shield to the starter assembly and tighten the retaining nuts to 106 inch lbs. (12 Nm), if removed
   - Wiring to the solenoid
   - Transmission mount and remove the supports
   - Secure the engine mounts, then remove the lifting device
   - Wheel assembly, if removed
7. Connect the negative battery cable.

Oil Pan

REMOVAL & INSTALLATION

2.2L Engine

1. Before servicing the vehicle, refer to the precautions in the beginning of this section.
2. Drain the engine oil.
3. Remove or disconnect the following:
   - Engine
   - Clutch pressure plate and disc, if equipped
   - Flywheel
   - Oil pan retainers and the pan

To install:

4. Clean the gasket mating surfaces
5. Install or connect the following:
   - New gasket and seal onto the oil pan using a thin bead of sealant at either side of the seal
   - Oil pan. Torque the bolts to 89 inch. lbs. (10 Nm).

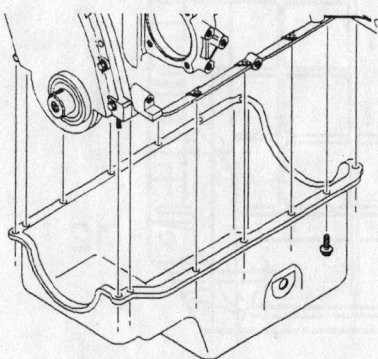

Oil pan mounting—2.2L engine

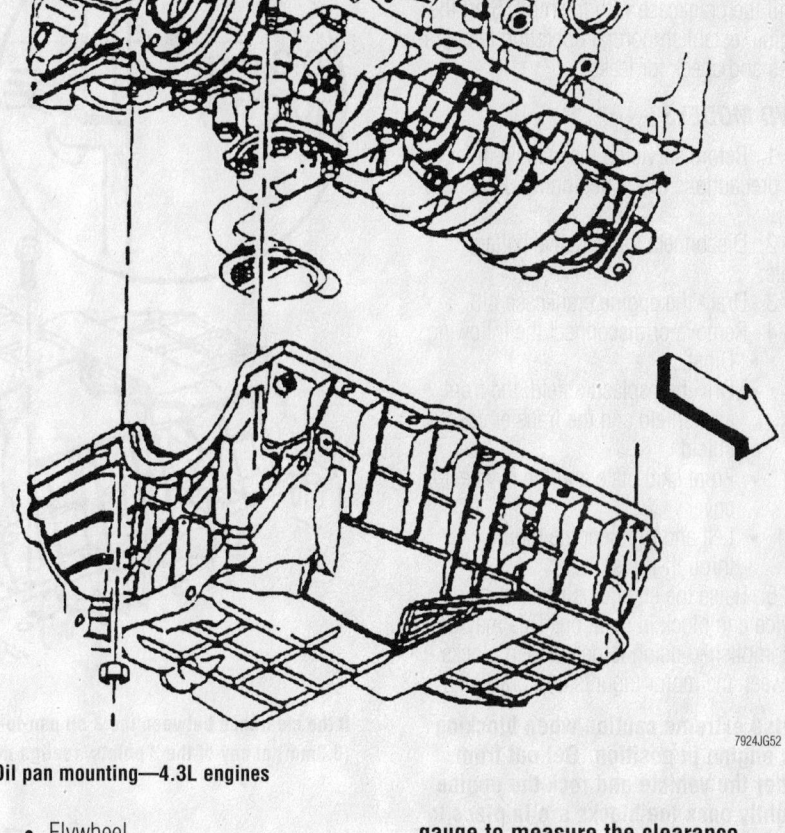

7924JG52

Oil pan mounting—4.3L engines

- Flywheel
- Pressure plate and disc, if equipped
- Engine

4.3L Engines

2WD MODELS

1. Before servicing the vehicle, refer to the precautions in the beginning of this section.
2. Drain the engine oil.
3. Remove or disconnect the following:
   - Engine
   - Oil level sensor, if equipped, and discard
   - Oil pan retainers (nuts, studs and/or bolts) and rail reinforcements, if equipped
   - Oil pan
   - Rubber bell housing plugs and gasket

To install:

4. Clean the gasket mounting surfaces.

➡**The alignment between the rear of the oil pan and the rear of the block is critical. The oil pan must be flush or slightly forward of the rear of the block to allow for proper alignment with the transmission housing. Use a feeler gauge to measure the clearance between the 3 oil pan-to-transmission contact points. If the clearance exceeds 0.011 in. (0.3mm) at any of the 3 points, realign the oil pan.**

5. Apply sealant to the oil pan rail where it contacts the timing cover-to-block joint (front) and the crankshaft rear seal retainer-to-block joint (rear). Continue the bead of sealant about 1 in. (25mm) in both directions from each of the 4 corners.

6. Install or connect the following:
   - Rubber bell housing plugs, if equipped
   - Oil pan using a new gasket

➡**The alignment between the rear of the pan and rear of the block is critical. The two surfaces must be flush to allow for proper alignment with the transmission housing.**

7. Use a feeler gauge to check the clearance between the oil pan-to-transmission contacts. If clearance exceeds 0.011 inch (0.3mm) at any of the three contact points, readjust the pan until the clearance is within specification.

8. Once the pan is in its correct position tighten the retainers to 18 ft. lbs. (25 Nm) using the proper sequence.

9. Install a new oil level sensor, if used and tighten to 115 inch lbs. (13 Nm).

10. Install the engine into the vehicle. Refill the crankcase with fresh oil. Start the engine, establish normal operating temperatures and check for leaks.

4WD MODELS

1. Before servicing the vehicle, refer to the precautions in the beginning of this section.

2. Disconnect the negative battery cable.

3. Drain the engine crankcase oil.

4. Remove or disconnect the following:
- Dipstick
- Drivebelt splash shield, the front axle shield and the transfer case shield
- Front skid plate and the flywheel cover
- Left and right engine mount through-bolts

5. Raise the engine using a lifting device and block in position. This may be accomplished using large wooden blocks between the motor mounts and brackets.

➡ Use extreme caution when blocking the engine in position. Get out from under the vehicle and rock the engine slightly once the blocks are in place to be sure the engine is properly supported.

6. Remove or disconnect the following:
- Oil cooler line
- Pitman arm bolt and pitman arm
- Idler arm bolts and idler arm
- Front differential through-bolts
- Front driveshaft, if necessary
- Differential assembly by rolling it forward for clearance
- Starter motor
- Oil pan bolts, nuts and reinforcements
- Oil pan and discard the gasket

To install:

7. Clean the gasket mounting surfaces.

➡ The alignment between the rear of the oil pan and the rear of the block is critical. The oil pan must be flush or slightly forward of the rear of the block to allow for proper alignment with the transmission housing. Use a feeler gauge to measure the clearance between the 3 oil pan-to-transmission contact points. If the clearance exceeds 0.011 in. (0.3mm) at any of the 3 points, realign the oil pan.

8. Apply sealant to the oil pan rail where it contacts the timing cover-to-block

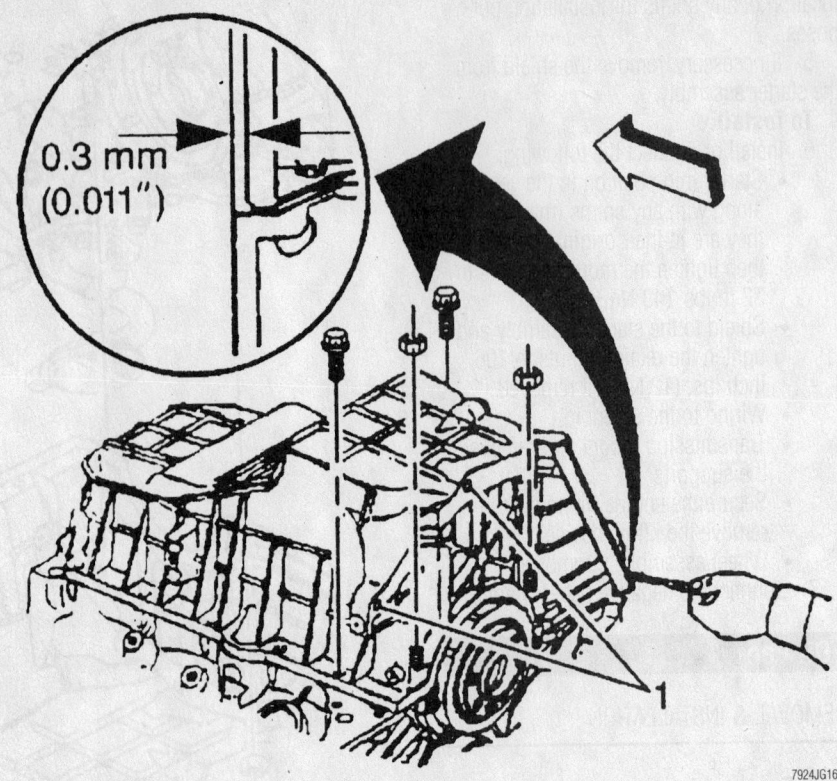

If the clearance between the 3 oil pan-to-transmission contact points exceeds 0.011 in. (0.3mm) at any of the 3 points, realign the oil pan—4.3L engine

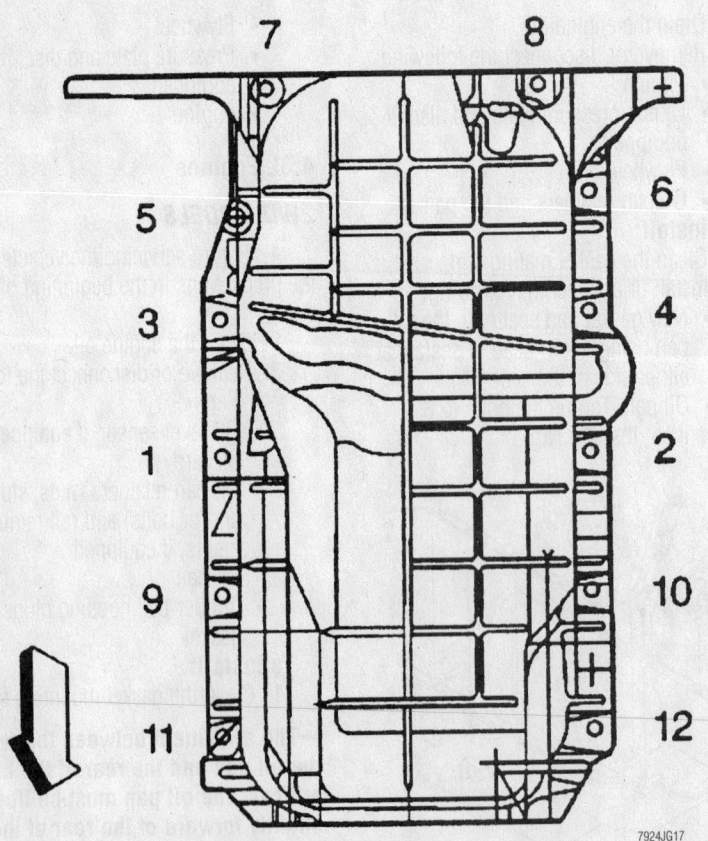

Tighten the bolts in sequence to prevent warping the sealing surface of the oil pan—4.3L vehicles

joint (front) and the crankshaft rear seal retainer-to-block joint (rear). Continue the bead of sealant about 1 in. (25mm) in both directions from each of the 4 corners.

9. Install or connect the following:
  • Oil pan, using a new gasket. Tighten the retainers, in sequence, to 18 ft. lbs. (25 Nm).
  • Starter motor
  • Differential by rolling it back into position
  • Front driveshaft
  • Front differential through-bolts
  • Idler arm and secure using the retaining bolts
  • Pitman arm and secure using the bolts
  • Transfer case shield
  • Flywheel cover
  • Front skid plate
  • Front axle shield
  • Drive belt splash shield
  • Dipstick
  • Negative battery cable
10. Refill the engine crankcase.
11. Start the engine and check for leaks.

Oil Pump

REMOVAL & INSTALLATION

1. Before servicing the vehicle, refer to the precautions in the beginning of this section.
2. Remove or disconnect the following:
  • Oil pan
  • Oil pump and the pickup tube/shaft, if equipped
  • Extension shaft and retainer from the pump, if necessary for the 2.2L engine

➡ **Be careful not to crack the retainer.**

To install:

3. For the 2.2L engine, if the extension shaft was removed, heat the extension shaft retainer in hot water, then install the shaft and retainer to the oil pump. Be sure the retainer does not crack during installation.
4. Ensure that the pump pickup tube is tight in the pump body. If the tube should come loose, oil pressure will be lost and oil starvation will occur. If the pickup tube is loose it should be replaced.
5. If the pump has been disassembled and is being replaced or for any reason oil has been removed, it must be primed. It can either be filled with oil before installing the cover plate and oil kept within the pump during handling or the entire pump cavity can be filled with petroleum jelly.

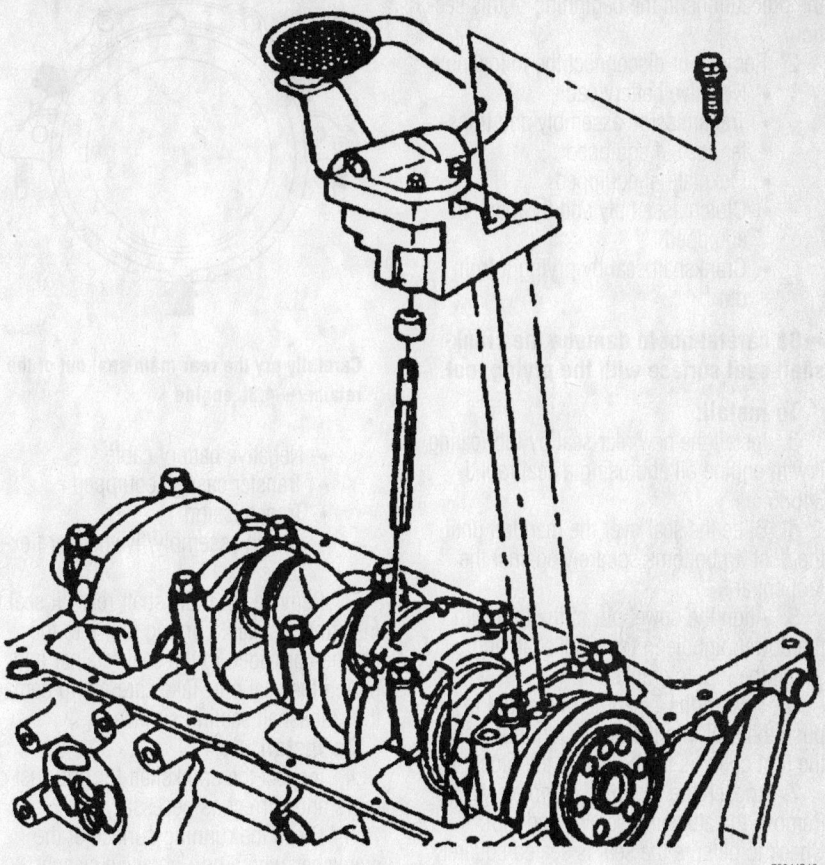

7924JG18

Exploded view of the oil pump mounting—2.2L engine

✳✳ WARNING

If the pump is not primed, the engine could be damaged upon start up.

6. Install or connect the following:
  • Oil pump. Tighten oil pump/pickup tube retainer(s) to 32 ft. lbs. (44 Nm) for the 2.2L engine or to 65 ft. lbs. (90 Nm), for the 4.3L engine.

➡ **If the oil pump does not build up oil pressure almost immediately, remove the pan and check for a loose oil pump-to-pickup tube attachment. If necessary dismantle the pump and pack the pump cavity with petroleum jelly.**

  • Oil pan
7. Refill the crankcase.
8. Disable the ignition system; crank engine for approximately 10 seconds to aid in priming the oil pump and reducing the risk of engine damage.

✳✳ WARNING

Running the engine without measurable oil pressure will cause extensive damage.

Rear Main Seal

REMOVAL & INSTALLATION

2.2L Engines

Please note that the transmission assembly and transfer case, if equipped, must be removed to perform this procedure.

1. Before servicing the vehicle, refer to

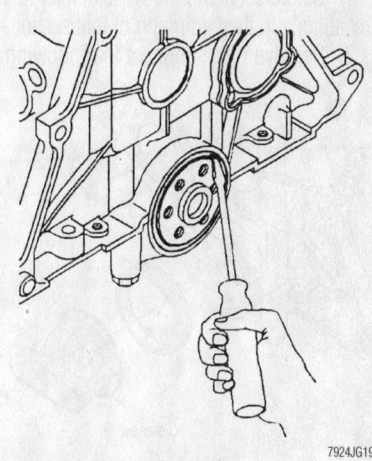

7924JG19

Carefully pry the rear main oil seal out of its bore—2.2L engine

the precautions in the beginning of this section.

2. Remove or disconnect the following:
- Negative battery cable
- Transmission assembly and transfer case, if equipped
- Flexplate, if equipped
- Clutch assembly and flywheel, if equipped
- Crankshaft seal by prying it from out

➡ Be careful not to damage the crankshaft seal surface with the prying tool.

To install:

3. Install the new rear seal by lubricating it with engine oil and using a seal tool J-34686.

4. Slide the seal over the mandrel until the dust lip bottoms squarely against the tool collar.

5. Align the dowel pin of the tool with the dowel pinhole in the crankshaft and attach the tool to crankshaft.

6. Tighten the T-handle of the tool to push the seal into the bore. Continue until the tool collar is flush against the block.

7. Loosen the T-handle completely. Remove the attaching screws and tool. Check to be sure the seal is seated squarely in the bore.

8. Install or connect the following:
- Flywheel/clutch assembly or flexplate
- Transmission assembly and transfer case, if equipped
- Negative battery cable

9. Start the engine and check for leaks.

4.3L Engines

Please note that the transmission assembly and transfer case, if equipped, must be removed to perform this procedure.

1. Before servicing the vehicle, refer to the precautions in the beginning of this section.

2. Remove or disconnect the following:

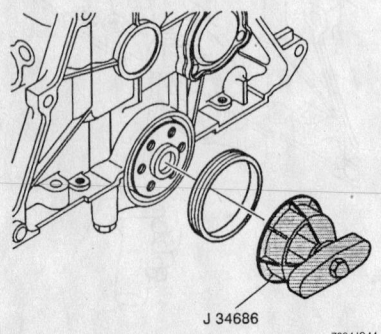

Rear main oil seal installation using tool
J-34686—2.2L engine

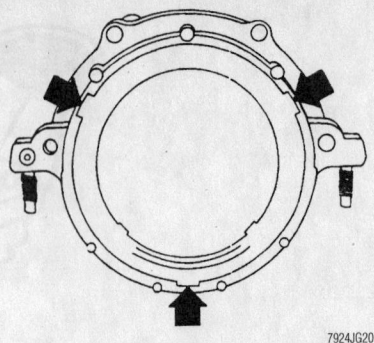

7924JG20

Carefully pry the rear main seal out of the retainer—4.3L engine

- Negative battery cable
- Transfer case, if equipped
- Transmission
- Clutch assembly/flywheel or flexplate

3. Remove the crankshaft rear oil seal by inserting a suitable prying tool into the notches provided in the seal retainer and prying the seal out. Take care not to damage the crankshaft sealing surface.

To install:

4. Inspect the crankshaft for grit, rust or burrs and correct as necessary.

5. Clean the running surface of the crankshaft with a non-abrasive cleaner.

6. Install or connect the following:
- New rear seal lubricated with engine oil and a seal installer
- Flywheel and clutch or flexplate
- Transmission
- Transfer case, if equipped
- Negative battery cable

7. Start the engine and verify no oil leaks.

Timing Chain, Sprockets, Front Cover and Seal

REMOVAL & INSTALLATION

Front Cover and Seal

2.2L ENGINES

1. Before servicing the vehicle, refer to the precautions in the beginning of this section.

2. Remove or disconnect the following:
- Negative battery cable
- Drive belt
- Cooling fan assembly and pulley
- Crankshaft pulley and hub
- Belt tensioner/idler pulley assembly
- Front oil pan-to-front cover nuts or studs
- Starter

- Alternator and brackets from the engine, then position them aside
- Oil pan bolts, loosen but do not remove
- Crankcase (timing) front cover bolts and the cover. Make sure all bolts are removed and be careful not to force and damage the cover.
- Old crankshaft seal from the cover using a suitable prytool. Be very careful not to distort the front cover or to score the end of the crankshaft.

To install:

3. Carefully remove all traces of gasket or sealant from the mating surfaces.

4. Lubricate the lips of a new seal with clean engine oil, then use a seal centering tool (such as J-35468) to install the seal to the front cover. Leave the tool in position in the seal until the cover is installed.

5. Apply a $\frac{3}{8}$ in. (10mm) wide by $\frac{5}{16}$ (5mm) thick bead of RTV sealer to the oil pan at the front crankcase cover sealing surface. Then apply a $\frac{1}{4}$ in. (6mm) by $\frac{1}{8}$ in. (3mm) thick bead of RTV to the crankcase front cover at the block sealing surface.

6. Install or connect the following:
- Crankcase front cover to the engine using the seal tool to assure it is properly centered and prevent damage to the hub. Tighten the cover retaining bolts to 97 inch lbs. (11 Nm), then remove the seal centering tool.
- Oil pan bolts
- Starter
- Alternator with brackets
- Belt tensioner/idler pulley assembly
- Belt assembly
- Crankshaft pulley and hub
- Cooling fan assembly and pulley
- Negative battery cable

4.3L ENGINES

1. Before servicing the vehicle, refer to the precautions in the beginning of this section.

2. Remove or disconnect the following:
- Negative battery cable
- Drain the engine cooling system.
- Crankshaft pulley and damper

✳✳ WARNING

The outer ring (weight) of the torsional damper is bonded to the hub with rubber. The damper must be removed with a puller that acts on the inner hub only. Pulling on the outer portion of the damper will break the rubber bond or destroy the tuning of the unit.

- Water pump assembly
- Crankshaft Position (CKP) sensor

➡**Depending upon the year of your truck, you may just need to loosen the oil pan bolts, or you may need to remove it completely.**

- Oil pan or loosen bolts, as applicable
- Crankshaft Position (CKP) sensor, if equipped
- Front cover bolts and the reinforcements, if equipped
- Front cover from the engine

3. Pry the seal out of the front cover using a small prytool. Be very careful not to distort the front cover or to score the end of the crankshaft.

To install:

➡**Anytime the front cover is removed, the cover must be replaced upon reassembly. If you reuse the old cover, oil leaks may develop.**

4. Clean the gasket mating surfaces of the engine and cover of all remaining gasket or sealer material. Be careful not to score or damage the surfaces.

➡**The manufacturer suggests you wait until the front cover is mounted to the engine before you install the replacement crankshaft oil seal. This assures the cover is properly supported.**

5. Install or connect the following:
- New front cover gasket to the engine or cover using gasket cement to hold it in position. Lubricate the front of the oil pan seal with engine oil to aid in reassembly.
- Front cover to the engine. Take care while engaging the front of the oil pan seal with the bottom of the cover. Apply sealer 12346141 to the oil pan rail where it contacts the timing cover-to-block joint (front) and the crankshaft rear seal retainer-to-block joint (rear). Continue the bead of sealant about 1 in. (25mm) in both directions from each of the four corners.
- Front cover retaining bolts and tighten to 106 inch. lbs. (12 Nm)

6. Lightly coat the lips of the replacement crankshaft seal with clean engine oil, then position the seal with the open end facing inward the engine. Use a suitable seal installation driver to position the seal in the front cover.
- CKP sensor O-ring and the sensor, if equipped

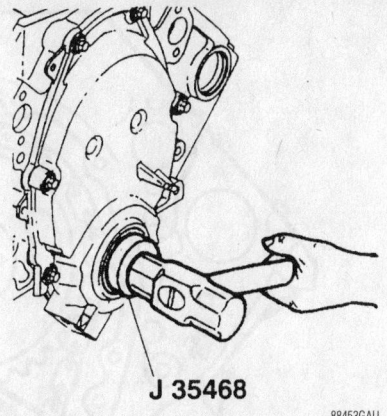

J 35468

88453GAU

Installing the crankshaft front oil seal—4.3L engines

- Oil pan, if removed
- Tighten the oil pan bolts
- Water pump
- Crankshaft damper and pulley
- Negative battery cable

7. Properly refill the engine cooling system.

8. Run the engine until normal operating temperature has been reached, then check for leaks.

Timing Chain and Sprockets

2.2L ENGINES

1. Before servicing the vehicle, refer to the precautions in the beginning of this section.

2. Remove or disconnect the following:
- Negative battery cable
- Crankcase (timing) front cover from the engine

3. Turn the crankshaft until the timing marks on the sprockets are in alignment. The marks should also be in alignment with the tabs on the tensioner.
- Tensioner retaining bolts
- Camshaft sprocket retaining bolts

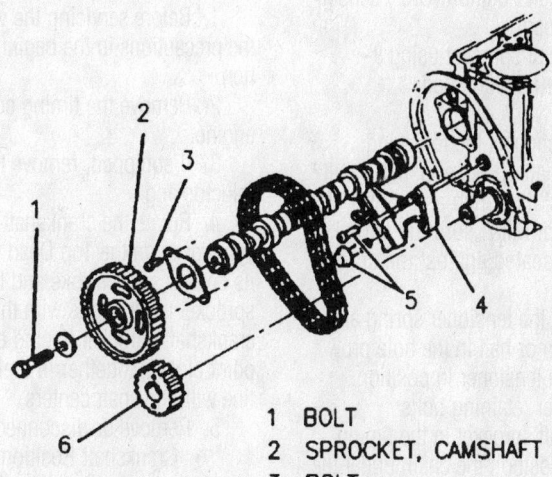

1   BOLT
2   SPROCKET, CAMSHAFT
3   BOLT
4   TENSIONER
5   BOLTS
6   SPOCKET, CRANKSHAFT

A ALIGN TABS ON TENSIONER WITH MARKS ON CAMSHAFT & CRANKSHAFT SPROCKETS.

85383289

Timing chain, sprocket and camshaft mounting—2.2L engine

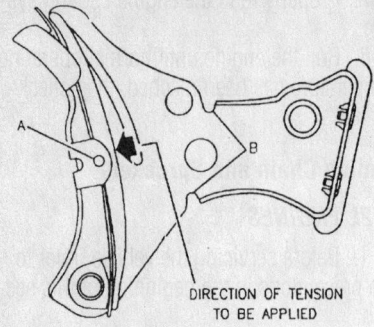

DIRECTION OF TENSION
TO BE APPLIED

A INSERT PIN AFTER TENSION HAS BEEN APPLIED
B TABS, USED FOR CAMSHAFT AND CRANKSHAFT
ALIGNMENT

85383290

Locking the timing chain tensioner into position for chain installation—2.2L engine

- Camshaft sprocket and timing chain at the same time
- Tensioner assembly
- Crankshaft Position (CKP) sensor, if equipped
- Crankshaft sprocket using J-22888-20, if equipped

To install:

4. Install or connect the following:
- Crankshaft sprocket using a suitable installer such as J-5590, if removed. Make sure the sprocket is fully seated against the crankshaft.

5. Compress the tensioner spring and insert a cotter pin or nail in the hole provided to hold the tensioner in position.
- Tensioner retaining bolts
- Camshaft sprocket in the timing chain, position the chain under the crankshaft sprocket and the camshaft sprocket to the camshaft

6. Verify that the timing marks are all properly aligned, then loosely install the camshaft sprocket bolt.
- CKP sensor, if equipped
- Tighten the tensioner bolts to 18 ft. lbs. (24 Nm), then tighten the camshaft sprocket bolt to 96 ft. lbs. (130 Nm)

7. Remove the cotter pin or nail holding the tensioner in position off the chain.
- Timing cover to the engine

4.3L ENGINES

➡ **The following procedure requires the use of the Crankshaft Sprocket Removal tool No. J-5825-A and the Crankshaft Sprocket Installation tool No. J-5590.**

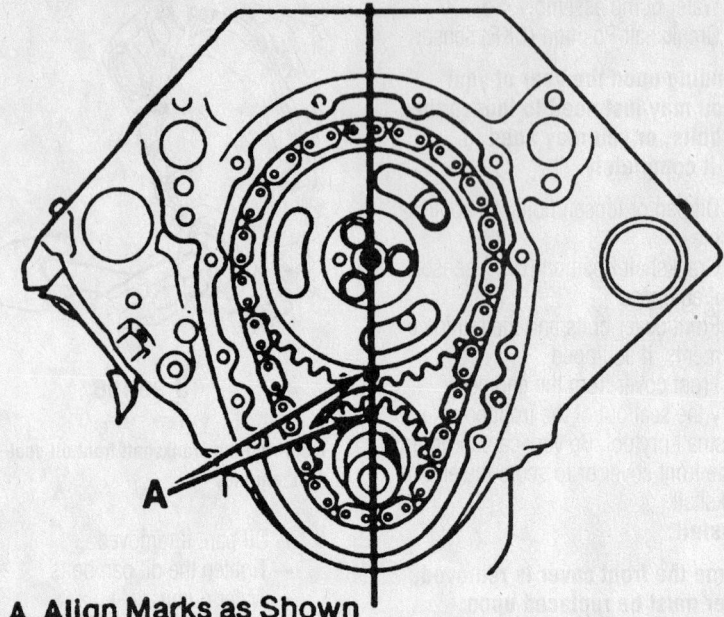

A. Align Marks as Shown

Timing mark alignment—4.3L engine

1. Before servicing the vehicle, refer to the precautions in the beginning of this section.

2. Remove the timing cover from the engine.

3. If equipped, remove the crankshaft reluctor ring.

4. Rotate the crankshaft until the No. 4 cylinder is on the Top Dead Center (TDC) of its compression stroke and the camshaft sprocket mark aligns with the mark on the crankshaft sprocket (facing each other at a point closest together in their travel) and in line with the shaft centers.

5. Remove or disconnect the following:
- Crankshaft Position (CKP) sensor reluctor ring, if equipped
- Camshaft sprocket-to-camshaft nut and/or bolts
- Camshaft sprocket (along with the timing chain). If the sprocket is difficult to remove, use a plastic mallet to bump the sprocket from the camshaft.

➡ **The camshaft sprocket (located by a dowel) is lightly pressed onto the camshaft and should come off easily. The chain comes off with the camshaft sprocket.**

6. If necessary use J-5825-A crankshaft sprocket removal tool to free the timing sprocket from the crankshaft.

7. If necessary, remove the crankshaft sprocket key.

To install:

8. Inspect the timing chain and the tim-

ing sprockets for wear or damage, replace the damaged parts as necessary.

9. Using a putty knife, clean the gasket mounting surfaces. Using solvent, clean the oil and grease from the gasket mounting surfaces.

10. Install or connect the following:
- Crankshaft sprocket key, if removed

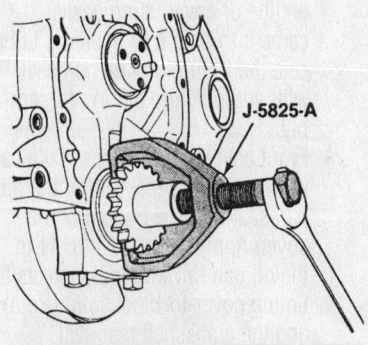

J-5825-A

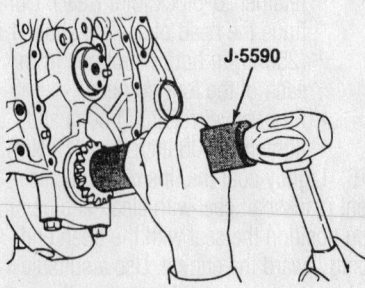

J-5590

85383293

Removal (top) and installation of the crankshaft timing gear

- Crankshaft sprocket onto the crankshaft using J-5590 crankshaft sprocket installation tool and a hammer without disturbing the position of the engine

→**During installation, coat the thrust surfaces lightly with Molykote• or an equivalent pre-lube.**

- Timing chain over the camshaft sprocket. Arrange the camshaft sprocket in such a way that the timing marks will align between the shaft centers and the camshaft locating dowel will enter the dowel hole in the cam sprocket.
- Timing chain under the crankshaft sprocket, then place the cam sprocket, with the chain still mounted over it, in position on the front of the camshaft
- Camshaft sprocket-to-camshaft retainers to 18 ft. lbs. (25 Nm)

11. With the timing chain installed, turn the crankshaft two complete revolutions, then check to make certain that the timing marks are in correct alignment between the shaft centers.

- CKP sensor reluctor ring, if equipped
- Timing cover

Piston and Ring

POSITIONING

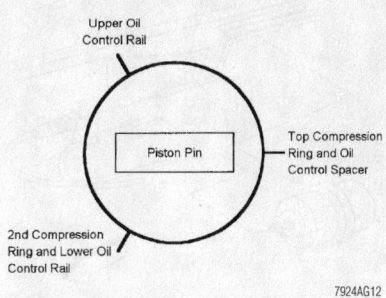

Piston ring end-gap spacing—GM/Isuzu 2.2L engine

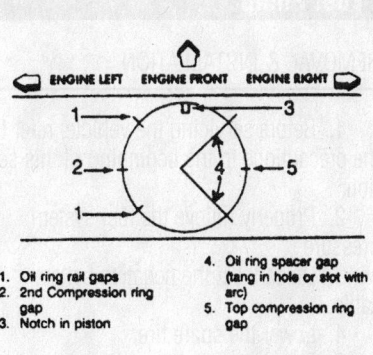

1. Oil ring rail gaps
2. 2nd Compression ring gap
3. Notch in piston
4. Oil ring spacer gap (tang in hole or slot with arc)
5. Top compression ring gap

Piston ring end-gap spacing—GM 4.3L engines

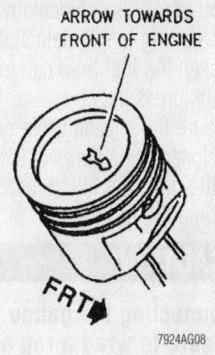

Piston/connecting rod-to-engine positioning—GM/Isuzu 2.2L engines

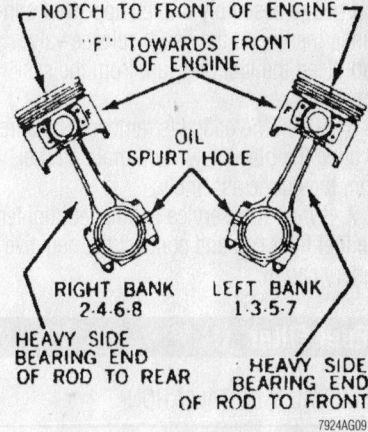

Piston and connecting rod assembly positioning—GM 4.3L engine

FUEL SYSTEM

Fuel System Service Precautions

Safety is the most important factor when performing not only fuel system maintenance but also any type of maintenance. Failure to conduct maintenance and repairs in a safe manner may result in serious personal injury or death. Maintenance and testing of the vehicle's fuel system components can be accomplished safely and effectively by adhering to the following rules and guidelines.

- To avoid the possibility of fire and personal injury, always disconnect the negative battery cable unless the repair or test procedure requires that battery voltage be applied.
- Always relieve the fuel system pressure prior to disconnecting any fuel system component (injector, fuel rail, pressure regulator, etc.), fitting or fuel line connection. Exercise extreme caution whenever relieving fuel system pressure, to avoid exposing skin, face and eyes to fuel spray. Please be advised that fuel under pressure may penetrate the skin or any part of the body that it contacts.

- Always place a shop towel or cloth around the fitting or connection prior to loosening to absorb any excess fuel due to spillage. Ensure that all fuel spillage (should it occur) is quickly removed from engine surfaces. Ensure that all fuel soaked cloths or towels are deposited into a suitable waste container.
- Always keep a dry chemical (Class B) fire extinguisher near the work area.
- Do not allow fuel spray or fuel vapors to come into contact with a spark or open flame.
- Always use a back-up wrench when loosening and tightening fuel line connection fittings. This will prevent unnecessary stress and torsion to fuel line piping. Always follow the proper torque specifications.
- Always replace worn fuel fitting O-rings with new. Do not substitute fuel hose or equivalent where fuel pipe is installed.

Fuel System Pressure

RELIEVING

Multi-Port Fuel Injection and Central Port Injection Systems

The fuel systems operate under high fuel pressures. It is very important that the pressure be properly relieved prior to servicing the system or any of its components.

A Schrader valve is provided on these fuel systems to conveniently test or release the system pressure. A fuel pressure gauge and adapter will be necessary to connect the gauge to the fitting. Most of the MFI systems utilize a service valve on one end of the fuel rail assembly.

1. Before servicing the vehicle, refer to the precautions in the beginning of this section.

2. Disconnect the negative battery cable to assure the prevention of fuel spillage if

the ignition switch is accidentally turned **ON** while a fitting is still detached.

3. Loosen the fuel filler cap to release the fuel tank pressure.

4. Be sure the release valve on the fuel gauge is closed, then connect the fuel gauge to the pressure fitting located on the inlet fuel pipe fitting.

✳✳ CAUTION

When connecting the gauge to the fitting, be sure to wrap a rag around the fitting to avoid spillage. After repairs, place the rag in an approved container.

5. Install the bleed hose portion of the fuel gauge assembly into an approved container, then open the gauge release valve and bleed the fuel pressure from the system.

6. When the gauge is removed, be sure to open the bleed valve and drain all fuel from the gauge assembly.

7. When fuel service is finished, tighten the fuel filler cap and connect the negative battery cable.

Fuel Filter

REMOVAL & INSTALLATION

1. Before servicing the vehicle, refer to the precautions in the beginning of this section.

2. Properly relieve the fuel system pressure.

3. Remove or disconnect the following:
- Negative battery cable
- Fuel filler cap
- Quick connect fittings from the filter
- Filter feed nut and the clamp bolt
- Filter and the clamp from the vehicle

To install:

4. Install or connect the following:
- Filter and clamp with the directional arrow facing away from the fuel tank, towards the throttle body

➥**The filter has an arrow (fuel flow direction) on the side of the case, be sure to install it correctly in the system, the with arrow facing away from the fuel tank.**

- Tighten the fuel feed nut
- Tighten the filter clamp assembly bolt
- Fuel quick disconnect fittings to the filter

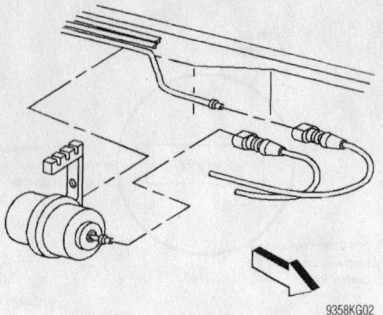

Typical fuel filter location along frame rail

9358KG02

- Fuel filler cap
- Negative battery cable

5. Start the engine and check for leaks.

Fuel Pump

REMOVAL & INSTALLATION

1. Before servicing the vehicle, refer to the precautions in the beginning of this section.

2. Properly relieve the fuel system pressure.

3. Disconnect the negative battery cable.

4. Lower the spare tire.

5. Remove or disconnect the following:
- Rear tail lamp assemblies
- Frame-to-pickup box bolts
- Wiring harness ground wire from the frame
- License plate lamp
- Fuel filler neck-to-pickup box ground wire and screws
- Pickup box
- Fuel sender electrical connectors
- Fuel sender and Evaporative Emission (EVAP) pipes

✳✳ WARNING

The fuel sender assembly may spring up from the fuel tank. When removing the sender from the fuel tank, keep in mind that the reservoir bucket is full of fuel, so you must tip the sender slightly during removal to avoid damaging the float. Discard the fuel sender O-ring and replace with a new on during installation.

6. While holding the module fuel sender down, remove the snapring from the designated slots (1) found on the retainer.

To install:

7. Install a new O-ring on the fuel sender to the tank.

8. Align the tab on the front of the

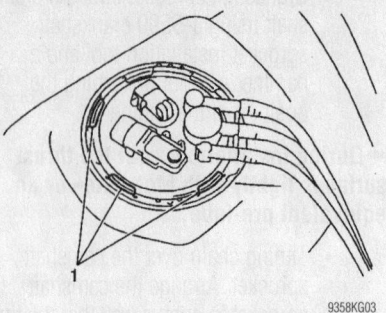

Hold the fuel sender down, and remove the snapring from the designated slots (1) found on the retainer

9358KG03

sender with the slot on the front of the retainer snapring.

9. Slowly apply pressure to the top of the spring-loaded sender until it aligns flush with the retainer on the tank.

➥**Make sure that the snapring is properly and fully seated in the tab slots.**

10. Install or connect the following:
- Snapring into the proper slots
- Fuel and EVAP pipes
- Electrical connectors
- Negative battery cable

11. Check for fuel leaks as follows:
a. Turn the ignition **ON** for 2 seconds.
b. Turn the ignition **OFF** for 10 seconds.
c. Turn the ignition **ON**.
d. Check for leaks.

12. Install or connect the following:
- Pickup box to the truck
- Filler neck-to-pickup box screws and ground wire
- License plate lamp
- Wiring harness ground wire to the frame
- Frame-to-pickup box bolts and tighten to 52 ft. lbs. (70 Nm)
- Rear tail lamp assemblies

Fuel Injector

REMOVAL & INSTALLATION

2.2L Engines

1. Before servicing the vehicle, refer to the precautions in the beginning of this section.

2. Relieve the fuel system pressure.

3. Remove or disconnect the following:
- Negative battery cable
- Intake manifold, if necessary
- Fuel injector electrical connections by pushing in the wire connector

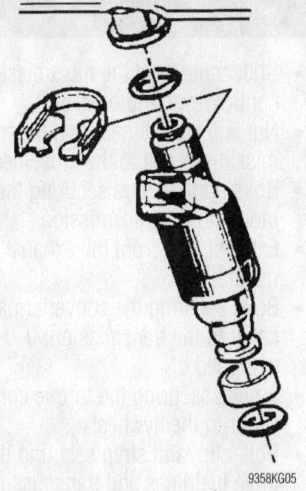

9358KG05

**Exploded view of a typical fuel injector—
2.2L engine**

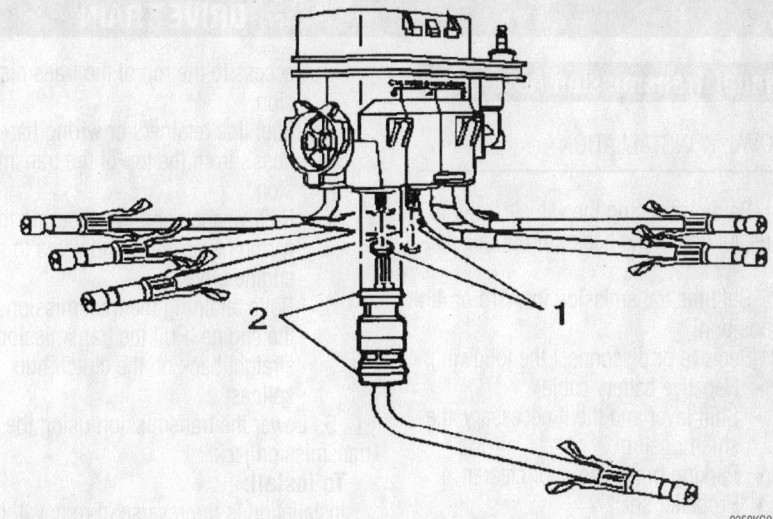

9358KG04

Exploded view of the fuel meter assembly, including the injectors

clip and gently pulling on the connector
• Fuel feed inlet pipe from the rail

➡**Use a back-up wrench on the fuel rail
return fitting to prevent it from turning.**

• Fuel return pipe from the fuel pres-
sure regulator
• Fuel pressure regulator
• Fuel rail attaching bolts and lift the
fuel rail assembly from the cylinder
head
• Fuel rail by moving the rail towards
the front of the engine
• Fuel injector retaining clip
• Fuel injector

➡**Because each injector is calibrated
for a specific flow rate, make sure you
only replace fuel injectors using an
IDENTICAL part number to the old
injectors.**

To install:

➡**When installing the injector care
should be taken not to tear or misalign
O-rings.**

4. Lubricate the injector O-ring seals with
clean engine oil and install them injector.
5. Install or connect the following:
• Upper O-ring, lower back-up O-
ring and lower O-ring
• Fuel injector to the fuel rail
• Fuel injector retaining clip
• Fuel rail and insert it into the cylin-
der head
• Fuel rail retaining bolts and tighten
to 18 ft. lbs. (25 Nm)
• Fuel pressure regulator

➡**Use a back-up wrench on the fuel rail
return fitting to prevent it from turning.**

• Return pipe to the fuel pressure
regulator. Tighten the fuel pipe nut
to 22 ft. lbs. (30 Nm).
• Fuel feed inlet pipe to rail

➡**Rotate the fuel injectors as necessary to avoid stretching the wire harness.**

• Injector electrical connections
• Intake manifold, if removed
• Negative battery cable
6. Inspect for leaks as follows:
a. Turn the switch to the **ON** position
for 2 seconds.
b. Turn the ignition switch **OFF** for 10
seconds.
c. Turn the ignition switch to the **ON**
position and check for leaks.

4.3L Engines

1. Before servicing the vehicle, refer to
the precautions in the beginning of this section.
2. Relieve the fuel system pressure.
Refer to the fuel system relief procedure in
this section.
3. Relieve the fuel system pressure.
4. Remove or disconnect the following:
• Negative battery cable
• Fuel meter body electrical connection and the fuel feed and return
hoses from the engine fuel pipes
• Upper manifold assembly
• Poppet nozzle out of the casting
socket
• Fuel meter body by releasing the
locktabs

➡**Each injector is calibrated. When
replacing the fuel injectors, be sure to
replace it with the correct injector.**

• Lower hold-down plate and nuts
5. While pulling the poppet nozzle tube
downward, push with a small prytool down
between the injector terminals and remove
the injectors.
To install:
6. Lubricate the new injector O-ring
seats with engine oil.
7. Install or connect the following:
• O-rings on the injector
• Fuel injector into the fuel meter
body injector socket.
• Lower hold-down plate and nuts.
Torque the nuts to 27 inch lbs. (3
Nm).
• Fuel meter body assembly into the
intake manifold. Torque the fuel
meter bracket retainer bolts to 88
inch. lbs. (10 Nm).

✳✳ **CAUTION**

**To reduce the risk of fire or injury
ensure that the poppet nozzles are
properly seated and locked in their
casting sockets**

• Fuel meter body into the bracket
and lock all the tabs in place
• Poppet nozzles into the casting
sockets
• Electrical connections
• New o-ring seals on the fuel return
and feed hoses.
• Fuel feed and return hoses and
tighten the fuel pipe nuts to 22 ft.
lbs. (30 Nm).
• Negative battery cable
8. Turn the ignition **ON** for 2 seconds
and then turn it **OFF** for 10 seconds. Again
turn the ignition **ON** and check for leaks.
9. Install the manifold plenum.

Manual Transmission Assembly

REMOVAL & INSTALLATION

1. Before servicing the vehicle, refer to the precautions in the beginning of this section.

2. Shift the transmission into 3rd or 4th gear position.

3. Remove or disconnect the following:
- Negative battery cable
- Shift lever and the if necessary, the shift housing
- Parking brake cable for clearance
- Propeller shaft
- Sid plate, if equipped
- Transfer case and shift lever, on 4WD models
- All wiring harness that would interfere with transmission removal
- Fuel line retainers from the rear crossmember
- Muffler from the catalytic converter
- Exhaust pipes from the exhaust manifold
- Catalytic converter hanger, if necessary
- Exhaust section
- Bolts and nuts attaching any transmission braces to the engine and transmission
- Hydraulic clutch quick-connect from the concentric slave cylinder following 1 of the 2 steps:

a. Use 2 small prytools at 180 degrees from each other to depress the white plastic sleeve on the quick connect to separate the clutch line from the concentric slave cylinder quick connect.

b. Use special tool J–36221 to depress the white plastic sleeve on the quick connect to separate the clutch line end from the concentric slave cylinder quick connect.

4. Remove or disconnect the following:
- Bolts securing the clutch housing cover to the transmission, if equipped
- Clutch plate and clutch cover, if necessary

5. Support the transmission with a suitable jack.
- Rear crossmember from the frame rail
- Wiring harness from the front crossmember, if equipped. Move the wiring harness away from the transmission oil pan. Lower the transmission enough to gain access to the top of the transmission.
- Fuel line retainers or wiring harness's from the top of the transmission
- Bolt, washer, and nut securing the wiring harness ground wires to the engine block
- Bolts retaining the transmission to the engine. Pull the transmission straight back on the clutch hub splines.

6. Lower the transmission using the transmission jack.

To install:

Installation is the reverse of removal, but please note the following important steps.

7. Place a THIN coat of high-temperature grease on the main drive gear (input shaft) splines.

8. Secure the transmission to the floor jack and raise the transmission into position.

➡**On some models, it may be necessary to rotate the transmission clockwise while inserting it into the clutch hub.**

9. Slowly insert the input shaft through the clutch. Rotate the output shaft slowly to engage the splines of the input shaft into the clutch while pushing the transmission forward into place. Do not force the transmission into position, the transmission should easily fall into place once everything is properly aligned.

10. Tighten the transmission mounting bolts to 35 ft. lbs. (47 Nm).

11. Do not remove the transmission jack until the crossmembers have been installed.

12. Check the transmission fluid level and replenish as necessary.

Automatic Transmission Assembly

REMOVAL & INSTALLATION

1. Before servicing the vehicle, refer to the precautions in the beginning of this section.

2. Remove or disconnect the following:
- Negative battery cable

3. Drain the transmission fluid.
- Driveshaft from the transmission (2WD) and transfer case, if equipped (4WD)

4. Support the transmission with a suitable transmission jack.
- Shift cable from the transmission control lever and bracket
- Nut and washer securing the transmission mount to the crossmember
- Bolts and washers securing the mount to the transmission
- Exhaust pipe from the exhaust manifold(s)
- Bolts securing the converter pan cover to the transmission, if equipped
- 3 bolts securing the torque converter to the flywheel
- Bolt, clip, and strap securing the three fuel lines and transmission vent hose to the transmission case
- Bolts and nut securing the transmission to the engine
- Oil filler tube and seal from the transmission
- Transmission cooler lines from the transmission. Plug the lines and the ports in the transmission.
- Wiring harness connectors from the transmission.

5. Inspect for any other wiring, brackets etc. which may interfere with the removal of the transmission.

6. Since the transmission acts as a rear engine mount, properly support the rear of the engine with an underbody support or other suitable support before attempting to remove the transmission. Otherwise the rear of the engine may pitch downward and components on the rear of the engine and on the firewall may be damaged.

7. Remove the transmission from the engine by pulling the transmission rearward to disengage it from the locator dowel pins on the back of the block. Carefully lower the transmission from the vehicle. Use care that the torque converter does not fall out of the front of the transmission.

➡**Use converter holding strap tool No. J-21366, to secure the torque converter to the transmission during removal and installation procedures.**

To install:

Installation is the reverse of removal, but please note the following important steps.

8. Make sure the torque converter is fully seated in the pump drive. If not, the transmission will not fit tightly to the rear of the engine block.

9. Raise the transmission into position and remove the torque converter holding

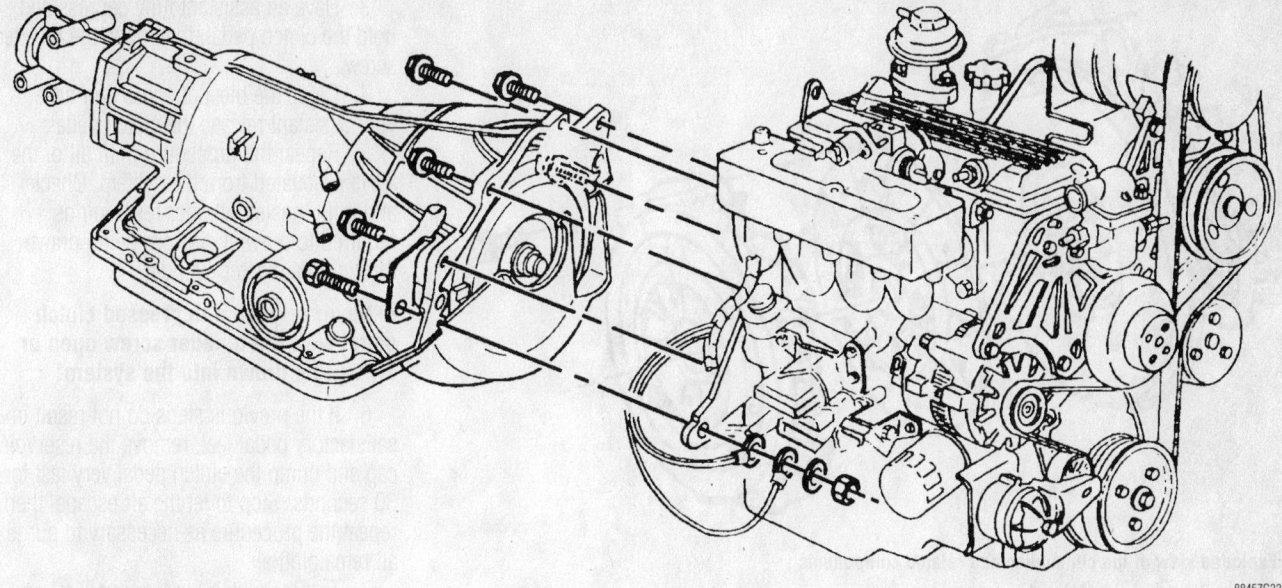

Transmission mounting on 2.2L engines

strap and carefully. Slide the transmission forward until the dowel pins are engaged.

10. The torque converter should be flush with the flywheel and turn freely by hand.

11. Install the transmission–to–engine bolts. Tighten the bolts to 34 ft. lbs. (47 Nm).

12. Tighten the torque converter–to–flywheel bolts to 46 ft. lbs. (63 Nm).

13. If equipped, tighten the converter pan cover to the transmission bolts to 37 ft. lbs. (50 Nm)

14. Tighten the bolts and washers securing the transmission mount to 35 ft. lbs. (47 Nm).

15. Tighten the nut and washer securing the transmission mount to the crossmember to 38 ft. lbs. (52 Nm).

16. Refill the transmission with the proper amount and type of fluid.

17. Connect the negative battery cable. Start the vehicle and allow to warm while checking for leaks. Road test the vehicle to check for shift quality.

Clutch

REMOVAL & INSTALLATION

1. Before servicing the vehicle, refer to the precautions in the beginning of this section.

2. Remove or disconnect the following:
   - Negative battery cable
   - Transmission

3. Install a clutch alignment tool or a used transmission input shaft to support the clutch.

4. If the clutch assembly is going to be reused, mark the flywheel, clutch cover and a pressure plate lug for alignment when installing.

5. Remove or disconnect the following:
   - Clutch cover bolts and washers
   - Clutch cover assembly and the clutch plate
   - Clutch alignment tool

6. Clean all parts and inspect for damage.

To install:

7. Install or connect the following:
   - Clutch alignment tool, to support the clutch
   - Clutch cover by aligning the matchmarks or, if new, align the lightest part of the cover, identified by a yellow dot, with the heaviest part identified by an **X**.
   - Clutch plate/clutch cover assembly to the flywheel. Tighten the bolts to 33 ft. lbs. (45 Nm) for 2.2L engines or to 29 ft. lbs. (40 Nm) for 4.3L engines.

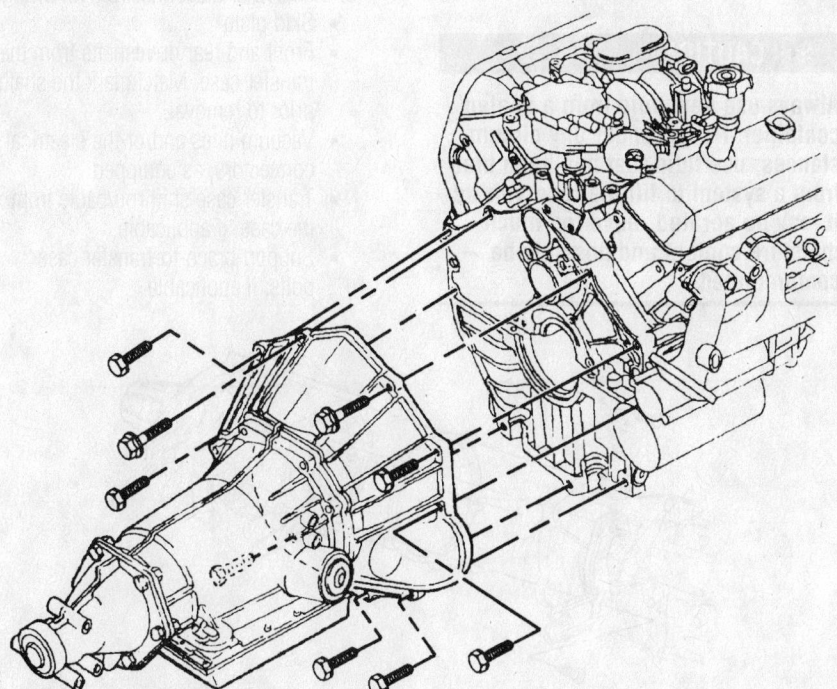

Transmission mounting on 4.3L engines

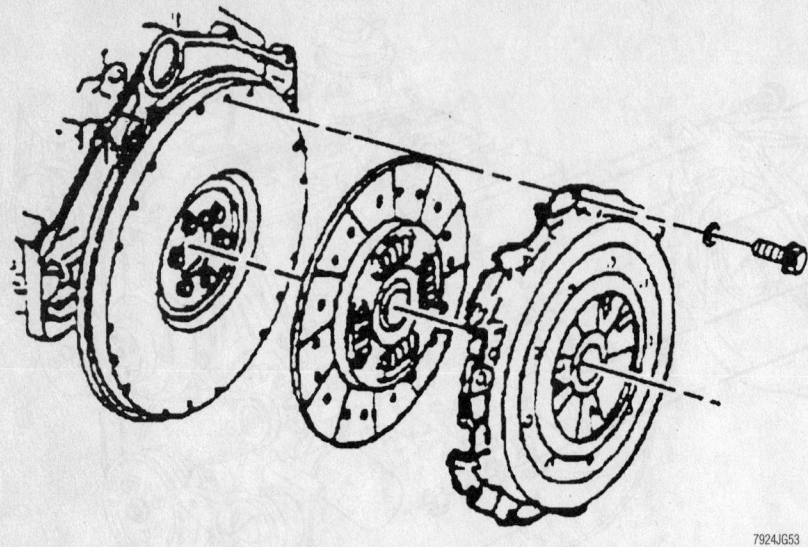

7924JG53

Exploded view of the clutch disc and related components

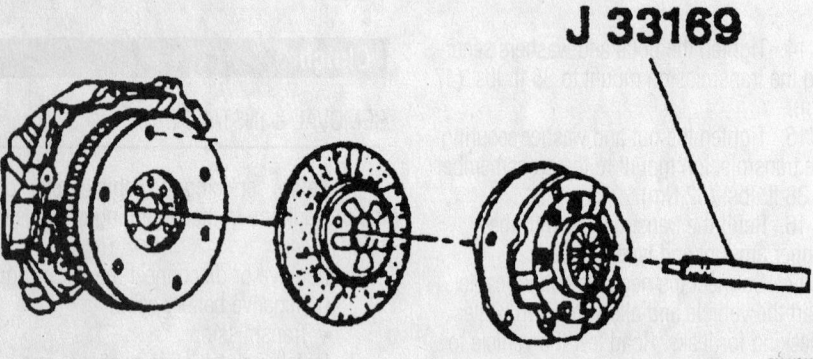

J 33169

7924JG22

Use the clutch alignment tool to center and support the clutch disc during installation

➡️ **Tighten each screw 1 turn at a time to avoid warping the clutch cover.**

8. Remove the clutch alignment tool.
9. Install or connect the following:
   • Transmission
   • Negative battery cable

Hydraulic Clutch System

Bleeding air from the hydraulic clutch system is necessary whenever any part of the system has been disconnected or the fluid level (in the reservoir) has been allowed to fall so low, that air has been drawn into the master cylinder.

BLEEDING

1. Before servicing the vehicle, refer to the precautions in the beginning of this section.
2. Fill master cylinder reservoir with new brake fluid conforming to DOT 3 specifications.

3. Have an assistant fully depress and hold the clutch pedal, then open the bleeder screw.
4. Close the bleeder screw and have your assistant release the clutch pedal.
5. Repeat the procedure until all of the air is evacuated from the system. Check and refill master cylinder reservoir as required to prevent air from being drawn through the master cylinder.

➡️ **Never release a depressed clutch pedal with the bleeder screw open or air will be drawn into the system.**

6. If the previous steps do not result in satisfactory pedal feel, remove the reservoir cap and pump the clutch pedal very fast for 30 seconds. Stop to let the air escape, then repeat the procedure as necessary to purge all remaining air.
7. Test the clutch for proper operation.

Transfer Case Assembly

REMOVAL & INSTALLATION

1. Before servicing the vehicle, refer to the precautions in the beginning of this section.
2. Disconnect the negative battery cable.
3. Shift the transfer case into the **4HI** range.
4. Drain the transfer case fluid.
5. Support the transfer case.
6. Remove or disconnect the following:
   • Skid plate
   • Front and rear driveshafts from the transfer case. Matchmark the shafts prior to removal.
   • Vacuum lines and/or the electrical connectors, as equipped
   • Transfer case shift rod/cable from the case, if applicable
   • Support brace-to-transfer case bolts, if applicable

✳✳ CAUTION

Always use new fluid from a sealed container. Never, under any circumstances, use fluid that has been bled from a system to fill the reservoir as it may be aerated, have too much moisture content and possibly be contaminated.

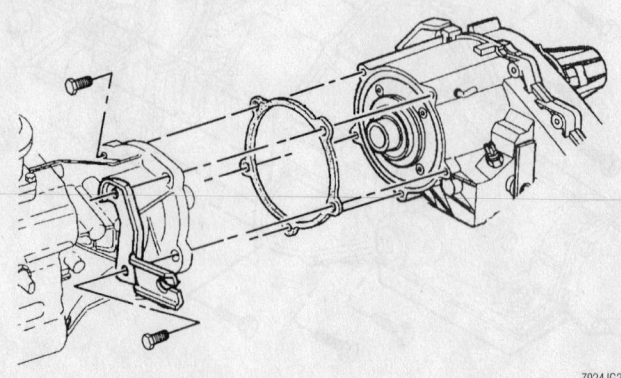

7924JG27

Transfer case-to-manual transmission mounting—Typical

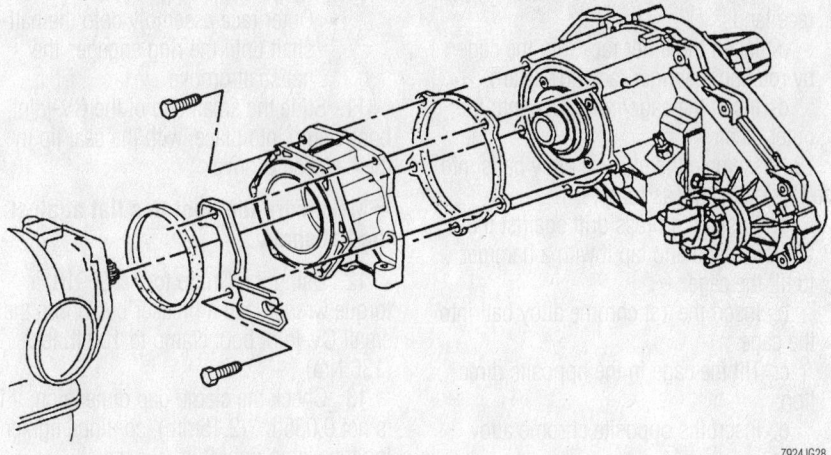

Transfer case-to-automatic transmission mounting—Typical

7924JG28

• Transfer case
7. Remove all traces of old gasket material from the mating surfaces.
   To install:
8. Install or connect the following:
   • New gasket using sealer to hold it in position
   • Transfer case. Torque the bolts to 33–35 ft. lbs. (45–47 Nm).
   • Support brace bolts. Torque the bolts to 35–37 ft. lbs. (47–50 Nm), if equipped
   • Shift rod to the case, if equipped
   • Vacuum lines and/or electrical connections, as necessary
   • Front and rear driveshafts by aligning the matchmarks
9. Refill the transfer case.
10. Install or connect the following:
    • Skid plate, if equipped
    • Negative battery cable

Halfshaft

REMOVAL & INSTALLATION

1. Before servicing the vehicle, refer to the precautions in the beginning of this section.
2. Unlock the steering column so the steering linkage is free to move.
3. Remove or disconnect the following:
   • Negative battery cable
   • Front wheels
   • Skid plate
   • 6 bolts and the flange
   • Snapring
   • Spindle washer

❋❋ CAUTION

The coil spring is under extreme pressure. Make sure the control arm is firmly supported with a hydraulic jack before removing the lower ball joint nut. After the lower ball joint nut has been removed, lower the hydraulic jack slowly to relieve coil spring pressure. If this precaution is not observed, serious bodily injury may result.

4. Support the lower control arm with a hydraulic jack.
5. Remove the cotter pin and nut attaching the ball joint to the lower control arm.
6. Remove the nuts and bolts connecting the strut to the steering knuckle. Separate the steering knuckle from the strut and lower control arm.
7. Slowly lower the hydraulic jack until coil spring pressure is relieved.
8. Remove the outer CV-joint from the steering knuckle.
9. If removing the right side halfshaft, place tool J 37780 or equivalent, between the front axle housing and the inner CV-joint. Gently tap the inner CV-joint away and out of the front axle housing.
10. If removing the left side halfshaft, scribe a reference mark on the left inner axle shaft flange and the inner CV-joint flange to ensure correct installation. Remove the three bolts and three nuts and separate the inner CV-joint from the left inner axle shaft.
11. Remove the halfshaft from the vehicle.
    To install:
12. If installing the right halfshaft, install the inner CV-joint into the axle housing, making sure the snapring seats in the differential side gear.
13. If installing the left halfshaft, install the left inner axle shaft flange to the inner CV-joint flange, aligning the reference marks made during removal. Install the three bolts and three nuts and tighten to 41 ft. lbs. (55 Nm).
14. Install the outer CV-joint into the steering knuckle.
15. Support the lower control arm with the hydraulic jack.
16. Attach the steering knuckle and lower ball joint to the lower control arm. Tighten the strut bolts and nuts to 65 ft. lbs. (90 Nm). Tighten the ball joint nut to 63 ft. lbs. (85 Nm) and install a new cotter pin.
17. Remove the hydraulic jack from the lower control arm.
18. Install the spindle washer and snapring to the end of the halfshaft.
19. Apply sealer, GM part no. 1052366, or equivalent, to the flange. Install the flange. Torque the flange bolts to 35 ft. lbs. (48 Nm).
20. If equipped, install the locking hub.
21. Install the front wheel.
22. Install the skid plate, if equipped.
23. Lower the vehicle.

CV-Joints

OVERHAUL

Outer CV-Joint

1. Before servicing the vehicle, refer to the precautions in the beginning of this section.
2. Remove or disconnect the following:
   • Front wheel
   • Halfshaft and position it in a vise
   • Large CV-joint boot clamp and discard it
   • Small CV-joint boot clamp and discard it
   • CV-joint boot and slide it back on the shaft
   • Outer race from the halfshaft, by spreading the outer race-to-halfshaft retaining ring, using Snapring Pliers J-8059
   • Retaining ring from the halfshaft and discard it
   • CV-joint boot from the halfshaft and discard it, if damaged
3. Disassemble the chrome alloy balls from the CV-joint cage as follows:
   a. Position a brass drift against the CV-joint cage and tap it with a hammer to tilt the cage.
   b. Remove the 1st chrome alloy ball from the cage.
   c. Tilt the cage in the opposite direction.

d. Remove the opposite chrome alloy ball.

e. Repeat the procedure until all 6 balls are removed.

4. Disassemble the CV-joint cage and inner race as follows:

a. Pivot the cage and race 90 degrees to the center line of the outer race.

b. Align the cage windows with outer race lands.

c. Remove the cage from the outer race.

d. Rotate the inner race upward and remove it from the cage.

5. Thoroughly clean and inspect all parts.

To install:

6. Lubricate the parts with a light coat of grease.

7. Assemble the CV-joint cage and inner race, as follows:

a. Rotate the inner race 90 degrees to the cage centerline.

b. Align the cage windows with inner race lands.

c. Insert the inner race into the cage by rotating the inner race downward.

d. Insert the cage/inner race into the outer race.

8. Assemble the chrome alloy balls into the CV-joint cage, as follows:

a. Position a brass drift against the CV-joint cage and tap it with a hammer to tilt the cage.

b. Insert the 1st chrome alloy ball into the cage.

c. Tilt the cage in the opposite direction.

d. Insert the opposite chrome alloy ball.

e. Repeat the procedure until all 6 balls are inserted.

9. Install ½ kit grease into the CV-joint.

10. Install or connect the following:
- Small ring clamp on the CV boot
- New retaining ring on the halfshaft
- Large ring clamp on the CV boot
- Outer race assembly onto the halfshaft until the ring engages the halfshaft groove

11. Slide the small end of the CV-joint boot/clamp into place, with the seal lip in the halfshaft groove

➡ **Make sure the boot lies flat against the halfshaft.**

12. Using the Crimp tool J-35910, a torque wrench and a breaker bar, crimp the small CV-joint boot clamp to 100 ft. lbs. (136 Nm).

13. Check the clamp gap dimension; if it is not 0.085 in. (2.15mm), continue tightening the clamp until it is.

14. Install ½ kit grease into the CV-joint boot.

15. Measure approximately 0.687 in. (17.5mm) up from the bottom edge of the outer CV-joint assembly.

16. Slide the large end of the CV boot/clamp into place, with the seal lip in place over the outer race.

➡ **Make sure the boot lies flat against the outer race.**

17. Using the Crimp tool J-35910, a torque wrench and a breaker bar, crimp the large CV-joint boot clamp to 130 ft. lbs. (176 Nm).

18. Check the clamp gap dimension; if it is not 0.102 in. (2.60mm), continue tightening the clamp until it is.

19. Install the halfshaft and the front wheel.

Inner (Tri-Pot) Joint

1. Before servicing the vehicle, refer to the precautions in the beginning of this section.

2. Remove or disconnect the following:
- Front wheel
- Halfshaft and place it in a vise
- Snapring from the stub shaft and discard it
- Small CV-joint boot clamp, cut and discard it
- Large CV-joint boot clamp, cut and discard it
- CV-joint boot by sliding it away from the tri-pot joint

3. Install a Stub Shaft Removal tool J-38868-A to the stub shaft snapring groove.

4. Using a slide hammer puller, press the stub shaft from the tri-pot housing.

5. Remove or disconnect the following:
- Tri-pot housing from the tri-pot spider

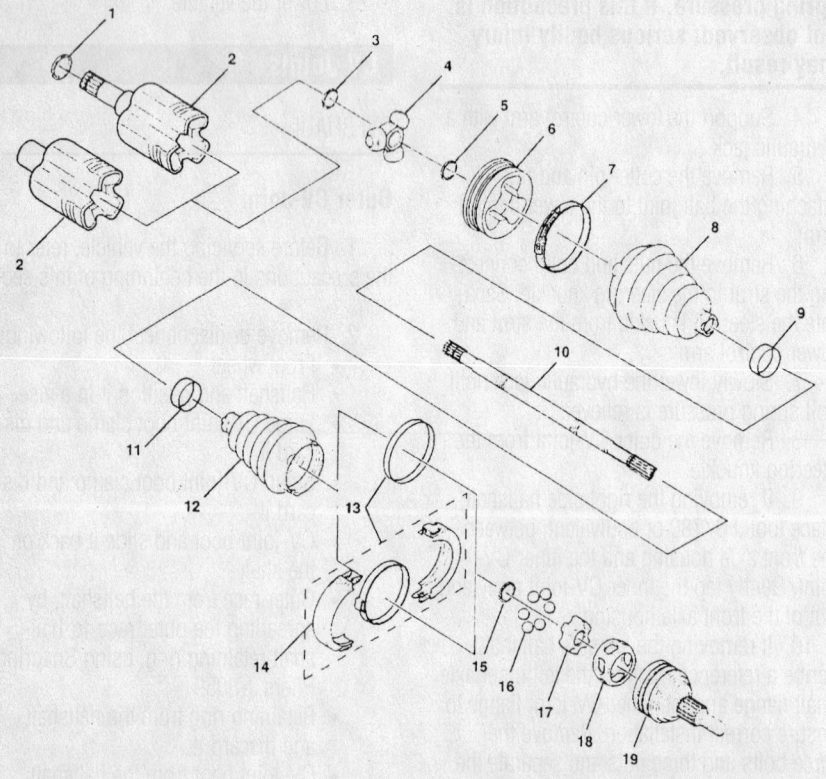

(1) Differential Shaft Ring
(2) Tripot Housing Assembly
(3) Spacer Ring
(4) Tripot Joint Spider Assembly
(5) Spacer Ring
(6) Tripot Bushing
(7) Boot Retaining Clamp
(8) Tripot Joint Boot
(9) Halfshaft Swage Ring
(10) Halfshaft Bar

(11) Halfshaft Swage Ring
(12) CV Joint Boot
(13) Swage Ring
(14) Clamp Protector
(15) Race Retaining Ring
(16) Ball
(17) CV Joint Inner Race
(18) CV Joint Cage
(19) CV Joint Outer Race

9308JG09

Exploded view of the CV-Joint Assembly

- Inboard spacer ring slide it rearward on the shaft using Snapring Pliers tool J-8059
- Outboard retaining ring using Snapring Pliers tool J-8059 and discard it
- Tri-pot joint spider assembly
- Inboard spacer ring and discard it
- CV-joint boot
- Trilobal tri-pot bushing from the housing

6. Thoroughly clean and inspect all parts.

To install:

7. Install or connect the following:

- New snapring onto the stub shaft
- Small boot clamp
- CV-joint boot

8. Using the Crimp tool J-35910, a torque wrench and a breaker bar, crimp the small CV-joint boot clamp to 100 ft. lbs. (136 Nm).

9. Install or connect the following:

- Inboard spacer ring slide it rearward on the shaft using Snapring Pliers tool J-8059, past the 2nd groove
- Tri-pot joint spider assembly onto the shaft until it passes the 2nd groove
- Outboard retaining ring into the axle shaft groove using Snapring Pliers tool J-8059
- Tri-pot joint spider assembly, slide it against the outboard retaining ring
- Inboard spacer ring, seat it in the groove
- ½kit grease into the boot
- ½kit grease into the tri-pot housing
- Trilobal tip-pot bushing flush with the tri-pot housing face
- New large seal clamp onto the CV-joint boot
- Tri-pot housing, slide it over the tri-pot joint spider assembly
- CV-joint boot/clamp, slide it into place, over the trilobal tri-pot bushing with the seal lip in the groove

➡ **Make sure the boot lies flat against the trilobal bushing.**

10. Position the CV-joint boot so it measures 4.9 in. (125mm).

11. Using the Crimp tool J-35566, latch the large CV-joint boot clamp.

12. Install the halfshaft and the front wheel.

Axle Shaft, Bearing and Seal

REMOVAL & INSTALLATION

For the Axle Shaft, Bearing and Seal, Removal and Installation, please refer to Wheel Bearing procedure located in the section.

Pinion Seal

REMOVAL & INSTALLATION

1. Before servicing the vehicle, refer to the precautions in the beginning of this section.

➡ **The following procedure requires the use of the Pinion Holding tool J-8614-10, the Pinion Flange Removal tool J-8614-1, J-8614-2, J-8614-3 and the Pinion Seal Installation tool J-23911.**

2. Remove or disconnect the following:
- Driveshaft from the pinion flange. Matchmark the driveshaft prior to removal.
- Driveshaft from the rear axle pinion flange and support the shaft up in body tunnel by wiring it to the exhaust pipe.

➡ **If the U-joint bearings are not retained by a retainer strap, use a piece of tape to hold bearings on their journals.**

3. Mark the position of the pinion stem, flange and nut for reference.

4. Use an inch lbs. torque wrench to measure the amount of torque necessary to turn the pinion, then note this measurement as it is the combined pinion bearing, seal, carrier bearing, axle bearing and seal preload.

5. Remove or disconnect the following:

- Pinion flange nut and washer, using a Pinion Holding tool J-8614-10 and a Pinion Flange Removal tool J-8614-1, J-8614-2, J-8614-3, as applicable
- Pinion flange
- Pinion oil seal by driving it out of the differential with a blunt chisel; DO NOT damage the carrier

To install:

6. Examine the seal surface of pinion flange for tool marks, nicks or damage, such as a groove worn by the seal. If damaged, replace flange.

7. Examine the carrier bore and remove any burrs that might cause leaks around the O.D. of the seal.

8. Apply GM seal lubricant 1050169 to the outside diameter of the pinion flange and sealing lip of new seal.

9. Install or connect the following:

- New pinion oil seal using a seal installer tool
- Pinion flange and tighten nut to the same position as marked earlier. Tighten the nut a little at a time and turn the pinion flange several times after each tightening in order to set the rollers.

10. Measure the torque necessary to turn the pinion and compare this to the reading taken during removal. Tighten the nut addi-

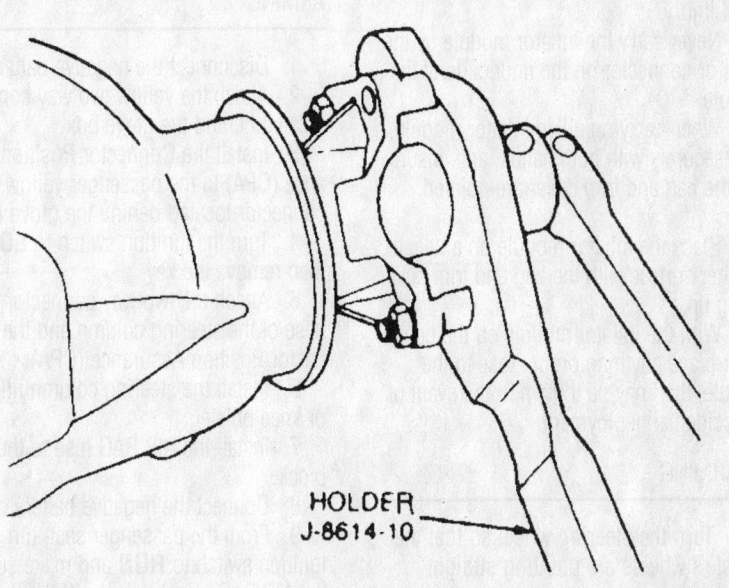

HOLDER
J-8614-10

88457G82

Removing the pinion nut using a pinion holding fixture tool

tionally, as necessary to achieve the same preload as measured earlier.

➡ **If fluid was lost from the differential housing during this procedure, be sure to check and add additional fluid, as necessary.**

11. Remove the support then align and secure the driveshaft assembly to the pinion flange.

➡ **The original matchmarks MUST be aligned to assure proper shaft balance and prevent vibration.**

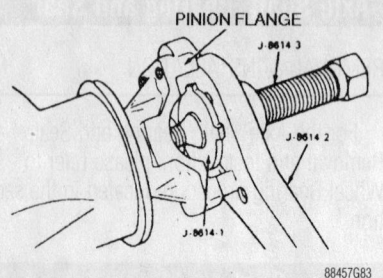

88457G83

A puller and adapter should be used to withdraw the pinion from the housing

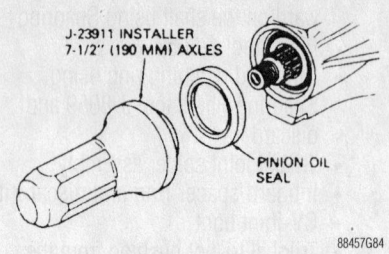

88457G84

Use the appropriately sized installation tool to drive the new seal into position.

STEERING AND SUSPENSION

Air Bag

※ CAUTION

Some vehicles are equipped with an air bag system, also known as the Supplemental Inflatable Restraint (SIR) system. The system must be disabled before performing service on or around system components, steering column, instrument panel components, wiring and sensors. Failure to follow safety and disabling procedures could result in accidental air bag deployment, possible personal injury and unnecessary system repairs.

PRECAUTIONS

Several precautions must be observed when handling the inflator module to avoid accidental deployment and possible personal injury.

• Never carry the inflator module by the wires or connector on the underside of the module.

• When carrying a live inflator module, hold securely with both hands, and ensure that the bag and trim cover are pointed away.

• Place the inflator module on a bench or other surface with the bag and trim cover facing up.

• With the inflator module on the bench, never place anything on or close to the module, that may be thrown in the event of an accidental deployment.

DISARMING

1. Turn the steering wheel so that the vehicle's wheels are pointing straight ahead.
2. Turn the ignition switch to **LOCK**,

remove the key, then disconnect the negative battery cable.

3. Remove the AIR BAG fuse from the fuse block.
4. Remove the steering column filler panel or knee bolster.
5. Unplug the Connector Position Assurance (CPA) and yellow two way connector at the base of the steering column.
6. Remove the Connector Position Assurance (CPA) from the passenger yellow two way connector located behind the glove box.
7. Unplug the yellow two way connector located behind the glove box.
8. Connect the negative battery cable.

➡ **With the AIR BAG fuse removed, the battery cable connected and the ignition in the ON position, the AIR BAG warning lamp will be ON. This is normal and does not indicate a system malfunction.**

ARMING

1. Disconnect the negative battery cable.
2. Attach the yellow two way connector located behind the glove box.
3. Install the Connector Position Assurance (CPA) to the passenger yellow two way connector located behind the glove box.
4. Turn the ignition switch to **LOCK**, then remove the key.
5. Attach the two way connector at the base of the steering column and the Connector Position Assurance (CPA).
6. Install the steering column filler panel or knee bolster.
7. Install the AIR BAG fuse to the fuse block.
8. Connect the negative battery cable.
9. From the passenger seat, turn the ignition switch to **RUN** and make sure that the AIR BAG warning lamp flashes seven times and then shuts off. If the warning

lamp does not shut off, make sure that the wiring is properly connected. If the light remains on, take the vehicle to a reputable repair facility for service.

Power Steering Gear

REMOVAL & INSTALLATION

Two Wheel Drive

1. Before servicing the vehicle, refer to the precautions in the beginning of this section.
2. Position a fluid catch pan under the power steering gear.

※ WARNING

Do NOT rotate the steering shaft after the steering column has been removed.

3. Lock the steering column through the access hole in the steering column lower trim cover using steering column anti-rotation pin J 42640.
4. Remove or disconnect the following:
   • Air cleaner assembly
   • Intermediate shaft from the steering gear

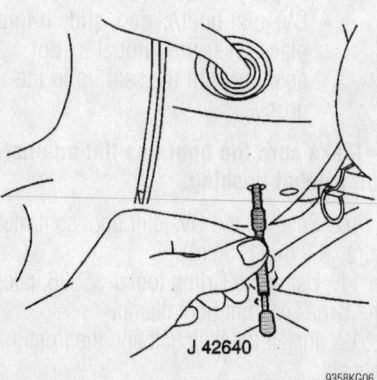

J 42640

9358KG06

Use J 42640 to lock the steering column

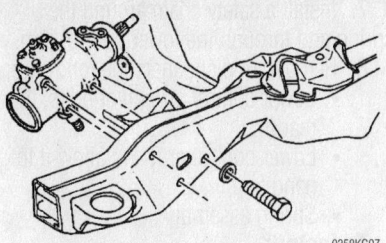

Power steering gear mounting

9358KG07

- Feed and return fluid hoses from the steering gear. Immediately cap or plug all openings to prevent system contamination or excessive fluid loss.
- Intermediate shaft lower coupling shield, if equipped
- Lower intermediate shaft coupling bolt
- Matchmark the lower intermediate shaft coupling and the steering shaft
- Lower intermediate shaft coupling from the steering shaft
- Pitman arm from the gear pitman shaft
- Power steering gear-to-frame bolts and washers, then carefully remove the steering gear from the vehicle

To install:
5. Install or connect the following:
- Steering gear to the vehicle and tighten the bolts to 55 ft. lbs. (75 Nm)
- Pitman arm
- Intermediate shaft to the power steering, making sure the matchmarks line up. Tighten the bolt to 26 ft. lbs. (35 Nm).
- Pressure and return hoses to the power steering gear. Tighten the pressure hose to 18 ft. lbs. (25 Nm) for 2.2L engines and 22 ft. lbs. (30 Nm) for 4.3L engines. Tighten the return hose to 18 ft. lbs. (25 Nm).
- Shield over the intermediate shaft lower coupling, if equipped
- Air cleaner assembly
6. Remove the steering column lock pin
7. Bleed the power steering system

Four Wheel Drive

This procedure requires the use of the following special tools: J 24319-B Steering Linkage and Tie Rod Puller, J 42640 Steering Column Anti-Rotation Pin, J 29193 Steering Linkage Installer (12mm), J 29194 Steering Linkage Installer (14mm).
1. Before servicing the vehicle, refer to

the precautions in the beginning of this section.
2. Position a fluid catch pan under the power steering gear.

※※ WARNING

Do NOT rotate the steering shaft after the steering column has been removed.

3. Lock the steering column through the access hole in the steering column lower trim cover using steering column anti-rotation pin J 42640.
4. Remove or disconnect the following:
- Air cleaner assembly
- Intermediate shaft lower coupling shield, if equipped
- Wiring harness clip from the power steering return hose at the power steering gear
- Feed and return fluid hoses from the steering gear. Immediately cap or plug all openings to prevent system contamination or excessive fluid loss.
- Lower intermediate shaft coupling bolt
- Matchmark the lower intermediate shaft coupling and the steering shaft
- Lower intermediate shaft coupling from the steering shaft
5. Raise the vehicle.
- Steering linkage shield
- Differential carrier shield mounting bolts
- Differential carrier shield
- Pitman arm ball stud cotter pin and nut at the relay rod
- Pitman arm from relay rod using a suitable puller
- Steering gear mounting bolts and the washers from the frame
- Steering gear
- Pitman arm

To install:
6. Install or connect the following:
- Pitman arm
- Steering gear
- Power steering gear to the frame washers and mounting bolts. Tighten the bolts to 55 ft. lbs. (75 Nm)
- Relay rod to the pitman arm ball stud. Ensure the seal is on the stud
7. Seat the taper using a J 29193 J 29194 and tighten the tool to 48 ft. lbs. (62 Nm).
8. Remove the special tool from the pitman arm ball stud.

9. Install or connect the following:
- New nut and cotter pin to the pitman arm ball stud at the relay rod and tighten the pitman arm ball stud nut at the relay rod to 61 ft. lbs. (83 Nm)
- Differential carrier shield
- Differential carrier shield mounting bolts
- Steering linkage shield
10. Lower the vehicle.
- Intermediate shaft to the power steering, making sure the matchmarks line up. Tighten the bolt to 26 ft. lbs. (35 Nm).
- Pressure and return hoses to the power steering gear. Tighten the pressure hose to 22 ft. lbs. (30 Nm) and the return hose to 18 ft. lbs. (25 Nm).
- Wiring harness clip to the power steering return hose at the power steering gear
- Shield over the intermediate shaft lower coupling, if equipped
- Air cleaner assembly
11. Remove the steering column lock pin
12. Bleed the power steering system

Shock Absorbers

REMOVAL & INSTALLATION

Front

2WD MODELS

1. Before servicing the vehicle, refer to the precautions in the beginning of this section.
2. Remove or disconnect the following:
- Wheel
- Mounting nut

➡**Hold the shock absorber stem with a wrench while backing the nut off.**

- Retaining nut and grommet
- Shock absorber-to-lower control arm bolts
- Shock absorber
- Replace the parts, as necessary.

To install:
3. Fully extend the shock absorber stem, then push it up through the lower control arm and spring so that the upper stem passes through the mounting hole in the upper control arm frame bracket.
4. Install or connect the following:
- Retaining nut and grommet on the stem. Tighten the nut to 106 inch lbs. (12 Nm).
- Shock absorber-to-lower control

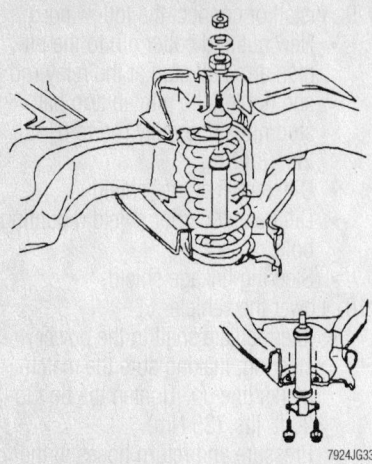

Front shock absorber mounting—2WD

arm bolts and tighten to 22 ft. lbs. (30 Nm)
- Wheel

4WD MODELS

1. Before servicing the vehicle, refer to the precautions in the beginning of this section.
2. Remove or disconnect the following:
   - Wheel
   - Lower nut/bolt and collapse the shock absorber

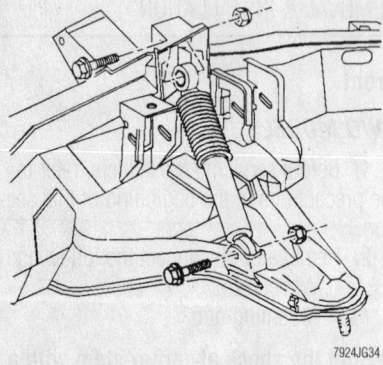

Front shock absorber mounting—4WD vehicles

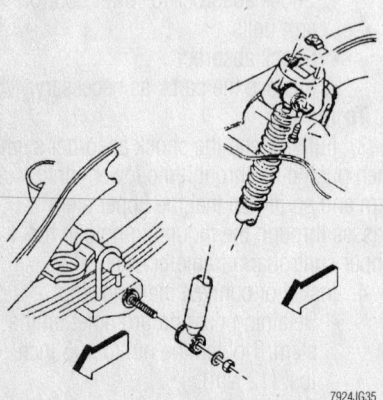

Rear shock absorber mounting

- Shock absorber upper nut and bolt
- Shock absorber

To install:

3. Install or connect the following:
   - Shock absorber to the bracket. Tighten the nuts/bolts to 54 ft. lbs. (73 Nm).
   - Wheel

Rear

1. Before servicing the vehicle, refer to the precautions in the beginning of this section.
2. Properly support the rear axle assembly.
3. Remove or disconnect the following:
   - Automatic level control air lines from the shock absorber, if equipped
   - Shock absorber-to-frame retainers at the top of the shock
   - Shock-to-axle retainers at the bottom of the shock
   - Shock absorber

To install:

4. Install the shock in the vehicle and loosely install the upper mounting fasteners to retain it
5. Align the lower-end of the shock absorber with the axle mounting, then loosely install the retainers.
6. Tighten the upper shock retainers to 18 ft. lbs. (25 Nm). Tighten the lower shock retainers to 62 ft. lbs. (84 Nm) on pick-up and two door utility models and 74 ft. lbs. (100 Nm) on four door utility models.
7. If equipped, attach the automatic level control air lines to the shock absorber.

Coil Springs

REMOVAL & INSTALLATION

1. Before servicing the vehicle, refer to the precautions in the beginning of this section.
2. Remove or disconnect the following:
   - Wheel
   - Shock absorber lower bolts
3. Push the shock absorber through the control arm and into the spring.
4. With the vehicle supported so the control arms hang free, install tool J-23028, onto a support and into the lower control arm bushings.
5. Remove or disconnect the following:
   - Stabilizer bar from the control arm
   - Stabilizer from the lower control arm
6. Raise and remove the tension on the lower control arm bolts.

7. Install a safety chain around the spring and through the lower control arm.
8. Remove or disconnect the following:
   - Lower control arm pivot bolts, the rear first
   - Lower control arm and allow it to hang free
   - Spring assembly

To install:

➡ **When positioning the spring in the lower control arm, be sure the spring insulator is in the proper position before lifting the control arm in place.**

9. Install or connect the following:
   - Spring assembly
   - Lower control arm
   - Lower control arm pivot bolts
   - Stabilizer to the lower control arm

Leaf Springs

REMOVAL & INSTALLATION

1. Before servicing the vehicle, refer to the precautions in the beginning of this section.

➡ **The following procedure requires the use of two sets of jackstands.**

2. Support the rear axle with jackstands, support the axle and the body separately in order to relieve the load on the rear spring.
3. Remove or disconnect the following:
   - Wheel
   - Shock absorber
   - U-bolt nuts, washers, anchor plate and bolts
   - Spare tire, if equipped
   - Rear exhaust hangers and lower the rear exhaust, if necessary
   - Shackle-to-frame bolt, washers and nut
   - Fuel tank, if necessary
   - Front bracket nut, washers and bolt
   - Spring
   - Shackle from the spring, if necessary

To install:

4. Install or connect the following:
   - Shackle to the rearward spring eye using the bolt, washers and nut, but do not fully tighten at this time.
   - Spring assembly
   - Spring to the front bracket using the bolt, washers and nut, but do not fully tighten at this time.
   - Fuel tank, if removed
   - Shackle-to-frame bolt, washers and nut, but do not fully tighten at this

time. If used, remove the spring support.
- U-bolts, anchor plate, washers and U-bolt nuts. Torque the nuts using 2 passes of a diagonal sequence:

a. Step 1: Torque to 18 ft. lbs. (25 Nm).

b. Step 2: Torque to 73 ft. lbs. (100 Nm) in the sequence.

5. Position the axle to achieve an approximate gap of 6.46–6.94 in. (164–176mm) between the axle housing tube and the metal surface of the rubber frame bumper bracket. Measure from the housing between the U-bolts to the metal part of the rubber bump stop on the frame.

6. While supporting the axle in this position, tighten the front and rear spring mounting fasteners to 89 ft. lbs. (122 Nm).

7. Install or connect the following:
- Rear exhaust in position and tighten the hangers
- Spare tire
- Shock absorber

Torsion Bar

Instead of the coil spring used on the front suspension of 2WD vehicles, the 4WD vehicles are equipped with a torsion bar.

REMOVAL & INSTALLATION

1. Before servicing the vehicle, refer to the precautions in the beginning of this section.

➡️**The following procedure requires the use of the Torsion Bar Unloader tool J-36202.**

2. Remove or disconnect the following:
- Transmission shield, if equipped
- Torsion bar unloader tool to relax the tension on the torsion bar adjusting arm screw; record the number of turns necessary to properly install the tool. Remove the adjusting screw and the unloader tool.
- Lower link mount nut from one side
- Torsion bars by disengaging them

➡️**Note the direction of the forward end and side of the torsion bar being removed**

- Lower link nut from the opposite side
- Lower link mount, upper link mount nut
- Upper link mount
- Torsion bar from the frame

To install:
3. Install or connect the following:
- Torsion bar and support
- Upper link mount. Torque the nut to 48 ft. lbs. (68 Nm).

4. Place a jack under the torsion bar to release tension.

5. Install or connect the following:
- Lower link mount bushing and nut. Torque the nut to 37 ft. lbs. (50 Nm).
- Torsion bar unloader tool. Tighten the tool against the adjusting arm the same number turns recorded earlier and remove the tool. This loads the torsion bars.
- Transmission shield, if removed

Ball Joints

REMOVAL & INSTALLATION

2WD Vehicles

UPPER

1. Before servicing the vehicle, refer to the precautions in the beginning of this section.

➡️**The following procedure requires the use of a ball joint separator tool such as J-23742 and J-9519-E ball joint remover and installer set.**

2. Raise and support the front of the vehicle safely by placing stands securely under the lower control arms. Because the vehicle's weight is used to relieve spring tension on the upper control arm, the stands must be positioned between the spring seats and the lower control arm ball joints for maximum leverage.

✸✸ CAUTION

With components unbolted, the stand is holding the lower control arm in place against the coil spring. Make sure the stand is firmly positioned and cannot move, or personal injury could result.

3. Remove or disconnect the following:
- Tire and wheel assembly
- Brake caliper and support it from the vehicle using a coat hanger or wire. Make sure the brake line is not stretched or damaged and that

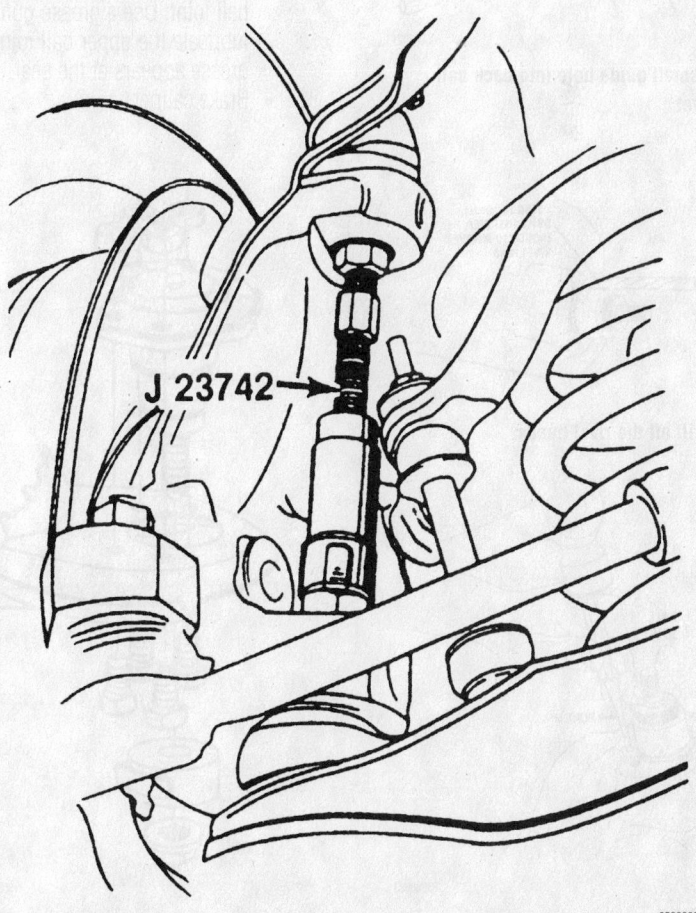

85388020

Use a ball joint separator tool to drive the upper ball joint from the steering knuckle

the caliper's weight is not supported by the line.
- Cotter pin and retaining nut from the upper ball joint
- Anti-lock brake sensor wire bracket, if equipped
- Upper ball joint from the steering knuckle using tool J-23742 and pull the steering knuckle free of the ball joint

➡ **After separating the steering knuckle from the upper ball joint, be sure to support the steering knuckle/hub assembly to prevent damaging the brake hose.**

4. Remove the riveted upper ball joint from the upper control arm as follows:

a. Drill a⅛in. (3mm) hole, about¼in. (6mm) deep into each rivet.

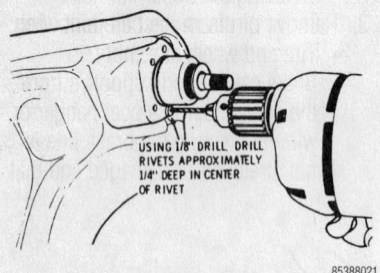

Drill a small guide hole into each ball joint rivet

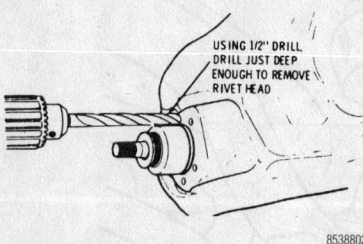

Then drill off the rivet heads

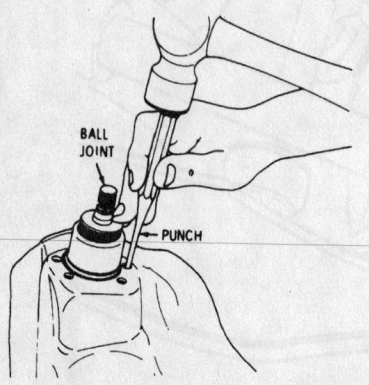

Punch the rivets out and remove the ball joint

b. Then use a½in. (13mm) drill bit, to drill off the rivet heads.

c. Using a pin punch and the hammer, drive out the rivets in order to free the upper ball joint from the upper control arm assembly, then remove the upper ball joint.

5. Clean and inspect the steering knuckle hole. Replace the steering knuckle if the hole is out of round.

To install:

6. Install or connect the following:
- Ball joint in the upper control arm
- Ball joint retaining nuts and bolts. Position the bolts threaded upward from under the control arm. Tighten the ball joint retainers to 17 ft. lbs. (23 Nm).
- Anti-lock brake sensor wire bracket, if removed
- Ball joint to the knuckle. Make sure the joint is seated, then install the stud nut and tighten to 61 ft. lbs. (83 Nm). Insert a new cotter pin.

➡ **When installing the cotter pin, never loosen the castle nut to expose the cotter pin hole.**

- Thread the grease fitting into the ball joint. Use a grease gun to lubricate the upper ball joint until grease appears at the seal.
- Brake caliper

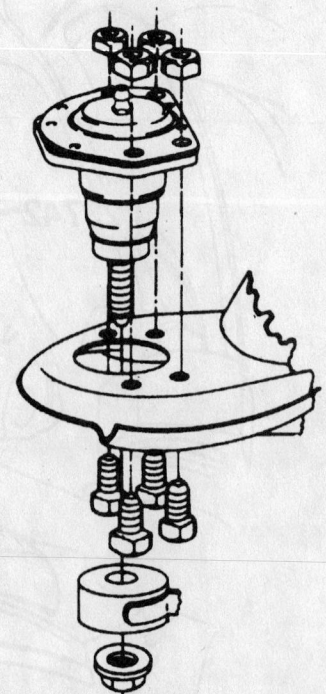

Service ball joints are bolted to the control arm

- Tire and wheel assembly

7. Check and adjust the front end alignment, as necessary.

LOWER

1. Before servicing the vehicle, refer to the precautions in the beginning of this section.

➡ **The following procedure requires the use of a ball joint remover/installer set (the particular set may vary upon application but must include a clamping-type tool with the appropriately sized adapters) and a ball joint separator tool, such as J-23742.**

- Tire and wheel assembly

2. Position a jack under the spring seat of the lower control arm, then raise the jack to support the arm.

✳✳ CAUTION

The jack MUST remain under the lower control arm, during the removal and installation procedures, to retain the arm and spring positions. Make sure the jack is securely positioned and will not slip or release during the procedure or personal injury may result.

3. Remove or disconnect the following:
- Brake caliper and support it aside using a hanger or wire. Make sure the brake line is not stressed or damaged.
- Lower ball joint cotter pin and discard
- Ball joint stud nut
- Lower ball joint from the steering knuckle using tool J-23742

4. Carefully guide the lower control arm out of the opening in the splash shield using a putty knife. Position a block of wood between the frame and upper control arm to keep the knuckle out of the way.
- Grease fitting
- Ball joint from the control arm using the ball joint remover set along with the appropriate adapters

To install:

5. Clean the tapered hole in the steering knuckle of any dirt or foreign matter, then check the hole to see if it is out of round, deformed or otherwise damaged. If a problem is found, then knuckle must be replaced.

6. Install or connect the following:
- Press the new ball joint (with grease fitting pointing inward) until it bottoms in the control arm using a suitable installation set. Make

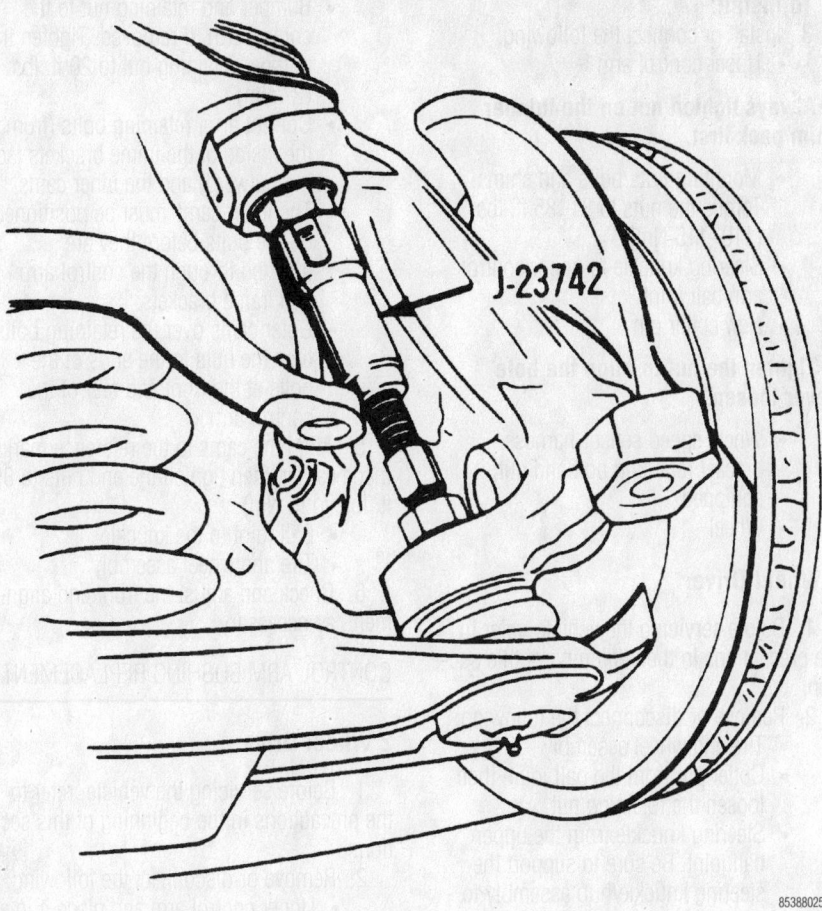

85388025

Use a ball joint separator to drive the lower joint from the knuckle

sure the grease seal is facing inboard.

- Ball joint stud into the steering knuckle
- Ball joint retaining nut and tighten to 79 ft. lbs. (108 Nm)

➡️**When installing the cotter pin, never loosen the castle nut to expose the cotter pin hole.**

- Grease fitting into the ball joint, if not already installed

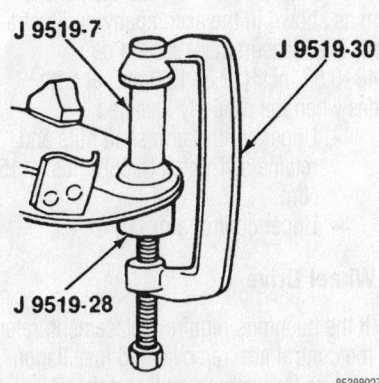

J 9519-7   J 9519-30

J 9519-28

85388027

Driving the lower joint from the control

7. Use a grease gun to lubricate the joint until grease appears at the seal.
- Brake caliper
- Tire and wheel assembly

8. Check and adjust the front end alignment, as necessary.

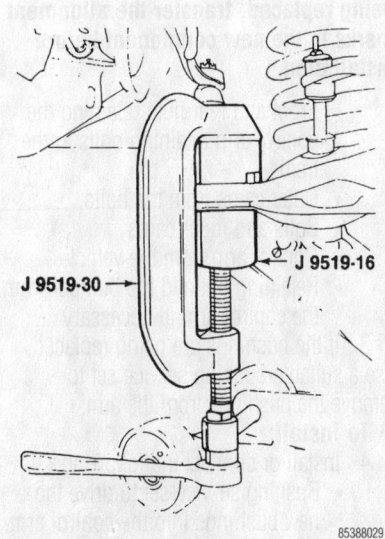

J 9519-30 ➡️   J 9519-16

85388029

Installing a new ball joint

4WD Vehicles

1. Before servicing the vehicle, refer to the precautions in the beginning of this section.

On 4WD vehicles both the upper and lower ball joints are removed in the same manner. Once the joint is separated from the steering knuckle the rivets are drilled and punched to free the joint from the control arm. Service joints are bolted into position with the retaining bolts threaded upward from beneath the control arm. In this manner, the joint is replaced in an almost identical fashion to the upper joints on 2WD vehicles.

2. Remove or disconnect the following:
- Tire and wheel assembly
- Wheel speed sensor wiring connector from the upper control arm, if removing the upper ball joint
- Cotter pin from the ball joint, then loosen the retaining nut

3. Position a suitable ball joint separator tool such as J-36607, then carefully loosen the joint in the steering knuckle. Remove the tool and the retaining nut, then separate the joint from the knuckle.

➡️**After separating the steering knuckle from the upper ball joint, be sure to support the steering knuckle/hub assembly to prevent damaging the brake hose.**

4. Remove the riveted ball joint from the control arm:

  a. Drill a ⅛ in. (3mm) hole, about ¼ in. (6mm) deep into each rivet.

  b. Then use a ½ in. (13mm) drill bit, to drill off the rivet heads.

  c. Using a pin punch and the hammer, drive out the rivets in order to free the ball joint from the control arm assembly, then remove the ball joint.

To install:

5. Install or connect the following:
- Ball joint in the control arm
- Ball joint retaining nuts and bolts. Position the bolts threaded upward from under the control arm. Tighten the ball joint retainers to 17 ft. lbs. (23 Nm).
- Ball joint to the knuckle. Make sure the joint is seated, tighten the lower nut to 79 ft. lbs. (108 Nm) and the upper nut to 61 ft. lbs. (83 Nm). Install a new cotter pin.

➡️**When installing the cotter pin, never loosen the castle nut to expose the cotter pin hole, but DO NOT tighten more than an additional ⅛ turn.**

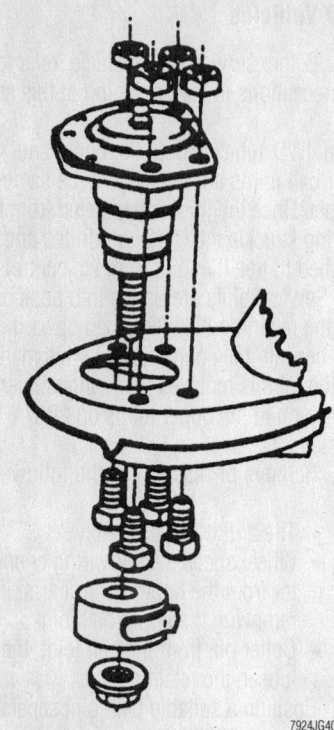

7924JG40

The replacement ball joint comes with nuts and bolts for installation

6. Use a grease gun to lubricate the upper ball joint.
   - Wheel speed sensor wiring connector to the upper control arm, if the upper ball joint was removed
   - Tire and wheel assembly
7. Check and adjust the front end alignment, as necessary.

Upper Control Arm

REMOVAL & INSTALLATION

2 Wheel Drive

1. Before servicing the vehicle, refer to the precautions in the beginning of this section.
2. Remove or disconnect the following:
   - Negative battery cable
   - Wheel
   - Wheel speed sensor harness bracket retaining bolt and nut, if equipped
   - Steering knuckle from upper control arm ball joint
   - Mounting nuts/bolts and shims

➡**Make sure to note the location of the control arm shims prior to removal so that they may be installed in their original positions.**
   - Upper control arm

To install:
3. Install or connect the following:
   - Upper control arm

➡**Always tighten nut on the thinner shim pack first.**
   - Mounting nuts/bolts and shims. Torque the nuts to 81–85 ft. lbs. (110–115 Nm).
   - Steering knuckle to upper control arm ball joint
   - New cotter pin

➡**Tighten the nut to align the hole never loosen.**
   - Wheel speed sensor harness bracket retaining bolt and nut, if equipped
   - Wheel

4 Wheel Drive

1. Before servicing the vehicle, refer to the precautions in the beginning of this section.
2. Remove or disconnect the following:
   - Tire and wheel assembly
   - Cotter pin from the ball joint, then loosen the retaining nut
   - Steering knuckle from the upper ball joint. Be sure to support the steering knuckle/hub assembly to prevent damaging the brake hose.

➡**The 4WD vehicles do not use shims to adjust the front wheel alignment. Instead, the upper control arm bolts are equipped with cams, which are rotated to achieve caster and camber adjustments. In order to preserve adjustment and ease installation, matchmark the cams to the control arm before removal. If the control arm is being replaced, transfer the alignment marks to the new component before installation.**
   - Front and rear nuts retaining the control arm retaining bolts to the frame
   - Outer cams from the bolts
   - Bolts and inner cams
   - Control arm from the vehicle
   - Retaining nut and the bumper from the control arm, if necessary
3. If the bushings are being replaced, use a suitable bushing service set to remove the bushings from the arm.

To install:
4. Install or connect the following:
   - Bushing service set to drive the new bushings into the control arm, if removed

   - Bumper and retaining nut to the control arm, if removed. Tighten the bumper retaining nut to 20 ft. lbs. (27 Nm).
   - Control arm, retaining bolts (from the inside of the frame brackets facing outward) and the inner cams. The inner cams must be positioned on the bolts before they are inserted through the control arm and frame brackets.
   - Outer cams over the retaining bolts, then the nuts to the ends of the bolts at the front and rear of the control arm
5. Align the cams to the reference marks made earlier, then tighten the end nuts to 85 ft. lbs. (115 Nm).
   - Ball joint to the knuckle
   - Tire and wheel assembly
6. Check and adjust the front end alignment, as necessary.

CONTROL ARM BUSHING REPLACEMENT

2 Wheel Drive

1. Before servicing the vehicle, refer to the precautions in the beginning of this section.
2. Remove or disconnect the following:
   - Upper control arm and place it in a vice
   - Upper control arm shaft nuts and retainers
   - Upper control arm bushings using tool J 22269-1, a slotted washer and a short piece if pipe that is slightly larger than the bushing
   - Upper control arm shaft

To install:
   - Upper control arm shaft
   - Upper control arm bushings using tool J 22269-1, a slotted washer and a short piece if pipe that is slightly larger than the bushing
3. Tighten J 22269-1 until the bushing is positioned on the shaft and the control arm as shown in the accompanying illustration. The measurement should be 0.48–0.52 inch (12.8–13.8mm) at both sides when the properly installed.
   - Upper control arm shaft nuts and retainers. Tighten to 85 ft. lbs. (115 Nm).
   - Upper control arm

4 Wheel Drive

If the bushings require replacement, refer to the control arm removal and installation procedure for bushing replacement.

Lower Control Arm

REMOVAL & INSTALLATION

2 Wheel Drive

1. Before servicing the vehicle, refer to the precautions in the beginning of this section.
2. Remove or disconnect the following:
   • Coil spring
   • Lower ball joint from the steering knuckle
   • Lower control arm from the vehicle

To install:

3. Install or connect the following:
   • Lower control arm
   • Lower ball joint stud into the steering knuckle
   • Ball joint-to-steering knuckle nut and tighten to specification
   • New cotter pin to the lower ball joint stud
   • Coil spring
4. Align the vehicle.

4 Wheel Drive

1. Before servicing the vehicle, refer to the precautions in the beginning of this section.

➡ **Tools Needed: universal tie rod separator J–24319–01, torsion bar unloader**

J–36202, lower control arm bushing service kit J–36618 (if the control arm bushing are being replaced) and ball joint C-clamp J–9519–23. Whether or not the control arm or bushing are being replaced, NEW control arm retaining nut should be used once the old ones have been loosened and removed.

2. Remove or disconnect the following:
   • Front wheels
   • 2 bolts from the front splash shield and pivot it in order to gain access to the tie rod
   • Stabilizer bar from the control arm (keeping all of the link hardware sorted for proper installation). If necessary, completely remove the bar from the vehicle for access.
   • Shock absorber
   • Inner tie rod from the relay rod using a tie rod separator
   • Outer halfshaft nut and washer
   • Bolts from the hub and bearing kit
3. Unload the torsion bar using the unloading tool J–36202. First, mark the adjuster for installation.
   • Adjustment arm. Slide the bar forward and the adapter out of the rear to remove the adjusting arm.
   • Lower ball joint cotter pin, nut and ball joint from the control arm using a ball joint separator

• Nuts and bolts and lower control arm with the torsion bar assembly. Note the direction which the control arm retaining bolts are facing for installation purposes.

To install:

4. Install or connect the following:
   • Torsion bar to the lower control arm and place the assembly into the vehicle. Position the front leg of the lower control arm into the crossmember before installing the rear leg into the frame bracket.
   • Control arm bolts (facing in the direction as noted during removal or shown in the accompanying illustration) with NEW nuts.

➡ **The control arm retainers MUST be tightened with the vehicle suspension at normal ride height. This can either be accomplished by starting the nuts now, then installing the remaining components along with the wheels and lowering the vehicle, or by moving jackstands under the ends of the lower control arms and resting the vehicle on them. If the latter solution is tried, make sure front suspension is at actual ride height compression. If you are unsure, it is best to start the nuts now and tighten them to specification once the vehicle is lowered.**

   • Ball joint stud in the knuckle
5. With the suspension at the correct height, tighten the control arm retaining nuts to 98 ft. lbs. (133 Nm).

➡ **The lower ball joint retaining nut MUST be tightened with the vehicle suspension at normal ride height. This can either be accomplished by starting the nut now, then installing the remaining components along with the wheels and lowering the vehicle, or by moving jackstands under the ends of the lower control arms and resting the vehicle on them. If the latter solution is tried, make sure the FULL WEIGHT of the vehicle front end is on the suspension.**

6. Install or connect the following:
   • Joint-to-control arm nut, then tighten the nut to 92 ft. lbs. (125 Nm) with the suspension at normal ride height and compression.
   • New cotter pin to the castellated nut. Tighten the nut (but no more than an additional ⅙ turn) in order to align the cotter pin. DO NOT loosen the nut from the specified torque.
   • Adjuster arm by sliding the adapter

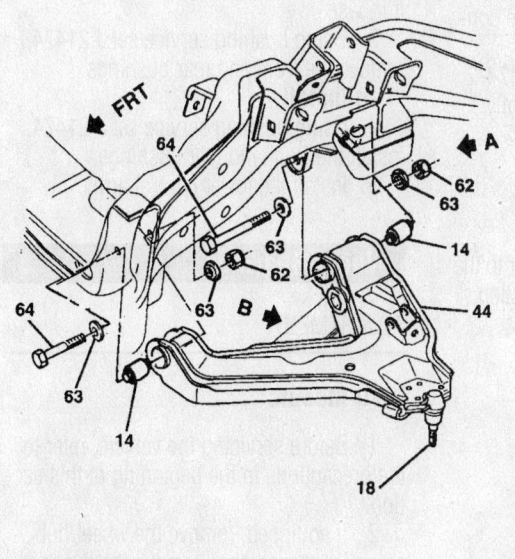
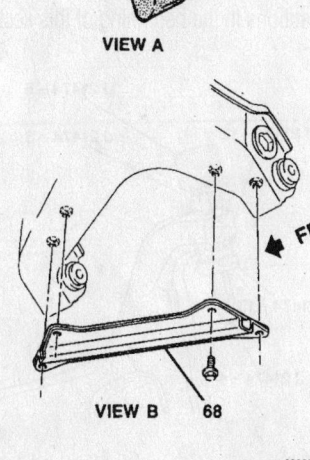

14. BUSHING
18. BALL JOINT, LOWER
44. ARM KIT, LOWER
62. NUT
63. WASHER
64. BOLT
66. NUT
67. BUMPER
68. BRACE

VIEW A

VIEW B 68

88268GB1

Exploded view of the lower control arm assembly mounting

forward, over the torsion bar to install the sides of the nut. Load the torsion bar and install the adjuster bolt aligning the installation mark.

- Drive axle through the hub and bearing assembly
- Tighten the hub and bearing assembly retaining bolts
- Drive axle shaft nut and washer
- Inner tie rod end to the relay rod
- Shock absorber
- Stabilizer bar, if removed
- Stabilizer link(s) to the control arm(s)
- Splash shield
- Front wheels

7. Recheck all fasteners for proper torque and installation before road testing.

8. Refill the differential if any fluid was lost.

9. Check and adjust the front end alignment, as necessary.

CONTROL ARM BUSHING REPLACEMENT

2 Wheel Drive

1. Before servicing the vehicle, refer to the precautions in the beginning of this section.

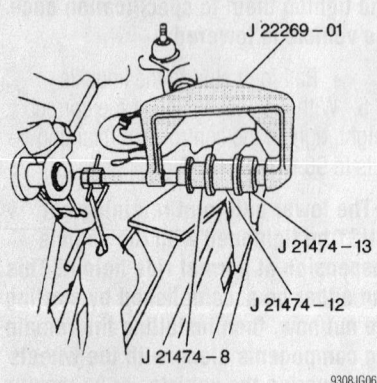

Removing the lower control arm rear bushing—all 2 wheel drive

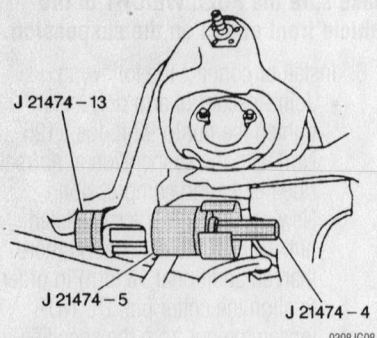

Installing the lower control arm front bushing—all 2 wheel drive

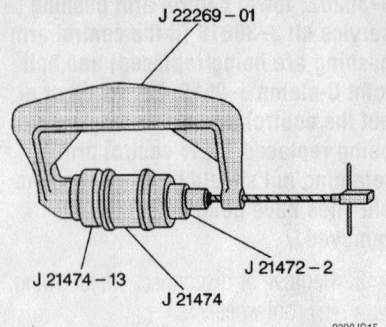

Installing the lower control arm rear bushing—all 2 wheel drive

2. Remove lower control arm and place it in vise.

3. Install tools J 22269-01, 21474-8, 12 and 13 on the rear bushing and tighten until the bushing is removed.

4. Using a blunt chisel, drive the front bushing flare flush with the rubber part of the bushing.

5. Place a wedge or a spacer between the bushing housing to keep the housing from bending while removing or installing the bushing.

6. Install tools J 21474-3, 4, 5 and 6 on the front bushing and tighten until the bushing is removed.

To install:

7. Install the front bushing into the control arm.

8. Install tools J 21474-4, 5 and 13. Tighten until the bushing is fully seated.

9. Install the rear bushing into the control arm

10. Install tools J 22269-01, J 21474-2 and 13. Tighten until the bushing is fully seated.

11. Install the lower control arm.

4 Wheel Drive

1. Before servicing the vehicle, refer to the precautions in the beginning of this section.

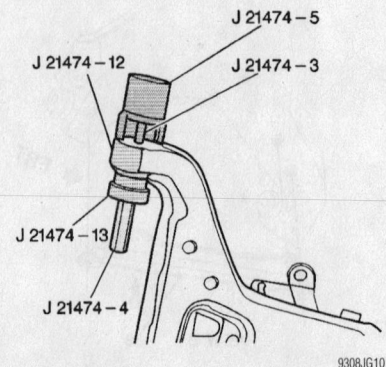

Removing the lower control arm front bushing—4 wheel drive

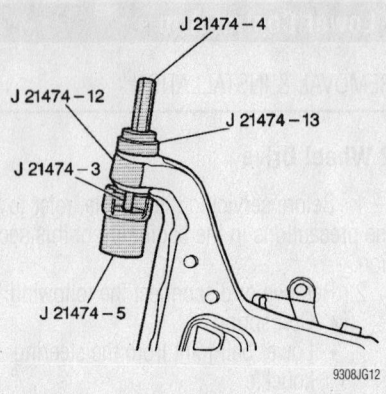

Installing the lower control arm front bushing—4 wheel drive

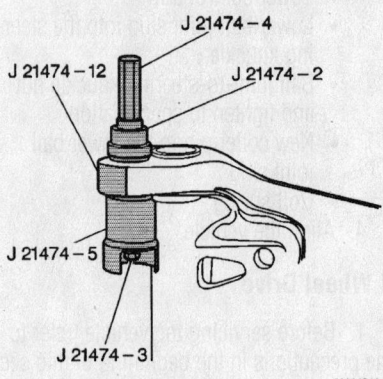

Installing the lower control arm rear bushing—4 wheel drive

2. Remove lower control arm and place it in vise.

3. Using bushing service set J 21474, remove the front and rear bushings.

To install:

4. Using bushing service set J 21474, install the front and rear bushings.

5. Install the lower control arm.

Wheel Bearings

ADJUSTMENT

2WD Models

1. Before servicing the vehicle, refer to the precautions in the beginning of this section.

2. If equipped, remove the wheel/hub cover for access, then remove the dust cap from the hub.

3. Remove the cotter pin and loosen the spindle nut.

4. Spin the wheel forward by hand and torque the nut to 12 ft. lbs. (16 Nm) in order to fully seat the bearings and remove any burrs from the threads.

5. Back off the nut until it is just loose, then finger-tighten the nut.

6. Loosen the nut ¼–½ turn until either hole in the spindle lines up with a slot in the nut, then install a new cotter pin. This may appear to be too loose, but it is the proper adjustment.

7. Proper adjustment creates 0.001–0.005 in. (0.025–0.127mm) end-play.

4WD Models

The front wheel bearings on the 4-wheel drive vehicles are not adjustable. If the bearings become loose or make noise, they must be replaced.

REMOVAL & INSTALLATION

Front

2WD MODELS

1. Before servicing the vehicle, refer to the precautions in the beginning of this section.

2. Remove or disconnect the following:
 • Wheel
 • Brake caliper with the pads without disconnecting the brake line
 • Grease cap
 • Cotter pin, spindle nut and washer
 • Hub

✳✳ WARNING

Be careful not to drop the outer wheel bearing. As the hub is pulled forward, the outer wheel bearings will often fall forward and they may easily be removed at this time.

 • Outer roller bearing assembly
 • Inner seal by prying it out of the hub and discard it
 • Inner bearing assembly

To install:

3. Clean all parts in solvent and allow to air dry, then check for excessive wear or damage. Inspect all of the parts for scoring, pitting or cracking and replace if necessary.

➡**DO NOT remove the bearing races from the hub, unless they show signs of damage.**

4. If it is necessary to remove the wheel bearing races, use the GM front bearing race removal tool J-29117 to drive the races from the hub/disc assembly. A hammer and brass drift may also be used to drive the races from the hub, but the race removal tool is quicker.

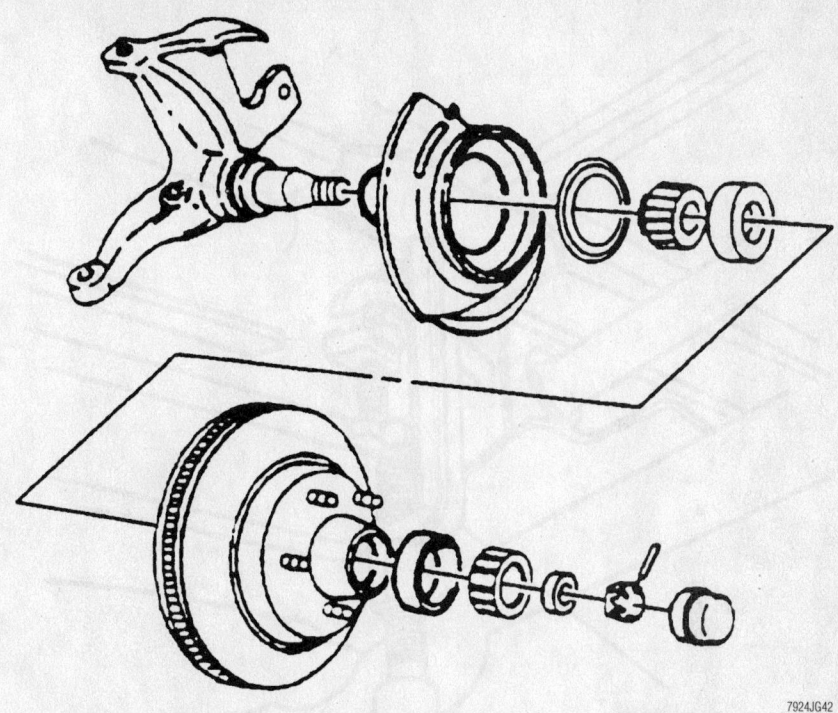

Wheel bearings, races and related components—2WD vehicles

7924JG42

5. If the bearing races were removed, position the replacement races in the freezer for a few minutes and then install them to the hub:

 a. Lightly lubricate the inside of the hub/disc assembly using wheel bearing grease.

 b. Using the GM seal installation tools J-8092 and J-8850, drive the inner bearing race into the hub/disc assembly until it seats. Be sure the race is properly seated against the hub shoulder and is not cocked.

➡**When installing the bearing races, be sure to support the hub/disc assembly with GM tool J-9746-02.**

 c. Using the GM seal installation tools J-8092 and J-8457, drive the outer race into the hub/disc assembly until it seats.

6. Using a high melting point wheel bearing grease, lubricate the bearings, races and spindle; be sure to place a gob of grease (inside the hub/disc assembly) between the races to provide an ample supply of lubricant.

➡**To lubricate each bearing, place a gob of grease in the palm of the hand, then scoop the bearing through the grease until it is well lubricated.**

7. Place the inner bearing in the hub, then apply a thin coating of grease to the sealing lip and install a new inner seal,

making sure the seal flange faces the bearing cup.

➡**Although a seal installation tool is preferable, a section of pipe with a smooth edge or a suitably sized socket may be used to drive the seal into position. Be sure the seal is flush with the outer surface of the hub assembly.**

8. Install or connect the following:
 • Wheel hub over the spindle
 • Outer bearing into the hub by hand
 • Spindle washer and nut
 • Brake caliper
 • Wheel
9. Properly adjust the wheel bearings
10. Install or connect the following:
 • New cotter pin
 • Dust cap
 • Wheel cover

4WD MODELS

1. Before servicing the vehicle, refer to the precautions in the beginning of this section.

2. Install Torsion Bar Unloading tool J 36202 on the torsion bar adjusting bolt and remove the bolt. To aid during installation, count the number of turns required to remove the bolt.

3. Remove the wheel.

4. Install an axle shaft boot seal protector to the Tri-pot axle joint.

5. Remove or disconnect the following:
 • Cotter pin and retainer

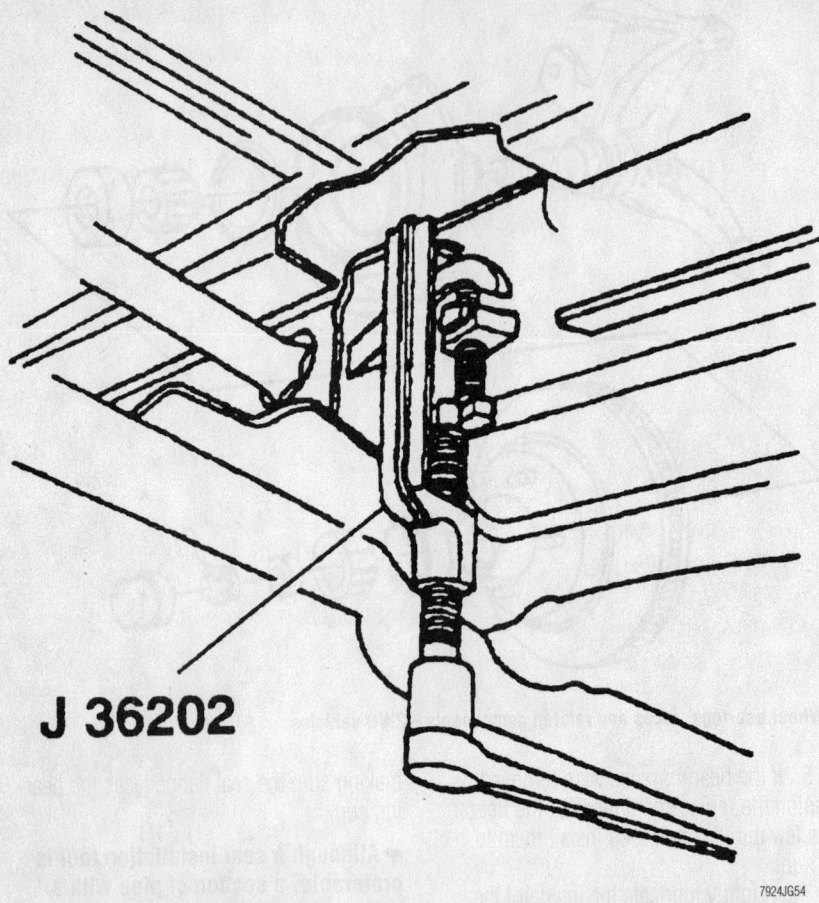

J 36202

Use Torsion Bar Unloading tool J 36202 to remove the adjusting bolt and unload the torsion bar

- Castle nut and the thrust washer
- Brake caliper and support it aside using wire or a coat hanger

→Be sure the brake line is not stretched or damaged.

- Brake disc from the wheel hub
- Halfshaft from the hub/bearing assembly, using a Spindle Remover tool J-28733-A to prevent damage to the shaft or hub/bearing assembly
- Hub/bearing assembly from the knuckle

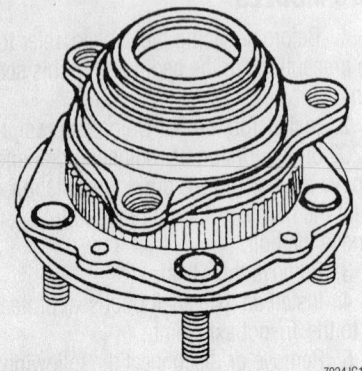

7924JG15

Hub and bearing assembly—4WD vehicles

6. Clean and inspect the parts for nicks, scores and/or damage, then replace them as necessary.

To install:

7. Install or connect the following:
- Hub and bearing assembly by aligning the threaded holes. Torque the bolts to 77 ft. lbs. (105 Nm).
- Tie rod end to the steering knuckle using the retaining nut
- New cotter pin
- Brake assembly
- Halfshaft nut. Tighten the nut to 180 ft. lbs. (245 Nm).
- Retainer and a new cotter pin but DO NOT back off specification in order to insert the cotter pin.

8. Remove the torsion bar unloader tool and the drive axle boot protector.

9. Install the wheel.

10. Check and/or adjust the vehicle trim height, as necessary.

Rear

A new pinion shaft lockbolt should be installed whenever either of the axle shafts is removed.

The axle shaft and seal may be removed and replaced without disturbing the bearing or seal but it is highly recommended to replace the seals when removing the axle shaft.

1. Before servicing the vehicle, refer to the precautions in the beginning of this section.

2. Remove or disconnect the following:
- Rear wheels
- Brake drums

3. Using a wire brush, clean the dirt/rust from around the rear axle cover.

4. Drain the fluid.

5. Remove or disconnect the following:
- Rear pinion shaft lockbolt and the pinion shaft
- C-lock from the button end of the axle shaft by pushing the axle shaft inward
- Axle shaft from the axle housing

→Be careful not to damage the oil seal.

※※ WARNING

If equipped with an Anti-Lock Brake System (ABS), be careful not to damage the reflector ring on the axle shaft or the speed sensor bolted to the backing plate, immediately adjacent to the shaft.

6. Remove or disconnect the following:
- Oil seal by prying the it from the end of the rear axle housing

※※ WARNING

DO NOT damage the housing oil seal surface.

- Wheel bearing using the GM Slide Hammer tool J-2619, the GM Adapter tool J-2619-4 and the GM Axle Bearing Puller tool J-22813-01

To install:

7. Clean and inspect the components for excessive wear or damage and replace them, if necessary.

8. Install or connect the following:
- New or reused bearing, coated with gear lubricant, using the Axle Shaft Bearing Installer tool J-34974 to drive the bearing in until it bottoms against the seat

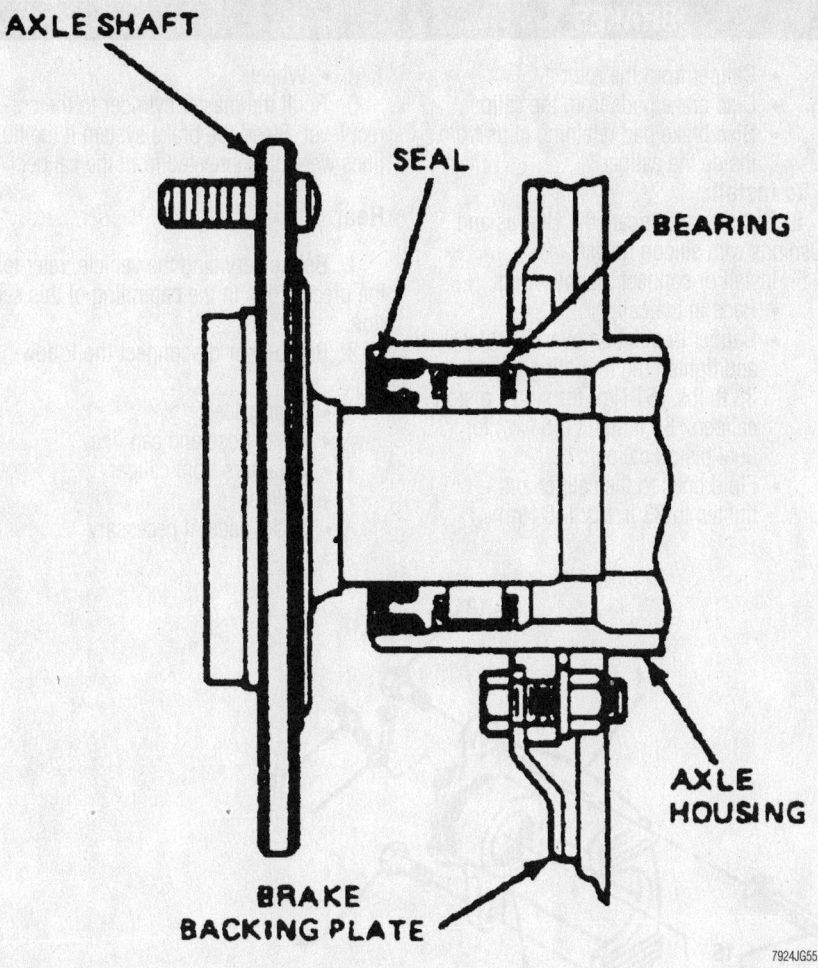

Cross-sectional view of the rear axle, bearing and seal assembly

7924JG55

Be sure the bearing installer does not contact and damage the speed sensor on ABS equipped vehicles.

- New seal lubricated with gear oil using the GM Axle Shaft Seal Installer tool J-33782 to seat it in the housing until it is flush with the axle tube

➡ **Be sure the seal installer does not contact and damage the speed sensor on ABS equipped vehicles.**

- Axle shaft into the housing by engaging the splines
- C-lock retainer on the axle shaft button end

BE CAREFUL not to damage the wheel bearing seal.

- Axle shaft by pulling it outward to seat the C-lock retainer in the counterbore of the side gears
- Pinion shaft through the case and the pinions. Tighten the new lock-bolt to 27 ft. lbs. (36 Nm).
- New rear axle cover gasket
- Housing cover
- Brake drums
- Wheels

9. Refill the housing.

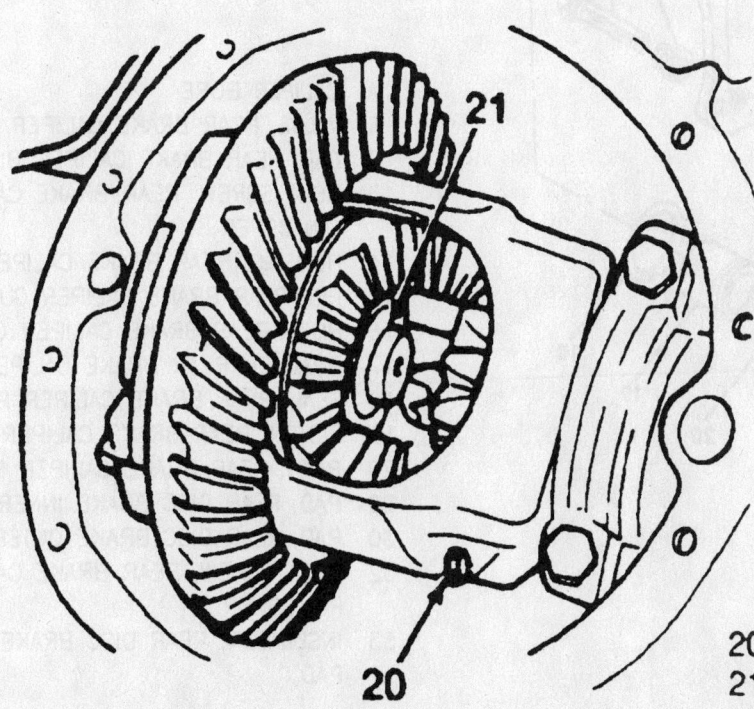

20. Lock bolt
21. "C" lock

7924JG56

Pinion shaft lockbolt and axle C-lock locations, inside the differential

BRAKES

Brake Caliper

REMOVAL & INSTALLATION

Front

1. Before servicing the vehicle, refer to the precautions in the beginning of this section.

2. Remove ⅔ of the brake fluid from the master cylinder reservoir.

3. Remove or disconnect the following:
- Wheel
- Caliper fluid line and plug to avoid contamination
- Bolts retaining the caliper to the rotor
- Caliper from the rotor
- Disc brake pads from the caliper
- Disc brake pad retaining clips from inside the caliper

To install:

4. Clean and lubricate the sleeves and bushings with silicon grease.

5. Install or connect the following:
- Pads in the caliper
- Caliper in position over the rotor and tighten the mounting bolts to 38 ft. lbs. (51 Nm) for single piston calipers; 85 ft. lbs. (115 Nm) for dual piston calipers
- Fluid lines to the caliper and tighten to 33 ft. lbs. (45 Nm)
- Wheel

6. Refill the master cylinder to the correct level. Bleed the brake system if the fluid lines were disconnected from the caliper.

Rear

1. Before servicing the vehicle, refer to the precautions in the beginning of this section.

2. Remove or disconnect the following:
- Rear wheels
- Brake hose and cap line
- Retainers from caliper
- Caliper
- Brake pads, if necessary

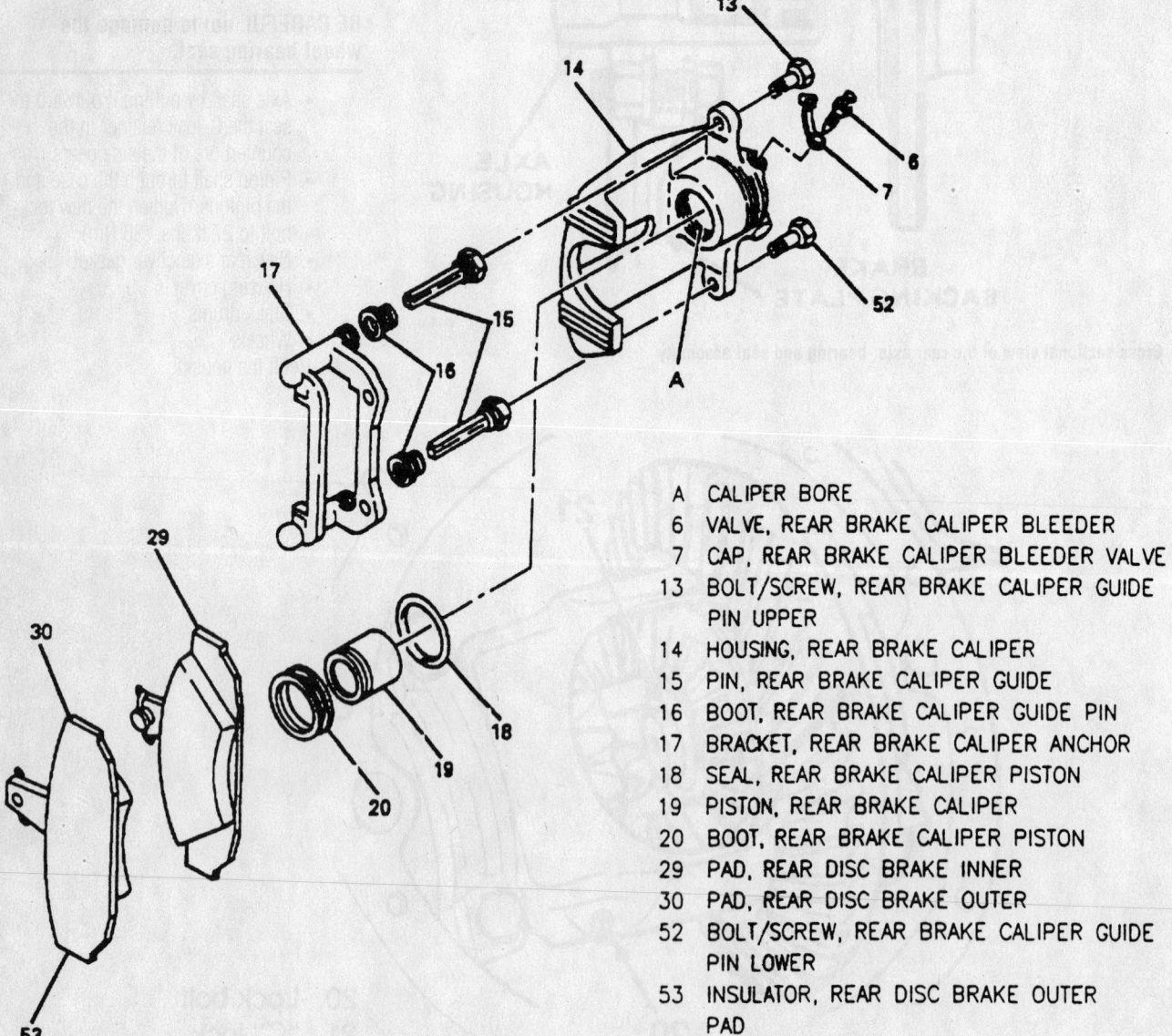

A   CALIPER BORE
6   VALVE, REAR BRAKE CALIPER BLEEDER
7   CAP, REAR BRAKE CALIPER BLEEDER VALVE
13   BOLT/SCREW, REAR BRAKE CALIPER GUIDE PIN UPPER
14   HOUSING, REAR BRAKE CALIPER
15   PIN, REAR BRAKE CALIPER GUIDE
16   BOOT, REAR BRAKE CALIPER GUIDE PIN
17   BRACKET, REAR BRAKE CALIPER ANCHOR
18   SEAL, REAR BRAKE CALIPER PISTON
19   PISTON, REAR BRAKE CALIPER
20   BOOT, REAR BRAKE CALIPER PISTON
29   PAD, REAR DISC BRAKE INNER
30   PAD, REAR DISC BRAKE OUTER
52   BOLT/SCREW, REAR BRAKE CALIPER GUIDE PIN LOWER
53   INSULATOR, REAR DISC BRAKE OUTER PAD

93026G44

Rear brake caliper

To install:

3. Install or connect the following:
    *   Brake pads, if removed
    *   Caliper and retainers, and tighten to 23 ft. lbs. or (31 Nm)
    *   Brake hose, and tighten to 20 ft. lbs. (27 Nm)
4. Bleed brake system and install the wheels.
5. Refill the master cylinder and pump pedal to attain full brake pedal before Road-testing the vehicle.

Disc Brake Pads

REMOVAL & INSTALLATION

Front

1. Before servicing the vehicle, refer to the precautions in the beginning of this section.
2. Remove ⅔ of the brake fluid from the master cylinder.
3. Place a C-clamp around the outer pad and caliper; tighten the C-clamp until the piston is fully compressed in the caliper.
4. Remove or disconnect the following:
    *   Top caliper retainer, and rotate caliper away from rotor
    *   Inboard pad and retaining spring from the caliper
    *   Outboard pad from the caliper
    *   Sleeves and bushings

To install:

5. Clean and lubricate the sleeves and bushing with silicone lubricant and install them in the caliper.
6. Install or connect the following:
    *   Retaining spring onto the inboard pad and the pad in the caliper
    *   Outboard pad into the caliper
    *   Caliper in position and tighten the

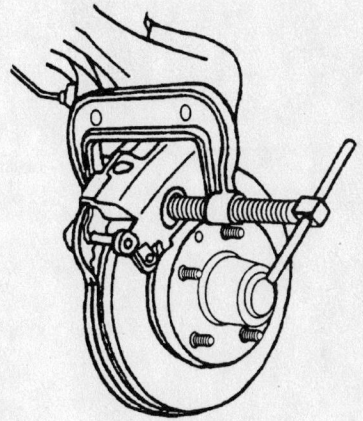

Compressing the caliper piston with a C-clamp

93026G47

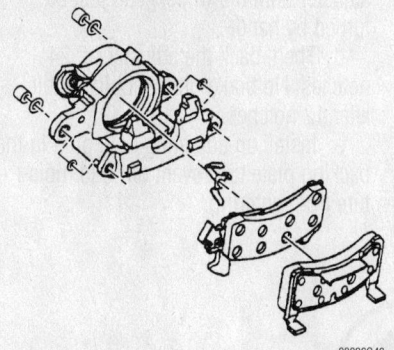

Exploded view of the disc brake assembly

93026G48

mounting bolts. Bend the tabs, on the outboard brake pad, over the caliper.
    *   Wheels
7. Refill the master cylinder and pump pedal to attain full brake pedal before Road-testing the vehicle.

Rear

1. Before servicing the vehicle, refer to the precautions in the beginning of this section.
2. Remove ⅔ of the brake fluid from the master cylinder.
3. Remove or disconnect the following:
    *   Wheel
4. Place a C-clamp around the outer pad and caliper; tighten the C-clamp until the piston is fully compressed in the caliper.
    *   Top caliper retainer, and rotate caliper away from rotor
    *   Inboard pad and retaining spring from the caliper
    *   Outboard pad from the caliper

To install:

5. Clean and lubricate the sleeves and bushing with silicone lubricant and install them in the caliper.
6. Install or connect the following:
    *   Retaining spring onto the inboard pad and the pad in the caliper
    *   Outboard pad into the caliper
    *   Caliper in position and tighten the mounting bolts
    *   Wheels
7. Refill the master cylinder and pump pedal to attain full brake pedal before Road-testing the vehicle.

Brake Drums

REMOVAL & INSTALLATION

1. Before servicing the vehicle, refer to the precautions in the beginning of this section.

2. Remove or disconnect the following:
    *   Wheel
    *   Brake drum. If the drum will not pull of the axle, use a rubber mallet and tap it around the edge.

To install:

3. Install or connect the following:
    *   Drum on the axle
    *   Wheel
4. Refill the master cylinder and pump pedal to attain full brake pedal before road-testing the vehicle.

Brake Shoes

REMOVAL & INSTALLATION

1. Before servicing the vehicle, refer to the precautions in the beginning of this section.
2. Remove or disconnect the following:
    *   Wheel
    *   Brake drum. If the drum will not pull of the axle, use a rubber mallet and tap it around the edge.
    *   Return springs from the brake shoes
    *   Shoe guide
    *   Hold-down springs and pins
    *   Actuator lever and pivot
    *   Lever return spring
    *   Actuator link
    *   Parking brake strut and spring
    *   Parking brake lever
    *   Brake shoes and the adjuster assembly

To install:

3. Lubricate the contact points on the backing plate and the adjuster with lithium grease.
4. Install or connect the following:
    *   Parking brake lever, adjusting screw and spring assembly
    *   Shoe assembly onto the backing plate
    *   Parking brake lever, strut and strut spring
    *   Actuator lever and lever pivot
    *   Actuator link
    *   Lever spring, the hold-down pins and springs
    *   Shoe guide
    *   Return springs
    *   Brake drum
5. Adjust the brakes as follows:
    a. Remove the knockout area in the backing plate, behind the adjuster assembly.

b. Ensure the parking brake system is adjusted properly with no tension on the cables or parking brake lever. The tops of the shoes should be firmly seated against the upper spring retaining anchor, if not as specified, loosen the parking brake cables.

c. Install the drum and turn the brake adjuster until the wheels can just be turned by hand.

d. Then, back the adjuster off 24 notches. No brake drag should be felt after 12 notches.

e. Install an adjusting hole plug in the backing plate to prevent dirt and moisture from entering.

f. Readjust the parking brake cable as necessary.

6. Install the wheels.

7. Refill the master cylinder and pump pedal to attain full brake pedal before Road-testing the vehicle.

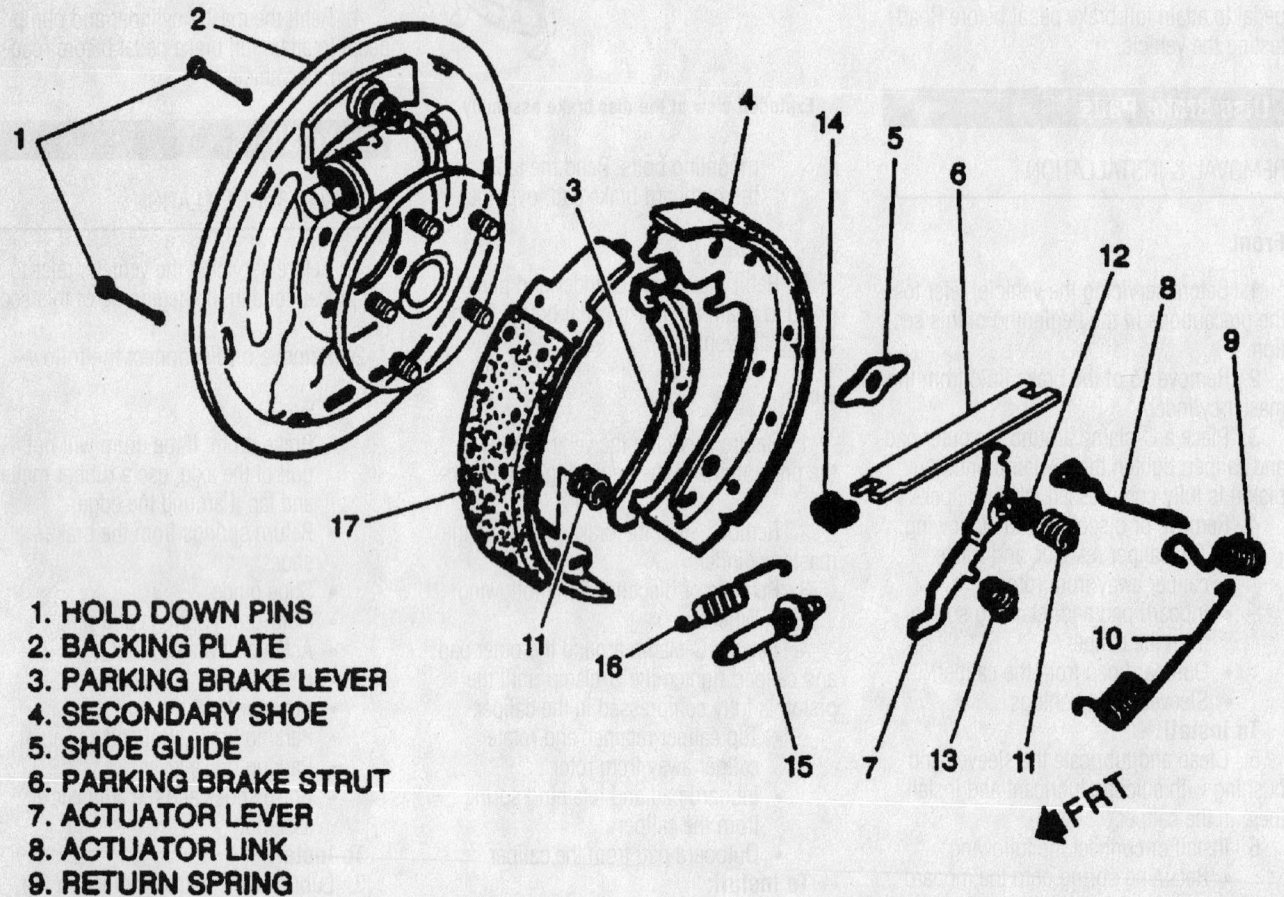

1. HOLD DOWN PINS
2. BACKING PLATE
3. PARKING BRAKE LEVER
4. SECONDARY SHOE
5. SHOE GUIDE
6. PARKING BRAKE STRUT
7. ACTUATOR LEVER
8. ACTUATOR LINK
9. RETURN SPRING
10. RETURN SPRING
11. HOLD DOWN SPRING
12. LEVER PIVOT
13. LEVER RETURN SPRING
14. STRUT SPRING
15. ADJUSTING SCREW ASSEMBLY
16. ADJUSTING SCREW SPRING
17. PRIMARY SHOE

Exploded view of the drum brake components

93026G51

ISUZU AND HONDA

SPECIFICATION CHARTS

ENGINE AND VEHICLE IDENTIFICATION

Engine							Model Year	
Code	Liters (cc)	Cu. In.	Cyl.	Fuel Sys.	Engine Type	Eng. Mfg.	Code ①	Year
D	2.2 (2198)	134	4	MFI	DOHC	Isuzu	Y	2000
W	3.2 (3165)	193	6	MFI	DOHC	Isuzu	1	2001
X	3.5 (3494)	213	6	MFI	DOHC	Isuzu	2	2002
							3	2003
							4	2004

NA: Not available

MFI: Multi-port Fuel Injection

DOHC: Double Overhead Camshaft

① 10th position of VIN

42356-ISUZ-C01

GENERAL ENGINE SPECIFICATIONS

Year	Model	Engine Displacement Liters (cc)	Engine Series VIN)	Fuel System	Net Horsepower @ rpm	Net Torque @ rpm (ft. lbs.)	Bore x Stroke (in.)	Compression Ratio	Oil Pressure @ rpm
2000	Amigo	2.2 (2198)	D	MFI	130@5200	144@4000	3.39x3.72	9.6:1	22@800
		3.2 (3165)	W	MFI	205@5400	214@3000	3.68x3.03	9.1:1	60-80@3000
	Rodeo	2.2 (2198)	D	MFI	130@5200	144@4000	3.39x3.72	9.6:1	22@800
		3.2 (3165)	W	MFI	205@5400	214@3000	3.68x3.03	9.1:1	60-80@3000
	Passport	3.2 (3165)	W	MFI	205@5400	214@3000	3.68x3.03	9.1:1	57-80@3000
	Trooper	3.5 (3494)	X	MFI	215@5400	230@3000	3.68x3.35	9.1:1	60-80@3000
	VehiCROSS	3.5 (3494)	X	MFI	215@5400	230@3000	3.68x3.35	9.1:1	60-80@3000
2001	Rodeo	2.2 (2198)	D	MFI	130@5200	144@4000	3.39x3.72	10.0:1	22@800
		3.2 (3165)	W	MFI	205@5400	214@3000	3.68x3.03	9.1:1	60-80@3000
	Rodeo Sport	2.2 (2198)	D	MFI	130@5200	144@4000	3.39x3.72	10.0:1	22@800
		3.2 (3165)	W	MFI	205@5400	214@3000	3.68x3.03	9.1:1	60-80@3000
	Passport	3.2 (3165)	W	MFI	205@5400	214@3000	3.68x3.03	9.1:1	57-80@3000
	Trooper	3.5 (3494)	X	MFI	215@5400	230@3000	3.68x3.35	9.1:1	60-80@3000
	VehiCROSS	3.5 (3494)	X	MFI	215@5400	230@3000	3.68x3.35	9.1:1	60-80@3000
2002	Rodeo	2.2 (2198)	D	MFI	130@5200	144@4000	3.39x3.72	10.0:1	22@800
		3.2 (3165)	W	MFI	205@5400	214@3000	3.68x3.03	9.1:1	60-80@3000
	Rodeo Sport	2.2 (2198)	D	MFI	130@5200	144@4000	3.39x3.72	10.0:1	22@800
		3.2 (3165)	W	MFI	205@5400	214@3000	3.68x3.03	9.1:1	60-80@3000
	Passport	3.2 (3165)	W	MFI	205@5400	214@3000	3.68x3.03	9.1:1	57-80@3000
	Axiom	3.5 (3494)	X	MFI	230@5400	230@3000	3.68x3.35	9.1:1	60-80@3000
	Trooper	3.5 (3494)	X	MFI	215@5400	230@3000	3.68x3.35	9.1:1	60-80@3000

MFI: Multiport fuel injection

42356-ISUZ-C02

ENGINE TUNE-UP SPECIFICATIONS

Year	Engine Displacement Liters (cc)	Engine VIN	Spark Plug Gap (in.)	Ignition Timing (deg.)		Fuel Pump (psi)	Idle Speed (rpm)		Valve Clearance	
				MT	AT		MT	AT	In.	Ex.
2000	2.2 (2198)	D	0.040	①	①	41-55	800	800	HYD	HYD
	3.2 (3165)	W	0.040	①	①	48-55	750	750	0.009-0.013	0.010-0.014
	3.5 (3494)	X	0.040	①	①	48-55	750	750	0.009-0.013	0.010-0.014
2001	2.2 (2198)	D	0.040	①	①	41-55	800	800	HYD	HYD
	3.2 (3165)	W	0.040	①	①	48-55	750	750	0.009-0.013	0.010-0.014
	3.5 (3494)	X	0.040	①	①	48-55	750	750	0.009-0.013	0.010-0.014
2002	2.2 (2198)	D	0.040	①	①	41-55	800	800	HYD	HYD
	3.2 (3165)	W	0.040	①	①	48-55	750	750	0.009-0.013	0.010-0.014
	3.5 (3494)	X	0.040	①	①	48-55	750	750	0.009-0.013	0.010-0.014

NOTE: The Vehicle Emission Control Information label figures must be used if they differ from those in this chart.

B: Before top dead center

HYD: Hydraulic

① Controlled by the PCM

42356-ISUZ-C03

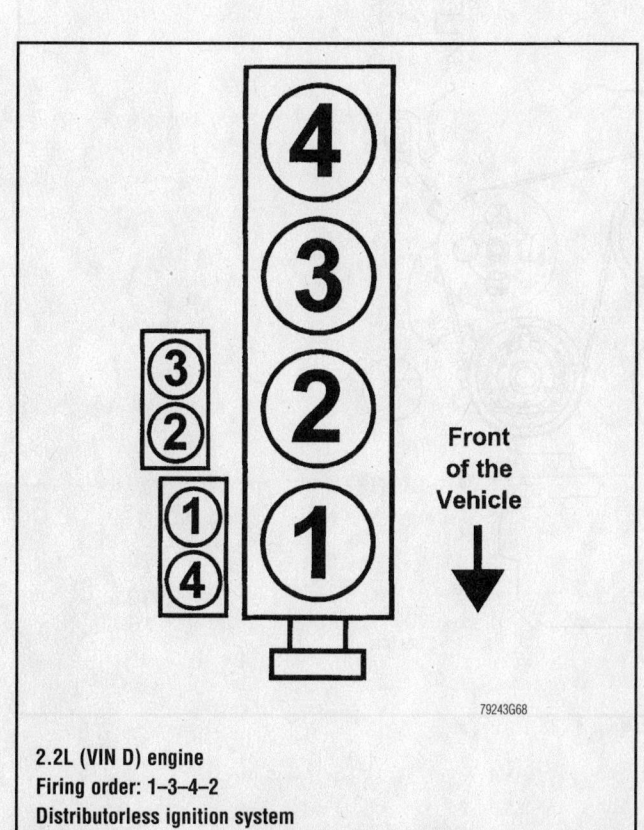

2.2L (VIN D) engine
Firing order: 1–3–4–2
Distributorless ignition system

79243G68

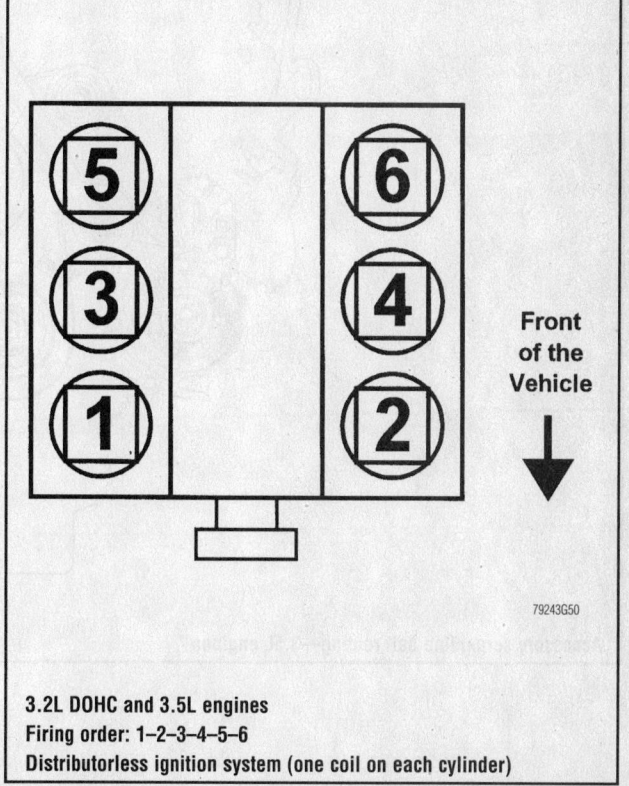

3.2L DOHC and 3.5L engines
Firing order: 1–2–3–4–5–6
Distributorless ignition system (one coil on each cylinder)

79243G50

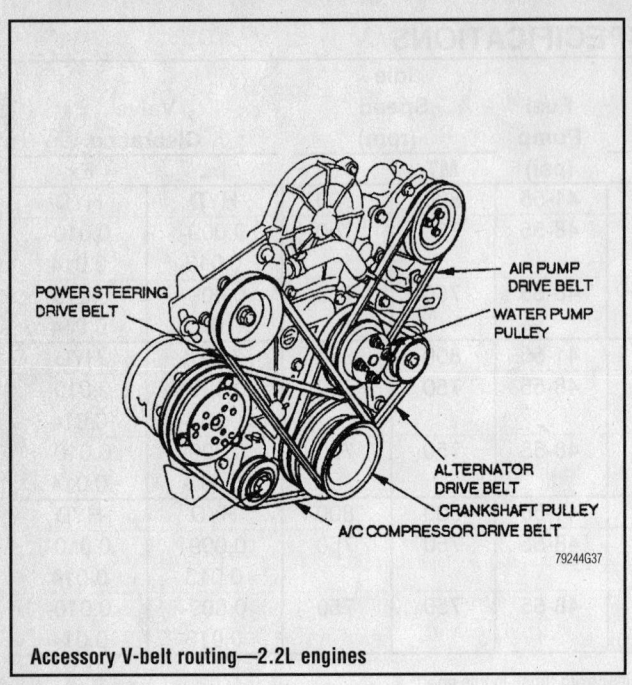

POWER STEERING
DRIVE BELT

AIR PUMP
DRIVE BELT

WATER PUMP
PULLEY

ALTERNATOR
DRIVE BELT

CRANKSHAFT PULLEY

A/C COMPRESSOR DRIVE BELT

79244G37

Accessory V-belt routing—2.2L engines

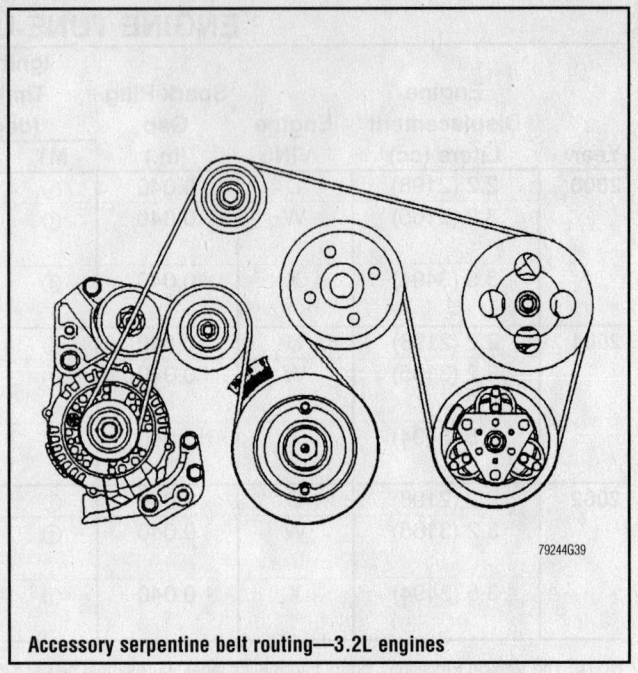

79244G39

Accessory serpentine belt routing—3.2L engines

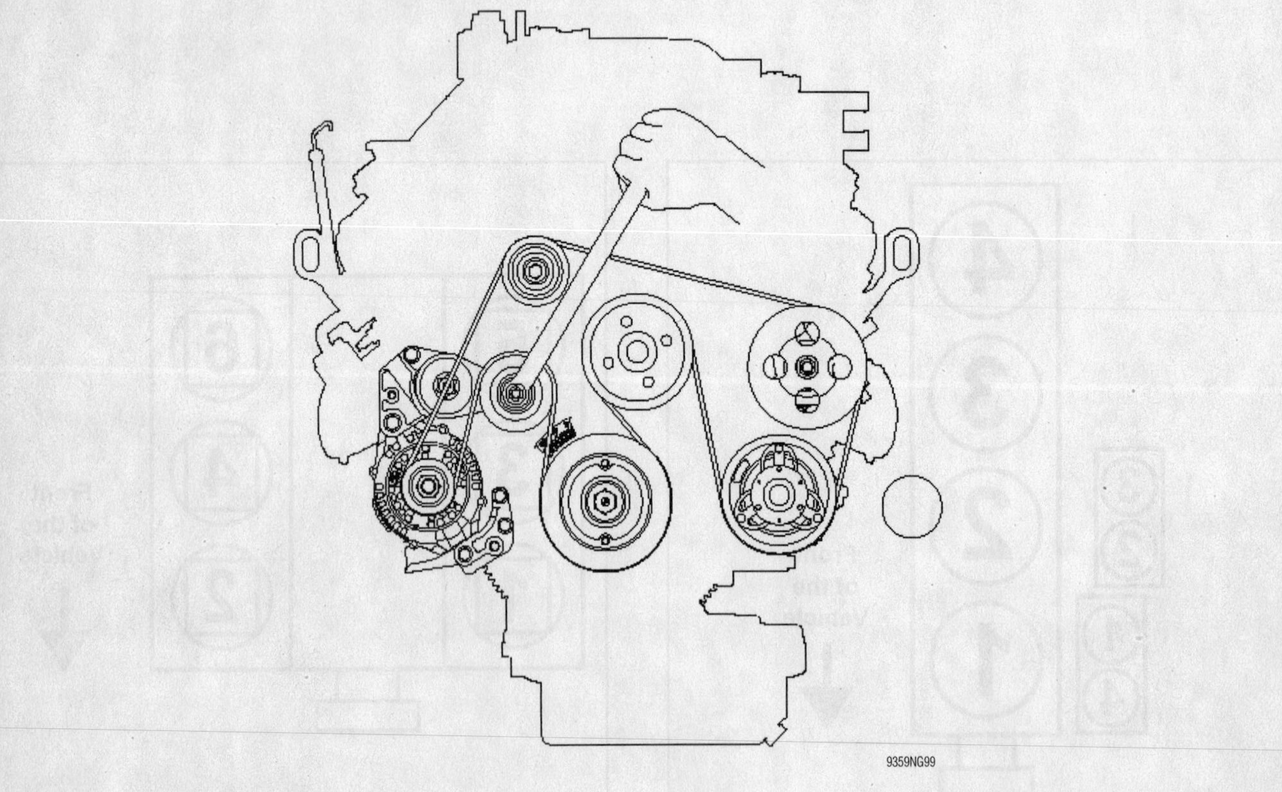

9359NG99

Accessory serpentine belt routing—3.5L engines

CAPACITIES

Year	Model	Engine Displacement Liters (cc)	Engine VIN	Oil with Filter (qts.)	Engine Transmission (pts.) Man.	Auto.	Transfer Case (pts.)	Drive Axle Front (pts.)	Rear (pts.)	Fuel Tank (gal.)	Cooling System (qts.)
2000	Amigo	2.2 (2198)	D	4.8	4.5	—	3.0	3.0	3.7	17.7	7.3
		3.2 (3165)	W	5.0	6.2	—	3.0	3.0	3.7	17.7	11.2
	Rodeo	2.2 (2198)	D	4.8	4.5	18.2	3.0	2.2	3.7	21.1	7.3
		3.2 (3165)	W	5.0	6.2	18.2	3.0	2.6	3.7	21.1	11.6
	Passport	3.2 (3165)	W	5.0	6.2	18.2	3.0	2.6	3.74	21.1	11.6
	Trooper	3.5 (3494)	X	5.0	5.8	18.2	3.0①	3.0	6.4	22.5	②
	VehiCROSS	3.5 (3494)	X	5.0	—	18.2	3.0①	3.0	6.4	22.5	②
2001	Rodeo Sport	2.2 (2198)	D	4.8	6.2	—	3.0	2.2	3.7	17.7	7.3
		3.2 (3165)	W	5.0	6.2	18.2	3.0	2.6	3.7	17.7	11.2
	Rodeo	2.2 (2198)	D	4.8	4.5	—	3.0	2.2	3.7	20.0	7.3
		3.2 (3165)	W	5.0	6.2	18.2	3.0	2.6	3.7	20.0	11.6
	Passport	3.2 (3165)	W	5.0	6.2	18.2	3.0	2.6	3.74	19.5	11.6
	Trooper	3.5 (3494)	X	5.0	5.8	18.2	3.0①	3.0	6.4	22.5	②
	VehiCROSS	3.5 (3494)	X	5.0	—	18.2	4.0	3.0	4.6	22.5	7.4
2002	Rodeo Sport	2.2 (2198)	D	4.8	6.2	—	3.0	2.2	3.7	17.7	7.3
		3.2 (3165)	W	5.0	6.2	18.2	3.0	2.6	3.7	17.7	11.2
	Rodeo	2.2 (2198)	D	4.8	6.2	—	3.0	2.2	3.7	19.5	7.3
		3.2 (3165)	W	5.0	6.2	18.2	3.0	2.6	3.7	19.5	11.6
	Passport	3.2 (3165)	W	5.0	6.2	18.2	3.0	2.6	3.74	19.5	11.6
	Axiom	3.5 (3494)	X	5.0	—	18.2	2.8	2.66	3.64	19.5	11.7
	Trooper	3.5 (3494)	X	5.0	5.8	18.2	3.0①	3.0	6.4	22.5	②

NOTE: All capacities are approximate. Add fluid gradually and check to ensure a proper level has been reached.

① 4.0 pts. if equipped with Torque On Demand
② A/T: 7.4 qts.
  M/T: 7.0 qts.

42356-ISUZ-C04

CRANKSHAFT AND CONNECTING ROD SPECIFICATIONS

All measurements are given in inches.

Year	Engine Displacement Liters (cc)	Engine VIN	Crankshaft				Connecting Rod		
			Main Brg. Journal Dia.	Main Brg. Oil Clearance	Shaft End-play	Thrust on No.	Journal Diameter	Oil Clearance	Side Clearance
2000	2.2 (2198)	D	2.2590-2.2610	0.0007-0.0016	0.0004-0.0008	2	1.9090-1.9100	0.0002-0.0012	0.0050-0.0150
	3.2 (3165)	W	2.5165-2.5170	0.0007-0.0017	0.0024-0.0094	3	2.1229-2.1235	0.0010-0.0023	0.0050-0.0150
	3.5 (3494)	X	2.5165-2.5170	0.0007-0.0017	0.0024-0.0094	3	2.1229-2.1235	0.0010-0.0023	0.0050-0.0150
2001	2.2 (2198)	D	2.2590-2.2610	0.0007-0.0016	0.0004-0.0008	2	1.9090-1.9100	0.0002-0.0012	0.0050-0.0150
	3.2 (3165)	W	2.5165-2.5170	0.0007-0.0017	0.0024-0.0094	3	2.1229-2.1235	0.0010-0.0023	0.0050-0.0150
	3.5 (3494)	X	2.5165-2.5170	0.0007-0.0017	0.0024-0.0094	3	2.1229-2.1235	0.0010-0.0023	0.0050-0.0150
2002	2.2 (2198)	D	2.2590-2.2610	0.0007-0.0016	0.0004-0.0008	2	1.9090-1.9100	0.0002-0.0012	0.0050-0.0150
	3.2 (3165)	W	2.5165-2.5170	0.0007-0.0017	0.0024-0.0094	3	2.1229-2.1235	0.0010-0.0023	0.0050-0.0150
	3.5 (3494)	X	2.5165-2.5170	0.0007-0.0017	0.0024-0.0094	3	2.1229-2.1235	0.0010-0.0023	0.0050-0.0150

42356-ISUZ-C05

VALVE SPECIFICATIONS

Year	Engine Displacement Liters (cc)	Engine VIN	Seat Angle (deg.)	Face Angle (deg.)	Spring Test Pressure (lbs. @ in.)	Spring Installed Height (in.)	Stem-to-Guide Clearance (in.)		Stem Diameter (in.)	
							Intake	Exhaust	Intake	Exhaust
2000	2.2 (2198)	D	NA	NA	NA	NA	0.0012-0.0022	0.0016-0.0026	NA	NA
	3.2 (3165)	W	45	45	41-44@1.38	1.38	0.0002-0.0009	0.0012-0.0025	0.2346-0.2353	0.2343-0.2350
	3.5 (3494)	X	45	45	41-44@1.38	1.38	0.0002-0.0009	0.0012-0.0025	0.2346-0.2353	0.2343-0.2350
2001	2.2 (2198)	D	NA	NA	NA	NA	0.0012-0.0022	0.0016-0.0026	NA	NA
	3.2 (3165)	W	45	45	41-44@1.38	1.38	0.0002-0.0009	0.0012-0.0025	0.2346-0.2353	0.2343-0.2350
	3.5 (3494)	X	45	45	41-44@1.38	1.38	0.0002-0.0009	0.0012-0.0025	0.2346-0.2353	0.2343-0.2350
2002	2.2 (2198)	D	NA	NA	NA	NA	0.0012-0.0022	0.0016-0.0026	NA	NA
	3.2 (3165)	W	45	45	41-44@1.38	1.38	0.0002-0.0009	0.0012-0.0025	0.2346-0.2353	0.2343-0.2350
	3.5 (3494)	X	45	45	41-44@1.38	1.38	0.0002-0.0009	0.0012-0.0025	0.2346-0.2353	0.2343-0.2350

NA: Not Available

42356-ISUZ-C06

PISTON AND RING SPECIFICATIONS
All measurements are given in inches.

Year	Engine Displacement Liters (cc)	Engine VIN	Piston Clearance	Ring Gap			Ring Side Clearance		
				Top Compression	Bottom Compression	Oil Control	Top Compression	Bottom Compressior	Oil Control
2000	2.2 (2198)	D	NA	0.0118-0.0195	0.0118-0.0195	0.016-0.055	0.0008-0.0546	0.0008-0.0546	NA
	3.2 (3165)	W	NA	0.0118-0.0157	0.0177-0.0236	0.0060-0.018	0.0006-0.002	0.0006-0.0015	NA
	3.5 (3494)	X	NA	0.0118-0.0157	0.0177-0.0236	0.006-0.018	0.0006-0.0015	0.0006-0.0015	NA
2001	2.2 (2198)	D	NA	0.0118-0.0195	0.0118-0.0195	0.016-0.055	0.0008-0.0546	0.0008-0.0546	NA
	3.2 (3165)	W	NA	0.0118-0.0157	0.0177-0.0236	0.0060-0.018	0.0006-0.002	0.0006-0.0015	NA
	3.5 (3494)	X	NA	0.0118-0.0157	0.0177-0.0236	0.006-0.018	0.0006-0.0015	0.0006-0.0015	NA
2002	2.2 (2198)	D	NA	0.0118-0.0195	0.0118-0.0195	0.016-0.055	0.0008-0.0546	0.0008-0.0546	NA
	3.2 (3165)	W	NA	0.0118-0.0157	0.0177-0.0236	0.0060-0.018	0.0006-0.002	0.0006-0.0015	NA
	3.5 (3494)	X	NA	0.0118-0.0157	0.0177-0.0236	0.006-0.018	0.0006-0.0015	0.0006-0.0015	NA

NA: Not Available

42356-ISUZ-C07

TORQUE SPECIFICATIONS
All readings in ft. lbs.

Year	Engine Displacement Liters (cc)	Engine VIN	Cylinder Head Bolts	Main Bearing Bolts	Rod Bearing Bolts	Crankshaft Damper Bolts	Flywheel Bolts	Manifold		Spark Plugs	Lug Nuts
								Intake	Exhaust		
2000	2.2 (2198)	D	①	②	③	④	⑤	16	⑥	18	87
	3.2 (3165)	W	⑦	29	40	123	40	18	42	13	87
	3.5 (3494)	X	⑦	⑧	40	123	40	18	38	13	87
2001	2.2 (2198)	D	①	②	③	④	⑤	16	⑥	18	87
	3.2 (3165)	W	⑦	29	40	123	40	18	42	13	87
	3.5 (3494)	X	⑦	⑧	40	123	40	18	38	13	87
2002	2.2 (2198)	D	①	②	③	④	⑤	16	⑥	18	87
	3.2 (3165)	W	⑦	29	40	123	40	18	42	13	87
	3.5 (3494)	X	⑦	⑧	40	123	40	18	38	13	87

① Step 1: 18 ft. lbs.
Step 2: Plus 90 degrees
Step 3: Plus 90 degrees
Step 4: Plus 90 degrees

② Step 1: 37 ft. lbs.
Step 2: Plus 45 degrees
Step 3: Plus 15 degrees

③ Step 1: 25 ft. lbs.
Step 2: Plus 45 degrees
Step 3: Plus 15 degrees

④ Crankshaft sprocket:
Step 1: 94 ft. lbs.
Step 2: Plus 45 degrees
Crankshaft balancer:
Step 1: 14 ft. lbs.
Step 2: Plus 45 degrees

⑤ Step 1: 48 ft. lbs.
Step 2: Plus 30 degrees
Step 3: Plus 15 degrees

⑥ Step 1: 112 inch lbs.
Step 2: 14 ft. lbs.
Step 3: 14 ft. lbs.

⑦ Step 1: 21 ft. lbs.
Step 2: 47 ft. lbs.

⑧ Step 1: 22 ft. lbs.
Step 2: Plus 55-65 degrees
Step 3: Crankcase side bolts to 29 ft. lbs.

42356-ISUZ-C08

WHEEL ALIGNMENT

Year	Model		Caster Range (+/-Deg.)	Caster Preferred Setting (Deg.)	Camber Range (+/-Deg.)	Camber Preferred Setting (Deg.)	Toe-in (in.)	Steering Axis Inclination (Deg.)
2000	Amigo	F	0.75	+2.50	0.30	0	0+/-0.08	—
		R	—	—	1.00	0	0+/-0.20	—
	Rodeo	F	1.00	+2.50	0.50	0	0+/-0.08	—
		R	—	—	1.00	0	0+/-0.20	—
	Passport	F	1.00	+2.50	0.50	0	0+/-0.08	—
		R	—	—	1.00	0	0+/-0.20	—
	Trooper		0.75	+2.10	0.50	0	0+/-0.08	—
	VehiCROSS		0.75	+2.10	0.50	0	0+/-0.08	—
2001	Rodeo Sport	F	1.00	+2.50	0.50	0	0+/-0.08	—
		R	—	—	1.00	0	0+/-0.20	—
	Rodeo	F	1.00	+2.50	0.50	0	0+/-0.08	—
		R	—	—	1.00	0	0+/-0.20	—
	Passport	F	1.00	+2.50	0.50	0	0+/-0.08	—
		R	—	—	1.00	0	0+/-0.20	—
	Trooper		0.75	+2.10	0.50	0	0+/-0.08	—
	VehiCROSS		0.75	+2.10	0.50	0	0+/-0.08	—
2002	Rodeo Sport	F	1.00	+2.50	0.50	0	0+/-0.08	—
		R	—	—	1.00	0	0+/-0.20	—
	Rodeo	F	0.75	+2.50	0.50	0	0.04+/-0.04	—
		R	—	—	1.00	0	0+/-0.20	—
	Passport	F	1.00	+2.50	0.50	0	0+/-0.08	—
		R	—	—	1.00	0	0+/-0.20	—
	Axiom		1.00	+2.50	0.50	0	0+/-0.08	—
	Trooper		0.75	+2.10	0.50	0	0+/-0.08	—

42356-ISUZ-C09

TIRE, WHEEL AND BALL JOINT SPECIFICATIONS

Year	Model	OEM Tires		Tire Pressures (psi)		Wheel Size	Ball Joint Inspection
		Standard	Optional	Front	Rear		
2000	Amigo	P245/70R16	none	26	26	7JJ	U: 4-28 ① L: 4-55
	Rodeo 2wd/4wd	P245/70R16	none	29	29	7JJ	U: 4-28 ① L: 4-55
	Passport LX 2wd	P225/75R16	None	29	29	6.5-JJ	NS
	Passport LX 4wd	P245/70R16	None	29	29	7J	NS
	Passport EX 2wd	P245/70R16	None	29	29	7J	NS
	Passport EX 4wd	P245/70R16	None	29	29	7-JJ	NS
	Trooper	P245/70R15	none	26	26	7JJ	U: 4-28 ① L: 4-55
	VehiCROSS	P245/70R16	none	26	26	7JJ	NS
2001	Rodeo Sport	P245/70R16	none	26	26	7JJ	U: 4-28 ① L: 4-55
	Rodeo 2wd/4wd	P225/70R16	P245/70R16	29	29	7JJ	U: 4-28 ① L: 4-55
	Passport LX 2wd	P225/75R16	None	29	29	6.5-JJ	NS
	Passport LX 4wd	P245/70R16	None	29	29	7J	NS
	Passport EX 2wd	P245/70R16	None	29	29	7J	NS
	Passport EX 4wd	P245/70R16	None	29	29	7-JJ	NS
	Trooper	P245/70R15	none	26	26	7JJ	U: 4-28 ① L: 4-55
	VehiCROSS	P245/60R18	none	26	26	7JJ	NS
2002	Rodeo Sport	P245/70R16	none	26	26	7JJ	U: 4-28 ① L: 4-55
	Axiom	P235/65R17	none	26	26	7JJ	U: 4-28 ① L: 4-55
	Rodeo 2wd/4wd	P225/75R16	P245/70R16 P245/60R18	26	26	7JJ	U: 4-28 ① L: 4-55
	Passport LX 2wd	P225/75R16	None	29	29	6.5-JJ	NS
	Passport LX 4wd	P245/70R16	None	29	29	7J	NS
	Passport EX 2wd	P245/70R16	None	29	29	7J	NS
	Passport EX 4wd	P245/70R16	None	29	29	7-JJ	NS
	Trooper	P245/70R15	none	26	26	7JJ	U: 4-28 ① L: 4-55

L: Lower

U: Upper

NA: Not available

NS: Not specified by manufacturer

① Torque required in inch lbs. to rotate ball joint when removed from the knuckle

42356-ISUZ-C10

BRAKE SPECIFICATIONS
All measurements in inches unless noted

Year	Model		Brake Disc Original Thickness	Brake Disc Machine Thickness	Brake Disc Minimum Thickness	Maximum Runout	Brake Drum Diameter Original Inside Diameter	Brake Drum Diameter Max. Wear Limit	Brake Drum Diameter Maximum Machine Diameter	Minimum Lining Thickness	Brake Caliper Bracket Bolts (ft. lbs.)	Brake Caliper Mounting Bolts (ft. lbs.)
2000	Amigo	F	1.020	0.983	0.969	0.005	—	—	—	0.039	115	54
		R	0.710	0.668	0.654	0.005	11.6	11.67	NA	0.039	76	32
	Rodeo	F	1.020	0.983	0.969	0.005	—	—	—	0.039	115	54
		R	0.710	0.668	0.654	0.005	11.6	11.67	NA	0.039	76	32
	Passport	F	1.020	0.983	0.969	0.005	—	—	—	0.039	115	54
		R	0.710	0.668	0.654	0.005	11.6	11.67	NA	0.039	76	32
	Trooper	F	1.024	0.983	0.969	0.005	—	—	—	0.039	115	54
		R	0.710	0.668	0.654	0.005	—	—	NA	0.039	76	32
	VehiCROSS	F	1.024	0.983	0.969	0.005	—	—	—	0.039	115	54
		R	0.710	0.668	0.654	0.005	—	—	—	0.039	76	32
2001	Rodeo Sport	F	1.020	0.983	0.969	0.005	—	—	—	0.039	115	54
		R	0.710	0.668	0.654	0.005	11.6	11.67	NA	0.039	76	32
	Rodeo	F	1.020	0.983	0.969	0.005	—	—	—	0.039	115	54
		R	0.710	0.668	0.654	0.005	11.6	11.67	NA	0.039	76	32
	Passport	F	1.020	0.983	0.969	0.005	—	—	—	0.039	115	54
		R	0.710	0.668	0.654	0.005	—	—	—	0.039	76	32
	Trooper	F	1.024	0.983	0.969	0.005	—	—	—	0.039	115	54
		R	0.710	0.668	0.654	0.005	—	—	—	0.039	76	32
	VehiCROSS	F	1.024	0.983	0.969	0.005	—	—	—	0.039	115	54
		R	0.710	0.668	0.654	0.005	—	—	—	0.039	76	32
2002	Rodeo Sport	F	1.020	0.983	0.969	0.005	—	—	—	0.039	115	54
		R	0.710	0.668	0.654	0.005	11.6	11.67	NA	0.039	76	32
	Rodeo	F	1.020	0.983	0.969	0.005	—	—	—	0.039	115	54
		R	0.710	0.668	0.654	0.005	11.6	11.67	NA	0.039	76	32
	Axiom	F	1.020	0.983	0.969	0.005	—	—	—	0.039	115	54
		R	0.710	0.668	0.654	0.005	11.6	11.67	NA	0.039	76	32
	Passport	F	1.020	0.983	0.969	0.005	—	—	—	0.039	115	54
		R	0.710	0.668	0.654	0.005	—	—	—	0.039	76	32
	Trooper	F	1.024	0.983	0.969	0.005	—	—	—	0.039	115	54
		R	0.710	0.668	0.654	0.005	—	—	—	0.039	76	32

NA: Not Available

42356-ISUZ-C11

SCHEDULED MAINTENANCE INTERVALS
Isuzu—Amigo, Axiom, Rodeo, Rodeo Sport, Trooper & vehiCROSS; Honda—Passport

TO BE SERVICED	TYPE OF SERVICE	7.5	15	22.5	30	37.5	45	52.5	60	67.5	75	82.5	90	97.5	105	112.5	120
Accelerator linkage ①	L	✓	✓	✓	✓	✓	✓	✓	✓	✓	✓	✓	✓	✓	✓	✓	✓
Accessory drive belts ②	S/I				✓				✓				✓				✓
Air cleaner filter	R				✓				✓				✓				✓
Auto cruise control linkage & hose ③	S/I		✓		✓		✓		✓		✓		✓		✓		✓
Automatic transmission fluid level ③	S/I	✓		✓		✓		✓		✓		✓		✓		✓	
Battery fluid level ③	S/I	✓	✓	✓	✓	✓	✓	✓	✓	✓	✓	✓	✓	✓	✓	✓	✓
Body and chassis ①	L	✓	✓	✓	✓	✓	✓	✓	✓	✓	✓	✓	✓	✓	✓	✓	✓
Brake fluid level ③	S/I	✓	✓	✓	✓	✓	✓	✓	✓	✓	✓	✓	✓	✓	✓	✓	✓
Brake lines & hoses ③	S/I	✓	✓	✓	✓	✓	✓	✓	✓	✓	✓	✓	✓	✓	✓	✓	✓
Brake pedal play ③	S/I		✓		✓		✓		✓		✓		✓		✓		✓
Clutch fluid level ③	S/I	✓	✓	✓	✓	✓	✓	✓	✓	✓	✓	✓	✓	✓	✓	✓	✓
Clutch lines & hose ③	S/I				✓				✓				✓				✓
Clutch pedal free-play ③	S/I		✓		✓		✓		✓		✓		✓		✓		✓
Clutch pedal spring, bushing and clevis pin ①	S/I		✓		✓		✓		✓		✓		✓		✓		✓
Cooling and heating system hoses ③	S/I		✓		✓		✓		✓		✓		✓		✓		✓
Driveshaft flange torque ③	S/I	✓		✓		✓		✓		✓		✓		✓		✓	
Drum and disc brakes ③	S/I		✓		✓		✓		✓		✓		✓		✓		✓
Engine coolant	R				✓				✓				✓				✓
Engine coolant level ③	S/I	✓	✓	✓	✓	✓	✓	✓	✓	✓	✓	✓	✓	✓	✓	✓	✓
Engine oil & filter ③	R	✓	✓	✓	✓	✓	✓	✓	✓	✓	✓	✓	✓	✓	✓	✓	✓
Exhaust system ③	S/I	✓	✓	✓	✓	✓	✓	✓	✓	✓	✓	✓	✓	✓	✓	✓	✓
Front and rear axle lubricant	R		✓		✓				✓				✓				✓
Front and rear driveshafts ①	S/I	✓	✓	✓	✓	✓	✓	✓	✓	✓	✓	✓	✓	✓	✓	✓	✓
Front wheel bearings	S/I & L				✓				✓				✓				✓
Fuel lines & tank cap ③	S/I								✓								✓
Inspect for fluid leaks ③	S/I	✓	✓	✓	✓	✓	✓	✓	✓	✓	✓	✓	✓	✓	✓	✓	✓
Key lock cylinder ③	L		✓		✓		✓		✓		✓		✓		✓		✓
Manual transmission and transfer case fluid ④	R		✓		✓				✓				✓				✓
Parking brake system ③	S/I		✓		✓		✓		✓		✓		✓		✓		✓
Power steering fluid	R				✓				✓				✓				✓
Radiator core and A/C condenser	S/I & C								✓								✓
Rotate tires	S/I	✓	✓	✓	✓	✓	✓	✓	✓	✓	✓	✓	✓	✓	✓	✓	✓
Shift-on-the-fly system gear fluid ③	S/I		✓		✓		✓		✓		✓		✓		✓		✓
Spark plug wires ⑤	S/I								✓								✓
Spark plugs	R	Every 100,000 miles
Starter safety switch ③	S/I	✓	✓	✓	✓	✓	✓	✓	✓	✓	✓	✓	✓	✓	✓	✓	✓
Steering operation ③	S/I	✓	✓	✓	✓	✓	✓	✓	✓	✓	✓	✓	✓	✓	✓	✓	✓
Suspension & steering ③	S/I	✓	✓	✓	✓	✓	✓	✓	✓	✓	✓	✓	✓	✓	✓	✓	✓
Throttle linkage ③	S/I		✓		✓		✓		✓		✓		✓		✓		✓
Timing belt	R										✓						

42356-ISUZ-C12

SCHEDULED MAINTENANCE INTERVALS
Isuzu—Amigo, Axiom, Rodeo, Rodeo Sport, Trooper & vehiCROSS; Honda—Passport

TO BE SERVICED	TYPE OF SERVICE	VEHICLE MILEAGE INTERVAL (x1000)														
		7.5	15	22.5	30	37.5	45	52.5	60	67.5	75	82.5	90	97.5	105	112.5	120
Tires and wheels ③	S/I	✓	✓	✓	✓	✓	✓	✓	✓	✓	✓	✓	✓	✓	✓	✓	✓
Valve clearance ④	A								✓								✓

R: Replace   S/I: Service or Inspect   L: Lubricate   A: Adjust   C: Clean

① Perform this at the mileage indicated or every 6 months, whichever occurs first.
② Perform this at the mileage indicated or every 24 months, whichever occurs first.
③ Perform this at the mileage indicated or every 12 months, whichever occurs first.
④ 3.2L V6 engine.
⑤ 2.2L 4 cyl. engine.

FREQUENT OPERATION MAINTENANCE (SEVERE SERVICE)

If a vehicle is operated under any of the following conditions it is considered severe service:

- Towing a trailer or using a camper or car-top carrier.
- Repeated short trips of less than 5 miles in temperatures below freezing.
- Extensive idling or low-speed driving for long distances as in heavy commercial use, such as delivery, taxi or police cars.
- Operating on rough, muddy or salt-covered roads.
- Operating on unpaved or dusty roads.

Air cleaner element: replace every 15,000 miles

Engine oil and filter: replace every 3000 miles or 3 months, whichever occurs first.

Automatic transmission fluid: replace every 20,000 miles.

Rear axle lubricant: replace every 15,000 miles.

42356-ISUZ-C13

PRECAUTIONS

Before servicing any vehicle, please be sure to read all of the following precautions, which deal with personal safety, prevention of component damage and important points to take into consideration when servicing a motor vehicle:

• Never open, service or drain the radiator or cooling system when the engine is hot; serious burns can occur from the steam and hot coolant.

• Observe all applicable safety precautions when working around fuel. Whenever servicing the fuel system, always work in a well-ventilated area. Do not allow fuel spray or vapors to come in contact with a spark, open flame, or excessive heat (a hot drop light, for example). Keep a dry chemical fire extinguisher near the work area. Always keep fuel in a container specifically designed for fuel storage; also, always properly seal fuel containers to avoid the possibility of fire or explosion. Refer to the additional fuel system precautions later in this section.

• Fuel injection systems often remain pressurized, even after the engine has been turned **OFF**. The fuel system pressure must be relieved before disconnecting any fuel lines. Failure to do so may result in fire and/or personal injury.

• Brake fluid often contains polyglycol ethers and polyglycols. Avoid contact with the eyes and wash your hands thoroughly after handling brake fluid. If you do get brake fluid in your eyes, flush your eyes with clean, running water for 15 minutes. If eye irritation persists, or if you have taken brake fluid internally, seek medical assistance IMMEDIATELY.

• The EPA warns that prolonged contact with used engine oil may cause a number of skin disorders, including cancer. You should make every effort to minimize your exposure to used engine oil. Protective gloves should be worn when changing oil. Wash your hands and any other exposed skin areas as soon as possible after exposure to used engine oil. Soap and water, or waterless hand cleaner should be used.

• All new vehicles are now equipped with an air bag system, often referred to as a Supplemental Restraint System (SRS) or Supplemental Inflatable Restraint (SIR) system. The system must be disabled before performing service on or around system components, steering column, instrument panel components, wiring and sensors. Failure to follow safety and disabling procedures could result in accidental air bag deployment, possible personal injury and unnecessary system repairs.

• Always wear safety goggles when working with, or around, the air bag system. When carrying a non-deployed air bag, be sure the bag and trim cover are pointed away from your body. When placing a non-deployed air bag on a work surface, always face the bag and trim cover upward, away from the surface. This will reduce the motion of the module if it is accidentally deployed. Refer to the additional air bag system precautions later in this section.

• Clean, high quality brake fluid from a sealed container is essential to the safe and proper operation of the brake system. You should always buy the correct type of brake fluid for your vehicle. If the brake fluid becomes contaminated, completely flush the system with new fluid. Never reuse any brake fluid. Any brake fluid that is removed from the system should be discarded. Also, do not allow any brake fluid to come in contact with a painted surface; it will damage the paint.

• Never operate the engine without the proper amount and type of engine oil; doing so WILL result in severe engine damage.

• Timing belt maintenance is extremely important. Many models utilize an interference-type, non-freewheeling engine. If the timing belt breaks, the valves in the cylinder head may strike the pistons, causing potentially serious (also time-consuming and expensive) engine damage. Refer to the maintenance interval charts in the front of this section for the recommended replacement interval for the timing belt.

• Disconnecting the negative battery cable on some vehicles may interfere with the functions of the on-board computer system(s) and may require the computer to undergo a relearning process once the negative battery cable is reconnected.

• When servicing drum brakes, only disassemble and assemble one side at a time, leaving the remaining side intact for reference.

• Only an MVAC-trained, EPA-certified automotive technician should service the A/C system or its components.

ENGINE REPAIR

➡**Disconnecting the negative battery cable on some vehicles may interfere with the functions of the on board computer system. The computer may undergo a relearning process once the negative battery cable is reconnected.**

Distributor

REMOVAL

These engines are equipped with a Distributorless Ignition System (DIS).

Alternator

REMOVAL

2.2L Engine

1. Before servicing the vehicle, refer to the precautions in the beginning of this section.
2. Remove or disconnect the following:
   • Negative battery cable
   • Accessory drive belt
   • Alternator harness connectors
   • Alternator

3.2L Engine

1. Before servicing the vehicle, refer to the precautions in the beginning of this section.
2. Remove or disconnect the following:
   • Negative battery cable
   • Accessory drive belt
   • Alternator harness connectors
   • Alternator

3.5L Engine

1. Before servicing the vehicle, refer to the precautions in the beginning of this section.

2. Remove or disconnect the following:
   - Negative battery cable
   - Accessory drive belt
   - Alternator wiring connectors
   - Alternator

INSTALLATION

2.2L Engine

Install or connect the following:
   - Alternator. Tighten the long bolt to 26 ft. lbs. (35 Nm) and the short bolt to 15 ft. lbs. (20 Nm).
   - Alternator harness connectors
   - Accessory drive belt
   - Negative battery cable

3.2L Engine

Install or connect the following:
   - Alternator. Tighten the 10mm bolt to 30 ft. lbs. (41 Nm) and the 8mm bolt to 15 ft. lbs. 21 Nm).
   - Alternator harness connectors
   - Accessory drive belt
   - Negative battery cable

3.5L Engine

1. Before servicing the vehicle, refer to the precautions in the beginning of this section.
2. Install or connect the following:
   - Alternator. Tighten the 10mm bolts to 30 ft. lbs. (41 Nm) and the 8mm bolts to 15 ft. lbs. (21 Nm).
   - Alternator wiring connectors
   - Accessory drive belt
   - Negative battery cable

Ignition Timing

ADJUSTMENT

These engines are equipped with a Distributorless Ignition System (DIS). No adjustment is possible.

Engine Assembly

REMOVAL & INSTALLATION

2.2L engines

1. Before servicing the vehicle, refer to the precautions in the beginning of this section.
2. Drain the cooling system.
3. Relieve the fuel system pressure.
4. Remove or disconnect the following:
   - Battery

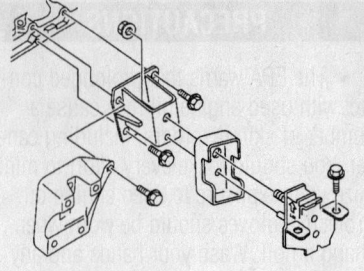

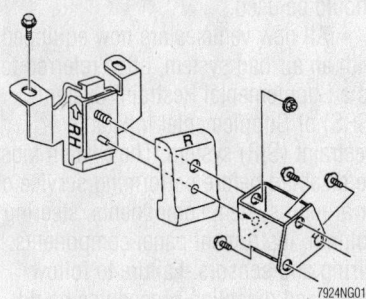

Left and right motor mounts—2.2L engine

7924NG01

   - Hood
   - Accessory drive belt
   - Accelerator cable
   - Air intake assembly
   - Engine wiring harness connectors at left rear of the engine compartment
   - Brake booster vacuum line
   - Engine ground cables
   - Clutch fluid line bracket and slave cylinder
   - Fuel lines and bracket
   - Exhaust front pipe
   - Transmission
   - A/C compressor
   - Power steering pump
   - Chassis harness connectors at right rear of the engine compartment
   - Frame ground cable
   - Radiator hoses
   - Heater hoses
   - Cooling fan connector
   - Cooling fan and shroud
   - Radiator
   - Left and right engine mounts
5. Lift the engine from the vehicle.

To install:

6. Position the engine in the engine compartment.
7. Install or connect the following:
   - Left and right engine mounts. Tighten the fasteners to 30 ft. lbs. (41 Nm).
   - Radiator
   - Cooling fan and shroud
   - Cooling fan connector
   - Heater hoses
   - Radiator hoses
   - Frame ground cable
   - Chassis harness connectors at right rear of the engine compartment
   - Power steering pump
   - A/C compressor
   - Transmission
   - Exhaust front pipe
   - Fuel lines and bracket
   - Clutch fluid line bracket and slave cylinder
   - Engine ground cables
   - Brake booster vacuum line
   - Engine wiring harness connectors at left rear of the engine compartment
   - Air intake assembly
   - Accelerator cable
   - Accessory drive belts
   - Hood
   - Battery
8. Fill the cooling system.
9. Start the engine and check for leaks.

3.2L Engines

1. Before servicing the vehicle, refer to the precautions in the beginning of this section.
2. Drain the cooling system.
3. Relieve the fuel system pressure.
4. Remove or disconnect the following:
   - Battery
   - Hood
   - Accelerator cable
   - Cruise control cable
   - Air intake assembly
   - Canister vacuum hose
   - Brake booster vacuum hose
   - Engine wiring harness connectors
   - Front axle harness connector, if equipped
   - Transmission harness connector and bracket
   - Frame ground cable
   - Firewall ground cable
   - Starter harness connectors
   - Alternator harness connectors
   - Coolant overflow reservoir hose
   - Radiator hoses
   - Cooling fan and shroud
   - Accessory drive belt
   - Power steering pump
   - A/C compressor
   - Heated Oxygen (HO$_2$S) sensor connectors
   - Exhaust front pipes
   - Heater hoses
   - Fuel lines
   - Transmission
   - Left and right engine mounts

5. Lift the engine from the vehicle.

To install:

6. Lower the engine into the vehicle.
7. Install or connect the following:
   - Left and right engine mounts. Tighten the bolts to 30 ft. lbs. (41 Nm) and the nuts to 37 ft. lbs. (50 Nm).
   - Transmission
   - Fuel lines
   - Heater hoses
   - Exhaust front pipes
   - Heated Oxygen (HO$_2$S) sensor connectors
   - A/C compressor
   - Power steering pump
   - Accessory drive belt
   - Cooling fan and shroud
   - Radiator hoses
   - Coolant overflow reservoir hose
   - Alternator harness connectors
   - Starter harness connectors
   - Firewall ground cable
   - Frame ground cable
   - Transmission harness connector and bracket
   - Front axle harness connector, if equipped
   - Engine wiring harness connectors
   - Brake booster vacuum hose
   - Canister vacuum hose
   - Air intake assembly
   - Cruise control cable
   - Accelerator cable
   - Hood
   - Battery
8. Fill the cooling system.
9. Start the engine and check for leaks.

3.5L Engine

EXCEPT AXIOM

1. Before servicing the vehicle, refer to the precautions in the beginning of this section.
2. Drain the cooling system.
3. Remove or disconnect the following:
   - Battery
   - Hood
   - Air cleaner assembly
   - Accelerator cable
   - Cruise control cable
   - Canister vacuum line
   - Brake booster vacuum line
   - Engine wiring harness connectors
   - Transmission harness connectors and bracket
   - Engine ground cable
   - Starter harness connector
   - Alternator harness connector
   - Coolant reservoir tank hose
   - Radiator hoses
   - Heater hoses
   - Upper fan shroud
   - Radiator
   - Cooling fan
   - Accessory drive belt
   - Power steering pump
   - A/C compressor
   - Heated Oxygen (HO$_2$S) sensor connectors
   - Left and right exhaust front pipes
   - Fuel lines
   - Flywheel dust cover
   - Transmission. Refer to the transmission procedure in this section.
   - Left and right engine mounts
   - Engine

To install:

4. Install or connect the following:
   - Engine
   - Left and right engine mounts. Tighten the bolts to 30 ft. lbs. (41 Nm).
   - Transmission
   - Flywheel dust cover
   - Fuel lines
   - Left and right exhaust front pipes
   - HO$_2$S sensor connectors
   - A/C compressor
   - Power steering pump
   - Accessory drive belt
   - Cooling fan. Tighten the nuts to 16 ft. lbs. (22 Nm).
   - Radiator
   - Upper fan shroud
   - Heater hoses
   - Radiator hoses
   - Coolant reservoir tank hose
   - Alternator harness connector
   - Starter harness connector
   - Engine ground cable
   - Transmission harness connectors and bracket
   - Engine wiring harness connectors
   - Brake booster vacuum line
   - Canister vacuum line
   - Cruise control cable
   - Accelerator cable
   - Air cleaner assembly
   - Hood
   - Battery
5. Fill the cooling system. Check all fluid levels and adjust as necessary.
6. Start the engine and check for leaks.

AXIOM

1. Before servicing the vehicle, refer to the precautions in the beginning of this section.
2. Drain the cooling system.
3. Remove or disconnect the following:
   - Battery
   - Hood
   - Air cleaner assembly
   - Canister vacuum line
   - Brake booster vacuum line
   - Engine wiring harness connectors
   - Transmission harness connectors and bracket
   - Engine ground cable
   - Bonding cable connectors
   - Starter harness connector
   - Alternator harness connector
   - Coolant reservoir tank hose
   - Radiator hoses
   - Upper fan shroud
   - Cooling fan
   - Accessory drive belt
   - Power steering pump
   - A/C compressor
   - Heated Oxygen (HO$_2$S) sensor connectors
   - Left and right exhaust front pipes
   - Flywheel dust cover
   - Heater hoses
   - Fuel lines
   - Transmission. Refer to the transmission procedure in this section.
   - Accelerator cable
   - Cruise control cable
   - Left and right engine mounts
   - Engine

To install:

4. Install or connect the following:
   - Engine
   - Left and right engine mounts. Tighten the bolts to 30 ft. lbs. (41 Nm).
   - Transmission
   - Flywheel dust cover
   - Fuel lines
   - Left and right exhaust front pipes
   - HO$_2$S sensor connectors
   - A/C compressor
   - Power steering pump
   - Accessory drive belt
   - Cooling fan. Tighten the nuts to 16 ft. lbs. (22 Nm).
   - Radiator
   - Upper fan shroud
   - Heater hoses
   - Radiator hoses
   - Coolant reservoir tank hose
   - Alternator harness connector
   - Starter harness connector
   - Engine ground cable
   - Transmission harness connectors and bracket
   - Engine wiring harness connectors
   - Brake booster vacuum line
   - Canister vacuum line
   - Cruise control cable
   - Accelerator cable
   - Air cleaner assembly
   - Hood

- Battery

5. Fill the cooling system. Check all fluid levels and adjust as necessary.

6. Start the engine and check for leaks.

Water Pump

REMOVAL & INSTALLATION

2.2L Engine

1. Before servicing the vehicle, refer to the precautions in the beginning of this section.

2. Drain the cooling system.

3. Remove or disconnect the following:
- Negative battery cable
- Radiator hose
- Accessory drive belt
- Front cover
- Timing belt. Refer to the Timing Belt unit repair section.
- Water pump

To install:

4. Install a new O-ring and coat the water pump sealing surface with silicone grease.

5. Install or connect the following:
- Water pump. Tighten the bolts to 18 ft. lbs. (25 Nm).
- Timing belt
- Front cover
- Accessory drive belt
- Radiator hose
- Negative battery cable

6. Fill the cooling system.

7. Start the engine and check for leaks.

3.2L Engines

1. Before servicing the vehicle, refer to the precautions in the beginning of this section.

2. Drain the cooling system.

3. Remove or disconnect the following:
- Negative battery cable
- Radiator hose
- Accessory drive belt
- Front cover
- Timing belt. Refer to the Timing Belt unit repair section.
- Timing belt idler pulley
- Water pump

To install:

4. Install or connect the following:
- Water pump. Tighten the bolts in sequence to 18 ft. lbs. (25 Nm).
- Timing belt idler pulley. Tighten the bolt to 38 ft. lbs. (52 Nm).
- Timing belt
- Front cover
- Accessory drive belt

- Radiator hose
- Negative battery cable

5. Fill the cooling system.

6. Start the engine and check for leaks.

3.5L Engine

1. Before servicing the vehicle, refer to the precautions in the beginning of this section.

2. Drain the cooling system.

3. Remove or disconnect the following:
- Negative battery cable
- Upper radiator hose

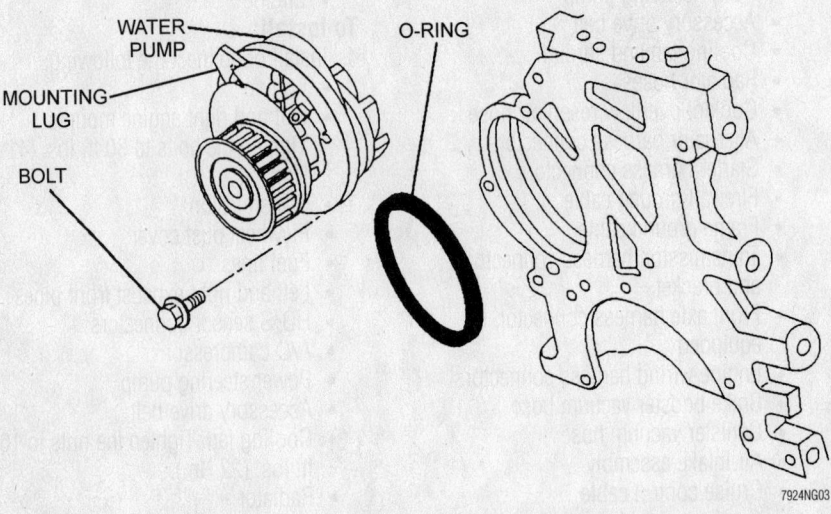

Exploded view of the water pump mounting, showing the location of the mounting lug—2.2L engine

- Timing belt. Refer to the Timing Belt Unit Repair Section.
- Idler pulley
- Water pump

To install:

➡ **Apply Loctite® 262 to bolt number 3 prior to installation.**

4. Install or connect the following:
- Water pump. Tighten the bolts in two passes, in sequence, to 13 ft. lbs. (18 Nm) for 3.2L engines or to 18 ft. lbs. (25 Nm) for 3.5L engines.

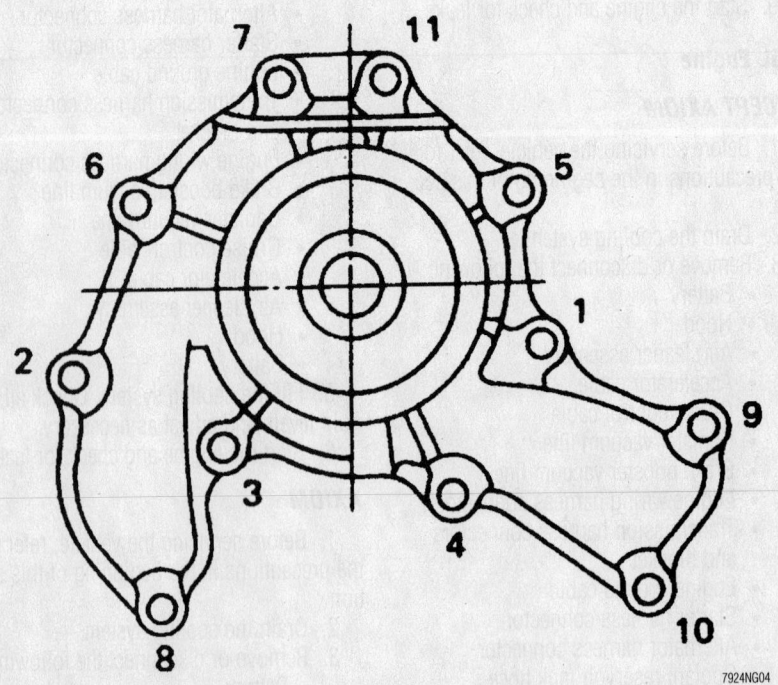

Water pump bolt tightening sequence—3.2L engines

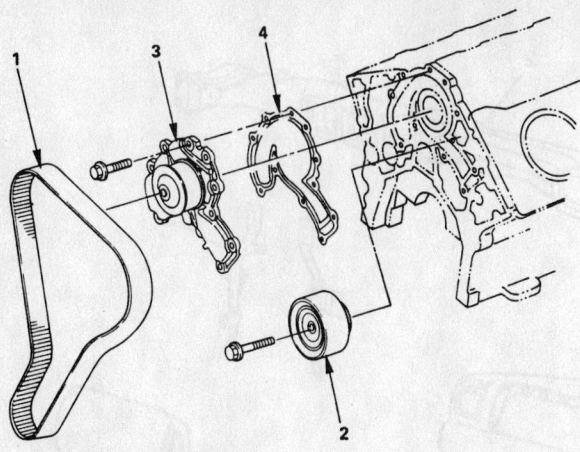

1. Timing belt
2. Idle pulley
3. Water pump assembly
4. Gasket

7924BG41

Exploded view of the water pump mounting

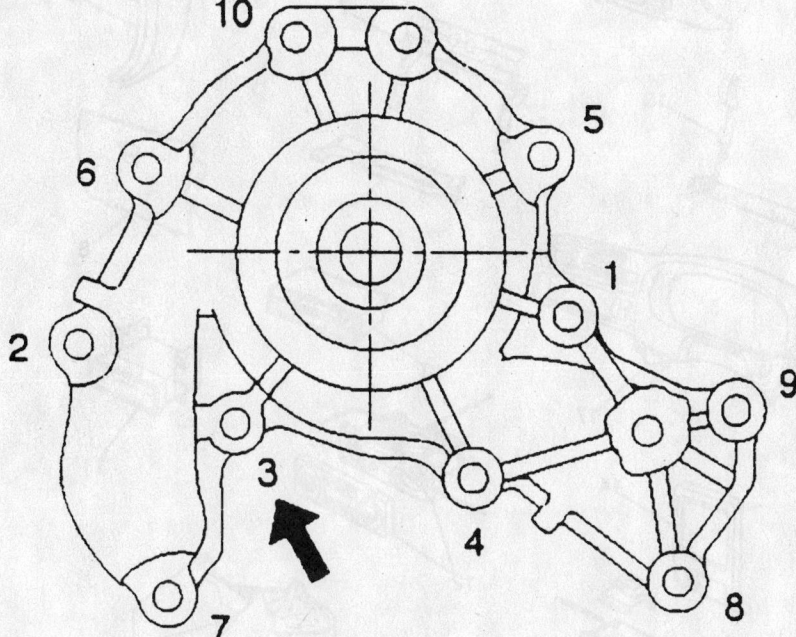

9302BG01

Water pump torque sequence. Apply LOCTITE® 262 to bolt number 3 (arrow)—3.5L engine

- Idler pulley
- Timing belt
- Upper radiator hose
- Negative battery cable

5. Fill the cooling system.

6. Start the engine and check for leaks.

Heater Core

REMOVAL & INSTALLATION

Amigo

1. If equipped with an air bag, perform the following procedure:

a. Turn the ignition to the LOCK position and remove the key.

b. From the lower left dash side fuse block, remove the SRS-1 fuse.

c. Disconnect the 2-pin yellow connector located at the base of the steering column.

d. Remove the glove box assembly.

e. Disconnect the 2-pin yellow connector located behind the glove box.

2. Disconnect the negative battery cable.

3. If equipped, discharge and recover the air conditioning system refrigerant.

4. Remove the evaporator lines at the firewall. Plug the air conditioning lines to minimize contamination.

5. Disconnect the cooling system hoses and drain the coolant into a clean container for reuse. Plug the cooling system hoses.

6. Remove the instrument panel by performing the following procedure:

a. Remove the lower center cover screw and pull it out at the clip positions; then, disconnect the cigarette lighter connector.

b. Remove both the rear and front console.

c. Remove the dash side trim panel sill plates and the panels.

d. Remove the 2 glove box screws and the glove box.

e. Remove the 2 hood release screws, the 6 instrument panel driver's lower cover assembly screws and the cover assembly.

f. Remove 5 instrument cluster screws, the 2 clips; then, disconnect the 8 switch connectors and remove the instrument cluster assembly.

g. Remove the 6 driver's knee bolster assembly bolts and screws and the knee bolster assembly.

h. Remove the 4 control lever assembly bolts; then, disconnect the 3 control cables (unit side) and the 3 harness connectors.

i. Remove the 4 radio/audio sub box assembly screws and the radio/audio sub box assembly.

j. Disconnect or remove the following instrument panel harness connectors or items:

- The 6 driver's side connectors
- The 3 passenger's side connectors
- The 2 center connectors
- Passenger's inflator module connector
- Radio antenna cable plug
- Ground cable bolt on the left dash side panel
- The 8 instrument panel-to-chassis bolts and the 3 nuts.

k. Remove the instrument panel assembly.

7. Remove the instrument panel bracket by performing the following procedure:

a. Remove the 2 passenger's inflator module bolts and 4 nuts.

b. Remove the 4 meter assembly screws; then, disconnect the meter wiring harness connectors and remove the meter assembly.

c. Remove the 5 vent duct assembly screws and the assembly.

d. Remove the 3 lower passenger bracket screws and the bracket.

e. Remove the 9 passenger knee bol-

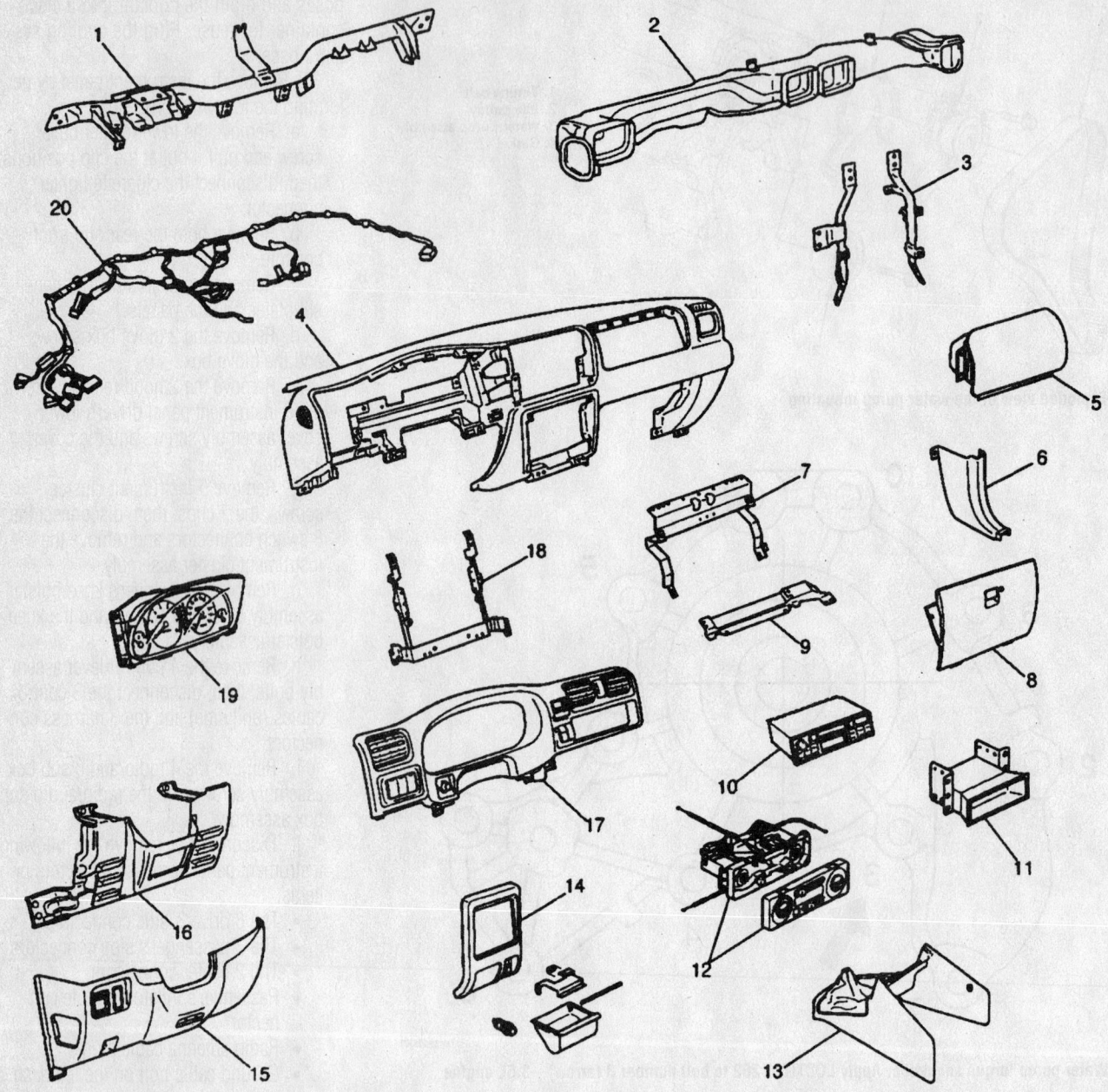

1	Cross Beam	11	Audio Sub Box
2	Vent Duct Assembly	12	Control Lever Assembly
3	Instrument Panel Bracket	13	Front Console Assembly
4	Instrument Panel Assembly	14	Lower Center Cover
5	Passenger Inflator Module	15	Instrument Panel Driver Lower Cover Assembly
6	Dash Side Trim Panel	16	Driver Knee Bolster Assembly
7	Passenger Knee Bolster Reinforcement Assembly	17	Meter Cluster Assembly
8	Glove Box	18	Instrument Panel Center Reinforcement
9	Passenger Lower Bracket	19	Meter Assembly
10	Radio Assembly	20	Instrument Harness Assembly

93113GB8

Exploded view of the instrument panel—Isuzu Amigo

ster reinforcement screws and the reinforcement.

f. Remove the 6 instrument panel
center reinforcement screws and the reinforcement.

g. Remove the instrument panel
wiring harness assembly clips and the
wiring harness.

h. Remove the 2 instrument panel
bracket nuts and 2 bolts for each bracket;
then, remove the bracket(s).

8. Remove the 5 cross beam assembly
nuts, 2 bolts and the 6 lower bolts; then,
remove the crossbeam.

9. Disconnect the resistor wiring connector.

10. Remove the duct from the heater
assembly.

11. If equipped with air conditioning,
remove the evaporator assembly.

12. Remove the driver's lap vent.

13. Remove the lower ventilation
duct.

14. Remove the footrest, the carpet, the 3
clips and the rear heater duct.

15. Remove the heater assembly.

16. Remove the mode control case-to-
temperature control case screws and

remove the mode control case; do not
remove the link unit.

17. Remove the temperature control case
screws and separate the cases.

18. Remove the heater core from the
case.

To install:

19. Install the heater core to the case.

20. Assemble the temperature control
cases and install the case screws.

21. Install the mode control case and the
mode control case-to-temperature control
case screws.

22. Install the heater assembly.

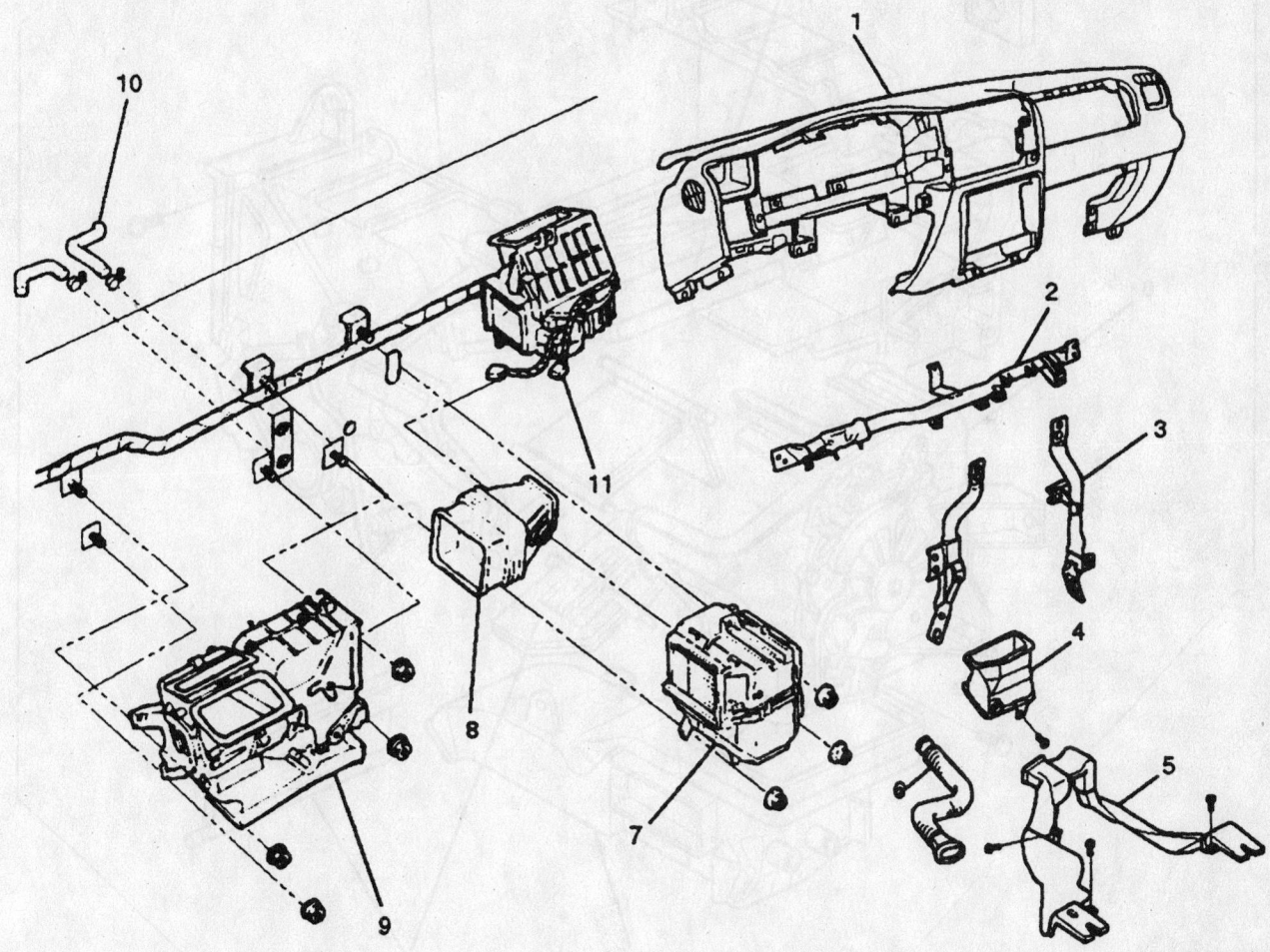

1	Instrument Panel Assembly	6	Driver Lap Vent Duct
2	Cross Beam Assembly	7	Evaporator Assembly (A/C only)
3	Instrument Panel Bracket	8	Duct
4	Ventilation Lower Duct	9	Heater Unit Assembly
5	Rear Heater Duct	10	Heater Hose
		11	Resistor Connector

View of the heater and air conditioning housing assemblies and related components—Isuzu Amigo

93113GB9

23. Install the rear heater duct, the footrest, the carpet, and the 3 clips.

24. Install the lower ventilation duct.

25. Install the driver's lap vent.

26. If equipped with air conditioning, install the evaporator assembly.

27. Install the duct to the heater assembly.

28. Connect the resistor wiring connector.

29. Install the crossbeam, the 5 crossbeam assembly nuts, 2 bolts and the 6 lower bolts.

30. Install the instrument panel bracket by performing the following procedure:

   a. Install the instrument panel bracket and the 2 nuts and 2 bolts for each bracket.

   b. Install the instrument panel wiring

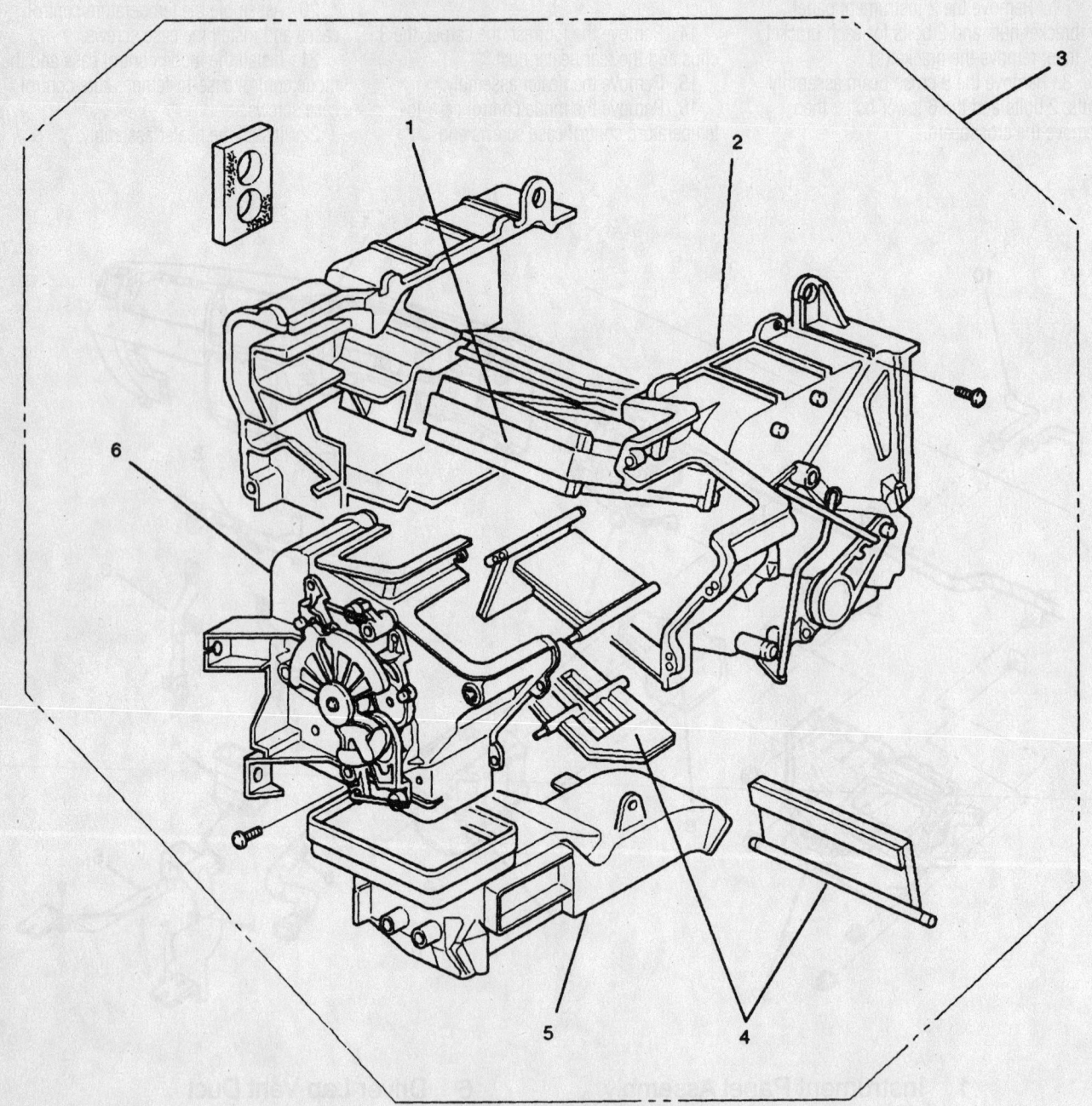

1	Heater Core	4	Mode Door
2	Case (Temperature Control)	5	Duct
3	Heater Unit	6	Case (Mode Control)

93113GB0

Exploded view of the heater housing assembly—Isuzu Amigo

harness assembly and the wiring harness clips.

c. Install the instrument panel center reinforcement and the 6 reinforcement screws.

d. Install the passenger knee bolster reinforcement and the 9 reinforcement screws.

e. Install the lower passenger bracket and the 3 bracket screws.

f. Install the vent duct assembly and the 5 vent duct assembly screws.

g. Install the meter assembly and the 4 meter assembly screws; then, connect the meter wiring harness connectors.

h. Install the 2 passenger's inflator module bolts and 4 nuts.

31. Install the instrument panel by performing the following procedure:

a. Install the instrument panel assembly.

b. Connect or install the following instrument panel harness connectors or items:

- The 6 driver's side connectors
- The 3 passenger's side connectors
- The 2 center connectors
- Passenger's inflator module connector
- Radio antenna cable plug
- Ground cable bolt on the left dash side panel
- The 8 instrument panel-to-chassis bolts and the 3 nuts.

c. Install the radio/audio sub box assembly and the 4 radio/audio sub box assembly screws.

d. Connect the 3 control cables (unit side) and the 3 harness connectors. Install the 4 control lever assembly bolts.

e. Install the knee bolster assembly and the 6 driver's knee bolster assembly bolts and screws.

f. Install the instrument cluster assembly. Connect the 8 switch connectors. Install 5 instrument cluster screws and the 2 clips.

g. Install the 2 hood release screws, the 6 instrument panel driver's lower cover assembly screws and the cover assembly.

h. Install the glove box and the 2 glove box screws.

i. Install the dash side trim panel sill plates and the panels.

j. Install both the rear and front console.

k. Connect the cigarette lighter connector and install the lower center cover screw.

32. Connect the cooling system hoses.

33. Refill the cooling system.

34. Install the evaporator lines at the firewall.

35. If equipped, evacuate and charge the air conditioning system.

36. Connect the negative battery cable.

37. If equipped with an air bag, perform the following procedure:

a. Turn the ignition to the LOCK position and remove the key.

b. Connect the 2-pin yellow connector located behind the glove box.

c. Install the glove box assembly.

d. Connect the 2-pin yellow connector located at the base of the steering column.

e. At the lower left dash side fuse block, install the SRS-1 fuse.

f. Turn the ignition switch to ON and verify that the AIR BAG warning light flashes 7 times and turns OFF.

38. Run the engine to normal operating temperatures; then, check the climate control operation and check for leaks.

Axiom

1. Disconnect the battery ground.
2. Drain the coolant.
3. Discharge and recover the refrigerant.
4. Remove the heater unit.
5. Remove the duct.
6. Remove the mix actuator.
7. Remove the mode actuator.
8. Remove the mode control case.
9. Remove the heater case and separate the halves.
10. Remove the core.
11. Installation is the reverse of removal.

Rodeo

1. If equipped with an air bag, perform the following procedure:

a. Turn the ignition to the LOCK position and remove the key.

b. From the lower left dash side fuse block, remove the SRS-1 fuse.

c. Disconnect the 2-pin yellow connector located at the base of the steering column.

d. Remove the glove box assembly.

e. Disconnect the 2-pin yellow connector located behind the glove box.

2. Disconnect the negative battery cable.

3. If equipped, discharge and recover the air conditioning system refrigerant.

4. Remove the evaporator lines at the firewall. Plug the air conditioning lines to minimize contamination.

5. Disconnect the cooling system hoses and drain the coolant into a clean container for reuse. Plug the cooling system hoses.

6. Remove the instrument panel by performing the following procedure:

a. Remove the lower center cover screw and pull it out at the clip positions; then, disconnect the cigarette lighter connector.

b. Remove both the rear and front console.

c. Remove the dash side trim panel sill plates and the panels.

d. Remove the 2 glove box screws and the glove box.

e. Remove the 2 hood release screws, the 6 instrument panel driver's lower cover assembly screws and the cover assembly.

f. Remove 5 instrument cluster screws and the 2 clips. Then, disconnect the 8 switch connectors and remove the instrument cluster assembly.

g. Remove the 6 driver's knee bolster assembly bolts and screws and the knee bolster assembly.

h. Remove the 4 control lever assembly bolts; then, disconnect the 3 control cables (unit side) and the 3 harness connectors.

i. Remove the 4 radio/audio sub box assembly screws and the radio/audio sub box assembly.

j. Disconnect or remove the following instrument panel harness connectors or items:

- The 6 driver's side connectors
- The 3 passenger's side connectors
- The 2 center connectors
- Passenger's inflator module connector
- Radio antenna cable plug
- Ground cable bolt on the left dash side panel
- The 8 instrument panel-to-chassis bolts and the 3 nuts.

k. Remove the instrument panel assembly.

7. Remove the instrument panel bracket by performing the following procedure:

a. Remove the 2 passenger's inflator module bolts and 4 nuts.

b. Remove the 4 meter assembly screws. Then, disconnect the meter wiring harness connectors and remove the meter assembly.

c. Remove the 5 vent duct assembly screws and the assembly.

d. Remove the 3 lower passenger bracket screws and the bracket.

e. Remove the 9 passenger knee bolster reinforcement screws and the reinforcement.

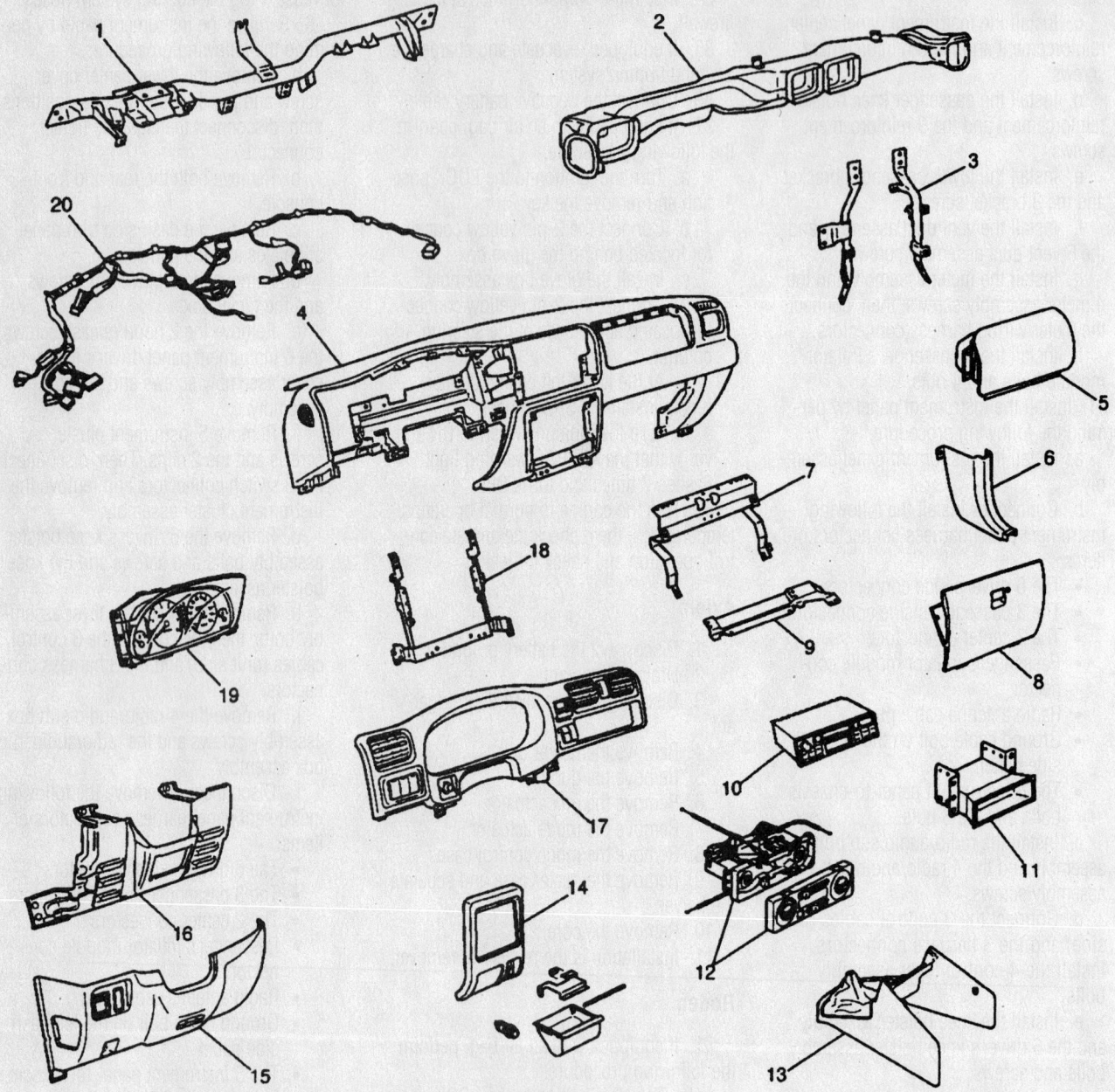

1	Cross Beam	11	Audio Sub Box
2	Vent Duct Assembly	12	Control Lever Assembly
3	Instrument Panel Bracket	13	Front Console Assembly
4	Instrument Panel Assembly	14	Lower Center Cover
5	Passenger Inflator Module	15	Instrument Panel Driver Lower Cover Assembly
6	Dash Side Trim Panel	16	Driver Knee Bolster Assembly
7	Passenger Knee Bolster Reinforcement Assembly	17	Meter Cluster Assembly
8	Glove Box	18	Instrument Panel Center Reinforcement
9	Passenger Lower Bracket	19	Meter Assembly
10	Radio Assembly	20	Instrument Harness Assembly

Exploded view of the instrument panel—Isuzu Rodeo

93113GB8

f. Remove the 6 instrument panel center reinforcement screws and the reinforcement.

g. Remove the instrument panel wiring harness assembly clips and the wiring harness.

h. Remove the 2 instrument panel bracket nuts and 2 bolts for each bracket; then, remove the bracket(s).

8. Remove the 5 cross beam assembly nuts, 2 bolts and the 6 lower bolts; then, remove the crossbeam.

9. Disconnect the resistor wiring connector.

10. Remove the duct from the heater assembly.

11. If equipped with air conditioning, remove the evaporator assembly.

12. Remove the driver's lap vent.

13. Remove the lower ventilation duct.

14. Remove the footrest, the carpet, the 3 clips and the rear heater duct.

15. Remove the heater assembly.

16. Remove the mode control case-to-temperature control case screws and remove the mode control case; do not remove the link unit.

17. Remove the temperature control case screws and separate the cases.

18. Remove the heater core from the case.

To install:

19. Install the heater core to the case.

20. Assemble the temperature control cases and install the case screws.

21. Install the mode control case and the mode control case-to-temperature control case screws.

22. Install the heater assembly.

23. Install the rear heater duct, the footrest, the carpet, and the 3 clips.

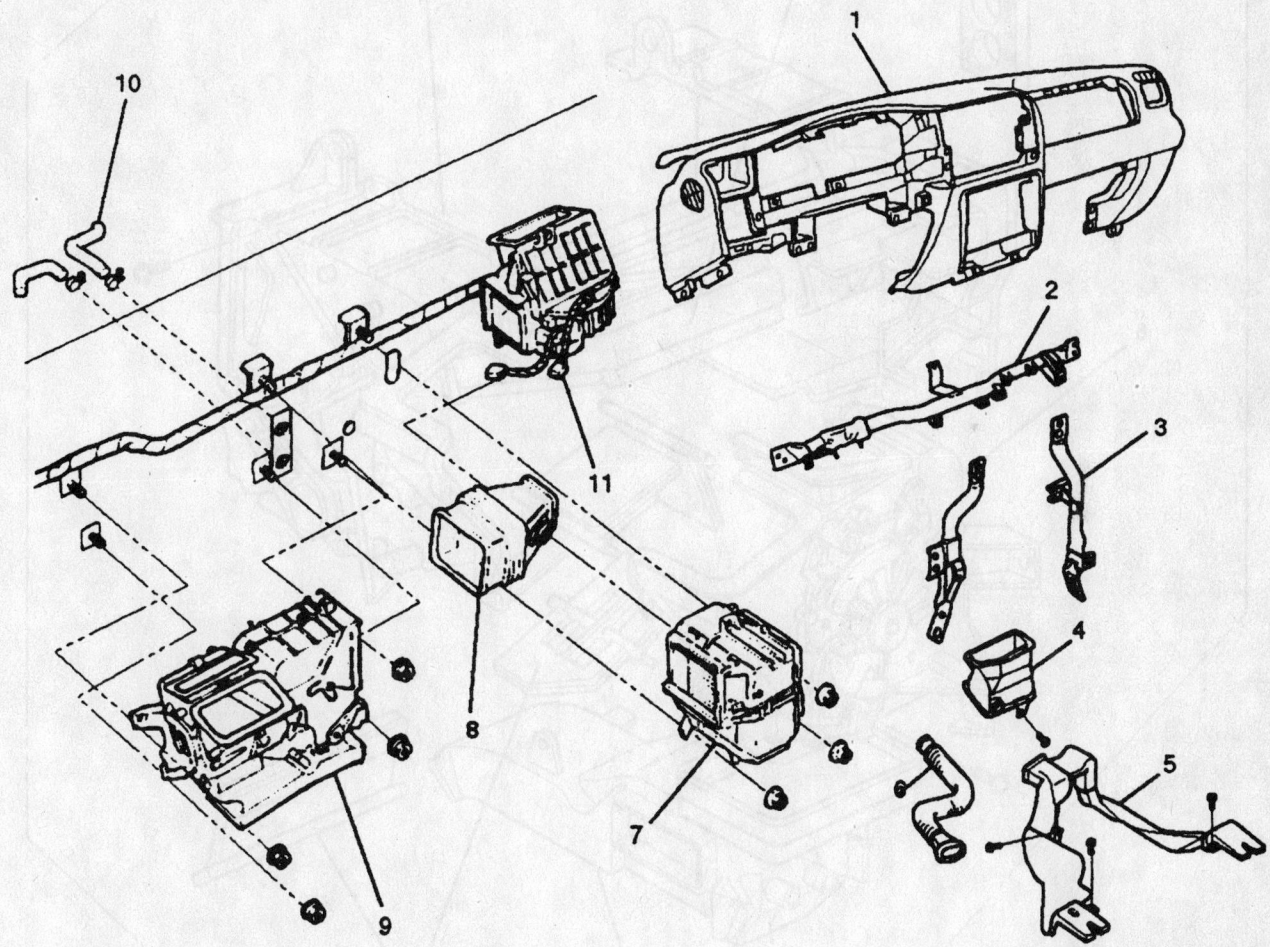

1	Instrument Panel Assembly	6	Driver Lap Vent Duct
2	Cross Beam Assembly	7	Evaporator Assembly (A/C only)
3	Instrument Panel Bracket	8	Duct
4	Ventilation Lower Duct	9	Heater Unit Assembly
5	Rear Heater Duct	10	Heater Hose
		11	Resistor Connector

93113GB9

View of the heater and air conditioning housing assemblies and related components—Isuzu Rodeo

24. Install the lower ventilation duct.
25. Install the driver's lap vent.
26. If equipped with air conditioning, install the evaporator assembly.
27. Install the duct to the heater assembly.
28. Connect the resistor wiring connector.
29. Install the crossbeam, the 5 cross-beam assembly nuts, 2 bolts and the 6 lower bolts.

30. Install the instrument panel bracket by performing the following procedure:

    a. Install the instrument panel bracket and the 2 nuts and 2 bolts for each bracket.

    b. Install the instrument panel wiring harness assembly and the wiring harness clips.

    c. Install the instrument panel center reinforcement and the 6 reinforcement screws.

    d. Install the passenger knee bolster reinforcement and the 9 reinforcement screws.

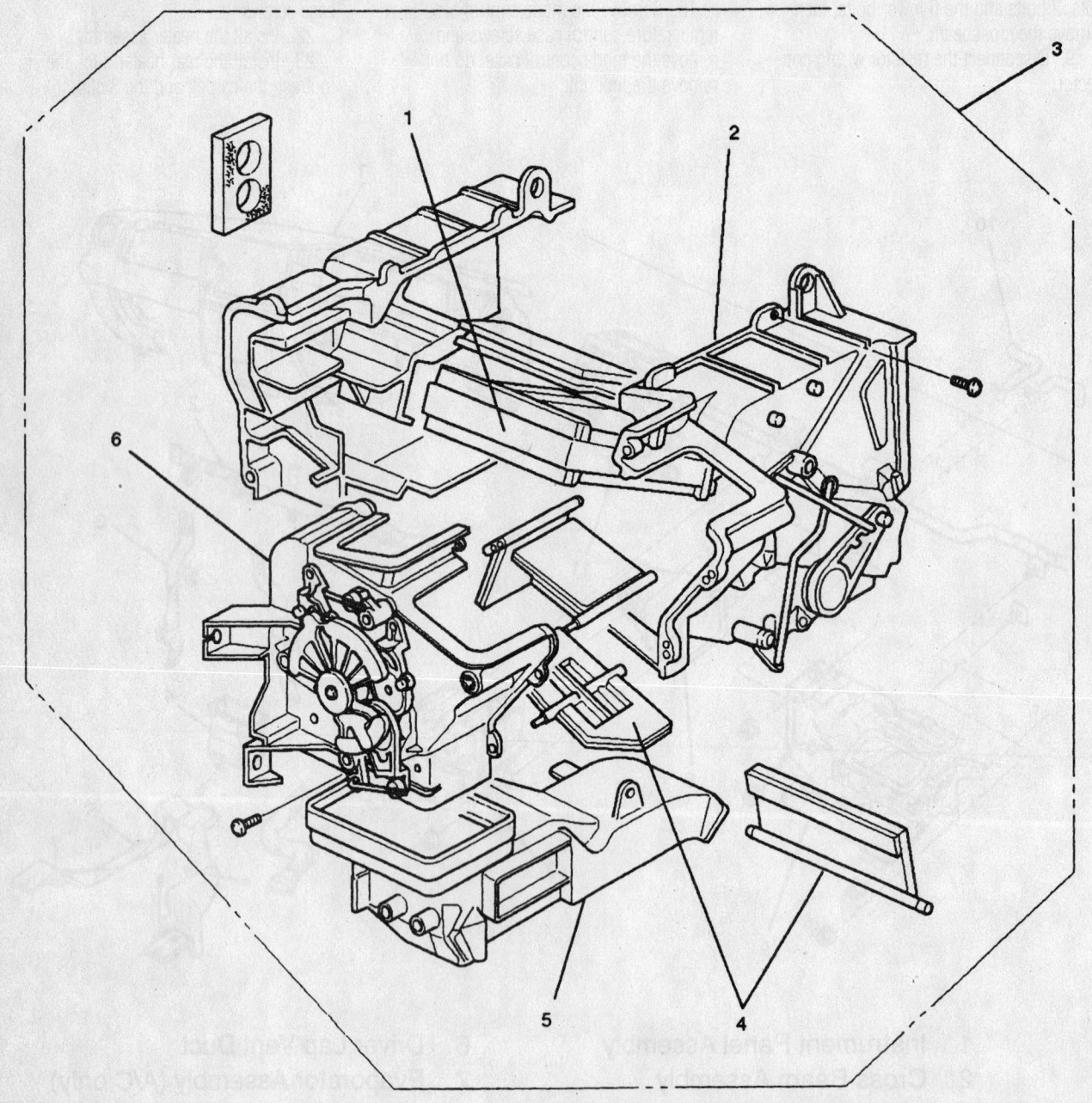

1	Heater Core	4	Mode Door
2	Case (Temperature Control)	5	Duct
3	Heater Unit	6	Case (Mode Control)

93113GB0

Exploded view of the heater housing assembly—Isuzu Rodeo

e. Install the lower passenger bracket and the 3 bracket screws.

f. Install the vent duct assembly and the 5 vent duct assembly screws.

g. Install the meter assembly and the 4 meter assembly screws. Then, connect the meter wiring harness connectors.

h. Install the 2 passenger's inflator module bolts and 4 nuts.

31. Install the instrument panel by performing the following procedure:

a. Install the instrument panel assembly.

b. Connect or install the following instrument panel harness connectors or items:

- The 6 driver's side connectors
- The 3 passenger's side connectors
- The 2 center connectors
- Passenger's inflator module connector
- Radio antenna cable plug
- Ground cable bolt on the left dash side panel
- The 8 instrument panel-to-chassis bolts and the 3 nuts.

c. Install the radio/audio sub box assembly and the 4 radio/audio sub box assembly screws.

d. Connect the 3 control cables (unit side) and the 3 harness connectors. Install the 4 control lever assembly bolts.

e. Install the knee bolster assembly and the 6 driver's knee bolster assembly bolts and screws.

f. Install the instrument cluster assembly. Connect the 8 switch connectors. Install 5 instrument cluster screws and the 2 clips.

g. Install the 2 hood release screws, the 6 instrument panel driver's lower cover assembly screws and the cover assembly.

h. Install the glove box and the 2 glove box screws.

i. Install the dash side trim panel sill plates and the panels.

j. Install both the rear and front console.

k. Connect the cigarette lighter connector and install the lower center cover screw.

32. Connect the cooling system hoses.

33. Refill the cooling system.

34. Install the evaporator lines at the firewall.

35. If equipped, evacuate and charge the air conditioning system.

36. Connect the negative battery cable.

37. If equipped with an air bag, perform the following procedure:

a. Turn the ignition to the LOCK position and remove the key.

b. Connect the 2-pin yellow connector located behind the glove box.

c. Install the glove box assembly.

d. Connect the 2-pin yellow connector located at the base of the steering column.

e. At the lower left dash side fuse block, install the SRS-1 fuse.

f. Turn the ignition switch to ON and verify that the AIR BAG warning light flashes 7 times and turns OFF.

38. Run the engine to normal operating temperatures; then, check the climate control operation and check for leaks.

Trooper

✳✳ CAUTION

The vehicle is equipped with a driver's side and a passenger's side air bag. Before starting service procedures on components, especially under the instrument panel and/or near the steering column, disable the air bag systems. There is sufficient voltage in the system to cause a deployment for up to 15 seconds after the battery has been disconnected, the ignition turned OFF or fuse C-21 is removed from the fuse panel.

1. If equipped with an air bag, perform the following procedures:

a. Disconnect the negative battery cable, then disconnect the positive battery cable.

b. Disconnect the yellow 2-pin connector located at the base of the steering column.

c. Remove the glove box and disconnect the yellow 2-pin connector located behind the glove box.

2. Disconnect the negative battery cable.

3. Drain the cooling system.

4. If equipped with air conditioning, discharge and recover the refrigerant.

5. Remove the instrument panel assembly by performing the following procedure:

a. At the front console assembly, disconnect the switch connectors; then, remove the console-to-chassis screws and the console.

b. At the lower cluster assembly, remove the cluster-to-instrument panel screws, disconnect the cigarette lighter and light connectors and remove the lower cluster.

c. Remove the glove box and the instrument panel lower cover and the passenger knee bolster reinforcement.

d. At the left side, remove the instrument panel lower cover and the knee bolster assembly.

e. At the top of the instrument panel, pry the 8 claws on the front side toward you, raise the defroster grille and remove it.

f. At the SRS adjust bracket and cross beam under the passenger air bag module, remove the 2 fixing bolts and remove the instrument panel assembly.

g. Disconnect the air conditioning control cables from the unit.

h. Remove the instrument panel harness connectors (5 on the driver's side and 3 on the passenger's side), the passenger air bag module connector, the radio antenna plug and the center bracket ground cable bolt.

i. Remove the passenger's air bag module nuts, disconnect the connectors and remove the module.

j. Remove the instrument panel cluster assembly screws, disconnect the switch connectors and the instrument panel assembly.

6. Disconnect the heater hoses from the heater unit.

7. Disconnect the heater resistor connector and the electro thermo connector (if equipped with air conditioning).

8. Remove the heater duct.

9. If equipped with air conditioning, remove the evaporator assembly by performing the following procedure:

a. Disconnect the drain hose.

b. Using a backup wrench, disconnect the refrigerant lines from the evaporator.

c. Plug or cap the refrigerant lines.

d. Remove the evaporator assembly.

10. Remove the instrument panel center bracket (crossbeam assembly) by performing the following procedure:

a. Remove the side support bracket bolts and brackets from both sides of the vehicle.

b. Remove the crossbeam center bracket nuts, disconnect the electrical connectors and the center bracket.

11. Remove the rear heater duct and heater assembly.

12. Disassemble the heater unit assembly by performing the following procedure:

a. Remove the lower air duct; do not remove the link unit.

b. Remove the temperature control case screws and lift the case from the heater unit.

c. Remove the heater core.

To install:

13. Assemble the heater unit assembly by performing the following procedure:

    a. Install the heater core into the heater unit.

    b. Install the temperature control case onto the unit and secure with screws.

    c. Install the lower air duct.

14. Install the heater unit assembly into the vehicle.

15. Install the rear heater duct.

16. Install the instrument panel cross beam assembly by reversing the removal procedures.

17. If equipped with air conditioning, install the evaporator assembly by performing the following procedures:

    a. If installing a new evaporator assembly, add 1.7 fl. oz. (50mL) of refrigerant oil to the evaporator.

    b. Using new O-rings and a backup wrench, install the refrigerant lines and torque the outlet line to 18 ft. lbs. (25

Nm) and the inlet line to 11 ft. lbs. (15 Nm).

18. Install the heater duct.

19. Connect the heater resistor connector and the electro-thermo connector (if equipped with air conditioning).

20. Connect the heater hoses to the heater unit.

21. Install the instrument panel assembly by reversing the removal procedures.

22. If equipped with air conditioning,

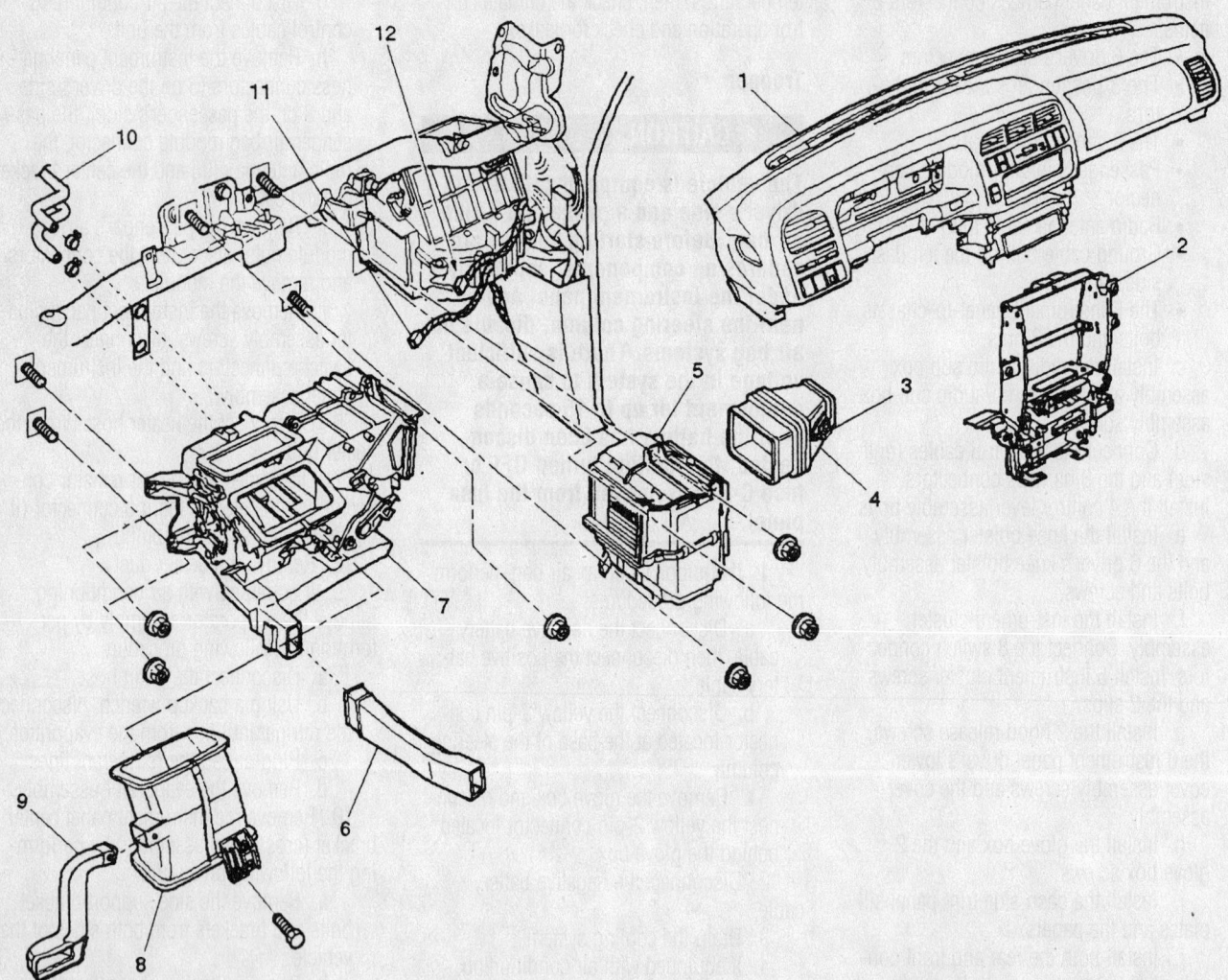

(1) Instrument Panel Assembly	(7) Heater Unit Assembly
(2) Instrument Panel Center Bracket	(8) Center Ventilation Lower Duct
(3) Resistor	(9) Driver Lap Vent Nozzle
(4) Duct	(10) Water Hose
(5) Evaporator Assembly (A/C only)	(11) Electro Thermo Connector (With A/C)
(6) Rear Heater Duct	(12) Resistor Connector

93113G01

Exploded view of the heater unit and related components—Isuzu Trooper

evacuate and charge and leak-test the system.

23. Refill the cooling system.
24. Connect the negative battery cable.

✳✳ CAUTION

Never use an air bag assembly from another vehicle and/or different model year. Starting in 1999, the air bag assemblies are equipped with identification colors on the bar code label as follows: YELLOW for the driver's air bag assembly, WHITE for the passenger's air bag assembly.

25. Enable the air bags by performing the following procedure:

a. Connect the passenger's side air bag yellow 2-pin connector.

b. Install the glove box.

c. At the base of the steering column, connect the yellow 2-pin connector.

d. Install the air bag fuse C-21 (if removed) or connect the negative battery cable.

e. Turn the ignition switch ON and

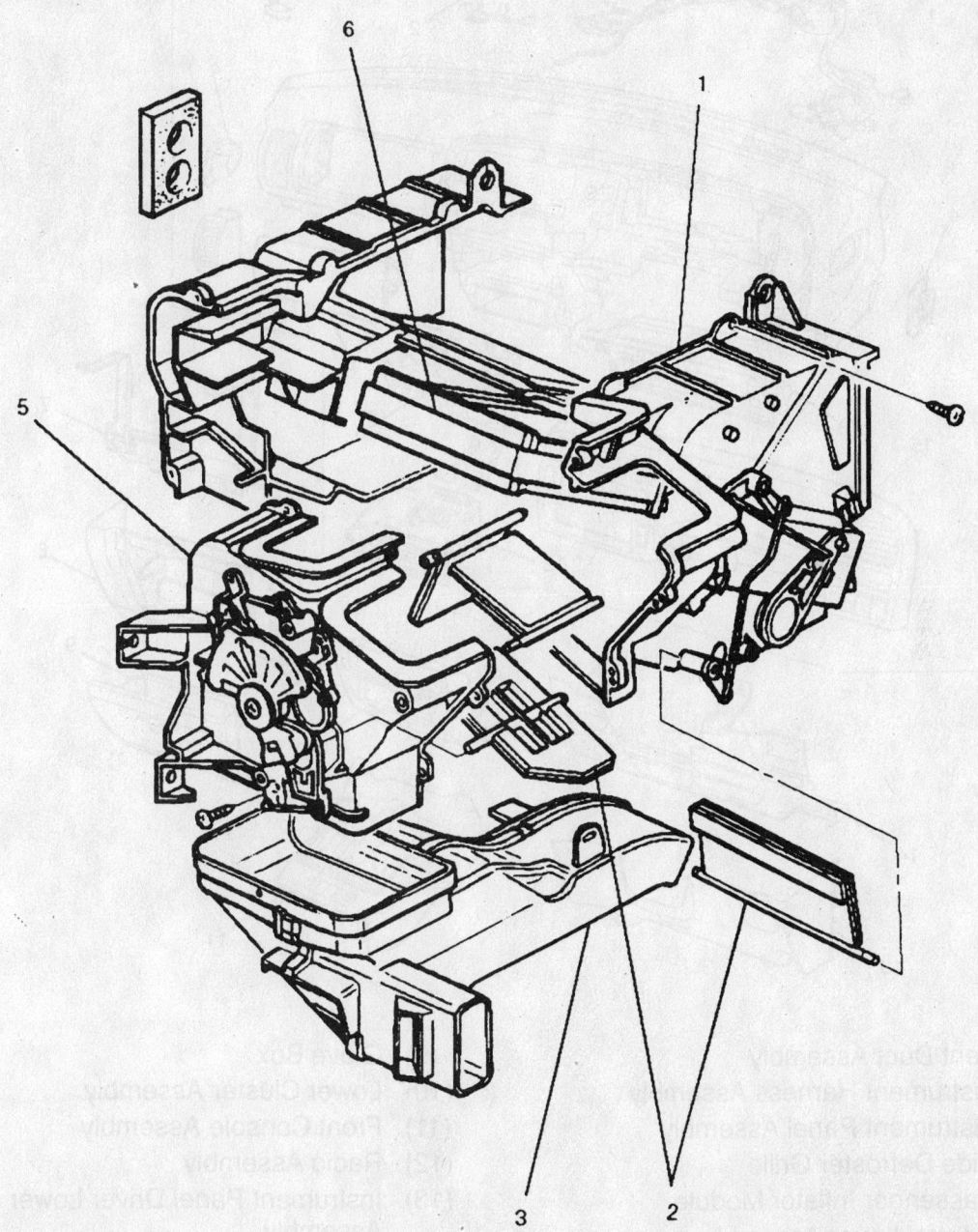

(1)	Case (Temperature Control)	(5)	Case (Mode Control)
(2)	Mode Door	(6)	Heater Core
(3)	Duct		

Exploded view of the heater unit—Isuzu Trooper

93113G02

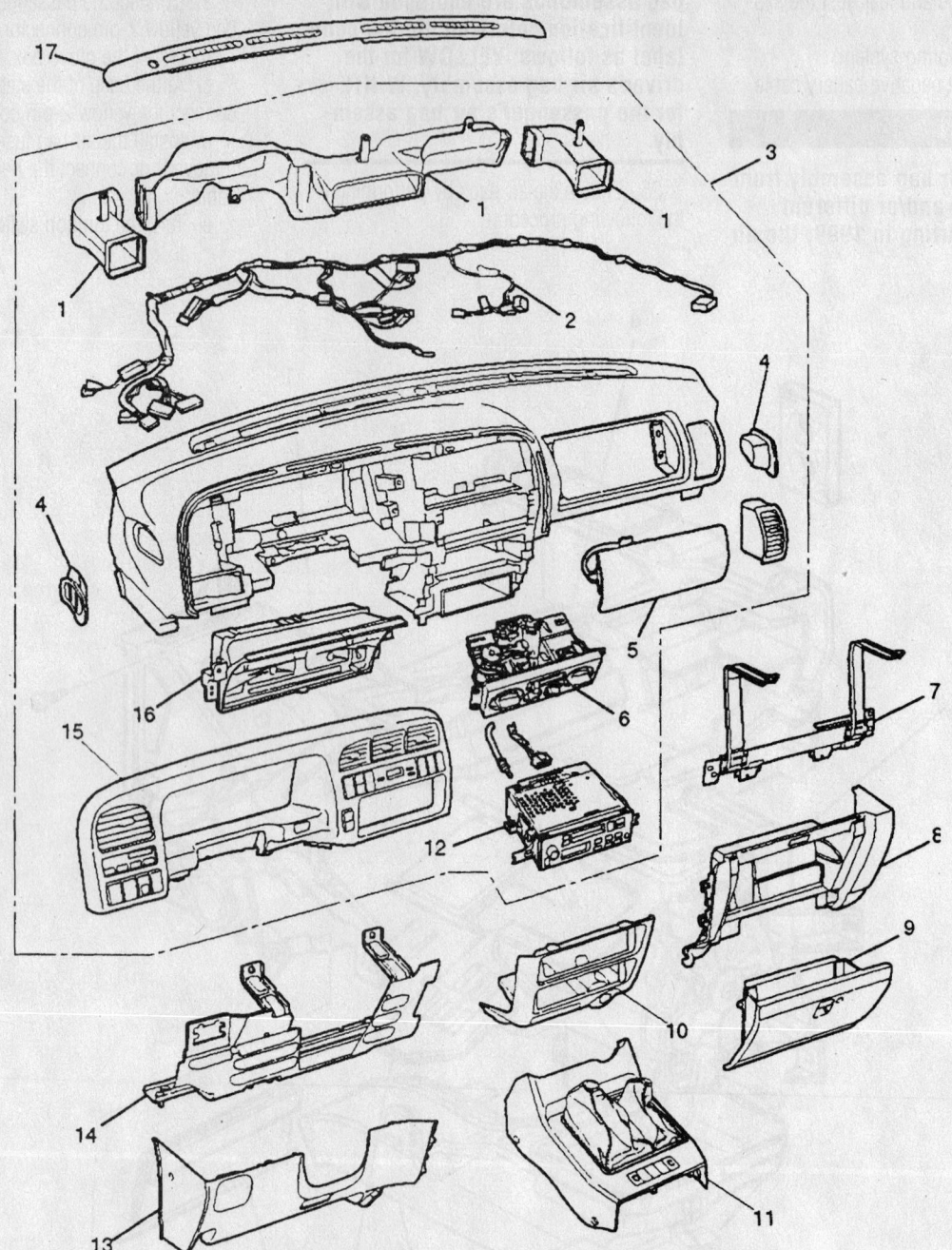

(1) Vent Duct Assembly
(2) Instrument Harness Assembly
(3) Instrument Panel Assembly
(4) Side Defroster Grille
(5) Passenger Inflator Module
(6) Control Lever Assembly
(7) Passenger Knee Bolster Reinforcement Assembly
(8) Instrument Panel Passenger Lower Cover Assembly

(9) Glove Box
(10) Lower Cluster Assembly
(11) Front Console Assembly
(12) Radio Assembly
(13) Instrument Panel Driver Lower Cover Assembly
(14) Driver Knee Bolster Assembly
(15) Instrument Panel Cluster Assembly
(16) Meter Assembly
(17) Front Defroster Grille

93113G03

Exploded view of the instrument panel and accessories—Isuzu Trooper

verify that the AIR BAG warning light flashes 7 times and then turns OFF.

26. Run the engine to normal operating temperatures and check for leaks. Check the systems for correct operation.

VehiCROSS

❄❄ CAUTION

The vehicle is equipped with a driver's side and a passenger's side air bag. Before starting service procedures on components, especially under the instrument panel and/or near the steering column, disable the air bag systems. There is sufficient voltage in the system to cause a deployment of the air bags for up to 15 seconds after the battery has been disconnected.

1. If equipped with an air bag, perform the following procedures:
   a. Disconnect the negative battery cable, then disconnect the positive battery cable.
   b. Disconnect the yellow 3-pin connector located at the base of the steering column.
2. Disconnect the negative battery cable.
3. Drain the cooling system.
4. If equipped with air conditioning, discharge and recover the refrigerant.
5. Remove the instrument panel assembly by performing the following procedure:
   a. Remove the front lower console cover screws and cover.
   b. Remove the glove box door and box.
   c. At the passenger's side, remove the instrument panel lower cover screws and panel.
   d. At the driver's side, disconnect the accelerator cable from the pedal and remove the instrument panel lower cover screws and panel.
   e. Remove the lower cluster.
   f. At the meter cluster assembly, disconnect the switch connectors, then remove the screws, clips and the meter assembly.
   g. At the driver's side, disconnect the data link connector, then remove the bolts and the knee bolster.
   h. At the lower cluster cover, remove the cover-to-instrument panel screws, disconnect the cigarette lighter and remove the lower cover.
   i. Disconnect the air conditioning control cables from the unit.
   j. Remove the instrument panel cluster assembly screws, disconnect the switch connectors and the instrument panel assembly.
   k. Remove the instrument harness connectors, the radio antenna plug.
   l. Remove the side defroster grille. Remove the instrument panel nuts, bolts and screws and the instrument panel.

6. Remove the passenger's air bag reinforcement screws and the reinforcement.
7. At the meter assembly, disconnect the electrical connector, then remove the screws and the meter assembly.
8. Remove the radio and the vent duct assembly.
9. Remove the passenger's knee bolster screws and bolster.
10. Remove the instrument panel center bracket bolts, nuts and bracket.
11. Remove the heater resistor connectors and the electro thermo connector (if equipped with air conditioning).
12. Remove the blower motor assembly.
13. If equipped with air conditioning, remove the evaporator assembly by performing the following procedure:
   a. Disconnect the drain hose.
   b. Using a backup wrench, disconnect the refrigerant lines from the evaporator.
   c. Plug or cap the refrigerant lines.
   d. Remove the evaporator assembly.
14. Remove the driver's side lap vent duct.
15. Remove the center and lower vent ducts.
16. Remove the heater assembly.
17. Disassemble the heater unit assembly by performing the following procedure:
   a. Remove the lower air duct; do not remove the link unit.
   b. Remove the temperature control

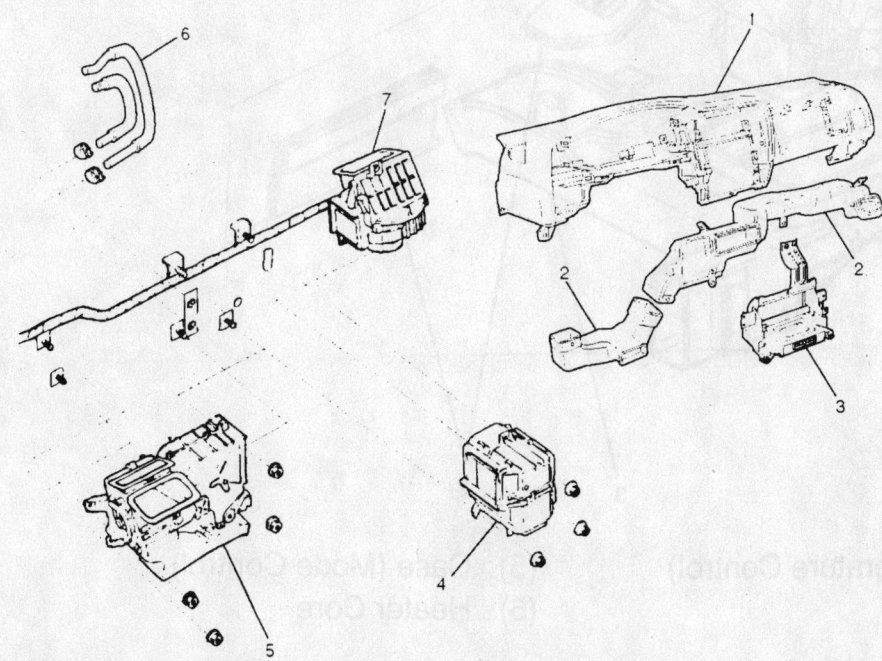

1 Instrument Panel Assembly

2 Center & Lower Vent Duct

3 Instrument Panel Center Bracket

4 Evaporator Assembly

5 Heater Unit Assembly

6 Heater Hose

7 Blower Unit

Exploded view of the heater unit and related components—Isuzu VehiCROSS

case screws and lift the case from the heater unit.

   c. Remove the heater core.

To install:

18. Assemble the heater unit assembly by performing the following procedure:

   a. Install the heater core into the heater unit.

   b. Install the temperature control case onto the unit and secure with screws.

   c. Install the lower air duct.

19. Install the heater unit assembly into the vehicle.

20. Install the center and lower vent ducts.

21. Install the driver's side lap vent duct.

22. Install the instrument panel cross beam assembly by reversing the removal procedures.

23. If equipped with air conditioning, install the evaporator assembly by performing the following procedures:

   a. If installing a new evaporator assembly, add 1.7 fl. oz. (50mL) of refrigerant oil to the evaporator.

   b. Using new O-rings and a backup wrench, install the refrigerant lines and torque the outlet line to 18 ft. lbs. (25 Nm) and the inlet line to 11 ft. lbs. (15 Nm).

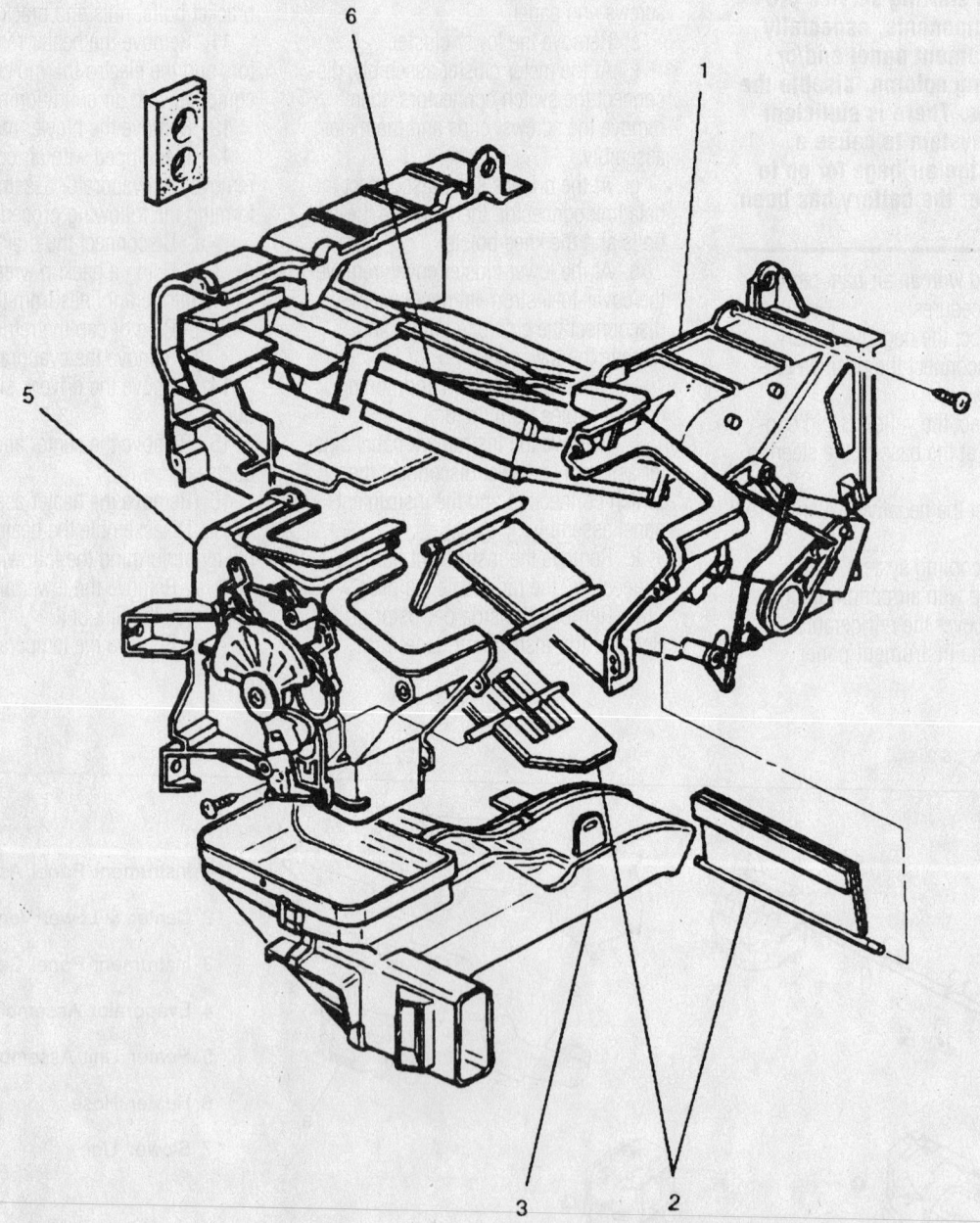

(1) Case (Temperature Control)	(5) Case (Mode Control)
(2) Mode Door	(6) Heater Core
(3) Duct	

93113G02

Exploded view of the heater unit—Isuzu VehiCROSS

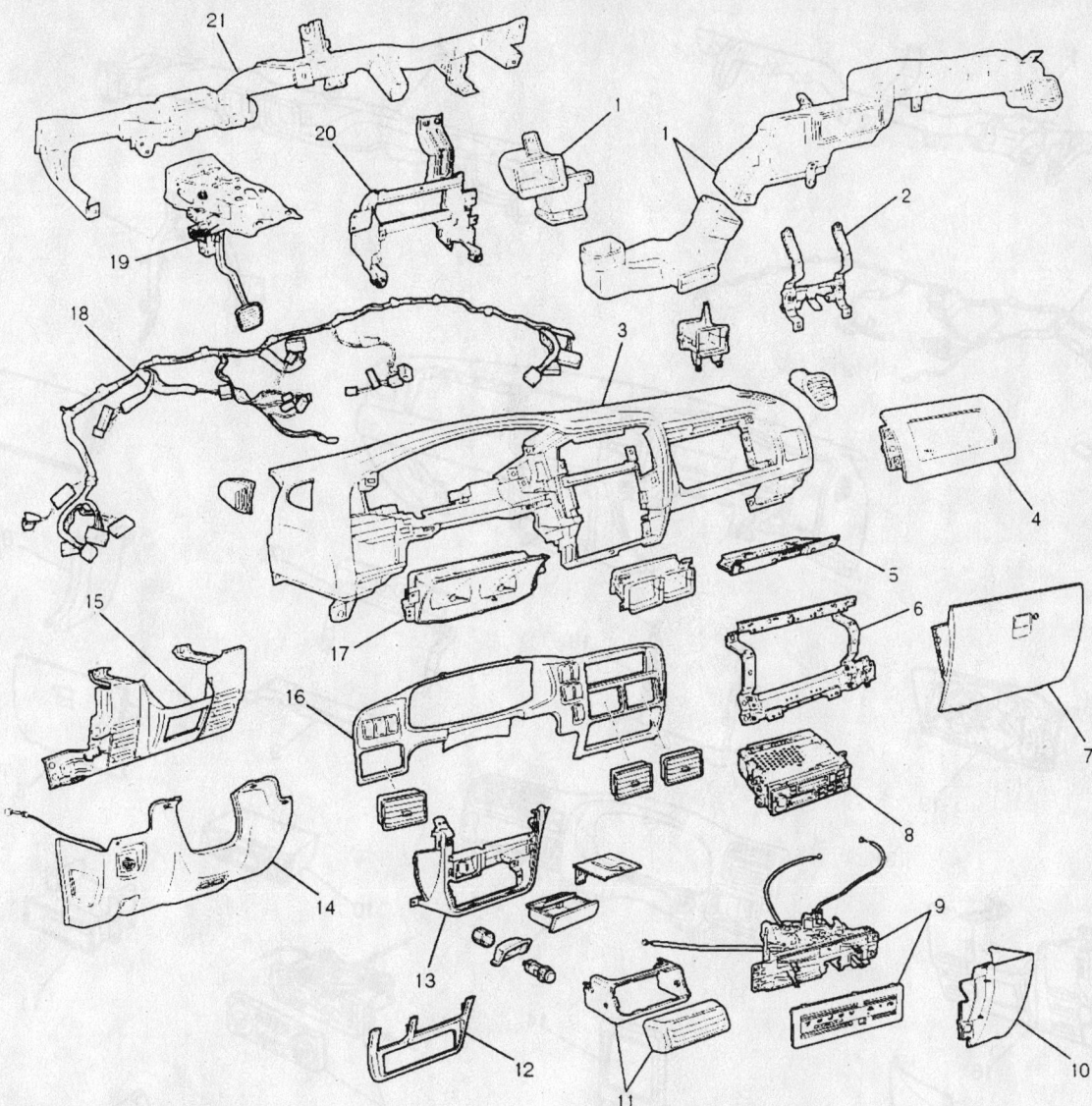

1 Vent Duct Assembly	12 Lower Cluster
2 Passenger Air Bag Reinforcement	13 Instrument Panel Lower Center Cover
3 Instrument Panel Assembly	14 Instrument Panel Driver Lower Cover
4 Passenger Air Bag Assembly	15 Driver Knee Bolster
5 Glove Box Cover	16 Meter Cluster Assembly
6 Passenger Knee Bolster	17 Meter Assembly
7 Glove Box Assembly	18 Instrument Harness Assembly
8 Radio Assembly	19 Brake Pedal & Bracket Assembly
9 Air Conditioner Control Lever Assembly	20 Instrument Panel Center Bracket
10 Instrument Panel Passenger Lower Cover	21 Cross Beam Assembly
11 Front Lower Console Cover	

Exploded view of the instrument panel and accessories—Isuzu VehiCROSS

93113G05

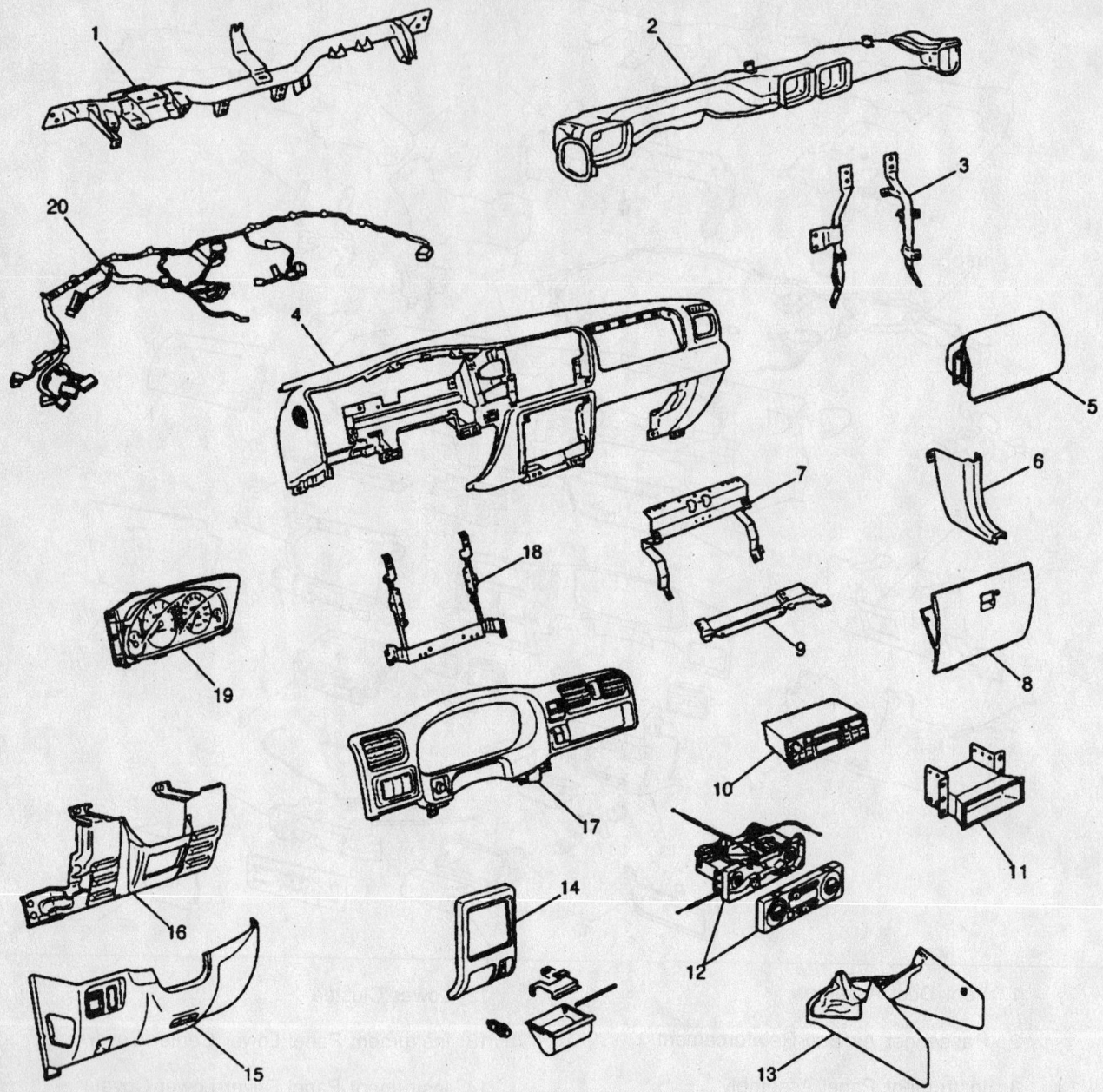

1	Cross Beam	11	Audio Sub Box
2	Vent Duct Assembly	12	Control Lever Assembly
3	Instrument Panel Bracket	13	Front Console Assembly
4	Instrument Panel Assembly	14	Lower Center Cover
5	Passenger Inflator Module	15	Instrument Panel Driver Lower Cover Assembly
6	Dash Side Trim Panel	16	Driver Knee Bolster Assembly
7	Passenger Knee Bolster Reinforcement Assembly	17	Meter Cluster Assembly
8	Glove Box	18	Instrument Panel Center Reinforcement
9	Passenger Lower Bracket	19	Meter Assembly
10	Radio Assembly	20	Instrument Harness Assembly

93113GB8

Exploded view of the instrument panel—Honda Passport

24. Install the blower motor.
25. Connect the heater resistor connectors and the electro-thermo connector (if equipped with air conditioning).
26. Connect the heater hoses to the heater unit.
27. Install the instrument panel assembly by reversing the removal procedures.
28. If equipped with air conditioning, evacuate and recharge the system.
29. Refill the cooling system.
30. Connect the negative battery cable.

✳✳ CAUTION

Never use an air bag assembly from another vehicle and/or different model year.

31. Enable the air bag by performing the following procedure:
   a. At the base of the steering column, connect the yellow 3-pin connector.
   b. Connect the negative battery cable.
   c. Turn the ignition switch ON and verify that the AIR BAG warning light flashes 7 times and then turns OFF.
32. Run the engine to normal operating temperatures and check for leaks. Check the systems for correct operation.

Passport

1. If equipped with an air bag, perform the following procedure:
   a. Turn the ignition to the LOCK position and remove the key.
   b. From the lower left dash side fuse block, remove the SRS-1 fuse.

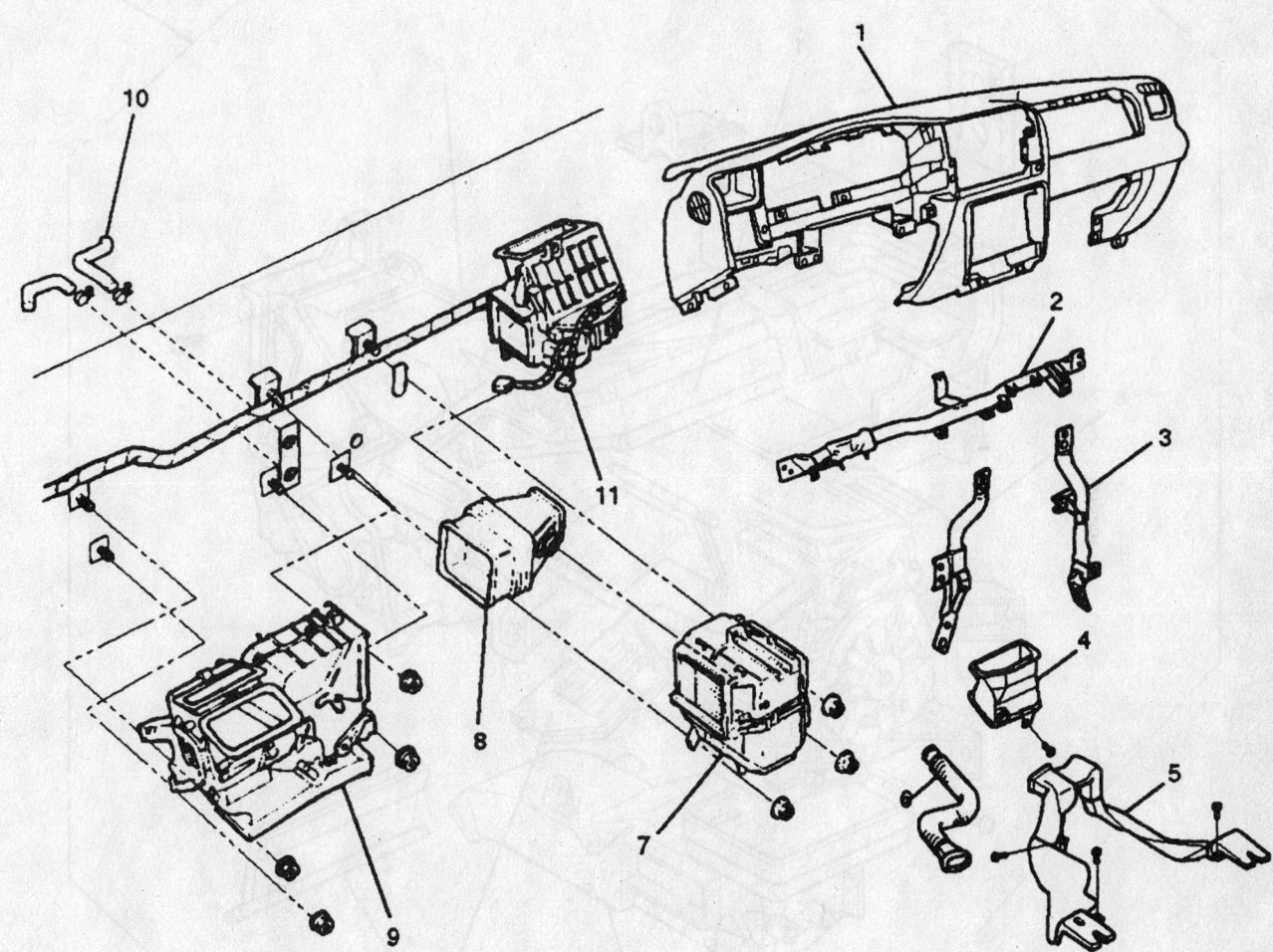

1	Instrument Panel Assembly	6	Driver Lap Vent Duct
2	Cross Beam Assembly	7	Evaporator Assembly (A/C only)
3	Instrument Panel Bracket	8	Duct
4	Ventilation Lower Duct	9	Heater Unit Assembly
5	Rear Heater Duct	10	Heater Hose
		11	Resistor Connector

93113GB9

View of the heater and air conditioning housing assemblies and related components—Honda Passport

c. Disconnect the 2-pin yellow connector located at the base of the steering column.

d. Remove the glove box assembly.

e. Disconnect the 2-pin yellow connector located behind the glove box.

2. Disconnect the negative battery cable.

3. If equipped, discharge and recover the air conditioning system refrigerant.

4. Remove the evaporator lines at the firewall. Plug the air conditioning lines to minimize contamination.

5. Disconnect the cooling system hoses and drain the coolant into a clean container for reuse. Plug the cooling system hoses.

6. Remove the instrument panel by performing the following procedure:

a. Remove the lower center cover screw and pull it out at the clip positions; then, disconnect the cigarette lighter connector.

b. Remove both the rear and front console.

c. Remove the dash side trim panel sill plates and the panels.

d. Remove the 2 glove box screws and the glove box.

e. Remove the 2 hood release screws, the 6 instrument panel driver's lower cover assembly screws and the cover assembly.

f. Remove the 5 instrument cluster

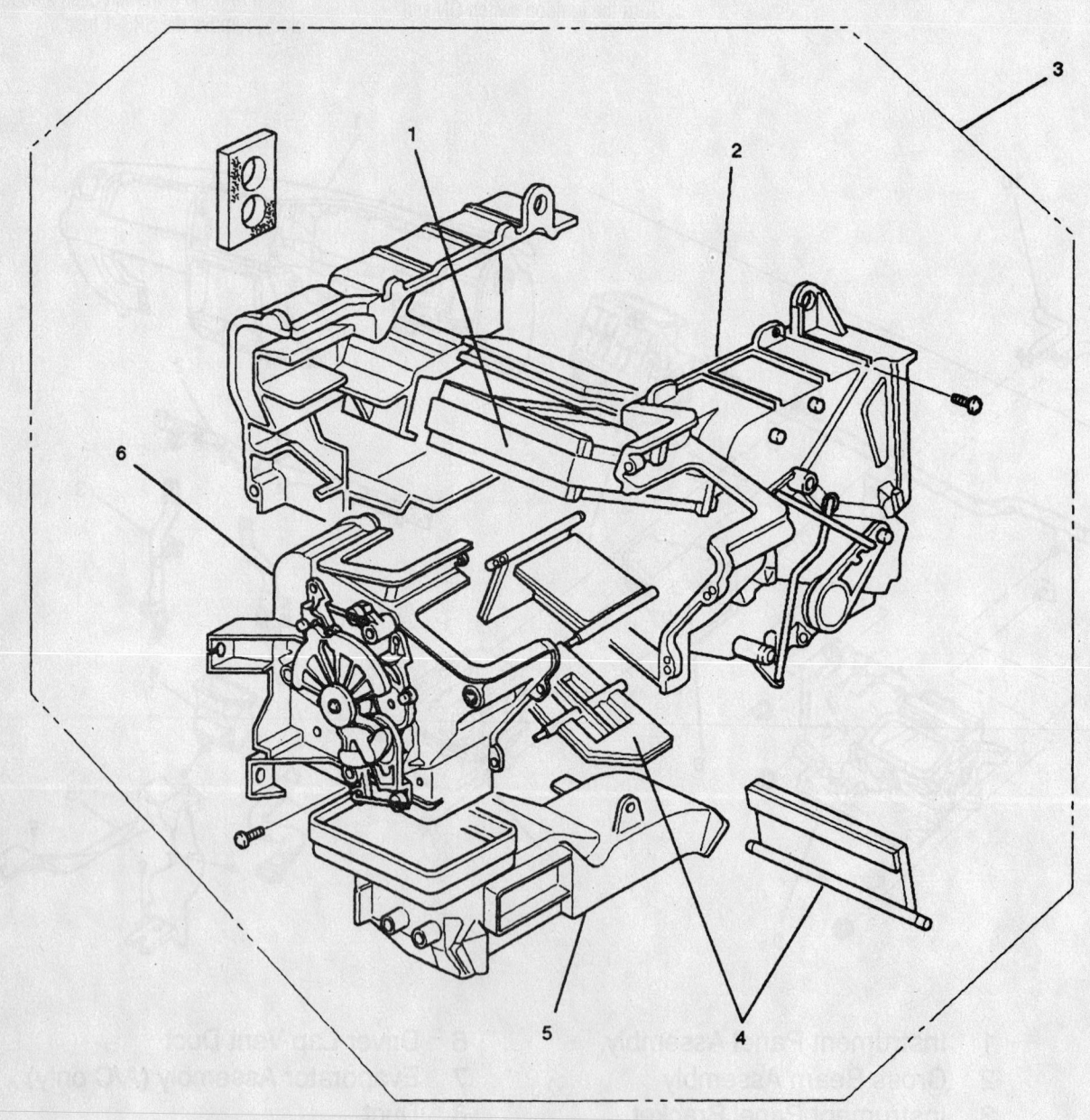

1	Heater Core	4	Mode Door
2	Case (Temperature Control)	5	Duct
3	Heater Unit	6	Case (Mode Control)

93113GB0

Exploded view of the heater housing assembly—Honda Passport

screws and the 2 clips. Disconnect the 8 switch connectors and remove the instrument cluster assembly.

g. Remove the 6 driver's knee bolster assembly bolts and screws and the knee bolster assembly.

h. Remove the 4 control lever assembly bolts; then, disconnect the 3 control cables (unit side) and the 3 harness connectors.

i. Remove the 4 radio/audio sub box assembly screws and the radio/audio sub box assembly.

j. Disconnect or remove the following instrument panel harness connectors or items:

- The 6 driver's side connectors
- The 3 passenger's side connectors
- The 2 center connectors
- Passenger's inflator module connector
- Radio antenna cable plug
- Ground cable bolt on the left dash side panel
- The 8 instrument panel-to-chassis bolts and the 3 nuts.

k. Remove the instrument panel assembly.

7. Remove the instrument panel bracket by performing the following procedure:

a. Remove the 2 passenger's inflator module bolts and 4 nuts.

b. Remove the 4 meter assembly screws. Then, disconnect the meter wiring harness connectors and remove the meter assembly.

c. Remove the 5 vent duct assembly screws and the assembly.

d. Remove the 3 lower passenger's bracket screws and the bracket.

e. Remove the 9 passenger's knee bolster reinforcement screws and the reinforcement.

f. Remove the 6 instrument panel center reinforcement screws and the reinforcement.

g. Remove the instrument panel wiring harness assembly clips and the wiring harness.

h. Remove the 2 instrument panel bracket nuts and 2 bolts for each bracket; then, remove the bracket(s).

8. Remove the 5 cross beam assembly nuts, 2 bolts and the 6 lower bolts; then, remove the crossbeam.

9. Disconnect the resistor wiring connector.

10. Remove the duct from the heater assembly.

11. If equipped with air conditioning, remove the evaporator assembly.

12. Remove the driver's lap vent.

13. Remove the lower ventilation duct.

14. Remove the footrest, the carpet, the 3 clips and the rear heater duct.

15. Remove the heater assembly.

16. Remove the mode control case-to-temperature control case screws and remove the mode control case; do not remove the link unit.

17. Remove the temperature control case screws and separate the cases.

18. Remove the heater core from the case.

To install:

19. Install the heater core to the case.

20. Assemble the temperature control cases and install the case screws.

21. Install the mode control case and the mode control case-to-temperature control case screws.

22. Install the heater assembly.

23. Install the rear heater duct, the footrest, the carpet, and the 3 clips.

24. Install the lower ventilation duct.

25. Install the driver's lap vent.

26. If equipped with air conditioning, install the evaporator assembly.

27. Install the duct to the heater assembly.

28. Connect the resistor wiring connector.

29. Install the crossbeam, the 5 cross beam assembly nuts, 2 bolts and the 6 lower bolts.

30. Install the instrument panel bracket by performing the following procedure:

a. Install the instrument panel bracket and the 2 nuts and 2 bolts for each bracket.

b. Install the instrument panel wiring harness assembly and the wiring harness clips.

c. Install the instrument panel center reinforcement and the 6 reinforcement screws.

d. Install the passenger knee bolster reinforcement and the 9 reinforcement screws.

e. Install the lower passenger bracket and the 3 bracket screws.

f. Install the vent duct assembly and the 5 vent duct assembly screws.

g. Install the meter assembly and the 4 meter assembly screws; then, connect the meter wiring harness connectors.

h. Install the 2 passenger's inflator module bolts and 4 nuts.

31. Install the instrument panel by performing the following procedure:

a. Install the instrument panel assembly.

b. Connect or install the following instrument panel harness connectors or items:

- The 6 driver's side connectors
- The 3 passenger's side connectors
- The 2 center connectors
- Passenger's inflator module connector
- Radio antenna cable plug
- Ground cable bolt on the left dash side panel
- The 8 instrument panel-to-chassis bolts and the 3 nuts.

c. Install the radio/audio sub box assembly and the 4 radio/audio sub box assembly screws.

d. Connect the 3 control cables (unit side) and the 3 harness connectors. Install the 4 control lever assembly bolts.

e. Install the knee bolster assembly and the 6 driver's knee bolster assembly bolts and screws.

f. Install the instrument cluster assembly. Connect the 8 switch connectors. Install 5 instrument cluster screws and the 2 clips.

g. Install the 2 hood release screws, the 6 instrument panel driver's lower cover assembly screws and the cover assembly.

h. Install the glove box and the 2 glove box screws.

i. Install the dash side trim panel sill plates and the panels.

j. Install both the rear and front console.

k. Connect the cigarette lighter connector and install the lower center cover screw.

32. Connect the cooling system hoses.

33. Refill the cooling system.

34. Install the evaporator lines at the firewall.

35. If equipped, evacuate and charge the air conditioning system.

36. Connect the negative battery cable.

37. If equipped with an air bag, perform the following procedure:

a. Turn the ignition to the LOCK position and remove the key.

b. Connect the 2-pin yellow connector located behind the glove box.

c. Install the glove box assembly.

d. Connect the 2-pin yellow connector located at the base of the steering column.

e. At the lower left dash side fuse block, install the SRS-1 fuse.

f. Turn the ignition switch to ON and verify that the AIR BAG warning light flashes 7 times and turns OFF.

38. Run the engine to normal operating temperatures; then, check the climate control operation and check for leaks.

Cylinder Head

REMOVAL & INSTALLATION

2.2L Engine

1. Before servicing the vehicle, refer to the precautions in the beginning of this section.
2. Drain the cooling system.
3. Relieve the fuel system pressure.
4. Remove or disconnect the following:
   - Negative battery cable
   - Intake Air Temperature (IAT) sensor connector
   - Positive Crankcase Ventilation (PCV) valve and hose
   - Air intake assembly
   - Upper radiator hose
   - Accessory drive belt
   - Exhaust front pipe
   - Alternator and brackets
   - Crankshaft Position (CKP) sensor connector
   - Knock sensor connector
   - Heater hoses
   - Water bypass hose
   - Fuel lines
   - Evaporative Emissions (EVAP) valve connector
   - Canister hose
   - Intake manifold

- Engine wiring harness connectors at left rear of the engine compartment
- Power steering pump pressure switch connector
- Front cover
- Spark plugs and wires
- Camshaft Position (CMP) sensor
- Valve cover
- Timing belt. Refer to the Timing Belt unit repair section.
- Timing belt idler pulleys
- Timing belt rear cover
- Oil pressure switch connector
- Camshafts
- Cylinder head. Remove the bolts in reverse of the tightening sequence.

To install:

➡**Use new cylinder head bolts for assembly.**

5. Install the cylinder head with a new gasket. Tighten the bolts in sequence as follows:
   a. Step 1: 18 ft. lbs. (25 Nm)
   b. Step 2: Plus 90 degrees
   c. Step 3: Plus 90 degrees
   d. Step 4: Plus 90 degrees
6. Install or connect the following:
   - Camshafts
   - Oil pressure switch connector
   - Timing belt rear cover

- Timing belt idler pulleys. Tighten the bolts to 18 ft. lbs. (25 Nm).
- Timing belt
- Valve cover
- CMP sensor
- Spark plugs and wires
- Front cover
- Power steering pump pressure switch connector
- Engine wiring harness connectors at left rear of the engine compartment
- Intake manifold
- Canister hose
- EVAP valve connector
- Fuel lines
- Water bypass hose
- Heater hoses
- Knock sensor connector
- CKP sensor connector
- Alternator and brackets
- Exhaust front pipe
- Accessory drive belt
- Upper radiator hose
- Air intake assembly
- PCV valve and hose
- IAT sensor connector
- Negative battery cable
7. Fill the cooling system.
8. Start the engine and check for leaks.

3.2L Engines

1. Before servicing the vehicle, refer to the precautions in the beginning of this section.
2. Drain the cooling system.
3. Relieve the fuel system pressure.
4. Remove or disconnect the following:
   - Negative battery cable
   - Hood
   - Engine cover
   - Mass Air Flow (MAF) sensor connector
   - Intake Air Temperature (IAT) sensor connector
   - Positive Crankcase Ventilation (PCV) valve and hose
   - Air cleaner assembly
   - Manifold Absolute Pressure (MAP) sensor connector
   - Vacuum Switching Valve (VSV) connector and vacuum line
   - Fuel injector connectors
   - Throttle Position (TP) sensor connector
   - Idle Air Control (IAC) valve connector
   - Ignition coils
   - Brake booster vacuum line
   - Canister purge vacuum line
   - Duty solenoid valve

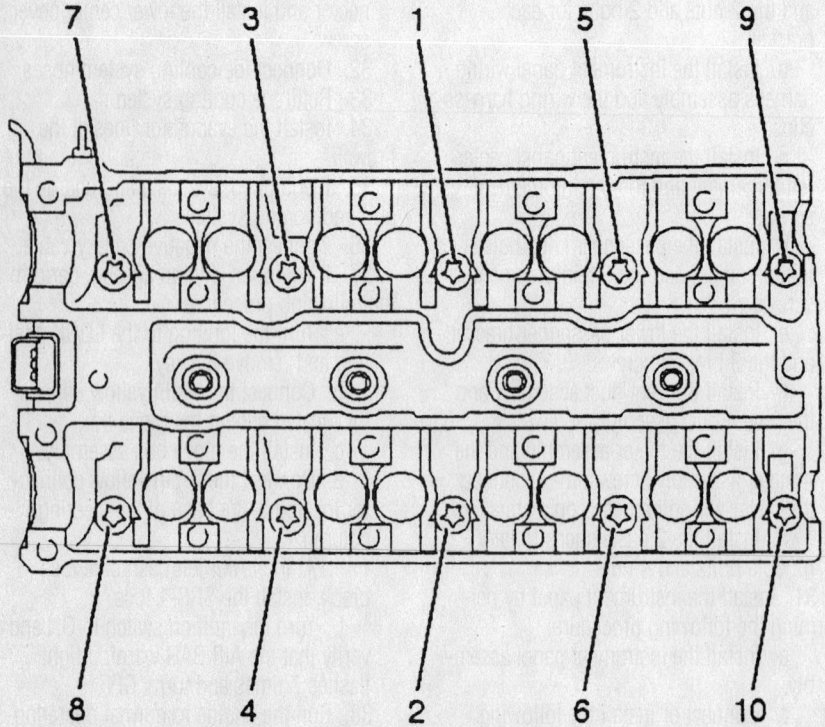

Cylinder head torque sequence—2.2L (VIN D) engine

7924NG05

- Fuel lines
- Intake manifold
- Radiator hoses
- Engine coolant manifold
- Upper fan shroud
- Accessory drive belt and tensioner
- Cooling fan and pulley
- Alternator
- Idler pulley
- Power steering pump and bracket
- A/C compressor
- Crankshaft pulley
- Oil cooler hoses
- Timing belt cover
- Valve covers
- Timing belt. Refer to the Timing Belt Unit Repair Section.
- Left and right exhaust front pipes
- Oil dipstick tube
- Cylinder heads

To install:

➡**Use new head bolts when installing the cylinder head.**

➡**The left and right cylinder head gaskets are not interchangeable.**

5. Install the cylinder heads with new gaskets. Tighten the bolts in sequence as follows:

   a. Step 1: 21 ft. lbs. (29 Nm)
   b. Step 2: 47 ft. lbs. (64 Nm)

6. Install or connect the following:

- Oil dipstick tube
- Left and right exhaust front pipes
- Timing belt
- Valve covers
- Timing belt cover
- Oil cooler hoses
- Crankshaft pulley. Tighten the pulley bolt to 123 ft. lbs. (167 Nm).
- A/C compressor
- Power steering pump and bracket. Tighten the bolts to 34 ft. lbs. (46 Nm).
- Idler pulley
- Alternator
- Cooling fan and pulley
- Accessory drive belt and tensioner
- Upper fan shroud
- Engine coolant manifold
- Radiator hoses
- Intake manifold
- Fuel lines
- Duty solenoid valve
- Canister purge vacuum line
- Brake booster vacuum line
- Ignition coils
- IAC valve connector
- TP sensor connector
- Fuel injector connectors
- VSV connector and vacuum line

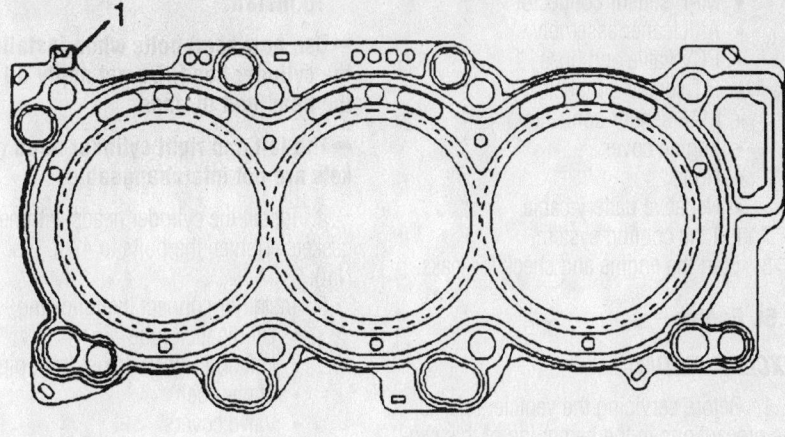

Right (1) and left (2) head gasket identification mark locations—3.2L DOHC engine

7924NG11

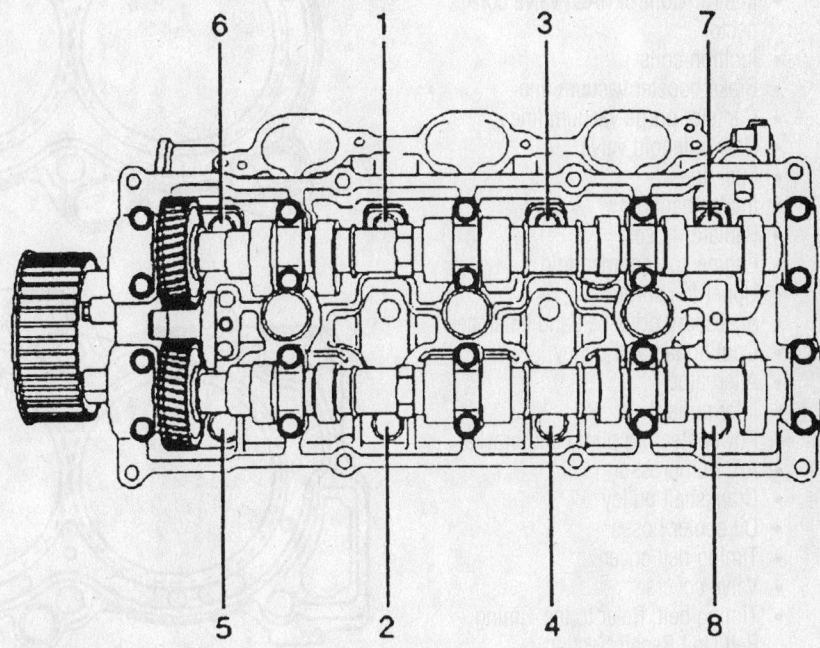

Cylinder head torque sequence—3.2L DOHC and 3.5L engines

7924NG12

- MAP sensor connector
- Air cleaner assembly
- PCV valve and hose
- IAT sensor connector
- MAF sensor connector
- Engine cover
- Hood
- Negative battery cable

7. Fill the cooling system.
8. Start the engine and check for leaks.

3.5L Engine

EXCEPT AXIOM

1. Before servicing the vehicle, refer to the precautions in the beginning of this section.
2. Drain the cooling system.
3. Remove or disconnect the following:
- Negative battery cable
- Hood
- Engine cover
- Mass Air Flow (MAF) sensor connector
- Intake Air Temperature (IAT) sensor connector
- Positive Crankcase Ventilation (PCV) valve and hose
- Air cleaner assembly
- Manifold Absolute Pressure (MAP) sensor connector
- Vacuum Switching Valve (VSV) connector and vacuum line
- Fuel injector connectors
- Throttle Position (TP) sensor connector
- Idle Air Control (IAC) valve connector
- Ignition coils
- Brake booster vacuum line
- Canister purge vacuum line
- Duty solenoid valve
- Fuel lines
- Intake manifold
- Radiator hoses
- Engine coolant manifold
- Upper fan shroud
- Accessory drive belt and tensioner
- Cooling fan and pulley
- Alternator
- Idler pulley
- Power steering pump and bracket
- A/C compressor
- Crankshaft pulley
- Oil cooler hoses
- Timing belt cover
- Valve covers
- Timing belt. Refer to the Timing Belt Unit Repair Section.
- Left and right exhaust front pipes
- Oil dipstick tube
- Cylinder heads

To install:

→Use new head bolts when installing the cylinder head. Do not apply oil to the head bolt threads.

→The left and right cylinder head gaskets are not interchangeable.

4. Install the cylinder heads with new gaskets. Tighten the bolts to 47 ft. lbs. (64 Nm).
5. Install or connect the following:
- Oil dipstick tube
- Left and right exhaust front pipes
- Timing belt
- Valve covers
- Timing belt cover
- Oil cooler hoses
- Crankshaft pulley. Tighten the pulley bolt to 123 ft. lbs. (167 Nm).
- A/C compressor
- Power steering pump and bracket. Tighten the bolts to 34 ft. lbs. (46 Nm).
- Idler pulley
- Alternator
- Cooling fan and pulley
- Accessory drive belt and tensioner
- Upper fan shroud

- Engine coolant manifold
- Radiator hoses
- Intake manifold
- Fuel lines
- Duty solenoid valve
- Canister purge vacuum line
- Brake booster vacuum line
- Ignition coils
- IAC valve connector
- TP sensor connector
- Fuel injector connectors
- VSV connector and vacuum line
- MAP sensor connector
- Air cleaner assembly
- PCV valve and hose
- IAT sensor connector
- MAF sensor connector
- Engine cover
- Hood
- Negative battery cable

6. Fill the cooling system.
7. Start the engine. Check for leaks and proper operation.

AXIOM

1. Before servicing the vehicle, refer to the precautions in the beginning of this section.

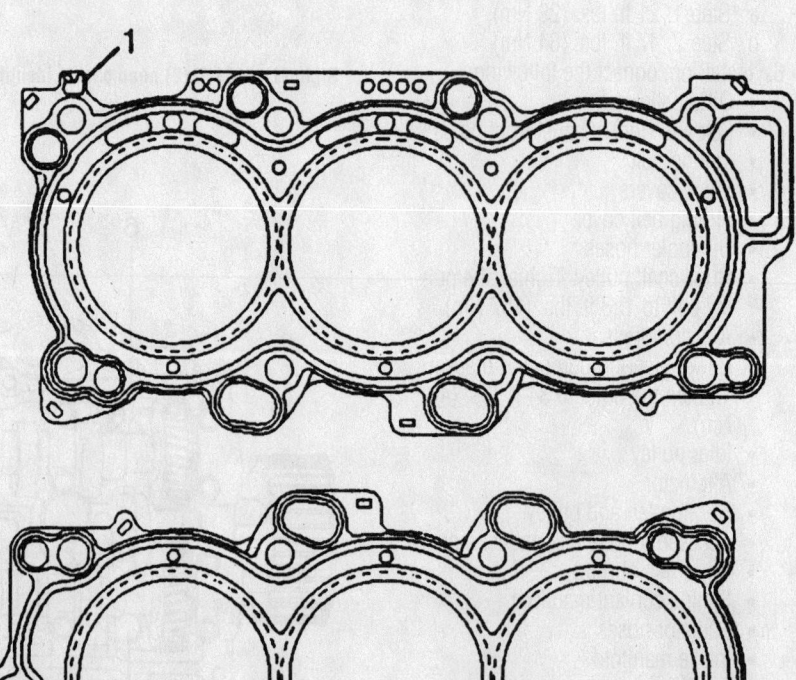

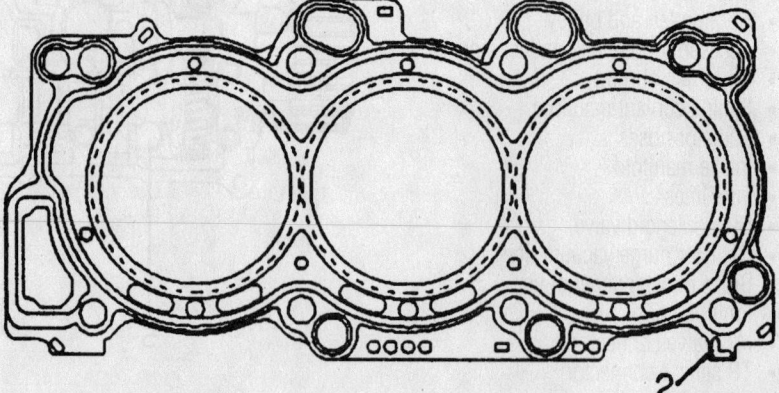

Right (1) and left (2) head gasket identification mark locations—3.5L engine

7924BG04

2. Drain the cooling system.
3. Remove or disconnect the following:
   - Negative battery cable
   - Hood
   - Engine cover
   - Mass Air Flow (MAF) sensor connector
   - Intake Air Temperature (IAT) sensor connector
   - Positive Crankcase Ventilation (PCV) valve and hose
   - Air cleaner assembly
   - Manifold Absolute Pressure (MAP) sensor connector
   - Vacuum Switching Valve (VSV) connector and vacuum line
   - Fuel injector connectors
   - Throttle Position (TP) sensor connector
   - Idle Air Control (IAC) valve connector
   - Ignition coils
   - Brake booster vacuum line
   - Canister purge vacuum line
   - Duty solenoid valve
   - Fuel lines
   - Intake manifold
   - Radiator hoses
   - Engine coolant manifold
   - Upper fan shroud
   - Accessory drive belt and tensioner
   - Cooling fan and pulley
   - Alternator
   - Idler pulley
   - Power steering pump and bracket
   - A/C compressor
   - Crankshaft pulley
   - Oil cooler hoses
   - Timing belt cover
   - Valve covers
   - Timing belt. Refer to the Timing Belt Unit Repair Section.
   - Left and right exhaust front pipes
   - Oil dipstick tube
   - Cylinder heads

To install:

➡**Use new head bolts when installing the cylinder head. Do not apply oil to the head bolt threads.**

➡**The left and right cylinder head gaskets are not interchangeable.**

4. Install the cylinder heads with new gaskets. Tighten the bolts to 22 ft. lbs. in sequence, then to 47 ft. lbs. (64 Nm) in sequence.
5. Install or connect the following:
   - Oil dipstick tube
   - Left and right exhaust front pipes
   - Timing belt
   - Valve covers

   - Timing belt cover
   - Oil cooler hoses
   - Crankshaft pulley. Tighten the pulley bolt to 123 ft. lbs. (167 Nm).
   - A/C compressor
   - Power steering pump and bracket. Tighten the bolts to 34 ft. lbs. (46 Nm).
   - Idler pulley
   - Alternator
   - Cooling fan and pulley
   - Accessory drive belt and tensioner
   - Upper fan shroud
   - Engine coolant manifold
   - Radiator hoses
   - Intake manifold
   - Fuel lines
   - Duty solenoid valve
   - Canister purge vacuum line
   - Brake booster vacuum line
   - Ignition coils
   - IAC valve connector
   - TP sensor connector
   - Fuel injector connectors
   - VSV connector and vacuum line
   - MAP sensor connector
   - Air cleaner assembly
   - PCV valve and hose
   - IAT sensor connector
   - MAF sensor connector
   - Engine cover
   - Hood
   - Negative battery cable
6. Fill the cooling system.
7. Start the engine. Check for leaks and proper operation.

Rocker Arms/Shafts

REMOVAL & INSTALLATION

➡**These engines are not equipped with rocker arms. The camshaft lobes act directly on the valve shims.**

Intake Manifold

REMOVAL & INSTALLATION

2.2L Engine

1. Before servicing the vehicle, refer to the precautions in the beginning of this section.
2. Drain the cooling system.
3. Relieve the fuel system pressure.
4. Remove or disconnect the following:
   - Negative battery cable
   - Accessory drive belt
   - Positive Crankcase Ventilation (PCV) valve and hose

   - Air intake duct
   - Throttle body water hoses
   - Throttle Position (TP) sensor connector
   - Idle Air Control (IAC) valve connector
   - Fuel lines
   - Fuel injector connectors
   - Fuel pressure regulator vacuum line
   - Fuel supply manifold
   - Accelerator cable
   - Alternator and brackets
   - Water pipe
   - Intake manifold bracket
   - Ignition coil and bracket
   - Brake booster vacuum line
   - Intake manifold

To install:

5. Install or connect the following:
   - Intake manifold. Use a new gasket and tighten the bolts to 16 ft. lbs. (22 Nm).
   - Brake booster vacuum line
   - Ignition coil and bracket
   - Intake manifold bracket. Tighten the bolts to 16 ft. lbs. (22 Nm).
   - Water pipe
   - Alternator and brackets. Tighten the short bolts to 14 ft. lbs. (20 Nm) and the long bolts to 25 ft. lbs. (35 Nm).
   - Accelerator cable
   - Fuel supply manifold
   - Fuel pressure regulator vacuum line
   - Fuel injector connectors
   - Fuel lines
   - IAC valve connector
   - TP sensor connector
   - Throttle body water hoses
   - Air intake duct
   - PCV valve and hose
   - Accessory drive belt
   - Negative battery cable
6. Fill the cooling system.
7. Start the engine and check for leaks.

3.2L Engine

1. Before servicing the vehicle, refer to the precautions in the beginning of this section.
2. Remove or disconnect the following:
   - Negative battery cable
   - Engine cover
   - Air cleaner assembly
   - Accelerator cable
   - Cruise control cable
   - Brake booster vacuum line
   - Manifold Absolute Pressure (MAP) sensor connector

- Idle Air Control (IAC) valve connector
- Throttle Position (TP) sensor connector
- Canister purge solenoid connector
- Electronic Vacuum Sensing Valve (EVSV) connector and vacuum line
- Exhaust Gas Recirculation (EGR) valve
- Positive Crankcase Ventilation (PCV) valve and hose
- Pressure regulator vacuum line
- Ventilation hose
- Throttle body
- Fuel lines
- Fuel injector connectors
- Intake manifold

To install:

3. Install or connect the following:
- Intake manifold. Tighten the fasteners to 18 ft. lbs. (25 Nm).
- Fuel injector connectors
- Fuel lines
- Throttle body. Tighten the bolts to 88 inch lbs. (10 Nm).
- Ventilation hose
- Pressure regulator vacuum line
- PCV valve and hose
- EGR valve
- EVSV connector and vacuum line
- Canister purge solenoid connector
- TP sensor connector
- IAC valve connector
- MAP sensor connector
- Brake booster vacuum line
- Cruise control cable
- Accelerator cable
- Air cleaner assembly
- Engine cover
- Negative battery cable

4. Start the engine and check for proper operation.

3.5L Engine

1. Before servicing the vehicle, refer to the precautions in the beginning of this section.
2. Remove or disconnect the following:
- Negative battery cable
- Engine cover
- Air cleaner assembly
- Accelerator cable
- Cruise control cable
- Brake booster vacuum line
- Manifold Absolute Pressure (MAP) sensor connector
- Idle Air Control (IAC) valve connector
- Throttle Position (TP) sensor connector
- Canister purge solenoid connector

- Electronic Vacuum Sensing Valve (EVSV) connector and vacuum line
- Exhaust Gas Recirculation (EGR) valve
- Positive Crankcase Ventilation (PCV) valve and hose
- Pressure regulator vacuum line
- Ventilation hose
- Throttle body
- Fuel lines
- Fuel injector connectors
- Intake manifold

To install:

3. Install or connect the following:
- Intake manifold. Tighten the fasteners to 18 ft. lbs. (25 Nm).
- Fuel injector connectors
- Fuel lines
- Throttle body. Tighten the bolts to 88 inch lbs. (10 Nm).
- Ventilation hose
- Pressure regulator vacuum line
- PCV valve and hose
- EGR valve
- EVSV connector and vacuum line
- Canister purge solenoid connector
- TP sensor connector
- IAC valve connector
- MAP sensor connector
- Brake booster vacuum line
- Cruise control cable
- Accelerator cable
- Air cleaner assembly
- Engine cover
- Negative battery cable

4. Start the engine and check for proper operation.

Exhaust Manifold

REMOVAL & INSTALLATION

2.2L Engine

1. Before servicing the vehicle, refer to the precautions in the beginning of this section.
2. Remove or disconnect the following:
- Negative battery cable
- Air intake duct
- Exhaust front pipe
- Exhaust manifold heat shield
- Exhaust manifold

To install:

3. Install the exhaust manifold. Tighten the nuts in sequence as follows:
   - a. Step 1: 10 ft. lbs. (14 Nm)
   - b. Step 2: 14 ft. lbs. (20 Nm)
   - c. Step 3: 14 ft. lbs. (20 Nm)
4. Install or connect the following:
- Exhaust manifold heat shield.

Tighten the bolts to 71 inch lbs. (8 Nm).
- Exhaust front pipe. Tighten the bolts to 18 ft. lbs. (25 Nm).
- Air intake duct
- Negative battery cable

5. Start the engine and check for leaks.

3.2L Engines

1. Before servicing the vehicle, refer to the precautions in the beginning of this section.
2. Remove or disconnect the following:
- Negative battery cable
- Air cleaner assembly
- Heated Oxygen (HO2S) sensor connectors
- Right torsion bar
- Exhaust Gas Recirculation (EGR) pipe and bracket
- Left and right exhaust front pipes
- Heat shields
- Accessory drive belt
- A/C compressor and bracket
- Exhaust manifolds

To install:

3. Install or connect the following:
- Exhaust manifolds. Tighten the bolts to 42 ft. lbs. (57 Nm).
- A/C compressor and bracket
- Accessory drive belt
- Heat shields
- Left and right exhaust front pipes
- EGR pipe and bracket
- Right torsion bar
- HO2S sensor connectors
- Air cleaner assembly
- Negative battery cable

4. Start the engine and check for leaks.

3.5L Engine

1. Before servicing the vehicle, refer to the precautions in the beginning of this section.
2. Remove or disconnect the following:
- Negative battery cable
- Air cleaner assembly
- Heated Oxygen (HO2S) sensor connectors
- Right torsion bar
- Exhaust Gas Recirculation (EGR) pipe and bracket
- Left and right exhaust front pipes
- Heat shields
- Accessory drive belt
- A/C compressor and bracket
- Exhaust manifolds

To install:

3. Install or connect the following:
- Exhaust manifolds. Tighten the bolts to 38 ft. lbs. (52 Nm).
- A/C compressor and bracket

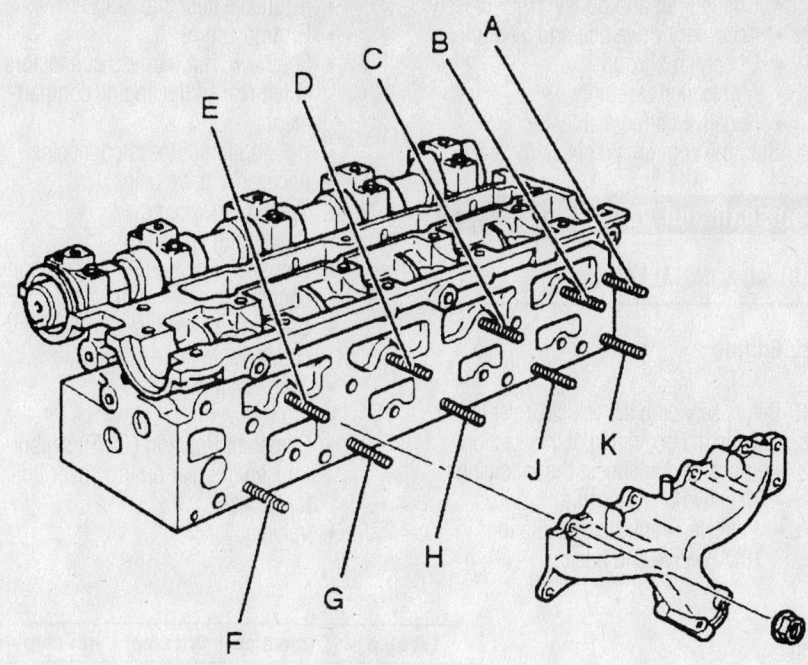

7924NG17

Tightening sequence:
Step1: J G H B D C J G B D
Step2: A B C D E F G H J K
Step3: A B C D E F G H J K

Tightening torque:
Step1: 14 N·m (10 lb ft)
Step2: 20 N·m (14 lb ft)
Step3: 20 N·m (14 lb ft)

Exhaust manifold torque sequence—2.2L engine

- Accessory drive belt
- Heat shields
- Left and right exhaust front pipes
- EGR pipe and bracket
- Right torsion bar
- HO$_2$S sensor connectors
- Air cleaner assembly
- Negative battery cable
4. Start the engine and check for leaks.

Front Crankshaft Seal

REMOVAL & INSTALLATION

2.2L Engines

1. Before servicing the vehicle, refer to the precautions in the beginning of this section.

2. Remove or disconnect the following:
- Negative battery cable
- Accessory drive belts
- Cooling fan
- A/C belt tensioner, if equipped
- Water pump pulley
- Power steering pump
- Crankshaft pulley
- Front cover
- Timing belt. Refer to the Timing Belt unit repair section.
- Crankshaft timing sprocket
- Rear timing cover
- Crankshaft oil seal

To install:
3. Install or connect the following:
- Crankshaft oil seal
- Rear timing cover
- Crankshaft timing sprocket. Tighten

the bolt to 94 ft. lbs. (130 Nm) plus 45 degrees.
- Timing belt. Refer to the Timing Belt unit repair section.
- Front cover
- Crankshaft pulley
- Power steering pump
- Water pump pulley
- A/C belt tensioner, if equipped
- Cooling fan
- Accessory drive belts
- Negative battery cable
4. Start the engine and check for leaks.

3.2L Engines

1. Before servicing the vehicle, refer to the precautions in the beginning of this section.

2. Remove or disconnect the following:
- Negative battery cable
- Air cleaner assembly
- Upper fan shroud
- Accessory drive belt and tensioner
- Cooling fan and pulley
- Idler pulley
- Power steering pump
- Crankshaft pulley
- Timing belt cover
- Timing belt. Refer to the Timing Belt Unit Repair Section.
- Crankshaft timing sprocket
- Oil seal

To install:
3. Install or connect the following:
- Oil seal so that it is flush with the oil pump housing
- Crankshaft timing sprocket
- Timing belt
- Timing belt cover
- Crankshaft pulley. Tighten the bolt to 123 ft. lbs. (167 Nm).
- Power steering pump
- Idler pulley
- Cooling fan and pulley
- Accessory drive belt and tensioner
- Upper fan shroud
- Air cleaner assembly
- Negative battery cable
4. Start the engine and check for leaks.

3.5L Engine

1. Before servicing the vehicle, refer to the precautions in the beginning of this section.

2. Remove or disconnect the following:
- Negative battery cable
- Air cleaner assembly
- Upper fan shroud
- Accessory drive belt and tensioner
- Cooling fan and pulley
- Idler pulley

- Power steering pump and move it aside
- Crankshaft pulley
- Timing belt cover
- Timing belt. Refer to the Timing Belt Unit Repair Section.
- Crankshaft timing sprocket
- Oil seal

To install:

3. Install or connect the following:
- Oil seal so that it is flush with the oil pump housing
- Crankshaft timing sprocket
- Timing belt
- Timing belt cover
- Crankshaft pulley. Tighten the bolt to 123 ft. lbs. (167 Nm).
- Power steering pump
- Idler pulley

- Cooling fan and pulley
- Accessory drive belt and tensioner
- Upper fan shroud
- Air cleaner assembly
- Negative battery cable
4. Start the engine and check for leaks.

Camshaft and Valve Lifters

REMOVAL & INSTALLATION

2.2L Engine

1. Before servicing the vehicle, refer to the precautions in the beginning of this section.
2. Remove or disconnect the following:
- Negative battery cable
- Positive Crankcase Ventilation (PCV) valve and hose

- Air intake duct and bracket
- Ground cables
- Engine wiring harness connectors at left rear of the engine compartment
- Cooling fan harness connector
- Accessory drive belt
- Spark plug wire cover
- Spark plug wires
- Camshaft Position (CMP) sensor connector
- Crankshaft Position (CKP) sensor connector
- Crankshaft pulley
- Front cover
- Camshaft Position (CMP) sensor. Loosen the rear timing cover bolt for access.
- Valve cover

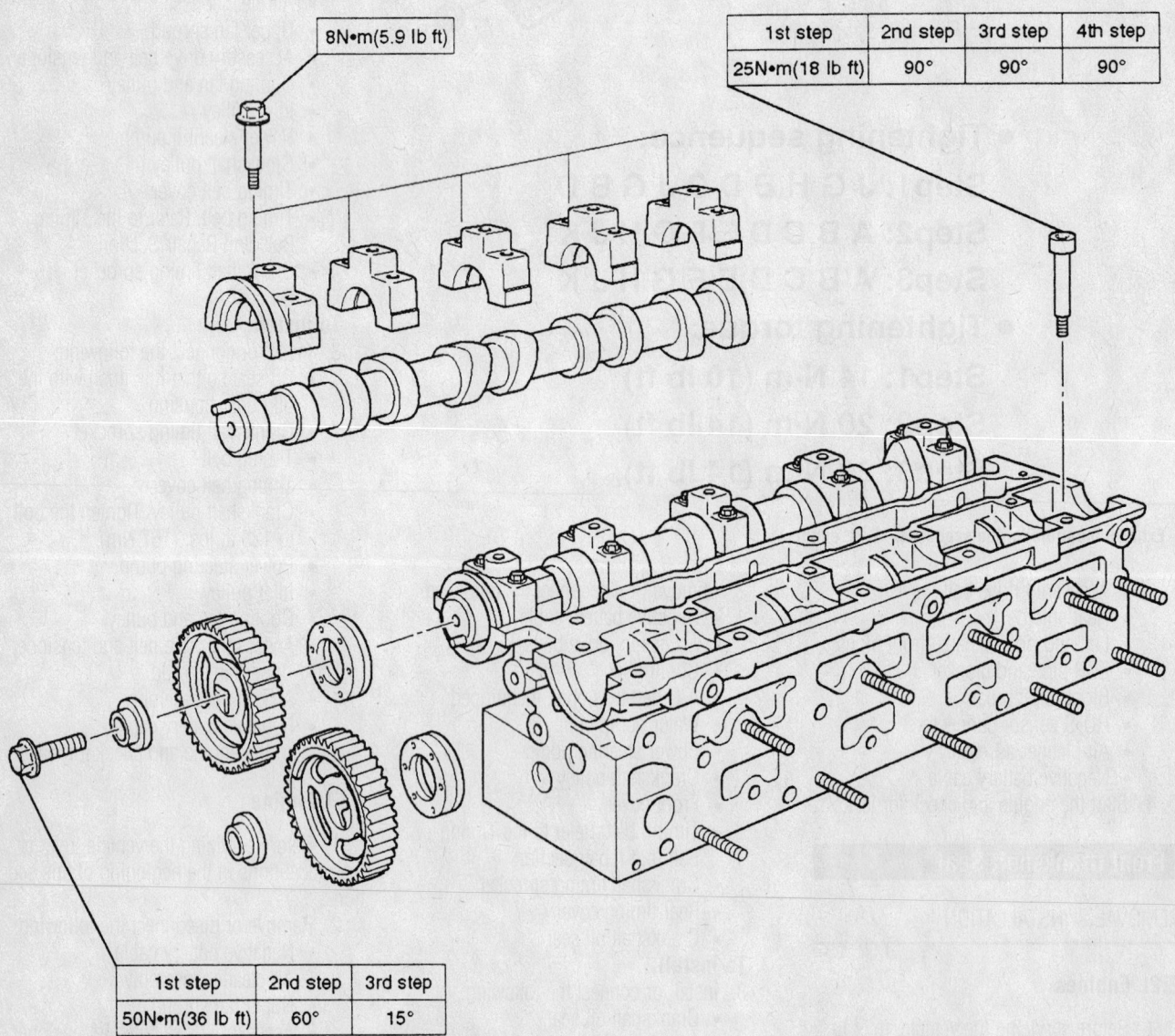

8N•m(5.9 lb ft)

	1st step	2nd step	3rd step	4th step
	25N•m(18 lb ft)	90°	90°	90°

	1st step	2nd step	3rd step
	50N•m(36 lb ft)	60°	15°

7924NG41

Exploded view of the cylinder head and camshaft components—2.2L engine

- Timing belt. Refer to the Timing Belt unit repair section.
- Camshaft sprockets
- Camshaft bearing caps
- Camshaft seals
- Camshafts
- Hydraulic tappets

➡ **Keep all valvetrain components in order for assembly.**

To install:

3. Install or connect the following:
- Hydraulic tappets in their original locations
- Camshafts
- Camshaft bearing caps. Tighten the bolts in sequence to 71 inch lbs. (8 Nm).
- Camshaft seals
- Camshaft sprockets

4. Tighten the camshaft sprocket bolts as follows:
   a. Step 1: 36 ft. lbs. (50 Nm)
   b. Step 2: Plus 60 degrees
   c. Step 3: Plus 15 degrees

5. Install or connect the following:
- Timing belt
- Valve cover
- CMP sensor. Loosen the rear timing cover bolt for access.
- Front cover
- Crankshaft pulley. Tighten the bolts to 14 ft. lbs. (20 Nm).
- CKP sensor connector
- CMP sensor connector
- Spark plug wires

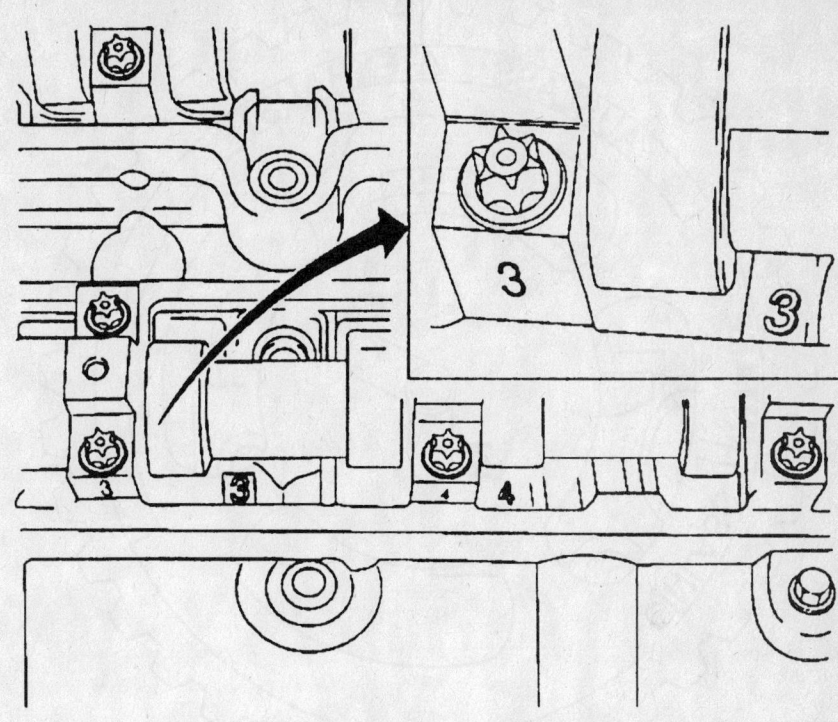

7924NG19

Camshaft bearing cap identification locations—2.2L engine

- Spark plug wire cover
- Accessory drive belt
- Cooling fan harness connector
- Engine wiring harness connectors at left rear of the engine compartment
- Ground cables
- Air intake duct and bracket
- PCV valve and hose
- Negative battery cable

3.2L Engine

1. Before servicing the vehicle, refer to the precautions in the beginning of this section.

2. Remove or disconnect the following:

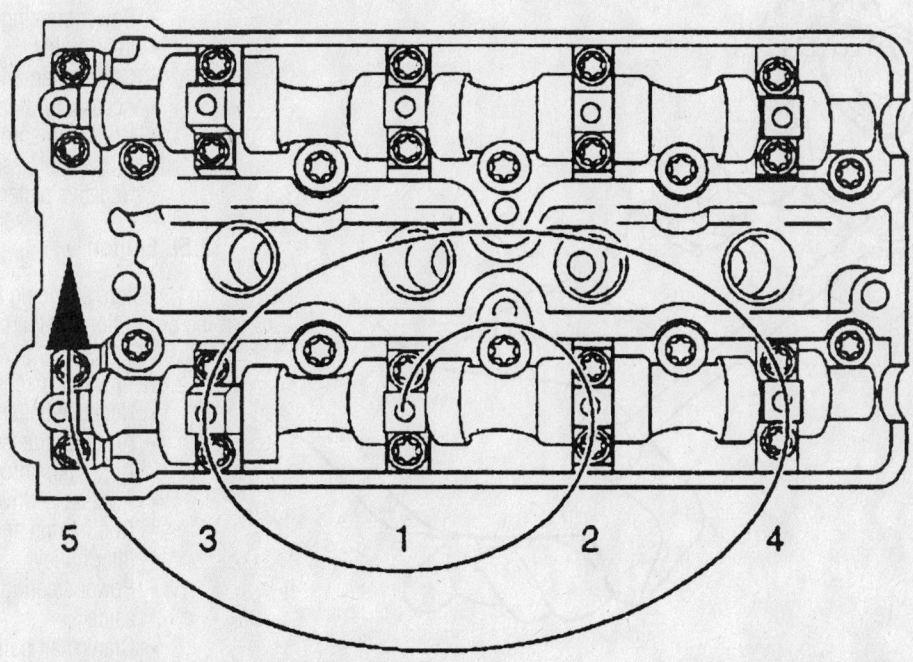

7924NG20

Camshaft bearing cap tightening sequence—2.2L engine

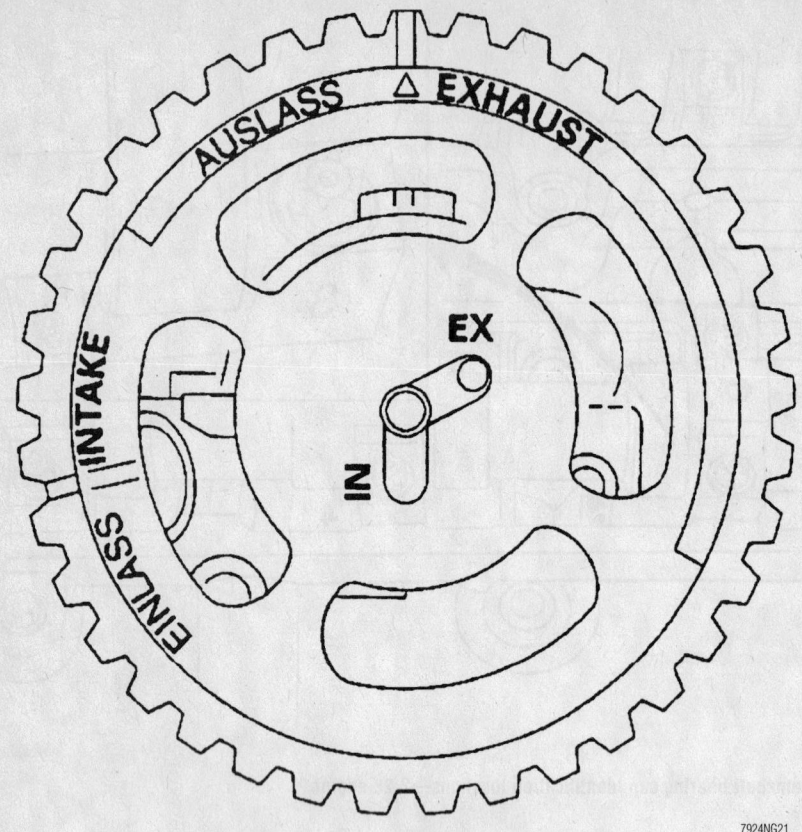

Guide pin location for the exhaust cam gear—2.2L engine

7924NG21

M5x0.8

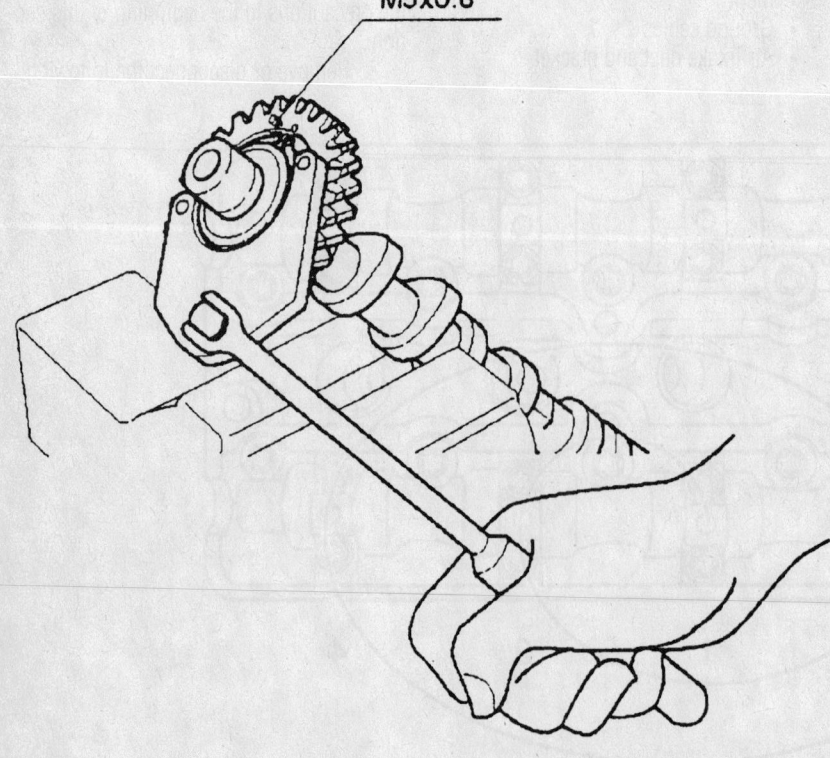

7924NG45

Aligning the sub gear with the Gear Spring Lever J-42686—3.2L DOHC engine

- Negative battery cable
- Air cleaner assembly
- Upper fan shroud
- Accessory drive belt and tensioner
- Cooling fan and pulley
- Idler pulley
- Power steering pump and move it aside
- Crankshaft pulley
- Timing belt cover
- Timing belt. Refer to the Timing Belt Unit Repair Section.
- Ignition coils
- Valve covers
- Camshafts
- Valve shims and tappets

➡ Keep the valve shims and tappets in order for installation.

To install:

3. Install the valve tappets and shims in their original locations.

4. Using Gear Spring Lever J-42686, turn the sub gear clockwise to align the 5mm bolt holes in the sub gear and the camshaft driven gear. Tighten the 5mm bolt.

5. Install or connect the following:
- Camshafts by aligning the timing marks as shown. Tighten the bolts in sequence to 89 inch lbs. (10 Nm).
- Valve covers
- Ignition coils
- Timing belt
- Timing belt cover
- Crankshaft pulley. Tighten the bolt to 123 ft. lbs. (167 Nm).
- Power steering pump
- Idler pulley
- Cooling fan and pulley
- Accessory drive belt and tensioner
- Upper fan shroud
- Air cleaner assembly
- Negative battery cable

3.5L Engine

1. Before servicing the vehicle, refer to the precautions in the beginning of this section.

2. Remove or disconnect the following:
- Negative battery cable
- Air cleaner assembly
- Upper fan shroud
- Accessory drive belt and tensioner
- Cooling fan and pulley
- Idler pulley
- Power steering pump and move it aside
- Crankshaft pulley
- Timing belt cover

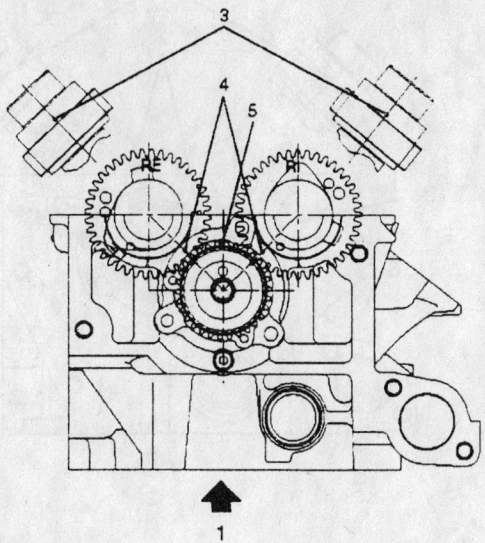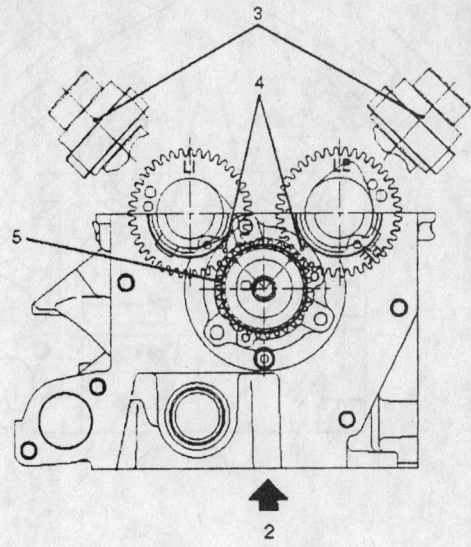

Legend
(1) Right Bank
(2) Left Bank

(3) Alignment Mark on Camshaft Drive Gear
(4) Alignment Mark on Camshaft
(5) Alignment Mark on Retainer

7924NG46

Camshaft alignment marks for the left and right cylinder heads—3.2L DOHC engine

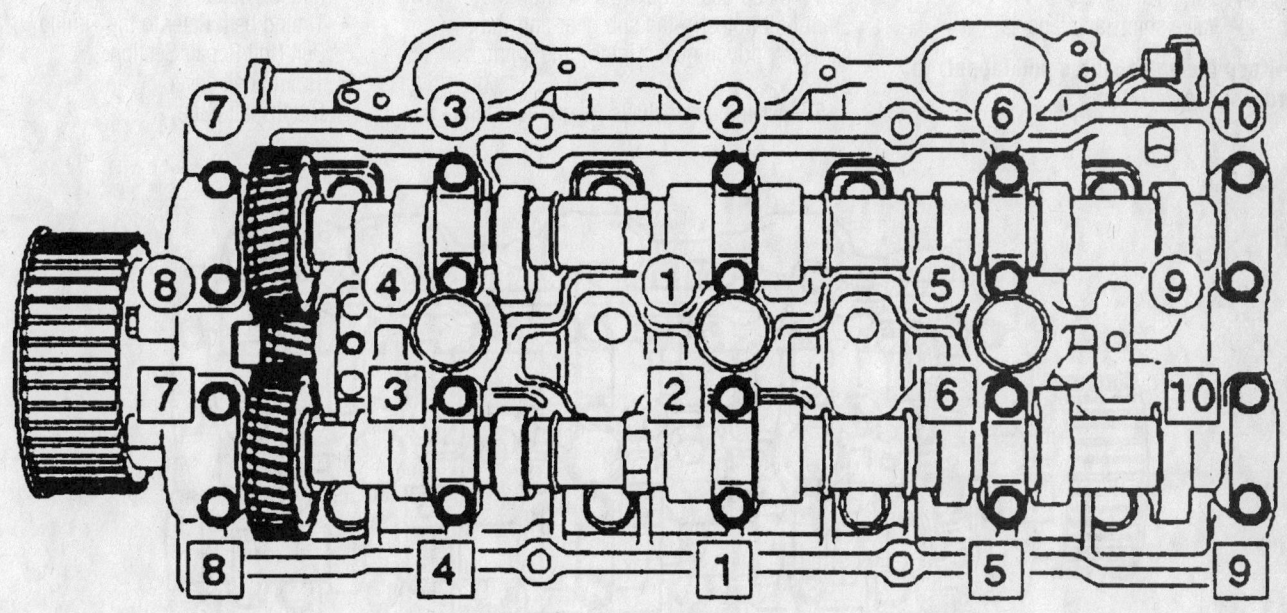

◯ : Intake   ▢ : Exhaust

7924NG47

Camshaft retaining bracket tightening sequence—3.2L DOHC engine

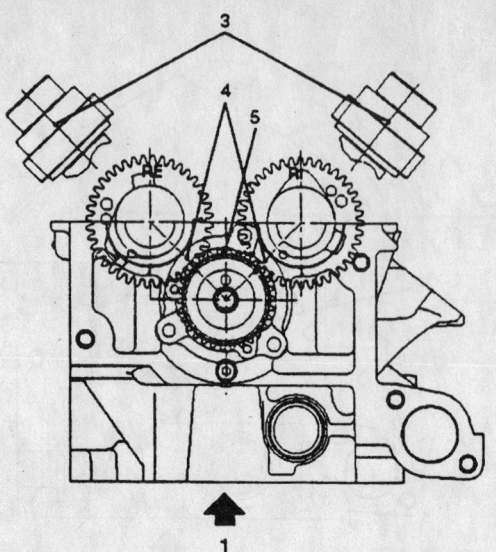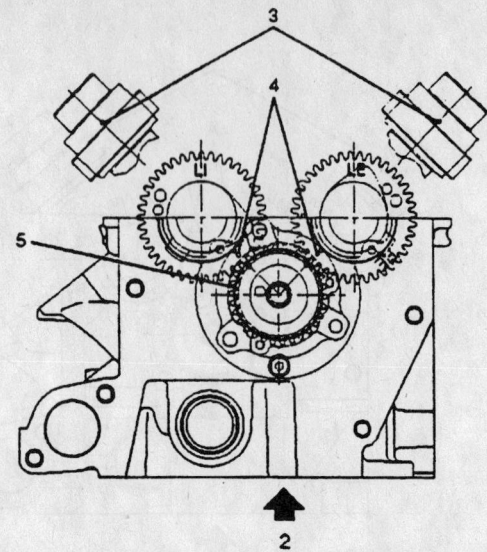

1   Right Bank
2   Left Bank

3   Alignment Mark on Camshaft Drive Gear
4   Alignment Mark on Camshaft
5   Alignment Mark on Retainer

7924BG11

Camshaft alignment marks for the left and right cylinder heads—3.5L engine

- Timing belt. Refer to the Timing Belt Unit Repair Section.
- Ignition coils
- Valve covers
- Camshafts
- Valve shims and tappets

➡ **Keep the valve shims and tappets in order for installation.**

To install:

3. Install the valve tappets and shims in their original locations.

4. Using Gear Spring Lever J-42686, turn the sub gear clockwise to align the 5mm bolt holes in the sub gear and the camshaft driven gear. Tighten the 5mm bolt.

5. Install the camshafts. Align the timing marks as shown. Tighten the bolts in sequence to 89 inch lbs. (10 Nm).

6. Install or connect the following:
- Valve covers
- Ignition coils
- Timing belt. Refer to the Timing Belt Unit Repair Section.
- Timing belt cover
- Crankshaft pulley

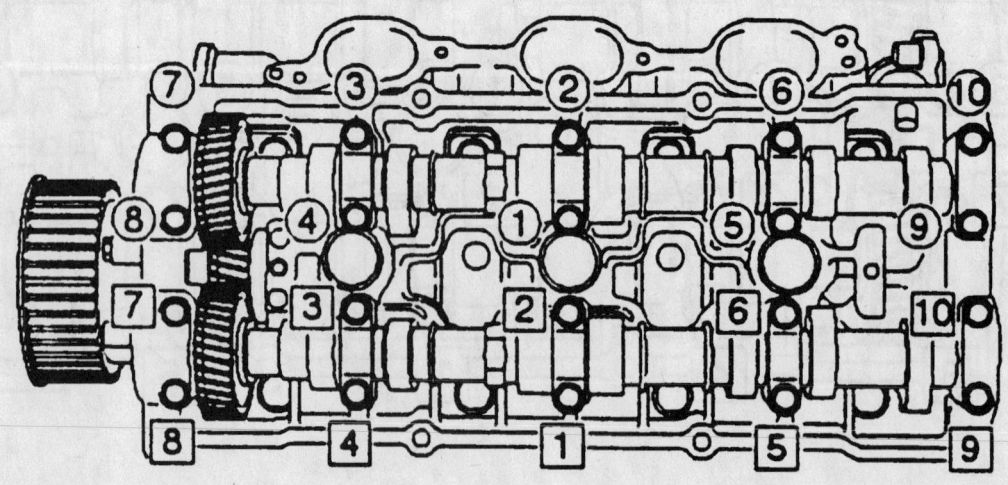

○ : Intake   □ : Exhaust

7924BG12

Camshaft retaining bracket tightening sequence—3.5L engine

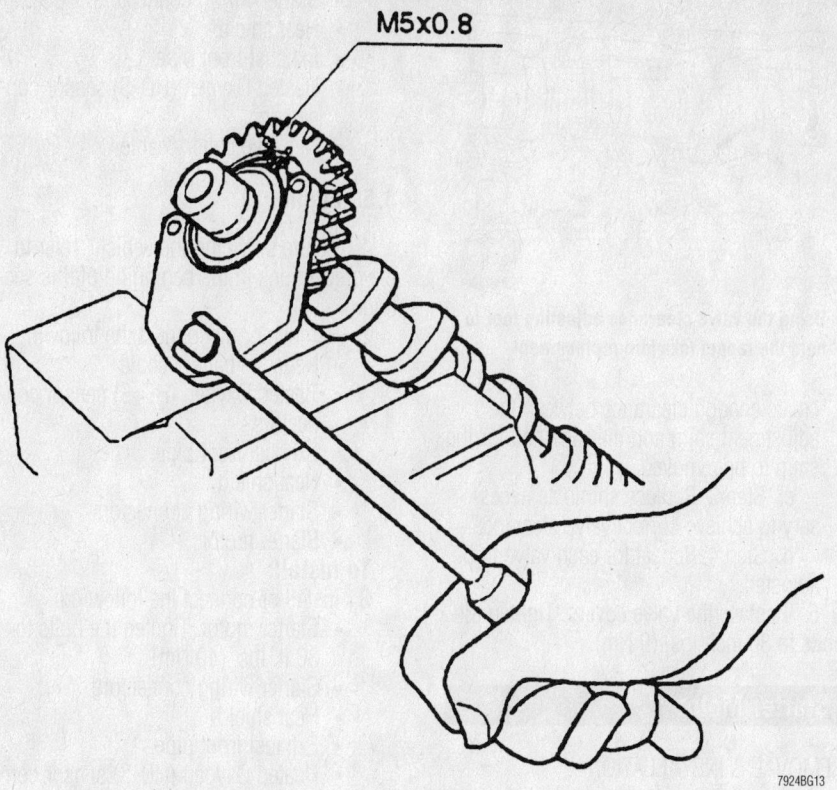

M5x0.8

7924BG13

Aligning the sub gear with the Gear Spring Lever J-42686—3.5L engine

- Power steering pump
- Idler pulley
- Cooling fan and pulley
- Accessory drive belt and tensioner
- Upper fan shroud
- Air cleaner assembly
- Negative battery cable

Valve Lash

ADJUSTMENT

2.2L Engine

The 2.2L DOHC engine is equipped with hydraulic lash adjusters. No valve adjustment is necessary.

3.2L Engines

➡**Measure valve clearance with the engine cold.**

1. Before servicing the vehicle, refer to the precautions in the beginning of this section.

2. Remove the valve covers.

3. Check the valve clearance with the camshafts positioned as shown. Intake valve clearance should be 0.0091–0.0130 in. (0.2311–0.3302mm). Exhaust valve clearance should be 0.0098–0.0138 in. (0.2489–0.3505mm).

4. If adjustment is required, replace the shims as follows:

   a. Step 1: Position special tool J-42689 on the edge of the tappet.

   b. Step 2: Rotate the crankshaft until the maximum lift portion of the camshaft lobe contacts the upper edge of the special tool and presses the tappet down to create enough clearance between the adjustment shim and the camshaft for the shim to be removed.

   c. Step 3: Replace shims as necessary to achieve correct valve clearance.

   d. Step 4: Repeat for each valve to be adjusted.

5. Replace the valve covers. Tighten the bolts to 80 inch lbs. (9 Nm).

3.5L Engine

➡**Measure valve clearance with the engine cold.**

1. Before servicing the vehicle, refer to the precautions in the beginning of this section.

2. Remove the valve covers.

3. Check the valve clearance with the camshafts positioned as shown. Intake valve clearance should be 0.0091–0.0130 inches. Exhaust valve clearance should be 0.0098–0.0138 inches.

4. If adjustment is required, replace the shims as follows:

   a. Step 1: Position special tool J–42689 on the edge of the tappet.

   b. Step 2: Rotate the crankshaft until the maximum lift portion of the camshaft lobe contacts the upper edge of the special tool and presses the tappet down to

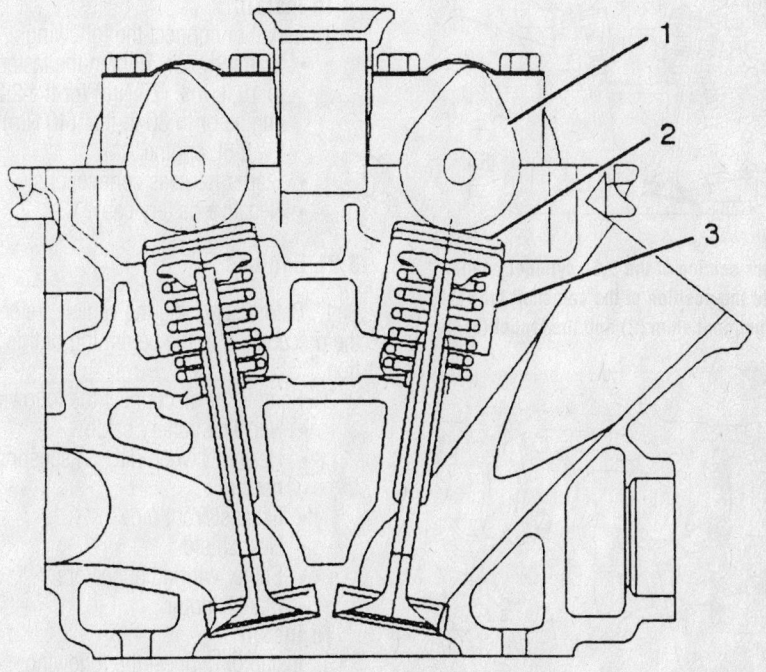

9302NG01

Cross section of the 3.2L DOHC cylinder head. Note the position of the camshaft lobe (1), adjustment shim (2) and the tappet (3)

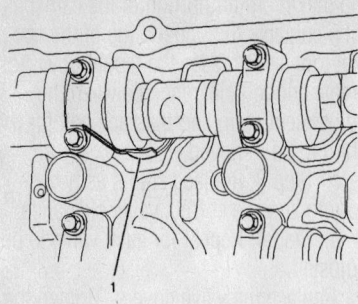

Insert special tool J-42689 (1) and use the camshaft to press the tappet down—3.2L DOHC engine

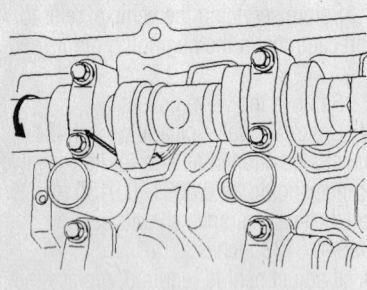

Rotate the camshaft to depress the valve with special tool J-42689—3.2L DOHC engine

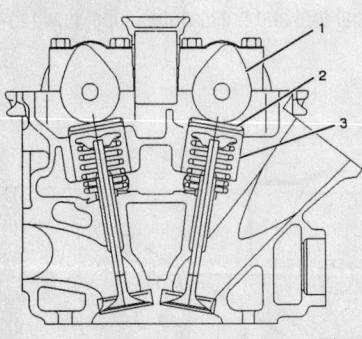

Cross section of the 3.5L cylinder head. Note the position of the camshaft lobe (1), adjustment shim (2) and the tappet (3)

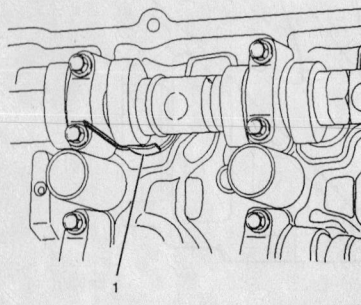

Valve clearance adjusting tool J–42689 (1)

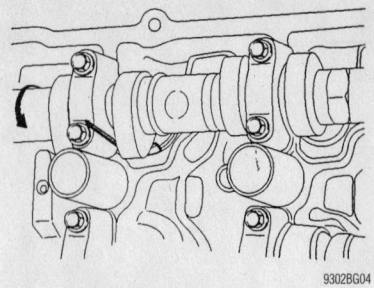

Using the valve clearance adjusting tool to hold the tappet for shim replacement

create enough clearance between the adjustment shim and the camshaft for the shim to be removed.

c. Step 3: Replace shims as necessary to achieve correct valve clearance.

d. Step 4: Repeat for each valve to be adjusted.

5. Replace the valve covers. Tighten the bolts to 80 inch lbs. (9 Nm).

Starter Motor

REMOVAL & INSTALLATION

2.2L Engines

1. Before servicing the vehicle, refer to the precautions in the beginning of this section.

2. Remove or disconnect the following:
   - Negative battery cable
   - Starter harness connections
   - Starter motor

To install:

3. Install or connect the following:
   - Starter motor. Tighten the fasteners to 18 ft. lbs. (25 Nm) for the 2.2L engine or to 30 ft. lbs. (40 Nm) for the 2.6L engine.
   - Starter harness connections
   - Negative battery cable

3.2L Engines

1. Before servicing the vehicle, refer to the precautions in the beginning of this section.

2. Remove or disconnect the following:
   - Negative battery cable
   - Heated Oxygen (HO2S) sensor connectors
   - Exhaust front pipe
   - Heat shield
   - Starter wiring connectors
   - Starter motor

To install:

3. Install or connect the following:
   - Starter motor. Tighten the bolts to 30 ft. lbs. (40 Nm).
   - Starter wiring connectors
   - Heat shield
   - Exhaust front pipe
   - Heated Oxygen (HO2S) sensor connectors
   - Negative battery cable

3.5L Engine

1. Before servicing the vehicle, refer to the precautions in the beginning of this section.

2. Remove or disconnect the following:
   - Negative battery cable
   - Heated Oxygen (HO2S) sensor connectors
   - Exhaust front pipe
   - Heat shield
   - Starter wiring connectors
   - Starter motor

To install:

3. Install or connect the following:
   - Starter motor. Tighten the bolts to 30 ft. lbs. (40 Nm).
   - Starter wiring connectors
   - Heat shield
   - Exhaust front pipe
   - Heated Oxygen (HO2S) sensor connectors
   - Negative battery cable

Oil Pan

REMOVAL & INSTALLATION

2.2L Engines

1. Before servicing the vehicle, refer to the precautions in the beginning of this section.

2. Drain the engine oil.

3. Remove or disconnect the following:
   - Flywheel dust cover
   - Left and right engine mounts. Raise the engine for access.
   - Oil pan
   - Oil pan support, for 2.2L engine

To install:

4. Perform the following:

   a. Install the oil pan support and tighten the bolts to 14 ft. lbs. (20 Nm).

   b. Install the oil pan and tighten the bolts to 70 inch lbs. (8 Nm) plus 30 degrees.

5. Install or connect the following:
   - Left and right engine mounts. Tighten the nuts to 41 ft. lbs. (55 Nm).
   - Flywheel dust cover

6. Fill the crankcase to the correct level.

7. Start the engine and check for leaks.

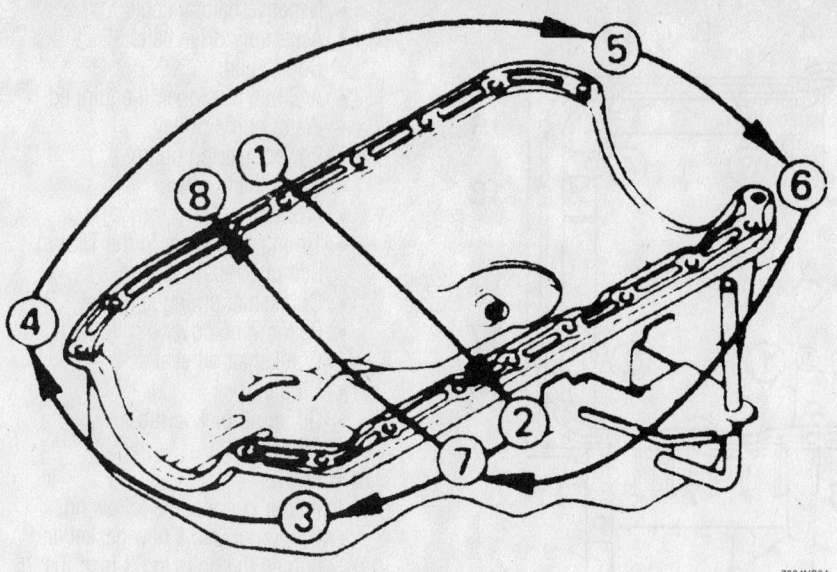

Oil pan bolt tightening sequence—2.2L engines

3.2L Engines

1. Before servicing the vehicle, refer to the precautions in the beginning of this section.
2. Drain the engine oil.
3. Remove or disconnect the following:
   • Negative battery cable
   • Front wheels
   • Oil level dipstick
   • Stone guard
   • Radiator under fan shroud
   • Suspension crossmember
   • Flywheel dust cover
   • Pitman arm
   • Idler arm

4. If equipped with 4 wheel drive, unbolt and lower the front axle housing assembly for clearance.
5. Remove or disconnect the following:
   • Oil pan
   • Lower crankcase

To install:

6. Apply a bead of silicone sealant to the crankcase flange and install the crankcase. Tighten the fasteners in sequence to 89 inch lbs. (10 Nm).
7. Apply a bead of silicone sealant to the oil pan flange and install the oil pan.

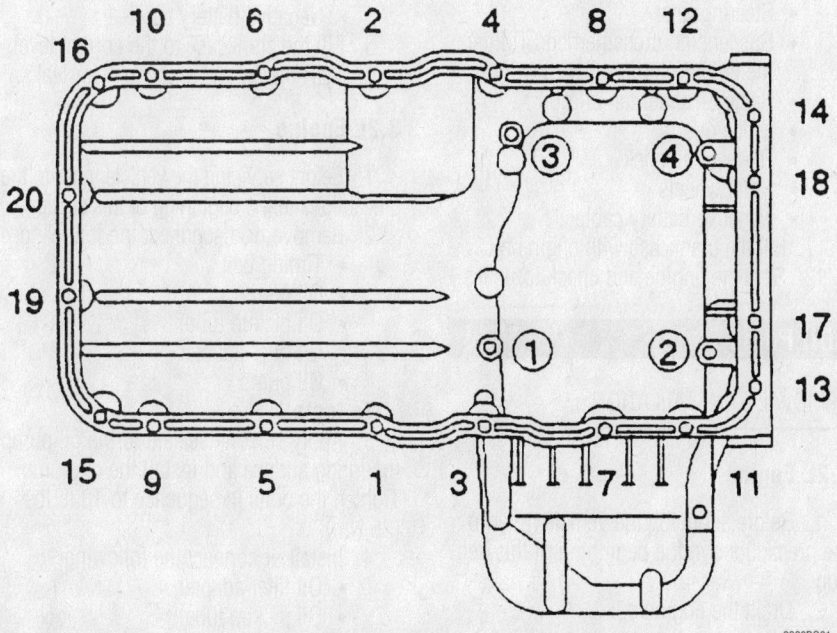

Lower crankcase torque sequence—3.2L DOHC engine

Tighten the fasteners to 89 inch lbs. (10 Nm).

8. If equipped, raise the axle housing assembly into position. Tighten the axle case bolts to 61 ft. lbs. (82 Nm) and the mounting bolts to 112 ft. lbs. (152 Nm).
9. Install or connect the following:
   • Pitman arm. Tighten the nut to 159 ft. lbs. (216 Nm).
   • Idler arm. Tighten the bolt to 33 ft. lbs. (44 Nm).
   • Flywheel dust cover
   • Suspension crossmember. Tighten the bolts to 58 ft. lbs. (78 Nm).
   • Radiator under fan shroud
   • Stone guard
   • Oil level dipstick
   • Front wheels
   • Negative battery cable
10. Fill the crankcase with engine oil.
11. Start the engine and check for leaks.

3.5L Engine

EXCEPT AXIOM

1. Before servicing the vehicle, refer to the precautions in the beginning of this section.
2. Drain the engine oil.
3. Remove or disconnect the following:

   • Negative battery cable
   • Front wheels
   • Oil level dipstick
   • Stone guard
   • Radiator under fan shroud
   • Suspension crossmember
   • Flywheel dust cover
   • Pitman arm
   • Idler arm

4. If equipped with 4 wheel drive, unbolt and lower the front axle housing assembly for clearance.
5. Remove or disconnect the following:
   • Oil pan
   • Lower crankcase

To install:

6. Apply a bead of silicone sealant to the crankcase flange and install the crankcase. Tighten the fasteners in sequence to 89 inch lbs. (10 Nm).
7. Apply a bead of silicone sealant to the oil pan flange and install the oil pan. Tighten the fasteners to 89 inch lbs. (10 Nm).
8. If equipped, raise the axle housing assembly into position. Tighten the axle case bolts to 61 ft. lbs. (82 Nm) and the mounting bolts to 112 ft. lbs. (152 Nm).
9. Install or connect the following:
   • Pitman arm. Tighten the nut to 159 ft. lbs. (216 Nm).

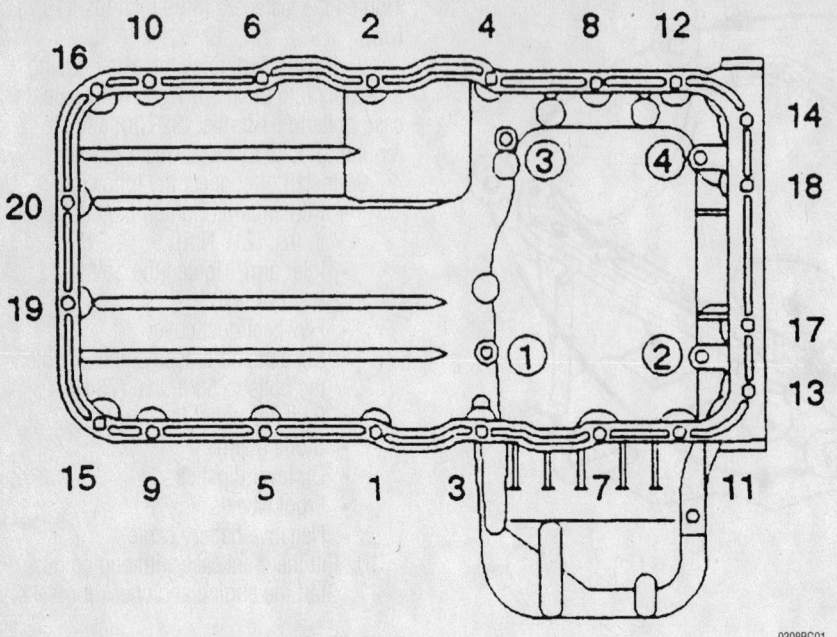

Lower crankcase torque sequence—3.5L engine

9308BG01

- Idler arm. Tighten the bolt to 33 ft. lbs. (44 Nm).
- Flywheel dust cover
- Suspension crossmember. Tighten the bolts to 58 ft. lbs. (78 Nm).
- Radiator under fan shroud
- Stone guard
- Oil level dipstick
- Front wheels
- Negative battery cable
10. Fill the crankcase with engine oil.
11. Start the engine and check for leaks.

AXIOM

1. Before servicing the vehicle, refer to the precautions in the beginning of this section.
2. Drain the engine oil.
3. Remove or disconnect the following:
   - Negative battery cable
   - Front wheels
   - Oil level dipstick
   - Stone guard
   - Radiator under fan shroud
   - Shift-on-the-fly from the axle
   - Suspension crossmember
4. If equipped with 4 wheel drive, unbolt and lower the front axle housing assembly for clearance.
5. Remove or disconnect the following:
   - Steering gear
   - Starter
   - Oil pan
   - Lower crankcase

To install:
6. Apply a bead of silicone sealant to the crankcase flange and install the

crankcase. Tighten the fasteners in sequence to 89 inch lbs. (10 Nm).

7. Apply a bead of silicone sealant to the oil pan flange and install the oil pan. Tighten the fasteners to 89 inch lbs. (10 Nm).

8. Install or connect the following:
   - Starter. Torque to 30 ft. lbs. (40 Nm)

9. If equipped, raise the axle housing assembly into position. Tighten the axle case bolts to 61 ft. lbs. (82 Nm) and the mounting bolts to 112 ft. lbs. (152 Nm).

10. Install or connect the following:
   - Steering gear
   - Suspension crossmember. Tighten the bolts to 58 ft. lbs. (78 Nm).
   - Radiator under fan shroud
   - Stone guard
   - Oil level dipstick
   - Front wheels
   - Negative battery cable
11. Fill the crankcase with engine oil.
12. Start the engine and check for leaks.

Oil Pump

REMOVAL & INSTALLATION

2.2L Engine

1. Before servicing the vehicle, refer to the precautions in the beginning of this section.
2. Drain the engine oil.
3. Remove or disconnect the following:

- Negative battery cable
- Accessory drive belts
- Cooling fan
- A/C belt tensioner, if equipped
- Water pump pulley
- Power steering pump
- Crankshaft pulley
- Front cover
- Timing belt. Refer to the Timing Belt unit repair section.
- Crankshaft timing sprocket
- Rear timing cover
- Crankshaft oil seal
- Oil pan
- Oil pump pickup tube
- Oil pump

To install:
4. Install or connect the following:
   - Oil pump. Use a new gasket and tighten the bolts to 53 inch lbs. (6 Nm).
   - Oil pump pickup tube. Tighten the bolts to 70 inch lbs. (8 Nm).
   - Oil pan
   - Crankshaft oil seal
   - Rear timing cover
   - Crankshaft timing sprocket. Tighten the bolt to 94 ft. lbs. (130 Nm) plus 45 degrees.
   - Timing belt. Refer to the Timing Belt unit repair section.
   - Front cover
   - Crankshaft pulley. Tighten the bolts to 14 ft. lbs. (20 Nm).
   - Power steering pump
   - Water pump pulley
   - A/C belt tensioner, if equipped
   - Cooling fan
   - Accessory drive belts
   - Negative battery cable
5. Fill the crankcase to the correct level.
6. Start the engine and check for leaks.

3.2L Engine

1. Before servicing the vehicle, refer to the precautions in the beginning of this section.
2. Remove or disconnect the following:
   - Timing belt
   - Oil pan
   - Oil pickup tube
   - Oil filter adapter
   - Oil pump

To install:
3. Apply silicone sealant to the oil pump mounting surface and install the oil pump. Tighten the bolts in sequence to 18 ft. lbs. (25 Nm).

4. Install or connect the following:
   - Oil filter adapter
   - Oil pickup tube
   - Oil pan

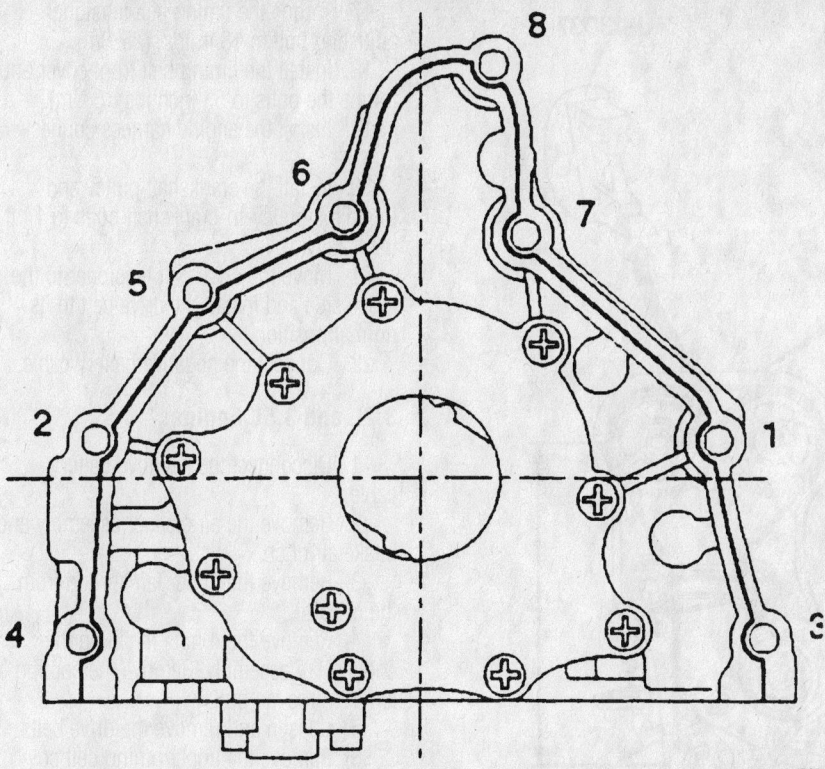

Oil pump torque sequence—3.2L and 3.5L engines

- Timing belt
5. Fill the crankcase to the correct level.
6. Start the engine and check for leaks.

3.5L Engine

1. Before servicing the vehicle, refer to the precautions in the beginning of this section.
2. Remove or disconnect the following:
   - Timing belt
   - Oil pan
   - Oil pick-up tube
   - Oil filter adapter
   - Oil pump

To install:

3. Apply silicone sealant to the oil pump mounting surface and install the oil pump. Tighten the bolts in sequence to 18 ft. lbs. (25 Nm).
4. Install or connect the following:
   - Oil filter adapter
   - Oil pickup tube
   - Oil pan
   - Timing belt

Rear Main Seal

REMOVAL & INSTALLATION

1. Before servicing the vehicle, refer to the precautions in the beginning of this section.

2. Remove or disconnect the following:
   - Negative battery cable
   - Transmission
   - Clutch assembly, if equipped with a manual transmission
   - Flywheel by loosening the flywheel bolts in a 2-step crisscross sequence
   - Rear main seal, using a seal puller

➡**Do not damage the crankshaft sealing surface.**

To install:
3. Install or connect the following:
   - New rear main seal, by lubricating it with engine oil
   - Flywheel, using new flywheel bolts. Tighten the bolts, in a 2-step criss-

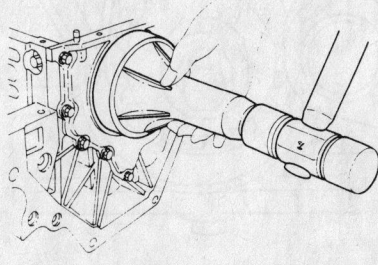

Installing a one-piece rear crankshaft oil seal

cross pattern, to 40 ft. lbs. (54 Nm).
   - Clutch assembly, if removed
   - Transmission
   - Negative battery cable
4. Check the oil and refill as necessary.

Timing Belt and Cover

REMOVAL & INSTALLATION

2.2L Engine

1. Disconnect the negative battery cable.
2. Using a box-end wrench on the drive belt adjuster, turn the adjuster clockwise and remove the drive belt.
3. From the left rear of the engine compartment, disconnect the 3 electrical connectors from the chassis harness.
4. Remove the crankshaft pulley-to-crankshaft bolts and remove the pulley.
5. From the front of the engine, remove the nut and the engine harness cover.
6. Remove the timing belt cover.
7. Rotate the crankshaft to position the timing marks at Top Dead Center (TDC) of the No. 1 cylinder's compression stroke.

➡**Mark the rotational direction of the timing belt for reinstallation purposes.**

8. Remove the timing belt tensioner adjusting bolt and the tensioner from the engine.
9. Remove the timing belt.
To install:
10. Install the timing belt tensioner and finger-tighten the tensioner bolt.
11. Inspect the timing marks to be sure that the engine is positioned at TDC of the No. 1 cylinder's compression stroke.
12. Using tool J-43037, or equivalent, place it between the intake and exhaust sprockets to prevent the camshaft gear from moving during the timing belt installation.
13. Install the timing belt.
14. Position the timing belt to ensure that the tension side of the belt is taut and move the timing belt tension adjusting lever clockwise until the tensioner pointer is flowing.
15. If installing a used timing belt (used over 60 min. from new), the pointer should be positioned approximately 0.16 in. (4mm) to the left of the "V" notch when viewed from the front of the engine.
16. If installing a new timing belt, the pointer should be positioned at the center of the "V" notch when viewed from the front of the engine.

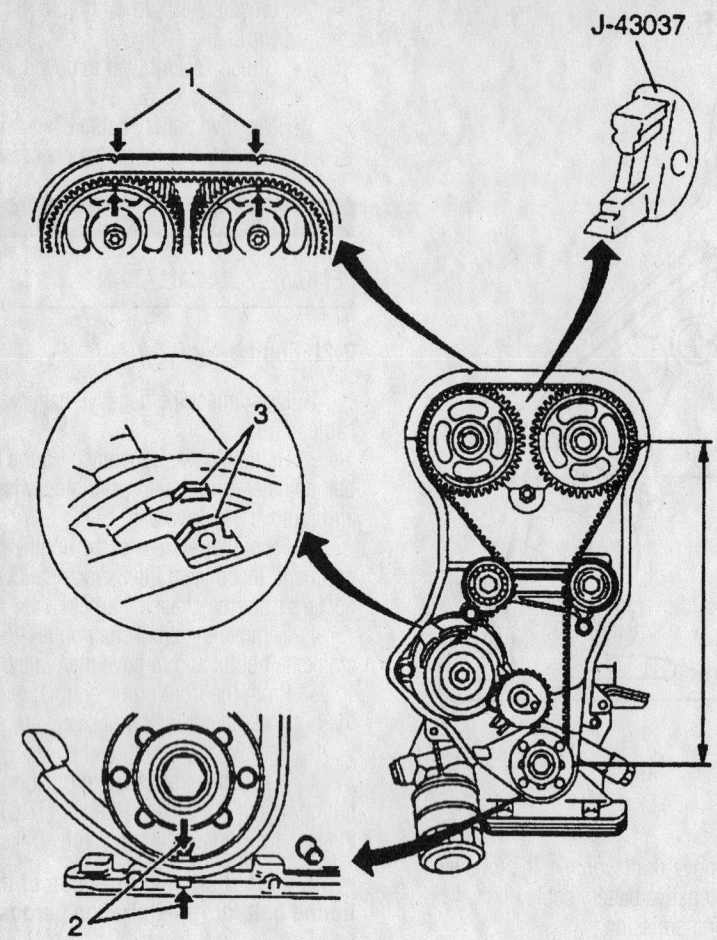

Aligning the timing marks and installing the timing belt —2.2L engine

93025G12

17. Torque the timing belt tensioner adjusting bolt to 18 ft. lbs. (25 Nm).

18. Install the timing belt front cover and torque the bolts to 53 inch lbs. (6 Nm).

19. Install the engine harness connectors.

20. Install the crankshaft pulley and toque the pulley-to-crankshaft bolts to 14 ft. lbs. (20 Nm).

21. Move the drive belt tensioner to the loose side and install the drive belt to its normal position.

22. Connect the negative battery cable.

3.2L and 3.5L Engines

1. Disconnect the negative battery cable.

2. Remove the air cleaner assembly and intake air duct.

3. Remove the upper fan shroud from the radiator.

4. Remove the 4 nuts retaining the cooling fan assembly. Remove the cooling fan from the fan pulley.

5. Loosen and remove the drive belts.

6. Remove the upper timing belt covers.

7. Remove the fan pulley assembly.

8. Rotate the crankshaft to align the camshaft timing marks with the pointer dots on the back covers. Verify that the pointer on the crankshaft aligns with the mark on the lower timing cover.

➡ When the timing marks are aligned, the No. 2 piston is at Top Dead Center (TDC) compression.

❋❋ WARNING

Align the camshaft and crankshaft sprockets with their alignment marks before removing the timing belt. Failure to align the belt and sprocket marks may result in valve damage.

9. Use tool No. J-8614-01, or a suitable pulley holding tool to remove the crankshaft pulley center bolt. Remove the crankshaft pulley.

10. If present, disconnect the 2 oil cooler hose bracket bolts on the timing cover. Move the oil cooler hoses and bracket off of the lower timing cover.

11. Remove the lower timing belt cover.

12. Remove the pusher assembly (tensioner) from below the belt tensioner pulley. The pusher rod must always face upward to prevent oil leakage. Depress the pusher rod, and insert a wire pin into the hole to keep the pusher rod retracted.

13. Remove the timing belt.

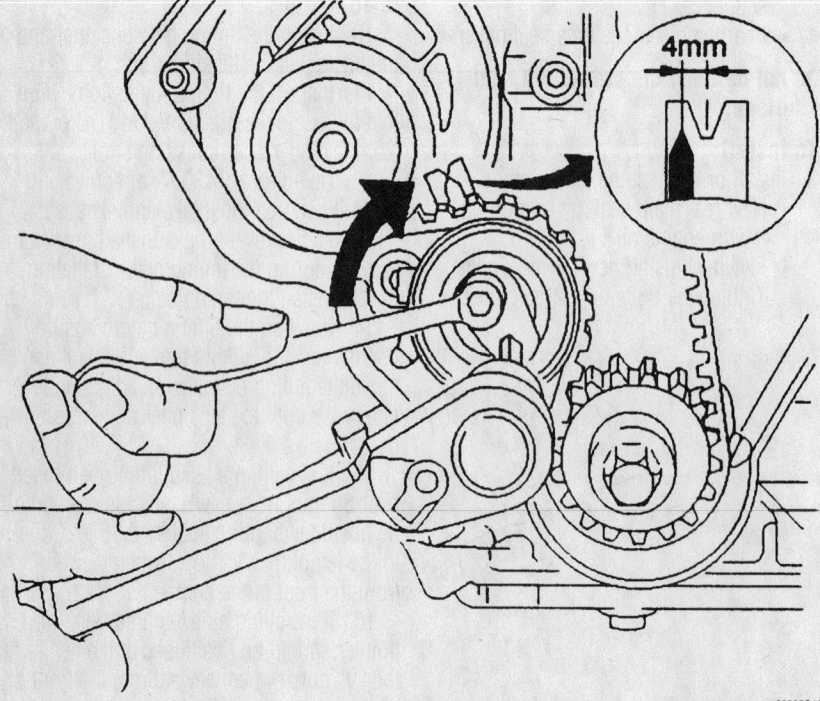

Tensioning the timing belt for a used timing belt —2.2L engine

93025G13

14. Inspect the water pump and replace it if there is any doubt about its condition.

15. Repair any oil or coolant leaks before installing a new timing belt. If the timing belt has been contaminated with oil or coolant, or is damaged, it must be replaced.

To install:

16. Verify that the sprocket timing marks are still aligned and that the groove and the keyway on the crankshaft timing sprocket align with the mark on the oil pump. The white pointers on the camshaft timing sprockets should align with the dots on the front plate.

17. Install the timing belt. Use clips to secure the belt onto each sprocket until the installation is complete. Align the dotted marks on the timing belt with the timing mark opposite the groove on the crankshaft sprocket.

➡The arrows on the timing belt must follow the belt's direction of rotation. The manufacturer's trademark on the belt's spine should be readable left-to-right when the belt is installed.

18. Align the white line on the timing belt with the alignment mark on the right bank camshaft timing pulley. Secure the belt with a clip.

⁂ WARNING

If any binding is felt when adjusting the timing belt tension by turning the crankshaft, STOP turning the engine, because the pistons may be hitting the valves.

19. Rotate the crankshaft counterclockwise to remove the slack between the crankshaft sprocket and the right camshaft timing belt sprocket.

20. Install the belt around the water pump pulley.

21. Install the belt on the idler pulley.

22. Align the white alignment mark on the timing belt with the alignment mark on the left bank camshaft timing belt sprocket.

23. Install the crankshaft pulley and tighten the center bolt by hand. Rotate the crankshaft pulley clockwise to give slack between the crankshaft timing belt pulley and the right bank camshaft timing belt pulley.

24. Insert a 1.4mm piece of wire through the hole in the pusher to hold the rod in. Install the pusher assembly while pushing the tension pulley toward the belt.

25. Pull the pin out from the pusher to release the rod.

26. Remove the clamps from the sprockets. Rotate the crankshaft pulley clockwise 2

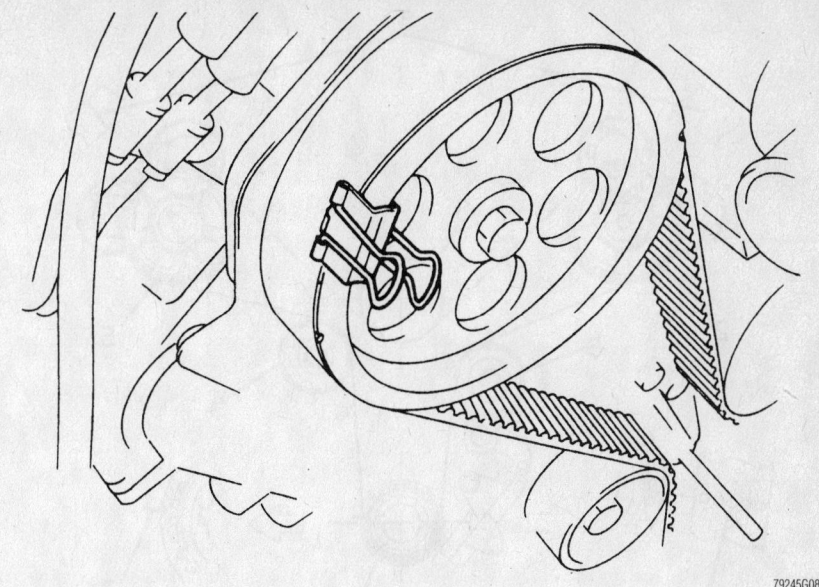

79245G08

Using a double clip to hold the belt in place—3.2L and 3.5L engines

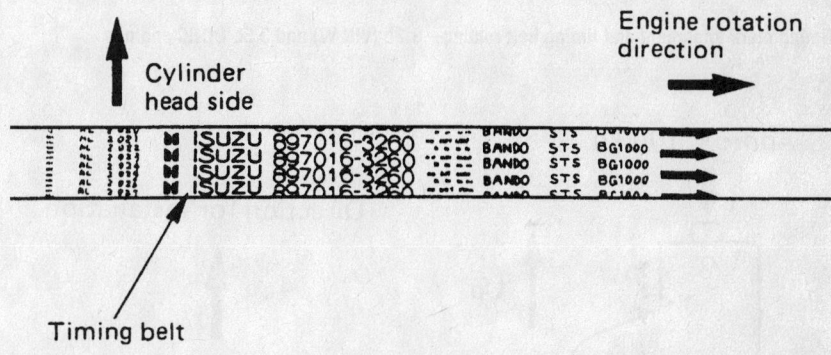

Engine rotation direction

Cylinder head side

Timing belt

79245G07

For maximum timing belt life, install the belt as shown—3.2L and 3.5L engines

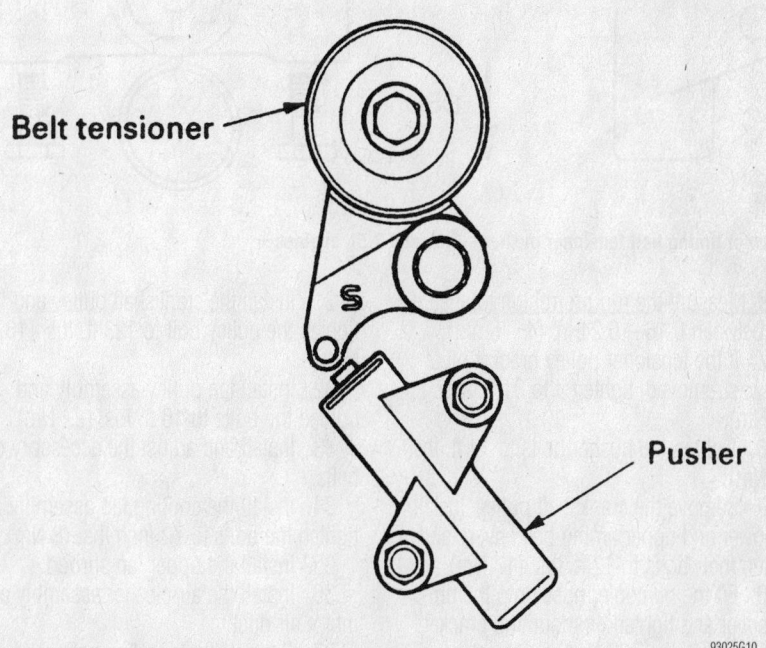

Belt tensioner

Pusher

93025G10

View of timing belt tensioner and pusher—3.2L and 3.5L engines

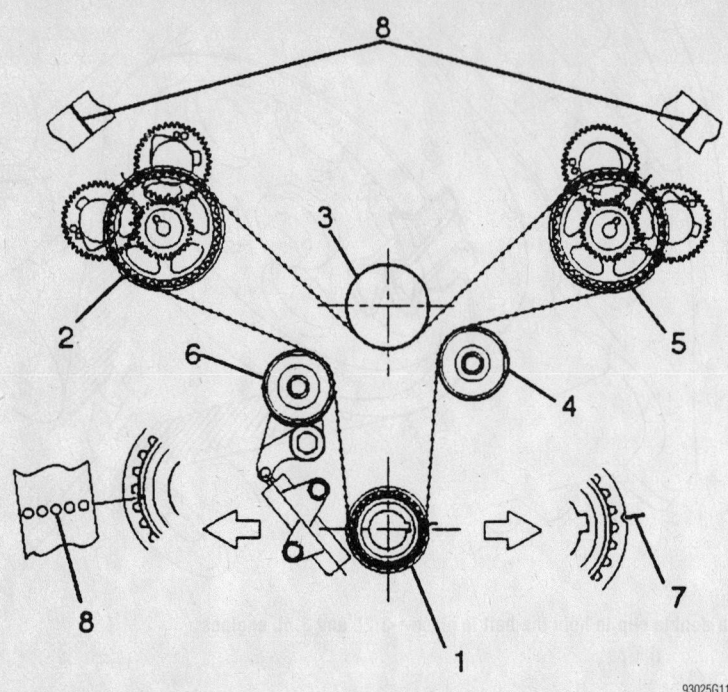

Timing mark alignment and timing belt routing—3.2L (VIN W) and 3.5L DOHC engines

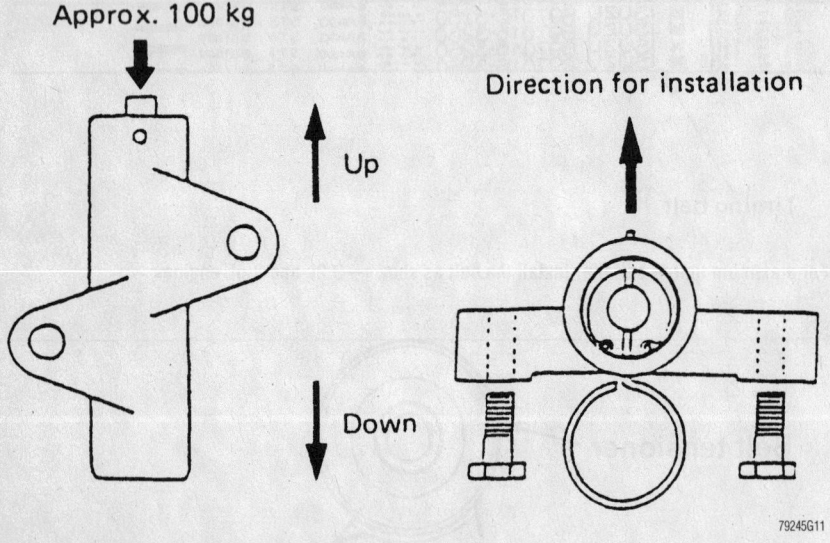

View of timing belt tensioner pusher—3.2L and 3.5L engines

Piston and Ring

POSITIONING

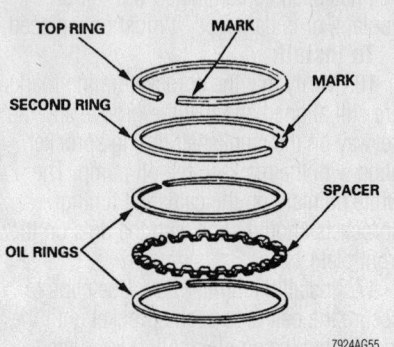

Piston ring positioning and top mark locations—all engines

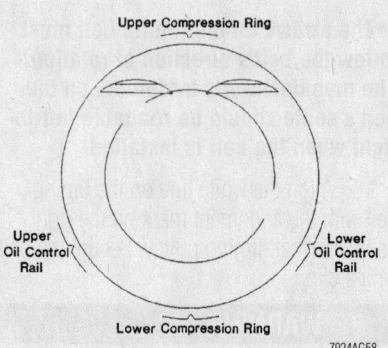

Piston ring end-gap spacing—2.2L engine

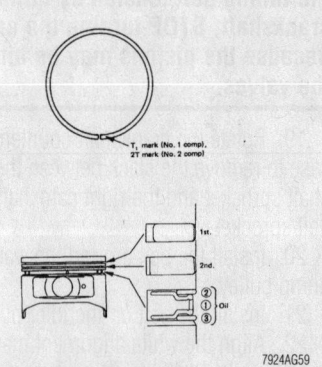

Piston ring positioning—3.2L and 3.5L engines

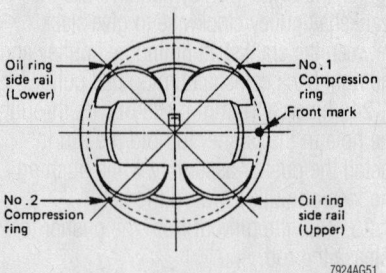

Piston ring end-gap spacing—3.2L and 3.5L engines

turns. Measure the rod protrusion to ensure it is between 0.16 – 0.24 in. (4 – 6mm).

27. If the tensioner pulley bracket pivot bolt was removed, tighten it to 31 ft. lbs. (42 Nm).

28. Tighten the pusher bolts to 14 ft. lbs. (19 Nm).

29. Remove the crankshaft pulley. Install the lower and upper timing belt covers and tighten their bolts to 12 ft. lbs. (17 Nm).

30. Fit the oil cooler hose onto the timing cover and tighten its mounting bracket bolts to 16 ft. lbs. (22 Nm).

31. Install the crankshaft pulley and tighten the pulley bolt to 123 ft. lbs. (167 Nm).

32. Install fan pulley assembly and tighten the bolts to 16 ft. lbs. (22 Nm).

33. Install and adjust the accessory drive belts.

34. Install the cooling fan assembly and tighten the bolts to 72 inch lbs. (8 Nm).

35. Install the upper fan shroud.

36. Install the air cleaner assembly and intake air duct.

37. Connect the negative battery cable.

FUEL SYSTEM

Fuel System Service Precautions

Safety is the most important factor when performing not only fuel system maintenance but any type of maintenance. Failure to conduct maintenance and repairs in a safe manner may result in serious personal injury or death. Maintenance and testing of the vehicle's fuel system components can be accomplished safely and effectively by adhering to the following rules and guidelines:

• To avoid the possibility of fire and personal injury, always disconnect the negative battery cable unless the repair or test procedure requires that battery voltage be applied.

• Always relieve the fuel system pressure prior to disconnecting any fuel system component (injector, fuel rail, pressure regulator, etc.), fitting or fuel line connection. Exercise extreme caution whenever relieving fuel system pressure, to avoid exposing skin, face and eyes to fuel spray. Please be advised that fuel under pressure may penetrate the skin or any part of the body that it contacts.

• Always place a shop towel or cloth around the fitting or connection prior to loosening to absorb any excess fuel due to spillage. Ensure that all fuel spillage (should it occur) is quickly removed from engine surfaces. Ensure that all fuel soaked cloths or towels are deposited into a suitable waste container.

• Always keep a dry chemical (Class B) fire extinguisher near the work area.

• Do not allow fuel spray or fuel vapors to come into contact with a spark or open flame.

• Always use a backup wrench when loosening and tightening fuel line connection fittings. This will prevent unnecessary stress and torsion to fuel line piping. Always follow the proper tightening specifications.

• Always replace worn fuel fitting O-rings with new. Do not substitute fuel hose or equivalent, where fuel pipe is installed.

Fuel System Pressure

RELIEVING

1. Before servicing the vehicle, refer to the precautions in the beginning of this section.
2. Remove the fuel filler cap.

3. Remove the fuel pump relay from the underhood relay box.
4. Start the engine and let it run until it stalls, then crank the engine for an additional 30 seconds.
5. Turn the ignition switch to the **OFF** position and remove the key. Disconnect the negative battery cable.
6. When service is completed, install the fuel pump relay and connect the negative battery cable.

Fuel Filter

REMOVAL & INSTALLATION

1. Before servicing the vehicle, refer to the precautions in the beginning of this section.
2. Relieve the fuel system pressure.
3. Remove or disconnect the following:
   • Fuel lines from the fuel filter
   • Fuel filter

To install:
4. Install or connect the following:
   • Fuel filter and tighten the bracket bolt. Note the fuel flow directional arrow.
   • Fuel lines to the fuel filter
   • Negative battery cable
5. Start the engine and inspect the fuel filter connections for leaks.

Fuel Pump

REMOVAL & INSTALLATION

1. Before servicing the vehicle, refer to the precautions in the beginning of this section.
2. Relieve fuel system pressure.
3. Drain the fuel tank.
4. Remove or disconnect the following:

   • Negative battery cable
   • Fuel filler and vent hoses
   • Fuel tank skid plate

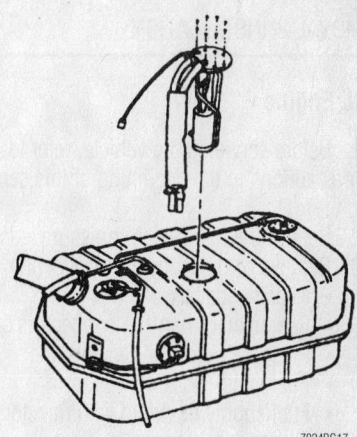

7924BG17

Fuel pump assembly mounting

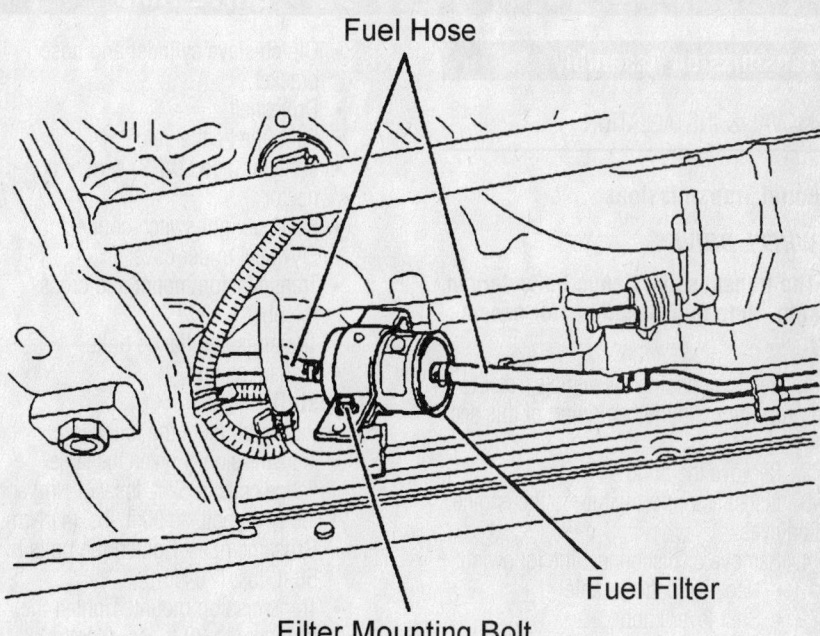

Fuel Hose

Fuel Filter

Filter Mounting Bolt

Fuel filter mounting location under the vehicle

7924NG25

- Fuel tank wiring connectors
- Fuel supply and return lines
- Fuel tank
- Fuel pump assembly

To install:

5. Install or connect the following:
- Fuel pump assembly
- Fuel tank. Tighten the bolts to 27 ft. lbs. (36 Nm).
- Fuel supply and return lines
- Fuel tank wiring connectors
- Fuel tank skid plate
- Fuel filler and vent hoses
- Negative battery cable

6. Start the engine and check for leaks.

Fuel Injector

REMOVAL & INSTALLATION

2.2L Engine

1. Before servicing the vehicle, refer to the precautions in the beginning of this section.
2. Relieve the fuel system pressure.
3. Remove or disconnect the following:
- Negative battery cable
- Fuel injector harness connectors
- Pressure regulator vacuum line
- Fuel lines
- Fuel supply manifold with injectors attached
- Fuel injector retaining clips

- Fuel injectors

To install:

4. Install or connect the following:
- Fuel injectors. Use new O-ring seals.
- Fuel injector retaining clips
- Fuel supply manifold with injectors attached. Tighten the fasteners to 14 ft. lbs. (19 Nm).
- Fuel lines
- Pressure regulator vacuum line
- Fuel injector harness connectors
- Negative battery cable

5. Start the engine and check for leaks.

3.2L Engines

1. Before servicing the vehicle, refer to the precautions in the beginning of this section.
2. Relieve fuel system pressure.
3. Remove or disconnect the following:
- Negative battery cable
- Engine cover
- Fuel injector wiring connectors
- Fuel lines
- Fuel supply manifold with injectors attached
- Fuel injector retaining clips
- Fuel injectors

To install:

4. Install or connect the following:
- Fuel injectors. Use new O-ring seals.
- Fuel injector retaining clips

- Fuel supply manifold with injectors attached. Tighten the bolts to 60 inch lbs. (6.5 Nm).
- Fuel lines
- Fuel injector wiring connectors
- Engine cover
- Negative battery cable

5. Start the engine and check for leaks.

3.5L Engine

1. Before servicing the vehicle, refer to the precautions in the beginning of this section.
2. Relieve fuel system pressure.
3. Remove or disconnect the following:
- Negative battery cable
- Engine cover
- Fuel injector wiring connectors
- Fuel lines
- Fuel supply manifold with injectors attached
- Clips
- Injectors from the supply manifold

To install:

4. Install or connect the following:
- New O-rings on the fuel injectors
- Fuel injectors
- Fuel supply manifold with injectors attached. Tighten the bolts to 60 inch lbs. (6.5 Nm).
- Fuel lines
- Fuel injector wiring connectors
- Engine cover
- Negative battery cable

5. Start the engine and check for leaks.

DRIVE TRAIN

Transmission Assembly

REMOVAL & INSTALLATION

Manual Transmissions

2 WHEEL DRIVE

➡ **The transmission flange bolts vary in length. Note their locations for assembly.**

1. Before servicing the vehicle, refer to the precautions in the beginning of this section.
2. Remove the hood.
3. Install a support fixture to the engine lifting eyes.
4. Remove or disconnect the following:
- Negative battery cable
- Shift lever knob
- Rear console assembly
- Grommet assembly
- Shift lever

- Clutch slave cylinder and hose bracket
- Driveshaft
- Fuel line heat shield
- Vehicle Speed (VSS) sensor connector
- Reverse light switch connector
- Flywheel under cover
- Transmission mount and crossmember
- Transmission flange bolts
- Transmission

To install:

5. Install or connect the following:
- Transmission. Tighten the large flange bolts to 52 ft. lbs. (71 Nm) and the small bolts to 30 ft. lbs. (41 Nm).
- Crossmember. Tighten the bolts to 56 ft. lbs. (76 Nm).
- Transmission mount. Tighten the fasteners to 30 ft. lbs. (41 Nm).
- Flywheel under cover. Tighten the bolts to 69 inch lbs. (8 Nm).

- Reverse light switch connector
- Vehicle Speed (VSS) sensor connector
- Fuel line heat shield
- Driveshaft. Tighten the bolts to 37 ft. lbs. (50 Nm).
- Clutch slave cylinder and hose bracket
- Shift lever
- Grommet assembly
- Rear console assembly
- Shift lever knob
- Negative battery cable

6. Remove the engine support fixture and install the hood.

4 WHEEL DRIVE

➡ **The transmission flange bolts vary in length. Note their locations for assembly.**

1. Before servicing the vehicle, refer to the precautions in the beginning of this section.

2. Remove the hood.

3. Install a support fixture to the engine lifting eyes.

4. Remove or disconnect the following:
- Negative battery cable
- Shift lever knob
- Console assembly
- Grommet assembly
- Shift lever
- Transfer case control lever
- Transfer case skid plate
- Front and rear driveshafts
- Reverse lamp switch connector
- Indicator switch connectors
- Vehicle Speed (VSS) sensor connector
- 4WD actuator connector
- Transmission harness clamps
- Fuel pipe bracket
- Clutch slave cylinder and heat shield
- Transmission mount and crossmember
- Heated Oxygen (HO2S) sensor connectors
- Right exhaust front pipe
- Wiring harness heat shield
- Flywheel under cover

5. Release the throw out bearing from the pressure plate as shown.

6. Remove the transmission flange bolts and remove the transmission.

To install:

7. Install the transmission. Tighten the large bolts to 56 ft. lbs. (76 Nm) and the small bolts to 52 inch lbs. (6 Nm).

8. Apply 13–18 lbs. (59–78 N) of force to the clutch fork to engage the throw out bearing to the pressure plate.

9. Install or connect the following:
- Flywheel under cover
- Wiring harness heat shield

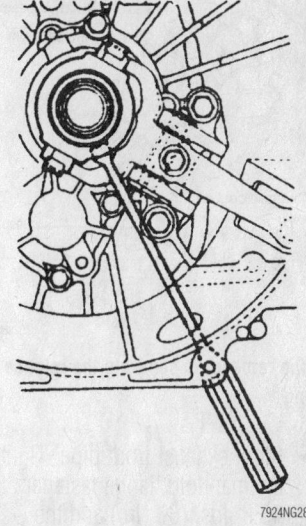

7924NG28

Insert the tool between the wedge collar and the release bearing

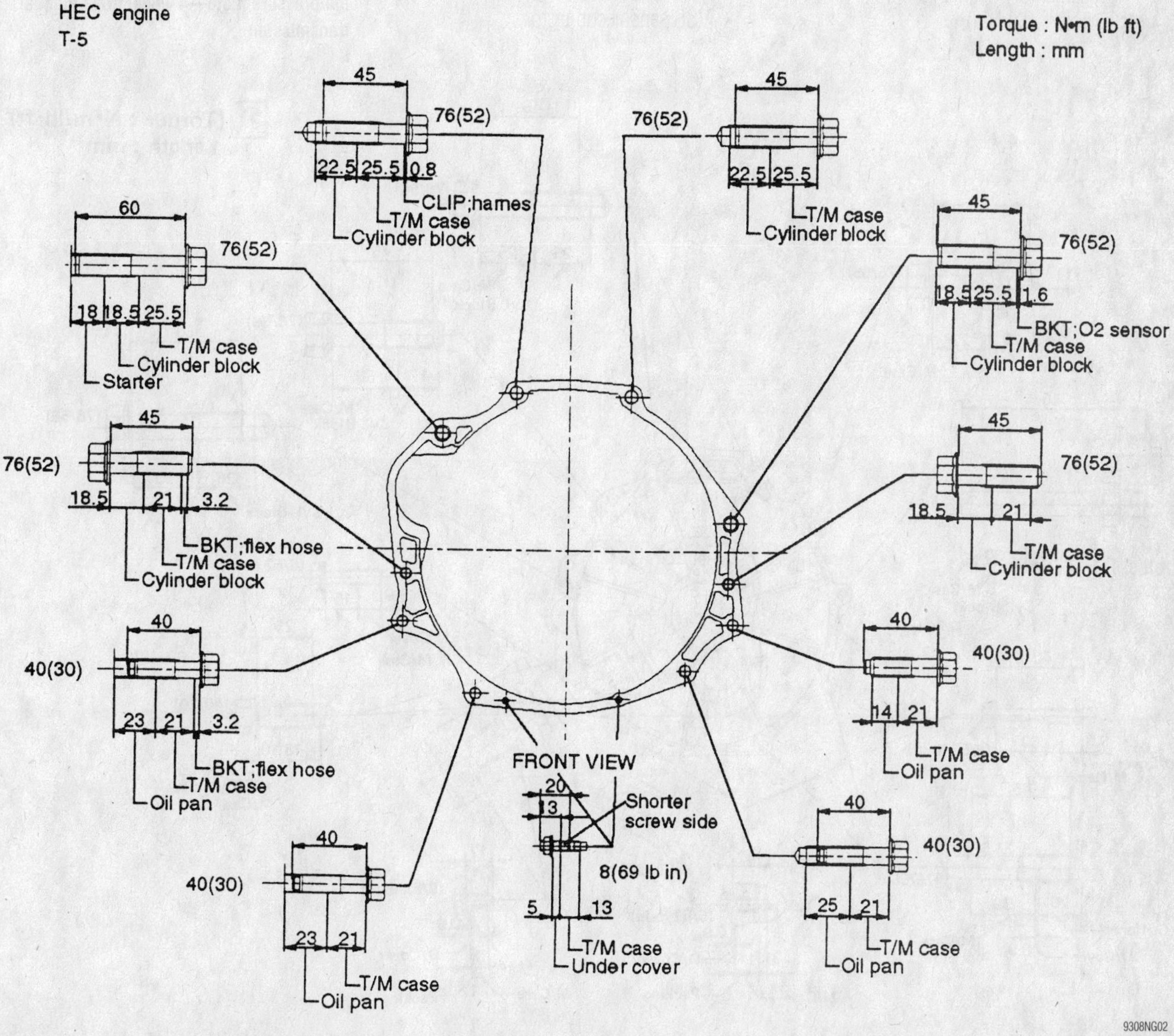

HEC engine
T-5

Torque : N•m (lb ft)
Length : mm

45
76(52)
22.5 25.5 0.8
CLIP;harnes
T/M case
Cylinder block

45
76(52)
22.5 25.5
T/M case
Cylinder block

60
76(52)
18 18.5 25.5
T/M case
Cylinder block
Starter

45
76(52)
18.5 25.5 1.6
BKT;O2 sensor
T/M case
Cylinder block

45
76(52)
18.5 21 3.2
BKT;flex hose
T/M case
Cylinder block

45
76(52)
18.5 21
T/M case
Cylinder block

40
40(30)
23 21 3.2
BKT;flex hose
T/M case
Oil pan

40
40(30)
14 21
T/M case
Oil pan

40
40(30)
23 21
T/M case
Oil pan

FRONT VIEW
20
13
Shorter screw side
8(69 lb in)
5 13
T/M case
Under cover

40
40(30)
25 21
T/M case
Oil pan

Transmission flange bolt identification and torque—2 wheel drive transmission

9308NG02

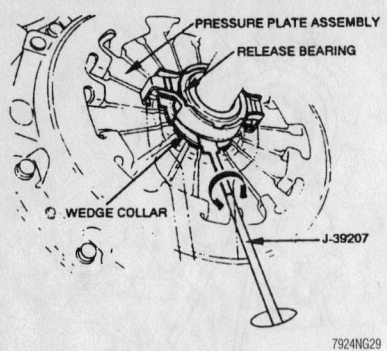

7924NG29

Turn the remover to separate the release bearing

- Right exhaust front pipe. Tighten the manifold flange fasteners to 49 ft. lbs. (67 Nm) and the exhaust flange bolts to 32 ft. lbs. (43 Nm).
- HO2S sensor connectors
- Crossmember. Tighten the bolts to 37 ft. lbs. (50 Nm).

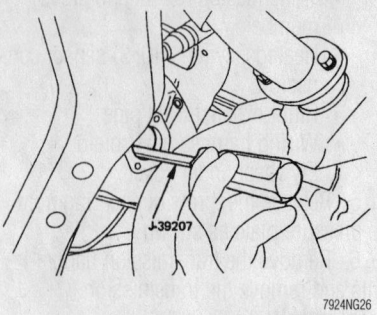

7924NG26

Insert the Release Bearing Remover tool J-39207 through the bell housing—4 wheel drive manual transmission

- Transmission mount. Tighten the bolts to 30 ft. lbs. (41 Nm).
- Clutch slave cylinder and heat shield
- Fuel pipe bracket
- Transmission harness clamps
- 4WD actuator connector
- VSS sensor connector

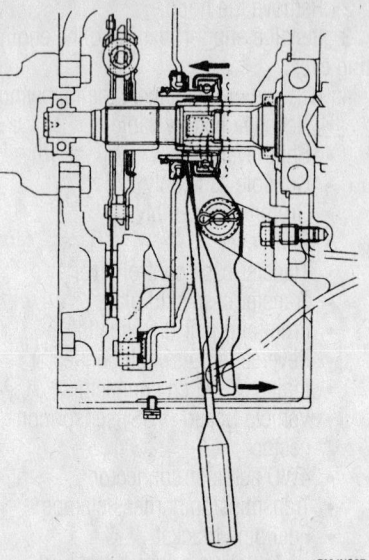

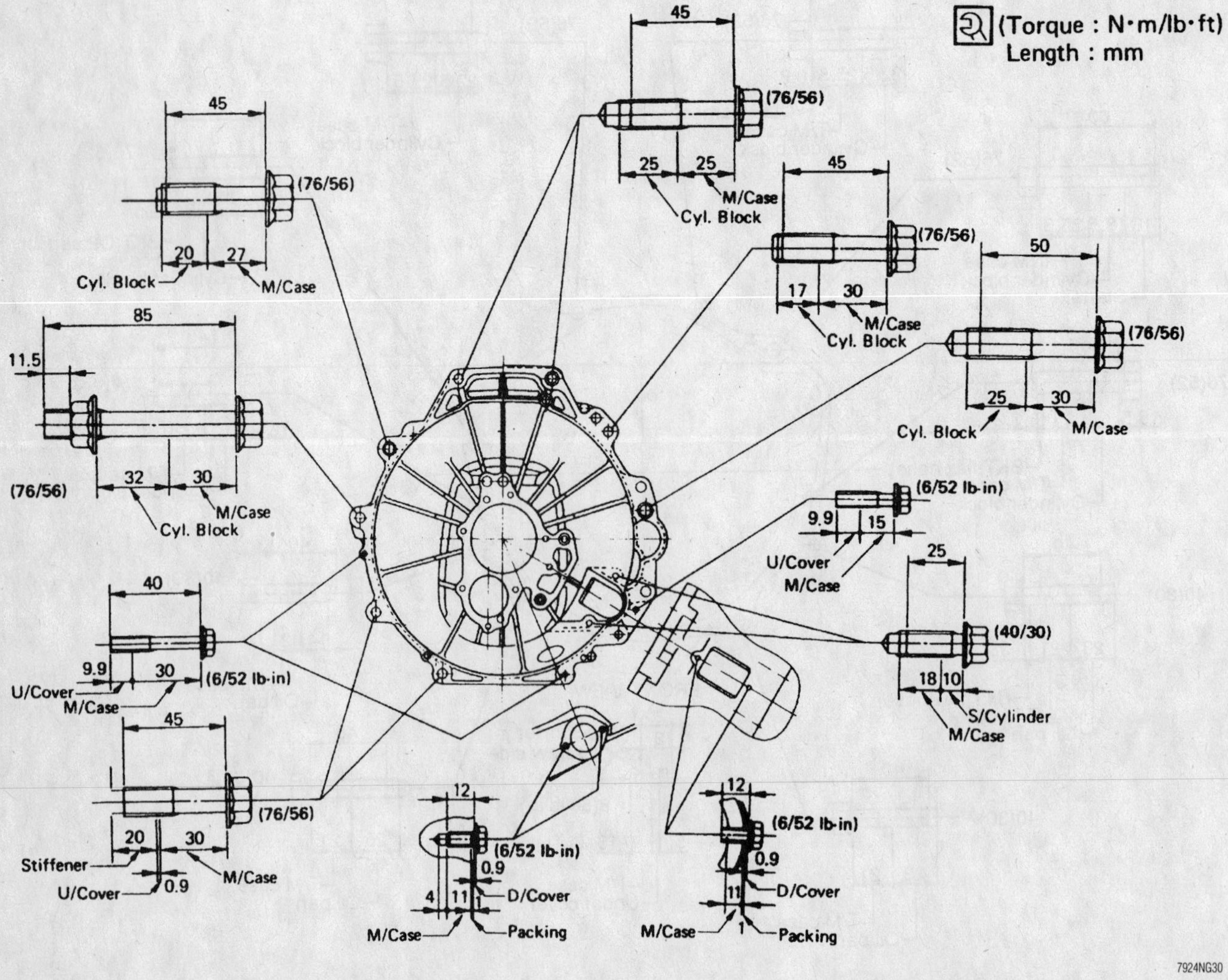

7924NG27

Push the release bearing fork toward the transmission to release the bearing from the pressure plate—4 wheel drive manual transmission

Transmission mounting bolt identification and torque specifications—V6 engine with 4 wheel drive transmission shown

7924NG30

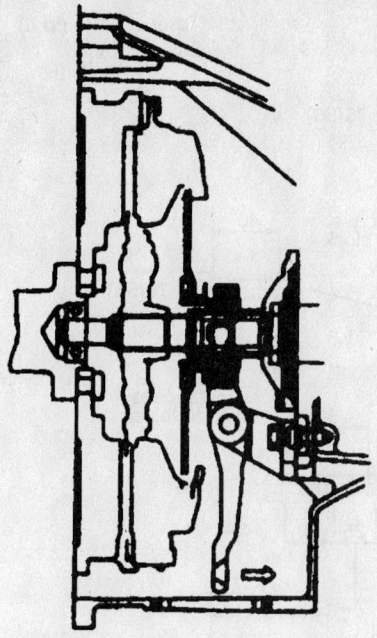

7924BG21

Push the release bearing fork toward the transmission to engage the release bearing with the pressure plate

- Indicator switch connectors
- Reverse lamp switch connector
- Front and rear driveshafts. Tighten the flange bolts to 46 ft. lbs. (63 Nm).
- Transfer case skid plate. Tighten the bolts to 27 ft. lbs. (37 Nm).
- Transfer case control lever
- Shift lever
- Grommet assembly
- Console assembly
- Shift lever knob
- Negative battery cable

10. Remove the engine support fixture and install the hood.

Automatic Transmissions with 2-Wheel Drive

EXCEPT AXIOM

1. Before servicing the vehicle, refer to the precautions in the beginning of this section.
2. Remove the hood.
3. Install a support fixture to the engine lifting eyes.
4. Remove or disconnect the following:
   - Negative battery cable
   - Front console assembly and wiring connectors
   - Shift lock cable
   - Shift control rod
   - Selector lever assembly
   - Driveshaft
   - Wiring harness heat shield
   - Transmission mount and crossmember
   - Heated Oxygen (HO2S) sensor connectors
   - Left and right exhaust front pipes
   - Transmission oil cooler lines
   - Starter motor
   - Fuel line bracket
   - Transmission harness connectors
   - Flywheel under covers
   - Torque converter
   - Transmission flange bolts
   - Transmission

To install:

➡ **Use new torque converter bolts.**

5. Install or connect the following:
   - Transmission. Tighten the large bolts to 56 ft. lbs. (76 Nm) and the small bolts to 69 inch lbs. (8 Nm).
   - Torque converter. Tighten the bolts to 40 ft. lbs. (54 Nm).
   - Flywheel under covers
   - Transmission harness connectors

- Fuel line bracket
- Starter motor. Tighten the bolts to 30 ft. lbs. (40 Nm).
- Transmission oil cooler lines
- Left and right exhaust front pipes. Tighten the manifold flange fasteners to 49 ft. lbs. (67 Nm) and the exhaust flange bolts to 32 ft. lbs. (43 Nm).
- HO2S sensor connectors
- Crossmember. Tighten the bolts to 37 ft. lbs. (50 Nm).
- Transmission mount. Tighten the bolts to 30 ft. lbs. (41 Nm).
- Wiring harness heat shield
- Driveshaft. Tighten the flange bolts to 46 ft. lbs. (63 Nm).
- Selector lever assembly
- Shift control rod
- Shift lock cable
- Front console assembly and wiring connectors
- Negative battery cable

6. Remove the engine support fixture and install the hood.

AXIOM

1. Before servicing the vehicle, refer to the precautions in the beginning of this section.
2. Remove or disconnect the following:
   - Negative battery cable
   - Driveshaft
   - Fuel line bracket from the transmission
   - Wiring harness heat shield
   - Transmission harness connectors
3. Support the transmission with a jack.
4. Remove or disconnect the following:
   - Transmission mount and crossmember
   - Transmission oil cooler lines
   - Selector cable
   - Starter motor
   - Undercovers
   - Torque converter bolts
   - Transmission flange bolts
   - Transmission

To install:

➡ **Use new torque converter bolts.**

5. Install or connect the following:
   - Transmission. Tighten the large bolts to 56 ft. lbs. (76 Nm) and the small bolts to 69 inch lbs. (8 Nm).
   - Torque converter. Tighten the bolts to 40 ft. lbs. (54 Nm).
   - Under covers
   - Starter motor. Tighten the bolts to 30 ft. lbs. (40 Nm).
   - Selector cable
   - Transmission oil cooler lines

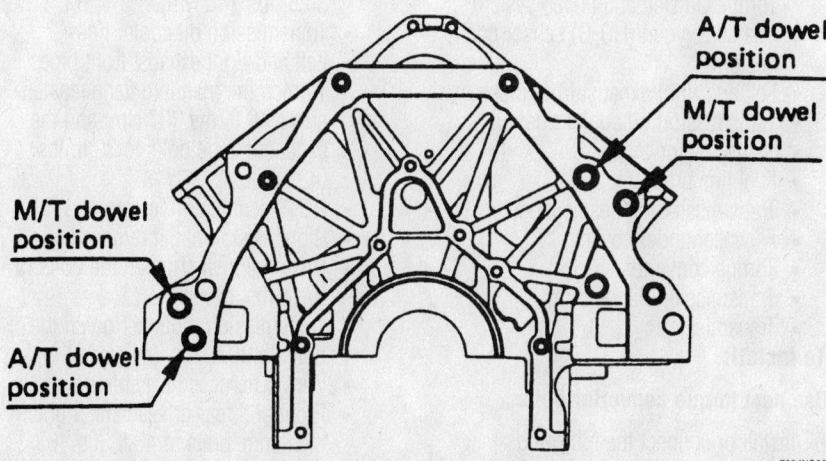

7924NG02

Dowel pin locations for automatic and manual transmissions—3.2L engine

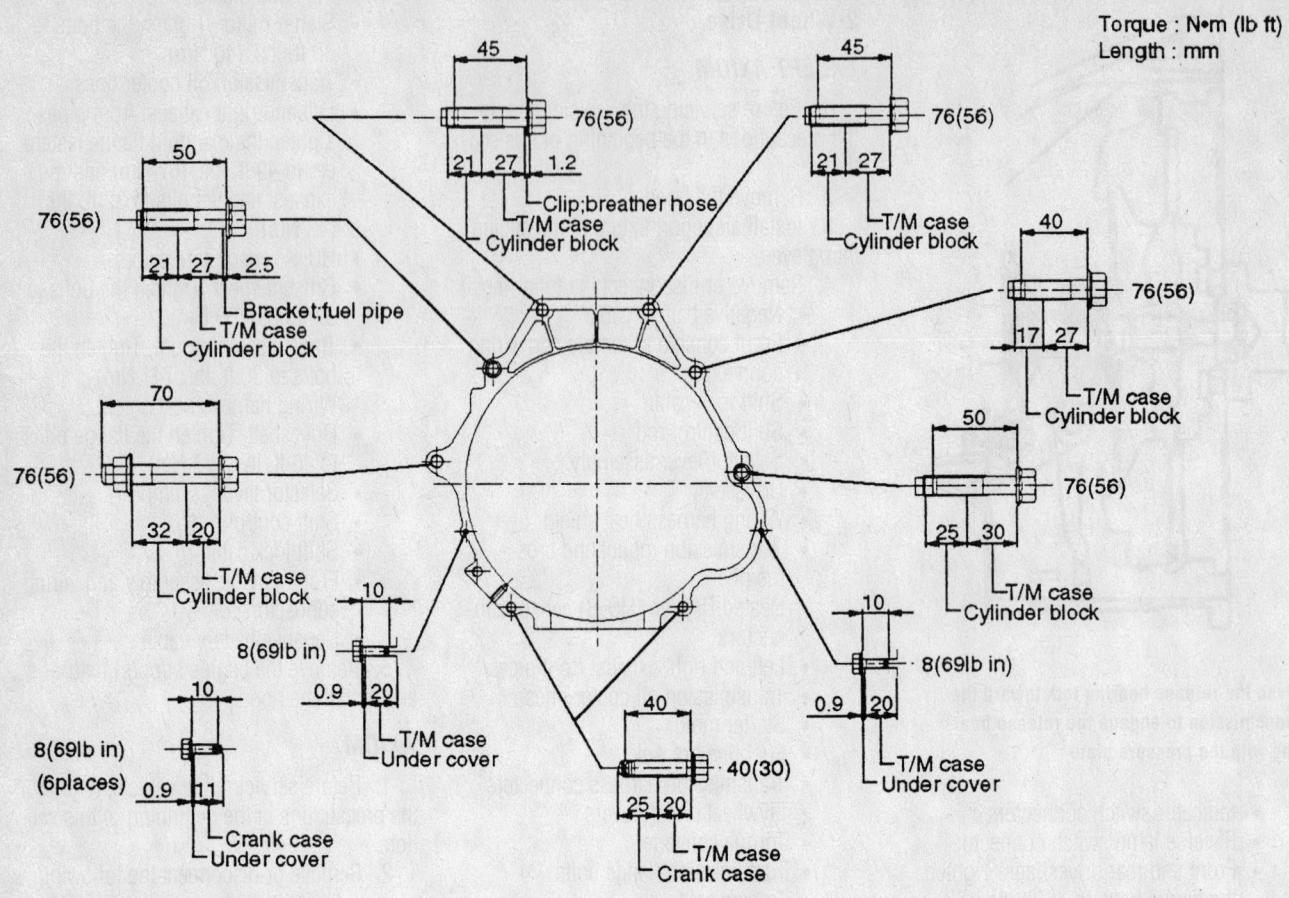

Torque : N•m (lb ft)
Length : mm

Automatic transmission mounting bolt locations and torque specifications

- Crossmember. Tighten the bolts to 85 ft. lbs. (116 Nm).
- Transmission mount. Tighten the bolts to 37 ft. lbs. (50 Nm).
- Transmission harness connectors
- Wiring harness heat shield
- Fuel line bracket
- Driveshaft. Tighten the flange bolts to 46 ft. lbs. (63 Nm).
- Negative battery cable

Automatic Transmissions with 4-Wheel Drive

EXCEPT AXIOM

1. Before servicing the vehicle, refer to the precautions in the beginning of this section.
2. Remove the hood.
3. Install a support fixture to the engine lifting eyes.
4. Remove or disconnect the following:
   - Negative battery cable
   - Transfer case shift lever knob
   - Front console assembly and wiring connectors
   - Shift lock cable
   - Shift control rod
   - Selector lever assembly
   - Transfer case shift lever
   - Transfer case skid plate
   - Front and rear driveshafts
   - Wiring harness heat shield
   - Transmission mount and crossmember
   - Right torsion bar, if equipped with Torque On Demand (TOD) system
   - Heated Oxygen (HO2S) sensor connectors
   - Left and right exhaust front pipes
   - Transmission oil cooler lines
   - Starter motor
   - Fuel line bracket
   - Transmission harness connectors
   - Flywheel under covers
   - Torque converter
   - Transmission flange bolts
   - Transmission

To install:

→ **Use new torque converter bolts.**

5. Install or connect the following:
   - Transmission. Tighten the large bolts to 56 ft. lbs. (76 Nm) and the small bolts to 30 ft. lbs. (40 Nm).
   - Torque converter. Tighten the bolts to 40 ft. lbs. (54 Nm).
   - Flywheel under covers
   - Transmission harness connectors
   - Fuel line bracket
   - Starter motor. Tighten the bolts to 30 ft. lbs. (40 Nm).
   - Transmission oil cooler lines
   - Left and right exhaust front pipes. Tighten the manifold flange fasteners to 49 ft. lbs. (67 Nm) and the exhaust flange bolts to 32 ft. lbs. (43 Nm).
   - HO2S sensor connectors
   - Right torsion bar, if removed
   - Crossmember. Tighten the bolts to 37 ft. lbs. (50 Nm).
   - Transmission mount. Tighten the bolts to 30 ft. lbs. (41 Nm).
   - Wiring harness heat shield
   - Front and rear driveshafts. Tighten the flange bolts to 46 ft. lbs. (63 Nm).

- Transfer case skid plate. Tighten the bolts to 27 ft. lbs. (37 Nm).
- Transfer case shift lever
- Selector lever assembly
- Shift control rod
- Shift lock cable
- Front console assembly and wiring connectors
- Transfer case shift lever knob
- Negative battery cable

6. Remove the engine support fixture and install the hood.

AXIOM

1. Before servicing the vehicle, refer to the precautions in the beginning of this section.
2. Remove or disconnect the following:
   - Negative battery cable
   - Skid plate
   - Driveshafts
   - Center exhaust pipe
   - Fuel line bracket from the transmission
   - Wiring harness heat shield
   - Transmission harness connectors
3. Support the transmission with a jack.
4. Remove or disconnect the following:
   - Transmission mount and crossmember
   - Transmission oil cooler lines
   - Selector cable
   - Starter motor
   - Undercovers
   - Torque converter bolts
   - Transmission flange bolts
   - Transmission

To install:

➡**Use new torque converter bolts.**

5. Install or connect the following:
   - Transmission. Tighten the large bolts to 56 ft. lbs. (76 Nm) and the small bolts to 69 inch lbs. (8 Nm).
   - Torque converter. Tighten the bolts to 40 ft. lbs. (54 Nm).
   - Under covers
   - Starter motor. Tighten the bolts to 30 ft. lbs. (40 Nm).
   - Selector cable
   - Transmission oil cooler lines
   - Crossmember. Tighten the bolts to 85 ft. lbs. (116 Nm).
   - Transmission mount. Tighten the bolts to 37 ft. lbs. (50 Nm).
   - Transmission harness connectors
   - Wiring harness heat shield
   - Fuel line bracket
   - Center exhaust pipe
   - Driveshafts. Tighten the flange bolts to 46 ft. lbs. (63 Nm).
   - Skid plate
   - Negative battery cable

Clutch

ADJUSTMENTS

➡**This vehicle is equipped with a hydraulic clutch linkage. No adjustment is necessary.**

REMOVAL & INSTALLATION

1. Before servicing the vehicle, refer to the precautions in the beginning of this section.
2. Remove the transmission.
3. Loosen the pressure plate mounting bolts in a 2-step crisscross sequence until the spring tension is relieved.
4. Remove the pressure plate and the clutch disc.

To install:

5. Install a new wedge collar and wire snapring into the pressure plate.
6. Using a clutch alignment tool, assemble the clutch disc and pressure plate onto the flywheel.
7. Tighten the pressure plate bolts in sequence and in two passes to 13 ft. lbs. (8 Nm).
8. Install the transmission.
9. Road test the vehicle and check for proper clutch operation.

Hydraulic Clutch System

BLEEDING

1. Before servicing the vehicle, refer to the precautions in the beginning of this section.
2. Have an assistant pump the clutch pedal slowly several times and hold it depressed.
3. Open the slave cylinder bleeder screw and allow air to escape.
4. Close the bleeder screw before releasing the clutch pedal.
5. Repeat until all air is purged from the clutch hydraulic system.
6. Refill the reservoir to the full mark.

Transfer Case Assembly

REMOVAL & INSTALLATION

Except Axiom

1. Before servicing the vehicle, refer to the precautions in the beginning of this section.
2. Remove or disconnect the following:
   - Negative battery cable
   - Transfer case skid plate

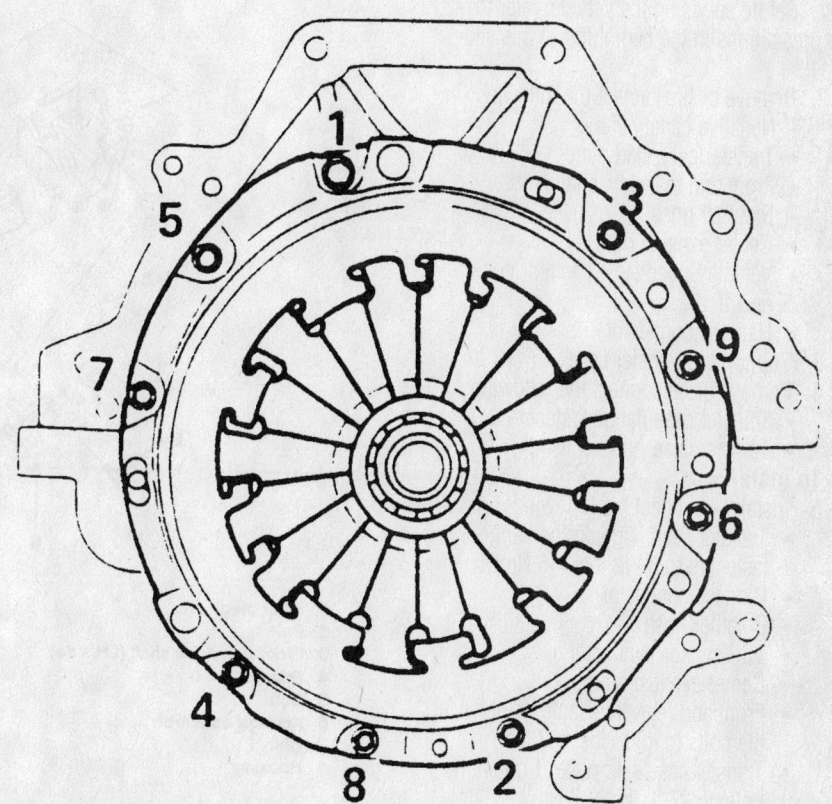

Pressure plate tightening sequence

7924NG49

- Front and rear driveshafts
- Heated Oxygen (HO2S) sensor connectors
- Left and right exhaust front pipes
- Transfer case control lever knob
- Selector lever assembly
- Transfer case control lever
- Vehicle Speed (VSS) sensor connector
- 4 wheel drive switch connector
- 4 wheel drive actuator connector
- Transfer case flange fasteners
- Transfer case

To install:

3. Install or connect the following:
- Transfer case. Tighten the flange fasteners to 34 ft. lbs. (46 Nm).
- 4 wheel drive actuator connector
- 4 wheel drive switch connector
- VSS sensor connector
- Transfer case control lever
- Selector lever assembly
- Transfer case control lever knob
- Left and right exhaust front pipes
- HO2S sensor connectors
- Front and rear driveshafts. Tighten the bolts to 46 ft. lbs. (63 Nm).
- Transfer case skid plate. Tighten the bolts to 27 ft. lbs. (37 Nm).
- Negative battery cable

Axiom

1. Before servicing the vehicle, refer to the precautions in the beginning of this section.
2. Remove or disconnect the following:
- Negative battery cable
- Transfer case skid plate
- Front and rear driveshafts
- Breather hose
- Center exhaust pipe
- Vehicle Speed (VSS) sensor connector
- Harness connector
3. Support the transfer case
4. Remove or disconnect the following:
- Transfer case flange fasteners
- Transfer case

To install:

5. Install or connect the following:
- Transfer case. Tighten the flange fasteners to 34 ft. lbs. (46 Nm).
- Harness connector
- Breather hose
- VSS sensor connector
- Center exhaust pipe
- Front and rear driveshafts. Tighten the bolts to 46 ft. lbs. (63 Nm).
- Transfer case skid plate. Tighten the bolts to 27 ft. lbs. (37 Nm).
- Negative battery cable

Halfshaft

REMOVAL & INSTALLATION

1. Before servicing the vehicle, refer to the precautions in the beginning of this section.
2. Remove or disconnect the following:
- Negative battery cable
- Front wheel
- Radiator skid plate
- Transfer case skid plate

- Brake calipers and mounting bracket
- Brake rotor
- Wheel speed sensor
- Steering knuckle

3. Support the axle housing with a jack. Unbolt the axle mounting bracket and remove the halfshaft/bracket assembly.

To install:

4. Install or connect the following:
- Axle/bracket assembly. Tighten the bracket flange bolts to 85 ft. lbs. (116 Nm) and the bracket mounting bolts to 112 ft. lbs. (152 Nm).

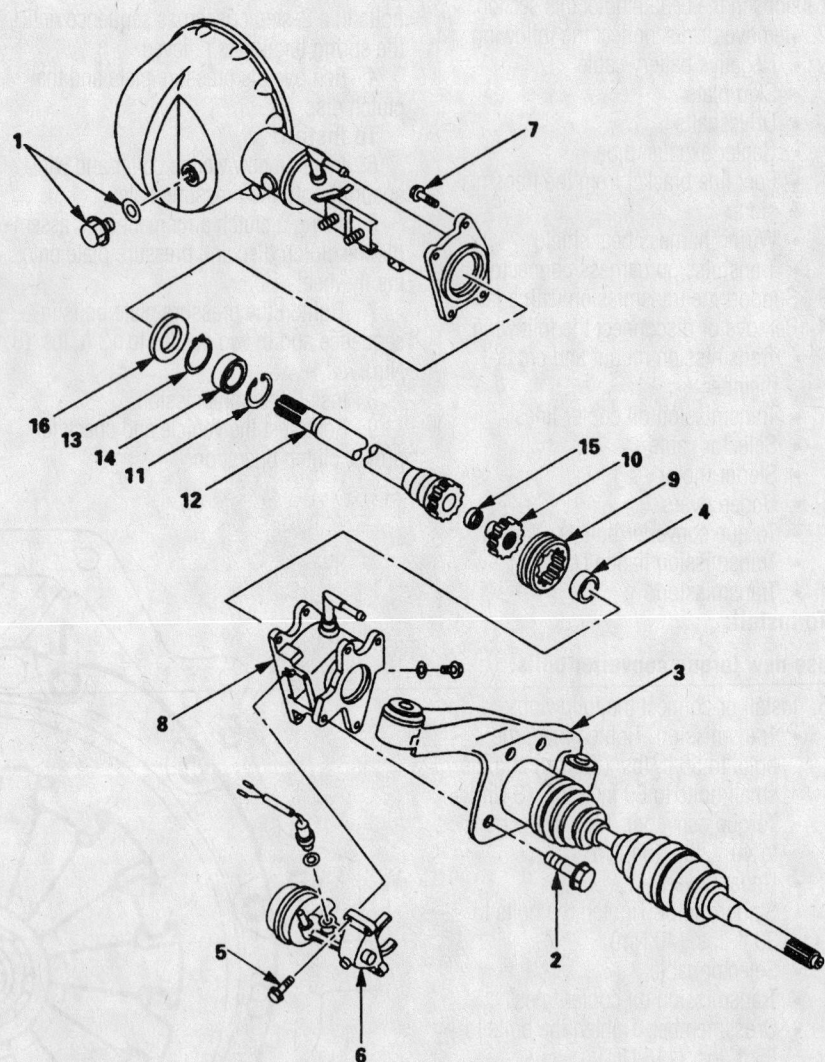

1. Filler plug	9. Sleeve
2. Bolt	10. Clutch gear
3. Front axle drive shaft (LH side)	11. Snap ring
4. Spacer	12. Inner shaft
5. Bolt	13. Snap ring
6. Actuator assembly	14. Inner shaft bearing
7. Bolt	15. Needle bearing
8. Housing	16. Oil seal

7924BG26

Exploded view of the left halfshaft, axle shaft and axle disconnect

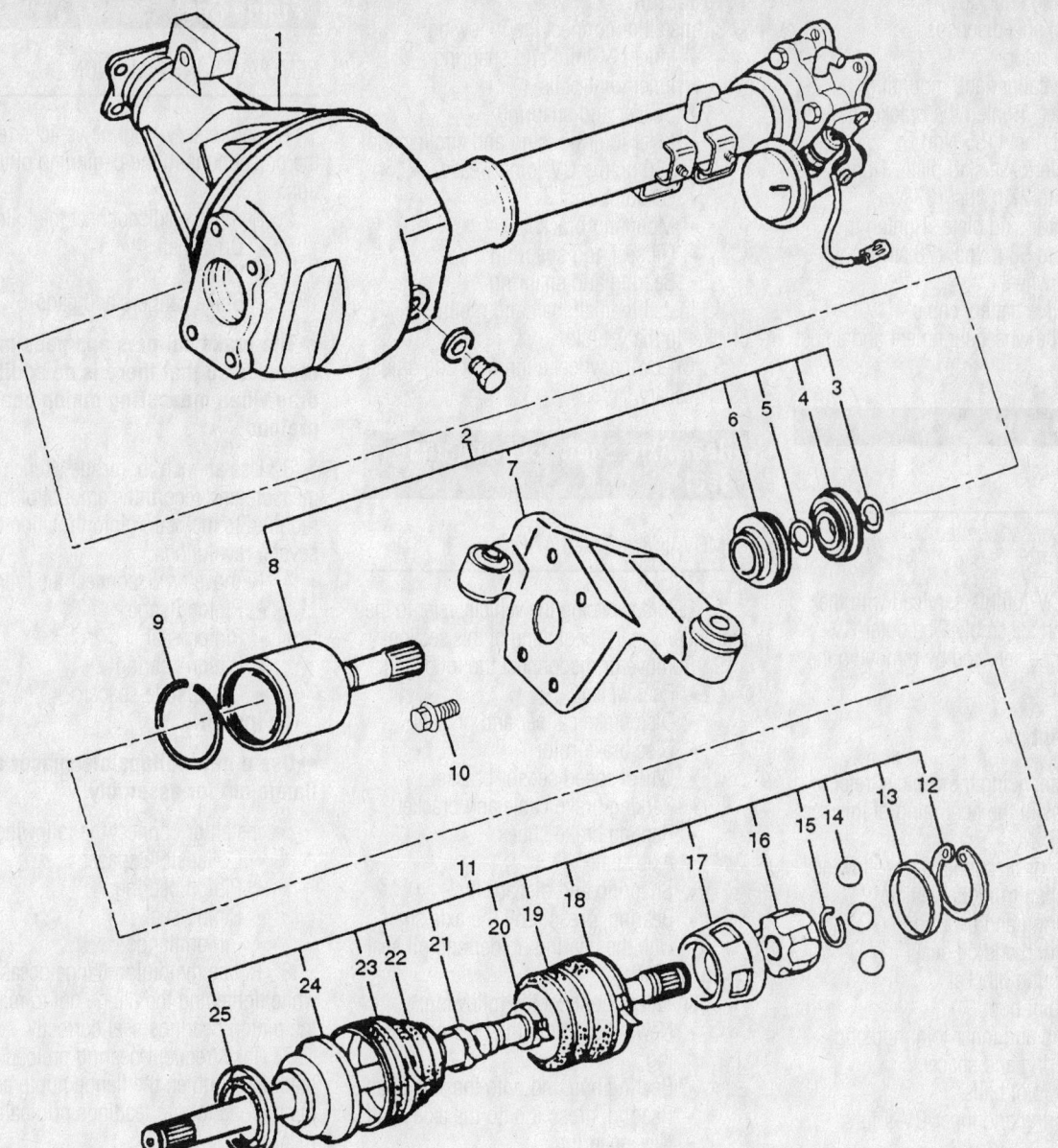

1 Axle Case and Differential
2 DOJ Case Assembly
3 Snap Ring
4 Bearing
5 Snap Ring
6 Oil Seal
7 Bracket
8 DOJ Case
9 Circlip
10 Bolt
11 Drive Shaft Joint Assembly
12 Snap Ring
13 Spacer
14 Ball
15 Snap Ring
16 Ball Retainer
17 Ball Guide
18 Band
19 Bellows
20 Band
21 Band
22 Bellows
23 Band
24 BJ Shaft
25 Dust Seal

9308BG03

Exploded view of the right halfshaft and mounting bracket

- Steering knuckle
- Wheel speed sensor
- Brake rotor
- Brake caliper and mounting bracket. Tighten the bracket bolts to 115 ft. lbs. (155 Nm).
- Transfer case skid plate. Tighten the bolts to 27 ft. lbs. (37 Nm).
- Radiator skid plate. Tighten the bolts to 58 ft. lbs. (78 Nm).
- Front wheel
- Negative battery cable

5. Check the wheel alignment and adjust as necessary.

CV-Joints

OVERHAUL

Outer CV-Joint

The outer CV-joint is serviced with the axle shaft as an assembly. The outer CV-joint boot can be serviced by removing the inner CV-joint.

Inner CV-Joint

1. Before servicing the vehicle, refer to the precautions in the beginning of this section.
2. Remove or disconnect the following:
   - Halfshaft from the vehicle
   - Snapring and bearing
   - Snapring and oil seal
   - Mounting bracket
   - CV-joint boot
   - Circlip and inner joint housing
   - Snapring and spacer
   - Inner joint balls
   - Snapring and inner CV-joint

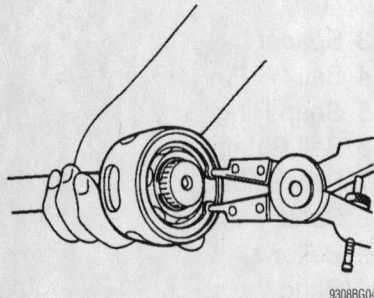

9308BG04

CV-joint spacer snapring—Inner CV-Joint

To install:
3. Install or connect the following:
   - Inner CV-joint and snapring
   - Inner joint balls
   - Spacer and snapring
   - Inner joint housing and circlip. Add 150 grams CV-joint grease.
   - CV-joint boot
   - Mounting bracket
   - Oil seal and snapring
   - Bearing and snapring
4. Install the halfshaft and mounting bracket to the vehicle.
5. Check the wheel alignment and adjust as necessary.

Rear Axle Shaft, Bearing and Seal

REMOVAL & INSTALLATION

1. Before servicing the vehicle, refer to the precautions in the beginning of this section.
2. Remove or disconnect the following:
   - Rear wheel
   - Disc brake caliper and bracket
   - Disc brake rotor
   - Wheel speed sensor bracket
   - Parking brake cable and bracket
   - Parking brake shoes
   - Axle shaft
   - Snapring and discard it
   - Bearing, press it off the axle shaft with the bearing holder and oil seal

To install:
3. Install or connect the following:
   - New oil seal into the bearing housing
   - Bearing housing onto the axle shaft
   - Bearing, press it onto the axle shaft
   - New snapring
   - Axle shaft. Use new lockwashers and tighten the bearing holder nuts to 54 ft. lbs. (74 Nm).
   - Parking brake shoes
   - Parking brake cable and bracket
   - Wheel speed sensor bracket
   - Disc brake rotor
   - Disc brake caliper and bracket. Tighten the bracket bolts to 76 ft. lbs. (103 Nm).
   - Rear wheel
4. Check the rear axle oil level and adjust as necessary.

Pinion Seal

REMOVAL & INSTALLATION

1. Before servicing the vehicle, refer to the precautions in the beginning of this section.
2. Remove or disconnect the following:
   - Driveshaft
   - Wheels
   - Brake calipers and pads

➡ **The brake calipers and pads must be removed so that there is no additional drag when measuring pinion bearing preload.**

3. Use an inch lb. torque wrench and measure and record the amount of torque required to maintain pinion rotation through several revolutions.
4. Remove or disconnect the following:
   - Pinion flange
   - Pinion seal
   - Pinion bearing
   - Collapsible spacer

To install:

➡ **Use a new collapsible spacer and flange nut for assembly.**

5. Install or connect the following:
   - Collapsible spacer
   - Pinion bearing
   - Pinion seal
   - Pinion flange
6. Rotate the pinion flange occasionally while tightening the flange nut to make sure the pinion bearings seat correctly.
7. Take frequent bearing preload torque readings. Tighten the flange nut to achieve the preload torque readings originally recorded.

✳✳ CAUTION

Never loosen the pinion nut to reduce bearing preload. If it is necessary to reduce bearing preload, install a new collapsible spacer and pinion nut.

8. Install or connect the following:
   - Driveshaft
   - Brake calipers and pads
   - Wheels
9. Fill the differential with gear lubricant and check for leaks.

STEERING AND SUSPENSION

Air Bag

✳✳ CAUTION

Some vehicles are equipped with an air bag system. The system must be disarmed before performing service on, or around, system components, the steering column, instrument panel components, wiring and sensors. Failure to follow the safety precautions and the disarming procedure could result in accidental air bag deployment, possible injury and unnecessary system repairs.

PRECAUTIONS

Several precautions must be observed when handling the inflator module to avoid accidental deployment and possible personal injury.

• Never carry the inflator module by the wires or connector on the underside of the module.

• When carrying a live inflator module, hold securely with both hands, and ensure that the bag and trim cover are pointed away from you.

• Place the inflator module on a bench or other surface with the bag and trim cover facing up.

• With the inflator module on the bench, never place anything on or close to the module which may be thrown in the event of an accidental deployment.

DISARMING

1. Before servicing the vehicle, refer to the precautions in the beginning of this section.
2. Turn the ignition switch to the **LOCK** position. Remove the key.
3. Disconnect the negative battery cable. Wait 1 minute before working around the air bags.
4. Disconnect the yellow 2-pin connector at the base of the steering column.
5. Disconnect the yellow 2-pin connector behind the glove box assembly.
6. When repairs are completed, connect the yellow 2-pin connectors.
7. Connect the negative battery cable.
8. Turn the ignition to the **ON** position, but don't start the engine. The AIR BAG warning light should turn on and flash on and off 7 times, and then turn off. This light sequence indicates that the SRS system is functioning normally. If the AIR BAG light doesn't come on, or stays on longer than 7 seconds, the system must be diagnosed.

Recirculating Ball Steering Gear

REMOVAL & INSTALLATION

1. Before servicing the vehicle, refer to the precautions in the beginning of this section.
2. Disable the air bag system.
3. Remove or disconnect the following:
   • Skid plates
   • Lower fan shroud
   • Stabilizer bar
   • Power steering pressure and return lines
   • Pitman arm
   • Steering column intermediate shaft
   • Steering gear

To install:
4. Install or connect the following:
   • Steering gear. Tighten the bolts to 33 ft. lbs. (44 Nm).
   • Steering column intermediate shaft. Tighten the pinch bolt to 18 ft. lbs. (25 Nm).
   • Pitman arm. Tighten the nut to 159 ft. lbs. (216 Nm).
   • Power steering pressure and return lines. Tighten the fittings to 33 ft. lbs. (44 Nm).
   • Stabilizer bar
   • Lower fan shroud
   • Skid plates
5. Fill the power steering fluid reservoir.
6. Check the wheel alignment and adjust as necessary.

Rack and Pinion Steering Gear

REMOVAL & INSTALLATION

2-Wheel Drive

1. Before servicing the vehicle, refer to the precautions in the beginning of this section.
2. Disable the air bag system.
3. Remove or disconnect the following:
   • Stone guard
   • Tie rod from knuckle
   • Power steering pressure and return lines
   • Steering gear

To install:
4. Install or connect the following:
   • Steering gear. Torque the mounting bolts to 85 ft. lbs. (116 Nm)
   • Fluid lines. Torque the fitting to 18 ft. lbs. (25 Nm)
   • Tie rod ends. 87 ft. lbs. (118 Nm)
   • Stone guard
5. Fill and bleed the system.

4-Wheel Drive

1. Before servicing the vehicle, refer to the precautions in the beginning of this section.
2. Disable the air bag system.
3. Remove or disconnect the following:
   • Stone guard
   • Tie rod from knuckle
   • Power steering pressure and return lines
   • Torsion bar
   • Lower control arm frame side bolt
   • Crossmember bolts
   • Steering gear with crossmember
   • Steering gear

To install:
4. Install or connect the following:
   • Steering gear. Torque the mounting bolts to 85 ft. lbs. (116 Nm)
   • Crossmember Torque the bolts to 128 ft. lbs. (173 Nm)
   • Lower control arm bolt
   • Torsion bar
   • Fluid lines. Torque the fitting to 18 ft. lbs. (25 Nm)
   • Tie rod ends. 87 ft. lbs. (118 Nm)
   • Stone guard
5. Fill and bleed the system.

Shock Absorber

REMOVAL & INSTALLATION

Front

EXCEPT AXIOM

1. Before servicing the vehicle, refer to the precautions in the beginning of this section.
2. Support the lower control arm with a jackstand.
3. Remove or disconnect the following:
   • Front wheels
   • Upper shock retaining nut and rubber bushing
   • Suspension bump stops
   • Shock absorber

To install:
4. Install or connect the following:
   • Shock absorber. Tighten the lower bolt to 60–61 ft. lbs. (82–84 Nm).

- Bump stop. Tighten the bolts to 30 ft. lbs. (41 Nm).
- Upper shock retaining nut and rubber bushing. Tighten the nut to 14–15 ft. lbs. (19–20 Nm).
- Front wheels

AXIOM

1. Before servicing the vehicle, refer to the precautions in the beginning of this section.
2. Support the lower control arm with a jackstand.
3. Remove or disconnect the following:
- Front wheels
- Upper shock retaining nut and rubber bushing
- ISC actuator and bracket
- Suspension bump stops
- Shock absorber

To install:

4. Install or connect the following:
- Shock absorber. Tighten the lower bolt to 69 ft. lbs. (93 Nm).
- ISC actuator and bracket
- Bump stop. Tighten the bolts to 30 ft. lbs. (41 Nm).
- Upper shock retaining nut and rubber bushing. Tighten the nut to 14 ft. lbs. (20 Nm).
- Front wheels

Rear

EXCEPT AXIOM

1. Before servicing the vehicle, refer to the precautions in the beginning of this section.
2. Support the rear axle with jackstands.
3. Remove the rear shock absorbers.

To install:

4. Install the rear shock absorbers. Tighten the upper bolt to 70 ft. lbs. (95 Nm). Tighten the lower bolt to 58 ft. lbs. (78 Nm).
5. Remove the jackstands.

AXIOM

1. Before servicing the vehicle, refer to the precautions in the beginning of this section.
2. Support the rear axle with jackstands.
3. Remove the rear shock absorbers.

To install:

4. Install the rear shock absorbers. Tighten the upper bolt to 14 ft. lbs. (20 Nm). Tighten the lower bolt to 58 ft. lbs. (78 Nm).

➡ **With ISC, do not use any grease on or near the bushings.**

5. Remove the jackstands.

Coil Spring

REMOVAL & INSTALLATION

Except Axiom

1. Before servicing the vehicle, refer to the precautions in the beginning of this section.
2. Support the vehicle under the frame.
3. Support the rear axle with a jack.
4. Remove or disconnect the following:
- Rear wheels
- Stabilizer bar links
- Parking brake cable brackets
- Shock absorbers
5. Lower the rear axle with the jack to release the coil spring tension. Remove the coil springs and insulators.

To install:

6. Place the coil springs on the axle assembly and the insulators on top of the springs.
7. Raise the axle assembly into position.
8. Install or connect the following:
- Shock absorbers
- Parking brake cable brackets
- Stabilizer bar links. Tighten the nuts to 37 ft. lbs. (50 Nm).
- Rear wheels

Axiom

1. Before servicing the vehicle, refer to the precautions in the beginning of this section.
2. Support the vehicle under the frame.
3. Support the rear axle with a jack.

4. Remove or disconnect the following:
- Rear wheels
- Breather hose
- Upper link fixing bolt, left side only
- Stabilizer bar links
- Shock absorbers
5. Lower the rear axle with the jack to release the coil spring tension. Remove the coil springs and insulators.

To install:

6. Place the coil springs on the axle assembly and the insulators on top of the springs.
7. Raise the axle assembly into position.
8. Install or connect the following:
- Shock absorbers. Torque to 58 ft. lbs. (78 Nm).
- Stabilizer bar links. Tighten the nuts to 23 ft. lbs. (31 Nm).
- Upper link. Torque to 101 ft. lbs. (137 Nm)
- Rear wheels

Torsion Bar

REMOVAL & INSTALLATION

1. Before servicing the vehicle, refer to the precautions in the beginning of this section.
2. Matchmark the adjusting bolt and end piece, then remove the bolt, end piece, and seat.
3. Matchmark the height control arm to the torsion bar, then remove the height control arm.
4. Matchmark the torsion bar to the lower control arm, then remove the torsion bar.

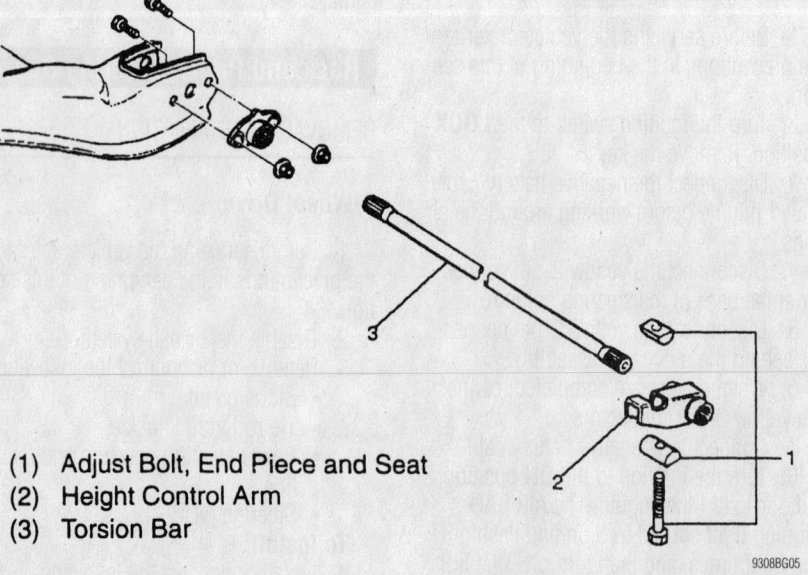

(1) Adjust Bolt, End Piece and Seat
(2) Height Control Arm
(3) Torsion Bar

Exploded view of the torsion bar assembly

9308BG05

To install:

5. Apply grease to the torsion bar splines.

6. Apply grease to the contact points of the height control arm, adjusting bolt end piece and seat.

7. Align the matchmarks and install the torsion bar.

8. Align the matchmarks and install the height control arm.

9. Install the adjusting bolt, seat and end piece.

10. Tighten the adjusting bolt to align the matchmarks.

Upper Ball Joint

REMOVAL & INSTALLATION

1. Before servicing the vehicle, refer to the precautions in the beginning of this section.

2. Support the lower control arm with a floor jack.

3. Remove or disconnect the following:
- Front wheel
- Wheel speed sensor
- Upper ball joint

To install:

➡ **Use new nuts, bolts and split pins for assembly.**

4. Install or connect the following:
- Upper ball joint. Tighten the mounting bolts to 42 ft. lbs. (57 Nm) and the nut to 72 ft. lbs. (96 Nm).
- Wheel speed sensor
- Front wheel

Lower Ball Joint

REMOVAL & INSTALLATION

1. Before servicing the vehicle, refer to the precautions in the beginning of this section.

2. Support the lower control arm with a jackstand.

3. Remove or disconnect the following:
- Front wheel
- Disc brake caliper and support
- Brake rotor and backing plate
- Wheel speed sensor
- Outer tie rod end
- Upper ball joint
- Steering knuckle
- Lower ball joint

To install:

➡ **Use new nuts, bolts and split pins for assembly.**

4. Install or connect the following:
- Lower ball joint. Tighten the mounting bolts to 76 ft. lbs. (103 Nm) on 1999–01 models; 85 ft. lbs. (116 Nm) on 2002 models.
- Steering knuckle. Tighten the lower ball joint nut to 108 ft. lbs. (147 Nm).
- Upper ball joint. Tighten the nut to 72 ft. lbs. (96 Nm).
- Outer tie rod end. Tighten the nut to 72 ft. lbs. (98 Nm).
- Wheel speed sensor
- Brake rotor and backing plate
- Disc brake caliper and support. Tighten the support bolts to 115 ft. lbs. (155 Nm).
- Front wheel

5. Check the wheel alignment and adjust as necessary.

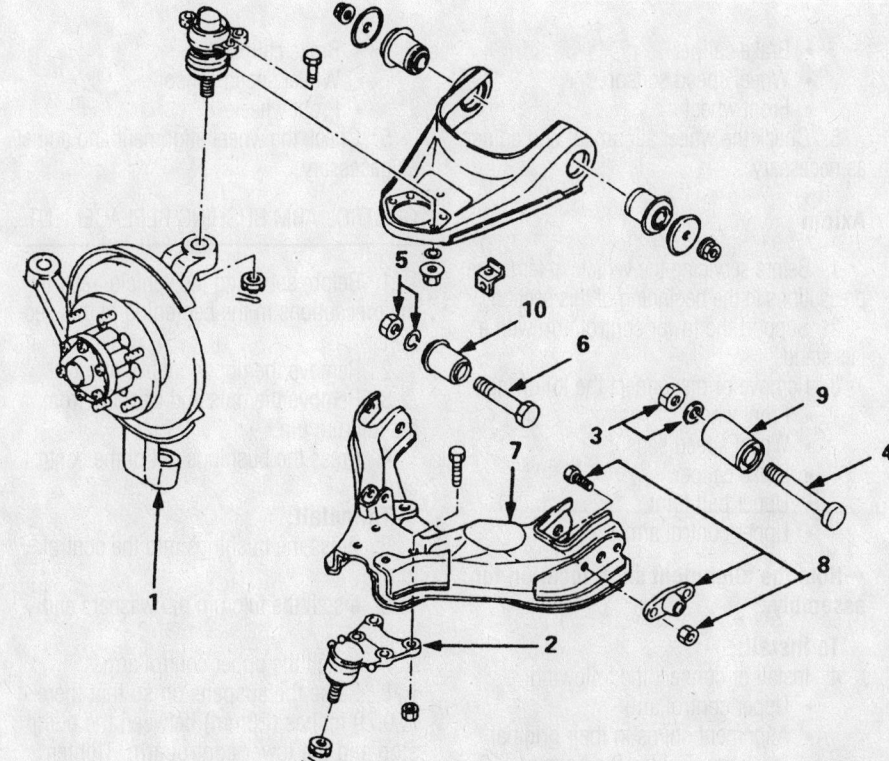

1. Knuckle
2. Lower end
3. Nut and washer, rear
4. Bolt, rear
5. Nut and washer, front
6. Bolt, front
7. Lower control arm assembly
8. Torsion bar arm bracket
9. Bushing, rear
10. Bushing, front

7924BG34

Exploded view of the control arm and ball joint components

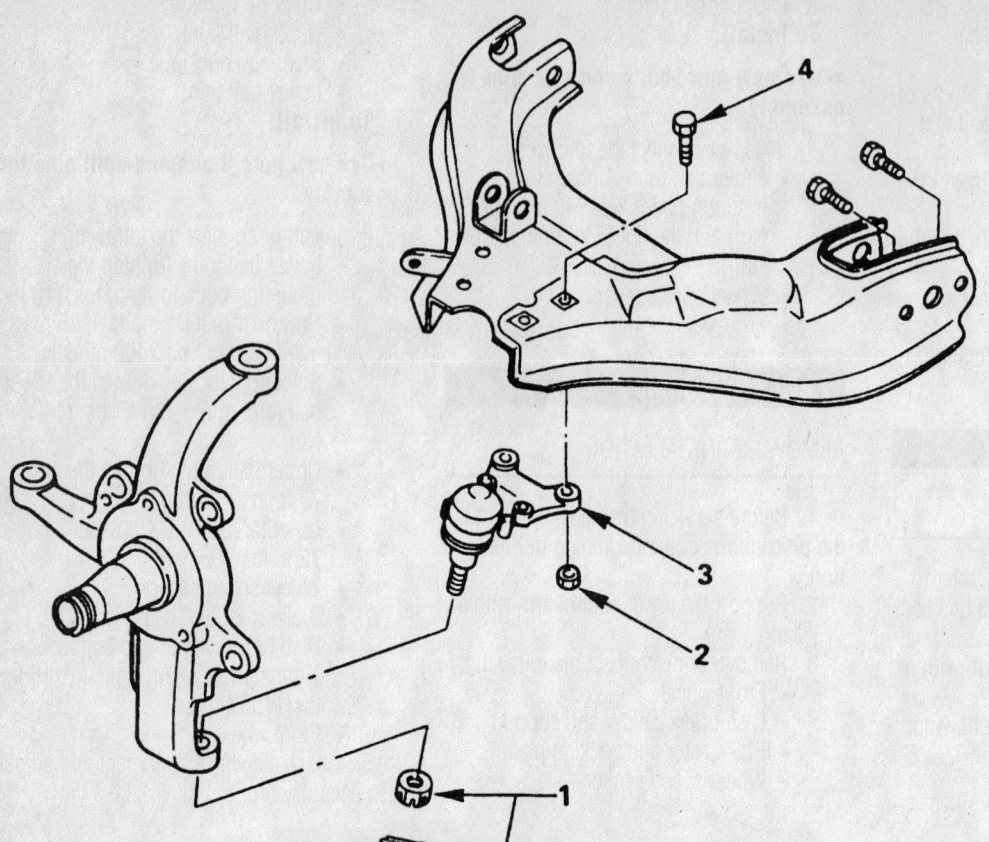

1. Nut and cotter pin
2. Nut
3. Lower ball joint
4. Bolt

7924BG35

Exploded view of the lower ball joint mounting and related components

Upper Control Arm

REMOVAL & INSTALLATION

Except Axiom

1. Before servicing the vehicle, refer to the precautions in the beginning of this section.
2. Support the lower control arm with a jackstand.
3. Remove or disconnect the following:
   • Front wheel
   • Wheel speed sensor
   • Brake caliper
   • Upper ball joint
   • Upper control arm

➡Note the alignment shim location for assembly.

To install:
4. Install or connect the following:
   • Upper control arm
   • Alignment shims in their original locations. Tighten the bolts to 112 ft. lbs. (152 Nm).
   • Upper ball joint. Tighten the nut to 72 ft. lbs. (98 Nm).

   • Brake caliper
   • Wheel speed sensor
   • Front wheel
5. Check the wheel alignment and adjust as necessary.

Axiom

1. Before servicing the vehicle, refer to the precautions in the beginning of this section.
2. Support the lower control arm with a jackstand.
3. Remove or disconnect the following:
   • Front wheel
   • Wheel speed sensor
   • Brake caliper
   • Upper ball joint
   • Upper control arm

➡Note the alignment shim location for assembly.

To install:
4. Install or connect the following:
   • Upper control arm
   • Alignment shims in their original locations. Tighten the bolts to 112 ft. lbs. (152 Nm).
   • Upper ball joint. Tighten the nut to 72 ft. lbs. (98 Nm).

   • Brake caliper
   • Wheel speed sensor
   • Front wheel
5. Check the wheel alignment and adjust as necessary.

CONTROL ARM BUSHING REPLACEMENT

1. Before servicing the vehicle, refer to the precautions in the beginning of this section.
2. Remove the upper control arm.
3. Remove the nuts and washers from the fulcrum pin.
4. Press the bushings out of the control arm.

To install:
5. Press the bushings into the control arm.
6. Install the fulcrum pin washers and nuts.
7. Install the upper control arm.
8. Raise the suspension so that there is 0.79 inches (20mm) between the bump stop and the lower control arm. Tighten the fulcrum pin nuts to 80 ft. lbs. (108 Nm).
9. Check the wheel alignment and adjust as necessary.

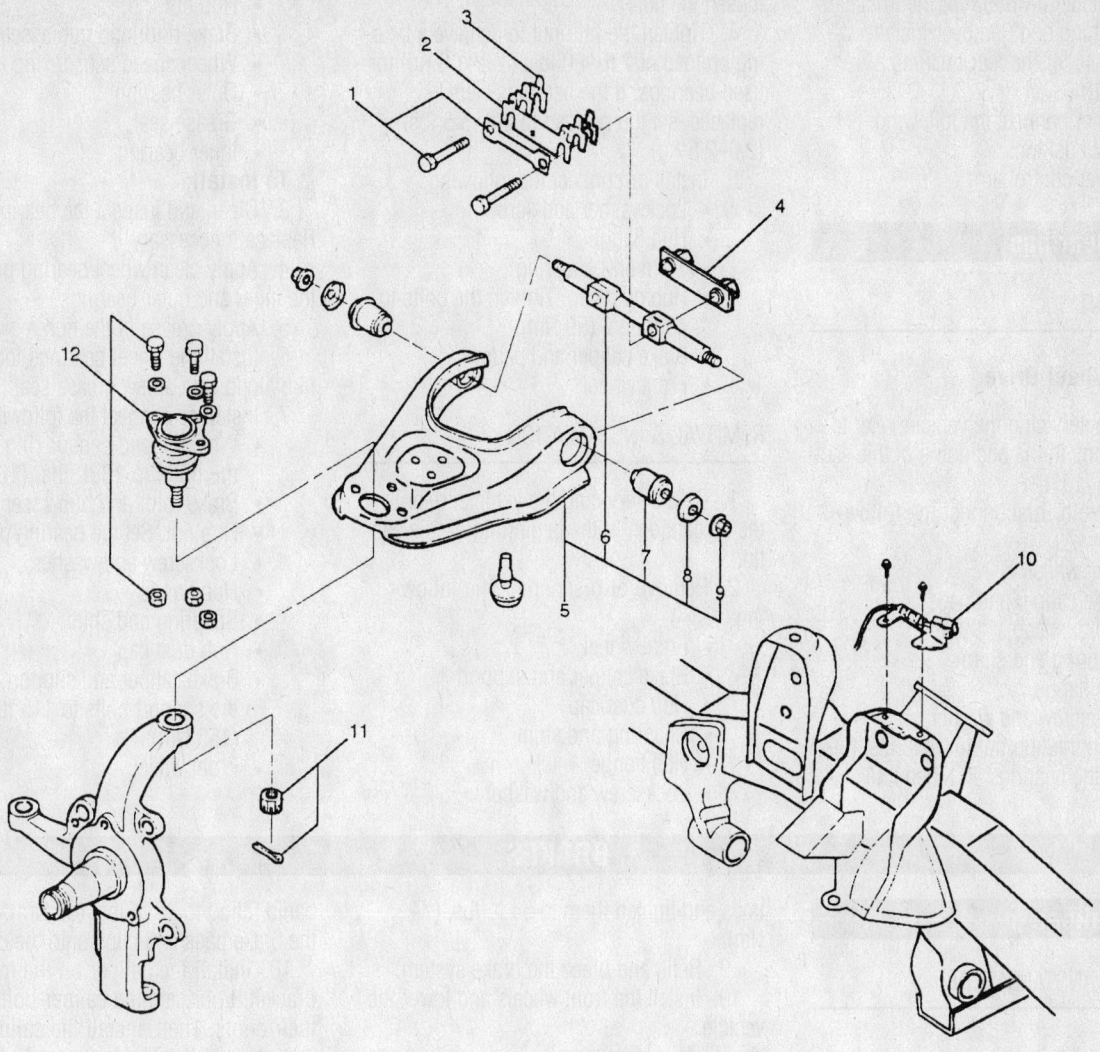

1	Bolt and Plate	7	Bushing
2	Camber Shims	8	Plate
3	Caster Shims	9	Nut
4	Nut Assembly	10	Speed Sensor Cable
5	Upper Control Arm Assembly	11	Nut and Cotter Pin
6	Fulcrum Pin	12	Upper Ball Joint

9308BG06

Upper control arm and related parts

Lower Control Arm

REMOVAL & INSTALLATION

1. Before servicing the vehicle, refer to the precautions in the beginning of this section.

2. Remove or disconnect the following:
   - Front wheel
   - Torsion bar
   - Lower ball joint
   - Stabilizer bar link
   - Shock absorber
   - Lower control arm

To install:

3. Install or connect the following:
   - Lower control arm
   - Shock absorber
   - Stabilizer bar link
   - Lower ball joint
   - Torsion bar
   - Front wheel

4. Raise the suspension so that there is 0.79 inches (20mm) between the bump stop and the lower control arm. For 1999–01, tighten the front control arm bolt to 116 ft. lbs. (157 Nm) and tighten the rear control arm bolt to 145 ft. lbs. (196 Nm). For 2002,

tighten the rear nut to 174 ft. lbs. (235 Nm); the rear nut to 137 ft. lbs. (186 Nm).

5. Check the wheel alignment and adjust as necessary.

CONTROL ARM BUSHING REPLACEMENT

1. Before servicing the vehicle, refer to the precautions in the beginning of this section.

2. Remove or disconnect the following:
   - Lower control arm
   - Bushings, press them from the control arm, using Remover

/Installer J-36833 for the front bushing and Remover/Installer J-36834 for the rear bushing.

To install:

3. Install or connect the following:
   - New bushings
   - Lower control arm

Wheel Bearings

ADJUSTMENT

2- and 4-Wheel Drive

1. Before servicing the vehicle, refer to the precautions in the beginning of this section.

2. Remove or disconnect the following:
   - Front wheel
   - Brake caliper and pads
   - Hub dust cap
   - Snapring and shim
   - Hub flange
   - Lockscrew and washer

3. Tighten the hub nut to 22 ft. lbs. (29 Nm) to seat the bearings and then fully loosen the nut.

4. Tighten the hub nut to achieve a bearing preload of 2.6–4.0 lbs. (1.2–1.8 Kg) for used bearings. If the bearings were replaced, set the preload to 4.4–5.5 lbs. (2.0–2.5 Kg).

5. Install or connect the following:
   - Lockwasher and screw
   - Hub flange
   - Shim and snapring
   - Hub dust cap. Tighten the bolts to 43 ft. lbs. (59 Nm).
   - Brake caliper and pads
   - Front wheel

REMOVAL & INSTALLATION

1. Before servicing the vehicle, refer to the precautions in the beginning of this section.

2. Remove or disconnect the following:
   - Front wheel
   - Brake caliper and support
   - Hub dust cap
   - Snapring and shim
   - Hub flange
   - Lockscrew and washer
   - Hub nut
   - Brake rotor and hub assembly
   - Wheel speed sensor ring
   - Outer bearing
   - Grease seal
   - Inner bearing

To install:

3. Clean and inspect the bearings. Replace if necessary.

4. Apply clean wheel bearing grease to the inner and outer bearings.

5. Apply grease in the hub.

6. Install the wheel bearings into the hub along with a new grease seal.

7. Install or connect the following:
   - Wheel speed sensor ring. Tighten the bolts to 13 ft. lbs. (18 Nm).
   - Brake rotor and hub assembly
   - Hub nut. Set the bearing preload.
   - Lockscrew and washer
   - Hub flange
   - Snapring and shim
   - Hub dust cap
   - Brake caliper and support. Tighten the support bolts to 115 ft. lbs. (155 Nm).
   - Front wheel

BRAKES

Brake Caliper

REMOVAL & INSTALLATION

Trooper

FRONT

1. Raise and safely support the vehicle.
2. Remove some brake fluid from the reservoir.
3. Remove the front wheels.
4. Disconnect the brake fluid line from the caliper. Plug the line to prevent fluid loss.
5. Loosen the brake caliper mounting bolt and guide bolt. Remove the caliper from the mount.
6. Remove the brake pads and clips from the caliper. Inspect the brake pads for wear and replace them if necessary.

To install:

7. Fill the brake caliper with clean brake fluid and connect the fluid line to the caliper using new washers. Tighten the brake line banjo fitting to 26 ft. lbs. (35 Nm). Install the brake pads and clips onto the caliper.

8. Install the caliper onto the mounting bracket. Lubricate the caliper bolts and their boots. Then, install the caliper mounting bolts and tighten them to 54 ft. lbs. (74 Nm).

9. Refill and bleed the brake system.

10. Install the front wheels and lower the vehicle.

REAR

1. Raise and safely support the vehicle.
2. Remove some brake fluid from the reservoir.
3. Remove the rear wheels.
4. Disconnect the brake fluid line from the caliper. Plug the line to prevent fluid loss.
5. Loosen the brake caliper mounting bolt and guide bolt. Remove the caliper from the mount bracket.
6. Remove the brake pads and clips from the caliper. Inspect the brake pads for wear; replace them if necessary.
7. If necessary for servicing, unbolt the caliper mounting bracket from the backing plate.

To install:

8. If removed, install the caliper mounting bracket and tighten its bolts to 76 ft. lbs. (103 Nm).

9. Fill the brake caliper with clean brake fluid and connect the fluid line to the caliper using new washers. Tighten the brake line banjo fitting to 26 ft. lbs. (35 Nm). Install the brake pads and clips onto the caliper.

10. Install the caliper on the mounting bracket. Lubricate the caliper bolts and their boots. Then, install the caliper mounting bolts. Tighten them to 32 ft. lbs. (44 Nm).

11. Refill and bleed the brake system.

12. Install the rear wheels and lower the vehicle.

Amigo, Axiom, Passport and Rodeo

FRONT

1. Raise and safely support the vehicle. Remove the front wheels.

2. Remove some brake fluid from the master cylinder reservoir.

3. Disconnect and plug the brake fluid line from the caliper.

4. Remove the brake caliper mounting bolt and guide bolt and remove the caliper from the mount. The brackets can be remove for additional work space.

5. Remove the brake pads and clips from the caliper. Inspect the brake pads for wear; replace them, if necessary.

To install:

6. Install the brake pads and clips onto the caliper.

7. If the caliper bracket was removed,

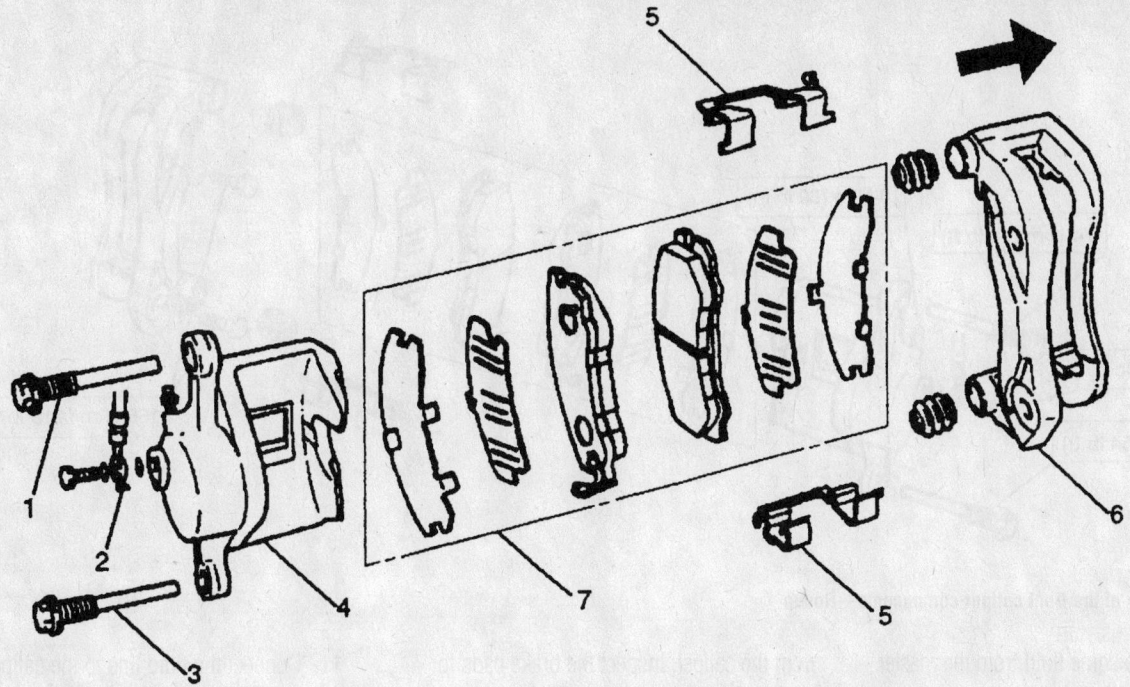

(1) Guide Bolt
(2) Brake Flexible Hose
(3) Lock Bolt

(4) Caliper Assembly
(5) Clip
(6) Support Bracket with Pad Assembly
(7) Pad Assembly

93026G02

Front caliper assembly—Trooper

tighten the bolts to 103–126 ft. lbs. (139–171 Nm).

8. Install the caliper on the mounting bracket. Torque the caliper-to-mounting bracket bolts to:
- 4-cylinder models: 20–27 ft. lbs. (27–37 Nm)
- 6-cylinder models: 54 ft. lbs. (74 Nm)

9. Connect the fluid line to the caliper using new washers. Torque the brake line banjo fitting to 26 ft. lbs. (35 Nm).

✳✳ WARNING

Be sure the hook end of the flexible brake line is positioned in the anti-rotation cavity.

10. Refill the master cylinder reservoir and bleed the brake system.

11. Install the front wheels and lower the vehicle.

REAR

1. Raise and safely support the vehicle. Remove the rear wheels.

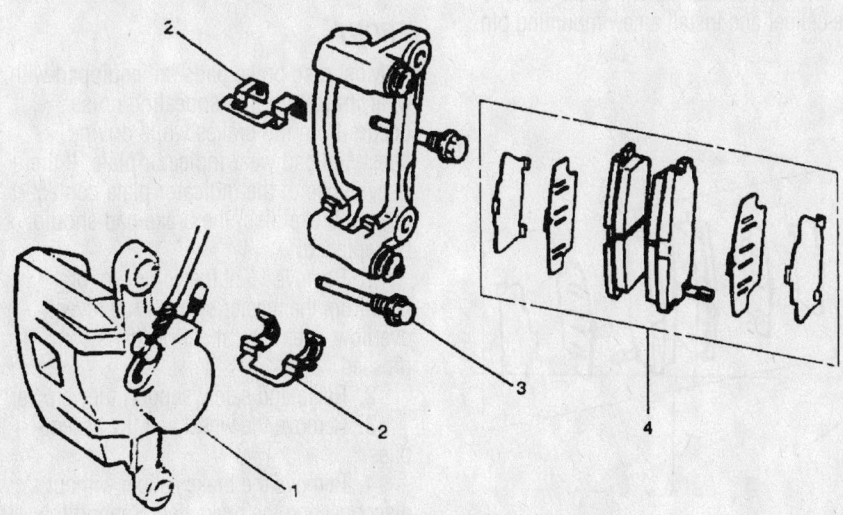

(1) Caliper Assembly
(2) Clip

(3) Lock Bolt
(4) Pad Assembly

93026G01

Rear caliper assembly—Trooper

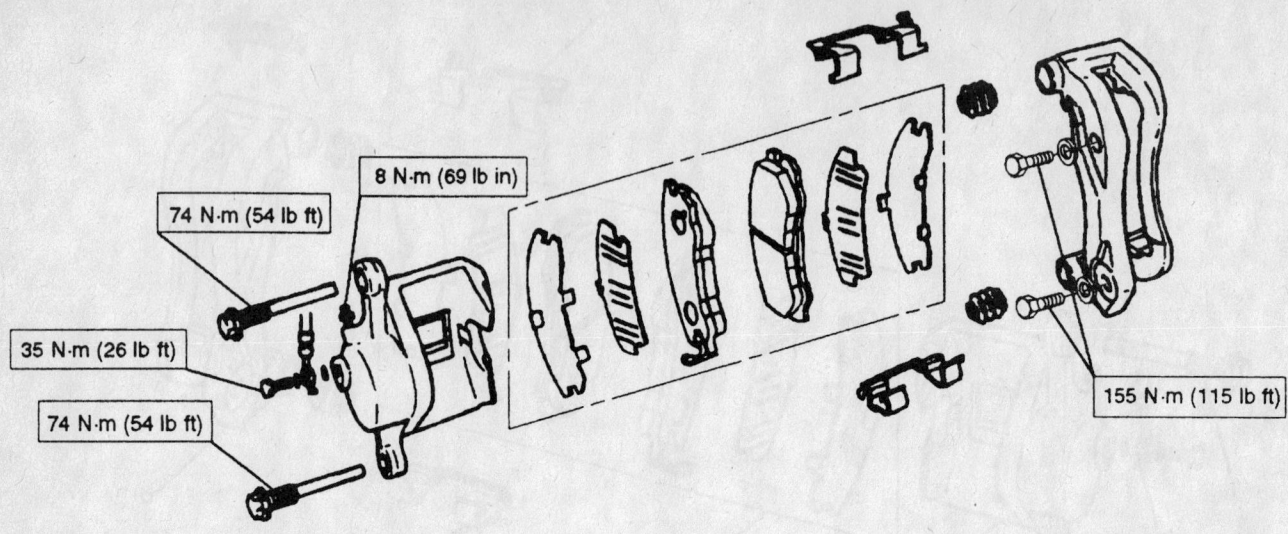

93026G54

Exploded view of the front caliper components—Rodeo

2. Remove some fluid from the master cylinder reservoir.

3. Disconnect and plug the brake fluid line from the caliper.

➡**Discard the parking brake cable mounting pin after removal.**

4. If equipped with caliper actuated parking brakes; remove the mounting pin from the parking brake cable and disconnect the parking cable from the disc caliper.

5. Remove the brake caliper mounting bolt and guide bolt and remove the caliper from the mount.

6. Remove the brake pads and clips from the caliper. Inspect the brake pads for wear; replace them, if necessary.

To install:

7. Install the brake pads and clips onto the caliper.

8. If the mounting bracket was removed, tighten the bolts to 69–84 ft. lbs. (93–114 Nm).

9. Install the caliper on the mounting bracket. Torque the caliper-to-mounting bracket bolts to 12–17 ft. lbs. (16–24 Nm), or 32 ft. lbs. (43 Nm) on vehicles with shoe-type parking brakes.

10. Connect the parking brake cable to the caliper and install a new mounting pin.

11. Connect the fluid line to the caliper using new washers. Torque the brake line banjo fitting to 26 ft. lbs. (35 Nm).

12. Refill the master cylinder reservoir and bleed the brake system.

13. Install the rear wheels and lower the vehicle.

Disc Brake Pads

REMOVAL & INSTALLATION

Amigo, Axiom, Passport and Rodeo

FRONT

Most disc brake pads are equipped with wear indicators. If a squealing noise occurs from the brakes while driving, check the pad wear indicator plate. If there is evidence of the indicator plate contacting the brake disc, the brake pad should be replaced.

1. Remove ½ of the volume of brake fluid from the master cylinder to prevent overflow when the caliper piston is compressed.

2. Raise and safely support the vehicle.

3. Remove the wheel and tire assemblies.

4. Remove the brake caliper without disconnecting the brake line. Support the caliper with a length of wire. Do not let the caliper hang from the brake hose.

➡**On some disc brake systems it is not necessary to remove the caliper when installing new brake pads. Remove the lower slide bolt and rotate the caliper upward to remove the pads.**

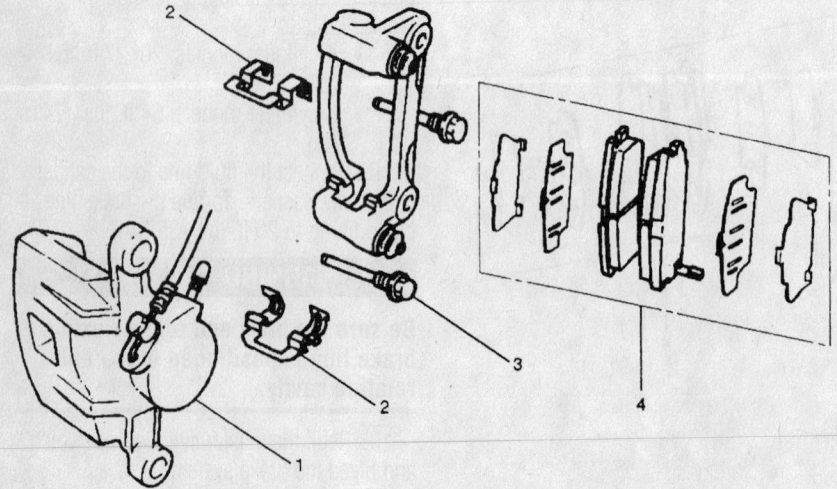

(1) Caliper Assembly
(2) Clip
(3) Lock Bolt
(4) Pad Assembly

93026G55

Exploded view of the rear caliper components—Rodeo

5. Remove the brake pads and shims. Inspect the brake rotor and machine or replace as necessary. Check the minimum thickness (specification is cast into the rotor) before machining.

To install:

6. Use a suitable tool to push the caliper piston into its bore.

7. Apply a thin coat of grease to the rear face of the brake pad and install the shim. Install the brake pads.

8. Install the calipers. Lubricate the caliper bolts and boots. If equipped with a 4-cylinder engine, tighten the caliper mounting bolts to 24 ft. lbs. (33 Nm). If equipped with a 6-cylinder engine, tighten the caliper mounting bolts to 54 ft. lbs. (74 Nm).

9. Install the wheel and tire assemblies and lower the vehicle.

10. Apply the brakes several times to seat the pads before moving the vehicle. Check the fluid in the master cylinder and add as necessary.

REAR

1. Raise and safely support the vehicle. Remove the rear wheels.

2. Remove the brake caliper mounting bolts and remove the caliper without disconnecting the brake fluid line. Support the caliper so it does not hang on the brake line.

3. Remove the brake pads and retaining clips from the caliper.

4. If equipped with caliper-activated parking brakes; use tool J-37617 or equivalent to rotate the piston clockwise until it retracts into the bore. Align the notches of the piston face so the centerline of the notches is perpendicular to the centerline of the mounting bosses.

To install:

5. Install the new brake pads and clips in the caliper and install the caliper in the mounting bracket.

6. Tighten the caliper mounting bolts to 12–17 ft. lbs. (16–24 Nm), or 32 ft. lbs. (43 Nm) on vehicles with shoe-type parking brakes.

7. Install the rear wheels. Check the brake fluid level.

8. Pump the brake pedal until pressure is felt before moving the vehicle.

Trooper

FRONT

1. Remove about ½ of the brake fluid from the master cylinder reservoir to prevent overflow when the caliper piston is compressed.

2. Raise and safely support the vehicle.

3. Remove the front wheels.

4. Remove the brake caliper from the caliper bracket without disconnecting the brake line. Support the caliper with a length of wire. Do not let the caliper hang from the brake hose.

5. Remove the brake pads and shims. Inspect the brake rotor and machine or replace as necessary. Check the minimum thickness (specification is cast into the rotor) before machining.

To install:

6. Use a large C-clamp or brake piston tool to push the caliper piston into its bore.

7. Apply a thin coat of brake grease to both sides of both inner shims. Assemble the pads and shims, then install them into the caliper. The wear indicator on the inner pad must face down.

8. Install the calipers. Clean and lubricate the caliper mounting bolts and lubricate the mounting bolt boots. Install the mounting bolts and tighten them to 54 ft. lbs. (74 Nm).

9. Install the front wheels and lower the vehicle.

10. Apply the brakes several times to seat the pads before moving the vehicle. Check the fluid level in the master cylinder reservoir and add as necessary.

REAR

1. Use a vacuum pump to remove some brake fluid from the master cylinder reservoir to prevent overflow when the caliper piston is compressed.

2. Raise and safely support the vehicle.

3. Remove the rear wheels.

4. Remove the brake caliper from the caliper bracket without disconnecting the brake line. Support the caliper with a length of wire. Do not let the caliper hang from the brake hose.

5. Remove the brake pads and shims. Inspect the brake rotor and machine or replace as necessary. Check the minimum thickness (specification is cast into the rotor) before machining.

To install:

6. Use a large C-clamp or brake piston tool to push the caliper piston into its bore.

7. Apply a thin coat of brake grease to both sides of both inner shims. Assemble the pads and shims, then install them into the caliper. The wear indicator on the inner pad must face down.

8. Install the calipers. Clean and lubricate the caliper mounting bolts and lubricate the mounting bolt boots. Install the mounting bolts and tighten them to 32 ft. lbs. (44 Nm).

9. Install the rear wheels and lower the vehicle.

10. Apply the brakes several times to seat the pads before moving the vehicle. Check the fluid level in the master cylinder reservoir and add as necessary.

Brake Drums

REMOVAL & INSTALLATION

Amigo, Axiom, Passport and Rodeo

1. Raise and safely support the vehicle. Release the parking brake.

2. Remove the rear wheels.

3. Use chalk to mark the brake drum to one of the wheel studs as an index mark for reinstallation.

4. Remove the retaining screw that holds the brake drum to the axle flange.

5. Pull the brake drum from the axle flange.

To install:

6. Align the index mark and install the brake drum to the axle flange.

7. Install the retaining screw to secure the brake drum to the axle flange.

8. Install the rear wheels.

Rear Brake Shoes

REMOVAL & INSTALLATION

Amigo, Axiom, Passport and Rodeo

1. Raise and safely support the vehicle.

2. Remove the rear wheels.

3. Remove the brake drums.

4. Remove the brake return springs.

5. Remove the leading shoe holding pin and spring, and then the leading shoe.

6. Remove the self-adjuster and the adjuster lever.

7. Remove the trailing shoe holding pin and spring.

8. Disconnect the parking brake cable from the trailing shoe and remove the trailing shoe. Remove the parking brake lever from the trailing shoe.

To install:

9. Attach the parking brake lever to the trailing shoe.

10. Connect the parking brake cable to the parking brake lever.

11. Apply a thin coat of high temperature grease to the shoe contact points on the brake backing plate (locations A and C in the accompanying illustration), piston contact surface (B), and self-adjuster (D).

12. Position the trailing shoe on the backing plate and install the hold-down pin, spring, and retainer. Don't stretch the return spring when fitting the shoes onto the backing plate.

13. Connect the upper return spring and the leading shoe to the trailing shoe and position the leading brake shoe on the backing plate.

14. Install the adjuster assembly and the hold-down pin, spring, and retainer.

15. Use a brake spring tool to install the lower return spring.

16. Install the self-adjuster lever and adjuster spring.

17. Adjust the shoe-to-drum clearance to 0.0098–0.0157 in. (0.25–0.40mm) and install the brake drum.

18. Check the brake drum for scoring or other wear. Machine or replace as necessary. Check the maximum brake drum diameter specification when machining.

19. Install the rear wheels. Lower the vehicle.

20. Road-test the vehicle.

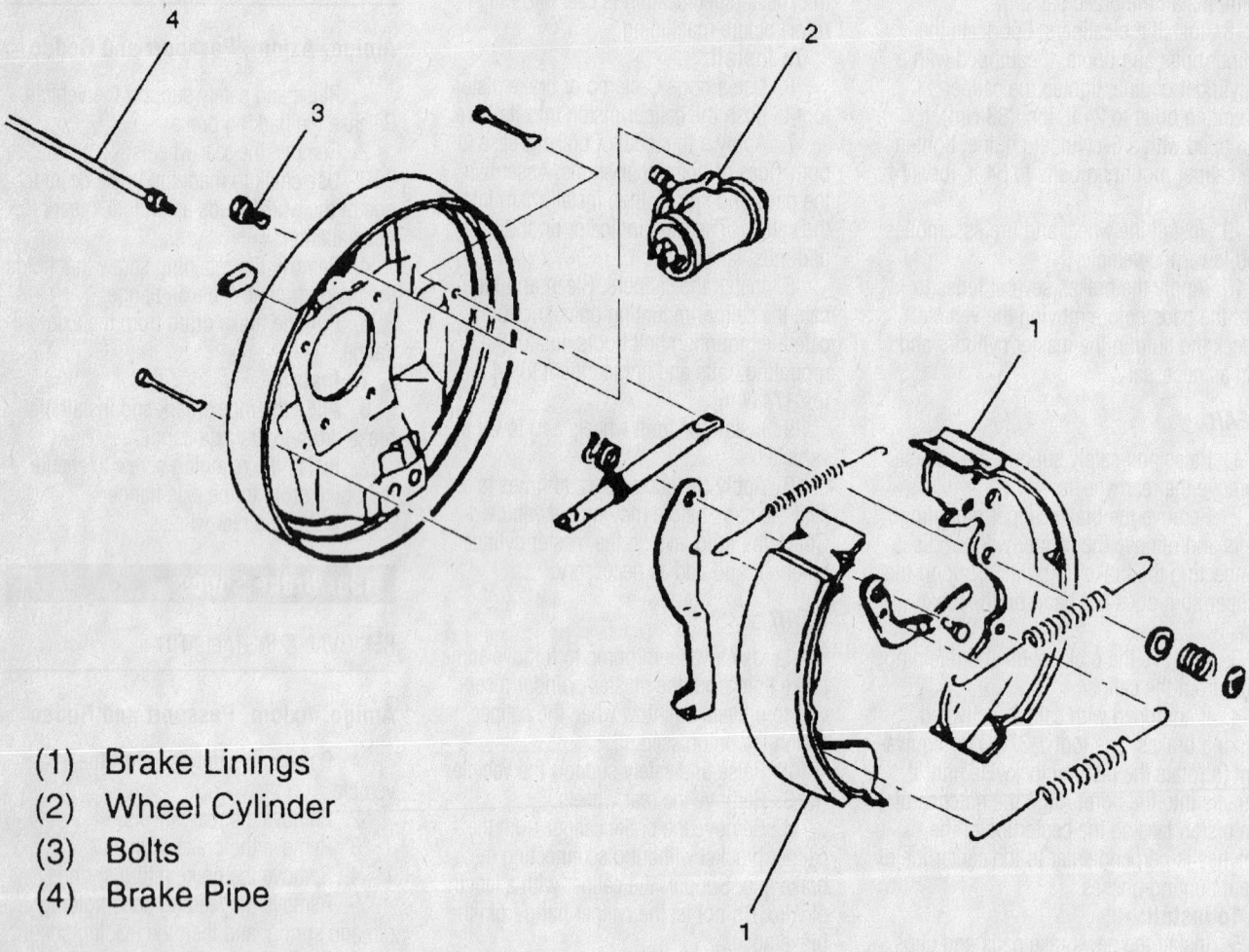

(1) Brake Linings
(2) Wheel Cylinder
(3) Bolts
(4) Brake Pipe

Exploded view of the rear drum brakes—Amigo and Rodeo

93026G56

SPECIFICATION CHARTS

ENGINE AND VEHICLE IDENTIFICATION

Engine							Model Year	
Code ①	Liters (cc)	Cu. In.	Cyl.	Fuel Sys.	Engine Type	Eng. Mfg.	Code ②	Year
1	1.8 (1793)	109	4	EGI	DOHC	KIA	Y	2000
3	1.5 (1493))	91	4	EGI	DOHC	KIA	1	2001
S	2.4 (2351)	144	4	EGI	DOHC	KIA	2	2002
8	2.7 (2656)	163	6	EGI	DOHC	KIA	3	2003
							4	2004

EGI: Electronic Gasoline Injection

DOHC: Double Overhead Camshafts

① 8th digit of VIN

② 10th digit of VIN

42356-KIAC-C01

GENERAL ENGINE SPECIFICATIONS

Year	Model	Engine Displacement Liters (cc)	Engine VIN	Fuel System Type	Net Horsepower @ rpm	Net Torque @ rpm (ft. lbs.)	Bore x Stroke (in.)	Compression Ratio	Oil Pressure @ rpm
2000	Sephia	1.8 (1793)	1	EGI	125@6000	120@4500	3.19x3.43	9.5:1	43-57@3000
2001	Rio	1.5 (1493)	3	EGI	96@5800	98@4500	NA	9.5:1	43-57@3000
	Sephia	1.8 (1793)	1	EGI	125@6000	120@4500	3.19x3.43	9.5:1	43-57@3000
	Spectra	1.8 (1793)	1	EGI	125@6000	120@4500	3.19x3.43	9.5:1	43-57@3000
	Optima	2.4 (2351)	S	EGI	149@6000	159@4500	NA	10.0:1	43-57@3000
		2.7 (2656)	8	EGI	178@6000	181@4000	NA	10.0:1	43-57@3000
2002	Rio	1.5 (1493)	3	EGI	96@5800	98@4500	NA	9.5:1	43-57@3000
	Spectra	1.8 (1793)	1	EGI	125@6000	120@4500	3.19x3.43	9.5:1	43-57@3000
	Optima	2.4 (2351)	S	EGI	149@6000	159@4500	NA	10.0:1	43-57@3000
		2.7 (2656)	8	EGI	178@6000	181@4000	NA	10.0:1	43-57@3000
2003	Rio	1.5 (1493)	3	EGI	96@5800	98@4500	NA	9.5:1	43-57@3000
	Spectra	1.8 (1793)	1	EGI	125@6000	120@4500	3.19x3.43	9.5:1	43-57@3000
	Optima	2.4 (2351)	S	EGI	149@6000	159@4500	NA	10.0:1	43-57@3000
		2.7 (2656)	8	EGI	178@6000	181@4000	NA	10.0:1	43-57@3000

EGI: Electronic Gasoline Injection

MFI: Multi-port Fuel Injection

42356-KIAC-C02

ENGINE TUNE-UP SPECIFICATIONS

Year	Engine Displacement Liters (cc)	Engine VIN	Spark Plug Gap (in.)	Ignition Timing (deg.) MT	Ignition Timing (deg.) AT	Fuel Pump (psi)	Idle Speed (rpm) MT	Idle Speed (rpm) AT	Valve Clearance Intake	Valve Clearance Exhaust
2000	1.8 (1793)	1	0.028-0.032	3-13B	3-13B	64	750-850	750-850	HYD	HYD
2001	1.5 (1493)	3	0.028-0.032	1-11B	1-11B	64	700-800	700-800	HYD	HYD
	1.8 (1793)	1	0.028-0.032	3-13B	3-13B	64	750-850	750-850	HYD	HYD
	2.4 (2351)	S	0.028-0.032	3-13B	3-13B	64	750-850	750-850	HYD	HYD
	2.7 (2656)	8	0.028-0.032	3-13B	3-13B	64	750-850	750-850	HYD	HYD
2002	1.5 (1493)	3	0.028-0.032	1-11B	1-11B	64	700-800	700-800	HYD	HYD
	1.8 (1793)	1	0.028-0.032	3-13B	3-13B	64	750-850	750-850	HYD	HYD
	2.4 (2351)	S	0.028-0.032	3-13B	3-13B	64	750-850	750-850	HYD	HYD
	2.7 (2656)	8	0.028-0.032	3-13B	3-13B	64	750-850	750-850	HYD	HYD
2003	1.5 (1493)	3	0.028-0.032	1-11B	1-11B	64	700-800	700-800	HYD	HYD
	1.8 (1793)	1	0.028-0.032	3-13B	3-13B	64	750-850	750-850	HYD	HYD
	2.4 (2351)	S	0.028-0.032	3-13B	3-13B	64	750-850	750-850	HYD	HYD
	2.7 (2656)	8	0.028-0.032	3-13B	3-13B	64	750-850	750-850	HYD	HYD

NOTE: The Vehicle Emission Control Information label often reflects specification changes made during production. The label figures must be used if they differ from those in this chart

B: Before top dead center

HYD: Hydraulic

42356-KIAC-C03

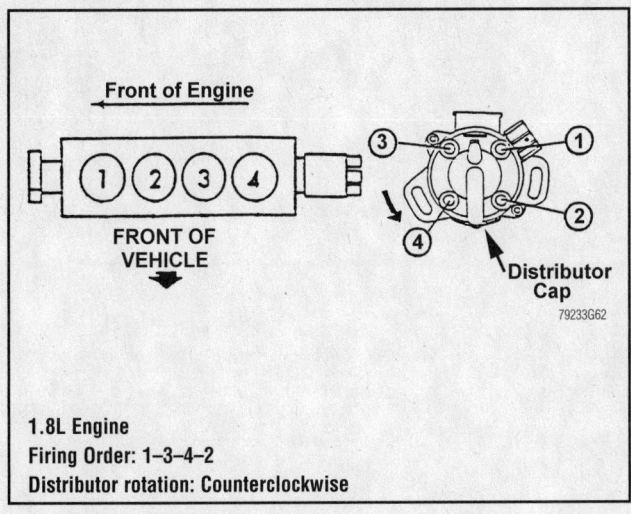

1.8L Engine
Firing Order: 1–3–4–2
Distributor rotation: Counterclockwise

Front of Engine

FRONT OF VEHICLE

Distributor Cap

79233G62

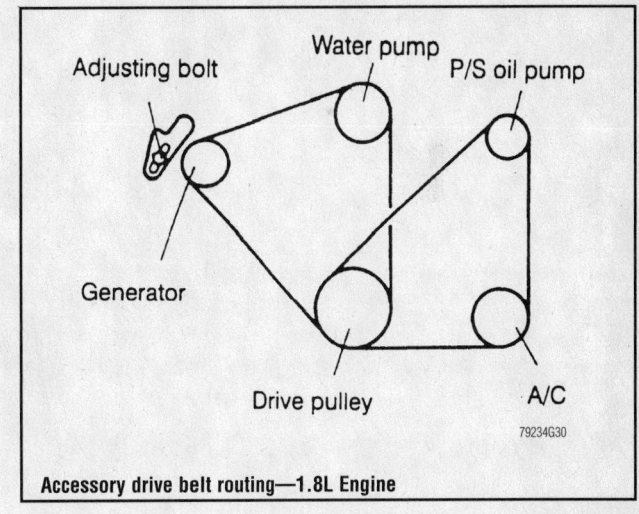

Adjusting bolt
Water pump
P/S oil pump
Generator
Drive pulley
A/C

Accessory drive belt routing—1.8L Engine

79234G30

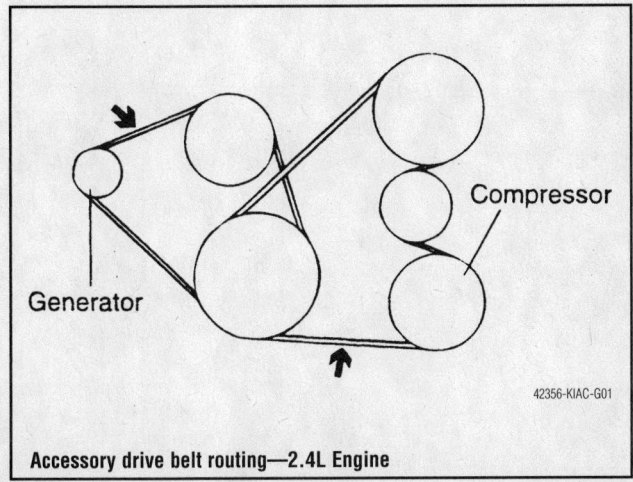

Compressor

Generator

Accessory drive belt routing—2.4L Engine

42356-KIAC-G01

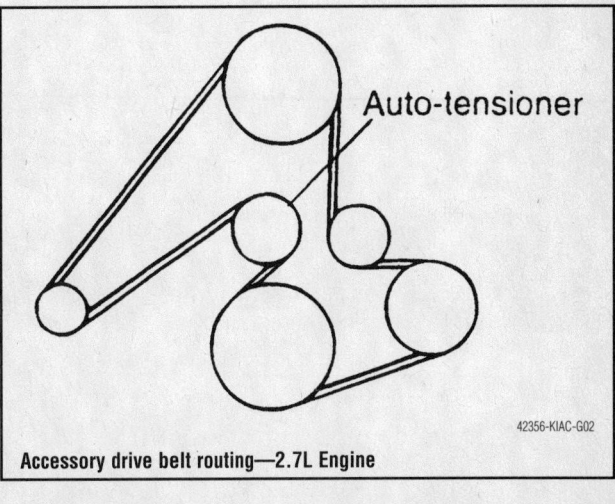

Auto-tensioner

Accessory drive belt routing—2.7L Engine

42356-KIAC-G02

CAPACITIES

Year	Model	Engine Displacement Liters (cc)	Engine VIN	Engine Oil with Filter	Transaxle (pts.) Manual	Auto.	Fuel Tank (gal.)	Cooling System (qts.)
2000	Sephia	1.8 (1793)	1	4.0	5.6	11.4	13.2	6.3
2001	Rio	1.5 (1493)	3	3.4	5.8	12.4	11.9	6.3
	Sephia	1.8 (1793)	1	4.0	5.6	11.4	13.2	6.3
	Spectra	1.8 (1793)	1	4.0	5.6	11.4	13.2	6.3
	Optima	2.4 (2351)	S	4.5	4.4	16.4	17.2	7.2
		2.7 (2656)	8	4.75	—	16.4	17.2	9.1
2002	Rio	1.5 (1493)	3	3.4	5.8	12.4	11.9	6.3
	Spectra	1.8 (1793)	1	4.0	5.6	11.4	13.2	6.3
	Optima	2.4 (2351)	S	4.5	4.4	16.4	17.2	7.2
		2.7 (2656)	8	4.75	—	16.4	17.2	9.1
2003	Rio	1.5 (1493)	3	3.4	5.8	12.4	11.9	6.3
	Spectra	1.8 (1793)	1	4.0	5.6	11.4	13.2	6.3
	Optima	2.4 (2351)	S	4.5	4.4	16.4	17.2	7.2
		2.7 (2656)	8	4.75	—	16.4	17.2	9.1

NOTE: All capacities are approximate. Add fluid gradually and ensure a proper level is obtained.

42356-KIAC-C04

VALVE SPECIFICATIONS

Year	Engine Displacement Liters (cc)	Engine VIN	Seat Angle (deg.)	Face Angle (deg.)	Maximum out of Square (in.)	Spring Free Length (in.)	Stem-to-Guide Clearance (in.)		Stem Diameter (in.)	
							Intake	Exhaust	Intake	Exhaust
2000	1.8 (1793)	1	45	45	0.0638	1.840	①	②	0.2350-0.2356	0.2348-0.2354
2001	1.5 (1493)	3	45	45	0.433	1.653	0.0007-0.0019	0.0019-0.0033	0.2131-0.2137	0.2117-0.2125
	1.8 (1793)	1	45	45	0.0638	1.840	①	②	0.2350-0.2356	0.2348-0.2354
	2.4 (2351)	S	44-44.5	45-45.5	0.0638	1.804	0.0008-0.0019	0.0020-0.0033	0.2585-0.2591	0.2571-0.2579
	2.7 (2656)	8	45	45-45.5	0.0638	1.673	0.0009-0.0020	0.0014-0.0026	0.2348-0.2354	0.2343-0.2348
2002	1.5 (1493)	3	45	45	0.433	1.653	0.0007-0.0019	0.0019-0.0033	0.2131-0.2137	0.2117-0.2125
	1.8 (1793)	1	45	45	0.0638	1.840	①	②	0.2350-0.2356	0.2348-0.2354
	2.4 (2351)	S	44-44.5	45-45.5	0.0638	1.804	0.0008-0.0019	0.0020-0.0033	0.2585-0.2591	0.2571-0.2579
	2.7 (2656)	8	45	45-45.5	0.0638	1.673	0.0009-0.0020	0.0014-0.0026	0.2348-0.2354	0.2343-0.2348
2003	1.5 (1493)	3	45	45	0.433	1.653	0.0007-0.0019	0.0019-0.0033	0.2131-0.2137	0.2117-0.2125
	1.8 (1793)	1	45	45	0.0638	1.840	①	②	0.2350-0.2356	0.2348-0.2354
	2.4 (2351)	S	44-44.5	45-45.5	0.0638	1.804	0.0008-0.0019	0.0020-0.0033	0.2585-0.2591	0.2571-0.2579
	2.7 (2656)	8	45	45-45.5	0.0638	1.673	0.0009-0.0020	0.0014-0.0026	0.2348-0.2354	0.2343-0.2348

NA: Not Available

① Standard range: 0.0010-0.0023 in.

   Maximum value: 0.0080 in.

② Standard range: 0.0012-0.0025 in.

   Maximum value: 0.0080 in.

42356-KIAC-C05

TORQUE SPECIFICATIONS
All readings in ft. lbs.

Year	Engine Displacement Liters (cc)	Engine VIN	Cylinder Head Bolts	Main Bearing Bolts	Rod Bearing Bolts	Crankshaft Damper Bolts	Flywheel Bolts	Manifold Intake	Manifold Exhaust	Spark Plugs	Lug Nuts
2000	1.8 (1793)	1	①	②	35-37	9-13 ③	71-76	14-19	28-34	11-17	65-87
2001	1.5 (1493)	3	④	40-43	22-24	28-38	71-76	11-14	11-14	11-17	65-87
	1.8 (1793)	1	①	②	35-37	9-13 ③	71-76	14-19	28-34	11-17	65-87
	2.4 (2351)	S	⑤	⑥	⑦	80-94	94-101	⑧	⑨	15-21	65-87
	2.7 (2656)	8	⑩	⑪	⑫	130-138	53-55	14-15	12-22	15-21	65-87
2002	1.5 (1493)	3	④	40-43	22-24	28-38	71-76	11-14	11-14	11-17	65-87
	1.8 (1793)	1	①	②	35-37	9-13 ③	71-76	14-19	28-34	11-17	65-87
	2.4 (2351)	S	⑤	⑥	⑦	80-94	94-101	⑧	⑨	15-21	65-87
	2.7 (2656)	8	⑩	⑪	⑫	130-138	53-55	14-15	12-22	15-21	65-87
2003	1.5 (1493)	3	④	40-43	22-24	28-38	71-76	11-14	11-14	11-17	65-87
	1.8 (1793)	1	①	②	35-37	9-13 ③	71-76	14-19	28-34	11-17	65-87
	2.4 (2351)	S	⑤	⑥	⑦	80-94	94-101	⑧	⑨	15-21	65-87
	2.7 (2656)	8	⑩	⑪	⑫	130-138	53-55	14-15	12-22	15-21	65-87

① Step 1: 36 ft. lbs.
  Step 2: Loosen fully
  Step 3: 29 ft. lbs.
  Step 4: Tighten 90 degrees
  Step 5: Additional 90 degrees

② Step 1: 29 ft. lbs.
  Step 2: Loosen fully
  Step 3: 14.5 ft. lbs.
  Step 4: Tighten 90 degrees
  Step 5: Tighten 60 degrees

③ Crankshaft pulley

④ Step 1: 36 inch lbs.
  Step 2: Loosen fully
  Step 3: 18 ft. lbs.
  Step 4: Tighten 90 degrees
  Step 5: Tighten 60 degrees

⑤ Step 1: 14 ft. lbs.
  Step 2: Loosen fully
  Step 3: 14 ft. lbs.
  Step 4: Tighten 90 degrees
  Step 5: Tighten 90 degrees

⑥ 18 ft. lbs., plus 90-94 degrees

⑦ 13-16 ft. lbs., plus 90-94 degrees

⑧ M10 bolts: 13-18 ft. lbs.
  M8 bolts: 11-14 ft. lbs.
  Nut: 22-30 ft. lbs.

⑨ M10 bolts: 25-40 ft. lbs.
  M8 bolts: 18-22 ft. lbs.

⑩ Step 1: 18 ft. lbs., plus 60-65 degrees
  Step 2: Tighten 45-49 degrees
  Step 3: 26-32 ft. lbs.

⑪ M10 bolts: 19.5-23.9 ft. lbs., plus 90-95 degrees
  M8 bolts: 9.6-14 ft. lbs., plus 90-95 degrees

⑫ 11.6-14 ft. lbs., plus 90-94 degrees

42356-KIAC-C06

PISTON AND RING SPECIFICATIONS

All measurements are given in inches.

Year	Engine Displacement Liters (cc)	Engine VIN	Piston Clearance	Ring Gap			Ring Side Clearance		
				Top Compression	Bottom Compression	Oil Control	Top Compression	Bottom Compression	Oil Control
2000	1.8 (1793)	1	0.0015-0.0021	0.006-0.011	0.012-0.018	0.008-0.027	0.0020-0.0030	0.0010-0.0030	SNUG
2001	1.5 (1493)	3	0.0015-0.0021	0.006-0.011	0.0016-0.0021	0.008-0.027	0.0020-0.0030	0.0010-0.0030	SNUG
	1.8 (1793)	1	0.0015-0.0021	0.006-0.011	0.012-0.018	0.008-0.027	0.0020-0.0030	0.0010-0.0030	SNUG
	2.4 (2351)	S	0.0008-0.0120	0.0098-0.0138	0.0157-0.0216	0.0039-0.0157	0.0012-0.0028	0.0008-0.0024	SNUG
	2.7 (2656)	8	0.0004-0.0012	0.0079-0.0138	0.0146-0.0205	0.0079-0.0276	0.0016-0.0031	0.0012-0.0028	SNUG
2002	1.5 (1493)	3	0.0015-0.0021	0.006-0.011	0.0016-0.0021	0.008-0.027	0.0020-0.0030	0.0010-0.0030	SNUG
	1.8 (1793)	1	0.0015-0.0021	0.006-0.011	0.012-0.018	0.008-0.027	0.0020-0.0030	0.0010-0.0030	SNUG
	2.4 (2351)	S	0.0008-0.0120	0.0098-0.0138	0.0157-0.0216	0.0039-0.0157	0.0012-0.0028	0.0008-0.0024	SNUG
	2.7 (2656)	8	0.0004-0.0012	0.0079-0.0138	0.0146-0.0205	0.0079-0.0276	0.0016-0.0031	0.0012-0.0028	SNUG
2003	1.5 (1493)	3	0.0015-0.0021	0.006-0.011	0.0016-0.0021	0.008-0.027	0.0020-0.0030	0.0010-0.0030	SNUG
	1.8 (1793)	1	0.0015-0.0021	0.006-0.011	0.012-0.018	0.008-0.027	0.0020-0.0030	0.0010-0.0030	SNUG
	2.4 (2351)	S	0.0008-0.0120	0.0098-0.0138	0.0157-0.0216	0.0039-0.0157	0.0012-0.0028	0.0008-0.0024	SNUG
	2.7 (2656)	8	0.0004-0.0012	0.0079-0.0138	0.0146-0.0205	0.0079-0.0276	0.0016-0.0031	0.0012-0.0028	SNUG

42356-KIAC-C07

CRANKSHAFT AND CONNECTING ROD SPECIFICATIONS

All measurements are given in inches.

Year	Engine Displacement Liters (cc)	Engine VIN	Crankshaft				Connecting Rod		
			Main Brg. Journal Dia.	Main Brg. Oil Clearance	Shaft End-play	Thrust on No.	Journal Diameter	Oil Clearance	Side Clearance
2000	1.8 (1793)	1	2.1629-2.1636	0.0010-0.0017	0.0032-0.0111	3	1.7693-1.7700	0.0008-0.0019	0.0044-0.0019
2001	1.5 (1493)	3	1.9365-1.9562	0.0007-0.0014	0.0032-0.0111	3	1.7693-1.7700	0.0008-0.0019	0.0044-0.0103
	1.8 (1793)	1	2.1629-2.1636	0.0010-0.0017	0.0032-0.0111	3	1.7693-1.7700	0.0008-0.0019	0.0044-0.0019
	2.4 (2351)	S	2.2434-2.2442	①	0.0020-0.0098	3	1.7709-1.7717	0.0008-0.0020	0.0040-0.0098
	2.7 (2656)	8	2.4402-2.4409	0.0002-0.0009	0.0028-0.0098	3	1.8891-1.8898	0.0007-0.0014	0.0039-0.0098
2002	1.5 (1493)	3	1.9365-1.9562	0.0007-0.0014	0.0032-0.0111	3	1.7693-1.7700	0.0008-0.0019	0.0044-0.0103
	1.8 (1793)	1	2.1629-2.1636	0.0010-0.0017	0.0032-0.0111	3	1.7693-1.7700	0.0008-0.0019	0.0044-0.0019
	2.4 (2351)	S	2.2434-2.2442	①	0.0020-0.0098	3	1.7709-1.7717	0.0008-0.0020	0.0040-0.0098
	2.7 (2656)	8	2.4402-2.4409	0.0002-0.0009	0.0028-0.0098	3	1.8891-1.8898	0.0007-0.0014	0.0039-0.0098
2003	1.5 (1493)	3	1.9365-1.9562	0.0007-0.0014	0.0032-0.0111	3	1.7693-1.7700	0.0008-0.0019	0.0044-0.0103
	1.8 (1793)	1	2.1629-2.1636	0.0010-0.0017	0.0032-0.0111	3	1.7693-1.7700	0.0008-0.0019	0.0044-0.0019
	2.4 (2351)	S	2.2434-2.2442	①	0.0020-0.0098	3	1.7709-1.7717	0.0008-0.0020	0.0040-0.0098
	2.7 (2656)	8	2.4402-2.4409	0.0002-0.0009	0.0028-0.0098	3	1.8891-1.8898	0.0007-0.0014	0.0039-0.0098

① Journal Nos. 1, 2, 4 & 5: 0.0007-0.0014 in.
 Journal Nos. 3: 0.0009-0.0016 in.

42356-KIAC-C08

WHEEL ALIGNMENT

Year	Model		Caster Range (+/-Deg.)	Caster Preferred Setting (Deg.)	Camber Range (+/-Deg.)	Camber Preferred Setting (Deg.)	Toe-in (in.)	Steering Axis Inclination (Deg.)
2000	Sephia	F	0.75	+2.45	0.50	0	0.11 +/- 0.12	12.58
		R	—	—	0.50	-0.52	0.07 +/- 0.12	—
2001	Rio	F	0.45	+1.41	0.45	+0.6	0.12 +/- 0.12	14.70
		R	—	—	0.18	-0.50	0.20 +/- 0.24	—
	Sephia	F	0.75	+2.45	0.50	0	0.11 +/- 0.12	—
		R	—	—	0.50	-0.52	0.07 +/- 0.12	—
	Spectra	F	0.75	+2.45	0.50	0	0.11 +/- 0.12	12.58
		R	—	—	0.50	-0.52	0.07 +/- 0.12	—
	Optima	F	1.00	+3.15	0.30	0	0.11 +/- 0.12	8.29
		R	—	—	0.30	-0.30	0.07 +/- 0.12	—
2002	Rio	F	0.45	+1.41	0.45	+0.6	0.12 +/- 0.12	14.70
		R	—	—	0.18	-0.50	0.20 +/- 0.24	—
	Spectra	F	0.75	+2.45	0.50	0	0.11 +/- 0.12	12.58
		R	—	—	0.50	-0.52	0.07 +/- 0.12	—
	Optima	F	1.00	+3.15	0.30	0	0.11 +/- 0.12	8.29
		R	—	—	0.30	-0.30	0.07 +/- 0.12	—
2003	Rio	F	0.45	+1.41	0.45	+0.6	0.12 +/- 0.12	14.70
		R	—	—	0.18	-0.50	0.20 +/- 0.24	—
	Spectra	F	0.75	+2.45	0.50	0	0.11 +/- 0.12	12.58
		R	—	—	0.50	-0.52	0.07 +/- 0.12	—
	Optima	F	1.00	+3.15	0.30	0	0.11 +/- 0.12	8.29
		R	—	—	0.30	-0.30	0.07 +/- 0.12	—

42356-KIAC-C09

TIRE, WHEEL AND BALL JOINT SPECIFICATIONS

Year	Model	OEM Tires		Tire Pressures (psi)		Wheel Size	Ball Joint Inspection
		Standard	Optional	Front	Rear		
2000	Sephia	P175/70SR13	P185/60HR14	29	29	Std: 5-JJ Opt: 6-JJ	①
2001	Rio	P155/80SR13	P175/70R13	②	②	Std: 5-JJ Opt: 5-JJ	①
	Sephia	P175/70SR13	P185/60HR14	29	29	Std: 5-JJ Opt: 6-JJ	①
	Spectra	P175/70SR13	P185/60HR14	29	29	Std: 5-JJ Opt: 6-JJ	①
	Optima	P175/70R14	P205/60R15	30	30	Std: 5.5-JJ Opt: 6-JJ	①
2002	Rio	P155/80SR13	P175/70R13	②	②	Std: 5-JJ Opt: 5-JJ	①
	Spectra	P175/70SR13	P185/60HR14	29	29	Std: 5-JJ Opt: 6-JJ	①
	Optima	P175/70R14	P205/60R15	30	30	Std: 5.5-JJ Opt: 6-JJ	①
2003	Rio	P155/80SR13	P175/70R13	②	②	Std: 5-JJ Opt: 5-JJ	①
	Spectra	P175/70SR13	P185/60HR14	29	29	Std: 5-JJ Opt: 6-JJ	①
	Optima	P175/70R14	P205/60R15	30	30	Std: 5.5-JJ Opt: 6-JJ	①

OEM: Original Equipment Manufacturer

PSI: Pounds Per Square Inch

STD: Standard

OPT: Optional

① Replace if any measurable movement is found.

② Standard front and rear tire pressure: 32 psi

   Optional front and rear tire pressure: 29 psi

42356-KIAC-C10

BRAKE SPECIFICATIONS
All measurements in inches unless noted

Year	Model		Brake Disc Original Thickness	Brake Disc Minimum Thickness	Brake Disc Maximum Run-out	Brake Drum Original Inside Diameter	Brake Drum Max. Wear Limit	Brake Drum Maximum Machine Diameter	Minimum Lining Thickness	Brake Caliper Bracket Bolts (ft. lbs.)	Brake Caliper Mounting Bolts (ft. lbs.)
2000	Sephia	F	0.940	0.710	0.0040	—	—	—	0.080	33-49	19-21
		R	0.400	0.320	0.0039	7.87	7.91	7.91	0.079	33-49	22-29
2001	Rio	F	0.870	0.710	0.0040	—	—	—	0.080	33-49	19-21
		R	—	—	—	7.87	7.91	7.91	0.079	33-49	22-29
	Sephia	F	0.940	0.710	0.0040	—	—	—	0.080	33-49	19-21
		R	0.400	0.320	0.0039	7.87	7.91	7.91	0.079	33-49	22-29
	Spectra	F	0.940	0.710	0.0040	—	—	—	0.080	33-49	19-21
		R	0.400	0.320	0.0039	7.87	7.91	7.91	0.079	33-49	22-29
	Optima	F	0.965	0.880	0.0012	—	—	—	0.079	51-63	16-24
		R	0.390	0.320	0.0039	9.00	9.08	9.08	0.079	51-63	16-24
2002	Rio	F	0.870	0.710	0.0040	—	—	—	0.080	33-49	19-21
		R	—	—	—	7.87	7.91	7.91	0.079	33-49	22-29
	Spectra	F	0.940	0.710	0.0040	—	—	—	0.080	33-49	19-21
		R	0.400	0.320	0.0039	7.87	7.91	7.91	0.079	33-49	22-29
	Optima	F	0.965	0.880	0.0012	—	—	—	0.079	51-63	16-24
		R	0.390	0.320	0.0039	9.00	9.08	9.08	0.079	51-63	16-24
2003	Rio	F	0.870	0.710	0.0040	—	—	—	0.080	33-49	19-21
		R	—	—	—	7.87	7.91	7.91	0.079	33-49	22-29
	Spectra	F	0.940	0.710	0.0040	—	—	—	0.080	33-49	19-21
		R	0.400	0.320	0.0039	7.87	7.91	7.91	0.079	33-49	22-29
	Optima	F	0.965	0.880	0.0012	—	—	—	0.079	51-63	16-24
		R	0.390	0.320	0.0039	9.00	9.08	9.08	0.079	51-63	16-24

F: Front

R: Rear

42356-KIAC-C11

SCHEDULED MAINTENANCE INTERVALS
Kia—Sephia, Spectra, Rio & Optima

TO BE SERVICED	TYPE OF SERVICE	VEHICLE MILEAGE INTERVAL (x1000)																
		7.5	15	22.5	30	37.5	45	52.5	60	67.5	75	82.5	90	97.5	105	112.5	120	127
Accessory drive belts	S/I				✓				✓				✓				✓	✓
Air cleaner element	R				✓				✓				✓				✓	✓
Air conditioner system	S/I	Inspect the system operation and refrigerant amount annually.																
Brake lines, hoses and connections	S/I				✓				✓				✓				✓	✓
Chassis and body fasteners	T				✓				✓				✓				✓	✓
Clutch pedal height, freeplay and operation	S/I				✓				✓				✓				✓	✓
Cooling system hoses and coolant level	S/I				✓				✓				✓				✓	✓
CV-joint boots	S/I				✓				✓				✓				✓	✓
Engine coolant	R				✓				✓				✓				✓	✓
Engine oil and filter	R	✓	✓	✓	✓	✓	✓	✓	✓	✓	✓	✓	✓	✓	✓	✓	✓	✓
Exhaust system heat shields	S/I				✓				✓				✓				✓	✓
Front and rear brakes	S/I				✓				✓				✓				✓	✓
Front ball joints S/I	S/I				✓				✓				✓				✓	✓
Fuel filter	R								✓								✓	✓
Fuel lines and hoses	S/I				✓				✓				✓				✓	✓
Idle speed	A				✓				✓				✓				✓	✓
Locks and hinges	L	✓	✓	✓	✓	✓	✓	✓	✓	✓	✓	✓	✓	✓	✓	✓	✓	✓
Spark plugs	R				✓				✓				✓				✓	✓
Steering operation and linkage	S/I				✓				✓				✓				✓	✓
Timing belt (California models)	R														✓			
Timing belt (California models)	S/I								✓				✓					
Timing belt (except California models)	R								✓								✓	✓

R: Replace   S/I: Inspect and service, if needed   L: Lubricate   A: Adjust   T: Tighten

FREQUENT OPERATION MAINTENANCE (SEVERE SERVICE)

If a vehicle is operated under any of the following conditions it is considered severe service

- Towing a trailer or using a camper or car-top carrier.

- Repeated short trips of less than 5 miles in temperatures below freezing, or trips of less than 10 miles in any temperature.

- Extensive idling or low-speed driving for long distances as in heavy commercial use, such as delivery, taxi or police cars.

- Operating on rough, muddy or salt-covered roads.

- Operating on unpaved or dusty roads.

- Driving in extremely hot (over 90°F) conditions.

Engine oil and filter: replace every 5000 miles or 5 months, whichever occurs first.

Air cleaner element: inspect ever 15,000 miles or 15 months and replace every 30,000 miles or 30 months, whichever occurs first.

Fuel system hoses (California models only): replace every 105,000 miles.

Emission system hoses (non-California models): inspect every 55,000 miles or 55 months, whichever occurs first.

Emission system hoses (California models): inspect every 60,000 miles or 60 months, whichever occurs first.

Front and rear disc brakes: inspect every 15,000 miles or 15 months, whichever occurs first.

Chassis and body fasteners: tighten every 15,000 miles or 15 months, whichever occurs first.

Locks and hinges: lubricate every 5000 miles or 5 months, whichever occurs first.

PRECAUTIONS

Before servicing any vehicle, please be sure to read all of the following precautions, which deal with personal safety, prevention of component damage, and important points to take into consideration when servicing a motor vehicle:

• Never open, service or drain the radiator or cooling system when the engine is hot; serious burns can occur from the steam and hot coolant.

• Observe all applicable safety precautions when working around fuel. Whenever servicing the fuel system, always work in a well-ventilated area. Do not allow fuel spray or vapors to come in contact with a spark, open flame, or excessive heat (a hot drop light, for example). Keep a dry chemical fire extinguisher near the work area. Always keep fuel in a container specifically designed for fuel storage; also, always properly seal fuel containers to avoid the possibility of fire or explosion. Refer to the additional fuel system precautions later in this section.

• Fuel injection systems often remain pressurized, even after the engine has been turned **OFF**. The fuel system pressure must be relieved before disconnecting any fuel lines. Failure to do so may result in fire and/or personal injury.

• Brake fluid often contains polyglycol ethers and polyglycols. Avoid contact with the eyes and wash your hands thoroughly after handling brake fluid. If you do get brake fluid in your eyes, flush your eyes with clean, running water for 15 minutes. If eye irritation persists, or if you have taken brake fluid internally, IMMEDIATELY seek medical assistance.

• The EPA warns that prolonged contact with used engine oil may cause a number of skin disorders, including cancer! You should make every effort to minimize your exposure to used engine oil. Protective gloves should be worn when changing oil. Wash your hands and any other exposed skin areas as soon as possible after exposure to used engine oil. Soap and water, or waterless hand cleaner should be used.

• All new vehicles are now equipped with an air bag system. The system must be disabled before performing service on or around system components, steering column, instrument panel components, wiring and sensors. Failure to follow safety and disabling procedures could result in accidental air bag deployment, possible personal injury and unnecessary system repairs.

• Always wear safety goggles when working with, or around, the air bag system. When carrying a non-deployed air bag, be sure the bag and trim cover are pointed away from your body. When placing a non-deployed air bag on a work surface, always face the bag and trim cover upward, away from the surface. This will reduce the motion of the module if it is accidentally deployed. Refer to the additional air bag system precautions later in this section.

• Clean, high quality brake fluid from a sealed container is essential to the safe and proper operation of the brake system. You should always buy the correct type of brake fluid for your vehicle. If the brake fluid becomes contaminated, completely flush the system with new fluid. Never reuse any brake fluid. Any brake fluid that is removed from the system should be discarded. Also, do not allow any brake fluid to come in contact with a painted surface; it will damage the paint.

• Never operate the engine without the proper amount and type of engine oil; doing so WILL result in severe engine damage.

• Timing belt maintenance is extremely important! Many models utilize an interference-type, non-freewheeling engine. If the timing belt breaks, the valves in the cylinder head may strike the pistons, causing potentially serious (also time-consuming and expensive) engine damage. Refer to the maintenance interval charts in the front of this section for the recommended replacement interval for the timing belt.

• Disconnecting the negative battery cable on some vehicles may interfere with the functions of the on-board computer system(s) and may require the computer to undergo a relearning process once the negative battery cable is reconnected.

• When servicing drum brakes, only dissemble and assemble one side at a time, leaving the remaining side intact for reference.

• Only an MVAC-trained, EPA-certified automotive technician should service the air conditioning system or its components.

ENGINE REPAIR

Alternator

REMOVAL

1. Before servicing the vehicle, refer to the precautions in the beginning of this section.
2. Remove or disconnect the following:
   • Negative battery cable
   • Front air intake inlet pipe bolts, 1.8L engine only
   • Top hose from the air intake inlet pipe, 1.8L engine only
   • Air intake inlet pipe clamp and the pipe, 1.8L engine only
   • Alternator **B** terminal cover cap
   • Alternator **B** terminal nut and the **B** terminal lead
   • Alternator **L** and **S** electrical connections
   • Alternator pivot bolt and the tensioner mounting bolt, loosen but do not remove
   • Drive belt(s); relieve tension on the belt by rotating the adjustment bolt
   • Alternator tensioner mounting bolt and the belt tensioner
   • Alternator pivot bolt
3. Loosen the bolt at the base of the adjusting bracket and rotate the bracket up.
   • Alternator

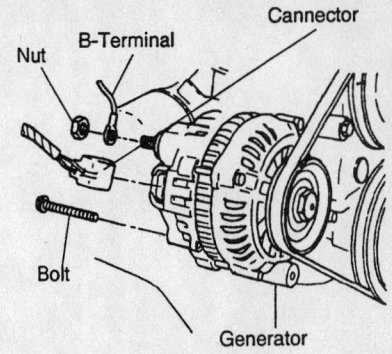

Alternator mounting—1.5L engine shown

9357NG01

2.4L DOHC ENGINE

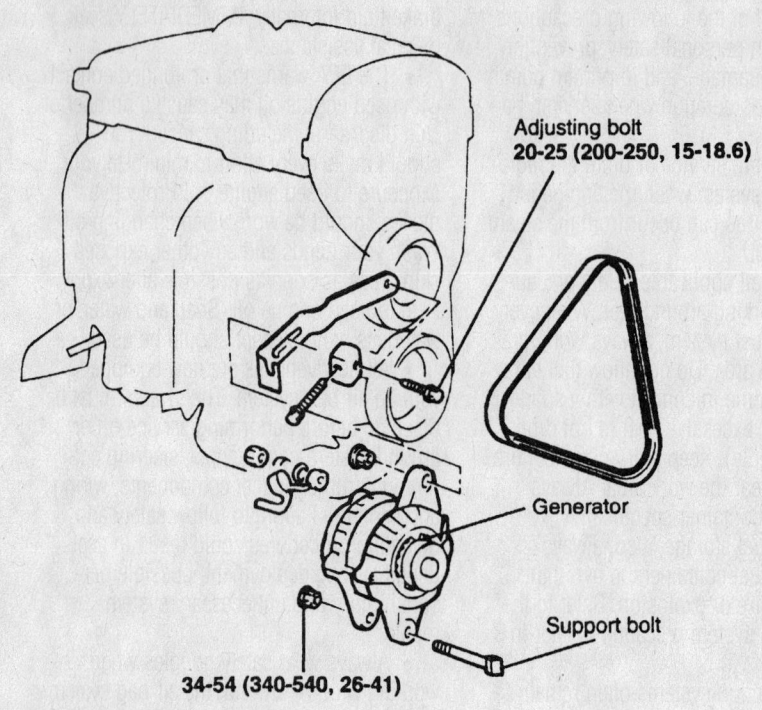

Adjusting bolt
20-25 (200-250, 15-18.6)

Generator

Support bolt

34-54 (340-540, 26-41)

V6 ENGINE

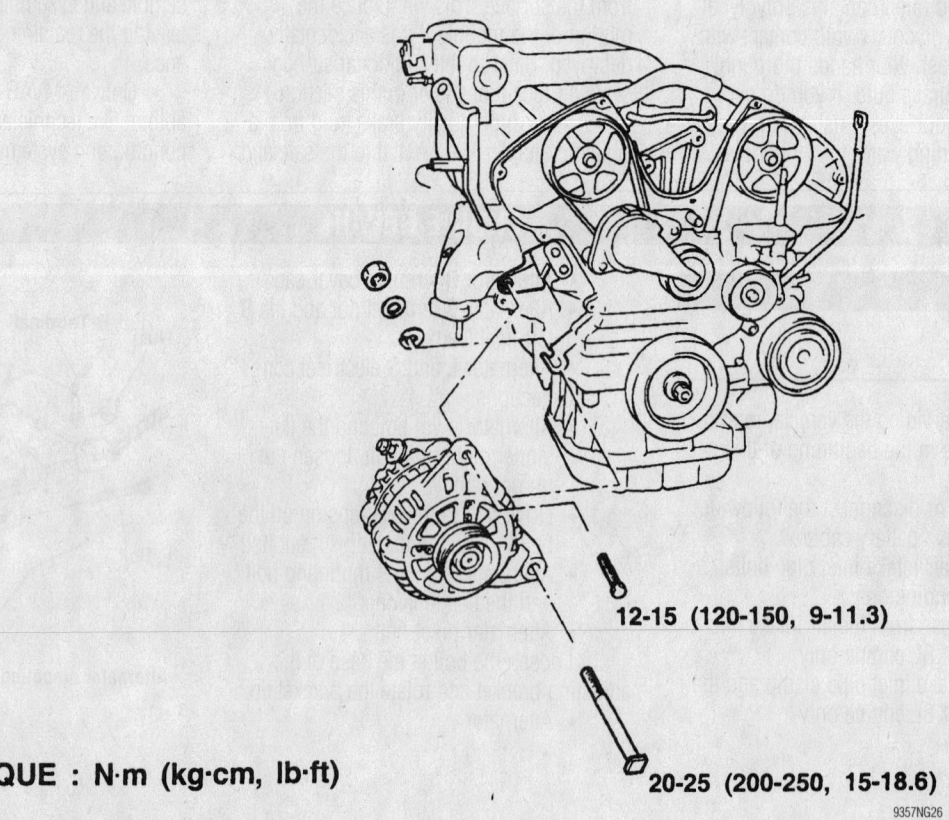

12-15 (120-150, 9-11.3)

TORQUE : N·m (kg·cm, lb·ft)

20-25 (200-250, 15-18.6)

9357NG26

Alternator mounting—2.4L and 2.7L engine

INSTALLATION

1. Before servicing the vehicle, refer to the precautions in the beginning of this section.
2. Install or connect the following:
   - Alternator
   - Alternator pivot bolt and hand-tighten at this time
3. Rotate the adjusting bracket into position, place the belt tensioner into position. And hand-tighten the mounting bolt.
   - Drive belt
4. Adjust the belt tension by rotating the adjustment bolt.
5. The belt deflection for 1.5L, 2.4L and 2.7L engines, is as follows:
   a. New belt: 0.22–0.28 in. (5.5–7mm).

b. Used belt: 0.24–0.28 in. (6–7mm).
6. The belt deflection for 1.8L engine, is as follows:
   a. New belt: 0.23–0.31 in. (6–8mm).
   b. Used belt: 0.28–0.35 in. (7–9mm).
7. Tighten the tensioner bolt to 14–19 ft. lbs. (19–26 Nm) and the pivot bolt to 28–38 ft. lbs. (38–51 Nm).
8. Install or connect the following:
   - Alternator **L** and **S** electrical connections
   - Alternator **B** terminal lead and nut
   - Alternator **B** terminal cover cap
   - Air intake inlet pipe and fasten the clamp, 1.8L engine only
   - Top hose to the air intake inlet pipe, 1.8L engine only

   - Front air intake inlet pipe bolts, 1.8L engine only
   - Negative battery cable

Ignition Timing

This vehicle is equipped with a Distributorless Ignition System (DIS). No adjustment is necessary or possible.

Engine Assembly

REMOVAL & INSTALLATION

1.5L Engine

1. Before servicing the vehicle, refer to the precautions in the beginning of this section.

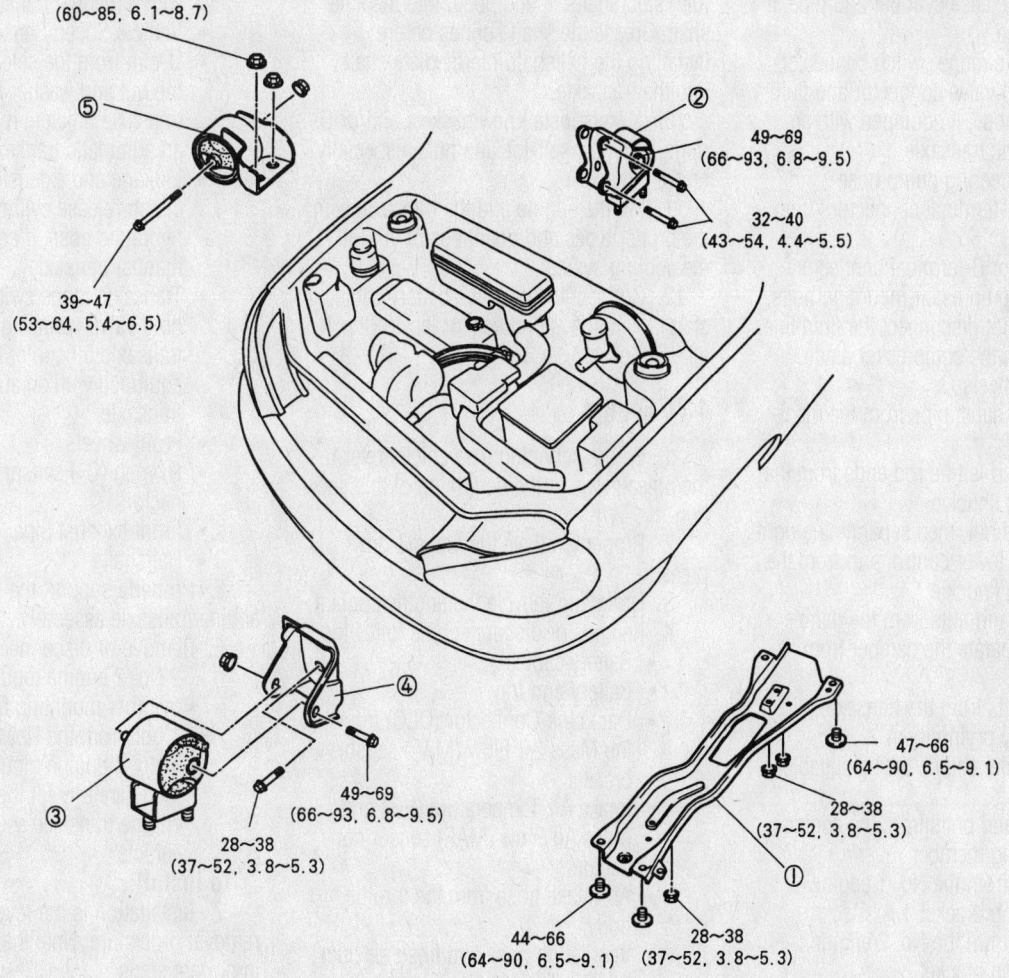

(1) Engine Mounting Member
(2) No.1 Engine Mounting Bracket
(3) No.2 Engine Mounting Rubber
(4) No.2 Engine Mounting Bracket
(5) No.3 Engine Mounting Bracket

lb-ft(N·m, kg-m)

9357NG02

Exploded view of the engine mount and brackets—1.5L engine

2. Properly relieve the fuel system pressure.

3. Drain and recycle the engine coolant.

4. Remove or disconnect the following:
- Battery cables
- Battery and tray
- Fresh air intake duct
- Upper and lower radiator hoses
- Accelerator cable
- Fuel hose from the fuel rail
- Heater hose
- Brake vacuum hose and purge control hose from the dynamic chamber
- Injector connectors
- Electrical connectors, tag before removing
- Transaxle linkage, if equipped with an automatic transaxle
- Manual transaxle linkage and extension bar, if equipped
- Clutch release cylinder and pipe, if equipped
- Transaxle range switch connector, solenoid valve connector and fluid cooler hose, if equipped with an automatic transaxle
- Power steering pump hose
- B and S-terminal connectors from starter
- Alternator B-terminal connector
- 4 A/C compressor mounting bolts, but do not disconnect the fluid line. Position the compressor aside.
- Front wheels
- Front exhaust pipe from the manifold
- Right and left tie rod ends from the steering knuckle
- Bolt and nut, then separate the right and left lower control arm from the steering knuckle
- 2 bolts and nuts from the damper, then separate the damper from the knuckle
- Halfshafts from the transaxle, by carefully prying them

5. Support the engine with a suitable hoist.
- 4 nuts and bolts from the engine mounting member
- 2 bolts from the No. 1 engine mounting bracket
- 4 bolts from the No. 2 engine mounting bracket
- 2 nuts from the No. 3 engine mounting rubber.
- Engine and transaxle assembly by lifting it out of the engine compartment as a unit

To install:
6. Installation is the reverse of the removal procedure. Note the following important steps.

7. When possible, leave the engine mounting nuts/bolts loose (hand tight) until all mounts are aligned and bolted. This may help in aligning the engine/transaxle assembly in the vehicle.

8. Tighten the engine mount bolts/nuts as follows:
- No. 3 engine mounting rubber (insulator) bolts: 49–68 ft. lbs. (68–93 Nm)
- No. 2 engine mounting nuts: 49–68 ft. lbs. (68–93 Nm)
- No. 2 engine mounting bolts: 28–38 ft. lbs. (38–51 Nm)
- Engine mounting member nuts: 28–38 ft. lbs. (39–52 Nm)
- Engine mounting member bolts: 47–66 ft. lbs. (65–91 Nm)

9. Install new circlips on the inner CV-joint stub shafts, if equipped, intermediate shaft. Grease the shaft splines before installing the halfshaft/intermediate shaft into the transaxle.

10. Always install new gaskets and/or O-rings. Use new self-locking nuts, especially on the exhaust.

11. Fill the engine and the transaxle with the proper types and quantities of oil. Fill the cooling system.

12. Connect the negative battery cable, start the engine and check for leaks. Check all fluid levels.

1.8L Engine

1. Before servicing the vehicle, refer to the precautions in the beginning of this section.

2. Properly relieve the fuel system pressure.

3. Drain and recycle the engine coolant.

4. Remove or disconnect the following:
- Battery cables
- Battery and tray
- Data Link Connector (DLC) from the Mass Air Flow (MAF) sensor bracket
- Intake Air Temperature (IAT) and Mass Air Flow (MAF) sensor connectors
- Air intake hose from the throttle body
- Ventilation hose and fresh air duct
- Accelerator and, if equipped, cruise control cables
- Air cleaner assembly
- Brake booster and purge control vacuum hoses from the intake manifold
- Upper and lower radiator hoses
- Fuel hose from the fuel injector rail
- Heater hoses
- Idle Air Control (IAC) and Throttle Position (TP) sensor connectors
- Fuel injector electrical connectors
- Starter and generator electrical connectors
- Engine ground strap
- Left and right splash shields
- Exhaust manifold heat shield and 3 power steering pump bracket bolts
- Air conditioning compressor mounting bolts and position it aside leaving the hoses attached
- Ground strap from the top of the transaxle
- No. 4 engine mount
- Input/turbine speed sensor connector, if equipped with an automatic transaxle
- Back-up light switch, if equipped with a manual transaxle
- Vehicle Speed Sensor (VSS)
- U-clip from the selector cable and the nut and washer from the transaxle linkage, if equipped with an automatic transaxle
- Linkage and extension bar, then the clutch release cylinder and hydraulic hose, if equipped with a manual transaxle
- Transaxle range switch and solenoid valve connectors, then the 2 transaxle oil cooler hoses, if equipped with an automatic transaxle
- Front wheels
- Oxygen (O_2) sensor electrical connectors
- Front exhaust pipe
- Halfshafts

5. Properly support the engine/transaxle assembly.

6. Remove or disconnect the following:
- 2 No. 2 engine mount-to-mounting member mounting bolts
- 1 bolt from the No. 3 engine mount
- 3 No. 1 engine mounting bolts, then carefully lift the engine/transaxle assembly from the vehicle

To install:
7. Installation is the reverse of the removal procedure. Note the following important steps.

8. When possible, leave the engine mounting nuts/bolts loose (hand tight) until all mounts are aligned and bolted. This may help in aligning the engine/transaxle assembly in the vehicle.

9. Tighten the engine mount bolts/nuts as follows:

- No. 1 mounting bolts: 50–70 ft. lbs. (67–93 Nm)
- No. 3 mounting bolts: 63–86 ft. lbs. (85–116 Nm)
- No. 2 mounting nuts: 28–38 ft. lbs. (38–51 Nm)
- No. 4 mounting bolts: 47–66 ft. lbs. (64–89 Nm)
- No. 4 mounting nuts: 49–68 ft. lbs. (68–93 Nm)

10. Install new circlips on the inner CV-joint stub shafts, if equipped, intermediate shaft. Grease the shaft splines before installing the halfshaft/intermediate shaft into the transaxle.

11. Always install new gaskets and/or O-rings. Use new self-locking nuts, especially on the exhaust.

12. Fill the engine and the transaxle with the proper types and quantities of oil. Fill the cooling system.

13. Connect the negative battery cable, start the engine and check for leaks. Check all fluid levels.

2.4L Engine

1. Before servicing the vehicle, refer to the precautions in the beginning of this section.

2. Properly relieve the fuel system pressure.

3. Drain and recycle the engine coolant, transaxle fluid and engine oil.

4. Remove or disconnect the following:

- Battery
- Air cleaner
- Back-up lamp and engine harness connectors
- Select control valve connector, if equipped with a manual transaxle
- Alternator harness and oil pressure gauge wiring
- Transaxle oil cooler hoses, if equipped with an automatic transaxle
- Upper and lower radiator hoses from the engine
- Engine ground
- Brake booster vacuum hose
- Main fuel line and return and vapor hoses from the engine side
- Inlet and outlet heater hoses from the engine
- Accelerator cable from the engine
- Clutch cable, shift control rod and extension rod if equipped with a manual transaxle
- Control cable from the transaxle, if equipped with an automatic transaxle

- Vehicle Speed Sensor (VSS) connector from the transaxle
- Column shaft from the steering gear box
- Engine and transaxle mounting insulators
- Wheel
- Caliper. Unbolt and support with a piece of wire. Do not disconnect the brake fluid line.
- Lower arm and fork and separate the upper arm and knuckle
- Front exhaust pipe from the manifold. Use wire to hang the exhaust pipe from the bottom of the vehicle.

5. Place a suitable supporter under the sub-frame. Make sure all hoses, vacuum lines and connectors are detached from the engine.

6. Loosen the sub-frame bolts, then slowly lift the vehicle. The engine/transaxle, sub-frame, steering gear and halfshafts are removed as an assembly.

7. Remove the halfshafts, raise the engine/transaxle assembly with a hoist, and separate the engine and transaxle from the sub-frame.

8. To separate the engine/transaxle from the sub-frame, remove the engine mounting bracket and transaxle mounting bracket, then lift up the engine hoist.

To install:

9. To place the engine and transaxle assembly on the sub-frame, install the front roll stopper and rear roll stopper among the sub-frame and engine/transaxle assembly.

10. Installation is the reverse of the removal procedure, noting the following steps:

a. Fill the engine and the transaxle with the proper types and quantities of oil.

b. Fill the cooling system.

c. Start the engine and check for leaks. Check all fluid levels.

2.7L Engine

1. Before servicing the vehicle, refer to the precautions in the beginning of this section.

2. Properly relieve the fuel system pressure.

3. Drain and recycle the engine coolant and engine oil.

4. Remove or disconnect the following:

- Battery and engine cover
- Air cleaner
- Engine harness connectors

- Alternator harness, oil pressure switch and oil pressure sender connectors
- Transaxle oil cooler hoses
- Upper and lower radiator hoses from the engine
- Radiator
- Spark plug wires
- Engine ground
- Brake booster vacuum hose
- Main fuel line and return and vapor hoses from the engine side
- Inlet and outlet heater hoses from the engine
- Accelerator and cruise control cables from the engine
- Control cable from the transaxle
- Speedometer cable from the transaxle
- Power steering pump hose
- Oil pan shield
- Front exhaust pipe from the manifold. Use a wire to support the exhaust pipe from the bottom of the vehicle.
- Lower control arm ball joint bolts and upper arm bolt from the steering knuckle
- Halfshafts from the transaxle. Plug the openings in the transaxle case and discard the halfshaft circlips.

5. Attach a cable or chain to the engine and use a chain hoist to lift the engine enough to pull the cable tight.

- Front and rear roll stoppers
- Connector from the starter motor harness
- Engine mount bolts
- Bolts and nuts that fasten the engine mount bracket to the body

6. Slowly raise the engine and transaxle assembly and temporarily hold in the raised position. Make sure that all hoses, cables, vacuum lines and connectors are detached from the engine.

- Transaxle mounting bracket bolts
- Left mount insulator bolt
- Engine and transaxle assembly. While directing the transaxle side down, lift the engine and transaxle up and out of the vehicle.

To install:

7. Installation is the reverse of the removal procedure, noting the following steps:

a. Fill the engine and the transaxle with the proper types and quantities of oil.

b. Fill the cooling system.

c. Start the engine and check for leaks. Check all fluid levels.

Water Pump

REMOVAL & INSTALLATION

1.5L Engine

1. Before servicing the vehicle, refer to the precautions in the beginning of this section.
2. Disconnect the negative battery cable.
3. Drain and recycle the engine coolant.
4. Remove or disconnect the following:
   - Drive belt
   - Timing belt

➡ **The power steering pump must be removed to access the water inlet pipe. Do not disconnect the pump fluid lines.**

   - Power steering pump and position aside
   - Water inlet pipe and gasket
   - Water bypass pipe and O-ring
   - Water pump bolts, pump and gasket
5. Clean all gasket mating surfaces.

To install:

6. Install or connect the following:
   - Water pump, using a new gasket and tighten the mounting bolts to 14–19 ft. lbs. (19–26 Nm)
   - Water pump bypass pump, with a new O-ring
   - Water inlet pipe, with a new O-ring
   - Power steering pump
   - Timing belt
   - Drive belt
7. Fill the engine coolant.
8. Connect the negative battery cable.
9. Start the engine and check for leaks.

1.8L Engine

1. Before servicing the vehicle, refer to the precautions in the beginning of this section.

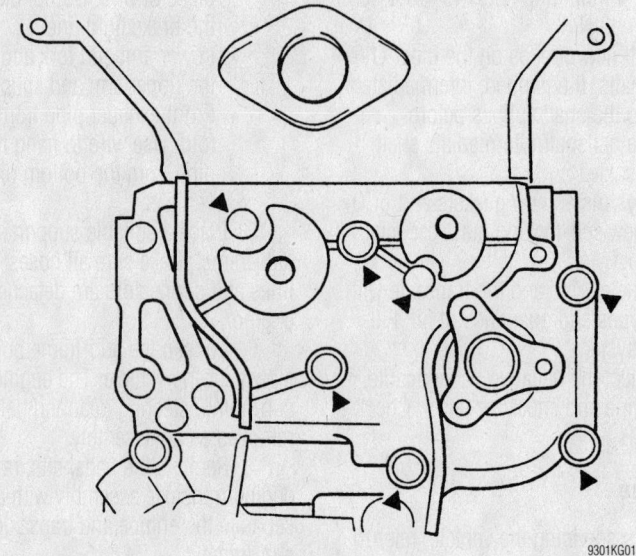

9301KG01

Water pump mounting bolt locations (arrows)—1.8L engine

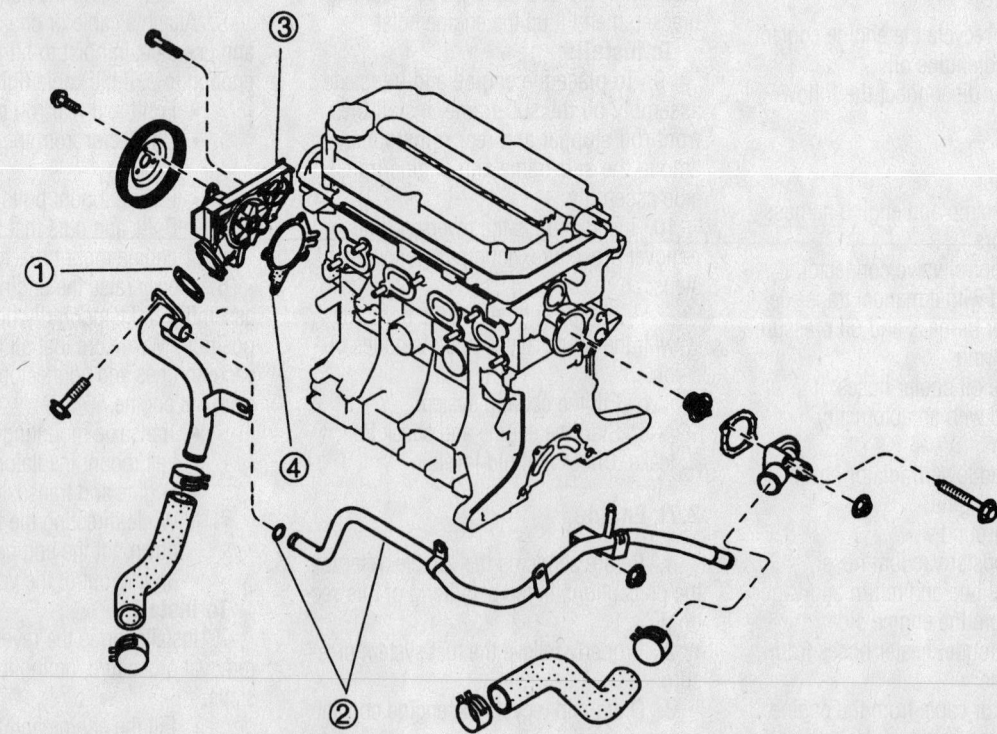

(1) Water inlet pipe and gasket	**(3) Water pump assembly**
(2) Water bypass pipe and 0-ring	**(4) Water pump gasket**

9357NG03

Water pump mounting—1.5L engine

2. Disconnect the negative battery cable.

3. Drain and recycle the engine coolant.

4. Remove or disconnect the following:
- Timing belt, tensioner and idler pulleys
- Water pump mounting bolts, then the pump

5. Clean all gasket mating surfaces.

To install:

6. Install or connect the following:
- Water pump, using a new gasket and tighten the mounting bolts to 14–19 ft. lbs. (19–26 Nm)
- Tensioner and idler pulleys
- Timing belt

7. Fill the engine coolant.

8. Connect the negative battery cable.

9. Start the engine and check for leaks.

2.4L and 2.7L Engines

1. Before servicing the vehicle, refer to the precautions in the beginning of this section.

2. Disconnect the negative battery cable.

3. Drain and recycle the engine coolant.

4. Remove or disconnect the following:
- Radiator outlet hose and coolant bypass hose from the water pump
- Drive belt
- Water pump pulley
- Timing belt covers and timing belt tensioner

- Water pump mounting bolts
- Alternator brace
- Water pump from the cylinder block

5. Clean all gasket mating surfaces.

To install:

6. Install or connect the following:
- Water pump, using a new gasket and tighten the mounting bolts to 14–20 ft. lbs. (19–27 Nm) for 2.4L engines and 11–16 ft. lbs. (15–22 Nm) for 2.7L engines
- Timing belt tensioner and timing belt
- Timing belt covers
- Coolant pump pulley and drive belt

7. Fill the engine coolant.

8. Connect the negative battery cable.

9. Start the engine and check for leaks.

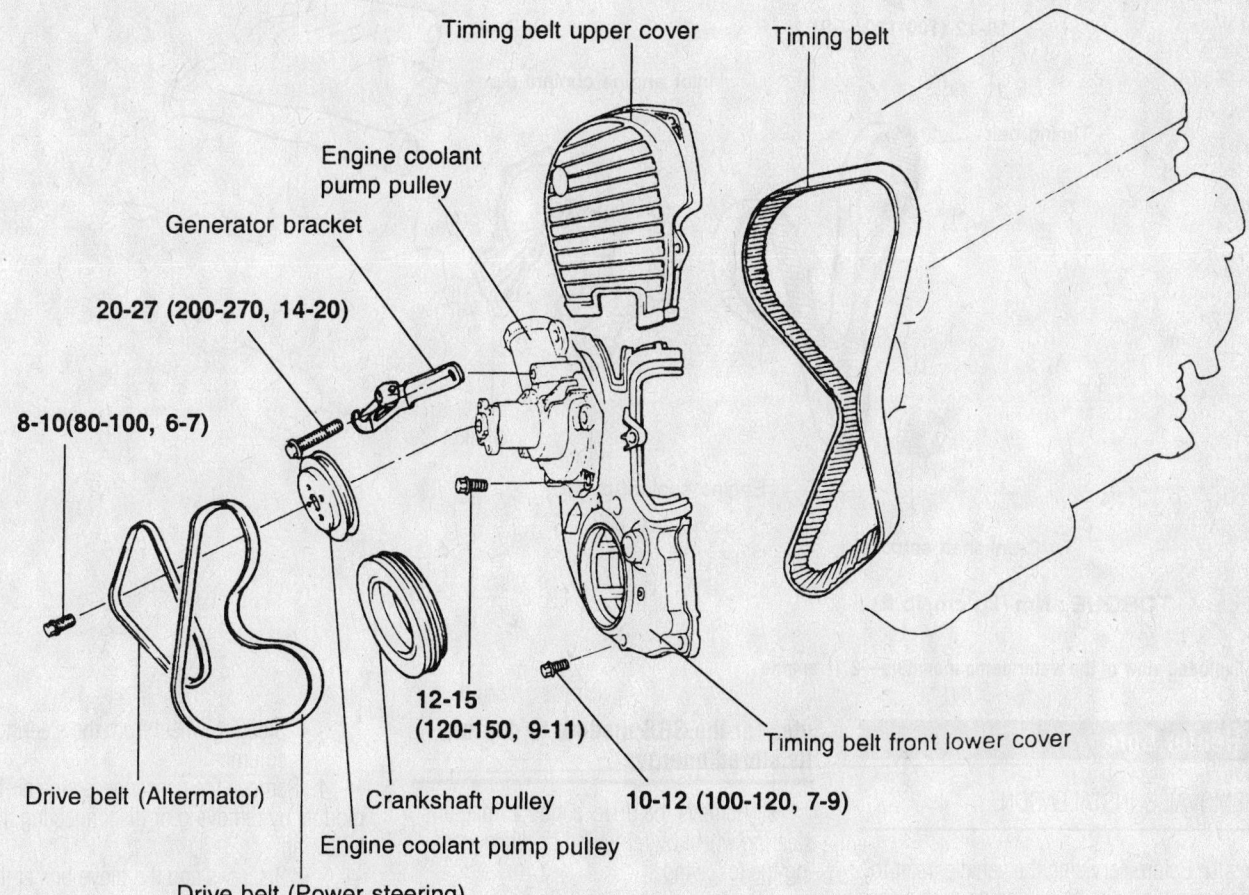

TORQUE : Nm (kg.cm, lb.ft)

9357NG27

Exploded view of the water pump mounting—2.4L engine

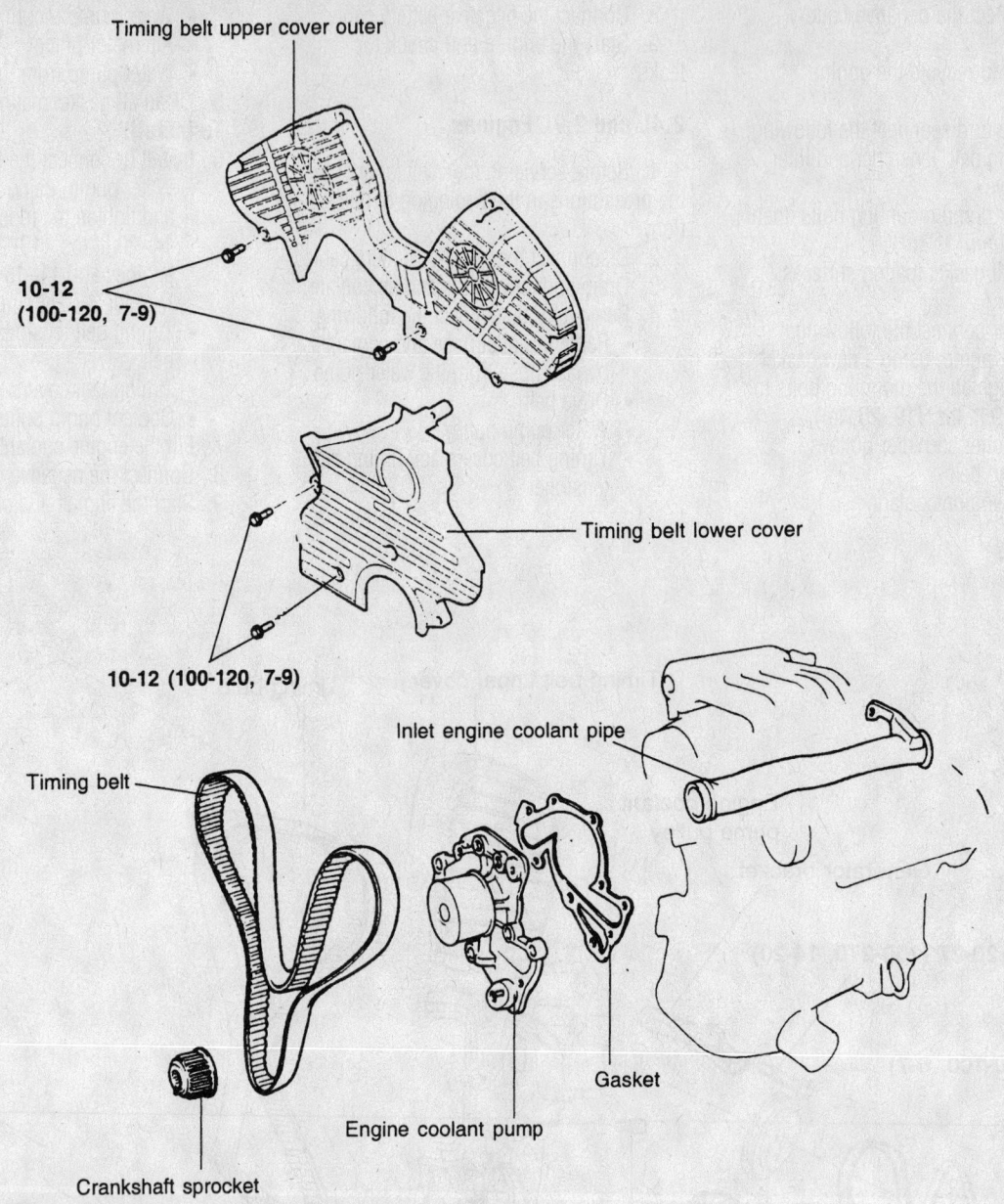

Timing belt upper cover outer

10-12
(100-120, 7-9)

Timing belt lower cover

10-12 (100-120, 7-9)

Inlet engine coolant pipe

Timing belt

Gasket

Engine coolant pump

Crankshaft sprocket

TORQUE : Nm (kg.cm, lb.ft)

9357NG28

Exploded view of the water pump mounting—2.7L engine

Heater Core

REMOVAL & INSTALLATION

1. Before servicing the vehicle, refer to the precautions in the beginning of this section.

2. Disconnect the negative battery cable.

✳✳ CAUTION

After disconnecting the negative battery cable, wait for at least 10 minutes for the SRS module to deplete its stored energy.

3. Remove the driver's side air bag and steering wheel by removing or disconnecting the following:

- Position the front wheels in the straight-ahead position
- 4 steering wheel-to-air bag module bolts
- Air bag module and disconnect the electrical connector
- Steering wheel-to-steering column nut

- Steering wheel from the steering column

4. Remove the passenger's side air bag module by removing or disconnecting the following:

- 2 screws and the glove box at the bottom of the glove box
- Side cover and pull the connector from the "T" bar side bracket
- 4 air bag module-to-instrument panel bolts
- Air bag module, disconnect the electrical connector and remove the air bag module

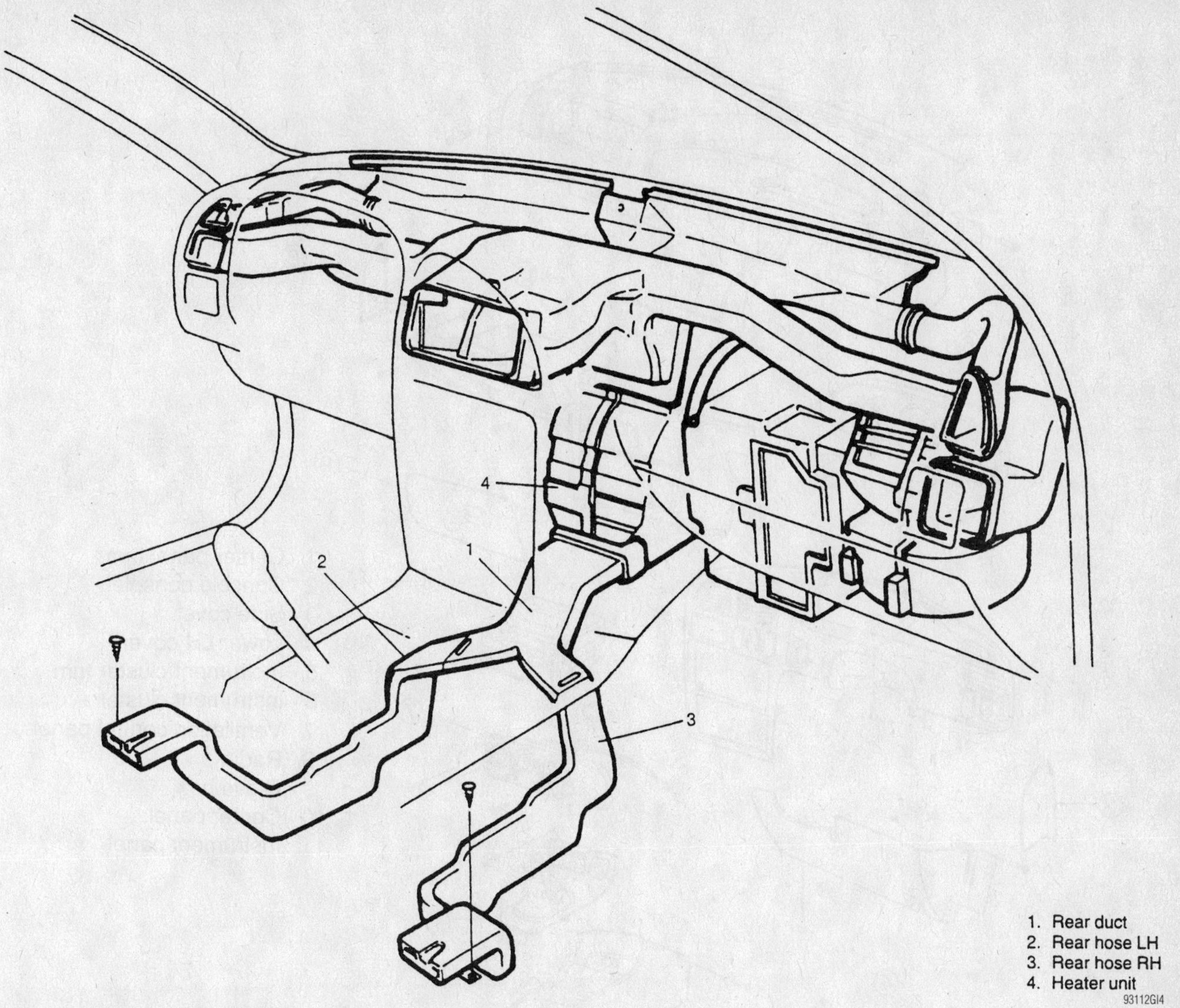

1. Rear duct
2. Rear hose LH
3. Rear hose RH
4. Heater unit

93112GI4

Exploded view of the steering wheel and air bag module assembly—Sephia

5. Drain the cooling system into a clean container for reuse.

6. Discharge and recover the air conditioning system refrigerant.

7. Remove the instrument panel by removing or disconnecting the following:

- Center panel trim
- Console
- Side cover
- Lower left side cover
- Steering column-to-instrument panel bolts and lower the steering column
- Instrument cluster trim
- Instrument cluster and disconnect the electrical connectors
- Ventilation control panel and disconnect the electrical connectors
- Radio and disconnect antenna and electrical connector
- Center panel

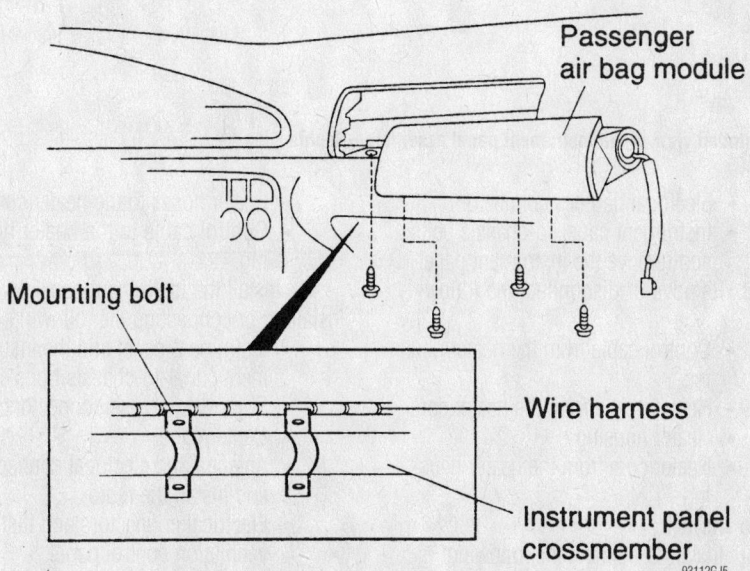

Passenger air bag module

Mounting bolt

Wire harness

Instrument panel crossmember

93112GJ5

Exploded view of the passenger's side air bag module assembly—Sephia

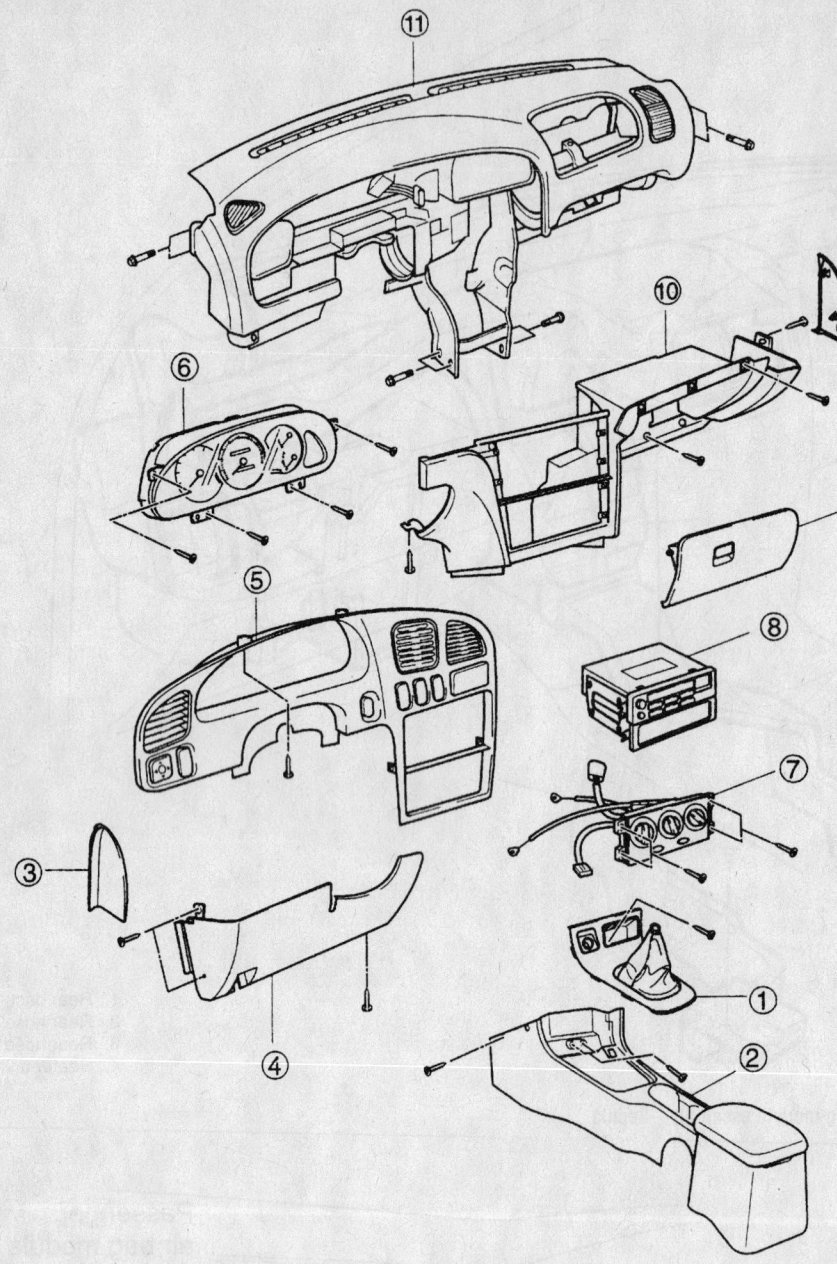

1 Center panel trim
2 Console console
3 Side cover
4 Lower LH cover
5 Instrument cluster trim
6 Instrument cluster
7 Ventilation control panel
8 Radio
9 Glove box
10 Center panel
11 Instrument panel

93112GJ6

Exploded view of the instrument panel assembly—Sephia

- Electrical harness connectors
- Instrument panel-to-chassis bolts and remove the instrument panel

8. Remove or disconnect the following:

- Control cable from the heater housing
- Heater hoses from the heater core
- Heater housing
- Heater core from the heater housing

To install:

9. Install or connect the following:

- Heater core to the heater housing
- Heater housing

- Heater hoses to the heater core
- Control cable to the heater housing

10. Install the instrument panel by installing or connecting the following:

- Instrument panel and the instrument panel-to-chassis bolts
- Electrical harness connectors
- Center panel
- Antenna and electrical connector and install the radio
- Electrical connectors and install the ventilation control panel
- Electrical connectors and install the instrument cluster

- Instrument cluster trim
- Steering column-to-instrument panel bolts
- Lower left side cover
- Side cover
- Console
- Center panel trim

11. Install the passenger's side air bag module by installing or connecting the following

- Air bag module and connect the electrical connector
- 4 air bag module-to-instrument panel bolts and torque the bolts to 18–32 ft. lbs. (24–43 Nm)

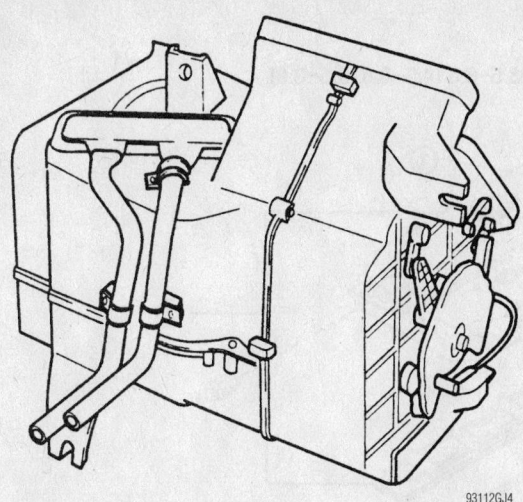

View of the heater housing—Sephia

- Connector to the "T" bar side bracket and the side cover
- Glove box and the 2 screws at the bottom of the glove box

12. Install the driver's side air bag and steering wheel by installing or connecting the following:
- Steering wheel to the steering column
- Steering wheel-to-steering column nut and torque the nut to 33 ft. lbs. (45 Nm)
- Air bag module and connect the electrical connector
- 4 steering wheel-to-air bag module bolts and torque to 72–106 inch lbs. (8–12 Nm)

13. Refill the cooling system.
14. Connect the negative battery cable.
15. Evacuate, charge and leak test the air conditioning system.
16. Operate the engine to normal operating temperatures; then, check the climate control operation and check for leaks.

Cylinder Head

REMOVAL & INSTALLATION

1.5L Engine

1. Before servicing the vehicle, refer to the precautions in the beginning of this section.
2. Disconnect the negative battery cable.
3. Properly relieve the fuel system pressure.
4. Drain and recycle the engine coolant.
5. Drain and recycle the engine oil.
6. Remove or disconnect the following:
- Upper radiator hose
- Breather hose from the between the air cleaner and rocker arm (valve) cover
- Air intake hose
- Vacuum hose, fuel hose and coolant hose
- Spark plug wires, tag before disconnecting
- Ignition coil
- Power steering pump and bracket and position aside. Do not disconnect the fluid lines.
- Intake manifold
- Heat shield from the exhaust manifold
- Exhaust manifold
- Surge tank
- Water pump and crankshaft pulleys
- Timing belt cover, belt tensioner and timing belt
- Cylinder head cover and cam carrier assembly
- Cylinder head bolts, in several steps in the proper sequence
- Cylinder head and gasket. Discard the gasket.

To install:

7. Thoroughly, clean the cylinder head and the block contact surfaces. Examine the head gasket and check the cylinder head for cracks. Check the cylinder head for warpage using a feeler gauge and straightedge. The maximum allowable distortion is 0.006 in. (0.15mm).
8. Clean the cylinder head bolts and the threads in the block. Be sure the bolts turn freely in the block.
9. Install or connect the following:
- New head gasket on the engine block
- Cylinder head
- Cylinder head bolts

10. Tighten the head bolts in the following step using the proper sequence:
   a. Step 1: Tighten to 36 ft. lbs. (49 Nm).
   b. Step 2: Loosen the bolts in the reverse order shown.
   c. Step 3: Tighten to 18 ft. lbs. (25 Nm).
   d. Step 4: Tighten 90° (¼ turn).
   e. Step 5: Tighten 90° (¼ turn).

11. Install or connect the following:
- Timing belt tensioner pulley and timing belt. Make sure all timing marks are aligned.
- Cylinder head cover. Tighten the bolts to 3.6–6.5 ft. lbs. (5–9 Nm).
- Timing belt cover
- Intake manifold, with a new gasket. Torque the bolts to 11–14 ft. lbs. (15–20 Nm).
- Exhaust manifold with a new gas-

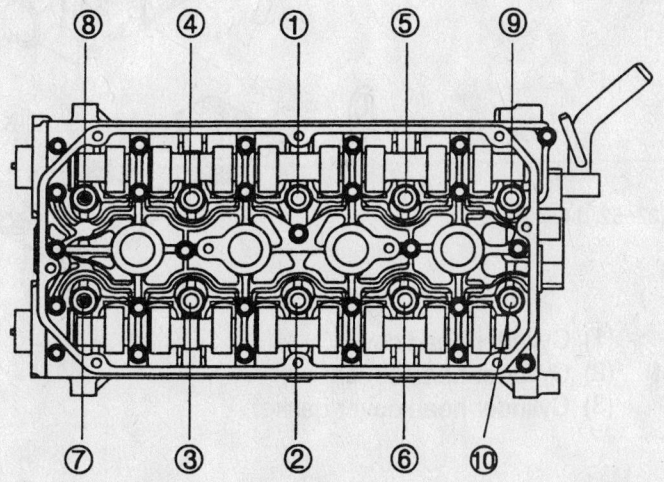

Cylinder head bolt tightening (loosen the reverse order)—1.5L engine

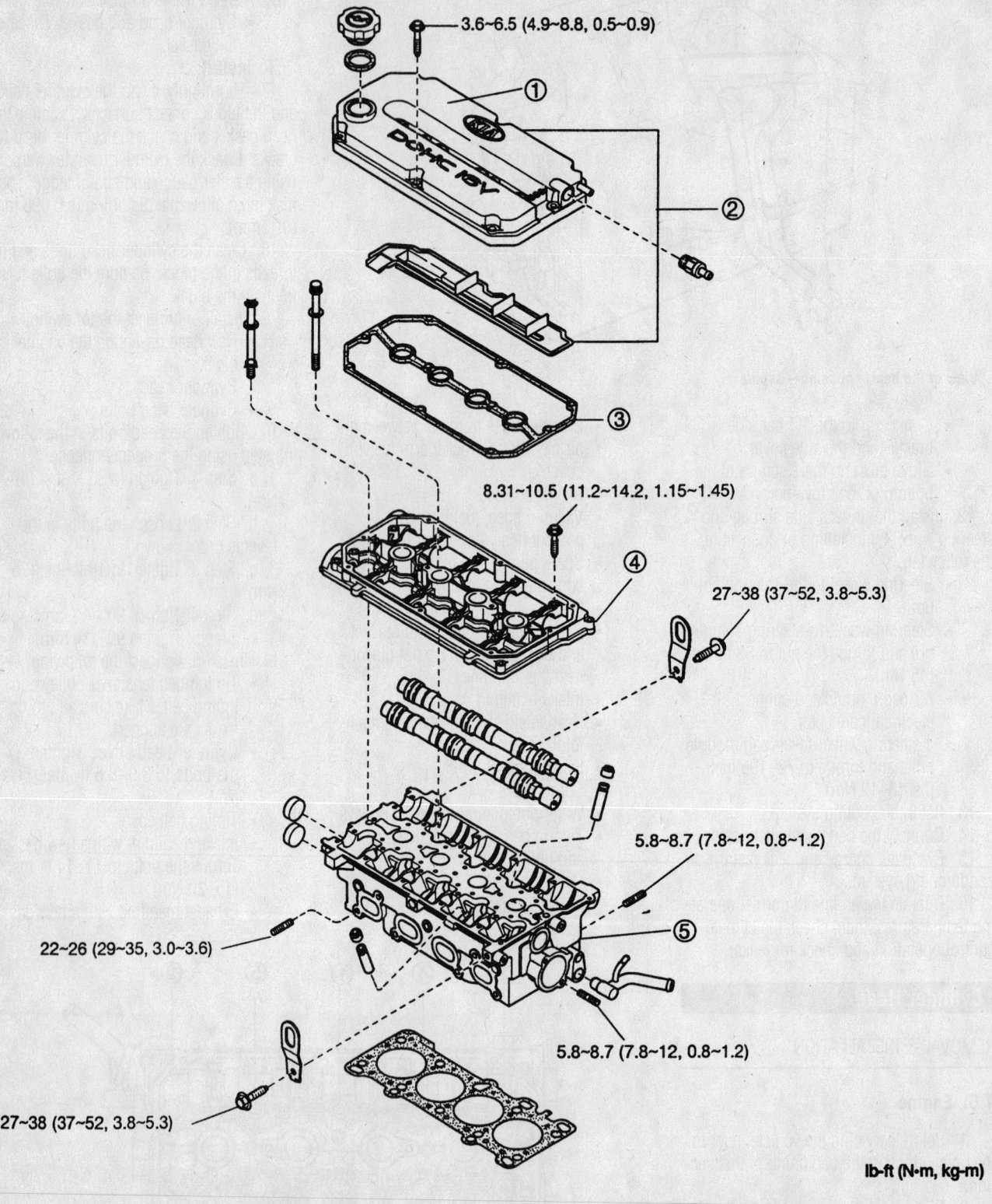

3.6~6.5 (4.9~8.8, 0.5~0.9)

8.31~10.5 (11.2~14.2, 1.15~1.45)

27~38 (37~52, 3.8~5.3)

5.8~8.7 (7.8~12, 0.8~1.2)

22~26 (29~35, 3.0~3.6)

5.8~8.7 (7.8~12, 0.8~1.2)

27~38 (37~52, 3.8~5.3)

lb-ft (N•m, kg-m)

(1) Cylinder head cover
(2) Cylinder head cover
(3) Cylinder head cover gasket

(4) Cam carrier assembly
(5) Cylinder head

9357NG05

Exploded view of the cylinder head and related components—1.5L engine

ket. Torque the bolts to 11–14 ft. lbs. (15–20 Nm).

- Exhaust manifold heat shield
- Surge tank and tighten the bolts to 11–14 ft. lbs. (15–20 Nm)
- Power steering pump and bracket
- Ignition coil and spark plug wires
- Air intake hose
- Vacuum hose, fuel hose and water hose
- Breather hose
- Fill the engine oil and coolant
- Start the vehicle and check for leaks

1.8L and 2.4L Engines

1. Before servicing the vehicle, refer to the precautions in the beginning of this section.

2. Disconnect the negative battery cable.

3. Properly relieve the fuel system pressure.

4. Drain and recycle the engine coolant.

5. Drain and recycle the engine oil.

6. Remove or disconnect the following:

- Positive Crankcase Ventilation (PCV) and crankcase ventilation hoses
- Accelerator and, if equipped, the cruise control cables
- Air intake hose from the throttle body
- Brake vacuum hose and the purge control vacuum hose
- Ventilation hose and fresh air duct
- Upper radiator hose
- Fuel hose from the fuel injector rail
- Heater hoses
- Idle Air Control (IAC) and Throttle Position (TP) sensor connectors
- Fuel injector electrical connectors
- Engine ground strap
- Exhaust manifold heat shield
- Front exhaust pipe from the manifold
- Exhaust manifold
- Coolant bypass pipe from the cylinder head
- Intake manifold support bracket
- Camshaft and Hydraulic Lash Adjusters (HLA's)
- Cylinder head bolts in several steps, in the order illustrated
- Cylinder head bolts, then the cylinder head

To install:

7. Thoroughly, clean the cylinder head and the block contact surfaces. Examine the head gasket and check the cylinder head for cracks. Check the cylinder head for warpage using a feeler gauge and straightedge. The

maximum allowable distortion is 0.006 in. (0.15mm).

8. Clean the cylinder head bolts and the threads in the block. Be sure the bolts turn freely in the block.

9. Install or connect the following:

- New head gasket on the engine block
- Cylinder head
- Cylinder head bolts

10. For 1.8L engines, tighten the head

bolts in the following step using the proper sequence:

a. Step 1: Tighten to 36 ft. lbs. (49 Nm).

b. Step 2: Loosen the bolts in the reverse order shown.

c. Step 3: Tighten to 29 ft. lbs. (39 Nm).

d. Step 4: Tighten 90° (¼ turn).

e. Step 5: Tighten 90° (¼ turn).

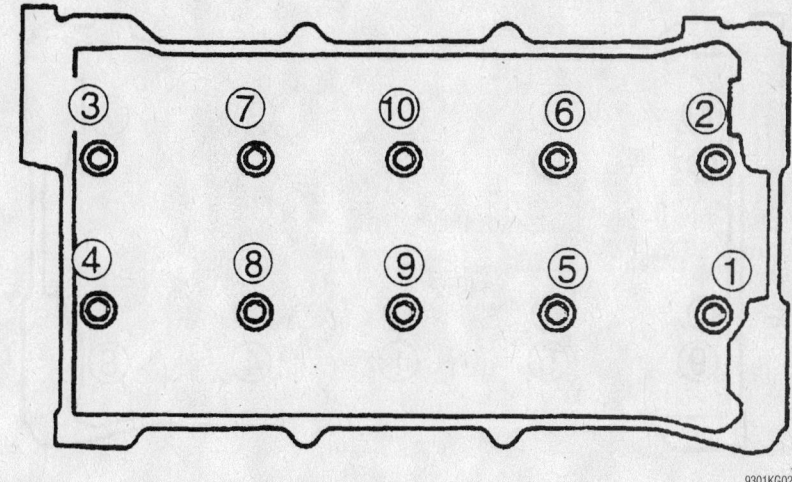

Cylinder head bolt removal sequence—1.8L engine

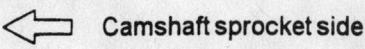

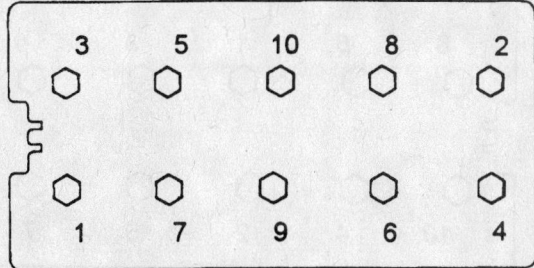

Cylinder head bolt removal sequence—2.4L engine

11. For 2.4L engines, tighten the head bolts in the following step using the proper sequence:

 a. Step 1: Tighten to 14 ft. lbs. (20 Nm).

 b. Step 2: Tighten 90° (¼ turn).

 c. Step 3: Loosen the bolts in the reverse order shown.

 d. Step 4: Tighten to 14 ft. lbs. (20 Nm).

 e. Step 5: Tighten 90° (¼ turn).

 f. Step 6: Tighten 90° (¼ turn).

12. Install or connect the following:
• Camshaft and HLA's
• Intake manifold support bracket and tighten the mounting bolts to 28–38 ft. lbs. (37–52 Nm)
• Coolant bypass pipe and tighten the mounting bolt to 66–86 ft. lbs. (89–117 Nm)

• Exhaust manifold, tighten the manifold-to-cylinder head mounting nuts to 28–34 ft. lbs. (38–46 Nm) and the manifold-to-exhaust pipe mounting nuts to 16–21 ft. lbs. (22–28 Nm)
• Exhaust manifold heat shield and tighten the mounting bolts to 13–22 ft. lbs. (19–30 Nm)

13. The remaining steps of the installation procedure is the reverse of the removal, while keeping in mind the following:

 a. Attach all hoses and connectors.

 b. Fill the engine oil and coolant.

 c. Start the vehicle and check for leaks.

2.7L Engine

1. Before servicing the vehicle, refer to the precautions in the beginning of this section.

2. Disconnect the negative battery cable.

3. Properly relieve the fuel system pressure.

4. Drain and recycle the engine coolant.

5. Drain and recycle the engine oil.

6. Remove or disconnect the following:
• Upper radiator hose
• Breather hose and air intake hose
• Vacuum hose, fuel hose and coolant hose
• Intake manifold
• Spark plug wires from the spark plugs
• Ignition coil
• Upper and lower timing belt covers
• Timing belt and camshaft sprockets
• Heat protector and exhaust manifold
• Water pump pulley and head cover
• Intake and exhaust camshafts
• Cylinder head bolts, loosening in the reverse of the torque sequence, in several steps

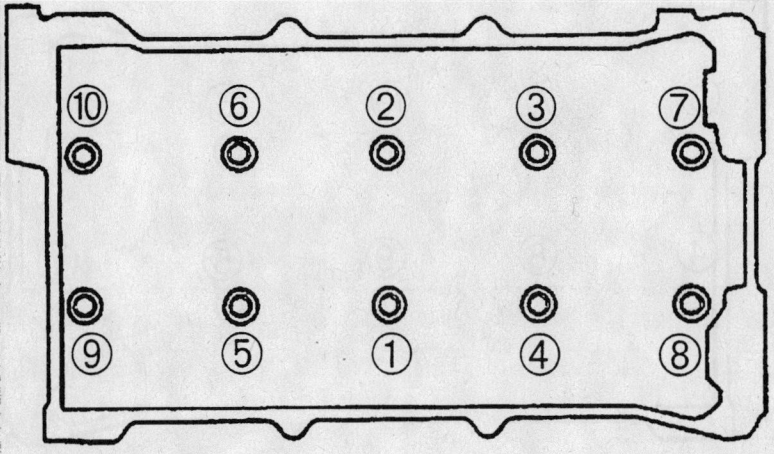

Cylinder head torque sequence—1.8L engine

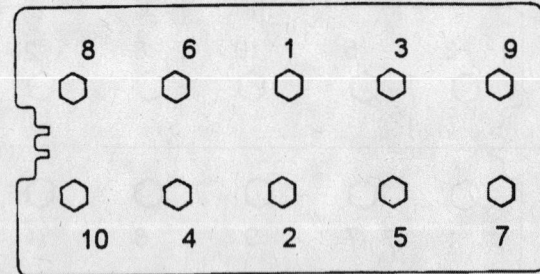

Cylinder head torque sequence—2.4L engine

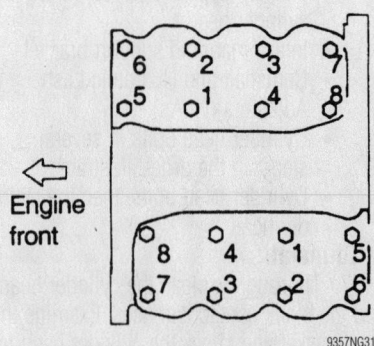

Cylinder head tightening sequence (use the reverse for removal)—2.7L engine

- Cylinder head
- Gaskets and discard

To install:

7. Thoroughly, clean the cylinder head and the block contact surfaces. Examine the head gasket and check the cylinder head for cracks. Check the cylinder head for warpage using a feeler gauge and straightedge. The maximum allowable distortion is 0.006 in. (0.15mm).

8. Clean the cylinder head bolts and the threads in the block. Be sure the bolts turn freely in the block.

9. Install or connect the following:
- New head gasket on the engine block

- Cylinder head
- Cylinder head bolts

10. Tighten the head bolts in the following step using the proper sequence:
   a. Step 1: Tighten to 18 ft. lbs. (25 Nm).
   b. Step 2: Tighten 60° (⅙ turn).
   c. Step 3: Tighten 45° (⅛ turn).

11. The remainder of installation is the reverse or the removal procedure, noting the following steps:
   a. Attach all hoses and connectors.
   b. Fill the engine oil and coolant.
   c. Start the vehicle and check for leaks.

Rocker Arms/Shafts

REMOVAL & INSTALLATION

The engines covered in this section are not equipped with rocker arms/shafts. The camshafts directly actuate the valve through a bucket type follower.

Intake Manifold

REMOVAL & INSTALLATION

1. Before servicing the vehicle, refer to the precautions in the beginning of this section.

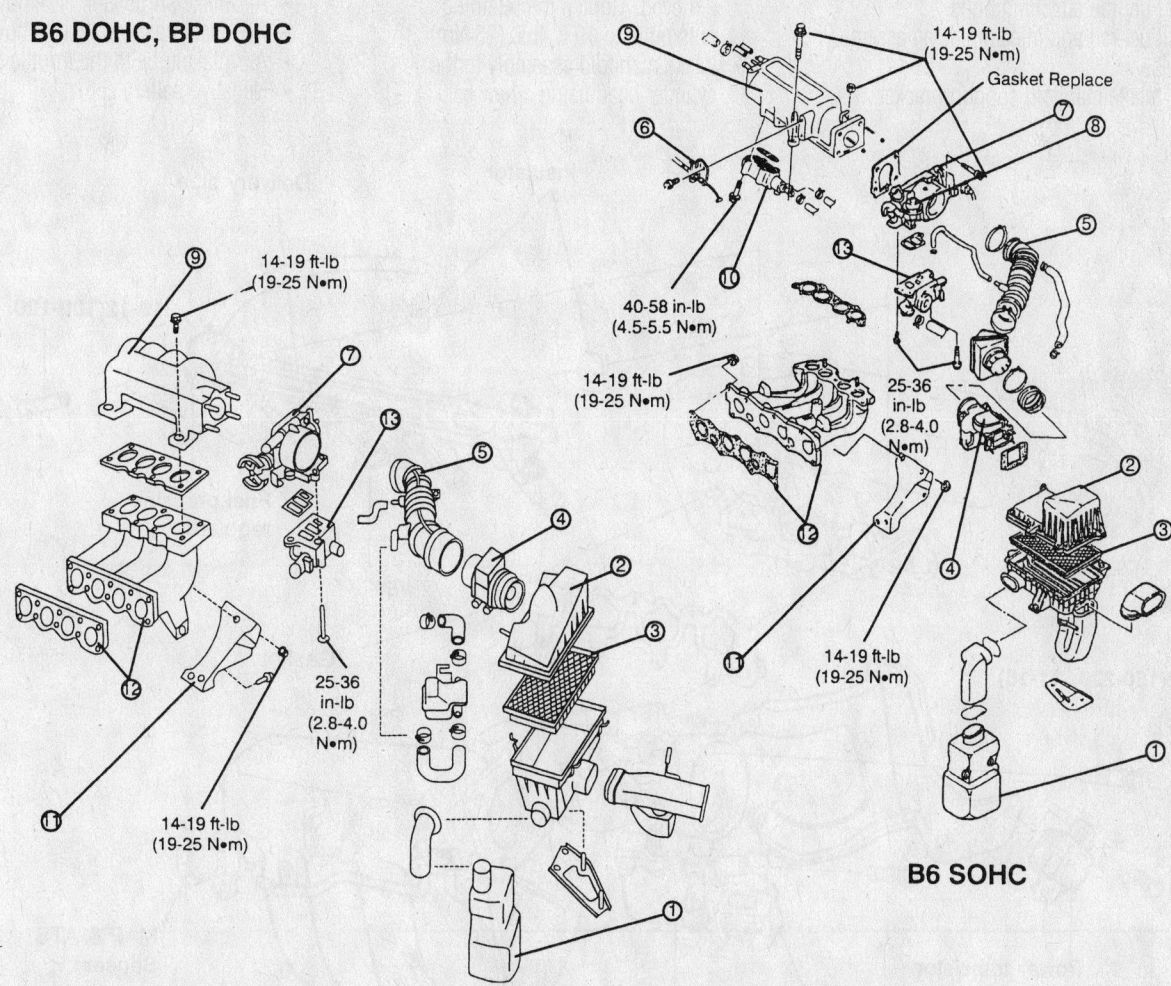

B6 DOHC, BP DOHC

B6 SOHC

1. Resonance chamber
2. Upper air filter housing
3. Air filter (B6 SOHC)
4. Mass air flow (MAF) sensor (B6 DOHC, BP DOHC)/ Volume air flow (VAF) sensor (B6 SOHC)
5. Intake air hose
6. Throttle cable
7. Throttle body
8. Dashpot (B6 SOHC)
9. Dynamic chamber
10. Air valve (B6 SOHC)
11. Intake manifold support bracket
12. Intake manifold and gasket (Replace)
13. Idle air control valve (B6 SOHC)/ Bypass air control (BAC) valve (B6 DOHC, BP DOHC)

7923KG17

Exploded view of the intake manifold assembly—1.8L engine shown

2. Properly relieve the fuel system pressure. Disconnect the negative battery cable and drain the cooling system.

3. Remove or disconnect the following:
- Air intake hose from the throttle body
- Air intake hose and air cleaner assembly, if necessary
- Air intake surge tank, if necessary
- Accelerator cable
- Fuel lines. Plug the lines to avoid contamination.
- All necessary vacuum hoses and electrical connectors
- Coolant hoses
- Exhaust Gas Recirculation (EGR) tube, if equipped
- Air valve, if equipped
- Fuel rail attaching bolts
- Fuel rail and injectors as an assembly
- Intake manifold support bracket

- Bolt retaining the dipstick tube bracket to the intake manifold, if necessary
- Intake manifold-to-cylinder bolts/nuts and the intake manifold assembly
- Throttle body, if necessary and separate the intake manifold upper and lower halves

To install:

4. Clean all gasket mating surfaces.
5. Install or connect the following:
- Upper and lower intake manifolds using a new gasket, if separated. Tighten the nuts/bolts to 19 ft. lbs. (25 Nm).
- Throttle body using a new gasket, if removed. Tighten the retaining nuts/bolts to 19 ft. lbs. (25 Nm).
- Intake manifold assembly to the cylinder head using a new gasket.

Tighten the nuts/bolts to 11–14 ft. lbs. (15–20 Nm) for 1.5L and 2.7L engines, or to 19 ft. lbs. (25 Nm) for 1.8L and 2.4L engines.

➡ **Tighten the bolts in the center of the manifold first and work outward toward the ends.**

- Bolt retaining the dipstick tube to the intake manifold, if equipped
- Intake manifold bracket. Tighten the attaching nuts/bolts to 19 ft. lbs. (25 Nm).
- EGR tube, if equipped
- Coolant and vacuum hoses, electrical connectors and fuel lines
- Accelerator cable
- Air intake surge tank, if removed
- Air cleaner assembly, if removed
- Air intake tube to the throttle body
- Negative battery cable

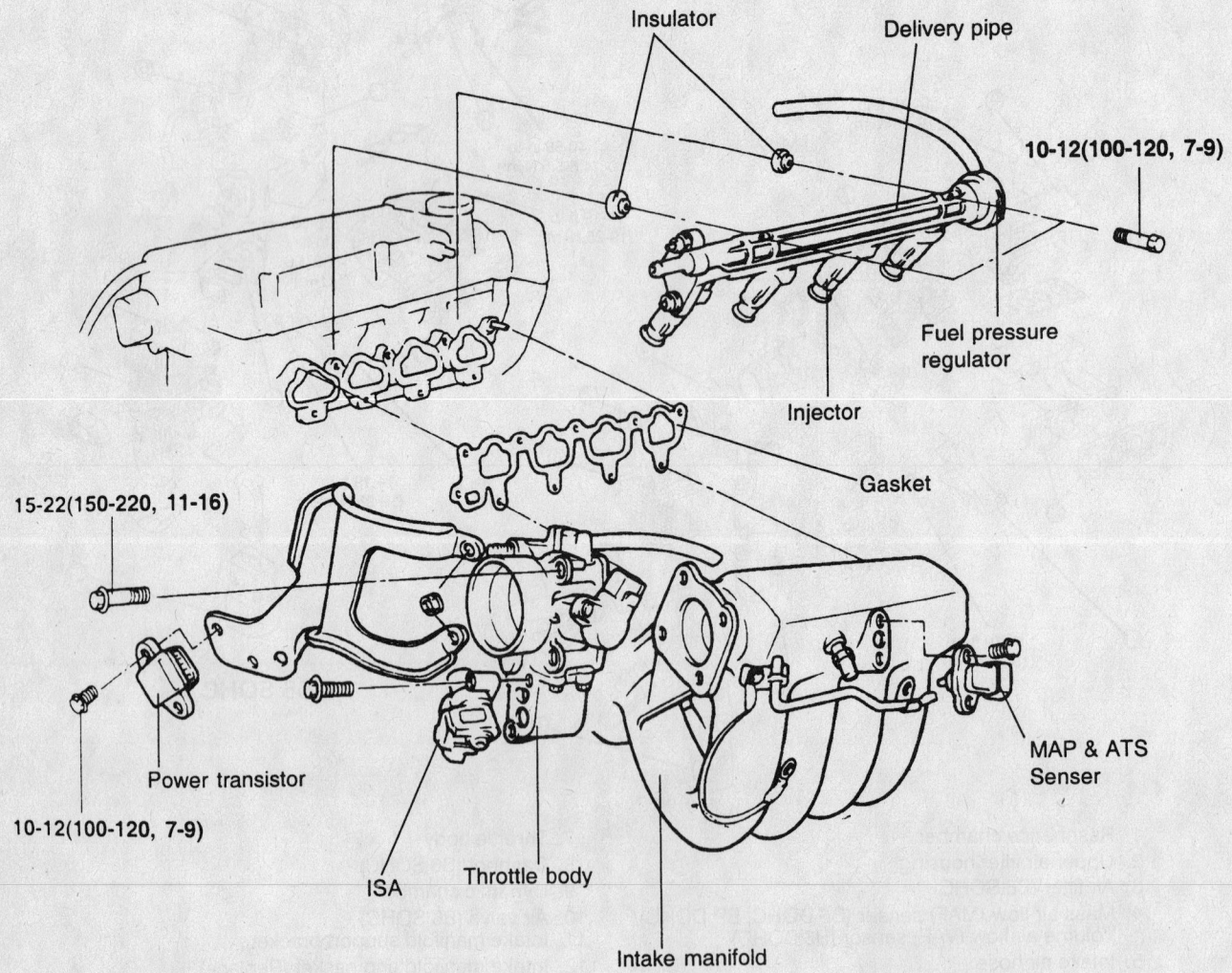

Insulator
Delivery pipe
10-12(100-120, 7-9)
Fuel pressure regulator
Injector
Gasket
15-22(150-220, 11-16)
Power transistor
10-12(100-120, 7-9)
ISA
Throttle body
Intake manifold
MAP & ATS Senser

TORQUE : Nm (kg.cm, lb.ft)

Exploded view of the intake manifold—2.4L engine

9357NG32

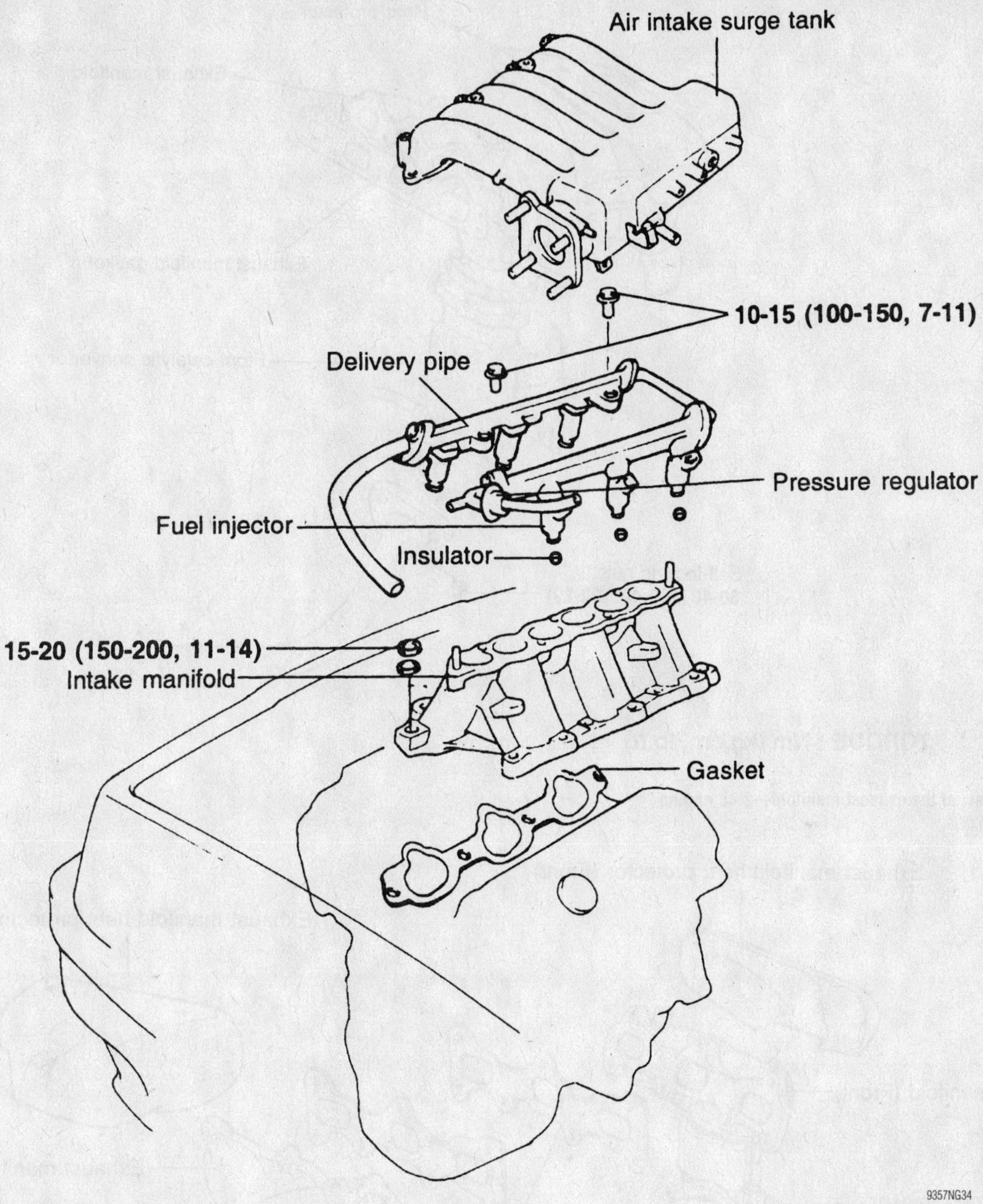

Air intake surge tank

10-15 (100-150, 7-11)

Delivery pipe

Pressure regulator

Fuel injector

Insulator

15-20 (150-200, 11-14)
Intake manifold

Gasket

9357NG34

Exploded view of the intake manifold—2.7L engine

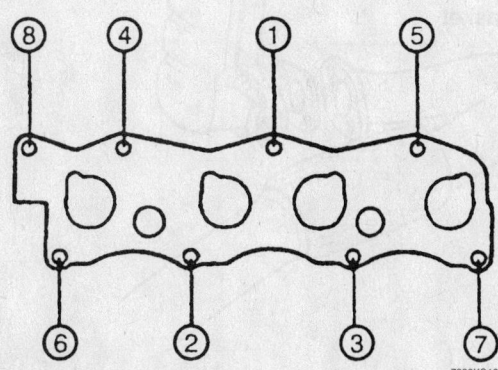

7923KG18

Intake manifold torque sequence—1.8L engine

6. Fill the cooling system.
7. Start the engine and bring to normal operating temperature. Check for leaks. Check the idle speed.

Exhaust Manifold

REMOVAL & INSTALLATION

1. Before servicing the vehicle, refer to the precautions in the beginning of this section.
2. Remove or disconnect the following:
   • Negative battery cable
   • Air cleaner

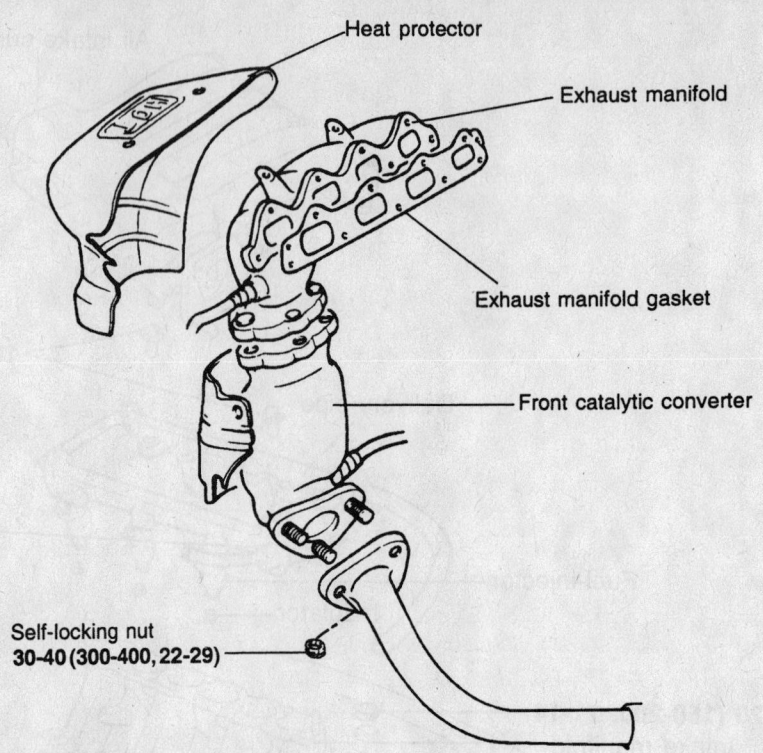

Heat protector

Exhaust manifold

Exhaust manifold gasket

Front catalytic converter

Self-locking nut
30-40 (300-400, 22-29)

TORQUE : Nm (kg.cm, lb.ft)

9357NG33

Exploded view of the exhaust manifold—2.4L engine

Exhaust manifold heat protector (Front)

Exhaust manifold heat protector (Rear)

Exhaust manifold (Front)

Exhaust manifold (Rear)

Oxygen sensor

Gasket

Front exhaust pipe

9357NG35

Exploded view of the exhaust manifolds—2.7L engine

- Air hose
- Water bypass pipe bolt
- Exhaust manifold heat shield bolts and the heat shield
- Oxygen Sensor (O_2S) electrical connector
- Exhaust pipe-to-exhaust manifold nuts and discard the nuts. Suspend the exhaust system with wire.
- Exhaust Gas Recirculation (EGR) pipe from the exhaust manifold
- Nuts, bolts and the exhaust manifold. Discard the nuts.

To install:

3. Clean all gasket mating surfaces.
4. Install or connect the following:
   - New exhaust manifold gasket and the exhaust manifold. Tighten the mounting nuts and bolts to 11–14 ft. lbs. (15–20 Nm) for 1.5L and 2.7L engines, to 29–34 ft. lbs. (39–47 Nm) for 1.8L engines or to 18–22 ft. lbs. (25–30 Nm) for 2.4L engines.
   - Exhaust pipe to the manifold. Install new nuts and tighten to 38 ft. lbs. (52 Nm).
   - O_2S sensor connector
   - EGR pipe to the back of the exhaust manifold and tighten to 34 ft. lbs. (47 Nm)
   - Heat shield and tighten the bolts to 88 inch lbs. (10 Nm)
   - Water bypass pipe bolt, and tighten to 48–65 ft. lbs. (64–89 Nm)
   - Air hose

- Air cleaner
- Negative battery cable

Front Crankshaft Seal

REMOVAL & INSTALLATION

1. Before servicing the vehicle, refer to the precautions in the beginning of this section.
2. Remove or disconnect the following:
   - Negative battery cable
   - Timing belt covers and belt
   - Timing belt pulley using a puller
   - Oil pump bolts and the pump
3. Wrap a suitable prytool with a rag and work the old seal from the oil pump housing.

To install:

4. Lubricate the seal lip with clean engine oil and push the seal slightly in by hand.
5. Install or connect the following:
   - Seal using a seal installer. Install the seal until it is flush with the oil pump body.
   - New O-ring seal
6. Apply a 0.04–0.08 in. (1–2mm) bead of silicone to the oil pump body as shown in the accompanying illustration.
7. Install or connect the following:
   - Oil pump
   - Timing belt pulley. Align the keyway groove on the pulley to the keyway on the crankshaft.

- Woodruff key with the tapered side towards the oil pump body
- Remaining components in the reverse order of removal

Camshaft

REMOVAL & INSTALLATION

1.5L Engine

1. Before servicing the vehicle, refer to the precautions in the beginning of this section.
2. Remove or disconnect the following:
   - Negative battery cable
   - Breather hose and Positive Crankcase Ventilation (PCV) hose
   - Water pump and crankshaft pulleys
   - Timing belt cover
3. Loosen the timing bolt tensioner pulley and temporarily secure it.
   - Timing belt from the camshaft sprocket
   - Center cover bolts and cover
   - Ignition coil
   - Cylinder head cover
   - Camshaft pulley (sprocket)
   - Cam carrier assembly and timing belt
   - Camshafts
   - Camshaft oil seal using a seal removal tool

➡ **The Hydraulic Lash Adjusters (HLA's) must be installed in the location from which they were removed.**

4. Mark the HLA's to identify their original positions.
5. Remove the HLA's from the cylinder head using a magnet, and store upside-down in a oil-filled container

To install:

6. Apply a coat of the clean engine oil to the sides of the HLA's.

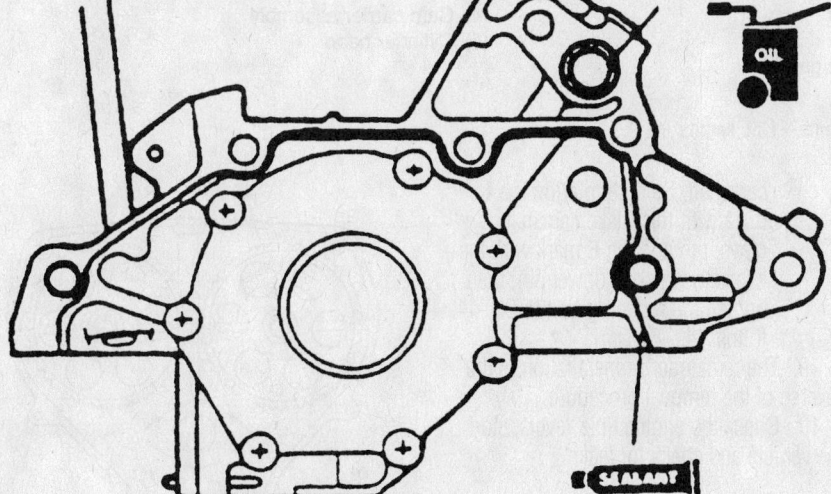

9307KG08

Apply a 0.04–0.08 in. (1–2mm) bead of silicone to the oil pump body

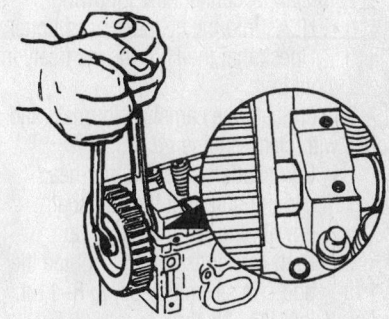

9357NG06

Remove the camshaft pulley (sprocket)—1.5L engine

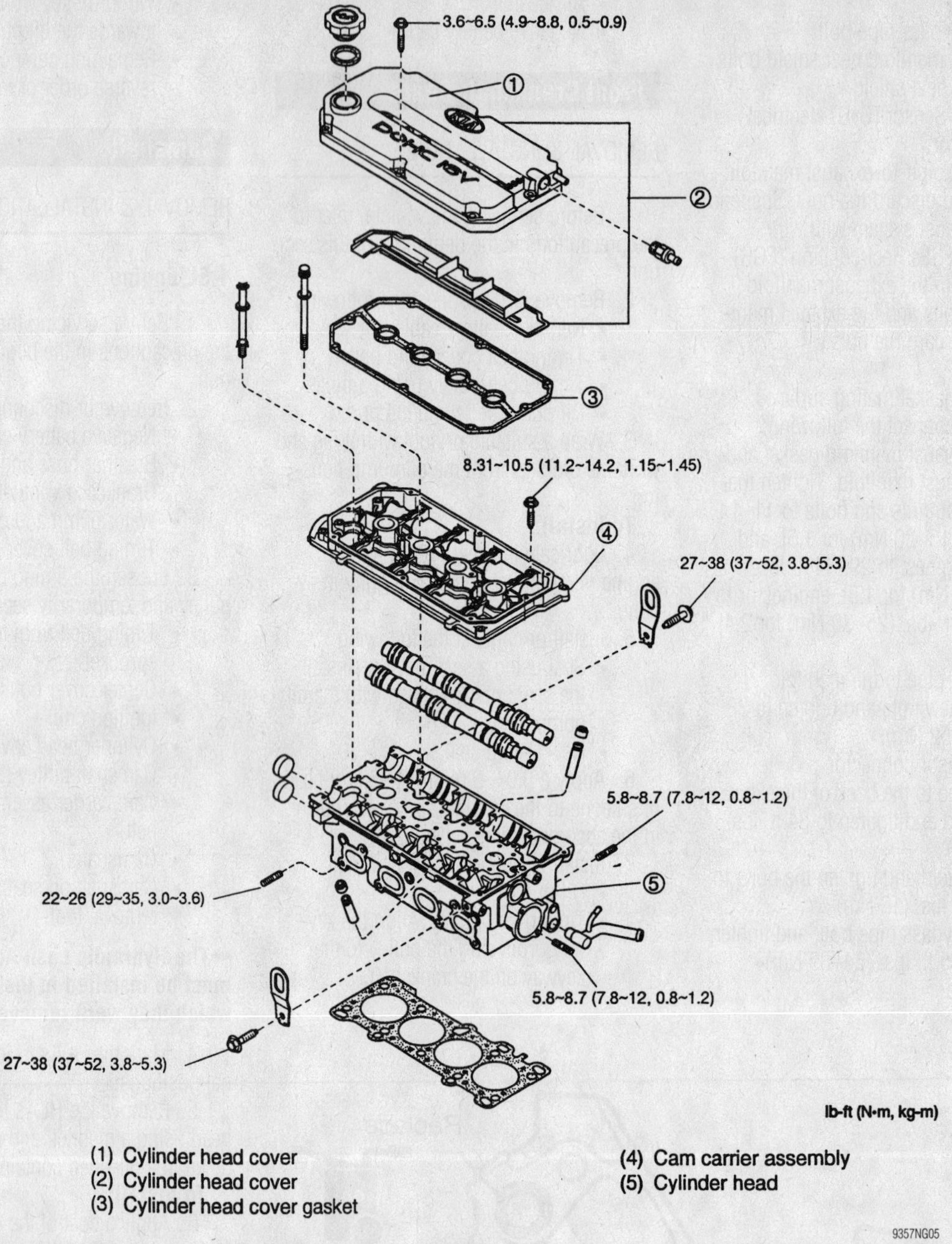

3.6~6.5 (4.9~8.8, 0.5~0.9)

① ②

8.31~10.5 (11.2~14.2, 1.15~1.45)

③

④

27~38 (37~52, 3.8~5.3)

5.8~8.7 (7.8~12, 0.8~1.2)

22~26 (29~35, 3.0~3.6)

⑤

5.8~8.7 (7.8~12, 0.8~1.2)

27~38 (37~52, 3.8~5.3)

lb-ft (N•m, kg-m)

(1) Cylinder head cover
(2) Cylinder head cover
(3) Cylinder head cover gasket
(4) Cam carrier assembly
(5) Cylinder head

9357NG05

Exploded view of the camshafts and related components—1.5L engine

7. Install or connect the following:
- HLA's into the cylinder head bore. Check that the HLA moves freely in its bore.

8. Lubricate the camshaft journals and lobes with clean engine oil.
- Camshafts in the cylinder head making sure that the camshaft dowel pins point straight up
- Cam carrier assembly. Tighten the bolts, in several steps, to 8–11 ft. lbs. (11–15 Nm).
- Camshaft oil seal, using a suitable seal installer
- Camshaft sprockets onto the

camshaft. Be sure to align the **I** mark with the intake camshaft dowel pin and the **E** mark with the exhaust camshaft dowel pin, then tighten the retaining bolt to 36–44 ft. lbs. (49–61 Nm).

9. The remainder of installation is the reverse of the removal procedure.

10. Check the engine fluid levels, start the vehicle and check for leaks.

1.8L Engine

1. Before servicing the vehicle, refer to the precautions in the beginning of this section.

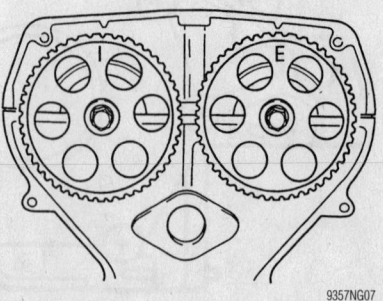

9357NG07

Proper alignment, prior to camshaft installation—1.8L engine

2. Remove or disconnect the following:
- Negative battery cable
- Ignition coils and high-tension cords
- Positive Crankcase Ventilation (PCV) valve and ventilation hoses

3. Position the engine so that No. 1 cylinder is at Top Dead Center (TDC).
- Timing belt
- Cylinder head cover
- Camshaft Position (CMP) sensor
- Camshaft sprockets

4. Loosen the camshaft bearing cap retaining bolts in several steps, following the proper sequence.
- Camshafts
- Camshaft oil seal using a seal removal tool

➡ **The Hydraulic Lash Adjusters (HLA's) must be installed in the location from which they were removed.**

5. Mark the HLA's to identify their original positions.

6. Remove the HLA's from the cylinder head using a magnet, and store upside-down in a oil-filled container

To install:

7. Apply a coat of the clean engine oil to the sides of the HLA's.

8. Install or connect the following:
- HLA's into the cylinder head bore

9. Check that the HLA moves freely in its bore.

10. Lubricate the camshaft journals and lobes with clean engine oil.

11. Install or connect the following:
- Camshafts in the cylinder head making sure that the camshaft dowel pins point straight up
- Camshaft caps in their original positions. Loosely install the cap bolts.
- Camshaft cap bolts in 5–6 steps to

13–20 ft. lbs. (18–27 Nm) in the proper sequence

12. Oil the lip of the new camshaft oil seal and, using a seal installer. Drive the seal into the cylinder head until it is flush with the edge of the camshaft bearing cap.

13. Install or connect the following:
- Camshaft sprockets onto the camshaft. Be sure to align the **I** mark with the intake camshaft dowel pin and the **E** mark with the exhaust camshaft dowel pin, then tighten the retaining bolt to 36–44 ft. lbs. (49–61 Nm).
- CMP sensor
- Cylinder head cover
- Timing belt
- PCV and ventilation hoses
- Ignition coils and high-tension cords
- Negative battery cable

14. Check the engine fluid levels, start the vehicle and check for leaks.

2.4L Engine

1. Before servicing the vehicle, refer to the precautions in the beginning of this section.

2. Drain the engine coolant.

3. Remove or disconnect the following:
- Negative battery cable
- Breather hose from between the air cleaner and rocker arm cover
- Air cleaner
- Timing belt cover
- Rocker arm cover and crank angle sensor
- Camshaft sprocket bolts and sprockets

➡ **Keep all valvetrain components in order as you remove them. The components must be installed in their original locations.**

- Bearing cap bolts, bearing caps, camshafts, rocker arms and valve adjusters

To install:

4. Lubricate the camshaft journals and lobes with clean engine oil.

5. Install or connect the following:
- Camshafts on the cylinder head. The intake camshaft has a slit on its rear side to drive the Crankshaft Position (CKP) sensor.
- Camshaft bearing caps. Make sure the cam can be easily turned by hand, then remove the bearing caps and install the rocker arms.
- Camshafts and bearings caps. Torque the bearing caps, in sequence, to 14–15 ft. lbs. (19–21 Nm).

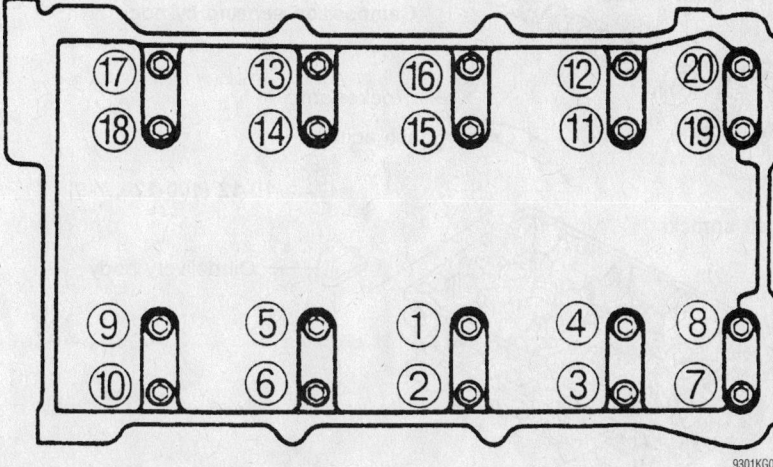

9301KG04

Camshaft bearing cap mounting bolt removal sequence—1.8L engine

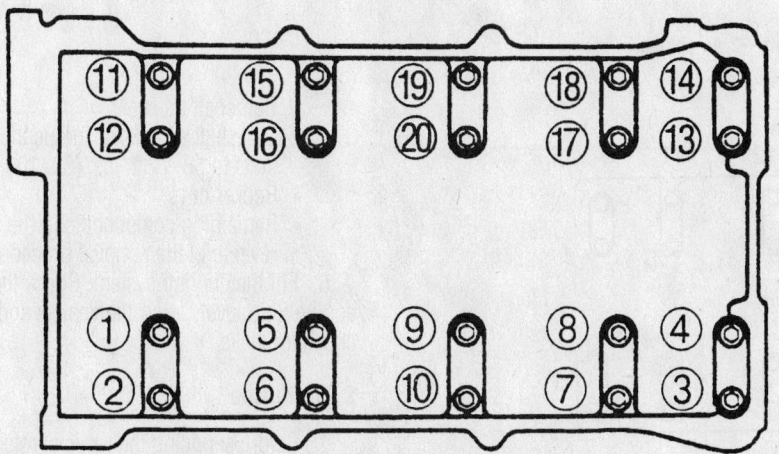

9301KG05

Camshaft bearing cap mounting bolt tightening sequence—1.8L engine

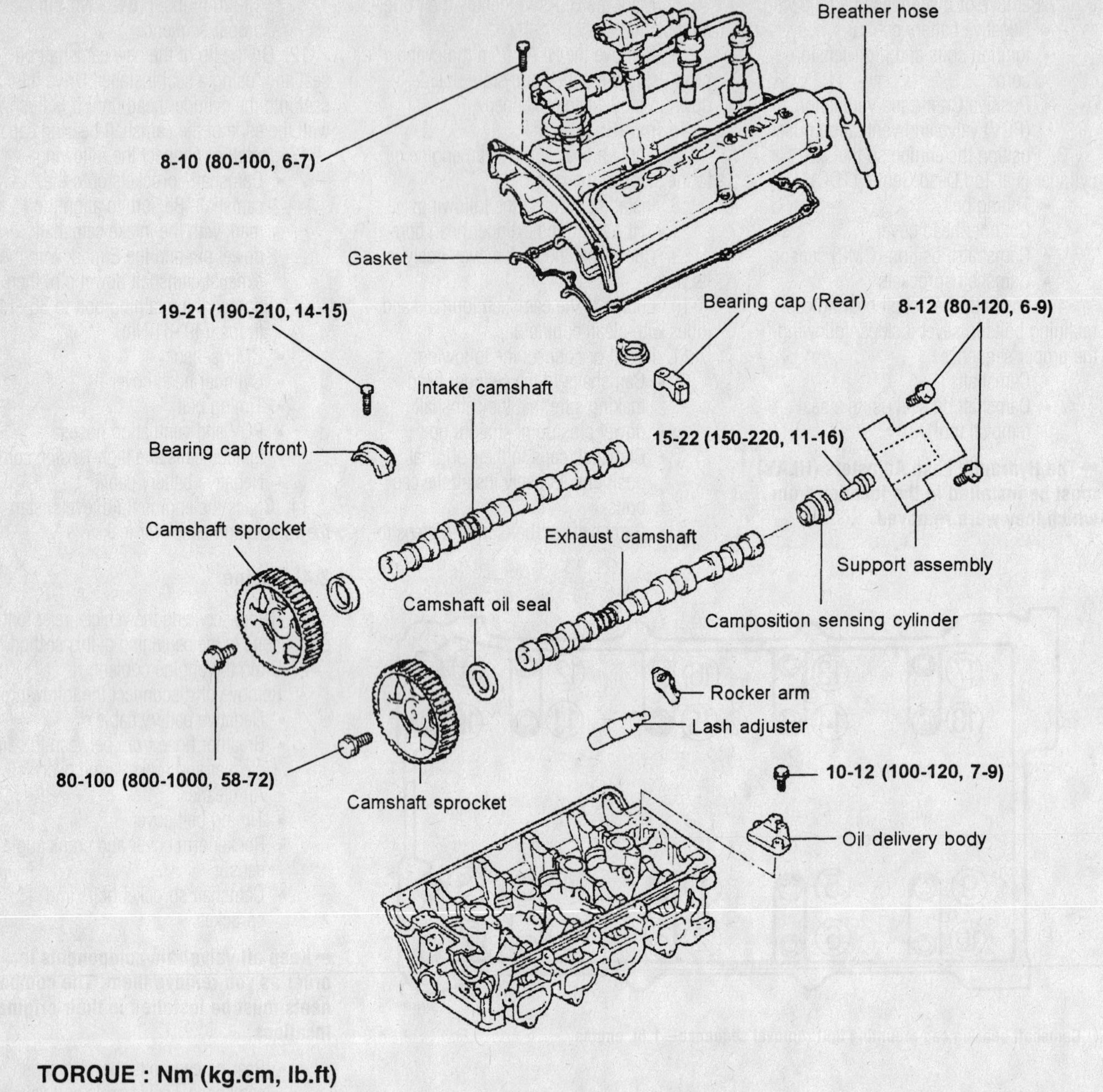

Breather hose

8-10 (80-100, 6-7)

Gasket

Bearing cap (Rear)

8-12 (80-120, 6-9)

19-21 (190-210, 14-15)

Intake camshaft

Bearing cap (front)

15-22 (150-220, 11-16)

Camshaft sprocket

Exhaust camshaft

Camshaft oil seal

Support assembly

Camposition sensing cylinder

Rocker arm

Lash adjuster

80-100 (800-1000, 58-72)

Camshaft sprocket

10-12 (100-120, 7-9)

Oil delivery body

TORQUE : Nm (kg.cm, lb.ft)

9357NG36

Exploded view of the camshafts and related components—2.4L engine

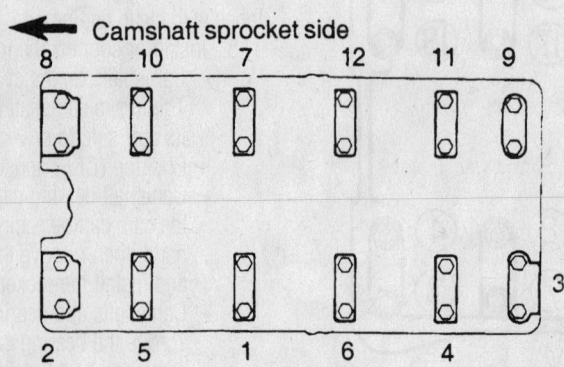

Camshaft sprocket side

| 8 | 10 | 7 | 12 | 11 | 9 |

| 2 | 5 | 1 | 6 | 4 | 3 |

9357NG37

Camshaft bearing cap tightening sequence—2.4L engine

- Camshaft oil seal
- Camshaft sprockets. Torque the bolts to 58–72 ft. lbs. (80–100 Nm).
- Rocker cover
- Remaining components in the reverse of the removal procedure

6. Fill the cooling system. Check the engine fluid levels, start the vehicle and check for leaks.

2.7L Engine

1. Before servicing the vehicle, refer to the precautions in the beginning of this section.

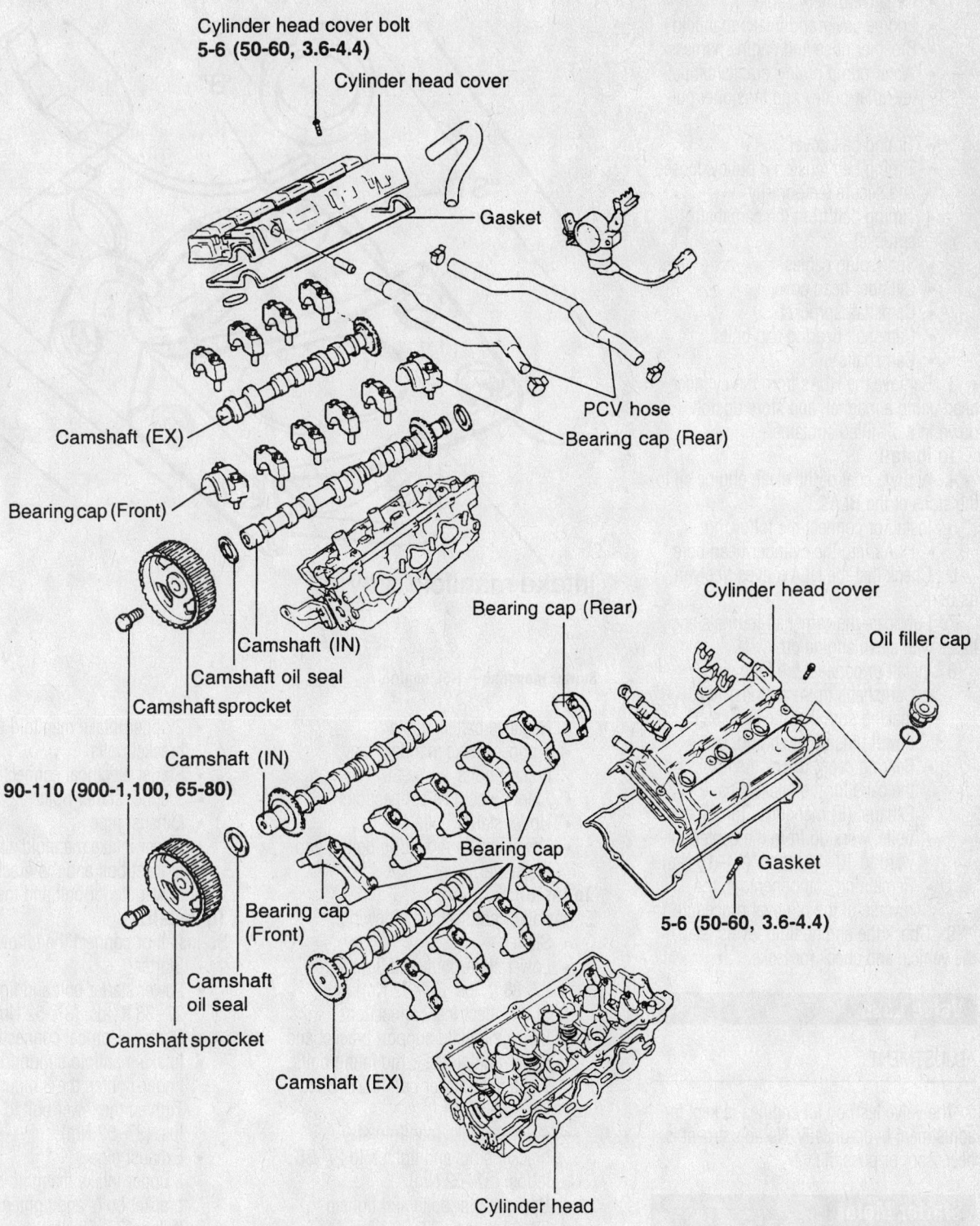

Cylinder head cover bolt
5-6 (50-60, 3.6-4.4)
Cylinder head cover
Gasket
PCV hose
Bearing cap (Rear)
Camshaft (EX)
Bearing cap (Front)
Camshaft (IN)
Camshaft oil seal
Camshaft sprocket
Cylinder head cover
Bearing cap (Rear)
Oil filler cap
90-110 (900-1,100, 65-80)
Camshaft (IN)
Bearing cap
Gasket
Bearing cap (Front)
5-6 (50-60, 3.6-4.4)
Camshaft oil seal
Camshaft sprocket
Camshaft (EX)
Cylinder head

TORQUE : Nm (kg.cm, lb.ft)

9357NG38

Exploded view of the camshafts and related components—2.7L engine

2. Remove or disconnect the following:
- Negative battery cable
- Engine cover and intake manifold
- Breather hose and engine harness
- Water pump pulley, crankshaft pulley, idler pulley and tensioner pulley
- Timing belt cover
- Timing belt tensioner pulley, loosen and secure temporarily
- Timing belt from the camshaft sprocket
- Spark plug cables
- Cylinder head cover
- Camshaft sprocket
- Camshaft bearing cap bolts
- Camshafts

3. Remove the HLA's from the cylinder head using a magnet, and store upside-down in a oil-filled container

To install:

4. Apply a coat of the clean engine oil to the sides of the HLA's.

5. Install or connect the following:
- HLA's into the cylinder head bore

6. Check that the HLA moves freely in its bore.

7. Lubricate the camshaft journals and lobes with clean engine oil.

8. Install or connect the following:
- Camshafts in the cylinder head making sure that the camshaft dowel pins point straight up
- Bearing caps. Check the marks on the caps for the Intake (I) and Exhaust (E) markings. Torque the bolts, working from the center outward to 10–11 ft. lbs. (14–15 Nm).
- Remaining components in the reverse of the removal procedure.

9. Check the engine fluid levels, start the vehicle and check for leaks.

Valve Lash

ADJUSTMENT

The valve lash on all engines is kept in adjustment hydraulically. No adjustment is necessary or possible.

Starter Motor

REMOVAL & INSTALLATION

1.5L Engine

1. Before servicing the vehicle, refer to the precautions in the beginning of this section.

2. Remove or disconnect the following:

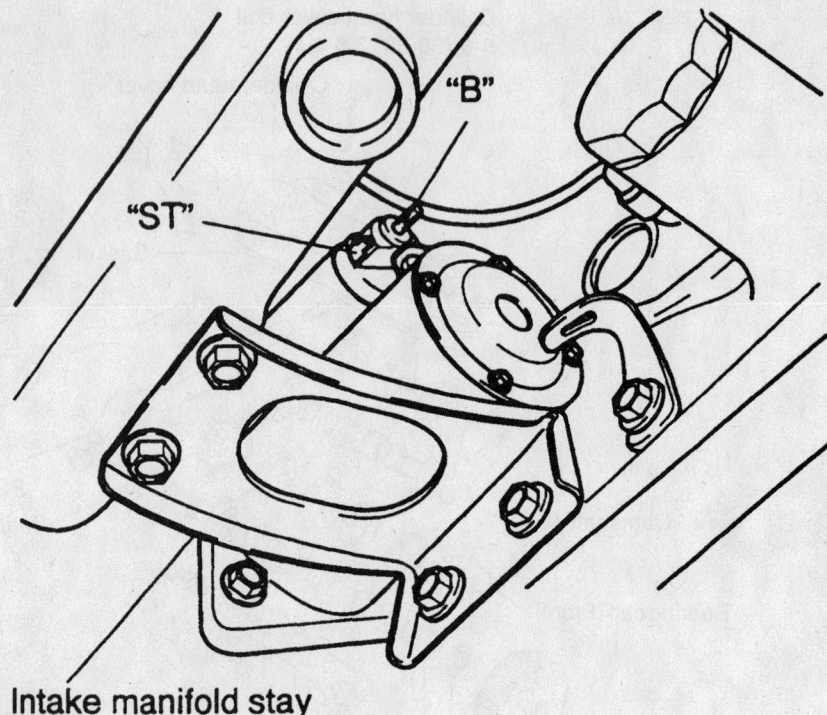

Intake manifold stay

9357NG08

Starter mounting—1.5L engine

- Negative battery cable
- 4 upper intake manifold stay bracket bolts
- Starter electrical connections
- Upper starter bolts
- Lower starter mounting bolt(s) and the starter

To install:

3. Install or connect the following:
- Starter
- Lower starter bolt and tighten to 27–38 ft. lbs. (37–52 Nm)
- Starter electrical connections
- Intake manifold support bracket and finger-tighten the 3 mounting bolts. Tighten the lower bolt to 27–38 ft. lbs. (37–52 Nm).
- 4 upper intake manifold stay bracket bolts and tighten to 27–38 ft. lbs. (37–52 Nm)
- Upper starter bolts and tighten to 27–38 ft. lbs. (37–52 Nm)
- Negative battery cable

1.8L Engine

1. Before servicing the vehicle, refer to the precautions in the beginning of this section.

2. Remove or disconnect the following:
- Negative battery cable

- 2 upper intake manifold support bracket bolts
- Starter electrical connections
- 2 upper starter bolts
- Exhaust pipe
- Lower intake manifold support bracket bolt and the bracket
- Lower starter bolt and the starter

To install:

3. Install or connect the following:
- Starter
- Lower starter bolt and tighten to 27–38 ft. lbs. (37–52 Nm)
- Starter electrical connections
- Intake manifold support bracket and finger-tighten the 3 mounting bolts. Tighten the lower bolt to 27–38 ft. lbs. (37–52 Nm).
- Exhaust pipe
- 2 upper intake manifold support bracket bolts and tighten to 27–38 ft. lbs. (37–52 Nm)
- 2 upper starter bolts and tighten to 27–38 ft. lbs. (37–52 Nm)
- Negative battery cable

2.4L and 2.7L Engines

1. Before servicing the vehicle, refer to the precautions in the beginning of this section.

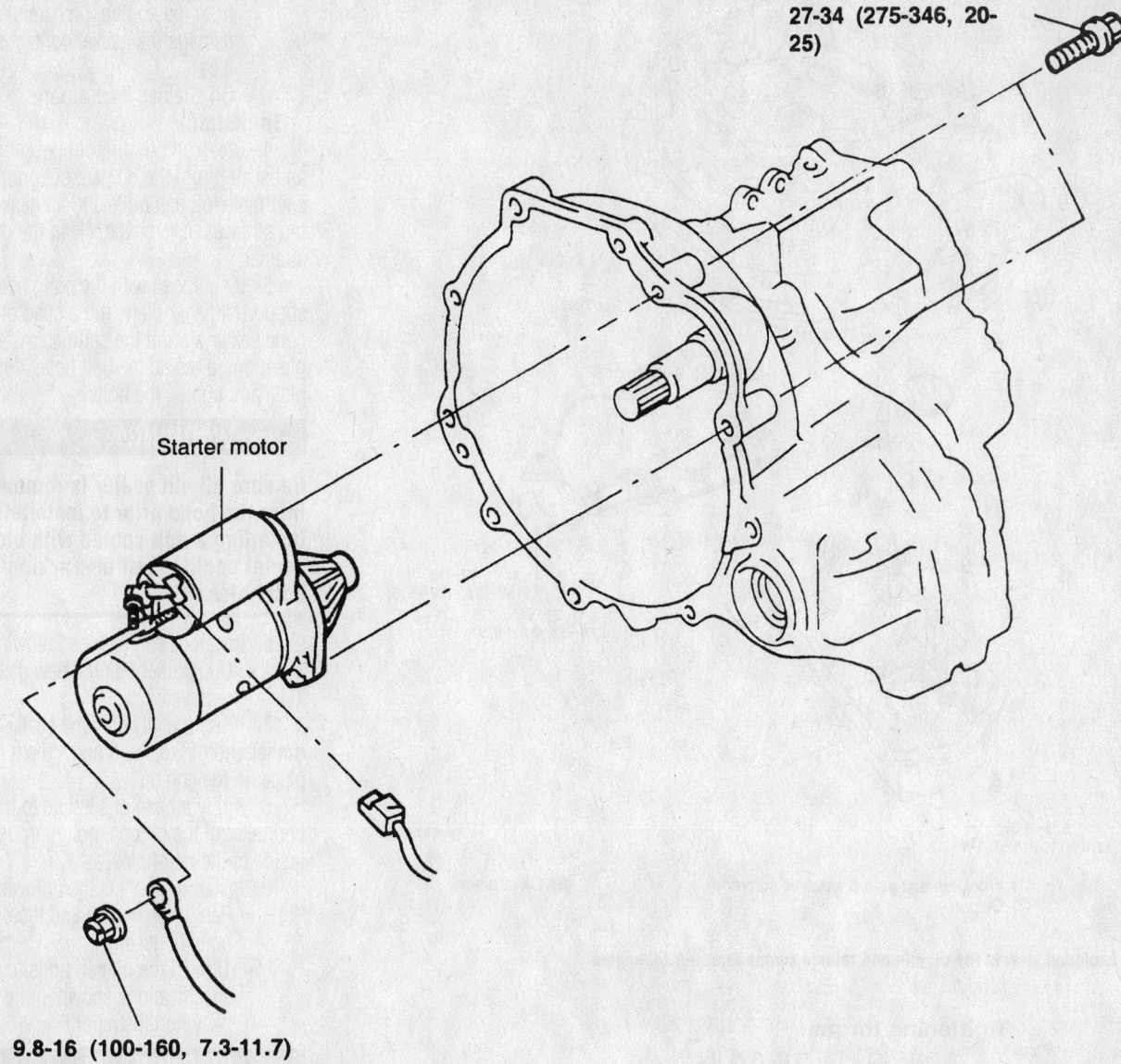

27-34 (275-346, 20-25)

Starter motor

9.8-16 (100-160, 7.3-11.7)

TORQUE : N·m (kg·cm, lb·ft)

9357NG39

Exploded view of the starter mounting—2.4L and 2.4L engines

2. Remove or disconnect the following:
   • Negative battery cable
   • Speed meter and shift cable
   • Starter connector and terminal connection
   • Starter mounting bolt(s)
   • Starter

To install:

3. Install or connect the following:
   • Starter and mounting bolt(s) and tighten to 20–25 ft. lbs. (27–34 Nm)
   • Terminal connection and electrical connection. Torque the terminal nut to 7.3–11.7 ft. lbs. (9.8–16 Nm).

   • Shift cable and speed meter
   • Negative battery cable

Oil Pan

REMOVAL & INSTALLATION

1.5L and 1.8L Engines

1. Before servicing the vehicle, refer to the precautions in the beginning of this section.

2. Remove or disconnect the following:
   • Negative battery cable

   • Engine undercover, if equipped
3. Drain the engine oil.
   • Exhaust pipe from the exhaust manifold and from the catalytic converter
   • Exhaust pipe bracket from the engine block, if necessary
   • Integrated stiffener from the engine block and transaxle, if equipped
   • Main bearing support/stiffener plate that is installed between the oil pan and engine block, if equipped
   • Bolts and the oil pan. It may be

necessary to pry the pan away from the engine; be careful not to damage the gasket contact surfaces.

- Oil strainer, if necessary

To install:

4. Clean all oil, dirt, old gasket material and sealer from the oil pan, support/stiffener plate, oil pan bolts and all gasket mating surfaces. If removed, clean the oil strainer.

5. If equipped with the main bearing support/stiffener plate, run a bead of silicone sealer around the perimeter of the plate, going inside the bolt holes. Install the plate and tighten the bolts.

❋❋ WARNING

Be sure all old sealer is removed from the bolts prior to installation. Installing a bolt coated with old sealer could result in cracking of the bolt holes.

6. Install or connect the following:
- Oil strainer using a new gasket, if removed

7. If used, apply silicone sealer to new rubber end gaskets and press them into place on the engine.

8. Apply a bead of silicone to the perimeter of the oil pan, going around the inside of the bolt holes.

9. Install or connect the following:
- Pan to the engine and the oil pan bolts finger-tight
- Tighten the oil pan bolts to the specifications shown in the accompanying illustrations

❋❋ WARNING

Be sure all old sealer is removed from the bolts prior to installation. Installing a bolt coated with old sealer could result in cracking of the bolt holes.

- Integrated stiffener to the engine block and transaxle, if removed. Tighten the bolts to 38 ft. lbs. (52 Nm).
- Transverse member, if removed. Tighten the bolts to 93 ft. lbs. (126 Nm).
- Front exhaust pipe bracket, if equipped
- Front exhaust pipe, using new gaskets. Tighten the exhaust manifold flange nuts to 27–38 ft. lbs. (37–52 Nm) for 1.5L engines, or 34 ft. lbs. (46 Nm) for 1.8L engines.

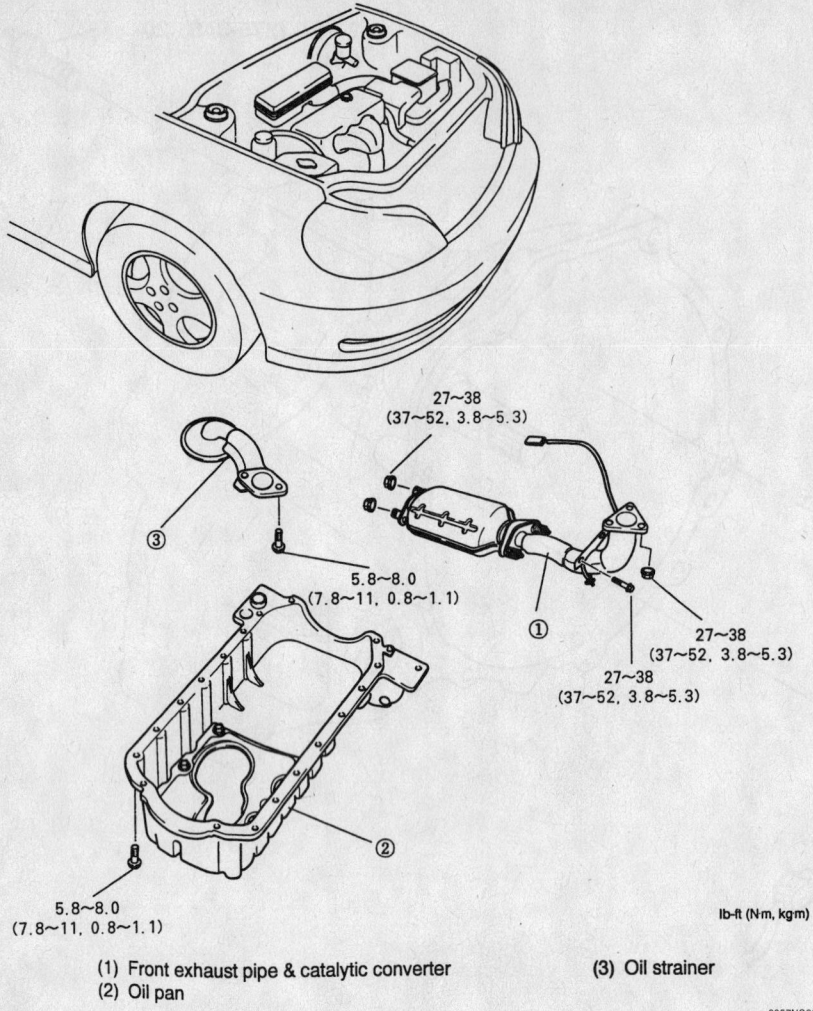

27~38
(37~52, 3.8~5.3)

5.8~8.0
(7.8~11, 0.8~1.1)

27~38
(37~52, 3.8~5.3)

27~38
(37~52, 3.8~5.3)

5.8~8.0
(7.8~11, 0.8~1.1)

lb-ft (N·m, kg·m)

(1) Front exhaust pipe & catalytic converter
(2) Oil pan
(3) Oil strainer

9357NG09

Exploded view of the oil pan and related components—1.5L engine

Tightening torque:

Ⓐ Ⓑ Ⓒ Ⓓ : 5.8~8.0 lb-ft
(7.8~10.8 N•m, 0.8~1.1 kg-m)

Ⓔ : 28~38 lb-ft (38~51 N•m, 3.8~5.3 kg-m)

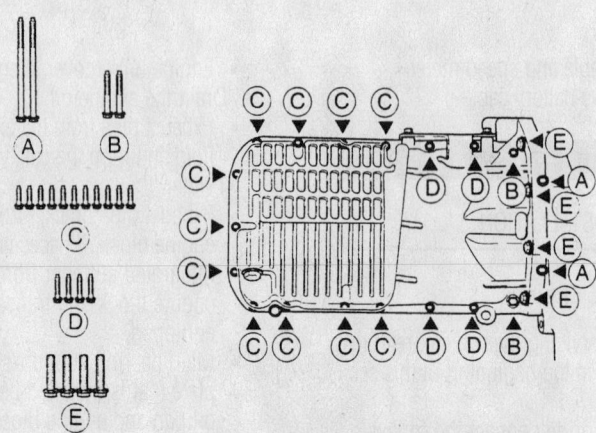

9307KG14

Oil pan mounting bolt locations and torque specifications—1.8L engine

- Oil pan drain plug using a new gasket. Tighten the drain plug to 30 ft. lbs. (41 Nm).
- Engine undercover

10. Fill the engine with the proper type and quantity of oil.

11. Connect the negative battery cable. Start the engine and bring to normal operating temperature. Check for leaks.

2.4L and 2.7L Engines

1. Before servicing the vehicle, refer to the precautions in the beginning of this section.

2. Drain the engine oil.

3. Remove or disconnect the following:
- Negative battery cable
- Timing belt
- Oil pan bolts

- Oil pan
- Oil pan screen and gasket, if necessary

To install:

4. Install or connect the following:
- Oil pump screen. Torque the bolts to 11–16 ft. lbs. (15–22 Nm).
- A 0.16 in. (44mm) bead of sealant into the groove of the oil pan

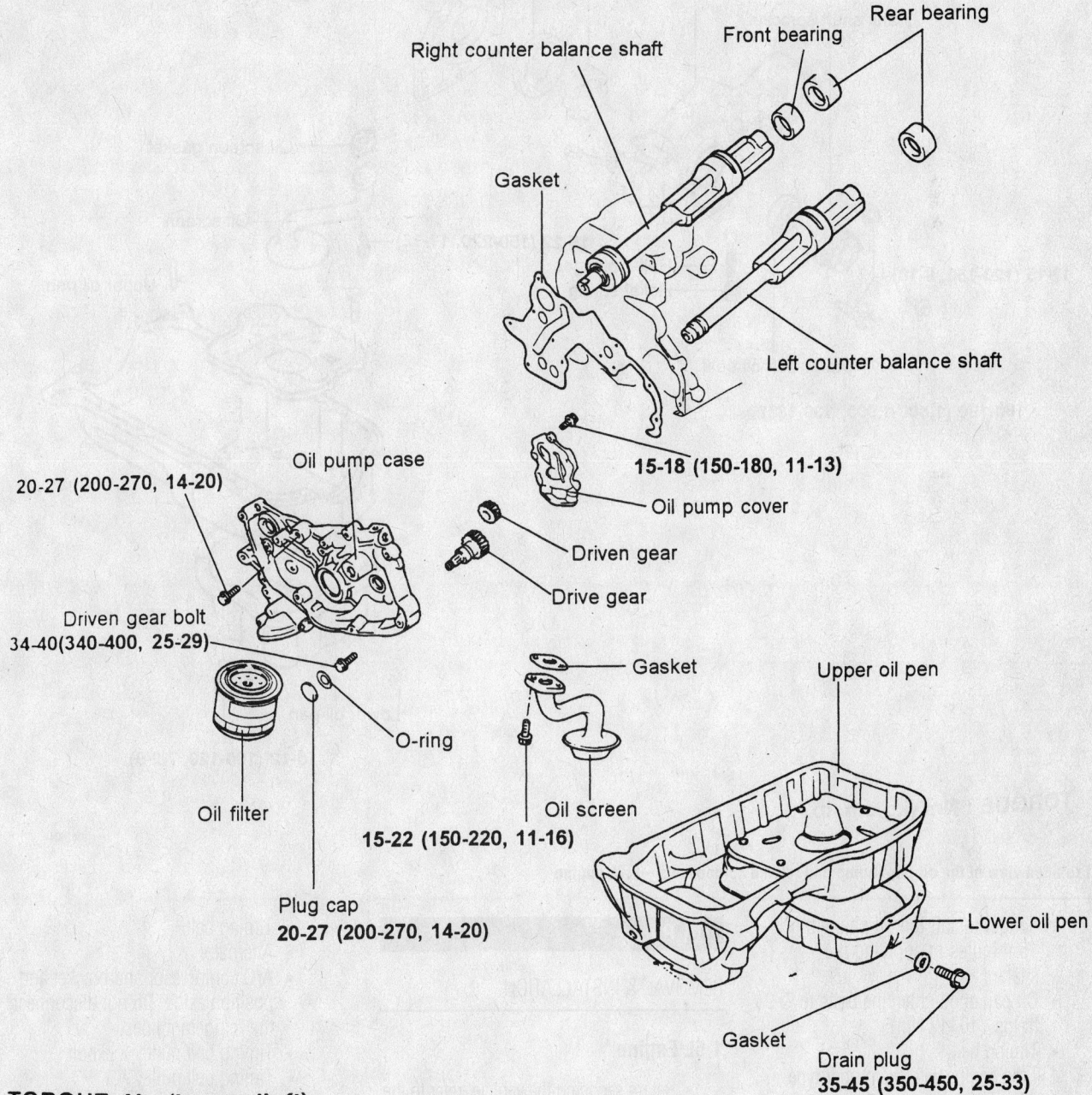

TORQUE : Nm(kg.cm, lb.ft)

9357NG40

Exploded view of the oil pan, pump and related components—2.4L engine

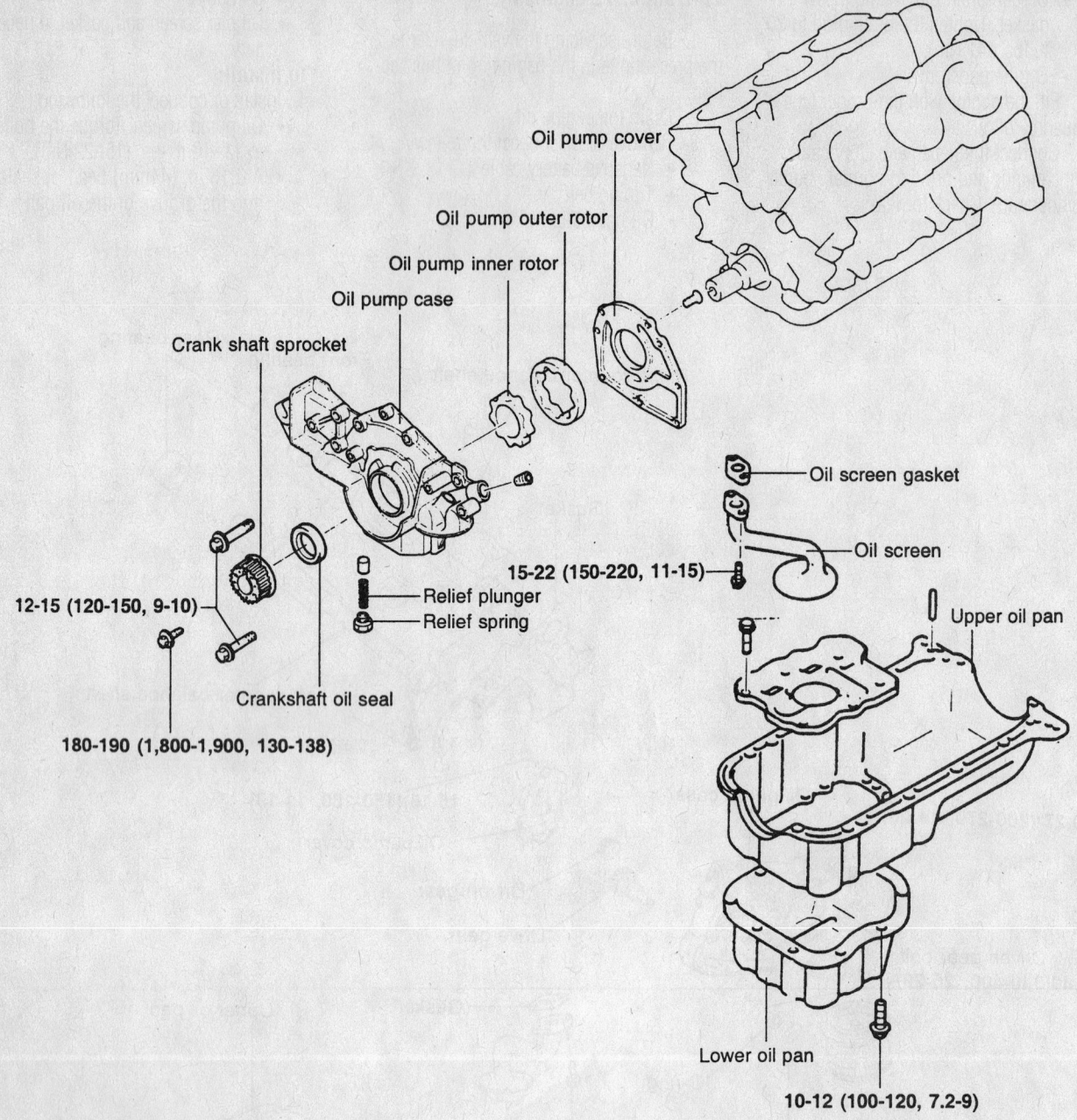

Oil pump cover

Oil pump outer rotor

Oil pump inner rotor

Oil pump case

Crank shaft sprocket

Oil screen gasket

Oil screen

15-22 (150-220, 11-15)

12-15 (120-150, 9-10)

Relief plunger

Relief spring

Crankshaft oil seal

180-190 (1,800-1,900, 130-138)

Upper oil pan

Lower oil pan

10-12 (100-120, 7.2-9)

TORQUE : Nm (kg.cm, lb.ft)

9357NG41

Exploded view of the oil pan, pump and related components—2.7L engine

flange. Install the oil pan within 15 minutes of applying the sealant.
- Oil pan and tighten the bolts to 7–9 ft. lbs. (10–12 Nm).
- Timing belt

5. Fill the engine with the proper type and quantity of oil.

6. Connect the negative battery cable. Start the engine and bring to normal operating temperature. Check for leaks.

Oil Pump

REMOVAL & INSTALLATION

1.5L Engine

1. Before servicing the vehicle, refer to the precautions in the beginning of this section.

2. Drain the engine oil.

3. Remove or disconnect the following:
- Negative battery cable
- Drive belt
- Timing belt
- Alternator
- A/C compressor and bracket and position aside. Do not disconnect the refrigerant lines.
- Timing belt pulley lockbolt
- Timing belt pulley
- Oil strainer
- Oil pump

4. Installation is the reverse of the removal procedure. Tighten the retainers to the specifications shown in the illustration.

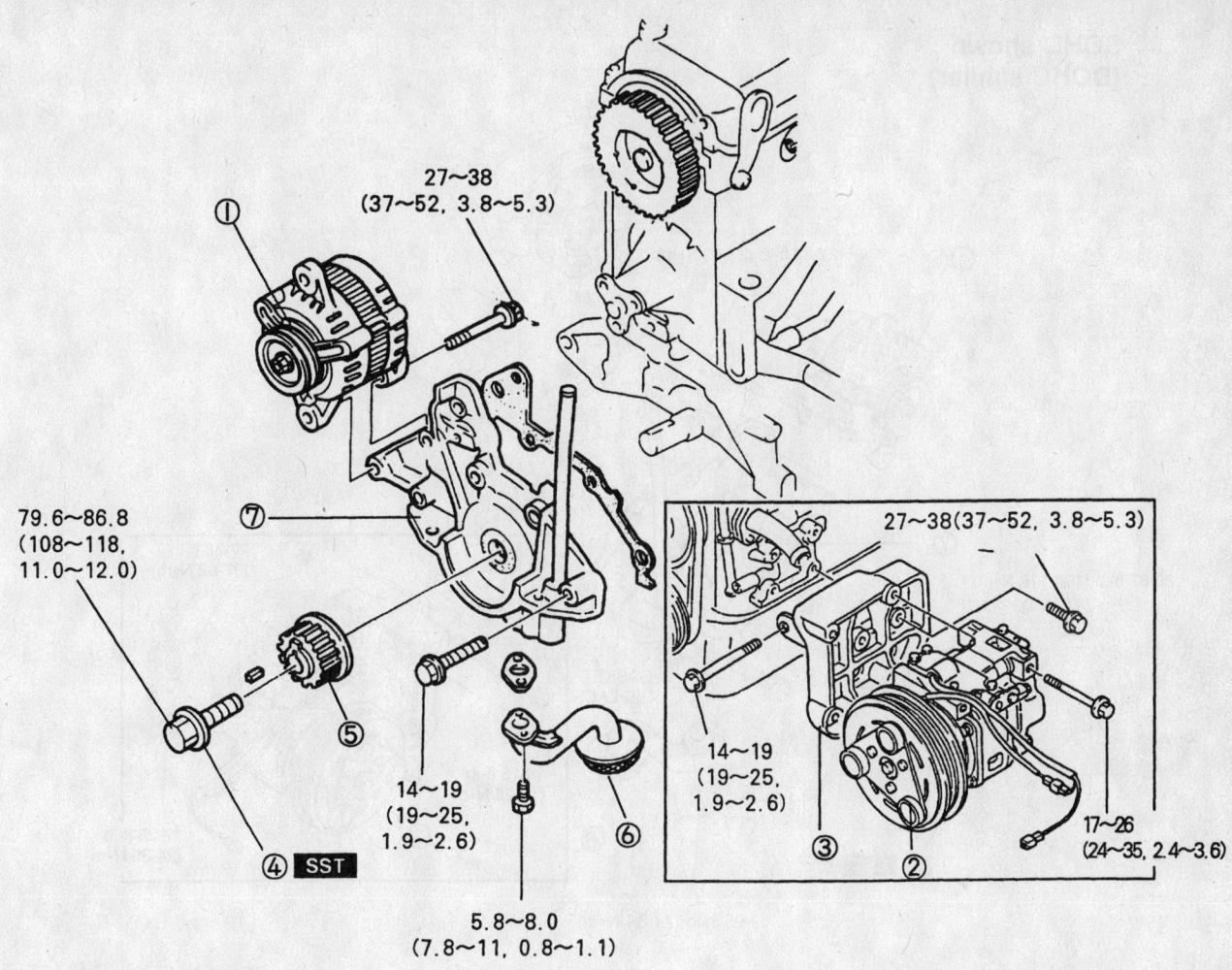

27~38
(37~52, 3.8~5.3)

79.6~86.8
(108~118,
11.0~12.0)

27~38(37~52, 3.8~5.3)

14~19
(19~25,
1.9~2.6)

14~19
(19~25,
1.9~2.6)

17~26
(24~35, 2.4~3.6)

5.8~8.0
(7.8~11, 0.8~1.1)

④ SST

lb-ft (N·m, kg·m)

(1) Alternator
(2) Compressor assembly
(3) Compressor bracket
(4) Timing belt pulley lock bolt

(5) Timing belt pulley
(6) Oil strainer
(7) Oil pump assembly

9357NG10

Exploded view of the oil pump mounting—1.5L engine

1.8L Engine

1. Before servicing the vehicle, refer to the precautions in the beginning of this section.
2. Drain the engine oil.
3. Remove or disconnect the following:
   - Negative battery cable
   - Crankshaft pulley, timing belt cover, belt and the crankshaft sprocket
   - Oil pan
   - Oil pick-up tube and discard the gasket
   - Oil pump attaching bolts
   - Oil pump
   - Front crankshaft seal from the oil pump if the pump is being replaced

To install:

4. Clean the oil, dirt and old sealant from all contact surfaces.
5. Install or connect the following:
   - New O-rings on the oil pump
6. If the oil seal was removed from the oil pump, apply clean engine oil to the lip of the seal. Push the seal in lightly by hand. Press the seal, with a protrusion of 0.02–0.04 in. (0.5–1.0mm), into the oil pump with a suitable tool (49 B014 401).
7. Apply a bead of silicone to the oil pump at the cylinder block contact surface, going inside the bolt holes.

8. Install or connect the following:
   - Oil pump and tighten the bolts to 14–18 ft. lbs. (19–25 Nm)
   - New gasket and the oil pump pick-up tube. Tighten the mounting bolts to 70–95 inch lbs. (8–11 Nm).
   - Oil pan
   - Crankshaft sprocket, timing belt and cover
   - Crankshaft pulley
   - Negative battery cable
9. Fill the engine with the proper type and quantity of oil. Run the engine and check for leaks.

**SOHC shown
(DOHC similar)**

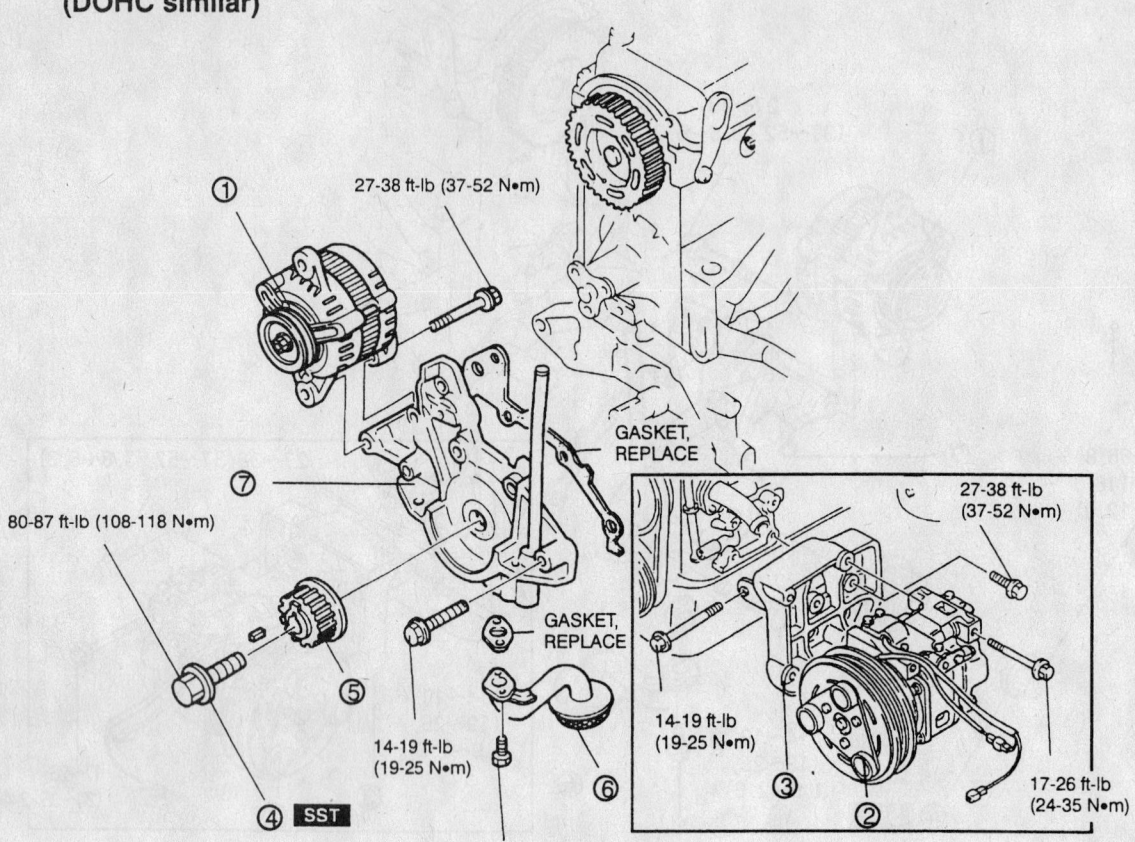

27-38 ft-lb (37-52 N•m)

GASKET,
REPLACE

80-87 ft-lb (108-118 N•m)

GASKET,
REPLACE

14-19 ft-lb
(19-25 N•m)

69-95 in-lb (7.8-11 N•m)

27-38 ft-lb
(37-52 N•m)

14-19 ft-lb
(19-25 N•m)

17-26 ft-lb
(24-35 N•m)

SST

1. Generator
2. A/C compressor (if equipped)
3. A/C compressor bracket (if equipped)
4. Crankshaft pulley lock bolt
5. Timing belt pulley
6. Oil strainer
7. Oil pump

7923KG26

Exploded view of the oil pump—1.8L engine

2.4L and 2.7L Engines

1. Before servicing the vehicle, refer to the precautions in the beginning of this section.
2. Drain the engine oil.
3. Remove or disconnect the following:
   - Negative battery cable
   - Oil pan, screen and gasket
   - Relief plunger and gasket
   - Relief spring and relief valve from the oil filter bracket
   - Oil pressure switch
   - Oil filter bracket and gasket
   - Plug cap from the oil pump portion of the front case

   - Left side of the cylinder block plug and insert a screwdriver with a 0.32 in. (8mm) diameter shaft in to the plug hole. The screwdriver must be inserted more than 2.4 in. (60mm).
   - Oil pump driven gear and left counter balance shaft retaining bolt
   - Front case and gasket
   - 2 counter balance shafts from the cylinder block
   - Oil pump cover and gears from the front case

To install:

4. Apply engine oil to the oil pump gears.

5. Install or connect the following:
   - Oil pump gears, aligning the 2 timing marks
   - Crankshaft front oil seal, using a proper tool
   - Special tool 09214-32100 on the front end of the crankshaft and apply a thin coat of oil to the outer circumference of the special tool to install the front case
   - Front case with a new gasket and temporarily tighten the flange bolts, except the bolts for the filter bracket. Torque the front case bolts to 14–20 ft. lbs. (20–27 Nm).

6. Insert a screwdriver through the plug hole in the left side of the cylinder block. After checking that the shaft is in position, tighten.

7. Install the remaining components in the reverse of the removal procedure.

8. Fill the engine with the proper type and quantity of oil. Run the engine and check for leaks.

Rear Main Seal

REMOVAL & INSTALLATION

1. Before servicing the vehicle, refer to the precautions in the beginning of this section.

2. Remove or disconnect the following:
   • Negative battery cable

   • Transaxle assembly
   • Clutch and flywheel assembly, if equipped with a manual transaxle
   • Flexplate-to-crankshaft bolts, the flexplate and shim plates, if equipped with an automatic transaxle

3. Cut the oil seal lip with a knife. Install a rag to the housing and using a screwdriver, carefully pry the oil seal from the oil seal housing. Clean the gasket mounting surfaces.

To install:

4. Clean the oil seal housing. Coat the oil seal and the housing with clean engine oil.

5. Install or connect the following:
   • Oil seal into the housing and tap it evenly into place with a hammer

and a large diameter piece of pipe. The seal must be flush with the edge of the rear cover.
   • Flywheel assembly or the flexplate, as applicable, and tighten the mounting bolts to 71–76 ft. lbs. (97–102 Nm)
   • Clutch assembly, if applicable
   • Transaxle
   • Negative battery cable

Timing Belt

REMOVAL & INSTALLATION

1. Before servicing the vehicle, refer to the precautions in the beginning of this section.

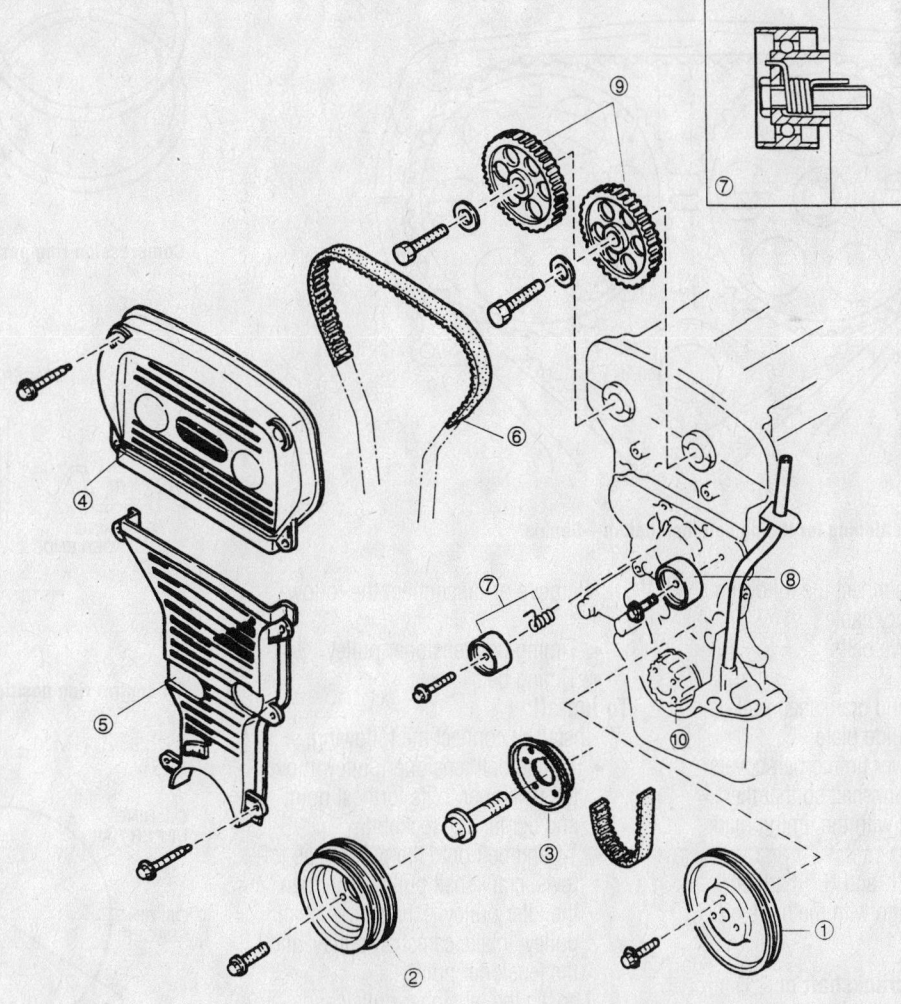

(1) Water pump pulley
(2) Crankshaft pulley
(3) Timing belt guide plate
(4) Timing belt cover (Upper)
(5) Timing belt cover (Lower)
(6) Timing belt
(7) Timing belt tensioner pulley & spring
(8) Idler pulley
(9) Camshaft pulley
(10) Timing belt pulley

93015G01

Exploded view of the timing belt cover mounting and related components–Sephia

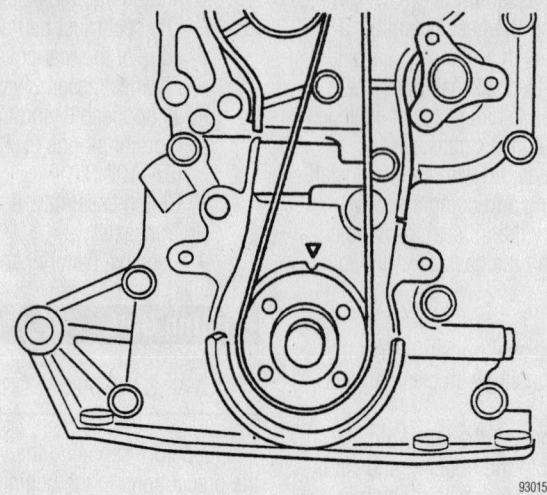

Crankshaft sprocket timing belt alignment mark—Sephia

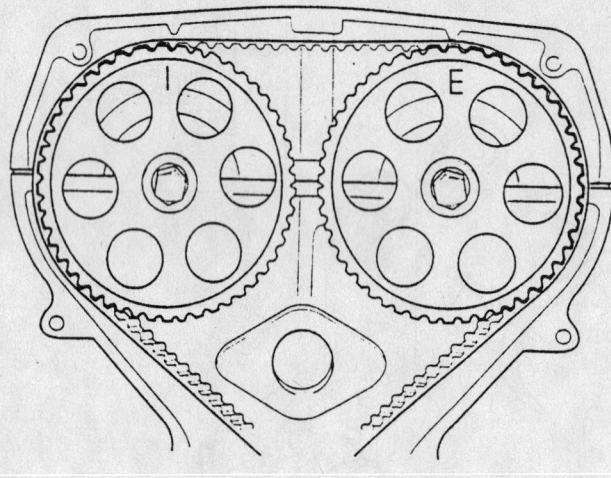

Camshaft sprocket positioning for timing belt installation—Sephia

2. Remove or disconnect the following:
- Negative battery cable
- Accessory drive belts
- Generator
- Water pump and crankshaft pulleys
- Timing belt guide plate
- Upper and lower timing belt covers

3. Position the crankshaft so that the timing mark is aligned with the timing mark on the engine.

4. Verify that the "I" and "E" mark on the camshaft pulley align with the mark on the cylinder head.

➡ **Do not move the crankshaft or camshaft once the timing marks have been correctly positioned.**

5. If the timing belt is to be reused, mark the direction of rotation on the timing belt.

6. Remove or disconnect the following:
- Timing belt tensioner pulley
- Timing belt

To install:

7. Install or connect the following:
- Timing belt tensioner pulley, move the tensioner to its furthest point and tighten the lockbolt.
- Timing belt onto the pulleys, as follows, crankshaft pulley first, then the idler pulley, exhaust camshaft pulley, intake camshaft pulley, and the tensioner pulley

8. Loosen the tensioner pulley and allow the tensioner spring to apply tension on the belt, then tighten the lockbolt to 28–38 ft. lbs. (38–51 Nm).

9. Rotate the crankshaft clockwise 2 turns and be sure all marks are still correctly aligned.

10. Install the remaining components in the revere order of the removal noting the following torque specifications:
- Crankshaft pulley: 9–13 ft. lbs. (12–17 Nm)
- Water pump pulley: 9–13 ft. lbs. (12–17 Nm)

11. Connect the negative battery cable.

Piston and Ring

POSITIONING

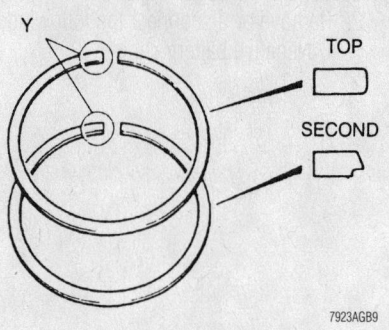

Compression ring positioning

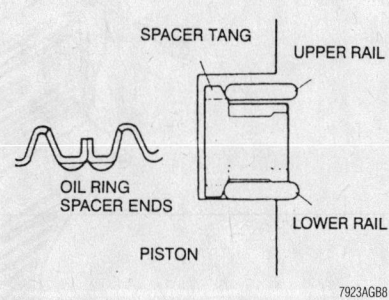

Oil control ring positioning

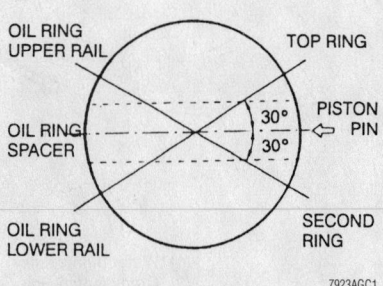

Piston ring end-gap spacing

FUEL SYSTEM

Fuel System Service Precautions

Safety is the most important factor when performing not only fuel system maintenance but any type of maintenance. Failure to conduct maintenance and repairs in a safe manner may result in serious personal injury or death. Maintenance and testing of the vehicle's fuel system components can be accomplished safely and effectively by adhering to the following rules and guidelines.

• To avoid the possibility of fire and personal injury, always disconnect the negative battery cable unless the repair or test procedure requires that battery voltage be applied.

• Always relieve the fuel system pressure prior to disconnecting any fuel system component (injector, fuel rail, pressure regulator, etc.), fitting or fuel line connection. Exercise extreme caution whenever relieving fuel system pressure, to avoid exposing skin, face and eyes to fuel spray. Please be advised that fuel under pressure may penetrate the skin or any part of the body that it contacts.

• Always place a shop towel or cloth around the fitting or connection prior to loosening to absorb any excess fuel due to spillage. Ensure that all fuel spillage

(should it occur) is quickly removed from engine surfaces. Ensure that all fuel soaked cloths or towels are deposited into a suitable waste container.

• Always keep a dry chemical (Class B) fire extinguisher near the work area.

• Do not allow fuel spray or fuel vapors to come into contact with a spark or open flame.

• Always use a back-up wrench when loosening and tightening fuel line connection fittings. This will prevent unnecessary stress and torsion to fuel line piping.

• Always replace worn fuel fitting O-rings with new. Do not substitute fuel hose or equivalent, where fuel pipe is installed.

Fuel System Pressure

RELIEVING

1. Before servicing the vehicle, refer to the precautions in the beginning of this section.
2. Release the rear seat retainers (clips or catches) and remove rear seat cushion.
3. Remove the fuel pump cover.
4. Detach the fuel pump electrical connector.

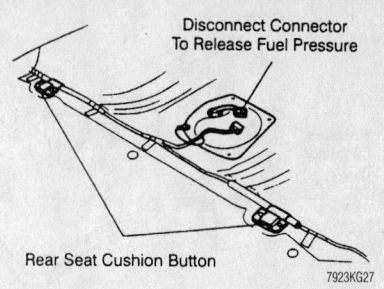

Fuel pump connector—Sephia shown

5. Start the engine, allowing it to idle until it runs out of fuel.
6. After the engine stalls, reattach the fuel pump connector and turn the ignition switch **OFF**.

Fuel Filter

REMOVAL & INSTALLATION

1.5L, 1.8L and 2.4L Engines

1. Before servicing the vehicle, refer to the precautions in the beginning of this section.
2. Relieve the fuel system pressure.

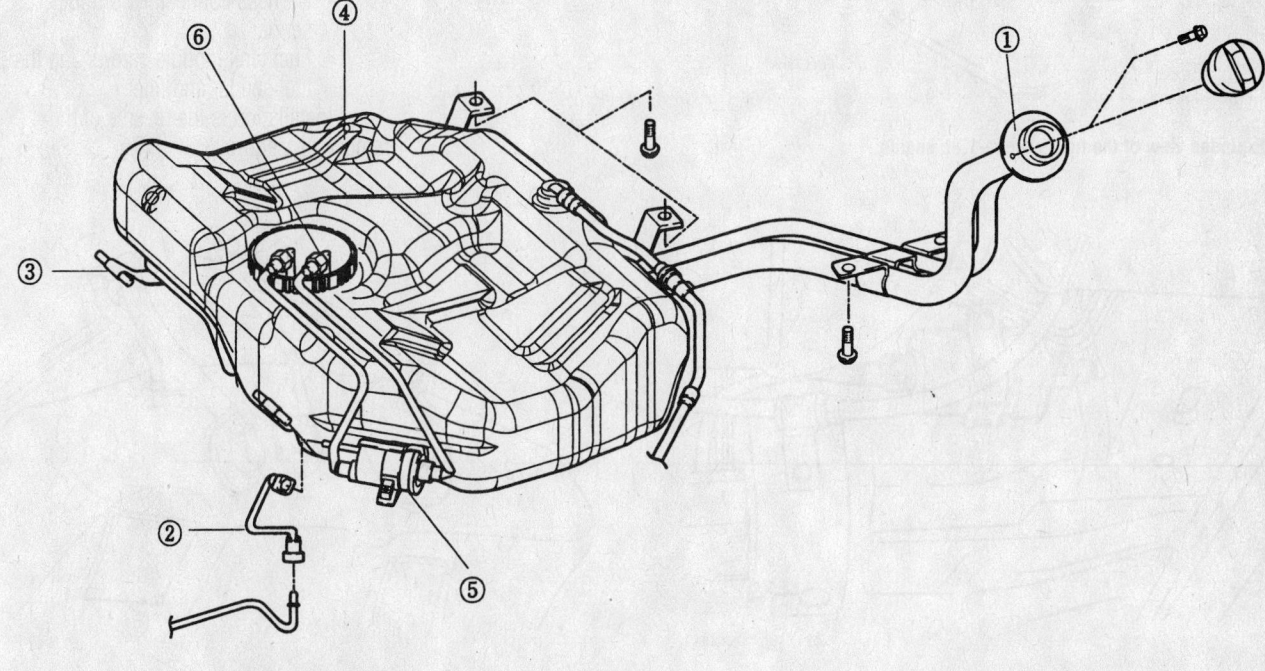

(1) Filler pipe
(2) Fuel tube
(3) Strap
(4) Fuel tank
(5) Fuel filter
(6) Fuel pump

Exploded view of the fuel system—1.5L engine

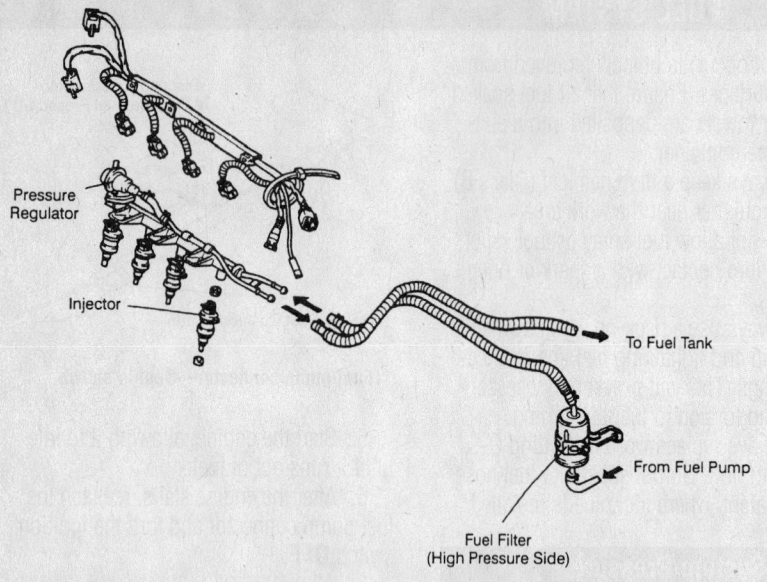

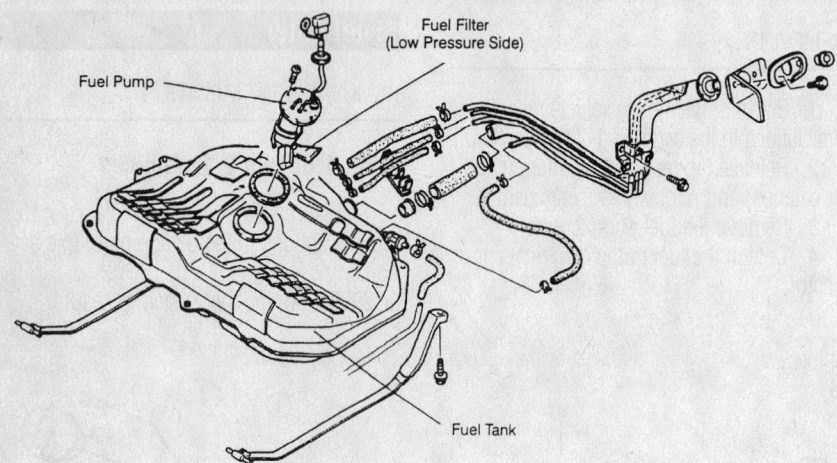

Exploded view of the fuel system—1.8L engine

7923KG28

3. Remove or disconnect the following:

- Negative battery cable
- Fuel lines from both ends of the fuel filter. Plug the lines to prevent leakage.
- Bracket retaining nuts and bolts, as necessary
- Filter and the mounting bracket

To install:

4. Install or connect the following:

- Filter in the mounting bracket
- Fuel lines to the filter
- Bracket nuts and tighten to 9.5 ft. lbs. (13 Nm) for 1.5L and 1.8L engines. For 2.4L engines, torque the nuts to 18–25 ft. lbs. (25–35 Nm).
- Negative battery cable

5. Run the engine and check for any fuel leaks.

2.7L Engine

1. Before servicing the vehicle, refer to the precautions in the beginning of this section.

2. Relieve the fuel system pressure.

3. Remove or disconnect the following:

- Negative battery cable
- Rear seat retainers (clips or catches) and rear seat cushion
- Harness connector from fuel sender
- Fuel tank module screws and the fuel sender and filter

4. Installation is the reverse of the remove procedure.

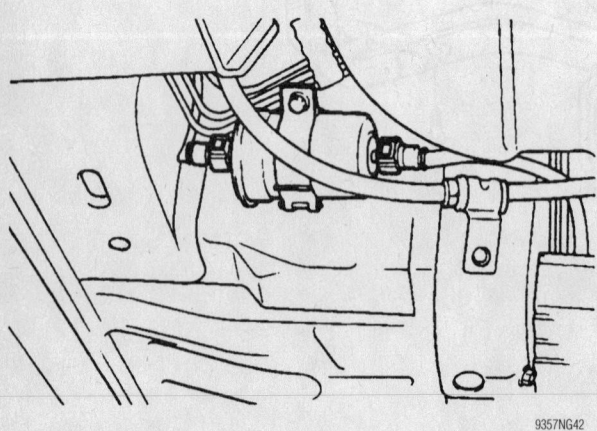

Fuel filter location—2.4L engine

9357NG42

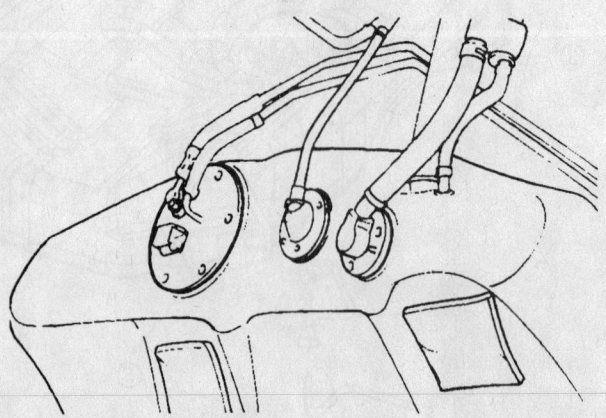

Fuel filter location—2.7L engine

9357NG43

Fuel Pump

REMOVAL & INSTALLATION

1. Before servicing the vehicle, refer to the precautions in the beginning of this section.

2. Relieve the fuel system pressure.

3. Remove or disconnect the following:
- Negative battery cable
- Rear seat cushion from the vehicle

4. Clean any dirt that has accumulated around the fuel pump cover so it will not enter the tank during pump removal and installation.

5. Remove or disconnect the following:
- Fuel pump cover
- Fuel gauge connector, hoses, and the gauge
- Fuel pump electrical connector
- Fuel pump from the bracket assembly
- Seal ring and discard

To install:

6. Clean the fuel pump mounting flange, fuel tank mounting surface and seal ring groove.

7. Apply a light coating of grease on a new seal ring to hold it in place during assembly

8. Install or connect the following:
- Seal ring
- Fuel pump to the bracket assembly carefully to ensure the filter is not damaged. Be sure the seal ring remains in the groove.

9. Hold the pump assembly in place, and pull the fuel pump down so that it is tight against the bracket.

10. Install or connect the following:
- Fuel pump electrical connector
- Fuel gauge, hoses, and gauge connector
- Fuel pump cover
- Rear seat cushion
- Negative battery cable

11. Start the engine and check for proper system operation and for fuel leaks.

Fuel Injector

REMOVAL & INSTALLATION

1. Before servicing the vehicle, refer to the precautions in the beginning of this section.

2. Relieve the fuel system pressure.

3. Remove or disconnect the following:
- Negative battery cable
- Injector electrical connectors
- Fuel line from the fuel rail
- Accelerator cable bracket and the cable, if necessary
- Fuel rail retainers and the fuel rail
- Injector retaining clips and the injectors
- Injector O-rings and discard

To install:

4. Apply a small amount of clean engine oil to the new O-rings and install them.

5. Install or connect the following:
- Injectors to the fuel rail and the retaining clips
- Fuel rail and the fuel rail retainers
- Accelerator cable and bracket, if removed
- Fuel line to the fuel rail
- Injector electrical connectors
- Negative battery cable

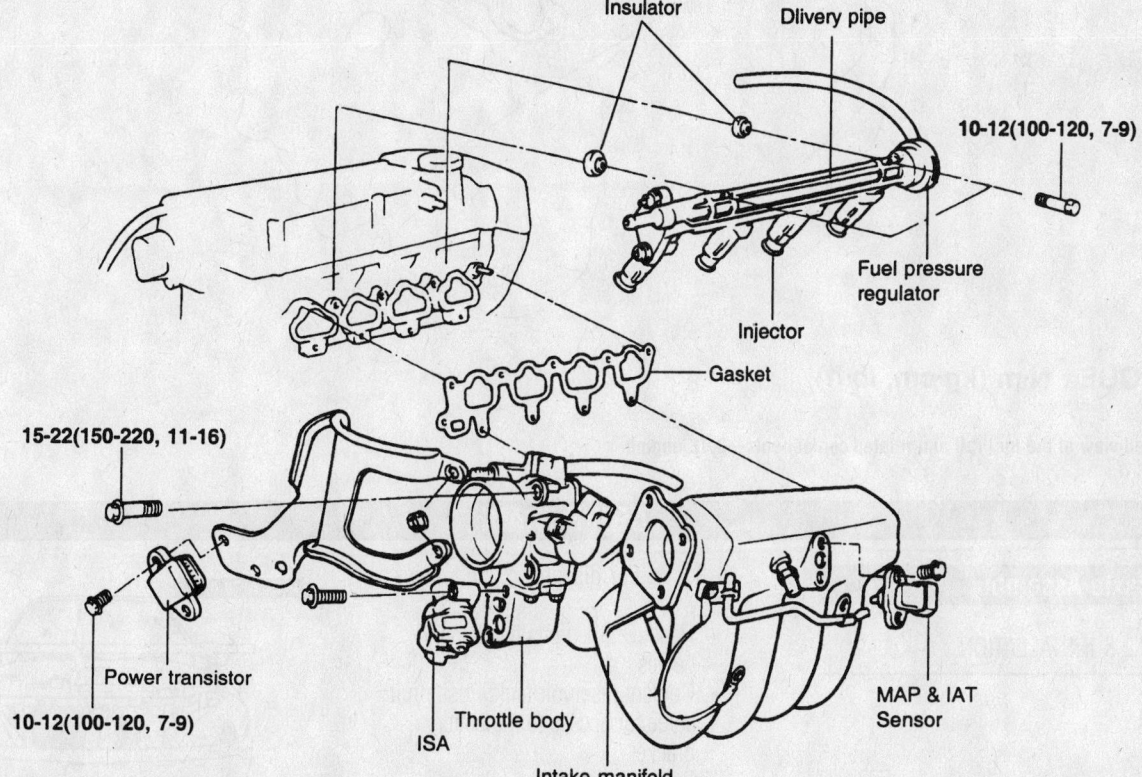

TORQUE : N·m (kg·cm, lb·ft)

9357NG44

Exploded view of the fuel rail and related components—2.4L engine

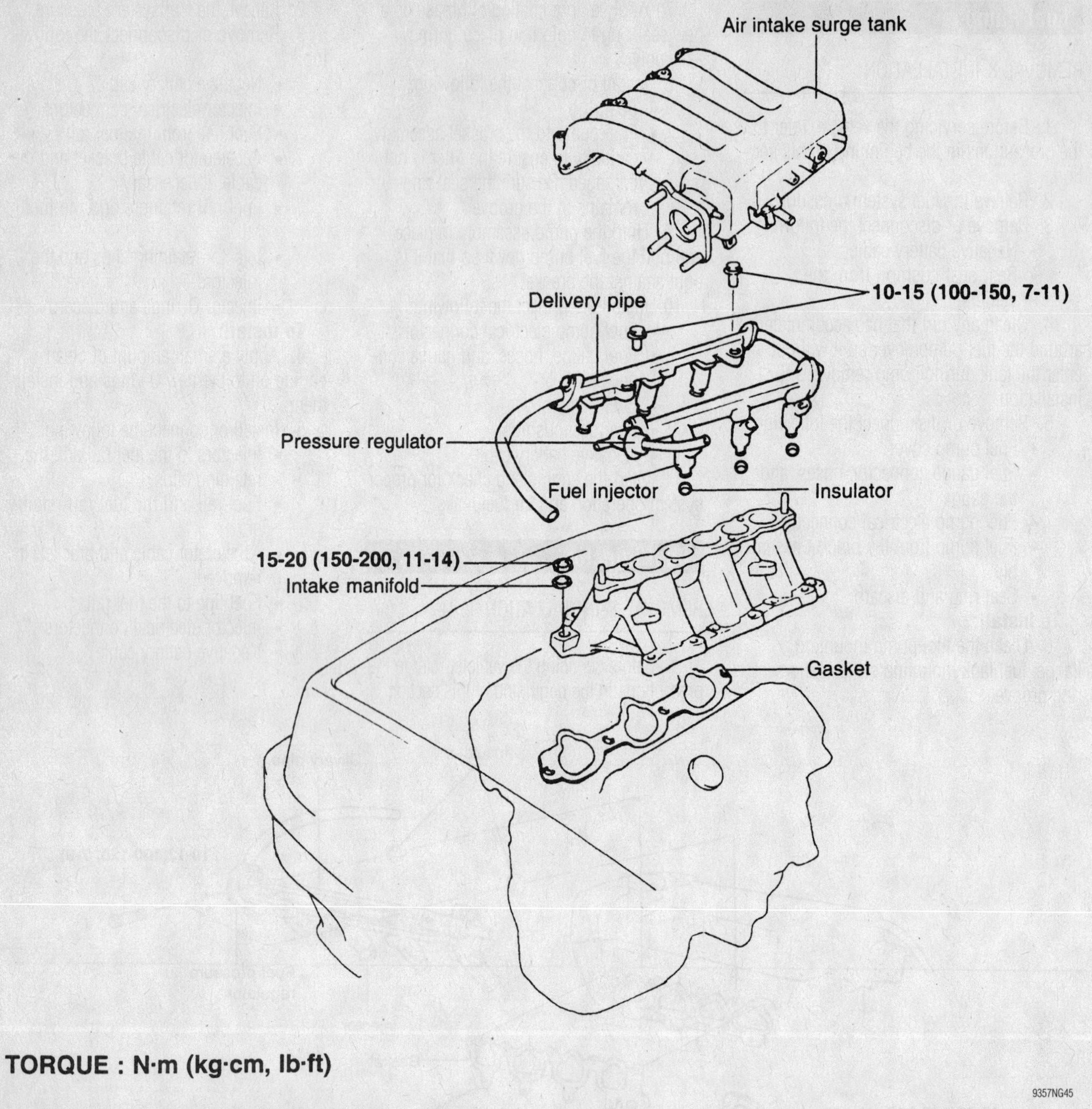

Air intake surge tank

Delivery pipe

10-15 (100-150, 7-11)

Pressure regulator

Fuel injector — Insulator

15-20 (150-200, 11-14)
Intake manifold

Gasket

TORQUE : N·m (kg·cm, lb·ft)

9357NG45

Exploded view of the fuel rail and related components—2.7L engine

---

DRIVE TRAIN

Transaxle Assembly

REMOVAL & INSTALLATION

Manual

RIO

1. Before servicing the vehicle, refer to the precautions in the beginning of this section.
2. Drain the transaxle oil.

3. Remove or disconnect the following:

- Negative, then positive battery cables
- Coolant reservoir tank. Position it aside for access to the battery bracket.
- Battery bracket
- Battery, battery tray and fixing holder
- Fresh air duct and air cleaner assembly

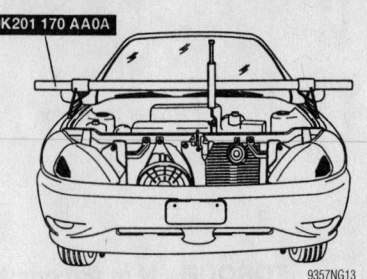

0K201 170 AA0A

9357NG13

Use the proper tool to support the engine—Rio

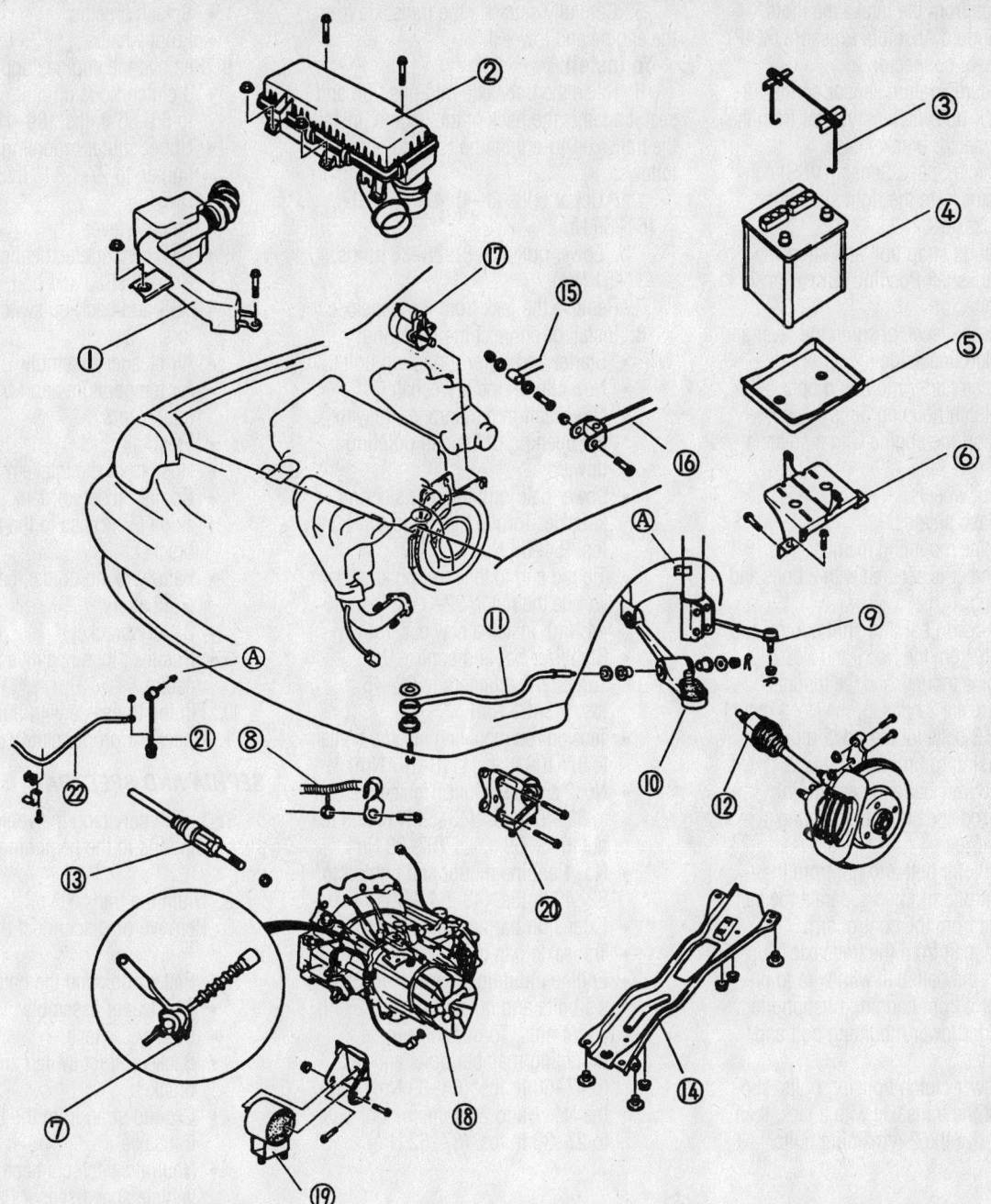

(1) Fresh air duct
(2) Air cleaner assembly
(3) Battery fixing holder
(4) Battery
(5) Battery tray
(6) Battery bracket
(7) Clutch cable
(8) Ground
(9) Tire-rod end
(10) Lower arm
(11) Torsion bar

(12) Driveshaft (Right side)
(13) Driverhaft (Left side)
(14) Engine mounting member
(15) Shift control rod
(16) Extension bar
(17) Starter
(18) Manual transaxle
(19) Engine mounting NO.2
(20) Engine mounting NO.1
(21) Shabilizer control link
(22) Stabilizer bar

9357NG12

Exploded view of the manual transaxle mounting and related components—Rio

- Hose from the intake manifold
- Manifold Absolute Pressure (MAP) sensor connector
- Air temperature sensor connector
- Back-up switch connector from the connector bracket case
- Vehicle Speed Sensor (VSS) connector from the right side of the transaxle
- Ground strap bolt and strap
- Crankshaft Position (CKP) sensor connector
- Release lever, position the lever and clutch line aside
- Upper starter mounting bolt
- 3 clutch housing bolts

4. Support the engine with a suitable support bar.

- Front wheels
- Splash shields
- Engine mounting member. The member is secured with 2 bolts and 4 nuts.
- Extension bar and shift control rod
- Bolts from the No. 1 and No. 2 engine mounts and the mounts. There are 2 bolts for the No. 1 mount and 3 bolts for the No. 2 mount.
- Tension rod mounting nuts
- Stabilizer bar and control link
- Tie rod ends from the steering knuckle
- Ball joint bolt and nut from the control arm, then separate the ball joint from the control arm
- Halfshaft from the transaxle. Support the halfshaft with wire to prevent it from hanging unsupported.
- Starter lower mounting bolt and starter
- 3 lower clutch housing bolts, support the transaxle with a jack, then remove the 2 remaining bolts

5. Carefully separate the transaxle from the engine and lower it.

To install:

6. Raise the transaxle into position and seat it against the back of the engine. Install the transaxle-to-engine bolts and tighten as follows:

    a. Upper bolts (1–4): 47–66 ft. lbs. (64–89 Nm).

    b. Lower bolts (5–8): 28–38 ft. lbs. (37–51 Nm).

7. Remove the jack from the transaxle.

8. Install or connect the following:

- Starter and lower mounting bolt
- New clip on the driveshaft
- Driveshaft into the transaxle with the opening of the clip pointing upward
- Lower ball joint into the steering knuckle. Torque the nut to 43–50 ft. lbs. (54–68 Nm).
- Tie rod end to the steering knuckle. Torque the nut to 22–33 ft. lbs. (30–44 Nm). Install a new cotter pin.
- Stabilizer bar and control link. Torque the retainers to 32–45 ft. lbs. (43–61 Nm).
- Tension rod mounting nut and tighten to 87–108 ft. lbs. (118–147 Nm).
- No. 2 engine mount. Tighten the bolt to 32–40 ft. lbs. (43–54 Nm) and the nut to 49–69 ft. lbs. (67–93 Nm).
- No. 1 engine mount and tighten to 32–40 ft. lbs. (43–54 ft. lbs.)
- Extension bar and shift control rod
- Transaxle pan drain plug
- Engine mounting member-to-chassis bolts and nuts. There are 2 bolt and 4 nuts. Torque the engine mounting member bolts and nuts to 47–66 ft. lbs. (64–89 Nm) and the No. 1 and 2 engine mount nuts to 27–38 ft. lbs. (37–52 Nm).

- Splash shields
- Front wheels

9. Remove the engine support bar

- 3 clutch housing bolts and torque to 66–86 ft. lbs. (89–116 Nm)
- Upper starter mounting bolts and tighten to 27–38 ft. lbs. (37–52 Nm)
- Release lever
- CKP sensor electrical connector
- Ground strap and bolt
- VSS and back-up switch connectors
- Air cleaner assembly
- Air temperature and MAF sensor connectors
- Fresh air duct
- Hose from the intake manifold
- Coolant reservoir tank. Position it aside for access to the battery bracket.
- Battery fixing holder, battery tray and battery
- Battery bracket
- Positive, then negative battery cables

10. Fill the transaxle with fluid.

11. Check for proper clutch operation.

SEPHIA AND SPECTRA

1. Before servicing the vehicle, refer to the precautions in the beginning of this section.

2. Drain the transaxle oil.

3. Remove or disconnect the following:

- Battery box and the battery
- Air cleaner assembly
- Battery carrier
- Back-up light switch and the bracket
- Ground strap from the top of the transaxle
- Neutral switch connector and the vehicle speed sensor (VSS) connector
- Wire harness bracket and ground cable
- Crankshaft Position (CKP) sensor
- Wheels
- Splash shield
- Transverse member
- Extension bar and the change control rod
- Tie-rod ends
- Stabilizer control link
- Halfshaft and the joint shaft
- Intake manifold bracket
- Starter
- Front exhaust pipe

4. Support the engine and remove the engine mounting member.

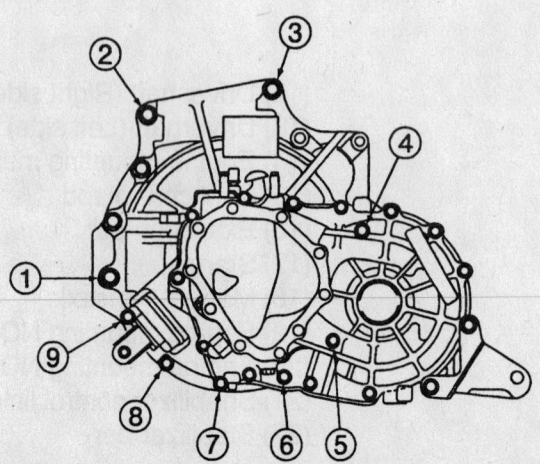

Location of the manual transaxle-to-engine mounting bolts—Rio

9357NG14

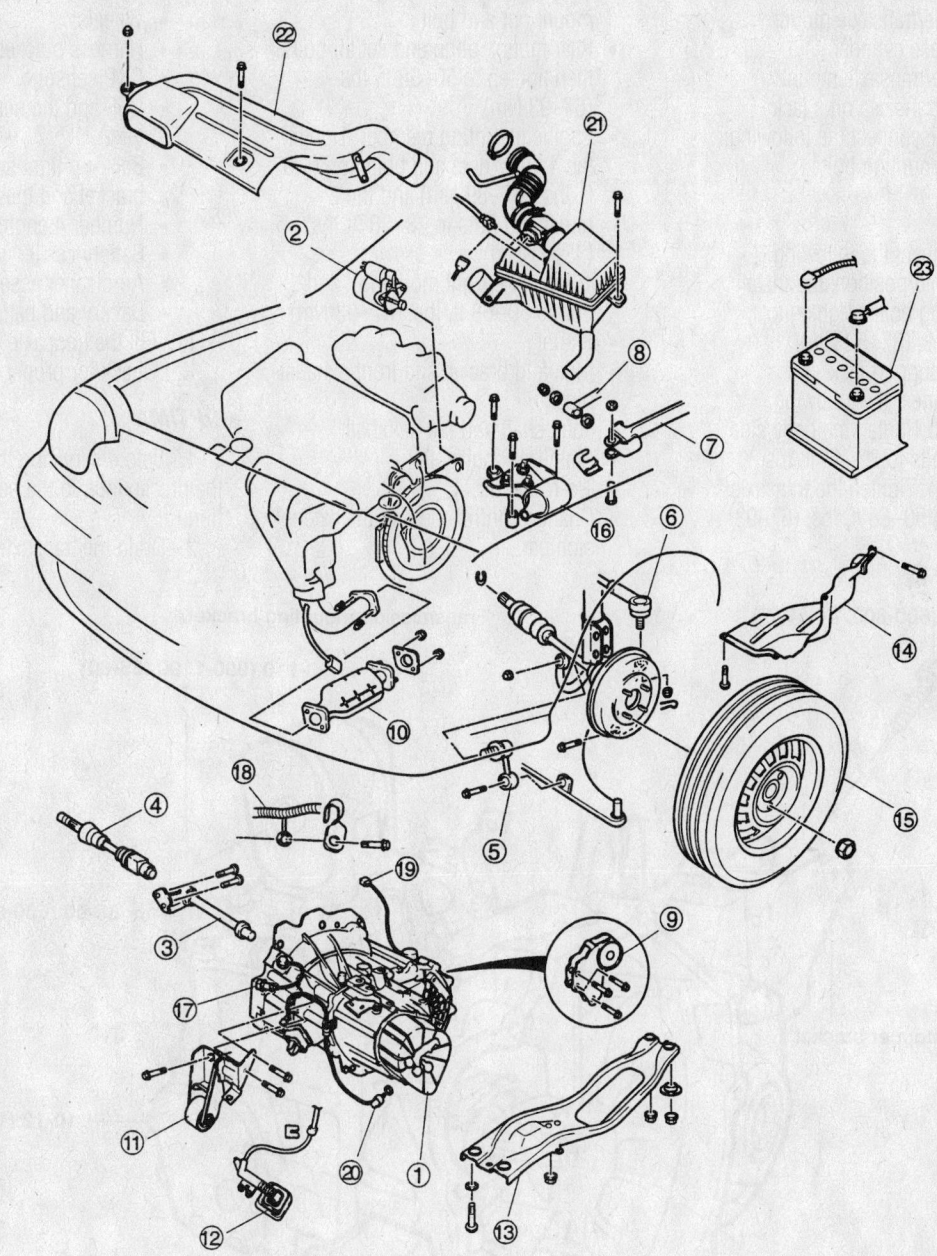

(1) Transaxle
(2) Starter
(3) Joint shaft
(4) Driveshaft
(5) Stabilizer control link
(6) Tie-rod end
(7) Change control rod
(8) Extension bar

(9) Engine mount No. 1
(10) Catalytic converter
(11) Engine mount No. 2
(12) Clutch release cylinder
(13) Engine mount member
(14) Splash shield
(15) Front wheel and tire
(16) Engine mount No. 4

(17) Crankshaft position sensor
(18) Ground
(19) Vehicle speed sensor connector
(20) Back-up switch connector
(21) Air cleaner assembly
(22) Fresh air duct
(23) Battery

9301KG07

Exploded view of the manual transaxle assembly mounting and related components—Sephia

5. Remove or disconnect the following:
- Rear engine/transaxle mount
- Front engine/transaxle mount
- Clutch release cylinder
- Side engine/transaxle mount
6. Support the transaxle on a jack
7. Remove or disconnect the following:
- Transaxle mounting bolts
- Transaxle

To install:

8. Install or connect the following:
- Transaxle into position and install the mounting bolts. Tighten to 28–38 ft. lbs. (37–52 Nm).
9. Remove the support jack.
10. Install or connect the following:
- Side mount. Tighten the body side nuts and bolts to 32–44 ft. lbs. (44–60 Nm). Tighten the transaxle side nuts to 50–68 ft. lbs. (67–93 Nm).

- Clutch release cylinder
- Front mount, loosely tighten the mount nut and bolt
- Rear mount, align and set all bolts, then tighten to 50–68 ft. lbs. (67–93 Nm)
- Engine mounting member. Tighten the 4 outer nuts and bolts to 50–65 ft. lbs. (67–89 Nm) and the 2 remaining nuts to 28–38 ft. lbs. (38–51 Nm).
- Tighten the front mount nut and bolt to 50–68 ft. lbs. (67–93 Nm)
- Starter
- Manifold bracket and front exhaust pipe
- Joint shaft and the halfshaft
- Stabilizer control link
- Tie-rod ends
- Change control rod and the extension bar

- Transverse member
- Splash shield
- Wheels
- Harness bracket
- CKP sensor
- VSS and the neutral switch connectors
- Back-up light switch connector bracket and the switch
- Number 4 engine mount
- Battery carrier
- Air cleaner assembly
- Battery and battery box
11. Fill the transaxle with fluid.
12. Check for proper clutch operation.

OPTIMA

1. Before servicing the vehicle, refer to the precautions in the beginning of this section.
2. Drain the transaxle oil.

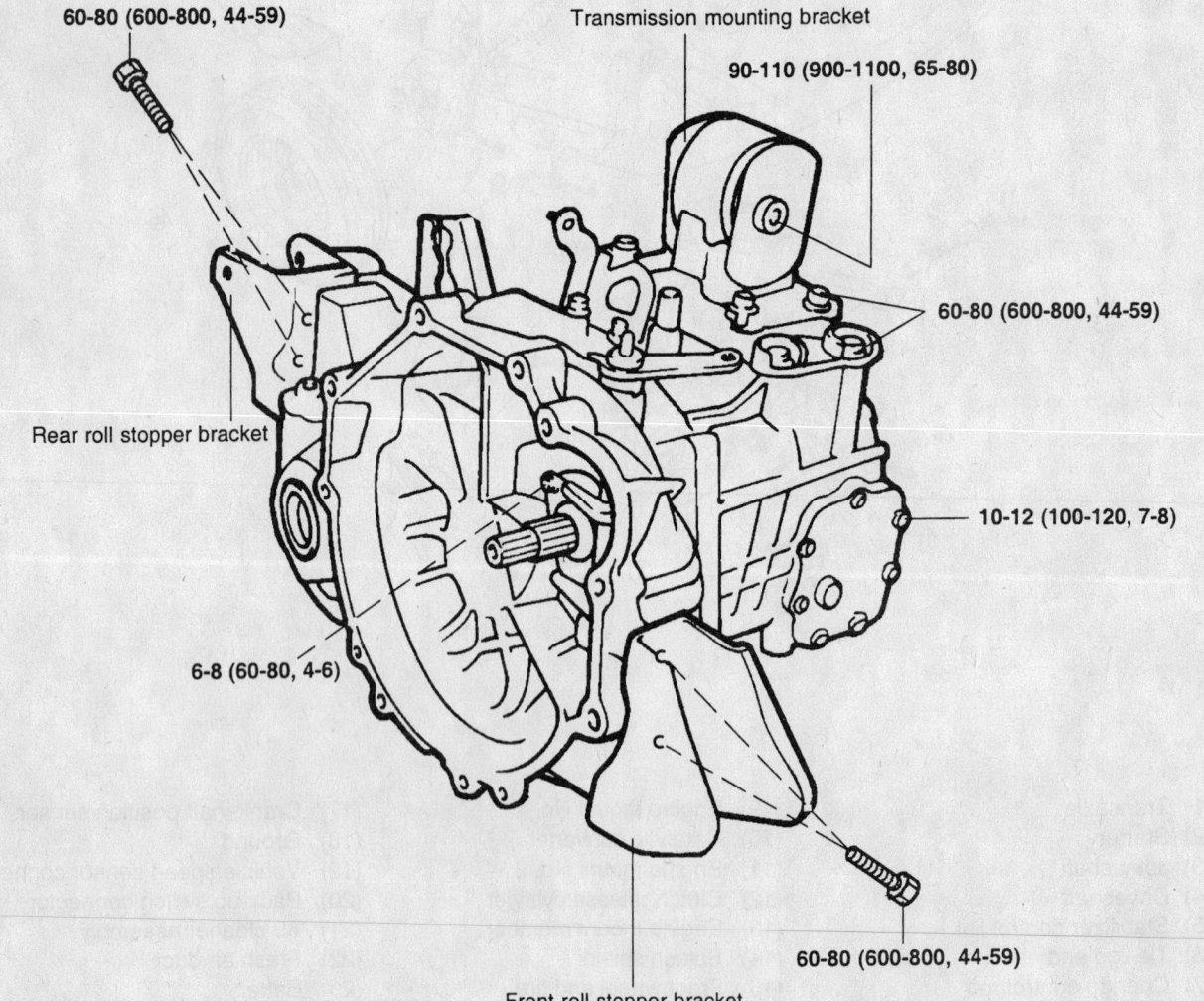

60-80 (600-800, 44-59)

Transmission mounting bracket

90-110 (900-1100, 65-80)

60-80 (600-800, 44-59)

Rear roll stopper bracket

10-12 (100-120, 7-8)

6-8 (60-80, 4-6)

60-80 (600-800, 44-59)

Front roll stopper bracket

TORQUE : Nm (kg·cm, lb·ft)

9357NG46

Manual transaxle mounting and tightening specifications—Optima

3. Remove or disconnect the following:
- Negative battery cable
- Air duct
- Air cleaner and air flow assembly
- Back-up light switch connector
- Clutch tube and clip
- Clutch release cylinder
- Speedometer cable
- Select cable and shift cable
- Starter mounting bolts
- Upper transaxle mounting bolts

4. Install engine hooks and support the engine with a suitable support bar.
- Transaxle mounting bracket and insulator
- Front wheels
- Steering gear box U-joint bolt and return tube mounting bolts
- Muffler
- Sub-frame mounting bolts and sub-frame
- Transaxle front and rear mounting bracket
- Transaxle side mounting bolts. Support the transaxle with a jack.
- Transaxle from the vehicle

5. Installation is the reverse of the removal procedure. Refer to the illustration for tightening specifications.

6. Fill the transaxle with fluid.

7. Check for proper clutch operation.

Automatic

RIO

1. Before servicing the vehicle, refer to the precautions in the beginning of this section.

2. Drain the transaxle fluid.

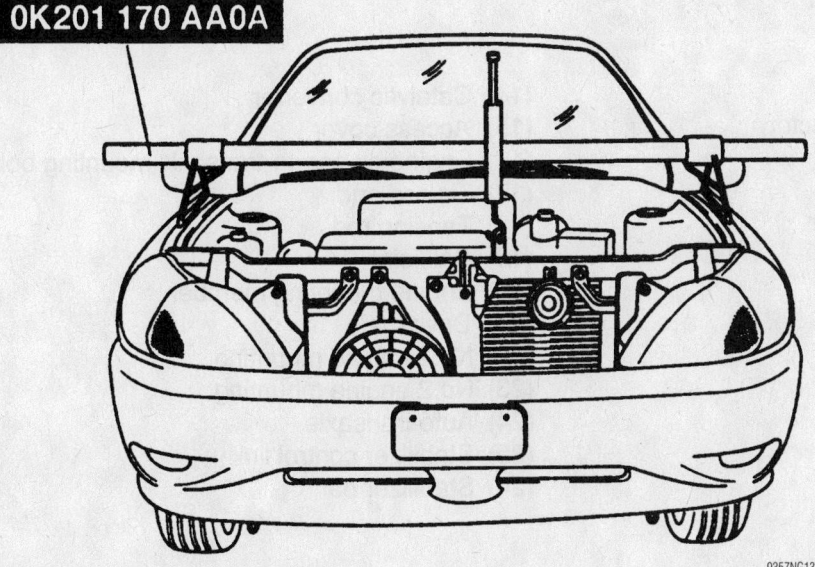

Use the proper tool to support the engine—Rio

3. Remove or disconnect the following:
- Negative, then positive battery cables
- Fresh air duct
- Input/turbine speed sensor, solenoid and transaxle range connector
- Ground strap bolt and ground strap from the top of the transaxle
- Vehicle Speed Sensor (VSS) connector from the right side of the transaxle
- U-clip connecting the selector cable to the linkage
- Nut and washer from the transaxle linkage
- Crankshaft Position (CKP) and Oxygen (O_2) sensor connectors
- 2 transaxle fluid cooler hoses
- Top 2 upper converter housing bolts

4. Support the engine with a suitable support bar.
- Front wheels
- Splash shield
- 2 nuts from the U-bolt
- Catalytic converter
- Converter housing access cover and 4 drive plate-to-torque converter mounting nuts. You will need to rotate the engine using the crank pulley to get to all of the bolts.
- Lower starter bolt
- 4 oil pan-to-transaxle mounting bolts
- Left tie rod end from the steering knuckle
- Tension rod
- Lower ball joint from the control arm
- No. 1 and 2 mounting nuts from the engine mounting member

- Engine mounting member-to-chassis bolts and nuts and the member
- Stabilizer bar and control link
- Left halfshaft from the transaxle

5. Install special tool 0K201 270 014 to prevent the side gear from becoming misaligned.
- No. 2 engine mount from the transaxle

6. Support and secure the transaxle with a suitable jack.
- 2 remaining converter housing bolts from the front and rear sides of the transaxle
- Transaxle. Slowly lower the drivetrain, letting the transaxle tilt toward the ground. Carefully separate the transaxle from the engine and pull the unit out through the wheel well.

To install:

7. Place the transaxle on a jack and place under the vehicle. Raise the transaxle into place and align with the engine.

8. Install or connect the following:
- Transaxle to engine, using 4 converter housing bolts (2 on top and 2 on bottom) to pull the components together. Torque the bolts to 47–66 ft. lbs. (64–89 Nm).

9. Remove the jack.
- No. 2 engine mounting to the transaxle. Tighten the nut to 49–69 ft. lbs. (67–93 Nm) and the bolts to 32–40 ft. lbs. (43–54 Nm).
- Starter and ground strap
- 3 No. 1 engine mounting-to-chassis bolts and tighten to 32–46 ft. lbs. (43–52 Nm)
- New clip on the left driveshaft
- Driveshaft into the transaxle with the opening of the clip facing up
- Left lower ball joint to the spindle. Torque the pinch bolt to 40–50 ft. lbs. (54–68 Nm).
- Stabilizer bar and control link and tighten to 32–45 ft. lbs. (43–61 Nm)
- Tension rod. Install the front part of the tension rod to the body after inserting the bushing into the tensioner rod. Tighten to 87–108 ft. lbs. (118–147 Nm).
- Engine-to-oil pan bolts and access cover. Torque the bolts to 27—38 ft. lbs. (37–52 Nm).
- Drive plate-to-torque converter mounting nuts and torque to 25–36 ft. lbs. (34–49 Nm). Rotate the engine with the crank pulley to get to all of the nuts.
- Engine mounting member-to-chas-

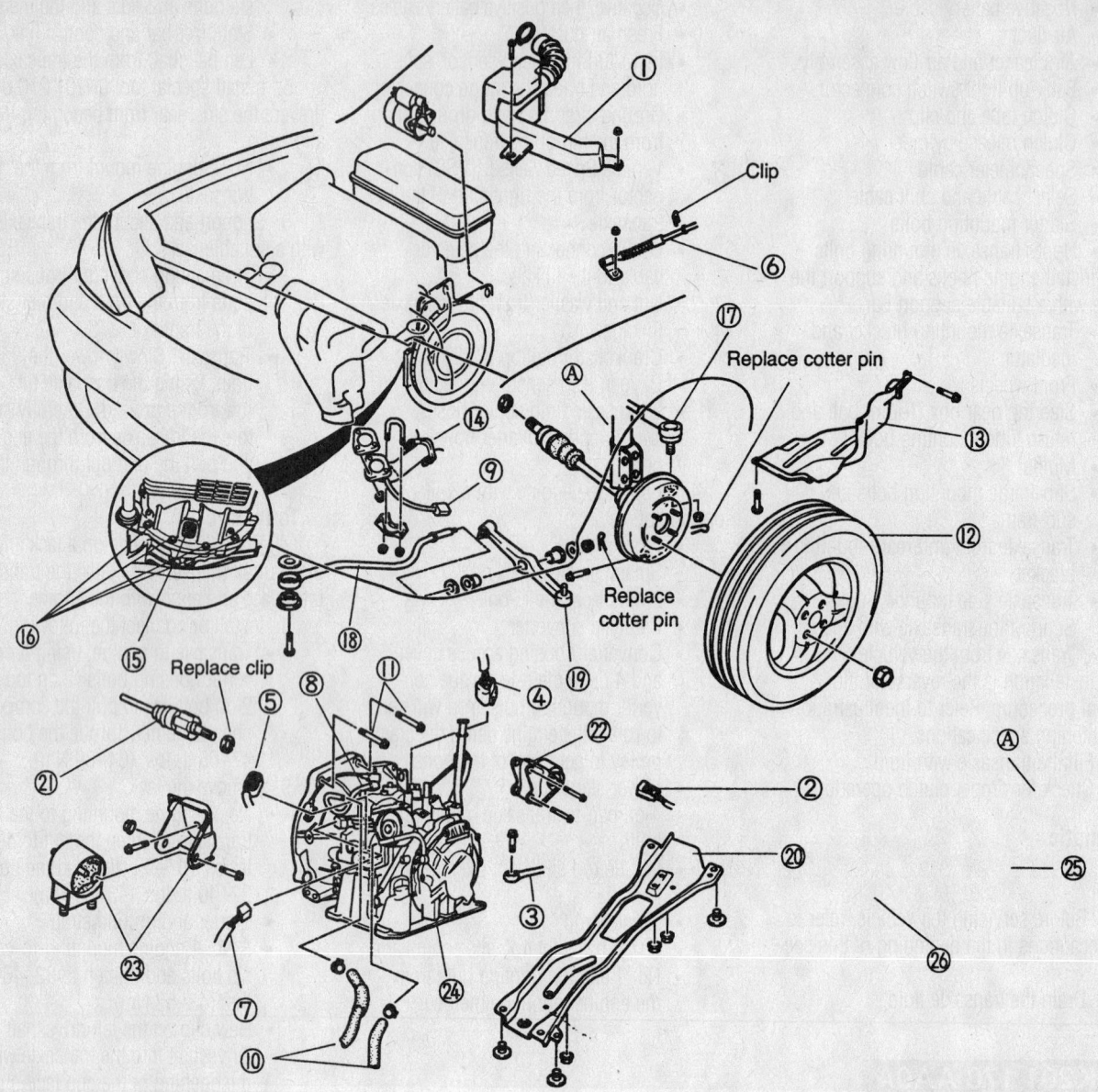

(1) Air cleaner assembly
(2) Input/turbine speed sensor connector
(3) Ground strap bolt
(4) Vehicle speed sensor connector
(5) Transaxle range switch connector
(6) Selector cable
(7) Solenoid valve connector
(8) Crankshaft position connector
(9) Oxygen sensor connector
(10) ATF cooler hose
(11) Upper converter housing bolts
(12) Wheel and tire
(13) Splash shield

(14) Catalytic converter
(15) Access cover
(16) Engine oil pan-to-transaxle mounting bolt
(17) Tie-rod end
(18) Tension rod
(19) Ball joint
(20) Engine mounting member
(21) Driveshaft
(22) No.1 engine mounting
(23) No.2 engine mounting
(24) Auto transaxle
(25) Stabilizer control link
(24) Stabilizer bar

9357NG15

Exploded view of the automatic transaxle mounting—Rio

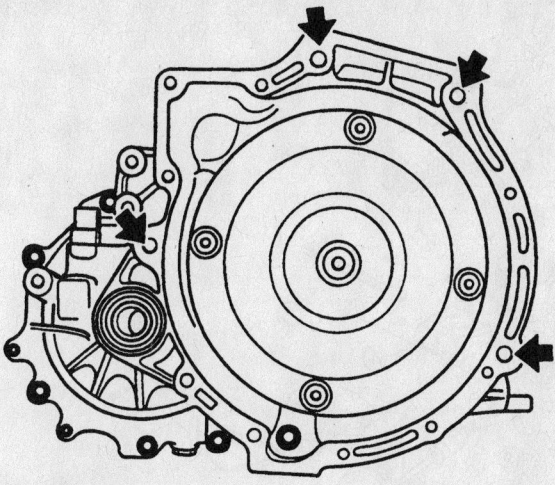

9357NG16

Location of the 4 converter housing bolts—Rio

sis nuts and bolts and tighten to 48–65 ft. lbs. (64–89 Nm)
- 2 nuts to the No. 1 and 2 engine mount-to-mounting member and tighten to 28–38 ft. lbs. (38–51 Nm).
- Catalytic converter and U-clip. Torque to 27–38 ft. lbs. (37–52 Nm).
- Splash shield.
- Front wheels
- Transaxle fluid hoses onto the cooler

10. Remove the engine support bar.
- CKP and O_2 sensor connectors
- Solenoid valve connector

11. Install the remaining components in the reverse of the removal procedure.

12. Fill the transaxle with the proper type and amount of fluid.

13. Test drive the vehicle. Check for proper operation in all gear ranges.

SEPHIA AND SPECTRA

1. Before servicing the vehicle, refer to the precautions in the beginning of this section.

2. Drain the transaxle fluid.

3. Remove or disconnect the following:
- Battery and battery cover
- Air cleaner assembly
- Battery carrier
- Solenoid and the Transaxle Range Switch (TRS) connectors
- Selector cable
- Vehicle Speed Sensor (VSS) connector
- Harness bracket
- Throttle cable
- Front wheels
- Splash shields
- Front exhaust pipe

- Transverse member, if equipped
- Tie-rod ends
- Stabilizer control links
- Lower arm by removing the cinch bolt from the lower arm ball joints. Pry the lower arm out of the knuckle.

4. Support the engine

5. Remove or disconnect the following:
- Engine mounting member
- Left and right halfshaft. Install Differential Side Gear holder K49A-4208-AT to hold the side gears.
- Joint shaft
- Intake manifold bracket
- Starter
- Front engine/transaxle mount
- Rear engine/transaxle mount
- Inner and outer oil hoses
- Side engine/transaxle mount

6. Hold the drive plate and remove the converter nuts.

7. Support the transaxle on a jack.

8. Remove or disconnect the following:
- Transaxle mounting bolts
- Transaxle

To install:

9. Support the transaxle on a jack and lift it into place. Align the transaxle with the engine and install the mounting bolts. Tighten to 65–86 ft. lbs. (89–116 Nm).

10. Hold the driveplate and install the torque converter mount nuts. 26–36 ft. lbs. (35–49 Nm).

11. Install or connect the following:
- Side engine/transaxle mount. Loosely tighten the nuts of the transaxle side. Tighten the nuts and bolts of the body side to 32–44 ft. lbs. (44–60 Nm). Tighten the nuts of the transaxle side to 50–68 ft. lbs. (67–93 Nm).

- Inner and outer oil hoses
- Rear engine/transaxle mount. Tighten the bolts to 50–68 ft. lbs. (67–93 Nm).
- Front engine/transaxle mount. Tighten the mount bracket to the transaxle to 28–38 ft. lbs. (38–51 Nm). Loosely tighten the nuts and bolts of the engine mount rubber, then tighten to 50–68 ft. lbs. (67–98 Nm).
- Starter
- Manifold bracket
- Joint shaft into the transaxle
- Joint shaft to the cylinder block and tighten the bolts in sequence (counterclockwise). Tighten to 32–46 ft. lbs. (42–62 Nm).
- Halfshafts, be sure that the shafts are properly installed and do not pull out
- Engine mounting member, the mounting nuts/bolts and tighten the nuts/bolts at the far corners to 48–65 ft. lbs. (64–89 Nm), tighten the remaining 2 nuts to 28–38 ft. lbs. (38–51 Nm).
- Front exhaust pipe
- Lower arm to the knuckle
- Stabilizer control link
- Tie-rod ends
- Transverse member, if removed
- Splash shields
- Wheels
- Throttle cable
- Harness bracket
- VSS connector
- Selector cable
- TRS connector
- Solenoid connector
- Battery carrier
- Air cleaner assembly
- Battery and battery cover

12. Fill the transaxle with the proper type and amount of fluid.

13. Test drive the vehicle. Check for proper operation in all gear ranges.

OPTIMA

1. Before servicing the vehicle, refer to the precautions in the beginning of this section.

2. Drain the transaxle fluid.

3. Remove or disconnect the following:
- Air cleaner assembly
- Control cable
- Vehicle Speed Sensor (VSS) connector
- Transaxle range switch connector, solenoid connector and oil temperature sensor connector
- Oil cooler hose

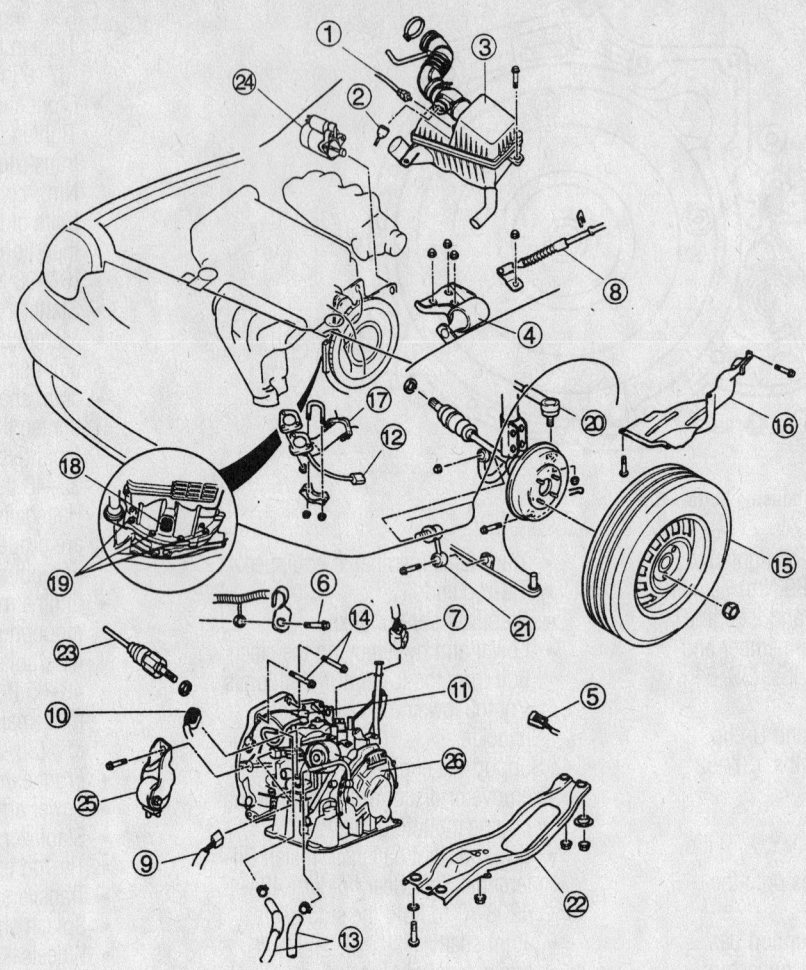

(1) Air temperature sensor connector
(2) MAF sensor connector
(3) Air cleaner assembly
(4) No. 4 mounting
(5) Input/turbine speed sensor connector
(6) Ground strap bolt
(7) Vehicle speed sensor connector
(8) Selector cable
(9) Transaxle range switch connector
(10) Solenoid valve connector
(11) Crankshaft position connector
(12) Oxygen sensor connector
(13) ATF cooler hose
(14) Upper converter housing bolts
(15) Wheel and tire

(16) Gravel shield
(17) Catalytic converter
(18) Converter housing
(19) Engine oil pan-to-transaxle mounting bolt
(20) Tie-rod end
(21) Stabilizer control link
(22) Engine mounting member
(23) Driveshaft
(24) Starter
(25) No. 2 engine mounting
(26) Auto transaxle

9301KG08

Exploded view of the automatic transaxle assembly mounting and related components—Sephia

4. Install a suitable engine support to the engine.
- Stabilizer bar, tie rod end, lower control arm ball joint and halfshafts
- Steering gear box U-joint bolt and return tube mounting bolts

- Sub-frame mounting bolts and sub-frame
- Starter
- Automatic transaxle mounting bolts
- Engine-to-transaxle bolts. Place a jack under the transaxle.

- Transaxle from the vehicle
5. Installation is the reverse of the removal procedure. Refer to the illustration for tightening specifications.
6. Fill the transaxle with fluid.
7. Check for proper clutch operation.

Lubricate all internal parts with automatic transmission fluid during reassembly

10-12 (100-120, 7-8)

20-26 (200-260, 14-18)

9

N

8

N

10

11

27-33 (270-330, 19-23)

10-12 (100-120, 7-8)

6

N

7

N

4-6 (40-60, 3-4)

3

4

12

20-26 (200-260, 14-18)

60-80 (600-800, 43-58)

2

15 14

5-7 (50-70, 4-5)

5

13

2

60-80 (600-800, 43-58)

8-10 (80-100, 6-7)

1

1. Torque converter
2. Roll stopper bracket
3. Harness bracket
4. Control cable support bracket
5. Oil level gauge
6. Eye bolt
7. Oil cooler feed tube
8. Input shaft speed sensor
9. Output shaft speed sensor
10. Manual control lever
11. Transaxle range switch
12. Vehicle speed sensor
13. Valve body cover
14. Manual control shaft detent spring
15. Manual control shaft detent

TORQUE : Nm (kg·cm, lb·ft)

9357NG47

Automatic transaxle mounting and tightening specifications—Optima

Clutch

ADJUSTMENTS

Pedal Height

1. Before servicing the vehicle, refer to the precautions in the beginning of this section.

2. Measure the distance from the upper surface of the pedal pad to the carpet.

3. The distance should be as follows:
   a. Sephia, Spectra and Optima: 7.83–8.15 in. (199–207mm).
   b. Rio: 7.67 in. (195mm).

4. If the distance is not as specified, loosen the locknut on the stopper bolt or switch.

5. Turn the switch or bolt until the distance is correct, then tighten the locknut.

Free-Play

1. Before servicing the vehicle, refer to the precautions in the beginning of this section.

2. Depress the clutch pedal by hand until resistance is felt. The free-play should be as follows:
   a. Sephia and Spectra: 0.12–0.20 in. (3.0–5.0mm).
   b. Rio: 0.35–0.59 in. (9–15mm).
   c. Optima: 0.2–0.5 in. (6–13mm).

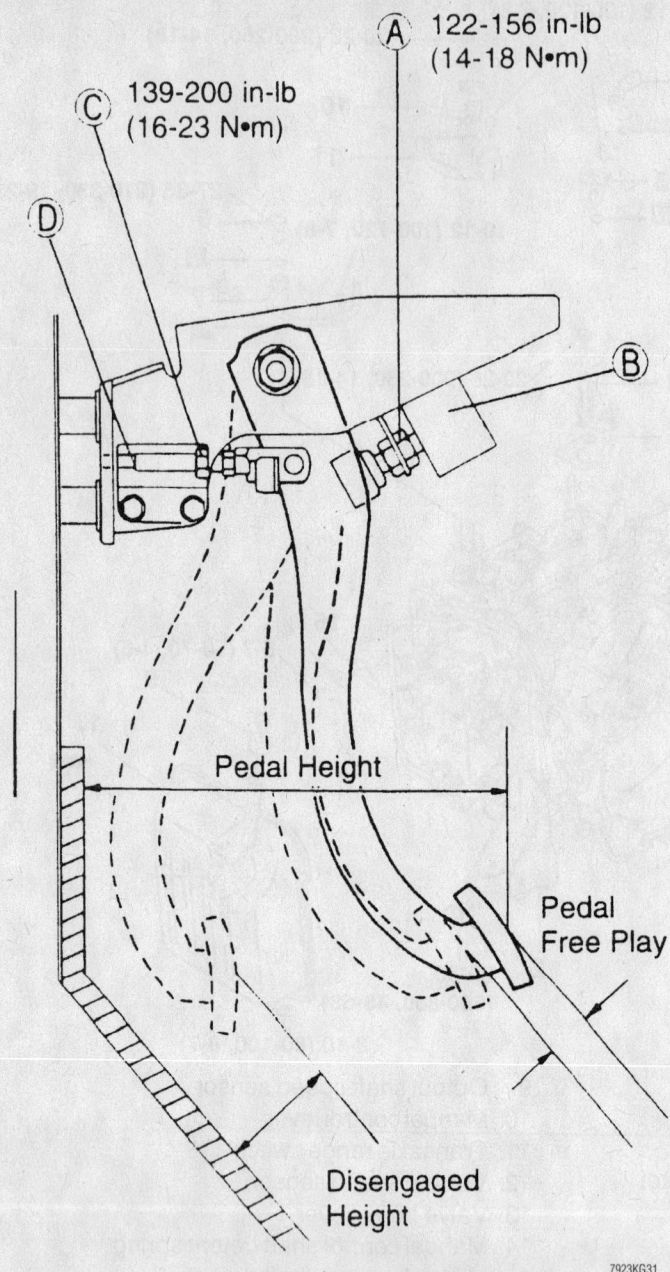

A 122-156 in-lb (14-18 N•m)

C 139-200 in-lb (16-23 N•m)

D

B

Pedal Height

Pedal Free Play

Disengaged Height

7923KG31

Clutch pedal measurement and adjustment points. (A) and (B) are for adjusting the pedal height, while (C) and (D) are for the free-play adjustment

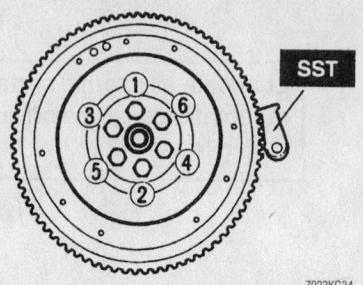

7923KG34

Flywheel tightening sequence

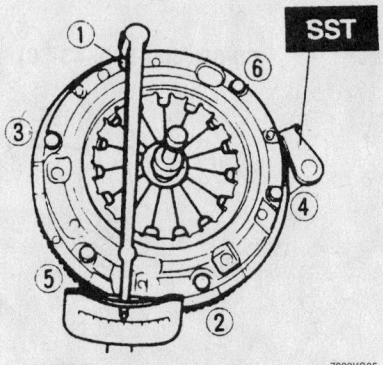

7923KG35

Pressure plate tightening sequence

3. If the free-play is not correct, loosen the clutch master cylinder pushrod locknut and turn the pushrod to adjust.

REMOVAL & INSTALLATION

1. Before servicing the vehicle, refer to the precautions in the beginning of this section.

2. Remove or disconnect the following:
   • Negative battery cable
   • Transaxle

3. Gradually loosen the clutch pressure plate bolts, in a crisscross pattern. Support the pressure plate and remove the bolts. Remove the pressure plate and clutch disc.

4. Inspect the pilot bearing. If it is worn or damaged and does not turn easily by hand, remove it using a puller/slide hammer.

5. Check the flywheel surface for scoring, cracks or burning and machine or replace, as necessary.

6. Install a flywheel holder to keep the flywheel from turning. Loosen the flywheel bolts evenly and gradually in a crisscross pattern. Remove the flywheel.

7. Inspect the clutch release bearing for wear. Replace it if it sticks or does not turn easily.

8. Inspect the release fork for wear or damage and replace as necessary.

To install:

9. Lubricate the release fork fingers and pivot with molybdenum grease and install in the release fork boot.

10. Install or connect the following:
    • Clutch release bearing on the release fork
    • New pilot bearing in the flywheel, if removed

11. Be sure the flywheel mounting surface and the crankshaft or eccentric shaft mounting surfaces are clean. Remove any old sealant from the flywheel bolt hole threads and the flywheel bolts.

12. Install or connect the following:
    • Flywheel

13. Apply sealant to the flywheel bolt threads and install them hand-tight. Install the flywheel holding tool. Tighten the bolts, in a crisscross pattern, to 71–76 ft. lbs. (96–103 Nm).

14. Apply a small amount of molybdenum grease to the clutch disc splines and install the clutch disc on the flywheel, spring side toward the transaxle. Install a suitable alignment tool in the pilot bearing to position the clutch disc.

15. Install or connect the following:
    • Clutch pressure plate, aligning the dowel holes with the flywheel dowels

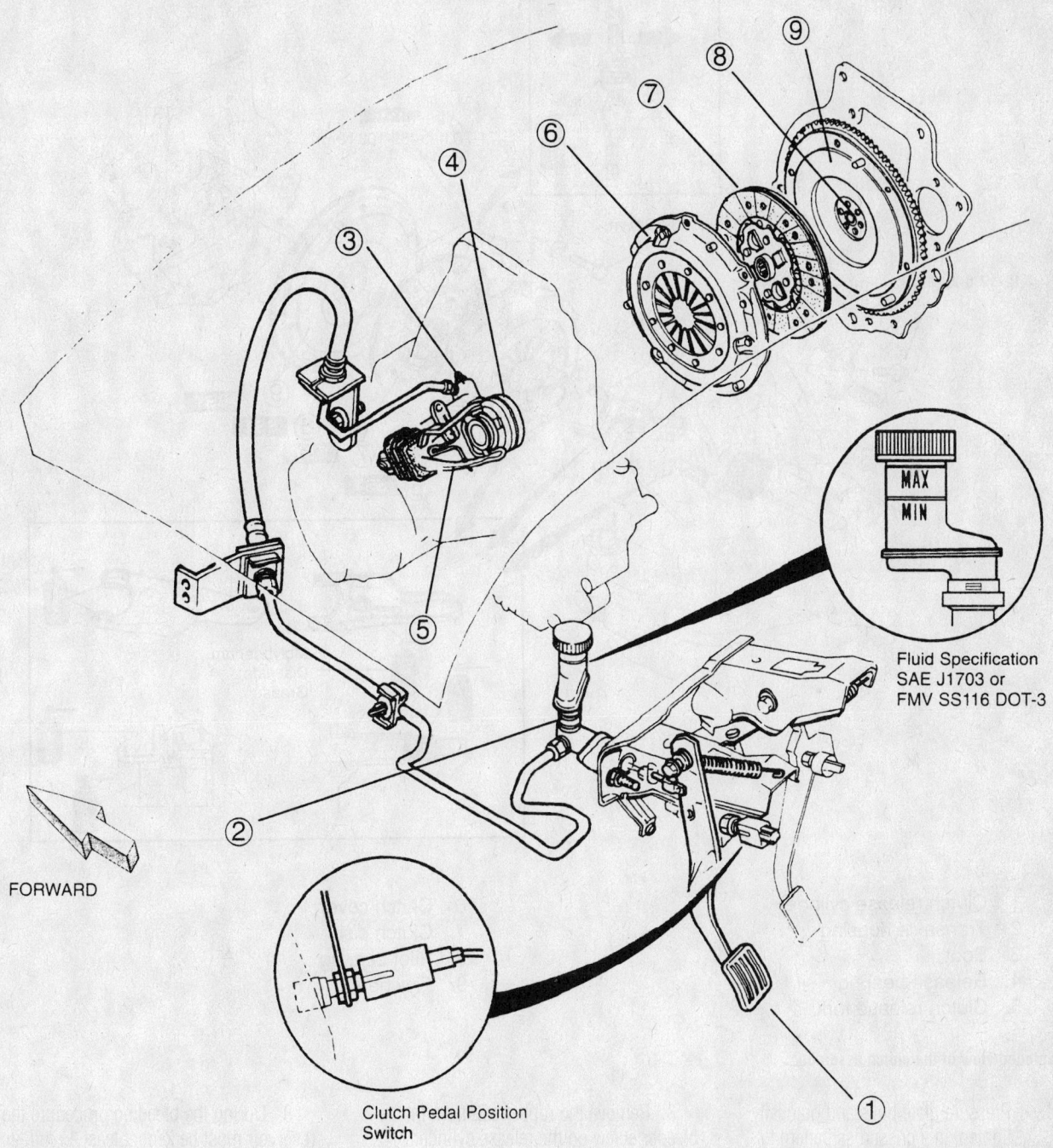

FORWARD

Fluid Specification
SAE J1703 or
FMV SS116 DOT-3

Clutch Pedal Position
Switch

1. Clutch pedal
2. Clutch master cylinder
3. Clutch release cylinder
4. Release bearing
5. Clutch release fork

6. Clutch cover
7. Clutch disc
8. Pilot bearing
9. Flywheel

7923KG32

Structural view of the hydraulic clutch system

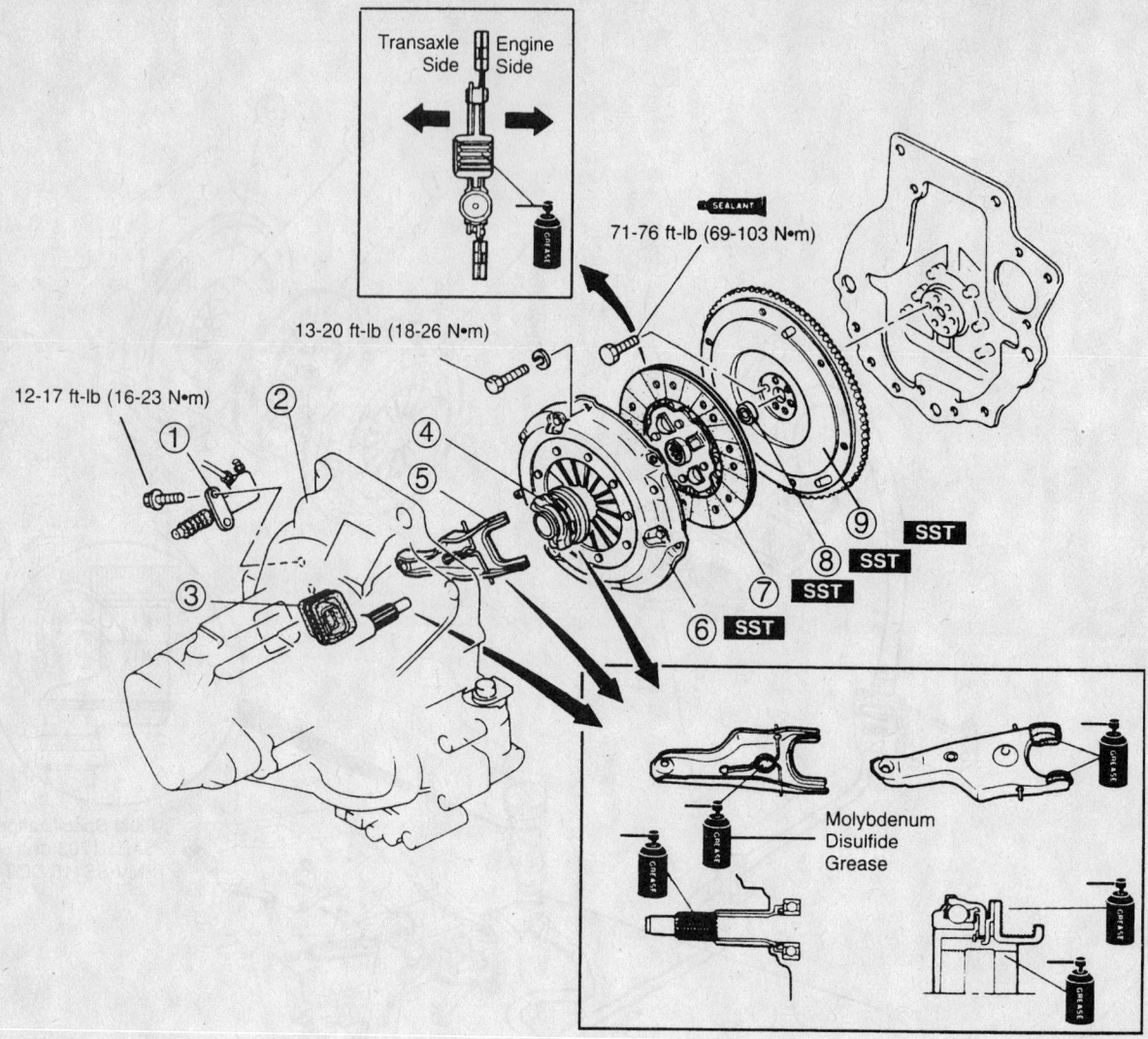

Transaxle Side   Engine Side

SEALANT

71-76 ft-lb (69-103 N•m)

13-20 ft-lb (18-26 N•m)

12-17 ft-lb (16-23 N•m)

① ② ③ ④ ⑤

⑨ SST
⑧ SST
⑦ SST
⑥ SST

Molybdenum
Disulfide
Grease

1. Clutch release cylinder
2. Transaxle housing
3. Boot
4. Release bearing
5. Clutch release fork

6. Clutch cover
7. Clutch disc
8. Pilot bearing
9. Flywheel

7923KG33

Exploded view of the clutch assembly

- Pressure plate bolts and gradually tighten, in a crisscross pattern to 20 ft. lbs. (26 Nm). Remove the alignment tool.
- Transaxle

Hydraulic Clutch System

BLEEDING

1. Before servicing the vehicle, refer to the precautions in the beginning of this section.
2. If necessary, remove the gravel shield from the drivers side.

3. Remove the rubber cap from the bleeder screw on the release cylinder.
4. Place a bleeder tube over the end of the bleeder screw.
5. Submerge the other end of the tube in a jar half filled with hydraulic brake fluid.
6. Slowly pump the clutch pedal fully and allow it to return slowly, several times.
7. While pressing the clutch pedal to the floor, loosen the bleeder screw until the fluid starts to run out. Then, close the bleeder screw. Keep repeating this Step, while watching the hydraulic fluid in the jar. As soon as the air bubbles disappear, close the bleeder screw.

8. During the bleeding procedure the reservoir must be kept at least ¾ full.

Halfshaft

REMOVAL & INSTALLATION

1. Before servicing the vehicle, refer to the precautions in the beginning of this section.
2. Remove or disconnect the following:
   - Wheel and tire assemblies
   - Splash shield, if equipped
3. Drain the transaxle.

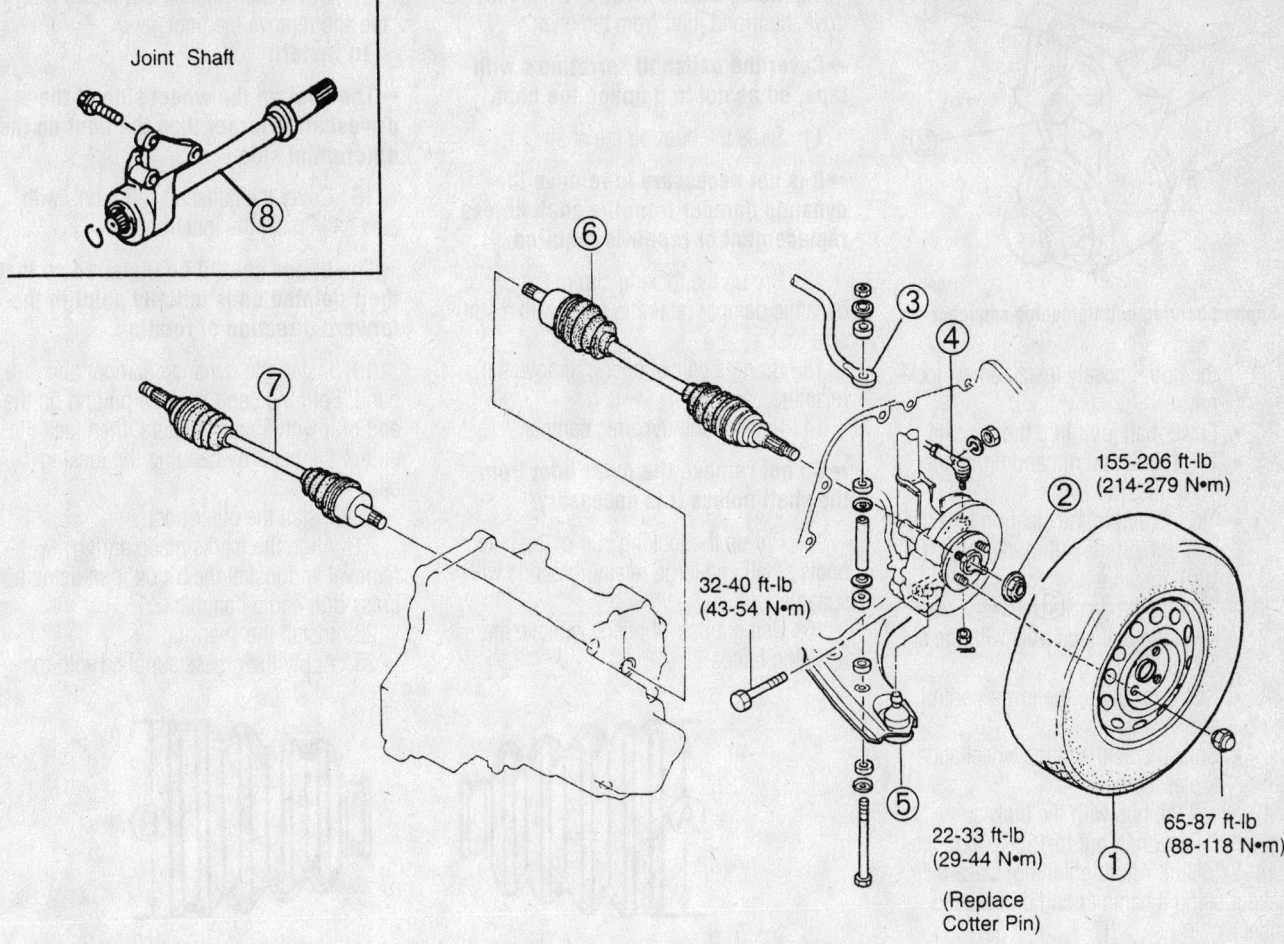

Joint Shaft

1. Wheel and tire
2. Locknut
3. Stabilizer bar
4. Tie-rod end
5. Ball joint
6. Left driveshaft
7. Right driveshaft
8. Joint shaft

155-206 ft-lb
(214-279 N•m)

32-40 ft-lb
(43-54 N•m)

22-33 ft-lb
(29-44 N•m)

(Replace
Cotter Pin)

65-87 ft-lb
(88-118 N•m)

7923KG36

Exploded view of the halfshafts and related components

4. Raise the staked portion of the hub locknut with a hammer and chisel. Lock the hub by applying the brakes and loosen the nut.

5. Remove or disconnect the following:
- Stabilizer bar from the lower control arm
- Cotter pin and nut from the tie rod end ball stud.
- Tie rod end from the knuckle, using a suitable tool
- Lower ball joint pinch bolt and nut
- Ball joint from the knuckle, using a prybar on the control arm

6. Position a prybar between the inner CV-joint and transaxle case. Carefully pry the halfshaft from the transaxle being careful not damage the oil seal. If equipped with a right side intermediate shaft, insert the pry-bar between the halfshaft and intermediate shaft and tap on the bar to uncouple them.
- Hub locknut and discard

7. Pull outward on the hub/knuckle assembly, push the outer CV-joint stub shaft through the hub, and remove the halfshaft. If the halfshaft is stuck in the hub, install the old hub nut to protect the stub shaft threads. Tap on the nut, using only a soft mallet, to remove the halfshaft.

➡**Install Differential Side Gear holder K49A-4208-AT, into the transaxle after removing the halfshaft, to keep the differential side gear in position. If the gear becomes out of position the differential may have to be removed to realign the gear.**

8. Remove the intermediate shaft, if necessary, by removing the support bearing

bolts and pulling the shaft from the transaxle.
To install:

9. Install or connect the following:
- New circlip on the end of the intermediate shaft, with the end gap facing upward, if removed
- Intermediate shaft in the transaxle, being careful not to damage the oil seals
- Support bearing bolts and tighten, in sequence, to 45 ft. lbs. (61 Nm)
- New circlip on the end of the halfshaft, with the end gap facing upward
- Halfshaft into the transaxle, being careful not to damage the oil seal
- Halfshaft into the intermediate shaft, if equipped
- Other end of the halfshaft through

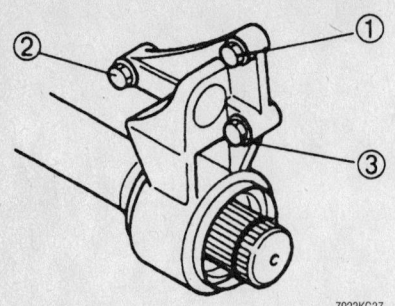

Support bearing bolt tightening sequence

7923KG37

the hub. Loosely install a new lock-nut.

- Lower ball joint into the knuckle
- Pinch bolt and nut and tighten to 40 ft. lbs. (54 Nm)
- Tie rod end to the steering knuckle and tighten the nut to 42 ft. lbs. (57 Nm). Install a new cotter pin. Tighten the nut, if necessary, to align the ball stud hole with the nut castellation.
- Stabilizer bar to the lower control arm
- Splash shield and the wheel and tire assemblies

10. Lock the hub with the brakes. Tighten the new hub nut to 155–206 ft. lbs. (214–279 Nm). After tightening, stake the locknut using a hammer and dull bladed chisel.

11. Fill the transaxle with the proper type and quantity of fluid.

CV-Joints

OVERHAUL

1. Before servicing the vehicle, refer to the precautions in the beginning of this section.

2. Remove the halfshaft.

3. Pry up the locking clip of the transaxle side boot retaining band with a suitable tool.

4. Using a pair of pliers, remove the retaining band.

5. Slide the boot away to access the CV-joint.

6. Matchmark the CV-joint housing and the shaft to ensure proper positioning during assembly.

7. Remove the outer ring.

8. Matchmark the shaft and tripod assembly to ensure proper positioning during assembly.

9. Remove the snapring.

➡**Be careful not to damage the needle bearings.**

10. Using a brass drift and a hammer, drive the tripod joint from the shaft.

➡**Cover the halfshaft serration's with tape, so as not to damage the boot.**

11. Slide the boot off the shaft.

➡**It is not necessary to remove the dynamic damper from the shaft unless replacement or repair is required.**

12. Pry up the locking clip of the dynamic damper retaining band with a suitable tool.

13. Using a pair of pliers, remove the retaining band.

14. Remove the dynamic damper.

➡**Do not remove the outer boot from the shaft unless it is necessary.**

15. Pry up the locking clip of the outer boots small and large retaining bands with a suitable tool.

16. Using a pair of pliers, remove the retaining bands.

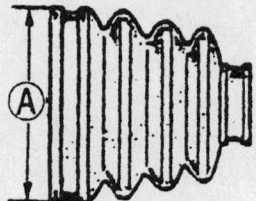 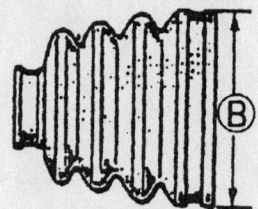

Transaxle End   **Wheel End**

9307KG12

Measure the boots as shown to ensure the larger boot is placed on the wheel side of the halfshaft.

17. Cover the halfshaft serration's with tape and remove the boot.

To install:

➡**The boot on the wheel side of the driveshaft is larger than the boot on the differential side.**

18. Cover the halfshaft serration's with tape and install the inner boot.

➡**The bands should be installed so that their pointed ends initially point in the forward direction of rotation.**

19. Install the dynamic damper and band. Fold the band back by pulling on the end of it with a pair of pliers, then lock the end of the band by bending the locking clip.

20. Install the outer boot.

21. Align the marks made during removal and install the tripod joint using a brass drift and a hammer.

22. Install the snapring.

23. Apply the grease supplied with the

Length of driveshaft   in (mm)

Item	Right side		Left side	
	MTX	ATX	MTX	ATX
TED	24.93 (633.3)	25.54 (648.7)	25.15 (638.9)	25.16 (639)

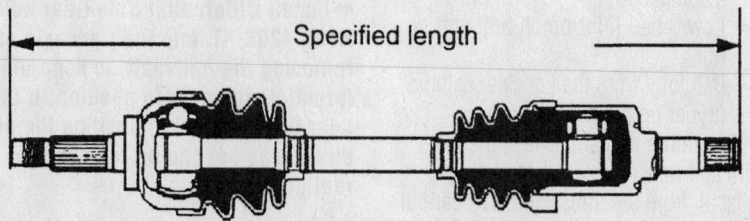

Specified length

Make sure the driveshaft's specified length is correct

9307KG13

joint rebuild kit to the tripod joint, outer ring and the boot.

24. Install the outer ring.

25. If the outer boot was removed, fill it with the correct amount of grease as follows:

a. Transaxle side: 4.94 oz. (140g).

b. Wheel side: 4.58 oz. (130g).

26. Make sure the boots are not damaged, then carefully up the small end of the boots to release any trapped air.

27. Measure the length of the driveshaft to ensure it is the correct length. Refer to the accompanying illustration for the drive-shaft measuring points and the correct specifications.

28. Install the boot retaining bands. Fold the bands back by pulling on the ends with a pair of pliers, then lock the end of the bands by bending the locking clips.

29. Install the halfshaft.

STEERING AND SUSPENSION

Air Bag

✳✳ CAUTION

Some vehicles are equipped with an air bag system. The system must be disabled before performing service on or around system components, steering column, instrument panel components, wiring and sensors. Failure to follow safety and disabling procedures could result in accidental air bag deployment, possible personal injury and unnecessary system repairs.

PRECAUTIONS

Several precautions must be observed when handling the inflator module to avoid accidental deployment and possible personal injury.

• Never carry the inflator module by the wires or connector on the underside of the module.

• When carrying a live inflator module, hold securely with both hands, and ensure that the bag and trim cover are pointed away.

• Place the inflator module on a bench or other surface with the bag and trim cover facing up.

• With the inflator module on the bench, never place anything on or close to the module which may be thrown in the event of an accidental deployment.

• An air bag is an explosive device. Handle with extreme caution.

• Always disconnect the battery and the air bag connector before removing the steering wheel or beginning work on the air bag system.

• Air bag components must not be repaired or opened. Always use new parts, including the wiring harness.

• Always place a removed air bag unit with the horn pad facing up. Put it in a safe place where it will not be disturbed.

• The air bag unit must not be exposed to grease, fluids, or cleaning agents.

• The air bag unit must not be exposed to temperatures above 194° F (90° C) at any time. Even the heat of a soldering iron can damage or ignite the charge.

• Storage and transport of air bags is subject to rules governing explosive devices and should be done only in the original package.

• Failure to follow proper safety precautions may result in personal injury through accidental firing of the air bag, or through failure of the air bag in an accident.

DISARMING

1. Before servicing the vehicle, refer to the precautions in the beginning of this section.

2. Turn the ignition switch to the **LOCK** position.

3. Disconnect the negative battery cable.

4. Wait 10 minutes for the battery back-up power to discharge.

ARMING

1. Before servicing the vehicle, refer to the precautions in the beginning of this section.

2. Connect the negative battery cable.

3. Turn the ignition switch **ON**.

4. Verify that the air bag indicator illuminates for 4–8 seconds, then goes off.

Manual Rack and Pinion Steering Gear

REMOVAL & INSTALLATION

Sephia

1. Before servicing the vehicle, refer to the precautions in the beginning of this section.

2. Remove or disconnect the following:

• Negative battery cable
• Front wheels
• Cotter pins from both steering tie rod ends and the nuts
• Tie rod out of the knuckle arm, using a suitable tool
• Set plate from the firewall
• Fixing bolt from the steering shaft to steering gear pinion shaft
• Steering shaft from the steering gear
• Steering gear mounting nuts

3. Move the steering gear to the right of the vehicle.

To install:

4. Install or connect the following:

• Steering gear to the vehicle

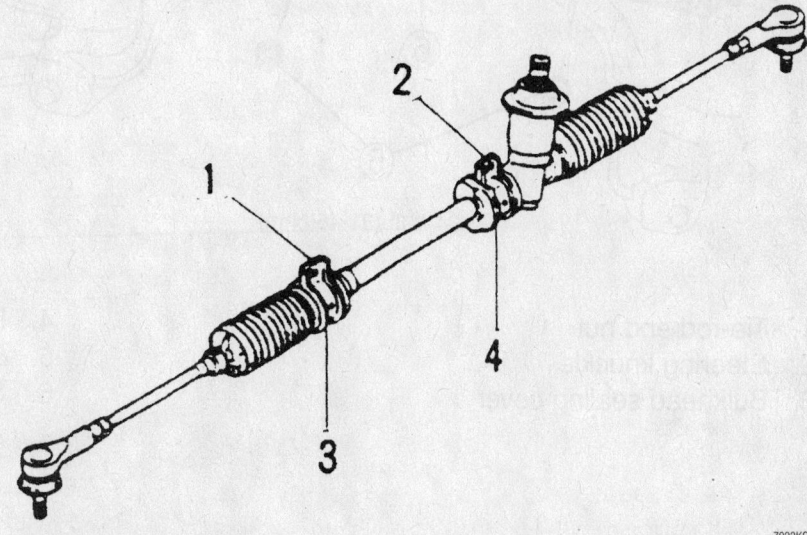

Rack mounting tightening sequence

7923KG39

- Mounting nuts in the order shown. Tighten the nuts to 23–34 ft. lbs. (31–46 Nm).
- Steering shaft to the steering gear pinion shaft. Tighten the bolt/nut to 13–20 ft. lbs. (18–27Nm).
- Set plate to the firewall
- Tie rod ends to the knuckle arm and tighten the nuts to 25–29 ft. lbs. (34–39 Nm)
- New cotter pins
- Wheels
- Negative battery cable

5. Check the front end alignment.

Power Rack and Pinion Steering Gear

REMOVAL & INSTALLATION

1. Before servicing the vehicle, refer to the precautions in the beginning of this section.

2. Remove or disconnect the following:
   - Negative battery cable
   - Front wheels
   - Cotter pins from both steering tie rod ends and the nuts
   - Tie rod out of the knuckle arm, using a suitable tool
   - Catalytic converter, if necessary for access
   - Pressure line and return pipe from the steering gear
   - Set plate from the firewall, if necessary
   - Fixing bolt from the steering shaft to steering gear pinion shaft
   - Steering shaft from the steering gear
   - Manual transaxle shifter linkage, if necessary

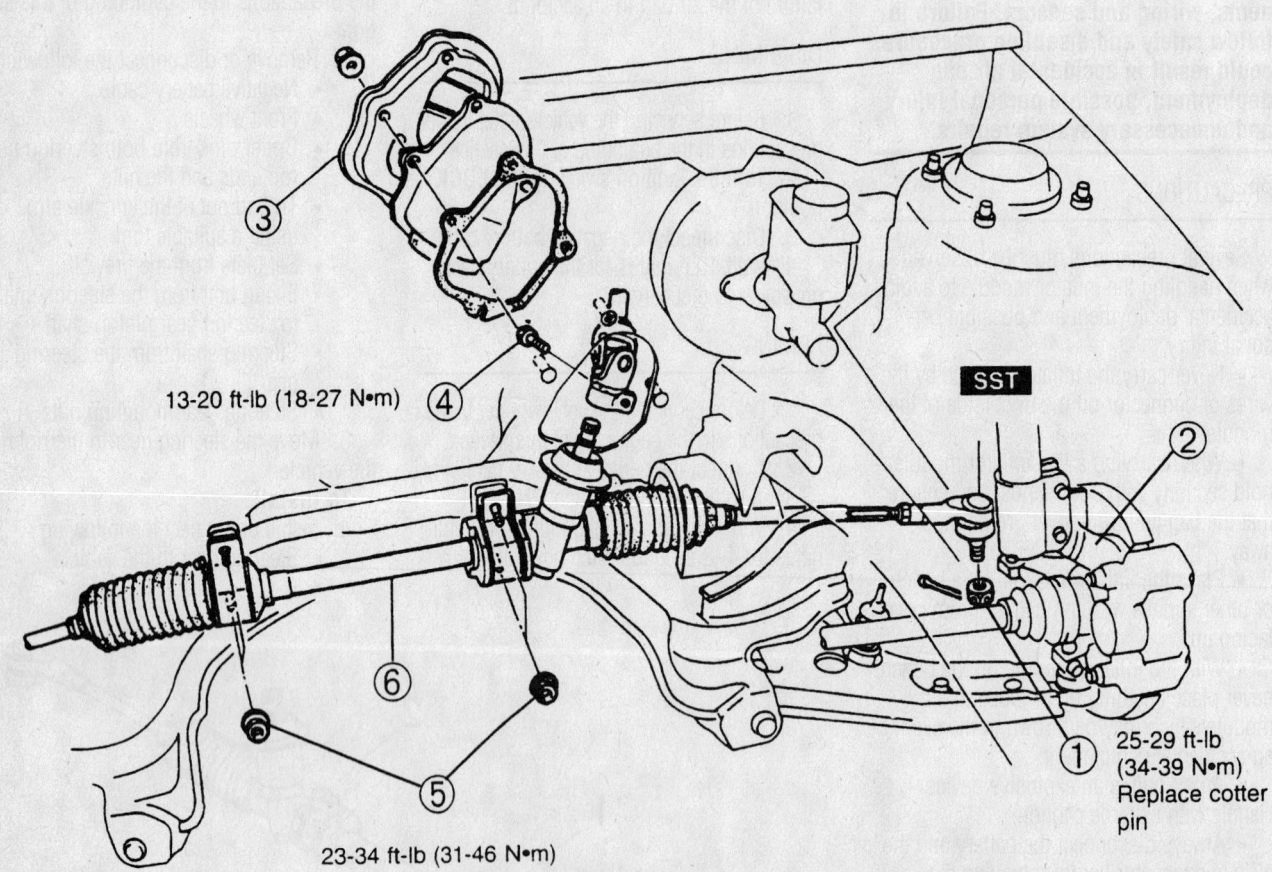

13-20 ft-lb (18-27 N•m)

SST

25-29 ft-lb (34-39 N•m) Replace cotter pin

23-34 ft-lb (31-46 N•m)

1. Tie-rod end nut
2. Steering knuckle
3. Bulkhead sealing cover
4. Pinch bolts
5. Steering rack nuts
6. Steering rack

7923KG38

Exploded view of the manual steering gear assembly mounting

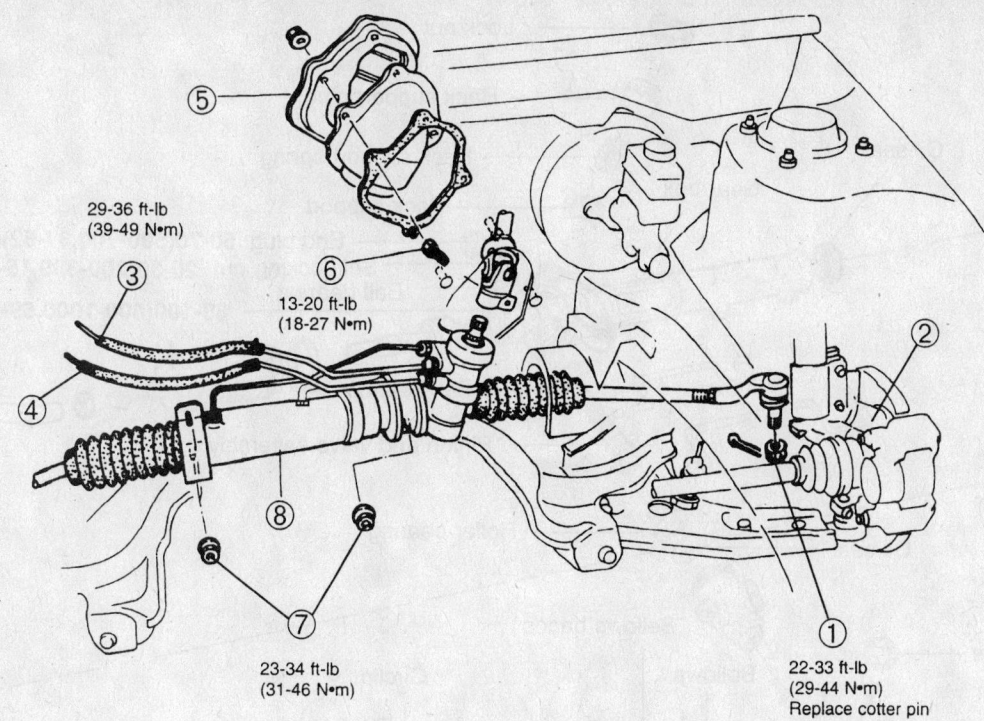

29-36 ft-lb
(39-49 N•m)

13-20 ft-lb
(18-27 N•m)

23-34 ft-lb
(31-46 N•m)

22-33 ft-lb
(29-44 N•m)
Replace cotter pin

1. Tie-rod end nut
2. Steering knuckle
3. High-pressure line
4. Return line

5. Sealing cover
6. Pinch bolt
7. Steering rack nuts
8. Steering rack and linkage

7923KG40

Exploded view of the power steering gear assembly mounting—Sephia

- Steering gear mounting nuts
- Steering gear to the right of the vehicle

To install:

3. Install or connect the following:
- Steering gear to the vehicle and the mounting nuts/bolts. Tighten the nuts to 23–34 ft. lbs. (31–46 Nm) for the Sephia and Spectra and 27–38 ft. lbs. (37–52 Nm) for Rio and Optima.
- Steering shaft to the steering gear pinion shaft. Tighten the bolt/nut to the specified torque.
- Manual transaxle shift linkage, if disconnected
- Set plate to the firewall, if removed
- Pressure line and return hose to the steering gear. Torque to 29–43 ft. lbs. (39–59 Nm).
- Catalytic converter, if removed
- Tie rod ends to the knuckle arm. Tighten the nuts to 22–33 ft. lbs. (29–44 Nm) for Sephia and Spectra and to 27–38 ft. lbs. (37–52 Nm) for Rio and Optima.
- New cotter pins

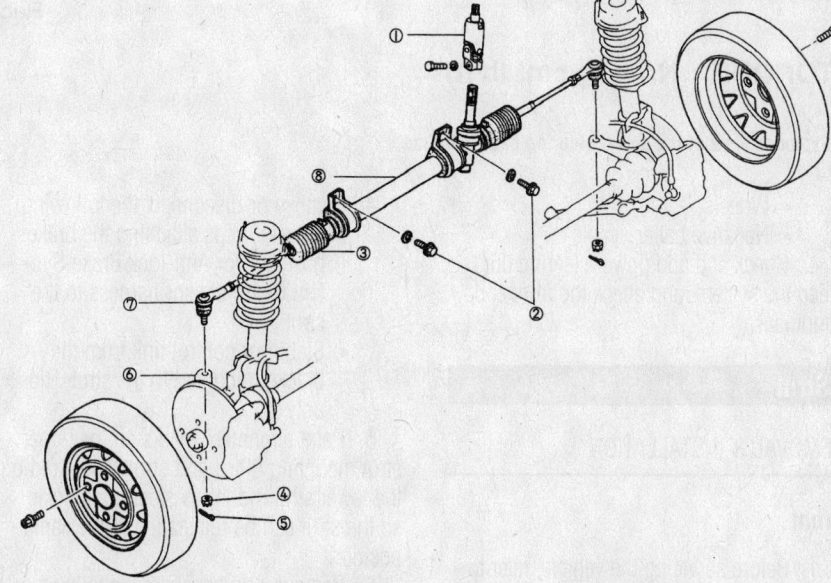

(1) Steering column intermediate shaft
(2) Steering gear mounting bolt
(3) Tie-rod boot
(4) Tie-rod nut

(5) Cotter pin
(6) Steering knuckle
(7) Tie-rod end
(8) Steering gear & Linkage

9357NG52

Power steering gear mounting—Rio

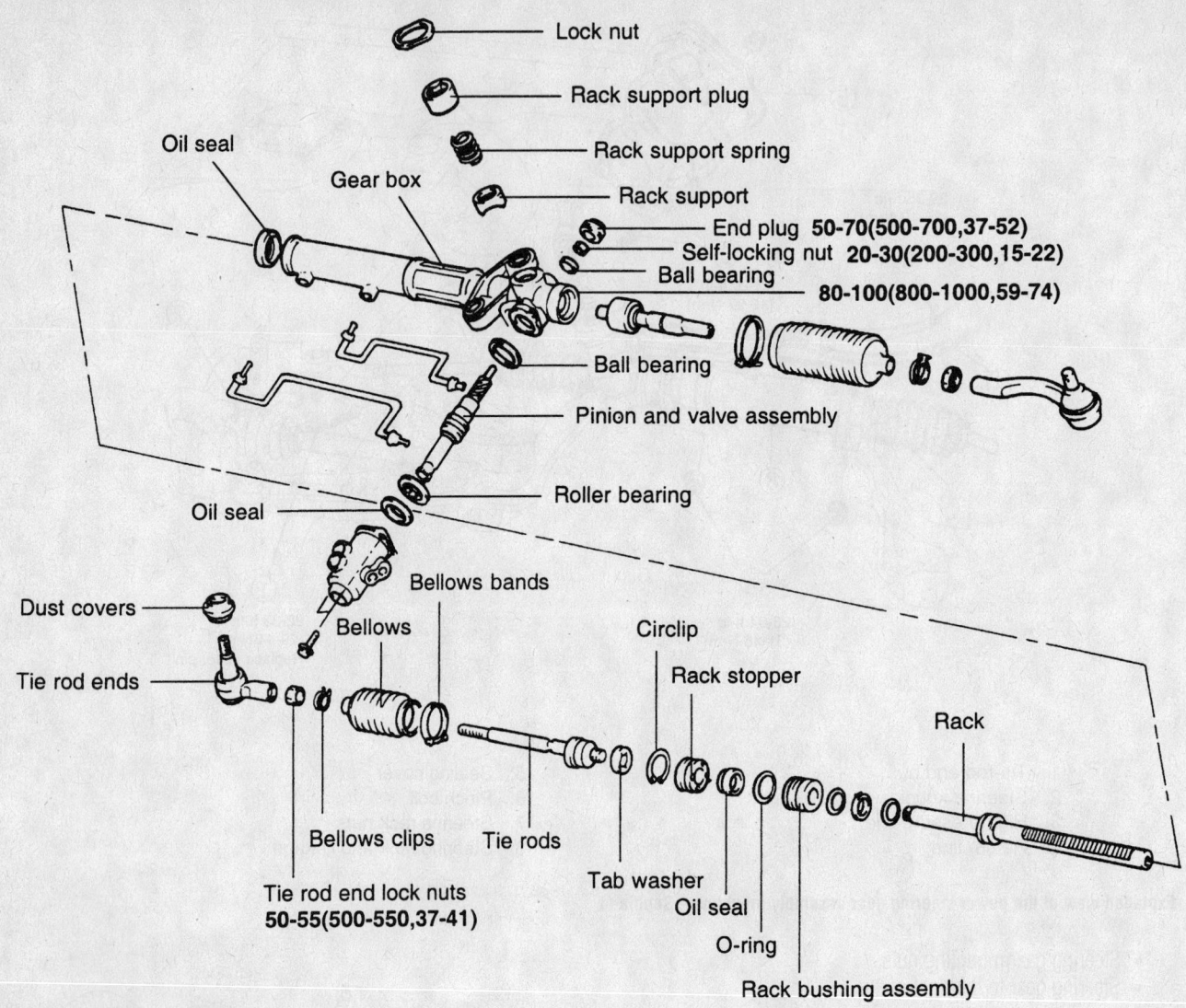

Lock nut

Rack support plug

Rack support spring

Oil seal

Gear box

Rack support

End plug **50-70(500-700,37-52)**

Self-locking nut **20-30(200-300,15-22)**

Ball bearing

80-100(800-1000,59-74)

Ball bearing

Pinion and valve assembly

Oil seal

Roller bearing

Bellows bands

Dust covers

Bellows

Circlip

Rack stopper

Tie rod ends

Rack

Bellows clips

Tie rods

Tie rod end lock nuts
50-55(500-550,37-41)

Tab washer

Oil seal

O-ring

Rack bushing assembly

TORQUE : Nm (kg·cm, lb·ft)

9357NG48

Exploded view of the power steering gear—Optima

- Wheels
- Negative battery cable

4. Check and add power steering fluid, bleed the system, and check the front end alignment.

Strut

REMOVAL & INSTALLATION

Front

1. Before servicing the vehicle, refer to the precautions in the beginning of this section.

2. Remove or disconnect the following:
- Wheel and tire assembly

3. Support the lower control arm with a jack.

4. Remove or disconnect the following:
- Bolts or clips attaching the brake hose and/or Anti-lock Brake System (ABS) sensor harness to the strut
- Stabilizer control link from the bracket mounted to the strut, Rio only

5. Paint alignment marks on the upper strut mounting block and strut tower, and on the lower strut mount-to-steering knuckle so the strut can be reinstalled in the same position.

6. Remove or disconnect the following:
- Lower strut-to-knuckle bolts
- Upper strut mounting block nuts. There may be 2 or 3 nuts, depending on model.
- Strut assembly

To install:

7. Install or connect the following:
- Strut into the strut tower, aligning the paint marks made during removal
- Strut mounting nuts. Tighten to 17–22 ft. lbs. (23–29 Nm) for Sephia and Spectra, to 34–46 ft. lbs. (43–61 Nm) for Rio or to 30–36 ft. lbs. (40–50 Nm) for Optima.
- Strut-to-knuckle bolts and tighten to 69–86 ft. lbs. (93–116 Nm) for Sephia and Spectra, to 76–90 ft. lbs. (103–123 Nm) for Rio or 44–59 ft. lbs. (60–80 Nm) for Optima.
- Stabilizer control link to the strut mounted bracket. Torque the

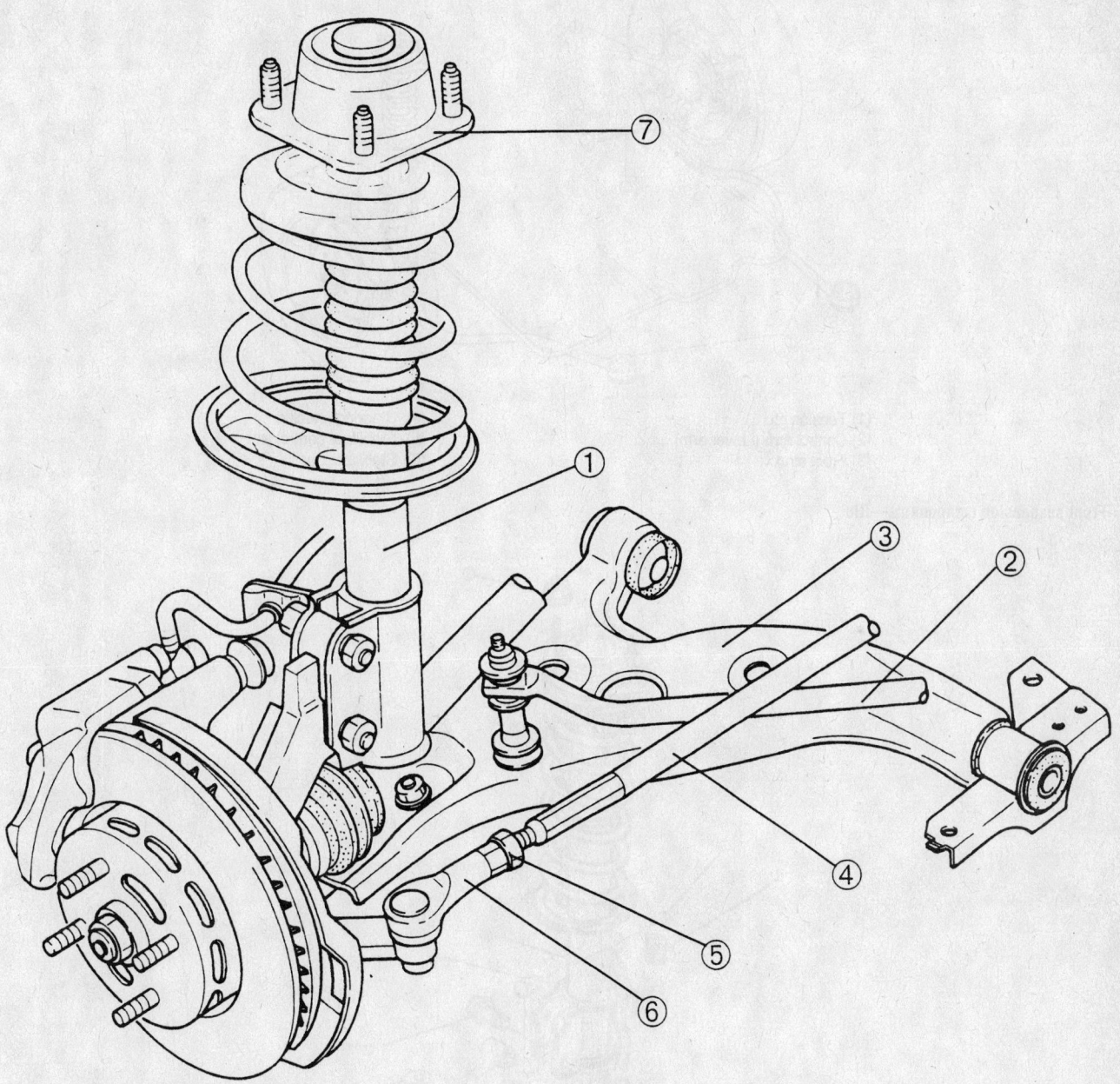

1. Front strut
2. Front stabilizer
3. Lower control arm
4. Tie-rod

5. Jam nut
6. Tie-rod end
7. Mounting block

7923KG41

Front suspension component identification

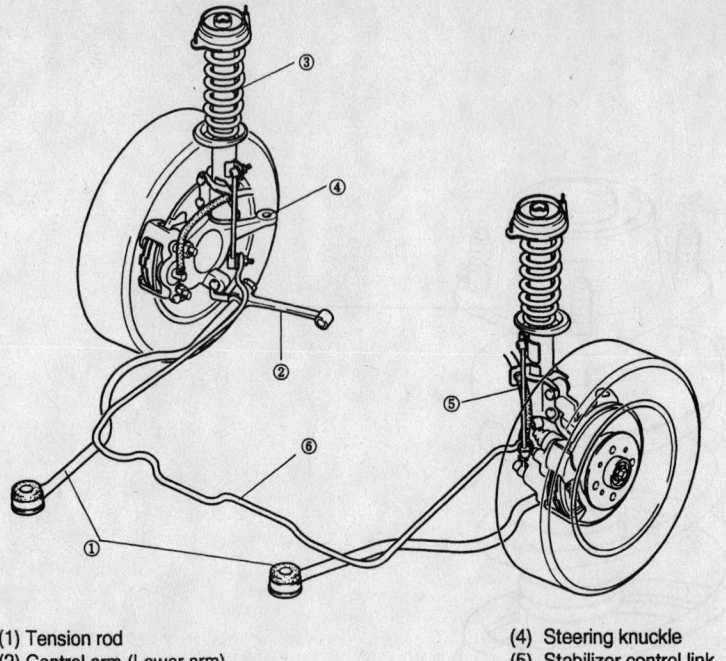

(1) Tension rod
(2) Control arm (Lower arm)
(3) Front strut

(4) Steering knuckle
(5) Stabilizer control link
(6) Stabilizer bar

9357NG17

Front suspension components—Rio

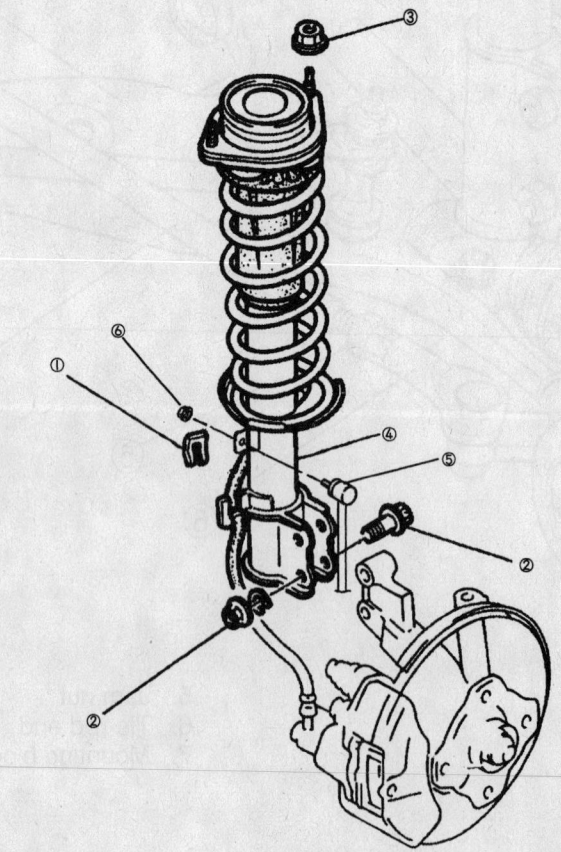

(1) Hose clip
(2) Bolt and nut
(3) Mounting block nut

(4) MacPherson strut
(5) Stabilizer control link
(6) Stabilizer control link nut

9357NG18

Exploded view of the front strut mounting—Rio

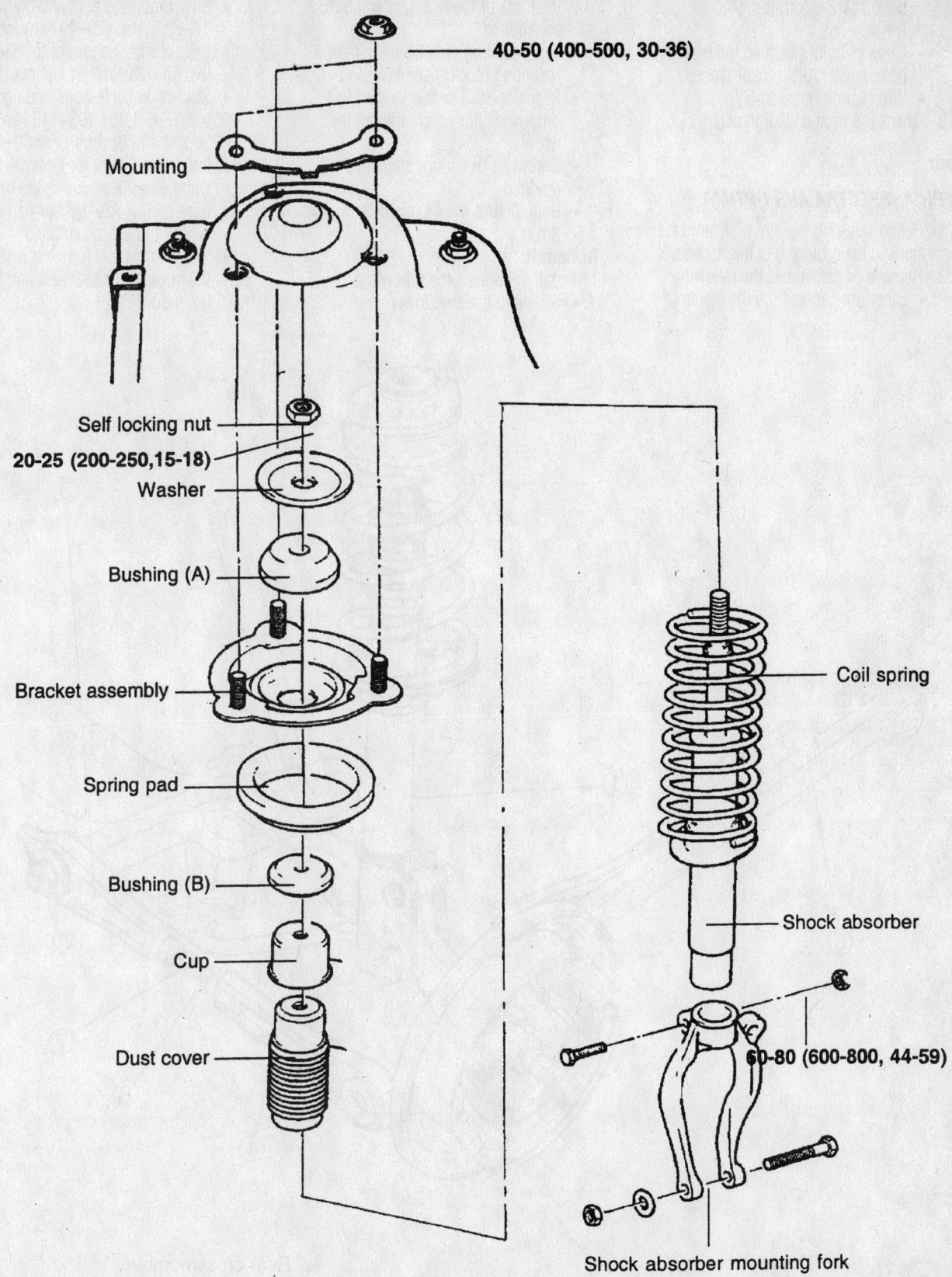

40-50 (400-500, 30-36)

Mounting

Self locking nut

20-25 (200-250,15-18)

Washer

Bushing (A)

Bracket assembly

Spring pad

Bushing (B)

Cup

Dust cover

Coil spring

Shock absorber

60-80 (600-800, 44-59)

Shock absorber mounting fork

TORQUE : Nm (kg·cm, lb·ft)

Exploded view of the front strut mounting—Optima

9357NG49

bolts to 32–45 ft. lbs. (43–61 Nm).
- Clips or bolts attaching the brake hose and/or ABS sensor harness
- Wheel and tire assembly

8. Check the front end alignment.

Rear

SEPHIA, SPECTRA AND OPTIMA

1. Before servicing the vehicle, refer to the precautions in the beginning of this section.
2. Remove or disconnect the following:
- Side trim panels from the inside of the trunk or the rear seat and trim, as required
- Top mounting nuts from the strut mounting block assembly
- Rear wheels. The suspension will drop when the weight lifts off the wheels.
- Brake line or wiring retainers as required
- Bottom strut mount retainers
- Strut

To install:

3. Install or connect the following:
- Strut into the strut tower
- Strut mounting nuts and tighten to 17–22 ft. lbs. (23–29 Nm) for Sephia and Spectra, or to 15–18 ft. lbs. (20–25 Nm) for Optima
- Strut-to-knuckle bolts and tighten to 69–86 ft. lbs. (93–116 Nm) for Sephia and Spectra or to 59–66 ft. lbs. (80–90 Nm) for Optima
- Clips or bolts attaching the brake hose and/or ABS sensor harness
- Wheel and tire assembly
- Side trim panels from the inside of the trunk or the rear seat and trim, if removed

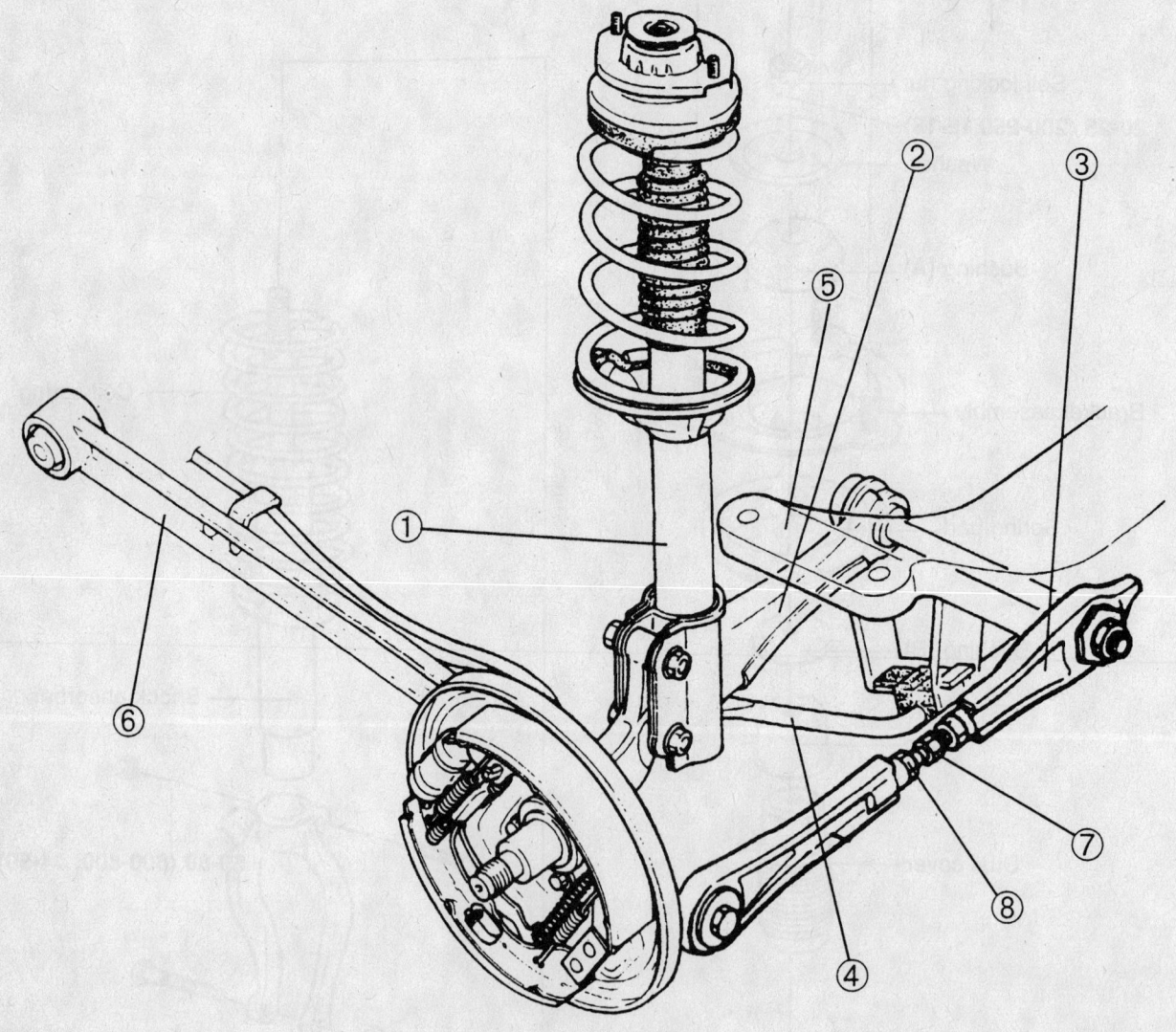

1. Rear strut		5. Rear crossmember
2. Front lateral link		6. Trailing link
3. Rear lateral link		7. Adjuster
4. Rear stabilizer		8. Jam nuts

Rear suspension component identification—Sephia and Spectra

7923KG42

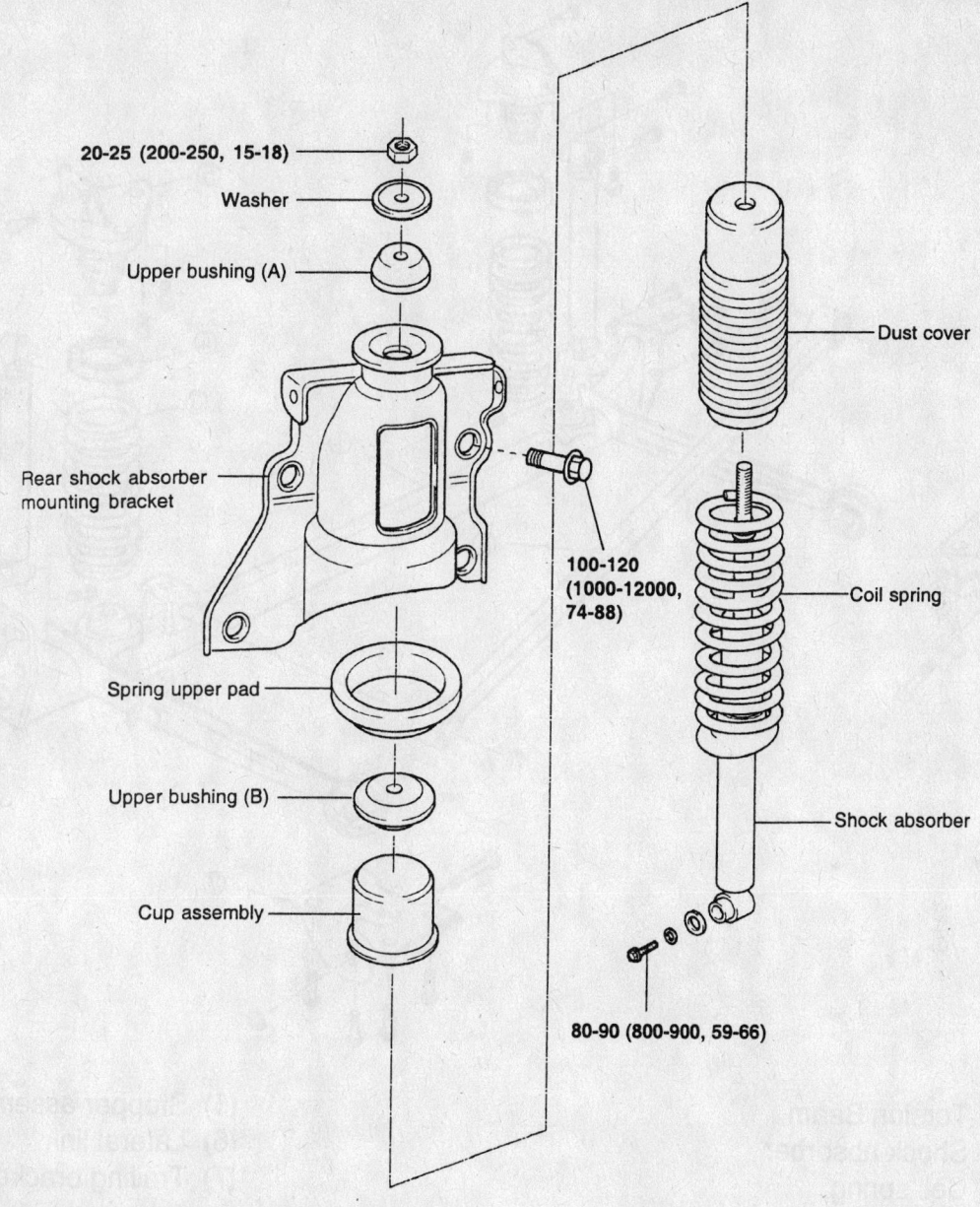

20-25 (200-250, 15-18)

Washer

Upper bushing (A)

Rear shock absorber
mounting bracket

100-120
(1000-12000,
74-88)

Spring upper pad

Upper bushing (B)

Cup assembly

Dust cover

Coil spring

Shock absorber

80-90 (800-900, 59-66)

TORQUE : Nm (kg·cm, lb·ft)

9357NG50

Exploded view of the rear strut—Optima

Shock Absorber and Coil Spring

REMOVAL & INSTALLATION

Rear

RIO

1. Before servicing the vehicle, refer to the precautions in the beginning of this section.

2. Remove or disconnect the following:
   - Rear wheels. Place a jack under the torsion beam axle.
   - Shock absorber lower mounting bolt
   - Coil spring from the torsion beam, after lowering the jack slightly
   - Shock absorber upper mounting bolt
   - Shock absorber

3. Installation if the reverse of the removal procedure, noting the following steps:
   a. Tighten the shock absorber upper bolt to 27–40 ft. lbs. (36–54 Nm).
   b. Tighten the shock absorber lower bolt to 58–72 ft. lbs. (78–98 Nm).

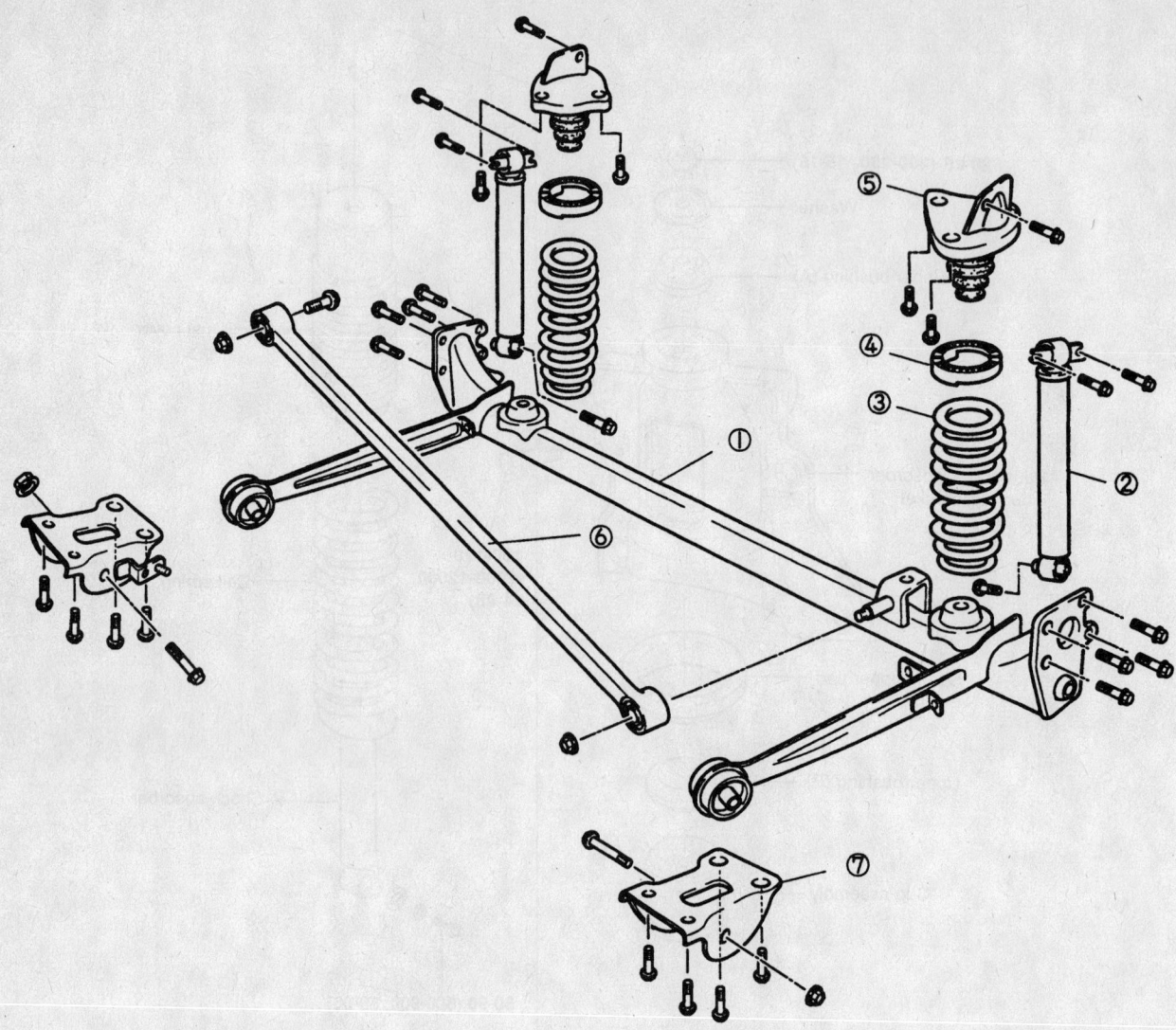

(1) Torsion Beam
(2) Shock absorber
(3) Coil spring
(4) Rubber seat

(5) Stopper assembly
(6) Lateral link
(7) Trailing bracket

9357NG19

Exploded view of the rear suspension components—Rio

Coil Spring

REMOVAL & INSTALLATION

Front and Rear

EXCEPT RIO—REAR SPRING

1. Before servicing the vehicle, refer to the precautions in the beginning of this section.
2. Remove the strut from the vehicle.
3. Install the strut securely in a vise with either aluminum or copper plates to protect the strut.

4. Loosen the piston rod upper nut several turns but DO NOT REMOVE IT.
5. Install the lower end of the strut in the vise and install a coil spring compressor. Compress the coil spring and remove the upper nut.

✳✳ CAUTION

Failure to fully compress the spring and hold it securely can be extremely dangerous.

6. Slowly release the coil spring tension.

7. Remove the suspension support, dust seal, spring seat, spring insulators, coil spring and bumper.
8. While pushing on the piston rod, be sure that the pull stroke is even and that there is no unusual noise or resistance. Also inspect for any oil leakage around the piston rod.
9. Push the piston rod in, then release it. Be sure that the return rate is constant.
10. If the shock absorber does not operate as described, replace it.

To assemble:

11. Install the strut assembly into a vise.

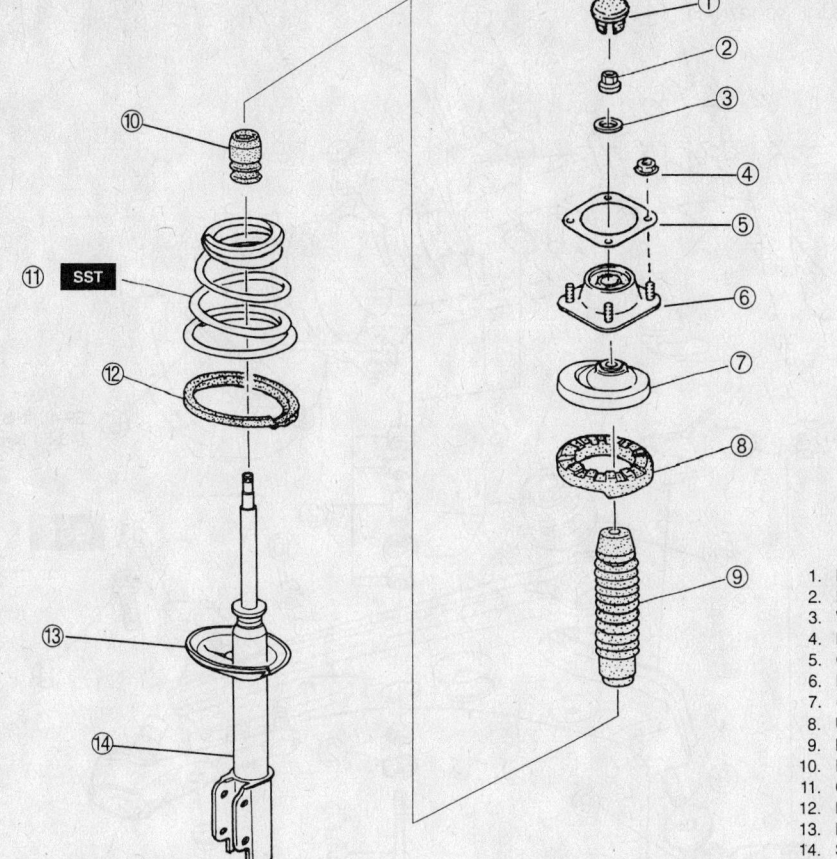

Exploded view of the front strut assembly

1. Dust cap
2. Piston retaining nut
3. Washer
4. Mounting nut
5. Gasket
6. Mounting block
7. Upper spring seat
8. Upper spring isolator
9. Dust boot
10. Rebound stopper
11. Coil spring
12. Lower spring isolator
13. Lower spring seat
14. Shock absorber

7932KG43

Exploded view of the rear strut assembly

1. Dust cap
2. Piston retaining nut
3. Washer
4. Mounting nut
5. Gasket
6. Mounting block/upper spring seat
7. Upper spring isolator
8. Dust boot
9. Coil spring
10. Lower spring isolator
11. Rebound stopper
12. Strut

7923KG44

12. Install the bound stopper and dust boot onto the piston rod.

13. Install the coil spring and compress the coil spring with the spring compressor.

14. Install the rubber seat, the spring upper seat, the bearing and the mounting block. Be sure that the spring upper seat notched portion is facing inward and tighten the piston rod upper nut.

15. Remove the spring compressor from the strut. Secure the upper mounting block in the vise. Tighten the nut to 41–50 ft. lbs. (55–68 Nm) for the front strut and 47–59 ft. lbs. (64–80 Nm) for the rear strut.

16. Be sure that the spring is well seated in the upper seats.

17. Install the strut to the vehicle.

Lower Ball Joint

REMOVAL & INSTALLATION

1. Before servicing the vehicle, refer to the precautions in the beginning of this section.

2. Remove or disconnect the following:
   • Wheel and tire assembly

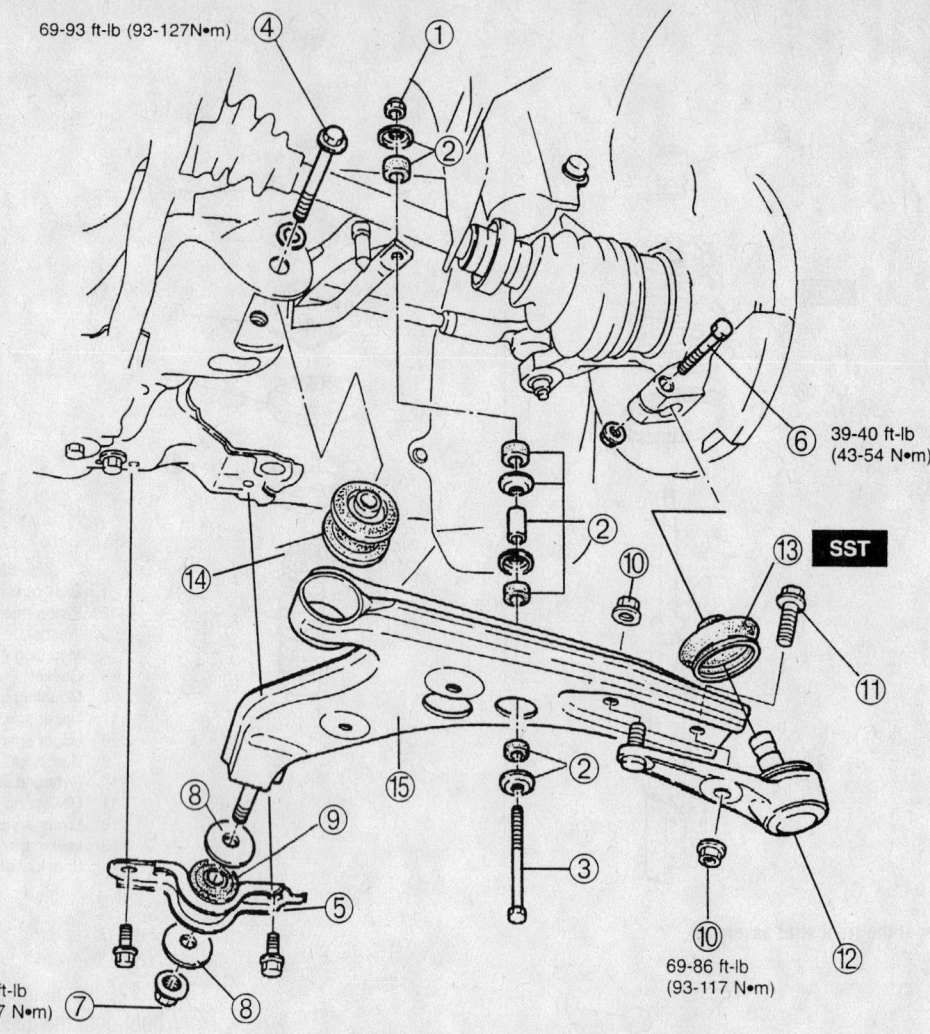

1. Stabilizer retaining nut
2. Stabilizer hardware - spacer, retainers, bushings
3. Stabilizer bolt
4. Pivot bolt
5. Mounting bolts
6. Pinch bolt,
7. Retaining nut
8. Washers
9. Control arm bushing - rear
10. Ball joint mounting nuts
11. Ball joint mounting bolt
12. Ball joint
13. Ball joint dust boot (Replace)
14. Control arm bushing - front
15. Control arm

7923KG45

Exploded view of the lower control arm with replaceable ball joint

- Ball joint stud pinch bolt and nut from the steering knuckle
- Ball joint from the knuckle, using a prytool
- Bolt, nut and the ball joint from the lower control arm

To install:

3. Installation is the reverse of the removal procedure. Tighten the ball joint-to-lower control arm bolt and nut to 86 ft. lbs. (117 Nm). Tighten the ball joint pinch bolt and nut to 43 ft. lbs. (59 Nm). Check the front wheel alignment.

Lower Control Arm

REMOVAL & INSTALLATION

Rio

1. Before servicing the vehicle, refer to the precautions in the beginning of this section.

2. Remove or disconnect the following:
- Front wheel
- Tension rod attaching bolt from the frame bracket
- Tensioner rod bushing cotter pin

and nut from the rear of the lower arm
- Washer and bushing
- Tension rod from the lower arm
- Front bushing and washer
- Lower arm, by lowering it and prying the ball joint from the steering knuckle after loosening the bolt and nut.
- Lower arm after loosening the lower arm attaching bolt from the frame bracket

To install:

3. Raise the inner end of the lower arm

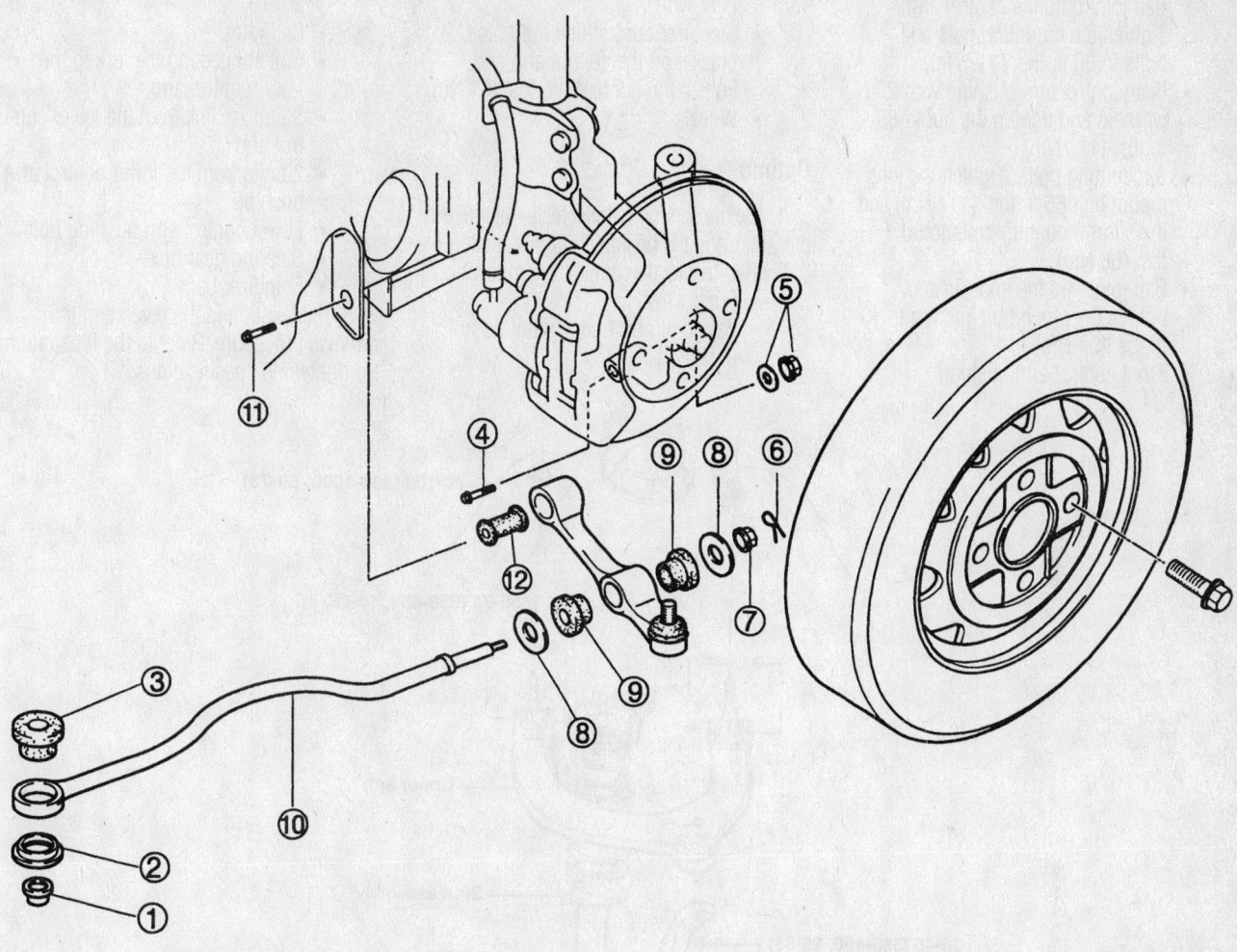

(1) Nut
(2) Stopper bushing
(3) Rubber
(4) Bolt
(5) Nut & Washer
(6) Cotter pin

(7) Nut
(8) Washer
(9) Bushing
(10) Tension rod
(11) Bolt
(12) Bushing

9357NG20

Exploded view of the lower control arm and tension rod—Rio

into the pivot bracket on the frame, and loosely tighten the arm frame bracket pivot to hold in place.

4. Install or connect the following:
- Lower control arm ball joint in the clamp bore of the knuckle. Tighten the bolt and nut to 40–50 ft. lbs. (54–68 Nm).

5. Torque the lower arm frame bracket pivot bolt to 40–50 ft. lbs. (54–68 Nm).
- Tension rod bolt into the lower arm, tighten washer and bushing and with a new cotter pin
- Bushing and insulator on the tensioner rod

6. Tighten the tension rod bolt to 87–108 ft. lbs. (118–147 Nm).

Sephia and Spectra

1. Before servicing the vehicle, refer to the precautions in the beginning of this section.
2. Remove or disconnect the following:
- Front wheel
- Stabilizer control link nut from the bracket on the lower control arm
- Pivot bolt
- Lower control arm ball joint bolt and nut from the steering knuckle
- 3 mounting bolts

- Retaining nut, washer and the rear control arm bushing
- 2 ball joint mounting nuts and bolts from the lower control arm

3. Place the ball joint in a vise and use a chisel top carefully remove the dust boot from the ball joint.
4. Remove or disconnect the following:
- Control arm front bushing
- Control arm

To install:

5. Apply a suitable general purpose grease to the new tire dust boot and press the boot onto the ball joint using a suitable tool.

6. Install or connect the following:
  • Ball joint onto the control arm. Tighten the mounting nuts and bolts to 86 ft. lbs. (117 Nm).
  • Rear control arm bushing with 2 washers and tighten the nut to 86 ft. lbs. (117 Nm)
  • 3 mounting bolts. Tighten the long mount bolt 86 ft. lbs. (117 Nm) and the short mounting bolts to 50 ft. lbs. (68 Nm).
  • Ball joint into the knuckle and tighten the pinch bolt and nut to 40 ft. lbs. (54 Nm)
  • Front control arm bushing

  • Pivot bolt and tighten to 86 ft. lbs. (117 Nm)
  • Stabilizer control link nut to the bracket on the control arm and tighten the nut to 45 ft. lbs. (61 Nm)
  • Wheels

Optima

1. Before servicing the vehicle, refer to the precautions in the beginning of this section.
2. Remove or disconnect the following:
  • Front wheel
  • Ball joint nut. Loosen, but do not remove it.

  • Ball joint from the lower control arm
  • Ball joint
  • Bolt connecting the fork to the lower control arm
  • Stabilizer link from the lower control arm
  • 2 bolts from the lower control arm bushing
  • Lower control arm bushing bolt
  • Steering gear box
  • Stabilizer bar

3. Installation is the reverse of the removal procedure. Refer to the illustration for tightening specifications.

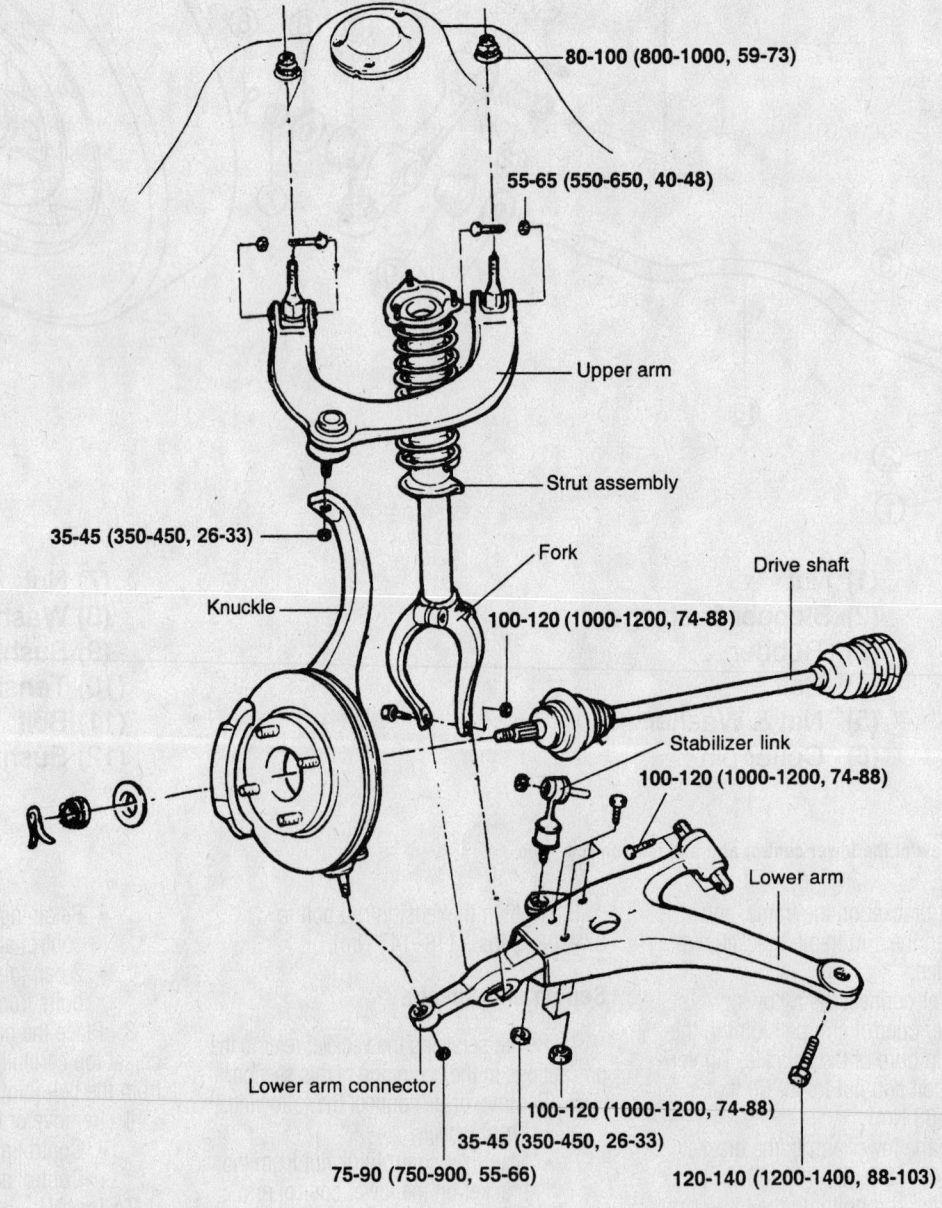

TORQUE : Nm (kg·cm, lb·ft)

9357NG51

Exploded view of the lower control arm—Optima

Wheel Bearings

ADJUSTMENT

Except Rio Rear Wheel Bearings

The wheel bearings on these vehicles are not adjustable. To check if the bearing requires service, remove the wheel and tire assembly, brake caliper and disc brake rotor. Install a dial indicator with the indicator foot resting on the wheel hub. Try to move the hub in and out. If there is more than 0.002 in. (0.05mm) bearing play, check the wheel hub nut torque or replace the hub and bearing assembly.

Rio Rear Wheel Bearings

BEARING PRELOAD

1. Make sure the parking brake is fully released.
2. Remove the wheel and hub (grease) cap.
3. Rotate the brake drum to be sure there is no drag.
4. Seat the bearings by tightening the nut after raising the nut tap. Tighten to 18–22 ft. lbs. (25–29 Nm).

5. Loosen the nut slightly until it can be turned by hand.

➡Before the bearing preload can be set, the amount of seal drag must be measured and added to the required preload. Use a pull scale to measure the oil seal drag.

6. Pull the scale squarely. Take the oil seal drag value when the wheel hub starters to turn and record it.
7. Add the oil seal drag value in the last step to the specified value to 0.6–1.9 lbs. (2.6–8.5 N). This is the standard bearing preload.
8. Turn the nut slowly to adjust the standard bearing preload while checking with the pull scale.
9. Firmly fix the lock nut into the groove.
10. Install the hub cap and wheel.

REMOVAL & INSTALLATION

Front

1. Before servicing the vehicle, refer to the precautions in the beginning of this section.

2. Remove or disconnect the following:
   • Front wheels
   • Center locknut. Discard the old locknut.
3. Remove or disconnect the following:
   • Caliper assembly from the knuckle. Do not disconnect the brake lines. Support the caliper with a piece of wire. Do not allow the caliper to hang by the hose at any time. Remove the brake disc.
   • Anti-lock Brake System (ABS) speed sensor, if equipped
   • Tie rod end cotter pin and nut
   • Tie rod end out of the knuckle assembly
   • Outer lower arm to ball joint mounting bolt and nut
   • Lower arm from the knuckle assembly
   • Knuckle assembly free of the half-shaft, using a plastic mallet
   • Knuckle assembly
4. Clamp the knuckle in a vise with protected jaws.
5. Remove or disconnect the following:
   • Inner oil seal from the knuckle
   • Front wheel hub from the knuckle

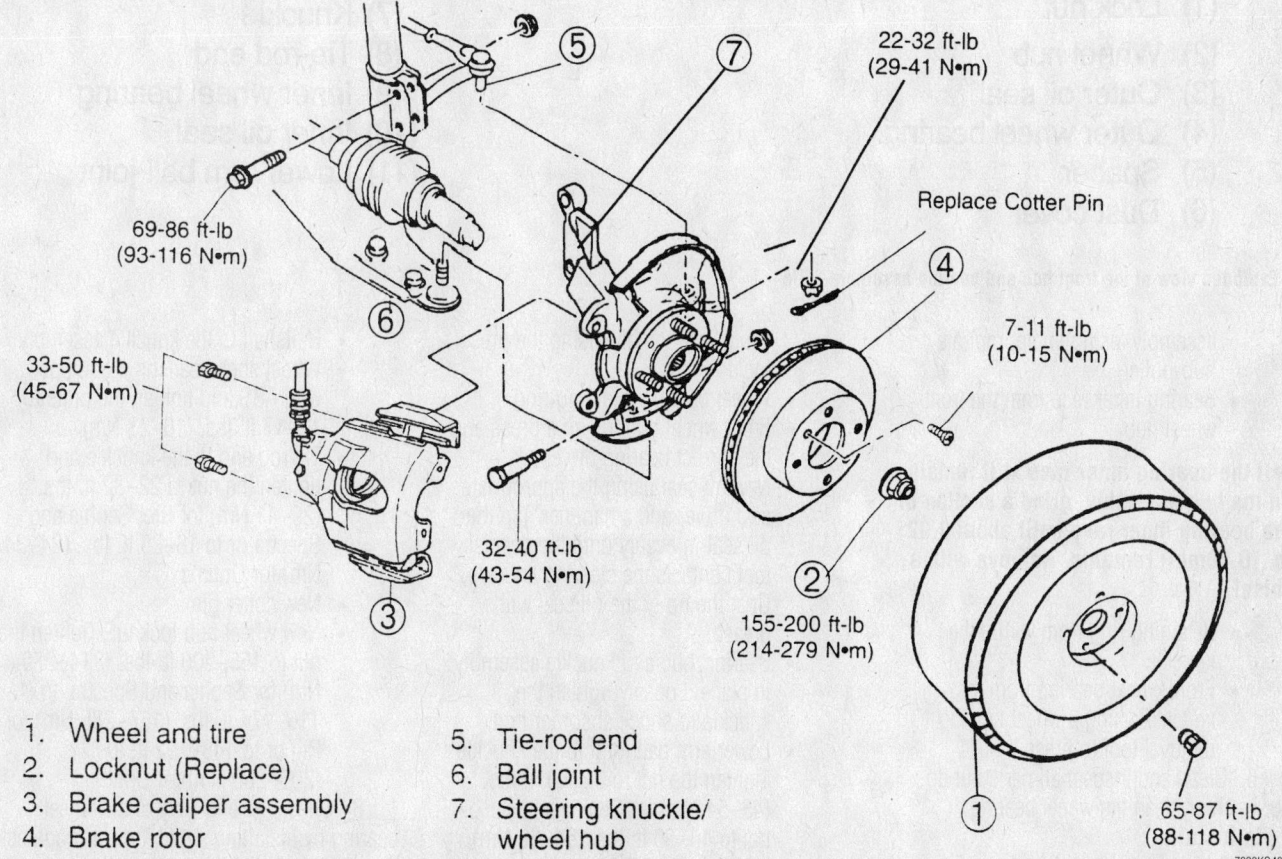

22-32 ft-lb (29-41 N•m)

Replace Cotter Pin

69-86 ft-lb (93-116 N•m)

7-11 ft-lb (10-15 N•m)

33-50 ft-lb (45-67 N•m)

32-40 ft-lb (43-54 N•m)

155-200 ft-lb (214-279 N•m)

65-87 ft-lb (88-118 N•m)

1. Wheel and tire
2. Locknut (Replace)
3. Brake caliper assembly
4. Brake rotor
5. Tie-rod end
6. Ball joint
7. Steering knuckle/ wheel hub

7923KG46

Exploded view of the front steering knuckle and related components—Sephia and Spectra

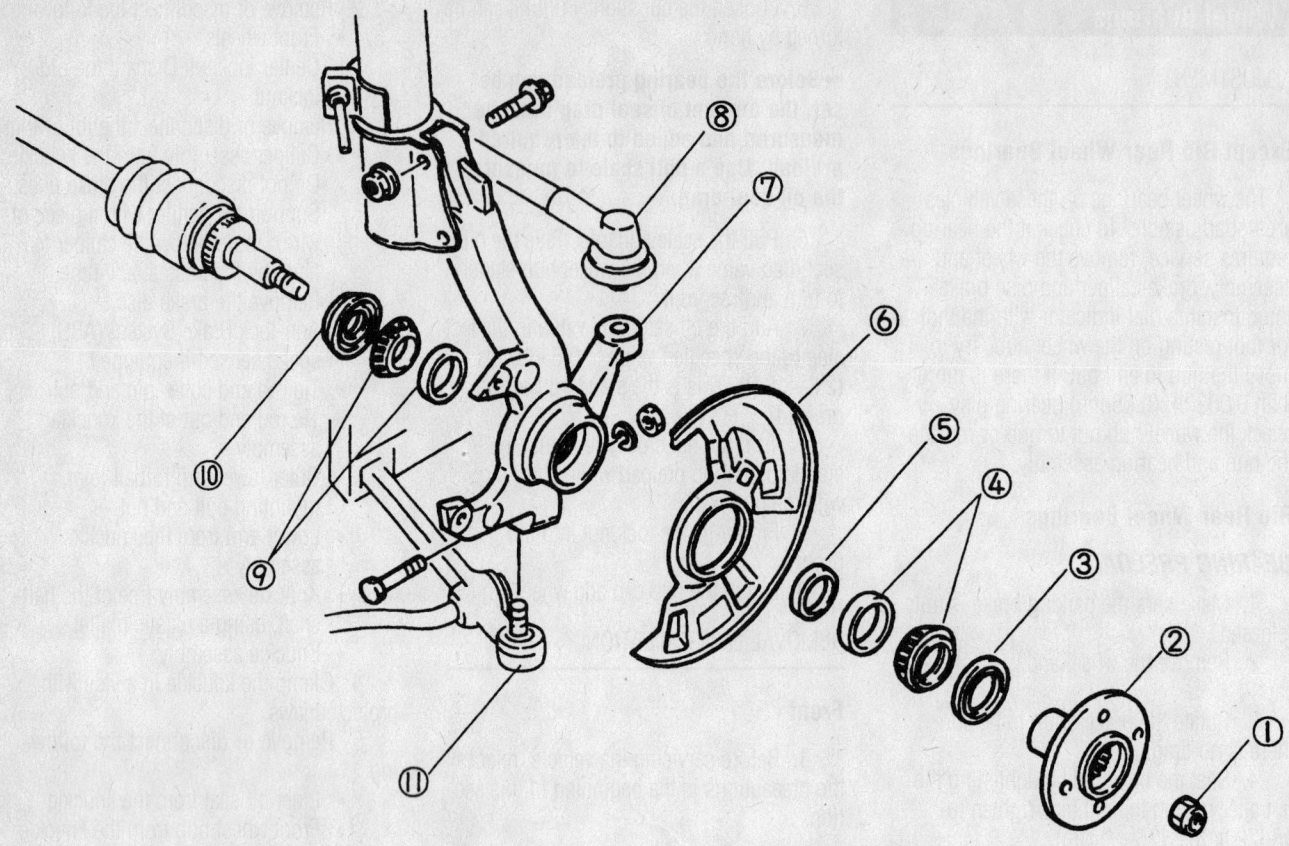

(1) Lock nut
(2) Wheel hub
(3) Outer oil seal
(4) Outer wheel bearing
(5) Spacer
(6) Dust cover

(7) Knuckle
(8) Tie-rod end
(9) Inner wheel bearing
(10) Inner oil seal
(11) Lower arm ball-joint

9357NG21

Exploded view of the front hub and bearing assembly—Rio

assembly, using an appropriate hub-puller
• Bearing inner race from the front wheel hub

➡If the bearing inner race still remains on the hub assembly, grind a section of the bearing inner race until about 0.02 in. (0.50mm) remains. Remove with a chisel.

• Retaining ring from within the knuckle
• Front wheel bearing from the knuckle, using a wheel bearing removal tool to press it out
6. Clean and inspect all parts but do not wash or clean the wheel bearing.
To install:
7. Install or connect the following:
• New wheel bearing into the

knuckle assembly, using the press tools
• Wheel bearing retaining ring
• Front wheel hub, using a press and the correct bearing driver
• New oil seal using the appropriate seal driver and a hammer. Tap the oil seal in evenly until the special tool contacts the steering knuckle. Coat the lip of the oil seal with grease.
• Bearing/hub and knuckle assembly in place. Loosely tighten the knuckle to shock absorber bolt.
• Lower arm ball joint to the knuckle. Tighten the nut to 32–40 ft. lbs. (43–54 Nm) for Sephia and Spectra, to 40–50 ft. lbs. (54–68 Nm) for Rio or to 74–88 ft. lbs. (100–120 Nm) for Optima.

• Halfshaft to the knuckle assembly
• Wheel speed sensor, if equipped with ABS and tighten the bolts to 12–17 ft. lbs. (16–23 Nm)
• Tie rod end to the knuckle and tighten the nut to 22–32 ft. lbs. (29–41 Nm) for Rio, Sephia and Spectra or to 18–25 ft. lbs. (24–34 Nm) for Optima
• New cotter pin
• New wheel hub locknut. Tighten the nut to 155–200 ft. lbs. (214–279 Nm) for Sephia and Spectra, to 116–174 ft. lbs. (157–235 Nm) for Rio or to 148–192 ft. lbs. (200–260 Nm) for Optima.
8. Check the end-play of the wheel bearing by installing a dial indicator against the wheel hub and tire to move the brake disc back and forth. There should be no

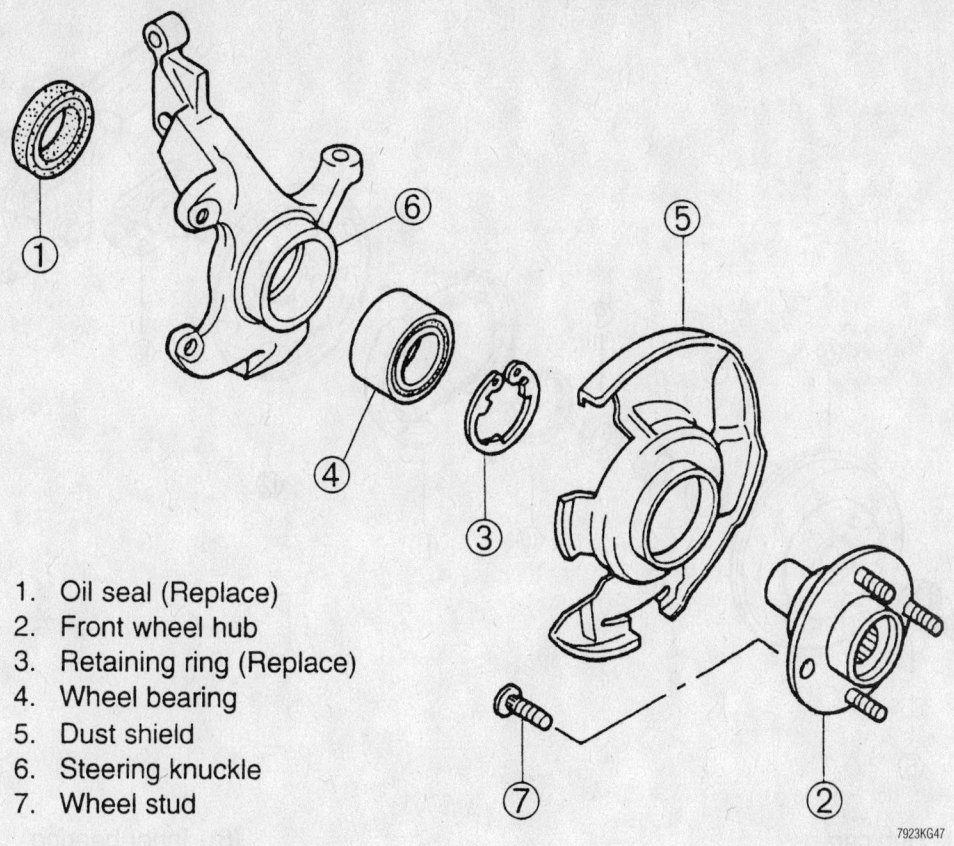

1. Oil seal (Replace)
2. Front wheel hub
3. Retaining ring (Replace)
4. Wheel bearing
5. Dust shield
6. Steering knuckle
7. Wheel stud

7923KG47

Exploded view of the front hub and bearing assembly—Sephia and Spectra

more than 0.002 in. (0.05mm) of free-play present.

9. Stake the locknut into place by bending it into the groove.

10. Install or connect the following:
- Brake caliper(s) and tighten the bolts to 33–50 ft. lbs. (45–67 Nm)
- Front wheels and lower the vehicle

11. With the vehicle lowered check all of the bolts and retighten as necessary.

12. Inspect the front end alignment and adjust as is necessary.

Rear

RIO

1. Before servicing the vehicle, refer to the precautions in the beginning of this section.

2. Remove or disconnect the following:
- Rear wheels
- Hub cap

3. Use a small chisel to raise the staked edge of the hub lock nut.
- Drum, washer and bearings as an assembly
- Anti-lock Brake System (ABS) sensor rotor using a suitable puller
- Bearing oil seal and discard

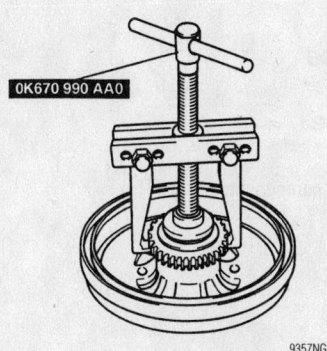

0K670 990 AA0

9357NG23

Removing the ABS rotor, using a puller—Rio

➡**If the bearings will be reused, they must be matchmarked so they can be installed in their original positions.**

- Inner bearings from the bearing hub
- Inner and outer bearing outer races using special tool 0K68A 173 002

To install:

4. Install or connect the following:
- Outer bearing race using a hammer. Tap the race in until it is fully seated in the hub.
- Inner bearing in the bearing hub
- Oil seal after lubricating it with

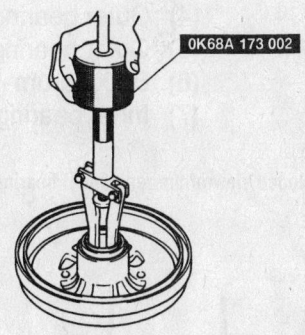

0K68A 173 002

9357NG24

View of the special tool need to remove the inner and outer bearing outer races—Rio

lithium grease, using a hammer and seal installation tool
- ABS sensor rotor using a flat plate to press it into place

5. Refer to the accompanying illustration and completely fill in the shaded area with lithium grease.
- Brake drum bearings and hub on the spindle. Keep the drum centered on the spindle to prevent damage to the oil seal and spindle threads.
- Outer bearing and washer
- New hub lock nut

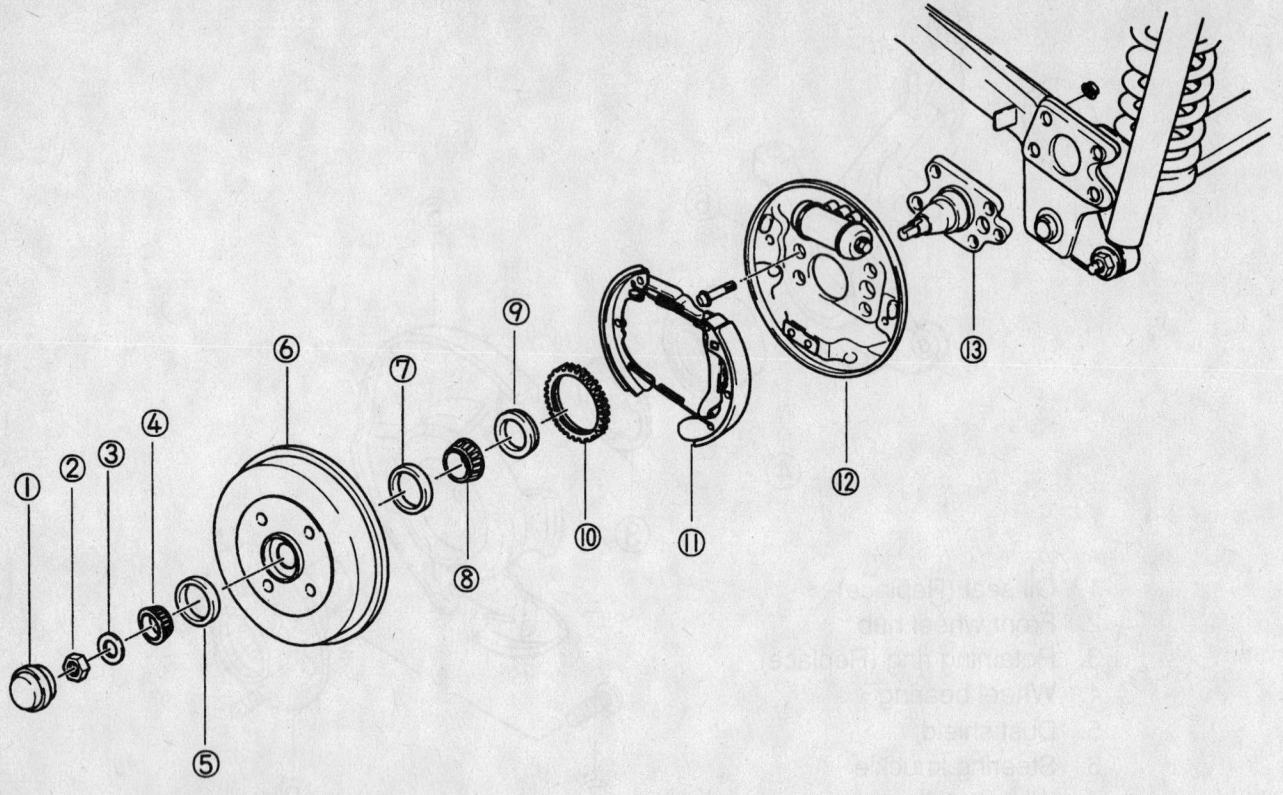

(1) Hub cap
(2) Lock nut
(3) Washer
(4) Outer bearing
(5) Outer bearing outer race
(6) Brake drum
(7) Inner bearing outer race

(8) Inner bearing
(9) Oil seal
(10) Sensor rotor(for ABS)
(11) Brake assembly
(12) Back plate
(13) Spindle

Exploded view of the rear wheel bearings and related components—Rio

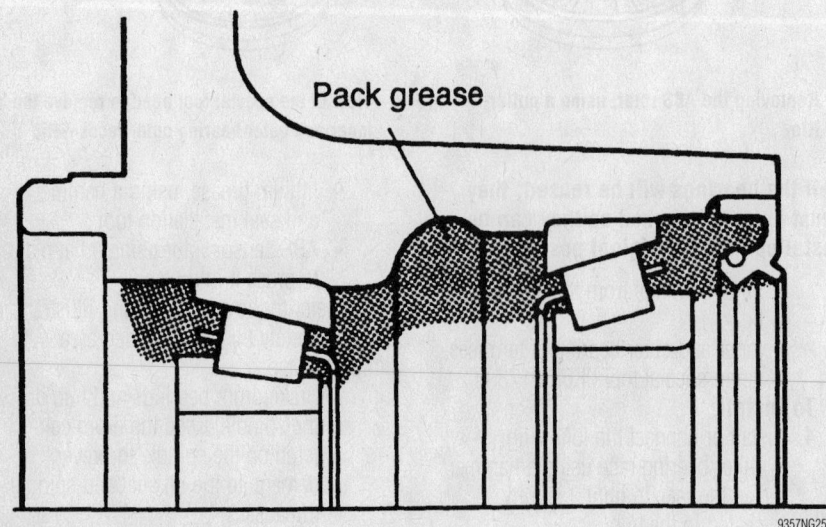

Fill in the shaded areas with lithium grease—Rio

6. Adjust the bearing preload to 5.6–8.6 ft. lbs. (0.63–0.98 Nm).
  • Hub lock nut to the groove firmly
  • Hub cap

SEPHIA, SPECTRA AND OPTIMA

1. Before servicing the vehicle, refer to the precautions in the beginning of this section.

2. Remove or disconnect the following:
  • Rear wheels
  • Hubcap. Hold the brake to remove the center axle nut.
  • Drum, if equipped with drum brakes
  • Disc brake caliper, if equipped with disc brakes without disconnecting the hydraulic hose and hang it from the body. Do not let it hang by the hose. Slide the disc off the spindle.
  • Hub and bearing assembly off the spindle. The hub and bearing can-

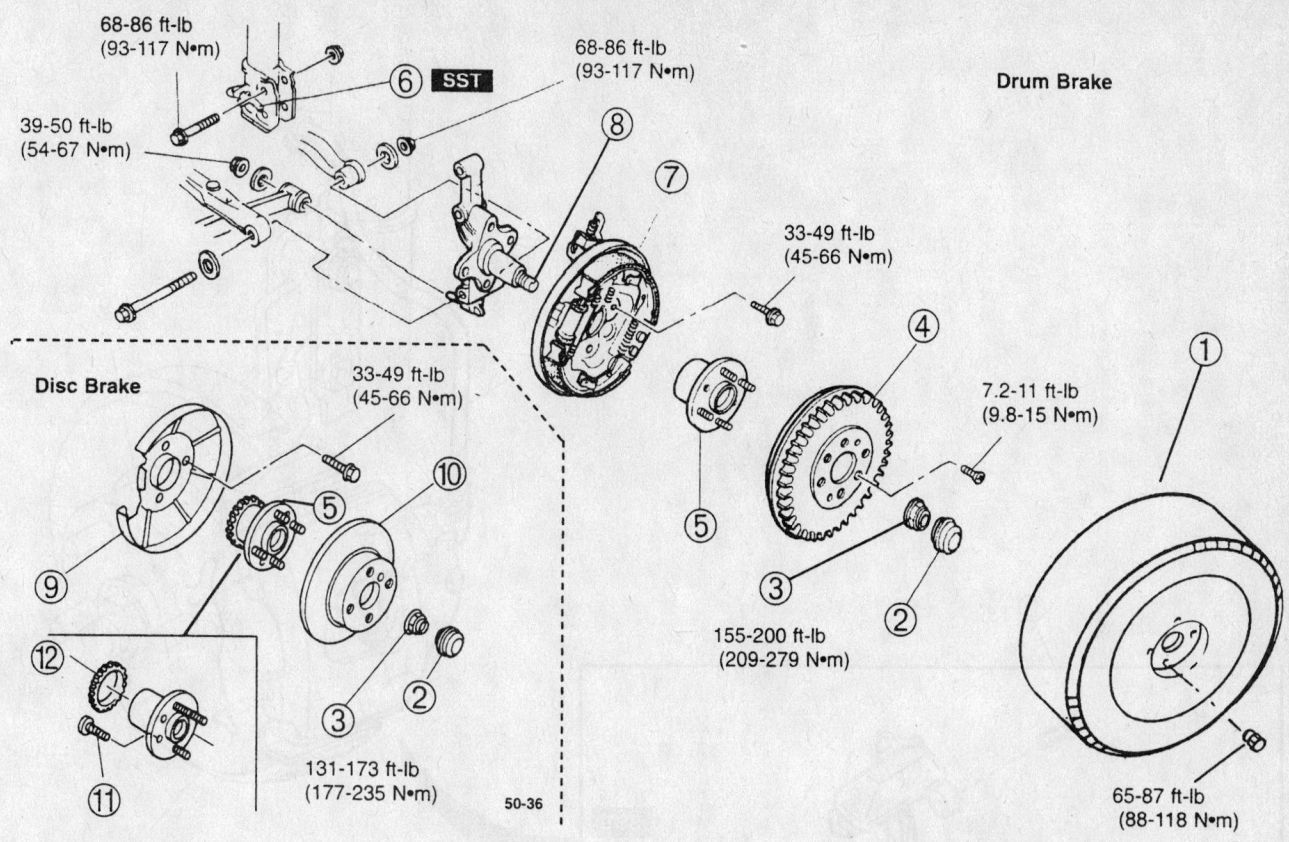

Drum Brake

68-86 ft-lb
(93-117 N•m)

68-86 ft-lb
(93-117 N•m)

39-50 ft-lb
(54-67 N•m)

33-49 ft-lb
(45-66 N•m)

7.2-11 ft-lb
(9.8-15 N•m)

Disc Brake

33-49 ft-lb
(45-66 N•m)

155-200 ft-lb
(209-279 N•m)

131-173 ft-lb
(177-235 N•m)

65-87 ft-lb
(88-118 N•m)

50-36

7923KG48

1. Wheel and tire
2. Dust cap
3. Locknut (Replace)
4. Brake drum (or disc)
5. Hub with bearing assembly
6. Brake line
7. Rear brake assembly (drum or disc)
8. Spindle
9. Dust cover
10. Brake rotor
11. Hub bolt
12. ABS sensor rotor

Exploded view of the rear axle assembly—Sephia and Spectra

not be separated and must be replaced as 1 piece.

To install:

3. Install or connect the following:
   • Hub and drum or rotor

   • Brake caliper, if equipped
   • New spindle nut and tighten to 131–173 ft. lbs. (177–235 Nm) for disk brakes and 155–200 ft. lbs. (209–279 Nm) for drum brakes.

Stake the nut into place. Replace the hubcap.
   • Wheel and tire assembly

BRAKES

Brake Caliper

REMOVAL & INSTALLATION

Front

1. Before servicing the vehicle, refer to the precautions in the beginning of this section.
2. Remove or disconnect the following:
   • Wheels
   • Brake hose from the caliper
   • Caliper guide pin bolts and caliper

To install:

3. Install or connect the following:
   • Caliper and tighten the guide pin bolts to 19–21 ft. lbs. (26–28 Nm)
   • Brake hose and tighten to 9–13 ft. lbs. (13–18 Nm)
4. Bleed the brakes and fill the master cylinder with clean brake fluid.
   • Wheels

Rear

1. Before servicing the vehicle, refer to the precautions in the beginning of this section.
2. Remove or disconnect the following:

   • Wheels
   • Parking brake cable and clip
   • Brake hose banjo bolt
   • Caliper lock bolts and the caliper

To install:

3. Install or connect the following:
   • Caliper and tighten the lock bolts to 22–29 ft. lbs. (29–39 Nm)
   • Brake hose and tighten the banjo bolt to 16–22 ft. lbs. (22–29 Nm)
   • Park brake cable and clip
4. Bleed the brakes and fill the master cylinder with clean brake fluid.
   • Wheels

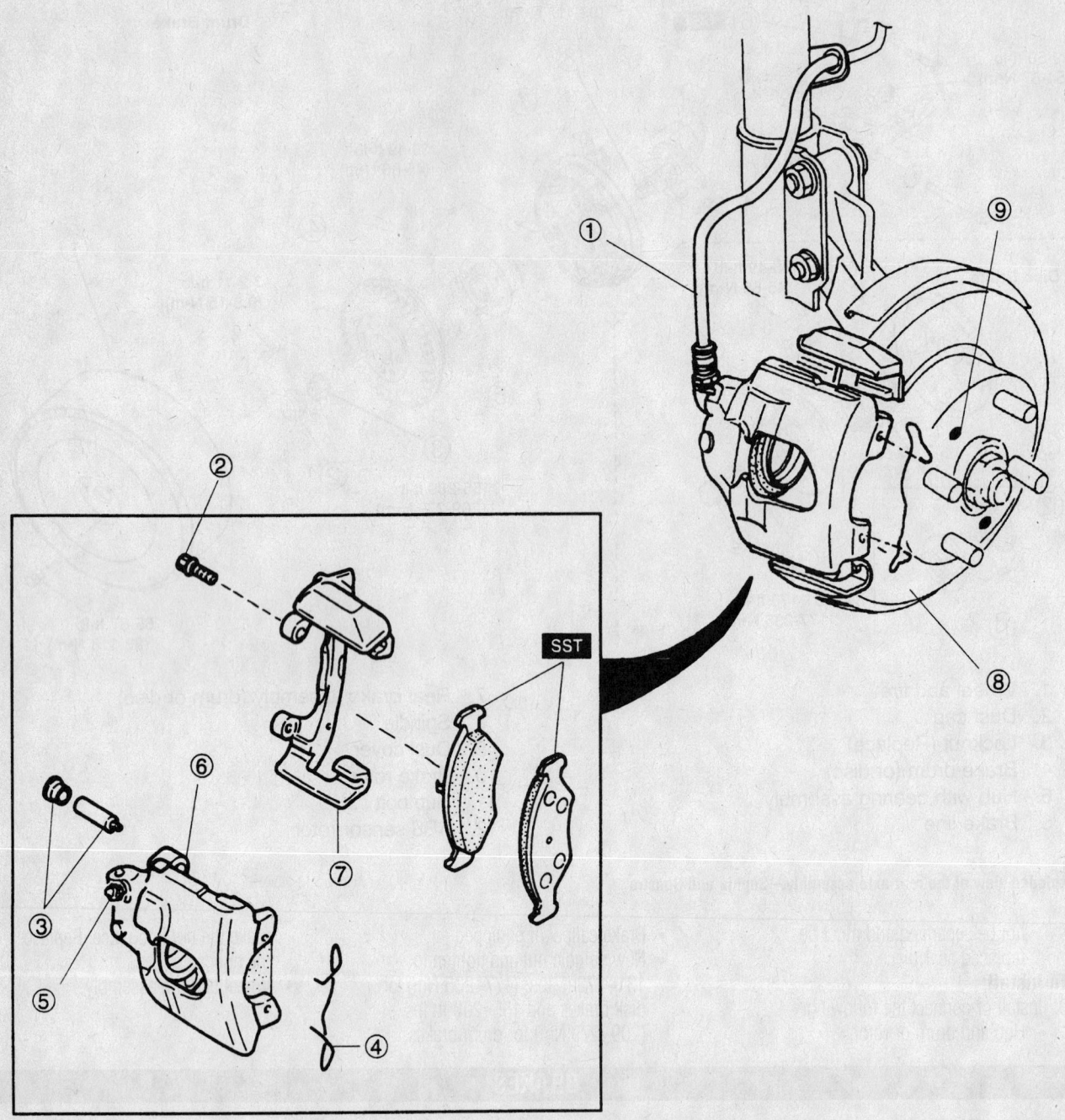

SST

(1) Flexible hose
(2) Bolt
(3) Cap, bolt (square head) and bushing
(4) Spring
(5) Cap and bleeder screw

(6) Caliper
(7) Supporting plate
(8) Brake rotor (Disc)
(9) Mounting screws

93016G15

Exploded view of the front brakes

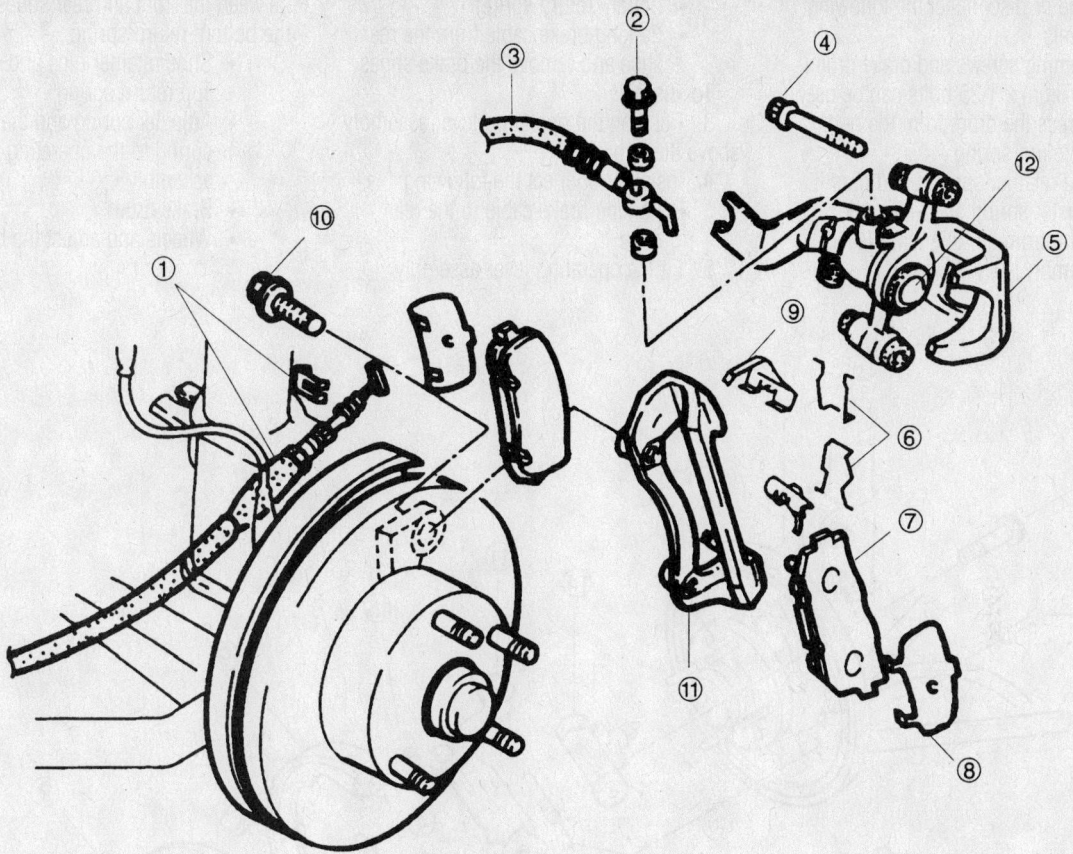

(1) Parking cable, clip
(2) Connecting bolt
(3) Brake hose
(4) Lock bolt

(5) Caliper
(6) V-spring
(7) Disc pad
(8) Shim

(9) Guide plate
(10) Bolt
(11) Mounting support
(12) Caliper piston

93016G16

Exploded view of the rear disc brakes

Disc Brake Pads

REMOVAL & INSTALLATION

Front

1. Before servicing the vehicle, refer to the precautions in the beginning of this section.
2. Remove or disconnect the following:
   - Wheels
   - Caliper guide pin bolts and lift the caliper away from the rotor
   - Brake pads and retainer spring from the caliper

To install:
3. Compress the caliper piston into the bore.
4. Install or connect the following:
   - Brake pads and retainer spring to the caliper
   - Caliper in position on the caliper support bracket
   - Guide pin bolts
   - Wheels

Rear

1. Before servicing the vehicle, refer to the precautions in the beginning of this section.
2. Remove or disconnect the following:
   - Wheels
   - Parking brake cable and clip
   - Caliper lock bolts and remove the caliper
   - V-springs from the brake pads
   - Brake pads and shims

To install:
3. Compress the caliper piston into the bore by rotating the piston with special tool OK9A4 263 001.
4. Install or connect the following:
   - New pads and shims
   - V-springs
   - Caliper and tighten the lock bolts to 22–29 ft. lbs. (29–39 Nm)
   - Park brake cable and clip
   - Wheels

Brake Drums

REMOVAL & INSTALLATION

1. Before servicing the vehicle, refer to the precautions in the beginning of this section.
2. Remove or disconnect the following:
   - Wheels
   - Retaining screws and brake drum. Two 8mm x 1.25 bolts can be used to press the drum from the hub.

To install:
3. Install or connect the following:
   - Brake drum to the hub
   - Wheels

Brake Shoes

REMOVAL & INSTALLATION

1. Before servicing the vehicle, refer to the precautions in the beginning of this section.

2. Remove or disconnect the following:
   - Wheels
   - Retaining screws and brake drum. Two 8mm x 1.25 bolts can be used to press the drum from the hub.
   - Top return spring
   - Shoe retainer springs and pins
   - Adjuster spring and anti-rattle spring from the operating lever assembly
   - Bottom return spring
   - Parking brake cable from the rear shoe and remove the brake shoes

To install:

3. Position the operating lever assembly above the hub.
4. Install or connect the following:
   - Parking brake cable to the rear shoe
5. Fit the operating lever assembly between the front and rear shoes and install the bottom return spring.
   - Shoe retainer pins and springs
   - Top return spring
   - Adjuster spring and the anti-rattle spring to the operating lever assembly
   - Brake drum
   - Wheels and adjust the brakes.

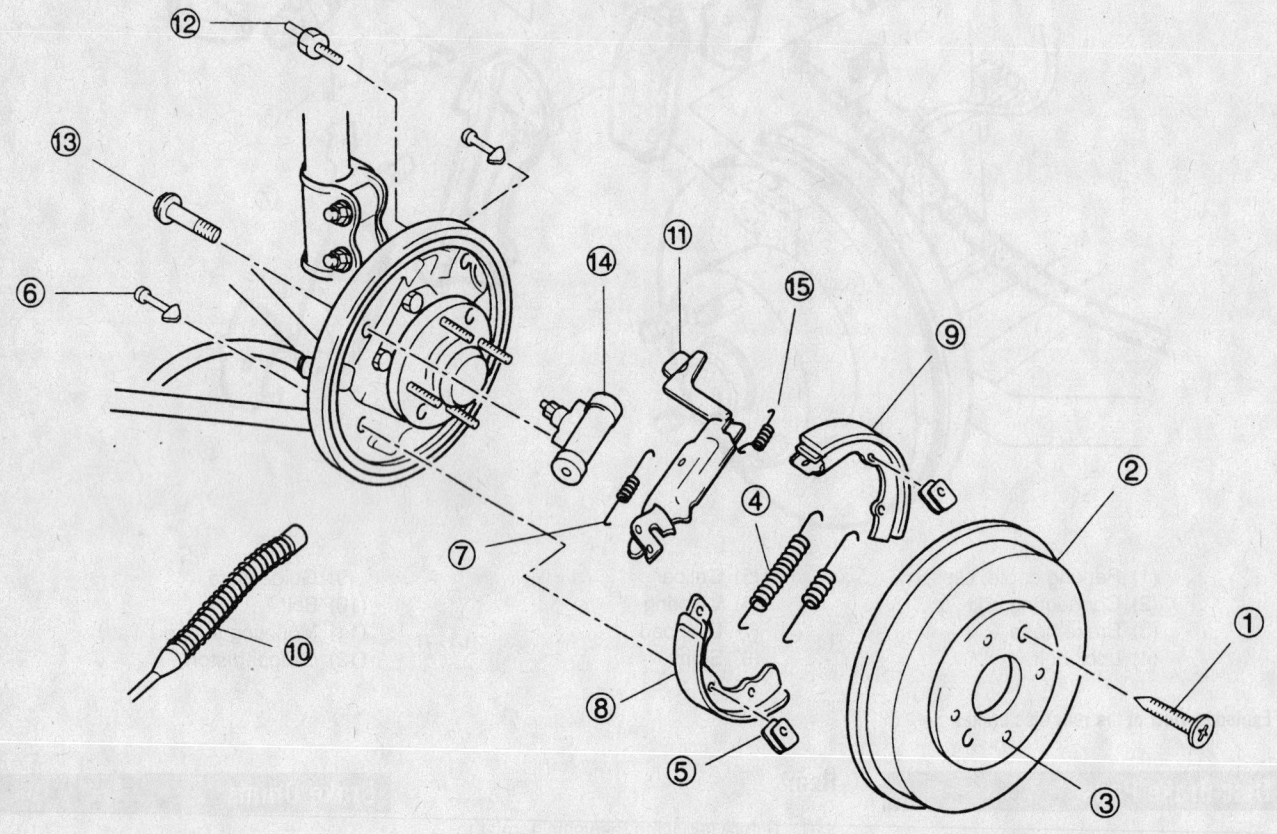

(1) Mounting screws
(2) Brake drum
(3) Drum pulling threads
(4) Return springs
(5) Spring clips
(6) Hold down pins
(7) Adjuster spring
(8) Brake shoe-leading
(9) Brake shoe-trailing
(10) Parking brake cable
(11) Operating lever assembly
(12) Brake line
(13) Bolts
(14) Wheel cylinder assembly
(15) Anti-rattle spring

93016G17

Exploded view of the rear drum brakes

KIA

Sedona

14

SPECIFICATION CHARTS

ENGINE AND VEHICLE IDENTIFICATION

Engine								Model Year	
Code ①	Liters (cc)	Cu. In.	Cyl.	Fuel Sys.	Engine Type	Eng. Mfg.		Code ②	Year
1	3.5 (3497)	213	6	EGI	DOHC	KIA		2	2002
								3	2003
								4	2004

EGI: Electronic Gasoline Injection

DOHC: Double Overhead Camshafts

① 8th digit of VIN

② 10th digit of VIN

42356-SEDO-C01

GENERAL ENGINE SPECIFICATIONS

Year	Model	Engine Displacement Liters (cc)	Engine VIN	Fuel System Type	Net Horsepower @ rpm	Net Torque @ rpm (ft. lbs.)	Bore x Stroke (in.)	Compression Ratio	Oil Pressure @ rpm
2002	Sedona	3.5 (3497)	1	EGI	195@5500	218@3500	3.66x3.88	10:01	43-57@3000
2003	Sedona	3.5 (3497)	1	EGI	195@5500	218@3500	3.66x3.88	10:01	43-57@3000

EGI: Electronic Gasoline Injection

MFI: Multi-port Fuel Injection

42356-SEDO-C02

ENGINE TUNE-UP SPECIFICATIONS

Year	Engine Displacement Liters (cc)	Engine VIN	Spark Plug Gap (in.)	Ignition Timing (deg.) MT	Ignition Timing (deg.) AT	Fuel Pump (psi)	Idle Speed (rpm) MT	Idle Speed (rpm) AT	Valve Clearance Intake	Valve Clearance Exhaust
2002	3.5 (3497)	1	0.039-0.043	—	①	46-49	—	600-800	HYD	HYD
2003	3.5 (3497)	1	0.039-0.043	—	①	46-49	—	600-800	HYD	HYD

NOTE: The Vehicle Emission Control Information label often reflects specification changes made during production. The label figures must be used if they differ from those in this chart

B: Before top dead center

HYD: Hydraulic

① Computer controled, no adjustment possible

42356-SEDO-C03

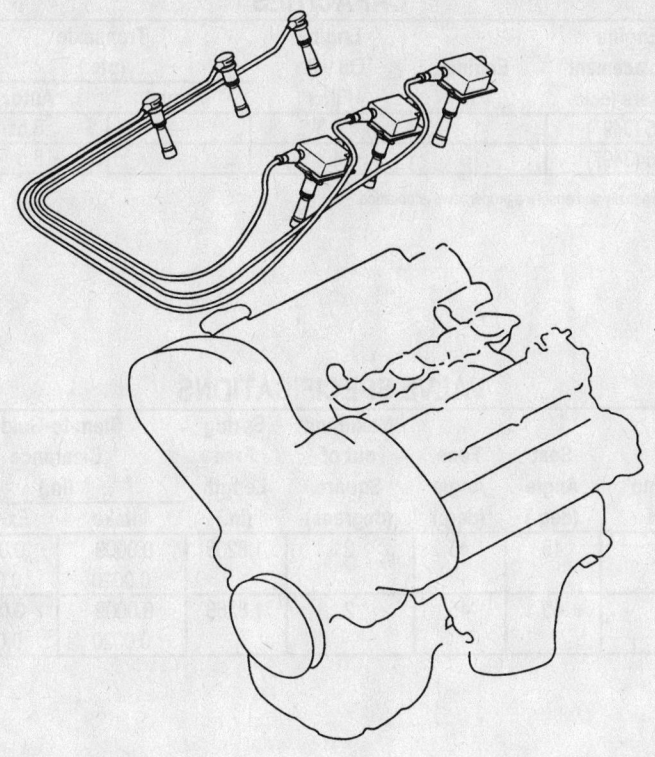

93581GY1

3.5L engine
Firing order: 1–2–3–4–5–6
Distributorless ignition system

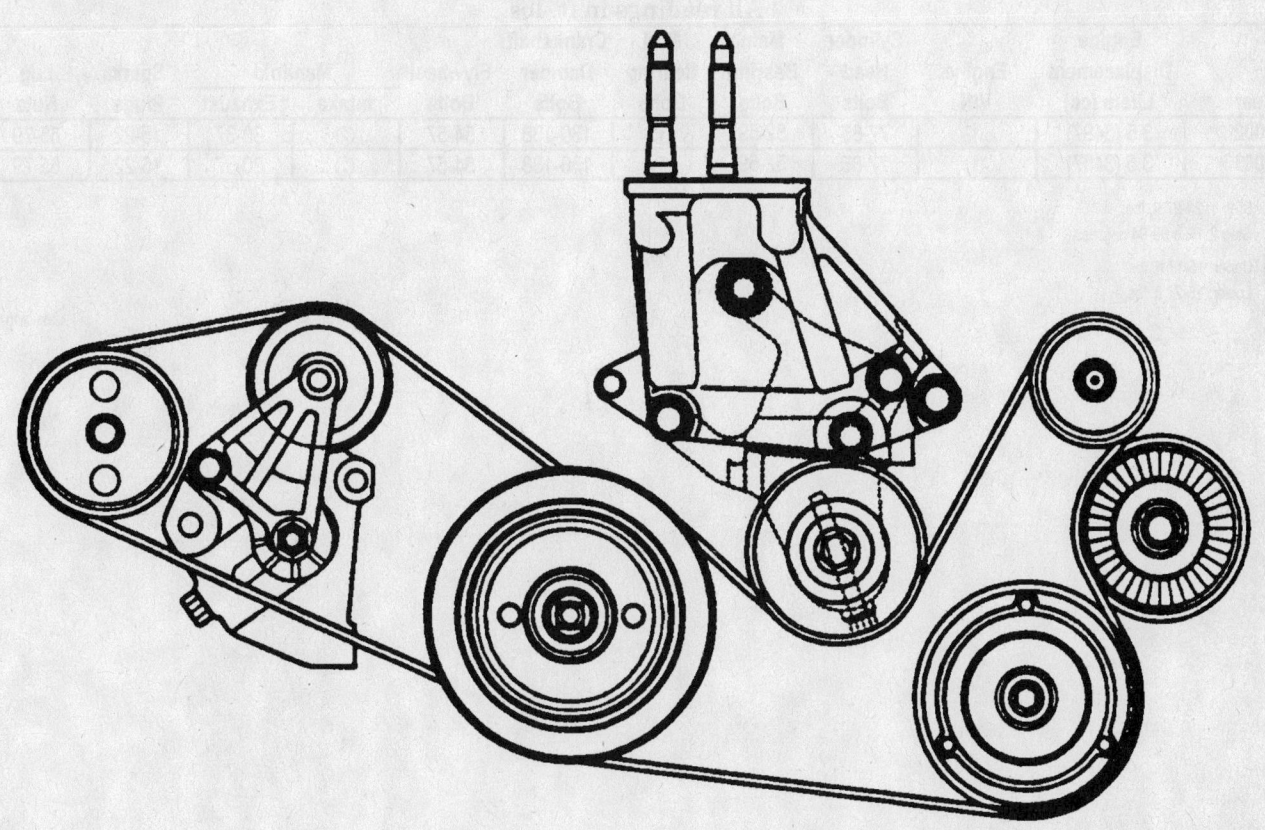

93581GY2

Accessory serpentine belt routing—3.5L engine

CAPACITIES

Year	Model	Engine Displacement Liters (cc)	Engine VIN	Engine Oil with Filter	Transaxle (pts.) Manual	Auto.	Fuel Tank (gal.)	Cooling System (qts.)
2002	Sedona	3.5 (3497)	1	4.3	—	8.5	19.8	8.2
2003	Sedona	3.5 (3497)	1	4.3	—	8.5	19.8	8.2

NOTE: All capacities are approximate. Add fluid gradually and ensure a proper level is obtained.

42356-SEDO-C04

VALVE SPECIFICATIONS

Year	Engine Displacement Liters (cc)	Engine VIN	Seat Angle (deg.)	Face Angle (deg.)	Maximum out of Square (degrees)	Spring Free Length (in.)	Stem-to-Guide Clearance (in.) Intake	Exhaust	Stem Diameter (in.) Intake	Exhaust
2002	3.5 (3497)	1	45	45	2	1.8268	0.0009-0.0020	0.0020-0.0039	0.258-0.259	0.257-0.258
2003	3.5 (3497)	1	45	45	2	1.8268	0.0009-0.0020	0.0020-0.0039	0.258-0.259	0.257-0.258

NA: Not Available

42356-SEDO-C05

TORQUE SPECIFICATIONS
All readings in ft. lbs.

Year	Engine Displacement Liters (cc)	Engine VIN	Cylinder Head Bolts	Main Bearing Bolts	Rod Bearing Bolts	Crankshaft Damper Bolts	Flywheel Bolts	Manifold Intake	Exhaust	Spark Plugs	Lug Nuts
2002	3.5 (3497)	1	77-85	52-59	①	130-138	54-57	②	30-37	15-22	65-79
2003	3.5 (3497)	1	77-85	52-59	①	130-138	54-57	②	30-37	15-22	65-79

① Step 1: 24-27 ft. lbs.
　Step 2: Plus 90-94 degrees

② Upper: 15-17 ft. lbs.
　Lower: 15-22 ft. lbs.

42356-SEDO-C06

PISTON AND RING SPECIFICATIONS

All measurements are given in inches.

Year	Engine Displacement Liters (cc)	Engine VIN	Piston Clearance	Ring Gap			Ring Side Clearance		
				Top Compression	Bottom Compression	Oil Control	Top Compression	Bottom Compression	Oil Control
2002	3.5 (3497)	1	0.0012-0.0020	0.0079-0.0118	0.0177-0.0236	0.0079-0.0276	0.0016-0.0315	0.0008-0.0024	SNUG
2003	3.5 (3497)	1	0.0012-0.0020	0.0079-0.0118	0.0177-0.0236	0.0079-0.0276	0.0016-0.0315	0.0008-0.0024	SNUG

42356-SEDO-C07

CRANKSHAFT AND CONNECTING ROD SPECIFICATIONS

All measurements are given in inches.

Year	Engine Displacement Liters (cc)	Engine VIN	Crankshaft				Connecting Rod		
			Main Brg. Journal Dia.	Main Brg. Oil Clearance	Shaft End-play	Thrust on No.	Journal Diameter	Oil Clearance	Side Clearance
2002	3.5 (3497)	1	2.5190-2.5197	0.0071-0.0140	0.0020-0.0100	3	2.1650-2.1654	0.0009-0.0016	0.0039-0.0098
2003	3.5 (3497)	1	2.5190-2.5197	0.0071-0.0140	0.0020-0.0100	3	2.1650-2.1654	0.0009-0.0016	0.0039-0.0098

42356-SEDO-C08

WHEEL ALIGNMENT

Year	Model		Caster		Camber		Toe-in (in.)	Steering Axis Inclination (Deg.)
			Range (+/-Deg.)	Preferred Setting (Deg.)	Range (+/-Deg.)	Preferred Setting (Deg.)		
2002	Sedona	F	0.50	+1.88	0.50	0.51	-0.04	—
		R	—	—	—	—	—	—
2003	Sedona	F	0.50	+1.88	0.50	0.51	-0.04	—
		R	—	—	—	—	—	—

42356-SEDO-C09

TIRE, WHEEL AND BALL JOINT SPECIFICATIONS

| Year | Model | OEM Tires | | Tire Pressures (psi) | | Wheel Size | Ball Joint Inspection |
		Standard	Optional	Front	Rear		
2002	Sedona	P215/70R15	—	35	35	6JJ	①
2003	Sedona	P215/70R15	—	35	35	6JJ	①

OEM: Original Equipment Manufacturer

PSI: Pounds Per Square Inch

STD: Standard

OPT: Optional

① Replace if any measurable movement is found.

42356-SEDO-C10

BRAKE SPECIFICATIONS
All measurements in inches unless noted

| Year | Model | | Brake Disc | | | Brake Drum | | | Minimum Lining Thickness | Brake Caliper | |
			Original Thickness	Minimum Thickness	Maximum Run-out	Original Inside Diameter	Max. Wear Limit	Maximum Machine Diameter		Bracket Bolts (ft. lbs.)	Mounting Bolts (ft. lbs.)
2002	Sedona	F	1.020	0.940	0.0020	—	—	—	0.100	7.2-9.4	18-26
		R	—	—	—	10.00	10.05	10.05	0.040	—	—
2003	Sedona	F	1.020	0.940	0.0020	—	—	—	0.100	7.2-9.4	18-26
		R	—	—	—	10.00	10.05	10.05	0.040	—	—

F: Front

R: Rear

42356-SEDO-C11

SCHEDULED MAINTENANCE INTERVALS
Kia—Sedona

TO BE SERVICED	TYPE OF SERVICE	VEHICLE MILEAGE INTERVAL (x1000)															
		7.5	15	22.5	30	37.5	45	52.5	60	67.5	75	82.5	90	97.5	105	112.5	120
Accessory drive belts	S/I			✔			✔			✔			✔			✔	
Air cleaner element	I/R		✔		✔		✔		✔		✔		✔		✔		✔
Air conditioner system	S/I	Inspect the system operation and refrigerant amount annually.															
Brake lines, hoses and connections	S/I		✔		✔		✔		✔		✔		✔		✔		✔
Chassis and body fasteners	T	✔	✔	✔	✔	✔	✔	✔	✔	✔	✔	✔	✔	✔	✔	✔	✔
Cooling system hoses and coolant level	S/I		✔		✔		✔		✔		✔		✔		✔		✔
CV-joint boots	S/I				✔				✔				✔				✔
Engine coolant	R				✔				✔				✔				✔
Engine oil and filter	R	✔	✔	✔	✔	✔	✔	✔	✔	✔	✔	✔	✔	✔	✔	✔	✔
Exhaust system heat shields	S/I				✔				✔				✔				✔
Front and rear brakes	S/I				✔				✔				✔				✔
Front ball joints S/I	S/I				✔				✔				✔				✔
Fuel filter	R								✔								✔
Fuel lines and hoses	S/I				✔				✔				✔				✔
Idle speed	A				✔				✔				✔				✔
Locks and hinges	L	✔	✔	✔	✔	✔	✔	✔	✔	✔	✔	✔	✔	✔	✔	✔	✔
Spark plugs	R				✔				✔				✔				✔
Steering operation and linkage	S/I				✔				✔				✔				✔
Timing belt	R								✔								✔

R: Replace   S/I: Inspect and service, if needed   L: Lubricate   A: Adjust   T: Tighten

FREQUENT OPERATION MAINTENANCE (SEVERE SERVICE)

If a vehicle is operated under any of the following conditions it is considered severe service

- Towing a trailer or using a camper or car-top carrier.

- Repeated short trips of less than 5 miles in temperatures below freezing, or trips of less than 10 miles in any temperature.

- Extensive idling or low-speed driving for long distances as in heavy commercial use, such as delivery, taxi or police cars.

- Operating on rough, muddy or salt-covered roads.

- Operating on unpaved or dusty roads.

- Driving in extremely hot (over 90°F) conditions.

Engine oil and filter: replace every 5000 miles or 5 months, whichever occurs first.

Air cleaner element: inspect ever 15,000 miles or 15 months and replace every 30,000 miles or 30 months, whichever occurs first.

Fuel system hoses (California models only): replace every 105,000 miles.

Emission system hoses (non-California models): inspect every 55,000 miles or 55 months, whichever occurs first.

Emission system hoses (California models): inspect every 60,000 miles or 60 months, whichever occurs first.

Front and rear disc brakes: inspect every 15,000 miles or 15 months, whichever occurs first.

Chassis and body fasteners: tighten every 15,000 miles or 15 months, whichever occurs first.

Locks and hinges: lubricate every 5000 miles or 5 months, whichever occurs first.

42356-SEDO-C12

PRECAUTIONS

Before servicing any vehicle, please be sure to read all of the following precautions, which deal with personal safety, prevention of component damage, and important points to take into consideration when servicing a motor vehicle:

• Never open, service or drain the radiator or cooling system when the engine is hot; serious burns can occur from the steam and hot coolant.

• Observe all applicable safety precautions when working around fuel. Whenever servicing the fuel system, always work in a well-ventilated area. Do not allow fuel spray or vapors to come in contact with a spark, open flame, or excessive heat (a hot drop light, for example). Keep a dry chemical fire extinguisher near the work area. Always keep fuel in a container specifically designed for fuel storage; also, always properly seal fuel containers to avoid the possibility of fire or explosion. Refer to the additional fuel system precautions later in this section.

• Fuel injection systems often remain pressurized, even after the engine has been turned **OFF**. The fuel system pressure must be relieved before disconnecting any fuel lines. Failure to do so may result in fire and/or personal injury.

• Brake fluid often contains polyglycol ethers and polyglycols. Avoid contact with the eyes and wash your hands thoroughly after handling brake fluid. If you do get brake fluid in your eyes, flush your eyes with clean, running water for 15 minutes. If eye irritation persists, or if you have taken

brake fluid internally, IMMEDIATELY seek medical assistance.

• The EPA warns that prolonged contact with used engine oil may cause a number of skin disorders, including cancer. You should make every effort to minimize your exposure to used engine oil. Protective gloves should be worn when changing oil. Wash your hands and any other exposed skin areas as soon as possible after exposure to used engine oil. Soap and water, or waterless hand cleaner should be used.

• All new vehicles are now equipped with an air bag system, often referred to as a Supplemental Restraint System (SRS) or Supplemental Inflatable Restraint (SIR) system. The system must be disabled before performing service on or around system components, steering column, instrument panel components, wiring and sensors. Failure to follow safety and disabling procedures could result in accidental air bag deployment, possible personal injury and unnecessary system repairs.

• Always wear safety goggles when working with, or around, the air bag system. When carrying a non-deployed air bag, be sure the bag and trim cover are pointed away from your body. When placing a non-deployed air bag on a work surface, always face the bag and trim cover upward, away from the surface. This will reduce the motion of the module if it is accidentally deployed. Refer to the additional air bag system precautions later in this section.

• Clean, high quality brake fluid from a sealed container is essential to the safe and

proper operation of the brake system. You should always buy the correct type of brake fluid for your vehicle. If the brake fluid becomes contaminated, completely flush the system with new fluid. Never reuse any brake fluid. Any brake fluid that is removed from the system should be discarded. Also, do not allow any brake fluid to come in contact with a painted surface; it will damage the paint.

• Never operate the engine without the proper amount and type of engine oil; doing so WILL result in severe engine damage.

• Timing belt maintenance is extremely important. Many models utilize an interference-type, non-freewheeling engine. If the timing belt breaks, the valves in the cylinder head may strike the pistons, causing potentially serious (also time-consuming and expensive) engine damage. Refer to the maintenance interval charts for the recommended replacement interval for the timing belt.

• Disconnecting the negative battery cable on some vehicles may interfere with the functions of the on-board computer system(s) and may require the computer to undergo a relearning process once the negative battery cable is reconnected.

• When servicing drum brakes, only disassemble and assemble one side at a time, leaving the remaining side intact for reference.

• Only an MVAC-trained, EPA-certified automotive technician should service the air conditioning system or its components.

ENGINE REPAIR

Alternator

REMOVAL

1. Before servicing the vehicle, refer to the precautions in the beginning of this section.

2. Remove or disconnect the following:
   • Negative battery cable
   • Accessory drive belt
   • Alternator mounting bolts
   • Alternator

To install:

3. Install or connect the following:
   • Alternator
   • Alternator electrical connectors. Tighten the battery terminal connector nut to 60 inch lbs. (7 Nm).

• Accessory drive belt
• Negative battery cable

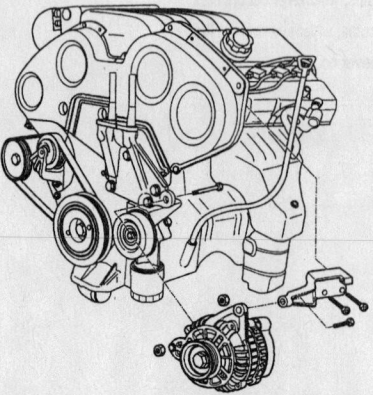

9358HG01

Alternator mounting exploded view

Ignition Timing

This vehicle is equipped with a Distributorless Ignition System (DIS). No adjustment is necessary or possible.

Engine Assembly

REMOVAL & INSTALLATION

1. Before servicing the vehicle, refer to the precautions in the beginning of this section.

2. Drain the cooling system.
3. Drain the transaxle fluid.
4. Relieve the fuel system pressure.
5. Remove or disconnect the following:

   • Battery

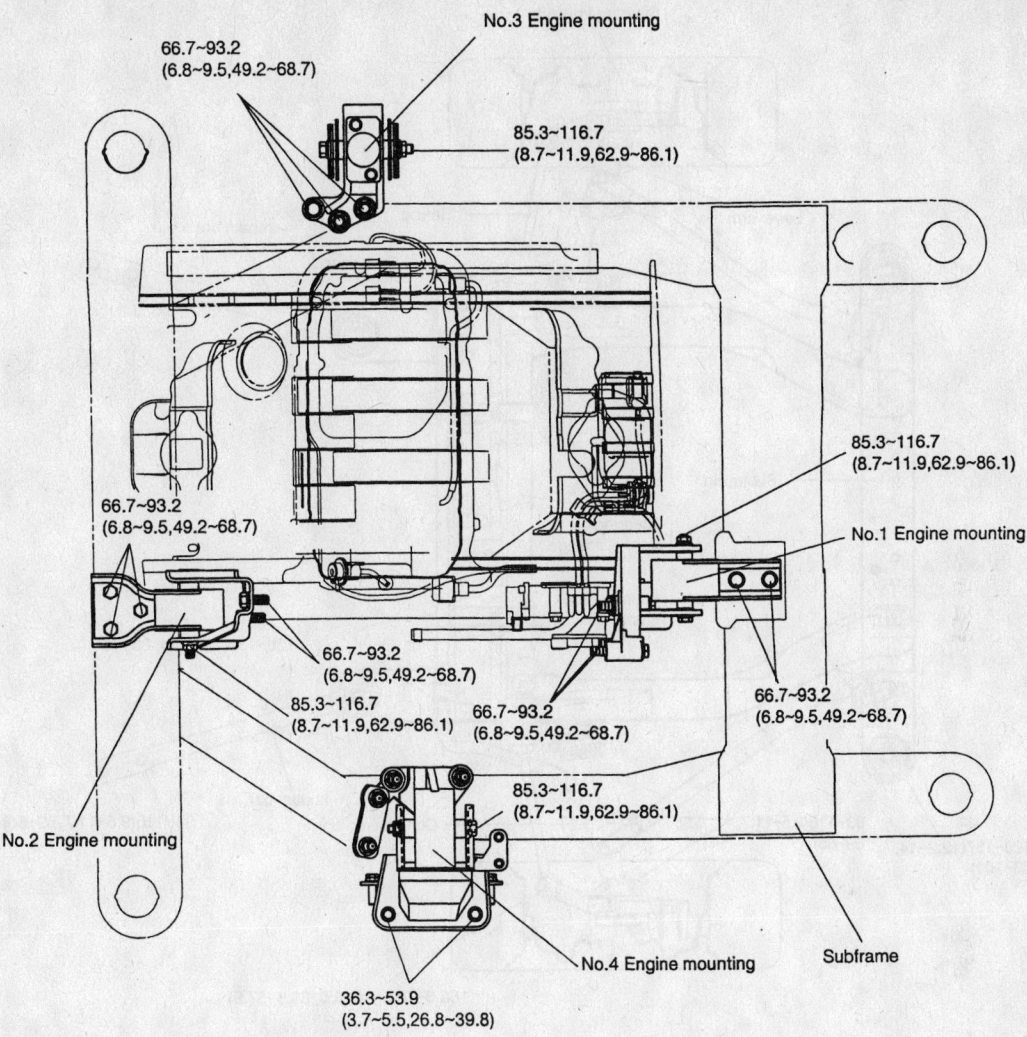

No.3 Engine mounting

66.7~93.2
(6.8~9.5,49.2~68.7)

85.3~116.7
(8.7~11.9,62.9~86.1)

85.3~116.7
(8.7~11.9,62.9~86.1)

No.1 Engine mounting

66.7~93.2
(6.8~9.5,49.2~68.7)

66.7~93.2
(6.8~9.5,49.2~68.7)

85.3~116.7
(8.7~11.9,62.9~86.1)

66.7~93.2
(6.8~9.5,49.2~68.7)

No.2 Engine mounting

85.3~116.7
(8.7~11.9,62.9~86.1)

No.4 Engine mounting

Subframe

36.3~53.9
(3.7~5.5,26.8~39.8)

TORQUE : N•m (kg•m, lb•ft)

9358HG02

Engine mount locations and torque specifications

- Engine cover
- Air cleaner assembly
- Engine wiring harness connectors
- Alternator wiring harness connectors
- Oil pressure switch connector
- Oil pressure sensor connector
- Starter motor wiring harness connectors
- Transaxle oil cooler hose
- Upper and lower radiator hoses
- Radiator
- Engine ground cable
- Brake booster vacuum hose

- EVAP canister hose
- Fuel lines
- Heater hoses
- Throttle and cruise control cables
- Transaxle shift cable
- Power steering pump hose
- Oil pan shield
- Exhaust front pipe
- Outer tie rod ends
- Sway bar links
- Lower ball joints
- Axle halfshafts
- Intermediate shaft bolt
- No. 3 and 4 engine mounts

6. Support the front subframe with a suitable powertrain jack.
- Subframe bolts
- Impact bar bolts

7. Lower the powertrain and subframe away from the vehicle.

To install:

8. Installation is the reverse of the removal procedure, while using the following torque values:
- Subframe bolts: 88-101 ft. lbs. (120-137 Nm)
- Impact bar bolts: 69-85 ft. lbs. (93-115 Nm)

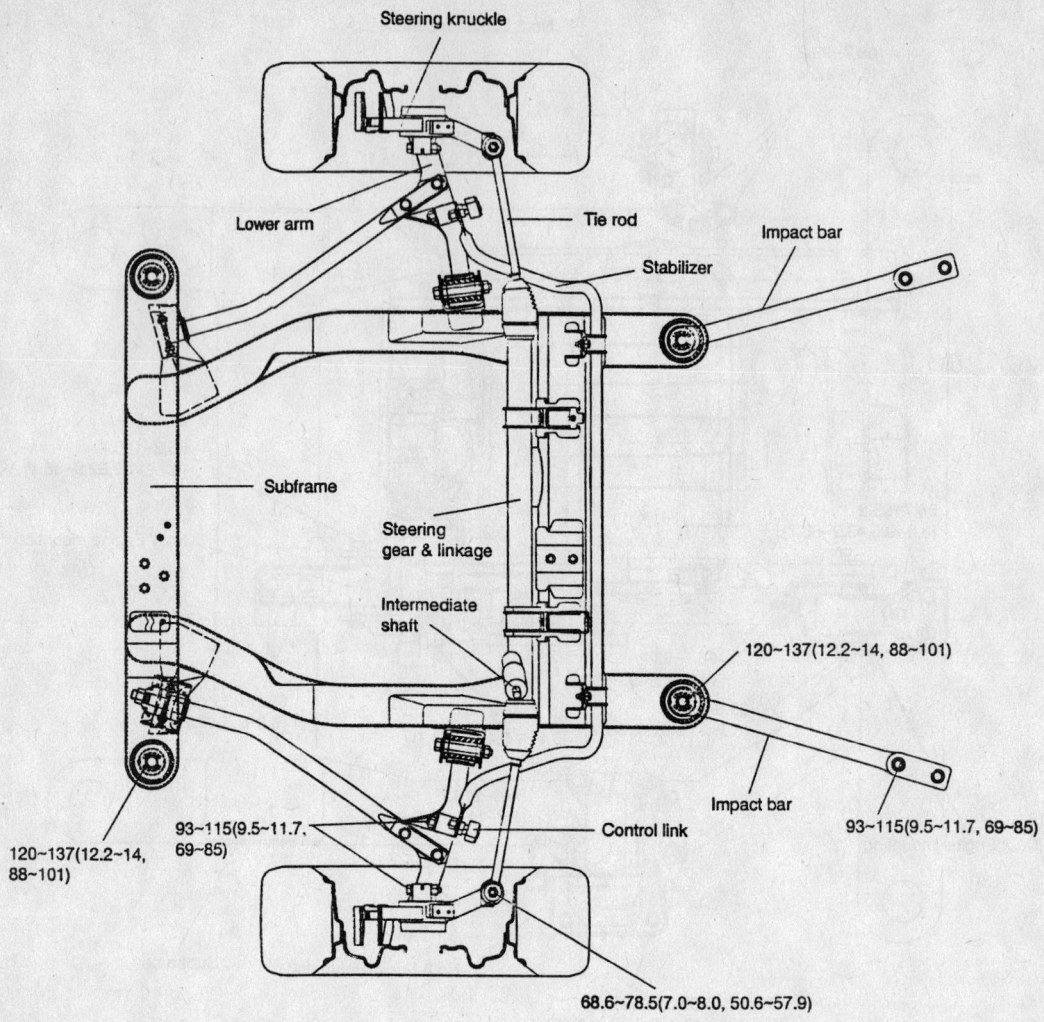

Steering knuckle

Lower arm

Tie rod

Impact bar

Stabilizer

Subframe

Steering gear & linkage

Intermediate shaft

120~137(12.2~14, 88~101)

Impact bar

Control link

93~115(9.5~11.7, 69~85)

120~137(12.2~14, 88~101)

93~115(9.5~11.7, 69~85)

68.6~78.5(7.0~8.0, 50.6~57.9)

TORQUE : N•m (kg•m, lb•ft)

9358HG03

Front subframe bolt locations and torque specifications

- Engine mount bolts: 49-69 ft. lbs. (67-93 Nm)
- Engine mount through-bolts: 63-86 ft. lbs. (85-117 Nm)
- Tie rod ends: 51-58 ft. lbs. (69-79 Nm)

Water Pump

REMOVAL & INSTALLATION

1. Before servicing the vehicle, refer to the precautions in the beginning of this section.
2. Drain the cooling system.
3. Remove or disconnect the following:
   - Negative battery cable
   - Engine cover
   - Accessory drive belt
   - Idler pulley
   - Crankshaft pulley

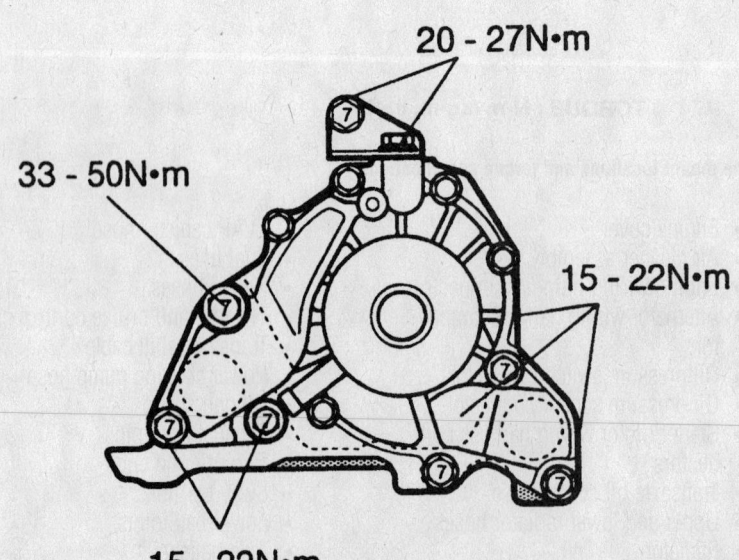

20 - 27N•m

33 - 50N•m

15 - 22N•m

15 - 22N•m

9358HG04

Water pump bolt locations and torque specifications

- Power steering pump pulley
- Tensioner pulley
- Upper and lower timing belt covers
- No. 3 engine mount
- Timing belt
- Timing belt tensioner
- Water pump

To install:
4. Installation is the reverse of the removal procedure, while using the following torque values:
- Water pump bolts: refer to the illustration
- Crankshaft pulley: 130-138 ft. lbs. (180-190 Nm)

REMOVAL & INSTALLATION

1. Before servicing the vehicle, refer to the precautions in the beginning of this section.

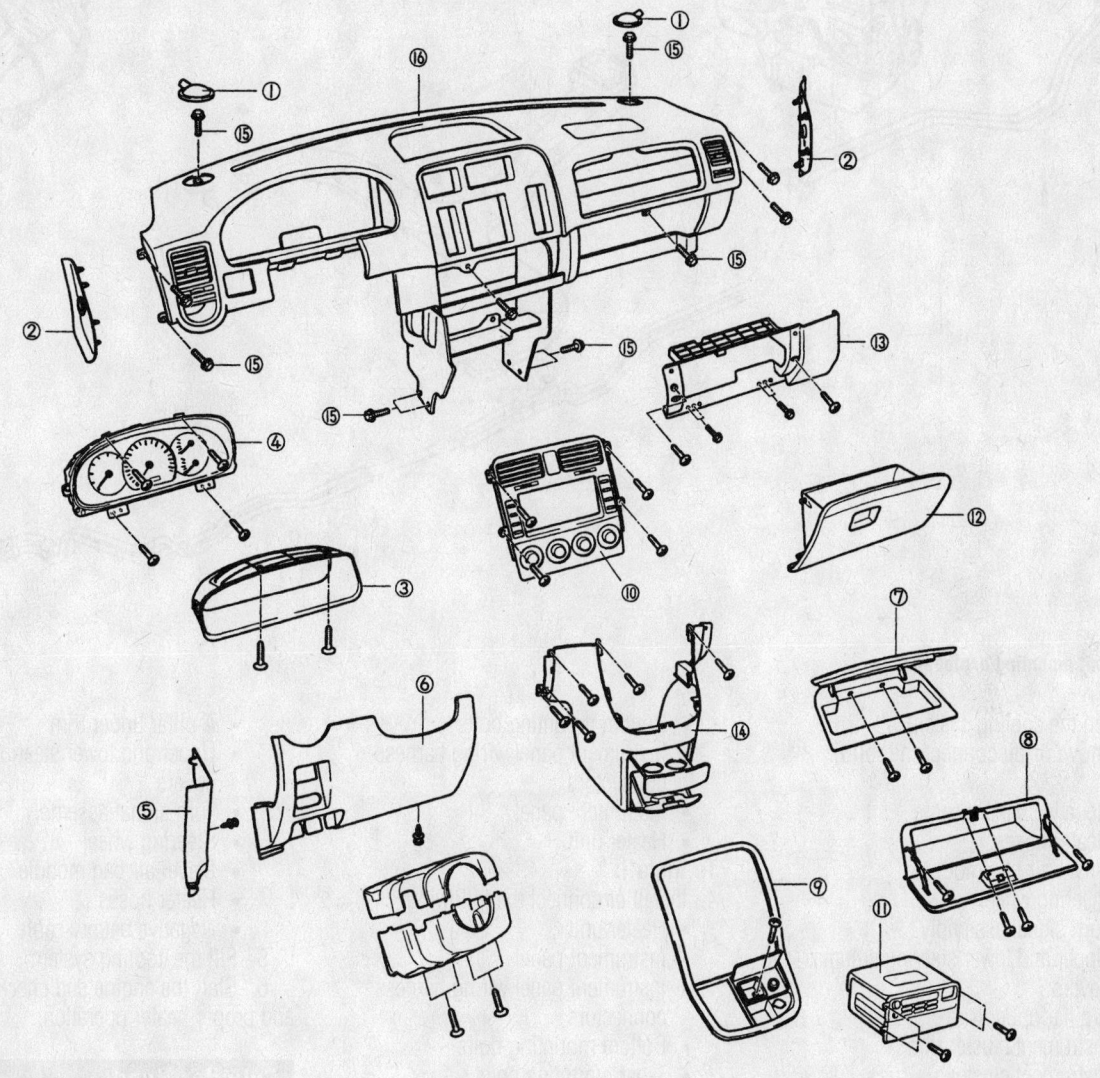

(1) Speaker assembly
(2) Side cover
(3) Instrument cluster trim
(4) Instrument cluster
(5) A-pillar lower trim
(6) Lower LH panel
(7) Center upper tray
(8) Multi box

(9) Audio panel
(10) Heater control panel
(11) Audio
(12) Glove box
(13) Lower RH panel
(14) Center console
(15) Mounting bolt
(16) Instrument panel

9358HG29

Instrument panel exploded view

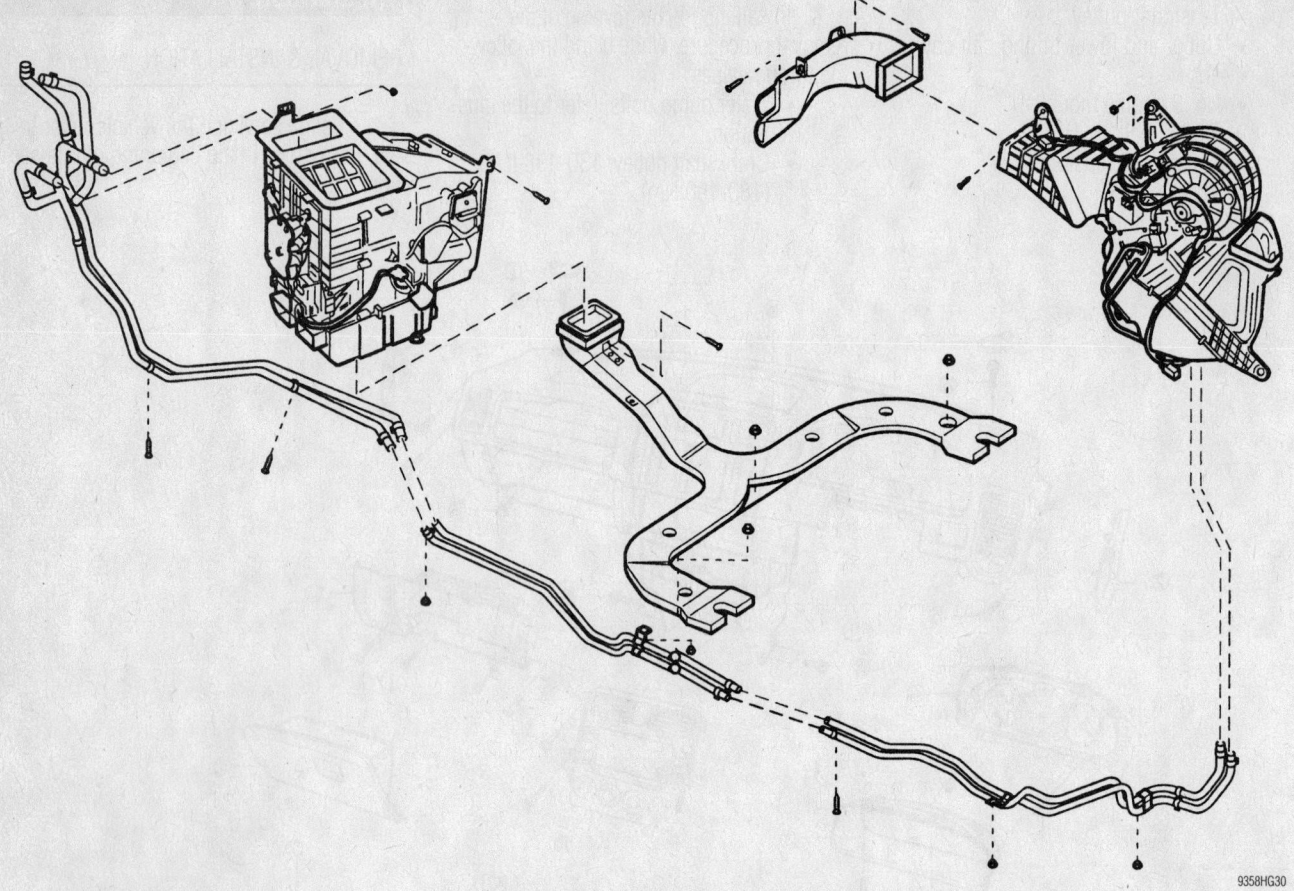

9358HG30

Heater unit mounting exploded view

2. Drain the cooling system.
3. Remove or disconnect the following:
  - Negative battery cable
  - Heater hoses
  - Driver air bag module
  - Steering wheel
  - Turn signal assembly
  - Upper and lower steering column covers
  - A pillar upper trim
  - Instrument cluster trim
  - Instrument cluster
  - A pillar lower trim
  - Hood release handle
  - Left lower trim panel
  - Multi box
  - Audio panel
  - Ventilation control panel
  - Radio
  - Glove box
  - Center console
  - Audio system speakers and mounting bolts from the top of the instrument panel
  - Side covers
  - T-bar mounting bolts

- Bottom mounting bolts
- Instrument panel wiring harness connectors
- Instrument panel
- Heater unit

To install:
4. Install or connect the following:
  - Heater unit
  - Instrument panel
  - Instrument panel wiring harness connectors
  - Bottom mounting bolts
  - T-bar mounting bolts
  - Side covers
  - Upper mounting bolts and audio system speakers
  - Center console
  - Glove box
  - Radio
  - Ventilation control panel
  - Audio panel
  - Multi box
  - Left lower trim panel
  - Hood release handle
  - A pillar lower trim
  - Instrument cluster
  - Instrument cluster trim

- A pillar upper trim
- Upper and lower steering column covers
- Turn signal assembly
- Steering wheel
- Driver air bag module
- Heater hoses
- Negative battery cable
5. Fill the cooling system.
6. Start the engine and check for leaks and proper heater operation.

Cylinder Head

REMOVAL & INSTALLATION

1. Before servicing the vehicle, refer to the precautions in the beginning of this section.
2. Drain the cooling system.
3. Relieve the fuel system pressure.
4. Remove or disconnect the following:
  - Negative battery cable
  - Engine cover
  - Timing belt
  - Intake manifold
  - Exhaust manifolds

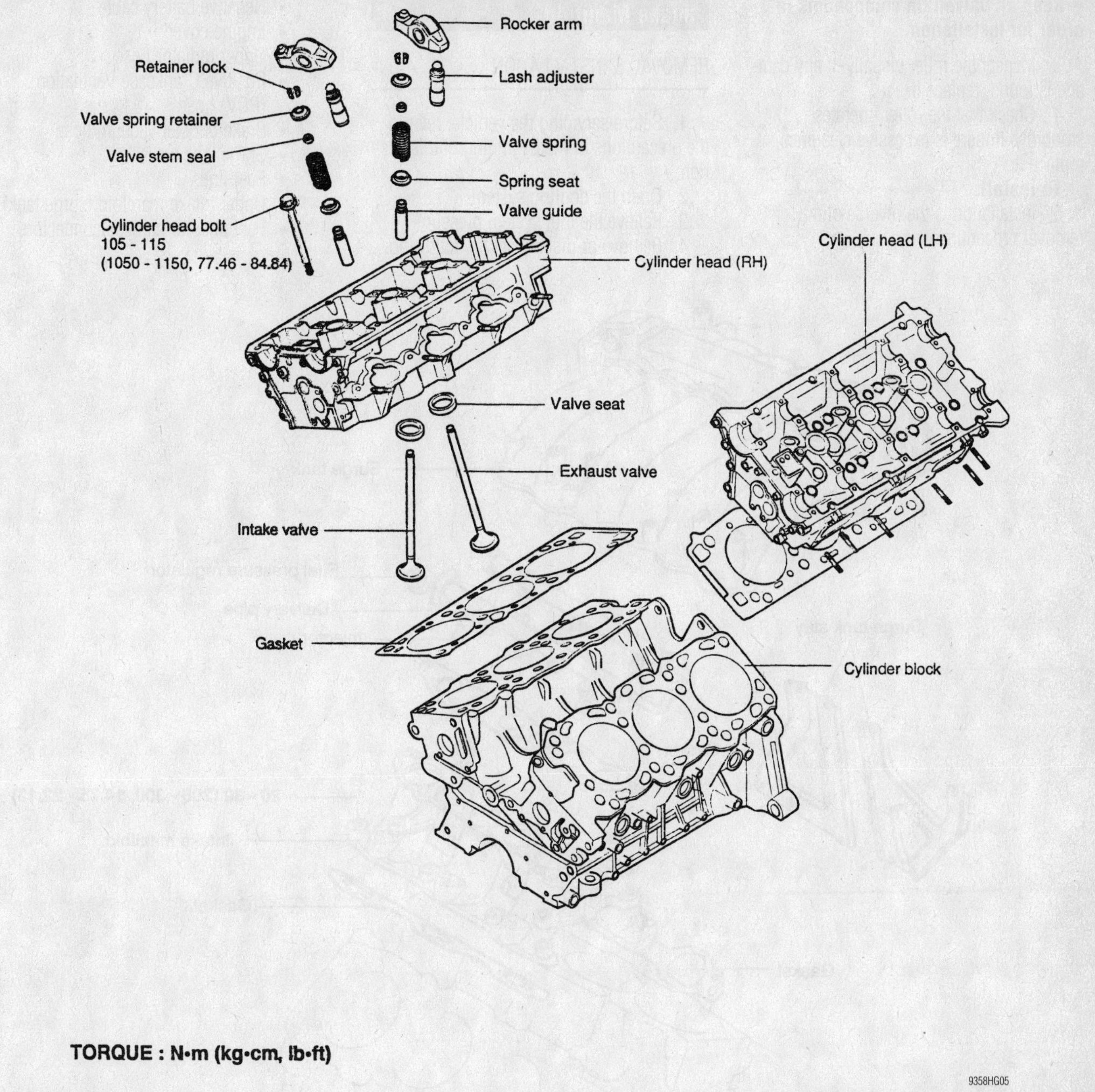

Rocker arm

Lash adjuster

Retainer lock

Valve spring retainer

Valve stem seal

Valve spring

Spring seat

Valve guide

Cylinder head bolt
105 - 115
(1050 - 1150, 77.46 - 84.84)

Cylinder head (RH)

Cylinder head (LH)

Valve seat

Exhaust valve

Intake valve

Gasket

Cylinder block

TORQUE : N•m (kg•cm, lb•ft)

9358HG05

Cylinder head exploded view

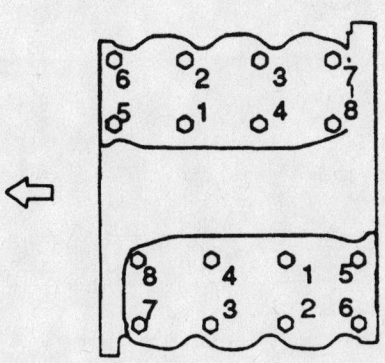

9358HG06

Cylinder head bolt torque sequence

- Cylinder head covers
- Camshafts
- Rocker arms and lash adjusters

➥**Keep all valvetrain components in order for assembly.**

- Cylinder head bolts
- Cylinder heads

To install:

5. Install the cylinder heads with new gaskets. Tighten the bolts in sequence to 78-85 ft. lbs. (105-115 Nm).

6. Installation is the reverse of the removal procedure.

Rocker Arms

REMOVAL & INSTALLATION

1. Before servicing the vehicle, refer to the precautions in the beginning of this section.

2. Remove or disconnect the following:
- Negative battery cable
- Engine cover
- Timing belt
- Cylinder head covers
- Camshafts
- Rocker arms and lash adjusters

➡**Keep all valvetrain components in order for installation.**

3. Inspect the roller visually. If any damage is found, replace it.

4. Check that the roller operates smoothly. If there is excessive clearance, replace it.

To install:

5. Installation is the reverse of the removal procedure.

Intake Manifold

REMOVAL & INSTALLATION

1. Before servicing the vehicle, refer to the precautions in the beginning of this section.

2. Drain the cooling system.

3. Relieve the fuel system pressure.

4. Remove or disconnect the following:

- Negative battery cable
- Engine cover
- Upper radiator hose
- Positive Crankcase Ventilation (PCV) hose
- Brake booster vacuum hose
- Surge tank stays
- Fuel lines
- Upper intake manifold (surge tank)
- Fuel injector harness connectors

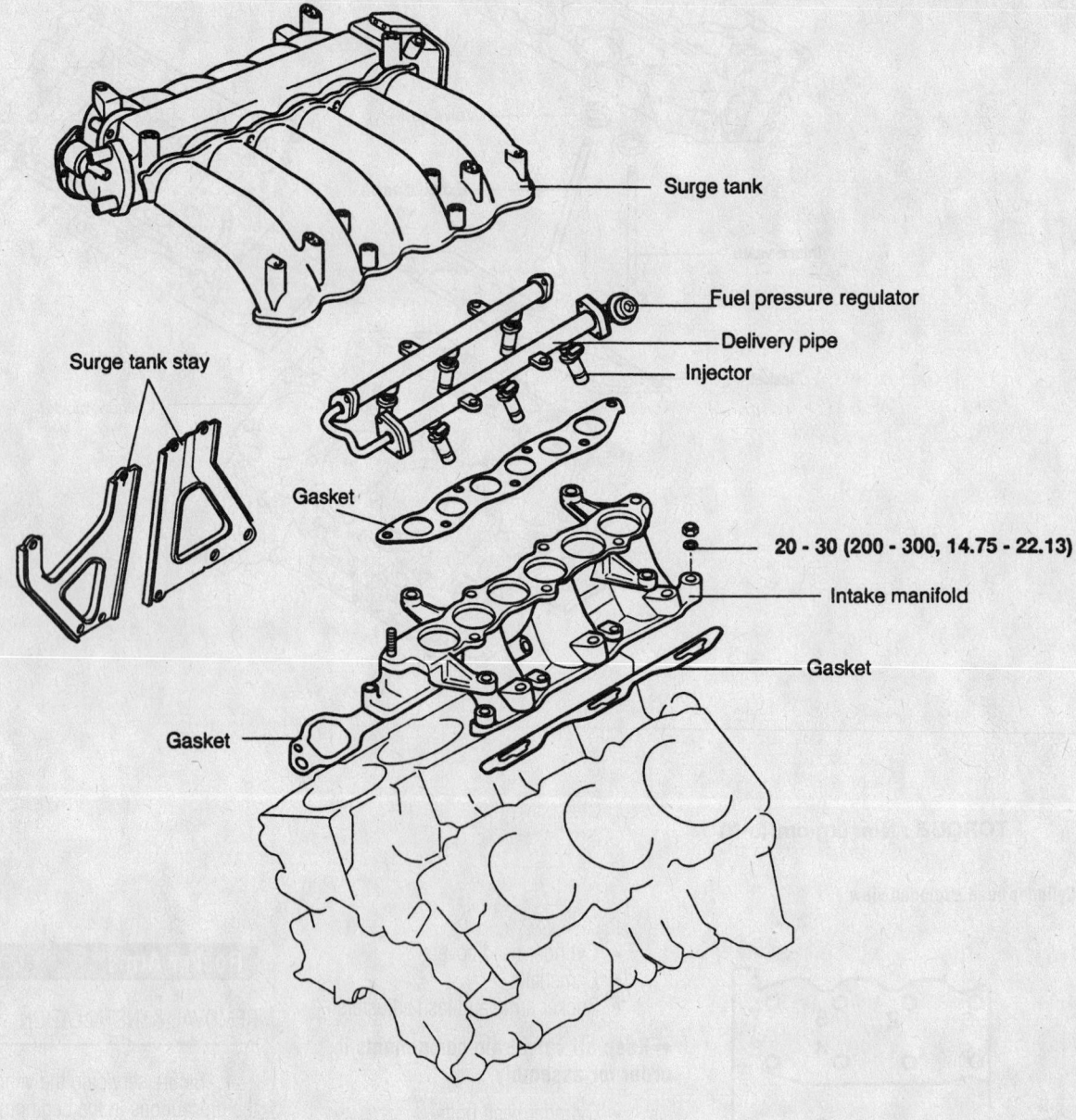

Surge tank

Fuel pressure regulator

Delivery pipe

Injector

Surge tank stay

Gasket

20 - 30 (200 - 300, 14.75 - 22.13)

Intake manifold

Gasket

Gasket

TORQUE : N•m (kg•cm, lb•ft)

Intake manifold exploded view

9358HG07

- Fuel supply manifold with injectors attached
- Engine Coolant Temperature (ECT) sensor connector
- Coolant temperature gauge sensor connector
- Thermostat
- Lower intake manifold

To install:

5. Installation is the reverse of the removal procedure, while using the following torque values:
- Lower intake manifold nuts: 15-22 ft. lbs. (20-30 Nm)

- Upper intake manifold bolts: 11-15 ft. lbs. (15-20 Nm)

Exhaust Manifold

REMOVAL & INSTALLATION

1. Before servicing the vehicle, refer to the precautions in the beginning of this section.
2. Remove or disconnect the following:
- Heated Oxygen (HO₂S) sensor connectors
- Exhaust Y pipe

- Exhaust manifold heat shields
- Exhaust manifolds

To install:

➡**Use only new gaskets and nuts for assembly.**

3. Install or connect the following:
- Exhaust manifolds with new gaskets. Tighten the fasteners to 30-37 ft. lbs. (40-50 Nm)
- Exhaust manifold heat shields. Tighten the bolts to 106-132 inch lbs. (12-15 Nm)
- Exhaust Y pipe

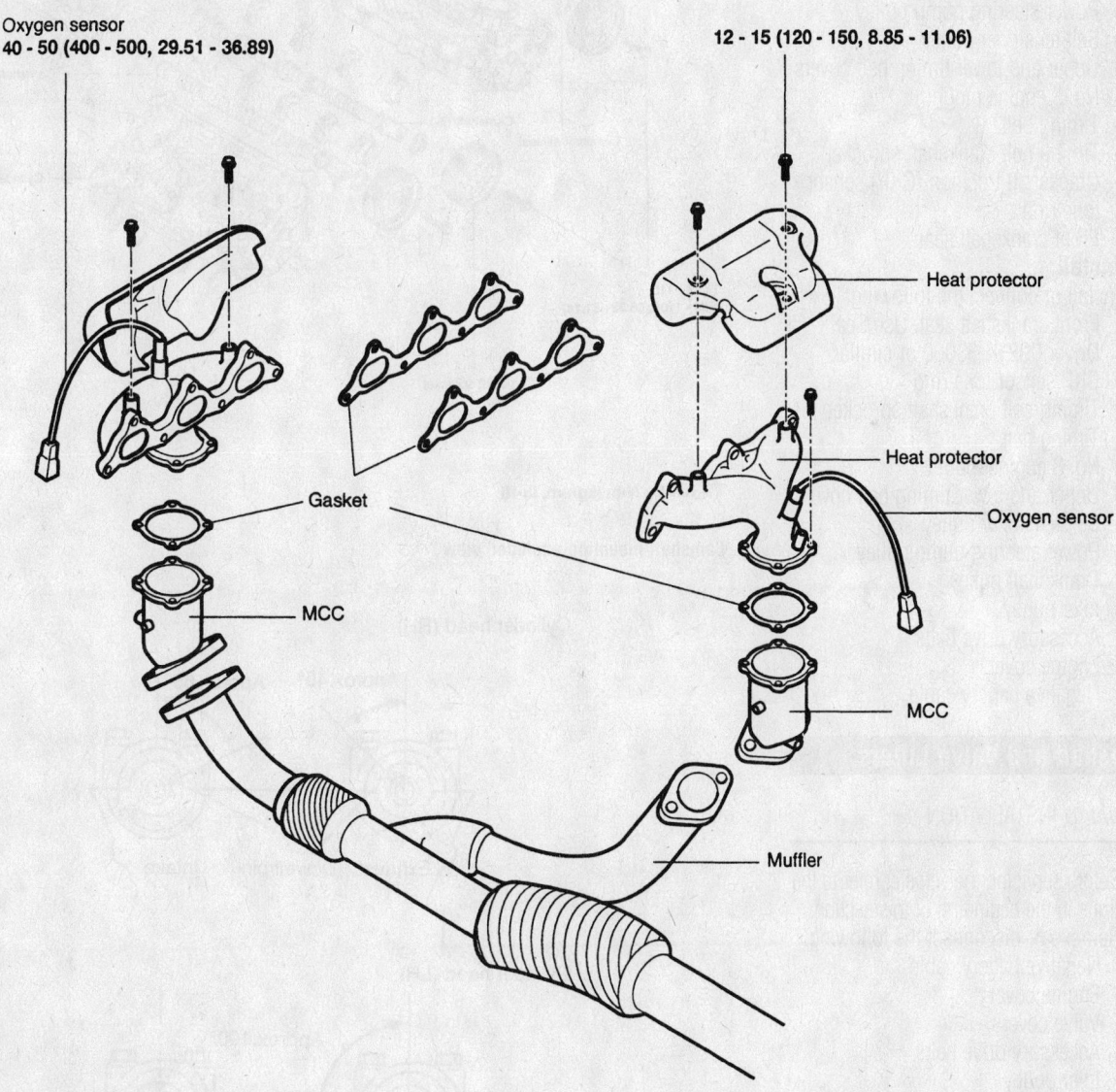

Oxygen sensor
40 - 50 (400 - 500, 29.51 - 36.89)

12 - 15 (120 - 150, 8.85 - 11.06)

Heat protector

Heat protector

Oxygen sensor

Gasket

MCC

MCC

Muffler

TORQUE : N•m (kg•cm, lb•ft)

Exhaust manifold mounting exploded view

9358HG08

- Heated Oxygen (HO2S) sensor connectors

Front Crankshaft Seal

REMOVAL & INSTALLATION

1. Before servicing the vehicle, refer to the precautions in the beginning of this section.
2. Remove or disconnect the following:
   - Negative battery cable
   - Engine cover
   - Accessory drive belts
   - Idler pulley
   - Crankshaft pulley
   - Power steering pump pulley
   - Belt tensioner pulley
   - Upper and lower timing belt covers
   - No. 3 engine mount
   - Timing belt
   - Timing belt crankshaft sprocket
   - Crankshaft Position (CKP) sensor tone ring
   - Front crankshaft seal

To install:
3. Install or connect the following:
   - Front crankshaft seal. Use Seal Driver 09214-33000 or similar.
   - CKP sensor tone ring
   - Timing belt crankshaft sprocket
   - Timing belt
   - No. 3 engine mount
   - Upper and lower timing belt covers
   - Belt tensioner pulley
   - Power steering pump pulley
   - Crankshaft pulley
   - Idler pulley
   - Accessory drive belts
   - Engine cover
   - Negative battery cable

Camshaft and Valve Lifters

REMOVAL & INSTALLATION

1. Before servicing the vehicle, refer to the precautions in the beginning of this section.
2. Remove or disconnect the following:
   - Negative battery cable
   - Engine cover
   - Valve covers
   - Accessory drive belts
   - Idler pulley
   - Crankshaft pulley
   - Power steering pump pulley
   - Belt tensioner pulley
   - Upper and lower timing belt covers
   - No. 3 engine mount
   - Timing belt
   - Camshaft sprockets

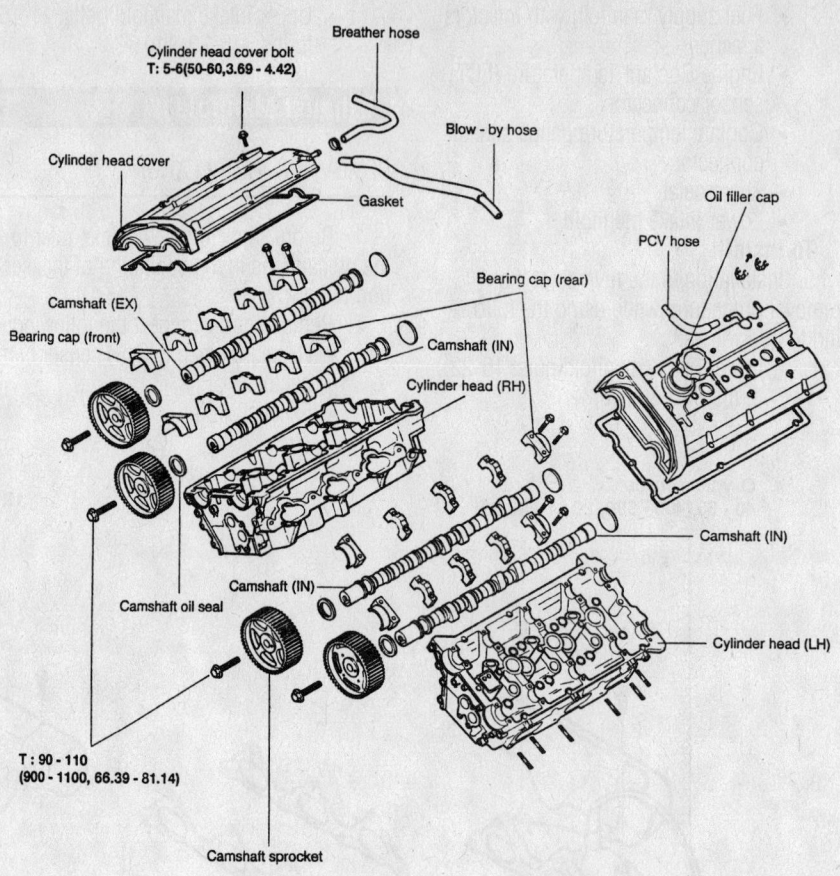

Cylinder head cover bolt
T: 5-6(50-60,3.69 - 4.42)

Breather hose
Blow - by hose
Oil filler cap
PCV hose
Cylinder head cover
Gasket
Camshaft (EX)
Bearing cap (rear)
Bearing cap (front)
Camshaft (IN)
Cylinder head (RH)
Camshaft oil seal
Camshaft (IN)
Camshaft (IN)
Cylinder head (LH)
T : 90 - 110
(900 - 1100, 66.39 - 81.14)
Camshaft sprocket

TORQUE : N·m (kg·cm, lb·ft)

Camshaft mounting exploded view

9358HG09

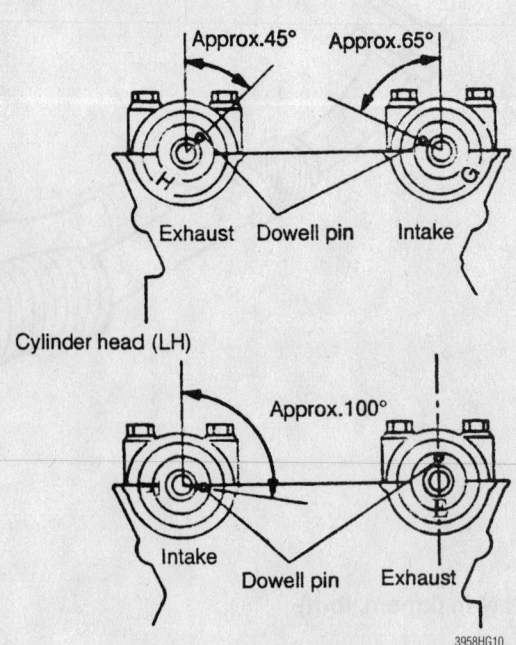

Cylinder head (RH)

Approx.45°   Approx.65°

H   Exhaust   Dowell pin   Intake   G

Cylinder head (LH)

Approx.100°

Intake   Dowell pin   Exhaust

3958HG10

Camshaft installation alignment

➡**Keep all valvetrain components in order for assembly.**

- Front bearing caps
- Rear bearing caps
- Center bearing caps
- Camshafts
- Rocker arms
- Hydraulic lifters

To install:

3. Set the No. 1 cylinder to Top Dead Center of the compression stroke.

4. Install the lifters and rocker arms in their original positions.

5. Install the camshafts aligned according to the illustration.

6. Install the bearing caps. Tighten the bolts evenly in several passes to the following torque specifications:

- Front and rear bearing caps: 16-19 ft. Lbs. (21-26 Nm)
- Center bearing caps: 88-106 inch lbs. (10-12 Nm)

7. Install or connect the following:

- Camshaft sprockets. Tighten the bolts to 67-81 ft. Lbs. (90-110 Nm).
- Timing belt
- No. 3 engine mount
- Upper and lower timing belt covers
- Belt tensioner pulley
- Power steering pump pulley
- Crankshaft pulley
- Idler pulley
- Accessory drive belts
- Valve covers. Tighten the bolts to 44-53 inch lbs. (5-6 Nm).
- Engine cover
- Negative battery cable

Valve Lash

ADJUSTMENT

This vehicle is equipped with hydraulic valve lifters. No adjustment is necessary.

Starter Motor

REMOVAL & INSTALLATION

1. Before servicing the vehicle, refer to the precautions in the beginning of this section.

2. Remove or disconnect the following:

- Negative battery cable
- Shift cable
- Starter motor electrical connectors
- Starter heat shield
- Starter motor

To install:

3. Install or connect the following:

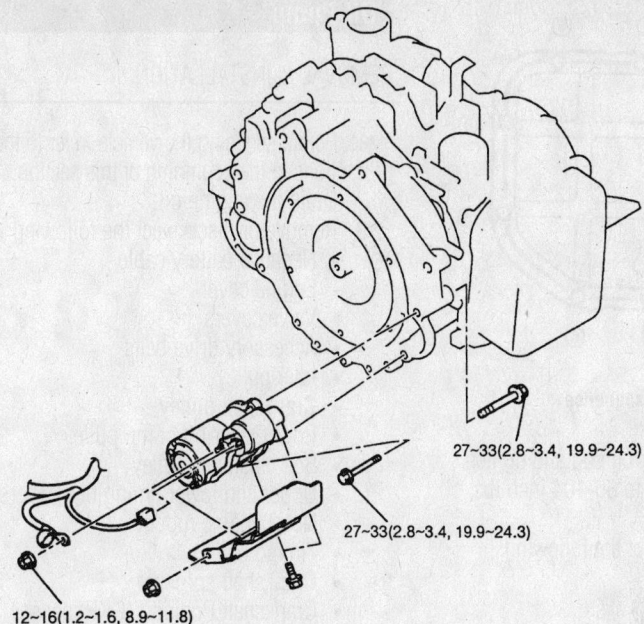

27~33(2.8~3.4, 19.9~24.3)

27~33(2.8~3.4, 19.9~24.3)

12~16(1.2~1.6, 8.9~11.8)

N•m(kg•m, lb•ft)

9358HG11

Starter motor mounting exploded view

- Starter motor. Tighten the bolts to 20-24 ft. Lbs. (27-33 Nm).
- Starter heat shield
- Starter motor electrical connectors. Tighten the battery terminal nut to 106-141 inch lbs. (12-16 Nm).
- Shift cable
- Negative battery cable

Oil Pan

REMOVAL & INSTALLATION

1. Before servicing the vehicle, refer to the precautions in the beginning of this section.

2. Drain the engine oil.

3. Remove or disconnect the following:

- Negative battery cable
- Starter motor
- Oil filter
- Lower oil pan
- Upper oil pan

To install:

4. Apply silicone sealant to the grove of the oil pan flange.

5. Install the upper oil pan and tighten the bolts in sequence as follows:

a. Bolts 1-14: 14-20 ft. lbs. (19-28 Nm)

b. Bolts 15 and 16: 44-62 inch lbs. (5-7 Nm)

c. Upper oil pan-to-transaxle mounting bolts: 22-33 ft. lbs. (30-42 Nm)

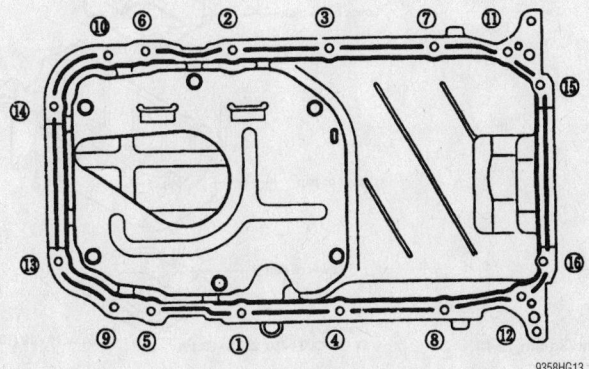

9358HG13

Upper oil pan torque sequence

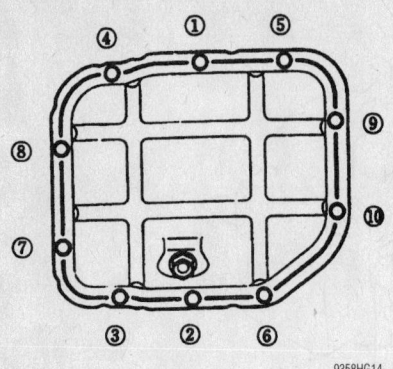

Lower oil pan torque sequence

9358HG14

6. Install the lower oil pan and tighten the bolts in sequence to 86-104 inch lbs. (10-12 Nm).

7. Install or connect the following:
- Oil filter
- Starter motor
- Negative battery cable

8. Fill the crankcase to the correct level with engine oil.

9. Start the engine and check for leaks.

Oil Pump

REMOVAL & INSTALLATION

1. Before servicing the vehicle, refer to the precautions in the beginning of this section.

2. Drain the engine oil.

3. Remove or disconnect the following:
- Negative battery cable
- Engine cover
- Valve covers
- Accessory drive belts
- Idler pulley
- Crankshaft pulley
- Power steering pump pulley
- Belt tensioner pulley
- Upper and lower timing belt covers
- No. 3 engine mount
- Timing belt
- Crankshaft sprocket
- Crankshaft Position (CKP) sensor tone ring
- Oil filter
- Starter motor

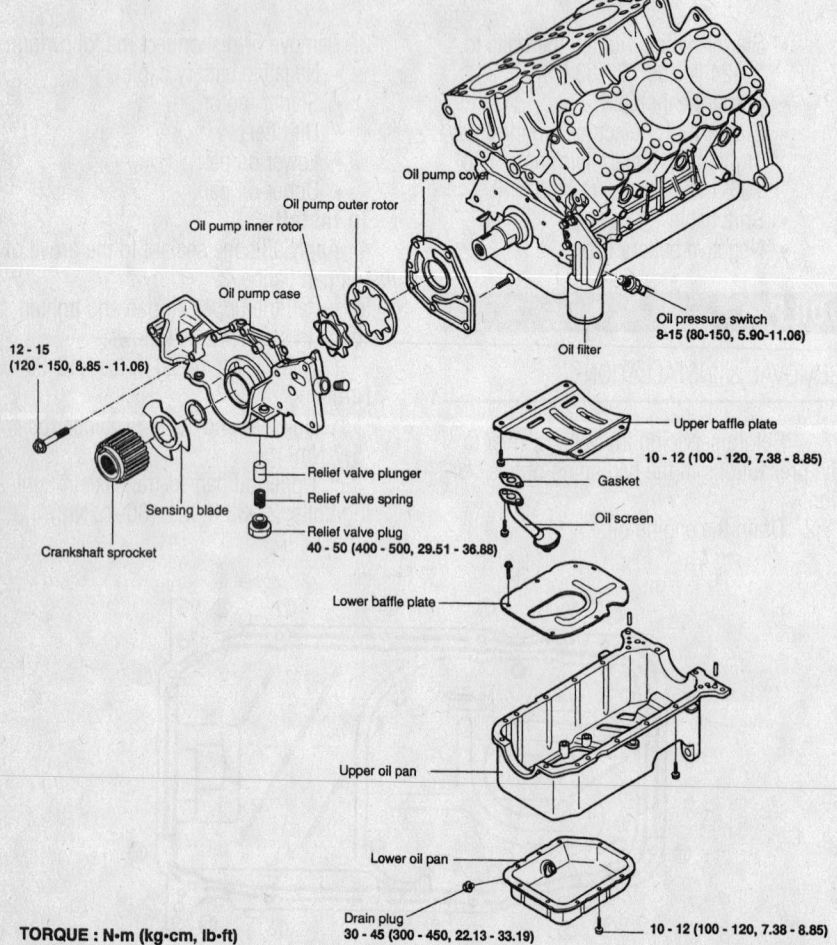

Oil pump and oil pan mounting exploded view

TORQUE : N·m (kg·cm, lb·ft)

9358HG12

- Lower oil pan
- Upper oil pan
- Oil pick-up tube
- Oil filter bracket
- Oil relief valve plug
- Oil pump

To install:

4. Install the oil pump with a new gasket. Tighten the oil pump case bolts to 106-133 inch lbs. (12-15 Nm). Tighten the oil pump cover screws to 71-106 inch lbs. (8-12 Nm).

5. Install or connect the following:
- Oil relief valve plug. Tighten the plug to 30-37 ft. lbs. (40-50 Nm).
- Oil filter bracket
- Oil pick-up tube. Tighten the bolts to 11-16 ft. lbs. (15-22 Nm).
- Upper oil pan
- Lower oil pan
- Starter motor
- Oil filter
- CKP sensor tone ring
- Crankshaft sprocket
- Timing belt
- No. 3 engine mount
- Upper and lower timing belt covers
- Belt tensioner pulley
- Power steering pump pulley
- Crankshaft pulley
- Idler pulley
- Accessory drive belts
- Valve covers
- Engine cover
- Negative battery cable

6. Fill the crankcase to the correct level with engine oil.

7. Start the engine and check for leaks.

Rear Main Seal

REMOVAL & INSTALLATION

1. Before servicing the vehicle, refer to the precautions in the beginning of this section.

2. Remove or disconnect the following:
- Negative battery cable
- Front wheels
- Starter motor
- Axle halfshafts
- Transaxle
- Flexplate
- Oil seal housing
- Oil seal

To install:

3. Install the oil seal to the seal housing using special tool 09231-33000 or similar seal driver.

4. Apply silicone sealant to the oil seal housing flange.

5. Install the seal housing and tighten the bolts to 94-106 inch lbs. (10-12 Nm).

6. Install or connect the following:
- Flexplate. Tighten the bolts to 54-57 ft. lbs. (73-77 Nm)
- Transaxle
- Axle halfshafts
- Starter motor
- Front wheels
- Negative battery cable

7. Fill the crankcase to the correct level with engine oil.

8. Start the engine and check for leaks.

Timing Belt

REMOVAL & INSTALLATION

1. Before servicing the vehicle, refer to the precautions in the beginning of this section.

2. Remove or disconnect the following:
- Negative battery cable
- Engine cover
- Accessory drive belts
- Idler pulley
- Crankshaft pulley
- Power steering pump pulley
- Belt tensioner pulley
- Upper and lower timing belt covers

3. Support the engine with a floor jack and remove the engine mount.

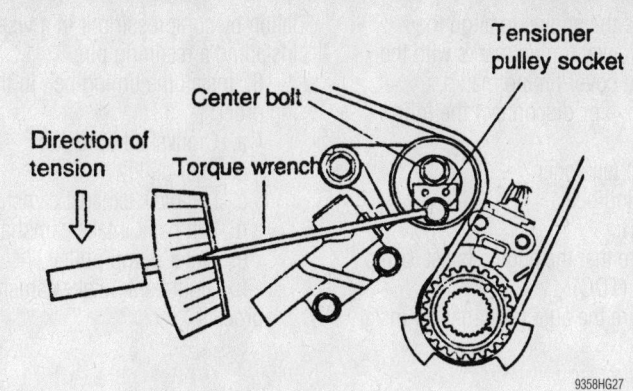

Adjusting the tensioner pulley-Kia 3.5L Engine

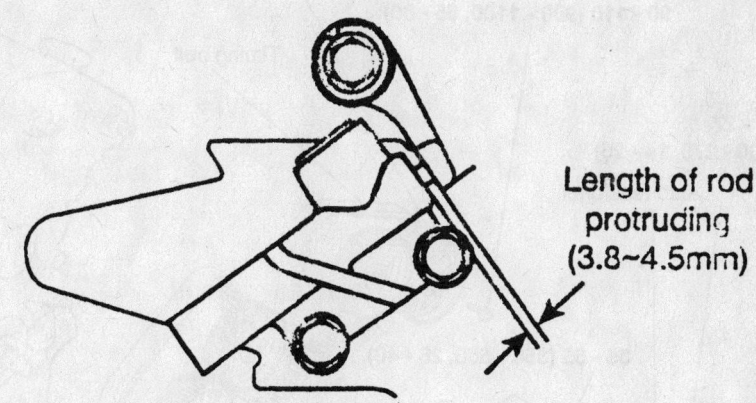

Measuring the auto tensioner rod-Kia 3.5L Engine

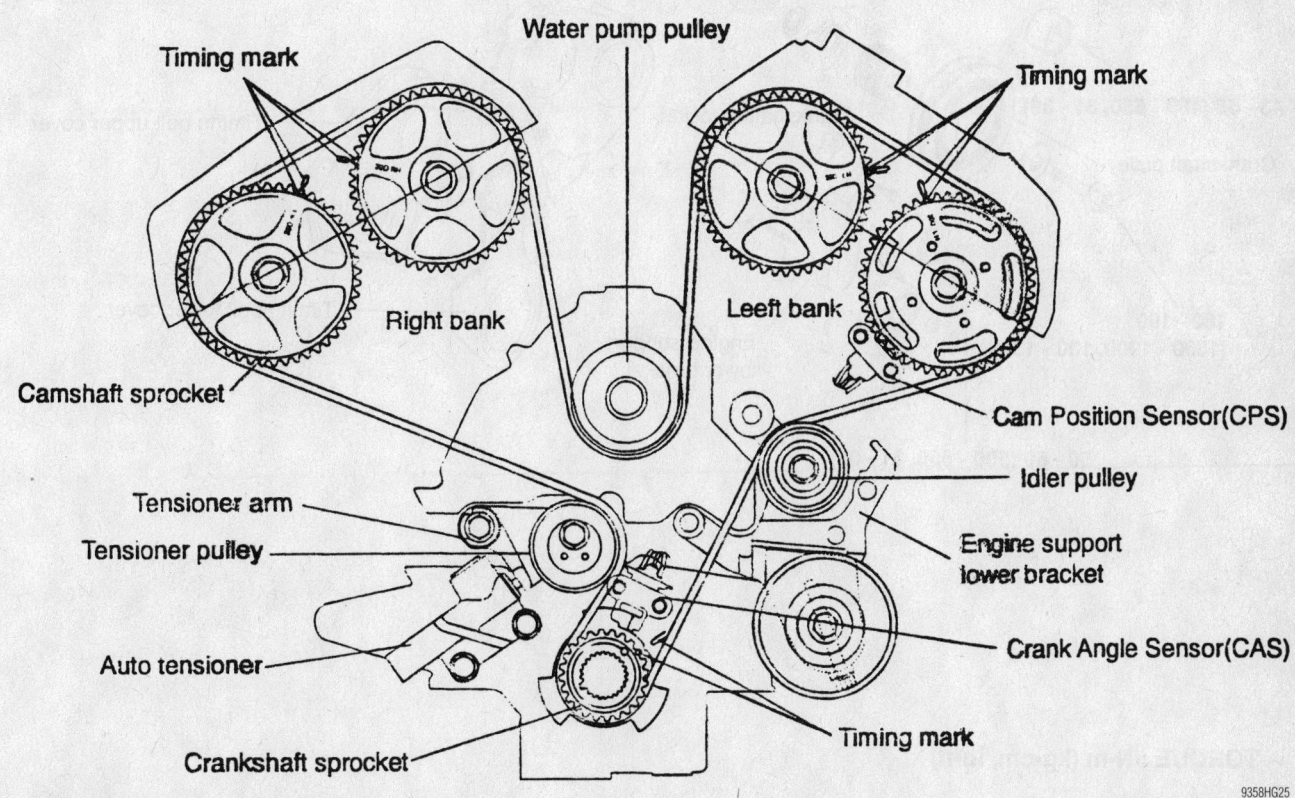

Timing belt routing and timing marks-Kia 3.5L Engine

4. Rotate the engine to align the camshaft sprocket timing marks with the cylinder head cover timing marks.

5. Remove or disconnect the following:
- Auto tensioner
- Timing belt

To install:

6. Ensure that the engine is set to Top Dead Center (TDC).

7. Prepare the auto tensioner for instal-lation by compressing it in a vise and installing a retaining pin.

8. Install the timing belt in the following order:
   a. Crankshaft sprocket
   b. Idler pulley
   c. Left bank exhaust camshaft sprocket
   d. Left bank intake camshaft sprocket
   e. Water pump pulley
   f. Right bank intake camshaft sprocket
   g. Right bank exhaust camshaft sprocket
   h. Tensioner pulley

9. Install the auto tensioner. Do not remove the retaining pin at this time.

10. Check that the crankshaft and camshaft timing marks are aligned correctly.

11. Rotate the crankshaft ¼ turn **Counterclockwise**.

12. Rotate the crankshaft ¼ turn **Clockwise** to return the engine to TDC.

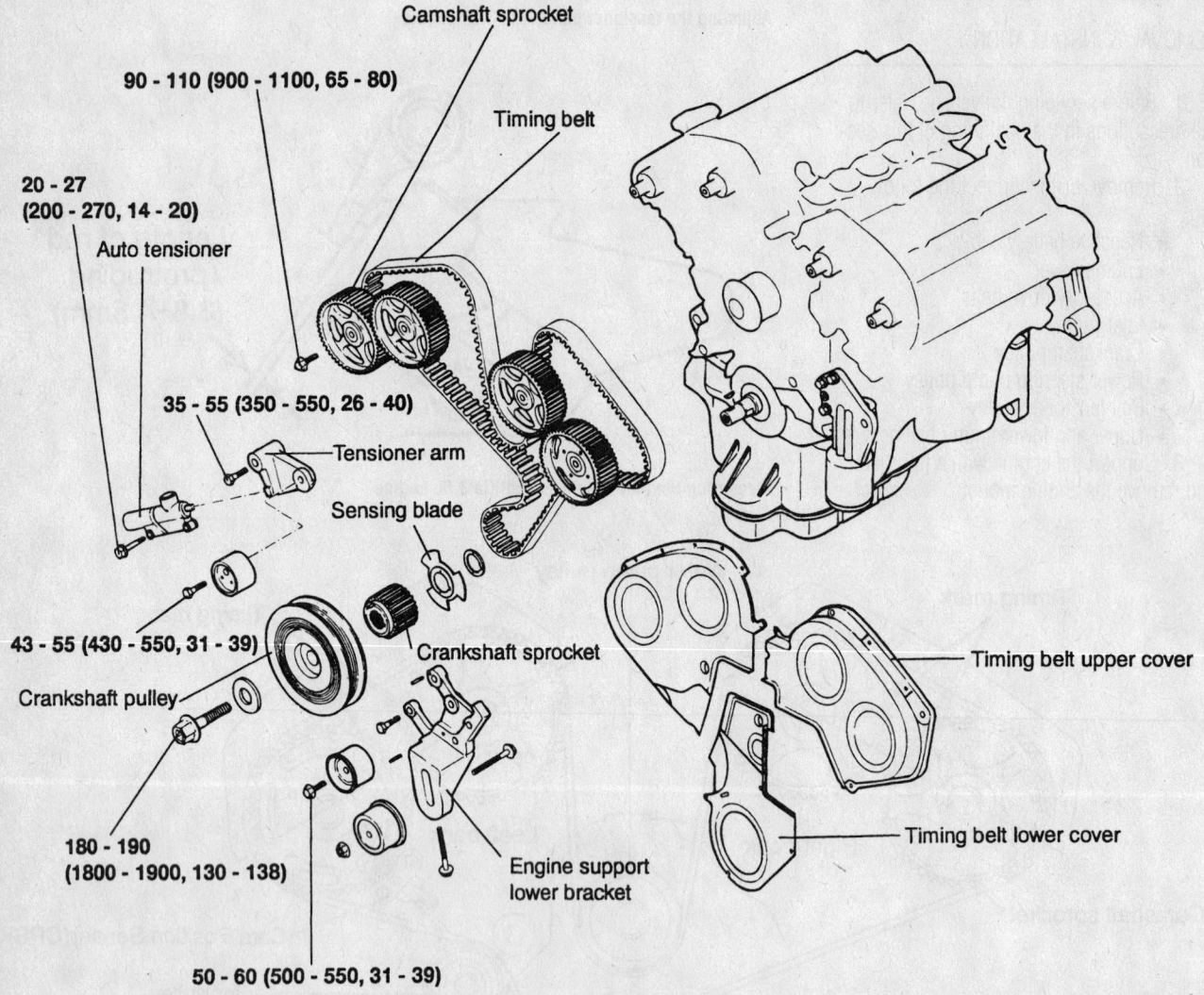

TORQUE : N•m (kg•cm, lb•ft)

Timing belt exploded view-Kia 3.5L Engine

9358HG26

13. Loosen the tensioner pulley center bolt.

14. Apply 44 inch lbs. (5 Nm) torque to the tensioner pulley as shown and tighten the center bolt to 32-41 ft. lbs. (43-55 Nm).

15. Remove the auto tensioner retaining pin.

16. Rotate the crankshaft 2 revolutions **Clockwise**, then wait 5 minutes for the auto tensioner to adjust.

17. Measure the auto tensioner rod as shown. If the measurement is not 3.8-4.5 mm, then repeat the belt tensioning procedure.

18. When the auto tensioner measurement is correct, install or connect the following:

- Engine mount
- Upper and lower timing belt covers
- Belt tensioner pulley
- Power steering pump pulley
- Crankshaft pulley
- Idler pulley
- Accessory drive belts
- Engine cover
- Negative battery cable

Piston and Ring

POSITIONING

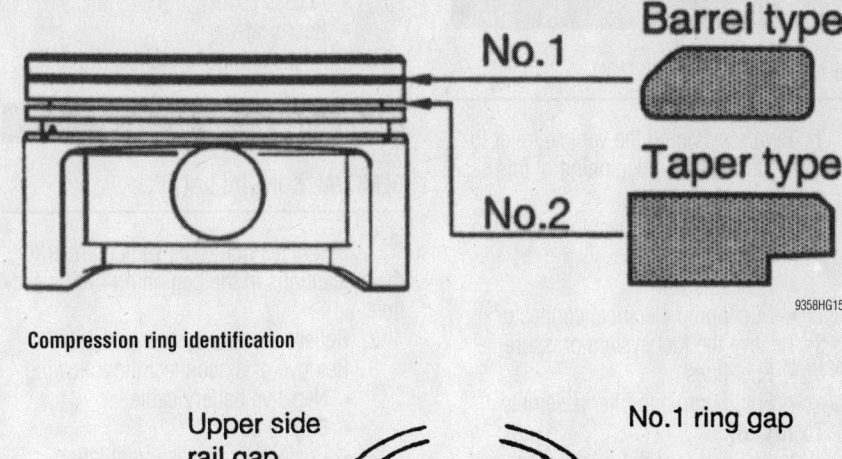

Compression ring identification

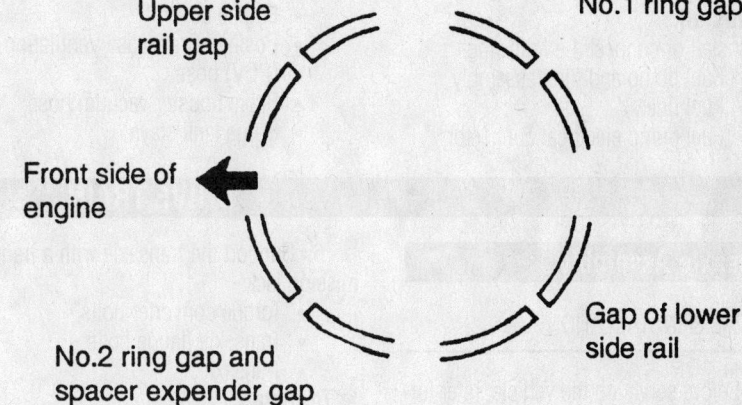

Piston ring end gap spacing

FUEL SYSTEM

Fuel System Service Precautions

Safety is the most important factor when performing not only fuel system maintenance, but any type of maintenance. Failure to conduct maintenance and repairs in a safe manner may result in serious personal injury or death. Maintenance and testing of the vehicle's fuel system components can be accomplished safely and effectively by adhering to the following rules and guidelines.

• To avoid the possibility of fire and personal injury, always disconnect the negative battery cable unless the repair or test procedure requires that battery voltage be applied.

• Always relieve the fuel system pressure prior to disconnecting any fuel system component (injector, fuel rail, pressure regulator, etc.), fitting or fuel line connection. Exercise extreme caution whenever relieving fuel system pressure, to avoid exposing skin, face and eyes to fuel spray. Please be advised that fuel under pressure may penetrate the skin or any part of the body that it contacts.

• Always place a shop towel or cloth around the fitting or connection prior to loosening to absorb any excess fuel due to spillage. Ensure that all fuel spillage (should it occur) is quickly removed from engine surfaces. Ensure that all fuel soaked cloths or towels are deposited into a suitable waste container.

• Always keep a dry chemical (Class B) fire extinguisher near the work area.

• Do not allow fuel spray or fuel vapors to come into contact with a spark or open flame.

• Always use a back-up wrench when loosening and tightening fuel line connection fittings. This will prevent unnecessary stress and torsion to fuel line piping. Always follow the proper torque specifications.

• Always replace worn fuel fitting O-rings with new. Do not substitute fuel hose or equivalent where fuel pipe is installed.

Fuel System Pressure

RELIEVING

1. Before servicing the vehicle, refer to the precautions in the beginning of this section.
2. Remove the rear seat.
3. Remove the access cover
4. Disconnect the fuel pump electrical connector.
5. Start the engine and allow it to run until it stalls.
6. Turn the ignition switch to the **OFF** position.
7. Restore the electrical connections after fuel system repairs are completed.

Fuel Filter

REMOVAL & INSTALLATION

The fuel filter is located inside the fuel tank and is replaced with the fuel pump as an assembly.

Fuel Pump

REMOVAL & INSTALLATION

1. Before servicing the vehicle, refer to the precautions in the beginning of this section.
2. Remove or disconnect the following:
   • Rear seat
   • Access cover
   • Fuel pump electrical connector
3. Relieve the fuel system pressure.
   • Fuel lines
   • Fuel pump and filter assembly

To install:

4. Install or connect the following:
   • Fuel pump and filter assembly
   • Fuel lines
   • Fuel pump electrical connector
   • Access cover
   • Rear seat
5. Start the engine and check for leaks.

Fuel Injector

REMOVAL & INSTALLATION

1. Before servicing the vehicle, refer to the precautions in the beginning of this section.
2. Relieve the fuel system pressure.
3. Remove or disconnect the following:
   • Negative battery cable
   • Engine cover
   • Positive Crankcase Ventilation (PCV) hose
   • Brake booster vacuum hose
   • Surge tank stays
   • Fuel lines
   • Upper intake manifold (surge tank)
   • Fuel injector harness connectors
   • Fuel supply manifold with injectors attached

To install:

4. Replace all injector seals.
5. Install or connect the following:
   • Fuel supply manifold with injectors attached
   • Fuel injector harness connectors
   • Upper intake manifold (surge tank)
   • Fuel lines
   • Surge tank stays
   • Brake booster vacuum hose
   • Positive Crankcase Ventilation (PCV) hose
   • Engine cover
   • Negative battery cable
6. Start the engine and check for leaks.

DRIVE TRAIN

Transaxle Assembly

REMOVAL & INSTALLATION

1. Before servicing the vehicle, refer to the precautions in the beginning of this section.
2. Drain the cooling system.
3. Drain the transaxle fluid.
4. Remove or disconnect the following:
   • Battery and tray
   • Engine wiring harness
   • Transaxle wiring harness
   • Air filter assembly
   • Shift cable
   • Transaxle cooler lines
   • Radiator hoses
   • Radiator
   • Heater hoses
   • Steering intermediate shaft
   • Power steering pressure and return lines from the steering gear
   • Engine upper roll stopper and bracket
   • Front wheels
   • Heated Oxygen (HO2S) sensor connector
   • Front muffler
   • Outer tie rod ends
   • Lower ball joints
   • Axle halfshafts
   • Steering tube mounting bolt
5. Support the engine from below.
   • Front subframe
   • Starter motor
   • Transaxle housing cover
6. Support the transaxle with a transmission jack.
   • Torque converter bolts
   • Transaxle flange bolts
   • Transaxle

To install:

7. Install or connect the following:
   • Transaxle. Tighten the flange bolts to 29-38 ft. lbs. (42-54 Nm)
   • Torque converter bolts
   • Transaxle housing cover
   • Starter motor
   • Front subframe
   • Steering tube mounting bolt
   • Axle halfshafts
   • Lower ball joints
   • Outer tie rod ends
   • Front muffler
   • Heated Oxygen (HO2S) sensor connector
   • Front wheels
   • Engine upper roll stopper and bracket
   • Power steering pressure and return lines from the steering gear
   • Steering intermediate shaft
   • Heater hoses
   • Radiator
   • Radiator hoses
   • Transaxle cooler lines
   • Shift cable
   • Air filter assembly
   • Transaxle wiring harness
   • Engine wiring harness
   • Battery and tray
8. Fill the transaxle to the correct level with the proper transmission fluid.
9. Fill the cooling system.
10. Start the engine and check for leaks.

Halfshaft

REMOVAL & INSTALLATION

1. Before servicing the vehicle, refer to the precautions in the beginning of this section.
2. Drain the transaxle fluid.
3. Remove or disconnect the following:
   • Front wheels
   • Hub retaining nuts
   • Sway bar links
   • Outer tie rod ends
   • Lower ball joints
4. Using a prybar, remove the left halfshaft from the transaxle. Separate the right halfshaft from the intermediate shaft.
5. Remove the axle halfshafts from the front hubs. It may be necessary to tap the stub shaft with a brass hammer to remove the axles.

To install:

➡Use new circlips for assembly.

6. Install the left halfshaft to the transaxle.
7. Install the right halfshaft to the intermediate shaft.
8. Install the halfshaft stub shafts to the wheel hubs.
9. Install or connect the following:
   • Lower ball joints. Tighten the pinch bolts to 69-85 ft. lbs. (93-115 Nm).
   • Outer tie rod ends. Tighten the nuts to 43-58 ft. lbs. (59-78 Nm).

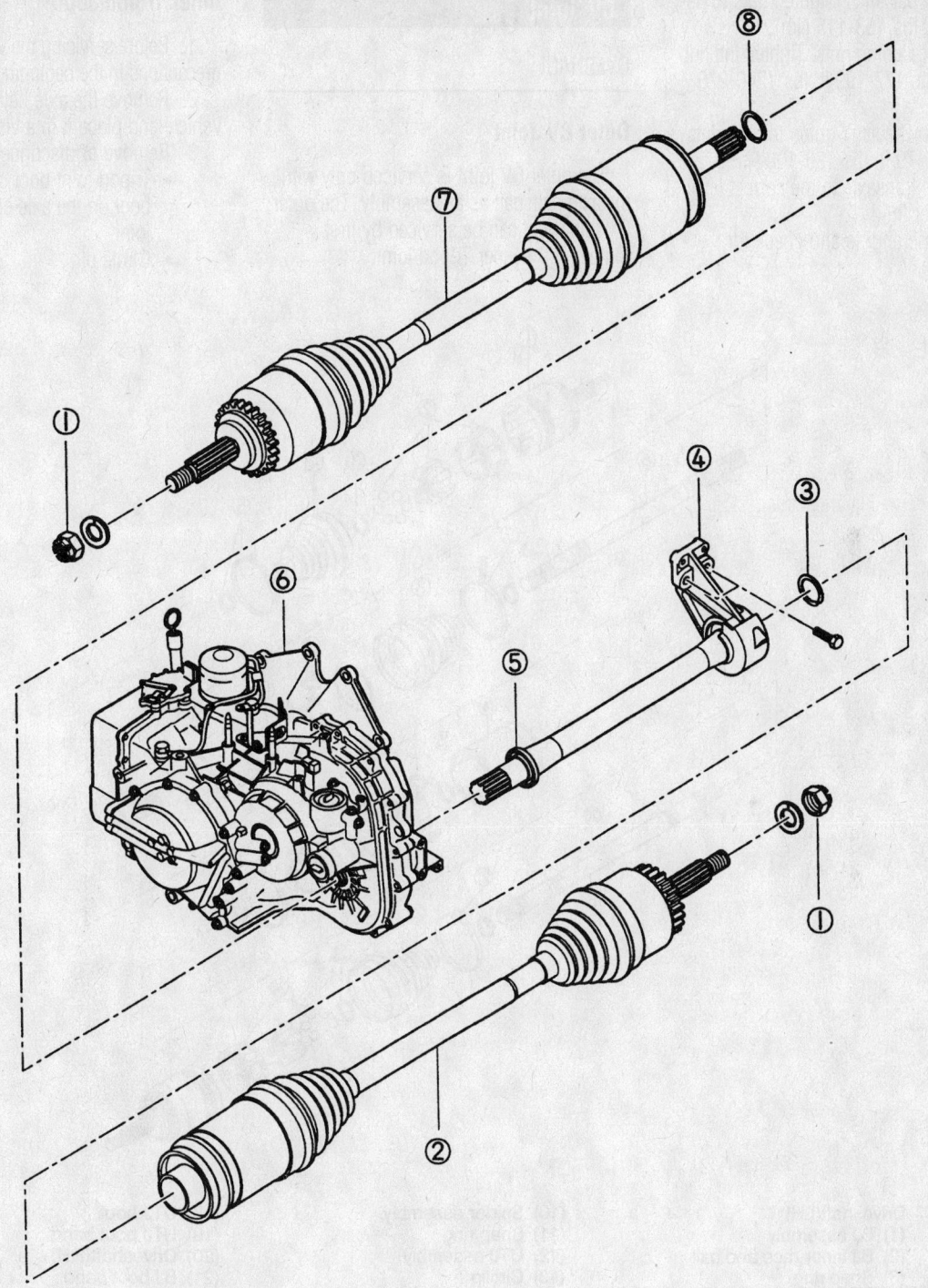

(1) Wheel nut
(2) Driveshaft(RH)
(3) Circlip
(4) Center bearing bracket

(5) constant velocity shaft
(6) Automatic transaxle
(7) Driveshaft(LH)
(8) Circle pin

9358HG17

Halfshaft mounting exploded view

- Sway bar links. Tighten nuts to 69-85 ft. lbs. (93-115 Nm).
- Hub retaining nuts. Tighten the hub nuts to 177-199 ft. lbs. (240-270 Nm).
- Front wheels. Tighten the lug nuts to 65-79 ft. lbs. (88-108 Nm).

10. Fill the transaxle to the correct level with the proper fluid.

11. Start the engine and check for leaks.

CV Joint

OVERHAUL

Outer CV Joint

The outer CV joint is serviced only with the axle halfshaft as an assembly. The outer CV joint boot can be serviced by first removing the inner Tripod joint.

Inner Tripod Joint

1. Before servicing the vehicle, refer to the precautions in the beginning of this section.

2. Remove the axle halfshaft from the vehicle and place it in a vise.

3. Remove or disconnect the following:
- Tripod joint boot clamps. Slide the boot on the axle shaft to expose the joint.
- Circle pin

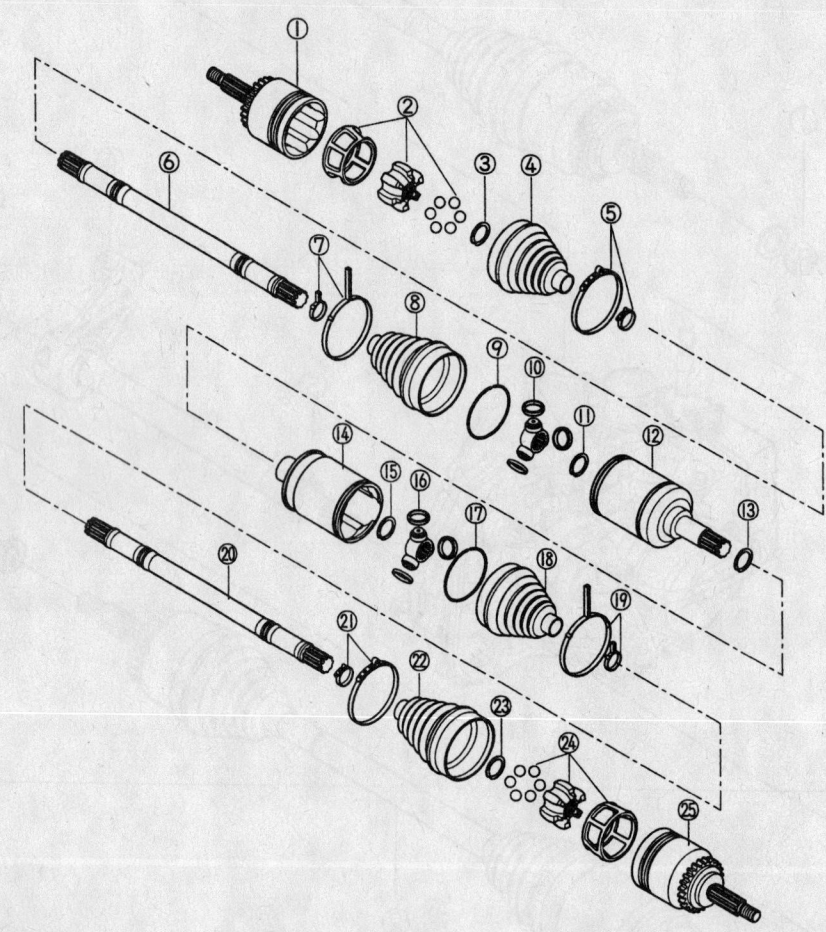

Driveshaft(LH)	(10) Spider assembly	(18) UTJ boot
(1) BJ assembly	(11) Snap ring	(19) UTJ boot band
(2) BJ inner race and ball	(12) UTJ assembly	(20) Driveshaft(RH)
(3) Snap ring	(13) Circlip	(21) BJ boot band
(4) BJ boot		(22) BJ boot
(5) BJ boot band	Driveshaft(RH)	(23) Snap ring
(6) Driveshaft(LH)	(14) UTJ assembly	(24) BJ inner race and ball
(7) UTJ boot band	(15) Snap ring	(25) BJ assembly
(8) UTJ boot	(16) Spider assembly	
(9) Circle pin	(17) Circle pin	

⚠ Caution
- a) *Install a protective material in a vice, and secure joint in the vice.*
- b) *Keep dust or foreign material from joint during procedure.*
- c) *Do not disassemble wheel side ball joint.*
- d) *Do not wash joint unless disassembling it.*

9358HG18

Inner and outer CV Joint exploded view and parts identification

- Tripod joint housing
- Tripod joint snapring
- Tripod joint
- Inner Tripod joint boot

To install:

4. Install or connect the following:

- Inner Tripod joint boot
- Tripod joint
- Tripod joint snapring
- Tripod joint housing. Fill with 7.5 ounces of CV Joint grease.
- Circle pin

- Tripod joint boot clamps. Pull on the clamp end with pliers and fold the locking tabs over to lock the clamp in place.

5. Install the axle halfshaft to the vehicle.

STEERING AND SUSPENSION

Air Bag

☀☀ CAUTION

Some vehicles are equipped with an air bag system. The system must be disarmed before performing service on, or around, system components, the steering column, instrument panel components, wiring and sensors. Failure to follow the safety precautions and the disarming procedure could result in accidental air bag deployment, possible injury and unnecessary system repairs.

PRECAUTIONS

Several precautions must be observed when handling the inflator module to avoid accidental deployment and possible personal injury.

- Never carry the inflator module by the wires or connector on the underside of the module.
- When carrying a live inflator module, hold securely with both hands, and ensure that the bag and trim cover are pointed away.
- Place the inflator module on a bench or other surface with the bag and trim cover facing up.
- With the inflator module on the bench, never place anything on or close to the module which may be thrown in the event of an accidental deployment.

DISARMING

1. Before servicing the vehicle, refer to the precautions in the beginning of this section.
2. Position the vehicle with the front wheels in a straight-ahead position.
3. Disconnect both battery cables.
4. Wait at least 1 minute for the air bag back-up power supply to deplete its stored energy before continuing.
5. Proceed with the repair.
6. Reconnect both battery cables once the repair is complete.

Rack and Pinion Steering Gear

REMOVAL & INSTALLATION

1. Before servicing the vehicle, refer to the precautions in the beginning of this section.
2. Drain the power steering fluid.
3. Remove or disconnect the following:

- Negative battery cable
- Front wheels
- Outer tie rod ends
- Power steering fluid pressure and return lines
- Steering intermediate shaft
- Steering rack brackets and fasteners

4. Remove the steering gear through the right wheel opening.

To install:

5. Install the steering gear through the right wheel opening.
6. Install or connect the following:

- Steering rack brackets and fasteners. Tighten the fasteners to 55-69 ft. lbs. (74-93 Nm).
- Steering intermediate shaft. Tighten the pinch bolt to 16-20 ft. lbs. (21-26 Nm).
- Power steering fluid pressure and return lines. Tighten the bolts to 17-26 ft. lbs. (24-35 Nm).
- Outer tie rod ends. Tighten the nuts to 43-58 ft. lbs. (59-78 Nm).
- Front wheels
- Negative battery cable

Strut

REMOVAL & INSTALLATION

Front

1. Before servicing the vehicle, refer to the precautions in the beginning of this section.
2. Remove or disconnect the following:
- Front wheel

- Brake hose from the bracket
- Wheel speed sensor cable
- Steering knuckle mounting bolts
- Upper strut mount nuts

3. Remove the strut from the vehicle.

To install:

4. Install the strut to the vehicle.
5. Install or connect the following:

- Upper strut mount nuts and tighten them to 33-46 ft. lbs. (46-62 Nm).
- Steering knuckle mounting bolts and tighten them to 88-101 ft. lbs. (119-137 Nm).
- Wheel speed sensor cable
- Brake hose to the bracket
- Front wheel

6. Check the front end alignment and adjust as necessary.

Shock Absorber

REMOVAL & INSTALLATION

Rear

1. Before servicing the vehicle, refer to the precautions in the beginning of this section.
2. Support the rear axle on jack stands.
3. Remove or disconnect the following:

- Rear wheel
- Shock absorber upper nut and washer
- Shock absorber lower nut and washer
- Shock absorber

To install:

4. Install or connect the following:

- Shock absorber
- Shock absorber lower nut and washer. Tighten the nut to 55-69 ft. lbs. (74-93 Nm).
- Shock absorber upper nut and washer. Tighten the nut to 41-47 ft. lbs. (55-64 Nm).
- Rear wheel

Coil Spring

REMOVAL & INSTALLATION

Front

1. Before servicing the vehicle, refer to the precautions in the beginning of this section.
2. Remove the strut from the vehicle and attach Service Tool 0K2A1 341 001 or other suitable spring compressor.
3. Compress the coil spring.
4. Remove or disconnect the following:
   - Upper mount retaining nut
   - Mounting block
   - Upper spring seat
   - Upper spring isolator
   - Coil spring
   - Dust boot
   - Bump stopper
   - Lower spring seat

To install:

5. Install or connect the following:
   - Lower spring seat
   - Bump stopper
   - Dust boot
   - Coil spring
   - Upper spring isolator
   - Upper spring seat
   - Mounting block
   - Upper mount retaining nut and tighten it to 88-101 ft. lbs. (120-137 Nm).
6. Install the strut to the vehicle.
7. Check the front end alignment and adjust as necessary.

Rear

1. Before servicing the vehicle, refer to the precautions in the beginning of this section.
2. Support the vehicle with jackstands forward of the lower control arm mounting.

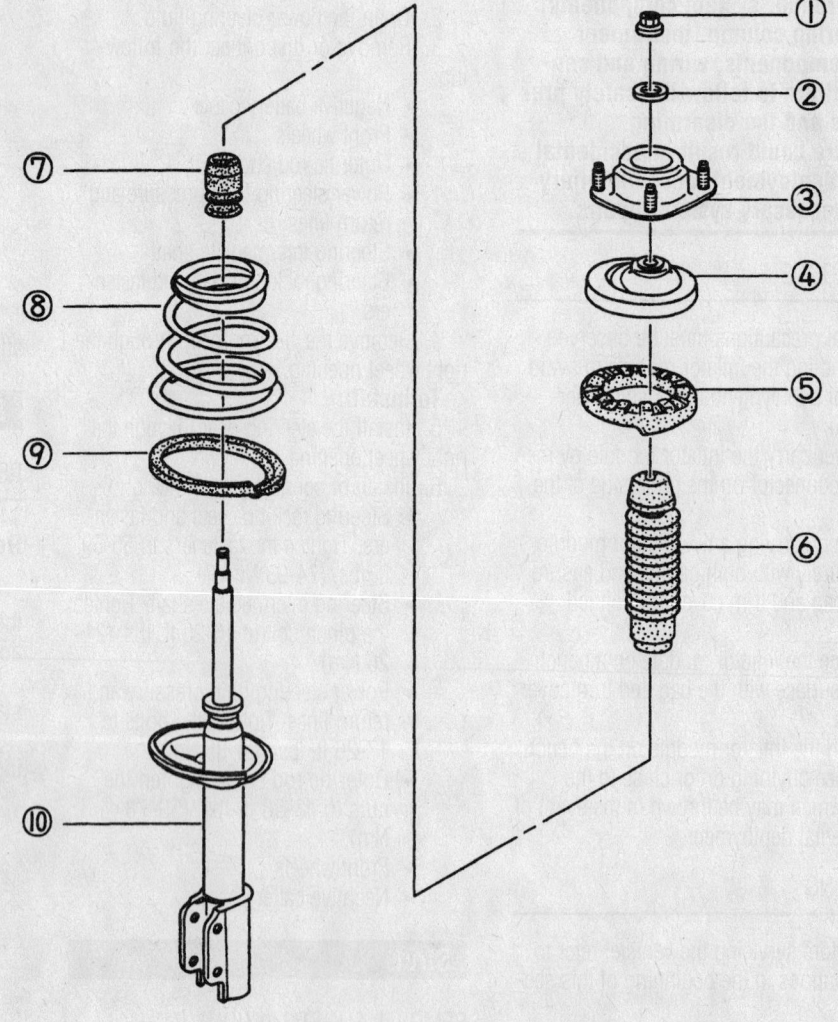

(1) **Piston retaining nut**
(2) **Washer**
(3) **Mounting block**
(4) **Upper spring seat**
(5) **Upper spring isolator**

(6) **Dust boot**
(7) **Bump stopper**
(8) **Coil spring**
(9) **Lower spring isolator**
(10) **Shock absorber**

9358HG19

Front strut and coil spring exploded view

3. Support the rear axle with jackstands.

4. Support the lower control arm with a floor jack.

5. Remove or disconnect the following:
   - Rear wheel
   - Sway bar link
   - Parking brake cable
   - Lower control arm bolts

6. Carefully lower the floor jack and remove the lower control arm, coil spring, and spring seats from the vehicle.

To install:

7. Install or connect the following:
   - Spring seats
   - Coil spring
   - Lower control arm. Tighten the

bolts to 87-101 ft. lbs. (118-137 Nm).
   - Parking brake cable. Tighten the bracket bolts to 14-19 ft. lbs. (16-23 Nm).
   - Sway bar link. Tighten the locknut to 17-20 ft. lbs. (24-28 Nm).
   - Rear wheel

Lower Ball Joint

REMOVAL & INSTALLATION

The ball joint is serviced with the lower control arm as an assembly.

Upper Control Arm

REMOVAL & INSTALLATION

Rear

1. Before servicing the vehicle, refer to the precautions in the beginning of this section.

2. Support the vehicle with jackstands forward of the lower control arm mounting.

3. Support the rear axle with jackstands.

4. Remove or disconnect the following:
   - Rear wheel
   - Upper control arm mounting bolts
   - Upper control arm

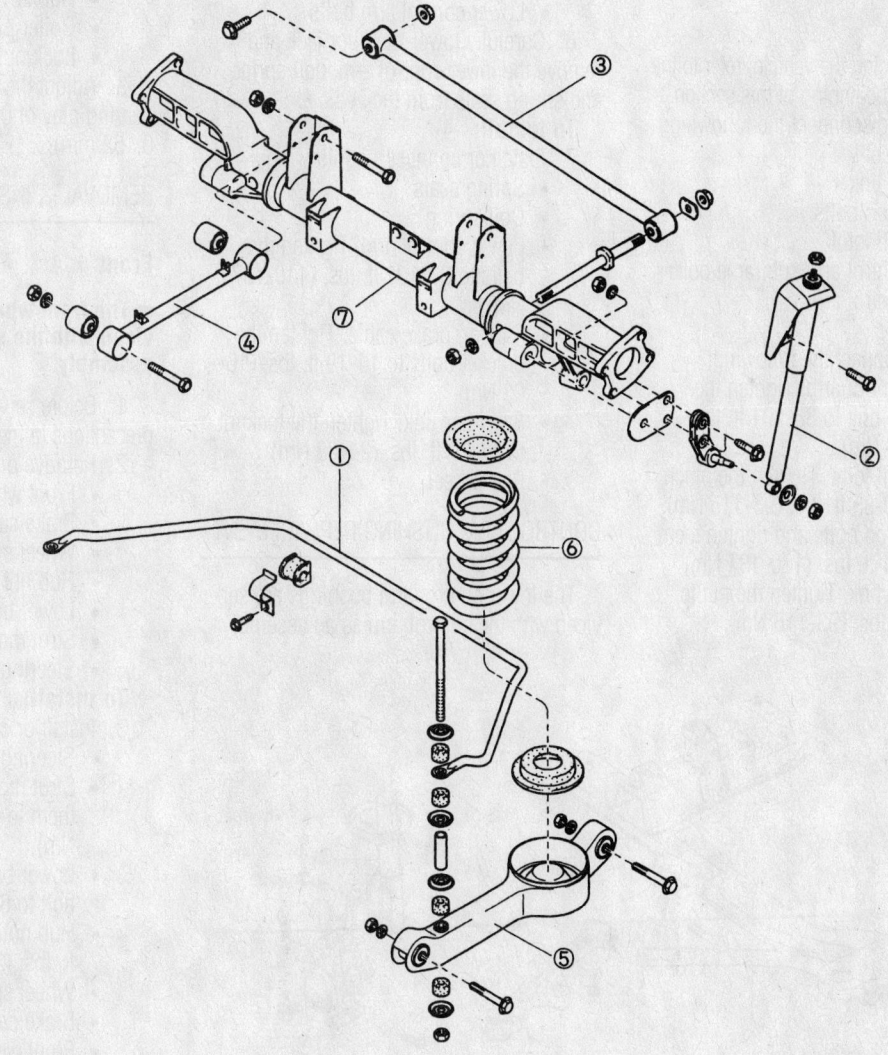

(1) Stabilizer bar
(2) Shock absorber
(3) Panhard rod
(4) Upper arm assembly
(5) Lower arm assembly
(6) Coil spring
(7) Rear casing

9358HG20

Rear suspension exploded view

To install:

5. Install or connect the following:
- Upper control arm
- Upper control arm mounting bolts. Tighten the bolts to 55-69 ft. lbs. (74-93 Nm).
- Rear wheel

CONTROL ARM BUSHING REPLACEMENT

The upper control arm bushings are serviced with the control arm as an assembly.

Lower Control Arm

REMOVAL & INSTALLATION

Front

1. Before servicing the vehicle, refer to the precautions in the beginning of this section.
2. Remove or disconnect the following:
- Front wheel
- Sway bar link
- Tension rod bolts
- Lower ball joint
- Lower control arm subframe bolt
- Lower control arm

To install:

3. Install or connect the following:
- Lower control arm. Tighten the subframe bolt to 88-101 ft. lbs. (120-137 Nm).
- Lower ball joint. Tighten the pinch bolt to 69-85 ft. lbs. (93-115 Nm).
- Tension rod bolts and tighten them to 88-101 ft. lbs. (120-137 Nm)
- Sway bar link. Tighten the nut to 69-85 ft. lbs. (93-115 Nm).

- Front wheel
4. Check the front end alignment and adjust as necessary.

Rear

1. Before servicing the vehicle, refer to the precautions in the beginning of this section.
2. Support the vehicle with jackstands forward of the lower control arm mounting.
3. Support the rear axle with jackstands.
4. Support the lower control arm with a floor jack.
5. Remove or disconnect the following:
- Rear wheel
- Sway bar link
- Parking brake cable
- Lower control arm bolts
6. Carefully lower the floor jack and remove the lower control arm, coil spring, and spring seats from the vehicle.

To install:

7. Install or connect the following:
- Spring seats
- Coil spring
- Lower control arm. Tighten the bolts to 87-101 ft. lbs. (118-137 Nm).
- Parking brake cable. Tighten the bracket bolts to 14-19 ft. lbs. (16-23 Nm).
- Sway bar link. Tighten the locknut to 17-20 ft. lbs. (24-28 Nm).
- Rear wheel

CONTROL ARM BUSHING REPLACEMENT

The lower control arm bushings are serviced with the control arm as an assembly.

Wheel Bearings

ADJUSTMENT

Front

The front wheel bearings are not adjustable. Replace the wheel bearings if play exceeds 0.002 inches (0.05 mm).

Rear

1. Before servicing the vehicle, refer to the precautions in the beginning of this section.
2. Remove or disconnect the following:
- Rear wheel
- Brake drum
- Hub cover
- Cotter pin
- Lock nut cover
3. Adjust the locknut to achieve wheel bearing play of 0.001-0.006 inches (0.025-0.152 mm).

REMOVAL & INSTALLATION

Front

➡ **The front wheel bearings are serviced with the steering knuckle as an assembly.**

1. Before servicing the vehicle, refer to the precautions in the beginning of this section.
2. Remove or disconnect the following:
- Front wheel
- Brake caliper and rotor
- Wheel speed sensor harness
- Hub nut
- Lower ball joint
- Strut mounting bolts
- Steering knuckle

To install:

3. Install or connect the following:
- Steering knuckle
- Strut mounting bolts and tighten them to 88-101 ft. lbs. (119-137 Nm).
- Lower ball joint. Tighten the pinch bolt to 69-85 ft. lbs. (93-115 Nm).
- Hub nut and tighten it to 177-199 ft. lbs. (240-270 Nm).
- Wheel speed sensor harness
- Brake caliper and rotor
- Front wheel
4. Check the alignment and adjust as necessary.

Rear

1. Before servicing the vehicle, refer to the precautions in the beginning of this section.
2. Remove or disconnect the following:

(1) Knuckle
(2) Shock absorber
(3) Tie rod end
(4) Stabilizer bar
(5) Stabilizer control link
(6) Tension rod
(7) Lower arm

9358HG21

Front suspension exploded view

- Rear wheel
- Brake drum
- Hub cap
- Cotter pin
- Lock nut cover
- Lock nut
- Wheel bearing retainer washer and the outer wheel bearing
- Hub assembly
- Grease seal

- Inner wheel bearing

3. Clean and inspect the wheel bearings and races for unusual wear or damage. Replace parts as necessary.

To install:

4. Pack the wheel bearings with grease for assembly.

5. Install or connect the following:
- Inner wheel bearing
- Grease seal

- Hub assembly
- Lock nut. Adjust the locknut to achieve wheel bearing play of 0.001-0.006 inches (0.025-0.152 mm).
- Lock nut cover
- Cotter pin
- Hub cap
- Brake drum
- Rear wheel

BRAKES

Brake Caliper

REMOVAL & INSTALLATION

1. Before servicing the vehicle, refer to the precautions in the beginning of this section.

2. Remove or disconnect the following:
- Front wheel
- Brake fluid hose
- Caliper mounting bolts
- Brake caliper

To install:

3. Install or connect the following:
- Brake caliper. Tighten the mounting bolts to 18–26 ft. lbs. (24–35 Nm).
- Brake fluid hose
- Front wheel

4. Bleed the brake system.

5. Before attempting to move the vehicle, pump the brake pedal to seat the pads against the rotors. Make sure the vehicle has a firm brake pedal. Check the level of the brake fluid and add DOT 3 or 4 brake fluid if necessary.

Disc Brake Pads

REMOVAL & INSTALLATION

1. Before servicing the vehicle, refer to the precautions in the beginning of this section.

2. Remove or disconnect the following:
- Front wheel
- Brake hose from the support bracket
- Outer brake pad retaining clip
- Brake caliper
- Inner and outer brake pads

To install:

3. Compress the caliper piston into the caliper bore.

4. Install or connect the following:
- Inner and outer brake pads
- Brake caliper. Tighten the mount-

ing bolts to 18–26 ft. lbs. (24–35 Nm).
- Brake hose to the support bracket
- Front wheel

Brake Drums

REMOVAL & INSTALLATION

1. Before servicing the vehicle, refer to the precautions in the beginning of this section.

2. Remove or disconnect the following:
- Rear wheel
- Brake drum retaining screws
- Brake drum

To install:

3. Install or connect the following:
- Brake drum
- Brake drum retaining screws
- Rear wheel

Brake Shoes

REMOVAL & INSTALLATION

1. Before servicing the vehicle, refer to the precautions in the beginning of this section.

2. Remove or disconnect the following:
- Rear wheels
- Brake drum
- Brake adjuster spring
- Adjuster lever
- Strut spring and retracting spring
- Upper return spring
- Hold-down spring and pins
- Leading (primary) brake shoe
- Trailing (secondary) brake shoe
- Parking brake cable

To install:

3. Transfer the parking brake lever to the new trailing (secondary) brake shoe.

4. Install or connect the following:

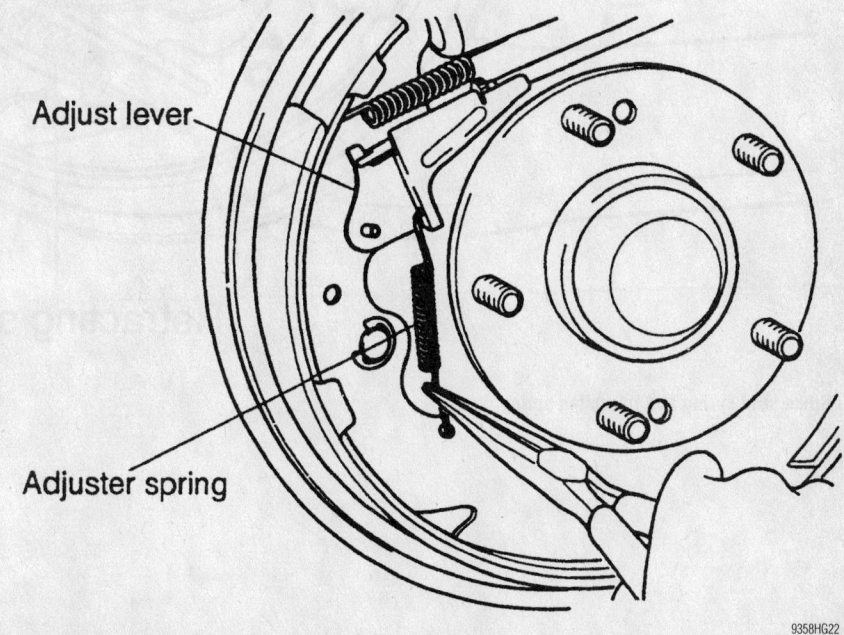

Adjust lever

Adjuster spring

Brake adjuster spring removal

9358HG22

- Parking brake cable
- Trailing (secondary) brake shoe
- Leading (primary) brake shoe
- Hold-down spring and pins
- Upper return spring
- Strut spring and retracting spring
- Adjuster lever
- Brake adjuster spring
- Brake drum
- Rear wheels

5. Work the parking brake control several times to complete the brake shoe adjustment and to check the parking brake adjustment as well.

6. Pump the brake pedal several times to assure a good pedal.

7. Road test the vehicle and check for proper brake system operation.

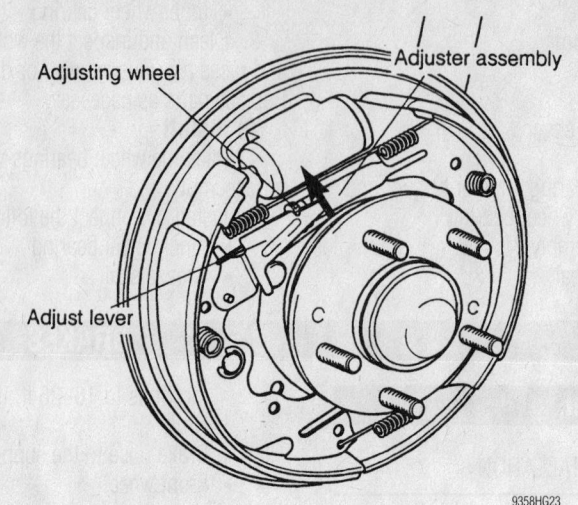

9358HG23

Brake adjuster assembly

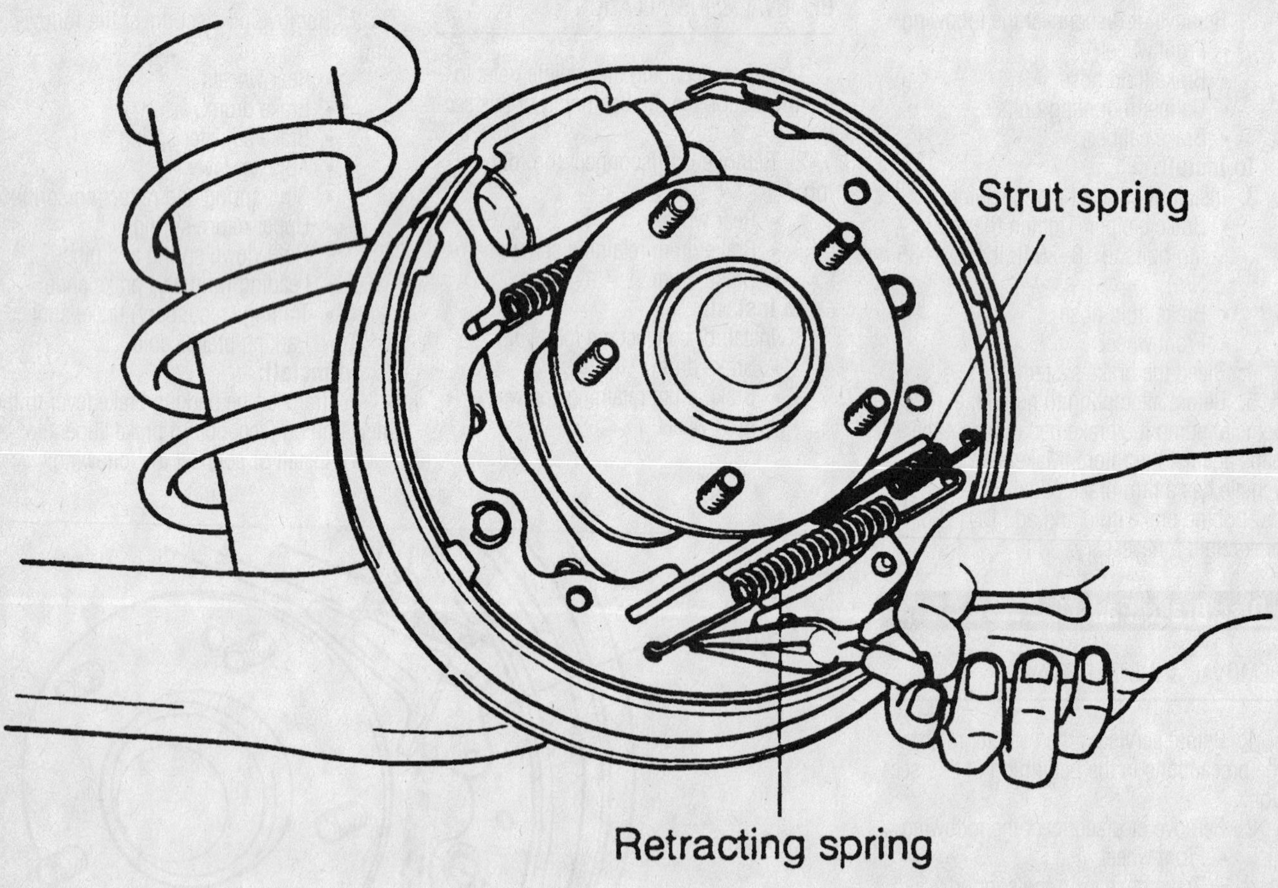

Brake strut spring and retracting spring

9358HG24

KIA

Sportage

SPECIFICATION CHARTS

ENGINE AND VEHICLE IDENTIFICATION

	Engine							Model Year	
Code ①	Liters (cc)	Cu. In.	Cyl.	Fuel Sys.	Engine Type	Eng. Mfg.		Code ②	Year
3	2.0 (1998)	122	4	MFI	DOHC	KIA		X	1999

MFI: Multi-port Fuel Injection

DOHC: Double Overhead Camshafts

① 8th digit of VIN

② 10th digit of VIN

Code ②	Year
X	1999
Y	2000
1	2001
2	2002
3	2003

42356-KSPO-C01

GENERAL ENGINE SPECIFICATIONS

Year	Model	Engine Displacement Liters (cc)	Engine VIN	Fuel System Type	Net Horsepower @ rpm	Net Torque @ rpm (ft. lbs.)	Bore x Stroke (in.)	Compression Ratio	Oil Pressure @ rpm
2000	Sportage	2.0 (1998)	3	MFI	130@5500	127@4000	3.39x3.39	9.2:1	43-57 ①
2001	Sportage	2.0 (1998)	3	MFI	130@5500	127@4000	3.39x3.39	9.2:1	43-57 ①
2002	Sportage	2.0 (1998)	3	MFI	130@5500	127@4000	3.39x3.39	9.2:1	43-57 ①
2003	Sportage	2.0 (1998)	3	MFI	130@5500	127@4000	3.39x3.39	9.2:1	43-57 ①

MFI: Multi-port Fuel Injection

① The manufacturer does not provide an engine speed specification for oil pump pressure.

42356-KSPO-C02

ENGINE TUNE-UP SPECIFICATIONS

Year	Engine Displacement Liters (cc)	Engine VIN	Spark Plug Gap (in.)	Ignition Timing (deg.) MT	Ignition Timing (deg.) AT	Fuel Pump (psi)	Idle Speed (rpm) MT	Idle Speed (rpm) AT	Valve Clearance Intake	Valve Clearance Exhaust
2000	2.0 (1998)	3	0.039-0.043	4B	4B	38	750-850	750-850	HYD.	HYD.
2001	2.0 (1998)	3	0.039-0.043	4B	4B	38	750-850	750-850	HYD.	HYD.
2002	2.0 (1998)	3	0.039-0.043	4B	4B	38	750-850	750-850	HYD.	HYD.
2003	2.0 (1998)	3	0.039-0.043	4B	4B	38	750-850	750-850	HYD.	HYD.

B: Before Top Dead Center

HYD: Hydraulic lash adjusters

42356-KSPO-C03

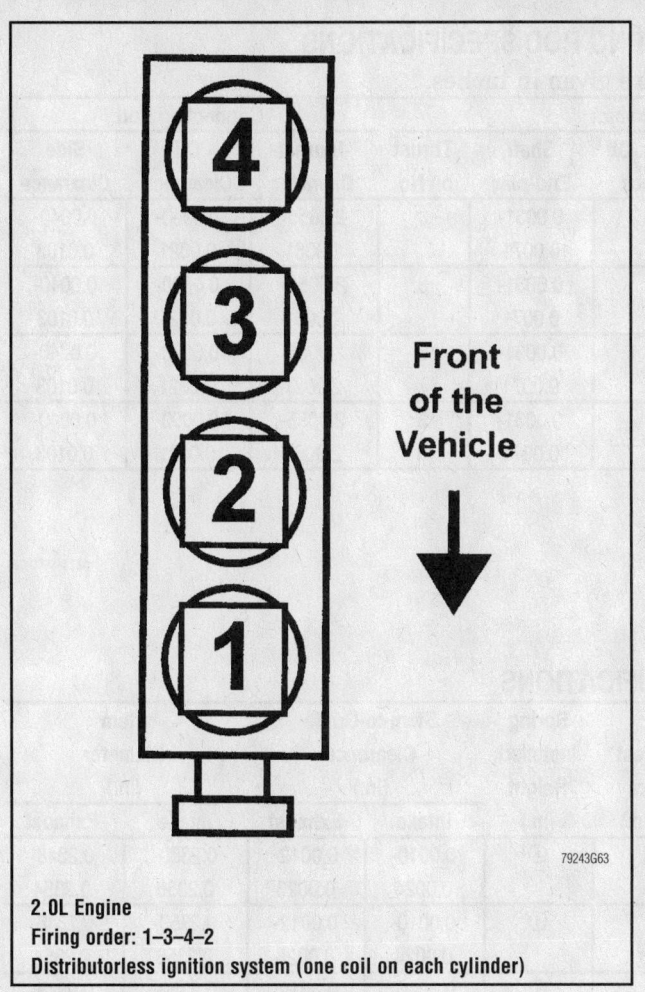

2.0L Engine
Firing order: 1-3-4-2
Distributorless ignition system (one coil on each cylinder)

Front
of the
Vehicle

79243G63

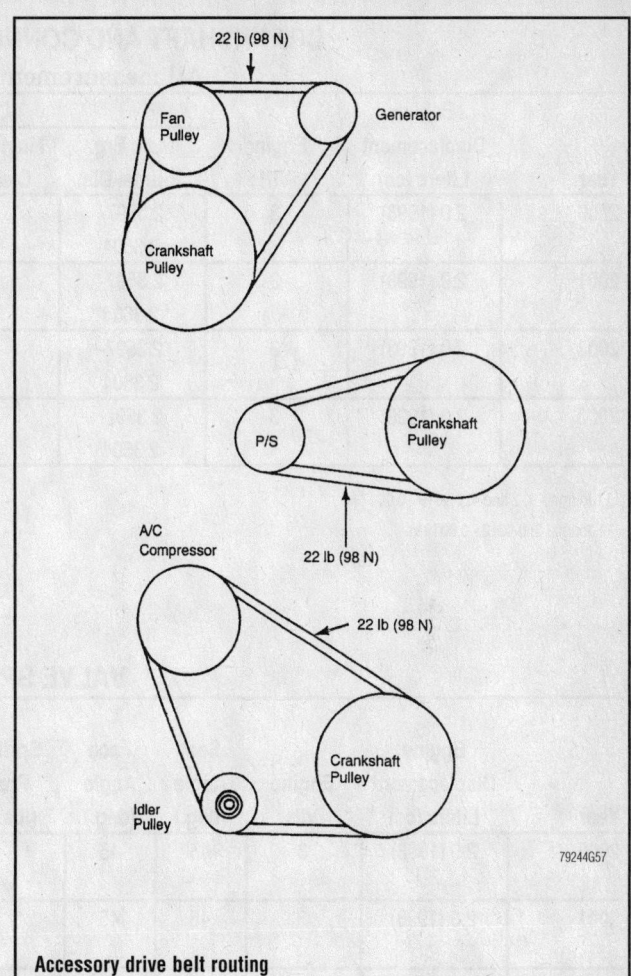

Accessory drive belt routing

79244G57

CAPACITIES

Year	Model	Engine Displacement Liters (cc)	Engine VIN	Engine Oil with Filter (qts.)	Transmission (pts.)		Transfer Case (pts.)	Drive Axle		Fuel Tank (gal.)	Cooling System (qts.)
					Manual	Auto.		Front (pts.)	Rear (pts.)		
2000	Sportage	2.0 (1998)	3	4.4	2.6	5.4	2.8	2.6	2.6	15.8	7.9
2001	Sportage	2.0 (1998)	3	4.4	2.6	5.4	2.8	2.6	2.6	15.8	7.9
2002	Sportage	2.0 (1998)	3	4.4	2.6	5.4	2.8	2.6	2.6	15.8	7.9
2003	Sportage	2.0 (1998)	3	4.4	2.6	5.4	2.8	2.6	2.6	15.8	7.9

NOTE: All capacities are approximate. Add fluid gradually and check to be sure a proper fluid level is obtained.

42356-KSPO-C04

CRANKSHAFT AND CONNECTING ROD SPECIFICATIONS

All measurements are given in inches.

Year	Engine Displacement Liters (cc)	Engine VIN	Crankshaft				Connecting Rod		
			Main Brg. Journal Dia.	Main Brg. Oil Clearance	Shaft End-play	Thrust on No.	Journal Diameter	Oil Clearance	Side Clearance
2000	2.0 (1998)	3	2.3597-2.3604	①	0.0031-0.0071	3	2.0055-2.0061	0.0090-0.0021	0.0040-0.0103
2001	2.0 (1998)	3	2.3597-2.3604	①	0.0031-0.0071	3	2.0055-2.0061	0.0090-0.0021	0.0040-0.0103
2002	2.0 (1998)	3	2.3597-2.3604	①	0.0031-0.0071	3	2.0055-2.0061	0.0090-0.0021	0.0040-0.0103
2003	2.0 (1998)	3	2.3597-2.3604	①	0.0031-0.0071	3	2.0055-2.0061	0.0090-0.0021	0.0040-0.0103

① Journals 1, 2 and 4: 0.0010 - 0.0017 in.
Journal 3: 0.0012 - 0.0019 in.

42356-KSPO-C05

VALVE SPECIFICATIONS

Year	Engine Displacement Liters (cc)	Engine VIN	Seat Angle (deg.)	Face Angle (deg.)	Spring Test Pressure (lbs. @ in.)	Spring Installed Height (in.)	Stem-to-Guide Clearance (in.)		Stem Diameter (in.)	
							Intake	Exhaust	Intake	Exhaust
2000	2.0 (1998)	3	45	45	①	①	0.0010-0.0024	0.0012-0.0026	0.2350-0.2356	0.2348-0.2354
2001	2.0 (1998)	3	45	45	①	①	0.0010-0.0024	0.0012-0.0026	0.2350-0.2356	0.2348-0.2354
2002	2.0 (1998)	3	45	45	①	①	0.0010-0.0024	0.0012-0.0026	0.2350-0.2356	0.2348-0.2354
2003	2.0 (1998)	3	45	45	①	①	0.0010-0.0024	0.0012-0.0026	0.2350-0.2356	0.2348-0.2354

① Spring test pressure or installed height not provided by the manufacturer.
Valve Spring Free Length:
Outer spring: 1.524-1.539 in.
Inner spring: 1.484-1.496 in.

42356-KSPO-C06

PISTON AND RING SPECIFICATIONS

All measurements are given in inches.

Year	Engine Displacement Liters (cc)	Engine VIN	Piston Clearance	Ring Gap			Ring Side Clearance		
				Top Compression	Bottom Compression	Oil Control	Top Compression	Bottom Compression	Oil Control
2000	2.0 (1998)	3	0.0019-0.0024	0.006-0.012	0.008-0.014	0.008-0.028	0.001-0.003	0.001-0.003	SNUG
2001	2.0 (1998)	3	0.0019-0.0024	0.006-0.012	0.008-0.014	0.008-0.028	0.001-0.003	0.001-0.003	SNUG
2002	2.0 (1998)	3	0.0019-0.0024	0.006-0.012	0.008-0.014	0.008-0.028	0.001-0.003	0.001-0.003	SNUG
2003	2.0 (1998)	3	0.0019-0.0024	0.006-0.012	0.008-0.014	0.008-0.028	0.001-0.003	0.001-0.003	SNUG

42356-KSPO-C07

TORQUE SPECIFICATIONS

All readings in ft. lbs.

Year	Engine Displacement Liters (cc)	Engine VIN	Cylinder Head Bolts	Main Bearing Bolts	Rod Bearing Bolts	Crankshaft Damper Bolts	Flywheel Bolts	Manifold Intake	Manifold Exhaust	Spark Plugs	Lug Nuts
2000	2.0 (1998)	3	62	63	50	11	73	16	31	11-17	73
2001	2.0 (1998)	3	62	63	50	11	73	16	31	11-17	73
2002	2.0 (1998)	3	62	63	50	11	73	16	31	11-17	73
2003	2.0 (1998)	3	62	63	50	11	73	16	31	11-17	73

42356-KSPO-C08

WHEEL ALIGNMENT

Year	Model		Caster Range (+/-Deg.)	Caster Preferred Setting (Deg.)	Camber Range (+/-Deg.)	Camber Preferred Setting (Deg.)	Toe-in (in.)	Steering Axis Inclination (Deg.)
2000	Sportage	F	0.75	+3.58	0.75	+0.44	0.10+/-0.10	—
		R	—	—	—	0	0	—
2001	Sportage	F	0.75	+3.58	0.75	+0.44	0.10+/-0.10	—
		R	—	—	—	0	0	—
2002	Sportage	F	0.75	+3.58	0.75	+0.44	0.10+/-0.10	—
		R	—	—	—	0	0	—
2003	Sportage	F	0.75	+3.58	0.75	+0.44	0.10+/-0.10	—
		R	—	—	—	0	0	—

42356-KSPO-C09

TIRE, WHEEL AND BALL JOINT SPECIFICATIONS

| Year | Model | OEM Tires | | Tire Pressures (psi) | | Wheel Size | Ball Joint Inspection |
		Standard	Optional	Front	Rear		
2000	Sportage	P205/75R15	none	26	26	6-JJ	①
2001	Sportage	P205/75R15	none	26	26	6-JJ	①
2002	Sportage	P205/75R15	none	26	26	6-JJ	①
2003	Sportage	P205/75R15	none	26	26	6-JJ	①

OEM: Original Equipment Manufacturer

PSI: Pounds Per Square Inch

STD: Standard

OPT: Optional

① Replace if any measurable movement is found.

42356-KSPO-C10

BRAKE SPECIFICATIONS

All measurements in inches unless noted

| Year | Model | Brake Disc | | | Brake Drum | | | Minimum Lining Thickness | | Brake Caliper Mounting Bolts (ft. lbs.) |
		Original Thickness	Minimum Thickness	Maximum Run-out	Original Inside Diameter	Max. Wear Limit	Maximum Machine Diameter	Front	Rear	
2000	Sportage	0.940	0.880	0.004	NA	9.89	NA	0.080	0.060	72
2001	Sportage	0.940	0.880	0.004	NA	9.89	NA	0.080	0.060	72
2002	Sportage	0.940	0.880	0.004	NA	9.89	NA	0.080	0.060	72
2003	Sportage	0.940	0.880	0.004	NA	9.89	NA	0.080	0.060	72

NA: Not Available

42356-KSPO-C11

SCHEDULED MAINTENANCE INTERVALS
Kia—Sportage

TO BE SERVICED	TYPE OF SERVICE	7.5	15	22.5	30	37.5	45	52.5	60	67.5	75	82.5	90	97.5	105	112.5	120
																	VEHICLE MILEAGE INTERVAL (x1000)
Accessory drive belt	S/I				✓				✓				✓				✓
Air cleaner filter	R				✓				✓				✓				✓
Automatic transmission fluid	R				✓		✓		✓		✓		✓		✓		✓
Ball joints	S/I				✓				✓				✓				✓
Brake lines & connections	S/I				✓				✓				✓				✓
Chassis/body fasteners	S/I				✓				✓				✓				✓
Cooling system	S/I				✓				✓				✓				✓
CV-joint boots	S/I		✓		✓		✓		✓		✓		✓		✓		✓
Disc brakes	S/I		✓		✓		✓		✓		✓		✓		✓		✓
Driveshaft U-joints	L		✓		✓		✓		✓		✓		✓		✓		✓
Drum brakes	S/I				✓				✓				✓				✓
Emission hoses & tubes	S/I								✓								✓
Emission hoses & tubes (Cal)	R															✓	
Engine coolant	R				✓				✓				✓				✓
Engine oil & filter	R	✓	✓	✓	✓	✓	✓	✓	✓	✓	✓	✓	✓	✓	✓	✓	✓
Exhaust system heat shields	S/I				✓				✓				✓				✓
Front differential fluid	R				✓				✓				✓				✓
	S/I	✓	✓	✓	✓	✓	✓	✓	✓	✓	✓	✓	✓	✓	✓	✓	✓
Fuel filter	R				✓				✓				✓				✓
Fuel lines & hoses	S/I				✓				✓				✓				✓
Idle speed	S/I				✓				✓				✓				✓
Locks & hinges	L	✓	✓	✓	✓	✓	✓	✓	✓	✓	✓	✓	✓	✓	✓	✓	✓
Manual transmission fluid	R				✓				✓				✓				✓
PCV valve	S/I								✓								✓
Rear differential fluid	R				✓				✓				✓				✓
	S/I	✓	✓	✓	✓	✓	✓	✓	✓	✓	✓	✓	✓	✓	✓	✓	✓
Spark plug wires	S/I								✓								✓
Spark plugs	R				✓				✓				✓				✓
Steering operation & linkage	S/I				✓				✓				✓				✓
Timing belt	R														✓		
	S/I								✓				✓				
Timing belt (non-California)	R								✓								✓
Transfer case fluid	R				✓				✓				✓				✓

42356-KSPO-C12

SCHEDULED MAINTENANCE INTERVALS
Kia—Sportage

TO BE SERVICED	TYPE OF SERVICE	VEHICLE MILEAGE INTERVAL (x1000)															
		7.5	15	22.5	30	37.5	45	52.5	60	67.5	75	82.5	90	97.5	105	112.5	120
Transfer case fluid	S/I		✓		✓		✓		✓		✓		✓		✓		✓
Transmission fluid	S/I	✓	✓	✓	✓	✓	✓	✓	✓	✓	✓	✓	✓	✓	✓	✓	✓

R: Replace   S/I: Inspect and service, if needed   L: Lubricate

FREQUENT OPERATION MAINTENANCE (SEVERE SERVICE)

If a vehicle is operated under any of the following conditions it is considered severe service:

- Towing a trailer or using a camper or car-top carrier.

- Repeated short trips of less than 5 miles in temperatures below freezing, or trips of less than 10 miles in any temperature.

- Extensive idling or low-speed driving for long distances as in heavy commercial use, such as delivery, taxi or police cars.

- Operating on rough, muddy or salt-covered roads.

- Operating on unpaved or dusty roads.

- Driving in extremely hot (over 90°) conditions.

Engine oil & filter: replace every 5000 miles or 5 months, whichever occurs first.

Air cleaner filter: inspect and replace if necessary, every 15,000 miles or 15 months, whichever occurs first.

Transfer case fluid: inspect the level every 5000 miles or 5 months, and replace every 15,000 miles or 15 months, whichever occurs first.

Transmission fluid: inspect the level every 5000 miles or 5 months, and replace every 15,000 miles or 15 months, whichever occurs first.

Front differential fluid: inspect the level every 5000 miles or 5 months, and replace every 15,000 miles or 15 months, whichever occurs first.

Rear differential fluid: inspect the level every 5000 miles or 5 months, and replace every 15,000 miles or 15 months, whichever occurs first.

42356-KSPO-C13

PRECAUTIONS

Before servicing any vehicle, please be sure to read all of the following precautions, which deal with personal safety, prevention of component damage, and important points to take into consideration when servicing a motor vehicle:

• Never open, service or drain the radiator or cooling system when the engine is hot; serious burns can occur from the steam and hot coolant.

• Observe all applicable safety precautions when working around fuel. Whenever servicing the fuel system, always work in a well-ventilated area. Do not allow fuel spray or vapors to come in contact with a spark, open flame, or excessive heat (a hot drop light, for example). Keep a dry chemical fire extinguisher near the work area. Always keep fuel in a container specifically designed for fuel storage; also, always properly seal fuel containers to avoid the possibility of fire or explosion. Refer to the additional fuel system precautions later in this section.

• Fuel injection systems often remain pressurized, even after the engine has been turned **OFF**. The fuel system pressure must be relieved before disconnecting any fuel lines. Failure to do so may result in fire and/or personal injury.

• Brake fluid often contains polyglycol ethers and polyglycols. Avoid contact with the eyes and wash your hands thoroughly after handling brake fluid. If you do get brake fluid in your eyes, flush your eyes with clean, running water for 15 minutes. If eye irritation persists, or if you have taken brake fluid internally, IMMEDIATELY seek medical assistance.

• The EPA warns that prolonged contact with used engine oil may cause a number of skin disorders, including cancer! You should make every effort to minimize your exposure to used engine oil. Protective gloves should be worn when changing oil. Wash your hands and any other exposed skin areas as soon as possible after exposure to used engine oil. Soap and water, or waterless hand cleaner should be used.

• All new vehicles are now equipped with an air bag system, often referred to as a Supplemental Restraint System (SRS) or Supplemental Inflatable Restraint (SIR) system. The system must be disabled before performing service on or around system components, steering column, instrument panel components, wiring and sensors. Failure to follow safety and disabling procedures could result in accidental air bag deployment, possible personal injury and unnecessary system repairs.

• Always wear safety goggles when working with, or around, the air bag system. When carrying a non-deployed air bag, be sure the bag and trim cover are pointed away from your body. When placing a non-deployed air bag on a work surface, always face the bag and trim cover upward, away from the surface. This will reduce the motion of the module if it is accidentally deployed. Refer to the additional air bag system precautions later in this section.

• Clean, high quality brake fluid from a sealed container is essential to the safe and proper operation of the brake system. You should always buy the correct type of brake fluid for your vehicle. If the brake fluid becomes contaminated, completely flush the system with new fluid. Never reuse any brake fluid. Any brake fluid that is removed from the system should be discarded. Also, do not allow any brake fluid to come in contact with a painted surface; it will damage the paint.

• Never operate the engine without the proper amount and type of engine oil; doing so WILL result in severe engine damage.

• Timing belt maintenance is extremely important! Many models utilize an interference-type, non-freewheeling engine. If the timing belt breaks, the valves in the cylinder head may strike the pistons, causing potentially serious (also time-consuming and expensive) engine damage. Refer to the maintenance interval charts in the front of this section for the recommended replacement interval for the timing belt, and to the timing belt procedure in this section for belt replacement and inspection.

• Disconnecting the negative battery cable on some vehicles may interfere with the functions of the on-board computer system(s) and may require the computer to undergo a relearning process once the negative battery cable is reconnected.

• When servicing drum brakes, only disassemble and assemble one side at a time, leaving the remaining side intact for reference.

• Only an MVAC-trained, EPA-certified automotive technician should service the A/C system or its components.

ENGINE REPAIR

Alternator

REMOVAL

1. Before servicing the vehicle, refer to the precautions in the beginning of this section.
2. Remove or disconnect the following:
   • Negative battery cable
   • Air cleaner inlet pipe front bolts
   • Top hose from the resonance chamber
   • Air cleaner inlet pipe
   • Alternator electrical connectors
   • Loosen the pivot and tensioner mounting bolts, do mot remove them

• Drive belt from the alternator pulley
• Drive belt tensioner
• Alternator pivot bolt
• Loosen the bolt at the base of the adjusting bracket and rotate the bracket up
• Alternator

To install:
3. Install or connect the following:
   • Alternator
   • Pivot bolt and hand tighten
   • Rotate the bracket down on top of the alternator
   • Belt tensioner on the adjustment bracket
   • Tensioner mounting bolt and hand tighten it
   • Drive belt

• Torque the tensioner bolt to 19 ft. lbs. (26 Nm).
• Torque the pivot bolt to 38 ft. lbs. (51 Nm).
• Alternator electrical connectors
• Air cleaner inlet pipe and tighten the clamp
• Hose to the resonance chamber
• Air inlet pipe bolts and tighten
• Negative battery cable

Ignition Timing

ADJUSTMENT

The 2.0L engine in the Sportage is equipped with a distributorless ignition system. The ignition timing is controlled by the Powertrain Control Module (PCM) through the input of engine control system sensors. The ignition timing is set at 4 degrees BTDC for vehicles equipped with manual or automatic transmissions. The ignition timing cannot be adjusted.

Engine Assembly

REMOVAL & INSTALLATION

1. Before servicing the vehicle, refer to the precautions in the beginning of this section .
2. Properly relieve the fuel system pressure.
3. Drain the cooling system.
4. Drain the engine oil.
5. Drain the transmission fluid.
6. Remove or disconnect the following:
   - Both battery cables
   - Windshield washer hose from the hood
   - Hood
   - 2 air duct mounting bolts from the top of the radiator. Loosen the clamp at the air intake housing and remove the duct
   - Accelerator cable by pulling the throttle back and rotating the cable until it aligns with the slot in the pulley
   - Transmission control cable
   - Resonance chamber mounting bolt, chamber bolt and air silencer
   - Idle Air Control (IAC) air hose, breather hose and vacuum line from the air intake tube
   - Manifold Air Flow (MAF) sensor connector
   - Loosen the air inlet hose clamp from the MAF sensor
   - 3 bolts from the air intake tube to the throttle body
   - Air intake hose and tube as an assembly
   - Upper radiator hose
   - Clutch fan nuts
   - Cooling fan shroud bolts
   - Fan and shroud at the same time
   - Alternator drive belt
   - Fan pulley
   - Alternator electrical connectors
   - Exhaust Gas Recirculation (EGR) solenoid valve connector on the intake manifold in front of the dynamic chamber
   - Both heater hoses from the pipes
   - Engine-to-body ground wire from the intake manifold and the harness bracket

- Brake booster vacuum hose from the dynamic chamber
- Fuel lines and fuel pressure regulator from the rear of the dynamic chamber
- Vacuum hose from the bottom of the EGR valve
- Purge solenoid valve vacuum hose from dynamic chamber
- Vacuum hoses from the top of the charcoal canister and slide the charcoal canister up and out of the bracket
- Lower radiator hose
- Cooling lines from the radiator, if equipped with an automatic transmission
- Radiator and raise and safely support the vehicle
- Lower splash panel
- Drive belt
- A/C pulley assembly
- A/C compressor mounting bolts and move the A/C compressor out of the way
- Power steering drive belt
- Intake manifold support bracket
- Starter wiring harness
- Starter
- Front exhaust pipe from the exhaust manifold
- Bracket bolt from the front exhaust pipe
- Exhaust-to-clutch (manual transmission) or converter (automatic transmission) housing bolts and the bracket
- Clutch housing (manual transmission) or converter (automatic transmission) housing bolts.
- Drive plate-to-torque converter bolts, if equipped

7. Support the transmission from underneath the vehicle.
8. Connect the engine hoist to the engine assembly.
   - Left and right side engine mounting bolts
9. Lift the engine up and forward slightly to provide access to three electrical connectors on the rear of the cylinder head.
   - Electrical connectors from the Camshaft Position (CMP) sensor, coil and condenser on the rear of the cylinder head
   - Engine from the vehicle

To install:

10. Lower the engine enough to connect the three electrical connectors to the CMP sensor, coil and condenser on the rear of the cylinder head.
11. Position the engine to the transmis-

sion. Install the transmission bolts and tighten the bolts. Torque according to bolt size:
- 14mm bolts to 80 ft. lbs. (108 Nm)
- 10mm bolts to 28 ft. lbs. (38 Nm)
- 6mm bolts to 60 inch lbs. (7 Nm)
- Right and left side engine mounting bolts. Torque the bolt s to 38 ft. lbs. (52 Nm).

12. Disconnect the engine hoist from the engine assembly.
13. Raise and safely support the vehicle.
   - Drive plate-to-torque converter bolts, if equipped.
   - Connect the front exhaust pipe to the exhaust manifold. Torque the flange bolts to 24 ft. lbs. (31 Nm).
   - Front exhaust pipe. Torque the bolts to 20 ft. lbs. (27 Nm).
   - Starter
   - Connect the starter wiring harness
   - Intake manifold support bracket bolts and the bracket.

14. Install the power steering pump lock and mounting bolts. Install the power steering drive belt. Torque the bolts to 30 ft. lbs. (42 Nm).
   - A/C compressor mounting bolts. Torque the bolts to 18 ft. lbs. (24 Nm).
   - A/C belt pulley assembly and drive belt. Install the two A/C idler pulley bracket bolts and torque to 24 ft. lbs. (32 Nm).
   - Lower splash panel
   - Radiator
   - Cooling lines to the radiator, if equipped
   - Lower radiator hose
   - Slide the charcoal canister in the bracket
   - Vacuum hoses to the top of the charcoal canister
   - Engine-to-body ground wire to the intake manifold and the harness bracket
   - Brake booster vacuum hose to the dynamic chamber
   - Fuel lines and fuel pressure regulator to the rear of the dynamic chamber
   - Vacuum hose to the bottom of the EGR valve
   - Purge solenoid valve vacuum hose to dynamic chamber
   - EGR solenoid valve connector on the intake manifold in front of the dynamic chamber
   - Heater hoses
   - Electrical terminal connectors to the alternator
   - Fan pulley

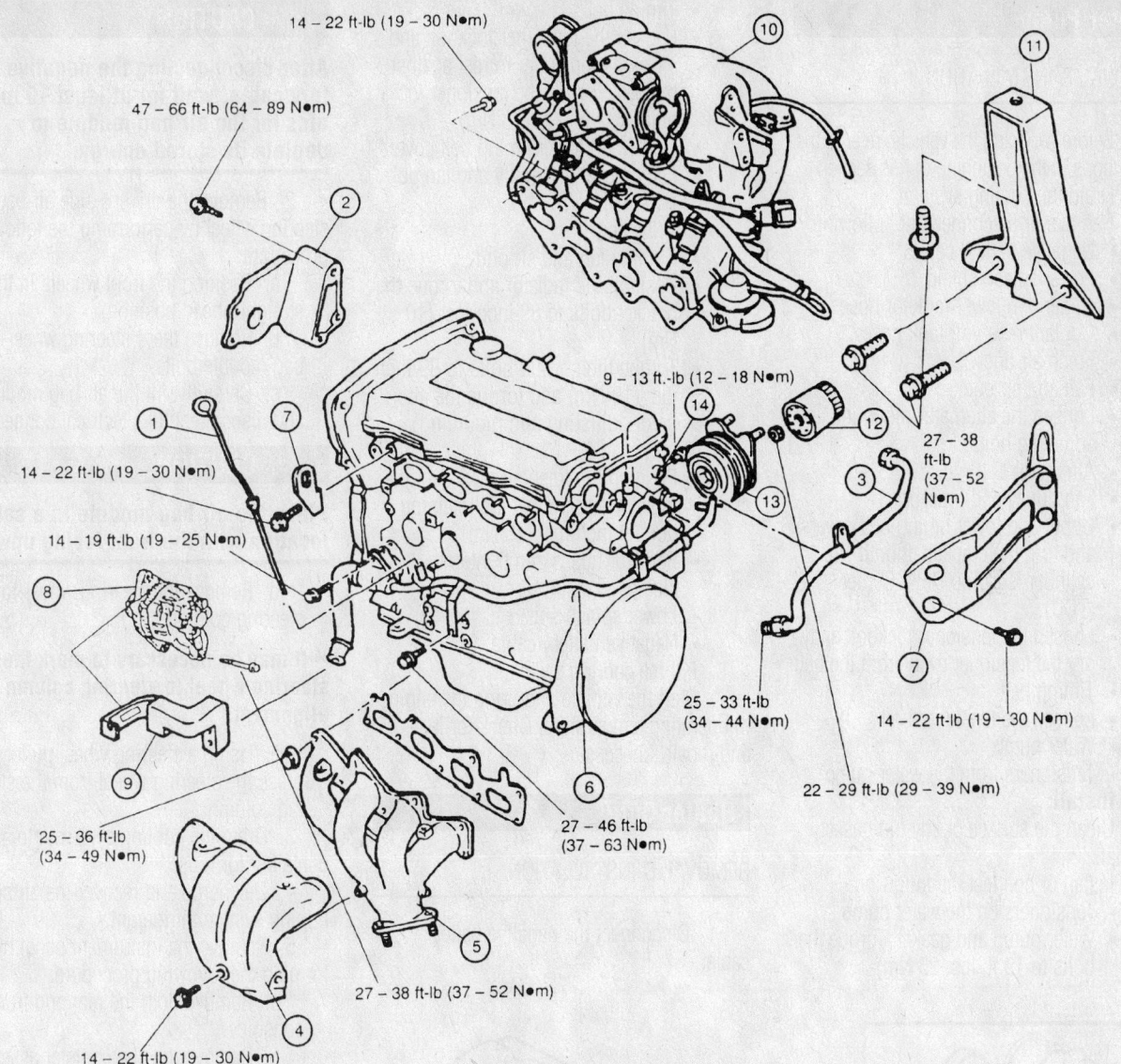

14 – 22 ft-lb (19 – 30 N•m)

47 – 66 ft-lb (64 – 89 N•m)

14 – 22 ft-lb (19 – 30 N•m)

14 – 19 ft-lb (19 – 25 N•m)

9 – 13 ft.-lb (12 – 18 N•m)

27 – 38 ft-lb (37 – 52 N•m)

25 – 33 ft-lb (34 – 44 N•m)

14 – 22 ft-lb (19 – 30 N•m)

22 – 29 ft-lb (29 – 39 N•m)

27 – 46 ft-lb (37 – 63 N•m)

25 – 36 ft-lb (34 – 49 N•m)

27 – 38 ft-lb (37 – 52 N•m)

14 – 22 ft-lb (19 – 30 N•m)

1. Oil Level Gauge	7. Engine Hanger	13. Oil Cooler
2. Thermo-Modulated Fan Bracket	8. Generator	14. Oil Pressure Switch
3. EGR Pipe	9. Generator Strap and Bracket	
4. Exhaust Manifold Heat Shield	10. Intake Manifold Assembly	
5. Exhaust Manifold	11. Intake Manifold Support Bracket	
6. Coolant Inlet Pipe and Bypass Pipe	12. Oil Filter	

7924QG36

Exploded view of some peripheral engine component mountings

- Alternator drive belt. Torque the adjusting bolt 16 ft. lbs. (22 Nm) and the mounting bolt to 32 ft. lbs. (45 Nm).
- Fan and shroud as an assembly. Torque the five cooling fan shroud bolts to 72 inch lbs. (8 Nm)
- Clutch fan nuts. Torque the nuts to 27 ft. lbs. (37 Nm).
- Upper radiator hose
- Air intake hose and tube as an assembly
- Air intake tube to the throttle body

and tighten the air inlet hose clamp to the MAF sensor
- MAF sensor electrical connector
- Resonance chamber mounting bolt, chamber bolt and air silencer
- IAC air hose, breather hose and vacuum line to the air intake tube
- Accelerator cable by pulling the throttle back and rotating the cable until it aligns with the slot in the pulley
- Transmission control cable
- Air duct mounting bolts to the top

of the radiator. Torque the clamp at the air intake housing
- Hood
- Windshield washer hose to the hood
15. Fill the engine with clean engine oil.
16. Connect the battery cables.
17. Fill the cooling system.
18. Fill the transmission fluid.
19. Recharge the A/C system.
20. Start the vehicle. Check for leaks, repair if necessary.
21. Road test the vehicle to check engine performance.

Water Pump

REMOVAL & INSTALLATION

1. Before servicing the vehicle, refer to the precautions in the beginning of this section .
2. Drain the cooling system.
3. Remove or disconnect the following:
   - Negative battery cable
   - Lower splash shield
   - Upper and lower radiator hoses
   - Coolant reservoir tank hose
   - Fresh air duct
   - Fan and shroud
   - Loosen the alternator mounting and adjusting bolts
   - Alternator belt
   - Fan pulley and bracket
   - Upper and lower timing belt covers and turn the crankshaft until No. 1 cylinder is at Top Dead Center (TDC).
   - Loosen the tensioner lockbolt and pry the tensioner away from the belt
   - Timing belt
   - Loosen the tensioner bolt
   - Water pump
   - Tensioners from the water pump

To install:

4. Clean the surface of any old gasket material.
5. Install or connect the following:
   - Tensioners on the water pump
   - Water pump and gasket. Torque the bolts to 19 ft. lbs. 25 Nm).
   - Timing belt
   - Loosen the tensioner lockbolt and allow the tensioner to rest against the belt. Torque the tensioner lockbolt to 32 ft. lbs. (43 Nm).
   - Upper and lower timing belt covers
   - Fan bracket assembly and fan pulley
   - Drive belt
   - Cooling fan and shroud
   - Position the radiator and torque the bracket bolts to 89 inch lbs. (10 Nm).
   - Torque the shroud bolts to 89 inch lbs. (10 Nm) and torque the alternator adjusting and mounting bolts
   - Position the fresh air duct over the radiator and tighten the retaining bolt 89 inch lbs. (10 Nm)
   - Radiator hoses and tighten the clamps
   - Lower splash shield
   - Negative battery cable
6. Fill the cooling system.
7. Start the vehicle and bring the engine to operating temperature. Check for leaks and repair if necessary.

Heater Core

REMOVAL & INSTALLATION

1. Disconnect the negative battery cable.

After disconnecting the negative battery cable, wait for at least 10 minutes for the air bag module to deplete its stored energy.

2. Remove the driver's side air bag and steering wheel by performing the following procedure:
   a. Position the front wheels in the straight-ahead position.
   b. Remove the 4 steering wheel-to-air bag module bolts.
   c. Carefully, lift the air bag module and disconnect the electrical connector.

Place the air bag module in a safe location with the front facing upward.

   d. Remove the steering wheel-to-steering column nut.

➡ If may be necessary to mark the steering wheel to steering column alignment.

   e. Using a steering wheel puller, press the steering wheel from the steering column.
3. Drain the cooling system into a clean container for reuse.
4. Discharge and recover the air conditioning system refrigerant.
5. Remove the instrument panel by performing the following procedure:
   a. Remove both the rear and front consoles.
   b. Remove the knee bolster assembly.
   c. Remove the "T" bar section.
   d. Remove the relay bracket.
   e. Remove the turn signal assembly and the upper/lower steering column covers.
   f. Remove the hood release handle lockscrew, the hood release handle and the cable assembly nut.
   g. Remove the left side front pillar trim and the lower left side cover.
   h. Remove the 2 left side of the "T" bar-to-chassis bolts.
   i. At the left side of the instrument panel, remove the 3 instrument panel-to-chassis bolts.
   j. Remove the ashtray.
   k. Remove the center panel trim.
   l. Remove the ventilation control panel.
   m. At the center of the windshield next to the windshield, remove the cap and the mounting bolt.
   n. Remove the right side front pillar trim and the lower right side cover.

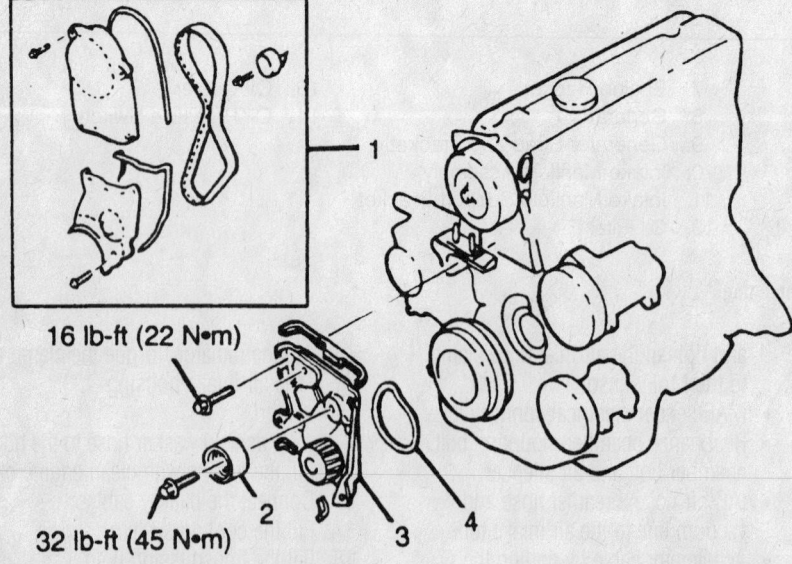

16 lb-ft (22 N•m)

32 lb-ft (45 N•m)

| 1 | TIMING BELT COVERS, GASKETS AND TIMING BELT | 3 | COOLANT PUMP |
| 2 | IDLER PULLEY | 4 | GASKET |

7924QG01

Exploded view of the water pump mounting

o. Remove the 2 right side of the "T" bar-to-chassis bolts.

p. At the right side of the instrument panel, remove the 3 instrument panel-to-chassis bolts.

q. Remove the steering column-to-instrument panel bolts and lower the steering column.

r. Disconnect the instrument panel electrical connectors.

s. Remove the instrument panel.

6. Remove the blower/evaporator housing by performing the following procedure:

a. Disconnect the air conditioning refrigerant lines from the evaporator core and discard the gaskets. Plug the openings to prevent contamination.

b. Disconnect the fresh air control cable from the blower/evaporator housing inlet duct.

c. Disconnect the 5 connectors from the bottom of the blower/evaporator housing.

d. Move the carpeting from the bulkhead to gain access to the hole cover plate.

e. Remove the 4 hole cover plate nuts and the plate.

f. Remove the 2 upper blower/evaporator housing bolts.

g. Remove the 2 lower blower/evaporator housing-to-bulkhead nuts.

h. Remove the blower/evaporator housing.

7. Disconnect the heater hoses from the heater core.

8. Remove the temperature control cable from the heater housing.

9. Remove the 2 lower heater housing nuts and the upper heater housing-to-bulkhead nut.

10. Remove the heater housing.

11. Disassemble the heater housing by performing the following procedure:

a. Remove the seal from the heater core tube connections.

b. Remove the vent seal.

c. Remove the 2 wiring harness-to-heater servo screws.

d. Remove the 8 heater housing clips located on the servo side (left side).

e. Remove the left side of the heater housing.

f. Remove the 6 heater housing assembly clips.

g. Remove the 4 heater core tube mounting bracket screws and the bracket.

h. Remove the 8 remaining heater housing clips and disassemble the housings.

i. Remove the heater core from the housing.

To install:

12. Assemble the heater housing by performing the following procedure:

a. Install the heater core to the housing.

b. Assemble the housings and install the 8 remaining heater housing clips.

c. Install the heater core tube mounting bracket and the 4 bracket screws.

d. Install the 6 heater housing assembly clips.

e. Install the left side of the heater housing.

f. Install the 8 heater housing clips located on the servo side (left side).

g. Install the 2 wiring harness-to-heater servo screws.

h. Install the vent seal.

i. Install the seal to the heater core tube connections.

13. Install the heater housing.

14. Install the 2 lower heater housing nuts and the upper heater housing-to-bulkhead nut.

15. Install the temperature control cable to the heater housing.

16. Connect the heater hoses to the heater core.

17. Install the blower/evaporator housing by performing the following procedure:

a. Install the blower/evaporator housing.

b. Install the 2 lower blower/evaporator housing-to-bulkhead nuts.

c. Install the 2 upper blower/evaporator housing bolts.

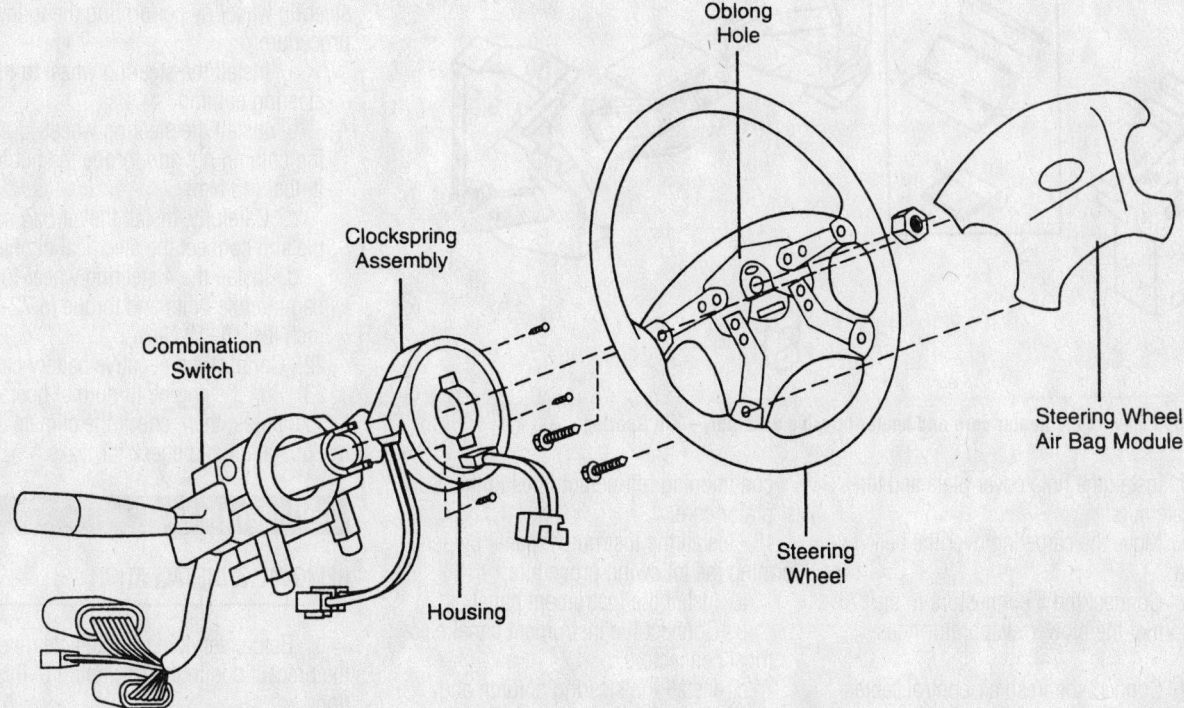

Exploded view of the steering wheel and air bag module assembly—Kia Sportage

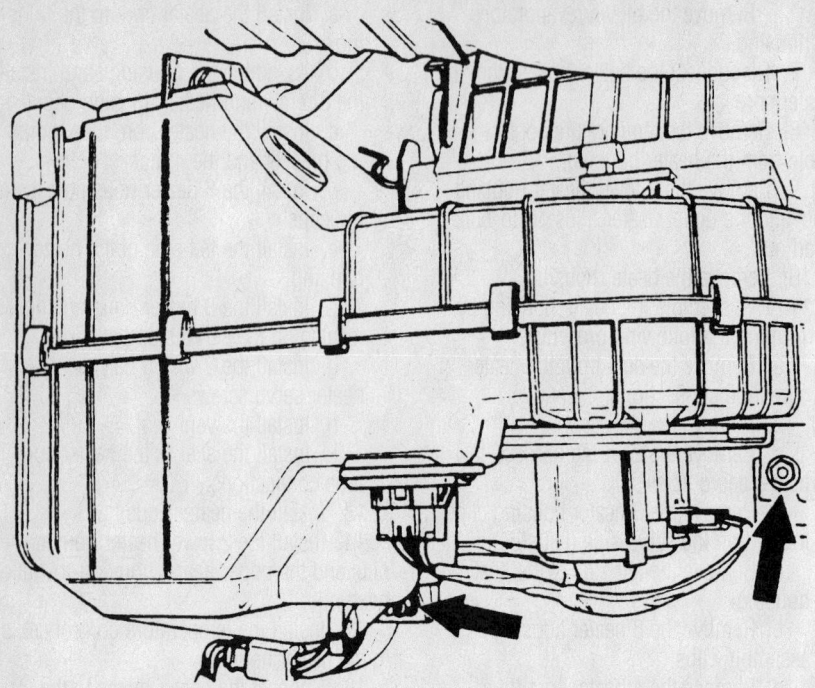

93113GI5

View of the blower/evaporator housing assembly—Kia Sportage

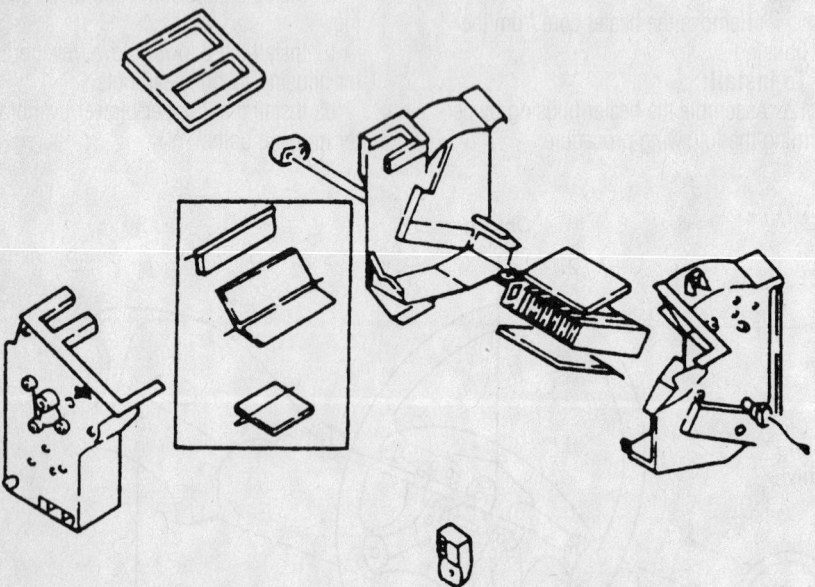

93113GI6

Exploded view of the heater core and heater housing assembly—Kia Sportage

d. Install the hole cover plate and the 4 plate nuts.

e. Move the carpeting over the bulkhead.

f. Connect the 5 connectors to the bottom of the blower/evaporator housing.

g. Connect the fresh air control cable to the blower/evaporator housing inlet duct.

h. Using new gaskets, connect the air conditioning refrigerant lines to the evaporator core.

18. Install the instrument panel by performing the following procedure:

a. Install the instrument panel.

b. Connect the instrument panel electrical connectors.

c. Install the steering column and lower the steering column-to-instrument panel bolts. Torque the bolts to 15 ft. lbs. (20 Nm).

d. At the right side of the instrument panel, install the 3 instrument panel-to-chassis bolts.

e. Install the 2 right side of the "T" bar-to-chassis bolts.

f. Install the right side front pillar trim and the lower right side cover.

g. At the center of the windshield next to the windshield, install the cap and the mounting bolt.

h. Install the ventilation control panel.

i. Install the center panel trim.

j. Install the ashtray.

k. At the left side of the instrument panel, install the 3 instrument panel-to-chassis bolts.

l. Install the 2 left side of the "T" bar-to-chassis bolts.

m. Install the lower left side cover and the left side front pillar trim.

n. Install the cable assembly nut, the hood release handle and the hood release handle lockscrew.

o. Install the turn signal assembly and the upper/lower steering column covers.

p. Install the relay bracket.

q. Install the "T" bar section.

r. Install the knee bolster assembly.

s. Install both the rear and front consoles.

19. Evacuate and charge the air conditioning system refrigerant.

20. Refill the cooling system.

21. Install the driver's side air bag and steering wheel by performing the following procedure:

a. Install the steering wheel to the steering column.

b. Install the steering wheel-to-steering column nut and torque the nut to 33 ft. lbs. (45 Nm).

c. Carefully, install the air bag module and connect the electrical connector.

d. Install the 4 steering wheel-to-air bag module bolts and torque to 72–106 inch lbs. (8–12 Nm).

22. Connect the negative battery cable.

23. Run the engine to normal operating temperatures; then, check the climate control operation and check for leaks.

Cylinder Head

REMOVAL & INSTALLATION

1. Before servicing the vehicle, refer to the precautions in the beginning of this section.

2. Properly relieve the fuel system pressure.

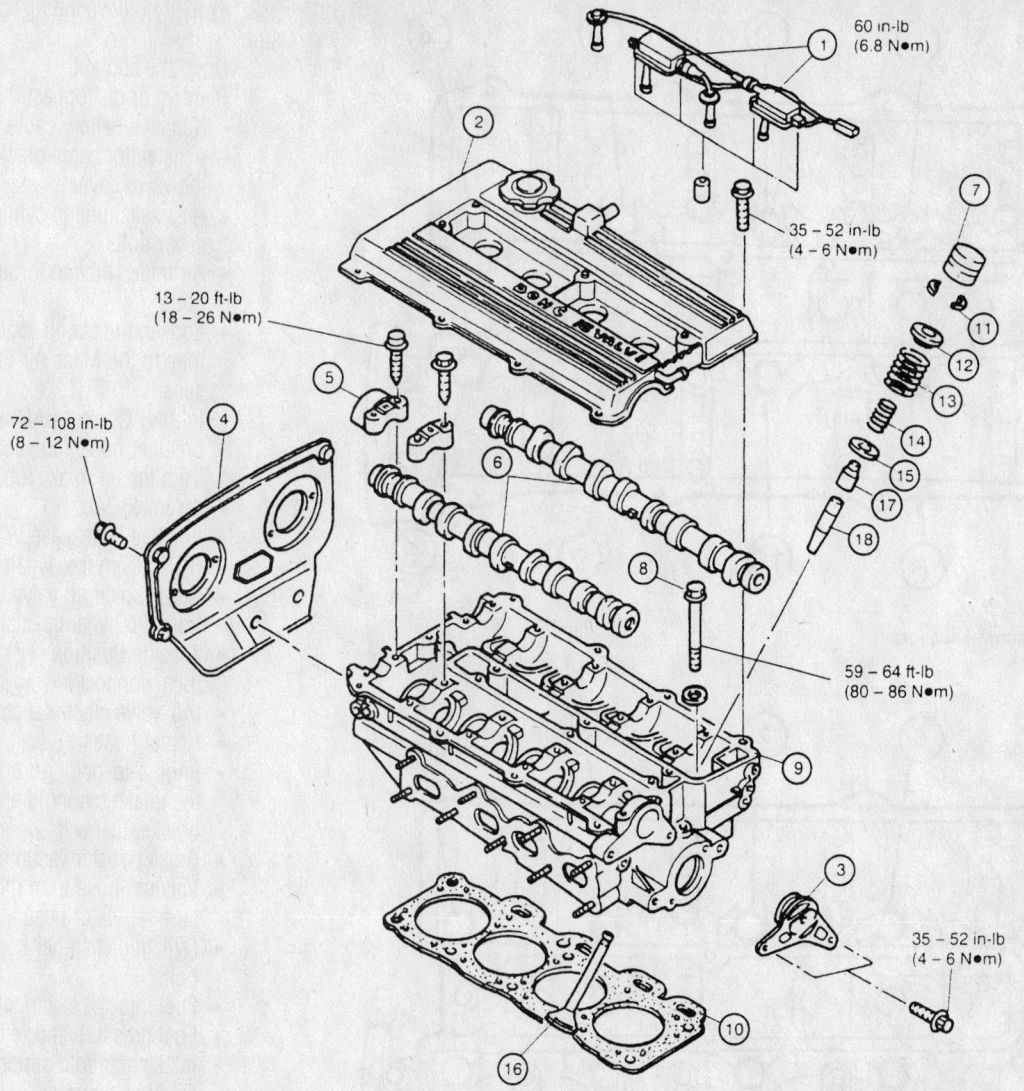

60 in-lb
(6.8 N•m)

35 – 52 in-lb
(4 – 6 N•m)

13 – 20 ft-lb
(18 – 26 N•m)

72 – 108 in-lb
(8 – 12 N•m)

59 – 64 ft-lb
(80 – 86 N•m)

35 – 52 in-lb
(4 – 6 N•m)

1. Ignition Coils and High Tension Leads
2. Cylinder Head Cover
3. Camshaft Position Sensor
4. Seal Plate
5. Camshaft Caps
6. Camshafts

7. Hydraulic Lash Adjuster
8. Cylinder Head Bolt
9. Cylinder Head
10. Cylinder Head Gasket
11. Valve Locks
12. Upper Spring Seat

13. Outer Valve Spring
14. Inner Valve Spring
15. Lower Spring Seat
16. Valve
17. Valve Stem Seal
18. Valve Guide

7924QG37

Exploded view of the cylinder head assembly

3. Drain the cooling system.
4. Remove or disconnect the following:
   - Negative battery cable
   - Brake booster vacuum hose from the dynamic chamber
   - Fuel line from the pressure regulator and the return line from the rear of the dynamic chamber
   - Ground wire from the intake manifold
   - Purge solenoid valve vacuum hose
   - Upper radiator hose
   - Intake manifold support bracket
   - Converter flange inlet pipe

   - Timing belt
   - Cylinder head cover
   - Cylinder head with the intake and exhaust manifolds attached
   - 3 wire harness connectors at the rear of the cylinder head

To install:

5. Place the new head gasket on the engine block.
6. Install or connect the following:
   - Cylinder head with the intake and exhaust manifolds attached
   - 3 wiring connectors at the rear of the cylinder head

   - Cylinder head bolts in the proper sequence. Torque the bolts to 64 ft. lbs. (87 Nm).
   - Cylinder head cover
   - Timing belt
   - Converter inlet pipe flange nuts. Torque the nuts to 24 ft. lbs. (33 Nm).
   - Upper radiator hose
   - Vacuum hose from the intake manifold to the charcoal canister
   - Purge solenoid vacuum hose
   - Ground wire and harness bracket to the intake manifold. Torque the bolts to 18 ft. lbs. (25 Nm).

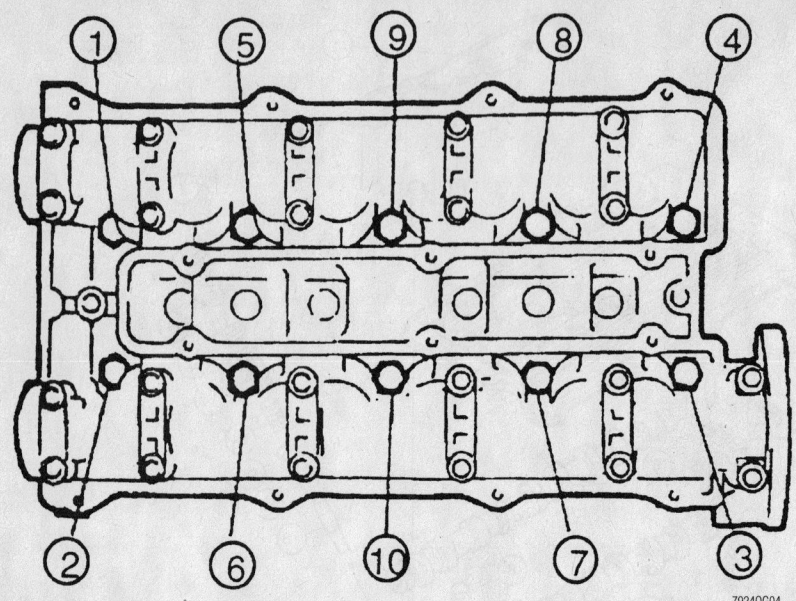

Cylinder head removal sequence

7924QG04

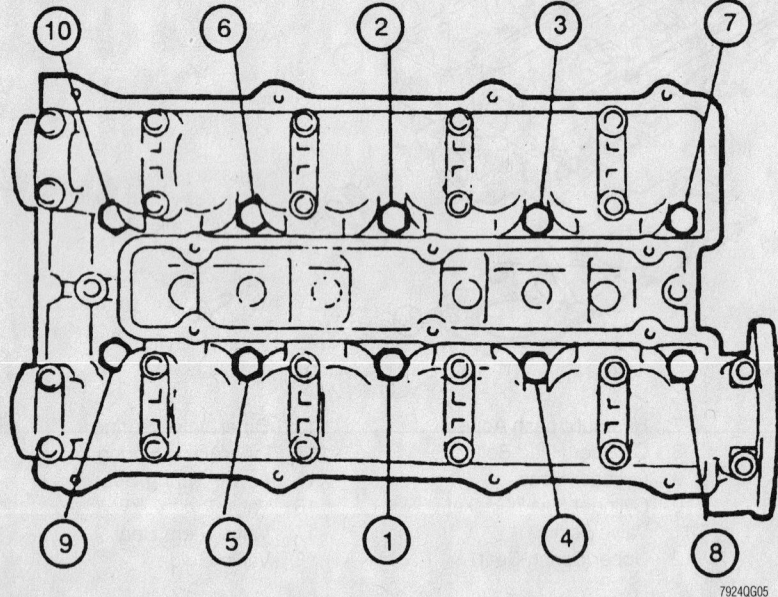

Cylinder head torque sequence—Kia Sportage

7924QG05

2. Properly relieve the fuel system pressure.
3. Drain the coolant.
4. Remove or disconnect the following:
   - Negative battery cable
   - Accelerator cable bracket bolts from the valve cover
   - Air intake tube to cylinder head cover bolts
   - Air intake tube to throttle body bolts
   - Loosen the clamp attaching the air tube to the Mass Air Flow (MAF) sensor
   - Idle Air Control (IAC) valve, breather hose and vacuum line from the air intake tube
   - Air intake tube
   - Positive Crankcase Ventilation hose (PCV) from the dynamic chamber
   - Purge solenoid valve vacuum hose from the dynamic chamber
   - Throttle Position (TP) sensor electrical connector
   - IAC valve electrical connector
   - Heater hoses
   - Engine-to-body ground strap from the intake manifold and the harness bracket below it
   - Brake booster vacuum line
   - Vacuum hose from the fuel pressure regulator hose
   - Dynamic chamber support bracket bolts
   - Fuel injector electrical connectors
   - Fuel pressure and return lines
   - Intake manifold support bracket
   - Oil filter
   - Intake manifold bolts
   - Bypass pipe from the heater hose
   - Intake manifold and discard the gasket

To install:
5. Install or connect the following:
   - Intake manifold with a new gasket to the cylinder head

   - Fuel line to the pressure regulator and the return line to the fuel rail
   - Brake booster vacuum hose
   - Negative battery cable
7. Properly fill the cooling system.
8. Start the engine and check for leaks, repair if necessary..

Intake Manifold

REMOVAL & INSTALLATION

1. Before servicing the vehicle, refer to the precautions in the beginning of this section .

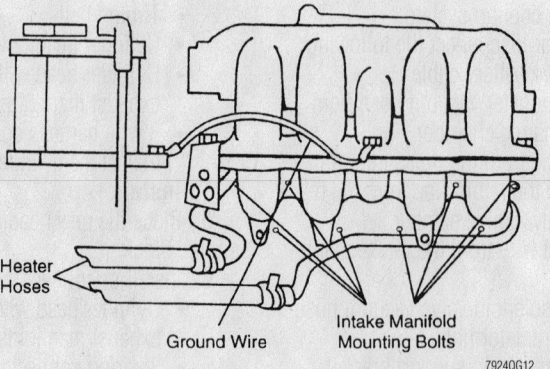

Heater Hoses

Ground Wire

Intake Manifold Mounting Bolts

7924QG12

Intake manifold mounting bolt locations. Be sure to connect the ground cable

- Heater hose to the bypass pipe
- Bolts and nuts attaching the intake manifold to the cylinder head. Torque the bolts to 14–22 ft. lbs. (19–30 Nm).
- New oil filter
- Intake manifold support bracket. Torque the bolts to 27–38 ft. lbs. (37–52 Nm)
- Fuel lines
- Fuel injector electrical connectors
- Engine to body ground wire. Torque the bolt to 18 ft. lbs. (25 Nm).
- Dynamic chamber support bracket. Torque the bolts to 18 ft. lbs. (25 Nm).
- Purge solenoid valve vacuum hose to the dynamic chamber

- Vacuum hose to the fuel pressure regulator
- Coolant hoses to the throttle body
- IAC valve electrical connector
- Brake booster vacuum hose
- Heater hoses
- TP sensor electrical connector
- Air intake tube and hose to the throttle body. Torque the bolts to 16 ft. lbs. (22 Nm).
- PCV hose to the dynamic chamber
- Accelerator cable to the throttle body pulley
- Accelerator cable bracket. Torque the bolts to 10 ft. lbs. (15 Nm).
- Air intake hose to the MAF sensor
- Resonance chamber

- IAC hose, breather hose and vacuum line to the intake manifold
- Negative battery cable
6. Properly fill the cooling system.
7. Start the engine and check for leaks, repair if necessary.

Exhaust Manifold

REMOVAL & INSTALLATION

1. Before servicing the vehicle, refer to the precautions in the beginning of this section .
2. Remove or disconnect the following:
- Negative battery cable
- Air intake hose
- Exhaust manifold heat shield

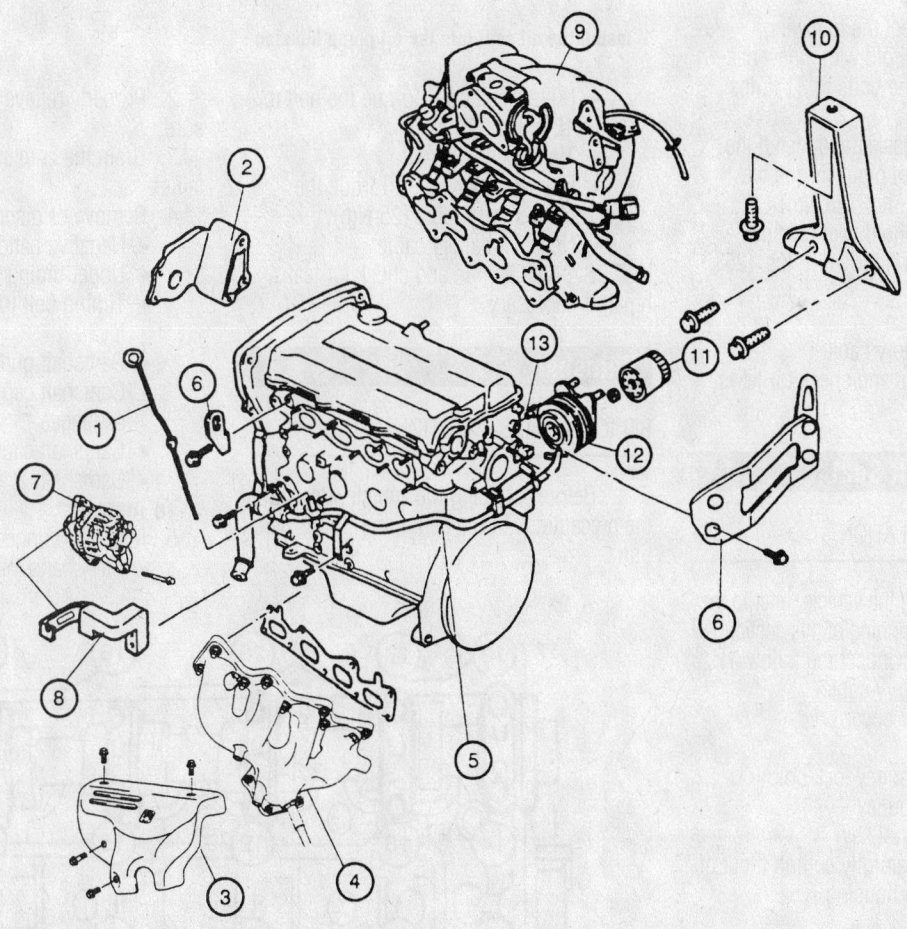

1. Oil Level Gauge
2. Thermo-Modulated Fan Bracket
3. Exhaust Manifold Heat Shield
4. Exhaust Manifold
5. Coolant Inlet Pipe and Bypass Pipe

6. Engine Hanger
7. Generator
8. Generator Strap and Bracket
9. Intake Manifold Assembly
10. Intake Support Bracket

11. Oil Filter
12. Oil Cooler
13. Oil Pressure Switch

9308QG01

Exploded view of the intake, exhaust manifold and related components

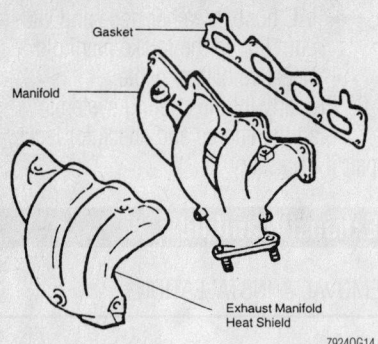

Exploded view of the exhaust manifold assembly

- Converter inlet pipe flange locknuts
- Exhaust manifold and discard the gasket
3. Clean the mating surfaces.

To install:

4. Install or connect the following:
- Exhaust manifold with a new gasket. Torque the bolts to 31 ft. lbs. (42 Nm).
- New flange gasket and install the converter inlet pipe. Torque the bolts to 24 ft. lbs. (33 Nm).
- Exhaust manifold heat shield. Torque the bolts to 18 ft. lbs. (25 Nm).
- Air intake hose
- Negative battery cable
5. Start the vehicle and check for leaks, repair if necessary.

Front Crankshaft Seal

REMOVAL & INSTALLATION

1. Before servicing the vehicle, refer to the precautions in the beginning of this section .
2. Remove or disconnect the following:
- Negative battery cable
- Engine under cover
- Timing belt
- Timing belt pulley lock bolt
- Timing belt pulley
- Pulley woodruff key
- Oil seal by carefully cutting it out of the oil pump housing

To install:

3. Lubricate the lip of the new seal with clean engine oil.
4. Install or connect the following:
- New oil seal into the oil pump housing by hand
- Press the oil seal into pump until it is flush with the edge of the oil pump body
- Timing belt pulley
- Pulley woodruff key

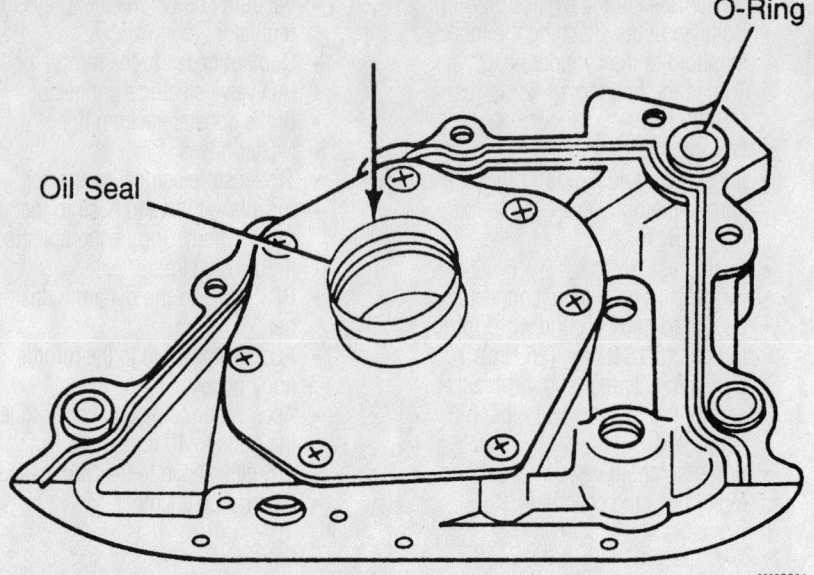

Install the oil seal into the oil pump housing

- Pulley lock bolt. Torque the bolt to 18 ft. lbs. (25 Nm).
- Timing belt
- Engine under cover. Torque the bolts to 18 ft. lbs. (25 Nm).
- Negative battery cable
5. Start the engine and check for leaks, repair if necessary.

Camshaft

REMOVAL & INSTALLATION

1. Before servicing the vehicle, refer to the precautions in the beginning of this section .

2. Properly relieve the fuel system pressure.

3. Drain the coolant into a suitable container.

4. Remove or disconnect the following:
- Negative battery cable
- Upper timing belt cover
- Timing belt from the camshaft pulley
- Camshaft pulleys
- Camshaft cap bolts in the proper sequence
- Camshaft caps
- Camshafts

To install:

5. Install or connect the following:
- Camshafts into the cylinder head.

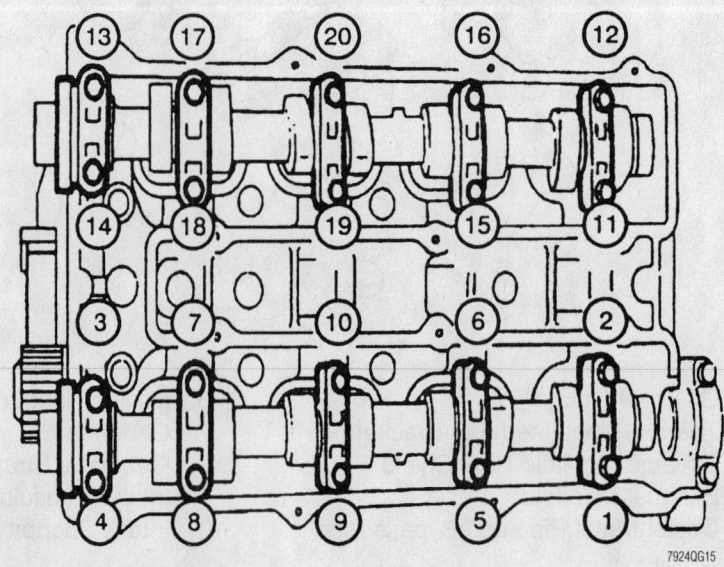

Camshaft cap bolt removal sequence

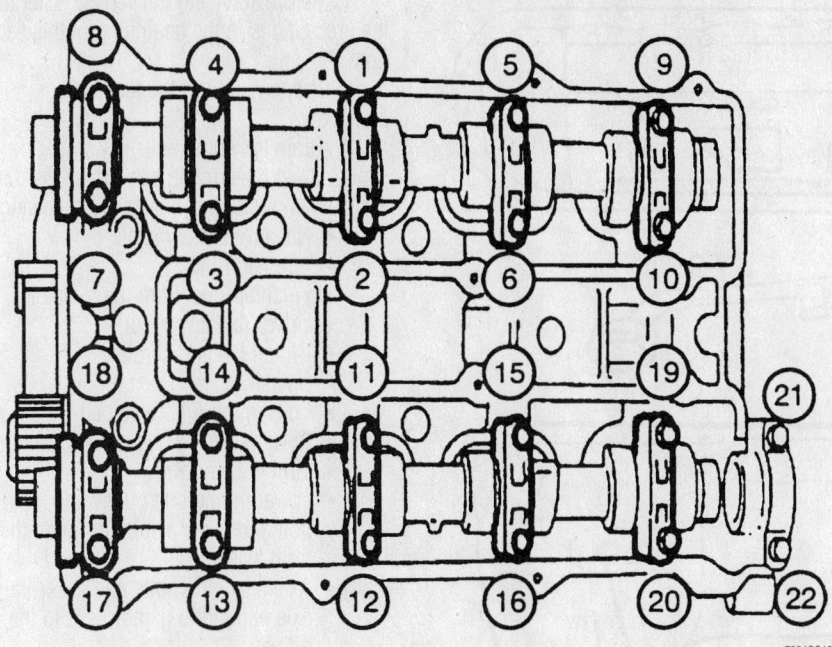

Camshaft journal bolt tightening sequence

The exhaust camshaft has a steel dowel pin at the rear for the camshaft position sensor

- Clean engine oil to the journals and bearings
- Camshaft oil seal
- Silicone sealant to the front camshaft cap and the camshaft position sensor mounting cap
- Camshaft caps in the proper sequence. Torque the bolts in three steps to 20 ft. lbs. (26 Nm).
- Camshaft pulleys
- Timing belt
- Timing belt cover
- Negative battery cable

Valve Lash

ADJUSTMENT

The DOHC engine uses Hydraulic Lash Adjusters (HLA's), which automatically maintain the proper amount of valve lash. Therefore, the DOHC engine does not need manual valve lash adjustment.

Starter

REMOVAL & INSTALLATION

1. Before servicing the vehicle, refer to the precautions in the beginning of this section .
2. Drain the engine oil.

3. Remove or disconnect the following:

- Negative battery cable
- Intake manifold bracket upper bolts
- Clutch release cylinder and move it aside, if equipped
- Intake manifold bracket lower bolts and remove the bracket
- Starter from the clutch, manual transmissions only
- Starter from the torque converter, automatic transmissions only
- Starter electrical connectors
- Move the transmission wire harness aside
- Starter

To install:
4. Install or connect the following:
- Starter to the engine well
- Starter electrical connectors
- Lower intake manifold bracket and install the upper bolts

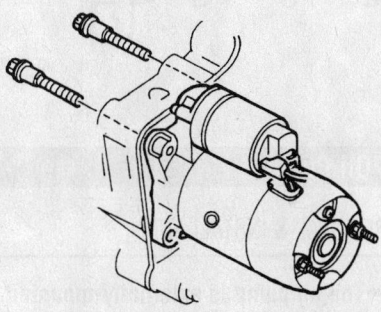

Exploded view of the starter

- Starter into position. When aligned properly torque the bolts to 40 ft. lbs. (54 Nm).
- Torque the intake manifold bracket bolts to 40 ft. lbs. (54 Nm).
- Properly position the clutch release cylinder, if equipped. Torque the bolts to 40 ft. lbs. (54 Nm).
- Torque the intake manifold bracket upper bolts to 40 ft. lbs. (54 Nm).
- Negative battery cable

Oil Pan

REMOVAL & INSTALLATION

1. Before servicing the vehicle, refer to the precautions in the beginning of this section .
2. Drain the engine oil.
3. Remove or disconnect the following:

- Negative battery cable
- 2 Top intake manifold bracket bolts
- Front 3 axle housing mounting bolts, 4WD only
- Left front bushing from the axle housing mount and lower the front axle housing
- Both gusset plates from the engine
- Transmission under cover
- Engine under cover
- Oil pan mounting bolts and using a scrapper tool separate the oil pan
- Oil pan
- Oil strainer assembly
- Oil baffle

To install:
4. Clean the engine block, oil pan and baffle pan surfaces of any gasket material.
5. Apply a continuous bead of Loctite Ultra Blue 587® silicone sealant around the baffle pan.
6. Install or connect the following:
- Oil baffle. Torque the bolt to 84 inch lbs. (9.5 Nm).
- Oil strainer. Torque the bolts to 84 inch lbs. (9.5 Nm).

7. Apply a continuous bead of Loctite Ultra Blue 587® silicone sealant around the oil pan.

- Oil pan. Torque the bolts to 84 inch lbs. (9.5 Nm).
- Transmission under cover. Torque the bolts to 84 inch lbs. (9.5 Nm).
- Gusset plates to the engine. Torque the bolts to 33 ft. lbs. (45 Nm).
- Engine under cover
- Front axle housing into position. When properly aligned, torque the bolts to 48 ft. lbs. (65 Nm).

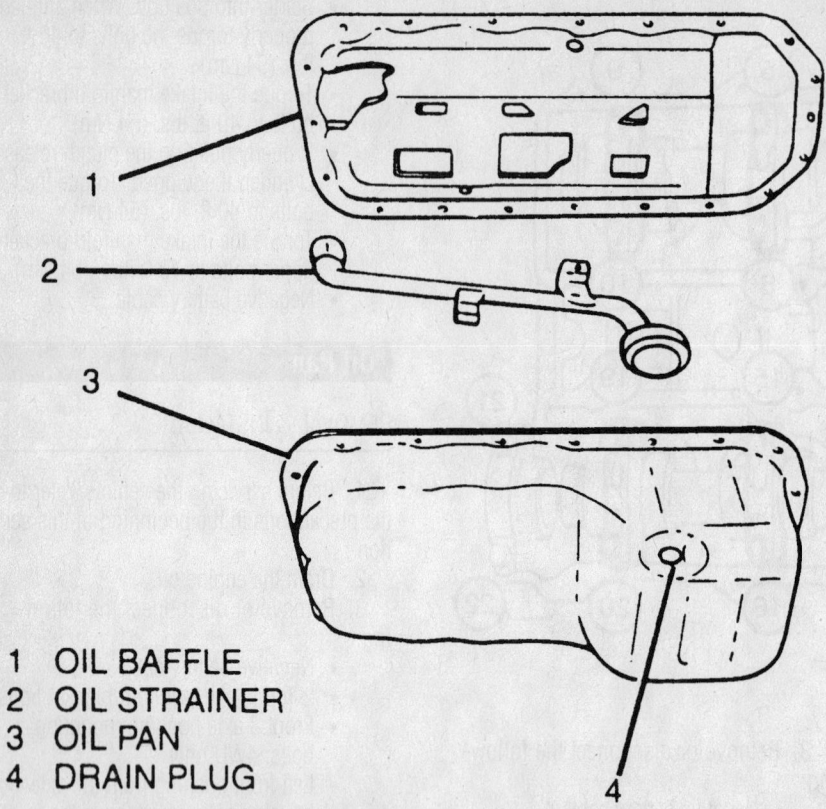

1 OIL BAFFLE
2 OIL STRAINER
3 OIL PAN
4 DRAIN PLUG

7924QG18

Exploded view of the oil pan assembly mounting

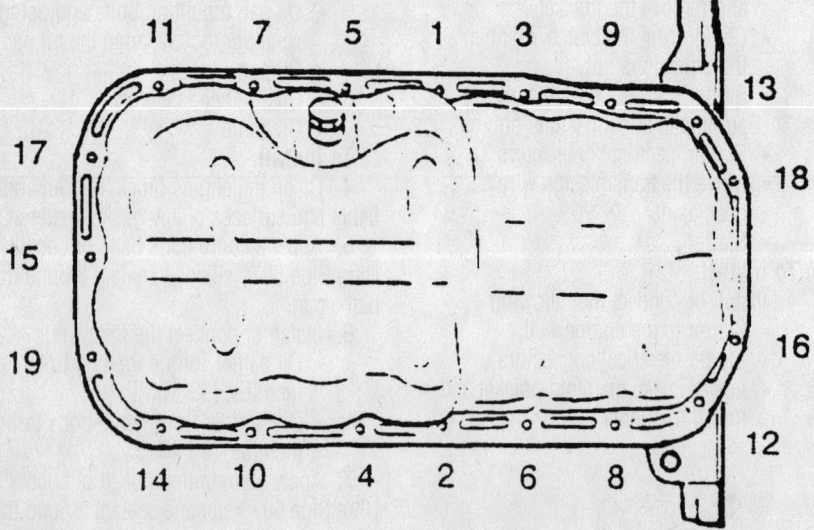

9308QG03

Tighten the oil pan bolts in sequence

- Intake manifold bracket bolts. Torque the bolts to 34 ft. lbs. (65 Nm).
- Negative battery cable
8. Fill the engine with clean oil.
9. Start the vehicle and check for leaks, repair if necessary.

1. Before servicing the vehicle, refer to the precautions in the beginning of this section .
2. Properly relieve the fuel system pressure.
3. Drain the engine oil.
4. Drain the cooling system.
5. Remove or disconnect the following:
- Negative battery cable
- Alternator belt
- Fresh air duct from the radiator
- Upper radiator hose
- Clutch fan and shroud
- Splash guard
- Loosen the A/C drive belt
- Power steering belt
- Timing belt covers
- Lower timing belt pulley and lock bolt and place a support under the front axle
- Axle attaching bolts and lower the axle enough to gain access to the oil pan
- Transmission under cover
- Oil pan
- Oil pump

To install:

6. Clean the engine block, oil pan and baffle pan surfaces of any gasket material.
7. Apply a continuous bead of silicone sealant around the oil pump.

➡**Do not allow sealant to get in the oil passages when applying sealant to the contact surface.**

8. Install or connect the following:
- New O-ring and mount the oil pump to the engine. Torque the "A" bolts to 16 ft. lbs. (22 Nm) and the "B" bolts to 33 ft. lbs. (45 Nm).
9. Remove the upper and lower A/C compressor mounting bolts.
10. Loosen the A/C compressor bracket.
- Power steering pump bracket and hand tighten the bolts
- A/C compressor bracket
- A/C compressor. Torque the

Oil Pump

REMOVAL & INSTALLATION

➡**The oil pump is externally-mounted, but still requires the removal of the oil pan to disconnect the oil pump strainer.**

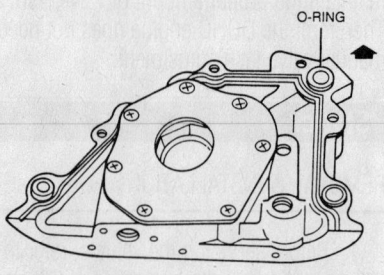

O-RING

7924QG19

Be sure the oil pump O-ring is in the proper location prior to installation

mounting bolts to 17 ft. lbs. (23 Nm).
- Torque the power steering pump bracket bolts to 24 ft. lbs. (33 Nm).
- Power steering pump. Torque the bolts to 43 ft. lbs. (58 Nm).
- Timing belt gear on the crankshaft. Torque the large crank bolt to 119 ft. lbs. (162 Nm).
- Oil baffle after applying sealant to the mating surface. Torque the bolt to 84 inch lbs. (9.5 Nm).
- Oil strainer. Torque the bolts to 84 inch lbs. (9.5 Nm).
- Oil pan
- Transmission under cover. Torque the bolts to 84 inch lbs. (9.5 Nm).
- Both gusset plates. Torque the bolts to 33 ft. lbs. (45 Nm).
- Raise the front axle into position. When aligned properly, torque the bolts to 123 ft. lbs. (167 Nm).
- Timing belt and cover
- Alternator belt
- A/C and power steering belt and adjust as needed
- Splash shield
- Upper radiator hose
- Clutch fan and shroud as an assembly
- Air duct to the top of the radiator
- Engine under cover. Torque the bolts to 18 ft. lbs. (25 Nm).
- Negative battery cable

11. Properly fill the cooling system.
12. Fill the engine with clean oil.
13. Start the engine, check for leaks, and repair if necessary.

Rear Main Seal

REMOVAL & INSTALLATION

1. Before servicing the vehicle, refer to the precautions in the beginning of this section.
2. Drain the transmission fluid.
3. Remove or disconnect the following:
- Negative battery cable
- Transmission
- Clutch cover and disc, if equipped
- Flywheel, if equipped
- Rear cover
- Rear main oil seal

To install:
4. Coat the new seal with clean oil and press the seal into the cover.

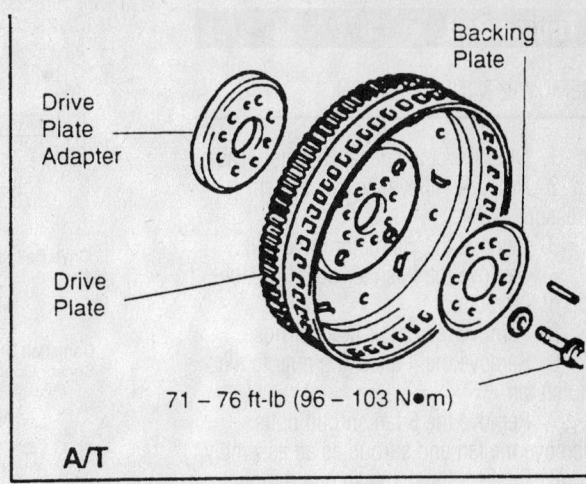

71 – 76 ft-lb (96 – 103 N•m)

A/T

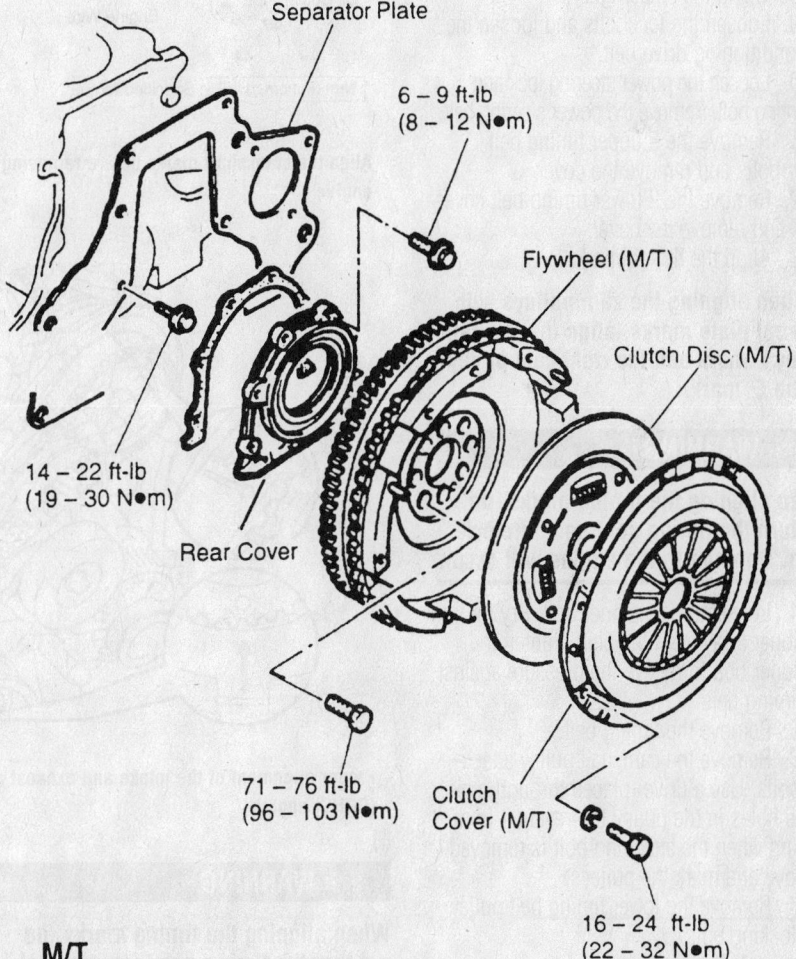

Separator Plate
6 – 9 ft-lb (8 – 12 N•m)
Flywheel (M/T)
Clutch Disc (M/T)
14 – 22 ft-lb (19 – 30 N•m)
Rear Cover
71 – 76 ft-lb (96 – 103 N•m)
Clutch Cover (M/T)
16 – 24 ft-lb (22 – 32 N•m)

M/T

Exploded view of the rear main seal and related components

5. Install or connect the following:
- Rear cover
- Flywheel onto the crankshaft. While holding the flywheel torque the bolts in sequence:
  a. Step 1: 30 ft. lbs. (41 Nm).
  b. Step 2: 60 ft. lbs. (81 Nm).
  c. Step 3: 73 ft. lbs. (99 Nm).
- Clutch disc and cover. Torque the bolts to 16 ft. lbs. (22 Nm).
- Transmission
- Negative battery cable
6. Fill the transmission to the proper level.
7. Start the vehicle and check for leaks, repair if necessary.

Timing Belt

REMOVAL & INSTALLATION

1. Disconnect the negative battery cable.
2. Properly relieve the fuel system pressure.
3. Remove the alternator drive belt.
4. Remove the fresh air duct from the top of the radiator.
5. Remove the upper radiator hose.
6. Remove the 4 attaching nuts to the clutch fan.
7. Remove the 5 fan shroud bolts. Remove the fan and shroud as an assembly.
8. Remove the 4 splash guard mounting bolts and the splash guard.
9. Loosen the lockbolts and loosen the air conditioning drive belt.
10. Loosen the power steering lock and mounting bolt. Remove the power steering belt.
11. Remove the 5 upper timing belt cover bolts and remove the cover.
12. Remove the 2 lower timing belt cover bolts and remove the cover.
13. Align the timing marks.

➡ When aligning the cam pulleys with the seal plate marks, align the left cam pulley I mark and the right cam pulley on the E mark.

❊❊ WARNING

When aligning the timing marks, do not turn the timing gear counterclockwise. Damage to the engine will occur.

14. Loosen the tensioner bolt. Pry the tensioner away from the belt. Tighten the tensioner bolt to relieve the pressure against the timing belt.
15. Remove the timing belt.
16. Remove the camshaft pulley attaching bolts. Use a driver placed through one of the holes in the pulley to prevent it from moving when the attaching bolt is removed. Remove and mark the pulleys.
17. Remove the lower timing belt pulley and locking bolt.

To install:
18. Install the camshaft pulleys. Tighten the bolts to 35 – 48 ft. lbs. (47 – 65 Nm).
19. Install the lower timing belt pulley and locking bolt. Tighten the bolt to 120 ft. lbs. (162 Nm).
20. If necessary, align the timing marks.

➡ When aligning the cam pulleys with the seal plate marks, align the left cam pulley "I" mark and the right cam pulley on the "E" mark.

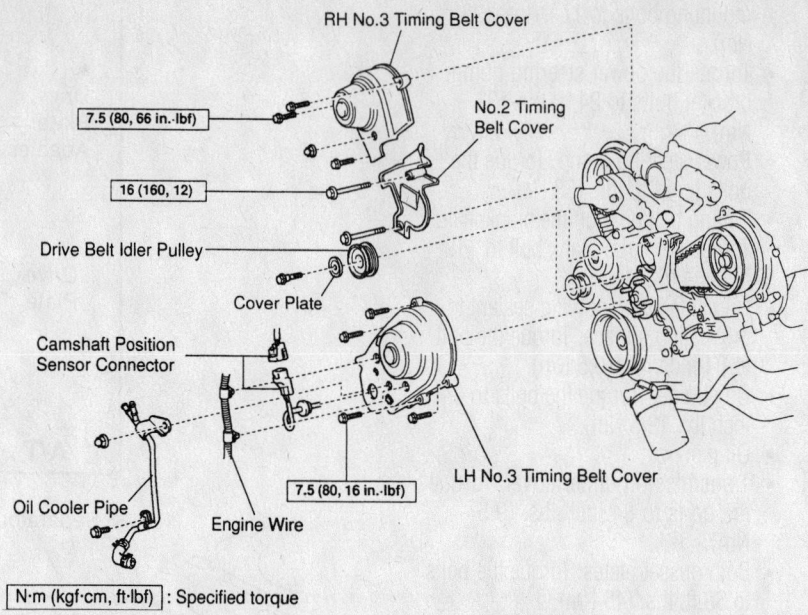

Align the crankshaft marks before removing the timing belt — KIA Sportage 2.0L (DOHC) engine

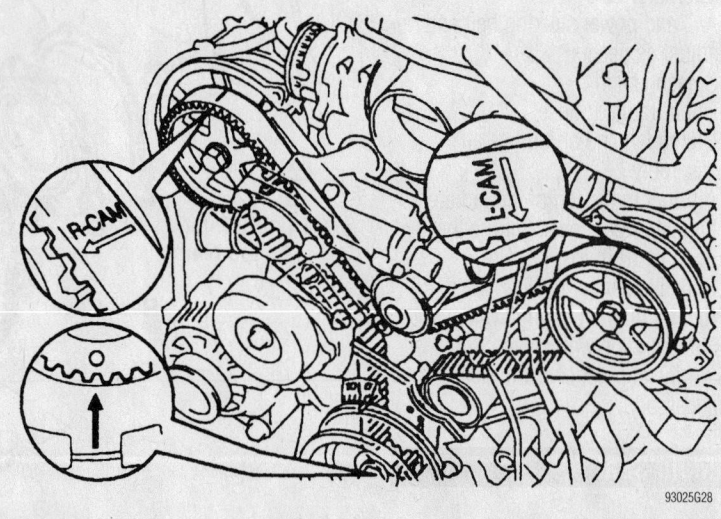

Proper alignment of the intake and exhaust camshaft pulley timing marks — KIA Sportage 2.0L (DOHC) engine

❊❊ WARNING

When aligning the timing marks, do not turn the timing gear counterclockwise. Damage to the engine will occur.

21. Loosen the tensioner bolt. Pry the tensioner away from the belt. Tighten tensioner bolt to relieve the pressure against the timing belt.
22. Install the timing belt.

❊❊ WARNING

If any binding is felt when adjusting the timing belt tension by turning the crankshaft, STOP turning the engine, because the pistons may be hitting the valves.

23. Loosen the tensioner bolt and allow the tensioner to tighten the timing belt. Tighten the tensioner bolt 27 – 38 ft. lbs. (37 – 52 Nm).
24. Check the timing belt deflection. If there is more than 0.30 – 0.33 in. (7.5 – 8.5mm) replace the tensioner spring.
25. Install the 2 lower timing belt cover bolts to the cover.
26. Install the 5 upper timing belt cover bolts to the cover.

27. Install and adjust the air conditioning and power steering drive belts.

28. Install the splash guard.

29. Install and tighten the alternator belt.

30. Install the upper radiator hose.

31. Install the fan and shroud as an assembly.

32. Install the 4 attaching nuts to the clutch fan.

33. Install the 5 fan shroud bolts.

34. Install the fresh air duct to the top of the radiator.

35. Properly fill the cooling system.

36. Connect the negative battery cable.

37. Start the engine and check for leaks.

38. Road test the vehicle.

Piston and Ring

POSITIONING

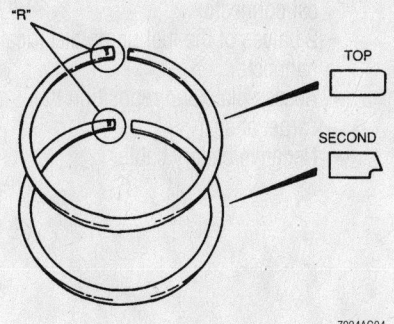

Kia 2.0L engine—compression ring positioning mark locations

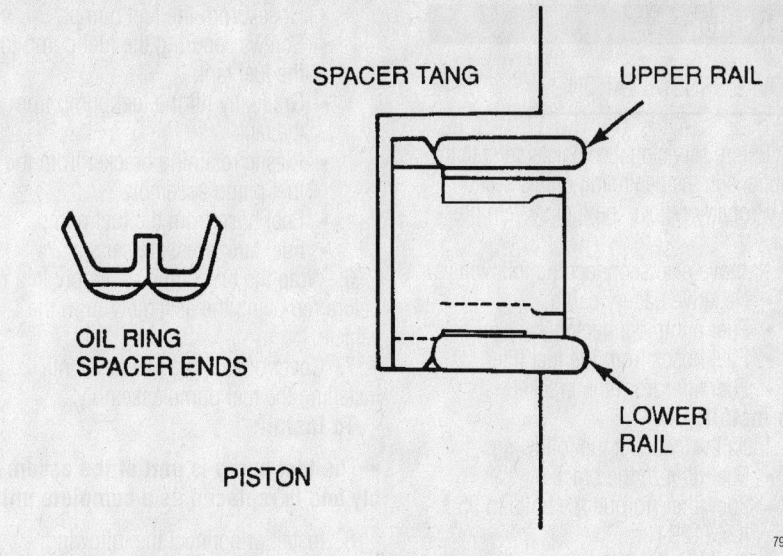

Kia 2.0L engine—oil control ring rail and spacer positioning

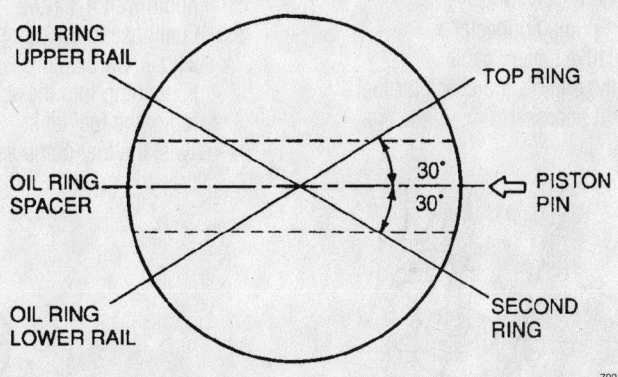

Kia 2.0L engine—piston ring end-gap spacing

FUEL SYSTEM

Fuel System Service Precautions

Safety is the most important factor when performing not only fuel system maintenance but any type of maintenance. Failure to conduct maintenance and repairs in a safe manner may result in serious personal injury or death. Maintenance and testing of the vehicle's fuel system components can be accomplished safely and effectively by adhering to the following rules and guidelines.

• To avoid the possibility of fire and personal injury, always disconnect the negative battery cable unless the repair or test procedure requires that battery voltage be applied.

• Always relieve the fuel system pressure prior to disconnecting any fuel system component (injector, fuel rail, pressure regulator, etc.), fitting or fuel line connection. Exercise extreme caution whenever relieving fuel system pressure to avoid exposing skin, face and eyes to fuel spray. Please be advised that fuel under pressure may penetrate the skin or any part of the body that it contacts.

• Always place a shop towel or cloth around the fitting or connection prior to loosening to absorb any excess fuel due to spillage. Ensure that all fuel spillage (should it occur) is quickly removed from engine surfaces. Ensure that all fuel soaked cloths or towels are deposited into a suitable waste container.

• Always keep a dry chemical (Class B) fire extinguisher near the work area.

• Do not allow fuel spray or fuel vapors to come into contact with a spark or open flame.

• Always use a back-up wrench when loosening and tightening fuel line connection fittings. This will prevent unnecessary stress and torsion to fuel line piping.

• Always replace worn fuel fitting O-rings with new. Do not substitute fuel hose or equivalent where fuel pipe is installed.

Fuel System Pressure

RELIEVING

1. Before servicing the vehicle, refer to the precautions in the beginning of this section.

2. Disconnect the fuel pump harness connector located behind the rear seat.

3. Start the engine and allow the engine to run out of fuel.

4. Once the engine has stalled, turn the key to the **OFF** position and connect the electrical connector.

5. Disconnect the negative battery cable so pressure cannot build up until work has been completed.

Fuel Filter

REMOVAL & INSTALLATION

1. Before servicing the vehicle, refer to the precautions in the beginning of this section.
2. Properly relieve the fuel system pressure.
3. Remove or disconnect the following:
    *   Negative battery cable
    *   Fuel pump connector
    *   Fuel hoses from the fuel filter
    *   Fuel filter from the bracket

To install:

4. Install or connect the following:
    *   Fuel filter in the bracket
    *   Fuel filter. Torque the bolts to 95 ft. lbs. (129 Nm).
    *   Fuel hoses on the filter and make certain that the hoses are seated properly
    *   Fuel pump connector
    *   Negative battery cable
5. Start the engine and check for fuel leaks, repair if necessary.

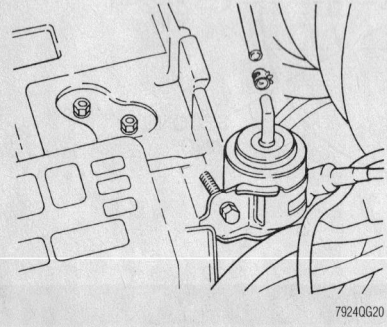

7924QG20

Fuel filter underhood mounting location

Fuel Pump

REMOVAL & INSTALLATION

1. Before servicing the vehicle, refer to the precautions in the beginning of this section .
2. Properly relieve the fuel system pressure.
3. Release the catch for the back seat and tilt the seat out of the way.
4. Move the carpet behind the seat that covers the fuel pump access panel.
5. Remove or disconnect the following:
    *   Negative battery cable
    *   Fuel pump electrical connectors
    *   Bolt securing the ground wire
    *   Fuel pump access panel
    *   Hose clamps connecting the fuel hoses to the fuel pump

*   Hoses from the fuel pump
*   Screws securing the fuel pump to the fuel tank
*   Gradually lift the fuel pump from the tank
*   Plastic retaining bracket from the fuel pump assembly
*   Fuel hose from the fuel pump
*   Fuel tank pressure sensor
6. Wrap the fuel pump assembly in a rag before removing the assembly from the vehicle.
7. Cover or seal the fuel tank until installing the fuel pump assembly.

To install:

➡ **The fuel pump is part of the assembly and is replaced as a complete unit.**

8. Install or connect the following:
    *   Plastic mounting bracket to the fuel pump and secure the bracket to the pump with 4 screws
    *   Fuel hose to the pump and secure with a new clamp
    *   Fuel pump into the access port on top of the fuel tank
    *   Twist the fuel pump as necessary to properly position it in the fuel tank

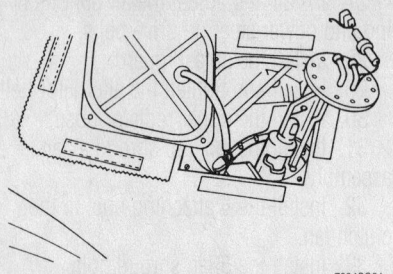

7924QG21

Removing the fuel pump through the access panel

*   8 screws around the top surface of the fuel pump
*   Fuel hoses to the pump and secure with clamps
*   Fuel tank pressure sensor
*   Fuel pressure sensor electrical connector
*   Ground wire and fuel pump electrical connector
*   2 halves of the fuel pump electrical connector
*   Access plate and reposition the carpet at seat
*   Negative battery cable

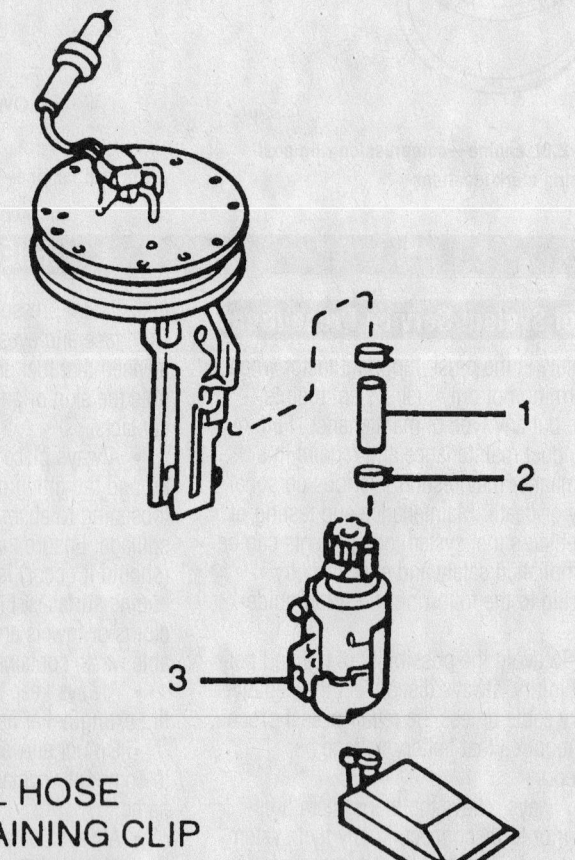

1 FUEL HOSE
2 RETAINING CLIP
3 FUEL PUMP

7924QG22

Exploded view of the fuel pump assembly

Fuel Injector

REMOVAL & INSTALLATION

1. Before servicing the vehicle, refer to the precautions in the beginning of this section.
2. Properly relieve the fuel system pressure.
3. Drain the cooling system.
4. Remove or disconnect the following:
   - Negative battery cable
   - Air intake hose assembly
   - Breather hoses from the air intake duct
   - Mass Air Flow (MAF) sensor electrical connector
   - Air intake hose bracket
   - Accelerator cable
   - Vacuum hose from the intake manifold to the vacuum pipe
   - Throttle Position (TP) sensor and Idle Air Control (IAC) valve electrical connectors
   - Coolant hoses from the throttle body
   - Clamp and hoses from the IAC valve
   - Brake booster vacuum hose from the dynamic chamber
   - Cruise control vacuum hose, if equipped
   - Bracket from the dynamic chamber
   - Manifold bracket
   - Heater inlet hose
   - Dynamic chamber
   - Fuel hose from the pressure regulator
   - Fuel injector rail clips
   - Fuel rail
   - Fuel injector insulators
   - Pressure regulator
   - Fuel injectors

To install:

5. Install or connect the following:
   - Fuel injectors to the fuel rail
   - New insulators
   - New injector clips
   - Fuel rail
   - Clamps and air hose to the fuel rail
   - Fuel hose to the pressure regulator
   - Dynamic chamber with a new gasket. Torque the bolts to 16 ft. lbs. (22 Nm).
   - IAC valve bracket bottom bolt
   - IAC valve and TP sensor electrical connectors
   - Heater inlet hose
   - Cruise control hose, if equipped
   - Vacuum hose to the pressure regulator
   - Air hose and clamp to the air rail
   - IAC valve vacuum hose
   - Manifold bracket bolts
   - Coolant hoses to the throttle body
   - Vacuum hose to the vacuum pipe
   - MAF sensor bracket
   - MAF sensor electrical connector
   - Accelerator cable
   - Breather hoses
   - Negative battery cable
6. Fill the cooling system.
7. Start the vehicle and check for leaks, repair if necessary.

DRIVE TRAIN

Transmission Assembly

REMOVAL & INSTALLATION

Manual

➡The removal of the manual transmission is virtually the same for 4WD and 2WD vehicles.

1. Before servicing the vehicle, refer to the precautions in the beginning of this section.
2. Drain the transmission fluid.
3. Drain the transfer case.
4. Remove or disconnect the following:
   - Negative battery cable
   - Rear portion of the center console
   - Shift lever and transfer lever knobs
   - Slide the boot cover over the shifters and remove the center console
   - Shift lever
   - Transfer lever
   - Front driveshaft by removing the bolts at the front differential and the bolts at the transfer case, if equipped
   - Bolts from the rear differential flange and the center support. Pull the driveshaft out of the tail shaft housing, if equipped with a 4 x 4
   - Bolts from the rear differential flange and center support. Pull the driveshaft out of the tail shaft housing, if equipped with a 4 x 2
   - Back-up light electrical connector and the Vehicle Speed Sensor (VSS) electrical connectors and move the wire harness aside
   - Crankshaft Position (CKP) sensor from the transmission housing
   - Clutch release cylinder and move it aside
   - Front exhaust pipe bracket
   - Front lower transmission housing bolts
   - Transfer case side mount, if equipped and properly support the transmission
   - Transmission crossmember
   - Transmission mount bolts
   - Starter from the front housing
   - Transmission and transfer case, if equipped

To install:

5. Install or connect the following:
   - Transmission into position at the rear of the engine
   - Wire harness along the right side of the transmission and route the VSS wire over the transmission to the rear of the control rod extension
   - Transmission to engine 14mm mounting bolts. Torque the bolts to 80 ft. lbs. (108 Nm).
   - Transmission to engine 10mm mounting bolts. Torque the bolts to 29 ft. lbs. (39 Nm).
   - Transmission to engine 6mm mounting bolts. Torque the bolts to 5 ft. lbs. (7 Nm).
   - Exhaust pipe to the bracket. Torque the bolts to 20 ft. lbs. (27 Nm).
   - Starter and ground wire. Torque the bolts to 29 ft. lbs. (39 Nm).
   - Transmission crossmember mount. Torque the bolts to 80 ft. lbs. (108 Nm).
   - Crossmember to the chassis. Torque the bolts to 32 ft. lbs. (44 Nm).
   - CKP sensor. Torque the bolt to 5 ft. lbs. (7 Nm).
   - 4WD indicator switch connector
   - Back-up light electrical connector
   - VSS electrical connector
   - Clutch release cylinder. Torque the bolts to 29 ft. lbs. (39 Nm).
   - Driveshaft to the rear differential flange, 4 x 4 only
   - Forward end of the driveshaft into the extension housing and attach the center support to the chassis
   - Shaft to the rear differential flange. Torque the bolts to 27 ft. lbs. (36 Nm).
   - Front driveshaft at the transfer case and install the bolts to the front differential, 4 x 4 only

- Transfer case side mount, if equipped. Torque the bolts to 38 ft. lbs. (52 Nm).
- Transfer case side mount to the chassis. Torque the bolts to 38 ft. lbs. (52 Nm).
- Shift lever assembly. Torque the shift lever bracket bolts to 18 ft. lbs. (25 Nm).
- Transfer lever assembly, if equipped. Torque the bolts to 18 ft. lbs. (25 Nm).
- Dust cover plate over the shifter lever handles. Torque the bolts to 15 ft. lbs. (20 Nm).
- Front console
- Shifter lever knobs
- Negative battery cable
- Fill the transfer case.

6. Fill the transmission assembly.
7. Start the vehicle and check for leaks, repair if necessary.

Automatic

1. Before servicing the vehicle, refer to the precautions in the beginning of this section.
2. Drain the transmission fluid.
3. Drain the cooling system.
4. Remove or disconnect the following:
- Negative battery cable
- Automatic transmission control cable from the throttle body

5. Slide the front seats forward and remove the 2 rear shift console mounting screws and set the parking brake.
- Rear console and slide the front seats rearward
- Front console mounting screws and untie the shifter boot draw strings
- Loosen the transfer case lock nut and remove the transfer case shifter lever knob
- Power/economy switch electrical connector
- Front console and shift the transfer lever to the 4L position
- transfer shift lever cover plate
- transfer case shift lever assembly and place the transmission in the park position
- Selector lever nuts
- Split pin from the shift selector lever
- Shift selector rod and spring washers
- Electrical connectors from the base of the selector lever

➡ **There is a fifth wire connection (Park Position). This wire is hard-wired, do not disconnect it.**

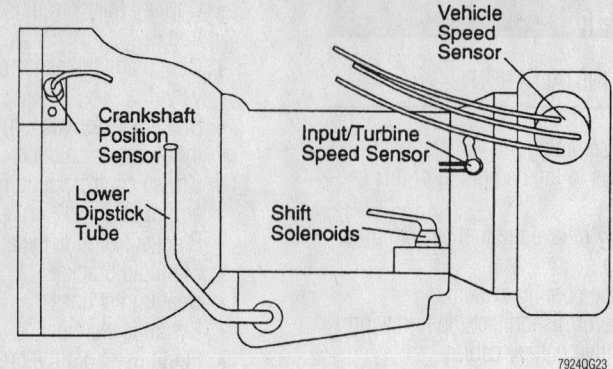

Automatic transmission wiring connections

Radiator Side

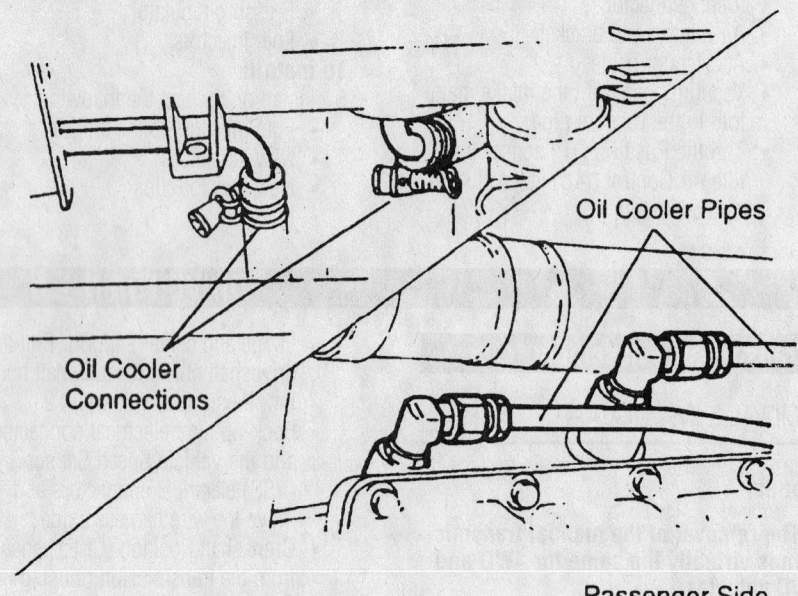

Exploded view of the oil cooler pipe connections

- Selector lever assembly
- Control cable from the throttle linkage and slide the cable from the bellcrank
- Upper dipstick tube from the lower dipstick tube
- Input/Turbine speed sensor from the top rear of the transmission
- Shift solenoids from the lower left side of the transmission
- Vehicle Speed Sensor (VSS) from the center of the transfer case
- Input speed sensor electrical connectors
- Matchmark the front and rear driveshafts. Remove the attaching bolts from both flanges and remove the driveshafts
- Oil cooler pipes at the transmission
- Starter
- Transfer case mounting bolts

- Transfer case nuts from the crossmember and support the transmission
- Crossmember
- Transmission to engine mounting bolts
- Front lower splash shield
- Left side lower gusset
- Torque converter inspection cover
- Torque converter to drive plate bolts
- Slide the transmission away from the engine and lower the transmission slightly
- Crankshaft Position (CKP) sensor attaching bolt and sensor
- 4WD, 4WD LOW indicator switches and the VSS from the transfer case

6. Lower the transmission. Be sure all wiring is clear and disconnected. Be sure the throttle cables come out without binding or attaching to anything.

To install:

7. Install or connect the following:

- Transmission into position to attach the sensor and indicator switch wiring. Be sure the throttle cables are guided into the engine compartment without binding or attaching to anything.
- 4WD, 4WD LOW indicator switches and the VSS to the transfer case
- Input/turbine speed sensor and CKP sensor. Torque the bolt to 12 ft. lbs. (16 Nm).
- Raise the transmission and position it to the engine. Install the upper housing bolts. Torque the bolts in the following sequence:

8. 10mm bolts to 38–60 ft. lbs. (57–81 Nm).

9. 12mm bolts to 51–65 ft. lbs. (69–88 Nm).

- Exhaust hanger and bracket. Torque the bolts to 38–60 ft. lbs. (57–81 Nm).
- Oil cooler pipe to the lower engine mount. Torque the bolt to 60 ft. lbs. (81 Nm).
- Crossmember. Torque the bolts to 23–34 ft. lbs. (31–46 Nm).
- Two transfer case mounting bolts located at the right center of the transfer case. Torque the bolts to 23–34 ft. lbs. (31–46 Nm).
- Four transfer case nuts to the crossmember. Torque the nuts to 23–34 ft. lbs. (31–46 Nm).
- Torque converter-to-drive plate bolts. Torque the bolts to 12–20 ft. lbs. (16–27 Nm).
- Torque converter inspection cover. Torque the bolts to 41–62 inch lbs. (5–7 Nm).
- Front splash guard. Torque the bolts to 41–62 inch lbs. (5–7 Nm).
- Starter. Torque bolts to 27–40 ft. lbs. (37–54 Nm).
- Left side lower gusset. Torque the bolts to 38–60 ft. lbs. (57–81 Nm).
- Right side lower gusset in 3 steps:

a. Install the bottom mounting bolts, but do not tighten.

b. Install the top bolt to the intake manifold support bracket and manifold. Tighten to 27–40 ft. lbs. (37–54 Nm).

c. Secure the manifold intake bracket by tightening the two attaching bolts to 38–60 ft. lbs. (57–81 Nm).

- Oil cooler pipes at the transmission. Torque the lines to 42–62 inch lbs. (5–7 Nm).
- Two oil cooling tube clamps to the lines

- Driveshafts with the matchmarks aligned. Torque the bolts to the differential flanges to 20–22 ft. lbs. (27–30 Nm) and the transfer case flange bolts to 36–43 ft. lbs. (49–59 Nm).
- Undercover splash shield. Torque the bolts to 42–62 inch lbs. (5–7 Nm).
- Upper dipstick tube to the lower tube and lower the vehicle

10. Provide automatic transmission control cable slack by gently pulling the cable to the left until the cable pin has rotated sufficiently to line the automatic transmission control cable up with the slot in the rear of the throttle body bellcrank.

11. Slide the automatic transmission control cable and cable pin into the bellcrank.

12. Tighten the locknut.

- Throttle kickdown cable to the mounting bracket
- Automatic transmission control cable to the throttle body
- Air silencer
- Shifter lever assembly to the transfer case. Torque the bolts to 72–102 inch lbs. (22–28 Nm).
- Four wiring connectors under the shift selector lever
- Shift selector lever, do not exert any force when installing the shift selector lever
- Four shift selector lever nuts. Tighten to 72–102 inch lbs. (22–28 Nm).
- Shift rod and washers to the selector lever and install the split pin to the shift selector lever
- Power/Economy switch wiring connector
- Rear console
- Front console and tie the shift boot draw strings
- Negative battery cable

13. Fill the transmission to the proper level.

14. Fill the cooling system.

15. Start the vehicle, check for leaks, and repair if necessary.

Clutch

ADJUSTMENT

Clutch Pedal Height

1. Before servicing the vehicle, refer to the precautions in the beginning of this section.

2. Pull back the carpet to measure the distance from the firewall to the top of the pedal. The standard height is 9.84 in. (250 mm).

3. If adjustment is required, loosen the locknut and turn the stopper bolt.

4. After adjustment is made tighten the locknut to 12 ft. lbs. (16 Nm).

Clutch Pedal Free-Play

1. Before servicing the vehicle, refer to the precautions in the beginning of this section.

2. Depress the clutch pedal gently by hand and measure the amount of free-play (distance the pedal travels before resistance is felt). The proper amount of free-play is 0.5 in. (12.7mm). If the free-play is not within the proper specifications, continue with the procedure.

3. Measure from the floor pan to the middle point of the clutch pedal when the pedal is in the fully released position. The proper clutch pedal height is 7.25 in. (184mm).

4. If the pedal height is incorrect, loosen locknut (A) and turn the pedal adjusting bolt (B) until the proper height is achieved, then retighten the locknut.

5. Remeasure the free-play. If it is still out of specification, loosen the clutch pushrod locknut (C) and turn the pushrod (D) until the proper free-play is achieved. Tighten locknut (C) securely.

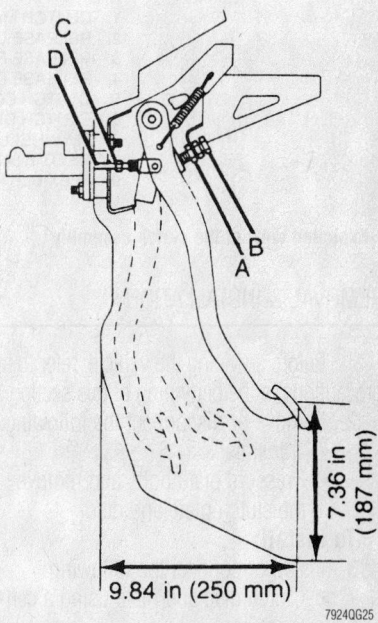

7.36 in (187 mm)

9.84 in (250 mm)

7924QG25

Clutch pedal height and free-play adjustment points

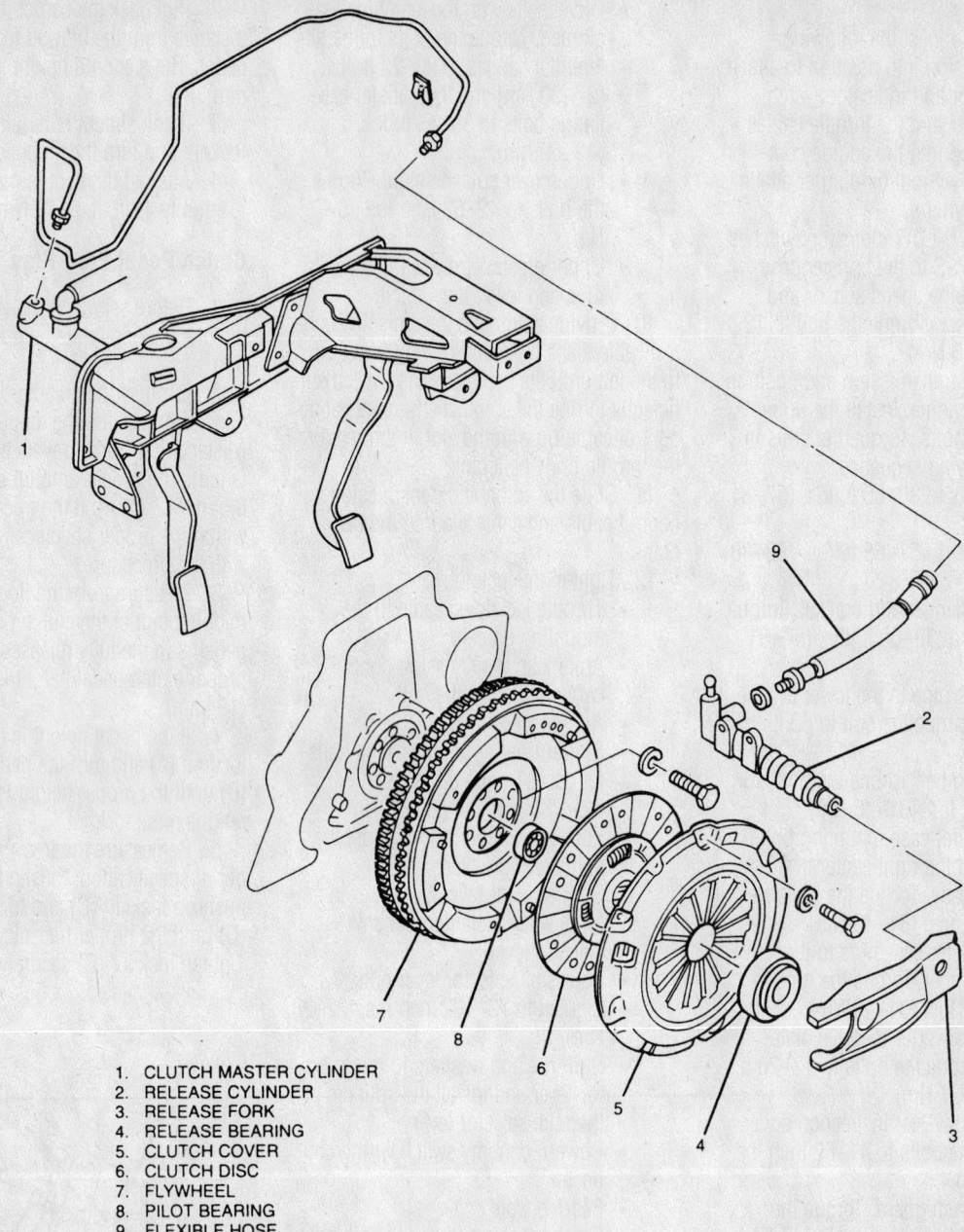

1. CLUTCH MASTER CYLINDER
2. RELEASE CYLINDER
3. RELEASE FORK
4. RELEASE BEARING
5. CLUTCH COVER
6. CLUTCH DISC
7. FLYWHEEL
8. PILOT BEARING
9. FLEXIBLE HOSE

79240G26

Exploded view of the clutch assembly

REMOVAL & INSTALLATION

1. Before servicing the vehicle, refer to the precautions in the beginning of this section .
2. Remove or disconnect the following:
   • Transmission
   • Pressure plate bolts and remove the clutch plate and disc

To install:
3. Install or connect the following:
   • Clutch disc and plate using a centering tool
   • Clutch cover. Torque the bolts to 73 ft. lbs. (99 Nm) and remove the centering tool

   • Check the release bearing condition and lubricate or replace as necessary
   • Transmission

Hydraulic Clutch System

BLEEDING

1. Before servicing the vehicle, refer to the precautions in the beginning of this section .
2. With an assistant in the vehicle, raise and safely support the vehicle.
3. Have your assistant pump the clutch pedal three times and hold the pedal to the floor.
4. Open the bleeder valve on the clutch slave cylinder until the air is purged from the cylinder.
5. Tighten the bleeder valve.
6. Have your assistant release the clutch pedal.
7. Fill the clutch master cylinder if below minimum.
8. Repeat Steps 2 through 6 until no air exits from the bleeder valve.
9. Lower the vehicle.
10. Fill the clutch master cylinder fluid reservoir.

Transfer Case Assembly

REMOVAL & INSTALLATION

1. Before servicing the vehicle, refer to the precautions in the beginning of this section .

2. Drain the transfer case.
3. Remove or disconnect the following:
   - Negative battery cable
   - Two rear console mounting screws. Slide the console forward to clear the parking brake handle and set aside

- Three mounting screws from the front console. Untie the shift boot draw strings and open the boot
- Loosen the transfer case shift lever locknut and remove the lever knob
- Pull the console up to access the Power/Economy switch wiring con-

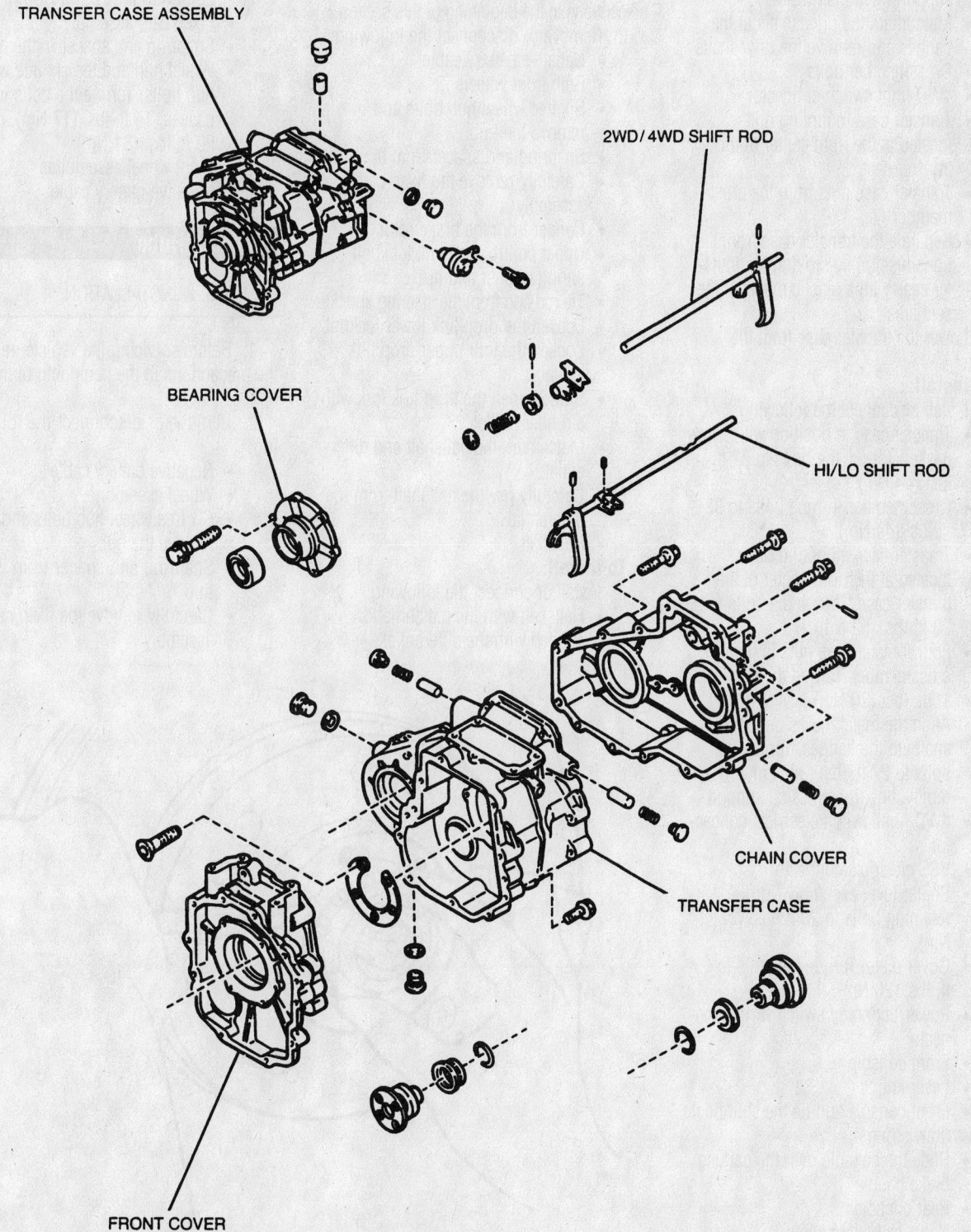

Exploded view of the transfer case assembly

9308QG05

nector. Unplug the connector and remove the console

4. Shift the transfer lever to the 4L position.

- Cover plate
- Retaining bolts from the transfer case and lift the shifter lever assembly straight out and properly support the transmission
- Matchmark the driveshafts at the flanges and remove the driveshafts
- Crossmember bolts
- 4WD light switch connector
- Transfer case mounting bolts located at the right center of the transfer case
- Transfer case nuts from the crossmember
- Separate the transfer case from the transmission by striking the transfer case with a plastic mallet at the seal area

5. Lower the transfer case from the vehicle.

To install:

6. Install or connect the following:

- Transfer case in position with a new gasket. Torque the bolts to 32 ft. lbs. (44 Nm).
- Crossmember. Torque bolts to 32 ft. lbs. (44 Nm).
- Transfer case mounting bolts located at the right center of the transfer case. Torque the bolts to 38 ft. lbs. (52 Nm).
- Four transfer case nuts to the crossmember. Torque the nuts to 15 ft. lbs. (20 Nm).
- Align the matchmarks on the driveshafts to the flanges. Torque the bolts to 27 ft. lbs. (36 Nm) and remove the transmission support
- 4WD light switch electrical connector
- VSS electrical connector
- Shifter lever assembly. Torque retaining bolts to 20 ft. lbs. (27 Nm).
- Cover plate. Torque the bolts to 15 ft. lbs. (20 Nm).
- Power/Economy switch wiring connector
- Front console
- Lever knobs
- Front console and tie the shift boot draw strings
- Slide the console over the parking brake handle
- Rear console
- Negative battery cable

7. Fill the transfer case to the proper level

8. Start the vehicle and check for leaks, repair if necessary.

Halfshaft

REMOVAL & INSTALLATION

1. Before servicing the vehicle, refer to the precautions in the beginning of this section .

2. Remove or disconnect the following:

- Negative battery cable
- Both front wheels
- Six free wheel hub bolts and remove the hub
- Snapring and spacer from the hub
- Carefully remove the fixed cam assembly
- Caliper from the brake rotor
- Upper control arm link lockbolt, spring washer and nut
- Tie rod end from the steering knuckle
- Loosen the drop link lower locknut
- Loosen the four upper drop link locknuts
- Spread open the drop link fork with a rubber mallet
- Matchmark the halfshaft and differential.
- Carefully pry the halfshaft from the differential
- Halfshaft

To install:

3. Install or connect the following:

- Halfshaft with the matchmarks aligned with the differential

- Torque the upper and lower drop link nuts to 36 ft. lbs. (49 Nm).
- Tie rod end to the steering knuckle. Torque the locknut to 27 ft. lbs. (36 Nm) and install a new cotter pin
- Upper control arm link lockbolt, spring washer and nut. Torque the bolt to 36 ft. lbs. (49 Nm).
- Fixed cam assembly
- Snapring and spacer in the hub
- Wheel hub and the six free wheel hub bolts. Torque the bolts in two passes, 14 ft. lbs. (17 Nm), then to 23 ft. lbs. (31 Nm).
- Front wheel assemblies
- Negative battery cable

Locking Hubs

REMOVAL & INSTALLATION

1. Before servicing the vehicle, refer to the precautions in the beginning of this section.

2. Remove or disconnect the following:

- Negative battery cable
- Wheel assembly
- Six free wheel hub bolts and remove the hub
- Snapring and spacer from the hub
- Carefully remove the fixed cam assembly

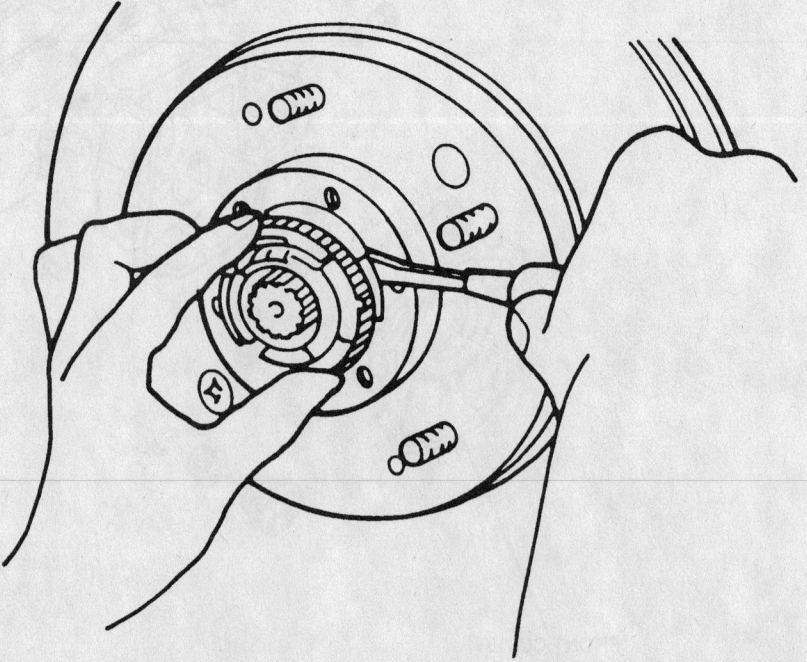

Removing the 4WD fixed cam assembly

7924QG27

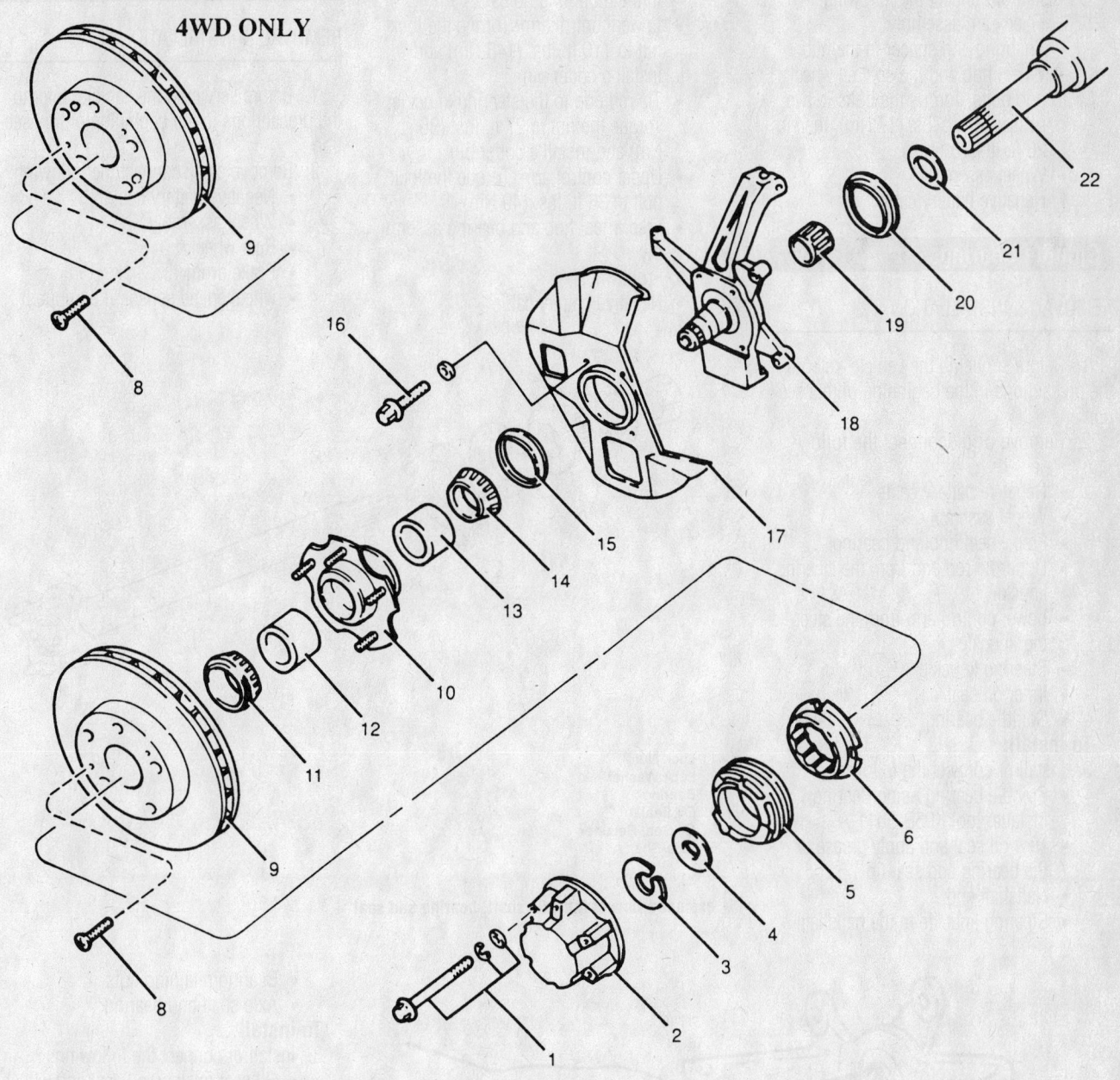

4WD ONLY

1	BOLT/WASHER	9	ROTOR	16	BOLT & SPRING WASHER
2	FREE WHEEL HUB BODY	10	WHEEL HUB	17	DUST COVER
3	SNAP RING	11	INNER BEARING INNER RACE	18	KNUCKLE
4	SPACER	12	INNER BEARING OUTER RACE	19	NEEDLE BEARING
5	FIXED CAM ASSEMBLY	13	OUTER BEARING OUTER RACE	20	OIL SEAL
6	LOCK NUT	14	OUTER BEARING INNER RACE	21	SPACER
8	SCREW	15	OIL SEAL	22	DRIVE SHAFT (LH)

7924QG28

Exploded view of the 4WD locking hub assembly

To install:
3. Install or connect the following:
- Fixed cam assembly
- Snapring and spacer in the hub
- Wheel hub and the six free wheel hub bolts. Torque the bolts in two passes, 14 ft. lbs. (17 Nm), then to 23 ft. lbs. (31 Nm).
- Wheel assembly
- Negative battery cable

Spindle Bearings

REMOVAL & INSTALLATION

1. Before servicing the vehicle, refer to the precautions in the beginning of this section .
2. Remove or disconnect the following:
- Negative battery cable
- Wheel assembly
- Free wheel hub and bearing
- Upper tie rod end from the steering knuckle
- Lower control arm from the steering knuckle
- Steering knuckle
- Inner oil seal
- Spindle bearing

To install:
3. Install or connect the following:
- Spindle bearing using Bearing Installer tool K95B-5011-A
- New oil seal and apply grease to the bearing and seal lip
- Halfshaft end
- Steering knuckle to the halfshaft

with the upper and lower ball joints in the mounting holes
- Lower control arm. Torque the lock nut to 110 ft. lbs. (148 Nm) and install a cotter pin
- Tie rod end to the steering knuckle. Torque the nut to 27 ft. lbs. (36 Nm) and install a cotter pin
- Upper control arm. Torque the lock bolt to 36 ft. lbs. (49 Nm).
- Free wheel hub and bearing assembly
- Front wheel
- Negative battery cable

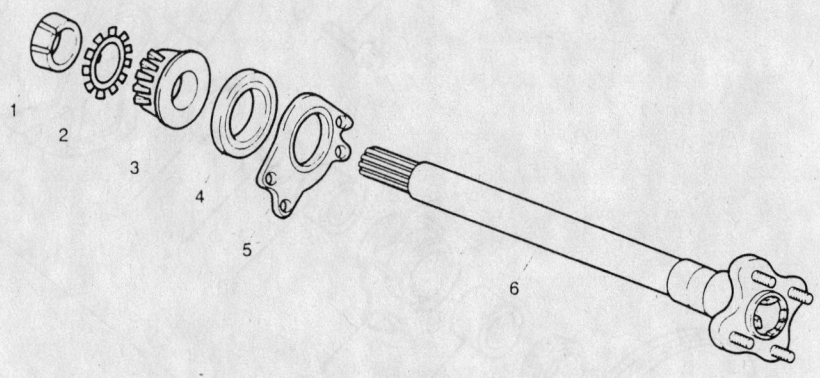

1. Lock Nut
2. Lock Washer
3. Bearing
4. Oil Seal
5. Oil Seal Retainer
6. Axle Shaft

9308QG07

exploded view of the axle shaft, bearing and seal

Axle Shaft Bearing and Seal

REMOVAL & INSTALLATION

1. Before servicing the vehicle, refer to the precautions in the beginning of this section .
2. Remove or disconnect the following:
- Negative battery cable
- Wheel assembly
- Rear wheel
- Brake drum
- wheel speed sensor, if equipped
- Bearing retaining nuts
- Axle shaft and bearing

To install:
3. Install or connect the following:
- Oil seal retainer, oil seal and wheel bearing to the axle shaft. Torque the new lock nut to 220 ft. lbs. (300 Nm).

➡ **Turn the right side halfshaft lock nut clockwise and the left side halfshaft lock nut counterclockwise.**

- Axle shaft assembly into the axle housing. Torque the nuts to 74 ft. lbs. (100 Nm).
- Wheel speed sensor, if equipped
- Brake drum
- Rear wheel assembly
- Negative battery cable
4. Check the fluid level and top off if necessary.

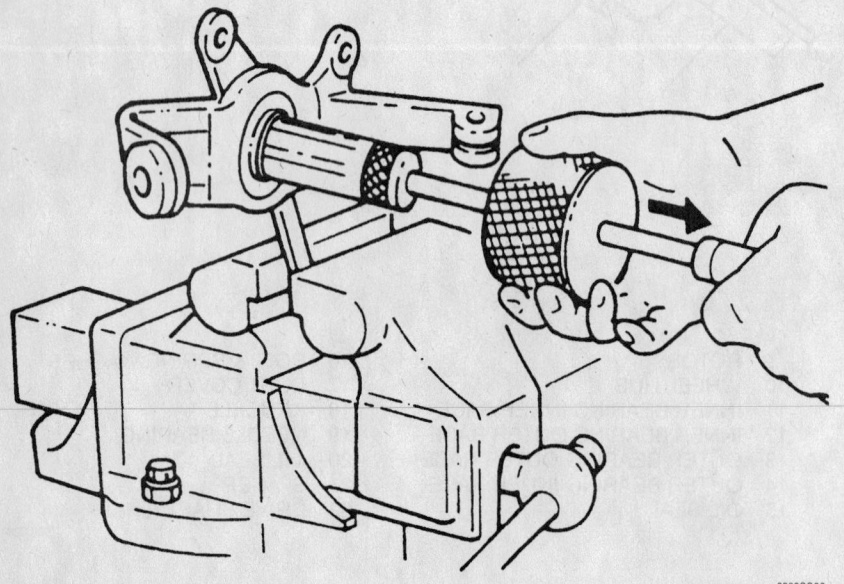

9308QG06

Remove the spindle bearing from the steering knuckle

Pinion Seal

REMOVAL & INSTALLATION

1. Before servicing the vehicle, refer to the precautions in the beginning of this section .
2. Drain the gear oil.
3. Remove or disconnect the following:

- Negative battery cable
- Wheel assemblies
- Brake drums
- driveshaft

➡**Use an inch lb. (Nm) torque wrench , measure and record the amount of torque required to maintain rotation of the pinion.**

- Pinion flange
- Pinion seal

To install:
4. Install or connect the following:

- New pinion seal lightly coated with clean gear oil
- Pinion flange

5. Rotate the pinion flange occasionally while tightening the flange nut and make certain that the pinion bearings are seated properly.

6. Take several bearing preload torque readings. Tighten the flange nut to achieve the preload torque reading. The maximum torque reading should not exceed 14 inch lbs. (1.6 Nm).

- Driveshaft after aligning the match-marks
- Brake drums
- Wheel assemblies
- Negative battery cable

7. Fill the gear oil to the proper level.
8. Start the vehicle and check for leaks, repair if necessary.

STEERING AND SUSPENSION

Air Bag

PRECAUTIONS

Several precautions must be observed when handling the inflator module to avoid accidental deployment and possible personal injury.

1. Never carry the inflator module by the wires or connector on the underside of the module.
2. When carrying a live inflator module, hold securely with both hands, and ensure that the bag and trim cover are pointed away from you.
3. Place the inflator module on a bench or other surface with the bag and trim cover facing up.
4. With the inflator module on the bench, never place anything on or close to the module which may be thrown in the event of an accidental deployment.

DISARMING

1. Before servicing the vehicle, refer to the precautions in the beginning of this section .
2. Turn the ignition switch to the **LOCK** position.
3. Disconnect the negative battery cable.
4. Wait 10 minutes for the back-up power to discharge.

ARMING

Assuming the system components (air bag control module, sensors, air bag, etc.) are installed correctly and are in good working order, the system is armed whenever the battery positive and negative battery cables are connected.

If you have disarmed the air bag system for any reason, to rearm, be sure no one is in the vehicle (as an added safety measure), then connect the negative battery cable.

Power Steering Gear

ADJUSTMENT

1. Before servicing the vehicle, refer to the precautions in the beginning of this section .
2. Place the steering gear in a vise with protective jaws.
3. Place a torque wrench on the Pitman arm end of the shaft.
4. Loosen the locknut on the adjusting bolt.
5. Slowly turn the adjusting bolt to until the breakaway torque is 65 ft. lbs. (88 Nm).
6. Hold the adjusting bolt in position and tighten the locknut to 25 ft. lbs. (34 Nm).

REMOVAL & INSTALLATION

1. Before servicing the vehicle, refer to the precautions in the beginning of this section .

2. Center the steering wheel.
3. Drain the power steering fluid.
4. Remove or disconnect the following:

- Negative battery cable
- Left front wheel
- Pitman arm-to-centerlink attaching nut
- Separate the Pitman arm from the centerlink with a ball joint puller
- Power steering hoses
- Coolant recovery tank and the power steering reserve tank
- Set bolt from the intermediate shaft
- Intermediate shaft from the steering gear
- Steering gear-to-frame bolts
- Steering gear

To install:
5. Install or connect the following:

- Position the steering gear to the frame. Torque the bolts to 159 ft. lbs. (215 Nm).
- Pitman arm to the centerlink. Torque the bolt to 36 ft. lbs. (49 Nm) and install a new cotter pin
- Intermediate shaft to the steering gear shaft. Torque the set bolt o 25 ft. lbs. (34 Nm).

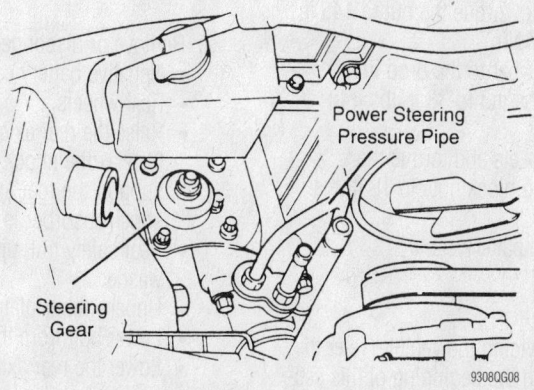

Power Steering
Pressure Pipe

Steering
Gear

9308QG08

Exploded view of the steering gear

- Power steering hoses/lines. Torque the fasteners to 29 ft. lbs. (34 Nm).
- Power steering reserve tank over the bracket and press until full engagement is reached
- Coolant recovery tank
- Left front wheel
- Negative battery cable

6. Fill the power steering fluid to the proper level and bleed the system.

7. Start the vehicle and check for leaks, repair if necessary.

8. Road test the vehicle to check that the steering wheel is straight.

Shock Absorber

REMOVAL & INSTALLATION

Front

The front shock absorber and coil spring are removed as a single unit.

1. Before servicing the vehicle, refer to the precautions in the beginning of this section.

2. Remove or disconnect the following:
- Negative battery cable
- Both front wheels
- Upper shock absorber mounting block nuts
- Stabilizer bar
- Drop link nut and allow the drop link to remain in place
- Both halves of the front fork
- Drop link
- Shock absorber and coil spring as an assembly

To install:

3. Install or connect the following:
- Coil spring to the shock absorber and position the assembly to the upper mounting block
- Upper mounting block nuts and hand tighten them
- Both front forks . Torque the bolts to 36 ft. lbs. (48 Nm).
- Drop link. Torque the nut to 145 ft. lbs. (197 Nm).
- Stabilizer bar to the drop link. Torque the nut to 36 ft. lbs. (48 Nm).
- Front wheels and torque the mounting block nuts to 18 ft. lbs. (25 Nm).
- Negative battery cable

Rear

1. Before servicing the vehicle, refer to the precautions in the beginning of this section.

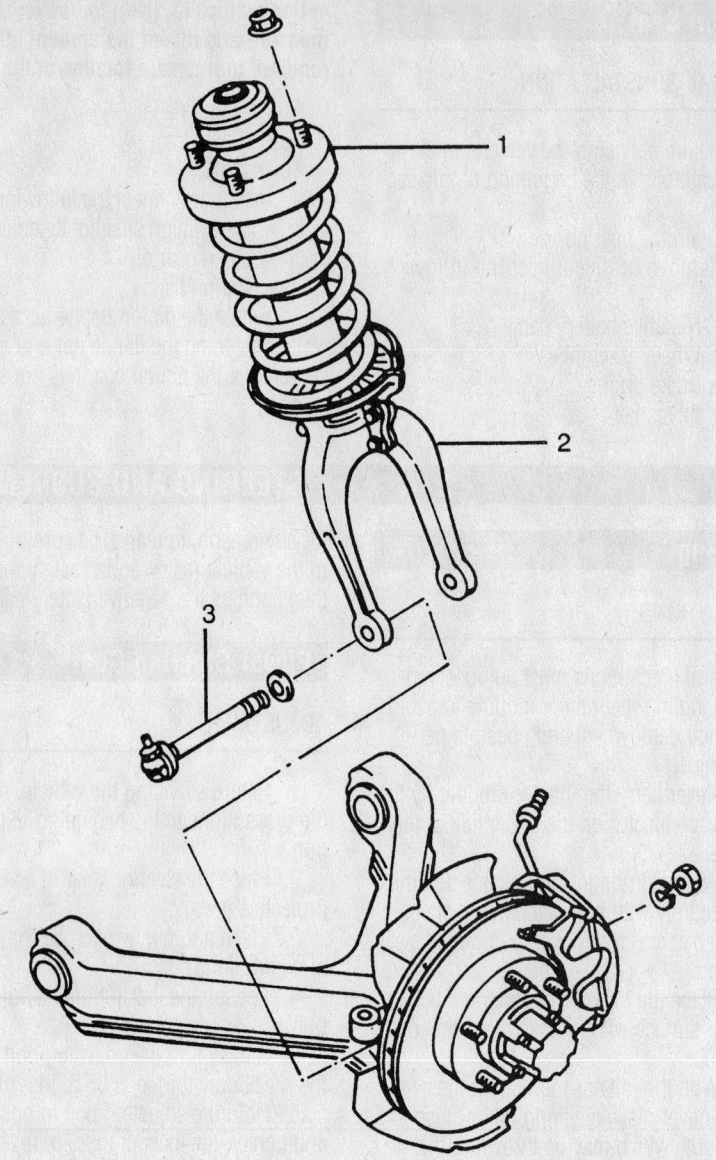

1 Front Shock Absorber & Coil Spring Assembly
2 Front Fork
3 Drop Link

9308QG09

Exploded view of the front shock absorber assembly

2. Remove or disconnect the following:
- Negative battery cable
- Rear wheels
- Raise the rear axle with a floor jack to relax the shock absorbers and support the rear axle when the shock absorber is removed.
- Rear safety nut, upper nut and washer
- Upper rubber plate
- Lower bolt from the shock absorber
- Lower the rear axle housing
- Shock absorber

3. Remove the lower mounting bolt and remove the shock absorber.

To install:

4. Install or connect the following:
- Bottom washer and rubber cushion on the top of the shock absorber. Position the shock absorber on the vehicle
- Lower bolt
- Rubber cushion, washer and nut. Tighten to 53 ft. lbs. (72 Nm).
- Safety nut. Torque the lower bolt to 62 ft. lbs. (84 Nm).

- Rear wheels
- Negative battery cable

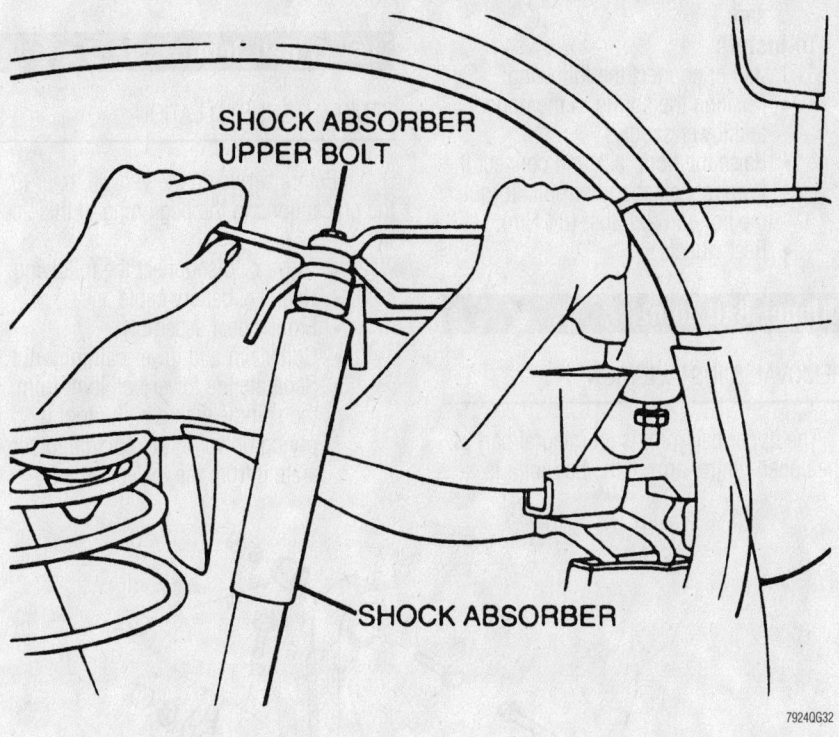

SHOCK ABSORBER
UPPER BOLT

SHOCK ABSORBER

79240G32

Upper shock absorber mounting nut

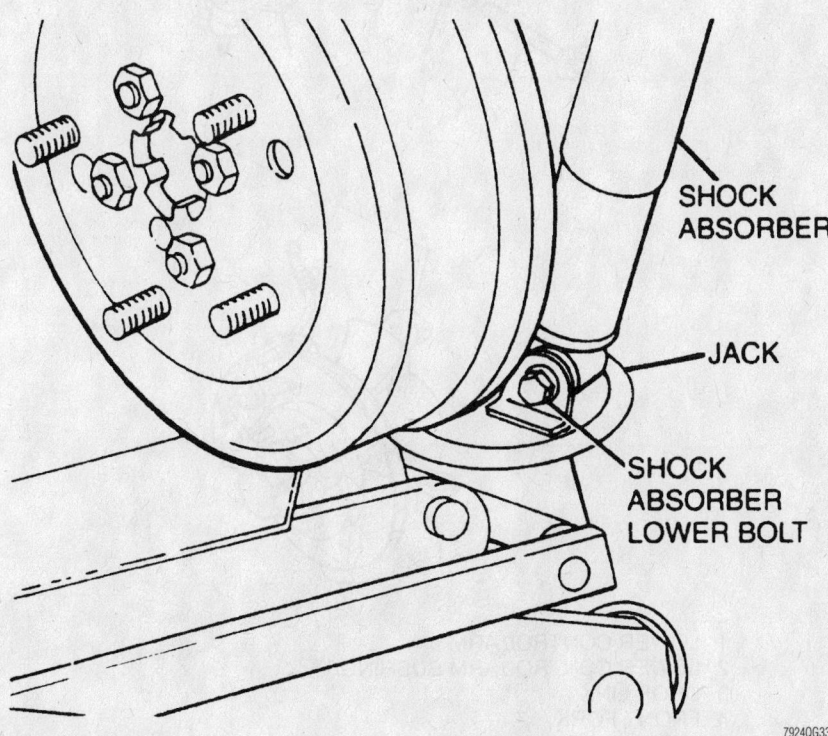

SHOCK
ABSORBER

JACK

SHOCK
ABSORBER
LOWER BOLT

79240G33

Lower shock absorber mounting bolt

Coil Spring

REMOVAL & INSTALLATION

Front

1. Before servicing the vehicle, refer to the precautions in the beginning of this section .
2. Remove or disconnect the following:
   - Negative battery cable
   - Shock absorber and coil spring as an assembly
   - Place the shock absorber in a vice
   - Loosen the pivot rod nut several turns
   - While still secured in a vise compress the coil spring
   - Piston rod nut and disassemble the coil spring as needed

To install:

3. Install or connect the following:
   - Bottom portion of the shock absorber in a vice and compress the coil spring
   - End of the coil spring to the rubber seat and install the spring
   - Assemble the dust boot and lower retainer, lower insulator, spring seat, boss, center washer, upper insulator and install the coil spring
   - Hand tighten the piston rod nut.
   - Carefully loosen the spring compressor tool and remove the tool
   - Torque the piston rod nut to 31 ft. lbs. (42 Nm).
   - Coil spring and shock absorber as an assembly

Rear

1. Before servicing the vehicle, refer to the precautions in the beginning of this section .
2. Remove or disconnect the following:
   - Both rear wheels
   - Raise the rear axle with a floor jack to relax the shock absorbers and support the rear axle when the shock absorber is removed.

➡**For easier installation, complete one side at a time.**

   - Lower mounting bolt, then the shock absorber
   - Lower the floor jack until the coil spring is fully expanded
   - Coil spring
   - Inspect the upper and lower rubber

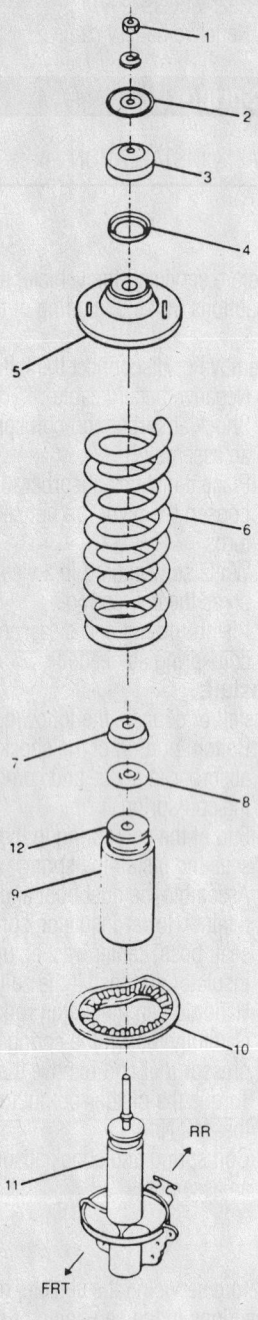

1. Nut
2. Upper retainer
3. Upper insulator
4. Centering washer
5. Spring seat
6. Coil spring
7. Lower insulator
8. Lower retainer
9. Dust boot
10. Rubber seat
11. Shock absorber
12. Front jounce stop

9308QG10

**Exploded view of the front shock absorber
and coil spring assembly**

spring seats and jounce stop for wear or damage, replace if necessary

To install:

3. Install or connect the following:
   • Position the spring in the upper and lower saddles
   • Raise the floor jack and connect the lower shock absorber bolt. Torque the bolt to 62 ft. lbs. (84 Nm).
   • Rear wheels

Upper Ball Joint

REMOVAL & INSTALLATION

The upper ball joint is an integral part of the upper control arm. If the ball joint is worn, replacement of the upper control arm is necessary.

Lower Ball Joint

REMOVAL & INSTALLATION

1. Before servicing the vehicle, refer to the precautions in the beginning of this section .

2. Remove or disconnect the following:
   • Negative battery cable
   • Front wheel assembly
   • Cotter pin and lower ball joint nut
   • Separate the lower ball joint from the spindle with a puller tool by prying down on the spindle to separate it from the lower ball joint

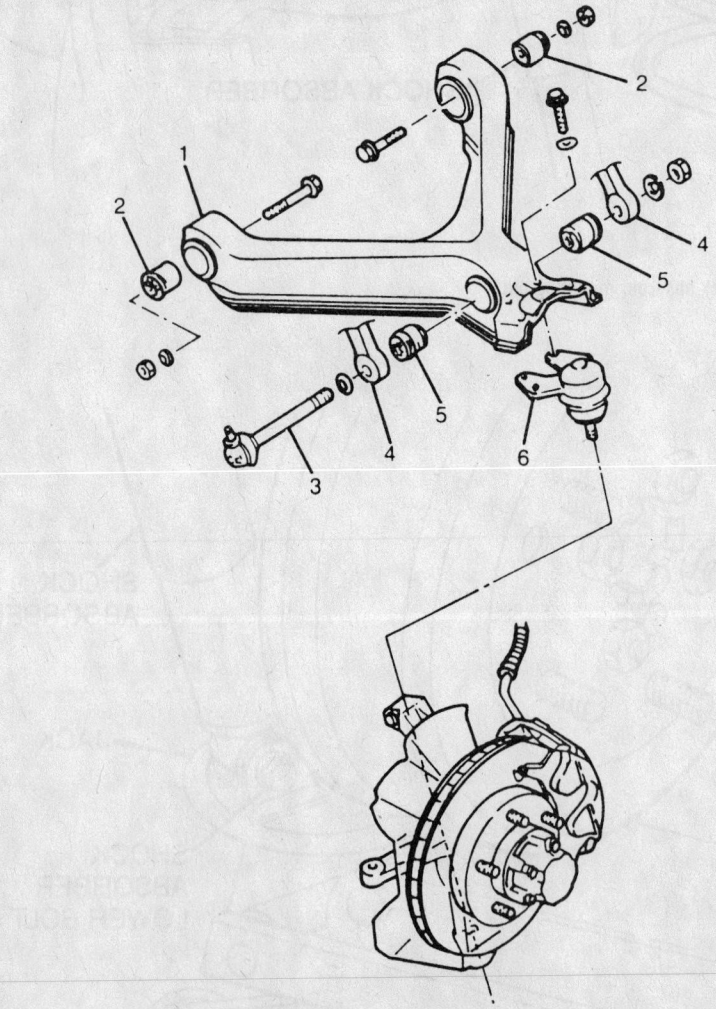

1 LOWER CONTROL ARM
2 LOWER CONTROL ARM BUSHING
3 DROP LINK
4 FRONT FORK
5 DROP LINK BUSHING
6 LOWER CONTROL ARM BALL JOINT

7924QG34

Exploded view of the lower control arm and ball joint assembly

- Lower ball joint attaching bolts
- Lower ball joint

To install:
3. Install or connect the following:
- Position the lower ball joint and install the attaching nuts and bolts. Torque the fasteners to 36 ft. lbs. (48 Nm).
- Pry down and guide the spindle onto the lower ball joint. Torque the nut 87 ft. lbs. (118 Nm) and install a new cotter pin
- Front wheel assembly.
- Negative battery cable

Upper Control Arm

REMOVAL & INSTALLATION

1. Before servicing the vehicle, refer to the precautions in the beginning of this section .
2. Remove or disconnect the following:
- Negative battery cable
- Front wheel assembly

- Bolt securing the upper ball joint to the steering knuckle

➡**Note the matchmark setting on the upper control arm mounting bolts before removal.**

- Upper control arm mounting bolts
- Upper control arm from the vehicle.

To install:
3. Install or connect the following:
- Position the upper control arms in the frame mounting. Hand tighten the bolts
- Position the ball joint in the spindle. Torque the through bolt to 36 ft. lbs. (48 Nm).

➡**Be sure the slot in the ball joint aligns with the through-bolt during installation.**

- Align the upper control arm bolts to the previous settings. Torque bolts to 62 ft. lbs. (108 Nm).
- Front wheel assembly
4. Check and adjust the alignment, if necessary.

UPPER CONTROL ARM BUSHING REPLACEMENT

1. Before servicing the vehicle, refer to the precautions in the beginning of this section .
2. Remove or disconnect the following:
- Upper control arm assembly
- Secure the control arm in a vise
- Using a standard press, remove the bushing

To install:
3. Install or connect the following:
- Lubricate the new bushing and press it into the upper control arm
- Upper control arm to the vehicle
4. Check the wheel alignment and adjust if necessary.

Lower Control Arm

REMOVAL & INSTALLATION

1. Before servicing the vehicle, refer to the precautions in the beginning of this section .
2. Remove or disconnect the following:
- Negative battery cable
- Front wheel assembly
- Stabilizer bar
- Driveshaft from the differential and steering knuckle
- Shock absorber and coil spring from the front half of the fork
- Cotter pin and castle nut from the lower control arm ball joint
- Lower control arm ball joint from the steering knuckle
- Lower control arm bushing bolts from the front frame crossmember brackets
- Lower control arm

To install:
3. Install or connect the following:
- Lower control arm to the front frame crossmember brackets and hand tighten the bushing bolts
- Lower control arm ball joint and bolt to the steering knuckle. Torque the bolt to 87 ft. lbs. (118 Nm).
- New cotter pin and castle nut
- Torque the lower control arm bushing bolts to 206 ft. lbs. (280 Nm).
- Front fork halves to the shock absorber and coil spring and position the lower portion over the lower control arm drop link holes
- Drop link. Torque the nut to 145 ft. lbs. (197 Nm).
- Driveshaft to the front differential and steering knuckle

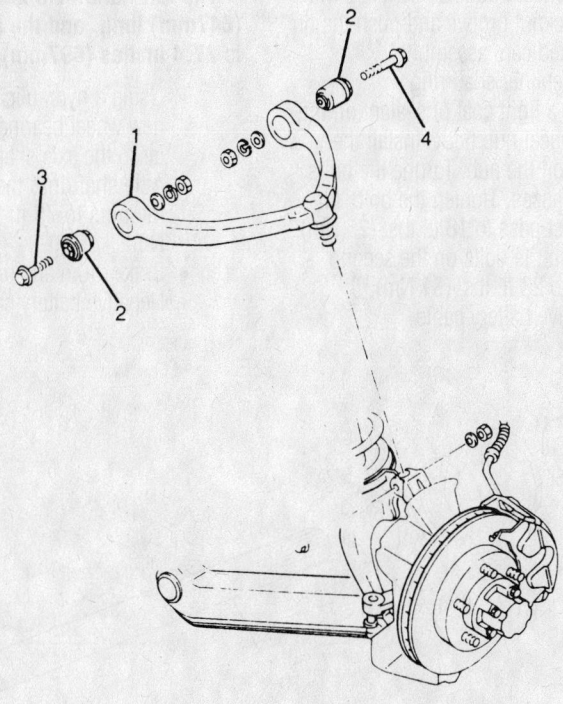

1 UPPER CONTROL ARM
2 UPPER CONTROL ARM BUSHING
3 FRONT SPINDLE
4 REAR SPINDLE

7924QG31

Exploded view of the upper control arm assembly

- Stabilizer bar
- Front wheel assembly
- Negative battery cable

4. Check and adjust the front wheel alignment if necessary.

LOWER CONTROL ARM BUSHING REPLACEMENT

1. Before servicing the vehicle, refer to the precautions in the beginning of this section .
2. Remove or disconnect the following:
   - Lower control arm assembly
   - Secure the control arm in a vise
   - Using a standard press, remove the bushing

To install:
3. Install or connect the following:
   - Lubricate the new bushing and press it into the lower control arm
   - lower control arm to the vehicle
4. Check the wheel alignment and adjust if necessary.

Wheel Bearings

ADJUSTMENT

Front

1. Remove or disconnect the following:
2. Before servicing the vehicle, refer to the precautions in the beginning of this section .
   - Negative battery cable
   - Front wheel
   - Brake caliper from the rotor and hang it out of the way
3. Attach a dial indicator to the axle hub and measure the bearing play
4. If the play exceeds .004 inch (.10mm), check and adjust locknut torque. The bolts should be tightened to 23 ft. lbs. (31 Nm).

Rear

The rear wheel bearings are not adjustable.

REMOVAL & INSTALLATION

Front

1. Before servicing the vehicle, refer to the precautions in the beginning of this section .
2. Remove or disconnect the following:
   - Negative battery cable
   - Front wheel
   - Free wheel hub body

- Brake caliper
- Brake rotor
- Wheel bearing and using a screw driver, pry out the oil seal
- Inner and outer bearings
- Using a drift punch, remove the inner and outer bearing race

To install:
3. Pack the new bearings with grease.
4. Install or connect the following:
   - Inner bearing and race and a new oil seal
   - Outer bearing and race and secure the dust cover with four screws
   - Apply grease to the new bearings and the lip of the oil seal
   - Hub assembly in the steering knuckle
   - Screw the locknut against the hub assembly until there is 10 inch lbs. (1.3 Nm) of preload on the hub
   - Brake rotor and retaining screws
   - Attach a run-out gauge to check the rotor run-out. The run-out should not exceed 0.004 inch (0.10mm)
   - Brake caliper
   - Free wheel hub fixed cam key with the locknut groove and push the on the fixed cam assembly
   - Axle retainer snap ring
   - Apply a light coat of sealant on the free wheel hub body. Install the body on the hub. Torque the bolts in 2 passes. Tighten the bolts on the first pass to 18 ft. lbs. (25 Nm). Tighten the bolts on the second pass to 23 ft. lbs. (31 Nm).
   - Negative battery cable

Rear

1. Before servicing the vehicle, refer to the precautions in the beginning of this section .
2. Remove or disconnect the following:
   - Negative battery cable
   - Rear wheels
   - Brake drum
   - Oil seal retainer flange

➡ **The axle shafts are different from side-to-side, mark the to ensure they are returned to the proper side.**

- Using a slide hammer, remove the axle shaft assembly
- Using a hydraulic press, remove the bearing collar and bearing from the axle
- oil seal from the differential

To install:
3. Install or connect the following:
   - Using the appropriate seal driver, install the new axle seal into the differential

➡ **The left-hand axle is 25.5 inches (647mm) long, and the right-hand axle is 27.4 inches (697mm) long.**

- Using a hydraulic press, install the new wheel bearing and retainer collar to the axle shaft
- Axle shaft into the carrier. Torque the nuts to 75 ft. lbs. (100 Nm).
- Brake drum and rear wheels
- Negative battery cable

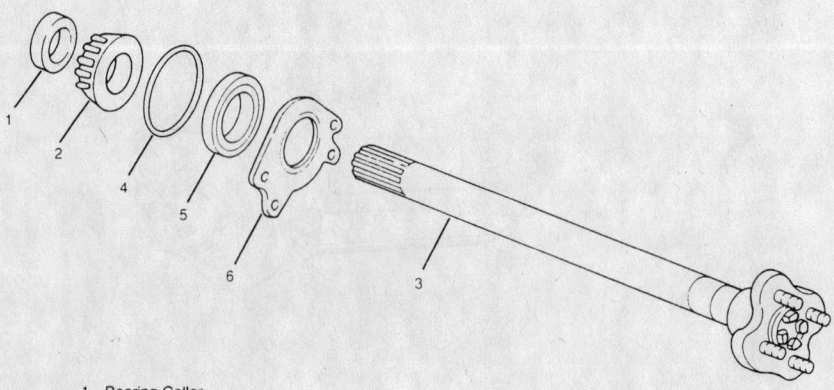

1   Bearing Collar
2   Bearing
3   Axle Shaft
4   Rib Ring
5   Oil Seal
6   Oil Seal Retainer

7924QG35

Rear axle bearing component identification

BRAKES

Brake Caliper

REMOVAL & INSTALLATION

Sportage

1. Raise and safely support the vehicle.
2. Remove the front wheels.
3. Remove the 2 caliper bolts and lift the caliper from the disc.
4. Disconnect the brake fluid flex line by removing the retaining bolt if the caliper is to be replaced.

To install:

5. Seat the caliper piston using a C-clamp.
6. Connect the brake fluid flex line bolt to the caliper. Tighten the bolt to 17 ft. lbs. (23.5 Nm).
7. Position the caliper over the disc assembly. Install the caliper bolts. Tighten to 72 ft. lbs. (98 Nm).
8. Install the front wheels. Tighten the lug nuts 77 ft. lbs. (99 Nm).
9. Bleed the hydraulic system if the flex hoses were removed.
10. Lower the vehicle.

Disc Brake Pads

REMOVAL & INSTALLATION

1. Raise and safely support the vehicle.
2. Remove the front wheels.
3. Remove the 2 caliper bolts and lift the caliper from the disc.
4. Slide the disc pads off the caliper bracket.

To install:

5. Clean the caliper bracket contact surface with a wire brush and lightly coat with assembly lube.
6. Position the disc pads.
7. Position the caliper over the disc assembly. Install the caliper bolts. Tighten to 72 ft. lbs. (98 Nm).
8. Install the front wheels. Tighten the lug nuts 77 ft. lbs. (99 Nm).
9. Bleed the hydraulic system if the flex hoses were removed.
10. Lower the vehicle.

Brake Drums

REMOVAL & INSTALLATION

1. Raise and safely support the vehicle.
2. Remove the rear wheels.
3. Apply the parking brake.
4. Remove the 4 attaching nuts.
5. Release the parking brake and remove the brake drum.
Installation is the reverse of the removal procedure.

Brake Shoes

REMOVAL & INSTALLATION

1. Raise and safely support the vehicle.
2. Remove the rear wheels.
3. Apply the parking brake.
4. Remove the 4 attaching nuts.
5. Release the parking brake and remove the brake drum.
6. Remove the spring from the secondary shoe to the adjusting lever.
7. Remove the adjusting lever.
8. Remove the return spring above the star adjusting wheel.
9. Turn the starwheel clockwise to relieve tension on the brake shoes.

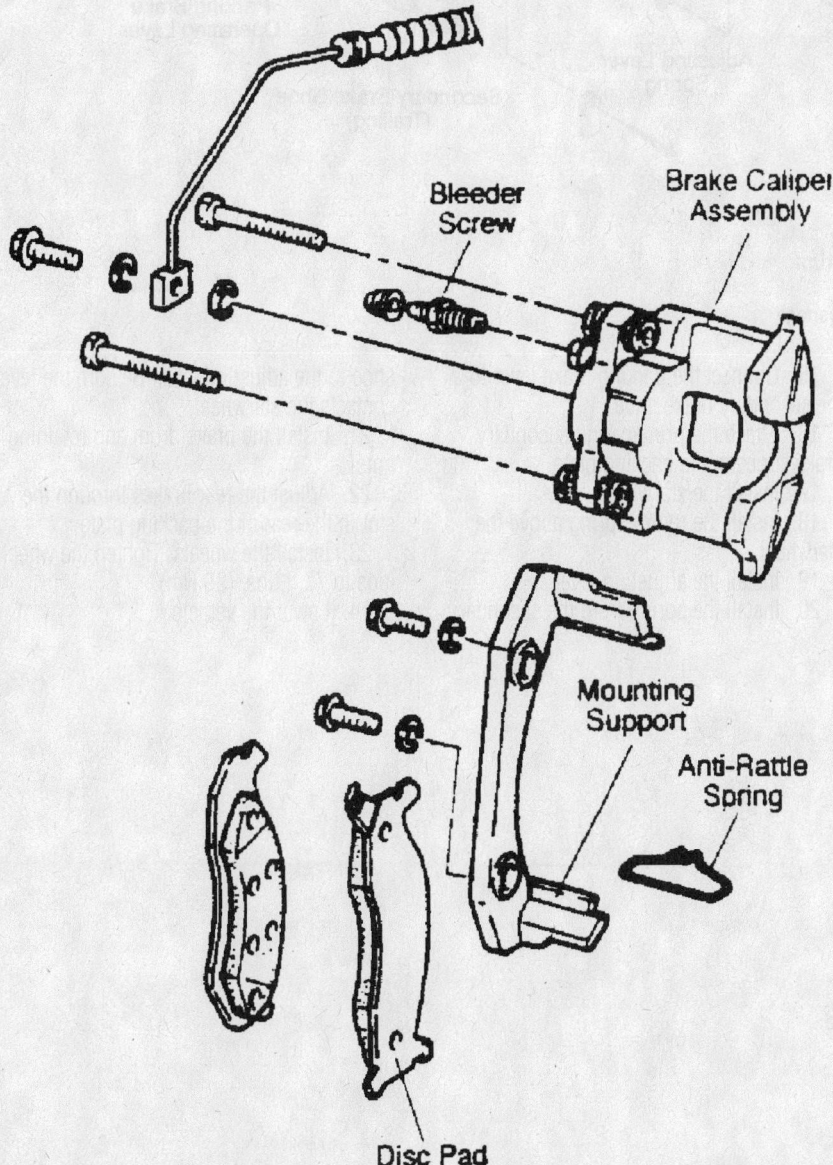

Bleeder Screw

Brake Caliper Assembly

Mounting Support

Anti-Rattle Spring

Disc Pad

93026G87

Exploded view of the front disc brake assembly—Sportage

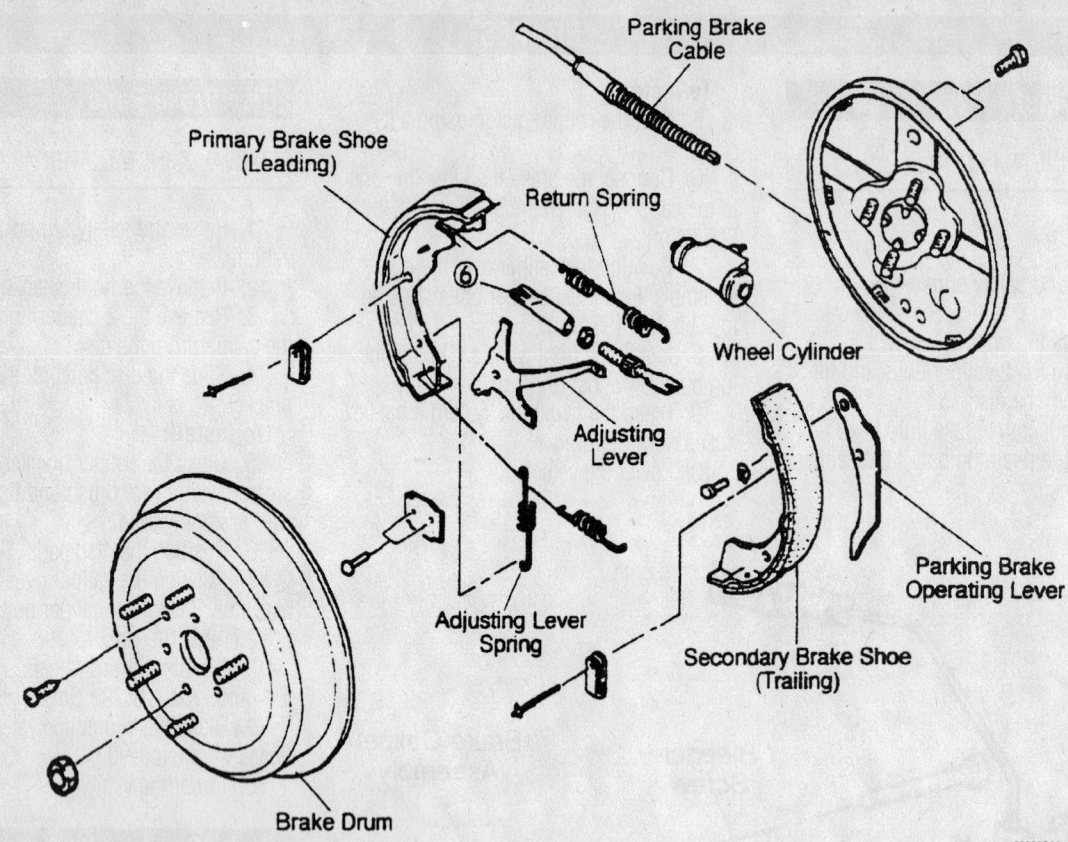

Parking Brake
Cable

Primary Brake Shoe
(Leading)

Return Spring

Wheel Cylinder

Adjusting
Lever

Parking Brake
Operating
Lever

Adjusting Lever
Spring

Secondary Brake Shoe
(Trailing)

Brake Drum

93026G88

Exploded view of the rear drum brake assembly—Sportage

10. Remove the starwheel.
11. Remove the hold-down pin clips.
12. Remove the primary shoe.
13. Disconnect the C-clip and pin attaching the parking brake lever to the secondary brake shoe.

To install:
14. Lubricate the backing plate contact points.

15. Connect the parking brake lever to the secondary brake shoe.
16. Attach the primary and secondary brake shoes to the backing plate.
17. Install the starwheel.
18. Install the return spring above the starwheel.
19. Install the adjusting lever.
20. Install the spring from the secondary

shoe to the adjusting lever. Be sure the lever contacts the starwheel.
21. Install the brake drum and retaining nuts.
22. Adjust the rear brakes through the slot in the rear of the backing plate.
23. Install the wheels. Tighten the wheel lugs to 77 ft. lbs. (99 Nm).
24. Lower the vehicle.

LEXUS

16

ES 300 • IS 300 • GS 300 • GS 400 • GS 430 • LS 400 • LS 430 • SC 300 • SC 400

SPECIFICATION CHARTS

ENGINE AND VEHICLE IDENTIFICATION

Code ①	Liters (cc)	Cu. In.	Cyl.	Fuel Sys.	Engine Type	Eng. Mfg.
1MZ-FE	3.0 (2995)	183	6	SFI	DOHC	Toyota
1UZ-FE	4.0 (3969)	242	8	SFI	DOHC	Toyota
2JZ-GE	3.0 (2997)	183	6	SFI	DOHC	Toyota
3UZ-FE	4.3 (4264)	262	8	SFI	DOHC	Toyota

SFI: Sequential Multi-port Fuel Injection

DOHC: Double Overhead Camshaft

① Located on the timing belt cover

② 10th digit of the VIN

Code ②	Year
Y	2000
1	2001
2	2002
3	2003
4	2004

42356-LEXC-C01

GENERAL ENGINE SPECIFICATIONS

All measurements are given in inches.

Year	Model	Engine Displacement Liters (cc)	Engine ID/VIN	Fuel System Type	Net Horsepower @ rpm	Net Torque @ rpm (ft. lbs.)	Bore x Stroke (in.)	Compression Ratio	Oil Pressure @ rpm
2000	ES 300	3.0 (2995)	1MZ-FE	SFI	188@5200	203@4400	3.44x3.27	10.5:1	43-78@3000
	GS 300	3.0 (2997)	2JZ-GE	SFI	220@5800	210@4800	3.39x3.39	10.0:1	47-84@3000
	GS 400	4.0 (3969)	1UZ-FE	SFI	260@5300	270@4500	3.45x3.25	10.4:1	43-85@3000
	LS 400	4.0 (3969)	1UZ-FE	SFI	260@5300	270@4500	3.45x3.25	10.4:1	43-85@3000
	SC 300	3.0 (2997)	2JZ-GE	SFI	220@5800	210@4800	3.39x3.39	10.0:1	47-84@3000
	SC 400	4.0 (3969)	1UZ-FE	SFI	290@5300	270@4500	3.45x3.25	10.4:1	43-85@3000
2001	ES 300	3.0 (2995)	1MZ-FE	SFI	210@5200	220@4400	3.44x3.27	10.5:1	43-78@3000
	GS 300	3.0 (2997)	2JZ-GE	SFI	220@5800	220@3800	3.39x3.39	10.0:1	47-84@3000
	GS 430	4.3 (4264)	3UZ-FE	SFI	300@5300	325@4500	3.58x3.25	10.4:1	43-85@3000
	IS 300	3.0 (2997)	2JZ-GE	SFI	215@5800	218@3800	3.39x3.39	10.0:1	47-84@3000
	LS 430	4.3 (4264)	3UZ-FE	SFI	290@5300	320@4500	3.58x3.25	10.4:1	43-85@3000
2002	ES 300	3.0 (2995)	1MZ-FE	SFI	210@5200	220@4400	3.44x3.27	10.5:1	43-78@3000
	GS 300	3.0 (2997)	2JZ-GE	SFI	220@5800	220@3800	3.39x3.39	10.0:1	47-84@3000
	GS 430	4.3 (4264)	3UZ-FE	SFI	300@5300	325@4500	3.58x3.25	10.4:1	43-85@3000
	IS 300	3.0 (2997)	2JZ-GE	SFI	215@5800	218@3800	3.39x3.39	10.0:1	47-84@3000
	LS 430	4.3 (4264)	3UZ-FE	SFI	290@5300	320@4500	3.58x3.25	10.4:1	43-85@3000
2003-04	ES 300	3.0 (2995)	1MZ-FE	SFI	210@5200	220@4400	3.44x3.27	10.5:1	43-78@3000
	GS 300	3.0 (2997)	2JZ-GE	SFI	220@5800	220@3800	3.39x3.39	10.0:1	47-84@3000
	GS 430	4.3 (4264)	3UZ-FE	SFI	300@5300	325@4500	3.58x3.25	10.4:1	43-85@3000
	IS 300	3.0 (2997)	2JZ-GE	SFI	215@5800	218@3800	3.39x3.39	10.0:1	47-84@3000
	LS 430	4.3 (4264)	3UZ-FE	SFI	290@5300	320@4500	3.58x3.25	10.4:1	43-85@3000

SFI : Sequential fuel injection

42356-LEXC-C02

ENGINE TUNE-UP SPECIFICATIONS

Year	Engine Displacement Liters (cc)	Engine ID/VIN	Spark Plug Gap (in.)	Ignition Timing (deg.)	Fuel Pump (psi)	Idle Speed (rpm)	Valve Clearance	
							Intake	Exhaust
2000	3.0 (2995)	1MZ-FE	0.043	8-12B①	44-50	650-750	0.006-0.010	0.010-0.014
	3.0 (2997)	2JZ-GE	0.043	8-12B②	44-50	650-750	0.006-0.010	0.010-0.014
	4.0 (3969)	1UZ-FE	0.043	8-12B②	44-50	700-800	0.006-0.010	0.010-0.014
2001	3.0 (2995)	1MZ-FE	0.043	8-12B①	44-50	650-750	0.006-0.010	0.010-0.014
	3.0 (2997)	2JZ-GE	0.043	8-12B②	44-50	650-750	0.006-0.010	0.010-0.014
	4.3 (4264)	3UZ-FE	0.043	8-12B③	44-50	700-800	0.006-0.010	0.010-0.014
2002	3.0 (2995)	1MZ-FE	0.039-0.043	8-12B①	44-50	650-750	0.006-0.010	0.010-0.014
	3.0 (2997)	2JZ-GE	0.039-0.043	8-12B②	44-50	650-750	0.006-0.010	0.010-0.014
	4.3 (4264)	3UZ-FE	0.043	8-12B③	44-50	700-800	0.006-0.010	0.010-0.014
2003-04	3.0 (2995)	1MZ-FE	0.039-0.043	8-12B①	44-50	650-750	0.006-0.010	0.010-0.014
	3.0 (2997)	2JZ-GE	0.039-0.043	8-12B②	44-50	650-750	0.006-0.010	0.010-0.014
	4.3 (4264)	3UZ-FE	0.043	8-12B③	44-50	700-800	0.006-0.010	0.010-0.014

NOTE: The Vehicle Emission Control Information label often reflects specification changes made during production. The label figures must be used if they differ from those in this chart.

B: Before top dead center

① Terminals TE1 and E1 of check connector must be connected

② Terminals TC and E1 of check connector must be connected

③ LS 430: Terminals TC and CG of check connector must be connected

    GS 430/300: Terminals TC and E1 of check connector must be connected

42356-LEXC-C03

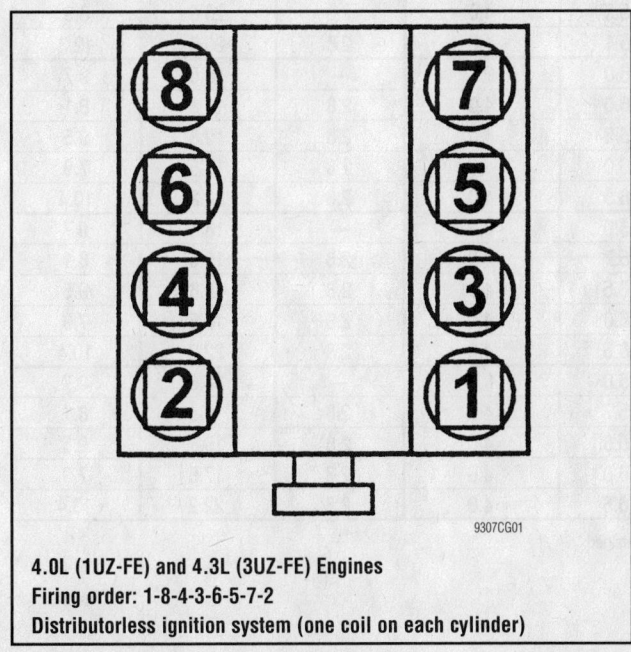

4.0L (1UZ-FE) and 4.3L (3UZ-FE) Engines
Firing order: 1-8-4-3-6-5-7-2
Distributorless ignition system (one coil on each cylinder)

9307CG01

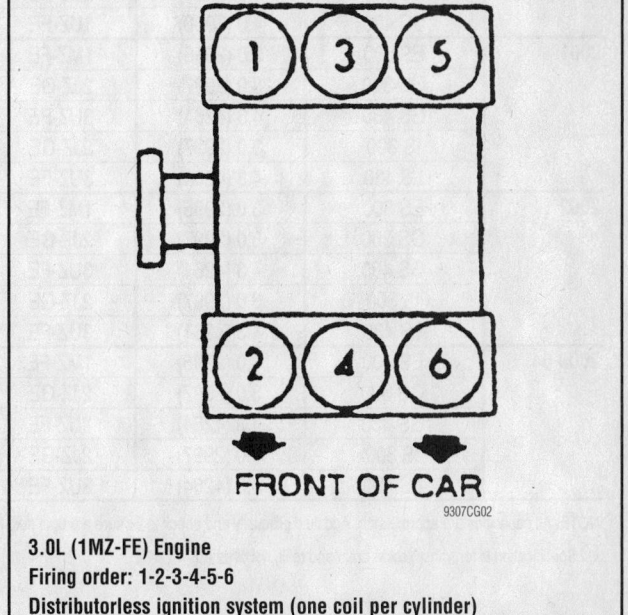

FRONT OF CAR

3.0L (1MZ-FE) Engine
Firing order: 1-2-3-4-5-6
Distributorless ignition system (one coil per cylinder)

9307CG02

3.0L (2JZ-GE) Engine
Firing order: 1-5-3-6-2-4
Distributorless ignition system (one coil on each cylinder)

Serpentine drive belt routing—3.0L (2JZ-GE) engine

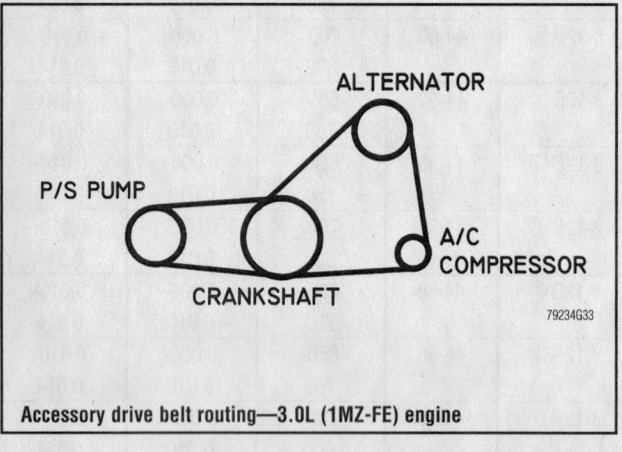

Accessory drive belt routing—3.0L (1MZ-FE) engine

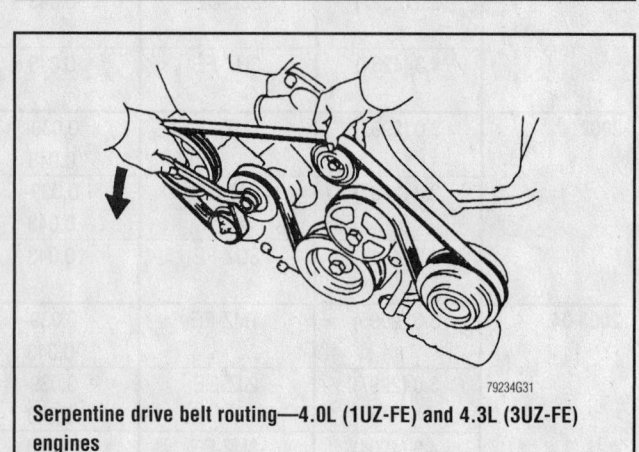

Serpentine drive belt routing—4.0L (1UZ-FE) and 4.3L (3UZ-FE) engines

CAPACITIES

Year	Model	Engine Displacement Liters (cc)	Engine ID/VIN	Engine Oil with Filter	Transmission (pts.) Auto. ①	Drive Axle (pts.)	Fuel Tank (gal.)	Cooling System (qts.)
2000	ES 300	3.0 (2995)	1MZ-FE	5.0	8.2	—	18.5	9.7
	GS 300	3.0 (2997)	2JZ-GE	6.0	4.0	2.8	19.8	8.1
	GS 400	4.0 (3969)	1UZ-FE	5.5	4.0	2.8	19.8	9.8
	LS 400	4.0 (3969)	1UZ-FE	6.5	4.0	2.8	22.5	11.6
	SC 300	3.0 (2997)	2JZ-GE	5.5	4.0	2.8	20.6	8.9
	SC 400	4.0 (3969)	1UZ-FE	5.1	4.0	2.8	20.6	12.3
2001	ES 300	3.0 (2995)	1MZ-FE	5.0	8.2	—	18.5	9.7
	GS 300	3.0 (2997)	2JZ-GE	6.0	4.0	2.8	19.8	8.1
	GS 430	4.3 (4264)	3UZ-FE	5.5	4.0	2.8	19.8	9.5
	IS 300	3.0 (2997)	2JZ-GE	6.0	4.0	2.8	17.5	7.9
	LS 430	4.3 (4264)	3UZ-FE	6.5	4.0	2.8	22.2	10.4
2002	ES 300	3.0 (2995)	1MZ-FE	5.0	8.2	—	18.5	9.7
	GS 300	3.0 (2997)	2JZ-GE	5.7	4.0	2.8	19.8	8.1
	GS 430	4.3 (4264)	3UZ-FE	5.5	4.0	2.8	19.8	9.5
	IS 300	3.0 (2997)	2JZ-GE	6.0	4.0	2.8	17.5	7.9
	LS 430	4.3 (4264)	3UZ-FE	6.5	4.0	2.8	22.2	10.4
2003-04	ES 300	3.0 (2995)	1MZ-FE	5.0	8.2	—	18.5	9.7
	GS 300	3.0 (2997)	2JZ-GE	5.7	4.0	2.8	19.8	8.1
	GS 430	4.3 (4264)	3UZ-FE	5.5	4.0	2.8	19.8	9.5
	IS 300	3.0 (2997)	2JZ-GE	6.0	4.0	2.8	17.5	7.9
	LS 430	4.3 (4264)	3UZ-FE	6.5	4.0	2.8	22.2	10.4

NOTE: All capacities are approximate. Add fluid gradually and check to be sure a proper fluid level is obtained.

① Specification is for transmission drain and refill, not overhaul.

VALVE SPECIFICATIONS

Year	Engine Displacement Liters (cc)	Engine ID/VIN	Seat Angle (deg.)	Face Angle (deg.)	Spring Test Pressure (lbs. @ in.)	Spring Free-Length (in.)	Stem-to-Guide Clearance (in.)		Stem Diameter (in.)	
							Intake	Exhaust	Intake	Exhaust
2000	3.0 (2995)	1MZ-FE	NA	44.5	41.9-46.3@ 1.331	1.791	0.0010- 0.0024	0.0012- 0.0026	0.2154- 0.2159	0.2152- 0.2157
	3.0 (2997)	2JZ-GE	NA	44.5	41.9-46.3@ 1.358	①	0.0010- 0.0024	0.0012- 0.0026	0.2350- 0.2356	0.2348- 0.2354
	4.0 (3969)	1UZ-FE	NA	44.5	②	2.130	0.0010- 0.0024	0.0012- 0.0026	0.2154- 0.2159	0.2152- 0.2157
2001	3.0 (2995)	1MZ-FE	NA	44.5	41.9-46.3@ 1.331	1.791	0.0010- 0.0024	0.0012- 0.0026	0.2154- 0.2159	0.2152- 0.2157
	3.0 (2997)	2JZ-GE	NA	44.5	41.9-46.3@ 1.358	①	0.0010- 0.0024	0.0012- 0.0026	0.2350- 0.2356	0.2348- 0.2354
	4.3 (4264)	3UZ-FE	45	44.5	45.9-50.7@ 1.3795	2.130	0.0010- 0.0024	0.0012- 0.0026	0.2154- 0.2159	0.2152- 0.2157
2002	3.0 (2995)	1MZ-FE	NA	44.5	41.9-46.3@ 1.331	1.791	0.0010- 0.0024	0.0012- 0.0026	0.2154- 0.2159	0.2152- 0.2157
	3.0 (2997)	2JZ-GE	NA	44.5	41.9-46.3@ 1.358	①	0.0010- 0.0024	0.0012- 0.0026	0.2350- 0.2356	0.2348- 0.2354
	4.3 (4264)	3UZ-FE	45	44.5	45.9-50.7@ 1.3795	2.130	0.0010- 0.0024	0.0012- 0.0026	0.2154- 0.2159	0.2152- 0.2157
2003-04	3.0 (2995)	1MZ-FE	NA	44.5	41.9-46.3@ 1.331	1.791	0.0010- 0.0024	0.0012- 0.0026	0.2154- 0.2159	0.2152- 0.2157
	3.0 (2997)	2JZ-GE	NA	44.5	41.9-46.3@ 1.358	①	0.0010- 0.0024	0.0012- 0.0026	0.2350- 0.2356	0.2348- 0.2354
	4.3 (4264)	3UZ-FE	45	44.5	45.9-50.7@ 1.3795	2.130	0.0010- 0.0024	0.0012- 0.0026	0.2154- 0.2159	0.2152- 0.2157

NA: Not Available

① Pink: 1.7209
   Yellow: 1.7362

② GS400: 45.9-50.7@1.3795
   SC400 & LS400: 45.9-50.7@1.378

42356-LEXC-C05

TORQUE SPECIFICATIONS
All readings in ft. lbs.

Year	Engine Displacement Liters (cc)	Engine ID/VIN	Cylinder Head Bolts	Main Bearing Bolts	Rod Bearing Bolts	Crankshaft Damper Bolts	Flywheel Bolts	Manifold		Spark Plugs	Lug Nuts
								Intake	Exhaust		
2000	3.0 (2995)	1MZ-FE	①	②	③	159	61	11	36	13	76
	3.0 (2997)	2JZ-GE	④	⑤	⑥	243	61	21	30	13	76
	4.0 (3969)	1UZ-FE	⑦	⑧	⑦	181	61	13	32	13	76
2001	3.0 (2995)	1MZ-FE	①	②	③	159	61	11	36	13	76
	3.0 (2997)	2JZ-GE	④	⑤	⑥	243	61	21	30	13	76
	4.3 (4264)	3UZ-FE	⑨	⑧	⑦	181	⑩	13	32	13	76
2002	3.0 (2995)	1MZ-FE	①	②	⑦	159	61	11	36	13	76
	3.0 (2997)	2JZ-GE	④	⑤	⑥	243	61	21	30	13	76
	4.3 (4264)	3UZ-FE	⑨	⑧	⑦	181	⑩	13	32	13	76
2003-04	3.0 (2995)	1MZ-FE	①	②	⑦	159	61	11	36	13	76
	3.0 (2997)	2JZ-GE	④	⑤	⑥	243	61	21	30	13	76
	4.3 (4264)	3UZ-FE	⑨	⑧	⑦	181	⑩	13	32	13	76

① Head bolt:
 Step 1: 40 ft. lbs.
 Step 2: Plus 90 degrees
 Recessed head bolt: 13 ft. lbs.

② 6-point bolts: 20 ft. lbs.
 12-point bolts:
 Step 1: 16 ft. lbs.
 Step 2: Plus an additional 90 degrees

③ Step 1: 18 ft. lbs.
 Step 2: Plus 90 degrees

④ Step 1: 25 ft. lbs.
 Step 2: Tighten an additional 90 degrees
 Step 3: Tighten an additional 90 degrees

⑤ Step 1: 33 ft. lbs.
 Step 2: Plus 90 degrees

⑥ Step 1: 22 ft. lbs.
 Step 2: Plus 90 degrees

⑦ Step 1: Tighten to 29 ft. lbs.
 Step 2: Plus 90 degrees

⑧ Nuts:
 Step 1: 20 ft. lbs.
 Step 2: Plus 90 degrees
 Bolts: 36 ft. lbs.

⑨ Step 1: 44 ft. lbs.
 Step 2: Plus 90 degrees

⑩ Step 1: 36 ft. lbs.
 Step 2: Plus 90 degrees

42356-LEXC-C06

PISTON AND RING SPECIFICATIONS

All measurements are given in inches.

| Year | Engine Displacement Liters (cc) | Engine ID/VIN | Piston Clearance | Ring Gap | | | Ring Side Clearance | | |
				Top Compression	Bottom Compression	Oil Control	Top Compression	Bottom Compression	Oil Control
2000	3.0 (2995)	1MZ-FE	0.0033-0.0042	0.0098-0.0138	0.0138-0.0177	0.0059-0.0157	0.0008-0.0028	0.0008-0.0024	SNUG
	3.0 (2997)	2JZ-GE	0.0014-0.0027	0.0118-0.0185	0.0138-0.0205	0.0051-0.0177	0.0004-0.0028	0.0012-0.0028	SNUG
	4.0 (3969)	1UZ-FE	0.0033-0.0041	0.0098-0.0177	0.0197-0.0276	0.0059-0.0197	0.0008-0.0028	0.0004-0.0020	SNUG
2001	3.0 (2995)	1MZ-FE	0.0033-0.0042	0.0098-0.0138	0.0138-0.0177	0.0059-0.0157	0.0008-0.0028	0.0008-0.0024	SNUG
	3.0 (2997)	2JZ-GE	0.0014-0.0027	0.0118-0.0185	0.0138-0.0205	0.0051-0.0177	0.0004-0.0028	0.0012-0.0028	SNUG
	4.3 (4264)	3UZ-FE	0.0033-0.0041	0.0118-0.0157	0.0157-0.0197	0.0059-0.0157	0.0012-0.0031	0.0008-0.0024	SNUG
2002	3.0 (2995)	1MZ-FE	0.0033-0.0042	0.0098-0.0138	0.0138-0.0177	0.0059-0.0157	0.0008-0.0028	0.0008-0.0024	SNUG
	3.0 (2997)	2JZ-GE	0.0022-0.0031	0.0118-0.0185	0.0138-0.0205	0.0051-0.0177	0.0004-0.0028	0.0012-0.0028	SNUG
	4.3 (4264)	3UZ-FE	0.0031-0.0041	0.0118-0.0197	0.0157-0.0236	0.0059-0.0197	0.0012-0.0031	0.0008-0.0024	SNUG
2003-04	3.0 (2995)	1MZ-FE	0.0033-0.0042	0.0098-0.0138	0.0138-0.0177	0.0059-0.0157	0.0008-0.0028	0.0008-0.0024	SNUG
	3.0 (2997)	2JZ-GE	0.0022-0.0031	0.0118-0.0185	0.0138-0.0205	0.0051-0.0177	0.0004-0.0028	0.0012-0.0028	SNUG
	4.3 (4264)	3UZ-FE	0.0031-0.0041	0.0118-0.0197	0.0157-0.0236	0.0059-0.0197	0.0012-0.0031	0.0008-0.0024	SNUG

42356-LEXC-C07

CRANKSHAFT AND CONNECTING ROD SPECIFICATIONS
All measurements are given in inches.

Year	Engine Displacement Liters (cc)	Engine ID/VIN	Crankshaft				Connecting Rod		
			Main Brg. Journal Dia.	Main Brg. Oil Clearance	Shaft End-play	Thrust on No.	Journal Diameter	Oil Clearance	Side Clearance
2000	3.0 (2995)	1MZ-FE	2.4011	①	0.0016- 0.0095	2	2.0863- 2.0866	0.0015- 0.0025	0.0059- 0.0118
	3.0 (2997)	2JZ-GE	2.4403- 2.4409	0.0010- 0.0016	0.0008- 0.0087	4	2.0465- 2.0472	0.0009- 0.0016	0.0098- 0.0158
	4.0 (3969)	1UZ-FE	2.6373- 2.6378	②	0.0008- 0.0087	3	2.0465- 2.0472	0.0011- 0.0021	0.0063- 0.0138
2001	3.0 (2995)	1MZ-FE	2.4011	①	0.0016- 0.0095	2	2.0863- 2.0866	0.0015- 0.0025	0.0059- 0.0118
	3.0 (2997)	2JZ-GE	2.4403- 2.4409	0.0010- 0.0016	0.0008- 0.0087	4	2.0465- 2.0472	0.0009- 0.0016	0.0098- 0.0158
	4.3 (4264)	3UZ-FE	2.6373- 2.6378	②	0.0008- 0.0087	3	2.0465- 2.0472	0.0008- 0.0019	0.0063- 0.0138
2002	3.0 (2995)	1MZ-FE	2.4011	①	0.0016- 0.0095	2	2.0863- 2.0866	0.0015- 0.0025	0.0059- 0.0118
	3.0 (2997)	2JZ-GE	2.4403- 2.4409	0.0010- 0.0016	0.0008- 0.0087	4	2.0465- 2.0472	0.0009- 0.0016	0.0098- 0.0158
	4.3 (4264)	3UZ-FE	2.6373- 2.6378	②	0.0008- 0.0087	3	2.0465- 2.0472	0.0008- 0.0019	0.0063- 0.0138
2003-04	3.0 (2995)	1MZ-FE	2.4011	①	0.0016- 0.0095	2	2.0863- 2.0866	0.0015- 0.0025	0.0059- 0.0118
	3.0 (2997)	2JZ-GE	2.4403- 2.4409	0.0010- 0.0016	0.0008- 0.0087	4	2.0465- 2.0472	0.0009- 0.0016	0.0098- 0.0158
	4.3 (4264)	3UZ-FE	2.6373- 2.6378	②	0.0008- 0.0087	3	2.0465- 2.0472	0.0008- 0.0019	0.0063- 0.0138

① Journal No. 1 and 4: 0.0006 - 0.0013 inch
Journal No. 2 and 3: 0.0010 - 0.0018 inch

② Journal No. 1 and 5: 0.0007 - 0.0013 inch
Remaning journals: 0.0011 - 0.0018 inch

42356-LEXC-C08

WHEEL ALIGNMENT

Year	Model		Caster Range (+/-Deg.)	Caster Preferred Setting (Deg.)	Camber Range (+/-Deg.)	Camber Preferred Setting (Deg.)	Toe-in (in.)	Steering Axis Inclination (Deg.)
2000	ES 300	F	0.75	+2.30	0.75	-0.62	0 +/- 0.08	13.06
		R	—	—	0.75	-0.80	0.16 +/- 0.08	—
	GS 300	F	0.50	+7.55	0.50	-0.27	0.06 +/- 0.08	8.83
		R	—	—	0.50	-0.78	0.06 +/- 0.08	—
	LS 400 ①	F	0.75	+7.00	0.75	+0.34	0.12 +/- 0.08	8.42
		R	—	—	0.75	-0.83	0.09 +/- 0.09	—
	LS 400 ②	F	0.75	+7.42	0.75	+0.08	0.04 +/- 0.08	8.75
		R	—	—	0.75	-1.42	0.12 +/- 0.08	—
	SC 300	F	0.75	+3.01	0.75	+2.00	0.04 +/- 0.08	9.03
		R	—	—	0.75	+1.08	0.08 +/- 0.24	—
	SC 400	F	0.75	+3.00	0.75	+2.00	0.04 +/- 0.08	9.03
		R	—	—	0.75	+1.08	0.08 +/- 0.24	—
2001	ES 300	F	0.75	+2.30	0.75	-0.62	0 +/- 0.08	13.06
		R	—	—	0.75	-0.80	0.16 +/- 0.08	—
	GS 300	F	0.50	+7.55	0.50	-0.27	0.06 +/- 0.08	8.83
		R	—	—	0.50	-0.78	0.06 +/- 0.08	—
	GS 430	F	0.50	+7.55	0.50	-0.27	0.06 +/- 0.08	8.83
		R	—	—	0.50	-0.78	0.06 +/- 0.08	—
	IS 300	F	0.50	+6.12	0.50	-0.50	0.04 +/- 0.08	9.42
		R	—	—	0.50	-0.07	0.08 +/- 0.08	—
	LS 430 ①	F	0.75	+6.75	0.75	-0.08	0.04 +/- 0.08	9.00
		R	—	—	0.75	-1.00	0.12 +/- 0.08	—
	LS 430 ②	F	0.75	+7.25	0.75	-0.25	0.04 +/- 0.08	9.25
		R	—	—	0.75	-1.55	0.12 +/- 0.08	—
2002	ES 300	F	0.75	+2.77	0.75	-0.72	0 +/- 0.08	11.45
		R	—	—	0.75	-1.38	0.16 +/- 0.08	—
	GS 300	F	0.50	+7.55	0.50	-0.27	0.06 +/- 0.08	8.83
		R	—	—	0.50	-0.78	0.06 +/- 0.08	—
	GS 430	F	0.50	+7.55	0.50	-0.27	0.06 +/- 0.08	8.83
		R	—	—	0.50	-0.78	0.06 +/- 0.08	—
	IS 300	F	0.50	+6.12	0.50	-0.50	0.04 +/- 0.08	9.42
		R	—	—	0.50	-0.07	0.08 +/- 0.08	—
	LS 430 ①	F	0.75	+6.75	0.75	-0.08	0.04 +/- 0.08	9.00
		R	—	—	0.75	-1.00	0.12 +/- 0.08	—
	LS 430 ②	F	0.75	+7.25	0.75	-0.25	0.04 +/- 0.08	9.25
		R	—	—	0.75	-1.55	0.12 +/- 0.08	—
2003-04	ES 300	F	0.75	+2.77	0.75	-0.72	0 +/- 0.08	11.45
		R	—	—	0.75	-1.38	0.16 +/- 0.08	—
	GS 300	F	0.50	+7.55	0.50	-0.27	0.06 +/- 0.08	8.83
		R	—	—	0.50	-0.78	0.06 +/- 0.08	—
	GS 430	F	0.50	+7.55	0.50	-0.27	0.06 +/- 0.08	8.83
		R	—	—	0.50	-0.78	0.06 +/- 0.08	—
	IS 300	F	0.50	+6.12	0.50	-0.50	0.04 +/- 0.08	9.42
		R	—	—	0.50	-0.07	0.08 +/- 0.08	—
	LS 430 ①	F	0.75	+6.75	0.75	-0.08	0.04 +/- 0.08	9.00
		R	—	—	0.75	-1.00	0.12 +/- 0.08	—
	LS 430 ②	F	0.75	+7.25	0.75	-0.25	0.04 +/- 0.08	9.25
		R	—	—	0.75	-1.55	0.12 +/- 0.08	—

① Except with air suspension
② With air suspension

42356-LEXC-C09

TIRE, WHEEL AND BALL JOINT SPECIFICATIONS

Year	Model	OEM Tires		Tire Pressures (psi)		Wheel Size	Ball Joint Inspection
		Standard	Optional	Front	Rear		
2000	GS 400	225/55VR16	235/45ZR17	Std: 32 Opt: 33	Std: 32 Opt: 33	Std: 7.5-JJ Opt: 8-JJ	U: 9-30 in. ①
	SC 300	225/55VR16	None	32	32	6.5-JJ	U: 9-30 in. ①
	SC 400	225/55VR16	None	32	32	6.5-JJ	U: 9-30 in. ①
	GS 300	P215/60VR16	225/55VR16	30	30	7.5-JJ	U: 9-30 in. ①
	LS 400	P225/60VR16	None	29	29	7-JJ	U: 9-30 in. ①
	ES 300	P205/65VR15	None	26	26	6-JJ	U: 9-30 in. ①
2001	GS 430	225/55VR16	235/45ZR17	Std: 32 Opt: 33	Std: 32 Opt: 33	Std: 7.5-JJ Opt: 8-JJ	U: 9-30 in. ①
	GS 300	P215/60VR16	225/55VR16	30	30	7.5-JJ	U: 9-30 in. ①
	LS 4300	P225/60VR16		29	29	7-JJ	U: 9-30 in. ①
			P225/55R17 95H	32	35		
			225/55R17 97W	35	36		
	IS 300	P205/55R16 89V	215/45ZR17	33	33	NA	U: 9-30 in. ①
	ES 300	P205/65VR15	None	26	26	6-JJ	U: 9-30 in. ①
2002	GS 430	225/55VR16	235/45ZR17	Std: 32 Opt: 33	Std: 32 Opt: 33	Std: 7.5-JJ Opt: 8-JJ	U: 9-30 in. ①
	GS 300	P215/60VR16	225/55VR16	30	30	7.5-JJ	U: 9-30 in. ①
	LS 4300	P225/60VR16		29	29	7-JJ	U: 9-30 in. ①
			P225/55R17 95H	32	35		
			225/55R17 97W	35	36		
	IS 300	P205/55R16 89V	215/45ZR17	33	33	NA	U: 9-30 in. ①
	ES 300	P205/65VR15	None	26	26	6-JJ	U: 9-30 in. ①
2003-04	GS 430	225/55VR16	235/45ZR17	Std: 32 Opt: 33	Std: 32 Opt: 33	Std: 7.5-JJ Opt: 8-JJ	U: 9-30 in. ①
	GS 300	P215/60VR16	225/55VR16	30	30	7.5-JJ	U: 9-30 in. ①
	LS 4300	P225/60VR16		29	29	7-JJ	U: 9-30 in. ①
			P225/55R17 95H	32	35		
			225/55R17 97W	35	36		
	IS 300	P205/55R16 89V	215/45ZR17	33	33	NA	U: 9-30 in. ①
	ES 300	P205/65VR15	None	26	26	6-JJ	U: 9-30 in. ①

NA: Not Available

OEM: Original Equipment Manufacturer

PSI: Pounds Per Square Inch

STD: Standard

OPT: Optional

L: Lower

U: Upper

① Torque required in inch lbs. to rotate ball joint when removed from the knuckle

42356-LEXC-C10

BRAKE SPECIFICATIONS
All measurements in inches unless noted

| Year | Model | Front Brake Disc | | | Rear Brake Disc | | | Minimum Lining Thickness | Brake Caliper | |
		Original Thickness	Minimum Thickness	Maximum Run-out	Original Thickness	Minimum Thickness	Maximum Run-out		Bracket Bolts (ft. lbs.)	Mounting Bolts (ft. lbs.)
2000	ES 300	1.102	1.024	0.0020	0.394	0.354	0.0059	0.0390	①	②
	GS 300	1.260	1.181	0.0020	0.472	0.413	0.0020	0.0390	③	25
	GS 400	1.260	1.181	0.0020	0.472	0.413	0.0020	0.0390	③	25
	LS 400	1.102	1.024	0.0020	0.630	0.591	0.0020	0.1180	③	25
	SC 300	1.102	1.024	0.0020	0.630	0.591	0.0020	0.0390	③	25
	SC 400	1.260	1.181	0.0020	0.630	0.591	0.0020	0.0390	③	25
2001	ES 300	1.102	1.024	0.0020	0.394	0.354	0.0059	0.0390	①	②
	GS 300	1.260	1.181	0.0020	0.472	0.413	0.0020	0.0390	③	25
	GS 430	1.260	1.181	0.0020	0.472	0.413	0.0020	0.0390	③	25
	IS 300	1.260	1.181	0.0020	0.472	0.413	0.0020	0.0390	③	25
	LS 430	1.181	1.102	0.0020	0.630	0.571	0.0020	0.0390	—	④
2002	ES 300	1.102	1.024	0.0020	0.472	0.413	0.0059	0.0390	①	②
	GS 300	1.260	1.181	0.0020	0.472	0.413	0.0020	0.0390	③	25
	GS 430	1.260	1.181	0.0020	0.472	0.413	0.0020	0.0390	③	25
	IS 300	1.260	1.181	0.0020	0.472	0.413	0.0020	0.0390	③	25
	LS 430	1.181	1.102	0.0020	0.630	0.571	0.0020	0.0390	—	④
2003-04	ES 300	1.102	1.024	0.0020	0.472	0.413	0.0059	0.0390	①	②
	GS 300	1.260	1.181	0.0020	0.472	0.413	0.0020	0.0390	③	25
	GS 430	1.260	1.181	0.0020	0.472	0.413	0.0020	0.0390	③	25
	IS 300	1.260	1.181	0.0020	0.472	0.413	0.0020	0.0390	③	25
	LS 430	1.181	1.102	0.0020	0.630	0.571	0.0020	0.0390	—	④

F: Front

R: Rear

① Front: 79 ft. lbs.
  Rear: 34 ft. lbs.

② Front: 25 ft. lbs.
  Rear: 14 ft. lbs.

③ Front: 87 ft. lbs.
  Rear: 77 ft. lbs.

④ Front: 81 ft. lbs.
  Rear: 58 ft. lbs.

42356-LEXC-C11

SCHEDULED MAINTENANCE INTERVALS
Lexus—ES300, IS300, GS300, GS400, GS430, SC300, SC400, LS400 & LS430

TO BE SERVICED	TYPE OF SERVICE	VEHICLE MILEAGE INTERVAL (x1000)												
		7.5	15	22.5	30	37.5	45	52.5	60	67.5	75	82.5	90	97.5
Engine oil & filter	R	✓	✓	✓	✓	✓	✓	✓	✓	✓	✓	✓	✓	✓
Air conditioning filter (LS 400) ①	S/I	✓	✓	✓	✓	✓	✓	✓	✓	✓	✓	✓	✓	✓
Automatic transaxle fluid & filter	S/I		✓		✓		✓		✓		✓		✓	
Ball joints & dust covers	S/I		✓		✓		✓		✓		✓		✓	
Bolts & nuts on chassis & body	S/I		✓		✓		✓		✓		✓		✓	
Brake fluid ②	S/I		✓		✓		✓		✓		✓		✓	
Brake line pipes & hoses	S/I		✓		✓		✓		✓		✓		✓	
Brake linings & drums	S/I		✓		✓		✓		✓		✓		✓	
Brake pads & discs (front & rear)	S/I		✓		✓		✓		✓		✓		✓	
Differential oil	S/I		✓		✓		✓		✓		✓		✓	
Driveshaft boots (ES 300)	S/I		✓		✓		✓		✓		✓		✓	
Steering gear housing oil	S/I		✓		✓		✓		✓		✓		✓	
Steering linkage	S/I		✓		✓		✓		✓		✓		✓	
Air filter	R				✓				✓				✓	
Exhaust pipes & mountings	S/I				✓				✓				✓	
Fuel lines & connections	S/I				✓				✓				✓	
Engine coolant	R						✓					✓		
Fuel tank cap gasket	R								✓					
Spark plugs	R								✓					
Charcoal canister	S/I								✓					
Drive belts	S/I								✓					
Valve clearance	S/I								✓					

R: Replace   S/I: Service or Inspect

① Replace at 15,000 miles.

② Replace at 30,000 miles (unless previously replaced).

FREQUENT OPERATION MAINTENANCE (SEVERE SERVICE)

If a vehicle is operated under any of the following conditions it is considered severe service

- **Extremely dusty areas.**
- **50% or more of the vehicle operation is in 32°C (90°F) or higher temperatures, or constant operation in temperatures below 0°C (32°F).**
- **Prolonged idling (vehicle operation in stop and go traffic).**
- **Frequent short running periods (engine does not warm to normal operating temperatures).**
- **Police, taxi, delivery usage or trailer towing usage.**

Oil & oil filter: change every 3750 miles.

Ball joints & dust covers: service or inspect every 7500 miles.

Bolts & nuts on chassis & body: service or inspect every 7500 miles.

Brake linings & drums: service or inspect every 7500 miles.

Brake pads & discs (front & rear): service or inspect every 7500 miles.

Driveshaft boots (ES 300): service or inspect every 7500 miles.

Brake linings & drums: service or inspect every 7500 miles.

Steering linkage: service or inspect every 7500 miles.

Air filter: service or inspect every 15,000 miles.

Automatic transmission fluid & filter: replace every 15,000 miles.

Differential oil: replace every 15,000 miles.

Exhaust pipes & mountings: service or inspect every 15,000 miles.

Drive belts: service or inspect at 60,000 miles & every 7500 miles thereafter.

Timing belts: replace every 60,000 miles.

PRECAUTIONS

Before servicing any vehicle, please be sure to read all of the following precautions that deal with personal safety, prevention of component damage, and important points to take into consideration when servicing a motor vehicle:

• Never open, service or drain the radiator or cooling system when the engine is hot; serious burns can occur from the steam and hot coolant.

• Observe all applicable safety precautions when working around fuel. Whenever servicing the fuel system, always work in a well-ventilated area. Do not allow fuel spray or vapors to come in contact with a spark, open flame or excessive heat (a hot drop light, for example). Keep a dry chemical fire extinguisher near the work area. Always keep fuel in a container specifically designed for fuel storage; also, always properly seal fuel containers to avoid the possibility of fire or explosion. Refer to the additional fuel system precautions later in this section.

• Fuel injection systems often remain pressurized, even after the engine has been turned **OFF**. The fuel system pressure must be relieved before disconnecting any fuel lines. Failure to do so may result in fire and/or personal injury.

• Brake fluid often contains polyglycol ethers and polyglycols. Avoid contact with the eyes and wash your hands thoroughly after handling brake fluid. If you do get brake fluid in your eyes, flush your eyes with clean, running water for 15 minutes. If eye irritation persists, or if you have taken

brake fluid internally, IMMEDIATELY seek medical assistance.

• The EPA warns that prolonged contact with used engine oil may cause a number of skin disorders, including cancer. You should make every effort to minimize your exposure to used engine oil. Protective gloves should be worn when changing oil. Wash your hands and any other exposed skin areas as soon as possible after exposure to used engine oil. Soap and water, or waterless hand cleaner should be used.

• All new vehicles are now equipped with an air bag system. The system must be disabled before performing service on or around system components, steering column, instrument panel components, wiring and sensors. Failure to follow safety and disabling procedures could result in accidental air bag deployment, possible personal injury and unnecessary system repairs.

• Always wear safety goggles when working with, or around, the air bag system. When carrying a non-deployed air bag, be sure the bag and trim cover are pointed away from your body. When placing a non-deployed air bag on a work surface, always face the bag and trim cover upward, away from the surface. This will reduce the motion of the module if it is accidentally deployed. Refer to the additional air bag system precautions later in this section.

• Clean, high quality brake fluid from a sealed container is essential to the safe and proper operation of the brake system. You should always buy the correct type of brake

fluid for your vehicle. If the brake fluid becomes contaminated, completely flush the system with new fluid. Never reuse any brake fluid. Any brake fluid that is removed from the system should be discarded. Also, do not allow any brake fluid to come in contact with a painted surface; it will damage the paint.

• Never operate the engine without the proper amount and type of engine oil; doing so WILL result in severe engine damage.

• Timing belt maintenance is extremely important. Many models utilize an interference-type, non-freewheeling engine. If the timing belt breaks, the valves in the cylinder head may strike the pistons, causing potentially serious (also time-consuming and expensive) engine damage. Refer to the maintenance interval charts in the front of this section for the recommended replacement interval for the timing belt, and to the timing belt procedure for belt replacement and inspection.

• Disconnecting the negative battery cable on some vehicles may interfere with the functions of the on-board computer system(s) and may require the computer to undergo a relearning process once the negative battery cable is reconnected.

• When servicing drum brakes, only dissemble and assemble one side at a time, leaving the remaining side intact for reference.

• Only an MVAC-trained, EPA-certified automotive technician should service the air conditioning system or its components.

ENGINE REPAIR

Alternator

REMOVAL

3.0L (1MZ-FE) Engine

1. Before servicing the vehicle, refer to the precautions in the beginning of this section.
2. Remove or disconnect the following:
   • Negative battery cable
   • Accessory drive belt
   • Alternator harness connectors
   • Alternator

3.0L (2JZ-GE) Engine

1. Before servicing the vehicle, refer to the precautions in the beginning of this section.

2. Remove or disconnect the following:
   • Negative battery cable
   • Engine under cover
   • Accessory drive belt
   • Alternator connector
   • Cap and nut
   • Alternator wire
   • Alternator wire clamp from the wire clip on the alternator
   • Bolt and pipe clamp
   • 2 automatic transmission oil cooler pipes from the alternator
   • Bolt, nut, pipe bracket and alternator

4.0L (1UZ-FE) and 4.3L (3UZ-FE) Engines

1. Before servicing the vehicle, refer to the precautions in the beginning of this section.

2. Remove or disconnect the following:
   • Negative battery cable
   • Air cleaner inlet
   • Accessory drive belt
   • Oil pan protector
   • Engine under cover
   • Power steering pump
   • Alternator harness connectors
   • Heated Oxygen (HO$_2$S) sensor wiring
   • Alternator

INSTALLATION

3.0L (1MZ-FE) Engine

1. Install or connect the following:
   • Alternator
   • Alternator harness connectors
   • Accessory drive belt. Tighten the adjusting lock bolt to 13 ft. lbs. (18

Nm) and the pivot bolt to 41 ft. lbs. (56 Nm).

- Negative battery cable

3.0L (2JZ-GE) Engine

1. Install or connect the following:
   - Bolt, nut, pipe bracket and alternator. Tighten the fasteners to 30 ft. lbs. (40 Nm).
   - 2 automatic transmission oil cooler pipes to the alternator
   - Bolt and pipe clamp
   - Alternator wire clamp to the wire clip on the alternator
   - Alternator wire
   - Cap and nut
   - Alternator connector
   - Accessory drive belt
   - Engine under cover
   - Negative battery cable

4.0L (1UZ-FE) and 4.3L (3UZ-FE) Engines

1. Install or connect the following:
   - Alternator. Tighten the fasteners to 29 ft. lbs. (39 Nm).
   - HO$_2$S sensor wiring
   - Alternator harness connectors
   - Power steering pump
   - Engine under cover
   - Oil pan protector
   - Accessory drive belt
   - Air cleaner inlet
   - Negative battery cable

Ignition Timing

ADJUSTMENT

The engines covered in this section are equipped with a Distributorless Ignition System (DIS). No timing adjustments are possible.

Engine Assembly

REMOVAL & INSTALLATION

ES 300

1. Before servicing the vehicle, refer to the precautions in the beginning of this section.
2. Release the fuel pressure.
3. Drain the engine coolant and engine oil.
4. Remove or disconnect the following:
   - Battery and tray
   - Hood

- Accelerator cable and the throttle cable
- Air cleaner cover,
- Volume air flow meter and air cleaner duct as an assembly
- Cruise control actuator, if equipped
- Radiator
- Engine relay box and 2 bolts
- 5 connections from the relay box
- 2 igniter connectors
- Left fender apron connector
- Noise filter connector
- 2 ground straps
- Engine wiring harness from the engine.
- Vacuum hoses from the following connections: intake air control valve vacuum tank, charcoal canister, brake booster vacuum hose from the intake chamber
- 2 heater hoses from the bulkhead
- Fuel feed and return lines
- Control cable from the transaxle
- Wiring harness from the PCM and route it through the bulkhead.
- Air conditioning compressor from the engine without disconnecting the lines and position it out of the way
- Front exhaust pipe
- Halfshafts
- 2 power steering air hoses from the engine
- Hydraulic cooling fan pressure hose
- Power steering pump without disconnecting the lines and position it out of the way
- Right and left lower engine mounts from the body
- Engine mounting shock absorber
- 3 front engine mounting bolts from the body

5. Attach a lifting device to the engine.
6. Remove or disconnect the following:
   - Coolant reservoir tank.
   - Right engine mounting bracket
   - Engine moving control rod and right No. 2 engine mounting bracket
   - Engine and transaxle as an assembly

✳✳ WARNING

Be careful not to hit the power steering or PNP switches.

- Engine mounting insulator below the oil filter
- Right rear engine-mounting insulator
- Front exhaust pipe stay

7. Label and detach the following connectors:
   - Overdrive solenoid
   - PNP switch
   - Speedometer
   - Starter terminal
   - Speed sensor.
   - 2 wire clamps from the transaxle
   - Oil dipstick and guide
   - Starter
   - Flywheel housing cover

8. Turn the crankshaft pulley to gain access to the 8 torque converter bolts. Secure the crankshaft and remove them as they become accessible.
9. Install or connect the following:
   - 2 exhaust manifold stays and plate
   - 2 bolts attaching the transaxle to the oil pan
   - 6 transaxle mounting bolts
   - Transaxle

To install:

10. Position the transaxle to the engine. Tighten the transaxle mounting bolts: 47 ft. lbs. (64 Nm).
11. Install or connect the following:
    - Bolts that attach the transaxle to the oil pan bolts and tighten them to 34 ft. lbs. (46 Nm)
    - Exhaust manifold support. Tighten the bolts to 14 ft. lbs. (20 Nm).
    - Bolts that attach the flywheel to the torque converter. Coat the threads with a locking compound. Rotate the engine and tighten the bolts alternately to 30 ft. lbs. (41 Nm).
    - Starter to the engine
    - Flywheel cover and tighten the bolts to 13 ft. lbs. (18 Nm)
    - Dipstick and tube with a new O-ring

12. Attach the clamps and following connectors:
    - Overdrive solenoid
    - PNP switch
    - Speedometer
    - Starter terminal
    - Speed sensor

13. Install or connect the following:
    - Exhaust pipe's stay. Tighten the bolts to 15 ft. lbs. (21 Nm).
    - Right rear insulator. Tighten the bolts to 47 ft. lbs. (64 Nm).
    - Front engine mounting insulator. Tighten the bolts to 47 ft. lbs. (64 Nm).

14. Lower the engine and transaxle into the engine compartment. Tilt the transaxle downward and clear the left mount.
15. Keep the engine level and align the right and left engine mounts.

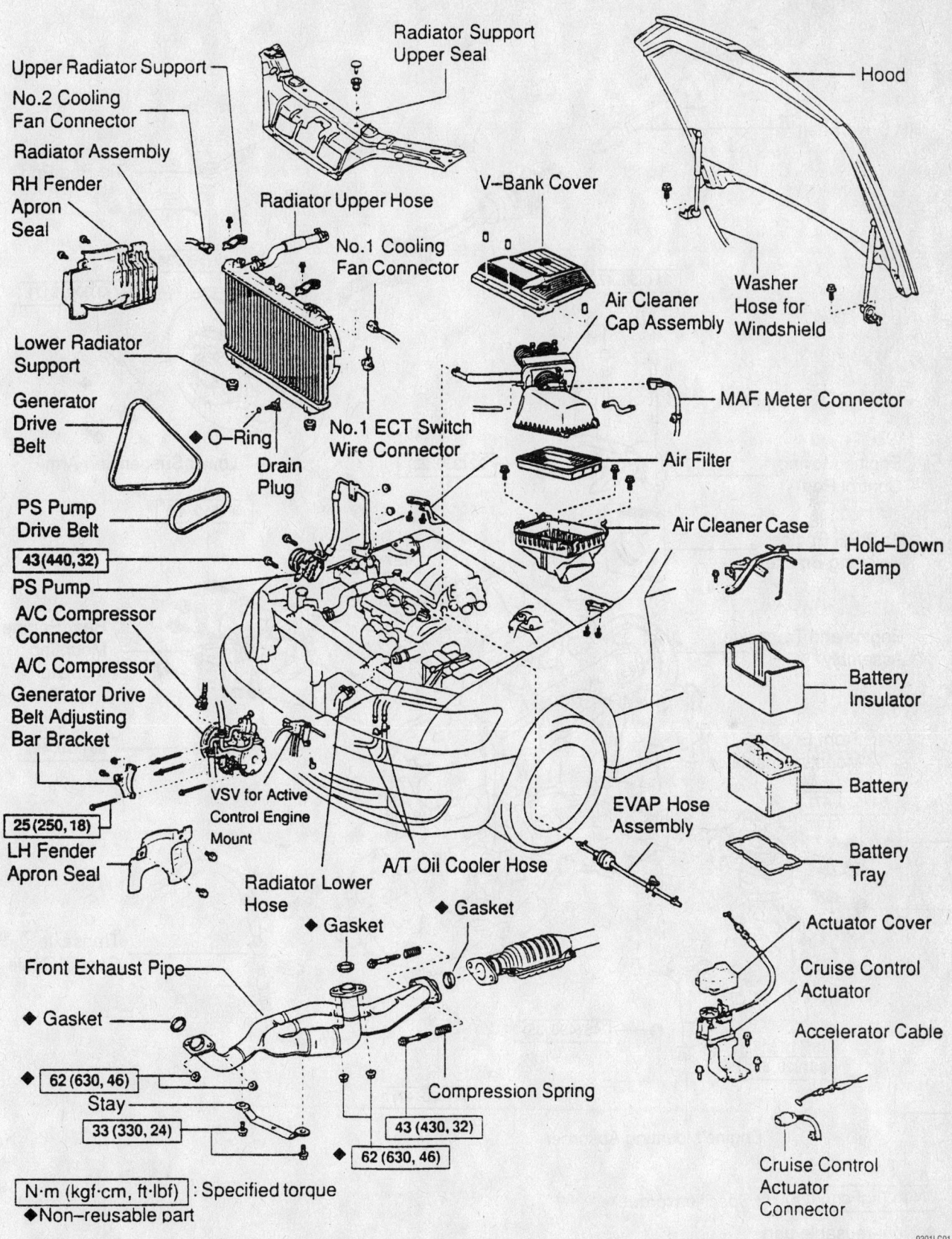

Upper Radiator Support

No.2 Cooling Fan Connector

Radiator Assembly

RH Fender Apron Seal

Radiator Support Upper Seal

Radiator Upper Hose

V–Bank Cover

Hood

No.1 Cooling Fan Connector

Air Cleaner Cap Assembly

Washer Hose for Windshield

MAF Meter Connector

Lower Radiator Support

No.1 ECT Switch Wire Connector

Generator Drive Belt

O–Ring

Drain Plug

Air Filter

Air Cleaner Case

Hold–Down Clamp

PS Pump Drive Belt

43 (440, 32)

PS Pump

A/C Compressor Connector

A/C Compressor

Generator Drive Belt Adjusting Bar Bracket

25 (250, 18)

LH Fender Apron Seal

VSV for Active Control Engine Mount

Radiator Lower Hose

A/T Oil Cooler Hose

EVAP Hose Assembly

Battery Insulator

Battery

Battery Tray

Actuator Cover

Cruise Control Actuator

Accelerator Cable

Front Exhaust Pipe

◆ Gasket

◆ Gasket

◆ Gasket

◆ Gasket

62 (630, 46)

Stay

33 (330, 24)

Compression Spring

43 (430, 32)

62 (630, 46)

Cruise Control Actuator Connector

N·m (kgf·cm, ft·lbf) : Specified torque
◆Non–reusable part

9301LG01

Exploded view of the engine removal and related components—ES 300

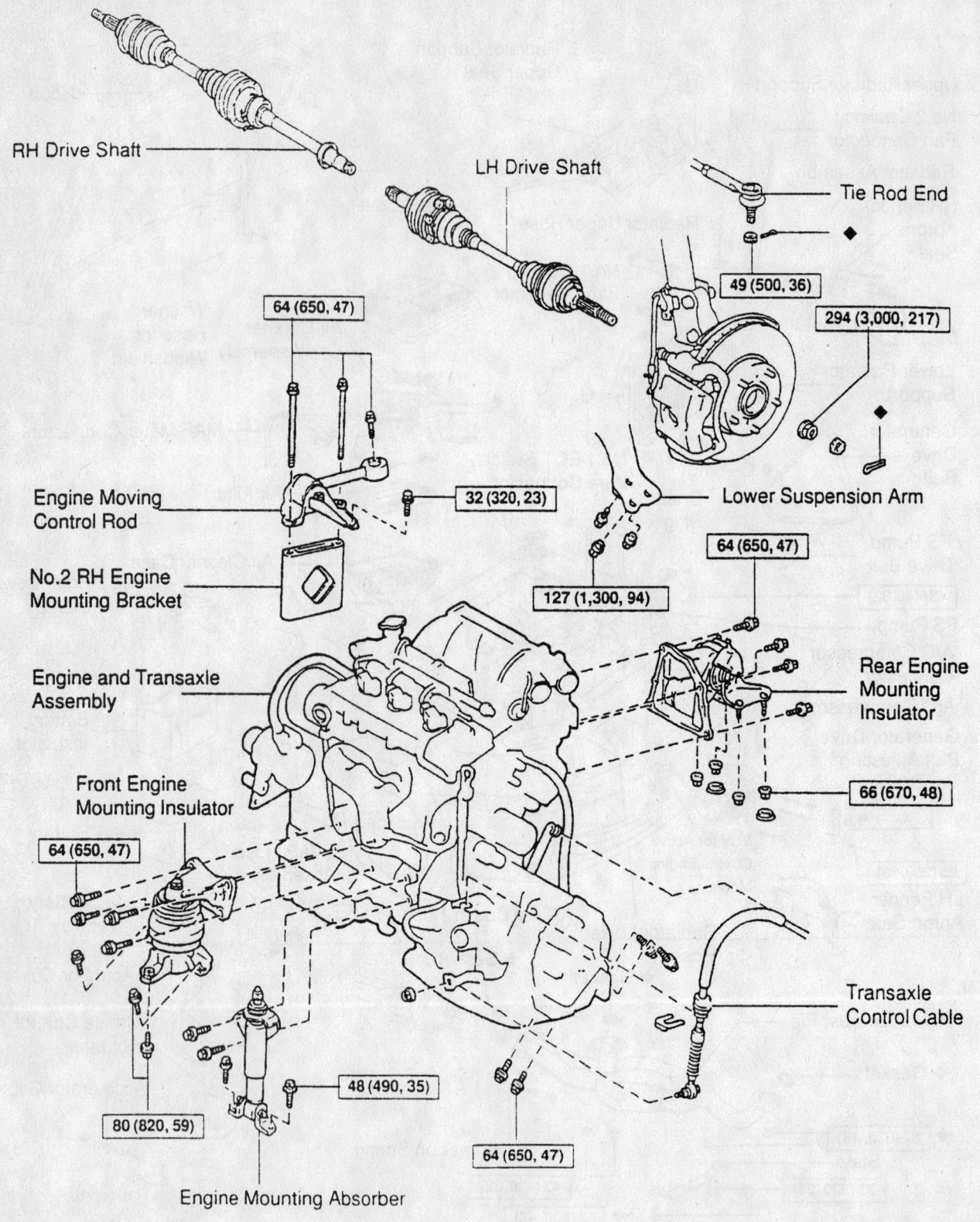

RH Drive Shaft

LH Drive Shaft

Tie Rod End

49 (500, 36)

294 (3,000, 217)

64 (650, 47)

Engine Moving
Control Rod

32 (320, 23)

Lower Suspension Arm

No.2 RH Engine
Mounting Bracket

64 (650, 47)

127 (1,300, 94)

Engine and Transaxle
Assembly

Rear Engine
Mounting
Insulator

Front Engine
Mounting Insulator

66 (670, 48)

64 (650, 47)

Transaxle
Control Cable

48 (490, 35)

80 (820, 59)

64 (650, 47)

Engine Mounting Absorber

N·m (kgf·cm, ft·lbf) : Specified torque

◆ Non–reusable part

9301LG02

Exploded view of the engine removal and related components (cont.)—ES 300

16. Install or connect the following:
- Engine mounting bracket and moving control rod. Tighten bolts to 47 ft. lbs. (64 Nm).
- Right engine stay. Tighten bolts to 23 ft. lbs. (32 Nm).
- Ground straps
- Coolant reservoir
- Front engine mounting insulator to the body. Tighten bolts to 59 ft. lbs. (81 Nm).
- Engine mounting shock absorber. Tighten bolts to 35 ft. lbs. (48 Nm).
- Right engine mount. Tighten bolts to 48 ft. lbs. (66 Nm).
- Left engine mount. Tighten bolts to 47 ft. lbs. (64 Nm).
- Engine lifting device
- Power steering pump and belt
- Hydraulic cooling fan pressure hose. Tighten the fitting to 33 ft. lbs. (44 Nm).
- Power steering air tube and hoses
- Halfshafts
- Front exhaust pipe with new gaskets
- Air conditioning compressor. Tighten bolts to 18 ft. lbs. (25 Nm).
- Harness to the Powertrain Control Module and assemble the instrument panel
- Control cable to the transaxle
- Fuel lines and tighten the fittings to 22 ft. lbs. (30 Nm)
- Heater hoses

17. Connect the vacuum hoses to the following connections:
- Intake air control valve vacuum tank
- Charcoal canister
- Air intake chamber from the brake booster

18. Install or connect the following:
- Engine wiring harness to the engine
- Engine relay box
- 2 bolts and attach the following connectors:
- 5 connections from the relay box
- 2 igniter connectors
- Left fender apron connector
- Noise filter connector
- 2 ground straps
- Radiator
- Cruise control actuator, if equipped
- Air cleaner cover
- Volume airflow meter and air cleaner duct assembly
- Throttle cable
- Accelerator cable

19. Fill the engine to the proper level with the recommended grade of oil.

20. Align the matchmarks and install the hood.
21. Fill the engine to the proper level with coolant.
22. Bleed the cooling system.
23. Install the battery and tray.
24. Check and/or adjust the ignition timing.
25. Start the engine and check for leaks.
26. Road test the vehicle.
27. Recheck the engine oil and coolant levels.

GS 300

1. Before servicing the vehicle, refer to the precautions in the beginning of this section.
2. Release the fuel pressure.
3. Drain the fuel from the tank.
4. Remove or disconnect the following:
- Negative battery cable. Wait at least 90 seconds before performing any other work.
- Hood insulator pad and the hood
- Engine undercover, then drain the engine coolant and oil
- Front suspension member brace, if equipped
- Accelerator cable, cruise control actuator cable and the automatic transmission throttle control cable from the throttle body
- Air cleaner assembly
- Volume air flow meter and the air intake hose
- Air cleaner duct
- Drive belt
- Radiator

5. Label and detach the following wires and electrical connectors:
- Igniter
- Ignition coil
- Wiring harness from the wire clamp and coolant tank
- Alternator and ground strap from the left engine mount
- Starter

6. Remove or disconnect the following:
- Fuel lines from the intake and return lines
- Power steering pump without disconnecting the lines and position it aside
- Air conditioning compressor without disconnecting the air conditioning lines and position it aside
- Brake booster vacuum hose
- Evaporative Emissions (EVAP) hose
- Heater hoses from the firewall
- Heater valve and engine wire from the firewall

- Electrical harness from the PCM and route it through the firewall
- Oxygen (O_2) sensor (if equipped) from the front exhaust pipe
- Front exhaust pipes and heat insulator
- Rear center floor crossmember brace
- Transmission control rod
- Driveshaft

7. Support the transmission with a jack. Use a piece of wood to prevent damage to the transmission oil pan.
8. Attach a lifting device to the engine.
9. Remove or disconnect the following:
- Rear transmission crossmember
- 2 hole plugs in the front crossmember
- 2 nuts holding the engine insulators to the front crossmember

10. Slowly and carefully remove the engine and transmission from the engine compartment as an assembly.

To install:
11. Install or connect the following:
- Engine and transmission into the engine compartment
- Stud bolts for the front engine mount into their bores in the front engine crossmember. Temporarily install the 2 nuts.

12. Remove the engine hoist
13. Install or connect the following:
- Temporarily, the rear engine support with the 4 nuts
- The 4 support bolts and tighten them to 19 ft. lbs. (25 Nm). Tighten the nuts to 10 ft. lbs. (13 Nm).
- Front engine crossmember to mount nuts to 54 ft. lbs. (74 Nm) and the hole plugs
- The driveshaft

14. Shift the transmission control shift rod into **N** (neutral) by shifting the lever all the way back and returning it 2 notches. Connect the shift rod to the lever and tighten it to 108 inch lbs. (13 Nm).

15. Install or connect the following:
- Rear center floor crossmember brace and tighten the bolts to 108 inch lbs. (13 Nm)
- Exhaust pipe heat insulator
- Front exhaust pipes
- Sub HO_2 sensor, if equipped
- Engine wiring harness to the PCM
- PCM and its cover
- The lower portion of the passenger side instrument panel, the vent, the carpet and the scuff panel
- Heater water valve and engine wire to the cowl panel

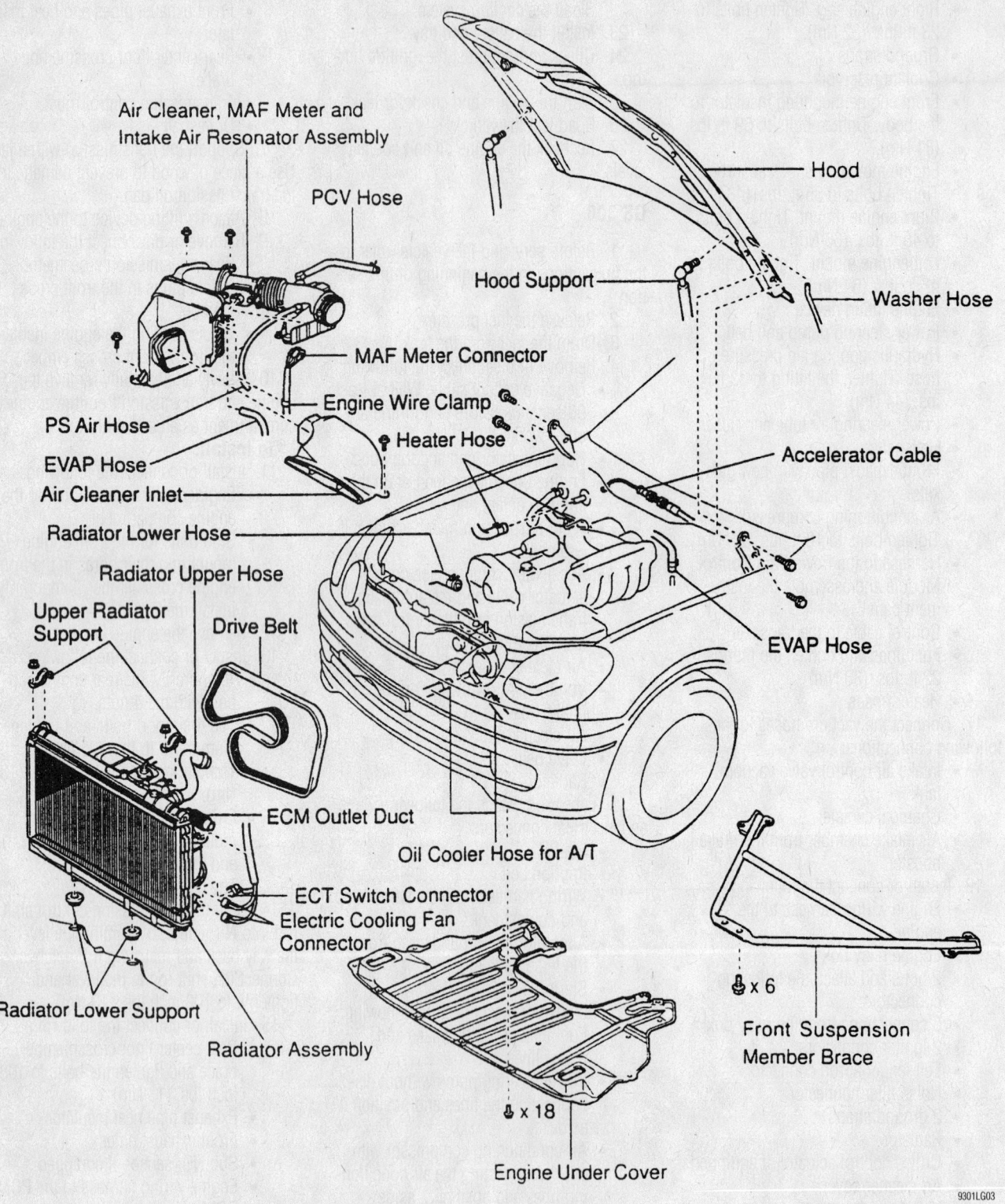

Air Cleaner, MAF Meter and
Intake Air Resonator Assembly

PCV Hose

Hood

Hood Support

Washer Hose

MAF Meter Connector

Engine Wire Clamp

Accelerator Cable

PS Air Hose

Heater Hose

EVAP Hose

Air Cleaner Inlet

Radiator Lower Hose

EVAP Hose

Radiator Upper Hose

Upper Radiator
Support

Drive Belt

ECM Outlet Duct

Oil Cooler Hose for A/T

ECT Switch Connector
Electric Cooling Fan
Connector

Radiator Lower Support

Radiator Assembly

Front Suspension
Member Brace

Engine Under Cover

9301LG03

Exploded view of the engine removal and related components—GS 300

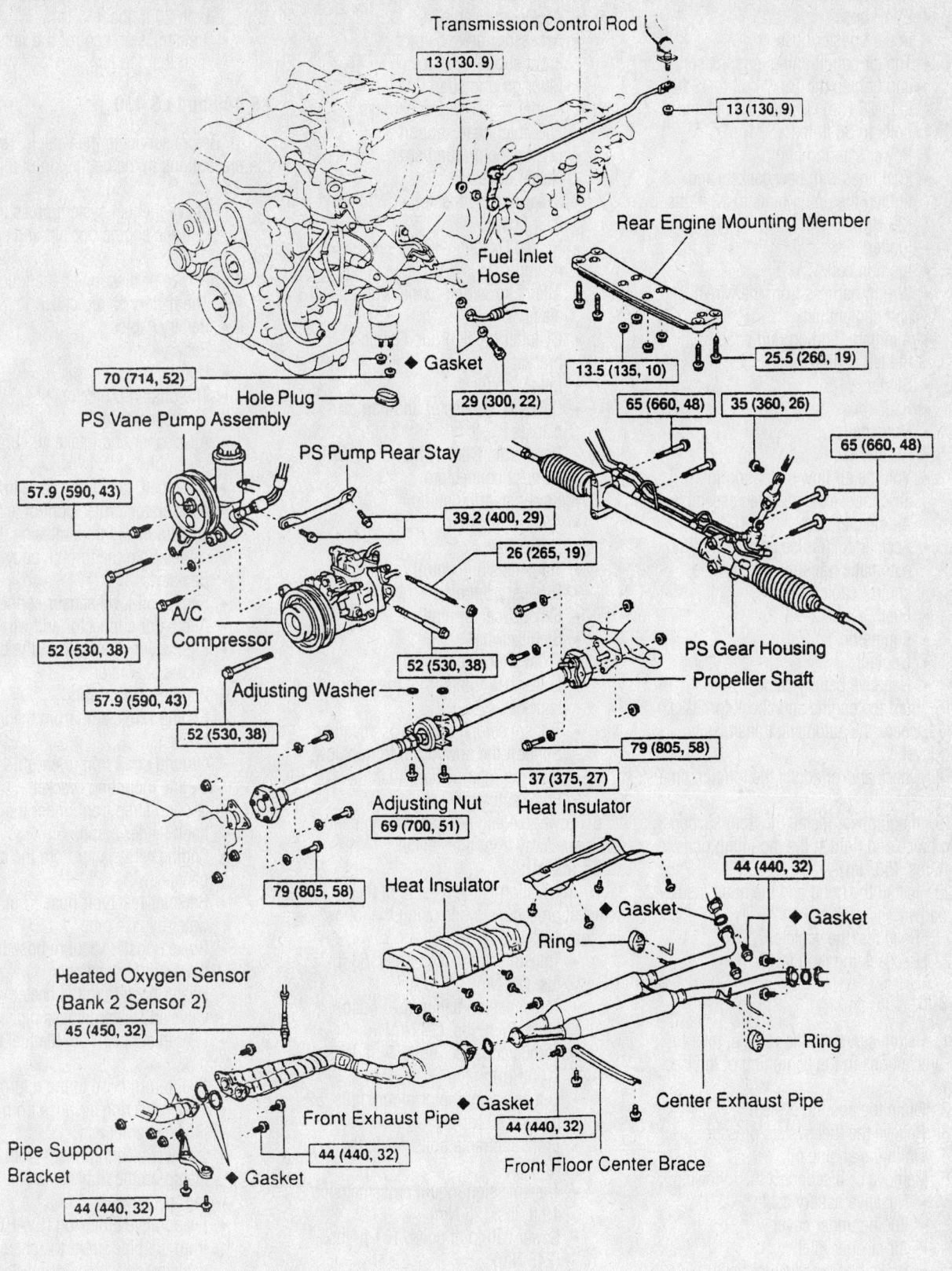

Transmission Control Rod

13 (130. 9)

13 (130, 9)

Rear Engine Mounting Member

Fuel Inlet Hose

◆ Gasket

70 (714, 52)

Hole Plug

29 (300, 22)

13.5 (135, 10)

25.5 (260, 19)

65 (660, 48)

35 (360, 26)

65 (660, 48)

PS Vane Pump Assembly

PS Pump Rear Stay

57.9 (590, 43)

39.2 (400, 29)

26 (265, 19)

52 (530, 38)

A/C Compressor

52 (530, 38)

PS Gear Housing

Propeller Shaft

57.9 (590, 43)

Adjusting Washer

52 (530, 38)

79 (805, 58)

37 (375, 27)

Adjusting Nut

Heat Insulator

69 (700, 51)

44 (440, 32)

79 (805, 58)

Heat Insulator

◆ Gasket

◆ Gasket

Ring

Heated Oxygen Sensor
(Bank 2 Sensor 2)

Ring

45 (450, 32)

Front Exhaust Pipe

◆ Gasket

Center Exhaust Pipe

Pipe Support Bracket

44 (440, 32)

44 (440, 32)

◆ Gasket

Front Floor Center Brace

44 (440, 32)

N·m (kgf·cm, ft·lbf) : Specified torque

◆ Non-reusable part

9301LG04

Exploded view of the engine removal and related components (cont.)—GS 300

- Heater hoses
- EVAP hose
- Brake booster hose
- The air conditioning compressor and tighten the Torx® bolt to 19 ft. lbs. (26 Nm). Tighten the nut and bolts to 38 ft. lbs. (52 Nm).
- Power steering pump
- Fuel lines with new gaskets and tighten the union bolts to 22 ft. lbs. (29 Nm)
- Igniter
- Ignition coil
- Wiring harness from the wire clamp and coolant tank
- Alternator and ground strap from the left engine mount
- Starter
- Radiator
- Drive belt
- Air cleaner
- Volume air flow meter and intake air connector pipe as an assembly
- Air cleaner duct
- Accelerator, cruise control and the automatic transmission throttle control cables
- Fuel
- Engine oil
- Coolant
- Negative battery cable

16. Start the engine and check for leaks.

17. Check the automatic transmission fluid level.

18. Check and/or adjust the ignition timing.

19. If equipped, install the front suspension brace and tighten the mounting bolts to 43 ft. lbs. (58 Nm).

20. Install the hood and the hood insulator pad.

21. Road test the vehicle.

22. Recheck the fluid levels.

IS 300

1. Before servicing the vehicle, refer to the precautions in the beginning of this section.

2. Drain the cooling system.

3. Relieve the fuel system pressure.

4. Drain the engine oil.

5. Remove or disconnect the following:
- Negative battery cable
- Engine under cover
- Air cleaner inlet
- Brake booster vacuum hose
- Radiator hoses
- Mass Air Flow (MAF) sensor connector
- Positive Crankcase Ventilation (PCV) hose

- Air intake resonator
- Accelerator cable
- Accessory drive belt
- Front subframe brace
- Floor ground strap
- Starter motor wiring harness
- Fuel inlet hose support
- Dash panel ground strap
- Heater hoses
- Evaporative Emissions (EVAP) canister hose
- Heated Oxygen (HO2S) sensor connectors
- Alternator wiring harness and clamp
- Cylinder block ground cable bracket
- Igniter connector
- Data link connector and harness clamps
- Powertrain Control Module (PCM) harness connectors
- Power steering pump
- A/C compressor
- Drive shaft
- Transmission control rod
- Exhaust front pipe
- Exhaust center pipe
- Stabilizer bar
- Front shock absorbers
- Lower ball joints from the steering knuckles
- Transmission mount crossmember. Support the powertrain from below.
- Steering intermediate shaft
- Front subframe

6. Lower the engine, transmission and subframe away from the vehicle.

To install:

7. Installation is the reverse of the removal procedure, while using the following torque values:
- Transmission flange bolts: 53 ft. lbs. (72 Nm)
- Transmission flange-to-oil pan bolts: 27 ft. lbs. (37 Nm)
- Torque converter bolts: 74 ft. lbs. (100 Nm)
- Left and right motor mount nuts: 52 ft. lbs. (70 Nm)
- Front subframe bolts: 52 ft. lbs. (70 Nm)
- Transmission mount crossmember: 19 ft. lbs. (26 Nm)
- Lower ball joint bolts: 181 ft. lbs. (245 Nm)
- Lower shock absorber bolts: 47 ft. lbs. (64 Nm)
- Stabilizer bar bolts: 13 ft. lbs. (18 Nm)
- Stabilizer bar nuts: 36 ft. lbs. (49 Nm)

- Steering intermediate shaft pinch bolt: 26 ft. lbs. (35 Nm)
- Transmission control rod nuts: 108 inch lbs. (13 Nm)

LS 400 and LS 430

1. Before servicing the vehicle, refer to the precautions in the beginning of this section.

2. Relieve the fuel system pressure

3. Drain the engine coolant and engine oil

4. Remove or disconnect the following:
- The battery clamp cover
- Battery cables
- Battery
- Hood
- The oil pan protector
- Air cleaner inlet
- Air cleaner and intake air connector assembly
- Drive belt, fan clutch and fan pulley
- Accelerator, cruise control actuator and automatic transmission throttle cables from the throttle body
- Radiator
- Engine oil level sensor connector
- Alternator connector and wire
- Engine wire clamp from the bracket on the alternator
- 2 igniter connectors
- Engine wire clamp from the igniter bracket
- Ground strap from the right-hand engine mounting bracket
- Ground strap from under the left-hand fender apron
- Engine wire clamp from the cowl panel
- Radiator reservoir hose from the water bypass pipe
- Brake booster vacuum hose from the air intake chamber
- Heater hose from the heater water valve and water bypass pipe
- Fuel inlet hose from the fuel inlet pipe
- Fuel return hose to the return pipe
- Power steering air hose from the air intake chamber
- 2 power steering hoses from the clamp on the right-hand No. 3 timing belt cover
- Evaporative Emission (EVAP) hose from the pipe (from the charcoal canister).
- Engine wire from the cabin as follows:
- Undercover from under the glove compartment
- Glove compartment

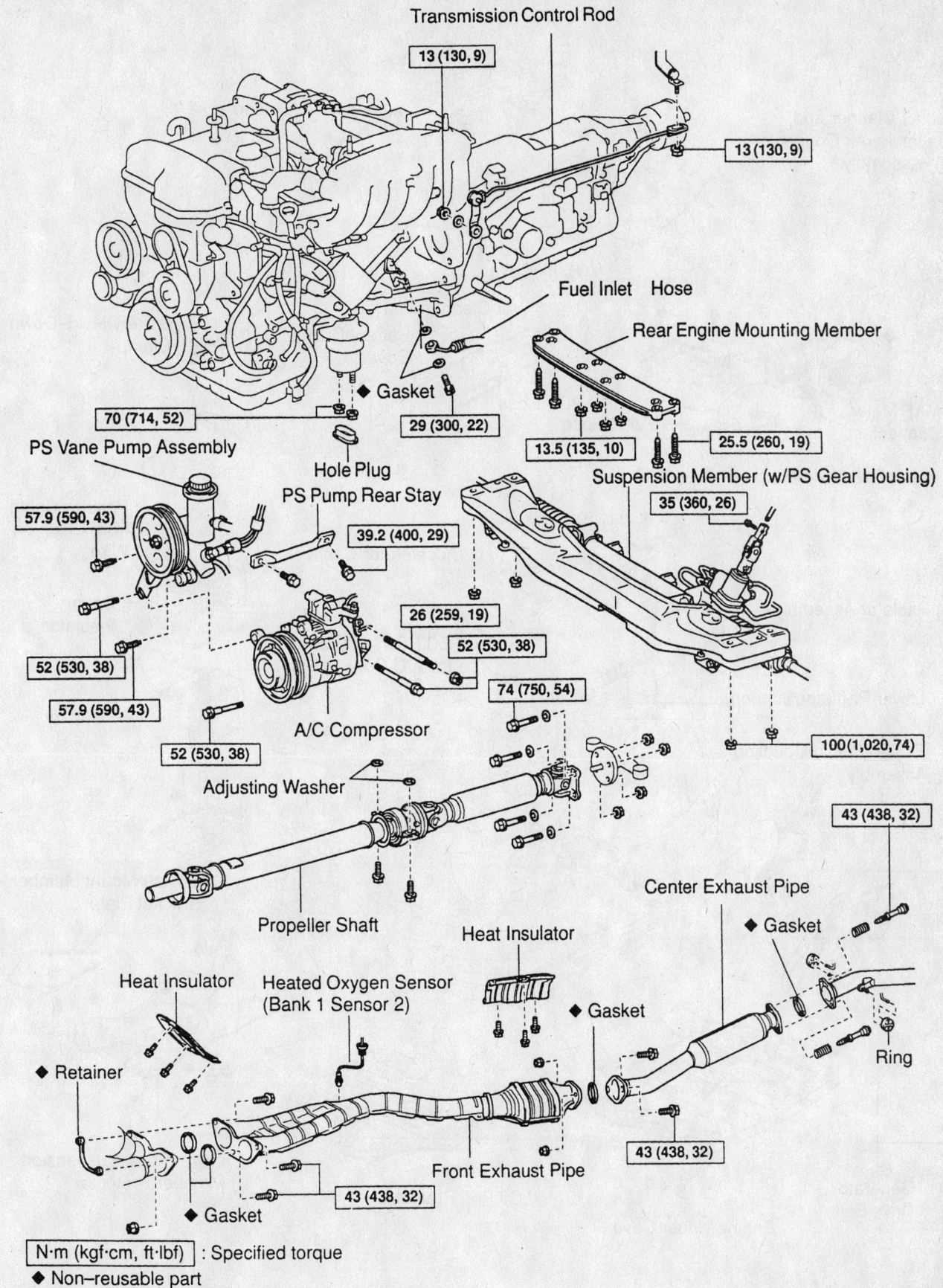

Transmission Control Rod

13 (130, 9)

13 (130, 9)

Fuel Inlet Hose

Rear Engine Mounting Member

◆ Gasket

70 (714, 52)

29 (300, 22)

13.5 (135, 10)

25.5 (260, 19)

PS Vane Pump Assembly

Hole Plug
PS Pump Rear Stay

Suspension Member (w/PS Gear Housing)

57.9 (590, 43)

39.2 (400, 29)

35 (360, 26)

52 (530, 38)

26 (259, 19)

57.9 (590, 43)

52 (530, 38)

A/C Compressor

52 (530, 38)

74 (750, 54)

100 (1,020, 74)

Adjusting Washer

43 (438, 32)

Center Exhaust Pipe

◆ Gasket

Propeller Shaft

Heat Insulator

◆ Gasket

Ring

Heat Insulator

Heated Oxygen Sensor
(Bank 1 Sensor 2)

Heat Insulator

◆ Gasket

◆ Retainer

43 (438, 32)

Front Exhaust Pipe

43 (438, 32)

◆ Gasket

N·m (kgf·cm, ft·lbf) : Specified torque

◆ Non–reusable part

Exploded view of the engine mounting—IS 300

9347LG03

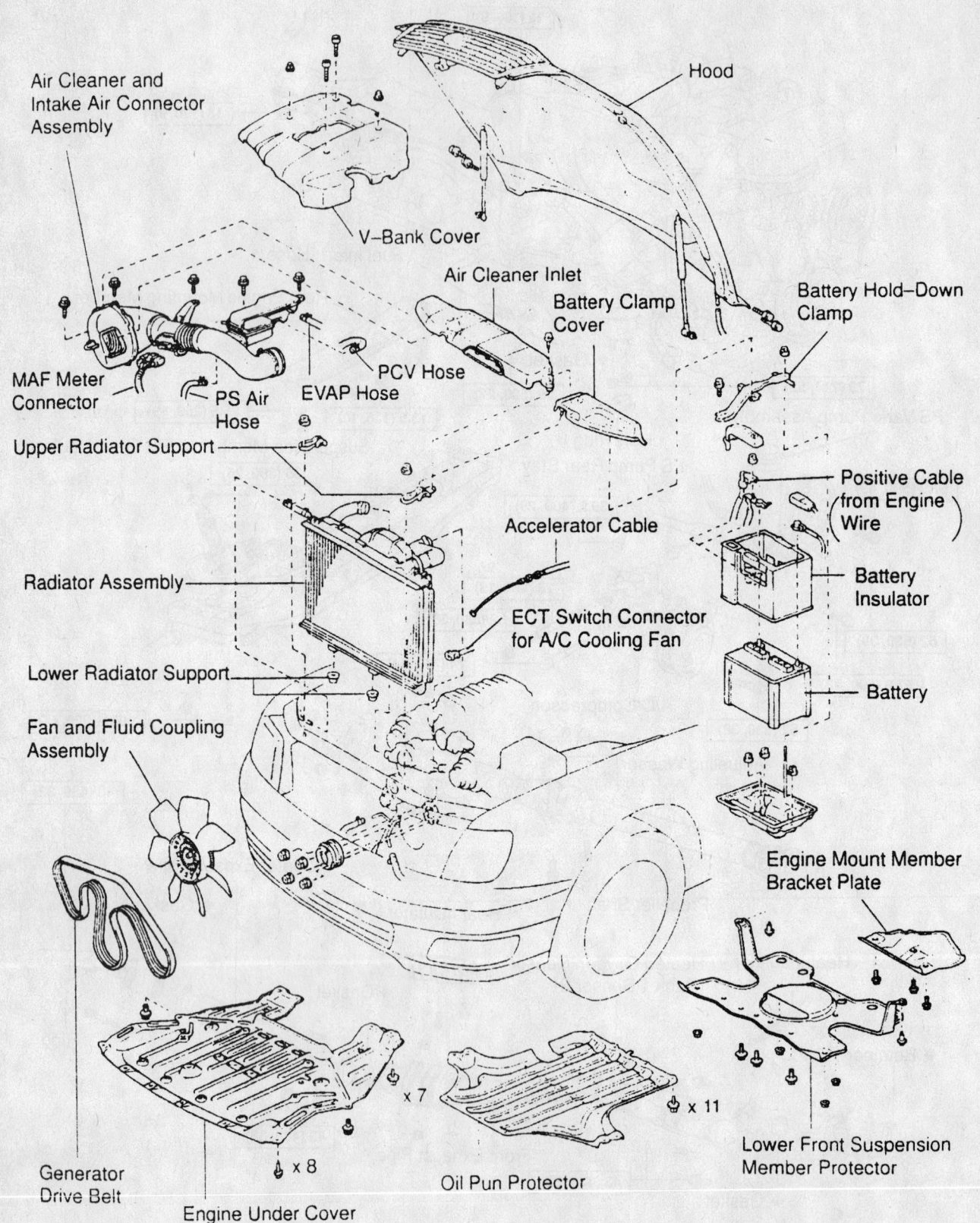

Hood

Air Cleaner and
Intake Air Connector
Assembly

V–Bank Cover

Air Cleaner Inlet

Battery Clamp
Cover

Battery Hold–Down
Clamp

MAF Meter
Connector

PS Air
Hose

PCV Hose

EVAP Hose

Upper Radiator Support

Positive Cable
from Engine
Wire

Accelerator Cable

Battery
Insulator

Radiator Assembly

ECT Switch Connector
for A/C Cooling Fan

Battery

Lower Radiator Support

Fan and Fluid Coupling
Assembly

Engine Mount Member
Bracket Plate

Generator
Drive Belt

Engine Under Cover

x 8

x 7

Oil Pun Protector

x 11

Lower Front Suspension
Member Protector

9301LG05

Exploded view of the engine removal and related components—LS 400

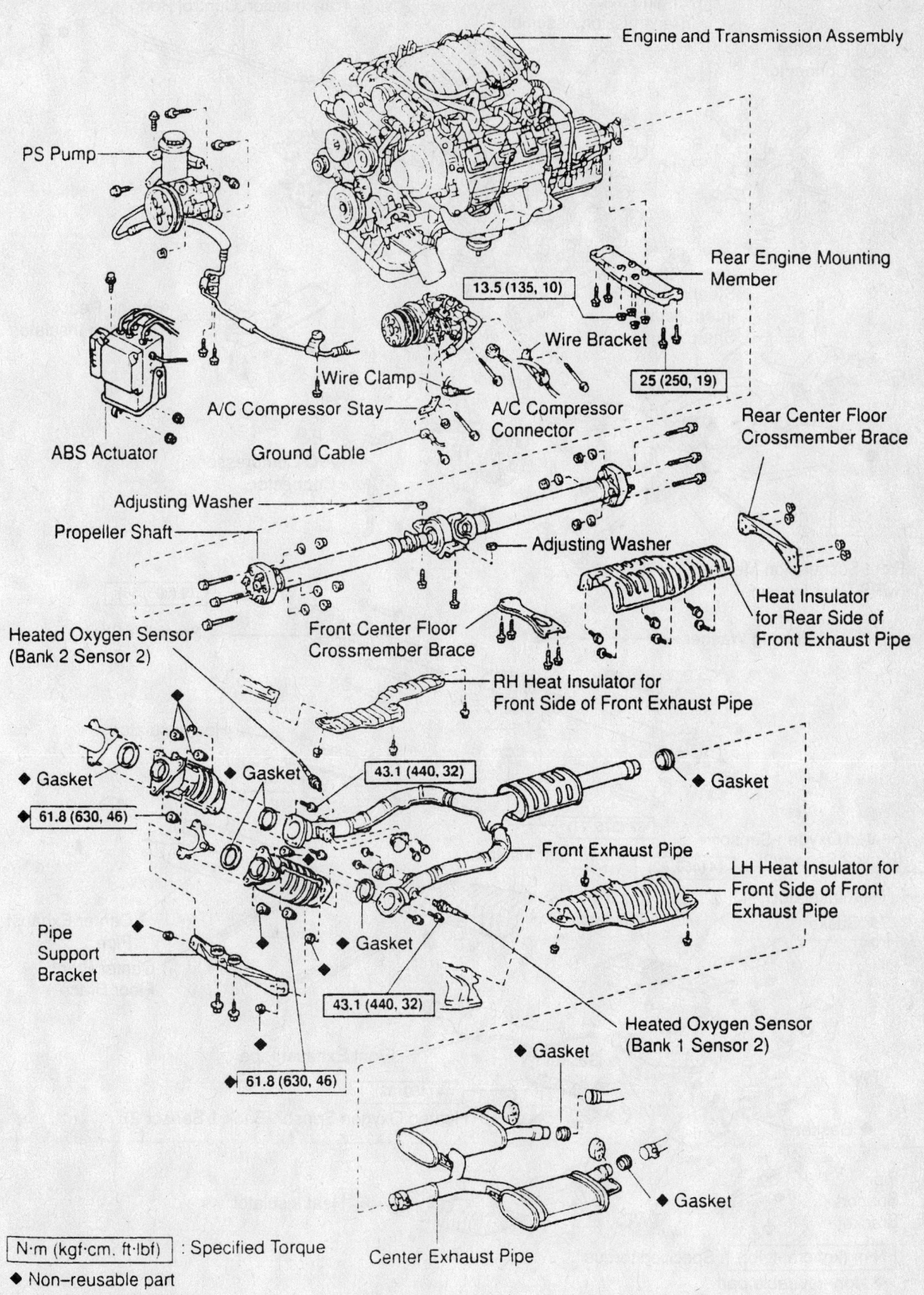

Engine and Transmission Assembly

PS Pump

13.5 (135, 10)

Rear Engine Mounting Member

Wire Bracket

25 (250, 19)

Wire Clamp

A/C Compressor Stay

A/C Compressor Connector

ABS Actuator

Ground Cable

Rear Center Floor Crossmember Brace

Adjusting Washer

Propeller Shaft

Adjusting Washer

Heat Insulator for Rear Side of Front Exhaust Pipe

Heated Oxygen Sensor (Bank 2 Sensor 2)

Front Center Floor Crossmember Brace

RH Heat Insulator for Front Side of Front Exhaust Pipe

◆ Gasket

◆ Gasket

43.1 (440, 32)

◆ Gasket

◆ 61.8 (630, 46)

Front Exhaust Pipe

LH Heat Insulator for Front Side of Front Exhaust Pipe

Pipe Support Bracket

◆ Gasket

Heated Oxygen Sensor (Bank 1 Sensor 2)

43.1 (440, 32)

◆ 61.8 (630, 46)

◆ Gasket

◆ Gasket

N·m (kgf·cm, ft·lbf) : Specified Torque

◆ Non-reusable part

Center Exhaust Pipe

9301LG06

Exploded view of the engine removal and related components (cont.)—LS 400

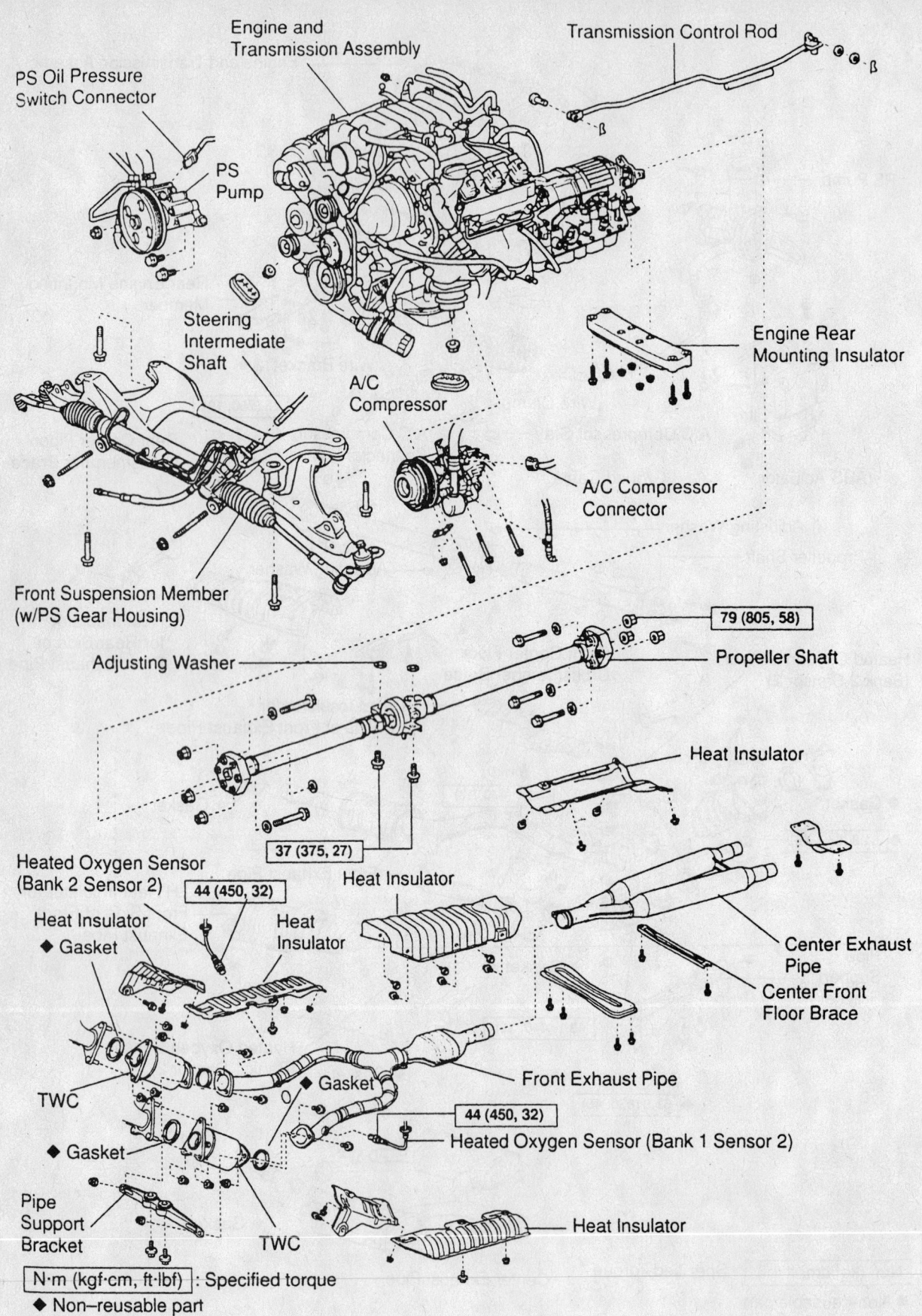

PS Oil Pressure Switch Connector

PS Pump

Engine and Transmission Assembly

Transmission Control Rod

Steering Intermediate Shaft

A/C Compressor

Engine Rear Mounting Insulator

A/C Compressor Connector

Front Suspension Member (w/PS Gear Housing)

Adjusting Washer

79 (805, 58)

Propeller Shaft

Heat Insulator

Heated Oxygen Sensor (Bank 2 Sensor 2)

37 (375, 27)

44 (450, 32)

Heat Insulator

Heat Insulator

Heat Insulator

◆ Gasket

Heat Insulator

Center Exhaust Pipe

Center Front Floor Brace

TWC

◆ Gasket

Front Exhaust Pipe

◆ Gasket

44 (450, 32)

Heated Oxygen Sensor (Bank 1 Sensor 2)

Pipe Support Bracket

TWC

Heat Insulator

N·m (kgf·cm, ft·lbf) : Specified torque

◆ Non-reusable part

9347LG04

Exploded view of the engine removal and related components—LS 430

- 3 Powertrain Control Module (PCM) connectors
- 2 cowl wire connectors from the connector on the bracket
- Wire clamp from the bracket
- Grommet from the cowl panel, then pull the engine wire out.
- Power steering oil cooler pipe from the oil pan
- Heated Oxygen (HO_2) sensors
- Front exhaust pipe
- 2 catalytic converters
- Center exhaust pipe
- Heat insulator from the rear side of the front exhaust pipe
- Front center floor and rear center floor crossmember braces
- Driveshaft
- Air conditioning compressor without disconnecting the air conditioning lines
- Power steering pump
- Heat insulators for the front side of the front exhaust pipe
- Heater water valve from the cowl panel by removing the 2 nuts

5. Attach the engine chain hoist to the engine hangers.

6. Remove or disconnect the following:
- Engine mounting insulators from the engine suspension crossmember by removing the 2 nuts
- Transmission control rod from the shift lever by removing the nut
- Rear engine mounting member by removing the 4 nuts and 4 bolts

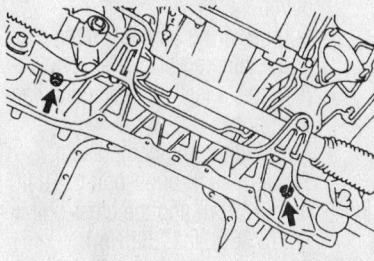

7923LG05

Engine mounting insulator fastener locations—LS 400

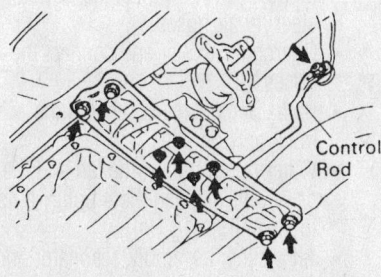

Control Rod

7923LG06

Engine mounting member bolt locations— LS 400

7. Lift the engine and transmission assembly out of the vehicle slowly and carefully.

8. Disconnect the engine from the transmission as follows:

9. Remove or disconnect the following:
- Vehicle Speed Sensor (VSS) connector
- Park/Neutral Position (PNP) switch connector
- Solenoid connector
- Direct clutch speed sensor connector
- 4 engine wire clamps from the brackets
- Oil dipstick and guide from the transmission
- Oil cooler pipes from the transmission and clamps
- The flywheel housing undercover by removing the 2 bolts
- The 6 torque converter bolts
- 10 bolts holding the transmission to the engine
- Transmission together with the torque converter clutch

To install:

10. Install the transmission to the engine and install the 10 bolts. Tighten the bolts as follows:
- 14mm: 27 ft. lbs. (37 Nm)
- 17mm: 53 ft. lbs. (72 Nm)

11. Install or connect the following:
- Torque converter clutch bolts. Apply adhesive to 2 or 3 threads of the bolt end. Tighten the bolts to 30 ft. lbs. (41 Nm).
- Flywheel housing undercover with the 2 bolts. Tighten bolts to 14 ft. lbs. (19 Nm).
- Oil cooler pipe for the transmission
- Dipstick guide and dipstick for the transmission
- Engine wire to the transmission
- VSS connector
- PNP switch connector
- Solenoid connector
- Direct clutch speed sensor connector
- 4 wire clamps to the brackets
- Engine and transmission assembly to the vehicle
- Rear engine mounting member to the vehicle and the 4 bolts and 4 nuts; bolts tightened to 19 ft. lbs. (25 Nm), nuts to 10 ft. lbs. (14 Nm)
- The transmission control rod to the shift lever with the nut. Tighten the nut to 108 inch lbs. (13 Nm).
- 2 nuts holding the engine mounting brackets to the front suspen-

sion crossmember. Tighten the 2 nuts to 52 ft. lbs. (70 Nm).
- Heater water valve to the cowl panel with the 2 nuts

12. Remove the engine hoist.

13. Install or connect the following:
- Heat insulators for the front side of the front exhaust pipe
- Power steering pump with the nut and 3 bolts-tighten the nut to 32 ft. lbs. (43 Nm); tighten the bolts to 29 ft. lbs. (39 Nm)
- The air conditioning compressor. Tighten the bolts to 36 ft. lbs. (49 Nm) and the nut to 22 ft. lbs. (29 Nm).
- Driveshaft to the vehicle
- Front center floor crossmember brace and tighten the bolts to 108 inch lbs. (13 Nm)
- Rear center floor crossmember brace and tighten the bolts to 108 inch lbs. (13 Nm)
- Heat insulator for the rear side of the front exhaust pipe
- Center exhaust pipe
- 2 front catalytic converters with 3 new nuts each. Tighten the nuts to 46 ft. lbs. (62 Nm).
- Front exhaust pipe. 4 bolts holding the pipe support bracket to the transmission: 32 ft. lbs. (44 Nm).
- HO_2 sensors. Tighten the sensors to 33 ft. lbs. (44 Nm).
- Power steering oil cooler pipe
- Engine wire harness to the passenger's compartment
- 3 PCM connectors
- 2 engine wire connectors to the connector on the bracket
- Engine wire clamp to bracket
- Glove compartment and the dash undercover
- All of the engine assembly connectors, wires, straps, clamps and hoses
- Radiator
- Accelerator and cruise control cables to the throttle body
- Throttle control cable to the throttle body, if equipped with automatic transmission
- Fan pulley
- Fan clutch and the drive belt. Tighten the 4 nuts for the fan to 16 ft. lbs. (21 Nm).
- Air cleaner and intake air connector assembly
- Air cleaner inlet
- Coolant
- Battery
- Engine oil

- Battery cables
- Battery cover
- Engine undercover
- Oil pan protector
- Hood

SC 300

1. Before servicing the vehicle, refer to the precautions in the beginning of this section.
2. Release the fuel system pressure.
3. Drain the fuel from the fuel tank.
4. Remove or disconnect the following:
- Negative battery cable. Wait at least 90 seconds before performing any other work.
- Battery and tray
- Hood
- Engine undercover, then drain the engine coolant and oil.
- Accelerator cable
- Cruise control cable
- Throttle control cable (automatic transmission only) from the throttle body
- Air cleaner assembly, resonator and the air intake hose.
- Drive belt, fan (with fluid coupling attached) and the water pump pulley
- Radiator
- Evaporative Emissions (EVAP) hoses (vacuum hose and air hose) from the charcoal canister
- Charcoal canister
- Power steering pump
- Air conditioning compressor
- All wires, electrical leads and vacuum hoses from the block
- Wiring clips or brackets
- Undercover beneath the glove box
- Lower instrument panel, the trim panel and the glove box door
- Right door sill trim (scuff plate). Lift the front edge of the carpet and remove the protective cover from the Powertrain Control Module (PCM).

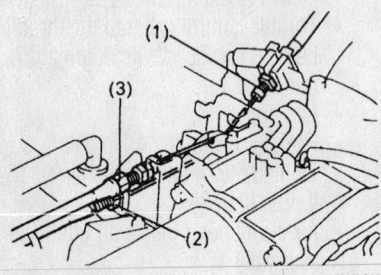

Accelerator cable (1), throttle control cable (2), cruise control cable (3) locations—SC 300

7923LG07

- PCM connectors, the cowl wire connectors and the control unit connectors behind the glove box
- 2 nuts holding the harness to the firewall and carefully pull the engine harness into the engine compartment
- Power steering pipe and 2 clamp bolts from the engine block
- Union bolt and 2 gaskets
- Fuel inlet hose
- Starter wiring and unhook the starter wiring from the clip
- Front exhaust pipe and heat shield
- Transmission control rod at the shift lever
- Intermediate shaft

➡ **Some vehicles are not equipped with adjusting washers.**

5. Attach an engine lift to the lift hooks
6. Remove or disconnect the following:
- The 2 nuts holding the engine to the front suspension crossmember
- The 4 bolts and 4 nuts holding the engine to the rear crossmember
- The rear engine mount
- Engine/transaxle assembly out of the engine compartment
7. Place the engine and transmission assembly onto a stand.
8. For vehicles with automatic transmission, remove or disconnect the following:
- Oil dipstick guide and dipstick for the transmission
- Oil cooler tubes
- Engine wire from the transmission
- Starter from the engine by removing the bolts
9. Separate the engine and transmission.

To install:

10. Install or connect the following:
- Engine and transmission
- Starter to the engine by installing the bolts
- Engine wire to the transmission
- With automatic transmission, the oil cooler tubes
- With automatic transmission, the oil dipstick guide and dipstick to the transmission
- Engine into the engine compartment
- The rear mount. Tighten the 4 nuts to 10 ft. lbs. (13 Nm). Tighten the 4 bolts to 19 ft. lbs. (25 Nm).
- Intermediate shaft to the rear differential and tighten the bolts and nuts to 54 ft. lbs. (74 Nm). Install the center support bearing set bolts

with the adjusting washers. Tighten the bolts to 36 ft. lbs. (49 Nm).
- Transmission control rod to the shift lever (automatic transmission only) by installing the nut.
- Exhaust heat insulator by installing the 4 nuts
- No. 2 front exhaust pipe, tighten the nuts to 46 ft. lbs. (62 Nm)
- The pipe support bracket to the transmission with the 2 bolts. Tighten the bolts to 32 ft. lbs. (43 Nm).
- No. 2 front exhaust pipe to the front exhaust pipe with the 2 bolts and nuts. Tighten the bolts to 32 ft. lbs. (43 Nm).
- With transmissions, the clutch release cylinder and tighten the bolts to 108 inch lbs. (12 Nm).
- Starter wiring and secure the harness in the clips
- Fuel inlet hose with 2 new gaskets and tighten the union bolt to 22 ft. lbs. (29 Nm)
- Power steering pipe below the engine
- Engine harness
- Connectors to the proper PCM, controller, or relay. Make certain each connector is square and secure.
- PCM and cover; connect the wiring harnesses. Refit the carpet and install the scuff plate.
- Lower instrument panel trim, the glove box door and the undercover.
- Engine wiring harness in the clips and retainers
- All wires, electrical leads and vacuum hoses
- Air conditioning compressor. Tighten the through-bolt to 19 ft. lbs. (26 Nm) and the other bolt and nut to 38 ft. lbs. (52 Nm).
- Power steering pump. Tighten the long bottom bolt to 43 ft. lbs. (58 Nm); tighten the others to 29 ft. lbs. (39 Nm). Connect the power steering air hoses.
- Charcoal canister and connect the hoses
- Radiator, coolant hoses and the transmission lines
- Water pump pulley, the fan and the drive belt. Tighten the 4 pulley nuts to 12 ft. lbs. (16 Nm).
- Air cleaner assembly, resonator and the air intake hose
- Accelerator, throttle control (automatic transmission only) and the

cruise control cables to the throttle body
- Battery tray and battery
- Battery cables
- All fluids, including fuel

11. Check the automatic transmission fluid level

12. Check the ignition timing

13. Shut the engine off and install the engine undercovers.

14. Install the hood

SC 400, GS 400 and GS 430

1. Before servicing the vehicle, refer to the precautions in the beginning of this section.

2. Relieve the fuel pressure from the fuel lines.

3. Drain the engine coolant from the cooling system.

4. Remove or disconnect the following:
- Battery cables and remove the battery. Wait at least 90 seconds before proceeding with any other work.
- Hood
- V-bank cover, if equipped
- Engine undercover and drain the engine oil. Lower the vehicle.
- Drive belt
- Throttle body
- Accelerator, transmission and cruise control cables from the throttle body.
- Air cleaner assembly
- Vacuum hose (from the power steering air control valve) from the air intake chamber.
- Intake air connector
- Coolant reservoir tank
- Radiator
- Igniter connectors and wire clamp
- Engine wires located next to the relay box, which is located next to the left strut tower
- Engine ground cable
- Power steering solenoid valve connector
- Alternator
- Power steering tubes from the suspension crossmember
- Power steering reservoir tank and bracket from the body by removing the 3 bolts
- Power steering pump by removing the pump mounting bolts and nut. Do not disconnect the power steering lines and place the pump off to the side.
- Air conditioning compressor from the engine. Do not remove the compressor pressure lines.

- Heater water hose from the water bypass hose
- Heater water hose from the heater water valve
- Brake booster hose from the union on the air intake chamber
- Vacuum hose from the Vacuum Switching Valve (VSV) for the heater water valve from the air intake chamber
- Ground strap from the bracket on the body
- Fuel inlet hose from fuel tube
- Charcoal canister from the engine
- Engine wire from the cabin as follows:
- Passenger's side lower instrument panel undercover
- 4 screws to the lower instrument panel finish panel and glove compartment door assembly
- Glove compartment and finish panel.
- Right scuff plate
- Take out the front side of the floor carpet
- 2 nuts and the Powertrain Control Module (PCM) protector
- Mounting nut and disconnect the PCM from the floor panel.
- 2 connectors from the PCM
- Connector from the Anti-lock brake system (ABS) and Traction control electronic control unit (TRAC ECU)
- 2 connectors from the TRAC ECU
- 4 connectors from connector cassette
- Connector from air conditioning control assembly
- Bolt holding the engine wire clamp to the heater water valve bracket.
- 2 bolts holding the engine wire clamp to the body.
- Engine wiring harness (through the cowl panel) from the vehicle cabin
- Oxygen (O$_2$) sensors from the front exhaust pipe
- Front exhaust pipe
- Front catalytic converter by removing the 3 nuts and gasket
- Tailpipes
- Center exhaust pipe by disconnecting the 2 hooks
- Heat insulator by removing the 4 nuts
- Center floor crossmember brace by removing the 4 bolts
- Driveshaft from the vehicle using the proper tools (2 of tool SST 09922–10010), loosen the adjusting nut on the driveshaft. Place

matchmarks on the transmission flange and the flexible coupling.
- The transmission control rod from the shift lever by removing the nut

5. Attach the engine chain hoist to the engine hangers.

6. Remove or disconnect the following:
- 2 nuts holding the engine mounting insulators to the front suspension crossmember.
- 4 bolts, 4 nuts and the rear engine mounting member
- Ground strap to the rear mounting member
- Engine out of the vehicle
- Oil dipstick guide and dipstick for transmission
- Oil cooler pipes for the transmission
- All the engine wiring
- Engine bolts holding the transmission to the engine
- Engine from the transmission

To install:

7. Install or connect the following:
- Transmission to the engine and install the bolts. Tighten the bolts to 42 ft. lbs. (57 Nm).
- Engine wiring
- Oil cooler pipe for the transmission. Tighten the unions on the pipes to 25 ft. lbs. (34 Nm).
- Engine oil dipstick guide and the dipstick for the transmission
- Engine and transmission to the vehicle
- Rear engine mounting member with the 4 bolts and 4 nuts. Tighten the bolts to 19 ft. lbs. (25 Nm) and the nuts to 10 ft. lbs. (13 Nm).
- 2 nuts holding the engine mounting brackets to the front suspension crossmember. Tighten the nuts to 43 ft. lbs. (59 Nm).
- Engine chain hoist
- Transmission control rod to the shift lever by installing the nut
- Driveshaft
- Center floor crossmember brace by installing the 4 bolts. Tighten the bolts to 108 inch lbs. (13 Nm).
- Heat insulator for the front exhaust pipe by installing the 4 bolts
- Center exhaust pipe by installing the 2 hooks
- Tailpipe and tighten the 2 bolts to 14 ft. lbs. (19 Nm)
- Front catalytic converter and tighten the nuts to 46 ft. lbs. (62 Nm)
- Front exhaust. Tighten the 4 bolts and nuts holding the catalytic converter to the front exhaust pipe to

32 ft. lbs. (43 Nm). Tighten the 2 bolts and nuts holding the front exhaust pipe to the center exhaust pipe. Tighten the bolts to 32 ft. lbs. (43 Nm). Tighten the 4 bolts holding the pipe support bracket to the transmission. Tighten the bolt to 32 ft. lbs. (43 Nm).

- O_2 sensors to the front exhaust and tighten the sensors to 33 ft. lbs. (44 Nm)
- Engine wiring harness in through the cowl panel
- Engine wire retainer with the 3 bolts
- Reattach the connectors under the dash panel
- PCM with the nut
- PCM protector with the 2 nuts
- Floor carpet
- Scuff plate
- Lower instrument panel finish panel and glove compartment door assembly with the 4 screws
- Instrument panel undercover with the 2 screws
- Charcoal canister
- All hoses and grounds
- Air conditioning compressor with the nut and 3 bolts. Tighten the bolts to 36 ft. lbs. (49 Nm) and the nut to 22 ft. lbs. (29 Nm).
- Power steering pump with the nut and 3 bolts. Tighten the bolts to 29 ft. lbs. (39 Nm) and the nut to 32 ft. lbs. (43 Nm).
- Power steering reservoir tank and bracket with the 3 bolts
- Power steering tubes with the clamp and bolt
- Alternator, tighten the nut and bolt to 27 ft. lbs. (37 Nm)
- Power steering solenoid valve connector
- Engine wire connectors
- Theft deterrent horn connector
- Ground cable to the body from the engine
- Igniter connectors
- Yellow taped connector to the igniter on the rear side
- Radiator assembly
- The reservoir tank and the inlet pipe to the fan shroud and tighten the 4 bolts to 43 inch lbs. (5 Nm)
- The 2 hydraulic lines for the fan motor and tighten the bolts to 47 ft. lbs. (64 Nm)
- Upper and lower radiator hoses to the radiator
- 2 oil cooler hoses for the transmission to the radiator

- Coolant tank
- Intake air connector
- Vacuum hose (from the power steering air control valve) to the air intake chamber
- Air cleaner
- Accelerator, transmission throttle control and the cruise control actuator cables to the engine
- Throttle cover and hose clamp with the cap nut and 2 bolts
- Evaporative emission control (EVAP) hose to the hose clamp
- Drive belt to the engine
- Battery to the engine compartment and attach the electrical connectors
- Engine coolant
- V-bank cover if it was removed
- Engine undercover
- Hood.

8. Fill the engine oil and check the transmission oil.

9. Start the engine, bleed the cooling system, and check for leaks.

Water Pump

REMOVAL & INSTALLATION

3.0L (1MZ-FE) Engine

1. Before servicing the vehicle, refer to the precautions in the beginning of this section.

2. Remove or disconnect the following:
- Negative battery terminal
- Engine coolant
- Timing belt
- Right and left camshaft pulleys
- No. 2 idler pulley by removing the bolt
- 3 clamps and engine wire from the rear timing belt cover
- 6 bolts holding the rear timing belt cover to the engine block
- 4 bolts and 2 nuts to the water pump
- Water pump
- All the old packing (sealant) and

Water pump mounting bolts—3.0L (1MZ-FE) engine

gasket material from the water pump and clean the mounting surfaces.
- All gasket material from the upper inner timing belt cover

To install:

3. Check that the water pump turns smoothly. Also, check the air hole for coolant leakage.

4. Using a new gasket, apply liquid sealer to the gasket, water pump and engine block.

5. Install or connect the following:
- Gasket and pump to the engine and install the 4 bolts and 2 nuts. Tighten the nuts and bolts to 53 inch lbs. (6 Nm).
- Rear timing belt cover and tighten the 6 bolts to 74 inch lbs. (9 Nm).
- Engine wire with the 3 clamps to the rear timing belt cover
- No. 2 idler pulley with the bolt. Tighten the bolt to 32 ft. lbs. (43 Nm). After tightening the bolt, be sure the idler pulley moves smoothly.
- Right-hand camshaft pulley, with the flange side **outward**. Be sure to align the knock pinhole on the camshaft pulley with the knock pin on the camshaft. Camshaft bolt to 65 ft. lbs. (88 Nm).
- Left-hand camshaft pulley with the flange side **inward**. Be sure to align the knock pin hole on the camshaft pulley with the knock pin on the camshaft. Camshaft bolt to 94 ft. lbs. (125 Nm).
- Timing belt
- Engine coolant
- Negative battery cable to the battery and start the engine.

3.0L (2JZ-GE) Engine

1. Before servicing the vehicle, refer to the precautions in the beginning of this section.

2. Disconnect the negative battery cable. Wait at least 90 seconds before performing any work.

3. Drain the engine coolant.

4. Remove or disconnect the following:
- Radiator assembly
- Air cleaner
- Mass Air Flow (MAF) meter
- Intake air connector pipe assembly
- Timing belt
- Idler pulley
- Water inlet and the thermostat
- 2 bolts, the water bypass outlet and the No. 1 water bypass pipe

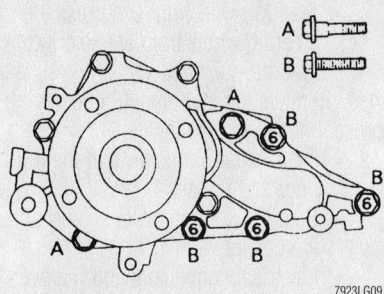

Water pump mounting bolt locations—3.0L (2JZ-GE) engine

Be sure to use new O-rings when installing the water bypass pipe—3.0L (2JZ-GE) engine

- 3 O-rings from the water bypass outlet and the No. 1 water bypass pipe
- Generator
- Bolt and engine wire bracket
- Bolt and clamp bracket for the Crankshaft Sensor (CKP) connector
- Nuts and the No. 2 water bypass pipe from the water pump
- 6 bolts and the water pump and gasket
- Drain hose and the O-ring from the cylinder block

To install:

5. Install or connect the following:
- New O-ring to the cylinder block
- Drain hose
- New gasket to the water pump
- Water pump to the water bypass pipe. Do not install the nut yet.
- Water pump with the 2 bolts (A) and the 4 bolts (B)

➡ **Hand-tighten the bolts (A) first. Tighten all 6 bolts to 15 ft. lbs. (21 Nm).**

- 2 nuts holding the No. 2 water bypass pipe to the water pump. Tighten the nuts to 15 ft. lbs. (21 Nm).
- Clamp bracket for the CKP sensor connector
- Engine wire bracket
- Generator
- O-rings to the No. 1 water bypass pipe.

- New O-ring and the water bypass outlet with the 2 bolts and tighten them to 78 inch lbs. (9 Nm)
- Thermostat and the water inlet
- Idler pulley
- Timing belt
- Air cleaner, the MAF meter and the intake air connector pipe assembly
- Radiator assembly
- Negative battery cable
- Coolant. Start the engine, check for leaks and bleed the cooling system.

4.0L (1UZ-FE) and 4.3L (3UZ-FE) Engines

1. Before servicing the vehicle, refer to the precautions in the beginning of this section.
2. Disconnect the negative battery cable.
3. Drain the cooling system.
4. Remove or disconnect the following:
- Timing belt
- No. 2 idler pulley
- Throttle body
- Bypass hose(s) from the water inlet housing
- 2 bolts holding the water inlet housing to water pump

- Water inlet housing and discard the gasket
- Mounting bolts, studs and the nut to the water pump
- Water pump by carefully prying between the pump and the cylinder block
- The old gasket and clean all mounting surfaces

To install:

5. Install new seal packing to the water pump groove and a new O-ring to the water bypass pipe end.
6. Install or connect the following:
- Water pump to the water bypass pipe end

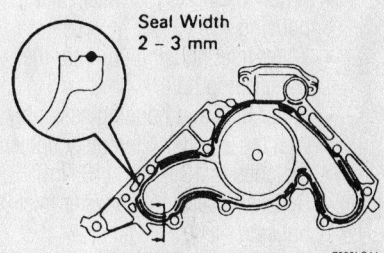

Apply silicone sealant to the water pump as shown—4.0L engine

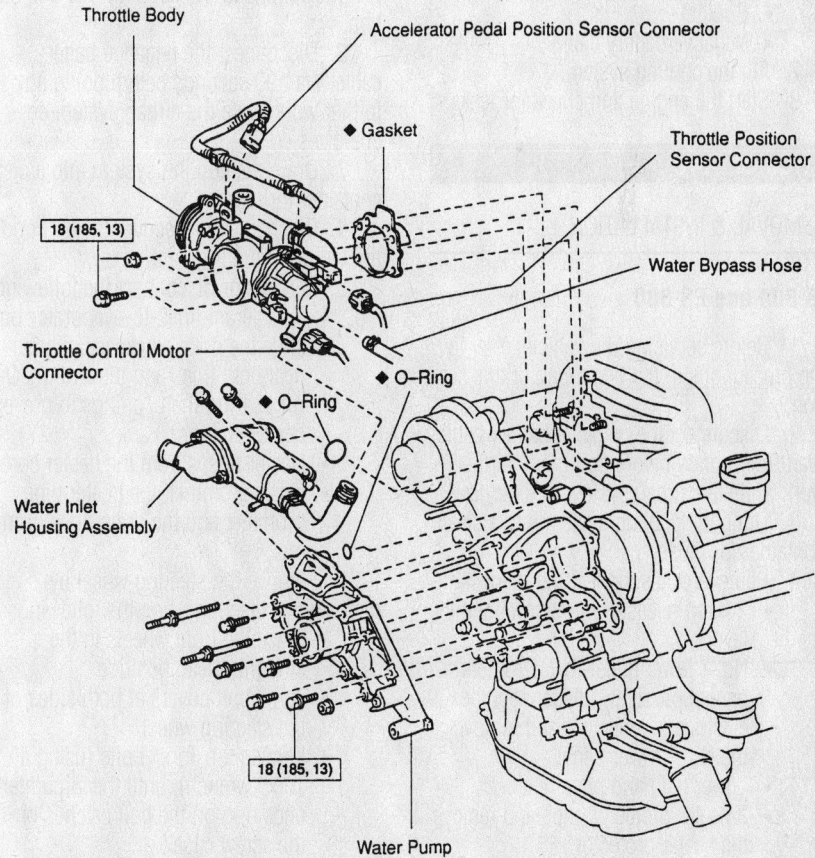

Exploded view of the water pump mounting and related components—4.0L engine shown

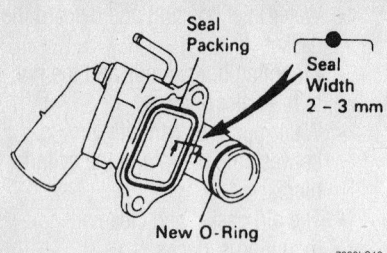

7923LG12

Apply sealant to the water inlet housing as shown—4.0L engine

- Water pump and tighten the mounting bolts and nut to13 ft. lbs. (18 Nm)
- Sealant to the groove of the water inlet housing
- A new O-ring to the water inlet housing
- Water inlet housing end into the water pump hole
- Water inlet and housing assembly with the 2 bolts. Alternately tighten the bolts to 13 ft. lbs. (18 Nm).
- Bypass hose(s) to the water inlet housing
- Throttle body. Tighten the mounting bolts to 13 ft. lbs. (18 Nm).
- No. 2 idler pulley
- Timing belt
- Coolant
- Negative battery cable
7. Fill the cooling system.
8. Start the engine and check for leaks.

Heater Core

REMOVAL & INSTALLATION

IS 300 and ES 300

1. Before servicing the vehicle, refer to the precautions in the beginning of this section.
2. Disconnect the negative battery cable. Wait 90 seconds before doing any further work while the airbag system de-energizes.
3. Drain the cooling system into a clean container for reuse.
4. Remove or disconnect the following:
- 2 hood release lever screws and the lever
- No. 1 lower panel-to-instrument panel bolt/screw, disconnect the electrical connectors and remove the No. 1 lower panel
- Lower left hand panel
- 3 heater protector clips and remove the heater protector
- 2 screws and the 2 clamps holding the heater core in place

- Heater core hoses and discard the O-rings
- Pull out heater core from the heater housing

To install:
5. Install or connect the following:
- Heater core to the heater housing
- Heater core hoses using new O-rings
- 2 clamps and the 2 screws holding the heater core in place
- Heater protector and connect the 3 heater protector clips
- Lower left hand panel
- No. 1 lower panel, connect the electrical connectors and install the No. 1 lower panel-to-instrument panel bolt/screw
- Hood release lever and the 2 lever screws
6. Refill the cooling system.
7. Connect the negative battery cable.
8. Operate the engine to normal operating temperatures; then, check the climate control operation and check for leaks.

GS 300, GS 400 and GS 430

1. Before servicing the vehicle, refer to the precautions in the beginning of this section.
2. Disconnect the negative battery cable. Wait 90 seconds before doing any further work while the airbag system de-energizes.
3. Drain the cooling system into a clean container for reuse.
4. Discharge and recover the air conditioning system refrigerant.
5. Remove or disconnect the following:
- Refrigerant lines-to-evaporator bolt, slide the plate clockwise, disconnect both lines and discard the O-rings. Plug the openings to prevent contamination
- Heater hoses from the heater core
- No. 1 grommet, the heater pipe grommet and the drain hose grommet
6. Remove the steering wheel by removing or disconnecting the following:
- Place the front wheels in the straight-ahead position.
- Torx® bolt covers at both sides of the steering wheel
- Loosen the Torx® bolts (using a Torx® wrench), until the circumference ring on the bolt catches on the screw case
- Lift the air bag, disconnect the electrical connector and remove it

- Steering wheel nut and press the steering wheel from the steering column
7. Remove the instrument by removing or disconnecting the following:
- Front pillar garnishes and the front door scuff plates
- Steering column cover screws and the covers
- Electrical connectors and remove the combination switch
- End pad
- 2 No. 1 undercover-to-instrument panel screws and the undercover
- 2 hood lock release screws and the release
- 4 No. 1 safety pad-to-instrument panel bolts, screw and the safety pad
- Parking brake handle and the No. 1 switch hole base
- 4 steering column-to-instrument panel nuts, disconnect the spring from the brake pedal and remove the steering column
- Instrument cluster finish panel using a suitable prytool
- 4 instrument cluster-to-instrument panel screws, disconnect the electrical connectors and remove the instrument cluster
- No. 2 undercover using a suitable prytool
- Plate and disconnect the air bag electrical connector inside the glove box
- Glove box-to-instrument panel 2 bolts, 3 screws, and the glove box
- 3 CD changer-to-instrument panel nuts, disconnect the electrical connectors and remove the CD changer
- Ashtray
- No. 2 register using a suitable prytool and disconnect the connector
- Audio unit-to-instrument panel 2 bolts, 2 screws and the audio unit
- Cluster finish panel using a suitable prytool, disconnect the connectors and remove the panel.
- Console box carpet
- Lower rear console box
- Rear console armrest
- No. 3 console box mounting bracket
- Console box
- No. 1 console box duct
- No. 7 heater-to-register
- 5 instrument panel-to-chassis bolts, the nut, the screw and the instrument panel
- No. 1 and No. 2 brace

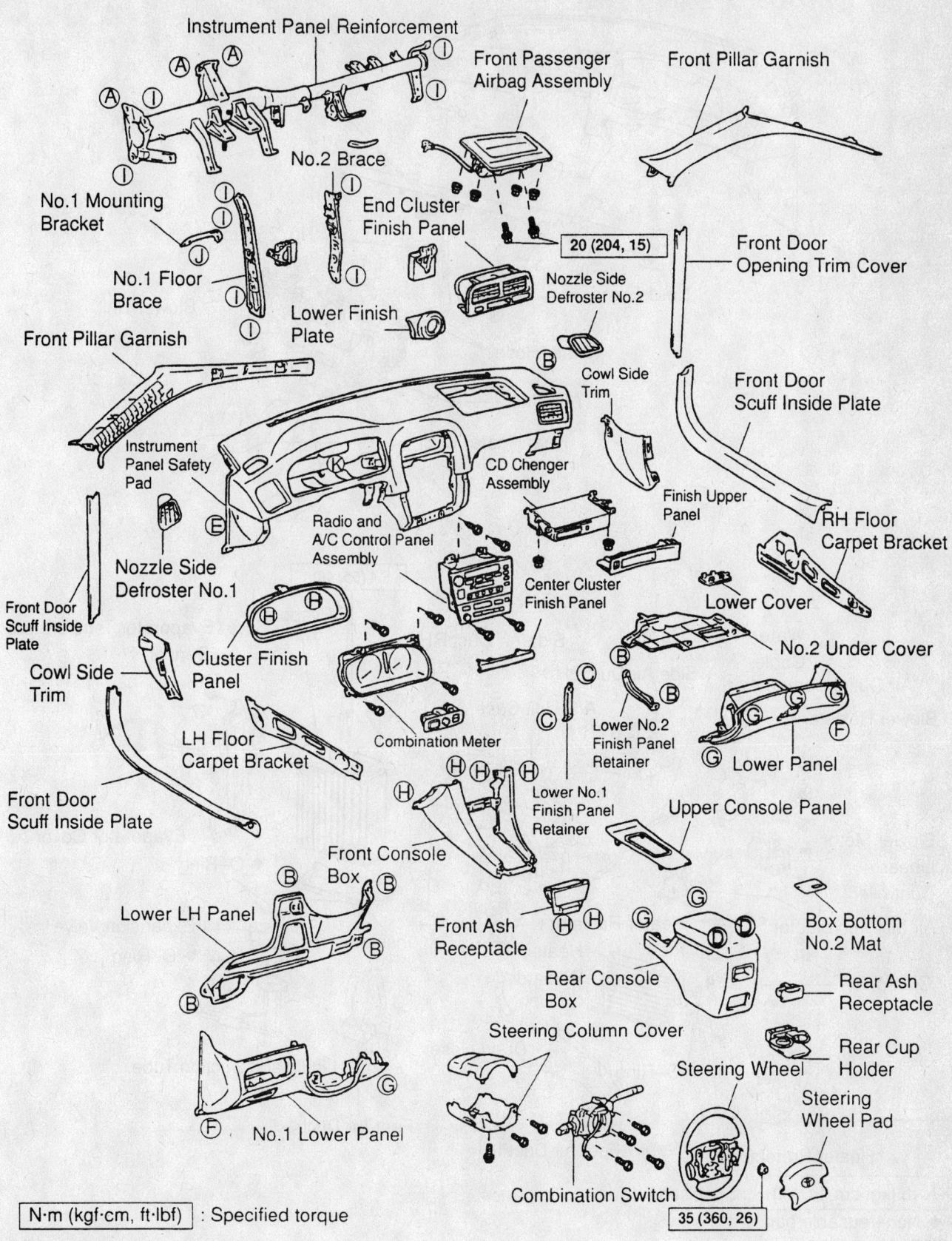

Instrument Panel Reinforcement

Front Passenger Airbag Assembly

Front Pillar Garnish

No.2 Brace

No.1 Mounting Bracket

End Cluster Finish Panel

20 (204, 15)

Front Door Opening Trim Cover

No.1 Floor Brace

Nozzle Side Defroster No.2

Lower Finish Plate

Front Door Scuff Inside Plate

Front Pillar Garnish

Cowl Side Trim

Front Door Scuff Inside Plate

Instrument Panel Safety Pad

Front Door Opening Trim Cover

CD Chenger Assembly

Finish Upper Panel

RH Floor Carpet Bracket

Nozzle Side Defroster No.1

Radio and A/C Control Panel Assembly

Center Cluster Finish Panel

Lower Cover

No.2 Under Cover

Cowl Side Trim

Cluster Finish Panel

Combination Meter

Lower No.2 Finish Panel Retainer

Lower Panel

LH Floor Carpet Bracket

Lower No.1 Finish Panel Retainer

Upper Console Panel

Front Door Scuff Inside Plate

Front Console Box

Lower LH Panel

Front Ash Receptacle

Rear Console Box

Box Bottom No.2 Mat

Rear Ash Receptacle

Rear Cup Holder

Steering Column Cover

Steering Wheel

Steering Wheel Pad

No.1 Lower Panel

Combination Switch

35 (360, 26)

N·m (kgf·cm, ft·lbf) : Specified torque

93112GS3

Exploded view of the instrument assembly—ES 300

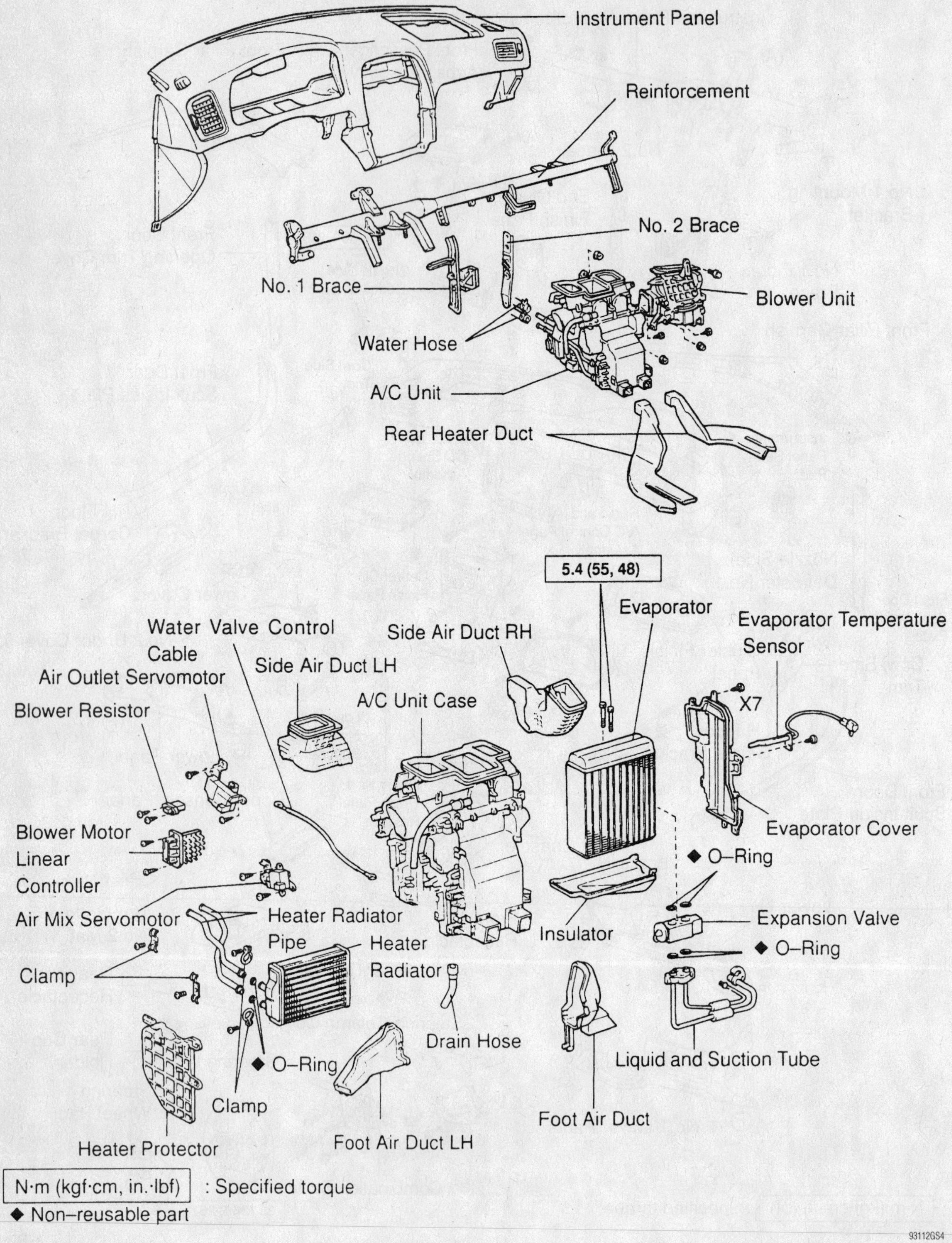

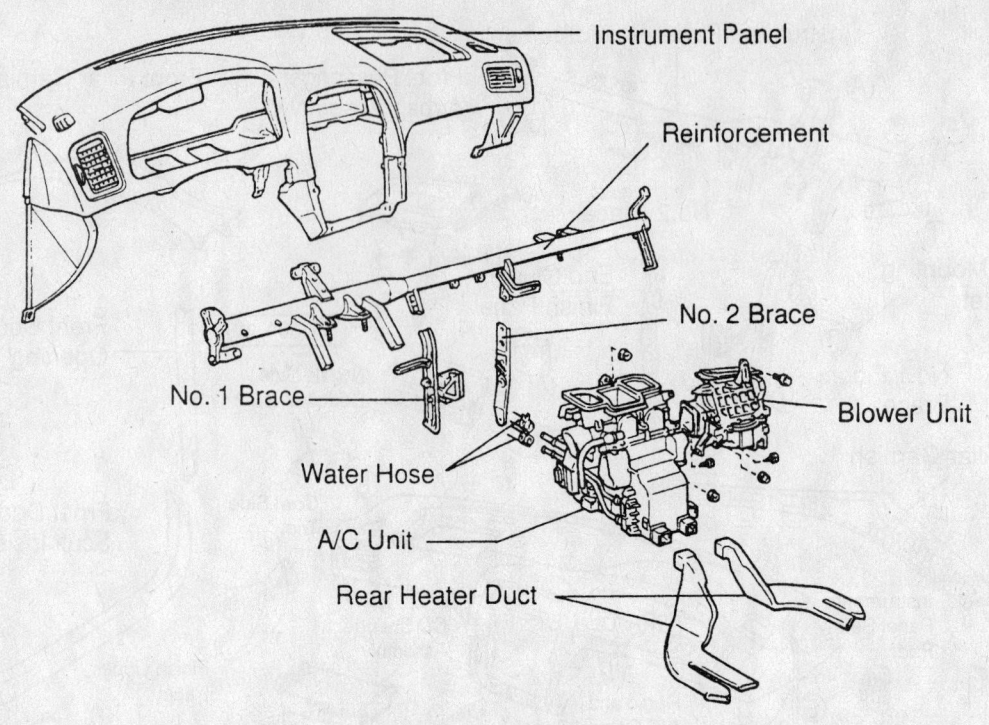

Instrument Panel

Reinforcement

No. 2 Brace

No. 1 Brace

Blower Unit

Water Hose

A/C Unit

Rear Heater Duct

5.4 (55, 48)

Evaporator

Water Valve Control Cable

Side Air Duct RH

Evaporator Temperature Sensor

Air Outlet Servomotor

Side Air Duct LH

Blower Resistor

A/C Unit Case

X7

Blower Motor Linear Controller

Evaporator Cover

◆ O–Ring

Air Mix Servomotor

Heater Radiator Pipe

Expansion Valve

Clamp

Heater Radiator

Insulator

◆ O–Ring

Clamp

Drain Hose

Liquid and Suction Tube

◆ O–Ring

Foot Air Duct

Heater Protector

Foot Air Duct LH

N·m (kgf·cm, in.·lbf) : Specified torque

◆ Non–reusable part

93112GS4

Exploded view of the heater core, heater/air conditioning housing and related components—ES 300

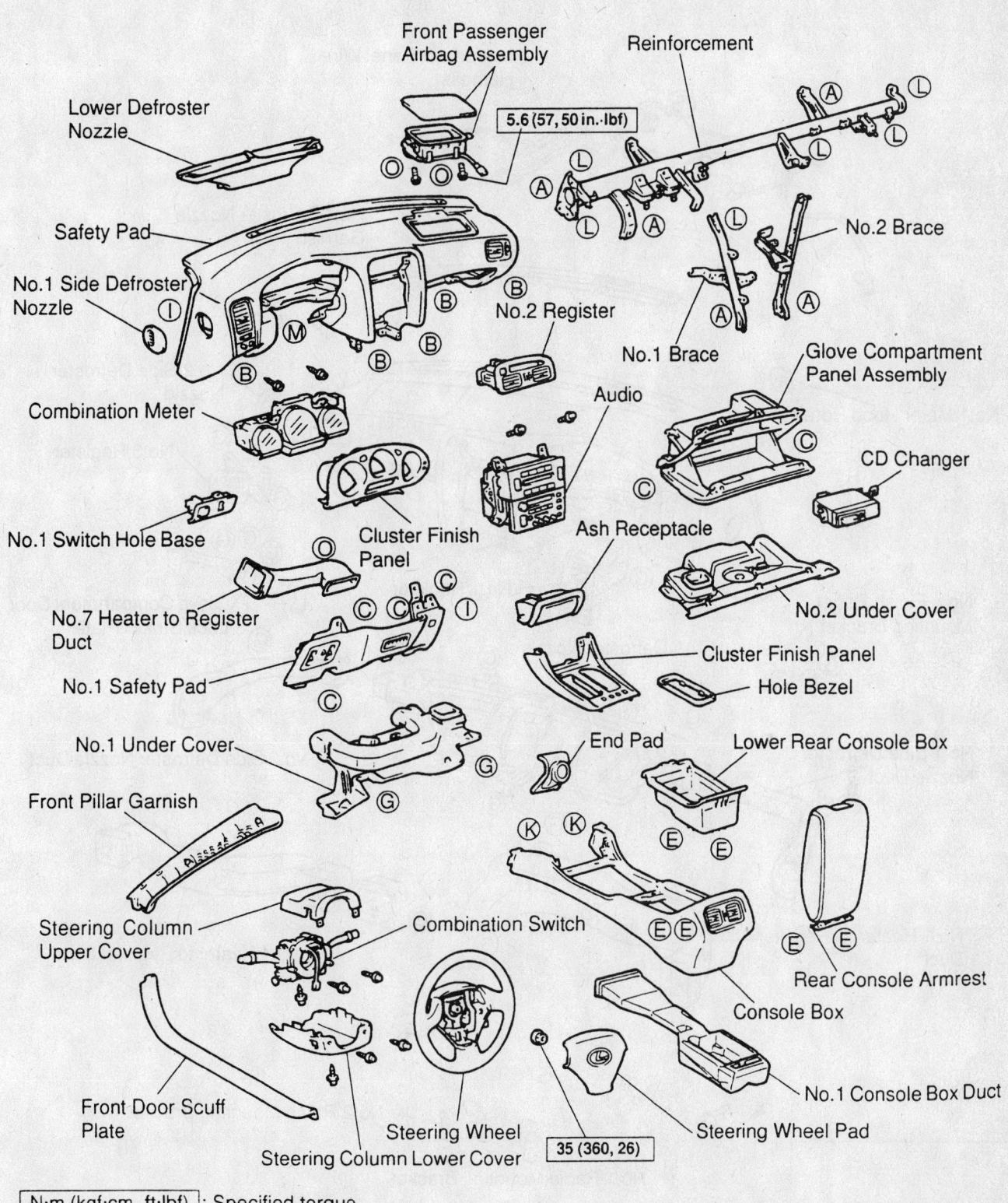

Front Passenger Airbag Assembly

Reinforcement

Lower Defroster Nozzle

5.6 (57, 50 in.·lbf)

No.2 Brace

Safety Pad

No.1 Side Defroster Nozzle

No.2 Register

No.1 Brace

Glove Compartment Panel Assembly

Audio

Combination Meter

CD Changer

No.1 Switch Hole Base

Cluster Finish Panel

Ash Receptacle

No.2 Under Cover

No.7 Heater to Register Duct

Cluster Finish Panel

Hole Bezel

No.1 Safety Pad

No.1 Under Cover

End Pad

Lower Rear Console Box

Front Pillar Garnish

Steering Column Upper Cover

Combination Switch

Rear Console Armrest

Console Box

Front Door Scuff Plate

No.1 Console Box Duct

Steering Wheel

Steering Wheel Pad

Steering Column Lower Cover

35 (360, 26)

N·m (kgf·cm, ft·lbf) : Specified torque

93112GT4

Exploded view of the instrument panel and related components—GS 300, GS 400 and GS 430

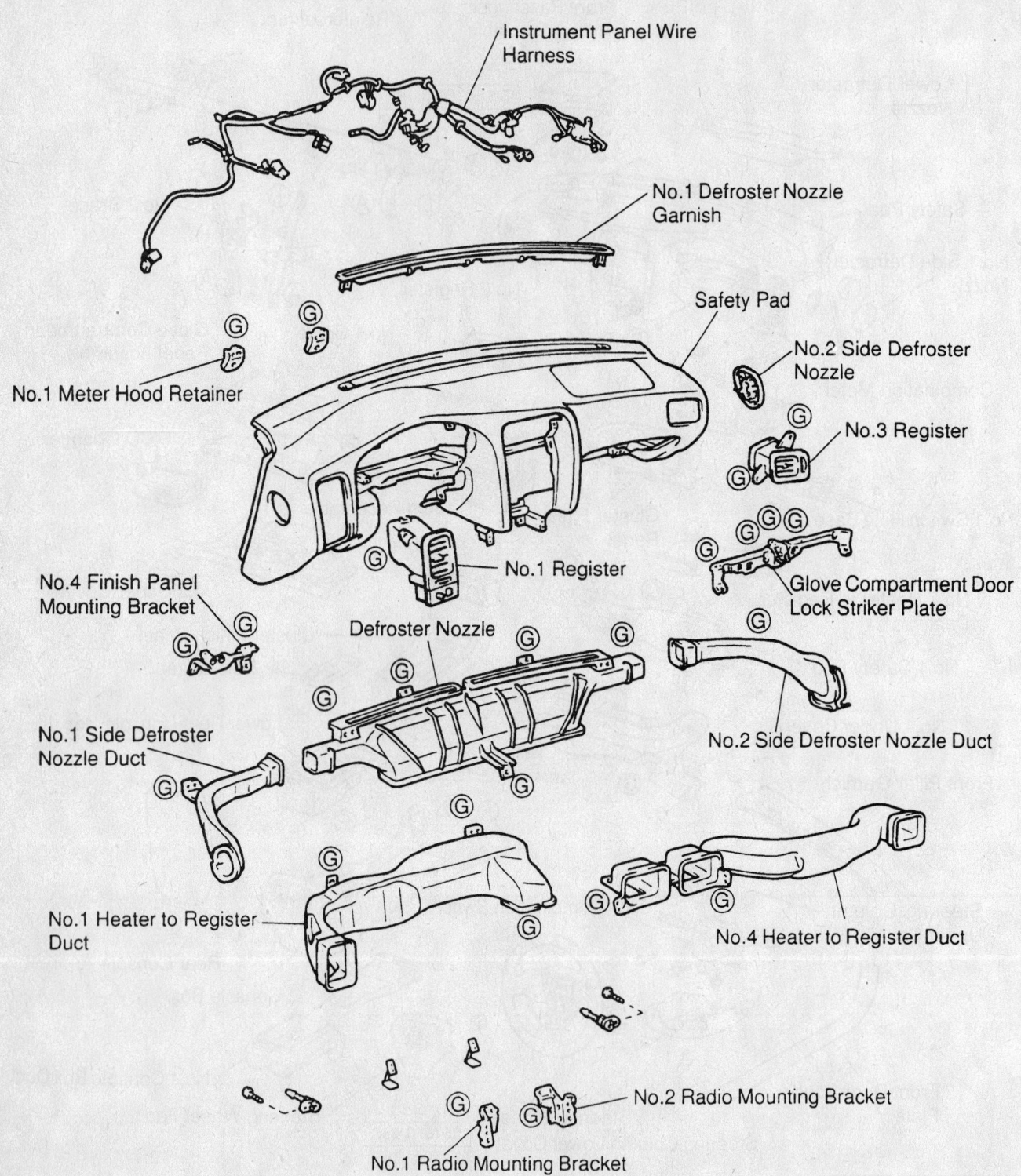

Instrument Panel Wire Harness

No.1 Defroster Nozzle Garnish

Safety Pad

No.2 Side Defroster Nozzle

No.3 Register

No.1 Meter Hood Retainer

Glove Compartment Door Lock Striker Plate

No.1 Register

No.4 Finish Panel Mounting Bracket

Defroster Nozzle

No.2 Side Defroster Nozzle Duct

No.1 Side Defroster Nozzle Duct

No.1 Heater to Register Duct

No.4 Heater to Register Duct

No.2 Radio Mounting Bracket

No.1 Radio Mounting Bracket

93112GT5

Exploded view of the instrument panel, ventilation ducts and related components—GS 300, GS 400 and GS 430

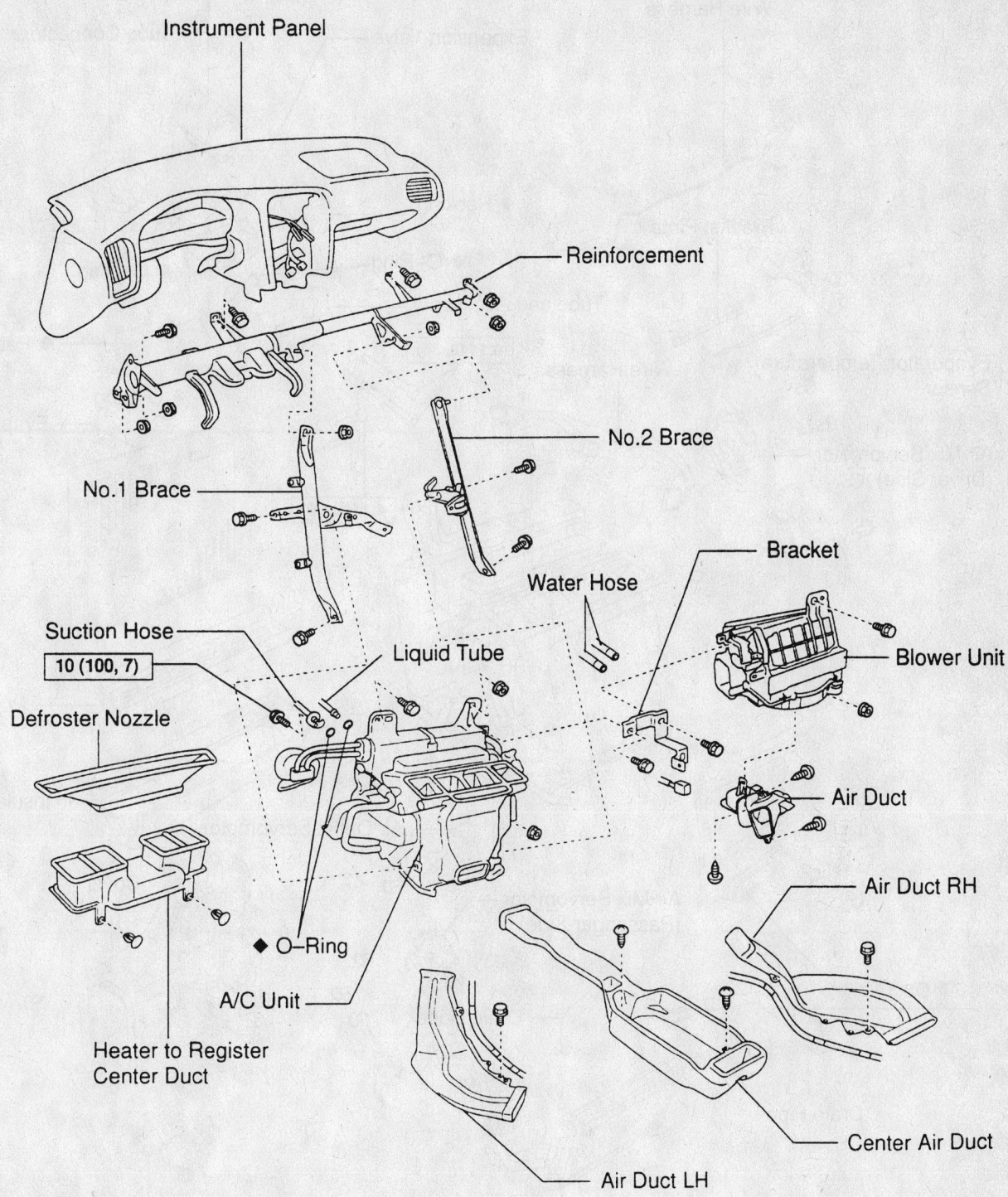

Instrument Panel

Reinforcement

No.2 Brace

No.1 Brace

Bracket

Water Hose

Blower Unit

Suction Hose

10 (100, 7)

Liquid Tube

Defroster Nozzle

Air Duct

Air Duct RH

♦ O–Ring

A/C Unit

Heater to Register
Center Duct

Air Duct LH

Center Air Duct

N·m (kgf·cm, ft·lbf) : Specified torque

♦ Non–reusable part

93112GT6

Exploded view of the heater/air conditioning housing and related components—GS 300, GS 400 and GS 430

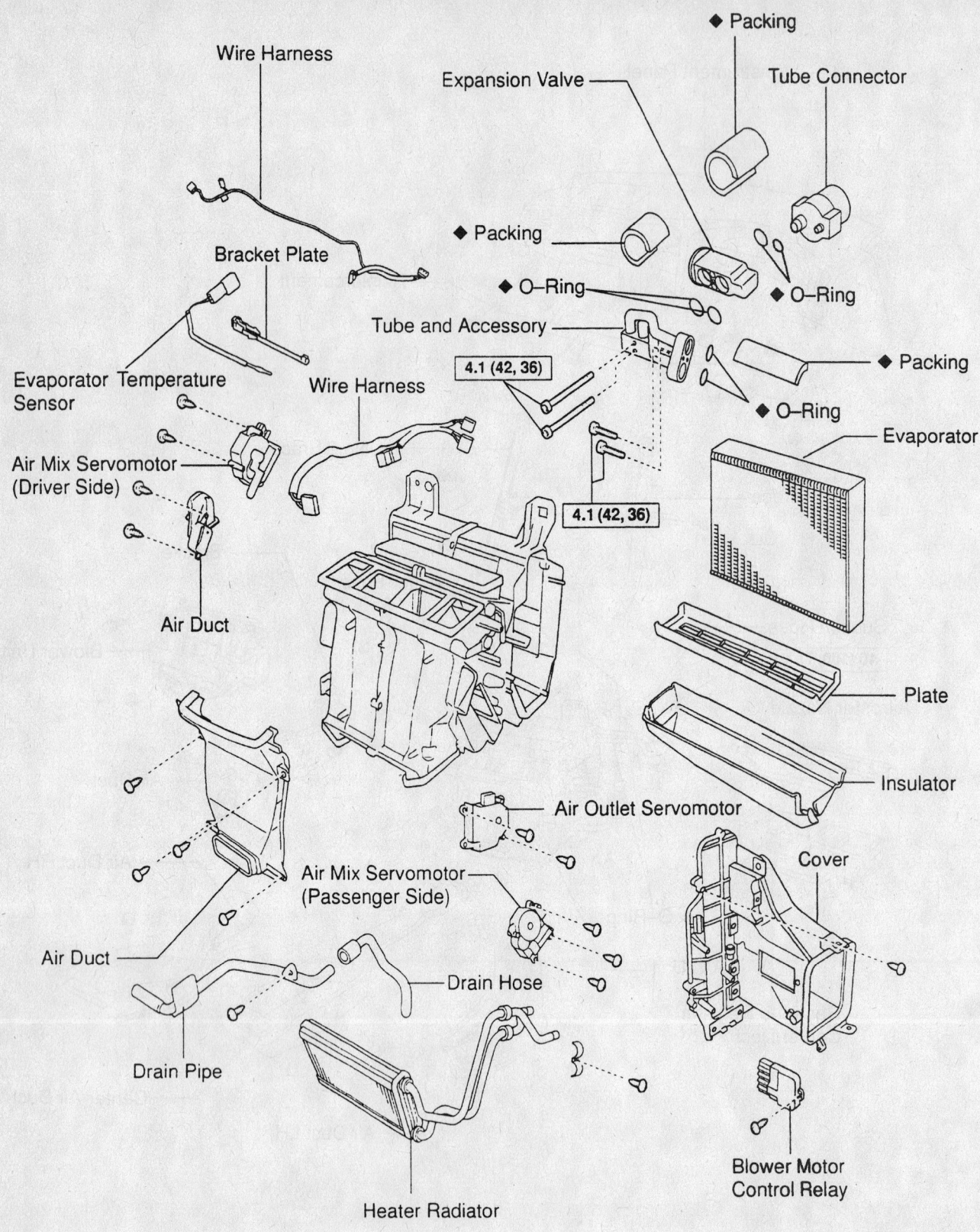

◆ Packing

Expansion Valve

Tube Connector

Wire Harness

Bracket Plate

◆ Packing

◆ O–Ring

◆ O–Ring

Tube and Accessory

◆ Packing

Evaporator Temperature Sensor

4.1 (42, 36)

◆ O–Ring

Wire Harness

Evaporator

Air Mix Servomotor (Driver Side)

4.1 (42, 36)

Air Duct

Plate

Insulator

Air Outlet Servomotor

Cover

Air Duct

Air Mix Servomotor (Passenger Side)

Drain Hose

Drain Pipe

Heater Radiator

Blower Motor Control Relay

N·m (kgf·cm, in.·lbf) : Specified torque

◆ Non–reusable part

93112GT7

Exploded view of the heater core, heater housing and related components—GS 300, GS 400 and GS 430

- Reinforcement-to-chassis 5 nuts, 4 bolts and the reinforcement
- Ventilation nozzles from the heater/air conditioning housing

8. Remove the blower unit by removing or disconnecting the following:
- Connector clamp
- 3 air duct-to-blower housing screws and the air duct
- Electrical connector bracket, the wiring harness clamps and the wiring harness
- Blower housing connectors
- 2 blower housing-to-bracket bolts and the bracket
- Blower housing-to-chassis bolt, screw, nut and the blower housing

9. Remove or disconnect the following:
- 2 center air duct-to-heater/air conditioning housing screws and the air duct
- Move the floor carpet rearward
- Wiring harness clamps
- 2 air duct bolts and the ducts at both sides

10. Remove the heater/air conditioning housing by removing or disconnecting the following:
- Electrical connector
- Wiring harness set nut
- Wiring harness clamp
- Heater/air conditioning housing 2 nuts and bolt
- Heater/air conditioning housing

11. Remove the heater core-to-heater/air conditioning housing clamp screw and the clamp.

12. Pull the heater core from the heater/air conditioning housing.

To install:

13. Install the heater core to the heater/air conditioning housing.

14. Install the heater core clamp and the clamp-to-heater/air conditioning housing screw.

15. Install the heater/air conditioning housing by installing or connecting the following:
- Heater/air conditioning housing
- Heater/air conditioning housing 2 nuts and bolt
- Wiring harness clamp
- Wiring harness set nut
- Electrical connector

16. Install or connect the following:
- Air duct and the 2 duct bolts on both sides
- Wiring harness clamps
- Move the floor carpet forward
- Center air duct and the 2 air duct-to-heater/air conditioning housing screws

17. Install the blower unit by installing or connecting the following:
- Blower housing and the blower housing-to-chassis bolt, screw and nut
- Blower housing and the 2 bracket-to-bracket bolts
- Blower housing connectors
- Electrical connector bracket, the wiring harness clamps and the wiring harness
- Air duct and the 3 air duct-to-blower housing screws
- Connector clamp

18. Install the instrument by installing or connecting the following:
- Ventilation nozzles to the heater/air conditioning housing
- Reinforcement and the reinforcement-to-chassis 5 nuts and 4 bolts
- No. 1 and No. 2 brace
- Instrument panel and the 5 instrument panel-to-chassis bolts, the nut and the screw

19. Install or connect the following:
- No. 7 heater-to-register
- No. 1 console box duct
- Console box
- No. 3 console box mounting bracket
- Rear console armrest
- Lower rear console box
- Console box carpet
- Connectors and install the cluster finish panel
- Audio unit and the audio unit-to-instrument panel 2 bolts and 2 screws
- Connector and install the No. 2 register
- Ashtray
- CD changer, connect the electrical connectors and install the 3 CD changer-to-instrument panel nuts
- Glove box and the glove box-to-instrument panel 2 bolts and 3 screws
- Air bag electrical connector and install the plate inside the glove box
- No. 2 undercover
- Instrument cluster, connect the electrical connectors and install the 4 instrument cluster-to-instrument panel screws
- Instrument cluster finish panel
- Steering column, connect the spring to the brake pedal and install the 4 steering column-to-instrument panel nuts
- Parking brake handle and the No. 1 switch hole base

- No. 1 safety pad and the 4 safety pad-to-instrument panel bolts and screw
- Hood lock release and the 2 release screws
- No. 1 undercover and the 2 under-cover-to-instrument panel screws
- End pad
- Combination switch and connect the electrical connectors
- Steering column covers and the cover screws
- Front pillar garnishes and the front door scuff plates

20. Install the steering wheel by installing or connecting the following:
- Steering wheel and torque the steering wheel nut to 26 ft. lbs. (35 Nm)
- Electrical connector and install the air bag
- Tighten the Torx® bolts to 80 inch lbs. (9.0 Nm) using a Torx® wrench
- Torx® bolt covers at both sides of the steering wheel
- No. 1 grommet, the heater pipe grommet and the drain hose grommet
- Heater hoses to the heater core
- Refrigerant lines (using new O-rings) and the refrigerant lines-to-evaporator bolt

21. Refill the cooling system.

22. Connect the negative battery cable.

23. Evacuate, charge and leak test the air conditioning system refrigerant.

24. Operate the engine to normal operating temperatures; then, check the climate control operation and check for leaks.

LS 400 and LS 430

1. Before servicing the vehicle, refer to the precautions in the beginning of this section.

2. Disconnect the negative battery cable. Wait 90 seconds before doing any further work while the airbag system de-energizes.

3. Disconnect the negative battery cable.

4. Drain the cooling system into a clean container for reuse.

5. Remove or disconnect the following:
- Undercover and the No. 1 safety pad-to-instrument panel screws and the panel at the driver's side
- No. 2 heater-to-register duct
- Heater core-to-heater housing screw and clamp
- Heater hoses from the heater core
- Heater core from the heater housing
- Discard the O-rings

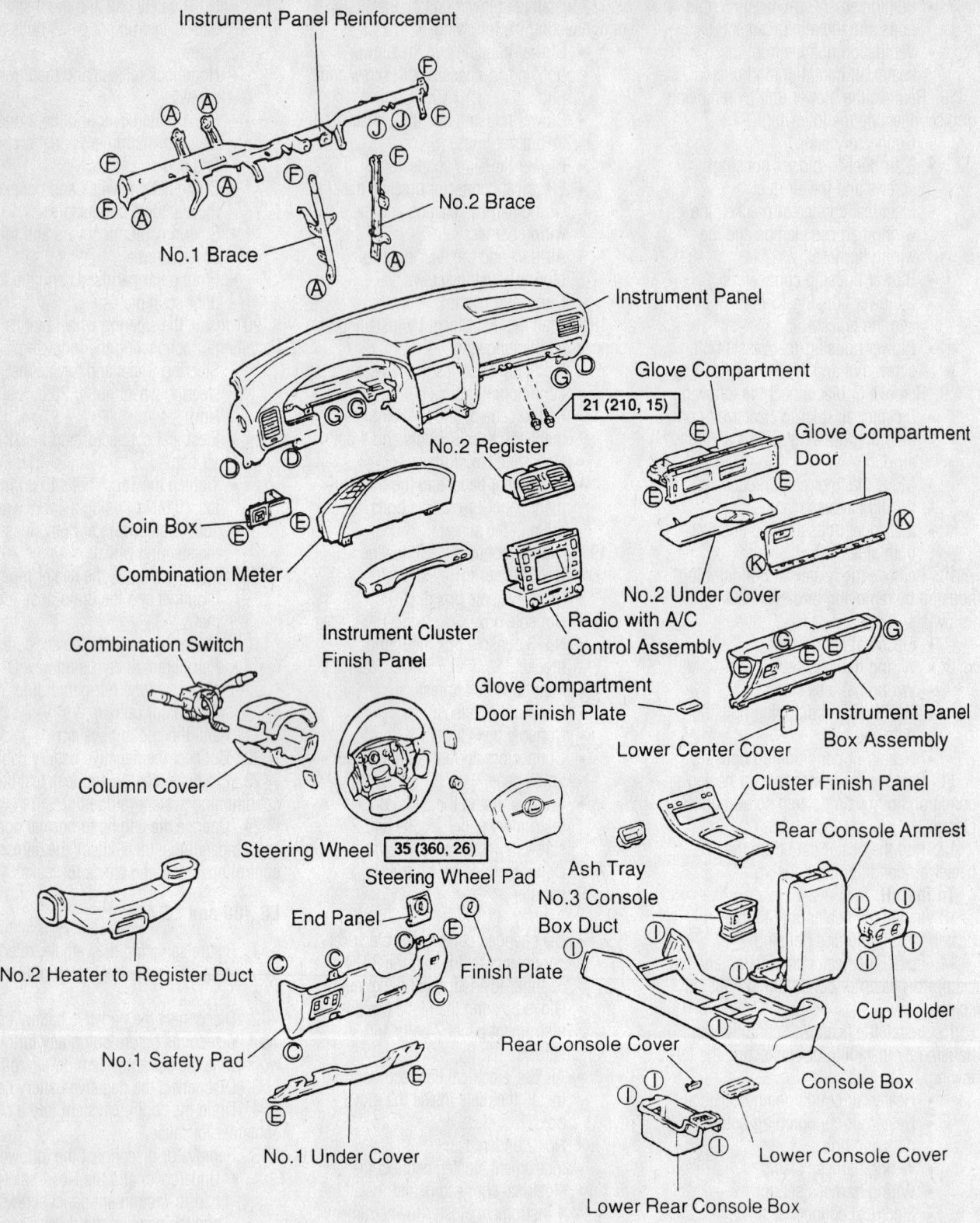

Instrument Panel Reinforcement

No.2 Brace

No.1 Brace

Instrument Panel

Glove Compartment

21 (210, 15)

Glove Compartment Door

No.2 Register

Coin Box

Combination Meter

No.2 Under Cover

Radio with A/C Control Assembly

Instrument Cluster Finish Panel

Glove Compartment Door Finish Plate

Lower Center Cover

Instrument Panel Box Assembly

Combination Switch

Column Cover

Cluster Finish Panel

Rear Console Armrest

Steering Wheel   35 (360, 26)

Steering Wheel Pad

End Panel

Finish Plate

Ash Tray

No.3 Console Box Duct

Cup Holder

No.2 Heater to Register Duct

Console Box

No.1 Safety Pad

Rear Console Cover

Lower Console Cover

No.1 Under Cover

Lower Rear Console Box

N·m (kgf·cm, ft·lbf) : Specified torque

93112GS1

Exploded view of the instrument panel—LS 400

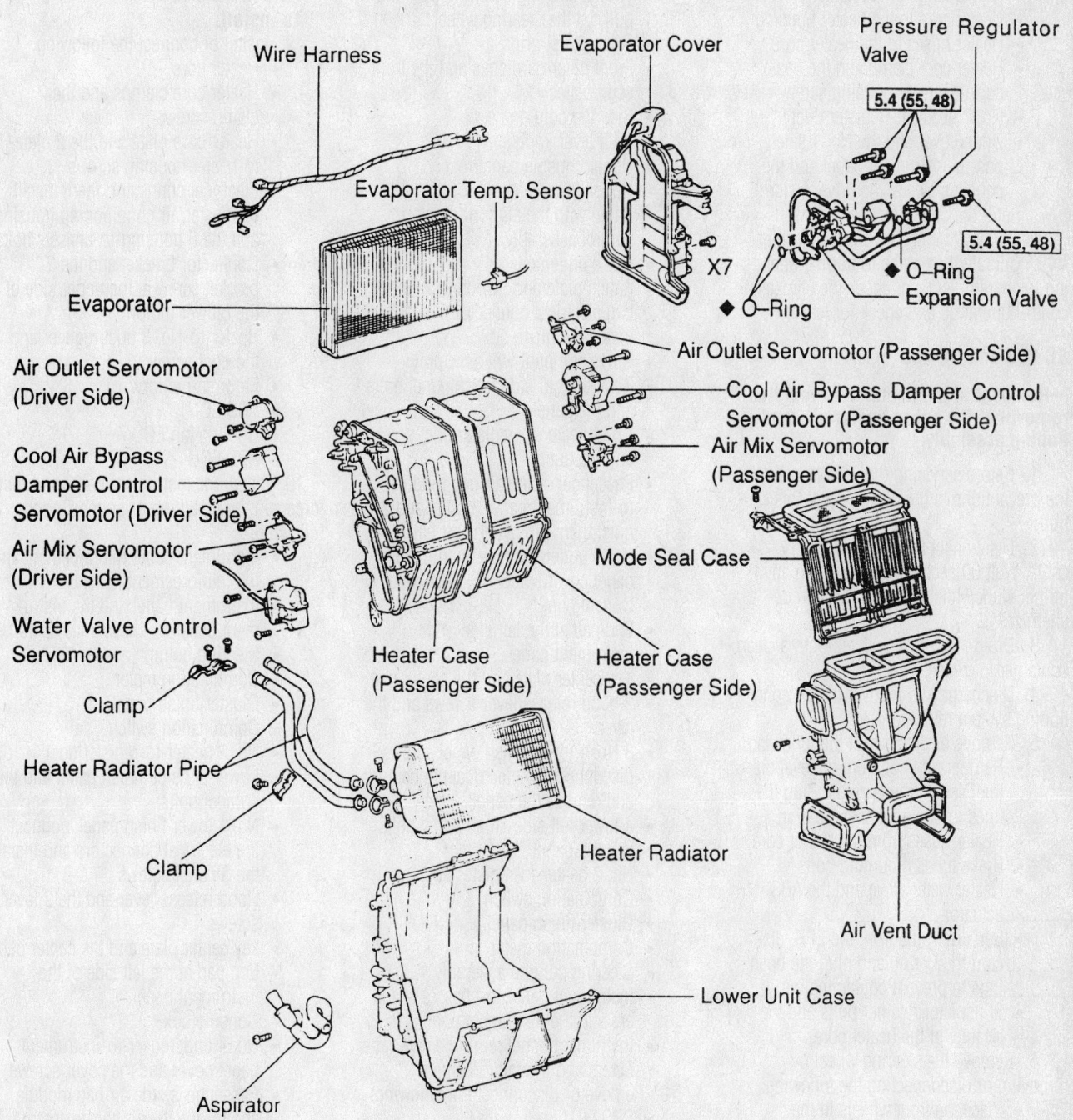

Wire Harness

Evaporator Cover

Pressure Regulator Valve

5.4 (55, 48)

Evaporator Temp. Sensor

Evaporator

Air Outlet Servomotor (Driver Side)

Cool Air Bypass Damper Control Servomotor (Driver Side)

Air Mix Servomotor (Driver Side)

Water Valve Control Servomotor

Clamp

Heater Radiator Pipe

Clamp

Aspirator

X7

◆ O–Ring

5.4 (55, 48)

◆ O–Ring

Expansion Valve

Air Outlet Servomotor (Passenger Side)

Cool Air Bypass Damper Control Servomotor (Passenger Side)

Air Mix Servomotor (Passenger Side)

Mode Seal Case

Heater Case (Passenger Side)

Heater Case (Passenger Side)

Heater Radiator

Air Vent Duct

Lower Unit Case

N·m (kgf·cm, in.·lbf) : Specified torque

◆ Non–reusable part

93112GS2

Exploded view of the heater core, heater/air conditioning housing and related components—LS 400

To install:

6. Install or connect the following:
   - New O-rings to the heater core
   - Heater core to the heater housing
   - Heater hoses to the heater core
   - Heater core clamp and the heater core-to-heater housing screw
   - No. 2 heater-to-register duct
   - Undercover and the No. 1 safety pad-to-instrument panel and the panel screws at the driver's side

7. Refill the cooling system.

8. Connect the negative battery cable.

9. Operate the engine to normal operating temperatures; then, check the climate control operation and check for leaks.

SC 300 and SC 400

➡ **Removal of the heater core requires removal of the entire heater air conditioning assembly.**

1. Before servicing the vehicle, refer to the precautions in the beginning of this section.

2. Disconnect the negative battery cable. Wait 90 seconds before doing any further work while the airbag system de-energizes.

3. Drain the cooling system into a clean container for reuse.

4. Discharge and recover the air conditioning system refrigerant.

5. Remove or disconnect the following:
   - Refrigerant lines from the evaporator. Discard the O-rings. Plug the lines to prevent contamination
   - Heater hoses from the heater core
   - Brake tubes mounting bolts
   - Heater water valve and the ABS actuator
   - Equalizer tube from the EPR, discard the O-ring and plug the openings to prevent contamination
   - 2 insulator retainer bolts and the retainer at the heater core

6. Remove the steering wheel by removing or disconnecting the following:
   - Place the front wheels in the straight-ahead position
   - Torx® bolt covers at both sides of the steering wheel
   - Loosen the Torx® bolts (using a Torx® wrench) until the circumference ring on the bolt catches on the screw case
   - Lift the air bag, disconnect the electrical connector and remove it
   - Steering wheel nut and press the steering wheel from the steering column

7. Remove the instrument panel and reinforcement by removing or disconnecting the following:
   - Tilt the steering column down and pull out the steering wheel
   - Front assist grips
   - Front pillar garnishes and the front scuff plates
   - Steering column cover
   - Shift lever knob
   - Upper console panel rear
   - Upper console panel
   - Radio with the air conditioning control assembly
   - No. 2 undercover
   - Finish plate and disconnect the air bag electrical connector located inside the glove box
   - Glove compartment assembly
   - 4 lower right side finish panel bolts and the panel
   - No. 4 heater-to-register duct screw and the duct
   - Passenger's side air bag module-to-instrument panel 3 bolts and 2 nuts; then, remove the air bag
   - No. 4 undercover-to-instrument panel cover screws and the cover
   - Console box
   - End pad at the left side of the instrument panel
   - Key center plate and the center pad
   - 2 hood release lever screws and the lever
   - 3 No. 1 lower finish panel screws, disconnect the electrical connectors and remove the panel
   - 4 lower left side finish panel bolts and the panel
   - No. 2 heater-to-register duct
   - Combination switch
   - Cluster finish panel
   - Combination meter
   - Steering column assembly
   - Instrument panel-to-chassis fasteners and the instrument panel
   - Instrument panel reinforcement fasteners and the reinforcement

8. Remove or disconnect the following:
   - PPS ECU
   - Cooling fan ECU
   - ABS ECU
   - Move the floor carpet rearward
   - Heater-to-No. 3 duct register screw and the duct
   - 2 connector bracket screws and the bracket at the under side of the blower motor
   - Electrical connector; then, remove the 6 heater/air conditioning housing-to-chassis bolts and the housing
   - 2 heater core plate-to-heater housing screws and the plate
   - 2 heater core clamp screws and the clamps
   - Heater core

To install:

9. Install or connect the following:
   - Heater core
   - Heater core clamps and the 2 clamp screws
   - Heater core plate and the 2 plate-to-heater housing screws
   - Electrical connector; then, install the heater/air conditioning housing and the 6 housing-to-chassis bolts
   - Connector bracket and the 2 bracket screw at the under side of the blower motor
   - Heater-to-No. 3 duct register and the duct screw
   - Floor carpet forward
   - ABS ECU
   - Cooling fan ECU
   - PPS ECU

10. Install the instrument panel and reinforcement by installing or connecting the following:
    - Instrument panel reinforcement and the reinforcement fasteners
    - Instrument panel and the instrument panel-to-chassis fasteners
    - Steering column assembly
    - Combination meter
    - Cluster finish panel
    - Combination switch
    - No. 2 heater-to-register duct
    - Lower left side finish panel and the 4 panel bolts
    - No. 1 lower finish panel, connect the electrical connectors and install the 3 panel screws
    - Hood release lever and the 2 lever screws
    - key center plate and the center pad
    - End pad at the left side of the instrument panel
    - Console box
    - No. 4 undercover-to-instrument panel cover and the cover screws
    - Passenger's side air bag module and torque the air bag-to-instrument panel 3 bolts and 2 nuts to 15 ft. lbs. (21 Nm)
    - No. 4 heater-to-register duct and the duct screw
    - Lower right side finish panel and the 4 panel bolts
    - Glove compartment assemble
    - Air bag electrical connector and install the finish plate inside the glove box
    - No. 2 undercover
    - Radio with the air conditioning control assembly

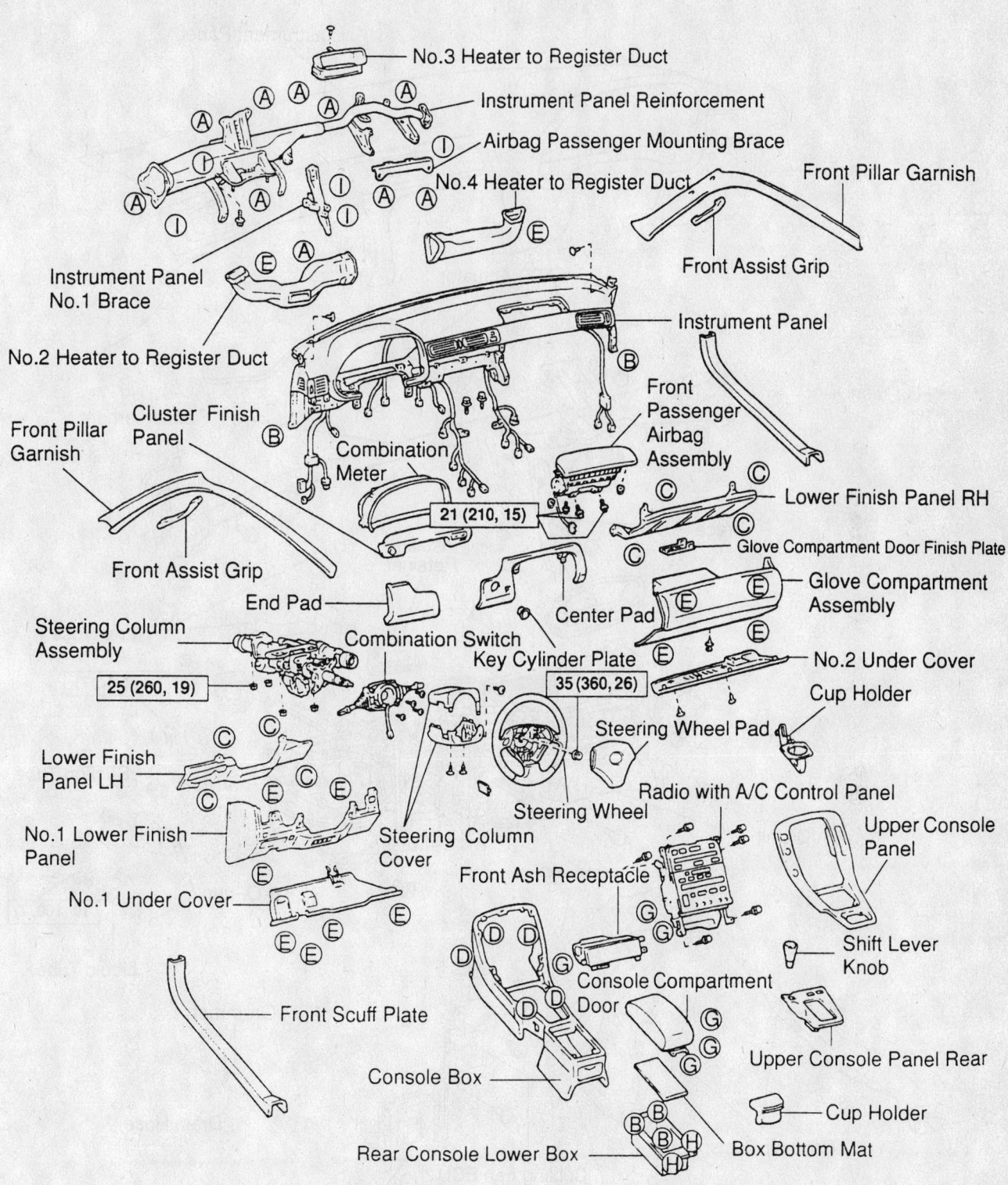

No.3 Heater to Register Duct

Instrument Panel Reinforcement

Airbag Passenger Mounting Brace

No.4 Heater to Register Duct

Front Pillar Garnish

Front Assist Grip

Instrument Panel Panel
No.1 Brace

No.2 Heater to Register Duct

Instrument Panel

Front Pillar Garnish

Cluster Finish Panel

Combination Meter

Front Passenger Airbag Assembly

Lower Finish Panel RH

21 (210, 15)

Glove Compartment Door Finish Plate

Front Assist Grip

End Pad

Center Pad

Glove Compartment Assembly

Steering Column Assembly

Combination Switch

Key Cylinder Plate

No.2 Under Cover

25 (260, 19)

35 (360, 26)

Cup Holder

Steering Wheel Pad

Lower Finish Panel LH

Steering Column Cover

Steering Wheel

Radio with A/C Control Panel

Upper Console Panel

No.1 Lower Finish Panel

Front Ash Receptacle

Shift Lever Knob

No.1 Under Cover

Console Compartment Door

Front Scuff Plate

Upper Console Panel Rear

Console Box

Cup Holder

Rear Console Lower Box

Box Bottom Mat

T N·m (kgf·cm, ft·lbf) : Specified torque

93112GS5

Exploded view of the instrument panel and related components—SC 300 and SC 400

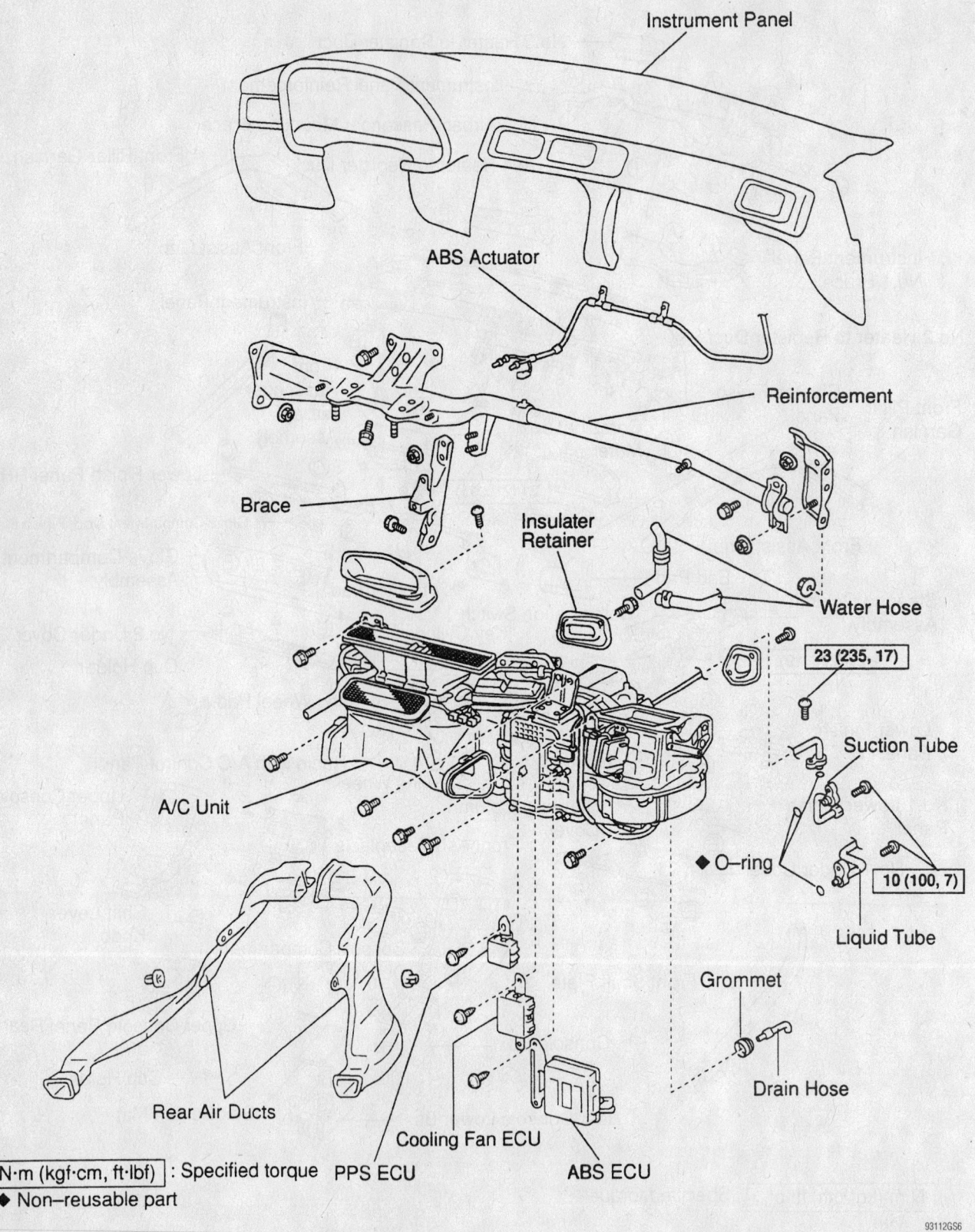

Instrument Panel

ABS Actuator

Reinforcement

Brace

Insulater Retainer

Water Hose

23 (235, 17)

Suction Tube

A/C Unit

◆ O-ring

10 (100, 7)

Liquid Tube

Grommet

Rear Air Ducts

Drain Hose

Cooling Fan ECU

PPS ECU

ABS ECU

N·m (kgf·cm, ft·lbf) : Specified torque
◆ Non–reusable part

93112GS6

Exploded view of the heater/air conditioning housing, reinforcement, ventilation ducts and related components—SC 300 and SC 400

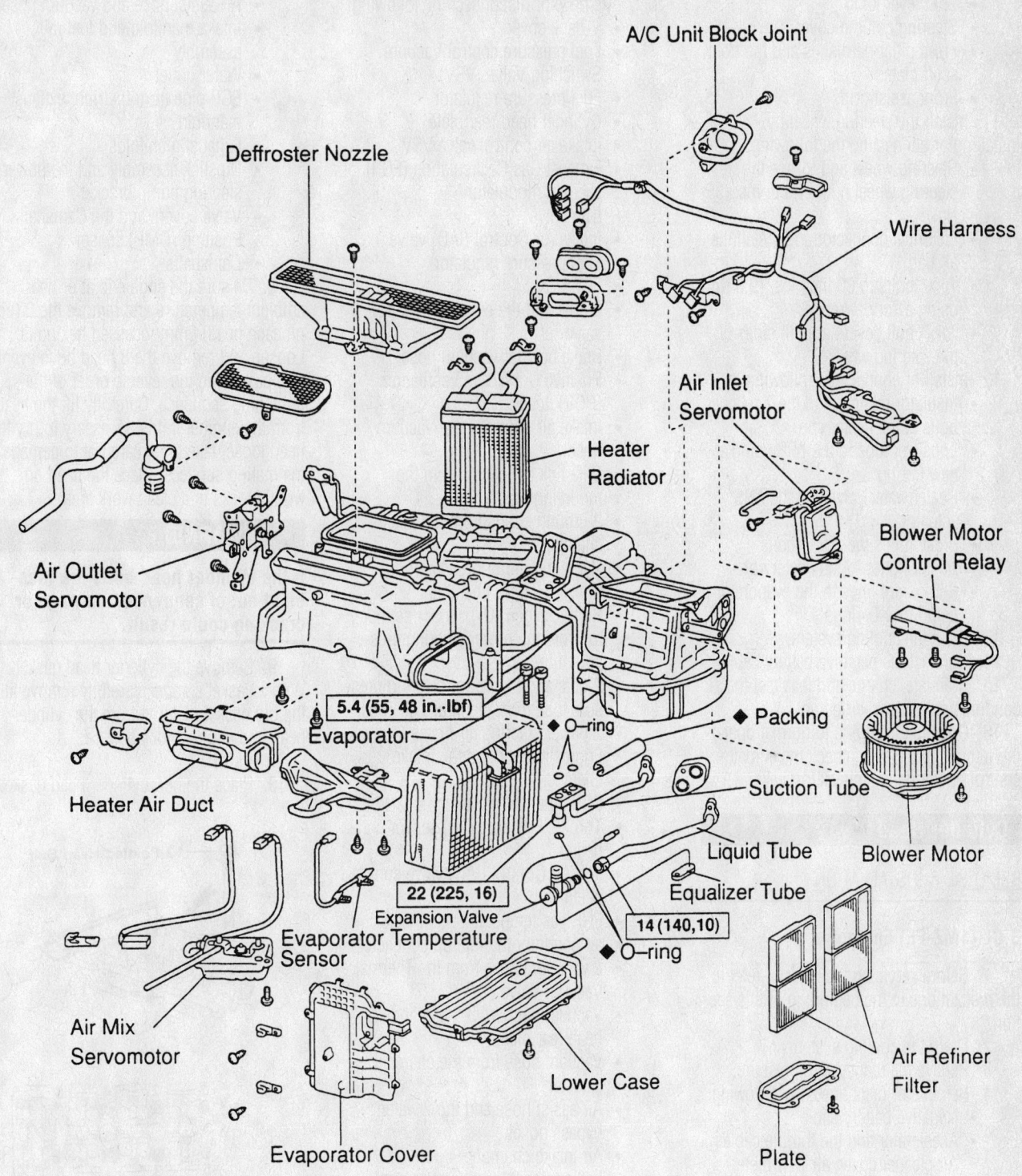

A/C Unit Block Joint

Deffroster Nozzle

Wire Harness

Air Inlet Servomotor

Heater Radiator

Blower Motor Control Relay

Air Outlet Servomotor

5.4 (55, 48 in.·lbf)

Evaporator

◆ Packing

◆ O–ring

Suction Tube

Blower Motor

Heater Air Duct

Liquid Tube

Equalizer Tube

22 (225, 16)

14 (140,10)

Expansion Valve

◆ O–ring

Evaporator Temperature Sensor

Air Mix Servomotor

Lower Case

Air Refiner Filter

Evaporator Cover

Plate

N·m (kgf·cm, ft·lbf) : Specified torque

◆ Non–reusable part

93112GS7

Exploded view of the heater core, heater/air conditioning housing and related components—SC 300 and SC 400

- Upper console panel
- Upper console panel rear
- Shift lever knob
- Steering column cover
- Front pillar garnishes and the front scuff plates
- Front assist grips

11. Install the steering wheel by installing or connecting the following:
- Steering wheel and torque the steering wheel nut to 26 ft. lbs. (35 Nm)
- Electrical connector and install the air bag
- Torx® bolts to 80 inch lbs. (9.0 Nm) using a Torx® wrench
- Torx® bolt covers at both sides of the steering wheel

12. Install or connect the following:
- Insulator retainer and the 2 retainer bolts at the heater core
- Equalizer tube to the EPR using a new O-ring
- Heater water valve and the ABS actuator
- Brake tubes mounting bolts
- Heater hoses to the heater core
- Refrigerant lines to the evaporator using new O-rings

13. Refill the cooling system.
14. Connect the negative battery cable.
15. Evacuate, charge and leak test the air conditioning system refrigerant
16. Operate the engine to normal operating temperatures; then, check the climate control operation and check for leaks.

Cylinder Head

REMOVAL & INSTALLATION

3.0L (1MZ-FE) Engine

1. Before servicing the vehicle, refer to the precautions in the beginning of this section.
2. Drain the cooling system.
3. Relieve the fuel system pressure.
4. Remove or disconnect the following:
- Negative battery cable
- Accelerator and the throttle cables
- Air cleaner cover, air flow meter and the air duct
- Cruise control actuator and bracket, if equipped
- 2 engine ground straps
- Right engine mounting support
- Radiator hoses
- 2 heater hoses

5. Plug the fuel feed and return lines from the fuel rail assembly.

6. Plug the pressure hose from the hydraulic motor
7. Remove or disconnect the following:
- V-bank cover
- Fuel pressure control Vacuum Switching Valve (VSV)
- Fuel pressure regulator
- Cylinder head rear plate
- Intake air control valve VSV
- Exhaust Gas Recirculation (EGR) vacuum modulator
- EGR valve
- Intake Air Control (IAC) valve
- Fuel pressure regulator
- EGR VSV
- 2 nuts and the emission control valve set
- Brake booster vacuum hose
- Positive Crankcase Ventilation (PCV) hose
- Intake air control valve vacuum hose
- Data link connector from the mounting bracket
- 2 ground straps from the intake chamber
- Hydraulic motor pressure hose from the intake chamber
- Right Oxygen (O₂) sensor connector from the power steering pressure tube
- 2 nuts and the power steering pressure tube from the intake chamber
- 2 power steering air hoses
- Engine hanger and the intake chamber support
- EGR pipe and gaskets
- Throttle Pressure (TP) sensor connector
- Idle Air Control (IAC) valve connector
- EGR gas temperature connector
- air conditioning idle up connector
- 2 vacuum hoses from the Thermal Vacuum Valve (TVV)
- Vacuum hose from the cylinder head rear plate
- Vacuum hose from the charcoal canister
- Air assist hose and the 2 water bypass hoses
- Air intake chamber
- Left engine wiring harness and position it out of the way
- Wiring harness from the rear of the engine
- Right engine wiring harness and position it out of the way
- Ignition coils and the spark plugs
- Timing belt
- Camshaft pulleys and the timing belt rear cover

- Cylinder head rear plate
- Water inlet pipe
- Air assist hose and vacuum hose
- Intake manifold and fuel rail assembly
- Water outlet
- EGR pipe from the right exhaust manifold
- Exhaust manifolds
- Dipstick assembly and the power steering pump bracket
- Valve covers and the Camshaft Position (CMP) sensor
- Camshafts

8. Be sure the engine is at or near ambient temperature and remove the 2 (one on each head) 8mm recessed hex bolts. Loosen and remove the 8 head bolts evenly, in 3 passes, in the reverse order of the tightening sequence. Carefully lift the head from the engine; if it is necessary to pry the head loose, take great care not to damage the mating surfaces. Place the head on wood blocks in a clean work area.

✳✳ WARNING

If the cylinder head bolts are loosened out of sequence, warpage or cracking could result.

9. Remove the cylinder head gasket. With a gasket scraper, carefully remove all the old gasket material from the cylinder head and engine block surfaces.

To install:
10. Place the new cylinder head gasket

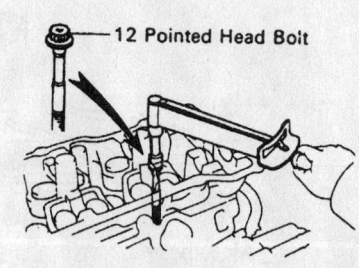

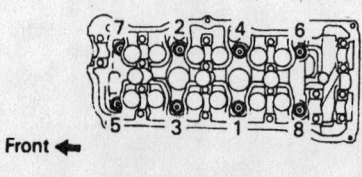

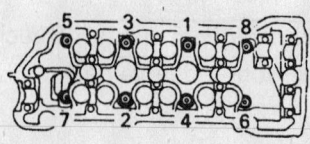

7923LG13

Cylinder head torque sequence—3.0L (1MZ-FE) engine

onto the cylinder block. Place the cylinder head onto the gasket.

11. Coat the threads of the 8 cylinder head bolts (12-sided) with clean engine oil and install the bolts into the cylinder head. Uniformly tighten the bolts in sequence in 3 steps to an ultimate tighten of 40 ft. lbs. (54 Nm). If any of the bolts does not meet the torque, replace it.

12. Mark the forward edge of each bolt with paint, then retighten each bolt, in proper sequence, an additional 90 degrees. Check that each painted mark is now at a 90 degrees angle to the front. The paint mark should have been applied to the bolt in the 9 o'clock position and should now be in the 12 o'clock position.

13. Coat the threads of the 2 remaining 8mm bolts with engine oil and install them. Tighten to 13 ft. lbs. (18 Nm).

14. Install or connect the following:
  • Camshafts and adjust the valves

➡**Apply sealant to the cylinder heads where the camshaft supports meet the cylinder heads.**

  • Cylinder head covers. Use new gaskets.
  • Dipstick and power steering pump bracket
  • Exhaust manifolds. Tighten the nuts to 36 ft. lbs. (49 Nm).
  • EGR pipe to the right exhaust manifold
  • Water outlet
  • Intake manifold and the fuel rail assembly. Tighten the intake manifold nuts and bolts to 11 ft. lbs. (15 Nm).
  • Air assist hose and the 2 water bypass hoses
  • Water inlet pipe and the cylinder head rear plate
  • Timing belt rear cover and the camshaft pulleys
  • Timing belt
  • Spark plugs and the ignition coils
  • Right engine wiring harness
  • Wiring harness to the rear of the engine
  • Left engine wiring harness
  • Air intake chamber
  • EGR pipe. Use new gaskets
  • 2 TVV vacuum hoses
  • Vacuum hose to the rear cylinder head plate
  • Charcoal canister vacuum hose
  • TP sensor connector
  • IAC valve connector
  • EGR gas temperature connector
  • Air conditioning idle up connector

  • Engine hanger and the intake chamber support
  • 2 power steering air hoses
  • Power steering pressure tube to the intake chamber
  • O_2 sensor connector to the pressure tube
  • 2 ground straps to the intake chamber
  • Data link connector to the bracket
  • Power brake booster vacuum hose
  • PCV hose
  • IAC valve vacuum hose
  • Emission control valve set and related vacuum hoses and connectors
  • V-bank cover
  • Pressure hose to the hydraulic motor
  • Fuel lines to the fuel rail assembly
  • Heater and radiator hoses
  • Right engine mounting support
  • 2 engine ground straps
  • Cruise control actuator and bracket
  • Air cleaner, air flow meter and air duct assembly
  • Accelerator and the throttle cables
  • Negative battery cable

15. Fill the cooling system to the proper level with coolant.

16. Start the engine and check for leaks. Bleed the air from the cooling system.

17. Adjust the ignition timing.

18. Road test the vehicle and check for unusual noise, shock, slippage, correct shift points and smooth operation.

19. Recheck the coolant and engine oil levels.

3.0L (2JZ-GE) Engine

1. Before servicing the vehicle, refer to the precautions in the beginning of this section.

2. Disconnect the negative battery cable. Wait at least 90 seconds before performing any other work.

3. Relieve the fuel pressure from the fuel lines.

4. Remove or disconnect the following:
  • Coolant
  • Undercovers
  • Accelerator, throttle control (automatic transmission only) and cruise control cables from the throttle body
  • Cleaner duct
  • Air cleaner, airflow meter and the intake air pipe
  • Drive belt, the fan and fluid coupling and the water pump pulley
  • No. 2 front exhaust pipe

  • Exhaust manifold cover
  • 2 Heated Oxygen (HO_2) sensor connector(s)
  • Exhaust manifolds and gaskets by removing the 8 bolts
  • Water bypass outlet and the No. 1 water bypass pipe
  • Power steering air hose from the No. 4 timing belt cover
  • Power steering hose from the air intake chamber
  • 2 bolts and the vane pump from the pump bracket
  • 2 bolts, the pump rear stay. Put aside the vane pump and suspend it.
  • Fuel return hose from the fuel return pipe. Plug the hose end.
  • Fuel return hose from the oil dipstick guide
  • Bolt and bracket
  • Engine wire from the intake manifold stay
  • Throttle body and intake air connector assembly
  • Bolt, pull out the oil dipstick guide with the dipstick and remove the O-ring from the dipstick guide
  • Transmission dipstick and guide, if equipped with automatic transmission
  • Connector from the No. 2 vacuum pipe
  • Exhaust Gas Recirculation (EGR) gas temperature sensor wiring harness
  • 2 nuts and the vacuum pipe from the air intake chamber and intake manifold
  • No. 2 vacuum pipe and Vacuum Switching Valve (VSV) assembly
  • Nuts and the vacuum tank from the intake manifold
  • VSV connector and hoses
  • Vacuum hose (from the air intake chamber) from port B of the vacuum tank
  • Vacuum hose (from actuator) from the VSV
  • Vacuum control valve set
  • Data Link Connector (DLC1) bracket and VSV assembly
  • Vacuum hose from the brake booster union and the Evaporative Emission (EVAP) hose from the No. 2 vacuum pipe
  • Bolt holding the engine wire protector to the air intake chamber
  • 5 bolts, nut, air intake chamber and gasket
  • No. 3 (top) timing belt cover by removing the oil filler cap and the 6

bolts using a 5mm hexagon wrench.
- 4 bolts, using a 5mm hexagon wrench, and the rear cylinder head cover
- Spark plugs
- Drive belt tensioner by removing the 3 bolts

5. Set the engine to Top Dead Center (TDC)/compression for cylinder No. 1 piston

6. Remove or disconnect the following:
- Timing belt tensioner and dust boot. Remove the timing belt from the camshaft pulleys. Support the belt so that it remains in contact with the crankshaft pulley.
- Wire clamp from the bracket.
- HO_2 and the Crankshaft Position (CKP) sensors.
- 2 ground straps from the intake manifold.
- Engine Coolant Temperature (ECT) sender gauge
- Knock Sensor (KS)
- Oil pressure switch
- Oil level sensor
- air conditioning compressor
- 6 injector electrical connectors
- 3 nuts and the engine wire protector from the intake manifold
- Water bypass hose from the clamp on the oil filter bracket
- Water outlet, 2 nuts, and the bolt with the water bypass hose
- 2 bolts and the intake manifold stay
- Fuel pressure pulsation damper
- Clamp bolt from the intake manifold
- Union bolt and gaskets
- Fuel inlet pipe
- 6 bolts, 2 nuts, the intake manifold
- Delivery pipe assembly and gasket
- Cylinder head covers (valve covers)
- Camshaft timing pulleys
- Rear (No. 4) timing belt cover
- Camshafts
- Cylinder head bolts in several passes and in the reverse order of the tightening sequence
- Head from the engine

7. Clean the head and block of all gasket material.

To install:

8. Install or connect the following:
- New gasket and cylinder head on the block
- Head bolts, lightly coated with engine oil and plate washers. Uniformly tighten the head bolts in several passes, in sequence to 26 ft. lbs. (35 Nm). Following the cor-

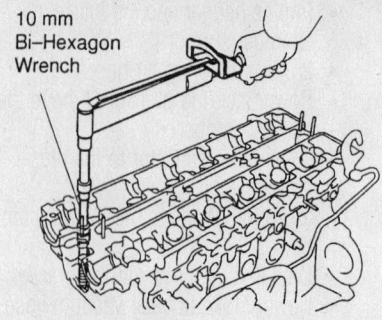

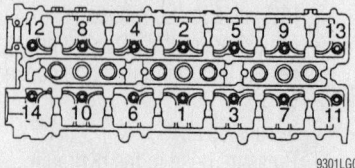

Cylinder head torque sequence—3.0L (2JZ-GE) engine

rect order, tighten each bolt an additional 90 degrees. Again following the correct order, tighten the bolts another 90 degrees of rotation.

> ❋❋ **WARNING**
>
> **Correct bolt torque must be achieved in 3 steps; do not attempt to shorten the procedure by combining the two 90 degree steps.**

- Camshafts. Coat the thrust portions of each with engine oil.
- No. 3 and No. 7 bearing caps in place. Coat the bolt threads with oil, then uniformly and alternately tighten them temporarily.
- New oil seals, coated with multipurpose grease, over the camshafts
- Seal packing to the No. 1 bearing cap
- Remaining bearing caps in their proper locations. Coat the threads of each bolt with clean oil, then

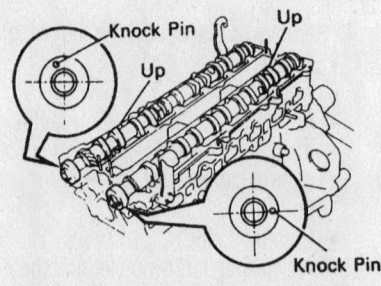

Position the knock pins as shown when installing the camshafts—3.0L (2JZ-GE) engine

Apply sealant to the areas indicated on the cylinder head before installing the cover— 3.0L (2JZ-GE) engine

tighten them, in several passes, in the correct sequence, to 14 ft. lbs. (20 Nm).
- The 2 oil seals in as far as it will go

9. Rotate each camshaft until the forward straight (knock) pin is straight up. Loosen the exhaust Nos. 1, 2 and 6 bearing cap bolts until they can be turned by hand; retighten the bolts, in several passes, to 14 ft. lbs. (20 Nm). Loosen the intake Nos. 1, 2 and 5 bearing cap bolts and retighten the bolts, in several passes, to 14 ft. lbs. (20 Nm).

10. Turn each camshaft ⅓ of a revolution (120 degrees). Loosen the exhaust Nos. 4 and 7 bearing cap bolts; retighten the bolts, in several passes, to 14 ft. lbs. (20 Nm). Loosen the intake Nos. 4 and 6 bearing cap bolts; retighten the bolts, in several passes, to 14 ft. lbs. (20 Nm).

11. Turn each camshaft an additional ⅓ of a revolution, loosen the exhaust bearing cap bolts Nos. 3 and 5, then retighten the bolts, in several passes, to 14 ft. lbs. (20 Nm). Loosen the intake bearing cap bolts Nos. 3 and 7, then retighten the bolts, in several passes, to 14 ft. lbs. (20 Nm).

12. Check and adjust the valve clearance.

13. Install or connect the following:
- Rear (No. 4) timing belt cover. Tighten the bolts to 78 inch lbs. (9 Nm).
- Camshaft timing pulleys. Align the shaft pin with the pulley groove and slide the pulley on. Install the bolt temporarily. Hold the hex portion of the camshaft with a wrench and tighten the pulley bolt to 59 ft. lbs. (79 Nm).
- Cylinder head covers
- Intake manifold and delivery pipe with a new gasket. Tighten the 6 bolts and 2 nuts to 20 ft. lbs. (27 Nm).
- Fuel inlet pipe to the fuel rail. Tighten the union bolt to 30 ft. lbs. (42 Nm).

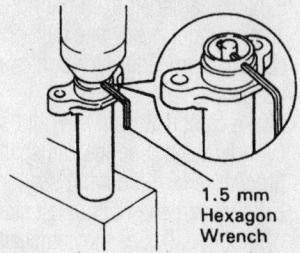

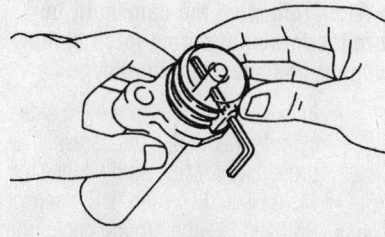

1.5 mm
Hexagon
Wrench

7923LG16

**Compressing the timing belt tensioner—
3.0L (2JZ-GE) engine**

- Clamp bolt to the intake manifold
- Fuel pressure pulsation damper
- Intake manifold stay and tighten the bolts to 29 ft. lbs. (39 Nm)
- Water outlet and the bypass hose. Tighten the bolts to 15 ft. lbs. (21 Nm).
- Engine wiring harness. Secure the wiring in all clamps and retainers.
- Wiring leads to the proper sender, sensor or switch
- Injector leads

14. Compress the timing belt tensioner in a vise and retain the pin with a 1.5mm hex wrench. Install the dust boot onto the tensioner.

15. Install the tensioner. Alternately tighten the bolts to 20 ft. lbs. (26 Nm). Remove the hex wrench with a pair of pliers, allowing the tensioner to be applied to the timing belt.

16. Turn the crankshaft 2 full revolutions clockwise. Check that all timing marks align as before. If the marks (cam and crankshaft) do not align, remove the timing belt and reinstall it.

17. Install the accessory drive belt tensioner. Take great care not to drop the bolts inside the lower timing cover. Tighten the bolts to 15 ft. lbs. (21 Nm).

18. Double check that the engine is still set to TDC/compression for cylinder No. 1. Check the alignment of both the crank and camshaft timing marks. Install the timing belt.

19. Install or connect the following:
- Spark plugs
- Wiring to the spark plugs
- No. 3 timing belt cover
- Cylinder head rear cover

- Air intake chamber with a new gasket. Tighten the bolts to 20 ft. lbs. (27 Nm). Install the bolt to hold the engine wire protector to the air intake chamber.
- Vacuum hose to the brake booster union and the EVAP hose to the No. 2 vacuum pipe.
- DLC connector and bracket and VSV connector
- Vacuum control set
- No. 2 vacuum pipe assembly and connect the hoses. Tighten the nuts to 20 ft. lbs. (27 Nm).
- EGR gas temperature sensor. Tighten it to 14 ft. lbs. (20 Nm).
- Vacuum hoses
- Dipstick tubes. Always use a new O-ring on each tube.
- Intake chamber supports and tighten the bolts to 13 ft. lbs. (18 Nm). The supports are marked **F** and **R** for the front and rear positions.
- Throttle body and intake air connector assembly
- Engine wire bracket
- Fuel return hose
- Vane pump to the pump bracket
- Power steering air hose to the No. 4 timing belt cover and intake chamber.
- Water bypass outlet and the bypass pipe. Always use new O-rings.
- Exhaust manifolds with new gaskets. Tighten the bolts to 29 ft. lbs. (39 Nm).
- O_2 sensor leads
- Front exhaust pipe. Tighten the bolts to 46 ft. lbs. (62 Nm).
- Manifold cover
- Water pump pulley
- Fan and coupling and the drive belt. Tighten the 4 nuts to 12 ft. lbs. (16 Nm).
- Air cleaner, airflow meter and the intake air connector pipe
- Air cleaner duct
- Control and accelerator cables to the throttle body
- Coolant
- Negative battery cable. Start the engine and check for leaks.
- Engine undercovers

4.0L (1UZ-FE) and 4.3L (3UZ-FE) Engines

1. Before servicing the vehicle, refer to the precautions in the beginning of this section.

2. Relieve the fuel system pressure.

3. Remove or disconnect the following:
- Negative battery cable. Wait at least 90 seconds before performing any other work.
- Oil pan protector
- Engine undercover
- Coolant
- Battery clamp cover
- Air cleaner inlet
- V bank cover by removing the bolt and 2 cap nuts
- Air cleaner and intake air connector assembly
- Drive belt, fluid coupling and the fan pulley. The drive belt tension may be slackened by turning the tensioner counterclockwise. The pulley bolt for the drive belt tensioner has a left-handed thread.
- Radiator
- Right-hand No. 3 timing belt cover
- Left-hand No. 3 timing belt cover
- Drive belt idler pulley by removing the pulley bolt and cover plate
- Right-hand No. 2 timing belt cover
- Left-hand No. 2 timing belt cover
- No. 1 ignition coil
- Air conditioning compressor from the engine
- Fan bracket by removing the 2 bolts and 2 nuts

4. Set the engine to Top Dead Center (TDC) on cylinder No. 1.

✷✷ WARNING

Since the thrust clearance of the camshaft is small, the camshaft must be kept level while it is being removed. If the camshaft is not kept level, the portion of the cylinder head receiving the shaft thrust may crack or be damaged, causing the camshaft to seize or break.

5. Turn the crankshaft pulley approximately 50 degrees clockwise and put the timing mark of the crankshaft pulley in line with the centers of the crankshaft pulley bolt and the idler pulley bolt.

✷✷ WARNING

If the timing belt is disengaged, having the crankshaft pulley at the wrong angle can cause the piston head and valve head to come into contact with each other when you remove the camshaft timing pulley. Always set the crankshaft pulley at the correct angle before removing the timing belt.

6. If the timing belt is to be reused, turn the crank pulley slowly; check that the 3 installation marks are present on the belt. If the marks are not present, make new installation marks before removing the belt. The marks should align with the timing marks on each camshaft pulley and the crank pulley.

7. Remove the timing belt tensioner. Alternately loosen the 2 bolts; remove the bolts, the tensioner and the dust protector.

8. Loosen the tension between the left side and the right side timing pulleys by slightly turning the left side camshaft clockwise.

9. Remove or disconnect the following:
- Timing belt from the camshaft timing pulleys. Using the proper tool, remove the bolt and the camshaft timing pulleys.
- Power steering pump from the engine. Do not disconnect the hoses or lines from the power steering pump. Support the power steering pump with a piece of wire. Do not allow the pump to hang.
- Front catalytic converter
- High tension spark plug wires, wire clamps and the wire cover assembly
- No. 2 ignition coil by removing the connector and the 2 bolts
- 2 bolts and the rear timing belt plate. Remove both plates
- Intake chamber assembly
- Throttle Position Sensor (TPS) connector
- With TRAC system, sub TP sensor connector
- With TRAC system, sub TP connector
- Idle Air Control (IAC) valve connector
- Exhaust Gas Recirculation (EGR) valve connector
- Vacuum Switching Valve (VSV) connector for fuel pressure control
- VSV connector for Evaporative Emissions (EVAP) system
- EGR gas temperature sensor connector
- Brake booster vacuum hose from the union on the air intake chamber
- Positive Crankcase Ventilation (PCV) hose from the PCV valve on the left-hand cylinder head
- Water bypass hose (from the EGR valve) from the rear water bypass joint
- Water bypass hose (from the throttle body) from the rear water bypass joint

- Vacuum hose (from the VSV for fuel pressure control) from the fuel pressure regulator
- EVAP hose (from charcoal canister) from the VSV for EVAP
- Heater hose from the water bypass pipe
- Fuel inlet hose from the delivery pipe
- Fuel return hose from the fuel return pipe
- Engine wire from the delivery pipes and rear water bypass joint
- Fuel hose from the fuel pressure regulator
- 2 bolts and fuel return pipe from the intake manifold
- 8 injector connectors
- 6 bolts, 4 nuts, the intake manifold assembly and the 2 gaskets
- Water inlet and inlet housing
- Front water bypass joint
- Rear water bypass joint and No. 1 EGR pipe assembly
- Oil dipstick and guide for the automatic transmission
- Oil dipstick and guide for the engine
- Engine hangers
- Right and left cylinder head covers by removing the 8 bolts, seal washers and gaskets
- Semi-circular plug, if necessary

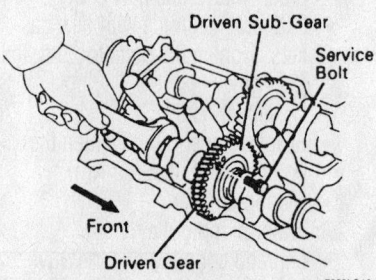

Securing the exhaust camshaft on the right cylinder head—4.0L (1UZ-FE) and 4.3L (3UZ-FE) Engines

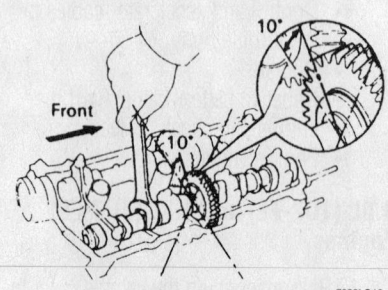

Turning the exhaust camshaft 10 degrees on the right cylinder head—4.0L (1UZ-FE) and 4.3L (3UZ-FE) Engines

- Exhaust camshaft from the right side cylinder head. See the camshaft procedure for tightening sequence.
- Intake camshaft from the right side cylinder head. See the camshaft procedure for tightening sequence.
- Exhaust camshaft of the left side cylinder head. See the camshaft procedure for tightening sequence.

➡When removing the camshaft, be sure the torsional spring force of the subgear has been eliminated.

- Intake camshaft from the left side cylinder head. See the camshaft procedure for tightening sequence.
- Main Heated Oxygen (HO$_2$) sensor
- Bolt and HO$_2$the ground cable from the right cylinder head
- Bolt and the ground strap from the left cylinder head
- Bolt and the engine wire protector from the left-hand cylinder head
- 2 bolts, seal washers, bearing cap and the camshaft housing plug from the right-hand cylinder head
- 10 cylinder head bolts and plate washers to each cylinder head. Loosen the bolts in the reverse order of the tightening sequence. Lift the heads from the dowels on the block with the exhaust manifolds attached. Place the heads on blocks of wood on the workbench.

➡Do not drop anything in the opening in the front of the right side cylinder head. The opening leads through the block and into the oil pan. If anything falls into the opening the oil pan will have to be removed in order to retrieve it.

✳✳ WARNING

If necessary to pry the head loose, take great care not to damage the contact surfaces of the head or block.

- 2 bolts, seal washers, bearing cap and camshaft housing plug from the right-hand cylinder head
- Right exhaust manifold from the cylinder head by removing the heat insulator, 8 nuts and the gasket
- Left exhaust manifold from the cylinder head by removing the heat insulator, 8 nuts and the gasket

To install:

10. Install or connect the following:
- Right exhaust manifold. The new gasket must be installed with the

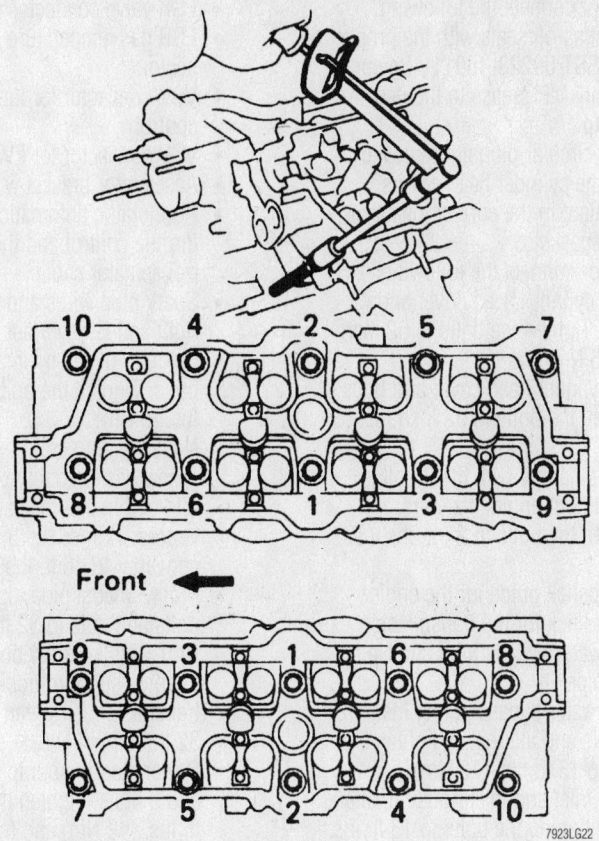

Cylinder head torque sequence—4.0L (1UZ-FE) engine

RH Bank

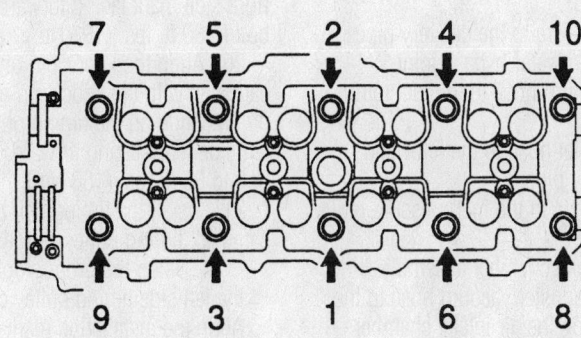

LH Bank

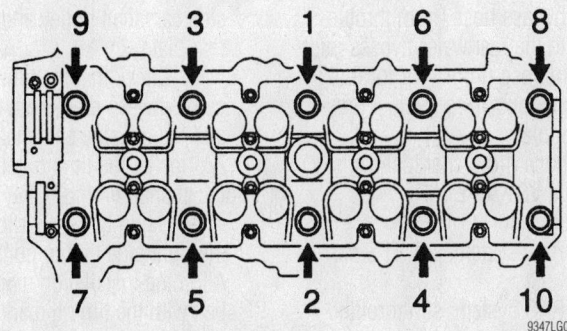

Cylinder head torque sequence—4.3L (3UZ-FE) engine

white marks facing the manifold side. Tighten the bolts to 33 ft. lbs. (44 Nm). Install the right O_2 sensor.

- Left exhaust manifold. The new gasket must be installed with the white marks facing the manifold side. Tighten the bolts to 33 ft. lbs. (44 Nm). Install the left O_2 sensor.
- The 2 new cylinder head gaskets in position on the engine block. Each gasket has a painted mark denoting the rear of the gasket. The gasket for the right bank has a white mark and the gasket for the left bank has a yellow mark. Double check the gasket position and placement.

11. For 4.0L engines, install the cylinder heads and tighten the bolts in sequence as follows:

   a. Step 1: 29 ft. lbs. (39 Nm)
   b. Step 2: Plus 90 degrees

12. For 4.3L engines, install the cylinder heads and tighten the bolts in sequence as follows:

   a. Step 1: 44 ft. lbs. (60 Nm)
   b. Step 2: Plus 90 degrees
- HO_2.
- Engine wire to the right-hand cylinder head with the 2 bolt

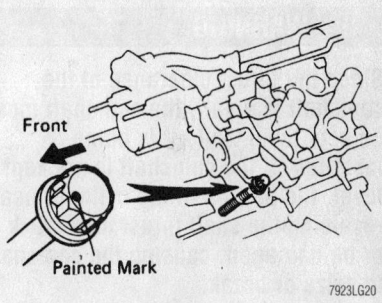

Apply a dot of paint at the front of each bolt—4.0L (1UZ-FE) and 4.3L (3UZ-FE) Engines

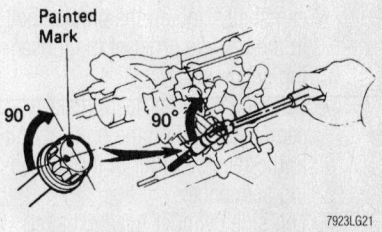

The paint mark must be 90 degrees from the starting point—4.0L (1UZ-FE) and 4.3L (3UZ-FE) Engines

- Ground cable to the right-hand cylinder head with the bolt
- The engine wire protector to the left-hand cylinder head with the bolt
- Ground cable to the left-hand cylinder head with the bolt
- Old packing and apply new seal packing to the bearing caps
- Bearing cap on the right side cylinder head, marked **I1**, in position with the arrow mark facing the rear. Install the bearing cap on the left side cylinder head, marked **I6**, in position with the arrow mark facing the front.
- Bearing cap bolts with new washers. Apply a light coat of oil on the threads of the cap bolts. Alternately tighten each bolt to 12 ft. lbs. (16 Nm).

➡**Use silver colored bolts 1.50 in. (38mm) in length.**

- Camshaft housing plugs on the cylinder heads. Be sure to face the cupped side forward.

13. Turn the crankshaft pulley clockwise or counterclockwise and put the timing mark of the crankshaft pulley in line with the centers of the crankshaft pulley bolt and the idler pulley bolt.

✳✳ WARNING

Since the thrust clearance of the camshaft is small, the camshaft must be kept level while it is being installed. If the camshaft is not kept level, the portion of the cylinder head receiving the shaft thrust may crack or be damaged, causing the camshaft to seize or break.

14. Install or connect the following:
- Right side cylinder head intake camshaft. Tighten the bracket bolt in the reverse order of the loosening sequence.
- Right side cylinder head exhaust camshaft. Tighten the bracket bolt in the reverse order of the loosening sequence.
- Left side cylinder head intake camshaft. Tighten the bracket bolt in the reverse order of the loosening sequence.
- Left side cylinder head exhaust camshaft. Tighten the bracket bolt in the reverse order of the loosening sequence.

15. Check and adjust the valve clearance.

16. Install or connect the following:
- Camshaft oil seals with the proper tool (SST 09223-46011). Be sure to apply MP grease to the new oil seal lip.
- Semi-circular plugs, if removed

17. Clean the cylinder head covers. Apply new sealant in the correct locations and install the gaskets.

18. Install or connect the following:
- Right cylinder head cover and bolts. Tighten the bolts to 52 inch lbs. (6 Nm).
- Left cylinder head cover and bolts. Tighten the bolts to 52 inch lbs. (6 Nm).
- Engine hanger with the 2 bolts. Install both engine hangers. Tighten the bolts to 27 ft. lbs. (37 Nm).
- Oil dipstick guide for the engine
- Oil dipstick for the transmission
- Rear water bypass joint and No. 1 EGR pipe
- Front water bypass joints. Install 2 gaskets and alternately tighten the nuts to 13 ft. lbs. (18 Nm).
- Water inlet and inlet housing, alternately tighten the bolts to 13 ft. lbs. (18 Nm)
- Delivery pipe and intake manifold
- Return pipe with 2 new gaskets. Tighten the union bolt to 26 ft. lbs. (35 Nm).
- Engine wire to the delivery pipes and rear water bypass joint
- Fuel return hose to the fuel return pipe
- Fuel inlet hose to the left-hand delivery pipe
- Fuel hose to the fuel pressure regulator
- Air intake chamber assembly
- Brake booster vacuum hose to the union on the air intake chamber
- PCV hose to the PCV valve on the left-hand cylinder head
- Water bypass hose (from EGR valve) to the rear water bypass joint
- Water bypass hose (from throttle body) to the rear water bypass joint
- Vacuum hose (from VSV for fuel pressure control) to the fuel pressure regulator
- EVAP hose (from charcoal canister) from the VSV for EVAP
- TPS connector
- With TRAC system, sub TPS connector
- With TRAC system, sub throttle actuator connector
- IAC valve connector

- EGR valve connector
- EGR gas temperature sensor connector
- VSV connector for fuel pressure control
- VSV connector for EVAP
- Accelerator bracket with the 2 bolts
- Accelerator, automatic transmission throttle control and the cruise control actuator cable
- Spark plug wires and clamps to the right and left cylinder head cover
- Belt rear plates by installing the bolts. Tighten the bolts to 66 inch lbs. (8 Nm).
- No. 2 ignition coil
- A new gasket to the exhaust manifold and install the catalytic converters. Tighten the 3 nuts to each converter to 46 ft. lbs. (62 Nm).
- Front exhaust pipe, tighten the bolts and nuts to 32 ft. lbs. (44 Nm). Tighten the 4 bolts holding the pipe support bracket to the transmission. Tighten the bolts to 32 ft. lbs. (44 Nm).
- Power steering pump with the nut and 3 bolts. Tighten the nut to 32 ft. lbs. (43 Nm) and the bolts to 29 ft. lbs. (39 Nm).

19. Align the knock pin on the right side camshaft with the knock pin of the timing pulley. Slide on the timing pulley with the right side mark facing forward. Tighten the bolt to 80 ft. lbs. (108 Nm).

20. Align the knock pin on the left side camshaft with the knock pin of the timing pulley. Slide on the timing pulley with the left side mark facing forward. Tighten the bolt to 80 ft. lbs. (108 Nm).

21. Install the timing belt to the left side camshaft timing pulley as follows:
a. Using the proper tool, slightly turn the left side timing pulley clockwise. Align the installation mark of the timing belt with the timing mark of the camshaft timing pulley and hang the timing belt on the left side camshaft pulley.
b. Align the timing marks of the left side camshaft pulley and the timing belt rear plate.
c. Check that the timing belt has tension between crankshaft timing pulley and the left side camshaft pulley.

22. Install the timing belt to the right side camshaft timing pulley as follows:
a. Using the proper tool, slightly turn the right side timing pulley clockwise. Align the installation mark of the timing belt with the timing mark of the camshaft timing pulley and hang the timing belt on the right side camshaft pulley.

b. Align the timing marks of the right side camshaft pulley and the timing belt rear plate.

c. Check that the timing belt has tension between crankshaft timing pulley and the right side camshaft pulley.

23. The timing belt tensioner must be set prior to installation. The tensioner can be set by:

a. Place a plate washer between the tensioner and a block. Using a press, press in the pushrod using 220–225 lbs. of pressure.

b. Align the holes of the pushrod and housing, pass a 0.05 inch (1.27mm) rod through the holes to keep the setting position of the pushrod.

c. Release the press and install the dust boot to the tensioner.

24. Loosely install the tensioner. Evenly and alternately tighten the bolts to 20 ft. lbs. (26 Nm). Remove the tool from the tensioner.

25. Turn the crankshaft pulley 2 complete revolutions from TDC to TDC. Always turn the crankshaft clockwise. Check that all belt and pulley marks align with their reference marks. If any mark is out of perfect alignment, the timing belt must be removed and reinstalled.

26. Install or connect the following:
• Drive belt tensioner and tighten the bolt and nuts to 12 ft. lbs. (16 Nm).

27. Install the fan bracket by installing the 2 bolts and 2 nuts. Tighten as follows:
a. 12mm: 12 ft. lbs. (16 Nm)
b. 14mm: 24 ft. lbs. (32 Nm)

28. Remove or disconnect the following:
• Air conditioning compressor. Tighten the bolts to 36 ft. lbs. (49 Nm) and the nut to 22 ft. lbs. (29 Nm).
• No. 1 ignition coil
• Right side No. 2 timing belt cover
• Left side No. 2 timing belt cover
• Drive belt idler pulley and cover plate. Tighten the bolt to 27 ft. lbs. (37 Nm).
• Secure the ignition wires. Make certain that all clips and retainers are securely engaged and that the wires are properly routed.
• Right side No. 3 timing belt
• Left-hand No. 3 timing belt cover
• Radiator assembly
• Fan pulley, fan, fluid coupling and the drive belt
• The air cleaner and intake air connector assembly
• V bank cover
• Coolant
• Negative battery cable to the battery
• Air cleaner inlet

• Battery clamp cover
• Engine undercover
• Oil pan protector

29. Start the engine and check for leaks
30. Bleed the cooling system and recheck the engine coolant level.
31. Make all the necessary engine adjustments.

Intake Manifold

REMOVAL & INSTALLATION

3.0L (1MZ-FE) Engine

1. Before servicing the vehicle, refer to the precautions in the beginning of this section.
2. Remove or disconnect the following:
• Negative battery cable
• Coolant
• Throttle/accelerator cable from the throttle body
• Air cleaner hose at the air intake chamber and remove it
• All lines and hoses. Tag them for installation.
• Idle Speed Control (ISC) valve and the throttle body
• Exhaust Gas Recirculation (EGR) valve and vacuum modulator
• Cylinder head rear plate
• Intake chamber stays, any wires, then, the air intake chamber
• Fuel injection delivery pipe and the injectors
• Water outlet and the bypass outlet
• 2 bolts and the No. 2 idler pulley bracket stay
• 8 bolts and 4 nuts, then lift out the intake manifold

To install:
3. Thoroughly clean the intake manifold and cylinder head surfaces. Using a machinist's straight edge and a feeler gauge, check the surface of the intake manifold for warpage. If the warpage is greater than 0.0039 in. (0.10mm), replace the intake manifold.
4. Place new gaskets onto the intake manifold and position the intake manifold between the cylinder heads. Tighten the nuts and bolts to 13 ft. lbs. (18 Nm). Tighten the No. 2 pulley bracket bolts to 13 ft. lbs. (18 Nm).
5. Install or connect the following:
• Water bypass outlet and tighten the bolts to 74 inch lbs. (8.3 Nm). Tighten the water outlet to 74 inch lbs. (8 Nm).
• Injectors and delivery pipe

• Air intake chamber and tighten the 2 bolts and 2 nuts to 32 ft. lbs. (43 Nm); use an 8mm hex wrench
• Chamber stays and tighten the mounting bolts to 29 ft. lbs. (39 Nm)
• Remaining components. Tighten the emission control valve set to 73 inch lbs. (8 Nm).
• All hoses
• Accelerator cable, if equipped with automatic transaxle
• Coolant
• Negative battery cable

3.0L (2JZ-GE) Engine

1. Before servicing the vehicle, refer to the precautions in the beginning of this section.
2. Remove or disconnect the following:
• Negative battery cable
• Coolant
• Spark plug wires at the spark plugs
• Spark plugs
• Radiator
• Water pump pulley
• Timing belt
• No. 2 front exhaust pipe
• 2 Oxygen (O_2) sensor leads
• 4 nuts, and the manifold heat shield
• Exhaust manifolds
• Water bypass outlet and the No. 1 bypass pipe. Remove the 3 O-rings.
• Water outlet
• No. 1 bypass hose
• Vacuum Control Valve (VCV) set and the No. 2 vacuum pipe
• Fuel return hose from the oil dipstick guide, remove the mounting bolt and pull the guide and dipstick from the pan. Plug the hole.
• Air intake chamber
• Fuel delivery pipe, then pull out the injectors
• No. 1 and 2 fuel pipes
• Engine harness from the intake manifold
• Intake manifold stay
• Loosen the 6 bolts and 2 nuts, then lift out the intake manifold

To install:
3. Install or connect the following:
• Install the intake manifold, with a new gasket, and tighten the bolts and nuts to 15 ft. lbs. (21 Nm)
• Mounting stay and tighten the bolts to 29 ft. lbs. (39 Nm)
• Engine harness to the manifold
• 2 fuel pipes and tighten the bolts to 78 inch lbs. (9 Nm)

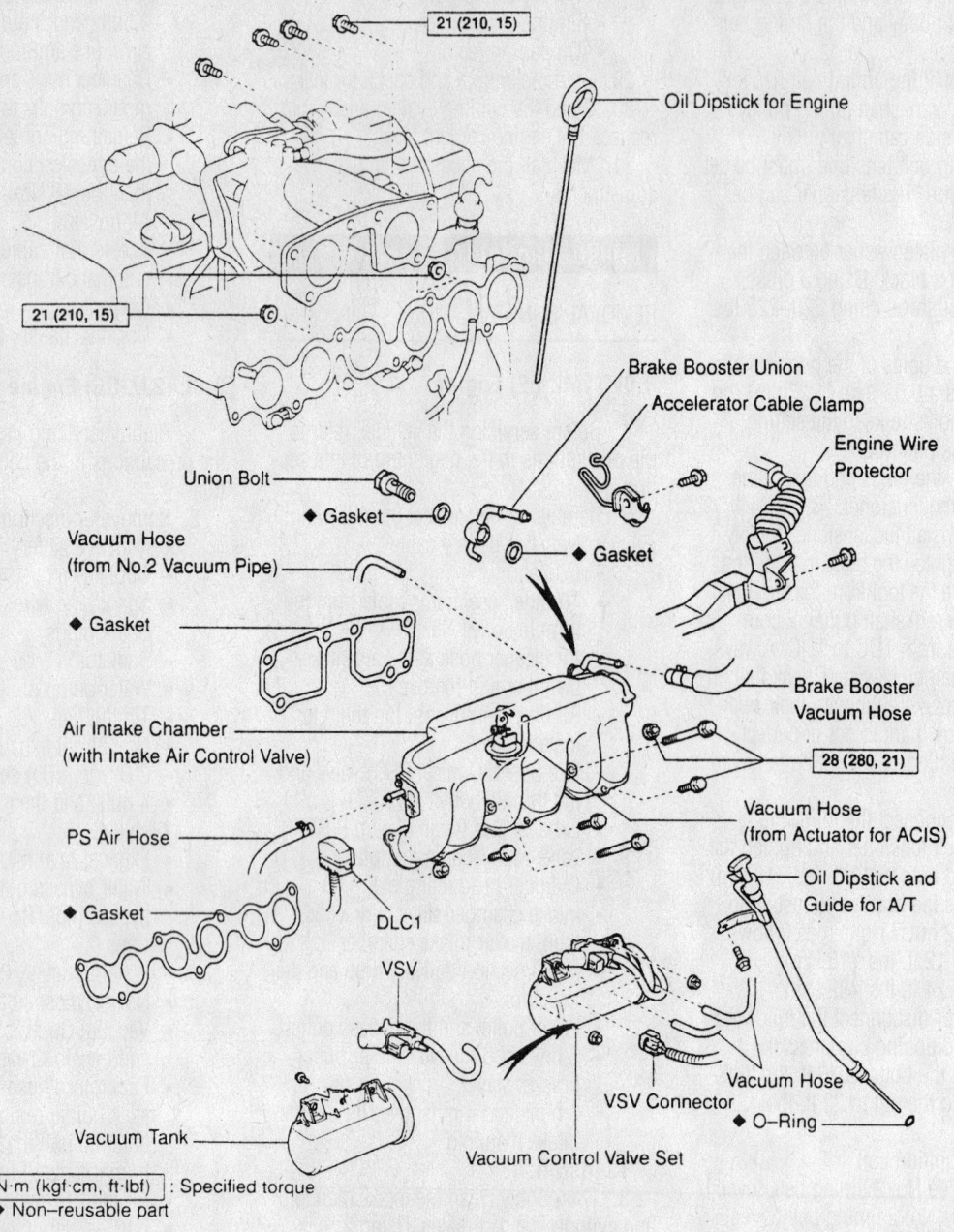

21 (210, 15)

Oil Dipstick for Engine

21 (210, 15)

Brake Booster Union

Accelerator Cable Clamp

Engine Wire Protector

Union Bolt

◆ Gasket

◆ Gasket

Vacuum Hose (from No.2 Vacuum Pipe)

◆ Gasket

Brake Booster Vacuum Hose

Air Intake Chamber (with Intake Air Control Valve)

28 (280, 21)

PS Air Hose

Vacuum Hose (from Actuator for ACIS)

◆ Gasket

DLC1

Oil Dipstick and Guide for A/T

VSV

Vacuum Hose

VSV Connector

Vacuum Tank

Vacuum Control Valve Set

◆ O-Ring

N·m (kgf·cm, ft·lbf) : Specified torque
◆ Non-reusable part

9301LG09

Exploded view of the intake manifold mounting and related components—3.0L (2JZ-GE) engine

- Delivery pipe and injectors. Tighten the pipe bolts to 15 ft. lbs. (21 Nm).
- Air intake chamber and tighten it to 15 ft. lbs. (21 Nm)
- 2 stays and tighten them to 13 ft. lbs. (18 Nm); The No. 1 stay is marked with an **F** and the No. 2 stay is marked with an **R**.
- The oil dipstick and guide, using a new O-ring
- VCV set and the vacuum pipe. Tighten the set mounting bolts to 15 ft. lbs. (21 Nm).

- Water bypass outlet and the pipe, tighten the bolts to 78 inch lbs. (9 Nm)
- Exhaust manifolds. Tighten the bolts to 29 ft. lbs. (39 Nm).
- Heat shield and tighten it to 13 ft. lbs. (18 Nm)
- No. 2 front pipe
- Timing belt
- Radiator and water pump pulley
- Plug wires to the plugs
- Coolant
- Negative battery cable

4.0L (1UZ-FE) and 4.3L (3UZ-FE) Engines

1. Before servicing the vehicle, refer to the precautions in the beginning of this section.

2. Properly relieve the fuel system pressure.

3. Remove or disconnect the following:
- Negative battery cable
- Coolant
- Accelerator cable
- Throttle and accelerator pedal position electrical connectors
- Throttle motor electrical connector

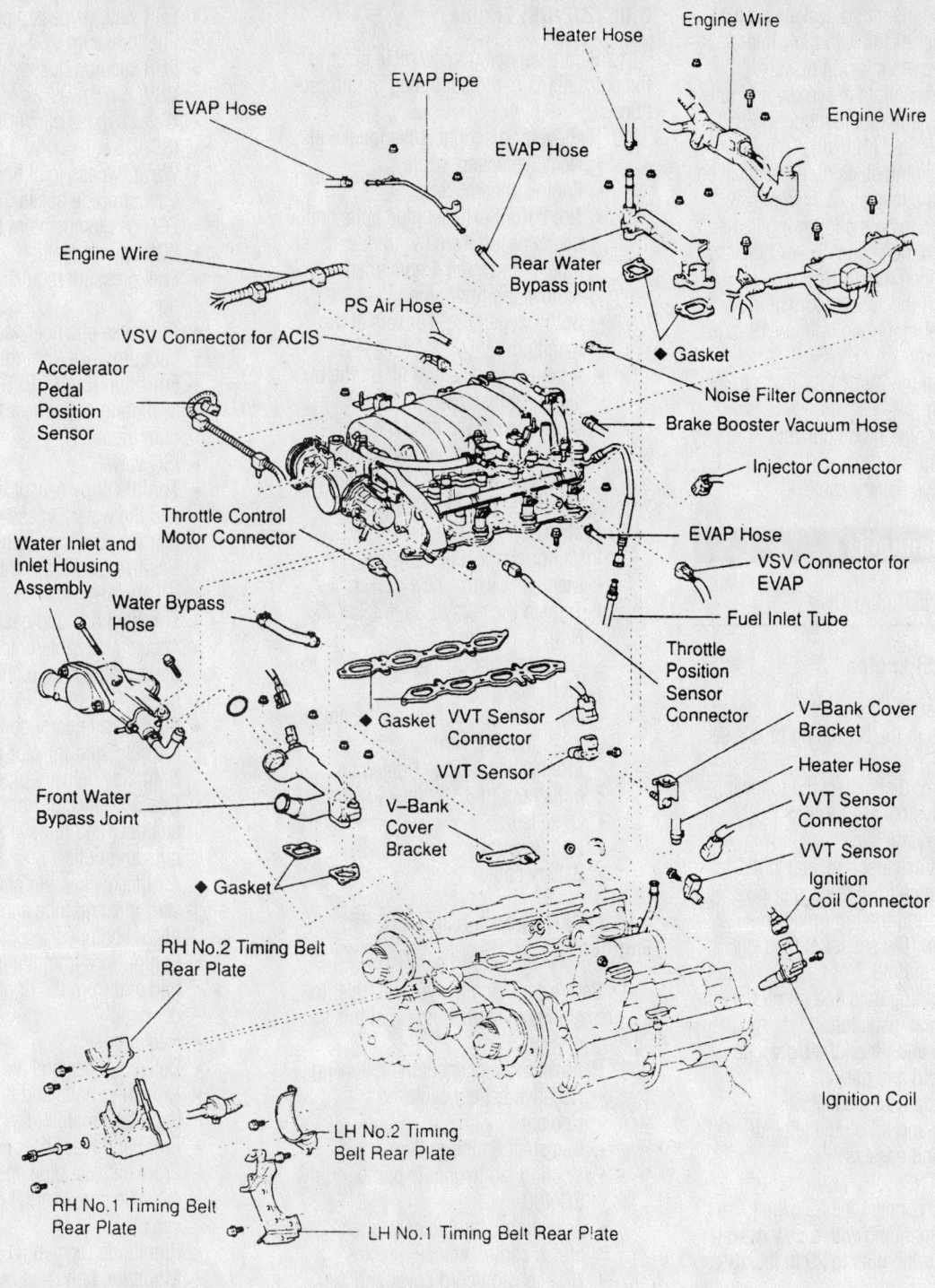

Heater Hose
Engine Wire
Engine Wire
Engine Wire
EVAP Pipe
EVAP Hose
EVAP Hose
Rear Water Bypass joint
PS Air Hose
VSV Connector for ACIS
Gasket
Noise Filter Connector
Accelerator Pedal Position Sensor
Brake Booster Vacuum Hose
Injector Connector
Throttle Control Motor Connector
EVAP Hose
VSV Connector for EVAP
Water Inlet and Inlet Housing Assembly
Fuel Inlet Tube
Water Bypass Hose
Throttle Position Sensor Connector
V–Bank Cover Bracket
Gasket
VVT Sensor Connector
Heater Hose
VVT Sensor
VVT Sensor Connector
Front Water Bypass Joint
V–Bank Cover Bracket
VVT Sensor
Gasket
Ignition Coil Connector
RH No.2 Timing Belt Rear Plate
LH No.2 Timing Belt Rear Plate
Ignition Coil
RH No.1 Timing Belt Rear Plate
LH No.1 Timing Belt Rear Plate

◆ Non–reusable part

9301LG10

Exploded view of the intake manifold mounting and related components—4.0L (1UZ-FE) and 4.3L (3UZ-FE) Engines

- Vacuum Switching Valve (VSV) electrical connectors
- Injector electrical connectors
- Noise filter electrical connector
- Brake booster vacuum hose from the intake manifold
- Positive Crankcase Ventilation (PCV) hose from the left-hand valve cover

- Evaporative Emission (EVAP) hoses and label for installation
- Power steering air hose from the intake manifold
- Coolant hoses from the throttle body
- 2 EVAP-to-intake manifold mounting bolts
- Accelerator cable bracket

- 3 V-bank cover brackets
- EVAP VSV
- Fuel supply and return lines
- 6 bolts and 4 nuts, then the intake manifold assembly

4. Clean the gasket mating surfaces of old gasket and sealant.
To install:
5. Install or connect the following:

- 2 intake manifold gaskets on the cylinder heads with the white painted mark facing upward.
- Intake manifold assembly and uniformly, tighten the mounting bolts to 13 ft. lbs. (18 Nm).

6. The remaining steps of the installation procedure are the reverse of the removal, while keeping in mind the following items:

- Tighten the V-bank cover brackets to 66 inch lbs. (8 Nm)
- Tighten the accelerator cable bracket mounting bolts to 13 ft. lbs. (18 Nm)
- Tighten the EVAP VSV to 13 ft. lbs. (18 Nm)
- All remaining components
- Coolant
- Negative battery cable

Exhaust Manifold

REMOVAL & INSTALLATION

3.0L (1MZ-FE) Engine

1. Before servicing the vehicle, refer to the precautions in the beginning of this section.

2. Remove or disconnect the following:

- Negative battery cable
- Engine undercovers
- 2 front exhaust pipe stay bolts
- Front pipe from the center pipe
- 3 nuts and the front pipe
- Oxygen (O_2) sensor at the right side manifold
- 3 mounting nuts and lift off the outside heat insulator
- 6 nuts and lift off the right side manifold and gasket
- Left side heat insulator
- 6 nuts and lift off the left side manifold and gaskets

To install:

3. Install or connect the following:

- Right manifold with a new gasket. Tighten the nuts to 29 ft. lbs. (39 Nm).
- Outer insulator
- Left manifold. Use a new gasket. Tighten the nuts to 29 ft. lbs. (39 Nm).
- Outer insulator
- Front exhaust pipe and tighten the manifold-to-pipe nuts to 46 ft. lbs. (62 Nm). Tighten the pipe-to-converter nuts to 32 ft. lbs. (43 Nm).
- O_2 sensor
- Undercovers
- Battery cable

3.0L (2JZ-GE) Engine

1. Before servicing the vehicle, refer to the precautions in the beginning of this section.

2. Remove or disconnect the following:

- Negative battery cable
- Engine undercovers
- No. 2 front exhaust pipe bolts and disconnect it from the front exhaust pipe. Loosen the 4 nuts, then remove the front pipe.
- Both Oxygen (O_2) sensors at the manifold
- 4 mounting nuts and lift off the outside heat insulator
- 4 nuts and disconnect the manifolds from the pipe. Loosen the mounting bolts and remove the 2 manifolds and the gasket.

To install:

3. Install or connect the following:

- Manifolds with a new gasket. Tighten the nuts to 30 ft. lbs. (40 Nm).
- Outer insulator. Tighten the nuts to 13 ft. lbs. (18 Nm).
- No. 2 front pipe. Tighten the nuts to 46 ft. lbs. (62 Nm).
- Front exhaust pipe. Tighten the bolts/nuts to 32 ft. lbs. (43 Nm).
- O_2 sensors
- Undercovers
- Battery cable

4.0L (1UZ-FE) and 4.3L (3UZ-FE) Engines

1. Before servicing the vehicle, refer to the precautions in the beginning of this section.

2. Remove or disconnect the following:

- Negative battery cable
- Coolant
- Camshaft timing pulleys
- Cooling fan hydraulic pump, on the SC 400
- Accelerator, throttle control and cruise control actuator cables
- High tension cord cover and the right side ignition coil
- Water inlet housing mounting bolts
- Water bypass hose from the Idle Speed Control (ISC) valve
- Water inlet and inlet housing assemblies
- O-ring from the water inlet housing
- Exhaust Gas Recirculation (EGR) pipe
- Vacuum Switching Valve (VSV) connector
- Vacuum pipe hose

- EGR water bypass pipe
- Fuel pressure VSV
- EGR vacuum hoses and the EGR VSV
- Water bypass pipe hose from the ISC valve
- Water bypass joint hose
- Vacuum pipe hoses
- EGR gas temperature sensor
- EGR valve adapter
- Fuel pressure regulator vacuum hose
- Air intake chamber vacuum hose
- Vacuum hose from the Evaporative Emission (EVAP) BVSV
- Mounting bolts, hoses and the vacuum pipe
- ISC valve
- Throttle body sensor connectors and the water bypass pipe from the rear water bypass joint
- Positive Crankcase Ventilation (PCV) valve hose
- Throttle body and gasket
- Accelerator cable bracket and the brake booster vacuum union and hose
- Cold start injector connector and the cold start injector tube from the right side delivery pipe, if equipped
- Check connector from the intake chamber and remove the mounting nuts and bolts
- Air intake chamber and the cold start injector, tube and wire assembly, if equipped
- Engine wire from the intake manifold and from the right side cylinder head
- Heater hoses
- Delivery pipes and the fuel injectors
- Mounting bolts and nuts. Lift up the intake manifold
- Front and rear water bypass joint
- Front exhaust pipe and the main catalytic converters. Lower the vehicle.
- Right side Oxygen (O_2) sensor
- Mounting bolts and nuts and remove the right side exhaust manifold.
- Oil dipstick and guide
- Left side O_2 sensor
- Mounting bolts and nuts
- Left side exhaust manifold

To install:

3. Install or connect the following:

- Right side exhaust manifold with a new gasket (the painted marks should face the manifold) and tighten the mounting bolts to 29 ft. lbs. (39 Nm).

- Right side O$_2$ sensor
- Left side exhaust manifold with a new gasket (the painted marks should face the manifold) and tighten the mounting bolts to 29 ft. lbs. (39 Nm)
- Left side O$_2$ sensor
- Oil dipstick and guide. Raise and safely support the vehicle.
- Catalytic converters and front exhaust pipe. Lower the vehicle.
- Front and rear water bypass joints. Tighten the mounting bolts to 13 ft. lbs. (18 Nm).
- Intake manifold, using new gaskets. Tighten the mounting nuts and bolts to 13 ft. lbs. (18 Nm).
- Delivery pipes and fuel injectors
- Fuel return pipe with new gaskets. Tighten the union bolt to 26 ft. lbs. (35 Nm).
- Fuel hoses and the injector connectors
- Engine wire to the delivery pipes
- Connectors on the left side delivery pipe
- ECT sensor
- Cold start injector time switch
- Water temperature sender gauge connectors
- Heater hoses and engine wire bracket
- Engine wire to the bracket.
- Cold start injector, tube and wire assembly. Tighten the mounting bolts to 69 inch lbs. (8 Nm), if equipped.
- Air intake chamber with new gaskets and tighten the mounting bolts to 13 ft. lbs. (18 Nm).
- Cold start injector tube to the right side delivery pipe and tighten the union bolt to 11 ft. lbs. (15 Nm), if equipped.
- Cold start injector connector, if necessary
- Accelerator cable bracket
- Brake booster union and connect the vacuum hose. Tighten the union bolt to 22 ft. lbs. (29 Nm).
- Water bypass hose to the throttle body and the PCV hose to the cylinder head cover.
- Throttle body, using a new gasket. Tighten the mounting bolts to 13 ft. lbs. (18 Nm).
- Water bypass pipe
- Sensor connectors
- Idle Speed Control (ISC) valve and tighten the mounting bolts to 13 ft. lbs. (18 Nm).
- Water bypass hose.

- Vacuum pipe and the assorted hoses
- Remaining components.
- Adjust the accelerator cable, the automatic transmission throttle cable and the cruise control actuator cable.
- Cooling fan hydraulic pump on the SC 400
- Camshaft timing pulleys
- Coolant
- Negative battery cable

Camshaft and Valve Lifters

REMOVAL & INSTALLATION

The following procedures have the valve lash adjuster removal and installation incorporated.

3.0L (1MZ-FE) Engine

1. Before servicing the vehicle, refer to the precautions in the beginning of this section.
2. Remove or disconnect the following:
   - Timing belt and idler pulley
   - Camshaft timing pulleys
   - Cylinder head covers

✳✳ WARNING

The thrust clearance on both the intake and exhaust camshafts is very small, the camshafts must be kept level during removal. If the camshafts are removed without being kept level, the camshaft may be caught in the cylinder head causing the head to break or the camshaft to seize.

3. To remove the exhaust and intake camshafts from the right side cylinder head:
   a. Turn the camshaft with a wrench until the 2 pointed marks drive and dri-

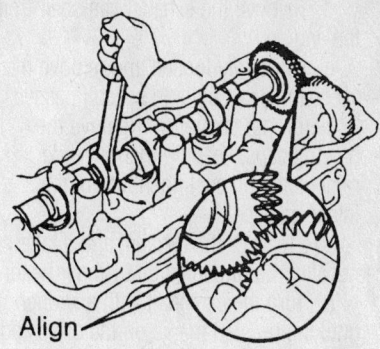

Aligning the right side camshaft timing marks—3.0L (1MZ-FE) engine

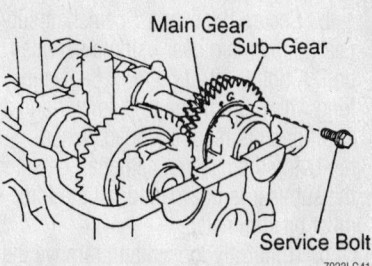

Securing the subgear and driven gear, right side—3.0L (1MZ-FE) engine

ven gears are aligned. (The right camshaft gears have 2 marks apiece; the left side camshaft gears have 1 mark each.)

   b. Secure the exhaust camshaft subgear to the main gear using a service bolt. A bolt 0.63–0.79 in. (16–20mm) long with a 6mm thread diameter and a 1mm pitch is recommended. When removing the exhaust camshaft be sure the subgear is not loaded; all the force must be eliminated.

   c. Uniformly loosen and remove the exhaust camshaft bearing cap bolts in several passes and in the proper sequence. Remove the 8 bearing cap bolts and remove the caps, keeping them in the correct order.

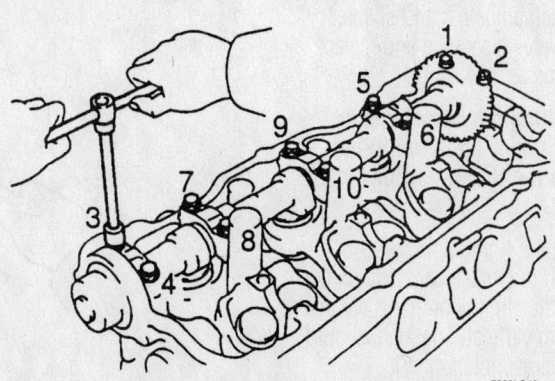

Right exhaust camshaft bearing loosening sequence—3.0L (1MZ-FE) engine

d. Remove the exhaust camshaft from the engine.

e. Uniformly loosen and remove the 10 bearing cap bolts in several passes, in the proper sequence. Remove the bearing caps, keeping them in order, remove the oil seal, then lift out the intake camshaft.

4. To remove the exhaust and intake camshafts from the left side cylinder head:

a. Turn the camshaft with a wrench until the pointed marks on the drive and driven gears are aligned. (The right camshaft gears have 2 marks apiece; the left side camshaft gears have 1 mark each.)

b. Secure the exhaust camshaft sub-gear to the main gear using a service bolt. A bolt 0.63–0.79 in. (16–20mm) long with a 6mm thread diameter and a 1mm pitch is recommended. When removing the exhaust camshaft be sure the subgear is not loaded; all the force must be eliminated.

c. Uniformly loosen and remove the exhaust camshaft bearing cap bolts in several passes and in the proper sequence. Remove the 8 bearing cap bolts and remove the caps. Keep the caps in the correct order.

d. Remove the exhaust camshaft from the engine.

e. Uniformly loosen and remove the 10 bearing cap bolts in several passes, in the proper sequence. Remove the bearing caps, keeping them in order, remove the oil seal, then lift out the intake camshaft.

5. Remove the valve lash adjuster shims and hydraulic lash adjusters. Identify each lash adjuster and shim as it is removed so it can be reinstalled in the same position. If the lash adjusters are to be reused, store them upside down in a sealed container.

To install:

6. Install or connect the following:
- Valve lash adjusters and shims. Check valve clearance and replace the shims as necessary.

➡ Before installing the camshafts in either cylinder head, apply multi-purpose grease to the thrust portions of each camshaft.

7. To install the right camshafts:

a. Position the intake camshaft on the head so that the alignment marks are at a 90° angle from vertical. The mark should be at the 3 o'clock position.

b. Apply sealant to the No. 1 bearing cap.

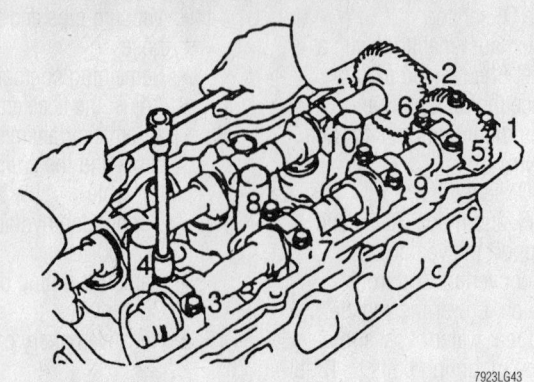

Right intake camshaft bearing loosening sequence—3.0L (1MZ-FE) engine

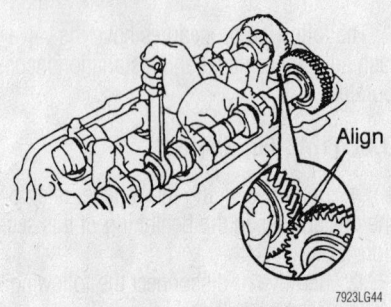

Aligning the left side camshaft timing marks—3.0L (1MZ-FE) engine

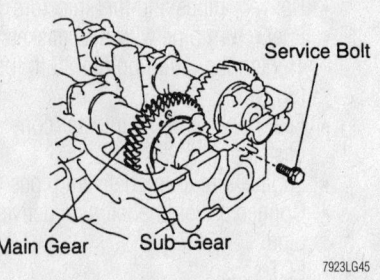

Securing the subgear and driven gear, left side—3.0L (1MZ-FE) engine

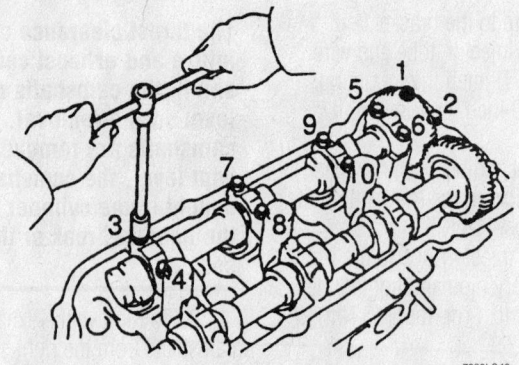

Left intake camshaft bearing cap bolt loosening sequence—3.0L (1MZ-FE) engine

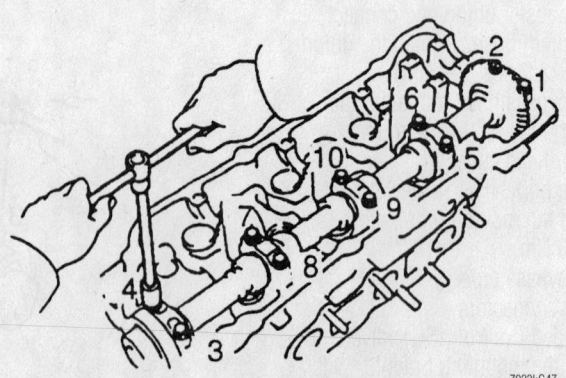

Left exhaust camshaft bearing cap bolt loosening sequence—3.0L (1MZ-FE) engine

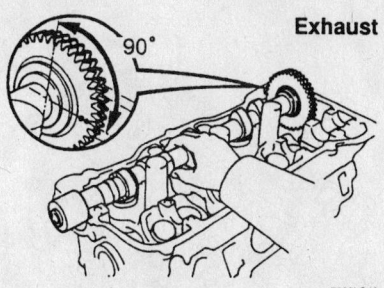

Exhaust camshaft installation position on the right cylinder head—3.0L (1MZ-FE) engine

7923LG48

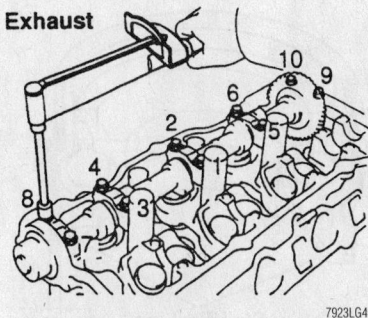

Exhaust camshaft bearing cap bolt tightening sequence on the right cylinder head—3.0L (1MZ-FE) engine

7923LG49

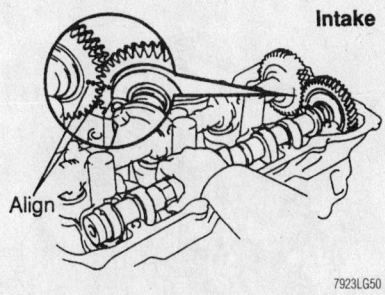

Intake camshaft installation position on the right cylinder head—3.0L (1MZ-FE) engine

7923LG50

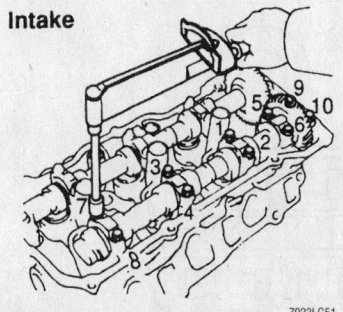

Intake camshaft bearing cap bolt tightening sequence on the right cylinder head—3.0L (1MZ-FE) engine

7923LG51

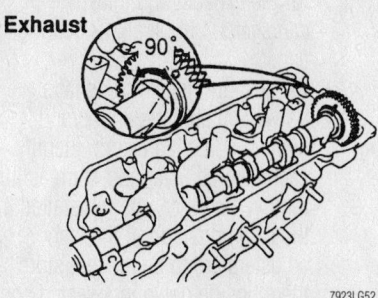

Exhaust camshaft installation position on the left cylinder head—3.0L (1MZ-FE) engine

7923LG52

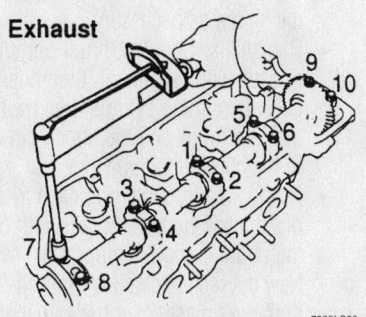

Exhaust camshaft bearing cap bolt tightening sequence on the right cylinder head—3.0L (1MZ-FE) engine

7923LG53

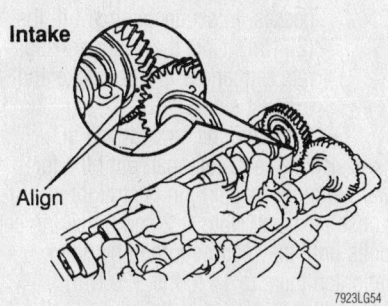

Intake camshaft installation position on the left cylinder head—3.0L (1MZ-FE) engine

7923LG54

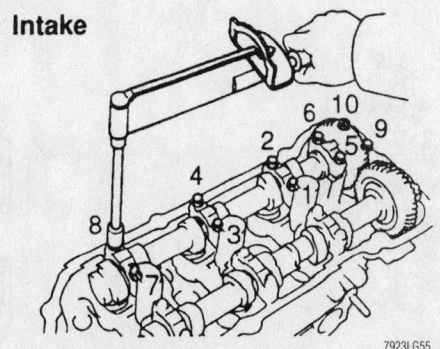

Intake camshaft bearing cap bolt tightening sequence on the right cylinder head—3.0L (1MZ-FE) engine

7923LG55

c. Apply a light coat of clean engine oil to the bolt threads and under the bolt head. Install the bearing caps to their proper position. Tighten the bolts evenly and in several passes in the reverse order of loosening to 12 ft. lbs. (16 Nm) in the proper sequence.

d. Position the exhaust camshaft on the head so that the alignment marks are at a 90° angle from vertical. The mark should be at the 9 o'clock position and must align with the marks on the other gear.

e. Apply a light coat of clean engine oil to the bolt threads and under the bolt head. Install the bearing caps to their proper position. Tighten the bolts evenly and in several passes in the reverse order of loosening to 12 ft. lbs. (16 Nm) in the proper sequence.

f. Remove the service bolt.

8. To install the left camshafts:

a. Position the intake camshaft on the head so that the alignment mark is at a 90 degree angle from vertical. The mark should be at the 9 o'clock position.

b. Apply sealant to the No. 1 bearing cap.

c. Apply a light coat of clean engine oil to the bolt threads and under the bolt head. Install the bearing caps to their proper position. Tighten the bolts evenly and in several passes to 12 ft. lbs. (16 Nm) in the proper sequence.

d. Position the exhaust camshaft on the head so that the alignment marks are at a 90 degree angle from vertical. The mark should be at the 3 o'clock position and must align with the marks on the other gear.

e. Apply a light coat of clean engine oil to the bolt threads and under the bolt head. Install the bearing caps to their proper position. Tighten the bolts evenly and in several passes to 12 ft. lbs. (16 Nm) in the proper sequence.

f. Remove the service bolt.

9. Apply multi-purpose grease to new camshaft oil seals. Install the seals.

10. Install or connect the following:
- No. 3 (rear) timing belt cover
- Camshaft timing gears
- Idler pulley, timing belt and covers

11. Check and adjust the valve clearance.

12. Install the cylinder head (valve) covers.

3.0L (2JZ-GE) Engine

1. Before servicing the vehicle, refer to the precautions in the beginning of this section.

2. Remove or disconnect the following:
- Negative battery cable from the battery
- Timing belt from the engine
- Cylinder head covers
- Camshaft sprocket
- 4 bolts and lift out the No. 4 (inner) timing belt cover.
- No. 1 camshaft bearing cap bolts. These are the bolts directly behind the sprockets.
- Bearing caps
- Remaining bearing cap bolts. Note that there are separate sequences

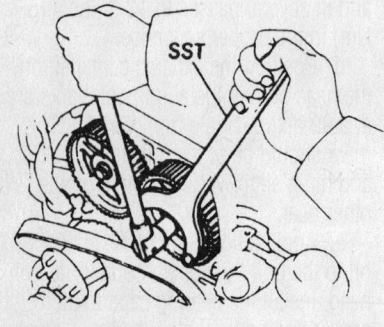

Removing the camshaft sprockets—3.0L (2JZ-GE) engine

for the exhaust and intake camshafts. Lift off all 12 bearing caps.
- Exhaust and intake camshafts
- Valve lash adjuster shims and hydraulic lash adjusters. Identify each lash adjuster and shim as it is removed so it can be reinstalled in the same position. If the lash adjusters are to be reused, store them upside down in a sealed container.

To install:

3. Install or connect the following:
- Valve lash adjusters and shims. Check valve clearance and replace the shims as necessary.
- Camshafts. Coat the thrust portions of each with engine oil, then position them in the cylinder head with the cam lobes and the knock pins in the correct position.
- No. 3 and No. 7 bearing caps in place, coat the bolt threads with oil, then tighten them temporarily
- New oil seals, coated with multi-purpose grease, over the camshafts
- No. 1 bearing caps, then apply some sealant. Install the bolts.
- All remaining bearing caps. Coat the threads of each bolt with clean oil, then tighten them, in several passes, in sequence, to 14 ft. lbs. (20 Nm). Note that there are separate sequences for the intake and exhaust sides.
- Oil seal in as far as it will go

4. Rotate each camshaft until the forward straight (knock) pin is straight up. Loosen exhaust Nos. 1, 2 and 6 bearing cap bolts until they can be turned by hand; retighten them to 14 ft. lbs. (20 Nm). Loosen intake Nos. 1 and 2 and retighten to 14 ft. lbs. (20 Nm).

5. Turn each camshaft ⅓ of a revolution (120 degrees). Loosen exhaust Nos. 4 and 7

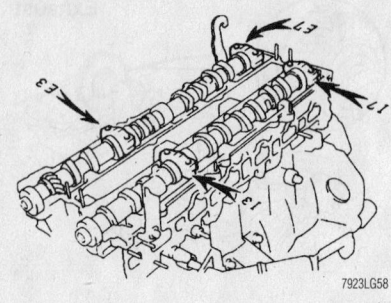

Installing No. 3 and 7 bearing caps—3.0L (2JZ-GE) engine

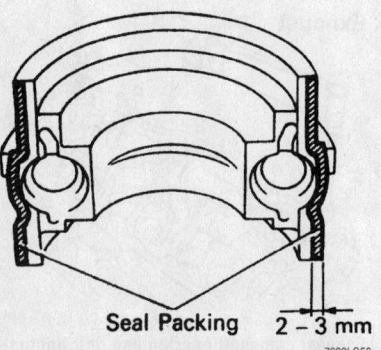

Applying sealant to the No. 1 bearing cap—3.0L (2JZ-GE) engine

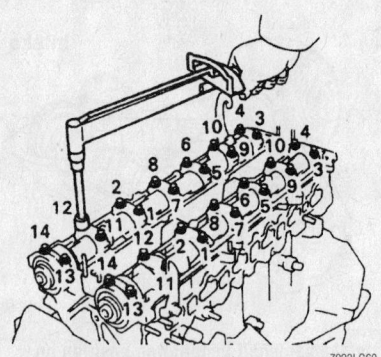

Camshaft bearing cap bolt tightening sequence—3.0L (2JZ-GE) engine

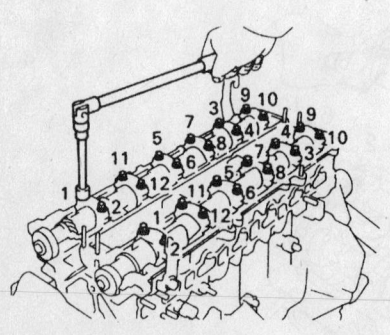

Camshaft bearing cap bolt removal sequence—3.0L (2JZ-GE) engine

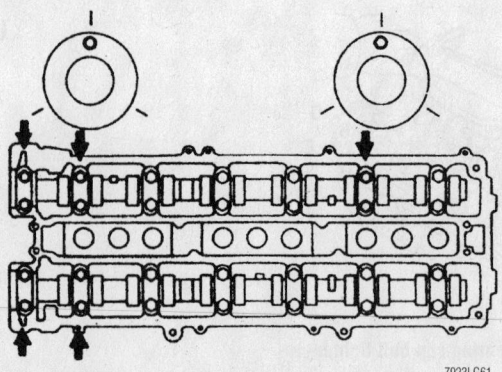

Tightening the camshafts (Step 1)—3.0L (2JZ-GE) engine

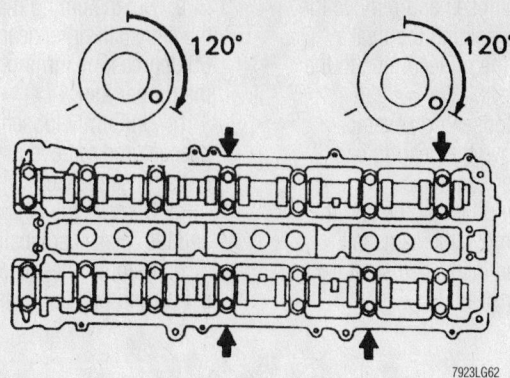

Tightening the camshafts (Step 2)—3.0L (2JZ-GE) engine

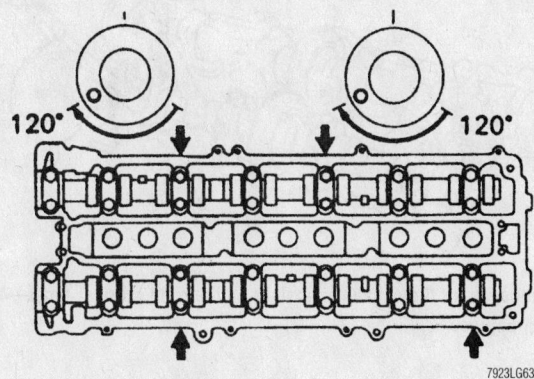

Tightening the camshafts (Step 3)—3.0L (2JZ-GE) engine

bearing cap bolts; retighten them to 14 ft. lbs. (20 Nm). Loosen intake Nos. 4 and 6 bearing cap bolts; retighten them to 14 ft. lbs. (20 Nm).

6. Turn each camshaft an additional ⅓ of a revolution, loosen exhaust bearing cap bolts Nos. 3 and 5, then retighten them to 14 ft. lbs. (20 Nm). Loosen intake bearing cap bolts Nos. 3 and 7, then retighten them to 14 ft. lbs. (20 Nm).

7. Check and adjust the valve clearance.

8. Install or connect the following:
   - No. 4 inside timing belt cover and the camshaft pulleys. Align the shaft pin with the pulley groove and slide the pulley on. Install the bolt temporarily. Hold the hex portion of the camshaft with a wrench; tighten the pulley bolt to 59 ft. lbs. (79 Nm).
   - Cylinder head covers
   - Timing belt to the engine
   - Negative battery cable to the battery

9. Check and/or adjust the ignition timing as necessary

4.0L (1UZ-FE) and 4.3L (3UZ-FE) Engines

1. Before servicing the vehicle, refer to the precautions in the beginning of this section.

2. Relieve the fuel pressure from the fuel lines.

3. Remove or disconnect the following:
   - Negative battery cable. Wait at least 90 seconds before performing any other work
   - Positive battery cable
   - Battery
   - Air cleaner inlet
   - V bank cover by removing the bolt and 2 cap nuts
   - Air cleaner and intake air connector assembly
   - Drive belt
   - Fluid coupling and the fan pulley. The drive belt tension may be slackened by turning the tensioner counterclockwise. The pulley bolt for the drive belt tensioner has a left-handed thread.
   - Radiator
   - The right-hand No. 3 timing belt cover
   - Left-hand No. 3 timing belt cover
   - Evaporative Emissions (EVAP) hose clamp from the timing belt cover. For all other vehicles, disconnect the EVAP hose from the hose clamp on the timing belt cover.

   - 4 bolts to the left-hand timing belt cover
   - Cord grommet from the timing belt cover and remove the timing belt cover
   - Drive belt idler pulley by removing the pulley bolt and cover plate
   - Right-hand and left No. 2 timing belt covers
   - No. 1 ignition coil
   - Air conditioning compressor from the engine. Do not disconnect the air conditioning pressure lines.
   - Fan bracket by removing the 2 bolts and 2 nuts
   - Alternator from the engine
   - Drive belt tensioner

4. Set the engine to Top Dead Center (TDC) on cylinder No. 1.

5. Turn the crankshaft pulley approximately 50° clockwise and put the timing mark of the crankshaft pulley in line with the centers of the crankshaft pulley bolt and the idler pulley bolt.

✳✳ WARNING

If the timing belt is disengaged, having the crankshaft pulley at the wrong angle can cause the piston head and valve head to come into contact with each other when you remove the camshaft timing pulley. Always set the crankshaft pulley at the correct angle before removing the timing belt.

6. If the timing belt is to be reused, turn the crank pulley slowly; check that the 3 installation marks are present on the belt. If the marks are not present, make new installation marks before removing the belt. The marks should align with the timing marks on each camshaft pulley and the crank pulley.

7. Remove or disconnect the following:
   - Timing belt tensioner
   - Timing belt from the camshaft timing pulleys
   - Camshaft timing pulleys
   - No. 2 ignition coil
   - Rear timing belt plates
   - Throttle body
   - Spark plug wires, wire clamps and the wire cover assembly from the right cylinder head
   - Right cylinder head cover by removing the 8 bolts and 8 washers
   - Transmission oil dipstick
   - Evaporative Emission (EVAP) control hose from the Vacuum Switching Valve (VSV)

- Engine wire clamp from the wire bracket on the delivery pipe
- Spark plug wires and clamps from the left-hand cylinder head cover
- Left cylinder head cover by removing the 8 bolts and 8 seal washers
- Semi-circular plugs, if necessary

✳✳ WARNING

Since the thrust clearance of the camshaft is small, the camshaft must be kept level while it is being removed. If the camshaft is not kept level, the portion of the cylinder head receiving the shaft thrust may crack or be damaged, causing the camshaft to seize or break.

8. To remove the exhaust camshaft from the right side cylinder head:

a. Position the service bolt hole of the drive subgear to the upright position. Secure the camshaft subgear to drive gear with a service bolt.

b. Set the timing mark (single dot) on the camshaft drive gear at approximately 10 degrees. Turn the camshaft with a wrench on the hexagonal flats.

c. Alternately loosen and remove the bearing cap bolts holding the intake camshaft side of the oil feed pipe to the cylinder head.

d. Uniformly loosen (in several passes) and remove the bearing cap bolts, in sequence.

e. Remove the oil feed pipe and the bearing caps. Remove the camshaft.

f. To remove the intake camshaft from the right side cylinder head:

g. Set the timing mark (single dot) on the camshaft drive gear at approximately 45 degrees. Turn the camshaft with a wrench on the hexagonal flats.

h. Uniformly loosen (in several passes) and remove the bearing cap bolts in the proper sequence.

i. Remove the bearing caps, oil seal and the intake camshaft.

9. To remove the exhaust camshaft of the left side cylinder head:

a. Position the service bolt hole of the drive subgear to the upright position. Secure the camshaft subgear to drive gear with a service bolt.

➡**When removing the camshaft, be sure the torsional spring force of the subgear has been eliminated.**

b. Set the timing mark (2 dots) on the camshaft drive gear at approximately 15 degrees, by turning the camshaft with the proper tool.

c. Alternately loosen and remove the bearing cap bolts holding the intake camshaft side of the oil feed pipe to the cylinder head.

d. Uniformly loosen (in several passes) and remove the bearing cap bolts in the proper sequence.

e. Remove the oil feed pipe and the bearing caps. Remove the camshaft.

10. To remove the intake camshaft from the left side cylinder head:

a. Set the timing mark (single dot) of the camshaft drive gear at approximately 60 degrees, by turning the camshaft with the proper tool.

b. Uniformly loosen (in several passes) and remove the bearing cap bolts, in sequence.

c. Remove the bearing caps, oil seal and the intake camshaft.

d. Remove the valve lash adjuster shims and hydraulic lash adjusters. Iden-

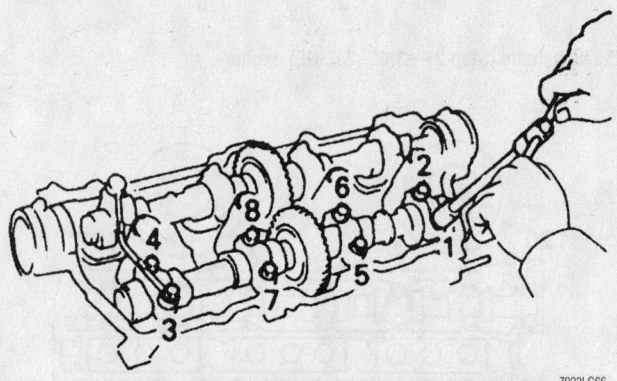

Bolt loosening sequence for the exhaust camshaft on the right cylinder head—4.0L (1UZ-FE) and 4.3L (3UZ-FE) Engines

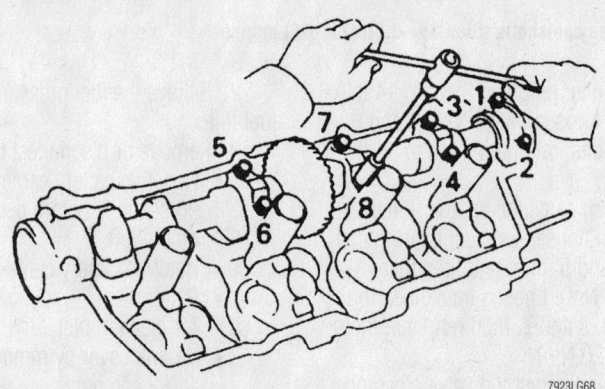

Loosen the bearing cap bolts for the intake camshaft on the right cylinder head in the sequence shown—4.0L (1UZ-FE) and 4.3L (3UZ-FE) Engines

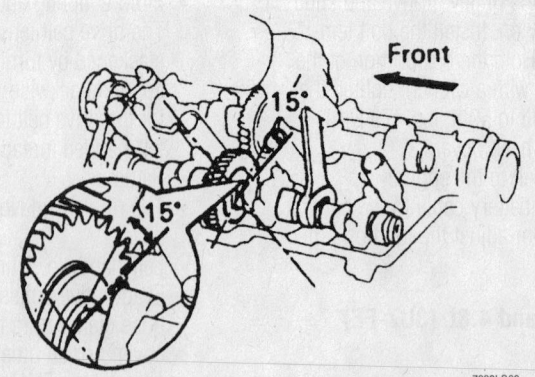

Turning the exhaust camshaft on the left cylinder head 15 degrees—4.0L (1UZ-FE) and 4.3L (3UZ-FE) Engines

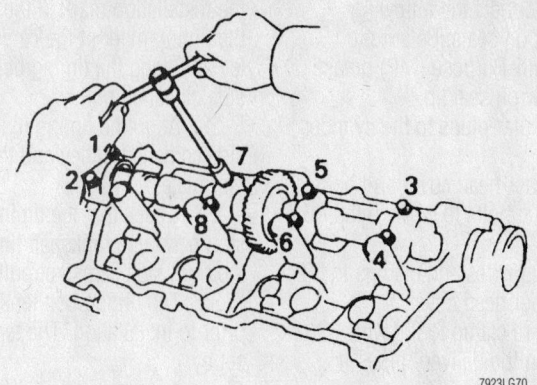

Bolt removal sequence for the intake camshaft on the left cylinder head—4.0L (1UZ-FE) and 4.3L (3UZ-FE) Engines

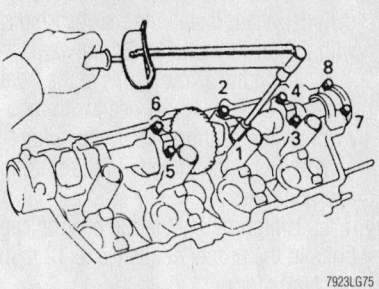

Right side intake camshaft bracket bolt tightening sequence—4.0L (1UZ-FE) and 4.3L (3UZ-FE) Engines

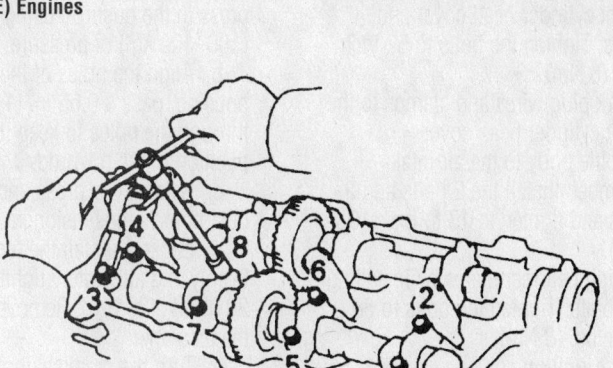

Bolt removal sequence for the exhaust camshaft on the left cylinder head—4.0L (1UZ-FE) and 4.3L (3UZ-FE) Engines

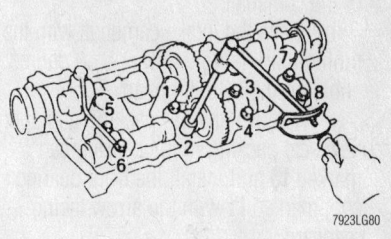

Right side exhaust camshaft bracket bolt tightening sequence—4.0L (1UZ-FE) and 4.3L (3UZ-FE) Engines

tify each lash adjuster and shim as it is removed so it can be reinstalled in the same position. If the lash adjusters are to be reused, store them upside down in a sealed container.

To install:

11. Install or connect the following:
- Valve lash adjusters and shims. Check the valve clearance and replace the shims as necessary.
- New seal packing to the bearing caps
- Bearing cap on the right side cylinder head, marked **I1**, in position with the arrow mark facing the rear. Install the bearing cap on the left side cylinder head, marked **I6**, in position with the arrow mark facing the front.
- Oil on the threads of the cap bolts. Install the bearing cap bolts with new washers and tighten to 12 ft. lbs. (16 Nm).

12. To install the right side cylinder head intake camshaft:

a. Apply grease to the thrust portion of the camshaft.

b. Place the intake camshaft at a 45

degree angle of the timing mark (single dot) on the cylinder head.

c. Remove any old packing and apply new seal packing to the bearing cap marked **I6** and install the front bearing cap, marked **I6** with the arrow facing rearward.

d. Align the arrows at the front and rear of the cylinder head with the bearing cap.

e. Install the remaining bearing caps in the proper sequence with the arrow mark facing rearward. Install the oil feed pipe and the mounting bolts.

f. Uniformly tighten the bearing cap bolts in the proper sequence to 12 ft. lbs. (16 Nm).

13. To install the right side cylinder head exhaust camshaft:

a. Set the timing mark (single dot) on the camshaft drive gear at a 10 degree angle by turning the intake camshaft with the proper tool.

b. Apply grease to the thrust portion of the camshaft.

c. Align the timing marks (single dots) on the camshaft drive and driven gears.

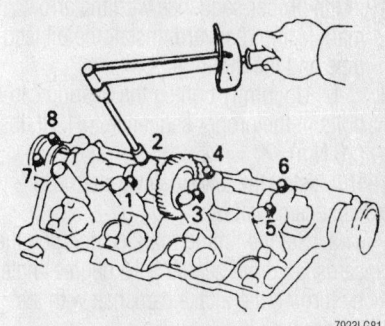

Left side intake camshaft bracket bolt tightening sequence—4.0L (1UZ-FE) and 4.3L (3UZ-FE) Engines

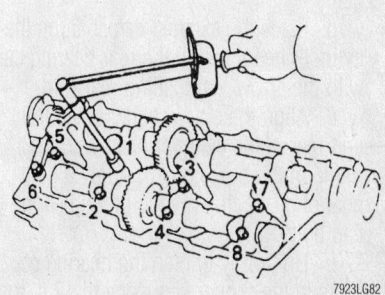

Left side exhaust camshaft bracket bolt tightening sequence—4.0L (1UZ-FE) and 4.3L (3UZ-FE) Engines

d. Place the exhaust camshaft in the cylinder head. Install the rear bearing cap with the arrow mark facing rearward.

e. Align the arrow marks at the front and rear of the cylinder head with the mark on the bearing cap. Apply a light coat of oil on the threads of the bearing cap bolts.

f. Uniformly tighten the bearing cap bolts in the proper sequence to 12 ft. lbs. (16 Nm).

g. Bring the service bolt installed upward by turning the camshaft with the proper tool. Remove the service bolt.

14. To install the left side cylinder head intake camshaft:

a. Apply grease to the thrust portion of the camshaft.

b. Place the intake camshaft with the timing mark (single dot) at a 60 degree angle on the cylinder head.

c. Remove any old packing and apply new seal packing to the bearing cap marked **I6** and install the front bearing cap, marked **I1** with the arrow facing rearward.

d. Align the arrows at the front and rear of the cylinder head with the bearing cap. Apply a light coat of oil on the threads of the bearing cap bolts.

e. Install the remaining bearing caps in the proper sequence with the arrow mark facing rearward. Install the oil feed pipe and the mounting bolts.

f. Uniformly tighten the bearing cap bolts in the proper sequence to 12 ft. lbs. (16 Nm).

15. Install the left side cylinder head exhaust camshaft by:

a. Set the timing mark (2 dots) on the camshaft drive gear at a 15 degree angle by turning the intake camshaft with the proper tool.

b. Apply grease to the thrust portion of the camshaft.

c. Align the timing marks (2 dots each) on the camshaft drive and driven gears.

d. Place the exhaust camshaft on the cylinder head. Install the rear bearing cap with the arrow mark facing rearward.

e. Align the arrow marks at the front and rear of the cylinder head with the mark on the bearing cap. Apply a light coat of oil on the threads of the bearing cap bolts.

f. Uniformly tighten the bearing cap bolts in the proper sequence to 12 ft. lbs. (16 Nm).

16. Bring the service bolt installed upward by turning the camshaft with the proper tool. Remove the service bolt.

17. Install or connect the following:

- Camshaft oil seals. Be sure to apply Multi-Purpose (MP) grease to the new oil seal lip.
- Semi-circular plugs to the cylinder heads
- Left cylinder head cover and bolts. Tighten the bolts to 52 inch lbs. (6 Nm).
- Spark plug wires and clamps to the left cylinder head cover
- Engine wire clamp to the wire bracket on the delivery pipe
- EVAP hose to the VSV
- Transmission oil dipstick
- Right cylinder head cover and bolts. Tighten the bolts to 52 inch lbs. (6 Nm).
- Spark plug wires and clamps to the right cylinder head cover
- Throttle body to the air intake chamber. Install the 2 bolts and 2 nuts and tighten to 13 ft. lbs. (18 Nm).
- Timing belt rear plates by installing the bolts. Tighten the bolts to 66 inch lbs. (8 Nm).
- No. 2 ignition coil

18. Align the knock pin on the right side camshaft with the knock pin of the timing pulley. Slide on the timing pulley with the right side mark facing forward. Tighten the bolt to 80 ft. lbs. (108 Nm).

19. Align the knock pin on the left side camshaft with the knock pin of the timing pulley. Slide on the timing pulley with the left side mark facing forward. Tighten the bolt to 80 ft. lbs. (108 Nm).

20. Turn the crankshaft pulley and align its groove with the **0** timing mark on the timing belt cover.

21. Turn each camshaft timing pulley and align the timing marks of the pulley with the timing belt rear plate.

22. Attach the timing belt to the left side camshaft timing pulley:

23. Using the proper tool, slightly turn the left side timing pulley clockwise. Align the installation mark of the timing belt with the timing mark of the camshaft timing pulley and hang the timing belt on the left side camshaft pulley.

24. Align the timing marks of the left side camshaft pulley and the timing belt rear plate.

25. Check that the timing belt has tension between crankshaft timing pulley and the left side camshaft pulley.

26. Install the timing belt to the right side camshaft timing pulley:

27. Using the proper tool, slightly turn the right side timing pulley clockwise. Align

the installation mark of the timing belt with the timing mark of the camshaft timing pulley and hang the timing belt on the right side camshaft pulley.

28. Align the timing marks of the right side camshaft pulley and the timing belt rear plate.

29. Check that the timing belt has tension between crankshaft timing pulley and the right side camshaft pulley.

30. The timing belt tensioner must be set prior to installation. The tensioner can be set by:

a. Place a plate washer between the tensioner and a block. Using a press, press in the pushrod using 220–225 lbs. (100–102 kg.) of pressure.

b. Align the holes of the pushrod and housing, pass a 0.05 in. (1.27mm) rod through the holes to keep the setting position of the pushrod.

c. Release the press and install the dust boot to the tensioner.

d. Loosely install the tensioner. Evenly and alternately tighten the bolts to 20 ft. lbs. (26 Nm). Remove the tool from the tensioner.

e. Turn the crankshaft pulley 2 complete revolutions from TDC to TDC. Always turn the crankshaft clockwise. Check that all belt and pulley marks align with their reference marks. If any mark is out of perfect alignment, the timing belt must be removed and reinstalled.

31. Install or connect the following:

- Remaining components
- V bank cover
- Coolant
- Battery and battery tray
- Battery cables, positive cable first
- Air cleaner inlet
- Battery clamp cover
- Engine undercover
- Oil pan protector

Valve Lash

ADJUSTMENT

3.0L (1MZ-FE) Engine

1. Before servicing the vehicle, refer to the precautions in the beginning of this section.

➡ **Adjust the valve clearance when the engine is cold.**

2. Remove or disconnect the following:

- Negative battery cable

- Accelerator/throttle cable from the throttle linkage
- Air intake chamber
- Cylinder head covers

3. Turn the crankshaft pulley and align it's groove with the timing mark **0** of the No. 1 timing cover.

4. Check that the valve lash adjusters on the No. 1 intake are loose and the exhaust are tight. If not, turn the crankshaft on complete revolution (360 degrees).

5. Measure the clearance between the valve lash adjuster and the camshaft. Record the measurements on valves No. 1, 2, 3 and 6.

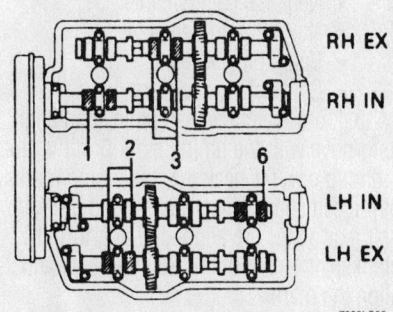

RH EX
RH IN

LH IN
LH EX

7923LG28

Adjust these valves FIRST—3.0L (1MZ-FE) engine

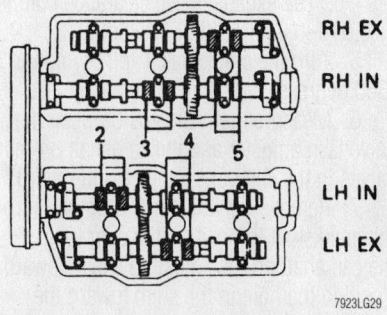

RH EX
RH IN

LH IN
LH EX

7923LG29

Adjust these valves SECOND—3.0L (1MZ-FE) engine

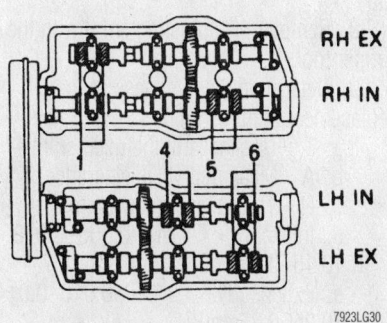

RH EX
RH IN

LH IN
LH EX

7923LG30

Adjust these valves THIRD—3.0L (1MZ-FE) engine

6. The intake valve clearance cold is 0.006–0.010 in. (0.15–0.25mm).

7. The exhaust valve clearance cold is 0.010–0.014 in. (0.25–0.35mm).

8. Turn the crankshaft ⅔ of a revolution (240 degrees) and check the clearance on valves No. 2, 3, 4 and 5 and record.

9. Turn the crankshaft another ⅔ of a revolution and check valves; No. 1, 4, 5 and 6 and record.

10. Remove or disconnect the following:

- Adjusting shim and turn the crankshaft to position the cam lobe of the camshaft on the adjusting valve upward. Press down the valve lash adjuster with the proper tool and place the proper tool between the camshaft and the valve lash adjuster. Remove the tool.
- Adjusting shim with the proper tool.

11. Determine the thickness of the replacement shim as follows:

a. T: Thickness of the used shim
b. A: Measured valve clearance
c. N: Thickness of new shim
d. Intake: $N = T + (A - 0.006-0.010$ in. $(0.15-0.25mm))$
e. Exhaust: $N = T + (A - 0.010-0.014$ in. $(0.25-0.35mm))$

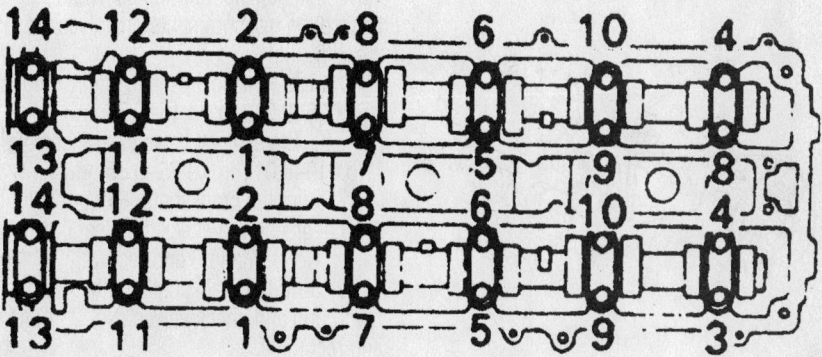

7923LG31

Camshaft bearing cap bolt tightening sequence—3.0L (2JZ-GE) engine

12. Install the specified valve shim on the valve lash adjuster

13. Recheck the valve clearance.

14. Install the cylinder head covers and intake chamber.

15. Connect the negative battery cable.

3.0L (2JZ-GE) Engine

➡**Adjust the valve clearance when the engine is cold.**

1. Before servicing the vehicle, refer to the precautions in the beginning of this section.

2. Remove or disconnect the following:

- Negative battery cable
- Accelerator/throttle cable from the throttle linkage
- Cylinder head covers

3. Turn the crankshaft pulley and align it's groove with the timing mark **0** of the No. 1 timing cover.

4. Check that the timing marks on the camshaft sprockets are in alignment with the marks on the No. 4 timing cover. If not, turn the crankshaft 1 complete revolution (360 degrees).

5. Uniformly tighten the camshaft bearing cap bolts in several passes, in the sequence, to 14 ft. lbs. (20 Nm).

6. Measure the clearance between the

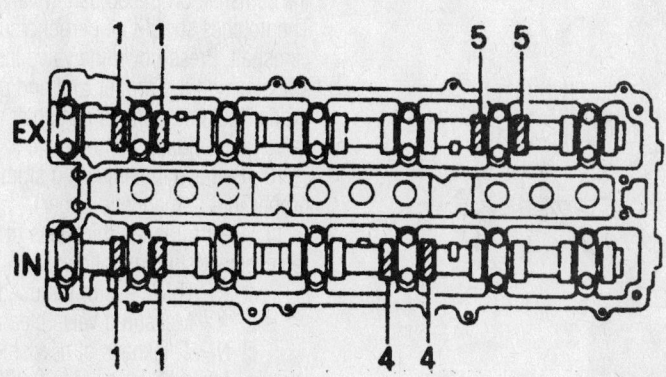

EX

IN

7923LG32

Adjust these valves FIRST—3.0L (2JZ-GE) engine

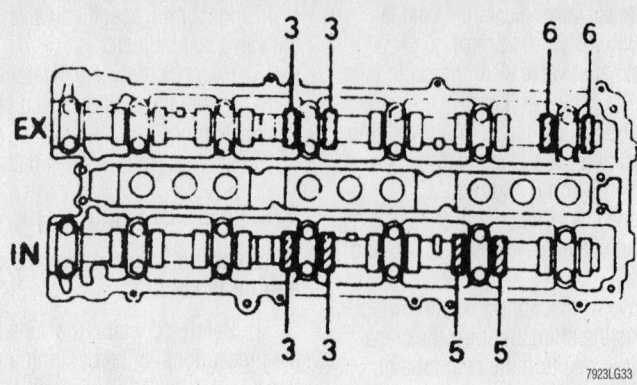

Adjust these valves SECOND—3.0L (2JZ-GE) engine

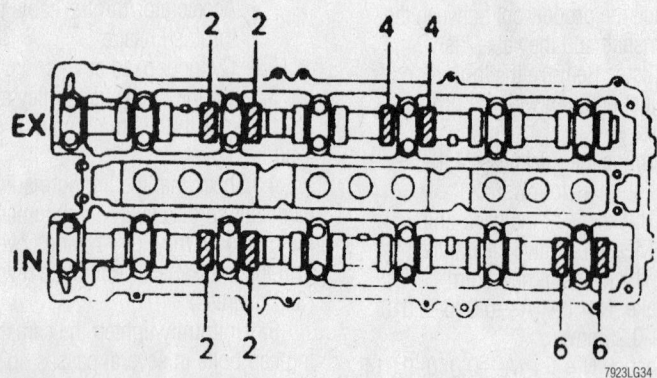

Adjust these valves THIRD—3.0L (2JZ-GE) engine

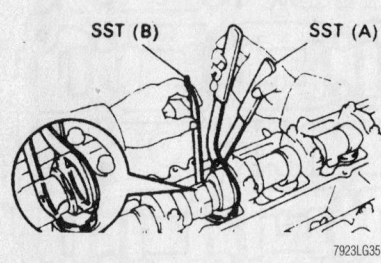

Press down the valve lash adjuster with a special tool—3.0L (2JZ-GE) engine

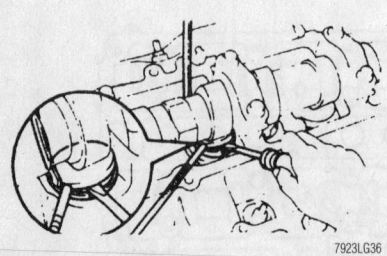

Removing the adjusting shim—3.0L (2JZ-GE) engine

valve lash adjuster and the camshaft. Record the measurements on valves No. 1, 4 and 5.

    a. The intake valve clearance cold is 0.006–0.010 in. (0.15–0.25mm).

    b. The exhaust valve clearance cold is 0.010–0.014 in. (0.25–0.35mm).

  7. Turn the crankshaft ⅔ of a revolution (240 degrees) and check the clearance on valves No. 3, 5 and 6 and record.

  8. Turn the crankshaft another ⅔ of a revolution and check valves: No. 2, 4 and 6 and record.

  9. Remove the adjusting shim and turn the crankshaft to position the cam lobe of the camshaft on the adjusting valve upward. The notches should be perpendicular to the camshaft. Press down the valve lash adjuster with the proper tool and place the proper tool between the camshaft and the valve lash adjuster. Remove the tool.

  10. Remove the adjusting shim with the proper tool (a magnetic finger).

  11. Determine the thickness of the replacement shim as follows:

    a. T = Thickness of the used shim

    b. A = Measured valve clearance

    c. N = Thickness of new shim

    d. Intake: N = T + (A—0.006–0.010 in. (0.15–0.25mm))

    e. Exhaust: N = T + (A—0.010–0.014 in. (0.25–0.35mm))

  12. Install the specified valve shim on the valve lash adjuster.

  13. Recheck the valve clearance.

  14. Install the cylinder head covers and intake chamber.

  15. Connect the negative battery cable.

4.0L (1UZ-FE) and 4.3L (3UZ-FE) Engines

  1. Before servicing the vehicle, refer to the precautions in the beginning of this section.

  2. Remove or disconnect the following:

- Negative battery cable
- No. 3 timing belt covers
- Spark plug wires
- Cylinder head covers

  3. Turn the crankshaft pulley and align its groove with the timing mark **0** of the No. 1 timing cover. Check that the timing marks of the camshaft timing pulleys and timing belt rear plates are aligned. If not, turn the crankshaft 1 revolution (360 degrees) and align the mark.

  4. Measure the clearance between the valve lash adjuster and the camshaft on the valves in the first sequence and record.

    a. The intake valve clearance cold is 0.006–0.010 in. (0.15–0.25mm).

    b. The exhaust valve clearance cold is 0.010–0.014 in. (0.25–0.35mm).

  5. Turn the crankshaft 1 full revolution (360 degrees) and align the mark.

  6. Measure the clearance between the valve lash adjuster and the camshaft on the valves in the second sequence and record.

  7. Remove the adjusting shim and turn the crankshaft to position the cam lobe of the camshaft on the adjusting valve upward. Position the hole in the shim toward the outside of the cylinder head. Press down the valve lash adjuster with the proper tool and place the proper tool between the camshaft and the valve lash adjuster. Remove the tool.

  8. Remove the adjusting shim with the proper tool.

  9. Determine the thickness of the replacement shim as follows:

    a. T = Thickness of the used shim

    b. A = Measured valve clearance

    c. N = Thickness of new shim

    d. Intake: N = T + (A—0.006–0.010 in. (0.15–0.25mm))

    e. Exhaust: N = T + (A—0.010–0.014 in. (0.25–0.35mm))

  10. Recheck the valve clearance. Install the cylinder head covers.

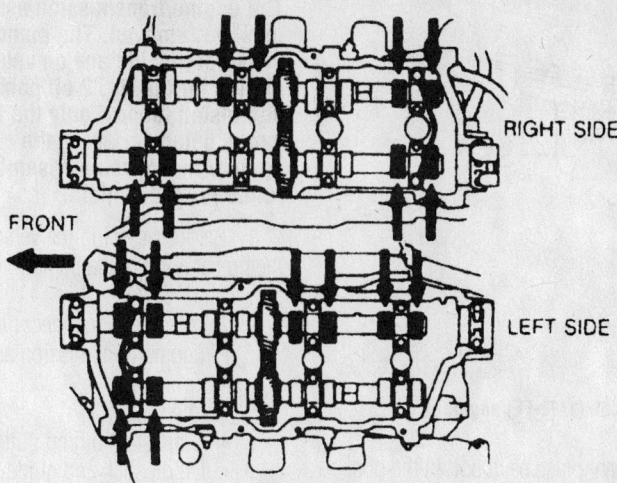

Adjust these valves FIRST—4.0L (1UZ-FE) and 4.3L (3UZ-FE) Engines

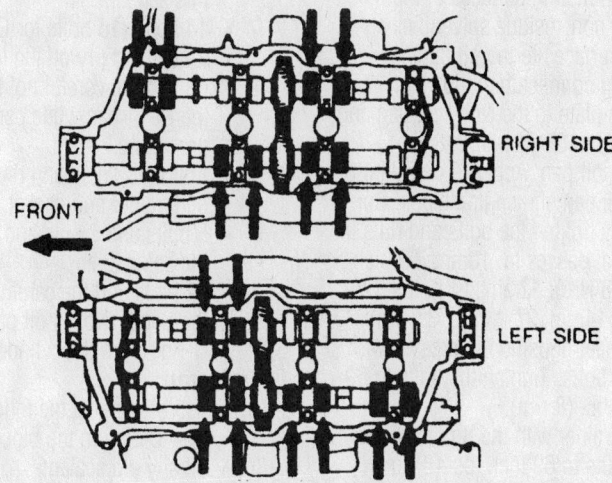

Adjust these valves SECOND—4.0L (1UZ-FE) and 4.3L (3UZ-FE) Engines

11. Connect the spark plug wires and install the No. 3 timing belt covers.

12. Connect the negative battery cable.

Starter Motor

REMOVAL & INSTALLATION

3.0L (1MZ-FE) Engine

1. Before servicing the vehicle, refer to the precautions in the beginning of this section.

2. Remove or disconnect the following:
- Negative cable
- Automatic transmission shift control cable
- Engine wire
- Starter connector
- Nut, and disconnect the starter wire
- 2 bolts, automatic transmission shift control cable clamp and starter

To install:

3. Installation is the reversal of the removal process.

4. Torque the starter bolts to 27 ft. lbs. (37 Nm).

3.0L (2JZ-GE) Engine

1. Before servicing the vehicle, refer to the precautions in the beginning of this section.

2. Remove or disconnect the following:
- Negative cable
- Starter connector
- Nut, and disconnect the starter wire
- 2 bolts and starter

To install:

3. Installation is the reversal of the removal procedure.

4. Tighten the starter bolts to 27 ft. lbs. (37 Nm).

4.0L (1UZ-FE) and 4.3L (3UZ-FE) Engines

1. Before servicing the vehicle, refer to the precautions in the beginning of this section.

2. Remove or disconnect the following:
- V-bank cover
- Accelerator cable
- Intake air connector
- Throttle Body
- Intake manifold assembly
- Rear water bypass joint
- Water bypass pipe
- 2 bolts
- Water bypass pipe from the water pump
- Wire clamp from the bracket on the water bypass pipe
- O-ring from the water bypass pipe
- Water bypass pipe bracket from the water bypass pipe
- Starter
- 2 bolts holding the starter to the cylinder block
- Starter from the cylinder block
- Starter connector
- Nut, and disconnect the starter wire
- Starter

To install:

3. Install or connect the following:
- Wire clamp to the wire bracket with the bolt. Tighten to 87 inch lbs.
- Starter wire with the nut. Tighten to 87 inch lbs.
- Starter connector
- Starter with the 2 bolts. Torque the bolts to 29 ft. lbs. (39 Nm).
- Water bypass pipe bracket to the water bypass pipe
- O-ring to the water bypass pipe
- Water bypass pipe
- Wire clamp to the bracket on the water bypass pipe
- Water bypass pipe bolts. Torque the bolts to 13 ft. lbs. (18 Nm).
- Rear water bypass joint
- Intake manifold assembly
- Throttle body
- Intake air connector
- Accelerator cable
- V-bank cover

Oil Pan

REMOVAL & INSTALLATION

3.0L (1MZ-FE) Engine

1. Before servicing the vehicle, refer to the precautions in the beginning of this section.

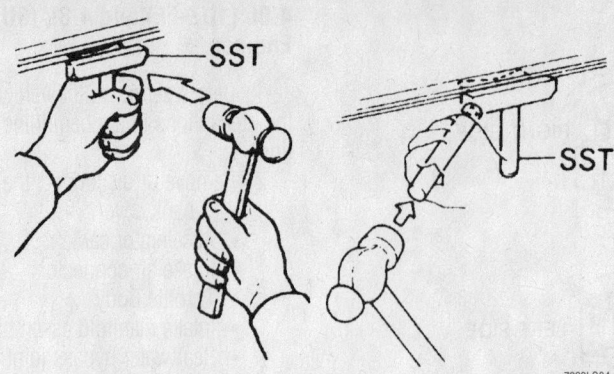

Use the special tool to break the seal and remove the oil pan—3.0L (1MZ-FE) engine

2. Drain the engine oil.
3. Remove or disconnect the following:
   - Negative battery cable from the battery.
   - Right front wheel
   - Fender apron seal
   - Engine undercover
   - Front exhaust pipe
   - Front exhaust pipe bracket from the No. 1 oil pan
   - Flywheel housing undercover
   - 10 bolts and 2 nuts to the No. 2 oil pan
4. Insert a blade between the No. 1 and No. 2 oil pans. Tap the tool sideways to break the seal and remove the pan. Clean the surfaces of the oil pans.
   - Oil strainer and gasket from the engine by removing the 3 nuts.
   - No. 1 oil pan as follows. Make a note of the position of the each bolt. When replacing the bolts into the oil pan, place each bolt in the position from which it was removed.
   - Baffle plate from the No. 1 oil pan

To install:
5. Clean all mating surfaces of the oil pans. Using a non-residue solvent, clean both sealing surfaces to the oil pan.
6. Install or connect the following:
   - Baffle plate to the No. 1 oil pan and tighten to 69 inch lbs. (8 Nm)
   - No. 1 oil pan. Apply RTV sealant to the oil pan and engine block. Uniformly tighten the bolts and nuts in several passes to: 10mm: 69 inch lbs. (8 Nm); 12mm: 14 ft. lbs. (20 Nm); 14mm: 27 ft. lbs. (37 Nm)
   - Flywheel housing undercover with the 2 bolts. Tighten the bolts to 69 inch lbs. (8 Nm).
   - Oil strainer with the 3 nuts. Tighten the nuts to 69 inch lbs. (8 Nm).
   - No. 2 oil pan. Apply RTV sealant to the oil pan and engine block. Uniformly tighten the bolts and nuts in several passes, to 69 inch lbs. (8 Nm).
   - Flywheel housing undercover
   - Front exhaust pipe bracket to the No. 1 oil pan. Tighten the bolts to 15 ft. lbs. (21 Nm).
   - Front exhaust pipe. 4 pipe-to-pipe nuts: 46 ft. lbs. (62 Nm); front exhaust pipe-to-the center exhaust pipe bolts and nuts: 41 ft. lbs. (56 Nm); bracket bolts: 14 ft. lbs. (19 Nm); support stay bolts: 22 ft. lbs. (30 Nm).
   - Engine undercover
   - Right fender apron seal
   - The right front wheel and lower the vehicle
   - Engine with oil

3.0L (2JZ-GE) Engine

➡The No. 1 oil pan can not be removed with the engine in the vehicle.

The engine/transmission assembly must be removed. The manufacturer does not provide any on vehicle information for the No. 2 oil pan removal and installation. If only the No. 2 oil pan is being serviced, the engine/transmission assembly can remain in the vehicle.

1. Before servicing the vehicle, refer to the precautions in the beginning of this section.
2. Remove or disconnect the following:
   - Engine/transmission assembly
   - Timing belt
   - Idler pulley
   - Crankshaft timing pulley
   - Oil dipstick and guide
   - Oil sensor lead
   - 4 attaching bolts and lift off the oil level sensor. Be careful not to drop this sensor.
   - 14 bolts (16 bolts for GS 300) and 2 nuts and pry off the lower (No. 2) oil pan. Be careful not to damage the No. 1 pan while performing this procedure.
   - Bolt and 2 nuts and drop down the oil strainer and gasket
   - 5 bolts and 2 nuts and drop down the baffle plate
   - 22 bolts and the carefully pry off the upper (No. 1) oil pan
   - O-ring from the cylinder block

To install:
3. Install or connect the following:
   - New O-ring in the block and scrape off any old sealant
   - A ⅛ inch (3–4mm) bead of RTV sealant to the pan mating surface
   - Upper pan. 12mm bolts: 15 ft. lbs. (21 Nm); 14mm bolts to 29 ft. lbs. (39 Nm)
   - Baffle plate and oil strainer. Tighten them both to 78 inch lbs. (9 Nm).
   - Lower pan in the same manner as the upper pan and tighten the bolts to 78 inch lbs. (9 Nm)

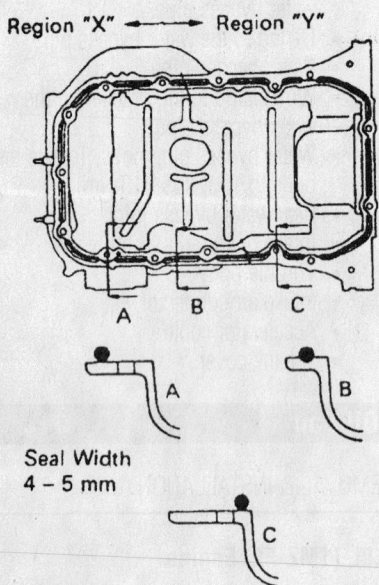

Region "X" ⟵ ⟶ Region "Y"

A   B   C

A

B

C

Seal Width
4 – 5 mm

Apply sealant as shown to the No. 1 (upper) oil pan—3.0L (1MZ-FE) engine

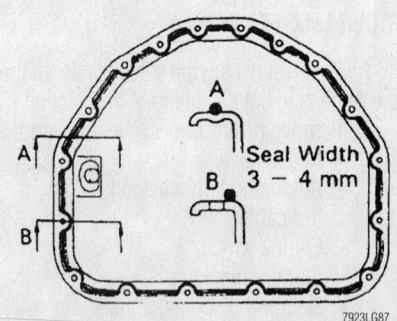

A

A

C

B

B

Seal Width
3 – 4 mm

Lower oil pan sealant application—3.0L (2JZ-GE) engine

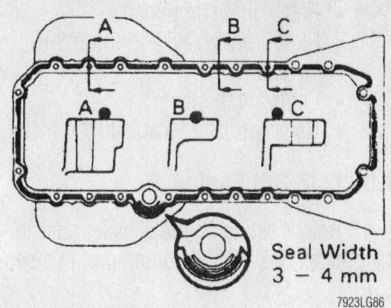

Upper oil pan sealant application—3.0L (2JZ-GE) engine

- Oil level sensor and tighten it to 48 inch lbs. (5 Nm)
- Oil dipstick and guide
- Timing pulleys and belt
- Transmission to the engine
- Engine and transmission
- All fluids

4.0L (1UZ-FE) and 4.3L (3UZ-FE) Engines

LS 400, LS 430

1. Before servicing the vehicle, refer to the precautions in the beginning of this section.
2. Remove or disconnect the following:
   - Engine/transmission assembly
   - Remove the timing belt
   - Idler pulleys
   - Crankshaft timing pulley
   - Oil dipstick and guide
   - Oil level sensor lead
   - 4 bolts and lift off the oil level sensor. Be careful not to drop this sensor.
   - Oil filter and the bracket assembly by removing the stud bolt and 2 nuts
   - Engine Crankshaft Position (CKP) sensor connector
   - Sensor by removing the bolt
   - 12 bolts and 2 nuts to the No. 2 oil pan. Use a gasket cutting tool to separate the No. 2 (lower) oil pan. Be careful not to damage the No. 1 pan while performing this procedure.
   - 2 bolts and 3 nuts and drop down the baffle plate
   - Oil strainer by removing the bolts and nuts
   - Bolts, then carefully pry off the No. 1 oil pan. There are slots for inserting the prybar.

To install:
3. Install or connect the following:
   - No. 1 pan. Apply a ⅛ inch (3–4mm) bead sealant to the pan mating surface. Bolts: 10mm: 66

inch lbs. (8 Nm); 12mm: 21 ft. lbs. (28 Nm)
   - Oil strainer. Bolts and nuts: 66 inch lbs. (8 Nm)
   - Baffle plate. Bolts and nuts: 66 inch lbs. (8 Nm)
   - No. 2 pan in the same manner as the No. 1 oil pan and tighten the bolts to 66 inch lbs. (8 Nm). Be sure the bolts are 14mm in length.
   - CKP sensor. Tighten the bolt to 56 inch lbs. (6 Nm).
   - New O-ring in position on the oil filter bracket
   - Bracket and tighten the bolt and nuts to 13 ft. lbs. (18 Nm)
   - Wiring to the pressure switch
   - Oil level sensor and tighten the 4 bolts to 48 inch lbs. (5 Nm). Use a new gasket.
   - Dipstick and guide
   - Timing belt pulleys and the timing belt components
   - Transaxle to the engine
   - Engine and transaxle
   - All fluids

SC 400, GS 400, GS 430

➡ **The No. 1 oil pan cannot be removed with the engine in the vehicle. The**

engine and transmission must be removed as a unit, then separated. It may be possible to remove the No. 2 oil pan from the vehicle while the engine is still in the vehicle.

1. Before servicing the vehicle, refer to the precautions in the beginning of this section.
2. Remove or disconnect the following:
   - Engine/transmission assembly
   - Oil dipstick and guide
   - 12 bolts and 2 nuts. Use a gasket-cutting tool to separate the No. 2 (lower) oil pan. Be careful not to damage the No. 1 pan while performing this procedure.
   - 6 bolts and 2 nuts; remove the baffle plate
   - 16 bolts, then carefully pry off the No. 1 oil pan

➡ **There are slots for inserting the pry-bar.**

To install:
3. Install or connect the following:
   - No. 1 pan. Apply a ⅛ inch (3–4mm) bead on RTV sealant to the pan mating surface. Bolts: 12mm: 69 inch lbs. (8mm); 14mm: 20 ft. lbs. (28 Nm)

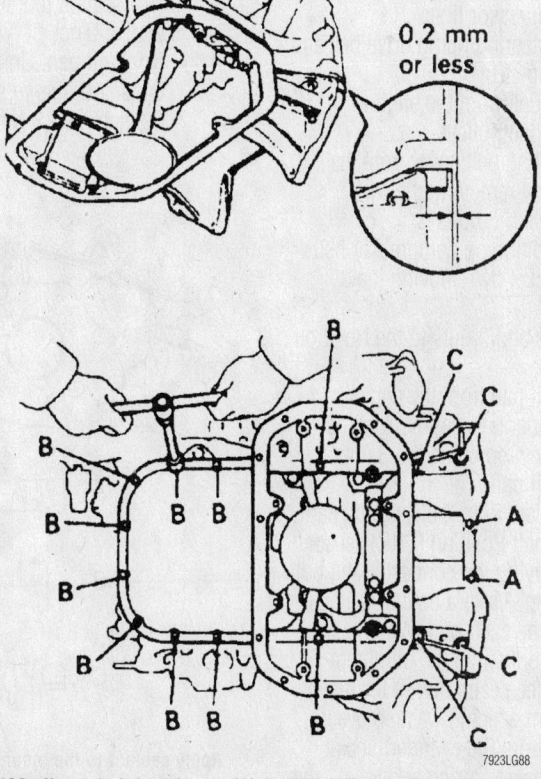

SC 400 and GS 400 oil pan bolt locations—(A) 0.78 in (20mm), (B) 1.38 in. (35mm), (C) 0.78 in (20mm) with 12mm bolt heads

- Baffle plate. Bolts and nuts: 69 inch lbs. (8 Nm).
- RTV sealant to the pan mating surface
- No. 2 oil pan. Bolts: 69 inch lbs. (8 Nm)
- Dipstick and guide
- Engine/transaxle assembly
- All fluids

Oil Pump

REMOVAL & INSTALLATION

3.0L (1MZ-FE) Engine

1. Before servicing the vehicle, refer to the precautions in the beginning of this section.
2. Remove or disconnect the following:
   - Negative battery cable from the battery
   - Right front wheel
   - Fender apron seal
   - Engine undercover
   - Engine oil
   - Front exhaust pipe
   - Front exhaust pipe bracket from the No. 1 oil pan
   - Alternator drive belt from the engine
   - Air conditioning compressor from the engine, without disconnecting the compressor lines
   - Power steering pump drive belt and adjusting strut
   - Timing belt from the engine
   - Timing belt pulleys
   - Rear timing belt cover from the engine by removing the wire clamps and 6 bolts
   - Air conditioning compressor housing bracket by removing the 3 bolts.
   - 10 bolts and 2 nuts to the No. 2 oil pan
   - No. 2 oil pan from the engine
   - Oil strainer and gasket from the engine by removing the 3 nuts
   - No. 1 oil pan
   - Baffle plate from the No. 1 oil pan
   - Crankshaft Position (CKP) sensor by removing the connector and bolt
   - Oil pump. Make a note of the position of the each bolt. When replacing the bolts into the oil pump body, place each bolt in the position from which it was removed.
   - O-ring from the cylinder block
   - Plug, gasket, spring and relief valve from the oil pump body

To install:

3. Install or connect the following:
   - Driven rotors, drive, pump body cover, then install the 9 screws
   - Relief valve, spring, gasket and the plug to the oil pump body
   - New O-ring on the cylinder block
   - RTV sealant to the oil pump as shown
   - Pump on the engine block. Be sure to engage the spline teeth of the oil pump drive gear with the large teeth of the crankshaft.
   - The 9 bolts to the oil pump and uniformly tighten the bolts in several passes. Tighten the bolts to: 10mm: 69 inch lbs. (8 Nm); 12mm: 14 ft. lbs. (20 Nm).
   - CKP sensor and bolt. Tighten the bolt to 69 inch lbs. (8 Nm).
   - Baffle plate to the No. 1 oil pan and tighten to 69 inch lbs. (8 Nm).
   - No. 1 oil pan Uniformly tighten the bolts and nuts in several passes: 10mm–69 inch lbs. (8 Nm); 12mm–14 ft. lbs. (20 Nm); 14mm–27 ft. lbs. (37 Nm).
   - Flywheel housing undercover with the 2 bolts. Tighten the bolts to 69 inch lbs. (8 Nm).
   - Oil strainer with the 3 nuts. Tighten the nuts to 69 inch lbs. (8 Nm).
   - No. 2 oil pan
   - RTV sealant to the oil pan and engine block
   - No. 2 oil pan. Uniformly tighten the bolts and nuts in several passes to 69 inch lbs. (8 Nm).

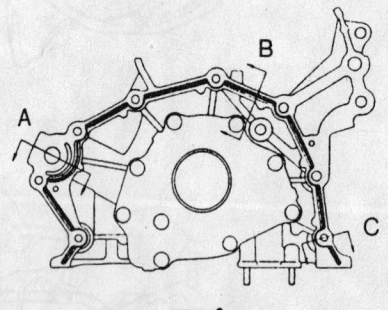

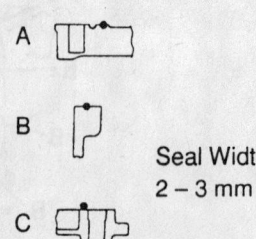

Seal Width 2 – 3 mm

7923LG89

Apply sealant to the mounting surface of the oil pump in the areas shown—3.0L (1MZ-FE) engine

- Remaining components
- Right front wheel and lower the vehicle
- Engine with oil
- Negative battery cable to the battery

3.0L (2JZ-GE) Engine

1. Before servicing the vehicle, refer to the precautions in the beginning of this section.
2. Remove or disconnect the following:
   - Engine and transmission
   - Timing belt
   - Idler pulley
   - Crankshaft timing pulley
   - Oil dipstick and tube
   - Oil level sensor
   - No. 2 (lower) oil pan
   - Oil strainer by removing the bolt and 2 nuts
   - Oil baffle plate by removing the 6 bolts
   - No. 1 (upper) oil pan by removing the 22 bolts. Take note of bolt size and placement for correct re-installation.
   - 9 mounting bolts to the oil pump body. Carefully drive the pump off the cylinder block using a brass drift.
   - 2 O-rings

To install:

3. Install or connect the following:
   - 2 new O-rings in the cylinder block
   - A ⅛ inch (3–4mm) bead of RTV sealant around the pump mating surface, taking great care around the oil passages
   - Pump and tighten the bolts to 15 ft. lbs. (21 Nm)
   - A new O-ring on the block
   - RTV sealant around the No. 1 oil pan
   - No. 1 oil pan. Bolts: 12mm–15 ft. lbs. (21 Nm); 14mm–29 ft. lbs. (39 Nm)
   - Oil baffle plate and tighten the nuts and bolts to 78 inch lbs. (9 Nm)
   - Oil strainer and tighten the nuts and bolts to 78 inch lbs. (9 Nm)
   - RTV sealant around the No. 2 oil pan
   - No. 2 oil pan and tighten the bolts to 78 inch lbs. (9 Nm)
   - Oil lever sensor with a new gasket and tighten the bolts to 48 inch lbs. (6 Nm)
   - Oil dipstick with a new O-ring
   - Remaining components
   - All fluids
   - Negative battery cable

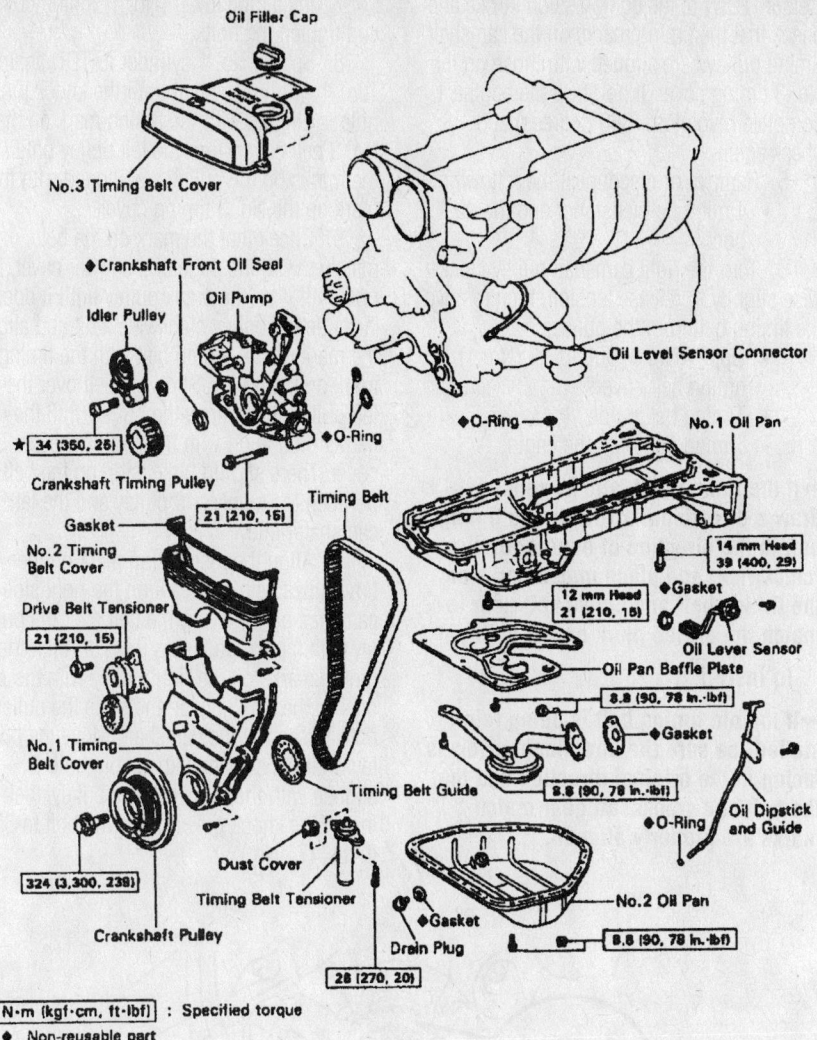

Exploded view of the oil pump and related component mountings—3.0L (2JZ-GE) engine

1. Before servicing the vehicle, refer to the precautions in the beginning of this section.
2. Remove or disconnect the following:
   - Engine/transmission assembly
   - Timing belt
   - Idler pulleys
   - Crankshaft timing pulley
   - Oil dipstick and guide
   - Oil level sensor lead
   - 4 bolts and lift off the oil level sensor. Be careful not to drop this sensor.
   - Main Oxygen (O₂) sensor bracket, if necessary
   - Oil filter and filter bracket assembly by removing the stud bolt and 2 nuts
   - Engine Crankshaft Position (CKP) sensor. Remove the sensor by removing the bolt.
   - 12 bolts and 2 nuts from the No. 2 oil pan
   - No. 2 (lower) oil pan. Use a gasket-cutting tool
   - 2 bolts and 3 nuts and drop down the baffle plate
   - Oil strainer
   - No. 1 oil pan. There are slots for inserting the prybar.
   - 8 bolts holding the oil pump to the engine

➡ **Make certain to observe bolt position during removal. The bolts are different lengths and sizes. Record their position for proper reassembly.**

   - Oil pump from the engine block
   - O-ring from the block

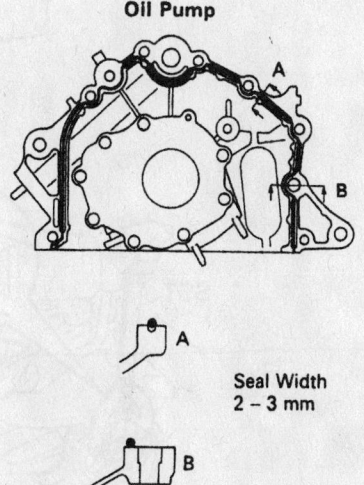

Oil pump mounting bolt installation locations—3.0L (2JZ-GE) engine

4.0L (1UZ-FE) and 4.3L (3UZ-FE) Engines

➡The oil pump cannot be removed with the engine in the vehicle. The engine and transmission must be removed as a unit, then separated.

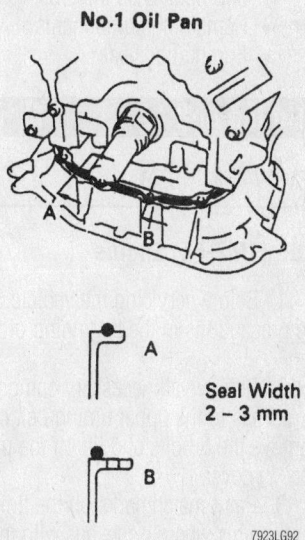

Apply sealant to the oil pump and the No. 1 oil pan, as shown, before installing the oil pump—4.0L (1UZ-FE) and 4.3L (3UZ-FE) Engines

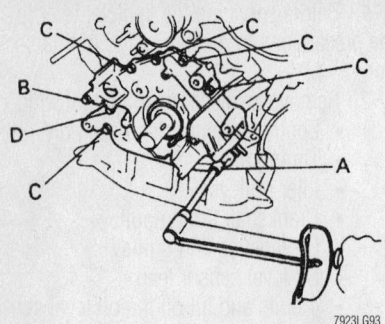

7923LG93

4.0L engine oil pump mounting bolt locations, according to bolt lengths—(A) 1.97 in. (50mm), (B) 4.17 in. (106mm), (C) 1.18 in. (30mm) and (D) 1.57 in. (40mm)

To install:

➡ **Prior to installing the oil pump, lubricate the gears with clean engine oil.**

3. Install or connect the following:
- A 2–3mm wide (0.08–0.12 in.) bead of RTV sealant to the oil pump
- New O-ring in position on the block
- Oil pump on the engine
- The 8 bolts in their correct locations. Tighten the bolts with 12mm heads to 12 ft. lbs. (16 Nm) and the bolts with 14mm heads to 22 ft. lbs. (30 Nm).
- A ⅛ inch (3–4mm) bead of RTV sealant to the pan mating surface.
- No. 1 pan. Bolts–10mm: 66 inch lbs. (8 Nm); 12mm: 21 ft. lbs. (28 Nm)
- Oil strainer and tighten the bolts to 66 inch lbs. (8 Nm)
- Baffle plate and tighten the bolts and nuts to 66 inch lbs. (8 Nm)
- Remaining components
- Engine/transaxle

Timing Belt

REMOVAL & INSTALLATION

3.0L (1MZ-FE) Engine

1. Before servicing the vehicle, refer to the precautions in the beginning of this section.

2. Remove all necessary components for access to the upper timing belt cover. Remove the 8 bolts and lift off the upper (No. 2) cover.

3. Paint matchmarks on the timing belt at all points where it meshes with the pulleys and the lower timing cover.

4. Set the No. 1 cylinder to Top Dead Center (TDC) of the compression stroke and check that the timing marks on the camshaft timing pulleys are aligned with those on the No. 3 timing cover. If not, turn the engine 1 complete revolution (360 degrees) and check again.

5. Remove or disconnect the following:
- Timing belt tensioner and the dust boot

6. Turn the right camshaft pulley clockwise slightly to release tension, then remove the timing belt from the pulleys.
- Upper (No. 3) and lower (No. 1) timing belt covers
- Timing belt guide
- Timing belt from the engine

➡ **If the timing belt is to be reused, draw a directional arrow on the timing belt in the direction of engine rotation (clockwise) and place matchmarks on the timing belt and crankshaft gear to match the drilled mark on the pulley.**

To install:

➡ **If the old timing belt is being reinstalled, be sure the directional arrow is facing in the original direction and that the belt and crankshaft gear matchmarks are properly aligned.**

7. Install the lower (No. 1) timing cover and tighten the bolts.

8. Set the No. 1 cylinder to TDC again. Turn the right camshaft until the knock pin hole is aligned with the timing mark on the No. 3 belt cover. Turn the left pulley until the marks on the pulley are aligned with the mark on the No. 3 timing cover.

9. Check that the mark on the belt matches with the edge of the lower cover. If not, shift it on the crank pulley until it does. Turn the left pulley clockwise a bit and align the mark on the timing belt with the timing mark on the pulley. Slide the belt over the left pulley. Now move the pulley until the marks on it align with the one on the No. 3 cover. There should be tension on the belt between the crankshaft pulley and the left camshaft pulley.

10. Align the installation mark on the timing belt with the mark on the right side camshaft pulley. Hang the belt over the pulley with the flange facing inward. Align the timing marks on the right pulley with the one on the No. 3 cover and slide the pulley onto the end of the camshaft. Move the pulley until the camshaft knock pin hole is aligned with the groove in the pulley, then install the knock pin. Tighten the bolt to 55 ft. lbs. (75 Nm).

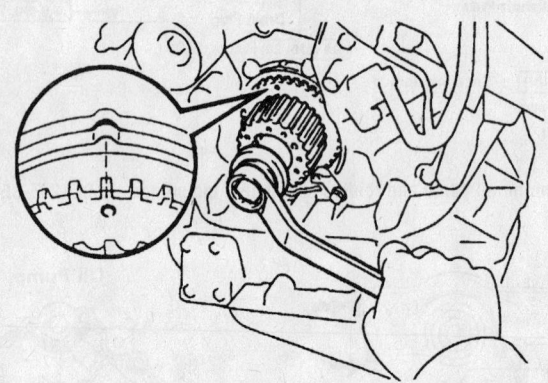

SST

79235G45

Camshaft and crankshaft pulley positioning for timing belt installation—3.0L (1MZ-FE) engine

11. Position a plate washer between the timing belt tensioner and the block, then press in the pushrod until the holes are aligned between it and the housing. Slide a 0.05 in. Allen wrench through the hole to keep the pushrod set. Install the dust boot, then install the tensioner. Tighten the bolts to 20 ft. lbs. (26 Nm). Don't forget to pull out the Allen wrench.

12. Turn the crankshaft clockwise 2 complete revolutions and check that all marks are still in alignment. If they aren't, remove the timing belt and start over again.

13. Install the remaining components.

3.0L (2JZ-GE) Engine

1. Before servicing the vehicle, refer to the precautions in the beginning of this section.

2. Remove all necessary components for access to the upper timing belt covers. Using a 5mm Allen wrench, remove the 9 bolts and lift off the two upper (No. 2 and No. 3) timing belt covers.

3. Rotate the crankshaft pulley clockwise so its groove is aligned with the **0** mark in the No. 1 (lower) timing cover. Check that the timing marks on the camshaft timing sprockets are aligned with the marks on the No. 4 (inner) cover. If not, rotate the crankshaft 1 complete revolution (360 degrees).

4. Alternately loosen the 2 tensioner mounting bolts and remove them, the tensioner and the dust boot. Slide the timing belt off of the 2 camshaft sprockets. Its a good idea to matchmark the belt to the pulleys.

5. Ensuring the timing belt is securely supported, hold the crankshaft pulley with a spanner wrench and loosen the mounting bolt. Remove the bolt and the pulley.

6. Remove or disconnect the following:

   • 5 bolts, then lift off the lower No. 1 timing belt cover

   • Timing belt guide
   • Timing belt

➡**If the timing belt is to be reused, draw a directional arrow on the timing belt in the direction of engine rotation (clockwise) and place matchmarks on the timing belt and crankshaft gear to match the drilled mark on the pulley.**

To install:

7. Install the timing belt on the crankshaft timing pulley and the idler pulleys.

➡**If the old timing belt is being reinstalled, be sure the directional arrow is facing in the original direction and that the belt and crankshaft gear matchmarks are properly aligned.**

8. Install the timing belt guide. Install the lower (No. 1) timing cover and tighten the bolts.

9. Align the crankshaft pulley set key with the key groove on the pulley and slide the pulley on. Tighten the bolt to 239 ft. lbs. (324 Nm).

10. Set the No. 1 cylinder to TDC again. Turn the camshaft until the sprocket timing marks are aligned with the timing marks on the No. 4 belt cover.

11. Check that the marks on the belt matches with those on the sprockets, then slide it over the sprockets. If not, shift it on the crank pulley until it does.

12. Position a plate washer between the timing belt tensioner and the block, then press in the pushrod until the holes are aligned between it and the housing. Slide a 1.5mm Allen wrench through the hole to keep the pushrod set. Install the dust boot, then install the tensioner. Tighten the bolts to 20 ft. lbs. (26 Nm). Don't forget to pull out the Allen wrench.

13. Turn the crankshaft clockwise two complete revolutions and check that all marks are still in alignment. If they aren't, remove the timing belt and start over again.

14. Position new gaskets, then install the upper (No. 2 and No. 3) timing covers.

4.0L (1UZ-FE) and 4.3L (3UZ-FE) Engines

1. Remove all necessary components for access to the right-hand side No. 3 and No. 2, and left-hand side No. 2 timing belt covers, then remove the covers.

2. Turn the crankshaft pulley and align it's groove with the timing mark **0** of the No. 1 timing cover. Check that the timing marks of the camshaft timing pulleys and

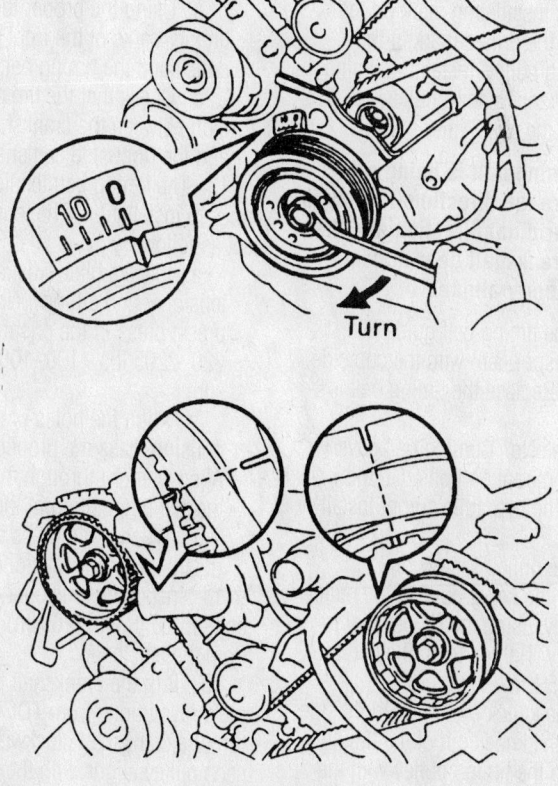

Turn

Set the engine to TDC by aligning the marks before removing the lower timing cover—Lexus 3.0L (2JZ-GE) engine

79235G46

Timing belt sprocket mark alignment for belt installation—Lexus 4.0L (1UZ-FE) engine

79235G47

timing belt rear plates are aligned. If not, turn the crankshaft 1 full revolution (360 degrees).

3. Remove or disconnect the following:

- Timing belt tensioner. Using the proper tool, loosen the tension between the left side and right side timing pulleys by slightly turning the left side camshaft clockwise.
- Timing belt from the camshaft timing pulleys
- Bolt and timing pulleys, using the proper tool
- Bolt and the crankshaft pulley with the proper tool.
- Fan bracket
- Hydraulic pump, on SC 400
- Mounting bolts and the No. 1 timing belt cover
- 2 upper and lower timing belt covers
- Timing belt guide (No. 1 crank position sensor plate)
- Timing belt

➡ If the timing belt is to be reused, draw a directional arrow on the timing belt in the direction of engine rotation (clockwise) and place matchmarks on the timing belt and crankshaft gear to match the drilled mark on the pulley.

To install:

4. Align the installation mark on the timing belt with the drilled mark of the crankshaft timing pulley. Install the timing belt on the crankshaft timing pulley, No. 1 idler pulley and the No. 2 idler pulley.

➡ If the old timing belt is being reinstalled, be sure the directional arrow is facing in the original direction and that the belt and crankshaft gear matchmarks are properly aligned.

5. Install the timing belt guide (No. 1 crank angle sensor plate) with the cup side facing forward. Replace the timing belt cover spacer.

6. Install the No. 1 timing belt cover and tighten the mounting bolts. On the SC400, install the hydraulic pump. Install the fan bracket.

7. Align the pulley set key on the crankshaft with the key groove of the pulley. Install the pulley, using the proper tool to tap in the pulley. Tighten the pulley bolt to 181 ft. lbs. (245 Nm).

8. Align the knock pin on the right side camshaft with the knock pin of the timing pulley. Slide on the timing pulley with the right side mark facing forward. Tighten the bolt to 80 ft. lbs. (108 Nm).

9. Align the knock pin on the left side camshaft with the knock pin of the timing pulley. Slide on the timing pulley with the left side mark facing forward. Tighten the bolt to 80 ft. lbs. (108 Nm).

10. Turn the crankshaft pulley and align its groove with the **0** timing mark on the No. 1 timing belt cover. Using the proper tool, turn the crankshaft timing pulley and align the timing marks of the camshaft timing pulley and the timing belt rear plate.

11. Install the timing belt to the left side camshaft timing pulley by:

a. Using the proper tool, slightly turn the left side timing pulley clockwise. Align the installation mark of the timing belt with the timing mark of the camshaft timing pulley and hang the timing belt on the left side camshaft pulley.

b. Using the proper tool, align the timing marks of the left side camshaft pulley and the timing belt rear plate.

c. Check that the timing belt has tension between crankshaft timing pulley and the left side camshaft pulley.

12. Install the timing belt to the right side camshaft timing pulley by:

a. Using the proper tool, slightly turn the right side timing pulley clockwise. Align the installation mark of the timing belt with the timing mark of the camshaft timing pulley and hang the timing belt on the right side camshaft pulley.

b. Using the proper tool, align the timing marks of the right side camshaft pulley and the timing belt rear plate.

c. Check that the timing belt has tension between the crankshaft timing pulley and the right side camshaft pulley.

13. The timing belt tensioner must be set prior to installation. The tensioner can be set as follows:

a. Place a plate washer between the tensioner and a block. Using a suitable press, press in the pushrod using 220–2205 lbs. (100–1000kg) of pressure.

b. Align the holes of the pushrod and housing, pass the proper tool (0.05 in. Allen wrench) through the holes to keep the setting position of the pushrod.

c. Release the press and install the dust boot on the tensioner.

14. Install the tensioner and tighten the bolts to 20 ft. lbs. (26 Nm). Remove the tool from the tensioner.

15. Turn the crankshaft pulley two complete revolutions from TDC-to-TDC. Always turn the crankshaft clockwise. Check that each pulley aligns with the timing marks.

16. Install all remaining components in the reverse order of removal.

Piston and Ring

POSITIONING

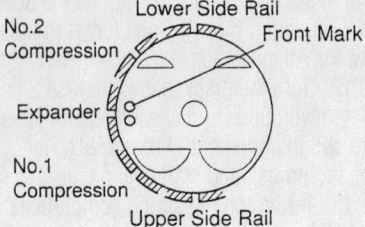

Piston ring positioning—3.0L (1MZE-FE) engine

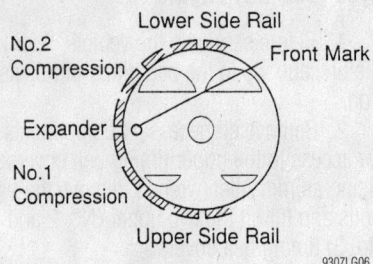

Piston ring positioning—3.0L (2JZ-GE) engine

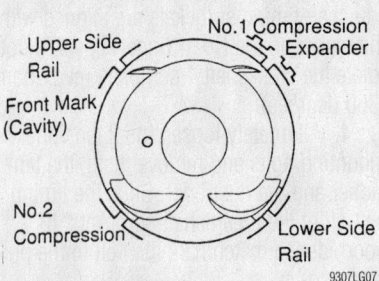

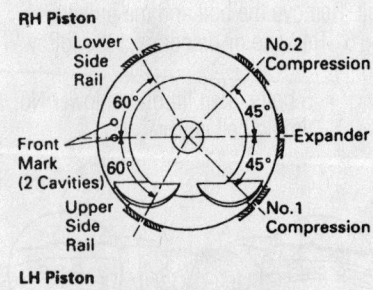

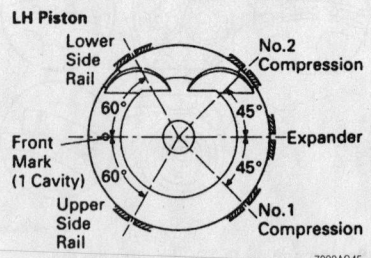

Piston ring positioning—4.0L (1UZ-FE) and 4.3L (3UZ-FE) engines

FUEL SYSTEM

Fuel System Service Precautions

Safety is the most important factor when performing not only fuel system maintenance but any type of maintenance. Failure to conduct maintenance and repairs in a safe manner may result in serious personal injury or death. Maintenance and testing of the vehicle's fuel system components can be accomplished safely and effectively by adhering to the following rules and guidelines.

• To avoid the possibility of fire and personal injury, always disconnect the negative battery cable unless the repair or test procedure requires that battery voltage be applied.

• Always relieve the fuel system pressure prior to disconnecting any fuel system component (injector, fuel rail, pressure regulator, etc.), fitting or fuel line connection. Exercise extreme caution whenever relieving fuel system pressure, to avoid exposing skin, face and eyes to fuel spray. Please be advised that fuel under pressure may penetrate the skin or any part of the body that it contacts.

• Always place a shop towel or cloth around the fitting or connection prior to loosening to absorb any excess fuel due to spillage. Ensure that all fuel spillage (should it occur) is quickly removed from engine surfaces. Ensure that all fuel soaked cloths or towels are deposited into a suitable waste container.

• Always keep a dry chemical (Class B) fire extinguisher near the work area.

• Do not allow fuel spray or fuel vapors to come into contact with a spark or open flame.

• Always use a back-up wrench when loosening and tightening fuel line connection fittings. This will prevent unnecessary stress and torsion to fuel line piping.

• Always replace worn fuel fitting O-rings with new. Do not substitute fuel hose or equivalent, where a fuel pipe is installed.

Fuel System Pressure

RELIEVING

1. Before servicing the vehicle, refer to the precautions in the beginning of this section.

2. Remove the fuse for the electronic fuel pump.

3. Start the engine until the engine stalls.

4. Disconnect the negative battery terminal.

5. Place a catch-pan under the joint to be disconnected. A large quantity of fuel may be released when the joint is opened.

6. Wear eye or full-face protection.

7. Place a shop towel over the area and slowly release the joint using a wrench of the correct size.

8. Allow the any fuel left in the line to bleed off slowly before fully disconnecting the joint.

9. Plug the opened lines immediately to prevent fuel spillage or the entry of dirt.

10. Dispose of the released fuel properly.

11. After connecting fuel lines, install the fuse for the fuel pump and start the engine.

12. Check for leaks and repair as needed.

Fuel Filter

REMOVAL & INSTALLATION

The fuel filter on the ES 300 is located under the hood, on the driver's side, by the fenderwell. The SC 300 fuel filter is located under the vehicle, on the driver's side, in front of the rear axle.

The fuel filter for the LS 400, LS 430 and SC 400 is located under the vehicle on the left side before the rear axle. The fuel filter on the GS 300, GS 400 and GS 430 is located under the vehicle, next to the left rear exhaust resonator.

1. Before servicing the vehicle, refer to the precautions in the beginning of this section.

2. Disconnect the negative battery cable. Wait at least 90 seconds before performing any other work.

3. On the GS 300, remove the rear body protector.

4. Slowly loosen the lower flare nut fitting until all the pressure is relieved and all the fuel is collected.

5. Loosen the union bolt on the upper portion of the filter and remove the banjo fitting and 2 metal gaskets. Discard the gaskets.

6. Loosen the fuel filter bracket bolt, remove the fuel line with the flared nut from the filter and pull the filter from the mounting bracket.

To install:

7. Install or connect the following:
   • A new fuel filter to the vehicle and tighten the bracket bolt

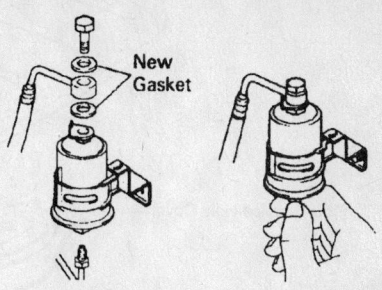

Exploded view of a typical fuel line connection at the filter

   • Banjo fitting with a new metal gasket on each side
   • Union bolt. Tighten the union bolt to 22 ft. lbs. (30 Nm).
   • Flare nut to the lower connection. Tighten the flare nut to 22 ft. lbs. (30 Nm).

8. On the GS 300, install the body protector.

9. Lower the vehicle if raised.

10. Remove the drain pan and/or rags and connect the negative battery cable.

11. Start the engine and visually inspect the upper and lower connections for leaks.

Fuel Pump

REMOVAL & INSTALLATION

ES 300 and IS 300

1. Before servicing the vehicle, refer to the precautions in the beginning of this section.

2. Relieve the fuel system pressure.

3. With the ignition switch in the **LOCK** position, disconnect the negative battery terminal.

4. Remove or disconnect the following:
   • Rear seat cushion
   • Fuel pump connector
   • Floor service hole cover

➡ **Do not lift the fuel pump assembly up using the wiring harness.**

   • Fuel filler cap
   • Fuel outlet pipe and the return hose from the pump bracket
   • 8 screws and lift out the pump/bracket assembly with gasket
   • Fuel pump lead wire
   • Lower end of the pump off the bracket
   • Fuel hose from the pump and remove the pump
   • Rubber cushion from the pump

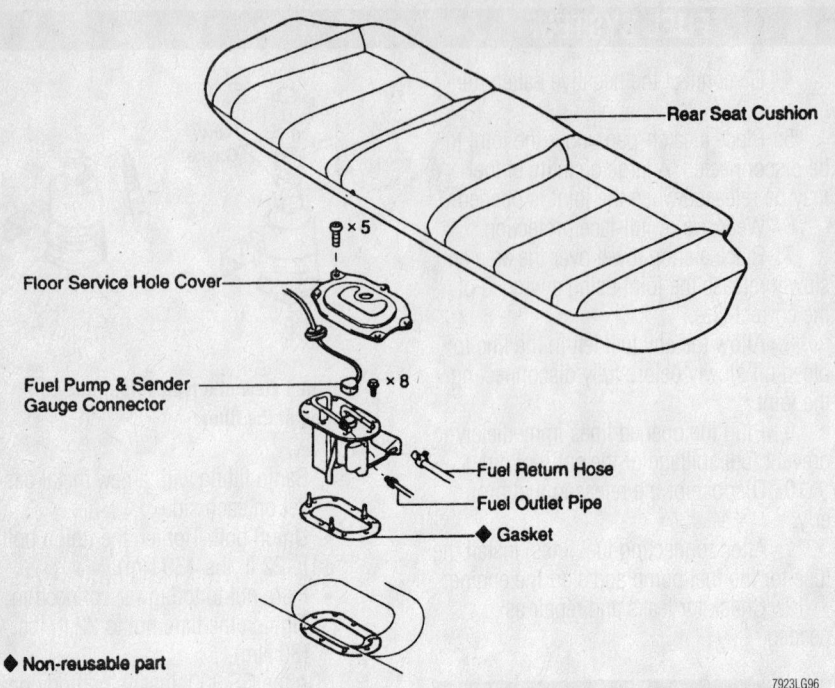

Rear Seat Cushion

Floor Service Hole Cover

× 5

Fuel Pump & Sender
Gauge Connector

× 8

Fuel Return Hose

Fuel Outlet Pipe

◆ Gasket

◆ Non-reusable part

Exploded view of the fuel pump assembly—ES 300

7923LG96

To install:

5. Install or connect the following:
   - Filter and rubber cushion on the new pump
   - Pump on the bracket
   - Fuel hose and the wire connector on the pump
   - Pump, using a new gasket and tighten the 8 screws to 35 inch lbs. (4 Nm)
   - Fuel pipe and return hose to the pump and tighten the bolts to 22 ft. lbs. (29 Nm)
   - Wire
   - Service cover, and replace the rear seat
   - Negative battery cable
6. Start the engine and check for leaks.

Except ES 300 and IS 300

1. Before servicing the vehicle, refer to the precautions in the beginning of this section.
2. Remove or disconnect the following:
   - Negative battery cable. Wait at least 90 seconds before performing any other work.
   - Trunk floor mat
   - Trunk trim cover
   - Fuel pump electrical connector
   - Rear seat bottom and seat back
   - Partition cover
   - Mounting bolts and remove the fuel pump set plate
   - 3 nuts and disconnect the fuel pump bracket from the tank
   - Fuel hose from the bracket
   - Pump, bracket and set plate as an assembly

To install:

3. Install or connect the following:
   - A new gasket on the set plate
   - Fuel hose to the pump and bracket
   - Pump and bracket assembly with the 3 nuts; tighten the nuts to 48 inch lbs. (5 Nm). Install the set plate and tighten the bolts to 26 inch lbs. (3 Nm).
   - Panel partition
   - Rear seat cushion and back
   - Fuel pump electrical connector
   - Trim panel
   - Spare tire and the trunk floor mat
   - Negative battery cable
4. Start the engine; check the fuel system for leaks

Fuel Injector

REMOVAL & INSTALLATION

3.0L (1MZ-FE) Engine

1. Before servicing the vehicle, refer to the precautions in the beginning of this section.
2. Remove or disconnect the following:
   - 3 cap nuts, using a 5mm hexagon

wrench. Loosen the V-bank cover fastener counterclockwise.
   - V-Bank cover
   - Air cleaner hose with resonator
   - air intake chamber assembly
   - Injector connectors
   - Air assist hoses and pipe
   - No. 1 fuel pipe and remove the fuel hose clamp
   - No. 1 fuel pipe (fuel tube connector) from the fuel filter outlet
   - 5 bolts and delivery pipes together with the 6 injectors and No. 1 fuel pipe.
   - 4 spacers from the intake manifold
   - 6 injectors form he delivery pipes
   - 2 O-rings and 2 grommets from each injector

To install:

3. Install or connect the following:
   - New insulator and grommet to each injector
   - New O-rings, coated with gasoline, to each injector
   - A light coat of gasoline on the place where a delivery pipe touches an O-ring of the injector
   - Injector, while turning it clockwise, into the delivery pipe

➡**Position the injector connector outward.**

   - The 4 spacers in position on the intake manifold
   - A light coat of gasoline on the place where the intake manifold touches an O-ring
   - The delivery pipe and fuel pipe together with the 6 injectors in position on the intake manifold. Position the injector connector outward.
   - 4 bolts holding the delivery pipes to the intake manifold, temporarily
   - Bolt holding the No. 1 fuel pipe to the intake manifold, temporarily

➡**Check that the injectors rotate smoothly. If the injectors do not rotate smoothly, the probable cause is incorrect installation of the O-rings. Replace the O-rings.**

   - 4 bolts holding the delivery pipes to the intake manifold. Torque to 14 ft. lbs. (19.5 Nm)
   - No. 1 fuel pipe (fuel tube connector) to the fuel filter
   - Fuel hose clamp to the fuel filter with a "click" sound. After installing the clamp, check that the

clamp is fixed by pulling up the clamp.
- Air assist hoses
- Injector connectors
- Air intake chamber assembly
- Air cleaner hose with resonator
- V-bank cover, using a 5mm hexagon wrench with the 3 cap nuts
- Press down the V-bank cover.

3.0L (2JZ-GE) Engine

1. Before servicing the vehicle, refer to the precautions in the beginning of this section.
2. Remove or disconnect the following:
   - Air intake chamber
   - Fuel pressure pulsation damper
   - Engine wire from intake manifold
   - Bolt holding the engine wire protector to the body
   - 6 injector connectors
   - Camshaft Position (CMP) sensor connector
   - Throttle Position (TP) sensor connector
   - Vacuum Switching Valve (VSV) connector for Evaporative Emission (EVAP) control
   - VSV connector for Acoustic Control Induction System (ACIS)
   - 3 nuts holding the engine wire protector to the intake manifold

➡**Be careful not to drop the injectors when removing the delivery pipe.**

- 3 bolts and delivery pipe together with the 6 injectors
- The injectors from the delivery pipe
- O-rings, insulator and grommet from each injector
- 3 spacers from the intake manifold

To install:

3. Install or connect the following:
   - A new insulator and grommet to each injector
   - A light coat of gasoline on the place where a delivery pipe touches an O-ring of the injector
   - Injector, while turning clockwise and counterclockwise, into the delivery pipe

➡**Position the injector connector outward.**

- The 3 spacers in position on the intake manifold
- A light coat of gasoline on the place where an intake manifold touches an O-ring
- Injectors together with the delivery pipe and 3 bolts in position on the intake manifold. Check that the injectors rotate smoothly. Position the injector connector upward.

➡**If the injectors do not rotate smoothly, the probable cause is incorrect installation of the O-rings. Replace the O-rings.**

4. Tighten the 3 bolts holding the delivery pipe to the intake manifold. Tighten the bolts to 15 ft. lbs. (21 Nm).
5. Install or connect the following:
   - Engine wire protector with the 3 nuts
   - 6 injector connectors

➡**The No. 1, No. 3 and No. 5 injector connectors are dark gray, and the No. 2, No. 4, and the No. 6 injectors connectors are brown.**

6. Install or connect the following:
   - Camshaft position sensor connector
   - Throttle position sensor connector
   - VSV connector for EVAP
   - Bolt holding the engine wire protector to the body

4.0L (1UZ-FE) and 4.3L (3UZ-FE) Engines

1. Before servicing the vehicle, refer to the precautions in the beginning of this section.
2. Remove or disconnect the following:
   - V-bank cover
   - Intake air connector
   - Accelerator cable
   - Fuel pressure pulsation dampers.
   - VVT sensor connectors
   - Vacuum Switching Valve (VSV) for Evaporative Emissions (EVAP)
   - 2 nuts and accelerator cable bracket
   - 2 nuts and accelerator cable bracket
   - 3 V-bank cover brackets
   - VSV connector for Acoustic Control

Induction System (ACIS) from the No. 1 V-bank cover bracket
   - 4 bolts and 3 V-bank cover brackets
   - Engine wire from the delivery pipe
   - 2 wire clamps from the wire clamp bracket on the right-hand deliver pipe
   - 8 injector connectors
   - 4 nuts holding the delivery pipe to the intake manifold
   - 2 delivery pipes and 8 injectors assembly and 4 spacers
   - 2 O-rings, grommet and insulator from each injector

To install:

3. Install or connect the following:
   - A new insulator and grommet to each injector
   - A light coat of gasoline to new O-rings and install them to each injector
   - A light coat of gasoline on the place where a delivery pipe touches an O-ring of the injector
   - Injector, while turning the clockwise and counterclockwise, into the delivery pipe

➡**Position the injector connector outward.**

- The 4 spacers in position on the intake manifold
- A light coat of gasoline on the place where an intake manifold touches an O-ring
- The delivery pipes in position on the intake manifold
- Temporarily, the 3 bolts holding the delivery pipe to the intake manifold

➡**Check that the injectors rotate smoothly. If the injectors do not rotate smoothly, the probable cause is incorrect installation of the O-rings. Replace the O-rings.**

4. Tighten the 3 bolts holding the delivery pipe to the intake manifold. Tighten the bolts to 15 ft. lbs. (21 Nm).
5. Install or connect the following:
   - Engine wire protector with the 3 nuts
   - Injector connectors
   - Remaining components

DRIVE TRAIN

Transmission Assembly

REMOVAL & INSTALLATION

Automatic

GS 300

1. Before servicing the vehicle, refer to the precautions in the beginning of this section.

2. Turn the ignition switch to the **LOCK** position and disconnect the negative battery cable. Wait at least 90 seconds or longer before doing any work on the vehicle.

3. Remove or disconnect the following:
   - Transmission level gauge
   - Transmission dipstick and tube
   - Throttle cable from the throttle body
   - Oxygen (O_2) sensor from the exhaust system.
   - Left and right tail pipes
   - Front and center exhaust pipe
   - Exhaust heat insulator
   - Rear center floor crossmember brace
   - Shift control rod from the shift lever
   - Driveshaft
   - Overdrive and direct clutch speed sensor
   - No. 1 Vehicle Speed Sensor (VSS)
   - No. 2 VSS
   - Solenoid wire
   - Park/Neutral Position (PNP) switch
   - Wiring from the starter
   - 2 oil cooler union nuts
   - Oil cooler hoses from the oil cooler pipes
   - Front oil cooler pipe bracket
   - Center and rear oil cooler pipe brackets
   - 2 oil cooler pipes
   - Torque converter inspection plate
   - Torque converter bolts

4. Support the transmission with a suitable jack.

5. Support the engine with a jack and a block of wood.

6. Remove or disconnect the following:
   - Rear transmission mount
   - Wiring harness clamps
   - Starter
   - 9 transmission mounting bolts and transmission

To install:

7. Install or connect the following:
   - Transmission and tighten the bolts to 53 ft. lbs. (52 Nm)

   - Starter and tighten the bolts to 27 ft. lbs. (37 Nm)
   - Rear transmission mount and tighten the bolts to 19 ft. lbs. (25 Nm)
   - And tighten the torque converter bolts to 30 ft. lbs. (41 Nm) while rotating the crankshaft
   - Converter inspection plate
   - 2 oil cooler pipes
   - Center and rear oil cooler pipe brackets
   - Front oil cooler pipe bracket and tighten to 49 inch lbs. (5.5 Nm)
   - Oil cooler hoses to the oil cooler pipes
   - 2 oil cooler union nuts and tighten to 32 ft. lbs. (44 Nm)
   - Wiring to the starter
   - Transmission electrical connectors
   - Driveshaft
   - Remaining components
   - Negative battery cable
   - Transmission level gauge

8. Fill the transmission to the proper level with Dexron®II or equivalent.

IS 300

1. Before servicing the vehicle, refer to the precautions in the beginning of this section.

2. Drain the cooling system.
   - Negative battery cable
   - Transmission oil dipstick and tube
   - Air cleaner
   - Mass Air Flow (MAF) sensor
   - Exhaust manifold
   - Engine under covers
   - Upper radiator hose
   - Exhaust front pipe
   - Exhaust center pipe
   - Shift control rod
   - Drive shaft
   - Oil cooler lines
   - Torque converter
   - Rear transmission mount crossmember. Support the transmission with a jack.
   - Transmission wiring harness connectors
   - Starter motor
   - Transmission flange bolts
   - Transmission

To install:

3. Installation is the reverse of the removal procedure, while using the following torque values:
   - 17mm transmission flange bolts: 53 ft. lbs. (72 Nm)

   - 14mm transmission flange bolts: 27 ft. lbs. (37 Nm)
   - Starter motor bolts: 27 ft. lbs. (37 Nm)
   - Rear transmission mount crossmember bolts: 19 ft. lbs. (25 Nm)
   - Torque converter bolts: 35 ft. lbs. (48 Nm)
   - Oil cooler lines: 33 ft. lbs. (44 Nm)
   - Drive shaft center support bolts: 36 ft. lbs. (49 Nm)
   - Drive shaft U-joint flange bolts: 54 ft. lbs. (74 Nm)
   - Shift control rod nuts: 108 inch lbs. (13 Nm)
   - Exhaust manifold nuts: 29 ft. lbs. (39 Nm)

LS 400 AND LS 430

1. Before servicing the vehicle, refer to the precautions in the beginning of this section.

2. Remove or disconnect the following:
   - Negative battery cable. Wait at least 90 seconds before performing any other work.
   - Transmission dipstick and tube
   - Throttle cable
   - Driveshaft
   - Engine undercover
   - Shift control rod
   - Exhaust pipe support bracket by removing the 2 bolts
   - Catalytic converters by removing the 6 nuts
   - Both side heat insulators
   - Oil cooler tube clamps and disconnect the tubes
   - Torque converter inspection plate by removing the 2 bolts
   - Torque converter bolts

3. Support the transmission with a suitable jack.

4. Remove or disconnect the following:
   - Rear transmission mount
   - Overdrive direct clutch speed sensor connector
   - Vehicle Speed Sensor (VSS) connector
   - Park/Neutral Position (PNP) switch connector
   - Solenoid connector
   - 3 wiring harness clamps from the bracket on the transmission.
   - 10 transmission mounting bolts and the transmission.

To install:

5. Install or connect the following:
   - Transmission and tighten the bolts

as follows: 14mm–27 ft. lbs. (37 Nm); 17mm–53 ft. lbs. (72 Nm)
- 3 wiring harness clamp to the bracket on the transmission
- Solenoid connector

- PNP switch connector
- VSS connector
- Overdrive direct clutch speed sensor connector
- Rear transmission mount and

tighten the bolts to 19 ft. lbs. (20 Nm) and the nuts to 10 ft. lbs. (13 Nm)
- Torque converter bolts to 30 ft. lbs. (41 Nm)

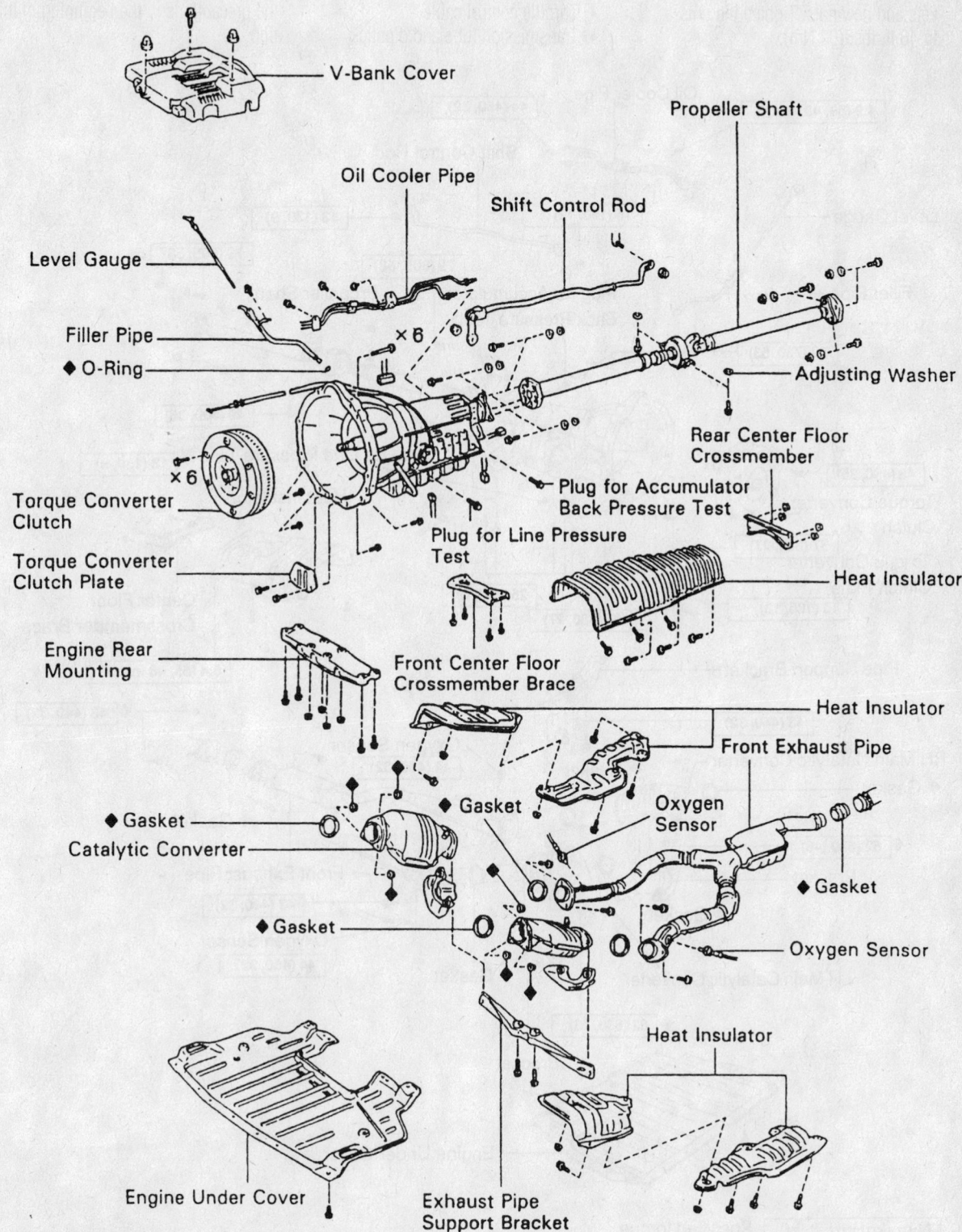

V-Bank Cover

Propeller Shaft

Oil Cooler Pipe

Shift Control Rod

Level Gauge

Filler Pipe

◆ O-Ring

Adjusting Washer

×6

Rear Center Floor Crossmember

Torque Converter Clutch

×6

Plug for Accumulator Back Pressure Test

Torque Converter Clutch Plate

Plug for Line Pressure Test

Heat Insulator

Engine Rear Mounting

Front Center Floor Crossmember Brace

Heat Insulator

Front Exhaust Pipe

◆ Gasket

◆ Gasket

Oxygen Sensor

Catalytic Converter

◆ Gasket

◆ Gasket

Oxygen Sensor

Heat Insulator

Engine Under Cover

Exhaust Pipe Support Bracket

◆   Non-reusable part

Exploded view of the transmission mounting—LS 400

7923LG97

- Converter inspection plate
- Support from the transmission
- Oil cooler pipes and tighten the union nuts to 32 ft. lbs. (44 Nm)
- Side heat insulators
- Catalytic converters with new gaskets and new nuts. Tighten the nuts to 46 ft. lbs. (62 Nm).

- Exhaust pipe support bracket with the 2 bolts and tighten the bolts to 32 ft. lbs. (44 Nm)
- Shift control rod
- Engine undercover
- Driveshaft
- Throttle control cable
- Transmission tube and dipstick

- Negative battery cable
6. Fill the transmission to the proper level with Dexron®II, or equivalent.

SC 300, SC 400, GS 400 AND GS 430

1. Before servicing the vehicle, refer to the precautions in the beginning of this section.

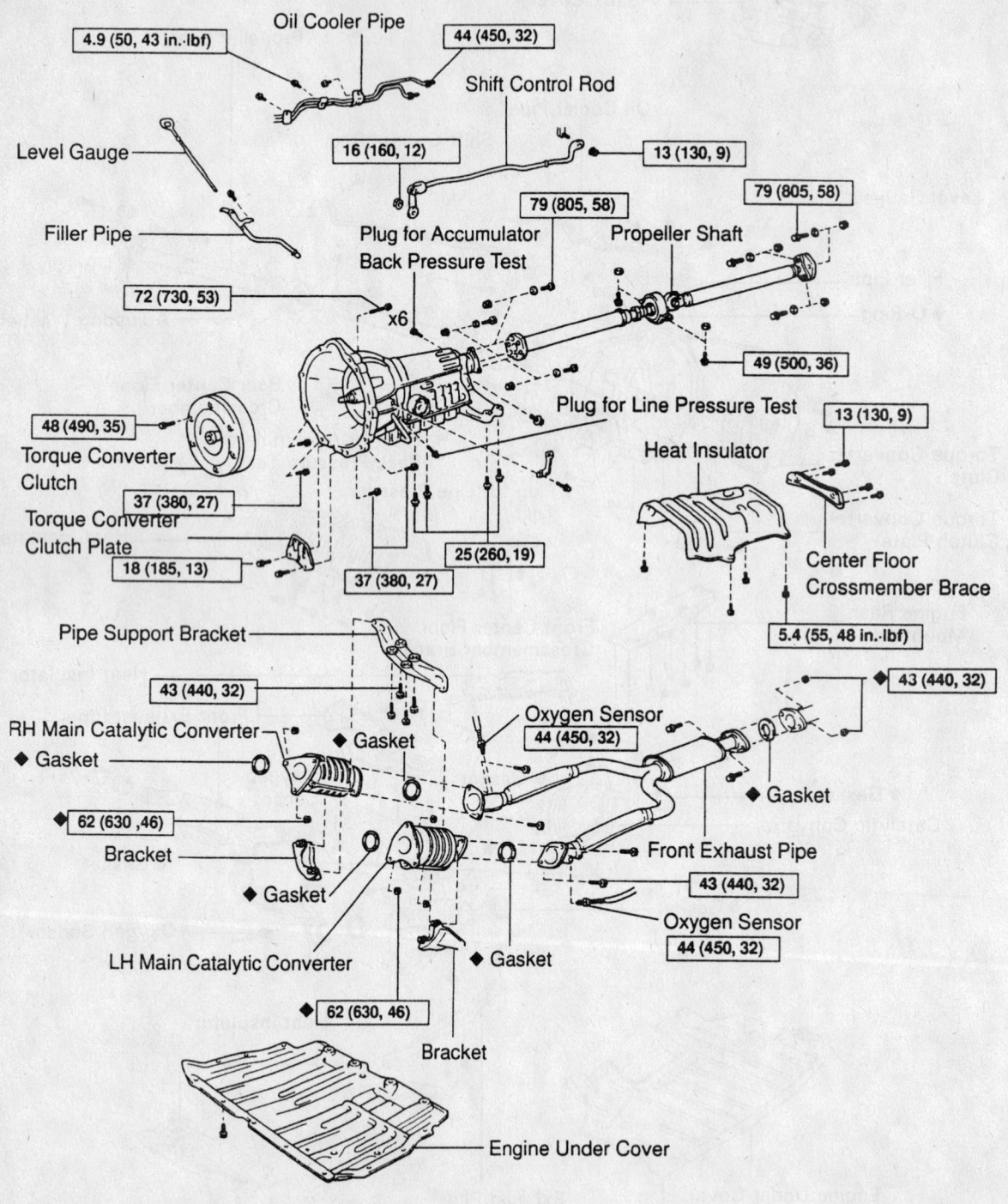

N·m (kgf·cm, ft·lbf) : Specified torque
◆ Non–reusable part

7923LG98

Exploded view of the transmission mounting—SC 300, SC 400 and GS 400

2. Remove or disconnect the following:
- Negative battery cable.
- V-bank cover, if equipped
- Automatic transmission oil level gauge if equipped
- Transmission dipstick and tube
- Throttle cable and clamps
- Exhaust pipe and converters
- Exhaust heat insulator
- Rear center floor crossmember brace
- Shift control rod
- Driveshaft

➡**The bolts inserted from the drive-shaft side should not be removed.**

- The electrical harness from the transmission.
- Oil cooler tube clamp and disconnect the tubes
- Lower engine cover
- Torque converter inspection plate
- Torque converter bolts

3. Support the transmission with a suitable jack.
4. Remove or disconnect the following:
- Starter, if necessary
- Rear transmission mount
- Transmission mounting bolts and the transmission.

To install:
5. Before installing the transmission, use calipers and a straightedge to check the distance between the installed surface of the torque converter and the front edge of the transmission case. Correct distance is 0.673 in. (17.1mm). If this distance is not correct, check the torque converter installation.
6. Install or connect the following:
- Transmission and tighten the bolts to 14mm: 29 ft. lbs. (39 Nm); 17mm: 42 ft. lbs. (57 Nm)
- If removed, the starter. Tighten the bolts to 27 ft. lbs. (37 Nm).
- Rear transmission mount and tighten the bolts to 19 ft. lbs. (20 Nm)
- And tighten the torque converter bolts to 25 ft. lbs. (33 Nm) while rotating the crankshaft
- Converter inspection plate
- Lower engine cover
- Oil cooler lines and tighten the lines to 25 ft. lbs. (34 Nm)
- Oil cooler pipe bracket and tighten the bolt
- Transmission electrical connectors
- Shift control rod and adjust the shift linkage. Tighten the nut to 12 ft. lbs. (16 Nm).
- Rear center floor crossmember

brace. Tighten the bolts to 108 inch lbs. (13 Nm).
- Heat insulator
- Transmission filler tube and dipstick
- Front exhaust pipe and converters with new gaskets
- Throttle control cable
- Driveshaft. Flange bolts: 58 ft. lbs. (79 Nm). Center bearing support bolts: 36 ft. lbs. (49 Nm). Adjusting nut: 35 ft. lbs. (48 Nm).
- Crossmember brace and tighten to 96 inch lbs. (13 Nm)
- Automatic transmission oil level gauge
- V-bank cover
- Negative battery cable
7. Adjust the PNP switch
8. Fill the transmission with Dexron®II.

Transaxle Assembly

REMOVAL & INSTALLATION

Automatic

ES 300

1. Before servicing the vehicle, refer to the precautions in the beginning of this section.
2. Turn the ignition switch to the **LOCK** position and disconnect the negative battery cable. Wait at least 90 seconds or longer before doing any work on the vehicle.
3. Remove or disconnect the following:
- Battery
- Air cleaner assembly
- Throttle cable from the throttle body
- Cruise control actuator cover and detach the connector
- Ground wire
- Starter
- Vehicle Speed Sensor (VSS) connectors
- Direct clutch speed sensor
- Park/Neutral Position (PNP) switch connector on the transaxle
- Solenoid connector on the transaxle
- Shift control cable
- Oil cooler hoses
- 2 front side transaxle mounting bolts
- 2 front engine mounting bolts
- Oil cooler line mounting bolts from the front frame
- 3 upper transaxle to engine mounting bolts
4. Install an engine support fixture. Tie

steering gear housing to engine support fixture.
5. Raise and safely support the vehicle.
6. Drain the transaxle/differential fluid.
7. Remove or disconnect the following:
- Front wheels
- Front exhaust pipe
- Engine side covers and undercovers
- Both halfshafts
- Front side engine mounting nut
- Rear side engine mounting bolts (remove hole plugs)
- 4 left side transaxle mounting bolts
- Steering gear housing
- Front frame assembly
8. Properly support the transaxle assembly.
9. Remove or disconnect the following:
- Rear end plate mounting bolts
- Torque converter cover
- Torque converter retaining bolts
- Remaining transaxle mounting bolts
10. Carefully remove the transaxle assembly from the vehicle.

To install:
11. Install or connect the following:
- Transaxle aligning the 2 dowel pins on the block with the converter housing. Tighten the bolts as follows: 10mm–34 ft. lbs. (46 Nm); 12mm–47 ft. lbs. (64 Nm).
- Torque converter bolts. Coat the threads of the with sealer. Install the bolts starting with the green bolt followed by the rest and tighten the bolts evenly to 20 ft. lbs. (27 Nm).
- End plate and tighten the bolts to 27 ft. lbs. (37 Nm)
- Front frame assembly and tighten the fasteners as follows: 12mm–24 ft. lbs. (32 Nm); 19mm–134 ft. lbs. (181 Nm); nut–27 ft. lbs. (36 Nm).
- 2 fender liner set screws
- Steering gear to the frame and tighten the bolts and nuts to 134 ft. lbs. (181 Nm)
- Sway bar brackets and toque the bolts to 14 ft. lbs. (19 Nm)
- Left transaxle mounting bolts and tighten them to 38 ft. lbs. (52 Nm)
- Rear side mounting bolts and nuts and tighten them to 48 ft. lbs. (66 Nm). Install the plugs.
- Front engine mounting nut and tighten it to 59 ft. lbs. (80 Nm)
- Halfshafts
- Right and left engine side covers
- Lower engine cover
- Exhaust pipe to the engine with

new gaskets and tighten the nuts to 46 ft. lbs. (62 Nm). Connect the exhaust pipe to the converter with a new gasket and tighten the nuts and bolts to 32 ft. lbs. (43 Nm).

- Wheel
- Engine support
- 4 upper transaxle mounting bolts and tighten them to 47 ft. lbs. (64 Nm)
- Oil cooler clamping bolts to the front frame
- 2 front side engine mounting bolts and tighten them to 59 ft. lbs. (80 Nm)
- 2 front side transaxle mounting bolts and tighten them to 59 ft. lbs. (80 Nm)
- Remaining components
- Battery and connect the battery cables

12. Fill the transaxle/differential to the proper level with Dexron®II, or equivalent.

13. Check the transaxle/differential fluid level.

14. Check the front wheel alignment.

Halfshaft

REMOVAL & INSTALLATION

ES 300

1. Before servicing the vehicle, refer to the precautions in the beginning of this section.

2. Remove or disconnect the following:
- Negative battery cable
- Front wheel(s)
- Front fender apron seal
- Transaxle fluid
- Tie rod end from the steering knuckle by removing the cotter pin and nut. Separate the tie rod from the steering knuckle.
- Stabilizer bar link from the lower control arm. Make note of the washers and cushions positions.
- Lower ball joint from the steering knuckle by removing the bolt and 2 nuts. Push down on the lower control arm and separate the steering knuckle from the ball joint.
- Cotter pin, lock cap and locknut holding the halfshaft to the steering knuckle
- Left halfshaft from the steering knuckle
- Halfshaft from the transaxle
- Snapring from the halfshaft
- Right halfshaft bearing lockbolt.

The lockbolt is located in the center of the halfshaft, near the dampener.
- Snapring and pull the halfshaft from the transaxle

To install:

3. Install or connect the following:
- Right halfshaft to the transaxle. Coat the side gear shaft and differential case sliding surface with gear oil.
- Snapring to the halfshaft
- Bearing lockbolt. Tighten the lockbolt to 24 ft. lbs. (32 Nm).
- New snapring to the inner spline of the left halfshaft. Coat the side gear shaft and differential case sliding surface with gear oil. Install the halfshaft to the transaxle with the snapring opening facing down. The halfshaft should click into place when installing.
- Halfshaft to the steering knuckle, then install the locknut. Tighten the locknut to 217 ft. lbs. (294 Nm).
- Lock cap and a new cotter pin to the halfshaft
- Steering knuckle to the lower ball joint. Install the 2 nuts and bolt. Tighten the nuts and bolt to 94 ft. lbs. (127 Nm).
- Stabilizer bar link to the lower control arm. Tighten the nut to 29 ft. lbs. (39 Nm).
- Tie rod to the steering knuckle and tighten the nut to 36 ft. lbs. (49 Nm). Install a new cotter pin to the tie rod end.
- Front fender apron seal
- Wheel(s) and lower the vehicle. Tighten the lug nuts to 76 ft. lbs. (103 Nm).
- Transaxle fluid
- Negative battery cable

Except ES 300

1. Before servicing the vehicle, refer to the precautions in the beginning of this section.

2. Remove or disconnect the following:
- Negative battery cable
- Rear tire and wheel assembly
- Cotter pin, locknut cap, and locknut
- Height control sensor, if equipped
- 2 exhaust pipe support brackets, if necessary

3. Place matchmarks on the halfshaft and the side gear shaft. Remove the 6 hex bolts and 2 washers.

4. Hold the inboard joint side of the halfshaft so the outboard joint side does not

bend too much. Tap the end of the halfshaft with a rubber mallet to loosen it from the axle hub and remove the halfshaft.

To install:

5. Insert the outboard joint side of the halfshaft through the axle hub. Align the matchmarks on the side gear shaft and the halfshaft.

6. Coat the threads with clean oil and install the hex bolts. Tighten the bolts to 61 ft. lbs. (83 Nm).

7. Install or connect the following:
- Exhaust pipe support brackets, if removed, and tighten to 14 ft. lbs. (19 Nm)
- Bearing locknut, if removed, and have a helper apply the brakes. Tighten the locknut to 213 ft. lbs. (289 Nm).
- Lockcap and a new cotter pin
- Height control sensor, if removed
- Rear tire and wheel assembly
- Negative battery cable

CV-Joints

OVERHAUL

ES 300

1. Before servicing the vehicle, refer to the precautions in the beginning of this section.

2. Once the driveshaft is removed from the vehicle, place matchmarks on the outboard and inboard joints and the shaft. Do not use a punch to make the marks.

3. Remove or disconnect the following:
- Boot clamps. Use a side cutter or pliers
- Outboard joint shaft expanding the snapring
- 2 boots
- Dust cover on the left-hand driveshaft, using a hammer and suitable chisel
- Dust cover from the inboard joint shaft, using a press
- Dust cover
- Snapring. Use a snapring expander.
- Bearing, using the press
- Snapring
- No. 2 dust deflector, using a hammer and suitable chisel

➡ **Be careful not to damage the Anti-lock Brake System (ABS) speed sensor rotor.**

To install:

4. Install or connect the following:
- No. 2 dust deflector

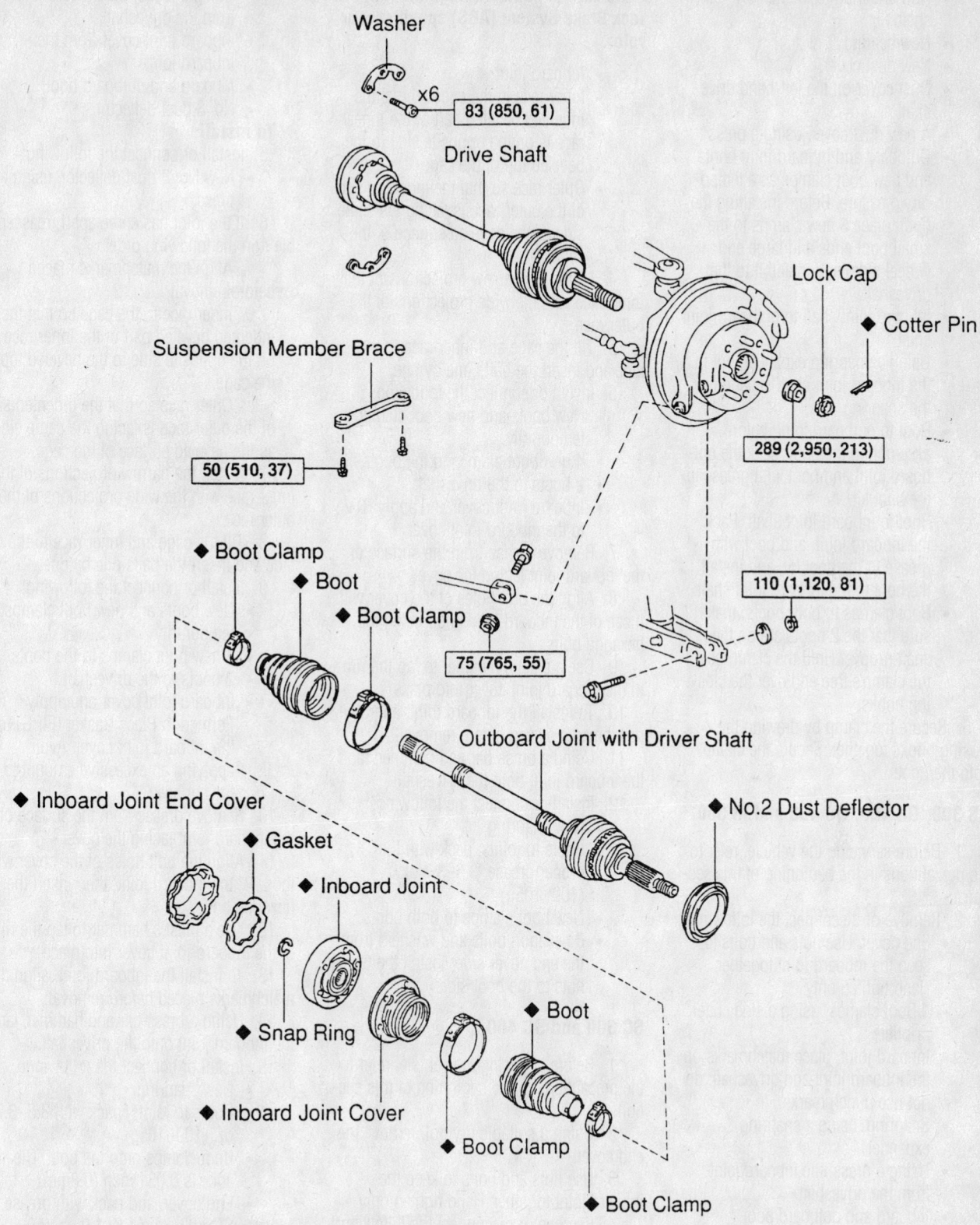

Washer

x6

83 (850, 61)

Drive Shaft

Lock Cap

◆ Cotter Pin

Suspension Member Brace

289 (2,950, 213)

50 (510, 37)

◆ Boot Clamp

◆ Boot

◆ Boot Clamp

110 (1,120, 81)

75 (765, 55)

Outboard Joint with Driver Shaft

◆ Inboard Joint End Cover

◆ Gasket

◆ No.2 Dust Deflector

◆ Inboard Joint

◆ Snap Ring

◆ Boot

◆ Inboard Joint Cover

◆ Boot Clamp

◆ Boot Clamp

N·m (kgf·cm, ft·lbf) : Specified torque
◆ Non–reusable part

9301LG12

Exploded view of the rear halfshaft and related components—except ES 300

- New snapring to the inboard joint shaft
- New bearing
- New dust cover
- Dust cover on the left-hand driveshaft
- A new dust cover, using a press
- Outboard and inboard joint boots and new boot clamps as a temporary measure. Before installing the boot, place 3 new clamps to the small boot ends and large end (wheel side) and install it to the driveshaft.
- Inboard joint shaft to outboard joint shaft
- Using a snapring expander, put in the inboard joint shaft expanding the snapring
- Boot to outboard joint, before assembling the boot, pack the outboard joint and boot with grease in the boot kit
- Boot to inboard joint shaft. Pack the inboard joint, and boot with grease in the boot kit, and install the boot to the inboard joint shaft.
- Boot clamps to both boots, make sure that the 2 boots are on the shaft groove. Hold the clamp near the clamp's free end over the closing hooks.

5. Secure the clamp by drawing the closing hooks together. Secure the clamp onto the boot.

GS 300, GS 400, GS 430 and IS 300

1. Before servicing the vehicle, refer to the precautions in the beginning of this section.

2. Remove or disconnect the following:
- End cover. Use nuts and bolts to keep the inboard joint together. Hand tighten only.
- 4 Boot clamps, using a side cutter or pliers
- Inboard joint, place matchmarks on the inboard joint and driveshaft; do not use punch marks.
- Snapring, using a snapring expander
- Using a press, the inboard joint from the driveshaft
- Inboard and outboard boot
- Inboard joint cover from the inboard joint

To install:

3. Install or connect the following:
- Inboard and outboard joint boots
- New No. 2 Dust deflector, using a press

➡Be careful not to damage the Antilock Brake System (ABS) speed sensor rotor.

- Inboard joint
- Inner race to the cage so that the indented beveled part of the inner race is on the opposite side to the beveled top of the cage
- Outer race so that the indented side of the outer race is facing the same side as the beveled surface of the cage

4. Match the narrow projections of the inner race with the wide projections of the outer race.

5. Tilt the cage and inner race to the side and insert the balls one by one.

6. Install or connect the following:
- New boots and new boot clamps, temporarily
- 4 new boot clamps to the boots
- 2 boots to the driveshaft
- Inboard joint cover and apply RTV to the inboard joint cover

7. Remove grease from the surface of the inboard joint facing the cover.

8. Align the bolt holes of the cover with those of the inboard joint, then insert the hexagon bolts.

9. Use a plastic hammer to tap the rim of the inboard joint cover into place

10. To install the inboard joint, align the matchmarks placed before removal.

11. Using a brass bar and hammer, tap the inboard joint onto the driveshaft.

12. Install or connect the following:
- New snapring
- Boots to joints, pack with the proper grease. 3.5–3.7 oz. (100–105g)
- New boot clamps to both boots
- 6 hexagon bolts and washers from the end cover side, install the 6 nuts to the boot side.

SC 300 and SC 400

1. Before servicing the vehicle, refer to the precautions in the beginning of this section.

2. Using a suitable prytool, remove the end cover.

3. Use nuts and bolts to keep the inboard joint together. Hand tighten only.

4. Remove or disconnect the following:
- 4 Boot clamps, using a side cutter or pliers
- Inboard joint, place matchmarks on the inboard joint and driveshaft; do not use punch marks
- Snapring, using a snapring expander

- Using a press, the inboard joint from the driveshaft
- Inboard joint cover from the inboard joint
- Inboard and outboard boot
- No. 3 dust deflector

To install:

5. Install or connect the following:
- New No. 3 dust deflector, using a press.

6. If the joint has come apart, reassemble it in the following order:
   a. Align the matchmarks placed before removal
   b. Inner race to the cage so that the indented beveled part of the inner race is on the opposite side to the beveled top of the cage.
   c. Outer race so that the indented side of the outer race is facing the same side as the beveled surface of the cage.

7. Match the narrow projections of the inner race with the wide projections of the outer race.

8. Tilt the cage and inner race to the side and insert the balls one by one.

9. Install or connect the following:
- New boots and new boot clamps, temporarily
- 4 new boot clamps to the boots
- 2 boots to the driveshaft
- Inboard joint cover and apply Formed In Place Gasket (FIPG) to the inboard joint cover. Avoid applying an excessive amount to the surface.

10. Remove grease from the surface of the inboard joint facing the cover

11. Align the bolt holes of the cover with those of the inboard joint, then insert the hexagon bolts.

12. Use a plastic hammer to tap the rim of the inboard joint cover into place

13. To install the inboard joint, align the matchmarks placed before removal.

14. Using a brass bar and hammer, tap the inboard joint onto the driveshaft.

15. Install or connect the following:
- New snapring
- Boots to joints, pack with 3.5–3.7 oz. (100–105g)
- Boot clamps onto the boot. Clearance is 0.031 inch (0.8mm)
- End cover, and pack with grease. SC 400 uses 1.8–1.9 oz. (50–55g); SC 300 uses 1.5–1.7 oz. (42–47g)

16. Remove grease from the surface of the inboard joint facing the cover

17. Glue on a new gasket, with the side with adhesive on it facing toward the outer race side of the inboard joint

18. Align the bolt holes of the cove with those of the inboard joint

19. Install or connect the following:
- 6 hexagon bolts an washer from the end cover side
- 6 nuts to the boot side

20. Check that the claw of the end cover touches the inboard joint

LS 400 and LS 430

1. Before servicing the vehicle, refer to the precautions in the beginning of this section.

2. Using a suitable prytool, remove the end cover.

3. Use nuts and bolts to keep the inboard joint together. Hand tighten only.

4. Remove or disconnect the following:
- 4 Boot clamps, using a side cutter or pliers
- Inboard joint, place matchmarks on the inboard joint and driveshaft; do not use punch marks.
- Snapring, using a snapring expander
- Using a press, the inboard joint from the driveshaft
- Inboard joint cover from the inboard joint
- Inboard and outboard boot
- No. 2 dust deflector, using a suitable chisel and hammer
- New No. 3 dust deflector, using a press

To install:

5. If the joint has come apart, reassemble it in the following order.

a. Align the matchmarks placed before removal

b. Inner race to the cage so that the indented beveled part of the inner race is on the opposite side to the beveled top of the cage.

c. Outer race so that the indented side of the outer race is facing the same side as the beveled surface of the cage.

6. Match the narrow projections of the inner race with the wide projections of the outer race.

7. Tilt the cage and inner race to the side and insert the balls one by one.

8. Install or connect the following:
- New boots and new boot clamps, temporarily
- 4 new boot clamps to the boots
- 2 boots to the driveshaft
- Inboard joint cover and apply Formed In Place Gasket (FIPG) to the inboard joint cover. Avoid applying an excessive amount to the surface

9. Remove grease from the surface of the inboard joint facing the cover

10. Align the bolt holes of the cover with those of the inboard joint, then insert the hexagon bolts.

11. Use a plastic hammer to tap the rim of the inboard joint cover into place

12. To install the inboard joint, align the matchmarks placed before removal.

13. Using a brass bar and hammer, tap the inboard joint onto the driveshaft.

14. Install or connect the following:
- New snapring
- Boots to joints, pack with 3.5–3.7 oz. (100–105g) of grease
- A new gasket, with the side with adhesive on it facing toward the outer race side of the inboard joint
- 6 hexagon bolts an washer from the end cover side
- 6 nuts to the boot side

15. Check that the claw of the end cover touches the inboard joint

Axle Shaft, Bearing and Seal

REMOVAL & INSTALLATION

GS 300, GS 400, GS 430 and IS 300

1. Drain the gear oil.

2. Remove or disconnect the following:
- Rear driveshaft
- Side gear shaft
- Snapring from the side gear, using a suitable tool
- Side gear shaft oil seal

To install:
- Side gear shaft oil seal
- New oil seal

3. Check installation of side gear shaft. There should be 0.08–0.12 inch (2–3mm) of play in the axial direction. Check that the side gear shaft will not come out by pulling on it.

SC 300 and SC 400

1. Remove or disconnect the following:
- Gear oil
- Rear driveshaft
- Side gear shaft
- Snapring from the side gear, using a suitable tool
- Side gear shaft oil seal

To install:

2. Install or connect the following:
- New oil seal until it is flush with the carrier end surface
- Multipurpose grease to the oil seal lip

- Side gear shaft, and install a new snapring
- Side gear shaft to the differential

➡ **Check that the side gear shaft does not come out by trying to pull it out by hand.**

- Reconnect the rear driveshaft
- Gear oil

LS 400 and LS 430

1. Remove or disconnect the following:
- Gear oil
- Rear driveshaft
- Side gear shaft
- Snapring from the side gear
- Side gear shaft oil seal

To install:

2. Install or connect the following:
- Side gear shaft oil seal
- New oil seal
- Multi-Purpose (MP) grease to the oil seal lip
- Side gear shaft
- Snapring from the side gear, using a suitable tool
- Side gear shaft oil seal
- Gear oil

Pinion Seal

REMOVAL & INSTALLATION

GS 300, GS 400 and GS 430

1. Drain the gear oil.

2. Remove the driveshaft and the companion flange

3. Remove the oil seal

4. Check oil slinger

To install:

5. Installation is the reversal of the removal procedure.

SC 300, SC 400, LS 400 and LS 430

1. Remove or disconnect:
- Gear oil
- Driveshaft
- Companion flange
- Oil seal and slinger

To install:

2. Install or connect:
- Oil slinger
- New oil seal

➡ **Oil seal drive-in depth: 0.079 inch (2.0mm).**

- Multi-Purpose (MP) grease to the oil seal lip
- Companion flange on the shaft

- Gear oil on the threads of a new nut. Torque to 80 ft. lbs. (108 Nm).
3. Adjust the drive pinion preload as necessary, stake drive the pinion nut, and install the driveshaft.
4. Fill the differential with hypoid gear oil

IS 300

1. Before servicing the vehicle, refer to the precautions in the beginning of this section.
2. Remove or disconnect the following:
   - Driveshaft
   - Rear wheels
   - Rear brake calipers
   - Pinion flange
   - Pinion seal

➡ **The rear brake calipers must be removed so that there is no additional drag when measuring pinion bearing preload.**

To install:
3. Install or connect the following:
   - Pinion seal and flange
   - New pinion flange nut
4. Rotate the pinion flange occasionally while tightening the flange nut to make sure the pinion bearings seat correctly. Do not exceed 249 ft. lbs. (338 Nm).
5. Take frequent bearing preload torque readings.
6. The pinion bearing preload specifications are as follows:

a. Used bearings: 4.3–6.9 inch lbs. (0.49–0.78 Nm).
b. New bearings: 8.7–13.9 inch lbs. (0.98–1.57 Nm).

✳✳ CAUTION

Never loosen the pinion nut to reduce bearing preload. If it is necessary to reduce bearing preload, install a new collapsible spacer and pinion nut.

7. Install or connect the following:
   - Driveshaft
   - Brake calipers
   - Rear wheels
8. Fill the differential with gear lubricant and check for leaks.

STEERING AND SUSPENSION

Air Bag

✳✳ CAUTION

These vehicles are equipped with an air bag system. The system must be disabled before performing service on or around system components, steering column, instrument panel components, wiring and sensors. Failure to follow safety and disabling procedures could result in accidental air bag deployment, possible personal injury and unnecessary system repairs.

PRECAUTIONS

Several precautions must be observed when handling the inflator module to avoid accidental deployment and possible personal injury.
- Never carry the inflator module by the wires or connector on the underside of the module.
- When carrying a live inflator module, hold securely with both hands and ensure that the bag and trim cover are pointed away.
- Place the inflator module on a bench or other surface with the bag and trim cover facing up.
- With the inflator module on the bench, never place anything on or close to the module which may be thrown in the event of an accidental deployment.

DISARMING

To avoid personal injury when working on vehicles equipped with an air bag, the negative battery cable must be disconnected and at least 90 seconds must elapse before working on the system. Failure to do so may result in deployment of the air bag.

ARMING

To rearm the air bag system, simply reconnect the battery cable(s).

Power Rack and Pinion Steering Gear

REMOVAL & INSTALLATION

ES 300

1. Before servicing the vehicle, refer to the precautions in the beginning of this section.
2. Remove or disconnect the following:
   - Negative battery cable and wait at least 90 seconds before working on the vehicle to disarm the air bag.
   - Front wheels
   - Left and right front fender apron seals
   - Cotter pin and nut holding the steering knuckle to the tie rod end. Using a tie rod puller, disconnect the tie rod end from the steering knuckle.
3. Place matchmarks on the intermediate shaft and the control valve shaft.
4. Loosen the upper bolt and remove the lower bolt holding the control valve shaft to the intermediate shaft. Disconnect the intermediate shaft from steering rack housing.
5. Remove or disconnect the following:
   - Tube clamp

- Return line and the pressure line from the control valve housing
- 4 stabilizer bar bolts and 2 nuts. Position the stabilizer bar out of the way. Do not remove the sway bar from the vehicle.
- Heated Oxygen (HO$_2$) sensor (bank 1 sensor 1)
- 2 steering gear mounting bolts and nuts. Remove the steering gear through the left side of the vehicle.

To install:
6. Install or connect the following:
- Steering gear on the vehicle and install the 2 mounting bolts and nuts. Tighten the nuts and bolts to 134 ft. lbs. (181 Nm).
- HO$_2$ sensor. Tighten the sensor to 33 ft. lbs. (44 Nm).
- Stabilizer bar bolts and nuts and tighten as follows: Bolts: 14 ft. lbs. (19 Nm); Nuts: 29 ft. lbs. (39 Nm).
- Pressure and return lines and tighten the connectors to 18 ft. lbs. (25 Nm)
- Tube clamp and tighten the nut to 84 inch lbs. (10 Nm)
- Intermediate shaft to the steering rack and tighten the retaining bolts to 26 ft. lbs. (35 Nm)
- Tie rods to the steering knuckles with the castellated nuts. Tighten the nut to 36 ft. lbs. and install a new cotter pin. The prongs of the cotter pin should be firmly wrapped around the flats of the nut.
- Front fender apron seals by installing the 2 bolts
- Front wheels and lower the vehicle
- Power steering fluid
- Negative battery cable

7. Release the steering wheel
8. Bleed the system
9. Check for leaks, adjust the toe-in and check the steering wheel center point

GS 300, GS 400 and GS 430

1. Before servicing the vehicle, refer to the precautions in the beginning of this section.
2. Remove or disconnect the following:
   • Negative battery cable
   • Front wheels
3. Matchmark the steering column universal joint to the control valve shaft.
4. Loosen the upper bolt and remove the lower bolt to the intermediate shaft universal joint.
5. Remove or disconnect the following:
   • Intermediate shaft from the control valve shaft
   • Tie rod ends from the steering knuckle
   • Fluid lines from the rack and pinion and cap the lines
   • 2 tube clamps by removing the bolt
   • Mounting bolts and nuts. Remove the rack and pinion.

To install:

6. Center the rack and pinion to the following dimensions:

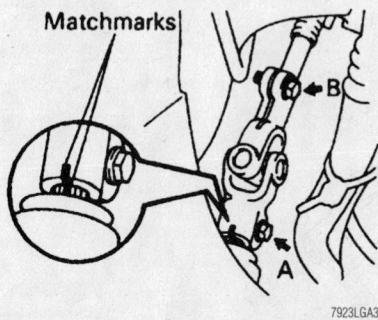

Matchmarking the intermediate shaft to the control valve shaft—GS 300, GS 400, GS 430

a. Dimension **A**: 1.14 in. (28.9mm)
b. Dimension **B**: 23.54 in. (589mm)
7. Install or connect the following:
   • Rack and tighten the bolts to 72 ft. lbs. (98 Nm)
   • 2 tube clamps and tighten the bolt to 12 ft. lbs. (17 Nm)
8. Align the matchmarks on the intermediate shaft and control valve shaft. Tighten the intermediate shaft bolts to 26 ft. lbs. (35 Nm).
9. Install or connect the following:
   • Fluid lines to the rack and pinion with new washers. Tighten the union bolts to 36 ft. lbs. (49 Nm).
   • Tie rod ends
   • Wheels
   • Negative battery cable
10. Lower the vehicle.
11. Check the steering wheel center point.
12. Check the front wheel alignment.

IS 300

1. Before servicing the vehicle, refer to the precautions in the beginning of this section.
2. Remove or disconnect the following:
   • Negative battery cable
   • Steering wheel
   • Front wheels
   • Brake calipers
   • Outer tie rod ends
   • Engine under cover
   • Intermediate shaft
   • Front subframe brace
   • Pressure and return lines
   • Steering gear

To install:

3. Installation is the reverse of the removal procedure, while using the following torque values:
   • Steering gear mounting bracket bolts: 54 ft. lbs. (74 Nm)
   • Fluid return line: 30 ft. lbs. (40 Nm)

   • Fluid pressure line: 31 ft. lbs. (42 Nm)
   • Front subframe brace large bolts: 88 ft. lbs. (119 Nm)
   • Front subframe brace small bolts: 43 ft. lbs. (58 Nm)
   • Tie rod end nuts: 40 ft. lbs. (54 Nm)
   • Steering wheel nut: 26 ft. lbs. (35 Nm)

LS 400 and LS 430

1. Before servicing the vehicle, refer to the precautions in the beginning of this section.
2. Remove or disconnect the following:
   • Wheel(s)
   • Engine undercover by removing the 8 bolts and 5 screws
   • Cotter pin and nut holding each tie rod to the steering knuckle
   • Tie rod end from the steering knuckle with a tie rod end puller
3. Place matchmarks on the sliding yoke and control valve shaft.
4. Loosen the top bolt holding the sliding yoke to the intermediate shaft. Remove the bottom bolt holding the sliding yoke to the steering rack.
5. Remove or disconnect the following:
   • Pressure feed and return lines to the rack and pinion
   • Power steering connector
   • 4 mount bolts and nuts to the power steering rack
   • 2 brackets and grommets
   • Power steering rack from the vehicle

To install:

6. Install or connect the following:
   • Power steering rack to the vehicle.
   • 2 brackets and grommets to the power steering rack.
   • 4 bolts and tighten the bolts to 56 ft. lbs. (76 Nm).
   • Power steering solenoid connector
   • Pressure feed and return tubes. Tighten the union bolt to 36 ft. lbs. (49 Nm).
7. Align the matchmarks on the sliding yoke and control valve shaft.
8. Tighten the bolt holding the sliding yoke to the steering rack to 26 ft. lbs. (35 Nm).
9. Tighten the bolt holding the sliding yoke to the intermediate shaft to 26 ft. lbs. (35 Nm).
10. Install or connect the following:
   • Tie rod end to the steering knuckle. Tighten the nut to 48 ft. lbs. (65 Nm). Install a new cotter pin.

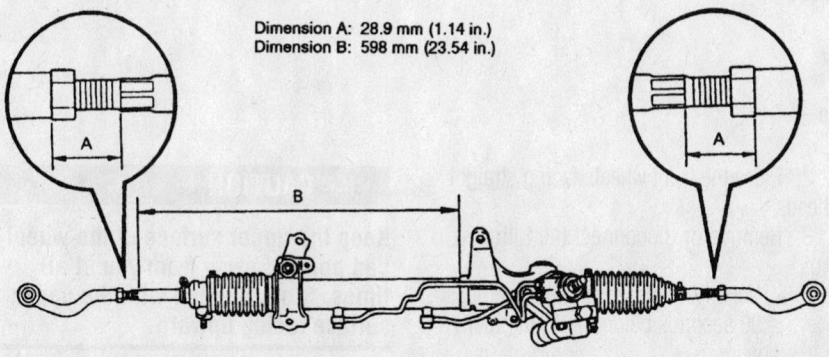

Dimension A: 28.9 mm (1.14 in.)
Dimension B: 598 mm (23.54 in.)

Centering the rack and pinion—GS 300, GS 400, GS 430

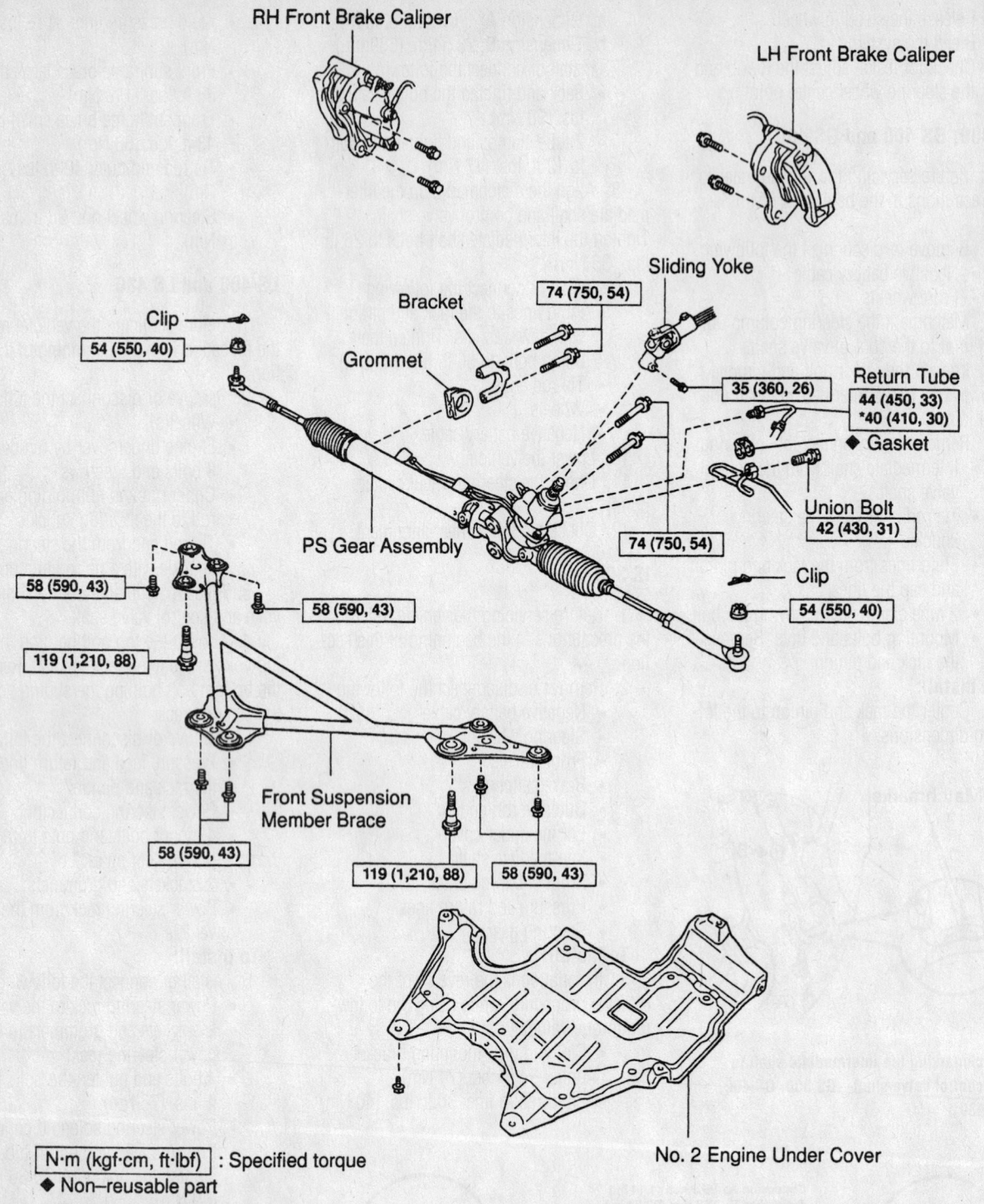

RH Front Brake Caliper

LH Front Brake Caliper

Sliding Yoke

Clip

Bracket

74 (750, 54)

Grommet

54 (550, 40)

35 (360, 26)

Return Tube
44 (450, 33)
*40 (410, 30)

◆ Gasket

PS Gear Assembly

Union Bolt
42 (430, 31)

74 (750, 54)

Clip

54 (550, 40)

58 (590, 43)

58 (590, 43)

119 (1,210, 88)

Front Suspension
Member Brace

58 (590, 43)

119 (1,210, 88)   58 (590, 43)

No. 2 Engine Under Cover

N·m (kgf·cm, ft·lbf) : Specified torque
◆ Non-reusable part
* For Use With SST

9347LG02

Exploded view of the steering gear mounting—IS 300

- Engine undercover
- Wheel(s)

11. Bleed the power steering system and check the front end alignment.

SC 300 and SC 400

1. Before servicing the vehicle, refer to the precautions in the beginning of this section.

2. Place the front wheels facing straight ahead.

3. Remove or disconnect the following:

- Negative battery cable. Wait at least 90 seconds before performing any work.
- Steering wheel pad

✳✳ CAUTION

Keep the upper surface of the wheel pad pointed away from you at all times. Store the pad with the upper surface facing upward.

- Intermediate shaft
- Right and left tie rod ends

- Union bolt and gasket and remove the pressure tube
- Union bolt and 2 gaskets; remove the return tube
- PPS solenoid connector
- On SC 400 models, the tube clamp
- 2 bolts and nuts and remove the bracket and grommet
- 2 bolts and nuts; remove the rack and pinion assembly

To install:

4. Install or connect the following:
- Rack and pinion assembly with the 2 set bolts and nuts. Tighten the bolts to 56 ft. lbs. (76 Nm).
- Bracket and grommet with the 2 bolts and nuts. Tighten the bolts to 56 ft. lbs. (76 Nm).
- On SC 400 models, the tube clamp
- PPS solenoid
- Return tube with the bolt and new gaskets. Tighten the union bolt to 36 ft. lbs. (49 Nm).
- Pressure tube with the union bolt and a new gasket. Tighten the union bolt to 36 ft. lbs. (49 Nm).
- Right and left tie rod ends
- Intermediate shaft
- The steering wheel. Tighten the steering wheel set nut to 26 ft. lbs. (35 Nm).
- Steering wheel pad
- Negative battery cable
- Steering fluid and bleed the steering system.

Strut and Coil Spring

REMOVAL & INSTALLATION

Front

ES 300

1. Before servicing the vehicle, refer to the precautions in the beginning of this section.
2. Remove or disconnect the following:

- Negative battery cable
- Tire and wheel assembly
- If equipped with an Anti-lock Brake System (ABS), the ABS speed sensor connector
- Brake line from the strut housing
- Strut assembly from the steering knuckle
- 3 upper mounting nuts from the strut tower
- Strut assembly

✻✻ CAUTION

Do not remove the center nut to the strut at this time. The spring on the strut is under high pressure and can cause serious injury.

3. Temporarily install the bolt and nuts to the lower bracket of the strut to support it and secure the strut in a vise.
4. Compress the coil spring
5. Remove or disconnect the following:

- Spring seat
- Upper strut retaining nut
- Suspension support
- Upper insulator
- Spring
- Bumper
- Insulator

To install:

6. Install or connect the following:
- Lower insulator
- Bumper to the piston rod
- Coil spring end into the gap of the lower seat
- Upper insulator
- Upper support to the piston rod, aligning it with the groove in the strut rod
7. Install or connect the following:
- Spring seat. Tighten the new upper

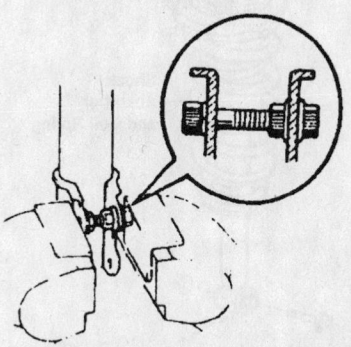

7923LGA5

Temporarily install the support nuts and bolt to the strut—ES 300

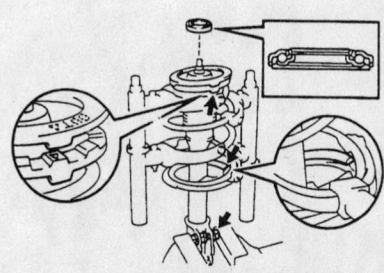

7923LGA6

Align the out mark of the upper spring seat with the mark on the upper insulator—ES 300 and LS 400

strut retaining nut to 36 ft. lbs. (49 Nm).

8. Remove the strut from the vise and disassemble the securing nuts and bolt.
9. Rotate the upper support so the lowest bolt on the support aligns with the projection part of the lower spring.
10. Install or connect the following:
- Strut and tighten the strut to body bolts to 59 ft. lbs. (80 Nm)
- Strut to the steering knuckle and tighten the bolts to 156 ft. lbs. (211 Nm)
11. Run the brake hose through the brake hose bracket and install the clip.
12. Install or connect the following:
- ABS speed sensor and tighten the mounting bolt to 48 inch lbs. (5 Nm)
- Brake line to the strut housing and tighten the bolt to 22 ft. lbs. (29 Nm)
- Wheel
- Negative battery cable
13. Check the front alignment.

GS 300, GS 400 AND GS 430

1. Before servicing the vehicle, refer to the precautions in the beginning of this section.
2. Remove or disconnect the following:
- Negative battery cable.
- Front wheel
- Brake caliper, leaving the line attached
3. Loosen the 3 upper strut mounting nuts.
4. Loosen, but do not remove, the upper strut rod nut.

✻✻ CAUTION

Do not remove the upper strut nut at this time.

5. Remove or disconnect the following:
- Anti-lock Brake System (ABS) speed sensor and harness
- Upper suspension arm from the steering knuckle
- Stabilizer bar from the link and remove the bracket
- Strut from the lower suspension arm.
- 3 upper strut mounting nuts and remove the strut
6. Compress the coil spring.
7. Remove or disconnect the following:
- Piston rod locknut
- Suspension support, coil spring and bumper
8. If disposing the strut, perform the following procedure:

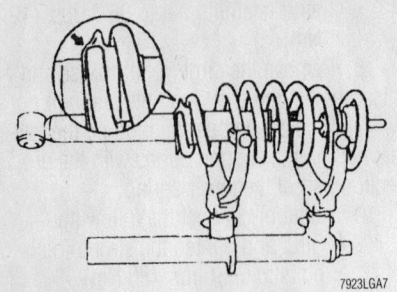

Matching the spring to the seat

a. Fully extend the strut rod.

b. Drill a hole near the bottom of the shock to remove the gas inside.

✳✳ CAUTION

The gas is harmless, but be careful of chips that may fly up when the gas is released.

To install:

9. Install or connect the following:
  • Spring bumper
  • Coil spring
  • Suspension support to the rod and temporarily install a new nut

10. Turn the suspension support so one of the bolts on the support faces the same direction as shown in the illustration.

→**Align the bolt so a line drawn between the rod and bolt would be at 90° to the direction of the lower bushing.**

11. Install or connect the following:
  • Spring compressor
  • Strut and tighten the upper retaining nuts to 41 ft. lbs. (56 Nm)
  • New upper strut rod nut to 20 ft. lbs. (27 Nm)
  • Strut to the lower arm and temporarily tighten the nut and bolt
  • Stabilizer bar bracket and tighten the bolts to 21 ft. lbs. (28 Nm)
  • The stabilizer bar to the link and tighten the bolts to 29 ft. lbs. (39 Nm)
  • Upper suspension arm to the steering knuckle. Tighten the nut to 64 ft. lbs. (87 Nm) and install a new cotter pin.
  • ABS speed sensor and tighten the bolt to 69 inch lbs. (8 Nm)
  • Caliper
  • Wheel

12. Bounce the vehicle several times to stabilize the suspension.

13. Tighten the lower strut bolt and nut to 116 ft. lbs. (157 Nm).

14. Check the front wheel alignment.

IS 300

1. Before servicing the vehicle, refer to the precautions in the beginning of this section.

2. Remove or disconnect the following:
  • Front wheel
  • Wheel speed sensor and harness clamp
  • Upper ball joint
  • Level control sensor link
  • Stabilizer bar link
  • Lower strut bolt
  • Upper strut mount cap
  • Upper strut mount nuts
  • Strut assembly

3. Install a suitable spring compressor and remove the center nut.

4. Remove the upper strut mount and the coil spring.

To install:

5. Installation is the reverse of the removal procedure, while using the following torque values:
  • Upper strut mount center nut: 25 ft. lbs. (34 Nm)
  • Upper strut mounting nuts: 26 ft. lbs. (35 Nm)
  • Lower strut mount bolt: 47 ft. lbs. (64 Nm)
  • Upper ball joint nut: 50 ft. lbs. (65 Nm)
  • Stabilizer bar link nut: 36 ft. lbs. (49 Nm)

LS 400 AND LS 430—WITHOUT AIR SUSPENSION

1. Before servicing the vehicle, refer to the precautions in the beginning of this section.

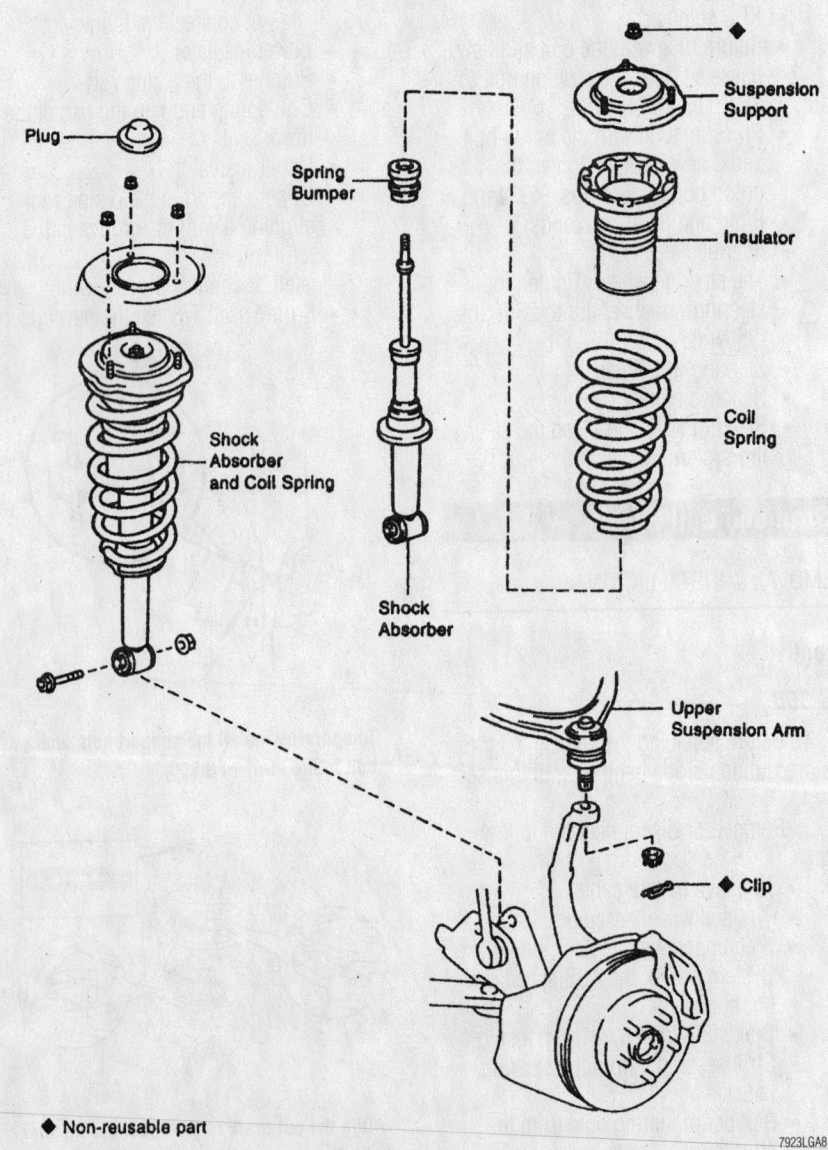

◆ **Non-reusable part**

Exploded view of the strut and spring mounting—LS 400 without air suspension shown

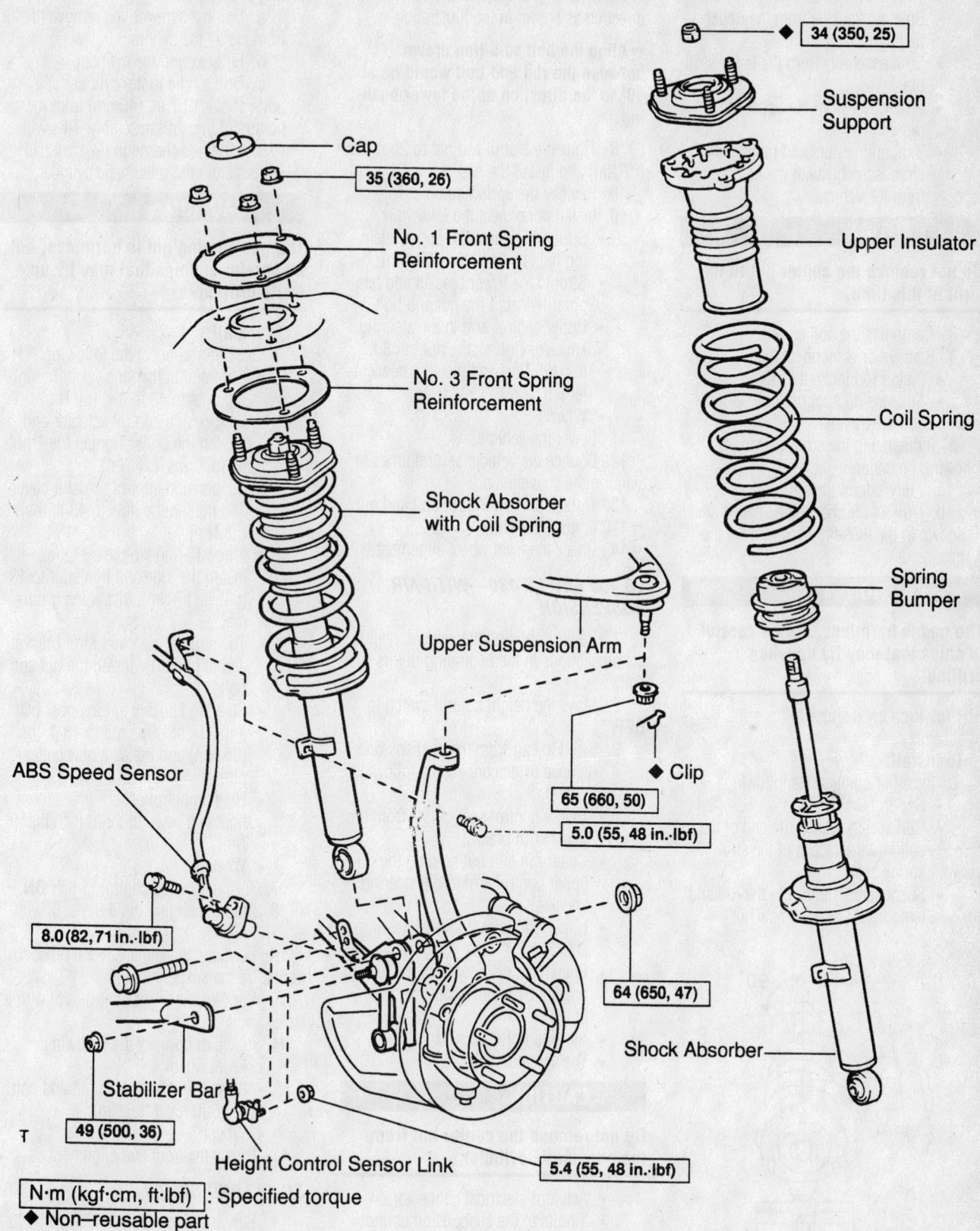

34 (350, 25)

Suspension Support

Cap

35 (360, 26)

No. 1 Front Spring Reinforcement

Upper Insulator

No. 3 Front Spring Reinforcement

Coil Spring

Shock Absorber with Coil Spring

Upper Suspension Arm

Spring Bumper

ABS Speed Sensor

◆ Clip

65 (660, 50)

5.0 (55, 48 in.·lbf)

8.0 (82, 71 in.·lbf)

64 (650, 47)

Shock Absorber

Stabilizer Bar

49 (500, 36)

Height Control Sensor Link

5.4 (55, 48 in.·lbf)

T

N·m (kgf·cm, ft·lbf) : Specified torque
◆ Non–reusable part

Exploded view of the front strut assembly mounting—IS 300

9347LG05

2. Remove or disconnect the following:
- Tire and wheel assembly
- Steering knuckle from the upper ball joint
- Strut assembly from the lower strut bracket
- Strut cover from the upper strut mount
- 3 mounting nuts and remove the strut assembly with the coil spring from the vehicle.

✳✳ CAUTION

Do not remove the center nut to the strut at this time.

3. Compress the coil spring.
4. Remove or disconnect the following:
- Piston rod locknut
- Suspension support, coil spring and the bumper

5. If disposing the strut, perform the following procedure:
    a. Fully extend the strut rod.
    b. Drill a hole within the shaded area shown in the illustration to remove the gas inside.

✳✳ CAUTION

The gas is harmless, but be careful of chips that may fly up when drilling.

    c. Properly dispose of the strut assembly.
To install:
6. Install or connect the following:
- Spring bumper
- Coil spring. Match the end of the coil into the recess of the strut spring seat.
- Suspension support to the rod and temporarily install a new nut

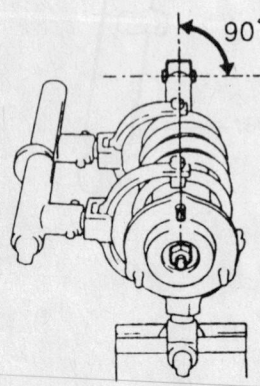

Be sure to align the suspension support with one of the upper mounting bolts as shown—LS 400 without air suspension

7. Turn the suspension support so one of the bolts on the support faces the same direction as shown in the illustration.

➡ **Align the bolt so a line drawn between the rod and bolt would be at 90° to the direction of the lower bushing.**

8. Tighten the strut rod nut to 20 ft. lbs. (27 Nm) and install the cap.
9. Remove the spring compressor
10. Install or connect the following:
- Strut and tighten the upper retaining nuts to 43 ft. lbs. (58 Nm)
- Strut to the lower bracket and temporarily install the nut and bolt
- Upper control arm to the steering knuckle. Tighten the nut to 48 ft. lbs. (65 Nm) and install a new cotter pin.
- Wheel

11. Lower the vehicle.
12. Bounce the vehicle several times to stabilize the suspension.
13. Tighten the lower strut bolt and nut to 116 ft. lbs. (157 Nm).
14. Check the front wheel alignment.

LS 400 AND LS 430—WITH AIR SUSPENSION

1. Before servicing the vehicle, refer to the precautions in the beginning of this section.
2. Move the height control switch to **OFF**.
3. Bleed the air from the suspension.
4. Remove or disconnect the following:
- Wheel
- Height control sensor link from the lower strut bracket
- Cotter pin and nut holding the upper control arm to the steering knuckle
- Upper ball joint from the steering knuckle
- Pneumatic cylinder from the lower bracket by removing the through-bolt
- Air tube from the strut
- The actuator cover

✳✳ CAUTION

Do not remove the center nut from the pneumatic cylinder.

- Actuator electrical connector
- 2 bolts to the suspension control actuator and position the actuator aside
- 3 upper mounting nuts and the strut from the vehicle

5. If disposing the strut perform the following procedure:
    a. Using a screwdriver, remove the air from inside the cylinder.
    b. Fully extend the cylinder.
    c. Drill a hole in the cylinder at a point above 1.57 in. (40mm) from the bottom of the strut assembly. This will release the gas charge in the strut. Do not puncture the pneumatic cylinder.

✳✳ CAUTION

The gas coming out is harmless, but be careful of chips that may fly up while drilling.

To install:
6. Install or connect the following:
- Strut and tighten the upper mounting nuts to 43 ft. lbs. (58 Nm)
- Suspension control actuator and tighten the bolts. Tighten the 2 nuts to 13 ft. lbs. (17 Nm).
- Suspension control actuator cover and tighten the nuts to 43 ft. lbs. (58 Nm)
- 2 new O-rings to the air tube. Install the tube and tighten it to 13 ft. lbs. (17 Nm). Install the grommet.
- The strut to the lower strut bracket and temporarily install the nut and bolt
- Steering knuckle to the upper ball joint. Tighten the nut to 48 ft. lbs. (65 Nm) and install a new cotter pin.
- Height control sensor link and tighten a new nut to 48 inch lbs. (5 Nm)
- Wheel

7. Turn the height control switch **ON**.
8. Start the engine to fill the strut with air.
9. Bounce the vehicle several times to normalize the suspension.
10. Support the lower control arm with a jack.
11. Install or connect the following:
- Front wheel
- Lower strut mounting nut and bolt to 76 ft. lbs. (106 Nm)
- Wheel

12. Check the front end alignment.

SC 300 AND SC 400

1. Before servicing the vehicle, refer to the precautions in the beginning of this section.
2. Remove or disconnect the following:
- Tire and wheel assembly

- Brake caliper support bracket
- Fender apron
- Engine undercover
- Front fender wheel opening molding
- If removing the left side strut, the windshield washer tank
- Anti-lock Brake System (ABS) speed sensor at the steering knuckle
- Wiring harness clamp in order to prevent the harness from being damaged when removing the through-bolt.
- Plug from the upper strut mount. Do not remove the center bolt.

✱✱ CAUTION

Do not remove the center bolt to the strut at this time. The spring on the strut is under high pressure and can cause serious injury or vehicle damage.

- Upper control arm through-bolt from the subframe
- Upper control arm and turn the control arm completely around. It is not necessary to remove the upper ball joint.
- Strut at the lower control arm by removing the nut and bolt
- 3 upper mounting nuts and remove the strut assembly with the coil spring from the vehicle
3. Compress the coil spring.
4. Remove or disconnect the following:
- Piston rod locknut
- Suspension support
- Coil spring
- Bumper
5. If disposing the strut, perform the following procedure:
  a. Fully extend the strut rod.
  b. Drill a hole within the shaded area shown in the illustration to remove the gas inside and dispose the old strut.

✱✱ CAUTION

The gas is harmless, but be careful of chips which may fly up when drilling.

To install:
6. Install or connect the following:
- Spring bumper
- Coil spring
- Suspension support to the rod and temporarily install a new nut
7. Turn the suspension support so one of the bolts on the support faces the same direction as shown in the illustration.

➡**Align the bolt so a line drawn between the rod and bolt would be at 90° to the direction of the lower bushing.**

8. Remove the spring compressor.
9. Install or connect the following:
- Strut and tighten the 3 upper strut mount nuts to 26 ft. lbs. (35 Nm). Tighten the middle nut to 22 ft. lbs. (29 Nm) and install the plug.
- Lower end of the strut to the lower control arm. Do not tighten the bolt at this time.
- Upper control arm and install the through-bolt and nut. Do not tighten the bolt at this time.
- VSS, wiring harness and the washer tank
- Fender apron and the engine undercover
- Caliper support bracket and tighten the bolts to 87 ft. lbs. (118 Nm)
- Tire and wheel assembly
10. Lower the vehicle.
11. Bounce the vehicle a few times to stabilize the suspension, then tighten the strut to lower arm bolt to 106 ft. lbs. (143 Nm). Tighten the upper arm to 121 ft. lbs. (164 Nm).
12. Check the front end alignment.

Rear

ES 300

1. Before servicing the vehicle, refer to the precautions in the beginning of this section.
2. Remove or disconnect the following:
- Tire and wheel assembly
- Load sensing proportioning valve spring assembly from the lower arm
- Anti-lock Brake System (ABS) speed sensor harness and brake line from the strut assembly
- Stabilizer bar link from the strut
3. Loosen the 2 nuts attaching the strut to the axle carrier.
4. Support the axle carrier.
5. Remove or disconnect the following:
- Rear seat back and package tray trim
- Upper mounting nuts
- 2 lower mounting bolts and remove the strut assembly
6. Compress the coil spring.
7. Temporarily install a bolt and 2 nuts on the bracket at the lower end of the strut and secure it in a vise.
8. Secure the upper support and remove the strut rod retaining nut.
9. Remove or disconnect the following:
- Upper suspension support

- Upper insulator
- Coil spring
- Spring bumper
- Lower insulator
10. If discarding the strut, perform the following:
  a. Fully extend the strut rod.
  b. Drill a hole in the side of the strut to release the gas.

✱✱ WARNING

The gas coming out is harmless, but be careful of chips which may fly up while drilling.

To install:
11. Install or connect the following:
- Lower insulator to the strut
- Spring bumper to the strut piston rod
- Compressed coil spring
- Coil spring with the end butted against the gap in the lower seat
- Upper insulator and support matching the bolt of the support with the cut-off part of the insulator
- Upper suspension support
- New strut piston rod nut to 36 ft. lbs. (49 Nm)
- Spring compressor
- Strut rod piston nut cap
- Strut and tighten the 3 nuts to 29 ft. lbs. (39 Nm)
- Strut to the axle carrier. Coat the nuts with engine oil and tighten the nuts and bolts to 188 ft. lbs. (255 Nm)
- ABS harness to the strut and tighten the bolt to 48 inch lbs. (6 Nm)
- Brake line to the strut and tighten the retaining nut to 22 ft. lbs. (29 Nm)
- Spring to the lower arm and tighten the nut to 10 ft. lbs. (13 Nm)
- LSPV to the lower arm and tighten the nut to 108 inch lbs. (12 Nm)
- Rear wheel
- Rear seat and package tray

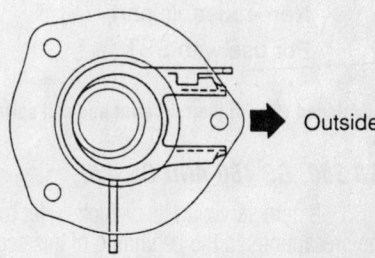

7923LGB1

Position the upper suspension support as shown when assembling the strut—ES 300

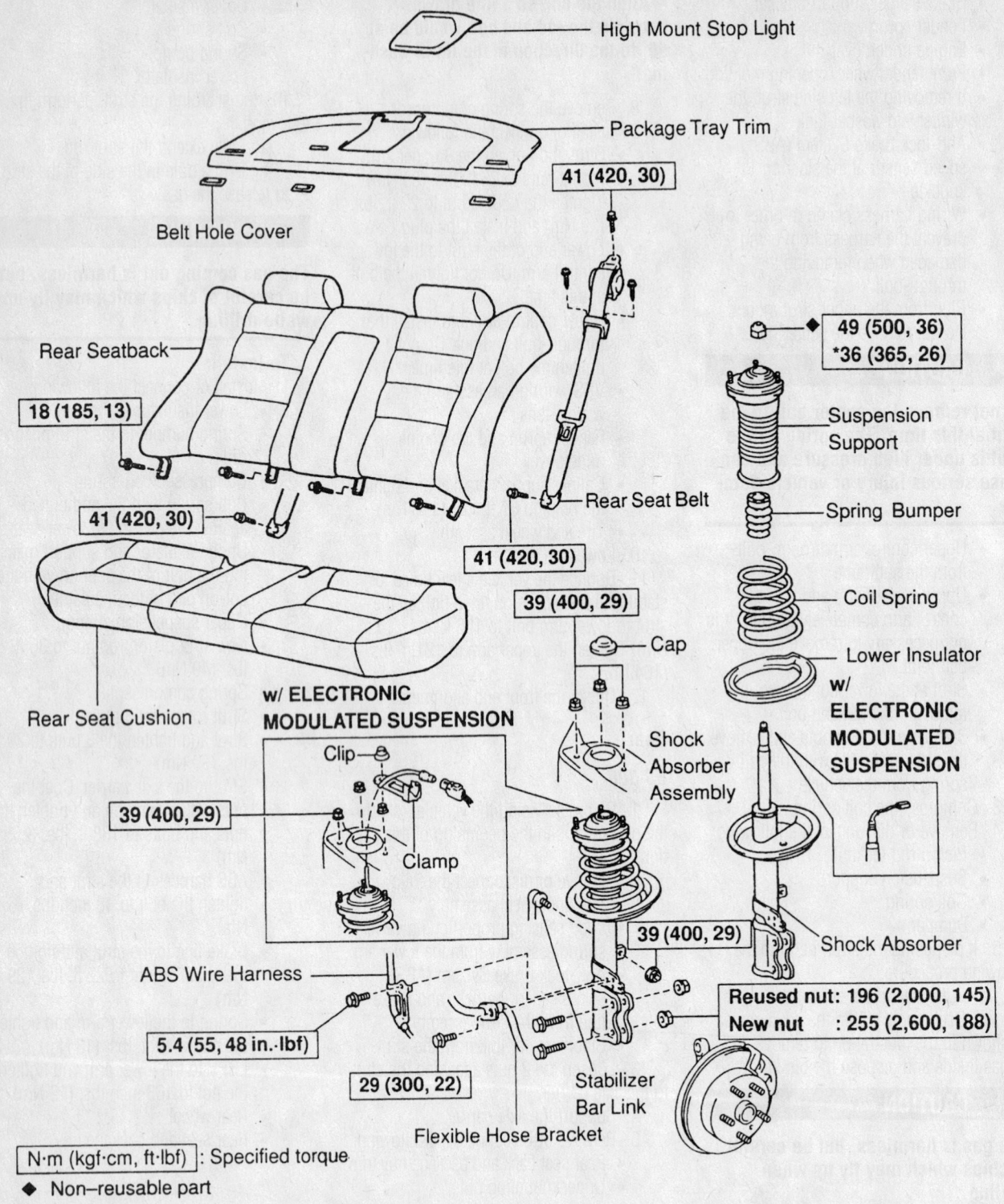

Exploded view of the rear strut and coil spring mounting—ES 300

GS 300, GS 400 AND GS 430

1. Before servicing the vehicle, refer to the precautions in the beginning of this section.
2. Remove or disconnect the following:
   • Front trunk compartment trim cover
   • Wheel(s)
   • Brake caliper support bracket from the rear axle carrier by removing the 2 bolts. Leave the brake line connected and position it out of the way.
   • Nut and disconnect the sway bar link from the lower control arm
   • Nut and bolt on the lower end of the strut
   • 3 upper nuts and lift out the strut. Do not remove the center nut

3. Compress the coil spring.

4. Secure the upper support and remove the strut rod retaining nut.

5. Remove or disconnect the following:
- Upper suspension support
- Upper insulator
- Coil spring
- Spring bumper
- Lower insulator

6. If discarding the strut, perform the following:

   a. Fully extend the strut rod.

   b. Drill a hole in the side of the strut to drain the gas inside

✲✲ CAUTION

The gas coming out is harmless, but be careful of chips that may fly up while drilling.

To install:

7. Install or connect the following:
- Lower insulator to the strut
- Spring bumper to the strut piston rod
- Compressed coil spring. Position the coil spring with the end butted against the gap in the lower seat.
- Upper insulator and suspension support
- Upper suspension support and tighten a new strut piston rod nut to 20 ft. lbs. (27 Nm)

8. Remove the spring compressor.

9. Install or connect the following:
- Strut to the vehicle and tighten the 3 upper mounting nuts to 14 ft. lbs. (20 Nm). Install the cap.
- And tighten the lower strut bolt and nut to 101 ft. lbs. (137 Nm)
- Sway bar link to the lower control arm. Tighten the nut to 33 ft. lbs. (44 Nm).
- Brake caliper to the rear axle carrier by installing the 2 bolts. Tighten the bolts to 77 ft. lbs. (104 Nm).
- Wheel(s)
- Trunk compartment cover trim

10. Check and adjust the vehicle alignment as necessary.

LS 400 AND LS 430—WITHOUT AIR SUSPENSION

1. Before servicing the vehicle, refer to the precautions in the beginning of this section.

2. Remove or disconnect the following:
- Rear seat cushion and seat back.
- Tray trim
- Tire and wheel assembly
- Rear halfshaft

- Stabilizer bar link from the stabilizer bar
- Anti-lock Brake System (ABS) speed sensor and wiring harness
- Brake caliper bracket from the axle carrier, leaving the brake line connected. Suspend the brake caliper aside with a piece of wire.
- Nut on the lower side of the strut. Do not remove the bolt.
- Rear axle assembly with a lifting device
- Strut cap by removing the 3 nuts
- 3 mounting nuts holding the strut assembly to the strut tower. Do not remove the center bolt.

✲✲ CAUTION

Do not remove the center nut to the strut at this time.

- Bolt on the lower side of the strut assembly
- Strut assembly with the coil spring

3. Compress the coil spring.

4. Secure the strut housing in a vise.

5. Remove or disconnect the following:
- Strut rod retaining nut
- Upper suspension support
- Upper insulator
- Coil spring
- Spring bumper
- Lower insulator

6. If discarding the strut, perform the following:

   a. Fully extend the strut rod.

   b. Drill a hole in the strut (about 1 in. above the strut lower mount) and drain the gas inside

✲✲ CAUTION

The gas coming out is harmless, but be careful of chips which may fly up while drilling.

To install:

7. Install or connect the following:
- Lower insulator to the strut.
- Spring bumper to the strut piston rod
- Coil spring
- Upper insulator and support
- Upper suspension support

8. Temporarily install the upper strut rod retaining nut.

9. Rotate the suspension support so that the rod and one of the bolts on the suspension support are aligned with the lower bushing.

10. Remove the spring compressor.

11. Install or connect the following:

- Strut assembly to the vehicle and tighten the 3 nuts to 47 ft. lbs. (64 Nm). Tighten the strut rod retaining nut to 20 ft. lbs. (27 Nm).
- Strut assembly cap and install the 3 nuts
- Strut to the rear axle carrier. Install the bolt from the rear of the vehicle and temporarily tighten the nut.
- Brake caliper and tighten the mounting bolts to 77 ft. lbs. (104 Nm)
- ABS speed sensor and wiring harness
- Stabilizer link to the stabilizer bar and tighten the nut to 48 ft. lbs. (65 Nm)
- Rear halfshaft
- Tire and wheel assembly

12. Bounce the vehicle up and down to stabilize the suspension.

13. Support the rear axle assembly with a lifting device. Tighten the lower strut bolt to 101 ft. lbs. (137 Nm).

14. Install or connect the following:
- Rear seat cushion and rear seat back.
- Package tray trim

15. Check the wheel alignment.

LS 400 AND LS 430—WITH AIR SUSPENSION

1. Before servicing the vehicle, refer to the precautions in the beginning of this section.

2. Bleed the air system from the suspension.

3. Remove or disconnect the following:
- Rear seat cushion and seat back
- Package tray trim
- Trunk trim panel. Move the height control switch, located in the trunk area, to the **OFF** position.
- Tire and wheel assembly
- Rear halfshaft
- Stabilizer links from the stabilizer bar
- Anti-lock Brake System (ABS) speed sensor and wiring harness
- Brake caliper bracket from the rear axle carrier. Do not disconnect the brake line.

4. Place matchmarks on the height control sensor link and bracket. Disconnect the height control sensor link from the No. 1 lower control arm.

5. Support the rear axle assembly with a lifting device.

6. Remove or disconnect the following:
- Nut on the lower side of the shock absorber. Do not remove the bolt.
- Grommet and disconnect the air tube from the shock absorber

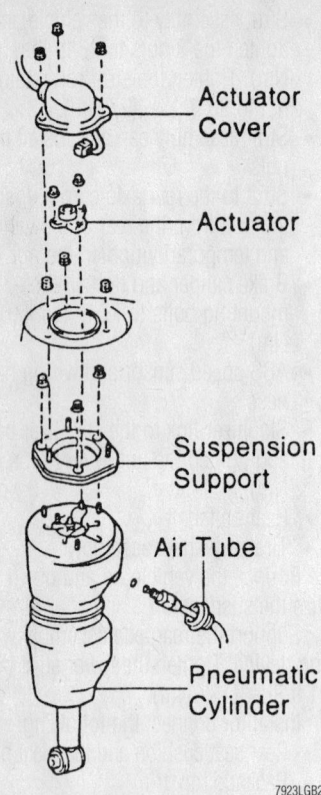

7923LGB2

Pneumatic cylinder (strut) component overview (air suspension)

- Actuator cover from the strut tower by removing the 3 nuts
- Actuator electrical connector from the top of the strut
- Actuator by removing the 2 nuts
- 3 upper mounting nuts holding the strut to the strut tower
7. Lower the rear axle assembly
8. Remove or disconnect the following:
- Bolt on the lower side of the shock absorber
- Pneumatic cylinder strut assembly from the vehicle
- Suspension support from the strut assembly by removing the 3 nuts
9. If discarding the pneumatic cylinder, perform the following:
   a. Using a screwdriver, depressurize the air from inside the cylinder.
   b. Drill a hole in the shaded area shown in the illustration and remove the gas inside.

✳✳ CAUTION

The gas coming out is harmless, but be careful of chips which may fly up when drilling.

To install:
10. Install or connect the following:
- Suspension support to the pneu-

matic cylinder (strut) and tighten the nuts to 27 ft. lbs. (36 Nm)
- Strut assembly to the vehicle and tighten the upper mounting nuts to 47 ft. lbs. (64 Nm)
11. Match the holes in the pneumatic cylinder with the holes in the suspension control actuator.
12. Install or connect the following:
- Actuator and tighten the mounting nuts to 69 inch lbs. (8 Nm)
- Actuator cover and tighten the 3 nuts to 18 ft. lbs. (25 Nm)
- New O-rings and connect the air line to the shock absorber. Tighten the fitting to 13 ft. lbs. (18 Nm)
- Strut to the rear axle carrier. Insert the bolt from the vehicle's rear and temporarily tighten the nut.
- Height control sensor link to the No. 1 lower control arm. Mounting nut: 48 inch lbs. (5 Nm).
- Rear brake caliper to the rear axle carrier and tighten the mounting bolts to 77 ft. lbs. (104 Nm)
- ABS speed sensor and wiring harness
- Stabilizer bar link and tighten the nut to 48 ft. lbs. (65 Nm)
- Halfshaft
- Actuator electrical connector to the top of the strut
- Tire
13. Move the height control switch to the **ON** position. Start the engine and fill the pneumatic cylinder with air.
14. Bounce the vehicle up and down several times to stabilize the suspension.
15. Turn the suspension height control to the **OFF** position.
16. Remove the tire and wheel assembly.
17. Support the rear axle carrier with a lifting device. Tighten the lower strut bolt to 101 ft. lbs. (137 Nm).
18. Install or connect the following:
- Package tray trim
- Rear seat cushion and seat back
19. Turn the suspension control switch to the **ON** position.
20. Check the wheel alignment.

SC 300 AND SC 400

1. Before servicing the vehicle, refer to the precautions in the beginning of this section.
2. Raise the rear of the vehicle and support it with safety stands.
3. Remove or disconnect the following:
- Wheel(s)
- Brake caliper support bracket by removing the 2 bolts. Leave the

brake line connected and position it aside.
- Nut and bolt on the lower end of the strut
- Cap nut on the upper end of the strut. Remove the 3 upper nuts and lift out the strut. Do not remove the center nut from the strut.

✳✳ CAUTION

Do not remove the center nut on the strut at this time. The spring on the strut is under high pressure and can cause serious injury.

4. Compress the coil spring.
5. Secure the strut housing with 2 nuts and a bolt as shown in the illustration and secure it in a vise.
6. Secure the upper support and remove the strut rod retaining nut.
7. Remove the upper suspension support, upper insulator, coil spring, spring bumper and the lower insulator.
8. If discarding the strut, perform the following:
   a. Fully extend the strut rod.
   b. Drill a hole in the strut in the shaded area shown in the illustration and drain the gas inside

✳✳ CAUTION

The gas coming out is harmless, but be careful of chips which may fly up while drilling.

To install:
9. Install or connect the following:
- Lower insulator to the strut
- Spring bumper to the strut piston rod
- The (compressed) coil spring
10. Position the coil spring with the end butted against the gap in the lower seat.
- Upper insulator and support matching the bolt of the support with the cut off part of the insulator
- Upper suspension support
11. Secure the upper suspension support and tighten the new strut piston rod nut to 20 ft. lbs. (27 Nm).
12. Remove the spring compressor.
- Strut rod piston nut cap
- Strut and the 3 nuts. Tighten the nuts to 19 ft. lbs. (25 Nm). Install the cap.
- Lower bolt to hold the strut to the lower control arm. Do not tighten the bolt at this time.
- Caliper support bracket and tighten the bolts to 77 ft. lbs. (104 Nm)
- Wheel(s)

13. Lower the vehicle.

14. Bounce the vehicle several times to normalize the suspension.

15. Support the lower arm.

16. Tighten the lower strut mounting bolt to 106 ft. lbs. (143 Nm).

17. Check the alignment and adjust as necessary.

Upper Ball Joint

REMOVAL & INSTALLATION

The upper ball joint is an integral part of the upper arm and is not replaced separately. The upper ball joint replacement is accomplished by replacing the upper arm.

Upper Control Arm

REMOVAL & INSTALLATION

GS 300

1. Before servicing the vehicle, refer to the precautions in the beginning of this section.

2. Remove or disconnect the following:
   - Negative battery cable
   - Wheel

3. Loosen the 3 upper strut mounting nuts.

4. Loosen, but do not remove, the upper strut rod nut.

✳✳ CAUTION

DO NOT completely remove the upper strut nut at this time.

5. Remove or disconnect the following:
   - Brake caliper, leaving the line attached and secure it out of the way
   - Anti-lock Brake System (ABS) speed sensor and harness
   - Cotter pin and nut from the upper control arm
   - Upper control arm from the steering knuckle
   - Stabilizer bar from the link and remove the bracket
   - Cotter pin and nut from the lower control arm
   - Strut from the lower suspension arm
   - 3 upper strut mounting nuts and remove the strut
   - Mounting bolts holding the upper control arm to the frame
   - Upper control arm from the vehicle

To install:

6. Install or connect the following:
   - Upper suspension arm and tighten the mounting bolts to 39 ft. lbs. (53 Nm)
   - Strut and tighten the upper retaining nuts to 41 ft. lbs. (56 Nm). Tighten the new upper strut rod nut to 20 ft. lbs. (27 Nm).
   - Strut to the lower arm and temporarily tighten the nut and bolt
   - Stabilizer bar bracket and tighten the bolts to 21 ft. lbs. (28 Nm)
   - Stabilizer bar to the link and tighten the bolts to 29 ft. lbs. (39 Nm)
   - Upper suspension arm to the steering knuckle. Tighten the nut to 64 ft. lbs. (87 Nm) and install a new cotter pin.
   - ABS speed sensor and tighten the bolt to 69 inch lbs. (8 Nm)
   - Caliper
   - Front wheel

7. Lower the vehicle.

8. Bounce the vehicle several times to stabilize the suspension.

9. Tighten the lower strut bolt and nut to 116 ft. lbs. (157 Nm).

10. Check the front wheel alignment.

IS 300

1. Before servicing the vehicle, refer to the precautions in the beginning of this section.

2. Remove or disconnect the following:
   - Front wheel
   - Strut and spring assembly
   - Inner bolts and the control arm

To install:

3. Install or connect the following:
   - Control arm and tighten the inner bolts to 44 ft. lbs. (59 Nm)
   - Strut and spring assembly
   - Front wheel

LS 400 and LS 430

1. Before servicing the vehicle, refer to the precautions in the beginning of this section.

2. Raise and safely support the vehicle.

3. Remove or disconnect the following:
   - Wheel
   - Strut or if equipped with air suspension, remove the pneumatic cylinder
   - Anti-lock Brake System (ABS) speed sensor wire harness from the upper control arm by removing the bolt.
   - Mounting bolts holding the upper control arm to the vehicle
   - Upper control arm

To install:

4. Install or connect the following:
   - Upper control arm and tighten the 2 mounting bolts to 83 ft. lbs. (113 Nm)
   - ABS speed sensor wire harness to the upper control arm with the attaching bolt
   - Strut, or if equipped with air suspension, install the pneumatic cylinder
   - Wheel

5. Lower the vehicle.

6. Check and adjust the wheel alignment as necessary.

SC 300 and SC 400

1. Raise the front of the vehicle and support it on safety stands.

2. Remove or disconnect the following:
   - Wheel
   - Caliper support bracket by removing the 2 bolts. Leave the brake line connected and suspend it aside.
   - Rotor
   - Front fender splash shield, fender liner and wheel opening molding
   - On the left side, the washer tank
   - Bolt and disconnect the Anti-lock Brake System (ABS) speed sensor from the steering knuckle. Remove the 3 bolts and disconnect the wire harness clamp.
   - Cotter pin and the nut from the upper ball joint; press the upper ball joint from the knuckle
   - Through-bolt, nut and the upper control arm

To install:

3. Install or connect the following:
   - Upper control arm. Connect the upper control arm to the subframe and install the through-bolt. Do not tighten the bolt at this time.

➡ **The upper control arm mounting bolts are not tightened until the suspension has been assembled and vehicle is on the ground.**

   - Ball joint to the knuckle and tighten the nut to 76 ft. lbs. (103 Nm). Install a new cotter pin.
   - Wire harness and ABS speed sensor. Tighten the speed sensor to knuckle bolt to 69 inch lbs. (8 Nm)
   - Washer tank, the fender liner, splash shield and molding
   - Rotor
   - Caliper support bracket and tighten the bolts to 87 ft. lbs. (118 Nm)
   - Wheel

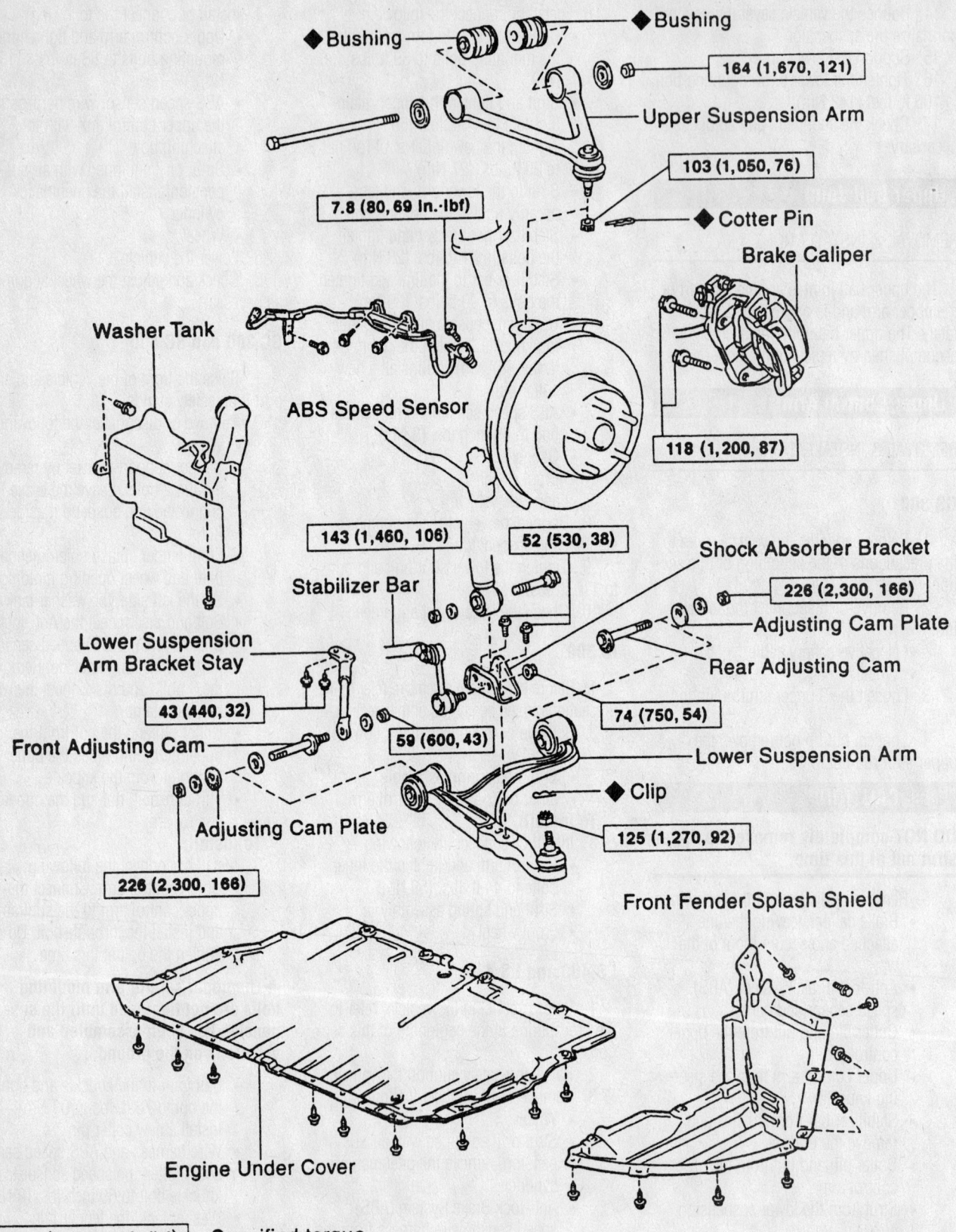

◆Bushing

◆Bushing

164 (1,670, 121)

Upper Suspension Arm

103 (1,050, 76)

7.8 (80, 69 In.·lbf)

◆Cotter Pin

Brake Caliper

Washer Tank

ABS Speed Sensor

118 (1,200, 87)

143 (1,460, 106)

52 (530, 38)

Shock Absorber Bracket

226 (2,300, 166)

Stabilizer Bar

Adjusting Cam Plate

Lower Suspension
Arm Bracket Stay

Rear Adjusting Cam

43 (440, 32)

74 (750, 54)

Front Adjusting Cam

59 (600, 43)

Lower Suspension Arm

◆Clip

125 (1,270, 92)

Adjusting Cam Plate

226 (2,300, 166)

Front Fender Splash Shield

Engine Under Cover

N·m (kgf·cm, ft·lbf) : Specified torque

◆ Non-reusable part

7923LGC7

Exploded view of the front suspension control arms and related components—SC 300 and SC 400 models

4. Lower the vehicle.

5. Bounce the suspension several times to set the suspension.

6. Support the lower arm and tighten the upper control arm through-bolt and nut to 121 ft. lbs. (164 Nm).

7. Check the front wheel alignment and adjust as necessary.

CONTROL ARM BUSHING REPLACEMENT

The control arm bushings are serviced with the control arm as an assembly.

Lower Control Arm

REMOVAL & INSTALLATION

ES 300

1. Before servicing the vehicle, refer to the precautions in the beginning of this section.

2. Remove or disconnect the following:
   • Negative battery cable
   • Front wheel(s)
   • Side fender apron seal
   • Steering knuckle with the axle hub, from the vehicle
   • Dust deflector from the knuckle
   • Cotter pin and the nut from the ball joint stud

3. Remove the lower ball joint from the steering knuckle.

To install:

4. Install the lower ball joint onto the steering knuckle and tighten nut to 90 ft. lbs. (123 Nm). Install new cotter pin.

5. Align the hole in the dust deflector

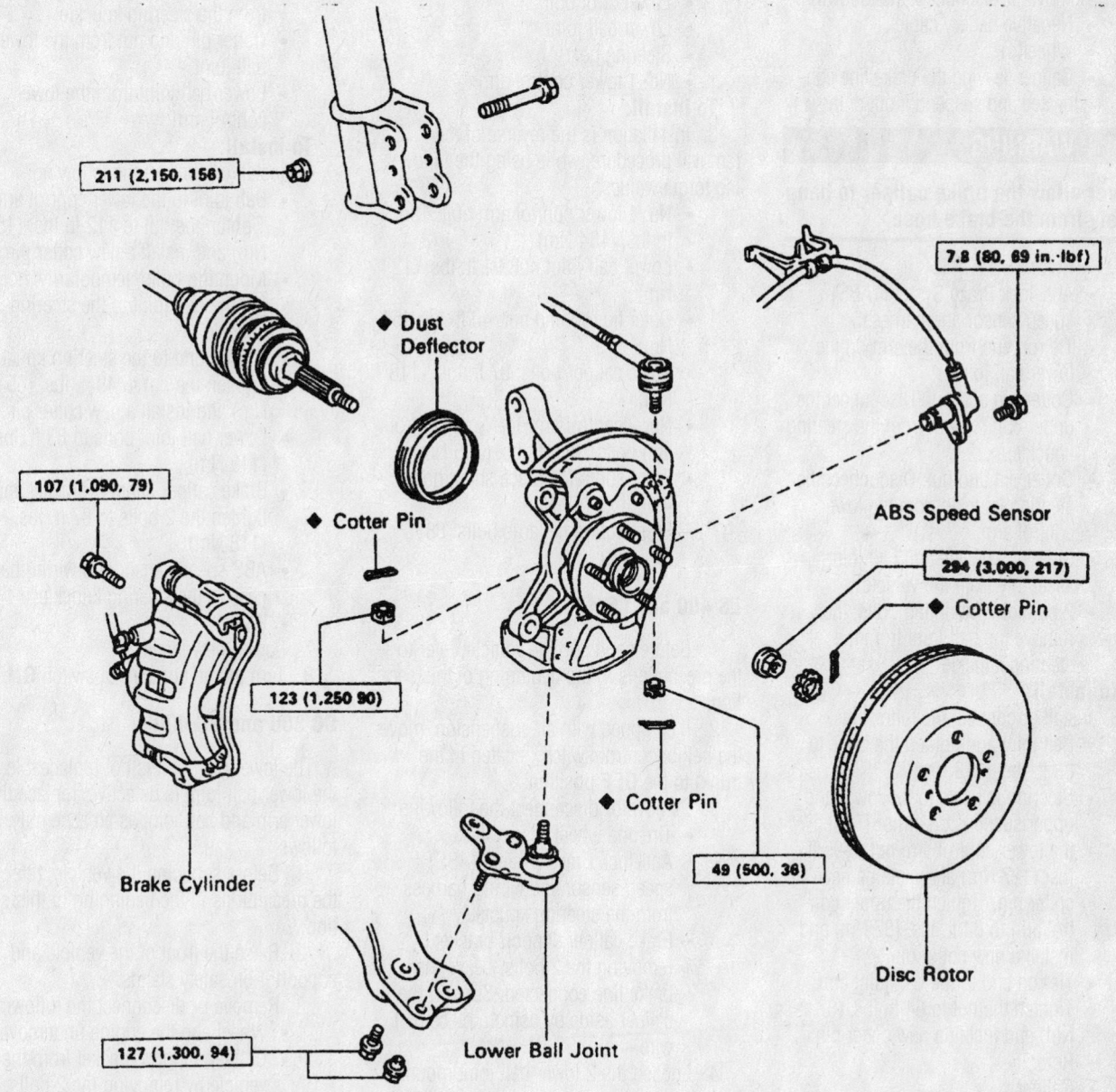

211 (2,150, 156)

7.8 (80, 69 in.·lbf)

♦ Dust Deflector

ABS Speed Sensor

107 (1,090, 79)

♦ Cotter Pin

294 (3,000, 217)

♦ Cotter Pin

123 (1,250 90)

♦ Cotter Pin

49 (500, 36)

Brake Cylinder

Disc Rotor

127 (1,300, 94)

Lower Ball Joint

N·m (kgf·cm, ft·lbf) : Specified torque

♦ Non-reusable part

Exploded view of the lower suspension—ES 300

7923LGB3

with the ABS speed sensor. Using the appropriate driver, install a new dust deflector.

6. Install or connect the following:
- Steering knuckle and hub onto the vehicle
- Fender apron seal
- Front wheel(s)
- Negative battery cable

GS 300, GS 400 and GS 430

1. Before servicing the vehicle, refer to the precautions in the beginning of this section.

2. Remove or disconnect the following:
- Negative battery cable
- Wheel(s)
- Caliper, leaving the brake line connected and suspend it out of the way

�֍ WARNING

Never allow the brake caliper to hang freely from the brake hose.

- Rotor
- Anti-lock Brake System (ABS) speed sensor and harness
- Tie rod end from the arm on the lower ball joint
- Cotter pin and nut. Disconnect the upper control arm from the steering knuckle.
- Cotter pin and nut. Disconnect the steering knuckle from the lower control arm.
- Steering knuckle and ball joint assembly from the vehicle
- 2 ball joint mounting bolts, then remove the ball joint from the steering knuckle

To install:
3. Install or connect the following:
- Ball joint and tighten the bolts to 83 ft. lbs. (113 Nm)
- Steering knuckle to the lower and upper suspension arms. Tighten the lower control arm nut to 95 ft. lbs. (127 Nm) and install a new cotter pin. Tighten the upper control arm to 64 ft. lbs. (87 Nm) and install a new cotter pin.
- Tie rod end to the ball joint arm. Tighten the nut to 64 ft. lbs. (87 Nm) and install a new cotter pin.
- Rotor
- Caliper
- ABS speed sensor and harness. Tighten the sensor retaining bolt to 69 inch lbs. (8 Nm).
- Wheel(s)
- Negative battery cable
4. Check the front wheel alignment.

IS 300

1. Before servicing the vehicle, refer to the precautions in the beginning of this section.

2. Remove or disconnect the following:
- Front wheel
- Engine under covers
- Level control sensor link
- Front subframe brace
- No. 2 lower control arm
- Brake caliper and rotor
- Outer tie rod end
- Stabilizer bar link
- Lower strut bolt
- Lower ball joint
- Steering gear
- No. 1 lower control arm

To install:
3. Installation is the reverse of the removal procedure, while using the following torque values:
- No. 1 lower control arm bolt: 136 ft. lbs. (184 Nm)
- Lower ball joint nut: 91 ft. lbs. (123 Nm)
- Outer tie rod end nut: 40 ft. lbs. (54 Nm)
- Brake caliper bolts: 87 ft. lbs. (118 Nm)
- No. 2 control arm-to-No. 1 control arm bolts: 181 ft. lbs. (245 Nm)
- Front subframe brace small bolts: 43 ft. lbs. (58 Nm)
- Front subframe large bolts: 88 ft. lbs. (119 Nm)

LS 400 and LS 430

1. Before servicing the vehicle, refer to the precautions in the beginning of this section.

2. If equipped with air suspension, move the height control switch (located in the trunk) to the **OFF** position.

3. Remove or disconnect the following:
- Tire and wheel assembly
- Anti-lock Brake System (ABS) speed sensor and wiring harness from the steering knuckle
- Brake caliper support bracket by removing the 2 bolts. Leave the brake line connected. Support the caliper aside by using a piece of wire.

4. Loosen the 2 lower ball joint mounting bolts.

➡ **Do not remove the bolts.**

5. Remove or disconnect the following:
- Clip and nut from the tie rod end
- Tie rod end from the steering arm with the proper tool

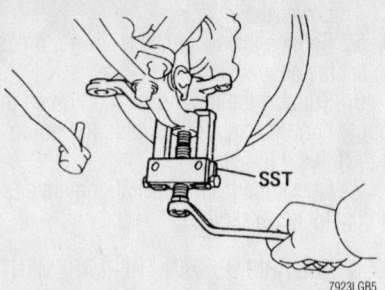

7923LGB5

Disconnecting the ball joint from the lower suspension arm—LS 400, LS 430

- Lower ball joint mounting bolts from the steering knuckle
- Cotter pin and nut from the lower ball joint
- Lower ball joint from the lower control arm

To install:
6. Install or connect the following:
- Ball joint to the lower control arm. Tighten the nut to 112 ft. lbs. (152 Nm) and install a new cotter pin.
- Mounting bolts, temporarily, holding the ball joint to the steering knuckle
- Tie rod end to the steering knuckle. Tighten the nut to 48 ft. lbs. (65 Nm) and install a new cotter pin.
- Lower ball joint bolts to 83 ft. lbs. (113 Nm)
- Brake caliper support bracket and tighten the 2 bolts to 87 ft. lbs. (118 Nm)
- ABS speed sensor and wiring harness to the steering knuckle
- Wheel
7. Lower the vehicle.
8. Turn the height control switch **ON**.

SC 300 and SC 400

The lower ball joint is not replaceable. If the lower ball joint is defective, replace the lower arm and ball joint as an assembly, as follows:

1. Before servicing the vehicle, refer to the precautions in the beginning of this section.

2. Raise the front of the vehicle and support it on safety stands.

3. Remove or disconnect the following:
- Wheel and the engine undercover
- Caliper support bracket from the vehicle by removing the 2 bolts. Support the caliper and bracket with a wire. Do not let the assembly hang from the brake line.
- Nut and disconnect the stabilizer bar from the lower control arm
- Cotter pin and nut from the lower

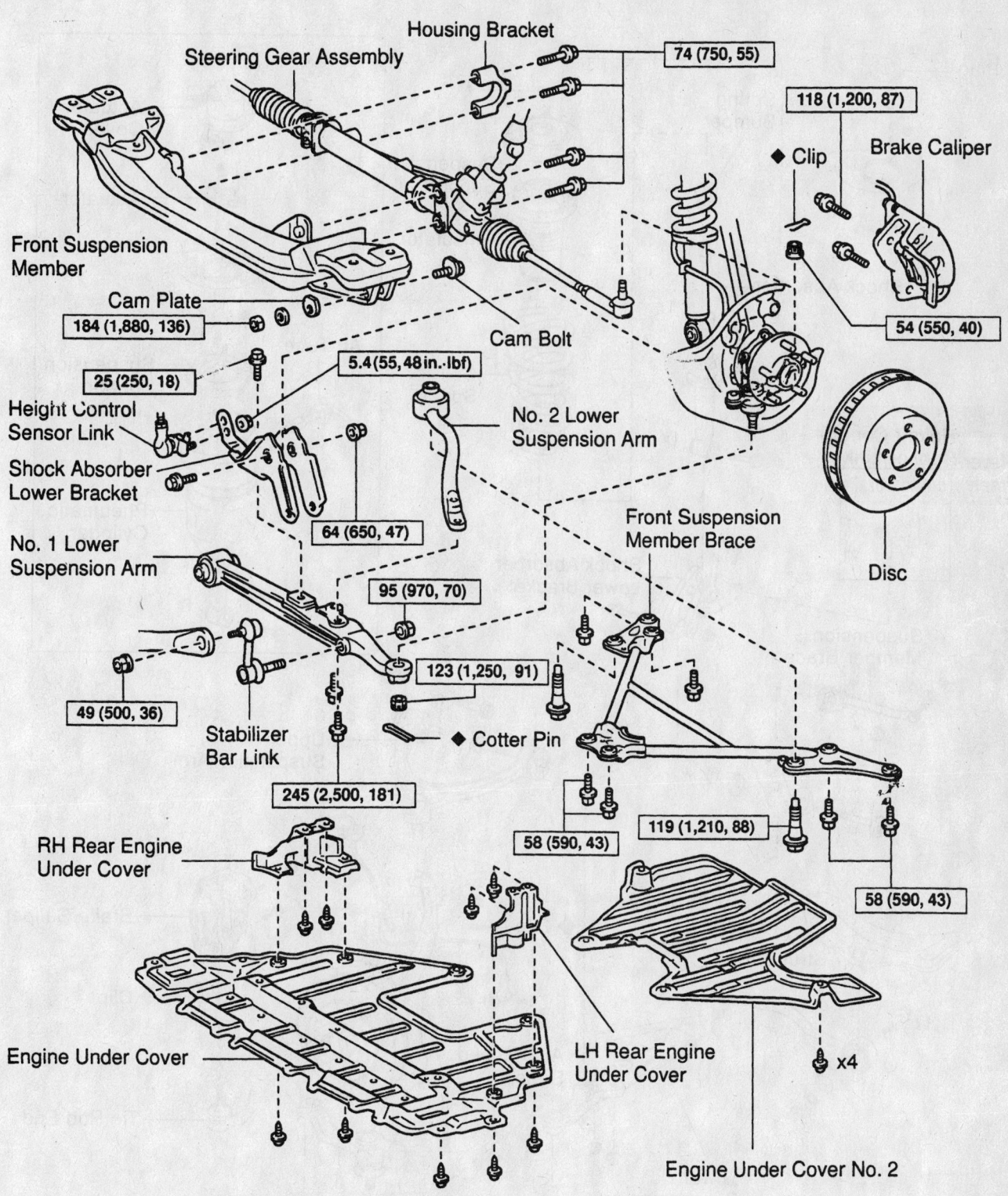

Housing Bracket

Steering Gear Assembly

74 (750, 55)

118 (1,200, 87)

◆ Clip

Brake Caliper

Front Suspension Member

Cam Plate

184 (1,880, 136)

Cam Bolt

54 (550, 40)

25 (250, 18)

5.4 (55, 48 in.·lbf)

No. 2 Lower Suspension Arm

Height Control Sensor Link

Shock Absorber Lower Bracket

64 (650, 47)

Front Suspension Member Brace

Disc

No. 1 Lower Suspension Arm

95 (970, 70)

49 (500, 36)

123 (1,250, 91)

Stabilizer Bar Link

◆ Cotter Pin

245 (2,500, 181)

119 (1,210, 88)

RH Rear Engine Under Cover

58 (590, 43)

58 (590, 43)

Engine Under Cover

LH Rear Engine Under Cover

x4

Engine Under Cover No. 2

N·m (kgf·cm, ft·lbf) : Specified torque

◆ Non–reusable part

9347LG06

Exploded view of the front suspension—IS 300

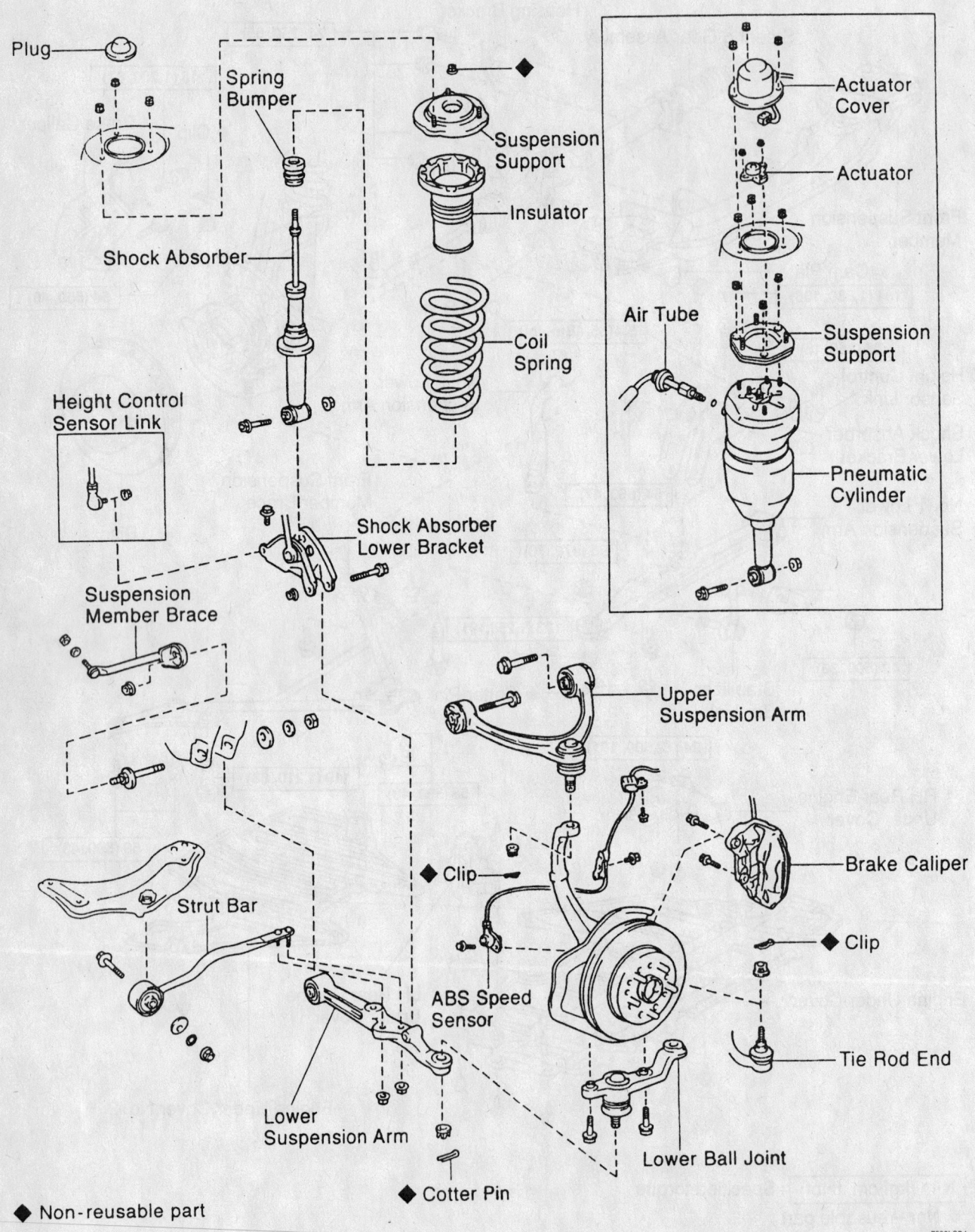

Plug

Spring Bumper

Suspension Support

Insulator

Shock Absorber

Height Control Sensor Link

Coil Spring

Air Tube

Actuator Cover

Actuator

Suspension Support

Pneumatic Cylinder

Shock Absorber Lower Bracket

Suspension Member Brace

Upper Suspension Arm

◆ Clip

Brake Caliper

◆ Clip

Strut Bar

ABS Speed Sensor

Tie Rod End

Lower Suspension Arm

◆ Cotter Pin

Lower Ball Joint

◆ Non-reusable part

7923LGB4

Exploded view of the lower ball joint mounting—LS 400, LS 430

ball joint. Press the lower ball joint out of the steering knuckle.
- Lower end of the strut by removing the nut and bolt
- Nut, 2 bolts and the front lower arm bracket stay

4. Matchmark the front and rear adjustment cams to the body and then remove the nuts and adjusting cams.

5. Lift out the lower control arm.

To install:

6. Install or connect the following:
- Bracket to the lower control arm by installing the 2 bolts. Tighten the bolts to 38 ft. lbs. (52 Nm).
- Lower control arm to the body and temporarily install the adjusting cams and nuts. Do not tighten the nuts at this time.
- Lower control arm to the knuckle and tighten the ball joint nut to 92 ft. lbs. (125 Nm). Install a new cotter pin.
- Strut to the arm and tighten the bolt and nut to 106 ft. lbs. (143 Nm)
- Stabilizer bar link and tighten the nut to 54 ft. lbs. (74 Nm)
- Brake caliper support bracket to the vehicle and tighten the bolts to 87 ft. lbs. (118 Nm)
- Wheel

7. Lower the vehicle.

8. Bounce it several times to set the suspension.

9. Support the lower arm, align the matchmarks on the adjusting cams and tighten the nuts to 166 ft. lbs. (226 Nm).

10. Check the wheel alignment.

CONTROL ARM BUSHING REPLACEMENT

The control arm bushings are serviced with the control arm as an assembly.

Wheel Bearings

ADJUSTMENT

Check the backlash in bearing shaft direction and the axle hub deviation. Maxi-

mum for backlash should be 0.0020 in. (0.05mm) and for axle hub deviation 0.020 in. (0.05mm).

➡ **The front and rear wheel bearings are non-adjustable. If the wheel bearing is out of specifications, replace the wheel bearing.**

REMOVAL & INSTALLATION

Front

ES 300

1. Before servicing the vehicle, refer to the precautions in the beginning of this section.

2. Remove or disconnect the following:
- Negative battery cable
- Front wheels
- Fender apron seal
- Cotter pin and lock cap from the end of the halfshaft
- Halfshaft locknut.
- Brake caliper and use a wire to support it out of the way

✲✲ WARNING

Never allow the caliper to hang freely from the brake hose.

- Rotor
- Anti-lock Brake System (ABS) speed sensor from the steering knuckle
- Nuts on the lower end of the strut
- Tie rod end from the steering knuckle
- Lower control arm from the ball joint
- Driveshaft from the axle hub
- 2 nuts on the lower end of the strut
- Steering knuckle

3. Clamp the steering knuckle in a vise with soft jaws to protect the knuckle.
- Dust deflector from the hub
- Ball joint from the steering knuckle
- Hub from the knuckle
- Inner race from the hub
- Dust cover

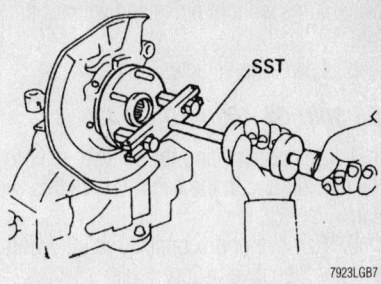

Removing the axle hub from the steering knuckle—ES 300

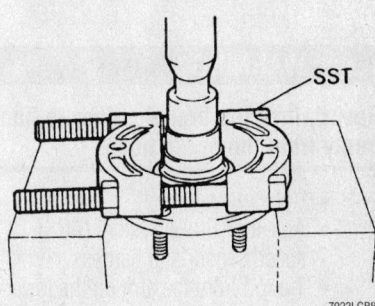

Remove the inner race from the hub—ES 300

- Snapring
- Bearing from the steering knuckle

To install:

4. Install or connect the following:
- Bearing into the knuckle
- Snapring
- Dust cover. Tighten the 4 bolts to 74 inch lbs. (8.3 Nm).
- Hub into the steering knuckle
- Lower ball joint to the steering knuckle. Tighten the nut to 90 ft. lbs. (123 Nm) and install a new cotter pin.
- Dust deflector
- Knuckle on the lower strut
- Lower ball joint to the lower arm. Tighten the bolts to 94 ft. lbs. (127 Nm).
- Tie rod end to the steering knuckle. Tighten the nut to 36 ft. lbs. (49 Nm).
- Nuts on the lower strut to 156 ft. lbs. (211 Nm)
- ABS speed sensor. Tighten the mounting bolt to 69 inch lbs. (8 Nm).
- Rotor
- Caliper. Tighten the mounting bolts to 79 ft. lbs. (107 Nm).
- Axle locknut. Tighten the nut to 217 ft. lbs. (294 Nm). Install the lock cap and a new cotter pin.
- Front fender apron seal
- Wheel

5. Turn the wheel by hand, verify that the

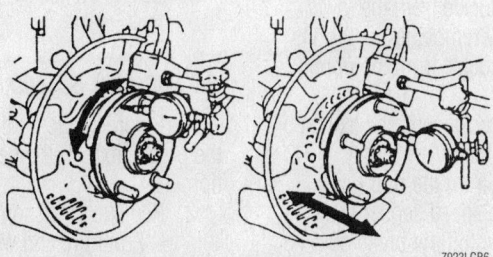

Checking wheel bearings for excessive play

wheel turns without noise and without binding.

6. Lower the vehicle.

GS 300, GS 400 AND GS 430

1. Before servicing the vehicle, refer to the precautions in the beginning of this section.

2. Remove or disconnect the following:
- Negative battery cable
- Front wheel
- Caliper, leaving the brake line connected and suspend it out of the way

❊ WARNING

Never allow the brake caliper to hang freely from the brake hose.

- Rotor
- Anti-lock Brake System (ABS) speed sensor and harness
- Tie rod from the arm on the lower ball joint
- Upper suspension arm from the steering knuckle
- Steering knuckle from the lower control arm
- Ball joint from the steering knuckle
- Front hub grease cap

3. Clamp the hub in a soft jaw vise.

4. Using a hammer and chisel, loosen the staked part of the locknut.

5. Remove or disconnect the following:
- Locknut
- ABS speed sensor rotor

➡**Do not scratch the serrations of the sensor rotor.**

- Brake dust cover bolts and shift the cover toward the outside.
- Hub from the steering knuckle
- Inner bearing race from the hub shaft
- Oil seal from the knuckle
- Bearing snapring from the steering knuckle
- Bearing from the steering knuckle

To install:

6. Install or connect the following:
- New bearing into the steering knuckle

➡**If the inner race and balls come loose from the bearing outer race, be sure to install them on the same side as before.**

- Snapring
- New outside inner race and tap in the new seal. Tap the seal until it is flush with the end surface of the steering knuckle.

- Brake dust cover to the knuckle and tighten the bolts to 74 inch lbs. (8 Nm)
- Hub into the steering knuckle
- ABS speed sensor rotor
- Axle hub locknut. Tighten the nut to 147 ft. lbs. (199 Nm) and stake it.
- Grease cap to the steering knuckle by tapping lightly around the circumference of the cap with a hammer
- Ball joint to the steering knuckle. Tighten the 2 bolts to 83 ft. lbs. (113 Nm).
- Steering knuckle to the upper and lower suspension arms. Tighten the upper nut to 64 ft. lbs. (87 Nm) and the lower nut to 95 ft. lbs. (127 Nm). Install a new cotter pin on the lower nut. Install the clip on the upper suspension arm nut.
- Tie rod end to the steering knuckle. Tighten the nut to 64 ft. lbs. (87 Nm) and install a new cotter pin.
- Rotor, disc brake pads and the brake caliper
- ABS speed sensor and harness. Tighten the sensor retaining bolt to 69 inch lbs. (8 Nm).
- Wheel

7. Lower the vehicle and connect the negative battery cable.

8. Check the front wheel alignment.

LS 400 AND LS 430

1. Before servicing the vehicle, refer to the precautions in the beginning of this section.

2. If equipped with air suspension, move the height control switch in the trunk area to the **OFF** position.

3. Remove or disconnect the following:
- Front tire and wheel assembly
- Brake caliper bracket from the steering knuckle, leaving the brake line connected. Support the caliper with a piece of wire.
- Brake rotor
- Anti-lock Brake System (ABS) speed sensor from the steering knuckle
- Steering knuckle from the lower ball joint by removing the 2 bolts
- Steering knuckle from the upper ball joint
- Steering knuckle with the axle hub from the vehicle
- Grease cap from the hub
- Nut and the speed sensor rotor
- 4 bolts and shift the brake dust cover towards the hub side
- Axle hub from the steering knuckle

- Outside inner race from the axle
- Oil seal from the steering knuckle
- Snapring and bearing from the steering knuckle

To install:

4. Install or connect the following:
- Bearing in the steering knuckle
- Snapring
- Inner race (outside)
- New oil seal until it is flush with the end surface of the steering knuckle
- Brake dust cover to the steering knuckle and tighten the bolts to 74 inch lbs. (8.4 Nm)
- Axle hub to the steering knuckle
- ABS speed sensor
- New nut on the axle shaft. Tighten the nut to 147 ft. lbs. (199 Nm). Stake the nut and install the grease cap.
- Steering knuckle to the lower ball joint and tighten the bolts to 83 ft. lbs. (113 Nm)
- Steering knuckle to the upper ball joint and tighten the nut to 48 ft. lbs. (65 Nm)
- Brake rotor
- Brake caliper and tighten the 2 bolts to 87 ft. lbs. (118 Nm)
- Speed sensor to the steering knuckle
- Front tire and wheel assembly

5. If equipped with air suspension, turn the height control switch to the **ON** position.

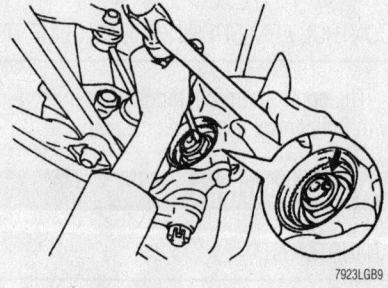

Axle hub nut is located on the inboard side of the knuckle assembly—LS 400, LS 430, SC 300 and SC 400

SC 300 AND SC 400

1. Before servicing the vehicle, refer to the precautions in the beginning of this section.

2. Remove or disconnect the following:
- Front tire and wheel assembly
- Brake caliper support bracket, leaving the brake line connected

LEXUS 16-103

ES 300 • IS 300 • GS 300 • GS 400 • GS 430 • LS 400 • LS 430 • SC 300 • SC 400

- Rotor by removing the 2 screws
- Anti-lock Brake System (ABS) speed sensor
- Cotter pin and nut and disconnect the tie rod from the steering knuckle
- Cotter pin and nut
- Steering knuckle from the upper control arm
- Clip and nut and press the knuckle off the lower control arm
- Steering knuckle from the vehicle
- Hub bearing cap from the steering knuckle
- Hub nut
- ABS sensor rotor
- 4 bolts and shift the brake dust shield toward the hub (outside)
- Axle hub from the knuckle
- Inner bearing race from the axle hub
- Oil seal
- Snapring and bearing

To install:

3. Press the bearing into the knuckle. If the inner race and balls come loose from the outer race, be sure to install them on the same side as before.

4. Install or connect the following:
- Snapring and inner race, then tap in a new oil seal until it is flush with the end surface of the knuckle
- Brake dust cover and tighten the bolts to 74 inch lbs. (8.3 Nm)
- Hub into the knuckle
- Speed sensor
- New locknut and tighten it to 147 ft. lbs. (199 Nm). Stake the nut with a chisel. Tap the bearing cap into place.
- Knuckle to the upper control arm and tighten the nut to 76 ft. lbs. (103 Nm). Install a new cotter pin
- Knuckle to the lower control arm and tighten the nut to 92 ft. lbs. (125 Nm). Install a new clip.
- Tie rod end to the steering knuckle with the nut. Tighten the nut to 36 ft. lbs. (49 Nm). Install a new cotter pin.
- Rotor by installing the 2 screws

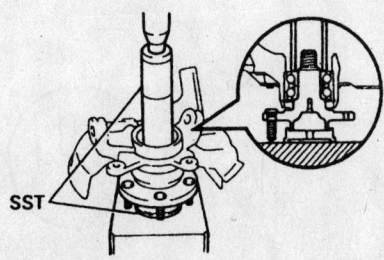

SST

7923LGB0

Pressing the hub into the knuckle—SC 300 and SC 400

- Caliper support bracket and tighten the bolt to 87 ft. lbs. (118 Nm)
- Speed sensor to the knuckle and tighten the bolt to 69 inch lbs. (8 Nm)
- Front wheel and tighten the lug nuts to 76 ft. lbs. (103 Nm)

5. Lower the vehicle.

6. Check the front end alignment and ABS speed sensor signal.

IS 300

1. Before servicing the vehicle, refer to the precautions in the beginning of this section.

2. Remove or disconnect the following:
- Front wheel
- Brake caliper and rotor
- Wheel speed sensor
- Upper and lower ball joints
- Steering knuckle from the vehicle
- Grease cap
- Hub locknut
- Brake dust cover
- Wheel speed sensor pulse ring

3. Press the hub out of the wheel bearing.

4. Remove the grease seal and the snapring, then press the wheel bearing out of the steering knuckle.

To install:

➡ **Use a new hub locknut for assembly.**

5. Installation is the reverse of the removal procedure, while using the following torque values:
- Hub locknut: 108 ft. lbs. (147 Nm)
- Brake dust cover bolts: 74 inch lbs. (8.3 Nm)
- Lower ball joint bolts: 83 ft. lbs. (113 Nm)
- Upper ball joint nut: 50 ft. lbs. (65 Nm)
- Brake caliper support bolts: 87 ft. lbs. (118 Nm)
- Wheel lug nuts: 76 ft. lbs. (103 Nm)

Rear

ES 300

1. Before servicing the vehicle, refer to the precautions in the beginning of this section.

2. Raise and safely support the vehicle.

3. Remove or disconnect the following:
- Rear tire and wheel assembly
- If equipped with rear disc brakes, the caliper mounting bolts. Leave the brake line connected and suspend the assembly out of the way.
- Brake rotor or drum
- 4 bolts and pull off the rear axle hub
- O-ring

➡ **If it is necessary to replace the hub or bearing, replace the components as an assembly.**

To install:

4. Install or connect the following:
- Hub on the carrier and tighten the bolts to 59 ft. lbs. (80 Nm)
- Rotor or drum
- Caliper, f equipped with rear disc brakes and tighten the bolts to 34 ft. lbs. (64 Nm)
- Wheel

GS 300, GS 400 AND GS 430

1. Before servicing the vehicle, refer to the precautions in the beginning of this section.

2. Remove or disconnect the following:
- Negative battery cable.
- Rear tire and wheel assembly
- Brake caliper support from the rear axle carrier and support it with a piece of wire

3. Place matchmarks on the disc brake rotor and the axle hub.

4. Remove or disconnect the following:
- Brake rotor
- Speed sensor
- Rear halfshaft
- Parking brake shoes
- Parking brake cable
- Strut rod

5. Place matchmarks on the adjusting cam and rear control crossmember.
- Nut, adjusting cam and the washer to the No. 1 control arm
- No. 1 lower control arm from the crossmember
- Loosen the nut holding the lower control arm to the axle carrier
- No. 2 lower control arm from the axle carrier
- Nut, then remove the No. 2 lower control arm from the axle carrier
- Nut holding the upper control arm to the axle carrier
- Axle carrier
- Nut holding the No. 1 control arm to the axle carrier
- No. 1 lower control arm from the axle carrier
- Dust deflector
- Axle hub from the carrier
- Backing plate
- Inner race (outside)
- Oil seal
- Snapring
- Bearing

To install:

6. Install or connect the following:
- Bearing to the axle carrier

➡If the inner races come loose from the bearing outer race, be sure to install them on the same side as before.

- Snapring. Install the inner race (outside) and a new oil seal.
- Backing plate. Install the inner race (inside) and press in the axle hub with the proper tools.
- Inner oil seal. Align the holes for the speed sensor in the dust deflector and axle carrier. Install the dust deflector.
- No. 1 lower arm to the axle carrier and install a new nut. Tighten the nut to 43 ft. lbs. (59 Nm).
- Upper control arm to the axle carrier. Tighten the new nut and bolt to 80 ft. lbs. (109 Nm).
- No. 2 lower control arm to the axle carrier and tighten a new nut to 110 ft. lbs. (150 Nm)
- No. 1 lower control arm to the rear crossmember. Tighten the nut to 136 ft. lbs. (184 Nm).
- Strut rod to the axle carrier. Tighten the nuts and bolts to 134 ft. lbs. (184 Nm).
- Parking brake cable and slide the backing plate to the inside. Install the hex bolt and tighten it to 132 ft. lbs. (180 Nm).
- Shoe guide plate set bolt. Tighten the bolt to 13 ft. lbs. (18 Nm).
- 4 hub bolts and tighten them to 19 ft. lbs. (26 Nm)

- Bolts at the speed sensor and tighten them to 69 inch lbs. (8 Nm)
- Parking brake shoes
- Halfshafts. Apply the brakes and tighten the locknut to 213 ft. lbs. (289 Nm).
- Brake rotor
- Brake caliper support to the rear axle carrier. Tighten the bolts to 77 ft. lbs. (104 Nm).
- Rear tire and wheel assembly
- Negative battery cable

7. Lower the vehicle and bounce it a few times to stabilize the suspension.

LS 400 AND LS 430

1. Before servicing the vehicle, refer to the precautions in the beginning of this section.

2. If equipped with air suspension, move the height control switch in the trunk area to the **OFF** position.

3. Remove or disconnect the following:
- Negative battery cable
- Rear wheel(s)
- Height control sensor link from the lower control arm
- Anti-lock Brake System (ABS) speed sensor and wiring harness
- Brake caliper bracket from the rear axle carrier by removing the 2 bolts. Support the caliper with a piece of wire.
- Brake rotor
- Parking brake shoes and cable

- Cotter pin, lock cap and the nut holding the halfshaft to the rear axle
- Suspension member brace by removing the 2 bolts
- Halfshaft bolts and washers
- Halfshaft from the vehicle
- Strut rod

4. Place matchmarks on the adjusting cam and body for the No. 1 control arm.

5. Remove or disconnect the following:
- Nut and adjusting cam
- Nut on the axle carrier side of the No. 1 lower control arm
- Separate the control arm from the axle carrier
- No. 1 lower control arm
- Stabilizer bar link from the No. 2 lower control arm.

6. Place matchmarks on the adjusting cam and body.

7. Remove or disconnect the following:
- Nut and adjusting cam from the No. 2 lower control arm
- Nut and bolt holding the No. 2 lower control arm to the axle carrier
- No. 2 control arm from the vehicle
- Nut and bolt on the lower side of the strut assembly
- 2 upper control arm set nuts and bolts
- Axle carrier with the upper control arm

8. Secure the axle carrier in a vise.

9. Remove or disconnect the following:
- Nut holding the upper control arm

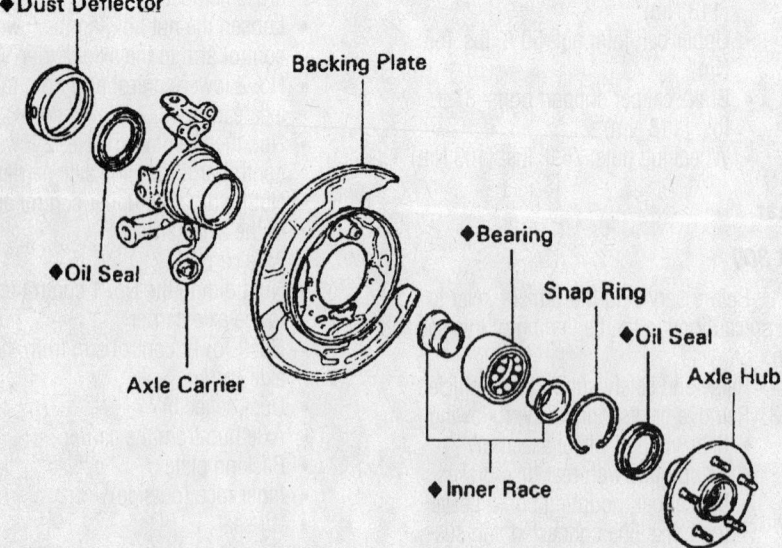

◆Dust Deflector

Backing Plate

◆Oil Seal

Axle Carrier

◆Bearing

Snap Ring

◆Oil Seal

Axle Hub

◆Inner Race

◆ Non-reusable part

Exploded view of the axle carrier — GS300

Exploded view of the axle carrier—GS 300

7923LGC1

SST

Removing the oil seal (inner)—LS 400, LS 430

7923LGC2

SST

SST

Removing the axle hub from the axle carrier—LS 400, LS 430

7923LGC3

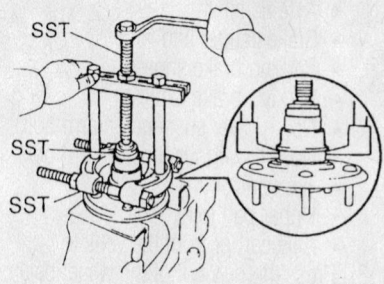

Removing the inner race (outside) from the axle hub—LS 400, LS 430

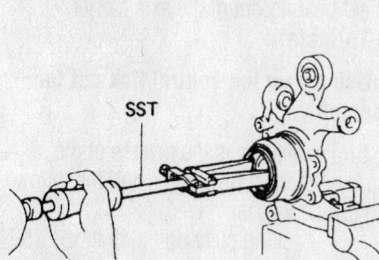

Removing the oil seal (outer)—LS 400, LS 430

to the axle carrier and remove the control arm
- Dust deflector. Use a suitable pry-tool.
- Oil seal

10. Remove the 2 bolts and nuts and shift the backing the plate towards the hub side (outside).

11. Remove or disconnect the following:
- Axle hub
- Backing plate.
- Inner race (outside) from the axle hub
- Oil seal (outer) from the axle
- Snapring from inside the axle housing
- Bearing from the axle housing

To install:

12. Install or connect the following:
- New bearing to the axle housing
- Snapring to the axle carrier, using snapring pliers

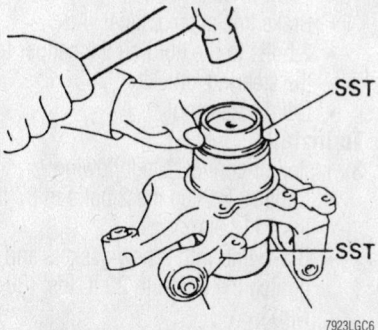

Installing the oil seal (outer)—LS 400, LS 430

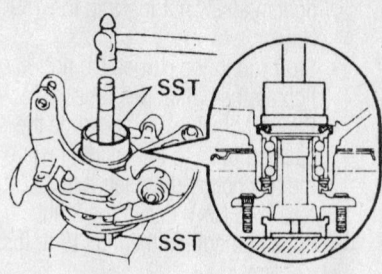

Installing the oil seal (inner)—LS 400, LS 430

- New outer oil seal. Coat the oil seal lip with multipurpose grease.
- Backing plate to the axle housing. Do not install the bolts or nuts at this time.
- Inner race (inside) to the axle housing
- Axle hub to the axle housing
- Backing plate in position. Tighten the bolts and nuts to 43 ft. lbs. (59 Nm).
- New oil seal (inner) to the axle housing. Coat the oil seal lip with multipurpose grease.
- New dust deflector. Be sure to align the hose for the ABS speed sensor in the dust deflector and axle carrier.
- Upper control arm to the axle carrier by installing the nut. Tighten the nut to 80 ft. lbs. (108 Nm).
- Axle carrier and upper control arm to the vehicle as an assembly
- 2 upper control arm set bolts and tighten the bolts to 121 ft. lbs. (164 Nm)
- Bolt and nut holding the strut to the axle carrier. Tighten to 101 ft. lbs. (137 Nm).
- Bolt and nut connecting the No. 2 lower control arm to the axle carrier. Tighten the bolt to 60 ft. lbs. (81 Nm).
- Nut and adjusting cam to hold the No. 2 lower control arm to the body. Align the adjusting cam marks and tighten the nut to 57 ft. lbs. (78 Nm).
- Stabilizer bar link to the No. 2 lower control arm and tighten the nut to 48 ft. lbs. (65 Nm)
- No. 1 lower control arm to the axle carrier and body. Install the nut to hold the No. 1 lower control arm to the axle carrier. Tighten the nut to 43 ft. lbs. (59 Nm).
- Nut and adjusting cam to hold the No. 1 lower control arm to the

body. Align the matchmarks and tighten the nut to 57 ft. lbs. (78 Nm).
- Strut rod to the axle carrier and body. Install the bolt and nut to hold the strut rod to the body. Tighten to 57 ft. lbs. (78 Nm).
- Bolt and nut to hold the strut rod to the axle carrier. Tighten to 136 ft. lbs. (184 Nm)
- Parking brake shoes and cable
- Outboard joint side of the halfshaft and align the matchmarks on the side gear shaft and the halfshaft. Coat the threads with clean oil and install the hexagon bolts. Tighten bolts to 61 ft. lbs. (83 Nm).
- Suspension member brace with the 2 bolts. Tighten the 2 bolts to 37 ft. lbs. (50 Nm).
- Nut to hold the halfshaft to the rear axle. Tighten the nut to 213 ft. lbs. (289 Nm).
- Lock cap and cotter pin
- Brake disc to the axle hub with the matchmarks aligned. Install the 2 screws and tighten the screws to 48 inch lbs. (5 Nm).
- Brake caliper to the vehicle and install the 2 bolts. Tighten the bolts to 77 ft. lbs. (104 Nm).
- ABS speed sensor and wiring harness
- Height control sensor link with the matchmarks aligned. Tighten the nut to 48 inch lbs. (5 Nm).
- Rear wheel(s)
- Negative battery cable

13. Lower the vehicle and turn **ON** the air suspension switch.

SC 300 AND SC 400

1. Before servicing the vehicle, refer to the precautions in the beginning of this section.

2. Install or connect the following:
- Rear tire and wheel assembly
- Brake caliper support bracket
- Brake rotor
- Speed sensor
- Rear halfshaft
- Parking brake shoes
- 2 bolts at the parking brake cable. Remove the 2 hub bolts and the hex bolt. Slide the backing plate to the outside and disconnect the parking brake cable.
- Strut rod at the axle carrier
- Nut, then press out the No. 1 lower suspension arm
- Nut, then press out the No. 2 lower suspension arm

- Nut, then press out the upper suspension arm.
- Axle carrier
- Dust deflector and pull out the oil seal
- Axle hub from the carrier
- Backing plate
- Inner race (outside) from the hub
- Oil seal
- Snapring
- Bearing and inner race (inside)

To install:

3. Install or connect the following:
- Bearing to the axle carrier

➥**If the inner races come loose from the bearing outer race, be sure to install them on the same side as before.**

- Snapring, the inner race (outside) and a new oil seal
- Backing plate. Install the inner race (inside) and press in the axle hub with the proper tools.
- Inner oil seal. Align the holes for the speed sensor in the dust deflector and axle carrier. Install the dust deflector.
- Upper arm to the axle carrier. Tighten the nut and bolt to 80 ft. lbs. (109 Nm).
- No. 2 lower arm to the carrier and tighten a new nut to 110 ft. lbs. (150 Nm)
- No. 1 lower arm to the carrier and

tighten a new nut to 43 ft. lbs. (59 Nm)
- Strut rod to the carrier. Do not tighten the bolt at this time.
- Parking brake cable and slide the backing plate to the inside. Install the hex bolt and tighten it to 132 ft. lbs. (180 Nm). Install the 2 hub bolts and tighten them to 19 ft. lbs. (26 Nm).
- 2 bolts at the parking brake cable and tighten them to 69 inch lbs. (8 Nm). Install the parking brake shoes and the ABS sensor.
- Halfshaft. Tighten the locknut to 213 ft. lbs. (289 Nm).
- Brake rotor
- Brake caliper to the rear axle carrier by installing the 2 bolts. Tighten the bolts to 77 ft. lbs. (104 Nm).
- Rear tire and wheel assembly. Lower the vehicle and bounce it a few times to stabilize the suspension. Raise the vehicle again, support the axle carrier and tighten the strut rod to 136 ft. lbs. (184 Nm).

IS 300

1. Before servicing the vehicle, refer to the precautions in the beginning of this section.
2. Remove or disconnect the following:
- Rear wheel
- Wheel speed sensor

- Axle halfshaft
- Brake caliper and rotor
- Parking brake shoes
- Parking brake cable
- No. 1 lower suspension arm bolt
- No. 2 lower suspension arm bolt
- Toe control link
- Upper ball joint
- Axle carrier from the vehicle

3. Press the hub out of the wheel bearing, then remove the backing plate.

4. Remove the snapring, then press the wheel bearing out of the axle carrier.

To install:

➥**Use a new toe control link nut for assembly.**

5. Installation is the reverse of the removal procedure, while using the following torque values:
- Backing plate bolts: 43 ft. lbs. (59 Nm)
- No. 1 lower suspension arm bolt: 55 ft. lbs. (75 Nm)
- No. 2 lower suspension arm bolt: 81 ft. lbs. (110 Nm)
- Toe control link nut: 36 ft. lbs. (49 Nm)
- Upper ball joint nut: 80 ft. lbs. (108 Nm)
- Brake caliper support bolts: 77 ft. lbs. (104 Nm)
- Rear wheel lug nuts: 76 ft. lbs. (103 Nm)

BRAKES

Brake Caliper

REMOVAL & INSTALLATION

ES 300

FRONT AND REAR

1. Before servicing the vehicle, refer to the precautions in the beginning of this section.
2. Remove or disconnect the following:
- Wheels
- Brake hose from the caliper
- Bolts that attach the caliper to the torque plate
- Caliper assembly by lifting the bottom

To install:

3. Grease the caliper slides and bolts with lithium grease.

4. Install or connect the following:
- Caliper. Torque the bolts to 25 ft. lbs. (34 Nm).
- Brake hose to the caliper using 2

new washers. Torque the union bolt to 21 ft. lbs. (29 Nm).

5. Fill the master cylinder to the proper level and bleed the brake system.

GS 300, GS 400 and GS 430

FRONT AND REAR

1. Before servicing the vehicle, refer to the precautions in the beginning of this section.
2. Remove or disconnect the following:
- Wheels
- Brake line at the caliper
- Mounting bolts, while holding the sliding pin with a wrench
- Caliper assembly

To install:

3. Install or connect the following:
- Caliper. Hold the sliding pin and tighten the mounting bolts to 25 ft. lbs. (34 Nm).
- Connect the brake line with 2 new gaskets and tighten the union bolt to 22 ft. lbs. (30 Nm)

4. Bleed the brake system.
- Wheels

5. Check and if necessary fill the master cylinder reservoir.

LS 400 and LS 430

FRONT

1. Before servicing the vehicle, refer to the precautions in the beginning of this section.
2. Remove or disconnect the following:
- Front wheels
- Brake line at the caliper
- 2 bolts to the holding the caliper to the steering knuckle
- Caliper assembly

To install:

3. Install or connect the following:
- Caliper. Tighten the 2 bolts to 87 ft. lbs. (118 Nm).
- Brake line with 2 new gaskets and tighten the union to 29 ft. lbs. (39 Nm)

4. Refill the reservoir as necessary and bleed the brake system.

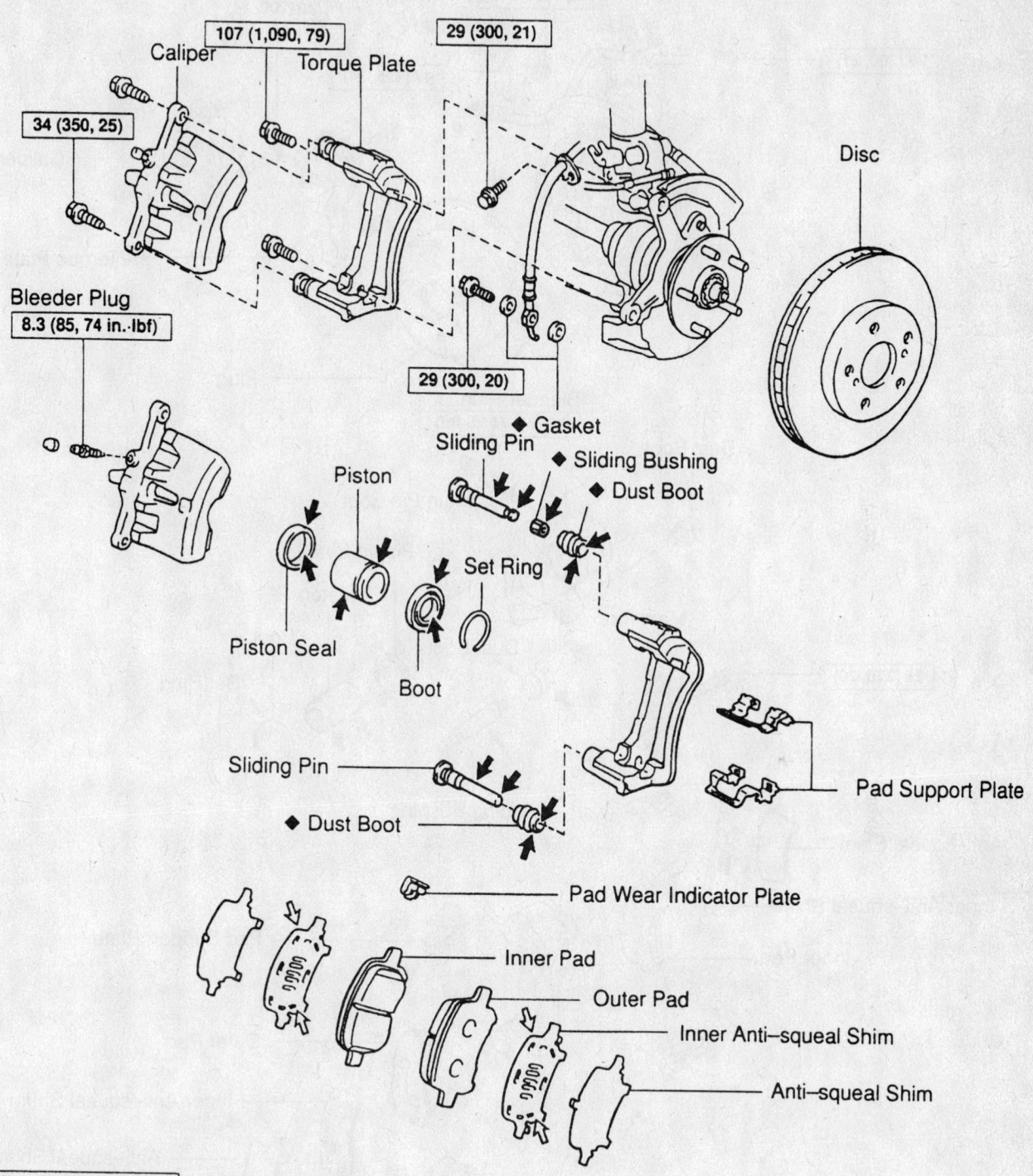

N·m (kgf·cm, ft·lbf) : Specified torque
◆ Non–reusable part
◄ Lithium soap base glycol grease
◁ Disc brake grease

93016G18

Front disc brakes—ES 300

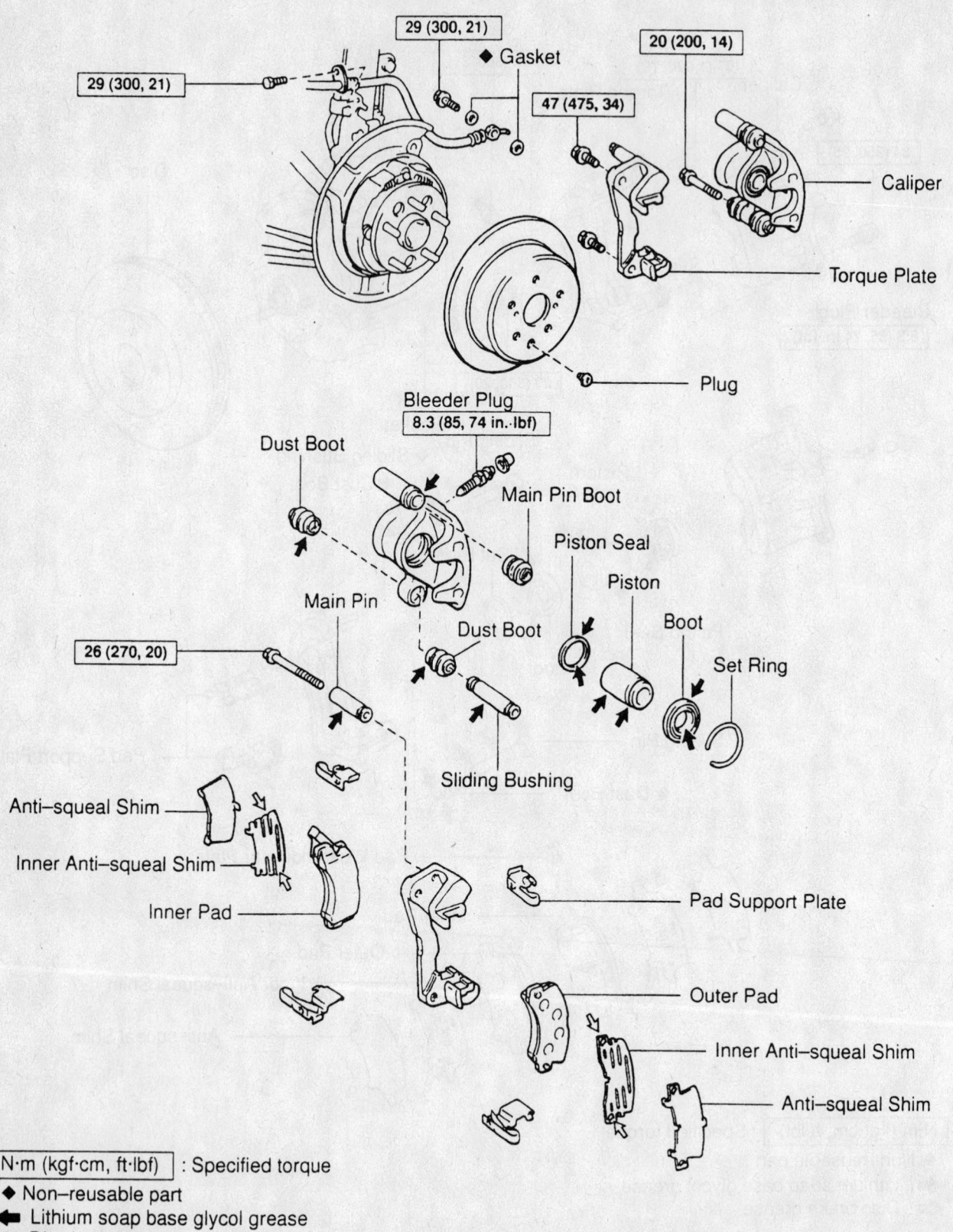

29 (300, 21)

29 (300, 21)

◆ Gasket

20 (200, 14)

47 (475, 34)

Caliper

Torque Plate

Plug

Bleeder Plug
8.3 (85, 74 in.·lbf)

Dust Boot

Main Pin Boot

Piston Seal

Piston

Boot

Set Ring

Main Pin

26 (270, 20)

Dust Boot

Sliding Bushing

Anti-squeal Shim

Inner Anti-squeal Shim

Inner Pad

Pad Support Plate

Outer Pad

Inner Anti-squeal Shim

Anti-squeal Shim

N·m (kgf·cm, ft·lbf) : Specified torque

◆ Non-reusable part

◀ Lithium soap base glycol grease

◁ Disc brake grease

93016G19

Rear disc brakes—ES 300

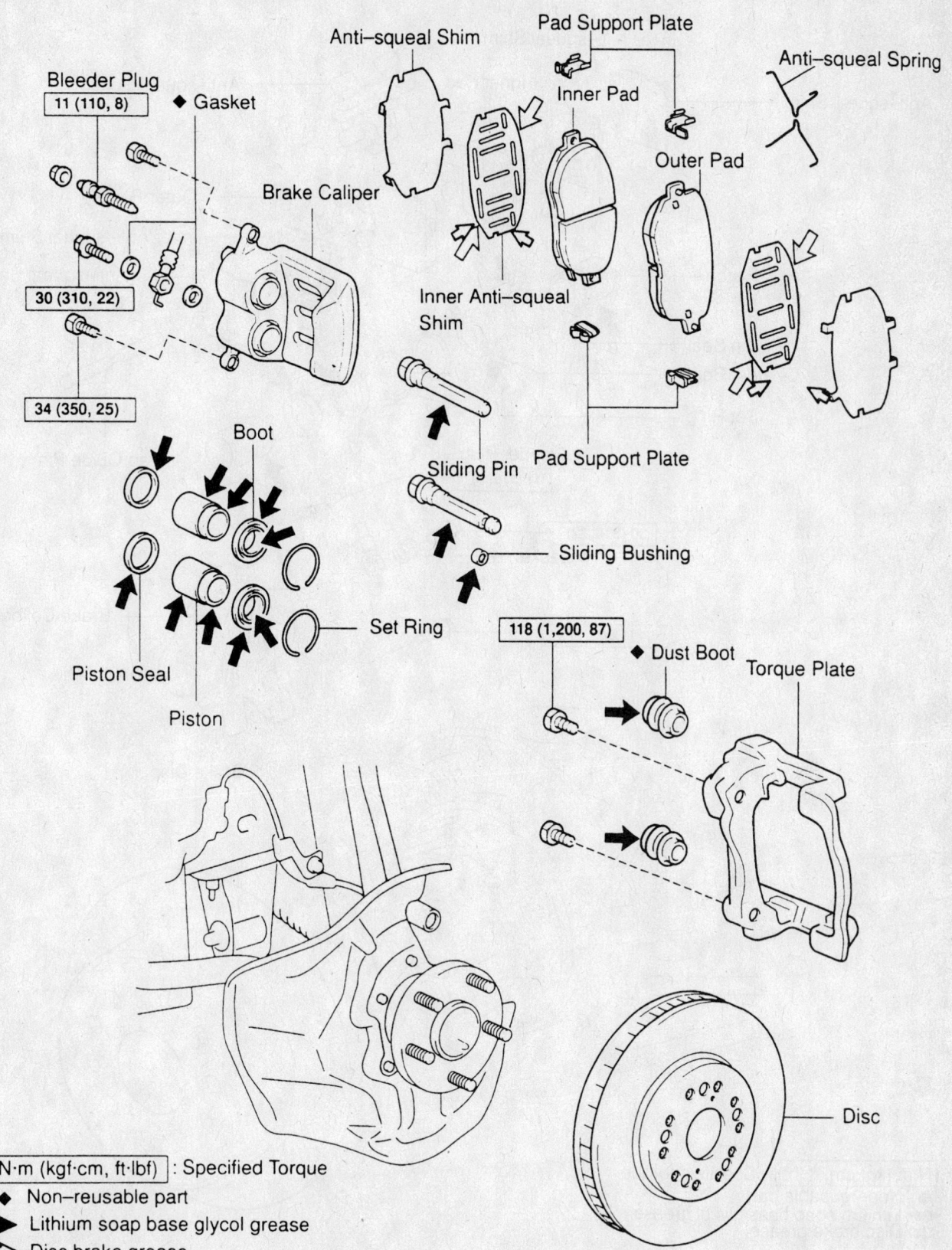

Anti–squeal Shim

Pad Support Plate

Anti–squeal Spring

Bleeder Plug

11 (110, 8)

◆ Gasket

Inner Pad

Brake Caliper

Outer Pad

Inner Anti–squeal Shim

30 (310, 22)

34 (350, 25)

Boot

Sliding Pin

Pad Support Plate

Sliding Bushing

Piston Seal

Set Ring

118 (1,200, 87)

◆ Dust Boot

Torque Plate

Piston

Disc

N·m (kgf·cm, ft·lbf) : Specified Torque

◆ Non–reusable part

➤ Lithium soap base glycol grease

⇨ Disc brake grease

93016G20

Front disc brakes—GS 300 and GS 400

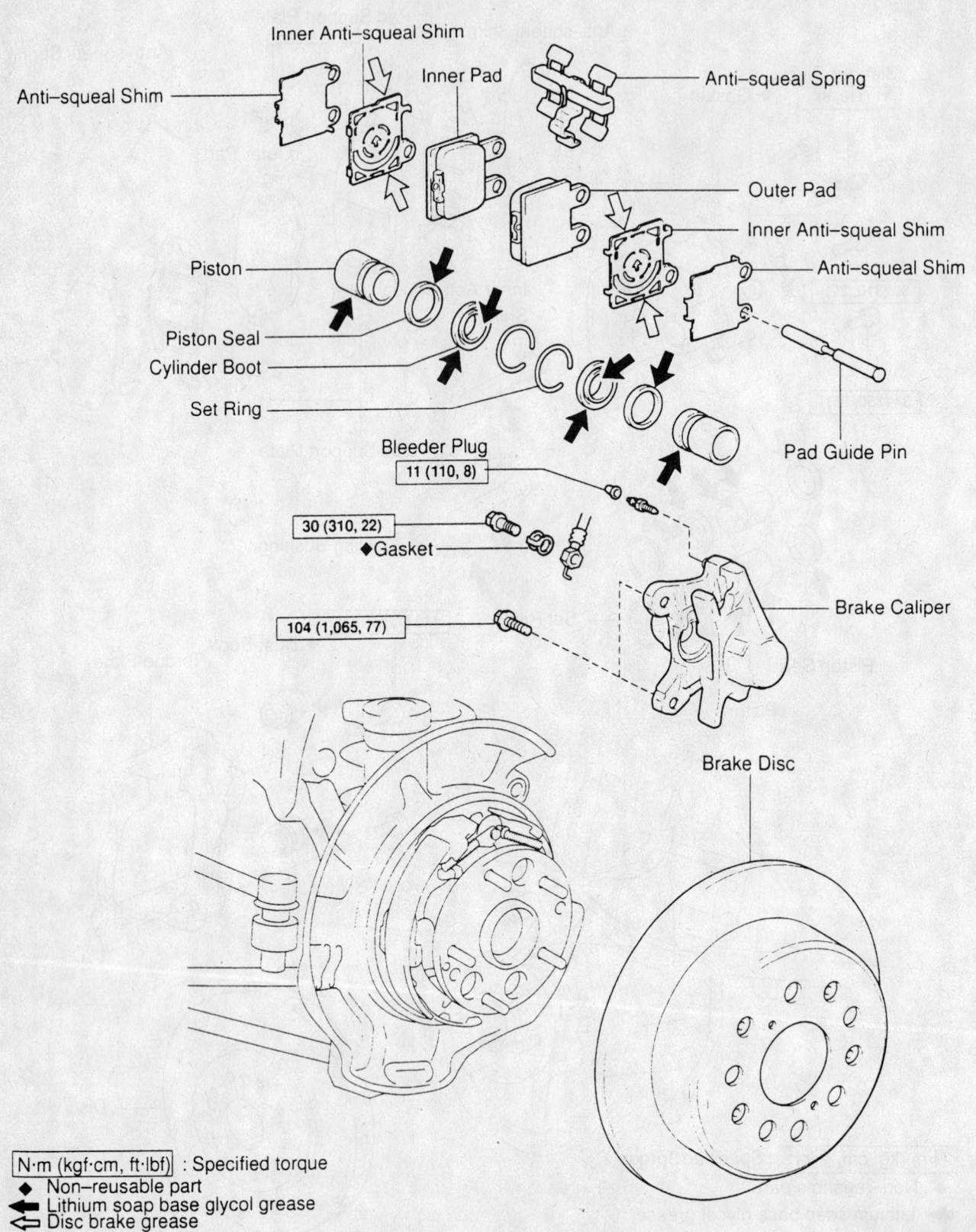

Inner Anti–squeal Shim

Anti–squeal Shim

Inner Pad

Anti–squeal Spring

Outer Pad

Inner Anti–squeal Shim

Anti–squeal Shim

Piston

Piston Seal

Cylinder Boot

Set Ring

Pad Guide Pin

Bleeder Plug
11 (110, 8)

30 (310, 22)
◆Gasket

104 (1,065, 77)

Brake Caliper

Brake Disc

N·m (kgf·cm, ft·lbf) : Specified torque
◆ Non–reusable part
◀ Lithium soap base glycol grease
◁ Disc brake grease

93016G21

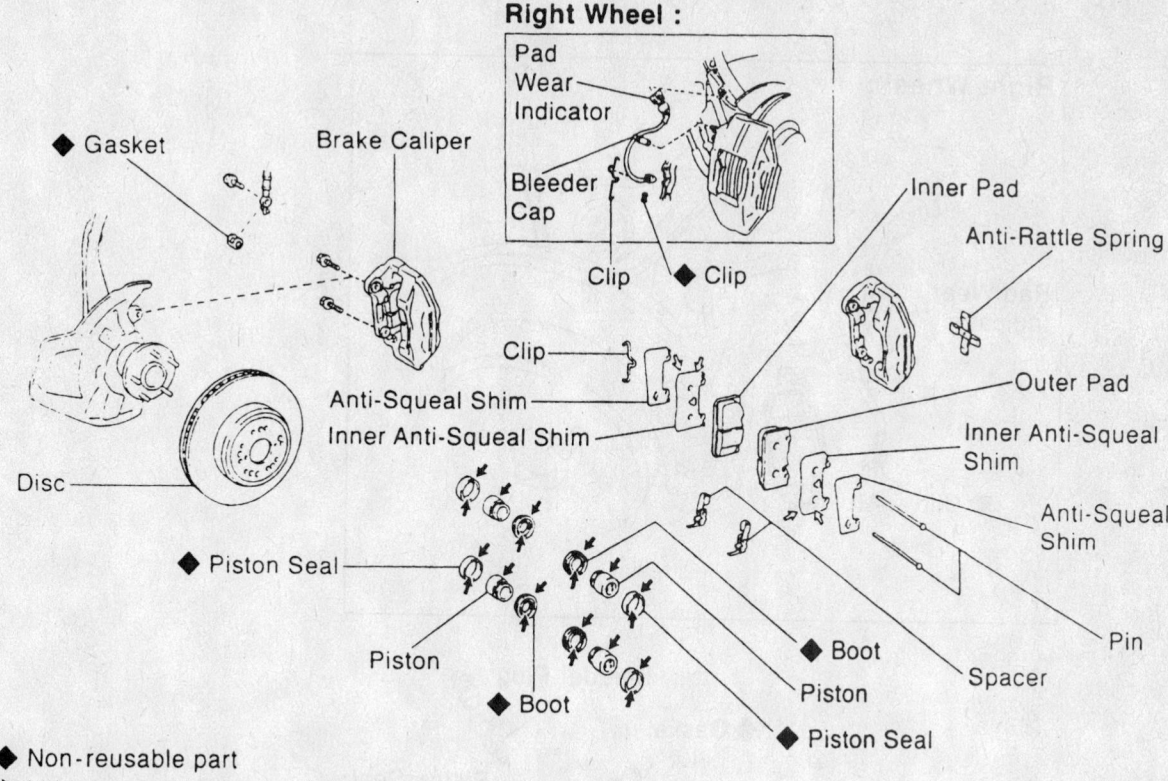

Right Wheel :

◆ Non-reusable part
➡ Lithium soap base glycol grease
⇨ Disc brake grease

93016G22

Front disc brakes—LS 400

REAR

1. Before servicing the vehicle, refer to the precautions in the beginning of this section.
2. Remove or disconnect the following:
   • Rear wheels
   • Brake line at the caliper, then plug it
   • Mounting bolts and the caliper assembly

To install:
3. Temporarily install the caliper on the torque plate with the 2 installation bolts.
4. Hold the sliding pin and tighten the mounting bolts to 25 ft. lbs. (34 Nm).
5. Connect the brake line with 2 new gaskets and tighten the union to 29 ft. lbs. (39 Nm).
6. Refill the reservoir as necessary and bleed the brake system.

SC 300 and SC400

FRONT AND REAR

1. Before servicing the vehicle, refer to the precautions in the beginning of this section.
2. Install or connect the following:
   • Wheels
   • Brake line at the caliper by removing the union bolt and 2 gaskets

• Mounting bolts, while holding the sliding pin with a wrench
• Caliper from the caliper support

To install:
3. Install or connect the following:
   • Caliper to the caliper support
   • Caliper bolts. Hold the sliding pin and tighten the mounting bolts to 25 ft. lbs. (34 Nm).
   • Brake line with (2) new gaskets and tighten the union to 22 ft. lbs. (30 Nm)
4. Bleed the brake system.
   • Wheels
5. Check and fill the master cylinder reservoir, if needed.

Disc Brake Pads

REMOVAL & INSTALLATION

ES 300

FRONT AND REAR

1. Before servicing the vehicle, refer to the precautions in the beginning of this section.
2. Remove or disconnect the following:
   • Wheels

• Lower installation bolt
• Caliper and suspend it securely. Do not disconnect the fluid line.
• Brake pads and retainers

To install:
3. Install or connect the following:
   • 2 pads so that the wear indicator plate is facing upward. Do not allow oil or grease to get in the rubbing face.
4. Draw out a small amount of brake fluid from the brake reservoir. Press in the caliper piston with a suitable tool.
   • Caliper. Torque the sliding main pin to 25 ft. lbs. (34 Nm).
   • Wheels
5. Check the fluid level in the master cylinder and add as necessary.

GS 300, SC 300 and SC 400

FRONT AND REAR

1. Before servicing the vehicle, refer to the precautions in the beginning of this section.
2. Remove or disconnect the following:
   • Wheels
3. Hold the sliding pin on the lower mounting bolt and remove the bolt. Swivel the caliper upward and out of the way.

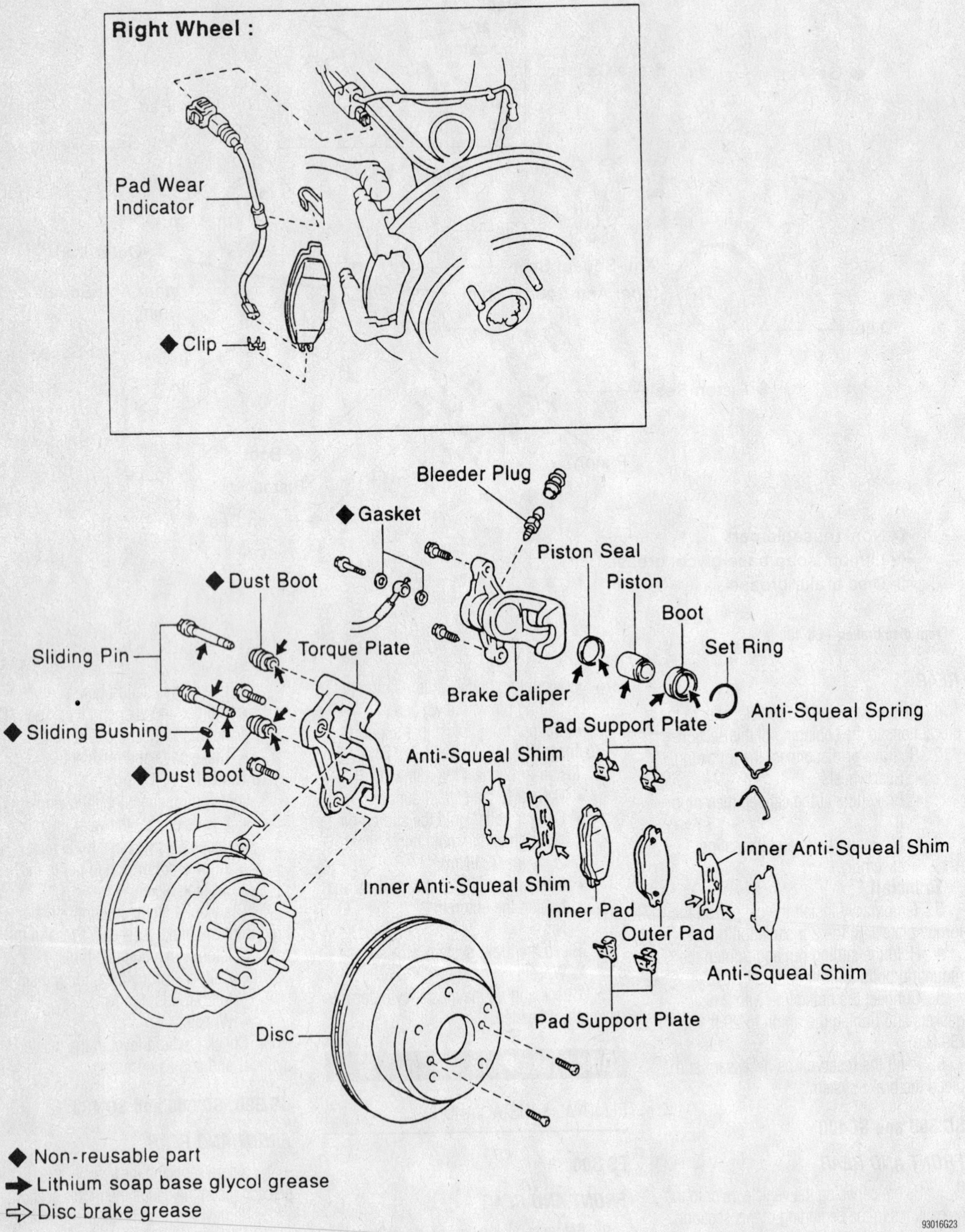

Right Wheel :
Pad Wear Indicator
◆ Clip
Bleeder Plug
◆ Gasket
Piston Seal
Piston
◆ Dust Boot
Boot
Set Ring
Sliding Pin
Torque Plate
Brake Caliper
Anti-Squeal Spring
Pad Support Plate
◆ Sliding Bushing
Anti-Squeal Shim
◆ Dust Boot
Inner Anti-Squeal Shim
Inner Anti-Squeal Shim
Inner Pad
Outer Pad
Anti-Squeal Shim
Pad Support Plate
Disc
◆ Non-reusable part
➡ Lithium soap base glycol grease
⇨ Disc brake grease
93016G23

Rear disc brakes—LS 400

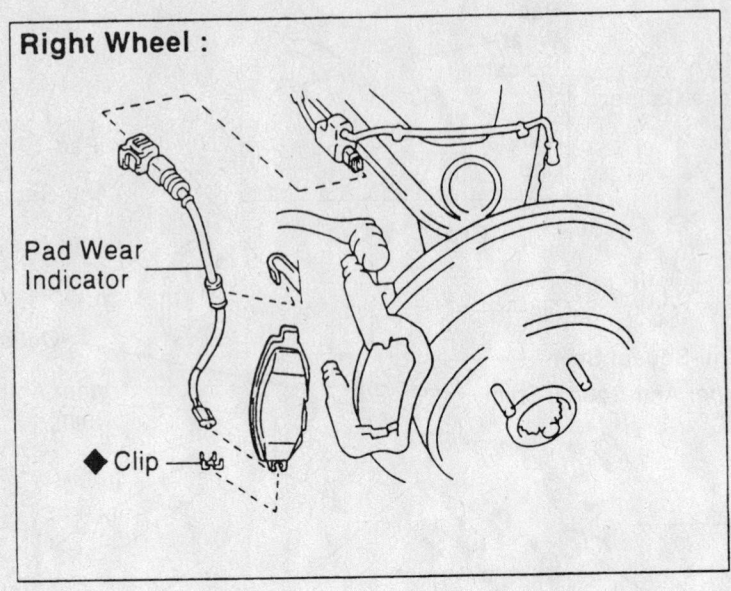

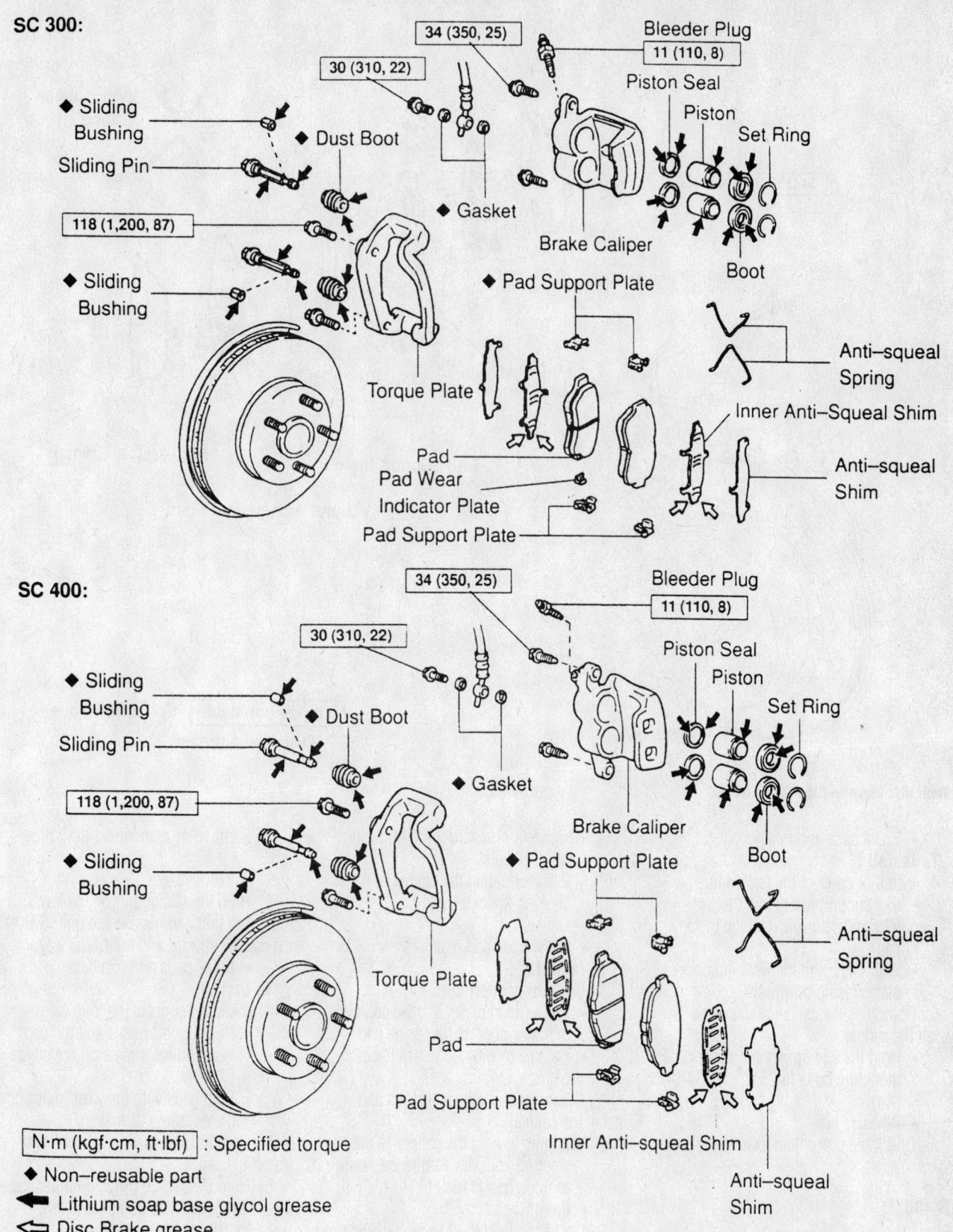

SC 300:

34 (350, 25)

30 (310, 22)

Bleeder Plug
11 (110, 8)

◆ Sliding Bushing

Sliding Pin

◆ Dust Boot

118 (1,200, 87)

◆ Sliding Bushing

◆ Gasket

Piston Seal

Piston

Set Ring

Brake Caliper

Boot

◆ Pad Support Plate

Torque Plate

Pad

Pad Wear Indicator Plate

Pad Support Plate

Anti–squeal Spring

Inner Anti–Squeal Shim

Anti–squeal Shim

SC 400:

34 (350, 25)

Bleeder Plug
11 (110, 8)

30 (310, 22)

◆ Sliding Bushing

Sliding Pin

◆ Dust Boot

118 (1,200, 87)

◆ Sliding Bushing

◆ Gasket

Piston Seal

Piston

Set Ring

Brake Caliper

Boot

◆ Pad Support Plate

Torque Plate

Pad

Pad Support Plate

Inner Anti–squeal Shim

Anti–squeal Spring

Anti–squeal Shim

N·m (kgf·cm, ft·lbf) : Specified torque

◆ Non–reusable part

◄ Lithium soap base glycol grease

◁ Disc Brake grease

93016G24

Front disc brakes—SC 300 and SC 400

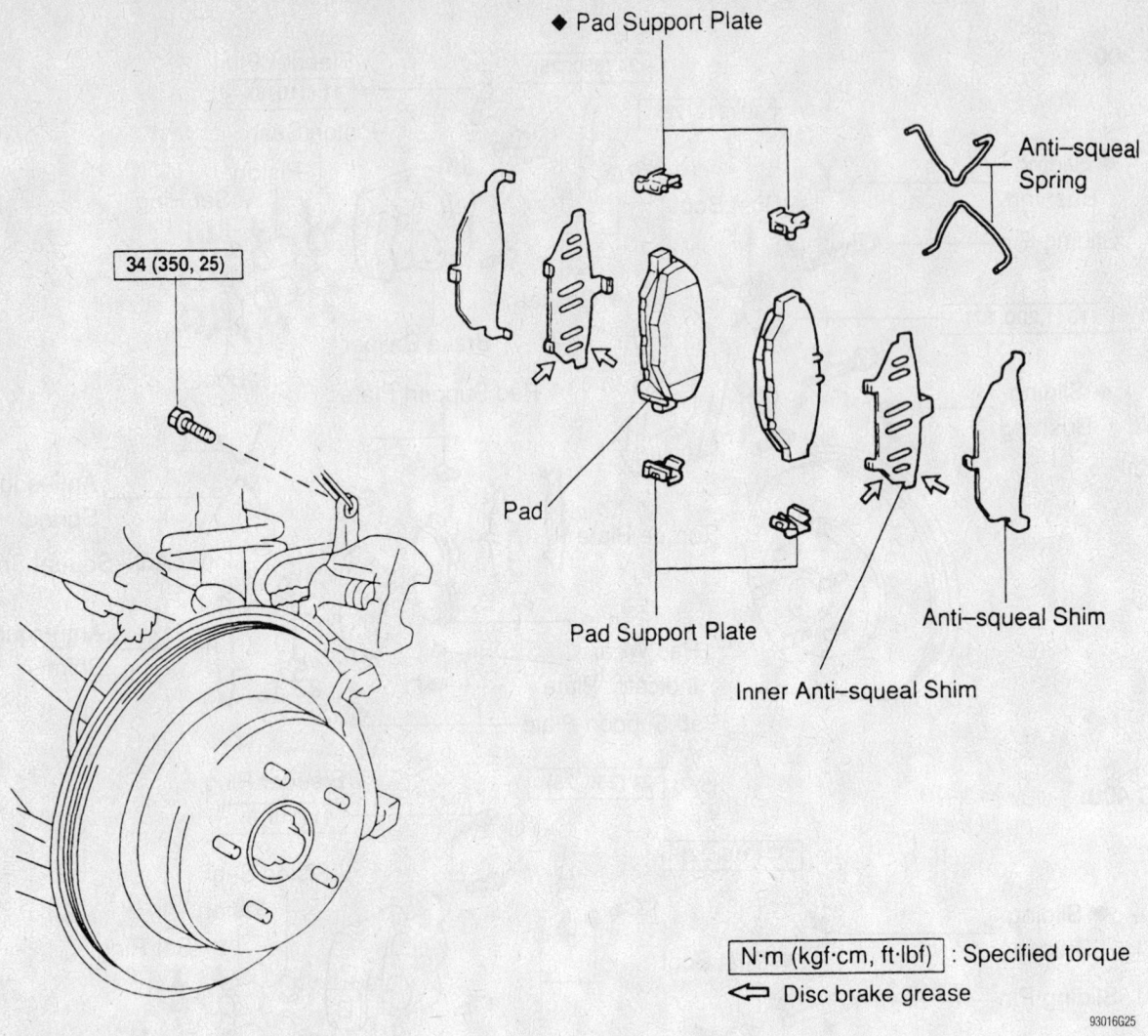

◆ Pad Support Plate

Anti-squeal Spring

34 (350, 25)

Pad

Pad Support Plate

Anti-squeal Shim

Inner Anti-squeal Shim

N·m (kgf·cm, ft·lbf) : Specified torque

⬅ Disc brake grease

93016G25

Rear disc brakes—SC 300 and SC 400

• Brake pads and retainers

To install:

4. Install or connect the following:
- Pad support plates and the pad wear indicator plate on the inside pad
- Both pads with the wear indicator plates facing downward

5. Compress the caliper pistons and install the caliper.
- Hold the sliding pin and tighten the mounting bolts to 25 ft. lbs. (34 Nm)
- Wheels

6. Check the brake fluid level in the reservoir.

LS 400

FRONT

1. Before servicing the vehicle, refer to the precautions in the beginning of this section.

2. Remove or disconnect the following:
- Front wheels
- 2 bolts holding the caliper to the steering knuckle
- Caliper
- Brake pads and retainers

To install:

3. Install or connect the following:
- Pad support plates and the pad wear indicator plate on the inside pad
- Both pads with the wear indicator plates

4. Compress the caliper pistons and install the caliper.
- 2 bolts to hold the caliper to the steering knuckle. Torque the bolts to 87 ft. lbs. (118 Nm).
- Front wheels

5. Check the fluid level in the reservoir.

REAR

1. Before servicing the vehicle, refer to the precautions in the beginning of this section.

2. Remove or disconnect the following:
- Rear wheels

3. Hold the sliding pin on the lower mounting bolt and remove the bolt. Swivel the caliper upward and out of the way.
- Brake pads and retainers

To install:

4. Install or connect the following:
- Pad support plates and the pad wear indicator plate on the inside pad
- Both pads with the wear indicator plates facing downward

5. Compress the caliper pistons and install the caliper.

6. Hold the sliding pin and tighten the mounting bolts to 25 ft. lbs. (34 Nm).

7. Install the rear wheels.

8. Check the brake fluid level in the reservoir.

SPECIFICATION CHARTS

ENGINE AND VEHICLE IDENTIFICATION

Code ①	Liters (cc)	Cu. In.	Cyl.	Fuel Sys.	Engine Type	Eng. Mfg.	Code ②	Year
BP	1.8 (1839)	112.2	4	MPFI	DOHC	Mazda	Y	2000
FP	1.8 (1839)	112.2	4	MPFI	DOHC	Mazda	1	2001
FS	2.0 (1991)	121.5	4	MPFI	DOHC	Mazda	2	2002
KJ	2.3 (2254)	137.2	6	MPFI	DOHC	Mazda	3	2003
KL	2.5 (2496)	152.3	6	MPFI	DOHC	Mazda	4	2004
ZM	1.6 (1597)	97.4	4	MPFI	DOHC	Mazda		

MPFI: Multi-Point Fuel Injection

DOHC: Double Over Head Cam

① Located above the starter

② 10th digit of the Vehicle Identification Number (VIN)

42356-MAZC-C01

GENERAL ENGINE SPECIFICATIONS

Year	Model	Engine Displacement Liters (cc)	Engine ID/VIN	Fuel System Type	Net Horsepower @ rpm	Net Torque @ rpm (ft. lbs.)	Bore x Stroke (in.)	Compression Ratio	Oil Pressure @ rpm
2000	626 ES	2.0 (1991)	FS	EFI	125@5500	127@3300	3.27x3.62	9.0:1	57-71@3000
	626 ES-V6	2.5 (2497)	KL	EFI	165@6000	161@5000	3.33x2.92	9.5:1	49-71@3000
	626 LX	2.0 (1991)	FS	EFI	125@5500	127@3300	3.27x3.62	9.0:1	57-71@3000
	626 LX-V6	2.5 (2497)	KL	EFI	165@6000	161@5000	3.33x2.92	9.5:1	49-71@3000
	Miata	1.8 (1839)	BP	EFI	①	②	3.27x3.35	9.5:1	43-57@3000
	Millenia	2.5 (2497)	KL	EFI	170@5800	160@4800	3.33x2.92	9.2:1	49-71@3000
	Millenia S	2.3 (2255)	KJ	EFI	210@5300	210@3500	3.16x2.92	10.0:1	44-66@3000
	Protege DX	1.6 (1597)	ZM	EFI	③	④	3.07x3.29	9.0:1	43-57@3000
	Protege ES	1.8 (1839)	FP	EFI	⑤	⑥	3.27x3.35	9.1:1	43-57@3000
	Protege LX	1.6 (1597)	ZM	EFI	③	④	3.07x3.29	9.0:1	43-57@3000
2001	626 ES	2.0 (1991)	FS	EFI	125@5500	127@3300	3.27x3.62	9.0:1	57-71@3000
	626 ES-V6	2.5 (2497)	KL	EFI	165@6000	161@5000	3.33x2.92	9.5:1	49-71@3000
	626 LX	2.0 (1991)	FS	EFI	125@5500	127@3300	3.27x3.62	9.0:1	57-71@3000
	626 LX-V6	2.5 (2497)	KL	EFI	165@6000	161@5000	3.33x2.92	9.5:1	49-71@3000
	Miata	1.8 (1839)	BP	EFI	①	②	3.27x3.35	9.5:1	43-57@3000
	Millenia	2.5 (2497)	KL	EFI	170@5800	160@4800	3.33x2.92	9.2:1	49-71@3000
	Millenia S	2.3 (2255)	KJ	EFI	210@5300	210@3500	3.16x2.92	10.0:1	44-66@3000
	Protege DX	1.6 (1597)	ZM	EFI	③	④	3.07x3.29	9.0:1	43-57@3000
	Protege ES	1.8 (1839)	FP	EFI	⑤	⑥	3.27x3.35	9.1:1	43-57@3000
	Protege LX	1.6 (1597)	ZM	EFI	③	④	3.07x3.29	9.0:1	43-57@3000
2002	626 ES	2.0 (1991)	FS	EFI	125@5500	127@3300	3.27x3.62	9.0:1	57-71@3000
	626 ES-V6	2.5 (2497)	KL	EFI	165@6000	161@5000	3.33x2.92	9.5:1	49-71@3000
	626 LX	2.0 (1991)	FS	EFI	125@5500	127@3300	3.27x3.62	9.0:1	57-71@3000
	626 LX-V6	2.5 (2497)	KL	EFI	165@6000	161@5000	3.33x2.92	9.5:1	49-71@3000
	Miata	1.8 (1839)	BP	EFI	142@7000	125@5500	3.3x3.4	10.0:1	43-56@3000
	Millenia	2.5 (2497)	KL	EFI	170@5800	160@4800	3.33x2.92	9.2:1	49-71@3000
	Millenia S	2.3 (2255)	KJ	EFI	210@5300	210@3500	3.16x2.92	10.0:1	44-66@3000
	Protege DX	1.6 (1597)	ZM	EFI	③	④	3.07x3.29	9.0:1	43-57@3000
	Protege LX	1.6 (1597)	ZM	EFI	③	④	3.07x3.29	9.0:1	43-57@3000
	Protege5	2.0 (1991)	FS	EFI	125@5500	127@3300	3.27x3.62	9.1:1	57-71@3000

EFI: Electronic Fuel Injection

① LEV states: 138@6500
Except LEV states: 140@6500

② LEV states: 117@5000
Except LEV states: 119@5500

③ LEV states: 103@5500
Except LEV states: 105@5500

④ LEV states: 106@4000
Except LEV states: 107@4000

⑤ LEV states: 120@6000
Except LEV states: 122@6000

⑥ LEV states: 119@4000
Except LEV states: 120@4000

42356-MAZC-C02

ENGINE TUNE-UP SPECIFICATIONS

Year	Engine Displacement Liters (cc)	Engine ID/VIN	Spark Plug Gap (in.)	Ignition Timing (deg.)		Fuel Pump (psi)	Idle Speed (rpm)		Valve Clearance	
				MT	AT		MT	AT	In.	Ex.
2000	1.6 (1597)	ZM	0.040-0.043	6-18B	6-18B	30-36	650-750	650-750	0.010-0.011	0.010-0.011
	1.8 (1839)	BP	0.040-0.043	6-18B	6-18B	53-61	750-850	750-850	0.008-0.009	0.012-0.013
	1.8 (1839)	FP	0.040-0.043	6-18B	6-18B	30-36	650-750	650-750	0.009-0.012	0.009-0.012
	2.0 (1991)	FS	0.040-0.043	6-18B	6-18B	37-45	550-850	500-800	0.008-0.0011	0.008-0.0011
	2.3 (2255)	KJ	28-31	6-8B	6-8B	41-48	600-700	600-700	0.011-0.012	0.011-0.012
	2.5 (2497)	KL	①	②	②	39-45	600-700	600-700	③	④
2001	1.6 (1597)	ZM	0.040-0.043	6-18B	6-18B	39-45	650-750	650-750	0.010-0.011	0.010-0.011
	1.8 (1839)	BP	0.040-0.043	6-18B	6-18B	53-61	750-850	750-850	0.008-0.009	0.012-0.013
	1.8 (1839)	FP	0.040-0.043	6-18B	6-18B	39-45	650-750	650-750	0.009-0.012	0.009-0.012
	2.0 (1991)	FS	0.040-0.043	6-18B	6-18B	37-45	550-850	500-800	0.008-0.0011	0.008-0.0011
	2.3 (2255)	KJ	28-31	6-8B	6-8B	41-48	600-700	600-700	0.011-0.012	0.011-0.012
	2.5 (2497)	KL	①	②	②	39-45	600-700	600-700	③	④
2002	1.6 (1597)	ZM	0.040-0.043	6-18B	6-18B	39-45	650-750	650-750	0.010-0.011	0.010-0.011
	1.8 (1839)	BP	0.040-0.043	6-18B	6-18B	53-61	750-850	750-850	0.008-0.0011	0.012-0.013
	2.0 (1991)	FS	0.040-0.043	6-18B	6-18B	37-45	550-850	500-800	0.008-0.0011	0.008-0.0011
	2.3 (2255)	KJ	28-31	6-8B	6-8B	41-48	600-700	600-700	0.011-0.012	0.011-0.012
	2.5 (2497)	KL	①	②	②	39-45	600-700	600-700	③	④

NOTE: The Vehicle Emission Control Information label often reflects specification changes made during production. The label figures must be used if they differ from those in this chart.

B: Before top dead center

HYD: Hydraulic

① 626 models: 0.028-0.031
   Millenia models: 0.039-0.043

① 626 models: 4-16
   Millenia models: 9-11

③ 626 models: 0.0097-0.012
   Millenia models: maintenace free

④ 626 models: 0.010-0.013
   Millenia models: maintenace free

42356-MAZC-C03

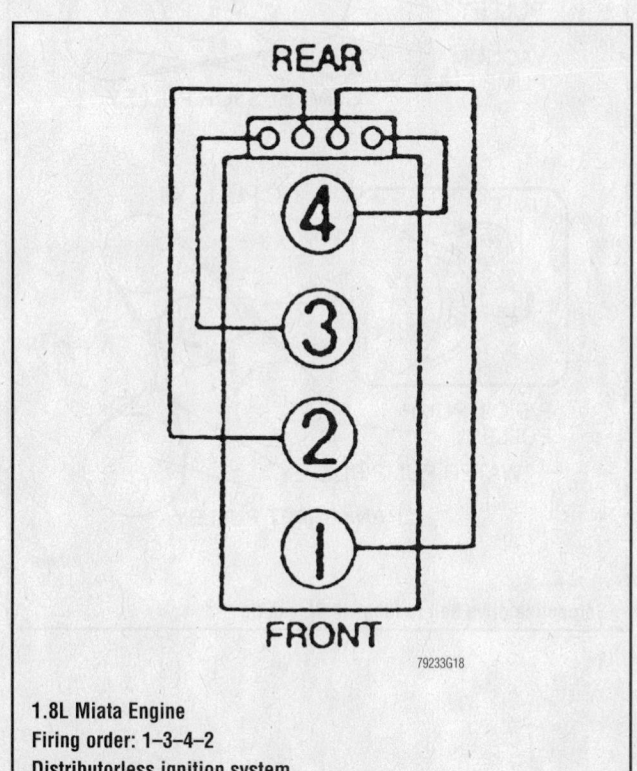

1.8L Miata Engine
Firing order: 1–3–4–2
Distributorless ignition system

79233G18

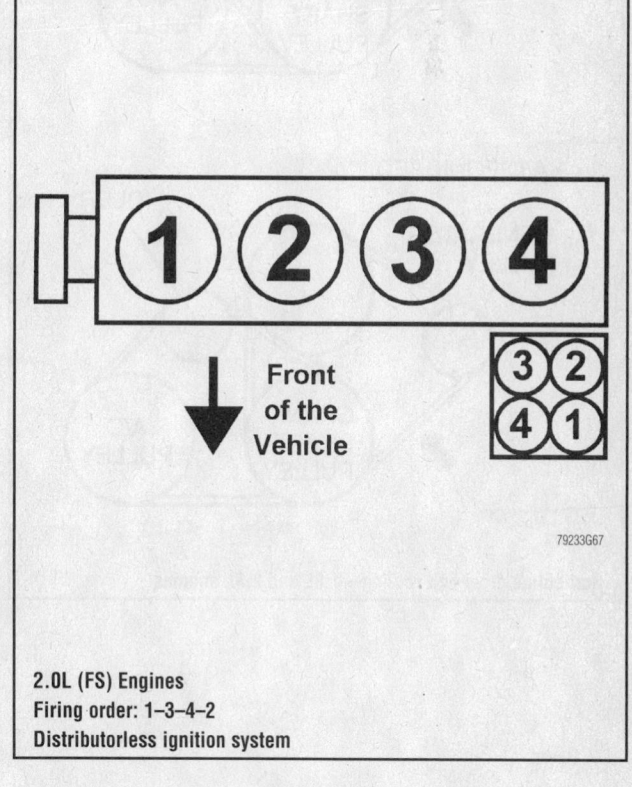

2.0L (FS) Engines
Firing order: 1–3–4–2
Distributorless ignition system

79233G67

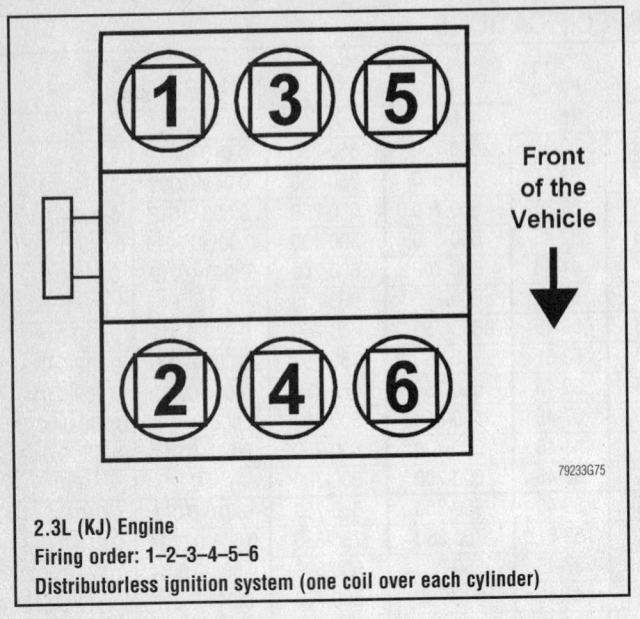

2.3L (KJ) Engine
Firing order: 1–2–3–4–5–6
Distributorless ignition system (one coil over each cylinder)

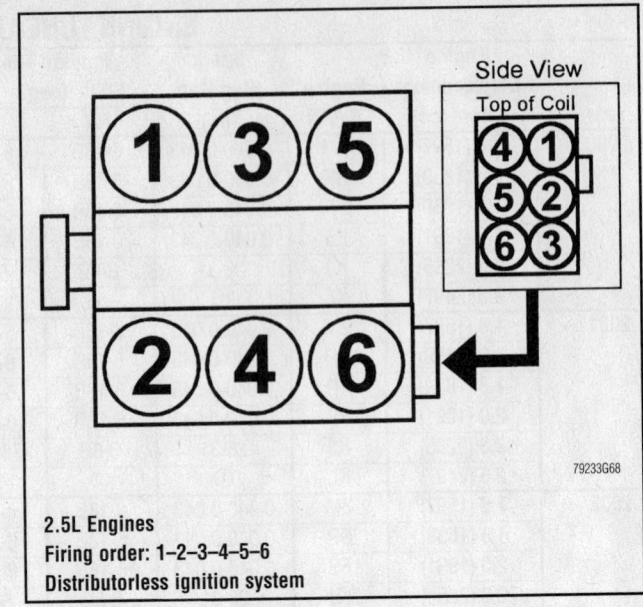

2.5L Engines
Firing order: 1–2–3–4–5–6
Distributorless ignition system

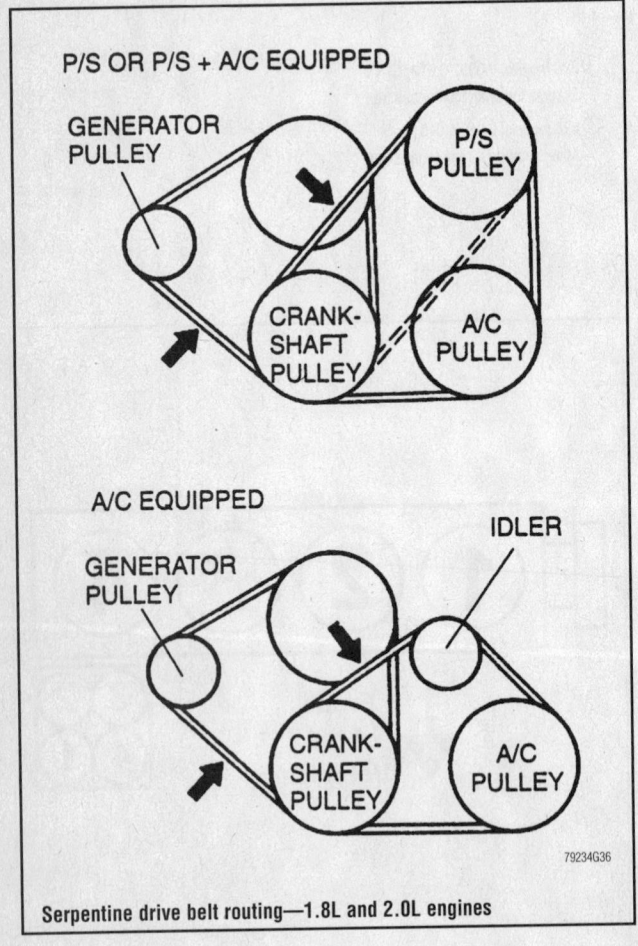

Serpentine drive belt routing—1.8L and 2.0L engines

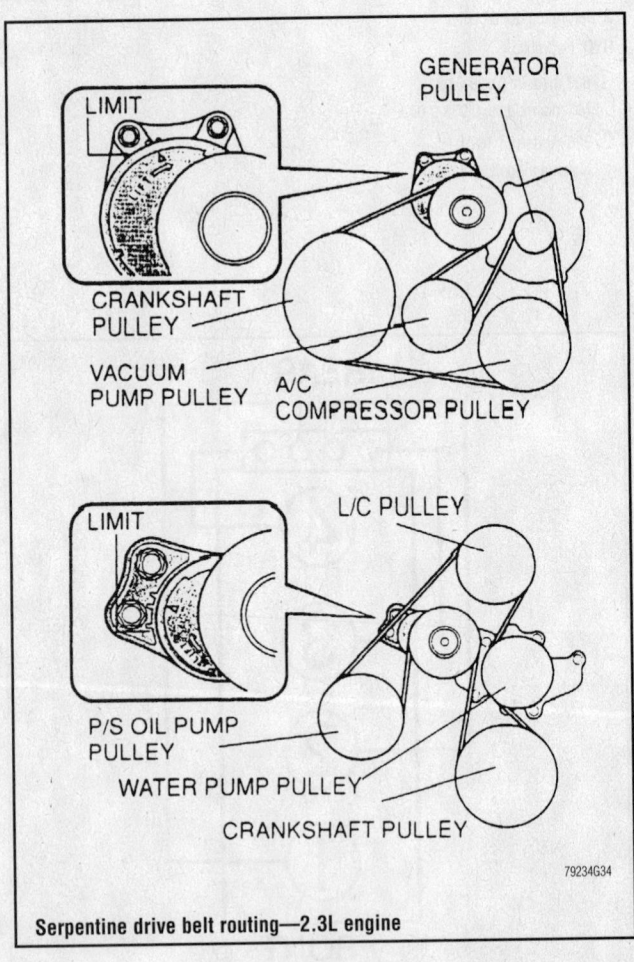

Serpentine drive belt routing—2.3L engine

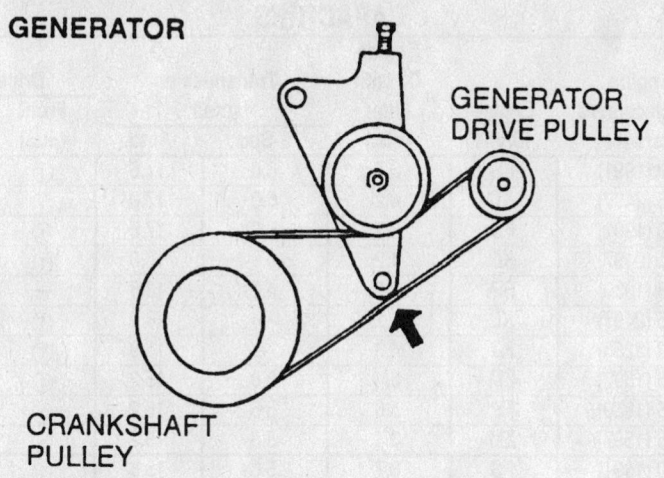

GENERATOR

GENERATOR
DRIVE PULLEY

CRANKSHAFT
PULLEY

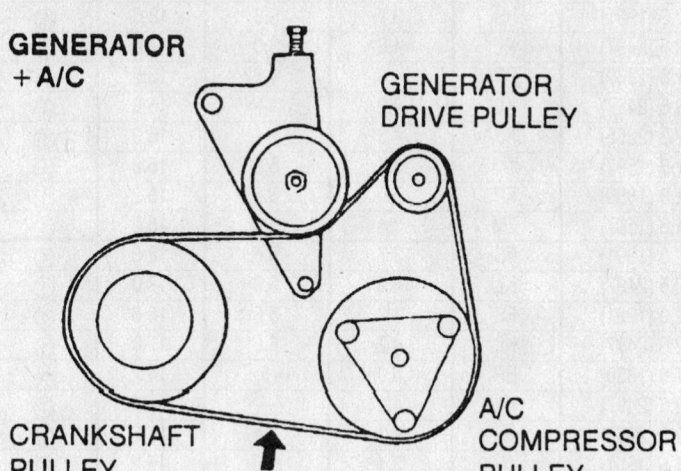

**GENERATOR
+ A/C**

GENERATOR
DRIVE PULLEY

CRANKSHAFT
PULLEY

A/C
COMPRESSOR
PULLEY

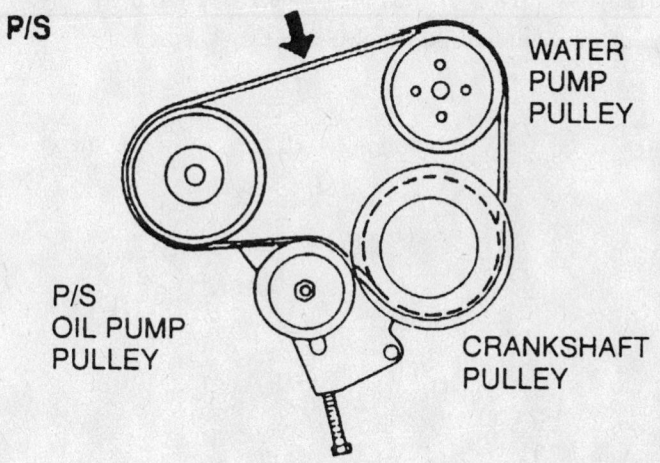

P/S

WATER
PUMP
PULLEY

P/S
OIL PUMP
PULLEY

CRANKSHAFT
PULLEY

79234G35

Serpentine drive belt routing—2.5L engines

CAPACITIES

Year	Model	Engine Displacement Liters (cc)	Engine ID/VIN	Engine Oil with Filter (qts.)	Transmission (pts.)		Drive Axle		Fuel Tank (gal.)	Cooling System (qts.)
					5-Spd	Auto.	Front (pts.)	Rear (pts.)		
2000	626 ES	2.0 (1991)	FS	3.7	6.0	17.6	①	—	16.9	7.9
	626 ES-V6	2.5 (2497)	KL	4.2	6.0	17.0	①	—	16.9	7.9
	626 LX	2.0 (1991)	FS	3.7	6.0	17.6	①	—	16.9	7.9
	626 LX-V6	2.5 (2497)	KL	4.2	6.0	17.0	①	—	16.9	7.9
	Miata	1.8 (1839)	BP	4.0	4.2	13.5	—	2.1	12.7	6.3
	Millenia	2.5 (2497)	KL	4.2	—	16.9	①	—	18.0	7.9
	Millenia S	2.3 (2255)	KJ	4.3	—	16.9	①	—	18.0	7.9
	Protege DX	1.6 (1597)	ZM	3.4	5.8	15.2	①	—	13.2	7.9
	Protege ES	1.8 (1839)	FP	5.0	5.6	15.2	①	—	13.2	9.9
	Protege LX	1.6 (1597)	ZM	3.4	5.8	15.2	①	—	13.2	7.9
2001	626 ES	2.0 (1991)	FS	3.7	5.8	18.6	①	—	16.9	7.9
	626 ES-V6	2.5 (2497)	KL	4.2	6.0	17.0	①	—	16.9	7.9
	626 LX	2.0 (1991)	FS	3.7	5.8	18.6	①	—	16.9	7.9
	626 LX-V6	2.5 (2497)	KL	4.2	6.0	17.0	①	—	16.9	7.9
	Miata	1.8 (1839)	BP	4.0	4.2	13.5	—	2.1	12.7	6.3
	Millenia	2.5 (2497)	KL	4.2	—	16.9	①	—	18.0	7.9
	Millenia S	2.3 (2255)	KJ	4.3	—	16.9	①	—	18.0	7.9
	Protege DX	1.6 (1597)	ZM	3.4	5.8	15.2	①	—	13.2	7.9
	Protege ES	1.8 (1839)	FP	5.0	5.6	15.2	①	—	13.2	9.9
	Protege LX	1.6 (1597)	ZM	3.4	5.8	15.2	①	—	13.2	7.9
2002	626 ES	2.0 (1991)	FS	3.7	5.8	18.6	①	—	16.9	7.9
	626 ES-V6	2.5 (2497)	KL	4.2	6.0	17.0	①	—	16.9	7.9
	626 LX	2.0 (1991)	FS	3.7	5.8	18.6	①	—	16.9	7.9
	626 LX-V6	2.5 (2497)	KL	4.2	6.0	17.0	①	—	16.9	7.9
	Miata	1.8 (1839)	BP	4.0	②	14.2	—	2.1	12.7	6.3
	Millenia	2.5 (2497)	KL	4.2	—	16.9	①	—	18.0	7.9
	Millenia S	2.3 (2255)	KJ	4.3	—	16.9	①	—	18.0	7.9
	Protege DX	1.6 (1597)	ZM	3.4	5.8	15.2	①	—	13.2	7.9
	Protege LX	2.0 (1991)	FS	3.7	5.8	15.2	①	—	14.5	7.9
	Protege5	2.0 (1991)	FS	3.7	6.0	17.6	①	—	14.5	7.9

NOTE: All capacities are approximate. Add fluid gradually and check to be sure a proper fluid level is obtained.

① Included in transaxle

② 5-speed: 4.2 qts. Included in transaxle

  6-apeed: 3.8 qts

42356-MAZC-C04

VALVE SPECIFICATIONS

Year	Engine Displacement Liters (cc)	Engine ID/VIN	Seat Angle (deg.)	Face Angle (deg.)	Maximum out of Square (in.)	Spring Free Length (in.)	Stem-to-Guide Clearance (in.)		Stem Diameter (in.)	
							Intake	Exhaust	Intake	Exhaust
2000	1.6 (1597)	ZM	NA	NA	NA	NA	NA	NA	NA	NA
	1.8 (1839)	BP	45	45	0.062	①	0.0010-0.0023	0.0012-0.0025	0.2351-0.2356	0.2349-0.2354
	1.8 (1839)	FP	NA	NA	NA	NA	NA	NA	NA	NA
	2.0 (1991)	FS	45	45	0.061	1.732	0.0010-0.0023	0.0012-0.0025	0.2351-0.2356	0.2349-0.2354
	2.3 (2255)	KJ	NA	45	0.062	1.413	0.0010-0.0023	0.0012-0.0025	0.2351-0.2356	0.2349-0.2354
	2.5 (2497)	KL	45	45	0.064	1.847	0.0010-0.0023	0.0012-0.0025	0.2351-0.2356	0.2349-0.2354
2001	1.6 (1597)	ZM	NA	NA	NA	NA	NA	NA	NA	NA
	1.8 (1839)	BP	45	45	0.062	①	0.0010-0.0023	0.0012-0.0025	0.2351-0.2356	0.2349-0.2354
	1.8 (1839)	FP	NA	NA	NA	NA	NA	NA	NA	NA
	2.0 (1991)	FS	45	45	0.061	1.732	0.0010-0.0023	0.0012-0.0025	0.2351-0.2356	0.2349-0.2354
	2.3 (2255)	KJ	NA	45	0.062	1.413	0.0010-0.0023	0.0012-0.0025	0.2351-0.2356	0.2349-0.2354
	2.5 (2497)	KL	45	45	0.064	1.847	0.0010-0.0023	0.0012-0.0025	0.2351-0.2356	0.2349-0.2354
2002	1.6 (1597)	ZM	NA	NA	NA	NA	NA	NA	NA	NA
	2.0 (1991)	FS	45	45	0.061	1.732	0.0010-0.0023	0.0012-0.0025	0.2351-0.2356	0.2349-0.2354
	2.3 (2255)	KJ	NA	45	0.062	1.413	0.0010-0.0023	0.0012-0.0025	0.2351-0.2356	0.2349-0.2354
	2.5 (2497)	KL	45	45	0.064	1.847	0.0010-0.0023	0.0012-0.0025	0.2351-0.2356	0.2349-0.2354

NA: Not Available

① Intake: 1.80 in.
   Exhaust: 1.903 in.

42356-MAZC-C05

CRANKSHAFT AND CONNECTING ROD SPECIFICATIONS

All measurements are given in inches.

Year	Engine Displacement Liters (cc)	Engine ID/VIN	Crankshaft				Connecting Rod		
			Main Brg. Journal Dia.	Main Brg. Oil Clearance	Shaft End-play	Thrust on No.	Journal Diameter	Oil Clearance	Side Clearance
2000	1.6 (1597)	ZM	NA	NA	NA	NA	NA	NA	NA
	1.8 (1839)	BP	1.9661-1.9667	0.0008-0.0014	0.0032-0.0111	4	1.7693-1.7699	0.0008-0.0017	0.0044-0.0103
	1.8 (1839)	FP	NA	NA	NA	NA	NA	NA	NA
	2.0 (1991)	FS	2.2022-2.2029	①	0.0031-0.0111	4	1.8874-1.8880	0.0005-0.0015	0.0043-0.0103
	2.3 (2255)	KJ	2.4385-2.4391	0.0015-0.0022	0.0032-0.0111	4	2.0843-2.0848	0.0010-0.0016	0.0071-0.0157
	2.5 (2497)	KL	2.4385-2.4391	0.0015-0.0022	0.0032-0.0111	4	2.0843-2.0848	0.0010-0.0016	0.0071-0.0157
2001	1.6 (1597)	ZM	NA	NA	NA	NA	NA	NA	NA
	1.8 (1839)	BP	1.9661-1.9667	0.0008-0.0014	0.0032-0.0111	4	1.7693-1.7699	0.0008-0.0017	0.0044-0.0103
	1.8 (1839)	FP	NA	NA	NA	NA	NA	NA	NA
	2.0 (1991)	FS	2.2022-2.2029	①	0.0031-0.0111	4	1.8874-1.8880	0.0005-0.0015	0.0043-0.0103
	2.3 (2255)	KJ	2.4385-2.4391	0.0015-0.0022	0.0032-0.0111	4	2.0843-2.0848	0.0010-0.0016	0.0071-0.0157
	2.5 (2497)	KL	2.4385-2.4391	0.0015-0.0022	0.0032-0.0111	4	2.0843-2.0848	0.0010-0.0016	0.0071-0.0157
2002	1.6 (1597)	ZM	NA	NA	NA	NA	NA	NA	NA
	1.8 (1839)	BP	1.9661-1.9667	0.0008-0.0014	0.0032-0.0111	4	1.7693-1.7699	0.0008-0.0017	0.0044-0.0103
	2.0 (1991)	FS	2.2022-2.2029	①	0.0031-0.0111	4	1.8874-1.8880	0.0005-0.0015	0.0043-0.0103
	2.3 (2255)	KJ	2.4385-2.4391	0.0015-0.0022	0.0032-0.0111	4	2.0843-2.0848	0.0010-0.0016	0.0071-0.0157
	2.5 (2497)	KL	2.4385-2.4391	0.0015-0.0022	0.0032-0.0111	4	2.0843-2.0848	0.0010-0.0016	0.0071-0.0157

NA: Not Avilable

① No. 1, 2, 4 & 5: 0.0009-0.0020 in.
  No. 3: 0.0012-0.0022 in.

42356-MAZC-C06

PISTON AND RING SPECIFICATIONS
All measurements are given in inches.

Year	Engine Displacement Liters (cc)	Engine ID/VIN	Piston Clearance	Ring Gap			Ring Side Clearance		
				Top Compression	Bottom Compression	Oil Control	Top Compression	Bottom Compression	Oil Control
2000	1.6 (1597)	ZM	NA	NA	NA	NA	NA	NA	NA
	1.8 (1839)	BP	0.0010-0.0014	0.006-0.011	0.006-0.011	0.008-0.027	0.0012-0.0026	0.0012-0.0027	0.0030-0.0060
	1.8 (1839)	FP	NA	NA	NA	NA	NA	NA	NA
	2.0 (1991)	FS	0.0015-0.0020	0.006-0.012	0.006-0.012	0.008-0.028	0.0014-0.0026	0.0014-0.0026	SNUG
	2.3 (2255)	KJ	0.0004-0.0014	0.006-0.010	0.010-0.014	0.008-0.030	0.0014-0.0025	0.0012-0.0025	0.0028-0.0062
	2.5 (2497)	KL	0.0012-0.0022	0.006-0.012	0.010-0.016	0.008-0.028	0.0008-0.0025	0.0012-0.0025	0.0008-0.0020
2001	1.6 (1597)	ZM	NA	NA	NA	NA	NA	NA	NA
	1.8 (1839)	BP	0.0010-0.0014	0.006-0.011	0.006-0.011	0.008-0.027	0.0012-0.0026	0.0012-0.0027	0.0030-0.0060
	1.8 (1839)	FP	NA	NA	NA	NA	NA	NA	NA
	2.0 (1991)	FS	0.0015-0.0020	0.006-0.012	0.006-0.012	0.008-0.028	0.0014-0.0026	0.0014-0.0026	SNUG
	2.3 (2255)	KJ	0.0004-0.0014	0.006-0.010	0.010-0.014	0.008-0.030	0.0014-0.0025	0.0012-0.0025	0.0028-0.0062
	2.5 (2497)	KL	0.0012-0.0022	0.006-0.012	0.010-0.016	0.008-0.028	0.0008-0.0025	0.0012-0.0025	0.0008-0.0020
2002	1.6 (1597)	ZM	NA	NA	NA	NA	NA	NA	NA
	1.8 (1839)	BP	0.0010-0.0014	0.006-0.011	0.006-0.011	0.008-0.027	0.0012-0.0026	0.0012-0.0027	0.0030-0.0060
	2.0 (1991)	FS	0.0015-0.0020	0.006-0.012	0.006-0.012	0.008-0.028	0.0014-0.0026	0.0014-0.0026	SNUG
	2.3 (2255)	KJ	0.0004-0.0014	0.006-0.010	0.010-0.014	0.008-0.030	0.0014-0.0025	0.0012-0.0025	0.0028-0.0062
	2.5 (2497)	KL	0.0012-0.0022	0.006-0.012	0.010-0.016	0.008-0.028	0.0008-0.0025	0.0012-0.0025	0.0008-0.0020

NA: Not Available

42356-MAZC-C07

TORQUE SPECIFICATIONS
All readings in ft. lbs.

Year	Engine Displacement Liters (cc)	Engine ID/VIN	Cylinder Head Bolts	Main Bearing Bolts	Rod Bearing Bolts	Crankshaft Damper Bolts	Flywheel Bolts	Manifold		Spark Plugs	Lug Nut
								Intake	Exhaust		
2000	1.6 (1597)	ZM	①	NA	NA	116-122	71-76	14-18	15-20	11-16	65-87
	1.8 (1839)	BP	56-60	40-43	35-36	116-130	71-76	17-21	29-33	11-16	65-87
	1.8 (1839)	FP	①	NA	NA	116-122	71-76	14-18	15-20	11-16	65-87
	2.0 (1991)	FS	①	②	③	116-122	71-76	14-18	⑦	11-16	65-87
	2.3 (2255)	KJ	④	⑤	③	116-122	45-49	14-18	14-18	11-16	65-87
	2.5 (2497)	KL	④	⑥	③	116-122	45-49	14-18	12-22	11-16	65-87
2001	1.6 (1597)	ZM	①	NA	NA	116-122	71-76	14-18	15-20	11-16	65-87
	1.8 (1839)	BP	56-60	40-43	35-36	116-130	71-76	17-21	29-33	11-16	65-87
	1.8 (1839)	FP	①	NA	NA	116-122	71-76	14-18	15-20	11-16	65-87
	2.0 (1991)	FS	①	②	③	116-122	71-76	14-18	⑦	11-16	65-87
	2.3 (2255)	KJ	④	⑤	③	116-122	45-49	14-18	14-18	11-16	65-87
	2.5 (2497)	KL	④	⑥	③	116-122	45-49	14-18	12-22	11-16	65-87
2002	1.6 (1597)	ZM	①	NA	NA	116-122	71-76	14-18	15-20	11-16	65-87
	1.8 (1839)	BP	56-60	40-43	35-36	116-130	71-76	17-21	29-33	11-16	65-87
	2.0 (1991)	FS	①	②	④	116-122	71-76	14-18	⑦	11-16	65-87
	2.3 (2255)	KJ	④	⑤	④	116-122	45-49	14-18	14-18	11-16	65-87
	2.5 (2497)	KL	④	⑥	④	116-122	45-49	14-18	12-22	11-16	65-87

NA: Not Available

① Step 1: 13-16 ft. lbs.
   Step 2: Tighten 85-95 degees
   Step 3: Repeat step 2

② Step 1: 16 ft. lbs.
   Step 2: Tighten each bolt 90 degrees

③ Nuts: 15-20 ft. lbs
   Bolts: 12-16 ft. lbs.

④ Step 1: 17-19 ft. lbs.
   Step 2: Tighten 85-95 degees
   Step 3: Repeat step 2

⑤ Step 1: Inner bolts: 17-19 ft. lbs.
   Step 2: Outer bolts: 13.5-15.5 ft. lbs.
   Step 3: Inner bolt Nos. 1-3: Tighten each bolt 70 degrees
   Step 4: Inner bolt No. 4: Tighten each bolt 80 degrees
   Step 5: Outer bolts: Tighten each bolt 60 degrees
   Step 6: Repeat Step 3-5

⑥ Step 1: Inner bolts: 17-18 ft. lbs.; Outer bolts: 13-15 ft. lbs.
   Step 2: Inner bolt Nos. 1-3: Tighten each bolt 70 degrees
   Step 3: Inner bolt No. 4: Tighten each bolt 80 degrees
   Step 4: Outer bolts: Tighten each bolt 60 degrees
   Step 5: Repeat Step 2

⑦ Nuts: 14-20 ft. lbs.
   Bolts: 12-22 ft. lbs.

42356-MAZC-C08

WHEEL ALIGNMENT

Year	Model		Caster Range (+/-Deg.)	Caster Preferred Setting (Deg.)	Camber Range (+/-Deg.)	Camber Preferred Setting (Deg.)	Toe-in (in.)	Steering Axis Inclination (Deg.)
2000	626 ①	F	1.00	+2.13	1.00	-0.70	0.12 +/- 0.16	
		R	—	—	1.00	-0.10	0.12 +/- 0.16	
	626 ②	F	1.00	+2.15	1.00	-0.70	0.12 +/- 0.16	
		R	—	—	1.00	-0.10	0.12 +/- 0.16	
	626 ③	F	1.00	+2.05	1.00	-0.70	0.12 +/- 0.16	
		R	—	—	1.00	-0.10	0.12 +/- 0.16	
	Miata	F	1.00	+5.75	1.00	+0.05	0.12 +/- 0.16	11.63
		R	—	—	1.00	-0.75	0.12 +/- 0.16	—
	Millenia	F	1.00	+2.23	0.75	-0.19	0.12 +/- 0.16	9.09
		R	—	—	1.00	-0.31	0.12 +/- 0.16	—
	Protégé	F	1.00	+1.88	1.00	-0.75	0.08 +/- 0.16	—
		R	—	—	1.00	-0.52	0.08 +/- 0.16	—
2001	626 ①	F	1.00	+2.13	1.00	-0.70	0.12 +/- 0.16	
		R	—	—	1.00	-0.10	0.12 +/- 0.16	
	626 ②	F	1.00	+2.15	1.00	-0.70	0.12 +/- 0.16	
		R	—	—	1.00	-0.10	0.12 +/- 0.16	
	626 ③	F	1.00	+2.05	1.00	-0.70	0.12 +/- 0.16	
		R	—	—	1.00	-0.10	0.12 +/- 0.16	
	Miata	F	1.00	+5.75	1.00	+0.05	0.12 +/- 0.16	11.63
		R	—	—	1.00	-0.75	0.12 +/- 0.16	—
	Millenia	F	1.00	+2.23	0.75	-0.19	0.12 +/- 0.16	9.09
		R	—	—	1.00	-0.31	0.12 +/- 0.16	—
	Protégé	F	1.00	+1.88	1.00	-0.75	0.08 +/- 0.16	—
		R	—	—	1.00	-0.52	0.08 +/- 0.16	—
2002	626 ①	F	1.00	+2.13	1.00	-0.70	0.12 +/- 0.16	
		R	—	—	1.00	-0.10	0.12 +/- 0.16	
	626 ②	F	1.00	+2.15	1.00	-0.70	0.12 +/- 0.16	
		R	—	—	1.00	-0.10	0.12 +/- 0.16	
	626 ③	F	1.00	+2.05	1.00	-0.70	0.12 +/- 0.16	
		R	—	—	1.00	-0.10	0.12 +/- 0.16	
	Miata	F	1.00	+5.75	1.00	+0.05	0.12 +/- 0.16	11.63
		R	—	—	1.00	-0.75	0.12 +/- 0.16	—
	Millenia	F	1.00	+2.23	0.75	-0.19	0.12 +/- 0.16	9.09
		R	—	—	1.00	-0.31	0.12 +/- 0.16	—
	Protégé	F	1.00	+1.88	1.00	-0.75	0.08 +/- 0.16	—
		R	—	—	1.00	-0.52	0.08 +/- 0.16	—

① With 14 in. wheels
② With 15 in. wheels
② With 16 in. wheels

42356-MAZC-C09

TIRE, WHEEL AND BALL JOINT SPECIFICATIONS

Year	Model	OEM Tires Standard	OEM Tires Optional	Tire Pressures (psi) Front	Tire Pressures (psi) Rear	Wheel Size	Ball Joint Inspection
2000	Millenia	P215/55R16	P215/50R17	32	29	Std: 6 1/2-JJ Opt:7-JJ	①
	Miata	P195/50R14	P205/45VR16	26	26	6-JJ	② ③
	Protégé 1.6L	P185/65R14	None	32	29	6-JJ	8-43 in. ②
	Protégé 1.8L	P185/65R14	P195/50VR15	32	32	Std: 5.5-JJ Opt: 6-JJ	8-43 in. ②
	626 2.0L	P205/60R15	P205/55R16	32	36	6-JJ	④
	626 2.5L	P205/60R15	P205/55R16	32	36	6-JJ	④
2001	Millenia	P215/55R16	P215/50R17	32	29	Std: 6 1/2-JJ Opt:7-JJ	①
	Miata	P195/50R14	P205/45VR16	26	26	6-JJ	② ③
	Protégé 1.6L	P185/65R14	None	32	29	6-JJ	8-43 in. ②
	Protégé 1.8L	P185/65R14	P195/50VR15	32	32	Std: 5.5-JJ Opt: 6-JJ	8-43 in. ②
	626 2.0L	P205/60R15	P205/55R16	32	36	6-JJ	④
	626 2.5L	P205/60R15	P205/55R16	32	36	6-JJ	④
2002	Millenia	P215/55R16	P215/50R17	32	29	Std: 6 1/2-JJ Opt:7-JJ	①
	Miata	P195/50R14	P205/45VR16	26	26	6-JJ	② ③
	Protégé 1.6L	P185/65R14	None	32	32	6-JJ	8-43 in. ②
	Protégé 2.0L	P185/65R14	P195/50VR15	32	32	Std: 5.5-JJ	8-43 in. ②
	626 2.0L	P205/60R15	P205/55R16	32	36	6-JJ	④
	626 2.5L	P205/60R15	P205/55R16	32	36	6-JJ	④

OEM: Original Equipment Manufacturer

PSI: Pounds Per Square Inch

STD: Standard

OPT: Optional

① Lower arm ball rotation torque: 0.7-7.7
Upper arm ball rotation torque: 0.7-8.8
Upper leading link rotation torque: 0.7-4.4

② Torque required in ft. lbs. to rotate ball joint using a pull scale

③ Lower arm ball rotation torque: 0.78-4.29
Upper arm ball rotation torque: 0.7-5.0

④ Lower arm ball rotation torque: 0.3-11

42356-MAZC-C10

BRAKE SPECIFICATIONS
All measurements in inches unless noted

| Year | Model | | Brake Disc | | | Brake Drum | | | Minimum Lining Thickness | Brake Caliper | |
			Original Thickness	Minimum Thickness	Maximum Runout	Original Inside Diameter	Max. Wear Limit	Maximum Machine Diameter		Bracket Bolts (ft. lbs.)	Mounting Bolts (ft. lbs.)
2000	626	F	0.940	0.870	0.002	—	—	—	0.080	58-75	22-28
		R	0.390	0.310	0.002	9.00	NA	9.059	0.040	34-49	26-28
	Miata	F	0.790	①	0.002	—	—	—	0.080	37-50	58-65
		R	0.350	0.310	0.002	—	—	—	0.040	37-50	26-28
	Millenia	F	1.100	1.000	0.002	—	—	—	0.080	47-62	47-62
		R	0.370	0.300	0.002	—	—	—	0.080	37-50	28-36
	Protege	F	NA	②	0.002	—	—	—	③	58-75	④
		R	—	—	—	7.87	7.993	NA	0.040	26-28	26-28
2001	626	F	0.940	0.870	0.002	—	—	—	0.080	58-75	22-28
		R	0.390	0.310	0.002	9.00	NA	9.059	0.040	34-49	26-28
	Miata	F	0.790	①	0.002	—	—	—	0.080	37-50	58-65
		R	0.350	0.310	0.002	—	—	—	0.040	37-50	26-28
	Millenia	F	1.100	1.000	0.002	—	—	—	0.080	47-62	47-62
		R	0.370	0.300	0.002	—	—	—	0.080	37-50	28-36
	Protege	F	NA	②	0.002	—	—	—	③	58-75	④
		R	—	—	—	7.87	7.993	NA	0.040	26-28	26-28
2002	626	F	0.940	0.870	0.002	—	—	—	0.080	58-75	22-28
		R	0.390	0.310	0.002	9.00	NA	9.059	0.040	34-49	26-28
	Miata	F	0.790	①	0.002	—	—	—	0.080	37-50	58-65
		R	0.350	0.310	0.002	—	—	—	0.040	37-50	26-28
	Millenia	F	1.100	1.000	0.002	—	—	—	0.080	47-62	47-62
		R	0.370	0.300	0.002	—	—	—	0.080	37-50	28-36
	Protege	F	NA	②	0.002	—	—	—	③	58-75	④
		R	NA	0.310	0.002	7.87	7.993	NA	0.040	26-28	26-28

NA: Not Avilable

F: Front

R: Rear

① With 15 inch wheel: 0.071 in.
   With 16 inch wheel: 0.079 in.

② With 1.6L engine: 0.780 in.
   With 1.8L engine: 0.870 in.
   With 2.0L engine: 0.870 in.

③ With 1.6L engine: 0.059 in.
   With 1.8L engine: 0.080 in.
   With 2.0L engine: 0.079 in.

④ Type A: 37-39
   Type B: 16-23

42356-MAZC-C11

SCHEDULED MAINTENANCE INTERVALS
Mazda Car

TO BE SERVICED	TYPE OF SERVICE	7.5	15	22.5	30	37.5	45	52.5	60	67.5	75	82.5	90	97.5
							VEHICLE MILEAGE INTERVAL (x1000)							
Engine oil & filter	R	✓	✓	✓	✓	✓	✓	✓	✓	✓	✓	✓	✓	✓
Air cleaner element	R				✓				✓				✓	
Engine coolant ①	R				✓				✓				✓	
Spark plugs	R				✓				✓				✓	
Automatic transaxle fluid	S/I				✓				✓				✓	
Bolts & nuts on chassis & body	S/I				✓				✓				✓	
Brake lines, hoses & connections	S/I				✓				✓				✓	
Cooling system	S/I				✓				✓				✓	
Disc brakes	S/I				✓				✓				✓	
Drive belts (Millenia ②)	S/I				✓				✓				✓	
Drive shaft dust boots	S/I				✓				✓				✓	
Exhaust system heat shield	S/I				✓				✓				✓	
Front & rear suspension ball joints	S/I				✓				✓				✓	
Fuel lines & hoses	S/I				✓				✓				✓	
Idle speed	S/I				✓				✓				✓	
Steering operation & linkages	S/I				✓				✓				✓	
Engine timing belt ③ ④	R								✓					
Fuel filter	R								✓					
Valve Clearance	I								✓					
Manual transmission	R								✓					
Hose & tube for emission	S/I								✓					

R: Replace   S/I: Service or Inspect

① Replace initially at 45,000 miles & every 30,000 miles thereafter except on 626. On 626, replace at 105,000 miles& every 30, 000 miles therafter

② (Millenia KJ engine): replace every 105,000 miles

③ Except Miata and Millenia KJ engine; inspect every 60,000 miles & replace at 105,000 miles (if not replaced previously).

④ Miata and Millenia KJ engine replace at 60,000 miles (if not replaced previously).

FREQUENT OPERATION MAINTENANCE (SEVERE SERVICE)

If a vehicle is operated under any of the following conditions it is considered severe service

- Extremely dusty areas.

- 50% or more of the vehicle operation is in 32°C (90°F) or higher temperatures, or constant operation in temperatures below 0°C (32°F).

- Prolonged idling (vehicle operation in stop and go traffic).

- Frequent short running periods (engine does not warm to normal operating temperatures).

- Police, taxi, delivery usage or trailer towing usage.

Oil & oil filter: change every 5000 miles.

Oil & oil filter (Puerto Rico): change every 3000 miles.

Air cleaner element: service or inspect every 15,000 miles

Automatic transaxle fluid: service or inspect every 15,000 miles.

Bolts & nuts on chassis & body: tighten every 15,000 miles.

Disc brakes: service or inspect every 15,000 miles.

42356-MAZC-C12

PRECAUTIONS

Before servicing any vehicle, please be sure to read all of the following precautions, which deal with personal safety, prevention of component damage, and important points to take into consideration when servicing a motor vehicle:

• Never open, service or drain the radiator or cooling system when the engine is hot; serious burns can occur from the steam and hot coolant.

• Observe all applicable safety precautions when working around fuel. Whenever servicing the fuel system, always work in a well-ventilated area. Do not allow fuel spray or vapors to come in contact with a spark, open flame or excessive heat (a hot drop light, for example). Keep a dry chemical fire extinguisher near the work area. Always keep fuel in a container specifically designed for fuel storage; also, always properly seal fuel containers to avoid the possibility of fire or explosion. Refer to the additional fuel system precautions later in this section.

• Fuel injection systems often remain pressurized, even after the engine has been turned **OFF**. The fuel system pressure must be relieved before disconnecting any fuel lines. Failure to do so may result in fire and/or personal injury.

• Brake fluid often contains polyglycol ethers and polyglycols. Avoid contact with the eyes and wash your hands thoroughly after handling brake fluid. If you do get brake fluid in your eyes, flush your eyes with clean, running water for 15 minutes. If eye irritation persists, or if you have taken brake fluid internally, IMMEDIATELY seek medical assistance.

• The EPA warns that prolonged contact with used engine oil may cause a number of skin disorders, including cancer! You should make every effort to minimize your exposure to used engine oil. Protective gloves should be worn when changing oil. Wash your hands and any other exposed skin areas as soon as possible after exposure to used engine oil. Soap and water, or waterless hand cleaner should be used.

• All new vehicles are now equipped with an air bag system. The system must be disabled before performing service on or around system components, steering column, instrument panel components, wiring and sensors. Failure to follow safety and disabling procedures could result in accidental air bag deployment, possible personal injury and unnecessary system repairs.

• Always wear safety goggles when working with, or around, the air bag system. When carrying a non-deployed air bag, be sure the bag and trim cover are pointed away from your body. When placing a non-deployed air bag on a work surface, always face the bag and trim cover upward, away from the surface. This will reduce the motion of the module if it is accidentally deployed. Refer to the additional air bag system precautions later in this section.

• Clean, high quality brake fluid from a sealed container is essential to the safe and proper operation of the brake system. You should always buy the correct type of brake fluid for your vehicle. If the brake fluid becomes contaminated, completely flush the system with new fluid. Never reuse any brake fluid. Any brake fluid that is removed from the system should be discarded. Also, do not allow any brake fluid to come in contact with a painted surface; it will damage the paint.

• Never operate the engine without the proper amount and type of engine oil; doing so WILL result in severe engine damage.

• Timing belt maintenance is extremely important! Many models utilize an interference-type, non-freewheeling engine. If the timing belt breaks, the valves in the cylinder head may strike the pistons, causing potentially serious (also time-consuming and expensive) engine damage.

• Disconnecting the negative battery cable on some vehicles may interfere with the functions of the on-board computer system(s) and may require the computer to undergo a relearning process once the negative battery cable is reconnected.

• When servicing drum brakes, only disassemble and assemble one side at a time, leaving the remaining side intact for reference.

• Only an MVAC-trained, EPA-certified automotive technician should service the air conditioning system or its components.

ENGINE REPAIR

Distributor

REMOVAL

1. Before servicing the vehicle, refer to the precautions in the beginning of this section.
2. Remove or disconnect the following:
   • Negative battery cable
   • Distributor cap and position it aside, leaving the ignition wires connected
   • Distributor electrical connector(s) from the side of the distributor
3. Using a wrench on the crankshaft pulley, rotate the crankshaft to position the No. 1 piston on Top Dead Center (TDC) of the compression stroke; the crankshaft pulley mark should align with the timing indicator and the distributor rotor should point towards the No. 1 spark plug wire tower position of the cap.
4. Using chalk or paint, mark the position of the distributor housing on the cylinder head. Also mark the position of the distributor rotor in relation to the distributor housing.
5. Remove distributor hold-down bolt(s).

6. On distributors attached to the end of the cylinder head (or inline with the camshaft), remove it by pulling it straight outward.
7. On distributors attached to the side of the cylinder head (or perpendicular with the camshaft), slowly pull it outward while watching the rotor. These distributors are

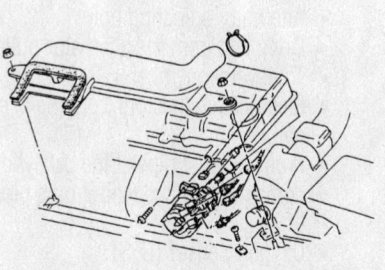

Exploded view of a typical side mounted distributor

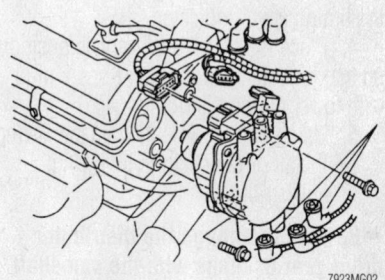

Exploded view of a typical end or inline mounted distributor gear driven and as you remove it, the gears will disengage inside the engine, causing the rotor to rotate. when the rotor stops moving, stop pulling outward. Re-align the distributor body-to-cylinder head match-mark (do not push it back in to do this, simply rotate the body to align the marks). Place a third mark indicating the new rotor position-to-distributor body relation. When installing the distributor, align this mark and the body-to-head mark to properly position the distributor.

8. Inspect the O-ring on the distributor housing and replace it, if it is damaged or worn.

INSTALLATION

Engine Not Disturbed

1. Using engine oil, lubricate the O-ring.
2. Install or connect the following:
   • Distributor

➡ Be sure to engage the distributor drive gear or tangs with the camshaft gear or slot. Align the mark that was made on the distributor housing with the mark that was made on the cylinder head.

   • Distributor hold-down bolt(s)
   • Electrical connector(s)
   • Distributor cap
   • Negative battery cable
3. Start the engine and check or adjust the ignition timing.

Engine Disturbed

1. Remove or disconnect the following:
   • Spark plug wire from the No. 1 cylinder spark plug
   • Spark plug from the No. 1 cylinder
2. Press a thumb over the spark plug hole.
3. Using a wrench on the crankshaft pulley, rotate the crankshaft until pressure is felt at the spark plug hole, indicating the piston is approaching TDC on the compression stroke. Continue rotating the crankshaft until the crankshaft pulley mark aligns with the timing cover indicator.
4. Place the distributor rotor in position so that it aligns with the No. 1 spark plug wire tower on the distributor cap.
5. Using engine oil, lubricate the O-ring.
6. Install or connect the following:
   • Distributor

➡ Be sure to engage the distributor drive gear or tangs with the camshaft gear or slot. Align the mark that was made on the distributor housing with

the mark that was made on the cylinder head.

   • Distributor hold-down bolt(s)
   • Electrical connector(s)
   • Distributor cap
   • Spark plug in the No. 1 cylinder and connect the spark plug wire
   • Negative battery cable
7. Start the engine and check or adjust the ignition timing.

Alternator

REMOVAL

Miata

1. Before servicing the vehicle, refer to the precautions in the beginning of this section.
2. Remove or disconnect the following:
   • Negative battery cable
   • Intake manifold bracket
   • Electrical connectors from the alternator
   • Alternator bolts
   • Alternator

Protégé

1. Before servicing the vehicle, refer to the precautions in the beginning of this section.
2. Remove or disconnect the following:
   • Negative battery cable
   • Electrical connectors from the alternator
   • Alternator drive belt
   • Alternator pivot and adjusting bar bolts
   • Alternator

626
2.0L (FS) ENGINES

1. Before servicing the vehicle, refer to the precautions in the beginning of this section.
2. Remove or disconnect the following:
   • Negative battery cable
   • Alternator upper mounting bolt
   • Alternator adjusting bolt
   • Drive belt from the alternator pulley
   • Transverse member
   • Electrical connectors from the alternator
   • Front exhaust pipe at the catalytic converter and suspend it on a piece of wire
   • Oxygen Sensor (O$_2$S)
   • 3 exhaust manifold flange nuts and the hold-down bracket clamp
   • Exhaust pipe

   • Alternator lower through-bolt
   • Alternator

2.5L (KL) ENGINES

1. Before servicing the vehicle, refer to the precautions in the beginning of this section.
2. Remove or disconnect the following:
   • Negative battery cable
   • Fresh air duct
   • Radiator upper bracket
   • Condenser fan, if equipped
   • Electrical connectors from the alternator
   • Loosen the belt tensioner locknut and tension adjusting bolt
   • Alternator upper mounting bolt
   • Right splash shield
   • Drive belt from the alternator pulley
   • A/C compressor and support it aside, leaving the refrigerant lines connected, if necessary
   • Alternator through-bolt
   • Alternator

Millenia

1. Before servicing the vehicle, refer to the precautions in the beginning of this section.
2. Disconnect the negative battery cable.
3. If equipped with the 2.3L (KJ) engine, remove the front charge air cooler, radiator upper seal board and the condenser fan assembly.
4. Remove or disconnect the following:
   • Intake air system on 2.4 (KL) engine
   • Electrical connectors from the alternator
   • Right splash shield
   • Drive belt from the alternator pulley
   • A/C compressor and support it aside, leaving the refrigerant lines connected
   • Upper and lower alternator mounting bolts
   • Alternator

INSTALLATION

Miata

1. Install or connect the following:
   • Alternator
   • Alternator bolts. Tighten the lower bolts to 38 ft. lbs. (51 Nm) and the upper bolt 18 ft. lbs. (25 Nm).
   • Electrical connectors to the alternator
   • Intake manifold bracket
   • Negative battery cable

Protégé

1. Install or connect the following:
   - Alternator with the pivot bolt
   - Alternator electrical connectors
   - Drive belt
   - Upper mounting bolt
2. Adjust the belt tension. Torque the lower through bolt to 28–38 ft. lbs. (38–51 Nm) and the upper mounting bolt to 14–18 ft. lbs. (19–26 Nm).
   - Negative battery cable

626

2.0L (FS) ENGINES

Install or connect the following:
- Alternator
- Alternator bolts. Tighten the lower bolt to 28–38 ft. lbs. (38–51 Nm) and the upper bolt to 14–18 ft. lbs. (19–25 Nm).
- Drive belt, check the belt deflection by applying moderate pressure 22 lbs. (98 N) between the alternator and the water pump pulley. The deflection on a new belt should be 6.5–7 inch (0.26–0.27mm) or on a used belt, 7.0–9.0 inch (0.28–0.35mm). Loosen the alternator mounting bolts and use the adjusting bolt to adjust the belt tension to the correct specification.
- Electrical connector and terminal wire
- Front exhaust pipe
- Transverse pipe
- Negative battery cable

2.5L (KL) ENGINES

1. Install or connect the following:
   - Alternator. Torque the lower bolt to 28–38 ft. lbs. (38–51 Nm) and the upper bolt to 14–18 ft. lbs. (19–25 Nm).
   - A/C compressor, if removed. Torque the bolts to 11–15 ft. lbs. (15–21 Nm).

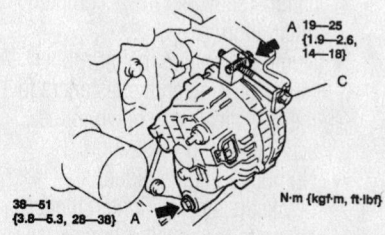

Adjust the alternator belt tension using the adjustment bolt C—626 4 cylinder models

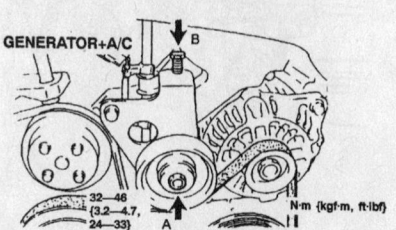

Adjust the alternator belt tension using the adjustment bolt B—626 6 cylinder models

- Drive belt, check the belt deflection by applying moderate pressure 22 lbf. (98 N) between the alternator and the crankshaft pulley to check the alternator belt or midway between the A/C compressor and the crankshaft pulley to check the Air conditioning/alternator belt. The deflection on a new belt should be 6.0–7 inch (0.24–0.27mm) or on a used belt, 7.0–8.0 inch (0.28–0.31mm) on the alternator belt. On models with a air conditioning/alternator belt; the deflection on a new belt should be 5.5–6.5 inch (0.22–0.25mm) or on a used belt, 6.5–7.5 inch (0.26–0.29mm) Loosen the alternator mounting bolts and use the adjusting bolt to adjust the belt tension to the correct specification.
- Negative battery cable

Millenia

1. Install or connect the following:
   - Alternator. Torque the upper bolt to 14–18 ft. lbs. (19–25 Nm) and the lower bolt to 28–38 ft. lbs. (38–51 Nm).
   - A/C compressor. Torque the bolts to 12–16 ft. lbs. (16–22 Nm).
   - Right splash shield
   - Drive belt
   - Electrical connectors to the alternator
   - Intake air system on 2.4 (KL) engine
2. If equipped with the 2.3L engine, install the condenser fan assembly, radiator upper seal board and the front charge air cooler using new O-rings. Torque the mounting bolts to 12–16 ft. lbs. (16–22 Nm).
3. Connect the negative battery cable.

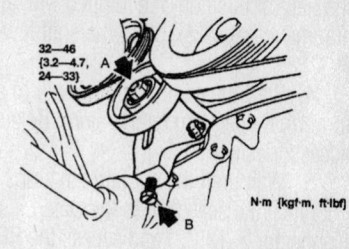

Ignition Timing

ADJUSTMENT

Except Millenia 2.5L (KL) Engines

1. Before servicing the vehicle, refer to the precautions in the beginning of this section.

➡ **If the information given in the following procedures differs from that on the emission information label located in the engine compartment, follow the directions given on the label. The label often reflects production changes made during the model year.**

The timing is controlled by the computer. Ignition timing adjustment is not possible or necessary.

2. If the timing is still not within specification. the following components may be defective:
   - Camshaft position (CMP) sensor
   - Crankshaft Position (CKP) sensor
   - Throttle Position (TP) sensor
   - Engine Coolant Temperature (ECT) sensor
   - Neutral switch if equipped with a manual transaxle
   - Clutch switch if equipped with a manual transaxle
   - Transaxle range switch if equipped with an automatic transaxle
3. If the above components are normal, replace the Powertrain Control Module (PCM).

Millenia 2.5L (KL) Engine

1. Before servicing the vehicle, refer to the precautions in the beginning of this section.
2. Let the engine warm to normal operating temperature.
3. Apply the parking brake. If equipped with a manual transaxle, place the shifter in

the neutral position. If equipped with an automatic transaxle, place the shift lever in **P**.

4. Start the engine and allow it to come to normal operating temperature. Be sure all accessories are **OFF**.

5. Wait until the cooling fan stops, then , connect the scan tool to the Data Link Connector 2 (DLC2) and access the RPM PID.

6. Locate the timing marks on the crankshaft pulley and timing belt lower cover. The engine may have to be cranked slightly to see the mark on the crankshaft pulley.

7. Check the idle speed and adjust, if necessary.

8. Connect a jumper wire between the TEN terminal and the GND terminal at the underhood diagnosis connector.

9. Connect the system selector tool to the DLC and set switch **A** to position **1**.

10. Set the test switch to **SELFT TEST**.

11. Connect an inductive timing light according to the manufacturer's instructions.

12. Start the engine and allow the idle to stabilize. Aim the timing light at the timing marks. The timing should be 9–11 degrees Before Top Dead Center (TDC).

13. Loosen the distributor lockbolts just enough to turn the distributor. While aiming the timing light at the timing marks, turn the distributor until the marks are aligned. Tighten the distributor lockbolts to 14–18 ft. lbs. (19–25 Nm) and recheck the timing.

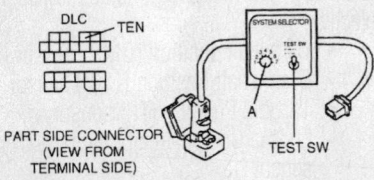

42356-MAZC-GFA

Jumper the connections shown on the data link connector and system selector tool

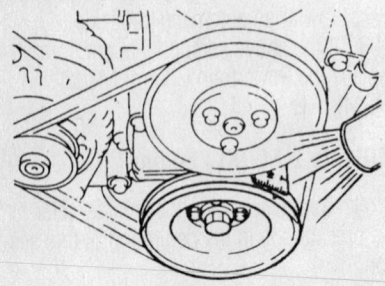

7923MG05

Connect an inductive timing light and aim it at the crankshaft pulley. Read the pulley mark against the scale

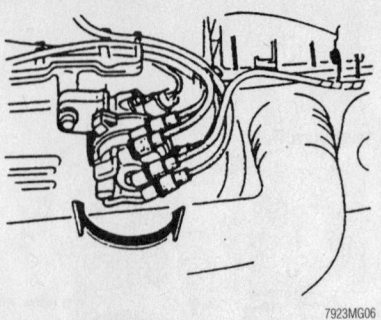

7923MG06

If adjustment is necessary, loosen the distributor lockbolts and rotate it until the mark is aligned

14. The ignition timing is now set. Disconnect the jumper wire from the DLC

15. Remove all test equipment.

Engine Assembly

REMOVAL & INSTALLATION

626

2.0L (FS) ENGINE

➡The procedure for pulling the engine requires removing the transaxle along with it. As a result, when the halfshafts are pulled from the transaxle, a special plug/side gear holding tool is recommended.

1. Before servicing the vehicle, refer to the precautions in the beginning of this section.

2. Properly relieve the fuel system pressure.

3. Drain the engine oil.

4. Drain the transaxle fluid.

5. Drain the cooling system.

6. Remove or disconnect the following:
- Negative battery cable
- Radiator
- Air cleaner assembly
- Accelerator cable
- Fuel hoses
- Front exhaust pipe
- Any rods, pipes or cables related to the transaxle that would hinder removal
- Battery
- Fuse box
- Power steering pump with the lines still attached and set aside
- A/C compressor with the lines still attached and position aside
- Cruise actuator connector, actuator retainers and the actuator
- Halfshafts
- Number 5 engine mount retainers. Refer to the accompanying illustration for location.
- Engine mount member (2) retainers. Refer to the accompanying illustration for location.
- Number 1 engine mount stay bracket retainers. Refer to the accompanying illustration for location.
- Number 3 engine mount rubber retainers. Refer to the accompanying illustration for location.
- Number 3 engine bracket retainers. Refer to the accompanying illustration for location.
- Number 1 engine mount nut and bolt. Refer to the accompanying illustration for location.
- Number 4 engine mount rubber retainers. Refer to the accompanying illustration for location.
- Number 4 engine bracket retainers. Refer to the accompanying illustration for location.
- Engine and transaxle assembly from the vehicle.

To install:

7. Installation is the reverse of removal. Tighten the fasteners to the specifications shown in the accompanying illustration.

8. When possible, leave the engine mounting nuts/bolts loose (hand tight) until all mounts are aligned and bolted. This may help in aligning the engine and transmission assembly in the vehicle.

9. Install or connect the following:
- Number 4 engine bracket retainers. Tighten the fasteners to the specifications shown in the accompanying illustration.
- Number 4 engine mount rubber retainers. Tighten the fasteners to the specifications shown in the accompanying illustration.
- Number 1 engine mount nut and bolt. Tighten the fasteners to the specifications shown in the accompanying illustration.
- Number 3 engine bracket retainers. Tighten the fasteners to the specifications shown in the accompanying illustration.
- Number 3 engine mount rubber retainers. Tighten the fasteners to the specifications shown in the accompanying illustration.
- Number 1 engine mount stay bracket retainers. Tighten the fasteners to the specifications shown in the accompanying illustration.
- Engine mount member (2) retainers. Tighten the fasteners to the

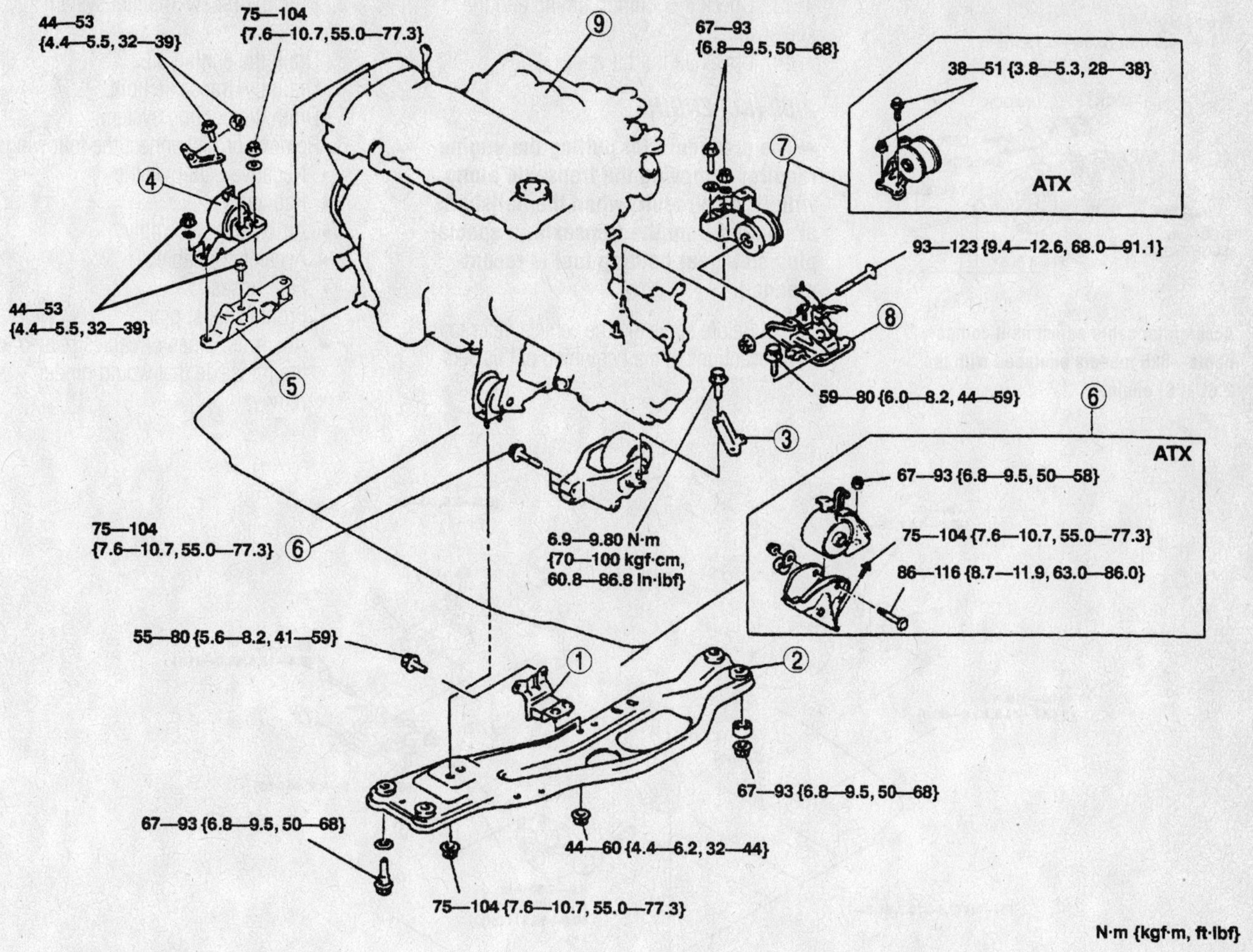

44—53
{4.4—5.5, 32—39}

75—104
{7.6—10.7, 55.0—77.3}

⑨

67—93
{6.8—9.5, 50—68}

38—51 {3.8—5.3, 28—38}

ATX

④

93—123 {9.4—12.6, 68.0—91.1}

⑦

44—53
{4.4—5.5, 32—39}

59—80 {6.0—8.2, 44—59}

⑧

⑥

⑤

③

ATX

67—93 {6.8—9.5, 50—58}

75—104 {7.6—10.7, 55.0—77.3}

86—116 {8.7—11.9, 63.0—86.0}

75—104
{7.6—10.7, 55.0—77.3} ⑥

6.9—9.80 N·m
{70—100 kgf·cm,
60.8—86.8 In·lbf}

55—80 {5.6—8.2, 41—59}

①

②

67—93 {6.8—9.5, 50—68}

67—93 {6.8—9.5, 50—68}

44—60 {4.4—6.2, 32—44}

75—104 {7.6—10.7, 55.0—77.3}

N·m {kgf·m, ft·lbf}

1	No.5 engine mount rubber	6	No.1 engine mount bolt and nut
2	Engine mount member	7	No.4 engine mount rubber
3	No.1 engine mount stay bracket	8	No.4 engine mount bracket
4	No.3 engine mount rubber	9	Engine, transaxle
5	No.3 engine bracket		

42356-MAZC-G03

Location of the engine mounting components and their torque specifications—626 models equipped with the 2.0L (FS) engine

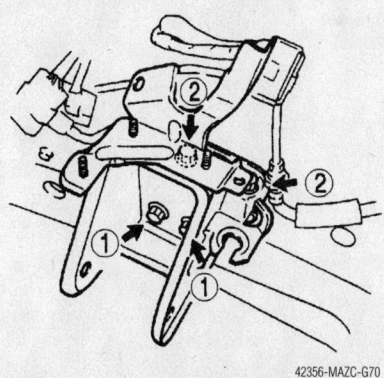

42356-MAZC-G70

Tighten the number 4 engine bracket retainers in the sequence shown—626 models equipped with the 2.0L (FS) engine

specifications shown in the accompanying illustration.
• Number 5 engine mount retainers. Tighten the fasteners to the specifications shown in the accompanying illustration.
10. When connecting the accelerator cable, perform the following adjustment:
  a. Move the white locking tab to the unlock position.
  b. Turn stopper B to the to the unlock position.

➡ If stopper B will not unlock, it may be necessary to carefully bend back tab C out a little using a suitable pry tool.

  c. Push or pull the cable housing directly behind the spring.
  d. Turn the stopper B to the lock position.
  e. Measure the free play which should be 0.06—0.15 inch (1.5—4mm) and make sure the cable free play is within specification.
  f. Move the white locking tab to the lock position and check for proper accelerator operation.
11. Fill the engine and the transaxle with the proper type and amount of fluids. Fill the cooling system.
12. Connect the negative battery cable, start the engine and check for leaks.

Free play
1.5—4.0 mm {0.06—0.15 in}

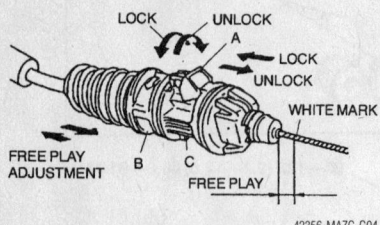

42356-MAZC-G04

Accelerator cable adjustment components—626 models equipped with the 2.0L (FS) engine

13. Check the ignition timing and the idle speed.
14. Check all fluid levels.

2.5L (KL) ENGINE

➡The procedure for pulling the engine requires removing the transaxle along with it. As a result, when the halfshafts are pulled from the transaxle, a special plug/side gear holding tool is recommended.

1. Before servicing the vehicle, refer to the precautions in the beginning of this section.

2. Properly relieve the fuel system pressure.
3. Drain the engine oil.
4. Drain the transaxle fluid.
5. Drain the cooling system.
6. Remove or disconnect the following:
   - Negative battery cable
   - Radiator
   - Air cleaner assembly
   - Accelerator cable
   - Fuel hoses
   - Front exhaust pipe
   - Any rods, pipes or cables related to the transaxle that would hinder removal

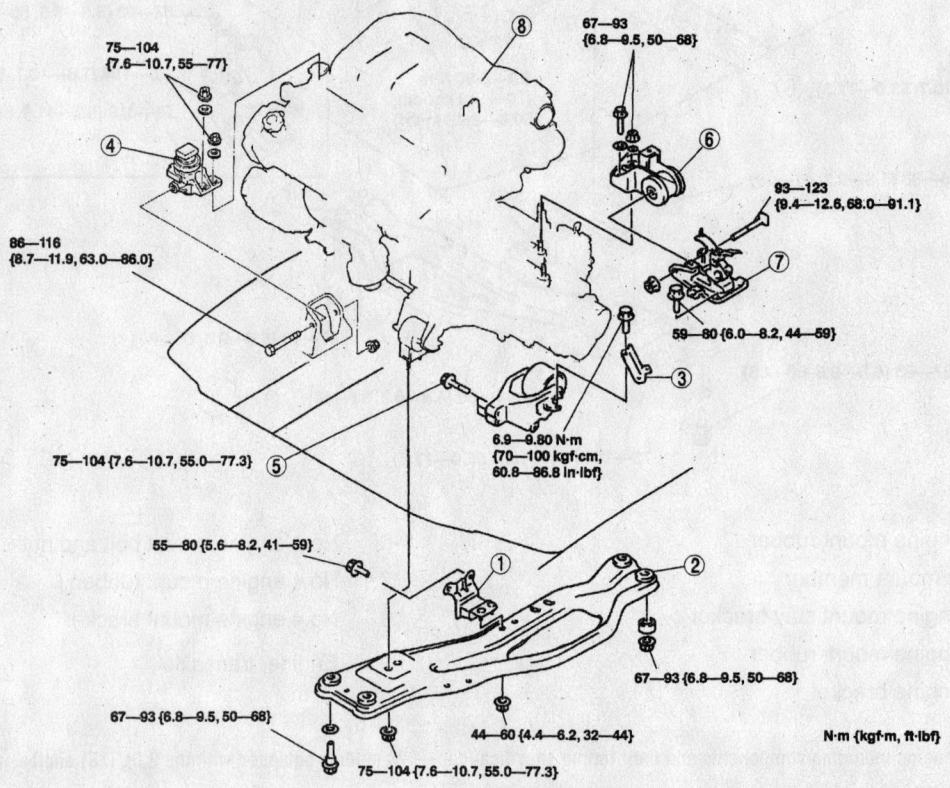

1	No.5 engine mount rubber
2	Engine mount member
3	No.1 engine mount stay bracket
4	No.3 engine mount rubber
5	No.1 engine mount bolt and nut

6	No.4 engine mount rubber
7	No.4 engine mount bracket
8	Engine, transaxle

No.4 Engine Mount Bracket Installation Note
- Tighten the bolt in the order shown.

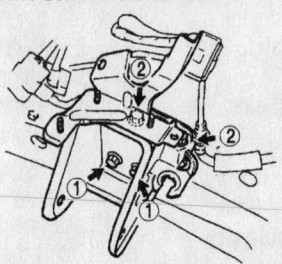

Location of the engine mounting components and their torque specifications—626 models equipped with the 2.5L (KL) engine

42356-MAZC-G05

- Battery
- Power steering pump with the lines still attached and set aside
- A/C compressor with the lines still attached and position aside
- Cruise actuator connector, actuator retainers and the actuator
- Halfshafts
- Number 5 engine mount retainers. Refer to the accompanying illustration for location.
- Engine mount member (2) retainers. Refer to the accompanying illustration for location.
- Number 1 engine mount stay bracket retainers. Refer to the accompanying illustration for location.
- Number 3 engine mount rubber retainers. Refer to the accompanying illustration for location.
- Number 1 engine mount nut and bolt. Refer to the accompanying illustration for location.
- Number 4 engine mount rubber retainers. Refer to the accompanying illustration for location.
- Number 4 engine bracket retainers. Refer to the accompanying illustration for location.
- Engine and transaxle assembly from the vehicle.

To install:

7. Installation is the reverse of removal. Tighten the fasteners to the specifications shown in the accompanying illustration.

8. When possible, leave the engine mounting nuts/bolts loose (hand tight) until all mounts are aligned and bolted. This may help in aligning the engine and transmission assembly in the vehicle.

9. Install or connect the following:
- Number 4 engine bracket retainers. Tighten the fasteners to the specifications shown in the accompanying illustration.
- Number 4 engine mount rubber

Free play
1.5—4.0 mm {0.06—0.15 in}

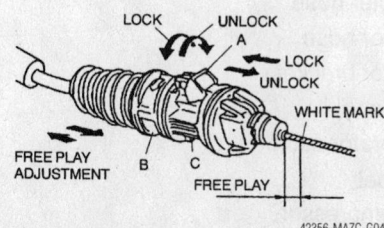

42356-MAZC-G04

Accelerator cable adjustment components—626 models equipped with the 2.5L (KL) engine

retainers. Tighten the fasteners to the specifications shown in the accompanying illustration.
- Number 1 engine mount nut and bolt. Tighten the fasteners to the specifications shown in the accompanying illustration.
- Number 3 engine mount rubber retainers. Tighten the fasteners to the specifications shown in the accompanying illustration.
- Number 1 engine mount stay bracket retainers. Tighten the fasteners to the specifications shown in the accompanying illustration.
- Engine mount member (2) retainers. Tighten the fasteners to the specifications shown in the accompanying illustration.
- Number 5 engine mount retainers. Tighten the fasteners to the specifications shown in the accompanying illustration.

10. When connecting the accelerator cable, perform the following adjustment:

a. Move the white locking tab to the unlock position.

b. Turn stopper B to the to the unlock position.

➡**If stopper B will not unlock, it may be necessary to carefully bend back tab C out a little using a suitable pry tool.**

c. Push or pull the cable housing directly behind the spring.

d. Turn the stopper B to the lock position.

e. Measure the free play which should be 0.06–0.15 inch (1.5–4mm) and make sure the cable free play is within specification.

f. Move the white locking tab to the lock position and check for proper accelerator operation.

11. Fill the engine and the transaxle with the proper type and amount of fluids. Fill the cooling system.

12. Connect the negative battery cable, start the engine and check for leaks.

13. Check the ignition timing and the idle speed.

14. Check all fluid levels.

Millenia

2.5L (KL) ENGINE

➡**The procedure for pulling the engine requires removing the transaxle along with it. As a result, when the halfshafts are pulled from the transaxle, a special plug/side gear holding tool is recommended.**

1. Before servicing the vehicle, refer to the precautions in the beginning of this section.

2. Properly relieve the fuel system pressure.

3. Drain the engine oil.

4. Drain the transaxle fluid.

5. Drain the cooling system.

6. Remove or disconnect the following:
- Hood
- Front wheels
- Splash shield
- Battery clamp, cover, battery, battery carrier and battery air duct
- Air cleaner assembly
- Upper seal board
- Radiator reservoir hose and the reservoir
- Condenser fan motor connector and the fan
- Cooling fan connector and fan
- Oil cooler hose
- Radiator hose, bracket and the radiator
- Accelerator cable
- Drive belt
- A/C compressor with the lines still attached and position aside
- Power steering pump with the lines still attached and position aside
- Selector cable
- Front exhaust pipe
- Tie rod end ball joint
- Upper and lower ball joints
- Halfshafts
- Joint shaft

7. Support the engine using a suitable support device.
- Number 1 engine mount stay. Refer to the accompanying illustration for location.
- Number 1 engine mount bracket. Refer to the accompanying illustration for location.
- Engine mount member retainers. Remove the bolts **A** first and then the bolts **B** and remove the member. Refer to the accompanying illustration for location.

✳✳ **CAUTION**

Engine load can damage the number 4 mount bolts holes when removing the bolts so make sure all weight is off the mount.

8. Attach a hoist to the engine, take the weight of the engine with the hoist and remove the engine support device, Lift up the engine/transaxle assembly slightly until

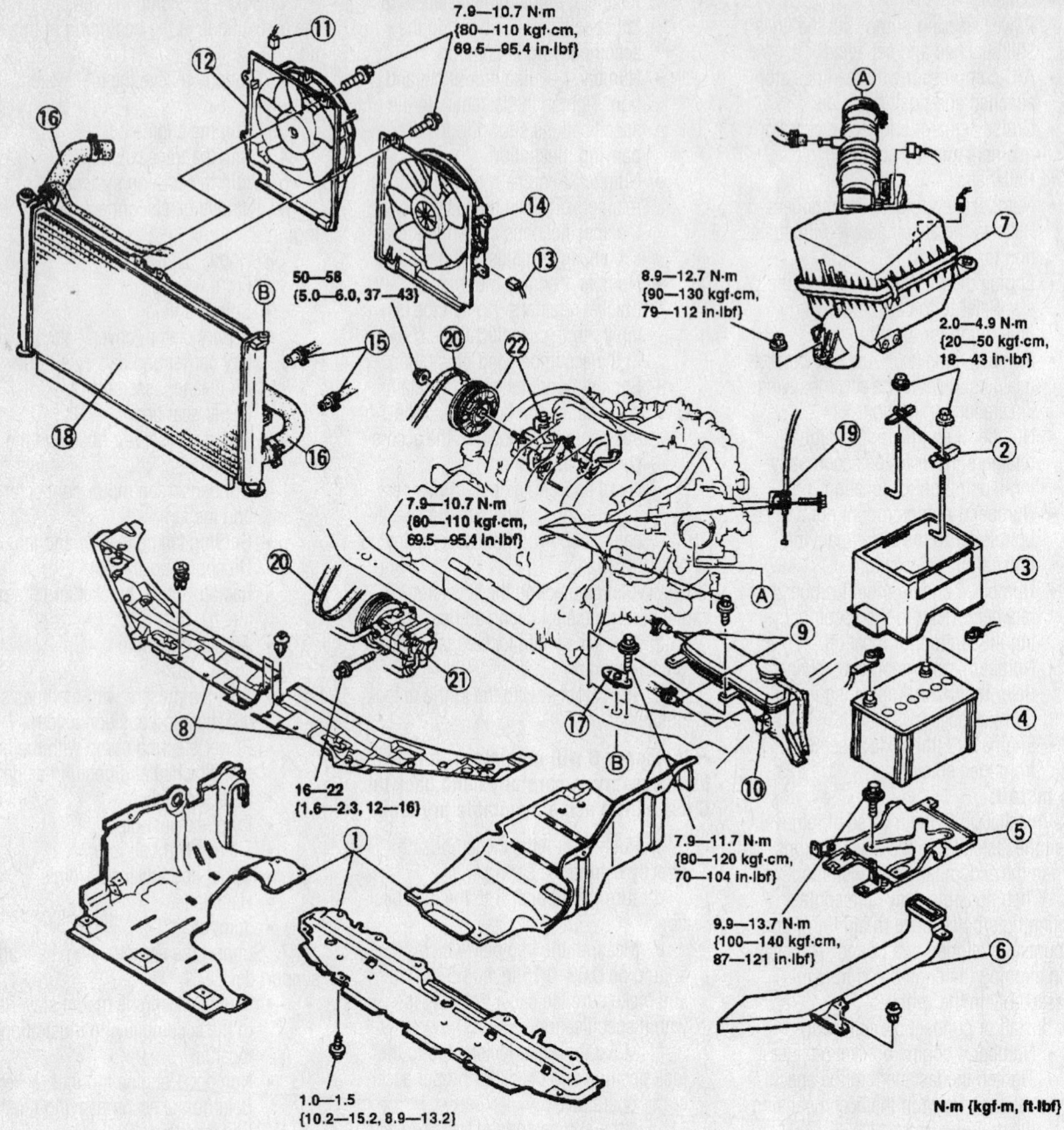

1	Splash shield	12	Condenser fan component
2	Battery clamp	13	Cooling fan motor connector
3	Battery cover	14	Cooling fan component
4	Battery	15	Oil cooler hose
5	Battery carrier	16	Radiator hose
6	Battery air duct	17	Radiator bracket
7	Air cleaner component	18	Radiator
8	Upper seal board	19	Accelerator cable
9	Radiator reservoir hose	20	Drive belt
10	Radiator reservoir	21	A/C compressor
11	Condenser fan motor connector	22	P/S oil pump

42356-MAZC-GDA

Location of the engine mounting components and their torque specifications (part 1 of 3)—Millenia models equipped with the 2.5L (KL) engine

the number 3 and 4 mounts are free from engine weight.

- Number 4 engine mount bracket. Refer to the accompanying illustration for location.
- Number 3 engine mount sub bracket. Refer to the accompanying illustration for location.
- Engine and transaxle assembly

from the vehicle. Be careful the powertrain assembly does not swing and strike the vehicle causing damage

To install:

9. Installation is the reverse of removal. Tighten the fasteners to the specifications shown in the accompanying illustration.

10. When possible, leave the engine mounting nuts/bolts loose (hand tight) until all mounts are aligned and bolted. This may help in aligning the engine and transmission assembly in the vehicle.

11. Install or connect the following:
- Number 4 engine bracket retainers and hand tighten the retainers

12. Using the hoist lower the powertrain

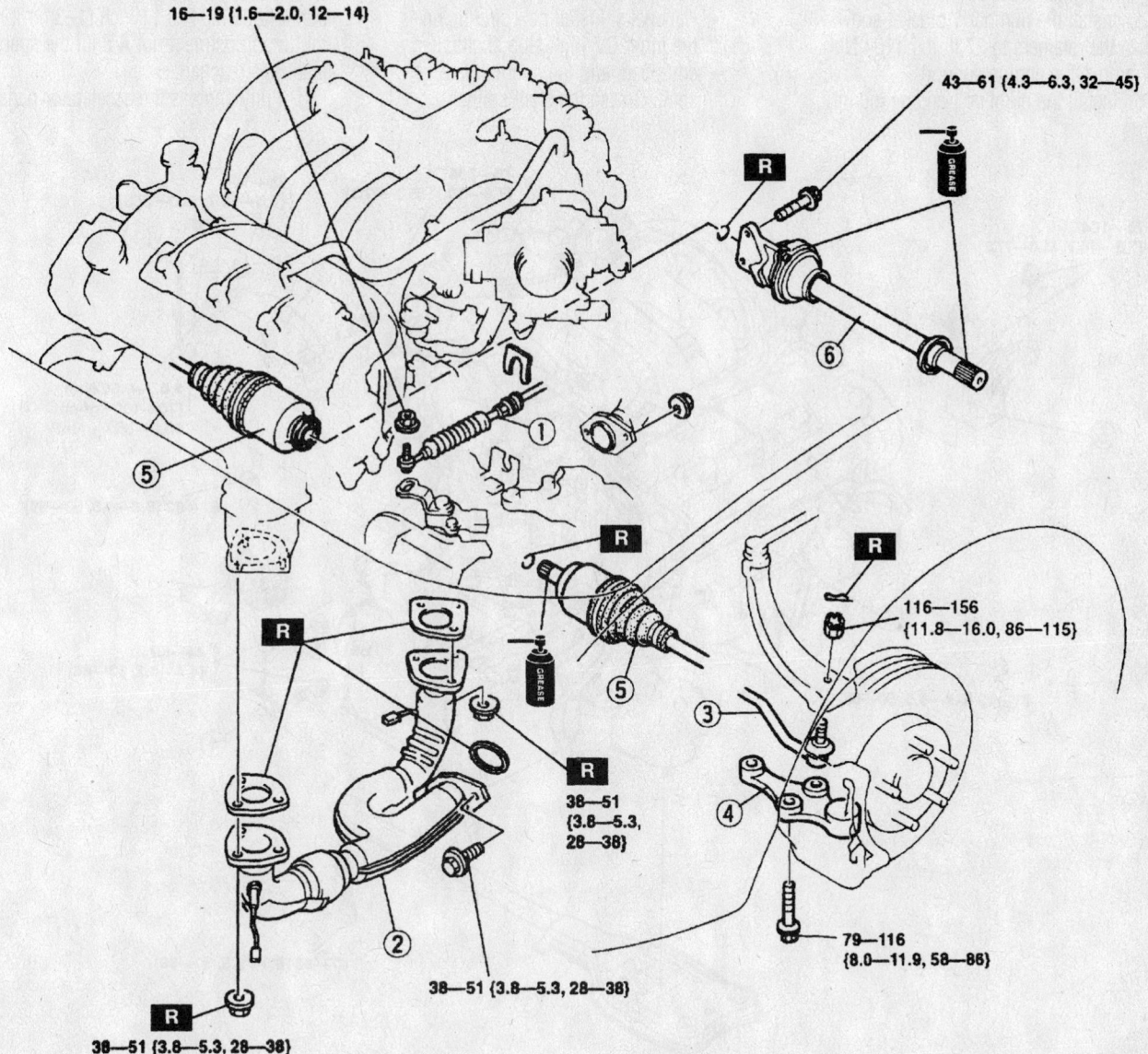

1   Selector cable
2   Front pipe
3   Upper lateral link ball joint
4   Lower ball joint
5   Drive shaft
6   Joint shaft

42356-MAZC-GDB

Location of the engine mounting components and their torque specifications (part 2 of 3)—Millenia models equipped with the 2.5L (KL) engine

assembly and hand tighten the number 3 mount sub bracket retainers.

• Number 4 engine bracket retainers and tighten the bolts **E** to 44 ft. lbs. (60 Nm) and hand tighten bolts **F**.

13. Remove the engine hoist and install the engine support device

14. Install the engine mount member and tighten bolts **B** to 68 ft. lbs. (93 Nm) and hand tighten bolts **A**.

15. Install the number 1 bracket and tighten the retainers to 77 ft. lbs. (104 Nm) and remove the engine support.

16. Install the number 1 engine mount

stay and tighten retainers **G** to 86 inch lbs. (10 Nm) and **H** to 77 ft. lbs. (104 Nm).

17. Tighten the number 2 engine mount bolts **A** to 77 ft. lbs. (104 Nm).

18. Tighten the number 3 engine mount sub bracket nuts **D** to 77 ft. lbs. (104 Nm).

19. Tighten the number 4 engine mount nuts and bolts **C** and **F** to 68 ft. lbs. (93 Nm)

• Joint shaft. Install a new circlip with the opening facing up. Tighten the bolts to 32–45 ft. lbs. (43–61 Nm).

• Halfshafts. Install new circlips on the inner CV-joint stub shafts, if equipped, and the intermediate shaft. Grease the shaft splines

before installing the halfshaft/intermediate shaft into the transaxle.

• Ball joints
• Front pipe
• Selector cable
• Power steering pump
• A/C Compressor
• Drive belt

20. When connecting the accelerator cable, perform the following adjustment:

a. Measure the free play of the cable, it should be 0.04–0.11 inch (1–3mm), if not turn adjustment nut **A** until the specification is reached.

b. Fully depress the accelerator pedal

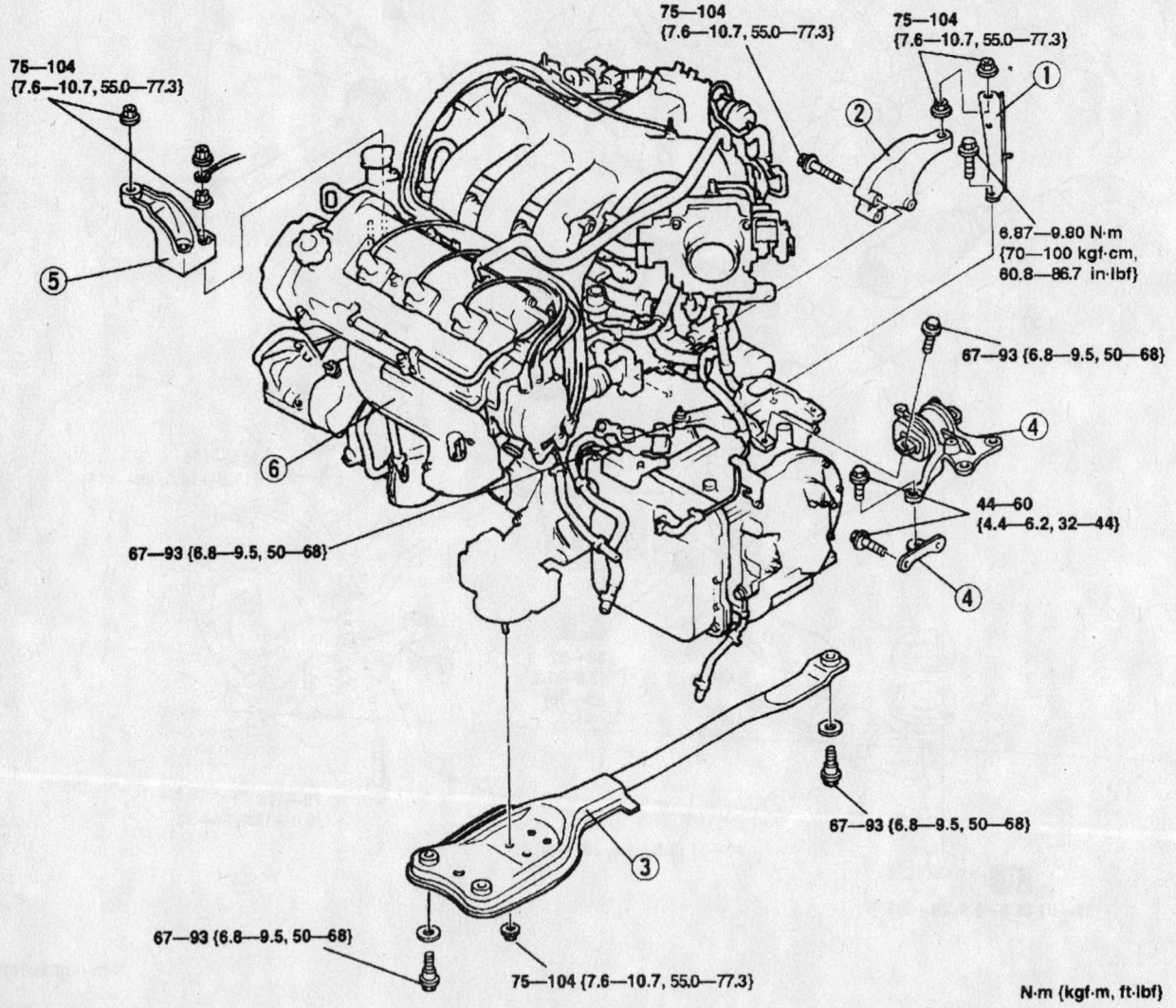

75—104 {7.6—10.7, 55.0—77.3}

75—104 {7.6—10.7, 55.0—77.3}

75—104 {7.6—10.7, 55.0—77.3}

6.87—9.80 N·m {70—100 kgf·cm, 60.8—86.7 in·lbf}

67—93 {6.8—9.5, 50—68}

44—60 {4.4—6.2, 32—44}

67—93 {6.8—9.5, 50—68}

67—93 {6.8—9.5, 50—68}

67—93 {6.8—9.5, 50—68}

75—104 {7.6—10.7, 55.0—77.3}

N·m {kgf·m, ft·lbf}

1	No.1 engine mount stay	4	No.4 engine mount bracket
2	No.1 engine mount bracket	5	No.3 engine mount sub bracket
3	Engine mount member	6	Engine and transaxle

42356-MAZC-GDC

Location of the engine mounting components and their torque specifications (part 3 of 3)—Millenia models equipped with the 2.5L (KL) engine

and check the throttle is wide open, if not adjust using nut **B**.

21. Install the remaining components in the reverse of removal.

22. Fill the engine and the transaxle with

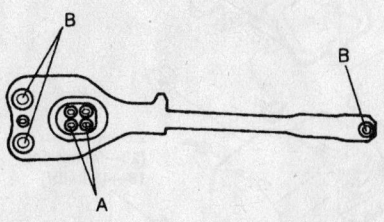

42356-MAZC-GDD

Remove the bolts A first and then the bolts B and remove the engine mount member—Millenia models equipped with the 2.5L (KL) engine

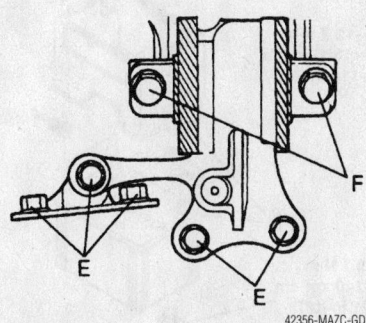

42356-MAZC-GDE

Tighten the number 4 engine bracket retainers E to specification and hand tighten bolts F—Millenia models equipped with the 2.5L (KL) engine

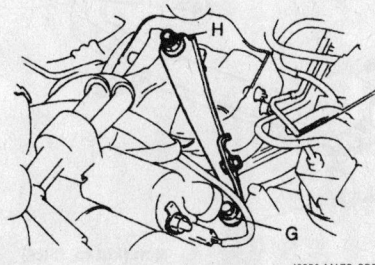

42356-MAZC-GDF

Tighten the number 1 engine mount stay retainers G and H to specification—Millenia models equipped with the 2.5L (KL) engine

42356-MAZC-GDG

Tighten the number 2 engine mount bolts A to specification—Millenia models equipped with the 2.5L (KL) engine

the proper type and amount of fluids. Fill the cooling system.

23. Connect the negative battery cable, start the engine and check for leaks.

24. Check the ignition timing and the idle speed.

25. Check all fluid levels.

2.3L (KJ) ENGINE

➡ **The procedure for pulling the engine requires removing the transaxle along with it. As a result, when the halfshafts are pulled from the transaxle, a special plug/side gear holding tool is recommended.**

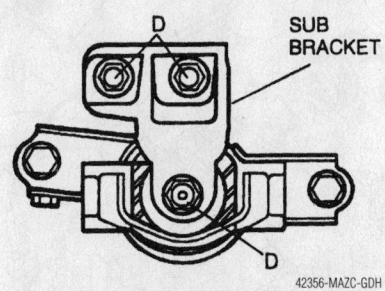

42356-MAZC-GDH

Tighten the number 3 engine mount sub bracket nuts D to specification—Millenia models equipped with the 2.5L (KL) engine

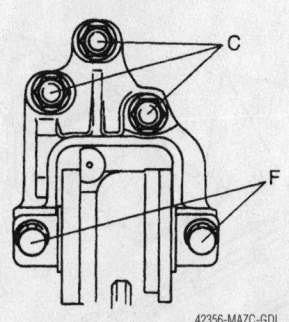

42356-MAZC-GDI

Tighten the number 4 engine mount nuts and bolts C and F to specification—Millenia models equipped with the 2.5L (KL) engine

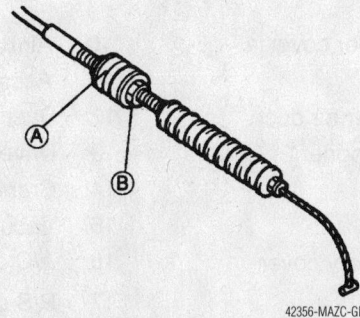

42356-MAZC-GDJ

Accelerator cable A and throttle cable B adjustment nuts—Millenia models equipped with the 2.5L (KL) engine

1. Before servicing the vehicle, refer to the precautions in the beginning of this section.

2. Properly relieve the fuel system pressure.

3. Drain the engine oil.

4. Drain the transaxle fluid.

5. Drain the cooling system.

6. Remove or disconnect the following:
   - Hood
   - Front wheels
   - Air cleaner assembly
   - Intake manifold cover
   - Splash shield
   - Charge air cooler duct
   - Battery clamp, cover, battery, battery carrier and battery air duct
   - Air duct
   - Accelerator cable
   - Dust cover
   - Drive belt
   - Crankshaft and vacuum pump pulleys
   - A/C compressor with the lines still attached and position aside
   - Power steering pump with the lines still attached and position aside
   - Radiator grille
   - Upper seal board
   - Radiator reservoir
   - Cooling fan connector and fan
   - Condenser fan motor connector and the fan
   - Oil cooler hose
   - Radiator hose, bracket and the radiator
   - Selector cable
   - Front exhaust pipe
   - Upper and lower ball joints
   - Halfshafts
   - Joint shaft

7. Support the engine using a suitable support device.
   - Number 1 engine mount bracket. Refer to the accompanying illustration for location.
   - Engine mount member retainers. Remove the bolts **A** first and then the bolts **B** and remove the member. Refer to the accompanying illustration for location.

✳✳ CAUTION

Engine load can damage the number 4 mount bolts holes when removing the bolts so make sure all weight is off the mount.

8. Attach a hoist to the engine, take the weight of the engine with the hoist and remove the engine support device, Lift up

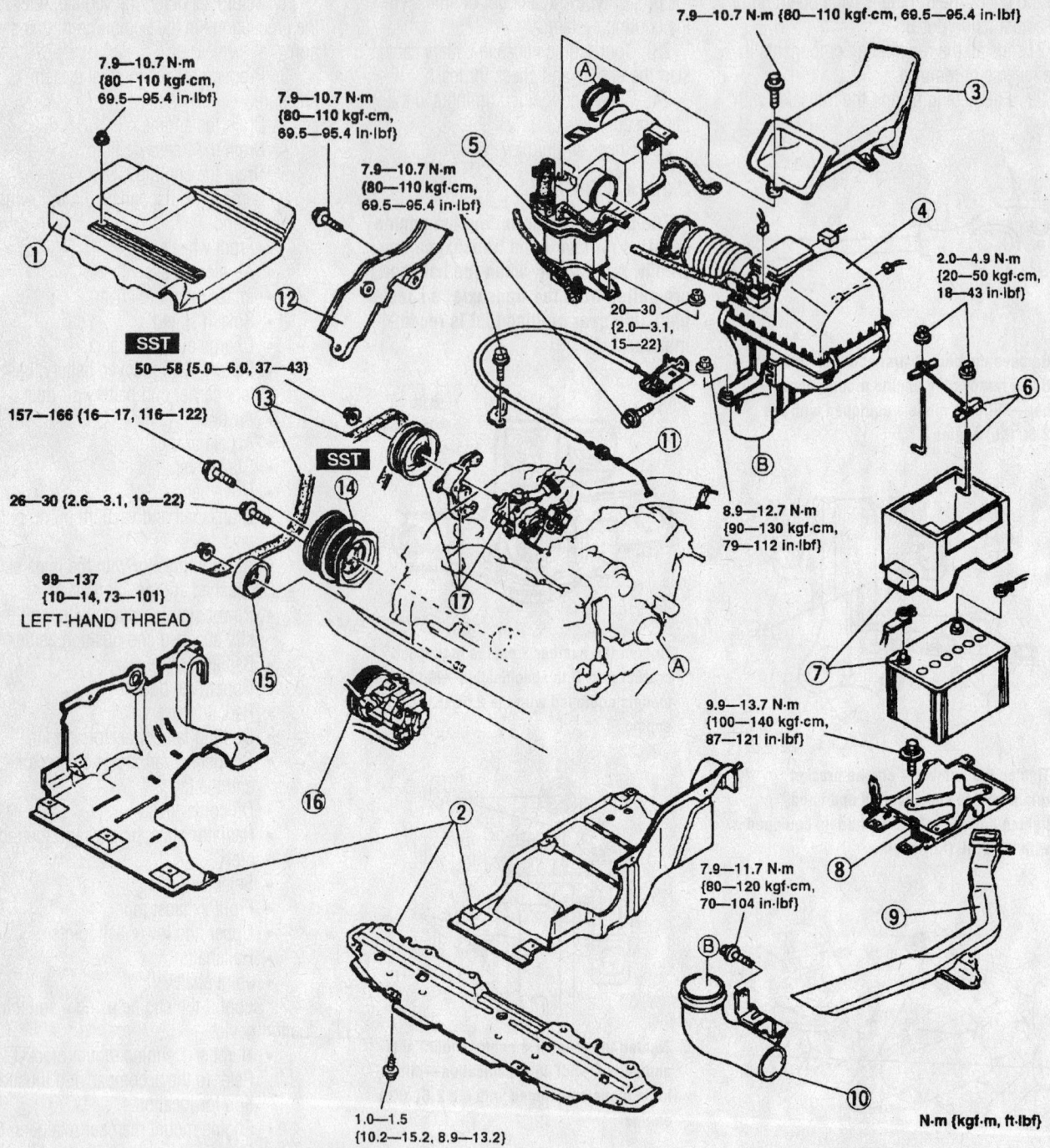

7.9—10.7 N·m {80—110 kgf·cm, 69.5—95.4 in·lbf}

7.9—10.7 N·m {80—110 kgf·cm, 69.5—95.4 in·lbf}

7.9—10.7 N·m {80—110 kgf·cm, 69.5—95.4 in·lbf}

7.9—10.7 N·m {80—110 kgf·cm, 69.5—95.4 in·lbf}

2.0—4.9 N·m {20—50 kgf·cm, 18—43 in·lbf}

SST
50—58 {5.0—6.0, 37—43}

157—166 {16—17, 116—122}

SST

26—30 {2.6—3.1, 19—22}

99—137 {10—14, 73—101}
LEFT-HAND THREAD

20—30 {2.0—3.1, 15—22}

8.9—12.7 N·m {90—130 kgf·cm, 79—112 in·lbf}

9.9—13.7 N·m {100—140 kgf·cm, 87—121 in·lbf}

7.9—11.7 N·m {80—120 kgf·cm, 70—104 in·lbf}

1.0—1.5 {10.2—15.2, 8.9—13.2}

N·m {kgf·m, ft·lbf}

1	Dynamic chamber cover	10	Air duct
2	Splash shield	11	Accelerator cable
3	Charge air cooler air duct	12	Dust cover
4	Air cleaner component	13	Drive belt
5	Resonator	14	Crankshaft pulley
6	Battery clamp	15	Vacuum pump pulley
7	Battery and battery cover	16	A/C compressor
8	Battery carrier	17	P/S oil pump
9	Battery air duct		

42356-MAZC-GDK

Location of the engine mounting components and their torque specifications (part 1 of 4)—Millenia models equipped with the 2.3L (KJ) engine

the engine/transaxle assembly slightly until the number 3 and 4 mounts are free from engine weight.

- Number 4 engine mount bracket. Refer to the accompanying illustration for location.
- Number 3 engine mount sub bracket. Refer to the accompanying illustration for location.
- Engine and transaxle assembly from the vehicle. Be careful the powertrain assembly does not

swing and strike the vehicle causing damage.

To install:

9. Installation is the reverse of removal. Tighten the fasteners to the specifications shown in the accompanying illustration.

10. When possible, leave the engine mounting nuts/bolts loose (hand tight) until all mounts are aligned and bolted. This may help in aligning the engine and transmission assembly in the vehicle.

11. Install or connect the following:

- Number 4 engine bracket retainers and hand tighten the retainers

12. Using the hoist lower the powertrain assembly and hand tighten the number 3 mount sub bracket retainers.

- Number 4 engine bracket retainers and tighten the bolts **E** to 44 ft. lbs. (60 Nm) and hand tighten bolts **F**.

13. Remove the engine hoist and install the engine support device

14. Install the engine mount member

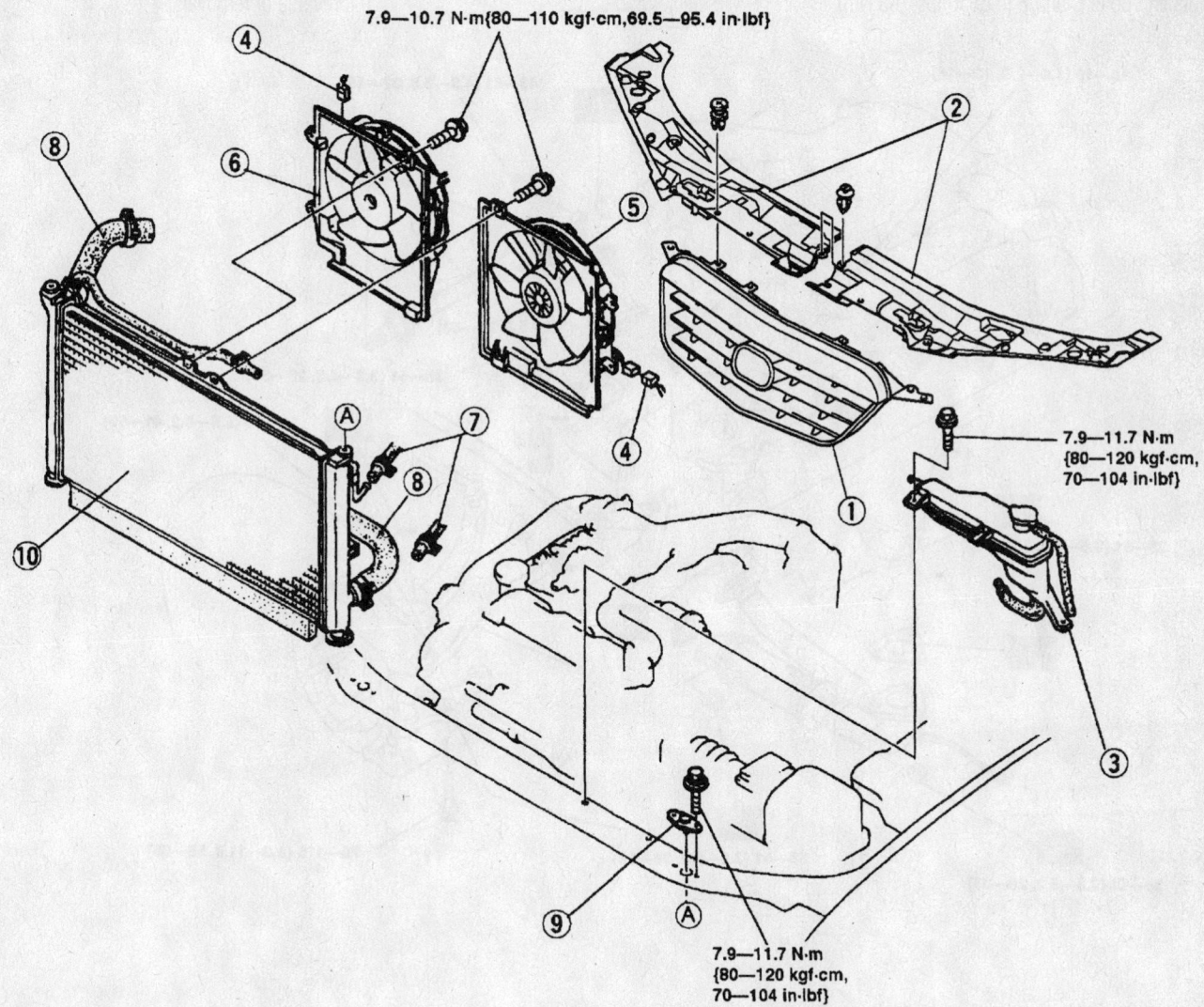

1	Radiator grille	6	Condenser fan component
2	Upper seal board	7	Oil cooler hose
3	Radiator reservoir	8	Radiator hose
4	Fan motor connector	9	Radiator bracket
5	Cooling fan component	10	Radiator

42356-MAZC-GDL

Location of the engine mounting components and their torque specifications (part 2 of 4)—Millenia models equipped with the 2.3L (KJ) engine

and tighten bolts **B** to 68 ft. lbs. (93 Nm) and hand tighten bolts **A**.

15. Install the number 1 bracket and tighten the retainers to 77 ft. lbs. (104 Nm) and remove the engine support.

16. Install the number 1 engine mount stay and tighten retainers **G** to 68 inch lbs. (10 Nm) and **H** to 77 ft. lbs. (104 Nm).

17. Tighten the number 2 engine mount bolts **A** to 77 ft. lbs. (104 Nm).

18. Tighten the number 3 engine mount sub bracket nuts **D** to 77 ft. lbs. (104 Nm).

19. Tighten the number 4 engine mount nuts and bolts **C** and **F** to 68 ft. lbs. (93 Nm)

20. Install or connect the following:
• Joint shaft. Install a new circlip with the opening facing up. Tighten the bolts to 32–45 ft. lbs. (43–61 Nm).
• Halfshafts. Install new circlips on the inner CV-joint stub shafts, if equipped, and the intermediate shaft. Grease the shaft splines before installing the halfshaft/inter-mediate shaft into the transaxle.
• Ball joints
• Front pipe
• Selector cable

21. Power steering pump
• A/C Compressor
• Drive belt

22. When connecting the accelerator cable, perform the following adjustment:
a. Measure the free play of the cable, it should be 0.04–0.11 inch (1–3mm), if not turn adjustment nut **A** until the speci-fication is reached.
b. Fully depress the accelerator pedal and check the throttle is wide open, if not adjust using nut **B**.

23. Install the remaining components in the reverse of removal.

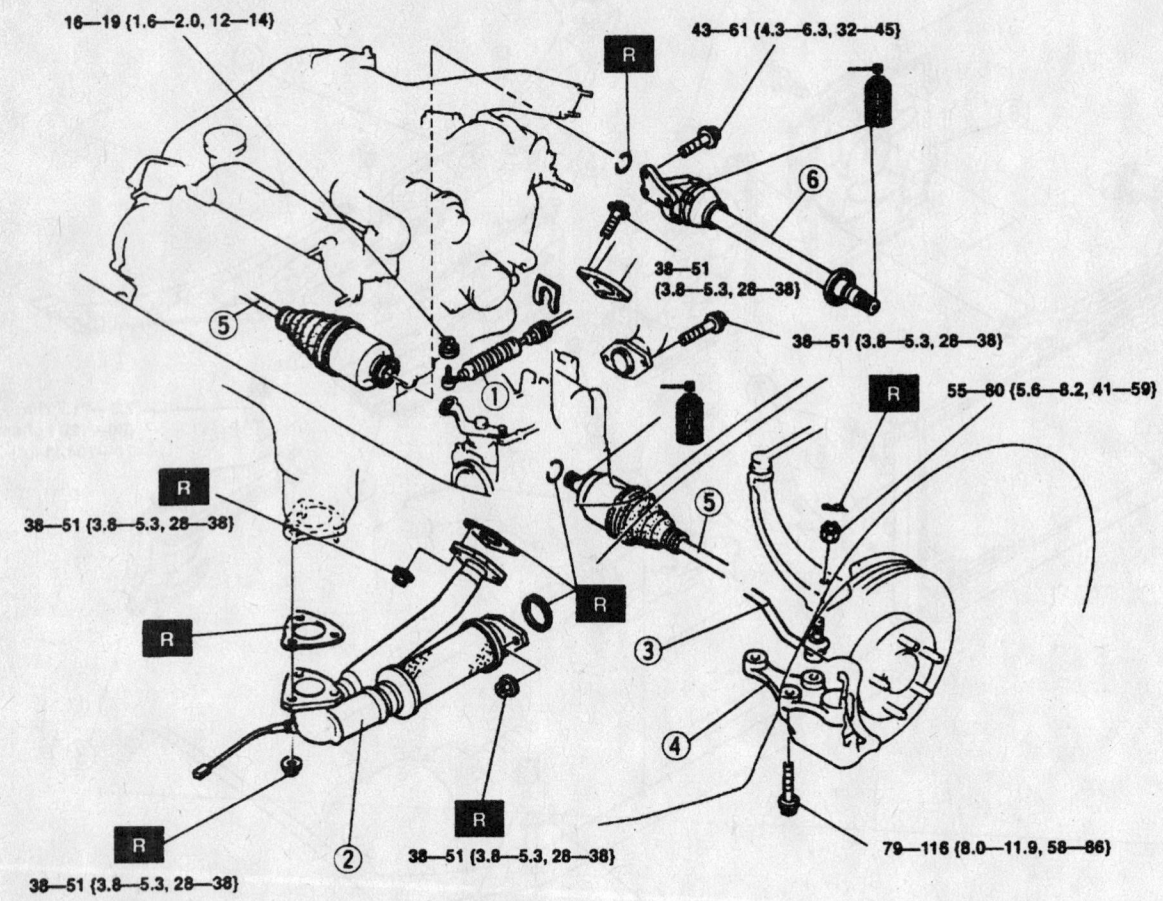

N·m {kgf·m, ft·lbf}

1	Selector cable	4	Lower ball joint
2	Front pipe	5	Drive shaft
3	Upper lateral link ball joint	6	Joint shaft

42356-MAZC-GDM

Location of the engine mounting components and their torque specifications (part 3 of 4)—Millenia models equipped with the 2.3L (KJ) engine

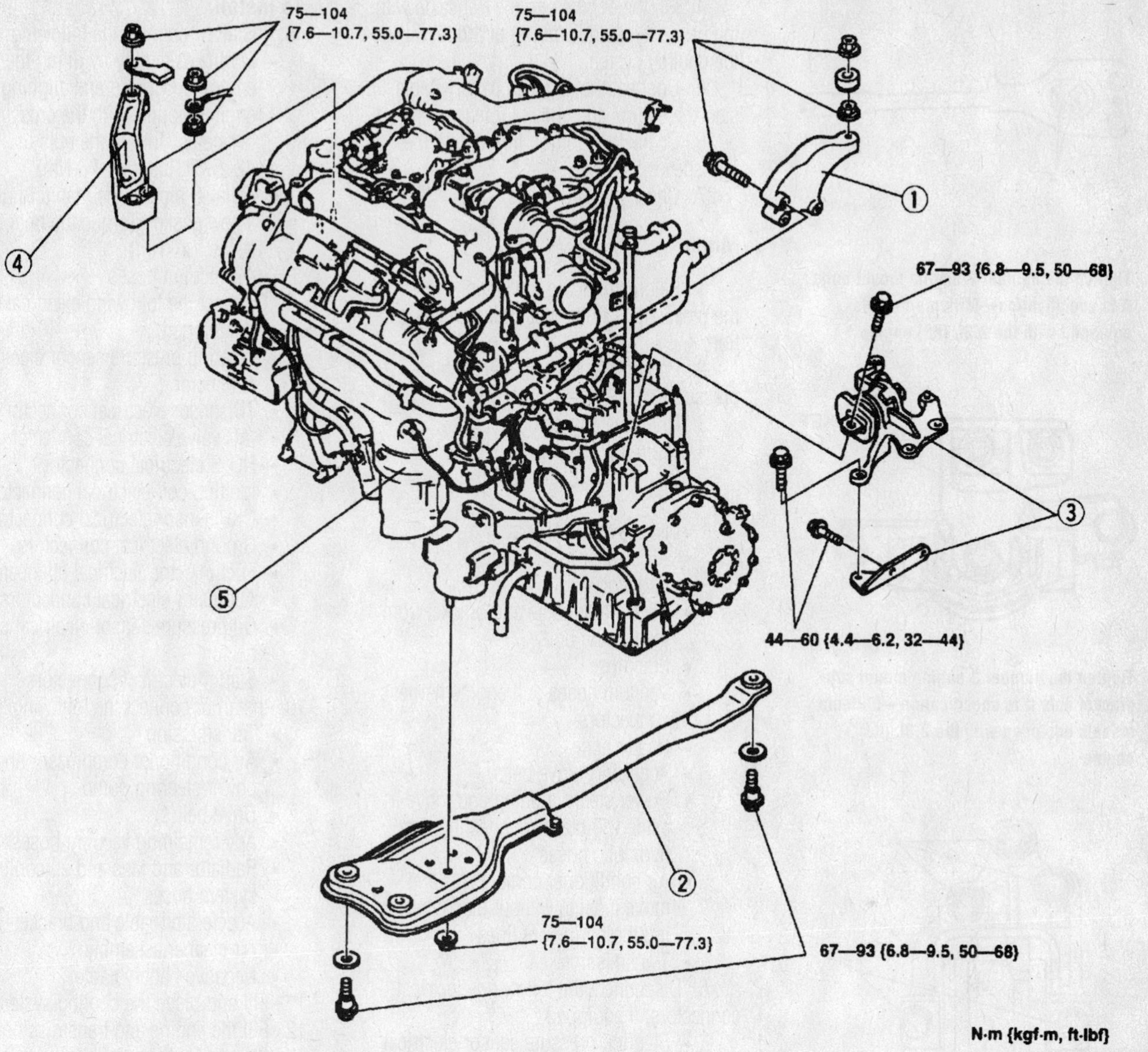

75—104
{7.6—10.7, 55.0—77.3}

75—104
{7.6—10.7, 55.0—77.3}

④

67—93 {6.8—9.5, 50—68}

③

44—60 {4.4—6.2, 32—44}

⑤

②

75—104
{7.6—10.7, 55.0—77.3}

67—93 {6.8—9.5, 50—68}

N·m {kgf·m, ft-lbf}

1	No.1 engine mount bracket	4	No.3 engine mount sub bracket
2	Engine mounting member	5	Engine and transaxle
3	No.4 engine mount bracket		

42356-MAZC-GDN

Location of the engine mounting components and their torque specifications (part 4 of 4)—Millenia models equipped with the 2.3L (KJ) engine

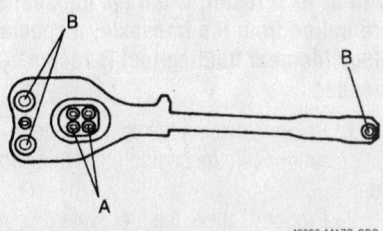

42356-MAZC-GDD

Remove the bolts A first and then the bolts B and remove the engine mount member—Millenia models equipped with the 2.3L (KJ) engine

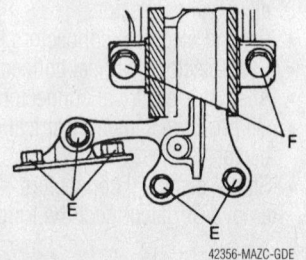

42356-MAZC-GDE

Tighten the number 4 engine bracket retainers E to specification and hand tighten bolts F—Millenia models equipped with the 2.3L (KJ) engine

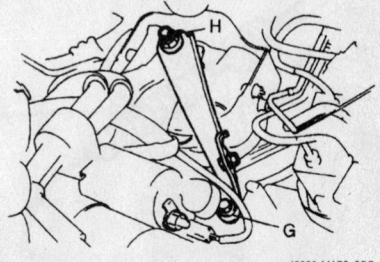

42356-MAZC-GDF

Tighten the number 1 engine mount stay retainers G and H to specification—Millenia models equipped with the 2.3L (KJ) engine

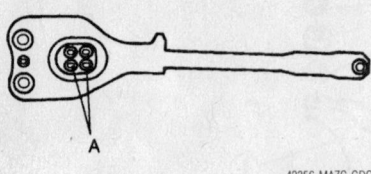

42356-MAZC-GDG

Tighten the number 2 engine mount bolts A to specification—Millenia models equipped with the 2.3L (KJ) engine

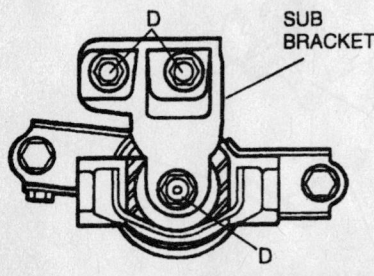

42356-MAZC-GDH

Tighten the number 3 engine mount sub bracket nuts D to specification—Millenia models equipped with the 2.3L (KJ) engine

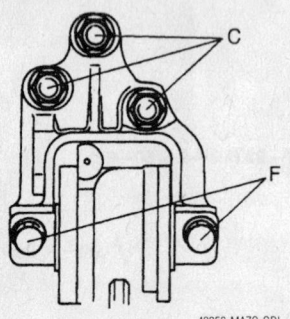

42356-MAZC-GDI

Tighten the number 4 engine mount nuts and bolts C and F to specification—Millenia models equipped with the 2.3L (KJ) engine

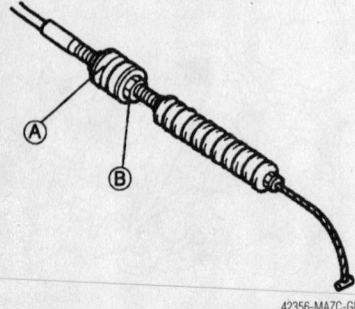

42356-MAZC-GDJ

Accelerator cable A and throttle cable B adjustment nuts—Millenia models equipped with the 2.3L (KJ) engine

24. Fill the engine and the transaxle with the proper type and amount of fluids. Fill the cooling system.

25. Connect the negative battery cable, start the engine and check for leaks.

26. Check the ignition timing and the idle speed.

27. Check all fluid levels.

Miata

1. Before servicing the vehicle, refer to the precautions in the beginning of this section.

2. Properly relieve the fuel system pressure.

3. Drain the engine oil.

4. Drain the cooling system.

5. Remove or disconnect the following:

- Negative battery cable
- Air cleaner assembly
- Radiator
- Accelerator cable and bracket from the throttle body
- Fuel hose
- Vacuum hoses and engine harness connectors
- Heater hose
- Accessory drive belts
- Power steering pump and move it aside without disconnecting the hydraulic hoses
- Air conditioner compressor and move it aside without disconnecting the refrigerant lines
- Transmission

6. Disconnect the following electrical connectors, if equipped:

- Steering pressure sensor electrical connector
- Throttle Position (TP) sensor electrical connector
- Idle Air Control (IAC) valve electrical connector
- Heated Oxygen (HO$_2$S) sensor electrical connector
- Ignition coil electrical connectors
- Crankshaft Position (CKP) sensor electrical connector
- Ground electrical connectors
- Fuel injector electrical connectors
- Alternator electrical connectors
- Oil pressure sensor electrical connector
- Starter electrical connectors

7. Remove or disconnect the following:

- exhaust pipe from the exhaust manifold and install suitable lifting equipment onto the engine.
- Engine from the vehicle

To install:

8. Install or connect the following:

- Engine assembly by tilting the engine downward and aligning the engine mounts with the crossmember holes. Torque the nuts to 42–57 ft. lbs. (57–78 Nm).
- Exhaust pipe to the manifold using a new gasket. Torque the nuts to 34 ft. lbs. (46 Nm).
- All vacuum hoses

9. Connect the following electrical connectors, if equipped:

- Steering pressure sensor electrical connector
- TP sensor electrical connector
- IAC valve electrical connector
- HO$_2$S electrical connector
- Ignition coil electrical connectors
- CKP sensor electrical connector
- Ground electrical connectors
- Fuel injector electrical connectors
- Alternator electrical connectors
- Oil pressure sensor electrical connector
- Starter electrical connectors

10. Install or connect the following:

- Transmission
- Air conditioner compressor and power steering pump
- Drive belt(s)
- Any remaining vacuum hoses
- Radiator and fans and all cooling system hoses
- Accelerator cable and bracket
- Air cleaner assembly
- Negative battery cable

11. Fill and bleed the cooling system.

12. Fill the engine and transmission.

13. Start the engine and check for leaks.

14. Check the ignition timing and idle speed.

Protégé

2.0L (FS) ENGINE

➡ The procedure for pulling the engine requires removing the transaxle along with it. As a result, when the halfshafts are pulled from the transaxle, a special plug/side gear holding tool is recommended.

1. Before servicing the vehicle, refer to the precautions in the beginning of this section.

2. Properly relieve the fuel system pressure.

3. Drain the engine oil.

4. Drain the transaxle fluid.

5. Drain the cooling system.

6. Remove or disconnect the following:

- Negative battery cable
- Radiator
- Air cleaner assembly
- Accelerator cable
- Fuel hoses
- Front exhaust pipe
- Any rods, pipes or cables related to the transaxle that would hinder removal
- Battery
- Fuse box
- Power steering pump with the lines still attached and set aside

- A/C compressor with the lines still attached and position aside
- Halfshafts
- Roll dampener

7. Support the engine with a suitable engine support device.

- Engine mount member retainers. Refer to the accompanying illustration for location.
- Number 3 engine mount. Refer to the accompanying illustration for location.
- Number 1 engine mount. Refer to

the accompanying illustration for location.
- Number 2 engine mount. Refer to the accompanying illustration for location.
- Number 4 engine mount. Refer to the accompanying illustration for location.
- Engine and transaxle assembly from the vehicle

To install:

8. Installation is the reverse of removal. Tighten the fasteners to the spec-

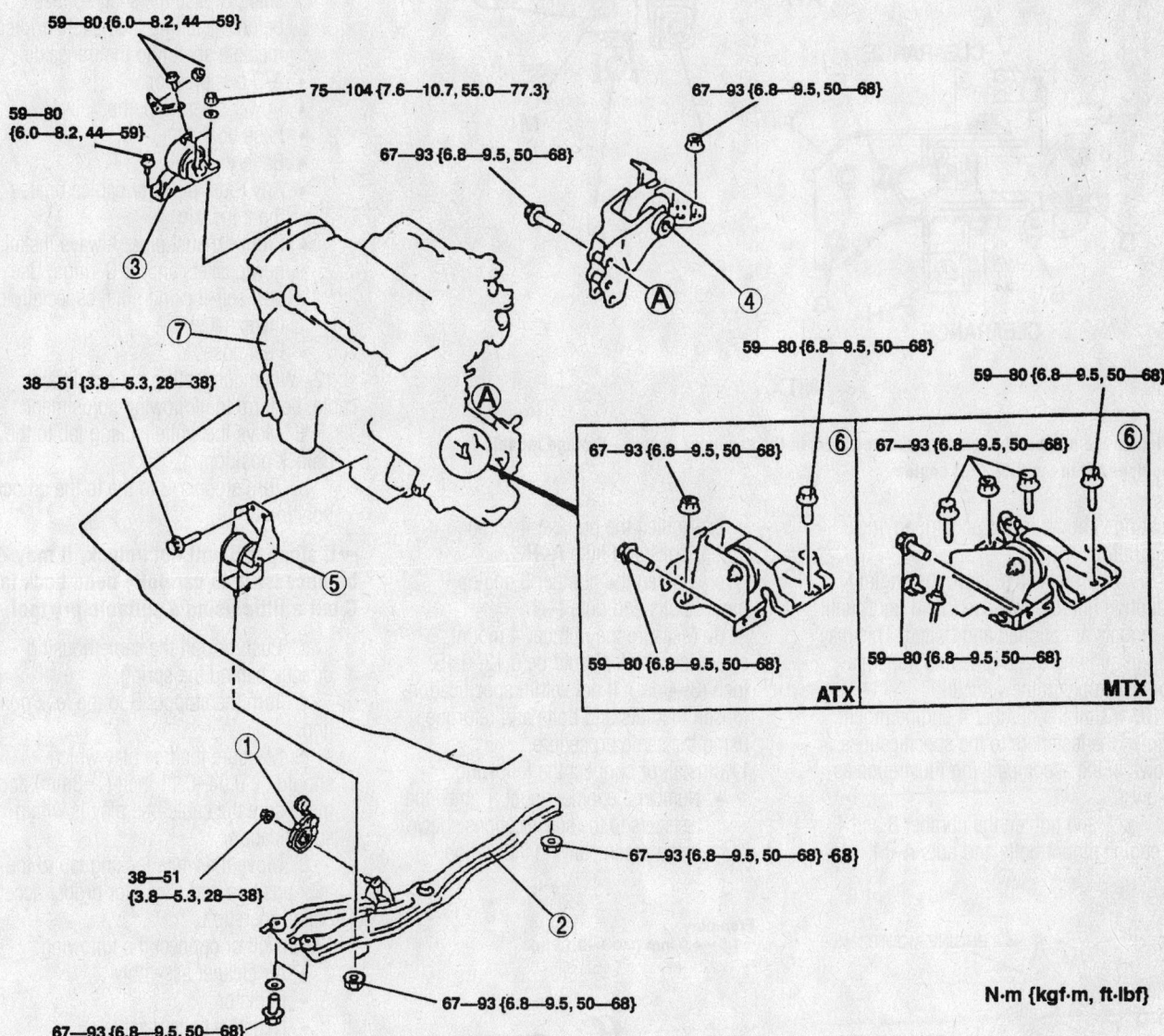

1 Roll damper
2 Engine mount member
3 No.3 Engine mount
4 No.1 Engine mount
5 No.2 Engine mount
6 No.4 Engine mount
7 Engine, transaxle

Location of the engine mounting components and their torque specifications—Protégé models equipped with the 2.0L (FS) engine

42356-MAZC-GGA

No.4 ENGINE MOUNT

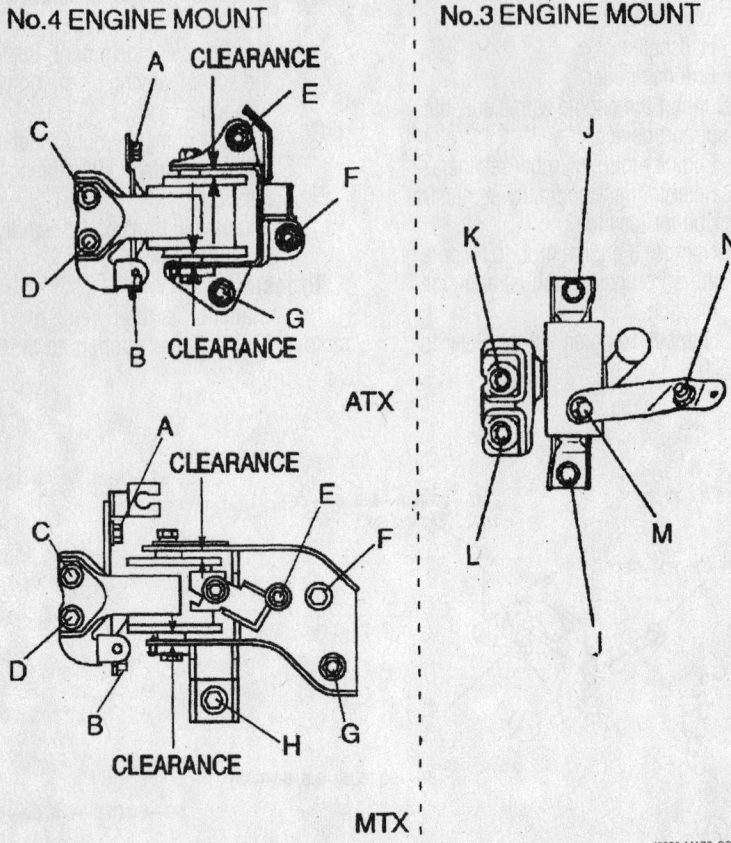

No.3 ENGINE MOUNT

ATX

MTX

42356-MAZC-GGB

Tighten the number 4 engine bracket retainers in the sequence shown—Protégé models equipped with the 2.0L (FS) engine

ifications shown in the accompanying illustration.

9. When possible, leave the engine mounting nuts/bolts loose (hand tight) until all mounts are aligned and bolted. This may help in aligning the engine and transmission assembly in the vehicle.

10. Install the number 4 engine mount. Tighten the fasteners to the specifications shown in the accompanying illustration as follows:

   a. Hand tighten the number 3 and 4 engine mount bolts and nuts **A–M**.

   b. Tighten the number 4 engine mount bolts and nuts **A–H**.

   c. Tighten the number 3 engine mount bolts and nuts **I–N**.

   d. Measure the number 4 mount clearance which should be 0.12–0.15 inch (3–4mm). If not within specification, loosen the nuts and bolts and retorque using the same procedure.

11. Install or connect the following:
   • Number 1 engine mount. Tighten the fasteners to the specifications shown in the accompanying illustration.

   • Number 3 engine mount, refer to the number 4 mount tightening sequence. Tighten the fasteners to the specifications shown in the accompanying illustration.
   • Engine mount member. Tighten the fasteners to the specifications shown in the accompanying illustration.
   • Roll damper, refer to the illustration or proper assembly.
   • Halfshafts. Install new circlips on the inner CV-joint stub shafts, if equipped, and the intermediate shaft. Grease the shaft splines before installing the halfshaft/intermediate shaft into the transaxle.
   • A/C compressor
   • Power steering pump
   • Fuse box
   • Battery
   • Any rods, pipes or cables related to the transaxle
   • Front exhaust pipe. Always install new gaskets and/or O-rings. Use new self-locking nuts, especially on the exhaust.
   • Fuel hoses

12. When connecting the accelerator cable, perform the following adjustment:
   a. Move the white locking tab to the unlock position.
   b. Turn stopper B to the to the unlock position.

➡ **If stopper B will not unlock, it may be necessary to carefully bend back tab C out a little using a suitable pry tool.**

   c. Push or pull the cable housing directly behind the spring.
   d. Turn the stopper B to the lock position.
   e. Measure the free play which should be 0.04–0.11 inch (1–3mm) and make sure the cable free play is within specification.
   f. Move the white locking tab to the lock position and check for proper accelerator operation.

13. Install or connect the following:
   • Air cleaner assembly
   • Radiator
   • Negative battery cable

14. Fill the engine and the transaxle with the proper type and amount of fluids. Fill the cooling system.

15. Connect the negative battery cable, start the engine and check for leaks.

16. Check the ignition timing and the idle speed.

17. Check all fluid levels.

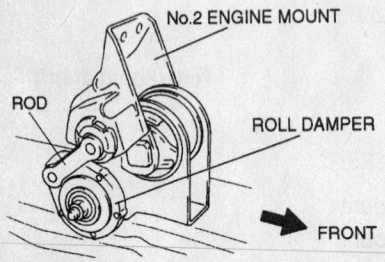

42356-MAZC-GGC

Assemble the roll damper as illustrated— Protégé models equipped with the 2.0L (FS) engine

**Free play
1.5—4.0 mm {0.06—0.15 in}**

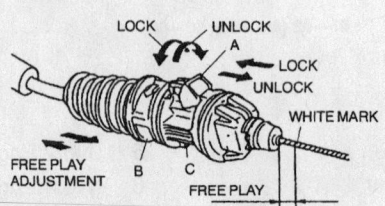

42356-MAZC-G04

Accelerator cable adjustment components—Protégé models equipped with the 2.0L (FS) engine

1.6L (ZM) ENGINE

➡The procedure for pulling the engine requires removing the transaxle along with it. As a result, when the halfshafts are pulled from the transaxle, a special plug/side gear holding tool is recommended.

1. Before servicing the vehicle, refer to the precautions in the beginning of this section.

2. Properly relieve the fuel system pressure.
3. Drain the engine oil.
4. Drain the transaxle fluid.
5. Drain the cooling system.
6. Remove or disconnect the following:

- Battery
- Air cleaner, air hose and resonance chamber
- Front exhaust pipe
- Accelerator cable and bracket
- Heater and vacuum hoses
- Radiator
- Drive belt
- Fuel hoses
- Any rods, pipes or cables related to the transaxle that would hinder removal
- Halfshafts
- Power steering pump with the lines still attached and set aside

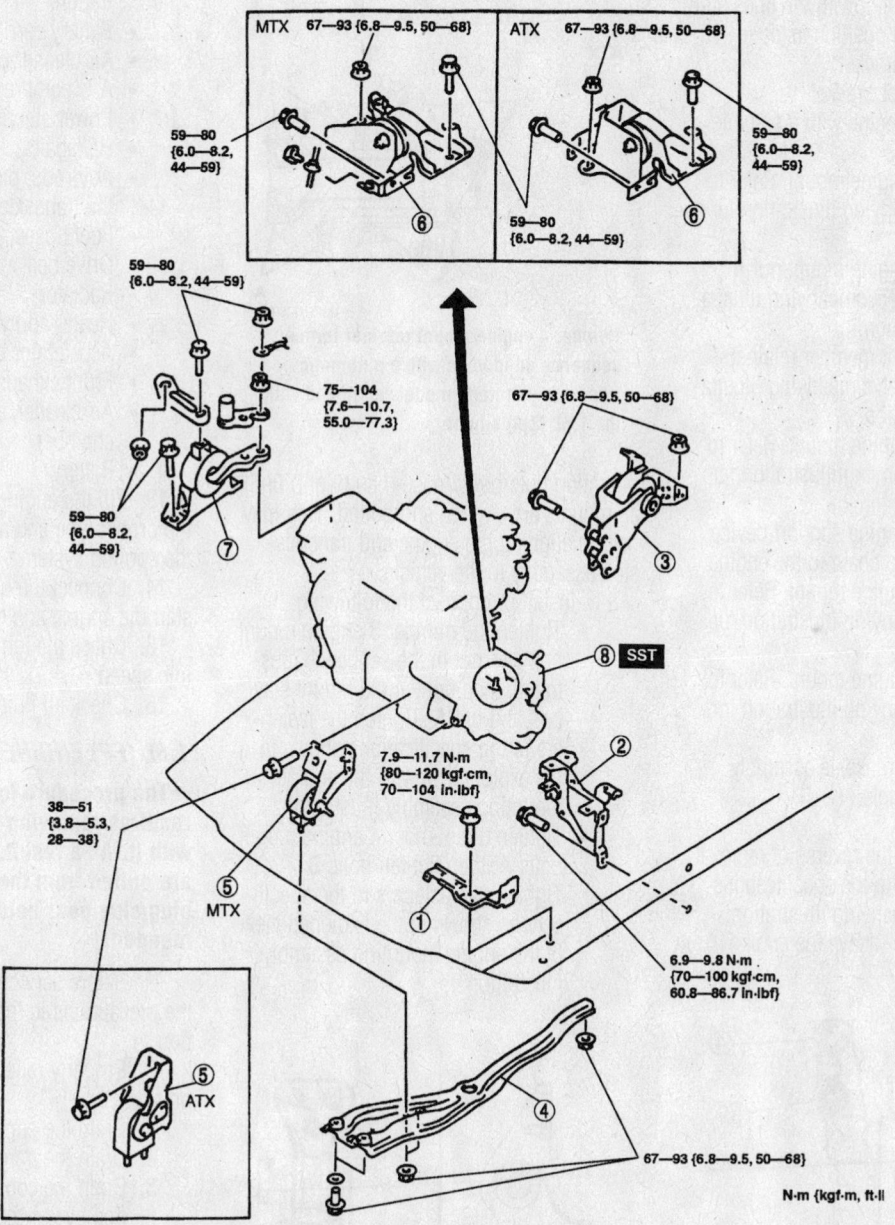

1	Air cleaner bracket	5	No.2 engine mount
2	Battery carrier bracket	6	No.4 engine mount
3	No.1 engine mount	7	No.3 engine mount
4	Engine mount member	8	Engine, transaxle

42356-MAZC-GHD

Location of the engine mounting components and their torque specifications—Protégé models equipped with the 1.6L (ZM) engine

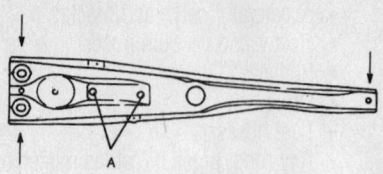

42356-MAZC-GHE

Remove the number 2 engine mount nut retainers A, then the engine mount retainers—Protégé models equipped with the 1.6L (ZM) engine

- A/C compressor with the lines still attached and position aside
- Air cleaner bracket
- Battery carrier bracket

7. Support the engine with a suitable engine support device.
- Number 1 engine mount. Refer to the accompanying illustration for location.
- Number 2 engine mount nut **A**. Refer to the accompanying illustration for location.
- Engine mount member retainers. Refer to the accompanying illustration for location.
- Number 2 engine mount. Refer to the accompanying illustration for location.

8. Remove the engine support device and attach a hoist and chain to the engine.
- Number 4 engine mount. Refer to the accompanying illustration for location.
- Number 3 engine mount. Refer to the accompanying illustration for location.
- Engine and transaxle assembly from the vehicle.

To install:

9. Installation is the reverse of removal. Tighten the fasteners to the specifications shown in the accompanying illustration.

10. When possible, leave the engine

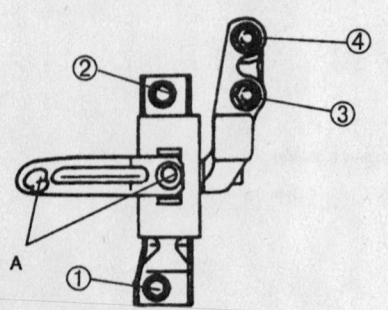

42356-MAZC-GHF

Tighten the number 3 engine mount bolt and nut in this sequence and the mount stay bolt A—Protégé models equipped with the 1.6L (ZM) engine

MTX

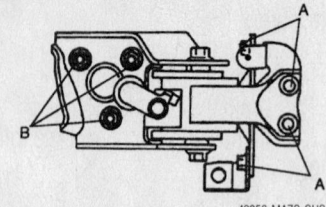

42356-MAZC-GHG

Number 4 engine mount retainer torque sequence on models with a manual transaxle—Protégé models equipped with the 1.6L (ZM) engine

ATX

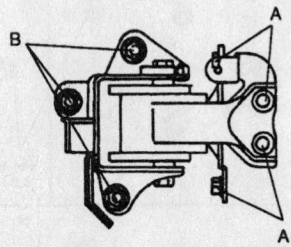

42356-MAZC-GHH

Number 4 engine mount retainer torque sequence on models with a automatic transaxle—Protégé models equipped with the 1.6L (ZM) engine

mounting nuts/bolts loose (hand tight) until all mounts are aligned and bolted. This may help in aligning the engine and transmission assembly in the vehicle.

11. Install or connect the following:
- Tighten the number 3 engine mount bolt and nut in the sequence illustrated, then tighten the mount stay bolt and nut **A**. Tighten the fasteners to the specifications shown in the exploded view of the engine mounting assembly illustration.
- Tighten the number 4 engine mount bolts **A**, then tighten bolts **B**. Tighten the fasteners to the specifications shown in the exploded view of the engine mounting assembly illustration.

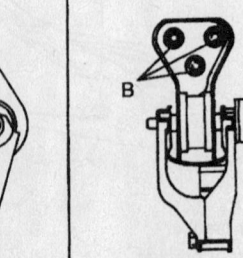

42356-MAZC-GHI

Number 1 engine mount retainer torque sequence —Protégé models equipped with the 1.6L (ZM) engine

12. Remove the hoist and chain and install the engine support device.
- Number 2 engine mount nut **A**, then tighten the mount member bolt and nut. Tighten the fasteners to the specifications shown in the exploded view of the engine mounting assembly illustration.
- Number 1 engine mount bolts **A** and then the bolt **B**. Tighten the fasteners to the specifications shown in the accompanying illustration.
- Battery carrier bracket
- Air cleaner bracket
- A/C compressor
- Power steering pump
- Halfshafts
- Any rods, pipes or cables related to the transaxle that were removal
- Fuel hoses
- Drive belt
- Radiator
- Heater and vacuum hoses
- Accelerator cable and bracket
- Front exhaust pipe
- Air cleaner, air hose and resonance chamber
- Battery

13. Fill the engine and the transaxle with the proper type and amount of fluids. Fill the cooling system.

14. Connect the negative battery cable, start the engine and check for leaks.

15. Check the ignition timing and the idle speed.

16. Check all fluid levels.

1.8L (FP) ENGINE

➡The procedure for pulling the engine requires removing the transaxle along with it. As a result, when the halfshafts are pulled from the transaxle, a special plug/side gear holding tool is recommended.

1. Before servicing the vehicle, refer to the precautions in the beginning of this section.

2. Properly relieve the fuel system pressure.

3. Drain the engine oil.

4. Drain the transaxle fluid.

5. Drain the cooling system.

6. Remove or disconnect the following:
- Negative battery cable
- Radiator
- Air cleaner
- Accelerator cable
- Fuel hoses
- Front exhaust pipe
- Any rods, pipes or cables related to

the transaxle that would hinder removal
- Battery
- Fuse box
- Drive belt
- Power steering pump with the lines still attached and set aside
- A/C compressor with the lines still attached and position aside
- Halfshafts

7. Support the engine with a suitable engine support device.
- Engine mount member retainers
- Number 3 engine mount. Refer to the accompanying illustration for location.
- Number 4 engine mount. Refer to the accompanying illustration for location.
- Number 1 engine mount. Refer to the accompanying illustration for location.
- Number 2 engine mount. Refer to the accompanying illustration for location.

8. Remove the engine support device and attach a hoist and chain to the engine.
- Engine and transaxle assembly from the vehicle.

To install:

9. Installation is the reverse of removal. Tighten the fasteners to the specifications shown in the accompanying illustration.

10. When possible, leave the engine mounting nuts/bolts loose (hand tight) until all mounts are aligned and bolted. This may help in aligning the engine and transmission assembly in the vehicle.

11. Install or connect the following:
- Engine and transaxle assembly

- Number 2 engine mount . Tighten the fasteners to the specifications shown in the accompanying illustration.
- Number 1 engine mount. Tighten the fasteners to the specifications shown in the accompanying illustration.
- Hand tighten the number 3 and 4 engine mount bolts **A–M**, tighten number 4 bolts and nuts **A–G**, then tighten the number 3 mount bolts and nuts **H–M**. Tighten the fasteners to the specifications shown in the exploded view of the engine mounting assembly illustration. Measure the number 4 mount clearance, if not 0.12–0.15 inch (3–4mm), repeat the tightening sequence.

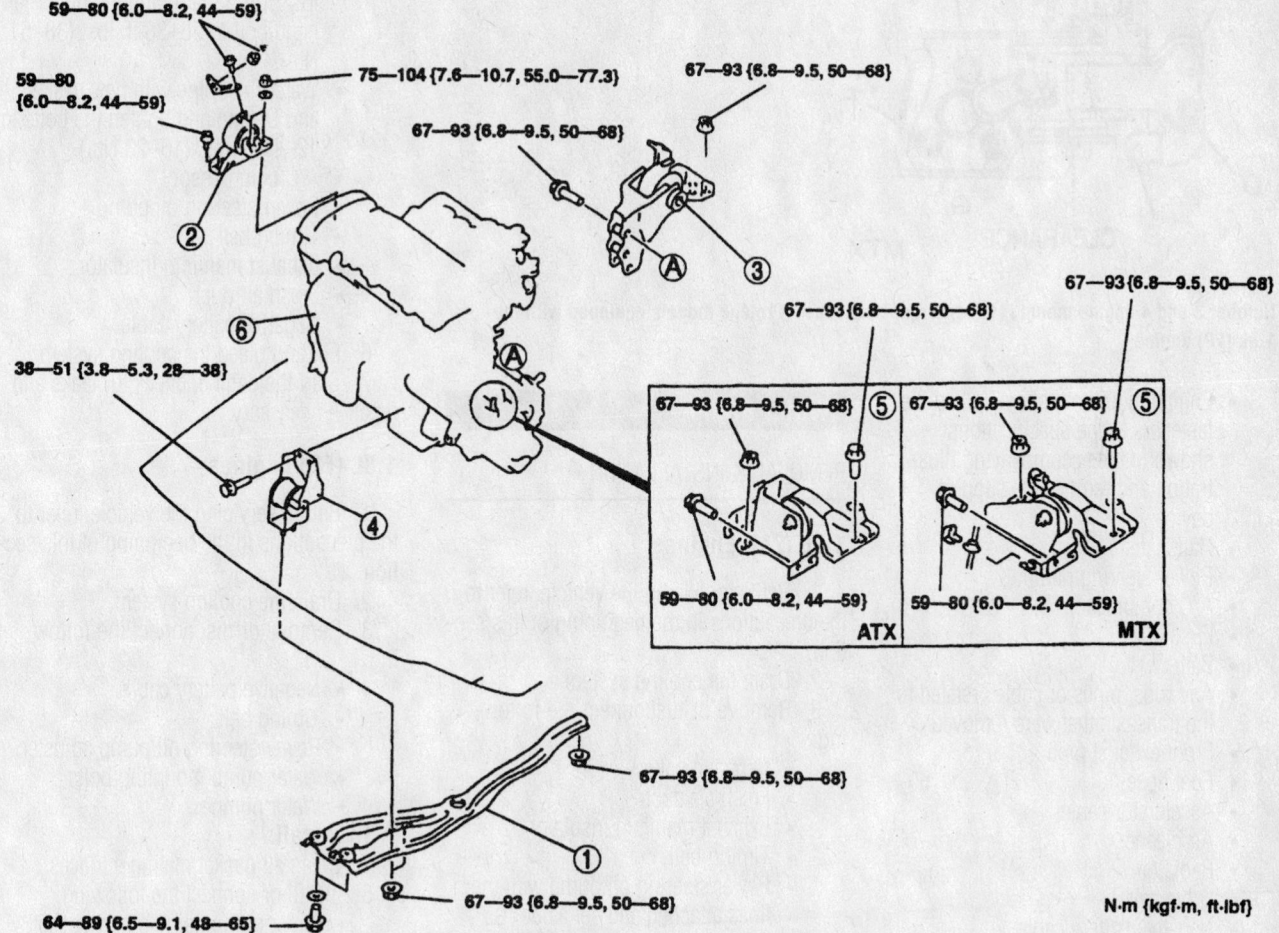

1	Engine mount member
2	No.3, No.4 Engine mount
3	No.1 Engine mount
4	No.2 Engine mount
5	Engine and transaxle

Location of the engine mounting components and their torque specifications—Protégé models equipped with the 1.8L (FP) engine

42356-MAZC-GJA

No.4 ENGINE MOUNT

ATX

No.3 ENGINE MOUNT

MTX

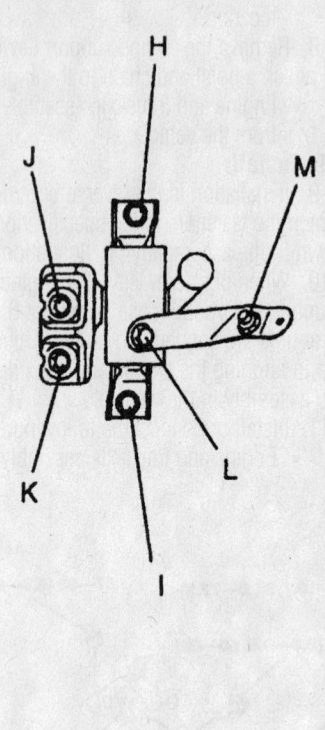

Number 3 and 4 engine mount retainer torque sequence—Protégé models equipped with the 1.8L (FP) engine

- Engine mount member. Tighten the fasteners to the specifications shown in the accompanying illustration and remove the support device.
- Halfshafts
- Power steering pump
- A/C compressor
- Fuse box
- Battery
- Any rods, pipes or cables related to the transaxle that were removed
- Front exhaust pipe
- Fuel hoses
- Accelerator cable
- Air cleaner
- Radiator
- Drive belt
- Negative battery cable

12. Fill the engine and the transaxle with the proper type and amount of fluids. Fill the cooling system.

13. Connect the negative battery cable, start the engine and check for leaks.

14. Check the ignition timing and the idle speed.

15. Check all fluid levels.

Water Pump

REMOVAL & INSTALLATION

1.6L (ZM) Engines

1. Before servicing the vehicle, refer to the precautions in the beginning of this section.

2. Drain the cooling system.

3. Remove or disconnect the following:

- Negative battery cable
- Fresh air duct
- Exhaust manifold insulator
- Timing belt
- Power steering oil pump with the lines attached and set aside
- A/C compressor with the lines attached and set aside
- Water inlet pipe
- Water pump mounting bolts
- Water pump

To install:

4. Clean all gasket mating surfaces.

5. Install or connect the following:

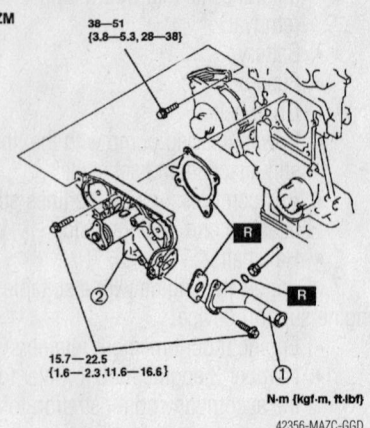

ZM

38—51
{3.8—5.3, 28—38}

15.7—22.5
{1.6—2.3, 11.6—16.6}

N·m {kgf·m, ft·lbf}

42356-MAZC-GGD

Exploded view of the water pump assembly—1.6L (ZM) engines

- Water pump using a new gasket. Torque the main bolts to 12–17 ft. lbs. (16–23 Nm) and the upper left hand bolt to 28–38 ft. lbs. (38–51 Nm).
- Water inlet pipe with new gasket and O–ring and tighten the bolts to 12–17 ft. lbs. (16–23 Nm).
- A/C compressor
- Power steering oil pump
- Timing belt
- Exhaust manifold insulator
- Fresh air duct
- Negative battery cable

6. Fill and bleed the cooling system.

7. Start the engine, check for leaks and repair if necessary.

1.8L (FP) Engines

1. Before servicing the vehicle, refer to the precautions in the beginning of this section.

2. Drain the cooling system.

3. Remove or disconnect the following:

- Negative battery cable
- Timing belt
- Power steering oil pump adjuster
- Water pump mounting bolts
- Water pump

To install:

4. Clean all gasket mating surfaces.

5. Install or connect the following:

- Water pump using a new gasket. Make sure the gasket is installed with the sealing ring facing the pump. Torque the bolts to 14–18 ft. lbs. (19–25 Nm).
- Power steering oil pump adjuster
- Timing belt
- Negative battery cable

6. Fill and bleed the cooling system.

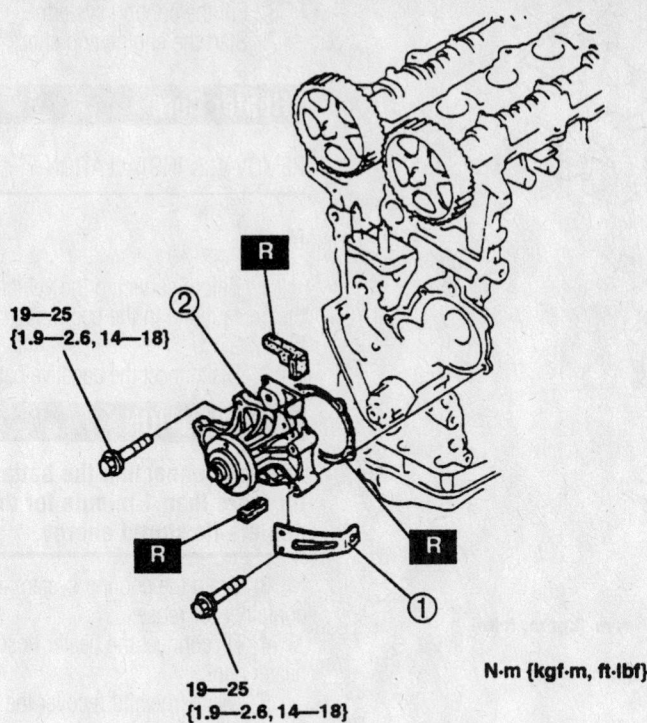

19—25
{1.9—2.6, 14—18}

19—25
{1.9—2.6, 14—18}

N·m {kgf·m, ft·lbf}

1   P/S oil pump adjuster
2   Water pump

42356-MAZC-GJC

Exploded view of the water pump assembly—1.8L (FP) engines

7. Start the engine, check for leaks and repair if necessary.

1.8L (BP) Engines

1. Before servicing the vehicle, refer to the precautions in the beginning of this section.
2. Disconnect the negative battery cable.
3. Drain the engine coolant.
4. Remove or disconnect the following:
   • Timing belt covers and timing belt

   • Power steering pump, leaving the lines attached and set aside
   • Idler, on models not equipped with power steering
   • Coolant inlet pipe and gasket
   • Water pump

To install:

5. Clean all gasket mating surfaces.
6. Using a new gasket, install the water pump on the engine. Torque the mounting

bolts to 14–18 ft. lbs. (19–25 Nm). Torque the bolt from the water pump to the alternator bracket to 28–38 ft. lbs. (38–51 Nm).
7. Install or connect the following:
   • Coolant inlet pipe with a new gasket. Torque the coolant inlet pipe bolts to 14–18 ft. lbs. (19–25 Nm).
   • Idler, on models not equipped with power steering
   • Power steering pump
   • Timing belt and the timing belt covers
   • Negative battery cable
8. Fill and bleed the cooling system.
9. Start the engine and bring to normal operating temperature. Check for leaks.

2.0L (FS) Engines

1. Before servicing the vehicle, refer to the precautions in the beginning of this section.
2. Drain the cooling system.
3. Remove or disconnect the following:
   • Negative battery cable
   • Timing belt
   • Power steering oil pump adjuster
   • Water pump mounting bolts
   • Water pump

To install:

4. Clean all gasket mating surfaces.
5. Install or connect the following:
   • Water pump using a new gasket. Torque the bolts to 14–18 ft. lbs. (19–25 Nm).
   • Power steering oil pump adjuster. Torque the bolts to 12–16 ft. lbs. (16–22 Nm).
   • Timing belt
   • Negative battery cable
6. Fill and bleed the cooling system.
7. Start the engine, check for leaks and repair if necessary.

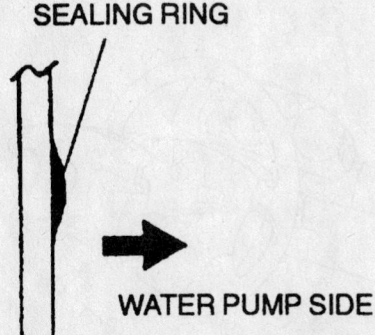

SEALING RING

WATER PUMP SIDE

42356-MAZC-GJD

Make sure the gasket is installed with the sealing ring facing the pump—1.8L (FP) engines

1. P/S oil pump
2. Idler (without P/S oil pump)
3. Water hose
4. Water pump
5. Water inlet pipe

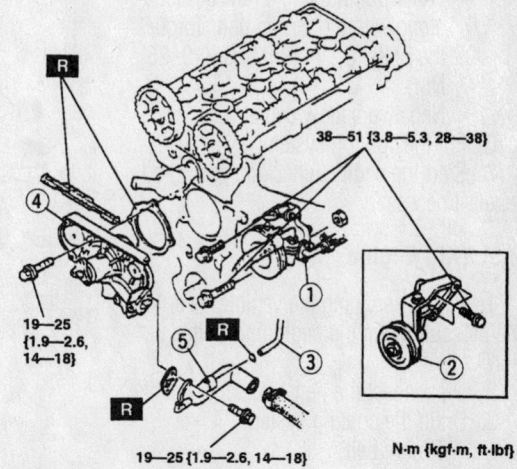

38—51 {3.8—5.3, 28—38}

19—25
{1.9—2.6, 14—18}

19—25 {1.9—2.6, 14—18}

N·m {kgf·m, ft·lbf}

42356-MAZC-G94

Exploded view of the water pump assembly—1.8L (BP) engines

FS

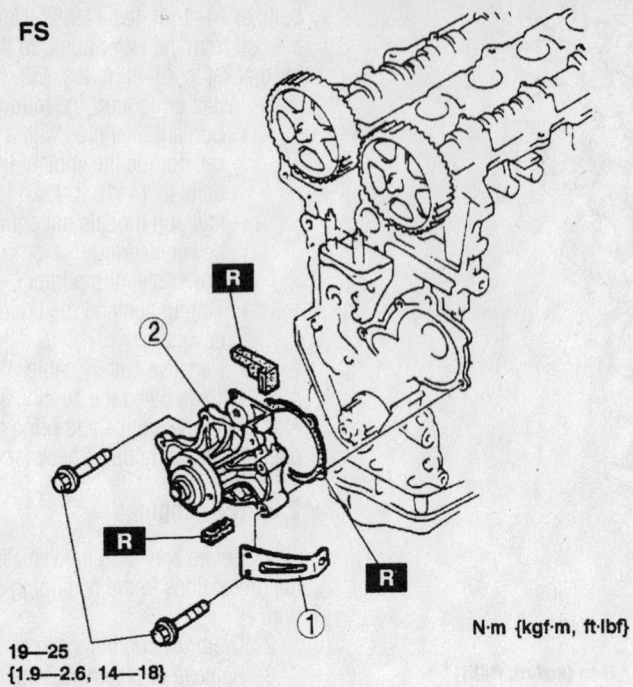

1   P/S oil pump adjuster

2   Water pump

N·m {kgf·m, ft·lbf}

19—25
{1.9—2.6, 14—18}

42356-MAZC-G5A

Exploded view of the water pump assembly—2.0L (FS) engines

2.3L (KJ) Engine

1. Before servicing the vehicle, refer to the precautions in the beginning of this section.
2. Drain the cooling system.
3. Remove or disconnect the following:
   - Negative battery cable
   - Timing belt and water pump together

To install:

4. Clean the mating surfaces.
5. Install or connect the following:
   - Water pump using a new gasket along with the timing belt. Torque the bolts to 14—18 ft. lbs. (19—25 Nm).
   - Negative battery cable
6. Fill the cooling system.
7. Start the engine, check for leaks and repair if necessary.

2.5L (KL) Engine

1. Before servicing the vehicle, refer to the precautions in the beginning of this section.
2. Disconnect the negative battery cable.
3. Drain the cooling system.
   - Timing belt
   - Number 3 engine mount bracket
   - Water pump bolts

- Water pump

To install:

4. Clean the mating surfaces.
5. Install or connect the following:
   - Water pump using a new gasket. Tighten the bolts to 14—18 ft. lbs. (19—25 Nm).
   - Number 3 engine mount bracket and tighten the bolts to 32—44 ft. lbs. (44—60 Nm).
   - Timing belt
   - Negative battery cable

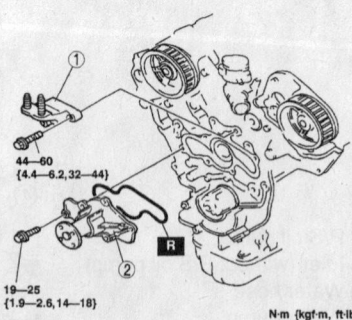

44—60
{4.4—6.2, 32—44}

19—25
{1.9—2.6,14—18}

N·m {kgf·m, ft·lbf}

1   No.3 engine mount bracket
2   Water pump

42356-MAZC-G06

Exploded view of the water pump assembly—2.5L (KL) engines

6. Fill the cooling system.
7. Start the engine and check for leaks.

Heater Core

REMOVAL & INSTALLATION

Miata

1. Before servicing the vehicle, refer to the precautions in the beginning of this section.
2. Disconnect the negative battery cable.

✱✱ CAUTION

After disconnecting the battery, wait for more than 1 minute for the SAS to deplete its stored energy.

3. Drain the cooling system into a clean container for reuse.
4. Disconnect the heater hoses from the heater core.
5. Discharge and recover the air conditioning system refrigerant.
6. At the driver's side, remove the SAS module and the steering wheel by removing or disconnecting the following:
   - Place the wheel in the straight-ahead position and turn the ignition switch to LOCK
   - Cover clips at both sides of the steering wheel
   - Steering wheel-to-SAS module bolts
   - SAS module from the steering wheel and disconnect the electrical connector
   - Steering wheel-to-column nut
   - Steering wheel from the steering column using a suitable puller
7. At the passenger's side, remove the SAS module by removing or disconnecting the following:
   - Glove compartment and the glove compartment cover
   - SAS module-to-dash bolts

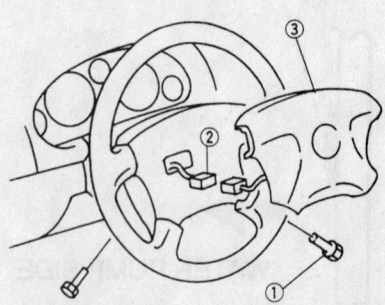

7.9—11.7 N·m {80—120 kgf·cm, 70—104 in·lbf}

93112GG0

View of the SAS module and the steering wheel—Miata

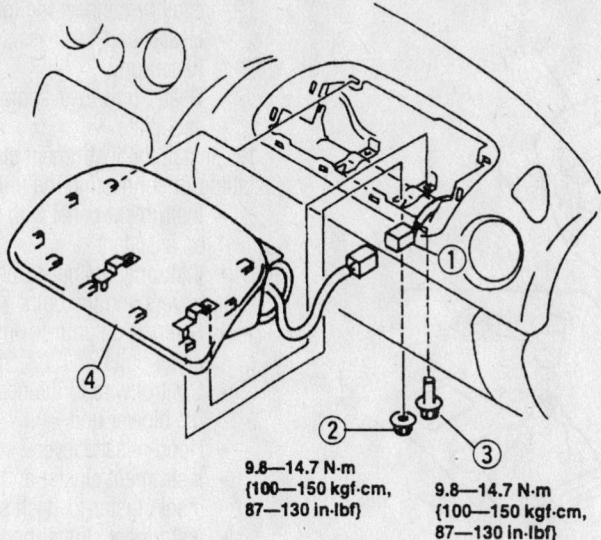

9.8—14.7 N·m
{100—150 kgf·cm,
87—130 in·lbf}

9.8—14.7 N·m
{100—150 kgf·cm,
87—130 in·lbf}

1   Connector
2   Nut
3   Bolt
4   Passenger-side air bag module

93112GH1

View of the passenger's side SAS module—Miata

- SAS module and disconnect the electrical connector
- Console

8. Remove the instrument cluster by removing or disconnecting the following:
- A-pillar trim at both sides
- Lower panel
- Instrument cluster hood
- Instrument cluster-to-dash screws and the instrument cluster
- Hood release lever
- Control wire from the heater unit and the blower unit
- Steering column-to-instrument panel bolts and lower the steering column
- Instrument panel-to-chassis bolt covers and the bolts
- Instrument panel with the help of an assistant

9. Remove or disconnect the following:
- Heater unit-to-evaporator housing seal plate
- Heater unit-to-chassis nuts and the heater unit

10. Separate the heater unit cases and remove the heater core.

To install:

11. Install the heater core and assemble the heater unit cases.

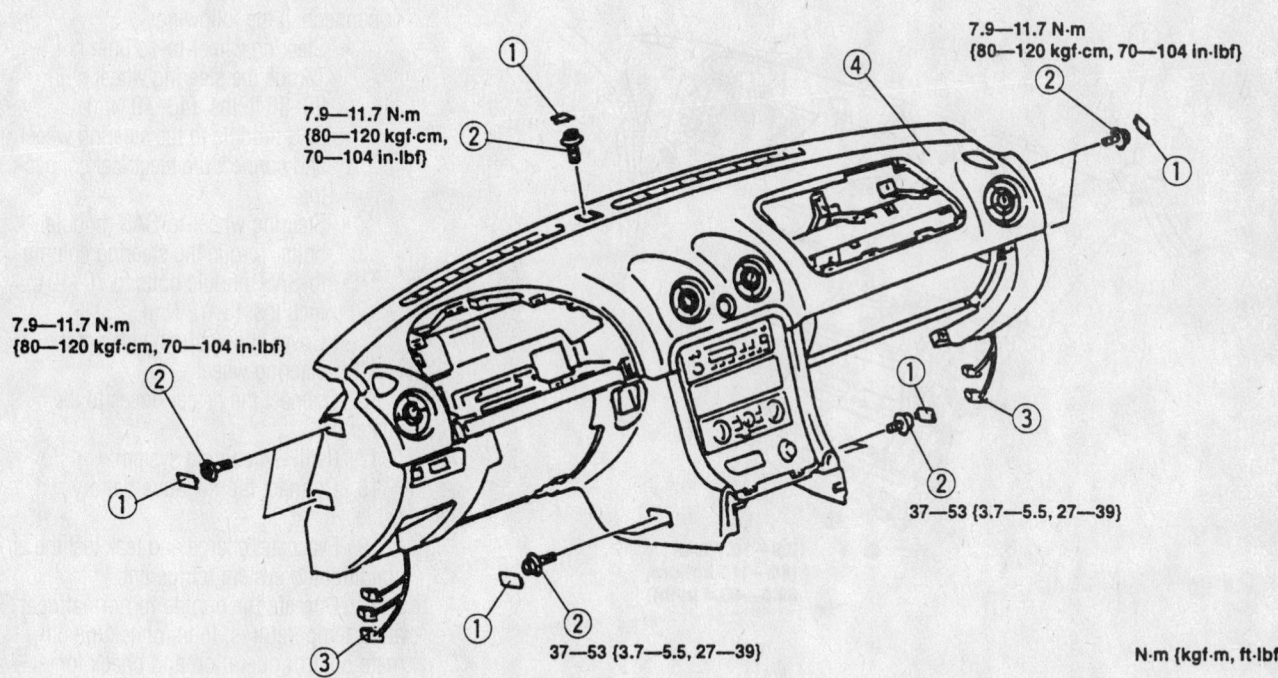

7.9—11.7 N·m
{80—120 kgf·cm, 70—104 in·lbf}

7.9—11.7 N·m
{80—120 kgf·cm,
70—104 in·lbf}

7.9—11.7 N·m
{80—120 kgf·cm, 70—104 in·lbf}

37—53 {3.7—5.5, 27—39}

37—53 {3.7—5.5, 27—39}

N·m {kgf·m, ft·lbf}

1   Cover
2   Bolt
3   Connector
4   Dashboard

93112GH2

View of the instrument panel—Miata

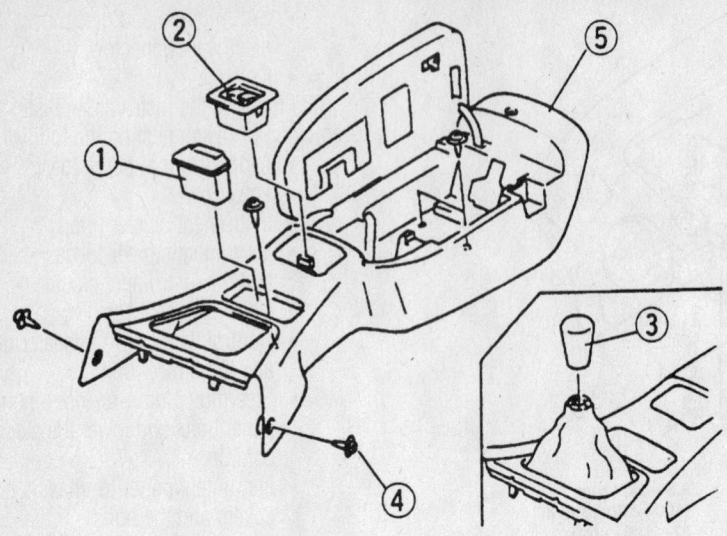

1　Ashtray

2　Power window switch

3　Shift lever knob (MT)

4　Screw

5　Console

93112GH3

View of the console—Miata

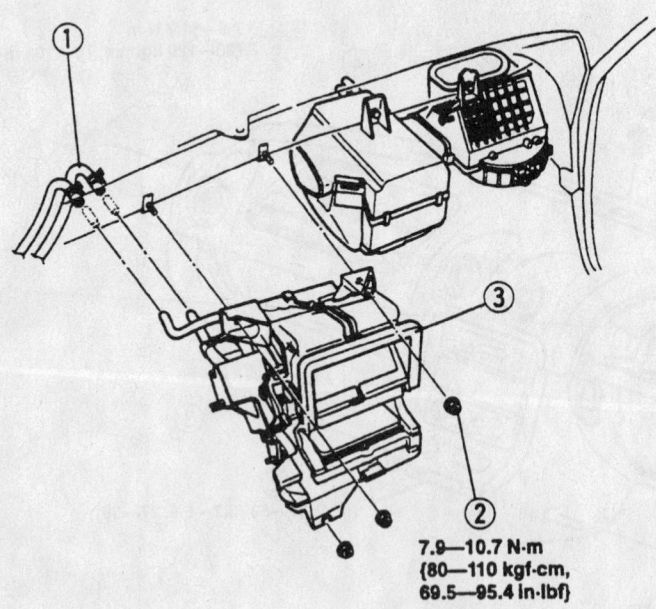

7.9—10.7 N·m
{80—110 kgf-cm,
69.5—95.4 in·lbf}

1　Heater hose

2　Nut

3　Heater unit

93112GH4

View of the heater unit and the heater unit-to-evaporator unit seal plate—Miata

12. Install or connect the following:
- Heater unit-to-chassis nuts and the heater unit
- Heater unit-to-evaporator housing seal plate

13. Install the instrument cluster by installing or connecting the following:
- Instrument panel with the help of an assistant
- Instrument panel-to-chassis bolt covers and the bolts
- Steering column-to-instrument panel bolts
- Control wire to the heater unit and the blower unit
- Hood release lever
- Instrument cluster and the instrument cluster-to-dash screws
- Instrument cluster hood
- Lower panel
- A-pillar trim at both sides
- Console

14. At the passenger's side, install the SAS module by installing or connecting the following:
- SAS module and connect the electrical connector
- SAS module-to-dash bolts
- Glove compartment cover and the glove compartment

15. At the driver's side, install the SAS module and the steering wheel by installing or connecting the following:
- Steering wheel-to-column nut. Torque the steering wheel nut to 29–36 ft. lbs. (40–49 Nm).
- SAS module to the steering wheel and connect the electrical connector
- Steering wheel-to-SAS module bolts. Torque the steering column-to-SAS module bolts to 70–104 inch lbs. (8–12 Nm).
- Cover clips at both sides of the steering wheel

16. Connect the heater hoses to the heater core.

17. Refill the cooling system.

18. Connect the negative battery cable.

19. Evacuate, charge and leak test the air conditioning system refrigerant.

20. Operate the engine to normal operating temperatures; then, check the climate control operation and check for leaks.

Millenia

1. Before servicing the vehicle, refer to the precautions in the beginning of this section.

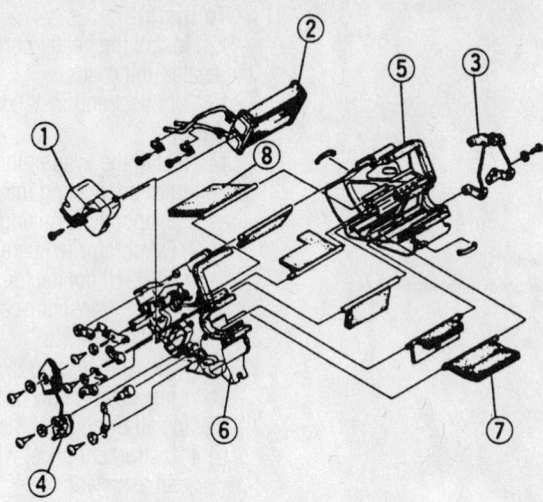

1   Cover
2   Heater core
3   Air mix link
4   Airflow mode link
5   Case (RH)
6   Case (LH)
7   Air mix door
8   Airflow mode door

93112GH5

Exploded view of the heater core and heater unit assembly—Miata

2. Disconnect the negative battery cable.

✲✲ CAUTION

After disconnecting the battery, wait for more than 1 minute for the SAS to deplete its stored energy.

3. Drain the cooling system into a clean container for reuse.

4. Discharge and recover the air conditioning system refrigerant.

5. At the driver's side, remove the SAS module and the steering wheel by removing or disconnecting the following:
- Place the wheel in the straight-ahead position and turn the ignition switch to LOCK
- Cover clips at both sides of the steering wheel
- Steering wheel-to-SAS module clips
- SAS module from the steering wheel and disconnect the electrical connector
- Steering wheel-to-column nut
- Steering wheel from the steering column using a suitable puller

6. At the passenger's side, remove the SAS module by removing or disconnecting the following:

- Glove compartment and the glove compartment cover
- SAS module-to-dash bolts
- SAS module and disconnect the electrical connector

7. Remove the rear console by removing or disconnecting the following:
- Rear console's box
- Bake boot
- Center panel
- Rear console

8. Remove or disconnect the following:
- A-pillar trim at both sides
- Undercover at the passenger's side
- Upper and lower steering column covers
- Electrical connectors and remove the combination switch from the steering column
- Meter hood
- Electrical connectors and remove the instrument cluster
- Steering column-to-chassis bolts and the steering column
- Hood release lever
- Both side panels
- Instrument panel with the help of an assistant

9. Remove the evaporator housing by removing or disconnecting the following:
- Air conditioning system refrigerant lines and discard the gaskets
- Aspirator hose
- Power transistor connector
- MAX-HI connector
- Evaporator temperature connector
- Evaporator housing assembly

10. Disconnect and remove the heater unit assembly.

11. Separate the heater unit cases and remove the heater core.

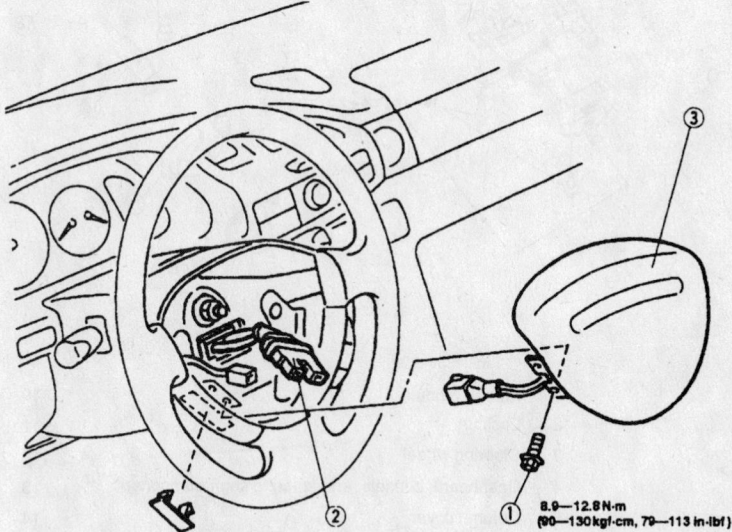

8.9—12.8 N·m
(90—130 kgf·cm, 79—113 in·lbf)

1   Bolt
2   Connector
3   Driver-side air bag module

93112GE5

Exploded view of the SAS module and steering wheel assembly—Millenia

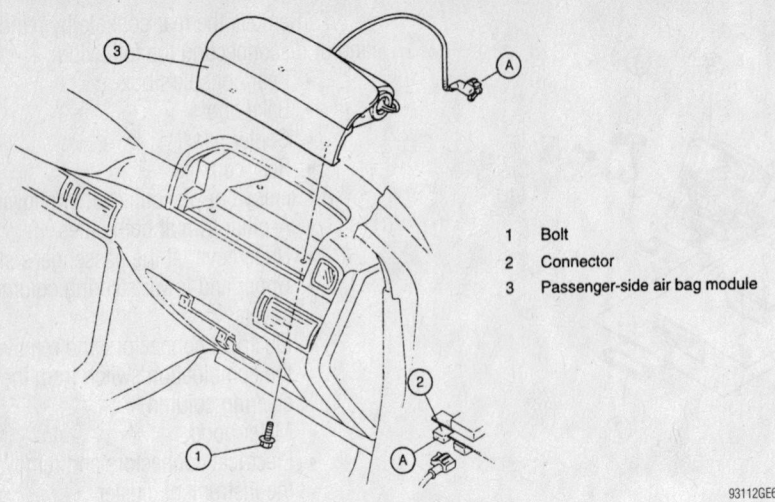

1	Bolt
2	Connector
3	Passenger-side air bag module

93112GE6

Exploded view of the passenger's side SAS module—Millenia

To install:

12. Install the heater core and assemble the heater unit cases.

13. Connect and install the heater unit assembly.

14. Install the evaporator housing by installing or connecting the following:

- Evaporator housing assembly
- Evaporator temperature connector
- MAX-HI connector
- Power transistor connector
- Aspirator hose
- Air conditioning system refrigerant lines using new gaskets

15. Install or connect the following:

- Instrument panel with the help of an assistant
- Both side panels

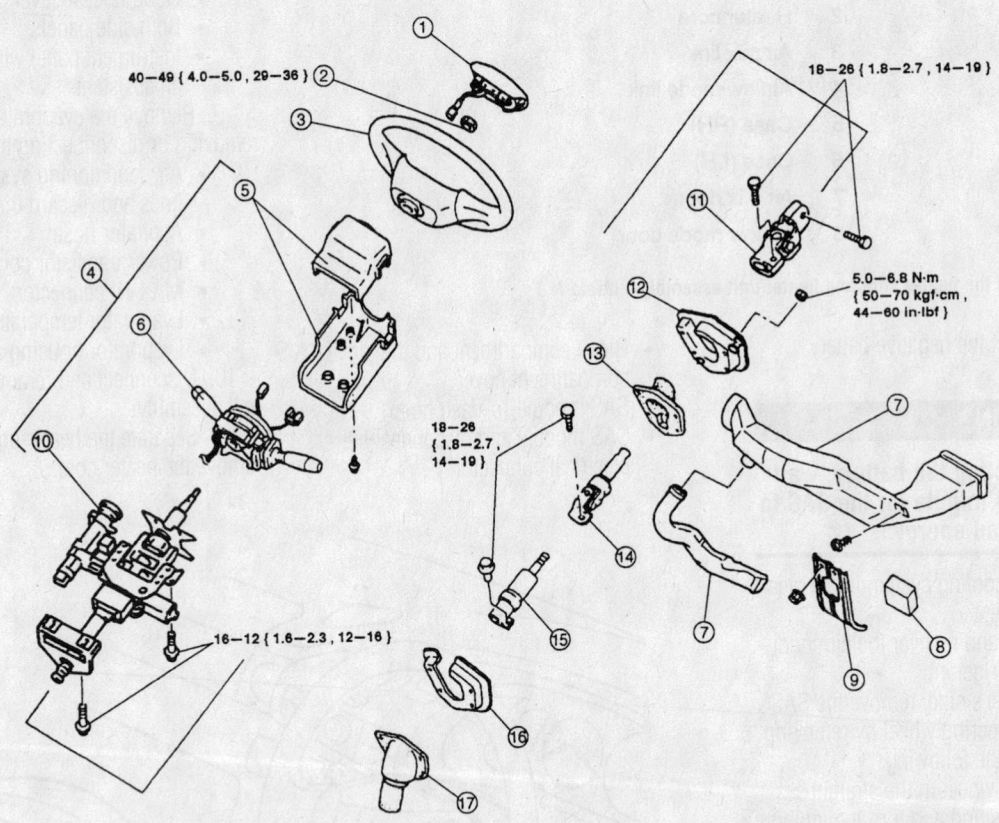

N·m { kgf·m , ft·lbf }

93112GE9

1	Air bag module	10	Steering shaft component
2	Locknut	11	Universal joint (intermediate shaft)
3	Steering wheel	12	Cover
4	Dashboard, console, and steering shaft component	13	Shaft seal
5	Column cover	14	Intermediate shaft
6	Combination switch	15	Collapsible shaft
7	Air duct	16	Set plate
8	Flasher unit	17	Dust cover
9	Bracket		

Exploded view of the steering wheel and steering column assembly—Millenia

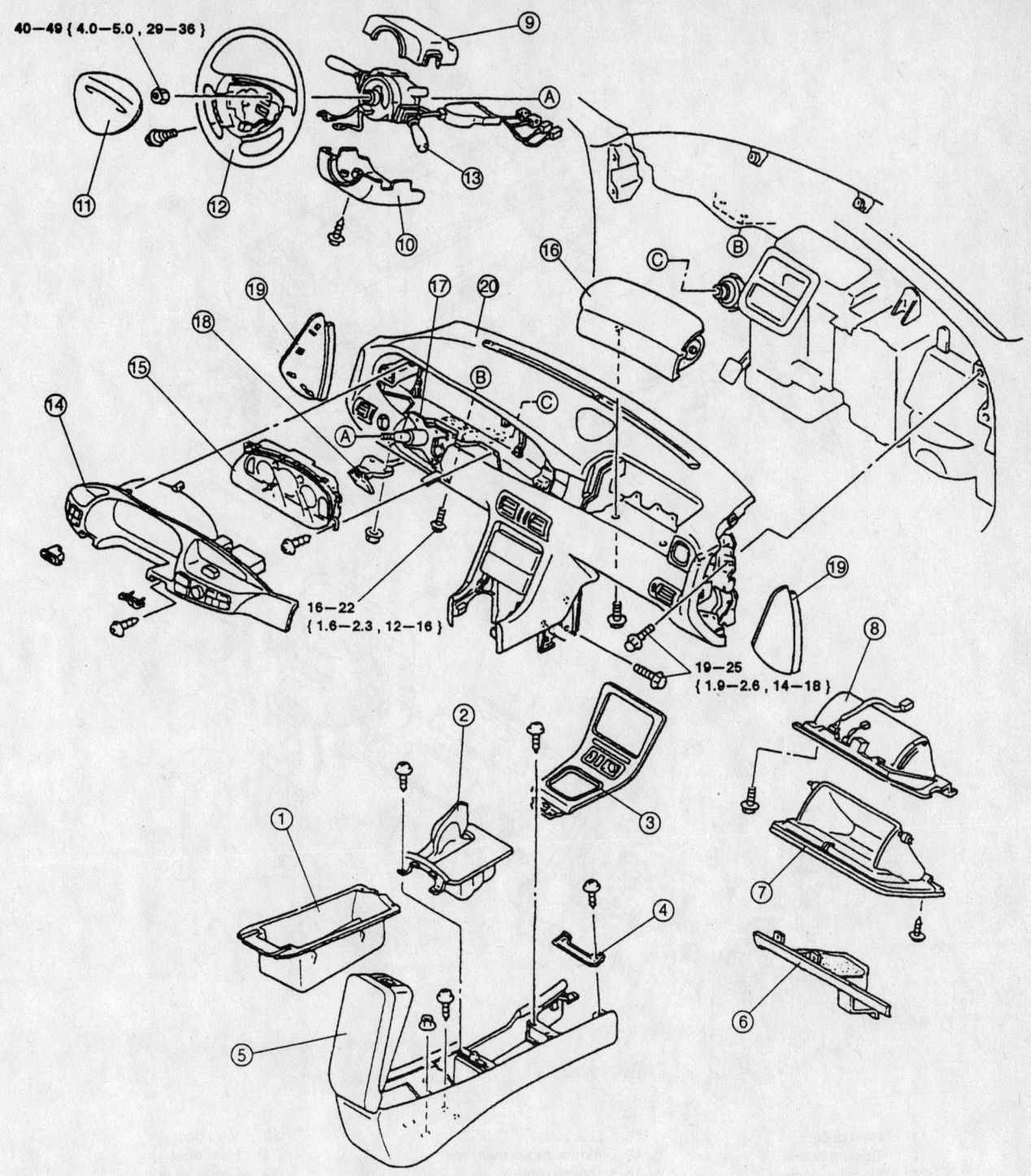

1	Rear console box	8	Glove compartment cover	15	Instrument cluster
2	Brake boot	9	Upper column cover	16	Passenger-side air bag module
3	Center panel	10	Lower column cover	17	Steering shaft
4	Bracket	11	Driver-side air bag module	18	Hood release lever
5	Rear console	12	Steering wheel	19	Side panel
6	Under cover	13	Combination switch	20	Dashboard
7	Glove compartment	14	Meter hood		

Exploded view of the instrument panel and rear console assemblies—Millenia

93112GE8

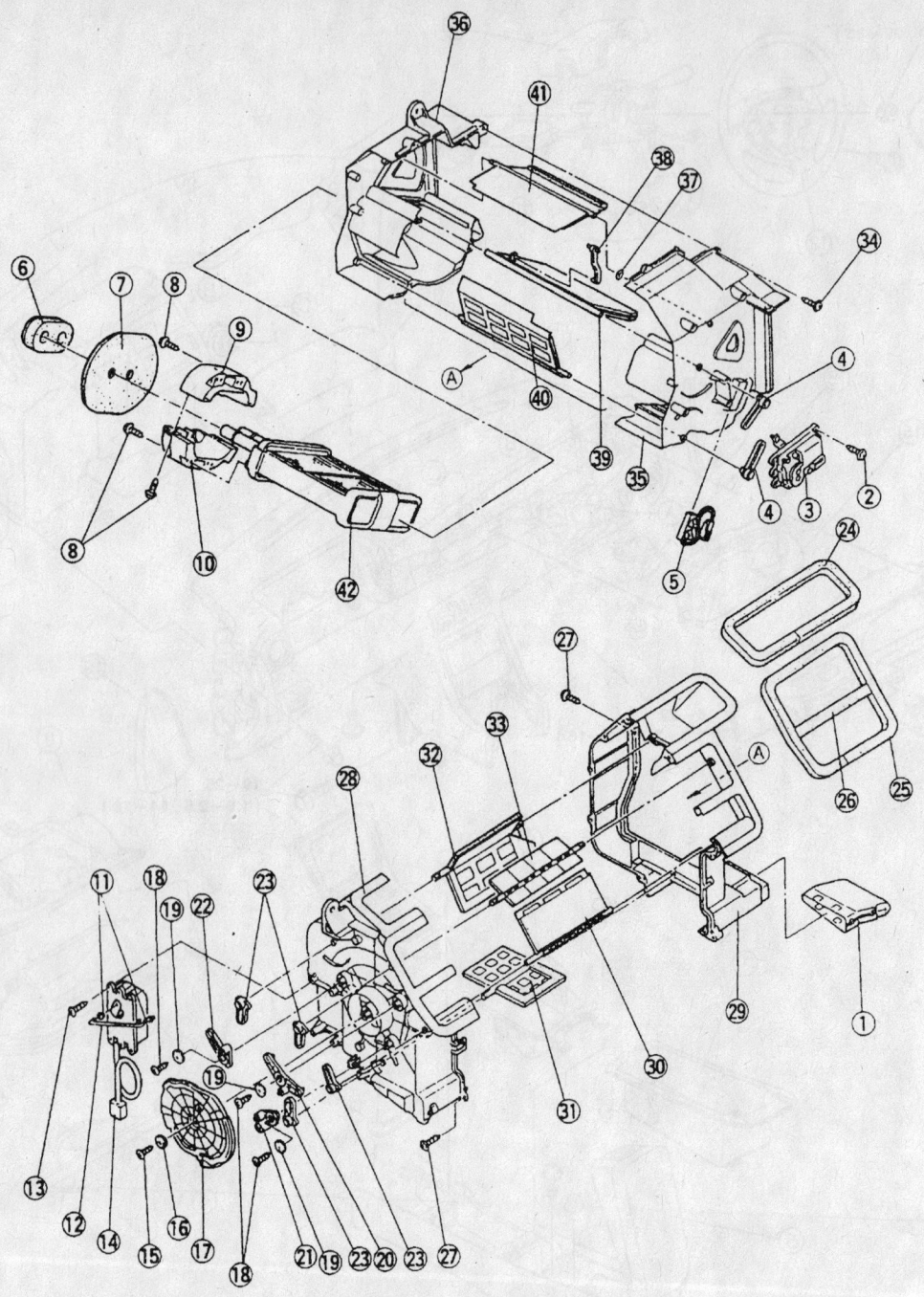

1	Heat duct	16	Link collar	30	Vent door
2	Tapping screw	17	Airflow mode main link	31	Heat door
3	Air mix actuator	18	Tapping screw	32	Defroster door
4	Air mix crank	19	Link collar	33	Side vent door
5	Water temperature sensor	20	Airflow mode sub link (VENT)	34	Tapping screw
6	Polyurethane foam (thick)	21	Airflow mode sub link (HEAT)	35	Heater case (1)
7	Polyurethane foam (thin)	22	Airflow mode sub link (DEFROSTER)	36	Heater case (2)
8	Tapping screw	23	Airflow mode crank	37	Collar
9	Heater core bracket (1)	24	Polyurethane protector (DEFROSTER)	38	Air mix rod
10	Heater core bracket (2)	25	Polyurethane protector (VENT)	39	Air mix main door
11	Rod stopper	26	Polyurethane protector (SIDE VENT)	40	Air mix sub door
12	Airflow mode rod	27	Tapping screw	41	Air mix guide door
13	Tapping screw	28	Heater case (4)	42	Heater core
14	Airflow mode actuator	29	Heater case (3)		
15	Tapping screw				

Exploded view of the heater core and heater case assembly—Millenia

93112GE7

- Hood release lever
- Steering column and the steering column-to-chassis bolts
- Instrument cluster and connect the electrical connectors
- Meter hood
- Combination switch to the steering column and connect the electrical connectors
- Upper and lower steering column covers
- Undercover at the passenger's side
- A-pillar trim at both sides

16. Install the rear console by installing or connecting the following:
- Rear console
- Center panel
- Brake boot
- Rear console's box

17. At the passenger's side, install the SAS module by installing or connecting the following:
- SAS module and connect the electrical connector
- SAS module-to-dash bolts
- Glove compartment cover and the glove compartment

18. At the driver's side, install the SAS module and the steering wheel by installing or connecting the following:
- Steering wheel to the steering column
- Steering wheel-to-column nut. Torque the nut to 29–36 ft. lbs. (40–49 Nm).
- SAS module to the steering wheel and connect the electrical connector. Torque the bolts to 79–113 inch lbs. (9–13 Nm).
- Steering wheel-to-SAS module clips
- Cover clips at both sides of the steering wheel

19. Refill the cooling system.
20. Connect the negative battery cable.
21. Evacuate, charge and leak test the air conditioning system refrigerant.
22. Operate the engine to normal operating temperatures; then, check the climate control operation and check for leaks.

Protégé

1. Before servicing the vehicle, refer to the precautions in the beginning of this section.
2. Disconnect the negative battery cable.

✳✳ CAUTION

After disconnecting the battery, wait for more than 1 minute for the SAS to deplete its stored energy.

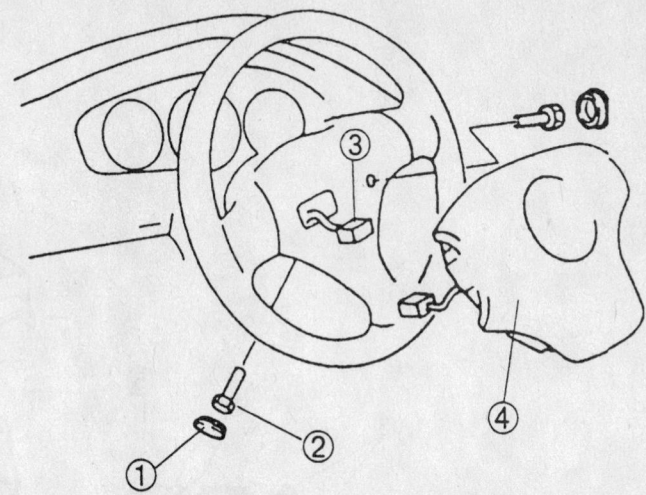

7.9—11.7 N·m {80—120 kgf·cm, 70—104 in·lbf}

1 Cap
2 Bolt
3 Connector
4 Driver-side air bag module

93112GG4

Exploded view of the SAS module and the steering wheel assembly—Protégé

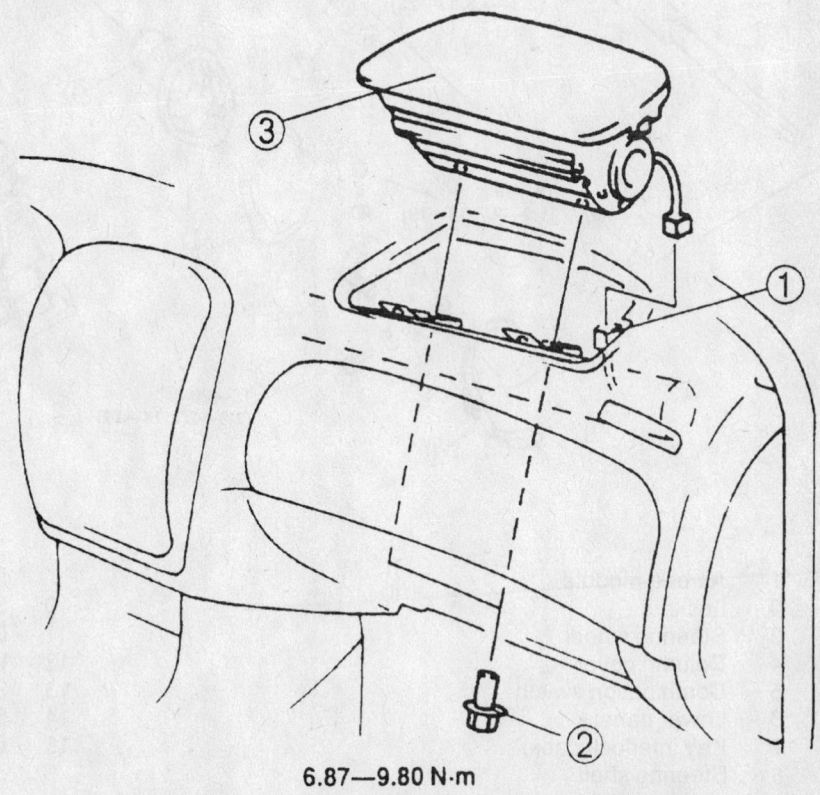

6.87—9.80 N·m
{70—100 kgf·cm, 60.8—86.7 in·lbf}

93112GG5

Exploded view of the passenger's side SAS module—Protégé

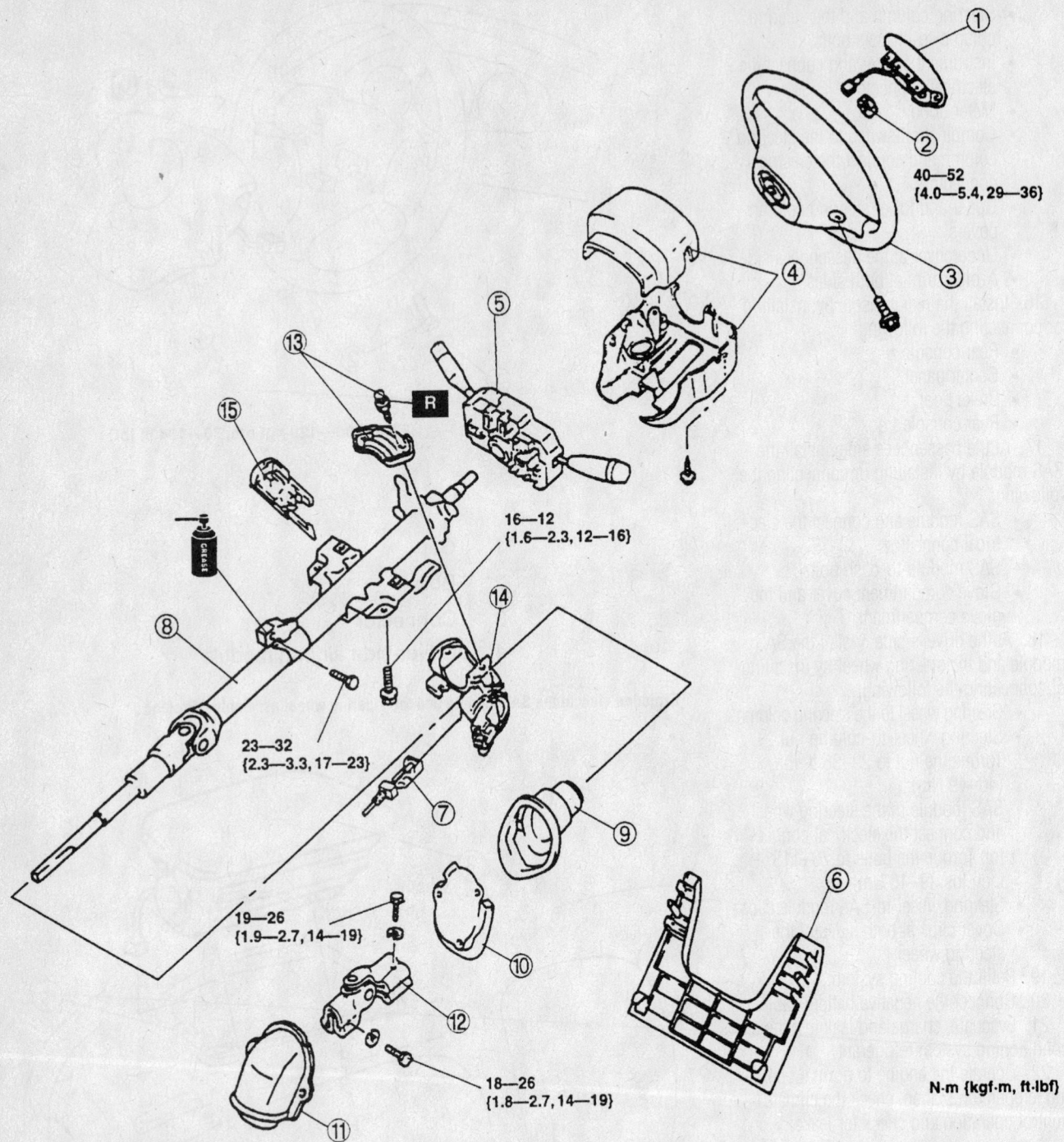

N·m {kgf·m, ft·lbf}

1	Air bag module	9	Shaft seal
2	Locknut	10	Set plate
3	Steering wheel	11	Dust cover
4	Column cover	12	Universal joint
5	Combination switch	13	Steering lock mounting bolts and bracket
6	Lower panel	14	Steering lock component
7	Key interlock cable	15	Cylinder outer component
8	Steering shaft		

Exploded view of the steering column assembly—Protégé

93112GG6

3. Drain the cooling system into a clean container for reuse.

4. Disconnect the heater hoses from the heater core.

5. Discharge and recover the air conditioning system refrigerant.

6. Place the wheel in the straight-ahead position and turn the ignition switch to LOCK.

7. At the driver's side, remove the SAS module and the steering wheel by removing or disconnecting the following:
- Cover clips at both sides of the steering wheel
- Steering wheel-to-SAS module bolts
- SAS module from the steering wheel and disconnect the electrical connector
- Steering wheel-to-column nut
- Steering wheel from the steering column using a suitable puller

8. At the passenger's side, remove the SAS module by removing or disconnecting the following:
- Glove compartment and the glove compartment cover
- SAS module-to-dash bolts
- SAS module and disconnect the electrical connector

9. Remove the console by removing or disconnecting the following:
- Shift lever knob, if equipped with a manual transmission
- Console's cover
- Console-to-chassis screws and console

10. Remove the combination switch by removing or disconnecting the following:
- Steering column cover
- Electrical connectors and remove the combination switch-to-steering

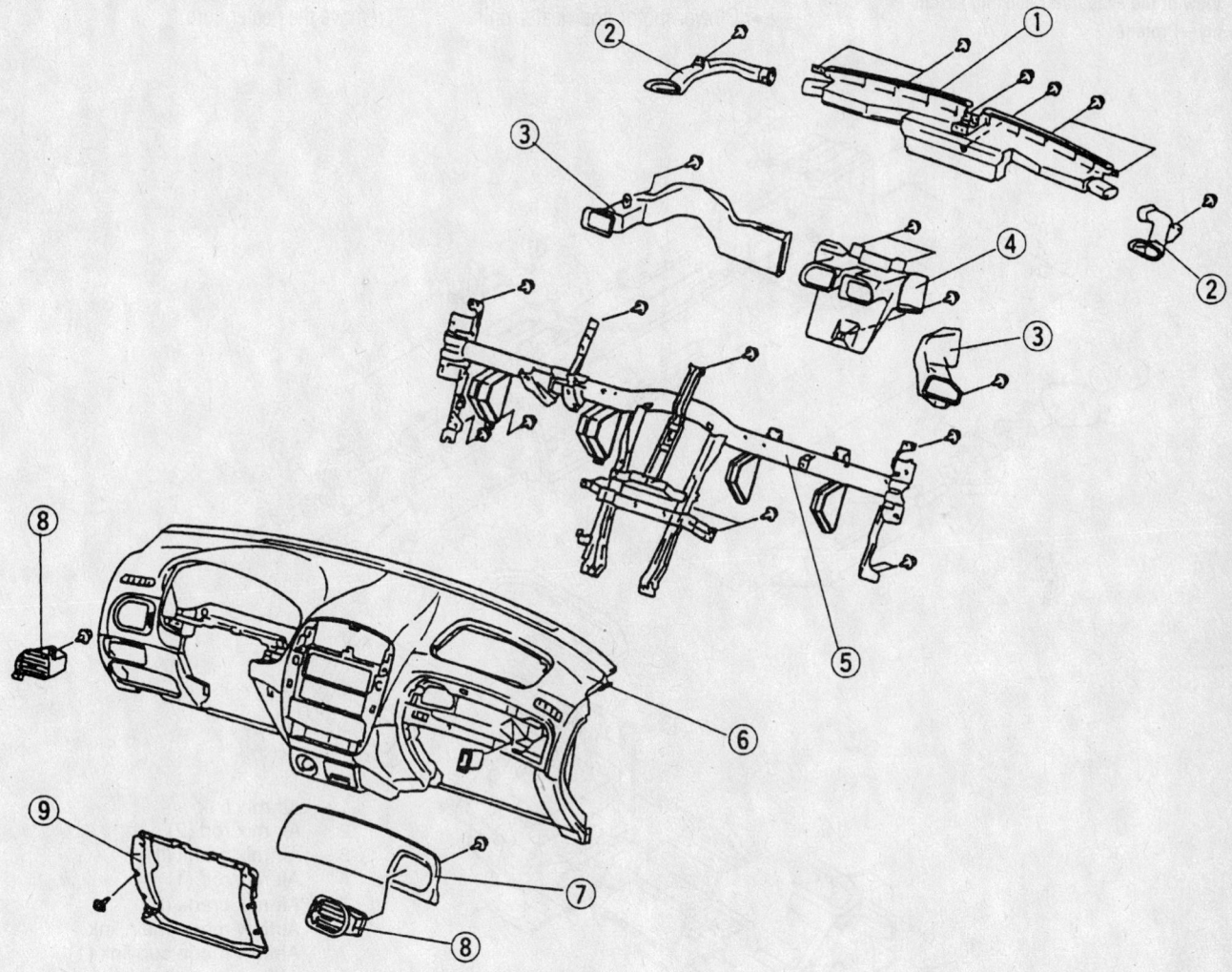

1 Defroster nozzle	6 Crush pad
2 Side demister duct	7 Pad
3 Duct	8 Ventilator grille
4 Center duct	9 Passenger-side lower panel
5 Dashboard member	

93112GG7

Exploded view of the dashboard assembly—Protégé

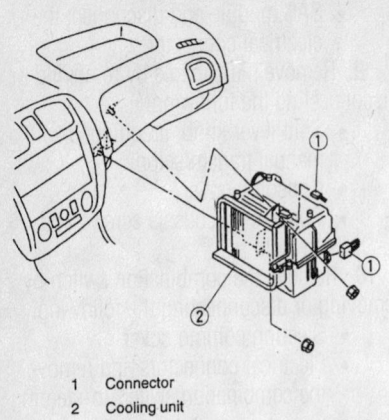

1 Connector
2 Cooling unit

93112GG8

View of the evaporator housing assembly—Protégé

column bolts and the combination switch
11. Remove the instrument cluster by:
- Meter hood
- Instrument cluster-to-dash panel screws
- Electrical connectors and remove the instrument cluster
12. Remove or disconnect the following:
- Lower panel
- Hood release cable installation nut
- Side wall trim
- "A" pillar trim at both sides
- Side panel
- Antenna connector
- Blower motor and heater unit

electrical connectors, if equipped with the wire-type climate control unit
- Electrical connectors and the bolts
- Dashboard-to-chassis bolts
- Dashboard from the vehicle with the help of an assistant
- Passenger's side lower panel
- Air intake wire from the climate control unit
- Air conditioning refrigerant lines from the evaporator and discard the O-rings
- Evaporator electrical connector(s)
- Evaporator housing
13. Disassemble the heater housing and remove the heater core.

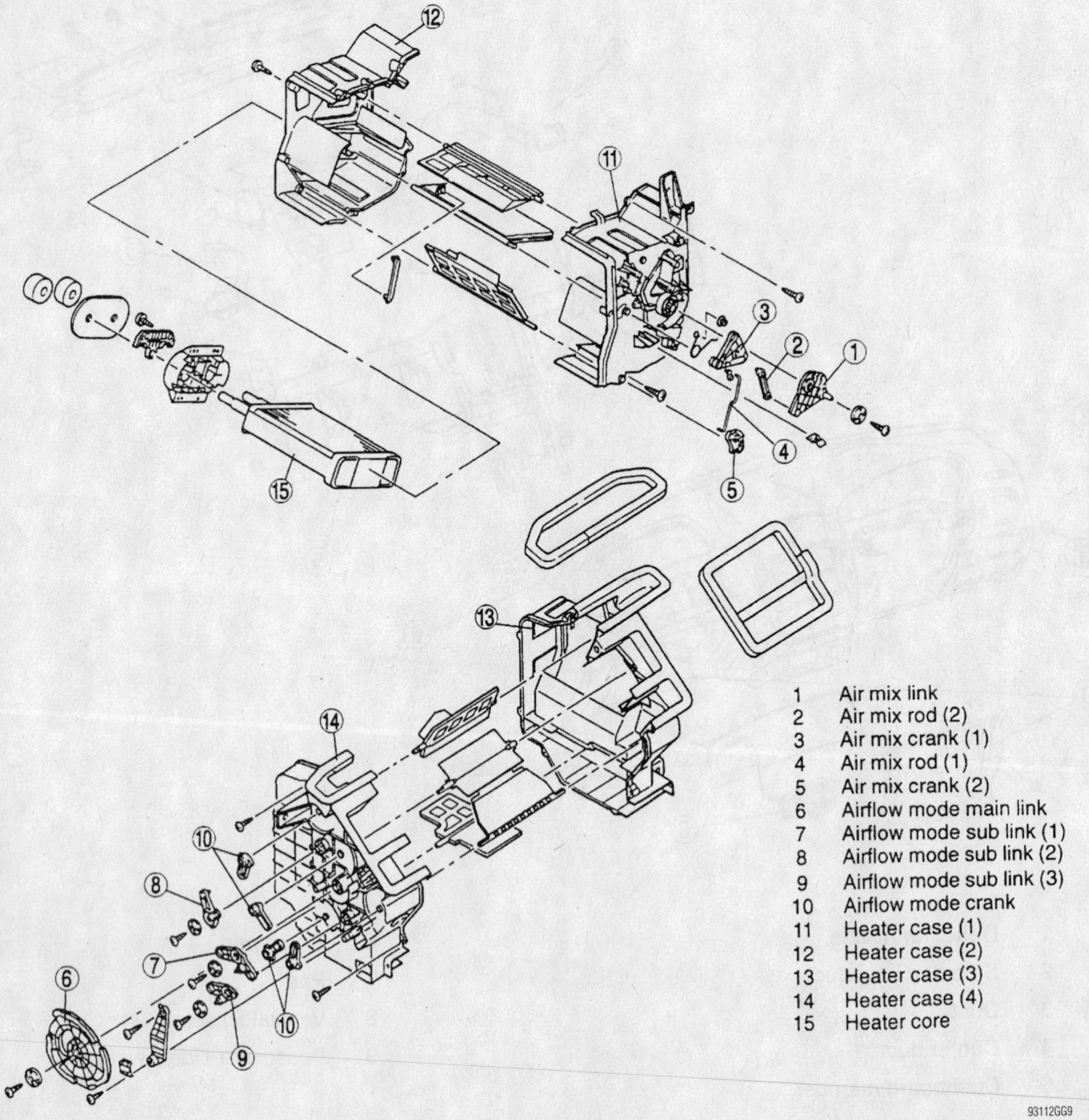

1	Air mix link
2	Air mix rod (2)
3	Air mix crank (1)
4	Air mix rod (1)
5	Air mix crank (2)
6	Airflow mode main link
7	Airflow mode sub link (1)
8	Airflow mode sub link (2)
9	Airflow mode sub link (3)
10	Airflow mode crank
11	Heater case (1)
12	Heater case (2)
13	Heater case (3)
14	Heater case (4)
15	Heater core

93112GG9

Exploded view of the heater housing assembly—Protégé

To install:

14. Install the heater core and assemble the heater housing.

15. Install or connect the following:
- Evaporator housing
- Evaporator electrical connector(s)
- Air conditioning refrigerant lines to the evaporator using new O-rings
- Air intake wire to the climate control unit
- Passenger's side lower panel
- Dashboard to the vehicle with the help of an assistant
- Dashboard-to-chassis bolts
- Electrical connectors and the bolts
- Blower motor and heater unit electrical connectors, if equipped with the wire-type climate control unit
- Antenna connector
- Side panel
- A-pillar trim at both sides
- Side wall trim
- Hood release cable installation nut
- Lower panel

16. Install the instrument cluster by installing or connecting the following:
- Instrument cluster and connect the electrical connectors
- Instrument cluster-to-dash panel screws
- Meter hood

17. Install the combination switch by installing or connecting the following:
- Electrical connectors and install the combination switch-to-steering column bolts and the combination switch
- Steering column cover

18. Install the console by installing or connecting the following:
- Console and the console-to-chassis screws
- Console's cover
- Shift lever knob, if equipped with a manual transmission

19. At the passenger's side, install the SAS module by installing or connecting the following:
- SAS module and connect the electrical connector
- SAS module-to-dash bolts
- Glove compartment and the glove compartment cover

20. At the driver's side, install the SAS module and the steering wheel by installing or connecting the following:
- Steering wheel-to-column nut
- SAS module from the steering wheel and connect the electrical connector
- Steering wheel-to-SAS module bolts

- Cover clips at both sides of the steering wheel

21. Connect the heater hoses to the heater core.

22. Refill the cooling system.

23. Connect the negative battery cable.

24. Evacuate, charge and leak test the air conditioning system.

25. Operate the engine to normal operating temperatures; then, check the climate control operation and check for leaks.

626

1. Before servicing the vehicle, refer to the precautions in the beginning of this section.

2. Disconnect the negative battery cable.

✳✳ CAUTION

After disconnecting the battery, wait for more than 1 minute for the SAS to deplete its stored energy.

3. Drain the cooling system into a clean container for reuse.

4. Discharge and recover the air conditioning system refrigerant.

5. Place the wheel in the straight-ahead position and turn the ignition switch to LOCK.

6. At the driver's side, remove the SAS module and the steering wheel removing or disconnecting the following:
- Cover clips at both sides of the steering wheel
- Steering wheel-to-SAS module bolts
- SAS module from the steering wheel and disconnect the electrical connector
- Steering wheel-to-column nut
- Steering wheel from the steering column using a suitable puller

7. At the passenger's side, remove the SAS module by removing or disconnecting the following:
- Glove compartment and the glove compartment cover
- SAS module-to-dash bolts
- SAS module and disconnect the electrical connector

8. Remove the instrument cluster by removing or disconnecting the following:
- Instrument cluster meter hood
- Instrument cluster-to-dash screws, the instrument cluster and disconnect the electrical connectors

9. Remove the climate control assembly

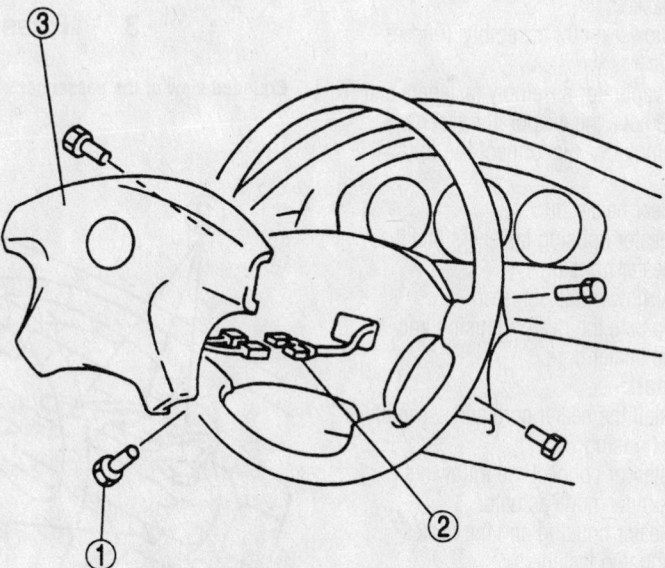

7.9—11.7 N·m {80—120 kgf·cm, 70—104 in·lbf}

1	Bolt
2	Connector
3	Driver-side air bag module

93112GH6

Exploded view of the steering wheel and SAS module—626

by removing or disconnecting the following:

- Climate control meter hood
- Climate control assembly screws
- Climate control assembly and disconnect the electrical connector and the assembly

10. Remove the audio unit by removing or disconnecting the following:

- Hole covers by inserting a small tape-wrapped flathead screwdriver into the slot and carefully pry off the hole covers
- Using 2 removal tools 49 UN01 050 or equivalent, insert them into sides of the audio unit.
- Slide audio unit outward and forward
- Electrical connectors and the antenna jack

11. Remove the instrument panel by removing or disconnecting the following:

- Upper center panel cover
- Dash panel-to-chassis bolts
- Dash panel with the help of an assistant

12. Remove the evaporator housing by removing or disconnecting the following:

- Center lower panel
- Refrigerant lines from the air conditioning evaporator and discard the gaskets
- Blower motor assembly, if necessary
- Evaporator assembly fasteners and remove the evaporator assembly

13. Remove or disconnect the following:

- Rear heater duct
- Heater housing fasteners and the heater housing
- Airflow mode actuator

14. Separate the heater housing and remove the heater core.

To install:

15. Install the heater core and assemble the heater housing.

16. Install or connect the following:

- Airflow mode actuator
- Heater housing and the heater housing fasteners
- Rear heater duct

17. Install the evaporator housing by installing or connecting the following:

- Center lower panel
- Refrigerant lines to the air conditioning evaporator using new gaskets
- Blower motor assembly, if necessary
- Evaporator assembly and the evaporator assembly fasteners

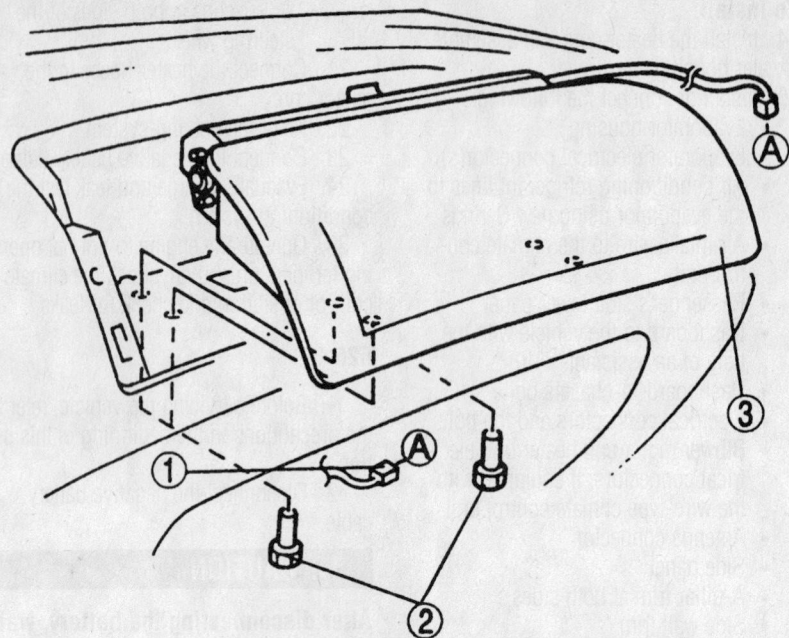

18—26 N·m {1.8—2.7 kgf·m, 14—19 ft·lbf}

1	Connector
2	Bolt
3	Passenger-side air bag module

Exploded view of the passenger's side SAS module—626

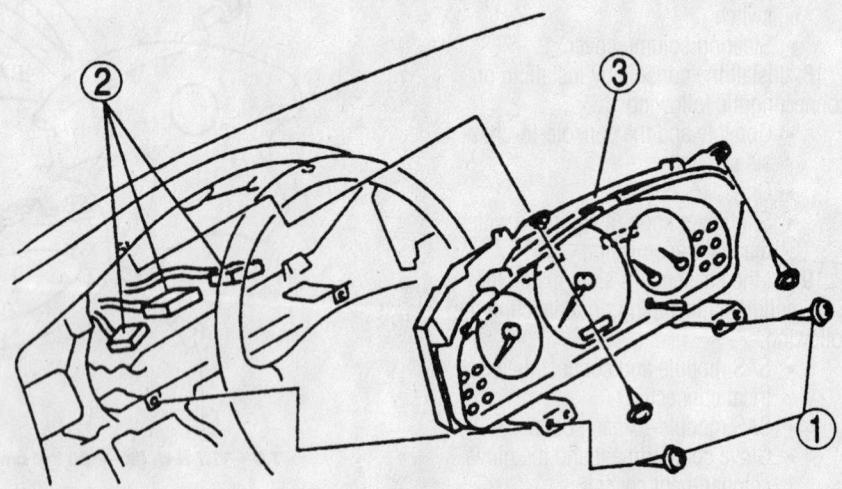

1	Screw
2	Connector
3	Instrument cluster

View of the instrument cluster assembly—626

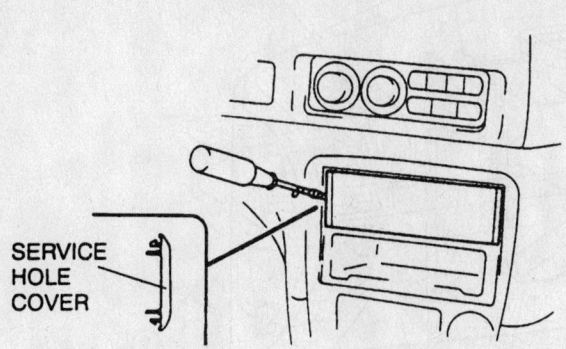

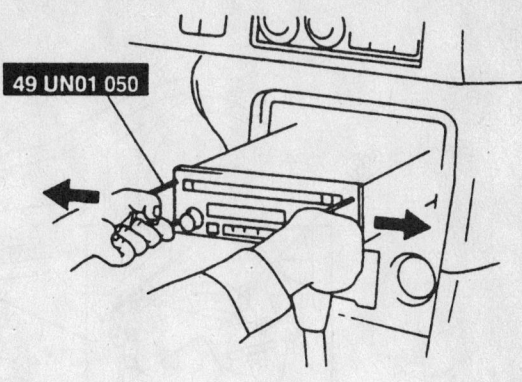

49 UN01 050

93112GH9

Removing the audio unit—626

18. Install the instrument panel by installing or connecting the following:
- Upper center panel cover
- Dash panel-to-chassis bolts
- Dash panel with the help of an assistant

19. Install the audio unit by installing or connecting the following:
- Hole covers

- Audio unit
- Electrical connectors and the antenna jack

20. Install the climate control assembly by installing or connecting the following:
- Climate control meter hood
- Climate control assembly screws
- Climate control assembly and connect the electrical connector

21. Install the instrument cluster by installing or connecting the following:
- Instrument cluster meter hood
- Instrument cluster-to-dash screws, the instrument cluster and connect the electrical connectors

22. At the passenger's side, install the SAS module by installing or connecting the following:

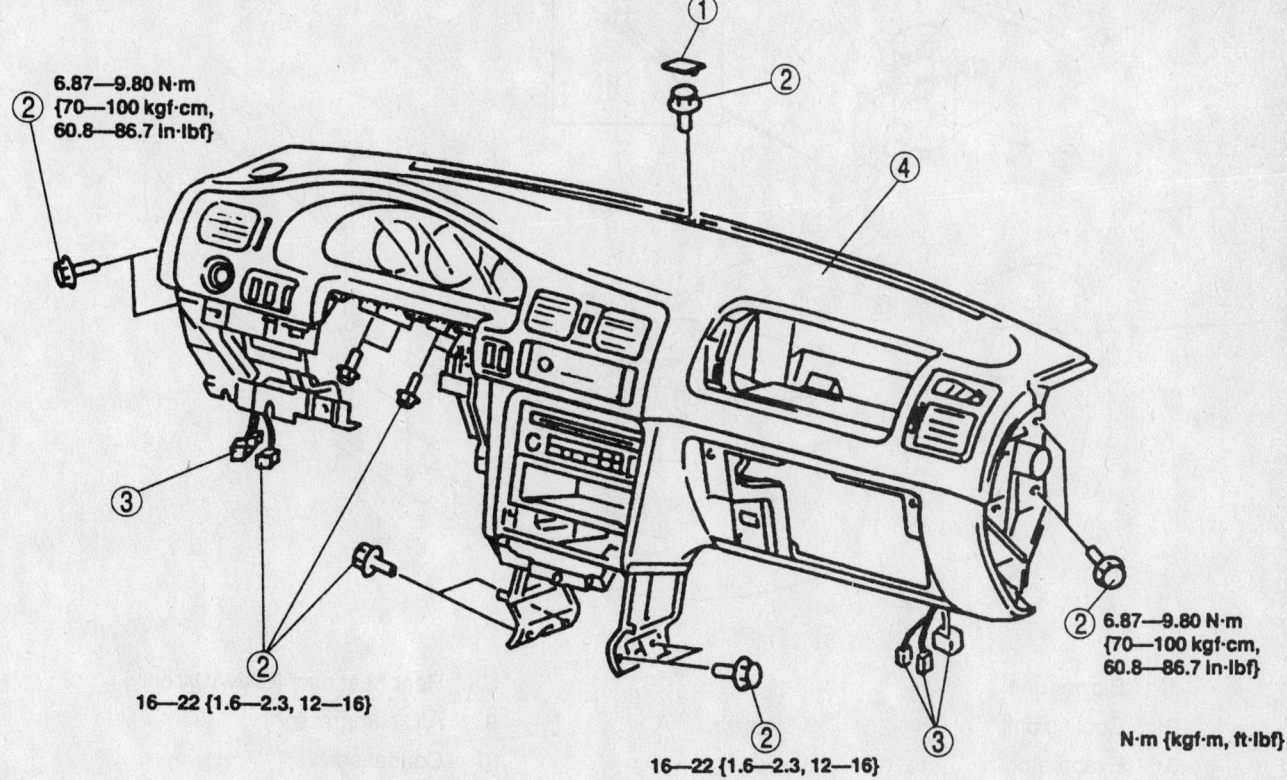

6.87—9.80 N·m
{70—100 kgf·cm,
60.8—86.7 In·lbf}

6.87—9.80 N·m
{70—100 kgf·cm,
60.8—86.7 In·lbf}

16—22 {1.6—2.3, 12—16}

16—22 {1.6—2.3, 12—16}

N·m {kgf·m, ft·lbf}

1	Cover	3	Connector
2	Bolt	4	Dashboard

View of the instrument panel assembly—626

93112GH0

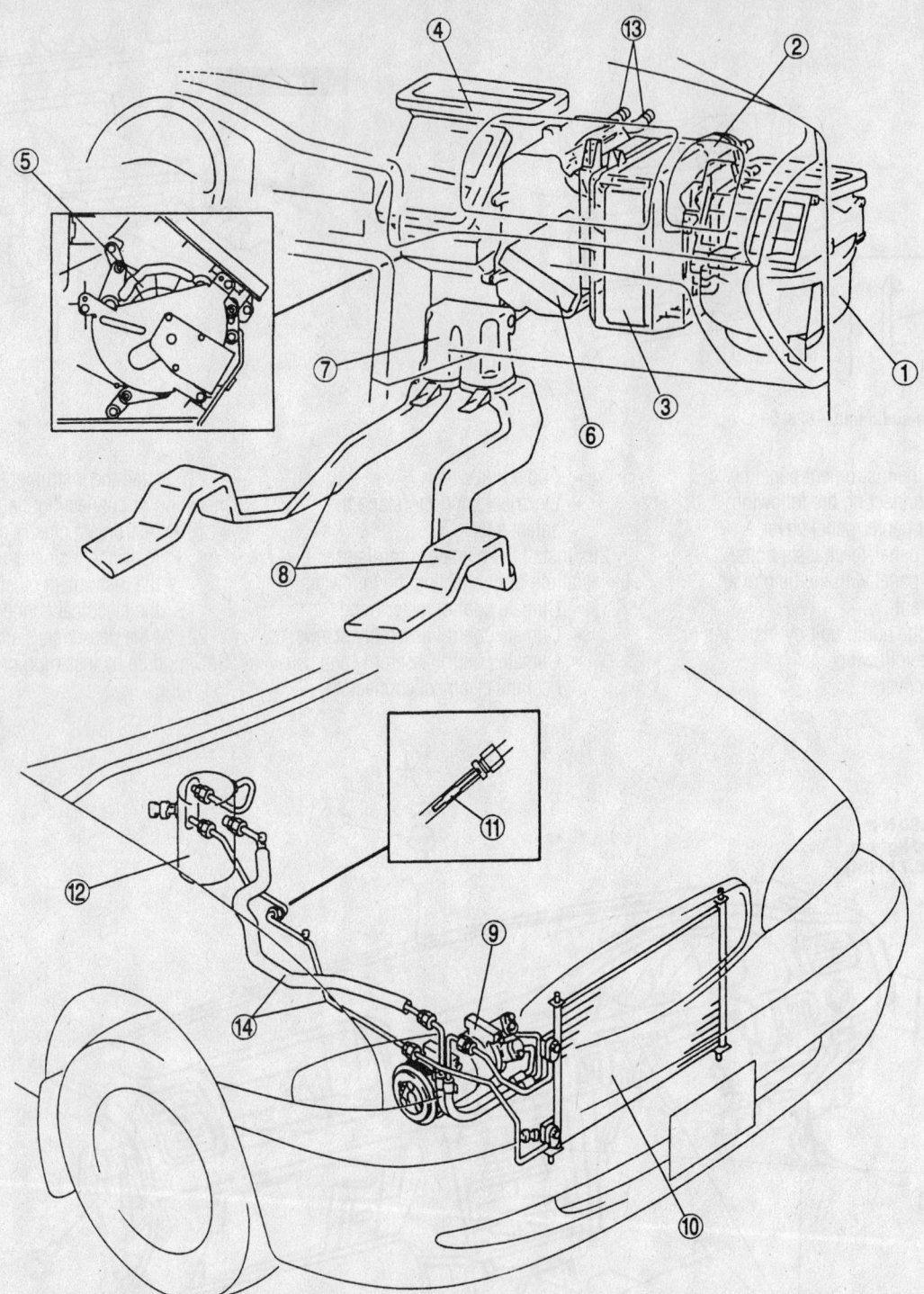

1	Blower unit	8	Rear heat duct (CANADA only)
2	Cooling unit	9	A/C compressor
3	Evaporator	10	Condenser
4	Heater unit	11	Orifice tube
5	Airflow mode main link	12	Accumulator tank
6	Heater core	13	Heater hose
7	Rear duct (CANADA only)	14	Refrigerant lines

93112GI1

View of the heater and air conditioning housing assemblies—626

- Glove compartment and the glove compartment cover
- SAS module-to-dash bolts
- SAS module and connect the electrical connector

23. At the driver's side, install the SAS module and the steering wheel by installing or connecting the following:

- Steering wheel and the steering wheel-to-column nut. Torque the steering wheel nut to 29–36 ft. lbs. (40–49 Nm).
- SAS module to the steering wheel and connect the electrical connector. Torque the steering column-to-SAS module bolts to 70–104 inch lbs. (8–12 Nm).
- Steering wheel-to-SAS module clips
- Cover clips, at both sides of the steering wheel

24. Refill the cooling system.
25. Connect the negative battery cable.
26. Evacuate, charge and leak test the air conditioning system refrigerant.
27. Operate the engine to normal operating temperatures; then, check the climate control system and check for leaks.

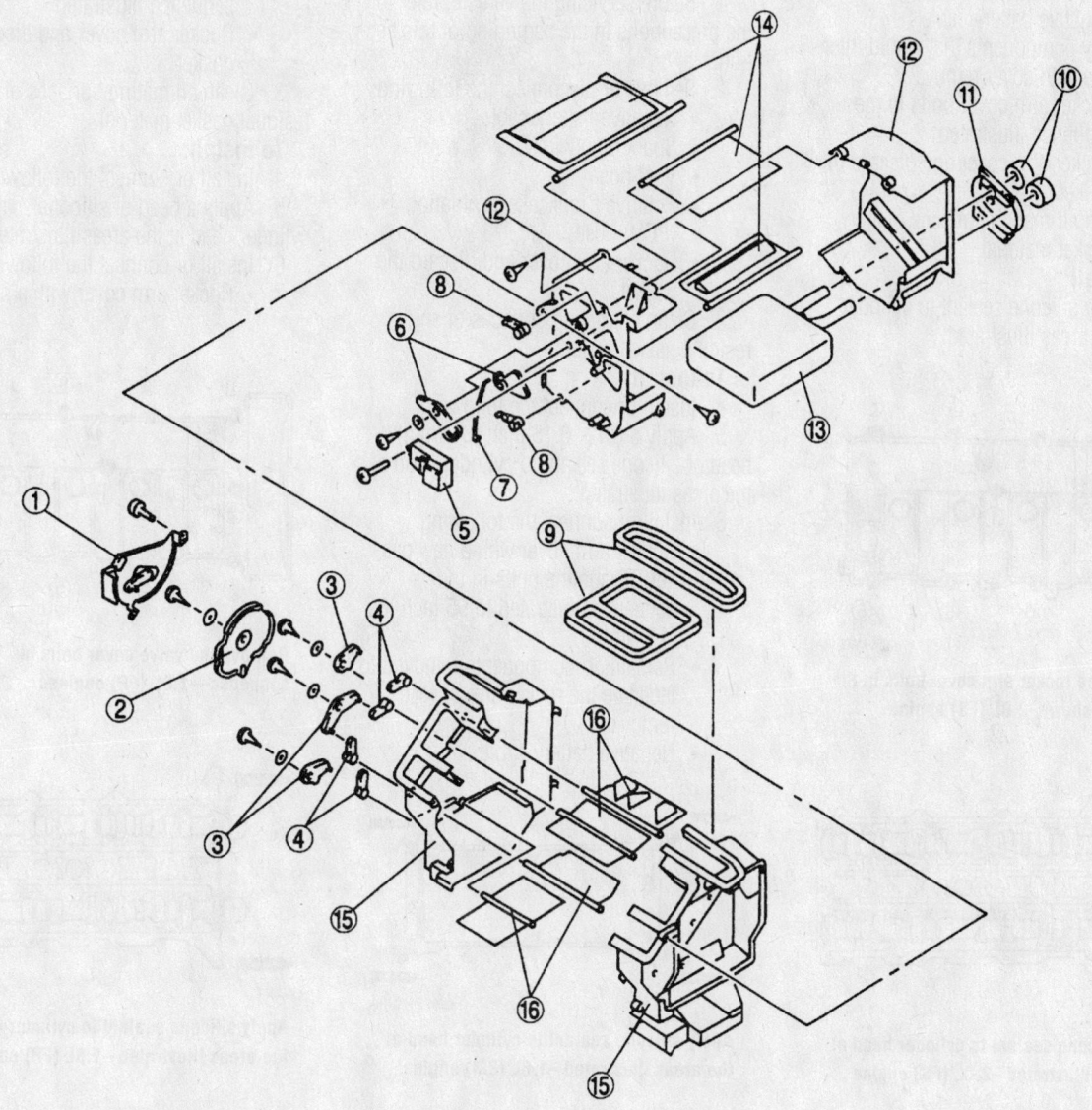

1	Airflow mode actuator	9	Polyurethane protector
2	Airflow mode main link	10	Seal
3	Airflow mode sub link	11	Cover
4	Airflow mode crank	12	Case
5	Air mix actuator	13	Heater core
6	Air mix link	14	Air mix door
7	Air mix rod	15	Case
8	Air mix crank	16	Airflow mode door

93112GI2

Exploded view of the heater core and heater housing assembly—626

Rocker Arm (Valve) Cover

REMOVAL & INSTALLATION

626 and Protégé

2.0L (FS) ENGINE

1. Before servicing the vehicle, refer to the precautions in the beginning of this section.
2. Remove or disconnect the following:
   - Negative battery cable
   - Any components that would interfere with cover removal
   - Rocker arm cover bolts in the sequence illustrated
   - Rocker arm cover and discard the gasket
3. Clean all mating surfaces of any residual gasket material.

To install:

4. Apply silicone sealant to cylinder head at the areas illustrated.

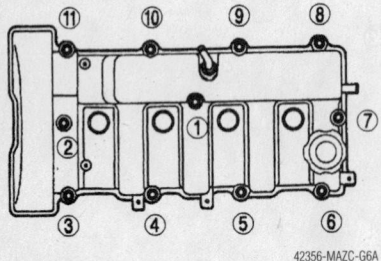

42356-MAZC-G6A

Remove the rocker arm cover bolts in the sequence shown–2.0L (FS) engine

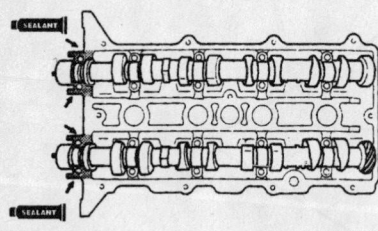

42356-MAZC-G07

Apply silicone sealant to cylinder head at the areas illustrated –2.0L (FS) engine

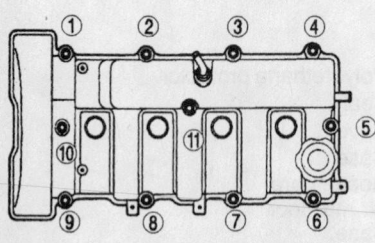

42356-MAZC-G08

Tighten the rocker arm cover bolts in the sequence shown–2.0L (FS) engine

5. Install or connect the following:
   - Rocker arm cover with a new gasket. Torque the bolts in the sequence illustrated to 95 inch lbs. (11 Nm).
   - Remaining components removed to facilitate the rocker arm cover removal
   - Negative battery cable

1.6L (ZM) Engines

1. Before servicing the vehicle, refer to the precautions in the beginning of this section.
2. Remove or disconnect the following:
   - Negative battery cable
   - Spark plug wires
   - Vent hose
   - Positive Crankcase Ventilation (PCV) hose
   - Rocker arm cover and discard the gasket
3. Clean all mating surfaces of any residual gasket material.

To install:

4. Install or connect the following:
5. Apply a 0.12–0.15 inch (3–4mm) bead of silicone sealant to cylinder head at the areas illustrated.
6. Install or connect the following:
   - Rocker arm cover with a new gasket. Torque the bolts in the sequence illustrated to 95 inch lbs. (11 Nm).
   - Remaining components removed to facilitate the rocker arm cover removal
   - Negative battery cable

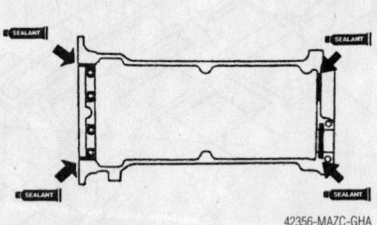

42356-MAZC-GHA

Apply silicone sealant to cylinder head at the areas illustrated –1.6L (ZM) engine

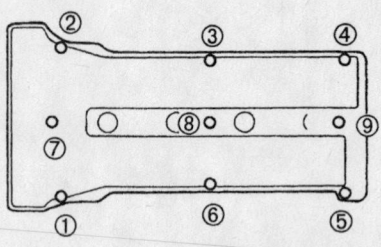

42356-MAZC-GHB

Tighten the rocker arm cover bolts in the sequence shown–1.6L (ZM) engine

1.8L (FP) Engines

1. Before servicing the vehicle, refer to the precautions in the beginning of this section.
2. Remove or disconnect the following:
   - Negative battery cable
   - Spark plug wires
   - Vent hose
   - Positive Crankcase Ventilation (PCV) hose
   - Rocker arm cover bolts in the sequence illustrated
   - Rocker arm cover and discard the gasket
3. Clean all mating surfaces of any residual gasket material.

To install:

4. Install or connect the following:
5. Apply a bead of silicone sealant to cylinder head at the areas illustrated.
6. Install or connect the following:
   - Rocker arm cover with a new gas-

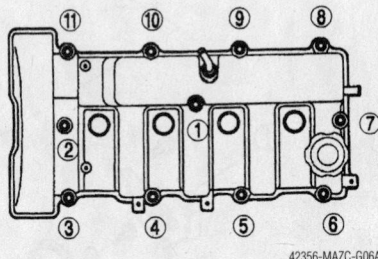

42356-MAZC-G06A

Remove the valve cover bolts in sequence—1.8L (FP) engines

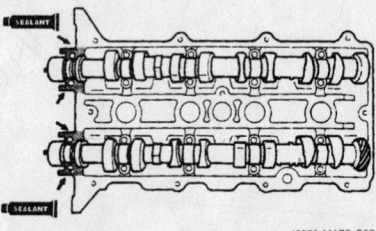

42356-MAZC-G07

Apply silicone sealant to cylinder head at the areas illustrated –1.8L (FP) engine

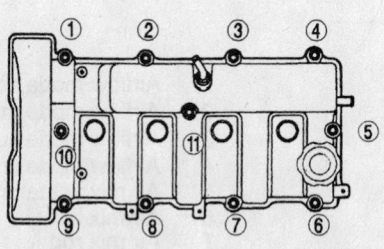

42356-MAZC-G08

Tighten the rocker arm cover bolts in the sequence shown–1.8L (FP) engine

**Thickness
1.5—2.5 mm {0.060—0.098 in}**

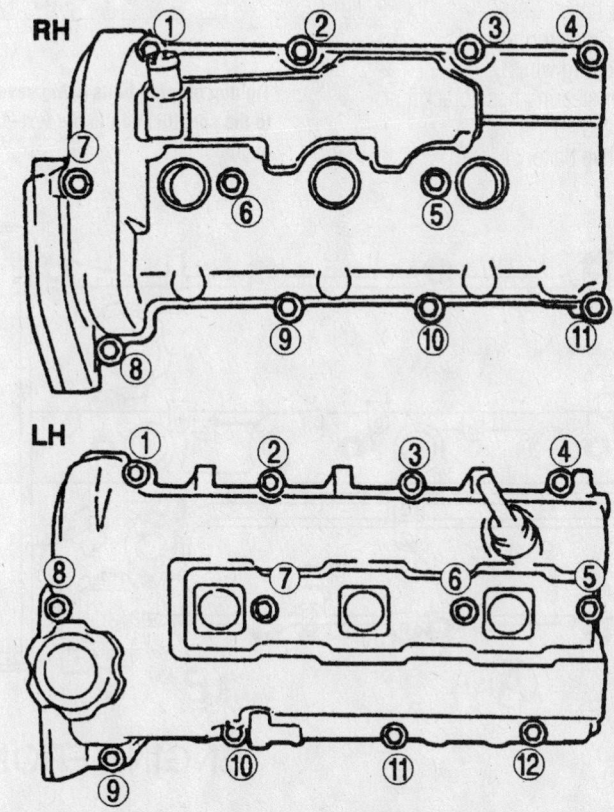

Apply silicone sealant to cylinder head at the areas illustrated –2.5L (KL) engine

42356-MAZC-G09

Tighten the rocker arm cover bolts in the sequence shown–2.5L (KL) engine

42356-MAZC-G10

ket. Torque the bolts in the sequence illustrated to 86 inch lbs. (10 Nm).
  • Remaining components removed to facilitate the rocker arm cover removal
  • Negative battery cable

2.5L (KL) ENGINE

1. Before servicing the vehicle, refer to the precautions in the beginning of this section.
2. Remove or disconnect the following:
  • Negative battery cable
  • Any components that would interfere with cover removal
  • Rocker arm cover bolts
  • Rocker arm cover and discard the gasket
3. Clean all mating surfaces of any residual gasket material.
To install:
4. Apply silicone sealant to cylinder head at the areas illustrated.
5. Install or connect the following:
  • Rocker arm cover with a new gasket. Torque the bolts in the sequence illustrated to 95 inch lbs. (11 Nm) using 2 or 3 steps. Retighten the right hand cover number 5 and 6 bolts and the left hand cover number 6 and 7 bolts.
  • Remaining components removed to facilitate the rocker arm cover removal
  • Negative battery cable

Miata

1.8L (BP) ENGINES

1. Before servicing the vehicle, refer to the precautions in the beginning of this section.
2. Remove or disconnect the following:
  • Negative battery cable
  • Upper radiator hose
  • Water hose
  • Oil pipe
  • Oil Control Valve (OCV)
  • Rocker arm cover and discard the gasket

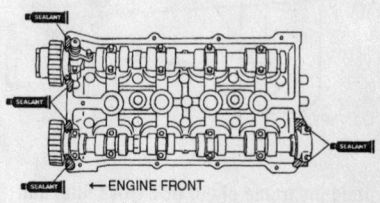

42356-MAZC-G95

Apply silicone sealant to cylinder head at the areas illustrated –1.8L (BP) engine

3. Clean all mating surfaces of any residual gasket material.

To install:

4. Install or connect the following:
   • Rocker arm cover with a new gasket.
   • Temporarily tighten the cover bolts **A**, refer to the illustration for loca-

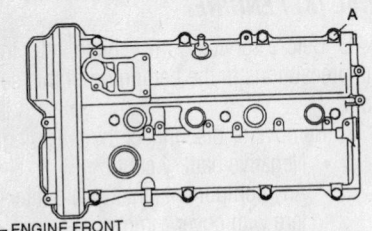

Temporarily tighten the cover bolts A–1.8L (BP) engine

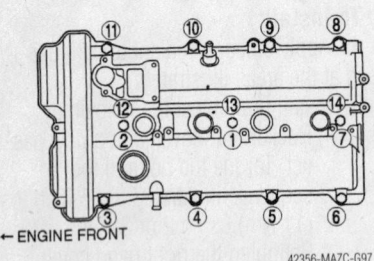

Tighten the rocker arm cover bolts in the sequence shown–1.8L (BP) engine

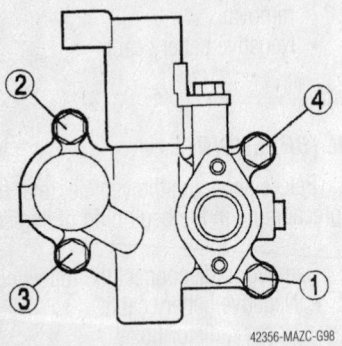

Tighten the OCV bolts in the sequence shown–1.8L (BP) engine

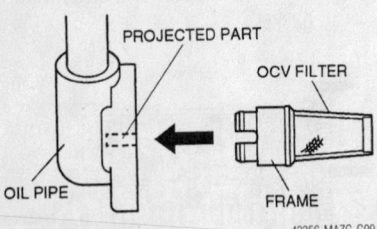

Hold the frame of the OCV valve filter and install the OCV so it is aligned with the projected part on the flange end of the oil pipe–1.8L (BP) engine

tion. Torque the bolts in sequence to 80 inch lbs. (9 Nm).

✳✳ CAUTION

When installing the OCV valve, be careful not to damage the O–ring, if damaged it may cause leaking.

   • OCV and tighten the bolts in the sequence illustrated
5. Install the oil pipe as follows:
   a. Oil pipe. Hold the frame of the OCV valve filter and install the OCV so it is aligned with the projected part on the flange end of the oil pipe. Coat the new washer with clean engine oil and temporarily install the upper and side oil pipe and position the pipe.
   b. **A** using several passes to 95 inch lbs. (11 Nm).
   c. Tighten oil pipe bolts **B** using several passes to 61 inch lbs. (7 Nm).
   d. Tighten oil pipe bolts **C** using several passes to 13 ft. lbs. (17 Nm).
   e. Tighten oil pipe bolts 1, 2, 3, 4 and 7 using several passes to 95 inch lbs. (11 Nm).
   f. Tighten oil pipe bolt 5 using several passes to 17 ft. lbs. (23 Nm).
   g. Tighten oil pipe bolt 6 using several passes to 34 ft. lbs. (47 Nm).
6. Install or connect the following:
   • Water hose
   • Upper radiator hose
   • Spark plug wires
   • Power steering hose bracket, if removed
   • Negative battery cable

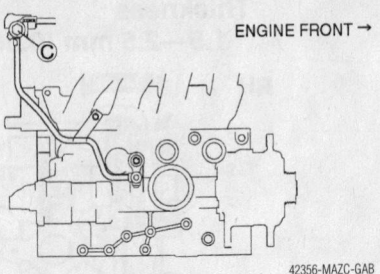

Tighten oil pipe bolts C using several passes to the specification in the text–1.8L (BP) engine

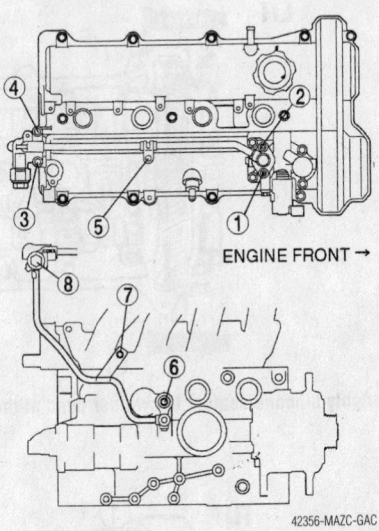

Tighten oil pipe bolts using several passes to the specification in the text–1.8L (BP) engine

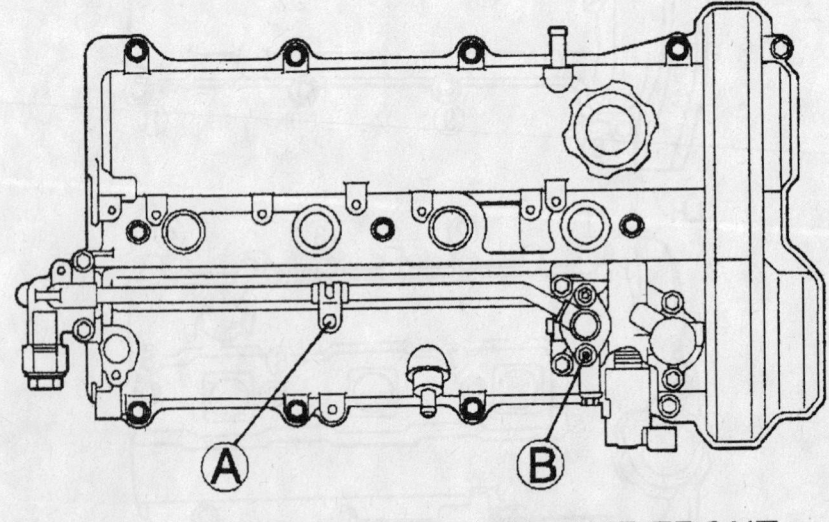

Tighten oil pipe bolts A and B using several passes to the specification in the text–1.8L (BP) engine

Millenia

2.3L (KJ) Engine

1. Before servicing the vehicle, refer to the precautions in the beginning of this section.

2. Remove or disconnect the following:
- Negative battery cable
- Ignition coils

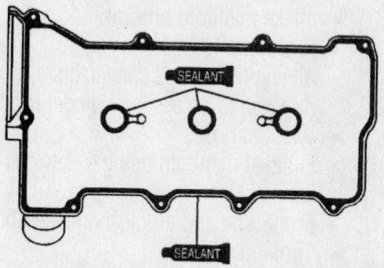

42356-MAZC-GDO

Apply a 0.004–0.07 inch (1–2mm) bead of sealant to rocker arm cover —Millenia models equipped with the 2.3L (KJ) engine

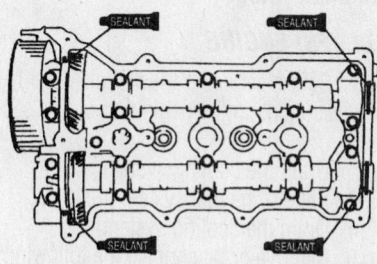

42356-MAZC-GDP

Apply a 0.059–0.098 inch (1.5–2.5mm) to the areas shown—Millenia models equipped with the 2.3L (KJ) engine

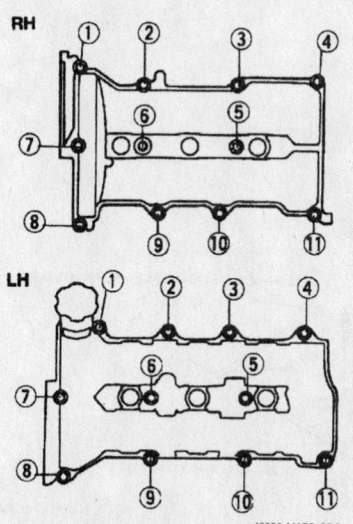

42356-MAZC-GDQ

Rocker arm cover bolt sequence—Millenia models equipped with the 2.3L (KJ) engine

- Vent hose
- Positive Crankcase Ventilation (PCV) hose
- Rocker arm cover and discard the gasket

3. Clean all mating surfaces of any residual gasket material.

To install:

4. Install or connect the following:

5. Apply a 0.004–0.07 inch (1–2mm) bead of sealant to rocker arm cover.
- New rocker arm gasket

6. Apply a 0.059–0.098 inch (1.5–2.5mm) to the areas illustrated
- Rocker arm cover. Torque the bolts in sequence to 86 inch lbs. (10 Nm), then retighten in sequence to 86 inch lbs. (10 Nm).
- PCV hose
- Vent hose
- Ignition coils
- Negative battery cable

2.5L (KL) ENGINE

1. Before servicing the vehicle, refer to the precautions in the beginning of this section.

2. Remove or disconnect the following:
- Negative battery cable
- Any components that would interfere with cover removal.
- Rocker arm cover bolts
- Rocker arm cover and discard the gasket

3. Clean all mating surfaces of any residual gasket material.

To install:

4. Apply silicone sealant to cylinder head at the areas illustrated.

5. Install or connect the following:
- Rocker arm cover with a new gasket. Torque the bolts in the sequence illustrated to 95 inch lbs. (11 Nm) using 2 or 3 steps. Retighten the right hand cover number 5 and 6 bolts and the left hand cover number 6 and 7 bolts.
- Remaining components removed to facilitate the rocker arm cover removal
- Negative battery cable

Thickness
1.5—2.5 mm {0.060—0.098 in}

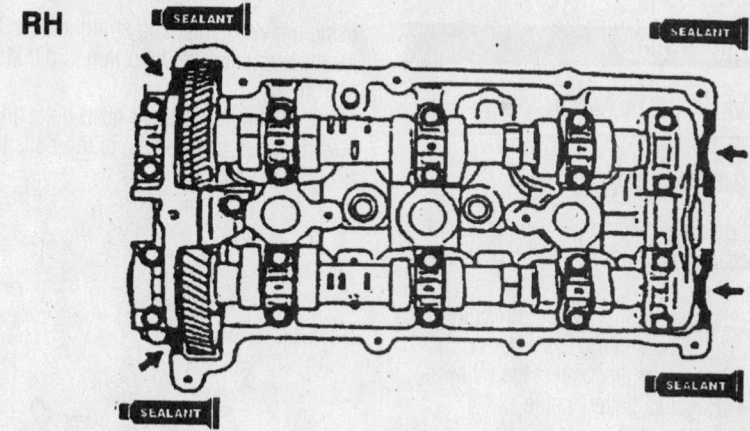

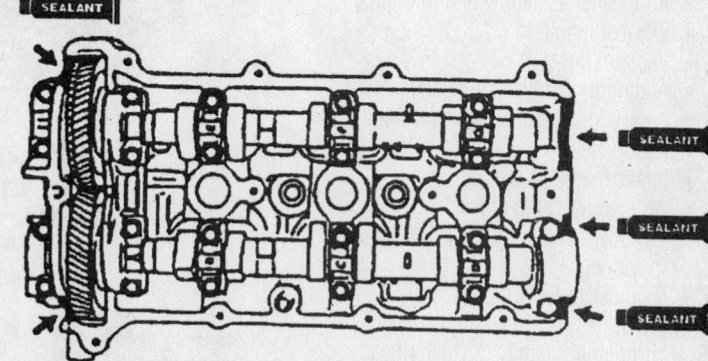

42356-MAZC-G09

Apply silicone sealant to cylinder head at the areas illustrated –2.5L (KL) engine

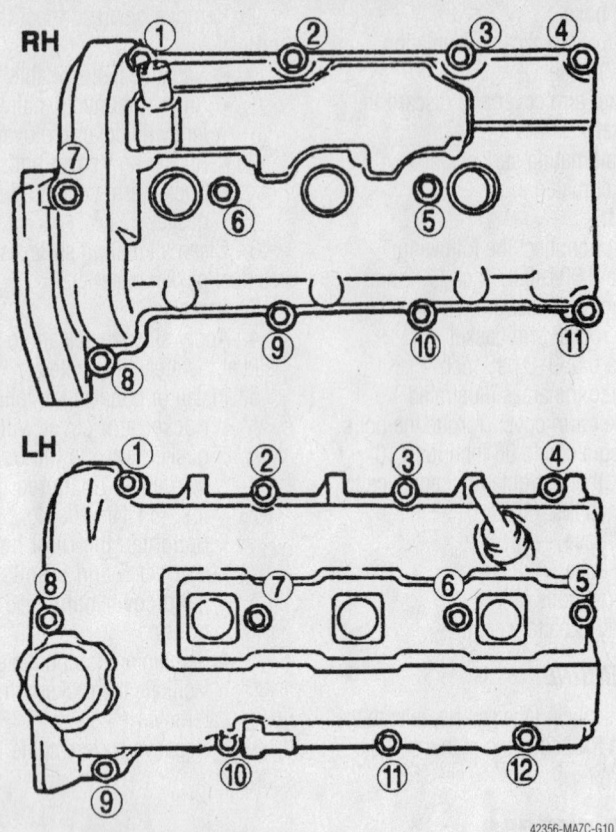

Tighten the rocker arm cover bolts in the sequence shown—2.5L (KL) engine

42356-MAZC-G10

Cylinder Head

REMOVAL & INSTALLATION

1.8L (BP) Engines

1. Before servicing the vehicle, refer to the precautions in the beginning of this section.
2. Relieve the fuel system pressure.
3. Drain the cooling system.
4. Remove or disconnect the following:
- Negative battery cable
- Timing belt
- Air cleaner assembly and front pipe
- Exhaust manifold
- Vacuum hoses
- All engine harness connectors necessary to access the cylinder head
- Fuel hose
- Intake manifold bracket
- Accelerator cable bracket
- Cylinder head bolts, in 2–3 steps, in sequence
- Cylinder head

To install:

5. Thoroughly, clean the cylinder head and the block contact surfaces. Examine the head gasket and check the cylinder head for cracks. Check the cylinder head for warpage using a feeler gauge and straightedge. The maximum allowable distortion is 0.004 inch (0.10mm).

6. Clean the cylinder head bolts and the threads in the block. Be sure the bolts turn freely in the block.

7. Install new head gasket on the engine block.
8. Install the cylinder head.
9. Lubricate the bolt threads and seat surfaces with clean engine oil and install them as follows:
   a. Torque the bolts in 2–3 steps to 56–60 ft. lbs. (75–81 Nm) in the proper sequence.
10. Install or connect the following:
- Accelerator cable bracket
- Intake manifold bracket
- Fuel hose
- All engine harness connectors removed to access the cylinder head
- Vacuum hoses
- Exhaust manifold using a new gasket
- Front pipe and air cleaner assembly
- Timing belt
- Negative battery cable
11. Fill and bleed the cooling system.
12. Change the oil and filter.
13. Run the engine and check for proper operation.

626 and Protégé

2.0L (FS) ENGINE

1. Before servicing the vehicle, refer to the precautions in the beginning of this section.
2. Drain the cooling system.
3. Relieve the fuel system pressure.
4. Drain the cooling system.
5. Remove or disconnect the following:
- Negative battery cable
- Timing belt

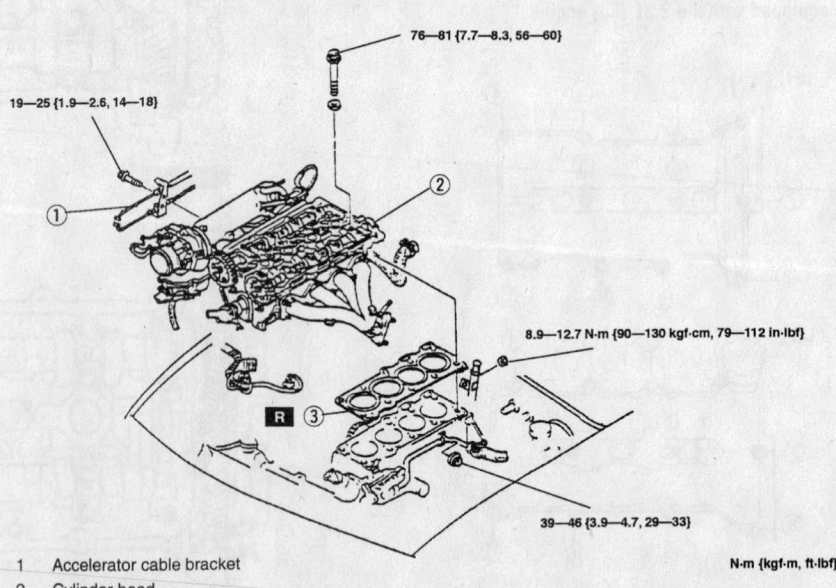

76—81 {7.7—8.3, 56—60}

19—25 {1.9—2.6, 14—18}

8.9—12.7 N·m {90—130 kgf·cm, 79—112 in·lbf}

39—46 {3.9—4.7, 29—33}

N·m {kgf·m, ft·lbf}

1	Accelerator cable bracket
2	Cylinder head
3	Cylinder head gasket

Exploded view of the cylinder head assembly—1.8L (BP) engines

42356-MAZC-GAE

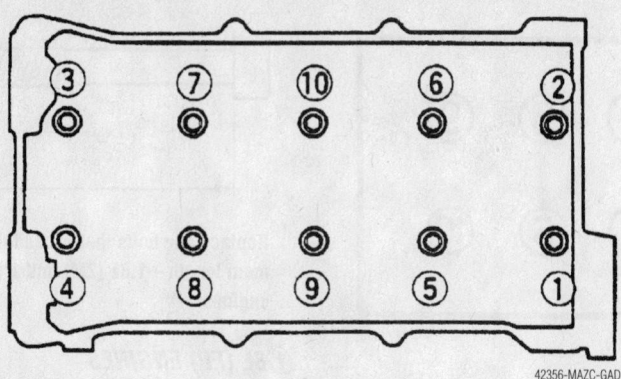

Cylinder head loosening sequence—1.8L (BP) engines

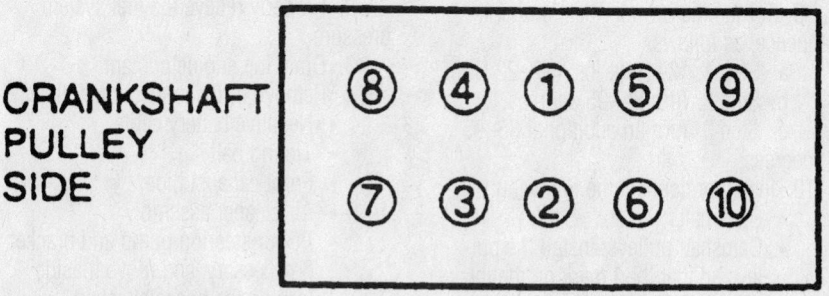

Cylinder head bolt tightening sequence—1.8L (BP) engines

- Front exhaust pipe
- Air cleaner assembly
- Power steering pump with the lines attached and set aside, if necessary on Protégé models
- Accelerator cable
- Fuel hose
- Ignition coil
- Camshaft pulley. Hold the camshaft pulley with a back–up wrench while loosening the pulley bolt.
- Camshaft
- Intake manifold bracket

6. Temporarily install the Number 3 engine mount to support the engine, on 626 models.

- Cylinder head bolts by loosening

them, in 2–3 steps, in the sequence illustrated
- Cylinder head

To install:

7. Install or connect the following:
- New cylinder head gasket
- Cylinder head

8. Apply clean engine oil to the bolt threads and seating faces.

9. Install new cylinder head bolts and torque in 2–3 steps, in sequence, to 13–16 ft. lbs. (17–22 Nm). If reusing old bolts, which is not recommended; make sure the maximum length of the bolt is 4.154 inch (105.5mm). The standard length of the bolt should be 4.103–4.125 inch recommended; (104.2–104.8mm).

Cylinder head bolt removal sequence—2.0L (FS) engines

Replace any bolts that exceed the maximum length of 4.154 inch (105.5mm)—2.0L (FS) engines

10. Paint a mark on the edge of each cylinder head bolt to use as a reference. Turn each bolt, in sequence, 85–95 degrees. Again, turn each bolt, in sequence, an additional 85–95 degrees.

11. Support the engine assembly with a suitable lifting device and remove the number 3 mount.

12. Install or connect the following:
- Intake manifold bracket
- Camshafts
- Camshaft pulley, make sure the camshaft sprocket pulleys are positioned with the pins facing up. Hold the camshaft pulley with a back–up wrench while tightening the pulley bolt to 37–44 ft. lbs. (50–60 Nm).
- Ignition coil
- Fuel hose
- Accelerator cable
- Power steering pump, if removed
- Air cleaner assembly
- Front exhaust pipe
- Timing belt
- Negative battery cable

13. Fill and bleed the cooling system.

14. Change the oil and filter.

15. Run the engine and check for proper operation.

1.6L (ZM) ENGINES

1. Before servicing the vehicle, refer to the precautions in the beginning of this section.

2. Properly relieve the fuel system pressure.

3. Drain the engine coolant.

4. Remove or disconnect the following:
- Negative battery cable
- Timing belt
- Front exhaust pipe
- Exhaust manifold insulator and the Exhaust Gas Recirculation (EGR) Pipe
- Fresh air duct and air cleaner Assembly
- Accelerator cable and bracket
- All vacuum hoses
- Engine wiring harness connectors
- Fuel supply and return hoses

CRANKSHAFT PULLEY SIDE

7923MG16

Cylinder head bolt tightening sequence—2.0L (FS) engines

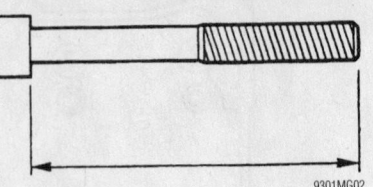

9301MG02

Replace any bolts that exceed the maximum length—1.6L (ZM) and 1.8L (FP) engines

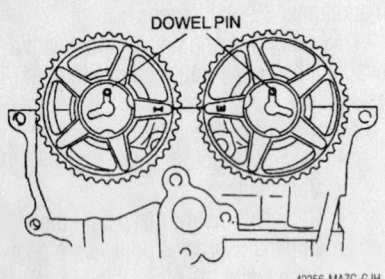

DOWEL PIN

42356-MAZC-GJH

Make sure the camshaft sprocket pulleys are positioned with the pins facing up—2.0L (FS) engine

- Intake manifold support bracket
- Heater hoses
- Camshaft pulleys by holding the them with a wrench
- Camshafts
- Cylinder head bolts in sequence
- Cylinder head

To install:

5. Thoroughly, clean the cylinder head and the block contact surfaces.

6. Clean the cylinder head bolts and the threads in the block. Be sure the bolts turn freely in the block.

7. Measure the length of the cylinder head bolts, as shown, maximum bolt length is 3.956 inch (100.5mm).

8. Install or connect the following:

- New head gasket
- Cylinder head

9. Torque the cylinder head bolts, in sequence, as follows.

a. Step 1: 13–16 ft. lbs. (17–22 Nm).

b. Step 2: Turn 85–95 degrees.

c. Step 3: Turn an additional 85–95 degrees.

10. Install or connect the following:

- Camshafts
- Camshaft pulleys, install the pulleys so that the **I** mark on the intake side or **E** mark on the exhaust side are facing up. Torque the bolts to 37–44 ft. lbs. (50–60 Nm).
- Heater hoses
- Intake manifold support bracket
- Fuel hoses
- Engine wiring harness connector
- Any vacuum hoses that were removed
- Accelerator cable and bracket
- Air cleaner assembly and fresh air duct
- EGR system
- Exhaust manifold insulator and front exhaust pipe
- Timing belt
- Negative battery cable

11. Fill the cooling system.

12. Start the vehicle, check for leaks and repair if necessary.

1.8L (FP) ENGINES

1. Before servicing the vehicle, refer to the precautions in the beginning of this section.

2. Properly relieve the fuel system pressure.

3. Drain the engine coolant.

4. Remove or disconnect the following:

- Negative battery cable
- Timing belt
- Front exhaust pipe
- Air cleaner assembly
- Power steering pump and bracket, if necessary, and move it aside leaving the hoses attached
- Accelerator cable
- Fuel supply and return hoses
- Ignition coil
- Camshaft pulleys by holding the them with a wrench
- Camshafts
- Cylinder head bolts in sequence
- Cylinder head

To install:

5. Thoroughly, clean the cylinder head and the block contact surfaces.

6. Clean the cylinder head bolts and the threads in the block. Be sure the bolts turn freely in the block.

7. Measure the length of the cylinder head bolts, as shown, maximum bolt length is 4.153 inch (105.5mm).

8. Install or connect the following:

- New head gasket
- Cylinder head

9. Torque the cylinder head bolts, in sequence, as follows.

a. Step 1: 13–16 ft. lbs. (17–22 Nm).

b. Step 2: Turn 85–95 degrees.

c. Step 3: Turn an additional 85–95 degrees.

10. Install or connect the following:

- Camshafts
- Camshaft pulleys, install the pulleys so that the dowels are facing up. Torque the bolts to 37–44 ft. lbs. (50–60 Nm).
- Ignition coil
- Fuel supply and return hoses

ENGINE FRONT

9301MG01

Cylinder head bolt removal sequence—1.6L (ZM) and 1.8L (FP) engines

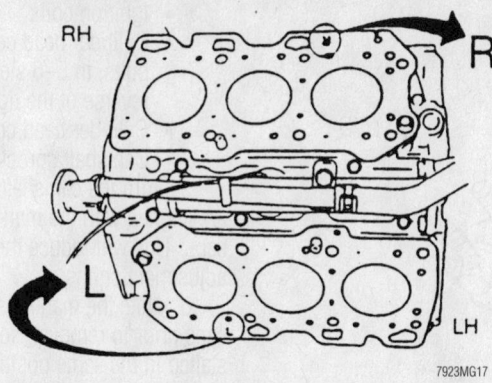

Cylinder head gasket positioning—6-cylinder engines

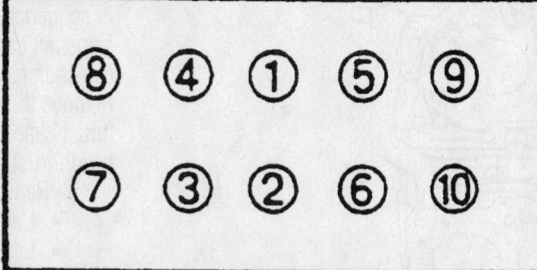

CRANKSHAFT PULLEY SIDE

⑧ ④ ① ⑤ ⑨

⑦ ③ ② ⑥ ⑩

7923MG16

Cylinder head bolt tightening sequence—1.6L (ZM) and 1.8L (FP) engines

11. When connecting the accelerator cable, perform the following adjustment:

a. Move the white locking tab A to the unlock position.

b. Turn stopper B to the to the unlock position.

➡**If stopper B will not unlock, it may be necessary to carefully bend back tab C out a little using a suitable pry tool.**

c. Push or pull the cable housing directly behind the spring.

d. Turn the stopper B to the lock position.

e. Measure the free play which should be 0.04–0.11 inch (1–3mm) and make sure the cable free play is within specification.

f. Move the white locking tab to the lock position and check for proper accelerator operation.

- Power steering pump and bracket
- Air cleaner assembly
- Front exhaust pipe
- Timing belt
- Negative battery cable

12. Fill the cooling system.

13. Start the vehicle, check for leaks and repair if necessary.

2.5L (KL) ENGINE

1. Before servicing the vehicle, refer to the precautions in the beginning of this section.

2. Drain the cooling system.

3. Relieve the fuel system pressure.

4. Drain the cooling system.

5. Remove or disconnect the following:

- Negative battery cable
- Timing belt
- Front exhaust pipe
- Air cleaner assembly
- Intake manifold
- Fuel hose
- Cylinder head cover
- Camshaft pulley. Hold the camshaft pulley with a back–up wrench while loosening the pulley bolt.
- Upper radiator hose
- Number 3 engine mount bracket
- Seal plate
- Water outlet
- Camshaft
- Alternator stay

6. Temporarily install the Number 3 engine mount to support the engine.

- Cylinder head bolts by loosening them, in 2–3 steps, in the sequence illustrated
- Cylinder head

To install:

7. Install or connect the following:

- New cylinder head gasket
- Cylinder head

8. Apply clean engine oil to the bolt threads and seating faces.

9. Install new cylinder head bolts and torque in 2–3 steps, in sequence, to 17–19 ft. lbs. (23–26 Nm). If reusing old bolts, which is not recommended; make sure the maximum length of the bolt is 5.217 inch (132.5mm). The standard length of the bolt should be 5.166–5.188 inch (131.2–131.8mm).

10. Paint a mark on the edge of each cylinder head bolt to use as a reference. Turn each bolt, in sequence, 85–95 degrees. Again, turn each bolt, in sequence, an additional 85–95 degrees.

11. Support the engine assembly with a suitable lifting device and remove the number 3 mount.

12. Install or connect the following:

- Alternator stay
- Camshaft
- Water outlet
- Seal plate
- Number 3 engine mount bracket
- Upper radiator hose
- Camshaft pulley. Hold the camshaft pulley with a back–up wrench while tightening the pulley bolt to 90–97 ft. lbs. (123–132 Nm).
- Cylinder head cover
- Fuel hose
- Intake manifold

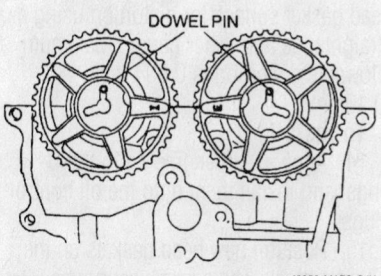

DOWEL PIN

42356-MAZC-GJH

Make sure the camshaft sprocket pulleys are positioned with the pins facing up— 1.8L (FP) engine

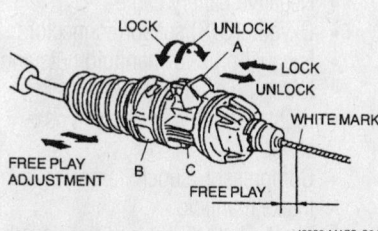

Free play
1.5—4.0 mm {0.06—0.15 in}

LOCK   UNLOCK
A
LOCK
UNLOCK
WHITE MARK

FREE PLAY ADJUSTMENT

B   C

FREE PLAY

42356-MAZC-G04

Accelerator cable adjustment components—Protégé models equipped with the 1.8L (FP) engine

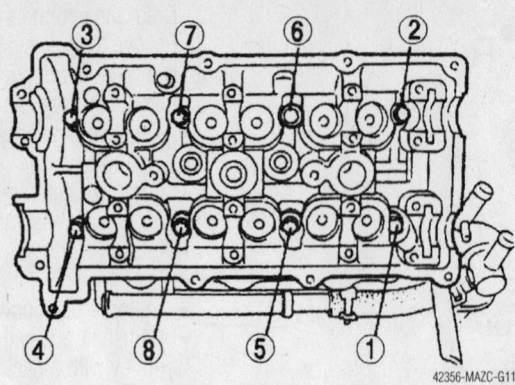

Cylinder head bolt removal sequence—2.5L (KL) engines

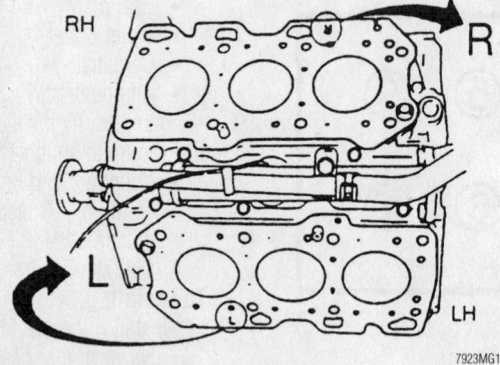

Cylinder head gasket positioning—2.5L (KL) engines

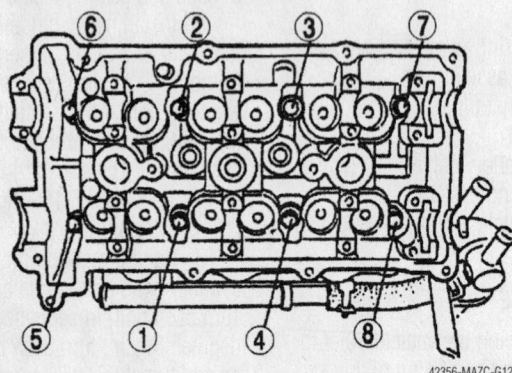

Cylinder head bolt tightening sequence—2.5L (KL) engines

- Air cleaner assembly
- Front exhaust pipe
- Timing belt
- Negative battery cable

13. Fill and bleed the cooling system.

14. Change the oil and filter.

15. Run the engine and check for proper operation.

Millenia

2.3L (KJ) ENGINES

1. Before servicing the vehicle, refer to the precautions in the beginning of this section.

2. Relieve the fuel system pressure.

3. Drain the engine coolant.

4. Remove or disconnect the following:
   - Negative battery cable
   - Oxygen (O$_2$S) sensor connectors
   - Exhaust pipe-to-manifold nuts and lower the exhaust pipes
   - Right-hand 3-way catalytic converter
   - Compressor (supercharger)
   - Intake manifold
   - Timing belt covers and timing belt
   - Spacer and O-ring from the front of the camshaft

- Ignition coils
- Cylinder head cover mounting bolts, in 5–6 steps, using the reverse of the tightening sequence
- Cylinder head cover
- Camshaft sprockets

5. Turn the camshafts so the knock pins are aligned with the marks on the camshaft caps. This will reduce the pressure on the adjustment shims.

6. Note the markings on the camshaft caps prior to removal, so they can be reinstalled in the same positions. The right-hand (rear) caps are marked with numbers and the left-hand (front) caps are marked with letters.

7. Loosen the front camshaft cap bolts in sequence, in 5–6 steps. Remove the front camshaft caps. remove the remaining camshaft cap bolts in the proper sequence. Remove the caps, being sure to remove the thrust caps last. Do not damage the cylinder head thrust bearing support.

8. Remove or disconnect the following:
   - Camshafts and oil seals
   - Lifters and adjustment shims

9. Identify and mark each lifter as it is removed so it can be reinstalled in the same position.

10. Remove or disconnect the following:
   - Lower radiator hose
   - Water inlet pipe
   - Compressor bracket
   - Alternator bracket bolt to gain additional clearance
   - Rubber insulator from the left-hand cylinder head

11. Temporarily install the No. 3 engine mount, which was removed with the timing belt, to support the engine.

12. Remove or disconnect the following:
   - Engine support device
   - Cylinder head bolts, in 2–3 steps, in the sequence illustrated
   - Cylinder heads
   - Oil control plug O-rings

13. Clean all gasket mating surfaces. Inspect the cylinder head for damage, cracks, and water and oil leakage. Check the head gasket surface for distortion using a straightedge and feeler gauge. Maximum allowable distortion is 0.004 inch (0.10mm).

To install:

14. Apply clean engine oil to the O-rings, and install them onto the oil control plugs.

15. Position new head gaskets on the cylinder block. The gaskets cannot be interchanged between sides and are marked **R** and **L** for right and left side.

16. Install the cylinder heads.

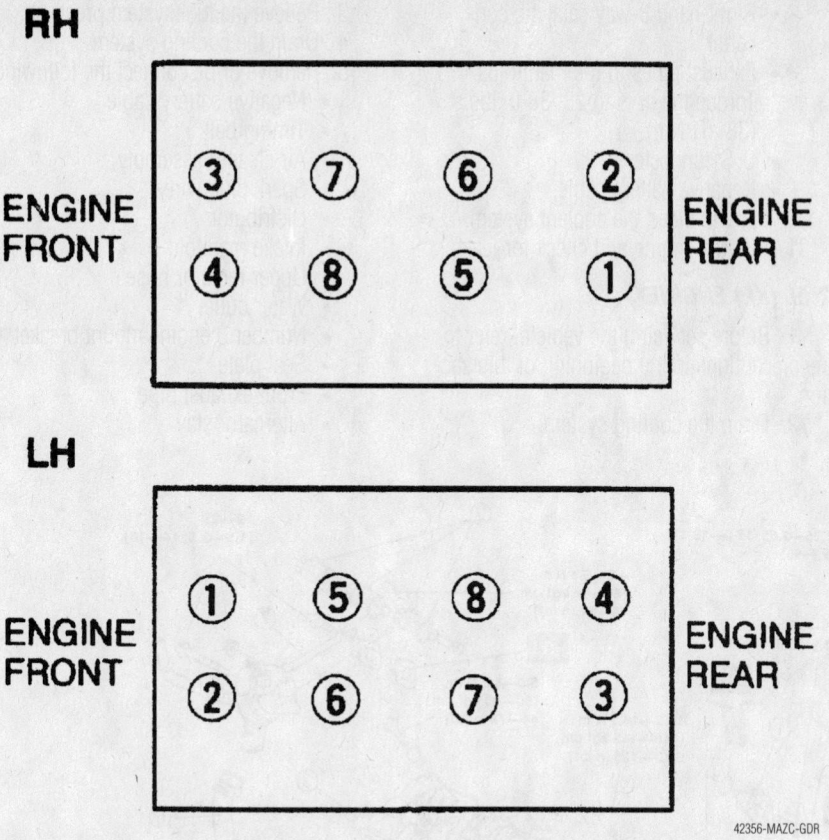

Loosen the cylinder head bolts in this sequence using several passes—Millenia models equipped with the 2.3L (KJ) engine

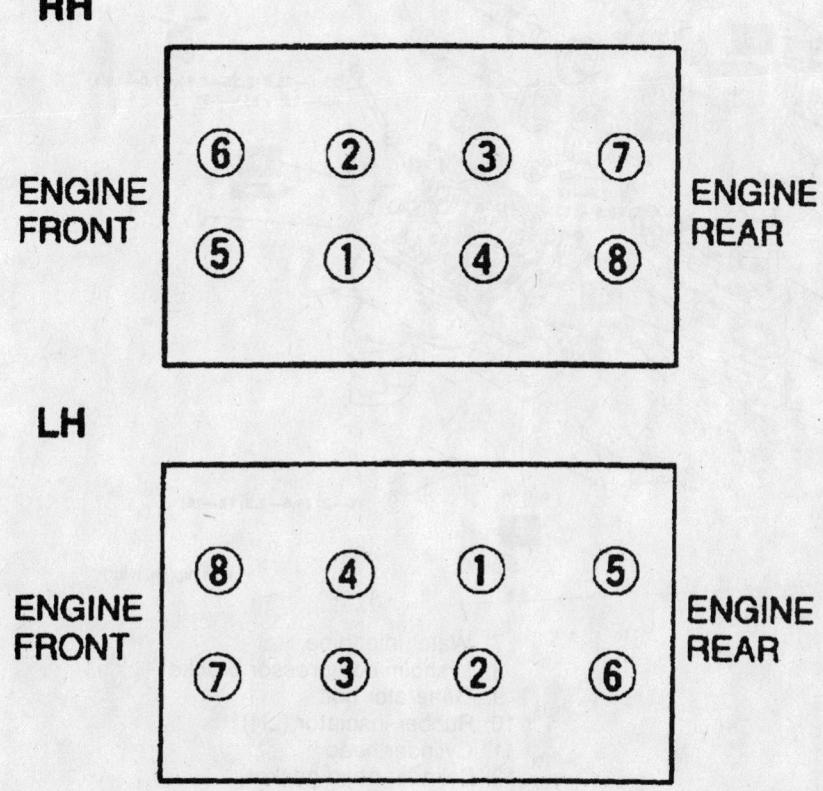

Cylinder head bolt torque sequence—Millenia models equipped with the 2.3L (KJ) engine

17. Apply clean engine oil to the threads of new cylinder head bolts and install. Torque the bolts in 2–3 steps, in sequence, to 17–19 ft. lbs. (23–26 Nm).

18. Paint a mark on the edge of each cylinder head bolt to use as a reference. Turn each bolt, in sequence, 85–95 degrees. Again, turn each bolt, in sequence, an additional 85–95 degrees.

19. Install the rubber insulator onto the left-hand cylinder head.

20. Fit the knock sensor harness into the drill hole on the cylinder block. Pass the harness under the rubber insulator.

21. Install or connect the following:
- Engine support device and remove the No. 3 engine mount
- Alternator bracket bolt. Torque the bolt to 12–16 ft. lbs. (16–22 Nm).
- Compressor bracket. Torque the bolts to 14–18 ft. lbs. (19–25 Nm).
- Water inlet pipe. Torque the bolts to 14–18 ft. lbs. (19–25 Nm).
- Lower radiator hose
- Lifters in their original positions by lubricating them with engine oil. Verify that they move smoothly in their bore
- New oil seals on the camshafts

22. Apply clean engine oil to the camshaft lobes, journals and supports.

23. Install or connect the following:
- Camshafts so the gear marks align
- Thrust caps. Torque the bolts, in 5–6 steps, until they are fully seated on the cylinder head

24. Apply silicone sealant, at a thickness of 0.06–0.09 inch (1.5–2.5mm), to the cylinder head surface in the area forward of the camshaft gear cavity.

25. Install or connect the following:
- Remaining camshaft caps in their original positions. Torque the caps, in sequence, in 5 equal steps, with the final step being 100–125 inch lbs. (11–14 Nm).
- New camshaft oil seal lubricated with engine oil. Tap the seal in evenly with a Seal Installer tool 49 F401 337A with a final protrusion of 0–0.02 inch (0–0.5mm). Tap in a new blind cap
- Camshaft sprockets. Torque the bolts to 91–103 ft. lbs. (123–140 Nm).

26. Measure and adjust the valve clearances.

27. Remove any sealant and gasket material from the cylinder head cover contact surfaces.

28. Apply silicone sealant to the cylinder

head in the area adjacent to the front and rear camshaft caps.

29. Install or connect the following:
- Cylinder head cover using a new gasket. Torque the bolts in 5–6 steps, in sequence, to 44–78 inch lbs. (5–9 Nm).
- Distributor using a new O-ring
- Ignition coils
- Spacer using a new O-ring. Torque the bolt to 14–18 ft. lbs. (19–25 Nm).
- Timing belt and timing belt cover
- Intake manifold
- Compressor (supercharger)

- Right-hand 3-way catalytic converter
- Exhaust pipes to the manifolds. Torque the nuts to 28–38 ft. lbs. (38–51 Nm).
- O_2S connectors
- Negative battery cable
30. Fill and bleed the coolant system.
31. Run the engine and check for leaks.

2.5L (KL) ENGINE

1. Before servicing the vehicle, refer to the precautions in the beginning of this section.
2. Drain the cooling system.

3. Relieve the fuel system pressure.
4. Drain the cooling system.
5. Remove or disconnect the following:
- Negative battery cable
- Timing belt
- Air cleaner assembly
- Spark plug wires
- Distributor
- Intake manifold
- Upper radiator hose
- Water outlet
- Number 3 engine mount bracket
- Seal plate
- Front exhaust pipe
- Alternator stay

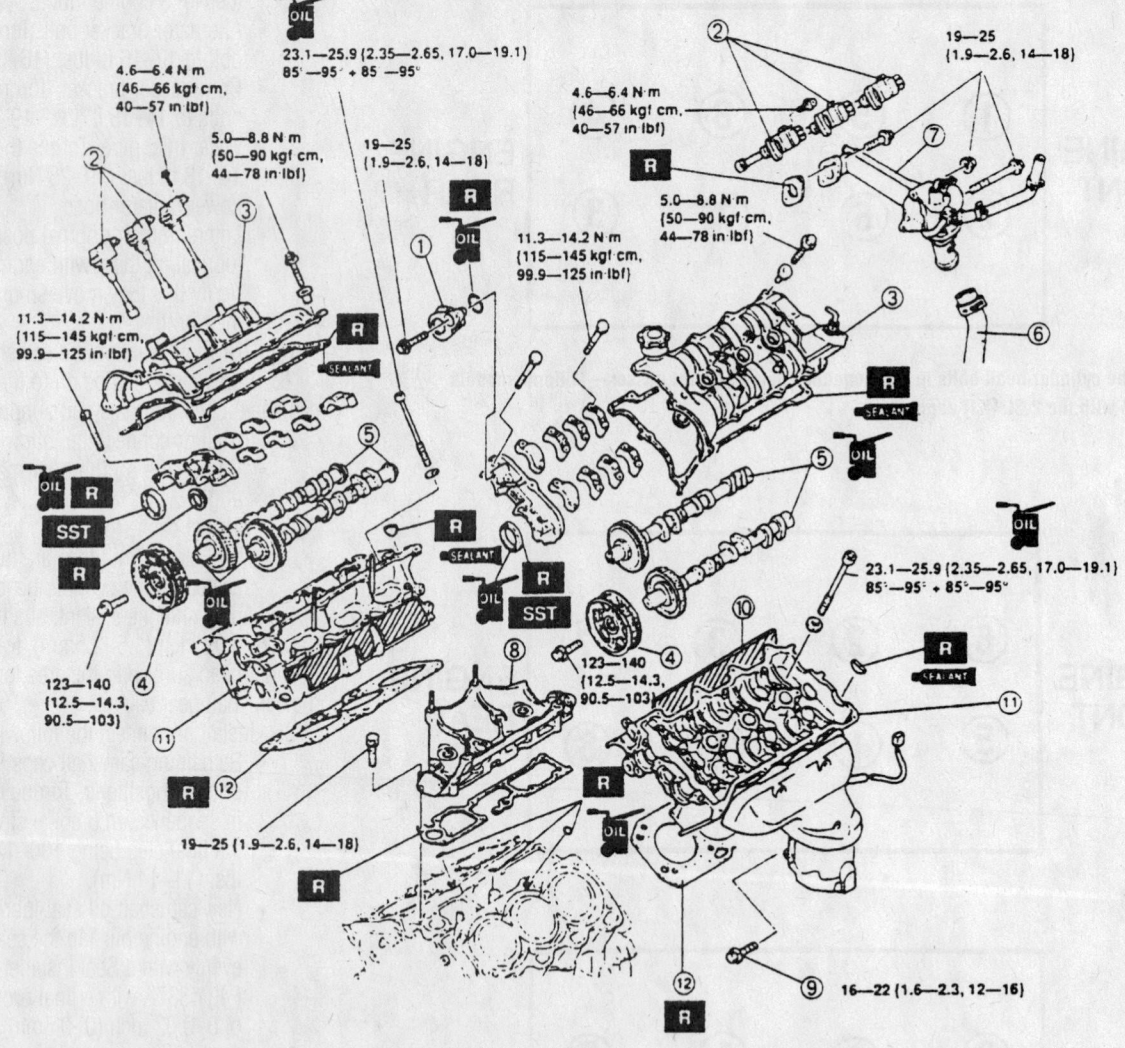

1. Spacer
2. Ignition coil
3. Cylinder head cover
4. Camshaft pulley
5. Camshaft
6. Lower radiator hose
7. Water inlet pipe
8. Lysholm compressor bracket
9. Generator bolt
10. Rubber insulator (LH)
11. Cylinder head
12. Cylinder head gasket

Exploded view of the cylinder head and related components—2.3L engine

7923MG24

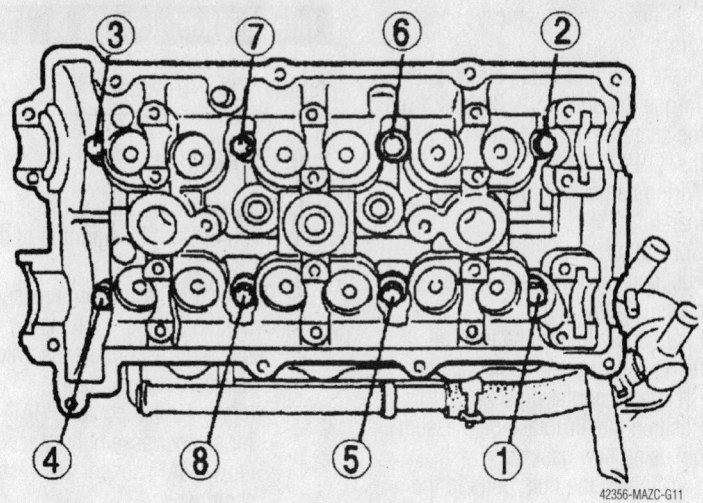

Cylinder head bolt removal sequence—2.5L (KL) engines

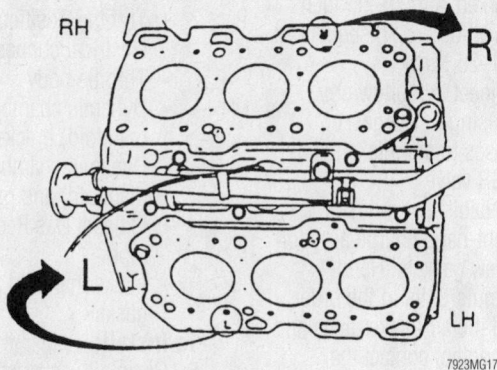

Cylinder head gasket positioning—2.5L (KL) engines

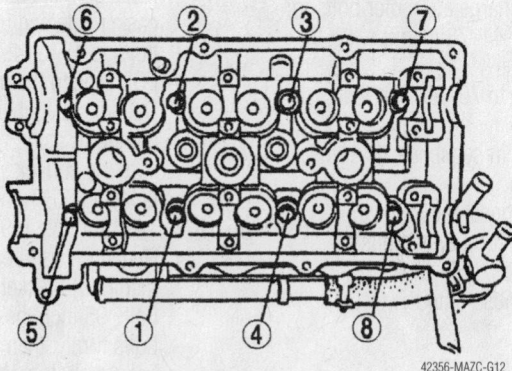

Cylinder head bolt tightening sequence—2.5L (KL) engines

- Cylinder head cover
- Camshaft pulley. Hold the camshaft pulley with a back–up wrench while loosening the pulley bolt.
- Camshaft
6. Temporarily install the Number 3 engine mount to support the engine.
  - Cylinder head bolts by loosening them, in 2–3 steps, in the sequence illustrated
- Cylinder head

To install:
7. Install or connect the following:
  - New cylinder head gasket
  - Cylinder head
8. Apply clean engine oil to the bolt threads and seating faces.
9. Install new cylinder head bolts and torque in 2–3 steps, in sequence, to 17–19 ft. lbs. (23–26 Nm). If reusing old bolts,

which is not recommended; make sure the maximum length of the bolt is 5.217 inch (132.5mm). The standard length of the bolt should be 5.166–5.188 inch (131.2–131.8mm).

10. Paint a mark on the edge of each cylinder head bolt to use as a reference. Turn each bolt, in sequence, 85–95 degrees. Again, turn each bolt, in sequence, an additional 85–95 degrees.

11. Support the engine assembly with a suitable lifting device and remove the number 3 mount.

12. Install or connect the following:
  - Camshaft
  - Camshaft pulley. Hold the camshaft pulley with a back–up wrench while tightening the pulley bolt to 90–97 ft. lbs. (123–132 Nm).
  - Cylinder head cover
  - Alternator stay
  - Front exhaust pipe
  - Seal plate
  - Number 3 engine mount bracket
  - Water outlet
  - Upper radiator hose
  - Intake manifold
  - Distributor and spark plug wires
  - Air cleaner assembly
  - Timing belt
  - Negative battery cable
13. Fill and bleed the cooling system.
14. Change the oil and filter.
15. Run the engine and check for proper operation.

Rocker Arms/Shafts

REMOVAL & INSTALLATION

All Mazda engines covered in this manual are not equipped with rocker arms/shafts, the camshafts directly actuate the valves through a bucket type cam follower.

Supercharger

REMOVAL & INSTALLATION

2.3L Engine

1. Before servicing the vehicle, refer to the precautions in the beginning of this section.
2. Relieve the fuel system pressure.
3. Drain the cooling system.
4. Remove or disconnect the following:
  - Negative battery cable
  - Dynamic chamber cover
  - Charge air cooler air duct

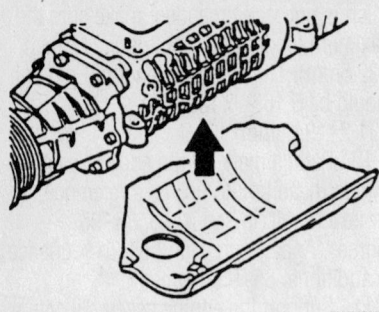

7923MG26

When installing the compressor, ensure that the rubber insulating pad is temporarily affixed to the compressor

- Vacuum hoses and electrical connectors from the air cleaner housing
- Air cleaner assembly
- Fresh air ducts
- Mass Air Flow (MAF) sensor and the air intake hose from the throttle body
- Resonator
- Right-hand charge air cooler
- Left-hand charge air cooler
- Accelerator cable
- Vacuum hoses from the rear of the intake manifold and Exhaust Gas Recirculation (EGR) valve
- EGR valve
- Air intake pipe assembly
- Charge air cooler pipe
- Fuel supply line at the fuel rails and discard the copper washers
- Fuel and vacuum lines from the fuel pressure regulator
- Coolant hoses
- Wiring harness from the intake manifold
- Intake manifold mounting nuts and bolts in 2–3 steps
- Intake manifold
- Fuel hoses and electrical connectors from the throttle body
- Throttle body
- Drive belt from the compressor (supercharger)
- Mounting bolts from the compressor
- Compressor

To install:

5. Clean all gasket mating surfaces.
6. Position the rubber shield for the compressor onto the compressor using double sided adhesive tape.
7. Install or connect the following:
- Compressor. Torque the nuts to 14–18 ft. lbs. (19–25 Nm).
- Compressor drive belt
- Throttle body. Torque the bolts to 14–18 ft. lbs. (19–25 Nm).

- Fuel hoses and electrical connectors
- Intake manifold using a new gasket. Torque the bolts in 2–3 steps, from the center to the ends, to 14–18 ft. lbs. (19–25 Nm).
- Wiring harness onto the intake manifold
- Coolant hoses
- Fuel and vacuum lines to the fuel pressure regulator
- Fuel supply line to the fuel rail using new copper crush washers
- Charge air cooler pipe
- Position the air intake pipe assembly using new gaskets

8. Hand-tighten the nuts/bolts in the order shown until the air intake pipe contacts the intake manifold. Verify that the rubber gaskets are not twisted or distorted. Torque the bolts marked **A** to 70–95 inch lbs. (8–11 Nm) and all others, in sequence, to 14–18 ft. lbs. (19–25 Nm).
9. Install or connect the following:
- EGR valve using a new gasket
- Vacuum hoses to the intake manifold and EGR valve
- Accelerator cable and adjust it
- Left and right-hand charge air coolers using new gaskets. Hand-tighten the nuts/bolts in the order shown until the air intake pipes and charge air coolers contact the intake manifold. Verify that the rubber gaskets are not twisted or distorted.
10. Torque the charge air cooler bolts to:
   a. Marked **A**: 44–78 inch lbs. (5–9 Nm).
   b. Marked **B**: to 70–95 inch lbs. (8–11 Nm).
   c. All others, in sequence: 14–18 ft. lbs. (19–25 Nm).
11. Install or connect the following:
- Resonator. Torque the bolts to 12–16 ft. lbs. (16–22 Nm).
- Air intake hose onto the throttle body
- MAF sensor
- Fresh air ducts
- Air cleaner assembly
- Vacuum hoses and electrical connectors to the air cleaner housing
- Charge air cooler air duct. Torque the bolts to 70–95 inch lbs. (8–11 Nm).
- Dynamic chamber cover
- Negative battery cable
12. Fill the cooling system.
13. Start the vehicle, check for leaks and repair if necessary.

REMOVAL & INSTALLATION

1.8L (BP) Engines

1. Before servicing the vehicle, refer to the precautions in the beginning of this section.
2. Relieve the fuel system pressure.
3. Drain the cooling system.
4. Remove or disconnect the following:
- Negative battery cable
- Intake Air Temperature (IAT) sensor
- Air cleaner assembly, Mass Air Flow (MAF) sensor and resonance chamber
- Throttle (automatic transaxle only) and accelerator cables
- Idle Air Control (IAC) valve and the Throttle Position Sensor (TPS) electrical connectors
- Throttle body
- Dynamic chamber (upper intake manifold) bracket
- Dynamic chamber (upper intake manifold) and gasket
- Exhaust Gas Recirculation (EGR) pipe
- Intake manifold and discard the gasket

To install:

5. Clean all gasket mating surfaces.
6. Install or connect the following:
- Intake manifold using a new gasket, make sure the convex side of the gasket is facing up. Torque the bolts to 17–21 ft. lbs. (19–25 Nm).
- EGR pipe to the manifold
- Dynamic chamber (upper intake manifold) using new gaskets, make sure the convex side of the gasket is facing up. Tighten the bolts to 14–18 ft. lbs. (19–25 Nm).
- Dynamic chamber (upper intake manifold) bracket and tighten the bolts 95 inch lbs. (11 Nm). Tighten bolts firmly, then tighten the chamber side bolt before tightening the fuel rail side bolt.
- Throttle body using a new gasket. Tighten the bolts to 14–18 ft. lbs. (19–25 Nm).
- Electrical connectors for the IAC valve and the TPS
- Connect and adjust the throttle and accelerator cables
- IAT sensor
- Air cleaner assembly, MAF sensor and ducts
- Negative battery cable

7. Fill the cooling system.
8. Run the engine and check for leaks.

626 and Protégé

1.6L (ZM) ENGINES

1. Before servicing the vehicle, refer to the precautions in the beginning of this section.

2. Relieve the fuel system pressure. Drain the cooling system.

3. Remove or disconnect the following:
- Negative battery cable
- Fuel hoses from the fuel rail

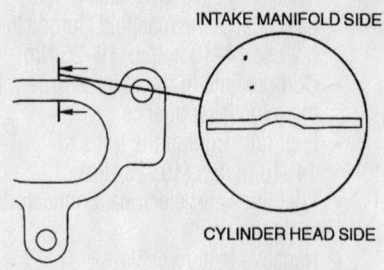

42356-MAZC-GAG

Make sure the convex side of the gasket is facing up when installing the intake manifold gasket—1.8L (BP) engine

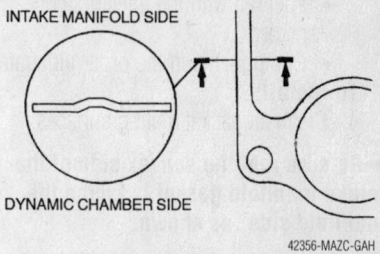

42356-MAZC-GAH

Make sure the convex side of the gasket is facing up when installing the upper intake manifold gasket—1.8L (BP) engine

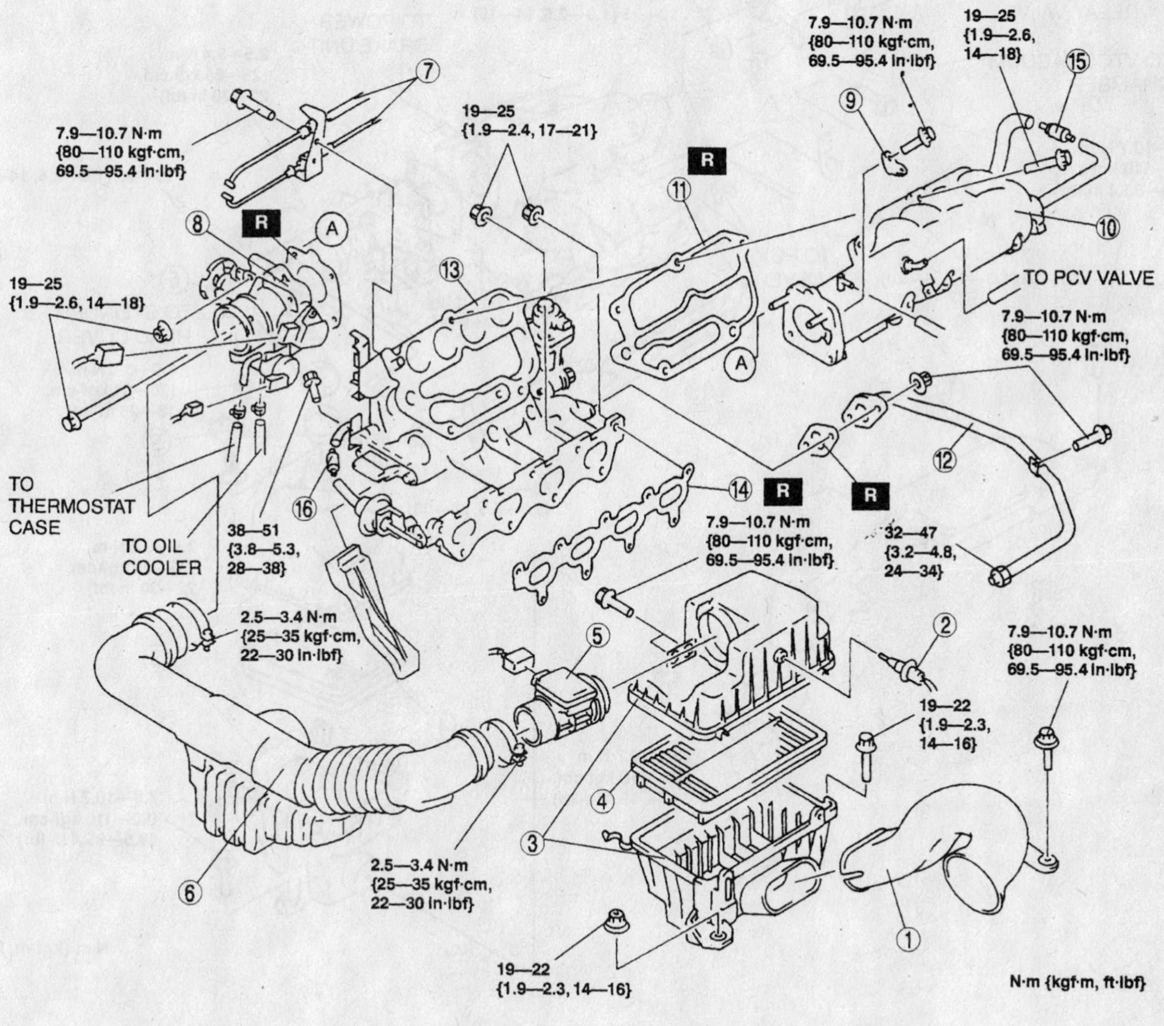

1	Fresh-air duct
2	IAT sensor
3	Air cleaner (ACL)
4	Air cleaner (ACL) element
5	MAF sensor
6	Air hose
7	Accelerator cable (and throttle cable (AT only))
8	Throttle body (TB)
9	Dynamic chamber bracket
10	Dynamic chamber
11	Dynamic chamber gasket
12	EGR pipe
13	Intake manifold
14	Intake manifold gasket
15	VTCS check valve (one-way)
16	Delay valve

42356-MAZC-GAF

Exploded view of the intake manifold and related components—1.8L (BP) engine

- Fuel injector electrical connectors
- Fuel rail with the injectors connected
- Components in the order illustrated

To install:

4. Clean all gasket mating surfaces.

➡ **Be sure that the convex side of the intake manifold gasket is facing the manifold side, as shown.**

5. Install or connect the following:
  - Intake manifold using a new gasket,

make sure the convex side of the gasket faces the manifold. Torque the bolts to 14–18 ft. lbs. (19–25 Nm).

- Components in the reverse order of the removal sequence
- Fuel rail. Tighten the bolts to 14–18 ft. lbs. (19–25 Nm).
- Fuel lines and electrical connectors to the fuel rail
- Negative battery cable

6. Fill the cooling system.
7. Run the engine and check for leaks.

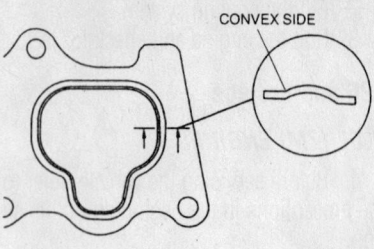

Cross-sectional view of the intake manifold gasket—1.6L (ZM) engines

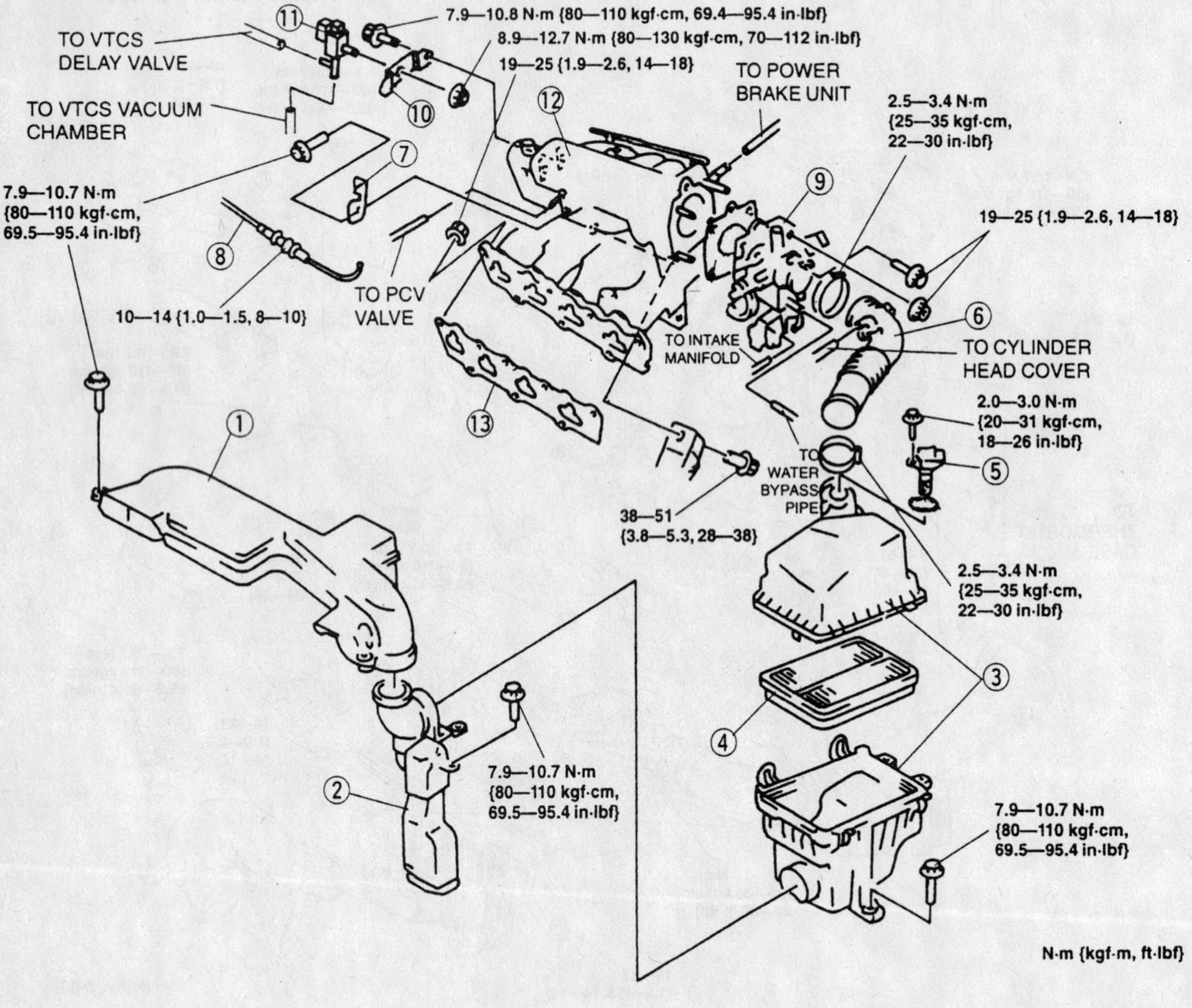

9301MG04

1	Fresh-air duct
2	Resonance chamber
3	Air cleaner
4	Air cleaner element
5	MAF sensor (Integrated with IAT sensor)
6	Air hose
7	Accelerator cable bracket
8	Accelerator cable
9	Throttle body
10	VTCS solenoid valve bracket
11	VTCS solenoid valve
12	Intake manifold
13	Intake manifold gasket

N·m {kgf·m, ft·lbf}

Exploded view of the intake manifold, illustrating the removal and installation components with tightening values—1.6L engine

9301MG03

1.8L (FP) ENGINES

1. Before servicing the vehicle, refer to the precautions in the beginning of this section.

2. Relieve the fuel system pressure. Drain the cooling system.

3. Remove or disconnect the following:
- Negative battery cable
- Fresh air duct and resonance chamber
- Intake Air Temperature (IAT) sensor
- Air cleaner and filter
- Mass Air Flow (MAF) sensor

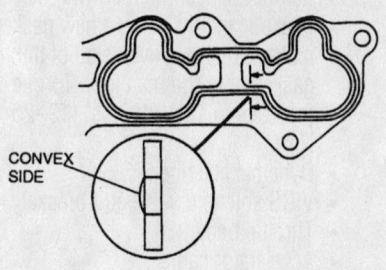

CONVEX SIDE

42356-MAZC-GJI

Be sure that the convex side of the intake manifold gasket is facing the manifold side—1.8L (FP) engine

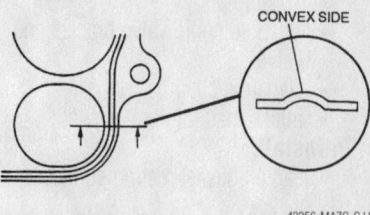

CONVEX SIDE

42356-MAZC-GJJ

Cross-sectional view of the dynamic chamber gasket—1.8L (FP) engines

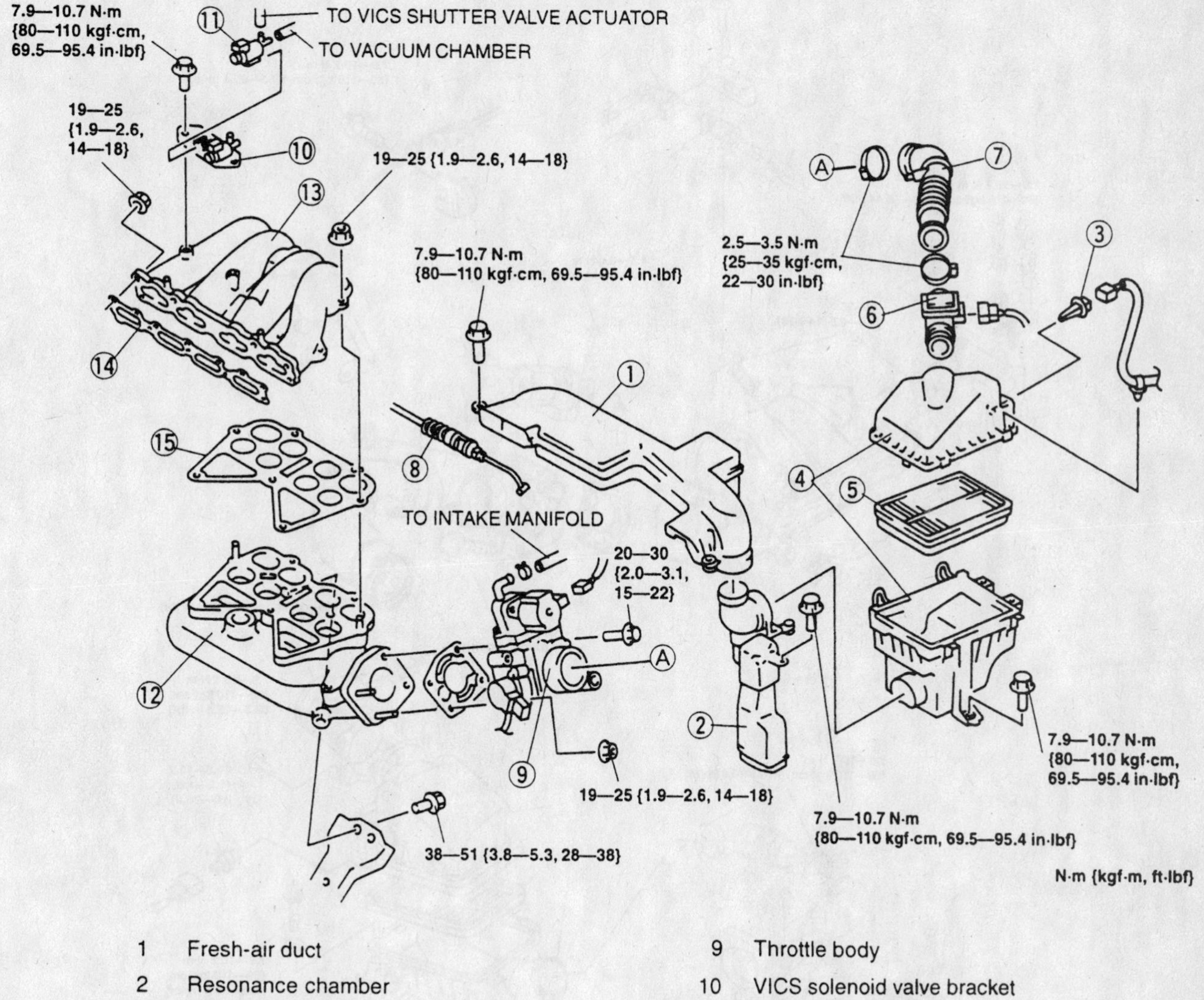

TO VICS SHUTTER VALVE ACTUATOR

TO VACUUM CHAMBER

7.9—10.7 N·m
{80—110 kgf·cm, 69.5—95.4 in·lbf}

19—25
{1.9—2.6, 14—18}

19—25 {1.9—2.6, 14—18}

7.9—10.7 N·m
{80—110 kgf·cm, 69.5—95.4 in·lbf}

2.5—3.5 N·m
{25—35 kgf·cm, 22—30 in·lbf}

TO INTAKE MANIFOLD

20—30
{2.0—3.1, 15—22}

19—25 {1.9—2.6, 14—18}

38—51 {3.8—5.3, 28—38}

7.9—10.7 N·m
{80—110 kgf·cm, 69.5—95.4 in·lbf}

7.9—10.7 N·m
{80—110 kgf·cm, 69.5—95.4 in·lbf}

N·m {kgf·m, ft·lbf}

1	Fresh-air duct	9	Throttle body
2	Resonance chamber	10	VICS solenoid valve bracket
3	IAT sensor	11	VICS solenoid valve
4	Air cleaner	12	Dynamic chamber
5	Air cleaner element	13	Intake manifold
6	MAF sensor	14	Intake manifold gasket
7	Air hose	15	Dynamic chamber gasket
8	Accelerator cable		

9301MG05

Exploded view of the intake manifold assembly components with tightening values—1.8L (FP) engine

- Air hose
- Accelerator cable
- Throttle body
- VICS solenoid valve bracket and valve
- Dynamic chamber
- Intake manifold and gasket

To install:

4. Clean all gasket mating surfaces.

➡**Be sure that the convex side of the intake manifold and dynamic gaskets are positioned as shown.**

5. Install or connect the following:
- Intake manifold using a new gasket, make sure the convex side of the gasket faces the manifold. Torque the bolts to 14–18 ft. lbs. (19–25 Nm).
- Dynamic chamber
- VICS solenoid valve and bracket
- Throttle body
- Accelerator cable
- Air hose
- MAF sensor
- Air cleaner and filter
- IAT sensor
- Fresh air duct and resonance chamber
- Negative battery cable

6. Fill the cooling system.
7. Run the engine and check for leaks.

626 WITH 2.0L (FS) ENGINE

1. Before servicing the vehicle, refer to the precautions in the beginning of this section.
2. Relieve the fuel system pressure.
3. Drain the cooling system.

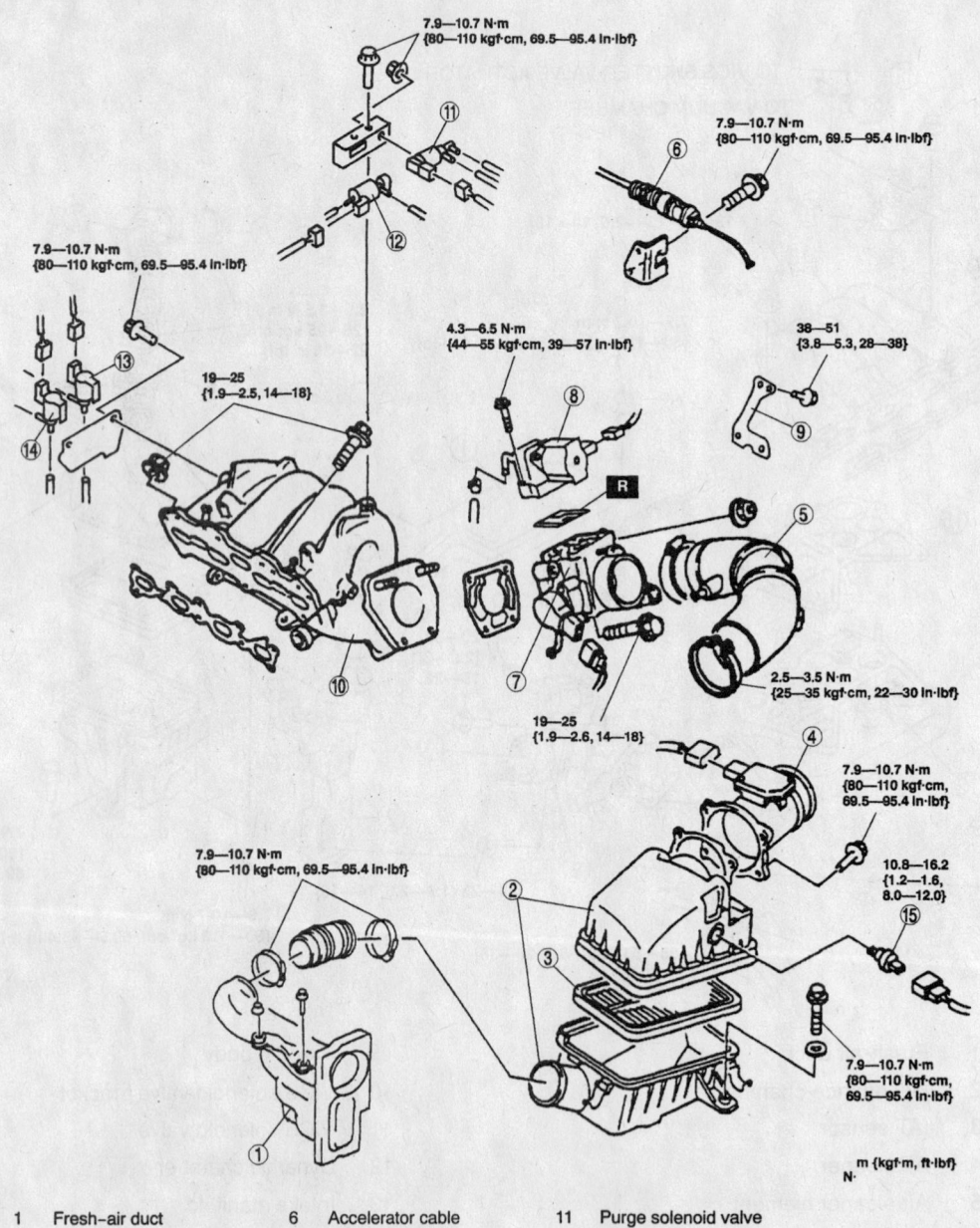

1	Fresh–air duct	6	Accelerator cable	11 Purge solenoid valve
2	Air cleaner	7	Throttle body	12 PRC solenoid valve
3	Air cleaner element	8	IAC valve	13 EGR boost solenoid valve
4	Mass air flow sensor	9	Intake manifold stay	14 VTCS solenoid valve
5	Air hose	10	Intake manifold	15 Intake air temperature sensor

42356-MAZC-G13

Exploded view of the intake manifold assembly–626 with 2.0L (FS) engine

Free play
1.5—4.0 mm {0.06—0.15 in}
LOCK
UNLOCK
A
LOCK
UNLOCK
WHITE MARK
FREE PLAY ADJUSTMENT
B
C
FREE PLAY
42356-MAZC-G04

Accelerator cable adjustment components—626 and Protégé models equipped with the 2.0L (FS) engine

4. Remove or disconnect the following:
   - Negative battery cable
   - Fresh air duct and air cleaner assembly
   - Mass Air Flow (MAF) sensor
   - Air hose
   - Accelerator cable
   - Throttle body assembly
   - Idle Air Control (IAC) valve
   - Intake manifold stay
   - Intake manifold and discard the gasket
5. Clean the mating surfaces of any gasket material

To install:

6. Install or connect the following:
   - Intake manifold with a new gasket. Make certain that the convex side of the new gasket faces the intake manifold. Torque the bolts to 18 ft. lbs. (25 Nm).
   - Intake manifold stay
   - IAC valve. Torque the bolt to 39–57 inch lbs. (4–6 Nm).
   - Throttle body. Torque the nuts to 18 ft. lbs. (25 Nm).
7. When connecting the accelerator cable, perform the following adjustment:

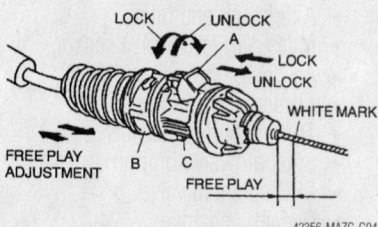

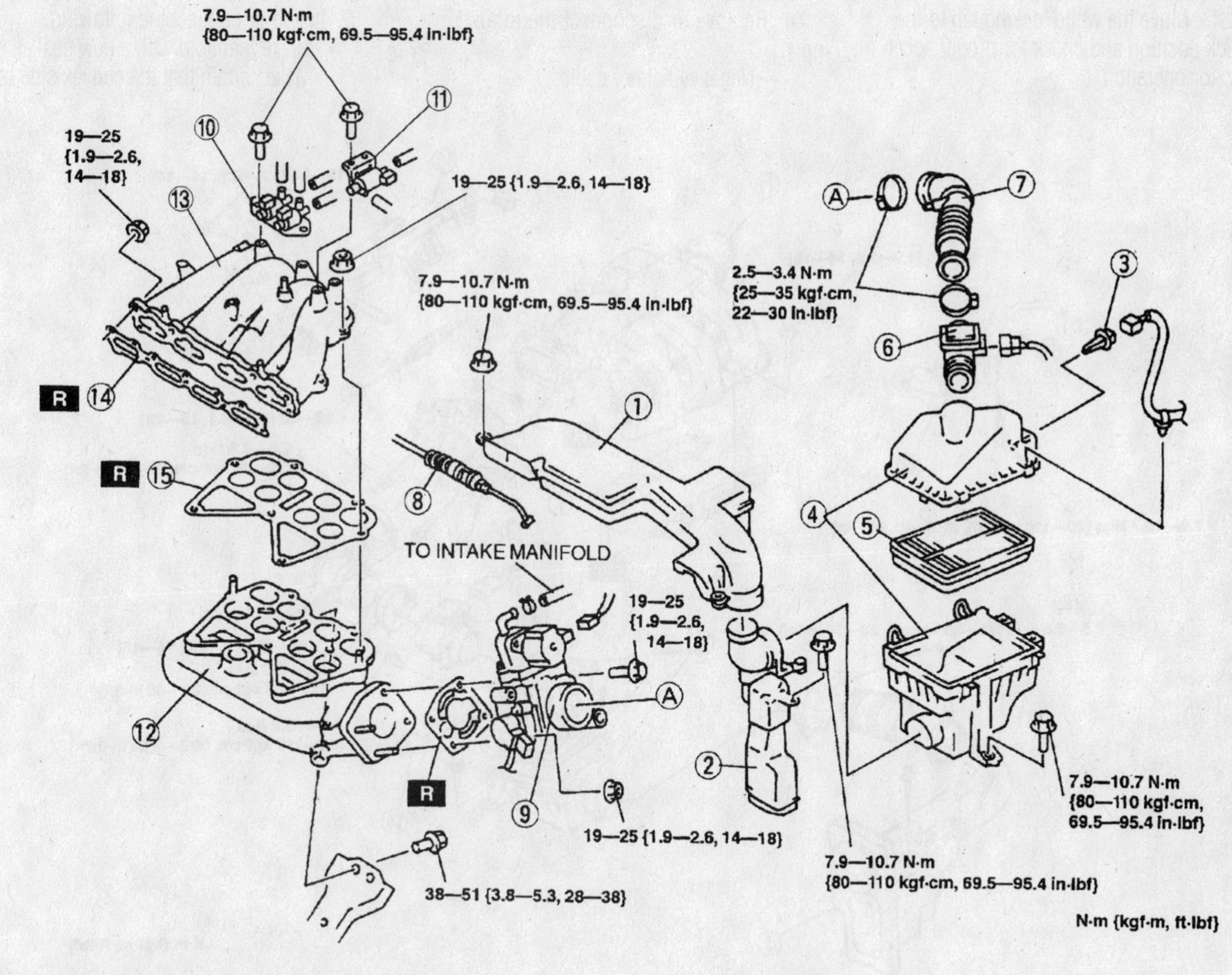

7.9—10.7 N·m
{80—110 kgf·cm, 69.5—95.4 in·lbf}
19—25
{1.9—2.6,
14—18}
19—25 {1.9—2.6, 14—18}
7.9—10.7 N·m
{80—110 kgf·cm, 69.5—95.4 in·lbf}
2.5—3.4 N·m
{25—35 kgf·cm,
22—30 in·lbf}
TO INTAKE MANIFOLD
19—25
{1.9—2.6,
14—18}
R
19—25 {1.9—2.6, 14—18}
38—51 {3.8—5.3, 28—38}
7.9—10.7 N·m
{80—110 kgf·cm,
69.5—95.4 in·lbf}
7.9—10.7 N·m
{80—110 kgf·cm, 69.5—95.4 in·lbf}
N·m {kgf·m, ft·lbf}

1	Fresh-air duct	9	Throttle body
2	Resonance chamber	10	Solenoid valve bracket
3	IAT sensor	11	PRC solenoid valve
4	Air cleaner	12	Dynamic chamber
5	Air cleaner element	13	Intake manifold
6	MAF sensor	14	Intake manifold gasket
7	Air hose	15	Dynamic chamber gasket
8	Accelerator cable		

42356-MAZC-GHC

Exploded view of the intake manifold assembly–Protégé with 2.0L (FS) engine

a. Move the white locking tab to the unlock position.

b. Turn stopper B to the to the unlock position.

➡ **If stopper B will not unlock, it may be necessary to carefully bend back tab C out a little using a suitable pry tool.**

c. Push or pull the cable housing directly behind the spring.

d. Turn the stopper B to the lock position.

e. Measure the free play which should be 0.06–0.15 inch (1.5–4mm) and make sure the cable free play is within specification.

f. Move the white locking tab to the lock position and check for proper accelerator operation.

8. Install or connect the following:
   • Air hose
   • MAF sensor. Torque the bolt to 95 inch lbs. (11 Nm).
   • Air cleaner/air duct assembly
   • Negative battery cable
9. Fill the cooling system.
10. Start the vehicle, check for leaks and repair if necessary.

PROTÉGÉ WITH 2.0L (FS) ENGINE

1. Before servicing the vehicle, refer to the precautions in the beginning of this section.
2. Relieve the fuel system pressure.
3. Drain the cooling system.
4. Remove or disconnect the following:

   • Negative battery cable
   • Fresh air duct and resonance assembly
   • Intake Air Temperature (IAT) sensor
   • Air cleaner and filter
   • Mass Air Flow (MAF) sensor
   • Air hose
   • Accelerator cable
   • Throttle body assembly
   • Solenoid valve bracket
   • PRC solenoid valve
   • Dynamic chamber
   • Intake manifold and discard the gasket
5. Clean the mating surfaces of any gasket material
   To install:
6. Install or connect the following:
   • Intake manifold with a new gasket. Make certain that the convex side of

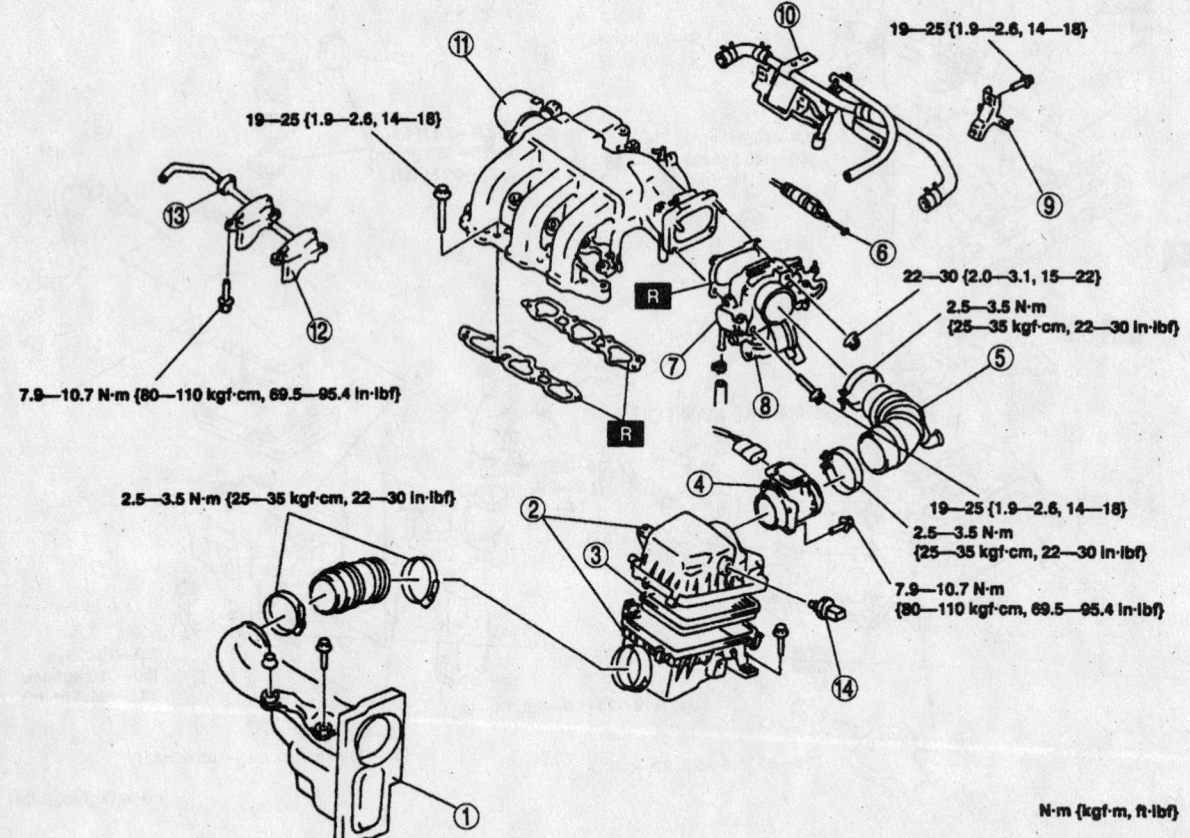

19—25 {1.9—2.6, 14—18}

7.9—10.7 N·m {80—110 kgf·cm, 69.5—95.4 in·lbf}

2.5—3.5 N·m {25—35 kgf·cm, 22—30 in·lbf}

22—30 {2.0—3.1, 15—22}

2.5—3.5 N·m {25—35 kgf·cm, 22—30 in·lbf}

19—25 {1.9—2.6, 14—18}

2.5—3.5 N·m {25—35 kgf·cm, 22—30 in·lbf}

7.9—10.7 N·m {80—110 kgf·cm, 69.5—95.4 in·lbf}

N·m {kgf·m, ft·lbf}

1	Fresh–air duct	8	IAC valve
2	Air cleaner	9	Intake manifold stay
3	Air cleaner element	10	Ventilation pipe
4	Mass air flow sensor	11	Intake manifold component
5	Air hose	12	Vacuum chamber
6	Accelerator cable	13	VRIS check valve (one–way)
7	Throttle body	14	Intake air temperature sensor

42356-MAZC-G14

Exploded view of the intake manifold assembly–626 with 2.5L (KL) engine

the new gasket faces the intake manifold. Torque the bolts to 18 ft. lbs. (25 Nm).

- Dynamic chamber
- PRC solenoid valve
- Solenoid valve bracket
- Throttle body assembly. Torque the bolts and nuts to 18 ft. lbs. (25 Nm).

7. When connecting the accelerator cable, perform the following adjustment:

   a. Move the white locking tab to the unlock position.

   b. Turn stopper B to the to the unlock position.

➡ **If stopper B will not unlock, it may be necessary to carefully bend back tab C out a little using a suitable pry tool.**

   c. Push or pull the cable housing directly behind the spring.

   d. Turn the stopper B to the lock position.

   e. Measure the free play which should be 0.06–0.15 inch (1.5–4mm) and make sure the cable free play is within specification.

   f. Move the white locking tab to the lock position and check for proper accelerator operation.

8. Install or connect the following:
- Air hose
- AF sensor
- Air filter and cleaner
- IAT sensor
- Fresh air duct and resonance assembly
- Negative battery cable

9. Fill the cooling system.

10. Start the vehicle, check for leaks and repair if necessary.

2.5L (KL) ENGINE

1. Before servicing the vehicle, refer to the precautions in the beginning of this section.

2. Relieve the fuel system pressure.
3. Drain the cooling system.
4. Remove or disconnect the following:
- Negative battery cable
- Fresh air duct and air cleaner assembly
- Mass Air Flow (MAF) sensor
- Air hose
- Accelerator cable
- Throttle body assembly
- Idle Air Control (IAC) valve
- Intake manifold stay
- Intake manifold and discard the gasket

5. Clean the mating surfaces of any gasket material

To install:

6. Install or connect the following:
- Intake manifold with a new gasket. Torque the bolts to 18 ft. lbs. (25 Nm).
- Intake manifold stay
- IAC valve.

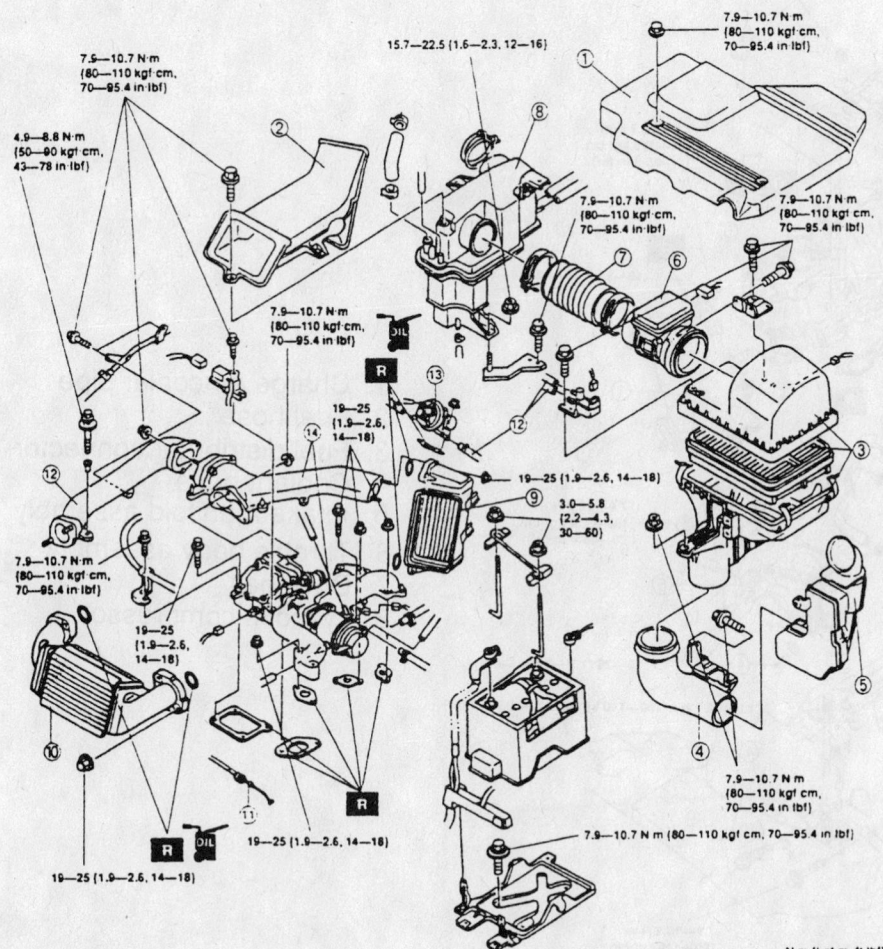

1. Dynamic chamber cover
2. Charge air cooler air duct
3. Air cleaner assembly
4. Air duct
5. Fresh air duct
6. Mass air flow sensor
7. Air intake hose
8. Resonator
9. Charge air cooler (RH)
10. Charge air cooler (LH)
11. Accelerator cable
12. Vacuum hose assembly
13. EGR control valve
14. Air intake pipe assembly

N·m (kgf·m, ft·lbf)

7923MG31

Exploded view of the intake manifold assembly (1 of 2)—2.3L engine

- Throttle body. Torque the nuts to 18 ft. lbs. (25 Nm).
- Accelerator cable
- Air hose
- MAF sensor. Torque the bolt to 95 inch lbs. (11 Nm).
- Air cleaner/air duct assembly
- Negative battery cable

7. Fill the cooling system.
8. Start the vehicle, check for leaks and repair if necessary.

Millenia

2.3L ENGINE

1. Before servicing the vehicle, refer to the precautions in the beginning of this section.
2. Relieve the fuel system pressure.
3. Drain the cooling system.
4. Remove or disconnect the following:
   - Negative battery cable
   - Dynamic chamber cover
   - Charge air cooler air duct
   - Vacuum hoses and electrical connectors from the air cleaner housing
   - Air cleaner assembly
   - Fresh air ducts
   - Mass Air Flow (MAF) sensor and the air intake hose from the throttle body
   - Resonator
   - Right-hand charge air cooler
   - Left-hand charge air cooler
   - Accelerator cable
   - Vacuum hoses from the rear of the intake manifold and Exhaust Gas Recirculation (EGR) valve
   - EGR valve
   - Air intake pipe assembly
   - Charge air cooler pipe
   - Fuel supply line at the fuel rails and discard the copper washers
   - Fuel and vacuum lines from the fuel pressure regulator
   - Coolant hoses
   - Wiring harness from the intake manifold
   - Intake manifold mounting nuts and bolts in 2–3 steps
   - Intake manifold

5. If necessary, label and disconnect the fuel hoses and electrical connectors from the throttle body. Remove the throttle body.

To install:

6. Clean all gasket mating surfaces.
7. Install or connect the following:
   - Throttle body, if removed. Tighten the nuts/bolts to 14–18 ft. lbs. (19–25 Nm).
   - Fuel hoses and electrical connectors.
   - Intake manifold using new gaskets. Tighten the nuts/bolts in 2–3 steps, from the center to the ends, to 14–18 ft. lbs. (19–25 Nm).
   - Wiring harness onto the intake manifold

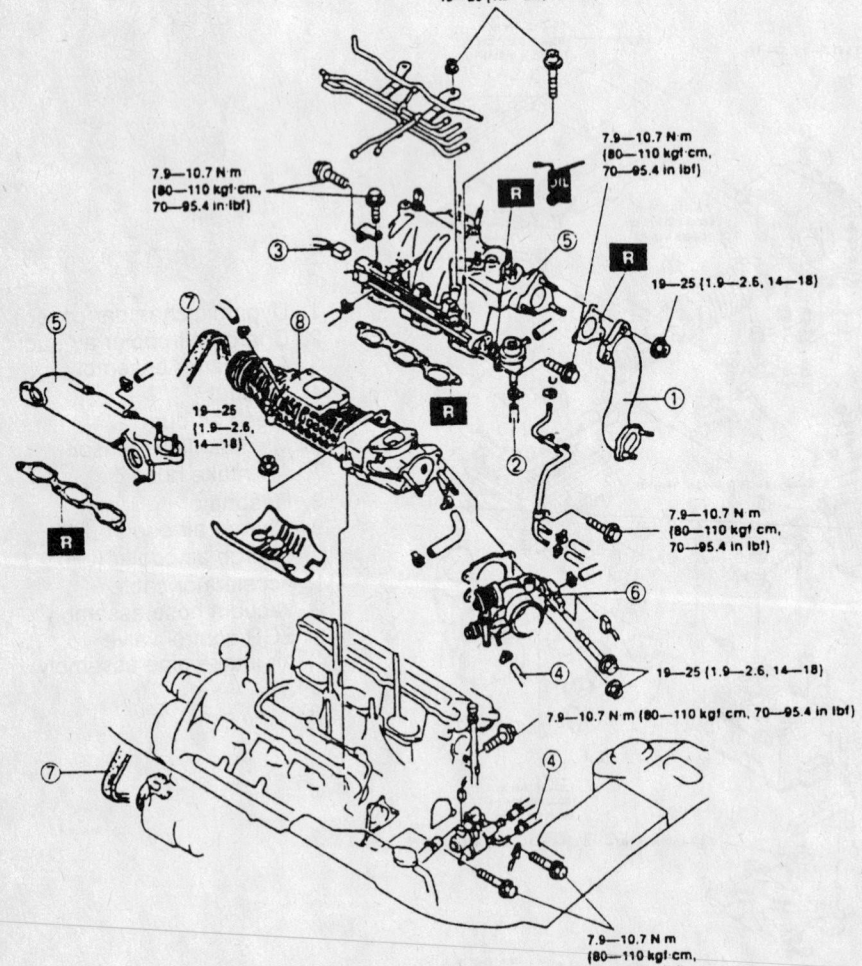

1. Charge air cooler pipe
2. Fuel hose
3. Fuel distributor connector
4. Coolant hose
5. Intake manifold assembly
6. Throttle body assembly
7. Drive belt
8. Lysholm compressor

Exploded view of the intake manifold assembly (2 of 2)—2.3L engine

- Coolant hoses
- Fuel and vacuum lines to the fuel pressure regulator
- Fuel supply line to the fuel rail using new copper washers
- Charge air cooler pipe
- Air intake pipe assembly using new gaskets. Verify that the rubber gaskets are not twisted or distorted. Tighten the bolts marked **A** to 70–95 inch lbs. (8–11 Nm) and all other bolts, in sequence, to 14–18 ft. lbs. (19–25 Nm).
- EGR valve using a new gasket

- Vacuum hoses to the intake manifold and EGR valve
- Accelerator cable, adjust as necessary
- Left and right-hand charge air coolers using new gaskets. Verify that the rubber gaskets are not twisted or distorted. Tighten the bolts marked **A** to 44–78 inch lbs. (5–9 Nm). Tighten the bolts marked **B** to 70–95 inch lbs. (8–11 Nm) and all other bolts, in sequence, to 14–18 ft. lbs. (19–25 Nm).

- Resonator. Tighten the nuts/bolts to 12–16 ft. lbs. (16–22 Nm).
- Air intake hose onto the throttle body
- MAF sensor
- Fresh air and air ducts
- Air cleaner assembly
- Vacuum hoses and electrical connectors to the air cleaner housing
- Charge air cooler air duct. Tighten the bolts to 70–95 inch lbs. (8–11 Nm).
- Dynamic chamber cover
- Negative battery cable

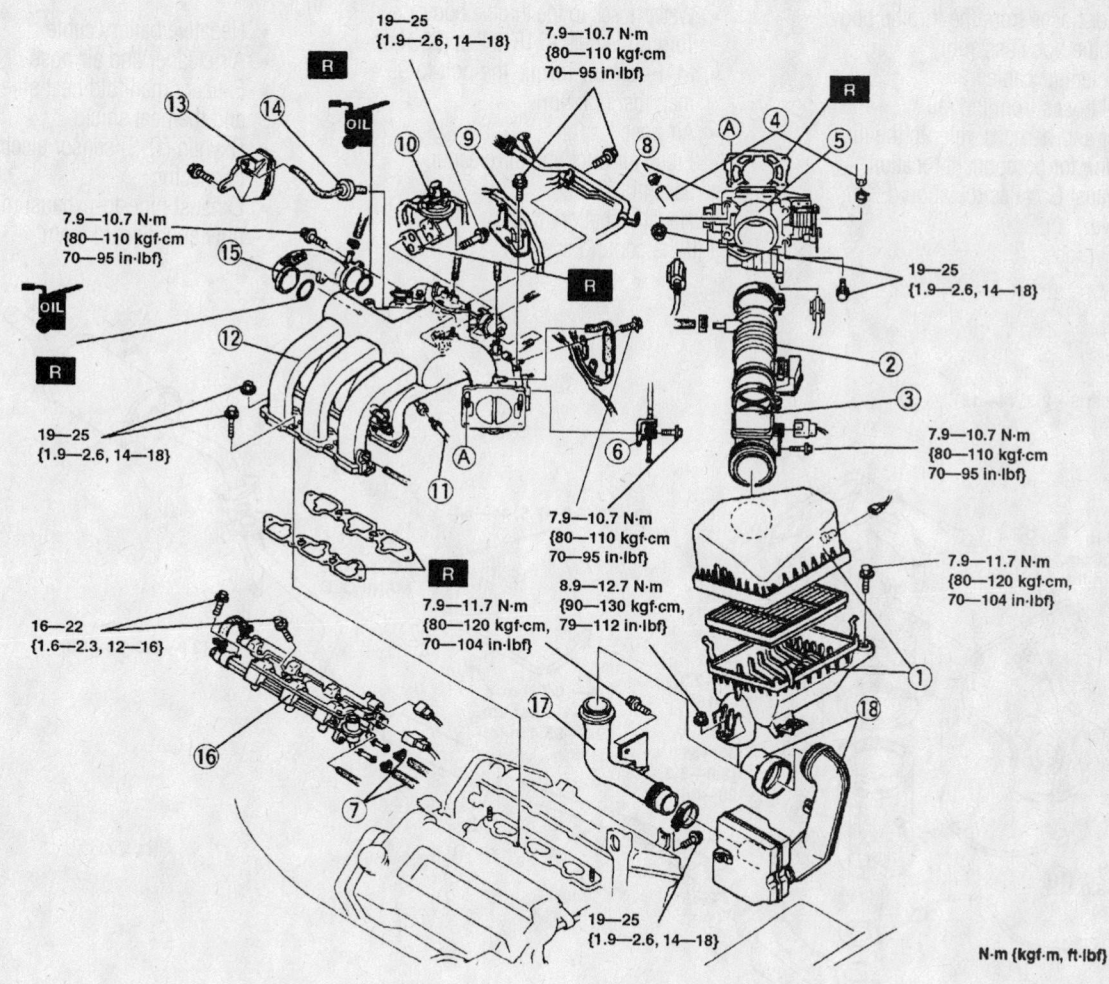

1 Air cleaner
2 Air intake hose
3 MAF sensor
4 Water hose
5 Throttle body component
6 Accelerator cable
7 Fuel hose
8 Pipe
9 Harness
10 EGR valve
11 EGR pipe
12 Intake manifold
13 Vacuum chamber
14 Check valve
15 Air intake pipe
16 Fuel distributor component
17 Air duct
18 Fresh-air duct

Exploded view of the intake manifold assembly–Millenia with 2.5L (KL) engine

8. Fill the cooling system.
9. Run the engine and check for leaks.

2.5L (KL) ENGINE

1. Before servicing the vehicle, refer to the precautions in the beginning of this section.
2. Relieve the fuel system pressure.
3. Drain the cooling system.
4. Remove or disconnect the following:
   - Negative battery cable
   - Fresh air duct and air cleaner assembly
   - Air hose
   - Mass Air Flow (MAF) sensor
   - Water hoses from the throttle body
   - Throttle body assembly
   - Accelerator cable
   - Fuel hoses from the rail
   - Pipe and harness, refer to the illustration for component location
   - Exhaust Gas Recirculation (EGR) Valve
   - EGR pipe

- Intake manifold and discard the gasket
5. Clean the mating surfaces of any gasket material

To install:
6. Install or connect the following:
   - Intake manifold with a new gasket. Torque the bolts to 18 ft. lbs. (25 Nm).
   - EGR pipe and valve
   - Pipe and harness, refer to the illustration for component location
   - Fuel hoses to the rail
   - Accelerator cable
   - Throttle body assembly
   - Water hoses to the throttle body. Torque the nuts to 18 ft. lbs. (25 Nm).
   - MAF sensor. Torque the bolt to 95 inch lbs. (11 Nm).
   - Air hose
   - Fresh air duct and air cleaner assembly
   - Negative battery cable
7. Fill the cooling system.

8. Start the vehicle, check for leaks and repair if necessary.

Exhaust Manifold

REMOVAL & INSTALLATION

Miata

1.8L (BP) ENGINES

1. Before servicing the vehicle, refer to the precautions in the beginning of this section.
2. Remove or disconnect the following:
   - Negative battery cable
   - Air cleaner and air hose
   - Exhaust manifold heat shield bolts and the heat shield
   - Oxygen (O2S) sensor electrical connector
   - Exhaust pipe-to-exhaust manifold nuts and discard them

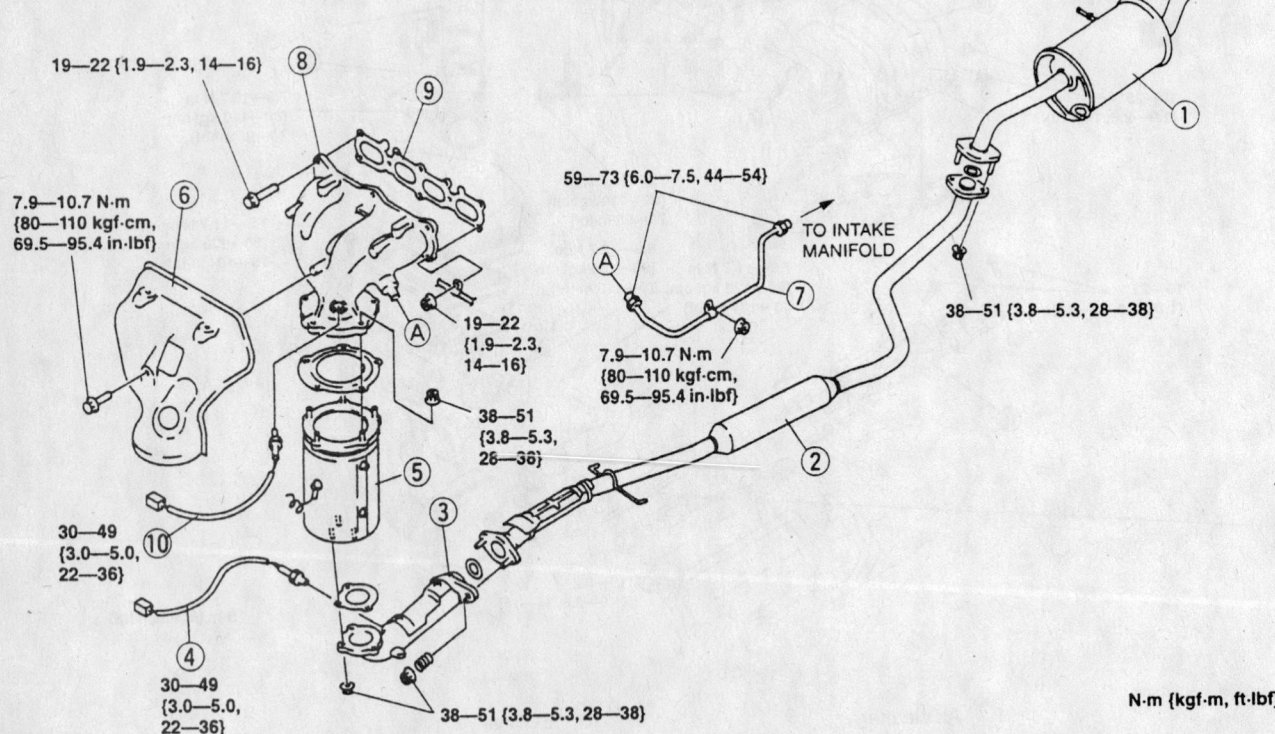

1	Main silencer	6	Exhaust manifold insulator
2	Presilencer	7	EGR Pipe
3	Front pipe	8	Exhaust manifold
4	HO2S (Rear)	9	Exhaust manifold gasket
5	WU-TWC	10	HO2S (Front)

Exploded view of the exhaust system—1.6L (ZM) engines

9301MG06

- Exhaust Gas Recirculation (EGR) pipe from the exhaust manifold
- Exhaust manifold nuts and bolts and discard the nuts
- Exhaust manifold

To install:

3. Clean all gasket mating surfaces.
4. Install or connect the following:
   - Exhaust manifold. Torque the bolts to 29–33 ft. lbs. (39–46 Nm).
   - Exhaust pipe. Torque the new nuts to 38 ft. lbs. (52 Nm).
   - O2S connector
   - EGR pipe. Torque the pipe to 34 ft. lbs. (47 Nm).

- Heat shield. Torque the bolts to 95 inch lbs. (11 Nm).
- Air hose and air cleaner
- Negative battery cable

626 and Protégé

1.6L (ZM) AND 1.8L (FP) ENGINES

1. Before servicing the vehicle, refer to the precautions in the beginning of this section.
2. Remove or disconnect the following:
   - Negative battery cable
   - Air cleaner and hose assembly
   - Water bypass pipe-to-engine block bolt, if equipped

- Exhaust Gas Recirculation (EGR) pipe
- Oxygen Sensor (O2S) from the exhaust system
- Front exhaust pipe from the Warm-Up Three Way Catalytic (WU-TWC) converter
- Exhaust manifold insulator
- WU-TWC converter from the manifold
- Exhaust manifold

To install:

3. Be sure all gasket mating surfaces are clean prior to assembly.
4. Tighten the components following the illustration.

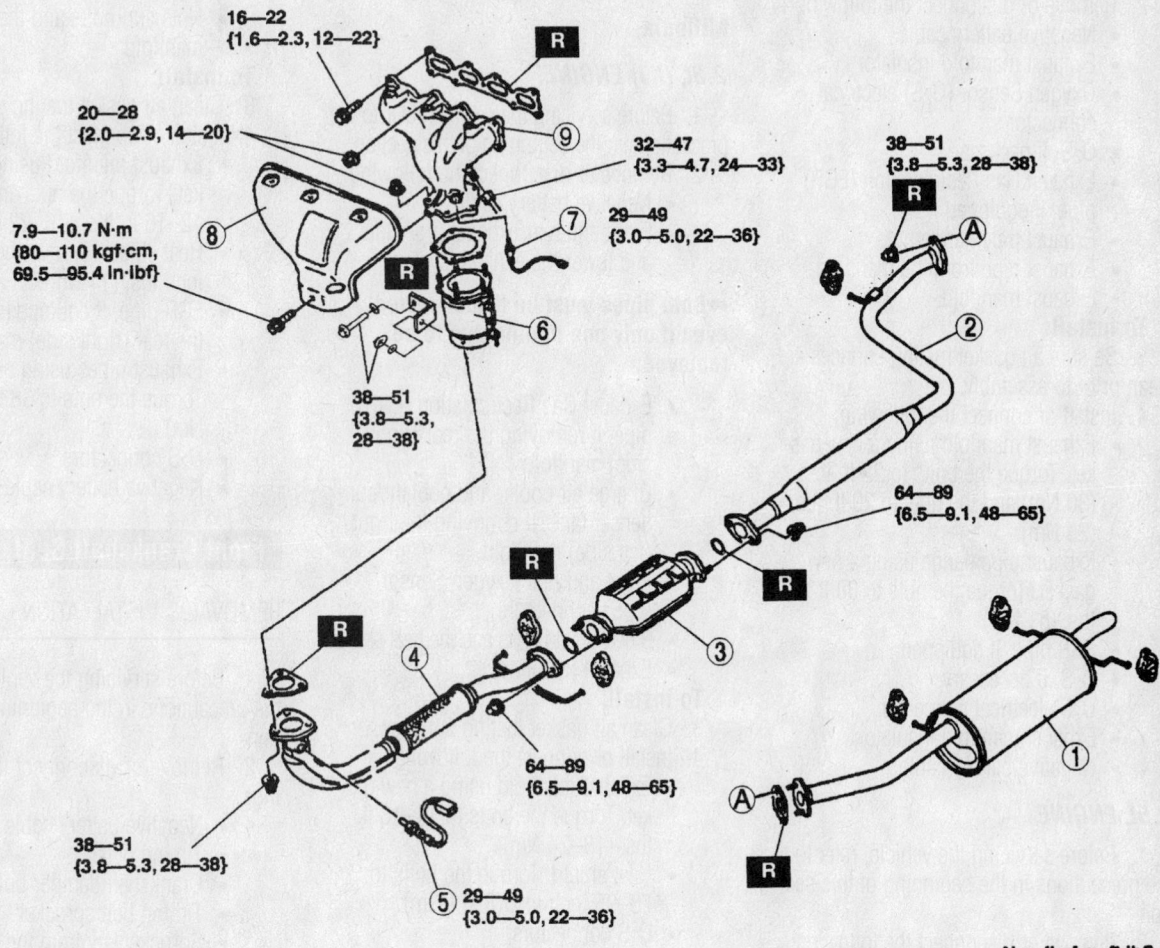

N·m {kgf·m, ft·lbf}

1	Main silencer	6	Warm up three way catalytic converter
2	Presilencer	7	Heated oxygen sensor (Front)
3	Three way catalytic converter	8	Exhaust manifold insulator
4	Front pipe	9	Exhaust manifold
5	Heated oxygen sensor (Rear)		

Exploded view of the exhaust system—626 shown, Protégé similar

42356-MAZC-G15

5. Install or connect the following:
- Exhaust manifold
- WU-TWC converter to the manifold
- Exhaust manifold insulator
- Front exhaust pipe from the WU-TWC converter
- O₂S to the exhaust system
- EGR pipe
- Water bypass pipe-to-engine block bolt
- Air cleaner and hose assembly
- Negative battery cable

2.0L (FS) ENGINE

1. Before servicing the vehicle, refer to the precautions in the beginning of this section.
2. Remove or disconnect the following:
- Negative battery cable
- Exhaust manifold insulator
- Oxygen Sensor (O₂S) electrical connector
- O₂S, if necessary
- Exhaust Gas Recirculation (EGR) pipe, if equipped
- Exhaust pipe flange nuts
- Exhaust pipe from the manifold
- Exhaust manifold

To install:
3. Be sure all gasket mating surfaces are clean prior to assembly.
4. Install or connect the following:
- Exhaust manifold using a new gasket. Torque the bolts to 22 ft. lbs. (30 Nm) and the nuts to 20 ft. lbs. (28 Nm).
- Exhaust pipe flange using a new gasket. Torque the nuts to 38 ft. lbs. (51 Nm).
- EGR pipe, if equipped
- O₂S, if necessary
- O₂S electrical connector
- Exhaust manifold insulator
- Negative battery cable

2.5L ENGINE

1. Before servicing the vehicle, refer to the precautions in the beginning of this section.
2. Remove or disconnect the following:
- Negative battery cable
- Oxygen (O₂S) sensor connectors
- Front and rear exhaust pipe nuts and lower the exhaust system

➡**Both pipes must be disconnected, even if only one manifold is to be removed.**

- Exhaust Gas Recirculation (EGR) pipe, if equipped; if removing the rear (right side) manifold
- Heat shield

- 2 nuts and 5 bolts and the exhaust manifold

To install:
3. Clean all gasket mating surfaces.
4. Install or connect the following:
- Exhaust manifold using a new gasket. Torque the nuts and bolts to 12–22 ft. lbs. (16–22 Nm).
- Heat shield. Torque the bolts to 95 inch lbs. (11 Nm).
- EGR pipe, if equipped; if installing the rear (right side) manifold
- Exhaust pipes using new gaskets. Torque the nuts to 38 ft. lbs. (51 Nm).
- O₂S connectors
- Negative battery cable

Millenia

2.3L (KJ) ENGINE

1. Before servicing the vehicle, refer to the precautions in the beginning of this section.
2. Remove or disconnect the following:
- Negative battery cable
- Front and rear exhaust pipe nuts and lower the exhaust system

➡**Both pipes must be disconnected, even if only one manifold is to be removed.**

- Exhaust Gas Recirculation (EGR) pipe, if removing the rear (right side) manifold
- Charge air cooler and coolant/condenser fans, If removing the front (left side) manifold
- Front and rear Oxygen Sensor (O₂S) connectors
- 3 heat shield bolts and the heat shield
- Exhaust manifold

To install:
3. Clean all gasket mating surfaces.
4. Install or connect the following:
- Exhaust manifold using a new gasket. Torque the bolts to 14–18 ft. lbs. (18–24 Nm).
- Heat shield. Torque the bolts to 70–95 inch lbs. (8–11 Nm).
- O₂S connectors
- Coolant/condenser fans and the charge air cooler, if installing the front (left side) manifold. Torque the bolts to 14–18 ft. lbs. (19–25 Nm).
- EGR pipe, if installing the rear (right side) manifold
- Exhaust pipes using new gaskets. Torque the nuts to 28–38 ft. lbs. (38–51 Nm).
- Negative battery cable

2.5L (KL) Engine

1. Before servicing the vehicle, refer to the precautions in the beginning of this section.
2. Remove or disconnect the following:
- Negative battery cable
- Oxygen Sensor (O₂S) connectors
- Front and rear exhaust pipe nuts and lower the exhaust system

➡**Both pipes must be disconnected, even if only one manifold is to be removed.**

- Exhaust Gas Recirculation (EGR) pipe, if equipped; if removing the rear (right side) manifold
- Heat shield
- Nuts and bolts and the exhaust manifold

To install:
3. Clean all gasket mating surfaces.
4. Install or connect the following:
- Exhaust manifold using a new gasket. Torque the nuts and bolts to 12–16 ft. lbs. (16–22 Nm).
- Heat shield. Torque the bolts to 95 inch lbs. (11 Nm).
- EGR pipe, if equipped; if installing the rear (right side) manifold
- Exhaust pipes using new gaskets. Torque the nuts to 38 ft. lbs. (51 Nm).
- O₂S connectors
- Negative battery cable

Front Crankshaft Seal

REMOVAL & INSTALLATION

1. Before servicing the vehicle, refer to the precautions in the beginning of this section.
2. Remove or disconnect the following:

- Negative battery cable
- Timing belt
- Crankshaft damper bolt and damper
- Timing belt sprocket
- Sprocket key from the crankshaft
- Oil seal from the engine block using a prybar

✸✸ WARNING

Be careful not to score the crankshaft or the seal seat.

3. Clean the seal bore.
To install:
4. Install or connect the following:
- New oil seal lubricated with clean

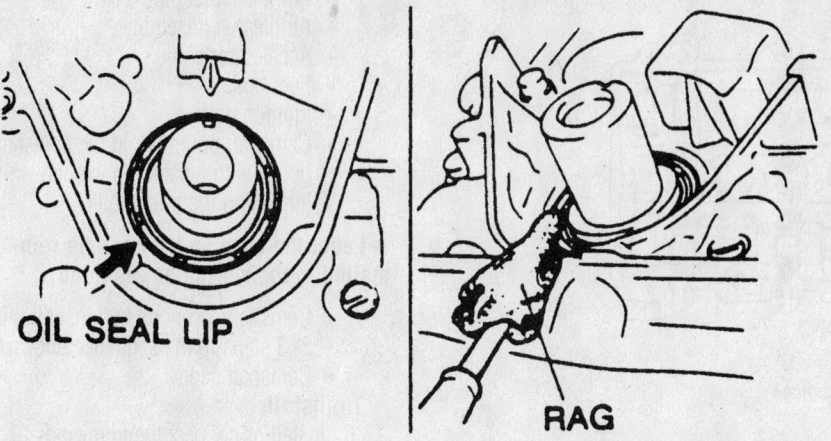

Remove the front engine seal by cutting the seal lip, then, so as not to damage the crankshaft, carefully pry the seal out with a prybar

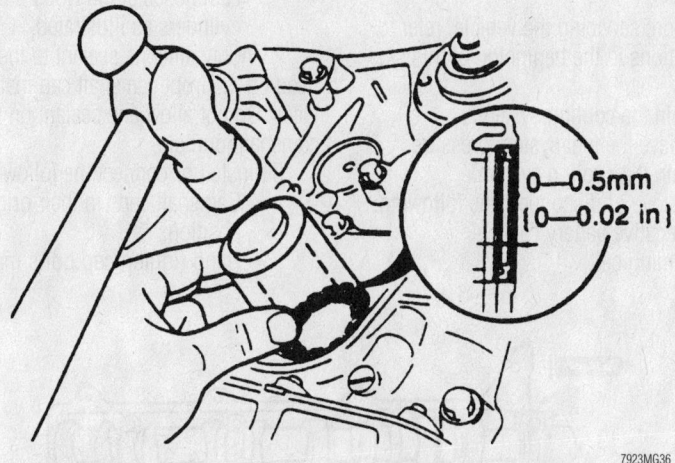

Install the seal using an appropriate driver, which fits over the crankshaft snout and presses on the outside edge of the seal

engine oil, drive it into the engine using an installation tool until it seats
- Sprocket key onto the crankshaft
- Timing belt sprocket
- Crankshaft damper
- Timing belt
- Negative battery cable

- Spark plugs
- Cylinder head cover hoses, if equipped
- Cylinder head cover bolts
- Cylinder head cover
- Timing belt

Camshaft

REMOVAL & INSTALLATION

1.6L (ZM) and 1.8L (FP) Engines

1. Before servicing the vehicle, refer to the precautions in the beginning of this section.
2. Remove or disconnect the following:
   - Negative battery cable
   - Spark plug wires

- Camshaft by holding it with a wrench on the hexagon cast into the camshaft
- Sprocket bolts
- Sprockets

➡Label the caps so they can be reinstalled in their original positions.

- Camshaft cap bolts by loosening in 2–3 steps in the sequence shown
- Camshaft caps

To install:

✷✷ CAUTION

Because there is little thrust clearance, the camshaft must be held in the horizontal position during installation. If not excessive force will be applied to the thrust area causing a burr on the receiving area of the cylinder head journal. Make sure to use the following procedure to avoid damage.

3. Install or connect the following:
   - Lubricate the camshaft journals and lobes with clean engine oil
   - Camshafts onto the head facing the cam noses at the no. 1 and 3 cylinders as illustrated. Make sure the camshaft sliding surface is free of sealant.
4. Apply a 0.04 inch (1mm) bead of silicone sealant to the areas illustrated in the illustration. Do not allow any sealant on the camshaft journals.
5. Install or connect the following:
   - Camshaft caps in their original positions
   - Cap bolts and hand tighten bolts 5, 7, 2 and 4
   - Cap bolts a few turns in the sequence illustrated
6. Make sure the camshaft settles horizontally when the bearing cap bolts at the

Camshaft cap bolt loosening sequence—1.6L (ZM) engines

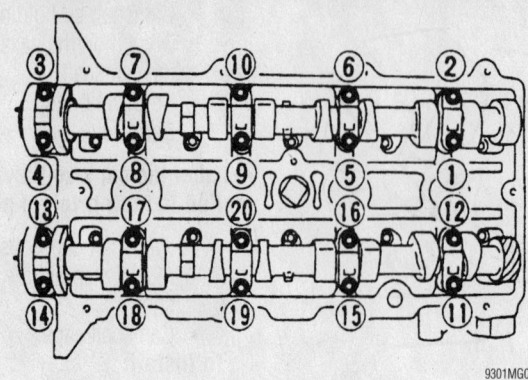

Camshaft cap bolt loosening sequence—1.8L (FP) engines

9301MG07

number 3 journals are tightened. Torque the bolts in 2–3 steps to 125 inch lbs. (14 Nm) in the proper sequence.

- New camshaft seal by lubricating it with engine oil. Tap the seal into position, using a seal installer, until it is recessed into the cylinder head 0.01 inch (0.4mm) on the 1.6L (ZM) engines and 0.012–0.027 inch (0.3–0.7mm) on1.8L (FP) engines.

7. Turn the camshafts until the dowel pins face straight up.

8. Install or connect the following:
- Camshaft sprockets. Torque the bolts to 44 ft. lbs. (60 Nm) by

holding the camshaft with the wrench on the cast hexagon.
- Remaining components

2.0L (FS) Engine

1. Before servicing the vehicle, refer to the precautions in the beginning of this section.

2. Drain the cooling system.

3. Relieve the fuel system pressure.

4. Drain the cooling system.

5. Remove or disconnect the following:
- Negative battery cable
- Timing belt

- Front exhaust pipe
- Air cleaner assembly
- Accelerator cable
- Fuel hose
- Ignition coil
- Camshaft pulley. Hold the camshaft pulley with a back–up wrench while loosening the pulley bolt.

➡**Label the caps so they can be rein-stalled in their original positions.**

- Camshaft cap bolts by loosening in 2–3 steps in the sequence shown
- Camshaft caps

To install:

6. Install or connect the following:
- Camshafts making sure to lubricate the journals and lobes with clean engine oil. Place the camshafts onto the cylinder head facing the cam noses at the No. 1 and No. 3 cylinders as illustrated.

7. Apply silicone sealant to the cylinder head on the front camshaft cap mating surfaces. Do not allow any sealant on the camshaft journals.

8. Install or connect the following:
- Camshaft caps in their original positions
- Hand tighten cap bolts marked 5,

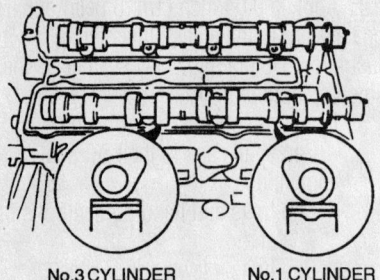

No.3 CYLINDER   No.1 CYLINDER

42356-MAZC-GHK

Install the camshafts onto the head facing the cam noses at the no. 1 and 3 cylinders as shown—1.6L (ZM) and 1.8L (FP) engines

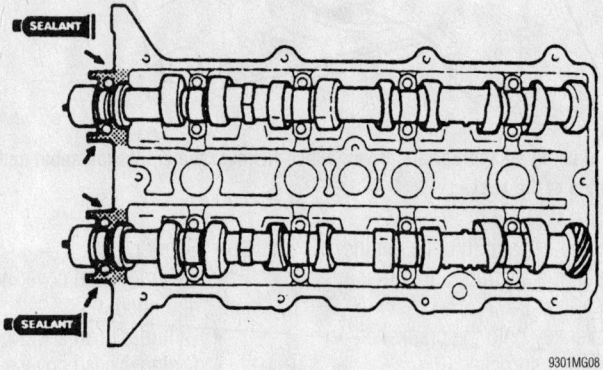

9301MG08

Apply silicone sealant to the cylinder head in the positions shown—1.8L (FP) engines

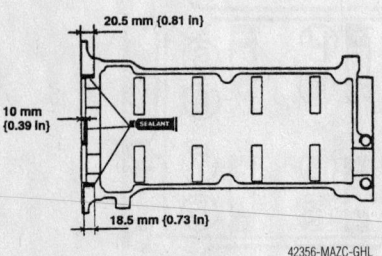

20.5 mm (0.81 in)

10 mm (0.39 in)

18.5 mm (0.73 in)

42356-MAZC-GHL

Apply silicone sealant to the cylinder head in the positions shown—1.6L (ZM) engines

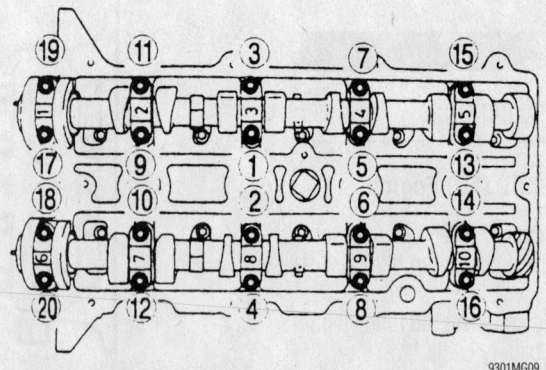

9301MG09

Camshaft cap bolt tightening sequence—1.6L (ZM) and 1.8L (FP) engines

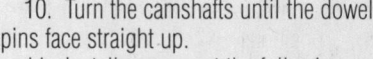
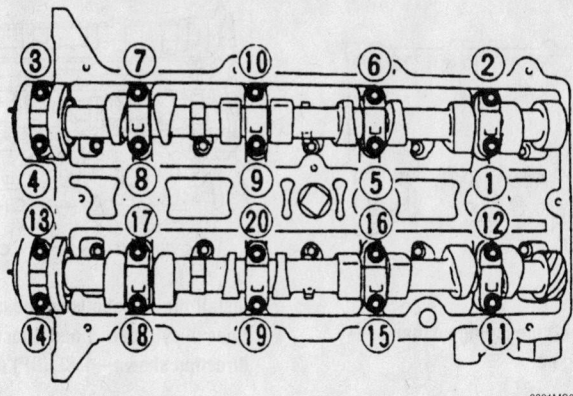

Camshaft cap bolt loosening sequence—2.0L (FS) engines

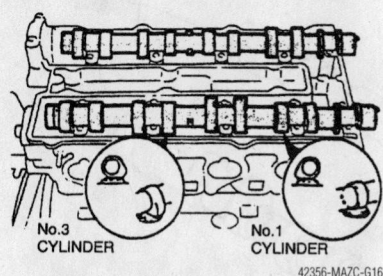

Install the camshafts so the cam noses are positioned at the No. 1 and No. 3 cylinders—2.0L (FS) engine

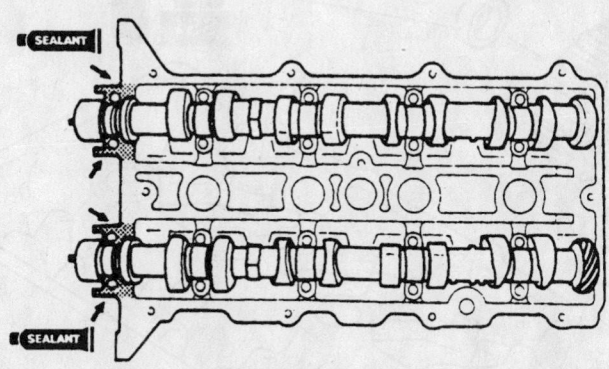

Apply silicone sealant to the cylinder head in the positions shown—2.0L (FS) engines

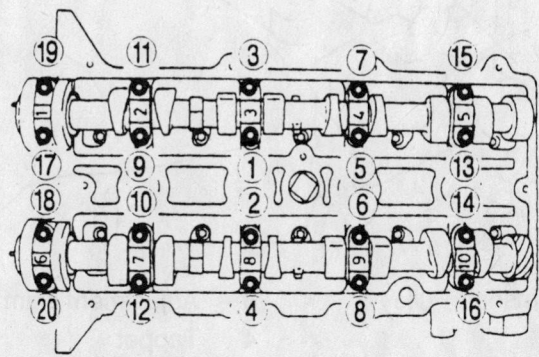

Camshaft cap bolt tightening sequence—2.0L (FS) engines

5, 2 and 4, refer to the torque sequence illustration for location
- Cap bolts. Torque the bolts in 2–3 steps to 125 inch lbs. (14 Nm) in the proper sequence.

9. Make sure the camshaft is settled horizontally when the 2 bearing cap bolts at the number 3 journal are tightened.
- New camshaft seal by lubricating it with engine oil. Tap the seal into position, using a seal installer, until it is recessed into the cylinder head 0.012–0.027 inch (0.3–0.7mm).

10. Turn the camshafts until the dowel pins face straight up.
11. Install or connect the following:
- Camshaft sprockets. Torque the bolts to 44 ft. lbs. (60 Nm) by holding the camshaft with the wrench on the cast hexagon.
- Ignition coil
- Fuel hose
- Accelerator cable
- Air cleaner assembly
- Front exhaust pipe
- Timing belt
- Negative battery cable
12. Fill and bleed the cooling system.
13. Change the oil and filter.
14. Run the engine and check for proper operation.

1.8L (BP) Engines

1. Before servicing the vehicle, refer to the precautions in the beginning of this section.
2. Remove or disconnect the following:
- Negative battery cable
- Timing belt
- Variable valve timing actuator and camshaft pulley. Hold the hexagonal part of the camshaft with a wrench to prevent the camshaft from turning and remove the actuator and pulley bolt.

➡**Label the caps so they can be reinstalled in their original positions.**

- Camshaft cap bolts by loosening in 2–3 steps in the sequence shown
- Camshaft caps
- Lifters and adjustment shims, if necessary

➡**Identify and mark each lifter as it is removed so it can be reinstalled in the same position.**

To install:

3. Install or connect the following:
- Lifters in their original positions by lubricating them with engine oil

➡**Verify that they move smoothly in their bore.**

4. Lubricate the camshaft journals and lobes with clean engine oil.
5. Install the camshafts so the cam projections of cylinders 1 and 3 face in the direction as illustrated.
6. Apply silicone sealant to the cylinder head on the front camshaft cap mating surfaces. Do not allow any sealant on the camshaft journals.
7. Install or connect the following:

- Camshaft caps in their original positions
- Cap bolts. Torque the bolts in 2–3 steps to 125 inch lbs. (14 Nm) in the proper sequence.

8. Make sure the camshaft is settled horizontally when the 2 bearing cap bolts at the number 3 journal are tightened.

- New camshaft seal by lubricating it with engine oil. Tap the seal into position, using a seal installer, until it is flush with the edge of the camshaft.

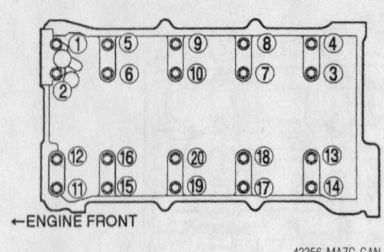

1
5
9
8
4
2
6
10
7
3
12
16
20
18
13
11
15
19
17
14
←ENGINE FRONT
42356-MAZC-GAN

Camshaft cap bolt loosening sequence—1.8L (BP) engine

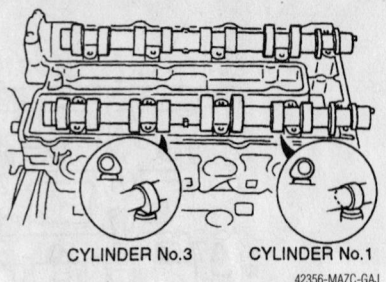

CYLINDER No.3
CYLINDER No.1
42356-MAZC-GAJ

Install the camshafts so the cam projections of cylinders 1 and 3 face in the direction shown—1.8L (BP) engine

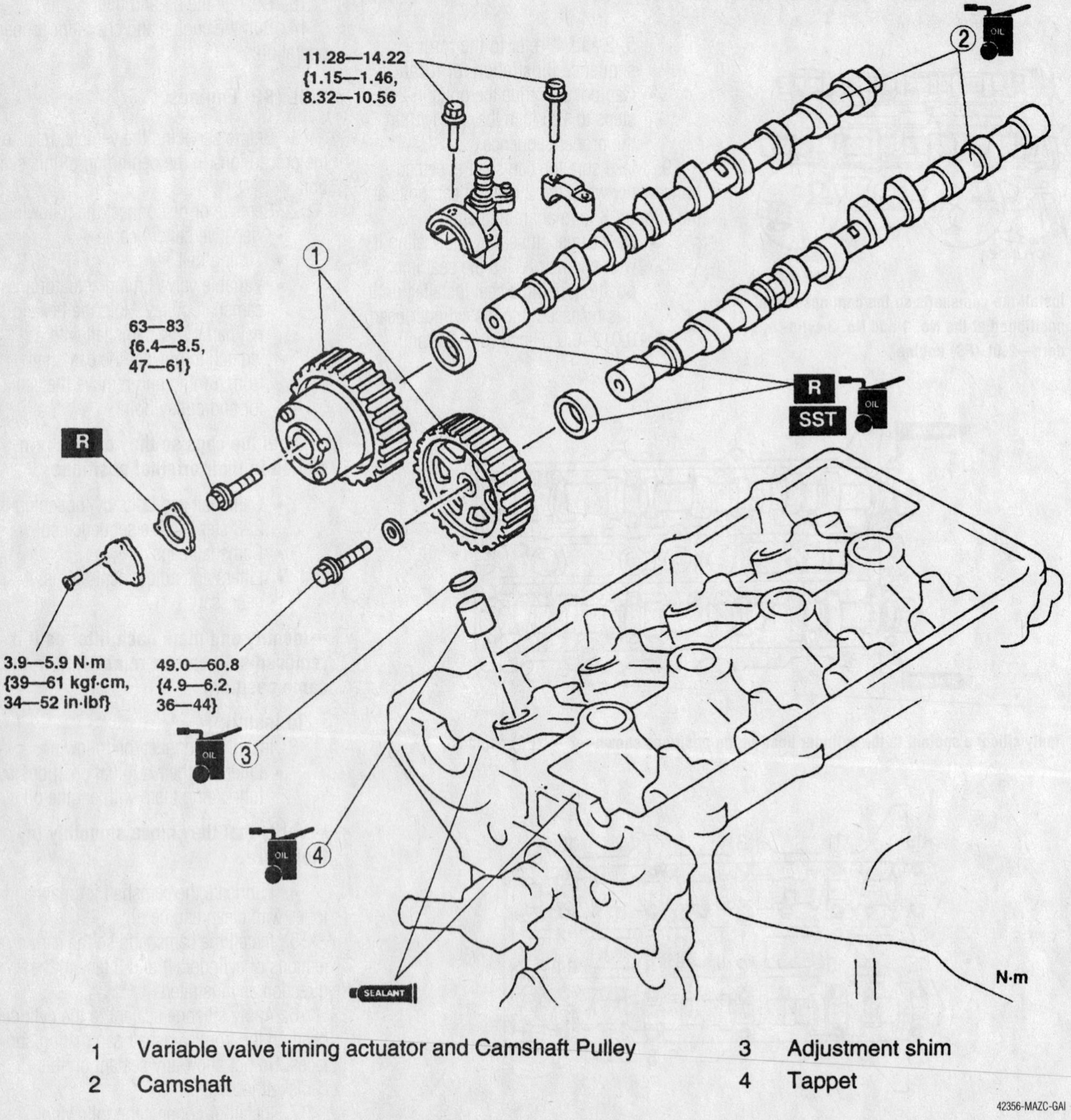

11.28—14.22
{1.15—1.46,
8.32—10.56
63—83
{6.4—8.5,
47—61}
R
R
SST
OIL
3.9—5.9 N·m
{39—61 kgf·cm,
34—52 in·lbf}
49.0—60.8
{4.9—6.2,
36—44}
SEALANT
N·m
1 Variable valve timing actuator and Camshaft Pulley
2 Camshaft
3 Adjustment shim
4 Tappet
42356-MAZC-GAI

Exploded view of the camshaft assembly and related components—1.8L (BP) engine

9. Install the variable valve timing actuator and camshaft pulley as follows:

a. Rotate the camshaft and face the knock pin to the position illustrated.

b. Install the camshaft so that the alignment mark of the pulley faces as illustrated.

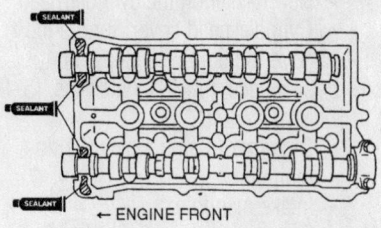

Apply silicone sealant to the cylinder head in the positions shown—1.8L (BP) engine

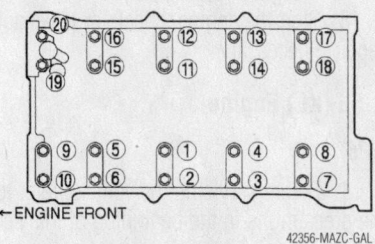

Camshaft cap bolt tightening sequence—1.8L (BP) engine

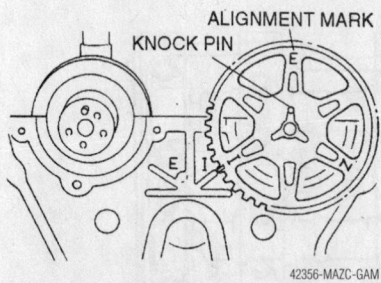

When installing the variable valve timing actuator and camshaft pulley, make sure the knock pin and alignment marks are aligned as shown—1.8L (BP) engine

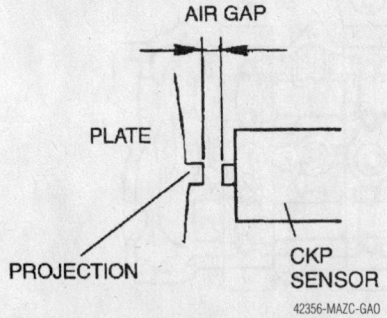

Check the Crankshaft position (CKP) sensor gap—1.8L (BP) engine

c. Install the camshaft so the knock pin of the camshaft is connected to the camshaft knock pin hole of the actuator.

d. Hold the hexagonal part of the camshaft with a wrench to prevent the camshaft from turning and tighten the actuator and pulley bolt. Refer to the illustration for torque specifications.

10. Install the remaining components in the reverse order of removal and inspect the Crankshaft position (CKP) sensor gap. Measure the gap between each 4 projections of the plate behind crankshaft pulley. The gap should be 0.020–0.059 inch (0.5–1.5mm). If not as specified, adjust.

2.3l (KJ) Engine

1. Before servicing the vehicle, refer to the precautions in the beginning of this section.

2. Relieve the fuel system pressure.

3. Drain the engine coolant.

4. Remove or disconnect the following:
- Negative battery cable
- Oxygen (O_2S) sensor connectors
- Exhaust pipe-to-manifold nuts and lower the exhaust pipes
- Right-hand 3-way catalytic converter

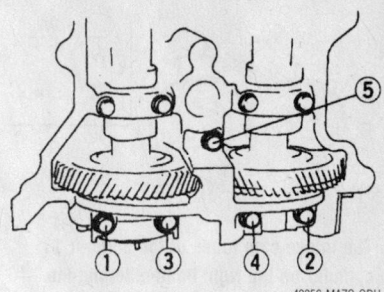

Loosen the front camshaft cap bolts using this sequence—2.3L (KJ) engine

- Compressor (supercharger)
- Intake manifold
- Timing belt covers and timing belt
- Spacer and O-ring from the front of the camshaft
- Ignition coils
- Cylinder head cover mounting bolts, in 5–6 steps, using the reverse of the tightening sequence
- Cylinder head cover
- Camshaft sprockets

5. Turn the camshafts so the knock pins are aligned with the marks on the camshaft caps. This will reduce the pressure on the adjustment shims.

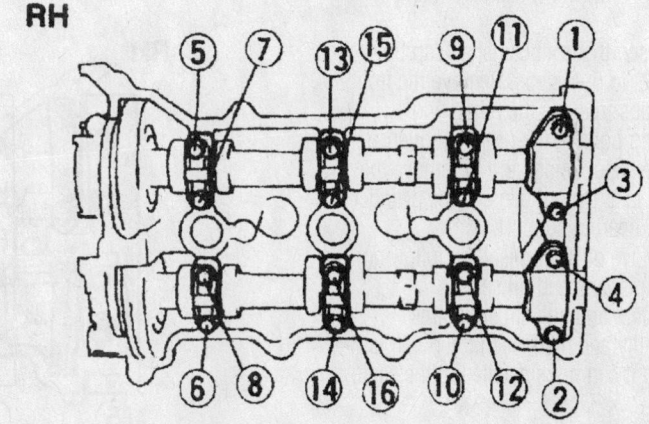

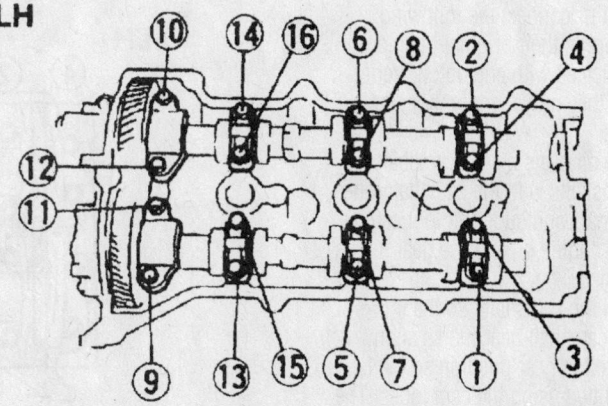

Loosen the remaining camshaft cap bolts using this sequence—2.3L (KJ) engine

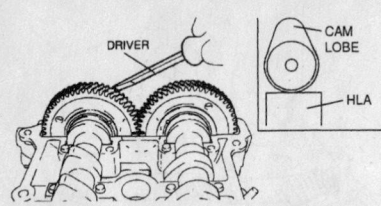

42356-MAZC-GDW

The intake cam lobes of the number 1 cylinder on the right hand side and the number 2 cylinder on the left hand side face straight up —2.3L (KJ) engine

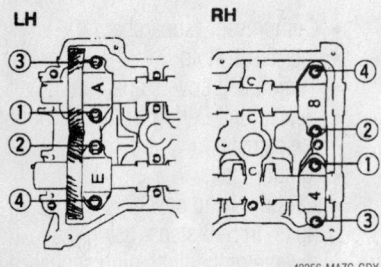

42356-MAZC-GDX

Tighten the front camshaft cap bolts in this sequence —2.3L (KJ) engine

6. Note the markings on the camshaft caps prior to removal, so they can be reinstalled in the same positions. The right-hand (rear) caps are marked with numbers and the left-hand (front) caps are marked with letters.

7. Loosen the front camshaft cap bolts in sequence, in 5–6 steps. Remove the front camshaft caps. remove the remaining camshaft cap bolts in the proper sequence. Remove the caps, being sure to remove the thrust caps last. Do not damage the cylinder head thrust bearing support.

8. Remove or disconnect the following:
   • Camshafts and oil seals
   • Lifters and adjustment shims

9. Identify and mark each lifter as it is removed so it can be reinstalled in the same position.

To install:

10. Install or connect the following:
   • Lifters in their original positions, lubricated with engine oil. Verify that they move smoothly in their bore.
   • New oil seals on the camshafts
   • Camshafts with the camshaft lobes, journals and supports lubricated with engine oil and the gear marks aligned

11. Install the camshafts so the intake and exhaust camshaft gear marks align. Adjust the friction gear position so the tappets are not lifted using the cam lobes. The intake cam lobes of the number 1 cylinder

on the right hand side and the number 2 cylinder on the left hand side face straight up.
   • Front trust caps. Torque the bolts, in 5–6 steps, until the caps are fully seated on the cylinder head.

12. Apply silicone sealant, at a thickness of 0.06–0.09 inch (1.5–2.5mm), to the cylinder head surface in the area forward of the camshaft gear cavity.

13. Install or connect the following:
   • Remaining camshaft caps in their original positions. Make sure the camshaft remains horizontal as the camshaft bolts marked **1** in the illustration are tightened. Torque the bolts, in sequence, in 5 equal steps, with the final step being 100–125 inch lbs. (11–14 Nm). Then retighten the bolts in the same sequence to 100–125 inch lbs. (11–14 Nm).
   • New camshaft oil seal lubricated with engine oil. Tap the seal in evenly with a Seal Installer tool 49 F401 337A with a final protrusion of 0–0.02 inch (0–0.5mm).
   • New blind cap by tapping it in
   • Camshaft sprockets. Torque the bolts to 91–103 ft. lbs. (123–140 Nm).

14. Measure and adjust valve clearances.

15. Remove any sealant and gasket material from the cylinder head cover contact surfaces.

16. Apply silicone sealant to the cylinder head in the area adjacent to the front and rear camshaft caps.

17. Install or connect the following:
   • New gasket on the cylinder head
   • Cylinder head cover. Torque the cover bolts in 5–6 steps, in sequence, to 44–78 inch lbs. (5–9 Nm).
   • Distributor using a new O-ring
   • Ignition coils
   • Intake manifold
   • Spacer using a new O-ring. Torque the bolt to 14–18 ft. lbs. (19–25 Nm).
   • Timing belt and timing belt cover
   • Negative battery cable

18. Start the engine, check for leaks and repair if necessary.

2.5L (KL) Engine

626

1. Before servicing the vehicle, refer to the precautions in the beginning of this section.

2. Drain the cooling system.

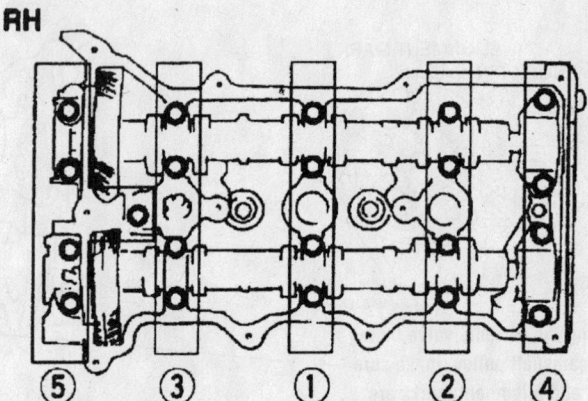

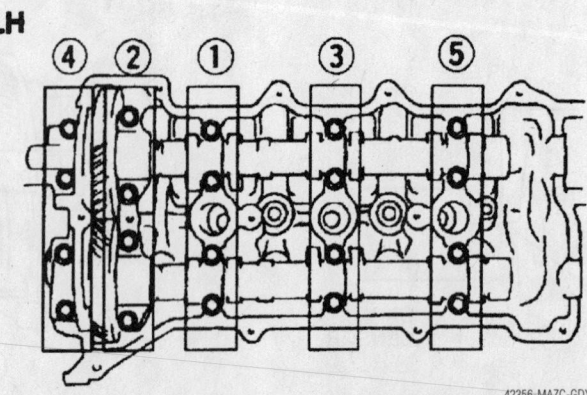

42356-MAZC-GDY

Tighten the camshaft cap bolts using this sequence—2.3L (KJ) engine

3. Relieve the fuel system pressure.
4. Drain the cooling system.
5. Remove or disconnect the following:
- Negative battery cable
- Timing belt
- Front exhaust pipe
- Air cleaner assembly
- Intake manifold
- Fuel hose
- Cylinder head cover
- Camshaft sprocket bolt by holding the camshaft with a wrench on the hexagon cast into the camshaft
- Upper radiator hose
- Number 3 engine mount bracket

- Seal plate
- Water outlet

6. Turn the camshaft, using a wrench on the cast hexagon, until the camshaft knock pin is aligned with the cylinder head marks.

➡ **Do not remove the camshaft caps when the camshaft lobe is pressing on a lifter, as the thrust journal support may become damaged.**

7. Loosen the front camshaft cap bolts in 5–6 steps, in the proper sequence. Bolt **A** is only on the right cylinder head. Remove the front camshaft cap.

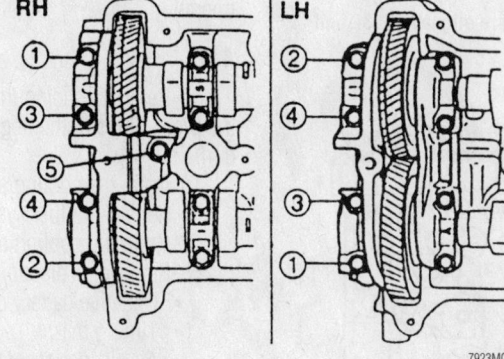

Front camshaft cap bolt loosening sequence—626 with 2.5L engines

7923MG48

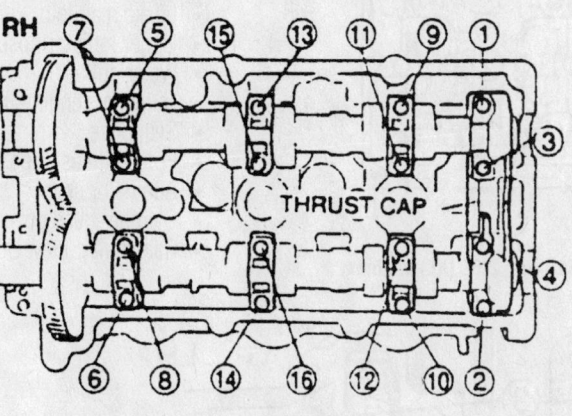

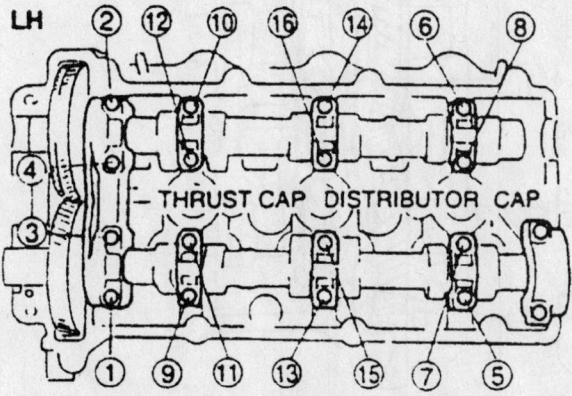

Camshaft cap bolt loosening sequence—626 with 2.5L (KL) engines

7923MG49

8. Mark the position of the camshaft caps so they can be reinstalled in their original locations. Loosen the remaining camshaft cap bolts in 5–6 steps, in the proper sequence, then remove the caps.
9. Remove or disconnect the following:
- Camshafts
- Lifters and adjustment shims, if necessary

➡ **Identify and mark each lifter as it is removed so it can be reinstalled in the same position.**

To install:

10. Install or connect the following:
- Lifters in their original positions lubricated with engine oil. Verify that they move smoothly in their bore.
- Camshafts by lubricating the camshaft journals, lobes and gears with clean engine oil and aligning the timing marks

➡ **The thrust plate positions for the right and left cylinder head camshafts are different.**

11. Be sure the camshaft cap and cylinder head surfaces are clean. Apply a small amount of sealant to the mating surface of the front camshaft cap on both cylinder heads and the rear exhaust camshaft cap on the left cylinder head. Do not get any sealant on the camshaft rotating surfaces.
12. Install or connect the following:
- Front camshaft caps and thrust plate caps. Torque the bolts until the cap seats fully to the cylinder head.
- Remaining caps in their original locations. Torque the bolts in 5–6 steps to 126 inch lbs. (14 Nm), in the proper sequence.
- New oil seals in the cylinder head using an installer
- New blind cap coated with sealant, tap it in place using a plastic hammer
- Camshaft sprockets

➡ **On the right cylinder head, install the sprocket so the R mark can be seen and the timing mark aligns with the camshaft knock pin. On the left cylinder head, install the sprocket so the L mark can be seen and the timing mark aligns with the camshaft knock pin.**

- Camshaft sprocket bolts lubricated with engine oil, by holding the camshaft with a wrench on the cast hexagon. Torque the bolt to 97 ft. lbs. (132 Nm).
- Water outlet

ALIGN THE MARKS

7923MG50

When installing the camshafts, ensure that the marks on the cam gears are aligned—2.3L and 2.5L engines

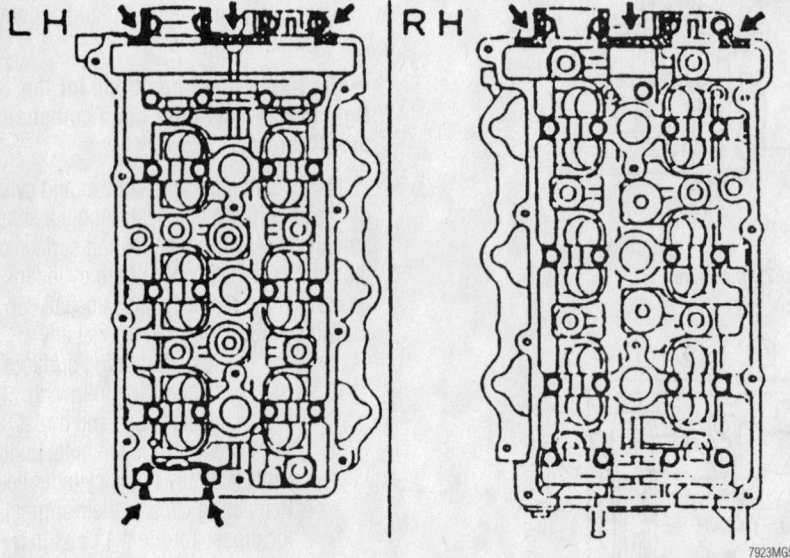

LH   RH

7923MG51

Put silicone sealant on the cylinder head at the positions shown—626 with 2.5L (KL) engines

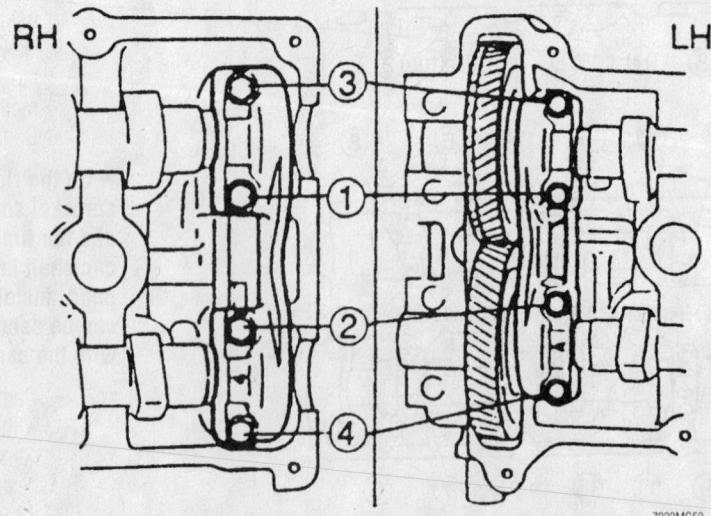

RH   LH

7923MG52

Front camshaft cap bolt tightening sequence—626 with 2.5L (KL) engines

- Seal plate
- Number 3 engine mount bracket
- Upper radiator hose
- Camshaft pulley. Hold the camshaft pulley with a back–up wrench while tightening the pulley bolt to 90–97 ft. lbs. (123–132 Nm).
- Cylinder head cover
- Fuel hose
- Intake manifold
- Air cleaner assembly
- Front exhaust pipe
- Timing belt
- Negative battery cable

13. Fill and bleed the cooling system.
14. Change the oil and filter.
15. Run the engine and check for proper operation.

MILLENIA

1. Before servicing the vehicle, refer to the precautions in the beginning of this section.
2. Drain the cooling system.
3. Relieve the fuel system pressure.
4. Drain the cooling system.
5. Remove or disconnect the following:
   - Negative battery cable
   - Timing belt
   - Air cleaner assembly
   - Spark plug wires
   - Distributor
   - Intake manifold
   - Upper radiator hose
   - Water outlet
   - Number 3 engine mount bracket
   - Seal plate
   - Front exhaust pipe
   - Alternator stay
   - Cylinder head cover
   - Camshaft pulley. Hold the camshaft

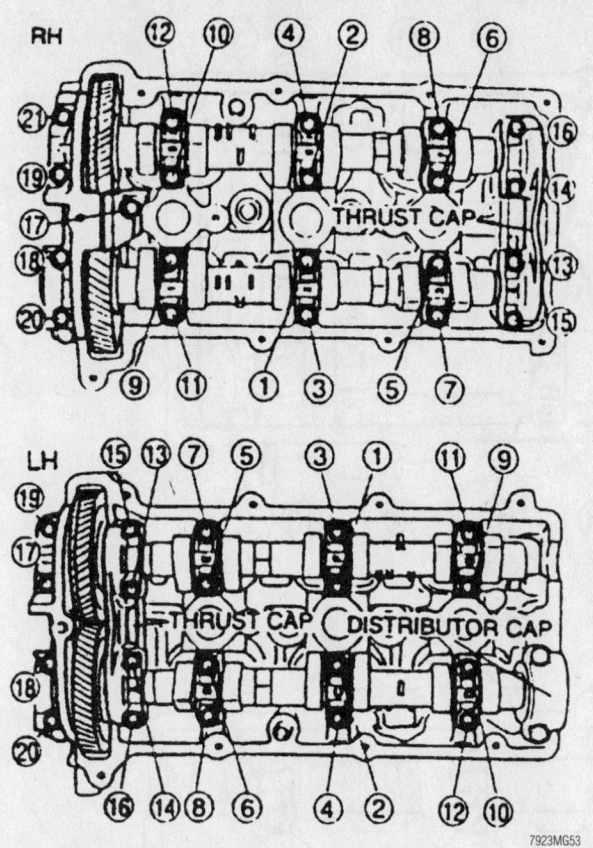

Camshaft cap bolt tightening sequence—626 with 2.5L (KL) engines

7923MG53

inal locations. Loosen the remaining camshaft cap bolts in 5–6 steps, in the proper sequence, then remove the caps.

9. Remove or disconnect the following:
   • Camshafts
   • Lifters and adjustment shims, if necessary

➡️**Identify and mark each lifter as it is removed so it can be reinstalled in the same position.**

To install:

10. Install or connect the following:
   • Lifters in their original positions lubricated with engine oil. Verify that they move smoothly in their bore.
   • Camshafts by lubricating the camshaft journals, lobes and gears with clean engine oil and aligning the timing marks

➡️**The thrust plate positions for the right and left cylinder head camshafts are different.**

11. Be sure the camshaft cap and cylinder head surfaces are clean. Apply a small amount of sealant to the mating surface of the front camshaft cap on both cylinder heads and the rear exhaust camshaft cap on the left cylinder head. Do not get any sealant on the camshaft rotating surfaces.

12. Install or connect the following:
   • Front camshaft caps and thrust plate caps. Torque the bolts until the cap seats fully to the cylinder head.
   • Remaining caps in their original locations. Torque the bolts in 5–6 steps to 126 inch lbs. (14 Nm), in the proper sequence.

pulley with a back–up wrench while loosening the pulley bolt.

6. Turn the camshaft, using a wrench on the cast hexagon, until the camshaft knock pin is aligned with the cylinder head marks.

➡️**Do not remove the camshaft caps when the camshaft lobe is pressing on** a lifter, as the thrust journal support **may become damaged.**

7. Loosen the front camshaft cap bolts in 5–6 steps, in the proper sequence. Bolt **A** is only on the right cylinder head. Remove the front camshaft cap.

8. Mark the position of the camshaft caps so they can be reinstalled in their orig-

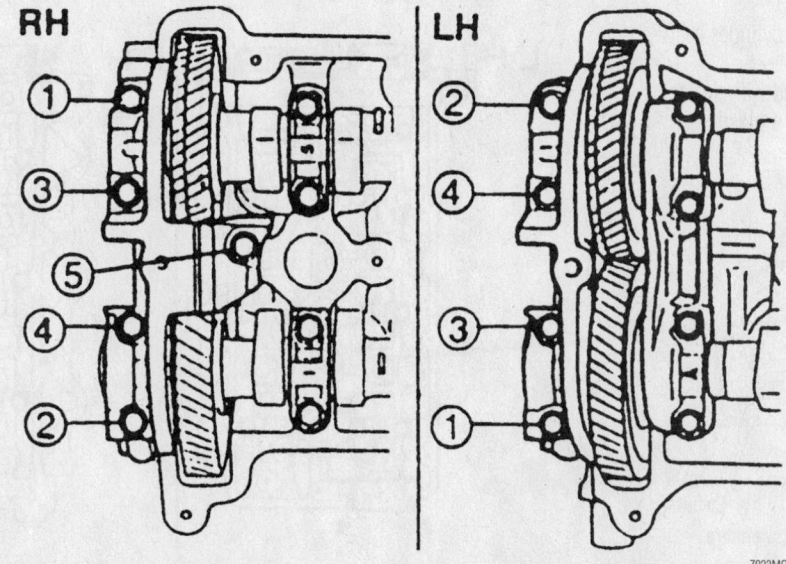

7923MG48

Front camshaft cap bolt loosening sequence—2.5L engines

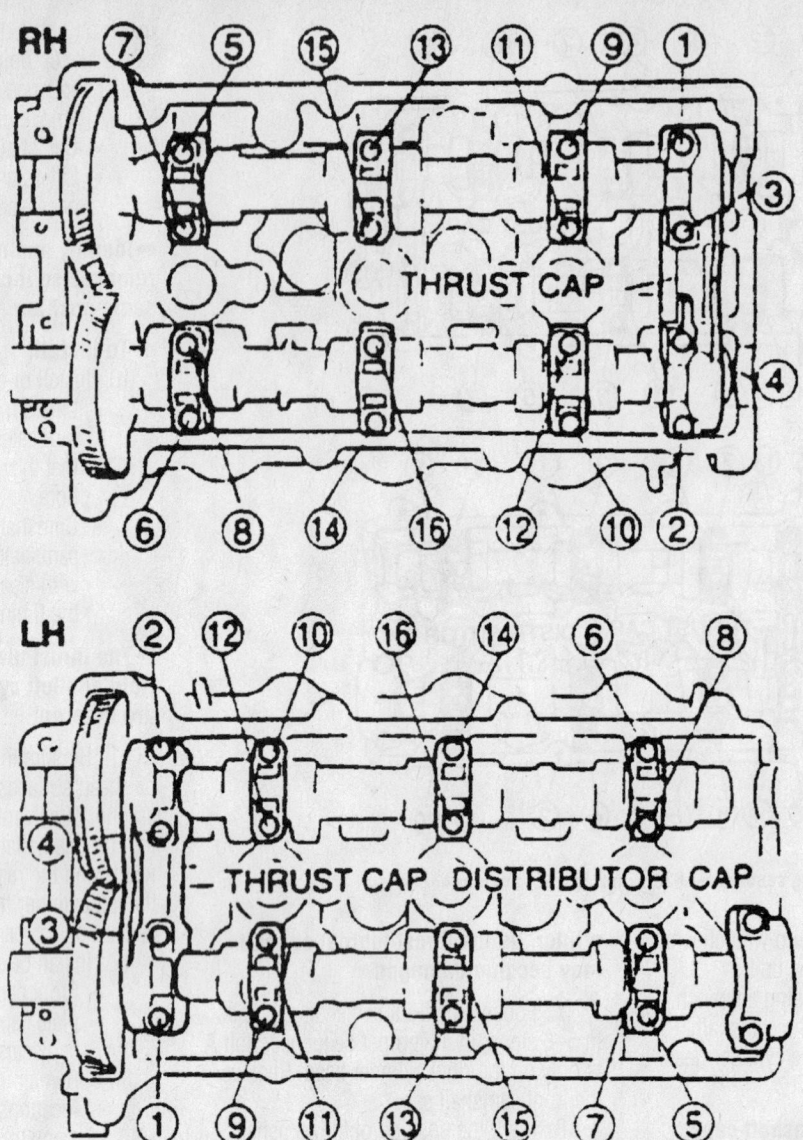

Camshaft cap bolt loosening sequence—2.5L (KL) engines

- New oil seals in the cylinder head using an installer
- New blind cap coated with sealant, tap it in place using a plastic hammer
- Camshaft sprockets

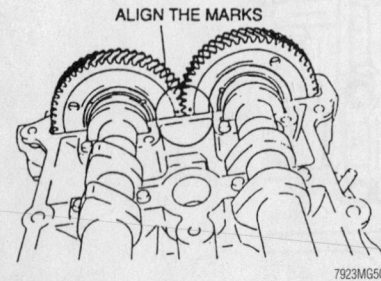

When installing the camshafts, ensure that the marks on the cam gears are aligned—2.5L (KL) engines

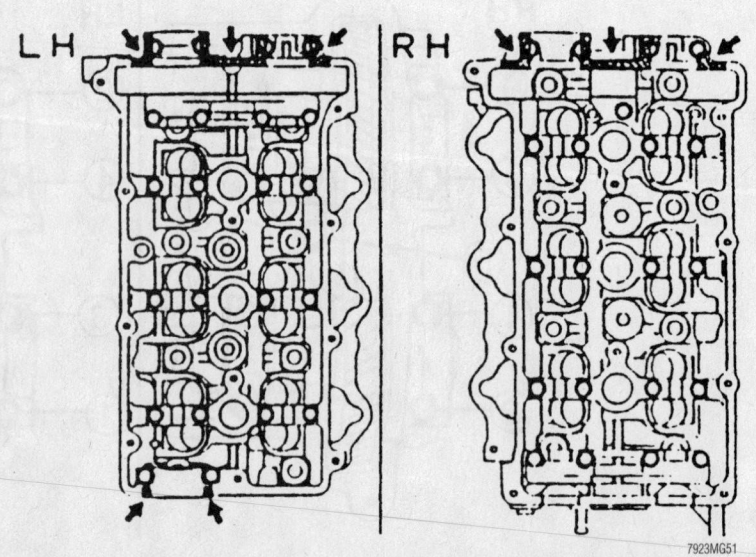

Put silicone sealant on the cylinder head at the positions shown—2.5L (KL) engines

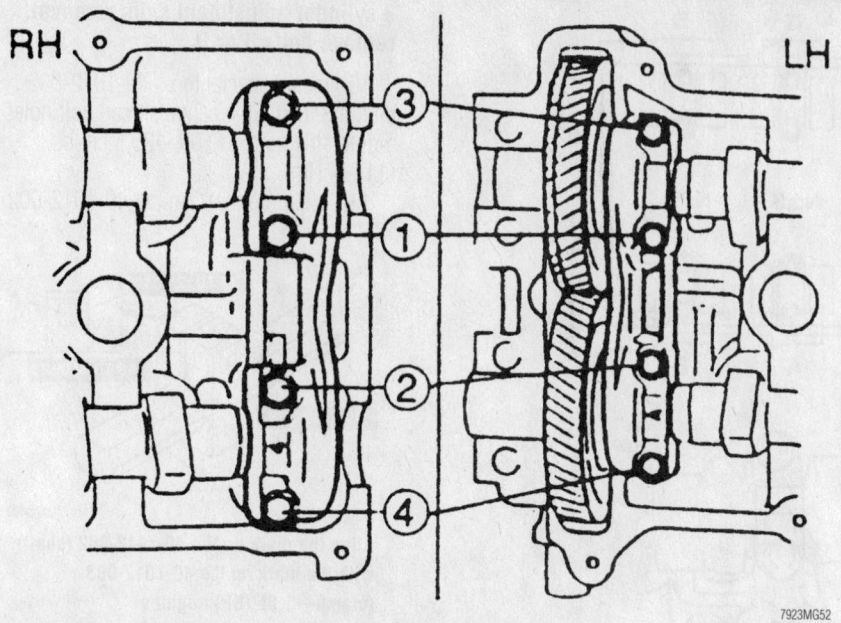

Front camshaft cap bolt tightening sequence—2.5L (KL) engines

7923MG52

➥On the right cylinder head, install the sprocket so the R mark can be seen and the timing mark aligns with the camshaft knock pin. On the left cylinder head, install the sprocket so the L mark can be seen and the timing mark aligns with the camshaft knock pin.

- Camshaft sprocket bolts lubricated with engine oil, by holding the camshaft with a wrench on the cast hexagon. Torque the bolt to 97 ft. lbs. (132 Nm).
- Cylinder head cover
- Alternator stay
- Front exhaust pipe
- Seal plate
- Number 3 engine mount bracket
- Water outlet
- Upper radiator hose
- Intake manifold
- Distributor and spark plug wires
- Air cleaner assembly
- Timing belt
- Negative battery cable

13. Fill and bleed the cooling system.
14. Change the oil and filter.
15. Run the engine and check for proper operation.

Valve Lash

ADJUSTMENT

These engines use solid cam followers with a removable adjustment shim. The valve lash clearance is measured with the original shim installed and checked against the specification. If adjustment is necessary, the original shim is removed, and a thicker or thinner shim is installed to obtain the proper clearance. Special tools are required in order to adjust the shim without removing the camshaft.

1.8L (BP) Engine

➥With the engine cold, standard valve clearance is 0.08–0.009 inch (0.18–0.24mm) on intake side and 0.012–0.013 inch (0.28–0.34mm) on the exhaust side.

1. Before servicing the vehicle, refer to the precautions in the beginning of this section.
2. Remove the cylinder head cover.
3. Measure the valve clearance by turning the crankshaft clockwise until the No. 1 piston is at Top Dead Center (TDC).
4. Measure the valve clearance at **A**. If the clearance exceeds specifications, replace the adjustment shim.

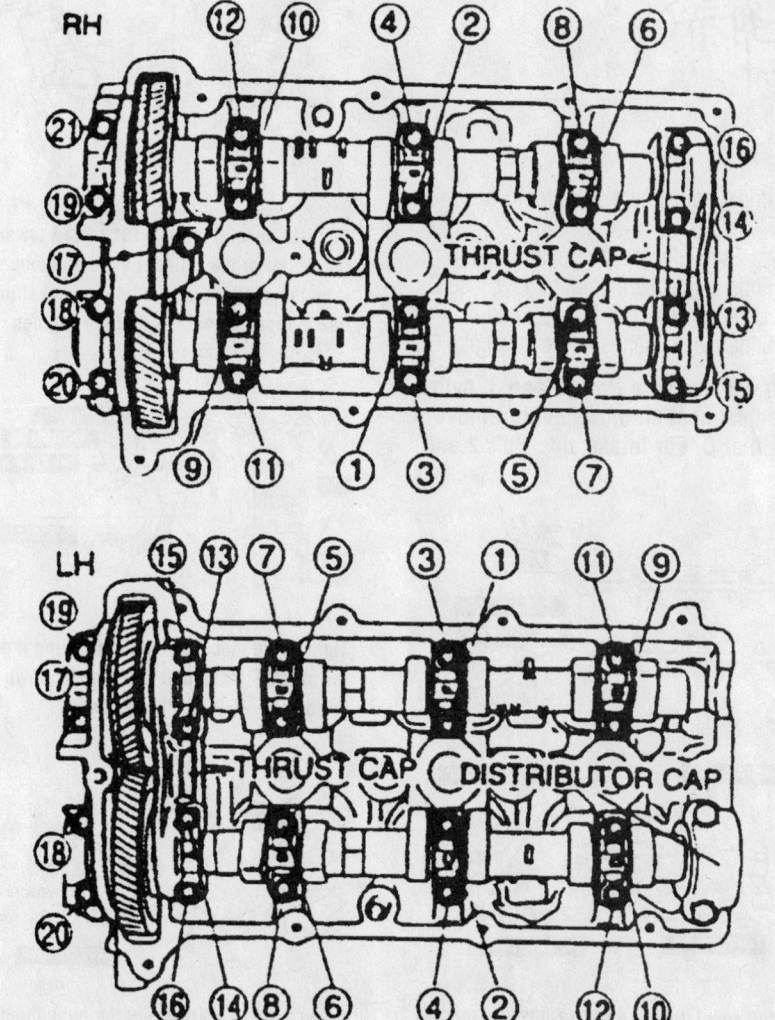

7923MG53

Camshaft cap bolt tightening sequence—2.5L (KL) engines

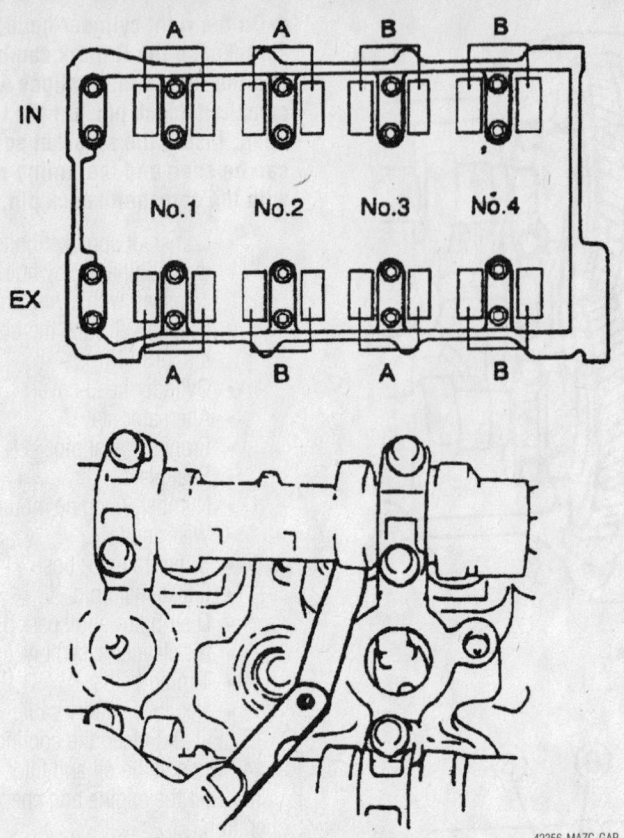

Valve clearance checking positions—1.8L (BP) engines

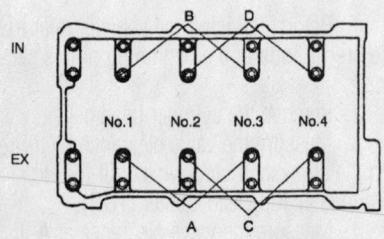

Cam bearing cap bolt removal positions—1.8L (BP) engines

42356-MAZC-GAR

5. Turn the crankshaft clockwise until the cam on the camshaft requiring the adjustment is positioned straight up.

6. Turn the crankshaft clockwise 360 degrees until the No. 4 piston is at TDC. Measure the valve clearance at **B**. If the clearance exceeds specifications, replace the adjustment shim.

7. Repeat this procedure for all the camshafts.

8. Turn the crankshaft clockwise until the cam on the camshaft requiring the adjustment is positioned straight up.

9. Remove the camshaft cap bolts one pair at a time as follows:

a. For exhaust side No. 1, 2 and 3 cylinder adjustment shim removal use **A**.

b. For intake side No. 1, 2 and 3 cylinder adjustment shim removal use **B**.

c. For exhaust side No. 2, 3 and 4 cylinder adjustment shim removal use **C**.

d. For intake side No. 2, 3 and 4 cylinder adjustment shim removal use **D**.

➡For exhaust side No's. 2 and 3, cylinder adjustment shim removal; remove bolts A or C. For intake side No's 2 and

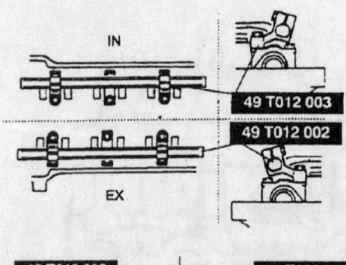

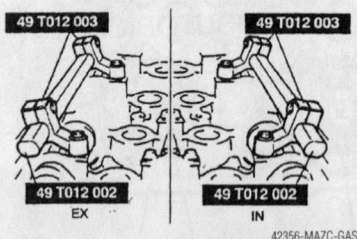

Install special tools 49-T012-002 and 003, using the camshaft cap bolt holes—1.8L (BP) engines

42356-MAZC-GAS

3 cylinder adjustment shim removal, remove bolts B or D.

10. Install special tools 49-T012-002 and 003, using the camshaft cap bolt holes. Tighten the bolts to 100–125 inch lbs. (11–14 Nm).

11. Align the mark on the 49-T012-002

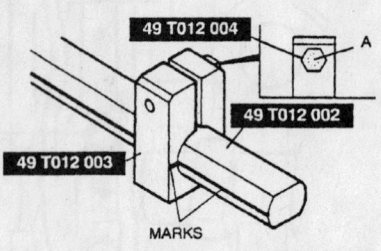

42356-MAZC-GAT

Align the mark on the 49-T012-002 (shaft) with the mark on the 49-T012-003 (clamp—1.8L (BP) engines

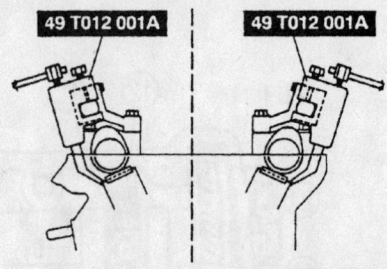

42356-MAZC-GAU

Position special tool 49-T012-001A toward the center of the cylinder head and mount it on the shaft where the adjustment shim needs replacement—1.8L (BP) engines

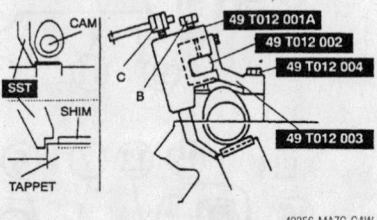

42356-MAZC-GAW

Tighten the mounting bolt B securing it on the shaft. Tighten bolt C and press down the tappet—1.8L (BP) engines

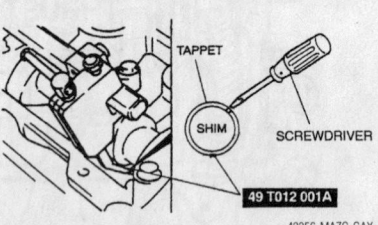

42356-MAZC-GAX

Using a small prytool, pry the adjustment shim upwards through the notch on the tappet—1.8L (BP) engines

(shaft) with the mark on the 49-T012-003 (clamp). Tighten special tool 49-T012-004 (bolt) to secure the shaft.

12. Position special tool 49-T012-001A toward the center of the cylinder head and mount it on the shaft where the adjustment shim needs replacement.

13. Position the notch of the tappet to allow a small prytool to be inserted.

14. Set the special tool on the tappet by its notch. Tighten the mounting bolt **B** securing it on the shaft.

15. Tighten bolt **C**, and press down the tappet.

16. Using a small prytool, pry the adjustment shim upwards through the notch on the tappet. Remove the shim with a magnet.

17. Select and install the proper adjustment shim. Loosen bolt **C** to allow the tappet to move up, and loosen bolt **B** to remove special tool 49-T012-001A.

18. Remove special tools 49-T012-002, 003 and 004, and torque the camshaft cap bolts to 100–125 inch lbs. (11–14 Nm).

19. Repeat the procedure for all necessary adjustment shims. Check the valve clearance.

2.0L (FS) Engine

➡ **With the engine cold, standard valve clearance is 0.089–0.0116 inch (0.225–0.295mm) on intake and exhaust sides.**

1. Before servicing the vehicle, refer to the precautions in the beginning of this section.

2. Remove the cylinder head cover.

3. Measure the valve clearance by turning the crankshaft clockwise until the No. 1 piston is at Top Dead Center (TDC).

4. Measure the valve clearance at **A**. If the clearance exceeds specifications, replace the adjustment shim.

5. Turn the crankshaft clockwise 360 degrees until the No. 4 piston is at TDC. Measure the valve clearance at **B**. If the

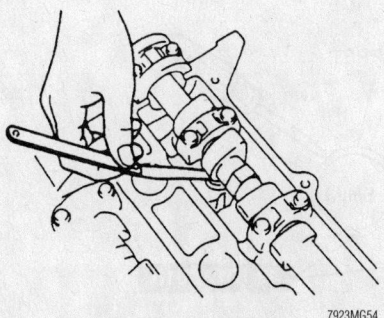

Ensure that the cam lobe faces away from the follower when checking the valve clearance

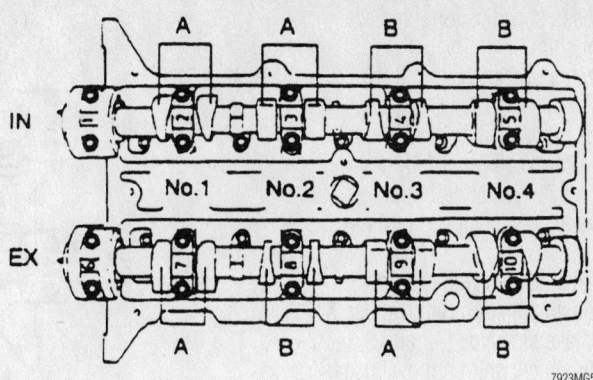

Valve clearance checking positions—2.0L (FS) engines

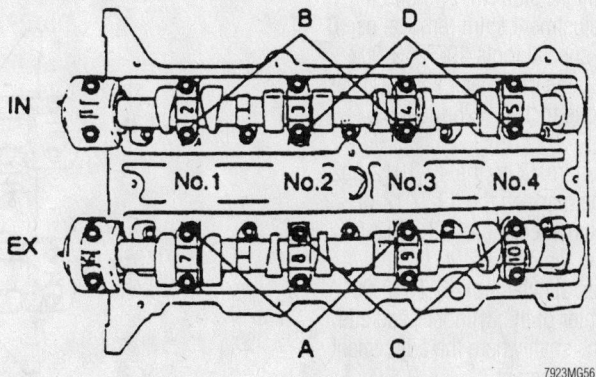

Cam bearing cap bolt removal positions—2.0L (FS) engines

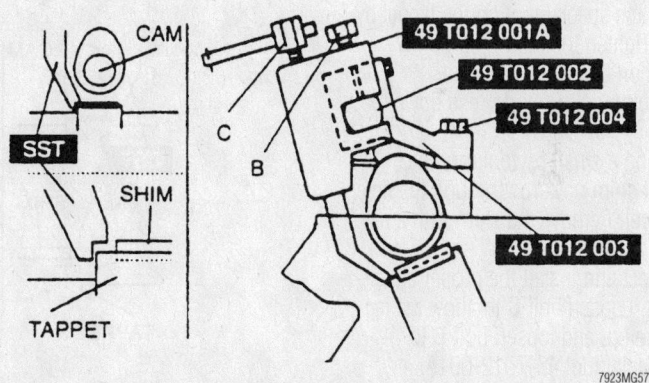

Mount the tappet depressor tool onto the shaft above the tappet that needs adjustment—2.0L (FS) engines

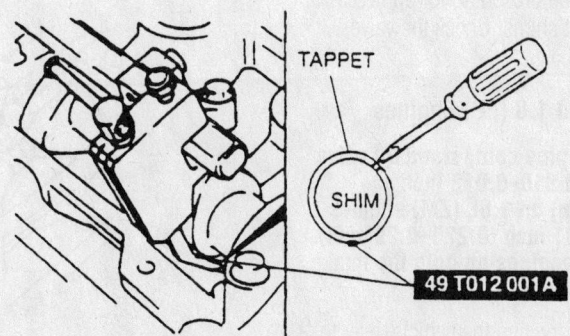

With the tappet depressed, use a small prytool to remove the adjustment shim—2.0L (FS) engines

clearance exceeds specifications, replace the adjustment shim.

6. Repeat this procedure for all the camshafts.

7. Turn the crankshaft clockwise until the cam on the camshaft requiring the adjustment is positioned straight up.

8. Remove the camshaft cap bolts as follows:

 a. For exhaust side No. 1, 2 and 3 cylinder adjustment shim removal use **A**.

 b. For intake side No. 1, 2 and 3 cylinder adjustment shim removal use **B**.

 c. For exhaust side No. 2, 3 and 4 cylinder adjustment shim removal use **C**.

 d. For intake side No. 2, 3 and 4 cylinder adjustment shim removal use **D**.

9. Install special tools 49-T012-002 and 003, using the camshaft cap bolt holes. Tighten the bolts to 100–125 inch lbs. (11–14 Nm).

10. Align the mark on the 49-T012-002 (shaft) with the mark on the 49-T012-003 (clamp). Tighten special tool 49-T012-004 (bolt) to secure the shaft.

11. Position special tool 49-T012-001A toward the center of the cylinder head and mount it on the shaft where the adjustment shim needs replacement.

12. Position the notch of the tappet to allow a small prytool to be inserted.

13. Set the special tool on the tappet by its notch. Tighten the mounting bolt **B** securing it on the shaft.

14. Tighten bolt **C**, and press down the tappet.

15. Using a small prytool, pry the adjustment shim upwards through the notch on the tappet. Remove the shim with a magnet.

16. Select and install the proper adjustment shim. Loosen bolt **C** to allow the tappet to move up, and loosen bolt **B** to remove special tool 49-T012-001A.

17. Remove special tools 49-T012-002, 003 and 004, and torque the camshaft cap bolts to 100–125 inch lbs. (11–14 Nm).

18. Repeat the procedure for all necessary adjustment shims. Check the valve clearance.

1.6L (ZM) and 1.8 (FP) Engines

➡ With the engine cold, standard valve clearance is 0.010–0.012 inch (0.25–0.31mm) on 1.6L (ZM) engines or 0.010–0.001 inch (0.225–0.295mm) on 1.8L (FP) engines on both the intake and exhaust sides.

1. Before servicing the vehicle, refer to the precautions in the beginning of this section.

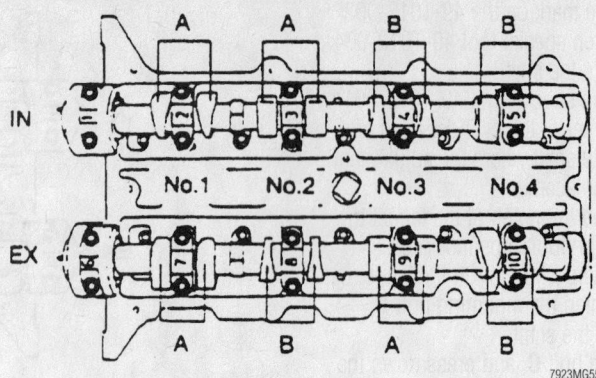

Valve clearance checking positions—4-cylinder engines

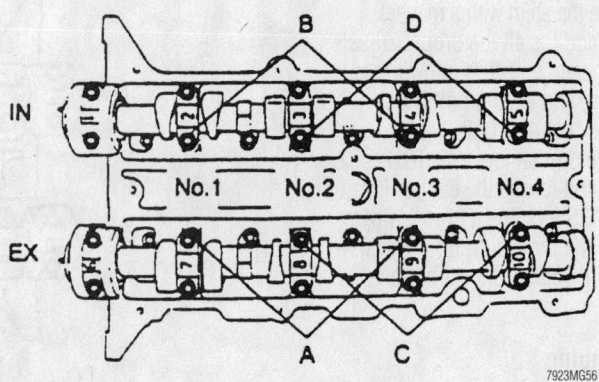

Cam bearing cap bolt removal positions—4-cylinder engine

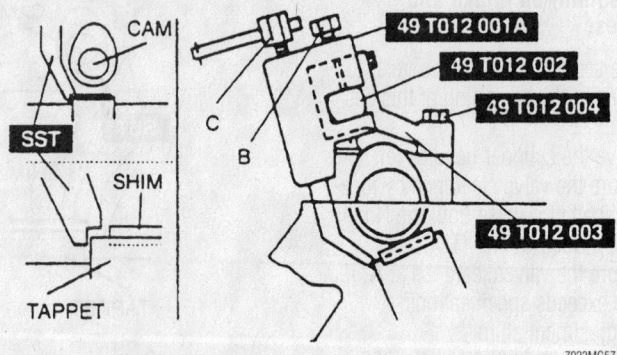

Mount the tappet depressor tool onto the shaft above the tappet that needs adjustment—4-cylinder engine

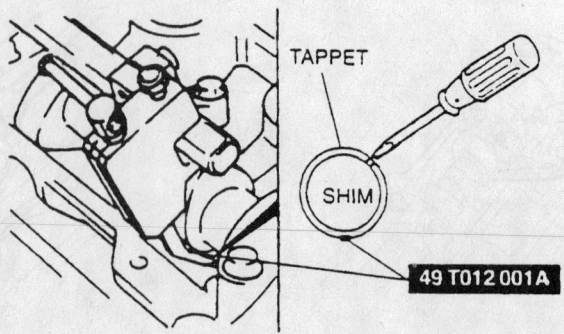

With the tappet depressed, use a small prytool to remove the adjustment shim—4-cylinder engine

2. Remove the cylinder head cover.

3. Measure the valve clearance by turning the crankshaft clockwise until the No. 1 piston is at Top Dead Center (TDC).

4. Measure the valve clearance at **A**. If the clearance exceeds specifications, replace the adjustment shim.

5. Turn the crankshaft clockwise 360 degrees until the No. 4 piston is at TDC. Measure the valve clearance at **B**. If the clearance exceeds specifications, replace the adjustment shim.

6. Repeat this procedure for all the camshafts.

7. Turn the crankshaft clockwise until the cam on the camshaft requiring the adjustment is positioned straight up.

8. Remove the camshaft cap bolts as follows:

    a. For exhaust side No. 1, 2 and 3 cylinder adjustment shim removal use **A**.

    b. For intake side No. 1, 2 and 3 cylinder adjustment shim removal use **B**.

    c. For exhaust side No. 2, 3 and 4 cylinder adjustment shim removal use **C**.

    d. For intake side No. 2, 3 and 4 cylinder adjustment shim removal use **D**.

9. Install special tools 49-T012-002 and 003, using the camshaft cap bolt holes. Tighten the bolts to 100–125 inch lbs. (11–14 Nm).

10. Align the mark on the 49-T012-002 (shaft) with the mark on the 49-T012-003 (clamp). Tighten special tool 49-T012-004 (bolt) to secure the shaft.

11. Position special tool 49-T012-001A toward the center of the cylinder head and mount it on the shaft where the adjustment shim needs replacement.

12. Position the notch of the tappet to allow a small prytool to be inserted.

13. Set the special tool on the tappet by its notch. Tighten the mounting bolt **B** securing it on the shaft.

14. Tighten bolt **C**, and press down the tappet.

15. Using a small prytool, pry the adjustment shim upwards through the notch on the tappet. Remove the shim with a magnet.

16. Select and install the proper adjustment shim. Loosen bolt **C** to allow the tappet to move up, and loosen bolt **B** to remove special tool 49-T012-001A.

17. Remove special tools 49-T012-002, 003 and 004, and torque the camshaft cap bolts to 100–125 inch lbs. (11–14 Nm).

18. Repeat the procedure for all necessary adjustment shims. Check the valve clearance.

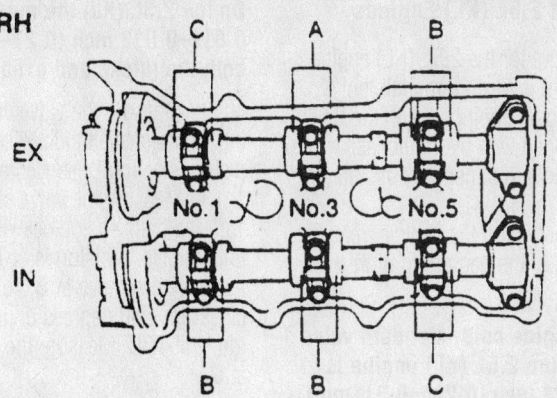

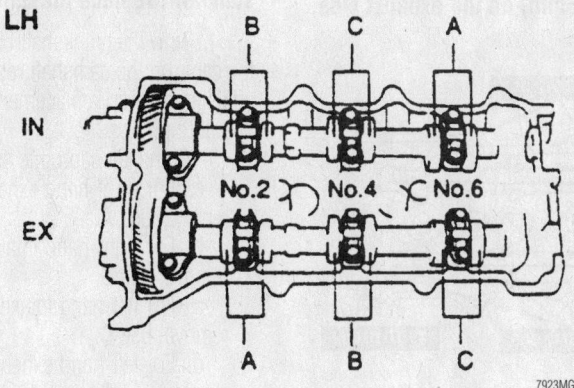

Valve clearance checking positions—6-cylinder engine

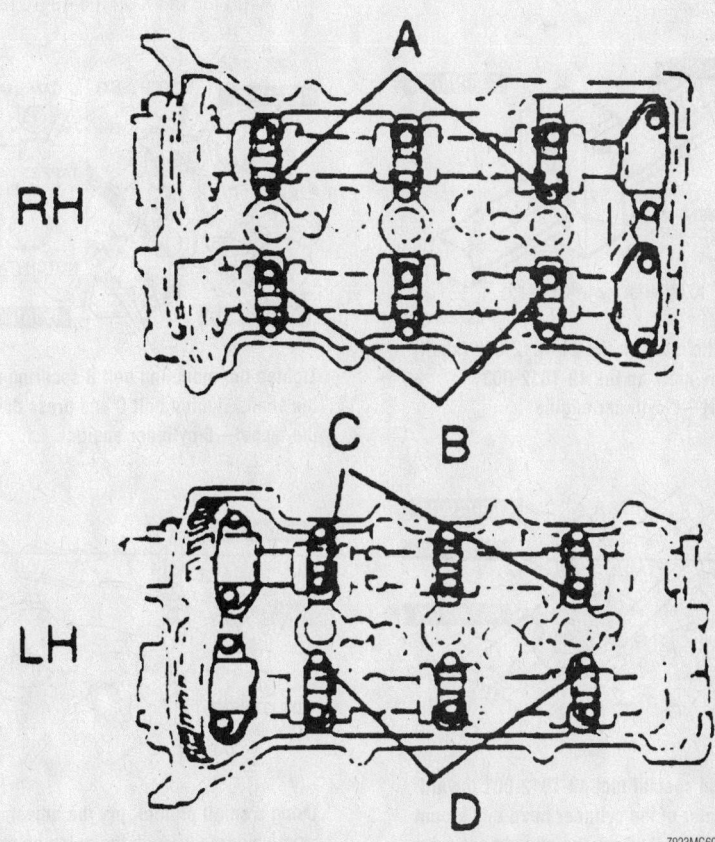

Camshaft cap bolt removal positions—6-cylinder engine—refer to text

2.3L (KJ) and 2.5L (KL) Engines

This procedure for the 2.5L (KL) engine valve adjustment is for 626 models only. The Millenia model equipped with the 2.5L engine is equipped with hydraulic lash adjusters and no adjustment is possible or necessary.

1. Before servicing the vehicle, refer to the precautions in the beginning of this section.

➡ With the engine cold, standard valve clearance on the 2.5L (KL) engine is 0.0097–0.0124 inch (0.245–0.311mm) on intake side and 0.0105–0.0131 inch (0.265–0.335mm) on the exhaust side.

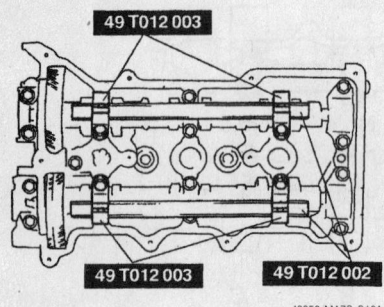

Install special tools 49-T012-002 and 003, using the camshaft cap bolt holes—6-cylinder engine

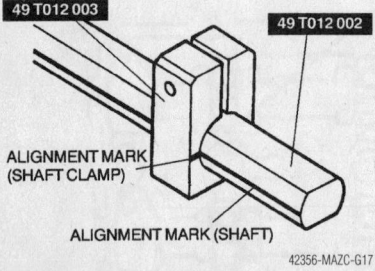

Align the mark on the 49-T012-002 (shaft) with the mark on the 49-T012-003 (clamp)—6-cylinder engine

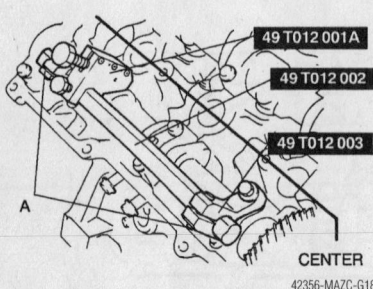

Position special tool 49-T012-001 toward the center of the cylinder head and mount it on the shaft where the adjustment shim needs replacement—6-cylinder engine

On the 2.3L (KJ) the measurement is 0.011–0.012 inch (0.27–0.33mm) on both the intake and exhaust sides.

2. Measure the valve clearance by turning the crankshaft clockwise until the No. 1 piston is at Top Dead Center (TDC).

3. Measure the valve clearance at **A**. Turn the crankshaft clockwise 240 degrees until the No. 3 piston is at TDC. Measure the valve clearance at **B**. Turn the crankshaft clockwise 240 degrees until the No. 5 piston is at TDC. Measure the valve clearance at **C**.

➡ If the valve clearance exceeds the standard, replace the adjustment shim.

4. Turn the crankshaft clockwise until the cam, on the camshaft requiring the adjustment shim replacement, is positioned straight up.

5. Camshaft cap bolts as follows:
    a. For right-hand exhaust side shim removal use **A**.
    b. For right-hand intake side shim removal use **B**.
    c. For left-hand intake side shim removal use **C**.
    d. For left-hand exhaust side shim removal use **D**.

6. Install special tools 49-T012-002 and 003, using the camshaft cap bolt holes.

7. Align the mark on the 49-T012-002

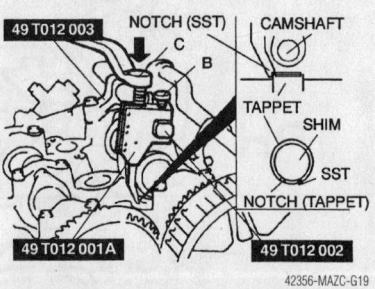

Tighten the mounting bolt B securing it on the shaft. Tighten bolt C and press down the tappet—6-cylinder engine

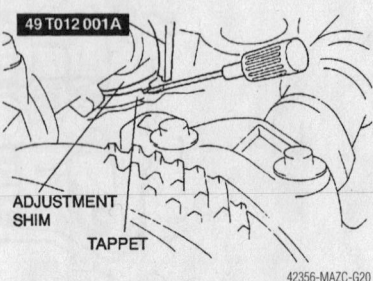

Using a small prytool, pry the adjustment shim upwards through the notch on the tappet—6-cylinder engine

(shaft) with the mark on the 49-T012-003 (clamp).

8. Position special tool 49-T012-001 toward the center of the cylinder head and mount it on the shaft where the adjustment shim needs replacement.

9. Position the notch of the tappet to allow a small prytool to be inserted.

10. Set the special tool on the tappet by its notch. Tighten the mounting bolt **B** securing it on the shaft.

11. Tighten bolt **C** and press down the tappet.

12. Using a small prytool, pry the adjustment shim upwards through the notch on the tappet. Remove the shim with a magnet.

13. Select and install the proper adjustment shim. Loosen bolt **C** to allow the tappet to move up and loosen bolt B to remove special tool 49-T012-001.

14. Remove special tools 49-T012-002, 003 and 004 and tighten the camshaft cap bolts to 100–125 inch lbs. (11–14 Nm).

15. Repeat the procedure for all necessary adjustment shims. Check the valve clearance.

Starter Motor

REMOVAL & INSTALLATION

Protégé

1. Remove or disconnect the following:
   • Negative battery cable
   • Air cleaner
   • Intake manifold support bracket bolts
   • Starter electrical connectors
   • Starter

To install:
2. Install or connect the following:
   • Starter and loosely tighten the lower starter mounting bolt
   • Starter electrical connectors
   • Intake manifold support bracket. Torque the bolts to 28–38 ft. lbs. (38–51 Nm).
   • Starter bolts. Torque the bolts 28–38 ft. lbs. (38–51 Nm). The upper mounting bolts must be tightened first.
   • Air cleaner
   • Negative battery cable

626

2.0L (FS) ENGINE

1. Remove or disconnect the following:

- Battery
- Air cleaner assembly
- Transverse member
- Intake manifold bracket
- Wiring at the starter
- Starter

To install:

2. Install or connect the following:
- Starter. Torque the bolts to 38 ft. lbs. (51 Nm).
- Wiring at the starter
- Intake manifold bracket
- Transverse member
- Air cleaner assembly
- Battery

2.5L (KL) ENGINE

1. Remove or disconnect the following:
- Battery
- Air cleaner assembly
- Transverse member
- Fuel filter with the hose still attached and set aside
- Oil filler pipe
- Transaxle selector cable from the automatic transaxle and remove the cable bracket
- Wiring at the starter
- Starter

To install:

2. Install or connect the following:
- Starter. Torque the bolts to 38 ft. lbs. (51 Nm).
- Wiring at the starter
- Cable bracket and connect the transaxle selector cable to the transaxle
- Oil filler pipe
- Fuel filter
- Transverse member
- Air cleaner assembly
- Battery

Miata

1. Remove or disconnect the following:
- Negative battery cable
- Air cleaner assembly
- Dipstick tube
- Intake manifold bracket
- Wiring at the starter
- Starter

To install:

2. Install or connect the following:
- Starter. Torque the bolts to 33 ft. lbs. (46 Nm).
- Wiring at the starter
- Intake manifold bracket. Torque the bolts to 38 ft. lbs. (51 Nm).
- Dipstick tube
- Air cleaner assembly
- Negative battery cable

Millenia

2.3L (KJ) ENGINE

1. Remove or disconnect the following:
- Negative battery cable
- Charge air cooler duct
- Battery clamp, box and battery
- Battery tray
- Rear charge air cooler
- Pipe bracket
- Starter electrical connectors
- Starter

To install:

2. Install or connect the following:
- Starter. Torque the bolts to 28–38 ft. lbs. (38–51 Nm).
- B-terminal wire
- S-terminal wire
- Pipe bracket. Torque the bolt to 14–18 ft. lbs. (19–25 Nm).
- Rear charge air cooler using new O-rings. Torque the nuts to 14–18 ft. lbs. (19–25 Nm).
- Battery tray
- Battery, box and clamp
- Charge air cooler duct
- Negative battery cable

2.5L (KL) ENGINE

1. Remove or disconnect the following:
- Negative battery cable
- Battery clamp, box and battery
- Battery tray
- Shift cable from the selector lever
- Cable from the bracket by squeeze the lock tabs
- Electrical connectors from the starter solenoid
- 2 selector cable bracket bolts and the bracket
- 2 nuts and the bolt from the starter bracket and the bracket
- Starter electrical connectors
- Starter

To install:

2. Install or connect the following:
- Starter. Torque the bolts to 28–38 ft. lbs. (38–51 Nm).
- B-terminal wire.
- S-terminal wire to the solenoid
- Starter bracket
- Selector cable bracket. Torque the bolts to 5–7 ft. lbs. (7–9 Nm).
- Starter solenoid electrical connectors
- Shift cable into the cable bracket and into the selector lever
- Battery tray
- Battery, box and clamp
- Negative battery cable

Oil Pan

REMOVAL & INSTALLATION

1.8L (BP) Engines

1. Before servicing the vehicle, refer to the precautions in the beginning of this section.

2. Drain the engine oil.

3. Remove or disconnect the following:
- Negative battery cable
- Air cleaner assembly
- Wheel speed sensor
- Dipstick tube
- Intermediate shaft

4. Attach an engine to the hoist, loosen the oil pan bolts, remove the engine mounting bolts and raise the engine slightly using a hoist.
- Stabilizer link nut
- Shock absorber-to-knuckle nut and bolt

✸✸ CAUTION

When removing the crossmember, be careful do to damage the brake hoses, A/C pipes and power steering pipes when lowering the crossmember.

5. Support the crossmember with a transmission jack, remove the crossmember retainers. Separate the steering intermediate shaft from the pinion shaft and lower the crossmember until the clearance between the oil pan and the steering gear exceeds 5.12 inch (130mm).

6. Remove or disconnect the following:
- Engine mount
- Oil pan bolts and the oil pan using a seal cutter, then insert a flat pry tool into the locations illustrated.

To install:

7. Clean the oil pan. Clean all dirt, oil, gasket and old sealant from the oil pan and cylinder block contact surfaces.

8. Apply a continuous bead of silicone sealant to the contact surfaces of the new oil pan gaskets as illustrated.

9. Install new gaskets into the oil pump body and rear cover facing the notches as illustrated.

10. Apply a 0.079 inch (2mm) continuous bead of silicone sealant on the oil pan as illustrated.

11. Apply a continuous bead of silicone sealant on the oil pan-to-block areas as illustrated.

12. Apply a 0.099–0.137 inch

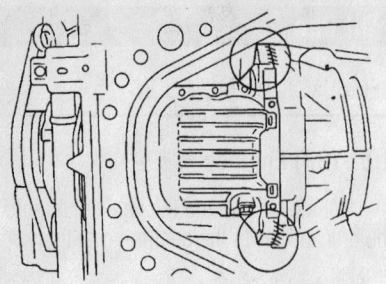

42356-MAZC-GAZ

Remove the oil pan by inserting a flat pry tool into the locations shown—1.8L (BP) engine

SEALANT
42356-MAZC-GBA

Apply a continuous bead of silicone sealant to the contact surfaces of the new oil pan gaskets as shown—1.8L (BP)

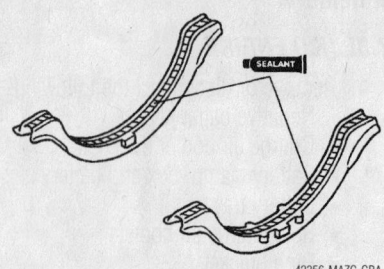

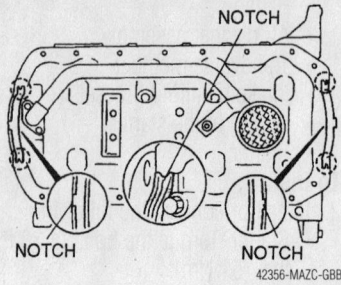

NOTCH
NOTCH
NOTCH
42356-MAZC-GBB

Install new gaskets into the oil pump body and rear cover facing the notches as shown—1.8L (BP) engine

18—26 {1.8—2.7, 14—19}
2
44—60 {4.4—6.0, 32—44}
4
6 118—137 {12.0—14.0, 87—101}
5 94—116 {9.5—11.9, 68.8—86.0}
37—53 {3.7—5.5, 27—39}
7
3
57—78 {5.8—8.0, 42—57}
SEALANT
R
10 16—21 {1.6—2.2, 12—15}
6 94—116 {9.5—11.9, 68.8—86.0}
9
7.9—10.7 N·m {80—110 kgf·cm, 69.5—95.4 in·lbf}
1
7.9—10.7 N·m {80—110 kgf·cm, 69.5—95.4 in·lbf}
SEALANT
R
SEALANT
R
R
8
64—86 {6.5—9.1, 48—65}
7.9—10.7 N·m {80—110 kgf·cm, 69.5—95.4 in·lbf}
N·m {kgf·m, ft·lbf}

1	Dipstick and pipe	6	Crossmember bolt and nut
2	Intermediate shaft	7	Engine mount
3	Engine mount nut	8	Oil pan
4	Stabilizer control link nut	9	Oil strainer
5	Shock absorber bolt and nut	10	MBSP

42356-MAZC-GAY

Exploded view of the oil pan and related components—1.8L (BP) engine

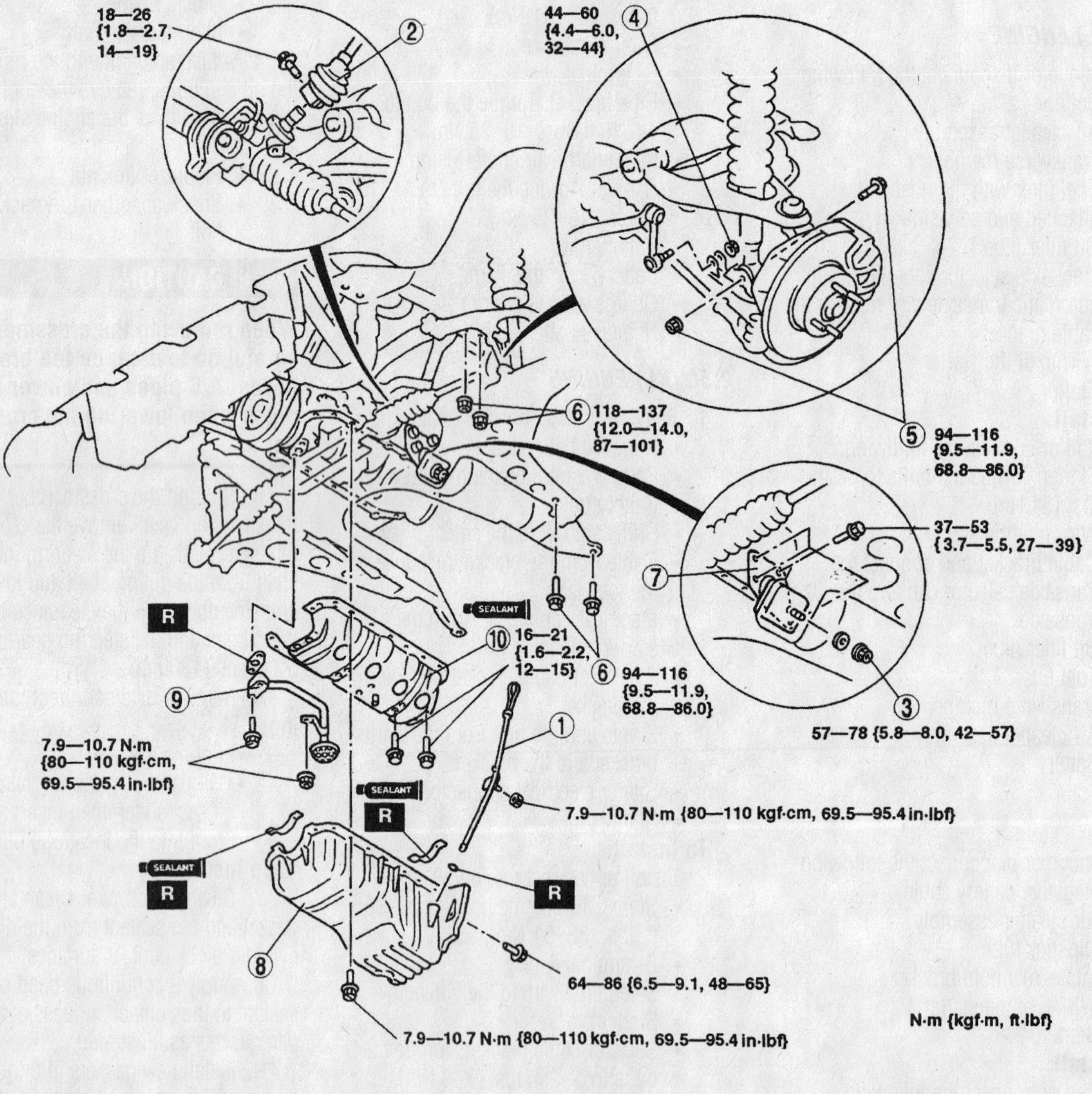

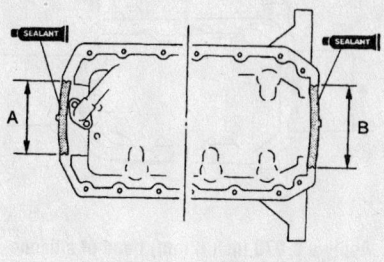

SEALANT
SEALANT
A
B
42356-MAZC-GBC

Apply a 0.079 inch (2mm) continuous bead of silicone sealant on the oil pan as shown—1.8L (BP) engine

SEALANT
SEALANT
SEALANT
SEALANT
42356-MAZC-GBD

Apply a continuous bead of silicone sealant on the oil pan-to-block areas as shown—1.8L (BP) engine

SEALANT
42356-MAZC-GBE

Apply a 0.099–0.137 inch (2.5–3.5mm) continuous bead of silicone sealant on the oil pan along the inside of the bolt holes and overlap the ends—1.8L (BP) engine

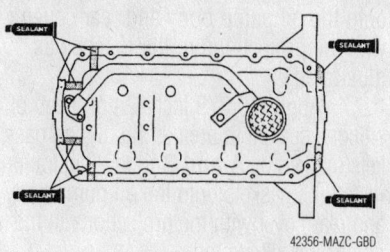

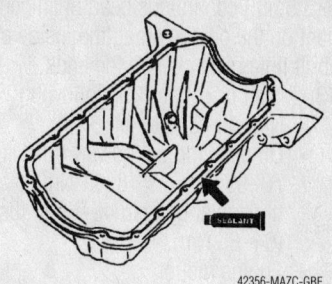

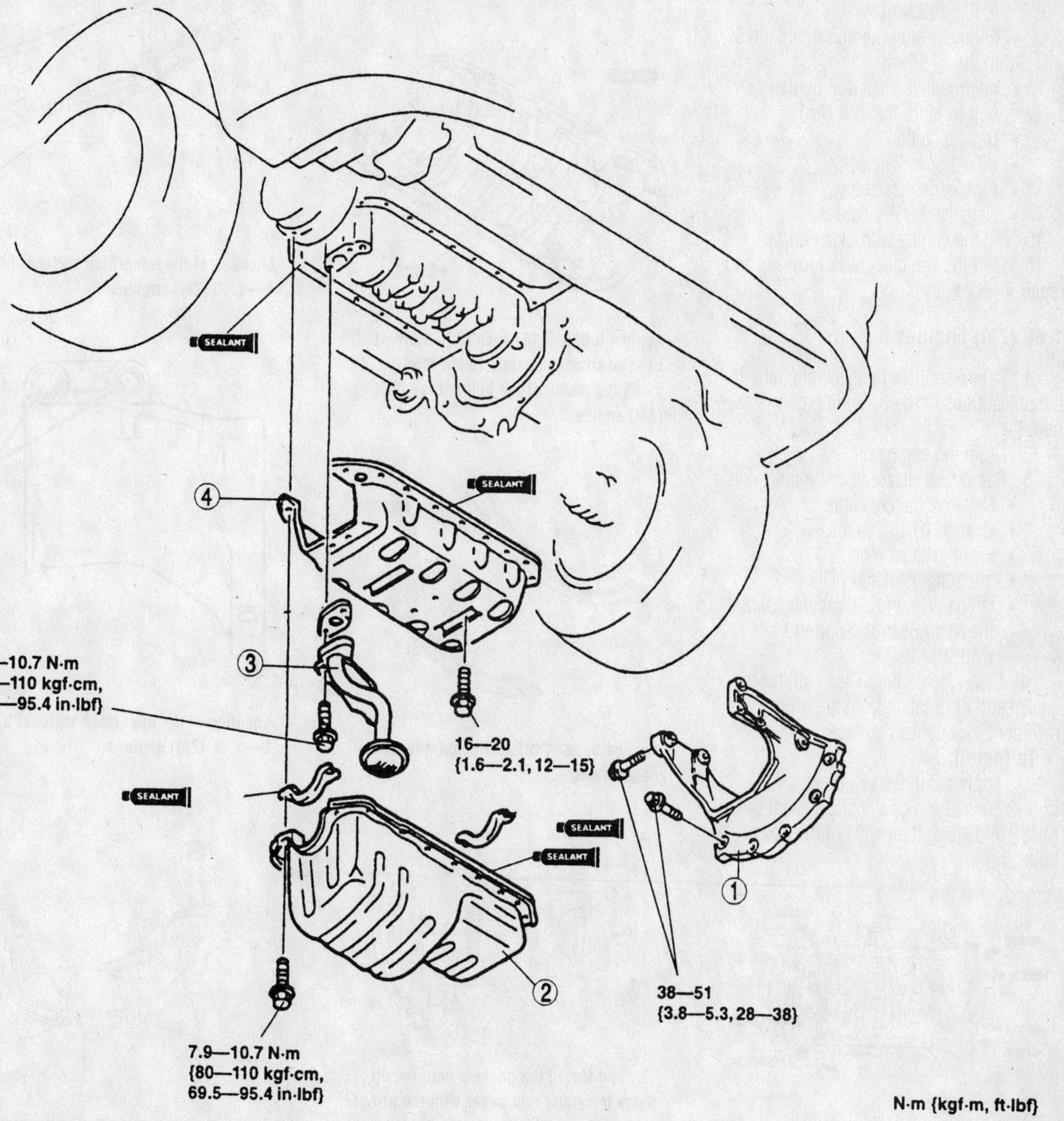

SEALANT
SEALANT
4
3
7.9—10.7 N·m
{80—110 kgf·cm,
69.5—95.4 in·lbf}
16—20
{1.6—2.1, 12—15}
SEALANT
SEALANT
SEALANT
1
2
38—51
{3.8—5.3, 28—38}
7.9—10.7 N·m
{80—110 kgf·cm,
69.5—95.4 in·lbf}
N·m {kgf·m, ft·lbf}
9301MG10

Exploded view of the oil pan and related components—1.6L engine

(2.5–3.5mm) continuous bead of silicone sealant on the oil pan along the inside of the bolt holes and overlap the ends.

13. Install or connect the following:
- New oil pan gaskets
- Oil pan. Torque the vertical bolts to 70–95 inch lbs. (8–11 Nm) and the horizontal bolts to 48–65 ft. lbs. (64–86 Nm).
- Engine mount

14. Raise the crossmember into position and tighten the bolts to 15 ft. lbs. (21 Nm) and the nuts to 101 ft lbs. (137 Nm).
- Shock absorber lower nut and bolt and tighten to 86 ft. lbs. (116 Nm)
- Stabilizer link nut and tighten to 44 ft. lbs. (60 Nm)
- Engine mount nut and tighten to 57 ft. lbs. (78 Nm)
- Intermediate shaft and tighten the bolt to 19 ft. lbs. (26 Nm)
- Dipstick tube
- Wheel speed sensor
- Air cleaner assembly
- Negative battery cable

15. Fill the engine with clean oil.

16. Start the vehicle, check for leaks and repair if necessary.

1.6L (ZM) Engines

1. Before servicing the vehicle, refer to the precautions in the beginning of this section.

2. Drain the engine oil.

3. Remove or disconnect the following:
- Negative battery cable
- Oxygen (O₂S) sensors
- Front exhaust pipe
- Integrated stiffener (1)
- Oil pan (2) and oil strainer (3) and the Main Bearing Support Plate (MBSP) (4)

4. Clean the oil pan. Clean all dirt, oil, gasket and old sealant from the oil pan and cylinder block contact surfaces.

To install:

5. Apply a 0.099–0.137 inch (2.5–3.5mm) bead of silicone sealant to the MBSP and along the inside of the bolt holes.

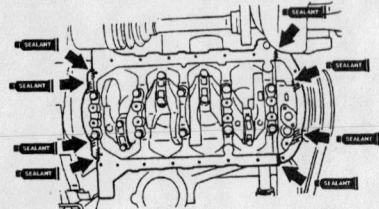

42356-MAZC-G00

Apply a bead of silicone sealant to area as illustrated—1.6L (ZM) engines

6. Apply a bead of silicone sealant to new oil pan gaskets. Install the gaskets onto the oil pump body and rear cover with the projections in the notches as illustrated.

7. Apply a 0.079 inch (2mm) bead of silicone sealant to area of the oil pan gaskets marked by **A and B** in the illustration. Install the gaskets onto the oil pump body and rear cover with the projections in the notches as illustrated.

8. Apply a 0.99–0.137 inch (2.5–3.5mm) bead of silicone sealant to the oil pan along the inside of the bolt holes and overlap the ends.

9. Install or connect the following:

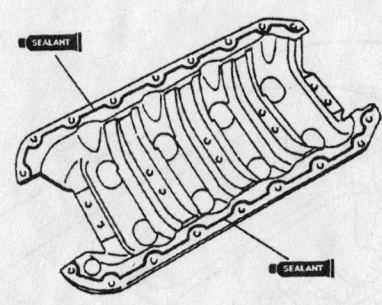

42356-MAZC-GHM

Apply a 0.099–0.137 inch (2.5–3.5mm) bead of silicone sealant to the MBSP and along the inside of the bolt holes—1.6L (ZM) engines

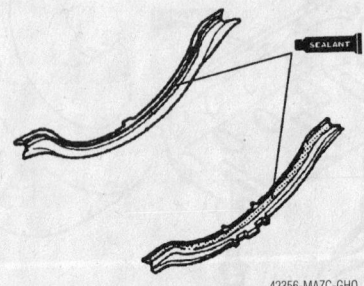

42356-MAZC-GHO

Apply a bead of silicone sealant to new oil pan gaskets . . .

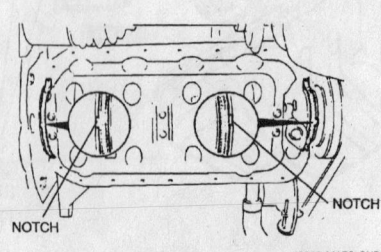

42356-MAZC-GHP

. . . and install the gaskets onto the oil pump body and rear cover with the projections in the notches as illustrated—1.6L (ZM) engines

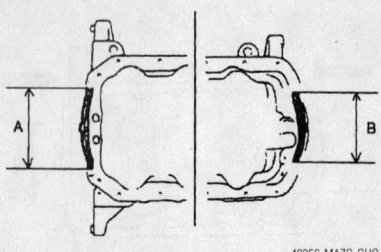

42356-MAZC-GHQ

Apply a 0.079 inch (2mm) bead of silicone sealant to area of the oil pan gaskets marked by A and B—1.6L (ZM) engines

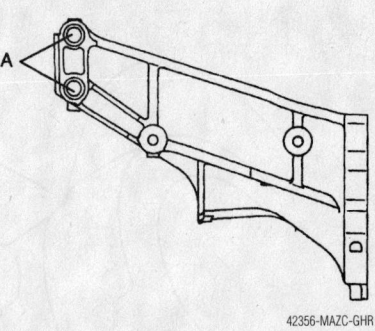

42356-MAZC-GHR

Location of the integrated stiffener bolt A—1.6L (ZM) engines

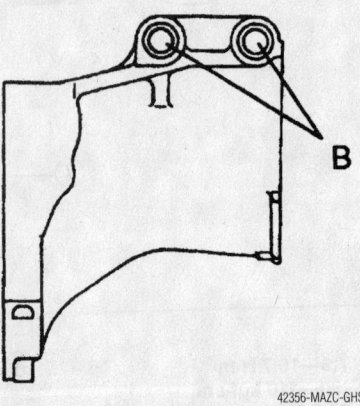

42356-MAZC-GHS

Location of the integrated stiffener bolt B—1.6L (ZM) engines

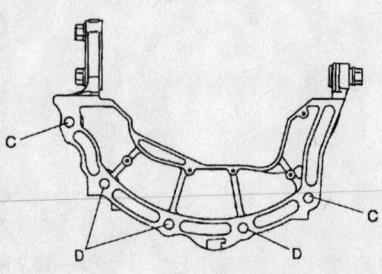

42356-MAZC-GHT

Location of the integrated stiffener bolts C and D—1.6L (ZM) engines

- Oil pan. Torque the bolts to 69.5–95.4 inch lbs. (7.9–10.7 Nm).
10. Install the integrated stiffener and hand tighten bolt **A**, then bolt **B**.
11. Tighten bolt **C** to 38 ft. lbs. (52 Nm).
12. Tighten bolt **D** to 38 ft. lbs. (52 Nm).
13. Tighten bolt **A** to 38 ft. lbs. (52 Nm).
14. Tighten bolt **B** to 38 ft. lbs. (52 Nm).
- Front exhaust pipe
- O$_2$S
- Negative battery cable
15. Fill the engine with clean oil.
16. Start the vehicle, check for leaks and repair if necessary.

1.8L (FP) Engines

1. Before servicing the vehicle, refer to the precautions in the beginning of this section.
2. Drain the engine oil.
3. Remove or disconnect the following:
- Negative battery cable
- Oxygen (O$_2$S) sensors
- Front exhaust pipe
- Oil pan
4. Clean the oil pan. Clean all dirt, oil, gasket and old sealant from the oil pan and cylinder block contact surfaces.
To install:
5. Apply a continuous bead of silicone sealant on the gaskets and around the oil pan, going on the inside of the bolt holes.
6. Install or connect the following:
- New oil pan gasket
- Oil pan. Torque the bolts to 14–18 ft. lbs. (19–25 Nm).
- Front exhaust pipe
- O$_2$S
- Negative battery cable
7. Fill the engine with clean oil.
8. Start the vehicle, check for leaks and repair if necessary.

2.0L (FS) Engine

1. Before servicing the vehicle, refer to the precautions in the beginning of this section.
2. Drain the engine oil.
3. Remove or disconnect the following:
- Negative battery cable
- Oxygen (O$_2$S) sensor
- Front exhaust pipe
- Oil pan bolts and the oil pan
To install:
4. Clean the oil pan. Clean all dirt, oil and old sealant from the oil pan and cylinder block contact surfaces.
5. Apply a continuous bead of silicone sealant around the oil pan, going on the inside of the bolt holes.
6. Install or connect the following:

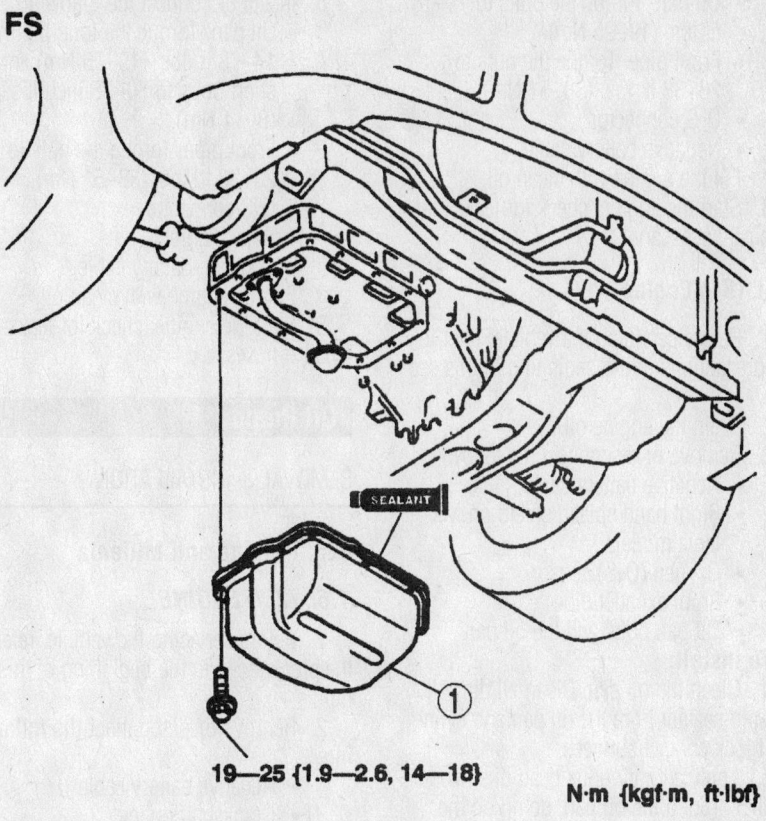

FS

19—25 {1.9—2.6, 14—18}

N·m {kgf·m, ft·lbf}

42356-MAZC-G21

Exploded view of the oil pan and related components for the 2.0L (FS) engines

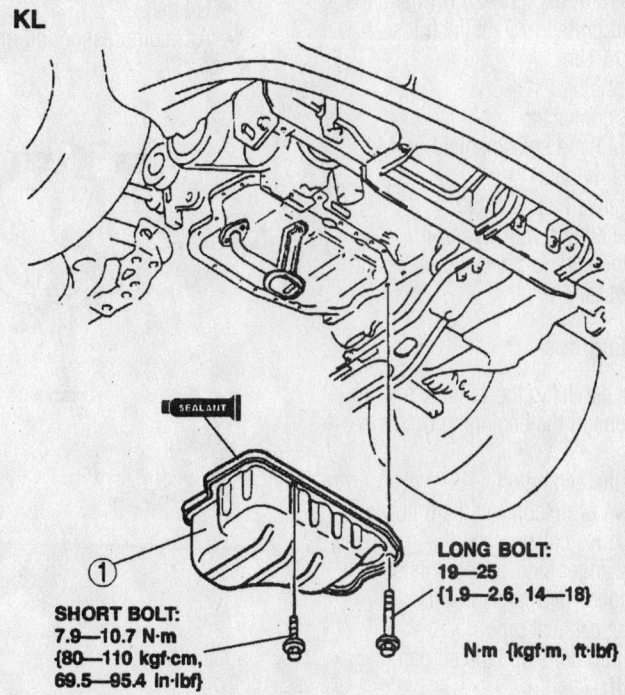

KL

SHORT BOLT:
7.9—10.7 N·m
{80—110 kgf·cm,
69.5—95.4 in·lbf}

LONG BOLT:
19—25
{1.9—2.6, 14—18}

N·m {kgf·m, ft·lbf}

1   Oil pan

42356-MAZC-G72

Exploded view of the oil pan and related components—2.5L (KL) engines, 626 model shown Millenia similar

- Oil pan. Torque the bolts to 14–18 ft. lbs. (19–25 Nm).
- Front pipe. Torque the nuts to 28–38 ft. lbs. (38–51 Nm).
- O2S connector
- Negative battery cable

7. Fill the engine with clean oil.

8. Start the engine, check for leaks and repair if necessary.

2.5L (KL) Engine

1. Before servicing the vehicle, refer to the precautions in the beginning of this section.

2. Drain the engine oil.

3. Remove or disconnect the following:
- Negative battery cable
- Right hand splash shield on Millenia models
- Oxygen (O2S) sensor
- Front exhaust pipe
- Oil pan bolts and the oil pan

To install:

4. Clean the oil pan. Clean all dirt, oil and old sealant from the oil pan and cylinder block contact surfaces.

5. Apply a continuous bead of silicone sealant around the oil pan, going on the inside of the bolt holes.

6. Install or connect the following:
- Oil pan. Torque the long bolts to 14–18 ft. lbs. (19–25 Nm) and the short bolts to 70–95 inch lbs. (8–11 Nm).
- Front pipe
- O2S connector
- Right hand splash shield on Millenia models
- Negative battery cable

7. Fill the engine with clean oil.

8. Start the engine, check for leaks and repair if necessary.

2.3L (KJ) Engines

1. Before servicing the vehicle, refer to the precautions in the beginning of this section.

2. Drain the engine oil.

3. Remove or disconnect the following:
- Negative battery cable
- Passenger side splash shield
- Oxygen (O2S) sensor
- Front exhaust pipe
- Oil pan bolts and the oil pan

To install:

4. Clean the oil pan. Clean all dirt, oil and old sealant from the oil pan and cylinder block contact surfaces.

5. Apply a continuous bead of silicone sealant around the oil pan, going on the inside of the bolt holes.

6. Install or connect the following:
- Oil pan. Torque the long bolts to 14–18 ft. lbs. (19–25 Nm) and the short bolts to 70–95 inch lbs. (8–11 Nm).
- Front pipe. Torque the nuts to 28–38 ft. lbs. (38–51 Nm).
- O2S connector
- Splash shield
- Negative battery cable

7. Fill the engine with clean oil.

8. Start the engine, check for leaks and repair if necessary.

Oil Pump

REMOVAL & INSTALLATION

626, Protégé and Millenia

1.6L (ZM) ENGINE

1. Before servicing the vehicle, refer to the precautions in the beginning of this section.

2. Remove or disconnect the following:

- Negative battery cable
- Crankshaft pulley
- Timing belt cover
- Timing belt
- Timing belt pulley
- Oil pan
- A/C compressor and move it aside, leaving the refrigerant lines attached
- A/C compressor mounting bracket
- Alternator
- Oil pump attaching bolts
- Front crankshaft seal from the oil pump, if the pump is being replaced

To install:

3. Clean the oil, dirt and old sealant from all contact surfaces.

4. If the oil seal was removed from the oil pump, apply clean engine oil to the lip of the seal. Push the seal in lightly be hand. Press the seal, with a protrusion of 0.02–0.04 inch (0.5–1.0mm) into the oil pump with a Seal Installer tool 49 B014 001.

5. Apply a bead of silicone to the oil pump at the cylinder block contact surface, going inside the bolt holes.

6. Install or connect the following:
- New O-rings on the oil pump
- Oil pump. Torque the bolts to 14–18 ft. lbs. (19–25 Nm).
- Alternator
- Air conditioning compressor bracket
- Air conditioning compressor
- Oil pan
- Timing belt pulley
- Timing belt
- Timing belt cover
- Crankshaft pulley
- Negative battery cable

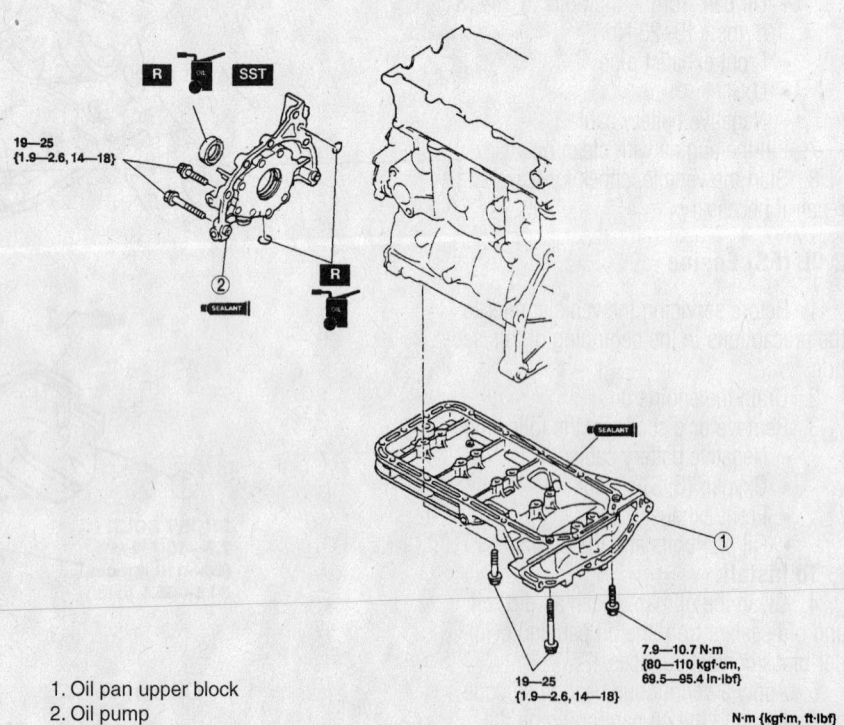

1. Oil pan upper block
2. Oil pump

19—25
{1.9—2.6, 14—18}

7.9—10.7 N·m
{80—110 kgf·cm, 69.5—95.4 in·lbf}

19—25
{1.9—2.6, 14—18}

N·m {kgf·m, ft·lbf}

42356-MAZC-G22

Exploded view of the oil pump and related components for the 1.8L (FP) and 2.0L (FS) engines

7. Fill the engine with clean oil.

8. Start the vehicle, check for leaks and repair if necessary.

1.8L (FP) AND 2.0L (FS) ENGINE

1. Before servicing the vehicle, refer to the precautions in the beginning of this section.

2. Remove or disconnect the following:
- Negative battery cable
- Crankshaft pulley
- Timing belt cover
- Timing belt
- Timing belt pulley
- Oil pan
- A/C compressor and move it aside, leaving the refrigerant lines attached
- A/C compressor mounting bracket
- Alternator
- Transaxle
- Two bolts at the rear of the oil pan upper block, on 1.8L (FP) engines or if equipped with a manual transaxle on 2.0L (FS) engines
- Rubber caps at the bottom surface, then remove the two bolts from the upper block through the holes uncovered by removing the caps, if equipped an automatic transaxle, on 2.0L (FS) engines
- Oil pan upper block bolts using 2–3 steps in the sequence illustrated
- Oil pan upper block using a rubber mallet and a suitable separator tool
- Oil pump body bolts and the oil pump body
- Seal from the oil pump, if the pump is being replaced

To install:

3. Clean the oil, dirt and old sealant from all contact surfaces.

4. If the oil seal was removed from the oil pump, apply clean engine oil to the lip of the seal. Push the seal in lightly be hand. Press the seal, with a protrusion of 0–0.019 inch (0–0.5mm) into the oil pump.

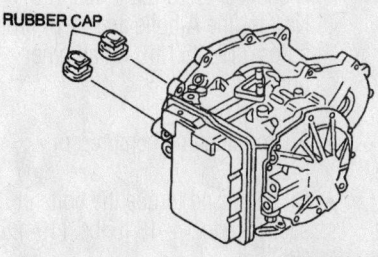

Remove the rubber caps at the bottom surface of the upper block, if equipped an automatic transaxle–2.0L (FS) engines

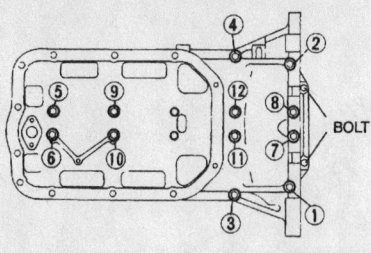

42356-MAZC-G24

Remove the oil pan upper block bolts using 2–3 steps in the sequence illustrated–1.8L (FP) and 2.0L (FS) engines

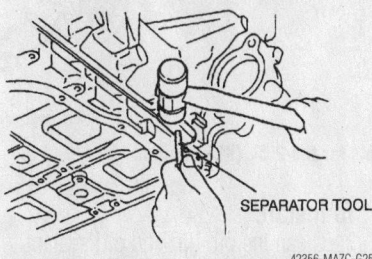

42356-MAZC-G25

Remove the oil pan upper block using a rubber mallet and a suitable separator tool–1.8L (FP) and 2.0L (FS) engines

5. Apply 0.04–0.07 inch (1–2mm) bead of sealant to the oil pump at the locations illustrated

6. Install or connect the following:
- Oil pump. Torque the bolts to 14–18 ft. lbs. (19–25 Nm).
- Oil pan

7. Apply a bead of sealant 0.08–0.11 inch (2–3mm) the mating surface of the upper block as illustrated, place the block assembly into position and tighten the bolts in sequence illustrated to 18 ft. lbs. (25 Nm) using 2–3 steps.
- Two bolts at the rear of the oil pan upper block, on 1.8L (FP) engines or if equipped with a manual transaxle on 2.0L (FS) engines and torque to 95 inch lbs. (11 Nm) and install the caps
- Two bolts to the upper block through the holes uncovered by

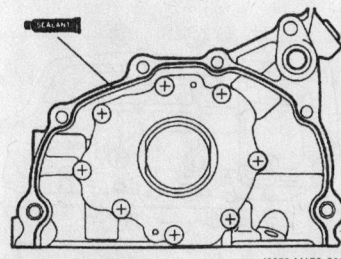

42356-MAZC-G26

Apply 0.04–0.07 inch (1–2mm) bead of sealant to the oil pump–1.8L (FP) and 2.0L (FS) engines

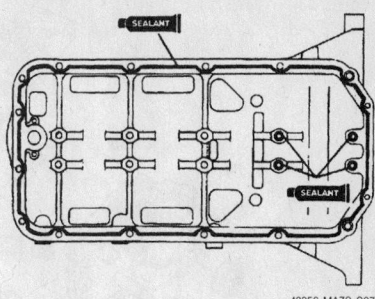

42356-MAZC-G27

Apply a bead of sealant 0.08–0.11 inch (2–3mm) the mating surface of the upper block–1.8L (FP) and 2.0L (FS) engines

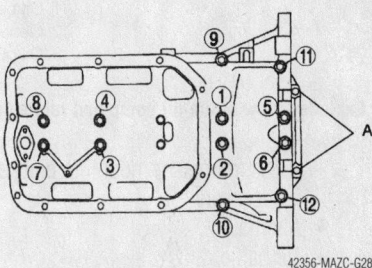

42356-MAZC-G28

Tighten the upper block bolts in this sequence–1.8L (FP) and 2.0L (FS) engines

removing the caps, if equipped an automatic transaxle, tighten the bolts to 95 inch lbs. (11 Nm)
- Transaxle, if equipped with a manual transaxle
- Air conditioning compressor bracket
- Air conditioning compressor
- Oil pan
- Timing belt pulley
- Timing belt
- Timing belt cover
- Crankshaft pulley
- Negative battery cable

8. Fill the engine with clean oil.

9. Start the vehicle, check for leaks and repair if necessary.

2.5L (KL) ENGINE

1. Before servicing the vehicle, refer to the precautions in the beginning of this section.

2. Remove or disconnect the following:
- Negative battery cable
- Crankshaft pulley
- Timing belt cover
- Timing belt
- Timing belt pulley
- Oil pan
- A/C compressor and move it aside, leaving the refrigerant lines attached

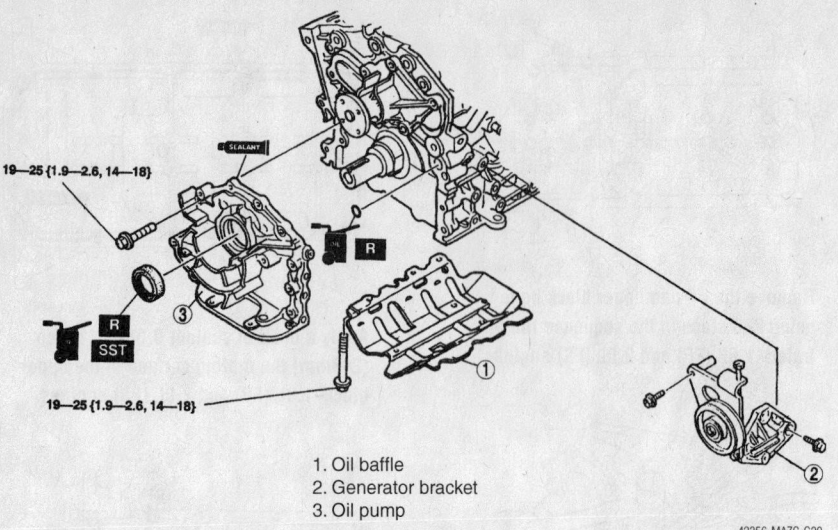

1. Oil baffle
2. Generator bracket
3. Oil pump

42356-MAZC-G29

Exploded view of the oil pump and related components for the 2.5L (KL) engines

- A/C compressor mounting bracket, if necessary
- Alternator
- Oil baffle
- Alternator bracket
- Oil pump body bolts and the oil pump body
- Seal from the oil pump, if the pump is being replaced

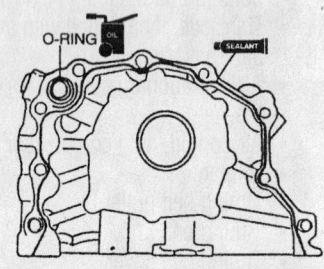

Apply clean engine oil to the NEW O–ring and install it in the pump–2.5L (KL) engines

42356-MAZC-G30

Bolt length
A: 40 mm {1.6 in}
B: 25 mm {1.0 in}

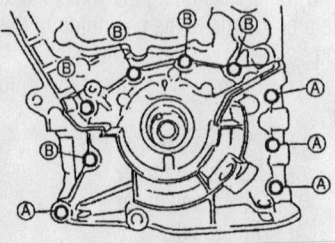

42356-MAZC-G31

The oil pump bolts are different lengths, make sure to install them where indicated. The long bolts A are 1.6 inch (40mm) long and the short bolts B are 1 inch (25mm) long –2.5L (KL) engines

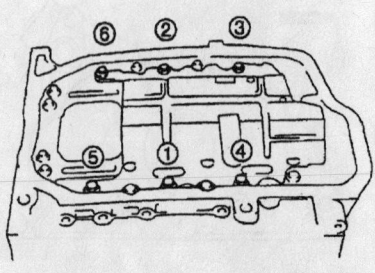

42356-MAZC-G32

Tighten the oil baffle and torque the bolts in sequence–2.5L (KL) engines

To install:

3. Clean the oil, dirt and old sealant from all contact surfaces.

4. If the oil seal was removed from the oil pump, apply clean engine oil to the lip of the seal. Push the seal in lightly be hand. Press the seal, with a protrusion of 0–0.019 inch (0–0.5mm) into the oil pump.

5. Apply clean engine oil to the NEW O–ring and install it in the pump

6. Install or connect the following:
- Oil pump bolts in the locations illustrated noting the location and length of the bolts. The long bolts **A** are 1.6 inch (40mm) long and the short bolts **B** are 1 inch (25mm) long. Torque the bolts to 14–18 ft. lbs. (19–25 Nm).
- Oil baffle and torque the bolts in sequence to 14–18 ft. lbs. (19–25 Nm)
- Alternator bracket
- Air conditioning compressor bracket
- Air conditioning compressor
- Oil pan
- Timing belt pulley

- Timing belt
- Timing belt cover
- Crankshaft pulley
- Negative battery cable

7. Fill the engine with clean oil.

8. Start the vehicle, check for leaks and repair if necessary.

2.3L (KJ) ENGINE

1. Before servicing the vehicle, refer to the precautions in the beginning of this section.

2. Remove or disconnect the following:
- Negative battery cable
- Crankshaft pulley
- Timing belt cover
- Timing belt
- Timing belt pulley
- Oil pan
- Alternator
- Power steering pump and bracket, move the pump aside leaving the refrigerant lines attached
- Oil baffle
- A/C compressor and move it aside, leaving the refrigerant lines attached
- A/C compressor mounting bracket, if necessary
- Vacuum pump
- Oil pump body bolts and the oil pump body
- Seal from the oil pump, if the pump is being replaced

To install:

3. Clean the oil, dirt and old sealant from all contact surfaces.

4. If the oil seal was removed from the oil pump, apply clean engine oil to the lip of the seal. Push the seal in lightly be hand. Press the seal, with a protrusion of 0–0.02 inch (0–0.7mm) into the oil pump.

5. Apply a 0.04–0.07 inch (1–2mm) bead of silicone sealant to the contact surface of the pump.

6. Apply clean engine oil to the NEW O–ring and install it in the pump

7. Install or connect the following:
- Oil pump. Torque the bolts, in sequence, to 15–22 ft. lbs. (20–30 Nm) for the **A** bolts and to 14–18 ft. lbs. (19–25 Nm) for all other bolts.
- Vacuum pump
- Air conditioning compressor bracket
- Oil baffle and torque the bolts in sequence to 14–18 ft. lbs. (19–25 Nm)
- Power steering pump and bracket
- Alternator
- Air conditioning compressor

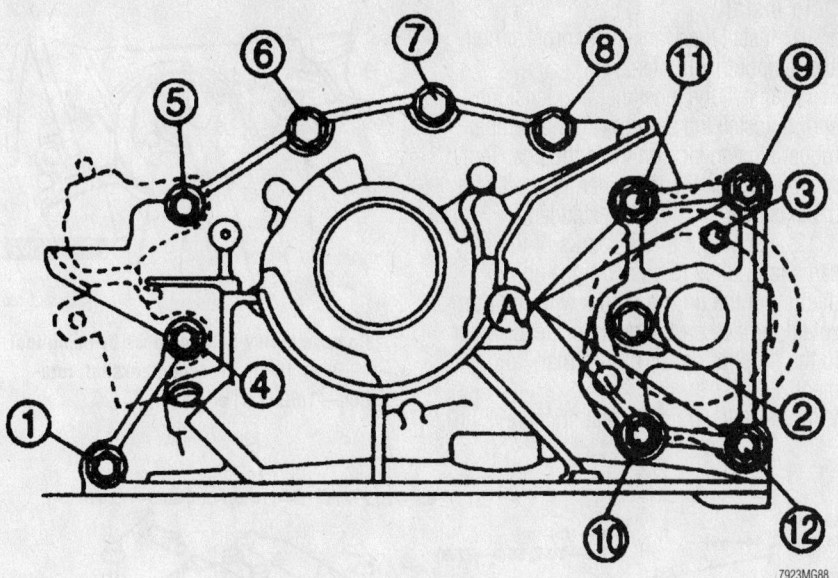

Tightening the oil pump mounting bolts in sequence—2.3L (KJ) engines

7923MG88

- Oil pan
- Timing belt pulley
- Timing belt
- Timing belt cover
- Crankshaft pulley
- Negative battery cable

8. Fill the engine with clean oil.

9. Start the vehicle, check for leaks and repair if necessary.

Miata

1.8L (BP) ENGINES

1. Before servicing the vehicle, refer to the precautions in the beginning of this section.

2. Remove or disconnect the following:
- Negative battery cable
- Crankshaft pulley
- Timing belt cover
- Timing belt
- Crankshaft sprocket
- A/C compressor and move it aside, leaving the refrigerant lines attached
- A/C compressor mounting bracket
- Alternator
- Oil pump bolts and the oil pump
- Front crankshaft seal from the oil pump, if the pump is being replaced

To install:

3. Clean the oil, dirt and old sealant from all contact surfaces.

4. If the oil seal was removed from the oil pump, apply clean engine oil to the lip of the seal. Push the seal in lightly be hand. Press the seal, with a protrusion of 0–0.02 inch (0–0.5mm) into the oil pump.

5. Apply a bead of silicone to the oil pump body-to-cylinder block contact surface, going inside the bolt holes.

6. Install or connect the following:
- Oil pump. Torque the bolts to 14–18 ft. lbs. (19–25 Nm).
- Air conditioning compressor bracket
- Air conditioning compressor
- Crankshaft sprocket
- Timing belt
- Timing belt cover
- Crankshaft pulley
- Negative battery cable

7. Fill the engine with clean oil.

8. Start the vehicle, check for leaks and repair if necessary.

Rear Main Seal

REMOVAL & INSTALLATION

1. Before servicing the vehicle, refer to the precautions in the beginning of this section.

2. Remove or disconnect the following:
- Negative battery cable
- Transaxle/transmission assembly
- Clutch/flywheel assembly, if equipped with a manual transaxle/transmission
- Flexplate/shim plates, if equipped with an automatic transaxle/transmission

3. Cut the oil seal lip with a knife. Install a rag to the housing and using a prytool, carefully pry the oil seal from the oil seal housing.

4. Clean the gasket mounting surfaces.

To install:

5. Clean the oil seal housing. Coat the oil seal and the housing with clean engine oil.

6. Install or connect the following:
- New oil seal into the housing by tapping it evenly into place with a hammer and a seal installer until it is flush with the edge of the rear cover
- Clutch/flywheel assembly or the flexplate, as applicable
- Transaxle/transmission
- Negative battery cable

Timing Belt

REMOVAL & INSTALLATION

626 and Protégé

1.6L (ZM) ENGINE

1. Before servicing the vehicle, refer to the precautions in the beginning of this section.

2. Drain the cooling system.

3. Remove or disconnect the following:
- Negative battery cable
- Camshaft Position (CMP) sensor
- Ignition coils
- Drive belt
- Crankshaft pulley and plate
- Water pump pulley
- Rocker arm cover and discard the gasket

4. Support the engine with a suitable support device and remove the number 3 engine mount.
- Timing belt cover
- Pulley using tool 49 D011 102 to prevent crankshaft rotation and remove the pulley boss

5. Install the pulley boss on the crankshaft and tighten the bolt.

6. Turn the crankshaft until the timing marks on the crankshaft and camshaft sprockets are aligned. Face the camshaft pulley marks **I** and **E** straight up, then align the timing marks with the horizontal surface on the cylinder head. The pin on the pulley boss must face upward. Hold the crankshaft pulley boss with a suitable tool and remove the pulley lockbolt, being careful not to rotate the crankshaft. Remove the crankshaft pulley boss.

➡**Protect the tensioner with a shop towel before prying on it. Do not rotate the crankshaft after the timing belt has been removed.**

7. Mark the direction of rotation on

the timing belt. Loosen the tensioner lock-bolt and pry the tensioner outward. Tighten the lockbolt with the tensioner spring fully extended. Remove the timing belt.

8. Remove the tensioner and spring. If necessary, remove the idler pulley.

9. Inspect the belt for wear, peeling, cracking, hardening or signs of oil contamination. Inspect the tensioner for free and smooth rotation. Check the tensioner spring free length; it should not exceed 2.43 inch (68mm). Inspect the sprocket teeth for wear or damage. Replace parts, as necessary.

To install:

10. Install the crankshaft sprocket bolt and temporarily tighten.

11. Install the tensioner and tensioner spring. Install the spring with the damper rubber closing face on the right side. Temporarily tighten the tensioner lockbolt with the tensioner spring fully extended.

12. Face the **I** and **E** marks of the camshaft pulley marks straight up, then align the timing marks with the horizontal surface on the cylinder head. Refer to the illustration for timing mark alignment.

13. Install the timing belt so there is no

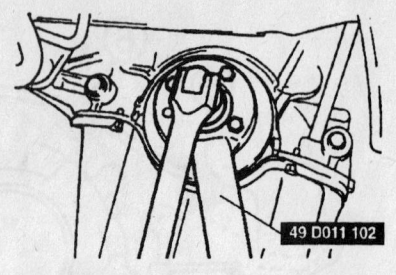

42356-MAZC-GBG

Remove pulley bolt and boss by using tool 49 D011 102 to prevent crankshaft rotation—1.6L (ZM) engines

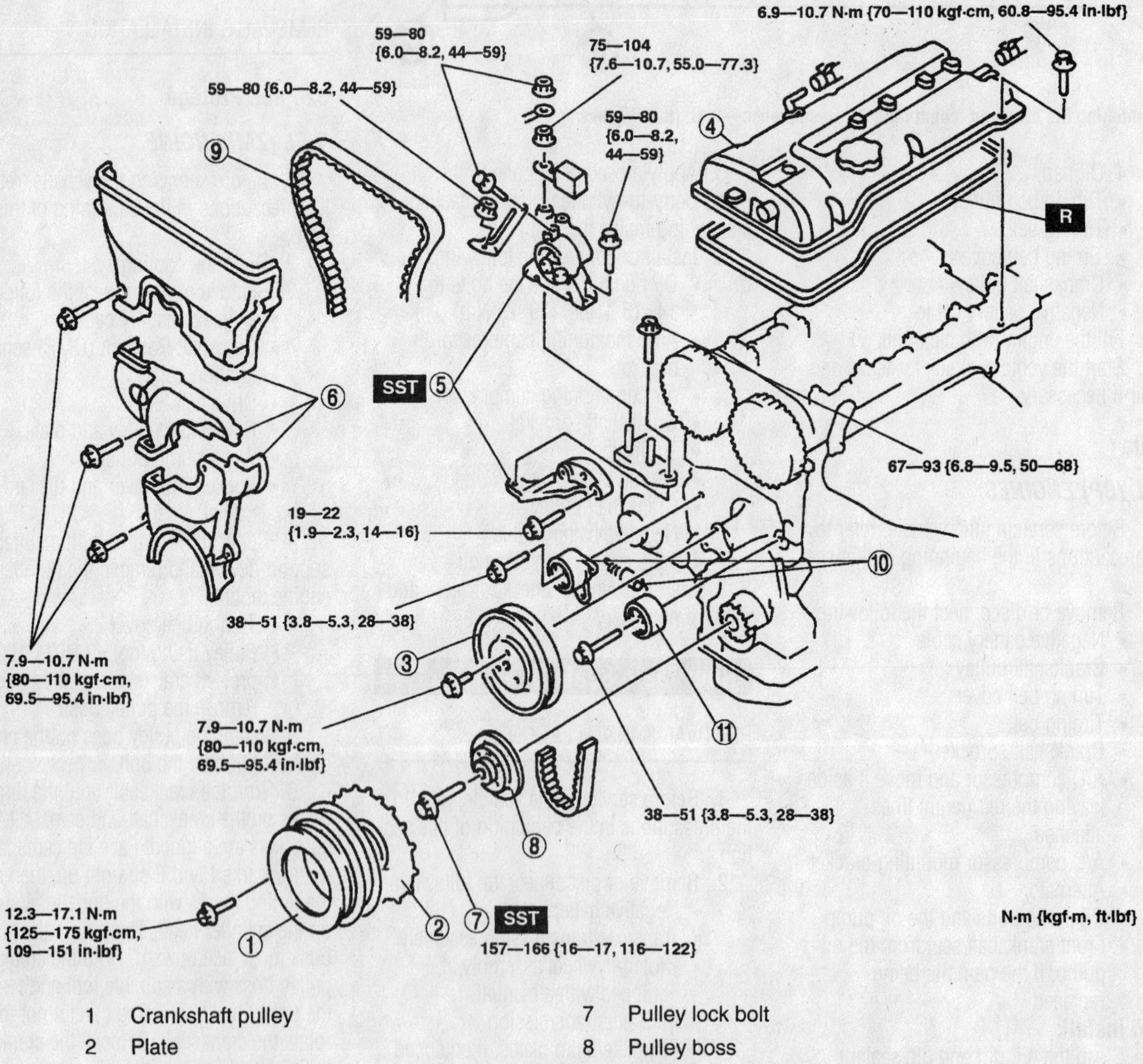

1	Crankshaft pulley
2	Plate
3	Water pump pulley
4	Cylinder head cover
5	No.3 engine mount
6	Timing belt cover

7	Pulley lock bolt
8	Pulley boss
9	Timing belt
10	Tensioner, tensioner spring (See 01–10A–11 Tensioner, Tensioner Spring Installation Note)
11	Idler

Exploded view of the timing belt assembly—1.6L (ZM) engines

42356-MAZC-GIA

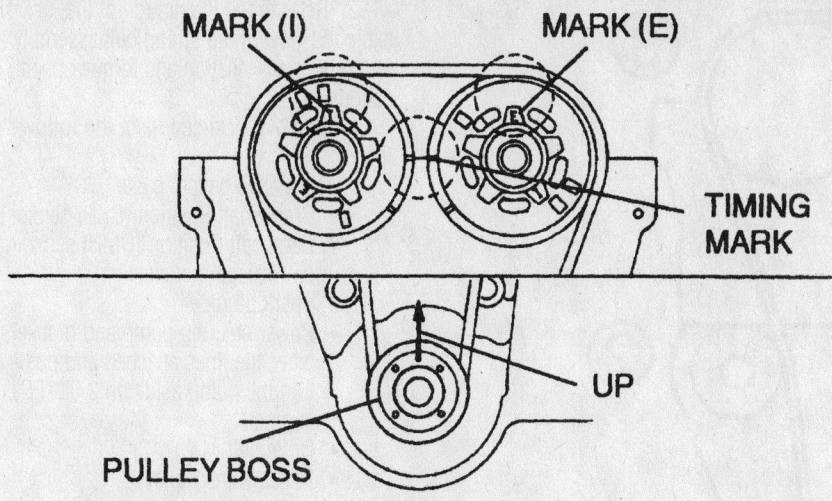

42356-MAZC-GIB

Turn the crankshaft until the timing marks on the crankshaft and camshaft sprockets are aligned. The pin on the pulley boss must face upward—1.6L (ZM) engines

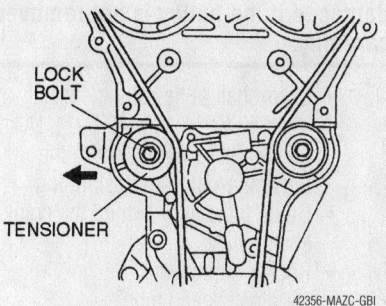

42356-MAZC-GBI

Loosen the tensioner lockbolt and pry the tensioner outward—1.6L (ZM) engines

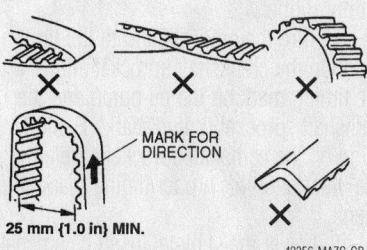

42356-MAZC-GBJ

Do not to bend, twist or get oil or other contaminates on the belt as this will damage the belt, if reusing the belt; mark the direction of rotation prior to removal—1.6L (ZM) engines

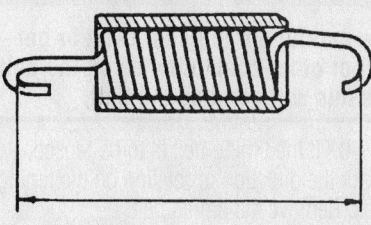

42356-MAZC-GBK

Check the tensioner spring free length; it should not exceed 2.43 inch (68mm)—1.6L (ZM) engines

looseness, refer to the illustration for location as follows:
- Crankshaft pulley
- Idler pulley
- Exhaust camshaft pulley
- Intake camshaft pulley
- Tensioner

14. Make sure no pressure other than the tensioner spring is applied to the belt. If reusing the old belt, be sure it is installed in the same direction of rotation.

15. Temporarily install the pulley boss and lockbolt.

16. Turn the crankshaft 1 ⅚ turns clockwise and align the crankshaft sprocket timing mark with the tension set mark for proper belt tension adjustment. Remove the lockbolt and pulley boss.
- Remove the pulley bolt and boss

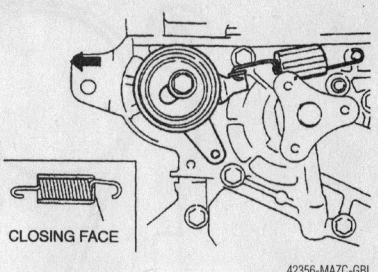

42356-MAZC-GBL

Install the tensioner and tensioner spring—1.6L (ZM) engines

and verify the timing belt pulley mark is still aligned with the tensioner set mark.

17. Tighten the tensioner lock bolt.
18. Temporarily install the pulley boss and lockbolt.
19. Turn the crankshaft 2 ⅙ turns clockwise and face the pin on the pulley boss upright. Be sure the camshaft sprocket timing marks are aligned. If they are not, repeat the alignment steps.

➡ The timing marks are aligned normally if the camshaft pulley marks I and E are facing straight up, the timing marks are aligned to the horizontal surface on the cylinder head.

20. Apply approximately 22 lbs. (10kg) pressure to the timing belt at a point midway between the camshaft sprockets. The belt should deflect 0.24–0.29 inch (6–7.5mm). If the deflection is not correct,

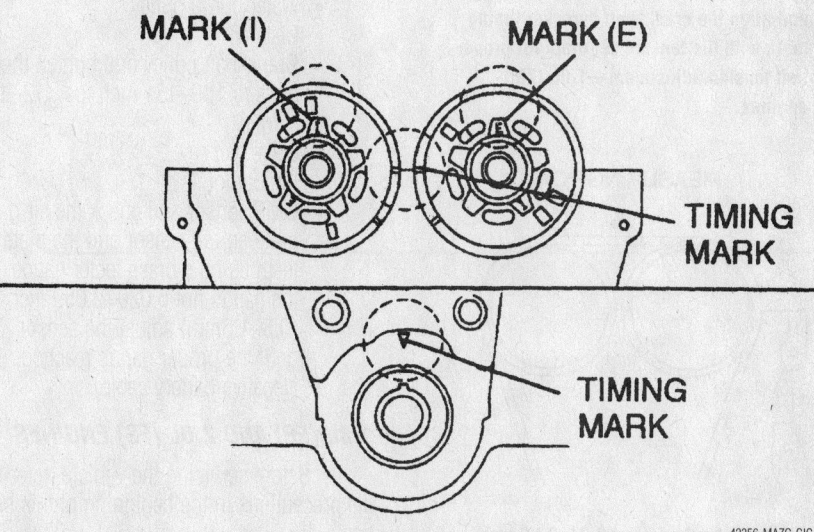

42356-MAZC-GIC

Verify the timing marks are aligned as shown before installing the belt—1.6L (ZM) engines

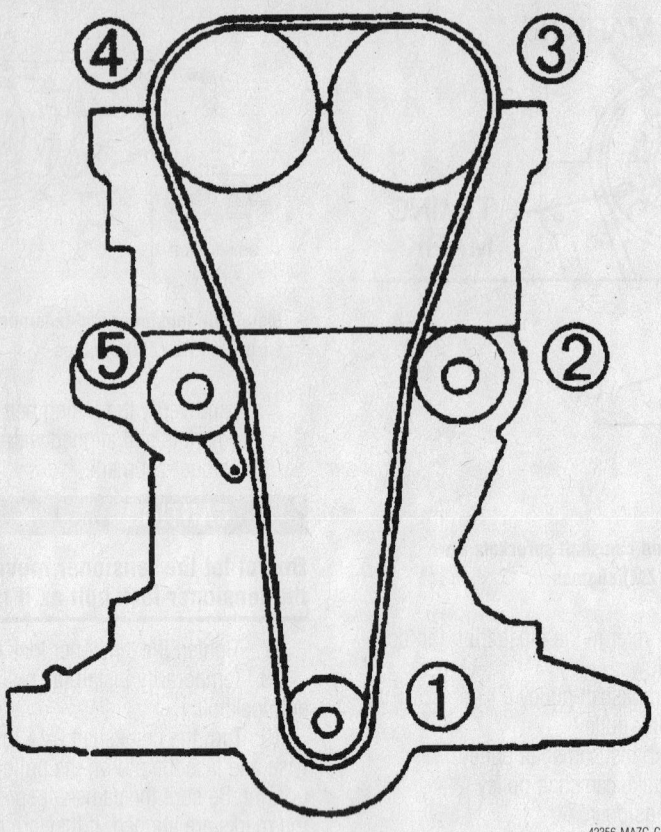

Install the timing belt in this order—1.6L (ZM) engines

42356-MAZC-GID

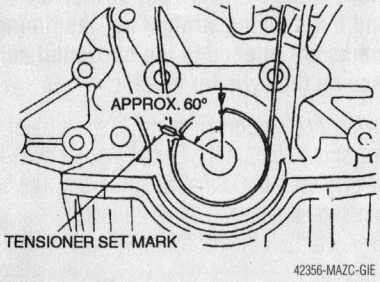

APPROX. 60°

TENSIONER SET MARK

42356-MAZC-GIE

Turn the crankshaft 1 ⅚ turns clockwise and align the crankshaft sprocket timing mark with the tension set mark for proper belt tension adjustment—1.6L (ZM) engines

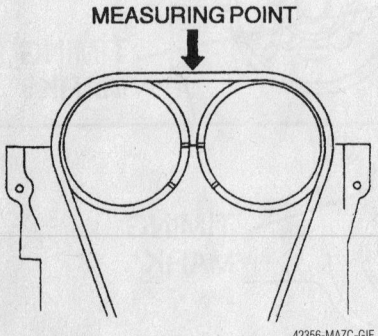

MEASURING POINT

42356-MAZC-GIF

The belt should deflect 0.24–0.29 inch (6–7.5mm)—1.6L (ZM) engines

repeat the alignment and tensioning procedure.

21. Install the pulley boss and lockbolt. Tighten the bolt to 116–122 ft. lbs. (157–166 Nm).

22. Install the timing belt covers and tighten the bolts to 95 inch lbs. (11 Nm).

23. Install the number 3 mount and remove the engine support device.

24. Install or connect the following:
- Rocker arm cover with a new gasket
- Water pump pulley
- Plate
- Crankshaft pulley and tighten the bolts to 109–151 inch lbs. (12–17 Nm)
- Drive belt
- Ignition coils
- CMP sensor and check the air gap between the sensor and the plate teeth using a brass feeler gauge. If the gap is not 0.020–0.059 inch (0.5–1.5mm) adjust the sensor until the proper gap is reached.
- Negative battery cable

1.8L (FP) AND 2.0L (FS) ENGINES

1. Before servicing the vehicle, refer to the precautions in the beginning of this section.

2. Refer to the illustration of the exploded view of the timing belt assembly for component location and torque specifications.

3. Remove or disconnect the following:
- Negative battery cable
- Crankshaft Position (CKP) sensor
- Camshaft Position (CMP) sensor on Protégé models
- Spark plugs
- Power steering pump and bracket, leave the lines attached and position the pump aside on 2.0L (FS) engines
- Drive belt
- Water pump pulley

✷✷ CAUTION

The CKP sensor rotor is on the rear of the crankshaft pulley and can be damaged if the pulley is not removed carefully.

- Crankshaft pulley using special tools 49 G011 103, 49 E011 1A1 and 49 S120 710. Refer to the illustration for tool positioning.
- Guide plate from behind the crankshaft pulley
- Rocker arm cover
- Oil dipstick and tube
- Timing belt cover

4. Support the engine assembly with a hoist and remove the number 3 engine mount rubber.

5. Turn the crankshaft until the timing mark on the crankshaft sprocket aligns with the timing mark on the oil pump and the camshaft sprocket timing marks E and I align on the camshaft sprockets. Refer to the illustration for proper timing mark alignment.

6. Lower the vehicle. Insert a camshaft sprocket holding tool between the camshaft sprockets.

7. Turn the timing belt tensioner with an Allen wrench and remove the tensioner spring from the hook pinch

✷✷ CAUTION

Be careful not to bend, twist or get oil or other contaminates on the belt as this will damage the belt.

8. If the timing belt is to be reused, mark the direction of rotation on the timing belt. Remove the timing belt.

9. Rotate the tensioner, if the tension rotates with no resistance or does not rotate; replace the tensioner

To install:

10. Install the crankshaft sprocket bolt. Install the flywheel locking tool, if equipped with automatic transaxle, or place the shift lever in **4th** gear and apply the parking brake, if equipped with manual transaxle. Tighten the bolt to 116–122 ft. lbs. (157–166 Nm).

11. Be sure the timing marks on the camshaft and crankshaft sprockets are still aligned.

12. Install the timing belt. If reusing the original timing belt, be sure it is installed in the same direction of rotation. Make sure there is no looseness on the idler side.

13. Rotate the crankshaft 2 turns in the normal direction of rotation and align the timing marks. Be sure all marks are still correctly aligned. If the marks are not aligned, remove the belt and then reinstall it and make sure the marks are properly aligned.

14. Check the tensioner spring length, if the free length is not 1.44 inch (36.5mm), replace the spring.

✳✳ CAUTION

Do not use tension other than that supplied by the tension spring or damage could occur

15. Turn the tensioner clockwise with an Allen wrench and install the tensioner spring. Remove the holding tool from between the camshaft sprockets.

16. Rotate the crankshaft 2 turns in the normal direction of rotation and align the timing marks. Be sure all marks are still correctly aligned.

17. Install the number 3 engine mount rubber and tighten the fasteners to the specifications shown in the illustration. Remove the engine hoist.

18. Install or connect the following:
  • Timing belt cover. Refer to the exploded view of the timing belt

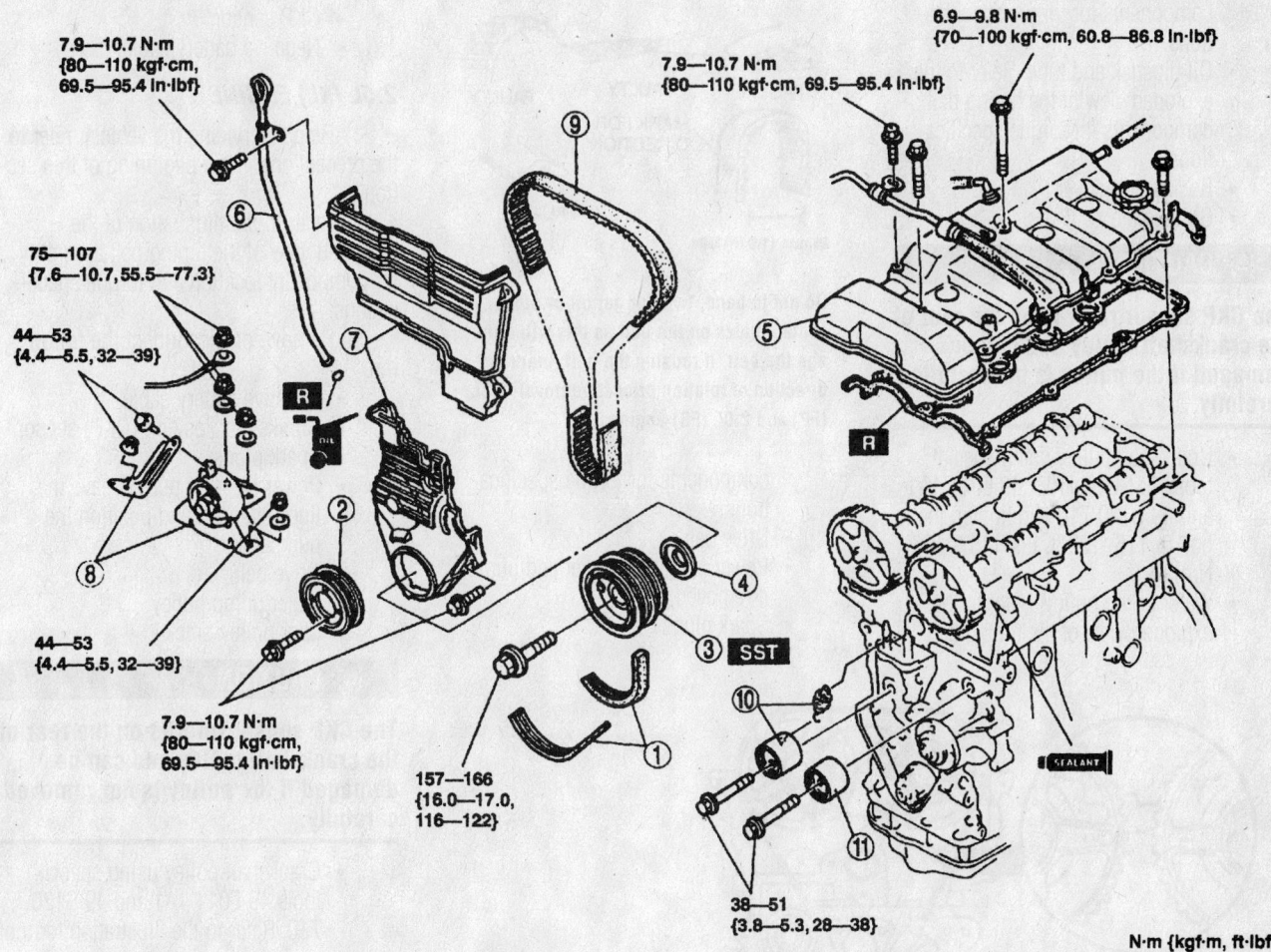

7.9—10.7 N·m
{80—110 kgf·cm, 69.5—95.4 in·lbf}

7.9—10.7 N·m
{80—110 kgf·cm, 69.5—95.4 in·lbf}

6.9—9.8 N·m
{70—100 kgf·cm, 60.8—86.8 in·lbf}

75—107
{7.6—10.7, 55.5—77.3}

44—53
{4.4—5.5, 32—39}

44—53
{4.4—5.5, 32—39}

7.9—10.7 N·m
{80—110 kgf·cm, 69.5—95.4 in·lbf}

157—166
{16.0—17.0, 116—122}

38—51
{3.8—5.3, 28—38}

N·m {kgf·m, ft·lbf}

1	Drive belt (See 01–10A–2 DRIVE BELT ADJUSTMENT [FS])	6	Dipstick and pipe
2	Water pump pulley	7	Timing belt cover
3	Crankshaft pulley	8	No.3 engine mount rubber
4	Guide plate	9	Timing belt
5	Cylinder head cover	10	Tensioner, tensioner spring
		11	Idler

Exploded view of the timing belt assembly–2.0L (FS) engines shown, 1.8L (FP) engines are similar

42356-MAZC-G33

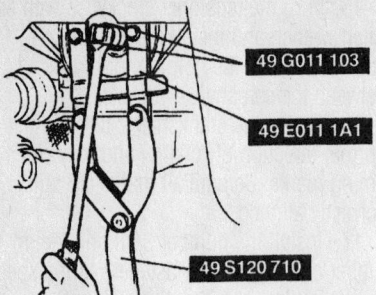

42356-MAZC-G34

Remove the crankshaft pulley using special tools 49 G011 v103, 49 E011 1A1 and 49 S120 710–1.8L (FP) and 2.0L (FS) engines

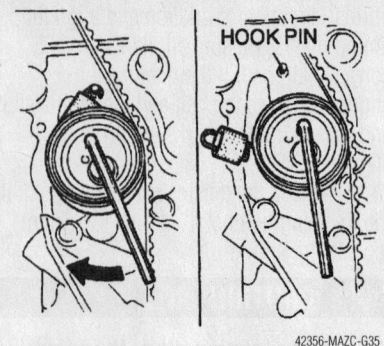

42356-MAZC-G35

Remove the tensioner spring from the hook pin–1.8L (FP) and 2.0L (FS) engines

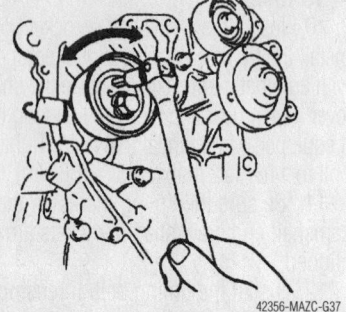

42356-MAZC-G37

Rotate the tensioner, if the tension rotates with no resistance or does not rotate; replace the tensioner–1.8L (FP) and 2.0L (FS) engines

components for torque specifications.
- Oil dipstick and tube. Refer to the exploded view of the timing belt components for torque specifications.
- Rocker arm cover
- Guide plate

❋❋ CAUTION

The CKP sensor rotor is on the rear of the crankshaft pulley and can be damaged if the pulley is not installed carefully.

- Crankshaft pulley using special tools 49 G011 v103, 49 E011 1A1 and 49 S120 710 and tighten the bolt to 116–122 ft. lbs. (157–166 Nm).
- Water pump pulley. Refer to the exploded view of the timing belt

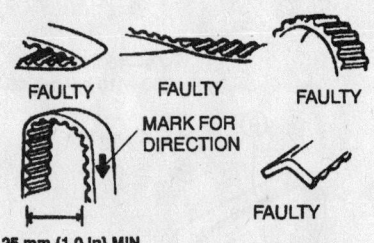

25 mm {1.0 In} MIN.

42356-MAZC-G36

Do not to bend, twist or get oil or other contaminates on the belt as this will damage the belt, if reusing the belt; mark the direction of rotation prior to removal–1.8L (FP) and 2.0L (FS) engines

components for torque specifications.
- Drive belt
- Power steering bracket and pump, if removed
- Spark plugs

- CKP sensor
- Negative battery cable

2.5L (KL) ENGINE

1. Before servicing the vehicle, refer to the precautions in the beginning of this section.

2. Refer to the illustration of the exploded view of the timing belt assembly for component location and torque specifications.

3. Remove or disconnect the following:

- Negative battery cable
- Crankshaft Position (CKP) sensor
- Spark plugs
- Power steering pump, leave the lines attached and position the pump aside
- Drive belt
- Water pump pulley
- Idler pulley bracket

❋❋ CAUTION

The CKP sensor rotor is on the rear of the crankshaft pulley and can be damaged if the pulley is not removed carefully.

- Crankshaft pulley using special tools 49 E011 1A1 and 49 S120 710. Refer to the illustration for tool positioning.
- Rocker arm cover
- Oil dipstick and tube
- Timing belt cover

4. Support the engine assembly with a hoist and remove the number 3 engine mount rubber.

5. Install the crankshaft pulley bolt and turn the crankshaft clockwise and align the timing marks as illustrated. The number 1 piston should be at Top Dead Center (TDC) of the compression stroke.

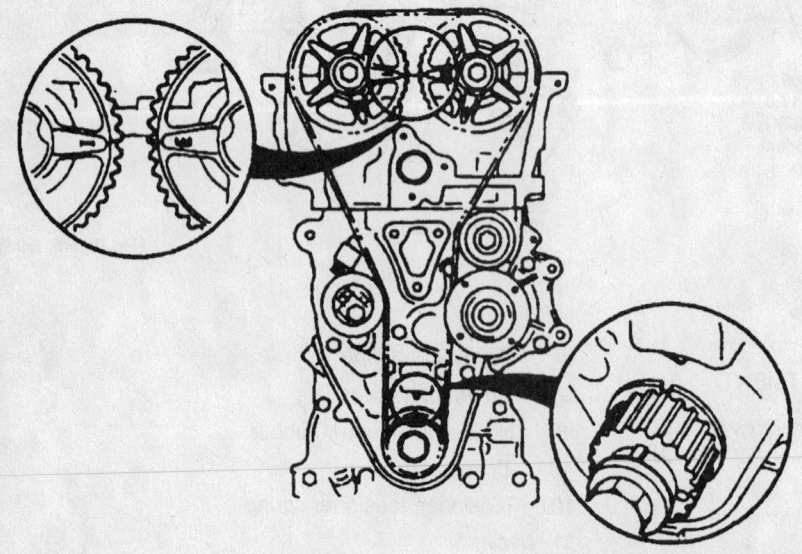

79235G52

When properly aligned for belt removal, the cam gear marks should face each other–1.8L (FP) and 2.0L (FS) engines

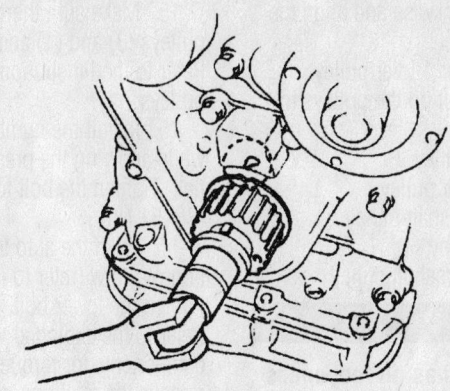

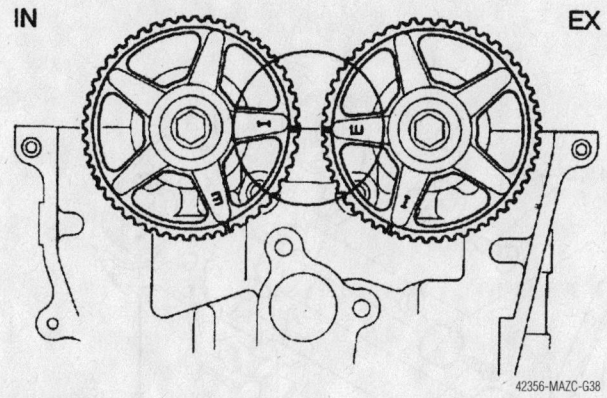

Be sure the timing marks on the camshaft and crankshaft sprockets are still aligned—1.8L (FP) and 2.0L (FS) engines

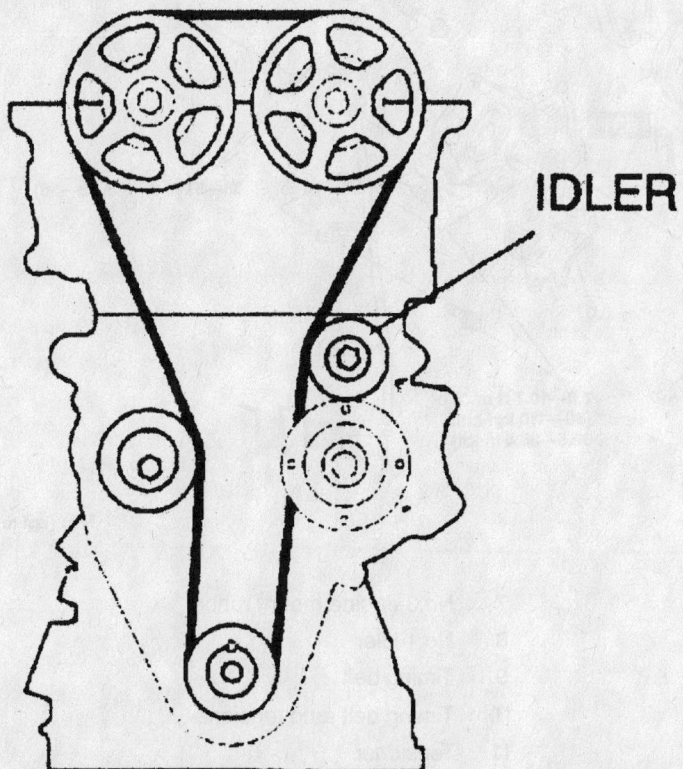

IDLER

Make sure there is no looseness on the idler side when the belt is installed—1.8L (FP) and 2.0L (FS) engines

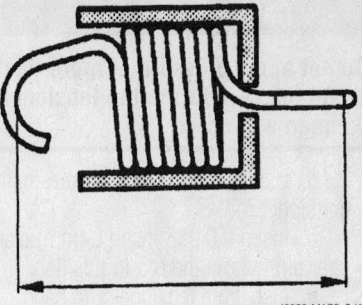

Check the tensioner spring length, if the free length is not 1.44 inch (36.5mm); replace the spring—1.8L (FP) and 2.0L (FS) engines

※※ CAUTION

When removing the number 1 idler pulley bolt, hold the pulley so that the threads are aligned or the threads can be damaged.

6. Remove the number 1 idler pulley

※※ CAUTION

When removing the auto tensioner bolts, hold the tensioner so that the threads are aligned or the threads can be damaged.

7. Loosen the auto tensioner bolts and remove the lower bolt.

※※ CAUTION

Be careful not to bend, twist or get oil or other contaminates on the belt as this will damage the belt.

8. If the timing belt is to be reused, mark the direction of rotation on the timing belt. Remove the timing belt.
9. Remove the belt.
10. If necessary, remove the auto tensioner.
To install:

※※ CAUTION

There are two type of auto tensioners and they are interchangeable.

11. If removed install the auto tensioner as follows:
   a. Measure the tensioner rod projection length. If the length exceeds 0.563–0.594 inch (14.3–15.1mm) on type **A** tensioners or 0.473–0.511 inch (12–13mm) on type **B** tensioners replace the tensioner. Refer to the illustration to distinguish tensioner type.
   b. Inspect the tensioner for leakage and replace if defective.

※※ **CAUTION**

Do not apply pressure of more than 2,200 lbf. (9.8 Kn) to the tensioner or damage will occur.

  c. Using a press, slowly press in the tensioner rod.

  d. Insert a 0.055 inch (1.4mm) diameter pin to hold the rod in position.

  e. Install the tensioner and hand tighten the bolt.

12. Turn the camshafts clockwise and align the timing marks as illustrated.

13. Install the crankshaft pulley bolt and turn the crankshaft clockwise and align the marks as illustrated.

14. With the number 1 idler pulley removed, install the belt on the pulleys in this order:

  • Timing belt pulley
  • Number 2 idler pulley
  • Left hand camshaft pulley
  • Tensioner pulley
  • Right hand camshaft pulley

※※ **CAUTION**

Make sure the belt has no looseness at the tension side.

15. Make sure there is tension between pulleys (3) and (1) and pulleys (1) and (5). Refer to the illustration for location of the pulleys.

16. Install the number 1 idler pulley while applying the pressure on the timing belt. Tighten the bolt to 28–38 ft. lbs. (38–51 Nm).

17. Push the auto tensioner in the direction of arrow (refer to illustration) and hand tighten the lower bolt, then tighten the bolts. Refer to the exploded view of the timing belt components for torque specifications. Then remove the retaining pin.

18. Turn the crankshaft clockwise and

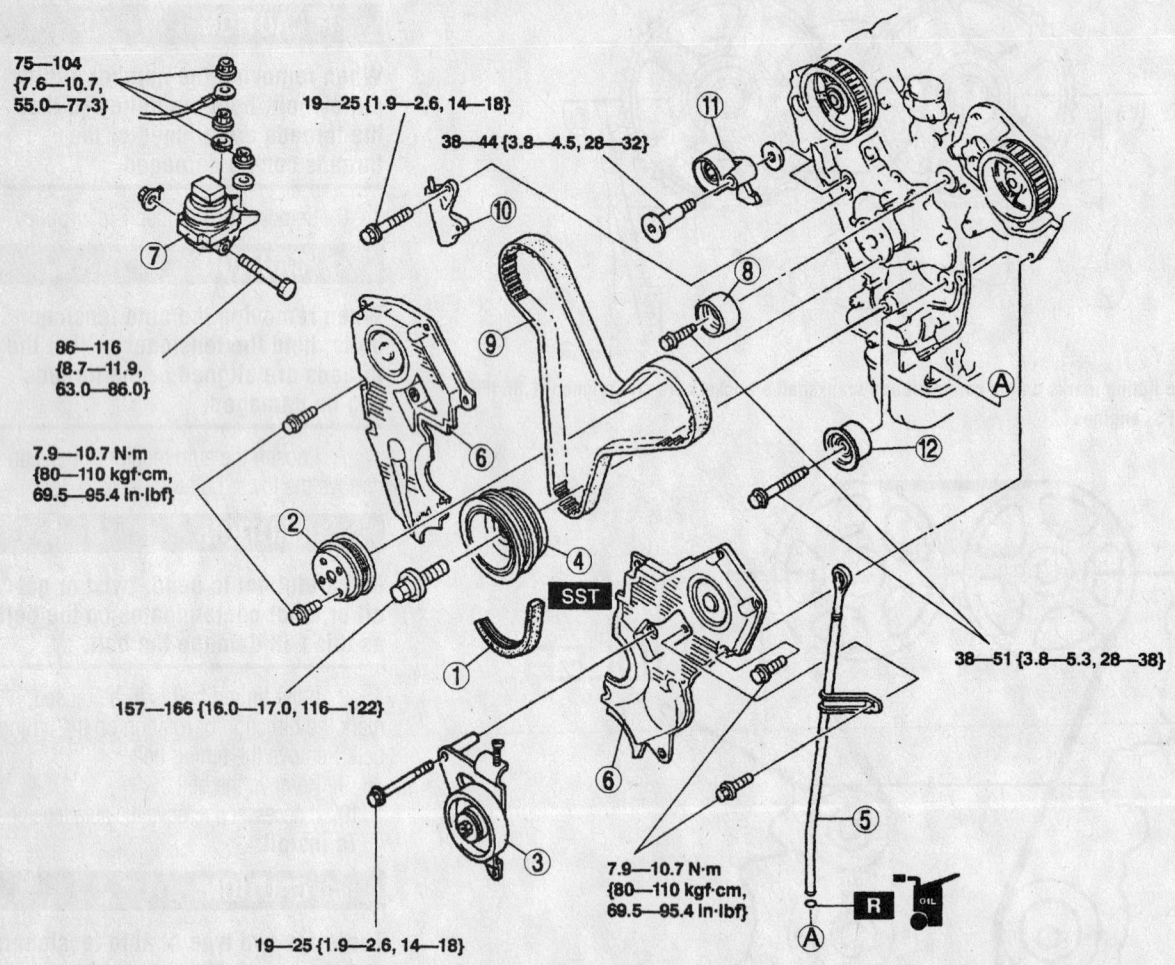

N·m {kgf·m, ft·lbf}

1	Drive belt	7	No.3 engine mount rubber
2	Water pump pulley	8	No.1 idler
3	Idler pulley bracket	9	Timing belt
4	Crankshaft pulley	10	Timing belt auto tensioner
5	Dipstick and pipe	11	Tensioner
6	Timing belt cover	12	No.2 idler

42356-MAZC-G41

Exploded view of the timing belt assembly–2.5L (KL) engines–626 models

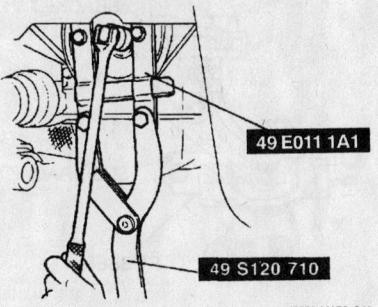

Remove the crankshaft pulley using special tools 49 E011 1A1 and 49 S120 710–2.5L (KL) engines

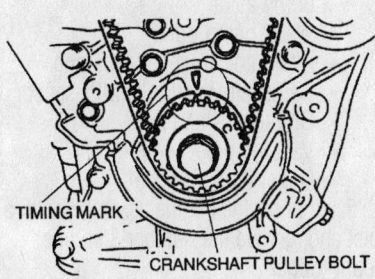

Turn the crankshaft clockwise and align the timing marks as illustrated. The number 1 piston should be at Top Dead Center (TDC) of the compression stroke–2.5L (KL) engines

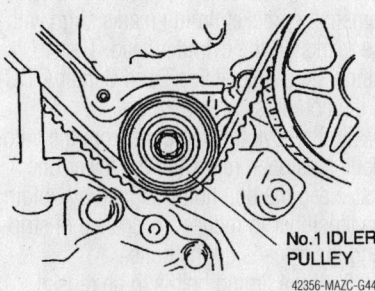

Remove the number 1 idler pulley–2.5L (KL) engines

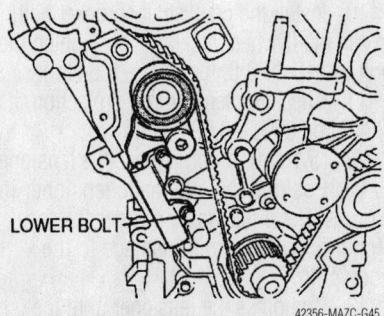

Loosen the auto tensioner bolts and remove the lower bolt–2.5L (KL) engines

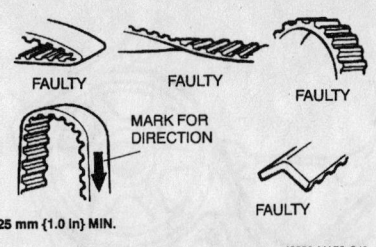

Do not to bend, twist or get oil or other contaminates on the belt as this will damage the belt, if reusing the belt; mark the direction of rotation prior to removal–2.5L (KL) engines

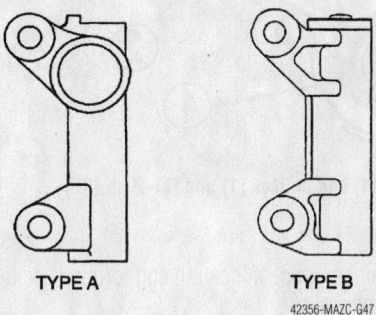

There are two type of auto tensioners and they are interchangeable–2.5L (KL) engines

Projection (Free length)
Type A: 14.3—15.1 mm {0.563—0.594 in}
Type B: 12.0—13.0 mm {0.473—0.511 in}

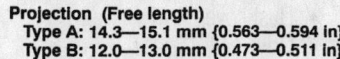
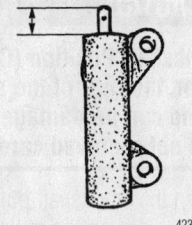

Measure the tensioner rod projection length and replace if it exceeds specification–2.5L (KL) engines

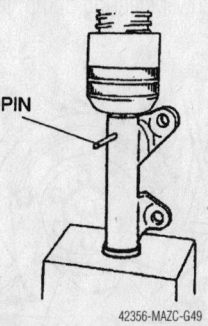

Using a press, slowly press in the tensioner rod and insert a 0.055 inch (1.4mm) diameter pin to hold the rod in position–2.5L (KL) engines

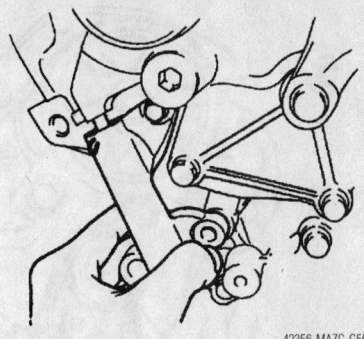

Install the tensioner and hand tighten the bolt–2.5L (KL) engines

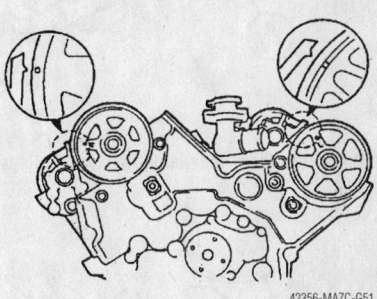

Turn the camshafts clockwise and align the timing marks–2.5L (KL) engines

align the timing marks. Make sure the timing marks are aligned, if not remove the belt and repeat the installation process.

19. Install or connect the following:
   • Number 3 engine mount rubber. Refer to the exploded view of the timing belt components for torque specifications. Remove the engine hoist.
   • Timing belt cover. Refer to the exploded view of the timing belt components for torque specifications.
   • Oil dipstick and tube. Refer to the exploded view of the timing belt components for torque specifications.
   • Rocker arm cover

✳✳ CAUTION

The CKP sensor rotor is on the rear of the crankshaft pulley and can be damaged if the pulley is not removed carefully.

   • Crankshaft pulley using special tools 49 E011 1A1 and 49 S120 710. Refer to the illustration for tool positioning. Tighten the bolt to 16–122 ft. lbs. (157–166 Nm).
   • Idler pulley bracket. Refer to the

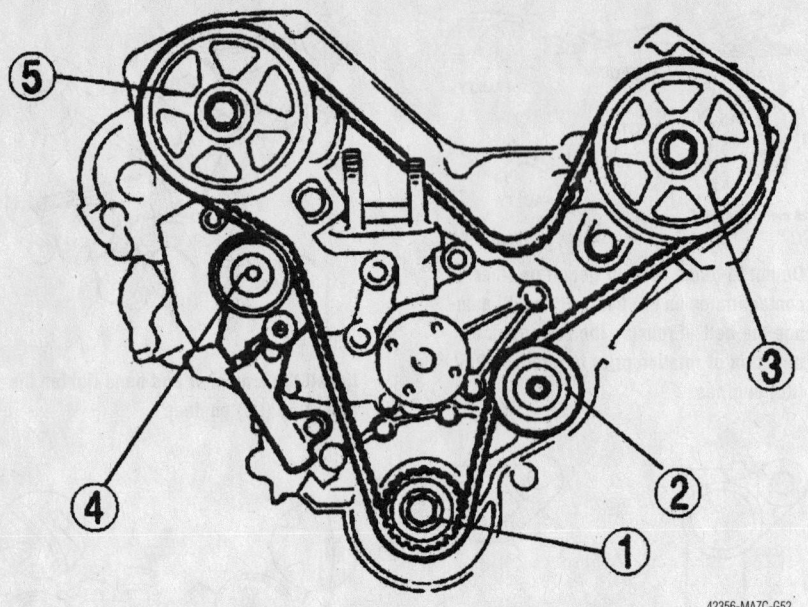

42356-MAZC-G52

Make sure there is tension between pulleys (3) and (1) and pulleys (1) and (5)–2.5L (KL) engines

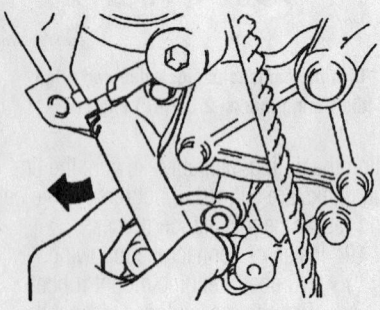

42356-MAZC-GEA

Push the auto tensioner in the direction of arrow and hand tighten the lower bolt–2.5L (KL) engine

exploded view of the timing belt components for torque specifications.
- Water pump pulley. Refer to the exploded view of the timing belt components for torque specifications.
- Drive belt
- Power steering pump
- Spark plugs
- CKP sensor
- Negative battery cable

Millenia

2.3L (KJ) ENGINE

1. Before servicing the vehicle, refer to the precautions in the beginning of this section.

2. Refer to the illustration of the exploded view of the timing belt assembly

for component location and torque specifications.

3. Remove or disconnect the following:
- Negative battery cable
- Right front wheel
- Splash shield from the right hand side
- Dust cover
- Drive belt

✳✳ CAUTION

The Crankshaft Position (CKP) sensor rotor is on the rear of the crankshaft pulley and can be damaged if the pulley is not removed carefully.

4. Turn the crankshaft pulley clockwise until the pin on the pulley is facing down.
- Crankshaft pulley using special tools 49 E120 710. Refer to the illustration for tool positioning.
- Power steering pump, leave the

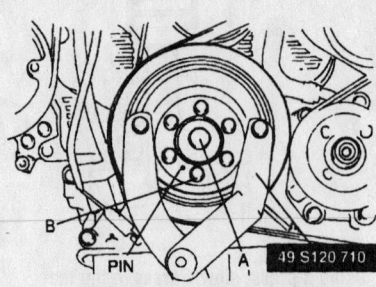

42356-MAZC-GED

Remove the crankshaft pulley using special tools 49 E120 710–2.3L (KJ) engines

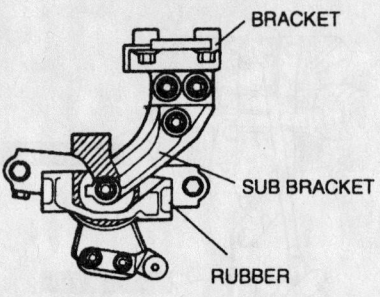

42356-MAZC-GEF

Remove the number 3 engine mount sub bracket, rubber and bracket–2.3L (KJ) engines

lines attached and position the pump aside
- Water pump pulley
- Alternator drive belt tensioner
- Camshaft Position (CMP) sensor
- Oil dipstick and tube
- Vacuum pipe
- Upper radiator hose

5. Support the engine assembly with a hoist and remove the number 3 engine mount rubber.
- Number 3 engine mount sub bracket, rubber and bracket
- Timing belt cover
- Power steering pump drive belt tensioner

6. Turn the crankshaft until the timing mark on the crankshaft sprocket aligns with the timing mark on the oil pump and the camshaft sprocket timing marks align with the marks on the cylinder head. The No. 1 piston should be at Top Dead Center (TDC) of the compression stroke.

7. Remove the two bolts from the automatic tensioner, removing the lower one first. Keep the bolt holes aligned by holding the tensioner to reduce the chance of stripping the threads on the bolts.

8. If the timing belt is to be reused, mark the direction of rotation on the timing belt.

9. Remove the timing belt.

To install:

10. Install the crankshaft sprocket bolt. Install the flywheel locking tool. Tighten the bolt to 116–122 ft. lbs. (157–166 Nm).

11. Check the tension rod projection, it should be 0.563–0.594 inch (14.3–15.1mm), if not, replace the tensioner

12. Position the automatic tensioner in a press. Set a flat washer under the tensioner body to prevent damage to the body plug.

13. Compress the tensioner until the hole in the piston is aligned with the 2nd hole in the tensioner case. Insert a 0.055

inch (1.4mm) diameter wire or pin through the 2nd hole to keep the piston compressed.

14. Be sure the camshaft sprocket timing marks are still aligned. Turn the crankshaft clockwise until the timing sprocket is aligned.

15. Install the timing belt. If the original belt is being reused, be sure it is installed in the same direction of rotation. The order of installation is: timing belt (crankshaft) sprocket, No. 2 idler pulley, left-hand camshaft sprocket, both No. 1 idler pulleys, right-hand camshaft sprocket and the tensioner pulley.

16. Install the automatic belt tensioner and tighten the bolts to 14–18 ft. lbs. (19–25 Nm). Remove the wire or pin from the tensioner.

17. Turn the crankshaft clockwise, until the crankshaft sprocket timing mark is again at TDC. This should place all of the belt slack in the automatic tensioner portion of the belt.

18. Rotate the crankshaft two turns in the normal direction of rotation and align the timing marks. Be sure all marks are still correctly aligned.

19. Inspect timing belt deflection, 0.24–0.31 inch (6–8mm), between the crankshaft sprocket and the tensioner pul-

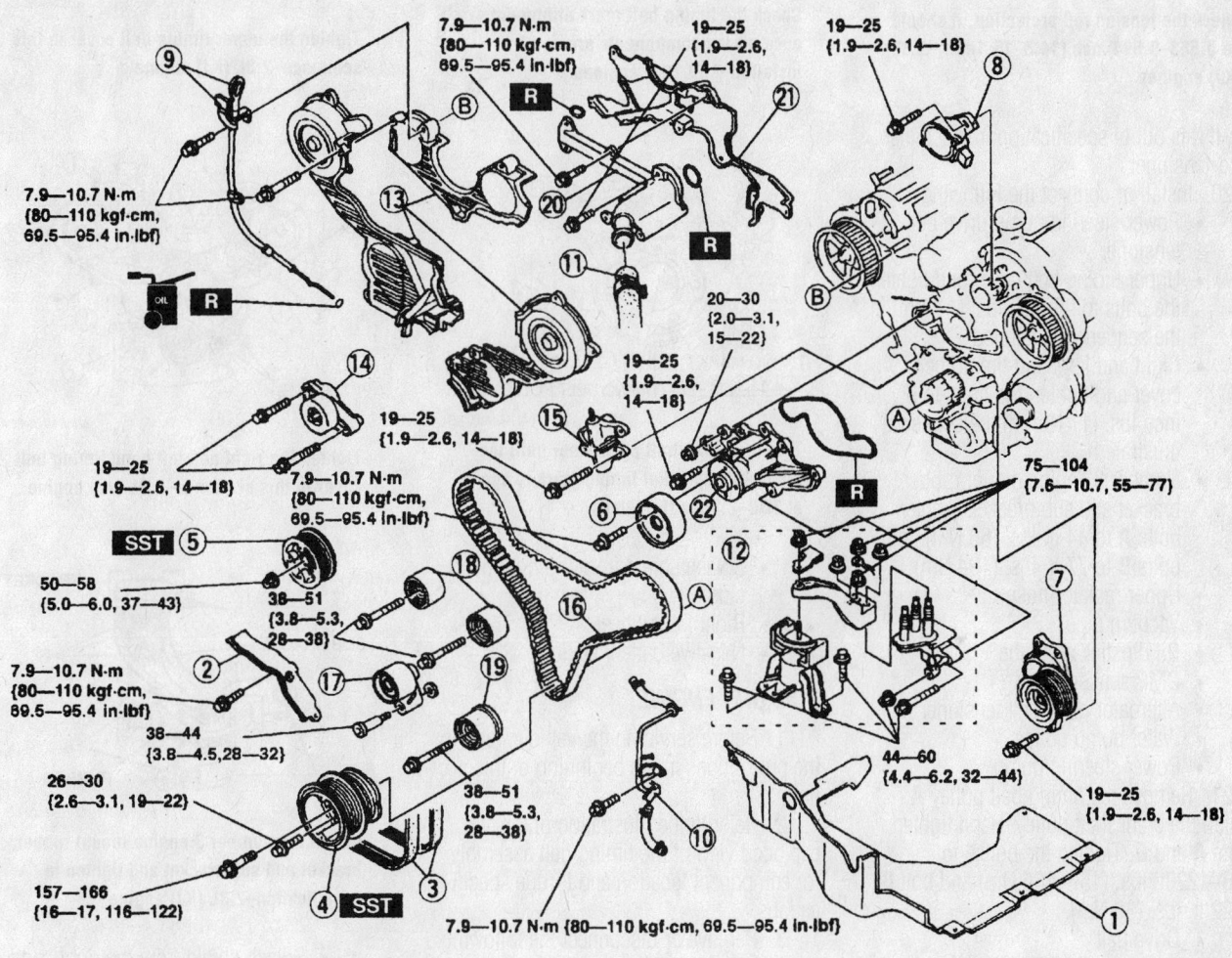

N·m {kgf·m, ft·lbf}

1	Splash shield (RH)	12	No.3 engine mount
2	Dust cover	13	Timing belt cover
3	Drive belt	14	Drive belt auto tensioner (P/S)
4	Crankshaft pulley	15	Timing belt auto tensioner
5	P/S oil pump pulley	16	Timing belt
6	Water pump pulley	17	Tensioner pulley
7	Drive belt auto tensioner (Generator)	18	No.1 idler pulley
8	CMP sensor	19	No.2 idler pulley
9	Dipstick and pipe	20	Water outlet pipe
10	Vacuum pipe	21	Seal plate
11	Upper radiator hose	22	Water pump

Exploded view of the timing belt assembly–2.3L (KJ) engines

42356-MAZC-GEC

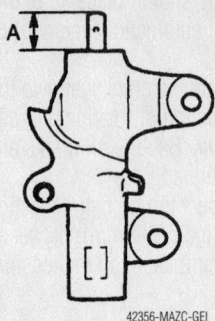

Check the tension rod projection, it should be 0.563–0.594 inch (14.3–15.1mm)–2.3L (KJ) engines

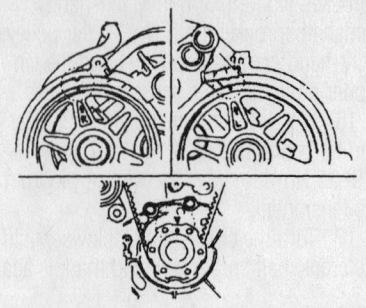

Check the timing belt mark alignment once all the components are installed–2.3L (KJ) engines

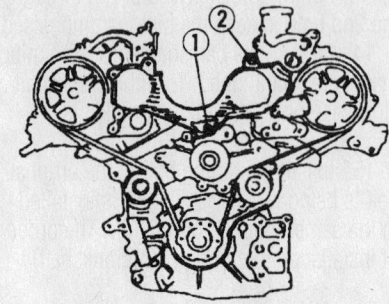

Tighten the upper timing belt cover in this sequence–2.3L (KJ) engine

ley. If it is out of specification, replace the auto-tensioner.

20. Install or connect the following:
- Power steering pump drive belt tensioner
- Upper timing belt cover and tighten the bolts to 95 inch lbs. (11 Nm) in the sequence illustrated
- Right and left hand timing belt cover and tighten the bolts to 95 inch lbs. (11 Nm) in the sequence illustrated
- Number 3 engine mount rubber, bracket and sub bracket. Tighten bolts **A** to 44 ft. lbs. (60 Nm) and bolts **B** to 77 ft. lbs. (104 Nm).
- Upper radiator hose
- Vacuum pipe
- Oil dipstick and tube
- CMP sensor
- Alternator drive belt tensioner
- Water pump pulley
- Power steering pump

21. Remove the timing belt pulley **A**, install the crankshaft pulley. Hand tighten bolts **A** and **B**. Tighten the bolt **A** to 116–122 ft. lbs. (157–166 Nm) and bolt **B** to 22 ft. lbs. (30 Nm).
- Drive belt
- Dust cover

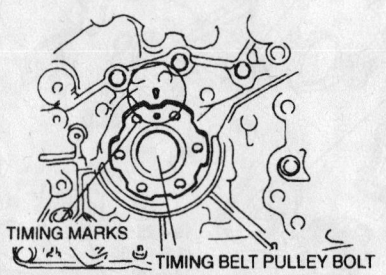

Turn the crankshaft clockwise, until the crankshaft sprocket timing mark is again at TDC–2.3L (KJ) engines

- Splash shield from the right hand side
- Right front wheel
- Negative battery cable

2.5L (KL) ENGINE

1. Before servicing the vehicle, refer to the precautions in the beginning of this section.

2. Refer to the illustration of the exploded view of the timing belt assembly for component location and torque specifications.

3. Remove or disconnect the following:
- Negative battery cable

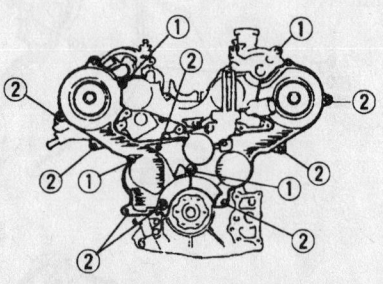

Tighten the right and left hand timing belt cover in this sequence–2.3L (KJ) engine

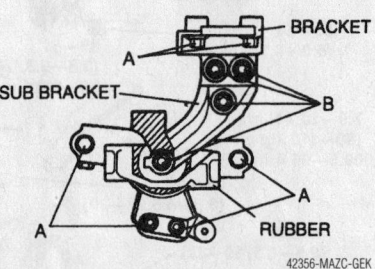

Install the number 3 engine mount rubber, bracket and sub bracket and tighten to specification–2.3L (KJ) engine

- Splash shield from the right hand side
- Drive belt
- Water pump pulley
- Idler pulley and bracket
- Power steering pump, leave the lines attached and position the pump aside

❊❊ CAUTION

The Crankshaft Position (CKP) sensor rotor is on the rear of the crankshaft pulley and can be damaged if the pulley is not removed carefully.

- Crankshaft pulley using special tools 49 E011 1A1 and 49 S120

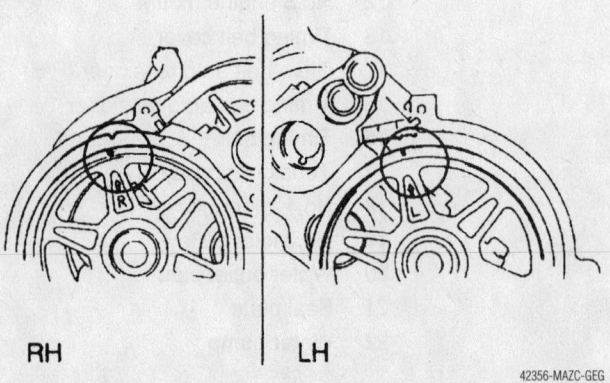

RH   LH

Be sure the camshaft sprocket timing marks are still aligned–2.3L (KJ) engines

710. Refer to the illustration for tool positioning.
- Rocker arm cover
- Oil dipstick and tube
- CKP sensor
- Timing belt cover

4. Support the engine assembly with a hoist and remove the number 3 engine mount rubber.

5. Install the crankshaft pulley bolt and turn the crankshaft clockwise and align the timing marks as illustrated. The number 1 piston should be at Top Dead Center (TDC) of the compression stroke.

※※ **CAUTION**

When removing the auto tensioner bolts, hold the tensioner so that the threads are aligned or the threads can be damaged.

6. Loosen the auto tensioner bolts and remove the lower bolt.

※※ **CAUTION**

When removing the number 1 idler pulley bolt, hold the pulley so that the threads are aligned or the threads can be damaged.

7. Remove the number 1 idler pulley

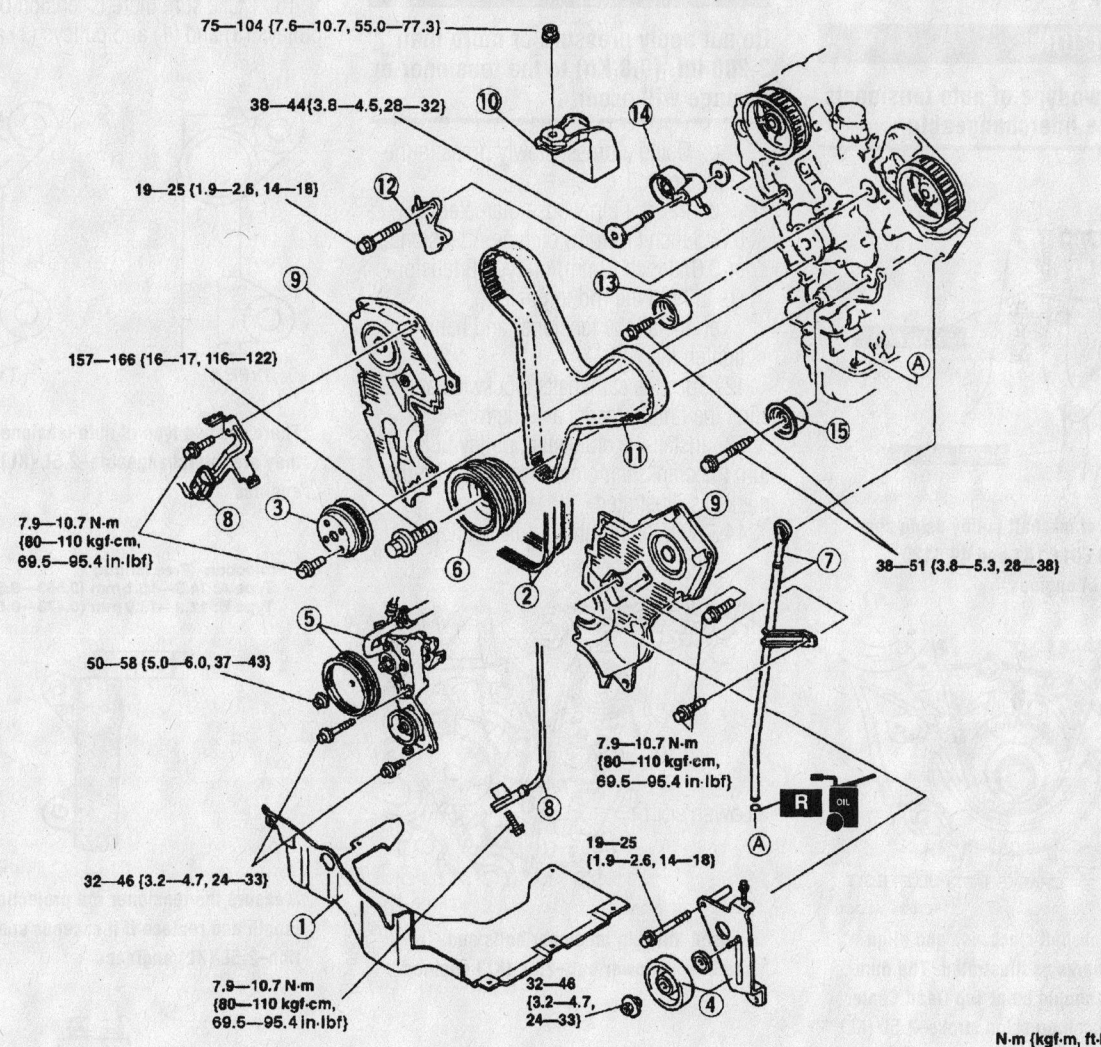

1 Splash shield (RH)	9 Timing belt cover
2 Drive belt	10 No.3 engine mount sub bracket
3 Water pump pulley	11 Timing belt
4 Idler pulley and bracket	12 Timing belt auto tensioner
5 P/S oil pump	13 No.1 idler pulley
6 Crankshaft pulley	14 Tensioner pulley
7 Dipstick and pipe	15 No.2 idler pulley
8 CKP sensor	

42356-MAZC-GEB

Exploded view of the timing belt assembly–2.5L (KL) engines–Millenia models

※※ CAUTION

Be careful not to bend, twist or get oil or other contaminates on the belt as this will damage the belt.

8. If the timing belt is to be reused, mark the direction of rotation on the timing belt. Remove the timing belt.

9. Remove the belt.

10. If necessary, remove the auto tensioner.

To install:

※※ CAUTION

There are two type of auto tensioners and they are interchangeable.

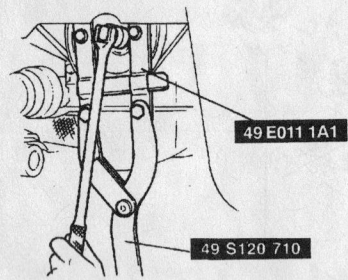

Remove the crankshaft pulley using special tools 49 E011 1A1 and 49 S120 710–2.5L (KL) engines

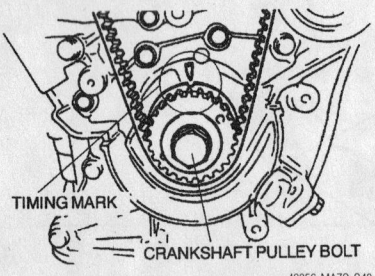

Turn the crankshaft clockwise and align the timing marks as illustrated. The number 1 piston should be at Top Dead Center (TDC) of the compression stroke–2.5L (KL) engines

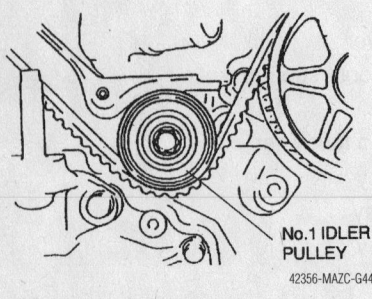

Remove the number 1 idler pulley–2.5L (KL) engines

11. If removed install the auto tensioner as follows:

a. Measure the tensioner rod projection length. If the length exceeds 0.0.512–0.551 inch (13–14mm) on type **A** tensioners or 0.563–0.594 inch (14.3–15.1mm) on type **B** tensioners replace the tensioner. Refer to the illustration to distinguish tensioner type.

b. Inspect the tensioner for leakage and replace if defective.

※※ CAUTION

Do not apply pressure of more than 2,200 lbf. (9.8 Kn) to the tensioner or damage will occur.

c. Using a press, slowly press in the tensioner rod.

d. Insert a pin whose diameter is 0.055 inch (1.4mm) on type A tensioners or 0.079 inch (2mm) on type B tensioners to hold the rod in position.

e. Install the tensioner and hand tighten the bolt.

12. Turn the camshafts clockwise and align the timing marks as illustrated.

13. Install the crankshaft pulley bolt and turn the crankshaft clockwise and align the marks as illustrated.

14. With the number 1 idler pulley

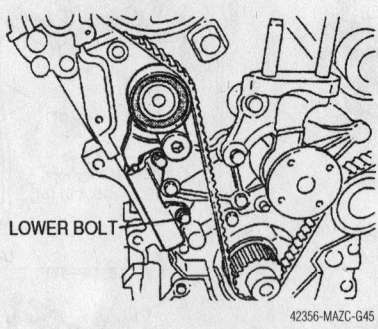

Loosen the auto tensioner bolts and remove the lower bolt–2.5L (KL) engines

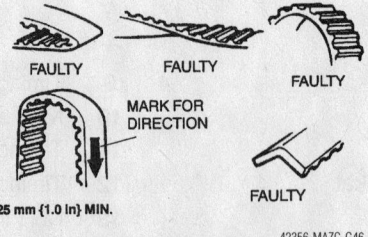

Do not to bend, twist or get oil or other contaminates on the belt as this will damage the belt, if reusing the belt; mark the direction of rotation prior to removal–2.5L (KL) engines

removed, install the belt on the pulleys in this order:

- Timing belt pulley
- Number 2 idler pulley
- Left hand camshaft pulley
- Tensioner pulley
- Right hand camshaft pulley

※※ CAUTION

Make sure the belt has no looseness at the tension side.

15. Make sure there is tension between pulleys (3) and (1) and pulleys (1) and (5).

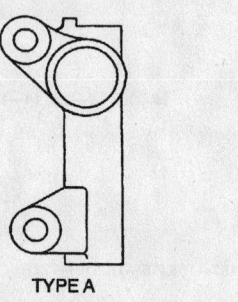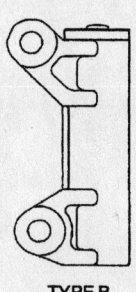

There are two type of auto tensioners and they are interchangeable–2.5L (KL) engines

Projection (Free length)
Type A: 14.3–15.1 mm {0.563–0.594 in}
Type B: 12.0–13.0 mm {0.473–0.511 in}

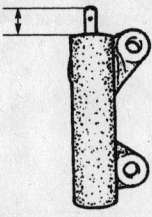

Measure the tensioner rod projection length and replace if it exceeds specification–2.5L (KL) engines

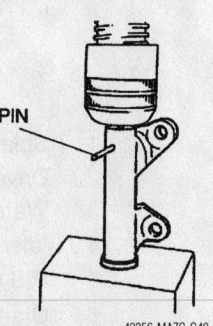

Using a press, slowly press in the tensioner rod and insert a 0.055 inch (1.4mm) diameter pin to hold the rod in position–2.5L (KL) engines

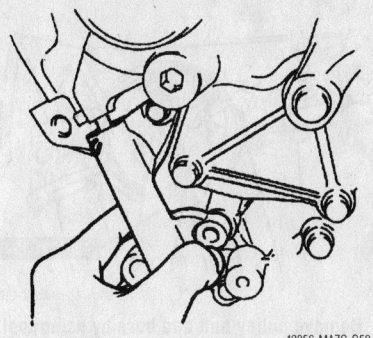

Install the tensioner and hand tighten the bolt–2.5L (KL) engines

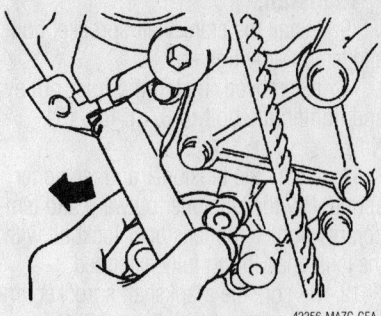

Push the auto tensioner in the direction of arrow and hand tighten the lower bolt–2.5L (KL) engine

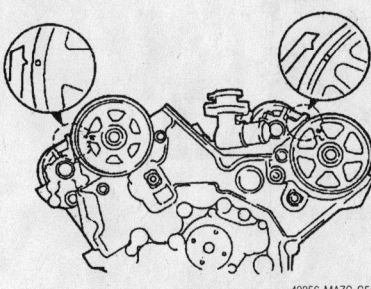

Turn the camshafts clockwise and align the timing marks–2.5L (KL) engines

Refer to the illustration for location of the pulleys.

16. Install the number 1 idler pulley while applying the pressure on the timing belt. Tighten the bolt to 28–38 ft. lbs. (38–51 Nm).

17. Push the auto tensioner in the direction of arrow (refer to illustration) and hand tighten the lower bolt, then tighten the bolts. Refer to the exploded view of the timing belt components for torque specifications. Then remove the retaining pin.

18. Turn the crankshaft clockwise and align the timing marks. Make sure the timing marks are aligned, if not remove the belt and repeat the installation process.

19. Install or connect the following:
- Number 3 engine mount rubber. Refer to the exploded view of the timing belt components for torque specifications. Remove the engine hoist.
- Timing belt cover. Refer to the exploded view of the timing belt components for torque specifications.
- Oil dipstick and tube. Refer to the exploded view of the timing belt components for torque specifications.
- CKP sensor
- Rocker arm cover

❋❋ CAUTION

The CKP sensor rotor is on the rear of the crankshaft pulley and can be damaged if the pulley is not removed carefully.

- Crankshaft pulley using special tools 49 E011 1A1 and 49 S120 710. Refer to the illustration for tool positioning. Tighten the bolt to 16–122 ft. lbs. (157–166 Nm).
- Power steering pump
- Idler pulley bracket. Refer to the exploded view of the timing belt components for torque specifications.
- Water pump pulley. Refer to the exploded view of the timing belt components for torque specifications.
- Drive belt
- Splash shield
- Negative battery cable

Miata

1.8L (BP) ENGINE

1. Before servicing the vehicle, refer to the precautions in the beginning of this section.

2. Drain the cooling system.

3. Remove or disconnect the following:
- Negative battery cable
- Front suspension lower bar
- Air pipe
- Drive belt
- Crankshaft Position (CKP) sensor
- High tension lead and the ignition coil
- Spark plugs
- Upper radiator hose
- Water hose
- Oil pipe
- Oil Control Valve (OCV)
- Rocker arm cover and discard the gasket
- Water pump pulley
- Crankshaft pulley and plate
- Pulley bolt and boss by using tool 49 D011 102 to prevent crankshaft rotation
- Timing belt cover

4. Install the pulley boss on the crankshaft and tighten the bolt.

5. Turn the crankshaft until the timing marks on the crankshaft and camshaft sprockets are aligned. The pin on the pulley boss must face upward. Hold the crankshaft pulley boss with a suitable tool and remove the pulley lockbolt, being careful not to rotate the crankshaft. Remove the crankshaft pulley boss.

➡**Protect the tensioner with a shop towel before prying on it. Do not rotate the crankshaft after the timing belt has been removed.**

6. Mark the direction of rotation on the timing belt. Loosen the tensioner lock-

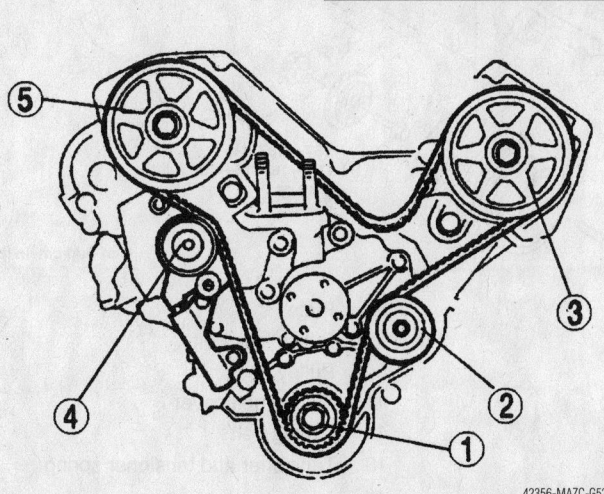

Make sure there is tension between pulleys (3) and (1) and pulleys (1) and (5)–2.5L (KL) engines

bolt and pry the tensioner outward. Tighten the lockbolt with the tensioner spring fully extended. Remove the timing belt.

7. Remove the tensioner and spring. If necessary, remove the idler pulley.

8. Inspect the belt for wear, peeling, cracking, hardening or signs of oil contamination. Inspect the tensioner for free and smooth rotation. Check the tensioner spring free length; it should not exceed 2.31 inch (59.2mm). Inspect the sprocket teeth for wear or damage. Replace parts, as necessary.

To install:

9. Install the crankshaft sprocket bolt and temporarily tighten.

10. If removed, install the idler pulley and tighten the bolt to 37 ft. lbs. (51 Nm).

11. Install the tensioner and tensioner spring. Pry the tensioner outward and temporarily tighten the tensioner lockbolt with the tensioner spring fully extended.

12. Be sure the crankshaft sprocket timing mark is aligned with the mark on the oil pump housing. Be sure the camshaft sprocket timing marks are aligned with the

49 D011 102

42356-MAZC-GBG

Remove pulley bolt and boss by using tool 49 D011 102 to prevent crankshaft rotation—1.8L (BP) engines

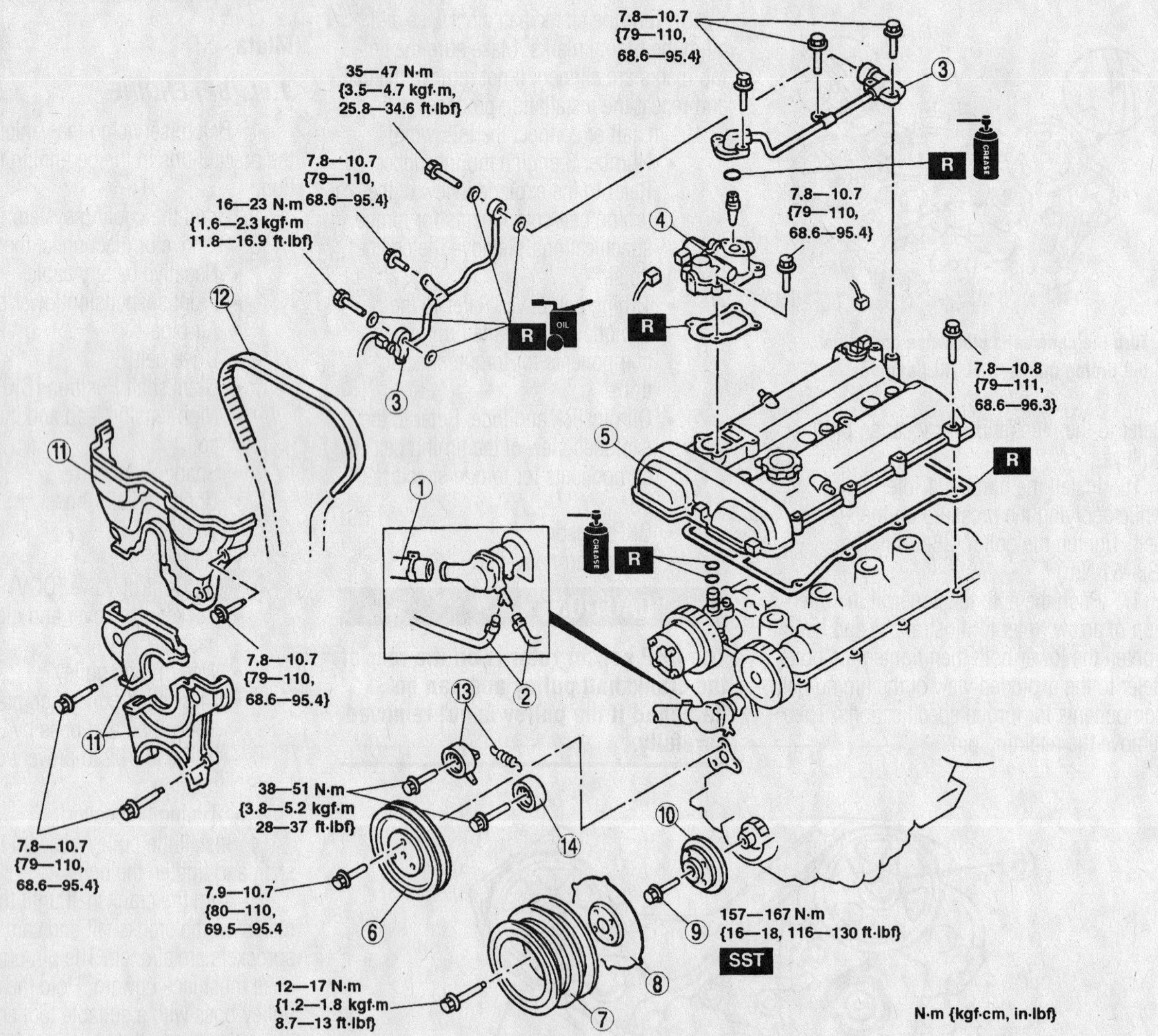

N·m {kgf·cm, in·lbf}

1	Upper radiator hose
2	Water hose
3	Oil pipe
4	Oil control valve (OCV) case
5	Cylinder head cover
6	Water pump pulley
7	Crankshaft pulley
8	Plate
9	Pulley lock bolt
10	Pulley boss
11	Timing belt cover
12	Timing belt
13	Tensioner and tensioner spring
14	Idler

42356-MAZC-GBF

Exploded view of the timing belt assembly—1.8L (BP) engines

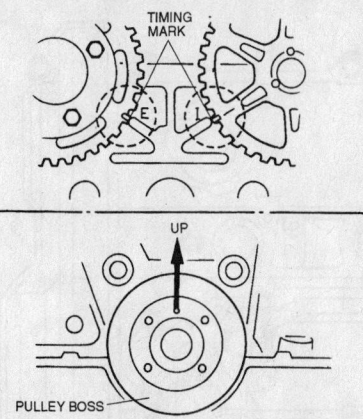

Turn the crankshaft until the timing marks on the crankshaft and camshaft sprockets are aligned. The pin on the pulley boss must face upward—1.8L (BP) engines

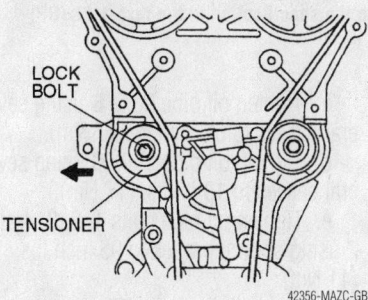

Loosen the tensioner lockbolt and pry the tensioner outward—1.8L (BP) engines

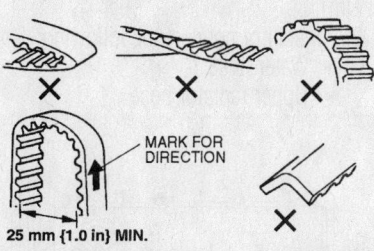

Do not to bend, twist or get oil or other contaminates on the belt as this will damage the belt, if reusing the belt; mark the direction of rotation prior to removal—1.8L (BP) engines

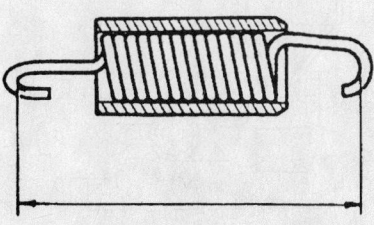

Check the tensioner spring free length; it should not exceed 2.31 inch (59.2mm)—1.8L (BP) engines

marks on the seal plate. Refer to the illustration for timing mark alignment.

13. Install the timing belt so there is no looseness at the idler pulley side or between the camshaft sprockets. If reusing the old belt, be sure it is installed in the same direction of rotation.

14. Temporarily install the pulley boss and lockbolt.

15. Turn the crankshaft 2 turns clockwise and align the crankshaft sprocket timing mark. Face the pin on the pulley boss

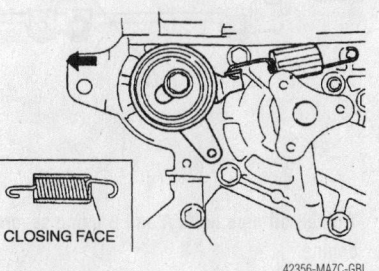

Install the tensioner and tensioner spring—1.8L (BP) engines

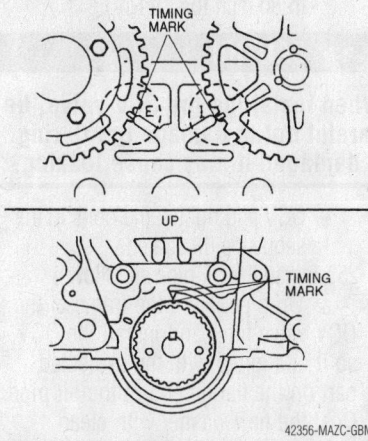

Be sure the crankshaft sprocket timing mark is aligned with the mark on the oil pump housing. Be sure the camshaft sprocket timing marks are aligned with the marks on the seal plate—1.8L (BP) engines

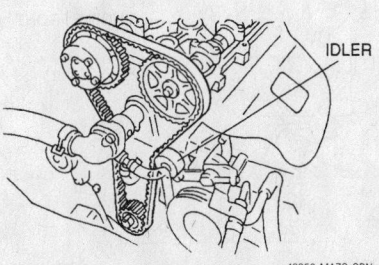

Install the timing belt so there is no looseness at the idler pulley side or between the camshaft sprockets—1.8L (BP)

upright. Be sure the camshaft sprocket timing marks are aligned. If they are not, repeat the alignment steps.

16. Turn the crankshaft 1 ⅝ turns clockwise and align the crankshaft sprocket timing mark with the tension set mark for proper belt tension adjustment. Remove the lockbolt and pulley boss.

17. Be sure the crankshaft sprocket timing mark is aligned with the tension set mark. Loosen the tensioner lockbolt, and allow the spring to apply tension to the belt.

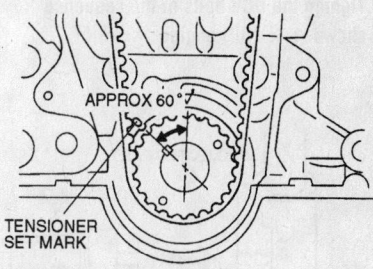

Turn the crankshaft 1 ⅝ turns clockwise and align the crankshaft sprocket timing mark with the tension set mark for proper belt tension adjustment—1.8L (BP) engines

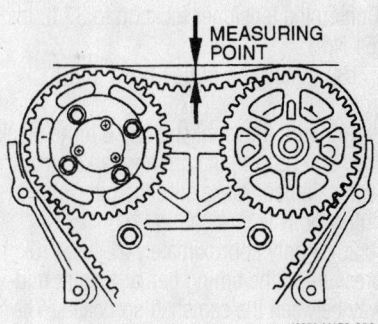

The belt should deflect 0.35–0.45 inch (9.0–11.5mm)—1.8L (BP) engines

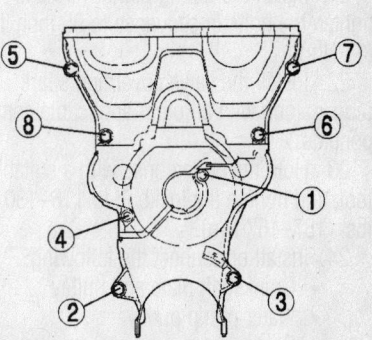

Tighten the timing belt cover using this sequence)—1.8L (BP) engines

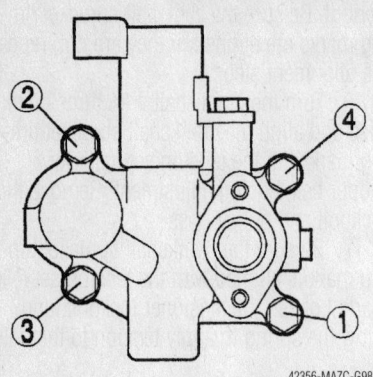

Tighten the OCV bolts in the sequence shown–1.8L (BP) engine

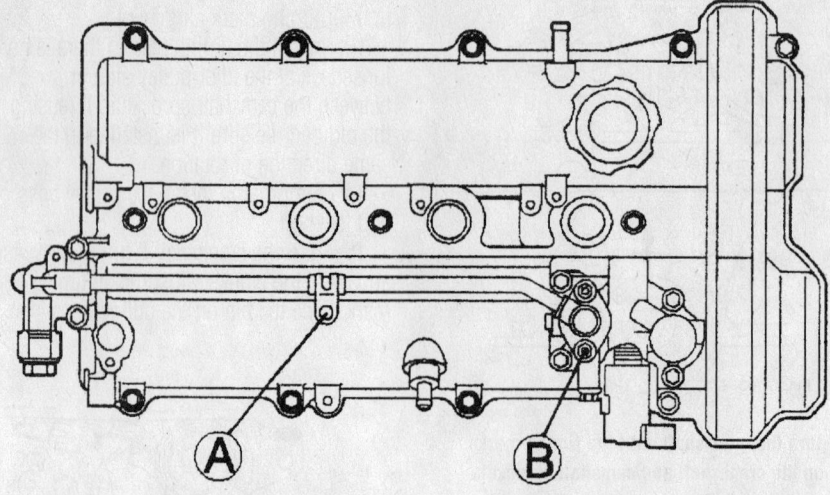

ENGINE FRONT →

42356-MAZC-GAA

Tighten oil pipe bolts A and B using several passes to the specification in the text–1.8L (BP) engine

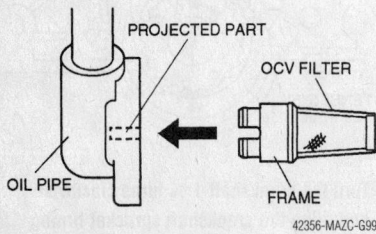

42356-MAZC-G99

Hold the frame of the OCV valve filter and install the OCV so it is aligned with the projected part on the flange end of the oil pipe–1.8L (BP) engine

Tighten the tensioner lockbolt to 37 ft. lbs. (51 Nm).

18. Install the pulley boss and lockbolt.

19. Turn the crankshaft 2 ⅙ turns clockwise and be sure the timing marks are correctly aligned. Make sure the pin on the pulley boss is straight up.

20. Apply approximately 22 lbs. (10kg) pressure to the timing belt at a point midway between the camshaft sprockets. The belt should deflect 0.35–0.45 inch (9.0–11.5mm). If the deflection is not correct, repeat the alignment and tensioning procedure.

21. Install the timing belt covers and tighten the bolts in sequence to 95 inch lbs. (11 Nm).

22. Install the valve cover and spark plugs, along with all other applicable components.

23. Hold the pulley boss with a suitable tool, and tighten the lockbolt to 116–130 ft. lbs. (157–167 Nm).

24. Install or connect the following:
- Crankshaft plate and pulley
- Water pump pulley
- Rocker arm cover with a new gasket
- Rocker arm cover with a new gasket.

- Temporarily tighten the cover bolts **A**, refer to the illustration for location. Torque the bolts in sequence to 80 inch lbs. (9 Nm).

✵✵ CAUTION

When installing the OCV valve, be careful not to damage the O–ring, if damaged it may cause leaking.

- OCV and tighten the bolts in the sequence illustrated

25. Install the oil pipe as follows:

a. Oil pipe. Hold the frame of the OCV valve filter and install the OCV so it is aligned with the projected part on the flange end of the oil pipe. Coat the new washer with clean engine oil and temporarily install the upper and side oil pipe and position the pipe.

b. **A** using several passes to 95 inch lbs. (11 Nm).

c. Tighten oil pipe bolts **B** using several passes to 61 inch lbs. (7 Nm).

d. Tighten oil pipe bolts **C** using several passes to 13 ft. lbs. (17 Nm).

e. Tighten oil pipe bolts 1, 2, 3, 4 and 7 using several passes to 95 inch lbs. (11 Nm).

f. Tighten oil pipe bolt 5 using several passes to 17 ft. lbs. (23 Nm).

g. Tighten oil pipe bolt 6 using several passes to 34 ft. lbs. (47 Nm).

26. Install or connect the following:
- Water hose
- Upper radiator hose

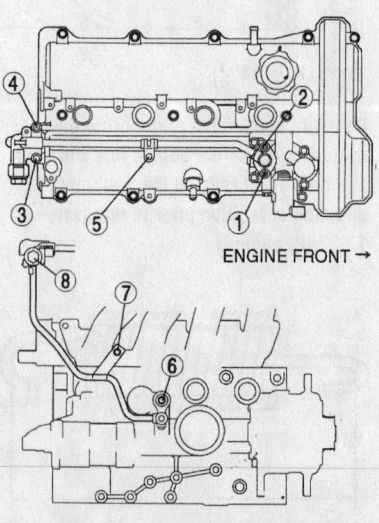

ENGINE FRONT →

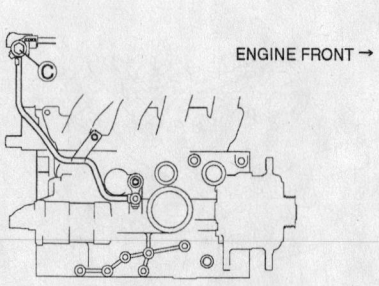

ENGINE FRONT →

42356-MAZC-GAB

Tighten oil pipe bolts C using several passes to the specification in the text–1.8L (BP) engine

42356-MAZC-GAC

Tighten oil pipe bolts using several passes to the specification in the text–1.8L (BP) engine

- Spark plugs
- High tension lead and the ignition coil
- CKP sensor
- Drive belt
- Air pipe
- Front suspension lower bar
- Negative battery cable

Piston and Ring

POSITIONING

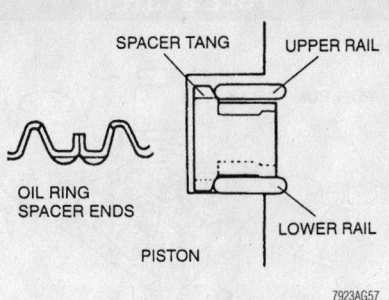

Upper, spacer and lower oil ring identification and positioning—Mazda engines

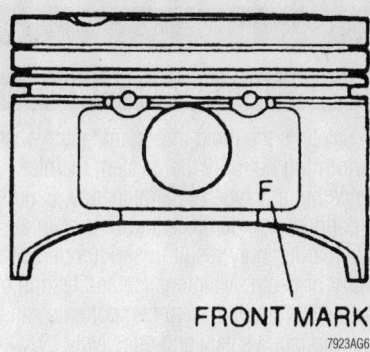

Piston-to-engine block mark location on the piston—Mazda 1.8L (BP) engines

Before removing the caps from the connecting rods, be sure to matchmark them—Mazda engines

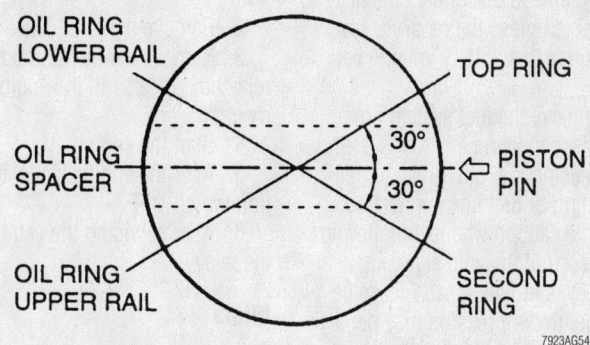

Piston ring end-gap spacing—Mazda engines

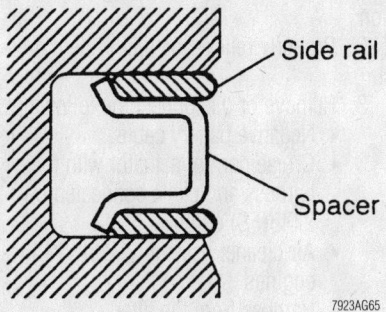

Compression ring identification and positioning—Mazda engines

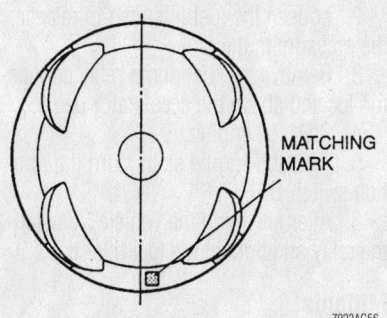

Piston-to-engine block mark location on the piston face—Mazda 2.0L (FS) and 2.5L (KL) engines

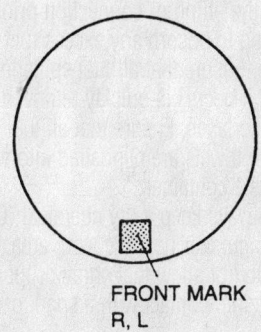

Piston-to-engine positioning mark location—Mazda 2.3L (KJ) engine

FUEL SYSTEM

Fuel System Service Precautions

Safety is the most important factor when performing not only fuel system maintenance but any type of maintenance. Failure to conduct maintenance and repairs in a safe manner may result in serious personal injury or death. Maintenance and testing of the vehicle's fuel system components can be accomplished safely and effectively by adhering to the following rules and guidelines.

1. To avoid the possibility of fire and personal injury, always disconnect the negative battery cable unless the repair or test procedure requires that battery voltage be applied.

2. Always relieve the fuel system pressure prior to disconnecting any fuel system component (injector, fuel rail, pressure regulator, etc.), fitting or fuel line connection. Exercise extreme caution whenever relieving fuel system pressure, to avoid exposing skin, face and eyes to fuel spray. Please be advised that fuel under pressure may penetrate the skin or any part of the body that it contacts.

3. Always place a shop towel or cloth around the fitting or connection prior to loosening to absorb any excess fuel due to spillage. Ensure that all fuel spillage (should it occur) is quickly removed from engine surfaces. Ensure that all fuel soaked cloths or towels are deposited into a suitable waste container.

4. Always keep a dry chemical (Class B) fire extinguisher near the work area.

5. Do not allow fuel spray or fuel vapors to come into contact with a spark or open flame.

6. Always use a back-up wrench when loosening and tightening fuel line connection fittings. This will prevent unnecessary stress and torsion to fuel line piping. Always follow the proper torque specifications.

7. Always replace worn fuel fitting O-rings with new. Do not substitute fuel hose where fuel pipe is installed.

Fuel System Pressure

RELIEVING

626 and Protégé

1. Before servicing the vehicle, refer to the precautions in the beginning of this section.

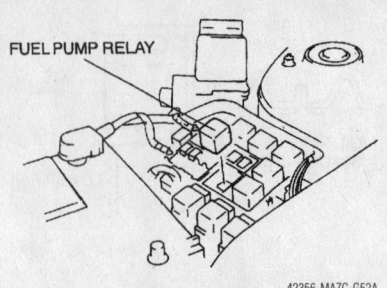

Fuel pump relay location—626 and with 2.0L and 2.5L engines

2. Remove the filler cap.

3. Remove the fuel pump relay from the relay box, located in the engine compartment.

4. Start the engine.

5. After the engine stalls, turn the ignition switch **OFF**.

6. After servicing the vehicle, reinstall the relay.

Miata

1. Before servicing the vehicle, refer to the precautions in the beginning of this section.

2. Loosen the fuel filler cap to release the pressure in the tank.

3. Remove the fuel pump relay connector, located above the accelerator pedal.

4. Start the engine.

5. After the engine stalls, turn the ignition switch **OFF**.

6. After servicing the vehicle, reinstall the relay and tighten the fuel filler cap.

Millenia

1. Before servicing the vehicle, refer to the precautions in the beginning of this section.

2. If necessary for clearance, remove the

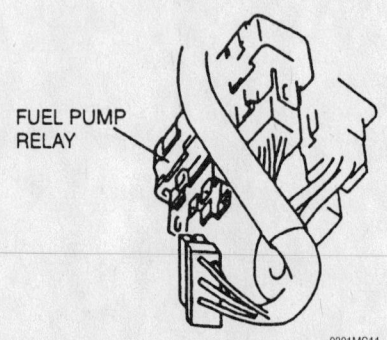

Fuel pump relay connector location—Miata

cruise control actuator and position aside on 2.3L (KJ) engines.

3. Remove the fuel pump relay from the relay box.

4. Start the engine.

5. After the engine stalls, turn the ignition switch **OFF**.

6. After servicing the vehicle, reinstall the relay and the cruise actuator, if necessary.

Fuel Filter

The fuel filter on all Mazda cars can be located on a bracket in the left rear of the engine compartment, next to or beneath the brake master cylinder fluid reservoir or as part of the fuel pump assembly.

On the Millenia, the fuel filter is located beneath an access cover in the trunk. Access to the cover is achieved by removing the trunk mat to expose the cover.

REMOVAL & INSTALLATION

626 and Protégé

1. Before servicing the vehicle, refer to the precautions in the beginning of this section.

2. Properly relieve the fuel system pressure.

3. Remove or disconnect the following:
- Negative battery cable
- Cruise control actuator with the harness and cable connected on 2.0L (FS) engines
- Air cleaner assembly, 2.5L (KL) engines
- Harness from the filter
- Fuel hoses from the filter
- Fuel filter

To install:

4. Installation is the reverse of removal.

5. Pressurize the fuel system and check all connections for leaks.

Miata

1. Before servicing the vehicle, refer to the precautions in the beginning of this section.

2. Properly relieve the fuel system pressure.

3. Remove or disconnect the following:
- Negative battery cable

4. Raise and safely support the rear of the vehicle.
- Fuel filter protector
- Fuel lines by squeezing the tabs.

Cover the lines to prevent leakage and contamination.
- Filter bracket and the filter. Mark the filter and bracket to aid for correct installation

To install:

5. Install or connect the following:
- Filter and bracket aligning the marks made during removal. Tighten the bracket bolts to 52 inch lbs. (6 Nm).
- Fuel lines making sure the tabs on the fittings are firmly engaged by pulling lightly on the lines
- Fuel filter protector
- Negative battery cable

6. Start the vehicle and check for leaks, then lower the vehicle.

Miata

1. Before servicing the vehicle, refer to the precautions in the beginning of this section.
2. Properly relieve the fuel system pressure.
3. Remove or disconnect the following:
- Negative battery cable
- Fuel lines by squeezing the tabs. Cover the lines to prevent leakage and contamination.
- Filter bracket and the filter. Mark the filter and bracket to aid for correct installation

To install:

4. Install or connect the following:
- Filter and bracket aligning the marks made during removal
- Fuel lines making sure the tabs on the fittings are firmly engaged by pulling lightly on the lines
- Negative battery cable

5. Start the vehicle and check for leaks, then lower the vehicle.

Millenia

1. Before servicing the vehicle, refer to the precautions in the beginning of this section.
2. Insure the ignition switch is **OFF**.
3. Relieve the fuel system pressure.
4. Remove or disconnect the following:
- Negative battery cable
- Trunk mat
- Service hole cover
- Fuel lines from both ends of the fuel filter
- Fuel filter and bracket
- Fuel filter from the mounting bracket

To install:

5. Install or connect the following:

- Filter in the bracket. Torque the nut to 70–95 inch lbs. (8–11 Nm).
- Fuel lines to the filter
- Service hole cover
- Trunk mat
- Negative battery cable

6. Run the engine and check for any fuel leaks.

Fuel Pump

REMOVAL & INSTALLATION

Protégé

1. Before servicing the vehicle, refer to the precautions in the beginning of this section.
2. Relieve the fuel pressure.
3. Remove or disconnect the following:
- Negative battery cable
- Rear seat cushion
- Fuel pump/sending unit electrical connector
- Fuel pump/sending unit access cover
- Fuel supply and return hoses from the fuel pump/sending unit
- Fuel pump/sending unit from the fuel tank
- Sending unit electrical connector
- Sending unit from the fuel pump assembly

To install:

4. Install or connect the following:
- Sending unit to the fuel pump assembly
- Sending unit electrical connector
- Fuel pump/sending unit into the fuel tank with a new gasket
- Fuel supply and return lines
- Access cover
- Sending unit electrical connector
- Rear seat cushion
- Negative battery cable

5. Start the engine and check fuel leaks.

Miata

1. Before servicing the vehicle, refer to the precautions in the beginning of this section.
2. Properly relieve the fuel pressure.
3. Remove or disconnect the following:
- Negative battery cable
- Rear package trim
- Service hole cover
- Fuel pump cover
- Fuel pump connector
- Fuel hoses
- Fuel pump and gauge sender unit as an assembly
- Fuel pump from the sender bracket

To install:

- New O-ring set
- Fuel pump to the sender bracket

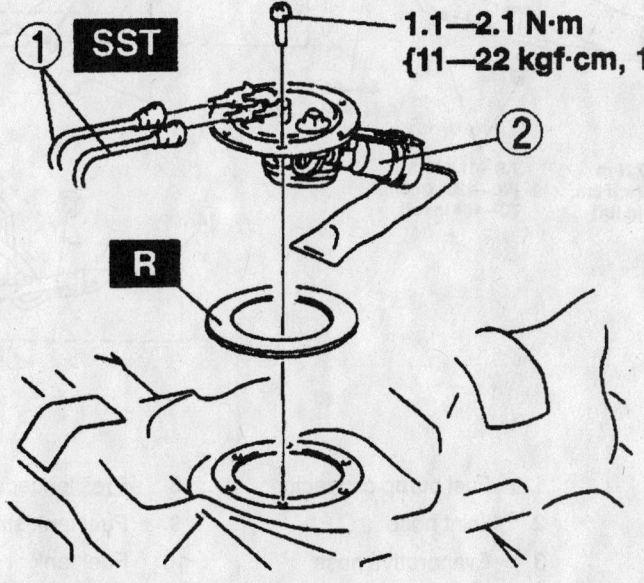

1   Fuel pipe
2   Fuel pump

Exploded view of the fuel pump assembly—626

42356-MAZC-G53

➡ **Pull the fuel pump down so that it is tight against the bracket.**

- Fuel pump and gauge sender unit as an assembly
- Fuel hoses
- Fuel pump connector
- Fuel pump cover
- Service hole cover
- Rear package trim
- Negative battery cable

626

1. Before servicing the vehicle, refer to the precautions in the beginning of this section.
2. Relieve the fuel system pressure.
3. Drain the fuel from the tank.
4. Remove or disconnect the following:
   - Negative battery cable
   - Fuel pump electrical connector
   - Hoses from the fuel tank

- Pressure control valve
- Fuel tank pressure sensor
- Fuel pipe
- Presilencer insulator
- Fuel tank strap while supporting the fuel tank
- Fuel tank
- All fuel hoses from the fuel pump unit
- Fuel pump ring retainers and the ring

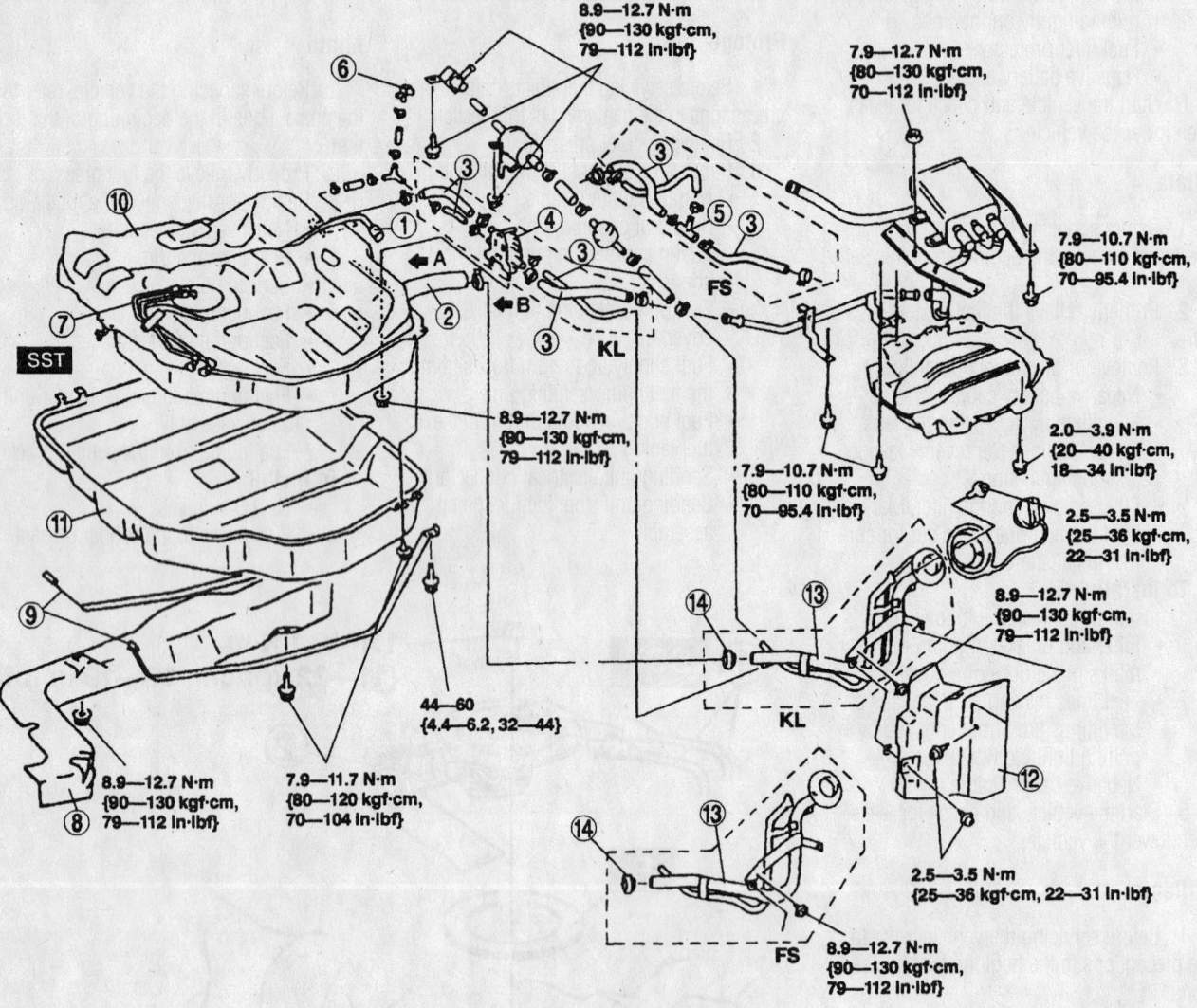

N·m {kgf·m, ft·lbf}

1	Fuel pump connector	8	Presilencer insulator
2	Joint hose	9	Fuel tank strap
3	Evaporative hose	10	Fuel tank
4	Pressure control valve	11	Fuel tank protector
5	Hose joint	12	Fuel–filler pipe insulator
6	Fuel tank pressure sensor	13	Fuel–filler pipe
7	Fuel pipe	14	Nonreturn valve

Exploded view of the fuel tank assembly—626

42356-MAZC-G54

- Fuel pump and gaskets from the fuel tank

To install:

5. Install or connect the following:
   - Fuel pump using a new gasket
   - Fuel pump ring and tighten the fasteners to 19 inch lbs. (2 Nm)
   - Fuel hoses to the fuel pump
   - Fuel tank
   - Fuel tank strap while supporting the fuel tank. Refer to the illustration for torque specifications.
   - Presilencer insulator
   - Fuel pipe
   - Fuel tank pressure sensor
   - Pressure control valve
   - Hoses to the fuel tank
   - Fuel pump electrical connector
   - Negative battery cable

6. Add a minimum of 10 gallons of fuel to the tank and check for leaks.

Millenia

1. Before servicing the vehicle, refer to the precautions in the beginning of this section.
2. Relieve the fuel system pressure.
3. Remove or disconnect the following:
   - Negative battery cable
   - Rear seat cushion
4. Drain the fuel from the tank.
   - Service hole cover
   - Fuel pump electrical connector
   - Hoses from the fuel tank
   - Fuel tank strap while supporting the fuel tank
   - Evaporative hose
   - Fuel tank
   - All fuel hoses from the fuel pump unit
   - Fuel pump ring using tool 49 T042 011
   - Fuel pump and gaskets from the fuel tank

To install:

5. Install or connect the following:
   - Fuel pump using a new gasket
   - Fuel pump ring and tighten the fasteners to 75 ft. lbs. (102 Nm)

- Fuel hoses to the fuel pump
- Fuel tank
- Evaporative hose
- Fuel tank strap while supporting the fuel tank. Tighten the strap bolts to 44 ft. lbs. (60 Nm).
- Hoses to the fuel tank
- Fuel pump electrical connector
- Service hole cover
- Rear seat cushion
- Negative battery cable

6. Add a minimum of 10 gallons of fuel to the tank and check for leaks.

Fuel Injector

REMOVAL & INSTALLATION

✶✶ CAUTION

Fuel injection systems remain under pressure after the engine has been turned OFF. Properly relieve fuel pressure before disconnecting any fuel lines. Failure to do so may result in fire or personal injury. Do not allow fuel spray or fuel vapors to come in contact with a spark or open flame. Keep a dry chemical fire extinguisher nearby. Never store fuel in an open container due to risk of fire or explosion.

626

2.0L (FS) ENGINE

1. Before servicing the vehicle, refer to the precautions in the beginning of this section.
2. Refer to the illustration for component location and torque specifications.
3. Relieve the fuel system pressure.
4. Remove or disconnect the following:
   - Negative battery cable
   - Fuel injector wiring harness
   - Fuel lines at the fuel rail
   - Hose from the pressure regulator
   - Fuel rail with the injectors attached
   - Fuel injectors, grommets and O-rings from the fuel rail
   - O-rings from the fuel injectors

To install:

5. Install or connect the following:
   - New O-rings and grommets lubricated with engine oil on the fuel injectors.
   - Insulators and injectors on the intake manifold
   - Grommets and the fuel rail onto the

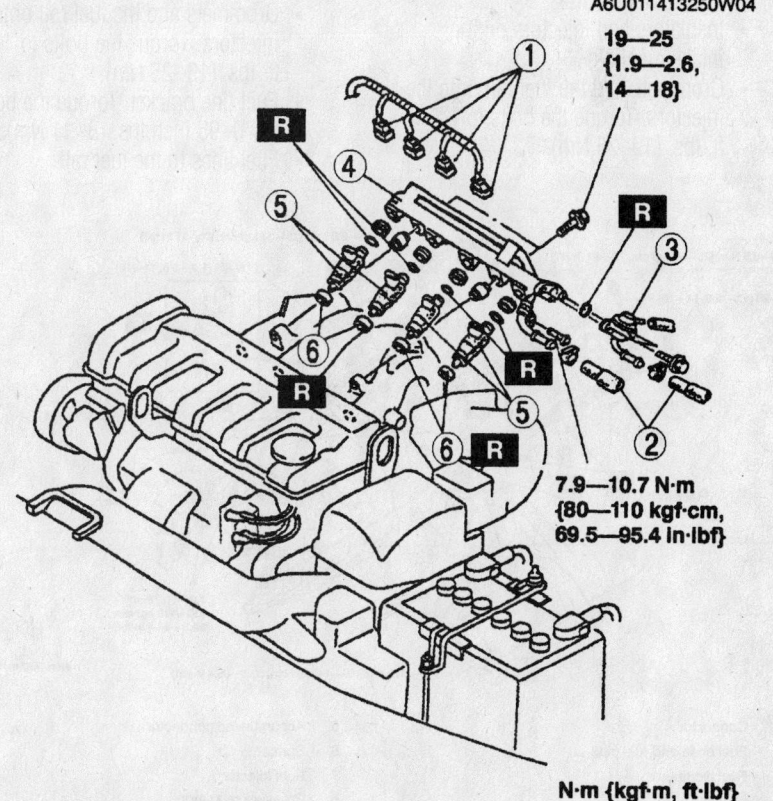

A6U011413250W04

19—25
{1.9—2.6,
14—18}

7.9—10.7 N·m
{80—110 kgf·cm,
69.5—95.4 in·lbf}

N·m {kgf·m, ft·lbf}

42356-MAZC-G56

Exploded view of the fuel rail and injector assembly—626 2.0L (FS)

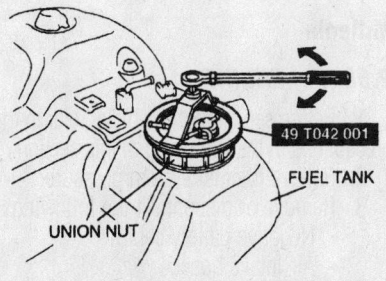

49 T042 001

FUEL TANK

UNION NUT

42356-MAZC-GEO

Remove the fuel pump ring—Millenia

injectors. Torque the bolts to 14–18 ft. lbs. (19–25 Nm).
- Fuel lines to the fuel rail
- Fuel injector wiring harness
- Negative battery cable

6. Turn the ignition switch **ON** to pressurize the fuel system.

7. Check for leaks and correct as necessary, before starting the engine.

2.5L (KL) ENGINE

1. Before servicing the vehicle, refer to the precautions in the beginning of this section.
2. Relieve the fuel system pressure.
3. Remove or disconnect the following:
   - Negative battery cable
   - Fuel injector wiring harness
   - Fuel lines at the fuel rail
   - Fuel rail
   - Accumulated connector
   - Spacer
   - Fuel injectors, grommets and O-rings from the fuel rail
   - O-rings from the fuel injectors

To install:
4. Install or connect the following:
   - New O-rings and grommets lubricated with engine oil on the fuel injectors. Fit the injector squarely into the rail while using a twisting motion. Fit the injector tab into the notch on the rail.
   - Insulators and injectors on the intake manifold
   - Grommets and the fuel rail onto the injectors. Torque the bolts to 14–18 ft. lbs. (19–25 Nm).

- Fuel lines to the fuel rail
- Fuel injector wiring harness
- Negative battery cable

5. Turn the ignition switch **ON** to pressurize the fuel system.

6. Check for leaks and correct as necessary, before starting the engine.

Miata

1. Before servicing the vehicle, refer to the precautions in the beginning of this section.
2. Relieve the fuel system pressure.
3. Remove or disconnect the following:
   - Negative battery cable
   - Upper intake manifold
   - Fuel lines at the fuel rail
   - Vacuum hose from the fuel pressure regulator
   - Fuel rail mounting bracket
   - Fuel rail mounting bolts, spacers and insulators
   - Fuel rail, with the injectors attached
   - Fuel injectors, grommets and O-rings from the fuel rail
   - O-rings from the fuel injectors

To install:
4. Install or connect the following:
   - New O-rings and grommets lubricated with engine oil on the fuel injectors
   - Insulators and injectors on the intake manifold
   - Grommets and the fuel rail onto the injectors. Torque the bolts to 14–18 ft. lbs. (19–25 Nm).
   - Fuel line bracket. Torque the bolts to 70–95 inch lbs. (8–11 Nm).
   - Fuel lines to the fuel rail

- Fuel injector wiring harness
- Upper intake manifold
- Negative battery cable

5. Turn the ignition switch **ON** to pressurize the fuel system.

6. Check for leaks and correct as necessary, before starting the engine.

Protégé

1. Before servicing the vehicle, refer to the precautions in the beginning of this section.
2. Relieve the fuel system pressure.
3. Remove or disconnect the following:
   - Negative battery cable
   - Accelerator cables and the cable bracket
   - Fuel injector connectors and wiring harness
   - Fuel lines at the fuel rail
   - Vacuum hose from the fuel pressure regulator
   - Fuel rail mounting bolts, spacers and insulators
   - Fuel rail, with the injectors attached
   - Fuel injectors, grommets and O-rings from the fuel rail
   - O-rings from the fuel injectors

To install:
4. Install or connect the following:
   - New O-rings and grommets lubricated with engine oil on the fuel injectors
   - Insulators and injectors on the intake manifold
   - Grommets and the fuel rail onto the injectors. Torque the bolts to 14–18 ft. lbs. (19–25 Nm).
   - Vacuum hose to the fuel pressure regulator
   - Fuel lines to the fuel rail
   - Fuel injector wiring harness
   - Cable bracket. Torque the bolt to 70–95 inch lbs. (8–11 Nm).
   - Accelerator cables, if removed, adjust as necessary
   - Negative battery cable

5. Turn the ignition switch **ON** to pressurize the fuel system.

6. Check for leaks and correct as necessary, before starting the engine.

Millenia

2.5L (KL) ENGINE

1. Before servicing the vehicle, refer to the precautions in the beginning of this section.
2. Relieve the fuel system pressure.
3. Remove or disconnect the following:
   - Negative battery cable
   - Air intake hose
   - Fuel injector wiring harness

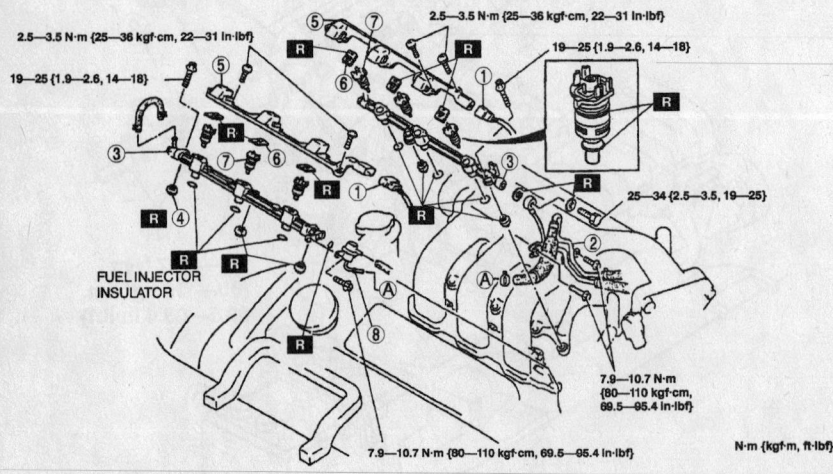

2.5—3.5 N·m {25—36 kgf·cm, 22—31 in·lbf}
19—25 {1.9—2.6, 14—18}
2.5—3.5 N·m {25—36 kgf·cm, 22—31 in·lbf}
19—25 {1.9—2.6, 14—18}
25—34 {2.5—3.5, 19—25}
7.9—10.7 N·m {80—110 kgf·cm, 69.5—95.4 in·lbf}
7.9—10.7 N·m {80—110 kgf·cm, 69.5—95.4 in·lbf}
N·m {kgf·m, ft·lbf}
FUEL INJECTOR INSULATOR

1	Connector	5	Accumulated connector
2	Fuel hose and fuel pipe	6	Spacer
3	Fuel distributor	7	Fuel injector
4	Insulator	8	Pressure regulator

42356-MAZC-G55

Exploded view of the fuel rail and injector assembly—626 2.5L (KL) engine

- Fuel lines at the fuel rail
- Fuel rail
- Insulator
- Accumulated connector
- Spacer
- Fuel injectors, grommets and O-rings from the fuel rail
- O-rings from the fuel injectors

To install:

4. Install or connect the following:
- New O-rings and grommets lubricated with engine oil on the fuel injectors. Fit the injector squarely into the rail while using a twisting motion. Fit the injector tab into the notch on the rail.
- Insulators and injectors on the intake manifold
- Grommets and the fuel rail onto the injectors. Torque the bolts to 14–18 ft. lbs. (19–25 Nm).
- Fuel lines to the fuel rail
- Fuel injector wiring harness
- Negative battery cable

5. Turn the ignition switch **ON** to pressurize the fuel system.

6. Check for leaks and correct as necessary, before starting the engine.

2.3L (KJ) ENGINE

1. Before servicing the vehicle, refer to the precautions in the beginning of this section.

2. Relieve the fuel system pressure.

3. Remove or disconnect the following:
- Negative battery cable
- Intake manifold cover
- Charge air cooler air duct
- Air cleaner assembly, as necessary, for clearance
- Resonator
- Left and right-hand charge air coolers
- Accelerator cable
- Air intake pipe assembly
- Vacuum hose assembly
- Fuel injector electrical connectors
- Fuel supply/return lines and discard the copper washers
- Fuel rail
- Insulators
- Distribution harness (accumulated connector) from the fuel rails
- Spacer from the top of each fuel injector and discard it
- Fuel injectors from the fuel rails by rotating back and forth
- Fuel pressure regulator, if necessary

To install:

4. Install or connect the following:
- Fuel pressure regulator, if removed. Torque the bolts to 61–86 inch lbs. (7–10 Nm).
- New O-rings lubricated with engine oil on the injectors
- Fuel injectors into the fuel rails
- New spacers on the injectors

- Distribution harness. Torque the screws to 22–31 inch lbs. (2.5–3.5 Nm).
- 6 insulators and the fuel rails. Torque the bolts to 14–18 ft. lbs. (19–25 Nm).
- Fuel supply and return lines using new copper washers
- Fuel injector electrical connectors
- Vacuum hose assembly. Torque the nuts to 70–95 inch lbs. (8–11 Nm).
- Air intake pipe assembly. Torque the nuts to 70–95 inch lbs. (8–11 Nm) and the bolts to 44–78 inch lbs. (5–9 Nm).
- Accelerator cable and adjust it. Torque the bolt to 70–95 inch lbs. (8–11 Nm).
- Both charge air coolers with new O-rings. Torque the bolts to 14–18 ft. lbs. (19–25 Nm).
- Resonator. Torque the nuts to 70–95 inch lbs. (8–11 Nm).
- Air cleaner assembly. Torque the nuts to 70–95 inch lbs. (8–11 Nm).
- Charge air cooler air duct. Torque the nuts to 70–95 inch lbs. (8–11 Nm).
- Negative battery cable

5. Turn the ignition switch **ON** to pressurize the fuel system.

6. Check for fuel leaks and correct as necessary before starting the engine.

DRIVE TRAIN

Manual Transaxle/Transmission Assembly

REMOVAL & INSTALLATION

Miata

1. Before servicing the vehicle, refer to the precautions in the beginning of this section.

2. Refer to the illustration for component location and torque specifications.

3. Drain the transaxle oil.

4. Remove or disconnect the following:
- Crossmember and bracket, on models equipped with 16 inch wheels
- Rear crossbar, on models equipped with 16 inch wheels
- Under cover
- Starter
- Front and middle exhaust pipes
- Shifter knob

- Rear console and insulator
- Shift lever assembly
- Dust boot
- Front crossbar
- Drive shaft
- Clutch release cylinder
- Back up light switch connector
- Neutral safety switch connector
- Speedometer sensor connector
- Wiring harness from the Power Plant Frame (PPF)

5. Support the transmission with a jack.
- PPF bracket
- Differential side bolts and pry out the spacer

✳✳ CAUTION

Removing the PPF spacers will reduce the performance of the PPF. If the spacers are removed, replace the PPF as an assembly.

- Differential mounting spacer
- Transmission side bolts and the PPF

- Transmission bolts and the transmission

To install:

6. Tilt the engine by pushing up on the front oil pan with a wooden block placed on a transmission jack.

7. Support the transmission with a jack.

8. Install or connect the following:
- Transmission into position and tighten the bolts to 48–65 ft. lbs. (64–89 Nm) on models equipped with the M15M–D transmission or 76–91 ft. lbs. (104–123 Nm) on models equipped with the Y16M–D transmission.
- Differential mounting spacer and tighten the bolts to 28–38 ft. lbs. (38–51 Nm).

9. Support the transmission with a jack until it is level.
- PPF in position and if removed install the sleeve
- Spacer and bolts and tighten the reamer bolt making sure the

threading is aligned correctly. The reamer bolt should be installed in the forward hole. Tighten the outer bolts making sure the threads are properly aligned. Tighten all bolts in the sequence illustrated to 91 ft .lbs. (123 (Nm).

• PPF bracket and tighten bolts **A** to 91 ft. lbs. (123 Nm) and bolts **B** to 39 ft. lbs. (53 Nm).

10. Remove the jack and connect the wiring harness.

11. Using a straightedge and Vernier caliper measure the distance **A** which should be 2.37–2.83 inch (60–72mm).

If the distance is not as specified, reposition the PPF to the transmission.

12. Install or connect the following:
• Speedometer sensor connector
• Neutral safety switch connector
• Back up light switch connector
• Clutch release cylinder
• Drive shaft
• Front crossbar
• Dust boot

➡The change control assembly must be filled with 4.9–5.8 cubic inch (80–95cc) of GL–4 or 5 oil if the extension housing was removed or the transmission overhauled.

13. Fill the change control case with the specified amount of the proper oil.

14. Apply grease to the shift lever components as illustrated.

15. Apply sealant to the contact surfaces of the shift lever and change control case.

16. Install the shift lever.

17. Install or connect the following:
• Rear console and insulator
• Shifter knob
• Starter
• Under cover
• Rear crossbar, on models equipped with 16 inch wheels
• Crossmember and bracket, on

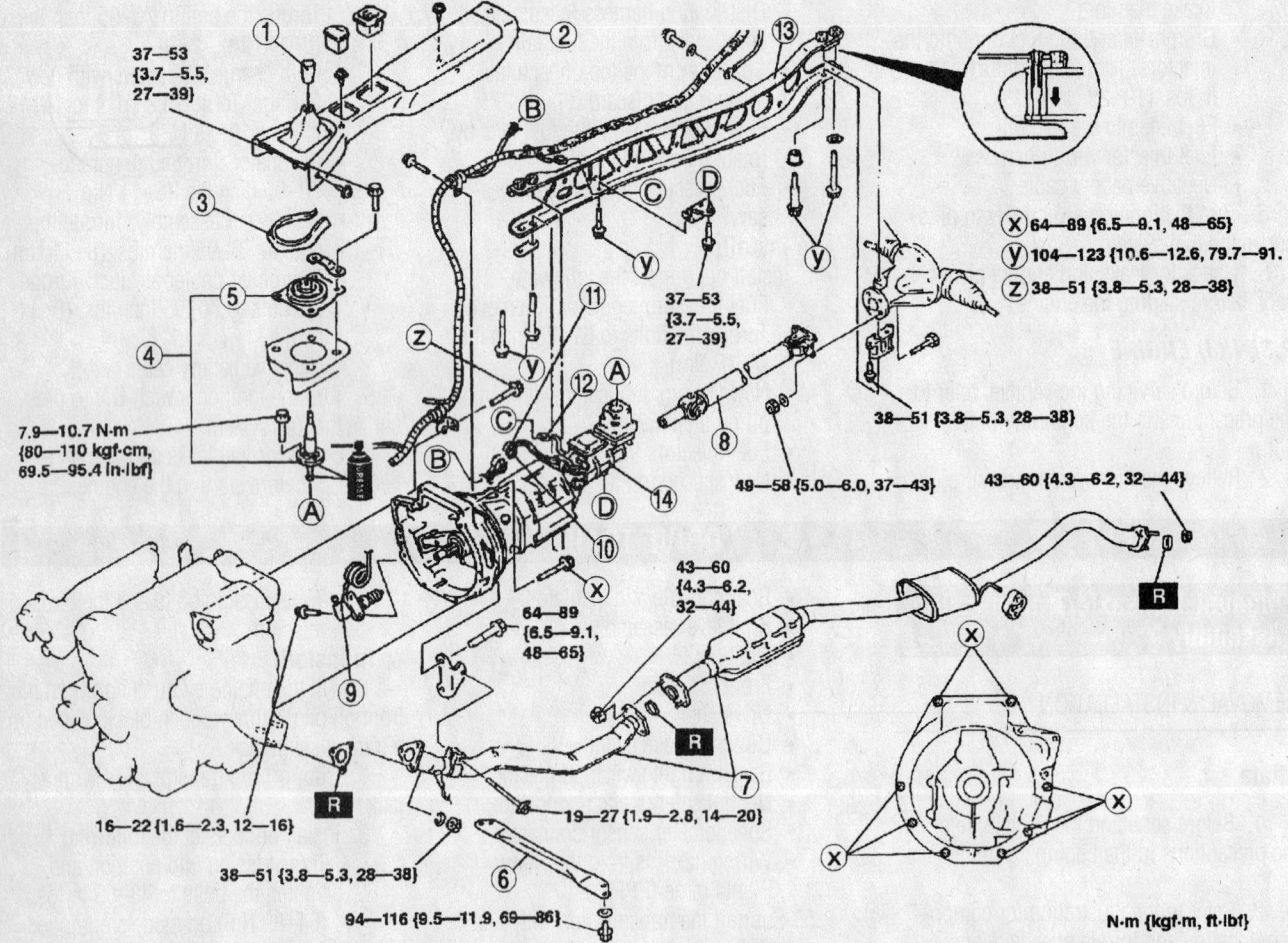

1	Shift lever knob	8	Propeller shaft
2	Rear console	9	Clutch release cylinder
3	Insulation	10	Back-up light switch connector
4	Shift lever component	11	Neutral switch connector
5	Dust boot	12	Speedometer sensor connector
6	Front crossbar	13	Power plant frame (PPF)
7	Front pipe and middle pipe	14	Transmission

Exploded view of the M15M–D manual transmission assembly—Miata

42356-MAZC-GBS

models equipped with 16 inch
wheels
- Front and middle exhaust pipes
18. Fill the transaxle with fluid. Road test
the vehicle and check for leaks. Top off all
fluids as needed.

626

2.0L (FS) ENGINE

1. Before servicing the vehicle, refer to the
precautions in the beginning of this section.
2. Refer to the illustration for component location and torque specifications.

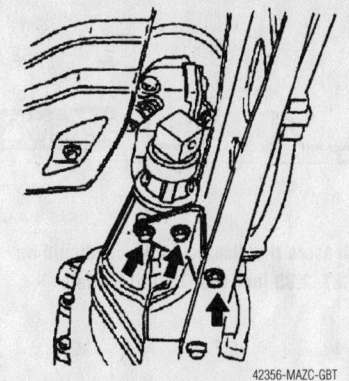

42356-MAZC-GBT

Remove the PPF bracket—Miata

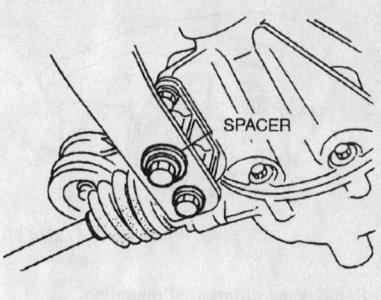

42356-MAZC-GBU

**Remove the differential side bolts and pry
out the spacer—Miata**

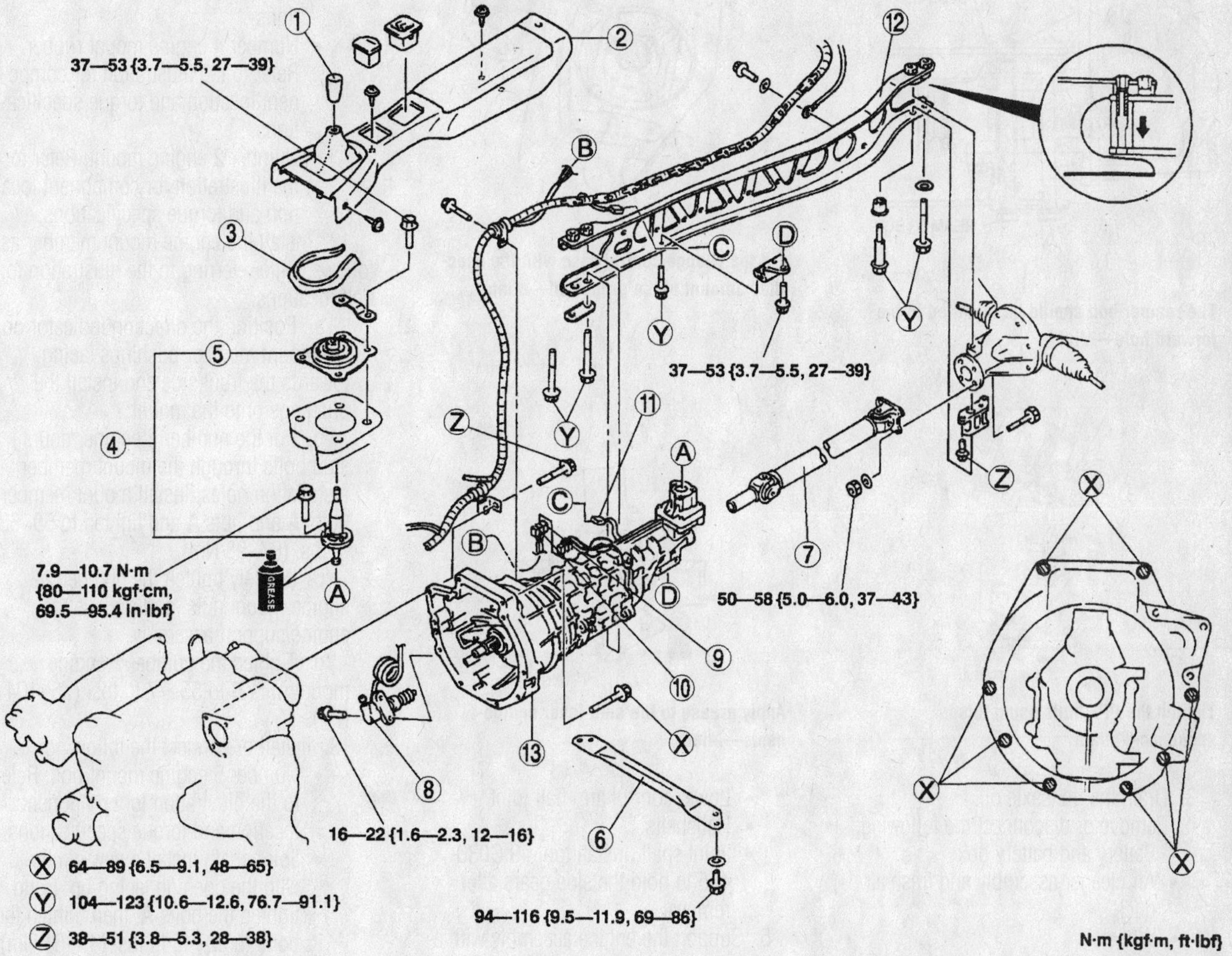

37—53 {3.7—5.5, 27—39}

37—53 {3.7—5.5, 27—39}

50—58 {5.0—6.0, 37—43}

7.9—10.7 N·m
{80—110 kgf·cm,
69.5—95.4 In·lbf}

16—22 {1.6—2.3, 12—16}

Ⓧ 64—89 {6.5—9.1, 48—65}
Ⓨ 104—123 {10.6—12.6, 76.7—91.1}
Ⓩ 38—51 {3.8—5.3, 28—38}

94—116 {9.5—11.9, 69—86}

N·m {kgf·m, ft·lbf}

1 Shift lever knob	8 Clutch release cylinder
2 Rear console	9 Back-up light switch connector
3 Insulator	10 Neutral switch connector
4 Shift lever component	11 Speedometer sensor connector
5 Dust boot	12 Power plant frame (PPF)
6 Front crossbar	13 Transmission
7 Propeller shaft	

Exploded view of the Y16M–D manual transmission assembly—Miata

42356-MAZC-GCD

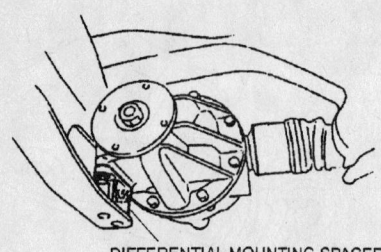

DIFFERENTIAL MOUNTING SPACER
42356-MAZC-GBV

Remove the differential mounting spacer—Miata

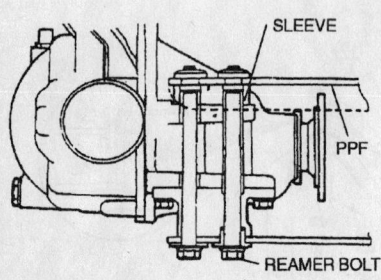

SLEEVE

PPF

REAMER BOLT

42356-MAZC-GBW

The reamer bolt should be installed in the forward hole—Miata

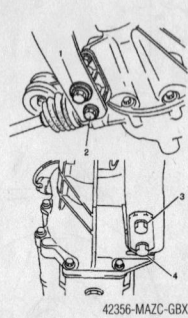

42356-MAZC-GBX

Tighten the PPF bolts using this sequence—Miata

3. Drain the transaxle oil.
4. Remove or disconnect the following:
   - Battery and battery box
   - Air cleaner assembly and fresh air duct
   - Wheels
   - Splash shields
   - Starter
   - Neutral safety switch connector
   - Back up light switch connector
   - Vehicle Speed Sensor (VSS) connector
   - Clutch release cylinder
   - Transverse member
   - Extension bar
   - Change control rod
   - Tie rod ends from the knuckle
   - Stabilizer bar link

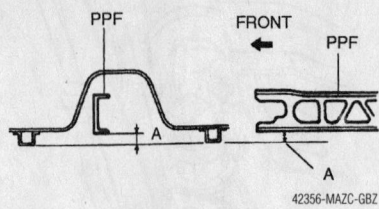

PPF   FRONT ←   PPF

A   A

42356-MAZC-GBZ

Measure the distance A which should be 2.37–2.83 inch (60–72mm)—Miata

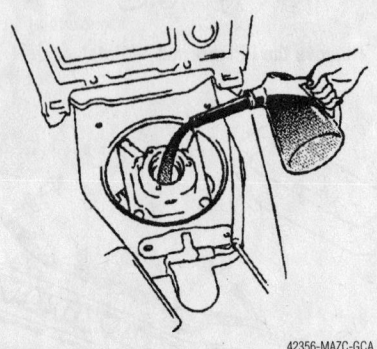

42356-MAZC-GCA

Fill the change control case with the specified amount of the proper oil—Miata

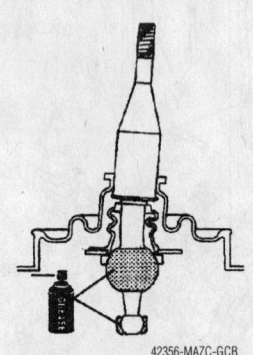

GREASE

42356-MAZC-GCB

Apply grease to the shift lever components—Miata

   - Lower control arm ball joint
   - Halfshafts
   - Joint shaft, install tool 49 G030 455 to hold the side gears after removal

5. Support the engine assembly with a engine support assembly such as 49 E017 5A0.
   - Number 5 engine mount bolt
   - Engine mount member
   - Number 2 engine mount
   - Number 4 engine mount rubber
   - Number 1 engine mount bracket

6. Loosen the engine support assembly and lean the engine towards the transaxle.

7. Support the transaxle with a jack, remove the transaxle bolts and the transaxle.

To install:

8. Place the transaxle onto a jack and raise into position.

9. Install the transaxle bolts and tighten the upper bolts to 66–86 ft. lbs. (90–116 Nm), the lower bolts (except very bottom bolt) to 28–38 ft. lbs. (38–51 Nm) and the very bottom bolt to 14–18 ft. lbs. (19–25 Nm). Refer to the exploded view illustration for bolt locations.

10. Lean the engine towards its normal position and tighten the support assembly.

11. Install or connect the following:
   - Number 1 engine mount bracket. Refer to the illustration for component location and torque specifications.
   - Number 4 engine mount rubber. Refer to the illustration for component location and torque specifications.
   - Number 2 engine mount. Refer to the illustration for component location and torque specifications.

12. Install the engine mount member as follows while referring to the illustration for bolt locations:

   a. Position the direction indicator on the mount member bushings facing towards the front side and install the bushings onto the mount.

   b. Put the number 2 engine mount stud bolts through the mount member installation holes. Install mount member bolts **A** and nuts **A** and tighten to 50–68 ft. lbs. (67–93 Nm).

   c. Loosely tighten the number 2 engine mount nuts and remove the engine support assembly.

   d. Tighten the number 2 engine mount nuts **B** to 55–77 ft. lbs. (75–104 Nm).

13. Install or connect the following:
   - Number 5 engine mount bolt. Refer to the illustration for component location and torque specifications.
   - Joint shaft. Install a new circlip with the opening facing up. Hand tighten the bolts **A**, then tighten the bolts to 32–45 ft. lbs. (43–61 Nm). Refer to the illustration for bolt location.
   - Halfshafts
   - Lower control arm ball joint
   - Stabilizer bar link
   - Tie rod ends to the knuckle and tighten the nut to 24–32 ft. lbs. (32–44 Nm) and install a new cotter pinch
   - Change control rod
   - Extension bar
   - Transverse member

FS ENGINE

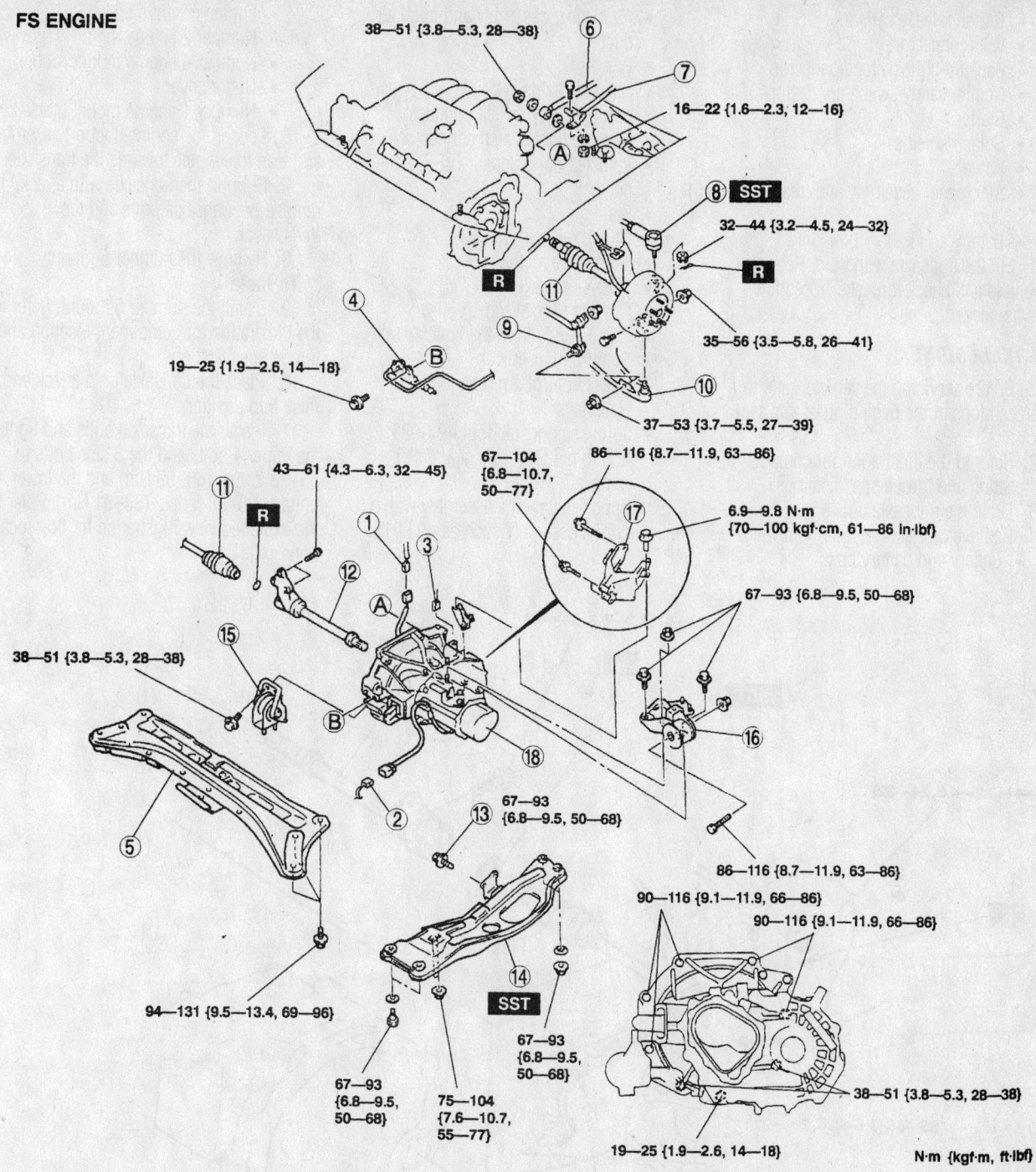

38—51 {3.8—5.3, 28—38}

16—22 {1.6—2.3, 12—16}

⑧ **SST**

32—44 {3.2—4.5, 24—32}

R

35—56 {3.5—5.8, 26—41}

19—25 {1.9—2.6, 14—18}

37—53 {3.7—5.5, 27—39}

43—61 {4.3—6.3, 32—45}

67—104 {6.8—10.7, 50—77}

86—116 {8.7—11.9, 63—86}

6.9—9.8 N·m {70—100 kgf·cm, 61—86 in·lbf}

67—93 {6.8—9.5, 50—68}

38—51 {3.8—5.3, 28—38}

67—93 {6.8—9.5, 50—68}

86—116 {8.7—11.9, 63—86}

90—116 {9.1—11.9, 66—86}

90—116 {9.1—11.9, 66—86}

94—131 {9.5—13.4, 69—96}

67—93 {6.8—9.5, 50—68}

SST

67—93 {6.8—9.5, 50—68}

75—104 {7.6—10.7, 55—77}

38—51 {3.8—5.3, 28—38}

19—25 {1.9—2.6, 14—18}

N·m {kgf·m, ft·lbf}

1	Neutral switch connector	10	Lower arm ball joint
2	Back-up light switch connector	11	Drive shaft
3	Vehicle speedometer sensor connector	12	Joint shaft
4	Clutch release cylinder	13	No.5 engine mount bolt
5	Transverse member	14	Engine mount member
6	Extension bar	15	No.2 engine mount
7	Change control rod	16	No.4 engine mount rubber
8	Tie-rod end ball joint	17	No.1 engine mount bracket
9	Stabilizer control link	18	Transaxle

Exploded view of the manual transaxle assembly mounting—626 with 2.0L (FS) engine

42356-MAZC-G57

- Clutch release cylinder
- VSS connector
- Back up light switch connector
- Neutral safety switch connector
- Starter
- Splash shields
- Wheels
- Air cleaner assembly and fresh air duct
- Battery and battery box

14. Fill the transaxle with fluid. Road test the vehicle and check for leaks. Top off all fluids as needed.

2.5L (KL) ENGINE

1. Before servicing the vehicle, refer to the precautions in the beginning of this section.

2. Refer to the illustration for component location and torque specifications.

3. Drain the transaxle oil.

4. Remove or disconnect the following:
- Battery and battery box
- Air cleaner assembly and fresh air duct
- Wheels
- Splash shields
- Neutral safety switch connector
- Back up light switch connector
- Vehicle Speed Sensor (VSS) connector
- Starter
- Clutch release cylinder
- Transverse member
- Extension bar
- Change control rod
- Tie rod ends from the knuckle
- Stabilizer bar link
- Lower control arm ball joint
- Halfshafts
- Joint shaft, install tool 49 G030 455 to hold the side gears after removal

5. Support the engine assembly with a engine support assembly such as 49 E017 5A0.

- Engine mount member
- Number 2 engine mount
- Number 4 engine mount rubber
- Under cover
- Number 1 engine mount bracket

6. Loosen the engine support assembly and lean the engine towards the transaxle.

7. Support the transaxle with a jack, remove the transaxle bolts and the transaxle.

8. Remove the number 5 mount

To install:

9. Install the number 5 mount. Refer to the illustration for component location and torque specifications.

10. Place the transaxle onto a jack and raise into position.

11. Install the transaxle bolts and tighten the upper 4 bolts and the 3 side bolts to 50–73 ft. lbs. (68–99 Nm) and the lower 4 bolts to 28–38 ft. lbs. (38–51 Nm). Refer to the exploded view illustration for bolt locations.

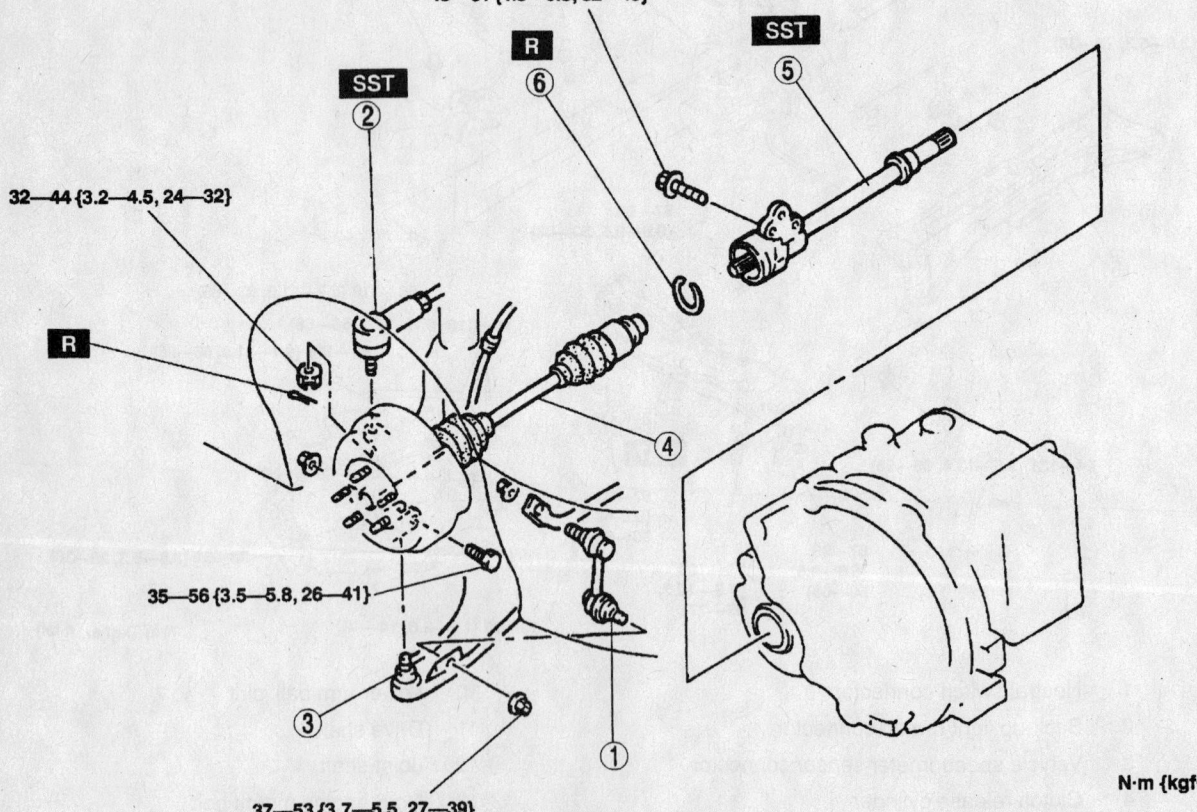

43—61 {4.3—6.3, 32—45}

32—44 {3.2—4.5, 24—32}

35—56 {3.5—5.8, 26—41}

37—53 {3.7—5.5, 27—39}

N·m {kgf·m, ft·lbf}

1	Stabilizer control link	4	Right drive shaft and axle
2	Tie-rod end ball joint	5	Joint shaft
3	Lower arm ball joint	6	Clip

42356-MAZC-G58

Exploded view joint shaft assembly—626

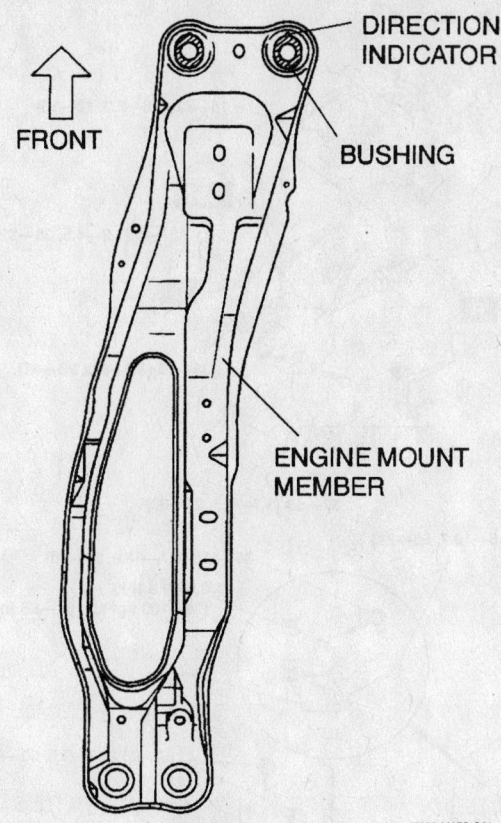

42356-MAZC-G61

Position the direction indicator on the mount member bushings facing towards the front side and install the bushings onto the mount—626

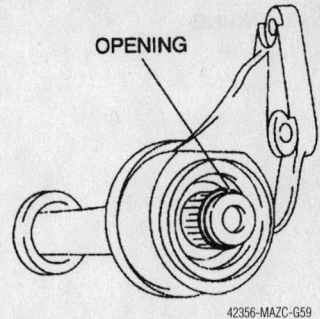

42356-MAZC-G59

When installing the joint shaft make sure to install a new clip with the opening facing up—626

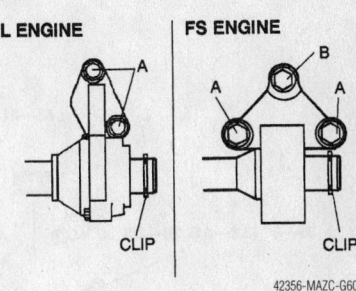

42356-MAZC-G60

Tighten the joint shaft bolts as shown—626

12. Lean the engine towards its normal position and tighten the support assembly.

13. Install or connect the following:
- Number 1 engine mount bracket. Refer to the illustration for component location and torque specifications.
- Under cover
- Number 4 engine mount rubber. Refer to the illustration for component location and torque specifications.
- Number 2 engine mount. Refer to the illustration for component location and torque specifications.

14. Install the engine mount member as follows while referring to the illustration for bolt locations:

a. Position the direction indicator on the mount member bushings facing towards the front side and install the bushings onto the mount.

b. Put the number 2 engine mount stud bolts through the mount member installation holes. Install mount member bolts **A** and nuts **A** and tighten to 50–68 ft. lbs. (67–93 Nm).

c. Loosely tighten the number 2 engine mount nuts and remove the engine support assembly.

d. Tighten the number 2 engine mount nuts **B** to 55–77 ft. lbs. (75–104 Nm).

15. Install or connect the following:
- Number 5 engine mount bolt. Refer to the illustration for component location and torque specifications.
- Joint shaft. Install a new circlip with the opening facing up. Hand tighten the bolts **A**, then tighten the bolts to 32–45 ft. lbs. (43–61 Nm). Refer to the illustration for bolt location.
- Halfshafts
- Lower control arm ball joint
- Stabilizer bar link
- Tie rod ends to the knuckle and tighten the nut to 24–32 ft. lbs. (32–44 Nm) and install a new cotter pinch
- Change control rod
- Extension bar

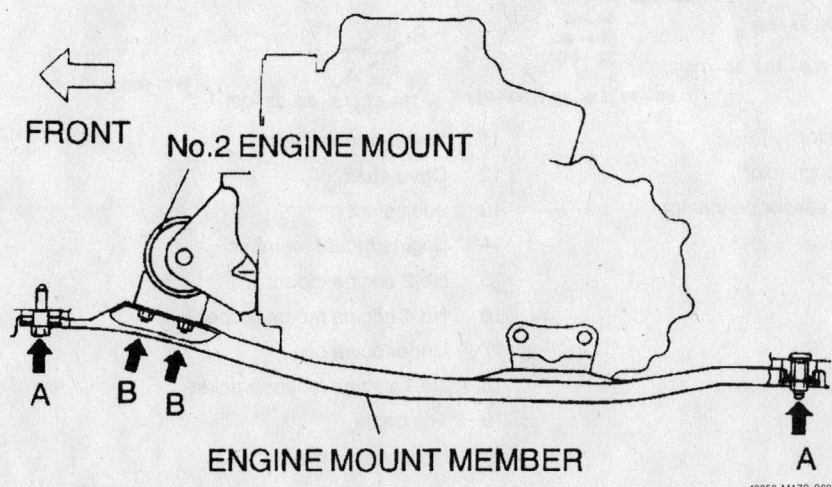

42356-MAZC-G62

Location of the front mount member bolt locations—626

KL ENGINE

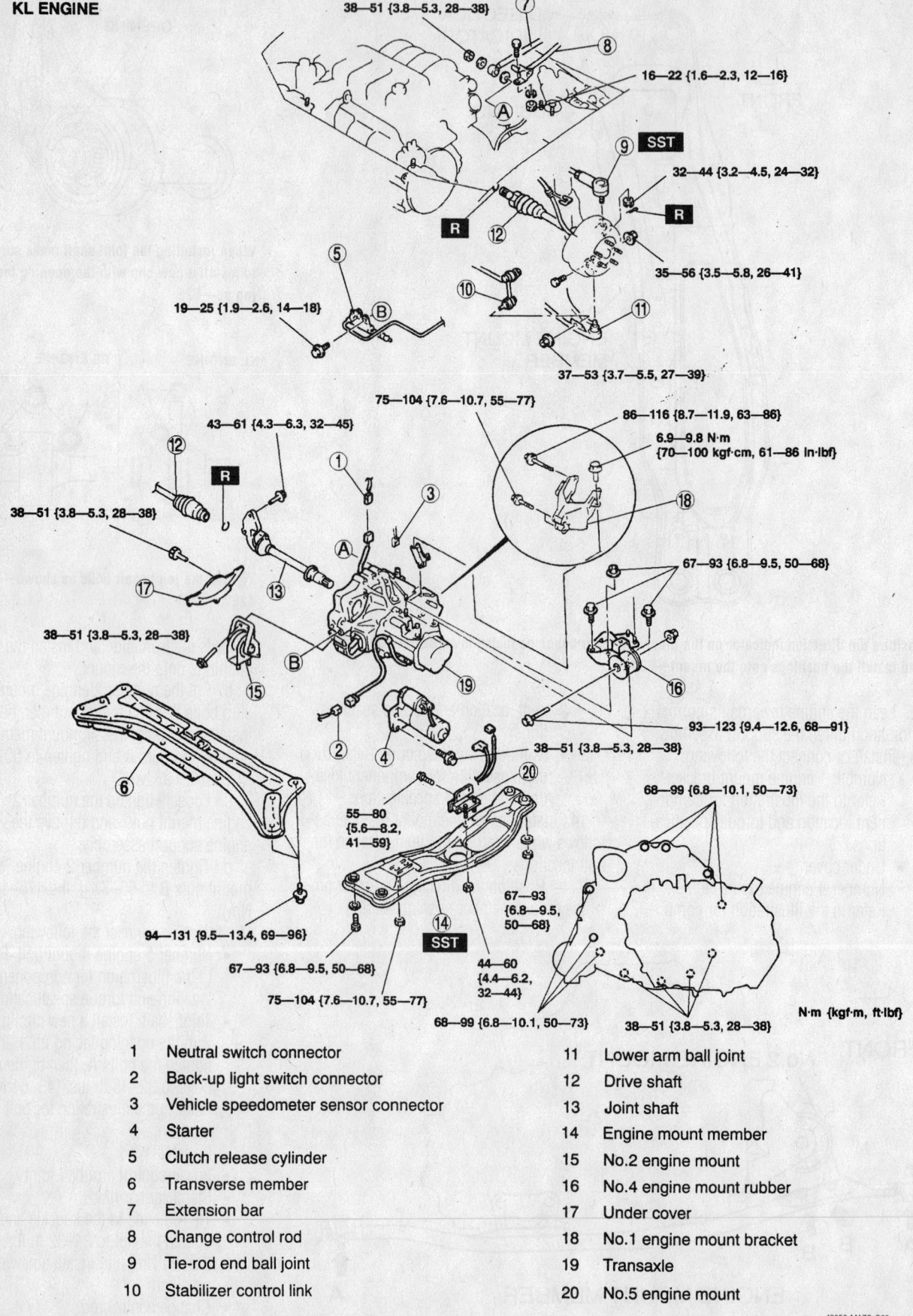

38—51 {3.8—5.3, 28—38}

16—22 {1.6—2.3, 12—16}

SST

32—44 {3.2—4.5, 24—32}

R

35—56 {3.5—5.8, 26—41}

19—25 {1.9—2.6, 14—18}

37—53 {3.7—5.5, 27—39}

75—104 {7.6—10.7, 55—77}

86—116 {8.7—11.9, 63—86}

6.9—9.8 N·m
{70—100 kgf·cm, 61—86 In·lbf}

43—61 {4.3—6.3, 32—45}

38—51 {3.8—5.3, 28—38}

R

67—93 {6.8—9.5, 50—68}

38—51 {3.8—5.3, 28—38}

93—123 {9.4—12.6, 68—91}

38—51 {3.8—5.3, 28—38}

68—99 {6.8—10.1, 50—73}

67—93
{6.8—9.5,
50—68}

55—80
{5.6—8.2,
41—59}

SST

94—131 {9.5—13.4, 69—96}

67—93 {6.8—9.5, 50—68}

44—60
{4.4—6.2,
32—44}

75—104 {7.6—10.7, 55—77}

68—99 {6.8—10.1, 50—73}

38—51 {3.8—5.3, 28—38}

N·m {kgf·m, ft·lbf}

1	Neutral switch connector	11	Lower arm ball joint
2	Back-up light switch connector	12	Drive shaft
3	Vehicle speedometer sensor connector	13	Joint shaft
4	Starter	14	Engine mount member
5	Clutch release cylinder	15	No.2 engine mount
6	Transverse member	16	No.4 engine mount rubber
7	Extension bar	17	Under cover
8	Change control rod	18	No.1 engine mount bracket
9	Tie-rod end ball joint	19	Transaxle
10	Stabilizer control link	20	No.5 engine mount

42356-MAZC-G63

Exploded view of the manual transaxle assembly mounting—626 with 2.5L (KL) engine

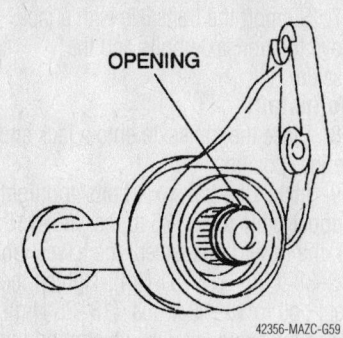

OPENING

42356-MAZC-G59

When installing the joint shaft make sure to install a new clip with the opening facing up—626

- Transverse member
- Clutch release cylinder
- Starter
- VSS connector
- Back up light switch connector
- Neutral safety switch connector
- Starter
- Splash shields
- Wheels
- Air cleaner assembly and fresh air duct
- Battery and battery box

16. Fill the transaxle with fluid. Road test the vehicle and check for leaks. Top off all fluids as needed.

Protégé

G15M–R TRANSAXLE

1. Before servicing the vehicle, refer to the precautions in the beginning of this section.

2. Refer to the illustration for component location and torque specifications.

3. Drain the transaxle oil.

4. Remove or disconnect the following:
- Battery and battery box
- Air cleaner assembly and fresh air duct
- Wheels
- Splash shields

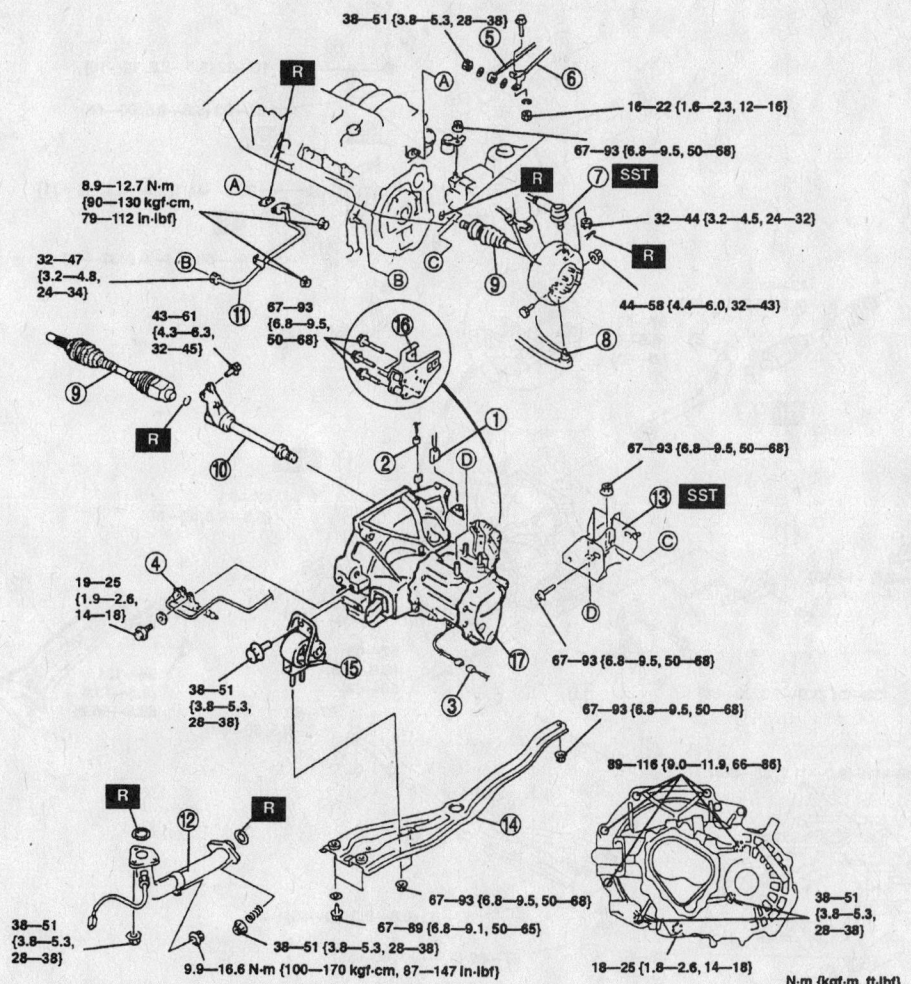

1	Speedometer sensor connector	
2	Neutral switch connector	
3	Back-up light switch connector	
4	Clutch release cylinder	
5	Extension bar	
6	Change control rod	
7	Tie-rod end ball joint	
8	Lower arm ball joint	
9	Drive shaft	
10	Joint shaft	
11	EGR pipe	
12	Front pipe	
13	No.4 engine mount bracket	
14	Engine mount member	
15	No.2 engine mount	
16	No.1	
17	MTX	

42356-MAZC-GQQ

Exploded view of the manual transaxle assembly mounting—2000–01 Protégé with G15M–R transaxle

- Exhaust Gas Recirculation (EGR) pipe
- Front exhaust pipe and three way catalytic converter
- Starter
- Speedometer sensor connector
- Neutral safety switch connector
- Back up light switch connector
- Clutch release cylinder
- Extension bar
- Change control rod
- Tie rod ends from the knuckle
- Lower control arm ball joint

- Halfshafts
- Joint shaft, install tool 49 B027 006 to hold the side gears after removal

5. Support the engine assembly with a engine support assembly such as 49 E017 5A0.

- Number 4 engine mount
- Engine mount member
- Number 2 engine mount
- Number 1 engine mount bracket

6. Loosen the engine support assembly and lean the engine towards the transaxle.

7. Support the transaxle with a jack, remove the transaxle bolts and the transaxle.

To install:

8. Place the transaxle onto a jack and raise into position.

9. Install the transaxle bolts and tighten the upper bolts to 65–86 ft. lbs. (89–116 Nm), the lower bolts except the lowest bolt to 28–38 ft. lbs. (38–51 Nm). Tighten the lowest bolt to 13–18 ft. lbs. (18–25 Nm). Refer to the exploded view illustration for bolt locations.

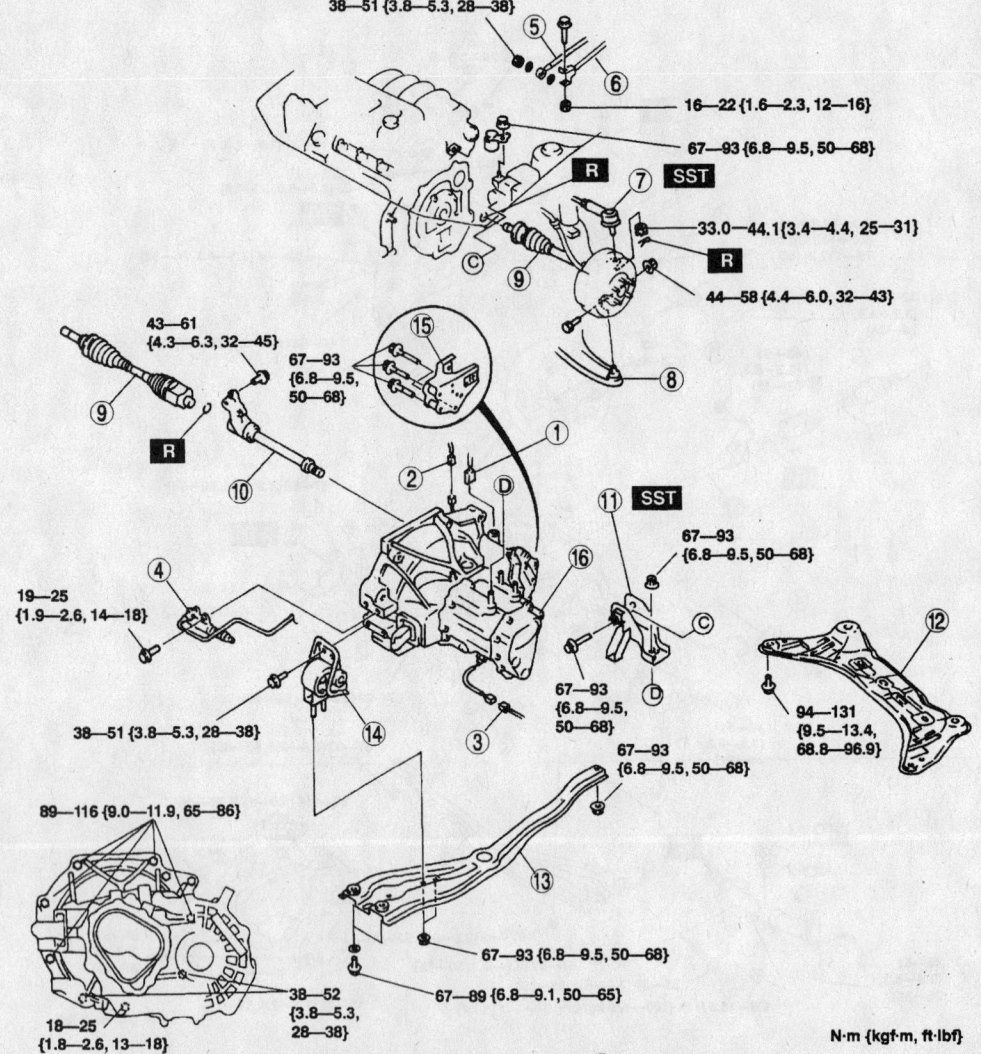

1	Speedometer sensor connector	9	Drive shaft
2	Neutral switch connector	10	Joint shaft
3	Back-up light switch connector	11	No.4 engine mount bracket
4	Clutch release cylinder	12	Transverse member
5	Extension bar	13	Engine mount member
6	Change control rod	14	No.2 engine mount
7	Tie-rod end ball joint	15	No.1 engine mount bracket
8	Lower arm ball joint	16	MTX

42356-MAZC-GIM

Exploded view of the manual transaxle assembly mounting—2002 Protégé with G15M–R transaxle

10. Install or connect the following:
  • Number 1 engine mount bracket. Refer to the illustration for component location and torque specifications.
  • Number 2 engine mount. Refer to the illustration for component location and torque specifications.

11. Install the engine mount member as follows while referring to the illustration for component locations:

  a. .Position the direction indicator on the mount member bushings as illustrated.

  b. Put the number 2 engine mount stud bolts through the mount member installation holes. Install mount member

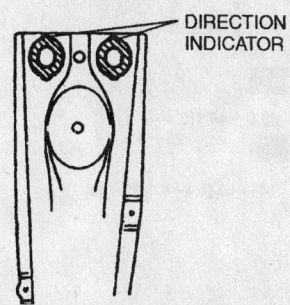

Position the direction indicator on the mount member and bushings—Protégé with both manual and automatic transaxles

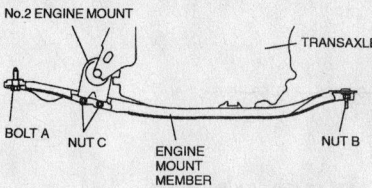

Location of the front mount member bolt locations—Protégé with both manual and automatic transaxles

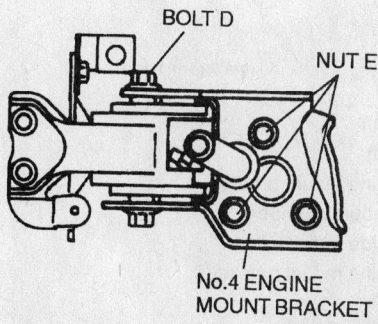

Number 4 mount fastener locations—2000—01 Protégé with G15M–R transaxle

bolts **A** and nuts **B** and tighten to 50–68 ft. lbs. (67–93 Nm).

  c. Tighten bolts **C** to 50–68 ft. lbs. (67–93 Nm).

  d. By aligning the holes on the stud bolts, install the number 4 mount bracket. Align hole of the bracket with the rubber and hand tighten bolt **D**, then nuts **E** and bolt **D** to 50–68 ft. lbs. (67–93 Nm).

12. Install or connect the following:
  • Joint shaft. Install a new circlip with the opening facing up. Hand tighten the bolts **A**, then tighten the bolts to 32–45 ft. lbs. (43–61 Nm). Refer to the illustration for bolt location.

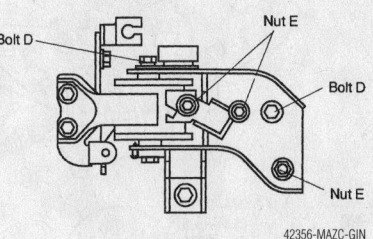

Number 4 mount fastener locations—2002 Protégé with G15M–R transaxle

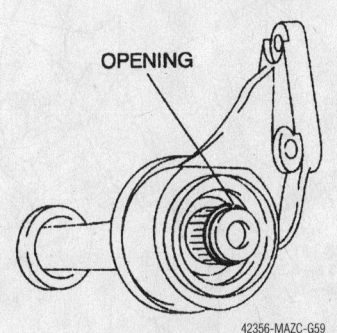

When installing the joint shaft make sure to install a new clip with the opening facing up—Protégé with both manual and automatic transaxles

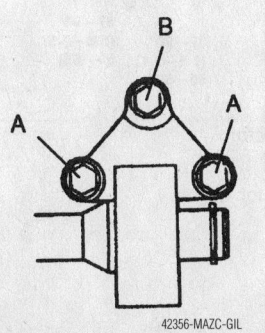

Tighten the joint shaft bolts as shown—Protégé with both manual and automatic transaxles

  • Halfshafts
  • Lower control arm ball joint
  • Tie rod ends to the knuckle
  • Change control rod
  • Extension bar
  • Clutch release cylinder
  • Neutral safety switch connector
  • Back up light switch connector
  • Speedometer connector
  • Starter
  • EGR pipe
  • Splash shields
  • Wheels
  • Air cleaner assembly and fresh air duct
  • Battery and battery box

13. Fill the transaxle with fluid. Road test the vehicle and check for leaks. Top off all fluids as needed.

F25M–R TRANSAXLE

1. Before servicing the vehicle, refer to the precautions in the beginning of this section.

2. Refer to the illustration for component location and torque specifications.

3. Drain the transaxle oil.

4. Remove or disconnect the following:
  • Battery and battery box
  • Air cleaner assembly and fresh air duct
  • Wheels
  • Splash shields
  • Exhaust Gas Recirculation (EGR) pipe
  • Front exhaust pipe and three way catalytic converter
  • Starter
  • Speedometer sensor connector
  • Neutral safety switch connector
  • Back up light switch connector
  • Clutch release cylinder
  • Extension bar
  • Change control rod
  • Tie rod ends from the knuckle
  • Lower control arm ball joint
  • Halfshafts
  • Joint shaft, install tool 49 B027 006 to hold the side gears after removal

5. Support the engine assembly with a engine support assembly such as 49 E017 5A0.
  • Number 4 engine mount
  • Engine mount member
  • Number 2 engine mount
  • Number 1 engine mount bracket

6. Loosen the engine support assembly and lean the engine towards the transaxle.

7. Support the transaxle with a jack, remove the transaxle bolts and the transaxle.

To install:

8. Place the transaxle onto a jack and raise into position.

9. Install the transaxle bolts and tighten the upper bolts to 48–65 ft. lbs. (64–89 Nm), the lower bolts to 28–38 ft. lbs. (38–51 Nm). Refer to the exploded view illustration for bolt locations.

10. Install or connect the following:
- Number 1 engine mount bracket. Refer to the illustration for component location and torque specifications.

- Number 2 engine mount. Refer to the illustration for component location and torque specifications.

11. Install the engine mount member as follows while referring to the illustration for bolt locations:

a. Position the direction indicator on the mount member bushings as illustrated.

b. Put the number 2 engine mount stud bolts through the mount member installation holes. Install mount member

bolts **A** and nuts **B** and tighten to 50–68 ft. lbs. (67–93 Nm).

c. Tighten bolts **C** to 50–68 ft. lbs. (67–93 Nm).

d. By aligning the holes on the stud bolts, install the number 4 mount bracket. Align hole of the bracket with the rubber and hand tighten bolt **D**, then nuts **E** and bolt **D** to 50–68 ft. lbs. (67–93 Nm).

12. Install or connect the following:
- Joint shaft. Install a new circlip with the opening facing up. Hand

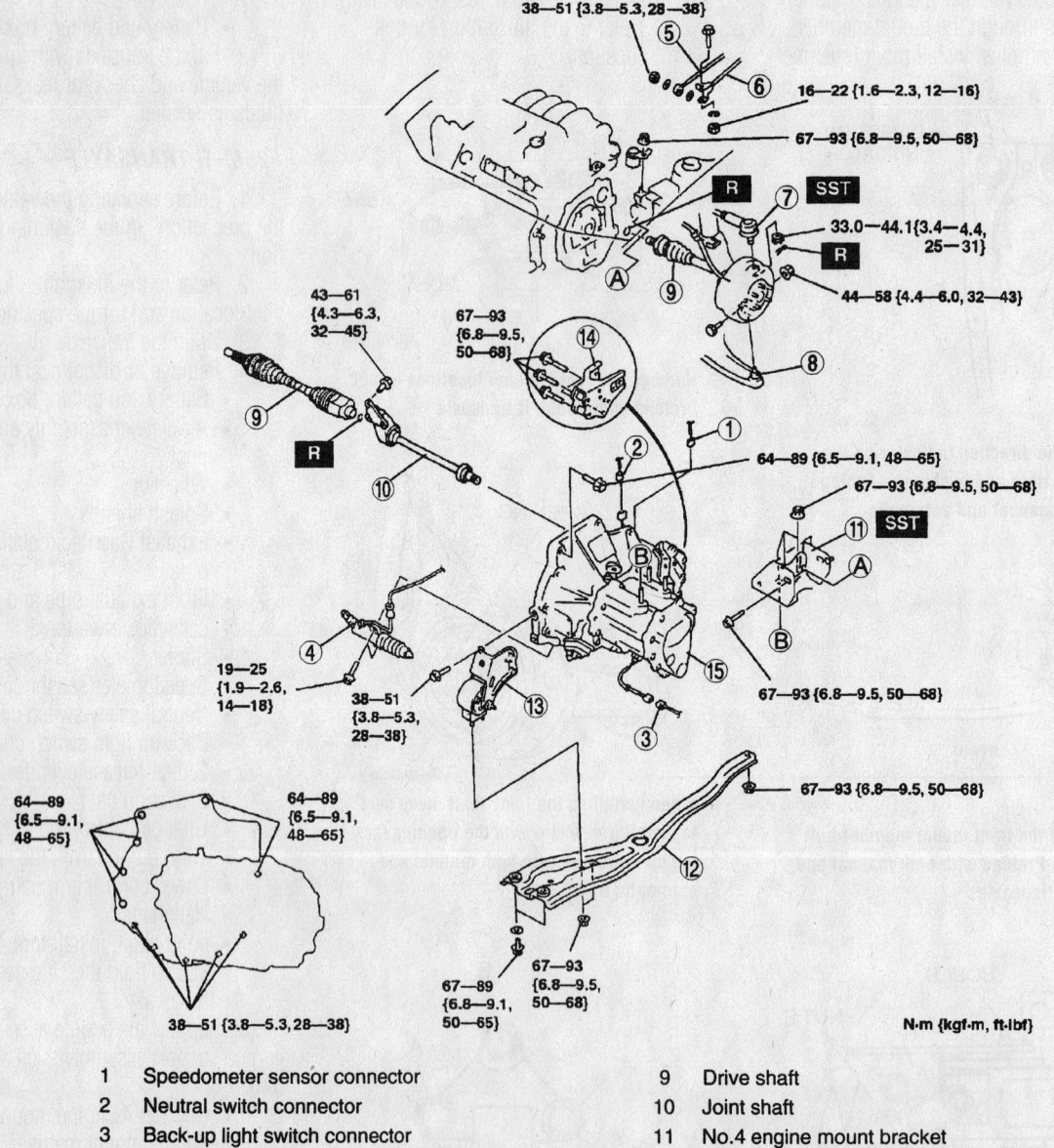

1	Speedometer sensor connector	9	Drive shaft
2	Neutral switch connector	10	Joint shaft
3	Back-up light switch connector	11	No.4 engine mount bracket
4	Clutch release cylinder	12	Engine mount member
5	Extension bar	13	No.2 engine mount
6	Change control rod	14	No.1 engine mount bracket
7	Tie-rod end ball joint	15	MTX
8	Lower arm ball joint		

Exploded view of the manual transaxle assembly mounting—Protégé with F25M–R transaxle

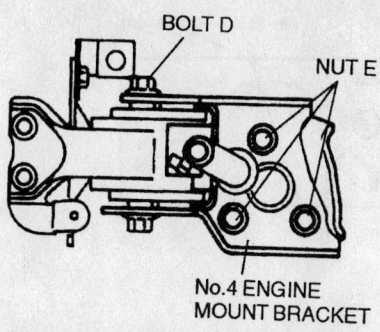

BOLT D

NUT E

No.4 ENGINE
MOUNT BRACKET

42356-MAZC-GIJ

Number 4 mount fastener locations—Protégé with F25M–R transaxle

tighten the bolts **A**, then tighten the bolts to 32–45 ft. lbs. (43–61 Nm). Refer to the illustration for bolt location.
- Halfshafts
- Lower control arm ball joint
- Tie rod ends to the knuckle
- Change control rod
- Extension bar
- Clutch release cylinder
- Neutral safety switch connector
- Back up light switch connector
- Speedometer connector
- Starter
- EGR pipe
- Splash shields
- Wheels
- Air cleaner assembly and fresh air duct
- Battery and battery box

13. Fill the transaxle with fluid. Road test the vehicle and check for leaks. Top off all fluids as needed.

Automatic Transaxle/Transaxle Assembly

REMOVAL & INSTALLATION

Protégé

1. Before servicing the vehicle, refer to the precautions in the beginning of this section.
2. Refer to the illustration for component location and torque specifications.
3. Drain the transaxle oil.
4. Remove or disconnect the following:
- Wheels
- Splash shields
- Battery and battery box
- Air cleaner assembly and fresh air duct
- Exhaust Gas Recirculation (EGR) pipe

- Front exhaust pipe and three way catalytic converter
- Speedometer sensor connector
- Transmission range switch connector
- Input/turbine speed sensor connector
- Transaxle connector
- Harness bracket
- Battery tray bracket
- Oil dipstick tube
- Oil hose
- Brake hose clip and ABS sensor bracket
- Tie rod ends from the knuckle
- Lower arm bolt
- Stabilizer link nut
- Lower arm
- Transverse member
- Halfshafts
- Joint shaft, install tool 49 G030 455 to hold the side gears after removal
- Selector cable
- Intake manifold stay
- Starter
- Torque converter nuts

5. Support the engine assembly with a engine support assembly such as 49 E017 5A0.
- Number 4 engine mount
- Number 1 engine mount bracket
- Roll damper on models with the 2.0L (FS) engine
- Engine mount member
- Number 2 engine mount

6. Loosen the engine support assembly and lean the engine towards the transaxle.
7. Support the transaxle with a jack, remove the transaxle bolts and the transaxle.

To install:

8. Place the transaxle onto a jack and raise into position.
9. Install the transaxle bolts and tighten the bolts as follows:
   a. Bolts **A** to 65 ft. lbs. (89 Nm).
   b. Bolts **B** to 86 ft. lbs. (116 Nm).
   c. Bolts **C** to 38 ft. lbs. (51 Nm).
   d. Bolts **D** to 18 ft. lbs. (25 Nm).
10. Install or connect the following:
- Number 2 engine mount. Refer to the illustration for component location and torque specifications.
11. Install the engine mount member as follows while referring to the illustration for bolt locations:
   a. Position the direction indicator on the mount member bushings as illustrated.
   b. Put the number 2 engine mount

stud bolts through the mount member installation holes. Install mount member bolts **A** and tighten to 50–68 ft. lbs. (67–93 Nm).
   c. Tighten the nuts **B** to 50–68 ft. lbs. (67–93 Nm) on 1.6L (ZM) and 1.8L (FP) engines or 63–86 ft. lbs. (86–116 Nm) on 2.0L (FS) engines.
   d. By aligning the holes on the stud bolts, install the number 4 mount bracket. Align hole of the bracket with the rubber and hand tighten bolt **A**. Tighten nut **B** and tighten bolt **A** to 50–68 ft. lbs. (67–93 Nm).
12. Install or connect the following:
- Roll damper, on 2.0L (FS) engines
- Number 1 engine mount
- Torque converter nuts
- Starter
- Intake manifold stay
- Selector cable
- Joint shaft. Install a new circlip with the opening facing up. Hand tighten the bolts **A**, then tighten the bolts to 32–45 ft. lbs. (43–61 Nm). Refer to the illustration for bolt location.
- Halfshafts
- Transverse member
- Lower arm
- Stabilizer link nut
- Lower arm bolt
- Tie rod ends to the knuckle
- ABS sensor bracket and brake hose clip
- Oil hose
- Oil dipstick tube
- Battery tray bracket
- Harness bracket
- Transaxle connector
- Input/turbine speed sensor connector
- Transmission range switch connector
- Speedometer sensor connector
- Front exhaust pipe and three way catalytic converter
- EGR pipe
- Air cleaner assembly and fresh air duct
- Battery and battery box
- Splash shields
- Wheels

13. Fill the transaxle with fluid. Road test the vehicle and check for leaks. Top off all fluids as needed.

Miata

1. Refer to the illustration for component location and torque specifications.
2. Before servicing the vehicle, refer to

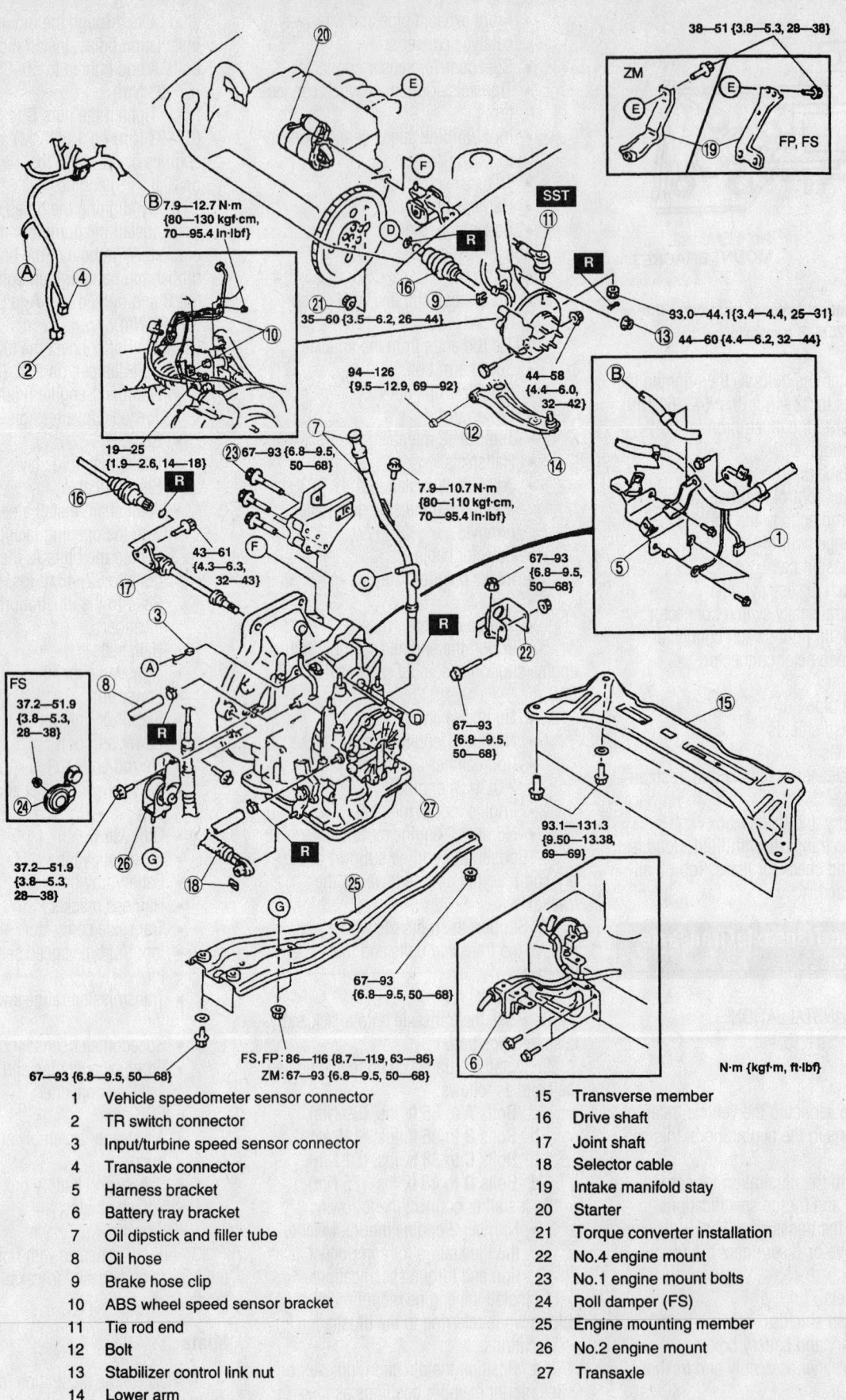

FS, FP : 86—116 {8.7—11.9, 63—86}
ZM : 67—93 {6.8—9.5, 50—68}

N·m {kgf·m, ft·lbf}

1	Vehicle speedometer sensor connector	15	Transverse member
2	TR switch connector	16	Drive shaft
3	Input/turbine speed sensor connector	17	Joint shaft
4	Transaxle connector	18	Selector cable
5	Harness bracket	19	Intake manifold stay
6	Battery tray bracket	20	Starter
7	Oil dipstick and filler tube	21	Torque converter installation
8	Oil hose	22	No.4 engine mount
9	Brake hose clip	23	No.1 engine mount bolts
10	ABS wheel speed sensor bracket	24	Roll damper (FS)
11	Tie rod end	25	Engine mounting member
12	Bolt	26	No.2 engine mount
13	Stabilizer control link nut	27	Transaxle
14	Lower arm		

Exploded view of the automatic transaxle assembly mounting—Protégé

42356-MAZC-GIP

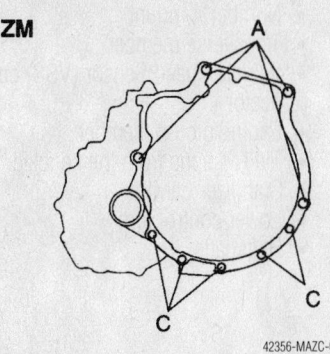

Automatic transaxle assembly bolt torque sequence—1.6L (ZM) engine

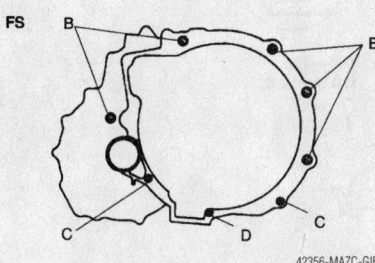

Automatic transaxle assembly bolt torque sequence—1.8L (FP) and 2.0L (FS) engines

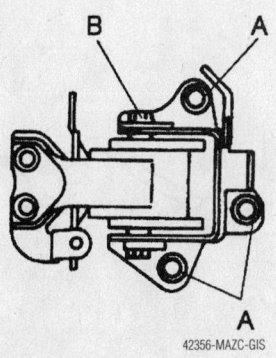

Number 4 mount fastener locations—Protégé with automatic transaxle

the precautions in the beginning of this section.

3. Refer to the illustration for component location and torque specifications.

4. Drain the transaxle oil, coolant and engine oil.

5. Remove or disconnect the following:
- Negative battery cable
- Crossmember and bracket, on models equipped with 16 inch wheels
- Rear crossbar, on models equipped with 16 inch wheels
- Exhaust system
- Drive shaft
- Throttle cable
- Dipstick tube
- Performance rod
- Exhaust bracket
- Shift rod
- Transmission range selector switch connector
- Output speed sensor connector
- Solenoid connector
- Input/turbine speed sensor
- Harness bracket
- Intake manifold bracket
- Oil filter
- Water hose and oil cooler
- Under cover
- Torque converter bolts
- Harness
- Wiring harness from the Power Plant Frame (PPF)

6. Support the transmission with a jack.
- PPF front bolts
- Differential side bolts and pry out the spacer

> ✳✳ **CAUTION**
>
> **Removing the PPF spacers will reduce the performance of the PPF. If the spacers are removed, replace the PPF as an assembly.**

- Differential mounting spacer
- Transmission side bolts and the PPF
- Transmission bolts and the transmission; keep the torque converter end slightly elevated during removal

To install:

7. Support the transmission with a jack.

8. Install or connect the following:
- Transmission into position, make sure the torque converter side is slightly elevated, once aligned, install and tighten the bolts to 48–65 ft. lbs.
- Differential mounting spacer and tighten the bolts to 28–38 ft. lbs. (38–51 Nm).

9. Support the transmission with a jack until it is level.
- PPF in position and if removed install the sleeve
- Spacer and bolts and tighten the reamer bolt making sure the threading is aligned correctly. The reamer bolt should be installed in the forward hole. Tighten the outer bolts making sure the threads are properly aligned. Tighten all bolts in the sequence illustrated to 91 ft. lbs. (123 (Nm).
- PPF front bolts and tighten to 91 ft. lbs. (123 Nm)

10. Remove the jack and connect the wiring harness.

11. Using a straightedge and Vernier caliper measure the distance **A** which should be 1.99–2.46 inch (50.5–62.5mm). If the distance is not as specified, reposition the PPF to the transmission.

12. Install or connect the following:
- Torque converter bolts. Align the holes while turning the converter, use a suitable tool to lock the flywheel and hand tighten the bolts in a criss–cross pattern. Once all the bolts are hand tight, tighten the bolts to 36 ft. lbs. (49 Nm).
- Under cover
- Oil cooler, tighten the oil cooler bolts to 28 ft. lbs. (39 Nm)
- Water hose and
- Oil filter
- Intake manifold bracket
- Harness bracket
- Input/turbine speed sensor
- Solenoid connector
- Output speed sensor connector
- Transmission range selector switch connector
- Shift rod
- Exhaust bracket
- Performance rod
- Dipstick tube
- Throttle cable. Measure the free-play on the cable, if the free-play is not 0.04–0.11 inch (1–3mm), then adjust by turning the lock nuts until the desired specification is reached and tighten the lock nuts to 10 ft. lbs. (14 Nm).
- Drive shaft
- Exhaust system
- Rear crossbar, on models equipped with 16 inch wheels
- Crossmember and bracket, on models equipped with 16 inch wheels
- Negative battery cable

13. Fill the transmission with fluid. Road test the vehicle and check for leaks. Top off all fluids as needed.

626

LA4A–EL TRANSXLE

1. Refer to the illustration for component location and torque specifications.

2. Before servicing the vehicle, refer to the precautions in the beginning of this section.

3. Refer to the illustration for component location and torque specifications.

4. Drain the transaxle oil.

5. Remove or disconnect the following:

- Negative battery cable
- Air cleaner assembly and fresh air duct
- Wheels
- Splash shields
- Ground cable from the transaxle
- Bracket
- Solenoid body connector
- Transaxle range switch connector

- Turbine speed shaft connector
- Oil filler tube
- Fuel filter nut
- Selector cable
- Number 1 engine mount nut, rubber and bracket
6. Support the engine assembly with a engine support assembly such as 49 E017 5A0.

- Number 4 mount
- Transverse member
- Vehicle Speed Sensor (VSS) connector
- Engine mount member
- Tie rod ends from the knuckle
- Stabilizer bar link
- Lower control arm
- Halfshafts

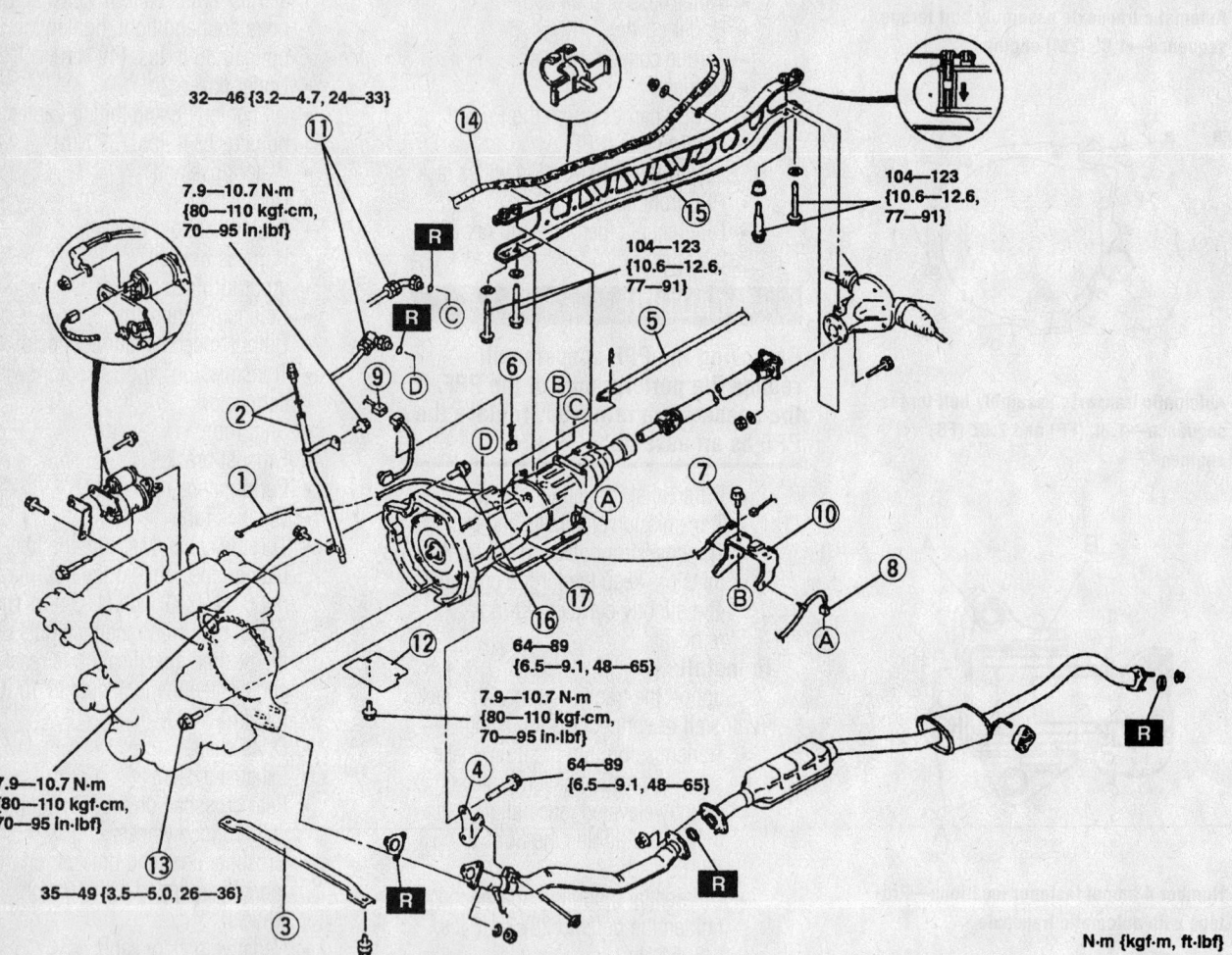

32—46 {3.2—4.7, 24—33}

7.9—10.7 N·m {80—110 kgf·cm, 70—95 in·lbf}

104—123 {10.6—12.6, 77—91}

104—123 {10.6—12.6, 77—91}

64—89 {6.5—9.1, 48—65}

7.9—10.7 N·m {80—110 kgf·cm, 70—95 in·lbf}

64—89 {6.5—9.1, 48—65}

7.9—10.7 N·m {80—110 kgf·cm, 70—95 in·lbf}

35—49 {3.5—5.0, 26—36}

N·m {kgf·m, ft·lbf}

1	Throttle cable	10	Harness bracket
2	Filler tube, dipstick	11	Oil pipe
3	Performance rod	12	Undercover
4	Exhaust bracket	13	Torque converter bolts
5	Shift rod	14	Harness
6	TR switch connector	15	Power plant frame
7	Output speed sensor connector	16	Transmission mount bolts
8	Solenoid connector	17	Transmission
9	Input/turbine speed sensor		

42356-MAZC-GXX

Exploded view of the automatic transmission assembly—Miata

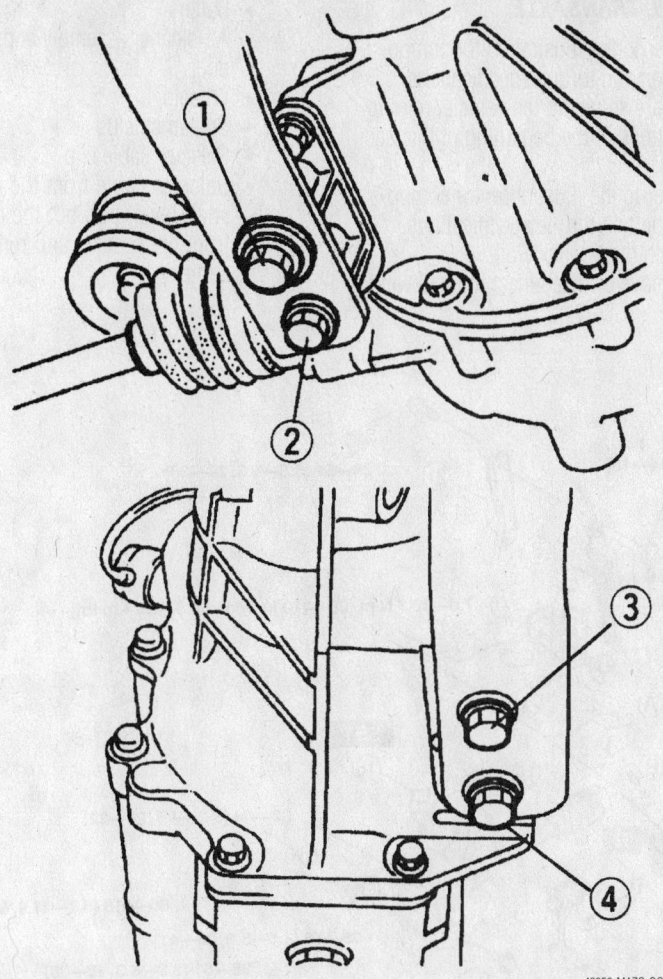

42356-MAZC-GCE

Tighten the differential bolts in the sequence illustrated—Miata with an automatic transmission

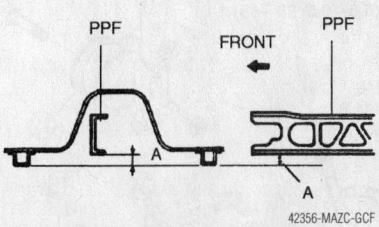

42356-MAZC-GCF

Measure the distance A which should be 1.99–2.46 inch (50.5–62.5mm)—Miata with an automatic transmission

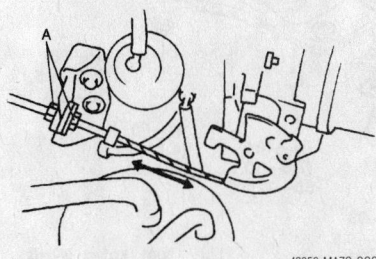

42356-MAZC-GCG

Adjust the free-play on the throttle and adjust using the lock nuts A—Miata with an automatic transmission

- Joint shaft, install tool 49 G030 455 to hold the side gears after removal
- Intake manifold bracket
- Starter
- Torque converter nuts
- Number 5 and 2 engine mounts
- Oil hose

7. Loosen the engine support assembly and lean the engine towards the transaxle.

8. Support the transaxle with a jack, remove the transaxle bolts and the transaxle.

To install:

9. Place the transaxle onto a jack and raise into position.

10. Install the transaxle bolts and tighten the bolts **A** to 66–86 ft. lbs. (90–116 Nm), bolts **B** to28–38 ft. lbs. (38–51 Nm) and bolts **C** to 14–18 ft. lbs. (19–25 Nm). Refer to the illustration for bolt locations.

- Hand tighten the number 4 mount retainers

11. Lean the engine towards its normal position and tighten the support assembly.

12. Install or connect the following:
- Oil hose
- Number 5 and 2 engine mounts. Refer to the illustration for component location and torque specifications.
- Torque converter nuts and tighten to 21–33 ft. lbs. (28–46 Nm)
- Starter
- Intake manifold bracket. Refer to the illustration for component location and torque specifications.
- Joint shaft. Install a new circlip with the opening facing up. Hand tighten the bolts **A**, then tighten the bolts to 32–45 ft. lbs. (43–61 Nm). Refer to the illustration for bolt location.
- Halfshafts
- Lower control arm
- Stabilizer bar link. Refer to the illustration for component location and torque specifications.
- Tie rod ends to the knuckle. Refer to the illustration for component location and torque specifications.

13. Install the engine mount member as follows while referring to the illustration for bolt locations:

a. Position the direction indicator on the mount member bushings facing towards the front side and install the bushings onto the mount.

b. Put the number 2 engine mount stud bolts through the mount member installation holes. Install mount member bolts **A** and nuts **A** and tighten to 50–68 ft. lbs. (67–93 Nm).

c. Loosely tighten the number 2 engine mount nuts and remove the engine support assembly.

d. Tighten the number 2 engine mount nuts **B** to 55–77 ft. lbs. (75–104 Nm).

14. Install or connect the following:
- VSS connector
- Transverse member. Refer to the illustration for component location and torque specifications.
- Number 4 mount. Refer to the illustration for component location and torque specifications.
- Number 1 engine mount nut, rubber and bracket. Refer to the illustration for component location and torque specifications.
- Selector cable
- Fuel filter nut
- Oil filler tube
- Turbine speed shaft connector
- Transaxle range switch connector
- Solenoid body connector

- Bracket
- Ground cable from the transaxle
- Splash shields
- Wheels
- Air cleaner assembly and fresh air duct
- Negative battery cable

15. Fill the transmission with fluid.
16. Start the engine and check for leaks and proper operation.

GFA4A–EL TRANSAXLE

1. Refer to the illustration for component location and torque specifications.
2. Before servicing the vehicle, refer to the precautions in the beginning of this section.
3. Refer to the illustration for component location and torque specifications.
4. Drain the transaxle oil.
5. Remove or disconnect the following:

- Battery
- Air cleaner assembly and fresh air duct
- Wheels
- Splash shields
- Selector cable clip
- Selector cable from the manual shaft lever, pull out the cable from the bracket and remove the cable

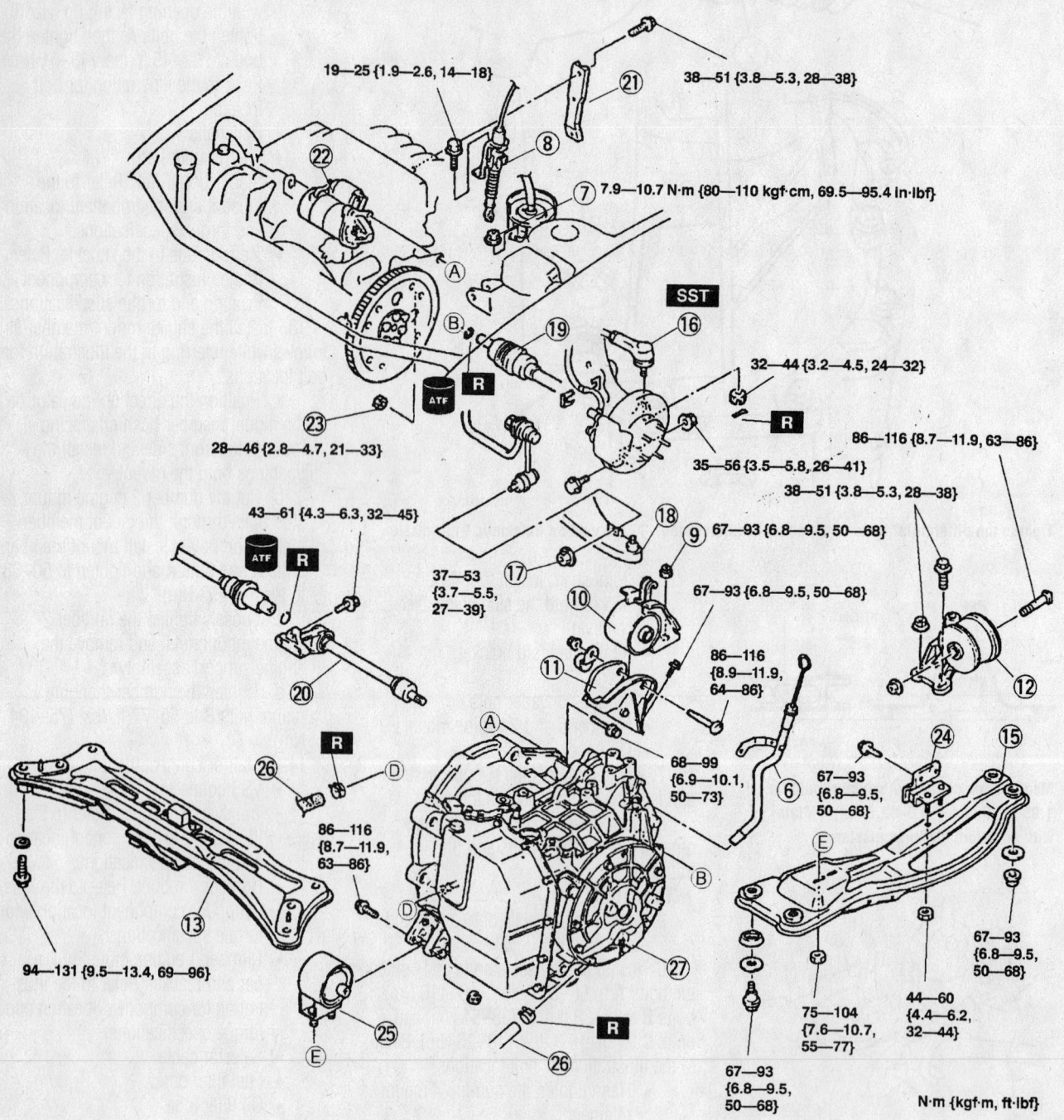

19—25 {1.9—2.6, 14—18}

38—51 {3.8—5.3, 28—38}

7.9—10.7 N·m {80—110 kgf·cm, 69.5—95.4 in·lbf}

32—44 {3.2—4.5, 24—32}

86—116 {8.7—11.9, 63—86}

35—56 {3.5—5.8, 26—41}

38—51 {3.8—5.3, 28—38}

67—93 {6.8—9.5, 50—68}

67—93 {6.8—9.5, 50—68}

28—46 {2.8—4.7, 21—33}

43—61 {4.3—6.3, 32—45}

37—53 {3.7—5.5, 27—39}

86—116 {8.9—11.9, 64—86}

68—99 {6.9—10.1, 50—73}

67—93 {6.8—9.5, 50—68}

86—116 {8.7—11.9, 63—86}

67—93 {6.8—9.5, 50—68}

94—131 {9.5—13.4, 69—96}

75—104 {7.6—10.7, 55—77}

67—93 {6.8—9.5, 50—68}

44—60 {4.4—6.2, 32—44}

N·m {kgf·m, ft·lbf}

42356-MAZC-G73

Exploded view of the LA4A–EL automatic transaxle assembly (1 OF 2)—626

- Transaxle range switch connector
- Solenoid valve connector
- Oxygen (O₂S) connector
- Vehicle Speed Sensor (VSS) connector
- Input/Turbine speed shaft connector
- Starter harness
- Engine mount stay
- Fuel filter bolts

- Oil hose and breather hose
- Lower control arm
- Tie rod ends from the knuckle
- Stabilizer bar link
- Transverse member
- Halfshafts
- Joint shaft, install tool 49 G030 455 to hold the side gears after removal
- Number 5 mount rubber

- Number 1 engine mount nut, rubber and bracket
6. Support the engine assembly with a engine support assembly such as 49 E017 5A0.

- Engine mount member
- Number 2 engine mount
- Under cover
- Torque converter nuts
- Number 4 engine mount

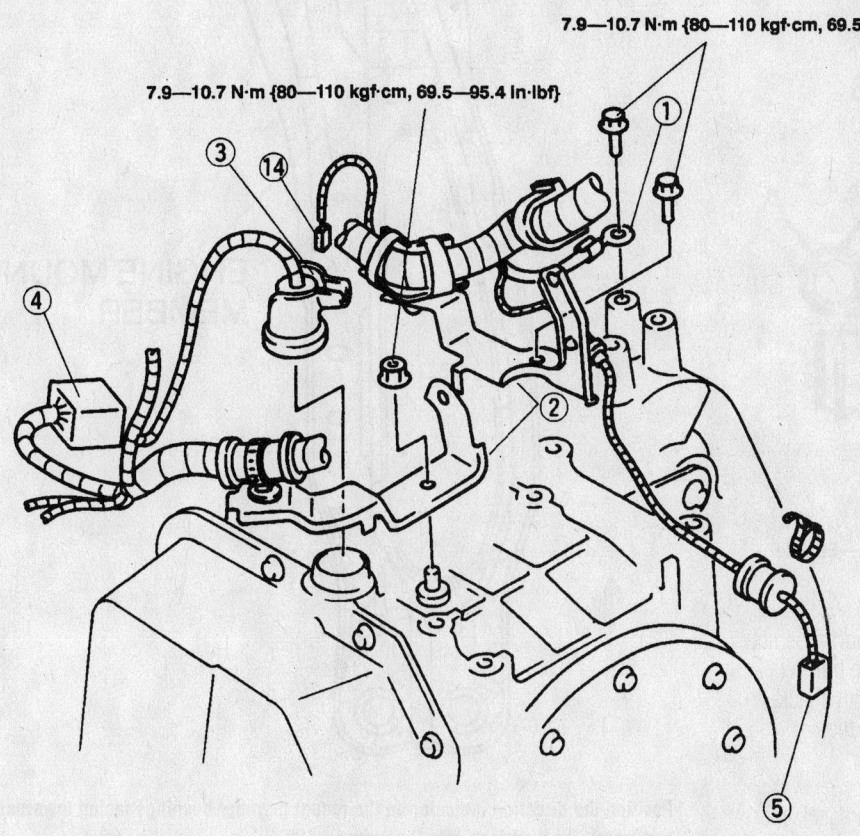

7.9—10.7 N·m {80—110 kgf·cm, 69.5—95.4 in·lbf}

7.9—10.7 N·m {80—110 kgf·cm, 69.5—95.4 in·lbf}

1	Ground	15	Engine mount member
2	Bracket	16	Tie-rod end
3	Solenoid body connector	17	Stabilizer control link
4	Transaxle range switch connector	18	Lower arm
5	Turbine shaft speed sensor connector	19	Drive shaft
6	Oil filler tube	20	Joint shaft
7	Fuel filter mounting nut	21	Intake manifold bracket
8	Selector cable	22	Starter
9	No.1 engine mount nut	23	Torque converter nut
10	No.1 engine mount rubber	24	No.5 engine mount
11	No.1 engine mount bracket	25	No.2 engine mount
12	No.4 engine mount	26	Oil hose
13	Transverse member	27	Transaxle
14	Vehicle speedometer sensor connector		

42356-MAZC-G74

Exploded view of the LA4A–EL automatic transaxle assembly (2 OF 2)—626

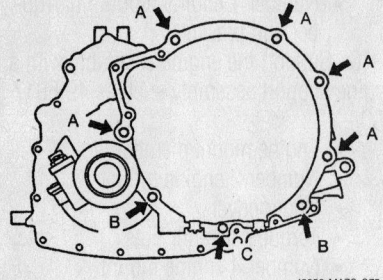

42356-MAZC-G75

LA4A–EL automatic transaxle assembly bolt locations—626

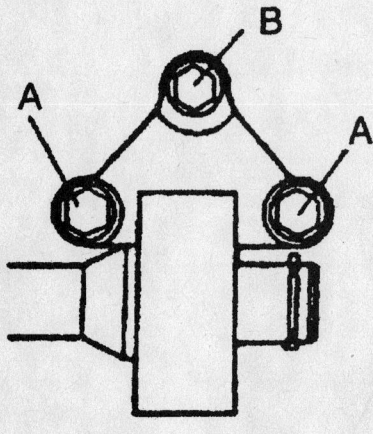

42356-MAZC-G76

Bolt locations on the joint shaft—EL automatic transaxle assembly

7. Loosen the engine support assembly and lean the engine towards the transaxle.

8. Support the transaxle with a jack, remove the transaxle bolts and the transaxle.

To install:

9. Place the transaxle onto a jack and raise into position.

10. Install the transaxle bolts and tighten the bolts to 50–73 ft. lbs. (68–99 Nm).

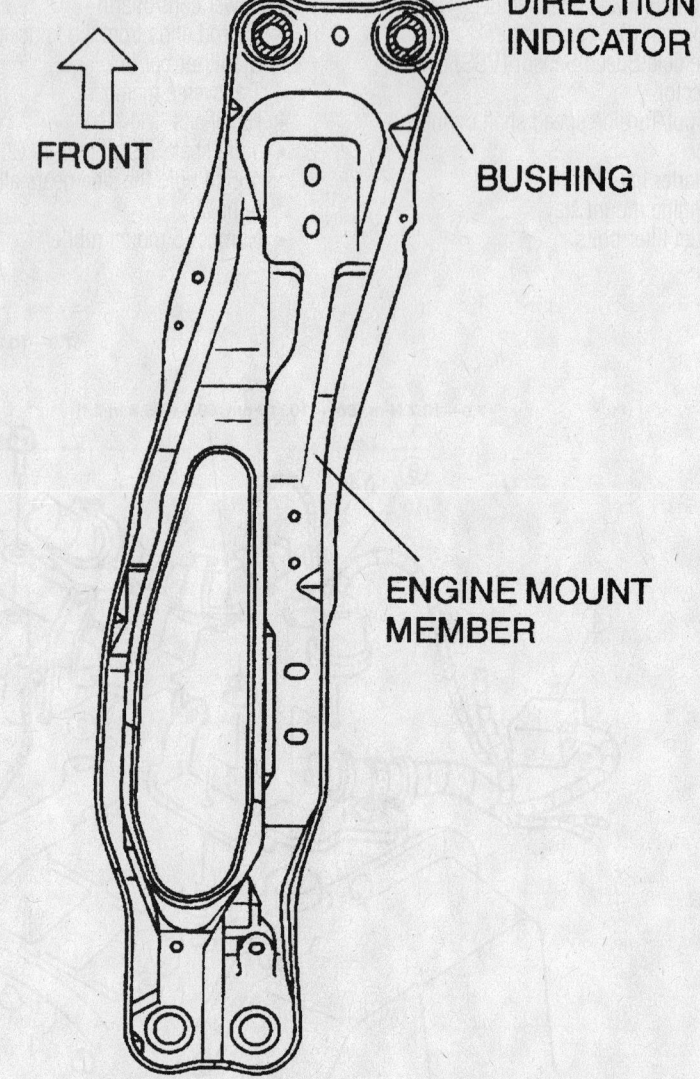

42356-MAZC-G61

Position the direction indicator on the mount member bushings facing towards the front side and install the bushings onto the mount—626

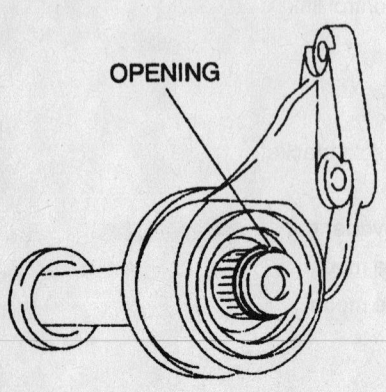

42356-MAZC-G59

When installing the joint shaft make sure to install a new clip with the opening facing up—626

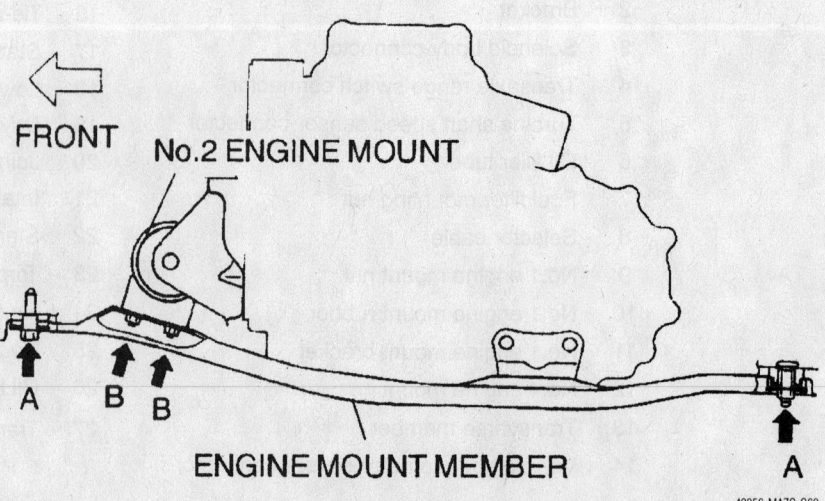

42356-MAZC-G62

Location of the front mount member bolt locations—626

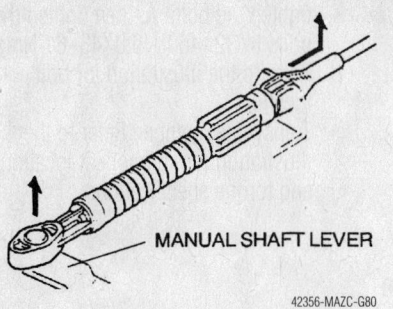

MANUAL SHAFT LEVER

42356-MAZC-G80

Remove the selector cable GF4A–EL automatic transaxle assembly (1 OF 2)—626

• Hand tighten the number 4 mount retainers, then tighten to 55–77 ft. lbs. (75–104 Nm)

11. Lean the engine towards its normal position and tighten the support assembly.

12. Install or connect the following:
• Torque converter nuts
• Under cover
• Number 2 engine mount

13. Install the engine mount member as follows while referring to the illustration for bolt locations:

a. Position the direction indicator on the mount member bushings facing towards the front side and install the bushings onto the mount.

b. Put the number 2 engine mount stud bolts through the mount member installation holes. Install mount member bolts **A** and nuts **A** and tighten to 50–68 ft. lbs. (67–93 Nm).

c. Loosely tighten the number 2 engine mount nuts and remove the engine support assembly.

d. Tighten the number 2 engine

6.9—9.80 N·m {70—100 kgf·cm, 60.7—86.7 in·lbf}

7.9—10.7 N·m {80—110 kgf·cm, 69.5—95.4 in·lbf}

7.9—10.7 N·m {80—110 kgf·cm, 69.5—95.4 in·lbf}

32—44 {3.2—4.5, 24—32}

35—56 {3.5—5.8, 26—41}

38—60 {3.8—6.2, 28—45}

37—53 {3.7—5.5, 27—39}

32—46 {3.2—4.7, 24—33}

43—61 {4.3—6.3, 32—45}

75—104 {7.6—10.7, 55—77}

67—93 {6.8—9.5, 50—68}

93—123 {9.4—12.6, 68—91}

55—80 {5.6—8.2, 41—59}

68—99 {6.9—10.1, 50—73}

94—131 {9.5—13.4, 69—96}

67—93 {6.8—9.5, 50—68}

44—60 {4.4—6.2, 32—44}

75—104 {7.6—10.7, 55—77}

38—51 {3.8—5.3, 28—38}

67—93 {6.8—9.5, 50—68}

N·m {kgf·m, ft·lbf}

42356-MAZC-G77

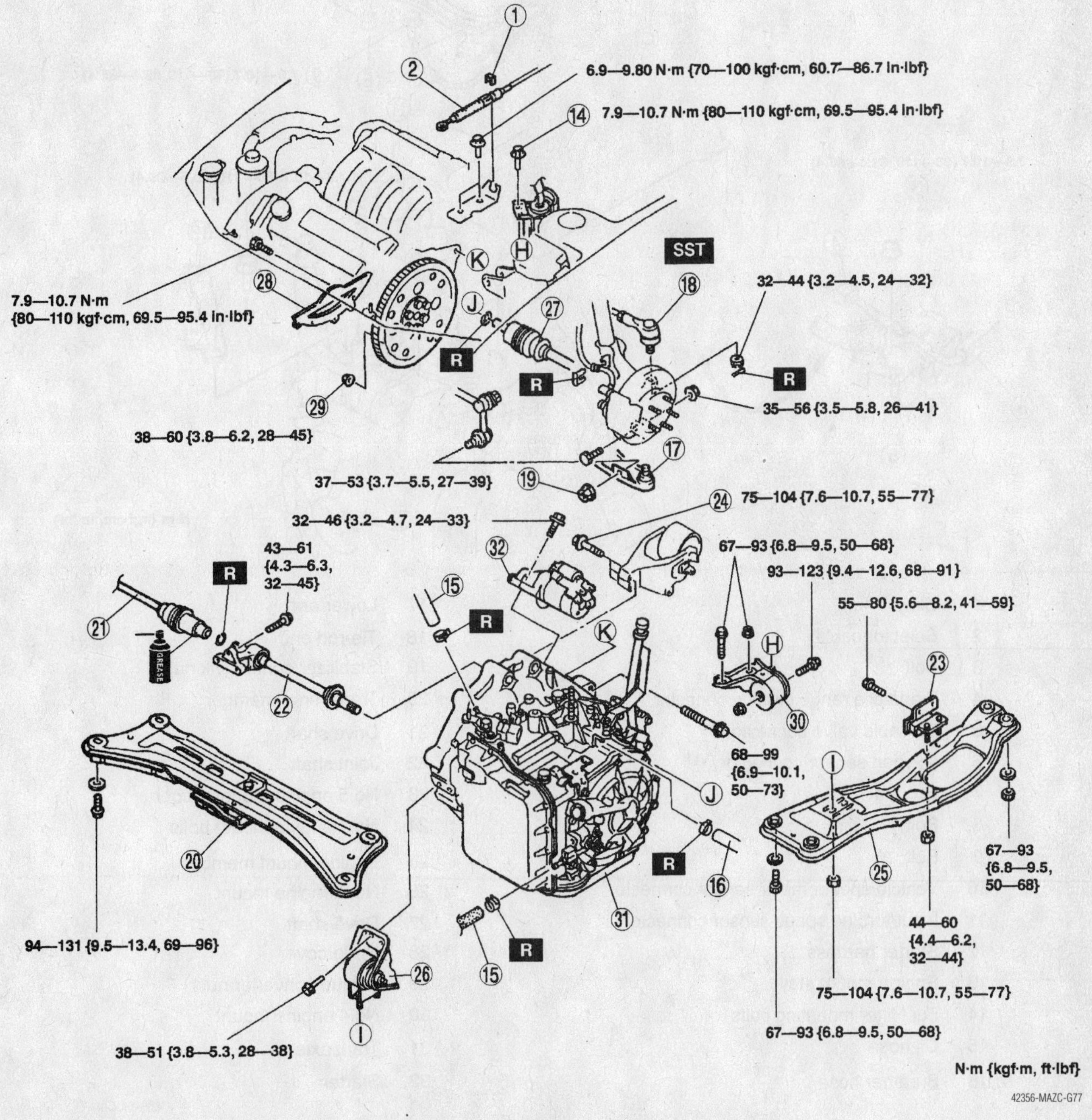

Exploded view of the GF4A–EL automatic transaxle assembly (1 OF 2)—626

mount nuts **B** to 55–77 ft. lbs. (75–104 Nm).

14. Install or connect the following:
- Number 1 engine mount nut, rubber and bracket. Refer to the illustration for component location and torque specifications.

- Number 5 mount rubber. Refer to the illustration for component location and torque specifications.
- Halfshafts
- Joint shaft. Install a new circlip with the opening facing up. Hand

tighten the bolts **A**, then tighten the bolts to 32–45 ft. lbs. (43–61 Nm). Refer to the illustration for bolt location.
- Transverse member. Refer to the illustration for component location and torque specifications.

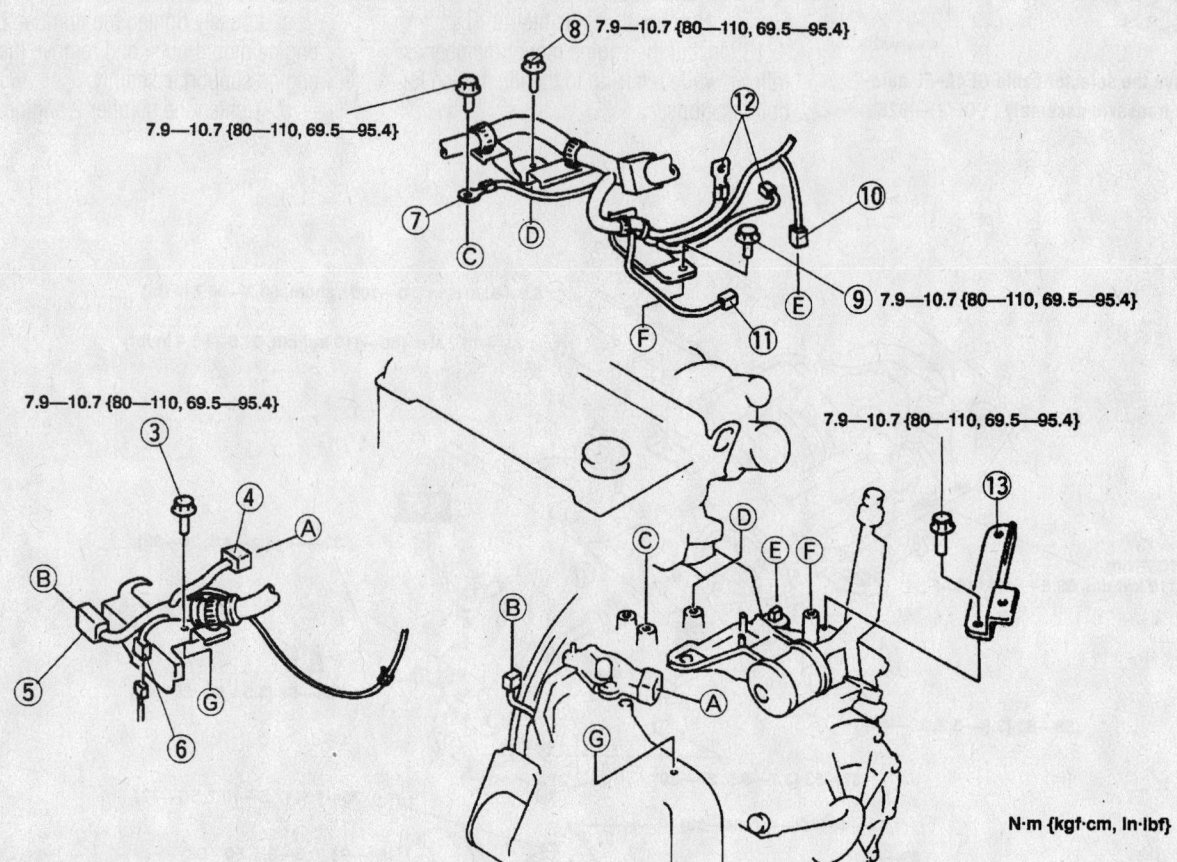

1	Clip	17	Lower arm
2	Selector cable	18	Tie-rod end
3	Bolt	19	Stabilizer control link nut
4	Transaxle range switch connector	20	Transverse member
5	Solenoid valve connector	21	Drive shaft
6	Oxygen sensor connector	22	Joint shaft
7	Ground	23	No.5 engine mount rubber
8	Bolt	24	No.1 engine mount bolts
9	Bolt	25	Engine mount member
10	Vehicle speedometer sensor connector	26	No.2 engine mount
11	Input/turbine speed sensor connector	27	Drive shaft
12	Starter harness	28	Undercover
13	Engine mount stay	29	Torque converter nuts
14	Fuel filter mounting bolts	30	No.4 engine mount
15	Oil hose	31	Transaxle
16	Breather hose	32	Starter

42356-MAZC-G78

Exploded view of the GF4A–EL automatic transaxle assembly (2 OF 2)—626

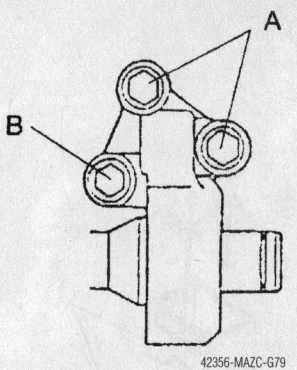

42356-MAZC-G79

Bolt locations on the joint shaft—GF4A–EL automatic transaxle assembly

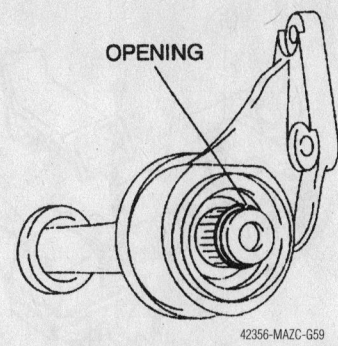

42356-MAZC-G59

When installing the joint shaft make sure to install a new clip with the opening facing up—626

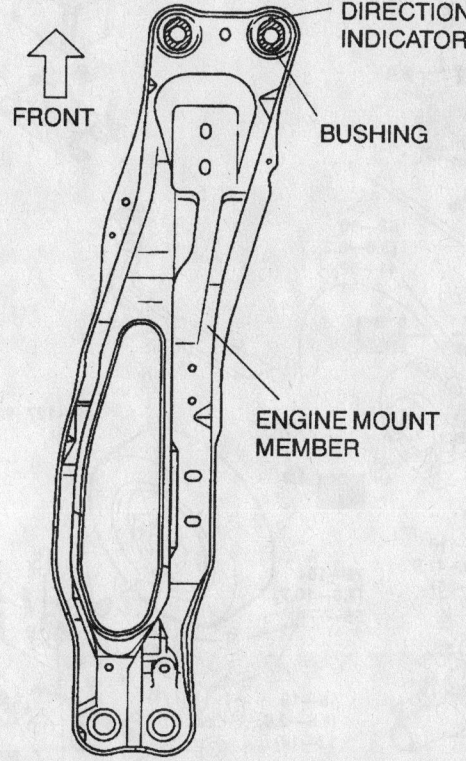

42356-MAZC-G61

Position the direction indicator on the mount member bushings facing towards the front side and install the bushings onto the mount—626

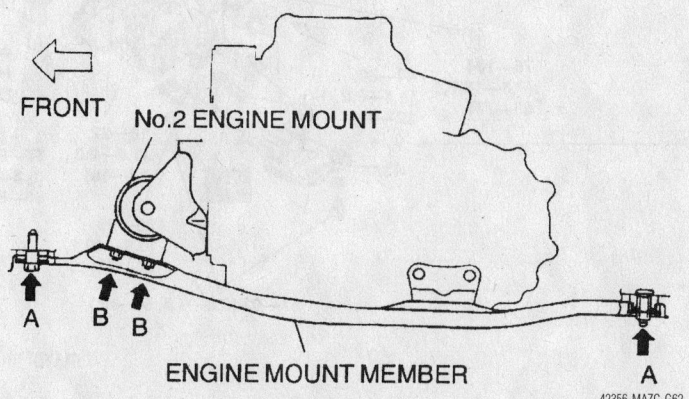

42356-MAZC-G62

Location of the front mount member bolt locations—626

- Stabilizer bar link. Refer to the illustration for component location and torque specifications.
- Tie rod ends to the knuckle. Refer to the illustration for component location and torque specifications.
- Lower control arm
- Oil hose and breather hose
- Fuel filter bolts
- Engine mount stay. Refer to the illustration for component location and torque specifications.
- Starter harness
- Input/Turbine speed shaft connector
- VSS connector
- O₂S connector
- Solenoid valve connector
- Transaxle range switch connector
- Selector cable to the bracket, install the clip and attach the cable to the manual shaft lever
- Splash shields
- Wheels
- Air cleaner assembly and fresh air duct
- Battery

15. Fill the transmission with fluid.
16. Start the engine and check for leaks and proper operation.

Millenia

LJ4A–EL TRANSAXLE

1. Refer to the illustration for component location and torque specifications.
2. Before servicing the vehicle, refer to the precautions in the beginning of this section.
3. Refer to the illustration for component location and torque specifications.
4. Drain the transaxle oil.
5. Remove or disconnect the following:
- Battery cover, battery and tray
- Rear air charge cooler duct
- Air cleaner assembly
- Selector cable clip and nut and the cable
- Output speed sensor connector
- Transaxle range switch connector
- Solenoid valve connector
- Harness bracket
- Selector cable bracket
- Rear intercooler
- Bracket and starter
- Front intercooler
- Electric fan
- Wheels
- Halfshaft locknut
- Splash shields
- Fresh air duct
- Exhaust pipe
- Lower ball joint

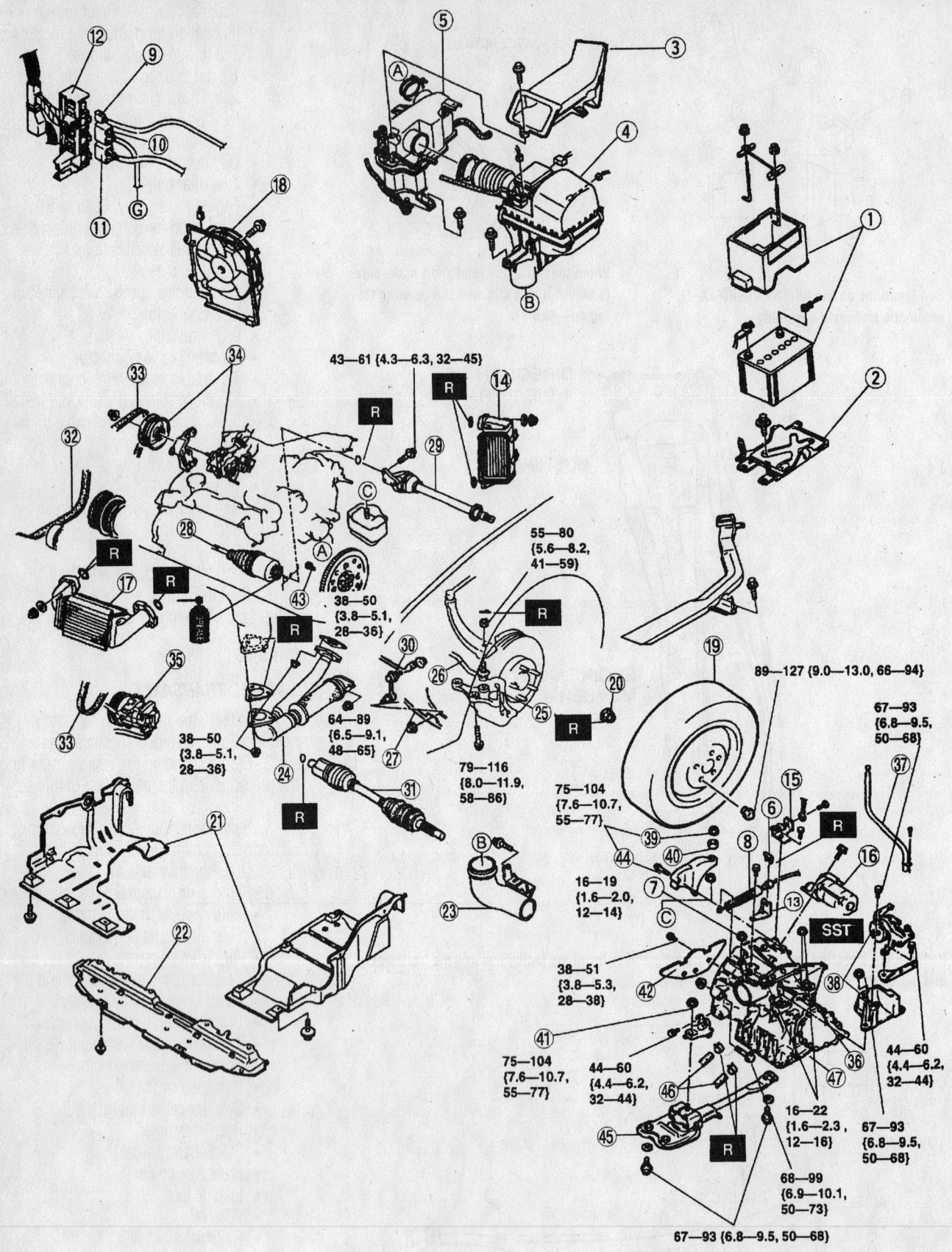

43—61 {4.3—6.3, 32—45}

55—80 {5.6—8.2, 41—59}

89—127 {9.0—13.0, 66—94}

67—93 {6.8—9.5, 50—68}

38—50 {3.8—5.1, 28—36}

38—50 {3.8—5.1, 28—36}

64—89 {6.5—9.1, 48—65}

79—116 {8.0—11.9, 58—86}

75—104 {7.6—10.7, 55—77}

16—19 {1.6—2.0, 12—14}

38—51 {3.8—5.3, 28—38}

75—104 {7.6—10.7, 55—77}

44—60 {4.4—6.2, 32—44}

44—60 {4.4—6.2, 32—44}

16—22 {1.6—2.3, 12—16}

67—93 {6.8—9.5, 50—68}

68—99 {6.9—10.1, 50—73}

67—93 {6.8—9.5, 50—68}

N·m {kgf·m, ft·lbf}

42356-MAZC-GEP

Exploded view of the LJ4A–EL automatic transaxle assembly (1 OF 2)—Millenia

1 Battery and battery cover
2 Battery carrier
3 Rear change air cooler air duct
4 Air cleaner component
5 Resonance chamber
6 Clip
7 Nut
8 Selector cable
9 Output speed sensor connector
10 Transaxle range switch connector
11 Solenoid valve connector
12 Harness bracket
13 Selector cable bracket
14 Rear intercooler
15 Bracket
16 Starter
17 Front intercooler
18 Electric coolant fan component
19 Wheel and tires
20 Locknut
21 Splash shield
22 Splash shield
23 Fresh air duct
24 Exhaust pipe

25 Lower ball joint
26 Upper lateral link (left side)
27 Lower arm
28 Drive shaft (right side)
29 Joint shaft
30 Stabilizer control link
31 Drive shaft (left side)
32 Timing belt
33 P/S, A/C Drive belt
34 P/S oil pump
35 A/C compressor
36 Selector rod
37 ATF filler tube
38 No.4 engine mount
39 No.1 engine mount nut
40 No.1 engine mount damper
41 No.2 engine mount nut
42 Under cover
43 Torque converter bolts
44 No.1 engine mount bolt
45 Engine mounting member
46 Oil hose
47 Transaxle

42356-MAZC-GEQ

Exploded view of the LJ4A–EL automatic transaxle assembly (2 OF 2)—Millenia

- Left side upper lateral link
- Lower arm
- Right side halfshaft
- Joint shaft
- Stabilizer bar link
- Left side halfshaft
- Timing belt
- Drive belts
- Power steering pump leaving the lines attached and set aside
- A/C compressor leaving the lines attached and set aside
- Selector rod
- Transmission dipstick tube

6. Support the engine assembly with a engine support assembly such as 49 E017 5A0.

- Loosen the number 3 engine mount bolt (refer to illustration) and remove the number 4 mount
- Number 1 engine mount nut, rubber and bracket
- Number 2 engine mount nut
- Under cover
- Torque converter bolts
- Number 1 engine mount bolt
- Engine mount member
- Oil hose

7. Loosen the engine support assembly and lean the engine towards the transaxle.

8. Support the transaxle with a jack, remove the transaxle bolts and the transaxle.

To install:

9. Place the transaxle onto a jack and raise into position.

10. Install the transaxle bolts and tighten the bolts to 50–73 ft. lbs. (68–99 Nm).

11. Install or connect the following:

- Oil hose
- Engine mount member and number 2 engine mount, make sure the number 2 mount stud bolt passes through the number 2 mount bracket installation hole and tighten the bolt **A** to 68 ft. lbs. (93 Nm). Tighten bolt **B** to 77 ft. lbs. (104 Nm). Refer to the illustration for component location.
- Number 3 engine mount bolt and tighten to 77 ft. lbs. (104 Nm)
- Torque converter bolts and tighten to 28–38 ft. lbs. (38–51 Nm)
- Under cover
- Number 2 engine mount nut
- Number 1 engine mount nut, rub-

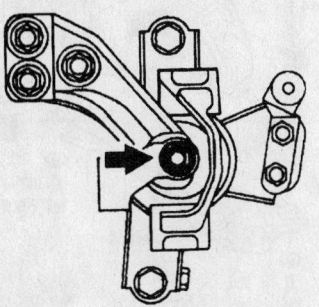

Loosen the number 3 engine mount bolt LJ4A–EL automatic transaxle assembly—Millenia

ber and bracket. Tighten the bolt to 77 ft. lbs. (104 Nm).
- Hand tighten the number 4 mount retainers. Tighten the bolts to 44 ft. lbs. (60 Nm), use the jack to ensure the mount holes and bracket holes are aligned and tighten the bolts to 68 ft. lbs. (93 Nm).

12. Remove the engine support assembly.

- Transmission dipstick tube

13. Move the selector lever to **P** and the manual shaft to **P** position. Install the selector rod and tighten the nut illustrated to 16 ft. lbs. (22 Nm).

- A/C compressor
- Power steering pump
- Drive belts
- Timing belt
- Left side halfshaft
- Stabilizer bar link
- Joint shaft
- Right side halfshaft
- Lower arm
- Left side upper lateral link
- Lower ball joint
- Exhaust pipe
- Fresh air duct
- Splash shields
- Halfshaft locknut
- Wheels
- Electric fan
- Front intercooler

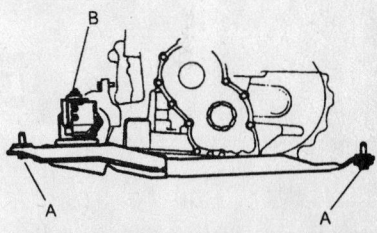

42356-MAZC-GES

Install the engine mount member bolts and tighten to specification— LJ4A–EL automatic transaxle assembly—Millenia with LJ4A–EL automatic transaxle

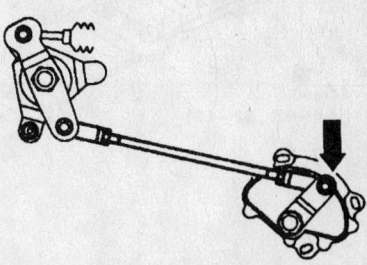

42356-MAZC-GET

Install the selector rod and tighten the nut—Millenia with LJ4A–EL automatic transaxle

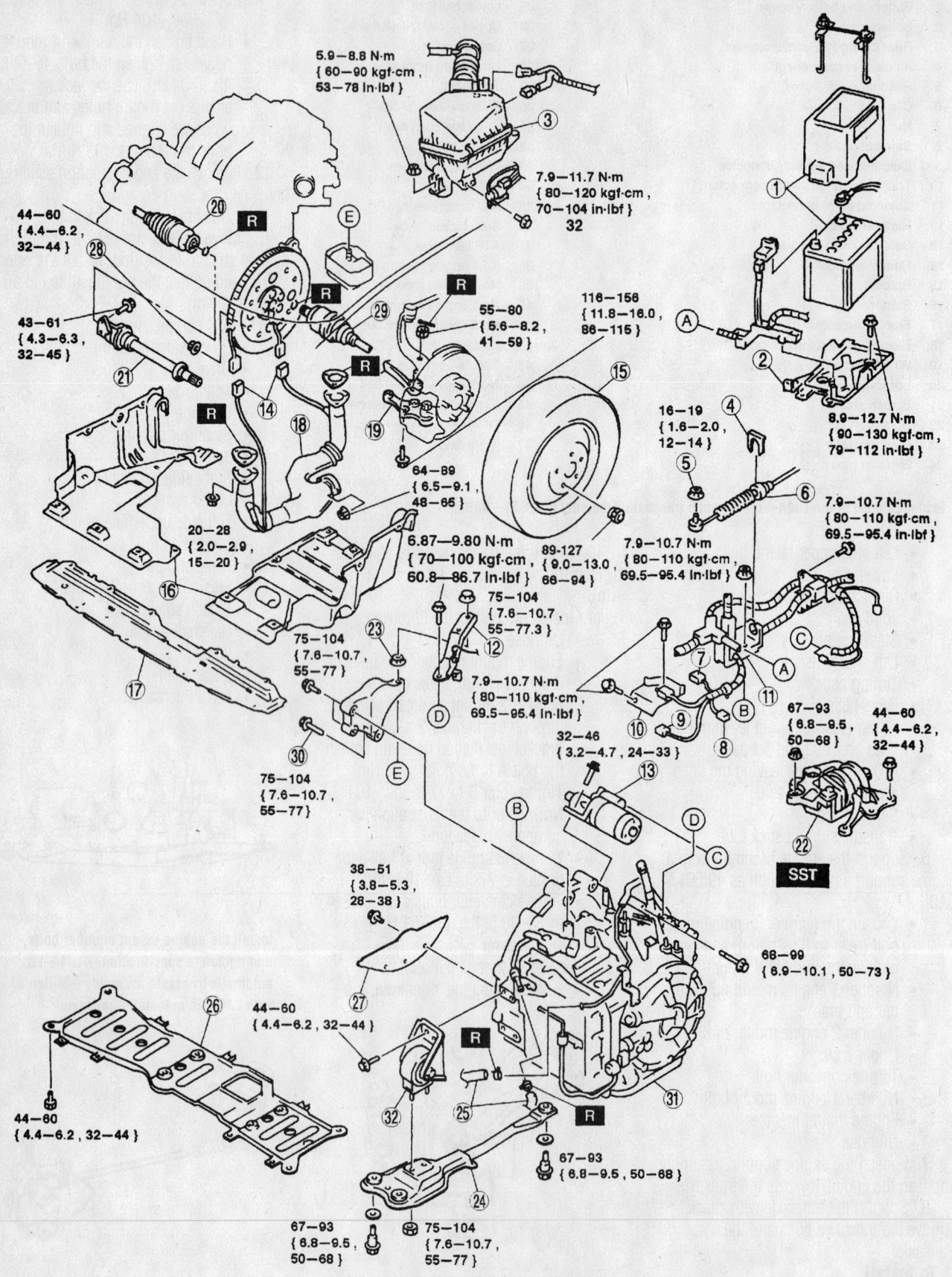

5.9—8.8 N·m
{ 60—90 kgf·cm,
53—78 In·lbf }

7.9—11.7 N·m
{ 80—120 kgf·cm,
70—104 in·lbf }
32

44—60
{ 4.4—6.2,
32—44 }

43—61
{ 4.3—6.3,
32—45 }

55—80
{ 5.6—8.2,
41—59 }

116—156
{ 11.8—16.0,
86—115 }

16—19
{ 1.6—2.0,
12—14 }

64—89
{ 6.5—9.1,
48—65 }

20—28
{ 2.0—2.9,
15—20 }

6.87—9.80 N·m
{ 70—100 kgf·cm,
60.8—86.7 In·lbf }

89—127
{ 9.0—13.0,
66—94 }

7.9—10.7 N·m
{ 80—110 kgf·cm,
69.5—95.4 In·lbf }

7.9—10.7 N·m
{ 80—110 kgf·cm,
69.5—95.4 in·lbf }

8.9—12.7 N·m
{ 90—130 kgf·cm,
79—112 In·lbf }

75—104
{ 7.6—10.7,
55—77.3 }

75—104
{ 7.6—10.7,
55—77 }

75—104
{ 7.6—10.7,
55—77 }

7.9—10.7 N·m
{ 80—110 kgf·cm,
69.5—95.4 In·lbf }

32—46
{ 3.2—4.7, 24—33 }

67—93
{ 6.8—9.5,
50—68 }

44—60
{ 4.4—6.2,
32—44 }

SST

68—99
{ 6.9—10.1, 50—73 }

38—51
{ 3.8—5.3,
28—38 }

44—60
{ 4.4—6.2, 32—44 }

44—60
{ 4.4—6.2, 32—44 }

67—93
{ 6.8—9.5, 50—68 }

67—93
{ 6.8—9.5,
50—68 }

75—104
{ 7.6—10.7,
55—77 }

42356-MAZC-GEU

Exploded view of the GF4A–EL automatic transaxle assembly (1 OF 2)—Millenia

1	Battery and battery cover	12	No.1 engine mount bracket stay	23	No.1 engine mount nut
2	Battery carrier	13	Starter	24	Engine mounting member
3	Air cleaner component	14	Heated oxygen sensor connector	25	Oil hose
4	Clip	15	Wheel and tire	26	Undercover
5	Nut	16	Splash shield	27	Undercover
6	Selector cable	17	Splash shield	28	Torque converter nut
7	Transaxle range switch connector	18	Front exhaust pipe	29	Drive shaft (left)
8	Input/turbine speed sensor connector	19	Lower ball joint	30	No.1 engine mount bolt
9	Solenoid valve connector	20	Drive shaft (right)	31	Transaxle
10	Harness bracket	21	Joint shaft	32	No.2 engine mount
11	Selector cable bracket	22	No.4 engine mount		

42356-MAZC-GEV

Exploded view of the GF4A–EL automatic transaxle assembly (2 OF 2)—Millenia

- Bracket and starter
- Rear intercooler
- Selector cable bracket
- Harness bracket
- Solenoid valve connector
- Transaxle range switch connector
- Output speed sensor connector
- Selector cable clip and nut and the cable
- Air cleaner assembly
- Rear air charge cooler duct
- Battery tray, battery and cover

14. Fill the transmission with fluid.

15. Start the engine and check for leaks and proper operation.

GF4A–EL TRANSAXLE

1. Refer to the illustration for component location and torque specifications.

2. Before servicing the vehicle, refer to the precautions in the beginning of this section.

3. Refer to the illustration for component location and torque specifications.

4. Drain the transaxle oil.

5. Remove or disconnect the following:
- Battery cover, battery and tray
- Air cleaner assembly
- Selector cable clip and nut and the cable
- Transaxle range switch connector
- Input/Turbine speed sensor connector
- Solenoid valve connector
- Harness bracket
- Selector cable bracket
- Number 1 engine mount stay bracket
- Starter
- Heated Oxygen (HO$_2$S) connector
- Wheels
- Splash shields
- Exhaust pipe
- Lower ball joint
- Right side halfshaft

- Joint shaft

6. Support the engine assembly with a engine support assembly such as 49 E017 5A0.

- Number 4 mount
- Number 1 engine mount nut
- Engine mount member
- Oil hose
- Under covers
- Torque converter bolts
- Left side halfshaft
- Number 1 engine mount bolt

7. Loosen the engine support assembly and lean the engine towards the transaxle.

8. Support the transaxle with a jack, remove the transaxle bolts and the transaxle.

To install:

9. Place the transaxle onto a jack and raise into position.

10. Install the transaxle bolts and tighten the bolts to 50–73 ft. lbs. (68–99 Nm).

11. Install or connect the following:
- Number 1 engine mount bolt. Make sure the holes align before tightening. Tighten the bolt to 77 ft. lbs. (104 Nm).
- Left side halfshaft
- Torque converter bolts and tighten to 32–44 ft. lbs. (44–60 Nm)
- Under covers
- Oil hose
- Engine mount member and number 2 engine mount, make sure the number 2 mount stud bolt passes through the number 2 mount bracket installation hole and tighten the bolt **A** to 68 ft. lbs. (93 Nm). Tighten bolt **B** to 77 ft. lbs. (104 Nm). Refer to the illustration for component location.
- Number 1 engine mount nut
- Number 4 mount retainers. Refer to the illustration for component location and torque specifications.

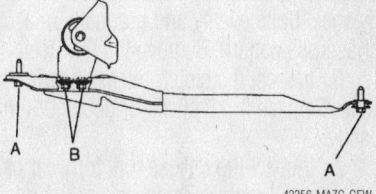

42356-MAZC-GEW

Install the engine mount member bolts and tighten to specification— LJ4A–EL automatic transaxle assembly—Millenia with GF4A–EL automatic transaxle

- Joint shaft
- Right side halfshaft
- Lower ball joint
- Exhaust pipe
- Splash shields
- Wheels
- HO$_2$S connector
- Starter
- Number 1 engine mount stay bracket
- Harness bracket
- Solenoid valve connector
- Input/Turbine speed sensor connector
- Transaxle range switch connector
- Selector cable nut and clip
- Air cleaner assembly
- Battery tray, battery and cover

12. Fill the transmission with fluid.

13. Start the engine and check for leaks and proper operation.

Clutch

REMOVAL & INSTALLATION

1. Before servicing the vehicle, refer to the precautions in the beginning of this section.

2. Remove or disconnect the following:
- Negative battery cable

- Clutch release cylinder
- Transaxle
- Rubber boot
- Clutch release collar
- Clutch release fork
- Pressure plate loosening the bolts one turn each in a criss–cross pattern
- Clutch disc

3. Inspect the pilot bearing. If it is worn or damaged and does not turn easily by hand, remove it using a puller/slide hammer.

4. Check the flywheel surface for scoring, cracks or burning and machine or replace, as necessary.

5. Install Holder tool 49 E011 1A0 to keep the flywheel from turning. Loosen the flywheel bolts evenly and gradually in a crisscross pattern. Remove the flywheel.

6. Inspect the clutch release bearing for wear. Replace it if it sticks or does not turn easily.

7. Inspect the release fork for wear or damage and replace as necessary.

To install:

8. Lubricate the release fork fingers and pivot with molybdenum grease and install in the release fork boot.

9. Install or connect the following:
- Clutch release bearing on the release fork
- New pilot bearing in the flywheel, if removed, using a installation tool

10. Be sure the flywheel mounting surface and the crankshaft or eccentric shaft mounting surfaces are clean. Remove any old sealant from the flywheel bolt hole threads and the flywheel bolts.
- Flywheel
- Sealant to the flywheel bolt threads and install them hand tight
- Flywheel holding tool. Tighten the bolts, in a crisscross pattern, to 71–75 ft. lbs. (97–102 Nm) for all engines except the 2.5L (KL) or to 45–49 ft. lbs. (61–66 Nm) for 2.5L (KL) engines.
- Small amount of molybdenum grease to the clutch disc splines
- Clutch disc on the flywheel with the spring side toward the transaxle
- An alignment tool in the pilot bearing to position the clutch disc
- Clutch pressure plate by aligning the dowel holes with the flywheel dowels
- Pressure plate. Gradually, torque the bolts, in a crisscross pattern, to 19 ft. lbs. (26 Nm).

11. Remove the alignment tool.
- Clutch release fork

- Clutch release collar
- Rubber boot
- Transaxle
- Clutch release cylinder
- Negative battery cable

Hydraulic Clutch System

BLEEDING

1. Before servicing the vehicle, refer to the precautions in the beginning of this section.

2. Remove the rubber cap from the bleeder screw on the release cylinder.

3. Place a bleeder tube over the end of the bleeder screw.

4. Submerge the other end of the tube in a jar half filled with hydraulic brake fluid.

5. Slowly pump the clutch pedal fully and allow it to return slowly, several times.

6. While pressing the clutch pedal to the floor, loosen the bleeder screw until the fluid starts to run out. Then, close the bleeder screw. Keep repeating this Step, while watching the hydraulic fluid in the jar. As soon as the air bubbles disappear, close the bleeder screw.

7. During the bleeding procedure the reservoir must be kept at least ¾ full.

Halfshafts

REMOVAL & INSTALLATION

1. Before servicing the vehicle, refer to the precautions in the beginning of this section.

2. Drain the transaxle oil.

3. Remove or disconnect the following:
- Wheels
- Splash shield, if equipped

4. Raise the staked portion of the hub locknut with a hammer and chisel.

5. Lock the hub by applying the brakes and remove the nut.

6. Remove or disconnect the following:
- Stabilizer bar from the lower control arm
- Cotter pin and nut from the tie rod end ball stud
- Tie rod end from the knuckle
- Transverse member, on the 626
- Lower ball joint pinch bolt and nut
- Lower ball joint from the knuckle

7. If removing the left side shaft on 626 with automatic transaxle, proceed as follows:

a. Suspend the engine using engine support tool 49 G017 5A0.

b. Remove the engine mount member.

8. Position a prybar between the inner CV-joint and transaxle case. Carefully pry the halfshaft from the transaxle being careful not to damage the oil seal. If equipped with a right side intermediate shaft, insert the prybar between the halfshaft and intermediate shaft and tap on the bar to uncouple them.

9. Pull outward on the hub/knuckle assembly, push the outer CV-joint stub shaft through the hub and remove the halfshaft. If the halfshaft is stuck in the hub, install the old hub nut to protect the stub shaft threads. Tap on the nut, using only a soft mallet, to remove the halfshaft.

➡Install plug tool 49 G030 455 into the transaxle after removing the halfshaft, to keep the differential side gear in position. If the gear becomes positioned incorrectly, the differential may have to be removed to realign the gear.

10. Remove the intermediate shaft, if necessary, by removing the support bearing bolts and pulling the shaft from the transaxle.

To install:

11. Install or connect the following:
- New circlip on the end of the intermediate shaft, if removed, with the end gap facing upward.
- Intermediate shaft in the transaxle, being careful not to damage the oil seals
- Intermediate shaft support bearing bolts. Torque in sequence, to 45 ft. lbs. (61 Nm).
- New circlip on the end of the halfshaft, with the end gap facing upward
- Halfshaft into the transaxle, being careful not to damage the oil seal

➡If equipped, push the halfshaft into the intermediate shaft.

- Other end of the halfshaft through the hub. Loosely install a new locknut

12. If installing the left side shaft on 626 with automatic transaxle, proceed as follows:

a. Install the engine mount member. Torque the mount member-to-body bolts to 66 ft. lbs. (89 Nm).

b. Torque the front mount-to-mount member nuts to 77 ft. lbs. (104 Nm) and the side mount bolts to 44 ft. lbs. (60 Nm).

c. Remove the engine support tool.

13. Install or connect the following:
- Lower ball joint into the knuckle.

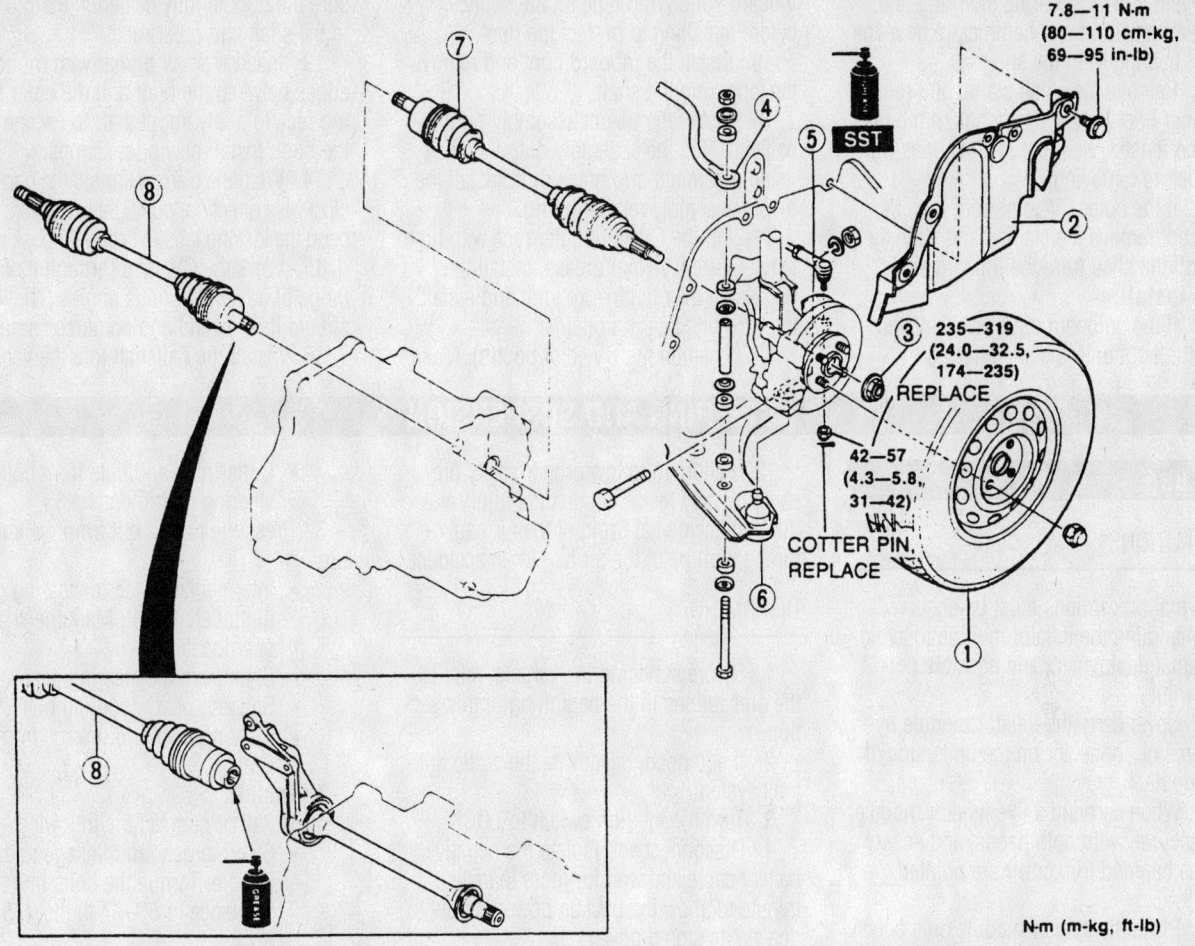

7.8—11 N·m
(80—110 cm-kg,
69—95 in-lb)

SST

235—319
(24.0—32.5,
174—235)
REPLACE

42—57
(4.3—5.8,
31—42)

**COTTER PIN
REPLACE**

N·m (m-kg, ft-lb)

1. Wheel and tire
2. Splash shield
3. Locknut
4. Stabilizer

5. Tie-rod end
6. Lower ball joint
7. Left driveshaft
8. Right driveshaft

7923MG76

Exploded view of a typical halfshaft mounting

Torque the pinch bolt to 40 ft. lbs. (54 Nm).
- Transverse member, on 626. Torque the bolts to 96 ft. lbs. (132 Nm).
- Tie rod end to the steering knuckle. Torque the nut to 42 ft. lbs. (57 Nm) on all except 626 and Millenia or to 32 ft. lbs. (44 Nm) for 626 and Millenia.
- New cotter pin. Tighten the nut, if necessary, to align the ball stud hole with the nut castellation.
- Stabilizer bar to the lower control arm
- Splash shield
- Wheels
- New hub nut. Torque it to 174–235 ft. lbs. (235–318 Nm). After tightening, stake the locknut using a hammer and dull bladed chisel.

14. Fill the transaxle.

CV-Joints

OVERHAUL

1. Before servicing the vehicle, refer to the precautions in the beginning of this section.

Two types of CV-joints are used. The inboard CV-joints are the tri-Pot type. All outboard CV-joints are Birfield type. The Birfield CV-joint cannot be disassembled; if an outboard CV-joint boot needs replacement, the inboard CV-joint must be removed. If the outboard CV-joint needs to be replaced, replace the entire halfshaft as an assembly.

2. Remove the halfshaft from the vehicle and clamp it in a vise equipped with jaw caps, to prevent damage to the machined surfaces. Do not allow the vise to contact the boot or its clamps.

3. Remove the large boot clamp from the inboard CV-joint, using side cutters. After removing the clamp, roll the boot back over the shaft.

➡**Check the grease for contamination by rubbing it between 2 fingers. Any gritty feeling indicates a contaminated CV-joint, in which case the entire CV-joint must be disassembled, cleaned and inspected. If the grease is not contaminated and the CV-joint has been operating satisfactorily, continue with the boot replacement procedure and add the required lubricant.**

4. Paint alignment marks on the outer race and shaft for assembly reference. Remove the wire ring bearing retainer and remove the outer race.

5. Paint alignment marks on the tri-pot bearing and shaft for assembly reference.

Remove the tri-pot bearing snapring and, using a brass drift and hammer, remove the tri-pot bearing from the shaft.

6. Remove the small clamp and remove the inner boot from the halfshaft. If the boot is to be reused, wrap the shaft splines with tape before removing.

7. If the outer CV-joint boot is to be replaced, remove the clamps and slide the boot off the shaft from the inboard side.

To install:

8. If the outboard boot was removed, slide the boot onto the shaft from the inboard side. Wrap tape on the splines before installing to protect the boot.

9. Install the inboard boot and remove the tape from the shaft.

10. Install the tri-pot assembly on the halfshaft. Tap the assembly onto the shaft using a hammer and brass drift. Install the tri-pot assembly retaining ring.

11. Fill the CV-joint outer race with high temperature CV-joint grease. Install the outer race over the tri-pot joint and install the wire ring bearing retainer.

12. Position the CV-joint boot(s). Make sure the boot is fully seated in the grooves in the shaft and outer race.

13. Insert a small prybar with rounded edges between the boot and the outer bearing race to allow trapped air to escape from the boot. Install new boot clamps.

14. Wrap the clamps around the boots in a clockwise direction, pull tight with pliers and bend the locking tabs to secure in position.

15. Work the CV-joint through its full range of travel at various angles. The joint should flex, extend and compress smoothly.

16. Install the halfshaft into the vehicle.

STEERING AND SUSPENSION

Air Bag

PRECAUTIONS

Several precautions must be observed when handling the inflator module to avoid accidental deployment and possible personal injury.

1. Never carry the inflator module by the wires or connector on the underside of the module.

2. When carrying a live inflator module, hold securely with both hands, and ensure that the bag and trim cover are pointed away.

3. Place the inflator module on a bench or other surface with the bag and trim cover facing up.

4. With the inflator module on the bench, never place anything on or close to the module which may be thrown in the event of an accidental deployment.

5. An air bag is an explosive device. Handle with extreme caution.

6. Always disconnect the battery and the air bag connector before removing the steering wheel or beginning work on the air bag system.

7. Air bag components must not be repaired or opened. Always use new parts, including the wiring harness.

8. Always place a removed air bag unit with the horn pad facing up. Put it in a safe place where it will not be disturbed.

9. The air bag unit must not be exposed to grease, fluids, or cleaning agents.

10. The air bag unit must not be exposed to temperatures above 194°F (90°C) at any time. Even the heat of a soldering iron can damage or ignite the charge.

11. Storage and transport of air bags is subject to rules governing explosive devices and should be done only in the original package.

12. Failure to follow proper safety precautions may result in personal injury through accidental firing of the air bag, or through failure of the air bag in an accident.

DISARMING

1. Before servicing the vehicle, refer to the precautions in the beginning of this section.

2. If equipped, deactivate the audio anti-theft system.

3. Turn the ignition switch to LOCK.

4. Disconnect and isolate the negative battery cable and wait for more than 1 minute to allow the backup power supply to deplete its stored power.

ARMING

1. Before servicing the vehicle, refer to the precautions in the beginning of this section.

2. Connect the negative battery cable, turn the ignition switch **ON** and verify the air bag warning light cones on for 6 seconds. If the light does not illuminate there are problems with the system.

3. If equipped, activate the audio anti-theft system.

Rack and Pinion Steering Gear

REMOVAL & INSTALLATION

Manual

MIATA

1. Before servicing the vehicle, refer to the precautions in the beginning of this section.

2. Remove or disconnect the following:
   • Negative battery cable
   • Front wheels

• Cotter pins and nuts from both steering tie rod ends

3. Press the tie rod out from the knuckle arm.

   • Intermediate shaft to steering gear pinion shaft bolt. Mark the shaft-to-gear location.
   • Shaft from the steering gear
   • Steering gear mounting nuts
   • Steering gear and linkage from the vehicle

To install:

4. Install or connect the following:
   • Steering gear and linkage to the vehicle. Torque the bolts in sequence to 55–77 ft. lbs. (75–104 Nm).
   • Steering shaft to the steering gear pinion shaft, align the marks made during removal and tighten the bolt to 19 ft. lbs. (26 Nm)
   • Tie rod ends to the knuckle arm. Torque the nuts to 32–41 ft. lbs. (43–56 Nm).
   • New cotter pins
   • Wheels
   • Negative battery cable

5. Check and/or adjust the front end alignment.

Power

MIATA

1. Before servicing the vehicle, refer to the precautions in the beginning of this section.

2. Remove or disconnect the following:
   • Negative battery cable
   • Front wheels
   • Cotter pins and nuts from both steering tie rod ends

3. Press the tie rod out from the knuckle arm.

   • Intermediate shaft to steering gear pinion shaft bolt. Mark the shaft-to-gear location.

- Shaft from the steering gear
- Pressure and return lines
- Steering gear mounting nuts
- Steering gear and linkage from the vehicle

To install:

4. Install or connect the following:
  - Steering gear and linkage to the vehicle. Torque the bolts in sequence to 55–77 ft. lbs. (75–104 Nm).
  - Steering shaft to the steering gear pinion shaft, align the marks made during removal and tighten the bolt to 19 ft. lbs. (26 Nm)
  - Pressure and return lines. Tighten the pressure line fitting to 34 ft. lbs. (47 Nm).
  - Tie rod ends to the knuckle arm. Torque the nuts to 32–41 ft. lbs. (43–56 Nm).
  - New cotter pins
  - Wheels
  - Negative battery cable

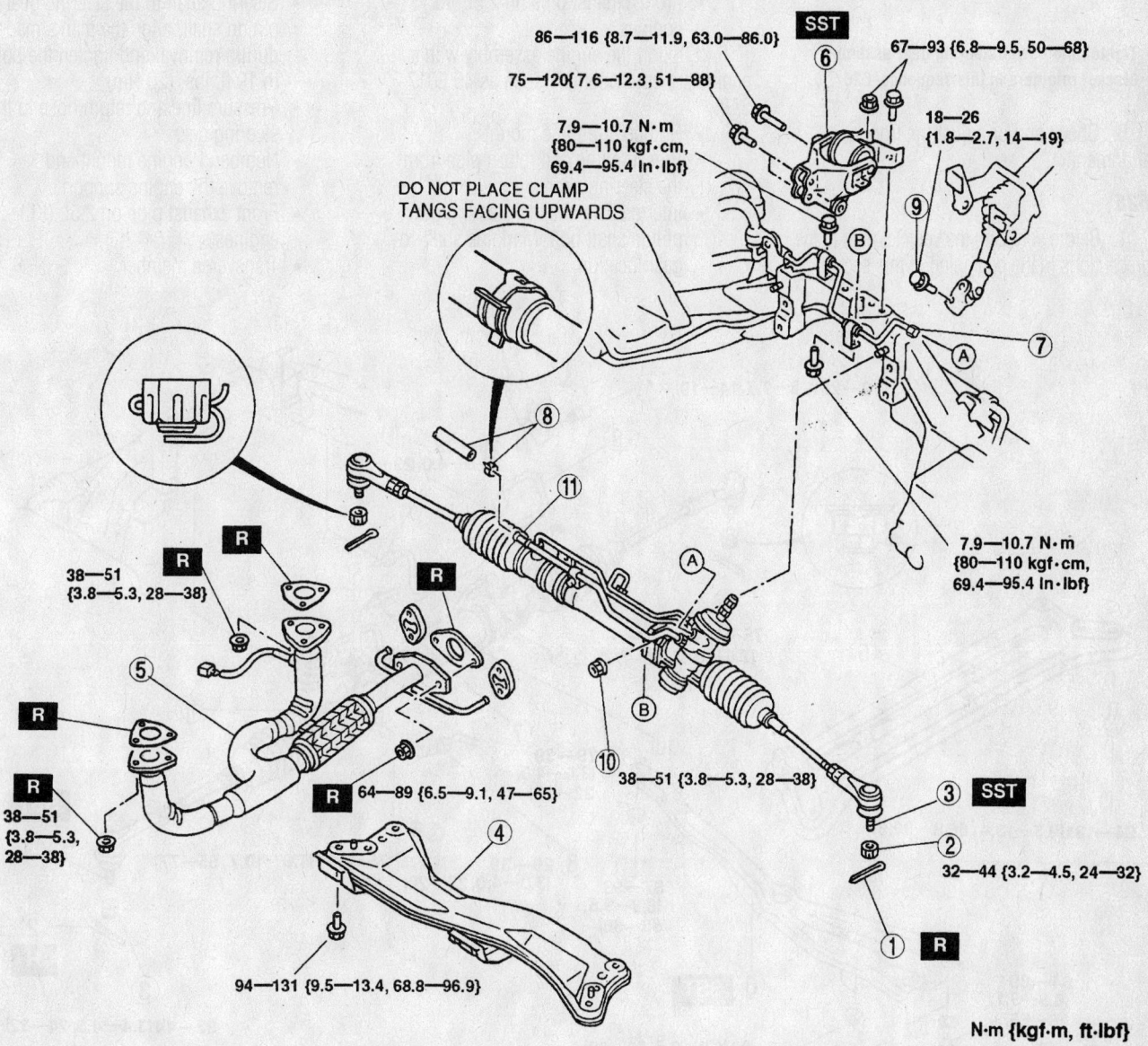

86—116 {8.7—11.9, 63.0—86.0}

75—120{ 7.6—12.3, 51—88}

7.9—10.7 N·m {80—110 kgf·cm, 69.4—95.4 In·lbf}

67—93 {6.8—9.5, 50—68}

18—26 {1.8—2.7, 14—19}

DO NOT PLACE CLAMP TANGS FACING UPWARDS

7.9—10.7 N·m {80—110 kgf·cm, 69.4—95.4 In·lbf}

38—51 {3.8—5.3, 28—38}

38—51 {3.8—5.3, 28—38}

64—89 {6.5—9.1, 47—65}

38—51 {3.8—5.3, 28—38}

94—131 {9.5—13.4, 68.8—96.9}

32—44 {3.2—4.5, 24—32}

N·m {kgf·m, ft·lbf}

1	Cotter pin	7	Pressure pipe
2	Nut	8	Return hose and clamp
3	Tie-rod end ball joint	9	Bolt (intermediate shaft)
4	Transverse member	10	Mounting bracket nut and bolt
5	Front exhaust pipe (KL engine)	11	Steering gear and linkage
6	No.1 engine mount component		

42356-MAZC-G81

Exploded view of the power steering gear assembly–626

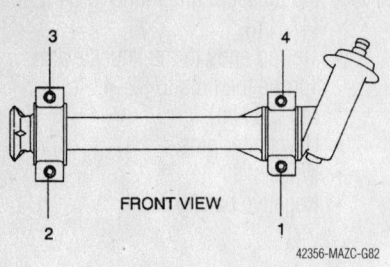

FRONT VIEW

42356-MAZC-G82

Tighten the power steering gear assembly bracket retainers in this sequence–626

5. Check and/or adjust the front end alignment.

626

1. Before servicing the vehicle, refer to the precautions in the beginning of this section.

2. Remove or disconnect the following:
• Negative battery cable
• Front wheels
• Cotter pins and nuts from both steering tie rod ends

3. Press the tie rod out from the knuckle arm.

4. Remove or disconnect the following:
• Transverse member
• Front exhaust pipe on 2.5L (KL) engines

5. Support the engine assembly with a engine support assembly such as 49 E017 5A0.

• Number 1 engine mount
• Pressure line and return pipe from the steering gear
• Intermediate shaft to steering gear pinion shaft bolt. Mark the shaft-to-gear location.

• Shaft from the steering gear
• Steering gear mounting nuts
• Steering gear from the right of the vehicle

To install:

6. Install or connect the following:
• Steering gear to the vehicle. Torque the nuts/bolts in sequence to 28–38 ft. lbs. (37–52 Nm).
• Steering shaft to the steering gear pinion shaft, align the marks made during removal and tighten the bolt to 19 ft. lbs. (26 Nm)
• Pressure line and return hose to the steering gear
• Number 1 engine mount and remove the engine support
• Front exhaust pipe on 2.5L (KL) engines
• Transverse member

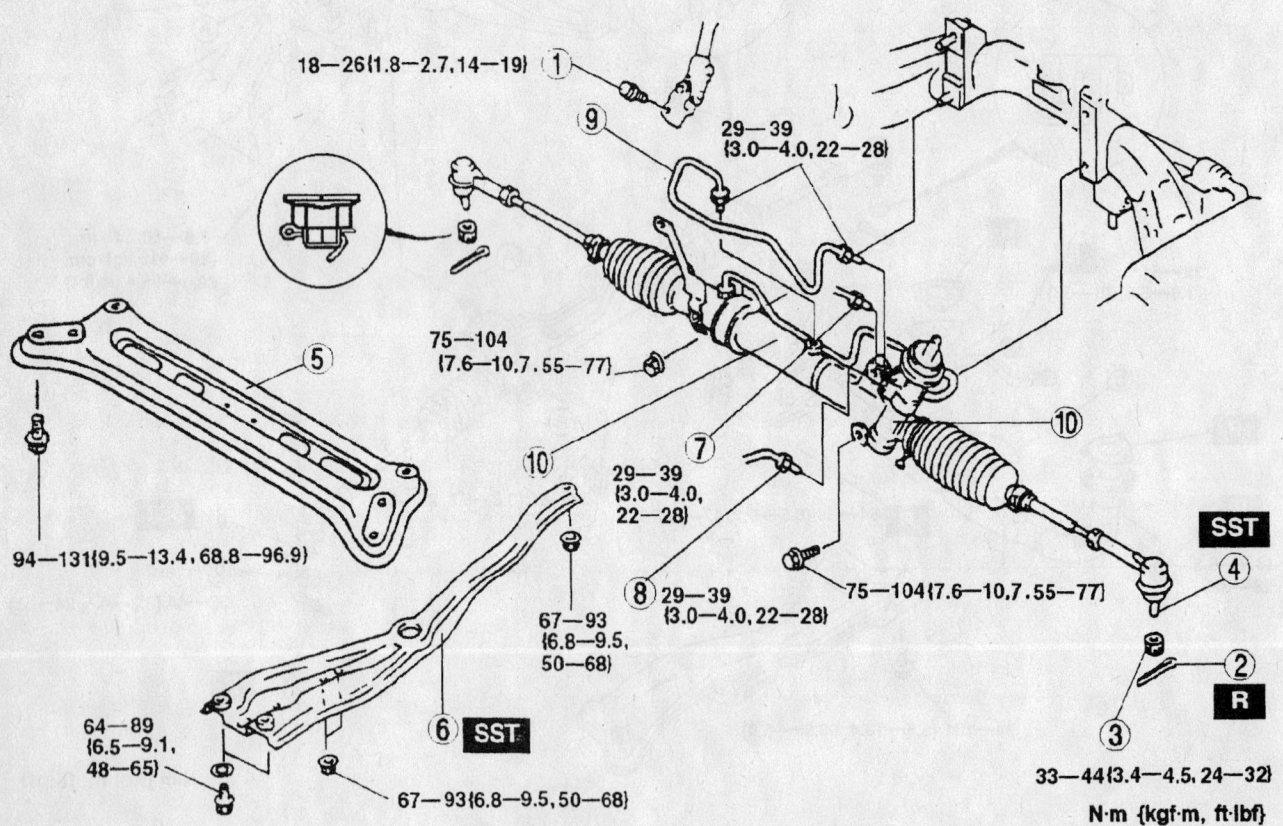

18—26 (1.8—2.7, 14—19)

29—39 (3.0—4.0, 22—28)

75—104 (7.6—10.7, 55—77)

94—131 (9.5—13.4, 68.8—96.9)

29—39 (3.0—4.0, 22—28)

29—39 (3.0—4.0, 22—28)

75—104 (7.6—10.7, 55—77)

SST

67—93 (6.8—9.5, 50—68)

64—89 (6.5—9.1, 48—65)

SST

67—93 (6.8—9.5, 50—68)

R

33—44 (3.4—4.5, 24—32)

N·m {kgf·m, ft·lbf}

1	Bolt (intermediate shaft)	6	Engine mount member
2	Cotter pin	7	Pressure pipe
3	Nut	8	Return pipe and clamp
4	Tie-rod end ball joint	9	Oil pipe
5	Transverse member (ZM (ATX), FS)	10	Steering gear and linkage

42356-MAZC-GKA

Exploded view of the power steering gear assembly–Protégé

- Tie rod ends to the knuckle arm. Torque the nuts to 24–32 ft. lbs. (32–44 Nm).
- New cotter pins and check the power steering fluid level
- Wheels
- Negative battery cable

7. Check and/or adjust the front end alignment.

PROTÉGÉ

1. Before servicing the vehicle, refer to the precautions in the beginning of this section.

2. Remove or disconnect the following:
- Negative battery cable
- Front wheels
- Intermediate shaft to steering gear pinion shaft bolt. Mark the shaft-to-gear location.
- Cotter pins and nuts from both steering tie rod ends

3. Press the tie rod out from the knuckle arm.

- Transverse member, if necessary on some models to facilitate removal

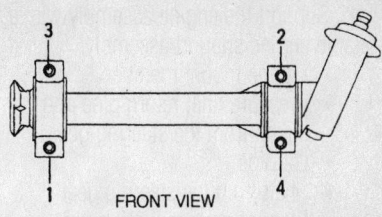

FRONT VIEW

42356-MAZC-GEZ

Tighten the power steering gear assembly bracket retainers in this sequence–Millenia

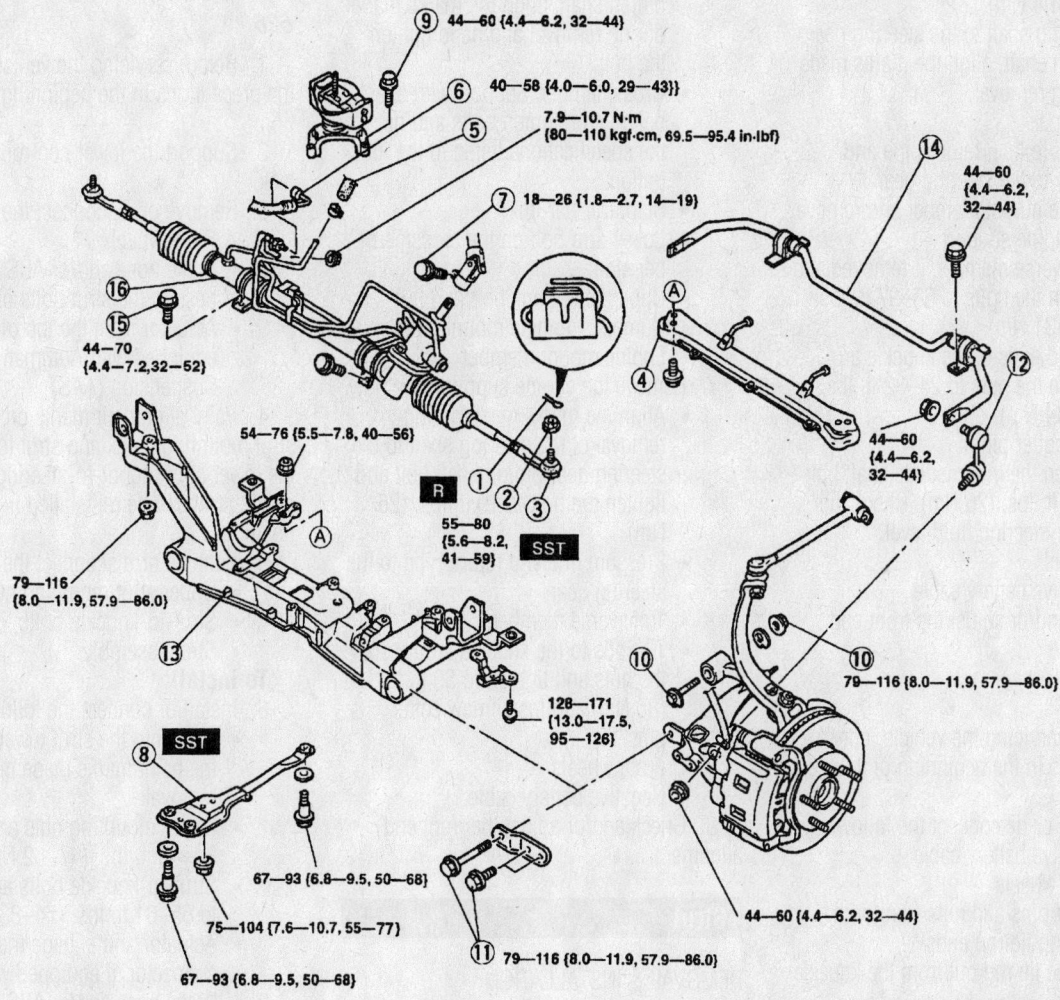

N·m (kgf·m, ft·lbf)

1	Cotter pin	9	Bolts (engine mount No.1)
2	Nut	10	Bolt and nut (upper lateral link)
3	Tie-rod end ball joint	11	Lower arm bolt (crossmember side)
4	Transverse member	12	Stabilizer control link
5	Return hose	13	Crossmember
6	Pressure pipe	14	Stabilizer
7	Bolt (intermediate shaft)	15	Mounting bracket bolts
8	Engine mount member	16	Steering gear and linkage

42356-MAZC-GEY

Exploded view of the power steering gear assembly–Millenia

4. Support the engine assembly with a suitable engine support assembly.
- Engine mount member
- Pressure line, return pipe and clamp from the steering gear
- Oil pipe
- Shaft from the steering gear
- Steering gear mounting nuts
- Steering gear from the right of the vehicle

To install:
5. Install or connect the following:
- Steering gear to the vehicle. Torque the nuts/bolts 55–77 ft. lbs. (75–104 Nm).
- Steering shaft to the steering gear pinion shaft, align the marks made during removal
- Oil pipe
- Pressure line, return pipe and clamp to the steering gear
- Engine mount member and remove the engine support
- Transverse member, if removed and tighten the bolts to 68–97 ft. lbs. (94–131 Nm)
- Tie rod ends to the knuckle arm. Torque the nuts to 24–32 ft. lbs. (32–44 Nm).
- New cotter pins
- Tighten the intermediate shaft bolt to 19 ft. lbs. (26 Nm). Check the power steering fluid level.
- Wheels
- Negative battery cable

6. Check and/or adjust the front end alignment.

MILLENIA

1. Before servicing the vehicle, refer to the precautions in the beginning of this section.
2. Remove or disconnect the following:
- Negative battery cable
- Front wheels
- Cotter pins and nuts from both steering tie rod ends

3. Press the tie rod out from the knuckle arm.
4. Remove or disconnect the following:
- Transverse member
- Pressure line and return pipe from the steering gear
- Intermediate shaft to steering gear pinion shaft bolt. Mark the shaft-to-gear location.

5. Support the engine assembly with a engine support assembly such as 49 E017 5A0.

- Engine mount member
- Number 1 engine mount bolts
- Upper lateral link bolt and nut

- Lower arm bolt on the crossmember side
- Stabilizer bar link
- Crossmember. Support with a jack before removing the bolts and nuts.
- Shaft from the steering gear
- Steering gear mounting nuts
- Steering gear

To install:
6. Install or connect the following:
- Steering gear to the vehicle. Torque the nuts/bolts in sequence to 23–52 ft. lbs. (44–70 Nm).
- Steering shaft to the steering gear pinion shaft, align the marks made during removal and hand tighten the bolt
- Crossmember. Support with a jack before and tighten bolts and nuts to the specifications listed in the illustration.
- Stabilizer bar link
- Lower arm bolt on the crossmember side
- Upper lateral link bolt and nut
- Number 1 engine mount bolts
- Engine mount member

7. Remove the engine support assembly.
- Align the marks made during removal of the steering shaft to the steering gear pinion shaft bolt and tighten the bolt to 19 ft. lbs. (26 Nm)
- Pressure line and return pipe to the steering gear
- Transverse member
- Tie rods to the knuckle arm. Install the nuts and tighten to 59 ft. lbs. (80 Nm) and install new cotter pins.
- Front wheels
- Negative battery cable

8. Check and/or adjust the front end alignment.

Strut

REMOVAL & INSTALLATION

Front

MIATA

1. Before servicing the vehicle, refer to the precautions in the beginning of this section.
2. Support the lower control arm with a jack.
3. Remove or disconnect the following:

- Front wheel
- Stabilizer bar nut

- Lower arm ball joint from the knuckle
- Lower arm bolts
- Lower the lower arm and remove the shock. Be careful not to lower the arms too much or it may cause damage.

To install:
4. Installation is the reverse of removal. Tighten the upper shock nuts to 26 ft. lbs. (36 Nm) and the lower shock nut and bolt to 86 ft. lbs. (116 Nm).
5. Check and/or adjust the front end alignment.

626

1. Before servicing the vehicle, refer to the precautions in the beginning of this section.
2. Support the lower control arm with a jack.
3. Remove or disconnect the following:
- Front wheel
- Brake hose and/or ABS sensor harness to the strut bolts or clips
- Actuator from the top of the strut, if equipped with Automatic Adjusting Suspension (AAS)

4. Paint alignment marks on the upper strut mounting block and strut tower, and on the lower strut mount-to-steering knuckle so the strut can be reinstalled in the same position.
5. Remove or disconnect the following:
- Upper strut mounting nuts
- Strut-to-knuckle bolts
- Strut assembly

To install:
6. Install or connect the following:
- Strut into the strut tower, aligning the paint marks made during removal
- Upper mounting nuts and tighten to 34–46 ft. lbs. (47–62 Nm)
- Strut-to-knuckle bolts and tighten to 55–61 ft. lbs. (74–83 Nm)
- Actuator and engage the electrical connector, if equipped with AAS
- Brake hose and/or ABS sensor harness clips or bolts
- Wheel

7. Check and/or adjust the front end alignment.

PROTÉGÉ

1. Before servicing the vehicle, refer to the precautions in the beginning of this section.
2. Support the lower control arm with a jack.
3. Remove or disconnect the following:
- Front wheel

- Brake hose and/or ABS sensor harness to the strut bolts or clips
- Stabilizer link nut
- Strut-to-knuckle bolt
- Upper strut mounting nuts
- Stiffener, if equipped
- Sheet
- Strut assembly

To install:

4. Install or connect the following:

- Face the mounting block direction indicator towards the rear outboard position and install the shock assembly.
- Sheet
- Stiffener, if equipped
- Upper mounting nuts and tighten to 34–46 ft. lbs. (47–62 Nm)
- Strut-to-knuckle bolts and tighten to 68–93 ft. lbs. (94–126 Nm)
- Stabilizer link nut and tighten to 32–44 ft. lbs. (47–60 Nm)
- Brake hose and/or ABS sensor harness clips or bolts
- Wheel

5. Check and/or adjust the front end alignment.

MILLENIA

1. Before servicing the vehicle, refer to the precautions in the beginning of this section.

2. Support the lower control arm with a jack.

3. Remove or disconnect the following:

- Front wheel
- Brake hose and/or ABS sensor harness to the strut bolts or clips

4. Paint alignment marks on the upper strut mounting block and strut tower, and on the lower strut mount-to-steering knuckle so the strut can be reinstalled in the same position.

5. Remove or disconnect the following:

- Upper strut mounting nuts
- Stiffener
- Strut-to-knuckle bolts
- Strut assembly

To install:

6. Install or connect the following:

- Stiffener. Make sure the word **LH or RH** faces up and the parting area faces the inside of the vehicle.
- Strut into the strut tower, aligning the paint marks made during removal
- Upper mounting nuts and tighten to 38–49 ft. lbs. (51–67 Nm)
- Strut-to-knuckle bolts and tighten to 73–101 ft. lbs. (99–137 Nm)
- Brake hose and/or ABS sensor harness clips or bolts

- Wheel

7. Check and/or adjust the front end alignment.

Rear

MIATA

1. Before servicing the vehicle, refer to the precautions in the beginning of this section.

2. Support the lower control arm with a jack.

3. Remove or disconnect the following:

- Front wheel
- Stabilizer bar nut
- Lower arm ball joint from the knuckle
- Lower arm bolts
- Lower the lower and upper arms and remove the shock. Be careful not to lower the arms too much or it may cause damage.

To install:

4. Installation is the reverse of removal. Tighten the upper shock nuts to 26 ft. lbs. (36 Nm) and the lower shock nut and bolt to 70 ft. lbs. (95 Nm).

5. Check and/or adjust the front end alignment.

626

1. Before servicing the vehicle, refer to the precautions in the beginning of this section.

2. Remove or disconnect the following:

- Rear wheel(s)
- Rear package trim
- Speed sensor harness
- Top strut nuts

➡**The suspension will drop when the weight lifts off the wheels.**

- Bottom strut mount bolt(s)
- Strut assembly

To install:

3. Install or connect the following:

- Strut assembly
- Bottom strut mount bolt(s) and tighten to 55–61 ft. lbs. (74–83 Nm)
- Upper mounting nuts and tighten to 33–46 ft. lbs. (47–62 Nm)
- Speed sensor harness
- Rear package trim
- Rear wheel(s)

PROTÉGÉ

1. Before servicing the vehicle, refer to the precautions in the beginning of this section.

2. Remove or disconnect the following:

- Rear wheel(s)

- Speed sensor harness
- Rear seat belt on 4SD models or trunk side trim on the 5HB.
- Clip and brake hose
- Stabilizer link nut
- Bottom strut mount bolt
- Cap
- Top strut nuts

➡**The suspension will drop when the weight lifts off the wheels.**

- Strut assembly

To install:

3. Install or connect the following:

- Strut assembly
- Upper mounting nuts and tighten to 34–46 ft. lbs. (47–62 Nm)
- Cap
- Bottom strut mount bolt and tighten to 68–93 ft. lbs. (94–126 Nm)
- Stabilizer link nut and tighten to 32–44 ft. lbs. (44–60 Nm)
- Brake hose and clip
- Rear seat belt on 4SD models or trunk side trim on the 5HB.
- Speed sensor harness
- Rear wheel(s)

MILLENIA

1. Before servicing the vehicle, refer to the precautions in the beginning of this section.

2. Remove or disconnect the following:

- Rear wheel(s)
- Rear package front trim
- Speed sensor harness
- Lower lateral link ball joint
- Top strut nuts
- Stiffener

➡**The suspension will drop when the weight lifts off the wheels.**

- Bottom strut mount bolt(s)
- Strut assembly

To install:

3. Install or connect the following:

- Stiffener. Make sure the word **OUT** faces up and the parting area faces the inside of the vehicle.
- Strut assembly
- Bottom strut mount bolt(s) and tighten to 76–101 ft. lbs. (102–137 Nm)
- Upper mounting nuts and tighten to 34–46 ft. lbs. (47–62 Nm)
- Lower lateral link ball joint, tighten the nut to 115 ft. lbs. (156 Nm) and install a new cotter pin
- Speed sensor harness
- Rear package front trim
- Rear wheel(s)

Coil Spring

REMOVAL & INSTALLATION

626 and Millenia

1. Before servicing the vehicle, refer to the precautions in the beginning of this section.
2. Remove or disconnect the following:
   - Strut from the vehicle
   - Cap from the top of the strut, if not equipped with Automatic Adjusting Suspension (AAS)
   - Piston rod upper nut 1 turn but do not remove it
3. Place the lower end of the strut in the vise.
4. Install a coil spring compressor and compress the coil spring.

❋❋ CAUTION

Failure to fully compress the spring and hold it securely can be extremely dangerous.

5. Remove or disconnect the following:
   - Upper strut nut
   - Slowly release the coil spring tension

1. Cap
2. Piston rod nut
3. Mounting rubber
4. Thrust bearing
5. Upper spring seat
6. Upper rubber spring seat
7. Dust cover
8. Bound stopper

9. Coil spring
10. Lower rubber spring seat
11. Shock absorber

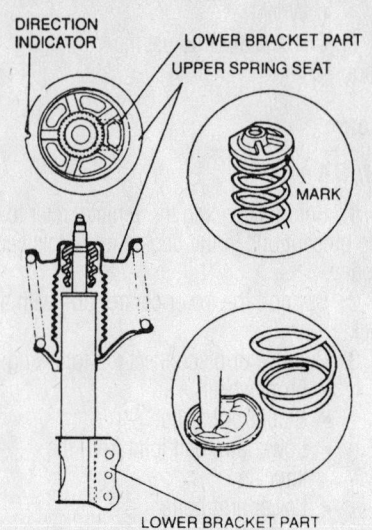

7923MG81

Be sure the end of the coil spring is in the step of the lower seat—626

   - Suspension support, dust seal, spring seat, spring insulators, coil spring and bumper
6. While pushing on the piston rod, be sure that the pull stroke is even and that there is no unusual noise or resistance. Also inspect for any oil leakage around the piston rod.

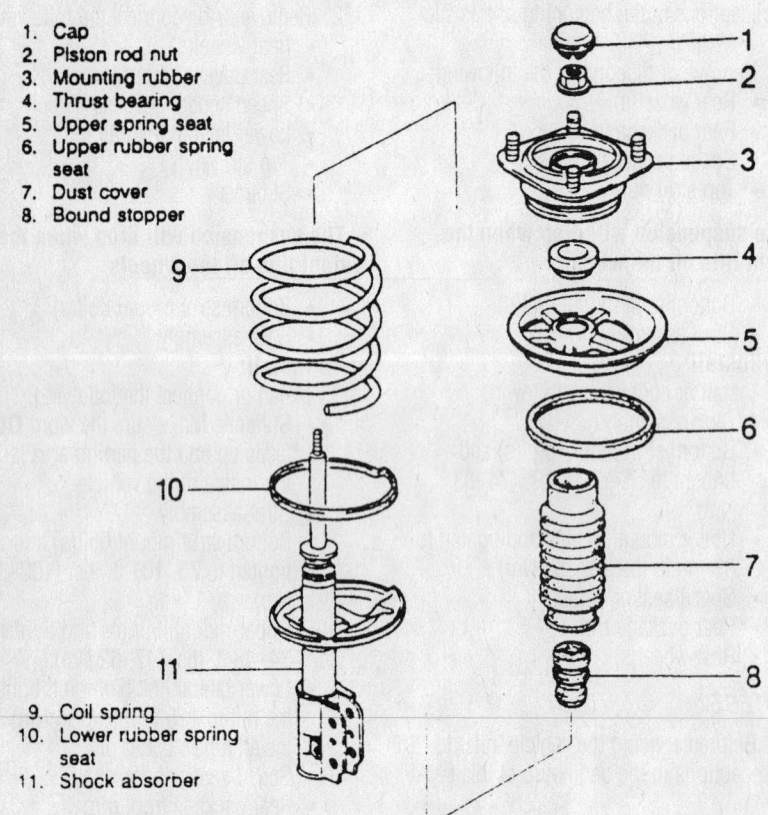

7923MG82

Exploded view of the front strut assembly—rear strut is similar

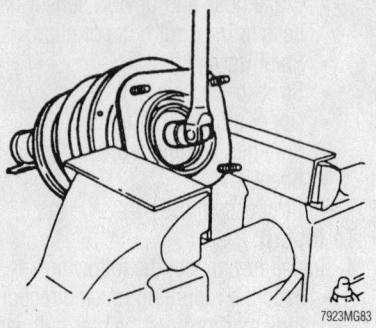

7923MG83

Secure the upper strut mount in a vise and loosen the piston rod nut one turn but do not remove it

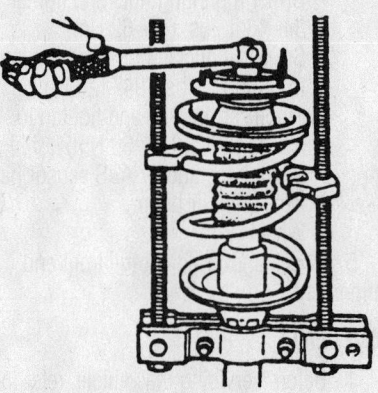

7923MG84

Use a coil spring compressor and relieve the spring tension from the upper mount, then remove the piston rod nut

7. Push the piston rod in, then release it. Be sure that the return rate is constant.
8. If the shock absorber does not operate as described, replace it.

To assemble:

9. Install or connect the following:
   - Strut assembly into a vise
   - Bump stopper and dust boot onto the piston rod
   - Temporarily install the upper spring seat, seat rubber and spring. Mark the seat, shock and spring assembly as illustrated for reassembly. Align the marks of the upper seat and coil spring. Protect the assembly with cloth and install the spring compressor.
   - Coil spring
10. Compress the coil spring with the spring compressor
11. Install or connect the following:
    - Rubber seat, the spring upper seat, the bearing and the mounting block
    - Piston rod upper nut
12. Be sure that the spring upper seat notched portion is facing inward and tighten the piston rod upper nut.

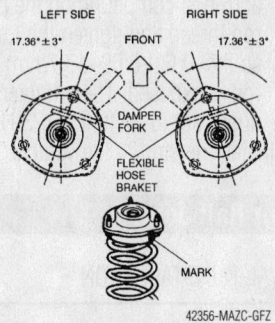

Mark the front seat, shock and spring assembly as illustrated for reassembly—Millenia

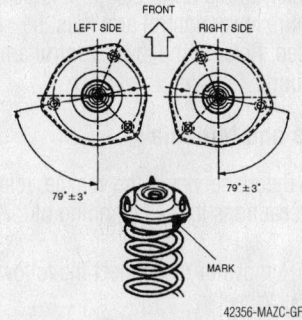

Mark the rear seat, shock and spring assembly as illustrated for reassembly—Millenia

13. Remove the spring compressor from the strut. Secure the upper mounting block in the vise and tighten the nut to 40–59 ft. lbs. (55–80 Nm) on 626 models and 24–33 ft. lbs. (32–46 ft. lbs. on Millenia models.

14. Be sure that the spring is well seated in the upper seats.

15. Install the strut to the vehicle.

Protégé

1. Before servicing the vehicle, refer to the precautions in the beginning of this section.

2. Remove or disconnect the following:
   • Strut from the vehicle
   • Piston rod upper nut 1 turn but do not remove it

3. Place the lower end of the strut in the vise.

4. Install a coil spring compressor and compress the coil spring.

✳✳ CAUTION

Failure to fully compress the spring and hold it securely can be extremely dangerous.

5. Remove or disconnect the following:
   • Upper strut nut

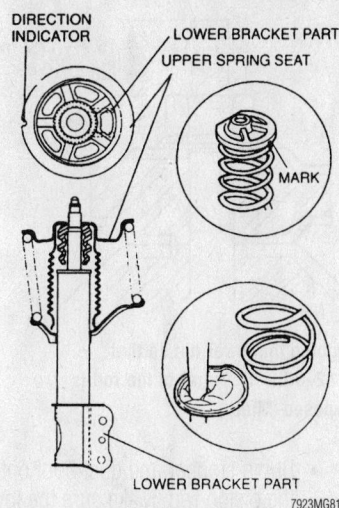

Be sure the end of the coil spring is in the step of the lower seat—Protégé

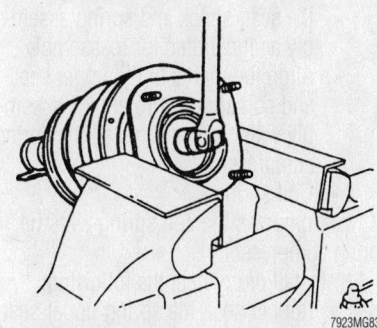

Secure the upper strut mount in a vise and loosen the piston rod nut one turn but do not remove it—Protégé

   • Slowly release the coil spring tension
   • Suspension support, dust seal, spring seat, spring insulators, coil spring and bumper

6. While pushing on the piston rod, be sure that the pull stroke is even and that there is no unusual noise or resistance. Also inspect for any oil leakage around the piston rod.

7. Push the piston rod in, then release it. Be sure that the return rate is constant.

8. If the shock absorber does not operate as described, replace it.

To assemble:

9. Install or connect the following:
   • Strut assembly into a vise
   • Bump stopper and dust boot onto the piston rod
   • Temporarily install the upper spring seat, seat rubber and spring so that the lower end of the spring is seated on the step of the lower spring seat. Mark the seat, shock

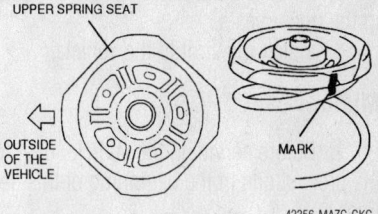

Mark the front seat, shock and spring assembly as illustrated for reassembly—Protégé

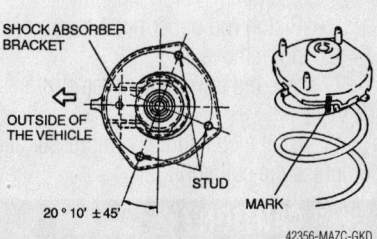

Mark the rear seat, shock and spring assembly as illustrated for reassembly—Protégé

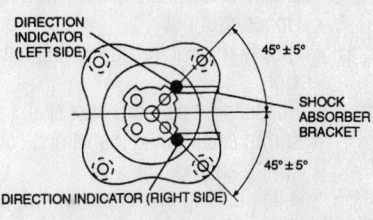

Make sure the bearing, rubber and nut are aligned as shown on the front suspension.—Protégé

and spring assembly as illustrated for reassembly. Align the marks of the upper seat and coil spring. Protect the assembly with cloth and install the spring compressor.
   • Coil spring

10. Compress the coil spring with the spring compressor

11. Install or connect the following:
   • Rubber seat, the spring upper seat, the bearing and the mounting block
   • Piston rod upper nut

12. Make sure that the marks on the shock absorber and upper spring seat are aligned. Tighten the piston rod upper nut.

13. Make sure the bearing, rubber and nut are aligned as shown on the front suspension.

14. Remove the spring compressor from the strut. Secure the upper mounting block in the vise and tighten the nut to 58–81 ft. lbs. (79–109 Nm) on the front strut and 41–49 ft. lbs. (55–67 ft. lbs. on the rear strut.

15. Be sure that the spring is well seated in the upper seats.

16. Install the strut to the vehicle.

Miata

1. Before servicing the vehicle, refer to the precautions in the beginning of this section.

2. Remove or disconnect the following:
- Strut from the vehicle
- Cap from the top of the strut, if equipped
- Piston rod upper nut 1 turn but do not remove it

3. Place the lower end of the strut in the vise.

4. Install a coil spring compressor and compress the coil spring.

※ CAUTION

Failure to fully compress the spring and hold it securely can be extremely dangerous.

5. Remove or disconnect the following:
- Upper strut nut
- Slowly release the coil spring tension
- Suspension support, dust seal, spring seat, spring insulators, coil spring and bumper

6. While pushing on the piston rod, be sure that the pull stroke is even and that there is no unusual noise or resistance. Also inspect for any oil leakage around the piston rod.

7. Push the piston rod in, then release it. Be sure that the return rate is constant.

8. If the shock absorber does not operate as described, replace it.

To assemble:

9. Install or connect the following:
- Strut assembly into a vise

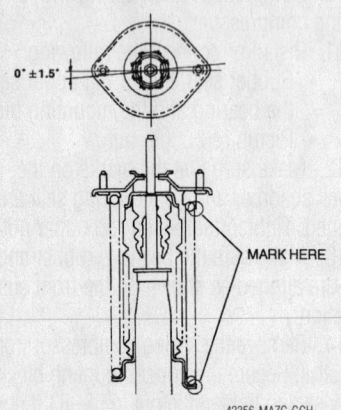

Mark the seat, shock and spring assembly as illustrated for reassembly–Miata

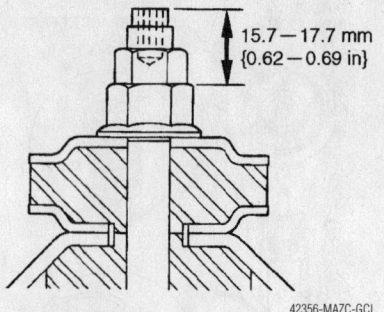

15.7 — 17.7 mm {0.62 — 0.69 in}

Tighten the lower nut so that 0.62–0.69–17.7mm) of the rod is exposed–Miata

- Bump stopper and dust boot onto the piston rod. Make sure the lower end of the stopper does not contact the cylinder.
- Temporarily install the upper spring seat, seat rubber and spring. Mark the seat, shock and spring assembly as illustrated for reassembly. Align the marks of the upper seat and coil spring. Protect the assembly with cloth and install the spring compressor.
- Coil spring

10. Compress the coil spring with the spring compressor.

11. Install or connect the following:
- Rubber seat, the spring upper seat, the bearing and the mounting block
- Piston rod upper nut. Apply an

1. Stabilizer nut
2. Retainer, bushing and spacer
3. Stabilizer bolt
4. Bolt, washer
5. Bolt
6. Bolt, nut
7. Nut
8. Washer
9. Lower control arm bushing (rear)
10. Nut
11. Bolt
12. Lower arm ball joint
13. Ball joint dust boot
14. Lower arm bushing (front)
15. Lower arm

anti–rust compound on the piston rod thread and tighten the lower nut so that 0.62–0.69–17.7mm) of the rod is exposed. Then, tighten the upper nut to 17 ft. lbs. (23 Nm)

12. Install the strut to the vehicle.

Lower Ball Joint

REMOVAL & INSTALLATION

Except Miata and Millenia

The lower ball joint is an integral part of the lower control and cannot be replaced separately. If the lower ball joint is defective, the entire lower control arm must be replaced. Refer to the lower control arm procedure.

Miata and Millenia

1. Before servicing the vehicle, refer to the precautions in the beginning of this section.

2. Remove or disconnect the following:
- Wheel
- Ball joint stud pinch bolt and nut from the steering knuckle
- Ball joint by prying it from the knuckle
- Ball joint to lower control arm bolt and nut

To install:

3. Install or connect the following:

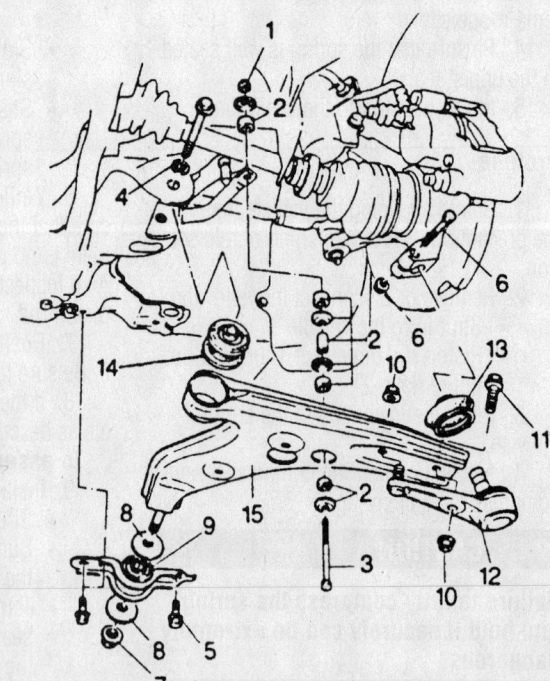

Exploded view of a common lower control arm with replaceable ball joint

- Ball joint to lower control arm. Torque the bolt to 86 ft. lbs. (116 Nm).
- Ball joint to the knuckle
- Ball joint to the steering knuckle. Torque the bolt to 57 ft. lbs. (77 Nm).
- Wheel

4. Check and/or adjust the front wheel alignment.

Upper Control Arm

REMOVAL & INSTALLATION

Miata

FRONT

1. Before servicing the vehicle, refer to the precautions in the beginning of this section.
2. Remove or disconnect the following:
   - Wheel
3. Support the lower control arm with a jack.
4. Remove or disconnect the following:
   - Cotter pin
   - Upper ball joint nut by loosening it
   - Ball joint by pressing it from the knuckle
   - Upper ball joint nut
   - Lower strut mounting bolt
   - Upper control arm bolt and nut
   - Upper control arm

To install:

5. Install the upper control arm.
6. Loosely tighten the bolt and nut.
7. Loosely install the lower strut mounting bolt.
   - Ball joint to the knuckle. Torque nut to 47–60 ft. lbs. (63–82 Nm).
   - New cotter pin to ball joint nut
   - Wheel
8. Torque upper control arm bolt to 87–101 ft. lbs. (118–137 Nm) and the lower strut mounting bolt to 69–86 ft. lbs. (94–116 Nm).
9. Check and/or adjust the front wheel alignment.

REAR

1. Before servicing the vehicle, refer to the precautions in the beginning of this section.
2. Remove or disconnect the following:
   - Wheel
3. Support the lower control arm with a jack.
4. Remove or disconnect the following:
   - Upper control arm bolts and nuts
   - Upper control arm

To install:

5. Install the upper control arm.
6. Loosely tighten the bolt and nut.
   - Wheel
7. Torque upper control arm nuts and bolts to 40–56 ft. lbs. (54–76 Nm).
8. Check and/or adjust the front wheel alignment.

CONTROL ARM BUSHING REPLACEMENT

All Mazda's use a pressed in control arm bushing, and the pressing can be done using two appropriately sized sockets (a press socket and a catch socket) and a large bench vise.

1. Position the control arm and the 2 sockets into a vise.
2. Position the press socket onto the control arm bushing.
3. Position the catch socket onto the control arm, opposite of the press socket.
4. Tighten the bench vise slowly and press the bushing into the catch socket.

To install:

5. Apply soapy water to the new control arm bushing.
6. Position the bushing against the control arm.
7. Using the same sockets, in the same positions, press the new bushing into the control arm.

Lower Control Arm

REMOVAL & INSTALLATION

626

FRONT

1. Before servicing the vehicle, refer to the precautions in the beginning of this section.
2. Remove or disconnect the following:
   - Wheel
   - Lower ball joint pinch bolt from the steering knuckle
   - Stabilizer bar link from the lower control arm
   - Lower control arm bolts and nuts
   - Lower control arm with the lower ball joint

To install:

3. Install or connect the following:
   - Lower control arm and loosely tighten the mounting nuts and bolts
   - Stabilizer link to the lower control arm. Torque the nut to 27–39 ft. lbs. (37–53 Nm).
   - Lower ball joint to the steering

knuckle. Torque the bolt to 26–41 ft. lbs. (35–56 Nm).
   - Wheel
4. With the vehicle at normal ride height, tighten the lower control arm mounting bolts. Torque the front bushing through-bolt to 40–59 ft. lbs. (55–80 Nm) and the rear bushing strap bolts to 69–96 ft. lbs. (94–131 Nm).
5. Check and/or adjust the front wheel alignment.

Protégé

FRONT

1. Before servicing the vehicle, refer to the precautions in the beginning of this section.
2. Remove or disconnect the following:
   - Wheel
   - Rear lower arm bolt
   - Lower ball joint pinch bolt from the steering knuckle
   - Bracket
   - Lower control arm with the lower ball joint

To install:

3. Install or connect the following:
   - Bracket and tighten the bolts to 68–97 ft. lbs. (94–126 Nm).
   - Lower control arm and loosely tighten the mounting nuts and bolts
   - Lower ball joint to the steering knuckle. Torque the bolt to 32–43 ft. lbs. (44–58 Nm).
   - Wheel
4. With the vehicle at normal ride height, tighten the lower control arm rear mounting bolts to 69–94 ft. lbs. (94–126 Nm).
5. Check and/or adjust the front wheel alignment.

Miata

1. Before servicing the vehicle, refer to the precautions in the beginning of this section.
2. Remove or disconnect the following:
   - Wheel
   - Undercover
   - Cotter pin and nut from the tie-rod end
   - Tie-rod end from the steering knuckle
   - Stabilizer bar link bolt and the lower strut mounting bolt
   - Cotter pin
   - Lower ball joint by loosening it
3. With the nut protecting the ball joint stud, separate the stud from the knuckle. Remove the nut.

- Dust boot, if necessary
- Lower control arm bolts, nuts and adjusting cams
- Lower control arm

To install:

4. Install or connect the following:
- Dust boot, if removed using a press. Always fill the inside of the new boot with grease prior to installation.
- Lower control arm by loosely tightening the bolts and nuts
- Lower ball joint to the knuckle. Torque the nut to 42–57 ft. lbs. (57–77 Nm).
- New cotter pin
- Lower strut and stabilizer link bolts by loosely tightening them
- Tie-rod end to the steering knuckle. Torque the nut to 22–32 ft. lbs. (30–44 Nm).
- New cotter pin
- Wheel and undercover

5. With the vehicle at normal ride height, torque the lower control arm bolts to 69–83 ft. lbs. (94–113 Nm) for the inner and 69–86 ft. lbs. (94–112 Nm) for the outer.

6. Torque the lower strut mounting bolt to 69–86 ft. lbs. (94–116 Nm) and the stabilizer link bolt to 32–44 lbs. (44–60 Nm).

7. Check and/or adjust the front wheel alignment.

Millenia

1. Remove or disconnect the following:
- Transverse member
- Power steering return hose and pressure pipe
- Intermediate steering shaft bolt

2. Support the engine from the top.
- Engine mount member
- Bolts for engine mount No. 1
- Lower strut mounting bolt
- Tie-rod end from the steering knuckle
- Upper lateral link ball joint
- Lower ball joint
- Stabilizer control link nut
- Gusset

3. Support the crossmember using a jack.
- Crossmember mounting nuts and lower the crossmember to gain clearance
- Lower arm assembly

To install:

4. Install or connect the following:
- Lower arm assembly to the vehicle
- Crossmember mounting bolts
- Gusset. Torque the bolts to 58–86 ft. lbs. (79–116 Nm).

- Stabilizer control link. Torque the nut to 32–44 ft. lbs. (44–60 Nm).
- Lower ball joint. Torque the bolts to 58–86 ft. lbs. (79–116 Nm) and the nut to 86–115 ft. lbs. (116–156 Nm).
- Upper lateral link nut and bolt. Torque to 58–86 ft. lbs. (79–116 Nm).
- Tie-rod end. Torque the nut to 41–59 ft. lbs. (55–80 Nm).
- Strut lower mounting bolt. Torque to 73–101 ft. lbs. (98–137 Nm).
- No. 1 engine mount. Torque the bolts to 32–44 ft. lbs. (44–60 Nm).
- Engine mount member. Torque the bolts to 50–68 ft. lbs. (67–93 Nm) and the nuts to 55–77.3 ft. lbs. (75–104 Nm).

5. Remove the engine support tool.
- Intermediate steering shaft. Torque the bolt to 14–19 ft. lbs. (18–26 Nm).
- Power steering pressure pipe and return hose
- Transverse member

6. Check the power steering fluid and fill to proper level, bleed if necessary.

7. Check and/or adjust the front end alignment.

REAR

Miata

1. Before servicing the vehicle, refer to the precautions in the beginning of this section.

2. Remove or disconnect the following:
- Wheel

3. Support the lower control arm with a jack.

4. Remove or disconnect the following:
- Stabilizer bar link nut
- Lower shock absorber bolt
- Lower control arm bolts and nuts
- Lower control arm

To install:

5. Install or connect the following:
- Lower control arm and loosely tighten the bolts and nuts
- Lower shock absorber bolt and tighten to 70 ft. lbs. (85 Nm)
- Stabilizer bar link nut and tighten to 32–44 ft. lbs. (44–60 Nm)
- Wheel

6. Torque lower control arm inner nuts and bolts to 54–70 ft. lbs. (73–95 Nm) and the outer nuts and bolts to 47–54 ft. lbs. (63–74 Nm).

7. Check and/or adjust the front wheel alignment.

CONTROL ARM BUSHING REPLACEMENT

Except Protégé

1. Before servicing the vehicle, refer to the precautions in the beginning of this section.

All Mazda's use a pressed in control arm bushing, and the pressing can usually be done using 2 appropriately sized sockets (a press socket and a catch socket) and a large vise.

2. Position the control arm and the 2 sockets into a vise.

3. Position the press socket onto the control arm bushing.

4. Position the catch socket onto the control arm, opposite of the press socket.

5. Tighten the vise slowly and press the bushing into the catch socket.

To install:

6. Apply soapy water to the new control arm bushing.

7. Position the bushing against the control arm.

8. Using the same sockets, in the same positions, press the new bushing into the control arm.

Protégé

1. Before servicing the vehicle, refer to the precautions in the beginning of this section.

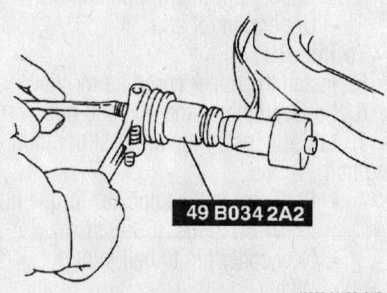

42356-MAZC-GKF

Use tool 49 B034 2A2 to remove and install the front bushing—Protégé

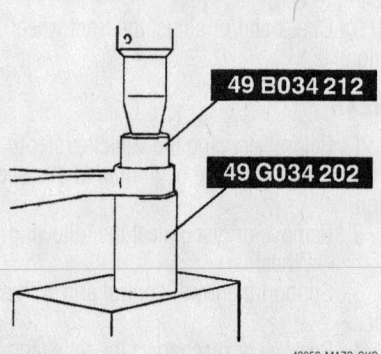

42356-MAZC-GKG

Use tools 49 B034 212 and 49 G034 202 to remove the rear bushing—Protégé

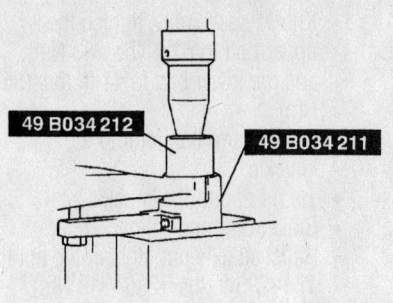

42356-MAZC-GKZ

Use tools 49 B034 212 and 49 G034 211 to install the rear bushing—Protégé

2. Position the control arm into a vise. Cut away the projecting rubber from the front bushing. Use tool 49 B034 2A2 to remove the bushing.

3. Remove the rear bushing using a

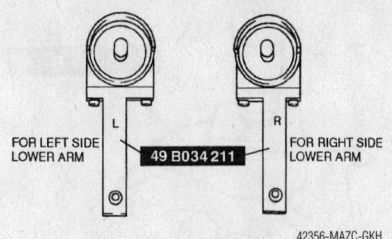

42356-MAZC-GKH

Align the mark of the lower arm and the small projection of the lower arm bushing (rear) when installing the rear bushing—Protégé

press and tools 49 B034 212 and 49 G034 202.

To install:

4. When installing the rear bushing on the lower arm, align the mark of the lower

arm and the small projection of the lower arm bushing (rear) as illustrated, set the arm onto the press and press the bushings into position.

5. Apply soapy water to the new control arm bushing.

6. Install the rear bushing using a press and tools 49 B034 212 and 49 G034 211.

7. Position the control arm into a vise. Use tool 49 B034 2A2 to pull the bushing into the arm bore.

Wheel Bearings

ADJUSTMENT

The front and rear wheel bearings are not adjustable. If the bearings become loose or make noise, they must be replaced.

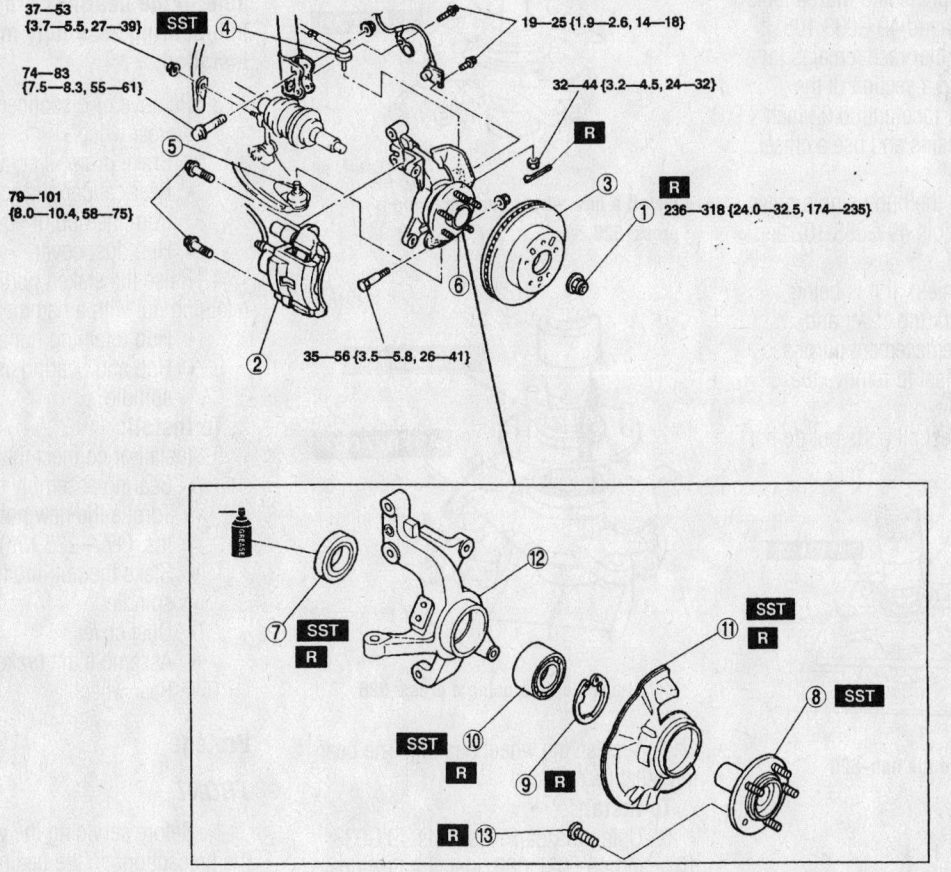

N·m {kgf·m, ft-lbf}

1	Locknut	8	Wheel hub component
2	Brake caliper component	9	Retaining ring
3	Disc plate	10	Wheel bearing
4	Tie-rod end ball joint	11	Dust cover
5	Lower arm ball joint	12	Steering knuckle
6	Wheel hub, steering knuckle, dust cover	13	Hub bolt
7	Oil seal		

Exploded view of the front wheel bearing and knuckle assembly–626

42356-MAZC-G83

REMOVAL & INSTALLATION

626

FRONT

1. Before servicing the vehicle, refer to the precautions in the beginning of this section.

2. Refer to the illustration for component location and torque specifications.

3. Remove or disconnect the following:
- Wheels
- Halfshaft axle nut, unstake the nut prior to removal
- Brake caliper and rotor
- Tie rid end from the knuckle
- Control arm ball joint from the knuckle
- Knuckle assembly
- Inner oil seal from the knuckle
- Hub using a press and Mazda tools 49 G033 103 and 49 G033 105. If the bearing inner race remains in the hub, grind a section of the bearing inner race until 0.02 inch (0.5mm) remains and use a chisel to remove it.
- Bearing from the hub using a press and Mazda tools 49 G033 102 and 49 G033 106
- Brake dust shield, if it is being replaced. Mark the cover and knuckle for replacement purposes and use a chisel to remove the shield.

4. Clean and inspect all parts but do not

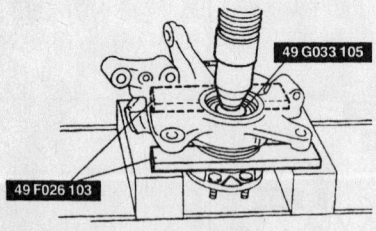

49 G033 105
49 F026 103

Use a press to remove the hub–626

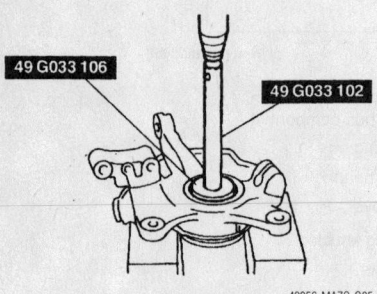

49 G033 106
49 G033 102

42356-MAZC-G85

Use a press to remove the wheel bearing–626

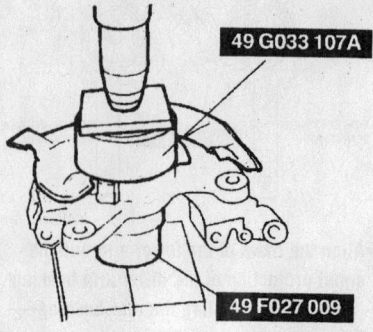

49 G033 107A
49 F027 009

42356-MAZC-G86

Install a new dust cover, if removed–626

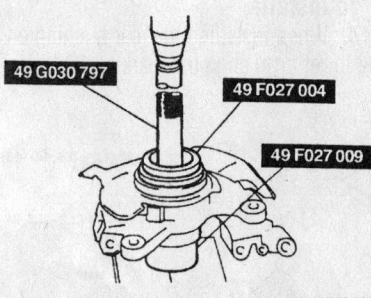

49 G030 797
49 F027 004
49 F027 009

42356-MAZC-G87

Install a new wheel bearing using a press–626

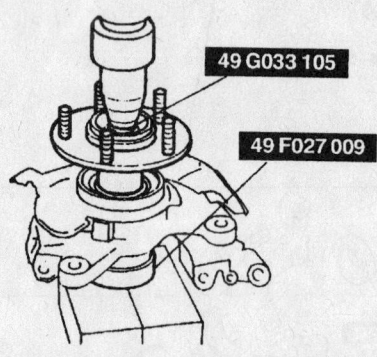

49 G033 105
49 F027 009

42356-MAZC-G88

Install a wheel hub using a press–626

wash or clean the wheel bearing. The bearing must be replaced.

To install:

5. Using Mazda Press tools 49 G033 107a and 49 F027 009, install a new dust shield cover assembly to the knuckle, if removed.

6. Using press tools 49 G033 797, 49 F027 004 and 49 F027 003; press a new wheel bearing into the knuckle assembly.

7. Install or connect the following:
- Wheel bearing retaining ring

8. Using press tools 49 G033 105, 49 F027 009; press in the hub assembly.
- New oil seal using installation tool 49 V001 795

- Knuckle assembly, tighten the upper bolt to 61 ft. lbs. (83 Nm) and the lower bolt to 41 ft. lbs. (56 Nm)
- Control arm ball joint to the knuckle
- Tie rid end to the knuckle
- Brake rotor and caliper
- Halfshaft axle nut, tighten the nut to 174–235 ft. lbs. (236–318 Nm) and stake the nut
- Wheels

REAR

1. Before servicing the vehicle, refer to the precautions in the beginning of this section.

2. Refer to the illustration for component location and torque specifications.

➡ **The wheel bearings are not serviceable. If the bearings are bad, a new hub/bearing assembly must be installed.**

3. Remove or disconnect the following:
- Rear wheels
- Brake drum, if equipped
- Rear caliper and rotor assembly from the hub, if equipped
- Hub dust cover

4. Raise the staked portion of the hub retaining nut with a hammer and chisel.
- Hub retaining nut and discard it
- Hub and bearing assembly from the spindle

To install:

5. Install or connect the following:
- Bearing assembly on the spindle. Torque the new nut to 131–173 ft. lbs. (177–235 Nm).
- Stake the nut into the groove in the spindle
- Dust cover
- Assemble the brakes
- Rear wheel

Protégé

FRONT

1. Before servicing the vehicle, refer to the precautions in the beginning of this section.

2. Refer to the illustration for component location and torque specifications.

3. Remove or disconnect the following:
- Wheels
- Halfshaft axle nut, Unstake the nut prior to removal
- Brake caliper and rotor
- Tie rid end from the knuckle
- Control arm ball joint from the knuckle

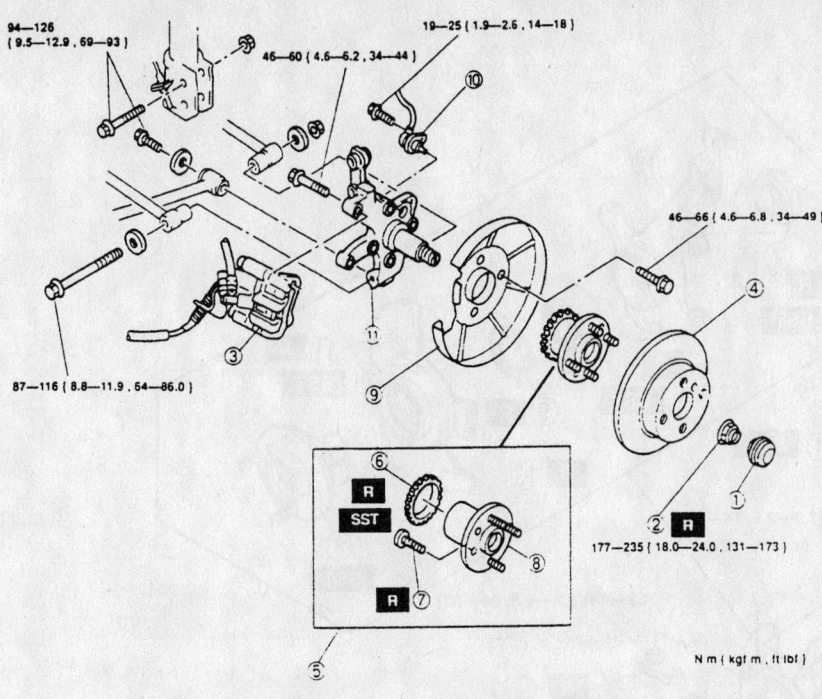

1. Hub cap
2. Locknut
3. Brake caliper assembly
4. Disc plate
5. Wheel hub assembly
   Inspect for damage
   Inspect bearing for damage and rough
   rotation
6. ABS sensor rotor
7. Hub bolt
8. Wheel hub
9. Dust cover
   Inspect for damage and cracks
10. ABS wheel-speed sensor
11. Hub spindle
   Inspect for damage and cracks

Exploded view of the rear wheel hub and bearing assembly (disc brake model shown, drum is similar)—626

- Knuckle assembly
- Inner oil seal from the knuckle
- Hub using a press and Mazda tools 49 G030 727, 49 G033 102 and 49 F026 103. If the bearing inner race remains in the hub, grind a section of the bearing inner race until 0.02 inch (0.5mm) remains and use a chisel to remove it.
- Bearing from the hub using a press and Mazda tools 49 F027 005, 49 F027 003 and 49 F026 103
- Brake dust shield, if it is being replaced. Mark the cover and knuckle for replacement purposes and use a chisel to remove the shield.

4. Clean and inspect all parts but do not wash or clean the wheel bearing. The bearing must be replaced.

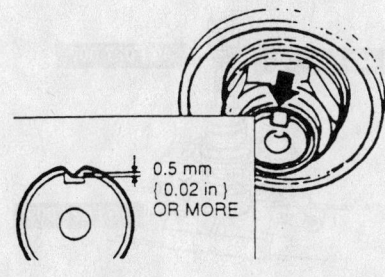

0.5 mm
{ 0.02 in }
OR MORE

When installing the new hub nut, be sure to stake it into the notch on the spindle

To install:

5. Using Mazda Press tools 49 E033 101 and 49 F027 009, install a new dust shield cover assembly to the knuckle, if removed.

6. Using press tools 49 F027 003, 49 F027 007 and 49 F027 009; press a new wheel bearing into the knuckle assembly.

7. Install or connect the following:
   - Wheel bearing retaining ring

8. Using press tool 49 F027 009 press in the hub assembly.
   - New oil seal using installation tool 49 V001 795
   - Knuckle assembly, tighten the upper bolt to 93 ft. lbs. (126 Nm) and the lower bolt to 43 ft. lbs. (58 Nm)
   - Control arm ball joint to the knuckle
   - Tie rid end to the knuckle
   - Brake rotor and caliper
   - Halfshaft axle nut, tighten the nut to 174–235 ft. lbs. (236–318 Nm) and stake the nut
   - Wheels

REAR—DRUM BRAKES

1. Before servicing the vehicle, refer to the precautions in the beginning of this section.

2. Refer to the illustration for component location and torque specifications.

➡**The wheel bearings are not service-able. If the bearings are bad, a new**

hub/bearing assembly must be installed.

3. Remove or disconnect the following:
   - Rear wheels
   - Wheel speed sensor

4. Raise the staked portion of the hub retaining nut with a hammer and chisel.
   - Hub retaining nut and discard it
   - Brake drum
   - Hub and bearing assembly from the spindle with the ABS sensor rotor, if equipped with ABS
   - ABS sensor rotor using a chisel

To install:

5. Position the ABS sensor rotor on the hub as illustrated.

6. Position tool 49 B026 103 so that the marking **A** faces the bottom.

7. Press the rotor onto the hub.

8. Install or connect the following:
   - Bearing assembly on the spindle. Torque the new nut to 131–173 ft. lbs. (177–235 Nm).
   - Stake the nut into the groove in the spindle
   - Dust cover
   - Assemble the brakes
   - Rear wheel

REAR—DISC BRAKES

1. Before servicing the vehicle, refer to the precautions in the beginning of this section.

2. Refer to the illustration for component location and torque specifications.

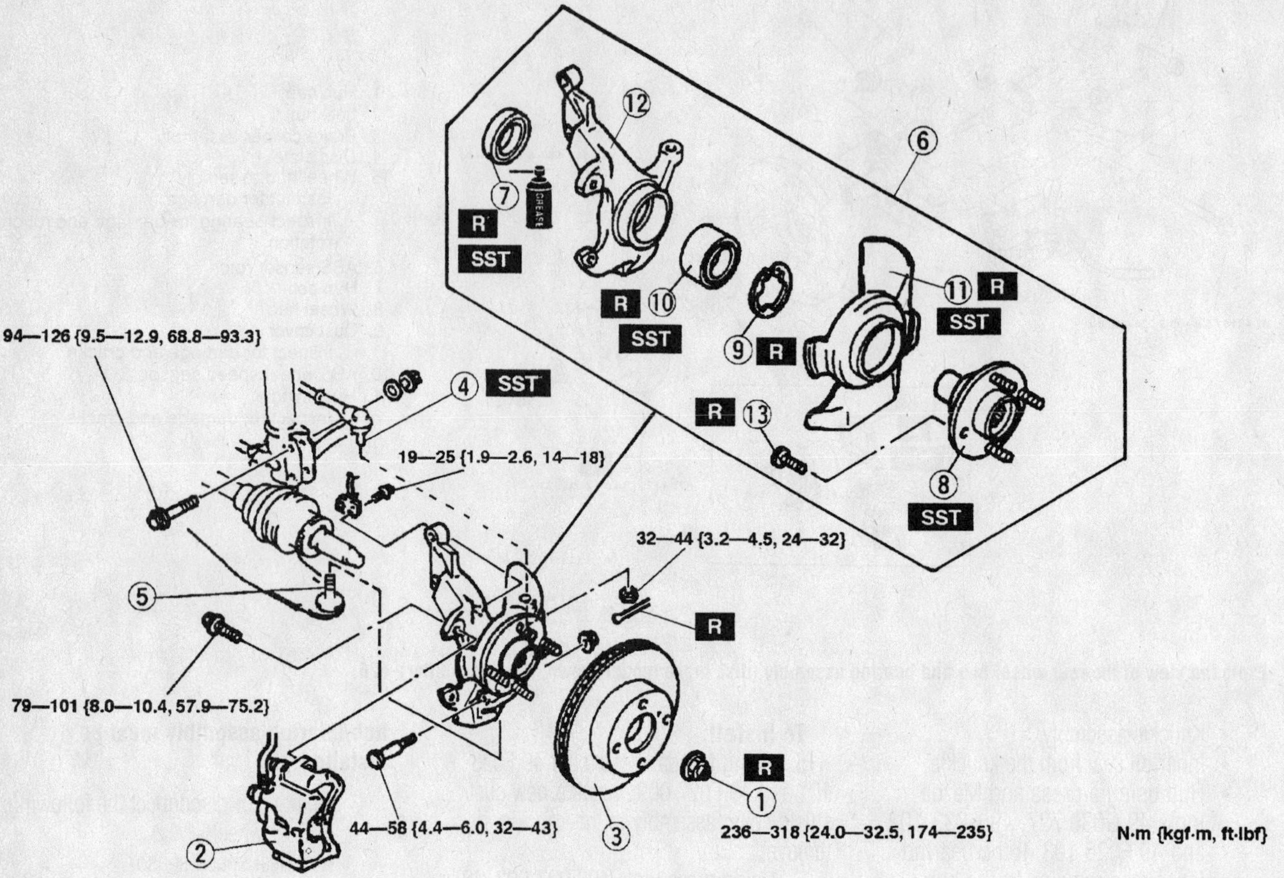

94—126 {9.5—12.9, 68.8—93.3}

19—25 {1.9—2.6, 14—18}

79—101 {8.0—10.4, 57.9—75.2}

32—44 {3.2—4.5, 24—32}

44—58 {4.4—6.0, 32—43}

236—318 {24.0—32.5, 174—235}

N·m {kgf·m, ft·lbf}

1	Locknut	8	Wheel hub component
2	Brake caliper component	9	Retaining ring
3	Disc plate	10	Wheel bearing
4	Tie-rod end	11	Dust cover
5	Lower arm ball joint	12	Steering knuckle
6	Wheel hub, steering knuckle, dust cover	13	Hub bolt
7	Oil seal		

42356-MAZC-GKI

Exploded view of the front wheel bearing and knuckle assembly–Protégé

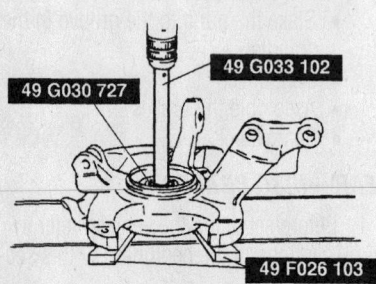

49 G033 102

49 G030 727

49 F026 103

42356-MAZC-GKJ

Use a press to remove the hub–Protégé

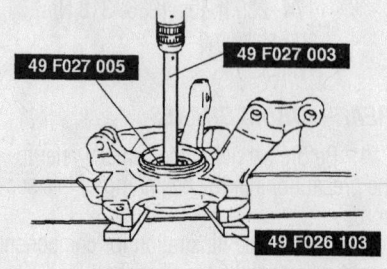

49 F027 005

49 F027 003

49 F026 103

42356-MAZC-GKK

Use a press to remove the wheel bearing–Protégé

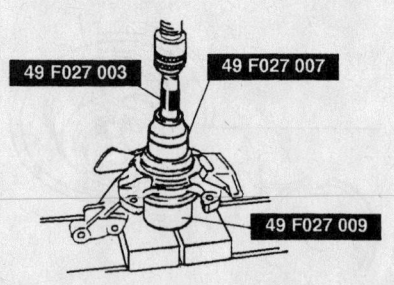

49 F027 003

49 F027 007

49 F027 009

42356-MAZC-GKL

Install a new wheel bearing using a press–Protégé

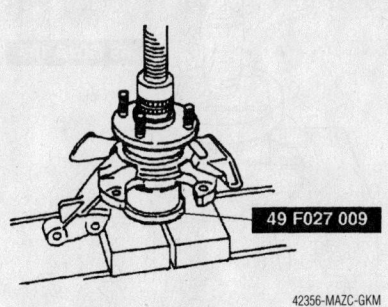

49 F027 009

42356-MAZC-GKM

Install a wheel hub using a press–Protégé

➡ The wheel bearings are not service-able. If the bearings are bad, a new hub/bearing assembly must be installed.

3. Remove or disconnect the following:
- Rear wheels
- Wheel speed sensor

4. Raise the staked portion of the hub retaining nut with a hammer and chisel.
- Hub retaining nut and discard it
- Brake caliper and rotor
- Hub and bearing assembly from the spindle with the ABS sensor rotor, if equipped with ABS

- ABS sensor rotor using a chisel

To install:

5. Position the ABS sensor rotor on the hub as illustrated.

6. Position tool 49 B026 105 so that the marking **B** faces the bottom.

7. Press the rotor onto the hub.

8. Install or connect the following:
- Bearing assembly on the spindle. Torque the new nut to 131–173 ft. lbs. (177–235 Nm).
- Stake the nut into the groove in the spindle
- Dust cover

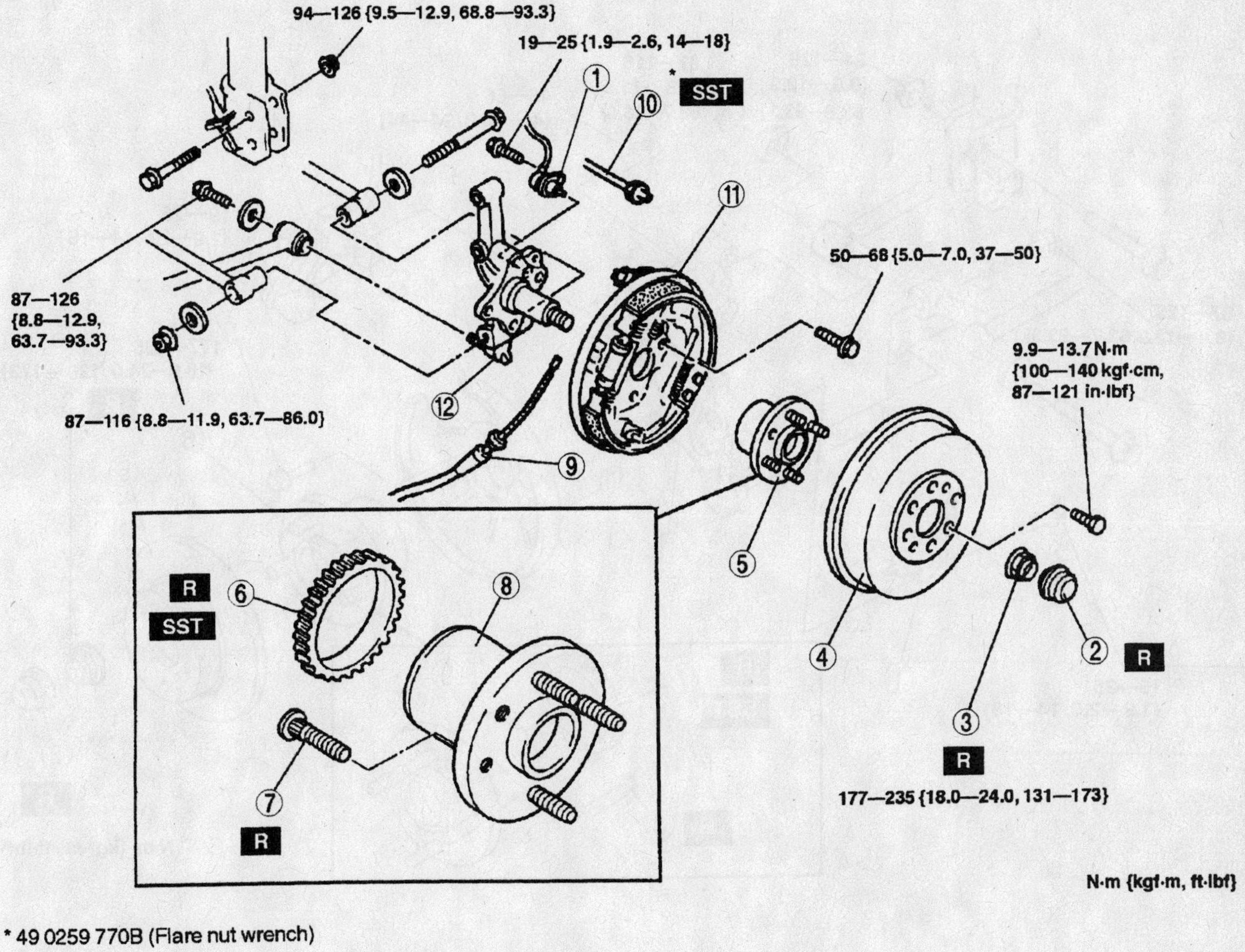

94—126 {9.5—12.9, 68.8—93.3}

19—25 {1.9—2.6, 14—18}

SST

50—68 {5.0—7.0, 37—50}

87—126 {8.8—12.9, 63.7—93.3}

9.9—13.7 N·m {100—140 kgf·cm, 87—121 in·lbf}

87—116 {8.8—11.9, 63.7—86.0}

177—235 {18.0—24.0, 131—173}

N·m {kgf·m, ft·lbf}

R SST

* 49 0259 770B (Flare nut wrench)

1	ABS-wheel speed sensor	7	Hub bolt
2	Hub cap	8	Wheel hub
3	Locknut	9	Parking brake cable
4	Brake drum	10	Brake pipe
5	Wheel hub and ABS sensor rotor (with ABS)	11	Rear brake component
6	ABS sensor rotor (with ABS)	12	Hub spindle

Exploded view of the rear wheel hub and bearing assembly–Protégé with drum brakes

42356-MAZC-GKO

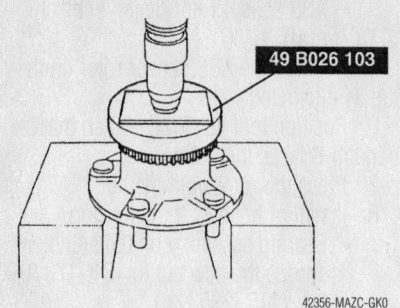

49 B026 103
42356-MAZC-GK0

Position the ABS sensor rotor on the hub–Protégé with drum brakes

WHEEL HUB
49 B026 103
3 mm {0.1 in}
A
ABS SENSOR ROTOR
42356-MAZC-GKP

Position tool 49 B026 103 so that the marking A faces the bottom–Protégé with drum brakes

49 G026 105
42356-MAZC-GKQ

Position the ABS sensor rotor on the hub–Protégé with disc brakes

94—126
{9.5—12.9,
68.8—93.3}
87—116
{8.8—11.9,
63.7—86.0}
46—66
{4.6—6.8, 34—44}
19—25
{1.9—2.6, 14—18}
87—126
{8.8—12.9, 63.7—93.3}
177—235
{18.0—24.0, 130—173}
R
19—25
{1.9—2.6, 14—18}
R
SST
R
R
N·m {kgf·m, ft·lbf}
1 ABS wheel-speed sensor
2 Hub cap
3 Locknut
4 Brake caliper component
5 Disc plate
6 Wheel hub and ABS sensor rotor
7 ABS sensor rotor
8 Hub bolt
9 Wheel hub component
10 Dust cover
11 Hub spindle
42356-MAZC-GKN

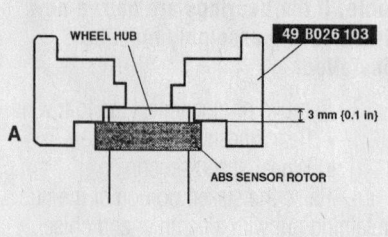

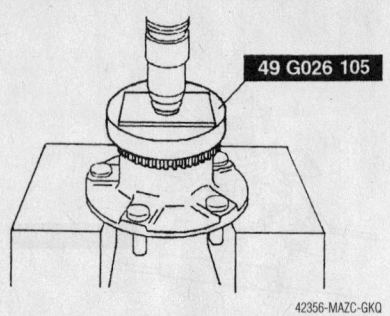

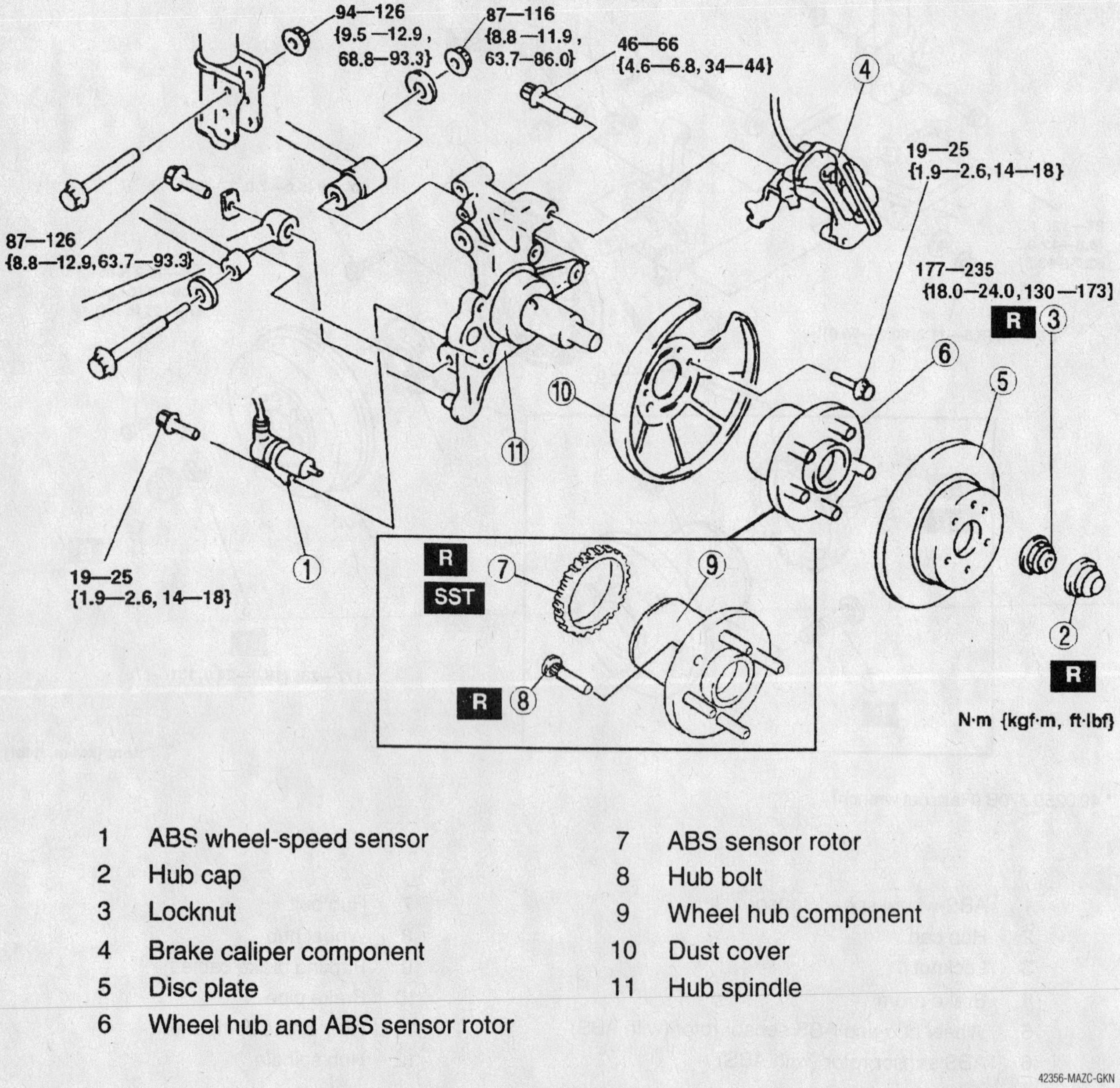

Exploded view of the rear wheel hub and bearing assembly–Protégé with disc brakes

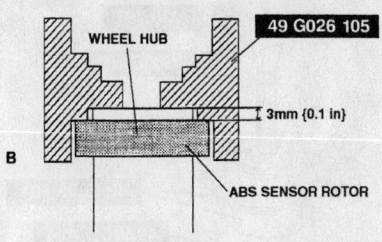

42356-MAZC-GKR

Position tool 49 B026 105 so that the marking B faces the bottom–Protégé with disc brakes

- Assemble the brakes
- Rear wheel

Miata

FRONT

1. Before servicing the vehicle, refer to the precautions in the beginning of this section.

➡**The wheel bearings are not serviceable. If the bearings are bad, a new hub/bearing assembly must be installed.**

2. Remove or disconnect the following:
   - Front wheels
   - Hub center dust cap

3. Raise the staked portion of the hub locknut with a hammer and chisel. Lock the hub by applying the brakes and loosen the nut.
   - Brake caliper
   - Brake rotor
   - Hub locknut and discard it
   - Hub from the spindle

To install:

4. Install or connect the following:
   - Hub to the spindle
   - Hub locknut. Torque the nut to 123–159 ft. lbs. (167–215 Nm).
   - Stake the new nut into the spindle groove using a dull chisel
   - Brake rotor
   - Brake caliper
   - Hub center dust cap
   - Front wheels

REAR

1. Before servicing the vehicle, refer to the precautions in the beginning of this section.

2. Remove the rear wheels.

3. Raise the staked portion of the hub locknut with a hammer and chisel. Lock the hub by applying the brakes and loosen the nut.

4. Remove or disconnect the following:
   - Brake caliper and position it aside

   - Brake rotor
   - Hub locknut and discard it
   - Anti-lock Brake System (ABS) wheel speed sensor, if equipped
   - Speed sensor bracket from the rear knuckle
   - Upper and lower knuckle mounting through bolts
   - Knuckle assembly from vehicle
   - Rear bearing oil seal by prying it from the knuckle
   - Wheel hub by pressing it from the knuckle
   - Retaining snapring from within the knuckle
   - Bearing assembly from the knuckle, once the wheel hub is removed

5. The inner race of the bearing may remain on the hub. Use a chisel and move the bearing race away from the rear hub flange. Once there is enough clearance, press the race from the hub.

To install:

6. Press the new bearing into the rear knuckle assembly.

7. Apply some grease to the wheel bearing inner race and press the rear hub into the bearing. Make sure to position a suitable support on the backside of the bearing.

8. Install or connect the following:
   - Retaining snapring to the knuckle
   - New rear oil seal by lubricating it with grease and pressing it into the rear knuckle
   - Knuckle assembly to the vehicle
   - Upper and lower knuckle mounting through bolts. Torque the upper bolt to 34–49 ft. lbs. (47–66 Nm) and the lower bolt to 47–54 ft. lbs. (63–74 Nm).
   - Speed sensor bracket to the rear knuckle
   - ABS wheel speed sensor, if equipped
   - New hub locknut. Torque the nut to 174–235 ft. lbs. (236–318 Nm).
   - Stake the new nut into the spindle groove using a dull chisel
   - Brake rotor
   - Brake caliper
   - Rear wheels

Millenia

FRONT

1. Before servicing the vehicle, refer to the precautions in the beginning of this section.

2. Refer to the illustration for component location and torque specifications.

3. Remove or disconnect the following:
   - Wheels
   - Halfshaft axle nut, Unstake the nut prior to removal.
   - Brake caliper and rotor
   - Tie rid end from the knuckle
   - Upper leading link ball joint
   - Upper lateral link ball joint
   - Lower ball joint
   - Knuckle assembly
   - Snap ring
   - Hub using a press and Mazda tools 49 F026 102 and 49 W017 101. If the bearing inner race remains in the hub, grind a section of the bearing inner race until 0.02 inch (0.5mm) remains and use a chisel to remove it.
   - Bearing from the hub using a press and Mazda tools 49 G0797, 49 G026 102, 49 E033 101 and 49 W017 101
   - Brake dust shield, if it is being replaced. Mark the cover and knuckle for replacement purposes and use a chisel to remove the shield.

4. Clean and inspect all parts but do not wash or clean the wheel bearing. The bearing must be replaced.

To install:

5. Using Mazda Press tools 49 G033 107a and 49 G026 103, install a new dust shield cover assembly to the knuckle, if removed.

6. Using press tools 49 G026 103 and 49 F026 102, press a new wheel bearing into the knuckle assembly.

7. Install or connect the following:
   - Wheel bearing retaining ring

8. Using press tools 49 G033 105, 49 F027 009; press in the hub assembly.
   - Snap ring
   - Knuckle assembly, tighten the upper bolt to the specifications shown in the illustration
   - Lower ball joint
   - Upper lateral link ball joint
   - Upper leading link ball joint
   - Tie rid end from the knuckle
   - Brake caliper and rotor
   - Halfshaft axle nut, tighten the nut to 174–235 ft. lbs. (236–318 Nm) and stake the nut
   - Wheels

REAR

1. Before servicing the vehicle, refer to the precautions in the beginning of this section.

2. Refer to the illustration for component location and torque specifications.

➡The wheel bearings are not serviceable. If the bearings are bad, a new hub/bearing assembly must be installed.

3. Remove or disconnect the following:
   - Rear wheels
   - Rear caliper and rotor assembly from the hub
   - Hub dust cover
4. Raise the staked portion of the hub retaining nut with a hammer and chisel.
   - Hub retaining nut and discard it

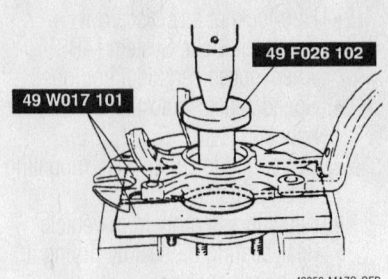

49 F026 102
49 W017 101
42356-MAZC-GFD

Use a press to remove the hub–Millenia

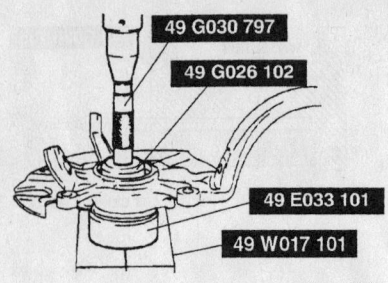

49 G030 797
49 G026 102
49 E033 101
49 W017 101
42356-MAZC-GFE

Use a press to remove the wheel bearing–Millenia

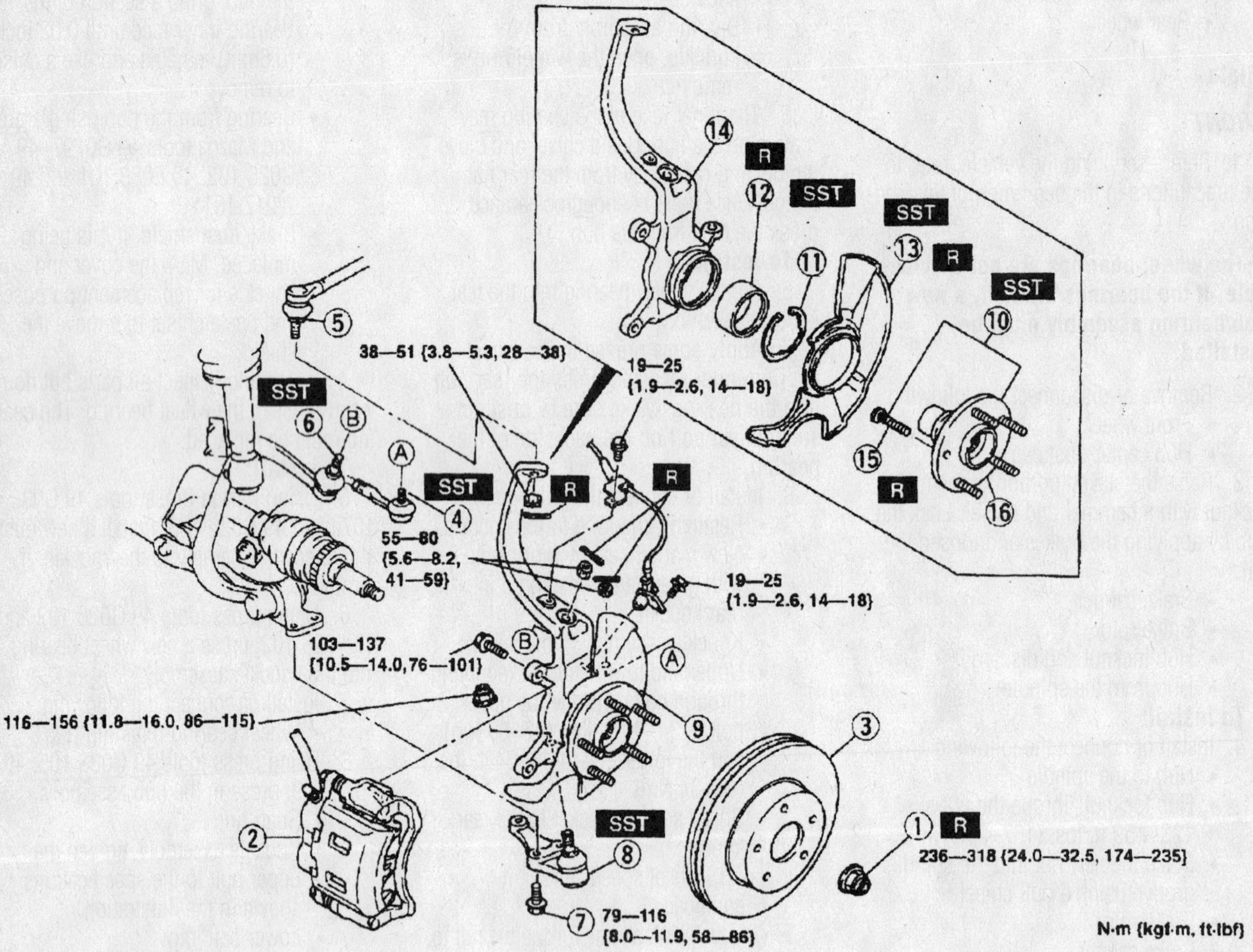

SST
R
38—51 {3.8—5.3, 28—38}
19—25
{1.9—2.6, 14—18}
55—80
{5.6—8.2,
41—59}
19—25
{1.9—2.6, 14—18}
103—137
{10.5—14.0,76—101}
116—156 {11.8—16.0, 86—115}
79—116
{8.0—11.9, 58—86}
236—318 {24.0—32.5, 174—235}
N·m {kgf·m, ft·lbf}

1	Locknut	9	Front wheel hub, steering knuckle
2	Brake caliper component	10	Wheel hub component
3	Disc plate	11	Snap ring
4	Tie-rod end ball joint	12	Wheel bearing
5	Upper leading link ball joint	13	Dust cover
6	Upper lateral link ball joint	14	Steering knuckle
7	Lower ball joint bolt	15	Hub bolt
8	Lower ball joint	16	Wheel hub

42356-MAZC-GFC

Exploded view of the front wheel bearing and knuckle assembly–Millenia

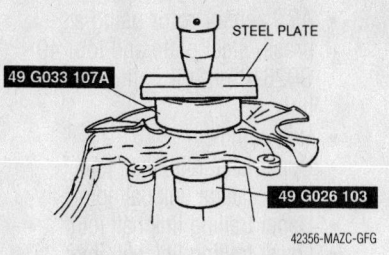

49 G033 107A
STEEL PLATE
49 G026 103
42356-MAZC-GFG

Install a new wheel bearing using a press–Millenia

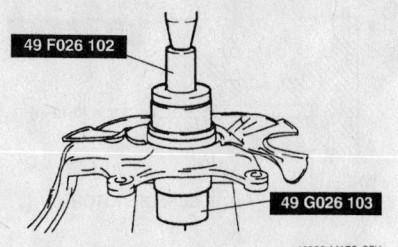

49 F026 102
49 G026 103
42356-MAZC-GFH

Install a wheel hub using a press–Millenia

- Parking brake shoe assembly
- Parking brake cable
- Backing plate
- Rear lower lateral link ball joint
- Lower trailing link ball joint
- Upper trailing link ball joint
- Lower lateral link ball joint
- Upper lateral link ball joint
- Hub spindle
- ABS sensor rotor using a chisel
- Hub bolts

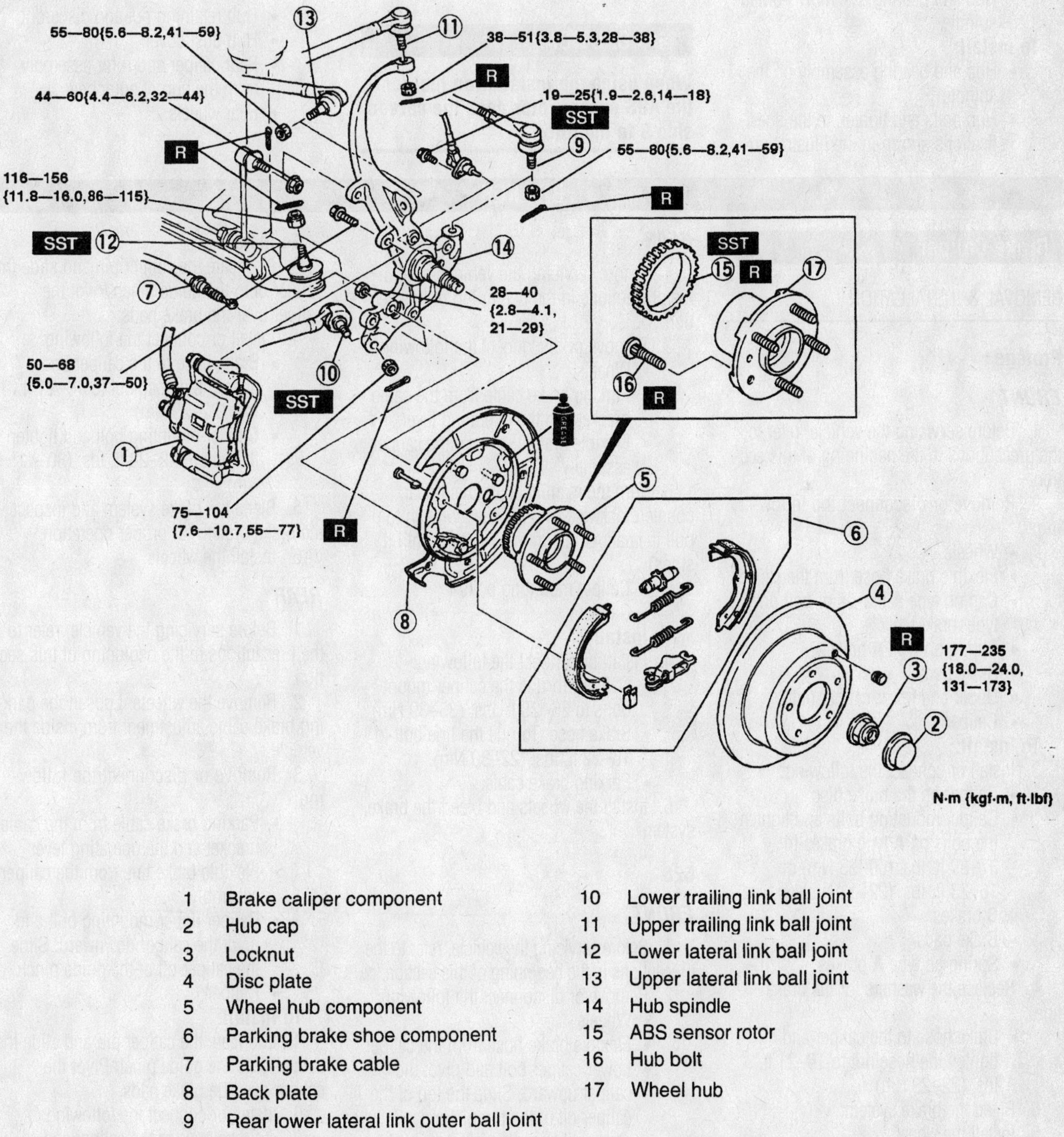

55—80{5.6—8.2,41—59}
44—60{4.4—6.2,32—44}
116—156 {11.8—16.0,86—115}
50—68 {5.0—7.0,37—50}
75—104 {7.6—10.7,55—77}
38—51{3.8—5.3,28—38}
19—25{1.9—2.6,14—18}
55—80{5.6—8.2,41—59}
28—40 {2.8—4.1, 21—29}
177—235 {18.0—24.0, 131—173}

N·m {kgf·m, ft·lbf}

1	Brake caliper component	10	Lower trailing link ball joint
2	Hub cap	11	Upper trailing link ball joint
3	Locknut	12	Lower lateral link ball joint
4	Disc plate	13	Upper lateral link ball joint
5	Wheel hub component	14	Hub spindle
6	Parking brake shoe component	15	ABS sensor rotor
7	Parking brake cable	16	Hub bolt
8	Back plate	17	Wheel hub
9	Rear lower lateral link outer ball joint		

Exploded view of the rear hub and spindle assembly–Millenia

42356-MAZC-GFI

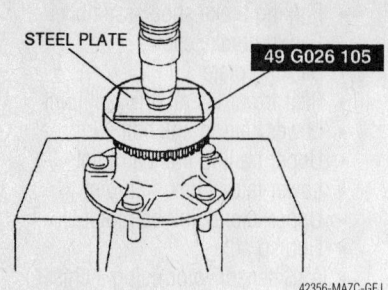

Install the ABS rotor—Millenia

- Hub and bearing assembly from the spindle

To install:

- Hub and bearing assembly on the spindle
- Hub bolts and tighten to the specifications shown in the illustration

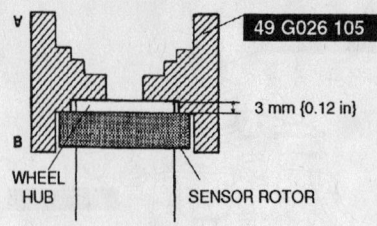

When using the installation tool for the ABS sensor rotor, face the carved side B to the rotor—Millenia

✳✳ CAUTION

When using the installation tool for the ABS sensor rotor, face the carved side B to the rotor.

- ABS sensor rotor using a press, steel plate and tool 49 G02610 until it is flush with the hub.
- Hub spindle
- Upper lateral link ball joint
- Lower lateral link ball joint
- Upper trailing link ball joint
- Lower trailing link ball joint
- Rear lower lateral link ball joint
- Backing plate
- Parking brake cable
- Parking brake shoe assembly
- Hub retaining nut and discard it
- Hub dust cover
- Rear caliper and rotor assembly from the hub, if equipped
- Rear wheels

BRAKES

Brake Caliper

REMOVAL & INSTALLATION

Protégé

FRONT

1. Before servicing the vehicle, refer to the precautions in the beginning of this section.
2. Remove or disconnect the following:

- Wheels
- Flexible brake hose from the caliper
- Cap on type **A** brakes or bolt on type brakes
- Spring on type **A** brakes
- Brake pads
- Upper and lower caliper bolts
- Caliper

To install:

3. Install or connect the following:

- Caliper on the brake disc
- Caliper mounting bolts and tighten the bolts on **A** type brakes to 37–39 ft. lbs. (50–53 Nm) or 16–23 ft. lbs. (22–31 Nm) on type **B** brakes
- Brake pads
- Spring on type **A** brakes

4. Replace the washers for the brake line.

- Brake hose to the caliper and tighten the hose nut to 16–21 ft. lbs. (22–29 Nm)

5. Bleed the brake system.
6. Install the wheels.

REAR

1. Before servicing the vehicle, refer to the precautions in the beginning of this section.
2. Remove or disconnect the following:

- Wheels
- Parking brake cable from the cable bracket and the operating lever
- Flexible brake line from the caliper assembly

3. Turn the manual adjustment gear counterclockwise with an Allen wrench to pull the caliper piston inward (turn until it stops).

- Caliper mounting bolts
- Caliper

To install:

4. Install or connect the following:

- Caliper. Torque the caliper mount bolts to 26–28 ft. lbs. (35–39 Nm).
- Brake hose. Torque the line bolt to 16–22 ft. lbs. (22–30 Nm).
- Parking brake cable

5. Install the wheels and bleed the brake system.

626

FRONT

1. Before servicing the vehicle, refer to the precautions in the beginning of this section.
2. Remove or disconnect the following:

- Wheels
- Flexible brake hose from the caliper
- Lower caliper bolt and pivot the caliper upward. Slide the top of the caliper off of the top pin and remove it from the vehicle.

To install:

3. Lubricate the caliper pin and slide the caliper onto the guide pinch Pivot the caliper over the brake pads.
4. Install or connect the following:

- Brake hose to the caliper and tighten the hose nut to 16–21 ft. lbs. (22–29 Nm)
- Caliper mounting bolt and tighten the bolt to 22–28 ft. lbs. (30–41 Nm)

5. Bleed the brake system and inspect the brake system for proper operation.
6. Install the wheels.

REAR

1. Before servicing the vehicle, refer to the precautions in the beginning of this section.
2. Remove the wheels. Loosen the parking brake cable adjustment from inside the vehicle.
3. Remove or disconnect the following:

- Parking brake cable from the cable bracket and the operating lever
- Flexible brake line from the caliper assembly
- Caliper upper mounting bolt and pivot the caliper downward. Slide the caliper off of the guide pinch
- Caliper

To install:

4. Lubricate the caliper pin and slide the caliper onto the guide pinch Pivot the caliper over the brake pads.
5. Install or connect the following:

- Brake hose to the caliper and

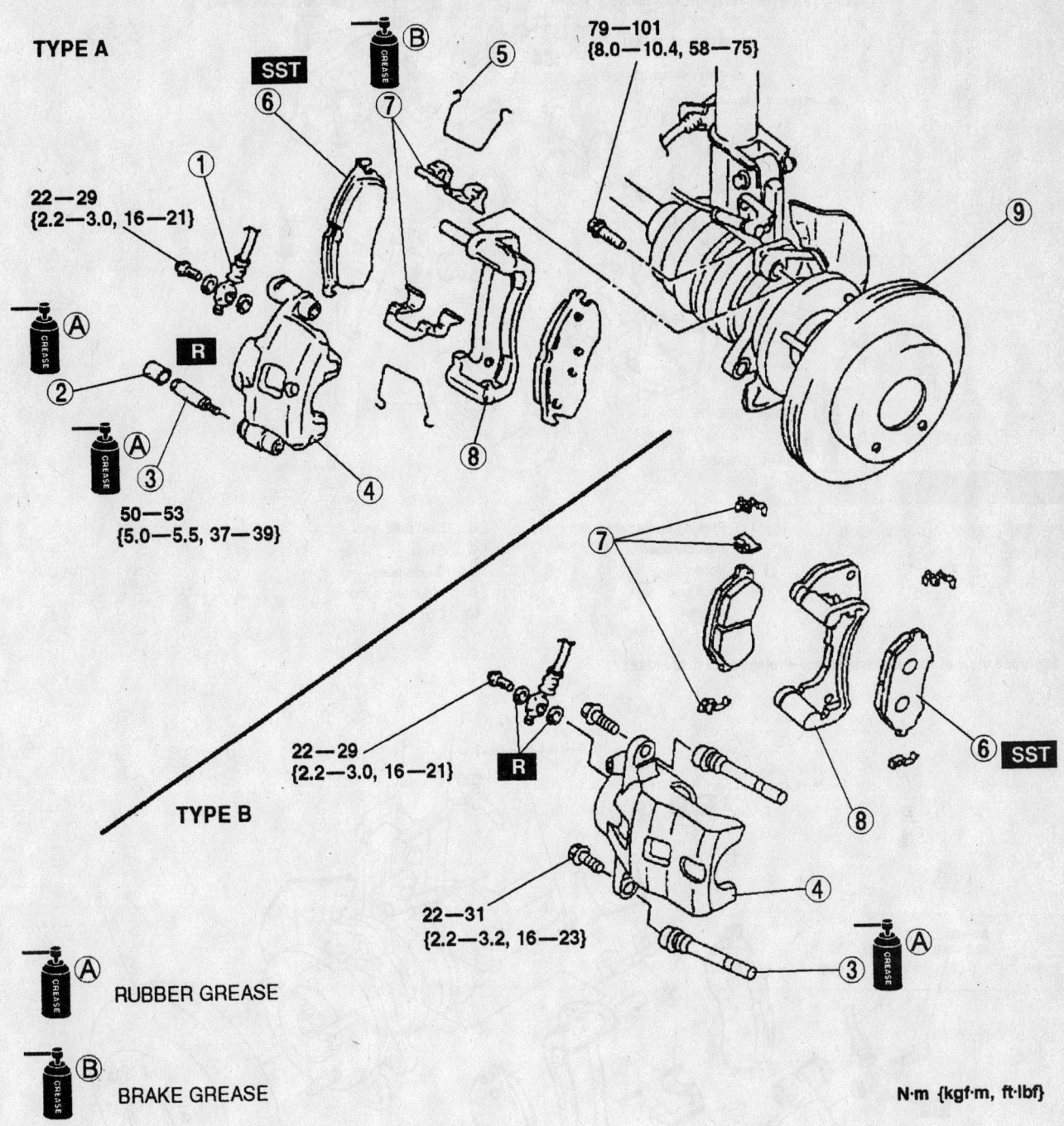

TYPE A

SST

22—29
{2.2—3.0, 16—21}

R

50—53
{5.0—5.5, 37—39}

79—101
{8.0—10.4, 58—75}

TYPE B

22—29
{2.2—3.0, 16—21}

R

22—31
{2.2—3.2, 16—23}

SST

Ⓐ RUBBER GREASE

Ⓑ BRAKE GREASE

N·m {kgf·m, ft·lbf}

1	Flexible hose	6	Disc pad
2	Cap (type A only)	7	Guide plate
3	Guide pin	8	Mounting support
4	Caliper	9	Disc plate
5	M-spring (type A only)		

42356-MAZC-GKT

Exploded view of the type A and B brake systems–Protégé with disc brakes

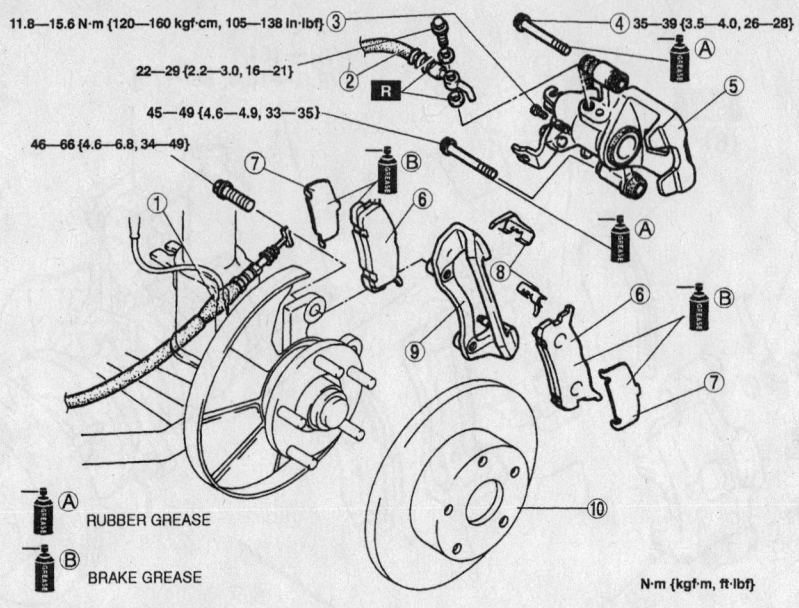

11.8—15.6 N·m {120—160 kgf·cm, 105—138 in·lbf}

22—29 {2.2—3.0, 16—21}

45—49 {4.6—4.9, 33—35}

46—66 {4.6—6.8, 34—49}

35—39 {3.5—4.0, 26—28}

RUBBER GREASE

BRAKE GREASE

N·m {kgf·m, ft·lbf}

1	Parking brake cable, clip	6	Disc pad
2	Flexible hose	7	Shim
3	Screw plug	8	Guide plate
4	Lock bolt	9	Mounting support
5	Caliper	10	Disc plate

42356-MAZC-GWW

Exploded view of rear brake systems—Protégé with disc brakes

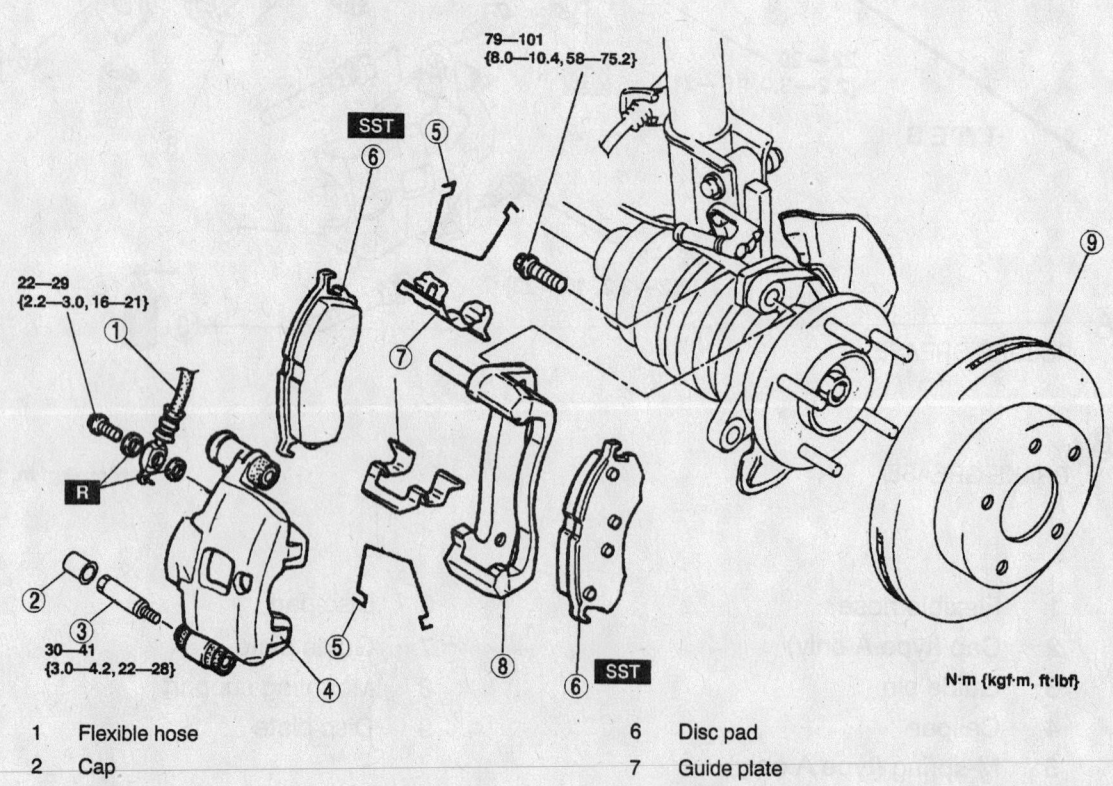

79—101
{8.0—10.4, 58—75.2}

SST

22—29
{2.2—3.0, 16—21}

R

30—41
{3.0—4.2, 22—28}

SST

N·m {kgf·m, ft·lbf}

1	Flexible hose	6	Disc pad
2	Cap	7	Guide plate
3	Lock bolt	8	Mounting support
4	Caliper	9	Disc plate
5	M-spring		

42356-MAZC-G89

Front disc brakes—626

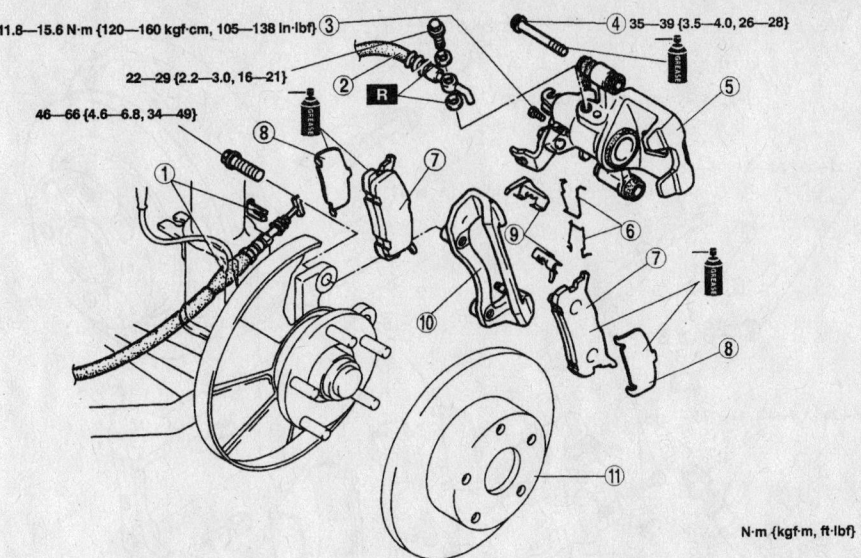

11.8—15.6 N·m {120—160 kgf·cm, 105—138 In·lbf} ③

④ 35—39 {3.5—4.0, 26—28}

22—29 {2.2—3.0, 16—21}

46—66 {4.6—6.8, 34—49}

N·m {kgf·m, ft·lbf}

1	Parking brake cable, clip	8	Shim
2	Flexible hose	9	Guide plate
3	Screw plug	10	Mounting support
4	Lock bolt	11	Disc plate
5	Caliper		
6	Spring		
7	Disc pad		

42356-MAZC-G91

Rear disc brakes—626

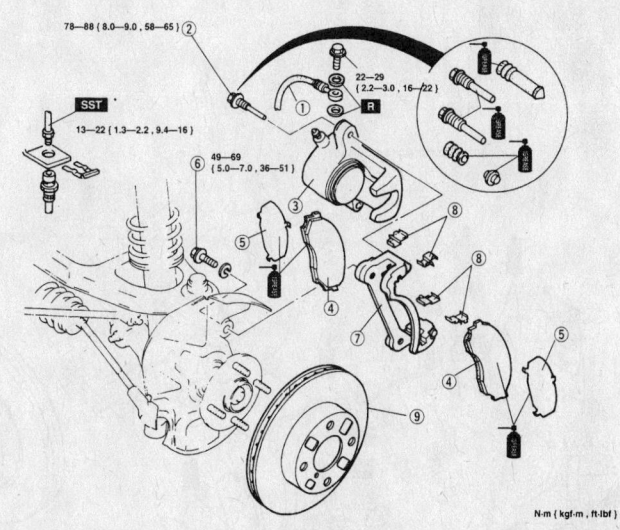

78—88 (8.0—9.0 , 58—65) ②

22—29 { 2.2—3.0 , 16—22 }

SST

13—22 { 1.3—2.2 , 9.4—16 }

49—69 { 5.0—7.0 , 36—51 }

N·m { kgf·m , ft·lbf }

1	Brake hose	6	Bolt
2	Connecting bolt	7	Mounting support
3	Caliper	8	Guide plate
4	Disc pad	9	Disc plate
5	Shim		

Disc Plate Removal Note
• Mark the wheel hub bolt and disc plate before removal for reference during installation.

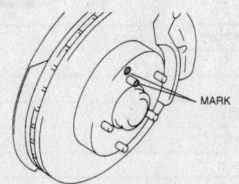

MARK

93016G30

Front disc brakes—Miata

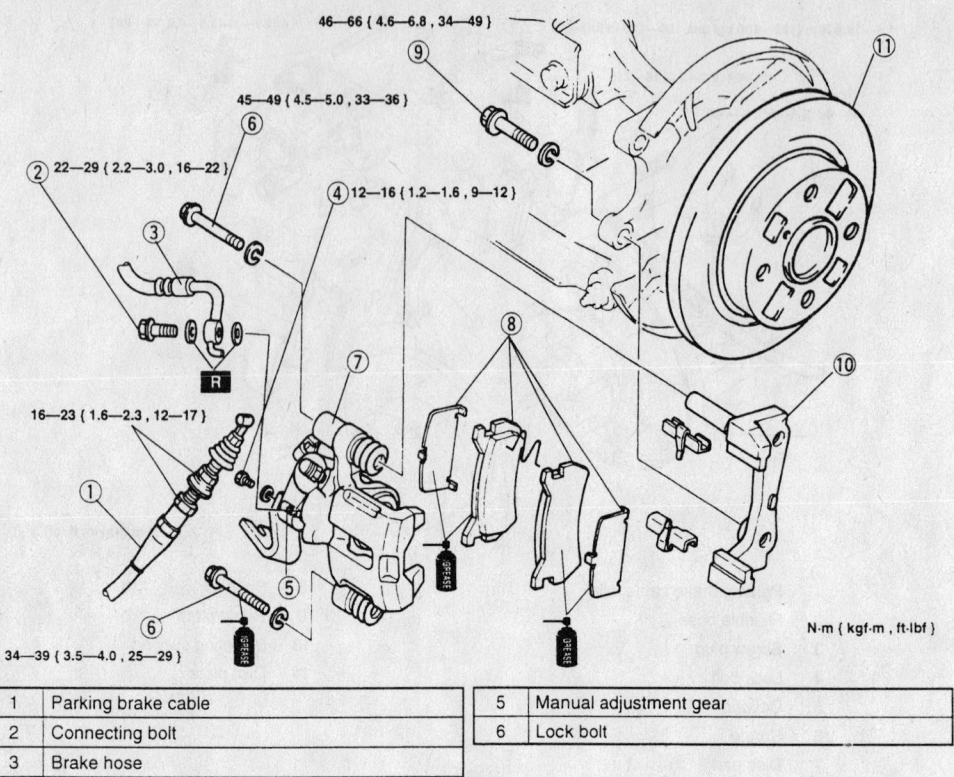

46—66 { 4.6—6.8 , 34—49 }

45—49 { 4.5—5.0 , 33—36 }

22—29 { 2.2—3.0 , 16—22 }

12—16 {1.2—1.6 , 9—12 }

16—23 { 1.6—2.3 , 12—17 }

34—39 { 3.5—4.0 , 25—29 }

N·m { kgf·m , ft·lbf }

1	Parking brake cable		5	Manual adjustment gear
2	Connecting bolt		6	Lock bolt
3	Brake hose			
4	Plug			

93016G31

Rear disc brakes—Miata

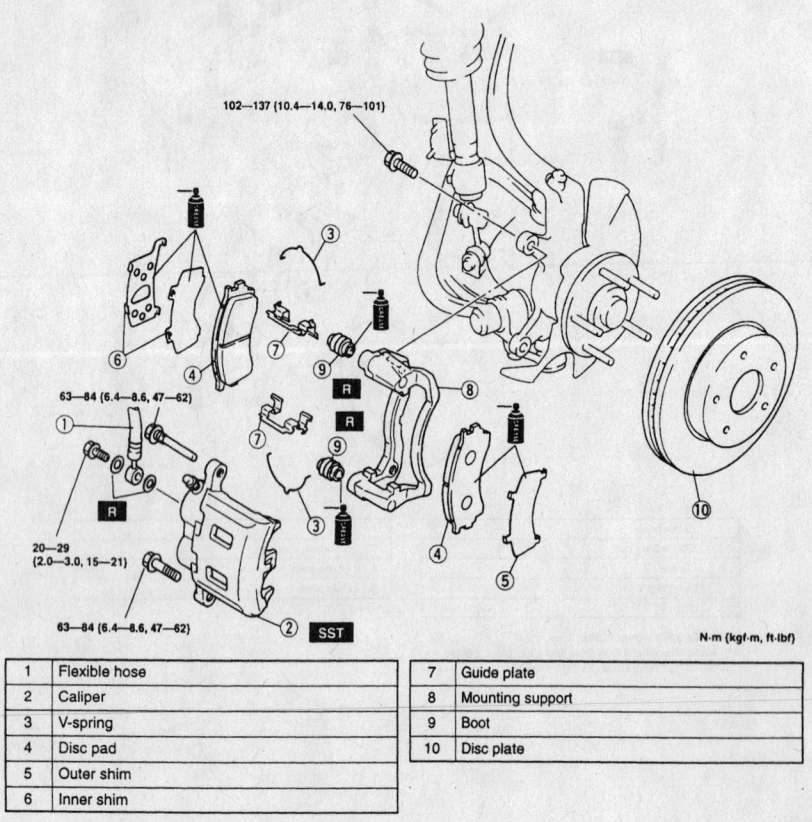

102—137 {10.4—14.0, 76—101}

63—84 {6.4—8.6, 47—62}

20—29
(2.0—3.0, 15—21)

63—84 {6.4—8.6, 47—62}

SST

N·m {kgf·m, ft·lbf}

1	Flexible hose		7	Guide plate
2	Caliper		8	Mounting support
3	V-spring		9	Boot
4	Disc pad		10	Disc plate
5	Outer shim			
6	Inner shim			

93016G32

Front disc brakes—Millenia

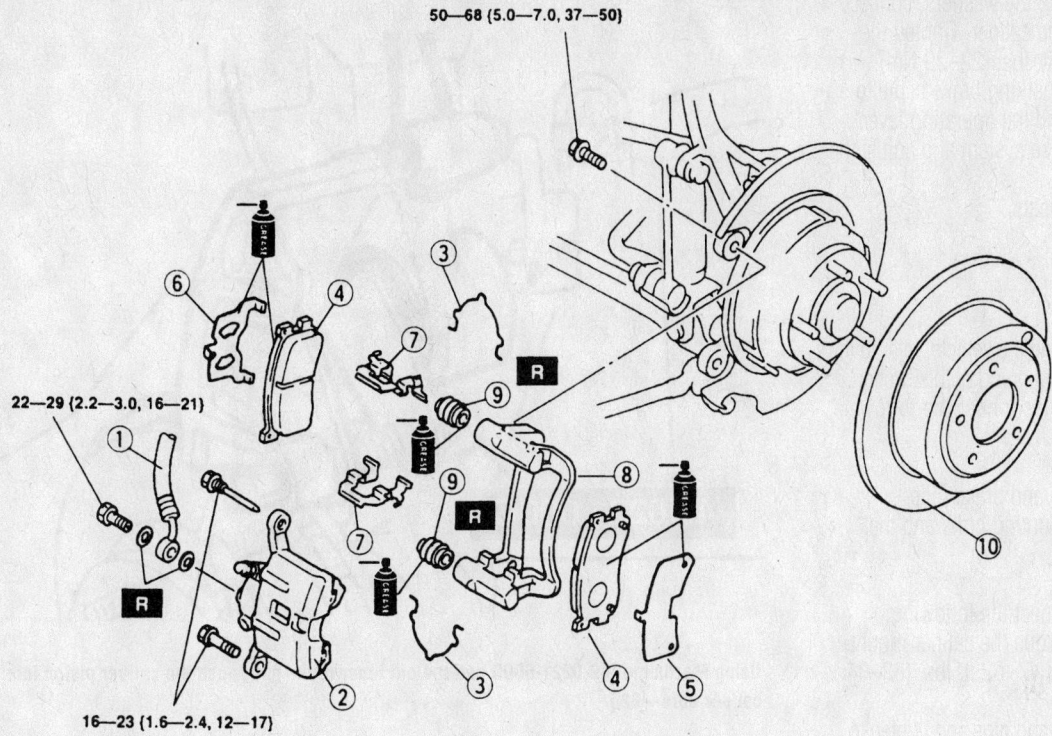

50—68 (5.0—7.0, 37—50)

22—29 (2.2—3.0, 16—21)

16—23 (1.6—2.4, 12—17)

N·m (kgf·m, ft·lbf)

93016G33

1	Flexible hose
2	Caliper
3	V-spring
4	Disc pad
5	Outer shim
6	Inner shim
7	Guide plate

8	Mounting support
9	Boot
10	Disc plate

Rear disc brakes—Millenia

tighten the hose nut to 16–21 ft. lbs. (22–29 Nm)
- Upper caliper mounting bolt and tighten the bolt to 26–28 ft. lbs. (35–39 Nm)
- Parking brake cable to the cable bracket and the operating lever
6. Bleed the brake system.
7. Install the wheels.

Miata

FRONT

1. Before servicing the vehicle, refer to the precautions in the beginning of this section.
2. Remove or disconnect the following:
- Wheels
- Flexible brake hose from the caliper
- Upper and lower caliper bolts
- Caliper from the vehicle

To install:
3. Install or connect the following:
- Caliper on the brake disc
- Caliper mounting bolts and tighten the bolts to 58–65 ft. lbs. (79–88 Nm)
4. Replace the washers for the brake line.
- Brake hose to the caliper and tighten the hose nut to 16–21 ft. lbs. (22–29 Nm)
5. Bleed the brake system and inspect the brake system for proper operation.
6. Install the wheels.

REAR

1. Before servicing the vehicle, refer to the precautions in the beginning of this section.
2. Loosen the parking brake cable adjustment from inside the vehicle.
3. Remove or disconnect the following:

- Wheels
- Parking brake cable from the cable bracket and the operating lever
- Flexible brake line from the caliper assembly
- Cover for the manual adjustment gear. Insert an Allen wrench and turn counterclockwise to retract the caliper piston.
- Caliper mounting bolts and remove the caliper from the vehicle

To install:
4. Install or connect the following:
- Caliper and install the mounting bolts. Tighten the upper bolt to 33–36 ft. lbs. (45–49 Nm), and the lower bolt to 26–28 ft. lbs. (35–39 Nm).
5. Turn the manual adjustment gear clockwise to return the caliper until the brake pads just touch the disc, then turn counterclockwise 1/3 of a turn. Replace the cover.

6. After replacing the washers, connect the brake hose to the caliper. Tighten the hose nut to 16–21 ft. lbs. (22–29 Nm).

7. Connect the parking brake cable to the cable bracket and the operating lever.

8. Bleed the brake system and adjust the parking brake.

9. Install the wheels.

Millenia

FRONT

1. Before servicing the vehicle, refer to the precautions in the beginning of this section.

2. Remove or disconnect the following:
- Wheels
- Brake hose and brake pipe.
- Caliper mounting bolts and the caliper

To install:

3. Install or connect the following:
- Caliper. Torque the caliper mounting bolts to 47–62 ft. lbs. (63–84 Nm).
- Brake hose and pipe and tighten to 15–21 ft. lbs. (22–29 Nm). Fill the master cylinder with clean brake fluid and bleed the hydraulic system.
- Wheels

REAR

1. Before servicing the vehicle, refer to the precautions in the beginning of this section.

2. Remove or disconnect the following:
- Wheels
- Brake hose
- Caliper bracket mounting bolts and remove the caliper

To install:

3. Install or connect the following:
- Caliper. Torque the caliper mounting bolts to 12–17 ft. lbs. (16–23 Nm).
- Brake hose and pipe and tighten to 16–21 ft. lbs. (22–29 Nm). Fill the master cylinder with clean brake fluid and bleed the hydraulic system.
- Wheels

Disc Brake Pads

REMOVAL & INSTALLATION

Protégé

FRONT

1. Before servicing the vehicle, refer to the precautions in the beginning of this section.

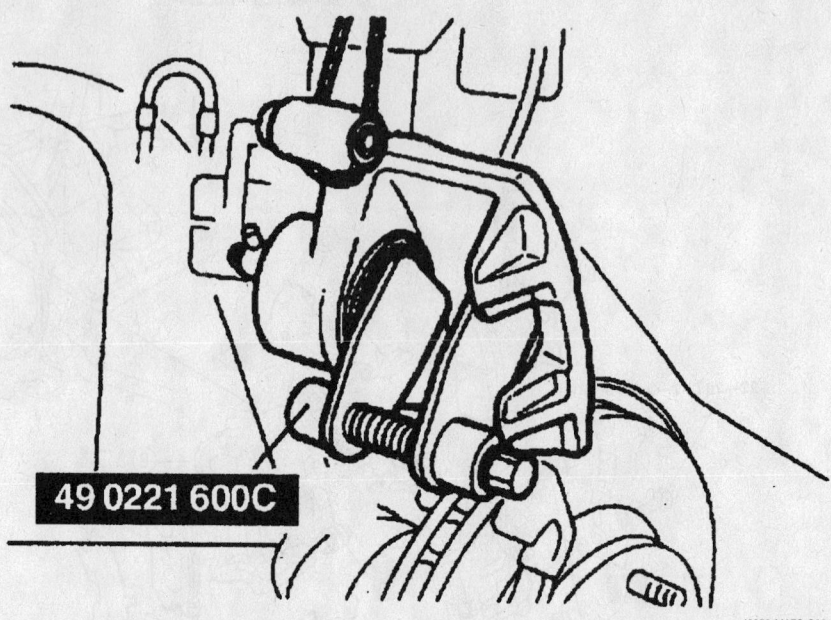

Using Mazda tool 49-0221-600C and the old inner brake pad, push the caliper piston into the caliper bore—626

2. Remove or disconnect the following:
- Wheels
- Cap on type **A** brakes or bolt on type brakes
- Caliper
- Spring on type **A** brakes
- Brake pads

To install:

3. Install or connect the following:
- Caliper on the brake disc

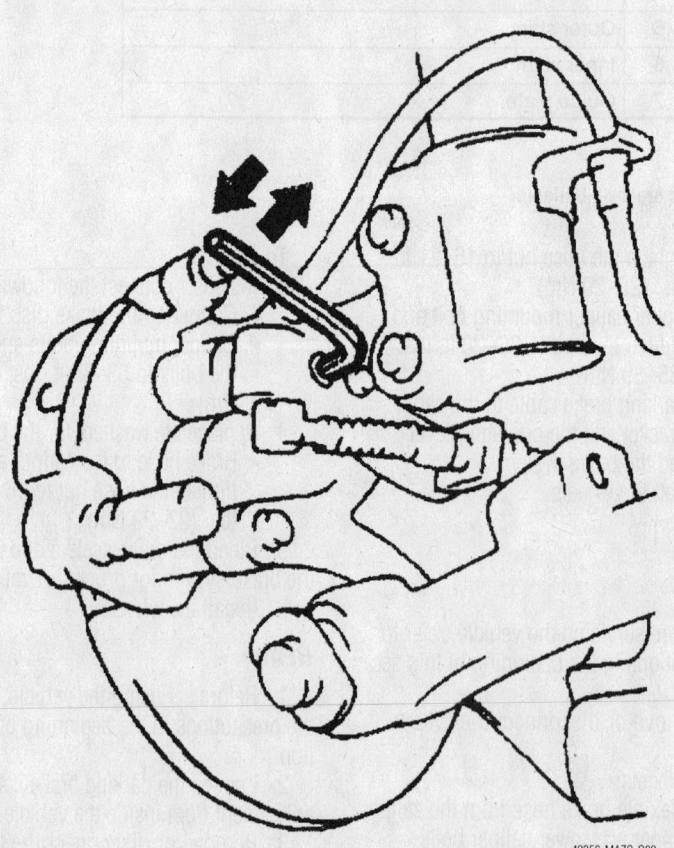

Location of the manual adjustment gear on the rear caliper—626

- Caliper mounting bolts and tighten the bolts on **A** type brakes to 37–39 ft. lbs. (50–53 Nm) or 16–23 ft. lbs. (22–31 Nm) on type **B** brakes
- Brake pads
- Spring on type **A** brakes

4. Install the wheels.

REAR

1. Before servicing the vehicle, refer to the precautions in the beginning of this section.
2. Remove or disconnect the following:
   - Wheels
   - Parking brake cable from the cable bracket and the operating lever
3. Turn the manual adjustment gear counterclockwise with an Allen wrench to pull the caliper piston inward (turn until it stops).
   - Caliper
   - Pads

To install:

4. Install or connect the following:
   - Pads
   - Caliper. Torque the caliper mount bolts to 26–28 ft. lbs. (35–39 Nm).
   - Parking brake cable
5. Install the wheels and bleed the brake system.

626

FRONT

1. Before servicing the vehicle, refer to the precautions in the beginning of this section.
2. Remove some of the brake fluid from the master cylinder reservoir.
3. Remove or disconnect the following:
   - Wheels
   - Caliper lower mounting bolt and pivot the caliper up and support it
   - Brake pads, shims and pin
4. Using Mazda tool 49-0221-600C and the old inner brake pad, push the caliper piston into the caliper bore.

To install:

5. Install or connect the following:
   - Brake pads and shims to the caliper support
   - Caliper over the brake pads
   - Caliper mounting bolt and torque to 22–28 ft. lbs. (30–41 Nm)
   - Wheels
6. Test the brakes for proper operation.

REAR

1. Before servicing the vehicle, refer to the precautions in the beginning of this section.

2. Remove some of the brake fluid from the master cylinder reservoir.
3. Loosen the parking brake cable adjustment from inside the vehicle.
4. Remove or disconnect the following:
   - Wheels.
   - Parking brake cable from the cable bracket and the operating lever
   - Upper caliper mounting bolt and pivot the caliper downward off of the pads
   - Brake pads and spring clips from the caliper support

To install:

5. Turn the manual adjustment gear counterclockwise using an Allen wrench to retract the caliper until it stops.
6. Install or connect the following:
   - Brake pads, shims and spring clips to the caliper support. Pivot the caliper over the brake pads.
7. Turn the adjustment gear clockwise until the pads start to touch the rotor and back of the gear ⅓ turn.
8. Lubricate the top caliper mounting bolt, then tighten the bolt to 26–28 ft. lbs. (35–39 Nm). Attach the parking brake cable to the operating lever.
9. Adjust the parking brake cable, as required.
10. Install the wheels.
11. Test the brakes for proper operation.

Miata

FRONT

1. Before servicing the vehicle, refer to the precautions in the beginning of this section.
2. Remove some of the brake fluid from the master cylinder reservoir.
3. Remove or disconnect the following:
   - Wheels
   - Lower lockbolt, and pivot the brake caliper upwards
   - Spring, if equipped with 15 inch wheels
   - Brake pads and shim
4. Using Mazda piston compressor tool 49-0221-600C and the old inner brake pad, push the caliper piston into the caliper bore.

To install:

5. Install or connect the following:
   - Brake pads and shims onto the caliper
   - Caliper
   - Spring, if equipped with 15 inch wheels
   - Lockbolt and tighten to 58–65 ft. lbs. (78–88 Nm)
   - Wheels
5. Test the brakes for proper operation.

REAR

1. Before servicing the vehicle, refer to the precautions in the beginning of this section.
2. Remove some of the brake fluid from the master cylinder reservoir.
3. Remove or disconnect the following:
   - Wheels
   - Plug for the manual adjustment gear. Using an Allen wrench turn the gear counterclockwise to retract the caliper piston.
   - Lower caliper mounting bolt and pivot the caliper upward off of the pads
   - Brake pads and spring clips from the caliper support

To install:

4. Install or connect the following:
   - Brake pads, shims and spring clips to the caliper support. Pivot the caliper over the brake pads.
   - Lubricate and install the lower caliper mounting bolt. Tighten the bolt to 26–28 ft. lbs. (35–39 Nm).
5. Turn the manual adjusting gear clockwise until the piston contacts the brake disc, then turn it clockwise ⅓ turn. Replace the plug.
6. Install the wheels.
7. Test the brakes for proper operation.

Millenia

FRONT

1. Before servicing the vehicle, refer to the precautions in the beginning of this section.
2. Remove or disconnect the following:
   - Wheels
   - Bottom caliper lock pin and swing the caliper upwards
   - V-springs the pads and shims

To install:

3. Install or connect the following:
   - Brake pads, shims and V-springs
4. Press the caliper pistons back into their cylinders.
   - Calipers. Torque the lock pin to 47–62 ft. lbs. (63–84 Nm).
   - Wheels

REAR

1. Before servicing the vehicle, refer to the precautions in the beginning of this section.
2. Remove or disconnect the following:
   - Wheels
   - Lock pin and rotate the caliper upwards
   - V-springs, pads and the shims from the pads

To install:

3. Press the caliper piston back into the cylinder.

4. Install or connect the following:
- Pads, shims, and V-springs
- Caliper and torque the lock pin to 37–50 ft. lbs. (50–68 Nm)
- Wheels

Brake Drums

REMOVAL & INSTALLATION

All Models

1. Before servicing the vehicle, refer to the precautions in the beginning of this section.

2. Remove or disconnect the following:
- Rear wheel
- Screw securing the rear brake drum, and pull the brake drum outward to remove

To install:

3. Install or connect the following:
- Drum and retaining screw
- Wheels and adjust the rear brakes as necessary

Brake Shoes

REMOVAL & INSTALLATION

1. Before servicing the vehicle, refer to the precautions in the beginning of this section.

2. Remove or disconnect the following:
- Wheel and the brake drum
- ABS speed sensor, if equipped
- Parking brake cable from the backside of the brake backing plate
- Upper return spring
- Hold pin and the spring from the lower shoe (leading side)
- Lower return spring and the anti-rattle spring and shoe from the lower shoe (leading side)
- Hold pin and spring and remove the upper brake shoe (trailing side)

To install:

3. Install or connect the following:
- Upper (trailing side) brake shoe to the operating lever and then to the wheel cylinder and backing plate
- Brake shoe hold spring and hold pin
- Anti-rattle spring
- Lower return spring to both brake shoes
- Leading side brake shoe to the operating lever and then to the wheel cylinder and anchor plate
- Hold spring and hold pin to the leading side brake shoe
- Upper return spring
- Brake drum
- ABS speed sensor, if equipped
- Wheels and adjust the brakes

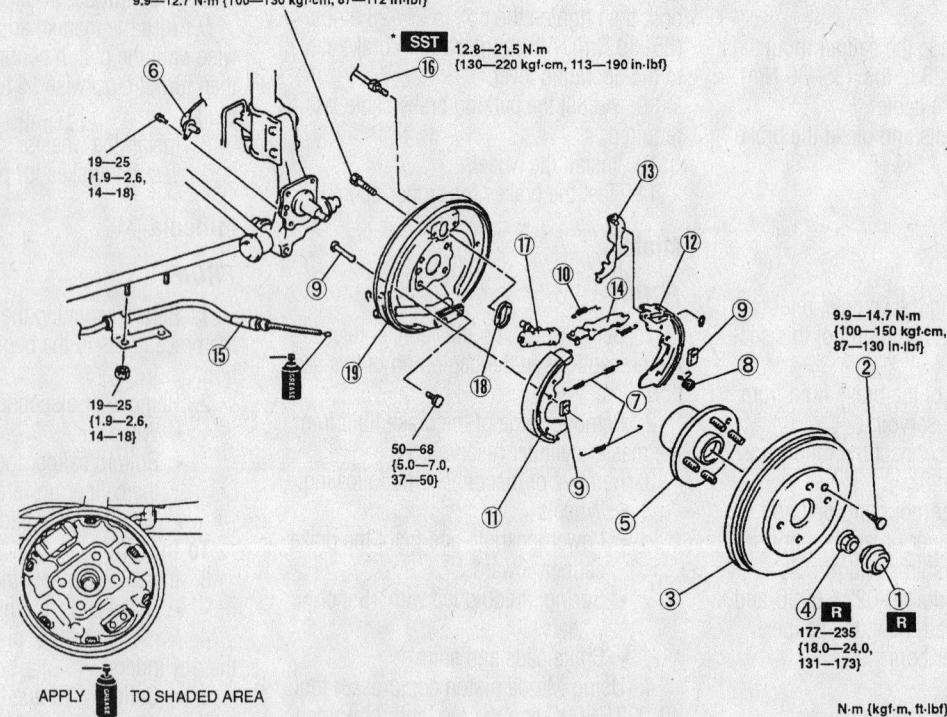

9.9–12.7 N·m {100—130 kgf·cm, 87—112 in·lbf}

* SST 12.8–21.5 N·m {130—220 kgf·cm, 113—190 in·lbf}

19—25 {1.9—2.6, 14—18}

19—25 {1.9—2.6, 14—18}

50—68 {5.0—7.0, 37—50}

9.9—14.7 N·m {100—150 kgf·cm, 87—130 in·lbf}

177—235 {18.0—24.0, 131—173}

N·m {kgf·m, ft·lbf}

APPLY TO SHADED AREA
*49 0259 770B

1	Hub cap	11	Leading shoe
2	Screw	12	Trailing shoe
3	Brake drum	13	Operating lever
4	Locknut	14	Adjuster
5	Wheel hub	15	Parking brake cable
6	ABS wheel-speed sensor (if equipped)	16	Brake pipe
7	Return spring	17	Wheel cylinder
8	Lever spring	18	O-ring
9	Hold pin and hold spring	19	Backing plate
10	Anti-rattle spring		

Typical rear drum brake assembly, Protégé shown others similar

42356-MAZC-GKU

MAZDA

B-Series Trucks

18

SPECIFICATION CHARTS

ENGINE AND VEHICLE IDENTIFICATION

Code ①	Liters (cc)	Cu. In.	Cyl.	Fuel Sys.	Type	Eng. Mfg.		Code ②	Year
					Engine			Model Year	
D	2.3 (2261)	138	4	MFI	DOHC	Ford		Y	2000
C	2.5 (2500)	152	4	MFI	SOHC	Ford		1	2001
U	3.0 (2999)	183	6	MFI	OHV	Ford		2	2002
X	4.0 (3998)	244	6	MFI	OHV	Ford		3	2003
E	4.0 (4000)	244	6	MFI	SOHC	Ford		4	2004

MFI: Multi-port Fuel Injection

OHV: Overhead Valve

DOHC: Dual Overhead Camshafts

SOHC: Single Overhead Camshaft

① 8th digit of the Vehicle Identification Number (VIN)

② 10th digit of the Vehicle Identification Number (VIN)

42356-BSER-C01

GENERAL ENGINE SPECIFICATIONS

Year	Model	Engine Displacement Liters (cc)	Engine Series (ID/VIN)	Fuel System Type	Net Horsepower @ rpm	Net Torque @ rpm (ft. lbs.)	Bore x Stroke (in.)	Com- pression Ratio	Oil Pressure @ rpm
2000	B2500	2.5 (2500)	C	MFI	119@5000	146@3000	3.78X3.90	9.1:1	40-60@2000
	B3000	3.0 (2982)	U	MFI	147@5000	162@3250	3.50x3.14	9.3:1	40-60@2500
	B4000	4.0 (3950)	X	MFI	160@4000	225@2500	3.81x3.39	9.0:1	40-60@2000
2001	B2300	2.3 (2261)	D	MFI	135@5050	153@3750	3.44x3.70	NA	29-39@3000
	B2500	2.5 (2500)	C	MFI	147@5000	147@5000	3.50x3.14	9.3:1	40-60@2500
	B3000	3.0 (2982)	U	MFI	147@5000	162@3250	3.50x3.14	9.3:1	40-60@2500
	B4000	4.0 (4000)	E	MFI	160@4000	225@2500	3.81x3.39	9.0:1	40-60@2000
2002	B2300	2.3 (2261)	D	MFI	135@5050	153@3750	3.44x3.70	NA	29-39@3000
	B3000	3.0 (2982)	U	MFI	147@5000	147@5000	3.50x3.14	9.3:1	40-60@2500
	B4000	4.0 (4000)	E	MFI	160@4000	225@2500	3.81x3.39	9.0:1	40-60@2000
2003	B2300	2.3 (2261)	D	MFI	135@5050	153@3750	3.44x3.70	NA	29-39@3000
	B3000	3.0 (2982)	U	MFI	147@5000	147@5000	3.50x3.14	9.3:1	40-60@2500
	B4000	4.0 (4000)	E	MFI	160@4000	225@2500	3.81x3.39	9.0:1	40-60@2000

MFI: Multi-port Fuel Injection

42356-BSER-C02

GASOLINE ENGINE TUNE-UP SPECIFICATIONS

Year	Engine Displacement Liters (cc)	Engine ID/VIN	Spark Plug Gap (in.)	Ignition Timing (deg.)		Fuel Pump (psi)	Idle Speed (rpm)		Valve Clearance	
				MT	AT		MT	AT	In.	Ex.
2000	2.5 (2500)	C	0.044	10B ①	10B ①	56-72	①	①	HYD	HYD
	3.0 (2982)	U	0.044	10B ①	10B ①	35-45	①	①	HYD	HYD
	4.0 (3950)	X	0.054	10B ①	10B ①	35-45	①	①	HYD	HYD
2001	2.3 (2261)	D	0.041-0.045	10B ①	10B ①	64-72	①	①	HYD	HYD
	2.5 (2500)	C	0.044	10B ①	10B ①	64-72	①	①	HYD	HYD
	3.0 (2982)	U	0.044	10B ①	10B ①	64-72	①	①	HYD	HYD
	4.0 (3950)	E	0.052-0.056	10B ①	10B ①	64-72	①	①	HYD	HYD
2002	2.3 (2261)	D	0.041-0.045	10B ①	10B ①	64-72	①	①	HYD	HYD
	3.0 (2982)	U	0.042-0.046	10B ①	10B ①	64-72	①	①	HYD	HYD
	4.0 (3950)	E	0.062-0.068	10B ①	10B ①	64-72	①	①	HYD	HYD
2003	2.3 (2261)	D	0.041-0.045	10B ①	10B ①	64-72	①	①	HYD	HYD
	3.0 (2982)	U	0.042-0.046	10B ①	10B ①	64-72	①	①	HYD	HYD
	4.0 (3950)	E	0.062-0.068	10B ①	10B ①	64-72	①	①	HYD	HYD

NOTE: The Vehicle Emission Control Information label often reflects specification changes changes made during production. The label figures must be used if they differ from those in this chart.

B: Before top dead center

HYD: Hydraulic

NA: Not Available

① Electronically controlled and cannot be adjusted

42356-BSER-C03

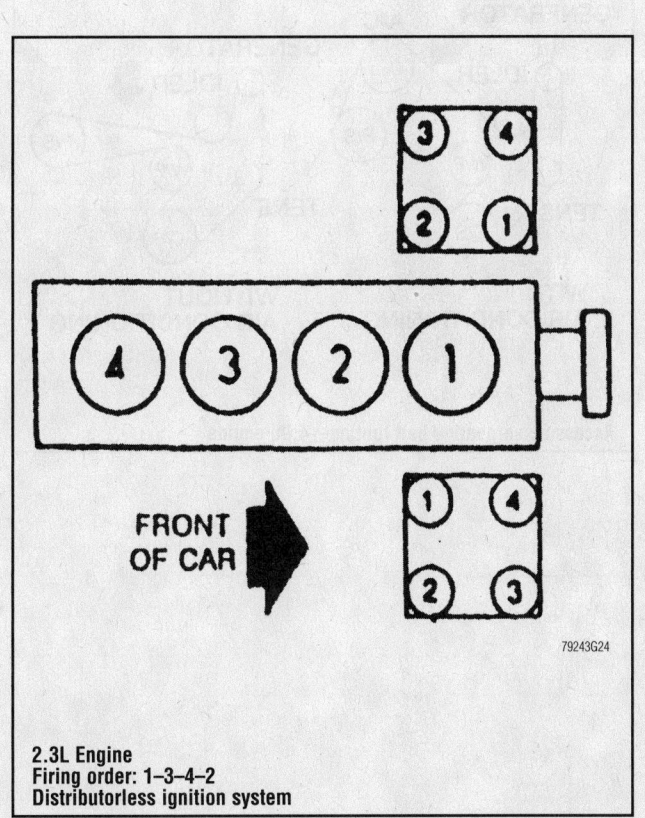

2.3L Engine
Firing order: 1-3-4-2
Distributorless ignition system

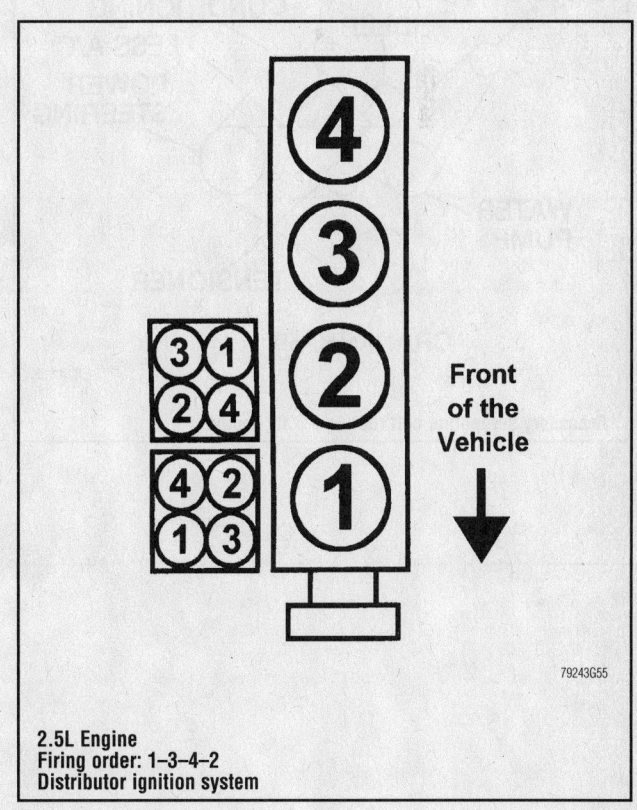

2.5L Engine
Firing order: 1-3-4-2
Distributor ignition system

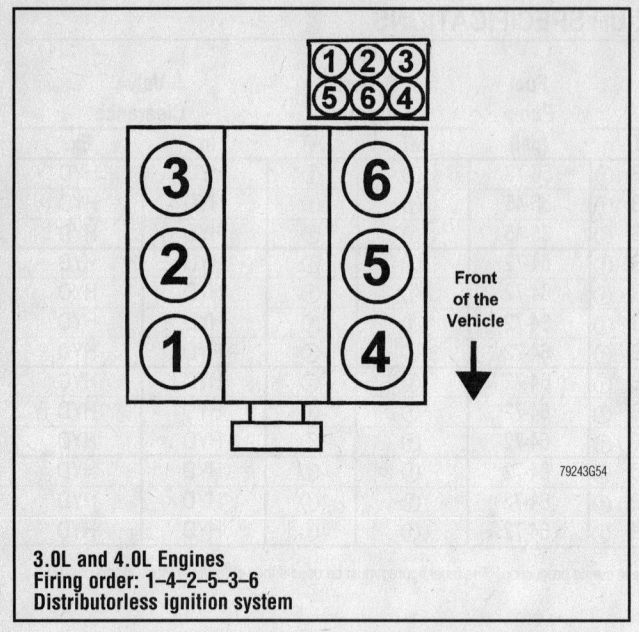

3.0L and 4.0L Engines
Firing order: 1–4–2–5–3–6
Distributorless ignition system

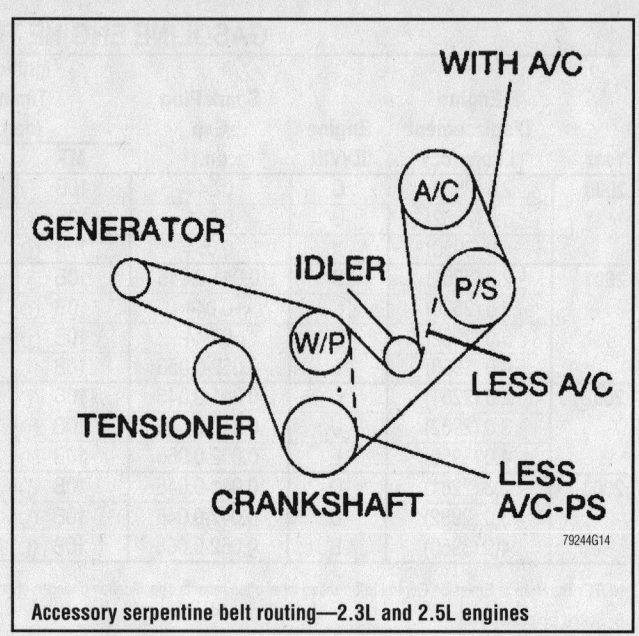

Accessory serpentine belt routing—2.3L and 2.5L engines

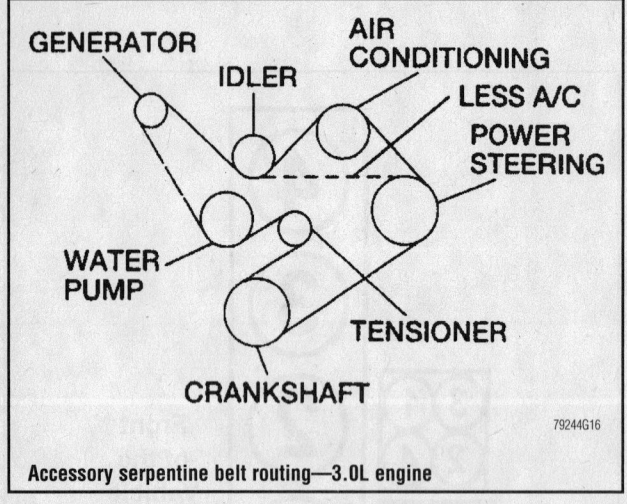

Accessory serpentine belt routing—3.0L engine

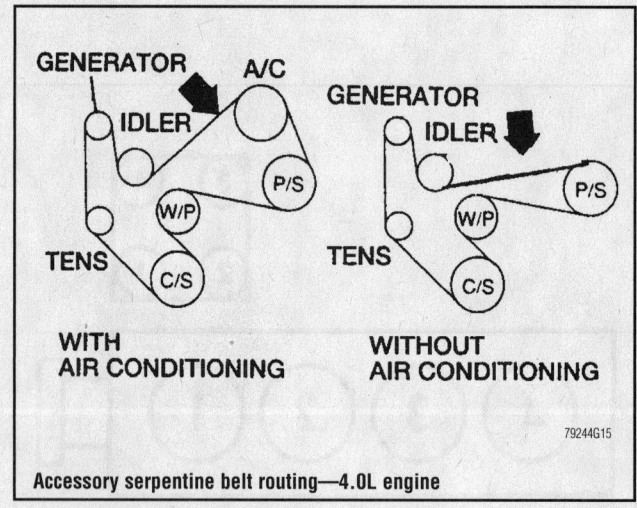

Accessory serpentine belt routing—4.0L engine

CAPACITIES

Year	Model	Engine Displacement Liters (cc)	Engine ID/VIN	Engine Oil with Filter (qts.)	Transmission (pts.)		Transfer Case (pts.)	Drive Axle		Fuel Tank (gal.)	Cooling System (qts.)
					5-Spd	Auto.		Front (pts.)	Rear (pts.)*		
2000	B2500	2.5 (2500)	C	5.0	5.6	19.8	—	—	5.0	①	②
	B3000	3.0 (2982)	U	4.5	5.6	③	2.5	3.0	5.0	①	④
	B4000	4.0 (3950)	X	5.0	5.6	③	2.5	3.6	5.0	①	⑤
2001	B2300	2.3 (2261)	D	4.0	3.0	19.8	—	—	5.0	①	⑥
	B2500	2.5 (2500)	C	5.0	3.0	19.8	—	—	5.0	①	⑥
	B3000	3.0 (2982)	U	4.5	3.0	③	2.5	3.3	5.0	①	⑦
	B4000	4.0 (4000)	E	5.0	3.0	③	2.5	3.25	5.0	①	⑧
2002	B2300	2.3 (2261)	D	4.0	3.0	19.8	—	—	5.0	①	⑥
	B3000	3.0 (2982)	U	4.5	3.0	③	2.5	2.7	5.0	①	⑦
	B4000	4.0 (4000)	E	5.0	3.0	③	2.5	2.7	5.0	①	⑧
2003	B2300	2.3 (2261)	D	4.0	3.0	19.8	—	—	5.0	①	⑥
	B3000	3.0 (2982)	U	4.5	3.0	③	2.5	2.7	5.0	①	⑦
	B4000	4.0 (4000)	E	5.0	3.0	③	2.5	2.7	5.0	①	⑧

NOTE: All capacities are approximate. Add fluid gradually and check to be sure a proper fluid level is obtained.

* For limited slip axles, add 4 oz. of friction modifier, exc. for 1-ton models

① Std: 16.5
　Long Wheelbase: 20.0
　Super Cab: 19.5
② MT: 10.5
　AT: 10.2
③ 2wd: 20.0
　4wd: 20.6
④ MT: 15.2
　AT: 14.8
⑤ MT: 13.5
　AT: 13.2

⑥ w/MT: 11.2
　w/AT: 10.9
⑦ w/MT: 15.2
　w/AT: 14.8
⑧ w/MT: 13.5
　w/AT: 13.2

42356-BSER-C04

VALVE SPECIFICATIONS

Year	Engine Displacement Liters (cc)	Engine ID/VIN	Seat Angle (deg.)	Face Angle (deg.)	Spring Test Pressure (lbs. @ in.)	Spring Installed Height (in.)	Stem-to-Guide Clearance (in.)		Stem Diameter (in.)	
							Intake	Exhaust	Intake	Exhaust
2000	2.5 (2500)	C	45	44	57-63@ 1.56	1.540- 1.580	0.0008- 0.0025	0.0018- 0.0037	0.2746- 0.2754	0.2736- 0.2744
	3.0 (2982)	U	45	44	185@1.16	1.580- 1.610	0.0010- 0.0027	0.0015- 0.0032	0.3126- 0.3134	0.3121- 0.3129
	4.0 (3950)	X	45	44	138@1.22	1.580- 1.610	0.0008- 0.0025	0.0018- 0.0035	0.3159- 0.3167	0.3149- 0.3156
2001	2.3 (2261)	D	44.5-45	45-45.5	①	1.492	0.0009	0.0011	0.2153- 0.2159	0.2151- 0.2157
	2.5 (2500)	C	44.75	44	118-132@1.16	1.540- 1.580	0.0008- 0.0027	0.0018- 0.0037	0.2746- 0.2754	0.2736- 0.2744
	3.0 (2982)	U	45	44	185@1.16	1.580- 1.610	0.0010- 0.0027	0.0015- 0.0032	0.3126- 0.3134	0.3121- 0.3129
	4.0 (4000)	E	45	45	202-225@ 1.413-1.445	1.569- 1.601	0.0010- 0.0020	0.0010- 0.0020	0.2740- 0.2748	0.2730- 0.2740
2002	2.3 (2261)	D	45	45	①	1.492	0.0009	0.0011	0.2153- 0.2159	0.2151- 0.2157
	3.0 (2982)	U	45	44	185@1.16	1.580- 1.610	0.0010- 0.0027	0.0015- 0.0032	0.3126- 0.3134	0.3121- 0.3129
	4.0 (4000)	E	45	45	202-225@ 1.413-1.445	1.569- 1.601	0.0010- 0.0020	0.0010- 0.0030	0.2742- 0.2748	0.2736- 0.2742
2003	2.3 (2261)	D	45	45	①	1.492	0.0009	0.0011	0.2153- 0.2159	0.2151- 0.2157
	3.0 (2982)	U	45	44	185@1.16	1.580- 1.610	0.0010- 0.0027	0.0015- 0.0032	0.3126- 0.3134	0.3121- 0.3129
	4.0 (4000)	E	45	45	202-225@ 1.413-1.445	1.569- 1.601	0.0010- 0.0020	0.0010- 0.0030	0.2742- 0.2748	0.2736- 0.2742

① Intake: 97.0@1.201
Exhaust: 93.3@1.201

42356-BSER-C05

PISTON AND RING SPECIFICATIONS

All measurements are given in inches.

Year	Engine Displacement Liters (cc)	Engine ID/VIN	Piston Clearance	Ring Gap			Ring Side Clearance		
				Top Compression	Bottom Compression	Oil Control	Top Compression	Bottom Compression	Oil Control
2000	2.5 (2500)	C	0.0010-0.0020	0.008-0.018	0.013-0.023	0.010-0.035	0.0014-0.0030	0.0014-0.0030	SNUG
	3.0 (2982)	U	0.0012-0.0023	0.010-0.020	0.010-0.020	0.010-0.049	0.0602-0.0612	0.0602-0.0612	SNUG
	4.0 (3950)	X	0.0008-0.0019	0.015-0.023	0.015-0.023	0.015-0.055	0.0020-0.0033	0.0020-0.0033	SNUG
2001	2.3 (2261)	D	0.0009-0.0017	0.006-0.0012	0.012-0.0180	0.007-0.0270	0.0008-0.0013	0.0004-0.0011	0.0025-0.0054
	2.5 (2500)	C	0.0010-0.0020	0.008-0.014	0.013-0.019	0.010-0.030	0.0014-0.0030	0.0014-0.0030	SNUG
	3.0 (2982)	U	0.0012-0.0023	0.010-0.020	0.010-0.020	0.010-0.049	0.0602-0.0612	0.0602-0.0612	SNUG
	4.0 (4000)	E	0.0008-0.0019	0.015-0.023	0.015-0.023	0.015-0.055	0.0010-0.0030	0.0010-0.0030	SNUG
2002	2.3 (2261)	D	0.0009-0.0017	0.006-0.0012	0.012-0.0180	0.007-0.0270	0.0008-0.0013	0.0004-0.0011	0.0025-0.0054
	3.0 (2982)	U	0.0012-0.0023	0.010-0.020	0.010-0.020	0.010-0.049	0.0602-0.0612	0.0602-0.0612	SNUG
	4.0 (4000)	E	0.0008-0.0019	0.015-0.023	0.015-0.023	0.015-0.055	0.0010-0.0030	0.0010-0.0030	SNUG
2003	2.3 (2261)	D	0.0009-0.0017	0.006-0.0012	0.012-0.0180	0.007-0.0270	0.0008-0.0013	0.0004-0.0011	0.0025-0.0054
	3.0 (2982)	U	0.0012-0.0023	0.010-0.020	0.010-0.020	0.010-0.049	0.0602-0.0612	0.0602-0.0612	SNUG
	4.0 (4000)	E	0.0008-0.0019	0.015-0.023	0.015-0.023	0.015-0.055	0.0010-0.0030	0.0010-0.0030	SNUG

42356-BSER-C06

CRANKSHAFT AND CONNECTING ROD SPECIFICATIONS
All measurements are given in inches.

Year	Engine Displacement Liters (cc)	Engine ID/VIN	Crankshaft				Connecting Rod		
			Main Brg. Journal Dia.	Main Brg. Oil Clearance	Shaft End-play	Thrust on No.	Journal Diameter	Oil Clearance	Side Clearance
2000	2.5 (2500)	C	2.2051-2.2059	0.0008-0.0015	0.0040-0.0080	3	2.0464-2.0472	0.0008-0.0015	0.0035-0.0115
	3.0 (2982)	U	2.5190-2.5198	0.0010-0.0014	0.0040-0.0080	3	2.1253-2.1261	0.0010-0.0014	0.0060-0.0140
	4.0 (3950)	X	2.2433-2.2441	0.0008-0.0015	0.0020-0.0120	3	2.1252-2.1260	0.0003-0.0024	0.0002-0.0025
2001	2.3 (2261)	D	2.0465-2.2059	0.0007-0.0013	0.0080-0.0160	3	1.9606-1.9685	0.0010-0.0020	0.0767-0.1200
	2.5 (2500)	C	2.2051-2.2059	0.0008-0.0015	0.0040-0.0080	3	2.0464-2.0472	0.0008-0.0015	0.0035-0.0115
	3.0 (2982)	U	2.5190-2.5198	0.0010-0.0014	0.0040-0.0080	3	2.1253-2.1261	0.0010-0.0014	0.0060-0.0140
	4.0 (4000)	E	2.2430-2.2440	0.0008-0.0015	0.0020-0.0125	3	2.7252-2.7260	0.0003-0.0024	0.0036-0.0106
2002	2.3 (2261)	D	2.0465-2.2059	0.0007-0.0013	0.0080-0.0160	3	1.9606-1.9685	0.0010-0.0020	0.0767-0.1200
	3.0 (2982)	U	2.5190-2.2059	0.0010-0.0015	0.0040-0.0080	3	2.1253-2.0472	0.0010-0.0015	0.0060-0.0115
	4.0 (4000)	E	2.2430-2.2440	0.0008-0.0015	0.0020-0.0125	3	2.7252-2.7260	0.0003-0.0024	0.0036-0.0106
2003	2.3 (2261)	D	2.0465-2.2059	0.0007-0.0013	0.0080-0.0160	3	1.9606-1.9685	0.0010-0.0020	0.0767-0.1200
	3.0 (2982)	U	2.5190-2.2059	0.0010-0.0015	0.0040-0.0080	3	2.1253-2.0472	0.0010-0.0015	0.0060-0.0115
	4.0 (4000)	E	2.2430-2.2440	0.0008-0.0015	0.0020-0.0125	3	2.7252-2.7260	0.0003-0.0024	0.0036-0.0106

42356-BSER-C07

TORQUE SPECIFICATIONS
All readings in ft. lbs.

	Engine Displacement Liters (cc)	Engine ID/VIN	Cylinder Head Bolts	Main Bearing Bolts	Rod Bearing Bolts	Crankshaft Damper Bolts	Flywheel Bolts	Manifold Intake *	Exhaust	Spark Plugs	Lug Nut
2000	2.5 (2500)	C	①	②	③	103-133	54-64	19-28	14-21	5-10	100
	3.0 (2982)	U	④	60	26	107	54-64	24	⑤	8-10	100
	4.0 (3950)	X	⑥	66-77	19-24	⑦	59	⑧	19	10-15	100
2001	2.3 (2261)	D	⑨	NA	NA	⑩	⑪	13	40	9	100
	2.5 (2500)	C	①	②	③	103-133	54-64	19-28	14-21	5-10	100
	3.0 (2982)	U	④	60	26	107	54-64	24	⑤	8-10	100
	4.0 (4000)	E	⑫	72	⑬	⑭	75-85	7	16	15	100
2002	2.3 (2261)	D	⑨	NA	NA	⑮	⑯	13	40	9	100
	3.0 (2982)	U	④	60	26	107	54-64	24	⑤	8-10	100
	4.0 (4000)	E	⑰	72	⑬	⑭	75-85	7	16	13	100
2003	2.3 (2261)	D	⑨	NA	NA	⑮	⑯	13	40	9	100
	3.0 (2982)	U	④	60	26	107	54-64	24	⑤	8-10	100
	4.0 (4000)	E	⑰	72	⑬	⑭	75-85	7	16	13	100

NA: Information not available

* NOTE: Applies to Lower Manifold only.

① Step 1: 52 ft. lbs.
Step 2: recheck at 52 ft. lbs.
Step 3: +90 degrees

② Step 1: 51-59 ft. lbs.
Step 2: 76-84 ft. lbs.

③ Step 1: 26-30 ft. lbs.
Step 2: 31-36 ft. lbs.

④ Step 1: 59 ft. lbs.
Step 2: Back off 1 full turn
Step 3: 34-40 ft. lbs.
Step 4: 63-73 ft. lbs.

⑤ Step 1: 89 inch lbs.
Step 2: 17 ft. lbs.

⑥ Step 1: 44 ft. lbs.
Step 2: 59 ft. lbs.
Step 3: Plus 85 degrees

⑦ Step 1: 30-37 ft. lbs.
Step 2: + 90 degrees

⑧ Step 1: 3-6 ft. lbs.
Step 2: 6-11 ft. lbs.
Step 3: 11-15 ft. lbs.
Step 4: 15-18 ft. lbs.

⑨ Step 1: 44 inch lbs.
Step 2: 11 ft. lbs.
Step 3: 33 ft. lbs.
Step 4: +90 degrees
Step 5: +90 degrees

⑩ Step 1: 30 ft. lbs.
Step 2: +90 degrees

⑪ Step 1: 51-59 ft. lbs.
Step 2: 50 ft. lbs.
Step 3: 83 ft. lbs.

⑫ Step 1: 24 ft. lbs.
Step 2: plus 90 degrees

⑬ Step 1: 15 ft. lbs.
Step 2: +90 degrees

⑭ Step 1: 37 ft. lbs.
Step 2: plus 90 degrees

⑮ Step 1: 74 ft. lbs.
Step 2: +90 degrees

⑯ Step 1: 37 ft. lbs.
Step 2: 50 ft. lbs.
Step 3: 83 ft. lbs.

⑰ 8mm bolts: 24 ft. lbs.
12mm bolts: 24 ft. lbs. +80 degrees, +80 degrees more

42356-BSER-C08

BRAKE SPECIFICATIONS

All measurements in inches unless noted

Year	Model		Brake Disc			Brake Drum Diameter			Minimum Lining Thickness	Brake Caliper	
			Original Thickness	Minimum Thickness	Maximum Runout	Original Inside Diameter	Max. Wear Limit	Maximum Machine Diameter		Bracket Bolts (ft. lbs.)	Mounting Bolts (ft. lbs.)
2000	B Series	①	0.850	0.810	0.0030	9.00	9.09	9.06	0.030	72-97	21-26
		②	0.850	0.810	0.0030	10.00	10.09	10.06	0.030	72-97	21-26
		③	0.850	0.810	0.0030	10.00	10.09	10.06	0.030	72-97	21-26
2001	B Series	①	0.850	0.810	0.0030	9.00	9.09	9.06	0.030	72-97	21-26
		②	0.850	0.810	0.0030	10.00	10.09	10.06	0.030	72-97	21-26
		③	0.850	0.810	0.0030	10.00	10.09	10.06	0.030	72-97	21-26
2002	B Series	①	0.850	0.810	0.0030	9.00	9.09	9.06	0.030	85	24
		②	0.850	0.810	0.0030	10.00	10.09	10.06	0.030	85	24
		③	0.850	0.810	0.0030	10.00	10.09	10.06	0.030	85	24
2003	B Series	①	0.850	0.810	0.0030	9.00	9.09	9.06	0.030	85	24
		②	0.850	0.810	0.0030	10.00	10.09	10.06	0.030	85	24
		③	0.850	0.810	0.0030	10.00	10.09	10.06	0.030	85	24

NOTE: Due to changes made during production, refer to manufacturer's specifications if they differ from those in this chart

NA: Not Available

F: Front

R: Rear

① With 9 inch brakes

② 4x2 with 10 inch brakes

③ 4x4 with 10 inch brakes

42356-BSER-C09

TIRE, WHEEL AND BALL JOINT SPECIFICATIONS

Year	Model	OEM Tires		Tire Pressures (psi)		Wheel Size	Ball Joint Inspection
		Standard	Optional	Front	Rear		
2000	B-Series 2wd	P205/75R14SL	P225/70R14SL	35	35	6-JJ	0.030 in. ①
	B-Series 4wd	P215/75R15SL	P235/75R15SL	35	35	6-JJ	0.030 in. ①
2001	B-Series 2wd	P205/75R14SL	P225/70R14SL	35	35	6-JJ	0.030 in. ①
	B-Series 4wd	P215/75R15SL	P235/75R15SL	35	35	6-JJ	0.030 in. ①
2002	B-Series 2wd	P235/75R15	none	②	②	7-JJ	0.030 in. ①
	B-Series 4wd	P245/75R16	none	②	②	7-JJ	0.030 in. ①
2003	B-Series 2wd	P235/75R15	none	②	②	7-JJ	0.030 in. ①
	B-Series 4wd	P245/75R16	none	②	②	7-JJ	0.030 in. ①

OEM: Original Equipment Manufacturer

PSI: Pounds Per Square Inch

STD: Standard

OPT: Optional

① Both upper and lower

② See placard on door post

42356-BSER-C10

WHEEL ALIGNMENT

Year	Model		Caster Range (+/-Deg.)	Caster Preferred Setting (Deg.)	Camber Range (+/-Deg.)	Camber Preferred Setting (Deg.)	Toe-in (in.)	Steering Axis Inclination (Deg.)
2000	B Series	2wd	1.0	①	0.70	-0.50	0.06+/-0.25	—
		4wd	1.0	②	0.70	-0.50	0.12+/-0.25	—
2001	B Series	2wd	1.0	④	0.70	-0.50	0.06+/-0.25	—
		4wd	1.0	②	0.70	-0.50	0.12+/-0.25	—
2002	B Series	2wd	1.0	④	0.70	-0.50	0.06+/-0.25	—
		4wd	1.0	②	0.70	-0.50	0.12+/-0.25	—
		Rear	—	—	0.75	0	0+/-0.30	—
2003	B Series	2wd	1.0	④	0.70	-0.50	0.06+/-0.25	—
		4wd	1.0	②	0.70	-0.50	0.12+/-0.25	—
		Rear	—	—	0.75	0	0+/-0.30	—

① Left: +4.1
   Right: +4.5

② Left: +3.9
   Right: +4.4

③ Left: +4.6
   Right: +5.0

④ Left: +4.0
   Right: +4.4

42356-BSER-C11

SCHEDULED MAINTENANCE INTERVALS
2000 MAZDA B-SERIES

TO BE SERVICED	TYPE OF SERVICE	VEHICLE MILEAGE INTERVAL (x1000)												
		5	10	15	20	25	30	35	40	45	50	55	60	65
Engine oil & filter	R	✓	✓	✓	✓	✓	✓	✓	✓	✓	✓	✓	✓	✓
Driveshaft fittings	L	✓	✓	✓	✓	✓	✓	✓	✓	✓	✓	✓	✓	✓
Exhaust system	I	✓	✓	✓	✓	✓	✓	✓	✓	✓	✓	✓	✓	✓
Cooling system hoses	I			✓			✓			✓			✓	
Coolant strength	I			✓			✓			✓			✓	
Brake caliper rails	L			✓			✓			✓			✓	
Air cleaner filter	R						✓						✓	
Front wheel bearings (2wd)	I/L						✓						✓	
Fuel filter ①	R										✓			
PCV valve	R												✓	
Accessory drive belts	S/I												✓	
Spark plugs 2.5L	R												✓	
Spark plugs 2.3, 3.0 & 4.0L	R	every 100,000 miles												
Transfer case fluid	R												✓	
Manual trans. fluid	R												✓	
Coolant ②	R										✓			
Differential fluid ③	R	every 100,000 miles												
Timing belt 2.5L	S/I	every 120,000 miles												
Clutch reservoir level	I	✓	✓	✓	✓	✓	✓	✓	✓	✓	✓	✓	✓	✓
Brake hoses	I			✓			✓			✓			✓	
Parking brake system	I						✓						✓	

R: Replace   S: Service   I: Inspect   L: Lubricate

① Recommended, but not required in Calif.

② Change at 50,000 miles, then every 30,000 miles or 36 months

③ Except synthetic

FREQUENT OPERATION MAINTENANCE (SEVERE SERVICE)

If a vehicle is operated under any of the following conditions it is considered severe service:

- Towing a trailer or using a camper or car-top carrier.

- Repeated short trips of less than 5 miles in temperatures below freezing, or trips of less than 10 miles in any temperature.

- Extensive idling or low-speed driving for long distance as in heavy commercial use, such as delivery, taxi or police cars.

- Operating on rough, muddy or salt-covered roads.

- Operating on unpaved or dusty roads.

- Driving in extremely hot (over 90°) conditions.

Engine oil & filter: replace every 3000 miles.

Air cleaner filter: service or inspect every 6000 miles.

Exhaust system: check every 6000 miles.

Rotate tires every 9000 miles. (City delivery vehicles & other unique applications that require constant turning may need frequent tire rotation.)

Automatic transmission fluid & filter: change every 21,000 miles.

42356-BSER-C12

SCHEDULED MAINTENANCE INTERVALS
2001-03 MAZDA B-SERIES

TO BE SERVICED	TYPE OF SERVICE	VEHICLE MILEAGE INTERVAL (x1000)												
		5	10	15	20	25	30	35	40	45	50	55	60	65
Engine oil & filter	R	✓	✓	✓	✓	✓	✓	✓	✓	✓	✓	✓	✓	✓
Tires	Rotate	✓	✓	✓	✓	✓	✓	✓	✓	✓	✓	✓	✓	✓
Auto trans. fluid	I			✓			✓			✓			✓	
Brake pads/shoes	I			✓			✓			✓			✓	
Coolant hoses	S/I			✓			✓			✓			✓	
Steering linkage	I			✓			✓			✓			✓	
Cabin air filter	R			✓			✓			✓			✓	
Ball joints (2wd)	L			✓			✓			✓			✓	
Exhaust system	I						✓						✓	
Engine air filter	R						✓						✓	
Fuel filter ①	R						✓						✓	
Auto trans fluid (4-speed)	R						✓						✓	
Green coolant ②	R									✓				
Wheel bearings (2wd)	L												✓	
Manual trans. fluid	R												✓	
Spark plugs	R	every 100,000 miles												
PCV valve	R	every 100,000 miles												
Orange coolant	R	every 150,000 miles												
Auto trans fluid (5-speed)	R	every 150,000 miles												
Differential fluid	R	every 150,000 miles												
Accessory drive belts	R	every 150,000 miles												
Transfer case fluid	R	every 150,000 miles												

R: Replace   S: Service   I: Inspect   L: Lubricate

① Recommended, but not required in Calif.

② Change every 30,000 miles or 36 months thereafter

FREQUENT OPERATION MAINTENANCE (SEVERE SERVICE)

If a vehicle is operated under any of the following conditions it is considered severe service:

- Towing a trailer or using a camper or car-top carrier.

- Repeated short trips of less than 5 miles in temperatures below freezing, or trips of less than 10 miles in any temperature.

- Extensive idling or low-speed driving for long distance as in heavy commercial use, such as delivery, taxi or police cars.

- Operating on rough, muddy or salt-covered roads.

- Operating on unpaved or dusty roads.

- Driving in extremely hot (over 90°) conditions.

Engine oil & filter: replace every 3000 miles.
Air cleaner filter: service or inspect every 6000 miles.
Exhaust system: check every 6000 miles.
Automatic transmission fluid & filter: change every 30,000 miles.
Transfer case fluid: change every 60,000 miles
Fule filter: change every 15,000 miles
Spark plugs: change every 60,000 miles
2wd front wheel bearings: lubricate every 30,000 miles

42356-BSER-C13

PRECAUTIONS

Before servicing any vehicle, please be sure to read all of the following precautions, which deal with personal safety, prevention of component damage, and important points to take into consideration when servicing a motor vehicle:

• Never open, service or drain the radiator or cooling system when the engine is hot; serious burns can occur from the steam and hot coolant.

• Observe all applicable safety precautions when working around fuel. Whenever servicing the fuel system, always work in a well-ventilated area. Do not allow fuel spray or vapors to come in contact with a spark, open flame, or excessive heat (a hot drop light, for example). Keep a dry chemical fire extinguisher near the work area. Always keep fuel in a container specifically designed for fuel storage; also, always properly seal fuel containers to avoid the possibility of fire or explosion. Refer to the additional fuel system precautions later in this section.

• Fuel injection systems often remain pressurized, even after the engine has been turned **OFF**. The fuel system pressure must be relieved before disconnecting any fuel lines. Failure to do so may result in fire and/or personal injury.

• Brake fluid often contains polyglycol ethers and polyglycols. Avoid contact with the eyes and wash your hands thoroughly after handling brake fluid. If you do get brake fluid in your eyes, flush your eyes with clean, running water for 15 minutes. If eye irritation persists, or if you have taken

brake fluid internally, IMMEDIATELY seek medical assistance.

• The EPA warns that prolonged contact with used engine oil may cause a number of skin disorders, including cancer! You should make every effort to minimize your exposure to used engine oil. Protective gloves should be worn when changing oil. Wash your hands and any other exposed skin areas as soon as possible after exposure to used engine oil. Soap and water, or waterless hand cleaner should be used.

• All new vehicles are now equipped with an air bag system, often referred to as a Supplemental Restraint System (SRS) or Supplemental Inflatable Restraint (SIR) system. The system must be disabled before performing service on or around system components, steering column, instrument panel components, wiring and sensors. Failure to follow safety and disabling procedures could result in accidental air bag deployment, possible personal injury and unnecessary system repairs.

• Always wear safety goggles when working with, or around, the air bag system. When carrying a non-deployed air bag, be sure the bag and trim cover are pointed away from your body. When placing a non-deployed air bag on a work surface, always face the bag and trim cover upward, away from the surface. This will reduce the motion of the module if it is accidentally deployed. Refer to the additional air bag system precautions later in this section.

• Clean, high quality brake fluid from a

sealed container is essential to the safe and proper operation of the brake system. You should always buy the correct type of brake fluid for your vehicle. If the brake fluid becomes contaminated, completely flush the system with new fluid. Never reuse any brake fluid. Any brake fluid that is removed from the system should be discarded. Also, do not allow any brake fluid to come in contact with a painted surface; it will damage the paint.

• Never operate the engine without the proper amount and type of engine oil; doing so WILL result in severe engine damage.

• Timing belt maintenance is extremely important! Many models utilize an interference-type, non-freewheeling engine. If the timing belt breaks, the valves in the cylinder head may strike the pistons, causing potentially serious (also time-consuming and expensive) engine damage. Refer to the maintenance interval charts in the front of this section for the recommended replacement interval for the timing belt, and to the timing belt section for belt replacement and inspection.

• Disconnecting the negative battery cable on some vehicles may interfere with the functions of the on-board computer system(s) and may require the computer to undergo a relearning process once the negative battery cable is reconnected.

• When servicing drum brakes, only disassemble and assemble one side at a time, leaving the remaining side intact for reference.

ENGINE REPAIR

Alternator

REMOVAL

2.3L Engines

1. Before servicing the vehicle, refer to the precautions in the beginning of this section.
2. Remove or disconnect the following:
   • Battery ground cable
   • Air cleaner outlet tube
   • Accessory drive belt
   • Mounting bolts and alternator
   • Nut and the electrical connectors.
3. To install, reverse the removal procedure. Torque all mounting bolts to 18 ft. lbs. (25Nm).

2.5L Engines

1. Before servicing the vehicle, refer to the precautions in the beginning of this section.
2. Remove or disconnect the following:
   • Negative battery cable
   • Drive belt
   • Electrical connections to the alternator
   • Alternator
To install:
3. Install or connect the following:
   • Alternator. Torque the bolts to 40 ft. lbs. (55 Nm).
   • Electrical connectors to the alternator
   • Drive belt
   • Negative battery cable

3.0L and 4.0L OHV Engines

1. Before servicing the vehicle, refer to the precautions in the beginning of this section.
2. Remove or disconnect the following:
   • Negative battery cable
   • Air cleaner outlet tube
   • Drive belt
   • Electrical connectors from the alternator
   • Wiring harness to alternator push pin
   • Alternator
To install:
3. Install or connect the following:
   • Alternator. Torque the bolts to 40 ft. lbs. (55 Nm).
   • Push pin for the alternator wiring harness

- Electrical connectors to the alternator
- Drive belt
- Air cleaner outlet tube
- Negative battery cable

4.0L OHC Engine

1. Before servicing the vehicle, refer to the precautions in the beginning of this section.
2. Remove or disconnect the following:
   - Negative battery cable
   - Air cleaner outlet tube
   - Accessory drive belt
   - Electrical connectors
   - Wiring harness-to-generator pin-type retainer
   - Stud bolt, the bolts and the alternator
3. To install, reverse the removal procedure. Torque all mounting bolts to 35 ft. lbs. (47Nm).

Ignition Timing

ADJUSTMENT

The ignition timing is preset to 10 degrees Before Top Dead Center (BTDC) and is not adjustable.

Engine Assembly

REMOVAL & INSTALLATION

2.3L Engines

1. Before servicing the vehicle, refer to the precautions in the beginning of this section.
2. Relieve the fuel system pressure.
3. Drain the cooling system.
4. Drain the engine oil.
5. Properly discharge the air conditioning system.
6. Remove or disconnect the following:
   - Hood
   - Accelerator control snow shield
   - Air cleaner tube
   - Upper radiator hose
   - Lower radiator hose
   - Fan and shroud
   - PCM electrical connector. Remove the retaining nut on the harness clamp. Position the harness on the engine.
   - Ground stud for the PCM
   - Heater hoses
   - All vacuum hoses
   - Coolant reservoir hoses

- Air conditioning compressor clutch
- MAF electrical connector
- Air conditioning compressor manifold, plug the lines and the compressor ports
- Accelerator and speed control cables
- Power steering return hose
- PSP switch electrical connector
- High pressure power steering hose
- Fuel supply hose
- 42-pin electrical connector
- VMV vacuum regulator solenoid supply hose
- Evaporative purge hose
- Brake booster vacuum hose and the engine ground strap
- Positive battery cable
- Solenoid control wire at the starter
- Starter wiring harness clamp bolt and position it out of the way.
- RH splash shield
- Alternator electrical connections
- Block heater electrical connector
- Front heated oxygen sensor electrical connector at the bell housing
- Oil pressure sensor electrical connector
- Engine wiring pushpins and position the engine wiring harnesses out of the way.
- Oil filter
- With automatic transmission, the bolt retaining the transmission cooling tubes to the engine. Remove the bracket.
- Transmission dust shield
- Starter motor
- Heated oxygen sensor electrical connector at the rear of the transmission
- Transmission wiring harness
- Vehicle speed sensor, transmission range sensor, backup light switch and the transmission electrical connectors. Disconnect the pushpins and position the harness forward to the engine.
- Oil filter adapter

➡ **Leave two side bolts in until the engine is ready to be removed.**

- Nine of the transmission-to-engine bolts
- With automatic transmission, the transmission fluid indicator and tube assembly
- Starter dust shield

➡ **Mark one stud and the flexplate for assembly reference.**

- With automatic transmission, the four torque converter nuts

7. Support the transmission with a floor jack.
8. Support the engine with a floor crane using a spreader bar.
9. Remove the two side transmission-to-engine bolts.
10. Remove the four engine support insulator.
11. Remove the engine from the vehicle.
12. Installation is the reverse of removal. Observe the following torques:
   - Torque converter bolts: 26 ft. lbs. (35Nm)
   - Nine transmission-to-engine bolts 35 ft. lbs. (48Nm)
   - Oil filter adapter: 18 ft. lbs. (25Nm)
   - Starter: 30 ft. lbs. (40Nm)
   - Engine support nuts: 75 ft. lbs. (102Nm)

2.5L Engines

1. Before servicing the vehicle, refer to the precautions in the beginning of this section.
2. Relieve the fuel system pressure.
3. Drain the cooling system.
4. Drain the engine oil.
5. Properly discharge the air conditioning system.
6. Remove or disconnect the following:
   - Hood
   - Air cleaner outlet tube
   - Accelerator control splash shield
   - Upper radiator hose
   - Lower radiator hose
   - The two bolts and position the fan shroud on the fan
   - Water pump pulley, fan and shroud
   - Radiator overflow tube
   - Transmission cooler lines
   - Mass air flow sensor
   - Heater hoses
   - Power connection from the alternator
   - Vacuum connection at the vacuum reservoir
   - Throttle body heater hose
   - Air conditioning cycling switch
   - Connector from the PCM
   - Ground wire stud from the powertrain control module
   - Power steering cut-out switch
   - Peanut fitting from the air conditioning condenser core
   - Air conditioning high pressure cut-out switch
   - Air conditioning manifold hose
   - Accelerator cable and speed control cable
   - Brake booster vacuum hose and

vacuum tube at the upper intake manifold assembly
- EVR vacuum supply hose and the vacuum reservoir vacuum line
- Fuel line
- Air conditioning compressor
- Power steering pressure and return hoses
- Nut retaining the wiring harness
- Block heater
- Engine ground cable
- With automatic transmission, the two transmission harness connectors
- With manual transmission, the transmission harness and the heated oxygen sensor
- Starter motor
- Nut retaining the starter harness and transmission cooler brackets
- Three way catalytic converter

➡ **The torque converter nuts are accessed through the starter motor hole.**

- With automatic transmission, remove the four torque converter nuts and six bolts

7. Support the transmission with a floor jack.

8. Remove the differential pressure feedback EGR sensor.

9. Support the engine with a floor crane.

10. Remove the two upper transmission-to-engine bolts.

11. On vehicles equipped with automatic transmission, disconnect then separate the heated exhaust gas oxygen sensor connector from the bracket located on the bell housing. Remove the four nuts.

Remove the engine from the vehicle.

12. Installation is the reverse of removal. Observe the following torques:
- Engine mount nuts: 85 ft. lbs. (115Nm)
- Transmission-to-engine bolts: 39 ft. lbs. (51 Nm)

3.0L Engines

1. Before servicing the vehicle, refer to the precautions in the beginning of this section.

2. Relieve the fuel system pressure.

3. Drain the cooling system.

4. Drain the engine oil.

5. Properly discharge the air conditioning system.

6. Remove or disconnect the following:
- Hood
- Air cleaner outlet tube
- Upper and the lower radiator hoses

✳✳ WARNING

The fan clutch has left-hand threads.

- The fan clutch and blade as an assembly
- Drive belt
- Fan shroud
- Radiator
- Air conditioning manifold and tube. Remove the nut and position the line aside.
- Air conditioning compressor wiring
- Air conditioning compressor and the air conditioning compressor mounting bracket
- Heater hoses
- Ground cable
- Fuel lines
- Snow shield
- Accelerator cable and the speed control actuator cable
- All vacuum lines
- 42-pin connector
- Powertrain control module connector
- Nut from the powertrain control module harness
- Stud bolt and the powertrain control module ground strap
- Alternator wiring and position aside
- Both heated oxygen sensors
- Transmission harness connectors
- MAF sensor
- LH heated oxygen sensor
- Dual converter Y pipe
- Starter motor and the starter grounding stud bolt
- Torque converter nuts
- 8 transmission-to-engine bolts

7. Install the lifting eyes.

8. Remove the four nuts.

9. Support the transmission.

10. Remove the engine from the vehicle.

11. Installation is the reverse of removal. Observe the following torques:
- Engine mount nuts: 80 ft. lbs. (109Nm)
- Transmission-to-engine bolts: 33 ft. lbs. (45Nm)
- Torque converter nuts: 26 ft. lbs. (35Nm)

4.0L OHV Engines

1. Before servicing the vehicle, refer to the precautions in the beginning of this section.

2. Relieve the fuel system pressure.

3. Drain the cooling system.

4. Drain the engine oil.

5. Properly discharge the air conditioning system.

6. Remove or disconnect the following:
- Both battery cables
- Mass Air Flow (MAF) sensor
- Air cleaner outlet tube
- Drive belt
- Accelerator control splash shield
- Upper and lower radiator hoses
- Fan guard/shroud
- Radiator overflow tube
- Radiator
- Transmission cooler lines, if equipped
- Heater hoses
- Alternator electrical connectors
- Vacuum reservoir connection
- Throttle body heater hose
- Air conditioning cycling switch
- Powertrain Control Module (PCM) connector
- Ground wire from the PCM
- Power steering cut-out switch
- Peanut fitting from the air conditioning condenser core
- Air conditioning high pressure cut-out switch
- Air conditioning manifold hose
- Accelerator and speed control cables
- Brake booster vacuum hose and tube from the intake manifold
- Vacuum reservoir line
- Fuel lines
- Power steering pressure and return hoses
- Block heater, if equipped
- Engine ground cable
- Automatic transmission harness connectors, if equipped
- Heated Oxygen (HO$_2$S) sensor, if equipped
- Starter
- Catalytic converter
- Torque converter bolts, if equipped and properly support the transmission
- Differential pressure feedback sensor and support the engine with a floor crane
- Exhaust Gas Recirculation (EGR) transducer
- Upper transmission-to-engine bolts
- Engine from the vehicle
- Clutch/ flywheel, if equipped

To install:

7. Install or connect the following:
- Flywheel. Torque the bolts to 64 ft. lbs. (87 Nm).
- Engine. Torque the mounting bolts to 85 ft. lbs. (115 Nm).

- HO$_2$S sensor
- Upper transmission-to-engine bolts. Torque the bolts to 38 ft. lbs. (51 Nm).
- EGR transducer and remove the floor jack from the transmission
- Torque converter-to-engine bolts. Torque the bolts to 38 ft. lbs. (51 Nm).
- Catalytic converter
- Transmission wiring harness connectors
- Engine ground cable. Torque the bolt to 106 inch lbs. (12 Nm).
- Block heater, if equipped
- Air conditioning high pressure cut out switch
- Air conditioning manifold to the evaporator core
- Power steering cut-out switch
- Power steering pressure and return lines
- Engine sensor control wiring harness to the air conditioning compressor
- Fuel lines
- 42 pin connector
- Vacuum reservoir vacuum line
- Brake booster vacuum hose and tube
- Accelerator cable
- Ground strap. Torque the bolt to 106 inch lbs. (12 Nm).
- Air conditioning manifold hose
- PCM ground strap. Torque the bolt to 106 inch lbs. (12 Nm).
- PCM wire harness bracket bolt. Torque the bolt to 61 inch lbs. (7 Nm).
- Air conditioning low pressure cut-out switch
- Throttle body heater hose
- Fan clutch and water pump pulley. Torque the bolt 17 ft. lbs. (23 Nm).
- Drive belt
- Alternator electrical connectors
- Inlet and outlet heater hoses
- Fuel charging wiring
- Radiator and fan shroud
- Transmission cooler lines, if equipped
- Upper, lower and overflow hoses to the radiator
- Accelerator control splash shield. Torque the bolts to 89 inch lbs. 10 Nm).
- Drive belt
- Air cleaner outlet tube
- MAF sensor
- Both battery cables

8. Recharge the air conditioning system.

9. Fill the cooling system.
10. Fill the engine with clean oil.
11. Run the engine and check for leaks and proper operation.
12. Check and adjust the front end alignment.

4.0L OHC Engines

✳✳ CAUTION

If the fuel supply manifold is used as a leverage device, damage may occur to the supply manifold. Care must be taken when working around the fuel supply manifold.

- Accelerator cable from engine
- Speed control cable from engine
- Radiator, the fan blade, and the fan shroud
- Accessory bracket bolts and position bracket aside
- Alternator wiring
- Wiring harness retainer and position generator wiring away from engine
- Engine electrical connector
- PCM connector
- PCM ground wire
- Engine ground wire
- Brake booster vacuum hose
- Air conditioning high pressure switch electrical connector
- Bolt and position the air conditioning lines aside

➡**Heater hose will be removed with engine.**

- Heater hoses
- Fuel line
- Starter motor
- Engine oil
- Oil drain plug
- Transmission portion of wiring harness
- RH and LH heated oxygen sensor connectors
- Transmission control connector
- Output shaft speed sensor connector
- Digital transmission range sensor connector
- Catalyst monitor sensor electrical connector
- Transmission/transfer case portion of the wiring harness from any routing clips or pushpins. Route transmission/transfer case portion of the wiring harness to top of engine.
- Bolt, and position the transmission cooling line bracket aside

- Air conditioning line bracket nut and position it aside
- Power steering return hose
- Power steering pressure hose
- Vapor management valve hose connector
- Eight bolts and the LH and the RH engine support insulator nuts

➡**The lifting eyes should be installed on the exhaust manifold studs for number three and number four cylinders.**

13. Install the lifting eyes.
14. Install the spreader bar to the lifting eyes.
15. Attach a floor crane to the spreader bar and remove the engine.
16. Installation is the reverse of removal. Observe the following torques:
- Left and right engine insulator nuts: 81 ft. lbs. (110 Nm)
- Engine mount nuts: 59 ft. lbs. (80Nm)
- Transmission-to-engine bolts: 35 ft. lbs. (47Nm)
- Torque converter nuts: 35 ft. lbs. (47Nm)

Water Pump

REMOVAL & INSTALLATION

2.3L Engine

1. Before servicing the vehicle, refer to the precautions in the beginning of this section.
2. Drain the cooling system.
3. Remove the drive belt.
4. Remove the water pump pulley.
5. Remove the water pump.

➡**Lubricate the water pump O-ring, with MERPOL®.**

6. To install, reverse the removal procedure. Torque the water pump mount bolts to 89 inch lbs. (10Nm). Torque the pulley bolts to 18 ft. lbs. (25Nm).

2.5L Engines

1. Before servicing the vehicle, refer to the precautions in the beginning of this section.
2. Drain the cooling system.
3. Remove or disconnect the following:
- Negative battery cable
- Drive belt
- Fan clutch and shroud
- Water pump pulley
- Heater hose from the water pump inlet tube

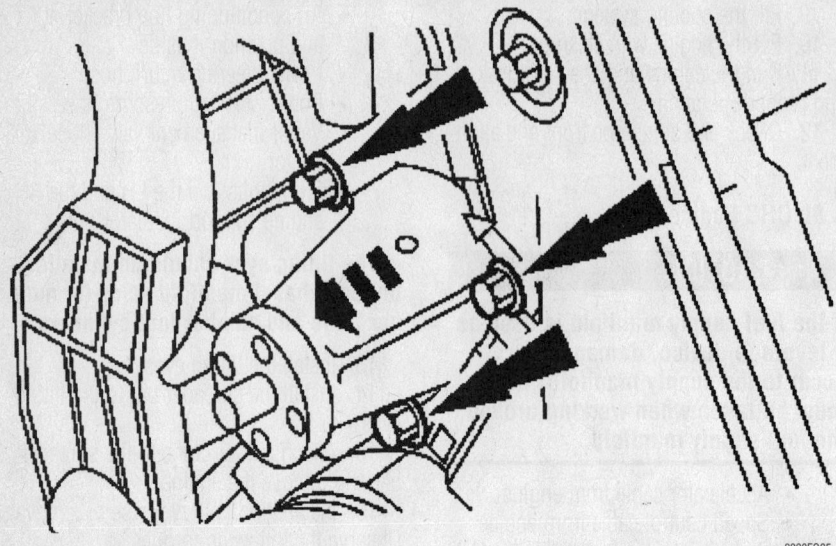

Exploded view of the water pump 2.5L engines

9308EG05

- Lower radiator hose
- Water pump inlet tube
- Water pump and discard the gasket

To install:

4. Clean the mating surface with the water pump connects to the engine.
5. Install or connect the following:
- Water pump with a new O-ring. Torque the bolts to 15 ft. lbs. (20 Nm).
- Inlet tube. Torque the bolts to 89 inch lbs. (10 Nm).
- Lower radiator hose
- Water pump pulley
- Heater hose from the water pump inlet tube
- Fan clutch and shroud
- Drive belt
- Negative battery cable

6. Fill the cooling system.
7. Start the vehicle and check for leaks, repair if necessary.

3.0L and 4.0L OHV Engines

1. Before servicing the vehicle, refer to the precautions in the beginning of this section.
2. Drain the cooling system.
3. Remove or disconnect the following:
- Negative battery cable
- Air cleaner outlet tube
- Fan and radiator shroud
- Water bypass tube
- Drive belt
- Heater hose
- Water pump pulley
- Lower radiator hose
- Air conditioning compressor and bracket assembly and move them aside

- Water pump

To install:

4. Clean the mating surfaces where the water pump attaches to the engine.
5. Install or connect the following:
- Water pump. Torque the bolts to 106 in lbs. (12 Nm).
- Air conditioning compressor mounting bracket. Torque the bolts to 44 ft. lbs. (61 Nm).
- Water pump pulley. Torque the bolts to 20 ft. lbs. (28 Nm).
- Drive belt
- Heater hose
- Lower radiator hose
- Fan and shroud
- Air cleaner outlet tube
- Negative battery cable

6. Fill the cooling system.
7. Start the vehicle and check for leaks, repair if necessary.

4.0L OHC Engine

1. Before servicing the vehicle, refer to the precautions in the beginning of this section.
2. Drain the cooling system.
3. Remove or disconnect the following:
- Fan shroud
- Accessory drive belt
- Idler pulley
- Water bypass hose
- Heater hose
- Lower radiator hose
- Water pump pulley
- Water pump

※※ WARNING

Use care when scraping the water pump-to-engine block mating sur-

faces. Gouges in the aluminum could form leak paths.

4. Clean all the sealing surfaces.
5. To install, reverse the removal procedure. Torque the water pump bolts to 89 inch lbs. (10Nm). Torque the pulley bolts to 18 ft. lbs. (25Nm).

Heater Core

REMOVAL & INSTALLATION

1. Disconnect the negative battery cable.

※※ CAUTION

After disconnecting the negative battery cable, wait for 1 minute for the SRS module to deplete its energy.

2. Drain the cooling system into a clean container for reuse.
3. Remove the steering column by performing the following procedure:
   a. Position the front wheels in the straight-ahead direction.
   b. At the both sides of the steering wheel, remove the cover plugs, the steering wheel-to-air bag module screws, disconnect the air bag electrical connector and carefully remove the air bag module.

※※ CAUTION

Safely store the air bag module with the front side facing upward.

   c. Remove the steering wheel-to-steering column nut.
   d. Using a steering wheel puller, press the steering wheel from the steering column.
   e. Remove the parking brake release handle screws and move the release handle aside.
   f. Remove the hood release screws and move the hood release aside.
   g. Remove the 2 instrument panel-to-steering column cover screws and the cover.
   h. Remove the instrument panel steering column opening reinforcement bolts and the reinforcement.
   i. Remove the ignition switch bolt and disconnect the ignition switch electrical connector.
   j. At the base of the steering column, disconnect the electrical connectors.
   k. If equipped with an automatic transmission, remove the transmission range indicator bolt and the cable.

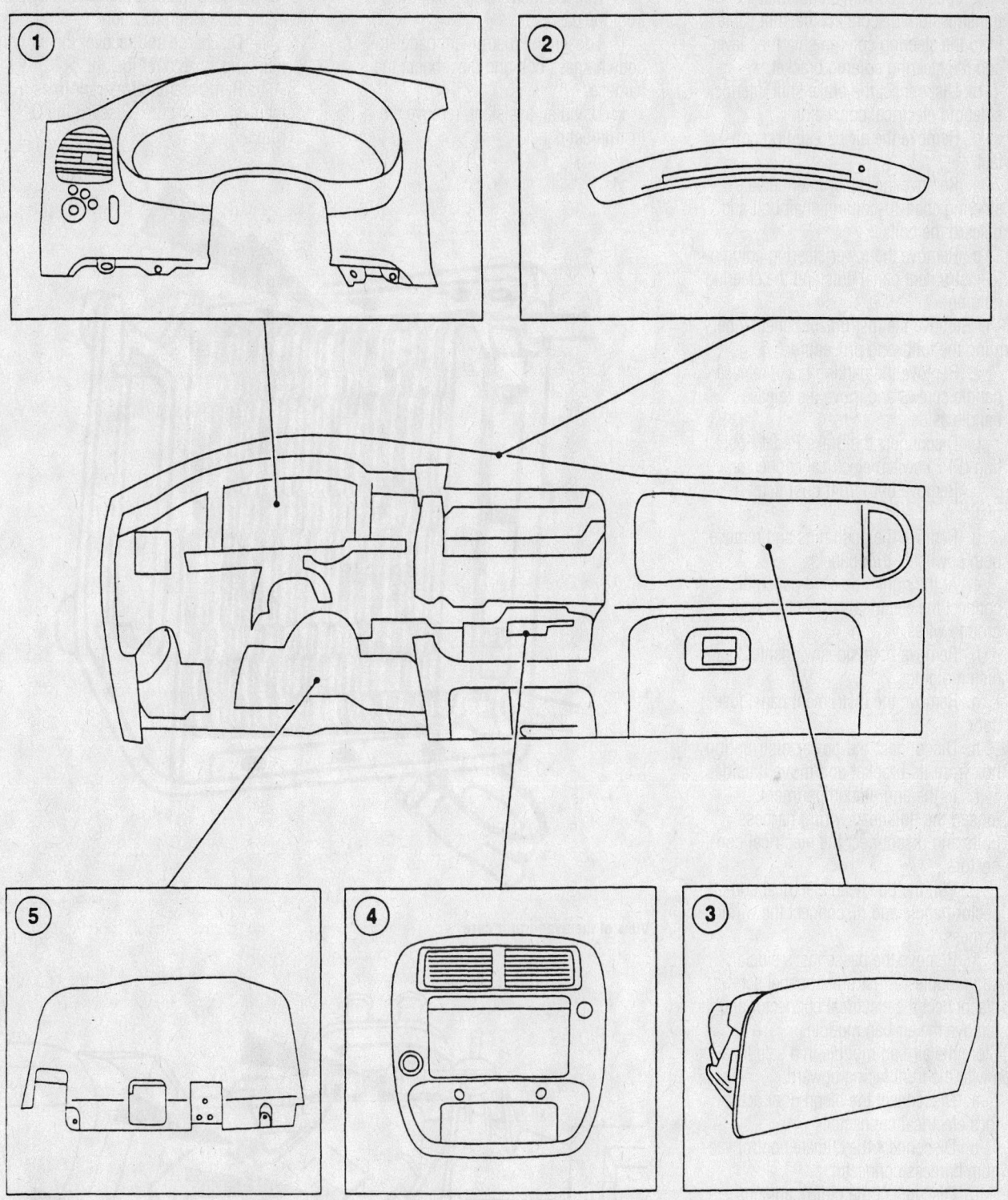

1 Instrument Panel Finish Panel
2 Instrument Panel Defroster
    Opening Grille Assembly
3 Passenger Side Air Bag Module
4 Instrument Panel Center Finish Panel
5 Instrument Panel Steering
    Column Cover

93113GL3

View of the instrument panel components

l. If equipped with an automatic transmission, disconnect the shift cable from the steering column shift tube lever and the steering column bracket.

m. Disconnect the brake shift interlock solenoid electrical connector.

n. Remove the air bag sliding contact.

o. Remove the upper intermediate steering shaft-to-column shaft bolt and discard the bolt.

p. Remove the lower steering column-to-instrument panel nuts and the steering column.

4. Remove the instrument panel by performing the following procedure:

a. Remove the parking brake release handle screws and move the release handle aside.

b. Disconnect the Brake Pedal Position (BPP) switch electrical connector.

c. Remove both front door scuff plates.

d. Remove the push pins and remove both cowl side trim panels.

e. At the right side cowl panel, disconnect the electrical connectors and ground wires.

f. Remove both sides windshield garnish moldings.

g. Remove the instrument panel fuse door.

h. Disconnect the power distribution box from its bracket and move it aside.

i. In the engine compartment, loosen the bulkhead wiring harness bolts and disconnect the electrical connectors.

j. Pull the bulkhead electrical connector handle and disconnect the wiring harness.

k. Remove the passenger's side air bag module-to-instrument panel screws, disconnect the electrical connector and remove the air bag module. Store the air bag module in a safe location with the front facing upward.

a. Disconnect the blend door actuator's electrical connector.

b. Disconnect the climate control vacuum harness connector.

c. Disconnect the radio's antenna connector.

d. Remove the glove compartment.

e. Remove the instrument panel defroster grille.

f. Remove the upper instrument panel bolts.

g. Under the steering column, remove the instrument panel brace bolt.

h. Remove both the right and left instrument panel-to-cowl bolts.

i. Pull the instrument panel away from the dash.

j. Loosen the instrument panel-to-body harness bolt and disconnect the harness.

k. Using an assistant, remove the instrument panel.

5. Remove the evaporator core by perform the following procedure:

a. Discharge and recover the air conditioning system refrigerant.

b. Remove the refrigerant lines from the evaporator core. Discard the O-rings.

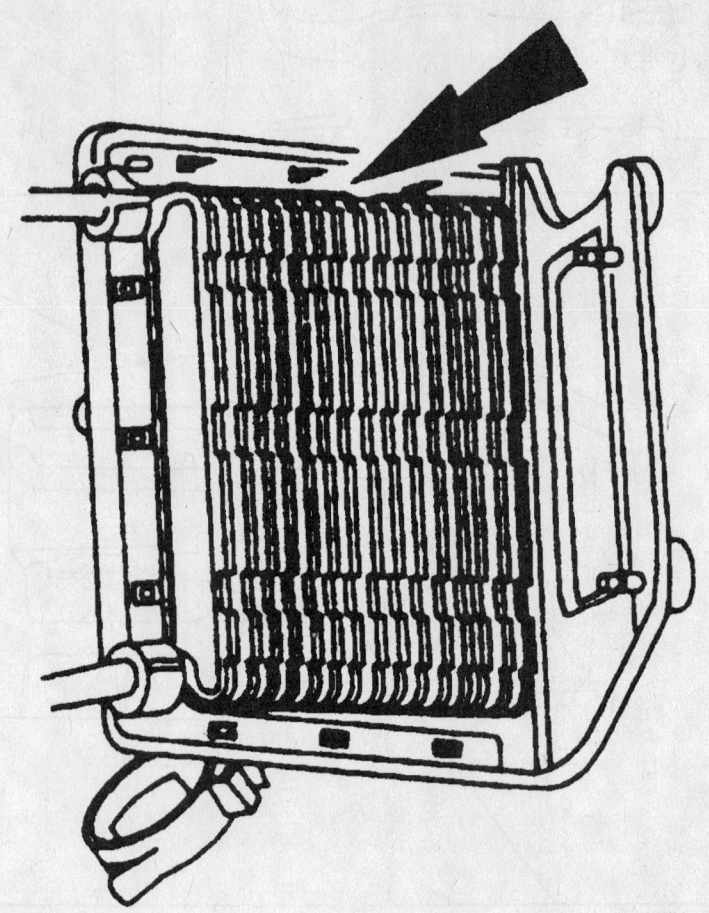

93113GL4

View of the evaporator core

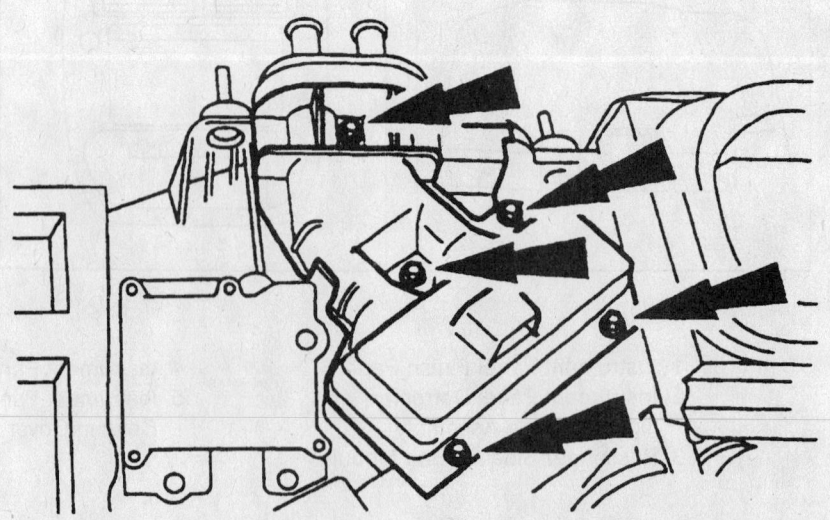

93113GL6

View of the heater core cover

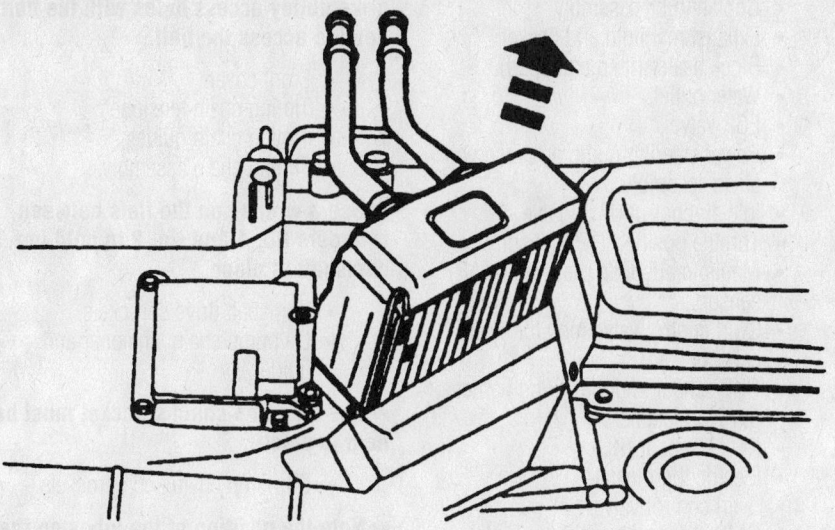

View of the heater core

93113GL5

c. If equipped, remove the air conditioning vacuum reservoir tank/bracket screws and reposition the tank.

d. If equipped, disconnect the speed control servo connector; then, remove the bolt and reposition the speed control servo.

e. If equipped with a 3.0L or 4.0L engine, remove the support bracket.

f. Disengage the windshield washer hose retainer and move it aside.

g. Disconnect the vacuum hose and the retainer; then, move the hose aside.

h. Remove the passenger's compartment nut.

i. At the back of the engine, remove the hose support bolts.

j. Remove the evaporator housing-to-chassis nuts.

k. Remove the air conditioning accumulator bracket screws.

l. Remove the evaporator housing cover screws, clips and the cover.

m. Remove the evaporator core from the housing.

6. Disconnect the heater hoses from the heater core.

7. Remove the heater housing plenum chamber nuts and the plenum chamber.

8. Remove the heater core-to-heater housing screws and the cover.

9. Remove the heater core.

To install:

10. Install the heater core.

11. Install the heater core-to-heater housing cover and the cover screws.

12. Install the heater housing plenum chamber and the plenum chamber nuts.

13. Connect the heater hoses to the heater core.

14. Install the evaporator core by performing the following procedure:

a. Install the evaporator core to the housing.

b. Install the evaporator housing cover, clips and the cover screws.

c. Install the air conditioning accumulator bracket screws.

d. Install the evaporator housing-to-chassis nuts.

e. At the back of the engine, install the hose support bolts.

f. Install the passenger's compartment nut.

g. Connect the vacuum hose and the retainer.

h. Engage the windshield washer hose retainer.

i. If equipped with a 3.0L or 4.0L engine, install the support bracket.

j. If equipped, install the speed control servo bolt and connect the connector.

k. If equipped, install the air conditioning vacuum reservoir tank and the bracket screws.

l. Using new O-rings, install the refrigerant lines to the evaporator core.

15. Install the instrument panel by performing the following procedure:

a. Using an assistant, install the instrument panel.

b. Connect the harness and tighten the instrument panel-to-body harness bolt.

c. Push the instrument panel toward the dash.

d. Install both the right and left instrument panel-to-cowl bolts.

e. Under the steering column, install the instrument panel brace bolt.

f. Install the upper instrument panel bolts.

g. Install the instrument panel defroster grille.

h. Install the glove compartment.

i. Connect the radio's antenna connector.

j. Connect the climate control vacuum harness connector.

k. Connect the blend door actuator's electrical connector.

l. Install the passenger's side air bag module, connect the electrical connector and torque the air bag module-to-instrument panel screws to 67–92 inch lbs. (7.6–10.4 Nm).

m. Connect the bulkhead electrical connector handle wiring harness.

n. In the engine compartment, connect the electrical connectors and tighten the bulkhead wiring harness bolts.

o. Connect the power distribution box to its bracket.

p. Install the instrument panel fuse door.

q. Install both sides windshield garnish moldings.

r. At the right side cowl panel, connect the electrical connectors and ground wires.

s. Install both cowl side trim panels and the push pins.

t. Install both front door scuff plates.

u. Connect the Brake Pedal Position (BPP) switch electrical connector.

v. Install the parking brake release handle and the release handle aside screws.

16. Install the steering column by performing the following procedure:

17. Install the steering column by performing the following procedure:

a. Install the lower steering column and the steering column-to-instrument panel nuts; then, torque the nuts to 10–13 ft. lbs. (13–17 Nm).

b. Using a new bolt, install the upper intermediate steering shaft-to-column shaft bolt and torque to 19–25 ft. lbs. (26–34 Nm).

c. Install the air bag sliding contact.

d. Connect the brake shift interlock solenoid electrical connector.

e. If equipped with an automatic transmission, connect the shift cable from the steering column shift tube lever and the steering column bracket.

f. If equipped with an automatic transmission, install the transmission range indicator cable and bolt.

g. At the base of the steering column, connect the electrical connectors.

h. Connect the ignition switch electrical connector and install the ignition switch bolt.

i. Install the instrument panel steering column opening reinforcement and the reinforcement bolts.

j. Install the instrument panel-to-steering column cover and the 2 cover screws.

k. Install the hood release and the hood release screws.

l. Install the parking brake release handle and the release handle screws.

m. Install the steering wheel to the steering column.

n. Install the steering wheel-to-steering column nut and torque the nut to 25–34 ft. lbs. (34–46 Nm).

o. At the both sides of the steering wheel, install the air bag module, connect the air bag electrical connector, install the steering wheel-to-air bag module screws and the cover plugs.

18. Refill the cooling system.

19. Connect the negative battery cable.

20. Evacuate and charge the air conditioning system.

21. Run the engine to normal operating temperatures; then, check the climate control operation and check for leaks.

Cylinder Head

REMOVAL & INSTALLATION

2.3L Engines

1. Before servicing the vehicle, refer to the precautions in the beginning of this section.

2. Relieve the fuel system pressure.

3. Drain the cooling system.

4. Properly discharge the air conditioning system.

5. Remove or disconnect the following:

- Negative battery cable
- Drive belt.
- Engine oil level indicator assembly.
- Engine oil level indicator.
- Engine oil level indicator tube.
- Water outlet tube.
- Water outlet tube.
- Air conditioning compressor.

➡ **The generator will be removed with the accessory bracket.**

- Accessory bracket.
- Right motor mount.
- Coolant hose from the thermostat.
- Coolant hose from the EGR valve.

- Coolant tube assembly.
- Exhaust manifold and gasket.
- Block heater (if so equipped).
- Water outlet.
- EGR valve.
- Power steering pump and reservoir as an assembly.
- Idle air control (IAC) valve.
- Throttle position (TP) sensor.
- Manifold absolute pressure (MAP) sensor.
- Swirl control valve monitor electrical connector.
- CKP sensor and the wiring harness pin-type retainers.
- Knock sensor (KS).
- Electric thermostat.
- Swirl control valve.
- CMP sensor electrical connector and disconnect the PCV hose from the intake manifold.
- Engine wiring harness pin-type retainers from the intake manifold.
- Engine wiring harness connector bracket. Position the engine wiring harness aside.
- EGR tube.
- Fuel supply line clip from the front of the intake manifold. Disconnect the vacuum hose from the intake manifold.
- Intake manifold assembly.
- Fuel injector electrical connectors. Detach the wiring harness pin-type retainers.
- Ignition coil and the cylinder head temperature (CHT) sensor electrical connectors.
- Engine wiring harness anchors from the valve cover studs. Remove the engine wiring harness.
- Ignition coil.
- Bypass hose.
- Thermostat housing.
- Knock sensor and the engine vent cover.
- Left motor mount.
- Fuel injector supply manifold with the injectors and the ground strap.
- Water pump pulley.
- Water pump.
- CMP sensor.
- CHT sensor.
- Spark plugs.
- Valve cover.
- CKP sensor.
- Crankshaft vibration damper

➡ **There is one front cover bolt behind the cooling fan drive pulley. To remove this bolt, align one of the cooling fan**

drive pulley access holes with the bolt head to access the bolt.

- Front cover.
- Timing chain tensioner.
- Timing chain guides.
- Timing chain assembly.

➡ **Use a wrench on the flats between cylinders No. 1 and No. 2 to hold the camshaft in place.**

- Camshaft drive sprockets.
- Oil pump chain tensioner and guide.

➡ **The oil pump chain sprocket must be held in place.**

- Oil pump chain and sprockets.

➡ **Note the position of the lobes on the No. 1 cylinder before removing the camshafts for assembly reference.**

6. Loosen the camshaft bearing cap bolts in sequence, one turn at a time. Repeat the first step until all tension is released from the camshaft bearing caps. Remove the camshaft bearing caps.

7. Remove or disconnect the following:
- Camshafts.
- Cylinder head bolts and the cylinder head.
- Cylinder head gasket.

8. Installation is the reverse of removal. Apply RTV sealer to the places shown. The head must be installed within 4 minutes of application. Observe the following torques:

a. Cylinder head:
- Step 1: Tighten the bolts to 5 Nm (44 inch lbs.)
- Step 2: Tighten the bolts to 15 Nm (11 ft. lbs.)
- Step 3: tighten the bolts to 45 Nm (33 ft. lbs.)
- Step 4: Tighten the bolts an additional 90 degrees (1/4 turn)
- Step 5: Tighten the bolts an additional 90 degrees (1/4 turn)

b. Camshafts:

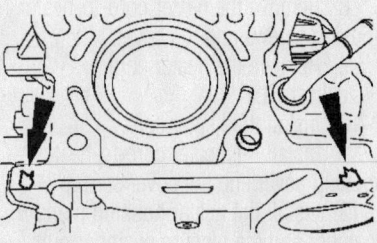

9348EG01

RTV sealer application—2.3L cylinder head

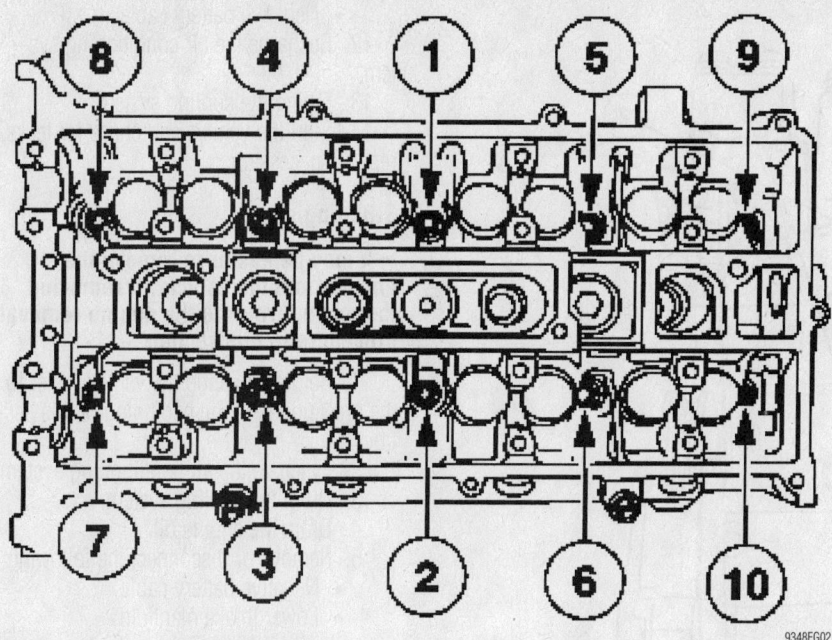

Head bolt torque sequence—2.3L

- Negative battery cable
- Loosen the water pump pulley bolts
- Drive belt
- Water pump pulley
- Fan and clutch assembly
- Intake manifolds
- Ignition wires from the spark plugs
- Spark plugs
- Oil level indicator tube
- Exhaust Gas Recirculation (EGR) valve to the exhaust manifold tube
- Valve cover
- Engine control wiring from the air conditioning compressor
- Air conditioning compressor mounting bracket with the power steering pump attached and move them aside
- Engine control sensor wiring from the alternator
- Lower radiator hose
- Water pump inlet tube
- Upper radiator hose
- Alternator
- Alternator mounting bracket
- Ignition wire and bracket
- Outer timing belt cover
- Timing belt

➡Install the camshafts with the alignment notches in the camshaft lined up so the camshaft alignment plate can be installed without rotating the camshafts. Make sure the lobes on the No. 1 cylinder are in the same position as noted in the disassembly procedure. Rotating the camshafts, or installing the camshafts 180 degrees out of position can cause severe damage to the valves and pistons. Lubricate the camshaft journals and bearing caps with clean engine oil. Install the camshafts and bearing caps. Tighten the bolts in the sequence shown in three stages.

- Step 1: Tighten the camshaft bearing caps one turn at a time until tight.
- Step 2: Tighten the bolts to 7 Nm . (62 inch lbs.)
- Step 3: Tighten the bolts to 16 Nm (12 ft. lbs.)
- c. Crankshaft vibration damper:

➡Do not reuse the crankshaft pulley bolt. Tighten the bolt in two stages.

- Step 1: Tighten the bolt to 40 Nm (30 ft. lbs.)
- Step 2: Tighten the bolt and additional 90 degrees (1/4 turn).

2.5L Engines

1. Before servicing the vehicle, refer to the precautions in the beginning of this section.

2. Relieve the fuel system pressure.
3. Drain the cooling system.
4. Properly discharge the air conditioning system.
5. Remove or disconnect the following:

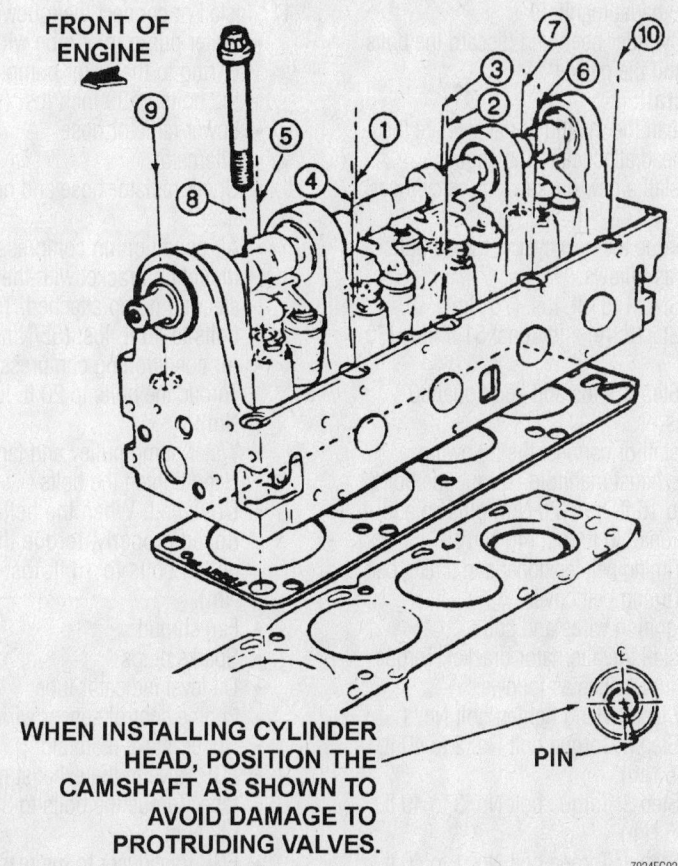

FRONT OF ENGINE

WHEN INSTALLING CYLINDER HEAD, POSITION THE CAMSHAFT AS SHOWN TO AVOID DAMAGE TO PROTRUDING VALVES.

PIN

Cylinder head bolt torque sequence—2.5L engines

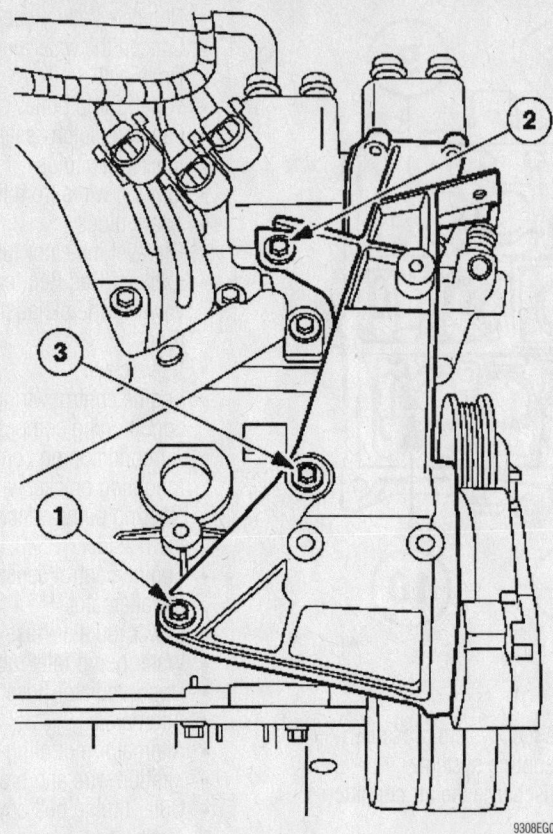

Alternator bracket bolt tightening sequence 2.5L engines

9308EG06

- Exhaust manifold
- Cylinder head and discard the bolts and the gasket

To install:

6. Clean the mating surface where the cylinder head attaches to the engine.

7. Install a new gasket and the cylinder head.

8. Torque the new cylinder head bolts in stages as follows:

    a. Step 1: 51 ft. lbs. (70 Nm).

    b. Step 2: An additional 51 ft. lbs. (70 Nm).

    c. Step 3: Plus and additional 90 degrees.

9. Install or connect the following:

- Exhaust manifold. Torque the bolts to 15 ft. lbs. (20 Nm) plus an additional 30 ft. lbs. (40 Nm).
- Timing belt tensioner and timing belt
- Timing belt cover
- Ignition wires and coil

10. Install the alternator bracket. Torque the bolts in 4 stages as follows:

    a. Step 1: Hand tighten bolt No. 1.

    b. Step 2: Torque bolt No. 2 to 40 ft. lbs. (55 Nm).

    c. Step 3: Torque bolt No. 3 to 40 ft. lbs. (55 Nm).

    d. Step 4: Torque bolt No. 1 to 40 ft. lbs. (55 Nm).

11. Install or connect the following:

- Water pump inlet tube with a new O-ring to the water pump. Torque the bolts to 89 inch lbs. (10 Nm).
- Lower radiator hose
- Alternator
- Upper radiator hose and heater hose
- Air conditioning compressor mounting bracket with the power steering pump attached. Torque the bolts to 40 ft. lbs. (55 Nm).
- Air conditioning compressor. Torque the bolts to 20 ft. lbs. (28 Nm).
- Water pump pulley and fan clutch. Hand tighten the bolts
- Drive belt. When the belt is positioned properly, torque the fan clutch bolts to 16 ft. lbs. (23 Nm).
- Fan shroud
- Sparks plugs
- Oil level indicator tube
- Engine control sensor wiring
- Upper intake manifold
- EGR valve to the exhaust manifold tube. Torque the bolts to 34 ft. lbs. (47 Nm).
- EGR transducer to the rear of the engine

- Negative battery cable

12. Recharge the air conditioning system.

13. Filling the cooling system.

14. Start the vehicle and check for leaks, repair if necessary.

3.0L Engine

➡ **It may be easier to remove the engine from the vehicle. If removing the engine, refer to the engine removal procedure in this section.**

1. Before servicing the vehicle, refer to the precautions in the beginning of this section.

2. Evacuate the air conditioning system.

3. Drain the cooling system.

4. Drain the engine oil.

5. Remove or disconnect the following:

- Negative battery cable
- Lower intake manifold
- Air conditioning compressor
- Alternator
- Power steering pump
- Alternator mounting bracket
- Air conditioning compressor mounting bracket
- Exhaust manifolds
- Cylinder head and discard the bolts and gasket

To install:

➡ **The "V" in the cylinder head gasket must face the front of the engine.**

6. Clean the mating surfaces where the head attaches to the engine.

7. Install a new cylinder head gasket and the cylinder head to the engine.

8. Torque the new cylinder head bolts in stages as follows:

    a. Step 1: 59 ft. lbs. (80 Nm).

    b. Step 2: Loosen the bolts one full turn.

    c. Step 3: 40 ft. lbs. (55 Nm).

    d. Step 4: 63 ft. lbs. (85 Nm).

9. Install or connect the following:

- Lower intake manifold
- Exhaust manifold
- Air conditioning compressor mounting bracket. Torque the bolts to 44 ft. lbs. (66 Nm).
- Alternator mounting bracket
- Power steering pump
- Alternator
- Air conditioning compressor
- Negative battery cable

10. Fill the engine with clean oil.

11. Fill the cooling system.

12. recharge the air conditioning system

13. Start the vehicle and check for leaks, repair if necessary.

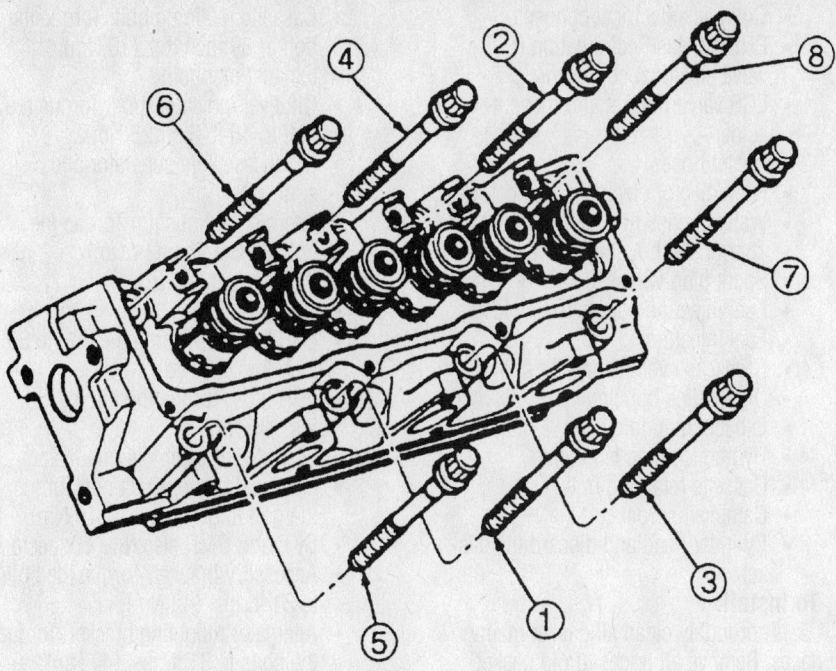

Cylinder head bolt torque sequence 3.0L engine

7924EG03

4.0L OHV Engine

➡️**New cylinder head bolts must be used when installing the cylinder head on the 4.0L engine.**

1. Before servicing the vehicle, refer to the precautions in the beginning of this section.
2. Relieve the fuel system pressure.
3. Evacuate the air conditioning system.
4. Drain the cooling system.
5. Drain the engine oil.
6. Remove or disconnect the following:
   - Negative battery cable
   - Drive belt
   - Separate the air conditioning manifold tube from the air conditioning compressor
   - Air conditioning compressor electrical connectors
   - Air conditioning compressor mounting bracket and move it aside
   - Alternator electrical connectors
   - Heater hose and move it aside
   - Alternator mounting bracket
   - Lower intake manifold
   - Both exhaust manifolds
7. Gradually loosen the rocker arm shafts and remove them.
   - Matchmark the position of the push rods and remove them
   - Cylinder head and gasket

To install:

8. Clean the mating surface where the cylinder head attaches to the engine.

9. Install a new cylinder head gasket and the cylinder head to the engine.
10. Torque the new cylinder head bolts in sequence as follows:
   a. Step 1: 25 ft. lbs. (34 Nm).
   b. Step 2: 53 ft. lbs. (72 Nm).
   c. Step 3: Plus an additional 90 degrees.
11. Install or connect the following:
   - Push rods
   - Rocker arm shafts gradually. Torque the bolts to 24 ft lbs. (33 Nm) plus an additional 90 degrees
   - Exhaust manifolds
   - Lower intake manifold
   - Alternator bracket. Torque the bolts to 35 ft. lbs. (47 Nm).
   - Heater hose retaining clip
   - Alternator electrical connectors
   - Air conditioning compressor mounting bracket. Torque the bolts to 35 ft. lbs. (47 Nm).
   - Air conditioning compressor electrical connectors
   - Air conditioning manifold tube to the air conditioning compressor
   - Drive belt
   - Negative battery cable
12. Fill the cooling system.
13. Fill the engine with clean oil. A filter replacement is also recommended.
14. Recharge the air conditioning system.

➡️**When the battery has been disconnected and reconnected, some abnormal drive symptoms may occur while the Powertrain Control Module (PCM) relearns its adaptive strategy. The vehicle may need to be driven about 10 miles (16 km) or more to relearn the strategy.**

15. Start the engine and check for leaks.

4.0L OHC Engine

➡️**If only one cylinder head is to be removed, only follow the procedures that apply. The following tools, or their equivalents are absolutely necessary to properly perform this procedure:**

- Cam Chain Tensioner tool T97T-6K254-A

Cylinder head bolt torque sequence 4.0L OHV engine

9308EG04

- Cam Gear Removal tool T97T-6256-F
- Cam Gear Torque adapter T97T-6256-G
- Camshaft Gear Positioning/Holding tool T97T-6256-B
- Camshaft Gear Positioning/Holding tool adapter T97T-6256-A
- Camshaft holding tool T97T-6256-C
- Crankshaft holding tool T97T-6303-A
- Camshaft holding tool adapter T97T-6256-D

1. Before servicing the vehicle, refer to the precautions in the beginning of this section.

2. Properly relieve the fuel system pressure.

3. Drain the cooling system.

4. Remove or disconnect the following:

- Negative battery cable
- Lower intake manifold
- Fan blade and shroud
- Valve cover
- Roller followers, if equipped
- Drive belt
- Upper radiator hose and tube
- Alternator electrical connectors
- Alternator mounting bracket
- Engine accessory bracket and move it aside
- Camshaft Position (CMP) electrical connector
- Crankshaft Position (CKP) sensor electrical connector
- Engine Coolant Temperature (ECT) sensor electrical connector
- Coil pack electrical connector
- Exhaust Gas Recirculation (EGR) valve electrical connector
- EGR valve bracket and move it aside
- Heater hoses
- Fuel injector electrical connectors
- Water bypass hose
- Thermostat housing
- Spark plug wires
- Fuel injection supply manifold
- Fuel injectors
- Crankcase vent separator spring
- Oil dipstick housing
- Exhaust manifold
- Hydraulic chain tensioner
- Cassette retaining bolt
- Camshaft sprocket
- Cylinder head and discard the gasket

To install:

5. Thoroughly clean all gasket mating surfaces. Remove all traces of old gasket material, oil, grease or dirt.

6. Insure that the rubber band is holding the right-hand chain to the cassette.

7. Install a new head gasket and the cylinder head.

8. Torque the new cylinder head bolts in sequence as follows:
   a. Step 1: 26 ft. lbs. (34 Nm).
   b. Step 2: Plus 90 degrees.
   c. Step 3: Plus an additional 90 degrees.

9. Install or connect the following:
- Camshaft sprocket in the cassette and make certain that the camshaft sprocket turns freely on the camshaft
- Cassette retaining bolt. Torque the bolt to 89 inch lbs. (10 Nm).
- Exhaust manifold
- Oil level indicator tube. Torque the bolt to 18 ft. lbs. (25 Nm).
- Crankcase vent separator and spring
- Thermostat housing. Torque the bolts to 8 ft. lbs. (11 Nm).
- Water bypass hose
- Heater hoses
- EGR bracket. Torque the bolt to 89 inch lbs. (10 Nm).
- EGR tube. Torque the nut to 30 ft. lbs. (40 Nm).
- ECT sensor electrical connector
- Electrical harness retainer. Torque the bolt to 89 inch lbs. (10 Nm).
- CKP and CMP electrical connectors
- Accessory bracket. Torque the bolts to 31 ft. lbs. (42 Nm).
- Alternator mounting bracket. Torque the bolts to 31 ft. lbs. (42 Nm).
- Alternator and electrical connectors
- Drive belt
- Fan shroud
- Roller followers
- Valve cover
- Lower intake manifold
- Negative battery cable

10. Change the engine oil and filter.

11. Refill the cooling system.

12. Start the engine and check for leaks, repair if necessary.

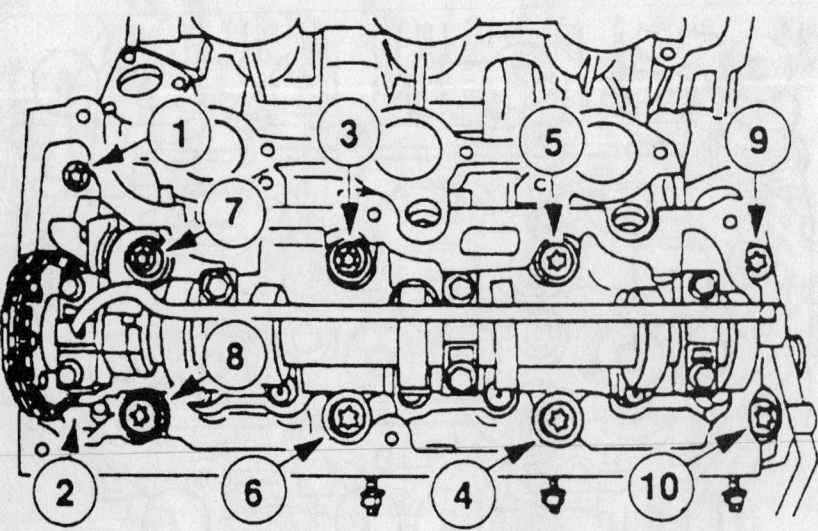

The correct cylinder head bolt loosening sequence must be used to prevent warpage–4.0L SOHC engine

7924EG04

Rocker Arms/Shafts

REMOVAL & INSTALLATION

2.3L Engines

This DOHC engine does not employ rocker arms. The camshafts bear directly on the lifters. For lifter removal, remove the camshafts.

2.5L Engines

➡ **A special tool is required to compress the valve spring.**

1. Before servicing the vehicle, refer to the precautions in the beginning of this section.

2. Disconnect the negative battery cable.

3. Remove the valve cover.

4. Rotate the camshaft so that the base circle of the cam is against the cam follower you intend to remove.

➡ **If removing more than one cam follower, label them so they can be returned to their original position.**

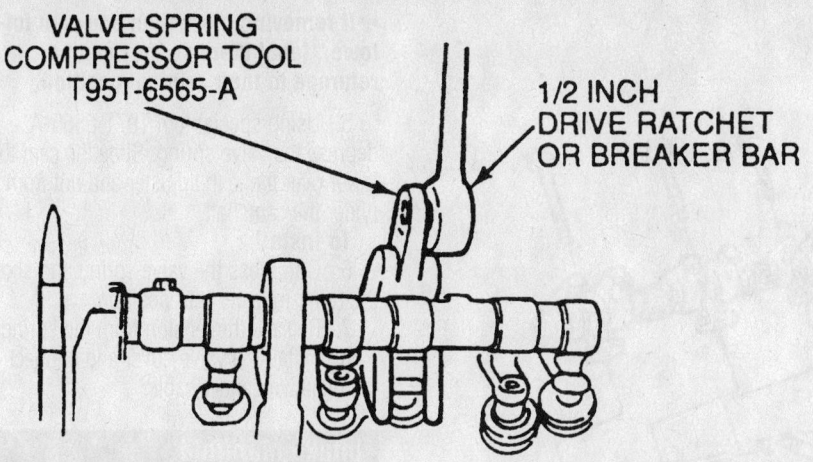

To remove the cam follower (rocker arm), use the special tool to depress the valve spring, then remove the cam follower—2.5L engines

5. Using special tool T88T-6565-BH depress the valve spring. Slide the cam follower over the lash adjuster and out from under the camshaft.

To install:

6. Compress the valve spring and slide the roller follower into position.

7. Release the tension from the spring.

8. Install the valve cover and connect the negative battery cable.

3.0L Engines

1. Before servicing the vehicle, refer to the precautions in the beginning of this section.

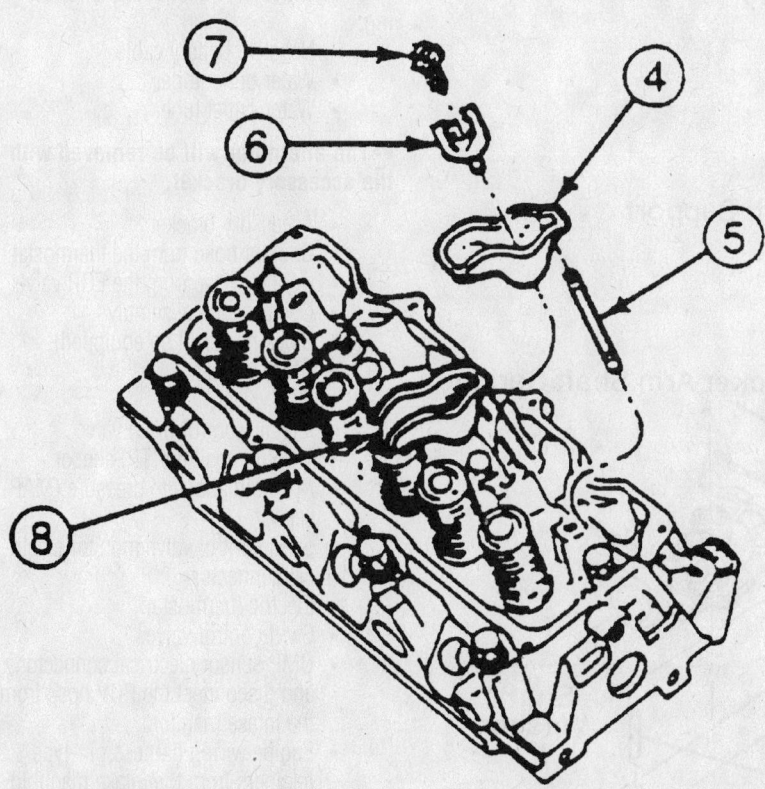

4. **Rocker arm**
5. **Pushrod**
6. **Fulcrum**
7. **Bolt**
8. **Assembled rocker arm**

Exploded view of the rocker arm assembly—3.0L engine

2. Remove or disconnect the following:
- Negative battery cable
- Rocker arm covers
- Retaining bolt at each rocker arm

3. The rocker arm and pushrod may then be removed from the engine. Keep all rocker arms and pushrods in order so they may be installed in their original locations.

To install:

4. Lubricate the rocker arm assemblies with SAE 50W engine oil.

5. Ensure that the fulcrums are properly seated into the cylinder head. Torque the rocker arm fulcrum bolts to 19 ft. lbs. (26 Nm).

6. Install the rocker arm covers and connect the negative battery cable.

4.0L OHV Engine

1. Before servicing the vehicle, refer to the precautions in the beginning of this section.

2. Remove or disconnect the following:
- Negative battery cable
- Rocker arm covers
- Rocker arm shaft stand attaching bolts by loosening the bolts 2 turns at a time, in sequence (from the end of the shaft to the middle of the shaft)
- Rocker arm and shaft assembly

To install:

3. If equipped, loosen the valve lash adjusting screws a few turns. Apply engine oil to the assembly to provide initial lubrication.

4. Install or connect the following:
- Rocker arm shaft assembly to the cylinder head and guide adjusting screws on to the pushrods
- Rocker arm stand. Torque the bolts to 46–52 ft. lbs. (62–70 Nm), 2 turns at a time, in sequence (from middle of shaft to the end of the shaft).
- Rocker arm covers
- Negative battery cable

4.0L SOHC Engines

➡A special tool is required to compress the valve spring.

1. Before servicing the vehicle, refer to the precautions in the beginning of this section.

2. Disconnect the negative battery cable.

3. Remove the valve cover.

4. Rotate the camshaft so that the base circle of the cam is against the cam follower you intend to remove.

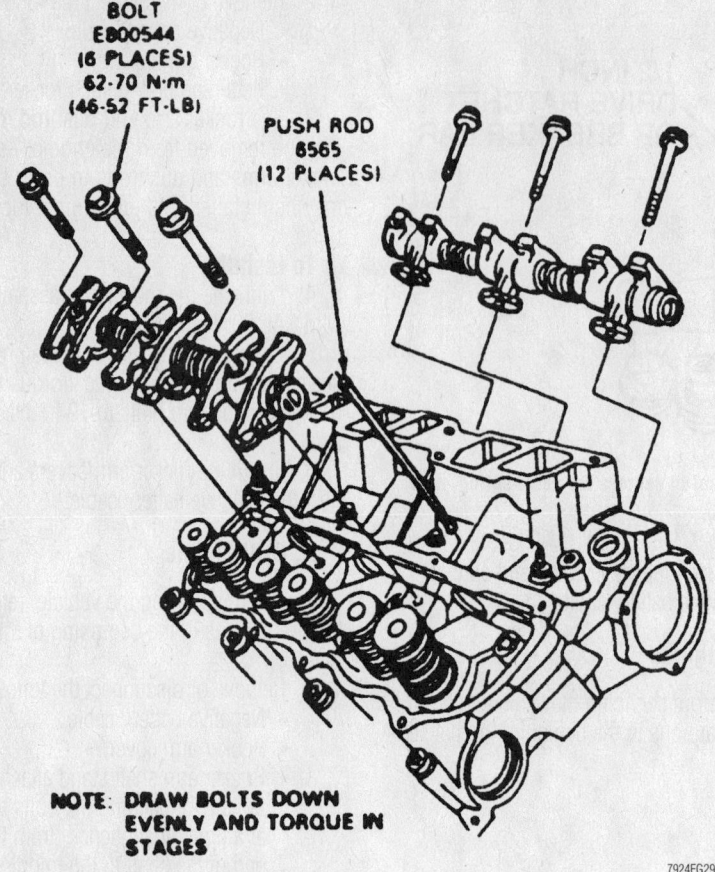

BOLT
E800544
(6 PLACES)
62-70 N·m
(46-52 FT-LB)

PUSH ROD
6565
(12 PLACES)

NOTE: DRAW BOLTS DOWN
EVENLY AND TORQUE IN
STAGES

7924EG29

Rocker arm and shaft assembly—4.0L engine

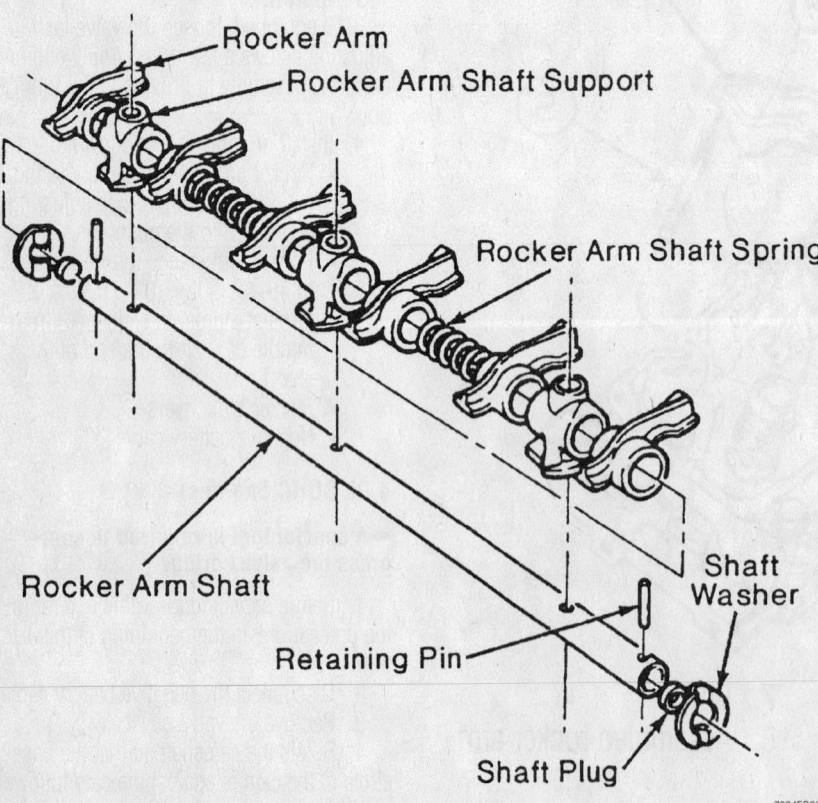

Rocker Arm
Rocker Arm Shaft Support
Rocker Arm Shaft Spring
Rocker Arm Shaft
Retaining Pin
Shaft Plug
Shaft Washer

7924EG26

Rocker arm and shaft assembly—4.0L SOHC engine

➡ **If removing more than one cam follower, label them so they can be returned to their original position.**

5. Using special tool T97T-6565-A depress the valve spring. Slide the cam follower over the lash adjuster and out from under the camshaft.

To install:

6. Compress the valve spring and slide the roller follower into position.

7. Release the tension from the spring.

8. Install the valve cover and connect the negative battery cable.

Intake Manifold

REMOVAL & INSTALLATION

2.3L Engine

1. Before servicing the vehicle, refer to the precautions in the beginning of this section.

2. Relieve the fuel system pressure.

3. Drain the cooling system.

4. Properly discharge the air conditioning system.

5. Remove or disconnect the following:

- Negative battery cable
- Water outlet tube.
- Water outlet tube.

➡ **The alternator will be removed with the accessory bracket.**

- Accessory bracket.
- Coolant hose from the thermostat.
- Coolant hose from the EGR valve.
- Coolant tube assembly.
- Block heater (if so equipped).
- Water outlet.
- EGR valve.
- Idle air control (IAC) valve.
- Throttle position (TP) sensor.
- Manifold absolute pressure (MAP) sensor.
- Swirl control valve monitor electrical connector.
- Electric thermostat.
- Swirl control valve.
- CMP sensor electrical connector and disconnect the PCV hose from the intake manifold.
- Engine wiring harness pin-type retainers from the intake manifold.
- Engine wiring harness connector bracket. Position the engine wiring harness aside.
- EGR tube.
- Fuel supply line clip from the front of the intake manifold. Disconnect

the vacuum hose from the intake manifold.
- Intake manifold assembly.

6. Installation is the reverse of removal. Torque the bolts to 13 ft. lbs. (18Nm). There is no special torque sequence.

2.5L Engine

1. Before servicing the vehicle, refer to the precautions in the beginning of this section.

2. Remove or disconnect the following:
- Negative battery cable
- Intake Air Temperature (IAT) sensor
- Air cleaner outlet tube
- Accelerator control splash shield

- Engine control sensor wiring from the Throttle Position (TP) sensor and the Idle Air Control (IAC) valve
- Accelerator cable and speed control cable, if equipped
- Accelerator cable bracket
- Crankcase vent hose from the valve cover
- Vacuum hoses from the intake manifold vacuum tee
- Heater hose from the intake manifold
- Exhaust Gas Recirculation (EGR) tube
- EGR valve
- Upper intake manifold and discard the gasket

To install:

3. Install a new upper intake manifold gasket.

4. Install the upper intake manifold. Torque the bolts in sequence as follows:
  a. Step 1: 89 inch lbs. (10 Nm).
  b. Step 2: 28 ft. lbs. (38 Nm).

5. Install or connect the following:
- EGR valve. Torque the bolts to 22 ft. lbs. (30 Nm).
- EGR valve tube. Torque the bolts to 34 ft. lbs. (47 Nm).
- Heater hoses to the intake manifold
- Vacuum hoses to the tee
- Crankcase vent hose to the valve cover
- Accelerator cable bracket. Torque the bolts to 20 ft. lbs. (28 Nm).
- Accelerator cable and speed control cable, if equipped
- IAC valve and TP sensor electrical connectors
- Air cleaner outlet tube
- IAT sensor
- Negative battery cable

6. Start the engine and check for leaks, repair if necessary.

3.0L Engine

1. Before servicing the vehicle, refer to the precautions in the beginning of this section.

2. Remove or disconnect the following:
- Negative battery cable
- Intake Air Temperature (IAT) sensor
- Air cleaner outlet tube
- Accelerator control splash shield
- Accelerator cable and speed control cable, if equipped
- Engine control sensor wiring from the Throttle Position (TP) sensor and the Idle Air Control (IAC) valve and Exhaust Gas Recirculation (EGR) transducer
- EGR tube from the valve
- EGR vacuum lines
- 42 pin connector bracket
- Throttle body and gasket
- Ignition coil and move it aside
- Evaporative Emissions (EVAP) hose
- Upper intake manifold bolts and discard them
- Crankcase vent hose
- Upper intake manifold and discard the gasket

To install:

3. Clean all mating surfaces.

4. Install or connect the following:
- New upper intake manifold
- Intake manifold
- Crankcase vent hose

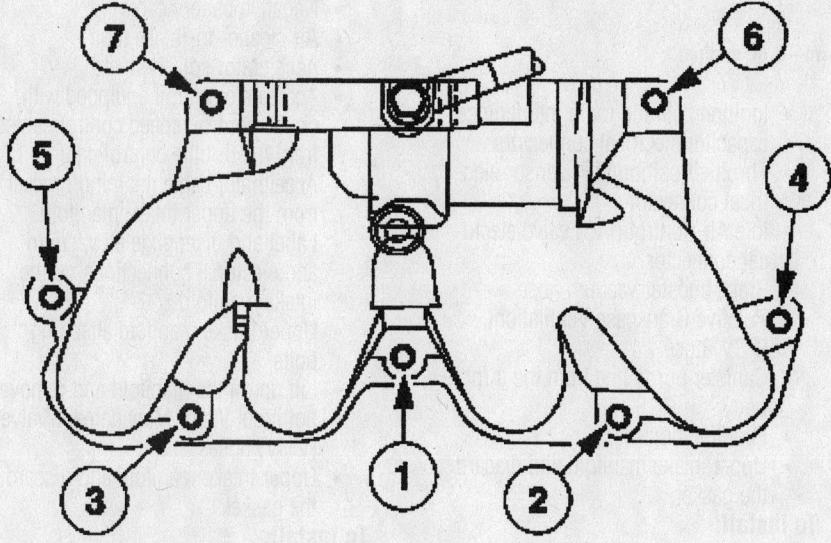

9308EG01

Tighten the lower manifold bolts in the sequence shown—2.5L engine

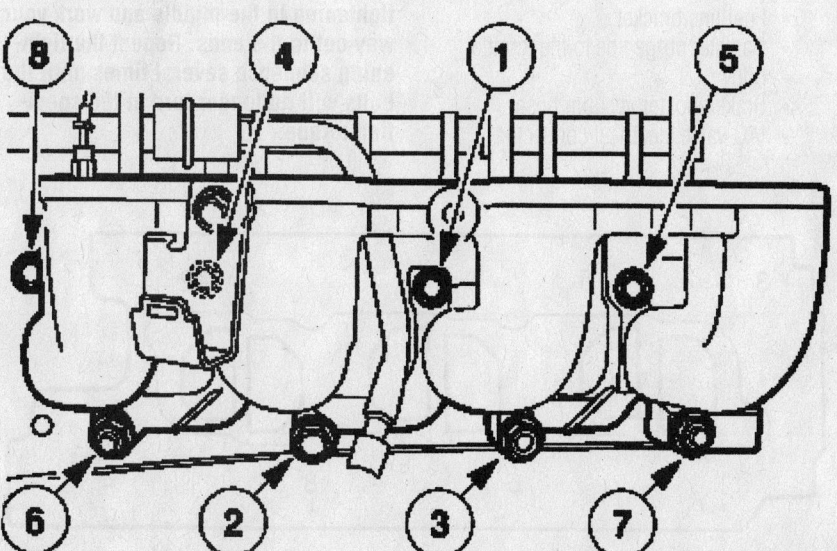

9308EG02

Tighten the upper manifold bolts in the sequence shown—2.5L engine

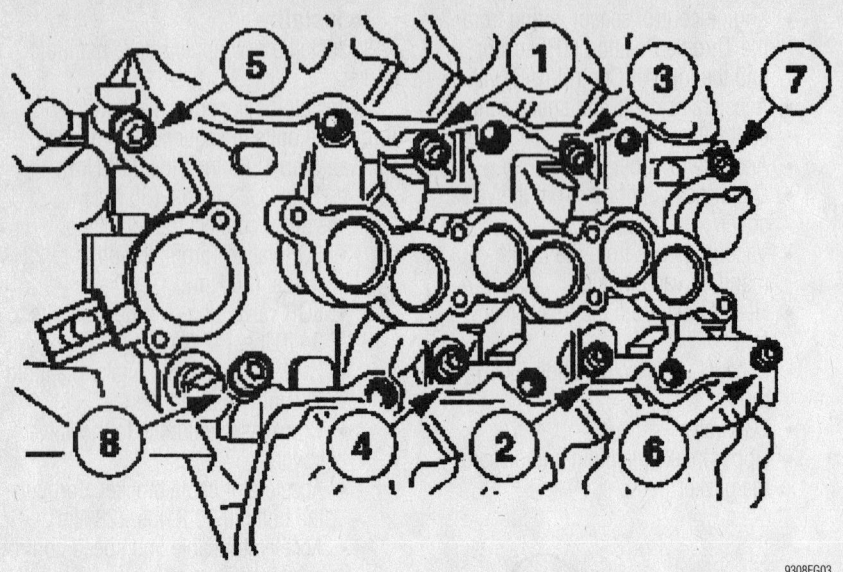

9308EG03

Tighten the lower manifold bolts in the sequence shown—3.0L engine

5. Torque the upper intake manifold bolts in sequence as follows:
   a. Step 1: 15 ft. lbs. (20 Nm).
   b. Step 2: 18 ft. lbs. (25 Nm).
6. Install or connect the following:
   - EVAP hose
   - EGR tube
   - EGR transducer. Torque the bolts to 89 inch lbs. (10 Nm).
   - Ignition coil. Torque the bolts to 15 ft. lbs. (20 Nm).
   - New gasket and throttle body. Torque the bolts to 22 ft. lbs. (30 Nm).
   - 42 pin connector
   - EGR vacuum lines
   - EGR transducer, IAC valve and TP sensor electrical connectors
   - Accelerator cable and speed control, if equipped
   - Accelerator cable splash shield
   - Air cleaner outlet tube
   - IAT sensor
   - Negative battery cable
7. Start the vehicle and check for leaks, repair if necessary.

4.0L OHV Engine

The intake manifold is a 4-piece assembly, consisting of the upper intake manifold, the throttle body, the fuel supply manifold, and the lower intake manifold.
1. Before servicing the vehicle, refer to the precautions in the beginning of this section.
2. Remove or disconnect the following:
   - Negative battery cable
   - Air cleaner outlet tube
   - Accelerator cable splash shield
   - Spark plug wires from the ignition coil

   - Ignition coil and radio interference capacitor electrical connectors
   - Throttle Position (TP) sensor electrical connector
   - Idle Air Control (IAC) valve electrical connector
   - Brake booster vacuum hose
   - Positive Crankcase Ventilation (PCV) hose
   - Canister purge line from the throttle body
   - Fuel line bracket
   - Upper intake manifold and discard the gasket

To install:
3. Install or connect the following:
   - New gasket and the upper manifold. Torque the nuts to 18 ft. lbs. (25 Nm).
   - Fuel line bracket
   - Canister purge line to the throttle body
   - Brake booster vacuum hose
   - IAC valve electrical connector

   - TP sensor electrical connector
   - Ignition coil and radio interference capacitor electrical connectors
   - Spark plug wires to the proper spark plug
   - Accelerator cable and speed control cable, if equipped
   - Accelerator cable splash shield
   - Air cleaner outlet tube
   - Negative battery cable
4. Start the vehicle and check for leaks, repair if necessary.

4.0L OHC Engine

1. Before servicing the vehicle, refer to the precautions in the beginning of this section.
2. Remove or disconnect the following:
   - Negative battery cable
   - Air cleaner-to-intake tube
   - Accelerator splash shield
   - Accelerator and, if equipped with cruise control, speed control cables from the throttle control cam
   - Accelerator cable retaining bracket from the upper intake manifold
   - Label and disengage all vacuum and electrical connections on the intake manifold.
   - Upper intake manifold attaching bolts
   - Lift up on the manifold and remove both fuel Vapor Management Valve (VMV) hoses
   - Upper intake manifold and discard the gasket

To install:

➡️ **Ford does not specify a sequence for either upper or lower intake manifolds, but it is recommended that you start tightening in the middle and work your way out to the ends. Repeat the tightening sequence several times until the bolts will no longer turn at the specified torque.**

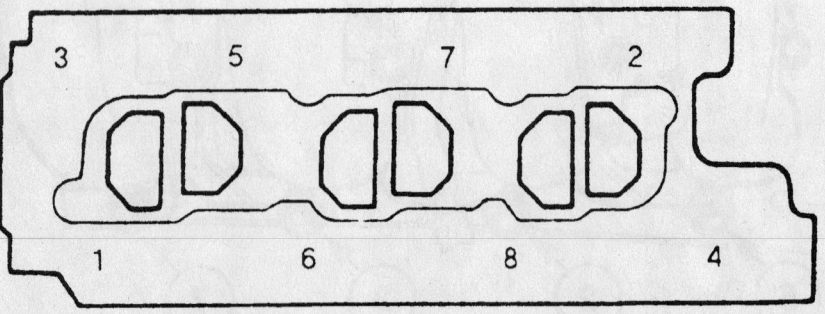

7924EG07

Tighten the manifold bolts in the sequence shown—4.0L OHV engine

3. Position the upper manifold on the lower manifold.

4. Install or connect the following:
- Attach both VMV hoses to the manifold
- Upper manifold attaching bolts. Torque the bolts to 62 inch lbs. (7 Nm).
- Attach any vacuum and electrical connections that were removed
- Accelerator cable bracket to the intake and the cable (or cables if equipped with cruise control) to the throttle cam
- Accelerator splash shield
- Air cleaner-to-intake supply tube
- Negative battery cable

5. Start the vehicle and check for leaks, repair if necessary.

Exhaust Manifold

REMOVAL & INSTALLATION

2.3L Engine

1. Before servicing the vehicle, refer to the precautions in the beginning of this section.

2. Remove or disconnect the following:
- Negative battery cable
- Exhaust flange nuts
- Drive belt
- Coolant
- Upper radiator hose and the engine reservoir hose
- Air conditioning compressor
- Heater hose
- Oil indicator and the upper bolt for the tube assembly
- Lower bolt and remove the oil indicator tube assembly
- Front radiator tube

3. Remove the pushpins and position the right inner fender splash shield out of the way.

4. Remove or disconnect the following:
- Alternator electrical connectors
- Lower front end accessory drive (FEAD) mounting bolts
- Upper mounting bolt and the FEAD assembly
- Two nuts and position the coolant tube out of the way
- Exhaust manifold
- Exhaust manifold gasket

To install:

5. Install or connect the following:
- Exhaust manifold gasket
- Exhaust manifold and the nuts
- Coolant tube and the nuts

- FEAD assembly and the upper mounting bolts
- Lower FEAD mounting bolts
- Alternator electrical connectors
- Right inner splash shield and pushpins
- Upper radiator tube and install the bolts
- Oil indicator tube assembly and the lower bolt
- Oil indicator tube upper bolt and the oil indicator
- Heater water hose
- Air conditioning compressor
- Upper radiator hose and the engine reservoir hose

6. Fill the cooling system.
7. Install the serpentine drive belt.
8. Install the exhaust flange nuts.
9. Connect the battery ground cable.

2.5L Engines

1. Before servicing the vehicle, refer to the precautions in the beginning of this section.

2. Remove or disconnect the following:
- Negative battery cable
- Intake Air Temperature (IAT) sensor
- Air cleaner outlet tube
- Differential Pressure Feedback (DPFE) sensor and move it aside
- Exhaust Gas Recirculation (EGR) transducer lines at the tube
- Loosen and remove the EGR valve-to-exhaust manifold tube

- Catalytic converter from the exhaust manifold
- Rear engine lifting eye nuts
- Exhaust manifold and discard the gasket

To install:

3. Clean the mating surfaces on the exhaust manifold and the cylinder head.

4. Install a new gasket and the exhaust manifold. Torque the bolts in sequence as follows:
   a. 16 ft. lbs. (23 Nm).
   b. 59 ft. lbs. (80 Nm).

5. Install or connect the following:
- Rear engine lifting eye. Torque the bolts to 15 ft. lbs. (20 Nm).
- Catalytic converter to the exhaust manifold
- EGR valve to the exhaust manifold tube
- EGR transducer lines
- DPFE sensor
- Air cleaner outlet tube
- IAT sensor
- Negative battery cable

6. Start the vehicle and check for leaks, repair if necessary.

3.0L Engine

LEFT SIDE

1. Before servicing the vehicle, refer to the precautions in the beginning of this section.

2. Install or connect the following:
- Negative battery cable

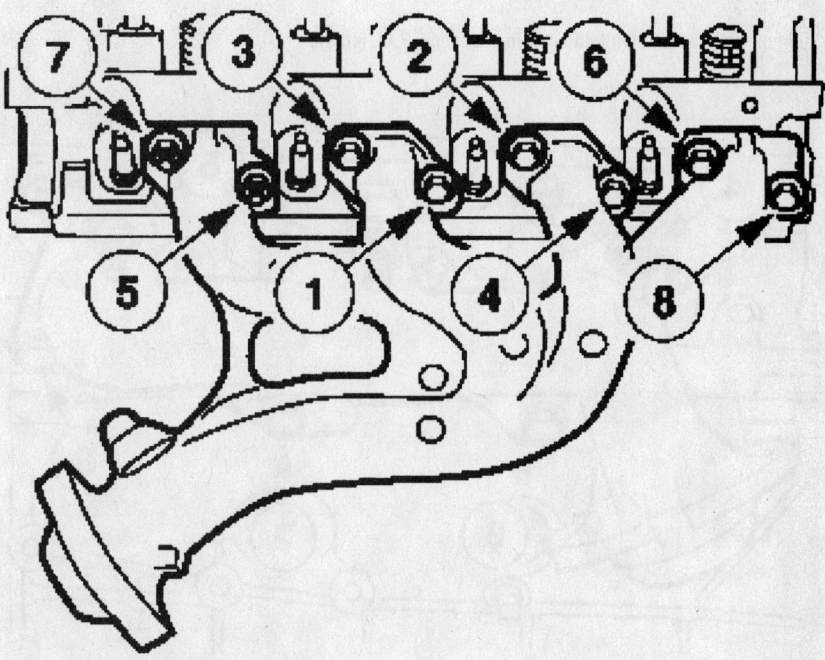

Tighten the exhaust manifold bolts in 2 stages

9308EG07

- Exhaust flange nuts
- Exhaust Gas Recirculation (EGR) valve from the exhaust manifold tube
- Oil lever indicator and bracket
- Exhaust manifold and discard the gasket

To install:

3. Clean the mating surfaces for the exhaust manifold and cylinder head.

4. Install a new gasket and the exhaust manifold. Torque the bolts in sequence to:
   a. 89 inch lbs. (10 Nm).
   b. 15 ft. lbs. (20 Nm).

5. Install or connect the following:
- Oil lever indicator tube and bracket. Torque the bolt to 12 ft. lbs. (16 Nm).
- EGR valve to the exhaust manifold

tube. Torque the fastener to 26 ft. lbs. (35 Nm).
- Exhaust flange. Torque the nuts to 25 ft. lbs. (34 Nm).
- Negative battery cable

6. Start the vehicle and check for leaks, repair if necessary.

RIGHT SIDE

1. Before servicing the vehicle, refer to the precautions in the beginning of this section.

2. Remove or disconnect the following:
- Negative battery cable
- Exhaust manifold flange
- Ignition coil support bracket
- Exhaust manifold and discard the gasket

To install:

3. Clean the mating surfaces for the exhaust manifold and cylinder head

4. Install a new gasket and the exhaust manifold. Torque the bolts is sequence to:
   a. 89 inch lbs. (10 Nm).
   b. 18 ft. lbs. (25 Nm).

5. Install or connect the following:
- Ignition coil support bracket. Torque the bolts to 15 ft. lbs. (20 Nm).
- Exhaust flange nuts. Torque the nuts to 33 ft. lbs. (46 Nm).
- Negative battery cable

6. Start the vehicle and check for leaks, repair if necessary.

4.0L OHV Engine

1. Before servicing the vehicle, refer to the precautions in the beginning of this section.

2. Remove or disconnect the following:
- Negative battery cable
- Oil level indicator tube and bracket
- Exhaust pipe-to-manifold bolts
- Power steering pump hoses, if removing the left-hand manifold
- Hot air intake shroud that is bolted around the manifold, if removing the right-hand manifold
- Exhaust manifold and gasket

To install:

3. Clean all mating surfaces for the exhaust manifold and cylinder head.

4. Install or connect the following:
- New gasket and the exhaust manifold. Torque the right side bolts to 18 ft. lbs. (25 Nm) and the left side bolts to 16 ft. lbs. (22 Nm).
- Exhaust pipe to the manifold
- Oil level bracket. Torque the bolt to 17 ft. lbs. (23 Nm).
- Negative battery cable

5. Start the vehicle and check for leaks, repair if necessary.

4.0L OHC Engine

1. Before servicing the vehicle, refer to the precautions in the beginning of this section.

2. Remove or disconnect the following:
- Negative battery cable
- Exhaust inlet pipe-to-manifold attaching bolts
- Differential Pressure Feedback EGR (DPFE) transducer hoses, left side manifold only
- Exhaust Gas Recirculation (EGR) tube from the manifold and valve, left side manifold only
- Exhaust manifold and discard the gasket

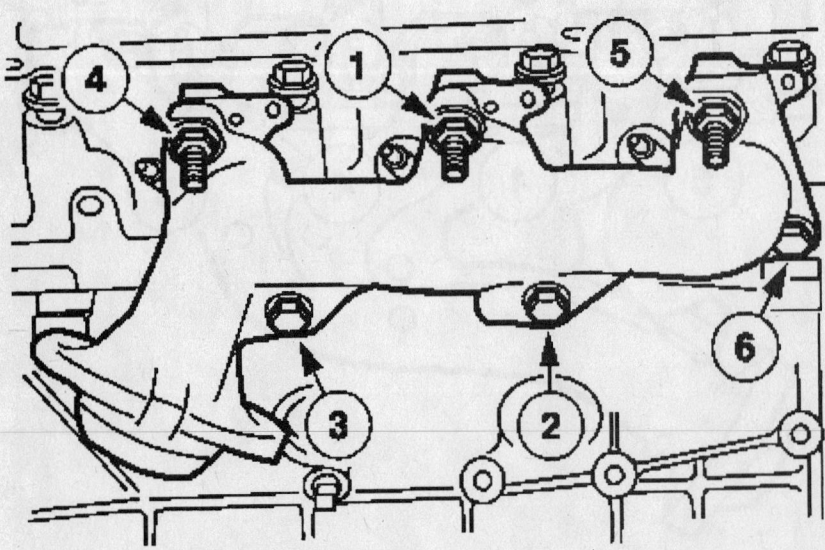

9308EG08

Tighten the exhaust manifold bolts in sequence—3.0L left side

9308EG09

Tighten the right side exhaust manifold bolts in the proper sequence—3.0L

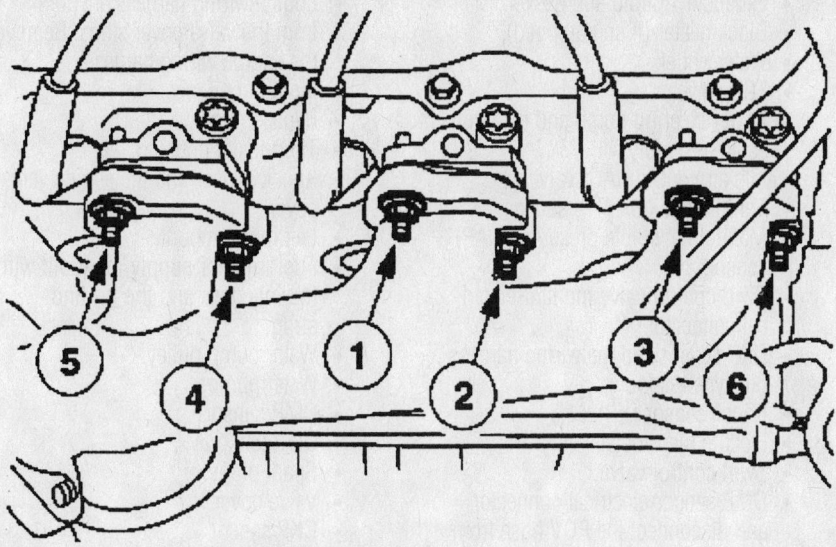

Tighten the right side exhaust manifold in sequence

9308EG10

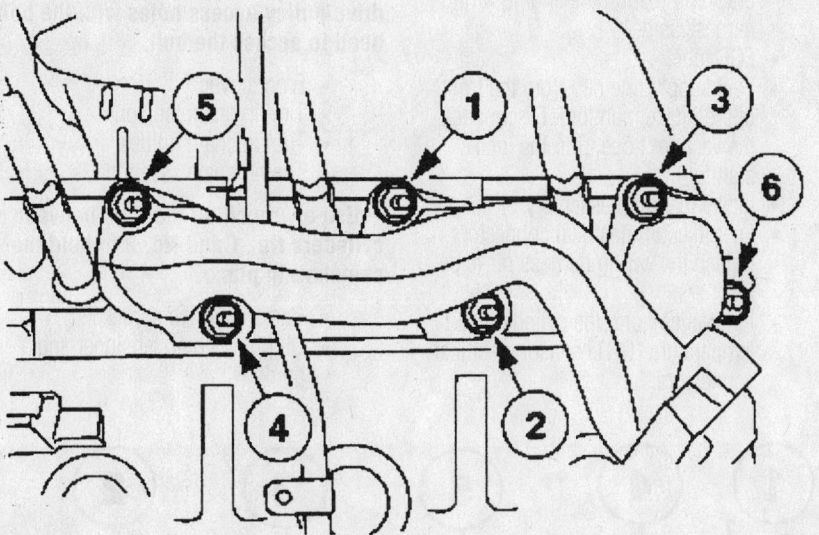

Tighten the left side exhaust manifold in sequence

9308EG11

To install:

3. Clean the gasket mating surfaces.
4. Install or connect the following:
  - New gasket and the exhaust manifold. Torque the bolts to 16 ft. lbs. (22 Nm).
  - EGR tube to the manifold. Torque the fastener to 30 ft. lbs. (40 Nm) left side manifold only
  - DPFE transducer hoses, left side manifold only
  - Exhaust inlet pipe-to-manifold attaching bolts. Torque the bolts to 30 ft. lbs. (40 Nm).
  - Negative battery cable
5. Start the vehicle and check for leaks, repair if necessary.

Front Crankshaft Seal

REMOVAL & INSTALLATION

➡️Only OHC engines will be covered here. For OHV engines, see the Timing Chain procedures later in this section. For front cover and timing belt procedures for the 2.5L engine, see the Timing Belt Section of this book.

2.3L Engines

1. Before servicing the vehicle, refer to the precautions in the beginning of this section.
2. Remove or disconnect the following:

  - Negative battery cable
  - Crankshaft pulley

Use care not to damage the engine front cover or the crankshaft when removing the seal.

  - Crankshaft front oil seal by prying the seal out of the front cover

To install:

3. Using the special tool, install the crankshaft front oil seal.
4. Install the crankshaft pulley.
5. Tighten the crankshaft damper in two stages:
  - Step 1: Tighten to 40 Nm (30 ft. lbs.)
  - Step 2: Tighten an additional 90 degrees

2.5L Engines

1. Before servicing the vehicle, refer to the precautions in the beginning of this section.
2. Remove or disconnect the following:
  - Negative battery cable
  - Timing belt cover
  - Drive belt
3. Align the crankshaft and camshaft timing marks and remove the timing belt.
  - Crankshaft pulley center bolt and slide the pulley off of the crankshaft
  - Crankshaft key

Do not damage the crankshaft sealing surface while removing the oil seal.

  - Crankshaft Seal Remover tool T74P-6700-B on the crankshaft and into the oil seal.
  - Oil seal and clean the seal journal

To install:

4. Apply clean engine oil to the rubber lip of the new seal to aid installation.
5. Using Cam Bearing Adapter Tube T72C-6250, or equivalent, and crankshaft center bolt, carefully install the new oil seal until flush with the engine.
6. Install or connect the following:
  - Key and crankshaft pulley, washer and bolt. Torque the bolt to 92–121 ft. lbs. (125–165 Nm).
  - Timing belt
  - Timing belt and cover
  - Drive belt
  - Negative battery cable
7. Start the vehicle and check for leaks, repair if necessary.

4.0L OHC Engine

1. Before servicing the vehicle, refer to the precautions in the beginning of this section.

2. Remove or disconnect the following:
   • Negative battery cable
   • Crankshaft pulley

3. Using a seal remover, remove the crankshaft front oil seal.

To install:

4. Lubricate the seal lip with clean engine oil.

5. Using a seal driver, install the crankshaft front oil seal.

6. Install the crankshaft pulley.

Camshaft and Valve Lifters

➡Although Ford suggests that this component is removable while the engine is installed in the vehicle, depending on the particular options with which your truck is equipped, working clearance may be extremely tight and this procedure may be much easier to perform with the engine removed. Before commencing, read through this procedure and make certain enough clearance, or working room, exists with the engine in the vehicle; if there is not enough space, the engine should be removed.

REMOVAL & INSTALLATION

2.3L Engine

1. Before servicing the vehicle, refer to the precautions in the beginning of this section.

2. Relieve the fuel system pressure.

3. Drain the cooling system.

4. Properly discharge the air conditioning system.

5. Remove or disconnect the following:
   • Negative battery cable
   • Drive belt.
   • Engine oil level indicator assembly.
   • Engine oil level indicator.
   • Engine oil level indicator tube.
   • Water outlet tube.
   • Water outlet tube.
   • Air conditioning compressor.

➡The generator will be removed with the accessory bracket.

   • Accessory bracket.
   • Right motor mount.
   • Coolant hose from the thermostat.
   • Coolant hose from the EGR valve.
   • Coolant tube assembly.

• Exhaust manifold and gasket.
• Block heater (if so equipped).
• Water outlet.
• EGR valve.
• Power steering pump and reservoir as an assembly.
• Idle air control (IAC) valve.
• Throttle position (TP) sensor.
• Manifold absolute pressure (MAP) sensor.
• Swirl control valve monitor electrical connector.
• CKP sensor and the wiring harness pin-type retainers.
• Knock sensor (KS).
• Electric thermostat.
• Swirl control valve.
• CMP sensor electrical connector and disconnect the PCV hose from the intake manifold.
• Engine wiring harness pin-type retainers from the intake manifold.
• Engine wiring harness connector bracket. Position the engine wiring harness aside.
• EGR tube.
• Fuel supply line clip from the front of the intake manifold. Disconnect the vacuum hose from the intake manifold.
• Intake manifold assembly.
• Fuel injector electrical connectors. Detach the wiring harness pin-type retainers.
• Ignition coil and the cylinder head temperature (CHT) sensor electrical connectors.

• Engine wiring harness anchors from the valve cover studs. Remove the engine wiring harness.
• Ignition coil.
• Bypass hose.
• Thermostat housing.
• Knock sensor and the engine vent cover.
• Left motor mount.
• Fuel injector supply manifold with the injectors and the ground strap.
• Water pump pulley.
• Water pump.
• CMP sensor.
• CHT sensor.
• Spark plugs.
• Valve cover.
• CKP sensor.
• Crankshaft vibration damper

➡There is one front cover bolt behind the cooling fan drive pulley. To remove this bolt, align one of the cooling fan drive pulley access holes with the bolt head to access the bolt.

   • Front cover.
   • Timing chain tensioner.
   • Timing chain guides.
   • Timing chain assembly.

➡Use a wrench on the flats between cylinders No. 1 and No. 2 to hold the camshaft in place.

   • Camshaft drive sprockets.
   • Oil pump chain tensioner and guide.

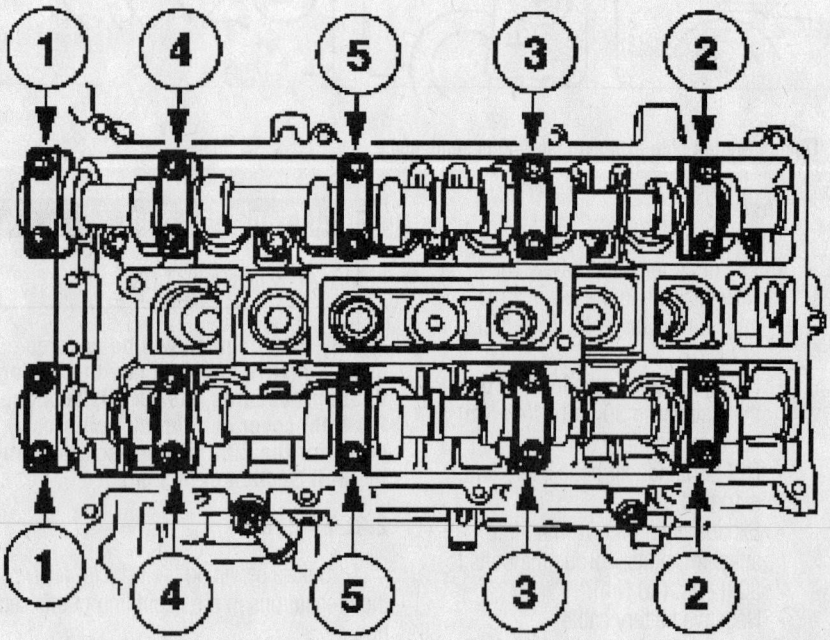

Camshaft cap loosening sequence—2.3L

9348EG03

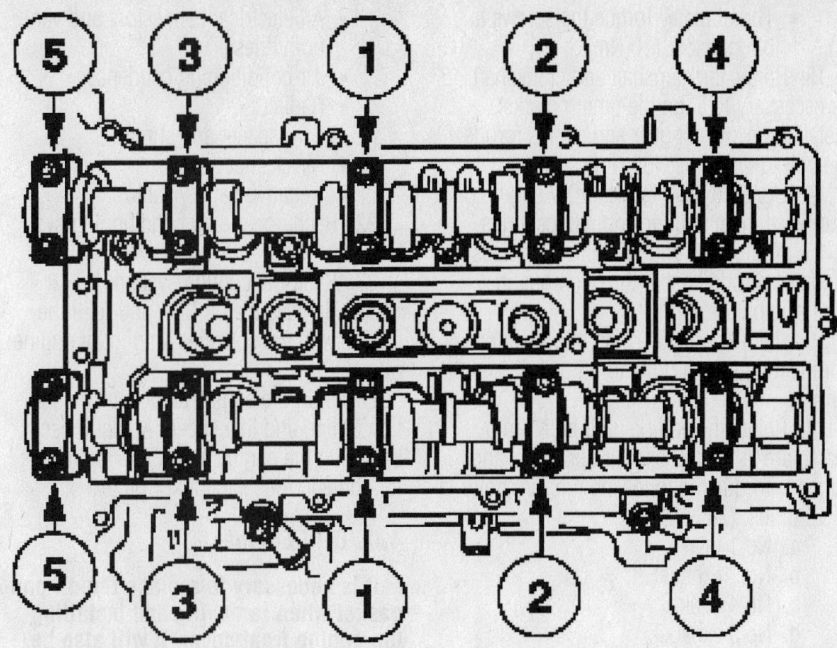

Camshaft cap torque sequence—2.3L

➡ **The oil pump chain sprocket must be held in place.**

- Oil pump chain and sprockets.

➡ **Note the position of the lobes on the No. 1 cylinder before removing the camshafts for assembly reference.**

6. Loosen the camshaft bearing cap bolts in sequence, one turn at a time. Repeat the first step until all tension is released from the camshaft bearing caps. Remove the camshaft bearing caps.
7. Remove the camshafts.
8. Installation is the reverse of removal.

➡ **Install the camshafts with the alignment notches in the camshaft lined up so the camshaft alignment plate can be installed without rotating the camshafts. Make sure the lobes on the No. 1 cylinder are in the same position as noted in the disassembly procedure. Rotating the camshafts, or installing the camshafts 180 degrees out of position can cause severe damage to the valves and pistons. Lubricate the camshaft journals and bearing caps with clean engine oil. Install the camshafts and bearing caps. Tighten the bolts in the sequence shown in three stages.**

- Step 1: Tighten the camshaft bearing caps one turn at a time until tight.
- Step 2: Tighten the bolts to 7 Nm (62 inch lbs.)

- Step 3: Tighten the bolts to 16 Nm (12 ft. lbs.)
  a. Crankshaft vibration damper:

➡ **Do not reuse the crankshaft pulley bolt. Tighten the bolt in two stages.**

- Step 1: Tighten the bolt to 40 Nm (30 ft. lbs.)
- Step 2: Tighten the bolt and additional 90 degrees (1/4 turn).

2.5L Engines

1. Drain the cooling system.
2. Before servicing the vehicle, refer to the precautions in the beginning of this section.
3. Remove or disconnect the following:

- Negative battery cable
- Air cleaner
- Spark plug wires and retainers
- Vacuum lines
- Drive belts
- Alternator and bracket
- Upper radiator hose
- Radiator shroud
- Fan blades
- Water pump pulley
- Fan shroud

4. Align the engine timing marks at Top Dead Center (TDC) for No. 1 cylinder. Remove the timing belt.

- Valve covers
- Rocker arms (camshaft followers)
- Camshaft drive gear and belt guide using a suitable puller. Remove the

front oil seal with Front Seal Replacer T74P-6150-A
- Camshaft retainer located on the rear mounting stand
- Front motor mount bolts
- Lower radiator hose from the radiator
- Automatic transmission cooler lines, if equipped

5. Position a piece of wood on a floor jack and raise the engine carefully as far as it will go. Place blocks of wood between the engine mounts and crossmember pedestals.
6. Remove camshaft by carefully withdrawing it toward the front of the engine. Caution should be used to prevent damage to cam bearings, lobes and journals.
7. Check the camshaft journals and lobes for wear. Inspect the cam bearings, if worn (unless the proper bearing installing tool is on hand), the cylinder head must be removed for new bearings to be installed by a machine shop.

To install:

8. Install or connect the following:

- Camshaft and lower the engine to its original position
- Transmission cooler lines, if equipped
- Lower radiator hose
- Front motor mount. Torque the bolts to 65 ft. lbs. (88 Nm).
- Camshaft retainer on the rear mounting stand
- Camshaft drive gear and belt guide
- Valve covers
- Timing belt. Make certain that the timing marks are properly aligned
- Fan shroud
- Water pump pulley
- Fan blades
- Upper radiator hose and radiator shroud
- Alternator and bracket
- Drive belts
- Vacuum lines
- Spark plugs wires and retainers
- Air cleaner
- Negative battery cable

9. Fill the cooling system.
10. Start the engine and check for leaks, repair if necessary.

3.0L Engine

1. Before servicing the vehicle, refer to the precautions in the beginning of this section.
2. Properly relieve the fuel system pressure.
3. Drain the cooling system.
4. Drain the engine oil.

5. Evacuate the air conditioning system.

6. Remove or disconnect the following:
- Negative battery cable
- Air cleaner hoses
- Fan, spacer and shroud
- Radiator

7. Rotate the crankshaft so that No. 1 piston is at Top Dead Center (TDC) on the compression stroke.
- Air conditioning condenser
- Fuel lines from the fuel supply manifold
- Vacuum hoses
- Electrical wiring
- Engine front cover
- Water pump
- Alternator
- Power steering pump. Do not disconnect the hoses
- Air conditioning compressor. Do not disconnect the hoses
- Throttle body
- Fuel injection wire harness

8. Turn the engine by hand to TDC of the power stroke on No. 1 cylinder.
- Spark plug wires from the plugs
- Distributor cap with the spark plug wires as an assembly, if equipped

9. Matchmark the rotor, distributor body and engine. Disconnect the distributor wiring harness and remove the distributor, if equipped.
- Rocker arm covers
- Intake manifold
- Loosen the rocker arm bolts enough to pivot the rocker arms out of the way and remove the pushrods. Identify them for installation
- Lifters and identify them for installation
- Crankshaft pulley/damper
- Starter
- Oil pan
- Camshaft gear attaching bolt and washer, then slide the gear off the camshaft
- Camshaft thrust plate

10. Carefully slide the camshaft out of the engine block, using caution to avoid any damage to the camshaft bearings.

To install:

11. Oil the camshaft journals and cam lobes with heavy SJ engine oil (50W). Install the spacer ring with the chamfered side toward the camshaft, then insert the camshaft key.

12. Install or connect the following:
- Camshaft using caution to avoid any damage to the camshaft bearings

- Thrust plate. Torque the screws to 84 inch lbs. (10 Nm).

13. Rotate the camshaft and crankshaft as necessary to align the timing marks. Install the camshaft gear and chain. Torque the bolt to 46 ft. lbs. (62 Nm).

14. Coat the tappets with 50W engine oil and place them in their original locations.

15. Apply 50W engine oil to both ends of the pushrods. Install the pushrods in their original locations.

16. Pivot the rocker arms into position. Torque the fulcrum bolts to 96 inch lbs. (11 Nm).

17. Rotate the engine until both timing marks are at the top of their sprockets and aligned. Torque the following fulcrum bolts to 18 ft. lbs. (24 Nm):
   a. No.1 intake.
   b. No.2 exhaust.
   c. No.4 intake.
   d. No.5 exhaust.

18. Rotate the engine until the camshaft timing mark is at the bottom of the sprocket and the crankshaft timing mark is at the top of the sprocket, and both are aligned. Torque the following fulcrum bolts to 18 ft. lbs. (24 Nm):
   a. No.1 exhaust.
   b. No.2 intake.
   c. No.3 intake and exhaust.
   d. No.4 exhaust.
   e. No.5 intake.
   f. No.6 intake and exhaust.

19. Torque all the bolts to 24 ft. lbs. (33 Nm).

20. Turn the engine by hand to 0 degrees Before Top Dead center (BTDC) of the power stroke on No. 1 cylinder.

21. Install or connect the following:
- Engine front cover and water pump assembly
- Oil pan
- Crankshaft damper/pulley and tighten the retaining bolt to 107 ft. lbs. (145 Nm).
- Intake manifold
- Starter
- Crankshaft pulley and damper
- Rocker arm covers
- Rotor and distributor cap, if equipped
- Spark plug wires
- Fuel lines to the fuel supply manifold
- Fuel injection wire harness
- Throttle body
- Air conditioning compressor
- Power steering pump
- Alternator
- Water pump
- Engine front cover

- All electrical connectors and vacuum lines
- Air conditioning condenser
- Radiator
- Fan, spacer and shroud
- Air cleaner hoses
- Negative battery cable

22. Recharge the air conditioning system.

23. Refill the cooling system.

24. Replace the oil filter and refill the engine with the specified amount of engine oil.

25. Start the engine and check the ignition timing and idle speed. Adjust if necessary. Run the engine at fast idle and check for coolant, fuel, vacuum or oil leaks.

4.0L OHV Engine

➡ **It is necessary to replace the oil pan gasket when removing and installing the engine front cover. It will also be necessary to remove the transmission to properly reseal the oil pan.**

1. Before servicing the vehicle, refer to the precautions in the beginning of this section.

2. Drain the engine oil.

3. Drain the cooling system.

4. Evacuate the air conditioning system.

5. Relieve fuel system pressure.

6. Remove or disconnect the following:
- Negative battery cable
- Radiator
- Air conditioning compressor. Do not disconnect the lines
- Air conditioning condenser
- Fan, spacer and shroud
- Air cleaner hoses
- Spark plug wires
- Ignition coil and bracket
- Crankcase pulley and damper
- Oil pump drive
- Alternator
- Fuel lines at the supply manifold
- Upper and lower intake manifold
- Rocker arm covers
- Rocker arm shafts
- Pushrods and identify them for installation
- Tappets and identify them for installation
- Oil pan
- Engine front cover
- Water pump

7. Turn the engine by hand until the timing marks align at Top dead Center (TDC) of the power stroke on No.1 piston.

8. Place the timing chain tensioner in the retracted position and install the retaining clip.

9. Check the camshaft end-play. If excessive, you'll have to replace the thrust plate.

10. Remove the camshaft gear attaching bolt and washer, then slide the gear off the camshaft.

11. Remove the camshaft thrust plate.

12. Carefully slide the camshaft out of the engine block, using caution to avoid any damage to the camshaft bearings.

To install:

13. Lubricate the camshaft using a good assembly lubricant.

14. Install or connect the following:
- Camshaft using caution to avoid any damage to the camshaft bearings
- Thrust plate. Make sure that it covers the main oil gallery. Torque the screws to 84–120 inch lbs. (9–13 Nm).

15. Rotate the camshaft and crankshaft, as necessary, to align the timing marks.
- Camshaft gear and chain. Torque the bolt to 44–50 ft. lbs. (60–68 Nm).

16. Remove the clip from the chain tensioner
- Engine front cover and water pump
- Crankshaft damper/pulley. Torque the bolt to 107 ft. lbs. (146 Nm).
- Oil pan

17. Coat the tappets with 50W engine oil and place them in their original locations.

18. Apply 50W engine oil to both ends of the pushrods. Install the pushrods in their original locations.
- Tappets
- Rocker arm shafts and covers
- Upper and lower intake manifolds
- Fuel lines to the fuel supply manifold
- Alternator and electrical connectors
- Oil pump drive
- Crankcase pulley and damper
- Ignition coil and bracket
- Spark plug wires
- Air cleaner hoses
- Fan, spacer and shroud
- Air conditioning condenser
- Air conditioning compressor
- Radiator
- Negative battery cable

19. Fill the cooling system.

20. Recharge the air conditioning system.

21. Replace the oil filter and refill the engine with clean oil.

22. Start the engine and check the ignition timing and idle speed; adjust if necessary. Run the engine at fast idle and check for coolant, fuel, vacuum or oil leaks.

4.0L OHC Engine

1. Before servicing the vehicle, refer to the precautions in the beginning of this section.

2. Remove or disconnect the following:
- Negative battery cable for safety
- Valve cover
- Hydraulic camshaft tensioner

➡**The right-hand camshaft sprocket bolt uses left-hand threads.**

3. For the right-hand camshaft use the Cam Gear Torque Adapter tool T97T-6256-F, to remove the camshaft sprocket bolt.

4. For the left-hand camshaft, remove the sprocket bolt.

➡**When removing the followers, label them so that they may be returned to their original positions.**

5. Using the Valve Spring Compressor tool ST1330-A, remove the camshaft roller followers.

6. Install or connect the following:
- Camshaft bearing cap bolts and the oil rail
- Camshaft

To install:

7. Lubricate all of the moving parts with SAE 50W engine oil.

8. Install camshaft onto the cylinder head.

9. Position the oil rail and install the bearing caps and bolts. Torque the bolts in 2 steps:
   a. Step 1—53.5 inch lbs. (6 Nm).
   b. Step 2—11–12.5 ft. lbs. (15–17 Nm).

10. Install or connect the following:
- Camshaft followers
- Camshaft sprocket bolt and hand tighten the bolt
- Camshaft Chain Tensioner T97T-6K254-A in the hole that the hydraulic chain tensioner was in

11. Turn the crankshaft one revolution clockwise until No. 1 piston is Top Dead Center (TDC).

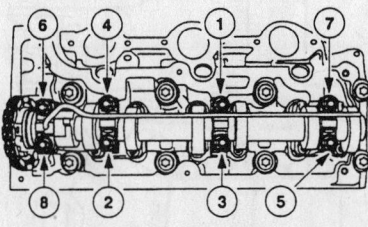

7924EG15

Use the proper sequence to prevent damage to the camshaft both when installing and removing the bearing caps—4.0L SOHC engine

12. Install or connect the following:
- Crankshaft Holding tool T97T-6303-A on the crankshaft to keep it from turning
- Position the timing slot on the rear of the camshaft to fit Camshaft Holding tool T97T-6256-C and install the holding tool on the rear of the head
- Camshaft Gear Holding tool T97T-6256-B and Camshaft Gear Holding tool T97T-6256-A on the front of the cylinder head to securely hold the camshaft gear
- Tighten the camshaft sprocket bolt to 63 ft. lbs. (85 Nm).

13. Remove the Camshaft Chain Tensioner tool and install the hydraulic chain tensioner, tighten the tensioner to 35–39 ft. lbs. (47–53 Nm).

14. Remove the special tools from the engine.

15. Install or connect the following:
- Valve cover
- Negative battery cable

16. Start the engine check for leaks and repair if necessary.

Oil Pan

REMOVAL & INSTALLATION

2.3L Engine

1. Before servicing the vehicle, refer to the precautions in the beginning of this section.

2. Drain the engine oil.

3. Remove or disconnect the following:
- Engine from the vehicle
- Engine oil level indicator assembly
- Engine oil pan bolts and oil pan

To install:

4. Clean and inspect all mating surfaces.

➡**The oil pan must be installed and the bolts tightened with four minutes of applying the silicone gasket and sealant.**

5. Apply a 2.5 mm bead of silicone gasket and sealant to the oil pan. Install the oil pan. Tighten the oil pan in the sequence shown.

6. Lubricate the O-ring with clean engine oil and install the engine oil level indicator assembly.

7. Install the engine into the vehicle.

2.5L Engines

1. Before servicing the vehicle, refer to the precautions in the beginning of this section.

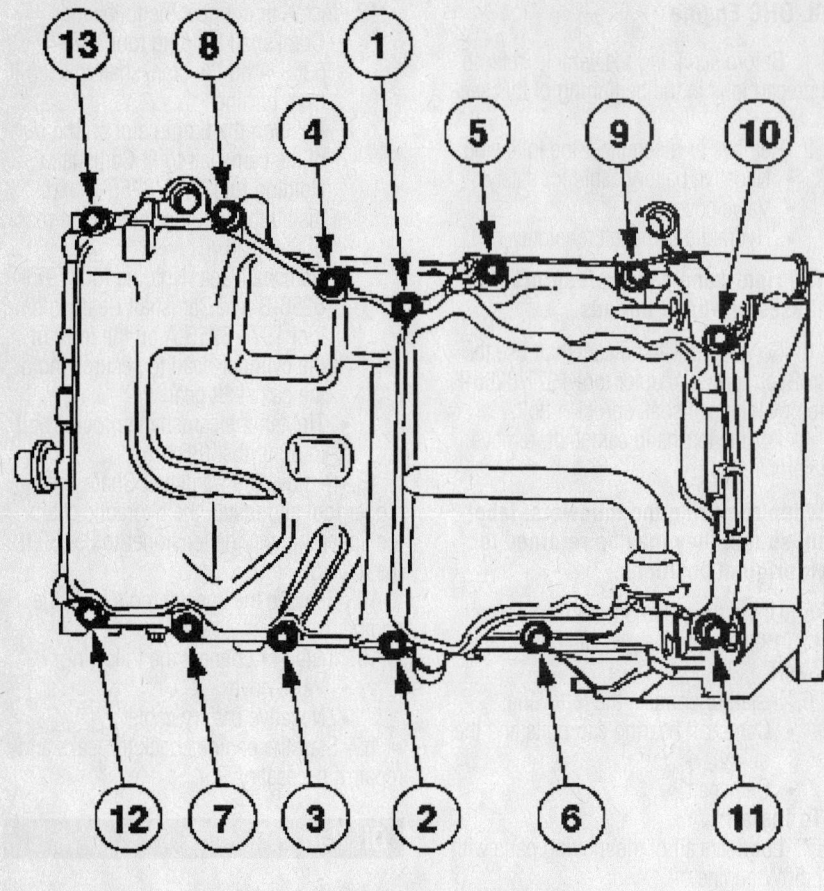

25 Nm (18 lb-ft)

9348EG05

Oil pan torque sequence—2.3L

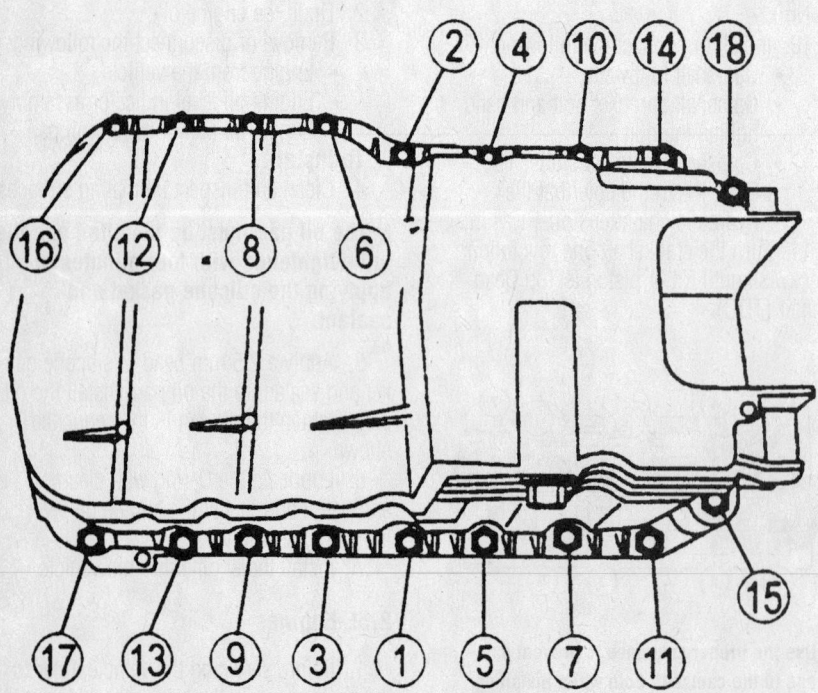

9308EG12

Tighten the oil pan bolts in sequence–2.5L engines

2. Drain the engine oil.
3. Remove or disconnect the following:
   - Negative battery cable
   - Engine from the vehicle and place it on a suitable engine stand
   - Oil pan and discard the gasket

To install:

4. Clean the mating surface on the oil pan.
5. Install or connect the following:
   - Oil pan gasket
   - Apply a bead of silicone sealant to the oil pan
   - Oil pan. Torque the bolts in sequence to 141 inch lbs. (16 Nm).
   - Engine
   - Negative battery cable
6. Fill the engine with clean oil.
7. Start the vehicle and check for leaks, repair if necessary.

3.0L Engine

2WD

1. Before servicing the vehicle, refer to the precautions in the beginning of this section.
2. Drain the engine oil.
3. Remove or disconnect the following:
   - Negative battery cable
   - Oil level dipstick tube
   - Fan shroud. Leave the fan shroud over the fan assembly
   - Motor mount nuts from the frame

✳✳ WARNING

On models equipped with distributor ignition, failure to remove the distributor will damage or break it when the engine is lifted.

   - Starter
   - Transmission inspection cover
   - Right hand axle I-beam. The brake caliper must be removed and secured out of the way.
   - Oil pan attaching bolts, using a suitable lifting device, raise the engine about 2 in. (5cm)
   - Oil pan and discard the gasket

➡ The oil pan fits tightly between the transmission spacer plate and oil pump pick-up tube. Use care when removing the oil pan from the engine.

4. Clean all gasket surfaces on the engine and oil pan. Remove all traces of old gasket and/or sealer.

To install:

5. Apply a 1/8 (4mm) bead of RTV sealer to the junctions of the rear main bearing cap and block, and the front cover and block.

The sealer sets in 15 minutes, so work quickly!

6. Apply adhesive to the gasket surfaces and install the oil pan gasket.

7. Install or connect the following:
- Oil pan on the engine block. Torque the bolts EVENLY to 9 ft. lbs. (12 Nm) working from the center to the end position on the oil pan.
- Right hand axle I-beam
- Brake caliper
- Transmission inspection cover
- Starter
- Fan shroud
- Motor mount retaining nuts
- Oil level dipstick tube
- Negative battery cable

8. Fill the engine with clean oil.

9. Start the vehicle and check for leaks, repair if necessary.

4WD

1. Before servicing the vehicle, refer to the precautions in the beginning of this section.

2. Drain the engine oil.

3. Remove or disconnect the following:
- Negative battery cable
- Engine from the vehicle and place it on a suitable engine stand
- Oil pan and discard the gasket

To install:

4. Install or connect the following:
- New oil pan gasket and secure the gasket with trim adhesive
- Oil pan. Torque the bolts to 9 ft. lbs. (12 Nm).
- Engine
- Negative battery cable

5. Fill the engine with clean oil.

6. Start the vehicle and check for leaks, repair if necessary.

4.0L OHV Engine

→Review the complete service procedure before starting this repair.

1. Before servicing the vehicle, refer to the precautions in the beginning of this section.

2. Drain the engine oil.

3. Remove or disconnect the following:
- Negative battery cable
- Engine from the vehicle and mount the engine on a suitable engine stand with the oil pan facing up
- Oil pan attaching bolts (note location of 2 spacers) and remove the pan from the engine block
- Oil pan gasket and crankshaft rear main bearing cap wedge seal

4. Clean all gasket surfaces on the

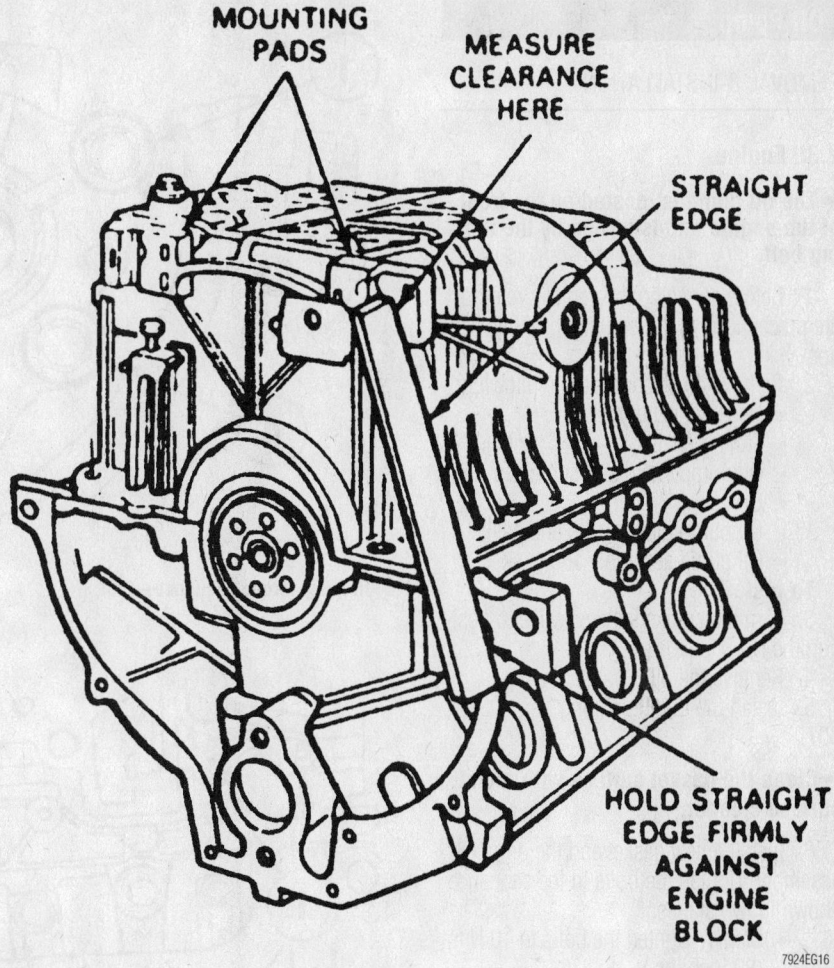

The correct spacer must be used to extend the mounting surface of the oil pan so it is flush with the mounting surface of the engine block—4.0L engine

engine and oil pan. Remove all traces of old gasket and/or sealer.

To install:

5. Install or connect the following:
- New crankshaft rear main bearing cap wedge seal. The seal should fit snugly into the sides of the rear main bearing cap
- New oil pan gasket to the engine block and place the oil pan in correct position on the 4 locating studs. Torque the bolts EVENLY to 60–84 inch lbs. (7–10 Nm).
- Transmission bolts to the engine and oil pan. There are 2 spacers on the rear of the oil pan to allow proper mating of the transmission and oil pan.
- Spacers to the mounting pads on the rear of the oil pan before bolting the engine and transmission together
- Engine to the vehicle
- Negative battery cable

6. Fill the engine with clean oil.

- Start the vehicle and check for leaks, repair if necessary.

4.0L OHC Engine

→The 4.0L OHC engine does not use an oil pan in the conventional sense. There is a separate access panel that unbolts from what would be considered the oil pan (which is now known as the ladder frame).

1. Before servicing the vehicle, refer to the precautions in the beginning of this section.

2. Drain the engine oil.

3. Remove or disconnect the following:
- Negative battery cable
- Oil pan and discard the gasket

To install:

4. Install or connect the following:
- New gasket and oil pan. Torque the bolts to 80 inch lbs. (9 Nm).
- Negative battery cable

5. Fill the engine with clean oil.

6. Start the vehicle and check for leaks, repair if necessary.

Oil Pump

REMOVAL & INSTALLATION

2.3L Engine

➡The oil pump is located on the front of the engine and is turned by the timing belt.

1. Before servicing the vehicle, refer to the precautions in the beginning of this section.
2. Remove or disconnect the following:
   - Negative battery cable
   - Timing chain
   - Oil pump chain and sprockets
   - Oil pan
   - Oil pump pickup tube and gasket
   - Oil pump assembly and gasket

To install:

3. Turn the crankshaft clockwise to position the No. 1 piston.
4. Remove the plug bolt.
5. Install the Engine Timing Peg 303-507.

➡Clean the gasket surface with metal surface cleaner.

6. Install a new gasket and the oil pump assembly. Tighten the bolts in the sequence shown in two stages.
   - Step 1: Tighten the bolts to 10 Nm (80 inch lbs.)
   - Step 2: Tight the bolts to 23 Nm (17 ft. lbs.)
7. Install a new oil pump pickup tube gasket and the pickup tube. Tighten the bolts in the sequence shown

2.5L Engines

➡The oil pump is located on the front of the engine and is turned by the timing belt.

1. Before servicing the vehicle, refer to the precautions in the beginning of this section.
2. Remove or disconnect the following:
   - Negative battery cable
   - Timing belt
   - Camshaft Position (CMP) sensor electrical connector
   - Oil pump sprocket
   - CMP sensor

➡Use a prybar or drift through one of the holes in the pump sprocket to keep it from turning while loosening the bolt.

   - 4 bolts retaining the oil pump to the engine block

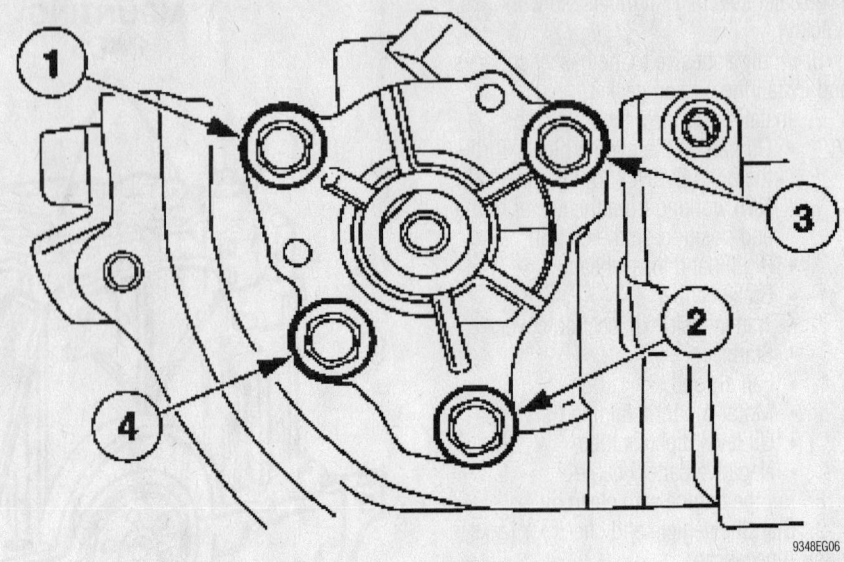

Oil pump torque sequence—2.3L

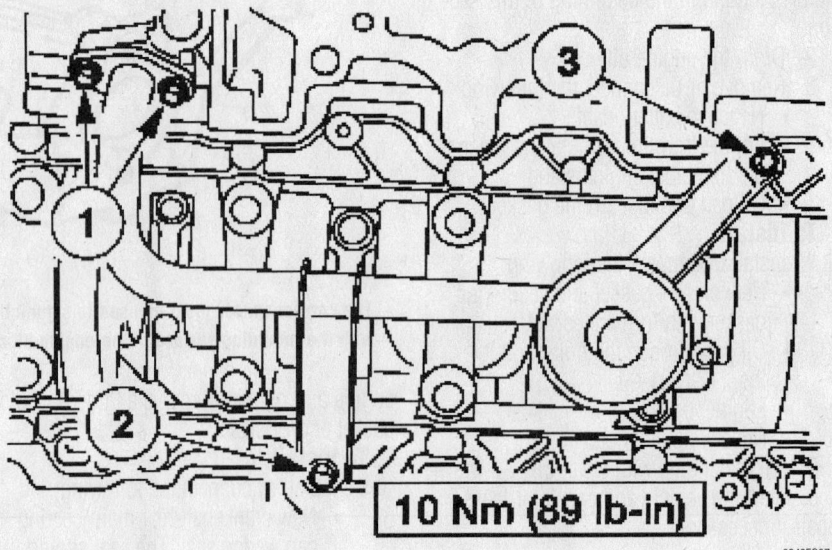

Oil pump pickup tube torque sequence—2.3L

10 Nm (89 lb-in)

   - Oil pump from the front of the engine and discard the gasket
3. Inspect the oil pump and O-rings and replace as necessary.
4. Clean all gasket mating surfaces thoroughly.

To install:

5. Prime the oil pump and with 8 ounces (236 ml) of new engine oil and lubricate the O-rings.
6. Install or connect the following:
   - New gasket on the oil pump
   - Oil pump. Torque the bolts to 89 inch lbs. (10 Nm).
   - CMP sensor. Torque the bolts to 61 inch lbs. (7 Nm).

   - Oil pump sprocket bolt. Torque the bolt to 40 ft. lbs. (55 Nm).
   - CMP sensor electrical connector
   - Timing belt
   - Negative battery cable
7. Fill the engine with clean oil.
8. Start the vehicle and check for leaks, repair if necessary.

3.0 L and 4.0L OHV Engines

➡On 4.0L it is necessary to remove the engine.

1. Before servicing the vehicle, refer to the precautions in the beginning of this section.
2. Drain the engine oil.

3. Remove or disconnect the following:
- Negative battery cable
- Oil pan
- Oil pick-up and tube assembly from the pump
- Oil pump retainer bolts and the oil pump

To install:

4. Prime the oil pump with clean engine oil by filling either the inlet or outlet port. Rotate the pump shaft to distribute the oil within the pump body.

5. Install the oil pump and tighten the mounting bolts to:

6. 3.0L: 30–40 ft. lbs. (41–54 Nm).

7. 4.0L: 13–15 ft. lbs. (18–20 Nm).

※※ WARNING

Do not force the oil pump if it does not seat readily. The oil pump driveshaft may be misaligned with the distributor or shaft assembly. If the pump is tightened down with the driveshaft misaligned, damage to the pump could occur. To align, rotate the intermediate driveshaft into a new position.

8. Install or connect the following:
- Oil pick-up and tube assembly
- Oil pan

9. Fill the engine with clean oil.

10. Start the vehicle and check for leaks, repair if necessary.

4.0L OHC Engines

➡**The oil pump cannot be removed with the engine in the vehicle.**

1. Before servicing the vehicle, refer to the precautions in the beginning of this section.

2. Drain the engine oil.

3. Remove or disconnect the following:
- Engine from the vehicle
- Oil pan
- Unbolt the oil pick-up tube
- The 8 ladder frame bolts that were under the oil pan
- The 2 rear outer ladder frame bolts
- The 7 left-hand and the 8 right-hand ladder frame bolts
- The ladder frame from the engine
- The 2 oil pump attaching bolts and the pump.

To install:

4. Submerge the pump in clean engine oil to prime it.

5. Install or connect the following:
- The ladder frame on the engine
- The 8 right-hand and 7 left-hand ladder frame bolts

- The 2 rear outer and the 8 frame bolts under the pan
- The oil pump. Torque the bolts to 13–15 ft. lbs. (17–21 Nm).
- Oil pick-up tube
- Oil pan
- Engine to the vehicle
- Negative battery cable

6. Fill the engine with clean oil.

7. Start the vehicle and check for leaks, repair if necessary.

Rear Main Seal

REMOVAL & INSTALLATION

2.3L Engine

- Flywheel or flexplate
- Bolts and the crankshaft rear oil seal

To install

- Rear oil seal on the Crankshaft Rear Main Oil Seal Installer
- Crankshaft Rear Main Oil Seal Installer and the crankshaft rear oil seal on the crankshaft

1. Tighten the bolts in the sequence shown to 10 Nm (89 inch lbs.)

2. Remove the Crankshaft Rear Main Oil Seal Installer.

3. Install the flywheel or flexplate.

2.5L Engine

1. Remove the flywheel or flexplate

➡**Clean the crankshaft rear oil seal and cylinder block prior to removing the rear oil seal.**

2. Screw in the Jet Plug Remover.

3. Remove the seal.

To install

➡**Apply 5W30 motor oil to seal and seal edge.**

4. Install crankshaft rear oil seal on Rear Main Seal Replacer.

5. Install the Rear Main Seal Replacer and the crankshaft rear oil seal on the crankshaft.

6. Alternate bolt tightening to crankshaft rear oil seal.

7. Install the flywheel or flexplate.

3.0L and 4.0L OHV Engines

1. Remove the flexplate or flywheel.

※※ WARNING

Use care to avoid scratching or damaging the oil seal surface or leakage may occur.

2. Using a sharp awl, punch one hole into the crankshaft rear oil seal metal surface between the seal lip and the cylinder block.

3. Screw the threaded end of the special tool into the oil seal. Use the special tool to remove the crankshaft rear oil seal.

To install:

4. Lubricate the outer lips and the inner seal on the crankshaft rear oil seal with clean engine oil.

5. Using the special tool, install the crankshaft rear oil seal. Alternate bolt tightening to correctly seat the crankshaft rear oil seal.

6. Install the flexplate or flywheel.

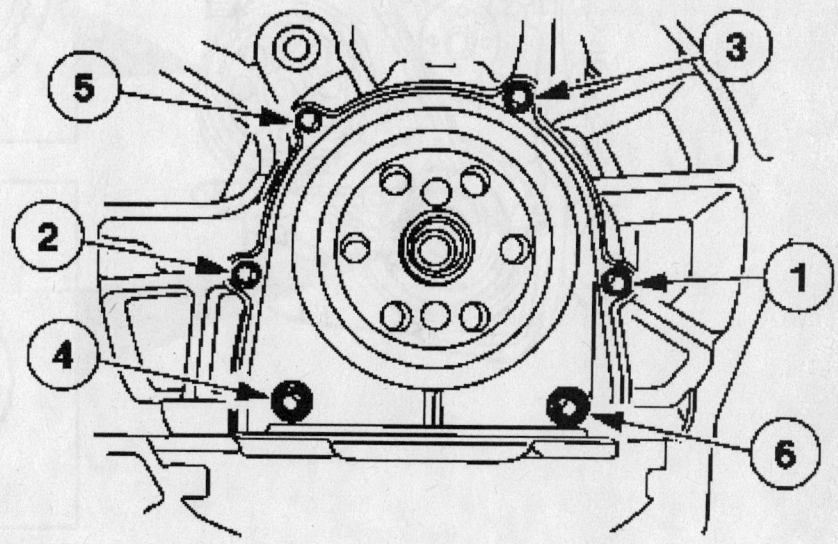

Rear main seal torque sequence—2.3L

9348EG08

4.0L OHC Engine

1. Remove the flexplate or flywheel.

❊❊ WARNING

Avoid scratching or damaging the oil crankshaft seal running surface during removal of the crankshaft rear oil seal.

2. Using the special tool, remove the crankshaft rear oil seal.

To install:

➡ Be sure the crankshaft rear sealing surface is clean and free of any rust or corrosion. To clean the crankshaft rear sealing surface, use extra-fine emery cloth or extra-fine 0000 steel wool with metal surface cleaner.

3. Lubricate the crankshaft rear oil seal with clean engine oil and install on the special tool.

4. Using the special tool, install the crankshaft rear oil seal.

5. Install the flexplate or flywheel.

Timing Belt and Cover

REMOVAL & INSTALLATION

2.5L Engine

1. Rotate the engine so that No. 1 cylinder is at Top Dead Center (TDC) on the compression stroke. Check that the timing marks are aligned on the camshaft and crankshaft pulleys. An access plug is provided in the cam belt cover so that the camshaft timing can be checked without removal of the cover or any other parts. Set the crankshaft to TDC by aligning the timing mark on the crank pulley with the TDC mark on the belt cover. Look through the access hole in the belt cover to be sure that the timing mark on the cam drive sprocket is aligned with the pointer on the inner belt cover.

➡ **Always turn the engine in the normal direction of rotation. Backward rotation may cause the timing belt to jump time, due to the arrangement of the belt tensioner.**

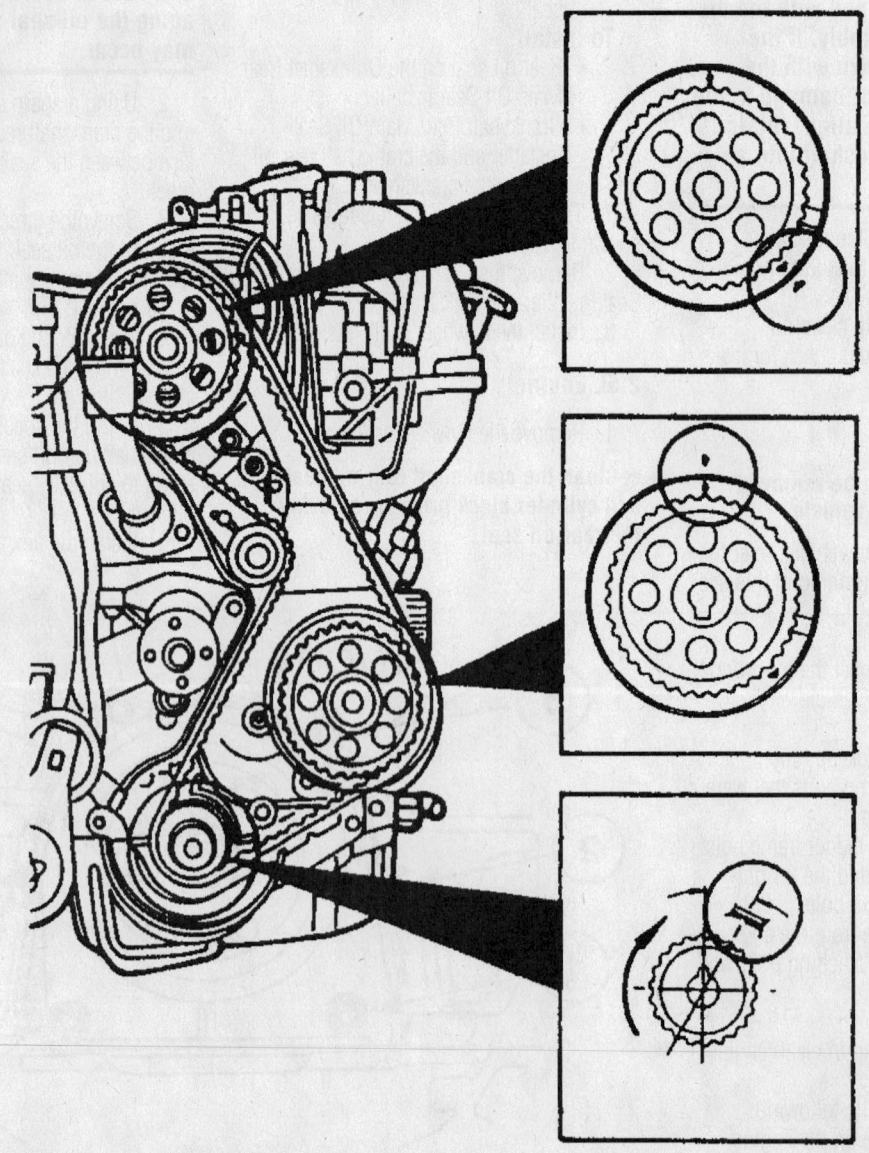

Camshaft, auxiliary shaft and crankshaft timing belt sprocket alignment mark locations—2.5L engines

79245G20

2. Drain cooling system. Remove the upper radiator hose as necessary. Remove the fan blade and water pump pulley bolts.

3. Loosen the alternator retaining bolts and remove the drive belt from the pulleys. Remove the water pump pulley.

4. Remove the power steering pump and set it aside.

5. Remove the 4 timing belt outer cover retaining bolts and remove the cover. Remove the crankshaft pulley and belt guide.

6. Loosen the belt tensioner pulley assembly, then position a camshaft belt adjuster tool T74P-6254-A, or equivalent, on the tension spring roll pin and retract the belt tensioner away from the timing belt. Tighten the adjustment bolt to lock the tensioner in the retracted position.

7. If the belt is to be reused, mark the direction of rotation on the belt for installation reference.

8. Remove the timing belt.

To install:

9. Install the new belt over the crankshaft sprocket and then counterclockwise over the auxiliary and camshaft sprockets, making sure the lugs on the belt properly engage the sprocket teeth on the pulleys. Be careful not to rotate the pulleys when installing the belt.

10. Release the timing belt tensioner pulley, allowing the tensioner to take up the belt slack. If the spring does not have enough tension to move the roller against the belt (belt hangs loose), it might be necessary to manually push the roller against the belt and tighten the bolt.

➡ **The spring cannot be used to set belt tension; a wrench must be used on the tensioner assembly.**

11. Rotate the crankshaft 2 complete turns by hand (in the normal direction of rotation) to remove slack from the belt. Tighten the tensioner adjustment to 26–33 ft. lbs. (35–45 Nm) and pivot bolts to 30–40 ft. lbs. (40–55 Nm). Be sure the belt is seated properly on the pulleys and that the timing marks are still in alignment when No. 1 cylinder is again at TDC/compression.

12. Install the crankshaft pulley and belt guide.

13. Install the timing belt cover.

14. Install the water pump pulley and fan blades. Install the upper radiator hose if necessary. Refill the cooling system.

15. Install the accessory drive belts.

16. Start the engine and check the ignition timing. Adjust the timing, if necessary.

Timing Chain, Sprockets, Front Cover and Seal

REMOVAL & INSTALLATION

2.3L Engine

1. Before servicing the vehicle, refer to the precautions in the beginning of this section.

2. Remove or disconnect the following:
- Negative battery cable
- Fan and shroud
- Drive belt
- Valve cover

3. Set No. 1 piston to TDC and install the Camshaft Alignment Plate 303-376.

4. Remove the plug for the crankshaft timing peg.

5. Install the Crankshaft Timing Peg 303-507.

6. Install an M6 bolt into the crankshaft pulley to verify the engine timing.

7. Remove or disconnect the following:

- Camshaft pulley
- Crankshaft position sensor
- Crankshaft position sensor
- Belt tensioner
- Water pump pulley
- Power steering high pressure hose. Remove the nylon O-ring.
- Power steering return hose
- Power steering pump

➡ **This step is needed only if a new front cover is being installed.**

8. Using a three-jaw puller, remove the fan drive pulley.

➡ **There is one bolt behind the cooling fan drive pulley. This bolt can be accessed by lining up one of the holes in the pulley with the bolt.**

9. Remove the bolts and the engine front cover.

10. Compress the timing chain tensioner and remove the tensioner.

11. Remove the right-hand timing chain guide.

12. Remove the timing chain.

13. Remove the bolts and the left-hand timing chain guide.

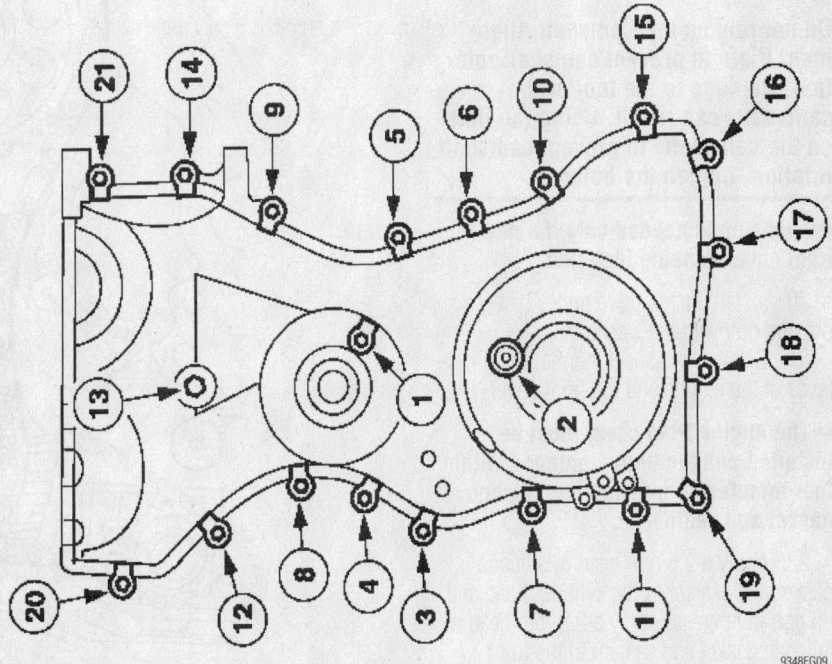

Front cover torque sequence—2.3L

9348EG09

※※ WARNING

Do not rely on the Camshaft Alignment Plate to prevent camshaft rotation. Damage to the tool or the camshaft can occur.

14. If necessary, remove the bolts and the camshaft sprockets. Use the flats on the camshaft to prevent camshaft rotation.

To install:

15. Remove the special tool.

※※ WARNING

Do not rotate the camshafts. Damage to the valves and pistons can occur.

If the camshaft sprockets were not removed, use the flats on the camshafts to prevent camshaft rotation and loosen the sprocket bolts.

16. If removed, install the camshaft sprockets and the bolts. Do not tighten the bolts at this time.

17. Install or connect the following:
  • Left-hand timing chain guide and bolts
  • Timing chain
  • Right-hand timing chain guide
  • Timing chain tensioner and release the piston
  • Timing chain tensioner and the bolts

18. Remove the drill rod to release the piston.

19. Install the special tool.

※※ WARNING

Do not rely on the Camshaft Alignment Plate to prevent camshaft rotation. Damage to the tool or the camshafts can result. Using the flats on the camshafts to prevent camshaft rotation, tighten the bolts.

➡**This step is needed only if a new front cover is being installed.**

20. Install the fan drive pulley using a nut and bolt with flat washers.

21. Clean and inspect the mounting surfaces of the engine and the front cover.

➡**The engine front cover must be installed and the bolts tightened within four minutes of applying the silicone gasket and sealant.**

22. Apply a 2.5 mm bead of silicone gasket and sealant to the cylinder head and oil pan joint areas. Apply a 2.5 mm bead of silicone gasket and sealant to the front cover.

23. Install the front cover. Tighten the bolts in the sequence shown, to the following specifications:
  • Step 1: 8 mm bolts to 10 Nm (89 inch lbs.)
  • Step 2: 10 mm bolts to 25 Nm (18 ft. lbs.)
  • Step 3: 13 mm bolts to 48 Nm (35 ft. lbs.)

24. Install or connect the following:
  • Power steering pump and lower retaining bolt
  • Power steering return hose
  • New nylon O-ring and install the high pressure line.
  • Water pump pulley
  • Belt tensioner

➡**Do not reuse the crankshaft damper bolt.**

  • Crankshaft pulley and hand-tighten the bolt

25. Install an M6 bolt in the crankshaft pulley. Tighten the crankshaft retaining bolt in two stages.
  • Step 1: 40 Nm (30 ft. lbs.)
  • Step 2: Rotate the bolt an additional 90 degrees.

26. Install the crankshaft position sensor, do not tighten the bolts at this time.

27. Adjust the crankshaft position sensor with the Alignment Tool, and tighten the mounting bolts.

28. Connect the crankshaft position sensor electrical connector.

29. Remove the M6 bolt from the crankshaft pulley.

30. Remove the Crankshaft Timing Peg 303-507.

31. Install the plug.

32. Remove the Camshaft Alignment Plate 303-376.

33. Install the valve cover.

34. Install the drive belt.

35. Install the fan and shroud.

36. Connect the battery ground cable.

3.0L and 4.0L OHV Engines

1. Before servicing the vehicle, refer to the precautions in the beginning of this section.

2. Remove or disconnect the following:
  • Negative battery cable
  • Engine front cover
  • Rotate the crankshaft and align the timing marks
  • Timing chain tensioner, 4.0L engine only
  • Sprocket bolt
  • Timing chain, camshaft sprocket

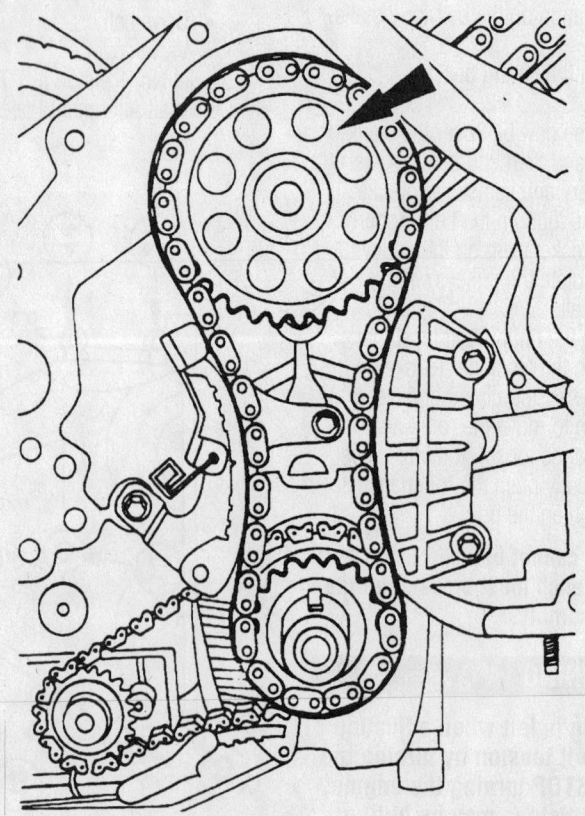

Remove the jackshaft sprocket–4.0L SOHC Engine

89683G51

and crankshaft sprocket as an assembly

To install:

3. Install or connect the following:
   - Timing chain, camshaft and crankshaft sprockets as an assembly
4. Align the timing marks.
5. Install or connect the following:
   - Timing chain tensioner
   - Sprocket bolt. Torque the bolt to 51 ft. lbs. (70 Nm).
   - Engine front cover
   - Negative battery cable

4.0L OHC Engine

1. Before servicing the vehicle, refer to the precautions in the beginning of this section.
2. Drain the engine oil.
3. Remove or disconnect the following:
   - Negative battery cable
   - Engine from the vehicle
   - Oil pan
   - Engine front cover
   - Cylinder heads
4. Lock the jackshaft tensioner by installing a pin.
   - Jackshaft sprocket and chain assembly
   - Left front cassette retaining bolt
   - Cassette chain and tensioner assembly
   - Rear jackshaft plug from the engine
   - Right rear cassette retaining bolt and spacer
   - Right rear cassette chain and tensioner
   - Timing chain (s)
5. Install or connect the following:
   - Timing chain(s)
   - Right rear cassette chain, tensioner and sprocket
   - Jackshaft sprocket and chain on the engine and remove the tensioner pin
6. Torque the jackshaft sprocket bolt in 2 stages:
   a. 32–35 ft. lbs. (43–47 Nm).
   b. Turn an additional 65 degrees.

7. Install or connect the following:
   - Cylinder heads
   - Front cover
   - Oil pan
   - Engine to the vehicle
   - Negative battery cable
8. Fill the engine with clean oil.
9. Start the vehicle, check for leaks and repair if necessary.

Piston and Ring

POSITIONING

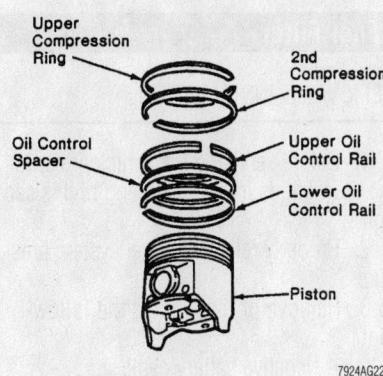

Piston ring positioning

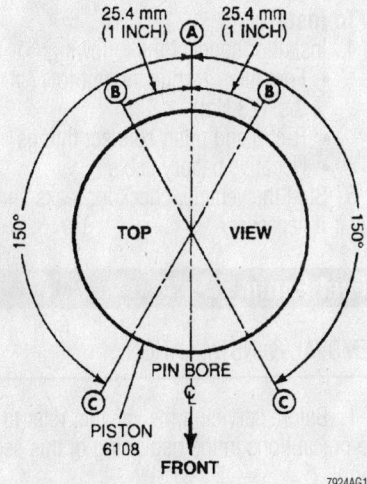

Piston ring end gap spacing

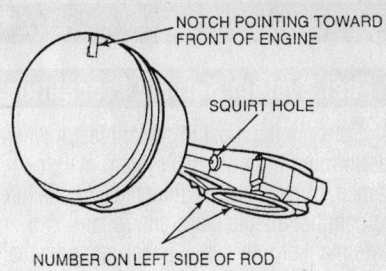

Piston and connecting rod positioning on 2.5L engines

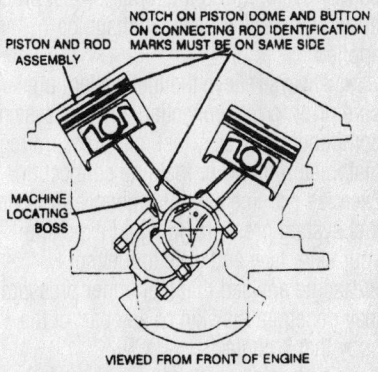

Piston and connecting rod positioning on 3.0L engines

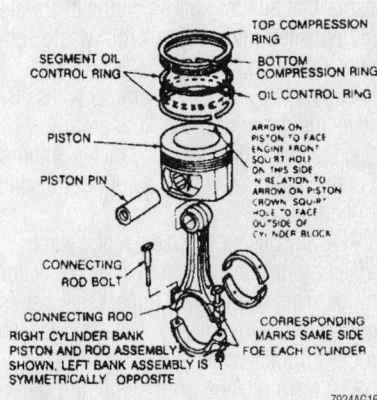

Piston and connecting rod positioning on 4.0L engines

FUEL SYSTEM

Fuel System Service Precautions

Safety is the most important factor when performing not only fuel system maintenance, but any type of maintenance. Failure to conduct maintenance and repairs in a safe manner may result in serious personal injury or death. Work on a vehicle's fuel system components can be accomplished safely and effectively by adhering to the following rules and guidelines.

- To avoid the possibility of fire and personal injury, always disconnect the negative battery cable unless the repair or test procedure requires that battery voltage be applied.
- Always relieve the fuel system pressure prior to disconnecting any fuel system component (injector, fuel rail, pressure regulator, etc.) fitting or fuel line connection. Exercise extreme caution whenever relieving fuel system pressure, to avoid exposing your skin, face and eyes to fuel spray. Please be advised that fuel under pressure may penetrate the skin or any part of the body that it contacts.
- Always place a shop towel or cloth around the fitting or connection prior to loosening to absorb any excess fuel due to spillage. Ensure that all fuel spillage is quickly remove from engine surfaces. Ensure that all fuel-soaked cloths or towels are deposited into a flame-proof waste container with a lid.
- Always keep a dry chemical (Class B) fire extinguisher near the work area.
- Do not allow fuel spray or fuel vapors to come into contact with a light bulb, spark or open flame.
- Always use a second wrench when loosening or tightening fuel line connection fittings. This will prevent unnecessary stress and torsion to fuel piping. Always follow the proper torque specifications.
- Always replace worn fuel fitting O-rings with new ones. Do not substitute fuel hose where rigid pipe is installed.

Relieving Fuel System Pressure

All Sequential Fuel Injection (SFI) fuel injected engines are equipped with a pressure relief valve located on the fuel supply manifold. Remove the fuel tank cap and attach fuel pressure gauge T80L-9974-B, to the valve to release the fuel pressure. Be sure to drain the fuel into a suitable container and to avoid gasoline spillage. If a pressure gauge is not available, disconnect the vacuum hose from the fuel pressure

regulator and attach a hand-held vacuum pump. Apply about 25 in. Hg (84 kPa) of vacuum to the regulator to vent the fuel system pressure into the fuel tank through the fuel return hose. Note that this procedure will remove the fuel pressure from the lines, but not the fuel. Take precautions to avoid the risk of fire and use clean rags to soak up any spilled fuel when the lines are disconnected.

An alternate method of relieving the fuel system pressure involves disconnecting the inertia switch.

Fuel Filter

REMOVAL & INSTALLATION

1. Before servicing the vehicle, refer to the precautions in the beginning of this section.
2. Properly relieve the fuel system pressure.
3. Remove or disconnect the following:

- Negative battery cable
- Push connect and R-clip fittings from the fuel filter
- Fuel filter

To install:
4. Install or connect the following:
- Fuel filter. Torque the nut to 17 ft. lbs. (23 Nm).
- R-clip and push connect fittings
- Negative battery cable
5. Start the vehicle, check for leaks and repair if necessary.

Fuel Pump

REMOVAL & INSTALLATION

1. Before servicing the vehicle, refer to the precautions in the beginning of this section.
2. Properly relieve the fuel system pressure.
3. Remove or disconnect the following:

- Negative battery cable
- Fuel tank
- Fuel tank pump locking retainer ring
- Fuel pump mounting gasket and discard the gasket
- Fuel pump

To install:
4. Install or connect the following:

- Fuel pump and a new mounting gasket
- Fuel tank pump locking retainer ring. Torque the ring to 66 ft. lbs. (90 Nm).
- Fuel tank
- Negative battery cable
5. Start the vehicle, check for leaks and repair if necessary.

Fuel Injectors

REMOVAL & INSTALLATION

2.3L Engine

1. Before servicing the vehicle, refer to the precautions in the beginning of this section.
2. Properly relieve the fuel system pressure.
3. Remove or disconnect the following:
- Negative battery cable
- Upper intake manifold
- Fuel injector connectors
- Fuel injector harness from the fuel injector supply manifold
- Fuel line spring lock
- Fuel line
- Fuel injection supply manifold
- Fuel injector retaining clip
- Fuel injector

✳✳ WARNING

Use O-ring seals that are made of special fuel-resistant material. Use of ordinary O-ring seals can cause the fuel system to leak. Do not reuse the O-ring seals.

4. Installation is the reverse of removal. Install new O-rings. Lubricate the O-rings with clean engine oil.

2.5L and 4.0L OHV Engines

1. Before servicing the vehicle, refer to the precautions in the beginning of this section.
2. Properly relieve the fuel system pressure.
3. Remove or disconnect the following:

- Negative battery cable
- Fuel injection supply manifold
- Fuel injectors by gently twisting them
- Inspect the O-rings and replace as needed

To install:

4. Install or connect the following:
   - Fuel injectors
   - Fuel injector supply manifold
   - Negative battery cable

5. Start the vehicle, check for leaks and repair if necessary.

3.0L and 4.0L OHC Engines

1. Before servicing the vehicle, refer to the precautions in the beginning of this section.

2. Properly relieve the fuel system pressure.

3. Remove or disconnect the following:
   - Negative battery cable
   - Upper intake manifold
   - Engine control sensor wiring from the fuel injectors
   - Fuel lines
   - Fuel injection supply manifold and injectors as an assembly
   - Vacuum line
   - Fuel injectors from the supply manifold
   - Inspect the O-rings and replace them as needed

To install:

4. Install or connect the following:
   - Fuel injectors
   - Vacuum line
   - Fuel injection supply manifold. Torque the bolts to 89 inch lbs. (10 Nm).
   - Fuel line
   - Engine control sensor wiring to the fuel injectors
   - Upper intake manifold
   - Negative battery cable

5. Start the vehicle, check for leaks and repair if necessary.

DRIVE TRAIN

Transmission Assembly

REMOVAL & INSTALLATION

Manual Transmission

1. Before servicing the vehicle, refer to the precautions in the beginning of this section.

2. Remove or disconnect the following:
   - Negative battery cable
   - Upper gearshift lever and the outer gearshift lever boot and console assembly as an assembly

3. If transmission disassembly is necessary, remove the drain plug, and drain the transmission fluid. Install the drain plug after draining all of the fluid.

4. Remove or disconnect the following:
   - Electrical connector from the reverse lamp switch
   - Electrical connector from the vehicle speed sensor (VSS)
   - Heated oxygen sensor (HO2S) electrical connector from the bracket
   - Wiring harness from the bracket
   - Electrical connectors from the heated oxygen sensors (HO2S)
   - Starter motor

➡The driveshaft centering socket yoke fits tightly on the rear axle pinion

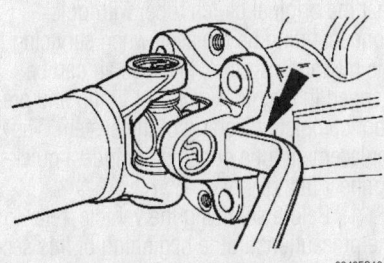

Pry here for driveshaft removal

9348EG10

flange pilot. Never hammer on the driveshaft or any of its components to disconnect the yoke from the flange. Pry only in the area shown, with a suitable tool, to disconnect the yoke from the flange.

➡If equipped, always disconnect the front driveshaft from the transfer case first. Otherwise, the weight of the driveshaft can cause the boot to tear.

   - Rear driveshaft, and the front driveshaft, if so equipped
   - Bolts retaining the exhaust inlet crossover pipe to the exhaust manifold
   - Bolts retaining the catalytic converter to the muffler. Discard the exhaust converter outlet gasket.
   - Exhaust hanger from the insulator. Position the exhaust assembly aside.
   - On 4-wheel drive vehicles, the transfer case
   - Clutch hydraulic line from the clutch slave cylinder

✳✳ WARNING

Secure the transmission to the jack with a suitable safety strap. Failure to follow these instructions may result in personal injury.

5. Using a suitable transmission jack, support the transmission. Secure the transmission to the jack with a suitable safety strap.

6. Loosen, but do not remove the nuts retaining the transmission insulator to the crossmember.

7. Remove the six bolts retaining the crossmember to the frame.

8. Remove the nuts and the crossmember.

➡Lower the transmission enough to gain access to the upper bolts retaining the transmission to the engine.

9. Remove the nine bolts retaining the transmission to the engine.

10. Remove the transmission from the vehicle.

11. Installation is the reverse of removal. Observe the following torques:
   - Transmission-to-engine bolts: 44 ft. lbs. (60Nm)
   - Crossmember-to-frame: 46 ft. lbs. (63Nm)
   - Transmission insulator-to-crossmember: 72 ft. lbs. (98Nm)

Automatic Transmission

4RE44 4-SPEED

1. Before servicing the vehicle, refer to the precautions in the beginning of this section.

2. Place the selector lever in NEUTRAL position.

3. Remove or disconnect the following:
   - Negative battery cable
   - Fluid level indicator
   - The two bolts retaining the fan shroud to the radiator.

➡If transmission disassembly is required, drain the transmission fluid. For additional information, refer to ®Fluid Pan, Gasket and Filter— in this section.

   - With 4wd, the transfer case

➡Mark the driveshaft yoke and axle flange, so they may be installed in their original alignment.

   - Rear driveshaft
   - Starter motor
   - Torque converter access cover

➡ Mark the torque converter and the flexplate for correct alignment at reinstallation.

- The four converter nuts
- Shift cable
- Transmission wiring harness
- Three way catalytic converter
- Left HO2S sensor
- Front exhaust crossover pipe
- Transmission cooler lines

4. Position a High-Lift Jack under the transmission. Raise and support the transmission.

5. Remove or disconnect the following:
- Crossmember.
- Transmission mount
- Transmission upper fill tube

➡ Lower the High-Lift Transmission Jack to gain access to screws.

- On 4x4 models, the vent tube assembly

❊❊ WARNING

Install the Torque Converter Holding Tool before lowering the transmission from the vehicle. Secure the transmission to the transmission jack with a safety chain. Failure to follow these instructions can result in personal injury.

6. Lower the transmission.
7. Installation is the reverse of removal. Observe the following torques:
- Transmission-to-engine bolts: 41 ft. lbs. (55Nm)
- Exhaust bracket bolts: 81 ft. lbs. (110Nm)
- Crossmember-to-frame: 87 ft. lbs. (118Nm)
- Transmission mount-to-crossmember: 81 ft. lbs. (110Nm)
- Converter-to-flexplate: 30 ft. lbs. (40Nm)
- Rear driveshaft-to-flange bolts: 95 ft. lbs. (129Nm)

Clutch

REMOVAL & INSTALLATION

1. Before servicing the vehicle, refer to the precautions in the beginning of this section.
2. Remove or disconnect the following:
- Negative battery cable
- Transmission

➡ If the clutch disc and pressure plate are to be reinstalled, bolts must be

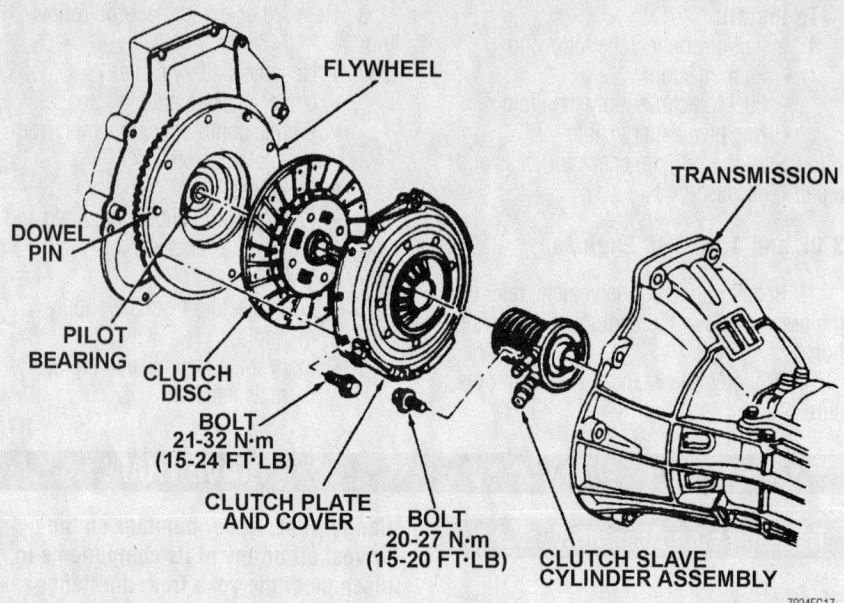

Clutch disc, pressure plate and bearing assembly

removed evenly or permanent damage to the diaphragm spring will occur resulting in complete clutch release.

- Bolts, clutch pressure plate and the clutch disc.

➡ If the parts are to be reused, index-mark the clutch pressure plate to the flywheel.

To installation:

3. Lubricate the transmission input shaft pilot bearing with front axle grease.
4. Using a suitable press, press downward on the pressure plate fingers until the adjusting ring moves freely.
5. Rotate the adjusting ring counterclockwise to compress the tension springs. Hold the adjusting ring in this position.
6. Release the pressure on the fingers. The adjusting ring will stay in the reset position.

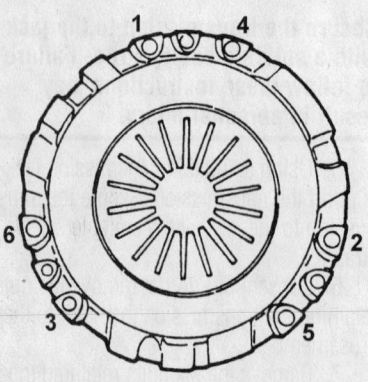

Tighten the bolts gradually in the correct sequence to avoid warping the pressure plate

7. Position the clutch disc on the flywheel.

➡ If reusing the clutch pressure plate and flywheel, align the marks made during removal.

8. Align the clutch disc and the clutch pressure plate. Install the bolts and tighten in a star pattern sequence to 24 ft. lbs. (35Nm).
- Install the transmission.

ADJUSTMENT

Because the clutch is hydraulically driven, there is no adjustment required.

In the event the clutch pedal develops a squeak or uneven feel when depressing, spray the pedal bushing assembly with penetrating oil and work the pedal back-and-forth.

Hydraulic Clutch System

BLEEDING

The following procedure is recommended for bleeding the clutch hydraulic system installed on the vehicle. It is recommended that the original clutch tube, with quick-connect fitting be replaced when servicing the hydraulic system, because air can be trapped in the quick-connect fitting and prevent complete bleeding of the system. The replacement tube does not include a quick-connect fitting.

1. Before servicing the vehicle, refer to the precautions in the beginning of this section.

2. Clean the dirt and grease from the dust cap.

3. Remove the cap and diaphragm and fill the reservoir to the top with approved brake fluid C6AZ-19542-AA or BA, (ESA-M6C25-A).

➡**To keep brake fluid from entering the clutch housing, route a suitable rubber tube of appropriate inside diameter from the bleed screw to a container.**

4. Loosen the bleed screw, located in the slave cylinder body, next to the inlet connection. Fluid will now begin to move from the master cylinder down the tube to the slave cylinder.

➡**The reservoir must be kept full at all times during the bleeding operation, to ensure no additional air enters the system.**

5. Observe the bleed screw outlet. When the slave cylinder is full, a steady stream of fluid will flow from the outlet port. Tighten the bleed screw.

6. Depress the clutch pedal to the floor and hold for 1–2 seconds. Release the pedal as rapidly as possible. The pedal must be released completely. Pause for 1–2 seconds. Repeat 10 times.

7. Check the fluid level in the reservoir.

The fluid should be level with the step when the diaphragm is removed.

8. Hold the pedal to the floor, slightly open the bleed screw to allow any additional air to escape. Close the bleed screw, then release the pedal.

9. Check the fluid in the reservoir. The hydraulic system should now be fully bled, and should actuate the clutch.

10. Check the vehicle by starting, pushing the clutch pedal to the floor and selecting reverse gear. There should be no grating of gears. If there is, and the hydraulic system still contains air; repeat the bleeding procedure.

Transfer Case Assembly

REMOVAL & INSTALLATION

1. Before servicing the vehicle, refer to the precautions in the beginning of this section.
2. Place the transmission in neutral.
3. Remove or disconnect the following:
   - Skid plate
   - Damper
   - Transfer case harness connector and position it aside
4. If transfer case (7005) disassembly is

necessary, remove the drain plug and drain the fluid.

➡**Index-mark the front output shaft assembly and the front driveshaft constant velocity (CV) joint. Always disconnect the front driveshaft from the transfer case first. Otherwise, the weight of the driveshaft can pinch the boot between the shaft and the boot can and cause the boot to tear.**

   - Front driveshaft from the transfer case and position the driveshaft aside. Remove and discard the bolts and washers.

➡**Index-mark the front flange on the rear driveshaft and the flange on the transfer case.**

   - Rear driveshaft

➡**Secure the transfer case to the jack with safety straps.**

5. Position a high lift jack under the transfer case.
6. Remove or disconnect the following:

   - Five bolts retaining the transfer case to the extension housing
   - Transfer case rearward and off of the transmission output shaft

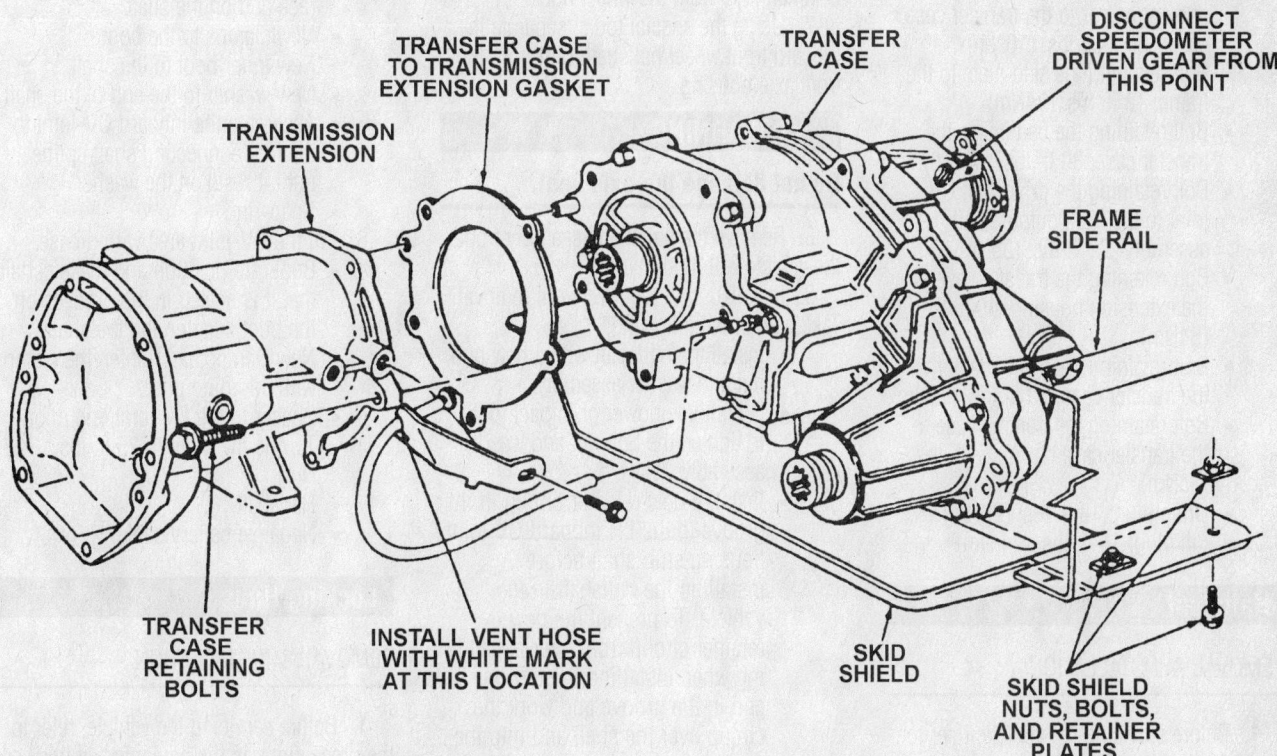

Exploded view of the 13-54 electronic shift transfer case-to-transmission mounting

7924EG20

7. Remove and discard the front extension housing gasket and clean the mating surfaces.

To install:

8. Installation is the reverse of removal. Take note of the following:

- Install the transfer case with a new gasket.
- Tighten the bolts that retain the transfer case to the extension housing in a clockwise direction beginning with the upper LH bolt.
- Install the front and the rear driveshafts with new bolts. If new bolts are not available, coat the threads of the original bolts with Threadlock and Sealer E0AZ-19554-AA or equivalent meeting Ford specification WSK-M2G351-A5.
- When installing the front driveshaft, always connect it to the axle first and then connect it to the transfer case.
- Align the index marks when installing the front and rear driveshafts.

9. Observe the following torques:

- Nut retaining the flange to the rear output shaft: 262 ft. lbs. (355Nm)
- Bolt retaining the rear driveshaft to the flange: 82 ft. lbs. (111Nm)
- Bolt retaining the motor assembly and connector to the transfer case cover: 89 inch lbs. (10Nm)
- Bolt retaining the skid plate to the frame: 18 ft. lbs. (24Nm)
- Bolt retaining the damper to the transfer case: 30 ft. lbs. (40Nm)
- Bolt retaining the driveshaft CV joint to the front output shaft assembly: 22 ft. lbs. (30Nm)
- Bolt retaining the transfer case to the extension housing: 40 ft. lbs. (54Nm)
- Bolt retaining the front adapter to the transfer case: 30 ft. lbs. (40Nm)
- Bolt retaining the transfer case to the transfer case cover: 27 ft. lbs. (36Nm)
- Drain plug: 18 ft. lbs. (24Nm)
- Fill plug: 18 ft. lbs. (24 Nm)

Halfshaft

REMOVAL & INSTALLATION

1. Before servicing the vehicle, refer to the precautions in the beginning of this section.
2. Place the transmission in NEUTRAL.

3. Remove or disconnect the following:

- Front wheel and tire assembly

※ WARNING

Do not reuse the torque prevailing design hub nut and washer assembly.

- Hub nut and washer assembly
- Front disc brake caliper, anchor plate, and pads as an assembly, and position the assembly aside
- Brake disc

※ WARNING

Do not use a hammer to separate the outboard front wheel halfshaft joint from the wheel hub. Damage to the outboard CV joint stub shaft threads and internal CV joint components may result.

- Outboard front wheel halfshaft joint from the wheel hub. Remove the special tool. Support the front suspension lower arm.
- Nut and bolt retaining the upper ball joint to the front wheel knuckle

4. Rotate the front wheel knuckle.
5. Compress the outboard front wheel halfshaft joint.
6. Remove the outboard front wheel halfshaft joint from the wheel hub.
7. Using the special tools, separate the inboard front wheel halfshaft joint from the front axle housing.

※ WARNING

Do not damage the axle seal.

8. Remove the halfshaft assembly from the vehicle with both hands.
9. Installation is the reverse of removal. Take note of the following:

- Install the halfshaft with a new hub nut and washer assembly.
- Do not use power or impact tools to tighten the hub nut and washer assembly.
- Install a new retainer circlip in the groove in the LH inboard CV joint housing stub shaft before installing the halfshaft in the vehicle. To prevent the new retainer circlip from over-expanding when installing it, start one end in the groove and work the circlip over the shaft and into the groove.
- Torque the hub nut to 162 ft. lbs. (220Nm).

CV-Joints

OVERHAUL

1. Before servicing the vehicle, refer to the precautions in the beginning of this section.
2. Remove or disconnect the following:

- Negative battery cable
- Halfshaft and place it in a vice with the inboard joint lower than the outboard joint

3. Cut the inner boot clamps with side cutters and remove the clamp from the boot.

- Larger boot end off the joint
- Inboard CV-joint bolts and separate the spacer and grease cap
- Snap-ring retaining the interconnecting shaft end to the CV-joint cage
- CV-joint and discard the washer

➡ **The outboard CV-joint is non-serviceable other than to replace the boot.**

To install:

4. Install or connect the following:

- Slide the boot over the shaft

5. Fill the CV-joint area with grease.

- Assemble the outer boot to the outboard CV-joint and interconnecting shaft. Make certain that the boot is seated in the grooves on the outer race and on the shaft
- New clamps to the boot
- New inner boot to the shaft
- New washer to the end of the shaft
- Assemble the inboard CV-joint to the interconnecting shaft spline until it rests on the washer
- Snap-ring

6. Fill the CV-joint area with grease.

- Boot into position and make certain that it is seated in the grooves on the boot adapter and the shaft
- New clamps and tighten the clamps with crimping pliers
- Spacer to the CV-joint end pilot. Torque the bolts to 25 ft. lbs. (34 Nm).
- Halfshaft
- Negative battery cable

Locking Hubs

REMOVAL & INSTALLATION

1. Before servicing the vehicle, refer to the precautions in the beginning of this section.
2. Remove or disconnect the following:

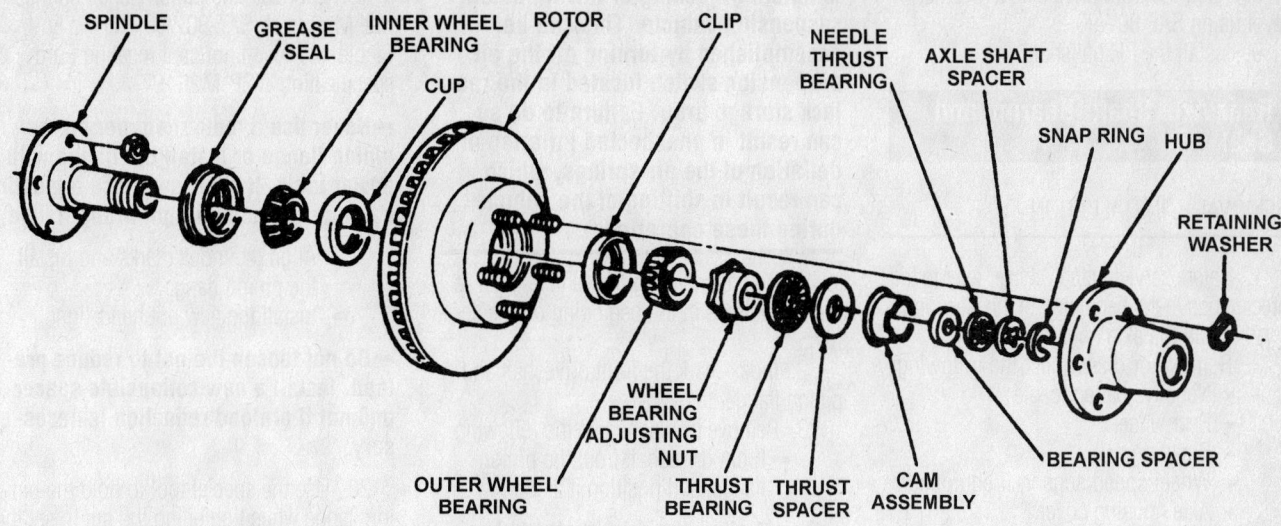

SPINDLE GREASE SEAL INNER WHEEL BEARING CUP ROTOR CLIP NEEDLE THRUST BEARING AXLE SHAFT SPACER SNAP RING HUB RETAINING WASHER

WHEEL BEARING ADJUSTING NUT

OUTER WHEEL BEARING THRUST BEARING THRUST SPACER CAM ASSEMBLY BEARING SPACER

7924EG23

Exploded view of the automatic locking hubs and related components

- Negative battery cable
- Wheel

➡**Some gentle tapping with a soft-faced hammer may help to loosen the locking hub cover if it seems stuck.**

- Automatic locking hub cover assembly from the rotor by pulling straight outward

3. Inspect the O-ring seal on the back-side of the hub assembly and, if damaged, replace it.

- Snap-ring from the end of the splined axle shaft
- Axle shaft spacer(s)

✳✳ WARNING

Do not pry on the locking cam or thrust spacers during removal. Prying may damage the cam or spacers.

- Pull the locking cam assembly and the 2 thrust spacers (behind cam assembly) from the wheel bearing adjusting nut.

To install:
4. Install or connect the following:

- 2 thrust spacers and locking cam into position. Make certain that the key in the cam assembly is aligned with the keyway of the front spindle
- Axle shaft spacer(s)
- Snap-ring to the end of the splined axle shaft
- Locking hub cover to the rotor
- Wheel
- Negative battery cable

Front Axle Tube Bearing

REMOVAL & INSTALLATION

1. Before servicing the vehicle, refer to the precautions in the beginning of this section.

2. Remove or disconnect the following:
- Right-hand halfshaft
- Right-hand axle shaft
- Axle seal, with a slide hammer
- Axle tube bearing, with a slide hammer

3. Clean the bearing and seal surfaces of any foreign debris.

To install:
4. Use an axle bearing replacer and the handle to replace the RH axle tube bearing.

5. Check the bearing depth as shown.

6. Use an axle seal replacer and the handle to replace the axle tube seal.

➡**Care should be taken not to damage the axle seal surface.**

7. Install the axle shaft.

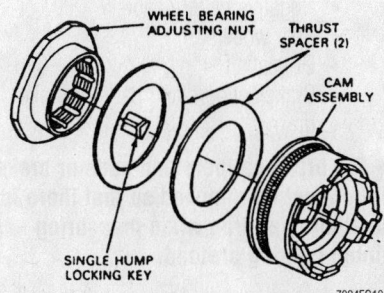

WHEEL BEARING ADJUSTING NUT THRUST SPACER (2) CAM ASSEMBLY

SINGLE HUMP LOCKING KEY

7924EG10

Exploded view of the locking cam assem-

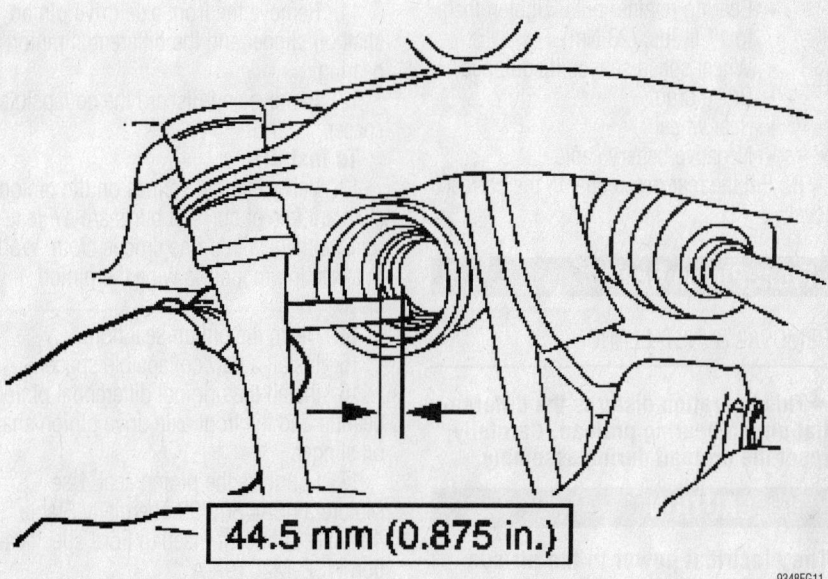

44.5 mm (0.875 in.)

9348EG11

Front axle tube bearing depth

8. Refill the front drive axle to proper level using SAE 80W90.

9. Install the RH halfshaft.

Rear Axle Shaft, Bearing and Seal

REMOVAL & INSTALLATION

1. Before servicing the vehicle, refer to the precautions in the beginning of this section.
2. Drain the axle housing fluid.
3. Remove or disconnect the following:
   - Negative battery cable
   - Rear wheel
   - Brake drum
   - Wheel speed sensor, if equipped
   - Axle housing cover
   - Bearing retainer nuts
   - Axle shaft and bearing
   - Axle shaft inner oil seal
4. If equipped with ABS, grind a flat spot on the wheel speed sensor tone ring, then split the ring with a chisel.
5. Press the wheel bearing off the axle shaft.
6. Remove the bearing retainer and the outer oil seal.

To install:

7. Install or connect the following:
   - Outer oil seal to the bearing retainer
   - Bearing retainer to the axle shaft
   - Bearing and retainer ring pressed onto the axle shaft
   - Wheel speed sensor tone ring pressed onto the axle shaft, if equipped
   - Axle shaft inner oil seal
   - Axle shaft and bearing
   - Bearing retainer nuts. Tighten them to 17 ft. lbs. (23 Nm).
   - Wheel speed sensor, if equipped
   - Brake drum
   - Rear wheel
   - Negative battery cable
8. Fill the rear differential to the correct level.

Front Pinion Seal

REMOVAL & INSTALLATION

➡**This operation disturbs the differential pinion bearing preload. Carefully reset the preload during assembly.**

✳✳ CAUTION

The electrical power to the air suspension system must be shut off prior to hoisting, jacking or towing an air suspension vehicle. This can be accomplished by turning off the air suspension switch located in the rear jack storage area. Failure to do so can result in unexpected inflation or deflation of the air springs, which can result in shifting of the vehicle during these operations.

1. Before servicing the vehicle, refer to the precautions in the beginning of this section.
2. Index-mark the front driveshaft and pinion flange.
3. Remove or disconnect the following:
   - Front driveshaft from the pinion flange, and position it aside

➡**Do not allow the driveshaft to hang unsupported.**

4. Using a Nm (inch-pound) torque wrench, measure the torque required to maintain pinion rotation. Record the measurement.
5. Index-mark the pinion flange and the pinion stem.
6. Hold the pinion flange while removing the nut.
7. Place a drain pan under the differential housing.
8. Using a puller, remove the pinion flange.
9. Inspect the pinion flange for burrs and damage. Inspect the end of the pinion flange that contacts the bearing cone, the nut counterbore, and the seal surface for nicks. Discard the pinion flange as necessary.
10. Using a seal remover and impact slide hammer, remove the pinion seal.
11. Remove the front axle drive pinion shaft oil slinger and the differential pinion bearing.
12. Remove and discard the collapsible spacer.

To install:

13. Verify that the splines on the pinion stem are free of burrs. If burrs are evident, remove them with a fine crocus cloth. Work in a rotating motion to wipe the pinion clean.
14. Clean the pinion seal bore.
15. Install a new collapsible spacer.
16. Install the original differential pinion bearing and the front axle drive pinion shaft oil slinger.
17. Lubricate the pinion seal. Use Motorcraft SAE 80W90 Thermally Stable 4x4 Axle Lubricant meeting Ford specification WSP-M2C197-A.
18. Install the pinion seal.

19. Lubricate the pinion flange splines. Use Motorcraft SAE 80W90 Thermally Stable 4x4 Axle Lubricant meeting Ford specification WSP-M2C197-A.

➡**Never use a metal hammer on the pinion flange or install the flange with power tools. If necessary, use a plastic hammer to tap on a tight fitting flange.**

- Align the index marks and install the pinion flange.
- Install the new nut hand-tight.

➡**Do not loosen the nut to reduce preload. Install a new collapsible spacer and nut if preload reduction is necessary.**

20. Use the special tool to hold the pinion flange while tightening the nut to set the preload.
21. Tighten the nut, rotating the pinion occasionally to ensure the differential pinion bearings are seating correctly. Take frequent differential pinion bearing preload readings by rotating the pinion with a Nm (inch-pound) torque wrench. The final reading must be 0.56 Nm (5 inch lbs.) more than the initial reading taken during removal.
22. Align the index marks and position the front driveshaft.
23. Install the universal joint spider retainers and bolts.
24. Check the fluid level and, if necessary, fill the axle to specification. Use Motorcraft SAE 80W90 Thermally Stable 4x4 Axle Lubricant meeting Ford specification WSP-M2C197-A.
25. Lower the vehicle.
26. If so equipped, reactivate the air suspension.

Rear Pinion Seal

REMOVAL & INSTALLATION

1. Before servicing the vehicle, refer to the precautions in the beginning of this section.
2. Drain the axle housing fluid.
3. Remove or disconnect the following:
   - Negative battery cable
   - Rear wheels
   - Driveshaft
   - Brake calipers and pads or brake drum

➡**The brake calipers and pads or brake drum must be removed so that there is no additional drag when measuring pinion bearing preload.**

4. Use an inch lb. torque wrench and

measure and record the amount of torque required to maintain pinion rotation through several revolutions.

5. Remove or disconnect the following:
- Pinion flange
- Pinion seal
- Pinion bearing
- Collapsible spacer

To install:

➥**Use a new collapsible spacer and flange nut for assembly.**

6. Install or connect the following:
- Collapsible spacer
- Pinion bearing
- Pinion seal
- Pinion flange

7. Rotate the pinion flange occasionally while tightening the flange nut to make sure the pinion bearings seat correctly.

8. Take frequent bearing preload torque readings. Tighten the flange nut to achieve the preload torque readings originally recorded.

✻✻ CAUTION

Never loosen the pinion nut to reduce bearing preload. If it is nec-essary to reduce bearing preload, install a new collapsible spacer and pinion nut.

9. Install or connect the following:
- Driveshaft
- Brake calipers and pads or brake drum
- Wheels
- Negative battery cable

10. Fill the differential with gear lubricant and check for leaks.

STEERING AND SUSPENSION

Air Bag

PRECAUTIONS

- Always wear safety glasses when servicing an air bag vehicle, and when handling an air bag.
- Never attempt to service the steering wheel or steering column on an air bag-equipped vehicle without first properly disarming the air bag system. The air bag system should be properly disarmed whenever ANY service procedure in this manual indicates that you should do so.
- When carrying a live air bag module, always make sure the bag and trim cover are pointed away from your body. In the unlikely event of an accidental deployment, the bag will then deploy with minimal chance of injury.
- When placing a live air bag on a bench or other surface, always face the bag and trim cover up, away from the surface. This will reduce the motion of the air bag if it is accidentally deployed.
- If you should come in contact with a deployed air bag, be advised that the air bag surface may contain deposits of sodium hydroxide, which is a product of the gas combustion and is irritating to the skin. Always wear gloves and safety glasses when handling a deployed air bag, and wash your hands with mild soap and water afterwards.

DISARMING THE SYSTEM

1. Before servicing the vehicle, refer to the precautions in the beginning of this section.
2. Disconnect the negative battery cable from the battery.
3. Disconnect the positive battery cable from the battery.
4. Wait 1 minute. This time is required

for the back-up power supply in the air bag diagnostic monitor to completely drain. The system is now disarmed.

ARMING THE SYSTEM

1. Before servicing the vehicle, refer to the precautions in the beginning of this section.
2. Connect the positive battery cable.
3. Connect the negative battery cable.
4. Stand outside the vehicle and carefully turn the ignition to the **RUN** position. Be sure that no part of your body is in front of the air bag module on the steering wheel, to prevent injury in case of an accidental air bag deployment.
5. Ensure the air bag indicator light turns off after approximately 6 seconds. If the light does not illuminate at all, does not turn off, or starts to flash, test the system.

Power Rack and Pinion Steering Gear

REMOVAL & INSTALLATION

✻✻ WARNING

If equipped, always turn off the Automatic Ride Control (ARC) service switch before lifting the vehicle off of the ground. Failure to do so could damage the ARC system components.

1. Before servicing the vehicle, refer to the precautions in the beginning of this section.
2. Raise and safely support the front of the vehicle, block the rear wheels and apply the parking brake.
3. Start the engine then rotate the steering wheel from lock-to-lock and record the number of rotations.

4. Divide the number of rotations by 2. This gives the number of rotations to achieve true center of the steering. Turn the wheel in one direction to the full lock.
5. Turn the wheel in the opposite direction the number of turns equal to true steering (lock-to-lock number divided by 2).

✻✻ WARNING

Do not rotate the steering wheel when the shaft is disconnected from the steering gear as damage to the clock spring could occur.

6. Drain the power steering fluid reservoir.
7. Remove or disconnect the following:
- Negative battery cable
- Bolt retaining the lower steering column shaft to the steering gear input shaft
- Stabilizer bar
- Quick-connect fittings for the power steering pressure and return hoses at the steering gear housing
- Nuts securing the power steering cooler and remove the cooler
- Outer tie rod ends
- Nuts, bolts and washer assemblies retaining the steering gear housing to the front crossmember
- Steering gear from the vehicle

To install:
8. Install or connect the following:
- Position the steering gear to the front crossmember and install the nuts, bolts and washer assemblies. Torque to 94–127 ft. lbs. (128–172 Nm).
- Power steering cooler retaining bolts
- Power steering lines to the steering gear housing and torque the fittings to 20–25 ft. lbs. (27–34 Nm).

- Outer tie rod ends and ensure that the steering shaft or gear input shaft has not been rotated
- Intermediate shaft-to-steering input shaft retaining (pinch) bolt and torque the bolt to 30–42 ft. lbs. (41–56 Nm)
- Negative battery cable

9. Fill the power steering pump reservoir.

10. Bleed the air from the power steering system.

11. Ensure that there are no leaks and the fluid is maintained at the proper level.

12. Check the alignment.

Shock Absorber

REMOVAL & INSTALLATION

Front

→Low pressure gas shocks are charged with Nitrogen gas. Do not attempt to open, puncture or apply heat to them. Prior to installing a new shock absorber, hold it upright and extend it fully. Invert it and fully compress and extend it at least 3 times. This will bleed trapped air.

1. Before servicing the vehicle, refer to the precautions in the beginning of this section.

2. Remove or disconnect the following:
- Negative battery cable
- Upper shock-to-frame attaching nut, washer and insulator assembly
- Lower shock-to-control arm attaching nuts
- Slightly compress the shock absorber by hand and remove it from the vehicle

To install:

3. Install or connect the following:
- Position the lower washer and insulator on the shock absorber rod and position the shock absorber to the upper frame bracket mount
- Position the upper insulator and washer on the shock absorber rod and install the attaching nut loosely.
- Position the lower shock absorber mounting studs into the control arm and install the attaching nuts loosely.
- Torque the lower shock attaching nuts to 15–21 ft. lbs. (21–29 Nm), and the upper shock attaching bolts to 30–40 ft. lbs. (40–55 Nm).
- Negative battery cable

Rear

→Low pressure gas shocks are charged with Nitrogen gas. Do not attempt to open, puncture or apply heat to them. Prior to installing a new shock absorber, hold it upright and extend it fully. Invert it and fully compress and extend it at least 3 times. This will bleed trapped air.

1. Before servicing the vehicle, refer to the precautions in the beginning of this section.

2. Remove or disconnect the following:
- Upper shock-to-frame attaching nut
- Lower shock nut
- Slightly compress the shock absorber by hand and remove it from the vehicle

To install:

3. Install or connect the following:
- Shock absorber upper end and nut
- Shock absorber lower end and nut
- Torque the upper and lower shock attaching nuts to 53 ft. lbs. (72Nm)

Coil Spring

REMOVAL & INSTALLATION

1. Before servicing the vehicle, refer to the precautions in the beginning of this section.

2. Remove or disconnect the following:
- Wheel and tire assembly
- Shock absorber
- Front stabilizer bar link nut

3. Use a coil spring compressor to compress the coil spring.

4. Remove the cotter pin and castellated nut.

5. Separate the lower ball joint from the front wheel spindle.

6. Position the front wheel spindle out of the way and remove the coil spring.

To install:

→The end of the coil spring must cover the first hole and should not be visible in the second hole.

7. Install the coil spring in the lower arm.

✳✳ WARNING

Always install the cotter pin into the lower ball joint castellated nut from outboard to inboard. Failure to do so will result in damage to the wheel and tire assembly.

8. Install the lower ball joint.

9. Install the front stabilizer bar link nut.

10. Remove the Coil Spring Compressor.

11. Install the front shock absorber and the two lower nuts.

12. Install the upper shock absorber bushing and nut/washer assembly.

13. Install the wheel and tire assembly.

Leaf Springs

REMOVAL & INSTALLATION

1. Before servicing the vehicle, refer to the precautions in the beginning of this section.

2. Remove or disconnect the following:
- Negative battery cable
- Rear wheels
- U-bolts from the rear spring plate
- Hardware from the spring to bracket at the front of the rear spring
- Upper and lower shackle bolts at the rear of the spring
- Spring and shackle from the bracket

To install:

3. Install or connect the following:
- Spring and shackle to the bracket
- Upper and lower shackle bolts at the rear of the spring. Torque the nuts to 87 ft. lbs. (118 Nm).
- U-bolts to the spring plate. Torque the nuts 87 ft. lbs. (113 Nm).
- Rear wheels
- Negative battery cable

Torsion Bar

REMOVAL & INSTALLATION

✳✳ CAUTION

The electrical power to the air suspension system must be shut off prior to hoisting, jacking or towing an air suspension vehicle. This can be accomplished by turning off the air suspension switch located in the rear jack storage area. Failure to do so can result in unexpected inflation or deflation of the air springs or shocks, which can result in shifting of the vehicle during these operations.

1. Before servicing the vehicle, refer to the precautions in the beginning of this section.

2. Remove or disconnect the following:

3. Remove the torsion bar cover plate

➡**Before relieving the torsion bar tension, measure and record the measurement of the torsion bar adjustment bolt. This measurement will be used as the preset depth for the new torsion bar adjustment bolt during installation.**

4. Relieve the torsion bar tension.
   a. Position the Torsion Bar Tool and adapters.
   b. Tighten the Torsion Bar Tool until the torsion bar adjuster lifts off the adjustment bolt.

✳✳ CAUTION

The torsion bar adjustment bolt is coated with dry adhesive; and must be replaced if it is backed off or removed. Failure to do so can cause the adjustment bolt to loosen during operation and cause a loss of vehicle alignment.

   c. Remove the torsion bar adjustment bolt and nut.
   d. ·Loosen the Torsion Bar Tool until the tension is removed from the torsion bar.
5. Mark the torsion bar and the adjuster for proper installation.
6. Remove the torsion bar insulator.
7. Grasp the torsion bar, and pull it free from the front suspension lower arm.
To install:
8. Position the torsion bar and the torsion bar adjuster.
9. Align the marks on the torsion bar and the torsion bar adjuster, then install the torsion bar adjuster.
10. Position the torsion bar insulator.
11. Install the Torsion Bar Tool and the adapters.
12. Tighten the Torsion Bar Tool until the new adjustment bolt and nut can be installed.
13. Turn the adjustment bolt until the preliminary adjustment measurement (recorded length of the old adjustment bolt) is reached.
14. Install the torsion bar cover plate. Torque the bolts to 46 ft. lbs. (63Nm).
15. If equipped with air suspension, reactivate the system by turning on the air suspension switch.
16. Lower the vehicle.
17. Adjust the ride height.
18. Check the alignment.

Upper Ball Joint

REMOVAL & INSTALLATION

The ball joints are integral with the control arm. If the ball joint is defective, the entire control arm must be replaced.

Lower Ball Joint

REMOVAL & INSTALLATION

The ball joints are integral with the control arm. If the ball joint is defective, the entire control arm must be replaced.

Upper Control Arm

REMOVAL & INSTALLATION

1. Before servicing the vehicle, refer to the precautions in the beginning of this section.
2. Remove or disconnect the following:
   • Wheel and tire assembly
   • Brake disc shield
3. Use a jack to support the front suspension lower arm.
4. Mark the position of the front suspension upper arm adjustment cams.
5. Remove the upper ball joint retaining nut and pinch bolt.
6. Separate the ball joint from the front wheel spindle (3105).
7. Remove the front suspension upper arm.
8. Installation is the reverse of removal. Align the marks made during removal on the front suspension upper arm adjustment cam. The forward front suspension upper arm nut must be tightened first while the arm is held at the curb position ride height. Observe the following torques:
   • Control arm attaching nuts: 98 ft. lbs. (133Nm)
   • Pinch bolt: 46 ft. lbs. (63Nm)

UPPER CONTROL ARM BUSHING REPLACEMENT

The control arm bushings are not serviceable. If they require service, the upper or lower arm must be replaced.

Lower Control Arm

REMOVAL AND & INSTALLATION

1. Before servicing the vehicle, refer to the precautions in the beginning of this section.
2. Remove or disconnect the following:
   • Negative battery cable
   • Front wheel
   • Brake rotor shield
   • Shock absorber
   • Stabilizer bar link hardware
3. Using a spring compressor tool, compress the coil spring.

   • Lower ball joint from the spindle
   • Lower control arm bolts
   • Lower control arm and coil spring
To install:
4. Install or connect the following:
   • Coil spring to the lower control arm

➡**The end of the coil spring must cover the first hole and should not be visible in the second hole.**

   • Lower arm and front coil spring
   • The two front suspension lower arm bolts and nuts. Do not tighten the nuts.

➡**On the RH front suspension lower arm, install the rear bolt adjustment cam, and nut in the center of the frame slot.**

✳✳ CAUTION

Always install the cotter pin into the lower ball joint castellated nut from outboard to inboard, with the fingers bent together at a right angle. Failure to do so will result in damage to the wheel and tire assembly.

   • Lower ball joint. Torque the nut to 113 ft. lbs. (153Nm).
5. Remove the Coil Spring Compressor.
6. Install or connect the following:
   • Front stabilizer bar link nut. Torque the nut to 21 ft. lbs. (29Nm).
   • Shock absorber and the two lower nuts
   • Upper shock absorber bushing and nut/washer assembly
   • Brake disc shield
   • Wheel and tire assembly
7. Inspect and adjust the front end alignment.

LOWER CONTROL ARM BUSHING REPLACEMENT

The control arm bushings are not serviceable. If they require service, the upper or lower arm must be replaced.

Wheel Bearings

ADJUSTMENT

2-Wheel Drive Vehicles

1. Before servicing the vehicle, refer to the precautions in the beginning of this section.
2. Remove the grease cap from the hub and wipe the excess grease from the end of

the spindle. Remove the cotter pin and retainer. Discard the cotter pin.

3. Loosen the adjusting nut 3 turns.

✳✳ WARNING

Obtain running clearance between the disc brake rotor surface and shoe linings by rocking the entire wheel assembly in and out several times in order to push the caliper and brake pads away from the rotor. An alternate method to obtain proper running clearance is to tap lightly on the caliper housing. Be sure not to tap on any other area that may damage the disc brake rotor or the brake lining surfaces. Do not pry on the phenolic caliper piston. The running clearance must be maintained throughout the adjustment procedure. If proper clearance cannot be maintained, the caliper must be removed from its mounting.

4. While rotating the wheel assembly, tighten the adjusting nut to 17–25 ft. lbs. (23–34 Nm) in order to seat the bearings. Loosen the adjusting nut a half turn. Retighten the adjusting nut 18–20 inch lbs. (2.0–2.2 Nm).

5. Place the retainer on the adjusting

nut. The castellations on the retainer must be in alignment with the cotter pin holes in the spindle. Once this is accomplished install a new cotter pin and bend the ends to insure its being locked in place.

6. Check for proper wheel rotation. If correct, install the grease cap.

7. Lower the vehicle and tighten the lug nuts to 100 ft. lbs., (136 Nm) if the wheel was removed. Before driving the vehicle, pump the brake pedal several times to restore normal brake pedal travel.

✳✳ CAUTION

If the wheel was removed, retighten the wheel lug nuts to specification after about 500 miles (804km) of driving. Failure to do this could result in the wheel coming off while the vehicle is in motion causing loss of vehicle control or collision.

4-Wheel Drive

1. Before servicing the vehicle, refer to the precautions in the beginning of this section.

2. Remove or disconnect the following:
   - Wheel assembly
   - Retainer washers from the lug nut studs and remove the automatic

locking hub assembly from the spindle
   - Snapring and spacer from the end of the spindle shaft
   - Pull the locking cam assembly and the 2 plastic spacers off of the wheel bearing adjusting nut

3. Use a magnet and remove the locking key from under the adjusting nut. If required, rotate the adjusting nut slightly to relieve pressure against the locking key.

✳✳ WARNING

To prevent damage to the adjusting nut and spindle threads on vehicles equipped with automatic hubs, look into the spindle keyway under the adjusting nut and remove the separate locking key before removing the adjusting nut.

4. Loosen the wheel bearing locknut using a 2⅜ inch (60.3mm) hex socket, such as Hex Locknut Wrench T70T-4252-B.

5. Tighten the inner locknut to 35 ft. lbs. (47 Nm) to seat the bearings.

6. Spin the rotor and back off the inner locknut ¼ turn (90°). Retighten the locknut to 16 inch lbs. (1.8 Nm).

7. Align the closest lug in the bearing adjusting nut with the center of the spindle keyway slot. Advance the nut to the next if required.

   To install:

8. Separate locking key in the spindle keyway under the adjusting nut.

✳✳ CAUTION

Extreme care must be taken when aligning the adjusting nut with the center of the spindle keyway slot to prevent damage to the separate locking key. The wheel and tire assembly may come off while the vehicle is in motion if the key is damaged.

9. Install or connect the following:
   - 2 plastic thrust spacers and push or press the cam assembly onto the adjusting nut by lining up the keyway in the cam assembly with the separate locking key

✳✳ WARNING

Do not damage the locking key when installing the cam assembly.

   - Axle shaft spacer
   - Clip the snapring onto the end of the spindle

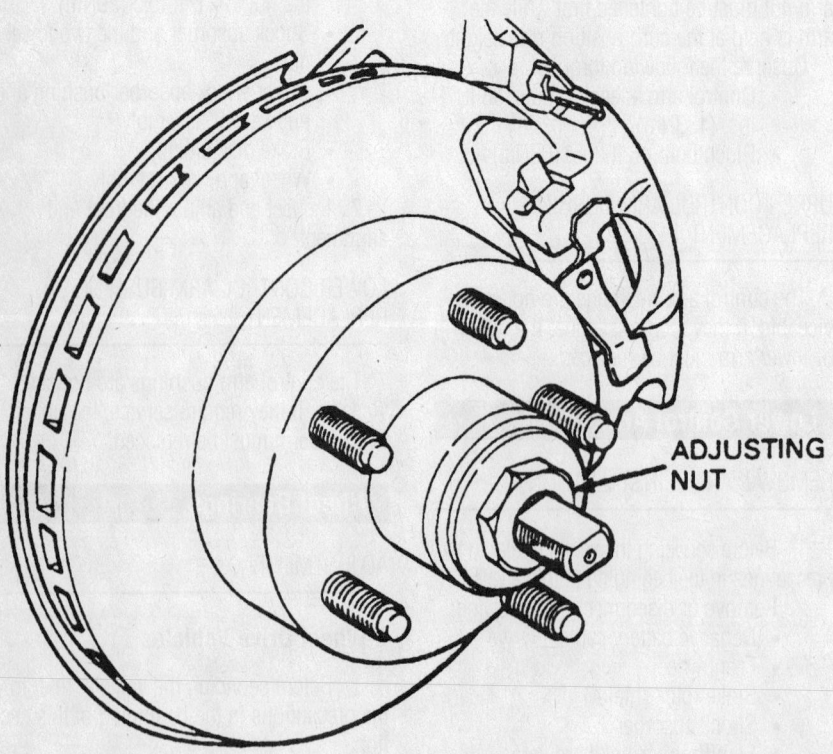

ADJUSTING NUT

7924EG34

Loosen the adjusting nut 3 turns, then rock the entire wheel assembly in-and-out to spread the brake pads before attempting to adjust the bearing—2wd vehicles

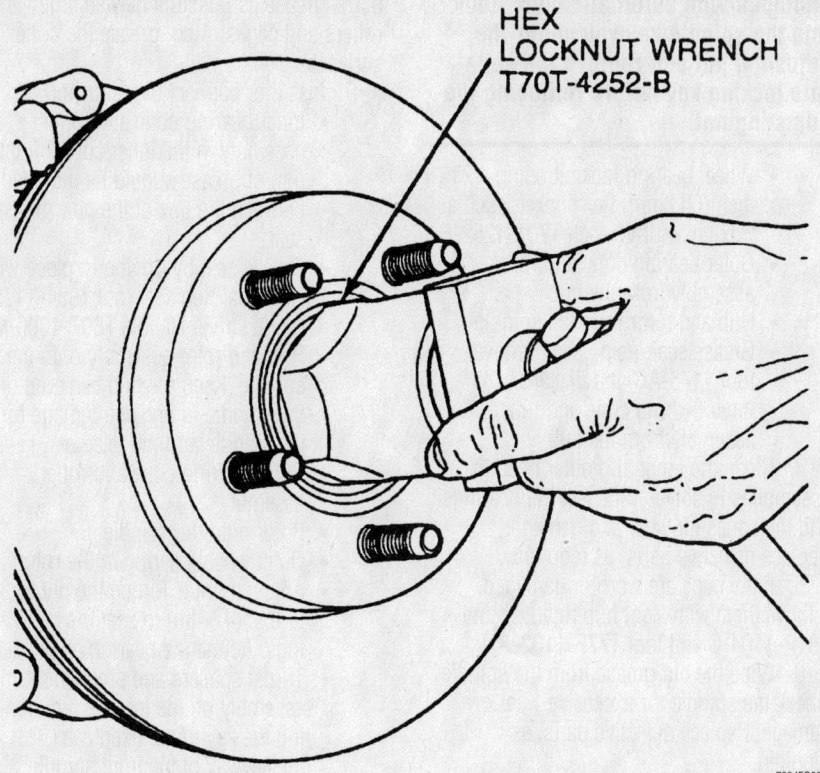

HEX
LOCKNUT WRENCH
T70T-4252-B

7924EG37

An oversize socket is needed to properly adjust the wheel bearing—automatic locking hub shown

- Manual hub assembly over the spindle. Install the retainer washers
- Wheel assembly

10. Check the end-play of the wheel and tire assembly on the spindle. End-play should be 0.001–0.003 in. (0.025–0.076mm) and the maximum torque to rotate the hub should be 25 inch lbs. (2.8 Nm).

REMOVAL & INSTALLATION

2-Wheel Drive

1. Before servicing the vehicle, refer to the precautions in the beginning of this section.

2. Remove or disconnect the following:
- Disc brake caliper anchor plate
- Hub grease cap
- Cotter pin
- Nut retainer
- Spindle nut
- Wheel outer bearing retainer washer
- Outer front wheel bearing
- Brake disc and hub
- Hub grease seal
- Inner wheel bearing

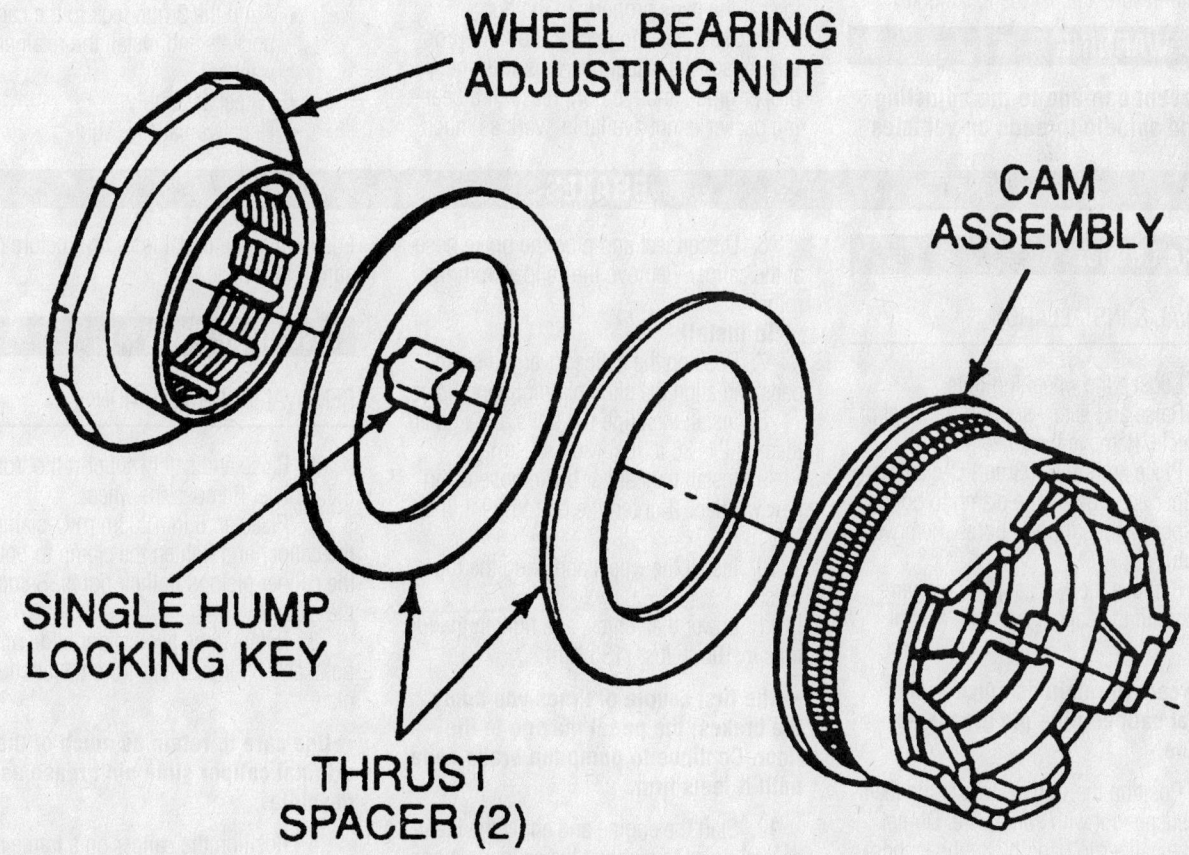

WHEEL BEARING
ADJUSTING NUT

CAM
ASSEMBLY

SINGLE HUMP
LOCKING KEY

THRUST
SPACER (2)

7924EG36

Exploded view of the wheel bearing adjusting nut and related components—automatic locking hub shown

To install:

3. Thoroughly clean and inspect the front wheel bearings and the brake disc and hub.

4. Lubricate the front wheel bearings.

5. Install the inner front wheel bearing.

6. Install a new wheel hub grease seal.

7. Position the brake disc and hub.

8. Assemble all parts and adjust the bearings.

4-Wheel Drive

1. Before servicing the vehicle, refer to the precautions in the beginning of this section.

2. Remove or disconnect the following:
- Negative battery cable
- Wheel assembly
- Retainer washers from the lug nut studs and remove the automatic locking hub assembly from the spindle
- Snapring and spacer from the end of the spindle shaft
- Pull the locking cam assembly and the 2 plastic spacers off of the wheel bearing adjusting nut

3. Use a magnet and remove the locking key from under the adjusting nut. If required, rotate the adjusting nut slightly to relieve pressure against the locking key

✳✳ WARNING

To prevent damage to the adjusting nut and spindle threads on vehicles

equipped with automatic hubs, look into the spindle keyway under the adjusting nut and remove the separate locking key before removing the adjusting nut.

- Wheel bearing locknut using a 2⅜ inch (60.3mm) hex socket, such as Hex Locknut Wrench T70T-4252-B
- Outer bearing cone and roller assembly from the hub
- Hub and rotor from the spindle
- Grease seal, using seal removal tool 1175-AC and discard
- Inner bearing cone and roller assembly from the hub

4. Clean the inner and outer bearing assemblies in solvent. Inspect the bearings and the cones for wear and damage. Replace defective parts, as required.

5. If the cups are worn or damaged, remove them with front hub remover tool T81P-1104-C and tool T77F-1102-A.

6. Wipe the old grease from the spindle. Check the spindle for excessive wear or damage. Replace defective parts, as required.

To install:

7. If the inner and outer cups were removed, use bearing driver handle tool T80-4000-W and replace the cups. Be sure to seat the cups properly in the hub.

8. Use a bearing packer tool and properly repack the wheel bearings with the proper grade and type of grease. If a bearing packer is not available, work as much

of the grease as possible between the rollers and cages. Also, grease the cone surfaces.

9. Install or connect the following:
- Inner bearing cone and roller assembly in the inner cup. A light film of grease should be included between the lips of the new grease seal.
- Grease seal by driving in place with Hub Seal Replacer tool T83T-1175-B and Driver Handle T80T-4000-W
- Hub and rotor assembly onto the spindle. Keep the hub centered on the spindle to prevent damage to the spindle and the retainer
- Outer bearing cone and roller assembly
- Rotor onto the spindle
- Outer wheel bearing in the rotor
- Adjusting nut. Torque the nut to 35 ft. lbs. (47 Nm) to seat the bearings. Adjust the bearing as needed.
- Thrust spacers and press the cam assembly on the locknut by aligning the key in the fixed cam with the keyway of the front spindle
- Axle shaft spacer
- Snapring on the end of the shaft
- Locking hub assembly over the front spindle
- Align the 3 hub legs to the cam pockets and install the retainer washers
- Wheel assembly
- Negative battery cable

BRAKES

Brake Caliper

REMOVAL & INSTALLATION

1. Loosen the wheel lug nuts.

2. Raise and safely support the front of the vehicle. Remove the wheel.

3. Place an 8 in. (203mm) C-clamp on the caliper and tighten the clamp to bottom the caliper pistons in their bores. Remove the clamp.

4. Remove the two caliper slide pin bolts and lift the caliper from the anchor plate.

➡**Use care to retain as much of the original caliper slide pin grease as possible.**

5. Position the caliper on a frame member or suspend it with some wire. Do not allow the caliper to hang by the brake hose.

6. Disconnect and plug the brake hose at the caliper. Remove the caliper from the rotor.

To install:

7. Position the caliper over the brake pads and align the slide pin mounting holes.

8. Install the slide pin bolts and tighten them to 21–26 ft. lbs. (30–36 Nm).

9. Install the caliper brake hose using new washers. Tighten the bolt to 29 ft. lbs. (40

10. Install the wheel and snug the lug nuts.

11. Lower the vehicle and tighten the lug nuts to 100 ft. lbs. (135 Nm).

➡**The first couple of times you apply the brakes, the pedal may go to the floor. Continue to pump the brake pedal until it feels firm.**

12. Start the engine and apply the brakes several times to readjust the caliper pistons.

Ensure that the pedal feels firm before operating the vehicle.

Disc Brake Pads

REMOVAL & INSTALLATION

1. Raise and safely support the front of the vehicle. Remove the wheel.

2. Place an 8 in. (203mm) C-clamp on the caliper and tighten the clamp to bottom the caliper pistons in their bores. Remove the clamp.

3. Remove the two caliper slide pin bolts and lift the caliper from the anchor plate.

➡**Use care to retain as much of the original caliper slide pin grease as possible.**

4. Position the caliper on a frame mem-

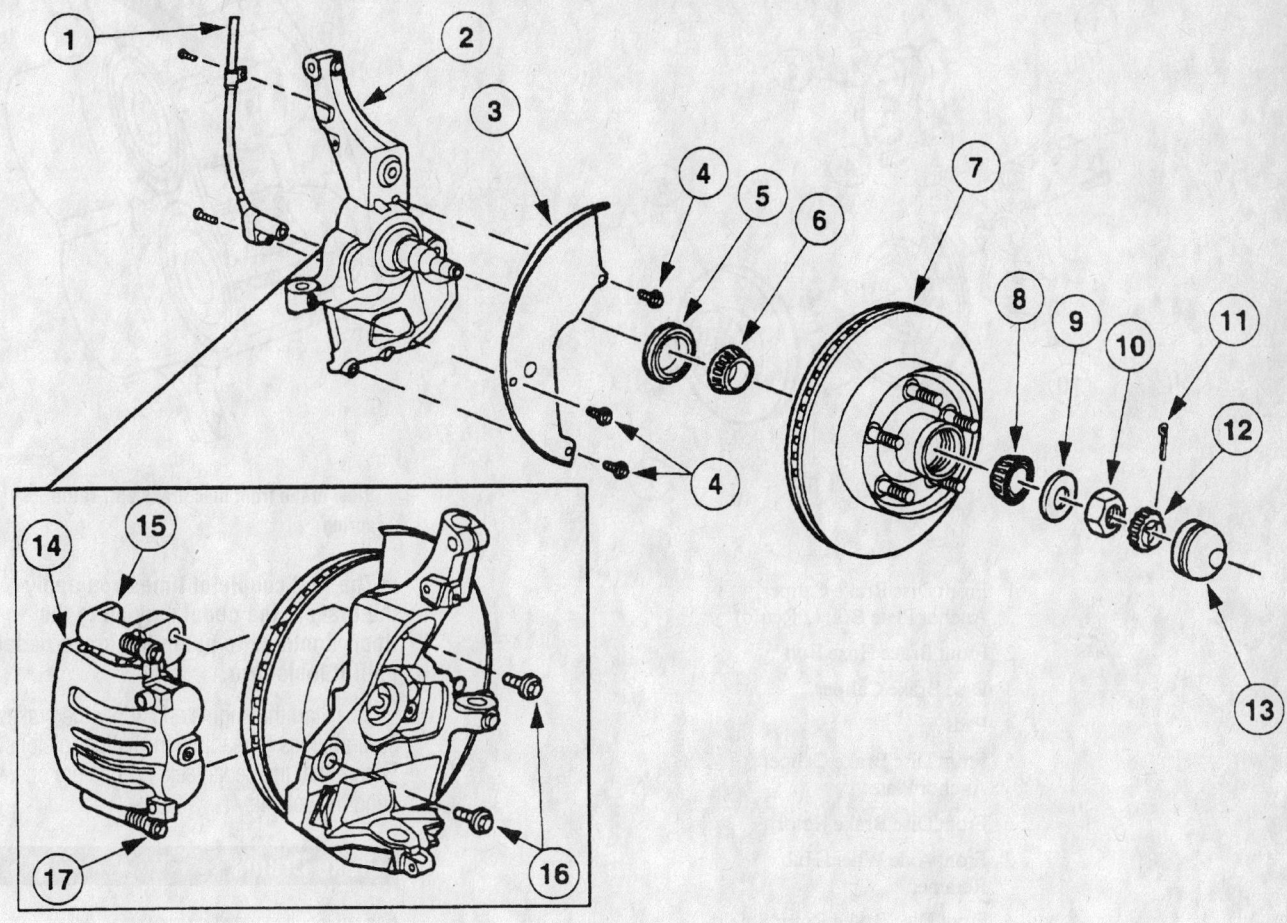

1 Front Brake Anti-Lock Sensor
2 Front Wheel Spindle
3 Front Disc Brake Rotor Shield
4 Rotor Shield Bolt
5 Grease Seal
6 Front Wheel Bearing

7 Front Disc Brake Hub and Rotor
8 Front Wheel Bearing
9 Front Wheel Outer Bearing Retainer Washer
10 Hub Spindle Nut
11 Cotter Pin

12 Nut Retainer
13 Hub Grease Cap
14 Disc Brake Caliper
15 Front Disc Brake Caliper Anchor Plate
16 Caliper Anchor Plate Bolts
17 Disc Brake Caliper Bolt

93026G22

Exploded view of the 2WD front disc brake assembly

ber or suspend it with some wire. Do not allow the caliper to hang by the brake hose.

5. Remove the brake pads and, if necessary, the anti-rattle clips from the anchor plate.

6. Remove the shims, if any, from the brake pads for re-use.

To install:

7. If removed, install the anti-rattle clips.

8. Install the brake pads to the anchor plate.

9. Position the caliper over the brake pads and align the slide pin mounting holes.

10. Install the slide pin bolts and tighten them to 21–26 ft. lbs. (30–36 Nm).

11. Install the wheel and snug the lug nuts.

12. Lower the vehicle and tighten the lug nuts to 100 ft. lbs. (135 Nm).

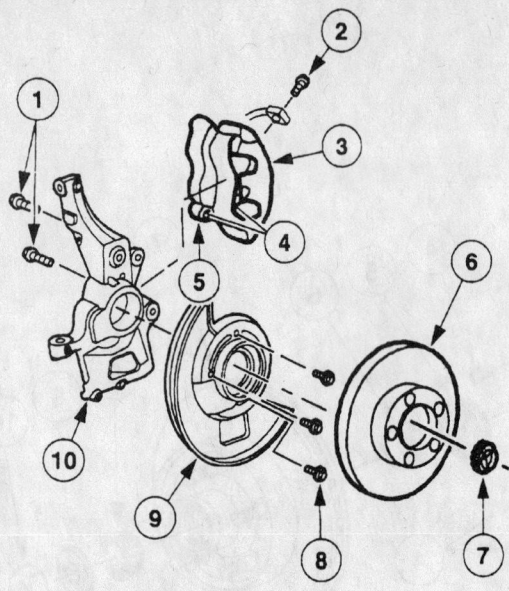

1   Front Disc Brake Caliper
    Anchor Plate Bolt (2 Req'd)
2   Front Brake Hose Bolt
3   Disc Brake Caliper
4   Pads
5   Front Disc Brake Caliper
    Anchor Plate
6   Front Disc Brake Rotor
7   Front Axle Wheel Hub
    Retainer
8   Front Disc Brake Rotor Shield
    Bolt (3 Req'd)
9   Front Disc Brake Rotor Shield
10 Front Wheel Knuckle

93026G23

Exploded view of the 4WD front disc brake assembly

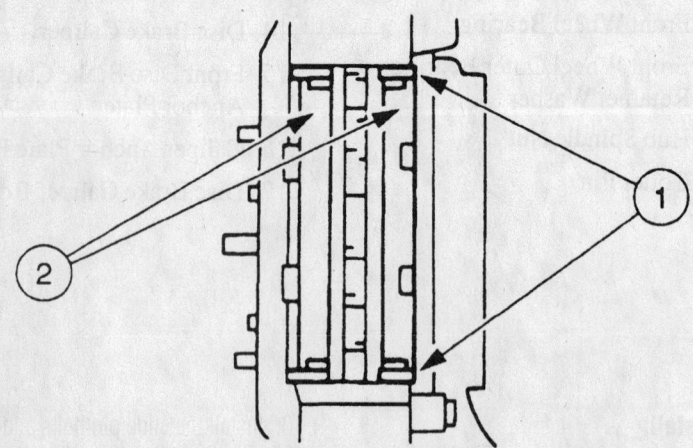

1   stainless slippers

2   pads

93026G24

Position of the front disc brake components

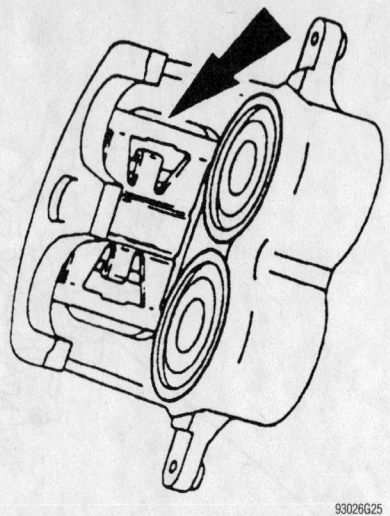

93026G25

View of the front disc brake anti-rattle spring

➡ **The first couple of times you apply the brakes, the pedal may go to the floor. Continue to pump the brake pedal until it feels firm.**

13. Start the engine and apply the brakes several times to readjust the caliper pistons. Ensure that the pedal feels firm before operating the vehicle.

Brake Drums

REMOVAL & INSTALLATION

1. Raise and safely support the vehicle. Remove the wheel and tire assembly.
2. Remove the retaining nuts, if equipped, and remove the brake drum.
3. Inspect the brake drum surface for wear, scoring and runout. Machine or replace, as necessary.
   To install:
4. Install the brake drum and secure in place with the retainer nuts, if equipped.
5. Adjust the rear brakes.
6. Install the wheel. Lower the vehicle.

Brake Shoes

REMOVAL & INSTALLATION

1. Raise and safely support the vehicle. Remove the wheel and tire assembly and the brake drum.
2. Pull backward on the adjusting lever cable to disengage the adjusting lever from the adjusting screw. Move the outboard side of the adjusting screw upward and back off the pivot nut as far as it will go.

3. Pull the adjusting lever, cable and automatic adjuster spring down and toward the rear to unhook the pivot hook from the large hole in the secondary shoe web. Do not pry the pivot hook from the hole.

4. Remove the automatic adjuster spring and adjusting lever.

5. Remove the secondary shoe-to-anchor spring using a suitable brake spring removal/installation tool. Using the tool,

remove the primary shoe-to-anchor spring and unhook the cable anchor. Remove the anchor pin plate, if equipped.

6. Remove the cable guide from the secondary shoe.

7. Remove the shoe hold-down springs, shoes, adjusting screw, pivot nut and socket. Note the color and position of each hold-down spring so they can be reassembled in the same position.

8. Remove the parking brake link and

spring. Disconnect the parking brake cable from the parking brake lever.

9. Remove the secondary brake shoe. On 9 in. (22.8cm) rear brakes, remove the parking brake lever from the shoe. On 10 in. (25.4cm) rear brakes, remove the retainer clip and spring washer and remove the parking brake lever.

To install:

10. Clean the backing plate ledge pads and sand lightly. Apply a light coating of

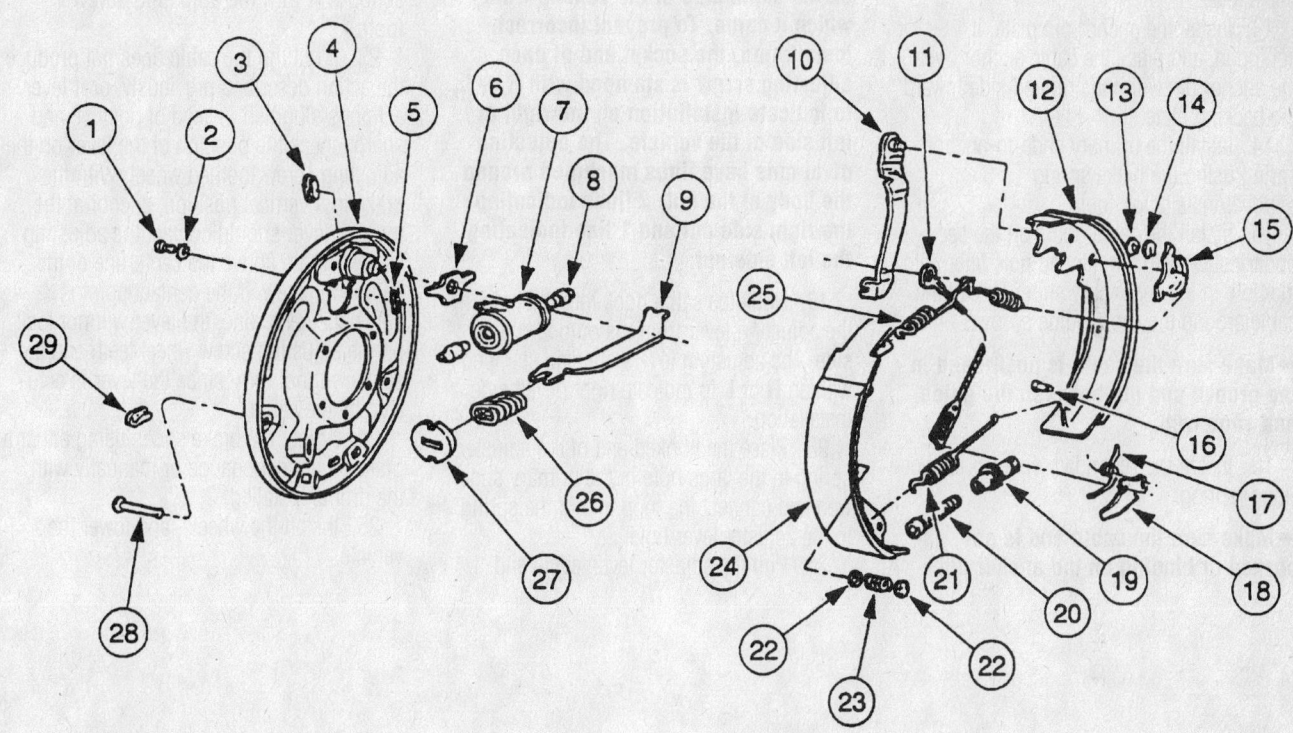

1 Wheel Cylinder-to-Backing Plate Bolt (2 Req'd)	12 Rear Brake Shoe and Lining, Secondary	22 Brake Shoe Hold-Down Spring Cup
2 Washer	13 Washer	23 Brake Shoe Hold-Down Spring
3 Inspection Hole Cover	14 Parking Brake Lever Pin Retainer	24 Rear Brake Shoe and Lining, Primary
4 Brake Backing Plate	15 Cable Guide	25 Brake Shoe Retracting Spring, Short
5 Lining Inspection Hole	16 Adjusting Lever Pin	26 Parking Brake Link Spring
6 Anchor Pin Guide Plate	17 Adjusting Lever Return Spring	27 Parking Brake Spring Retainer
7 Rear Wheel Cylinder	18 Brake Shoe Adjusting Lever	28 Brake Shoe Hold-Down Spring Pin
8 Wheel Cylinder Brake Shoe Link	19 Brake Shoe Adjusting Screw Nut	29 Brake Adjusting Hole Cover
9 Parking Brake Strut	20 Brake Adjuster Screw	
10 Parking Brake Lever	21 Brake Shoe Adjusting Screw Spring	
11 Brake Shoe Adjusting Lever Cable		

93026G21

Exploded view of the rear brake shoes and components

high temperature lithium grease to the points where the brake shoes touch the backing plate. Lubricate the adjusting cable eye and the anchor pin area.

11. Install the parking brake lever on the secondary shoe. On 10 in. (25.4cm) brakes, secure with the spring washer and retaining clip.

12. Position the brake shoes on the backing plate and install the hold-down spring pins, springs and cups. Install the parking brake link, spring and washer. Connect the parking brake cable to the parking brake lever.

13. Install the anchor pin plate, if equipped, and place the cable anchor over the anchor pin with the crimped side toward the backing plate.

14. Install the primary shoe-to-anchor spring using the brake spring removal/installation tool.

15. Install the cable guide on the secondary shoe with the flanged hole fitted into the hole in the secondary shoe. Thread the cable around the cable guide groove.

➡**Make sure the cable is positioned in the groove and not between the guide and shoe web.**

16. Install the secondary shoe-to-anchor (long) spring.

➡**Make sure the cable end is not cocked or binding on the anchor pin when installed. All parts should be flat on the anchor pin.**

17. Apply high temperature lithium grease to the threads and the socket end of the adjusting screw. Turn the adjusting screw into the adjusting pivot nut to the end of the threads and then loosen, ½ turn.

18. Place the adjusting socket on the screw and install the assembly between the shoe ends with the adjusting screw nearest the secondary shoe.

➡**Be sure to install the adjusting screw on the same side of the vehicle from which it came. To prevent incorrect installation, the socket end of each adjusting screw is stamped with R or L, to indicate installation on the right or left side of the vehicle. The adjusting pivot nuts have lines machined around the body of the nut, 2 lines indicating the right side nut and 1 line indicating the left side nut.**

19. Hook the cable hook into the hole in the adjusting lever from the outboard plate side. The adjusting levers are also stamped with an **R** or **L** to indicate right or left side installation.

20. Place the hooked end of the adjuster spring in the large hole in the primary shoe web and connect the loop end of the spring to the adjuster lever hole.

21. Pull the adjuster lever, cable and automatic adjuster spring down toward the rear to engage the pivot hook in the large hole in the secondary shoe web.

22. After installation, check the action of the adjuster by pulling the section of the cable between the cable guide and the adjusting lever toward the secondary shoe web far enough to lift the lever past a tooth on the adjusting screw wheel. The lever should snap into position behind the next tooth and releasing the cable should cause the adjuster spring to return the lever to its original position. This return action will turn the adjusting screw 1 tooth.

23. If pulling the cable does not produce the action described previously, or if lever action is sluggish instead of positive and sharp, check the position of the lever on the adjusting screw toothed wheel. With the brake in a vertical position, anchor at the top, the lever should contact the adjusting wheel 1 tooth above the centerline of the adjusting screw. If the contact point is below the centerline, the lever will not lock on the adjusting screw wheel teeth and the screw will not turn, since the lever is actuated by the cable.

24. Adjust the brake shoes using either a brake adjustment gauge or manually with the drums installed.

25. Install the wheels, and lower the vehicle.

MAZDA

19

MPV

SPECIFICATION CHARTS

ENGINE AND VEHICLE IDENTIFICATION

Code ①	Liters (cc)	Cu. In.	Cyl.	Fuel Sys.	Engine Type	Eng. Mfg.
GY	2.5 (2507)	153	6	SFI	DOHC	Mazda
AJ	3.0 (NA)	NA	6	SFI	DOHC	Mazda

Code ②	Year
Y	2000
1	2001
2	2002
3	2003
4	2004

MFI: Multi-port Fuel Injection

NA: Not available

SFI: Sequential Fuel Injection

SOHC: Single Overhead Camshaft

DOHC: Double Overhead Camshaft

OHV: Overhead Valve

① 8th digit of the Vehicle Identification Number (VIN)

② 10th digit of the Vehicle Identification Number (VIN)

42356-MMPV-C01

GENERAL ENGINE SPECIFICATIONS

Year	Model	Engine Displacement Liters (cc)	Engine ID	Fuel System Type	Net Horsepower @ rpm	Net Torque @ rpm (ft. lbs.)	Bore x Stroke (in.)	Compression Ratio	Oil Pressure @ rpm
2000	MPV	2.5 (2507)	GY	SFI	170@6250 ①	165@4250	3.25x3.13	9.7:1	20-45@1500
2001	MPV	2.5 (2507)	GY	SFI	170@6250 ①	165@4250	3.25x3.13	9.7:1	20-45@1500
2002	MPV	3.0 (NA)	AJ	SFI	200@6200	200@3000	NA	NA	20-45@1500

SFI: Sequential Fuel Injection

NA: Not available

① California LEV: 160@6250

42356-MMPV-C02

ENGINE TUNE-UP SPECIFICATIONS

Year	Engine Displacement Liters (cc)	Engine ID	Spark Plug Gap (in.)	Ignition Timing (deg.) MT	Ignition Timing (deg.) AT	Fuel Pump (psi)	Idle Speed (rpm) MT	Idle Speed (rpm) AT	Valve Clearance Intake	Valve Clearance Exhaust
2000	2.5 (2507)	GY	0.054	—	10B	37-41	—	①	HYD	HYD
2001	2.5 (2507)	GY	0.054	—	10B	37-41	—	①	HYD	HYD
2002	3.0 (NA)	AJ	NA	—	10B	61-66	—	650-750	HYD	HYD

NOTE: The Vehicle Emission Control Information label often reflects specification changes made during production. The label figures must be used if they differ from those in this chart

B: Before top dead center

HYD: Hydraulic

NA: Not available

① Refer to Vehicle's Emission Control Information label

42356-MMPV-C03

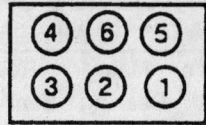

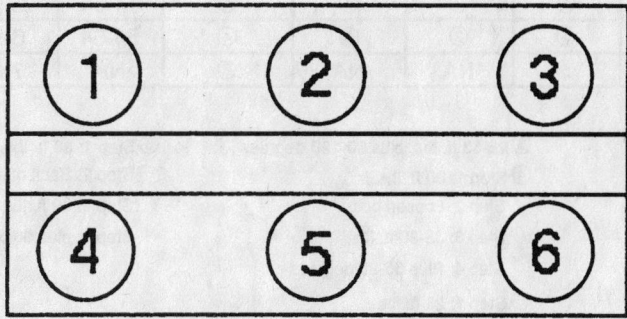

FRONT OF VEHICLE

↓

79223G07

2.5L Engine
Firing order 1–4–2–5–3–6
Distributorless ignition system

CAPACITIES

Year	Model	Engine Displacement Liters (cc)	Engine ID	Engine Oil with Filter (qts.)	Transmission (pts.)		Transfer Case (pts.)	Drive Axle		Fuel Tank (gal.)	Cooling System (qts.)
					Manual	Auto.		Front (pts.)	Rear (pts.)		
2000	MPV	2.5 (2507)	GY	6.0	—	20.6 ①	—	—	—	18.5	②
2001	MPV	2.5 (2507)	GY	6.0	—	20.6 ①	—	—	—	18.5	②
2002	MPV	3.0 (NA)	AJ	5.5	—	17.0	—	—	—	19.8	NA

NOTE: All capacities are approximate. Add fluid gradually and ensure a proper fluid level is obtained.

NA: Not available

① Transaxle

② With rear heater: 12.7
   Without rear heater: 10.8

42356-MMPV-C04

VALVE SPECIFICATIONS

Year	Engine Displacement Liters (cc)	Engine ID	Seat Angle (deg.)	Face Angle (deg.)	Spring Test Pressure (lbs. @ in.)	Spring Installed Height (in.)	Stem-to-Guide Clearance (in.)		Stem Diameter (in.)	
							Intake	Exhaust	Intake	Exhaust
2000	2.5 (2507)	GY	44.75	45.5	153@1.18	1.570	0.0007-0.0027	0.0017-0.0037	0.2350-0.2358	0.2343-0.2350
2001	2.5 (2507)	GY	44.75	45.5	153@1.18	1.570	0.0007-0.0027	0.0017-0.0037	0.2350-0.2358	0.2343-0.2350
2002	3.0 (NA)	AJ	NA	NA	NA	NA	NA	NA	NA	NA

NA: Not Available

① Intake:
   Inner: 21-22@1.77
   Outer: 31-33@1.73

Exhaust:
   Inner: 33-37@1.59
   Outer: 21-22@1.56

② Intake:
   Inner: 1.840
   Outer: 2.004

Exhaust:
   Inner: 2.092
   Outer: 2.296

42356-MMPV-C05

TORQUE SPECIFICATIONS
All readings in ft. lbs.

Year	Engine Displacement Liters (cc)	Engine ID	Cylinder Head Bolts	Main Bearing Bolts	Rod Bearing Bolts	Crankshaft Damper Bolts	Flywheel Bolts	Manifold Intake	Manifold Exhaust	Spark Plugs	Lug Nut
2000	2.5 (2507)	GY	①	②	③	④	54-64	6-9	13-16	14	66-86
2001	2.5 (2507)	GY	①	②	③	④	54-64	6-9	13-16	14	66-86
2002	3.0 (NA)	AJ	⑤	NA	NA	⑥	NA	7.5	18	8-14	108

NA: Not available

① Step 1: 28-31 ft. lbs.
Step 2: Plus 90 degrees
Step 3: Loosen bolts 1 turn
Step 4: 28-31 ft. lbs.
Step 5: Plus 90 degrees
Step 6: Plus 90 degrees

② Step 1: 12-43 inch lbs.
Step 2: Outer cap bolts: 16-21 ft. lbs.
Step 3: Inner cap bolts: 27-32 ft. lbs.
Step 4: All cap bolts: 85-95 degrees
Step 5: Remaining bolts: 15-22 ft. lbs.

③ 26-33 ft. lbs. plus 90-120 degrees
④ Step 1: 89 ft. lbs.
Step 2: Loosen bolt
Step 3: 35-39 ft. lbs.
Step 4: Plus 85-95 degrees
⑤ Step 1: 28 ft. lbs.
Step 2: plus 90 degrees
Step 3: back off one full turn
Step 4: 28 ft. lbs.
Step 5: plus 90 degrees
Step 6: plus an additional 90 degrees

⑥ Step 1: 88 ft. lbs.
Step 2: Back off one full turn
Step 3: 39 ft. lbs.
Step 4: plus 90 degrees

42356-MMPV-C06

PISTON AND RING SPECIFICATIONS
All measurements are given in inches.

Year	Engine Displacement Liters (cc)	Engine ID	Piston Clearance	Ring Gap Top Compression	Ring Gap Bottom Compression	Ring Gap Oil Control	Ring Side Clearance Top Compression	Ring Side Clearance Bottom Compression	Ring Side Clearance Oil Control
2000	2.5 (2507)	GY	0.0005-0.0009	0.004-0.010	0.011-0.017	0.006-0.026	0.0015-0.0029	0.0015-0.0033	SNUG
2001	2.5 (2507)	GY	0.0005-0.0009	0.004-0.010	0.011-0.017	0.006-0.026	0.0015-0.0029	0.0015-0.0033	SNUG
2002	3.0 (NA)	AJ	NA	NA	NA	NA	NA	NA	NA

NA: Not available

42356-MMPV-C07

CRANKSHAFT AND CONNECTING ROD SPECIFICATIONS
All measurements are given in inches.

Year	Engine Displacement Liters (cc)	Engine ID	Crankshaft Main Brg. Journal Dia.	Crankshaft Main Brg. Oil Clearance	Crankshaft Shaft End-play	Thrust on No.	Connecting Rod Journal Diameter	Connecting Rod Oil Clearance	Connecting Rod Side Clearance
2000	2.5 (2507)	GY	2.4670-2.4790	0.0009-0.0019	0.0040-0.0090	4	1.9670-1.9680	0.0010-0.0025	0.0039-0.0118
2001	2.5 (2507)	GY	2.4670-2.4790	0.0009-0.0019	0.0040-0.0090	4	1.9670-1.9680	0.0010-0.0025	0.0039-0.0118
2002	3.0 (NA)	AJ	NA	NA	NA	NA	NA	NA	NA

NA: Not available

42356-MMPV-C08

BRAKE SPECIFICATIONS
All measurements in inches unless noted

| Year | Model | Brake Disc | | | Brake Drum Diameter | | | Minimum Lining Thickness | | Brake Caliper | |
		Original Thickness	Minimum Thickness	Maximum Runout	Original Inside Diameter	Max. Wear Limit	Maximum Machine Diameter	Front	Rear	Bracket Bolts (ft. lbs.)	Mounting Bolts (ft. lbs.)
2000	MPV	1.100	1.030	0.002	10.000	①	10.050	0.080	0.04	66-79	62-68
2001	MPV	1.100	1.030	0.002	10.000	①	10.050	0.080	0.04	66-79	62-68
2002	MPV	NA	1.030	0.002	NA	NA	NA	0.080	NA	65-79	62-68

NA: Not Available

① Refer to the maximum diameter stamped on drum

42356-MMPV-C09

WHEEL ALIGNMENT

| Year | Model | | Caster ① | | Camber ① | | Toe-in (in.) | Kingpin Angle (Deg.) |
			Range (+/-Deg.)	Preferred Setting (Deg.)	Range (+/-Deg.)	Preferred Setting (Deg.)		
2000	MPV	F	1.0	+1.70	1.0	-0.90	0.08+/-0.16	11.09
		R	—	—	1.0	-1.00	0.12+/-0.16	—
2001	MPV	F	1.0	+1.70	1.0	-0.90	0.08+/-0.16	11.09
		R	—	—	1.0	-1.00	0.12+/-0.16	—
2002	MPV	F	1.0	+2.03	1.0	-0.23	0.08+/-0.16	11.18
		R	—	—	1.0	-1.00	0.12+/-0.16	—

① Empty vehicle

42356-MMPV-C10

TIRE, WHEEL AND BALL JOINT SPECIFICATIONS

| Year | Model | OEM Tires | | Tire Pressures (psi) | | Wheel Size | Ball Joint Inspection |
		Standard	Optional	Front	Rear		
2000	MPV	P215/70R15	None	32	32	6-JJ	18-30 in. ①
2001	MPV	P215/70R15	None	32	32	6-JJ	18-30 in. ①
2002	MPV	205/65R15	215/60R16 P215/60R17	②	②	std: 6-JJ opt: 6.5-JJ, 7-JJ	NA

NA: not available

OEM: Original Equipment Manufacturer

PSI: Pounds Per Square Inch

① Torque required in inch lbs. to rotate ball joint when removed from the knuckle

② See placard on vehicle

42356-MMPV-C11

SCHEDULED MAINTENANCE INTERVALS
MAZDA—2000-01 MPV

TO BE SERVICED	TYPE OF SERVICE	VEHICLE MILEAGE INTERVAL (x1000)												
		7.5	15	22.5	30	37.5	45	52.5	60	67.5	75	82.5	90	97.5
Engine oil & filter	R	✓	✓	✓	✓	✓	✓	✓	✓	✓	✓	✓	✓	✓
Air cleaner filter	R				✓				✓				✓	
Brake fluid	R				✓				✓				✓	
Spark plugs	R				✓				✓				✓	
Bolts & nuts on chassis & body	S/I				✓				✓				✓	
Cooling system	S/I				✓				✓				✓	
Disc brakes, brake lines, hoses & connections	S/I				✓				✓				✓	
Drive belt(s)	S/I				✓				✓				✓	
Driveshaft dust boots (4WD)	S/I				✓				✓				✓	
Exhaust system heat shields	S/I				✓				✓				✓	
Front suspension ball joints	S/I				✓				✓				✓	
Fuel lines & hoses	S/I				✓				✓				✓	
Idle speed	S/I		✓				✓				✓			
Steering operation & linkages	S/I				✓				✓				✓	
Engine coolant	R						✓				✓			
Timing belt (except Calif.)	R								✓					
Timing belt (Calif.)①	S/I								✓				✓	
Automatic transmission fluid & filter	R								✓					
Fuel filter & PCV valve	R								✓					
Emission hoses & tubes②	S/I								✓					
Ignition timing	S/I								✓					

R: Replace   S/I: Service or Inspect

① Timing belt (Calif.): replace at 105,000 miles, unless previously replaced.

② Emission hoses & tubes: replace at 80,000 miles.

FREQUENT OPERATION MAINTENANCE (SEVERE SERVICE)

If a vehicle is operated under any of the following conditions it is considered severe service:

- Extremely dusty areas.
- 50% or more of the vehicle operation is in 32°C (90°F) or higher temperatures, or constant operation in temperatures below 0°C (32°F).
- Prolonged idling (vehicle operation in stop and go traffic).
- Frequent short running periods (engine does not warm to normal operating temperatures).
- Police, taxi, delivery usage or trailer towing usage.

Air cleaner filter: service or inspect every 15,000 miles

Engine oil & filter: replace every 5000 miles.

Ball joints & dust covers: service or inspect every 7500 miles.

Bolts & nuts on chassis & body: tighten every 15,000 miles.

Spark plugs: replace every 15,000 miles.

Automatic transmission fluid & filter: replace every 30,000 miles.

Front & rear axle oil: replace every 30,000 miles.

Transfer case oil (4WD): replace every 30,000 miles.

42356-MMPV-C12

SCHEDULED MAINTENANCE INTERVALS
MAZDA—2002 MPV

TO BE SERVICED	TYPE OF SERVICE	VEHICLE MILEAGE INTERVAL (x1000)												
		7.5	15	22.5	30	37.5	45	52.5	60	67.5	75	82.5	90	97.5
Engine oil & filter	R	✓	✓	✓	✓	✓	✓	✓	✓	✓	✓	✓	✓	✓
Drive belt(s)	S/I				✓				✓				✓	
PCV valve	I								✓					
Spark plugs(platinum tip)	R	every 100,000 miles												
Air cleaner filter	R				✓				✓				✓	
Brake lines and hoses	I				✓				✓				✓	
Brake fluid	R				✓				✓				✓	
Brake pads/shoes	I				✓				✓				✓	
Bolts & nuts on chassis & body	S/I				✓				✓				✓	
Cooling system hoses	S/I				✓				✓				✓	
Driveshaft dust boots	S/I				✓				✓				✓	
Exhaust system heat shields	S/I				✓				✓				✓	
Front suspension ball joints	S/I				✓				✓				✓	
Fuel lines & hoses	S/I				✓				✓				✓	
Steering operation & linkages	S/I				✓				✓				✓	
Engine coolant	R						✓				✓			

R: Replace   S/I: Service or Inspect

FREQUENT OPERATION MAINTENANCE (SEVERE SERVICE)

If a vehicle is operated under any of the following conditions it is considered severe service:

- Extremely dusty areas.

- 50% or more of the vehicle operation is in 32°C (90°F) or higher temperatures, or constant operation in temperatures below 0°C (32°F).

- Prolonged idling (vehicle operation in stop and go traffic).

- Frequent short running periods (engine does not warm to normal operating temperatures).

- Police, taxi, delivery usage or trailer towing usage.

Air cleaner filter: service or inspect every 15,000 miles

Engine oil & filter: replace every 5000 miles.

Ball joints & dust covers: service or inspect every 7500 miles.

Bolts & nuts on chassis & body: tighten every 15,000 miles.

Automatic transmission fluid & filter: replace every 30,000 miles.

42356-MMPV-C13

PRECAUTIONS

Before servicing any vehicle, please be sure to read all of the following precautions, which deal with personal safety, prevention of component damage, and important points to take into consideration when servicing a motor vehicle:

• Never open, service or drain the radiator or cooling system when the engine is hot; serious burns can occur from the steam and hot coolant.

• Observe all applicable safety precautions when working around fuel. Whenever servicing the fuel system, always work in a well-ventilated area. Do not allow fuel spray or vapors to come in contact with a spark, open flame or excessive heat (a hot drop light, for example). Keep a dry chemical fire extinguisher near the work area. Always keep fuel in a container specifically designed for fuel storage; also, always properly seal fuel containers to avoid the possibility of fire or explosion. Refer to the additional fuel system precautions later in this section.

• Fuel injection systems often remain pressurized, even after the engine has been turned **OFF**. The fuel system pressure must be relieved before disconnecting any fuel lines. Failure to do so may result in fire and/or personal injury.

• Brake fluid often contains polyglycol ethers and polyglycols. Avoid contact with the eyes and wash your hands thoroughly after handling brake fluid. If you do get brake fluid in your eyes, flush your eyes with clean, running water for 15 minutes. If eye irritation persists, or if you have taken brake fluid internally, IMMEDIATELY seek medical assistance.

• The EPA warns that prolonged contact with used engine oil may cause a number of skin disorders, including cancer! You should make every effort to minimize your exposure to used engine oil. Protective gloves should be worn when changing oil. Wash your hands and any other exposed skin areas as soon as possible after exposure to used engine oil. Soap and water, or waterless hand cleaner should be used.

• All new vehicles are now equipped with an air bag system. The system must be disabled before performing service on or around system components, steering column, instrument panel components, wiring and sensors. Failure to follow safety and disabling procedures could result in accidental air bag deployment, possible personal injury and unnecessary system repairs.

• Always wear safety goggles when working with, or around, the air bag system. When carrying a non-deployed air bag, be sure the bag and trim cover are pointed away from your body. When placing a non-deployed air bag on a work surface, always face the bag and trim cover upward, away from the surface. This will reduce the motion of the module if it is accidentally deployed. Refer to the additional air bag system precautions later in this section.

• Clean, high quality brake fluid from a sealed container is essential to the safe and proper operation of the brake system. You should always buy the correct type of brake fluid for your vehicle. If the brake fluid becomes contaminated, completely flush the system with new fluid. Never reuse any brake fluid. Any brake fluid that is removed from the system should be discarded. Also, do not allow any brake fluid to come in contact with a painted surface; it will damage the paint.

• Never operate the engine without the proper amount and type of engine oil; doing so WILL result in severe engine damage.

• Timing belt maintenance is extremely important! Many models utilize an interference-type, non-freewheeling engine. If the timing belt breaks, the valves in the cylinder head may strike the pistons, causing potentially serious (also time-consuming and expensive) engine damage.

• Disconnecting the negative battery cable on some vehicles may interfere with the functions of the on-board computer system(s) and may require the computer to undergo a relearning process once the negative battery cable is reconnected.

• When servicing drum brakes, only disassemble and assemble one side at a time, leaving the remaining side intact for reference.

• Only an MVAC-trained, EPA-certified automotive technician should service the air conditioning system or its components.

ENGINE REPAIR

➡Disconnecting the negative battery cable on some vehicles may interfere with the functions of the on board computer system. The computer may undergo a relearning process once the negative battery cable is reconnected.

Distributor

This engine is equipped with a Distributorless Ignition System (DIS).

Alternator

REMOVAL & INSTALLATION

2.5L Engine

1. Before servicing the vehicle, refer to the precautions in the beginning of this section.

2. Remove or disconnect the following:

• Negative battery cable
• Accessory drive belt
• Subframe transverse section
• Exhaust front pipe
• Right axle halfshaft and center shaft assembly
• Alternator harness connectors
• Center shaft support bracket
• Alternator

To install:

3. Install or connect the following:

• Alternator. Tighten the bolts to 29–41 ft. lbs. (40–50 Nm).
• Center shaft support bracket. Tighten the bolts to 32–45 ft. lbs. (43–61 Nm).
• Alternator harness connectors. Tighten the battery terminal nut to 87–130 inch lbs. (10–15 Nm).

• Right axle halfshaft and center shaft assembly
• Exhaust front pipe
• Subframe transverse section. Tighten the bolts to 69–96 ft. lbs. (94–131 Nm).
• Accessory drive belt
• Negative battery cable

3.0L Engine

1. Before servicing the vehicle, refer to the precautions in the beginning of this section.

2. Remove or disconnect the following:

• Negative battery cable
• Accessory drive belt
• Exhaust front pipe
• Right axle halfshaft and center shaft assembly
• Alternator harness connectors
• Center shaft support bracket

- Alternator

To install:

3. Install or connect the following:
- Alternator. Tighten the bolts to 29–41 ft. lbs. (40–50 Nm).
- Center shaft support bracket. Tighten the bolts to 32–45 ft. lbs. (43–61 Nm).
- Alternator harness connectors. Tighten the battery terminal nut to 87–130 inch lbs. (10–15 Nm).
- Right axle halfshaft and center shaft assembly
- Exhaust front pipe
- Accessory drive belt
- Negative battery cable

Ignition Timing

ADJUSTMENT

2.5L Engine

This engine is equipped with a Distributorless Ignition System (DIS). No adjustment is necessary.

Engine Assembly

REMOVAL & INSTALLATION

2.5L Engine

1. Before servicing the vehicle, refer to the precautions in the beginning of this section.
2. Drain the cooling system.
3. Drain the engine oil.
4. Drain the transaxle.
5. Relieve the fuel system pressure.
6. Install a support fixture to the engine lifting eyes.
7. Remove or disconnect the following:
- Battery and tray
- Inner fender liners
- Axle halfshafts
- Air intake assembly
- Accelerator cable and bracket
- Gear select cable
- Transaxle dipstick tube
- Cruise control actuator
- Radiator
- Fuel line
- Brake booster vacuum line
- Powertrain Control Module (PCM) connector. Pull the harness through the firewall into the engine compartment.
- Exhaust front pipe
- Accessory drive belt
- Alternator and bracket

- Right engine mount bracket
- Power steering hoses
- Front engine mount
- Subframe center section
- Rear engine mount
- Left engine mount

8. Lower the powertrain from the vehicle.

To install:

9. Raise the powertrain into position.
10. Install or connect the following:
- Left engine mount
- Rear engine mount
- Subframe center section. Tighten the bolts to 48–65 ft. lbs. (64–89 Nm) and the nut to 50–67 ft. lbs. (67–93 Nm).
- Front engine mount
- Power steering hoses
- Right engine mount bracket

11. Check that the right engine mount stud is centered in the mount bracket with no tension applied to the stud by the bracket.

12. Tighten the engine mount fasteners as follows:
   a. Step 1: Left and rear engine mount through bolts to 63–86 ft. lbs. (85–116 Nm).
   b. Step 2: Front engine mount nuts to 50–67 ft. lbs. (67–93 Nm).
   c. Right engine mount bracket nuts to 56–76 ft. lbs. (75–104 Nm).

13. Install or connect the following:
- Alternator and bracket
- Accessory drive belt
- Exhaust front pipe
- PCM connector. Pull the harness through the firewall into the passenger compartment.
- Brake booster vacuum line
- Fuel line
- Radiator
- Cruise control actuator
- Transaxle dipstick tube
- Gear select cable
- Accelerator cable and bracket
- Air intake assembly
- Axle halfshafts
- Inner fender liners
- Battery and tray

14. Fill the crankcase and transaxle to the correct level.
15. Fill the cooling system.
16. Start the engine and check for leaks.

3.0L Engine

1. Before servicing the vehicle, refer to the precautions in the beginning of this section.
2. Drain the cooling system.

3. Drain the engine oil.
4. Drain the transaxle.
5. Relieve the fuel system pressure.
6. Disconnect the negative batter cable.
7. Install a support fixture to the engine lifting eyes.
8. Remove or disconnect the following:
- Timing chain plug hole plate
- Power steering hoses
- Splash shield
- Axle halfshafts
- Battery and tray
- Air intake assembly
- Accelerator cable and bracket
- Gear select cable
- Transaxle dipstick tube
- Cruise control actuator
- Radiator
- Fuel line
- Powertrain Control Module (PCM) connector. Pull the harness through the firewall into the engine compartment.
- Exhaust front pipe
- Accessory drive belt
- A/C compressor with lines still attached
- Alternator and bracket
- Engine mounts

To install:

9. Installation is the reverse of removal. Observe the following torques:
- Subframe center section. Tighten the bolts to 48–65 ft. lbs. (64–89 Nm) and the nut to 50–67 ft. lbs. (67–93 Nm).
- Left and rear engine mount through bolts to 63–86 ft. lbs. (85–116 Nm).
- Front engine mount nuts to 50–67 ft. lbs. (67–93 Nm).
- Right engine mount bracket nuts to 56–76 ft. lbs. (75–104 Nm).

Water Pump

REMOVAL & INSTALLATION

2.5L and 3.0L Engines

1. Before servicing the vehicle, refer to the precautions in the beginning of this section.
2. Drain the cooling system.
3. Remove or disconnect the following:
- Battery and tray
- Water pump drive belt
- Water pump drive pulley
- Thermostat housing
- Water pump belt tensioner
- Oil cooler hose

- Water outlet pipe
- Water pump

To install:

4. Install or connect the following:
   - Water pump. Tighten the bolts to 89 inch lbs. (10 Nm) plus 90 degrees.
   - Water outlet pipe
   - Oil cooler hose
   - Water pump belt tensioner
   - Thermostat housing
   - Water pump drive pulley
   - Water pump drive belt
   - Battery and tray
5. Fill the cooling system.
6. Start the engine and check for leaks.

Heater Core

REMOVAL & INSTALLATION

Front System

1. Disconnect the negative battery cable.

✳✳ CAUTION

After disconnecting the battery, wait for more than 1 minute for the air bag system to deplete its stored energy.

2. Drain the cooling system into a clean container for reuse.
3. Disconnect the heater hoses from the heater core.
4. Discharge and recover the air conditioning system refrigerant.
5. At the driver's side, remove the SAS module and the steering wheel by performing the following procedure:
   a. Place the wheel in the straight-ahead position and turn the ignition switch to LOCK.
   b. Remove the lower steering column cover.
   c. Disconnect the clock spring connector.
   d. Remove the steering wheel-to-SAS module bolts.
   e. Carefully, lift the SAS module from the steering wheel.

✳✳ CAUTION

Place the SAS in a safe place with the module facing upward.

   f. Remove the steering wheel-to-column nut.
   g. Using a steering wheel puller, press the steering wheel from the steering column.
6. At the passenger's side, remove the SAS module by performing the following procedure:
   a. Remove the glove compartment by sliding it to the left; then, pull the right side forward to remove the stopper and the pin, then, move it to the right to remove it.
   b. Remove the SAS module-to-dash bolts.
   c. Carefully, lift the SAS module and disconnect the electrical connector.

✳✳ CAUTION

Place the SAS in a safe place with the module facing upward.

7. Remove the instrument panel by performing the following procedure:
   a. Remove the A-pillar trim from both sides.
   b. Remove the front side trim and the side panel trim.
   c. Remove the hood release handle and the lower panel.
   d. Remove the meter hood and the instrument cluster.
   e. Remove the steering column-to-instrument panel bolts and lower the steering column.
   f. Disconnect the electrical connectors from the blower motor and the heater housing.

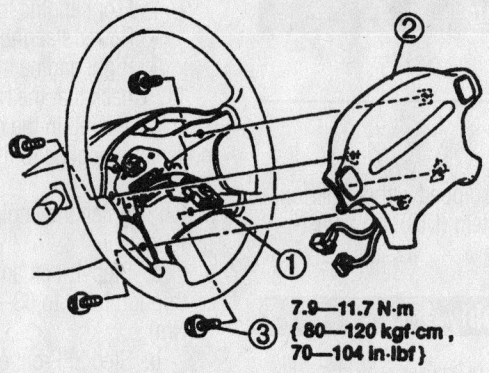

7.9—11.7 N·m
{ 80—120 kgf·cm , 70—104 in·lbf }

1	Connector
2	Driver-side air bag module
3	Bolt

93113GF1

Exploded view of the steering wheel and SAS module

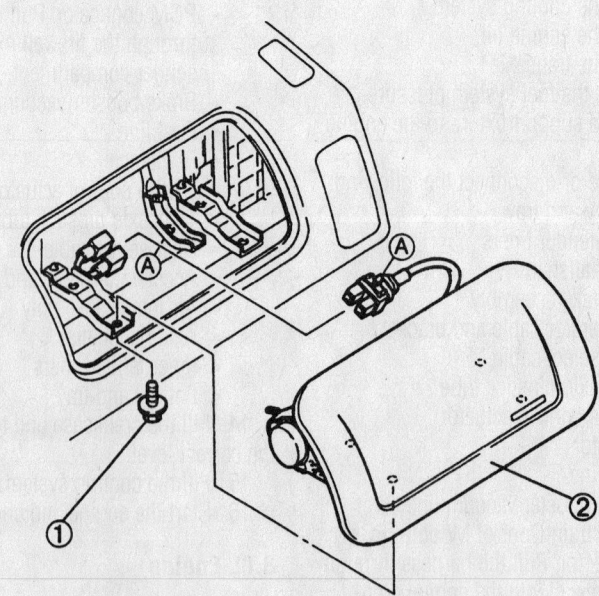

| 1 | Bolt |
| 2 | Passenger-side air bag module |

93113GF2

Exploded view of the passenger's side SAS module

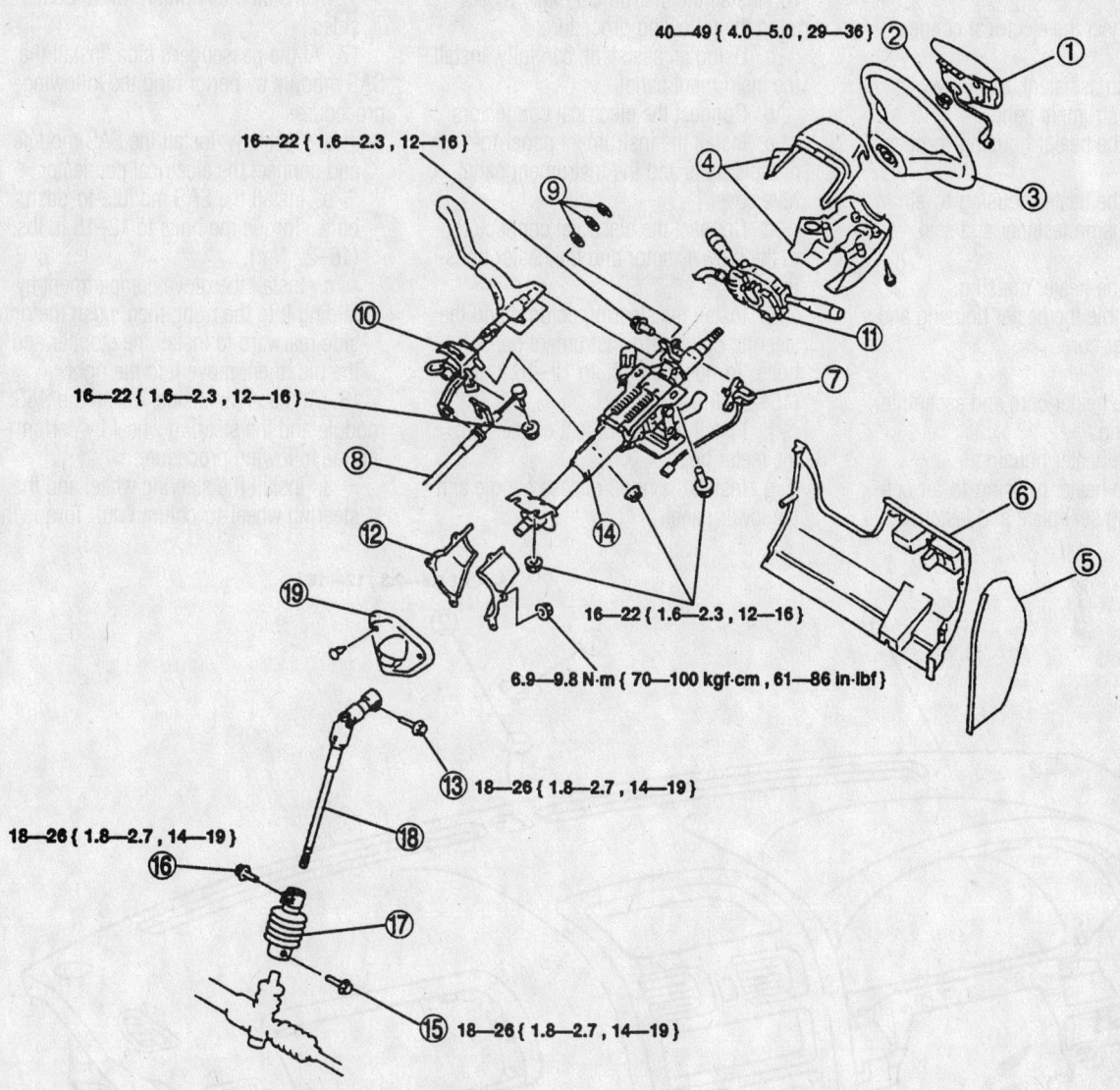

40—49 { 4.0—5.0 , 29—36 }

16—22 { 1.6—2.3 , 12—16 }

16—22 { 1.6—2.3 , 12—16 }

16—22 { 1.6—2.3 , 12—16 }

6.9—9.8 N·m { 70—100 kgf·cm , 61—86 in·lbf }

18—26 { 1.8—2.7 , 14—19 }

18—26 { 1.8—2.7 , 14—19 }

18—26 { 1.8—2.7 , 14—19 }

N·m { kgf·m , ft·lbf }

1	Air bag module	10	Selector lever component
2	Locknut	11	Combination switch
3	Steering wheel	12	Joint cover
4	Column cover	13	Fixing bolt (steering shaft/intermediate shaft)
5	Side panel	14	Steering shaft
6	Lower panel	15	Fixing bolt (universal joint/pinion shaft)
7	Shift-lock actuator	16	Fixing bolt (intermediate shaft/universal joint)
8	Selector cable	17	Universal joint
9	Retaining ring, wave washer, adjustment washer(s)	18	Intermediate shaft
		19	Dust cover

93113GF3

Exploded view of the steering column and related components

g. Remove the instrument panel hole cover and the instrument panel-to-chassis bolts.

h. Disconnect the electrical connectors.

i. Using an assistant, carefully remove the instrument panel.

8. Remove the heater housing-to-chassis nuts.

9. Remove the heater housing-to-air conditioning housing fastener and seal plate.

10. Remove the heater housing.

11. Disassemble the heater housing and remove the heater core.

To install:

12. Install the heater core and assemble the heater housing.

13. Install the heater housing.

14. Install the heater housing-to-air conditioning housing seal plate and fastener.

15. Install the heater housing-to-chassis nuts.

16. Install the instrument panel by performing the following procedure:

a. Using an assistant, carefully, install the instrument panel.

b. Connect the electrical connectors.

c. Install the instrument panel-to-chassis bolts and the instrument panel hole cover.

d. Connect the electrical connectors to the blower motor and the heater housing.

e. Install the steering column and the steering column-to-instrument panel bolts. Torque the bolts to 12–16 ft. lbs. (16–22 Nm).

f. Install the instrument cluster and the meter hood.

g. Install the hood release handle and the lower panel.

h. Install the front side trim and the side panel trim.

i. Install the A-pillar trim to both sides.

17. At the passenger's side, install the SAS module by performing the following procedure:

a. Carefully, install the SAS module and connect the electrical connector.

b. Install the SAS module-to-dash bolts. Torque the bolts to 12–16 ft. lbs. (16–22 Nm).

c. Install the glove compartment by sliding it to the right; then, push the right side rearward to install the stopper and the pin, then, move it to the right.

18. At the driver's side, install the SAS module and the steering wheel by performing the following procedure:

a. Install the steering wheel and the steering wheel-to-column nut. Torque the

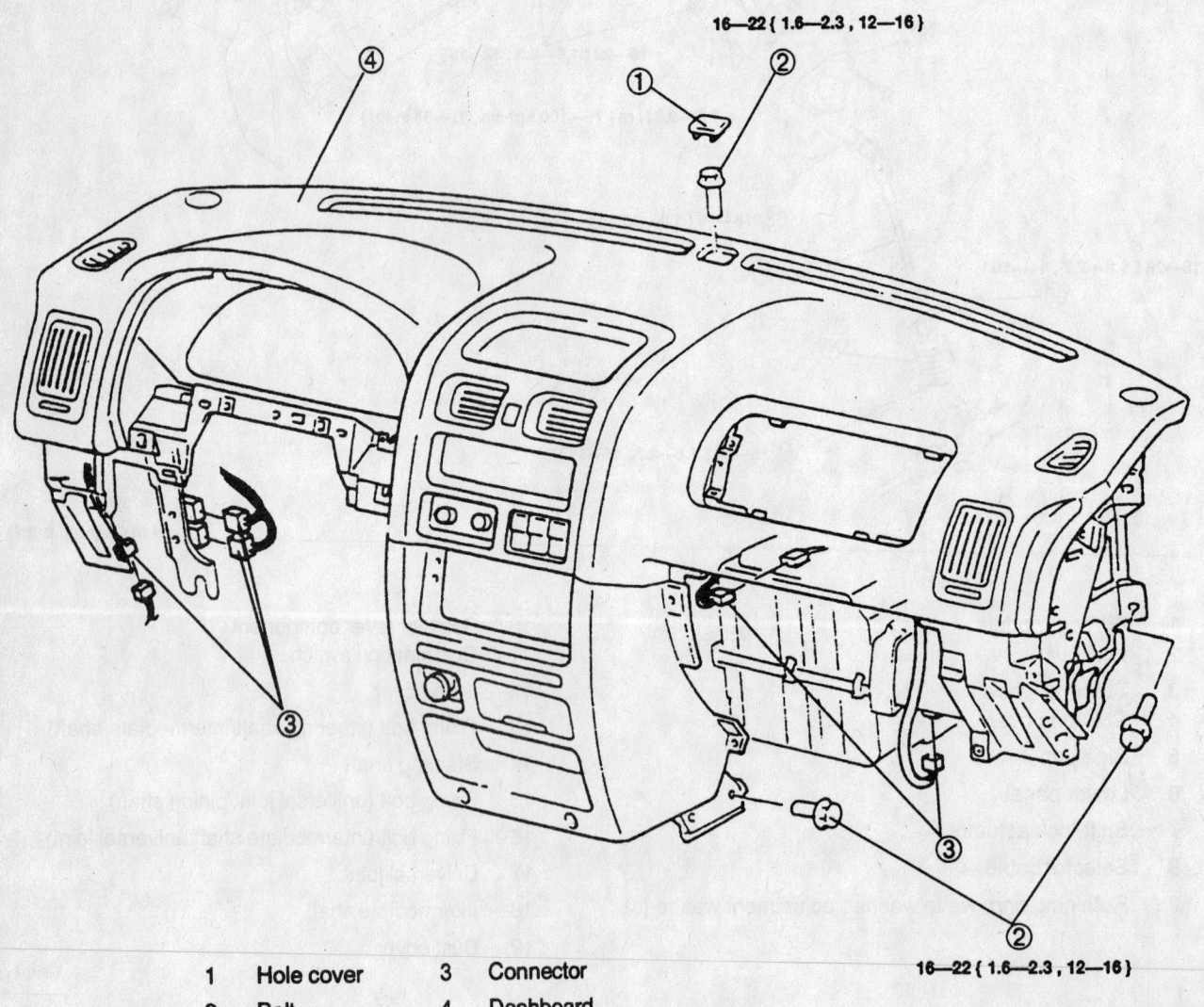

16—22 { 1.6—2.3 , 12—16 }

| 1 | Hole cover | 3 | Connector |
| 2 | Bolt | 4 | Dashboard |

16—22 { 1.6—2.3 , 12—16 }

N·m { kgf·m , ft·lbf }

93113GF4

View of the instrument panel and related components

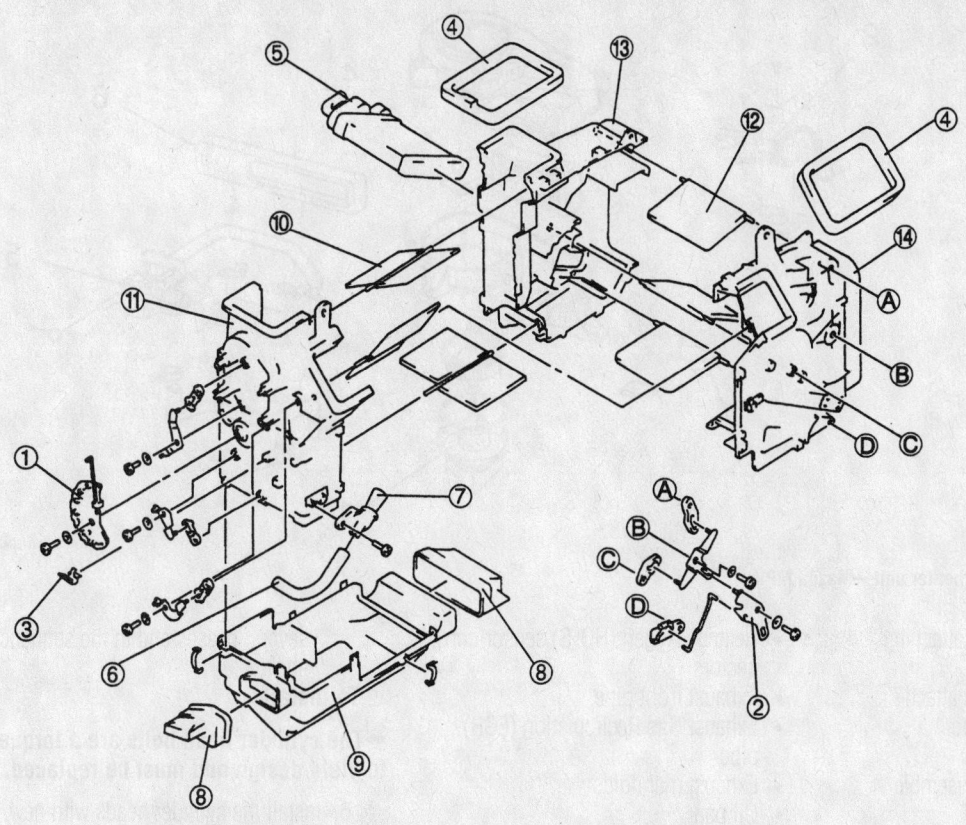

1	Air mix link
2	Airflow mode link
3	Wire clamp
4	Polyurethane protector
5	Front heater core
6	Drain hose
7	Joint
8	Duct
9	Case (bottom)
10	Airflow mode door
11	Case (left)
12	Air mix door
13	Case (right No.1)
14	Case (right No.2)

93113GF5

Exploded view of the heater core, heater housing and related components

steering wheel nut to 29–36 ft. lbs. (40–49 Nm).

b. Carefully, install the SAS module to the steering wheel.

c. Install the steering wheel-to-SAS module bolts. Torque the bolts to 70—104 inch lbs. (7.9–11.7 Nm).

d. Connect the clock spring connector.

e. Install the lower steering column cover.

19. Connect the heater hoses to the heater core.

20. Refill the cooling system.

21. Connect the negative battery cable.

22. Run the engine to normal operating temperatures; then, check the climate control operation and check for leaks.

Rear Auxiliary System

1. Disconnect the negative battery cable.

2. Set the rear heater control knob to the WARM position to open the water valve.

3. Drain the cooling system into a clean container for reuse.

4. Remove the driver's seat.

5. Disconnect the heater hoses.

6. Disconnect the rear heater unit wire connector.

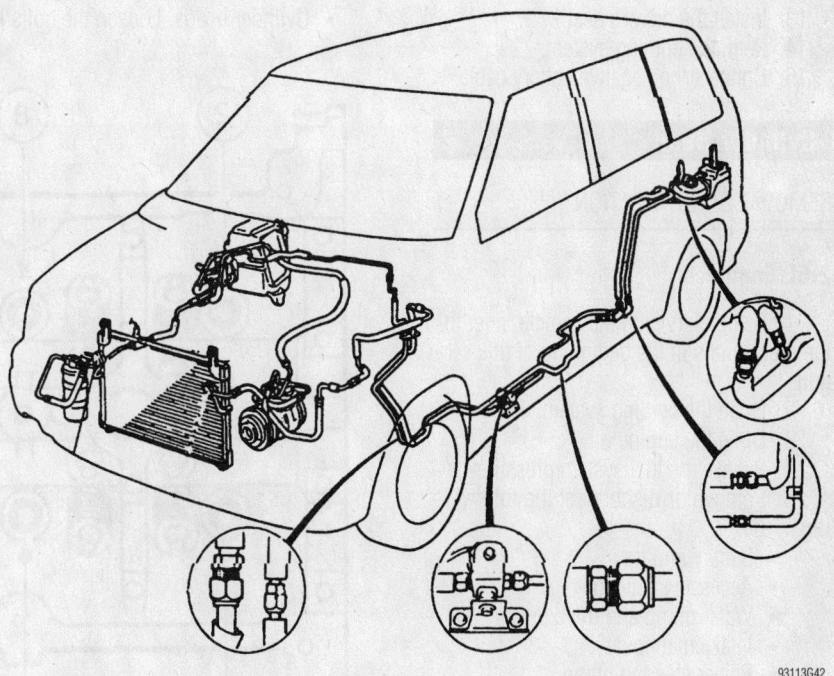

93113G42

View of the rear auxiliary heater assembly and related components—Mazda MPV

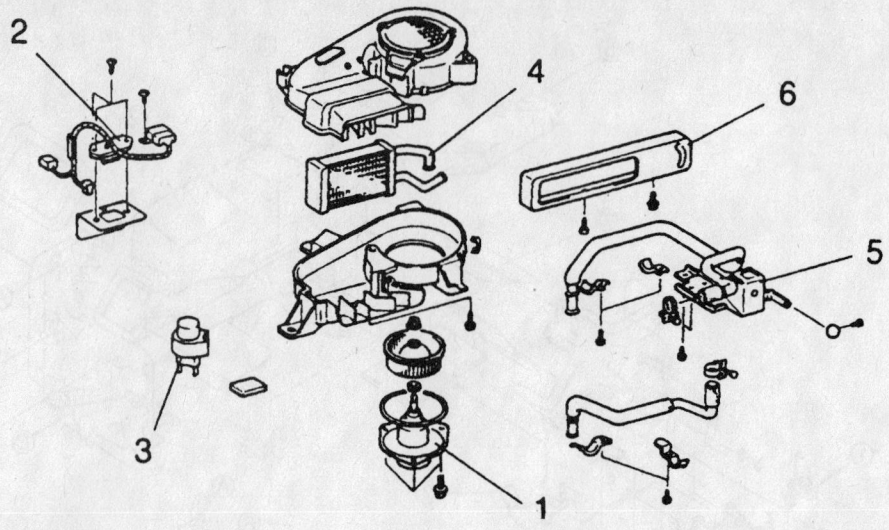

1. Rear heater blower motor
2. Resistor assembly
3. Rear heater relay
4. Heater core
5. Water valve
6. Switch panel

93113G43

Exploded view of the rear auxiliary heater unit—Mazda MPV

7. Remove the rear heater unit attaching bolts and remove the assembly.

8. Separate the rear heater unit attaching bolts and remove the heater core.

To install:

9. Install the heater core and assemble the rear heater unit.

10. Install the assembly and the rear heater unit.

11. Connect the rear heater unit wire connector.

12. Connect the heater hoses to the heater core.

13. Install the driver's seat.

14. Refill the cooling system.

15. Connect the negative battery cable.

Cylinder Head

REMOVAL & INSTALLATION

2.5L Engine

1. Before servicing the vehicle, refer to the precautions in the beginning of this section.

2. Drain the cooling system.

3. Drain the engine oil.

4. Relieve the fuel system pressure.

5. Remove or disconnect the following:

- Battery and tray
- Accessory drive belt
- Water pump and drive pulley
- Intake manifold
- Power steering pump
- Intake Manifold Runner Control (IMRC) actuator
- Spark plug wires
- Ignition coil
- Heated Oxygen (HO$_2$S) sensor connectors
- Exhaust front pipe
- Exhaust Gas Recirculation (EGR) pipe
- Exhaust manifolds
- Oil pan
- Alternator
- A/C compressor
- Valve covers
- Front cover
- Timing chains
- Camshafts
- Cylinder heads. Loosen the bolts in several passes and in the sequence shown.

To install:

➡ **The cylinder head bolts are a torque-to-yield design and must be replaced.**

6. Install the cylinder heads with new gaskets. Tighten the bolts in sequence as follows:

  a. Step 1: 28–31 ft. lbs. (37–43 Nm).

  b. Step 2: Plus 90 degrees.

  c. Step 3: Loosen one full turn.

  d. Step 4: 28–31 ft. lbs. (37–43 Nm).

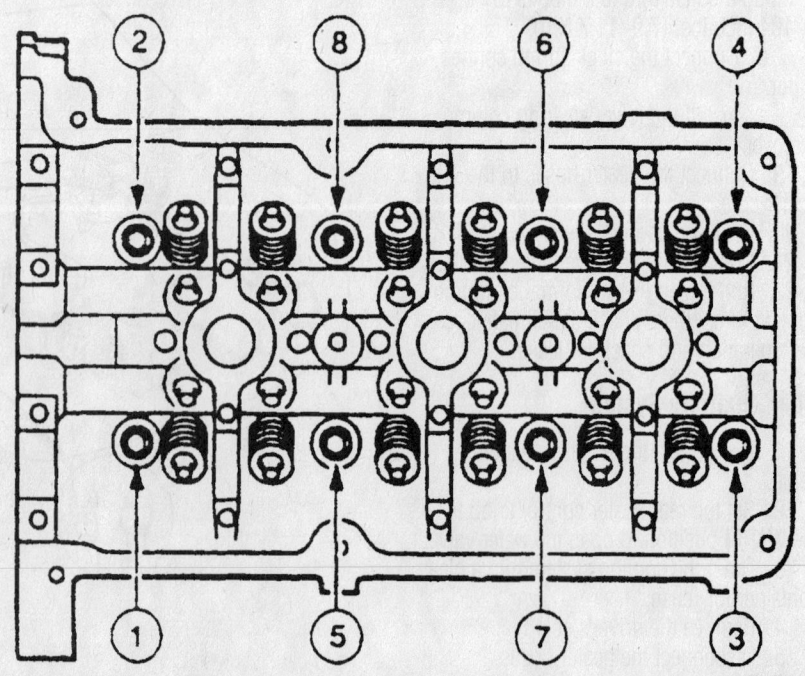

9308TG05

Cylinder head loosening sequence—2.5L and 3.0L engine

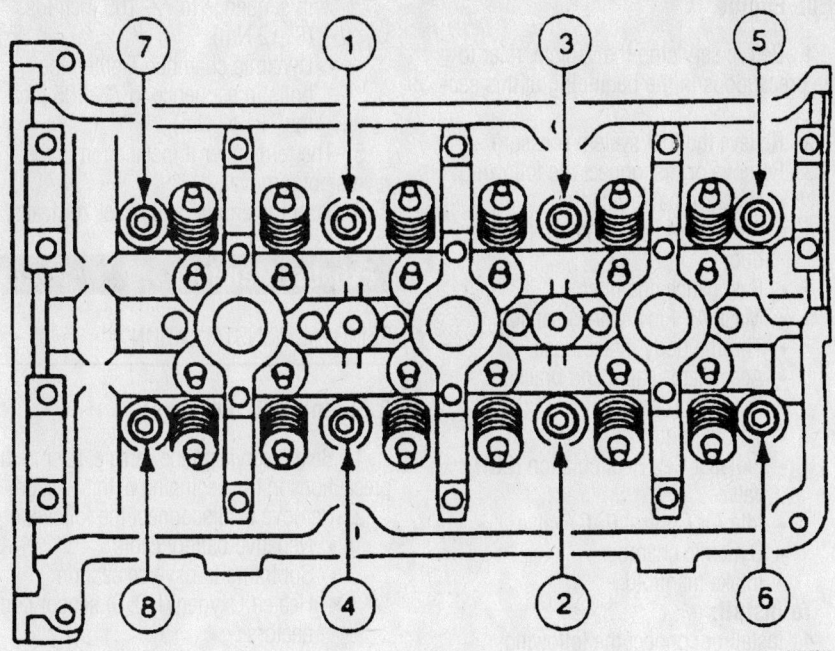

Cylinder head torque sequence—2.5L and 3.0L engines

e. Step 5: Plus 90 degrees.
f. Step 6: Plus 90 degrees.
7. Install or connect the following:
- Camshafts
- Timing chains
- Front cover
- Valve covers
- A/C compressor
- Alternator
- Oil pan
- Exhaust manifolds
- EGR pipe
- Exhaust front pipe
- HO2S sensor connectors
- Ignition coil
- Spark plug wires
- IMRC actuator
- Power steering pump
- Intake manifold
- Water pump and drive pulley
- Accessory drive belt
- Battery and tray
8. Fill the crankcase to the correct level.
9. Fill the cooling system.
10. Start the engine and check for leaks.

3.0L Engine

1. Before servicing the vehicle, refer to the precautions in the beginning of this section.
2. Drain the cooling system.
3. Drain the engine oil.
4. Relieve the fuel system pressure.
5. Remove or disconnect the following:
- Accessory drive belt
- Water pump and drive pulley
- Timing chains
- No.3 engine mount rubber and joint bracket
- Ventilation pipe
- Water bypass tube
- Camshafts

➡**Remove the Nos. 1 and 5 caps first. Don't loosen any other cap bolts until these caps are removed.**

- Rocker arms
- Cylinder heads. Loosen the bolts in several passes and in the sequence shown.

To install:
6. Installation is the reverse of removal. Observe the following torques:

➡**The cylinder head bolts are a torque-to-yield design and must be replaced.**

7. Install the cylinder heads with new gaskets. Tighten the bolts in sequence as follows:
   a. Step 1: 24–28 ft. lbs. (32–38 Nm).
   b. Step 2: Plus 90 degrees.
   c. Step 3: Loosen one full turn.
   d. Step 4: 24–28 ft. lbs. (32–38 Nm).
   e. Step 5: Plus 90 degrees.
   f. Step 6: Plus 90 degrees.

Rocker Arms/Shafts

REMOVAL & INSTALLATION

2.5L and 3.0L Engines

1. Before servicing the vehicle, refer to the precautions in the beginning of this section.
2. Relieve the fuel system pressure.
3. Drain the engine oil.
4. Remove or disconnect the following:
- Negative battery cable
- Intake manifold
- Accessory drive belt
- Power steering pump
- Intake Manifold Runner Control (IMRC) actuator
- Spark plug wires
- Ignition coil
- Water pump drive belt and pulley
- Camshaft seal housing
- Wiring harness connector bracket
- Valve covers
- Oil pan
- Front cover
- Timing chains
- Camshafts
- Rocker arms

➡**Keep all valvetrain components in order for assembly.**

To install:
5. Install or connect the following:
- Rocker arms
- Camshafts
- Timing chains
- Front cover
- Oil pan
- Valve covers
- Wiring harness connector bracket
- Camshaft seal housing
- Water pump drive belt and pulley
- Ignition coil
- Spark plug wires
- Intake Manifold Runner Control (IMRC) actuator
- Power steering pump
- Accessory drive belt
- Intake manifold
- Negative battery cable
6. Fill the crankcase to the correct level.
7. Start the engine and check for leaks.

Intake Manifold

REMOVAL & INSTALLATION

2.5L Engine

1. Before servicing the vehicle, refer to the precautions in the beginning of this section.
2. Relieve the fuel system pressure.
3. Remove or disconnect the following:
   - Negative battery cable
   - Air cleaner housing and fresh air duct
   - Mass Air Flow (MAF) sensor
   - Throttle body intake hose
   - Accelerator cable and bracket
   - Intake Manifold Runner Control (IMRC) cable at the IMRC housing
   - Throttle body
   - Exhaust Gas Recirculation (EGR) valve
   - Idle Air Control (IAC) valve
   - Pressure Regulator Control (PRC) solenoid
   - Intake manifold
   - Fuel lines
   - Fuel pressure regulator vacuum line
   - Fuel supply manifold
   - IMRC housing

To install:

4. Install or connect the following:
   - IMRC housing. Tighten the bolts in sequence to 72–105 inch lbs. (8–12 Nm).
   - Fuel supply manifold. Tighten the bolts to 72–105 inch lbs. (8–12 Nm).
   - Fuel pressure regulator vacuum line
   - Fuel lines
   - Intake manifold. Tighten the bolts in sequence to 72–105 inch lbs. (8–12 Nm).
   - PRC solenoid. Tighten the bolt to 45–61 inch lbs. (5–7 Nm).
   - IAC valve. Tighten the bolts to 72–105 inch lbs. (8–12 Nm).
   - EGR valve
   - Throttle body. Tighten the bolts to 72–105 inch lbs. (8–12 Nm).
   - IMRC cable
   - Accelerator cable and bracket. Tighten the bolt to 71–94 inch lbs. (8–11 Nm).
   - Throttle body intake hose
   - MAF sensor
   - Air cleaner housing and fresh air duct
   - Negative battery cable
5. Start the engine and check for leaks.

3.0L Engine

1. Before servicing the vehicle, refer to the precautions in the beginning of this section.
2. Relieve the fuel system pressure.
3. Remove or disconnect the following:
   - Negative battery cable
   - Air cleaner housing and fresh air duct
   - Resonance chamber
   - Mass Air Flow (MAF) sensor
   - Throttle body intake hose
   - Accelerator cable and bracket
   - IMCC actuator
   - Throttle body
   - Exhaust Gas Recirculation (EGR) valve
   - Idle Air Control (IAC) valve
   - Dynamic chamber
   - Intake manifold

To install:

4. Install or connect the following:
   - Intake manifold. Tighten the bolts in sequence to 72–105 inch lbs. (8–12 Nm).
   - Dynamic chamber. Tighten the bolts in sequence to 72–105 inch lbs. (8–12 Nm).
5. The remainder if installation is the reverse of removal.
6. Start the engine and check for leaks.

Exhaust Manifold

REMOVAL & INSTALLATION

2.5L and 3.0L Engines

1. Before servicing the vehicle, refer to the precautions in the beginning of this section.
2. Remove or disconnect the following:
   - Negative battery cable
   - Subframe transverse section
   - Heated Oxygen (HO$_2$S) sensor connectors
   - Exhaust front pipe

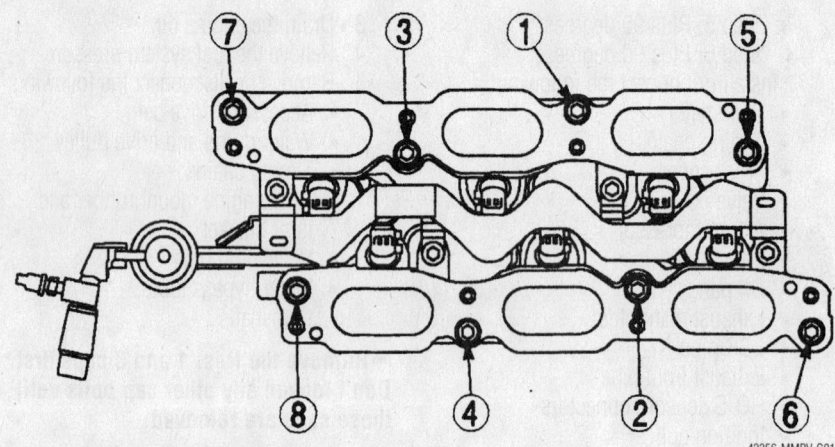

Intake manifold torque sequence—3.0L engine

42356-MMPV-G01

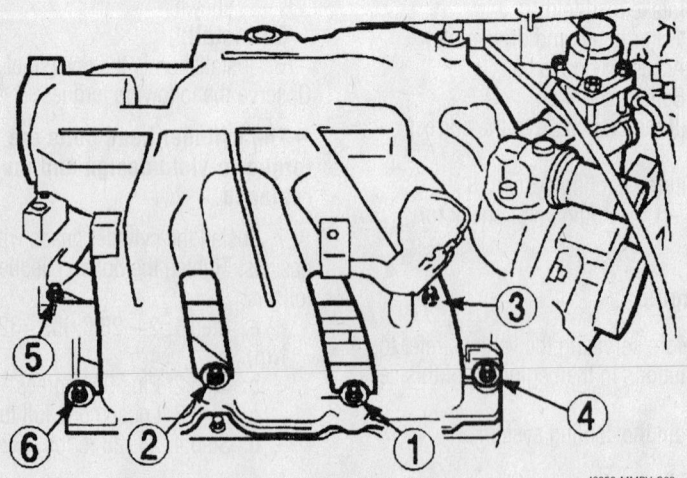

Dynamic chamber torque sequence—3.0L engine

42356-MMPV-G02

Right exhaust manifold torque sequence–2.5L and 3.0L engine

Left exhaust manifold torque sequence–2.5L and 3.0L engine

- Exhaust Gas Recirculation (EGR) pipe
- Exhaust manifolds

To install:
3. Install or connect the following:
- Exhaust manifolds. Tighten the nuts in sequence to 14 ft. lbs. (20 Nm).
- EGR pipe
- Exhaust front pipe
- HO2S sensor connectors
- Subframe transverse section. Tighten the bolts to 69–96 ft. lbs. (94–131 Nm).
- Negative battery cable
4. Start the engine and check for leaks.

Front Crankshaft Seal

REMOVAL & INSTALLATION

Refer to the Timing Chain, Sprockets, Front Cover and Seal procedure in this section.

Camshaft and Valve Lifters

REMOVAL & INSTALLATION

2.5L Engine

1. Before servicing the vehicle, refer to the precautions in the beginning of this section.
2. Relieve the fuel system pressure.
3. Drain the engine oil.
4. Remove or disconnect the following:
- Negative battery cable
- Intake manifold
- Accessory drive belt
- Power steering pump
- Intake Manifold Runner Control (IMRC) actuator
- Spark plug wires
- Ignition coil
- Exhaust front pipe
- Oil pan
- Alternator and bracket
- A/C compressor
- Water pump belt and drive pulley
- Camshaft oil seal housing
- Wiring harness connector bracket
- Valve covers
- Front cover
- Timing chains

➡️**Keep all valvetrain components in order for assembly**

➡️**Remove the camshaft thrust bearing caps before loosening any of the other bearing cap bolts.**

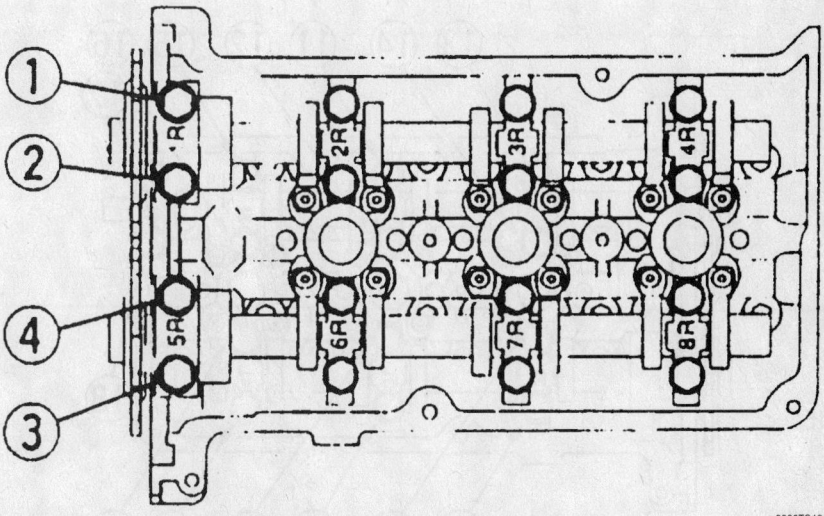

Right bank camshaft thrust cap loosening sequence—2.5L and 3.0L engine

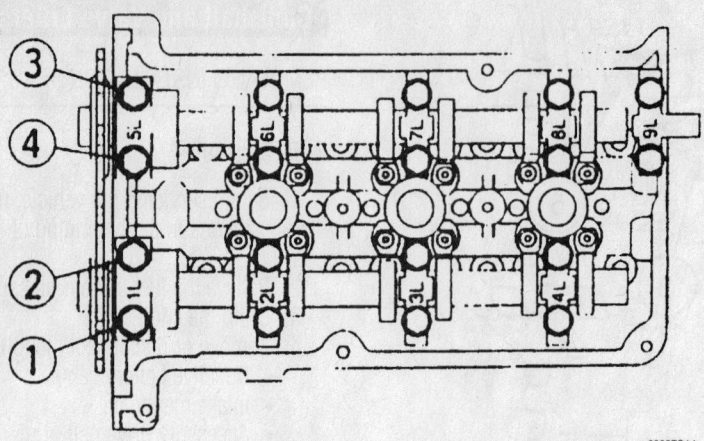

Left bank camshaft thrust cap loosening sequence—2.5L and 3.0L engine

9308TG14

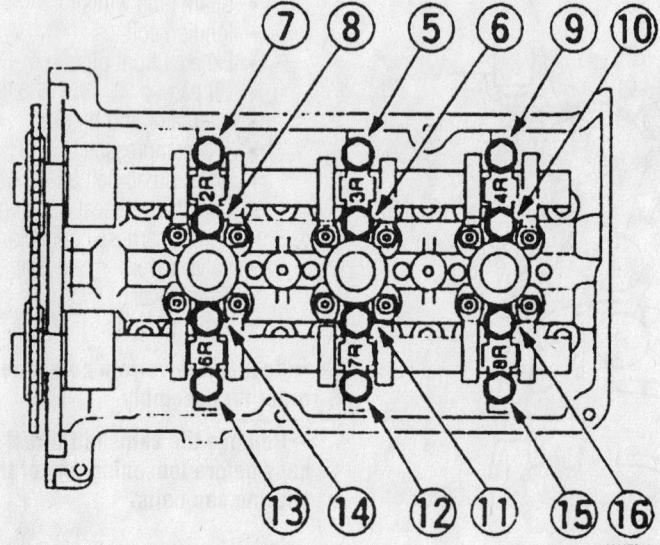

Right bank camshaft bearing cap loosening sequence—2.5L and 3.0L engine

9308TG13

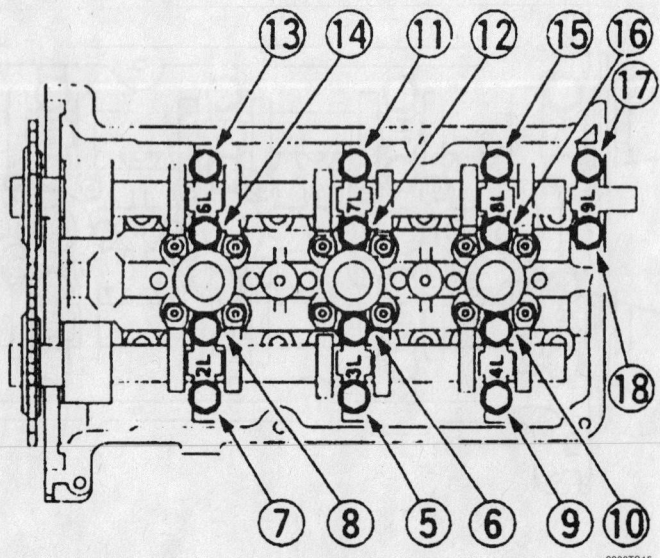

Left bank camshaft bearing cap loosening sequence—2.5L and 3.0L engine

9308TG15

- Camshaft thrust bearing caps. Loosen the bolts evenly in several passes.
- Remaining camshaft bearing caps. Loosen the bolts evenly in several passes.
- Camshafts
- Rocker arms
- Hydraulic lifters

To install:

❋❋ WARNING

The crankshaft keyway must be at the 11 o'clock position before reassembly. Failure to do so may lead to engine damage.

5. Rotate the crankshaft so that the keyway is at the 11 o'clock position for installation of the camshafts.

6. Install or connect the following:
- Hydraulic lifters
- Rocker arms.
- Camshafts. Align the sprocket timing marks.

➡**Do not install the camshaft journal thrust caps until the rocker arms and timing chains have been installed and the camshaft journal caps are secured into position.**

- All camshaft journal caps except the thrust caps.
- Timing chains. Tighten the camshaft journal cap bolts in reverse of the loosening order and in several steps to 71–106 inch lbs. (8–12 Nm).
- Thrust caps. Tighten the bolts to 71–106 inch lbs. (8–12 Nm).
- Front cover
- Valve covers
- Wiring harness connector bracket
- Camshaft oil seal housing
- Water pump belt and drive pulley
- A/C compressor
- Alternator and bracket
- Oil pan
- Exhaust front pipe
- Ignition coil
- Spark plug wires
- IMRC actuator
- Power steering pump
- Accessory drive belt
- Intake manifold
- Negative battery cable

7. Fill the crankcase to the correct level.

8. Fill the cooling system.

9. Start the engine and check for leaks.

3.0L Engine

1. Before servicing the vehicle, refer to the precautions in the beginning of this section.
2. Drain the cooling system.
3. Drain the engine oil.
4. Relieve the fuel system pressure.
5. Remove or disconnect the following:
   - Accessory drive belt
   - Water pump and drive pulley
   - Timing chains
   - No.3 engine mount rubber and joint bracket
   - Ventilation pipe
   - Water bypass tube
   - Camshafts

➡**Remove the Nos. 1 and 5 caps first. Don't loosen any other cap bolts until these caps are removed.**

   - Rocker arms

To install:

6. Installation is the reverse of removal. Observe the following torques:
7. Install the camshaft caps. Tighten the bolts in sequence to 71–106 inch lbs.

Valve Lash

ADJUSTMENT

The engine covered in this section are equipped with hydraulic lash adjusters. Valve clearance adjustments are not possible.

Starter Motor

REMOVAL & INSTALLATION

2.5L and 3.0L Engines

1. Before servicing the vehicle, refer to the precautions in the beginning of this section.
2. Remove or disconnect the following:
   - Battery and tray
   - Air intake assembly
   - Gear select cable
   - Starter harness connectors
   - Starter motor

To install:

3. Install or connect the following:
   - Starter motor. Tighten the bolts to 28–38 ft. lbs. (38–51 Nm).
   - Starter harness connectors. Tighten the battery cable nut to 87–104 inch lbs. (10–12 Nm).
   - Gear select cable
   - Air intake assembly
   - Battery and tray

Oil Pan

REMOVAL & INSTALLATION

2.5L Engine

1. Before servicing the vehicle, refer to the precautions in the beginning of this section.
2. Drain the engine oil.
3. Remove or disconnect the following:
   - Negative battery cable
   - Subframe transverse section
   - Exhaust front pipe
   - Flywheel access panel
   - Transaxle housing bolts
   - Oil pan bolts. Loosen the bolts in reverse of the tightening sequence and in several steps.
   - Oil pan

To install:

4. Apply a bead of silicone sealer to the gasket area where the pan meets the parting lines of the lower cylinder block and the front engine cover.

5. Install or connect the following:
   - Oil pan. Use a new gasket, tighten the pan bolts in several passes to 15–22 ft. lbs. (20–30 Nm), then tighten the transaxle case bolts to 28–38 ft. lbs. (38–51 Nm).
   - Flywheel access panel
   - Exhaust front pipe
   - Subframe transverse section. Tighten the bolts to 69–96 ft. lbs. (94–131 Nm).
   - Negative battery cable

6. Fill the crankcase to the correct level.
7. Start the engine and check for leaks.

3.0L Engine

1. Before servicing the vehicle, refer to the precautions in the beginning of this section.
2. Drain the engine oil.
3. Remove or disconnect the following:
   - Negative battery cable
   - Exhaust front pipe
   - Flywheel access panel
   - Transaxle housing bolts

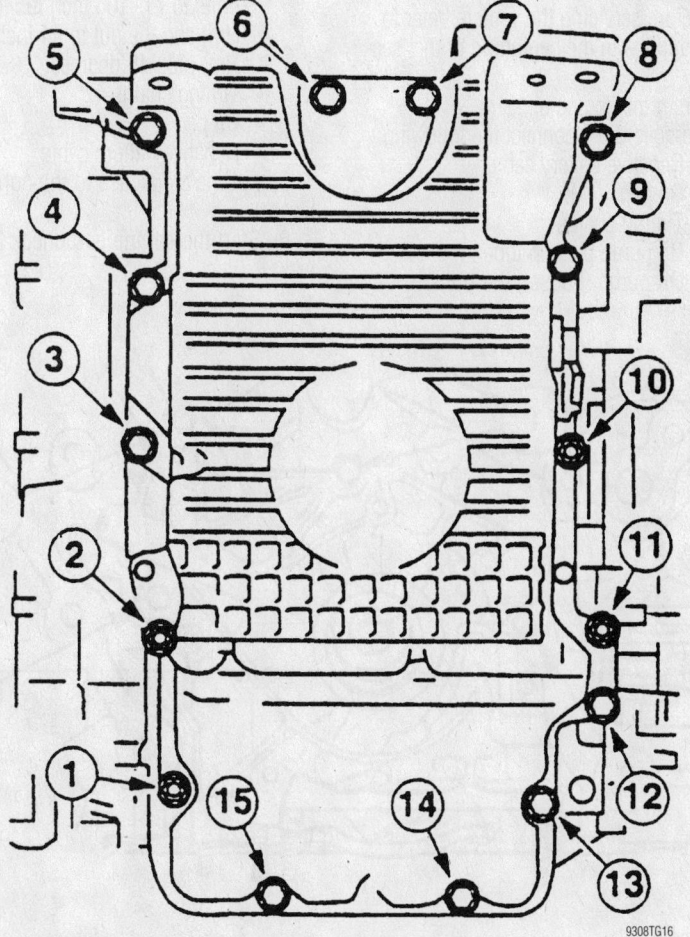

9308TG16

Oil pan torque sequence—2.5L and 3.0L engine

- Oil pan bolts. Loosen the bolts in the sequence shown.
- Oil pan

To install:

4. Apply a bead of silicone sealer to the gasket area where the pan meets the parting lines of the lower cylinder block and the front engine cover.

5. Install or connect the following:
- Oil pan. Use a new gasket, tighten the pan bolts in several passes to 15–22 ft. lbs. (20–30 Nm), then tighten the transaxle case bolts to 28–38 ft. lbs. (38–51 Nm).
- Flywheel access panel
- Exhaust front pipe
- Negative battery cable

6. Fill the crankcase to the correct level.

7. Start the engine and check for leaks.

Oil Pump

REMOVAL & INSTALLATION

2.5L and 3.0L Engine

1. Before servicing the vehicle, refer to the precautions in the beginning of this section.

2. Drain the engine oil.

3. Remove or disconnect the following:
- Negative battery cable
- Oil pan
- Timing chains
- Oil pump pick up tube
- Oil pump. Loosen the bolts in

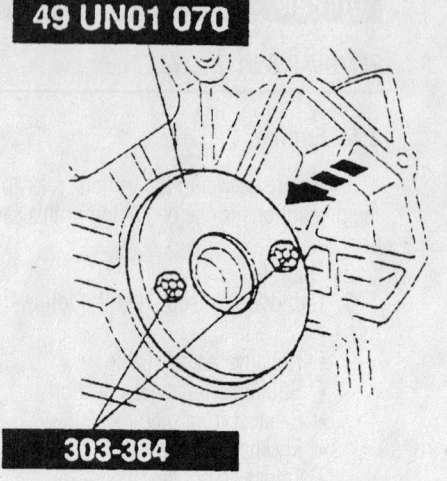
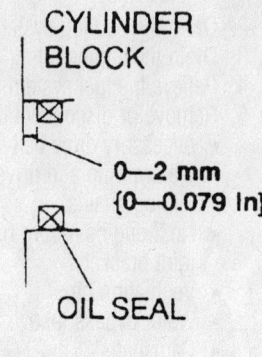

49 UN01 070

303-384

Rear main seal installation—2.5L and 3.0L engines

reverse of the tightening sequence.

To install:

4. Install or connect the following:
- Oil pump. Tighten the bolts in sequence to 71–106 inch lbs. (8–12 Nm).
- Oil pump pick up tube. Tighten the bolts to 71–106 inch lbs. (8–12 Nm) and the nut to 44 inch lbs. (5 Nm) plus 45 degrees.
- Timing chains
- Oil pan
- Negative battery cable

5. Fill the crankcase to the correct level.

6. Start the engine and check for leaks.

CYLINDER BLOCK

0—2 mm
{0—0.079 In}

OIL SEAL

9308TG18

Rear Main Seal

REMOVAL & INSTALLATION

2.5L and 3.0L Engines

1. Before servicing the vehicle, refer to the precautions in the beginning of this section.

2. Remove or disconnect the following:
- Negative battery cable
- Transaxle
- Flywheel
- Oil seal

To install:

3. Install or connect the following:
- Oil seal. Press the seal in evenly with Special Service Tools 49 UN01 070 and 303-384 as shown.
- Flywheel. Tighten the bolts to 54–64 ft. lbs. (73–87 Nm).
- Transaxle
- Negative battery cable

4. Start the engine and check for leaks.

Timing Chain, Sprockets, Front Cover and Seal

REMOVAL & INSTALLATION

2.5L and 3.0L Engines

1. Before servicing the vehicle, refer to the precautions in the beginning of this section.

2. Relieve the fuel system pressure.

3. Drain the engine oil.

4. Remove or disconnect the following:
- Negative battery cable
- Intake manifold

Oil pump torque sequence—2.5L and 3.0L engines

9308TG17

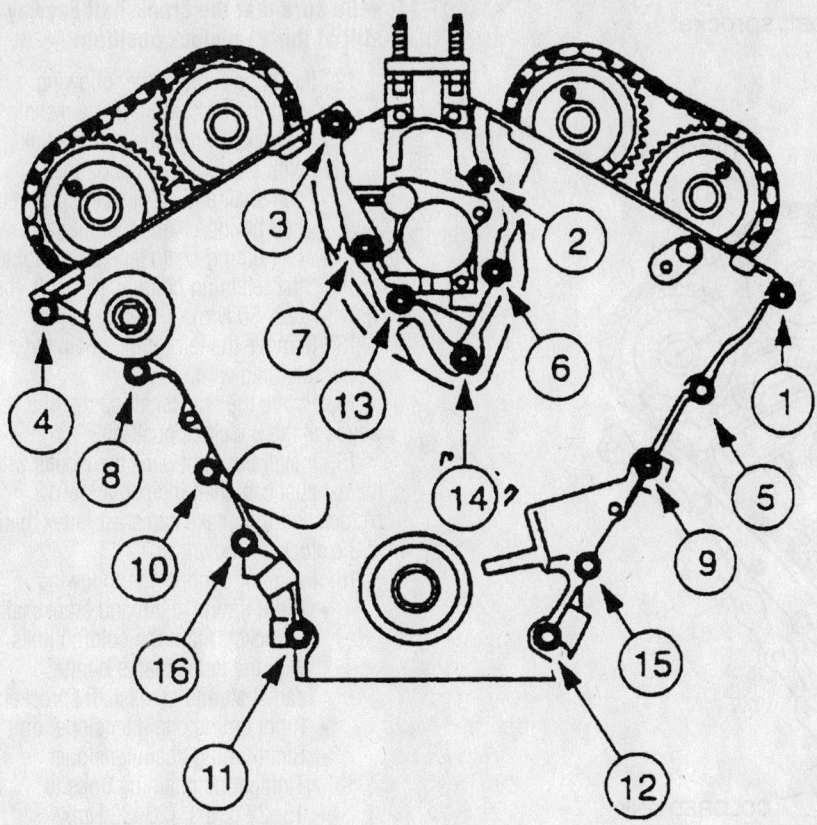

Front cover bolt removal sequence—2.5L and 3.0L engine

9308TG21

- Accessory drive belt
- Intake Manifold Runner Control (IMRC) actuator
- Spark plug wires
- Ignition coil
- Exhaust front pipe
- Oil pan
- Alternator and bracket
- A/C compressor
- Wiring harness connector bracket
- Water pump drive belt and pulley
- Camshaft oil seal housing
- Valve covers
- Right motor mount and bracket
- Crankshaft pulley
- Front crankshaft seal
- Front cover. Loosen the bolts in the sequence shown.
- Crankshaft Position (CKP) sensor pulse wheel

5. Rotate the crankshaft so that the keyway is at the 11 o'clock position to locate the crankshaft at TDC for No. 1 cylinder.

6. Verify that the alignment arrows on the camshafts are aligned. If not, rotate the crankshaft 1 complete revolution and recheck.

7. Rotate the crankshaft so that the keyway is at the 3 o'clock position. This positions the right cylinder head camshafts to the neutral position.

➡ **Keep all valvetrain components in order for assembly.**

8. Remove or disconnect the following:

- Right timing chain tensioner
- Right timing chain tensioner arm
- Right timing chain and crankshaft timing sprocket
- Right bank camshafts

9. Rotate the crankshaft 1 and ⅔ turns and set the crankshaft keyway at the 11

VIEW FROM BACK

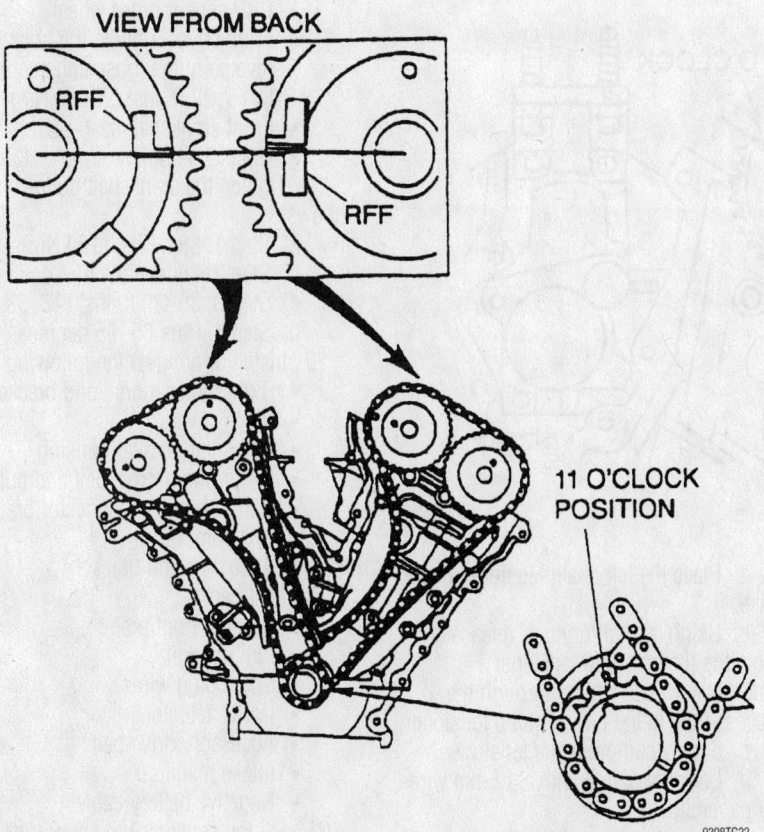

11 O'CLOCK POSITION

9308TG22

Camshaft alignment with the crankshaft in the 11 o'clock position—2.5L and 3.0L engine

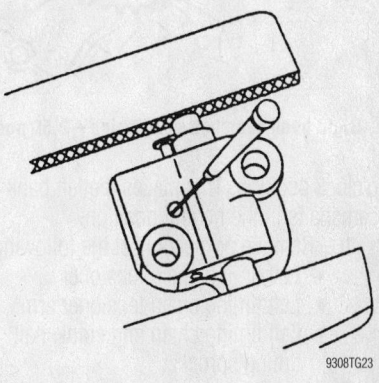

9308TG23

Using a thin prytool, release and hold the timing chain tensioner ratchet/pawl mechanism—2.5L and 3.0L engine

(1) Timing chain crankshaft sprocket
(2) Chain guide
(3) Timing chain
(4) Tensioner arm

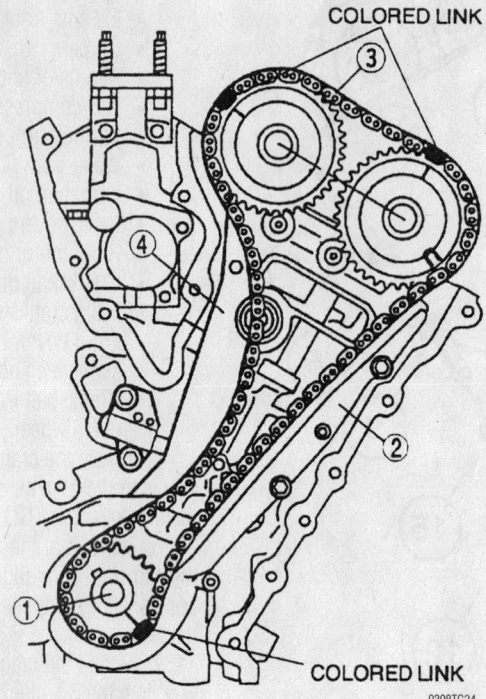

COLORED LINK
③

④

COLORED LINK
9308TG24

Left bank timing chain alignment—2.5L and 3.0L engine

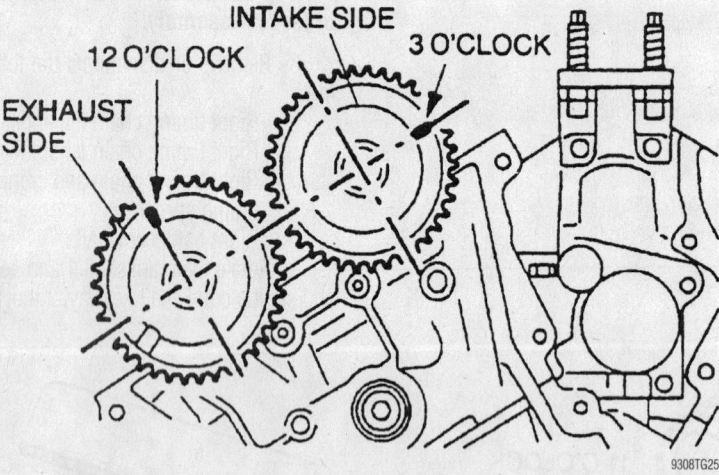

INTAKE SIDE

12 O'CLOCK   3 O'CLOCK

EXHAUST
SIDE

9308TG25

Right bank camshaft positioning—2.5L and 3.0L engine

o'clock position. This places the left bank camshafts in the neutral position.

10. Remove or disconnect the following:
- Left timing chain tensioner
- Left timing chain tensioner arm
- Left timing chain and crankshaft timing sprocket

To install:

11. Prepare the timing chain tensioners for installation as follows:

a. Place the left chain tensioner in a vise.

b. Using a small prytool, release and hold the timing chain tensioner ratchet/pawl mechanism through the access hole in the timing chain tensioner.

c. Slowly compress the tensioner.

d. Lock the piston with a 1.5mm wire or paperclip.

e. Repeat for the right chain tensioner.

➡ **Be sure that the crankshaft keyway is still at the 11 o'clock position.**

12. Install or connect the following:
- Left timing chain and crankshaft sprocket. Align the colored links with the index marks on the camshaft and crankshaft sprockets.
- Left timing chain tensioner arm
- Left timing chain tensioner. Tighten the retaining bolts to 15–22 ft. lbs. (20–30 Nm).

13. Remove the left timing chain tensioner retaining wire.

14. Rotate the crankshaft so that the keyway is at the 3 o'clock position.

15. Install the right bank camshafts with the exhaust camshaft index mark at 12 o'clock and the intake camshaft index mark at 3 o'clock as shown.

16. Install or connect the following:
- Right timing chain and crankshaft sprocket. Align the colored links with the index marks on the camshaft and crankshaft sprockets.
- Right timing chain tensioner arm
- Right timing chain tensioner. Tighten the retaining bolts to 15–22 ft. lbs. (20–30 Nm).

17. Remove the right timing chain tensioner retaining wire.

18. Install or connect the following:
- CKP sensor pulse wheel
- Front cover. Tighten the bolts in the reverse of the loosening sequence to 15–22 ft. lbs. (20–30 Nm).
- Front crankshaft seal
- Crankshaft pulley

19. Tighten the crankshaft pulley bolt as follows:
   a. Step 1: 88 ft. lbs. (120 Nm).
   b. Step 2: Loosen the bolt one turn.
   c. Step 3: 35–39 ft. lbs. (47–53 Nm).
   d. Step 4: Plus 85–95 degrees.

20. Install or connect the following:
- Right motor mount and bracket
- Valve covers
- Camshaft oil seal housing
- Water pump drive belt and pulley
- Wiring harness connector bracket
- A/C compressor
- Alternator and bracket
- Oil pan
- Exhaust front pipe
- Ignition coil
- Spark plug wires
- IMRC actuator
- Accessory drive belt
- Intake manifold
- Negative battery cable

21. Fill the crankcase to the correct level.

22. Start the engine and check for leaks.

(1) Chain guide
(2) Timing chain
(3) Tensioner arm

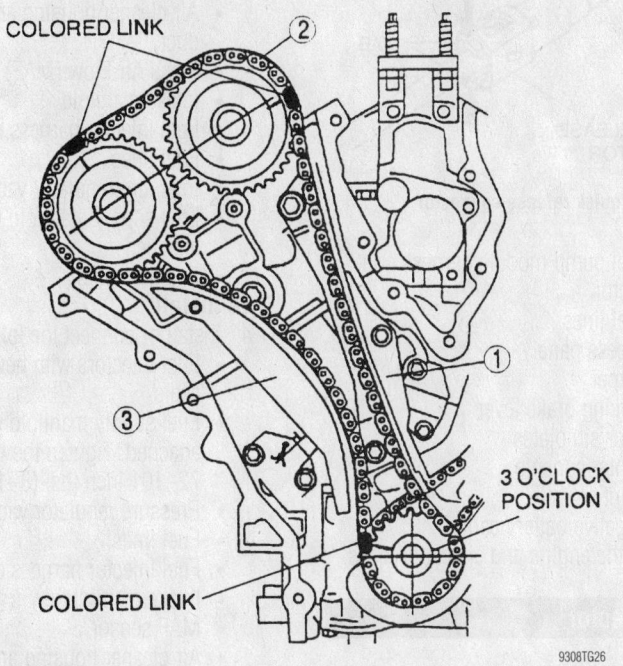

Right bank timing chain alignment—2.5L and 3.0L engine

Piston and Ring

POSITIONING

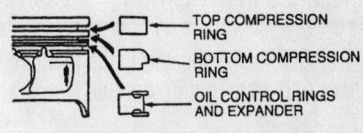
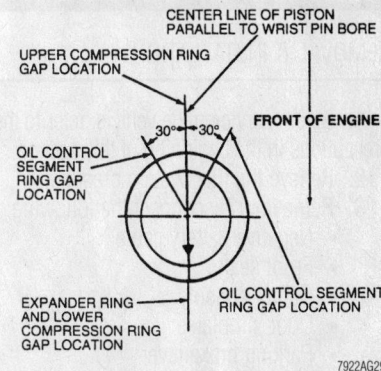

Piston ring positioning, end-gap spacing, and piston positioning. The small directional arrow must face the front of the engine—2.5L and 3.0L engine

FUEL SYSTEM

Fuel System Service Precautions

Safety is the most important factor when performing not only fuel system maintenance but any type of maintenance. Failure to conduct maintenance and repairs in a safe manner may result in serious personal injury or death. Maintenance and testing of the vehicle's fuel system components can be accomplished safely and effectively by adhering to the following rules and guidelines.

• To avoid the possibility of fire and personal injury, always disconnect the negative battery cable unless the repair or test procedure requires that battery voltage be applied.

• Always relieve the fuel system pressure prior to disconnecting any fuel system component (injector, fuel rail, pressure regulator, etc.), fitting or fuel line connection. Exercise extreme caution whenever relieving fuel system pressure, to avoid exposing skin, face and eyes to fuel spray. Please be advised that fuel under pressure may penetrate the skin or any part of the body that it contacts.

• Always place a shop towel or cloth around the fitting or connection prior to loosening to absorb any excess fuel due to spillage. Ensure that all fuel spillage

(should it occur) is quickly removed from engine surfaces. Ensure that all fuel soaked cloths or towels are deposited into a suitable waste container.

• Always keep a dry chemical (Class B) fire extinguisher near the work area.

• Do not allow fuel spray or fuel vapors to come into contact with a spark or open flame.

• Always use a back-up wrench when loosening and tightening fuel line connection fittings. This will prevent unnecessary stress and torsion to fuel line piping. Always follow the proper tighten specifications.

• Always replace worn fuel fitting O-rings with new. Do not substitute fuel hose or equivalent, where fuel pipe is installed.

Fuel System Pressure

RELIEVING

1. Before servicing the vehicle, refer to the precautions in the beginning of this section.

2. Disconnect the fuel pump relay, located at the ECM.

3. Start the engine.

4. After the engine stalls, crank the engine several times.

5. Turn the ignition switch **OFF**.

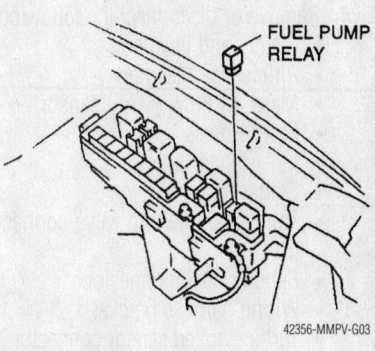

Fuel pump relay—2000-01

Fuel pump relay—2002

6. When repairs are complete, connect the fuel pump relay.

Fuel Filter

REMOVAL & INSTALLATION

The fuel filter is located in the fuel tank as part of the fuel pump module.

Fuel Pump

REMOVAL & INSTALLATION

1. Before servicing the vehicle, refer to the precautions in the beginning of this section.
2. Relieve the fuel system pressure.
3. Remove or disconnect the following:
   - Negative battery cable
   - Front seats
   - Center console
   - Door sill plates
   - Parking brake lever
   - Carpet
   - Access panel
   - Fuel lines
   - Fuel pump module harness connector
   - Fuel pump module

To install:
4. Install or connect the following:
   - Fuel pump module

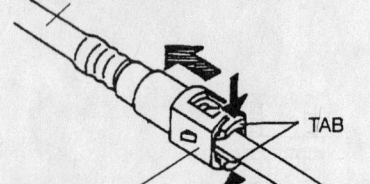

PLASTIC FUEL HOSE

TAB

QUICK RELEASE CONNECTOR

9308TG20

Fuel hose quick release connector

   - Fuel pump module harness connector
   - Fuel lines
   - Access panel
   - Carpet
   - Parking brake lever
   - Door sill plates
   - Center console
   - Front seats
   - Negative battery cable
5. Start the engine and check for leaks.

Fuel Injector

REMOVAL & INSTALLATION

2.5L and 3.0L Engine

1. Before servicing the vehicle, refer to the precautions in the beginning of this section.

2. Relieve the fuel system pressure.
3. Remove or disconnect the following:
   - Negative battery cable
   - Air cleaner housing and fresh air duct
   - Mass Air Flow (MAF) sensor
   - Intake manifold
   - Fuel injector harness connectors
   - Fuel lines
   - Pressure regulator vacuum hose
   - Fuel supply manifold with injectors attached
   - Fuel injectors

To install:
4. Install or connect the following:
   - Fuel injectors with new O-ring seals
   - Fuel supply manifold with injectors attached. Tighten the bolts to 72–101 inch lbs. (8–11 Nm).
   - Pressure regulator vacuum hose
   - Fuel lines
   - Fuel injector harness connectors
   - Intake manifold
   - MAF sensor
   - Air cleaner housing and fresh air duct
   - Negative battery cable
5. Start the engine and check for leaks.

DRIVE TRAIN

Automatic Transaxle Assembly

REMOVAL & INSTALLATION

1. Before servicing the vehicle, refer to the precautions in the beginning of this section.
2. Drain the transaxle fluid.
3. Attach a support fixture to the engine lifting eyes.
4. Remove or disconnect the following:
   - Battery and tray
   - Air cleaner assembly
   - Mass Air Flow (MAF) sensor
   - Front wheels
   - Inner fender liner
   - Starter motor
   - Transaxle solenoid valve connector
   - Range switch connector
   - Wiring harness bracket
   - Turbine speed sensor connector
   - Vehicle Speed (VSS) sensor connector

   - Shift cable
   - Transaxle oil cooler hoses
   - Subframe transverse section
   - Axle halfshafts
   - Flywheel access panel
   - Torque converter
   - Left engine mount bracket
   - Subframe center section
   - Rear engine mount
   - Transaxle flange bolts. Support the transaxle.
5. Lower the transaxle from the vehicle.

To install:
6. Install or connect the following:
   - Transaxle. Tighten the flange bolts to 28–38 ft. lbs. (38–51 Nm).
   - Rear engine mount. Tighten the bracket bolts to 50–68 ft. lbs. (67–93 Nm) and the through bolt to 63–86 ft. lbs. (86–116 Nm).
   - Subframe center section. Tighten the bolts to 48–65 ft. lbs. (64–89 Nm) and the nuts to 50–67 ft. lbs. (67–93 Nm).
   - Left engine mount bracket. Tighten

the bracket fasteners to 50–68 ft. lbs. (67–93 Nm) and the through bolt to 63–86 ft. lbs. (86–116 Nm).
   - Torque converter. Tighten the nuts to 26–36 ft. lbs. (35–49 Nm).
   - Flywheel access panel
   - Axle halfshafts
   - Subframe transverse section. Tighten the bolts to 69–97 ft. lbs. (94–131 Nm).
   - Transaxle oil cooler hoses
   - Shift cable
   - VSS sensor connector
   - Turbine speed sensor connector
   - Wiring harness bracket
   - Range switch connector
   - Transaxle solenoid valve connector
   - Starter motor
   - Inner fender liner
   - Front wheels
   - MAF sensor
   - Air cleaner assembly
   - Battery and tray
7. Fill the transaxle to the correct level.
8. Start the engine and check for leaks.

Halfshaft

REMOVAL & INSTALLATION

Left

1. Before servicing the vehicle, refer to the precautions in the beginning of this section.
2. Drain the transaxle fluid.
3. Remove or disconnect the following:
   - Front wheel
   - Wheel speed sensor
   - Hub locknut
   - Outer tie rod end
   - Lower ball joint
   - Stabilizer bar link
4. Separate the stub shaft from the hub and pry the inner joint from the transaxle.

To install:

➡**Use a new circlip, split pin and locknut for assembly.**

5. Insert the stub shaft into the wheel hub.
6. Lubricate the oil seal with transaxle fluid, then push the axle halfshaft into the transaxle. Pull on the inner joint to confirm that the circlip is seated.
7. Install or connect the following:
   - Stabilizer bar link
   - Lower ball joint. Tighten the pinch bolt to 32–43 ft. lbs. (44–58 Nm).
   - Outer tie rod end. Tighten the nut to 24–32 ft. lbs. (32–44 Nm).
   - Hub locknut. Tighten the nut to 174–235 ft. lbs. (236–318 Nm).
   - Wheel speed sensor
   - Front wheel

Right

✳✳ WARNING

Attempting to remove the right axle halfshaft while the center shaft support bracket is installed may result in damage to the center shaft support bracket.

1. Before servicing the vehicle, refer to the precautions in the beginning of this section.
2. Remove or disconnect the following:
   - Front wheel
   - Wheel speed sensor
   - Hub locknut
   - Brake caliper and rotor
   - Outer tie rod end

- Lower ball joint
- Strut bracket bolts
- Steering knuckle. Separate the stub shaft from the wheel hub.
- Center shaft support bracket
- Axle halfshaft and center shaft assembly

3. Separate the axle halfshaft and the center shaft as follows:

   a. Step 1: Place the center shaft in a vise.

   b. Step 2: Insert a pry tool between the center shaft and the axle halfshaft.

   c. Step 3: Tap on the pry tool to separate the axle halfshaft from the center shaft.

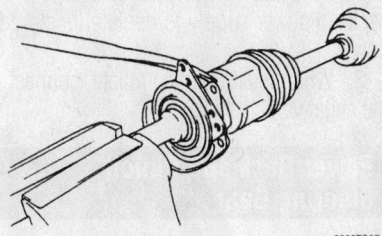

9308TG27

Separating the axle halfshaft from the center shaft

To install:

➡**Use a new split pin, locknut, and new circlips for assembly.**

4. Place the axle halfshaft in a vise and install the center shaft by tapping it with a plastic hammer as shown.

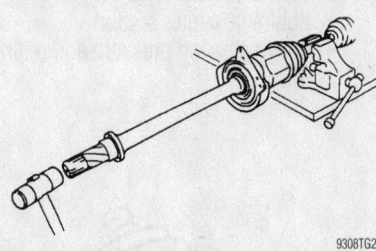

9308TG28

Installing the center shaft

5. Lubricate the oil seal with transaxle fluid, then push the center shaft into the transaxle. Pull on the inner joint to confirm that the circlip is seated.
6. Install or connect the following:
   - Center shaft support bracket. Tighten the nuts to 16–22 ft. lbs. (22–30 Nm).
   - Steering knuckle. Guide the stub shaft into the wheel hub.
   - Strut bracket bolts. Tighten the

bolts to 76–90 ft. lbs. (103–122 Nm).
   - Lower ball joint. Tighten the pinch bolt to 32–43 ft. lbs. (44–58 Nm).
   - Outer tie rod end. Tighten the nut to 24–32 ft. lbs. (32–44 Nm).
   - Hub locknut. Tighten the nut to 174–235 ft. lbs. (236–318 Nm).
   - Brake caliper and rotor. Tighten the caliper bracket bolts to 66–79 ft. lbs. (89–107 Nm).
   - Wheel speed sensor
   - Front wheel

7. Check the wheel alignment and adjust as necessary.

CV-Joint

OVERHAUL

Outer CV-Joint

The outer CV-joint is serviced with the axle halfshaft as an assembly. The outer CV-joint boot may be serviced by removing the inner joint.

Inner CV-Joint

1. Before servicing the vehicle, refer to the precautions in the beginning of this section.
2. Remove or disconnect the following:
   - Axle halfshaft from the vehicle
   - Inner CV-joint boot clamps
   - Housing retainer clip
   - CV-joint housing
   - CV-joint balls and cage
   - Snapring
   - CV-joint inner race
   - CV-joint boot

To install:

➡**Use new snaprings, clips, and boot clamps for assembly.**

3. Install or connect the following:
   - CV-joint boot
   - CV-joint inner race
   - Snapring
   - CV-joint balls and cage
   - CV-joint housing
   - Housing retainer clip
4. Fill the CV-joint housing and boot with CV-joint grease and tighten the boot clamps.
5. Install the axle halfshaft.

Inner Tri-Pot Joint

1. Before servicing the vehicle, refer to the precautions in the beginning of this section.

2. Remove or disconnect the following:
- Axle halfshaft from the vehicle
- Inner tri-pot joint boot clamps
- Tri-pot joint housing
- Snapring
- Tri-pot joint

To install:

➡**Use new snaprings, clips, and boot clamps for assembly.**

3. Install or connect the following:
- Tri-pot joint

- Snapring
- Tri-pot joint housing

4. Fill the tri-pot joint housing and boot with grease and tighten the boot clamps.

5. Install the axle halfshaft.

STEERING AND SUSPENSION

Air Bag

❋❋ CAUTION

Some vehicles are equipped with an air bag system. The system must be disarmed before performing service on, or around, system components, the steering column, instrument panel components, wiring and sensors. Failure to follow the safety precautions and the disarming procedure could result in accidental air bag deployment, possible injury and unnecessary system repairs.

PRECAUTIONS

Several precautions must be observed when handling the inflator module to avoid accidental deployment and possible personal injury.

• Never carry the inflator module by the wires or connector on the underside of the module.

• When carrying a live inflator module, hold securely with both hands, and ensure that the bag and trim cover are pointed away.

• Place the inflator module on a bench or other surface with the bag and trim cover facing up.

• With the inflator module on the bench, never place anything on or close to the

module which may be thrown in the event of an accidental deployment.

DISARMING

1. Turn the ignition switch to the **LOCK** position.

2. Disconnect the negative battery cable and wait at least 1 minute to allow the back-up power supply to deplete its stored power.

3. When repairs are complete, connect the negative battery cable.

Power Rack and Pinion Steering Gear

REMOVAL & INSTALLATION

1. Before servicing the vehicle, refer to the precautions in the beginning of this section.

2. Attach a support fixture to the engine lifting eyes.

3. Remove or disconnect the following:
- Front wheels
- Wheel speed sensors
- Steering shaft pinch bolt
- Outer tie rod ends
- Subframe transverse section
- Subframe center section
- Power steering pressure and return lines
- Steering gear

To install:

4. Install or connect the following:
- Steering gear. Tighten the fasteners in sequence to 55–77 ft. lbs. (75–104 Nm).
- Power steering pressure and return lines
- Subframe center section. Tighten the bolts to 48–65 ft. lbs. (64–89 Nm) and the nuts to 50–68 ft. lbs. (67–93 Nm).
- Subframe transverse section. Tighten the bolts to 69–96 ft. lbs. (94–131 Nm).
- Outer tie rod ends. Tighten the nuts to 24–32 ft. lbs. (32–44 Nm).
- Steering shaft pinch bolt. Tighten the bolt to 14–19 ft. lbs. (19–26 Nm).
- Wheel speed sensors
- Front wheels

5. Fill the power steering reservoir.

6. Check the wheel alignment and adjust as necessary.

Strut

REMOVAL & INSTALLATION

1. Before servicing the vehicle, refer to the precautions in the beginning of this section.

2. Remove or disconnect the following:
- Front wheel
- Brake hose clip
- Stabilizer bar link
- Steering knuckle bolts
- Upper strut mount nuts
- Strut assembly

To install:

3. Install or connect the following:
- Strut assembly. Tighten the upper strut mount nuts to 34–46 ft. lbs. (47–62 Nm) and the steering knuckle bolts to 76–90 ft. lbs. (103–122 Nm).
- Stabilizer bar link. Tighten the nut to 32–44 ft. lbs. (44–60 Nm).
- Brake hose clip
- Front wheel

4. Check the wheel alignment and adjust as necessary.

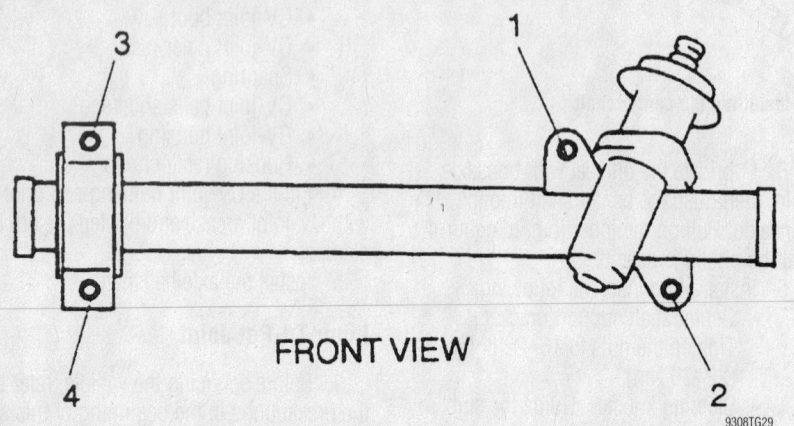

FRONT VIEW

9308TG29

Steering gear torque sequence

Shock Absorber

REMOVAL & INSTALLATION

1. Before servicing the vehicle, refer to the precautions in the beginning of this section.
2. Support the rear axle with a jack or stands.
3. Remove or disconnect the following:
   • Rear wheel
   • Shock absorber

To install:

4. Install or connect the following:
   • Shock absorber. Tighten the upper nut to 56–76 ft. lbs. (76–102 Nm) and the lower bolt to 71–94 ft. lbs. (97–127 Nm).
   • Rear wheel

Coil Spring

REMOVAL & INSTALLATION

Front

1. Before servicing the vehicle, refer to the precautions in the beginning of this section.
2. Remove the strut assembly from the vehicle.
3. Compress the coil spring and remove the piston rod nut.
4. Remove or disconnect the following:
   • Upper strut mount
   • Strut mount bearing
   • Spring upper seat
   • Coil spring

To install:

5. Install or connect the following:
   • Coil spring
   • Spring upper seat
   • Strut mount bearing
   • Upper strut mount. Tighten the piston rod nut to 66–94 ft. lbs. (90–127 Nm).
6. Remove the spring compressor and install the strut assembly to the vehicle.
7. Check the wheel alignment and adjust as necessary.

Rear

1. Before servicing the vehicle, refer to the precautions in the beginning of this section.
2. Support the vehicle at the frame and support the axle with a jack.
3. Remove or disconnect the following:
   • Rear wheels
   • Lateral rod
   • Shock absorber
4. Lower the rear axle and remove the coil springs.

To install:

5. Place the coil springs on the spring seats and raise the axle into position.
6. Install or connect the following:
   • Shock absorber. Tighten the upper nut to 56–76 ft. lbs. (76–102 Nm) and the lower bolt to 71–94 ft. lbs. (97–127 Nm).
   • Lateral rod. Tighten the fastener to 76–101 ft. lbs. (102–137 Nm).
   • Rear wheels

Lower Ball Joint

REMOVAL & INSTALLATION

The lower ball joint is serviced with the lower control arm as an assembly.

Lower Control Arm

REMOVAL & INSTALLATION

2000–01

1. Before servicing the vehicle, refer to the precautions in the beginning of this section.
2. Support the arm.
3. Remove or disconnect the following:
   • Front wheel
   • Lower ball joint
   • Front inner control arm bolt
   • Rear inner control arm bracket
   • Lower control arm

To install:

4. Install or connect the following:
   • Lower control arm. Tighten the front inner bolt to 69–93 ft. lbs. (94–126 Nm).
   • Rear inner control arm bracket. Tighten the nut to 69–97 ft. lbs. (94–131 Nm).
   • Lower ball joint. Tighten the pinch bolt to 32–43 ft. lbs. (44–58 Nm).
   • Front wheel

5. Check the wheel alignment and adjust as necessary.

2002

1. Before servicing the vehicle, refer to the precautions in the beginning of this section.
2. Support the arm.
3. Remove or disconnect the following:
   • Front wheel
   • Pivot bolt
   • Dynamic damper
   • Ball joint bolt
   • Ball joint bracket
   • Nut
   • Lower arm
4. Installation is the reverse of removal. Observe the following torques:
   • Nut: 76–97 ft. lbs. (103–131 Nm)
   • Bracket bolt: 32–43 ft. lbs. (43–59 Nm)
   • Pivot bolt: 74–101 ft. lbs. (101–137 Nm)

CONTROL ARM BUSHING REPLACEMENT

1. Before servicing the vehicle, refer to the precautions in the beginning of this section.
2. Remove the control arm from the vehicle.
3. Mark the control arm to indicate the alignment of the rear bushing as shown.
4. Remove the control arm bushings with a hydraulic press.

To install:

5. Lubricate the control arm bushings with liquid soap.
6. If replacing the rear bushing, align the direction marks as shown.
7. Press the bushings into the control arm until the bushing flange contacts the housing edge of the control arm.
8. Install the control arm to the vehicle.
9. Check the wheel alignment and adjust as necessary.

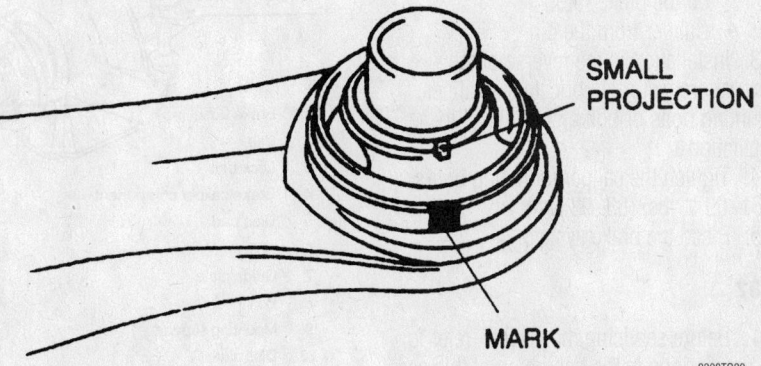

SMALL PROJECTION

MARK

9308TG30

Rear bushing alignment marks

Wheel Bearing

ADJUSTMENT

1. Before servicing the vehicle, refer to the precautions in the beginning of this section.
2. Remove or disconnect the following:
   - Front wheel
   - Brake caliper and rotor
3. Position a dial indicator gauge against the wheel hub. Push and pull the wheel hub in and out and measure the end-play of the wheel bearing.
4. End-play should not exceed 0.002 in. (0.05mm).
5. If end-play is excessive, replace the hub retainer locknut and tighten it to specification. Recheck the end-play.
6. If end-play is not within specification, replace the wheel bearing assembly.

REMOVAL & INSTALLATION

Front

1. Before servicing the vehicle, refer to the precautions in the beginning of this section.

2. Remove or disconnect the following:
   - Front wheel
   - Brake caliper and rotor
   - Wheel speed sensor
   - Outer tie rod end
   - Lower ball joint
   - Hub retainer locknut
   - Strut bracket bolts
   - Steering knuckle
   - Inner oil seal
   - Hub
   - Snapring
   - Wheel bearing cartridge

To install:

→ **Use new locknuts, split pins and oil seals for assembly.**

3. Install or connect the following:
   - Wheel bearing cartridge
   - Snapring
   - Hub
   - Inner oil seal
   - Steering knuckle. Tighten the strut bracket bolts to 76–90 ft. lbs. (103–122 Nm).
   - Hub retainer locknut. Tighten the nut to 174–235 ft. lbs. (236–318 Nm).

   - Lower ball joint. Tighten the pinch bolt to 32–43 ft. lbs. (44–58 Nm).
   - Outer tie rod end. Tighten the nut to 24–32 ft. lbs. (32–44 Nm).
   - Wheel speed sensor. Tighten the bolt to 14–18 ft. lbs. (19–25 Nm).
   - Brake caliper and rotor
   - Front wheel

Rear

1. Before servicing the vehicle, refer to the precautions in the beginning of this section.
2. Remove or disconnect the following:
   - Rear wheel
   - Brake drum
   - Dust cap
   - Hub retaining lock nut
   - Wheel bearing and hub assembly

To install:

3. Install or connect the following:
   - Wheel bearing and hub assembly. Tighten the locknut to 131–173 ft. lbs. (177–235 Nm).
   - Dust cap
   - Brake drum
   - Rear wheel

BRAKES

Brake Caliper

REMOVAL & INSTALLATION

2000–01

1. Before servicing the vehicle, refer to the precautions in the beginning of this section.
2. Remove or disconnect the following:
   - Wheel assembly
   - Banjo bolt and disconnect the brake hose from the caliper. Plug the hose to prevent fluid leakage.
   - Caliper mounting bolt and pivot the caliper about the mounting pin and off the brake rotor
   - Caliper from the pin
3. Installation is the reverse of the removal procedure. Lubricate the caliper mounting bolts or bolt and pin prior to installation.
4. Tighten the caliper mounting bolt(s) to 61–69 ft. lbs. (83–93 Nm).
5. Bleed the brake system.

2002

1. Before servicing the vehicle, refer to the precautions in the beginning of this section.

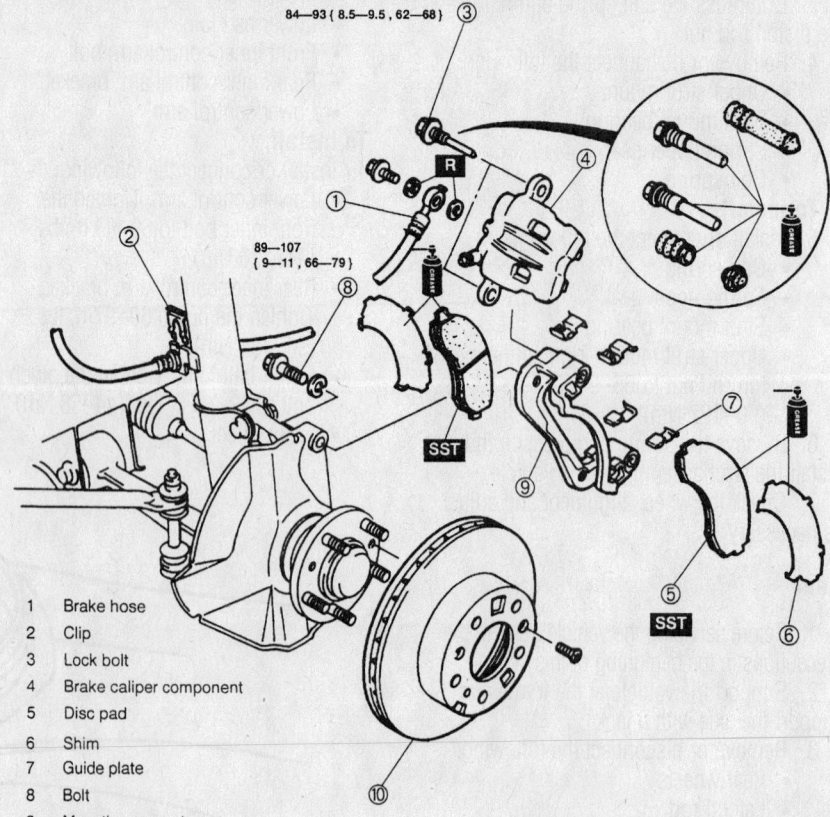

84—93 { 8.5—9.5 , 62—68 }

89—107 { 9—11 , 66—79 }

1	Brake hose
2	Clip
3	Lock bolt
4	Brake caliper component
5	Disc pad
6	Shim
7	Guide plate
8	Bolt
9	Mounting support
10	Disc plate

N·m { kgf·m , ft·lbf }

93026G35

Front disc brake assembly—2000–01

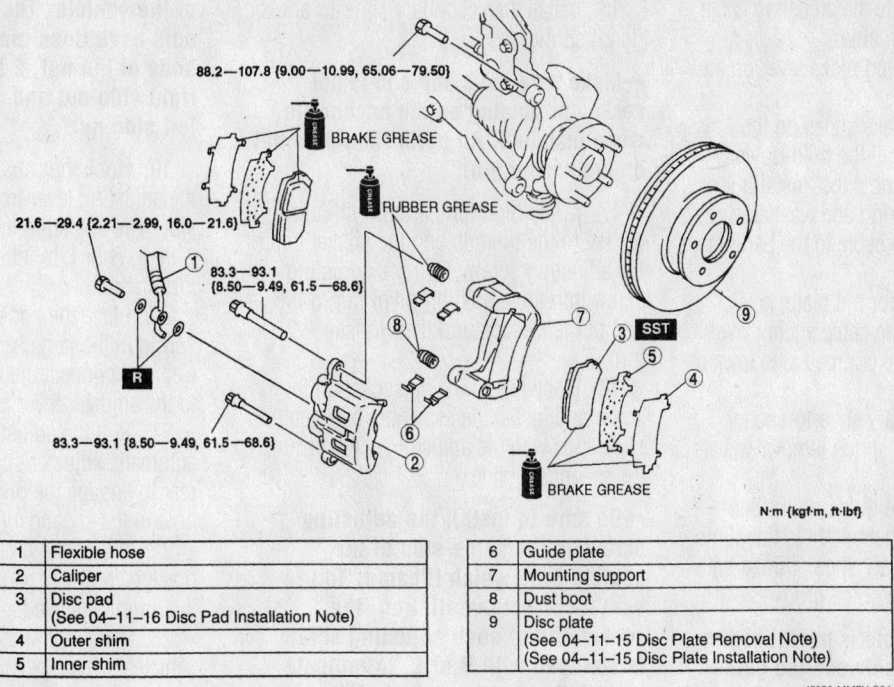

88.2—107.8 {9.00—10.99, 65.06—79.50}

BRAKE GREASE

RUBBER GREASE

21.6—29.4 {2.21—2.99, 16.0—21.6}

83.3—93.1
{8.50—9.49, 61.5—68.6}

R

83.3—93.1 {8.50—9.49, 61.5—68.6}

SST

BRAKE GREASE

N·m {kgf·m, ft·lbf}

1	Flexible hose
2	Caliper
3	Disc pad (See 04–11–16 Disc Pad Installation Note)
4	Outer shim
5	Inner shim

6	Guide plate
7	Mounting support
8	Dust boot
9	Disc plate (See 04–11–15 Disc Plate Removal Note) (See 04–11–15 Disc Plate Installation Note)

42356-MMPV-G04

Front disc brake assembly—2002

2. Remove or disconnect the following:
- Wheel assembly
- Banjo bolt and disconnect the brake hose from the caliper. Plug the hose to prevent fluid leakage.
- Caliper mounting bolts

3. Installation is the reverse of the removal procedure. Lubricate the caliper mounting bolts or bolt and pin prior to installation.

4. Tighten the caliper mounting bolts to 61–69 ft. lbs. (83–93 Nm).

5. Bleed the brake system.

Disc Brake Pads

REMOVAL & INSTALLATION

1. Before servicing the vehicle, refer to the precautions in the beginning of this section.

2. Remove or disconnect the following:
- Wheel assembly
- Lower lock pin bolt from the caliper

3. Rotate the caliper upward and remove the brake pads, shims, guide plates and if equipped, the springs.

To install:

4. Remove the master cylinder reservoir cap and remove about ½ of the fluid from the reservoir.

5. Using a large C-clamp and piece of wood, depress the caliper piston(s) until they bottom in their bores.

6. Install the shims, guide plates, new pads and if removed, the springs.

7. Reposition the caliper and install the lock pin bolt. Torque the lockbolt to 62–68 ft. lbs. 84–93 Nm).

8. Install the wheels, lower the vehicle, refill the master cylinder and depress the brake pedal a few times to restore pressure. Bleed the system if required.

Brake Drum

REMOVAL & INSTALLATION

1. Raise and safely support the vehicle. Remove the wheel and tire assembly.

2. Remove the screws, if equipped, and remove the brake drum.

3. Inspect the brake drum surface for wear, scoring and runout. Machine or replace, as necessary.

To install:

4. Install the brake drum and secure in place with the screws, if equipped. Torque the screws to 10 ft. lbs. (14 Nm).

5. Adjust the rear brakes.

6. Install the wheel. Lower the vehicle.

Brake Shoes

REMOVAL & INSTALLATION

1. Raise and safely support the vehicle. Remove the wheel and tire assembly and the brake drum.

2. Pull backward on the adjusting lever

cable to disengage the adjusting lever from the adjusting screw. Move the outboard side of the adjusting screw upward and back off the pivot nut as far as it will go.

3. Pull the adjusting lever, cable and automatic adjuster spring down and toward the rear to unhook the pivot hook from the large hole in the secondary shoe web. Do not pry the pivot hook from the hole.

4. Remove the automatic adjuster spring and adjusting lever.

5. Remove the secondary shoe-to-anchor spring using a suitable brake spring removal/installation tool. Using the tool, remove the primary shoe-to-anchor spring and unhook the cable anchor. Remove the anchor pin plate, if equipped.

6. Remove the cable guide from the secondary shoe.

7. Remove the shoe hold-down springs, shoes, adjusting screw, pivot nut and socket. Note the color and position of each hold-down spring so they can be reassembled in the same position.

8. Remove the parking brake link and spring. Disconnect the parking brake cable from the parking brake lever.

9. Remove the secondary brake shoe. Remove the retainer clip and spring washer and remove the parking brake lever.

To install:

10. Clean the backing plate ledge pads and sand lightly. Apply a light coating of high temperature lithium grease to the points where the brake shoes touch the

backing plate. Lubricate the adjusting cable eye and the anchor pin area.

11. Install the parking brake lever on the secondary shoe.

12. Position the brake shoes on the backing plate and install the hold-down spring pins, springs and cups. Install the parking brake link, spring and washer. Connect the parking brake cable to the parking brake lever.

13. Install the anchor pin plate, if equipped, and place the cable anchor over the anchor pin with the crimped side toward the backing plate.

14. Install the primary shoe-to-anchor spring using the brake spring removal/installation tool.

15. Install the cable guide on the secondary shoe with the flanged hole fitted into the hole in the secondary shoe. Thread the cable around the cable guide groove.

➡ Make sure the cable is positioned in the groove and not between the guide and shoe web.

16. Install the secondary shoe-to-anchor (long) spring.

➡ Make sure the cable end is not cocked or binding on the anchor pin when installed. All parts should be flat on the anchor pin.

17. Apply high temperature lithium grease to the threads and the socket end of the adjusting screw. Turn the adjusting screw into the adjusting pivot nut to the end of the threads and then loosen, ½ turn.

18. Place the adjusting socket on the screw and install the assembly between the shoe ends with the adjusting screw nearest the secondary shoe.

➡ Be sure to install the adjusting screw on the same side of the vehicle from which it came. To prevent incorrect installation, the socket end of each adjusting screw is stamped with R or L, to indicate installation on the right or left side

of the vehicle. The adjusting pivot nuts have lines machined around the body of the nut, 2 lines indicating the right side nut and 1 line indicating the left side nut.

19. Hook the cable hook into the hole in the adjusting lever from the outboard plate side. The adjusting levers are also stamped with an **R** or **L** to indicate right or left side installation.

20. Place the hooked end of the adjuster spring in the large hole in the primary shoe web and connect the loop end of the spring to the adjuster lever hole.

21. Pull the adjuster lever, cable and automatic adjuster spring down toward the rear to engage the pivot hook in the large hole in the secondary shoe web.

22. Adjust the brake shoes using either a brake adjustment gauge or manually with the drums installed.

23. Install the wheels, and lower the vehicle.

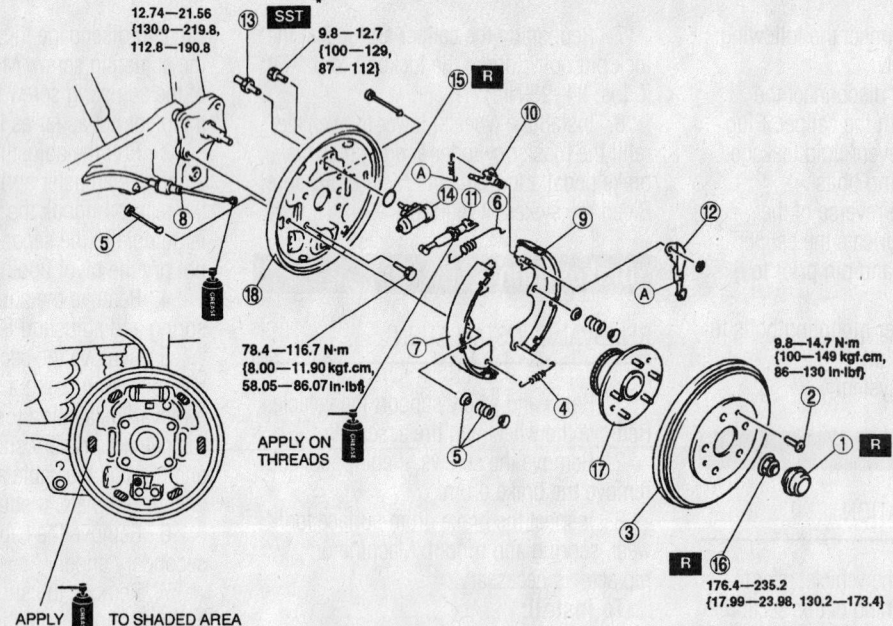

1	Hub cap		10	Ajuster lever
2	Screw		11	Adjuster component
3	Brake drum		12	Operating lever
4	Return spring		13	Brake pipe
5	Hold pin and hold spring		14	Wheel cylinder
6	Anti-rattle spring		15	O-ring
7	Leading shoe		16	Locknut
8	Parking brake cable		17	Wheel hub
9	Trailing shoe		18	Backing plate

Exploded view of the rear brake shoes and components

42356-MMPV-G05

MAZDA

Tribute

20

SPECIFICATION CHARTS

ENGINE AND VEHICLE IDENTIFICATION

		Engine						Model Year	
Code ①	Liters (cc)	Cu. In.	Cyl.	Fuel Sys.	Engine Type	Eng. Mfg.		Code ②	Year
B	2.0 (1998)	121	4	SFI	DOHC	Ford		1	2001
1	3.0 (3049)	182	6	SFI	DOHC	Ford		2	2002
SFI: Multi-port Fuel Injection								3	2003
DOHC: Double Overhead Camshafts								4	2004

① 8th digit of VIN

② 10th digit of VIN

42356-TRIB-C01

GENERAL ENGINE SPECIFICATIONS

Year	Model	Engine Displacement Liters (cc)	Engine ID/VIN	Fuel System Type	Net Horsepower @ rpm	Net Torque @ rpm (ft. lbs.)	Bore x Stroke (in.)	Compression Ratio	Oil Pressure @ rpm①
2001	Tribute	2.0 (1998)	B	SFI	135@5500	135@4500	3.34x3.46	9.6:1	54-80
	Tribute	3.0 (3049)	1	SFI	200@5500	200@4500	3.50x3.13	10.0:1	45
2002	Tribute	2.0 (1998)	B	SFI	135@5500	135@4500	3.34x3.46	9.6:1	54-80
	Tribute	3.0 (3049)	1	SFI	200@5500	200@4500	3.50x3.13	10.0:1	45
2003	Tribute	2.0 (1998)	B	SFI	135@5500	135@4500	3.34x3.46	9.6:1	54-80
	Tribute	3.0 (3049)	1	SFI	200@5500	200@4500	3.50x3.13	10.0:1	45

SFI: Multi-port Fuel Injection

① The manufacturer does not provide an engine speed specification for oil pump pressure.

42356-TRIB-C02

ENGINE TUNE-UP SPECIFICATIONS

Year	Engine Displacement Liters (cc)	Engine ID/VIN	Spark Plug Gap (in.)	Ignition Timing (deg.) MT	Ignition Timing (deg.) AT	Fuel Pump (psi)	Idle Speed (rpm) MT	Idle Speed (rpm) AT	Valve Clearance Intake	Valve Clearance Exhaust
2001	2.0 (1998)	B	0.039-0.043	10 BTDC	—	65	①	—	HYD.	HYD.
	3.0 (3049)	1	0.052-0.056	10 BTDC	10 BTDC	65	①	①	HYD.	HYD.
2002	2.0 (1998)	B	0.051	10 BTDC	—	65	①	—	HYD.	HYD.
	3.0 (3049)	1	0.052-0.056	10 BTDC	10 BTDC	65	①	①	HYD.	HYD.
2003	2.0 (1998)	B	0.051	10 BTDC	—	65	①	—	HYD.	HYD.
	3.0 (3049)	1	0.052-0.056	10 BTDC	10 BTDC	65	①	①	HYD.	HYD.

BTDC: Before Top Dead Center

HYD: Hydraulic lash adjusters

① Refer to Vehicle Emission Control Information Label

42356-TRIB-C03

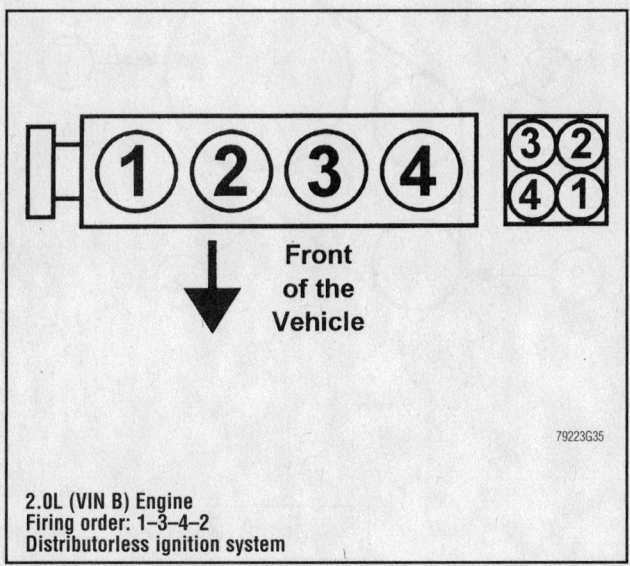

2.0L (VIN B) Engine
Firing order: 1–3–4–2
Distributorless ignition system

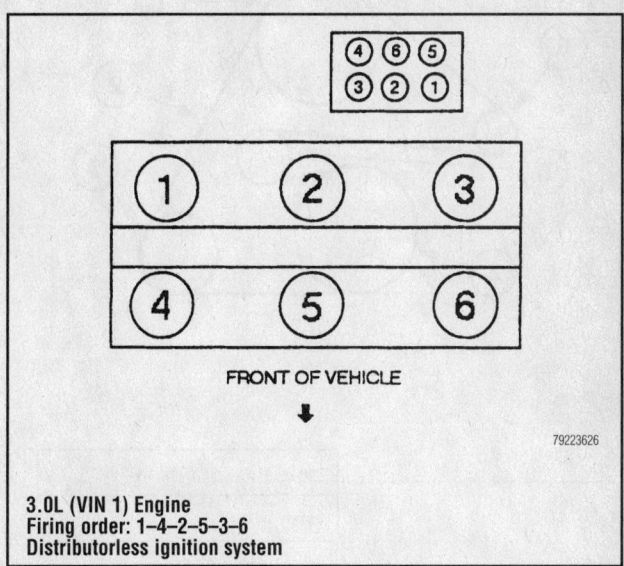

3.0L (VIN 1) Engine
Firing order: 1–4–2–5–3–6
Distributorless ignition system

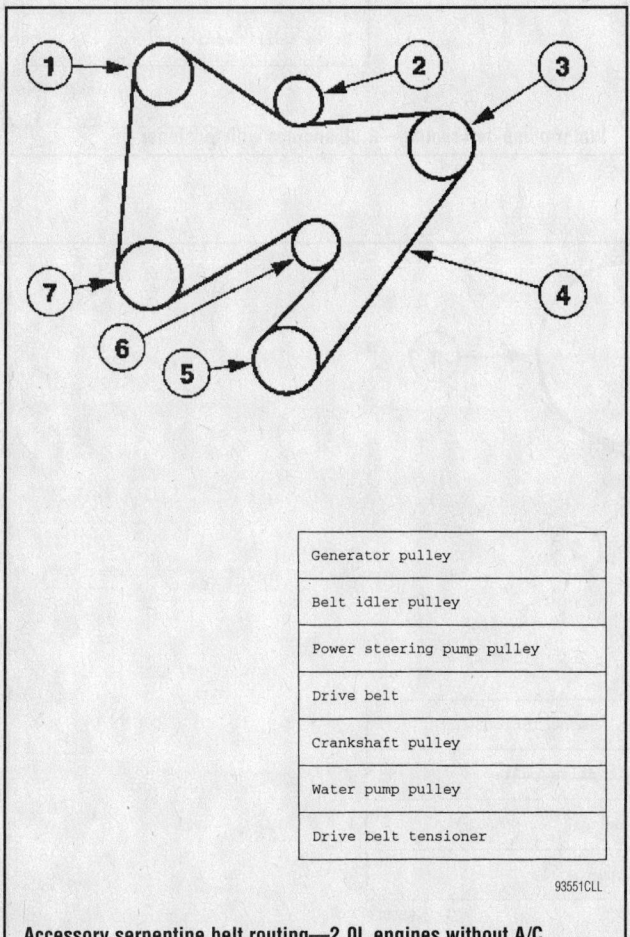

| Generator pulley |
| Belt idler pulley |
| Power steering pump pulley |
| Drive belt |
| Crankshaft pulley |
| Water pump pulley |
| Drive belt tensioner |

93551CLL

Accessory serpentine belt routing—2.0L engines without A/C

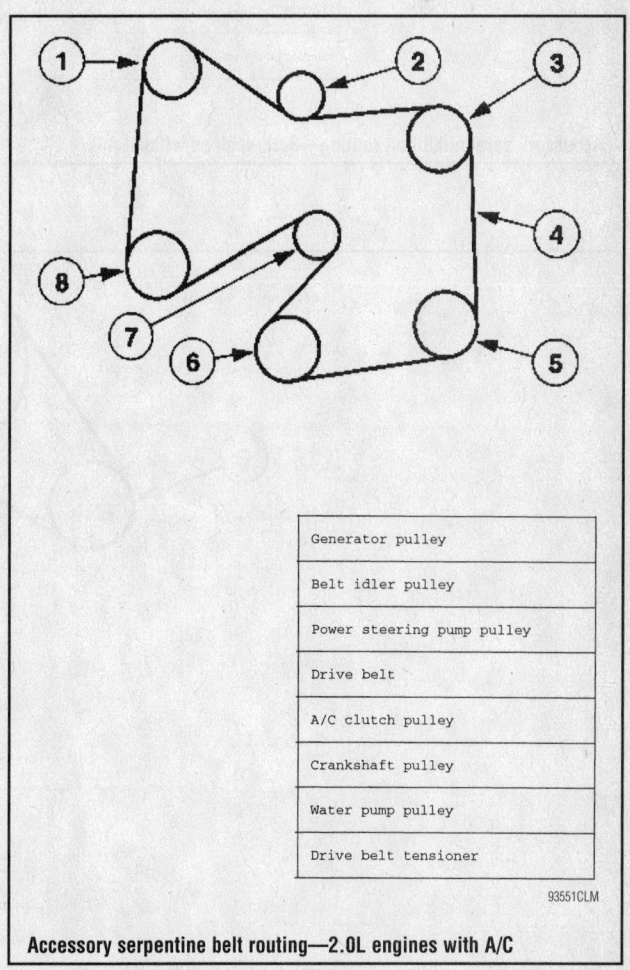

| Generator pulley |
| Belt idler pulley |
| Power steering pump pulley |
| Drive belt |
| A/C clutch pulley |
| Crankshaft pulley |
| Water pump pulley |
| Drive belt tensioner |

93551CLM

Accessory serpentine belt routing—2.0L engines with A/C

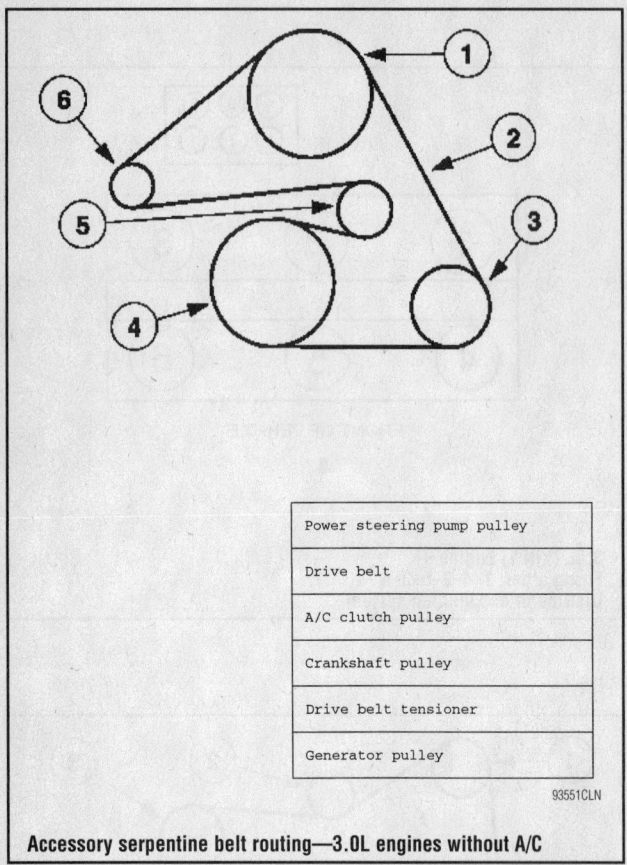

| Power steering pump pulley |
| A/C clutch pulley |
| Crankshaft pulley |
| Drive belt tensioner |
| Generator pulley |

93551CLN

Accessory serpentine belt routing—3.0L engines without A/C

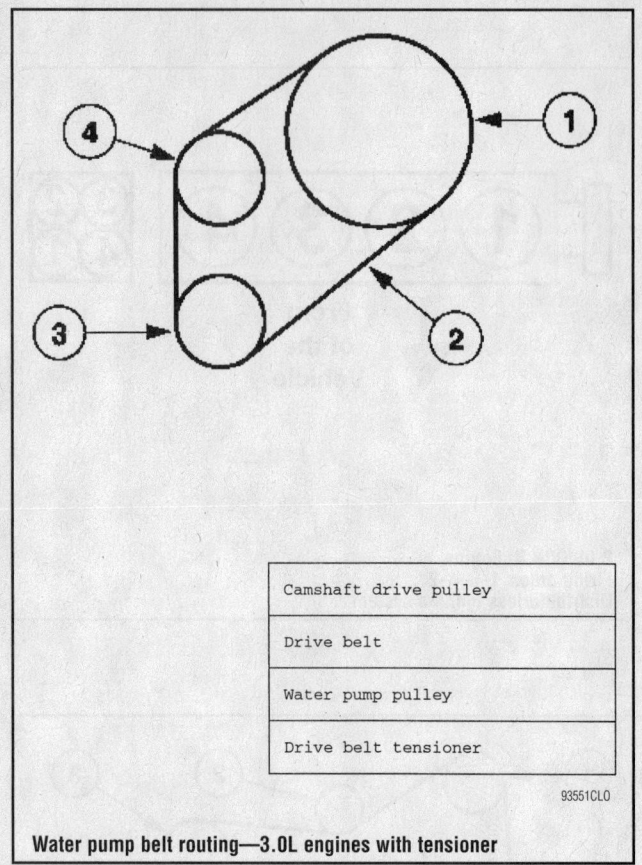

| Camshaft drive pulley |
| Drive belt |
| Water pump pulley |
| Drive belt tensioner |

93551CLO

Water pump belt routing—3.0L engines with tensioner

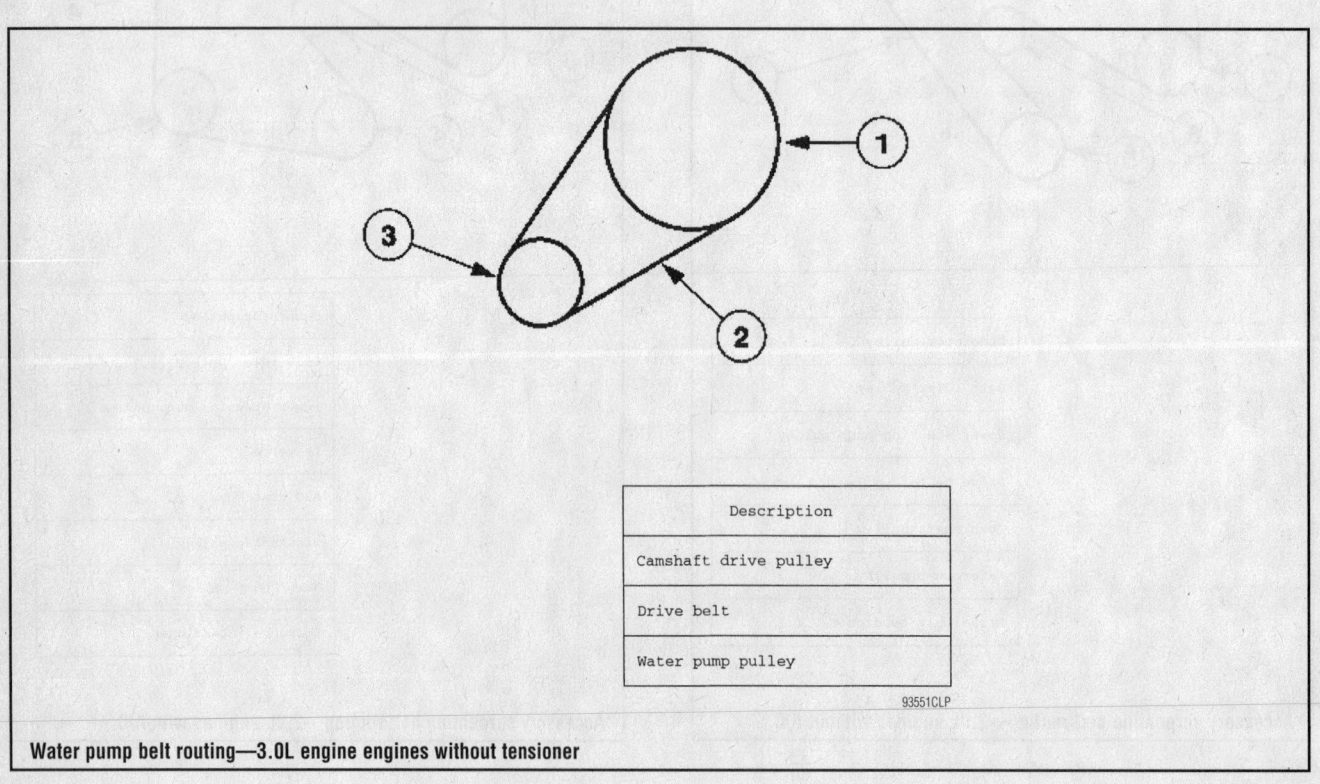

Description
Camshaft drive pulley
Drive belt
Water pump pulley

93551CLP

Water pump belt routing—3.0L engine engines without tensioner

CAPACITIES

Year	Model	Engine Displacement Liters (cc)	Engine ID/VIN	Engine Oil with Filter (qts.)	Transmission (pts.)		Transfer Case (pts.)	Drive Axle		Fuel Tank (gal.)	Cooling System (qts.)
					Manual	Auto.		Front (pts.)	Rear (pts.)		
2001	Tribute	2.0 (1998)	B	4.5	5.7	—	—	—	3.0	15.0	7.0
	Tribute	3.0 (3049)	1	5.8	5.7	20.0	①	2.6	3.0	16.0	10.5
2002	Tribute	2.0 (1998)	B	4.5	5.7	—	—	—	3.0	15.0	7.0
	Tribute	3.0 (3049)	1	5.5	5.7	20.0	①	2.6	3.0	16.0	10.5
2003	Tribute	2.0 (1998)	B	4.5	5.7	—	—	—	3.0	15.0	7.0
	Tribute	3.0 (3049)	1	5.5	5.7	20.0	①	2.6	3.0	16.0	10.5

NOTE: All capacities are approximate. Add fluid gradually and check to be sure a proper fluid level is obtained.

① The transfer case is lubricated for life and is not to be checked unless a leak is suspected or a repair is necessary.

42356-TRIB-C04

CRANKSHAFT AND CONNECTING ROD SPECIFICATIONS
All measurements are given in inches.

Year	Engine Displacement Liters (cc)	Engine ID/VIN	Crankshaft				Connecting Rod		
			Main Brg. Journal Dia.	Main Brg. Oil Clearance	Shaft End-play	Thrust on No.	Journal Diameter	Oil Clearance	Side Clearance
2001	2.0 (1998)	B	2.2827- 2.2835	①	0.0035- 0.0102	3	1.8460- 1.8468	0.0006- 0.0028	0.0040- 0.0110
	3.0 (3049)	1	2.4791- 2.4800	0.0010- 0.0018	0.0043- 0.0091	3	1.9673- 1.9681	0.0010- 0.0025	0.0039- 0.0118
2002	2.0 (1998)	B	2.2827- 2.2835	①	0.0035- 0.0102	3	1.8460- 1.8468	0.0006- 0.0028	0.0040- 0.0110
	3.0 (3049)	1	2.4791- 2.4800	0.0010- 0.0018	0.0043- 0.0091	3	1.9673- 1.9681	0.0011- 0.0026	0.0039- 0.0118
2003	2.0 (1998)	B	2.2827- 2.2835	①	0.0035- 0.0102	3	1.8460- 1.8468	0.0006- 0.0028	0.0040- 0.0110
	3.0 (3049)	1	2.4791- 2.4800	0.0010- 0.0018	0.0043- 0.0091	3	1.9673- 1.9681	0.0011- 0.0026	0.0039- 0.0118

① Journals 1, 2 and 4: 0.0010 - 0.0017 in.
  Journal 3: 0.0012 - 0.0019 in.

42356-TRIB-C05

VALVE SPECIFICATIONS

Year	Engine Displacement Liters (cc)	Engine ID/VIN	Seat Angle (deg.)	Face Angle (deg.)	Spring Test Pressure (lbs. @ in.)	Spring Installed Height (in.)	Stem-to-Guide Clearance (in.) Intake	Stem-to-Guide Clearance (in.) Exhaust	Stem Diameter (in.) Intake	Stem Diameter (in.) Exhaust
2001	2.0 (1998)	B	45	45	①	1.346	0.0007-0.0025	0.0007-0.0025	0.2374	0.2374
	3.0 (3049)	1	44.75	45.5	153@ 1.18	1.57	0.0008-0.0027	0.0018-0.0037	0.2352-0.2360	0.2343-0.2350
2002	2.0 (1998)	B	45	45	①	1.346	0.0007-0.0025	0.0007-0.0025	0.2374	0.2374
	3.0 (3049)	1	44.75	45.5	153@ 1.18	1.57	0.0008-0.0027	0.0018-0.0037	0.2352-0.2360	0.2343-0.2350
2003	2.0 (1998)	B	45	45	①	1.346	0.0007-0.0025	0.0007-0.0025	0.2374	0.2374
	3.0 (3049)	1	44.75	45.5	153@ 1.18	1.57	0.0008-0.0027	0.0018-0.0037	0.2352-0.2360	0.2343-0.2350

① Intake: 82.1@ 0.988

Exhaust: 95@ 1.0275

42356-TRIB-C06

PISTON AND RING SPECIFICATIONS

All measurements are given in inches.

Year	Engine Displ. Liters (cc)	Engine ID/VIN	Piston Clearance	Ring Gap Top Comp.	Ring Gap Bottom Comp.	Ring Gap Oil Control	Ring Side Clearance Top Comp.	Ring Side Clearance Bottom Comp.	Ring Side Clearance Oil Control
2001	2.0 (1998)	B	0.0004-0.0012	0.0100-0.0300	0.0100-0.0300	0.0160-0.0660	0.0015-0.0032	0.0015-0.0035	snug
	3.0 (3049)	1	0.0005-0.0009	0.0039-0.0098	0.0106-0.0165	0.0059-0.0256	0.0016-0.0030	0.0016-0.0033	snug
2002	2.0 (1998)	B	0.0004-0.0012	0.0100-0.0300	0.0100-0.0300	0.0160-0.0660	0.0015-0.0032	0.0015-0.0035	snug
	3.0 (3049)	1	0.0005-0.0009	0.0039-0.0098	0.0106-0.0165	0.0059-0.0256	0.0016-0.0030	0.0016-0.0033	snug
2003	2.0 (1998)	B	0.0004-0.0012	0.0100-0.0300	0.0100-0.0300	0.0160-0.0660	0.0015-0.0032	0.0015-0.0035	snug
	3.0 (3049)	1	0.0005-0.0009	0.0039-0.0098	0.0106-0.0165	0.0059-0.0256	0.0016-0.0030	0.0016-0.0033	snug

42356-TRIB-C07

TORQUE SPECIFICATIONS
All readings in ft. lbs.

Year	Engine Displacement Liters (cc)	Engine ID/VIN	Cylinder Head Bolts	Main Bearing Bolts	Rod Bearing Bolts	Crankshaft Damper Bolts	Flywheel Bolts	Manifold Intake	Manifold Exhaust	Spark Plugs	Lug Nuts
2001	2.0 (1998)	B	①	②	③	80-87	83	13	12	11	98
	3.0 (3049)	1	④	⑤	⑥	⑦	59	⑧	15	11	98
2002	2.0 (1998)	B	①	②	③	80-87	83	13	12	11	98
	3.0 (3049)	1	④	⑤	⑥	⑦	59	⑧	15	11	98
2003	2.0 (1998)	B	①	②	③	80-87	83	13	12	11	98
	3.0 (3049)	1	④	⑤	⑥	⑦	59	⑧	15	11	98

① Step 1: 15 ft. lbs. (20 Nm).

  Step 2: 30 ft. lbs. (40 Nm).

  Step 3: Plus an additional 90 degrees.

② Step 1: 18 ft. lbs.

  Step 2: +60 degrees

③ Step 1: 26 ft. lbs.

  Step 2: +90 degrees

④ Step 1: 30 ft. lbs. (40 Nm).

  Step 2: Tighten the bolts 90 degrees.

  Step 3: Loosen the bolts one full turn.

  Step 4: 30 ft. lbs. (40 Nm).

  Step 5: Tighten the bolts 90 degrees.

  Step 6: Tighten the bolts 90 degrees.

⑤ Step 1: Fasteners 1-8: 18 ft. lbs.

  Step 2: Fasteners 9-19: 30 ft. lbs.

  Step 3: Fasteners 1-16: +90 degrees

  Step 4: fasteners 17-22: 18 ft. lbs.

⑥ Step 1: 17 ft. lbs.

  Step 2: 32 ft. lbs.

⑦ Step 1: 89 ft. lbs.

  Step 2: Loosen 1 full turn

  Step 3: 37 ft. lbs.

  Step 4: 66 ft. lbs.

⑧ 89 inch lbs.

42356-TRIB-C08

WHEEL ALIGNMENT

Year	Model		Caster Range (+/-Deg.)	Caster Preferred Setting (Deg.)	Camber Range (+/-Deg.)	Camber Preferred Setting (Deg.)	Toe-in (in.)	Steering Axis Inclination (Deg.)
2001	ALL	F	NA	+1.93	NA	-0.84	0.12+/-0.12	11.40
		R	NA	NA	NA	-0.04	0.09+/-0.11	NA
2002	ALL	F	NA	+1.93	NA	-0.84	0.12+/-0.12	11.40
		R	NA	NA	NA	-0.04	0.09+/-0.11	NA
2003	ALL	F	NA	+1.93	NA	-0.84	0.12+/-0.12	11.40
		R	NA	NA	NA	-0.04	0.09+/-0.11	NA

42356-TRIB-C09

TIRE, WHEEL AND BALL JOINT SPECIFICATIONS

| Year | Model | OEM Tires | | Tire Pressures (psi) | | Wheel Size | Ball Joint Inspection |
		Standard	Optional	Front	Rear		
2001	Tribute	P225/70SR15	P235/70R16	NA	NA	NA	0.030 in.
2002	Tribute	P215/70R16	P225/70SR15 P235/70R16	NA	NA	NA	0.030 in.
2003	Tribute	P215/70R16	P225/70SR15 P235/70R16	NA	NA	NA	0.030 in.

OEM: Original Equipment Manufacturer

PSI: Pounds Per Square Inch

STD: Standard

OPT: Optional

NA: Not Available

42356-TRIB-C10

BRAKE SPECIFICATIONS

All measurements in inches unless noted

| Year | Model | Brake Disc | | | Brake Drum | | | Minimum Lining Thickness | Brake Caliper | |
		Original Thickness	Minimum Thickness	Maximum Run-out	Original Inside Diameter	Max. Wear Limit	Maximum Machine Diameter		Bracket Bolts (ft. lbs.)	Mounting Bolts (ft. lbs.)
2001	Tribute	0.940	0.860	0.002	9.06	0.06	8.92	0.039	111	26
2002	Tribute	0.940	0.860	0.004	9.06	0.06	8.92	0.039	111	26
2003	Tribute	0.940	0.860	0.004	9.06	0.06	8.92	0.039	111	26

42356-TRIB-C11

SCHEDULED MAINTENANCE INTERVALS
Mazda Tribute

TO BE SERVICED	TYPE OF SERVICE	VEHICLE MILEAGE INTERVAL (x1000)												
		5	10	15	20	25	30	35	40	45	50	55	60	65
Air cleaner filter	R						✓						✓	
Accessory drive belt	S/I												✓	
Brake system ①	S/I			✓			✓			✓			✓	
Clutch pedal operation	S/I						✓						✓	
Cooling fan operation	S/I		✓		✓		✓		✓		✓		✓	
Cooling system hoses	S/I			✓			✓			✓			✓	
CV-joint boots & axle seals	S/I						✓						✓	
Engine coolant	R	Ten years or 150,000 miles												
Engine oil & filter	R	✓	✓	✓	✓	✓	✓	✓	✓	✓	✓	✓	✓	✓
Exterior Lights	S/I	Check monthly												
PCV valve	S/I												✓	
Exhaust system & heat shields	S/I						✓						✓	
Parking brake system	S/I	Every 6 months												
Power steering fluid	S/I	Every 6 months												
Rotate tires	S/I	✓		✓		✓		✓		✓		✓		✓
Steering linkage	S/I						✓						✓	
Spark plugs	R	Change at 100,000 miles												
Suspension components	S/I						✓						✓	

R: Replace   S/I: Inspect and service, if necessary   L: Lubricate   A: Adjust   C: Clean

① Inspect the reservoir fluid level, rotor and or drum, brake lines, hoses, calipers and or wheel cylinders

FREQUENT OPERATION MAINTENANCE (SEVERE SERVICE)

If a vehicle is operated under any of the following conditions it is considered severe service:

- Extremely dusty areas.

- 50% or more of the vehicle operation is in 32°C (90°F) or higher temperatures, or constant operation in temperatures below 0°C (32°F).

- Prolonged idling (vehicle operation in stop and go traffic).

- Frequent short running periods (engine does not warm to normal operating temperatures).

- Police, taxi, delivery usage or trailer towing usage.

Oil & oil filter change: change every 3000 miles.

Air filter element: change every 15,000 miles.

42356-TRIB-C12

PRECAUTIONS

Before servicing any vehicle, please be sure to read all of the following precautions, which deal with personal safety, prevention of component damage, and important points to take into consideration when servicing a motor vehicle:

• Never open, service or drain the radiator or cooling system when the engine is hot; serious burns can occur from the steam and hot coolant.

• Observe all applicable safety precautions when working around fuel. Whenever servicing the fuel system, always work in a well-ventilated area. Do not allow fuel spray or vapors to come in contact with a spark, open flame, or excessive heat (a hot drop light, for example). Keep a dry chemical fire extinguisher near the work area. Always keep fuel in a container specifically designed for fuel storage; also, always properly seal fuel containers to avoid the possibility of fire or explosion. Refer to the additional fuel system precautions later in this section.

• Fuel injection systems often remain pressurized, even after the engine has been turned **OFF**. The fuel system pressure must be relieved before disconnecting any fuel lines. Failure to do so may result in fire and/or personal injury.

• Brake fluid often contains polyglycol ethers and polyglycols. Avoid contact with the eyes and wash your hands thoroughly after handling brake fluid. If you do get brake fluid in your eyes, flush your eyes with clean, running water for 15 minutes. If eye irritation persists, or if you have taken brake fluid internally, IMMEDIATELY seek medical assistance.

• The EPA warns that prolonged contact with used engine oil may cause a number of skin disorders, including cancer! You should make every effort to minimize your exposure to used engine oil. Protective gloves should be worn when changing oil. Wash your hands and any other exposed skin areas as soon as possible after exposure to used engine oil. Soap and water, or waterless hand cleaner should be used.

• All new vehicles are now equipped with an air bag system, often referred to as a Supplemental Restraint System (SRS) or Supplemental Inflatable Restraint (SIR) system. The system must be disabled before performing service on or around system components, steering column, instrument panel components, wiring and sensors. Failure to follow safety and disabling procedures could result in accidental air bag deployment, possible personal injury and unnecessary system repairs.

• Always wear safety goggles when working with, or around, the air bag system. When carrying a non-deployed air bag, be sure the bag and trim cover are pointed away from your body. When placing a non-deployed air bag on a work surface, always face the bag and trim cover upward, away from the surface. This will reduce the motion of the module if it is accidentally deployed. Refer to the additional air bag system precautions later in this section.

• Clean, high quality brake fluid from a sealed container is essential to the safe and proper operation of the brake system. You should always buy the correct type of brake fluid for your vehicle. If the brake fluid becomes contaminated, completely flush the system with new fluid. Never reuse any brake fluid. Any brake fluid that is removed from the system should be discarded. Also, do not allow any brake fluid to come in contact with a painted surface; it will damage the paint.

• Never operate the engine without the proper amount and type of engine oil; doing so WILL result in severe engine damage.

• Timing belt maintenance is extremely important! Many models utilize an interference-type, non-freewheeling engine. If the timing belt breaks, the valves in the cylinder head may strike the pistons, causing potentially serious (also time-consuming and expensive) engine damage. Refer to the maintenance interval charts in the front of this section for the recommended replacement interval for the timing belt, and to the timing belt section for belt replacement and inspection.

• Disconnecting the negative battery cable on some vehicles may interfere with the functions of the on-board computer system(s) and may require the computer to undergo a relearning process once the negative battery cable is reconnected.

• When servicing drum brakes, only disassemble and assemble one side at a time, leaving the remaining side intact for reference.

• Only an MVAC-trained, EPA-certified automotive technician should service the air conditioning system or its components.

ENGINE REPAIR

Distributor

The Tribute uses a Direct Ignition System (DIS). No distributor is used.

Alternator

REMOVAL

2.0L Engine

1. Remove or disconnect the following:
   • Negative battery cable
   • Drive belt
   • Alternator electrical connectors and loosen the upper alternator bolt

while moving the alternator to the rear of the engine
   • Alternator. Torque the mounting and adjusting bolts to 33 ft. lbs. (45Nm).

3.0L Engine

1. Remove or disconnect the following:
   • Negative battery cable
   • Right side intermediate axle shaft
   • Right side splash shield and retainers
   • Drive belt
   • Alternator electrical connectors
   • Alternator. Torque the mounting and adjusting bolts to 35 ft. lbs. (48Nm).

INSTALLATION

2.0L Engine

1. Install or connect the following:
   • Alternator with the upper bolt in the alternator before installation. Torque the bolts to 18 ft. lbs. (25 Nm).
   • Alternator electrical connectors
   • Drive belt
   • Negative battery cable

3.0L Engine

1. Install or connect the following:
   • Alternator. Torque the bolts to 35 ft. lbs. (48 Nm).

- Alternator electrical connectors
- Drive belt
- Negative battery cable

Ignition Timing

ADJUSTMENT

Ignition timing is controlled by the Powertrain Control Module (PCM). No adjustment is necessary or possible.

Engine Assembly

REMOVAL & INSTALLATION

2.0L Engine

MANUAL TRANSMISSION

1. Before servicing the vehicle, refer to the precautions in the beginning of this section.
2. Properly recover the air conditioning system refrigerant.
3. Properly relieve the fuel system pressure.
4. Drain the cooling system.
5. Drain the engine oil.
6. Remove or disconnect the following:

- Hood
- Battery and battery tray
- Air cleaner housing
- Fuel lines
- Throttle cable and speed control cable, if equipped
- Exhaust Gas Recirculation (EGR) vacuum valve regulator
- EGR electrical connectors and vacuum hoses
- Brake booster vacuum hose
- Powertrain Control Module (PCM) wire harness and ground
- Wire harness connector
- Power distribution board electrical connectors
- Evaporative emissions (EVAP) canister vacuum lines
- Upper radiator hose
- Power steering line bracket
- Upper power steering pump bolts
- Coolant hose
- Heater hoses
- Speed control unit, if equipped
- Catalytic converter
- A/C compressor
- Both halfshafts
- Shifter linkages
- Block heater electrical connector, if equipped

- Front transmission through bolt
- Engine-to-transmission bolts
- Lower radiator hose
- Power steering pump
- Clutch slave cylinder line from the bracket and move it aside
- Rear transmission mount
- Left side transmission mount
- Lower ground cable
- Engine mount upper bracket
- Engine and transmission as an assembly by using a proper lifting device
- Alternator electrical connectors
- Knock Sensor (KS) electrical connector
- Oil pressure sender electrical connector
- Starter electrical connector
- Vehicle Speed Sensor (VSS) electrical connector
- Park Neutral Position (PNP) electrical connector
- Fuel charging wire harness electrical connector
- PCM wire harness from the bracket
- PCM ground wire
- Back up lamp switch electrical connector
- Wire harness
- Differential Pressure Feedback (DPFEE) EGR sensor

7. Separate the engine from the transmission.
8. Lock the flywheel to the engine.
9. Clutch pressure plate and disc.
10. Flywheel and rear cover plates.

To install:

11. Install or connect the following:

- Flywheel. Torque the bolts to 83 ft. lbs. (112 Nm).
- Clutch disc to the flywheel
- Pressure plate to the flywheel. Torque the bolts in the proper sequence to 18 ft. lbs. (25 Nm).
- Transmission to the engine. Torque the bolts to 33 ft. lbs. (45 Nm).
- Starter. Torque the bolts to 18 ft. lbs. (25 Nm).
- Wire harness and attach it to the powertrain assembly
- DPFEE sensor electrical connector
- Reverse lamp switch electrical connector
- Ground wire. Torque the bolt to 80 inch lbs. (9 Nm).
- PCM wire harness to the bracket
- Fuel charging wire harness electrical connector
- PNP switch electrical connector
- VSS electrical connector
- KS, Oil pressure sender and starter electrical connector. Torque the fasteners to 9 ft. lbs. (12 Nm).
- Alternator electrical connectors. Torque the fasteners to 71 inch lbs. (8 Nm).
- Powertrain assembly in the vehicle
- Left side transmission mount. Torque the side bolts to 41 ft. lbs. (55 Nm) and the center bolt to 66 ft. lbs. (90 Nm).
- Engine mount upper bracket. Torque the side bolts to 72 ft. lbs. (98 Nm) and the center bolt to 57 ft. lbs. (77 Nm).
- Ground wire. Torque the bolt to 25 ft. lbs. (34 Nm).

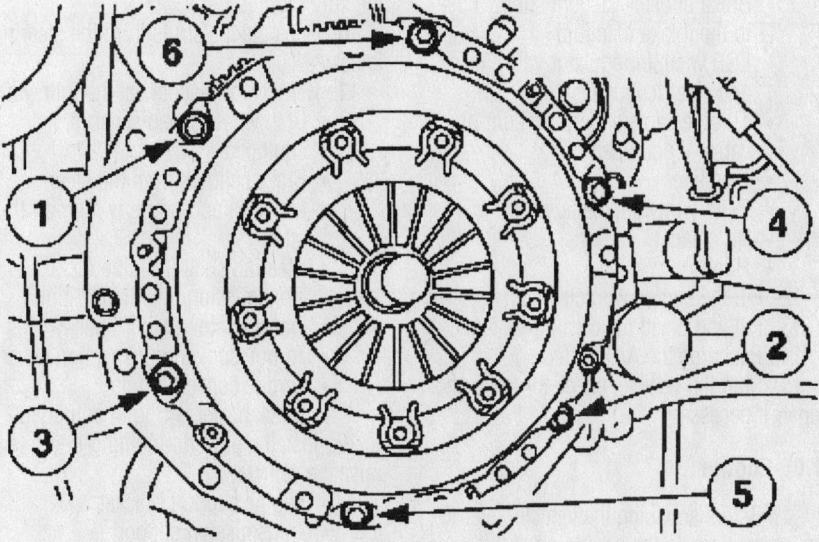

9308TG07

Tighten the pressure plate bolts in the proper sequence–2.0L engine

- Rear transmission mount. Torque the bolts to 41 ft. lbs. (55 Nm).
- Speed control unit, if equipped. Torque the bolts to 89 inch lbs. (10 Nm).
- Power steering pump and hand tighten the bolts
- Lower power steering line bracket. Torque the bolt to 89 inch lbs. (10 Nm).
- Upper power steering line bolt. Torque the bolt to 15 ft. lbs. (20 Nm).
- Slave cylinder line and clip. Torque the bolts to 16 ft. lbs. (22 Nm).
- Power steering lines. Torque the retaining bolts to 89 inch lbs. (10 Nm). Torque the power steering pump bolts to 18 ft. lbs. (25 Nm).
- Lower radiator hose
- Engine-to-transmission bolts. Torque the bolts to 33 ft. lbs. (45 Nm).
- Front transmission through bolt. Torque the bolt to 66 ft. lbs. (90 Nm).
- Block heater electrical connector
- Shifter linkages. Torque the upper bolt to 33 ft. lbs. (45 Nm) and the lower bolt to 15 ft. lbs. (20 Nm).
- Coolant hose
- Catalytic converter
- Heater hoses
- Upper radiator hose
- EVAP canister vacuum lines
- Power distribution box electrical connector. Torque the fastener to 9 ft. lbs. (12 Nm).
- Wire harness electrical connector
- Ground wires
- PCM wire harness and ground
- Brake booster vacuum supply hose to the intake manifold
- EGR vacuum regulator valve hoses and electrical connector
- Throttle cable and speed control cable, if equipped
- Fuel lines
- Battery tray and battery
- Air cleaner
- Hood

12. Fill the engine with clean oil.
13. Fill and bleed the cooling system.
14. Recharge the A/C system.
15. Start the vehicle, check for leaks and repair if necessary.

3.0L Engine

1. Before servicing the vehicle, refer to the precautions in the beginning of this section.

2. Properly recover the air conditioning system refrigerant.
3. Properly relieve the fuel system pressure.
4. Drain the cooling system.
5. Drain the engine oil.
6. Remove or disconnect the following:
- Hood
- Battery and battery tray
- Air cleaner outlet tube and housing
- Lower radiator air deflectors
- Fuel lines
- Water pump drive belt
- Accelerator cable and speed control cable, if equipped
- Vapor Management Valve (VMV)
- Powertrain Control Module (PCM)
- PCM ground wire
- Thermostat housing and hose assembly and move them aside
- Power distribution box electrical connector
- Power distribution box cover
- Nuts and cables from inside the power distribution box
- Transmission linkage
- Brake booster vacuum hose
- Heater hoses
- Power steering return line
- Power Steering Pressure (PSP) switch electrical connector
- Power steering supply line
- Oil level indicator
- Catalytic converter
- A/C compressor
- Both front wheels
- Intermediate drive shaft, if equipped

7. Separate both side ball joints.
8. Separate both side tie rod ends from the steering knuckles.
9. Separate both sway bar links from the strut mounts.
10. Separate the struts from the steering knuckles.
11. Remove or disconnect the following:
- Both wheel speed sensors, if equipped
- Brake calipers from the steering knuckles and properly support the struts
- Steering shaft from the rack
- Transmission line bracket bolt
- Transmission cooler lines
- Torque converter inspection cover
- Torque converter nuts
- Block heater wiring, if equipped

12. Install a powertrain lifting devise and raise the vehicle.
- Engine support bracket
- Transmission support
- 2 rear subframe bolts
- 2 subframe side bolts

- Motor mount support bolts
- Engine and transmission as an assembly
- Heated Oxygen (HO2S) sensor
- Transmission Range (TR) sensor
- Transmission harness electronic control switch
- Transmission control harness from the bracket
- Starter and wire harness
- Knock Sensor (KS) electrical connector
- Output Shaft Speed (OSS) sensor electrical connector
- HO2S sensor and Exhaust Gas Recirculation (EGR) tube from the exhaust manifold
- Alternator and electrical connectors
- Right side exhaust manifold and gasket
- Halfshaft support bracket and move it aside

13. Separate the engine from the transmission assembly

To install:

14. Install or connect the following:
- Powertrain assembly on the subframe
- Transmission-to-engine bolts. Torque the bolts to 30 ft. lbs. (40 Nm).
- Halfshaft bracket. Torque the bolts to 18 ft. lbs. (25 Nm).
- Right side exhaust manifold and new gasket. Torque the bolts to 15 ft. lbs. (25 Nm).
- Alternator. Torque the larger bolts to 18 ft. lbs. (25 Nm) and smaller bolt to 89 inch lbs. (10 Nm).
- EGR tube and HO2S sensor electrical connectors
- OSS sensor electrical connector
- KS jumper electrical connector
- Starter. Torque the bolts to 18 ft. lbs. (25 Nm).
- Transmission control harness to the bracket. Torque the bolt to 18 ft. lbs. (25 Nm).
- Transmission harness
- Transmission range sensor
- Powertrain assembly
- Motor mount support. Torque the bolts to 66 ft. lbs. (90 Nm).
- Subframe side nuts. Torque the nuts to 76 ft. lbs. (103 Nm). Raise the vehicle and support the powertrain assembly with a lifting device.
- Transmission mount. Torque the bolts to side bolts to 66 ft. lbs. (90 Nm) and the other bolts to 76 ft. lbs. (103 Nm).
- Motor mount. Torque the bolts to

side bolts to 66 ft. lbs. (90 Nm) and the other bolts to 76 ft. lbs. (103 Nm). Remove the powertrain lift.
- Block heater electrical connector, if equipped
- Torque converter. Torque the nuts to 27 ft. lbs. (37 Nm).
- Transmission cover plate and plug
- Transmission cooler lines
- Transmission cooler line bracket. Torque the bolt to 15 ft. lbs. (20 Nm).
- Steering shaft to the rack. Torque the bolt to 18 ft. lbs. (25 Nm).
- Struts to the steering knuckles. Torque the bolts to 75 ft. lbs. (102 Nm).
- Brake calipers to the steering knuckles
- Wheel speed sensors, if equipped. Torque the bolts to 89 inch lbs. (10 Nm).
- Sway bar links to the strut mount. Torque the bolts to 41 ft. lbs. (55 Nm).
- Tie rods to the steering knuckles. Torque the bolts to 41 ft. lbs. (55 Nm).
- Ball joints. Torque the bolts to 52 ft. lbs. (70 Nm).
- Intermediate drive shaft, if equipped
- Both front wheels
- A/C compressor
- Lower radiator air deflectors
- Catalytic converter
- Oil level indicator dipstick tube
- Power steering line and bracket. Torque the bolt to 13 ft. lbs. (17 Nm).
- PSP switch electrical connector
- Power steering return line
- Heater hoses
- Vacuum lines
- Transmission linkage
- Wire harness cables and nuts to the power distribution box. Torque the nuts to 89 inch lbs. (10 Nm).
- Power distribution box wire harness
- Thermostat housing and connect the hoses
- Ground wire. Torque the bolt to 89 inch lbs. (10 Nm).
- PCM electrical connector
- VMV electrical connector
- Accelerator cable and speed control cable, if equipped
- Air cleaner assembly
- Water pump drive belt
- Battery and tray
15. Fill and bleed the cooling system.

16. Fill the engine with clean oil.
17. Recharge the A/C system.
18. Inspect and top off the power steering fluid.
19. Start the vehicle, check for leaks and repair if necessary.

Water Pump

REMOVAL & INSTALLATION

2.0L Engine

1. Before servicing the vehicle, refer to the precautions in the beginning of this section.
2. Drain the cooling system.
3. Remove or disconnect the following:
   - Negative battery cable
   - Right front wheel
   - Splash shield
   - Drive belt
   - Water pump pulley
   - Water pump

To install:
4. Install or connect the following:
   - Water pump. Torque the bolts to 89 inch lbs. (10 Nm).
   - Water pump pulley. Torque the bolts to 89 inch lbs. (10 Nm).
   - Drive belt
   - Splash shield
   - Right front wheel
   - Negative battery cable
5. Refill the cooling system.
6. Start the vehicle and check for leaks, repair if necessary.

3.0L Engine

1. Before servicing the vehicle, refer to the precautions in the beginning of this section.
2. Drain the cooling system.
3. Remove or disconnect the following:
   - Negative battery cable
   - Air cleaner outlet tube

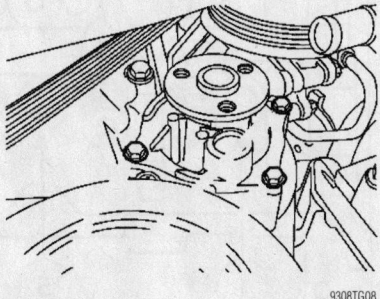

9308TG08

Exploded view of the water pump–2.0L engine

- Water pump belt tensioner
- Coolant hoses
- Water pump
- Water pump from the housing

To install:
- Water pump to the housing. Torque the bolts to 89 inch lbs. (10 Nm).
- Water pump. Torque the bolts to 89 inch lbs. (10 Nm).
- Coolant hoses
- Water pump belt tensioner
- Air cleaner outlet tube
- Negative battery cable
4. Refill the cooling system.
5. Start the vehicle and check for leaks, repair if necessary.

Heater Core

REMOVAL & INSTALLATION

1. Drain the engine coolant.
2. Disconnect the heater hoses from the heater core.
3. Remove the driver air bag module.
4. Remove the two front door scuff plates.
5. Remove the four pin-type retainers.
6. Remove the two front door scuff plates.
7. Remove the two A-pillar lower trim panels.
8. Remove the two pin-type retainers.
9. Remove the two A-pillar lower trim panels.
10. Disconnect the electrical connectors located by the LH cowl.
11. Position the hood latch release handle aside.
12. Remove the bolts.
13. Position the hood latch release handle aside.
14. Remove the utility compartment.
15. Remove the four pin-type retainers.
16. Remove the utility compartment.
17. Disconnect the electrical connector.
18. Remove the instrument panel steering column cover.
19. Release the upper clips and rotate the cover outward to release the lower pivot retainers.
20. Remove the steering column lower cover.
21. Remove the screws.
22. Remove the steering column lower cover.
23. If equipped, disconnect the shift cable.
24. Disconnect the shift cable.
25. Disconnect the shift cable from the retaining bracket.

26. Remove the steering column coupler access cover.

27. Disconnect the steering column coupler.

28. Remove the steering column coupler bolt and nut.

29. Disconnect the steering column coupler.

30. Remove the cover panel.

31. Remove the pin-type retainer.

32. Release the retaining clip.

33. Disconnect the electrical connectors.

34. Disconnect the climate control vacuum harness connector.

35. Disconnect the in-line electrical connector.

36. Remove the four instrument panel center brace bolts.

37. Remove the passenger air bag module.

38. Disconnect the vacuum harness connector.

39. Disconnect the temperature control cable.

40. Position the locator pin.

41. Release the locking tab.

42. Disconnect the temperature control cable from the blend door shaft.

43. Close the glove compartment.

44. Press the release tabs inward while raising the glove compartment.

45. Disconnect the electrical connectors at the blower motor.

46. Disconnect the antenna cable in-line connector.

47. Open the four A-pillar passenger assist handle covers.

48. Remove the two A-pillar passenger assist handles.

49. Remove the four bolts.

50. Remove the two A-pillar passenger assist handles.

51. Remove the two windshield side garnish mouldings.

52. Remove the instrument panel cowl top cover.

53. Remove the instrument panel cowl top bolt.

54. Loosen the tilt lever (if equipped) and lower steering column.

55. Position the transmission range selector lever (if equipped) down to provide access to the instrument cluster finish panel and instrument cluster.

56. Remove the screws and the instrument cluster finish panel.

57. Remove the screws.

58. Disconnect the electrical connectors and remove the instrument cluster.

59. Through the instrument cluster opening, remove the instrument panel nut.

60. Remove the two instrument panel finish end panels.

61. Remove the four instrument panel cowl side bolts.

➡**This step requires an assistant.**

62. Remove the instrument panel.

63. Remove the heater blending door levers.

64. Remove the screw for heater blending door.

65. Remove the levers for the blending door.

66. Remove the heater core.

67. Remove the three screws.

68. Remove the cover for the heater core and pull the heater core out of the housing.

➡**Before installing the temperature control cable, make sure the blend door, cable and temperature switch are correctly positioned.**

69. To install, reverse the removal procedure.

⁂ **CAUTION**

Electronic modules are sensitive to static electrical charges. If exposed to these charges, damage may result.

⁂ **CAUTION**

Once the new module is installed, it is necessary to download the module configuration information from the

scan tool into the new instrument cluster.

Cylinder Head

REMOVAL & INSTALLATION

2.0L Engine

1. Before servicing the vehicle, refer to the precautions in the beginning of this section.

2. Properly relieve the fuel system pressure.

3. Drain the engine oil.

4. Remove or disconnect the following:
- Negative battery cable
- Ignition coil bracket
- Thermostat housing
- Positive Crankcase Ventilation (PCV) tube
- Intake manifold
- Exhaust manifold
- Power steering bracket and move it aside
- Valve tappets
- Engine mount lower bracket
- Engine mount upper bracket
- Cylinder head bolts in the proper sequence and discard the gasket

To install:

5. Install a new head gasket and the cylinder head.

6. Lubricate the cylinder head bolt threads.

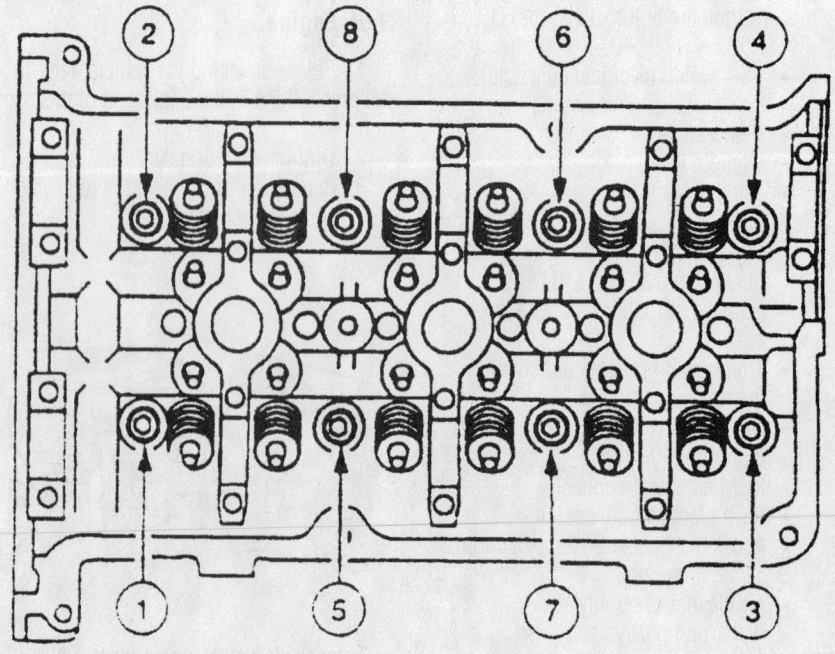

9308TG04

Cylinder head bolt torque sequence 2.0L engine

7. Torque the cylinder head bolts in the proper sequence as follows:

   a. Step 1: 15 ft. lbs. (20 Nm).

   b. Step 2: 30 ft. lbs. (40 Nm).

   c. Step 3: Plus an additional 90 degrees.

8. Install or connect the following:

- Engine mount upper bracket. Torque the 2 upper bolts to 72 ft. lbs. (98 Nm) and the center bolt to 57 ft. lbs. (77 Nm).
- Engine mount lower bracket. Torque the bolts to 37 ft. lbs. (50 Nm).
- Valve tappets
- Power steering pump bracket. Torque the bolts to 20 ft. lbs. (28 Nm).
- Exhaust manifold
- Intake manifold
- PCV tube
- Thermostat housing
- Ignition coil bracket
- Negative battery cable

9. Fill the engine with clean oil and replace the filter.

10. Start the vehicle and check for leaks, repair if necessary.

3.0L Engine

The procedure for the left side cylinder head and right side are similar. Changes in the procedure will be noted for either side cylinder head.

1. Before servicing the vehicle, refer to the precautions in the beginning of this section.

2. Properly relieve the fuel system pressure.

- Drain the cooling system.

3. Remove or disconnect the following:

- Negative battery cable
- Camshaft
- Exhaust Gas Recirculation (EGR) tube, right side only
- Exhaust manifold
- Camshaft followers
- Hydraulic lash adjusters and matchmark them for proper installation
- Cylinder head bolts in sequence and discard them
- Cylinder head and discard the gasket

To install:

4. Install a new head gasket and the cylinder head.

5. Lubricate the cylinder head bolt threads.

6. Torque the cylinder head bolts in the proper sequence as follows:

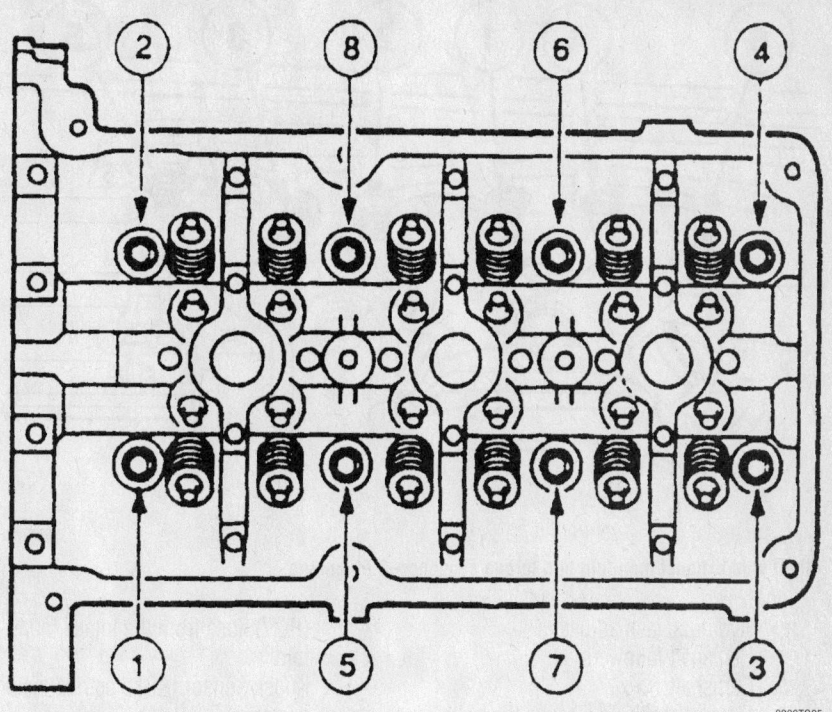

Left side cylinder head bolt torque sequence 3.0L engine

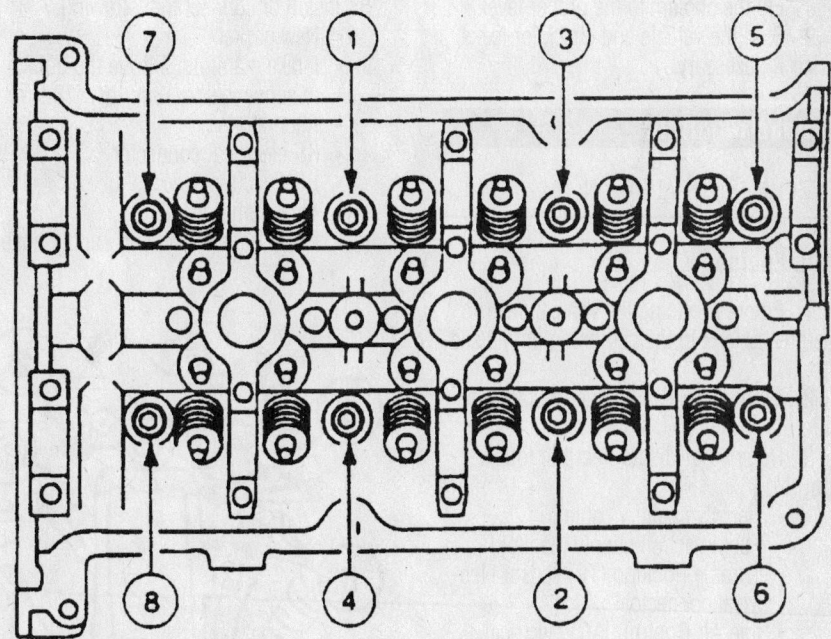

Right side cylinder head bolt torque sequence 3.0L engine

   a. Step 1: 30 ft. lbs. (40 Nm).

   b. Step 2: Additional 90 degrees.

   c. Step 3: Loosen the bolts one full turn.

   d. Step 4: 30 ft. lbs. (40 Nm).

   e. Step 5: Plus an additional 90 degrees.

   f. Step 6: Plus an additional 90 degrees.

7. Install or connect the following:

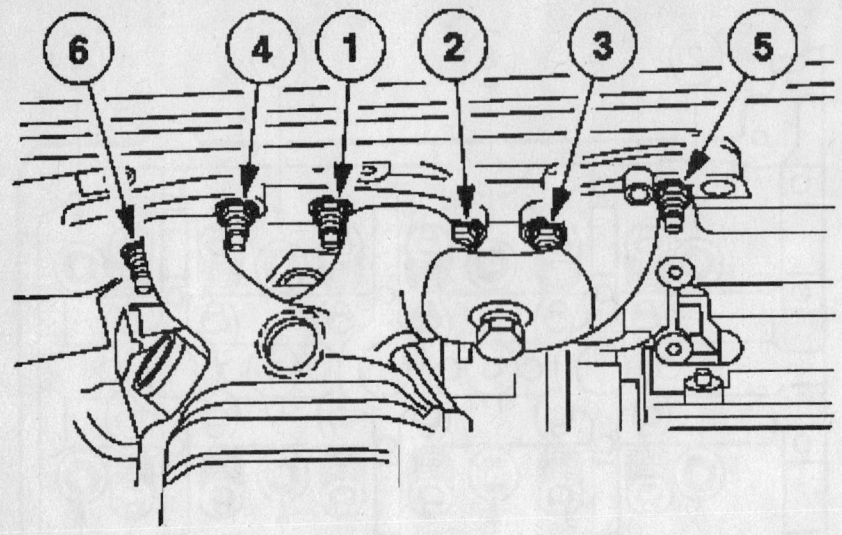

9308TG09

Right side exhaust manifold bolt torque sequence–3.0L engine

- Hydraulic lash adjusters
- Camshaft followers
- Camshaft
- Exhaust manifold. Torque the bolts in sequence to 15 ft. lbs. (20 Nm), right side only
- EGR tube, right side only
- Coolant bypass tube
- Negative battery cable

8. Fill the coolant to the proper level.

9. Start the vehicle and check for leaks, repair if necessary.

Intake Manifold

REMOVAL & INSTALLATION

2.0L Engine

1. Before servicing the vehicle, refer to the precautions in the beginning of this section.

2. Properly relieve the fuel system pressure.

3. Remove or disconnect the following:

- Negative battery cable
- Fuel injection supply manifold
- Throttle Position (TP) sensor electrical connector
- Idle Air Control (IAC) electrical connector and unclip the harness from the bracket
- Main engine control sensor wiring
- Connector from the bracket
- Powertrain Control Module (PCM) wire harness from the bracket
- Brake booster vacuum hose
- 4 additional vacuum lines
- Positive Crankcase Ventilation

(PCV) hose from the intake manifold
- Knock Sensor (KS) electrical connector
- Alternator
- Intake manifold and discard the gasket

4. Clean the mating surfaces.

To install:

5. Install or connect the following:

- New gasket
- Intake manifold. Torque the bolts, in sequence, to 13 ft. lbs. (18 Nm).
- Alternator
- KS electrical connector
- PCV vacuum line
- 4 vacuum lines
- Brake booster vacuum supply hose

- PCM wire harness to the bracket
- Main engine control sensor wiring
- IAC valve electrical connector and attach the harness to the bracket
- TP sensor electrical connector
- Fuel injection supply manifold
- Negative battery cable

6. Start the vehicle and check for leaks, repair if necessary.

3.0L Engine

UPPER

1. Before servicing the vehicle, refer to the precautions in the beginning of this section.

2. Properly relieve the fuel system pressure.

3. Drain the coolant system.

4. Remove or disconnect the following:

- Negative battery cable
- Air cleaner outlet tube
- Engine appearance cover
- Throttle cable
- Speed control cable, if equipped
- Throttle cable bracket
- Throttle Position (TP) sensor electrical connector
- Idle Air Control (IAC) valve electrical connector
- Exhaust Gas Recirculation (EGR) valve vacuum hose and tube
- EGR vacuum regulator valve electrical connector and hose
- Chassis vacuum hose
- Engine vacuum hose
- Positive Crankcase Ventilation (PCV) hose
- Vapor Management Valve (VMV) vacuum hose

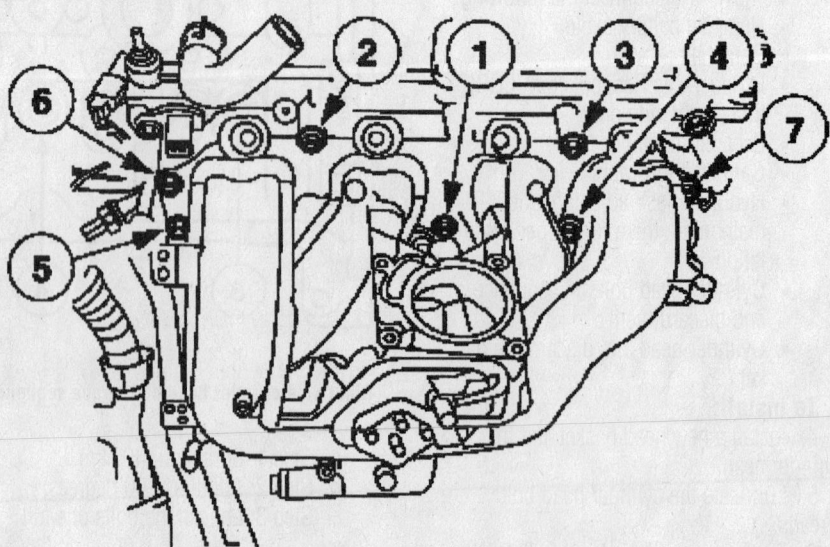

9308TG01

Tighten the intake manifold bolts in the sequence shown—2.0L engine

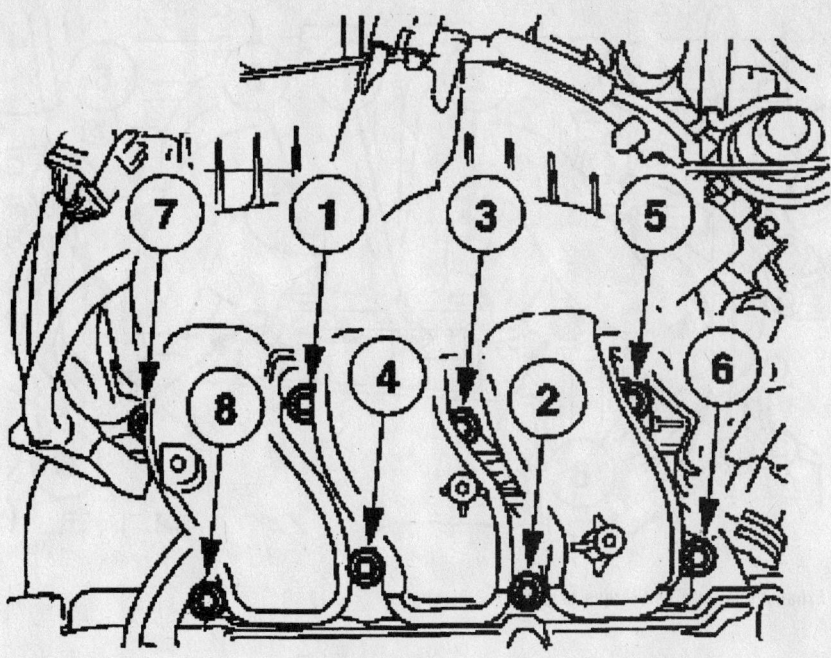

Tighten the lower intake manifold bolts in the sequence shown—3.0L engine

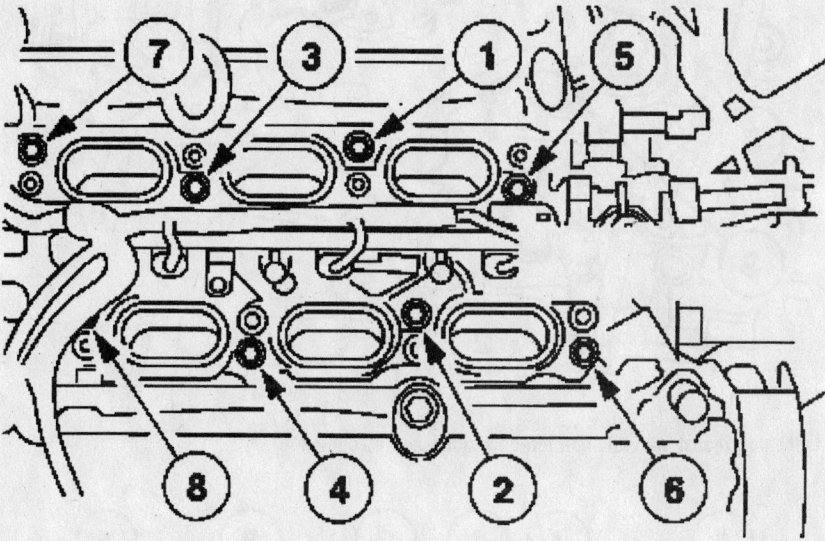

Tighten the upper intake manifold bolts in the sequence shown—3.0L engine

- Electrical connectors from the left side of the upper intake manifold
- Power Steering Pressure (PSP) sensor electrical connector
- Upper intake manifold and discard the gasket
5. Clean the mating surfaces.
To install:
6. Install or connect the following:
  - New gasket
  - Intake manifold. Torque the bolts, in sequence, to 89 inch lbs. (10 Nm).

- PSP electrical connector
- Electrical connectors on the left side of the upper intake manifold
- VMV vacuum hose
- Chassis, engine and PCV hoses
- EGR valve vacuum regulator
- EGR valve vacuum hose and tube. Torque the nut to 30 ft. lbs. (40 Nm).
- TP sensor electrical connector
- IAC valve electrical connector
- Throttle cable and speed control cable, if equipped. Torque the

bracket bolts to 89 inch lbs. (10 Nm).
- Air cleaner outlet tube
- Engine appearance cover. Torque the bolts to 53 inch lbs. (6 Nm).
- Negative battery cable
7. Fill the coolant system to the proper level.
8. Start the vehicle and check for leaks, repair if necessary.

LOWER

1. Before servicing the vehicle, refer to the precautions in the beginning of this section.
2. Properly relieve the fuel system pressure.
3. Remove or disconnect the following:
  - Negative battery cable
  - Fuel line spring lock coupling
  - Upper intake manifold
  - Fuel rail
  - Fuel injector electrical connectors
  - Fuel pressure damper vacuum line
  - Lower intake manifold
  - Lower intake manifold from the fuel rail
  - Fuel injectors from the manifold and discard the gasket
4. Clean the mating surfaces.
To install:
5. Inspect the fuel injector O-rings and replace if necessary.
6. Install or connect the following:
  - Fuel injectors into the lower intake manifold
  - Fuel rail. Torque the bolts to 89 inch lbs. (10 Nm).
  - New gasket
  - Intake manifold. Torque the bolts, in sequence, to 89 inch lbs. (10 Nm).
  - Fuel rail electrical connectors
  - Fuel injector electrical connectors
  - Fuel pressure damper vacuum line
  - Upper intake manifold
  - Fuel line spring lock coupling
  - Negative battery cable
7. Start the vehicle and check for leaks, repair if necessary.

Exhaust Manifold

REMOVAL & INSTALLATION

2.0L Engine

1. Before servicing the vehicle, refer to the precautions in the beginning of this section.
2. Remove or disconnect the following:

- Negative battery cable
- Catalytic converter
- Oil level indicator tube and bracket
- Exhaust manifold and discard the gasket

To install:

3. Clean the sealing surfaces of any old gasket material.

4. Install or connect the following:
- Exhaust manifold and new gasket. Torque the bolts to 12 ft. lbs. (16 Nm).
- Oil level indicator tube and bracket. Torque the bolt to 89 inch lbs. (10 Nm).
- Catalytic converter
- Negative battery cable

5. Start the vehicle and check for leaks, repair if necessary.

3.0L Engine

LEFT SIDE

1. Before servicing the vehicle, refer to the precautions in the beginning of this section.

2. Remove or disconnect the following:
- Negative battery cable
- Heated Oxygen (HO$_2$S) sensor and catalyst monitor
- Splash shield
- Exhaust crossover pipe
- Drive belt
- A/C compressor and move it aside
- Exhaust manifold and discard the gasket

To install:

3. Clean the sealing surfaces of any old gasket material.

4. Install or connect the following:
- Exhaust manifold and new gasket. Torque the bolts to 15 ft. lbs. (20 Nm).
- A/C compressor. Torque the bolts to 18 ft. lbs. (20 Nm).
- Drive belt
- Exhaust crossover pipe. Torque the bolts to 30 ft. lbs. (40 Nm).
- Splash shield. Torque the bolts to 80 inch lbs. (9 Nm).
- Left side HO$_2$S sensor and catalyst monitor
- Negative battery cable

5. Start the vehicle and check for leaks, repair if necessary.

RIGHT SIDE

1. Before servicing the vehicle, refer to the precautions in the beginning of this section.

2. Remove or disconnect the following:
- Negative battery cable

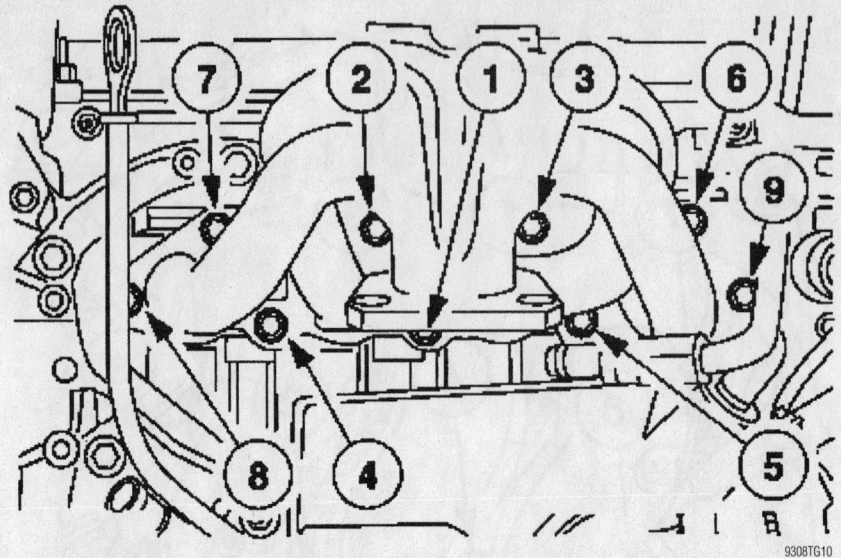

Exhaust manifold bolt torque sequence–2.0L engine

Left side exhaust manifold bolt torque sequence–3.0L engine

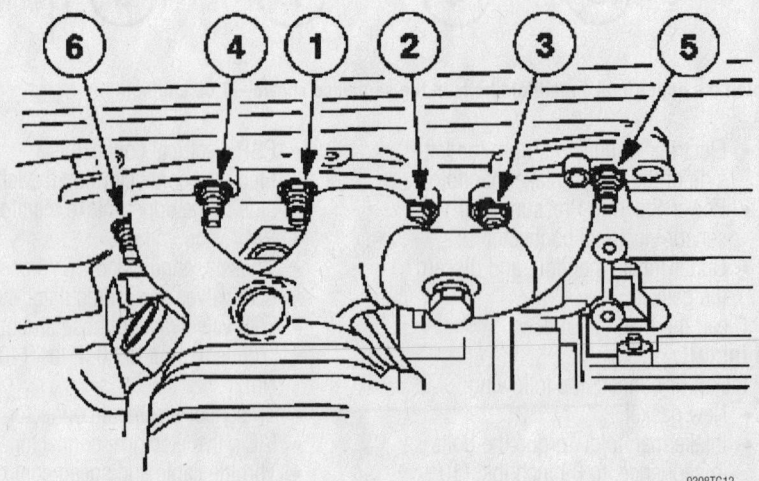

Right side exhaust manifold bolt torque sequence–3.0L engine

- Exhaust Gas Recirculation (EGR) tube
- Alternator
- Right side Heated Oxygen (HO2S) sensor
- Right side exhaust manifold and discard the gasket

To install:

3. Clean the sealing surfaces of any old gasket material.

4. Install or connect the following:
- Exhaust manifold and new gasket. Torque the bolts to 15 ft. lbs. (20 Nm).
- Right side HO2S sensor
- Alternator
- EGR tube
- Negative battery cable

5. Start the vehicle and check for leaks, repair if necessary.

Front Crankshaft Seal

REMOVAL & INSTALLATION

2.0L Engine

1. Before servicing the vehicle, refer to the precautions in the beginning of this section.

2. Remove or disconnect the following:

- Negative battery cable
- Timing belt
- Crankshaft sprocket and timing belt guide
- Crankshaft oil seal

➡ **Be careful not to damage the seal surface of the cover.**

To install:

3. Install or connect the following:
- New front crankshaft oil seal
- Timing belt guide and crankshaft sprocket
- Timing belt
- Negative battery cable

4. Start the vehicle and check for leaks, repair if necessary.

3.0L Engine

1. Before servicing the vehicle, refer to the precautions in the beginning of this section.

2. Remove or disconnect the following:
- Negative battery cable
- Crankshaft pulley
- Front oil seal

To install:

3. Install or connect the following:
- New front crankshaft oil seal

- Crankshaft pulley
- Negative battery cable

4. Start the vehicle and check for leaks, repair if necessary.

Camshaft and Lifters

REMOVAL & INSTALLATION

2.0L Engine

1. Before servicing the vehicle, refer to the precautions in the beginning of this section.

2. Remove or disconnect the following:

- Negative battery cable
- Camshaft timing sprocket and verify the valve clearance
- Camshaft journal cap bolts by loosening them in several passes in the proper sequence
- Camshafts

3. Inspect the camshaft for wear and discard the oil seals

To install:

4. Install or connect the following:
- Camshaft cam followers, lubricate

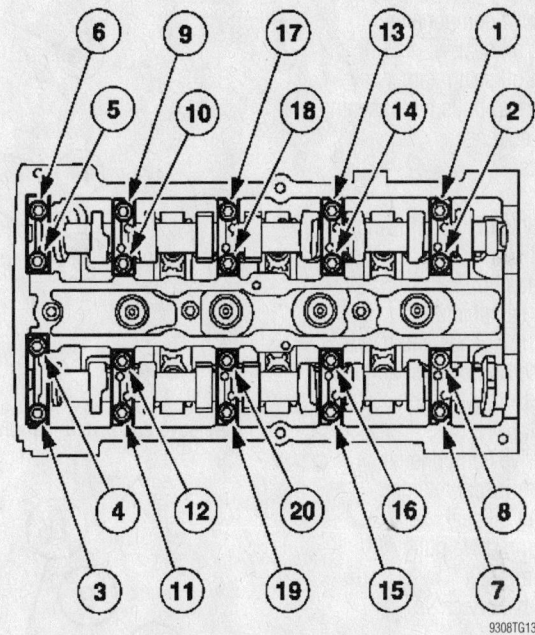

Remove the camshaft bearing caps in sequence–2.0L engine

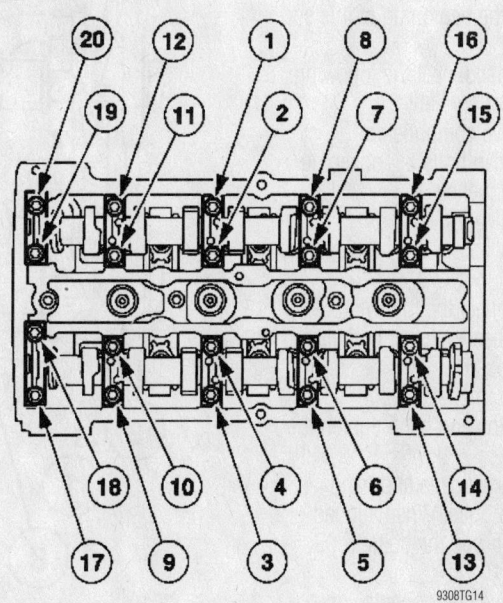

Camshaft bearing cap tightening sequence–2.0L engine

the bearing journals thoroughly. Torque the caps to 14 ft. lbs. (19 Nm).

- Exhaust camshaft oil seal
- Camshaft timing sprocket
- Negative battery cable

3.0L Engine

LEFT SIDE

1. Before servicing the vehicle, refer to the precautions in the beginning of this section.
2. Remove or disconnect the following:
   - Negative battery cable
   - Water pump belt
   - Timing drive components
   - Camshaft oil seal
   - Camshaft oil seal retainer
   - Camshaft cap bolts by loosening them in sequence
   - Camshafts

To install:

3. Install or connect the following:
   - Camshaft bearing caps in their original position
   - Align the camshafts
   - Bearing thrust caps and hand tighten the bolts. When aligned properly, torque the bolts to 89 inch lbs. (10 Nm).
   - Timing drive components
   - Camshaft oil seal retainer
   - Crankshaft oil seal
   - Water pump drive pulley
   - Water pump belt
   - Negative battery cable

RIGHT SIDE

1. Before servicing the vehicle, refer to the precautions in the beginning of this section.
2. Remove or disconnect the following:
   - Negative battery cable
   - Timing drive components
   - Camshaft cap bolts by loosening them in sequence
   - Camshafts caps
   - Camshafts

To install:

3. Install or connect the following:
   - Camshaft bearing caps in their original position
   - Align the camshafts
   - Bearing caps and hand tighten the bolts
   - Bearing thrust caps and hand tighten the bolts. When aligned properly, torque the bolts to 89 inch lbs. (10 Nm).
   - Timing drive components
   - Negative battery cable

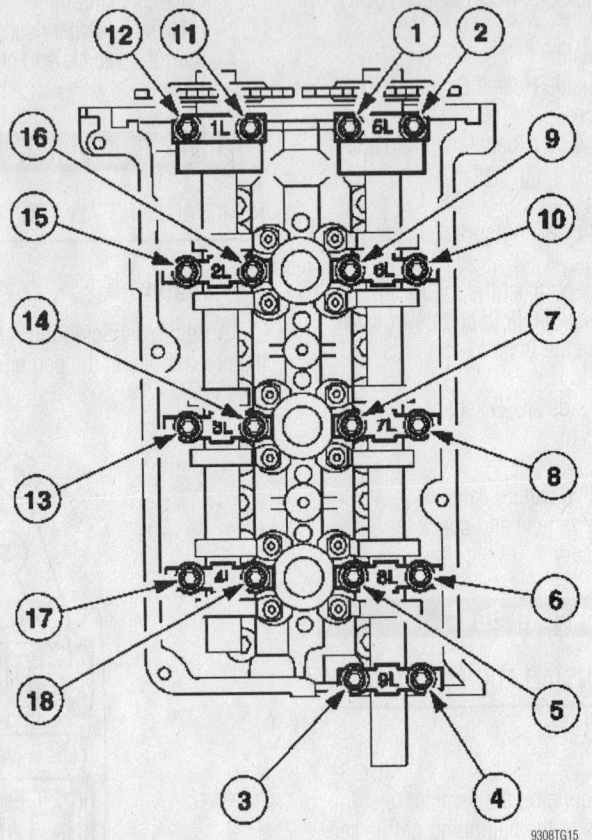

Remove and install the left side camshaft bearing caps in sequence–3.0L engine

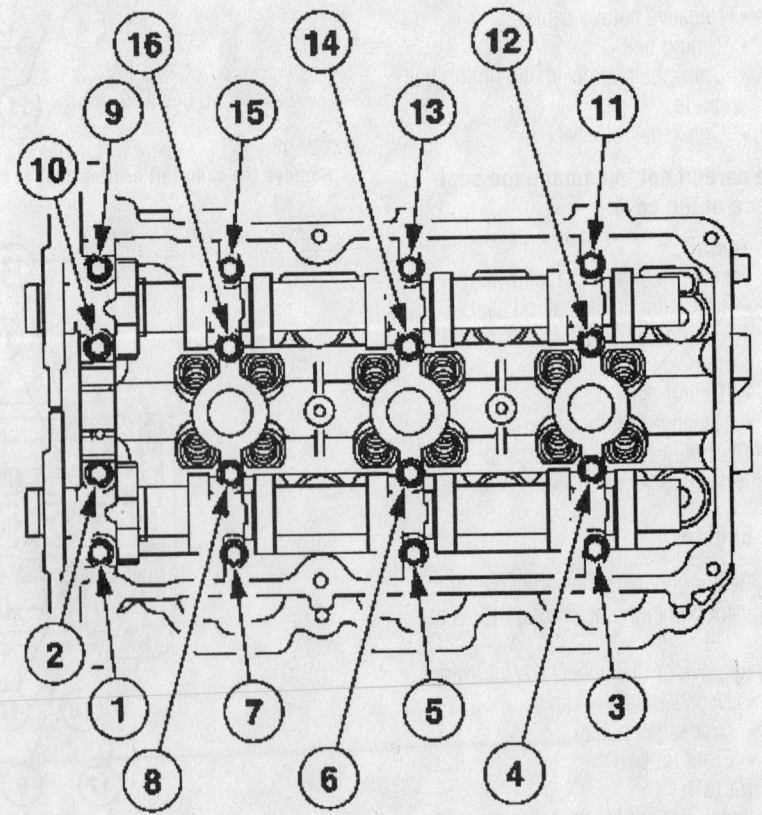

Remove and install the right side camshaft bearing caps in sequence–3.0L engine

Valve Lash

ADJUSTMENT

2.0L Engine

1. Before servicing the vehicle, refer to the precautions in the beginning of this section.
2. Remove or disconnect the following:
   - Negative battery cable
   - Timing belt
3. Measure each valve's clearance at the base circle with the lobe facing away from the tappet.
4. Use a feeler gauge to measure and record each valve's clearance
5. Remove or disconnect the following:
   - Camshafts
   - Valve tappets from the cylinder head
6. A mid range clearance is recommended as follows:
   a. Intake: 0.006 inch (0.15mm).
   b. Exhaust: 0.012 inch (0.3mm).

To install:

7. Install or connect the following:
   - Valve tappets after lubricating them with clean engine oil
   - Camshafts and verify each valve's clearance at the base circle with the lobe facing away from the tappet
   - Timing belt
   - Negative battery cable

3.0L Engine

1. Before servicing the vehicle, refer to the precautions in the beginning of this section.
2. Remove or disconnect the following:
   - Negative battery cable
   - Camshaft followers
   - Hydraulic lash adjusters

➡**Mark the position of the hydraulic lash adjusters to assure they are assembled in their original position**

3. Inspect the adjusters for scoring or uneven wear in the bore and replace them as required.

To install:

4. Install or connect the following:
   - Hydraulic lash adjusters after lubricating them with clean engine oil
   - Camshaft followers
   - Negative battery cable

Starter Motor

REMOVAL & INSTALLATION

2.0L Engine

1. Before servicing the vehicle, refer to the precautions in the beginning of this section.
2. Remove or disconnect the following:
   - Negative battery cable
   - Starter bolts
   - Exhaust system, AWD vehicles only
   - Halfshaft support bracket bolts
   - Starter electrical connectors
   - Starter

To install:

3. Install or connect the following:
   - Starter. Torque bolts to 20 ft. lbs. (27 Nm).
   - Starter electrical connectors
   - Halfshaft support bracket. Torque the bolts to 11 ft. lbs. (15 Nm).
   - Exhaust system on AWD vehicles. Torque the bolts to 18 ft. lbs. (25 Nm).
   - Negative battery cable

3.0L Engine

1. Before servicing the vehicle, refer to the precautions in the beginning of this section.
2. Drain the cooling system.
3. Remove or disconnect the following:
   - Negative battery cable
   - Air cleaner outlet tube
   - Coolant hoses and move the thermostat aside
   - Starter electrical connectors

- Starter

To install:

4. Install or connect the following:
   - Starter. Torque bolts to 20 ft. lbs. (27 Nm).
   - Starter electrical connectors and reposition the thermostat
   - Connect the 4 coolant hoses
   - Air cleaner outlet tube
   - Negative battery cable
5. Fill the cooling system to the proper level.
6. Start the vehicle and check for leaks, repair if necessary.

Oil Pan

REMOVAL & INSTALLATION

2.0L Engine

1. Before servicing the vehicle, refer to the precautions in the beginning of this section.
2. Drain the engine oil.
3. Support the powertrain assembly.
4. Remove or disconnect the following:
   - Negative battery cable
   - Catalytic converter
   - Oil pan and gasket
5. Thoroughly clean the gasket mating surfaces.

To install:

6. Apply silicone sealer to the oil pan.
7. Install a new gasket on the oil pan.
8. Oil pan. Torque the bolts in sequence to:
   a. Step 1: 53 inch lbs. (6 Nm).
   b. Step 2: 106 in lbs. (12 Nm).
9. Install or connect the following:

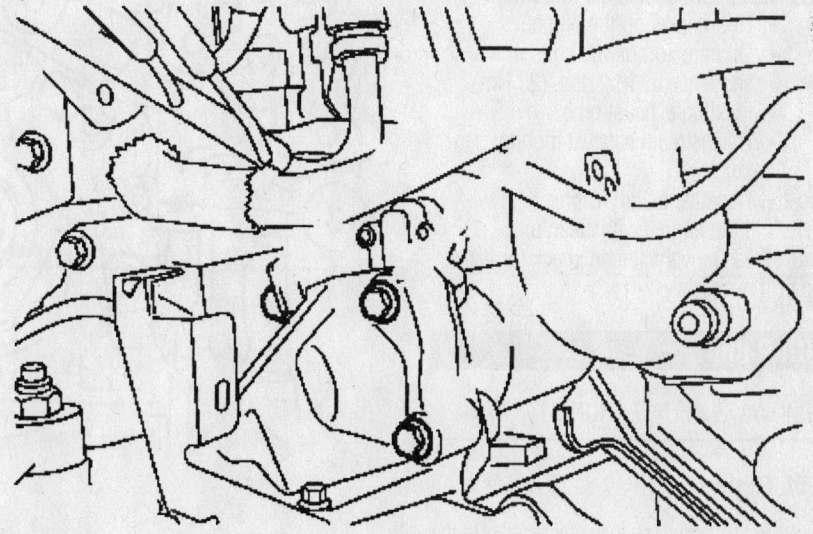

Removal of the starter motor–2.0L engine

9308TG17

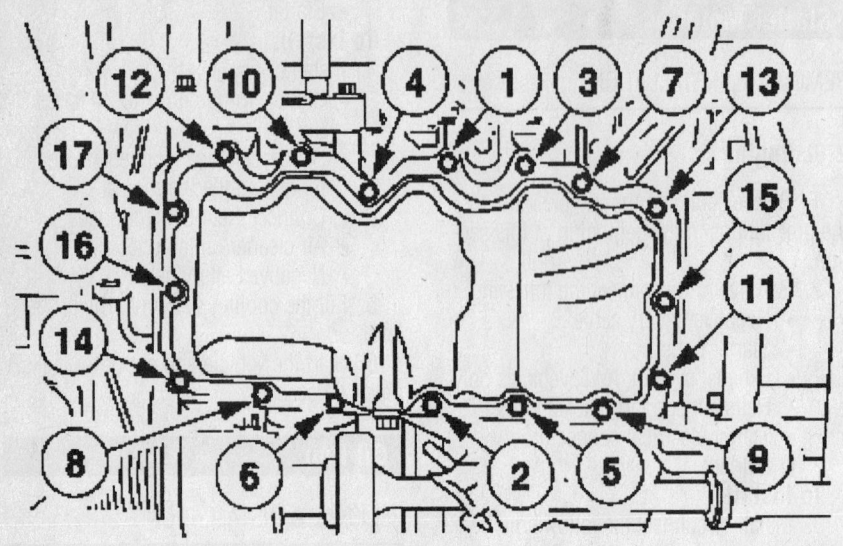

Tighten the oil pan bolts in sequence—2.0L engine

- Catalytic converter
- Negative battery cable

10. Fill the engine with clean oil.

11. Start the engine and check for leaks, repair if necessary.

3.0L Engine

1. Before servicing the vehicle, refer to the precautions in the beginning of this section.

2. Drain the engine oil.

3. Remove or disconnect the following:
- Negative battery cable
- Flexible exhaust pipe
- Downstream catalyst monitor sensor
- Oil pan and gasket

4. Thoroughly clean the gasket mating surfaces.

To install:

5. Apply silicone sealer to the oil pan.

6. Install or connect the following:
- New gasket on the oil pan
- Oil pan. Torque the bolts in sequence to 18 ft. lbs. (25 Nm).
- Flexible exhaust pipe
- Downstream catalyst monitor sensor
- Negative battery cable

7. Fill the engine with clean oil.

8. Start the vehicle and check for leaks, repair if necessary.

Oil Pump

REMOVAL & INSTALLATION

2.0L Engine

1. Before servicing the vehicle, refer to the precautions in the beginning of this section.

2. Drain the engine oil.

3. Remove or disconnect the following:
- Negative battery cable
- Oil pan
- Oil pump screen cover and tube
- Oil pump and discard the gasket

4. Thoroughly clean the gasket mating surfaces.

To install:

5. Install or connect the following:
- Oil pump screen cover and tube

with a new gasket. Torque the bolts to 89 inch lbs. (10 Nm).
- Oil pump to the oil pan
- Oil pan
- Negative battery cable

6. Refill the engine with clean oil.

7. Start the engine and check for leaks; repair if necessary.

3.0L Engine

1. Before servicing the vehicle, refer to the precautions in the beginning of this section.

2. Drain the engine oil.

3. Remove or disconnect the following:

- Negative battery cable
- Timing drive components
- Oil pump screen cover and tube
- Damper bolt and crankshaft sprockets
- Oil pump bolts in the proper sequence

4. Thoroughly clean the gasket mating surfaces.

To install:

5. Install or connect the following:
- Oil pump and bolts in the proper sequence. Torque the bolts to 89 inch lbs. (10 Nm).
- Crankshaft sprockets
- Oil pump screen cover and tube
- Timing drive components
- Negative battery cable

6. Refill the engine with clean oil.

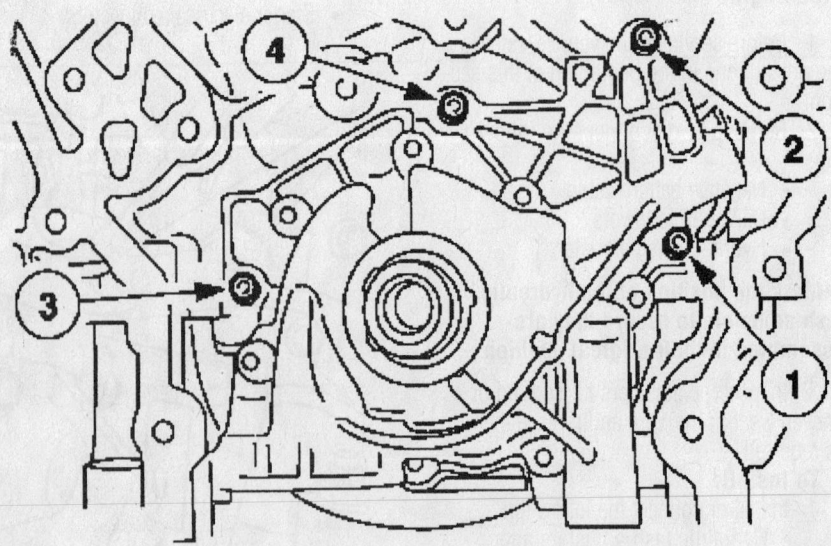

Remove the oil pump bolts in the proper sequence—3.0L engine

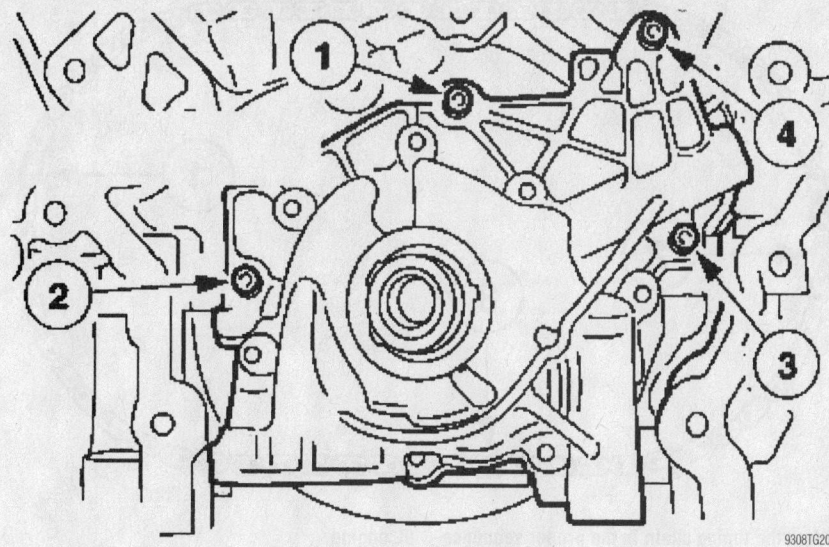

Install the oil pump bolts in the proper sequence–3.0L engine

7. Start the engine and check for leaks; repair if necessary.

Rear Main Seal

REMOVAL & INSTALLATION

2.0L Engine

1. Before servicing the vehicle, refer to the precautions in the beginning of this section.
2. Remove or disconnect the following:
   - Negative battery cable
   - Flywheel
   - Rear main seal

To install:

3. Coat the oil seal with clean engine oil.
4. Install or connect the following:
   - Crankshaft rear oil seal
   - Flywheel
   - Negative battery cable

3.0L Engine

1. Before servicing the vehicle, refer to the precautions in the beginning of this section.
2. Remove or disconnect the following:
   - Negative battery cable
   - Flexplate
   - Rear main oil seal

To install:

3. Coat the oil seal with clean engine oil.

4. Install or connect the following:
   - Crankshaft rear oil seal
   - Flywheel
   - Negative battery cable

Timing Belt and Covers

REMOVAL & INSTALLATION

2.0L Engine

1. Remove the valve cover.
2. Remove the spark plugs.
3. Remove the catalytic converter.
4. Remove the bolt, nut, and position the coolant tube aside.
5. Remove the right wheel and tire assembly.
6. Remove the right lower splash shield.
7. Rotate the crankshaft to just before top dead center (TDC) (No. 1 cylinder).
8. Remove the stud.
9. Install the special tool.

➡**Make sure the correct (second) notch in the pulley is indexed to the lower cylinder block.**

10. Rotate the crankshaft clockwise against the peg to bring it to TDC (No. 1 cylinder).
11. Loosen the water pump pulley bolts.
12. Disconnect the battery ground cable.
13. Remove the crankshaft pulley.
14. Remove the bolts and the lower timing belt cover.

15. Lower the vehicle.
16. Install the special tool 303-F072.
17. Remove the ground strap.
18. Remove the engine mount upper bracket.
19. Remove the studs.
20. Remove the knock sensor connector.
21. Remove the bolts and the upper timing belt cover.
22. Remove the water pump pulley.
23. Remove the accessory drive belt idler pulley.

➡**Installation of the alignment tool into the exhaust camshaft may require the camshafts to be rotated clockwise slightly.**

24. Install the special tool and align the camshafts.
25. Raise and support the vehicle.
26. Remove the bolts and the engine mount lower bracket.
27. Loosen the timing belt tensioner pulley and allow to slide down to the bottom of its travel.

✳✳ CAUTION

If the camshaft timing belt is to be reused, mark the direction of the camshaft timing belt to rotation of camshaft prior to removal or premature wear or failure may occur.

28. Slide the timing belt off of the camshaft and crankshaft sprockets. the timing belt for wear. Install a new belt if necessary.

To install:

➡**Make sure the correct (second) notch in the pulley is indexed to the lower cylinder block.**

29. Slide the crankshaft pulley onto the crankshaft and confirm the crankshaft position is at TDC (No. 1 cylinder) by rotating it clockwise against the alignment peg.
30. Remove the crankshaft pulley.
31. Lower the vehicle.
32. Confirm that the timing belt tensioner is installed correctly with the tab positioned in the slot in the inner timing cover.
33. Install the timing belt onto the timing belt sprockets.
34. Adjust the timing belt tensioner.
35. Using a 6 mm Allen wrench, rotate the adjuster counterclockwise and align the marks as shown.
36. Tighten the tensioner pulley bolt.
37. Raise the vehicle.
38. Install the front engine mount lower bracket.

39. Install the accessory drive belt idler pulley.

40. Install the water pump pulley.

41. Hand tighten the bolts.

42. Lower the vehicle.

43. Install the timing belt covers. Torque the cover bolts to 62 inch lbs. (7Nm).

44. Tighten the water pump pulley bolts.

45. Remove the special tool.

46. Install the stud. Torque to 25 ft. lbs. (34Nm).

47. Install the coolant tube.

48. Install the catalytic converter.

49. Remove the special tool.

50. Install the valve cover.

51. Install the spark plugs.

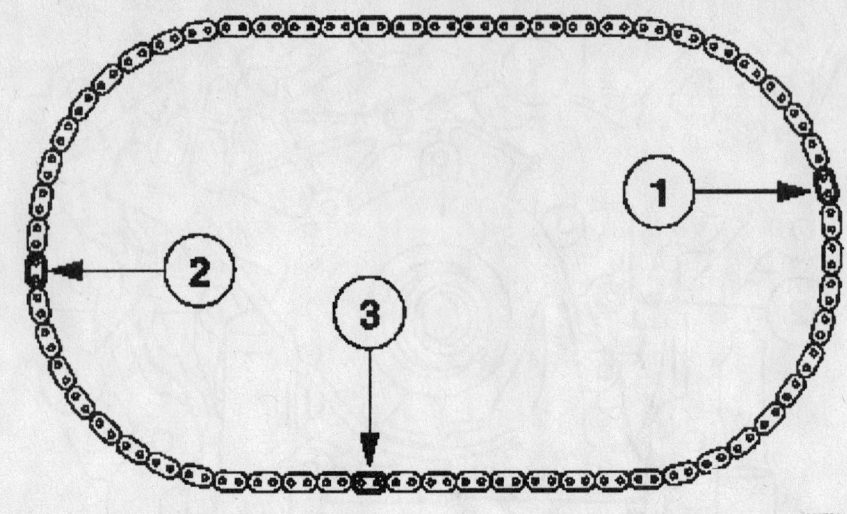

Mark the timing chain in the proper sequence–3.0L engine

9308TG21

Timing Gears, Front Cover and Seal

REMOVAL & INSTALLATION

3.0L Engine

1. Before servicing the vehicle, refer to the precautions in the beginning of this section.

2. Remove or disconnect the following:

- Negative battery cable
- Engine front cover
- Ignition pulse wheel and install a damper bolt
- Spark plugs

3. Rotate the crankshaft clockwise to position the keyway at the 11 o'clock position and the camshafts in the correct positions. The No. 1 cylinder will be at Top Dead Center (TDC).

4. Rotate the crankshaft clockwise 120 degrees to the 3 o'clock position to locate the right side camshafts in the neutral position.

5. Remove or disconnect the following:

- Right side timing chain and tensioner
- Tensioner arm and timing chain guide

6. Rotate the crankshaft clockwise 2 times to position the keyway at the 11 o'clock position. This will position the left side camshafts in the neutral position.

7. Verify that the left side crankshafts

are in the neutral position and mark the link position on the crankshaft sprocket.

8. Remove or disconnect the following:

- Left side timing chain and tensioner
- Tensioner arm and timing chain guide
- Damper bolt and crankshaft sprockets

To install:

9. Install the crankshaft sprockets.

10. Position the timing chain tensioner in a soft jaw vise. Hold the ratchet lock mechanism away from the ratchet stem and slowly compress the timing chain tensioner

11. If the timing marks on the chain are not visible, use a permanent marker to mark the left and right side timing chains. Mark the timing chains in the following sequence:

a. Mark any link to use as the crankshaft timing mark.

b. Count 29 links from the crankshaft timing mark and mark the link as the exhaust cam sprocket timing mark.

c. Continue counting to 42 and mark the link as the intake sprocket timing mark

12. Install the guide. Torque the bolts to 18 ft. lbs. (25 Nm).

13. Install the left side timing chain and align the chain in the following sequence:

a. Mark any link to use as the crankshaft timing mark.

b. Count 29 links from the crankshaft timing mark and mark the link as the exhaust cam sprocket timing mark.

c. Continue counting to 42 and mark the link as the intake sprocket timing mark

14. Install or connect the following:

- Left side timing chain and tensioner arm. Torque the bolts to 18 ft. lbs. (25 Nm).
- Crankshaft damper bolt and rotate the keyway to the 3 o'clock position

15. Verify that the right side camshafts are properly positioned and install the right side timing chain and guide. Torque the bolts to 18 ft. lbs. (25 Nm).

16. Make certain that the timing chain aligns with the marks on the camshaft and crankshaft sprockets

✳✳ CAUTION

Install the pulse wheel with the keyway in the slot stamped 20–25–34Y–30M (Color Blur).

17. Install or connect the following:

- Right side timing chain tensioner and arm. Torque the bolts to 18 ft. lbs. (25 Nm) and remove the damper bolt
- Ignition pulse wheel
- Spark plugs
- Engine front cover
- Negative battery cable

Piston and Ring

POSITIONING

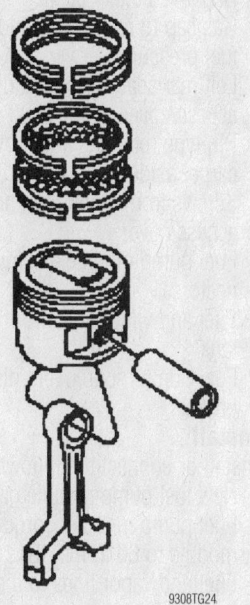

9308TG24

2.0L engine piston and connecting rod
positioning ring end-gap spacing

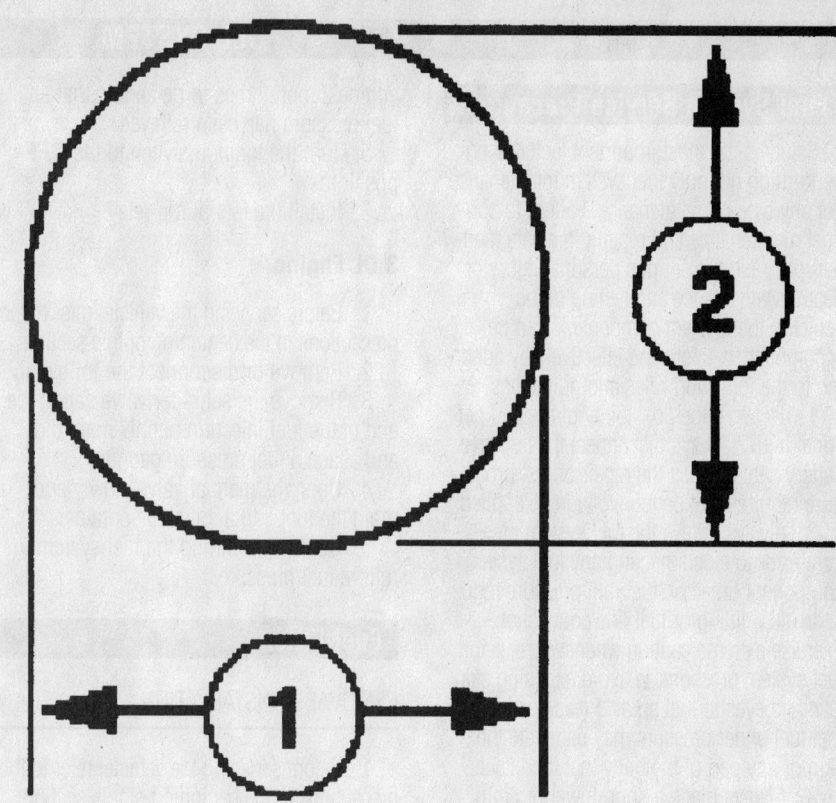

9308TG26

2.0L (VIN B) engine —piston ring end-gap spacing

9308TG25

3.0L (VIN 1) engine—piston ring end-gap spacing

FUEL SYSTEM

Fuel System Service Precautions

Safety is the most important factor when performing not only fuel system maintenance but any type of maintenance. Failure to conduct maintenance and repairs in a safe manner may result in serious personal injury or death. Maintenance and testing of the vehicle's fuel system components can be accomplished safely and effectively by adhering to the following rules and guidelines.

1. To avoid the possibility of fire and personal injury, always disconnect the negative battery cable unless the repair or test procedure requires that battery voltage be applied.

2. Always relieve the fuel system pressure prior to disconnecting any fuel system component (injector, fuel rail, pressure regulator, etc.), fitting or fuel line connection. Exercise extreme caution whenever relieving fuel system pressure, to avoid exposing skin, face and eyes to fuel spray. Please be advised that fuel under pressure may penetrate the skin or any part of the body that it contacts.

3. Always place a shop towel or cloth around the fitting or connection prior to loosening to absorb any excess fuel due to spillage. Ensure that all fuel spillage (should it occur) is quickly removed from engine surfaces. Ensure that all fuel soaked cloths or towels are deposited into a suitable waste container.

4. Always keep a dry chemical (Class B) fire extinguisher near the work area.

5. Do not allow fuel spray or fuel vapors to come into contact with a spark or open flame.

6. Always use a backup wrench when loosening and tightening fuel line connection fittings. This will prevent unnecessary stress and torsion to fuel line piping.

7. Always replace worn fuel fitting O-rings with new. Do not substitute fuel hose or equivalent, where fuel pipe is installed.

Before servicing the vehicle, make sure to refer to the precautions in the beginning of this section as well.

Fuel System Pressure

RELIEVING

2.0L Engine

1. Before servicing the vehicle, refer to the precautions in the beginning of this section.

2. Remove or disconnect the following:

3. Remove the fuel pump relay and start the engine.

4. After the engine stalls, crank the engine 2 more times to be certain that all fuel pressure has been relieved.

5. Turn the ignition switch to the **OFF**-position.

6. Install the fuel pump relay.

3.0L Engine

1. Before servicing the vehicle, refer to the precautions in the beginning of this section.

2. Remove or disconnect the following:

3. Remove the schrader valve cap at the end of the fuel injection supply manifold and attach a fuel pressure gauge.

4. Open the manual valve slowly and drain the fuel into a suitable container.

5. Continue draining the fuel system to relieve fuel pressure.

Fuel Filter

REMOVAL & INSTALLATION

1. Before servicing the vehicle, refer to the precautions in the beginning of this section.

2. Properly relieve the fuel system pressure.

3. Remove or disconnect the following:
   - Negative battery cable
   - Fuel line to the fuel filter

4. Loosen the clamp and remove the filter

To install:

5. Install or connect the following:
   - New clips to the fuel lines
   - Fuel filter and tighten the clamp
   - Fuel lines to the fuel filter
   - Negative battery cable

6. Start the vehicle and check for leaks, repair if necessary.

Fuel Pump

REMOVAL & INSTALLATION

1. Before servicing the vehicle, refer to the precautions in the beginning of this section.

2. Properly relieve the fuel system pressure.

3. Remove or disconnect the following:
   - Negative battery cable
   - Gas cap to relieve any additional fuel pressure
   - Left rear seat cushion and lift the access cover on the scuff plate
   - Pin type retainers and move the carpet aside
   - Screws from the fuel pump module access cover
   - Fuel pump module electrical connectors
   - Fuel and vapor lines from the fuel tank
   - Fuel pump module and discard the gasket

To install:

4. Install or connect the following:
   - New fuel pump module gasket
   - Fuel pump module. Torque the module to 60 ft. lbs. (81 Nm).
   - Fuel and vapor lines to the fuel tank
   - Fuel pump module electrical connectors
   - Fuel pump module access cover and tighten the screws securely
   - Pin type retainers and reposition the carpet
   - Left rear seat cushion
   - Gas cap
   - Negative battery cable

5. Start the engine and check for leaks, repair if necessary.

Fuel Injectors

REMOVAL & INSTALLATION

1. Before servicing the vehicle, refer to the precautions in the beginning of this section.

2. Release the fuel system pressure.

3. Remove or disconnect the following:

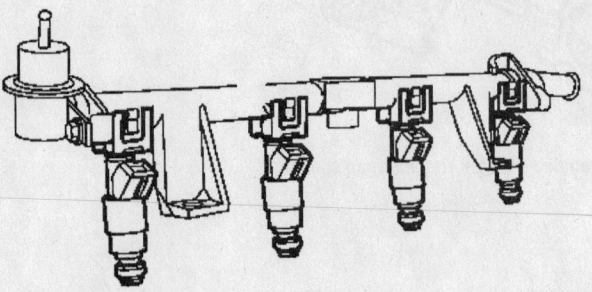

Remove the fuel injectors from the fuel supply manifold—2.0L engine

9308TG22

- Negative battery cable
- Fuel injection supply manifold
- Retaining clips and gently twist the fuel injector out of the manifold

4. Check the O-rings and replace if damaged.

To install:
5. Install or connect the following:
- Fuel injector(s) using new O-rings lubricated with clean engine oil
- Fuel injector into the supply manifold

- Retaining clips when the fuel injectors are seated properly
- Fuel injection supply manifold
- Negative battery cable

6. Start the vehicle and check for leaks, repair if necessary.

DRIVE TRAIN

Transmission Assembly

REMOVAL & INSTALLATION

Manual Transmission

1. Before servicing the vehicle, refer to the precautions in the beginning of this section.
2. Drain the transmission fluid.
3. Remove or disconnect the following:

- Battery cables
- Battery and tray
- Mass Air Flow (MAF) sensor electrical connector
- Accelerator cable from the air cleaner outlet tube
- Emission management tube and hose
- Crankcase ventilation hose
- Air cleaner outlet tube
- Air cleaner housing
- Back-up lamp switch electrical connector
- Front wire harness bracket and move it aside
- Front wire harness bracket spacer
- Wire harness from the rear harness bracket
- Park Neutral Position (PNP) electrical connector
- Rear wire harness bracket and move it aside
- Vehicle Speed Sensor (VSS) electrical connector
- Clutch slave cylinder line from the bracket and move it aside while properly supporting the engine
- Left side transmission support insulator and bracket
- Rear transmission support insulator
- Front transmission support insulator and bracket
- Starter and move it aside
- Top transmission flywheel housing bolts
- Front transmission flywheel housing bolts
- Transfer case, if equipped
- Left side halfshaft

- Rear transmission support insulator bracket
- Shifter linkage and stabilizer bar
- Transverse crossmember
- Front to aft crossmember
- Left side splash shield and properly support the transmission
- Remaining transmission flywheel housing bolts
- Transmission and separate the right side halfshaft from the transmission

To install:
4. Align the right side half shaft to the transmission and position the transmission to the engine.
5. Install or connect the following:
- Transmission flywheel housing bolts. Torque the bolts to 33 ft. lbs. (45 Nm) and remove the transmission support
- Left side splash shield
- Front-to-aft crossmember. Torque the bolts to 66 ft. lbs. (90 Nm).
- Transverse crossmember. Torque the bolts to 85 ft. lbs. (115 Nm).
- Shifter linkage. Torque the bolt to 15 ft. lbs. (20 Nm).
- Stabilizer bar. Torque the bolt to 30 ft. lbs. (40 Nm).
- Rear transmission support bracket. Torque the bolts to 66 ft. lbs. (90 Nm).
- Left side halfshaft
- Transfer case, if equipped
- Front transmission flywheel housing bolts. Torque the bolts to 33 ft. lbs. (45 Nm).
- Top transmission flywheel housing bolts. Torque the bolts to 33 ft. lbs. (45 Nm).
- Starter. Torque the bolts to 33 ft. lbs. (45 Nm).
- Front transmission support insulator and bracket. Torque the lower bolt to 66 ft. lbs. (90 Nm) and the 3 upper bolts to 41 ft. lbs. (55 Nm).
- Rear transmission support insulator bolt. Torque the bolt to 66 ft. lbs. (90 Nm).
- Left side transmission support insulator bracket. Torque the bolts to 66 ft. lbs. (90 Nm).

- Left side transmission support insulator. Torque the large bolt to 66 ft. lbs. (90 Nm) and the 3 remaining bolts to 41 ft. lbs. (55 Nm).
- Clutch slave cylinder. Torque the bolt to 15 ft. lbs. (20 Nm).
- Clutch slave cylinder line to the bracket and install the retaining clip
- VSS electrical connector
- Rear wire harness bracket. Torque the bolts to 80 inch lbs. (9 Nm).
- PNP switch electrical connector
- Wire harness to the rear bracket
- Front wire harness bracket spacer and bracket. Torque the bolt to 9 ft. lbs. (12 Nm).
- Back-up lamp switch electrical connector
- Air cleaner housing
- MAF sensor electrical connector
- Air cleaner outlet tube
- Crankcase ventilation hose
- Emission management tube and hose
- Accelerator cable to the air cleaner outlet tube
- Battery and tray
- Both battery cables

6. Fill the transmission to the proper level.
7. Start the vehicle and check for leaks, repair if necessary.

Automatic Transmission

1. Before servicing the vehicle, refer to the precautions in the beginning of this section.
2. Remove or disconnect the following:
- Battery cables
- Battery and tray
- Breather tube
- Mass Air Flow (MAF) sensor
- Intake tube and air cleaner cover
- Air cleaner assembly
- Transmission Range (TR) sensor
- Heated Oxygen (HO$_2$S) sensors
- Transmission harness connector and bracket
- Wire harness bracket spacer and move the bracket aside

- Shift cable
- Shift cable bracket and move the bracket aside
- Starter electrical connectors
- Starter
- Electrical connectors from the valve cover and install an engine support bar
- Upper transmission retaining bolts
- Left side upper transmission mounting plate
- Rear transmission mount
- Right side engine mount bolt and slightly raise the engine
- Both front wheels and splash shields
- Right side halfshaft and intermediate shaft assembly after match-marking them
- Cross brace
- Center exhaust pipe and rubber hanger
- Front exhaust pipe and flange
- Rear exhaust pipe flange
- Driveshaft
- PTU vent tube
- Lower transmission bracket
- Access cover
- Flexplate nuts
- Output Shaft Speed (OSS) sensor
- Turbine Shaft Speed (TSS) sensor
- Fluid cooler tube and move it aside
- Fluid cooler line and install a transmission jack
- Bolts from the PTU unit
- Transmission with the PTU unit attached

To install:

3. Install or connect the following:
- Transmission with the PTU unit. Torque the engine-to-transmission mounting bolts to 30 ft. lbs. (40 Nm).
- Fluid cooler line. Torque the fastener to 17 ft. lbs. (23 Nm) and remove the transmission jack.
- Fluid cooler tube. Torque the bolt to 17 ft. lbs. (23 Nm).
- OSS sensor
- TSS sensor
- Flexplate nuts. Torque the nuts to 27 ft. lbs. (36 Nm).
- Access cover
- Cross brace. Torque the bolts to 96 ft. lbs. (130 Nm).
- Transmission bracket. Torque the bolts to 30 ft. lbs. (40 Nm).
- PTU vent tube
- Driveshaft. Torque the bolts to 15 ft. lbs. (20 Nm).
- Exhaust pipe and flange. Torque the bolts to 21 ft. lbs. (29 Nm).

- Exhaust pipe and rubber hanger. Torque the bolts to 21 ft. lbs. (29 Nm).
- Left side halfshaft assembly
- Right side halfshaft and intermediate shaft assembly by aligning the matchmarks
- Splash shields
- Both front wheels and lower the engine on to the right side engine mount
- Right side engine mount bolt. Torque the bolt to 89 ft. lbs. (120 Nm).
- Rear transmission mount. Torque the upper bolt to 89 ft. lbs. (120 Nm) and the lower bolts to 35 ft. lbs. (45 Nm).
- Transmission mount assemble. Torque the bolts to 30 ft. lbs. (40 Nm) and remove the engine support bar.
- Electrical connectors to the valve cover
- Starter. Torque the bolts to 20 ft. lbs. (27 Nm).
- Starter electrical connectors
- Shifter cable and bracket. Torque the bolt to 14 ft. lbs. (19 Nm) and connect the shifter cable.
- Wire harness and install the harness bracket spacer
- Wire harness bracket. Torque the bolt to 89 inch lbs. (10 Nm).
- HO2S sensor
- TR sensor and make certain it is properly aligned
- Air cleaner assembly
- Intake tube and air cleaner cover

- Breather tube
- MAF sensor
- Battery tray
- Battery and cables

4. Fill the transmission with clean fluid to the proper level.

5. Start the vehicle and check for leaks, repair if necessary

Clutch

ADJUSTMENTS

The clutch is hydraulically driven and therefore no adjustment is required.

REMOVAL & INSTALLATION

1. Before servicing the vehicle, refer to the precautions in the beginning of this section.

2. Remove or disconnect the following:
- Negative battery cable
- Transmission and lock the flywheel to the engine with special tool 303–103
- Pressure plate bolts by loosening them evenly
- Clutch pressure plate and disc

3. Clean the pressure plate and inspect it for burn marks, scores, flatness or ridges, replace if damaged.

4. Inspect the pressure plate diaphragm finger for wear, replace if damaged.

5. Measure the depth of the rivet heads. Minimum depth is 0.012 inch (0.3mm).

6. Inspect the clutch disc for signs of wear and replace if needed.

7. Check the clutch disc runout. Replace

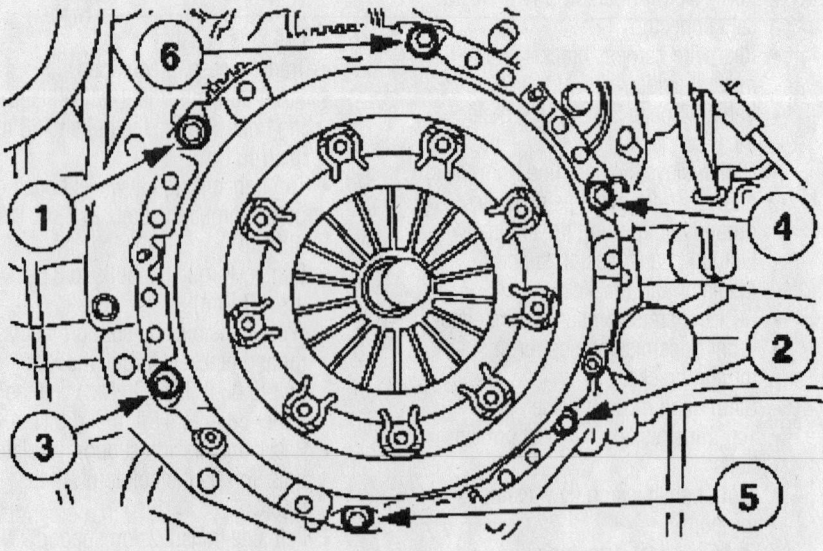

Torque the pressure plate bolts in the proper sequence

9308TG07

the disc if not with specification: 0.027 inch (0.7mm).

To install:

8. Install or connect the following:
   - Clutch disc to the flywheel
   - Pressure plate to the flywheel. Torque the bolts in sequence to 21 ft. lbs. (29 Nm).
   - Transmission
   - Negative battery cable
9. Check the transmission fluid level and top off if necessary.

Hydraulic Clutch System

BLEEDING

The following procedure is recommended for bleeding the clutch hydraulic system installed on the vehicle. It is recommended that the original clutch tube, with quick-connect fitting be replaced when servicing the hydraulic system, because air can be trapped in the quick-connect fitting and prevent complete bleeding of the system.

1. Before servicing the vehicle, refer to the precautions in the beginning of this section.
2. Clean the dirt and grease from the dust cap.
3. Remove the cap and diaphragm and fill the reservoir ¾ of the way with approved brake fluid C6AZ-19542-AB or DOT 3 equivalent fluid (ESA-M6C25-A).
4. Loosen the bleeder screw cover from the slave cylinder and attach a hose to the screw.
5. Place the hose in a container and slowly pump the clutch pedal several times.
6. With the clutch pedal depressed, loosen the bleeder screw to release the fluid and air.
7. Remove the hose and tighten the bleeder screw.
8. Repeat this procedure until all the air is removed from the hydraulic system.

Transfer Case

REMOVAL & INSTALLATION

3.0L Engine

1. Before servicing the vehicle, refer to the precautions in the beginning of this section.
   Place the transmission in the NEUTRAL position.
2. Remove or disconnect the following:
   - Negative battery cable
   - Driveshaft

- Intermediate shaft
- Exhaust crossover pipe
- Right side exhaust manifold
- Heat shield
- Lower bracket
- Crossmember brace
- Transfer case bolts and vent tube
- Transfer case

To install:

3. Install or connect the following:
   - Transfer case with a new driven gear seal. Torque the bolts to 33 ft. lbs. (45 Nm).
   - Transfer case vent tube
   - Crossmember brace. Torque the bolts to 30 ft. lbs. (40 Nm).
   - Lower bracket. Torque the bolts to 30 ft. lbs. (40 Nm).
   - Exhaust manifold
   - Heat shield. Torque the bolts to 10 ft. lbs. (14 Nm).
   - Exhaust crossover pipe
   - Intermediate shaft
   - Driveshaft
   - Negative battery cable
4. Check the transfer case fluid level and top off if necessary.
5. Start the vehicle and check for leaks, repair if necessary.

Halfshaft

REMOVAL & INSTALLATION

1. Before servicing the vehicle, refer to the precautions in the beginning of this section.
2. Place the transmission in the PARK position.
3. Remove or disconnect the following:
   - Negative battery cable
   - Front wheel
   - Front brake disc
   - Front axle wheel hub nut and discard the nut
   - Tie rod end and separate the lower ball from the steering knuckle
   - Halfshaft from the steering knuckle
   - Halfshaft

To install:

4. When seated properly, the halfshaft bearing retainer circlip will snap into the differential side gear groove.
5. Position the halfshaft and joint so that the splines align with differential side gear splines. Push the halfshaft into side gear.
6. Install or connect the following:
   - Halfshaft into the steering knuckle
   - Lower ball joint to steering knuckle.

Torque the pinch bolt to 52 ft. lbs. (70 Nm).
   - Tie rod end. Torque the nut to 41 ft. lbs. (55 Nm).
   - New front axle wheel hub nut. Torque the nut to 214 ft. lbs. (290 Nm).
   - Front brake disc
   - Front wheel
   - Negative battery cable
7. Check the fluid level and adjust as needed.

CV-Joints

OVERHAUL

1. Before servicing the vehicle, refer to the precautions in the beginning of this section.
2. Remove or disconnect the following:
   - Negative battery cable
   - Halfshaft and secure it in a soft-jawed vise
   - Inboard halfshaft boot clamp
   - Boot from the inboard CV-joint housing
   - Tripod joint from the CV-joint housing and matchmark the tripod joint to the halfshaft
   - Snapring and boot from the halfshaft
   - Outboard halfshaft boot clamps
   - Outboard boot back to expose the CV-joint and matchmark the joint to the halfshaft
   - Outboard CV-joint from the halfshaft
   - Halfshaft retainer circlip and discard it
   - Boot from the halfshaft

To install:

3. Install or connect the following:
   - Outboard CV-joint and boot
   - New halfshaft bearing circlip
   - Inboard CV-joint to the halfshaft
   - Outboard halfshaft boot forward on to the outboard CV-joint
   - New outboard halfshaft boot clamps
   - Inboard halfshaft boot
   - Tripod joint on the halfshaft by aligning the matchmarks
   - New snapring to the tripod joint and lubricate the needle bearings while filling the housing with CV-joint grease, E43Z–19590–A
   - Inboard halfshaft boot with new clamps
   - Halfshaft
   - Negative battery cable

Axle Housing

REMOVAL & INSTALLATION

1. Before servicing the vehicle, refer to the precautions in the beginning of this section.
2. Remove or disconnect the following:

- Negative battery cable
- Rotary blade coupling
- Rear halfshafts
- Axle assembly-to-front bracket bolts
- Rear axle-to-side bracket bolts
- Axle assembly

To install:

3. Install or connect the following:

- Axle assembly
- Rear axle-to-side-bearing bolts. Torque the bolts to 59 ft. lbs. (80 Nm).
- Axle assembly-to-front bracket bolts. Torque the bolts to 59 ft. lbs. (80 Nm).
- Rotary blade coupling
- Negative battery cable

STEERING AND SUSPENSION

Air Bag

✳✳ CAUTION

Some vehicles are equipped with an air bag system. The system MUST BE disabled before performing service on or around system components, steering column, instrument panel components, wiring and sensors. Failure to follow safety and disabling procedures could result in accidental air bag deployment, possible personal injury and unnecessary system repairs.

PRECAUTIONS

Several precautions must be observed when handling the inflator module to avoid accidental deployment and possible personal injury:

1. Never carry the inflator module by the wires or connector on the underside of the module.
2. When carrying a live inflator module, hold securely with both hands and ensure that the bag and trim cover are pointed away.
3. Place the inflator module on a bench or other surface with the bag and trim cover facing up.
4. With the inflator module on the bench, never place anything on or close to the module, which may be thrown in the event of an accidental deployment.

DISARMING

✳✳ CAUTION

The Supplemental Inflatable Restraint (SIR) system must be disarmed before performing service around SIR system components or SIR system wiring. Failure to do so may cause accidental deployment of the air bag, resulting in unnecessary SIR system repairs and/or personal injury.

The positive battery cable must be disconnected for a minimum of 1 minute before beginning any air bag work to de-energize the back-up power supply. It is a good idea to disengage both the positive and negative battery cables to ensure that the Air Bag system is definitely discharged.

ARMING THE SYSTEM

✳✳ WARNING

If the air bag simulators have been used, the air bag simulators must be removed and the air bags reconnected when the system is reactivated to avoid non-deployment in a collision resulting in possible personal injury.

1. Disconnect the positive battery cable.
2. Wait 1 minute, this is required for the back-up power supply in the air bag diagnostic monitor to deplete its stored energy.
3. Remove the air bag simulator from the air bag sliding contact connector at the top of the steering column. Reconnect the driver's side air bag module assembly. Position the driver's air bag module on the steering wheel and secure with the 2 bolts and washers. Tighten the bolt and washer assembly to 8–10 ft. lbs. (10–14 Nm).
4. Connect the positive battery cable.
5. Turn the ignition switch from the **OFF** to **RUN** and visually monitor the air bag warning indicator. The light will illuminate continuously for approximately 6 seconds and then turn off. If a fault occurs, the air bag indicator will either fail to light, remain lighted continuously or flash. The flashing may not occur until approximately 30 seconds after the ignition switch has been turned from **OFF** to **RUN**. This is the time needed for the air bag diagnostic monitor to complete testing the system. If the air bag indicator is inoperative, an air bag system fault exists, a tone will sound in a pattern of 5 sets of 5 beeps. If this occurs, the air bag indicator will need to be serviced before further diagnostics can be done.

Steering Gear

REMOVAL & INSTALLATION

1. Before servicing the vehicle, refer to the precautions in the beginning of this section.
2. Place the steering wheel in the straight-ahead position. Lock the steering wheel in place, using a steering wheel holder.

➡ **Locking the steering wheel keeps the clockspring in alignment position.**

3. Drain the power steering fluid.
4. Remove or disconnect the following:
- Negative battery cable
- Rear transmission insulator
- Rear transmission insulator bracket, if equipped with an automatic transmission
- Both front wheels
- Rear transmission insulator bracket, if equipped with a manual transmission
- Tie rod end cotter pin and nut
- Tie rod end from the steering knuckle and record the number of turns required to remove the tie rod end
- Steering gear coupling pinch bolt
- Power steering pressure and return lines and bracket
- Steering gear mounting bolts
- Steering gear and separate the steering coupling from the steering gear shaft
- Steering gear

To install:

- Slide the steering gear rearward to connect the steering coupling to the steering gear shaft

5. Install or connect the following:
- Steering gear mounting bolts. Torque the bolts to 93 ft. lbs. (126 Nm).
- Pressure and return lines and bracket. Torque the bracket bolts to 89 inch lbs. (10 Nm).

- Power steering pressure and return lines to the steering gear. Torque the bolt to 18 ft. lbs. (25 Nm).
- Steering gear pinch bolt and reposition the boot. Torque the bolt to 18 ft. lbs. (25 Nm).
- Tie rod end to the tie rod using the number of turns required to remove the tie rod end
- Jam nuts. Torque the nuts to 35 ft. lbs. (47 Nm).
- Tie rod end to the steering knuckle. Torque the nut to 41 ft. lbs. (57 Nm) and install a new cotter pin
- Rear transmission insulator bracket. Torque the bolts to 66 ft. lbs. (90 Nm).
- Both front wheels
- Rear transmission insulator bracket. Torque the bolts to 66 ft. lbs. (90 Nm).
- Rear transmission insulator. Torque the bolts to 66 ft. lbs. (90 Nm).
- Negative battery cable

6. Fill and bleed the power steering system.

7. Start the vehicle and check for leaks, repair if necessary.

8. Check and adjust the front end alignment.

Strut

REMOVAL & INSTALLATION

1. Before servicing the vehicle, refer to the precautions in the beginning of this section.

2. Install or connect the following:
- Negative battery cable
- Front wheel
- Brake hose grommet from the bracket
- Antilock Brake System (ABS) harness from the strut assembly and move the brake hose bracket aside
- Stabilizer bar link nut and move the bar aside
- Strut to steering knuckle bolts and support the strut assembly
- Upper strut nuts
- Strut and coil spring assembly

To install:
3. Install or connect the following:
- Strut and spring assembly. Torque the upper nuts to 59 ft. lbs. (80 Nm).
- Lower strut assembly to the steering knuckle. Torque the lower bolts to 85 ft. lbs. (115 Nm).
- Stabilizer bar into position.

Torque the bolts to 35 ft. lbs. (48 Nm).
- Brake hose bracket. Torque the bolts to 14 ft. lbs. (18 Nm).
- ABS harness to the strut assembly, if equipped
- Brake hose grommet to the bracket
- Front wheel
- Negative battery cable

Shock Absorber

REMOVAL & INSTALLATION

1. Before servicing the vehicle, refer to the precautions in the beginning of this section.

2. Remove or disconnect the following:
- Negative battery cable
- Rear quarter trim panel
- Upper shock absorber nut and raise the vehicle enough to relax the suspension
- Lower shock absorber nut
- Shock absorber

To install:
3. Install or connect the following:
- Shock absorber. Torque the lower nut to 85 ft. lbs. (115 Nm).
- Upper shock absorber nut. Toprque the nut to 13 ft. lbs. (18 Nm).

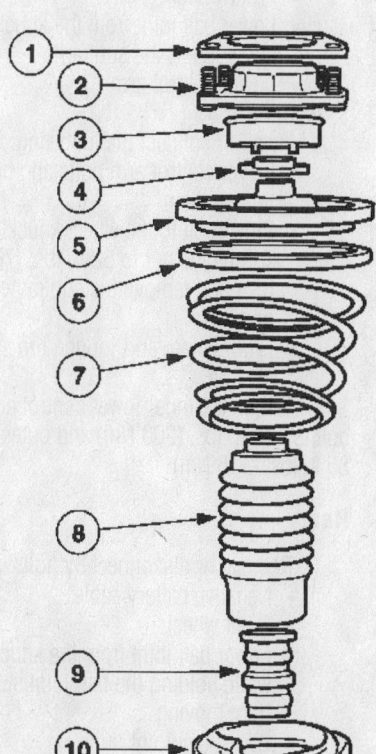

- Rear quarter trim panel
- Negative battery cable

Coil Spring

REMOVAL & INSTALLATION

Front

1. Before servicing the vehicle, refer to the precautions in the beginning of this section.

2. Install or connect the following:
- Negative battery cable
- Front wheel
- Strut and spring assembly and mount the strut assembly in a holding fixture and compress the coil spring using a suitable tool
- Strut piston rod nut

3. Coil spring by disassembling the strut in the following sequence:
   a. Step 1: Metal sheet plate.
   b. Step 2: Upper strut mount.
   c. Step 3: Thrust bearing plate.
   d. Step 4: Thrust bearing.
   e. Step 5: Upper spring seat.
   f. Step 6: Upper spring seat isolator.
   g. Step 7: Coil spring.
   h. Step 8: Dust boot.

1	Metal sheet plate
2	Upper strut mount
3	Thrust bearing plate
4	Thrust bearing
5	Upper spring seat
6	Upper spring seat isolator
7	Spring
8	Dust boot
9	Rubber bump stopper
10	Lower spring seat

9308TG23

Disassemble the strut assembly in the proper sequence

i. Step 9: Rubber bump stopper.
j. Step 10: Lower spring seat.

To install:
- Assemble the strut assembly in the reverse order of the removal procedure

4. Install or connect the following:
- Strut piston rod nut. Torque the nut to 76 ft. lbs. (103 Nm) and remove the assembly from the holding fixture
- Strut and spring assembly
- Front wheel
- Negative battery cable

Rear

1. Before servicing the vehicle, refer to the precautions in the beginning of this section.

2. Remove or disconnect the following:
- Wheel and install 1 lug nut to retain the brake drum
- Brake line from the wheel cylinder
- Brake line bracket
- Bolts from the Antilock Braking System (ABS) sensor bracket and move the sensor aside, if equipped
- Rear knuckle and loosen the inside upper and lower arm bolts
- Shock absorber lower nut
- Spring

To install:

3. Install or connect the following:
- Spring to the shock absorber
- Lower shock absorber nut. Torque the nut to 85 ft. lbs. (115 Nm).
- Inside upper and lower arm bolts. Torque the bolts to 85 ft. lbs. (115 Nm).
- ABS sensor bracket, if equipped. Torque the bolts to 80 inch lbs. (9 Nm).
- Brake line bracket. Torque the bolt to 15 ft. lbs. (20 Nm).
- Brake line to the wheel cylinder. Torque the fastener to 11 ft. lbs. (15 Nm) and remove the lug nut
- Wheel
- Negative battery cable

Upper Control Arm

REMOVAL & INSTALLATION

The upper and lower ball joints are an integral part of the control arms and are not a serviceable components. Replacement of the ball joint requires replacing the appropriate control arm.

1. Before servicing the vehicle, refer to the precautions in the beginning of this section.

2. Remove or disconnect the following:
- Negative battery cable
- Rear wheel
- Upper control arm from the knuckle while holding the ball joint stud from turning
- Upper ball joint nut
- Upper control arm
- Upper control arm inner bolt

To install:

3. Install or connect the following:
- Upper control arm inner bolt
- Upper control arm. Torque the bolts to 85 ft. lbs. (115 Nm).
- Upper ball joint nut
- Upper control arm to the knuckle. Torque the ball joint nut to 85 ft. lbs. (115 Nm).
- Rear wheel
- Negative battery cable

Lower Control Arm

REMOVAL & INSTALLATION

Front

1. Remove or disconnect the following:
- Negative battery cable
- Front wheel
- Lower ball joint from the knuckle and support the subframe
- Lower control arm

To install:

2. Install or connect the following:
- Lower control arm bolts and hand tighten them
- Pinch bolt to the wheel knuckle. Torque the nut to 52 ft. lbs. (70 Nm) and remove the subframe support
- Front wheel and jounce the vehicle

3. Torque the inner lower control arm bolt to 148 ft. lbs. (200 Nm) and outer bolt 85 ft. lbs. (115 Nm).

Rear

1. Remove or disconnect the following:
- Negative battery cable
- Front wheel
- Lower ball joint from the knuckle while holding the ball joint stud from moving
- Lower ball joint nut
- Lower control arm
- Lower control arm inner bolt

To install:

2. Install or connect the following:
- Lower control arm inner bolt
- Lower control arm. Torque the bolts to 85 ft. lbs. (115 Nm).
- Lower ball joint nut
- Lower ball joint the knuckle. Torque the ball joint nut to 85 ft. lbs. (115 Nm).
- Rear wheel
- Negative battery cable

Wheel Bearings

REMOVAL & INSTALLATION

Front

1. Before servicing the vehicle, refer to the precautions in the beginning of this section.

2. Remove or disconnect the following:
- Negative battery cable
- Front wheel
- Brake disc
- Wheel hub nut
- Tie rod end cotter pin and nut
- Tie rod end from the knuckle
- Antilock Brake System (ABS) sensor bolt and move the sensor aside, if equipped
- Lower ball joint from the knuckle
- Halfshaft from the knuckle and properly support the halfshaft
- Steering knuckle

3. Press the hub from the wheel bearing and knuckle

4. Press the inner wheel bearing race from the knuckle and remove the snapring

5. Press the outer wheel bearing race from the knuckle

To install:

6. Install or connect the following:
- Wheel bearing into the steering knuckle
- Snapring
- Wheel hub into the wheel bearing by using a press
- Steering knuckle. Torque the bolts to 85 ft. lbs. (115 Nm).
- Halfshaft into the wheel hub
- Pinch bolt to knuckle. Torque the nut to 52 ft. lbs. (70 Nm).
- Ball joint stud into the knuckle
- ABS sensor. Torque the bolt to 80 inch lbs. (9 Nm), if equipped
- Tie rod end to the knuckle. Torque the nut to 41 ft. lbs. (55 Nm).
- New cotter pin to the tie rod end nut
- Wheel hub. Torque the nut to 214 ft. lbs. (290 Nm).
- Brake disc
- Front wheel
- Negative battery cable

Rear

2WD VEHICLES

1. Before servicing the vehicle, refer to the precautions in the beginning of this section.
2. Remove or disconnect the following:
   - Negative battery cable
   - Rear wheel
   - Rear brake drum
   - Wheel hub nut
   - Wheel hub
   - Inner wheel bearing race from the hub
   - Snapring
   - Wheel bearing outer race from the knuckle

To install:

3. Install or connect the following:
   - Wheel bearing in to the knuckle
   - Snapring
   - Wheel hub into the wheel bearing
   - Wheel hub nut. Torque the nut to 214 ft. lbs. (290 Nm).
   - Brake drum
   - Rear wheel
   - Negative battery cable

4WD VEHICLES

1. Before servicing the vehicle, refer to the precautions in the beginning of this section.

2. Remove or disconnect the following:
   - Negative battery cable
   - Rear wheel
   - Rear brake shoes
   - Rear halfshaft nut and loosen the halfshaft from the hub
   - Wheel hub and place it in a vise
   - Inner wheel bearing race from the hub
   - Antilock Brake System (ABS) sensor bracket and move the sensor aside, if equipped
   - Parking brake cable from the steering knuckle
   - Brake line from the wheel cylinder and support the knuckle
   - Lower shock absorber nut
   - Lower ball joint by holding the ball joint stud
   - Upper ball joint
   - Coil spring while noting the location of the insulator
   - Steering knuckle cam
   - Steering knuckle
   - Snapring and press out the outer wheel bearing race from the knuckle

To install:

3. Install or connect the following:
   - New wheel bearing into the steering knuckle
   - Snapring to the knuckle

   - Wheel hub
   - Steering knuckle cam and hand tighten the bolt
   - Coil spring
   - Shock absorber lower nut. Torque the nut to 85 ft. lbs. (115 Nm).
   - Upper ball joint. Torque the nut to 85 ft. lbs. (115 Nm).
   - Lower ball joint. Torque the nut to 85 ft. lbs. (115 Nm). Align the steering knuckle cam and torque the bolt to 85 ft. lbs. (115 Nm).
   - Brake line to the wheel cylinder. Torque the brake line bracket bolt to 15 ft. lbs. (20 Nm) and the brake line fastener to 11 ft. lbs. (15 Nm).
   - Parking brake cable to the backing plate. Torque the bolt to 16 ft. lbs. (22 Nm).
   - ABS sensor bracket. Torque the bolt to 80 inch lbs. (9 Nm), if equipped
   - Halfshaft nut. Torque the nut to 214 ft. lbs. (290 Nm).
   - Brake shoes
   - Rear wheel
   - Negative battery cable
4. Fill and bleed the brake system.
5. Check and adjust the wheel alignment as needed.

BRAKES

Brake Caliper

REMOVAL & INSTALLATION

1. Remove the wheel and tire assembly.
2. Remove the brake caliper clip.
3. Remove the brake caliper bolt caps and bolts.
4. Position the caliper aside.
5. Disconnect and cap the brake line from the caliper and remove the caliper. To install, reverse the removal procedure. Torque the mounting bolts to 26 ft. lbs. (35Nm). Torque the brake line to 15 ft. lbs. (20Nm).
6. Bleed the brake system.

Brake Pads

REMOVAL & INSTALLATION

1. Remove the wheel and tire assembly.

2. Remove the brake caliper clip.
3. Position the caliper aside.
4. Remove brake caliper bolt caps and the bolts.
5. Position the caliper aside and support.
6. Remove the brake pads.
7. Remove the outer brake pad from the anchor.
8. Remove the inner brake pad from the caliper piston.
9. To install, reverse the removal procedure.

Brake Rotor

REMOVAL & INSTALLATION

1. Remove the brake caliper anchor plate.
2. Remove the brake disc retaining clips (if equipped) and the brake disc.
3. To install, reverse the removal procedure. Torque the anchor plate bolts to 111 ft. lbs. (150Nm).

Brake Drum

REMOVAL & INSTALLATION

1. Remove the tire and wheel assembly.

✳✳ CAUTION

Use of a brake drum puller or a torch is not recommended. Brake drum distortion can result.

➡**If the brake drum is rusted to the axle shaft pilot diameter, tap the center of the brake drum between the wheel studs.**

2. Remove the brake drum.
3. If equipped, remove the brake drum retaining clips.
4. If the brake drums will not come off, follow these steps.
5. Move the brake shoe adjusting lever off the brake adjuster screw.
6. Loosen the brake shoe adjuster screw nut by adjusting the nut upward.

7. Using the special tool, 134-R0191, measure the brake drum inside diameter.

8. Install a new brake drum if the maximum inside diameter exceeds the specification shown on the outside of the brake drum.

To install:

9. Clean the wheel hub mounting surface and wheel pilot.

10. Install the tire and wheel assembly.

Brake Shoes

REMOVAL & INSTALLATION

1. Remove the brake drum.

2. Use the Brake/Clutch/Service Vacuum to remove brake dust and dirt from the brake assemblies.

➡**If new rear brake shoes and linings are being installed, resurface the brake drums to remove glazing and to ensure an equal friction surface from side-to-side. Resurfacing will also correct out-of-round and bell conditions.**

3. Using the special tool, measure the braking surface diameter. If the inside diameter measures more than the maximum specification shown on the outside of the brake drum, install a new brake drum.

4. Remove the parking brake cable from the parking brake cable lever.

5. Remove the hold-down clips and pins.

6. Remove the lower spring.

7. Remove the rear brake shoes.

8. Pull the bottom of the brake shoe forward.

9. Release the upper return spring.

10. Remove both brake shoes together.

11. Remove the self adjuster lever.

12. Remove the self adjuster and spring assembly.

13. Return the self adjuster to the fully seated position.

14. Remove the parking brake lever.

15. Remove the horseshoe clip.

16. Remove the parking brake lever.

17. Inspect the rear brake shoes for minimum thickness above the backing plate, and install new as necessary.

18. To install, reverse the removal procedure.

MITSUBISHI

Diamante • Eclipse • Galant • Lancer • Mirage

21

SPECIFICATION CHARTS

ENGINE AND VEHICLE IDENTIFICATION

Engine							Model Year	
Code ①	Liters (cc)	Cu. In.	Cyl.	Fuel Sys.	Type	Eng. Mfg.	Code ②	Year
4G15/A	1.5 (1468)	87	4	MFI	SOHC	Mitsubishi	Y	2000
4G93/C	1.8 (1834)	112	4	MFI	SOHC	Mitsubishi	1	2001
4G94/E	2.0 (1999)	122	4	MFI	SOHC	Mitsubishi	2	2002
4G64/G	2.4 (2351)	143	4	MFI	SOHC	Mitsubishi	3	2003
6G72/H	3.0 (2972)	181	6	MFI	SOHC	Mitsubishi	4	2004
6G72/J	3.0 (2972)	181	6	MFI	DOHC	Mitsubishi		
6G72/K	3.0 (2972)	181	6	MFI-Turbo	DOHC	Mitsubishi		
6G72/L	3.0 (2972)	181	6	MFI	SOHC	Mitsubishi		
6G74/P	3.5 (3497)	213	6	MFI	SOHC	Mitsubishi		

MFI: Multiport fuel injection

SOHC: Single overhead camsha

DOHC: Double overhead camshafts

① Engine ID / 8th digit of the VIN

② 10th digit of the VIN

42356-GALA-C01

GENERAL ENGINE SPECIFICATIONS

Year	Model	Engine Displacement Liters (cc)	Engine ID/VIN	Fuel System Type	Net Horsepower @ rpm	Net Torque @ rpm (ft. lbs.)	Bore x Stroke (in.)	Compression Ratio	Oil Pressure @ rpm
2000	Mirage	1.5 (1468)	4G15/A	MFI	92@6000	93@3000	2.97x3.23	9.2:1	54@2000
	Mirage	1.8 (1834)	4G93/C	MFI	113@6000	116@4500	3.19x3.50	9.5:1	41@2000
	Eclipse	2.4 (2350)	4G64/G	MFI	③	148@3000	3.41x3.94	9.5:1	41@2000
	Eclipse	3.0 (2972)	6G72/L	MFI	175@5500	185@3000	3.59x2.99	8.9:1	30-80@2000
	Eclipse Spyder	2.4 (2350)	4G64/G	MFI	③	148@3000	3.41x3.94	9.5:1	41@2000
	Eclipse Spyder	3.0 (2972)	6G72/L	MFI	175@5500	185@3000	3.59x2.99	8.9:1	30-80@2000
	Galant	2.4 (2350)	4G64/G	MFI	③	148@3000	3.41x3.94	9.5:1	41@2000
	Galant	3.0 (2972)	6G72/L	MFI	175@5500	185@3000	3.59x2.99	8.9:1	30-80@2000
	Diamante	3.5 (3497)	6G74/P	MFI	214@5000	228@3000	3.66x3.38	9.5:1	30-80@2000
2001	Mirage	1.5 (1468)	4G15/A	MFI	92@6000	93@3000	2.97x3.23	9.2:1	54@2000
	Mirage	1.8 (1834)	4G93/C	MFI	113@6000	116@4500	3.19x3.50	9.5:1	41@2000
	Eclipse	2.4 (2350)	4G64/G	MFI	③	148@3000	3.41x3.94	9.5:1	41@2000
	Eclipse	3.0 (2972)	6G72/L	MFI	175@5500	185@3000	3.59x2.99	8.9:1	30-80@2000
	Eclipse Spyder	2.4 (2350)	4G64/G	MFI	③	148@3000	3.41x3.94	9.5:1	41@2000
	Eclipse Spyder	3.0 (2972)	6G72/L	MFI	175@5500	185@3000	3.59x2.99	8.9:1	30-80@2000
	Galant	2.4 (2350)	4G64/G	MFI	③	148@3000	3.41x3.94	9.5:1	41@2000
	Galant	3.0 (2972)	6G72/L	MFI	175@5500	185@3000	3.59x2.99	8.9:1	30-80@2000
	Diamante	3.5 (3497)	6G74/P	MFI	214@5000	228@3000	3.66x3.38	9.5:1	30-80@2000
2002	Mirage	1.5 (1468)	4G15/A	MFI	92@6000	93@3000	2.97x3.23	9.2:1	54@2000
	Mirage	1.8 (1834)	4G93/C	MFI	113@6000	116@4500	3.19x3.50	9.5:1	41@2000
	Lancer	2.0 (1999)	4G94/E	MFI	120@5500	130@4250)	3.21x3.77	9.5:1	43-100@3500
	Eclipse	2.4 (2350)	4G64/G	MFI	③	148@3000	3.41x3.94	9.5:1	41@2000
	Eclipse	3.0 (2972)	6G72/L	MFI	175@5500	185@3000	3.59x2.99	8.9:1	30-80@2000
	Eclipse Spyder	2.4 (2350)	4G64/G	MFI	③	148@3000	3.41x3.94	9.5:1	41@2000
	Eclipse Spyder	3.0 (2972)	6G72/L	MFI	175@5500	185@3000	3.59x2.99	8.9:1	30-80@2000
	Galant	2.4 (2350)	4G64/G	MFI	③	148@3000	3.41x3.94	9.5:1	41@2000
	Galant	3.0 (2972)	6G72/L	MFI	175@5500	185@3000	3.59x2.99	8.9:1	30-80@2000
	Diamante	3.5 (3497)	6G74/P	MFI	214@5000	228@3000	3.66x3.38	9.5:1	30-80@2000
2003-04	Mirage	1.5 (1468)	4G15/A	MFI	92@6000	93@3000	2.97x3.23	9.2:1	54@2000
	Mirage	1.8 (1834)	4G93/C	MFI	113@6000	116@4500	3.19x3.50	9.5:1	41@2000
	Lancer	2.0 (1999)	4G94/E	MFI	120@5500	130@4250)	3.21x3.77	9.5:1	43-100@3500
	Eclipse	2.4 (2350)	4G64/G	MFI	③	148@3000	3.41x3.94	9.5:1	41@2000
	Eclipse	3.0 (2972)	6G72/L	MFI	175@5500	185@3000	3.59x2.99	8.9:1	30-80@2000
	Eclipse Spyder	2.4 (2350)	4G64/G	MFI	③	148@3000	3.41x3.94	9.5:1	41@2000
	Eclipse Spyder	3.0 (2972)	6G72/L	MFI	175@5500	185@3000	3.59x2.99	8.9:1	30-80@2000
	Galant	2.4 (2350)	4G64/G	MFI	③	148@3000	3.41x3.94	9.5:1	41@2000
	Galant	3.0 (2972)	6G72/L	MFI	175@5500	185@3000	3.59x2.99	8.9:1	30-80@2000
	Diamante	3.5 (3497)	6G74/P	MFI	214@5000	228@3000	3.66x3.38	9.5:1	30-80@2000

MFI: Multiport fuel injection

① Manual transaxle: 210@6000
　Automatic transaxle: 205@6000

② Manual transaxle: 214@3000
　Automatic transaxle: 220@3000

③ California: 138@5500
　Except California: 141@5500

42356-GALA-C02

ENGINE TUNE-UP SPECIFICATIONS

Year	Engine Displacement Liters (cc)	Engine ID/VIN	Spark Plugs Gap (in.)	Ignition Timing (deg.) MT	AT	Fuel Pump (psi)	Idle Speed (rpm) MT	AT	Valve Clearance In.	Ex.
2000	1.5 (1468)	4G15/A	0.039-0.043	2-8B	2-8B	38	600-800	600-800	HYD	HYD
	1.8 (1834)	4G93/C	0.039-0.043	2-8B	2-8B	38	600-800	600-800	HYD	HYD
	2.4 (2350)	4G64/G	0.039-0.043	2-8B	2-8B	38	650-850	650-850	HYD	HYD
	3.0 (2972)	6G72/L	0.039-0.043	5B	5B	38	600-800	600-800	HYD	HYD
	3.5 (3497)	6G74/P	0.039-0.043	2-8B	2-8B	38	600-800	600-800	HYD	HYD
2001	1.5 (1468)	4G15/A	0.039-0.043	2-8B	2-8B	38	600-800	600-800	HYD	HYD
	1.8 (1834)	4G93/C	0.039-0.043	2-8B	2-8B	38	600-800	600-800	HYD	HYD
	2.4 (2350)	4G64/G	0.039-0.043	2-8B	2-8B	38	650-850	650-850	HYD	HYD
	3.0 (2972)	6G72/L	0.039-0.043	5B	5B	38	600-800	600-800	HYD	HYD
	3.5 (3497)	6G74/P	0.039-0.043	—	2-8B	38	600-800	600-800	HYD	HYD
2002	1.5 (1468)	4G15/A	0.039-0.043	2-8B	2-8B	38	600-800	600-800	HYD	HYD
	1.8 (1834)	4G93/C	0.039-0.043	2-8B	2-8B	38	600-800	600-800	HYD	HYD
	2.0 (1999)	4G94/E	0.039-0.043	2-8B	2-8B	38	600-800	600-800	HYD	HYD
	2.4 (2350)	4G64/G	0.039-0.043	2-8B	2-8B	38	650-850	650-850	HYD	HYD
	3.0 (2972)	6G72/L	0.039-0.043	5B	5B	38	600-800	600-800	HYD	HYD
	3.5 (3497)	6G74/P	0.039-0.043	—	2-8B	38	600-800	600-800	HYD	HYD
2003-04	1.5 (1468)	4G15/A	0.039-0.043	2-8B	2-8B	38	600-800	600-800	HYD	HYD
	1.8 (1834)	4G93/C	0.039-0.043	2-8B	2-8B	38	600-800	600-800	HYD	HYD
	2.0 (1999)	4G94/E	0.039-0.043	2-8B	2-8B	38	600-800	600-800	HYD	HYD
	2.4 (2350)	4G64/G	0.039-0.043	2-8B	2-8B	38	650-850	650-850	HYD	HYD
	3.0 (2972)	6G72/L	0.039-0.043	5B	5B	38	600-800	600-800	HYD	HYD
	3.5 (3497)	6G74/P	0.039-0.043	—	2-8B	38	600-800	600-800	HYD	HYD

NOTE: The Vehicle Emission Control Information label often reflects specification changes made during production. The label figures must be used if they differ from those in this chart.

B: Before top dead center

HYD: Hydraulic

42356-GALA-C03

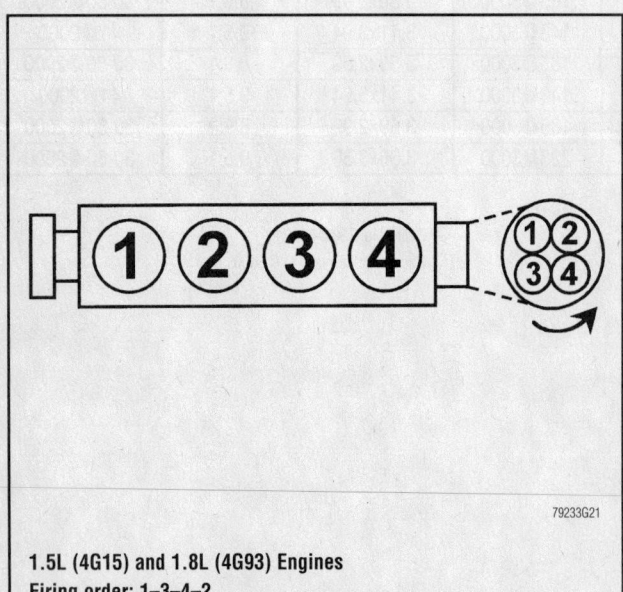

1.5L (4G15) and 1.8L (4G93) Engines
Firing order: 1–3–4–2
Distributor rotation: Counterclockwise

79233G21

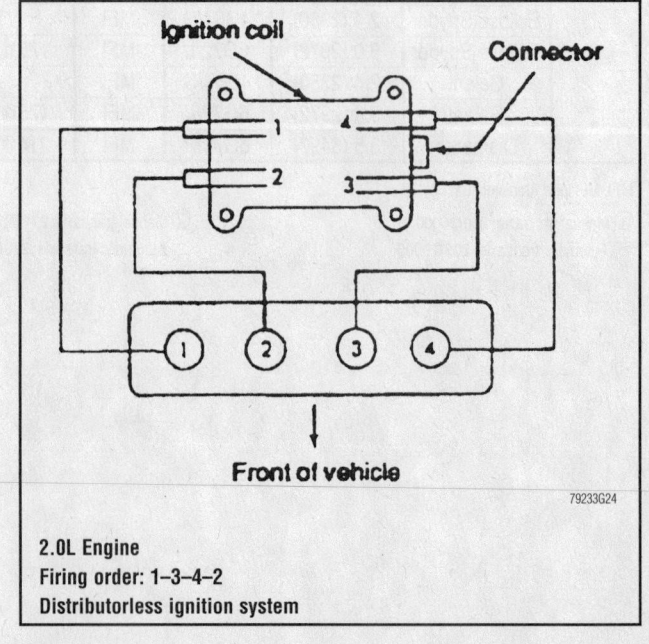

2.0L Engine
Firing order: 1–3–4–2
Distributorless ignition system

79233G24

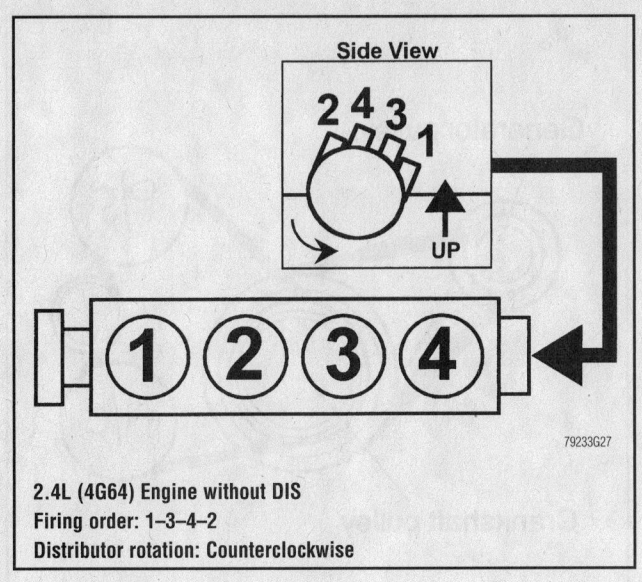

2.4L (4G64) Engine without DIS
Firing order: 1–3–4–2
Distributor rotation: Counterclockwise

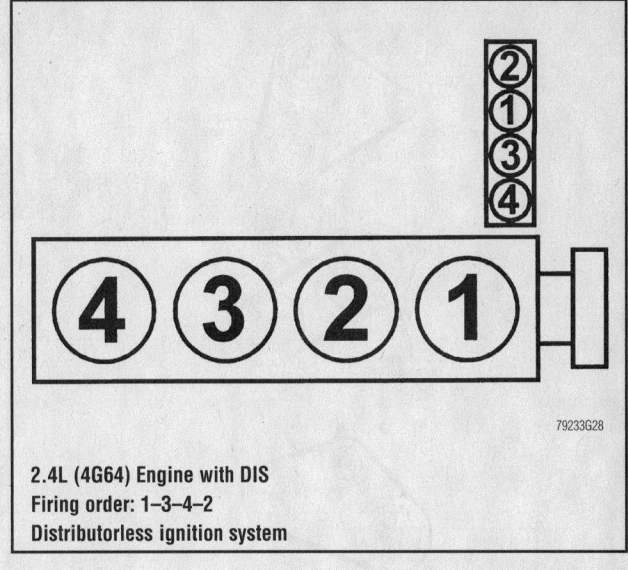

2.4L (4G64) Engine with DIS
Firing order: 1–3–4–2
Distributorless ignition system

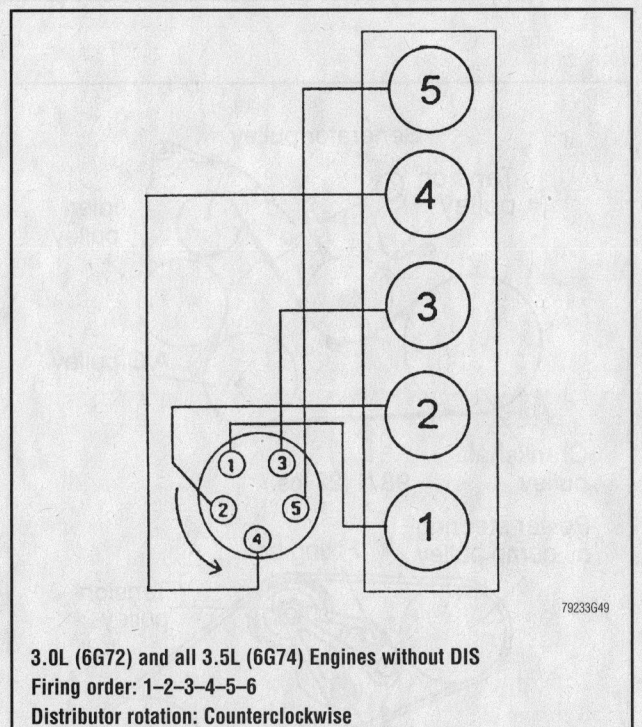

3.0L (6G72) and all 3.5L (6G74) Engines without DIS
Firing order: 1–2–3–4–5–6
Distributor rotation: Counterclockwise

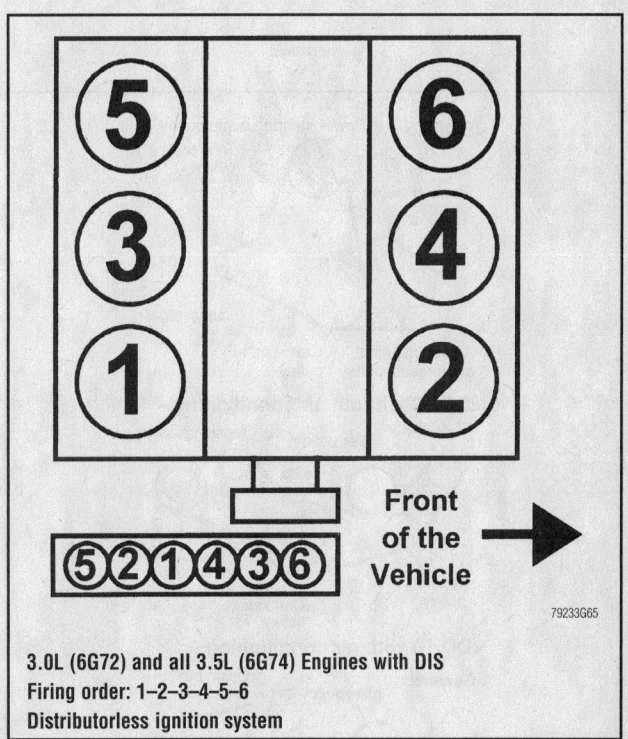

3.0L (6G72) and all 3.5L (6G74) Engines with DIS
Firing order: 1–2–3–4–5–6
Distributorless ignition system

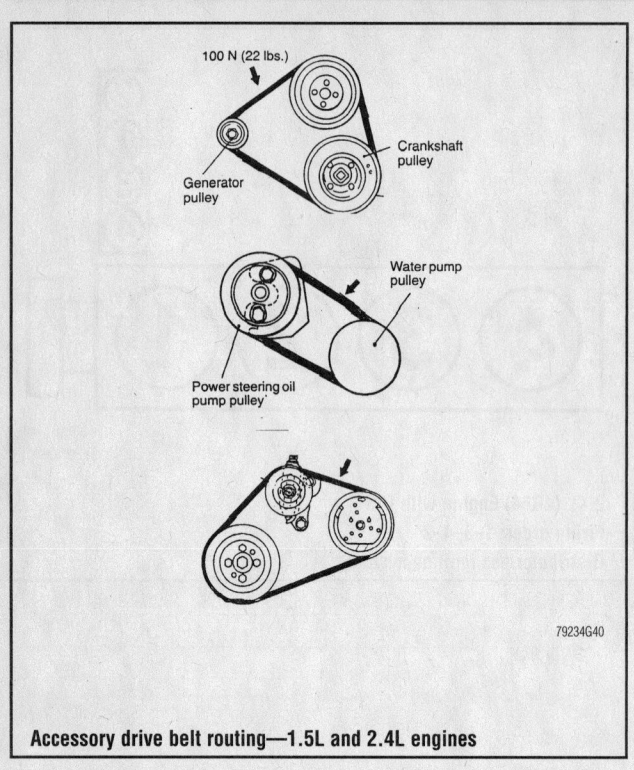

Accessory drive belt routing—1.5L and 2.4L engines

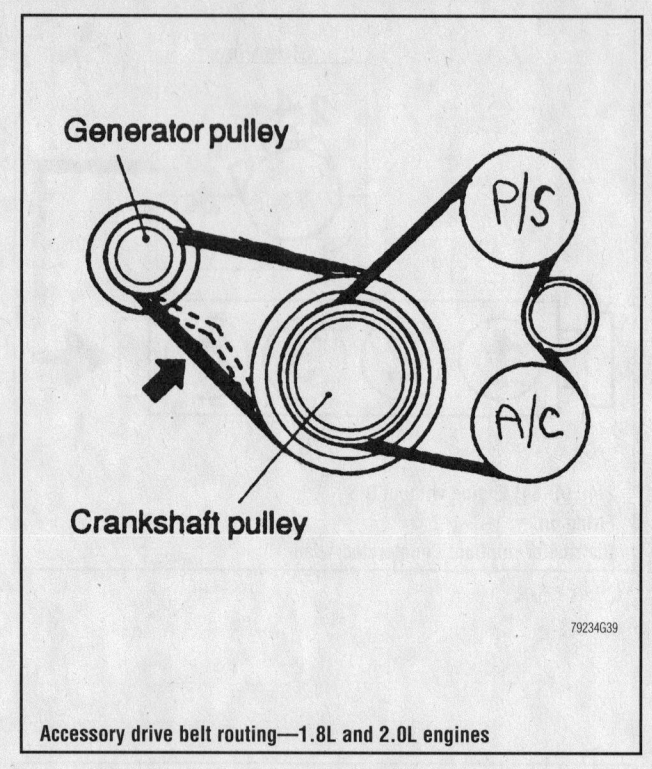

Accessory drive belt routing—1.8L and 2.0L engines

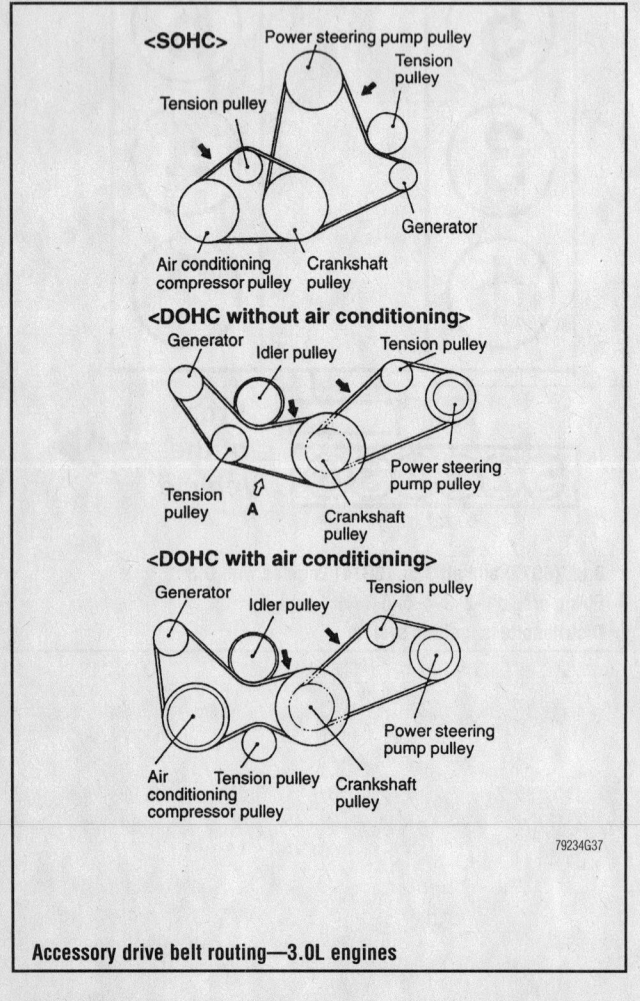

Accessory drive belt routing—3.0L engines

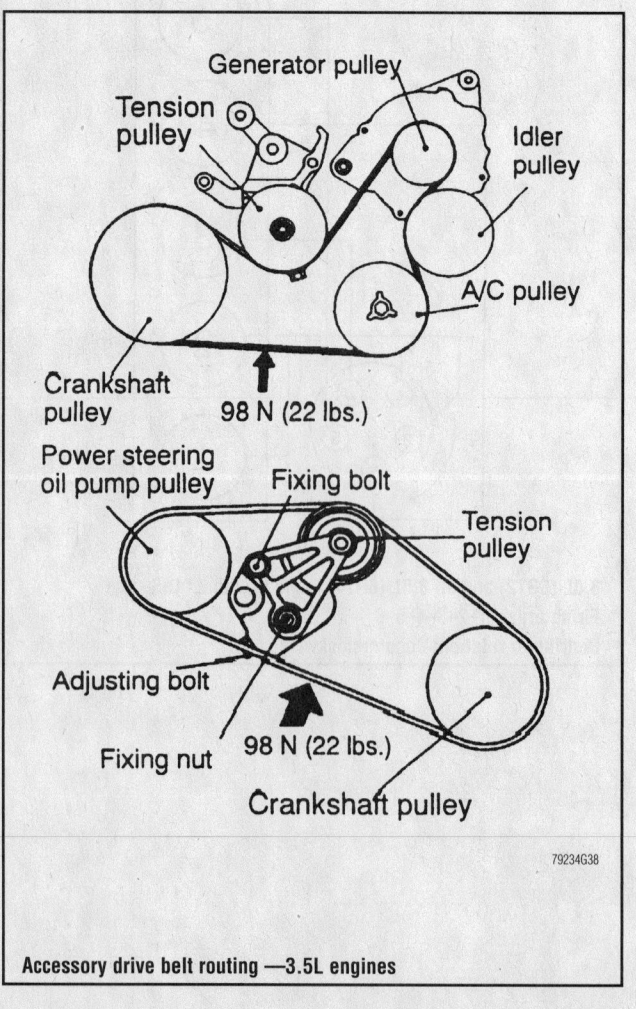

Accessory drive belt routing —3.5L engines

CAPACITIES

Year	Model	Engine Displacement Liters (cc)	Engine ID/VIN	Engine Oil with Filter	Transmission (pts.)		Transfer Case (pts.)	Drive Axle		Fuel Tank (gal.)	Cooling System (qts.)
					5 or 6-Spd	Auto.		Front (pts.)	Rear (pts.)		
2000	Mirage	1.5 (1468)	4G15/A	3.7	3.8	12.6	—	—	—	13.2	5.3
	Mirage	1.8 (1834)	4G93/C	4.2	3.8	12.6	—	—	—	13.2	6.3
	Eclipse	2.4 (2350)	4G64/G	4.5	4.6	16.4	—	—	—	16.3	7.4
	Eclipse	3.0 (2972)	6G72/L	4.5	6.0	17.8	—	—	—	16.3	8.5
	Eclipse Spyder	2.4 (2350)	4G64/G	4.5	4.6	16.4	—	—	—	16.3	7.4
	Eclipse Spyder	3.0 (2972)	6G72/L	4.5	6.0	17.8	—	—	—	16.3	8.5
	Galant	2.4 (2350)	4G64/G	4.5	4.6	12.8	—	—	—	17.0	14.8
	Galant	3.0 (2972)	6G72/L	4.5	4.6	15.8	—	—	—	19.8	8.5
	Diamante	3.5 (3497)	6G74/P	4.7	—	18.0	—	—	—	18.7	10.0
2001	Mirage	1.5 (1468)	4G15/A	3.3	4.4	16.2	—	—	—	12.4	5.3
	Mirage	1.8 (1834)	4G93/C	3.9	4.6	16.2	—	—	—	12.4	6.3
	Eclipse	2.4 (2350)	4G64/G	4.5	4.6	16.4	—	—	—	16.4	7.4
	Eclipse	3.0 (2972)	6G72/L	4.5	6.0	17.8	—	—	—	16.4	8.5
	Eclipse Spyder	2.4 (2350)	4G64/G	4.5	4.6	16.4	—	—	—	16.4	7.4
	Eclipse Spyder	3.0 (2972)	6G72/L	4.5	6.0	17.8	—	—	—	16.4	8.5
	Galant	2.4 (2350)	4G64/G	4.5	—	16.2	—	—	—	16.3	14.8
	Galant	3.0 (2972)	6G72/L	4.5	—	17.8	—	—	—	16.3	8.5
	Diamante	3.5 (3497)	6G74/P	4.5	—	17.8	—	—	—	18.7	10.0
2002	Mirage	1.5 (1468)	4G15/A	3.3	4.4	16.2	—	—	—	12.4	5.3
	Mirage	1.8 (1834)	4G93/C	3.9	4.6	16.2	—	—	—	12.4	6.3
	Lancer	2.0 (1999)	4G94/E	3.8	4.6	16.2	—	—	—	13.2	5.3
	Eclipse	2.4 (2350)	4G64/G	4.5	4.6	16.4	—	—	—	16.4	7.4
	Eclipse	3.0 (2972)	6G72/L	4.5	6.0	17.8	—	—	—	16.4	8.5
	Eclipse Spyder	2.4 (2350)	4G64/G	4.5	4.6	16.4	—	—	—	16.4	7.4
	Eclipse Spyder	3.0 (2972)	6G72/L	4.5	6.0	17.8	—	—	—	16.4	8.5
	Galant	2.4 (2350)	4G64/G	4.5	—	16.2	—	—	—	16.3	14.8
	Galant	3.0 (2972)	6G72/L	4.5	—	17.8	—	—	—	16.3	8.5
	Diamante	3.5 (3497)	6G74/P	4.5	—	17.8	—	—	—	18.7	10.0
2003-04	Mirage	1.5 (1468)	4G15/A	3.3	4.4	16.2	—	—	—	12.4	5.3
	Mirage	1.8 (1834)	4G93/C	3.9	4.6	16.2	—	—	—	12.4	6.3
	Lancer	2.0 (1999)	4G94/E	3.8	4.6	16.2	—	—	—	13.2	5.3
	Eclipse	2.4 (2350)	4G64/G	4.5	4.6	16.4	—	—	—	16.4	7.4
	Eclipse	3.0 (2972)	6G72/L	4.5	6.0	17.8	—	—	—	16.4	8.5
	Eclipse Spyder	2.4 (2350)	4G64/G	4.5	4.6	16.4	—	—	—	16.4	7.4
	Eclipse Spyder	3.0 (2972)	6G72/L	4.5	6.0	17.8	—	—	—	16.4	8.5
	Galant	2.4 (2350)	4G64/G	4.5	—	16.2	—	—	—	16.3	14.8
	Galant	3.0 (2972)	6G72/L	4.5	—	17.8	—	—	—	16.3	8.5
	Diamante	3.5 (3497)	6G74/P	4.5	—	17.8	—	—	—	18.7	10.0

NOTE: All capacities are approximate. Add fluid gradually and ensure a proper fluid level is obtained.

① FWD: 4.2 pts.
  AWD: 4.6 pts.

② FWD: 4.8 pts.
  AWD: 5.0 pts.

42356-GALA-C04

CRANKSHAFT AND CONNECTING ROD SPECIFICATIONS
All measurements are given in inches.

Year	Engine Displacement Liters (cc)	Engine ID/VIN	Crankshaft				Connecting Rod		
			Main Brg. Journal Dia.	Main Brg. Oil Clearance	Shaft End-play	Thrust on No.	Journal Diameter	Oil Clearance	Side Clearance
2000	1.5 (1468)	4G15/A	1.8900	0.0008-0.0040	0.0020-0.0120	3	1.6500	0.0008-0.0040	0.0039-0.0160
	1.8 (1834)	4G93/C	1.9678-1.9685	0.0008-0.0040	0.0020-0.0098	3	1.7709-1.7717	0.0008-0.0040	0.0039-0.0160
	2.4 (2351)	4G64/G	2.2436-2.2441	0.0008-0.0040	0.0020-0.0098	3	1.7709-1.7717	0.0008-0.0040	0.0039-0.0160
	3.0 (2972)	6G72/L	2.3614-2.3622	0.0008-0.0040	0.0020-0.0120	3	2.1646-2.1654	0.0008-0.0040	0.0039-0.0160
	3.5 (3497)	6G74/P	2.3614-2.3622	0.0008-0.0040	0.0020-0.0120	3	1.9700	0.0008-0.0040	0.0039-0.0160
2001	1.5 (1468)	4G15/A	1.8900	0.0008-0.0040	0.0020-0.0120	3	1.6500	0.0008-0.0040	0.0039-0.0160
	1.8 (1834)	4G93/C	1.9678-1.9685	0.0008-0.0040	0.0020-0.0098	3	1.7709-1.7717	0.0008-0.0040	0.0039-0.0160
	2.4 (2351)	4G64/G	2.2436-2.2441	0.0008-0.0040	0.0020-0.0098	3	1.7709-1.7717	0.0008-0.0040	0.0039-0.0160
	3.0 (2972)	6G72/L	2.3614-2.3622	0.0008-0.0040	0.0020-0.0120	3	2.1646-2.1654	0.0008-0.0040	0.0039-0.0160
	3.5 (3497)	6G74/P	2.3614-2.3622	0.0008-0.0040	0.0020-0.0120	3	1.9700	0.0008-0.0040	0.0039-0.0160
2002	1.5 (1468)	4G15/A	1.8900	0.0008-0.0040	0.0020-0.0120	3	1.6500	0.0008-0.0040	0.0039-0.0160
	1.8 (1834)	4G93/C	1.9678-1.9685	0.0008-0.0040	0.0020-0.0098	3	1.7709-1.7717	0.0008-0.0040	0.0039-0.0160
	2.0 (1999)	4G94/E	1.9700	0.0008-0.0012	0.0020-0.0098	3	NA	0.0008-0.0016	0.0039-0.0098
	2.4 (2351)	4G64/G	2.2436-2.2441	0.0008-0.0040	0.0020-0.0098	3	1.7709-1.7717	0.0008-0.0040	0.0039-0.0160
	3.0 (2972)	6G72/L	2.3614-2.3622	0.0008-0.0040	0.0020-0.0120	3	2.1646-2.1654	0.0008-0.0040	0.0039-0.0160
	3.5 (3497)	6G74/P	2.3614-2.3622	0.0008-0.0040	0.0020-0.0120	3	1.9700	0.0008-0.0040	0.0039-0.0160
2003-04	1.5 (1468)	4G15/A	1.8900	0.0008-0.0040	0.0020-0.0120	3	1.6500	0.0008-0.0040	0.0039-0.0160
	1.8 (1834)	4G93/C	1.9678-1.9685	0.0008-0.0040	0.0020-0.0098	3	1.7709-1.7717	0.0008-0.0040	0.0039-0.0160
	2.0 (1999)	4G94/E	1.9700	0.0008-0.0012	0.0020-0.0098	3	NA	0.0008-0.0016	0.0039-0.0098
	2.4 (2351)	4G64/G	2.2436-2.2441	0.0008-0.0040	0.0020-0.0098	3	1.7709-1.7717	0.0008-0.0040	0.0039-0.0160
	3.0 (2972)	6G72/L	2.3614-2.3622	0.0008-0.0040	0.0020-0.0120	3	2.1646-2.1654	0.0008-0.0040	0.0039-0.0160
	3.5 (3497)	6G74/P	2.3614-2.3622	0.0008-0.0040	0.0020-0.0120	3	1.9700	0.0008-0.0040	0.0039-0.0160

NA - Not Available

PISTON AND RING SPECIFICATIONS
All measurements are given in inches.

Year	Engine Displacement Liters (cc)	Engine ID/VIN	Piston Clearance	Ring Gap			Ring Side Clearance		
				Top Compression	Bottom Compression	Oil Control	Top Compression	Bottom Compression	Oil Control
2000	1.5 (1468)	4G15/A	0.0008-0.0016	0.0079-0.0310	0.0079-0.0310	0.0079-0.0390	0.0012-0.0040	0.0008-0.0040	NA
	1.8 (1834)	4G93/C	0.0008-0.0016	0.0098-0.0310	0.0157-0.0310	0.0078-0.0390	0.0012-0.0039	0.0008-0.0039	NA
	2.4 (2351)	4G64/G	0.0008-0.0016	0.0098-0.0310	0.0157-0.0310	0.0039-0.0390	0.0012-0.0040	0.0012-0.0040	NA
	3.0 (2972)	6G72/L	0.0008-0.0020	0.0118-0.0310	0.0177-0.0310	0.0079-0.0390	0.0012-0.0040	0.0008-0.0040	NA
	3.5 (3497)	6G74/P	0.0008-0.0020	0.0118-0.0310	0.0177-0.0310	0.0079-0.0390	0.0012-0.0040	0.0008-0.0040	NA
2001	1.5 (1468)	4G15/A	0.0008-0.0016	0.0079-0.0310	0.0079-0.0310	0.0079-0.0390	0.0012-0.0040	0.0008-0.0040	NA
	1.8 (1834)	4G93/C	0.0008-0.0016	0.0098-0.0310	0.0157-0.0310	0.0078-0.0390	0.0012-0.0039	0.0008-0.0039	NA
	2.4 (2351)	4G64/G	0.0008-0.0016	0.0098-0.0310	0.0157-0.0310	0.0039-0.0390	0.0012-0.0040	0.0012-0.0040	NA
	3.0 (2972)	6G72/L	0.0008-0.0020	0.0118-0.0310	0.0177-0.0310	0.0079-0.0390	0.0012-0.0040	0.0008-0.0040	NA
	3.5 (3497)	6G74/P	0.0008-0.0020	0.0118-0.0310	0.0177-0.0310	0.0079-0.0390	0.0012-0.0040	0.0008-0.0040	NA
2002	1.5 (1468)	4G15/A	0.0008-0.0016	0.0079-0.0310	0.0079-0.0310	0.0079-0.0390	0.0012-0.0040	0.0008-0.0040	NA
	1.8 (1834)	4G93/C	0.0008-0.0016	0.0098-0.0310	0.0157-0.0310	0.0078-0.0390	0.0012-0.0039	0.0008-0.0039	NA
	2.0 (1999)	4G94/E	0.0008-0.0016	0.0059-0.0118	0.0157-0.0217	0.0039-0.0138	0.0012-0.0028	0.0008-0.0024	NA
	2.4 (2351)	4G64/G	0.0008-0.0016	0.0098-0.0310	0.0157-0.0310	0.0039-0.0390	0.0012-0.0040	0.0012-0.0040	NA
	3.0 (2972)	6G72/L	0.0008-0.0020	0.0118-0.0310	0.0177-0.0310	0.0079-0.0390	0.0012-0.0040	0.0008-0.0040	NA
	3.5 (3497)	6G74/P	0.0008-0.0020	0.0118-0.0310	0.0177-0.0310	0.0079-0.0390	0.0012-0.0040	0.0008-0.0040	NA
2003-04	1.5 (1468)	4G15/A	0.0008-0.0016	0.0079-0.0310	0.0079-0.0310	0.0079-0.0390	0.0012-0.0040	0.0008-0.0040	NA
	1.8 (1834)	4G93/C	0.0008-0.0016	0.0098-0.0310	0.0157-0.0310	0.0078-0.0390	0.0012-0.0039	0.0008-0.0039	NA
	2.0 (1999)	4G94/E	0.0008-0.0016	0.0059-0.0118	0.0157-0.0217	0.0039-0.0138	0.0012-0.0028	0.0008-0.0024	NA
	2.4 (2351)	4G64/G	0.0008-0.0016	0.0098-0.0310	0.0157-0.0310	0.0039-0.0390	0.0012-0.0040	0.0012-0.0040	NA
	3.0 (2972)	6G72/L	0.0008-0.0020	0.0118-0.0310	0.0177-0.0310	0.0079-0.0390	0.0012-0.0040	0.0008-0.0040	NA
	3.5 (3497)	6G74/P	0.0008-0.0020	0.0118-0.0310	0.0177-0.0310	0.0079-0.0390	0.0012-0.0040	0.0008-0.0040	NA

NA - Not Available

42356-GALA-C06

VALVE SPECIFICATIONS

Year	Engine Displacement Liters (cc)	Engine ID/VIN	Seat Angle (deg.)	Face Angle (deg.)	Spring Test Pressure (lbs. @ in.)	Spring Installed Height (in.)	Stem-to-Guide Clearance (in.)		Stem Diameter (in.)	
							Intake	Exhaust	Intake	Exhaust
2000	1.5 (1468)	4G15/A	44.5-45	45-45.5	①	1.570	0.0008-0.0020	0.0020-0.0035	0.260	0.260
	2.4 (2350)	4G64/G	44.5-45	45-45.5	60@1.740	1.740	0.0008-0.0020	0.0008-0.0028	0.236	0.232
	1.8 (1834)	4G93/C	44.5-45	45-45.5	59@1.740	1.740	0.0008-0.0020	0.0020-0.0035	0.234	0.234
	3.0 (2972)	6G72/L	44.5-45	45-45.5	40.4@1.591	1.591	0.0012-0.0024	0.0020-0.0035	0.315	0.311
	3.5 (3497)	6G74/P	44-44.5	45-45.5	60@1.740	1.740	0.0008-0.0020	0.0016-0.0028	0.236	0.236
2001	1.5 (1468)	4G15/A	44.5-45	45-45.5	①	1.570	0.0008-0.0020	0.0020-0.0035	0.260	0.260
	2.4 (2350)	4G64/G	44.5-45	45-45.5	60@1.740	1.740	0.0008-0.0020	0.0008-0.0028	0.236	0.232
	1.8 (1834)	4G93/C	44.5-45	45-45.5	59@1.740	1.740	0.0008-0.0020	0.0020-0.0035	0.234	0.234
	3.0 (2972)	6G72/L	44.5-45	45-45.5	40.4@1.591	1.591	0.0012-0.0024	0.0020-0.0035	0.315	0.311
	3.5 (3497)	6G74/P	44-44.5	45-45.5	60@1.740	1.740	0.0008-0.0020	0.0016-0.0028	0.236	0.236
2002	1.5 (1468)	4G15/A	44.5-45	45-45.5	①	1.570	0.0008-0.0020	0.0020-0.0035	0.260	0.260
	1.8 (1834)	4G93/C	44.5-45	45-45.5	59@1.740	1.740	0.0008-0.0020	0.0020-0.0035	0.234	0.234
	2.0 (1999)	4G94/E	44.5-45	45-45.5	44.2@1.740	1.740	0.0008-0.0020	0.0016-0.0024	0.236	0.236
	2.4 (2350)	4G64/G	44.5-45	45-45.5	60@1.740	1.740	0.0008-0.0020	0.0008-0.0028	0.236	0.232
	3.0 (2972)	6G72/L	44.5-45	45-45.5	40.4@1.591	1.591	0.0012-0.0024	0.0020-0.0035	0.315	0.311
	3.5 (3497)	6G74/P	44-44.5	45-45.5	60@1.740	1.740	0.0008-0.0020	0.0016-0.0028	0.236	0.236
2003-04	1.5 (1468)	4G15/A	44.5-45	45-45.5	①	1.570	0.0008-0.0020	0.0020-0.0035	0.260	0.260
	1.8 (1834)	4G93/C	44.5-45	45-45.5	59@1.740	1.740	0.0008-0.0020	0.0020-0.0035	0.234	0.234
	2.0 (1999)	4G94/E	44.5-45	45-45.5	44.2@1.740	1.740	0.0008-0.0020	0.0016-0.0024	0.236	0.236
	2.4 (2350)	4G64/G	44.5-45	45-45.5	60@1.740	1.740	0.0008-0.0020	0.0008-0.0028	0.236	0.232
	3.0 (2972)	6G72/L	44.5-45	45-45.5	40.4@1.591	1.591	0.0012-0.0024	0.0020-0.0035	0.315	0.311
	3.5 (3497)	6G74/P	44-44.5	45-45.5	60@1.740	1.740	0.0008-0.0020	0.0016-0.0028	0.236	0.236

① Intake: 51@1.57
Exhaust: 64@1.57

TORQUE SPECIFICATIONS
All readings in ft. lbs.

Year	Engine Displacement Liters (cc)	Engine ID/VIN	Cylinder Head Bolts	Main Bearing Bolts	Rod Bearing Bolts	Crankshaft Damper Bolts	Flywheel Bolts	Manifold Intake	Manifold Exhaust	Spark Plugs	Lug Nut
2000	1.5 (1468)	4G15/A	①	37	12 ②	93	95	12	12	18	65-80
	1.8 (1834)	4G93/C	③	18 ②	15 ②	131	71	15	④	18	65-80
	2.4 (2350)	4G64/G	⑤	14.5 ②	14.5 ②	87	98	13	⑥	18	⑦
	3.0 (2972)	6G72/L	⑧	57	38	136	54	13	14	18	⑦
	3.5 (3497)	6G74/P	80	67	38	134	54	16	36	18	65-80
2001	1.5 (1468)	4G15/A	①	37	12 ②	93	95	12	12	18	65-80
	1.8 (1834)	4G93/C	③	18 ②	15 ②	131	71	15	④	18	65-80
	2.4 (2350)	4G64/G	⑤	14.5 ②	14.5 ②	87	98	13	⑥	18	⑦
	3.0 (2972)	6G72/L	⑧	57	38	136	54	13	14	18	⑦
	3.5 (3497)	6G74/P	80	67	38	134	54	16	36	18	65-80
2002	1.5 (1468)	4G15/A	①	37	12 ②	93	95	12	12	18	65-80
	1.8 (1834)	4G93/C	③	18 ②	15 ②	131	71	15	④	18	65-80
	2.0 (1999)	4G94/E	③	18 ②	15 ②	134	73	14	④	19	66-80
	2.4 (2350)	4G64/G	⑤	14.5 ②	14.5 ②	87	98	13	⑥	18	⑦
	3.0 (2972)	6G72/L	⑧	57	38	136	54	13	14	18	⑦
	3.5 (3497)	6G74/P	80	67	38	134	54	16	36	18	65-80
2003-04	1.5 (1468)	4G15/A	①	37	12 ②	93	95	12	12	18	65-80
	1.8 (1834)	4G93/C	③	18 ②	15 ②	131	71	15	④	18	65-80
	2.0 (1999)	4G94/E	③	18 ②	15 ②	134	73	14	④	19	66-80
	2.4 (2350)	4G64/G	⑤	14.5 ②	14.5 ②	87	98	13	⑥	18	⑦
	3.0 (2972)	6G72/L	⑧	57	38	136	54	13	14	18	⑦
	3.5 (3497)	6G74/P	80	67	38	134	54	16	36	18	65-80

① Step 1: Tighten all bolts to 35 ft. lbs.
　Step 2: Loosen all bolts to 0 ft. lbs.
　Step 3: Tighten all bolts to 15 ft. lbs.
　Step 4: Tighten all bolts 90 degrees.
　Step 5: Tighten all bolts an additional 90 degrees.

② Torque to specification plus
　an additional 90 degrees.

③ Step 1: Tighten all bolts to 54 ft. lbs.
　Step 2: Loosen all bolts to 0 ft. lbs.
　Step 3: Tighten all bolts to 15 ft. lbs.
　Step 4: Tighten all bolts 90 degrees.
　Step 5: Tighten all bolts an additional 90 degrees.

④ 8mm: 13 ft. lbs.
　10mm: 21 ft. lbs

⑤ Step 1: Tighten all bolts to 58 ft. lbs.
　Step 2: Loosen all bolts to 0 ft. lbs.
　Step 3: Tighten all bolts to 14.5 ft. lbs.
　Step 4: Tighten all bolts 90 degrees.
　Step 5: Tighten all bolts an additional 90 degrees.

⑥ 8mm: 20 ft. lbs.
　10mm: 21 ft. lbs.

⑦ Diamante/Galant: 65-80 ft. lbs.
　Eclipse/3000GT: 85-100 ft. lbs.

⑧ Step 1: Tighten all bolts in 3 steps to 80 ft. lbs.
　Step 2: Loosen all bolts to 0 ft. lbs.
　Step 3: Tighen all bolts in 3 steps to 80 ft. lbs.

42356-GALA-C08

WHEEL ALIGNMENT

Year	Model		Caster Range (+/-Deg.)	Caster Preferred Setting (Deg.)	Camber Range (+/-Deg.)	Camber Preferred Setting (Deg.)	Toe-in (in.)	Steering Axis Inclination (Deg.)
2000	Diamante ①	F	0.50	+3.00	0.50	0	0 +/- 0.13	—
		R	—	—	0.50	-0.69	0.13 +/- 0.13	—
	Diamante ②	F	0.50	+3.00	0.50	0	0 +/- 0.13	—
		R	—	—	0.50	-0.81	0.13 +/- 0.13	—
	Eclipse ③	F	1.50	+4.69	0.50	-0.09	0 +/- 0.13	7.19
		R	—	—	0.50	-1.31	0.13 +/- 0.13	—
	Eclipse ②	F	1.50	+4.69	0.50	-0.31	0 +/- 0.13	7.19
		R	—	—	0.50	-1.69	0.13 +/- 0.13	—
	Galant	F	0.50	+3.00	0.50	0	0 +/- 0.13	—
		R	—	—	0.50	-1.00	0 +/- 0.13	—
	Mirage	F	0.50	+2.84	0.50	0	0 +/- 0.13	—
		R	—	—	0.50	-0.69	0.13 +/- 0.10	—
2001	Diamante ①	F	0.50	+3.00	0.50	0	0 +/- 0.13	—
		R	—	—	0.50	-0.69	0.13 +/- 0.13	—
	Diamante ②	F	0.50	+3.00	0.50	0	0 +/- 0.13	—
		R	—	—	0.50	-0.81	0.13 +/- 0.13	—
	Eclipse ③	F	1.50	+4.69	0.50	-0.09	0 +/- 0.13	7.19
		R	—	—	0.50	-1.31	0.13 +/- 0.13	—
	Eclipse ②	F	1.50	+4.69	0.50	-0.31	0 +/- 0.13	7.19
		R	—	—	0.50	-1.69	0.13 +/- 0.13	—
	Galant	F	0.50	+3.00	0.50	0	0 +/- 0.13	—
		R	—	—	0.50	-1.00	0 +/- 0.13	—
	Mirage	F	0.50	+2.84	0.50	0	0 +/- 0.13	—
		R	—	—	0.50	-0.69	0.13 +/- 0.10	—
2002	Diamante ①	F	0.50	+3.00	0.50	0	0 +/- 0.13	—
		R	—	—	0.50	-0.69	0.13 +/- 0.13	—
	Diamante ②	F	0.50	+3.00	0.50	0	0 +/- 0.13	—
		R	—	—	0.50	-0.81	0.13 +/- 0.13	—
	Eclipse ③	F	1.50	+4.69	0.50	-0.09	0 +/- 0.13	7.19
		R	—	—	0.50	-1.31	0.13 +/- 0.13	—
	Eclipse ②	F	1.50	+4.69	0.50	-0.31	0 +/- 0.13	7.19
		R	—	—	0.50	-1.69	0.13 +/- 0.13	—
	Lancer	F	0.30	+2.50	0.30	0	0.09 +/- 0.13	—
		R	—	—	0.30	-0.40	0.12 +/- 0.08	—
	Galant	F	0.50	+3.00	0.50	0	0 +/- 0.13	—
		R	—	—	0.50	-1.00	0 +/- 0.13	—
	Mirage	F	0.50	+2.84	0.50	0	0 +/- 0.13	—
		R	—	—	0.50	-0.69	0.13 +/- 0.10	—

42356-GALA-C09

WHEEL ALIGNMENT

Year	Model		Caster		Camber		Toe-in (in.)	Steering Axis Inclination (Deg.)
------	-------	---	Range (+/-Deg.)	Preferred Setting (Deg.)	Range (+/-Deg.)	Preferred Setting (Deg.)		
2003-04	Diamante ①	F	0.50	+3.00	0.50	0	0 +/- 0.13	—
		R	—	—	0.50	-0.69	0.13 +/- 0.13	—
	Diamante ②	F	0.50	+3.00	0.50	0	0 +/- 0.13	—
		R	—	—	0.50	-0.81	0.13 +/- 0.13	—
	Eclipse ③	F	1.50	+4.69	0.50	-0.09	0 +/- 0.13	7.19
		R	—	—	0.50	-1.31	0.13 +/- 0.13	—
	Eclipse ②	F	1.50	+4.69	0.50	-0.31	0 +/- 0.13	7.19
		R	—	—	0.50	-1.69	0.13 +/- 0.13	—
	Lancer	F	0.30	+2.50	0.30	0	0.09 +/- 0.13	—
		R	—	—	0.30	-0.40	0.12 +/- 0.08	—
	Galant	F	0.50	+3.00	0.50	0	0 +/- 0.13	—
		R	—	—	0.50	-1.00	0 +/- 0.13	—
	Mirage	F	0.50	+2.84	0.50	0	0 +/- 0.13	—
		R	—	—	0.50	-0.69	0.13 +/- 0.10	—

① With 15 in. wheels
② With 16 in. wheels
③ With 14 in. wheels

42356-GALA-C10

TIRE, WHEEL AND BALL JOINT SPECIFICATIONS

| Year | Model | OEM Tires | | Tire Pressures (psi) | | Wheel Size | Ball Joint Inspection |
		Standard	Optional	Front	Rear		
2000	Diamante	205/65VR15	None	32	30	6-JJ	87-109 in. ①
	Galant LS/GT-Z/ES V6	205/55R16	None	32	29	6-JJ	3-13 in. ①
	Galant DE/ES 4-cyl	195/65R15	None	32	29	6-JJ	3-13 in. ①
	Mirage DE	P175/65R14	None	31	31	5.5-JJ	9-56 in. ①
	Mirage LS	P185/65R14	None	31	31	5.5-JJ	9-56 in. ①
	Eclipse RS	195/65R15	None	32	29	6-JJ	U: 3-13 in. ① L: 13 in.
	Eclipse GS	205/55R16	None	32	29	6-JJ	U: 3-13 in. ① L: 13 in.
	Eclipse GS-T	215/50R17	None	32	29	6.5-JJ	U: 3-13 in. ① L: 13 in.
	Eclipse GS-X	P205/55HR16	P215/50VR17	32	30	6-JJ	U: 3-13 in. ① L: 13 in.
2001	Diamante	205/65VR15	None	32	30	6-JJ	87-109 in. ①
	Galant LS/GT-Z/ES V6	205/55R16	None	32	29	6-JJ	3-13 in. ①
	Galant DE/ES 4-cyl	185/70HR14	None	29	26	5.5-JJ	3-13 in. ①
	Mirage DE	P175/70R13	None	31	31	5-J	9-56 in. ①
	Mirage LS	P185/65R14	None	31	31	5.5-JJ	9-56 in. ①
	Mirage GS Spyder	P215/50VR17	None	32	30	6.5-JJ	9-56 in. ①
	Eclipse RS	P195/70HR14	None	32	30	5.5-JJ	U: 3-13 in. ① L: 13 in.
	Eclipse GS	P195/70HR14	None	32	30	5.5-JJ	U: 3-13 in. ① L: 13 in.
	Eclipse GS-T	P205/55HR16	P215/50VR17	32	30	6-JJ	U: 3-13 in. ① L: 13 in.
2002	Diamante	205/65VR15	None	32	30	6-JJ	87-109 in. ①
	Galant LS/GT-Z/ES V6	205/55R16	None	32	29	6-JJ	3-13 in. ①
	Galant DE/ES 4-cyl	185/70HR14	None	29	26	5.5-JJ	3-13 in. ①
	Mirage DE	P175/70R13	None	31	31	5-J	9-56 in. ①
	Mirage LS	P185/65R14	None	31	31	5.5-JJ	9-56 in. ①
	Mirage GS Spyder	P215/50VR17	None	32	30	6.5-JJ	9-56 in. ①
	Lancer ES	P185/65 R14 85S	None	32	29	5-JJ	0-35 in. ①
	Lancer LS/OZ/Rally	P195/60R15 87T	None	32	29	6-JJ	0-35 in. ①
	Eclipse RS	P195/70HR14	None	32	30	5.5-JJ	U: 3-13 in. ① L: 13 in.
	Eclipse GS	P195/70HR14	None	32	30	5.5-JJ	U: 3-13 in. ① L: 13 in.
	Eclipse GS-T	P205/55HR16	P215/50VR17	32	30	6-JJ	U: 3-13 in. ① L: 13 in.

42356-GALA-C11

TIRE, WHEEL AND BALL JOINT SPECIFICATIONS

| Year | Model | OEM Tires | | Tire Pressures (psi) | | Wheel Size | Ball Joint Inspection |
		Standard	Optional	Front	Rear		
2003-04	Diamante	205/65VR15	None	32	30	6-JJ	87-109 in. ①
	Galant LS/GT-Z/ES V6	205/55R16	None	32	29	6-JJ	3-13 in. ①
	Galant DE/ES 4-cyl	185/70HR14	None	29	26	5.5-JJ	3-13 in. ①
	Mirage DE	P175/70R13	None	31	31	5-J	9-56 in. ①
	Mirage LS	P185/65R14	None	31	31	5.5-JJ	9-56 in. ①
	Mirage GS Spyder	P215/50VR17	None	32	30	6.5-JJ	9-56 in. ①
	Lancer ES	P185/65 R14 85S	None	32	29	5-JJ	0-35 in. ①
	Lancer LS/OZ/Rally	P195/60R15 87T	None	32	29	6-JJ	0-35 in. ①
	Eclipse RS	P195/70HR14	None	32	30	5.5-JJ	U: 3-13 in. ① L: 13 in.
	Eclipse GS	P195/70HR14	None	32	30	5.5-JJ	U: 3-13 in. ① L: 13 in.
	Eclipse GS-T	P205/55HR16	P215/50VR17	32	30	6-JJ	U: 3-13 in. ① L: 13 in.

OEM: Original Equipment Manufacturer

PSI: Pounds Per Square Inch

STD: Standard

OPT: Optional

L: Lower

U: Upper

① Torque required in inch lbs. to rotate ball joint when removed from the knuckle

42356-GALA-C12

BRAKE SPECIFICATIONS
All measurements in inches unless noted

Year	Model		Brake Disc Original Thickness	Brake Disc Minimum Thickness	Brake Disc Maximum Runout	Brake Drum Diameter Original Inside Diameter	Brake Drum Diameter Max. Wear Limit	Brake Drum Diameter Maximum Machine Diameter	Minimum Lining Thickness Front	Minimum Lining Thickness Rear	Brake Caliper Bracket Bolts (ft. lbs.)	Brake Caliper Mounting Bolts (ft. lbs.)
2000	Diamante	F	0.940	0.880	0.002	—	—	—	0.080	—	54	65
		R	0.410	0.330	0.0023	—	—	—	—	0.039	24	36-43
	Galant	F	0.940	0.880	0.0031	—	—	—	0.080	—	54	65
		R	—	—	—	8.976	9.078	9.078	—	0.040	—	—
	Mirage	F	0.710	0.650	0.0024	—	—	—	0.080	—	36	67-81
		R	—	—	—	8.00	8.10	8.10	—	0.039	—	—
	Eclipse	F	0.940	0.882	0.003	—	—	—	0.080	—	58-72	46-62
		R	—	—	—	9.000	—	9.100	—	0.039	—	—
	Eclipse w/rear disc	F	0.940	0.882	0.003	—	—	—	0.080	—	58-72	46-62
		R	0.390	0.331	0.003	—	—	—	—	0.080	36-43	54
2001	Diamante	F	0.940	0.880	0.002	—	—	—	0.080	—	54	65
		R	0.410	0.330	0.0023	—	—	—	—	0.039	24	36-43
	Galant	F	②	③	0.002	—	—	—	0.080	—	①	65
		R	0.390	0.331	0.002	9.000	—	9.100	—	0.040	36-43	54
	Mirage	F	0.940	0.882	0.0012	—	—	—	0.080	—	36	67-81
		R	—	—	—	8.00	8.10	8.10	—	0.039	—	—
	Eclipse	F	0.940	0.882	0.002	—	—	—	0.080	—	58-72	46-62
		R	—	—	—	9.000	—	9.100	—	0.039	—	—
	Eclipse w/rear disc	F	0.940	0.882	0.003	—	—	—	0.080	—	①	46-62
		R	0.390	0.331	0.003	—	—	—	—	0.080	36-43	54
2002	Diamante	F	0.940	0.880	0.002	—	—	—	0.080	—	54	65
		R	0.410	0.330	0.0023	—	—	—	—	0.039	24	36-43
	Galant	F	②	③	0.002	—	—	—	0.080	—	①	65
		R	0.390	0.331	0.002	9.000	—	9.100	—	0.040	36-43	54
	Mirage	F	0.940	0.882	0.0012	—	—	—	0.080	—	36	67-81
		R	—	—	—	8.00	8.10	8.10	—	0.039	—	—
	Lancer	F	0.900	0.880	0.002	—	—	—	0.080	—	37	62
		R	—	—	—	7.99	8.07	8.07	—	0.040	—	—
	Eclipse	F	0.940	0.882	0.002	—	—	—	0.080	—	58-72	46-62
		R	—	—	—	9.000	—	9.100	—	0.039	—	—
	Eclipse w/rear disc	F	0.940	0.882	0.003	—	—	—	0.080	—	①	46-62
		R	0.390	0.331	0.003	—	—	—	—	0.080	36-43	54

42356-GALA-C13

BRAKE SPECIFICATIONS
All measurements in inches unless noted

| Year | Model | | Brake Disc | | | Brake Drum Diameter | | | Minimum Lining Thickness | | Brake Caliper | |
			Original Thickness	Minimum Thickness	Maximum Runout	Original Inside Diameter	Max. Wear Limit	Maximum Machine Diameter	Front	Rear	Bracket Bolts (ft. lbs.)	Mounting Bolts (ft. lbs.)
2003-04	Diamante	F	0.940	0.880	0.002	—	—	—	0.080	—	54	65
		R	0.410	0.330	0.0023	—	—	—	—	0.039	24	36-43
	Galant	F	②	③	0.002	—	—	—	0.080	—	①	65
		R	0.390	0.331	0.002	9.000	—	9.100	—	0.040	36-43	54
	Mirage	F	0.940	0.882	0.0012	—	—	—	0.080	—	36	67-81
		R	—	—	—	8.00	8.10	8.10	—	0.039	—	—
	Lancer	F	0.900	0.880	0.002	—	—	—	0.080	—	37	62
		R	—	—	—	7.99	8.07	8.07	—	0.040	—	—
	Eclipse	F	0.940	0.882	0.002	—	—	—	0.080	—	58-72	46-62
		R	—	—	—	9.000	—	9.100	—	0.039	—	—
	Eclipse w/rear disc	F	0.940	0.882	0.003	—	—	—	0.080	—	①	46-62
		R	0.390	0.331	0.003	—	—	—	—	0.080	36-43	54

NA: Not Available

F: Front

R: Rear

① Lock pin (2.4L): 55 ft. lbs.
   Lock bolt (3.0L): 28 ft. lbs.

② 2.4L: 0.940
   3.0L: 1.020

③ 2.4L: 0.88
   3.0L: 0.96

42356-GALA-C14

SCHEDULED MAINTENANCE INTERVALS
Mitsubishi—Diamante, Eclipse, Galant, Mirage & Lancer

TO BE SERVICED	TYPE OF SERVICE	VEHICLE MILEAGE INTERVAL (x1000)												
		7.5	15	22.5	30	37.5	45	52.5	60	67.5	75	82.5	90	97.5
Engine oil & filter ①	R	✓	✓	✓	✓	✓	✓	✓	✓	✓	✓	✓	✓	✓
Automatic transaxle fluid & filter	S/I		✓		✓		✓		✓		✓		✓	
Brake hoses	S/I		✓		✓		✓		✓		✓		✓	
Disc brake pads	S/I		✓		✓		✓		✓		✓		✓	
Driveshaft boots	S/I		✓		✓		✓		✓		✓		✓	
Valve clearance (Mirage)	S/I		✓		✓		✓		✓		✓		✓	
Air cleaner element	R				✓				✓				✓	
Engine coolant	R				✓				✓				✓	
Spark plugs (except Diamante w/platinum tip)	R				✓				✓				✓	
Spark plugs (Diamante w/platinum tip)	R								✓					
Ball joints & steering linkage seals	S/I				✓				✓				✓	
Drive belt(s)	S/I				✓				✓				✓	
Exhaust system	S/I				✓				✓				✓	
Fuel hoses	S/I				✓				✓				✓	
Manual transaxle fluid (Mirage)	S/I				✓				✓				✓	
Manual transaxle fluid (including transfer) (Eclipse)	S/I				✓				✓				✓	
Manual transaxle fluid (Galant)	S/I				✓				✓				✓	
Rear axle oil (Eclipse AWD)	S/I				✓				✓				✓	
Rear drum brake linings & rear wheel cylinders (Eclipse, Galant & Mirage)	S/I				✓				✓				✓	
Ignition cables	R								✓					
Timing belt(s)	R								✓					
Distributor cap & rotor	S/I								✓					
EVAP system (except canister)	S/I								✓					

42356-GALA-C15

SCHEDULED MAINTENANCE INTERVALS
Mitsubishi—Diamante, Eclipse, Galant, Mirage & Lancer

TO BE SERVICED	TYPE OF SERVICE	VEHICLE MILEAGE INTERVAL (x1000)												
		7.5	15	22.5	30	37.5	45	52.5	60	67.5	75	82.5	90	97.5
Fuel system (tank, pipe line, connection & fuel tank filler tube cap)	S/I								✓					

R: Replace   S/I: Service or Inspect
① 3000GT turbo: replace every 5000 miles.

FREQUENT OPERATION MAINTENANCE (SEVERE SERVICE)
If a vehicle is operated under any of the following conditions it is considered severe service:

- Extremely dusty areas.

- 50% or more of the vehicle operation is in 32°C (90°F) or higher temperatures, or constant operation in temperatures below 0°C (32°F).

- Prolonged idling (vehicle operation in stop and go traffic).

- Frequent short running periods (engine does not warm to normal operating temperatures).

- Police, taxi, delivery usage or trailer towing usage.

Oil & oil filter: change every 3000 miles.

Disc brake pads: service or inspect ever 6000 miles.

Air filter element: service or inspect every 15,000 miles.

Automatic transaxle fluid & filter: replace every 15,000 miles.

Spark plugs (except Diamante w/platinum tip): replace every 15,000 miles.

Manual transaxle oil (including transfer (Galant, Mirage & 3000GT): replace every 30,000 miles.

42356-GALA-C16

PRECAUTIONS

Before servicing any vehicle, please be sure to read all of the following precautions, which deal with personal safety, prevention of component damage, and important points to take into consideration when servicing a motor vehicle:

• Never open, service or drain the radiator or cooling system when the engine is hot; serious burns can occur from the steam and hot coolant.

• Observe all applicable safety precautions when working around fuel. Whenever servicing the fuel system, always work in a well-ventilated area. Do not allow fuel spray or vapors to come in contact with a spark, open flame, or excessive heat (a hot drop light, for example). Keep a dry chemical fire extinguisher near the work area. Always keep fuel in a container specifically designed for fuel storage; also, always properly seal fuel containers to avoid the possibility of fire or explosion. Refer to the additional fuel system precautions in this section.

• Fuel injection systems often remain pressurized, even after the engine has been turned **OFF**. The fuel system pressure must be relieved before disconnecting any fuel lines. Failure to do so may result in fire and/or personal injury.

• Brake fluid often contains polyglycol ethers and polyglycols. Avoid contact with the eyes and wash your hands thoroughly after handling brake fluid. If you do get brake fluid in your eyes, flush your eyes with clean, running water for 15 minutes. If eye irritation persists, or if you have taken brake fluid internally, IMMEDIATELY seek medical assistance.

• The EPA warns that prolonged contact with used engine oil may cause a number of skin disorders, including cancer. You should make every effort to minimize your exposure to used engine oil. Protective gloves should be worn when changing oil. Wash your hands and any other exposed skin areas as soon as possible after exposure to used engine oil. Soap and water, or waterless hand cleaner should be used.

• All new vehicles are now equipped with an air bag system. The system must be disabled before performing service on or around system components, steering column, instrument panel components, wiring and sensors. Failure to follow safety and disabling procedures could result in accidental air bag deployment, possible personal injury, and unnecessary system repairs.

• Always wear safety goggles when working with, or around, the air bag system. When carrying a non-deployed air bag, be sure the bag and trim cover are pointed away from your body. When placing a non-deployed air bag on a work surface, always face the bag and trim cover upward, away from the surface. This will reduce the motion of the module if it is accidentally deployed. Refer to the additional air bag system precautions later in this section.

• Clean, high quality brake fluid from a sealed container is essential to the safe and proper operation of the brake system. You should always buy the correct type of brake fluid for your vehicle. If the brake fluid becomes contaminated, completely flush the system with new fluid. Never reuse any brake fluid. Any brake fluid that is removed from the system should be discarded. Also, do not allow any brake fluid to come in contact with a painted surface; it will damage the paint.

• Never operate the engine without the proper amount and type of engine oil; doing so will result in severe engine damage.

• Timing belt maintenance is extremely important. Many models utilize an interference-type, non-freewheeling engine. If the timing belt breaks, the valves in the cylinder head may strike the pistons, causing potentially serious (also time-consuming and expensive) engine damage. Refer to the maintenance interval charts in the front of this section for the recommended replacement interval for the timing belt, and to the timing belt procedure in this section for belt replacement and inspection.

• Disconnecting the negative battery cable on some vehicles may interfere with the functions of the on-board computer system(s) and may require the computer to undergo a relearning process once the negative battery cable is reconnected.

• When servicing drum brakes, only disassemble and assemble one side at a time, leaving the remaining side intact for reference.

ENGINE REPAIR

➡ **Disconnecting the negative battery cable on some vehicles may interfere with the functions of the on board computer systems and may require the computer to undergo a relearning process, once the negative battery cable is reconnected.**

Distributor

REMOVAL

Before removing the distributor, position No. 1 cylinder at TDC on the compression stroke and align the timing marks.

1. Before servicing the vehicle, refer to the precautions in the beginning of this section.

2. Remove or disconnect the following:
   • Negative battery cable

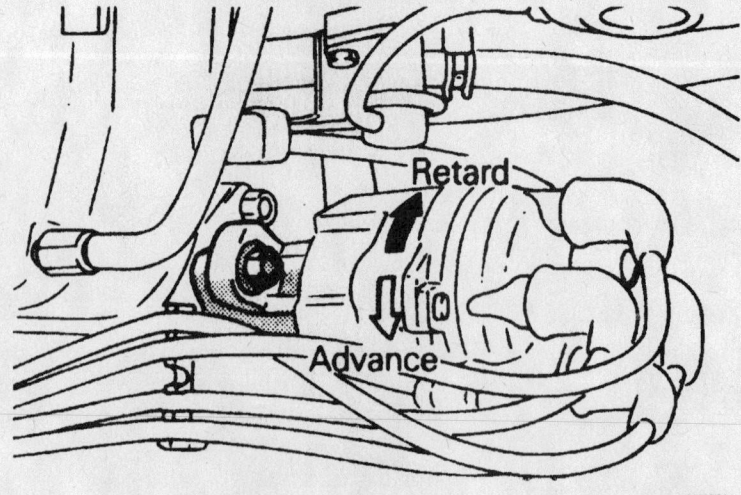

Adjusting the distributor—1.5L Mirage shown

7923PG01

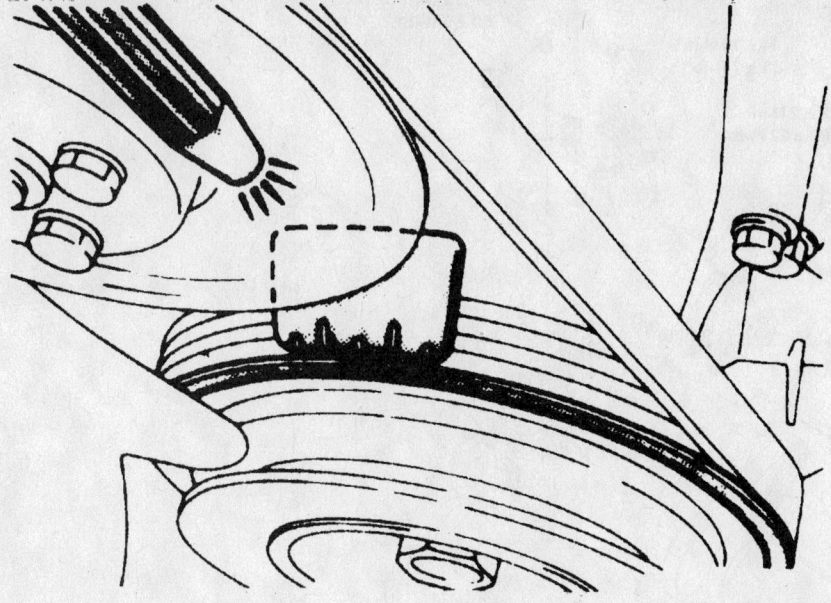

Checking the ignition timing—1.5L Mirage shown

7923PG02

- Ignition wire cover, if equipped
- Distributor harness connector
- Distributor cap with all ignition wires still connected
- Coil wire, if necessary

3. Matchmark the rotor to the distributor housing, and the distributor housing to the engine.

4. Remove or disconnect the following:
- Hold-down nut
- Distributor from the engine

INSTALLATION

Timing Not Disturbed

1. Install or connect the following:
- New distributor housing O-ring and lubricate with clean oil
- Distributor in the engine, matchmarks aligned
- Hold-down nut
- Distributor harness connectors
- Distributor cap
- Coil wire, if removed
- Negative battery cable

2. Adjust the ignition timing and tighten the hold-down nut to 96 inch lbs. (11 Nm).

Timing Disturbed

1. Install a new distributor housing O-ring and lubricate with clean oil.

2. Position the engine so the No. 1 piston is at TDC of its compression stroke and the mark on the vibration damper is aligned with **0** on the timing indicator.

3. Align the distributor housing and gear

mating marks. Install the distributor in the engine so the slot or groove of the distributor's installation flange aligns with the distributor installation stud in the engine block. Be sure the distributor is fully seated. Inspect alignment of the distributor rotor making sure the rotor is aligned with the position of the No. 1 ignition wire in the distributor cap.

4. Install or connect the following:
- Hold-down nut
- Distributor harness connectors
- Distributor cap
- Negative battery cable

5. Adjust the ignition timing and tighten the hold-down nut to 96 inch lbs. (11 Nm).

Alternator

REMOVAL

1.5L, 1.8L, and 2.4L Engines

1. Before servicing the vehicle, refer to the precautions in the beginning of this section.

2. Remove or disconnect the following:
- Negative battery cable
- Left side cover panel under the vehicle
- Air intake hose, on turbocharged Galant models
- Drive belts
- Water pump pulleys
- Alternator upper bracket/brace
- Alternator wiring connectors
- Alternator mounting bolts and remove the alternator

2.0L Engine

1. Before servicing the vehicle, refer to the precautions in the beginning of this section.

2. Remove or disconnect the following:
- Negative battery cable
- Drive belts
- Power steering hose clamp
- Alternator brace
- Alternator electrical connector
- Engine mount, on vehicles with Anti-lock Brake Systems (ABS)
- Alternator, tilting the engine as necessary

3.0L DOHC Engine

1. Remove or disconnect the following:
- Negative battery cable
- Headlamp washer reservoir tank
- Condenser fan and upper radiator insulator
- Alternator drive belt
- Alternator upper and lower mounting bolts
- Alternator support bracket mounting bolts
- Alternator support bracket from the vehicle
- Alternator wiring harness
- Alternator from the vehicle

3.0L SOHC and 3.5L Engines

1. Remove or disconnect the following:
- Negative battery cable
- Air intake hose
- Alternator drive belt

2. On California models, remove the rear bank converter assembly.

3. Remove the engine roll stopper stay bracket assembly.

4. On the 3.0L SOHC engine, disconnect the Exhaust Gas Recirculation (EGR) temperature sensor wire and remove the EGR pipe assembly.

5. On the 3.0L SOHC engine, remove the intake plenum stay bracket assembly.

6. Remove or disconnect the following:
- Alternator wiring harness connectors
- Alternator upper and lower mounting bolts

7. From beneath the vehicle, remove the alternator.

INSTALLATION

1.5L, 1.8L, and 2.4L Engines

1. Position the alternator on the lower mounting fixture and install the lower

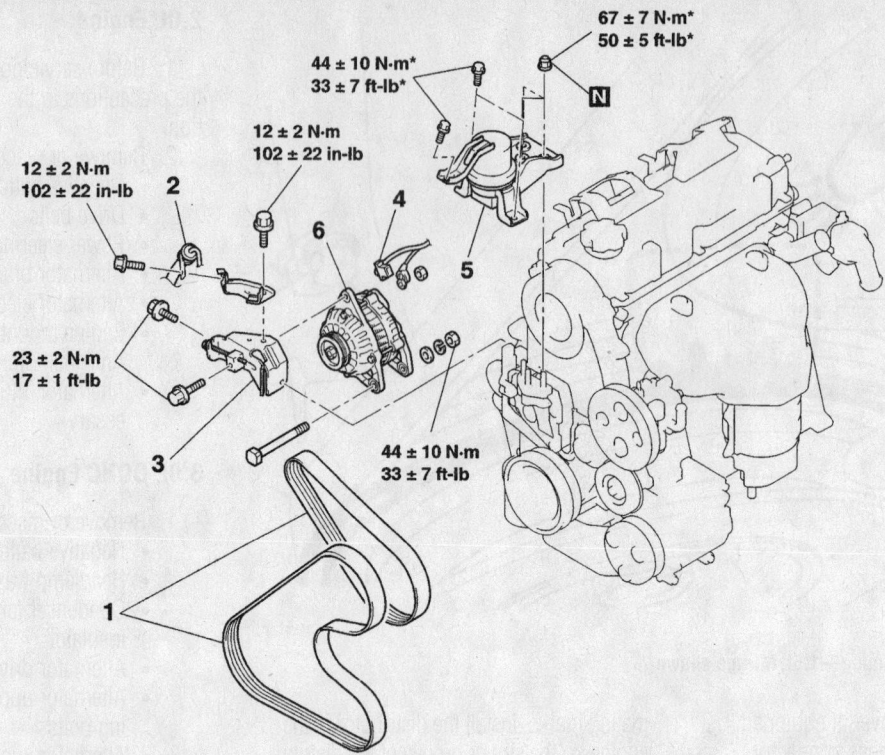

67 ± 7 N·m*
50 ± 5 ft-lb*

44 ± 10 N·m*
33 ± 7 ft-lb*

12 ± 2 N·m
102 ± 22 in-lb

12 ± 2 N·m
102 ± 22 in-lb

23 ± 2 N·m
17 ± 1 ft-lb

44 ± 10 N·m
33 ± 7 ft-lb

1.	DRIVE BELT	4.	GENERATOR CONNECTOR
2.	POWER STEERING HOSE CLAMP	5.	ENGINE MOUNT<VEHICLES WITH ABS>
3.	GENERATOR BRACE	6.	GENERATOR

93570G01

Exploded view of the alternator and related components—Lancer with 2.0L engine

mounting bolt and nut. Tighten nut just enough to allow for movement of the alternator.

2. Install or connect the following:
- Alternator upper bracket/brace and connect the alternator electrical harness
- Water pump pulleys
- Drive belts and adjust to the proper tension

3. On turbocharged Galant models, install the air intake hose.

4. Install or connect the following:
- Left side cover panel under the vehicle as required
- Negative battery cable and check for proper operation

2.0L Engines

1. Install or connect the following:
- Alternator
- Engine mount, if removed
- Alternator connector
- Alternator brace
- Power steering hose clamp
- Drive belts
- Negative battery cable

2. Adjust drive belts.

3.0L DOHC Engine

1. Install or connect the following:
- Alternator to the vehicle and connect the wiring harness
- Alternator support bracket to the vehicle and tighten the bracket mounting bolts to specifications

2. Position the alternator on the mounting bracket. Install and tighten the mounting bolt and nut to 17 ft. lbs. (24 Nm).

3. Install or connect the following:
- Drive belt and adjust the tensioner until the proper belt tension is achieved
- Upper radiator insulator and condenser fan
- Headlamp washer reservoir tank
- Negative battery cable and check the charging system for proper operation

3.0L SOHC and 3.5L Engines

1. Position the alternator on the lower mounting fixture. Install and tighten the mounting bolt and nut to 14–18 ft. lbs. (20–25 Nm).

2. Install or connect the following:
- Alternator wiring harness

3. On the 3.0L SOHC engine, install the intake plenum stay bracket and tighten the mounting bolt to 13 ft. lbs. (18 Nm).

4. On the 3.0L SOHC engine, install the EGR pipe and tighten the fitting connections to 43 ft. lbs. (60 Nm).

5. On the 3.0L SOHC engine, connect the EGR temperature sensor wire.

6. Install or connect the following:
- Engine roll stopper stay and tighten the mounting bolt to 35 ft. lbs. (45 Nm) and the nut to 36–43 ft. lbs. (50–60 Nm)
- Rear converter assembly, if removed
- Drive belt and adjust the tensioner until the proper belt tension is achieved
- Air intake hose
- Negative battery cable and check the charging system for proper operation

Ignition Timing

ADJUSTMENT

The ignition timing is controlled by the Electronic Control Module (ECM) and is not adjustable. However it can be inspected using a scan tool.

Engine Assembly

REMOVAL & INSTALLATION

Diamante

1. Before servicing the vehicle, refer to the precautions in the beginning of this section.
2. Remove the hood assembly.
3. Relieve fuel system pressure.
4. Remove or disconnect the following:
   - Negative, then the positive battery cable
   - Battery
   - Air cleaner assembly and all adjoining air intake duct work
5. Drain the engine coolant and remove the radiator assembly and coolant reservoir (and bracket).
6. Remove or disconnect the following:
   - Engine undercover, if equipped
   - Front exhaust pipe
   - Transaxle assembly
   - Accelerator cable from the throttle body
   - Vacuum hoses from the intake manifold, label for installation
   - High pressure fuel line and the fuel return line
   - Vacuum hoses from the solenoid valves
   - Vacuum hoses from the purge canister
   - Heater hose connections from the engine
   - Harness for the Exhaust Gas Recirculation (EGR) temperature sensor connection, if equipped
   - Engine drive belts
   - Power steering pump oil pressure switch connection from the pump
   - Power steering pump and secure away from the work area
   - Air conditioning compressor. Wire the compressor aside. Do not discharge or disconnect the air conditioning lines.
   - Wiring to the alternator
   - Harness plugs for the Barometer (BARO) sensor, Idle Speed Control (ISC) motor, Throttle Position (TP) sensor, fuel injectors and Knock (KS) sensor
   - Harness plugs for the Engine Coolant Temperature (ECT) switch, sensor and gauge
   - Harness plugs for the ignition coil, condenser and ignition power transistor
   - Harness plugs for the variable induction control motor and the Manifold Absolute Pressure (MAP) sensor
   - Harness plugs for the Crankshaft Position (CKP) and Camshaft Position (CMP) sensors
   - Radiator overflow tank and remove the mounting bracket
   - Ground cable connections
7. Attach a hoist to the engine and take up the engine weight. Remove the engine mount bracket. Remove any torque control brackets (roll stoppers).
8. Lift the engine slowly and remove from the engine compartment.

To install:

9. Install or connect the following:
   - Engine and secure all control brackets
   - Transaxle assembly
   - Engine ground cable connections
   - Harness plugs for the CKP and CMP sensors
   - Harness plugs for the variable induction control motor and the MAP sensor
   - Harness plugs for the ignition coil, condenser and ignition power transistor
   - Harness plugs for the ECT switch, sensor and gauge
   - Harness plugs for the BARO sensor, ISC motor, TP sensor, fuel injectors and KS sensor
   - Wiring to the alternator
   - Air conditioning compressor assembly
   - Power steering pump assembly
   - Power steering pump oil pressure switch harness plug to the pump
   - Engine drive belts, adjust
   - Harness for the EGR temperature sensor
   - Heater hose connections to the engine, using new hose clamps
   - Vacuum hoses to the purge canister
   - Vacuum hoses to the solenoid valves
   - High pressure fuel line and the fuel return line, using new clamps or O-rings
   - Vacuum hoses
   - Accelerator cable to the throttle body
   - Air cleaner assembly and all adjoining air intake duct work
   - Radiator and coolant reservoir assembly
   - Transaxle assembly
   - Exhaust system to the engine, using new gaskets
   - Battery to the vehicle
   - Positive, then the negative battery cables
   - Engine undercover, if equipped
10. Fill the engine with the proper amount of engine oil and coolant.
11. Install the hood.
12. Start the engine and check for leaks.

Eclipse

2.4L (4G64) ENGINES

1. Before servicing the vehicle, refer to the precautions in the beginning of this section.

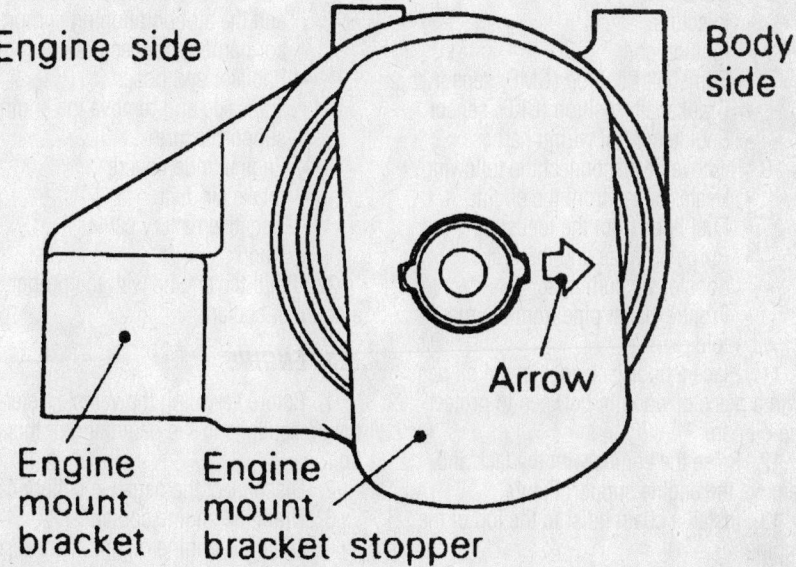

Alignment of the engine mount stopper bracket—Diamante shown

7923PG06

2. Relieve the fuel system pressure.
3. Remove or disconnect the following:
   - Negative battery cable
   - Hood
   - Intake air duct
4. Drain the engine coolant.
5. Remove or disconnect the following:
   - Hoses and remove the radiator
   - Engine undercover
6. Attach an engine lifting fixture to the engine and remove the transaxle assembly.
7. Disconnect the following connectors:
   - Air conditioning compressor
   - Power steering pressure switch
   - Heated Oxygen (HO2S) sensor
   - Engine Coolant Temperature (ECT) gauge sender
   - ECT sensor
   - Manifold Absolute Pressure (MAP) sensor
   - Intake Air Temperature (IAT) sensor
8. Remove or disconnect the following:
   - Power steering pump from the bracket and position the pump out of the way
   - Air conditioning compressor from the bracket and position it out of the way. Do not disconnect the hoses.
   - Accelerator cable from the throttle body and mounting bracket
9. Disconnect the following connectors:
   - Idle Air Control (IAC) motor
   - Knock (KS) sensor
   - Ignition module (power transistor)
   - Exhaust Gas Recirculation (EGR) solenoid
   - Oil pressure switch
   - Throttle Position (TP) sensor
   - Condenser
   - Injectors
   - Ignition coil
   - Camshaft Position (CMP) sensor
   - Crankshaft Position (CKP) sensor
   - Engine control wiring harness
10. Remove or disconnect the following:
    - Heater hoses from the engine
    - Fuel lines from the fuel supply rail
    - Purge air hose and the brake booster vacuum hose
    - Front exhaust pipe from the manifold
11. Place a floor jack against the oil pan with a piece of wood in between to protect the oil pan.
12. Raise the engine with the jack and remove the engine support fixture.
13. Install a chain hoist to the top of the engine.
14. Remove the engine mount bracket.
15. Lift the engine up slowly out of the engine compartment.

To install:

16. Slowly lower the engine assembly into the vehicle.
17. Position the floor jack under the oil pan with a piece of wood in between. Use the floor jack to adjust the height of the engine while installing the engine mount bracket.
18. Remove the chain hoist and install the engine support fixture.
19. Install or connect the following:
    - Front exhaust pipe to the manifold
    - Brake booster vacuum hose
    - New O-ring on the high pressure fuel line. Apply a small amount of clean engine oil to the O-ring and connect the fuel lines to the fuel supply rail.
20. Connect the following connectors:
    - IAC motor
    - KS sensor
    - Ignition module (power transistor)
    - EGR solenoid
    - Oil pressure switch
    - TP sensor
    - Condenser
    - Injectors
    - Ignition coil
    - CMP sensor
    - CKP sensor
    - Engine control wiring harness
21. Install or connect the following:
    - Accelerator cable, adjust
    - Air conditioning compressor and the power steering pump in their brackets
    - IAT sensor, MAP sensor, ECT sensor and gauge sender, HO2S sensor, power steering pressure switch and the air conditioning compressor harness connectors
    - Radiator and hoses
    - Transaxle and remove the engine support fixture
    - Engine undercovers
    - Intake air duct
    - Negative battery cable
    - Hood
22. Refill the engine with the proper amount of coolant.

3.0L ENGINE

1. Before servicing the vehicle, refer to the precautions in the beginning of this section.
2. Disconnect the negative battery cable.
3. Drain the engine coolant.
4. Drain the engine oil and the transmission oil.
5. Relieve the fuel system pressure.
6. Remove or disconnect the following:
   - All wires, cables and hoses connected to the engine
   - Hood
   - Air intake and breather hoses
   - Radiator hoses and remove the radiator
   - Front exhaust pipe
   - Power steering pump and position it aside
   - Air conditioning compressor drive belt
   - Compressor from its mount and hang it out of the way. Do not disconnect the hoses and do not allow the compressor to hang by the hoses.
   - Engine hoist equipment and make certain the attaching points on the engine are secure
7. Raise the hoist enough to support the engine.
8. Remove or disconnect the following:
   - Front and rear engine roll stoppers
   - Left engine mount and support bracket
9. Slowly lift the engine and remove it from the vehicle.

Galant

1. Before servicing the vehicle, refer to the precautions in the beginning of this section.
2. Disconnect the negative battery cable.
3. Drain the engine coolant.
4. Drain the engine oil and the transmission oil.
5. Relieve the fuel system pressure.
6. Remove or disconnect the following:
   - Hood
   - Transaxle assembly
   - Radiator hoses and remove the radiator
   - Accelerator cable and remove the bracket
   - Air intake and breather hoses
   - Heater hoses
   - Brake booster vacuum hose at the engine
   - Vacuum hoses at the throttle body, label
   - Fuel feed and return hoses
7. Disconnect the following:
   - Power steering pressure switch
   - Alternator
   - Oil pressure switch
   - Air conditioning compressor
   - Each injector
   - Power transistor
   - Ignition coil
   - Throttle Position (TP) sensor

- Idle Air Control (IAC) motor
- Engine Coolant Temperature (ECT) switch
- ECT sensor
- Exhaust Gas Recirculation (EGR) temperature sensor
- Engine control wiring harness
- Heated Oxygen (HO$_2$S) sensor
- Crankshaft Position (CKP) sensor
- Camshaft Position (CMP) sensor
- Refrigerant temperature switch
- Condenser connection

8. Remove or disconnect the following:
- Power steering pump and position it aside
- Air conditioning compressor drive belt
- Compressor from its mount and hang it out of the way. Do not disconnect the hoses and do not allow the compressor to hang by the hoses.
- Front exhaust pipe
- Engine hoist equipment and make certain the attaching points on the engine are secure

9. Raise the hoist enough to support the engine.

10. Remove or disconnect the following:
- Front and rear engine roll stoppers
- Left engine mount and support bracket

11. Slowly lift the engine and remove it from the vehicle.

To install:

12. Lower the engine into the vehicle.

13. Install the front and rear roll stoppers and the left engine mount. Do not torque the through-bolts at this time.

14. Remove the lifting apparatus from the engine.

15. Connect the exhaust system to the manifold, using a new gasket and new locking nuts. Tighten the nuts and the small bolt to 33 ft. lbs. (44 Nm).

16. Tighten the engine mount nuts and bolts. Correct torque values are:
- Upper mount to engine nuts: 42 ft. lbs. (57 Nm)
- Upper mount to engine bolt: 108 inch lbs. (12 Nm)
- Upper mount through-bolt: 72–87 ft. lbs. (98–118 Nm)
- Rear roll stopper through-bolt: 32 ft. lbs. (44 Nm)
- Front roll stopper through-bolt: 41 ft. lbs. (57 Nm)

17. Install or connect the following:
- Air conditioning compressor, tightening the mounting bolts to 18 ft. lbs. (25 Nm)
- Power steering pump, tightening

the front bolts to 21 ft. lbs. (28 Nm) and the rear bolt to 16 ft. lbs. (22 Nm)
- Accessory drive belts

18. Connect the following:
- Power steering pressure switch
- Alternator
- Oil pressure switch
- Air conditioning compressor
- Each injector
- Power transistor
- Ignition coil
- TP sensor
- IAC motor
- ECT switch
- ECT sensor
- EGR temperature sensor
- Engine control wiring harness
- HO$_2$S sensor
- CKP sensor
- CMP sensor
- Refrigerant temperature switch
- Condenser connection

19. Install or connect the following:
- Fuel return hose and secure with the retaining clamp
- New O-ring, connect the high pressure fuel line and tighten the bolts to 48 inch lbs. (6 Nm)
- Vacuum lines running to the throttle body
- Heater hoses
- Accelerator cable bracket, tightening the bolts to 48 inch lbs. (6 Nm), and connect the accelerator cable
- Radiator and connect the hoses
- Transaxle

20. Fill the coolant system.

21. Connect the negative battery cable.

22. Start the engine and check for leaks.

23. Install the hood.

Lancer

1. Before servicing the vehicle, refer to the precautions in the beginning of this section.

2. Relieve fuel system pressure.

3. Remove or disconnect the following:
- Negative battery cable
- Undercover, if equipped
- Hood assembly
- Air cleaner assembly and all adjoining air intake duct work

4. Drain the engine coolant and engine oil.

5. Remove or disconnect the following:
- Radiator
- Front exhaust pipe
- Battery and battery tray
- Accelerator cable

6. Detach the electrical connectors from the following components:
- Air Conditioning (A/C) compressor
- Power steering oil pressure switch
- Crank angle sensor
- Manifold differential pressure sensor
- Evaporative emission (EVAP) purge solenoid
- Exhaust Gas Recirculation (EGR) solenoid valve
- Ignition coil
- Fuel injectors
- Throttle Position (TP) sensor
- Idle Air Control (IAC) motor
- Engine Coolant Temperature (ECT) sensor
- Camshaft Position (CMP) sensor
- Knock Sensor (KS)
- ECT gauge unit
- Heated Oxygen Sensor (HO$_2$S)
- Starter and alternator
- Oil pressure switch

7. Remove or disconnect the following:
- Brake booster vacuum hose
- Power steering pump and A/C compressor drive belt
- Power steering pump and brace. Position the assembly aside, but do NOT disconnect the fluid line.
- A/C compressor, but do NOT disconnect the lines

➡ **Matchmark the installed position of the radiator hoses before disconnecting them.**

- Upper and lower radiator hoses
- Heater and purge hoses
- Fuel lines. Discard the O-rings
- Transaxle assembly. Do NOT remove the flywheel bolt indicated by the arrow in the accompanying illustration. Removal of this bolt will cause the flywheel to be out of balance.

8. Remove the engine mount insulator and bracket as follows:

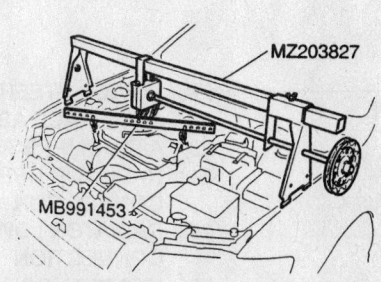

93570G02

These special tools are needed to support the engine while the transaxle is removed—Lancer with 2.0L engine

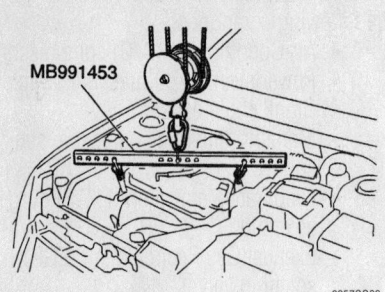

MB991453

9357QG03

View of the special tool needed to support the engine during mount removal—Lancer with 2.0L engine

a. Support the engine with a suitable floor jack.

b. Remove the special tools that were installed for transaxle removal.

c. Support the engine with special tool MB991453 attached to a chain block or engine hoist.

d. Place a jack under the oil pan with a block of wood in between to protect the pan. Jack up the engine to take the weight off the engine mount insulator and bracket, then remove the insulator and bracket.

9. Make sure that all cables, hoses and harnesses are disconnected from the engine, then use the engine hose to slowly lift the engine up and out of the engine compartment

To install:

10. Installation is the reverse of the removal procedure. Torque the engine mounting fasteners to the specifications shown in the accompanying illustration.

11. Fill the coolant system and engine crankcase.

12. Connect the negative battery cable.

13. Start the engine and check for leaks.

14. Install the hood.

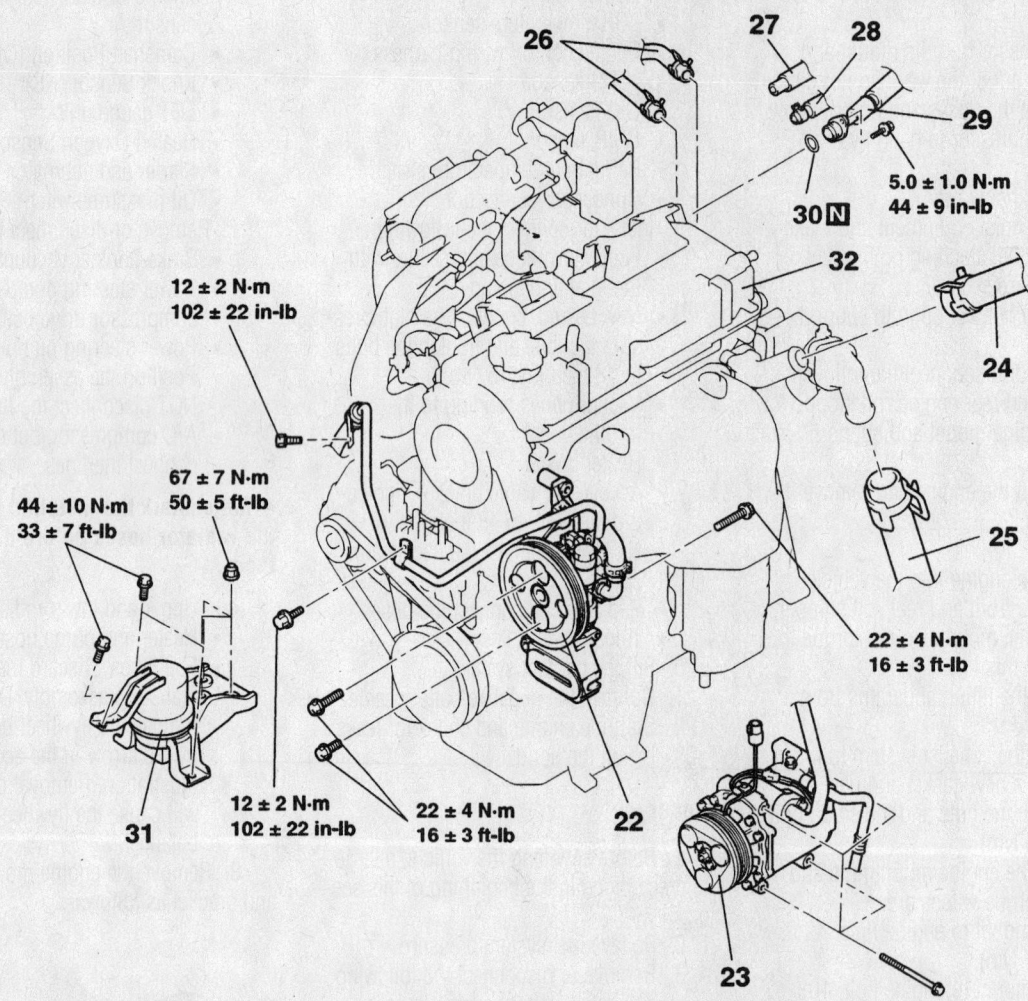

9357QG04

View of the engine mounts and related components—Lancer with 2.0L engine

22. POWER STEERING OIL PUMP AND BRACE ASSEMBLY
23. A/C COMPRESSOR
24. RADIATOR UPPER HOSE CONNECTION
25. RADIATOR LOWER HOSE CONNECTION
26. HEATER HOSE CONNECTION
27. PURGE HOSE CONNECTION
28. FUEL RETURN HOSE CONNECTION
29. HIGH-PRESSURE FUEL HOSE CONNECTION
30. O-RING
31. ENGINE MOUNT INSULATOR AND BRACKET ASSEMBLY
32. ENGINE ASSEMBLY

Mirage

1. Before servicing the vehicle, refer to the precautions in the beginning of this section.
2. Relieve fuel system pressure.
3. Remove or disconnect the following:
   - Negative battery cable
   - Undercover, if equipped
   - Hood assembly
   - Air cleaner assembly and all adjoining air intake duct work
4. Drain the engine coolant.
5. Remove or disconnect the following:
   - Radiator assembly and coolant reservoir
   - Transaxle assembly
   - Ground cable, accelerator cable, breather hose and heater hose connections from the engine
6. Note locations and remove vacuum hoses from engine.
7. Remove or disconnect the following:
   - Fuel feed and return hoses
   - Crankshaft Position (CKP) and Camshaft Position (CMP) sensor wiring
   - Heated Oxygen (HO$_2$S sensor), Engine Coolant (ECT) gauge and ECT sensor connections
   - Oil pressure switch
   - Thermo switch, with automatic transmissions
   - Harness connections for the Idle Speed Control (ISC) motor and Throttle Position (TP) sensor
   - Intake Air Temperature (IAT) sensor
   - Exhaust Gas Recirculation (EGR) temperature sensor (California)
   - Injector harness plugs
   - Power transistor and the ignition coil connections
   - Alternator and power steering switch wiring
   - Air conditioning compressor and hang it out of the way. Do NOT allow the compressor to hang by the hoses.
   - Power steering pump and hang it out of the way—Do not allow the pump to hang by the hoses.
   - Starter and alternator harness clamp, for 1.8L engines
   - Exhaust manifold to head pipe nuts
8. Attach a hoist to the engine and support the engine weight. Remove the engine mount bracket. Remove any torque control brackets (roll stoppers).
9. Remove the engine assembly from the vehicle.
   To install:
10. Install the engine and secure in position. The front lower mount through-bolt nut

should not be tightened until the full weight of the engine is on the mount. Tighten through-bolt to 72 ft. lbs. (100 Nm) and bracket mounting bolts to 42 ft. lbs. (58 Nm). Tighten bracket mounting nut to 38 ft. lbs. (53 Nm).
11. Using a new gasket, position exhaust pipe onto the manifold and tighten the flange nuts to 36 ft. lbs. (50 Nm).
12. Install or connect the following:
   - Power steering pump, alternator and air conditioning compressor
   - Accessory drive belts
   - Alternator and power steering wiring
   - Alternator and starter harness clamp for 1.8L engines
   - Ignition coil and power transistor connections
   - Fuel injector harness connections
   - EGR temperature sensor plug–California models
   - IAT sensor
   - IAC and TPS connectors
   - Thermo switch, automatic transmission
   - Oil pressure switch wiring
   - HO$_2$S sensor, ECT gauge and ECT sensor
   - CKP and CMP sensors
   - Fuel feed hose and tighten bolts to 44 inch lbs. (5 Nm), using new O-rings
   - Fuel return hose, using a new hose clamp
   - Vacuum hoses and the brake booster vacuum supply
   - Breather hose, heater hoses, accelerator cable and ground cables. Inspect accelerator cable for proper adjustment.
   - Transaxle assembly
   - Radiator assembly and refill the cooling system
   - Air cleaner and hood assembly
   - Negative battery cable

Water Pump

REMOVAL & INSTALLATION

Diamante

1. Before servicing the vehicle, refer to the precautions in the beginning of this section.
2. Drain the cooling system.
3. Remove or disconnect the following:
   - Negative battery cable
   - Engine undercover
   - Clamp bolt from the power steering hose

4. Support the engine with the appropriate equipment and remove the engine mount bracket.
5. Remove or disconnect the following:
   - Timing belt
   - Coolant hoses from the pump, if equipped
   - Alternator brace

➡ **The water pump bolts are different in size. Note their locations for installation.**

6. Remove the water pump, gasket and O-ring where the water inlet pipe joins the pump.
   To install:
7. Thoroughly clean both gasket surfaces of the water pump and block.
8. Install or connect the following:
   - New O-ring into the groove on the front end of the water inlet pipe. Do not apply oils or grease to the O-ring. Wet with water only.
   - Water pump assembly to the engine block, with new gasket. Torque the mounting bolts to 17 ft. lbs. (24 Nm)
   - Hoses to the pump
   - Timing belt
   - Engine drive belts
9. Fill the system with coolant.
10. Connect the negative battery cable, run the vehicle until the thermostat opens and fill the radiator completely.
11. Once the vehicle has cooled, recheck the coolant level.

Eclipse

1. Before servicing the vehicle, refer to the precautions in the beginning of this section.
2. Disconnect the negative battery cable.
3. Drain the engine coolant.
4. Remove or disconnect the following:
   - Timing belt
   - Alternator brace from the water pump
   - Timing belt rear cover
   - Water pump mounting bolts
   - Water pump, gasket and O-ring
   To install:
5. Install or connect the following:
   - New O-ring on the water inlet pipe. Coat the O-ring with water or coolant. Do not allow oil or other grease to contact the O-ring.
   - Water pump to the engine block, with new gasket. Torque the mounting bolts to 10 ft. lbs. (13 Nm)
   - Alternator brace on the water pump. Torque the brace pivot bolt to 17 ft. lbs. (24 Nm).

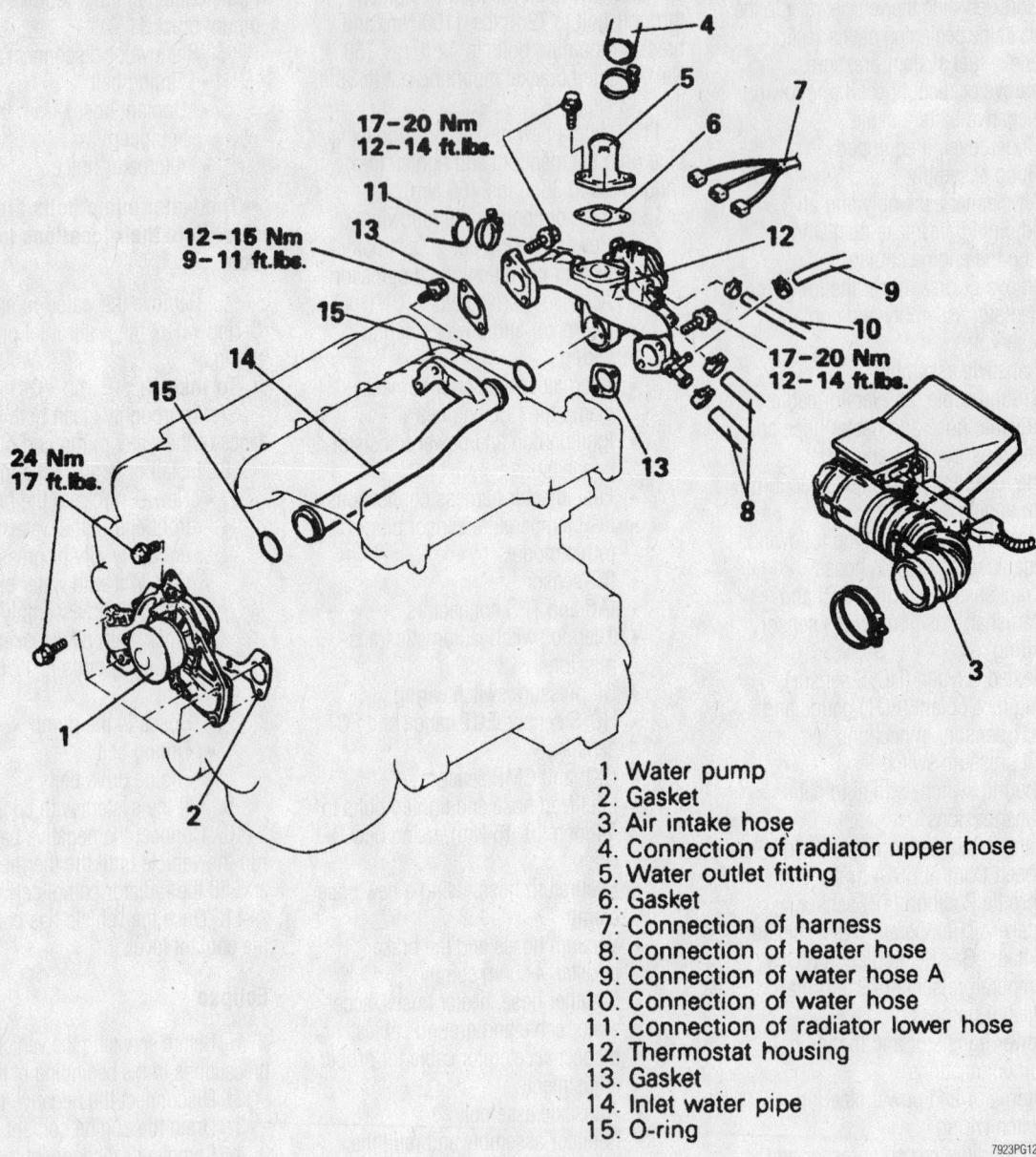

17–20 Nm
12–14 ft.lbs.

12–15 Nm
9–11 ft.lbs.

17–20 Nm
12–14 ft.lbs.

24 Nm
17 ft.lbs.

1. Water pump
2. Gasket
3. Air intake hose
4. Connection of radiator upper hose
5. Water outlet fitting
6. Gasket
7. Connection of harness
8. Connection of heater hose
9. Connection of water hose A
10. Connection of water hose
11. Connection of radiator lower hose
12. Thermostat housing
13. Gasket
14. Inlet water pipe
15. O-ring

7923PG12

Water pump and related components—DOHC Diamante

- Timing belt rear cover
- Timing belt
- Remaining components
6. Refill the engine with coolant.
7. Connect the negative battery cable, start the engine and check for leaks.

Galant

1. Before servicing the vehicle, refer to the precautions in the beginning of this section.
2. Disconnect the negative battery cable.
3. Drain the cooling system.

4. Remove or disconnect the following:
- Engine undercover
- Clamp bolt from the power steering hose
5. Support the engine with the appropriate equipment and remove the engine mount bracket.
6. Remove or disconnect the following:
- Engine drive belts and the air conditioning tensioner bracket
- Timing belt covers from the front of the engine
- Camshaft and silent shaft timing belts
- Alternator brace

- Water pump, gasket and O-ring where the water inlet pipe(s) joins the pump
To install:
7. Thoroughly clean both gasket surfaces of the water pump and block.
8. Install a new O-ring into the groove on the front end of the water inlet pipe and wet with clean antifreeze only. Do not apply oils or grease to the O-ring.
9. Using a new gasket, install the water pump assembly. Tighten bolts with the head mark **4** to 10 ft. lbs. (14 Nm) and bolts with the head mark **7** to 18 ft. lbs. (24 Nm).
10. Install or connect the following:

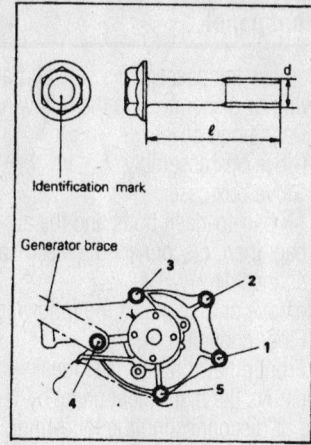

Water pump bolt identification—Galant

No.	Identification mark	Bolt diameter (d) x length (ℓ) mm (in.)	Torque Nm (ft.lbs.)
1	4	8 x 14 (.31 x .55)	
2	4	8 x 22 (.31 x .87)	12–15 (9–10)
3	4	8 x 30 (.31 x 1.18)	
4	7	8 x 65 (.31 x 2.56)	20–27 (15–19)
5	4	8 x 28 (.31 x 1.10)	12–15 (9–10)

7923PG11

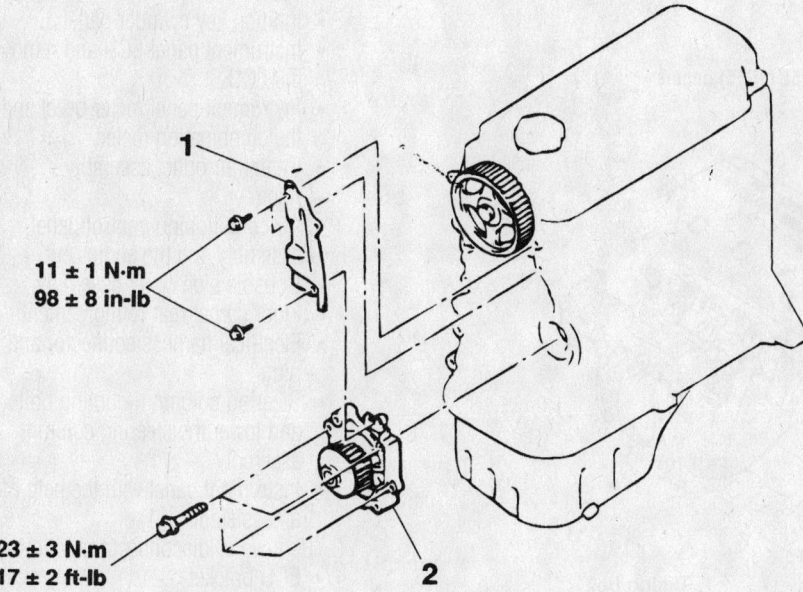

11 ± 1 N·m
98 ± 8 in-lb

23 ± 3 N·m
17 ± 2 ft-lb

REMOVAL STEPS
1. TIMING BELT REAR UPPER COVER CONNECTION
2. WATER PUMP

9357QG05

Exploded view of the water pump—Lancer with 2.0L engine

- Timing belts
- Engine drive belts
- Engine mount bracket
- Engine undercover
11. Fill the system with coolant.
12. Connect the negative battery cable, run the vehicle until the thermostat opens and fill the radiator completely.
13. Once the vehicle has cooled, recheck the coolant level.

Lancer

1. Before servicing the vehicle, refer to precautions in the beginning of this section.
2. Disconnect the negative battery cable.
3. Drain the cooling system.
4. Remove or disconnect the following:
- Timing belt
- Timing belt rear upper cover connection
- Water pump bolts(s) and pump

To install:
5. Install or connect the following:
- Water pump. Tighten the mounting bolt(s) to 15–19 ft. lbs. (20–26 Nm).
- Timing belt rear upper cover connection and tighten the bolts to 89–106 inch lbs. (10–12 Nm)
- Timing belt
6. Refill the cooling system and connect the negative battery cable.

Mirage

1. Before servicing the vehicle, refer to the precautions in the beginning of this section.
2. Disconnect the negative battery cable.
3. Drain the cooling system.
4. Remove or disconnect the following:
- Engine undercover
- Clamp bolt from the power steering hose
- Engine drive belts
5. Support the engine with the appropriate equipment and remove the engine mount bracket.
6. Remove or disconnect the following:
- Timing belt
- Power steering pump bracket
- Alternator brace

➡ **The water pump mounting bolts are different in length, note their positioning for reassembly.**

7. Remove the water pump, gasket and O-ring where the water inlet pipe(s) joins the pump.

To install:
8. Thoroughly clean both gasket surfaces of the water pump and block.
9. For 1.5L engines, install a new O-ring into the groove on the front end of the water inlet pipe. Do not apply oils or grease to the O-ring. Wet the O-ring with water only.
10. For 1.8L engines, apply a 0.09–0.12 in. (2.5–3.0mm) continuous bead of sealant to water pump and install the pump assembly. Install the water pump within 15 minutes of the application of the sealant. Wait 1 hour after installation of the water pump to refill the cooling system or starting the engine.
11. Install or connect the following:
- Gasket and pump assembly and tighten the bolts to 17 ft. lbs. (24 Nm)
- Remaining components in the reverse order of removal
12. Fill the system with coolant.

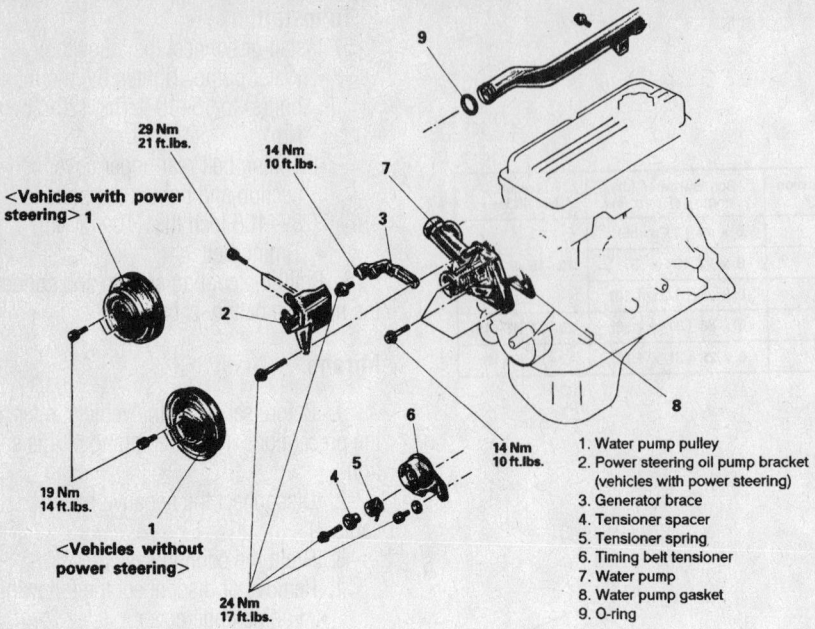

29 Nm
21 ft.lbs.

14 Nm
10 ft.lbs.

<Vehicles with power steering> 1

19 Nm
14 ft.lbs.

<Vehicles without power steering>

14 Nm
10 ft.lbs.

24 Nm
17 ft.lbs.

1. Water pump pulley
2. Power steering oil pump bracket (vehicles with power steering)
3. Generator brace
4. Tensioner spacer
5. Tensioner spring
6. Timing belt tensioner
7. Water pump
8. Water pump gasket
9. O-ring

7923PG07

Water pump and related components—Mirage with 1.5L (4G15) engine

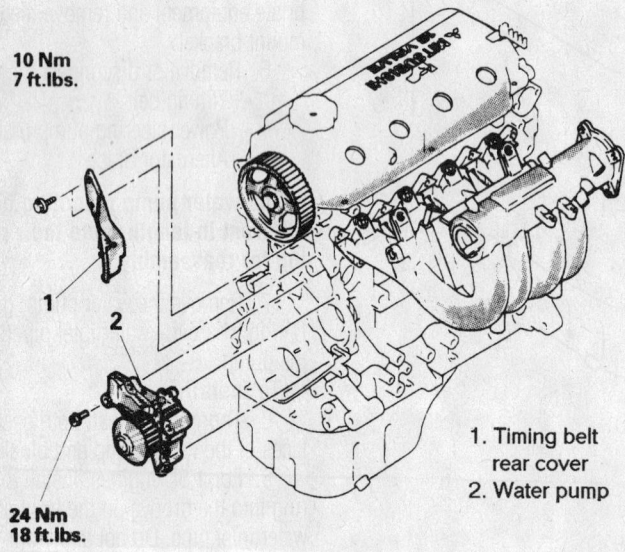

10 Nm
7 ft.lbs.

24 Nm
18 ft.lbs.

1. Timing belt rear cover
2. Water pump

7923PG08

Water pump and related components—Mirage with 1.8L (4G93) engines

13. Connect the negative battery cable, run the vehicle until the thermostat opens and fill the radiator completely.

14. Once the vehicle has cooled, recheck the coolant level.

Heater Core

REMOVAL & INSTALLATION

Diamante

1. Before servicing the vehicle, refer to the precautions in the beginning of this section.

2. Disconnect the negative battery cable.

3. Drain the cooling system into a clean container for reuse.

4. Discharge and recover the air conditioning system refrigerant.

5. Remove or disconnect the following:
- Heater hoses from the heater core
- Refrigerant lines from the evaporator core and discard the O-rings

✻✻ CAUTION

After disconnecting the negative battery cable, wait at least 60 seconds

before working on the SRS module or instrument panel.

6. Remove the passenger's side air bag by removing or disconnecting the following:
- Dash undercover
- Glove box assembly
- Glove box case
- Air bag-to-dash bolts and the air bag; then, disconnect the electrical connector

7. Remove or disconnect the following:
- Floor console
- Front pillar trim at both sides

8. Remove the instrument panel by removing or disconnecting the following:
- Steering column covers
- Hood lock release handle
- Parking brake release handle
- Lower left side instrument panel cover
- Ignition key cylinder panel
- Instrument panel ECU and remove the ECU
- Instrument panel meter bezel and the combination meter
- Center air outlet assembly
- Ashtray
- Air conditioning control panel assembly and the audio unit
- Console side cover assembly
- Floor carpet rear reinforcement
- Electrical harness connector and plug
- Steering column mounting bolts and lower the steering column assembly
- Instrument panel with the help of an assistant

9. Remove or disconnect the following:
- ECU bracket
- Center stay assembly
- Heater hose connection and the center duct assembly
- Foot distribution duct and the breather hose
- Refrigerant lines from the evaporator and discard the O-rings
- Air conditioning housing drain hose and remove the evaporator housing
- Heater housing unit
- Heater core support and the heater core

To install:

10. Install or connect the following:
- Heater core support and the heater core
- Heater housing unit
- Air conditioning housing drain hose and Install the evaporator housing

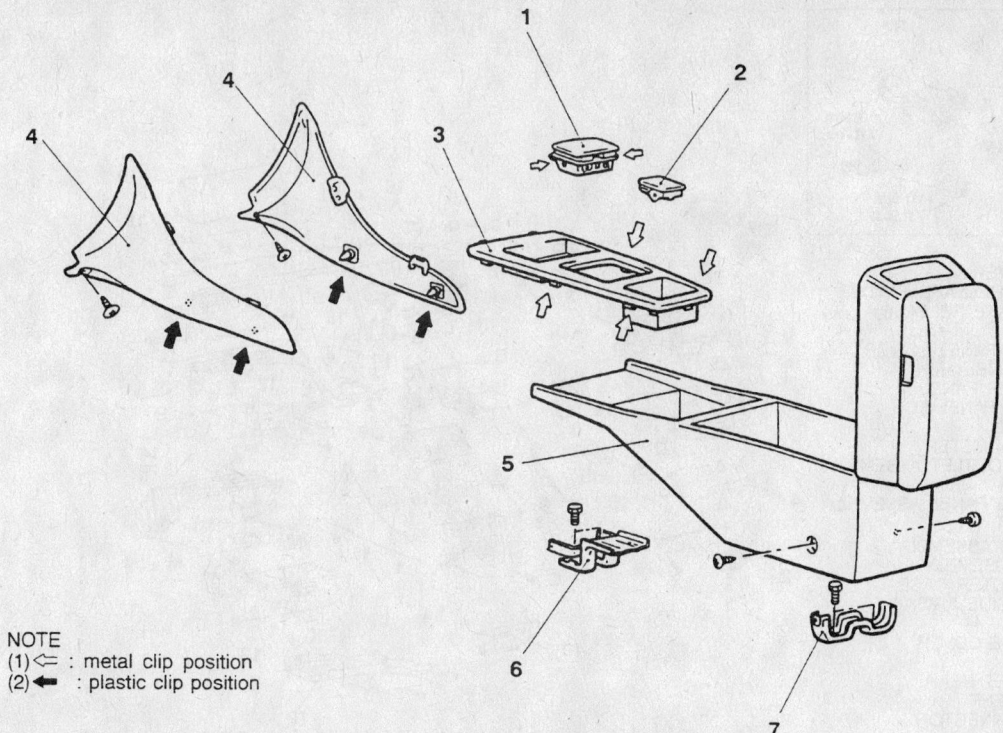

1. CUP HOLDER ASSEMBLY
2. COIN BOX ASSEMBLY
3. FLOOR CONSOLE PANEL
4. CONSOLE SIDE COVER ASSEMBLY
5. FLOOR CONSOLE BOX
6. CONSOLE BRACKET A
7. CONSOLE BRACKET C

NOTE
(1) ⇐ : metal clip position
(2) ⬅ : plastic clip position

Exploded view of the floor console and related components—Diamante

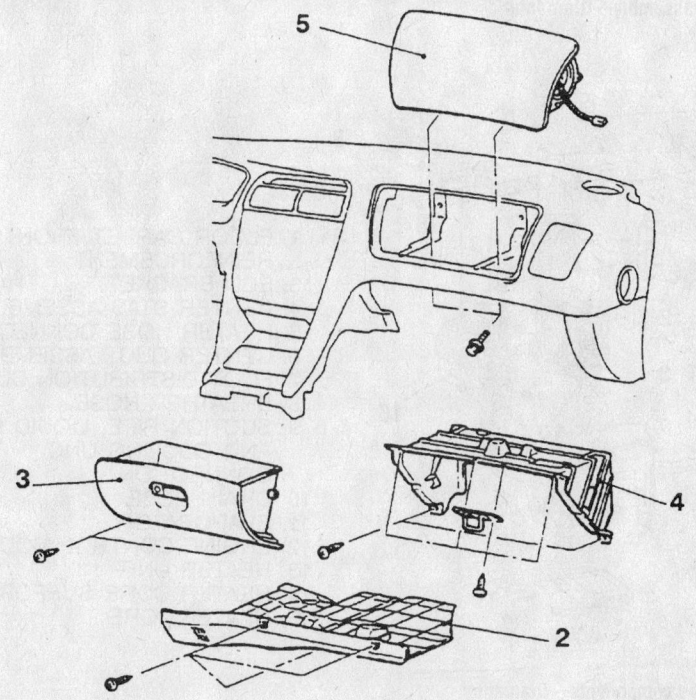

2. UNDERCOVER
3. GLOVE BOX ASSEMBLY
4. GLOVE BOX CASE
5. AIR BAG MODULE

Exploded view of the passenger's air bag module—Diamante

- Refrigerant lines to the evaporator using new O-rings
- Foot distribution duct and the breather hose
- Heater hose connection and the center duct assembly
- Center stay assembly
- ECU bracket

11. Install the instrument panel by installing or connecting the following:
- Instrument panel with the help of an assistant
- Steering column assembly and the steering column mounting bolts. Torque the bolts to 84 inch lbs. (10 Nm)
- Electrical harness connector and plug
- Floor carpet rear reinforcement
- Console side cover assembly
- Air conditioning control panel assembly and the audio unit
- Ashtray
- Center air outlet assembly
- Instrument panel meter bezel and the combination meter
- Instrument panel ECU and connect the ECU electrical connector
- Ignition key cylinder panel
- Lower left side instrument panel cover
- Parking brake release handle
- Hood lock release handle
- Steering column covers

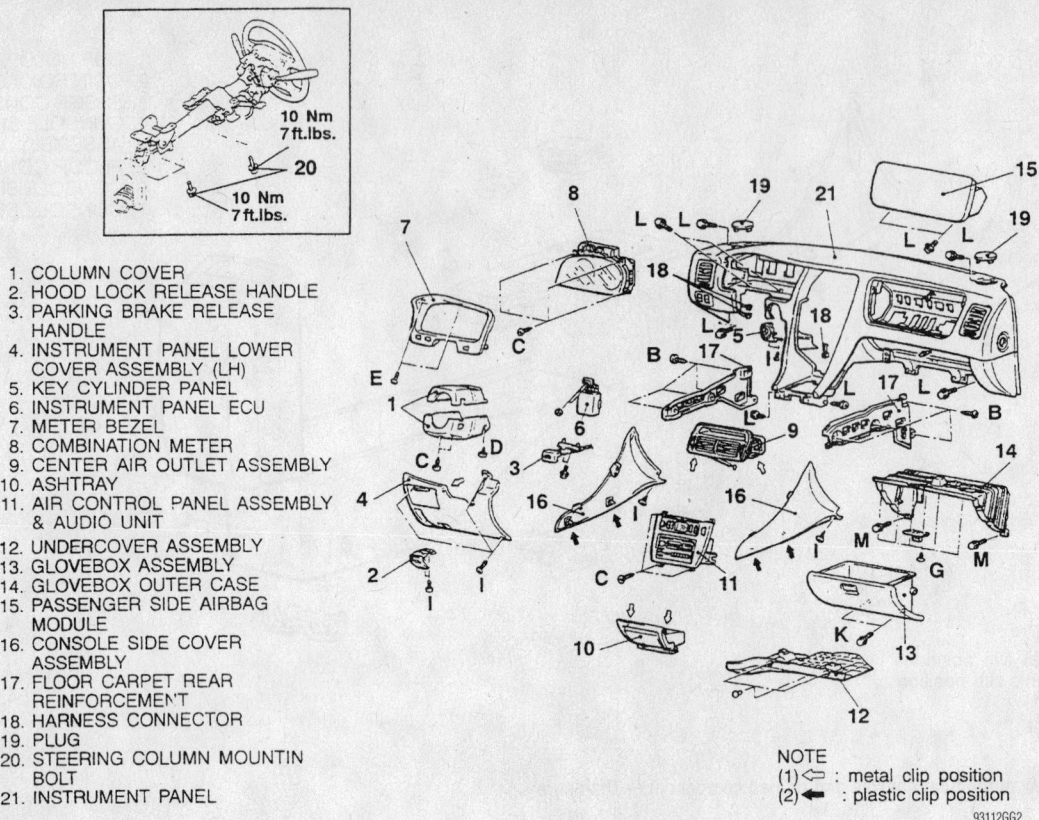

1. COLUMN COVER
2. HOOD LOCK RELEASE HANDLE
3. PARKING BRAKE RELEASE HANDLE
4. INSTRUMENT PANEL LOWER COVER ASSEMBLY (LH)
5. KEY CYLINDER PANEL
6. INSTRUMENT PANEL ECU
7. METER BEZEL
8. COMBINATION METER
9. CENTER AIR OUTLET ASSEMBLY
10. ASHTRAY
11. AIR CONTROL PANEL ASSEMBLY & AUDIO UNIT
12. UNDERCOVER ASSEMBLY
13. GLOVEBOX ASSEMBLY
14. GLOVEBOX OUTER CASE
15. PASSENGER SIDE AIRBAG MODULE
16. CONSOLE SIDE COVER ASSEMBLY
17. FLOOR CARPET REAR REINFORCEMEN'T
18. HARNESS CONNECTOR
19. PLUG
20. STEERING COLUMN MOUNTIN BOLT
21. INSTRUMENT PANEL

NOTE
(1)⇐ : metal clip position
(2)◀ : plastic clip position

93112GG2

Exploded view of the instrument panel and steering column assembly—Diamante

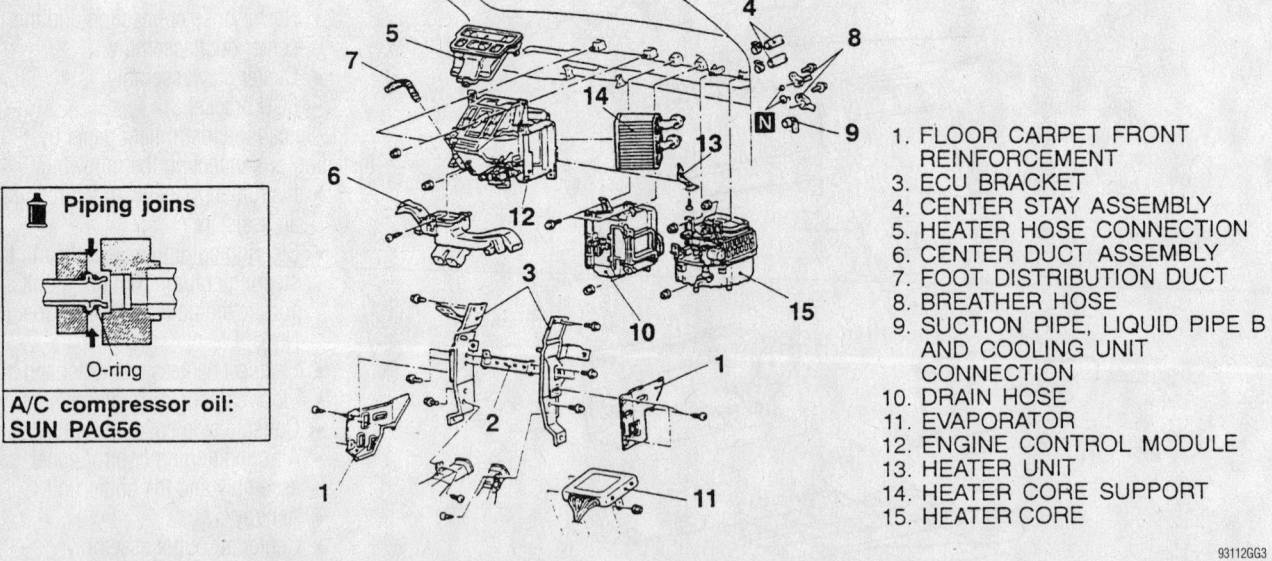

Piping joins

O-ring

A/C compressor oil: SUN PAG56

1. FLOOR CARPET FRONT REINFORCEMENT
3. ECU BRACKET
4. CENTER STAY ASSEMBLY
5. HEATER HOSE CONNECTION
6. CENTER DUCT ASSEMBLY
7. FOOT DISTRIBUTION DUCT
8. BREATHER HOSE
9. SUCTION PIPE, LIQUID PIPE B AND COOLING UNIT CONNECTION
10. DRAIN HOSE
11. EVAPORATOR
12. ENGINE CONTROL MODULE
13. HEATER UNIT
14. HEATER CORE SUPPORT
15. HEATER CORE

93112GG3

Exploded view of the heater core, heater housing and related components—Diamante

12. Front pillar trim at both sides
13. Floor console
14. Install the passenger's side air bag by installing or connecting the following:
- Air bag-to-dash bolts and the air bag; then, connect the electrical connector
- Glove box case
- Glove box assembly
- Dash undercover
- Refrigerant lines to the evaporator core using new O-rings
- Heater hoses to the heater core
15. Refill the cooling system.

16. Connect the negative battery cable.
17. Evacuate, charge and leak test the air conditioning system.
18. Operate the engine to normal operating temperatures; then, check the climate control operation and check for leaks.

Eclipse

✳✳ CAUTION

If equipped with an air bag, wait for 1 minute after disconnecting the negative battery cable before working inside the vehicle. The air bag system is set to deploy for a short period of time after the battery is disconnected.

1. Before servicing the vehicle, refer to the precautions in the beginning of this section.

2. Disconnect the negative battery cable.

3. Drain the cooling system into a clean container for reuse.

4. Disconnect the heater hoses from the heater core tubes at the firewall. Do not allow the coolant to damage the vehicle speed sensor located below the heater hoses on the non-turbo manual transmission vehicles.

✳✳ WARNING

To prevent damage to the air bag control unit during removal or installation of the floor console, avoid shocks or impact. Do not drop.

5. Remove the floor console by removing or disconnecting the following:
   • Center console trim panel
   • Ashtray and cup holder assembly
   • Shift lever knob on manual transmission
   • Retaining screws
   • Floor console assembly

6. Locate the rectangular plugs in the knee protector on either side of the steering column. Pry these plugs out and remove the screws.

7. Remove or disconnect the following:
   • Driver's side air bag assembly, the steering wheel and the passenger's side air bag assembly
   • Lap cooler duct and steering column covers
   • Instrument cluster bezel and then the instrument cluster
   • Radio
   • Glove box
   • Center air outlet assembly
   • Hood release handle and the lower cover
   • Heater control assembly
   • Front speakers and the instrument panel switch
   • Steering shaft support bolts and lower the steering column

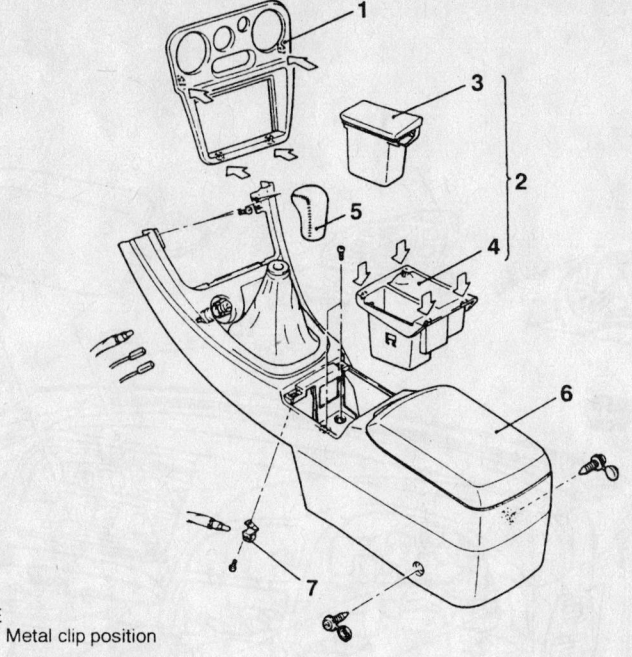

NOTE
⟵ : Metal clip position

1. Center console panel
2. Ashtray and cupholder assembly
3. Ashtray
4. Cup holder
5. Shift lever knob <M/T>
6. Floor console assembly
7. Ashtray illumination light bracket

93112G56

Exploded view of the floor console and related components—Eclipse

   • Instrument panel mounting hardware and remove the instrument panel from the vehicle
   • Stamped steel center reinforcement
   • Lower ductwork from the heater box
   • Evaporator case mounting bolt and nut to allow clearance for the heater unit removal
   • Heater unit
   • Heater core from the heater unit

To install:

8. Install or connect the following:
   • Heater core to the heater unit
   • Heater unit
   • Evaporator case mounting bolt and nut
   • Lower ductwork to the heater box
   • Stamped steel center reinforcement
   • Instrument panel and the instrument panel mounting hardware
   • Steering column and the steering shaft support bolts
   • Front speakers and the instrument panel switch
   • Heater control assembly
   • Hood release handle and the lower cover
   • Center air outlet assembly
   • Glove box
   • Radio
   • Instrument cluster and the instrument cluster bezel
   • Steering column covers and the lap cooler duct
   • Steering wheel, the driver's side air bag assembly and the passenger's side air bag assembly
   • Screws and the rectangular plugs in the knee protector on either side of the steering column

9. Install the floor console by installing or connecting the following:
   • Floor console assembly
   • Retaining screws
   • Shift lever knob on manual transmission
   • Ashtray and cup holder assembly
   • Center console trim panel
   • Center console trim panel

10. Connect the heater hoses to the heater core tubes at the firewall.

11. Refill the cooling system.

12. Connect the negative battery cable.

13. Operate the engine to normal operating temperatures; then, check the climate control operation and check for leaks.

Galant

1. Before servicing the vehicle, refer to the precautions in the beginning of this section.

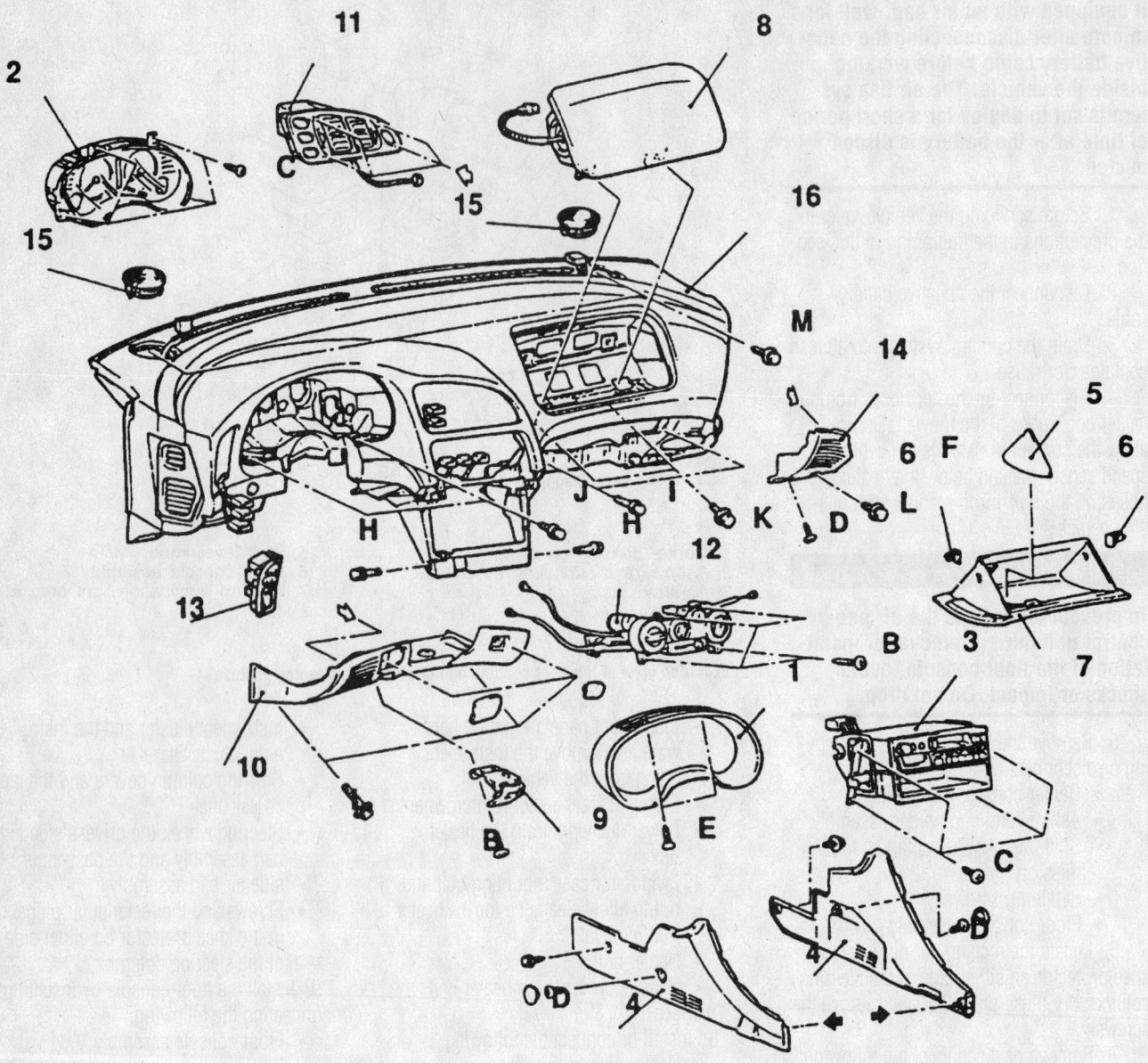

1. Meter bezel
2. Combination meter
3. Radio and tape player, and box
4. Console side cover
5. Sunglasses holder
6. Stopper
7. Glove box
8. Passenger's side air bag module assembly
9. Hood lock release handle
10. Instrument under cover L.H.
11. Center air outlet assembly
12. Heater control assembly
13. Instrument panel switch
14. Instrument under cover R.H.
15. Front speaker
16. Instrument panel assembly

93112G74

Exploded view of the instrument panel and related components—Eclipse

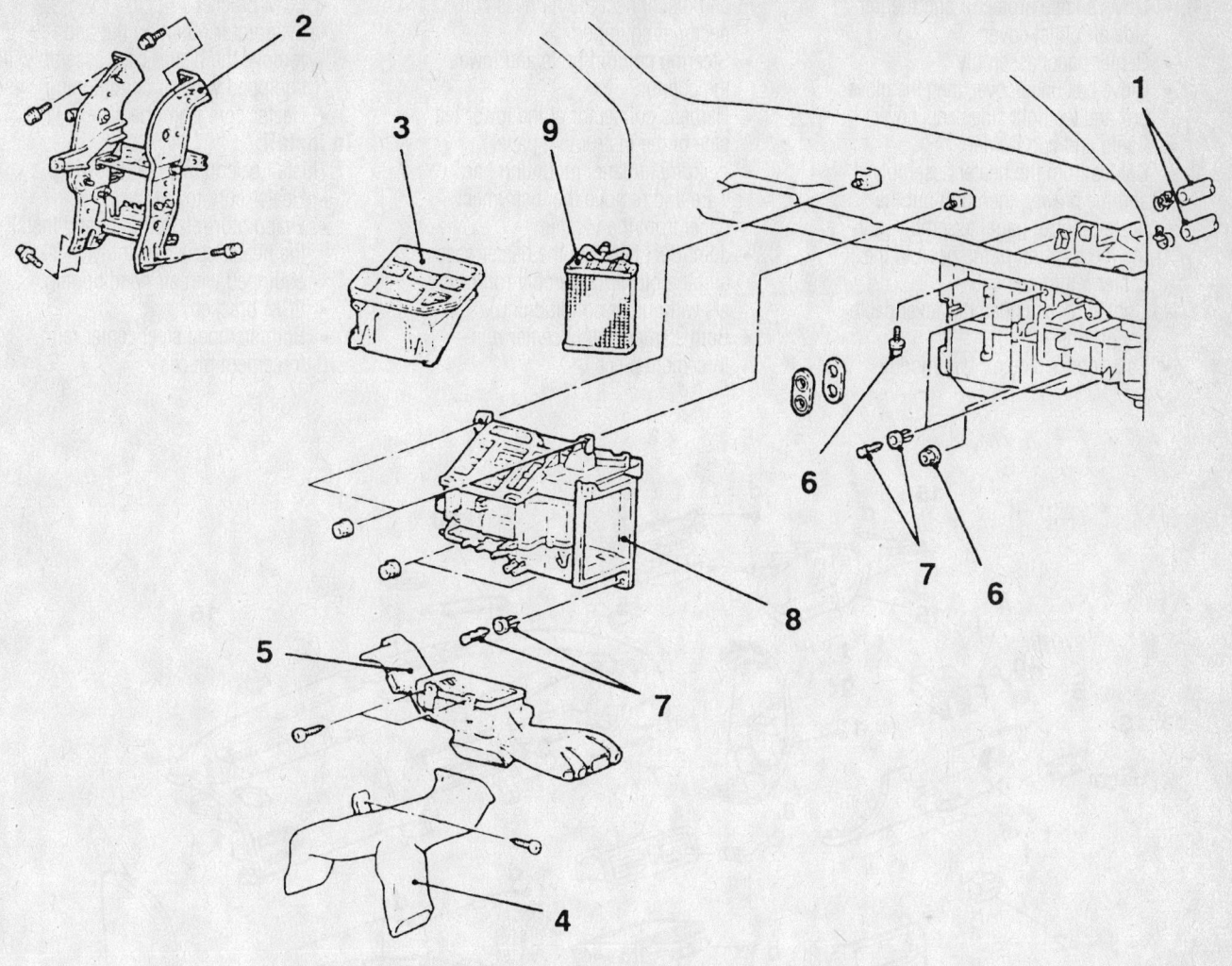

1. Heater hose connection
2. Center stay
3. Center duct
4. Semi rear heater duct
5. Foot distribution duct
6. Cooling unit installation bolt and nut
7. Clip
8. Heater unit
9. Heater core

93112GE0

Exploded view of the heater core, heater case and related components—Eclipse

2. Disconnect the negative battery cable.

❊❊ CAUTION

After disconnecting the negative battery cable, wait at least 60 seconds before working on the SRS module or instrument panel.

3. Drain the cooling system into a clean container for reuse.

4. Disconnect the heater hoses from the heater core at the firewall.

❊❊ WARNING

To prevent damage to the air bag control unit during removal or installation of the floor console, avoid shocks or impact. Do not drop.

5. Remove the floor console by removing or disconnecting the following:
- Shift lever knob on manual transmission vehicles or the shift indicator plate on automatic transmissions

- Coin holder behind the shifter, then the center console trim cover in front of the shifter
- Center console retaining bolt cover plugs, then remove the bolts
- Console assembly, then the brackets
6. Remove or disconnect the following:
- Steering column covers
- Instrument cluster bezel and then the instrument cluster
- Instrument panel switch, hood lock release handle and the lower duct work

- Driver's knee protector and the left side air outlet cover
- Center panel assembly
- Glove box undercover, then the glove box and the right side panel cover
- Radio and cup holder
- Cables from the heater assembly and the blower, then pull out the heater control panel assembly, noting the location of the boss in the center reinforcement
- Cool air bypass damper lever cable connection
- Passenger's side air bag module

and disconnect the harness connector, if equipped
- Steering column bolts and lower the column
- Harness connector at the lower left side of the instrument panel
- Instrument panel mounting hardware and remove the instrument panel from the vehicle
- Joint duct between the heater case and the blower assembly (on models without air conditioning)
- Both stamped steel center reinforcement piece

- ECM bracket
- Evaporator retaining nut and remove the heater case assembly, if equipped with air conditioning
- Heater core from the case

To install:
7. Install or connect the following:
- Heater core to the case
- Evaporator retaining nut and Install the heater case assembly, if equipped with air conditioning
- ECM bracket
- Both stamped steel center reinforcement pieces

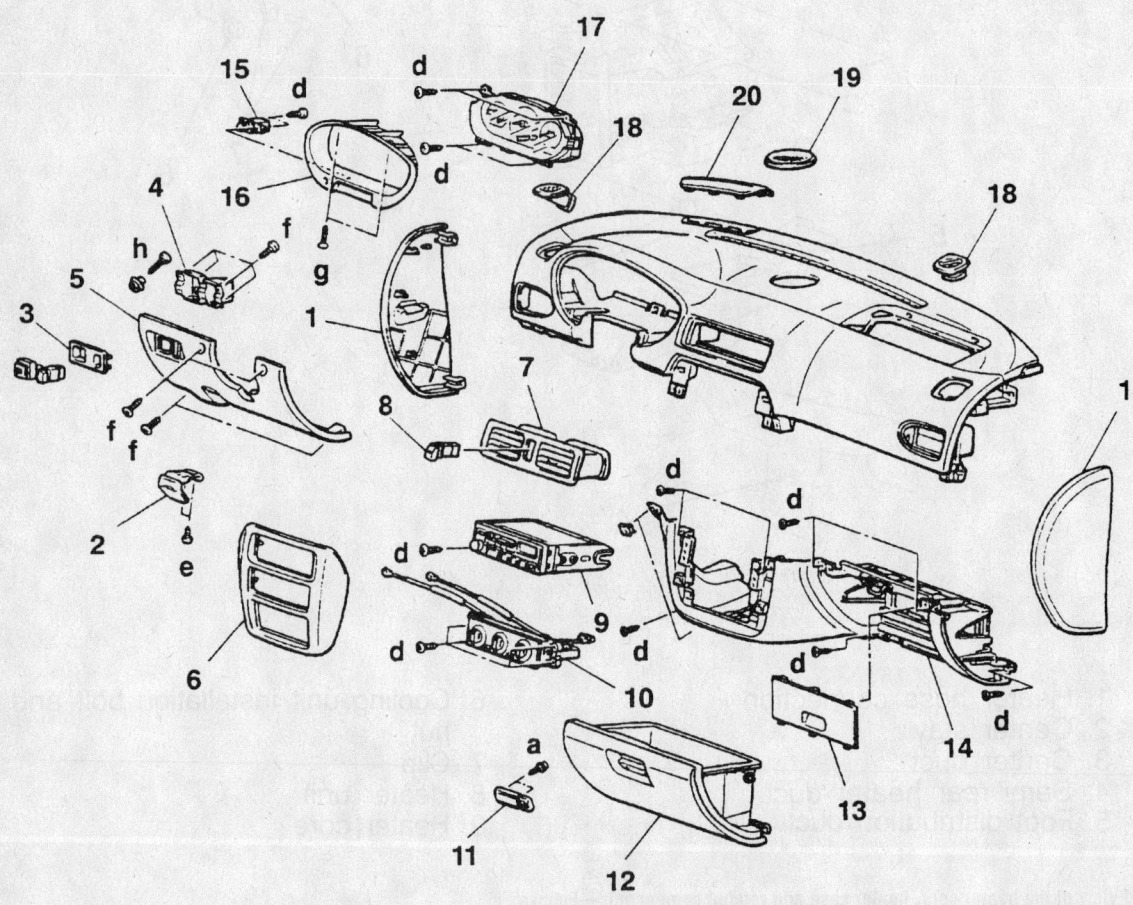

1. INSTRUMENT PANEL SIDE COVER
2. HOOD LOCK RELEASE HANDLE
3. SWITCH PANEL ASSEMBLY
4. CONNECTOR HOLDER
5. FRONT DRIVER'S SIDE UNDER COVER
6. CENTER PANEL ASSEMBLY
7. CENTER AIR OUTLET ASSEMBLY
8. HAZARD WARNING LIGHT SWITCH
9. RADIO AND TAPE PLAYER
10. HEATER CONTROL ASSEMBLY
11. GLOVE BOX STRIKER
12. GLOVE BOX
13. FRONT PASSENGER'S UNDER COVER PLUG
14. FRONT PASSENGER'S SIDE UNDER COVER
15. RHEOSTAT
16. METER BEZEL
17. COMBINATION METER
18. SIDE DEFROSTER GRILLE
19. SPEAKER GRILLE
20. INSTRUMENT PANEL UPPER PLUG

93112GF1

Exploded view of the instrument panel assembly—Galant

- Joint duct between the heater case and the blower assembly (on models without air conditioning)
- Instrument panel and install the instrument panel mounting hardware
- Harness connector at the lower left side of the instrument panel
- Steering column bolts
- Passenger's side air bag module and connect the harness connector, if equipped
- Cool air bypass damper lever cable connection

- Cables to the heater assembly and the blower and install the heater control panel assembly
- Radio and cup holder
- Glove box undercover, then the glove box and the right side panel cover
- Center panel assembly
- Left side air outlet cover and the driver's knee protector
- Lower duct work, the instrument panel switch and hood lock release handle
- Instrument cluster and the instrument cluster bezel

- Steering column covers
8. Install the floor console installing or connecting the following:
- Console assembly brackets and the console
- Center console retaining bolts, then the bolt cover plugs
- Coin holder behind the shifter, then the center console trim cover in front of the shifter
- Shift lever knob on manual transmission vehicles or the shift indicator plate on automatic transmissions

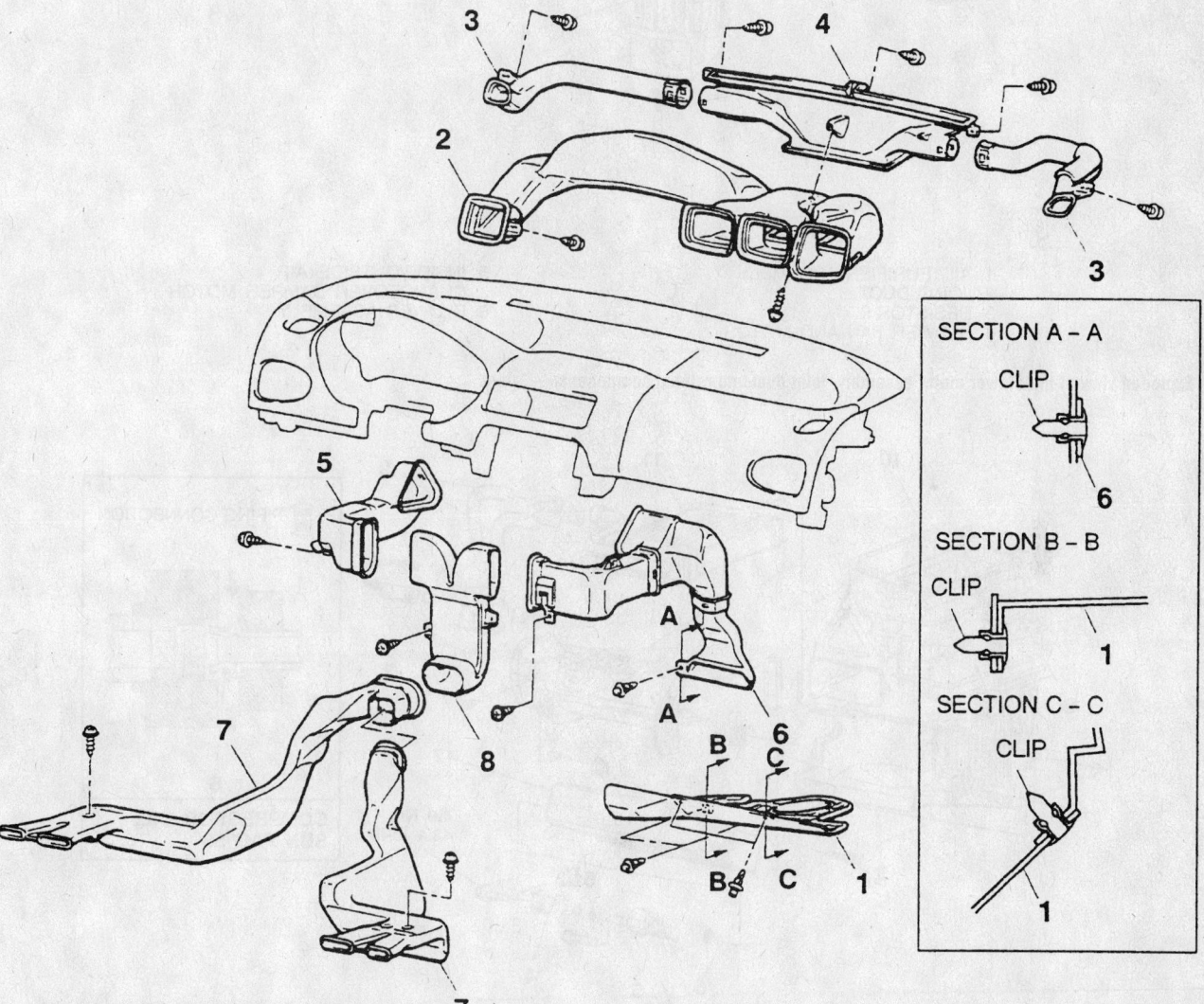

SECTION A – A
CLIP
6

SECTION B – B
CLIP
1

SECTION C – C
CLIP
1

1. UNDER COVER
2. DISTRIBUTION DUCT
3. SIDE DEFROSTER DUCT
4. DEFROSTER NOZZLE ASSEMBLY

5. FOOT DUCT (LH)
6. FOOT DUCT (RH)
7. REAR HEATER DUCT
8. FOOT CENTER DUCT

93112GF2

Exploded view of the ventilator assembly—Galant

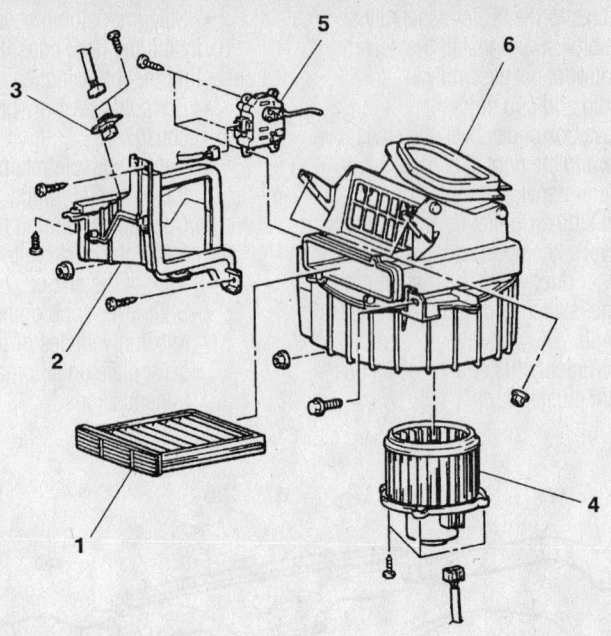

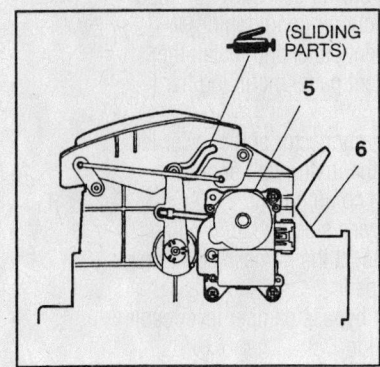

1. AIR PURIFIER ASSEMBLY
2. JOINT DUCT
3. RESISTOR
4. BLOWER FAN AND MOTOR

5. INSIDE/OUTSIDE AIR
   CHANGEOVER DAMPER MOTOR
6. BLOWER ASSEMBLY

93112GF3

Exploded view of the blower motor assembly, joint duct and related components—Galant

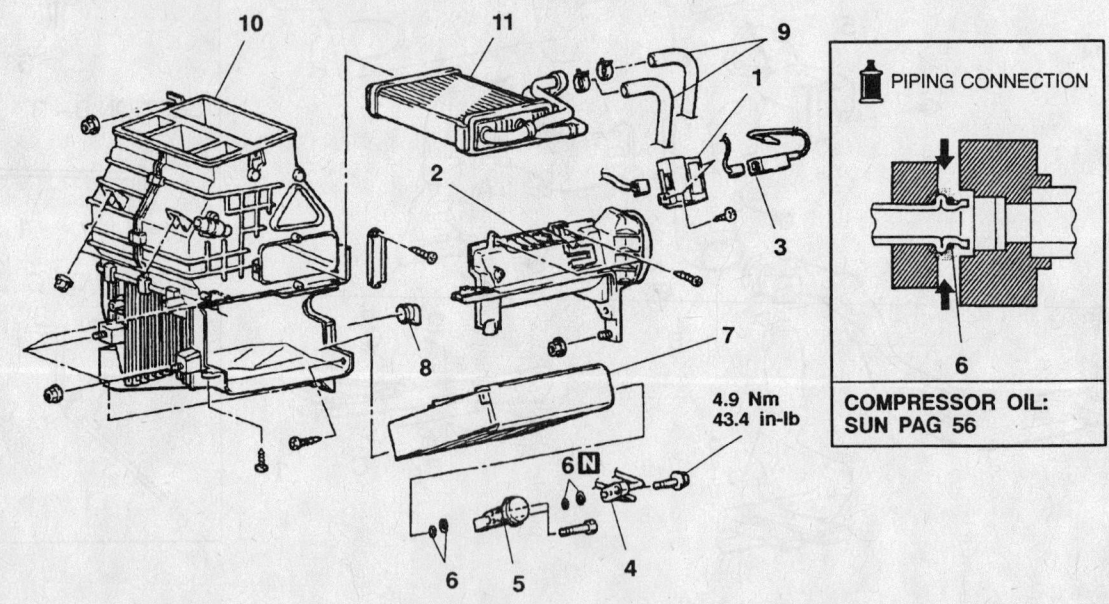

1. BELT LOCK CONTROLLER
2. COVER
3. AUTOMATIC COMPRESSOR
   CONTROLLER
4. A/C PIPE
5. EXPANSION VALVE

6. O-RING
7. EVAPORATOR
8. DRAIN HOSE
9. HEATER HOSE
10. HEATER/COOLER UNIT
11. HEATER CORE

93112GF4

Exploded view of the heater core, heater housing and related components—Galant

- Heater hoses to the heater core at the firewall
9. Refill the cooling system.
10. Connect the negative battery cable.
11. Operate the engine to normal operating temperatures; then, check the climate control operation and check for leaks.

Lancer

1. Before servicing the vehicle, refer to the precautions in the beginning of this section.
2. Disconnect the negative battery cable.

3. Drain the cooling system into a clean container for reuse.
4. Discharge and recover the air conditioning system refrigerant.
5. Remove or disconnect the following:
   - Instrument panel
   - Front seat assembly
   - Front console assembly
   - Front floor carpet
   - Steering shaft attachment bolt
   - Front deck crossmember
   - Heater hoses
   - Flexible suction hose
   - Liquid pipe B connection

   - Center duct
   - Heater unit
   - Intake duct
   - Blower assembly
6. Disassemble the heater unit as necessary for access to components,

To install:
7. Assemble the heater unit as necessary.
8. Install or connect the following:
   - Blower assembly
   - Intake duct
   - Heater unit
   - Center duct
   - Liquid pipe B connection

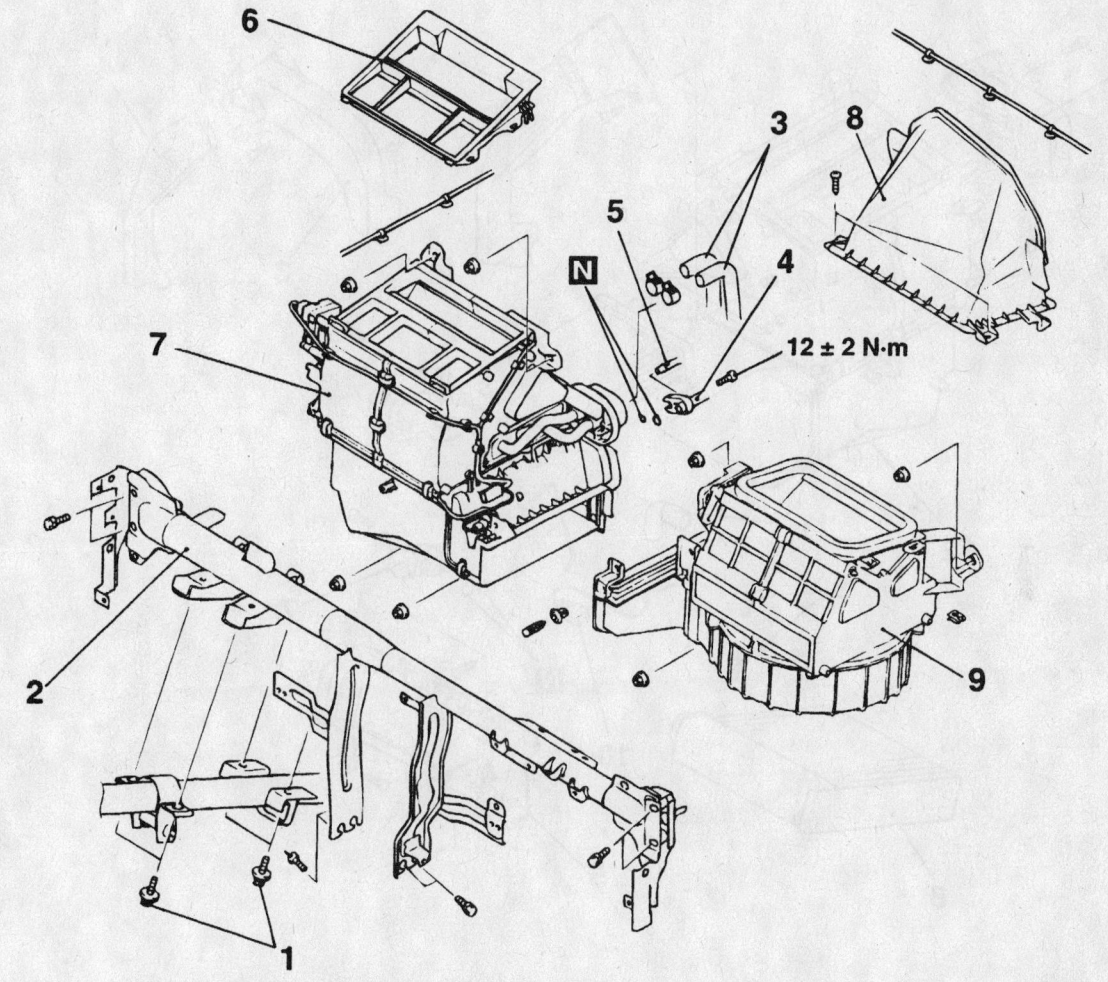

1. **STEERING SHAFT ATTACHMENT BOLT**
2. **FRONT DECK CROSSMEMBER**
3. **HEATER HOSE CONNECTION**
4. **FLEXIBLE SUCTION HOSE CONNECTION**
5. **LIQUID PIPE B CONNECTION**
6. **CENTER DUCT**
7. **HEATER UNIT**
8. **INTAKE DUCT**
9. **BLOWER ASSEMBLY**

42356-GALA-G01

Exploded view of the heater core and related components—Lancer

- Flexible suction hose
- Heater hoses
- Front deck crossmember
- Steering shaft attachment bolt
- Front floor carpet
- Front console assembly
- Front seat assembly
- Instrument panel

9. Refill the cooling system.

10. Connect the negative battery cable.

11. Evacuate, charge and leak test the air conditioning system.

12. Operate the engine to normal operating temperatures; then, check the climate control operation and check for leaks.

Mirage

1. Before servicing the vehicle, refer to the precautions in the beginning of this section.

2. Disconnect the negative battery cable.

3. Drain the cooling system into a clean container for reuse.

4. Remove the air cleaner cover and the air intake hose.

5. Disconnect the heater hoses from the heater core.

6. If equipped with air conditioning, discharge and recover the air conditioning system refrigerant.

7. If equipped with air conditioning, disconnect the refrigerant lines from the evaporator core and discard the O-rings.

✳✳ CAUTION

After disconnecting the negative battery cable, wait at least 60 seconds before working on the SRS module or instrument panel.

8. Remove the passenger's side air bag by removing or disconnecting the following:
- Glove box assembly
- Air bag-to-dash bolts and the air

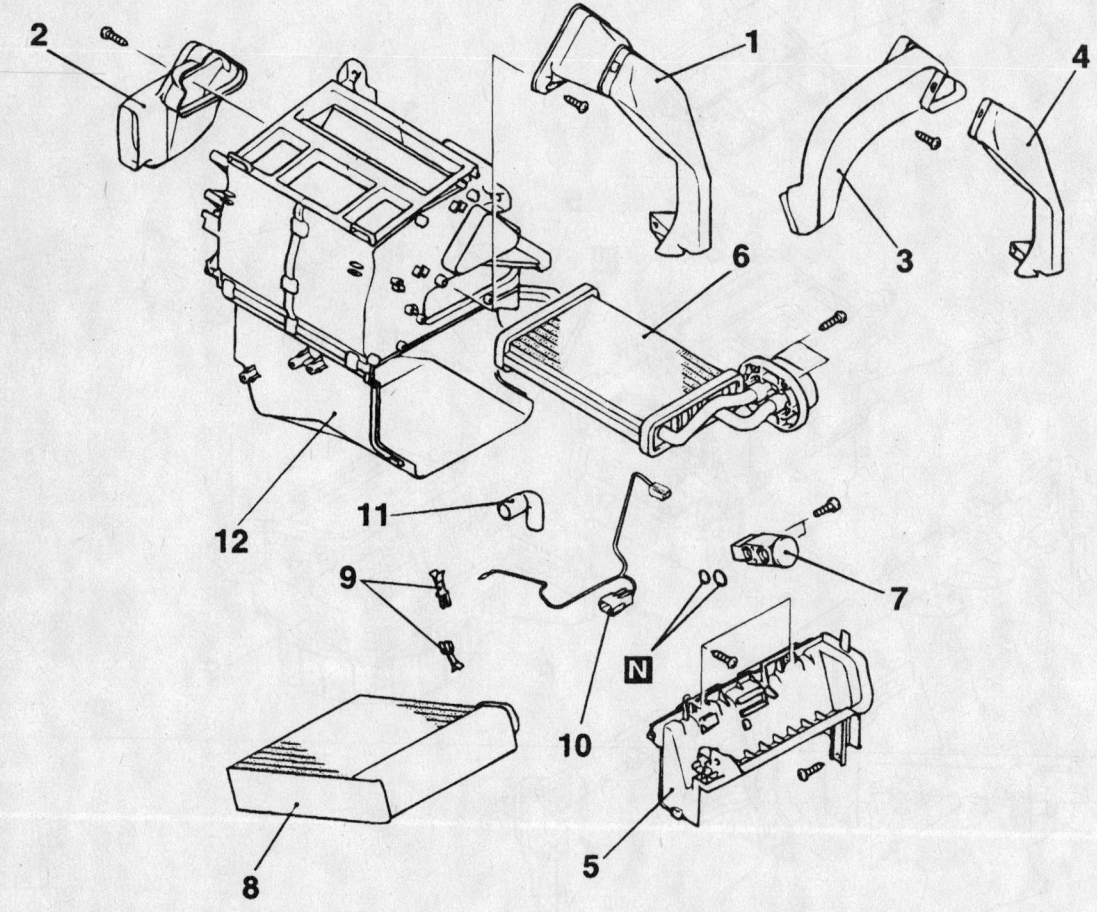

1. RIGHT-HAND FOOT DUCT
2. LEFT-HAND FOOT DUCT
3. LEFT-HAND FOOT DUCT <VEHICLE WITH REAR HEATER DUCT>
4. LEFT-HAND UPPER REAR HEATER DUCT A <VEHICLE WITH REAR HEATER DUCT>
5. EVAPORATOR COVER
6. HEATER CORE
7. EXPANSION VALVE
8. EVAPORATOR
9. AIR THERMO SENSOR CLIP
10. AIR THERMO SENSOR
11. DRAIN PLUG
12. HEATER CASE

42356-GALA-G02

Disassembly of the heater unit—Lancer

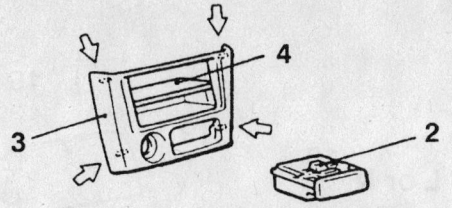

NOTE
⇦ : metal clip position

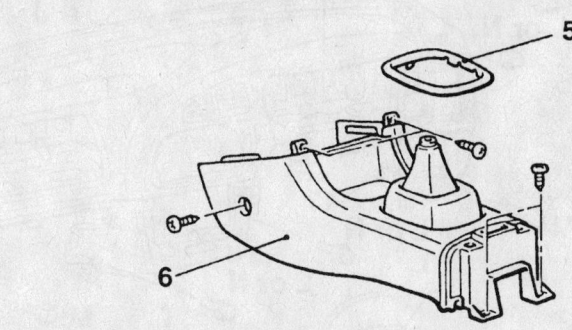

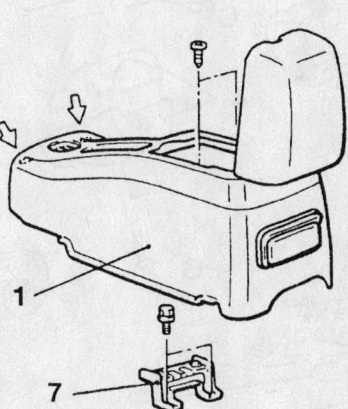

1. Rear floor console assembly
2. Ashtray
3. Audio panel
4. Box
● Shift lever knob
5. A/T panel
6. Front floor console assembly
7. Rear console bracket

93112GF5

Exploded view of the floor console and related components—Mirage

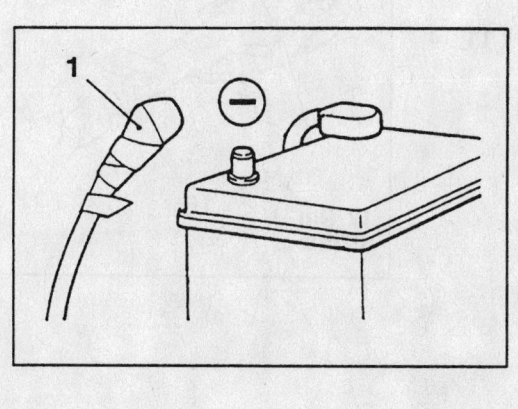

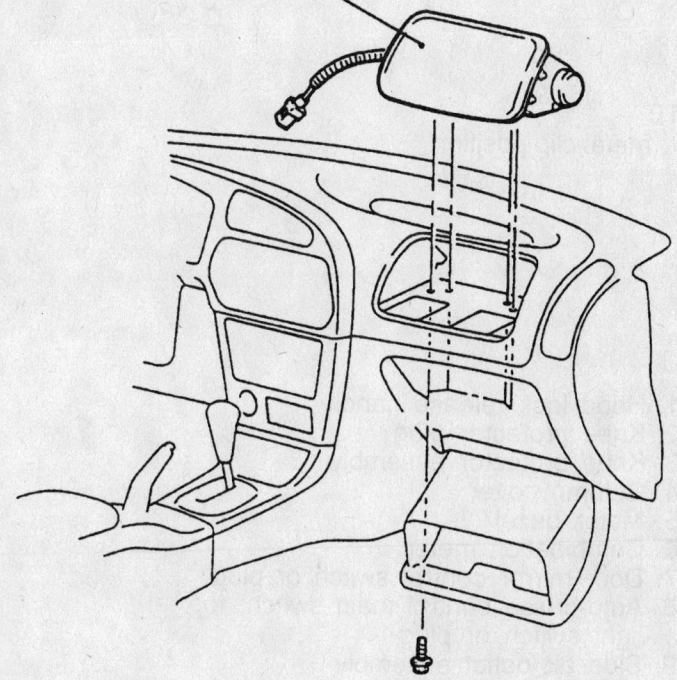

1. Negative (–) battery cable connection
2. Air bag module

93112GF6

Exploded view of the passenger's air bag module—Mirage

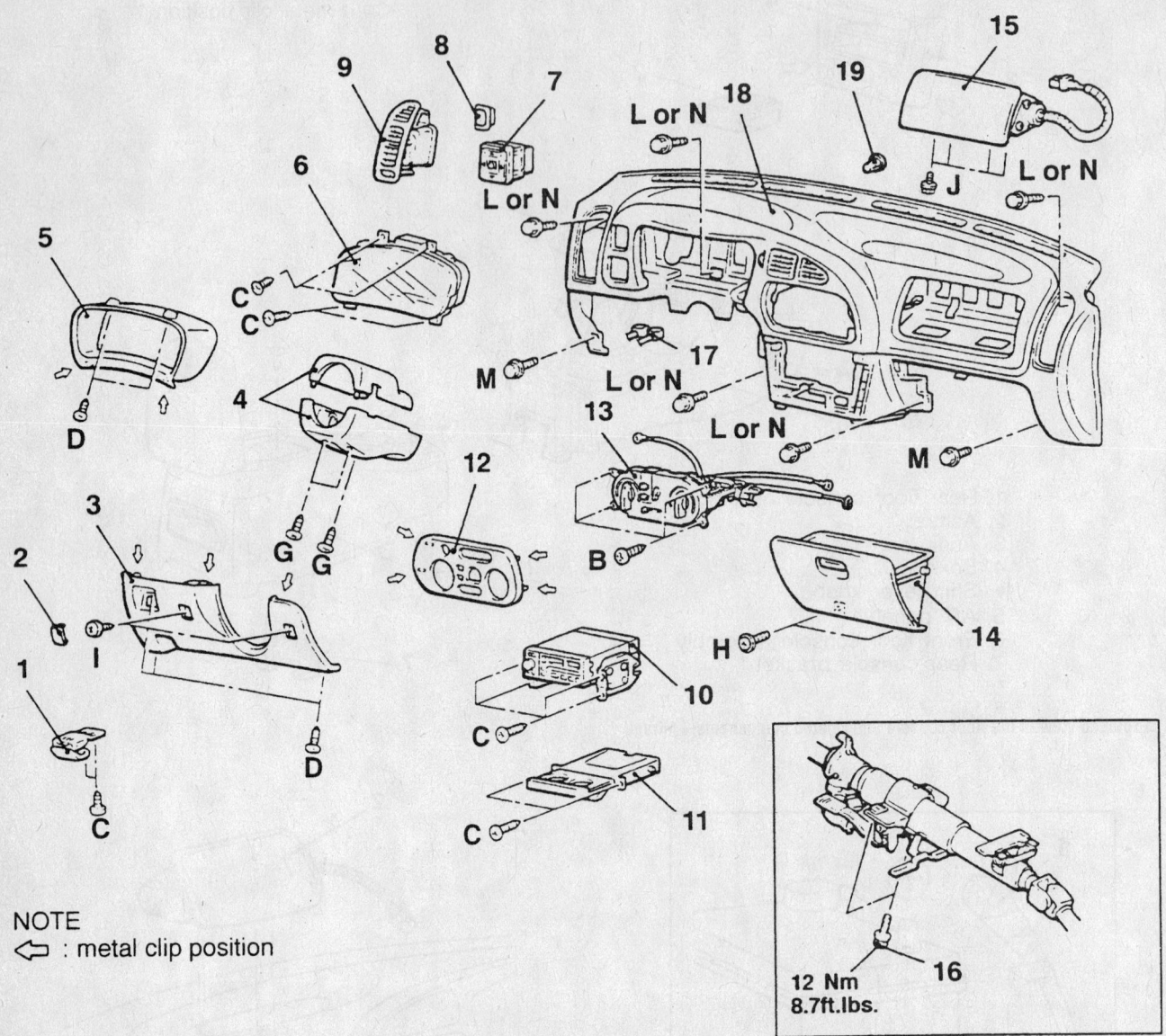

NOTE
⟸ : metal clip position

12 Nm
8.7ft.lbs.

1. Hood lock release handle
2. Knee protector plug
3. Knee protector assembly
4. Column cover
5. Meter bezel
6. Combination meter
7. Door mirror control switch or plug
8. Auto-cruise control main switch, fog light switch or plug
9. Side air outlet assembly
10. Radio and tape player
11. Cup holder
12. Heater control panel
13. Heater control assembly
14. Glove box
15. Front passenger's air bag module assembly
16. Steering column assembly installation bolt
17. Harness connector
18. Instrument panel assembly
19. Grommet

Exploded view of the instrument panel and steering column assembly—Mirage

93112GF7

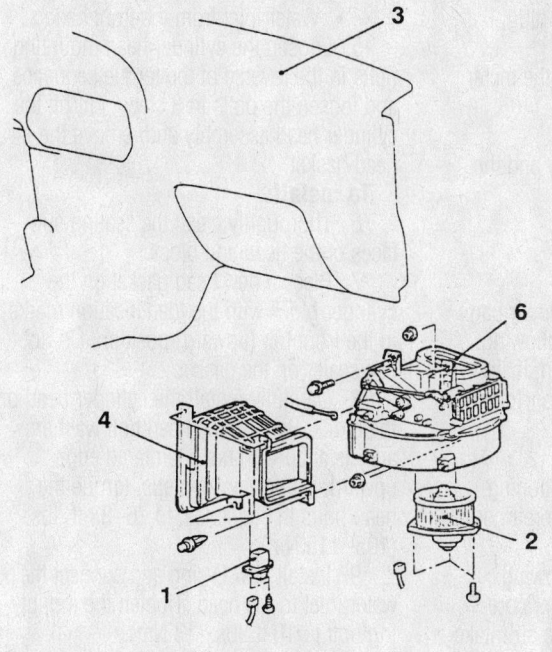

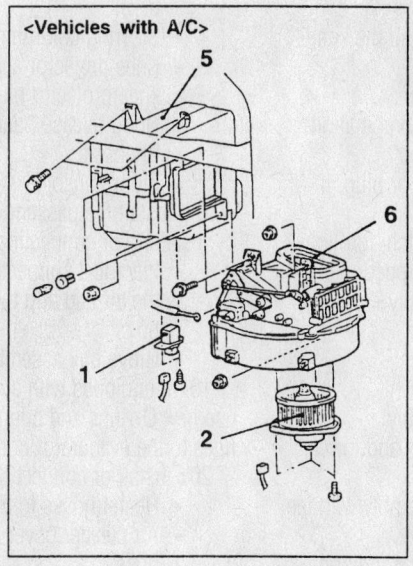

1. Resistor
2. Blower fan and motor
3. Instrument panel
4. Joint duct
5. Evaporator
6. Blower unit assembly

<Vehicles with A/C>

93112GF9

view of the joint duct and blower motor assembly—Mirage

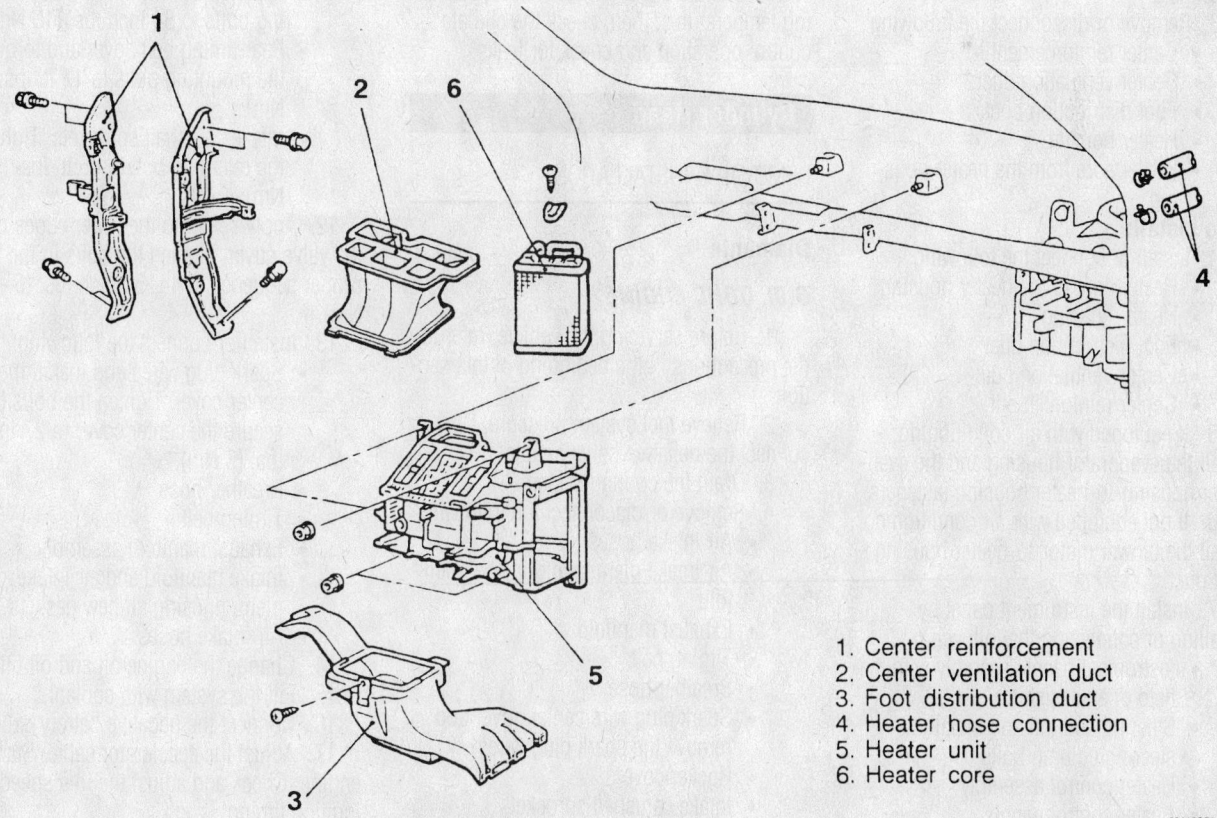

1. Center reinforcement
2. Center ventilation duct
3. Foot distribution duct
4. Heater hose connection
5. Heater unit
6. Heater core

93112GF8

Exploded view of the heater core, heater housing and related components—Mirage

bag; then, disconnect the electrical connector

9. Remove the floor console.

10. Remove the instrument panel by removing or disconnecting the following:
- Rheostat
- Hood release handle
- Knee protector plug and the knee protector assembly
- Steering column cover
- Meter bezel and the combination meter
- Mirror control switch or plug, if equipped
- Auto-cruise main switch, fog light switch or plug, if equipped
- Side air outlet assembly
- Radio and tape player
- Cup holder
- Heater control panel
- Heater control assembly
- Steering column bolts and lower the steering column
- Instrument panel assembly with the help of an assistant

11. If not equipped with air conditioning, remove the blower motor-to-heater housing joint duct.

12. If equipped with air conditioning, remove the evaporator housing-to-heater housing fasteners and remove the evaporator housing.

13. Remove or disconnect the following:
- Center reinforcement
- Center ventilation duct
- Foot distribution duct
- Heater housing
- Heater core from the heater housing

To install:

14. Install or connect the following:
- Heater core to the heater housing
- Heater housing
- Foot distribution duct
- Center ventilation duct
- Center reinforcement

15. If equipped with air conditioning, install the evaporator housing and the evaporator housing-to-heater housing fasteners.

16. If not equipped with air conditioning, install the blower motor-to-heater housing joint duct.

17. Install the instrument panel by installing or connecting the following:
- Instrument panel assembly with the help of an assistant
- Steering column and install the steering column bolts
- Heater control assembly
- Heater control panel
- Cup holder
- Radio and tape player

- Side air outlet assembly
- Auto-cruise main switch, fog light switch or plug, if equipped
- Mirror control switch or plug, if equipped
- Combination meter and the meter bezel
- Steering column cover
- Knee protector assembly and the knee protector plug
- Hood release handle
- Rheostat
- Floor console

18. Install the passenger's side air bag by installing or connecting the following:
- Electrical connector; then, install the air bag and the air bag-to-dash bolts
- Glove box assembly

19. If equipped with air conditioning, use new O-rings and connect the refrigerant lines to the evaporator core.

20. Install or connect the following:
- Heater hoses to the heater core
- Air cleaner cover and the air intake hose

21. Refill the cooling system.

22. Connect the negative battery cable.

23. If equipped with air conditioning, evacuate, charge and leak test the air conditioning system refrigerant.

24. Operate the engine to normal operating temperatures; then, check the climate control operation and check for leaks.

Cylinder Head

REMOVAL & INSTALLATION

Diamante

3.0L DOHC ENGINE

1. Before servicing the vehicle, refer to the precautions in the beginning of this section.

2. Relieve fuel system pressure. Disconnect the negative battery cable.

3. Drain the cooling system.

4. Remove or disconnect the following:
- Air intake hoses
- Air intake plenum and intake manifold
- Exhaust manifold
- Timing belt
- Breather hose
- Spark plug wire center cover and remove the spark plug wires
- Rocker covers
- Intake camshaft sprockets
- Rear timing belt cover
- Ignition coil assembly

- Water hoses from the thermostat housing
- Thermostat housing
- Water inlet from the front head

5. Loosen the cylinder head mounting bolts in the reverse of the torque sequence and loosen the bolts in 3 steps. Lift off the cylinder head assembly and remove the head gasket.

To install:

6. Thoroughly clean the sealing surfaces of the head and block.

7. Place a new head gasket on the cylinder block with the identification marks in the front top (upward) position. Do not use sealer on the gasket.

8. Carefully install the cylinder head on the block. Be sure the head bolt washers are installed with the chamfered edge upward. Using 3 even steps, torque the head bolts in sequence, to 76–83 ft. lbs. (105–115 Nm).

9. Install new O-ring and connect the water inlet to the head. Tighten the mounting bolt to 10 ft. lbs. (13 Nm).

10. Replace the gaskets and install the thermostat housing. Tighten the mounting bolts to 12–14 ft. lbs. (17–20 Nm).

11. Install or connect the following:
- Hoses to the thermostat housing, using new clamps
- Ignition coil and torque the mounting bolts to 84 inch lbs. (10 Nm)
- Rear timing belt cover and torque the mounting bolts to 17 ft. lbs. (24 Nm)
- Intake camshaft sprockets. Tighten the retaining bolt to 65 ft. lbs. (90 Nm).

12. Apply sealer to the lower edges of the valve cover. Tighten the bolts in the proper sequence to 44–51 inch lbs. (5–6 Nm).

13. Install or connect the following:
- Spark plug wires and install the center cover. Tighten the bolts that secure the center cover to 27 inch lbs. (3 Nm).
- Breather hose
- Timing belt
- Exhaust manifold assembly
- Intake manifold and air intake plenum, using all new gaskets
- Air intake hoses

14. Change the engine oil and oil filter.

15. Fill the system with coolant.

16. Connect the negative battery cable.

17. Adjust the accelerator cable. Start the engine. Check and adjust the idle speed and ignition timing.

18. Once the vehicle has cooled, recheck the coolant level.

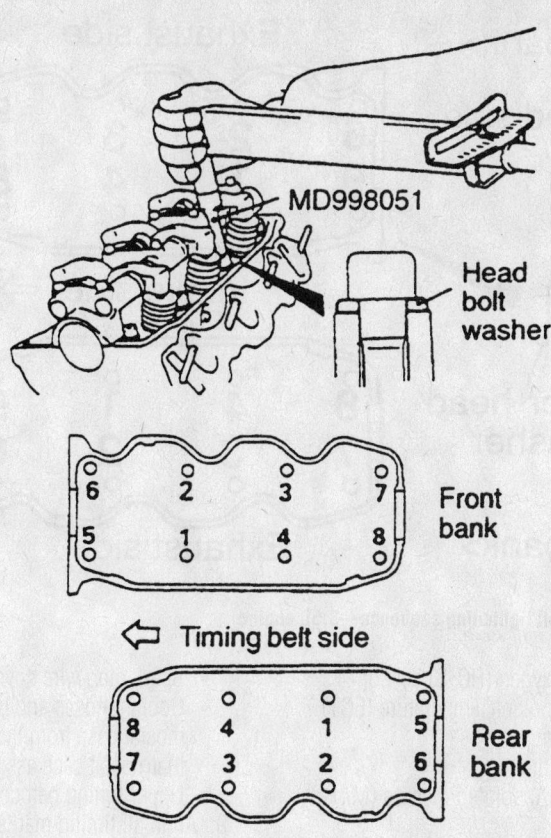

Front bank

⇐ Timing belt side

Rear bank

7923PG26

Tighten the cylinder head bolts according to the sequence shown—3.0L (SOHC and DOHC) engines

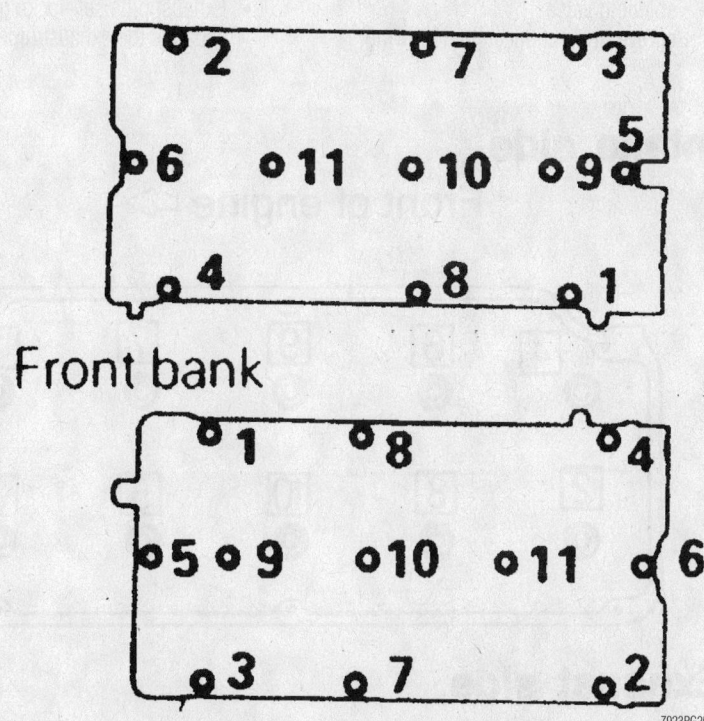

Rear bank

Front bank

7923PG25

Rocker cover bolt torque sequence—Diamante 3.0L DOHC engine

3.0L SOHC ENGINE

1. Before servicing the vehicle, refer to the precautions in the beginning of this section.

2. Relieve the fuel system pressure. Disconnect the negative battery cable.

3. Drain the cooling system.

4. Remove or disconnect the following:
- Air intake hose
- Exhaust manifold
- Air intake plenum and intake manifold
- Timing belt
- Camshaft sprockets and the rear timing belt cover
- Power steering pump bracket. If removing the rear head, remove the alternator brace.
- Water inlet pipe
- Purge pipe assembly
- Valve cover

5. Using the reverse sequence of the installation sequence, loosen the cylinder head mounting bolts in 3 steps. Lift off the cylinder head assembly and remove the head gasket.

To install:

6. Thoroughly clean the sealing surfaces of the head and block.

7. Place a new head gasket on the cylinder block making sure the identification mark on the cylinder head gasket is in the front top (upward) location. Do not use sealer on the gasket.

8. Carefully install the cylinder head on the block. Be sure the head bolt washers are installed with the chamfered edge upward. Using 3 even steps, torque the head bolts in sequence, to 76–83 ft. lbs. (105–115 Nm).

9. Apply sealer to the lower edges of the half-round portions and install the valve cover. Tighten valve cover bolts to 84 inch lbs. (9 Nm).

10. Install or connect the following:
- Purge pipe assembly
- Water inlet pipe
- Power steering pump bracket and alternator brace
- Rear timing belt cover and camshaft sprockets. Torque the retaining bolt to 65 ft. lbs. (90 Nm).
- Timing belt
- Intake manifold, air intake plenum and exhaust manifold, using new gaskets
- Air intake hose

11. Fill the system with coolant.

12. Connect the negative battery cable.

13. Start the engine.

14. Check and adjust the idle speed and ignition timing.

15. Once the vehicle has cooled, recheck the coolant level.

3.5L ENGINE

1. Before servicing the vehicle, refer to the precautions in the beginning of this section.

2. Disconnect the negative battery cable.

3. Drain the engine coolant

4. Remove or disconnect the following:
- Timing belt
- Intake and exhaust manifolds
- Spark plug wires
- Cylinder head covers
- Timing belt rear center cover

5. Loosen the cylinder head bolts gradually in 3 stages, in the opposite of the installation sequence.

6. Remove the cylinder head.

To install:

7. Clean the cylinder head and mounting surface on the engine block.

8. Install the cylinder head using a new gasket.

9. Tighten the bolts in sequence using 3 stages to 76–83 ft. lbs. (103–113 Nm).

10. Install or connect the following:
- Timing belt rear center cover
- Cylinder head covers using new gaskets. Tighten the bolts to 24–36 inch lbs. (3–4 Nm).
- Spark plug wires
- Intake and exhaust manifolds
- Timing belt
- Remaining components

11. Refill the cooling system.

12. Connect the negative battery cable.

Eclipse

1. Before servicing the vehicle, refer to the precautions in the beginning of this section.

2. Relieve the fuel system pressure.

3. Disconnect the negative battery cable.

4. Remove the air cleaner with all air intake hoses.

5. Drain the cooling system.

6. Remove or disconnect the following:
- Accelerator cable
- Cable mounting brackets and position the cable aside
- Breather hose
- Vacuum lines at the throttle body, label for identification
- High pressure fuel line, plug
- Fuel return hose, plug

7. Disconnect the following connectors:
- Air conditioning compressor
- Power steering pressure switch

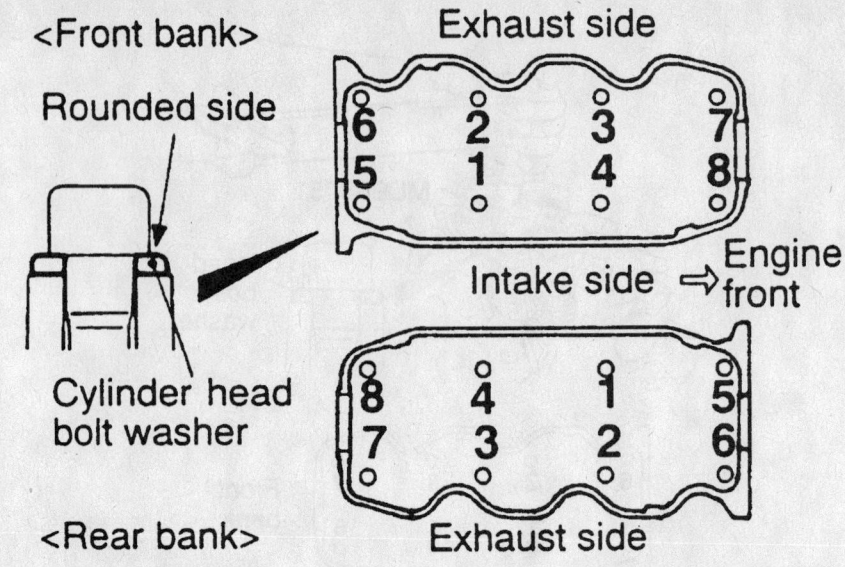

Cylinder head bolt tightening sequence—3.5L engine

- Heated Oxygen (HO2S) sensor
- Engine Coolant Temperature (ECT) gauge sender
- ECT sensor
- Manifold Absolute Pressure (MAP) sensor
- Intake Air Temperature (IAT) sensor
- Throttle Position (TP) sensor
- Idle Air Control (IAC) motor
- Injector harness
- Ignition coil
- Camshaft Position(CMP) sensor
- Exhaust Gas Recirculation (EGR) solenoid valve

8. Remove or disconnect the following:

- Spark plug wire cover and wires
- Coolant hoses and unbolt the thermostat case from the engine, at the thermostat case assembly
- Upper timing belt cover

9. Align all timing marks.

10. Secure the timing belt to the camshaft sprocket with cord or a wire tie.

11. Remove or disconnect the following:
- Camshaft sprocket
- Valve cover and the half-round seal
- Intake manifold stay bracket from the intake manifold
- Exhaust pipe self-locking nuts and separate the exhaust pipe from the

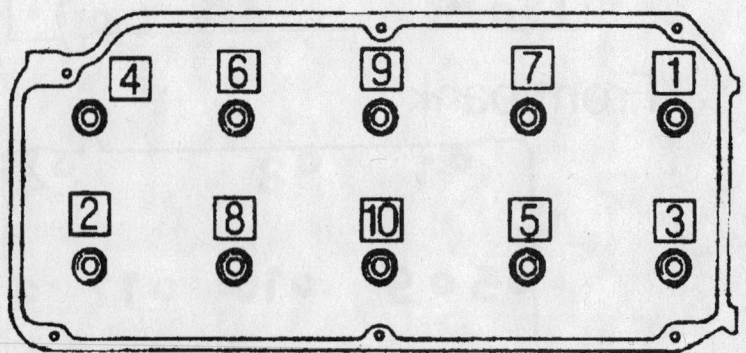

Cylinder head bolt removal sequence—2.4L (4G64) engine

exhaust manifold. Discard the gasket.

12. Loosen the cylinder head mounting bolts in 3 steps, starting from the outside and working inward. Lift off the cylinder head assembly and remove the head gasket.

To install:

13. Thoroughly clean the mating surfaces of the head and block.

14. Place a new head gasket on the cylinder block with the identification marks at the front top (upward) position. Do not use sealer on the gasket.

15. Inspect the cylinder head bolt length prior to installation. If the length exceeds 3.91 in. (99.4mm), the bolt must be replaced. Install the washer onto the bolt so the chamfer on the washer faces towards the head of the bolt.

16. Carefully install the cylinder head on the block and tighten the cylinder head bolts as follows:

    a. Following the proper tightening sequence, tighten the cylinder head bolts to 58 ft. lbs. (78 Nm).

    b. Loosen all bolts completely.

    c. Torque bolts to 15 ft. lbs. (20 Nm).

    d. Tighten bolts an additional ¼ turn.

    e. Tighten bolts an additional ¼ turn.

17. Install or connect the following:

- New exhaust pipe gasket and connect the exhaust pipe to the manifold. Tighten the bolts to 33 ft. lbs. (44 Nm).
- Thermostat case and tighten the mounting bolts to 18 ft. lbs. (24 Nm)
- Coolant hoses to the thermostat case

18. Apply sealer to the perimeter of the half-round seal and to the lower edges of the half-round portions of the belt-side of the new gasket. Install the valve cover.

19. Install or connect the following:

- Camshaft sprocket with the timing belt attached. Remove the cord or wire tie.
- Upper timing belt cover
- Intake manifold stay and tighten the mounting bolts to 22 ft. lbs. (30 Nm)

20. Connect the following connectors:

- Air conditioning compressor
- Power steering pressure switch
- HO₂S sensor
- ECT gauge sender
- ECT sensor
- MAP sensor
- IAT sensor
- TP sensor
- IAC motor
- Injector harness

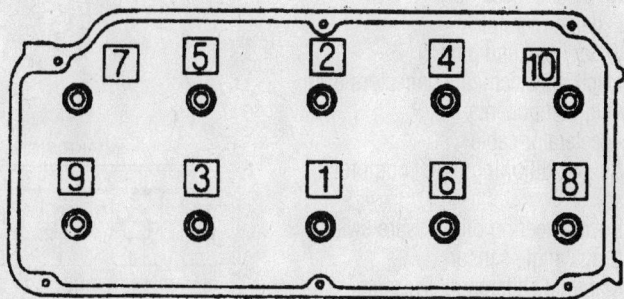

Intake side

Front of engine ⇨

Exhaust side

7923PG22

Cylinder head bolt installation sequence—2.4L (4G64) engine

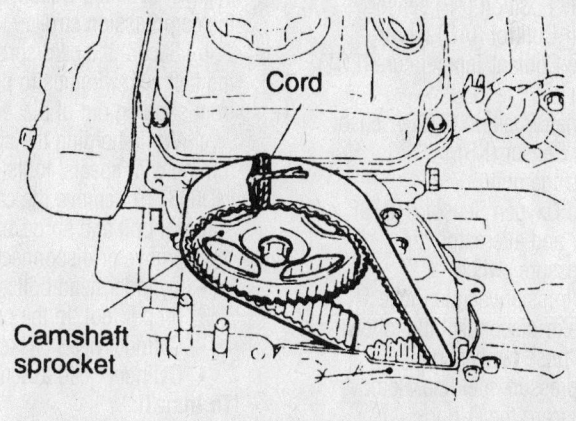

Cord

Camshaft sprocket

Secure the timing belt to the camshaft sprocket and remove the sprocket—2.4L (4G64) engine

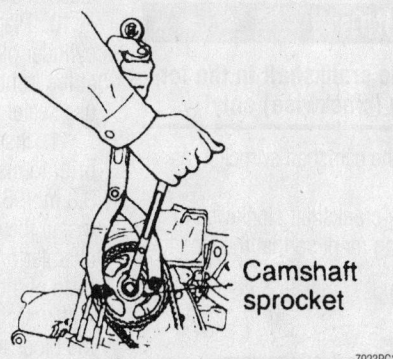

Camshaft sprocket

7923PG23

- Ignition coil
- CMP sensor
- EGR solenoid valve

21. Remove or disconnect the following:

- Spark plug wires and cover
- Fuel lines using new O-rings
- Air cleaner and intake hose

- Breather hose

22. Fill the cooling system.

23. Connect the negative battery cable

Lancer

1. Before servicing the vehicle, refer to the precautions in the beginning of this section.

2. Relieve the fuel system pressure. Disconnect the negative battery cable.

3. Drain the cooling system.

4. Remove or disconnect the following:
- Engine undercover
- Air cleaner assembly
- Exhaust manifold
- Water hose and pipe

5. Detach the electrical connectors from the following components:
- Accelerator cable
- Air Conditioning (A/C) compressor
- Power steering oil pressure switch
- Crank angle sensor
- Manifold Differential Pressure (MDP) sensor
- Evaporative emission (EVAP) purge solenoid
- Exhaust Gas Recirculation (EGR) solenoid valve
- Ignition coil
- Fuel injectors
- Throttle Position (TP) sensor
- Idle Air Control (IAC) motor
- Engine Coolant Temperature (ECT) sensor
- Camshaft Position (CMP) sensor
- Knock Sensor (KS)
- ECT gauge unit
- Heated Oxygen Sensor (HO2S)
- Starter and alternator
- Oil pressure switch
- Brake booster vacuum hose

6. Remove or disconnect the following:
- Purge hose connection
- High pressure fuel line. Remove and discard the O-rings.
- Timing belt front upper cover

❋❋ WARNING

Always turn the crankshaft in the forward direction (clockwise) only!

7. Remove the camshaft sprocket, as follows:

   a. Turn the crankshaft clockwise to align the timing mark so that the No. 1

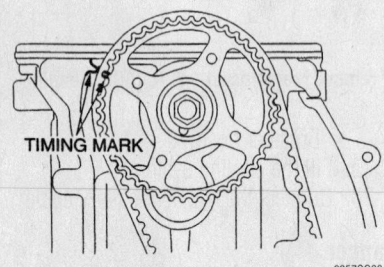

TIMING MARK

9357QG06

Proper timing mark alignment—Lancer with the 2.0L engine

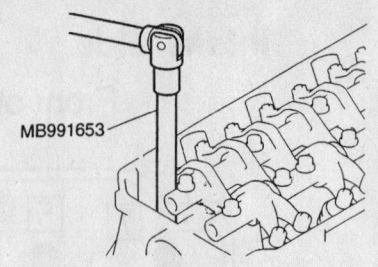

MB991653

INTAKE.SIDE

```
3   5   10   8   2
1   7   9    6   4
```

EXHAUST SIDE

9357QG08

Cylinder head bolt removal sequence— Lancer with 2.0L engine

cylinder is at Top Dead Center (TDC) of the compression stroke.

   b. Secure the cam sprocket and timing belt with wire ties to prevent them from slipping out of place.

   c. While holding the sprocket from turning with special tools MB990767 and MD998719, remove the camshaft sprocket bolt and sprocket.

8. Remove or disconnect the following:
- Cylinder head bolts, using the proper tool, in the proper sequence
- Cylinder head gasket
- Cylinder head assembly

To install:

9. Thoroughly clean the mating surfaces of the head and block.

10. Place a new head gasket on the cylinder block with the identification marks at the front top (upward) position. Do not use sealer on the gasket.

11. Inspect the cylinder head bolt length prior to installation. If the length exceeds 3.8 in. (96.4mm), the bolt must be replaced. Install the washer onto the bolt so the chamfer on the washer faces towards the head of the bolt.

12. Carefully install the cylinder head on the block and tighten the cylinder head bolts as follows:

   a. Following the proper tightening sequence, tighten the cylinder head bolts to 52–58 ft. lbs. (70–78 Nm).

   b. Loosen all bolts completely.

   c. Torque bolts to 15 ft. lbs. (20 Nm).

   d. Tighten bolts an additional ¼ turn.

   e. Tighten bolts an additional ¼ turn.

13. The remainder of installation is the reverse of the removal procedure.

14. Fill the system with coolant.

15. Connect the negative battery cable.

Mirage

1.5L ENGINE

1. Before servicing the vehicle, refer to the precautions in the beginning of this section.

2. Relieve the fuel system pressure. Disconnect the negative battery cable.

3. Drain the cooling system.

4. Remove or disconnect the following:
- Air intake hose and the air cleaner assembly
- Ground cable connection and the accelerator cable
- Positive Crankcase Ventilation (PCV) and the breather hose connection
- Vacuum hoses from the intake and throttle body, label for reference
- Vacuum line for the brake booster
- Upper radiator hose, throttle body hoses, bypass hose and heater hose connections
- Fuel feed and return lines
- Spark plug wires

5. Disconnect the electrical harness plugs from the following:
- Crankshaft Position (CKP) and Camshaft Position (CMP) sensors
- Heated Oxygen (HO2S) sensor
- Engine Coolant Temperature (ECT) sensor and gauge sender
- Idle Speed Control (ISC) motor
- Throttle Position (TP) sensor
- Intake Air Temperature (IAT) sensor
- Exhaust Gas Recirculation (EGR) temperature sensor

6. Remove or disconnect the following:
- Electrical harness plugs from the ignition distributor, fuel injectors, power transistor and ground cable
- Engine control wiring harness
- Clamp that holds the power steering pressure hose to the engine mounting bracket

7. Place a jack and wood block under the oil pan and carefully lift just enough to take the weight off the engine mounting bracket and remove the bracket.
- Valve cover
- Timing belt upper cover

8. Rotate the crankshaft clockwise and align the timing marks.

9. Attach the timing belt to the camshaft sprocket with cord or a wire tie.

10. Secure the camshaft from turning and remove the camshaft sprocket with the timing belt attached.

11. Remove the timing belt rear upper cover.

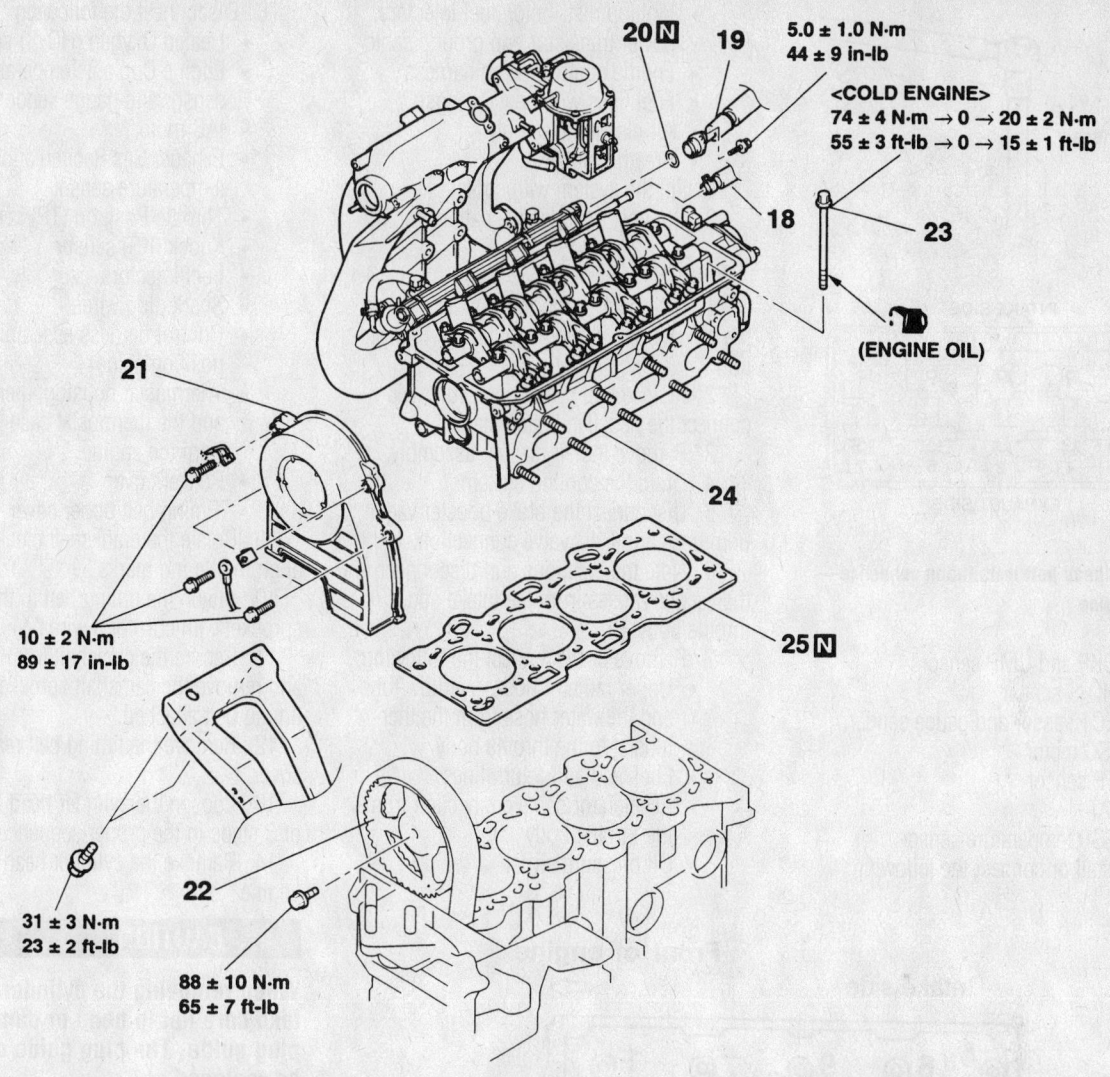

5.0 ± 1.0 N·m
44 ± 9 in-lb

<COLD ENGINE>
74 ± 4 N·m → 0 → 20 ± 2 N·m
55 ± 3 ft-lb → 0 → 15 ± 1 ft-lb

(ENGINE OIL)

10 ± 2 N·m
89 ± 17 in-lb

31 ± 3 N·m
23 ± 2 ft-lb

88 ± 10 N·m
65 ± 7 ft-lb

18. PURGE HOSE CONNECTION
19. HIGH-PRESSURE FUEL HOSE
    CONNECTION
20. O-RING
21. TIMING BELT FRONT UPPER
    COVER

22. CAMSHAFT SPROCKET
23. CYLINDER HEAD BOLTS
24. CYLINDER HEAD ASSEMBLY
25. CYLINDER HEAD GASKET

93570G09

Exploded view of the cylinder head and related components—Lancer with 2.0L engine

12. Remove the exhaust pipe from the exhaust manifold.

13. Loosen the cylinder head mounting bolts in sequence using 3 steps. Remove the cylinder head.

To install:

14. Thoroughly clean the mating surfaces of the head and block.

15. Place a new head gasket on the cylinder block with the identification marks facing upward. Do not use sealer on the gasket.

16. Carefully install the cylinder head on the block. Tighten the cylinder head bolts as follows:

a. 36 ft. lbs. (49 Nm) in the correct sequence.

b. Loosen the bolts completely in the reverse order.

c. Tighten the bolts in sequence to 14 ft. lbs. (20 Nm).

d. Tighten each bolt in sequence 90 degrees.

e. Tighten each bolt in sequence an additional 90 degrees.

17. Install or connect the following:

• New exhaust pipe gasket and connect the exhaust pipe to the manifold

• Upper rear timing cover

18. Align the timing marks and install the cam sprocket. Torque the retaining bolt to 51 ft. lbs. (70 Nm). Check the belt tension and adjust, if necessary. Install the outer timing cover.

19. Install or connect the following:

• Valve cover and torque the retaining bolts to 16 inch lbs. (1.8 Nm)

• Engine mount bracket and remove the support jack

• Clamp that holds the power steering pressure hose to the engine mounting bracket

20. Connect the following:

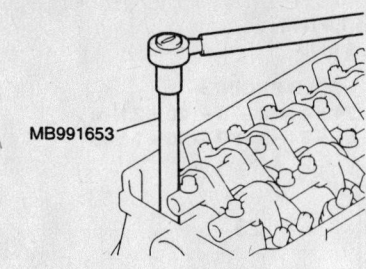

MB991653

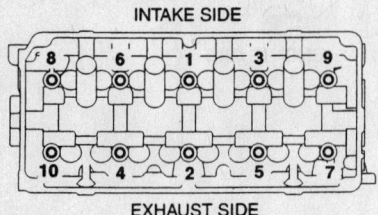

INTAKE SIDE

EXHAUST SIDE

93570G10

Cylinder head bolt installation sequence— 2.0L engine

- CKP and CMP sensors
- HO$_2$S sensor
- ECT sensor and gauge sender
- ISC motor
- TP sensor
- IAT
- EGR temperature sensor

21. Install or connect the following:

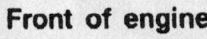

Front of engine ⇨

Intake side

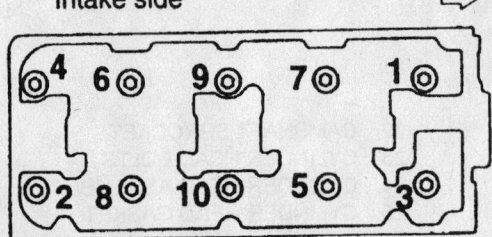

Exhaust side

7923PG13

Cylinder head bolt loosening sequence—Mirage with 1.5L (4G15) engine

Intake side ⇦ Front of engine

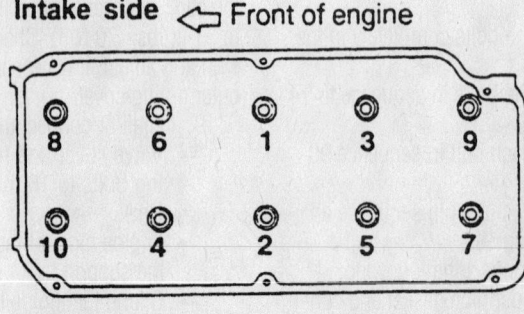

Exhaust side

7923PG14

Cylinder head bolt tightening sequence—Mirage with 1.5L (4G15) engine

- Ignition distributor, fuel injectors, power transistor and ground cable
- Engine control wiring harness
- Fuel lines with new O-rings
- Air cleaner assembly
- Breather hose

22. Fill the system with coolant.
23. Connect the negative battery cable.

1.8L ENGINE

1. Before servicing the vehicle, refer to the precautions in the beginning of this section.
2. Relieve fuel system pressure. Disconnect the negative battery cable.
3. Remove the air cleaner assembly.
4. Drain the cooling system.
5. Disconnect the brake booster vacuum hose and PVC valve connection.
6. Note the locations and disconnect the vacuum hoses from the intake and throttle body.
7. Remove or disconnect the following:
   - Upper radiator hose, overflow tube and the water hose from the thermostat to the throttle body
   - Fuel feed and return lines
   - Accelerator cable connection from the throttle body
   - Oil pressure switch

8. Disconnect the following:
   - Heated Oxygen (HO$_2$S) sensor
   - Engine Coolant Temperature (ECT) sensor and gauge sender
   - IAC motor
   - Exhaust Gas Recirculation (EGR) temperature sensor
   - Throttle Position (TP) sensor
   - Knock (KS) sensor
   - Fuel injectors
   - Spark plug wires
   - Control harness assembly and position aside
   - Thermostat housing, thermostat and the thermostat case with O-ring from the engine
   - Rocker cover
   - Timing belt upper cover

9. Rotate the crankshaft clockwise and align the timing marks.
10. Attach the timing belt to the camshaft sprocket with cord or a wire tie.
11. Secure the camshaft from turning and remove the camshaft sprocket with the timing belt attached.
12. Remove the timing belt rear upper cover.
13. Loosen the cylinder head bolts in 2 or 3 steps in the proper sequence.
14. Remove the cylinder head from the engine.

✳✳ CAUTION

When removing the cylinder head, take care not to bend or damage the plug guide. The plug guide can not be replaced.

To install:

15. Thoroughly clean the mating surfaces of the head and block.
16. Place a new head gasket on the cylinder block with the identification marks facing upward. Do not use sealer on the gasket.
17. Carefully install the cylinder head on the block.
18. Measure the cylinder head bolts prior to installation. Replace any that exceed 3.795 in. (96.4mm).
19. Apply a small amount of engine oil to the thread section of the bolt and install so the chamfer of the washer faces upward.
20. Tighten the cylinder head bolts as follows:
    a. In the proper tightening sequence, torque bolts to 54 ft. lbs. (75 Nm).
    b. In the reverse order of the tightening sequence, fully loosen all bolts.
    c. In the proper tightening sequence, torque bolts to 14 ft. lbs. (20 Nm).

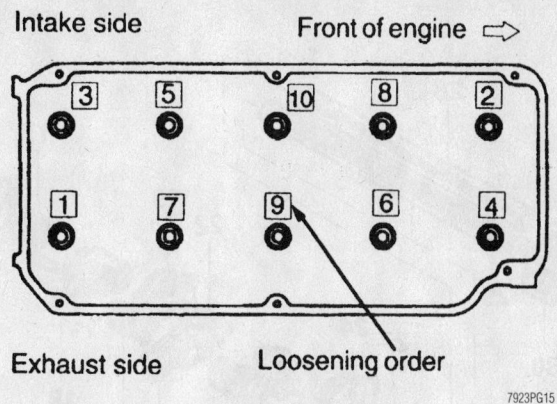

Cylinder head bolt loosening sequence—1.8L engine

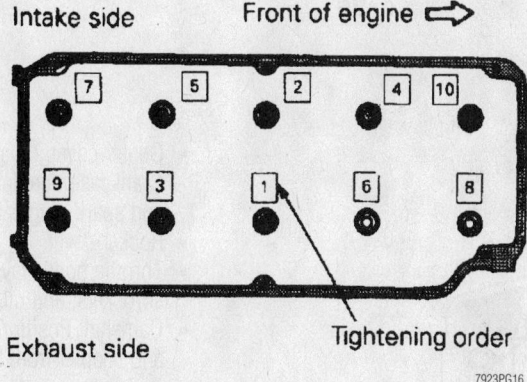

Cylinder head bolt torque sequence—1.8L engine

d. In the proper tightening sequence, tighten bolts ¼ turn (90 degrees).

e. In the proper tightening sequence, tighten bolts an additional ¼ turn (90 degrees).

21. Install the camshaft sprocket and tighten the bolt to 65 ft. lbs. (90 Nm), while holding the sprocket in place using the appropriate wrench. Confirm proper timing mark alignment.

22. Install the upper timing belt cover and rocker cover. Torque the rocker cover bolts to 29 inch lbs. (3.3 Nm).

23. Loosen the water pipe mounting bolt for ease of thermostat housing installation.

24. Apply a thin bead of RTV sealant to the water tube connection on the thermostat case.

25. Apply a small amount of water to the O-ring of the water inlet pipe and press the thermostat case assembly onto the water inlet pipe. Install the thermostat case assembly mounting bolt tightening to 16 ft. lbs. (22 Nm).

26. Tighten the water pipe mounting bolt.

27. Install the thermostat into the housing so the jiggle valve is located at the top.

Tighten the housing bolts to 10 ft. lbs. (14 Nm).

28. Connect the following:
- HO2S sensor
- ECT sensor and gauge sender
- IAC motor
- EGR temperature sensor
- TP sensor
- KS sensor
- Fuel injectors

29. Install or connect the following:
- Upper radiator hose to the thermostat housing
- Accelerator cable connection to the throttle body
- Oil pressure switch
- Spark plug wires
- Control harness assembly

30. Replace the O-ring for the high pressure hose and install a new clamp on the return hose and reconnect the fuel lines.
- Air intake hose
- Breather hose and air cleaner case cover
- Brake booster and the PCV vacuum hoses

31. Fill the system with coolant.

32. Connect the negative battery cable

Rocker Arms/Shafts

REMOVAL & INSTALLATION

Diamante

3.0L SOHC ENGINE

On this engine, the hydraulic lash adjusters are built into the rocker arms.

1. Before servicing the vehicle, refer to the precautions in the beginning of this section.

2. Disconnect the negative battery cable.

3. Remove the valve cover. Install lash adjuster retainer tools to prevent the auto-lash adjuster from falling out of the rocker arm.

4. Loosen rocker arm and shaft assembly evenly in several steps. Remove the rocker arm and shaft assembly as a complete unit.

5. Remove the rear camshaft bearing cap and slide the rocker arms, springs and washers from the shaft. If they are to be reused, note the location and positioning of all rocker shaft components. It is recommended that all lash adjusters and rockers be replaced as a complete set.

To install:

6. Immerse the lash adjusters in clean diesel fuel. Using a small wire, move the plunger of the lash adjuster up and down 4 or 5 times while pushing down lightly on the check ball in order to bleed out the air. Install the lash adjusters in the rocker arms.

7. Using a light coat of engine oil, assemble the rocker arms to the shaft. Install the rear camshaft bearing cap.

8. Lubricate the camshaft and rocker shaft with clean engine oil and position on the cylinder head.

9. Apply a drop of sealant to the rear edges of the end caps.

10. Install or connect the following:
- Assembly making sure the notches in the rocker shafts are facing up
- Cap bolts and tighten evenly and gradually to 14 ft. lbs. (20 Nm). Remove the lash adjuster retainers.
- Valve cover
- Negative battery cable

3.0L DOHC ENGINE

1. Before servicing the vehicle, refer to the precautions in the beginning of this section.

2. Relieve the fuel system pressure.

3. Remove or disconnect the following:
- Battery negative cable
- Timing belt cover and timing belt

13. Bearing cap
14. Rocker arm
15. Spring
16. Rocker arm
17. Spring
18. Bearing cap no. 3
19. Rocker arm
20. Spring
21. Rocker arm
22. Spring
23. Bearing cap no. 2
24. Rocker arm
25. Spring
26. Rocker arm
27. Spring
28. Rocker arm shaft
29. Rocker arm shaft
30. Bearing cap no. 1

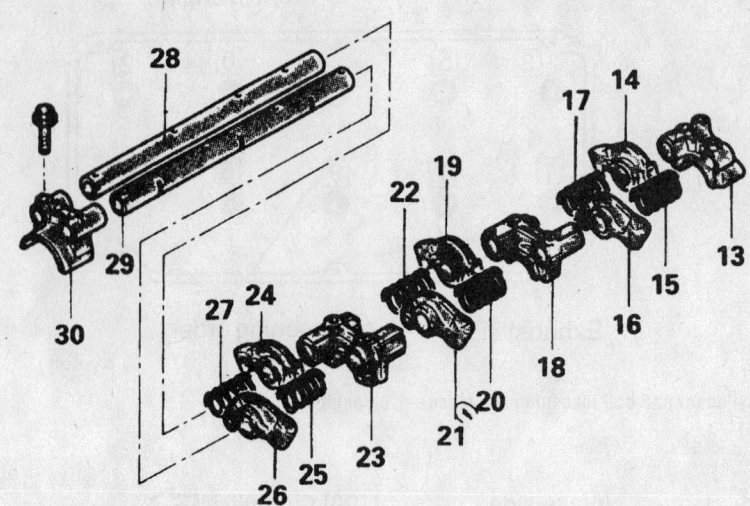

7923PG34

Rocker arm assembly—Diamante 3.0L SOHC engine

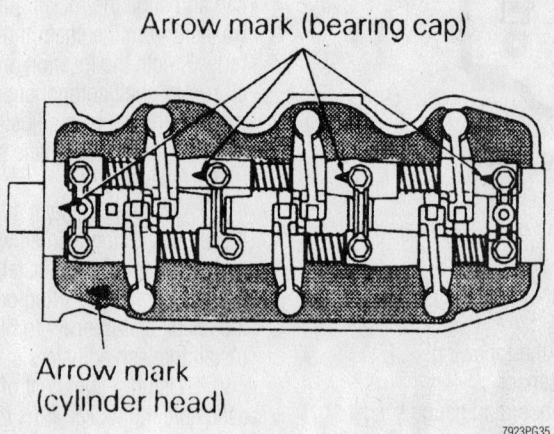

7923PG35

When installing the rocker arm/shaft assemblies, ensure that the arrow marks point in the same direction as the arrow stamped into the cylinder head—Diamante 3.0L SOHC engine

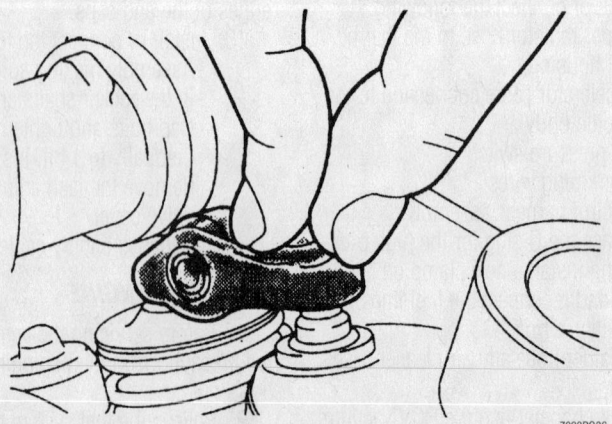

7923PG36

The rocker arms sit beneath the camshaft and are supported on one end by the valve stem and on the other end by the hydraulic lash adjuster—Diamante 3.0L DOHC engine

- Center cover, breather and Positive Crankcase Ventilation (PCV) hoses, and spark plug cables
- Rocker cover
- Throttle body stay, both camshaft sprockets, and oil seals
- Camshaft Position (CMP) sensor and adapter from the rear of the camshaft
- Intake and exhaust camshafts
- Rocker arms and lash adjusters from the head

➡ It is recommended that all lash adjusters and rockers be replaced as a complete set.

To install:

4. Immerse the lash adjusters in clean diesel fuel. Using a small wire, move the plunger of the lash adjuster up and down 4 or 5 times while pushing down lightly on the check ball in order to bleed out the air. Lubricate and install the lash adjusters in the cylinder head.

5. Lubricate the camshafts with clean engine oil and position the camshafts on the cylinder head.

6. Install the bearing caps. Tighten the caps in sequence and in 2 or 3 steps. Caps 2, 3 and 4 have a front mark. Install with the mark aligned with the front mark on the cylinder head. Intake caps have I stamped on the cap and exhaust caps have E. Also, be sure the rocker arm is correctly mounted on the lash adjuster and the valve stem end. Torque the front and rear retaining cap bolts to 14 ft. lbs. (20 Nm) and tighten the center 3 retaining cap bolts to 96 inch lbs. (11 Nm).

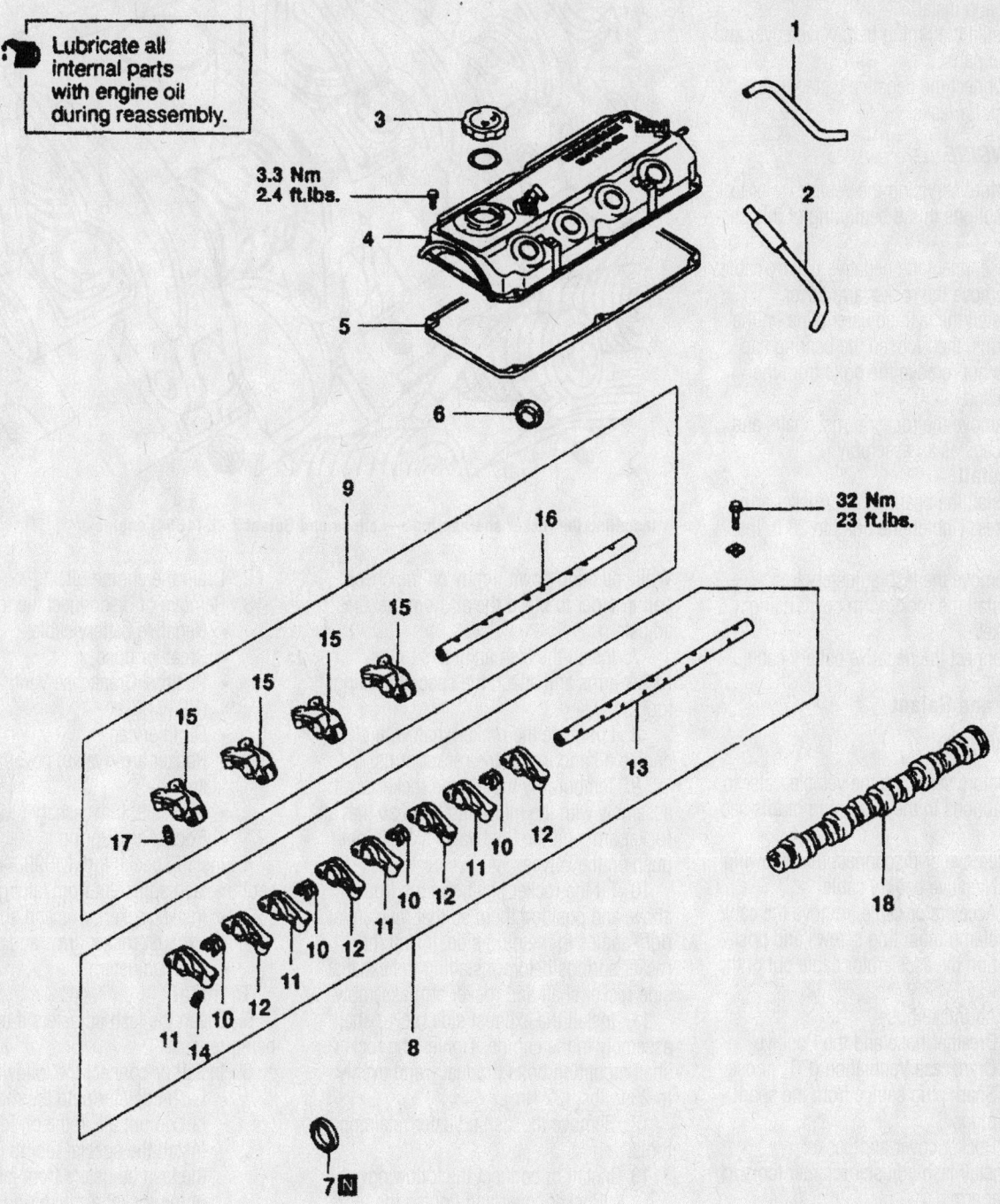

Lubricate all internal parts with engine oil during reassembly.

3.3 Nm
2.4 ft.lbs.

32 Nm
23 ft.lbs.

1. Breather hose
2. P.C.V. hose
3. Oil filler cap
4. Rocker cover
5. Rocker cover gasket
6. Oil seal
7. Oil seal
8. Rocker arms and rocker arm shaft
9. Rocker arms and rocker arm shaft
10. Rocker shaft spring
11. Rocker arm A
12. Rocker arm B
13. Rocker arm shaft (Intake side)
14. Lash adjuster
15. Rocker arm C
16. Rocker arm shaft (Exhaust side)
17. Lash adjuster
18. Camshaft

7923PG32

Rocker arm shafts and components—Eclipse and Galant 2.4L (4G64) engines

7. Apply a coating of engine oil to the oil seals and install.

8. Install the timing belt, valve cover and all related parts.

9. Connect the negative battery cable and check for leaks.

3.5L ENGINE

1. Before servicing the vehicle, refer to the precautions in the beginning of this section.

2. Disconnect the negative battery cable.

3. Remove the rocker arm cover.

4. Install the lash adjuster clips on the rocker arms, then loosen the bearing cap bolts. Do not remove the bolts from the bearing caps.

5. Remove the rocker arms, shafts and bearing caps as an assembly.

To install:

6. Install the bearing caps/rocker arm assemblies. Tighten the bolts to 23 ft. lbs. (31 Nm).

7. Remove the lash adjuster clips.

8. Install the rocker arm cover using a new gasket.

9. Connect the negative battery cable.

Eclipse and Galant

2.4L ENGINE

1. Before servicing the vehicle, refer to the precautions in the beginning of this section.

2. Remove or disconnect the following:
- Negative battery cable
- Accelerator cable, remove the cable clamp mounting screws and position the accelerator cable out of the way.
- Air intake hose
- Breather hose and the Positive Crankcase Ventilation (PCV) hose
- Spark plug cables from the spark plugs
- Rocker cover and gasket

3. Install lash adjuster retainer tools to the rocker arm.

4. Remove the rocker shaft hold-down bolts gradually and evenly and remove the rocker shaft/arm assemblies.

5. Disassemble the rockers and the rocker shaft springs from the rocker shafts. If they are to be reused, note the location and positioning of all rocker shaft components. It is recommended that all lash adjusters and rockers be replaced as a complete set.

To install:

6. Immerse the lash adjusters in clean diesel fuel, and using a small wire, move the plunger up and down 4 or 5 times.

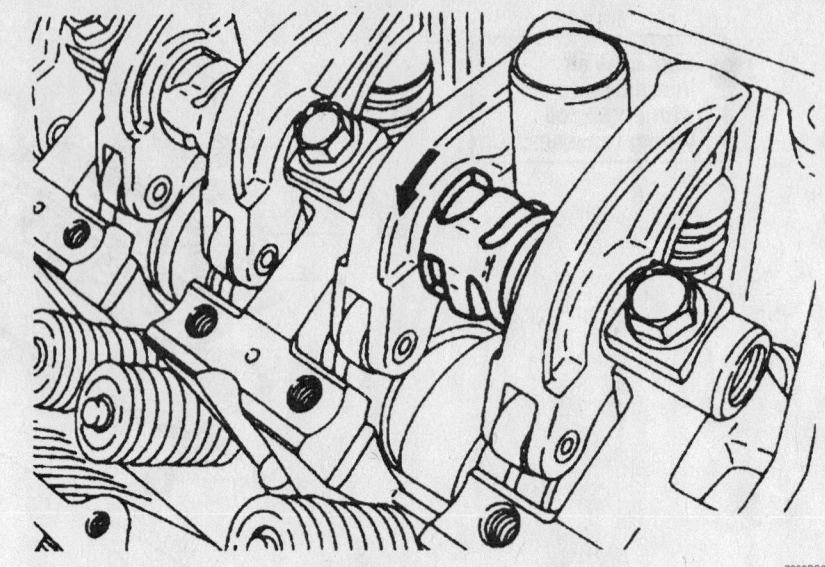

Installing the rocker shaft springs—Eclipse and Galant 2.4L (4G64) engines

7923PG33

While pushing down lightly on the check ball in order to bleed the air from the adjuster.

7. Install the lash adjusters to the rocker arms and attach the special holding tool.

8. Lubricate the rocker shaft with clean engine oil and install the rocker arms.

9. Temporarily tighten the rocker shaft assembly with the mounting bolts so that all rocker arms on the inlet valve side do not push on the valves.

10. Fit the rocker shaft springs from above and position them so that they are at right angles to the plug side. Install the rocker springs before installing the exhaust side rocker shaft and rocker arm assembly.

11. Install the exhaust side rocker shaft assembly in the engine. Tighten the rocker shaft mounting bolts gradually and evenly to 23 ft. lbs. (32 Nm).

12. Remove the lash adjuster retaining tools.

13. Install or connect the following:
- Rocker cover and tighten the mounting bolts to 29 inch lbs. (3.3 Nm)
- Spark plug wires to the spark plugs
- PCV and breather hoses
- Air intake hose
- Accelerator cable brackets and reconnect the accelerator cable
- Negative battery cable

Lancer

1. Before servicing the vehicle, refer to the precautions in the beginning of this section.

2. Drain the engine oil.

3. Remove or disconnect the following:
- Negative battery cable
- Breather hose
- Positive Crankcase Ventilation (PCV) hose
- Oil filler cap
- Rocker arm (valve) cover and gasket
- Oil seals, if necessary
- Rocker arm spring

4. Install special tool MD998443 to prevent the lash adjusters from falling out.
- Intake rocker arms and shafts
- Exhaust rocker arms and shafts
- Lash adjusters

To install:

5. Clean the lash adjusters if they are being reused.

6. Install or connect the following:
- Lash adjuster onto the rocker arm, but do not allow the oil to spill out. Install the special tool to prevent the lash adjusters from falling out of the rocker arms during installation.
- Rocker arm shafts, placing the end with the notched side toward the timing belt
- Rocker arms. Move the rocker arms from side to side before tightening the shaft bolts to 21–25 ft. lbs. (28–34 Nm).
- Rocker arm spring. Insert the spring at an angle to the spark plug guide, then install it so that it is at a right angle to the guide.

7. Remove the special tool from the rocker arms.

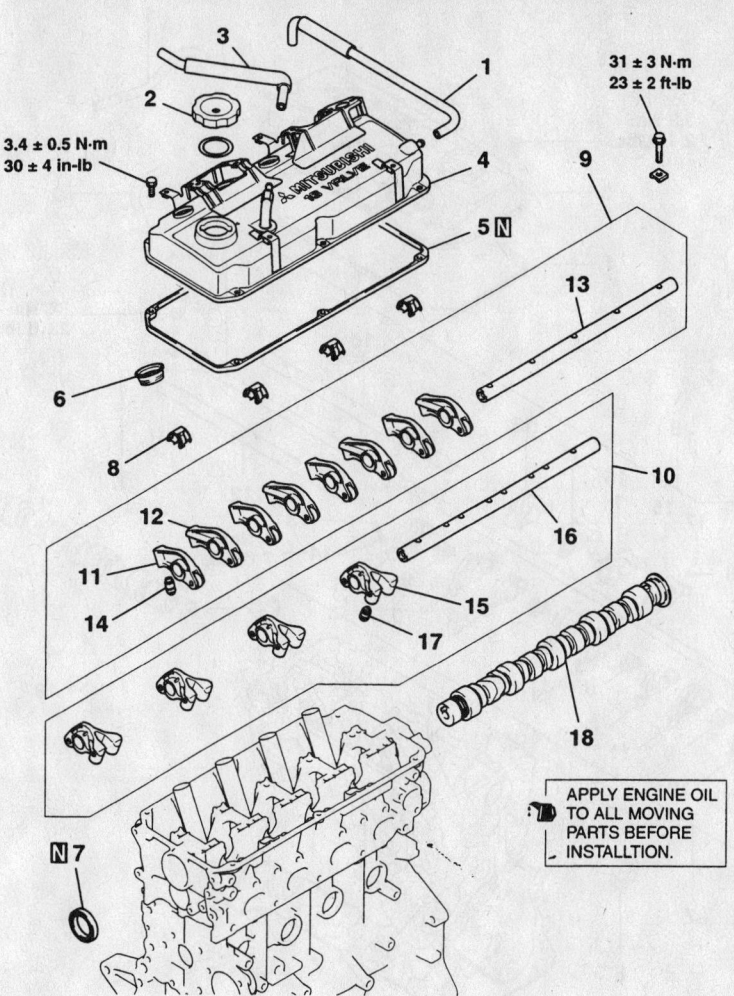

3.4 ± 0.5 N·m
30 ± 4 in-lb

31 ± 3 N·m
23 ± 2 ft-lb

APPLY ENGINE OIL TO ALL MOVING PARTS BEFORE INSTALLTION.

1. Breather hose
2. Positive crankcase ventilation hose
3. Oil hiller cap
4. Rocker cover
5. Rocker cover gasket
6. Oil seal
7. Oil seal
8. Rocker arm spring
9. Rocker arms and rocker arm shaft (Intake)
10. Rocker arms and rocker arm shaft (Exhaust)
11. Rocker arm B
12. Rocker arm A
13. Rocker arm shaft
14. Lash adjuster
15. Rocker arm C
16. Rocker arm shaft
17. Lash adjuster
18. Camshaft

9357QG11

Exploded view of the rocker arms and related components—Lancer 2.0L engine

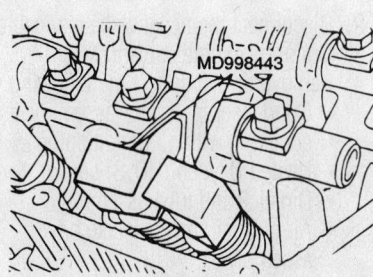

9357QG12

This special tool must be used to keep the lash adjusters in place

- Oil seals, if necessary
- Rocker arm cover gasket and cover. Tighten the bolts to 26–34 inch lbs. (2.9–3.9 Nm).
- Oil filler cap
- PCV hose
- Breather hose

8. Fill the crankcase with oil and connect the negative battery cable.

Mirage

1. Before servicing the vehicle, refer to the precautions in the beginning of this section.
2. Remove or disconnect the following:

- Negative battery cable
- Spark plug cables—for 1.8L engine
- Accelerator cable, breather hose and Positive Crankcase Ventilation (PCV) hose connections
- Rocker cover

3. Loosen both rocker arm shaft assemblies gradually and evenly and remove the rocket shafts from the vehicle.
4. If disassembly is required, keep all parts in the exact order of removal.

To install:

5. Lubricate the rocker shaft with clean engine oil and install the rockers and springs.

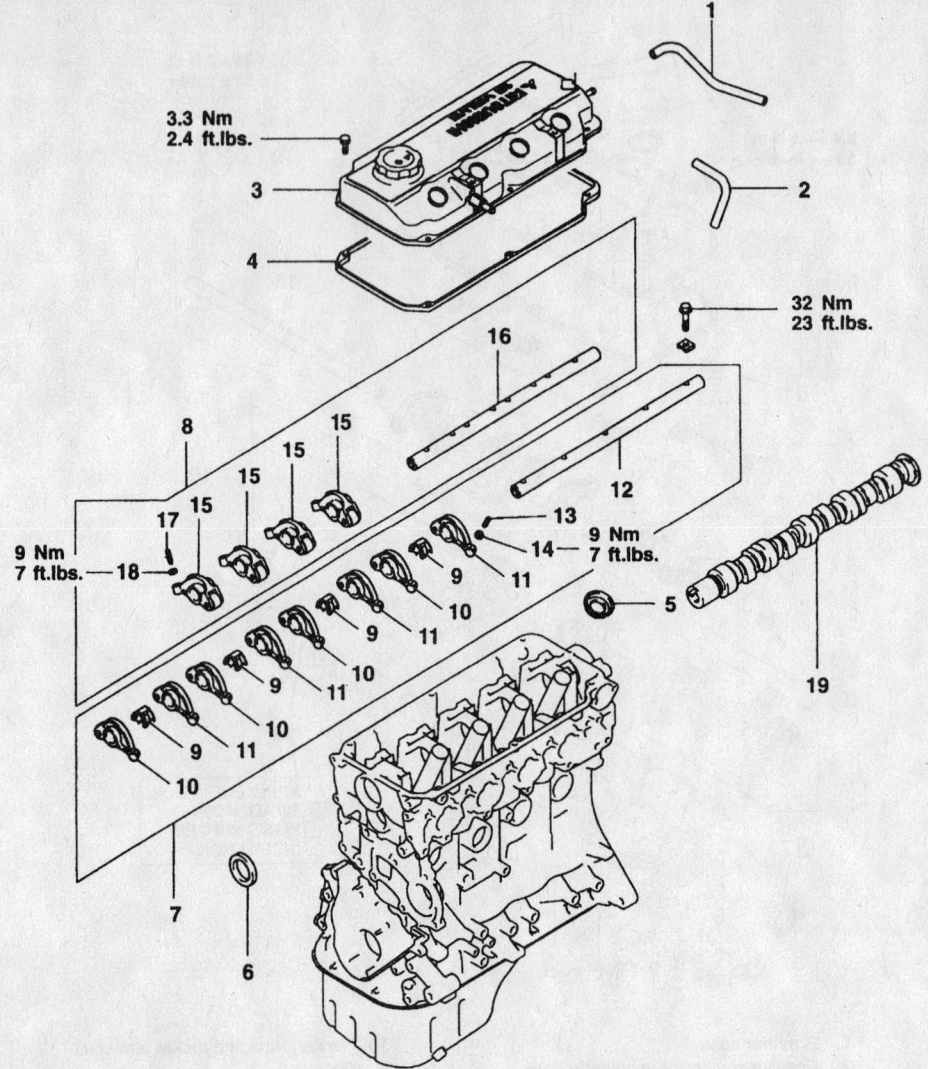

3.3 Nm
2.4 ft.lbs.

32 Nm
23 ft.lbs.

9 Nm
7 ft.lbs.

9 Nm
7 ft.lbs.

1. Breather hose
2. P.C.V. hose
3. Rocker cover
4. Rocker cover gasket
   Valve clearance pre-adjustment
5. Oil seal
6. Oil seal
7. Rocker arms and rocker arm shaft
8. Rocker arms and rocker arm shaft
9. Rocker shaft spring
10. Rocker arm A
11. Rocker arm B
12. Rocker arm shaft (Intake side)
13. Adjusting screw
14. Nut
15. Rocker arm C
16. Rocker arm shaft (Exhaust side)
17. Adjusting screw
18. Nut
19. Camshaft

7923PG27

Camshaft, rocker arm and shaft assemblies—Mirage 1.8L (4G93) engine

6. Install the rocker arm and shaft assemblies. Tighten the rocker arm shaft retainer bolts to 23 ft. lbs. (32 Nm).

7. Check valve adjustment and install the valve cover. Tighten the valve cover bolts to 16 inch lbs. (1.8 Nm) for the 1.5L engine or to 29 inch lbs. (3.3 Nm) for the 1.8L engine.

8. Install or connect the following:
• Spark plug cables, if detached
• Accelerator cable, breather hose and PCV hose
• Negative battery cable

Turbocharger

REMOVAL & INSTALLATION

Eclipse

✳✳ CAUTION

The air bag system must be disarmed before removing the turbocharger.

1. Before servicing the vehicle, refer to the precautions in the beginning of this section.

2. Disconnect the negative battery cable.
3. Drain the engine coolant.
4. Remove or disconnect the following:
• Condenser fan motor assembly, if equipped with air conditioning
• Heated Oxygen (HO2S) sensor
• Dipstick and tube assembly
• Air cleaner and air intake hose assembly
• Air intake hose from the turbocharger
• Engine coolant hoses from the turbocharger

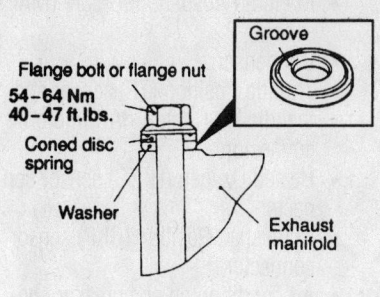

Flange bolt or flange nut
54–64 Nm
40–47 ft. lbs.
Coned disc spring
Washer
Groove
Exhaust manifold

7923PG37

Install the groove of the cone-shaped disc spring toward the flange bolt or nut—Eclipse

- Oil supply pipe connection
- Heat shields
- Engine hanger
- Front exhaust pipe from the turbocharger
- Oil return pipe and gaskets
- Flange bolts and nut that attach the turbo to the exhaust manifold. Take note of the positions of the coned disc springs and the washers.
- Turbocharger, gasket and ring

To install:

5. Use a new gasket and install the turbo to the exhaust manifold. Be sure the coned disc spring and the washers are installed in their original positions. Torque the bolts and nut to 20–23 ft. lbs. (27–31 Nm). Further tighten the bolts and nuts 60–70 degrees.

6. Install or connect the following:
- Exhaust pipe to the turbo—Torque the mounting bolts to 40–47 ft. lbs. (54–64 Nm).
- Oil return pipe
- Engine hanger
- Heat shields
- Oil supply pipe. Torque the flare nut fittings to 14 ft. lbs. (19 Nm).

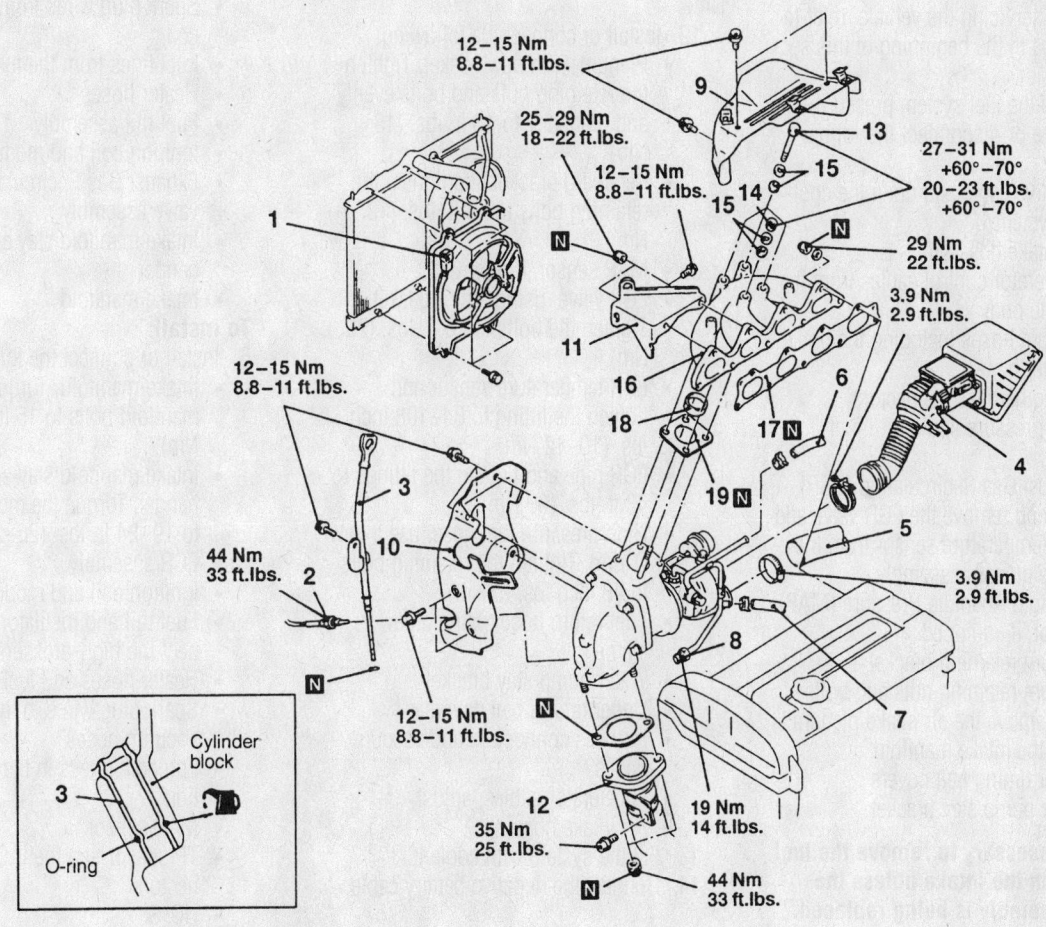

12–15 Nm
8.8–11 ft.lbs.

25–29 Nm
18–22 ft.lbs.

12–15 Nm
8.8–11 ft.lbs.

27–31 Nm
+60°–70°
20–23 ft.lbs.
+60°–70°

29 Nm
22 ft.lbs.

3.9 Nm
2.9 ft.lbs.

12–15 Nm
8.8–11 ft.lbs.

3.9 Nm
2.9 ft.lbs.

44 Nm
33 ft.lbs.

12–15 Nm
8.8–11 ft.lbs.

19 Nm
14 ft.lbs.

35 Nm
25 ft.lbs.

44 Nm
33 ft.lbs.

Cylinder block
O-ring

7923PG47

Removal steps

1. Condenser fan motor assembly <Vehicles with air conditioning>
2. Heated oxygen sensor <front>
3. Engine oil level gauge guide
4. Air cleaner and air intake hose assembly
5. Air hose (A) connection
6. Water hose connection
7. Water hose connection
8. Oil pipe (A) connection
9. Heat protector (A)

10. Heat protector (B)
11. Engine hanger
12. Front exhaust pipe connection
13. Flange bolts
14. Flange nut
15. Coned disc spring
16. Exhaust manifold
17. Exhaust manifold gasket
18. Ring
19. Gasket (A)

Exploded view of the turbocharger mounting—Eclipse

- Engine coolant hoses to the turbo
- Air hose
- Air cleaner and duct assembly
- Dipstick tube and dipstick, with new O-ring
- HO2S sensor
- Condenser fan assembly if removed
- Negative battery cable and refill the engine with coolant

Intake Manifold

REMOVAL & INSTALLATION

Diamante

1. Before servicing the vehicle, refer to the precautions in the beginning of this section.
2. Relieve the fuel system pressure.
3. Remove or disconnect the following:

- Negative cable and drain the cooling system
- Air intake hose(s)
- Accelerator control cables from the throttle body
- Vacuum hoses including the brake booster hose
- Wiring harness connectors
- High pressure and return fuel hoses
- Exhaust Gas Recirculation (EGR) pipe and remove the EGR valve and EGR temperature sensor from the intake plenum assembly
- Manifold Absolute Pressure (MAP) sensor, if equipped
- Plenum retaining bracket
- Plenum retaining nuts and bolts and remove the air intake plenum from the intake manifold
- Upper timing belt covers
- Water pump stay bracket

➡ **It is not necessary to remove the fuel injectors from the intake unless the manifold assembly is being replaced.**

- Fuel rail with the injectors attached
- Coolant hoses from the intake manifold
- Intake manifold mounting nuts and remove the intake manifold

4. Clean the gasket mounting surfaces.

To install:

5. Thoroughly clean the mating surfaces of the heads, intake manifold and air intake plenum.
6. Install new intake manifold gaskets to the cylinder heads with the adhesive side facing up.

7. Place the manifold on the cylinder heads.
8. Lubricate the studs lightly with oil and install the nuts.
9. Tighten the mounting nuts as follows:

- Front bank nuts: 48–72 inch lbs. (5–8 Nm).
- Rear bank nuts: 14–17 ft. lbs. (20–23 Nm).
- Front bank nuts: 14–17 ft. lbs. (20–23 Nm).

10. Connect the coolant hoses to the intake manifold.
11. Using new O-rings, install the fuel rail assembly, if removed. Tighten the mounting bolts to 84–108 inch lbs. (10–13 Nm).
12. Install or connect the following:

- Plenum, with new gasket. Tighten the retaining nuts and bolts evenly and gradually to 13 ft. lbs. (18 Nm).
- Retaining bracket and tighten the retaining bolts to 13 ft. lbs. (18 Nm)
- MAP sensor, if removed
- EGR valve, using a new gasket. Tighten the bolts to 16 ft. lbs. (22 Nm).
- EGR temperature sensor and tighten the fitting to 84–108 inch lbs. (10–12 Nm)
- EGR pipe and tighten the fittings to 43 ft. lbs. (60 Nm)
- High pressure fuel hose, use a new O-ring. Tighten the retaining bolts to 48 inch lbs. (5 Nm).
- Fuel return hose, using a new clamp
- Water pump stay bracket
- Upper timing belt covers
- Harness connectors and vacuum hoses
- Accelerator cables, adjust
- Air intake hose(s)

13. Fill the system with coolant.
14. Connect the negative battery cable.

Eclipse

2.4L (4G64) ENGINE

1. Before servicing the vehicle, refer to the precautions in the beginning of this section.
2. Relieve the fuel system pressure.
3. Remove the battery.
4. Drain the engine coolant.
5. Remove or disconnect the following:

- Accelerator cable
- Air intake hose
- Ignition coil and the module wiring connectors

- Manifold Absolute Pressure (MAP) sensor
- Condenser
- Throttle Position (TP) sensor and the Idle Air Control (IAC) motor connectors
- Heated Oxygen (HO2S) sensor connector
- Crankshaft Position (CKP) sensor connector
- Air conditioning compressor connector
- Engine control wiring harness bracket and position the harness out of the way
- Vacuum hoses, label for reference
- Spark plug wires from the ignition coil
- Fuel lines from the fuel rail
- Heater hoses
- Fuel rail assembly
- Ignition coil and module
- Exhaust Gas Recirculation (EGR) valve assembly
- Intake manifold stay and the engine hanger
- Intake manifold

To install:

6. Install or connect the following:

- Intake manifold. Torque the intake manifold bolts to 15 ft. lbs. (20 Nm).
- Intake manifold stay and the engine hanger. Torque the mounting bolts to 19–24 ft. lbs. (26–33 Nm).
- EGR assembly
- Ignition coil and module
- Fuel rail and insulators and reconnect the high-pressure fuel hose
- Heater hoses and fuel lines
- Spark plug wires to the coil towers
- Vacuum hoses
- Engine harness in the proper position
- MAP sensor
- TP sensor and the IAC motor connectors
- HO2S
- Ignition condenser
- Accelerator cable, adjust
- Battery

7. Refill the engine with coolant.

Galant

1. Before servicing the vehicle, refer to the precautions in the beginning of this section.
2. Relieve the fuel system pressure.
3. Disconnect battery negative cable.
4. Drain the cooling system.
5. Remove or disconnect the following:

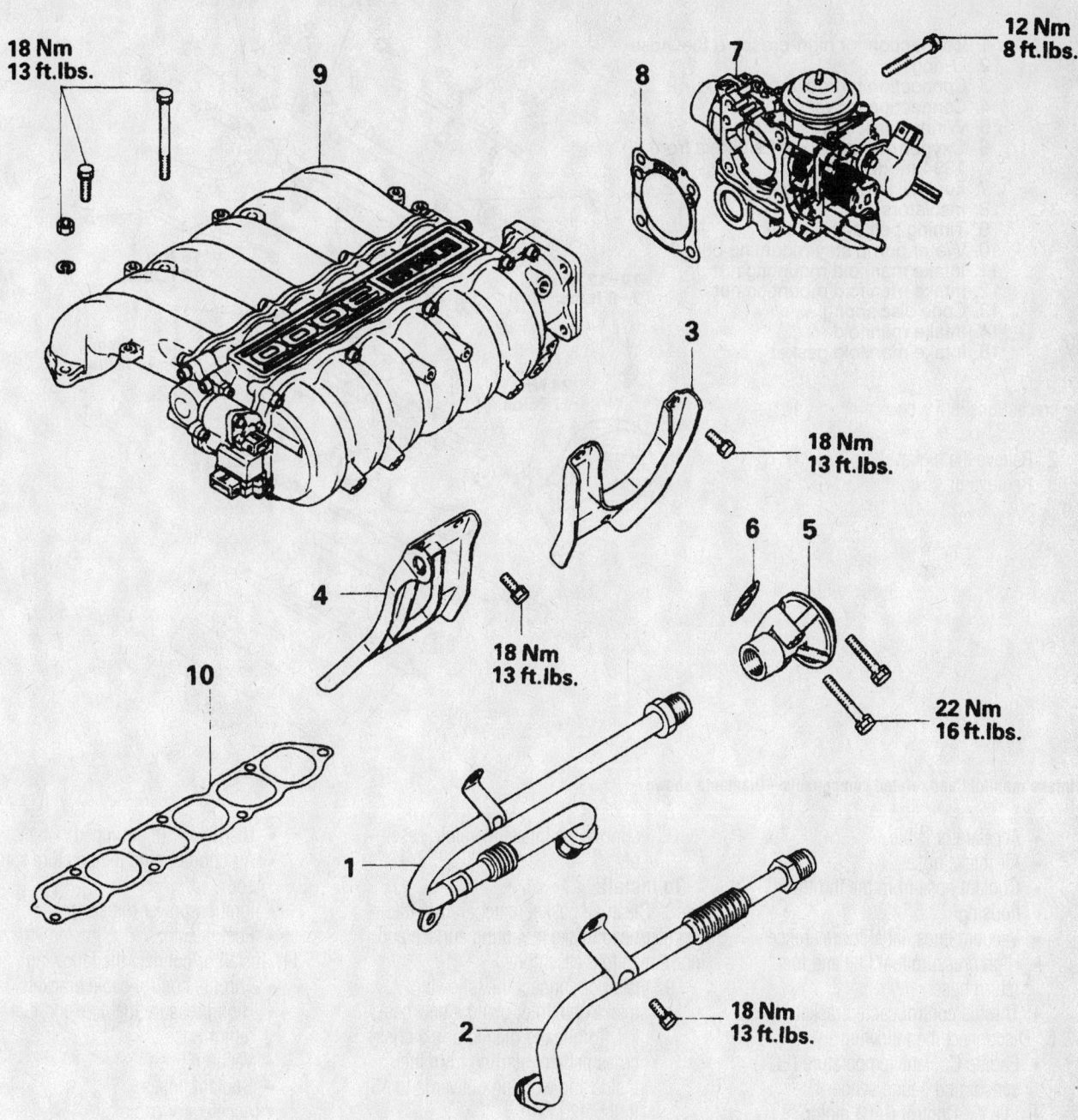

18 Nm
13 ft.lbs.

12 Nm
8 ft.lbs.

18 Nm
13 ft.lbs.

18 Nm
13 ft.lbs.

22 Nm
16 ft.lbs.

18 Nm
13 ft.lbs.

1. EGR pipe -- Up to 1993 <California> model
2. EGR pipe – From 1994 <California> model
3. Intake manifold plenum stay, rear
4. Intake manifold plenum stay, front
5. EGR valve
6. EGR valve gasket } <For California>
7. Throttle body
8. Throttle body gasket
9. Intake manifold plenum
10. Intake manifold plenum gasket

7923PG42

Exploded view of air intake plenum assembly—Diamante

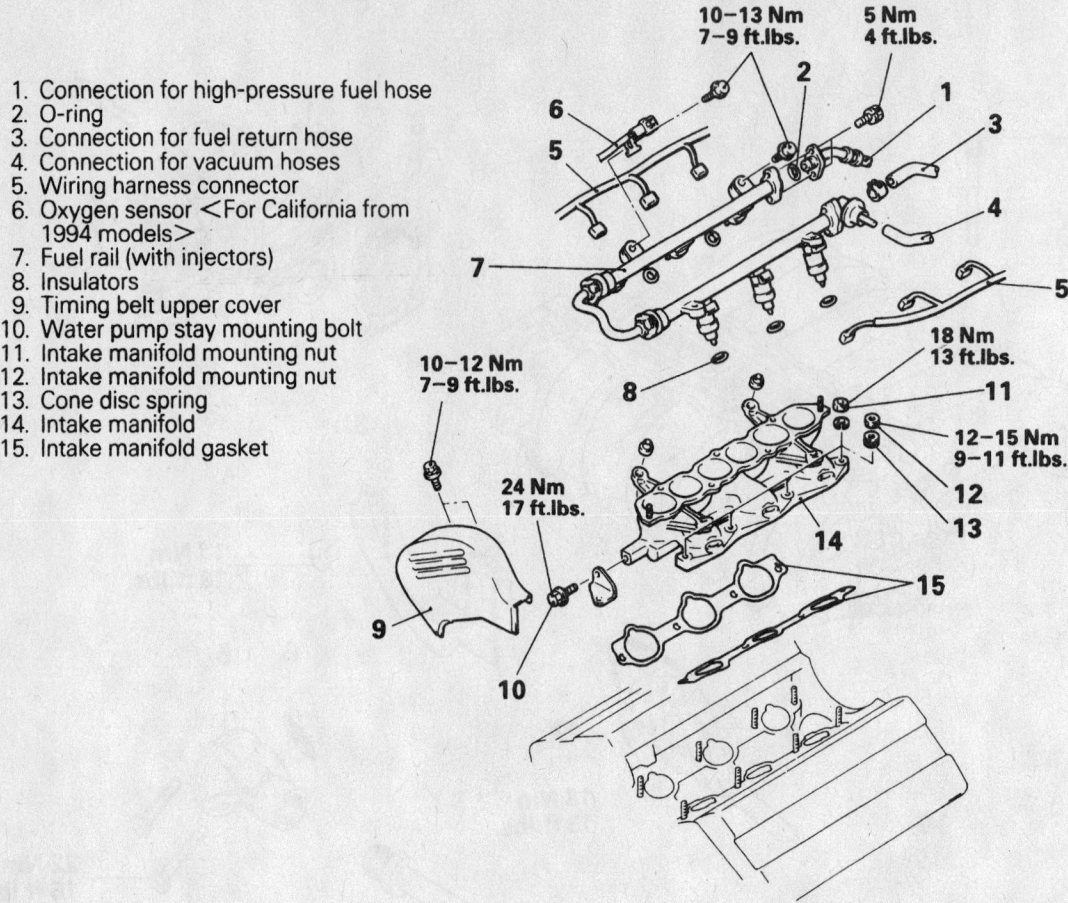

1. Connection for high-pressure fuel hose
2. O-ring
3. Connection for fuel return hose
4. Connection for vacuum hoses
5. Wiring harness connector
6. Oxygen sensor <For California from 1994 models>
7. Fuel rail (with injectors)
8. Insulators
9. Timing belt upper cover
10. Water pump stay mounting bolt
11. Intake manifold mounting nut
12. Intake manifold mounting nut
13. Cone disc spring
14. Intake manifold
15. Intake manifold gasket

Intake manifold and related components—Diamante shown

- Accelerator cable
- Air intake hose
- Coolant hose from the throttle housing
- Vacuum lines, label for reference
- High pressure fuel line and fuel return hose
- Throttle control cable brackets

6. Disconnect the following:
- Engine Coolant Temperature (ECT) sensor and gauge sender
- Idle Air Control (IAC) motor
- Exhaust Gas Recirculation (EGR) temperature sensor
- Ignition coil
- Knock (KS) sensor
- Heated Oxygen (HO$_2$S) sensor
- Throttle Position (TP) sensor
- Distributor (if equipped)
- Air conditioning temperature sensor
- Ignition power transistor
- Fuel injectors

7. Remove or disconnect the following:
- Spark plug wires
- Intake manifold stay bracket
- Intake manifold mounting bolts and

remove the intake manifold assembly

To install:

8. Clean all gasket material from the cylinder head intake mounting surface and intake manifold assembly.

9. Install or connect the following:
- Intake manifold, using a new gasket. Torque the manifold in a criss-cross pattern, starting from the inside and working outwards to 15 ft. lbs. (20 Nm).
- Fuel injectors, fuel rail and pressure regulator to the engine. Torque the retaining bolts to 48 inch lbs. (6 Nm).
- Intake manifold brace bracket and tighten bolts to 21 ft. lbs. (29 Nm)
- Spark plug wires

10. Connect the following:
- ECT sensor and gauge sender
- IAC motor
- EGR temperature sensor
- Ignition coil
- (KS) sensor
- HO$_2$S sensor
- TP sensor

- Distributor (if equipped)
- Air conditioning temperature sensor
- Ignition power transistor
- Fuel injectors

11. Install or connect the following:
- Throttle control cable brackets
- High pressure fuel line and fuel return hose
- Vacuum lines
- Coolant hoses
- Accelerator cable
- Air intake hose

12. Fill the system with coolant.
13. Connect the negative battery cable.

Lancer

1. Before servicing the vehicle, refer to the precautions in the beginning of this section.
2. Relieve the fuel system pressure.
3. Remove or disconnect the following:
- Battery negative cable and drain the cooling system
- Air cleaner assembly
- Throttle body
- Fuel rail assembly

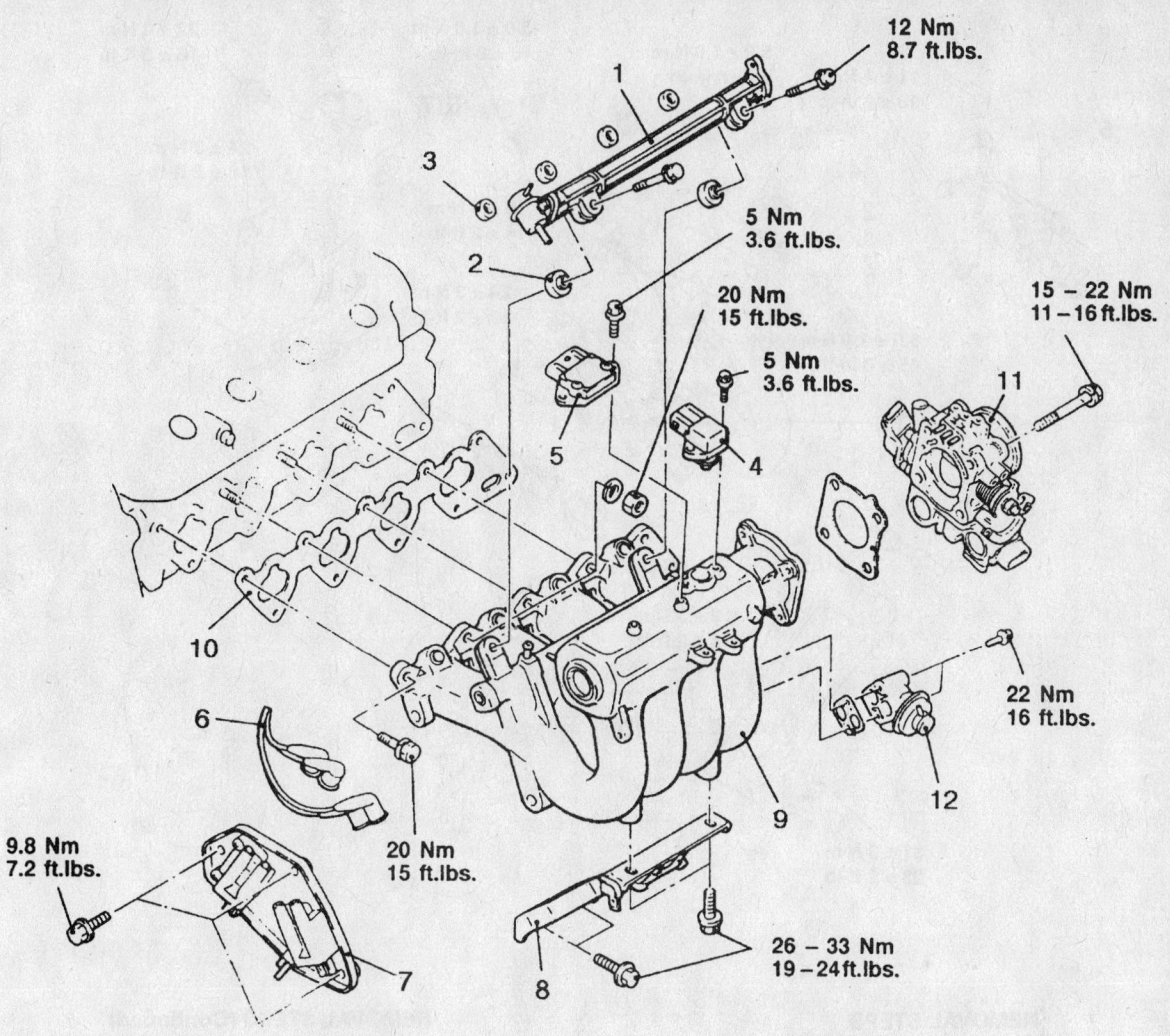

12 Nm
8.7 ft.lbs.

5 Nm
3.6 ft.lbs.

20 Nm
15 ft.lbs.

5 Nm
3.6 ft.lbs.

15 – 22 Nm
11 – 16 ft.lbs.

22 Nm
16 ft.lbs.

9.8 Nm
7.2 ft.lbs.

20 Nm
15 ft.lbs.

26 – 33 Nm
19 – 24 ft.lbs.

7923PG41

1. Fuel rail, fuel injector and pressure regulator assembly
2. Insulator
3. Insulator
4. Manifold differential pressure sensor
5. Ignition power transistor
6. Spark plug cable connection
7. Ignition coil
8. Intake manifold stay
9. Intake manifold
10. Intake manifold gasket
11. Throttle body
12. EGR valve assembly

Intake manifold and related components—Eclipse and Galant 2.4L (4G64) engines

- Manifold Differential Pressure (MDP) sensor
- Auto cruise vacuum hose connection
- Vacuum pipe
- Brake booster vacuum hose connection
- Vacuum hose and pipe assembly
- Exhaust Gas Recirculation (EGR) valve and gasket
- Engine hanger
- Throttle body stay
- Intake manifold stay

- Intake manifold bolt and manifold.
- Intake manifold gasket and discard

To install:

4. Clean all gasket material from the cylinder head intake mounting surface and intake manifold assembly.

5. Install or connect the following:
- Intake manifold, using a new gasket. Torque the manifold in a criss-cross pattern, starting from the inside and working outwards to 12–16 ft. lbs. (16–22 Nm).
- Intake manifold stay

- Throttle body stay
- Engine hanger
- EGR valve and gasket
- Vacuum hose and pipe assembly
- Brake booster vacuum hose connection
- Vacuum pipe
- Auto cruise vacuum hose connection
- MDP sensor
- Fuel rail assembly
- Throttle body
- Air cleaner assembly

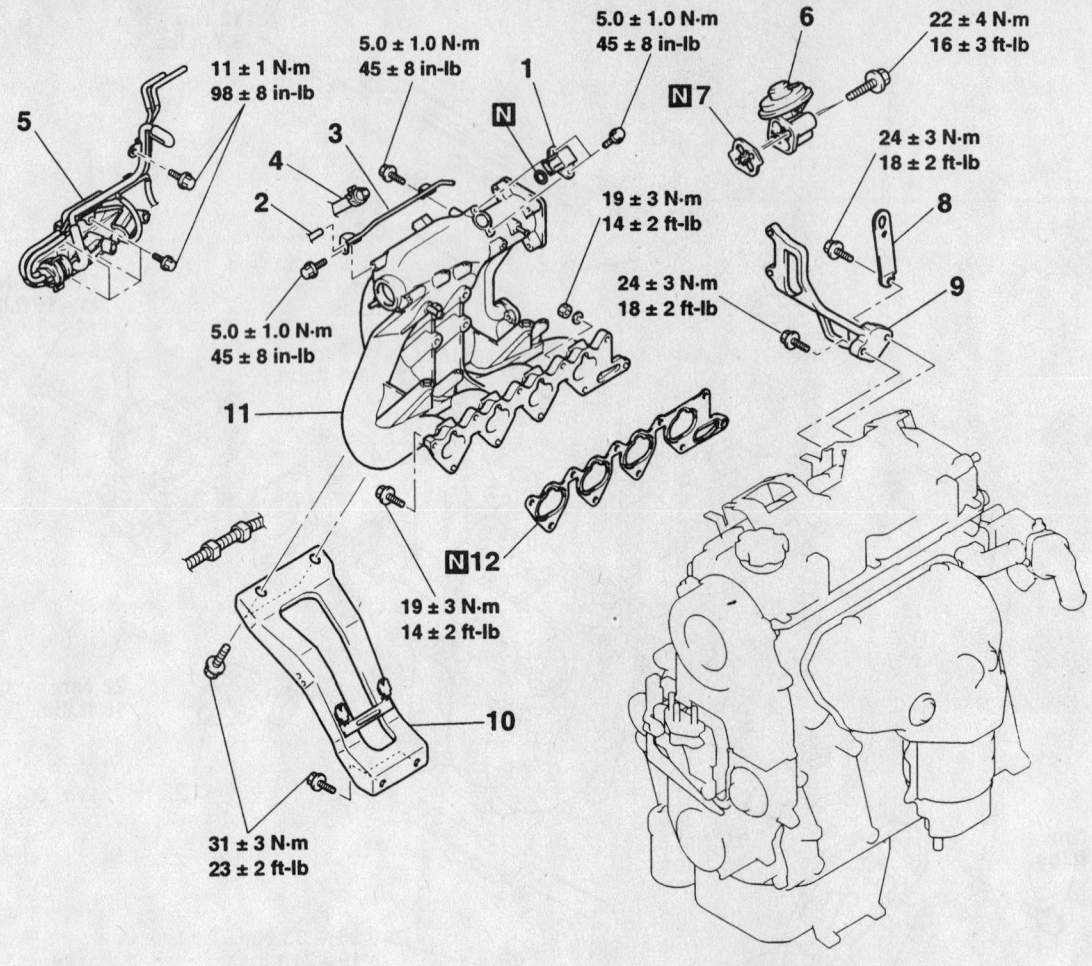

11 ± 1 N·m
98 ± 8 in-lb

5.0 ± 1.0 N·m
45 ± 8 in-lb

5.0 ± 1.0 N·m
45 ± 8 in-lb

5.0 ± 1.0 N·m
45 ± 8 in-lb

22 ± 4 N·m
16 ± 3 ft-lb

24 ± 3 N·m
18 ± 2 ft-lb

19 ± 3 N·m
14 ± 2 ft-lb

24 ± 3 N·m
18 ± 2 ft-lb

19 ± 3 N·m
14 ± 2 ft-lb

31 ± 3 N·m
23 ± 2 ft-lb

REMOVAL STEPS
1. MANIFOLD DIFFERENTIAL PRESSURE SENSOR
2. AUTO CRUISE VACUUM HOSE CONNECTION
3. VACUUM PIPE
4. BRAKE BOOSTER VACUUM HOSE CONNECTION
5. VACUUM HOSE AND PIPE ASSEMBLY

REMOVAL STEPS (Continued)
6. EGR VALVE
7. EGR VALVE GASKET
8. ENGINE HANGER
9. THROTTLE BODY STAY
10. INTAKE MANIFOLD STAY
11. INTAKE MANIFOLD
12. INTAKE MANIFOLD GASKET

93570G13

Exploded view of the intake manifold and related components—Lancer with 2.0L engine

6. Fill the system with coolant.
7. Connect the negative battery cable.

Mirage

1.5L (4G15) ENGINE

1. Before servicing the vehicle, refer to the precautions in the beginning of this section.
2. Relieve the fuel system pressure.
3. Remove or disconnect the following:
   - Battery negative cable and drain the cooling system
   - Upper radiator hose, heater hose and water bypass hose
   - Thermostat housing from intake manifold
   - Accelerator cable, breather hose and air intake hose
   - Vacuum hoses, label for reference
   - Throttle body assembly
   - High pressure fuel line and the fuel return hose
4. Disconnect the following:
   - Heated Oxygen (HO$_2$S) sensor
   - Engine Coolant Temperature (ECT) sensor
   - Idle Air Control (IAC) motor
   - Intake Air Temperature (IAT) sensor
   - Distributor (if equipped)
   - Exhaust Gas Recirculation (EGR) temperature sensor
5. Remove or disconnect the following:
   - Spark plug wires
   - Fuel rail, fuel injectors, pressure regulator and insulators
   - EGR valve from the intake manifold
   - Intake manifold support bracket and remove the engine mount support bracket
   - Intake manifold mounting bolts and remove the intake manifold assembly

To install:

6. Clean all gasket material from the cylinder head intake mounting surface and intake manifold assembly.

7. Install or connect the following:
- Intake manifold gasket, using a new gasket. Torque the manifold in a crisscross pattern, starting from the inside and working outwards to 13 ft. lbs. (18 Nm).
- Intake manifold support bracket and tighten the mounting bolts to 16 ft. lbs. (22 Nm)
- Engine mount support bracket and tighten the mounting bolts to 26 ft. lbs. (36 Nm)
- EGR valve and tighten the mounting bolts to 15 ft. lbs. (21 Nm)
- Install the fuel rail, fuel injectors and pressure regulator to the engine, using new insulators and O-rings. Torque the retaining bolts to 84–108 inch lbs. (10–13 Nm).
- Spark plug wires

8. Connect the following:
- HO2S sensor
- ECT sensor

- IAC motor
- IAT sensor
- Distributor (if equipped)
- EGR temperature sensor

9. Install or connect the following:
- Fuel feed and return lines
- Throttle body assembly
- Vacuum hoses and pipes as necessary, including the brake booster vacuum line
- Accelerator cable
- Breather and air intake hose
- Thermostat housing to the intake manifold and tighten the mounting bolts to 13 ft. lbs. (18 Nm)
- Upper radiator hose, heater hose and water bypass hose

10. Fill the system with coolant.
11. Connect the negative battery cable.

1.8L (4G93) ENGINE

1. Before servicing the vehicle, refer to the precautions in the beginning of this section.
2. Relieve the fuel system pressure.
3. Remove or disconnect the following:
- Battery negative cable and drain the cooling system

- Accelerator cable and the air intake hose

4. Disconnect the following:
- Heated Oxygen (HO2S) sensor
- Engine Coolant Temperature (ECT) sensor
- Idle Air Control (IAC) motor
- Exhaust Gas Recirculation (EGR) temperature sensor
- Throttle Position (TP) sensor
- Oil pressure switch
- Distributor (if equipped)
- Fuel injectors

5. Label and remove all vacuum hoses.
6. Remove or disconnect the following:
- Upper radiator hose, heater hose and water bypass hose
- High pressure fuel line and the fuel return hose
- Fuel rail, fuel injectors, pressure regulator and insulators
- Intake manifold support bracket
- Thermostat housing, if necessary for clearance
- Intake manifold mounting bolts/nuts and remove the intake manifold assembly

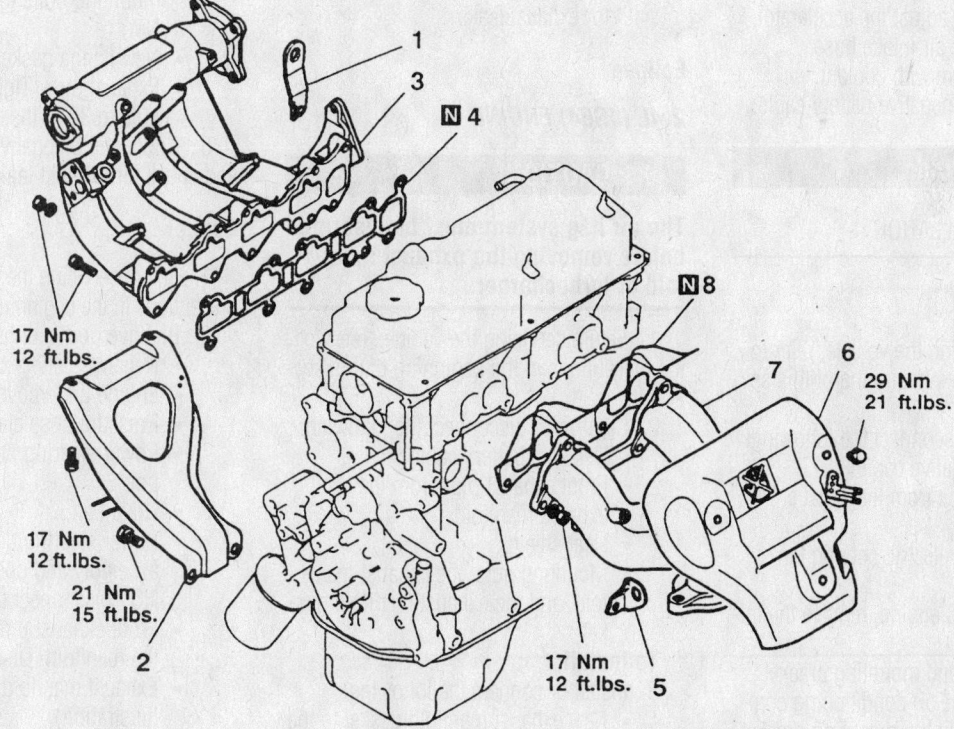

1. Engine hanger
2. Intake manifold stay
3. Intake manifold
4. Intake manifold gasket
5. Engine hanger
6. Exhaust manifold cover
7. Exhaust manifold
8. Exhaust manifold gasket

7923PG38

Exploded view of the intake and exhaust manifold mounting—Mirage 1.5L (4G15) engine

To install:

7. Clean all gasket material from the cylinder head intake mounting surface and intake manifold assembly.

8. Install or connect the following:
- Intake manifold, using a new gasket. Torque the manifold in a criss-cross pattern, starting from the inside and working outwards to 14 ft. lbs. (20 Nm).
- Thermostat housing
- Intake manifold brace bracket
- Fuel rail, fuel injectors and pressure regulator to the engine. Torque the retaining bolts to 108 inch lbs. (12 Nm).
- Fuel feed and return lines
- Upper radiator hose, heater hose and water bypass hoses
- Vacuum hoses

9. Connect the following:
- HO2S sensor
- ECT sensor
- IAC motor
- EGR temperature sensor
- TP sensor
- Oil pressure switch
- Distributor (if equipped)
- Fuel injectors

10. Connect and adjust the accelerator cable and install the air intake hose.

11. Fill the system with coolant.

12. Connect the negative battery cable.

Exhaust Manifold

REMOVAL & INSTALLATION

Diamante

1. Before servicing the vehicle, refer to the precautions in the beginning of this section.

2. Remove or disconnect the following:
- Battery negative cable
- Exhaust pipe from the exhaust manifold
- Condenser electric cooling fan assembly

3. For the DOHC engine, remove the following:
- Alternator and mounting bracket
- Separate the air conditioning compressor from the mounting bracket. Leaving the hoses connected, position the compressor aside.
- Heat protector

4. If removing the front manifold, remove the oil dipstick and tube from the engine.

5. If removing the rear manifold, disconnect the Exhaust Gas Recirculation (EGR) tube.

6. For the SOHC engine, if removing the rear manifold, remove the intake plenum stay and the roll stopper bracket.

7. Remove or disconnect the following:
- Electrical connector and remove the Heated Oxygen (HO2S) sensor
- Exhaust manifold mounting bolts the manifold

To install:

8. Clean all gasket material from the mating surfaces.

9. Install or connect the following:
- New gasket and install the manifold. Tighten the nuts in a criss-cross pattern to 21 ft. lbs. (30 Nm) for the J- engine or to 14 ft. lbs. (19 Nm) for the H- engine.
- Heat shields
- EGR tube and intake plenum stay and roll stopper bracket, if removed
- HO2S sensor
- Electric cooling fan assembly, air conditioning compressor, dipstick tube and alternator, as required
- New flange gasket and connect the exhaust pipe or converter assembly
- Negative battery cable and check for exhaust leaks

Eclipse

2.4L (4G64) ENGINE

※※ CAUTION

The air bag system must be disarmed before removing the exhaust manifold or turbocharger.

1. Before servicing the vehicle, refer to the precautions in the beginning of this section.

2. Remove or disconnect the following:
- Negative battery cable
- Front exhaust pipe from the exhaust manifold
- Heat shield
- Mounting nuts, the exhaust manifold, and the exhaust manifold gasket

To install:

3. Install or connect the following:
- New exhaust manifold gasket to the cylinder head
- Exhaust manifold. Torque the mounting nuts to 21 ft. lbs. (29 Nm).
- Heat shield and tighten the bolts to 10 ft. lbs. (13 Nm)
- New gasket between the exhaust manifold and the front exhaust pipe

and reconnect the pipe. Torque the nuts to 32 ft. lbs. (34 Nm).
- Negative battery cable

4. Start the engine and check for any exhaust leaks.

Galant

1. Before servicing the vehicle, refer to the precautions in the beginning of this section.

2. Remove or disconnect the following:
- Battery negative cable
- Exhaust pipe from the exhaust manifold
- Outer exhaust manifold heat shield and engine hanger
- Exhaust manifold mounting nuts and the exhaust manifold from the engine

To install:

3. Clean all gasket material from the mating surfaces.

4. Install or connect the following:
- New gasket and install the manifold. Tighten the nuts to in a criss-cross pattern to 18–21 ft. lbs. (25–29 Nm).
- Heat shields and tighten the mounting bolts to 10 ft. lbs. (14 Nm)
- New flange gasket
- Exhaust pipe. Tighten the mounting nuts to 32 ft. lbs. (44 Nm).

5. Connect the negative battery cable and check for exhaust leaks.

Lancer

1. Before servicing the vehicle, refer to the precautions in the beginning of this section.

2. Remove or disconnect the following:
- Negative battery cable
- Engine undercover
- Pressure hose clamp bolt
- Power steering pump bracket stay bolt
- Drive belt
- Power steering pump and bracket assembly and position it aside. Do NOT disconnect the fluid lines.
- Front exhaust pipe and gasket from the manifold. Discard the gasket.
- Exhaust manifold bracket "B" (see illustration)
- Heated Oxygen Sensor (HO2S)
- Heat shield
- Exhaust manifold retainers and manifold
- Exhaust manifold gasket, and discard
- Exhaust manifold bracket "A" (see illustration)

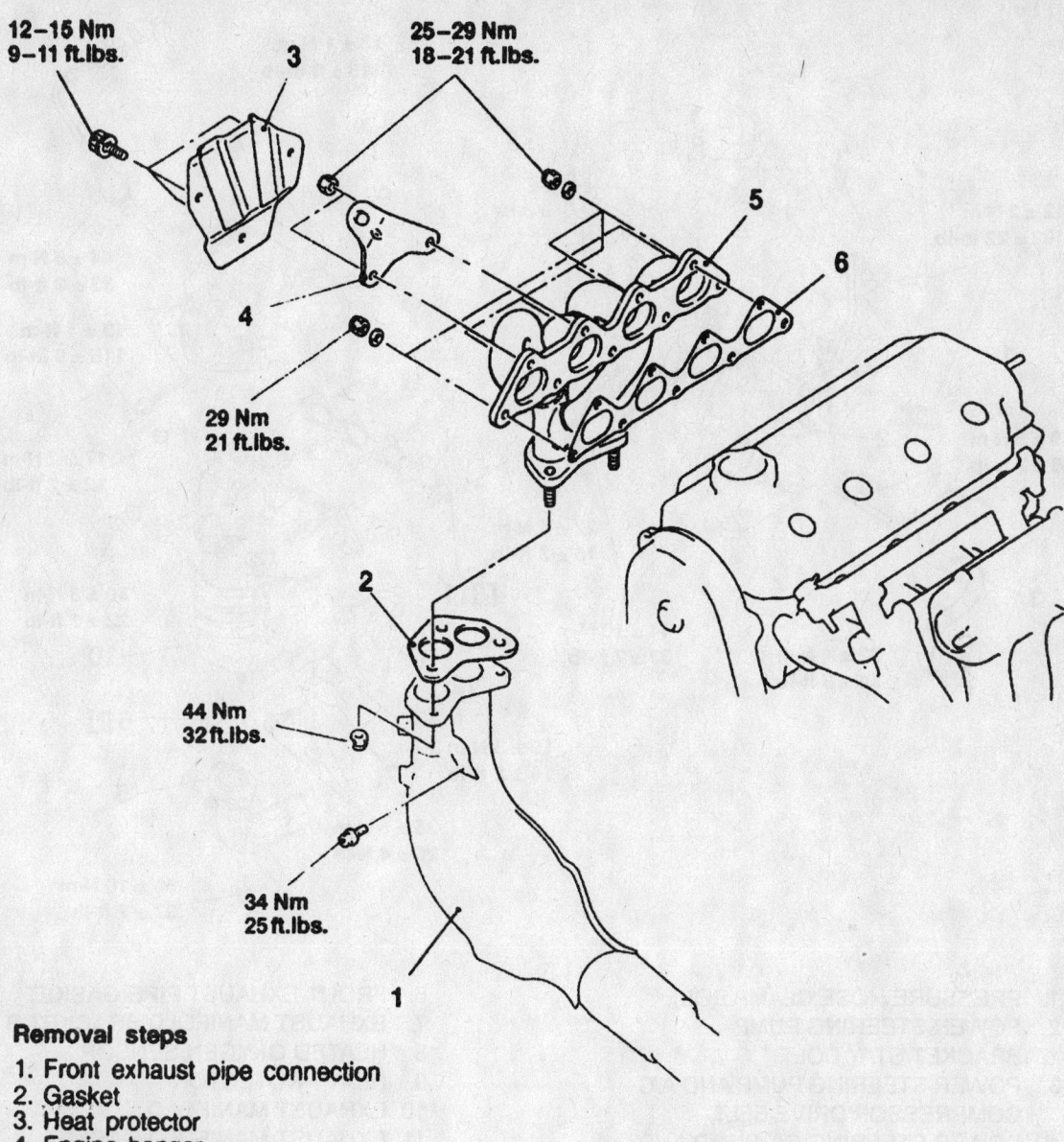

12–15 Nm
9–11 ft.lbs.

25–29 Nm
18–21 ft.lbs.

29 Nm
21 ft.lbs.

44 Nm
32 ft.lbs.

34 Nm
25 ft.lbs.

Removal steps
1. Front exhaust pipe connection
2. Gasket
3. Heat protector
4. Engine hanger
5. Exhaust manifold
6. Exhaust manifold gasket

7923PG46

Exploded view of the exhaust manifold mounting—Eclipse 2.4L (4G64) engine shown, Galant similar

To install:
3. Clean all gasket material from the mating surfaces.
4. Install or connect the following:
  • Exhaust manifold bracket "A"
  • New gasket and exhaust manifold. Tighten the upper retainers to 20–24 ft. lbs. (27–33 Nm) and the lower retainers to 10–14 ft. lbs. (15–19 Nm).
  • Heat shield
  • HO$_2$S
  • Exhaust manifold bracket "B"
  • Front exhaust pipe to the exhaust manifold, using a new gasket

  • Power steering pump assembly
  • Drive belt
  • Power steering pump bracket stay bolt
  • Pressure hose clamp bolt
  • Engine undercover
  • Negative battery cable

Mirage

1. Before servicing the vehicle, refer to the precautions in the beginning of this section.
2. Remove or disconnect the following:
  • Battery negative cable

  • Exhaust pipe from the exhaust manifold
  • Electric cooling fan assembly
  • Heated Oxygen (HO$_2$S) sensor
  • Exhaust Gas Recirculation (EGR) pipe
  • Outer exhaust manifold heat shield and engine hanger
  • Exhaust manifold mounting bolts, the inner heat shield and the exhaust manifold

To install:
3. Clean all gasket material from the mating surfaces.

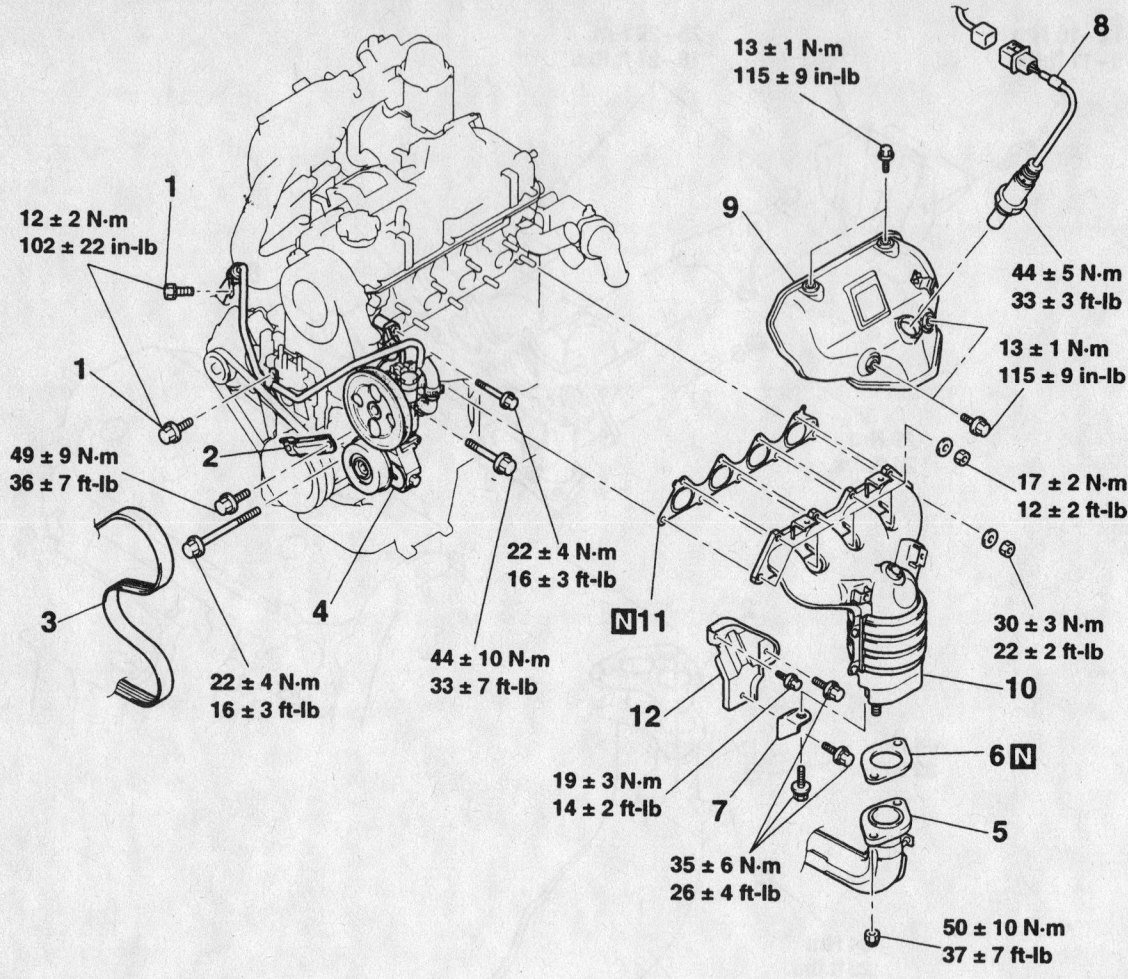

13 ± 1 N·m
115 ± 9 in-lb

44 ± 5 N·m
33 ± 3 ft-lb

13 ± 1 N·m
115 ± 9 in-lb

17 ± 2 N·m
12 ± 2 ft-lb

30 ± 3 N·m
22 ± 2 ft-lb

12 ± 2 N·m
102 ± 22 in-lb

49 ± 9 N·m
36 ± 7 ft-lb

22 ± 4 N·m
16 ± 3 ft-lb

22 ± 4 N·m
16 ± 3 ft-lb

44 ± 10 N·m
33 ± 7 ft-lb

19 ± 3 N·m
14 ± 2 ft-lb

35 ± 6 N·m
26 ± 4 ft-lb

50 ± 10 N·m
37 ± 7 ft-lb

1. PRESSURE HOSE CLAMP BOLT
2. POWER STEERING PUMP
   BRACKET STAY BOLT
3. POWER STEERING PUMP AND A/C
   COMPRESSOR DRIVE BELT
4. POWER STEERING OIL PUMP AND
   BRACKET ASSEMBLY
5. FRONT EXHAUST PIPE
   CONNECTION
6. FRONT EXHAUST PIPE GASKET
7. EXHAUST MANIFOLD BRACKET B
8. HEATED OXYGEN SENSOR
9. HEAT PROTECTOR
10. EXHAUST MANIFOLD
11. EXHAUST MANIFOLD GASKET
12. EXHAUST MANIFOLD BRACKET A

9357QG14

Exploded view of the exhaust manifold and related components—Lancer 2.0L engine

4. Using a new gasket and install the manifold. For 1.5L engines, tighten the nuts on a crisscross patter to 13 ft. lbs. (18 Nm). For 1.8L engines, tighten the inner nuts to in a crisscross pattern to 13 ft. lbs. (18 Nm) and tighten the 2 outer (larger) nuts to 22 ft. lbs. (30 Nm).

5. Install or connect the following:
   • Heat shields
   • EGR pipe
   • HO2S sensor
   • Electric cooling fan assembly
   • New flange gasket and connect the exhaust pipe

• Negative battery cable and check for exhaust leaks

Front Crankshaft Seal

REMOVAL & INSTALLATION

Diamante

1. Before servicing the vehicle, refer to the precautions in the beginning of this section.

2. Remove or disconnect the following:

• Negative battery cable
• Drive belts
• Crankshaft pulley
• Timing belt covers and the timing belt
• Crankshaft Position (CKP) sensor
• Crankshaft sprocket, the sensing blade, spacer and Woodruff® key

3. Pry the seal from the bore, using a suitable tool.

To install:

4. Using a seal driver, install the new crankshaft seal. Lubricate the lips of the seal with clean engine oil.

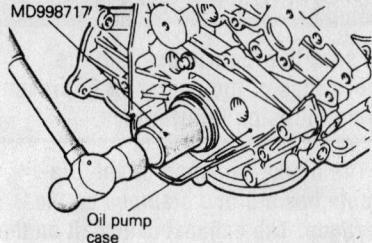

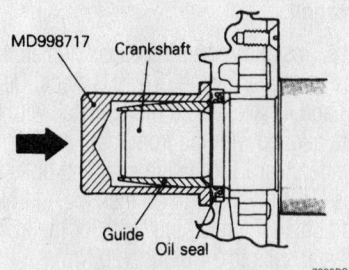

Crankshaft seal installation—Diamante

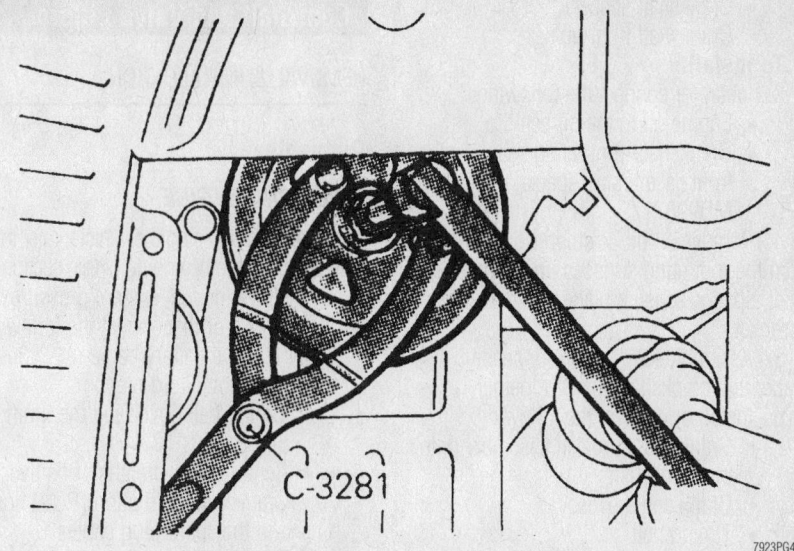

The crankshaft pulley pin spanner tool should be used to hold pulley while bolt is removed—Diamante

5. Install or connect the following:
- Woodruff® key, spacer, sensing blade and the crankshaft sprocket
- CKP sensor and tighten the retaining bolts to 84 inch lbs. (9 Nm)
- Timing belt and the timing belt covers
- Crankshaft pulley and retaining bolt. Torque the retaining bolt to 130–137 ft. lbs. (180–190 Nm) for the DOHC engine or to 108–116 ft. lbs. (150–160 Nm) for the SOHC engine.
- Drive belts, adjust
- Negative battery cable

Eclipse

2.4L ENGINE

1. Before servicing the vehicle, refer to the precautions in the beginning of this section.
2. Remove or disconnect the following:

- Negative battery cable
- Timing belt
- Crankshaft sprocket

3. Carefully pry the oil seal out of the front case. Be careful not to damage the oil seal bore or the crankshaft sealing surface.

To install:

4. Apply clean engine oil to the oil seal lip. Using a seal driver, install the oil seal.
5. Install or connect the following:
- Crankshaft sprocket. If equipped, tighten the crankshaft bolt to 87 ft. lbs. (118 Nm).
- Timing belt
- Negative battery cable

Lancer

1. Before servicing the vehicle, refer to the precautions in the beginning of this section.
2. Remove or disconnect the following:

- Negative battery cable
- Timing belt
- Crank angle sensor
- Crankshaft sprocket
- Spring pin
- Crankshaft sending blade

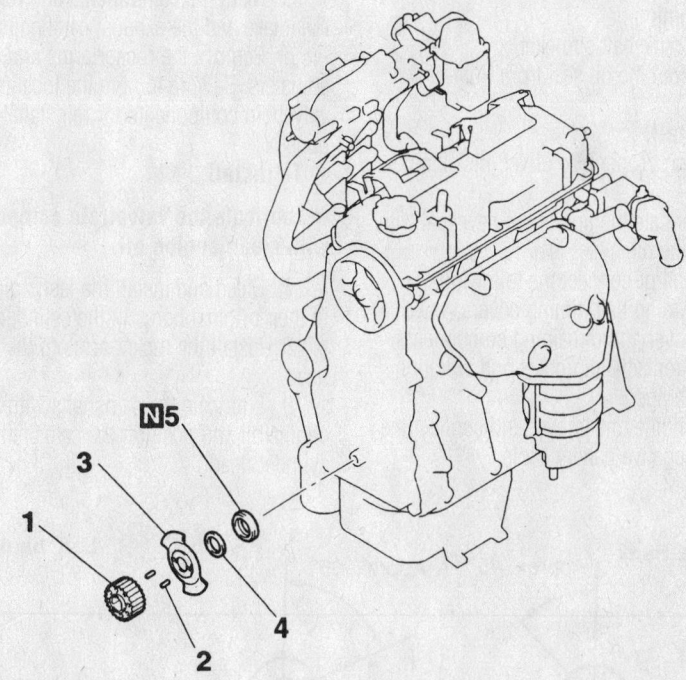

1.	CRANKSHAFT SPROCKET
2.	SPRING PIN
3.	CRANKSHAFT SENSING BLADE
4.	CRANKSHAFT SPACER
5.	CRANKSHAFT FRONT OIL SEAL

View of the front crankshaft seal—Lancer 2.0L engine

- Crankshaft spacer
- Crankshaft front oil seal

To install:

3. Install or connect the following:
- Engine oil to the oil seal lip
- Crankshaft front using seal into the front case, using special driver tool MD998717

4. Remove all oil or other lubricants from the mounting surfaces on the crankshaft, spacer, sensing blade and crankshaft sprocket.

5. Assemble the spring pin, sensing blade, and crankshaft spacer together.

6. Install or connect the following:
- Crankshaft sprocket assembly onto the crankshaft
- Crank angle sensor
- Timing belt
- Negative battery cable

Mirage

1. Before servicing the vehicle, refer to the precautions in the beginning of this section.

2. Remove or disconnect the following:
- Negative battery cable
- Crankshaft pulley retainer bolts and remove the pulley
- Vibration damper retainer bolt and washer and remove damper
- Timing belt
- Crankshaft sprocket

3. Pry out the oil seal from front of engine.

To install:

4. Using proper size driver, install a new front seal.

5. Lubricate the lips of the new seal with clean engine oil.

6. Install or connect the following:
- Timing belt, timing covers, valve cover and remaining components
- Crankshaft sprocket and vibration damper
- Engine undercover and connect the negative battery cable

Camshaft and Valve Lifters

REMOVAL & INSTALLATION

Diamante

3.0L DOHC ENGINE

1. Before servicing the vehicle, refer to the precautions in the beginning of this section.

2. Relieve the fuel system pressure.

3. Remove or disconnect the following:
- Negative battery cable
- Intake manifold plenum
- Timing belt cover and the timing belt
- Center cover, breather, Positive Crankcase Ventilation (PCV) hoses, and the spark plug cables
- Rocker cover and the semi-circular packing
- Camshaft Position (CMP) sensor
- Camshaft sprockets

➡**Be sure to keep the valvetrain components labeled and in proper order for reassembly.**

4. Loosen the bearing cap bolts in 2–3 steps. Label and remove all camshaft bearing caps.

5. Mark the components and remove the intake and the exhaust camshafts.

6. Remove the rocker arms and the lash adjusters. Be sure to note the location of the valvetrain components for reinstallation purposes.

To install:

➡**Lubricate the valvetrain components with clean engine oil.**

7. Bleed and install the lash adjusters in their original bores in the cylinder head.

8. Install the rocker arms to the cylinder head.

9. Lubricate the camshafts with clean engine oil and position the camshafts on the cylinder head.

☀☀ WARNING

Be sure to properly position the knock pins of the camshaft to prevent valve to piston interference.

➡**The intake camshaft on the Diamante has a B or J stamped on the hexagon. The exhaust camshaft on the Diamante has a D or K stamped on the hexagon.**

10. Install the bearing caps. Tighten the caps in sequence and in 2 or 3 steps. Caps 2, 3 and 4 have a front mark. Install with the mark aligned with the front mark on the cylinder head. Torque the retaining bolts for caps No. 2, 3 and 4 to 96 inch lbs. (11 Nm) and tighten the retaining bolts for the front and rear caps to 14 ft. lbs. (20 Nm).

11. Apply a coating of engine oil to the oil seals and install the oil seals to the front and rear of the camshafts.

12. Install or connect the following:
- Camshaft sprockets and tighten the sprocket bolts to 65 ft. lbs. (90 Nm)
- CMP sensor
- Timing belt
- Rocker cover and the semi-circular packing
- Intake manifold plenum
- Spark plug cables, center cover, breather and PCV hoses
- Negative battery cable and check for leaks

3.0L SOHC ENGINE

1. Before servicing the vehicle, refer to the precautions in the beginning of this section.

2. Remove or disconnect the following:
- Negative battery cable
- Intake manifold plenum stay bracket
- Camshaft Position (CMP) sensor
- Valve covers and the timing belt

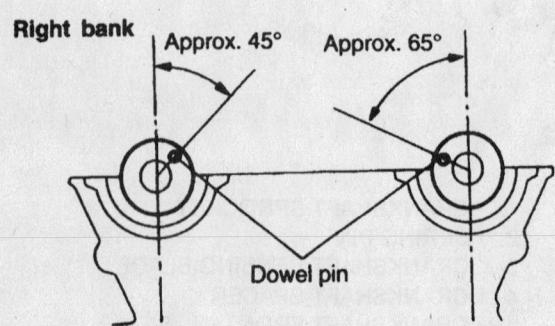

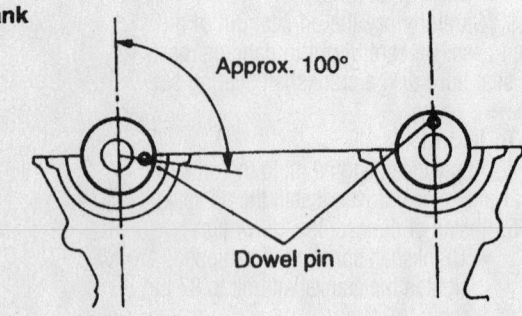

Proper positioning of the camshaft knock pins—Diamante 3.0L DOHC engine

7923PG58

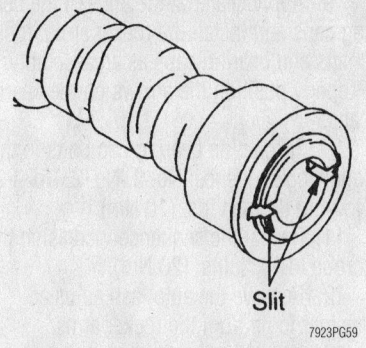

Slit

7923PG59

Right bank camshaft identification—Diamante 3.0L SOHC engine

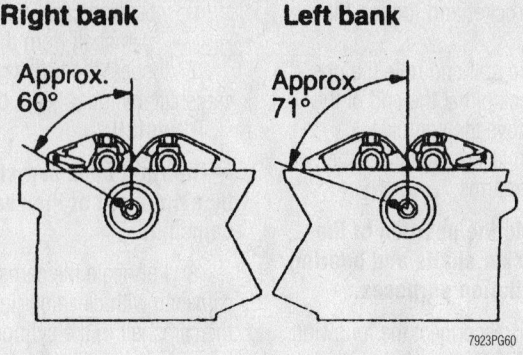

Right bank

Left bank

Approx. 60°

Approx 71°

7923PG60

Proper positioning of the camshafts—Diamante 3.0L SOHC engine

Arrow mark (bearing cap)

Timing belt side

Arrow mark (cylinder head)

Arrow mark (bearing cap)

7923PG61

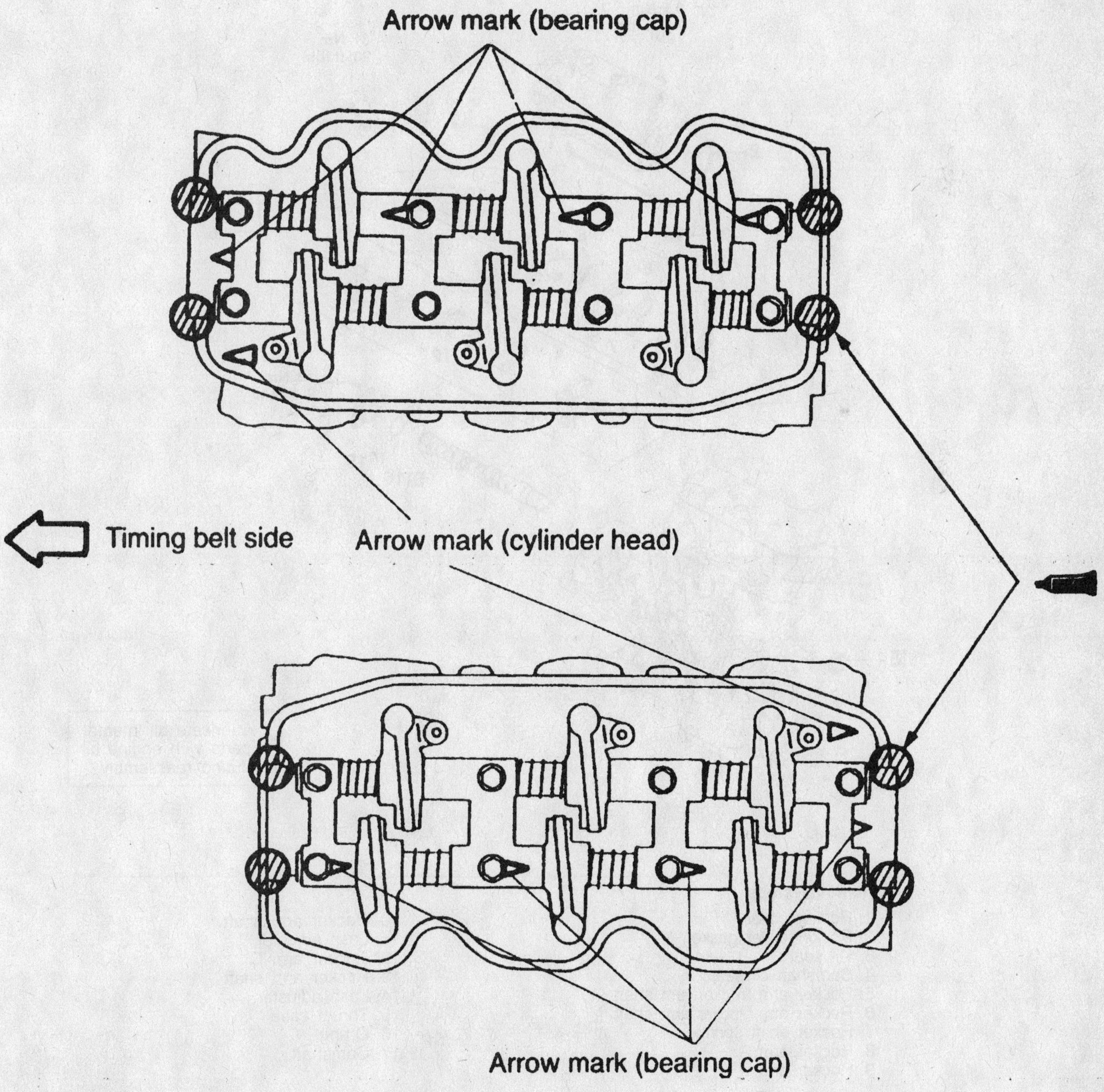

Alignment of the rocker shafts and application of sealant—Diamante 3.0L SOHC engine

3. Using a camshaft sprocket holding tool, hold the sprocket and loosen the bolt.

4. Remove the bolt and note the positioning of the knock pin at the end of the camshaft and remove the sprocket.

5. Install auto lash adjuster retainer tools on the rocker arms.

➡**Be sure to note the position of the rocker arms, rocker shafts and bearing caps for reinstallation purposes.**

6. Remove or disconnect the following:
- Camshaft bearing caps but do not remove the bolts from the caps
- Rocker arms, rocker shafts and bearing caps, as an assembly
- Camshaft from the cylinder head

7. Inspect the bearing journals on the camshaft, cylinder head, and bearing caps.

To install:

➡**The right bank camshaft is identified by a 4mm slit at the rear end of the camshaft.**

8. Lubricate the camshaft journals and camshaft with clean engine oil and install the camshaft in the cylinder head. Be sure to properly position the knock pin of the camshaft as noted during removal.

9. Apply sealer at the ends of the bearing caps and install the rocker arms, rocker shafts and bearing caps as an assembly. Properly position the arrows on the bearing caps.

10. Torque the bearing cap bolts in the following sequence: No. 3, No. 2, No. 1 and No. 4 to 85 inch lbs. (10 Nm).

11. Repeat the sequence increasing the torque to 14 ft. lbs. (20 Nm).

12. Remove the auto lash adjuster retainer tools from the rocker arms.

13. Install the camshaft sprocket and bolt.

14. Using a camshaft sprocket holding

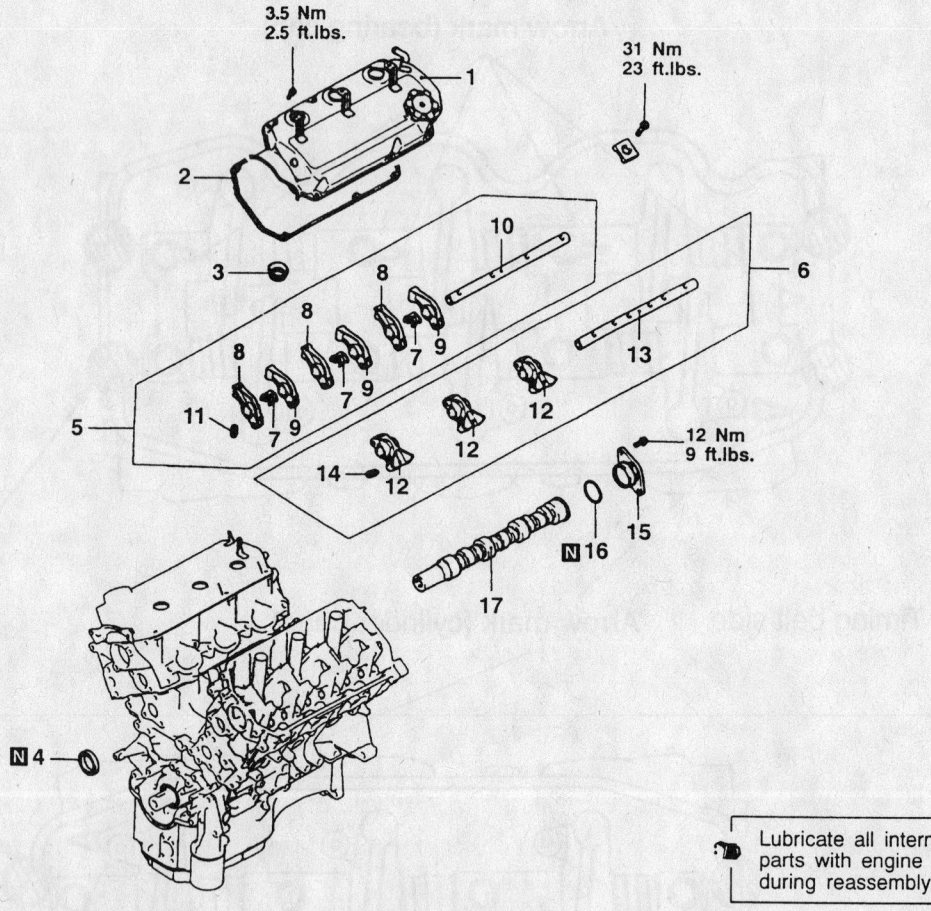

Removal steps
1. Rocker cover
2. Rocker cover gasket
3. Oil seal
4. Camshaft oil seal
5. Rocker arm, rocker arm shaft
6. Rocker arm, rocker arm shaft
7. Rocker shaft spring
8. Rocker arm A
9. Rocker arm B
10. Rocker arm shaft
11. Lash adjuster
12. Rocker arm C
13. Rocker arm shaft
14. Lash adjuster
15. Thrust case
16. O ring
17. Camshaft

Lubricate all internal parts with engine oil during reassembly.

Exploded view of the camshaft mounting—3.5L engine

7923PGD3

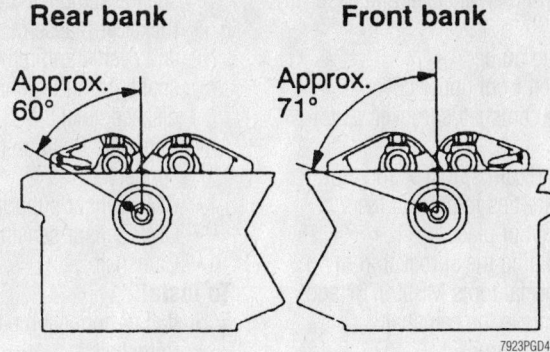

Rear bank Approx. 60°

Front bank Approx. 71°

7923PGD4

Camshaft dowel position during installation—3.5L engine

tool, hold the sprocket and tighten the bolt to 65 ft. lbs. (90 Nm).
15. Install or connect the following:
- Timing belt and valve covers
- CMP sensor
- Intake manifold plenum stay bracket
- Negative battery cable and check for leaks

3.5L ENGINE

1. Before servicing the vehicle, refer to the precautions in the beginning of this section.
2. Remove or disconnect the following:
- Negative battery cable
- Timing belt
- Rocker arm cover
- Lash adjuster clips on the rocker arms, then loosen the bearing cap bolts. Do not remove the bolts from the bearing caps.
- Rocker arms, shafts and bearing caps as an assembly
- Camshafts

To install:
3. Lubricate the camshafts with engine oil and position them on the cylinder heads.
4. Position the dowel pins as shown in the drawing.
5. Install or connect the following:
- Bearing caps/rocker arm assemblies. Tighten the bolts to 23 ft. lbs. (31 Nm).
- Rocker arm cover using a new gasket
- Timing belt and remaining components
- Negative battery cable

Eclipse

1. Before servicing the vehicle, refer to the precautions in the beginning of this section.
2. Remove or disconnect the following:
- Remove the battery

- Accelerator cable bracket and position the cable aside
- Air intake hose
- Breather hose and disconnect the Positive Crankcase Ventilation (PCV) hose
- Spark plug cables
- Rocker cover
3. Install lash adjuster retainer tools to the rocker arm.
- Timing belt covers and the timing belt
- Camshaft sprocket retainer bolt and remove the sprocket from the shaft
- Camshaft oil seal
- Both rocker arm shaft assemblies from the head
- Camshaft from the cylinder head

To install:
4. Lubricate the camshaft journals and camshaft with clean engine oil and install the camshaft in the cylinder head.
5. Install the rocker arm and shaft assemblies. Tighten the rocker arm shaft retainer bolts to 21–25 ft. lbs. (29–35 Nm).
6. Apply a coating of engine oil to the oil seal. Using the proper size driver, press-fit the seal into the cylinder head.
7. Install or connect the following:
- Camshaft sprocket and retainer bolt to 65 ft. lbs. (90 Nm)
- Timing belt and belt covers
8. Remove the lash adjuster retaining tools.
9. Install or connect the following:
- Rocker cover using new gasket material on mating surfaces
- Spark plug cables
- Air intake hose
- Breather hose and connect the PCV hose
- Battery
10. Run the engine at idle until normal operating temperature is reached. Check idle speed and ignition timing; adjust as required.

Galant

1. Before servicing the vehicle, refer to the precautions in the beginning of this section.
2. Relieve the fuel system pressure.
3. Remove or disconnect the following:
- Negative battery cable
- Accelerator cable, Positive Crankcase Ventilation (PCV) hoses, breather hoses, spark plug cables
- Valve cover
- Timing belt upper and lower covers
- Timing belt
- Camshaft sprockets
4. Loosen the bearing cap bolts in 2–3 steps. Label and remove all camshaft bearing caps.
- Intake and exhaust camshafts
- Rocker arms and lash adjusters

To install:
5. Install the lash adjusters and rocker arms into the cylinder head. Lubricate lightly with clean oil prior to installation.
6. Lubricate the camshafts with clean engine oil and position the camshafts on the cylinder head.
7. Be sure the dowel pin on both camshaft sprocket ends are located on the top.
8. Install the bearing caps. Tighten the caps in sequence and in 2 or 3 steps. No. 2 and 5 caps are of the same shape. Check the markings on the caps to identify the cap number and intake/exhaust symbol. Only **L** (intake) or **R** (exhaust) is stamped on No. 1 bearing cap. Also, be sure the rocker arm is correctly mounted on the lash adjuster and the valve stem end. Torque the retaining bolts to 15 ft. lbs. (20 Nm).
9. Apply a coating of engine oil to the oil seal. Using the proper size driver, press-fit the seal into the cylinder head.
10. Install or connect the following:
- Camshaft sprockets and tighten the sprocket bolts to 58–72 ft. lbs. (80–100 Nm)
- Timing belt, covers and related components

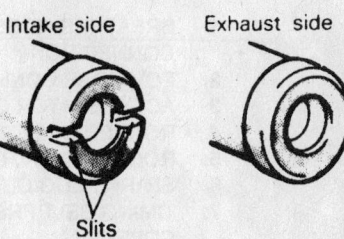

Intake side **Exhaust side**

Slits

7923PG57

Camshaft identification—2.4L (4G64) engine

- Valve cover and reconnect all related components
- Negative battery cable

Lancer

1. Before servicing the vehicle, refer to the precautions in the beginning of this section.
2. Remove or disconnect the following:
   - Negative battery cable
   - Air cleaner assembly
   - Ignition coil
   - Breather hose
   - Positive Crankcase Ventilation (PCV) hose
   - Rocker arm (valve) cover and gasket
   - Spark plug guide
   - Timing belt front upper cover
3. Remove the camshaft sprocket, as follows:

   a. Secure the cam sprocket and timing belt with wire ties to prevent them from slipping out of place.

   b. While holding the sprocket from turning with special tools MB990767 and MD998719, remove the camshaft sprocket bolt and sprocket.
4. Remove or disconnect the following:
   - Intake and exhaust rocker arm and shaft assemblies. Loosen both rocker arm assemblies gradually and evenly and remove the rocker shafts from the vehicle. Do NOT disassemble!
   - Camshaft Position (CMP) sensor connector
   - CMP sensor support
   - CMP sensor sensing cylinder
   - Camshaft

To install:
5. Install or connect the following:
   - Camshaft
   - CMP sensor sensing cylinder, support and connector

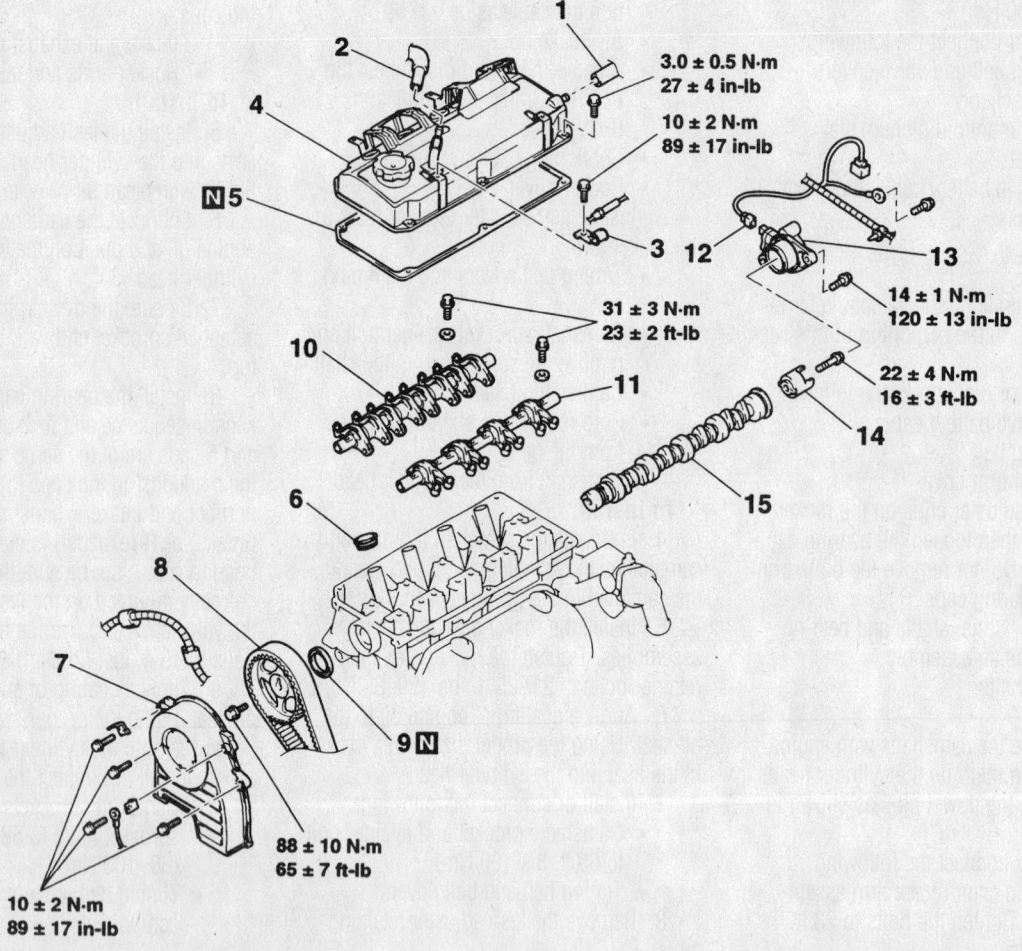

1.	BREATHER HOSE CONNECTION
2.	PCV HOSE CONNECTION
3.	ACCELERATOR CABLE CLAMP
4.	ROCKER COVER
5.	ROCKER COVER GASKET
6.	SPARK PLUG GUIDE
7.	TIMING BELT FRONT UPPER COVER
8.	CAMSHAFT SPROCKET
9.	CAMSHAFT OIL SEAL
10.	INTAKE ROCKER ARM AND SHAFT ASSEMBLY
11.	EXHAUST ROCKER ARM AND SHAFT ASSEMBLY
12.	CAMSHAFT POSITION SENSOR CONNECTOR
13.	CAMSHAFT POSITION SENSOR SUPPORT

Exploded view of the camshaft and related components—Lancer

93570G17

- Rocker arm and shaft assemblies
- Camshaft oil seal
- Camshaft sprocket. Tighten the bolt to 58–72 ft. lbs. (78–98 Nm).
- Timing belt front upper cover
- Spark plug guide
- Rocker arm cover, with a new gasket. Tighten the bolts to 23–31 inch lbs. (2.5–3.5 Nm).
- Accelerator cable clamp
- PCV and breather hoses
- Negative battery cable

Mirage

1.5L (4G15) ENGINE

1. Before servicing the vehicle, refer to the precautions in the beginning of this section.
2. Remove or disconnect the following:
   - Negative battery cable
   - Accelerator cable, breather hose and Positive Crankcase Ventilation (PCV) hose connections
   - Distributor, if equipped
   - Valve cover and discard the gasket
3. Loosen both rocker arm assemblies gradually and evenly and remove the rocker shafts from the vehicle.
4. Remove or disconnect the following:
   - Timing belt covers
   - Timing belt
   - Camshaft sprocket from the camshaft. Note the positioning of the dowel pin at the end of the camshaft.
   - Camshaft oil seal from the front of the cylinder head
   - Camshaft from the head

To install:

5. Lubricate the camshaft with clean engine oil and slide it into the head. Be sure to position the dowel pin at the 12 o'clock position.
6. Install or connect the following:
   - New camshaft oil seal. Be sure to lubricate the lips of the seal with clean engine oil.
   - Camshaft sprocket and install the mounting bolt. Tighten the bolt to 51 ft. lbs. (70 Nm)
   - Timing belt
   - Timing belt covers
   - Rocker shaft assemblies. Torque the bolts gradually and evenly to 23 ft. lbs. (32 Nm).
7. Install the valve cover with a new gasket. Tighten the valve cover bolt to 16 inch lbs. (1.8 Nm).
8. Install or connect the following:
   - Distributor, if equipped
   - Accelerator cable, breather hose and PCV hose

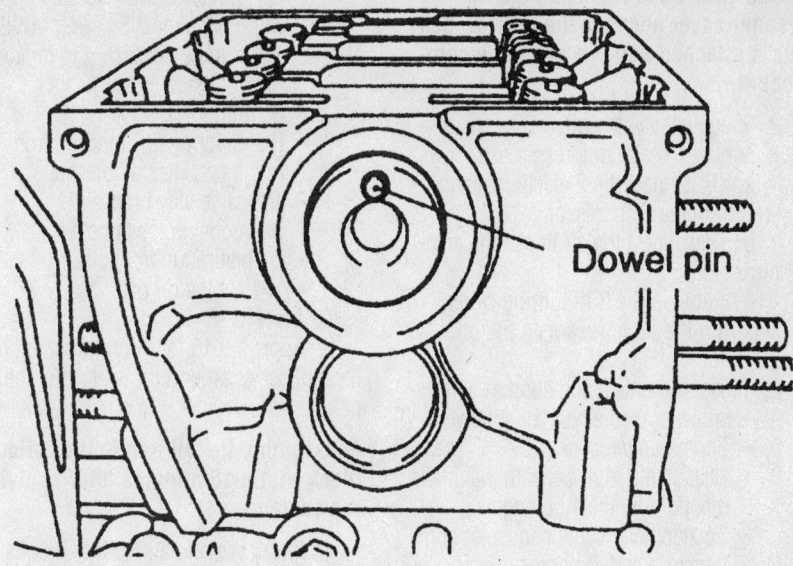

Positioning of the camshaft dowel pin—Mirage 1.5L (4G15) engine

7923PG51

   - Negative battery cable and check the ignition timing

1.8L (4G93) ENGINE

1. Before servicing the vehicle, refer to the precautions in the beginning of this section.
2. Remove or disconnect the following:
   - Negative battery cable
   - Spark plug cables
   - Manifold Absolute Pressure (MAF) sensor connector and remove the air cleaner case cover
   - Accelerator cable, breather hose and Positive Crankcase Ventilation (PCV) hose connections
   - Rocker cover and discard the gasket
3. Loosen both rocker arm shaft assemblies gradually and evenly and remove the rocket shafts from the vehicle.
4. Remove or disconnect the following:
   - Timing belt covers
   - Timing belt
   - Camshaft sprocket from the camshaft. Note the positioning of the dowel pin at the end of the camshaft.
   - Camshaft oil seal from the front of the cylinder head
   - Camshaft from the head

To install:

5. Lubricate the camshaft journals and camshaft with clean engine oil and install the camshaft in the cylinder head. Be sure to position the dowel pin at the end of the camshaft as noted during the removal procedure.

6. Install or connect the following:
   - New camshaft oil seal.
   - Camshaft sprocket and tighten the retainer bolt to 65 ft. lbs. (90 Nm)
   - Timing belt
   - Timing belt covers
   - Rocker arm and shaft assemblies. Tighten the rocker arm shaft retainer bolts to 23 ft. lbs. (32 Nm).
   - Valve cover with a new gasket. Tighten the valve cover bolts to 29 inch lbs. (3.3 Nm).
   - Spark plug cables
   - Accelerator cable, breather hose and PCV hose
   - MAP sensor connector and install the air cleaner case cover
   - Negative battery cable

Valve Lash

ADJUSTMENT

Valve clearance is not adjustable on these vehicles.

Starter Motor

REMOVAL & INSTALLATION

1. Remove or disconnect the following:
   - Negative battery cable
   - Air-flow sensor assembly connector and remove the breather hose
   - Resonator retaining nuts and remove the air intake hose and resonator assembly as required

➡ **Use care when removing the air cleaner cover because the air-flow sensor is attached and is a sensitive component.**

2. If equipped with Active-ECS suspension, remove the air compressor as follows:

    a. Disconnect the 2 electrical connectors, from the compressor.

    b. Disconnect the air line at the compressor.

    c. Remove the 3 mounting bolts, securing the compressor to the chassis.

3. Raise the vehicle and support safely.

4. Remove or disconnect the following:
- Engine undercover
- Heat shield from beneath the intake manifold on the 1.5L engine
- Speedometer cable connector at the transaxle end, if necessary
- Starter motor electrical connections
- Starter motor mounting bolts and remove the starter

5. Install or connect the following:
- Starter motor mounting bolts and remove the starter. Tighten the starter mounting bolts to 22 ft. lbs. (31 Nm).
- Starter motor electrical connections
- Speedometer cable connector at the transaxle end, if necessary
- Hat shield from beneath the intake manifold on the 1.5L engine
- Engine undercover

6. Lower the vehicle

7. Install or connect the following:
- Air compressor, if equipped with Active-ECS suspension
- Resonator retaining nuts and remove the air intake hose and resonator assembly as required
- Air-flow sensor assembly connector and remove the breather hose
- Negative battery cable and check the starter for proper operation

Oil Pan

REMOVAL & INSTALLATION

Diamante

3.0L ENGINES

1. Before servicing the vehicle, refer to the precautions in the beginning of this section.

2. Remove or disconnect the following:
- Negative battery cable
- Oil pan drain plug and drain the engine oil

- Left side crossmember. If equipped with Four Wheel Steering (4WS), it will also be necessary to remove the right side crossmember.
- Starter motor
- Roll stopper stay bracket, from the rear transaxle stay bracket
- Transaxle stay brackets
- Bell housing lower cover
- Oil pan mounting bolts
- The engine oil pan

To install:

3. Apply a 0.16 in. (4mm) continuous bead of sealer around the surface of the oil pan.

➡ **Assemble the oil pan to the cylinder block within 15 minutes after applying the sealant.**

4. Install the oil pan mounting bolts. Following proper sequence, tighten mounting bolts to 48 inch lbs. (6 Nm).

5. Install or connect the following:
- Lower bell housing cover and the starter motor
- Transaxle stay brackets and connect the roll stopper bracket
- Crossmember(s) and tighten the mounting bolts to 43–51 ft. lbs. (60–70 Nm)

6. Fill the engine with the proper amount of oil.

7. Connect the negative battery cable and check for leaks.

3.5L ENGINE

1. Before servicing the vehicle, refer to the precautions in the beginning of this section.

2. Disconnect the negative battery cable.

3. Drain the engine oil.

4. Remove the mounting bolts from the lower oil pan.

5. Place a block of wood against the side of the pan and tap the block with a hammer to break the seal and remove the lower pan.

6. Remove or disconnect the following:
- Starter
- Dipstick tube
- Upper oil pan

✳✳ WARNING

Do not pry or use seal breaker tool to remove the oil pan. Damage to the aluminum surface can result.

7. Screw a bolt into the threaded hole to force the oil pan from the engine block and remove the pan.

8. Remove the bolt used to remove the pan.

To install:

9. Clean and degrease the sealing surfaces of the upper oil pan and engine block.

10. Apply a bead of silicone sealant along the mounting surface of the upper oil pan.

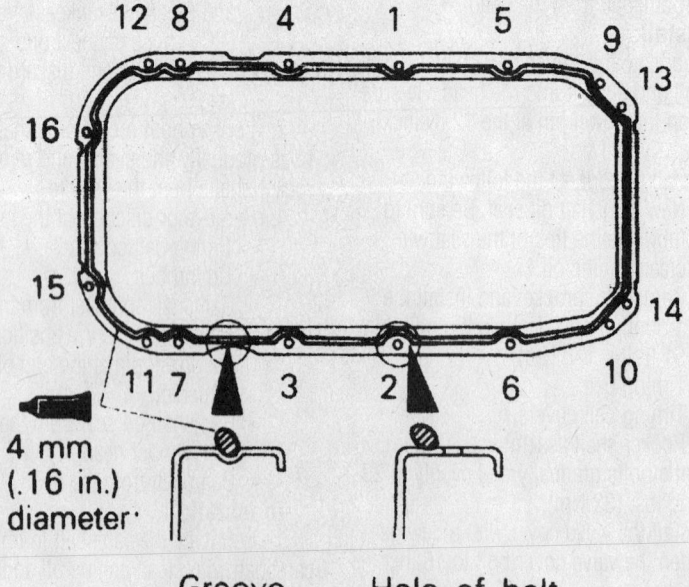

Oil pan bolt tightening sequence and application of sealant to the pan—Diamante 3.0L (J- and H-) engines

7923PG67

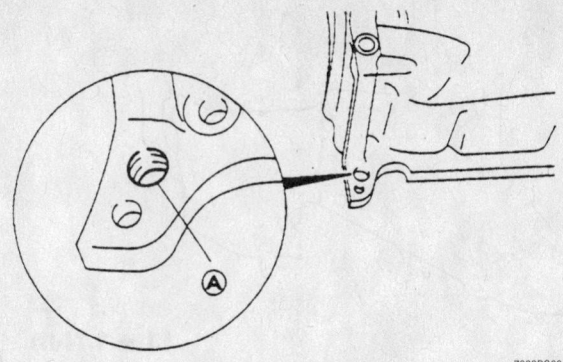

7923PG68

Install a bolt in the threaded hole to force the oil pan from the engine block—3.5L engine

11. Install or connect the following:
- Upper oil pan. Tighten the bolts in sequence to 48 inch lbs. (6 Nm).
- Dipstick tube using a new O-ring
- Starter assembly

12. Clean and degrease the sealing surface of the lower oil pan.

13. Place a bead of sealant on the mounting surface of the lower oil pan. Install the lower pan. Tighten the bolts in sequence to 84–108 inch lbs. (10–12 Nm).

14. Install the drain plug using a new washer. Tighten the drain plug to 29 ft. lbs. (39 Nm).

15. Lower the vehicle and fill the crankcase to the correct level.

16. Connect the negative battery cable.

17. Start the engine and check for leaks.

Eclipse

2.4L (4G64) ENGINE

1. Before servicing the vehicle, refer to the precautions in the beginning of this section.
2. Remove the negative battery cable.
3. Drain the engine oil.
4. Remove or disconnect the following:
- Engine dipstick and tube assembly
- Front exhaust pipe
- Bell housing inspection cover
- Bolts attaching the oil pan to the cylinder block
5. Remove the oil pan assembly.

To install:

6. Clean the sealing surface on the oil pan and engine block. Apply a continuous bead of sealant to the oil pan.

7. Install or connect the following:
- Oil pan to the cylinder block and tighten the bolts to 60 inch lbs. (7 Nm)
- Bell housing inspection cover. Torque the bolts to 84 inch lbs. (9 Nm).
- Front exhaust pipe
- Engine dipstick and tube assembly using a new O-ring

8. Refill the engine with oil. Connect the negative battery cable. Start the engine and check for leaks.

Galant

1. Before servicing the vehicle, refer to the precautions in the beginning of this section.
2. Remove or disconnect the following:
- Negative battery cable
- Oil pan drain plug and drain the engine oil
- Oil dipstick and tube assembly
- Heated Oxygen (HO$_2$S) sensor connector
- Front exhaust pipe from the vehicle
- Bell housing cover
- Oil pan retainer bolts
3. Tap in between the engine block and the oil pan.

➡ **Do not use a pry tool when removing the oil pan. Damage to engine components may occur.**

To install:

4. Apply sealant around the gasket surfaces of the oil pan.

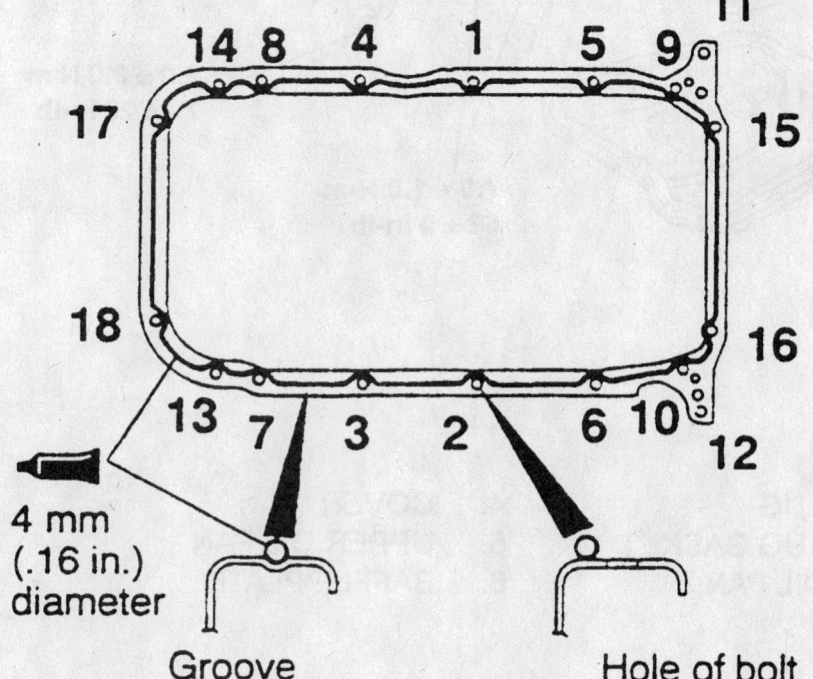

7923PG69

Apply sealant and tighten the bolts in the order shown—3.5L engine, upper oil pan

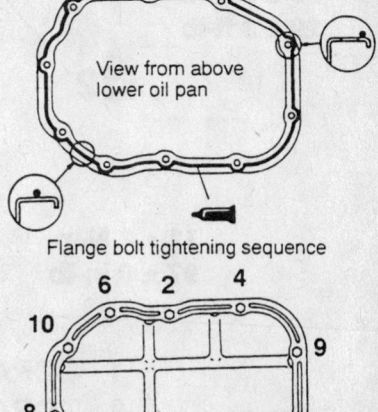

7923PG70

Apply sealant and tighten the bolts in the order shown—3.5L engine, lower oil pan

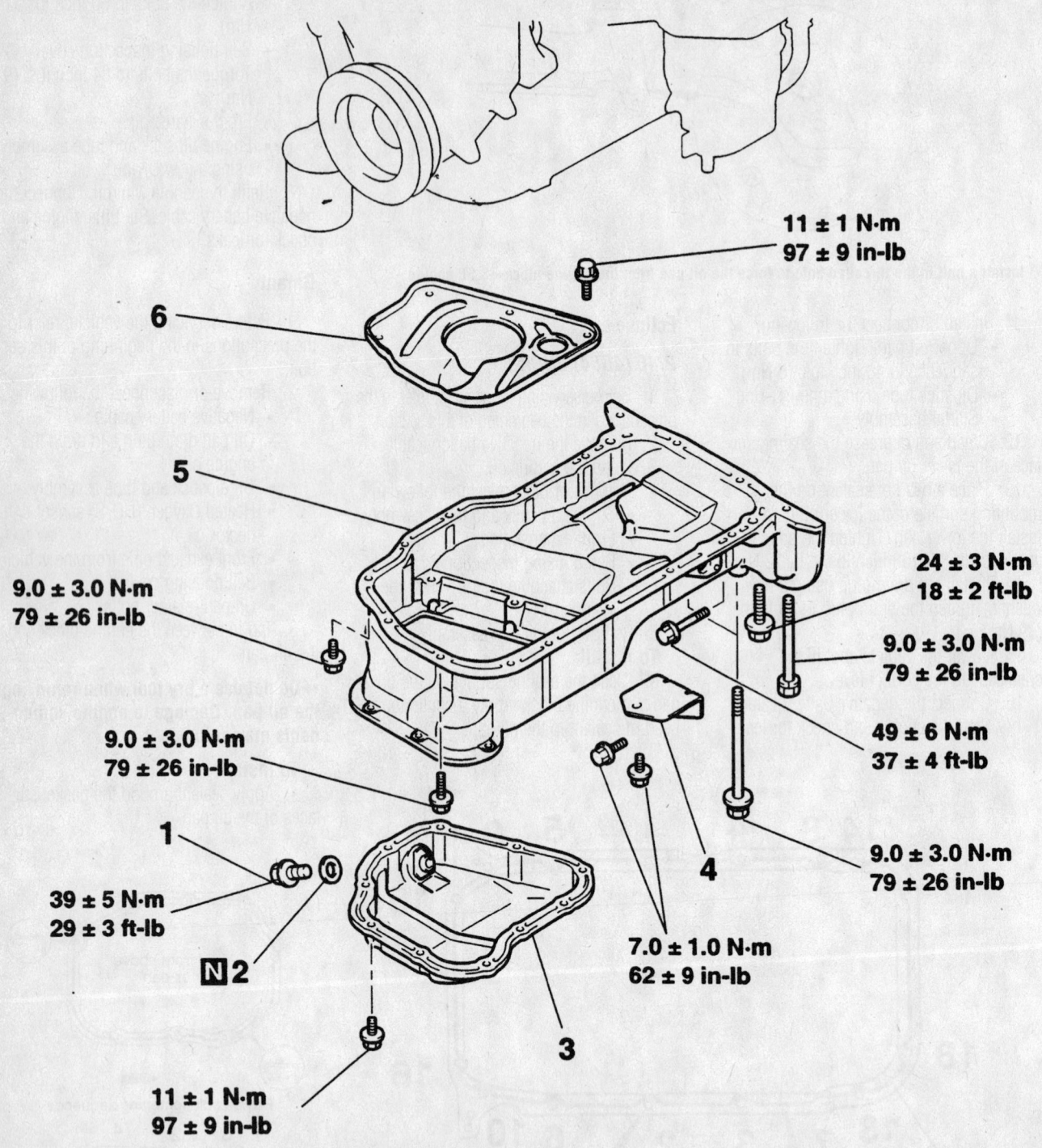

11 ± 1 N·m
97 ± 9 in-lb

24 ± 3 N·m
18 ± 2 ft-lb

9.0 ± 3.0 N·m
79 ± 26 in-lb

9.0 ± 3.0 N·m
79 ± 26 in-lb

49 ± 6 N·m
37 ± 4 ft-lb

9.0 ± 3.0 N·m
79 ± 26 in-lb

9.0 ± 3.0 N·m
79 ± 26 in-lb

39 ± 5 N·m
29 ± 3 ft-lb

7.0 ± 1.0 N·m
62 ± 9 in-lb

11 ± 1 N·m
97 ± 9 in-lb

1. DRAIN PLUG
2. DRAIN PLUG GASKET
3. LOWER OIL PAN
4. COVER
5. UPPER OIL PAN
6. BAFFLE PLATE

Exploded view of the oil pan mounting—Lancer with 2.0L engine

9357QG33

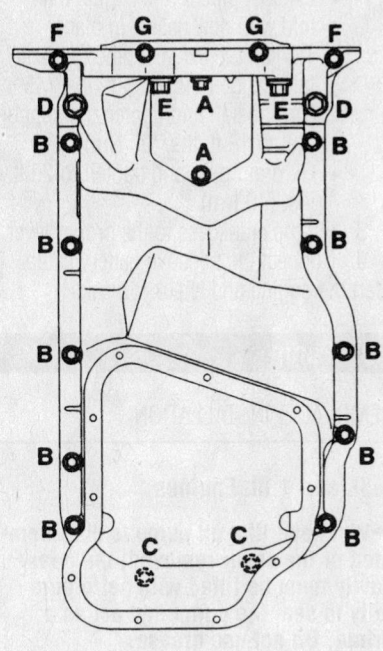

Designation	Symbol	Qty	Diameter × length mm (in)	Tightening torque
Flange Bolt	A	2	6 × 10 (0.2 × 0.4)	7.0 ± 1.0 N·m (62 ± 9 in-lb)
	B	10	6 × 18 (0.2 × 0.7)	9.0 ± 3.0 N·m (79 ± 26 in-lb)
	C	2	6 × 22 (0.2 × 0.9)	
	D	2	8 × 40 (0.3 × 1.6)	24 ± 3 N·m (18 ± 2 ft-lb)
	E	2	10 × 40 (0.4 × 1.6)	49 ± 6 N·m (37 ± 4 ft-lb)
Bolts with Washers	F	2	6 × 50 (0.2 × 2.0)	9.0 ± 3.0 N·m (79 ± 26 in-lb)
	G	2	6 × 127 (0.2 × 5.0)	

9357QG18

Upper oil pan bolt location and torque sequence—Lancer with 2.0L engine

5. Install or connect the following:
- Oil pan onto the cylinder block within 15 minutes after applying sealant. Tighten to 72 inch lbs. (8 Nm).
- Oil drain plug and tighten to 29 ft. lbs. (39 Nm)
- Bell housing cover. Tighten the mounting bolts to 84 inch lbs. (9 Nm).
- Front exhaust pipe and tighten the bolts at the catalytic converter to 36 ft. lbs. (49 Nm). Tighten the nuts at the exhaust manifold to 32 ft. lbs. (44 Nm).
- HO2S sensor connector

6. Fill the crankcase to the proper level.
7. Connect the negative battery cable. Start the engine and check for leaks.

Lancer

1. Before servicing the vehicle, refer to the precautions in the beginning of this section.
2. Disconnect the negative battery cable.
3. Drain the engine oil.
4. Remove or disconnect the following:
- Engine undercover
- Front exhaust pipe
- Lower oil pan bolts and lower pan
- Cover
- Upper oil pan bolt and upper pan
- Baffle plate

To install:
5. Clean all gasket surfaces of the cylinder block and the upper and lower oil pan.
6. Install or connect the following:
- Baffle plate

7. Apply a 0.16 in. (4mm) bead of sealant to the gasket surfaces of the upper oil pan.
- Upper oil pan onto the cylinder block within 15 minutes after applying sealant. Tighten the bolts as shown in the accompanying figure.

8. Apply 0.16 in. (4mm) bead of sealant to the gasket surfaces of the lower oil pan.
- Lower oil pan and tighten the bolts, in the sequence shown, to 88–106 inch lbs. (10–12 Nm).
- Front exhaust pipe
- Engine undercover
- Oil drain plug with a new gasket and tighten to 29 ft. lbs. (40 Nm)

9. Lower the vehicle and fill the crankcase to the proper level with clean engine oil.

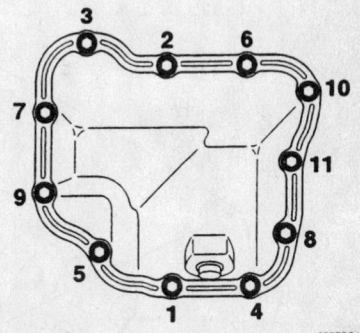

9357QG19

Lower oil pan bolt tightening sequence— Lancer with 2.0L engine

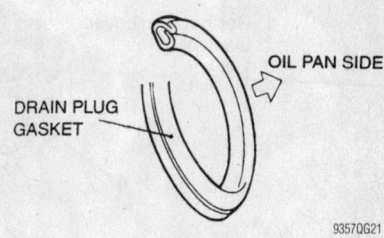

9357QG21

Make sure to install the new drain plug gasket as shown, or leaks will occur

10. Connect the negative battery cable. Start the engine and check for leaks.

Mirage

1.5L (4G15) ENGINE

1. Before servicing the vehicle, refer to the precautions in the beginning of this section.
2. Disconnect the negative battery cable.
3. Drain the engine oil.
4. Remove or disconnect the following:
   • Bell housing lower cover
   • Oil pan retainer bolts

➡ **Do not use a pry tool when removing the oil pan.**

To install:

5. Clean all gasket surfaces of the cylinder block and the oil pan.
6. Apply sealant to the gasket surfaces of the oil pan.
7. Install or connect the following:
   • Oil pan onto the cylinder block within 15 minutes after applying sealant. Tighten to 60 inch lbs. (7 Nm).
   • Bell housing cover
   • Oil drain plug with a new seal and tighten to 29 ft. lbs. (40 Nm)
8. Lower the vehicle and fill the

crankcase to the proper level with clean engine oil.

9. Connect the negative battery cable. Start the engine and check for leaks.

1.8L (4G93) ENGINE

1. Before servicing the vehicle, refer to the precautions in the beginning of this section.
2. Disconnect the negative battery cable.
3. Raise the vehicle and support safely.
4. Remove or disconnect the following:
   • Oil pan drain plug and drain the engine oil
   • Exhaust pipe from the engine manifold
   • Bell housing lower cover
   • Oil pan retainer bolts and remove the oil pan

➡ **Do not use a pry tool when removing the oil pan.**

To install:

5. Clean all gasket surfaces of the cylinder block and the oil pan.
6. Apply sealant around the gasket surfaces of the oil pan.
7. Install or connect the following:
   • Oil pan onto the cylinder block within 15 minutes after applying

sealant. Tighten to 60 inch lbs. (5 Nm).
   • Bell housing cover
   • Exhaust pipe to the engine manifold with new gasket in place. Tighten the exhaust pipe to manifold flange nuts to 33 ft. lbs. (45 Nm). Install and tighten the support bolt to 18 ft. lbs. (25 Nm).
   • Oil drain plug and tighten to 29 ft. lbs. (40 Nm)
8. Fill the crankcase to the proper level.
9. Connect the negative battery cable. Start the engine and check for leaks.

Oil Pump

REMOVAL & INSTALLATION

1.5L and 1.8L Engines

➡ **Whenever the oil pump is disassembled or the cover removed, the gear cavity must be filled with petroleum jelly to seal the pump and act as a prime. Do not use grease.**

1. Before servicing the vehicle, refer to the precautions in the beginning of this section.
2. Remove or disconnect the following:
   • Negative battery cable
   • Front engine mount bracket and accessory drive belts
   • Timing belt upper and lower covers
   • Timing belt and crankshaft sprocket
   • Oil pan and remove the oil screen
   • Front cover mounting bolts. Note the lengths of the mounting bolts as they are removed for proper installation.
   • Front case assembly and oil pump assembly
   • Oil pump cover
   • Inner and outer gears from the front case

To install

3. Remove all gasket material from the mating surfaces and clean all parts.
4. Thoroughly coat both oil pump gears with clean engine oil and install them in the correct direction of rotation.
5. Install the pump cover and tighten the bolts to 84 inch lbs. (10 Nm).
6. Coat the relief valve and spring with clean engine oil. Install them and tighten the plug to 33 ft. lbs. (45 Nm).
7. Install or connect the following:
   • New front crankshaft seal and coat the lips of the seal with clean engine oil
   • Front case and oil pump assembly

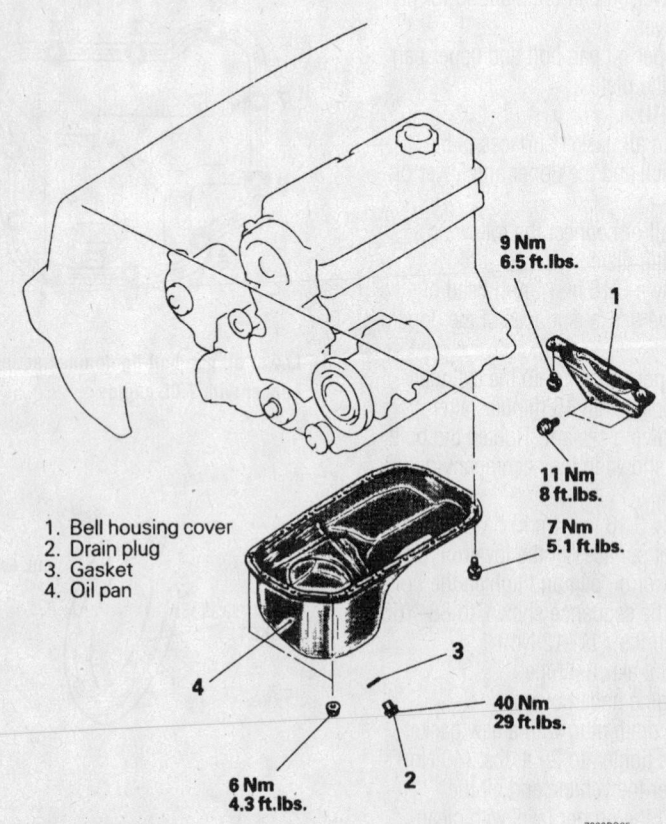

9 Nm
6.5 ft.lbs.

11 Nm
8 ft.lbs.

7 Nm
5.1 ft.lbs.

40 Nm
29 ft.lbs.

6 Nm
4.3 ft.lbs.

1. Bell housing cover
2. Drain plug
3. Gasket
4. Oil pan

7923PG65

Oil pan and related components—Mirage 1.5L (4G15) engine

to the engine block using a new gasket. Tighten the bolts to 10 ft. lbs. (14 Nm).

- Oil screen with new gasket. Torque the screen bolts to 14 ft. lbs. (19 Nm).
- Oil pan
- Crankshaft sprocket and timing belt

8. Fill the crankcase to the proper level.
9. Connect the negative battery cable.

2.0L (4G94) Engine

1. Before servicing the vehicle, refer to the precautions in the beginning of this section.

2. Drain the engine oil.
3. Remove or disconnect the following:

- Negative battery cable
- Oil pressure switch
- Oil filter
- Drain plug and gasket. Discard the gasket.
- Cover
- Upper oil pan. Remove the 5 in. (127mm) bolt, which is closest to the flywheel/flexplate first, then, the other bolts.
- Baffle plate

- Lower oil pan. Remove the 5 in. (127mm) bolt, which is closest to the flywheel/flexplate first, then, the other bolts.
- Oil screen and gasket
- Relief plug and spring
- Oil seal
- Front case
- O-ring
- Oil pump case cover

➡**Matchmark the installed position of the pump rotors before removing them.**

- Outer and inner oil pump rotors

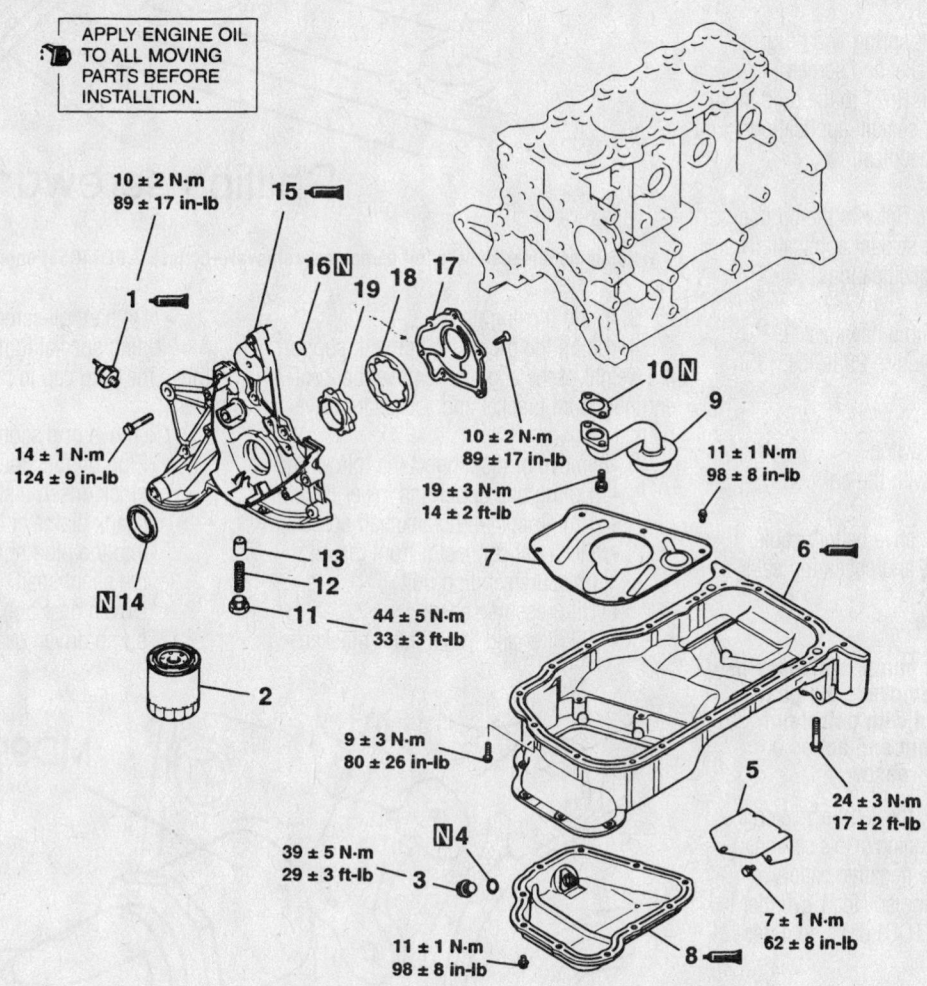

APPLY ENGINE OIL TO ALL MOVING PARTS BEFORE INSTALLTION.

1. Oil pressure switch	11. Relief plug
2. Oil filter	12. Relief spring
3. Drain plug	13. Relief plunger
4. Gasket	14. Oil seal
5. Cover	15. Front case
6. Upper oil pan	16. O-ring
7. Baffle plater	17. Oil pump case cover
8. Lower oil pan	18. Outer rotor
9. Oil screen	19. Inner rotor
10. Oil screen gasket	

93570G20

Exploded view of the oil pump and related components—2.0L (4G94) engine

To install:

4. Install or connect the following:
- Inner and outer rotors, making sure the alignment marks are matched up
- Oil pump case cover
- O-ring

➡**After installation or the front case, wait at least one hour before filling the crankcase with oil or starting the engine.**

- Front case. Apply a 0.12 inch (3mm) bead of sealant, then tighten the case bolts to 124 inch lbs. (14 Nm).
- Oil seal
- Relief plunger, spring and plug
- Oil screen gasket and screen
- Lower oil pan. Refer to the oil pan procedure for sealant application and torque specifications.
- Baffle plate
- Upper oil pan. Refer to the oil pan procedure for sealant application and torque specifications.
- Cover
- Drain plug with a new gasket. Tighten the plug to 29 ft. lbs. (35 Nm).
- Oil filter
- Oil pressure switch

5. Fill the engine with the correct amount of oil.
6. Connect the negative battery cable.
7. Start the engine and check for leaks.

2.4L (4G64) Engine

➡**Whenever the oil pump is disassembled or the cover removed, the gear cavity must be filled with petroleum jelly to seal the pump and act as a prime. Do not use grease.**

1. Before servicing the vehicle, refer to the precautions in the beginning of this section.
2. Disconnect the negative battery cable. Rotate the engine so No. 1 cylinder is on Top Dead Center (TDC) of its compression stroke.

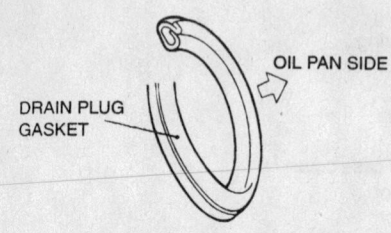

DRAIN PLUG GASKET
OIL PAN SIDE

9357QG21

Make sure to the install the new drain plug gasket as shown, or leaks will occur

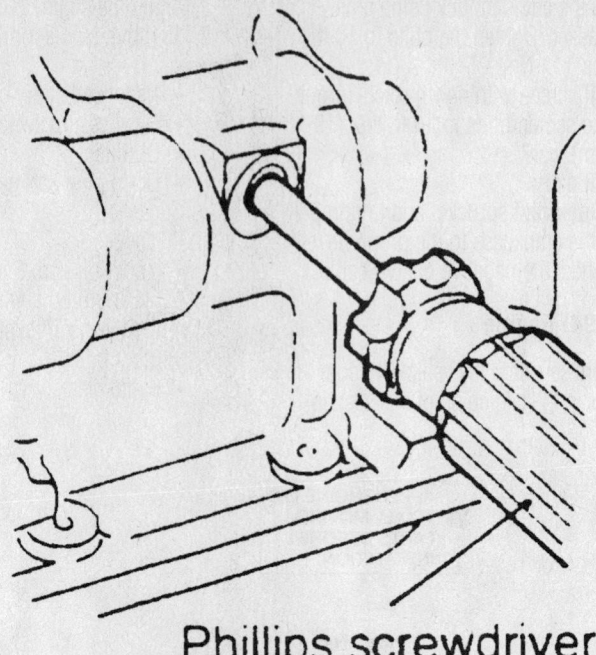

Phillips screwdriver

7923PG71

Holding the silent shaft for oil pump gear removal—Eclipse 2.0L (4G63) engine

3. Drain the engine oil.
4. Using the proper equipment, support the weight of the engine. Remove the front engine mount bracket and accessory drive belts.
5. Remove or disconnect the following:
- Timing belt upper and lower covers
- Timing belt and crankshaft sprocket
- Electrical connector from the oil pressure sending unit
- Oil pressure sensor
- Oil filter and the oil filter bracket

- Oil pan, oil screen and gasket
6. Using special tool MD998162, remove the plug cap in the engine front cover.
7. Remove or disconnect the following:
- Plug on the side of the engine block. Insert a steel rod with a shank diameter of 0.32 in. (8mm) into the plug hole. This will hold the silent shaft.
- Driven gear bolt that secures the oil pump driven gear to the silent shaft

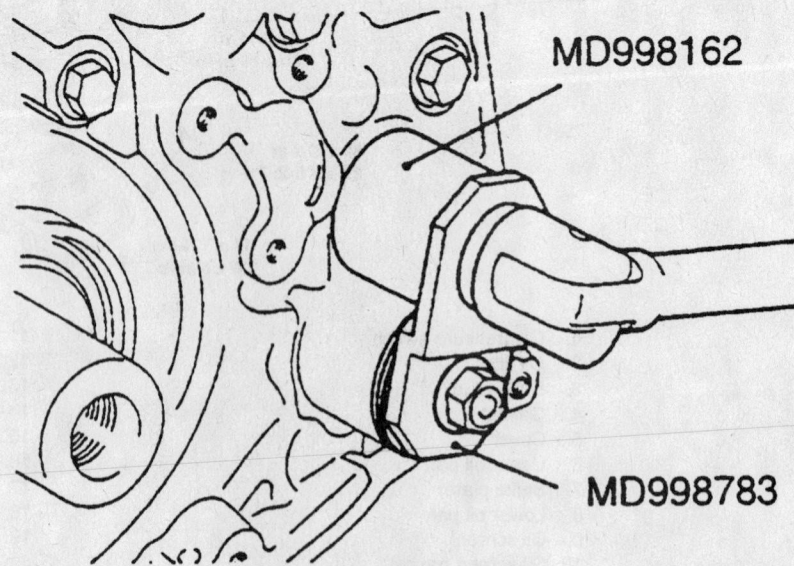

MD998162

MD998783

7923PG72

Use the special socket and holder to remove the balance shaft plug—Eclipse 2.0 (4G63) engine

- Front cover mounting bolts. Note the lengths of the mounting bolts as they are removed for proper installation.
- Front case cover and oil pump assembly. If necessary, the silent shaft can come out with the cover assembly.
- Oil pump cover, located on the back of the engine front cover. Remove the oil pump drive and driven gears.

8. After disassembling the oil pump, clean all components and remove gasket material from mating surfaces.

9. Assemble the oil pump gears into the front case and rotate it to ensure smooth rotation and no looseness. Be sure there is no ridge wear on the contact surface between the front case and the gear surface of the oil pump front cover.

To install

10. Align the timing mark on the oil pump drive gear with that on the driven gear and install them into the engine front case. Apply engine oil to the gears.

11. Install the oil pump cover and tighten the retainer bolts to 13 ft. lbs. (18 Nm) on Eclipse models and 17 ft. lbs. (24 Nm) on Galant models.

12. Using the appropriate driver, install a new crankshaft seal into the front case.

13. Position new front case gasket in place. Set seal guide tool MD998285 on the front end of the crankshaft to protect the seal from damage. Apply a thin coat of oil to the outer circumference of the seal pilot tool.

14. Install the front case assembly through a new front case gasket and temporarily tighten the flange bolts.

15. Mount the oil filter on the bracket with new oil filter bracket gasket in place. Install the bolts with washers and tighten to 14 ft. lbs. (19 Nm).

16. Insert a Phillips screwdriver into the hole in the left side of the engine block to lock the silent shaft in place.

17. Install or connect the following:
- Oil pump drive gear onto the left silent shaft. Tighten the driven gear bolt to 27 ft. lbs. (37 Nm).
- New O-ring to the groove in the front case and install the plug cap. Tighten the cap to 17 ft. lbs. (24 Nm).
- Oil screen in position with new gasket in place

18. Clean both mating surfaces of the oil pan and the cylinder block. Apply sealant in the groove in the oil pan flange.

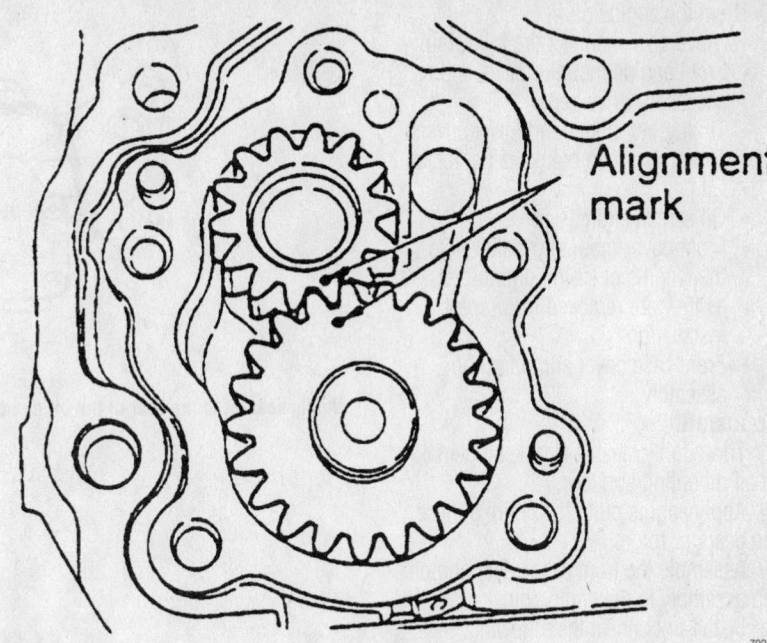

Aligning oil pump timing marks—Eclipse 2.0L (4G63) engine

7923PG73

L = 20 (.79)   L = 40 (1.57)   *L = 30 (1.18)   L = 20 (.79)

L = 40 (1.57)

L = 25 (.98)

L = 75 (2.95)   L = 55 (2.17)

Tighten together with belt tensioner

L = Bolt length below head [mm (in.)]

Front case bolt identification—Eclipse 2.0L (4G63) and 2.4L (4G64) engines

7923PG74

➡ **After applying sealant to the oil pan, do not exceed 15 minutes before installing the oil pan.**

19. Install or connect the following:
- Oil pan to the engine and secure with the retainers. Tighten bolts to 60 inch lbs. (7 Nm).
- Oil pressure gauge unit and the oil pressure switch
- Electrical harness connector
- Oil cooler. Oil cooler bolt to 31 ft. lbs. (43 Nm).

20. Refill the crankcase. Install new oil filter.

21. Connect the negative battery cable and start the engine. Verify correct oil pressure. Inspect for leaks.

3.0L Engines

➡ **Whenever the oil pump is disassembled or the cover removed, the gear cavity must be filled with petroleum jelly to seal the pump and act as a prime. Do not use grease.**

1. Before servicing the vehicle, refer to the precautions in the beginning of this section.

2. Disconnect the negative battery cable.

3. Drain the engine oil.
4. Remove or disconnect the following:
- Front engine mount bracket and accessory drive belts
- Timing belt upper and lower covers
- Timing belt and crankshaft sprocket
- Oil pan
- Oil screen and gasket
- Front cover mounting bolts. Note the lengths of the mounting bolts as they are removed for proper installation.
- Front case cover and oil pump assembly

To install:

5. Thoroughly clean all gasket material from all mounting surfaces.
6. Apply engine oil to the entire surface of the gears or rotors.
7. Assemble the front case cover and oil pump assembly to the engine block.
8. Install or connect the following:
- Oil screen with new gasket
- Oil pan
- Crankshaft sprocket and timing belt
- Timing belt covers
- Drive belts and the front engine mount bracket
- Negative battery cable, refill the crankcase and check for adequate oil pressure

3.5L Engine

1. Before servicing the vehicle, refer to the precautions in the beginning of this section.
2. Remove or disconnect the following:
- Negative battery cable
- Timing belt
3. Drain the engine oil.
4. Remove or disconnect the following:
- Splash shield from the wheel well
- Oil filter adapter
- Lower and upper oil pans
- Lower baffle, oil pump pick-up and upper baffle
- Oil pump case mounting bolts and the oil pump case
- Oil pump gear cover

5. Make matchmarks on the oil pump rotors before removing them.
- Crankshaft seal from the oil pump case

To install:

6. Install or connect the following:
- New crankshaft seal in the oil pump cover

7. Apply engine oil to the rotors, then align the matchmarks and install the rotors in the oil pump case.
- Rotor cover and tighten the bolts to 84 inch lbs. (10 Nm)

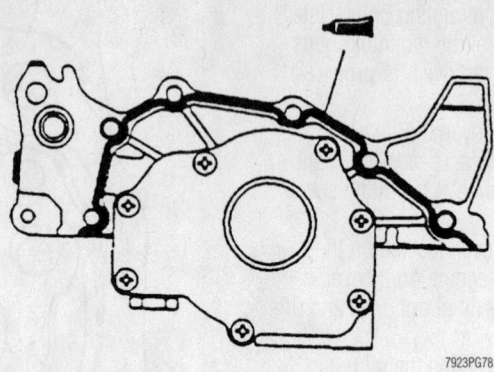

Apply sealant to the rear of the oil pump case—3.5L engine

7923PG78

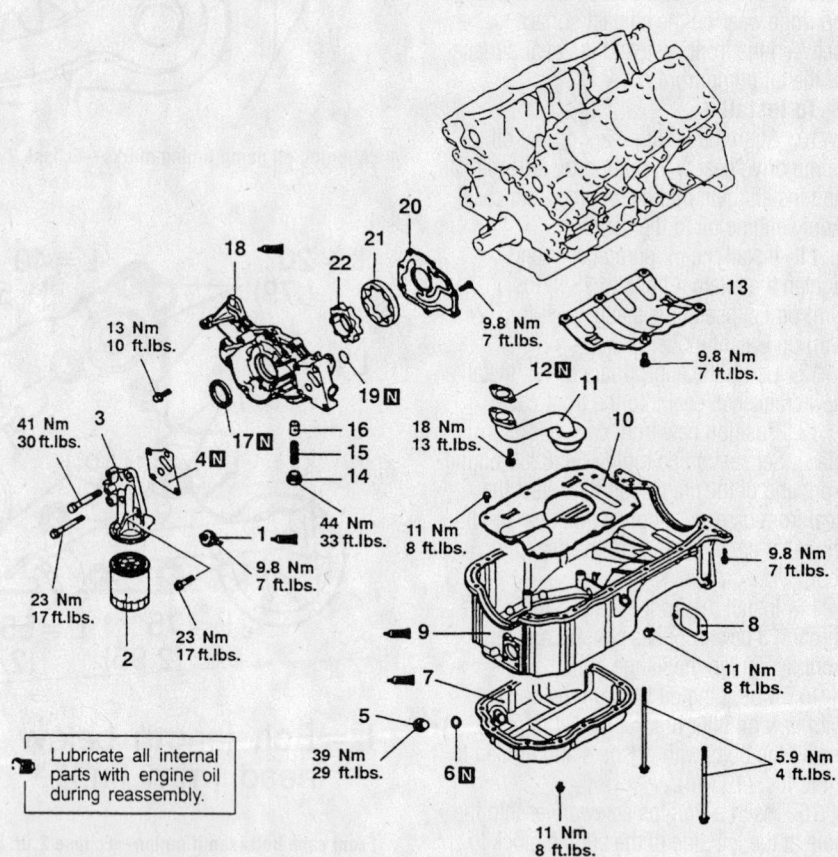

Lubricate all internal parts with engine oil during reassembly.

Removal steps

1. Oil pressure gauge unit
2. Oil filter
3. Oil filter bracket
4. Oil filter bracket gasket
5. Drain plug
6. Drain plug gasket
7. Oil pan, lower
8. Cover
9. Oil pan, upper
10. Baffle plate
11. Oil screen
12. Oil screen gasket
13. Baffle plate
14. Plug
15. Relief spring
16. Relief plunger
17. Crankshaft oil seal
18. Oil pump case
19. O-ring
20. Oil pump cover
21. Oil pump outer rotor
22. Oil pump inner rotor

Exploded view of the oil pump mounting—3.5L engine

7923PG77

8. Apply a 0.113 in. (3mm) bead of sealant to the back of the oil pump case. Install the case on the engine and tighten the bolts to 10 ft. lbs. (13 Nm).
- Upper baffle plate and oil pump pick-up using a new gasket–Tighten the baffle bolts to 84 inch lbs. (10 Nm) and the pick-up bolts to 13 ft. lbs. (18 Nm).
- Lower baffle in the upper oil pan. Tighten the bolts to 96 inch lbs. (11 Nm).
- Oil pans
- Oil filter adapter using a new gasket. Tighten the larger bolt to 30 ft.

lbs. (41 Nm) and the smaller bolt to 17 ft. lbs. (23 Nm).
- Timing belt and remaining components
9. Fill the engine with the correct amount of oil.
10. Connect the negative battery cable.
11. Start the engine and check for leaks.

Rear Main Seal

REMOVAL & INSTALLATION

1. Before servicing the vehicle, refer to the precautions in the beginning of this section.

2. Remove or disconnect the following:
- Transaxle
- Flywheel or flexplate from the crankshaft
3. Carefully pry the seal out of the oil seal case without damaging the sealing surface of the crankshaft.
To install:
4. Apply engine oil to the lip of the new seal and install the seal in the case using the proper size seal driver.
5. Install or connect the following:
- Flywheel or flexplate
- Transaxle

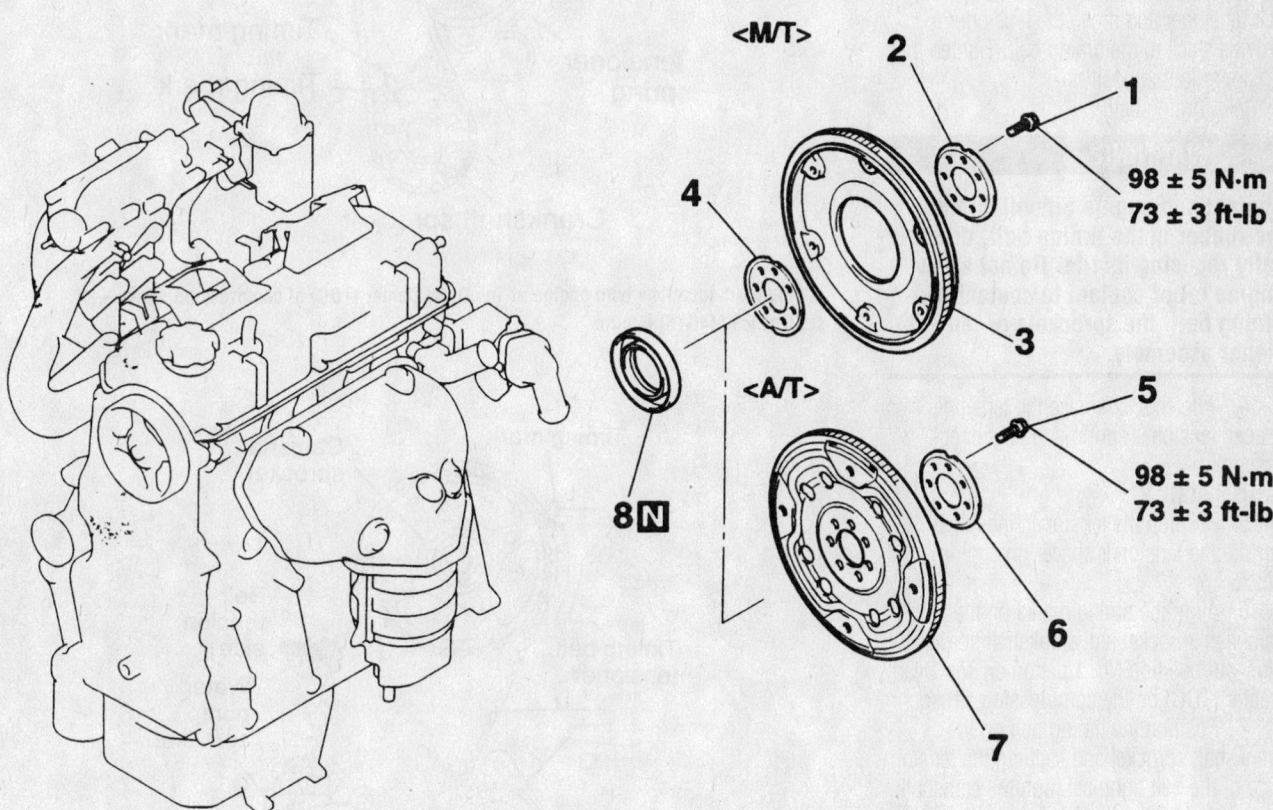

1. FLYWHEEL BOLT <M/T>
2. ADAPTER PLATE <M/T>
3. FLYWHEEL ASSEMBLY <M/T>
4. ADAPTER PLATE <M/T>
5. DRIVE PLATE BOLT <A/T>
6. ADAPTER PLATE <A/T>
7. DRIVE PLATE <A/T>
8. CRANKSHAFT REAR OIL SEAL

Exploded view of a typical rear main seal

Timing Belt

REMOVAL & INSTALLATION

1.5L and 1.8L Engines

1. Before servicing the vehicle, refer to the precautions in the beginning of this section.

2. Remove or disconnect the following:
   - Negative battery cable
   - Electrical connectors, tag before disconnecting
   - Timing belt upper and lower covers

3. Make a mark on the back of the timing belt indicating the direction of rotation so it may be reassembled in the same direction if it is to be reused. Loosen the timing belt tensioner and move the tensioner to provide slack to the timing belt. Tighten the tensioner in this position.
   - Timing belt

✳✳ WARNING

Coolant and engine oil will damage the rubber in the timing belt, drastically reducing its life. Do not allow engine oil or coolant to contact the timing belt, the sprockets or tensioner assembly.

4. If defective, replace the tensioner spacer, tensioner spring and tensioner assembly.

To install:

5. Position the tensioner, tensioner spring and tensioner spacer on engine block.

6. Align the timing marks on the camshaft sprocket and crankshaft sprocket. This will position No. 1 piston on Top Dead Center (TDC) on the compression stroke.

7. Position the timing belt on the crankshaft sprocket and keeping the tension side of the belt tight, set it on the camshaft sprocket, then the tensioner.

8. Apply slight counterclockwise force to the camshaft sprocket to give tension to the belt and be sure all timing marks are aligned.

9. Loosen the pivot side tensioner bolt and the slot side bolt. Allow the spring to remove the slack.

10. Tighten the slot side tensioner bolt, then the pivot side bolt. If the pivot side bolt is tightened first, the tensioner could turn with bolt, causing over tension.

11. For 1.5L engines, turn the crankshaft clockwise. Loosen the pivot side tensioner bolt, then the slot side bolt to allow the spring to take up any remaining slack.

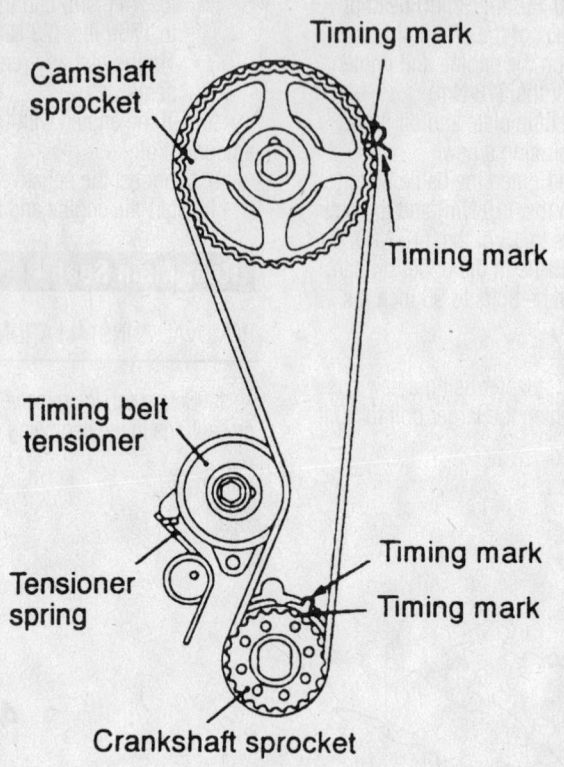

Timing mark locations with engine at Top Dead Center (TDC) of compression stroke—Mitsubishi 1.5L (4G15) engine

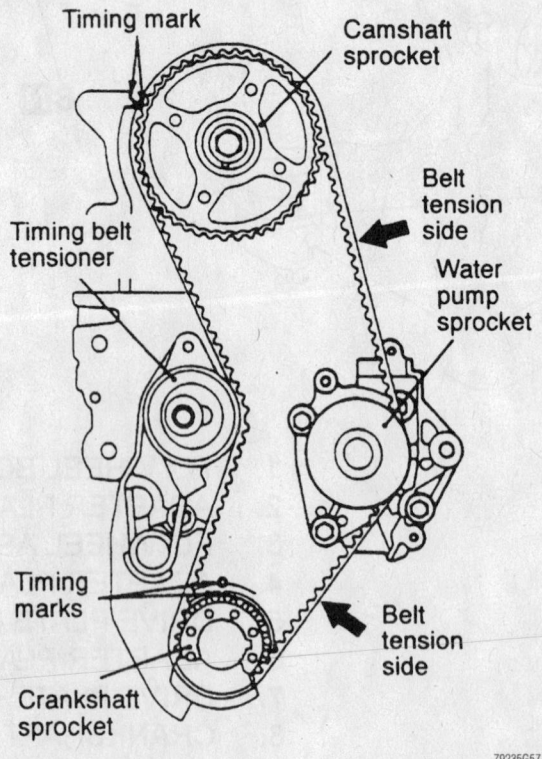

Timing mark locations with engine at Top Dead Center (TDC) of compression stroke—Mitsubishi 1.8L (4G93) engine

Tighten the slot bolt, then the pivot side bolt to 17 ft. lbs. (24 Nm).

12. For 1.8L engines, turn the crankshaft clockwise two rotations and tighten the adjuster bolt to 18 ft. lbs. (24 Nm) and the pivot (spring) bolt to 35 ft. lbs. (45 Nm).

13. Install the timing belt covers and tighten the cover bolts to 84–96 inch lbs. (10–11 Nm).

14. Install the remaining components in the reverse order of removal.

2.0L Engine

1. Before servicing the vehicle, refer to the precautions in the beginning of this section.

2. Remove or disconnect the following:
   • Front timing belt cover

➡ **If the timing belt is going to be reused, mark the direction of rotation on the belt with an arrow. Install the belt in the same direction.**

3. Rotate the crankshaft sprocket clockwise until the timing marks are aligned.

4. Place 8mm Allen wrench into the belt tensioner, then using the long end of a ⅛ in. (3mm) Allen wrench, rotate the tensioner counterclockwise until it slides into the locking hole.

5. Remove the belt.

✳✳ WARNING

Do not rotate the crankshaft or the camshafts while the belt is removed.

To install:

6. Using a vise, slowly compress the plunger into the body of the tensioner and install a pin through the body of the tensioner to retain the plunger.

7. Be sure the timing marks are still aligned, if not, align the camshaft sprocket timing marks facing each other. Align the crankshaft sprocket timing mark with the mark on the oil pump housing, then turn the crankshaft sprocket backward ½ notch.

8. Install the timing belt, starting at the crankshaft, then around the water pump sprocket, idler pulley, camshaft sprockets and the tensioner pulley.

9. Turn the crankshaft sprocket ½ notch to Top Dead Center (TDC) to take up the slack in the belt.

10. Install the tensioner on the engine, but do not tighten the bolts.

11. Place a torque wrench on the tensioner pulley and apply 21 ft. lbs. (28 Nm) of torque in the direction of the water pump. Push the tensioner up against the tensioner pulley and tighten the mounting bolts to 23 ft. lbs. (31 Nm).

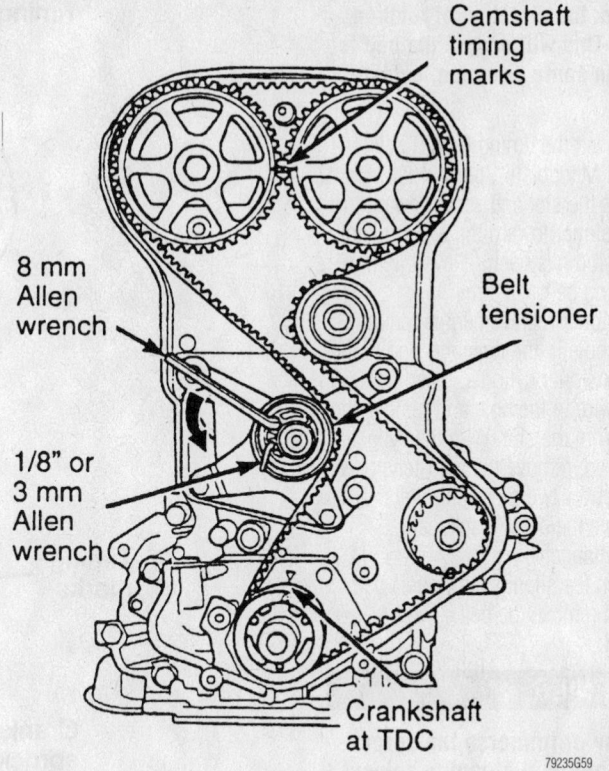

Timing belt sprocket mark alignment for belt service—Mitsubishi 2.0L non-turbo engines

12. Pull the pin out of the tensioner. Belt tension is correct when the pin can be removed and installed.

13. Rotate the crankshaft two revolutions and check the timing marks for alignment. Repeat the previous steps, if necessary.

14. Install the timing belt cover and all other applicable components.

2.4L Engine

1. Before servicing the vehicle, refer to the precautions in the beginning of this section.

2. Position the engine so that the No. 1 piston is at Top Dead Center (TDC).

3. Remove the timing belt covers.

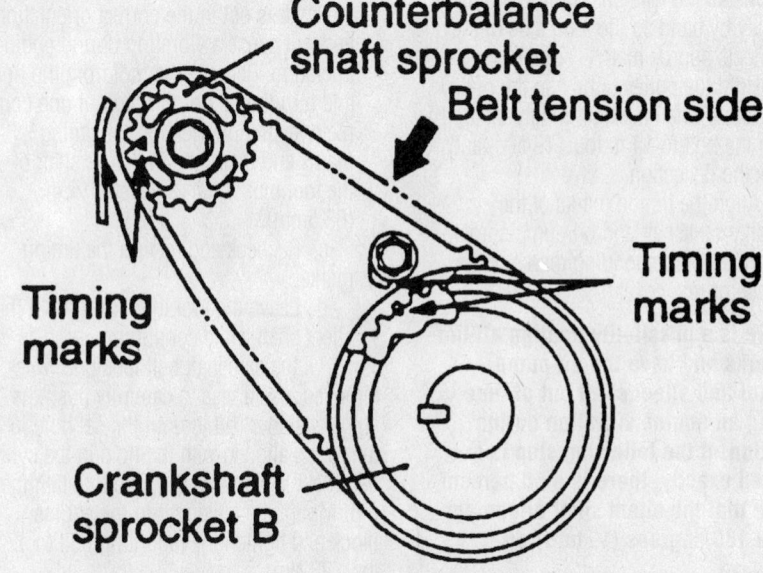

Timing belt "B" installation mark alignment—2.4L engines

➡ If the timing belts are going to be reused, mark the direction of rotation on the belt. This will ensure the belt is reinstalled in same direction, extending belt life.

4. To loosen the timing (outer) belt tensioner, install Mitsubishi Special tool MD998738 to the slot and screw inward to move the tensioner toward the water pump. Once the tension has been relieved, remove the outer timing belt.

5. If tensioner replacement is required, align the pin hole in the tensioner rod to the hole in the tensioner cylinder. Insert a 0.055 in. (1.4mm) wire in the hole and remove the special tool from the slot. With the cylinder tension relieved, remove the auto-tensioner cylinder assembly two mounting bolts.

6. Remove the outer crankshaft sprocket and flange.

7. Loosen the silent shaft (inner) belt tensioner and remove the belt.

To install:

✳✳ WARNING

Do not spray or immerse the sprockets or tensioners in cleaning solvent. The sprocket may absorb the solvent and transfer it to the belt. The tensioners are internally lubricated and the solvent will dilute or dissolve the lubricant.

8. Align the timing marks of the silent shaft sprockets and the crankshaft sprocket with the timing marks on the front case. Route the timing belt around the sprockets so there is no slack in the upper span of the belt and the timing marks are still aligned.

9. Install the tensioner pulley and move the pulley by hand so the long side of the belt deflects approximately ¼ in. (6mm).

10. Hold the pulley tightly so the pulley cannot rotate when the bolt is tightened. Tighten the bolt to 14 ft. lbs. (19 Nm) and recheck the deflection.

11. Align the timing marks of the camshaft, crankshaft and oil pump sprockets with their corresponding marks on the front case or rear cover.

➡ There is a possibility to align all timing marks and have the oil pump sprocket and silent shaft out of time, causing an engine vibration during operation. If the following step is not followed exactly, there is a 50 percent chance that the silent shaft alignment will be 180 degrees (½ turn) off.

12. Before installing the timing belt, ensure that the left side (rear) silent shaft

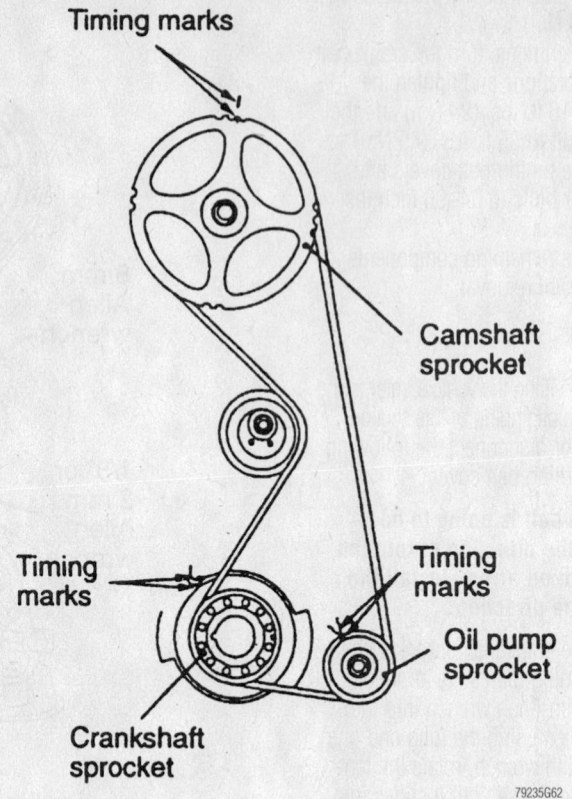

Proper alignment of the timing belt sprocket marks for belt service— 2.4L engines

(oil pump sprocket) is in the correct position as follows:

a. Remove the plug from the rear side of the block and insert a tool with shaft diameter of 0.31 in. (8mm) into the hole.

b. With the timing marks still aligned, the shaft of the tool must be able to go in at least 2 ½ in. (63.5mm). If the tool can only go in approximately 1 in. (25mm), the shaft is not in the correct orientation and will cause a vibration during engine operation. Remove the tool from the hole and turn the oil pump sprocket one complete revolution. Realign the timing marks and insert the tool. The shaft of the tool must go in at least 2 ¼ in. (63.5mm).

c. Recheck and realign the timing marks.

d. Leave the tool in place to hold the silent shaft while continuing.

13. If the camshaft belt tensioner was removed, use a vise to carefully push the auto-tensioner rod in until the set hole in the rod is aligned with the hole in the cylinder. Place a wire into the hole to retain the rod. Mount the tensioner to the engine block and tighten the mounting bolt to 17 ft. lbs. (23 Nm).

14. Install the belt to the crankshaft

sprocket, oil pump sprocket, then camshaft sprocket, in that order. While doing so, be sure there is no slack between the sprocket except where the tensioner is installed.

15. To adjust the timing (outer) belt perform the following steps:

a. Turn the crankshaft ¼ turn counterclockwise, then turn it clockwise to move No. 1 cylinder to TDC.

b. Loosen the center bolt. Using tool MD998752 and a torque wrench, apply a torque of 2.6 ft. lbs. (3.6 Nm) to the tensioner. Tighten the center bolt.

c. Screw the special tool into the engine left support bracket until its end makes contact with the tensioner arm. At this point, screw the special tool in some more and remove the set wire attached to the auto-tensioner, if the wire was not previously removed. Then, remove the special tool.

d. Rotate the crankshaft two complete turns clockwise and let it sit for approximately 15 minutes. Then, measure the auto-tensioner protrusion (the distance between the tensioner arm and auto-tensioner body) to ensure that it is within 0.15–0.18 in. (3.8–4.5mm). If out of specification, repeat substeps **a** through **d** until the specified value is obtained.

➡**Do not manually overtighten the belt or it will howl.**

16. Install the upper and lower timing belt covers.

3.0L (6G72) SOHC Engine

1. Before servicing the vehicle, refer to the precautions in the beginning of this section.

2. Position the engine so the No. 1 cylinder is at Top Dead Center (TDC) of its compression stroke.

✳✳ CAUTION

Wait at least 90 seconds after the negative battery cable is disconnected to prevent possible deployment of the air bag.

3. Remove all necessary components for access to the timing belt covers, then remove the covers from the engine.

4. If the same timing belt will be reused, mark the direction of the timing belt's rotation for installation in the same direction. Be sure the engine is positioned so the No. 1 cylinder is at the TDC of its compression stroke and the timing marks are aligned with the engine's timing mark indicators.

5. Loosen the timing belt tensioner bolt and remove the belt. If the tensioner is not being removed, position it as far away from the center of the engine as possible and tighten the bolt.

6. If the tensioner is being removed, mark the outside of the spring to ensure that it is not installed backwards. Unbolt the tensioner and remove it along with the spring.

✳✳ WARNING

Do not rotate the camshafts when the timing belt is removed from the engine. Turning the camshaft when the timing belt is removed could cause the valves to interfere with the pistons thus causing severe internal engine damage.

To install:

7. Install the tensioner, if removed, and hook the upper end of the spring to the water pump pin and the lower end to the tensioner in exactly the same position as originally installed.

8. Ensure both camshafts are still positioned so the timing marks align with those on the rear timing covers. Rotate the crankshaft so the timing mark aligns with the mark on the front cover.

9. Install the timing belt on the crankshaft sprocket and while keeping the belt tight on the tension side, install the belt on the front (left) camshaft sprocket.

10. Install the belt on the water pump pulley, then the rear (right) camshaft sprocket and the tensioner.

11. Loosen the bolt that secures the adjustment of the tensioner and lightly press the tensioner against the timing belt.

12. Check that the timing marks are in alignment.

13. Rotate the crankshaft 2 full turns in the clockwise direction only, then realign the timing marks.

14. Tighten the bolt that secures the tensioner to 19 ft. lbs. (26 Nm).

15. Install the lower and the upper timing belt covers, along with all other applicable components.

3.0L (6G72) DOHC Engine

1. Before servicing the vehicle, refer to the precautions in the beginning of this section.

2. Position the engine so the No. 1 cylinder is at Top Dead Center (TDC) of its compression stroke.

3. Remove all necessary components for access to the timing belt covers, then remove the covers from the engine.

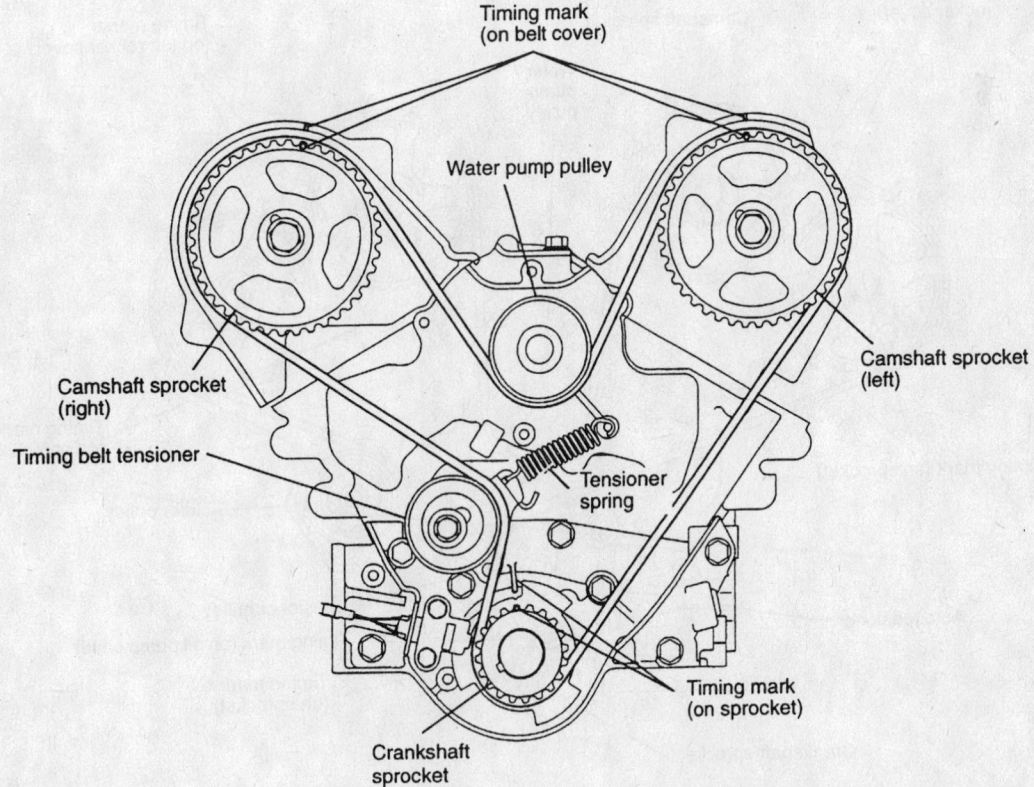

79235G64

Align the sprockets properly before removing or installing the timing belt—Diamante with 3.0L (6G72) SOHC engine

✸✸ CAUTION

Be sure to disconnect the negative battery cable. Wait at least 90 seconds after the negative battery cable is disconnected to prevent possible deployment of the air bag.

4. If the same timing belt will be reused, mark the direction of the timing belt's rotation for installation in the same direction. Be sure the engine is positioned so the No. 1 cylinder is at the TDC of its compression stroke and the timing marks are aligned with the engine's timing mark indicators on the rear timing covers.

5. Remove the timing belt.

✸✸ WARNING

Turning the camshaft sprocket when the timing belt is removed could cause the valves to contact with the pistons, resulting in severe engine damage.

6. Remove the bolts that secure the auto-tensioner to the engine block and remove the tensioner.

To install:

➡The auto-tensioner assembly must be reset to correctly adjust belt tension.

7. Loosen the center bolt of tensioner pulley to provide timing belt slack. Remove the timing belt assembly.

8. Position the auto-tensioner into a vise with soft jaws. The plug at the rear of tensioner protrudes, be sure to use a washer as a spacer to protect the plug from contacting vise jaws.

9. Slowly push the rod into the tensioner until the set hole in rod is aligned with set hole in the auto-tensioner.

10. Insert a 0.055 in. (1.4mm) wire into the aligned set holes. Unclamp the tensioner from the vise and install it on the engine. Tighten tensioner mounting bolts to 17 ft. lbs. (24 Nm).

✸✸ WARNING

DO NOT rotate or turn the camshafts when removing the sprockets or severe engine damage will result from internal component interference.

11. Align the mark on the crankshaft sprocket with the mark on the front case. Then, move the crankshaft sprocket 1 tooth counterclockwise.

12. Align the timing marks of the camshafts with the marks on the rear covers.

13. Using large paper clips to secure the timing belt to the sprockets, install the timing belt in the following order. Be sure camshafts-to-cylinder heads and crankshaft-to-front cover timing marks are aligned. Install the timing belt around the pulleys in the following order:

   a. Exhaust camshaft sprocket (front bank).

   b. Intake camshaft sprocket (front bank).

   c. Water pump pulley.

   d. Intake camshaft sprocket (rear bank).

   e. Exhaust camshaft sprocket (rear bank).

   f. Tensioner pulley.

   g. Crankshaft pulley.

   h. Idler pulley.

➡Since the camshaft sprockets turn easily, secure them with box wrenches when installing the timing belt.

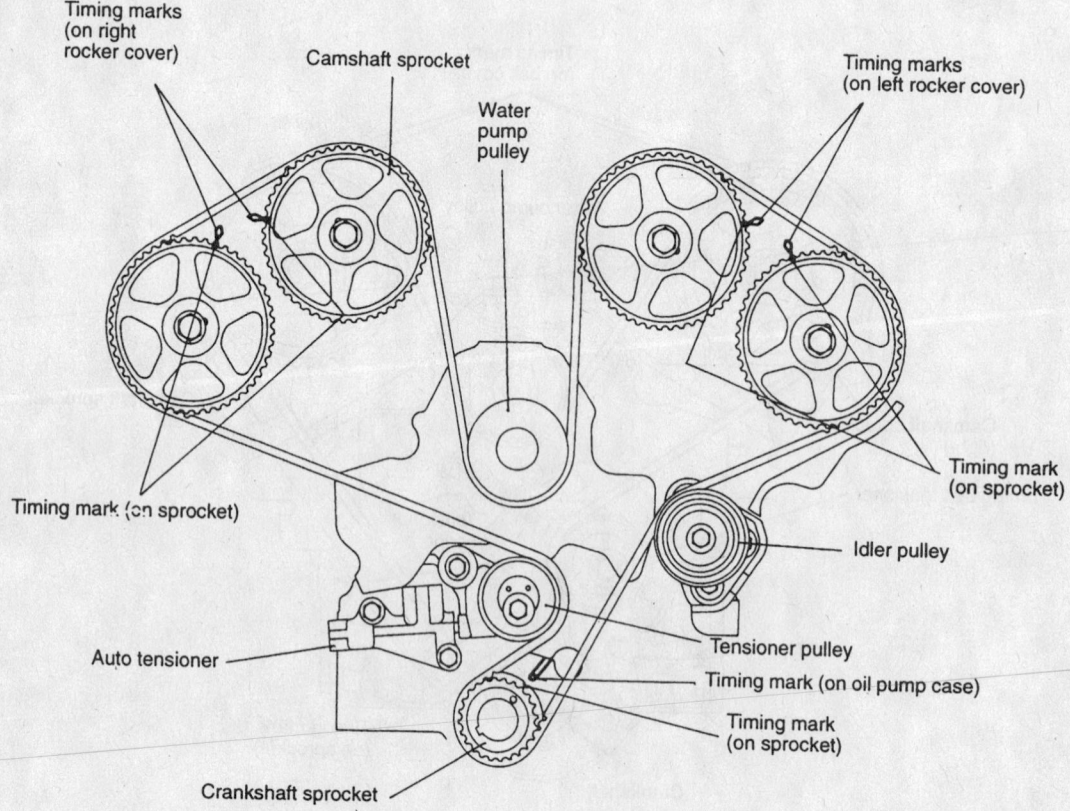

Sprocket alignment for timing belt installation—Diamante with the 3.0L (6G72) DOHC engine

14. Align all timing mark on the crankshaft and raise tensioner pulley against belt to remove slack, snug tensioner bolt.

15. Check the alignment of all the timing marks and remove the clips that secure the timing belt to the camshaft sprockets.

16. Rotate the engine ¼ turn counterclockwise, then rotate the engine clockwise to align the timing marks. Check that all the timing marks are in alignment.

17. Loosen the center bolt on the tensioner pulley. Using tool MD998752 and a torque wrench, apply 84 inch lbs. (10 Nm) to the tool on the tensioner. Tighten the tensioner bolt to 35 ft. lbs. (49 Nm) and be sure the tensioner does not rotate with the bolt.

18. Rotate the crankshaft two complete turns clockwise and let it sit for approximately five minutes. Then, check that the set pin can easily be inserted and removed from the hole in the auto-tensioner.

19. Remove the set wire attached to the auto-tensioner.

20. Measure the auto-tensioner protrusion (the distance between the tensioner arm and auto-tensioner body) to ensure that it is within 0.15–0.18 in. (3.8–4.5mm). If out of specification, repeat adjustment procedure until the specified value is obtained.

21. Check again that the timing marks on all sprockets are in proper alignment.

22. Install the timing belt covers and all other applicable components.

3.5L (6G74) SOHC Engine

1. Before servicing the vehicle, refer to the precautions in the beginning of this section.

2. Position the engine so the No. 1 cylinder is at Top Dead Center (TDC) of its compression stroke.

3. Remove all necessary components for access to the timing belt covers, then remove the covers from the engine.

❋❋ CAUTION

Be sure to disconnect the negative battery cable. Wait at least 90 seconds after the negative battery cable is disconnected to prevent possible deployment of the air bag.

4. If the same timing belt will be reused, mark the direction of the timing belt's rotation for installation in the same direction. Be sure the engine is positioned so the No. 1 cylinder is at the TDC of its compression stroke and the timing marks are aligned with the engine's timing mark indicators on the rear timing covers.

5. Remove the timing belt.

❋❋ WARNING

Turning the camshaft sprocket when the timing belt is removed could cause the valves to contact with the pistons, resulting in severe engine damage.

6. Remove the bolts that secure the auto-tensioner to the engine block and remove the tensioner.

To install:

➡The auto-tensioner assembly must be reset to correctly adjust belt tension.

7. Loosen the center bolt of tensioner pulley to provide timing belt slack. Remove the timing belt tensioner assembly.

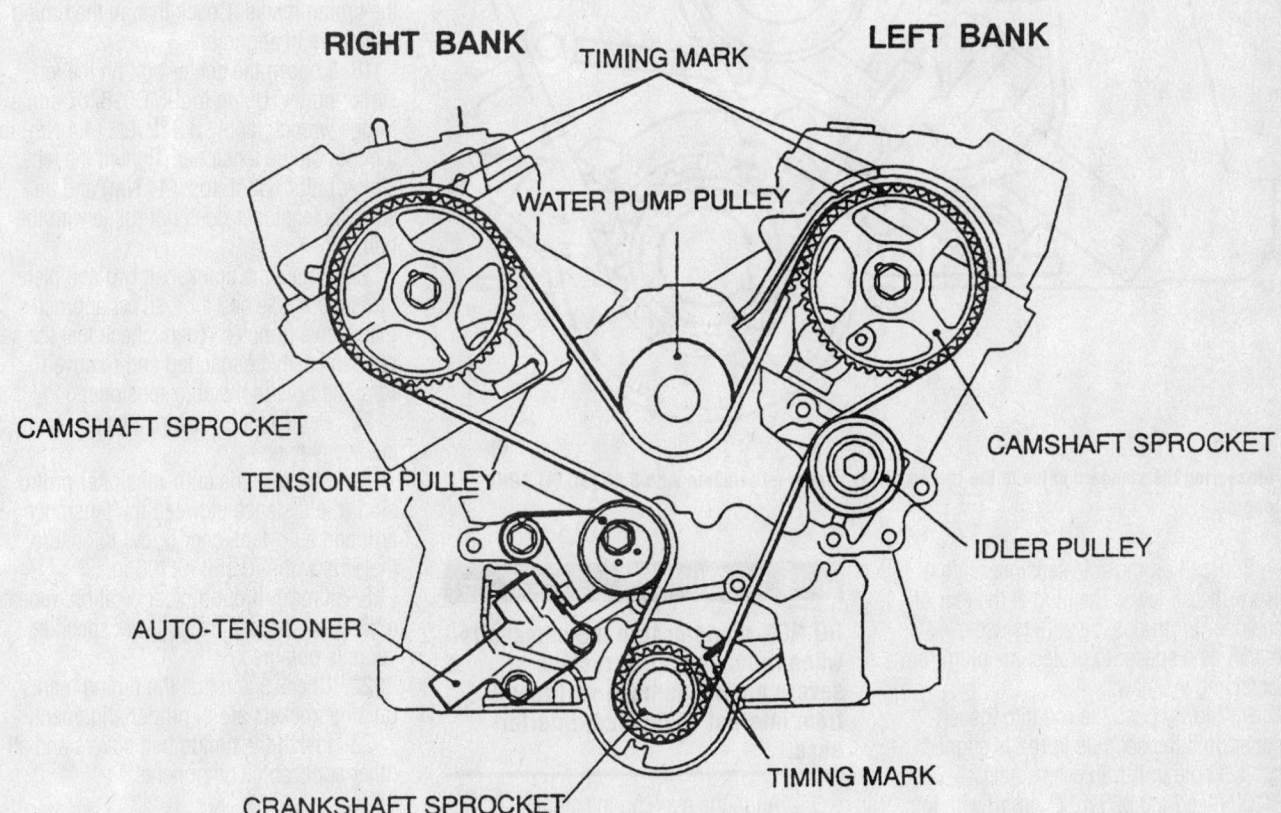

RIGHT BANK   TIMING MARK   LEFT BANK

WATER PUMP PULLEY

CAMSHAFT SPROCKET

TENSIONER PULLEY

AUTO-TENSIONER

CRANKSHAFT SPROCKET

TIMING MARK

CAMSHAFT SPROCKET

IDLER PULLEY

93015G27

Sprocket alignment for timing belt installation—Diamante with 3.5L (6G74) SOHC engine

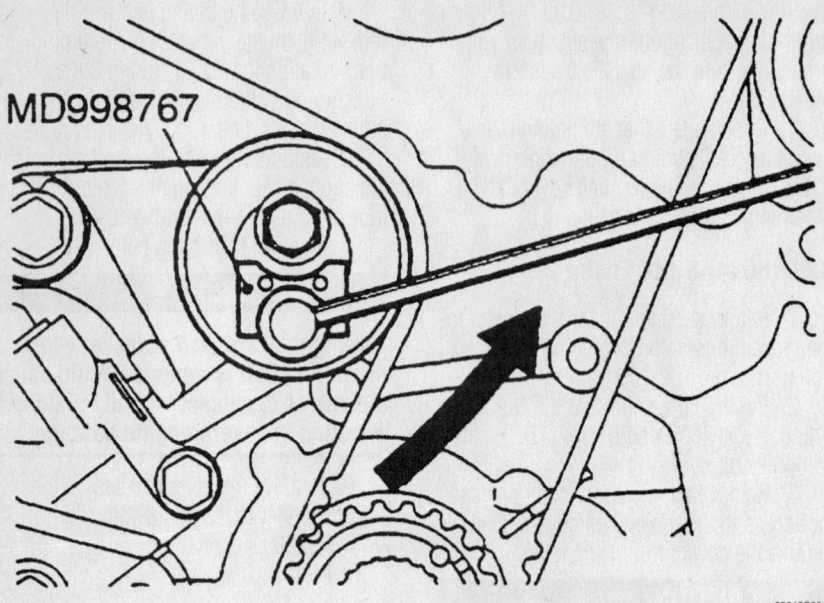

MD998767

93015G28

Special tool used for tightening timing belt—Diamante with 3.5L (6G74) SOHC engine

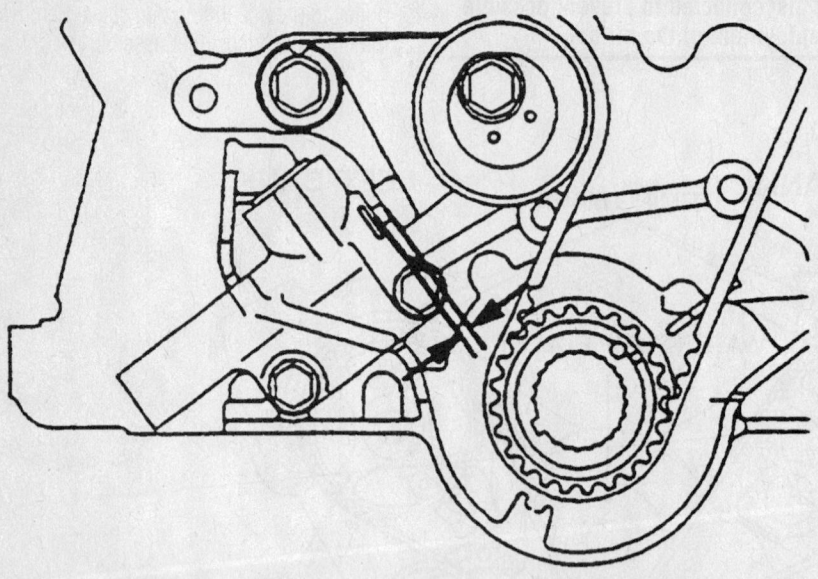

93015G29

Measuring the standard value of the timing belt tensioner—Diamante with 3.5L (6G74) SOHC engine

8. Position the auto-tensioner into a vise with soft jaws. The plug at the rear of tensioner protrudes, be sure to use a washer as a spacer to protect the plug from contacting vise jaws.

9. Slowly push the rod into the tensioner until the set hole in rod is aligned with set hole in the auto-tensioner.

10. Insert a 0.055 in. (1.4mm) wire into the aligned set holes. Unclamp the tensioner from the vise and install it on the engine. Tighten tensioner mounting bolts to 17 ft. lbs. (24 Nm).

❈❈ WARNING

DO NOT rotate or turn the camshafts when removing the sprockets or severe engine damage will result from internal component interference.

11. Align the mark on the crankshaft sprocket with the mark on the front case. Then, move the crankshaft sprocket 3 teeth counterclockwise.

12. Align the timing marks of the camshafts with the marks on the rear covers.

13. Realign the crankshaft pulley with timing mark on the housing.

➡ **Be sure camshafts-to-cylinder heads and crankshaft-to-front cover timing marks are aligned.**

14. Install the timing belt around the pulleys in the following order:
   a. Crankshaft pulley.
   b. Idler pulley.
   c. Left camshaft sprocket.
   d. Water pump pulley.
   e. Right camshaft sprocket.
   f. Tensioner pulley.

➡ **Since the camshaft sprockets turn easily because of spring action, be careful not to get your fingers caught.**

15. Align all timing mark on the crankshaft and raise tensioner pulley against belt to remove slack, snug tensioner bolt.

16. Check the alignment of all the timing.

17. Using special tool MD998769, rotate the crankshaft ¼ turn counterclockwise, then rotate the crankshaft clockwise to align the timing marks. Check that all the timing marks are in alignment.

18. Loosen the center bolt on the tensioner pulley. Using tool MD998767 and a torque wrench, apply 3.3 ft. lbs. (4.4 Nm) to the tool on the tensioner. Tighten the tensioner bolt to 33 ft. lbs. (44 Nm) and be sure the tensioner does not rotate with the bolt.

19. Rotate the crankshaft two complete turns clockwise and let it sit for approximately five minutes. Then, check that the set pin can easily be inserted and removed from the hole in the auto-tensioner.

20. Remove the set wire attached to the auto-tensioner.

21. Measure the auto-tensioner protrusion (the distance between the tensioner arm and auto-tensioner body) to ensure that it is within 0.150–0.196 in. (3.8–5.0mm). If out of specification, repeat adjustment procedure until the specified value is obtained.

22. Check again that the timing marks on all sprockets are in proper alignment.

23. Install the timing belt covers and all other applicable components.

Piston and Ring

POSITIONING

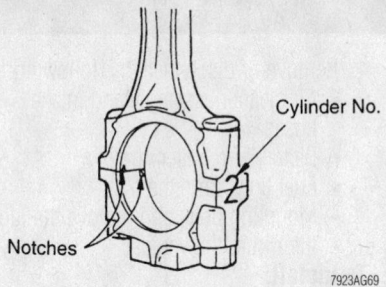

Before removing the caps from the connecting rods, be sure to matchmark them as shown

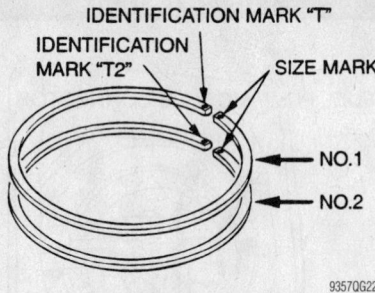

2.0L engines—compression ring identification mark locations

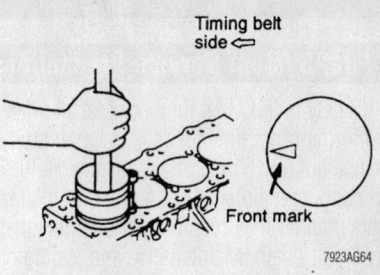

1.5L, 1.8L. 2.0L and 2.4L engines—piston-to-engine block mark location on the piston face

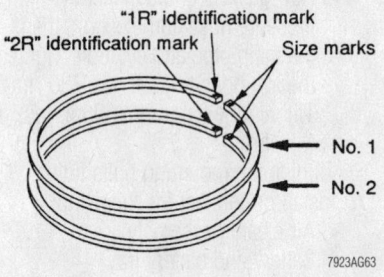

1.5L and 2.4L engines—compression ring identification mark locations

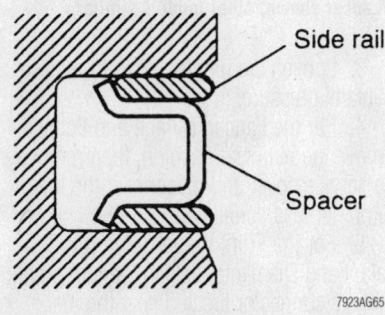

1.8L engine—oil side and spacer ring positioning

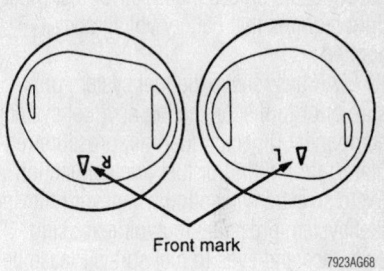

3.0L engine—piston-to-engine block mark locations

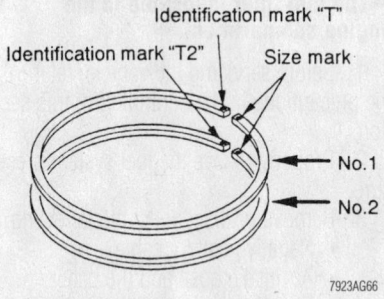

1.8L engine—compression ring identification mark locations

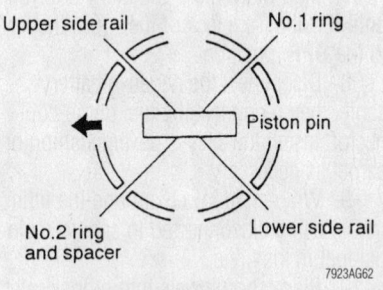

Piston ring end-gap spacing

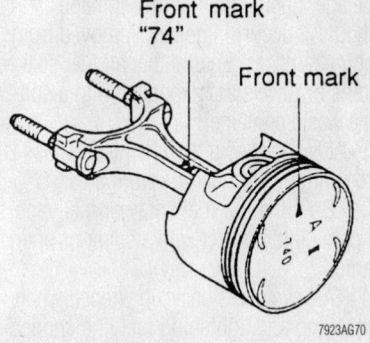

3.5L engine—piston and connecting rod assembly positioning

FUEL SYSTEM

Fuel System Service Precautions

Safety is the most important factor when performing not only fuel system maintenance but any type of maintenance. Failure to conduct maintenance and repairs in a safe manner may result in serious personal injury or death. Maintenance and testing of the vehicle's fuel system components can be accomplished safely and effectively by adhering to the following rules and guidelines.

• To avoid the possibility of fire and personal injury, always disconnect the negative battery cable unless the repair or test procedure requires that battery voltage be applied.

• Always relieve the fuel system pressure prior to disconnecting any fuel system component (injector, fuel rail, pressure regulator, etc.), fitting or fuel line connection. Exercise extreme caution whenever relieving fuel system pressure, to avoid exposing skin, face and eyes to fuel spray. Please be advised that fuel under pressure may penetrate the skin or any part of the body that it contacts.

• Always place a shop towel or cloth around the fitting or connection prior to loosening to absorb any excess fuel due to spillage. Ensure that all fuel spillage (should it occur) is quickly removed from engine surfaces. Ensure that all fuel soaked cloths or towels are deposited into a suitable waste container.

• Always keep a dry chemical (Class B) fire extinguisher near the work area.

• Do not allow fuel spray or fuel vapors to come into contact with a spark or open flame.

• Always use a back-up wrench when loosening and tightening fuel line connection fittings. This will prevent unnecessary stress and torsion to fuel line piping. Always follow the proper torque specifications.

• Always replace worn fuel fitting O-rings with new. Do not substitute fuel hose or equivalent, where fuel pipe is installed.

Fuel System Pressure

RELIEVING

1. Before servicing the vehicle, refer to the precautions in the beginning of this section.
2. Turn the ignition to the **OFF** position.

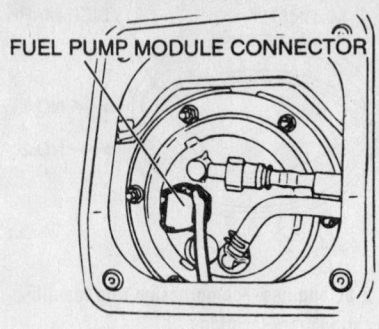

FUEL PUMP MODULE CONNECTOR

9357QG23

Location of the fuel pump connector—
Lancer shown, other models similar

3. Loosen the fuel filler cap to release fuel tank pressure.
4. For the Lancer, Mirage and Eclipse, remove the rear seat cushion, then remove the service cover and disconnect the fuel pump harness connector.
5. For the Front Wheel Drive (FWD) Galant and Diamante, detach the fuel pump harness connector located near the fuel tank. It may be necessary to raise the vehicle to access the connector.
6. For the All Wheel Drive (AWD) Galant, remove the carpet from the trunk, locate the fuel tank wiring at the pump access cover, then detach the wiring.
7. Start the vehicle and allow it to run until it stalls from lack of fuel. Turn the key to the **OFF** position.
8. Disconnect the negative battery cable, then reconnect the fuel pump connector. Install the access cover, cushion or carpet as necessary.
9. Wrap shop towels around the fitting that is being disconnected to absorb residual fuel in the lines.
10. Place shop towels into proper safety container.

Fuel Filter

REMOVAL & INSTALLATION

Diamante

1. Before servicing the vehicle, refer to the precautions in the beginning of this section.
2. Properly relieve the fuel pressure.
3. Disconnect the negative battery cable.

➡The filter is located in the engine compartment, mounted on the inner fender panel.

4. Remove or disconnect the following:
• Air cleaner assembly and intake hoses
• Battery and battery tray
• Fuel lines from the filter
• Mounting bolts and remove the fuel filter from the vehicle

To install:

➡Install new gaskets or O-rings whenever fuel connections have been disassembled.

5. Install or connect the following:
• Filter to its bracket finger-tight
• New gaskets and connect the high pressure hose and eye bolt, then the main pipe and eye bolt. Tighten the eye bolts to 22 ft. lbs. (30 Nm). Tighten the flare nut to 25 ft. lbs. (35 Nm).
6. Tighten the mounting bolts fully.
7. Install or connect the following:
• Air cleaner assembly
• Battery and battery tray
• Negative battery cable, install the fuel filler cap, turn the key to the **ON** position to pressurize the fuel system and check for leaks.

Lancer, Mirage and Galant

➡The fuel filter is located in the engine compartment.

1. Before servicing the vehicle, refer to the precautions in the beginning of this section.
2. Properly relieve the fuel system pressure.
3. Remove or disconnect the following:
• Negative battery cable
• Air intake hose and the battery
• Fuel lines from the filter
• Mounting bolts and the fuel filter from the vehicle

To install:

4. If equipped with flare fitting, tighten the fitting by hand before installing the filter to the vehicle.
5. Install or connect the following:
• Filter to its bracket finger-tight
• New gaskets and connect the high pressure hose and eye bolt, then the main pipe. Tighten the eye bolts to 22 ft. lbs. (30 Nm). Tighten the flare nut to 27 ft. lbs. (37 Nm).
6. Tighten the filter mounting bolts fully.
7. Install or connect the following:
• Air intake hose
• Battery
• Negative battery cable, install the

fuel filler cap, turn the key to the **ON** position to pressurize the fuel system and check for leaks.

Eclipse

1. Before servicing the vehicle, refer to the precautions in the beginning of this section.
2. Properly relieve the fuel system pressure.
3. Disconnect the negative battery cable.
4. On models equipped with the 2.4L engine, remove the battery and the air intake hose.
5. Remove the fuel lines from the filter.
6. Remove clamp and the hose from the fuel pressure regulator.
7. Remove or disconnect the following:
   - Fuel filter mounting bracket bolts and remove the fuel filter
   - Bracket screw and remove the fuel filter from the mounting bracket
8. On 2.0L non-turbo models, remove the following from the filter:
   a. The eye bolt and washer.
   b. The fuel connector and washer with the fuel pressure regulator.

To install:
9. On 2.0L non-turbo models, install the fuel connector with the fuel pressure regulator to the filter with 2 new washers and tighten the eye bolt to 22 ft. lbs. (36 Nm).
10. Install or connect the following:
   - Fuel filter to the mounting bracket with the screw
   - Fuel filter to the vehicle with the bracket mounting bolts
   - Main fuel pipe to the fuel filter connector or the filter itself. Torque the flare nut to 27 ft. lbs. (36 Nm).
11. On the 2.0L non-turbo engine models reconnect the hose and clamp to the fuel pressure regulator.
12. Reconnect the high pressure fuel hose to the fuel filter. Torque the eye bolt to 22 ft. lbs. (29 Nm).
13. On 2.4L engine models, install the battery and the air intake hose.
14. Reconnect the negative battery cable, start the engine and check for fuel leaks.

Fuel Pump

REMOVAL & INSTALLATION

Diamante

1. Before servicing the vehicle, refer to the precautions in the beginning of this section.

2. Properly relieve the fuel system pressure.
3. Disconnect the negative battery cable.
4. Remove the left rear wheel well liner, if equipped.
5. Disconnect the center exhaust system from the main muffler. Disconnect the rear exhaust hangers, lower the exhaust and secure aside.
6. Remove the tank drain plug and drain the fuel into an approved container.
7. On models equipped with 4WS, remove the mounting bolts and lower the rear steering gear.
8. Remove or disconnect the following:

- Fuel return hose, high pressure hose and vent hose from the sending unit
- Electrical connector
- Filler and vent hoses. Place a support under the tank and remove the retaining nuts.
9. Lower the tank from the vehicle.
10. Remove the fuel pump retaining nuts and remove the assembly from the tank.

To install:
11. Install or connect the following:
- Pump assembly to the tank and tighten the retaining nuts to 24 inch lbs. (3 Nm)

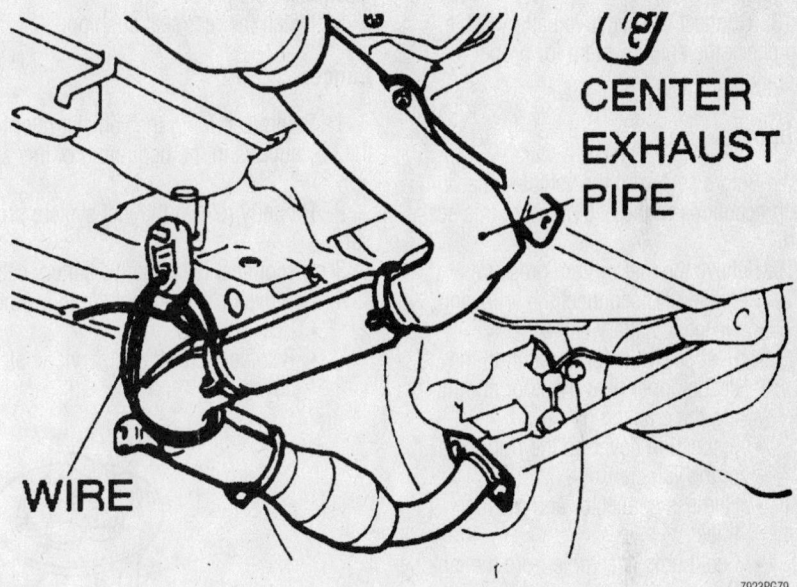

Proper method of supporting rear exhaust system—Diamante 3.0L engine

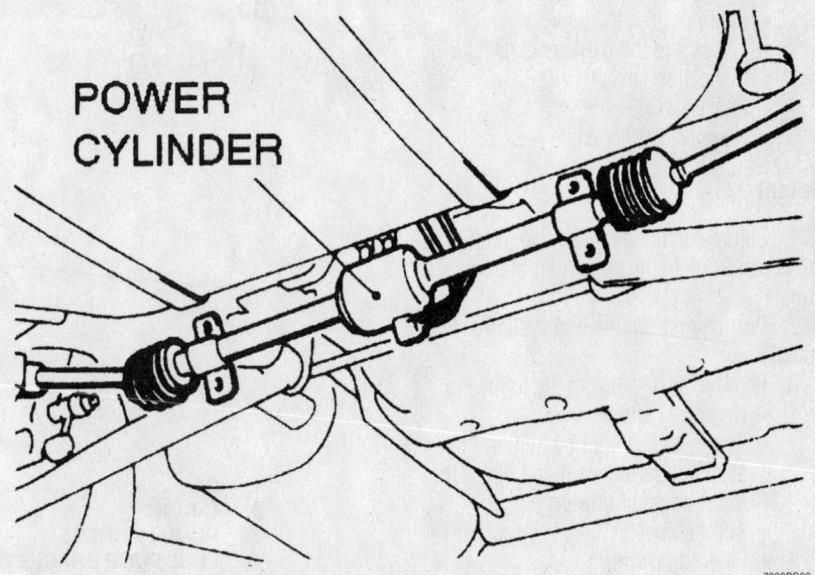

Power cylinder identification—Diamante 3.0L engine

- Fuel tank and connect the filler and vent hoses. Tighten the tank retaining nuts and bolts to 19 ft. lbs. (26 Nm).
- Return hose, high pressure hose and all other hoses and connectors connected to the pump/sending unit
- Power cylinder unit and tighten the mounting bolts to 31 ft. lbs. (43 Nm), if equipped with 4WS
- Exhaust pipe and secure the rear hangers
- Left rear wheel well liner, if removed

12. Lower the vehicle and return fuel to the gas tank.

13. Connect the negative battery cable and check the entire system for proper operation and leaks.

Eclipse

1. Before servicing the vehicle, refer to the precautions in the beginning of this section.

2. Relieve the fuel system pressure.

3. Remove or disconnect the following:
- Negative battery cable
- Rear seat cushion by pulling the seat stopper near the floor and lifting the cushion up
- Inspection cover on the right side of the vehicle
- Harness connector and the fuel lines
- Fuel pump assemble from the tank. Remove the locking ring on the All Wheel Drive (AWD) model.

To install:

4. Install or connect the following:
- Fuel pump in the tank
- Hoses and the harness connector
- Inspection cover
- Rear seat
- Negative battery cable

Galant

1. Before servicing the vehicle, refer to the precautions in the beginning of this section.

2. Properly relieve the fuel system pressure.

3. Remove or disconnect the following:
- Negative battery cable
- Rear seat cushion, by pulling the seat stopper outward and lifting the lower cushion upward
- Access cover
- Fuel pump wiring
- Return hose and the high pressure fuel hose

- Pump mounting nuts and remove the pump assembly

To install:

4. Install the fuel pump assembly to the tank and tighten the retaining nuts to 22 inch lbs. (2.5 Nm).

➡ Tilt the float to the left of the vehicle, when installing the pump assembly.

5. Install or connect the following:
- High pressure hose, return hose and the fuel tank wiring
- Negative battery cable

6. Check the fuel pump for proper pressure and inspect the entire system for leaks.

7. Apply sealant to the access cover and install the cover.

8. Install the rear seat cushion.

Lancer

1. Before servicing the vehicle, refer to the precautions in the beginning of this section.

2. Properly relieve the fuel system pressure.

3. Disconnect the negative battery cable.

4. Remove or disconnect the following:
- Rear seat assembly
- Retainer screws and service hole cover

- Harness electrical connector
- Fuel lines
- Fuel pump module mounting nuts
- Fuel pump module

5. Disassemble the fuel pump module as necessary. Refer to the accompanying figure.

6. Installation is the reverse of the removal procedure. Torque the fuel pump module nuts to 23 inch lbs. (2.5 Nm).

Mirage

1. Before servicing the vehicle, refer to the precautions in the beginning of this section.

2. Properly relieve the fuel system pressure.

3. Disconnect the negative battery cable.

4. Raise and safely support the vehicle.

5. Drain the fuel from the fuel tank into an approved container.

6. Disconnect the filler and vent hoses.

7. Support the tank with a transmission jack. Disconnect the retainer straps and lower the tank to gain access to the fitting on top of the tank.

8. Remove or disconnect the following:
- Return hose, high pressure hose and vapor hoses from the pump/sending unit

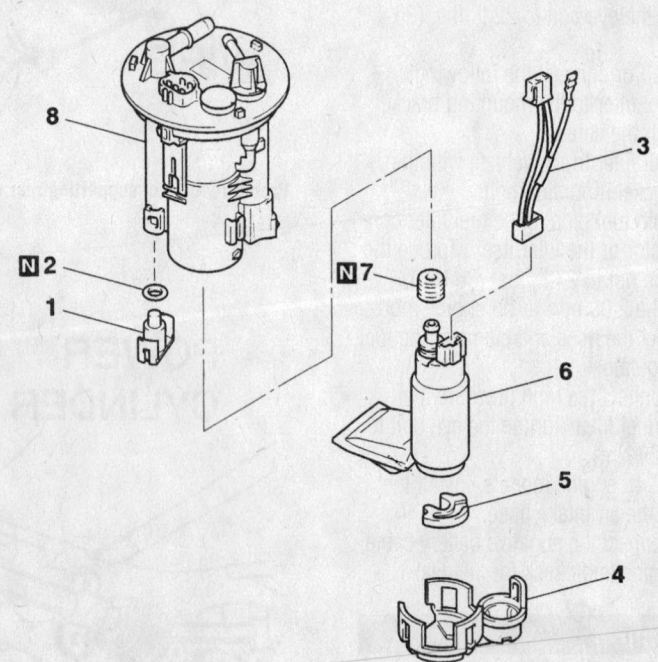

1.	CAP	5.	FUEL PUMP CUSHION
2.	O-RING	6.	FUEL PUMP
3.	PUMP HARNESS	7.	GROMMET
4.	FUEL PUMP BRACKET	8.	FUEL FILTER ASSEMBLY

9357QG24

Exploded view of the fuel pump module—Lancer

- Electrical connectors at the pump/sending unit
- Fuel tank from the vehicle
- Access plate to the fuel tank and remove the pump assembly

To install:

9. Install fuel pump into fuel tank, with new packing gasket, and tighten mounting nuts.

10. Raise the tank in position under the vehicle.

11. Attach all connections to the top of the tank.

12. Raise the tank completely and position the retainer straps around the fuel tank. Install new fuel tank self-locking nuts and tighten to 22 ft. lbs. (31 Nm).

13. Connect the return hose and high pressure hoses.

14. Install the vapor hose and the filler hose. Install the filler hose retainer screws to the fender, if removed.

15. Lower the vehicle and pour the drained fuel into the gas tank.

16. Connect the negative battery cable. Check the fuel pump for proper pressure and inspect the entire system for leaks.

Fuel Injector

REMOVAL & INSTALLATION

1.5L & 1.8L Engines

1. Relieve the fuel system pressure as described in this section.

2. Remove or disconnect the following:
   - Positive Crankcase Ventilation (PCV) hose from the valve cover
   - Breather hose at the opposite end of the valve cover
   - High pressure fuel line

❊❊ CAUTION

Observe all applicable safety precautions when working around fuel. Whenever servicing the fuel system, always work in a well ventilated area. Do not allow fuel spray or vapors to come in contact with a spark or open flame. Keep a dry chemical fire extinguisher near the work area. Always keep fuel in a container specifically designed for fuel storage; also, always properly seal fuel containers to avoid the possibility of fire or explosion.

3. Remove or disconnect the following:
   - Vacuum hose from the fuel pressure regulator

- Fuel return hose from the pressure regulator
- Electrical connector from each injector. Label for reference
- Bolt(s) holding the fuel rail to the manifold. Carefully lift the rail up and remove it with the injectors attached. Take great care not to drop an injector. Place the rail and injectors in a safe location on the workbench; protect the tips of the injectors from dirt and/or impact.
- Injector insulators from the intake manifold, discard. The insulators are not reusable.
- Injectors from the fuel rail by pulling gently in a straight outward motion. Make certain the grommet and O-ring come off with the injector.

To install:

4. Install a new insulator in each injector port in the manifold.

5. Remove the old grommet and O-ring from each injector. Install a new grommet and O-ring; coat the O-ring lightly with clean, thin oil.

6. If the fuel pressure regulator was removed, replace the O-ring with a new one and coat it lightly with clean, thin oil. Insert the regulator straight into the rail, then check that it can be rotated freely. If it does not rotate smoothly, remove it and inspect the O-ring for deformation or jamming. When properly installed, align the mounting holes and tighten the retaining bolts to 84 inch lbs. (9 Nm). This procedure must be followed even if the fuel rail was not removed.

7. Install or connect the following:
   - Injector into the fuel rail, constantly turning the injector left and right during installation. When fully installed, the injector should still turn freely in the rail. If it does not, remove the injector and inspect the O-ring for deformation or damage.
   - Delivery pipe and injectors to the engine. Make certain that each injector fits correctly into its port and that the rubber insulators for the fuel rail mounts are in position.
   - Fuel rail retaining bolts and tighten them to 108 inch lbs. (12 Nm)
   - Wiring harnesses to the appropriate injector
   - Fuel return hose to the pressure regulator, then connect the vacuum hose

8. Replace the O-ring on the high pressure fuel line, coat the O-ring lightly with clean, thin oil and install the line to the fuel rail. Tighten the mounting bolts.

- PCV hose and the breather hose
- Negative battery cable

9. Pressurize the fuel system and inspect all connections for leaks.

2.0L Engine

1. Relieve the fuel system pressure as described in this section.

2. Disconnect the negative battery cable.

3. Wrap the connection with a shop towel and disconnect the high pressure fuel line at the fuel rail.

❊❊ CAUTION

Observe all applicable safety precautions when working around fuel. Whenever servicing the fuel system, always work in a well ventilated area. Do not allow fuel spray or vapors to come in contact with a spark or open flame. Keep a dry chemical fire extinguisher near the work area. Always keep fuel in a container specifically designed for fuel storage; also, always properly seal fuel containers to avoid the possibility of fire or explosion.

4. Remove or disconnect the following:
   - Positive Crankcase Ventilation (PCV) hose
   - Exhaust Gas Recirculation (EGR) solenoid valve connector
   - Manifold Differential Pressure (MDP) sensor connector
   - Purge control solenoid valve connector
   - Throttle Position (TP) sensor connector
   - Idle Air Control (IAC) motor connector
   - Electrical connector from each injector, label for reference
   - High pressure fuel hose
   - Fuel return hose
   - Vacuum hose(s)
   - Fuel pressure regulator
   - Fuel hose
   - Fuel return pipe
   - Bolt(s) holding the fuel rail to the manifold. Carefully lift the rail up and remove it with the injectors attached. Take great care not to drop an injector. Place the rail and injectors in a safe location on the workbench; protect the tips of the injectors from dirt and/or impact.
   - Injector insulators from the intake manifold, discard. The insulators are not reusable.
   - Injectors from the fuel rail by

pulling gently in a straight outward motion. Make certain the grommet and O-ring come off with the injector.

To install:

5. Install a new insulator in each injector port in the manifold.

6. Remove the old grommet and O-ring from each injector. Install a new grommet and O-ring; coat the O-ring lightly with clean, thin oil.

7. If the fuel pressure regulator was removed, replace the O-ring with a new one and coat it lightly with clean, thin oil. Insert the regulator straight into the rail, then check that it can be rotated freely. If it does not rotate smoothly, remove it and inspect the O-ring for deformation or jamming. When properly installed, align the mounting

holes and tighten the retaining bolts to 84 inch lbs. (9 Nm). This procedure must be followed even if the fuel rail was not removed.

8. Install or connect the following:
   • Injector into the fuel rail, constantly turning the injector left and right during installation. When fully installed, the injector should still turn freely in the rail. If it does not, remove the injector and inspect the O-ring for deformation or damage.
   • Fuel rail and injectors to the engine. Make certain that each injector fits correctly into its port and that the rubber insulators for the fuel rail mounts are in position.
   • Fuel return pipe and fuel hose
   • Fuel return hose to the fuel rail

   • High pressure fuel line to the fuel rail
   • Fuel injector connectors
   • IAC motor connector
   • TP sensor connector
   • Purge control solenoid connector
   • MDP sensor connector
   • EGR solenoid valve connector
   • PCV hose connector
   • Negative battery cable

9. Pressurize the fuel system and inspect all connections for leaks.

2.4L Engine

1. Relieve the fuel system pressure as described in this section.

2. Label and disconnect the spark plug wires. Position the wires aside.

3. Remove or disconnect the following:

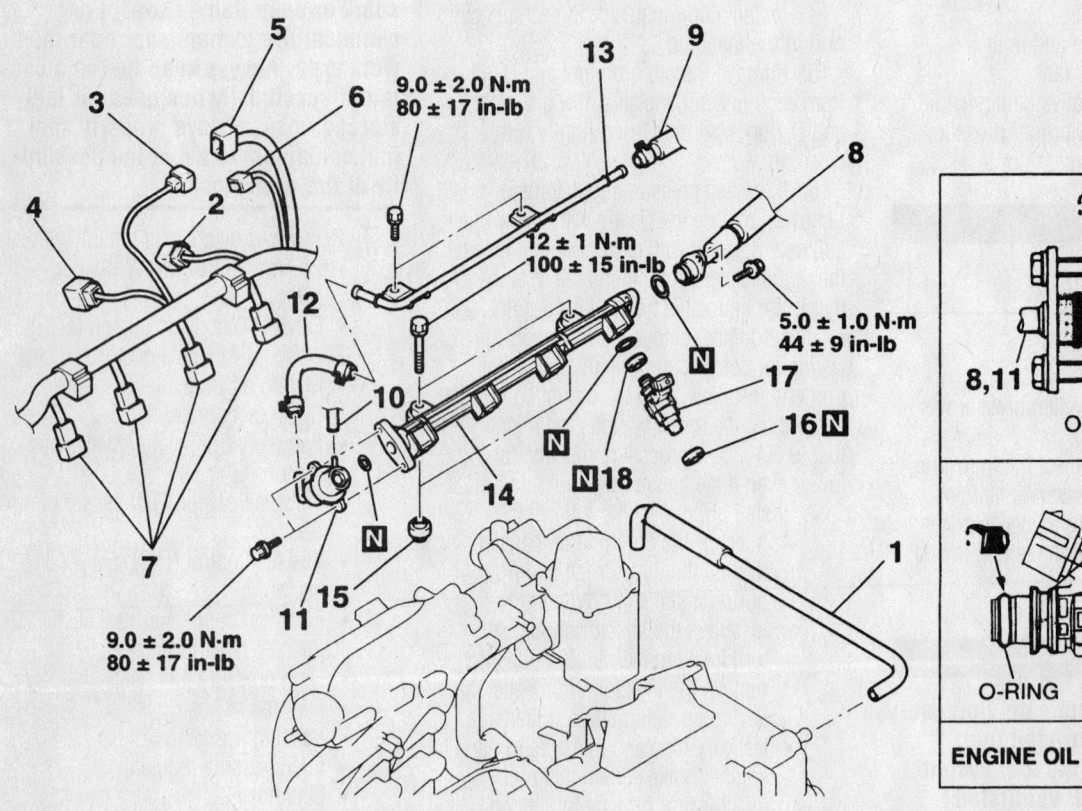

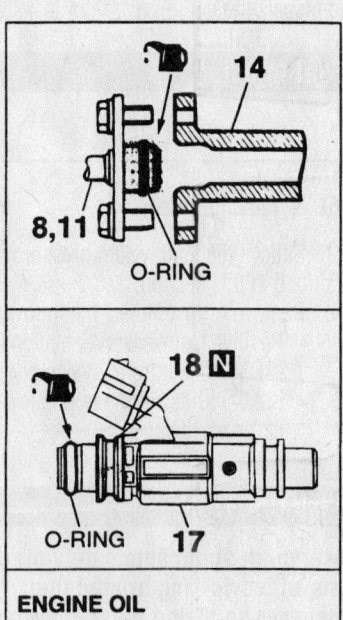

1. PCV HOSE CONNECTION
2. EGR SOLENOID VALVE CONNECTOR
3. MANIFOLD DIFFERENTIAL PRESSURESENSOR CONNECTOR
4. PURGE CONTROL SOLENOID VALVE CONNECTOR
5. THROTTLE POSITION SENSOR CONNECTOR
6. IDLE AIR CONTROL MOTOR CONNECTOR
7. INJECTOR CONNECTOR

Exploded view of the fuel rail and injectors—2.0L engine

9357QG25

- Positive Crankcase Ventilation (PCV) hose from the valve cover
- High pressure fuel line to the fuel rail and disconnect the line. Be prepared to contain fuel spillage; plug the line to keep out dirt and debris.

❊❊ CAUTION

Observe all applicable safety precautions when working around fuel. Whenever servicing the fuel system, always work in a well ventilated area. Do not allow fuel spray or vapors to come in contact with a spark or open flame. Keep a dry chemical fire extinguisher near the work area. Always keep fuel in a container specifically designed for fuel storage; also, always properly seal fuel containers to avoid the possibility of fire or explosion.

4. Remove or disconnect the following:
- Vacuum hose from the fuel pressure regulator
- Fuel return hose from the pressure regulator
- Electrical connector from each injector, label for reference
- Bolt(s) holding the fuel rail to the manifold. Carefully lift the rail up and remove it with the injectors attached. Take great care not to drop an injector. Place the rail and injectors in a safe location on the workbench; protect the tips of the injectors from dirt and/or impact.
- Injector insulators from the intake manifold, discard. The insulators are not reusable.
- Injectors from the fuel rail by pulling gently in a straight outward motion. Make certain the grommet and O-ring come off with the injector.

To install:

5. Install a new insulator in each injector port in the manifold.
6. Remove the old grommet and O-ring from each injector. Install a new grommet and O-ring; coat the O-ring lightly with clean, thin oil.
7. If the fuel pressure regulator was removed, replace the O-ring with a new one and coat it lightly with clean, thin oil. Insert the regulator straight into the rail, then check that it can be rotated freely. If it does not rotate smoothly, remove it and inspect

the O-ring for deformation or jamming. When properly installed, align the mounting holes and tighten the retaining bolts to 84 inch lbs. (9 Nm). This procedure must be followed even if the fuel rail was not removed.

8. Install or connect the following:
- Injector into the fuel rail, constantly turning the injector left and right during installation. When fully installed, the injector should still turn freely in the rail. If it does not, remove the injector and inspect the O-ring for deformation or damage.
- Delivery pipe and injectors to the engine. Make certain that each injector fits correctly into its port and that the rubber insulators for the fuel rail mounts are in position.
- Fuel rail retaining bolts and tighten them to 108 inch lbs. (12 Nm)
- Wiring harnesses to the appropriate injector
- Fuel return hose to the pressure regulator, then connect the vacuum hose
- O-ring on the high pressure fuel line, coat the O-ring lightly with clean, thin oil and install the line to the fuel rail. Tighten the mounting bolts to 48 inch lbs. (6 Nm).
- PCV hose and spark plug wires
- Negative battery cable
9. Pressurize the fuel system and inspect all connections for leaks.

3.0L and 3.5L Engines

1. Relieve the fuel system pressure.
2. Disconnect the negative battery cable.

❊❊ CAUTION

Work MUST NOT be started until at least 90 seconds after the ignition switch is turned to the LOCK position and the negative battery cable is disconnected from the battery. This will allow time for the air bag system backup power supply to deplete its stored energy preventing accidental air bag deployment which could result in unnecessary air bag system repairs and/or personal injury.

3. Drain the cooling system.
4. Disconnect all components from the air intake plenum and remove the plenum from the intake manifold.
5. Wrap the connection with a shop towel and disconnect the high pressure fuel line at the fuel rail.

❊❊ CAUTION

Observe all applicable safety precautions when working around fuel. Whenever servicing the fuel system, always work in a well ventilated area. Do not allow fuel spray or vapors to come in contact with a spark or open flame. Keep a dry chemical fire extinguisher near the work area. Always keep fuel in a container specifically designed for fuel storage; also, always properly seal fuel containers to avoid the possibility of fire or explosion.

6. Remove or disconnect the following:
- Fuel return hose and remove the O-ring
- Vacuum hose from the fuel pressure regulator
- Electrical connectors from each injector
- Fuel pipe connecting the fuel rails
- Injector rail retaining bolts. Make sure the rubber mounting bushings do not get lost.
7. Lift the rail assemblies up and away from the engine.
8. Remove the injectors from the rail by pulling gently. Discard the lower insulator.

To install:

➡ **Some of the vehicles may have a clip that secures the injector to the fuel rail. Be sure to remove or install the injector clip where necessary.**

9. Install or connect the following:
- New grommet and O-ring to the injector. Coat the O-ring with light oil.
- Injector to the fuel rail
10. Replace the seats in the intake manifold. Install the fuel rails and injectors to the manifold. Make sure the rubber bushings are in place before tightening the mounting bolts.
11. Tighten the retaining bolts to 84–108 inch lbs. (10–13 Nm). Install the fuel pipe with new gasket.
12. Install or connect the following:
- Electrical connectors to the injectors
- Fuel return hose
- O-ring, lightly lubricate it and connect the high pressure fuel line
- Intake plenum and all related items, using new gaskets
13. Fill the cooling system.
14. Connect the negative battery cable and check the entire system for proper operation and leaks.

DRIVE TRAIN

Transaxle assembly

REMOVAL & INSTALLATION

Manual

ECLIPSE

1. Before servicing the vehicle, refer to the precautions in the beginning of this section.
2. Remove or disconnect the following:
   - Battery and the air intake hoses
   - Battery tray and support
   - Auto-cruise actuator and bracket, if equipped with cruise control
   - Charcoal canister and bracket
   - Shift and select cables from the transaxle
   - Back-up light switch and the Vehicle Speed Sensor (VSS) connectors
   - Starter assembly
   - Engine support fixture to the engine and remove the transaxle mounting bolts
3. Remove or disconnect the following:
   - Rear roll stopper bracket mounting bolts
   - Transaxle mounting bracket mounting nuts
   - Engine undercovers
   - Transfer case assembly, if equipped with All Wheel Drive (AWD)
   - Axle shafts

- Slave cylinder from the bell housing without disconnecting the fluid line. Position it out of the way.
- Bell housing cover and the right-hand center member stay (support)
- Center member

4. Place a transmission jack under the transaxle and remove the transaxle mounting bolt.
5. Remove the transaxle mounting and lower the transaxle.

To install:

6. Raise the transaxle into position and install the transaxle mounting. Torque the through-bolt to 50 ft. lbs. (69 Nm).
7. Install or connect the following:
   - Transaxle assembly mounting bolt. Torque the bolt to 22–25 ft. lbs. (30–34 Nm).
   - Center member assembly and the right-hand stay
   - Bell housing cover and the slave cylinder
   - Axle shafts. Be sure to install the washer in the proper direction.
   - Engine undercovers and lower the vehicle
   - Transfer case assembly, if removed
   - Transaxle mounting bracket mounting nuts
   - Rear roll stopper bracket mounting bolts
   - Transaxle assembly mounting bolts. Torque the mounting bolts to 35 ft. lbs. (48 Nm).

8. Remove the engine support fixture.
9. Install or connect the following:
   - Starter assembly
   - VSS and the back-up light connectors
   - Cruise control actuator if removed
   - Battery tray support and the tray
   - Charcoal canister bracket and the canister
   - Air duct and the air cleaner assembly

GALANT

1. Before servicing the vehicle, refer to the precautions in the beginning of this section.
2. Remove or disconnect the following:
   - Negative battery cable
   - Air cleaner and intake hoses
   - Cotter pins and clips securing the select and shift cables and remove the cable ends from the transaxle
   - Air compressor, if equipped with Active Electronic Control Suspension (Active-ECS)
   - Back-up light switch harness and position aside
   - Speedometer electrical connector from the transaxle assembly
   - Starter motor and position aside
3. Support the engine assembly.
   - Rear roll stopper mounting bracket
   - Transaxle mount bracket
   - Upper transaxle mounting bolts
   - Front wheel assemblies
   - Right-hand undercover
   - Cotter pin and disconnect the tie rod end, from the steering knuckle
   - Stabilizer bar link from the damper fork
   - Damper fork from the lateral lower control arm
   - Lateral lower arm and the compression arm lower ball joints from the steering knuckle
   - Halfshafts from the transaxle, and secure aside
   - Clutch release cylinder without disconnecting the hydraulic line and secure aside
   - Cover from the transaxle bell housing
   - Engine front roll stopper through-bolt
   - Crossmember and the triangular right-hand stay
4. Support the transaxle with a transmission jack and remove the transaxle lower coupling bolt.

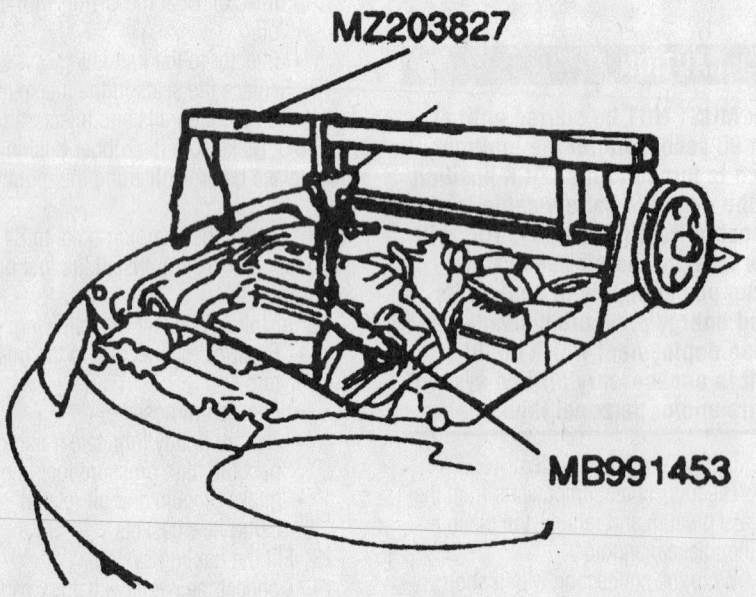

MZ203827

MB991453

7923PG82

Proper method of supporting the engine assembly for transaxle removal

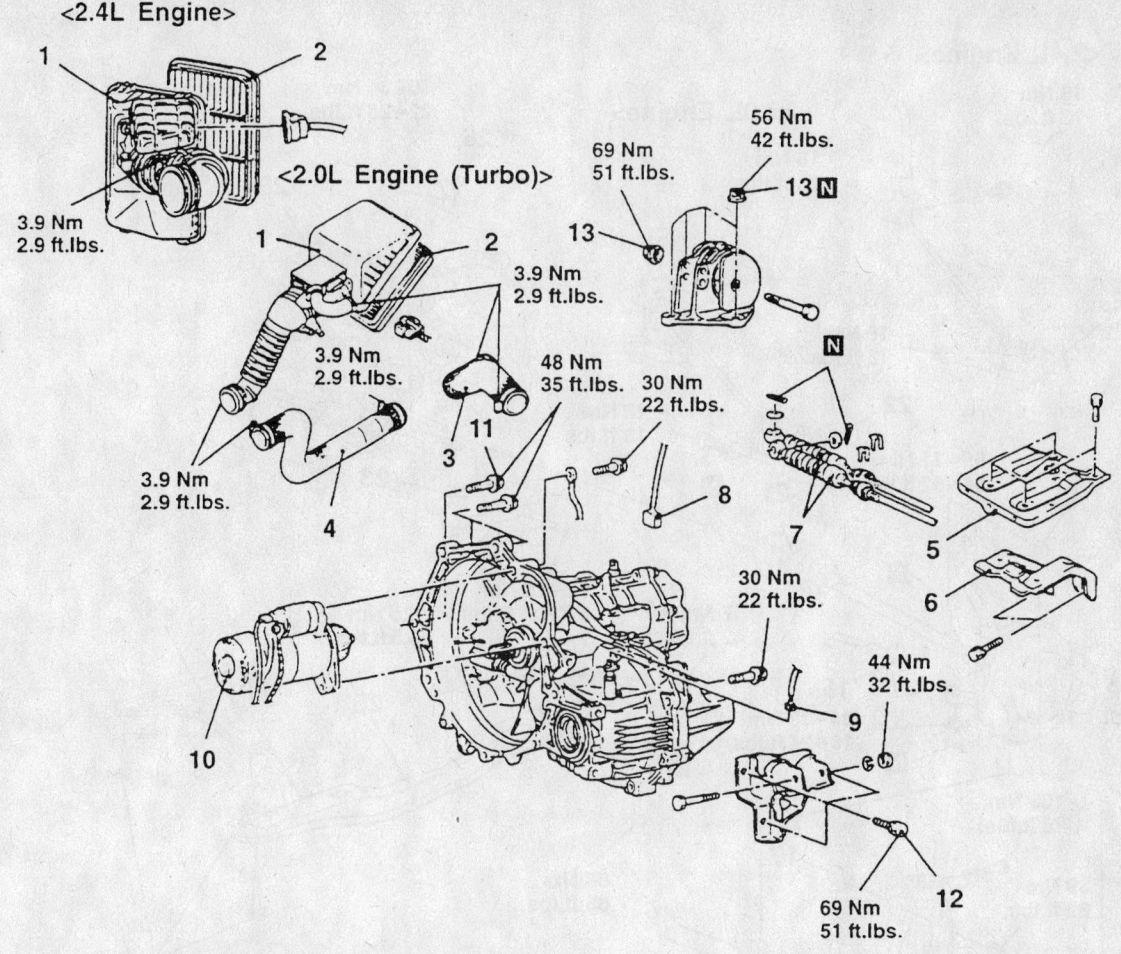

Removal steps
1. Air cleaner cover and air intake hose assembly
2. Air cleaner element
3. Air hose C <2.0L Engine (Turbo)>
4. Air hose A <2.0L Engine (Turbo)>
5. Battery tray
6. Battery tray stay
7. Shift cable and select cable connection
8. Backup light switch connector
9. Vehicle speed sensor connector
10. Starter motor
11. Transaxle assembly mounting bolts
12. Rear roll stopper bracket mounting bolts
13. Transaxle mounting bracket mounting nuts
● Supporting engine assembly

7923PG86

Exploded view of the manual transaxle mounting (1 of 2)—FWD Eclipse with 2.4L engine

➡**The coupling bolt threads from the engine side into the transaxle and is located just above the halfshaft opening.**

5. Slide the transaxle rearward and carefully lower it from the vehicle.
To install:
6. Install or connect the following:
● Transaxle to the engine and install the mounting bolts. Tighten to 35 ft. lbs. (48 Nm). Install the transaxle lower coupling bolt and tighten to 22–25 ft. lbs. (30–34 Nm).

● Cover to the transaxle bell housing and tighten the mounting bolts to 84 inch lbs. (9 Nm)
● Crossmember and tighten the front mounting bolts to 65 ft. lbs. (88 Nm) and the rear bolt to 54 ft. lbs. (73 Nm). Install the front engine roll stopper through-bolt and lightly tighten. Once the full weight of the engine is on the mounts, tighten the bolt to 42 ft. lbs. (57 Nm).
● Triangular stay bracket and tighten the mounting bolts to 65 ft. lbs. (88 Nm)

● Clutch release cylinder
● Halfshafts, using new circlips on the axle ends
● Tie rod and ball joints to the steering knuckle. Tighten the ball joint self-locking nuts to 48 ft. lbs. (65 Nm). Tighten the tie rod end nut to 21 ft. lbs. (28 Nm) and secure with a new cotter pin.
● Damper fork to the lower control arm and tighten the through-bolt to 65 ft. lbs. (88 Nm)
● Stabilizer link to the damper fork and tighten the self-locking nut to 29 ft. lbs. (39 Nm)

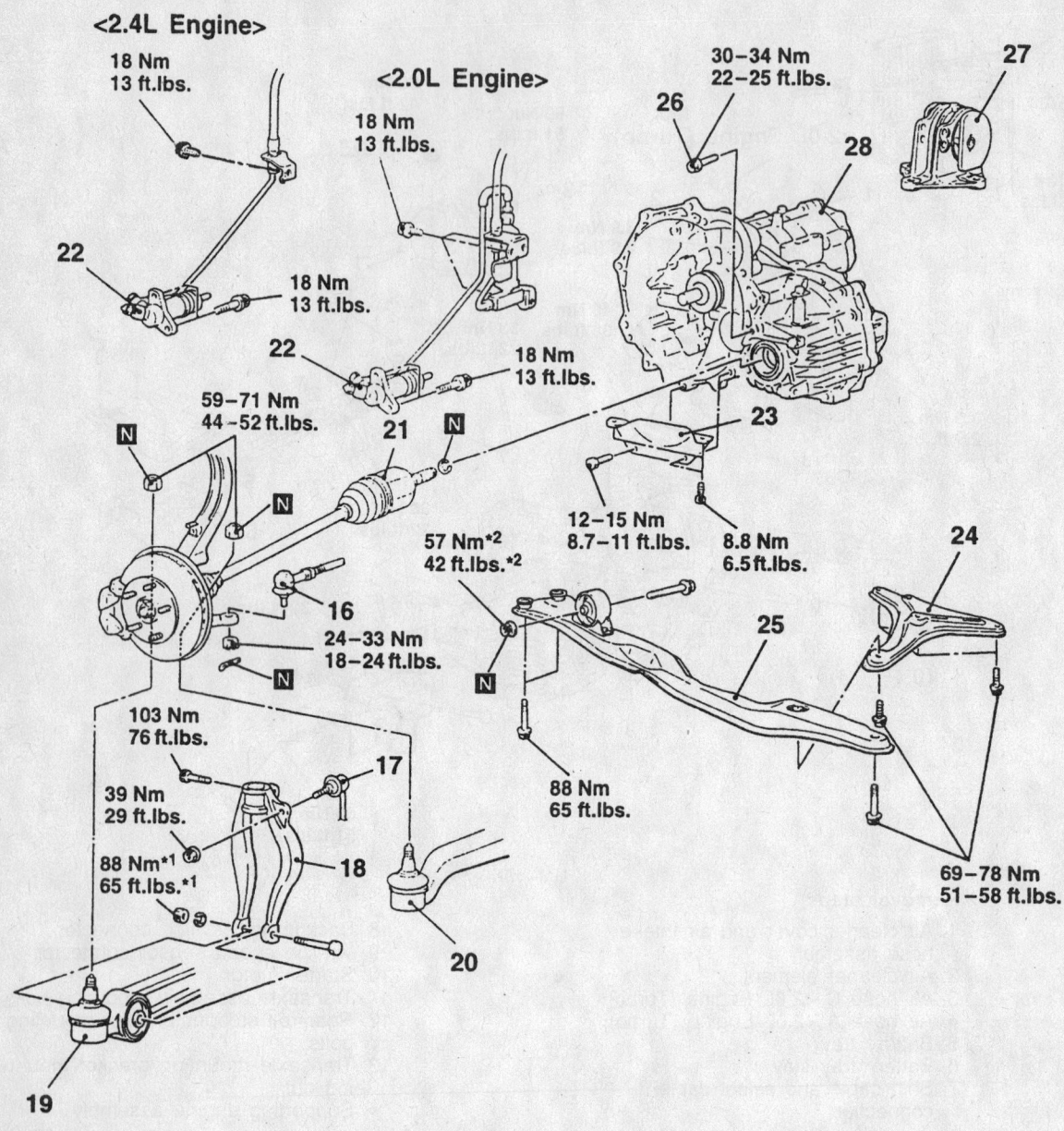

<2.4L Engine>

18 Nm
13 ft.lbs.

<2.0L Engine>

18 Nm
13 ft.lbs.

30–34 Nm
22–25 ft.lbs.

26

27

28

22

18 Nm
13 ft.lbs.

22

18 Nm
13 ft.lbs.

23

59–71 Nm
44–52 ft.lbs.

N

21

N

12–15 Nm
8.7–11 ft.lbs.

8.8 Nm
6.5 ft.lbs.

24

57 Nm*2
42 ft.lbs.*2

16

24–33 Nm
18–24 ft.lbs.

N

N

25

103 Nm
76 ft.lbs.

39 Nm
29 ft.lbs.

17

88 Nm*1
65 ft.lbs.*1

18

88 Nm
65 ft.lbs.

20

69–78 Nm
51–58 ft.lbs.

19

Lifiting up of the vehicle

16. Tie rod end ball joint and kunckle connection
17. Stabilizer link connection
18. Damper fork
19. Lateral lower arm ball joint and kunckle connection
20. Compression lower arm ball joint and kunckle connection
21. Drive shaft connection
22. Clutch release cylinder connection
23. Bell housing cover
24. Stay (R.H.)
25. Center member assembly
26. Transaxle assembly mounting bolt

27. Transaxle mounting
28. Transaxle assembly

Caution
*1: Indicates parts which should be temporarily tightened, and then fully tightened with the vehicle on the ground in the unladen condition.
*2: For tightening locations indicated by the symbol, first tighten temporarily, and then make the final tightening with the entire weight of the engine applied to the vehicle body.

7923PG87

Exploded view of the manual transaxle mounting (2 of 2)—FWD Eclipse with 2.4L engine

- Underpan
- Wheels and lower vehicle
- Transaxle mount bracket to the transaxle and tighten the mounting nuts to 32 ft. lbs. (43 Nm)
- Rear roll stopper mounting bracket

7. Remove the engine support. Tighten the transaxle mount through-bolt to 51 ft. lbs. (69 Nm) and tighten the front engine roll stopper through-bolt.

- Upper transaxle mounting bolts and tighten to 35 ft. lbs. (48 Nm)
- Starter motor
- Back-up light switch and the speedometer connector
- Air compressor, if equipped with Active Electronic Control Suspension (Active-ECS)
- Select and shift cables and install new cotter pins
- Air cleaner and the air intake hose
- Negative battery cable

8. Check the transaxle for proper operation.

LANCER

1. Before servicing the vehicle, refer to the precautions in the beginning of this section.
2. Drain the transaxle fluid.
3. Remove or disconnect the following:

- Negative battery cable
- Engine undercover
- Evaporative canister
- Positive battery cable, battery and battery tray
- Shifter cables
- Back-up light switch and Vehicle Speed Sensor (VSS) connector
- Starter motor
- Clutch hose
- Upper engine-to-transaxle bolts
- Transaxle mount
- Transaxle mount stopper

4. Install a suitable engine support assembly, then raise and safely support the vehicle.

- Stabilizer bar
- Wheel Speed Sensor (WSS) connector, if equipped with Anti-lock Brakes (ABS)
- Brake hose clamp
- Tie rod end
- Lower control arm
- Centermember
- Halfshafts by inserting a prybar between the transaxle case and the driveshaft and prying the shaft from the transaxle. Do not pull on the driveshaft.
- Bell housing lower cover
- Transaxle to engine bolts and lower the transaxle from the vehicle

To install:

5. Install or connect the following:

- Transaxle to the engine and install the lower mounting bolts
- Bell housing cover

➡**When installing the halfshafts, use new circlips on the axle ends.**

6. Install or connect the following:

- Halfshafts into the transaxle
- Centermember
- Lower control arm
- Tie rod end
- Brake hose clamp
- WSS connector, if equipped
- Stabilizer bar

7. Lower the vehicle, then remove the engine support.

- Transaxle mount bracket. Tighten the nuts to 35 ft. lbs. (47 Nm).
- Transaxle mount stopper. Tighten the nuts to 61 ft. lbs. (82 Nm).
- Transaxle mount
- Upper transaxle-to-engine bolts and torque to 36 ft. lbs. (48 Nm)
- Clutch line
- Starter motor
- VSS connector
- Back-up light switch connector
- Shifter cables, adjust
- Evaporative canister
- Battery and battery tray
- Engine undercover
- Positive and negative battery cables
- Transaxle with fluid

8. Bleed the clutch, check and adjust the front wheel alignment, then check the transaxle for proper operation.

MIRAGE

1. Before servicing the vehicle, refer to the precautions in the beginning of this section.
2. Remove or disconnect the following:

- Negative battery cable
- Front wheels and the inner wheel panels
- Air cleaner assembly and vacuum hoses

3. Note the locations and disconnect the shifter cables.

- Back-up lamp switch connector
- Speedometer cable and remove the starter motor
- Upper transaxle-to-engine mounting bolts

4. Remove the undercover and splash pan.
5. Drain the transaxle oil.
6. Support the engine and remove the crossmember.

7. Remove or disconnect the following:

- Upper transaxle mounting bolt and bracket
- Stabilizer bar, tie rod ends and the lower ball joint connections
- Clutch release cylinder and clutch oil line bracket. Disconnect the clutch cable, if equipped with cable controlled clutch system.
- Halfshafts by inserting a prybar between the transaxle case and the driveshaft and prying the shaft from the transaxle. Do not pull on the driveshaft.
- Bell housing lower cover
- Transaxle to engine bolts and lower the transaxle from the vehicle

To install:

8. Install or connect the following:

- Transaxle to the engine and install the mounting bolts
- Bell housing cover

➡**When installing the halfshafts, use new circlips on the axle ends.**

9. Install or connect the following:

- Halfshafts into the transaxle
- Slave cylinder or connect the clutch cable
- Ball joints, tie rod ends and stabilizer bar connections
- Upper transaxle mounting bracket and bolt
- Crossmember
- Undercover
- Upper transaxle-to-engine mounting bolts
- Starter motor
- Back-up light switch connector and speedometer cable
- Shifter cables, adjust
- Air cleaner assembly
- Front wheels
- Negative battery cable and check the transaxle for proper operation

Automatic

DIAMANTE

1. Before servicing the vehicle, refer to the precautions in the beginning of this section.
2. Properly disarm the Supplemental Restraint System (SRS) system.
3. Raise and safely support the vehicle.
4. Remove or disconnect the following:

- Front wheels
- Engine side cover and undercovers

5. Drain the transaxle assembly.
6. If equipped, remove the front catalytic converter.
7. Remove or disconnect the following:

- Exhaust pipe, main muffler and cat-alytic converter
- Tie rod end and ball joint from the steering knuckle
- Support bearing for the left side halfshaft
- Halfshafts by inserting a prybar between the transaxle case and the driveshaft and prying the shaft from the transaxle
- Air cleaner assembly and adjoining ductwork
- Engine harness connection
- Compressor assembly, if the vehicle is equipped with Active Electronic Controlled Suspension (Active-ECS)—suspend with wire—Do not allow the compressor to hang from the air hose.
- Roll stopper stay bracket, if equipped
- Speedometer cable from the transaxle
- The clip that secures the shifter
- Shifter control cable from the transaxle
- Plug the oil cooler hoses from the transaxle

8. Disconnect the following:
- Park/neutral switch electrical har-ness
- Kickdown servo switch
- Pulse generator
- Oil temperature sensor electrical harness
- Shift control solenoid valve har-ness.

9. Support the transaxle and remove the transaxle mounting bracket.

10. Remove the 3 upper transaxle-to-engine mounting bolts.

11. For vehicles with Four Wheel Steer-ing (4WS), remove the heat shield for the 4WS oil pump and remove the pump. Do not allow the pump to hang from the oil hoses.

12. For vehicles equipped with Active-ECS, disconnect the height sensor rod from the lower control arm.

13. Remove or disconnect the following:
- Bolt that secures the Heated Oxy-gen (HO2S) sensor harness to the right side crossmember
- Starter assembly
- Mounting brackets for access to the bell housing cover
- Bell housing/oil pan covers assem-bly
- Bolts holding the flexplate to the torque converter
- Lower transaxle to engine bolts and remove the transaxle assembly

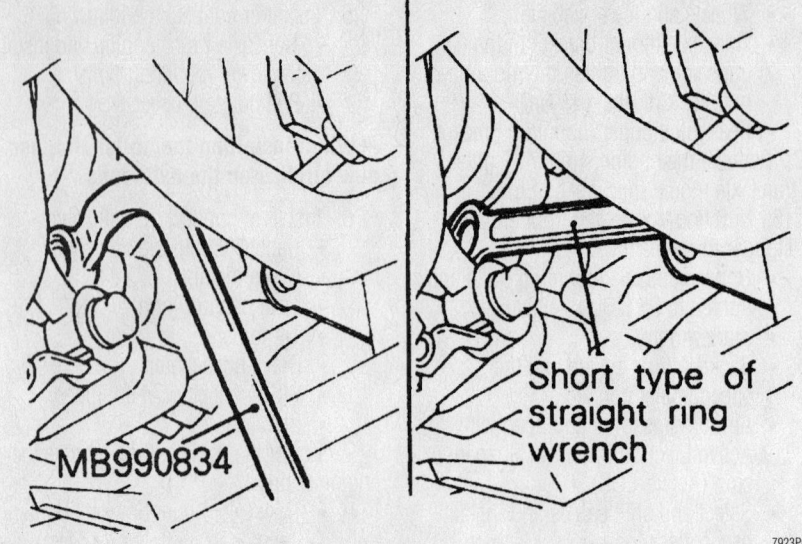

Location of 4-wheel steering oil pump mounting bolts—Diamante with a F4A33 automatic transaxle

To install:

14. Install or connect the following:
- Transaxle assembly to the engine block and install the mounting bolts
- Bolts that secure the torque con-verter to the driveplate. Tighten the bolts to 34–38 ft. lbs. (46–53 Nm).
- Bell housing/oil pan covers
- Transaxle stay brackets that were removed for access to the bell housing cover
- Starter assembly and connect the wiring
- Bolt that secures the HO2S sensor harness to the right side cross-member and tighten the bolt to 84–108 inch lbs. (10–12 Nm)

15. For vehicles equipped with Active-ECS, connect the height sensor rod from the lower control arm. Check the height sensor rod for a length (A) of 10.59–10.63 in. (269–270mm).

16. If removed, install the 4WS oil pump and tighten the mounting bolts to 17 ft. lbs. 24 Nm).

17. If removed, install the 4WS oil pump heat shield and tighten the mounting bolts to 17 ft. lbs. 24 Nm).

18. Install the 3 upper transaxle-to-engine mounting bolts. Tighten the mount-ing bolts to 54 ft. lbs. (75 Nm).

➡ One of the upper bolts has a ground-ing strap to secure under the bolt.

19. Install or connect the following:
- Transaxle mounting bracket. Tighten the mounting nut and bolts to 51 ft. lbs. (70 Nm).
- Shift control solenoid valve harness
- Kickdown servo switch, pulse gen-erator and oil temperature sensor electrical harness
- Park/neutral switch electrical har-ness

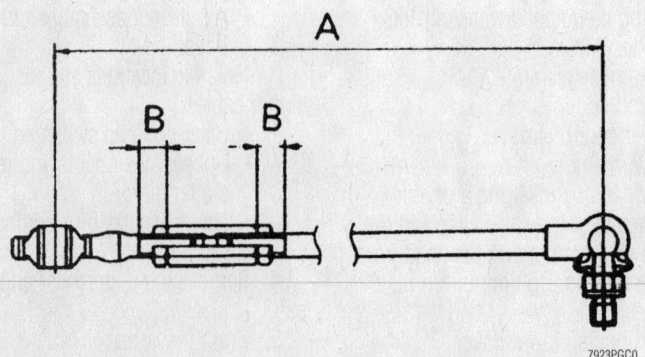

Height sensor rod adjustment—Diamante with a F4A33 automatic transaxle

- Oil cooler hoses to the transaxle, using new hose clamps
- Shifter control cable to the transaxle and secure the cable with clip
- Speedometer cable to the transaxle
- Roll stopper stay bracket and tighten the one through nut and bolt to 36–43 ft. lbs. (50–60 Nm), if removed. Tighten the 2 mounting bolts to 16 ft. lbs. (22 Nm).
- Active-ECS compressor assembly, if removed. Tighten the mounting bolts to 48 inch lbs. (5 Nm) and connect the electrical harness.

- Engine harness connection
- Air cleaner assembly and adjoining ductwork
- Halfshafts and seat halfshafts into the transaxle, using new circlips
- Bolt that secure the left side support bearing and tighten the bolts to 33 ft. lbs. (45 Nm)
- Ball joint and tie rod end to the steering knuckle. Using new nuts, tighten the ball joint castle nut to 43–52 ft. lbs. (60–72 Nm) and tighten the tie rod castle nut to 22 ft. lbs. (30 Nm). Install new cotter pins.

- Exhaust system, using new gaskets
- Front catalytic converter, if removed
- Engine undercovers
- Negative battery cable
20. Fill the transaxle to the correct level.
21. Start the engine and check for leaks.

ECLIPSE

1. Before servicing the vehicle, refer to the precautions in the beginning of this section.
2. Remove or disconnect the following:
- Battery and the air intake hoses
- Battery tray and support

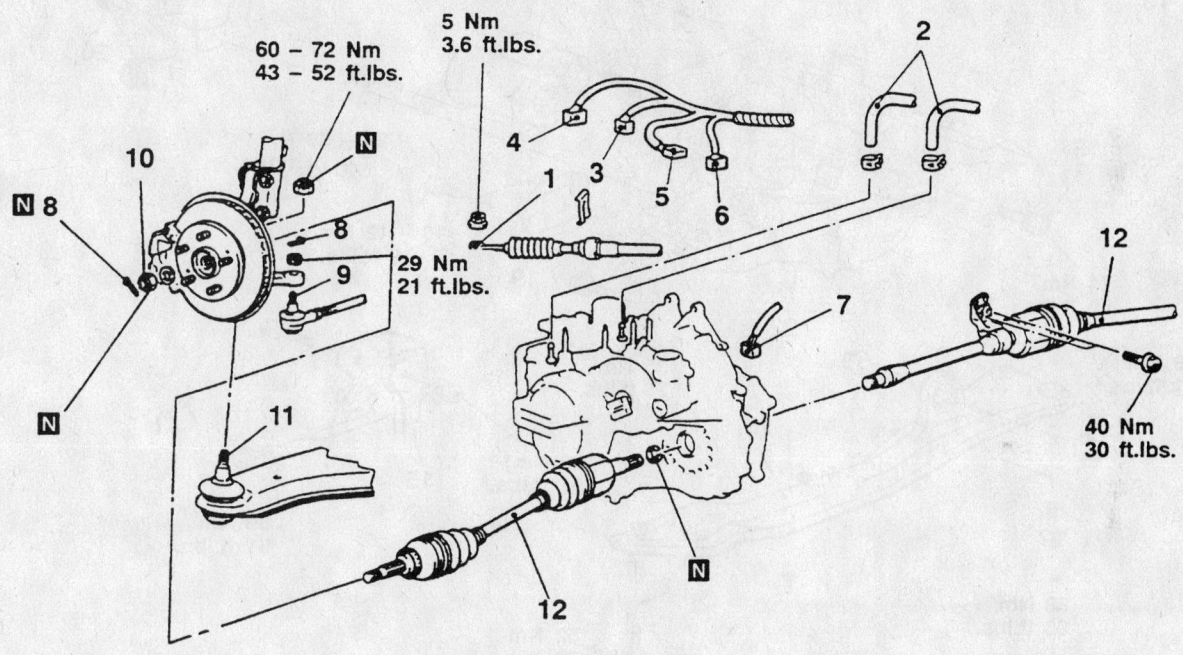

Removal steps

1. Transaxle control cable connection
2. Transaxle oil cooler hoses connection
3. PNP switch connector
4. A/T control solenoid valve connector
5. Input shaft speed sensor connector
6. Output shaft speed sensor connector
7. Vehicle speed sensor connector
8. Split pin
9. Connection of the tie rod end
10. Drive shaft nut

11. Connection for the lower arm ball joint
12. Drive shaft and inner shaft assembly (RH) and the drive shaft (LH)

Caution
Mounting locations marked by * should be provisionally tightened, and then fully tightened when the body is supporting the full weight of the engine.

7923PG84

- Auto-cruise actuator and bracket, if equipped with cruise control
- Charcoal canister and bracket
- Shift and select cables from the transaxle
- Back-up light switch and the vehicle speed sensor connectors

- Dipstick and tube assembly
- Starter assembly
- Park/neutral switch
- Oil temperature sensor
- Kick down servo switch
- Solenoid valve
- Pulse generator

- Speedometer connections

3. Attach an engine support fixture to the engine and remove the transaxle mounting bolts.

4. Remove or disconnect the following:
- Rear roll stopper bracket mounting bolts

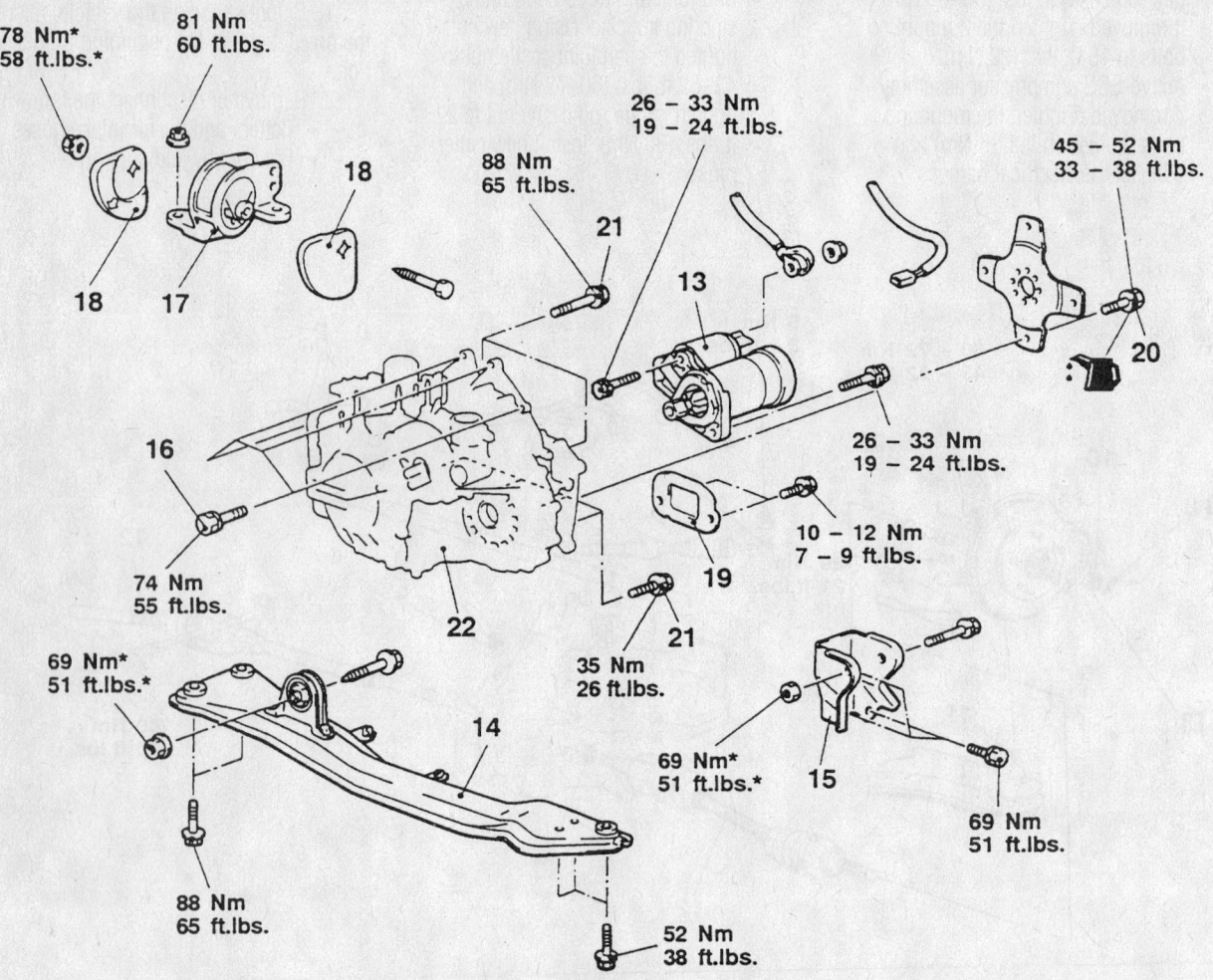

Lifting up of the vehicle

13. Starter motor
14. Center member assembly
15. Rear roll stopper bracket
16. Transaxle upper portion fixing bolt
17. Transaxle mounting bracket
18. Transaxle mount stopper
• Support the engine and transaxle assembly
19. Bell housing cover

20. Drive plate attaching bolt
21. Transaxle lower portion fixing bolt
22. Transaxle assembly

Caution
Mounting locations marked by * should be provisionally tightened, and then fully tightened when the body is supporting the full weight of the engine.

7923PG85

Transaxle removal—Diamante with F4A51 transaxle 2 of 2

- Transaxle mounting bracket mounting nuts

5. Raise the vehicle and remove the engine undercovers.

6. Remove or disconnect the following:
- Front exhaust pipe
- Transfer case assembly, if equipped with All Wheel Drive (AWD)
- Axle shafts
- Slave cylinder from the bell housing without disconnecting the fluid line. Position it out of the way.
- Bell housing cover and the right-hand center member stay (support)
- Center member
- Drive plate connecting bolts

7. Place a transmission jack under the transaxle and remove the transaxle mounting bolt.

8. Lower the transaxle.

To install:

9. Raise the transaxle into position and install the transaxle mounting bracket. Torque the through-bolt to 51 ft. lbs. (69 Nm).

10. Install or connect the following:
- Transaxle assembly mounting bolt. Torque the bolt to 22–25 ft. lbs. (29–34 Nm).
- Drive plate connecting bolts. Torque the bolts to 33–38 ft. lbs. (45–52 Nm).
- Center member assembly and the right-hand stay
- Bell housing cover and the slave cylinder
- Axle shafts.
- Front exhaust pipe
- Transfer case assembly if removed
- Engine undercovers and lower the vehicle
- Transaxle mounting bracket mounting nuts
- Rear roll stopper bracket mounting bolts
- Transaxle assembly mounting bolts. Torque the bolts to 35 ft. lbs. (48 Nm).

11. Remove the engine support fixture.
- Park/neutral switch
- Oil temperature sensor
- Kick down servo switch
- Solenoid valve
- Pulse generator
- Speedometer connections
- Starter assembly
- Dipstick and tube assembly
- Vehicle speed sensor and the back-up light connectors
- Cruise control actuator if removed
- Battery tray support and the tray
- Charcoal canister bracket and the canister

- Air duct and the air cleaner assembly

12. Refill the transaxle and the transfer case, if equipped, with the proper fluid.

GALANT

1. Before servicing the vehicle, refer to the precautions in the beginning of this section.

2. Remove or disconnect the following:
- Negative battery cable
- Air cleaner and intake hoses

3. Drain the transaxle into a suitable waste container.
- Nut securing the shifter lever to the transaxle
- Cable retaining clip and remove the cable from the transaxle
- Shifter cable mounting bracket
- Electrical connectors for the speedometer, neutral safety switch (inhibitor switch), the pulse generator, kickdown servo switch, and the oil temperature sensor
- Oil cooler lines, at the transaxle
- Dipstick and tube from the transaxle
- Starter motor and position it aside

4. Support the engine assembly.
- Rear roll stopper mounting bracket
- Transaxle mount bracket
- Upper transaxle mounting bolts

5. Raise and safely support the vehicle.

6. Remove or disconnect the following:
- Front wheel assemblies
- Right-hand undercover
- Tie rod end from the steering knuckle
- Stabilizer bar link from the damper fork
- Damper fork from the lateral lower control arm
- Later lower arm and the compression arm lower ball joints, from the steering knuckle
- Halfshafts from the transaxle and secure aside
- Cover from the transaxle bell housing
- Engine front roll stopper through-bolt
- Crossmember and the triangular right-hand stay
- Bolts holding the flexplate to the torque converter

7. Support the transaxle using a transmission jack, and remove the transaxle lower coupling bolt.

➡ **The coupling bolt threads from the engine side into the transaxle and is located just above the halfshaft opening.**

8. Slide the transaxle rearward and carefully lower it from the vehicle.

To install:

9. After the torque converter has been mounted on the transaxle, install the transaxle assembly to the engine. Install the mounting bolts and tighten to 35 ft. lbs. (48 Nm). Install the transaxle lower coupling bolt and tighten to 21–25 ft. lbs. (29–34 Nm).

10. Install or connect the following:
- Torque converter to the flexplate and tighten the bolts to 33–38 ft. lbs. (45–52 Nm)
- Cover to the transaxle bell housing and tighten the mounting bolts to 84 inch lbs. (9 Nm)
- Crossmember and tighten the front mounting bolts to 65 ft. lbs. (88 Nm) and the rear bolt to 54 ft. lbs. (73 Nm)
- Front engine roll stopper through-bolt and lightly tighten. Once the full weight of the engine is on the mounts, tighten the bolt to 42 ft. lbs. (57 Nm).
- Triangular stay bracket and tighten the mounting bolts to 65 ft. lbs. (88 Nm)
- Halfshafts, using new circlips on the axle ends
- Tie rod and ball joints to the steering knuckle. Tighten the ball joint self-locking nuts to 48 ft. lbs. (65 Nm). Tighten the tie rod end nut to 21 ft. lbs. (28 Nm) and secure with a new cotter pin.
- Damper fork to the lower control arm and tighten the through-bolt to 65 ft. lbs. (88 Nm)
- Stabilizer link to the damper fork, and tighten the self-locking nut to 29 ft. lbs. (39 Nm)
- Underpan
- Wheels and lower the vehicle
- Transaxle mount bracket to the transaxle, and tighten the mounting nuts to 32 ft. lbs. (43 Nm)
- Rear roll stopper mounting bracket
- Engine support. Tighten the transaxle mount through-bolt to 51 ft. lbs. (69 Nm) and tighten the front engine roll stopper through-bolt.
- Upper transaxle mounting bolts and tighten to 35 ft. lbs. (48 Nm)
- Starter motor
- Dipstick tube and the dipstick
- Shifter cable mounting bracket
- Shifter lever and tighten the retaining nut to 14 ft. lbs. (19 Nm)
- Oil cooler lines and secure with clamps

- Electrical connectors for the speedometer, neutral safety switch (inhibitor switch), pulse generator, kickdown servo switch and oil temperature sensor
- Air cleaner and the air intake hose
- Negative battery cable

11. Fill the transaxle to the correct level.

LANCER

1. Before servicing the vehicle, refer to the precautions in the beginning of this section.

2. Install or connect the following:
- Negative battery cable
- Engine undercover

3. Drain the transaxle oil and engine coolant.
- Front exhaust pipe
- Battery and battery tray
- Air cleaner assembly
- Transaxle control cable
- Vehicle Speed Sensor (VSS) connector
- Input and output shaft speed sensor connectors
- Inhibitor switch sensor connector
- A/C control solenoid valve connector
- Starter
- Transaxle oil cooler line
- Upper engine-to-transaxle bolts
- Transaxle mount
- Transaxle mount stopper

4. Install a suitable engine support assembly, then raise and safely support the vehicle.
- Wheel Speed Sensor (WSS)
- Brake hose clamp
- Stabilizer bar
- Lower control arm
- Tie rod end
- Lower control arm
- Halfshafts by inserting a prybar between the transaxle case and the driveshaft and prying the shaft from the transaxle. Do not pull on the driveshaft.
- Centermember
- Bell housing lower cover
- Transaxle to engine bolts and lower the transaxle from the vehicle
- Front roll stopper installation bolt
- Bell housing cover

5. Support the transaxle with a suitable jack.
- Driveplate bolts
- Lower transaxle-to-engine mounting bolts
- Transaxle from the vehicle

To install:

6. Installation is the reverse of the removal procedure, noting the following:

- Driveplate bolts: 36 ft. lbs. (49 Nm)
- Centermember bolts: 51 ft. lbs. (69 Nm)
- Hub nut: 181 ft. lbs. (245 Nm)
- Transaxle mount bracket nuts: 36 ft. lbs. (49 Nm)
- Transaxle mount stopper nuts: 60 ft. lbs. (81 Nm)
- Transaxle-to-engine upper mounting bolts: 36 ft. lbs. (49 Nm)

7. Fill the transaxle and the engine cooling system to the correct level.

8. Check and adjust the front wheel alignment.

9. Check the speedometer and gear selector for proper operation.

10. Start the engine and check for leaks.

MIRAGE

1. Before servicing the vehicle, refer to the precautions in the beginning of this section.

2. Remove or disconnect the following:
- Negative battery cable
- Battery and battery tray
- Air hose and air cleaner assembly
- Under guard pan

3. Drain the transaxle oil.
- Control cable and cooler lines
- Shift control solenoid valve connector
- Inhibitor switch, kickdown servo switch, the pulse generator and oil temperature sensor, if equipped
- Speedometer cable and remove the starter
- Transaxle mounting bolts and bracket
- Stabilizer bar from the lower control arm
- Steering tie rod end and the ball joint from the steering arm
- Halfshafts at the inboard side from the transaxle. Tie the joint assembly aside.

4. Support the engine and remove the center member.
- Bell housing cover and remove the driveplate bolts
- Transaxle assembly lower connecting bolt, located just over the halfshaft opening
- Transaxle

To install:

5. Install or connect the following:
- Transaxle assembly on the engine. Tighten the driveplate bolts to 33–38 ft. lbs. (46–53 Nm).
- Bell housing cover
- Center member
- Halfshafts to the transaxle, using new circlips

- Tie rods, ball joints and stabilizer links to the steering arm
- Transaxle mounting bracket and bolts
- Starter
- Speedometer cable
- Inhibitor switch, kickdown servo switch, the pulse generator and oil temperature sensor, if disconnected
- Shift control solenoid valve connector
- Control cables and oil cooler lines
- Air cleaner assembly
- Battery tray and battery. Connect the positive, then the negative terminal.

6. Fill the transaxle to the correct level.

7. Start the engine and check for leaks.

Clutch

ADJUSTMENT

➡The following adjustment is for cable actuated clutch systems. Hydraulic systems are self-adjusting.

Mirage

1. Before servicing the vehicle, refer to the precautions in the beginning of this section.

2. Measure the clutch pedal height (measurement A). The specification is 6.38–6.50 in. (162–165mm).

➡The clutch pedal height is not adjustable. If not within specifications, part replacement is required.

3. Depress clutch pedal several times and check the pedal free-play (measurement B).

4. If measurement is not 0.67–0.87 in. (17–22mm), adjustment is required.

5. To adjust turn the outer cable adjust-

Clutch pedal height

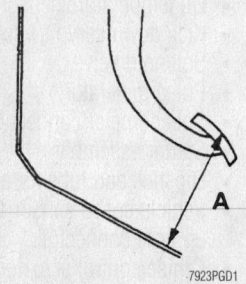

7923PGD1

Clutch pedal height (A) measurement— Mirage

ing nut, located at the firewall, until free-play is within range.

6. Depress clutch pedal several times and recheck measurement.

REMOVAL & INSTALLATION

Eclipse, Galant and Mirage

1. Before servicing the vehicle, refer to the precautions in the beginning of this section.
2. Remove or disconnect the following:
   - Negative battery cable
   - Transaxle assembly from the vehicle
   - Pressure plate attaching bolts, pressure plate and clutch disc. If the pressure plate is to be reused, loosen the bolts in a diagonal pattern, 1 or 2 turns at a time. This

will prevent warping the clutch cover assembly.
   - Return clip and the pressure plate release bearing. Do not use solvent to clean the bearing.
3. Inspect the clutch release fork and fulcrum for damage or wear. If necessary, remove the release fork and unthread the fulcrum from the transaxle.
4. Carefully inspect the condition of the clutch components and replace any worn or damaged parts.

To install:
5. Inspect the flywheel for heat damage or cracks. Resurface or replace the flywheel as required.
6. Install the fulcrum and tighten to 25 ft. lbs. (35 Nm). Install the release fork. Apply a coating of multi-purpose grease to the point of contact with the fulcrum and the

point of contact with the release bearing. Apply a coating of multi-purpose grease to the end of the release cylinder pushrod and the pushrod hole in the release fork.
7. Apply multi-purpose grease to the clutch release bearing. Pack the bearing inner surface and the groove with grease. Do not apply grease to the resin portion of the bearing. Place the bearing in position and install the return clip.
8. Using the proper alignment tool, install the clutch disc to the flywheel. Install the pressure plate assembly. Install the retainer bolts and tighten a little at a time, in a diagonal sequence. Tighten them to a final torque of 16 ft. lbs. (22 Nm). Remove the aligning tool.
9. Install the transaxle assembly.
10. Check for proper clutch operation.

Lancer

1. Before servicing the vehicle, refer to the precautions in the beginning of this section.
2. Remove or disconnect the following:
   - Negative battery cable
   - Transaxle assembly from the vehicle
   - Clutch fluid line bracket, insulator and washer
   - Clutch fluid line
   - Clutch slave (release) cylinder
   - Boot
   - Clutch cover (pressure plate) attaching bolts, cover plate and clutch disc. If the pressure plate is to be reused, loosen the bolts in a diagonal pattern, 1 or 2 turns at a time. This will prevent warping the clutch cover assembly.
3. Carefully inspect the condition of the clutch components and replace any worn or damaged parts.

To install:
4. Inspect the flywheel for heat damage or cracks. Resurface or replace the flywheel as required.
5. Apply multi-purpose grease to the clutch release bearing. Pack the bearing inner surface and the groove with grease. Do not apply grease to the resin portion of the bearing. Place the bearing in position and install the return clip.
6. Using the proper alignment tool, install the clutch disc to the flywheel. Install the clutch cover (pressure plate) assembly. Install the retainer bolts and tighten a little at a time, in a diagonal sequence. Tighten them to a final torque of 14 ft. lbs. (19 Nm). Remove the aligning tool.
7. Install or connect the following:
   - Boot

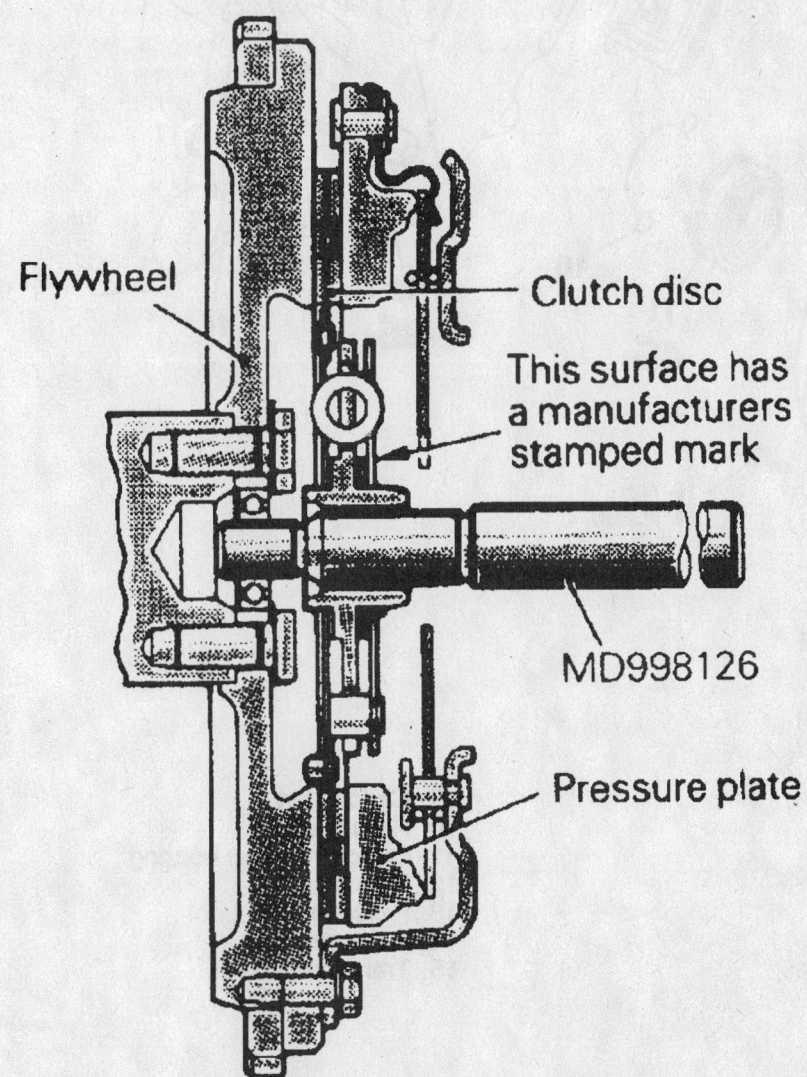

Flywheel

Clutch disc

This surface has a manufacturers stamped mark

MD998126

Pressure plate

7923PG88

Use the alignment dowel to center the disc on the flywheel—Mirage

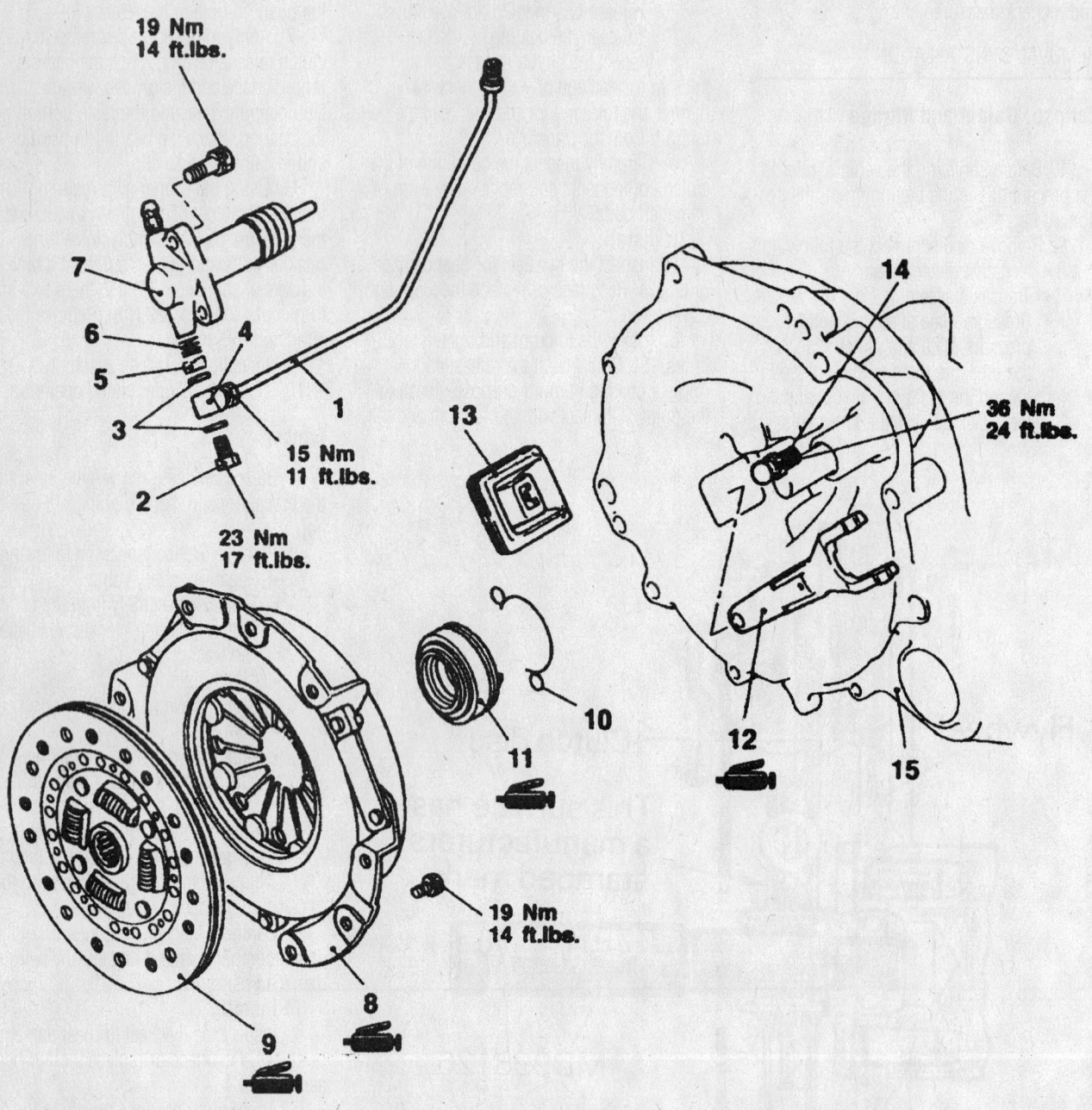

1. Clutch oil tube
2. Union bolt
3. Gasket
4. Union
5. Valve plate
6. Valve plate spring
7. Clutch release cylinder
8. Clutch cover
9. Clutch disc
10. Return clip
11. Clutch release bearing
12. Release fork
13. Release fork boot
14. Fulcrum
15. Transaxle

Exploded view of clutch assembly—Eclipse shown

7923PG89

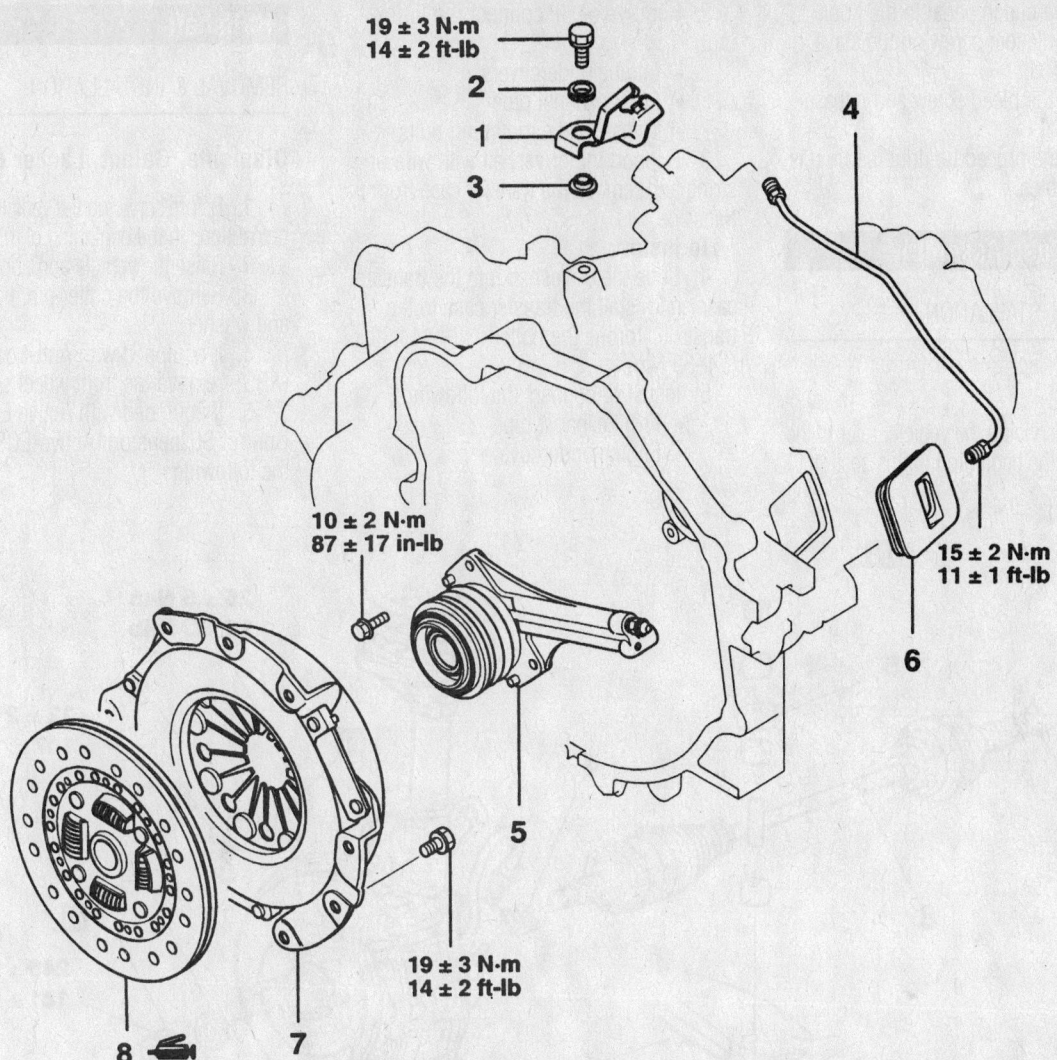

1. **CLUTCH FLUID LINE BRACKET**
2. **INSULATOR**
3. **WASHER**
4. **CLUTCH TUBE**
5. **CLUTCH RELEASE CONCENTRIC CYLINDER**
6. **BOOT**
7. **CLUTCH COVER**
8. **CLUTCH DISC**

9357QG26

Exploded view of the clutch components—Lancer shown

- Clutch line, washer, insulator and bracket
- Transaxle assembly
8. Check for proper clutch operation.

Hydraulic Clutch System

BLEEDING

1. Before servicing the vehicle, refer to the precautions in the beginning of this section.

2. Fill the reservoir with clean brake fluid meeting DOT 3 specifications.

7923PG91

Bleeding a typical clutch hydraulic system

3. Press the clutch pedal to the floor, then open the bleeder screw on the slave cylinder.

4. Tighten the bleed screw and release the clutch pedal.

5. Repeat the procedure until the fluid is free of air bubbles.

Transfer Case Assembly

REMOVAL & INSTALLATION

Eclipse

1. Before servicing the vehicle, refer to the precautions in the beginning of this section.

2. Remove or disconnect the following:
- Engine undercovers
- Front exhaust pipe
- Transfer case mounting bolts

3. Support the driveshaft with wire or string and remove the transfer case from the transaxle.

To install:

4. Slide the driveshaft into the transfer case and install the transfer case to the transaxle. Torque the bolts to 40–44 ft. lbs. (54–59 Nm).

5. Install or connect the following:
- Front exhaust pipe
- Engine undercover

Halfshaft

REMOVAL & INSTALLATION

Diamante, Galant, Lancer & Mirage

1. Before servicing the vehicle, refer to the precautions in the beginning of this section.

2. Raise the vehicle and support it safely.

3. Remove the cotter pin, halfshaft nut and washer.

4. If equipped with Anti-Lock Brake (ABS), remove the front wheel speed sensor.

5. If equipped with Active Electronic Control Suspension (Active-ECS) perform the following:

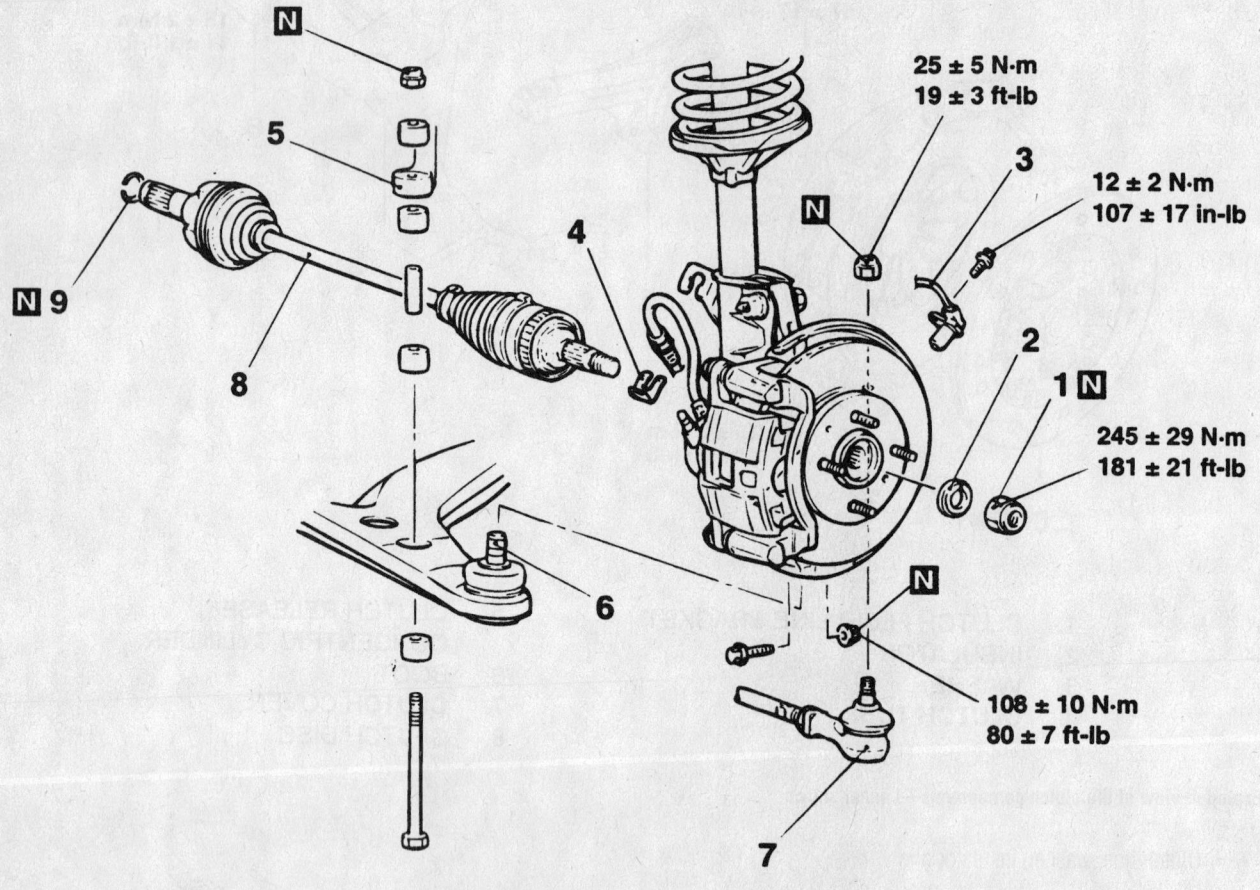

25 ± 5 N·m
19 ± 3 ft-lb

12 ± 2 N·m
107 ± 17 in-lb

245 ± 29 N·m
181 ± 21 ft-lb

108 ± 10 N·m
80 ± 7 ft-lb

1. **DRIVESHAFT NUT**
2. **WASHER**
3. **FRONT SPEED SENSOR <VEHICLES WITH ABS>**
4. **BRAKE HOSE CRAMP**
5. **STABILIZER BAR CONNECTION**
6. **LOWER ARM BALL JOINT CONNECTION**
7. **TIE ROD END CONNECTION**

9357QG27

Exploded view of the halfshaft mounting—Lancer shown, others similar

a. Loosen the nut that secures the air line to the to the top of the strut and discard the O-ring.

b. Remove the bolts that secure the actuator to the top of the strut and remove the component. Disconnect the wiring harness.

6. Disconnect the lower ball joint and the tie rod end from the steering knuckle.

7. If removing the left side axle with an inner shaft, remove the center support bearing bracket bolts and washers. Then, remove the halfshaft by setting up a puller on the outside wheel hub and pushing the halfshaft from the front hub. Tap the shaft union at the joint case with a plastic hammer to remove the halfshaft and inner shaft from the transaxle.

8. If removing right side axle shafts without an inner shaft, remove the halfshaft by setting up a puller on the outside wheel hub and pushing the halfshaft from the front hub. After pressing the outer shaft, insert a prybar between the transaxle case and the halfshaft and pry the shaft from the transaxle.

➡**Do not pull on the shaft; doing so damages the inboard joint.**

To install:

9. Replace the circlips on the ends of the halfshafts.

10. Insert the halfshaft into the transaxle. Be sure it is fully seated.

11. Pull the strut assembly out and install the other end to the hub.

12. Install the center bearing bracket bolts and tighten to 33 ft. lbs. (45 Nm).

13. Install the washer so the chamfered edge faces outward. Install the nut and tighten to 145–188 ft. lbs. (200–260 Nm), for all models except Lancer. For Lancer, tighten the nut to 181 ft. lbs. (245 Nm). Secure with a new cotter pin.

14. Connect the ball joint to the steering knuckle. Torque the new retaining nut to 43–52 ft. lbs. (60–72 Nm) and secure with a new cotter pin.

15. Connect the tie rod end to the steering knuckle. Torque the retaining nut to 21 ft. lbs. (29 Nm) and secure with a new cotter pin.

16. If equipped with ABS, install the front wheel speed sensor.

17. If equipped with Active-ECS, perform the following:

a. Install the air line with a new O-ring.

b. Install the actuator to the top of the strut. Connect the wiring harness.

18. Install the wheel and lower the vehicle to the floor.

Eclipse

FRONT

1. Before servicing the vehicle, refer to the precautions in the beginning of this section.

2. Raise and safely support the vehicle.

3. Remove or disconnect the following:
- Front wheel
- Halfshaft nut and washer
- Tie rod end from the knuckle
- Stabilizer link from the damper fork
- Compression and lateral arm ball joint studs from the knuckle

4. Mount a puller on the wheel studs and push the halfshaft through the hub assembly.

5. Detach the inner halfshaft from the transaxle by carefully prying the CV-joint housing out.

6. Pull the knuckle assembly outward and remove the halfshaft.

To install:

7. Place a new circlip on the inner halfshaft and install the halfshaft in the transaxle.

8. Push out on the knuckle assembly and install the halfshaft through the hub.

9. Using new nuts, install the lateral and compression arm ball joint studs in the knuckle. Tighten the nuts to 43–52 ft. lbs. (59–71 Nm). Install new cotter pins.

10. Install the damper fork on the knuckle. Do not tighten the nut at this time.

11. Attach the stabilizer link to the damper fork. Tighten the nut to 29 ft. lbs. (39 Nm).

12. Install the washer and nut on the halfshaft. Prevent the hub from turning and tighten the nut to 145–188 ft. lbs. (196–255 Nm).

13. Install the wheel and lower the vehicle to the floor. Tighten the damper fork nut to 65 ft. lbs. (88 Nm).

REAR

1. Before servicing the vehicle, refer to the precautions in the beginning of this section.

2. Raise and safely support the vehicle.

3. Remove or disconnect the following:
- Rear wheel
- Rear wheel speed sensor, if equipped
- Caliper and rotor, if equipped with disc brakes
- Brake drum and shoes, If equipped with drum brakes
- Brake hydraulic line from the wheel cylinder
- Parking brake cable from the rear brakes
- Lower end of the shock absorber from the knuckle
- Trailing and lower arms from the knuckle
- Toe control arm ball joint from the knuckle

4. Prevent the hub assembly from turning by using a tool such as MB990767 and remove the halfshaft nut and washer.

5. Remove the differential mount support.

✳✳ WARNING

Do not pull on the halfshaft to remove it from the differential. Damage to the inner CV-joint will occur.

6. Push the lower part of the knuckle outward and pry the inner halfshaft out of the differential.

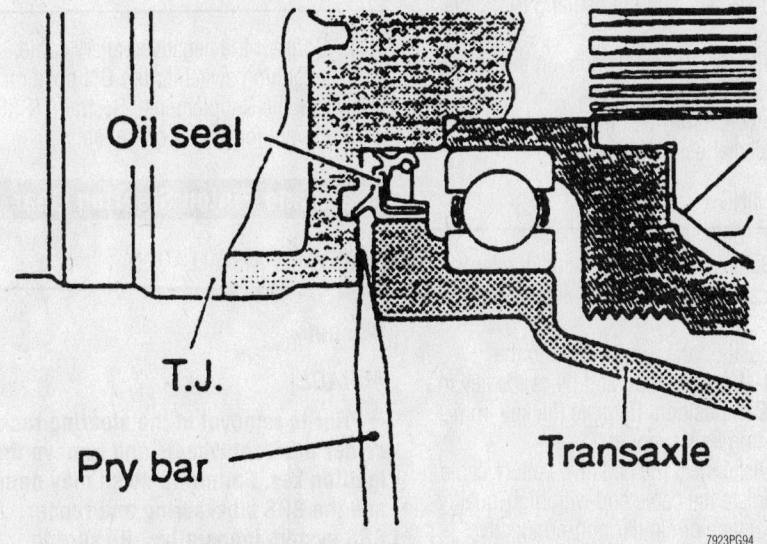

Oil seal

T.J.

Pry bar

Transaxle

7923PG94

Proper method for removing the inner halfshaft from the transaxle or differential

7. Push the outer end of the halfshaft through the hub/knuckle and remove it.

To install:

8. Install the outer end of the halfshaft through the hub/knuckle.

9. Place a new circlip on the inner halfshaft and install the halfshaft in the differential.

10. Install or connect the following:
   • Differential mount support

• Washer and a new nut on the end of the halfshaft. Tighten the nut to 145–188 ft. lbs. (196–255 Nm).

• Toe control arm to the knuckle. Tighten the new nut to 20 ft. lbs. (28 Nm).

• Lower and trailing arms to the knuckle. Do not tighten the fasteners at this time.

• Shock absorber. Tighten the bolt to 71 ft. lbs. (98 Nm).

11. Assemble the brake components.

12. Install or connect the following:
   • Rear wheel speed sensor
   • Rear wheel and lower the vehicle to the floor. Tighten the lower arm nut to 71 ft. lbs. (98 Nm) and the trailing arm nut to 85–99 ft. lbs. (118–137 Nm).

STEERING AND SUSPENSION

Air Bag

✳✳ CAUTION

All vehicles are equipped with an air bag system. The system must be disabled before performing service on or around system components, steering column, instrument panel components, wiring and sensors. Failure to follow safety and disabling procedures could result in accidental air bag deployment, possible personal injury and unnecessary system repairs.

PRECAUTIONS

Several precautions must be observed when handling the inflator module to avoid accidental deployment and possible personal injury.

• Never carry the inflator module by the wires or connector on the underside of the module.

• When carrying a live inflator module, hold securely with both hands, and ensure that the bag and trim cover are pointed away.

• Place the inflator module on a bench or other surface with the bag and trim cover facing up.

• With the inflator module on the bench, never place anything on or close to the module which may be thrown in the event of an accidental deployment.

DISARMING

1. Before servicing the vehicle, refer to the precautions in the beginning of this section.

2. Position the front wheels in the straight-ahead position and place the key in the **LOCK** position. Remove the key from the ignition lock cylinder.

3. Disconnect the negative battery cable and insulate the cable end with high-quality electrical tape or similar non-conductive wrapping.

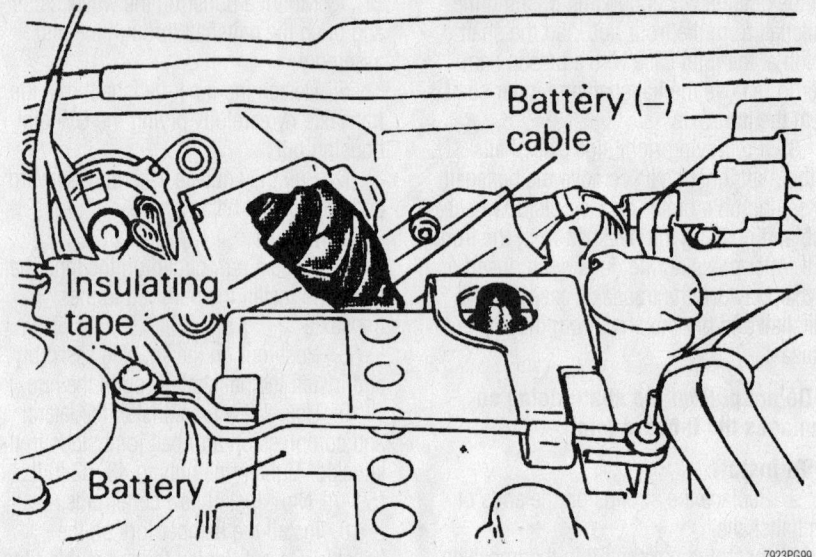

Insulate the negative battery cable to prevent accidental deployment of the air bag

4. Wait at least 1 minute before working on the vehicle. The air bag system is designed to retain enough voltage to deploy the air bag for a short period of time after the battery has been disconnected.

REARMING

1. Connect the negative battery cable, turn the ignition switch to the **ON** position and check the Supplemental Restraint (SRS) warning light for proper operation.

Rack and Pinion Steering Gear

REMOVAL & INSTALLATION

Manual

MIRAGE

➡**Prior to removal of the steering rack, center the front wheels and remove the ignition key. Failure to do so may damage the SRS clockspring and render SRS system inoperative. Be sure to properly disarm the air bag system.**

1. Before servicing the vehicle, refer to the precautions in the beginning of this section.

2. Remove or disconnect the following:
   • Battery negative cable
   • Wheels
   • Heated Oxygen (HO2S) sensor and remove the front exhaust pipe

3. Properly support the engine. Remove both roll stopper mounting bolts and the 4 center member installation bolts.
   • Center member

➡**Matchmark the pinion input shaft of the rack to the lower steering column joint for installation purposes.**

4. Remove or disconnect the following:
   • Pinch bolt holding the lower steering column joint to the rack and pinion input shaft
   • Cotter pins and disconnect the tie rod ends from the steering knuckle
   • Rack and pinion steering assembly and its rubber mounts from the right side of the vehicle

To install:

5. Align the matchmarks of the input shaft and install the rack to the vehicle.

6. Secure the rack using the retainer clamps and bolts. Torque the bolts to 51 ft. lbs. (70 Nm).

7. Torque the steering column pinch bolt to 13 ft. lbs. (18 Nm).

8. Install or connect the following:
- Center member
- Front exhaust pipe
- HO2S sensor
- Tie rod ends to the steering knuckles and tighten the castle nuts to 25 ft. lbs. (34 Nm). Install new cotter pins.
- Wheels and connect the negative battery cable

9. Perform a front end alignment.

Power

DIAMANTE

➡ **Prior to removal of the steering gear box, center the front wheels and remove the ignition key. Failure to do so may damage the Supplemental Restraint System (SRS) clock spring and render SRS system inoperative.**

1. Before servicing the vehicle, refer to the precautions in the beginning of this section.

2. Remove or disconnect the following:
- Negative battery cable.
- Front exhaust pipe.
- Transfer case assembly, if equipped with All Wheel Drive (AWD)
- Bolt holding the lower steering column joint to the rack and pinion input shaft
- Tie rod ends
- Left and right frame members
- Stabilizer bar bracket
- Lines going to the rear pump, if equipped with Four Wheel Steering (4WS)
- Rack and pinion steering assembly and its rubber mounts. Move the rack to the right to remove it from the crossmember.

To install:

3. Install or connect the following:
- Rack and install the mounting bolts, tightening bolts to 51 ft. lbs. (70 Nm). When installing the rubber rack mounts, align the projection of the mounting rubber with the indentation in the crossmember. Install the pinch bolt.
- Pressure and return lines to the rack and to the rear pump, if equipped

- Frame members and tighten the bolts to 43–51 ft. lbs. (60–70 Nm)
- Tie rods and install new cotter pins
- Transfer case and front exhaust pipe

4. Refill the reservoir and bleed the system.

5. Perform a front end alignment.

ECLIPSE (NON-TURBO)

➡ **Prior to removal of the steering gear box, center the front wheels and remove the ignition key. Failure to do so may damage the SRS clock spring and render SRS system inoperative.**

1. Before servicing the vehicle, refer to the precautions in the beginning of this section.

2. Install or connect the following:
- Negative battery cable

3. Drain the power steering fluid.
- Stabilizer bar
- Windshield washer reservoir
- Pinch bolt from the joint assembly
- Fluid lines from the steering rack
- Tie rod ends from the steering knuckles
- Left and right stays (supports)

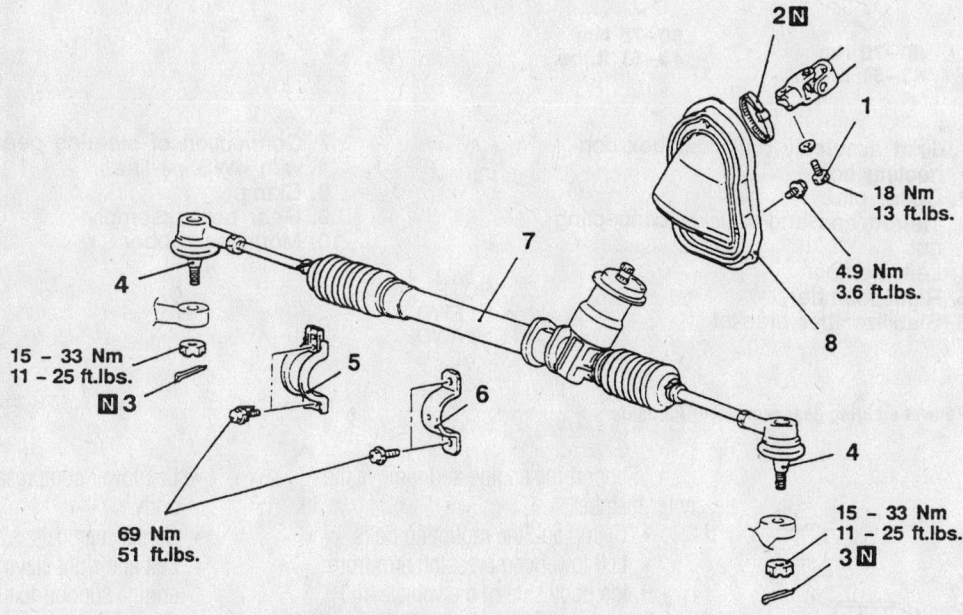

1. Steering shaft assembly and gear box connecting bolt
2. Band
3. Cotter pin
4. Tie-rod end and knuckle connection
5. Cylinder clamp
6. Gear housing clamp
7. Gear box assembly
8. Steering cover assembly

7923PG00

Exploded view of the manual steering gear mounting—Mirage

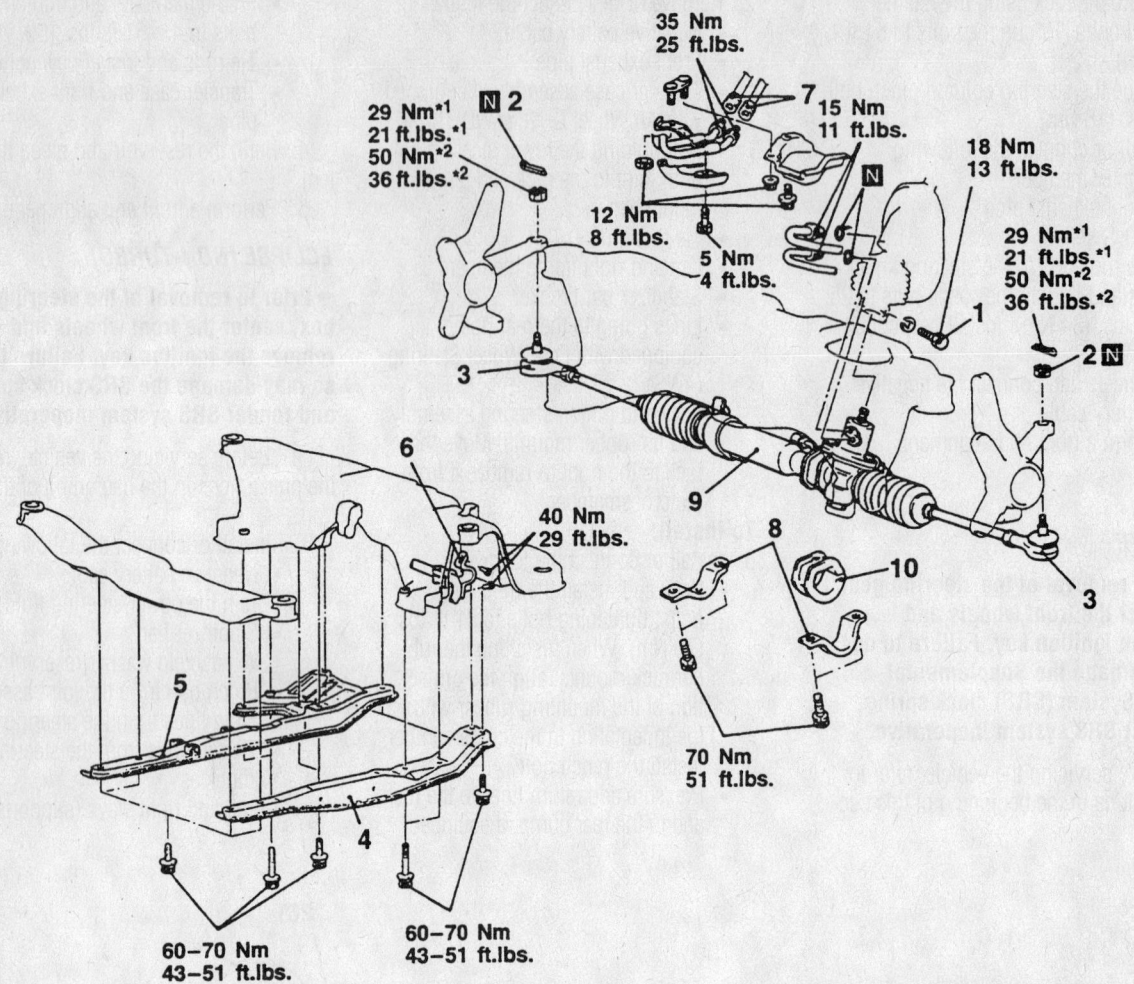

35 Nm
25 ft.lbs.

29 Nm*1
21 ft.lbs.*1
50 Nm*2
36 ft.lbs.*2

15 Nm
11 ft.lbs.

18 Nm
13 ft.lbs.

12 Nm
8 ft.lbs.

5 Nm
4 ft.lbs.

29 Nm*1
21 ft.lbs.*1
50 Nm*2
36 ft.lbs.*2

40 Nm
29 ft.lbs.

70 Nm
51 ft.lbs.

60—70 Nm
43—51 ft.lbs.

60—70 Nm
43—51 ft.lbs.

1. Joint assembly and gear box connecting bolt
2. Cotter pin
3. Tie-rod end and knuckle connecting nut
4. Left member
5. Right member
6. Stabilizer bar bracket

7. Connection of steering gear box with 4WS oil line
8. Clamp
9. Gear box assembly
10. Mounting rubber

NOTE
*1: FWD
*2: AWD

7923PGA5

Exploded view of the power steering gear removal—Diamante

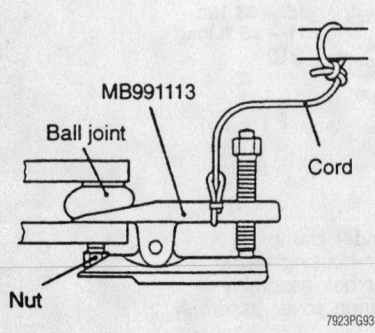

MB991113

Ball joint

Cord

Nut

7923PG93

Proper tie-rod end removal method

4. Support the engine and remove the center member.
- Clamp and the mounting bolts
- Left lower compression arm from the body side of the vehicle and support it with wire or string
- Steering rack from the joint assembly and remove the rack from the left side of the vehicle

To install:
5. Position the steering rack in the vehicle and install the clamp and the mounting bolts. Be sure the rack is centered before connecting it to the joint assembly.
6. Install or connect the following:

- Left lower compression arm to the body
- Center member
- Left and right stays and remove the engine support fixture or jack
- Tie rods to the steering knuckles
- Fluid lines to the steering rack
- Pinch bolt in the joint assembly
- Stabilizer bar and the windshield washer reservoir

7. Safely lower the vehicle.
8. Connect the negative battery cable.
9. Refill and bleed the power steering system.
10. Perform a front end alignment.

ECLIPSE (TURBO)

✳✳ WARNING

Prior to removal of the steering gear box, center the front wheels and remove the ignition key. Failure to do so may damage the Supplemental Restraint System (SRS) clock spring and render SRS system inoperative.

1. Before servicing the vehicle, refer to the precautions in the beginning of this section.
2. Install or connect the following:
   - Negative battery cable
3. Drain the power steering fluid.
   - Stabilizer bar
   - Fluid level sensor and remove the brake fluid reservoir and position it out of the way. Do not disconnect the brake hose.
   - Electrical connector from the air conditioning compressor
   - Air conditioning compressor from the bracket and position it out of the way. Do not disconnect the hoses.
   - Pinch bolt from the joint assembly
   - Fluid lines from the steering rack
   - Tie rod ends from the steering knuckles
   - Left and right stays (supports)
4. Support the engine and remove the center member assembly.
   - Clamp and the mounting bolts
   - Left lower compression arm from the body side of the vehicle and support it with wire or string
5. Disconnect the steering rack from the joint assembly and remove the rack from the left side of the vehicle.

To install:

6. Position the steering rack in the vehicle and install the clamp and the mounting bolts. Be sure the rack is centered before connecting it to the joint assembly.
7. Install or connect the following:
   - Left lower compression arm to the body
   - Center member assembly
   - Left and right stays
   - Tie rod ends to the steering knuckles
   - Fluid lines to the steering rack
   - Pinch bolt in the joint assembly
   - Stabilizer bar
   - Air conditioning compressor and connect the harness connector
   - Brake fluid reservoir and connect the fluid level sensor
   - Negative battery cable
8. Refill and bleed the power steering system.
9. Perform a front end alignment.

GALANT

✳✳ WARNING

Prior to removal of the steering gear box, center the front wheels and remove the ignition key. Failure to do so may damage the Supplemental Restraint System (SRS) clock spring and render SRS system inoperative.

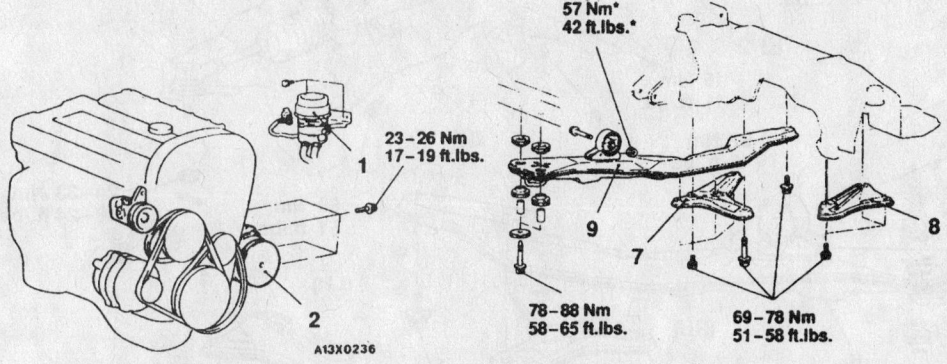

57 Nm*
42 ft.lbs.*

23–26 Nm
17–19 ft.lbs.

78–88 Nm
58–65 ft.lbs.

69–78 Nm
51–58 ft.lbs.

A13X0236

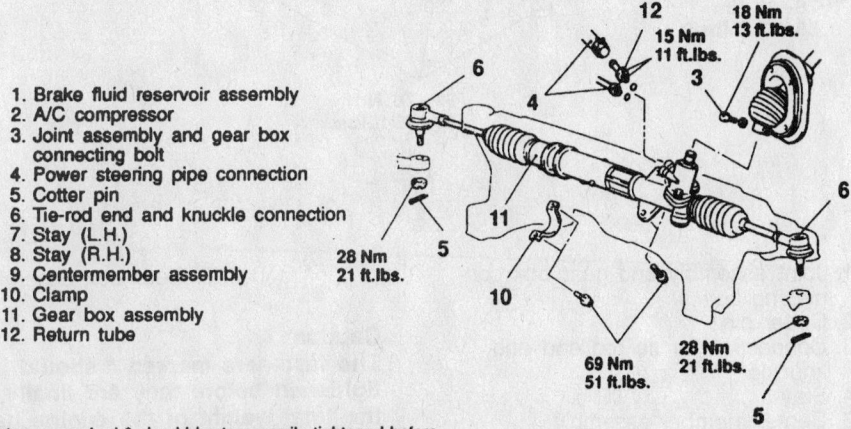

1. Brake fluid reservoir assembly
2. A/C compressor
3. Joint assembly and gear box connecting bolt
4. Power steering pipe connection
5. Cotter pin
6. Tie-rod end and knuckle connection
7. Stay (L.H.)
8. Stay (R.H.)
9. Centermember assembly
10. Clamp
11. Gear box assembly
12. Return tube

12
15 Nm
11 ft.lbs.

18 Nm
13 ft.lbs.

28 Nm
21 ft.lbs.

69 Nm
51 ft.lbs.

28 Nm
21 ft.lbs.

NOTE
The fasteners marked * should be temporarily tightened before they are finally tightened once the total weight of the engine has been placed on the vehicle body.

7923PGA2

Power steering rack assembly and related components—Eclipse

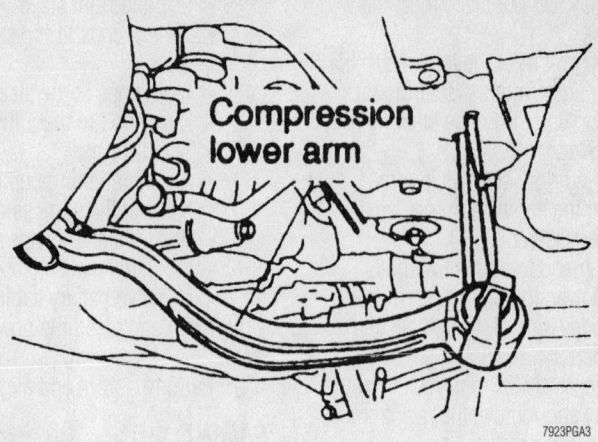

Compression lower arm

7923PGA3

Disconnect the lower compression arm from the body—Eclipse

1. Before servicing the vehicle, refer to the precautions in the beginning of this section.

2. Install or connect the following:
- Negative battery cable
- Both front wheel assemblies
- Bolt holding lower steering column joint to the rack and pinion input shaft
- Stabilizer bar
- Cotter pins and the tie rod ends from the steering knuckle

3. On vehicles equipped with Electronic Control Power steering (EPS), disconnect the wiring harness from the solenoid connector.

4. Locate the 2 triangular braces near the crossmember and remove both.

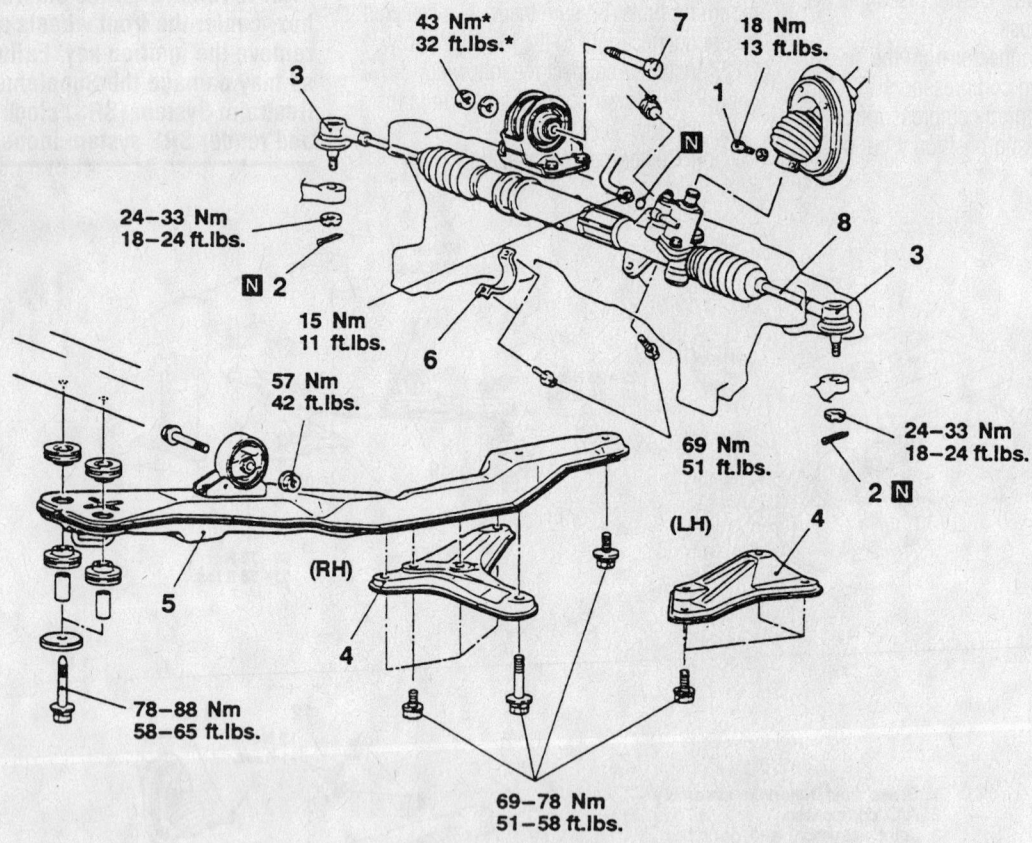

43 Nm*
32 ft.lbs.*

7

18 Nm
13 ft.lbs.

3

1

24–33 Nm
18–24 ft.lbs.

8

2

3

15 Nm
11 ft.lbs.

6

57 Nm
42 ft.lbs.

69 Nm
51 ft.lbs.

24–33 Nm
18–24 ft.lbs.

2

(LH)

4

(RH)

5

4

4

78–88 Nm
58–65 ft.lbs.

69–78 Nm
51–58 ft.lbs.

1. Joint assembly and gear box connecting bolt
2. Cotter pin
3. Connection for tie rod end and knuckle
4. Stay
5. Center member assembly
6. Clamp
7. Bolt

8. Gear box assembly

Caution
The fasteners marked * should be temporarily tightened before they are finally tightened once the total weight of the engine has been placed on the vehicle body.

7923PGA4

Exploded view of the power steering gear removal procedure—Galant

5. Support the center crossmember. Remove the through-bolt from the front round roll stopper and remove the bolts securing the center crossmember.

6. Remove the center crossmember.

7. Properly support the engine and remove the rear roll stopper through-bolt.

8. Remove or disconnect the following:
- Power steering fluid pressure pipe and return hose from the rack fittings. Plug the fittings to prevent excessive fluid leakage.
- Clamp bolts and the 2 bolts securing the rack assembly to the chassis
- Rack and pinion steering assembly and its rubber mounts

➡**When removing the rack and pinion assembly, tilt the assembly to the vehicle side of the compression lower arm and remove from the left side of the vehicle.**

To install:

9. Center the rack assembly and insert the pinion into the steering column shaft.

10. Install or connect the following:
- Rack and with the mounting bolts.

Torque the mounting bolts to 51 ft. lbs. (69 Nm).
- Pinch bolt and tighten the bolt to 13 ft. lbs. (18 Nm)
- Power steering fluid lines to the rack and tighten the pressure hose fitting to 11 ft. lbs. (15 Nm). Secure the return hose with the clamp.

11. Raise the engine into position. Install the rear roll stopper through-bolt and tighten to 32 ft. lbs. (43 Nm).

12. Raise the crossmember into position. Install the center member mounting bolts and tighten the front bolts to 58–65 ft. lbs. (78–88 Nm) and the rear bolt to 51–58 ft. lbs. (69–78 Nm).

13. Install or connect the following:
- Front roll stopper bolt and tighten the nut to 32 ft. lbs. (43 Nm)
- 2 triangular braces and tighten the mounting bolts to 50–56 ft. lbs. (69–78 Nm)
- Stabilizer bar
- Tie rod ends and tighten nuts to 20 ft. lbs. (27 Nm)

14. On vehicles equipped with EPS, connect the wiring harness to the solenoid connector.

15. Install the wheel assemblies and lower the vehicle.

16. Refill the reservoir with power steering fluid and bleed the system.

17. Perform a front end alignment.

LANCER & MIRAGE

➡**Prior to removal of the steering gear box, center the front wheels and remove the ignition key. Failure to do so may damage the Supplemental Restraint System (SRS) clockspring and render SRS system inoperative.**

1. Before servicing the vehicle, refer to the precautions in the beginning of this section.

2. Drain the power steering system.

3. Remove or disconnect the following:
- Battery negative cable. Raise the vehicle and support safely
- Heated Oxygen (HO$_2$S) sensor and remove the front exhaust pipe, if necessary

4. Properly support the engine.
- Both roll stopper mounting bolts and the 4 center member installation bolts. Remove the center member.
- Center member

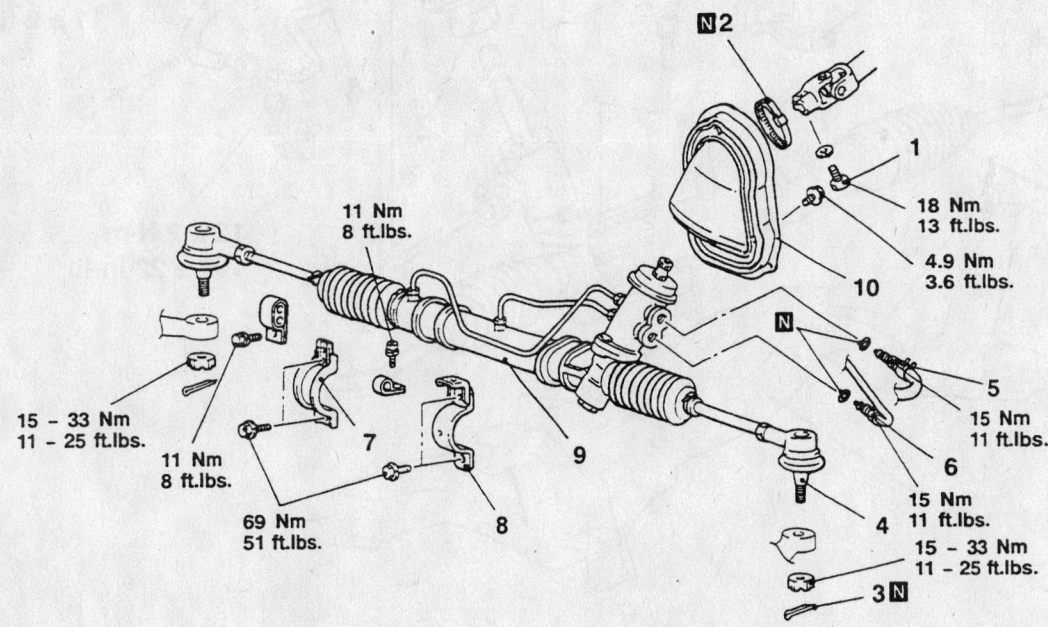

1. Steering shaft assembly and gear box connecting bolt
2. Band
3. Cotter pin
4. Tie-rod end and knuckle connection
5. Return tube connection

6. Pressure tube connection
7. Cylinder clamp
8. Gear housing clamp
9. Gear box assembly
10. Steering cover assembly

Exploded view of the power steering gear assembly—Mirage

7923PGA1

➡**Matchmark the pinion input shaft of the rack to the lower steering column joint for installation purposes.**

5. Remove or disconnect the following:
- Pinch bolt holding the lower steering column joint to the rack and pinion input shaft
- Cotter pins and disconnect the tie rod ends from the steering knuckle
- Power steering fluid pressure pipe and return hose from the rack fittings
- Rack and pinion steering assembly

and its rubber mounts from the right side of the vehicle

To install:

6. Align the matchmarks of the input shaft and install the rack to the vehicle.

7. Secure the rack using the retainer clamps and bolts. Torque the bolts to 52 ft. lbs. (70 Nm).

8. Torque the steering column pinch bolt to 13 ft. lbs. (18 Nm).

9. Using new O-rings, connect the power steering fluid lines to the rack fittings.

10. Install or connect the following:
- Center member
- Front exhaust pipe
- HO2S sensor
- Tie rod ends to the steering knuckles and tighten the castle nuts to 25 ft. lbs. (34 Nm). Install new cotter pins.
- Wheels and connect the negative battery cable

11. Refill the reservoir and bleed the system.

12. Perform a front end alignment.

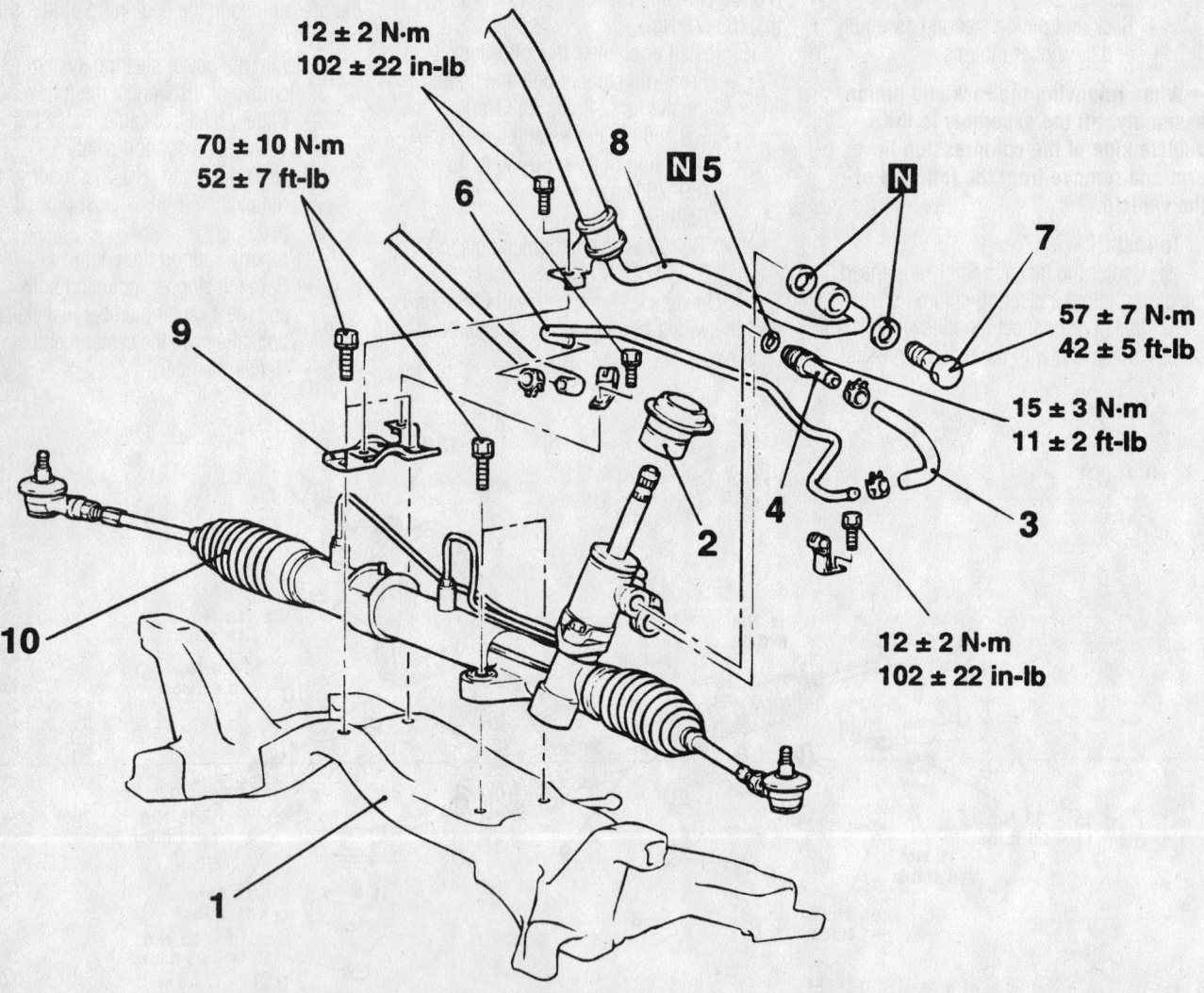

1. CROSSMEMBER
2. JOINT COVER GROMMET
3. RETURN HOSE
4. RETURN TUBE
5. O-RING
6. RETURN TUBE
7. EYE BOLT
8. PRESSURE HOSE ASSEMBLY
9. CLAMP
10. STEERING GEAR AND LINKAGE

9357QG28

Exploded view of the power steering gear—Lancer

Strut and Coil Spring

REMOVAL & INSTALLATION

Front

DIAMANTE

1. Before servicing the vehicle, refer to the precautions in the beginning of this section.

2. Disconnect the negative battery cable.

3. Raise and safely support the vehicle.

4. Remove the brake hose and the tube bracket.

➡ **Do not pry the brake hose and tube clamp away when removing it.**

5. If equipped with Anti-Lock Brake (ABS), disconnect the front speed sensor mounting clamp from the strut.

6. Support the lower arm and remove the strut to knuckle bolts. Use a piece of wire to suspend the knuckle to keep the weight off the brake hose.

7. If equipped with Active Electronic Control Suspension (Active-ECS) perform the following:

a. Loosen the nut that secures the air line to the to the top of the strut and discard the O-ring.

b. Remove the bolts that secure the actuator to the top of the strut and remove the component. Disconnect the wiring harness.

➡ **Before removing the top bolts, make matchmarks on the body and the strut insulator for proper reassembly.**

8. Remove the strut upper nuts and remove the strut assembly from the vehicle.

9. Compress the coil spring using a spring compressor until the spring just comes away from one of the seats.

10. Remove or disconnect the following:
   • Center nut from the strut and remove the upper mounting bracket and bushings
   • Coil spring

To install:

11. Install or connect the following:
   • Compressed spring on the strut assembly
   • Upper bushings and the mounting bracket
   • Nut and tighten it to 43 ft. lbs. (59 Nm)
   • Strut to the vehicle and tighten the upper mounting nuts to 33 ft. lbs. (45 Nm)

12. Align the strut to the knuckle and connect with the mounting bolts. Torque the mounting bolts to 70–76 ft. lbs. (90–105 Nm).

13. If equipped with Active-ECS, perform the following:

a. Install the air line with a new O-ring.

b. Install the actuator to the top of the strut. Connect the wiring harness.

14. Install or connect the following:
   • Brake hose bracket and the ABS clamp, if equipped
   • Wheel and tire assembly

15. Perform a front end alignment.

LANCER & MIRAGE

1. Before servicing the vehicle, refer to the precautions in the beginning of this section.

2. Disconnect the negative battery cable.

3. Raise and safely support vehicle.

4. Remove the brake hose and tube bracket retainer bolt and bracket from the front strut. Do not pry the brake hose and tube clamp away when removing.

5. If equipped with ABS, disconnect the front speed sensor mounting clamp from the strut.

6. Support the lower arm using floor jack. Remove the lower strut to knuckle bolts.

➡ **Before removing the top bolts, make matchmarks on the body and the strut insulator for proper reassembly.**

7. Remove the strut upper mounting bolts. Remove the strut assembly from the vehicle.

8. Compress the coil spring using a spring compressor until the spring just comes away from one of the seats.

9. Remove or disconnect the following:
   • Center nut from the strut and remove the upper mounting bracket and bushings
   • Coil spring

To install:

10. Install or connect the following:
   • Compressed spring on the strut assembly
   • Upper bushings and the mounting bracket
   • Nut and tighten it to 43 ft. lbs. (59 Nm)
   • Strut to the vehicle and install the top mounting bolts. Tighten the mounting bolts to 29 ft. lbs. (40 Nm) for Mirage and 33 ft. lbs. (44 Nm) for Lancer.

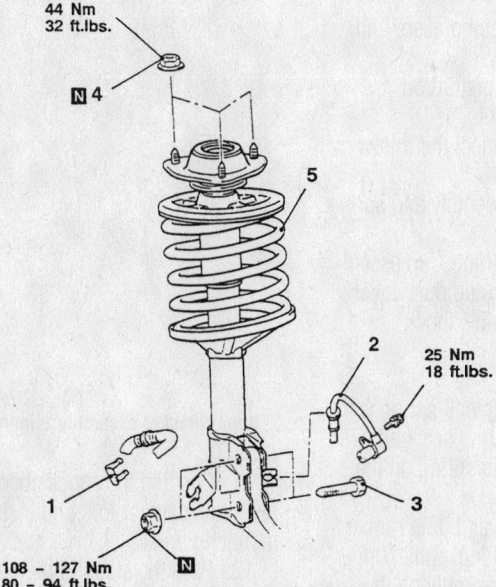

44 Nm
32 ft.lbs.

N 4

5

2

25 Nm
18 ft.lbs.

1

3

108 – 127 Nm
80 – 94 ft.lbs.

N

Removal steps
1. Brake hose clamp
2. Front speed sensor <Vehicles with ABS>
3. Bolts
4. Self-locking nut
5. Strut assembly

Caution
For vehicles with ABS, be careful when handling the pole piece at the tip of the speed sensor so as not to damage it by striking against other parts.

7923PGA6

Front strut assembly and related parts—Mirage shown, Lancer similar

11. Position the strut on the knuckle and install the mounting bolts. While holding the head of the lower mounting bolt, tighten the nuts to 80–94 ft. lbs. (110–130 Nm) for Mirage and 123 ft. lbs. (167 Nm) for Lancer.

12. Install or connect the following:
- Brake hose bracket and the ABS clamp, if equipped
- Wheel and tire assembly

13. Perform a front end alignment.

Shock Absorber and Coil Spring

REMOVAL & INSTALLATION

Front

ECLIPSE

1. Before servicing the vehicle, refer to the precautions in the beginning of this section.

2. Raise and safely support the vehicle.

3. Remove or disconnect the following:
- Front wheel
- 3 upper shock absorber mounting nuts. Do not remove the larger nut in the center of the strut at this time.
- Stabilizer link from the damper fork
- Damper fork mounting bolt
- Shock absorber assembly from the vehicle

4. Use a coil spring compressor and compress the coil spring.

5. While holding the piston rod, remove the self-locking nut.

6. Remove or disconnect the following:
- Upper bracket assembly and spring pad
- Collar, upper bushing, cup assembly, bump rubber and dust cover
- Coil spring from the shock absorber

To install:

7. Align the end of the coil spring with the stepped part of the spring seat and install the compressed coil spring on the shock.

8. Install the dust cover, bump rubber, cup assembly, upper bushing, collar, upper spring pad and bracket assembly on the strut.

9. Install or connect the following:
- Upper bushing and washer on the piston rod
- New self-locking nut on the piston rod. Temporarily tighten the nut.

10. Carefully remove the spring compressor from the spring. Torque the self-locking nut to 16 ft. lbs. (25 Nm).

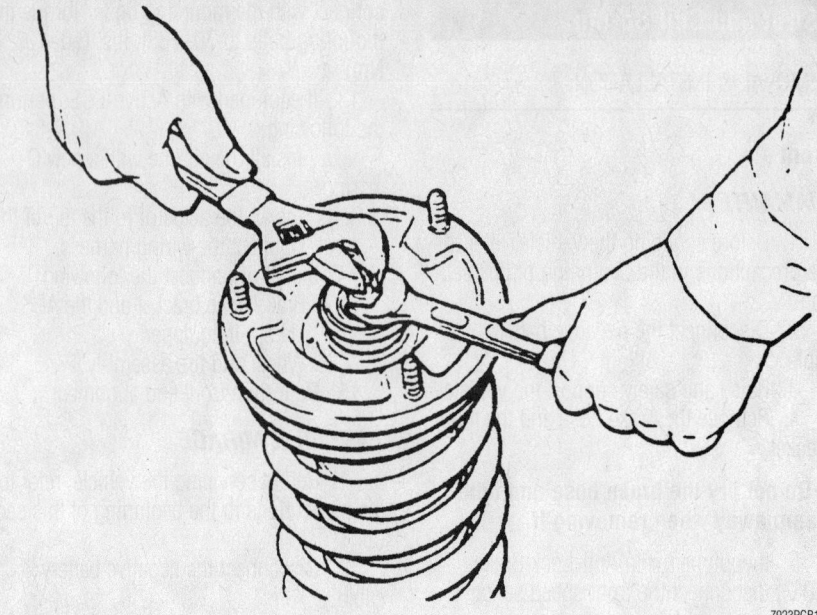

Removing the self-locking nut—Eclipse

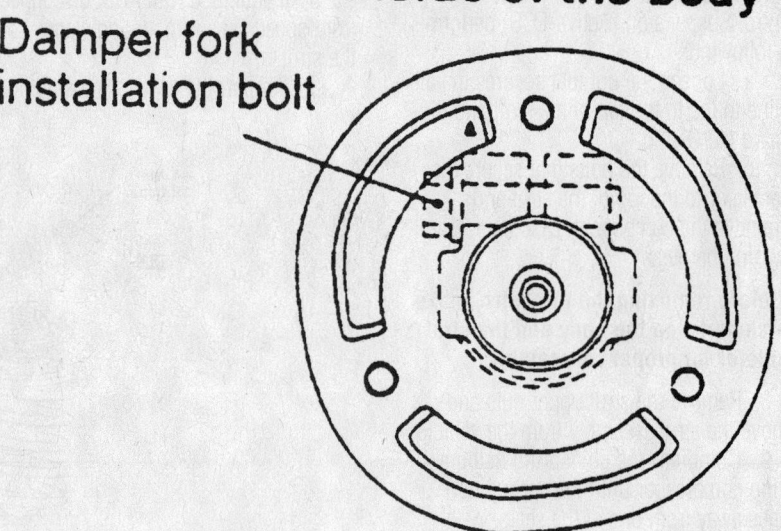

Upper bracket assembly alignment—Eclipse

11. Position the shock absorber assembly in the damper fork and install the mounting bolt.

12. Pass the studs in the upper bracket assembly through the holes in the inner fender and install the 3 mounting nuts.

13. Connect the stabilizer link to the damper fork.

14. Install the wheel assembly.

15. Safely lower the vehicle to the floor.

16. Check the front wheel alignment and adjust if necessary.

GALANT

1. Before servicing the vehicle, refer to the precautions in the beginning of this section.

2. Disconnect the negative battery cable.

3. Raise and safely support vehicle.

4. Remove or disconnect the following:
- Appropriate wheel assembly
- Sway bar link from the damper fork
- Damper fork lower through-bolt and upper pinch bolt
- Damper fork assembly

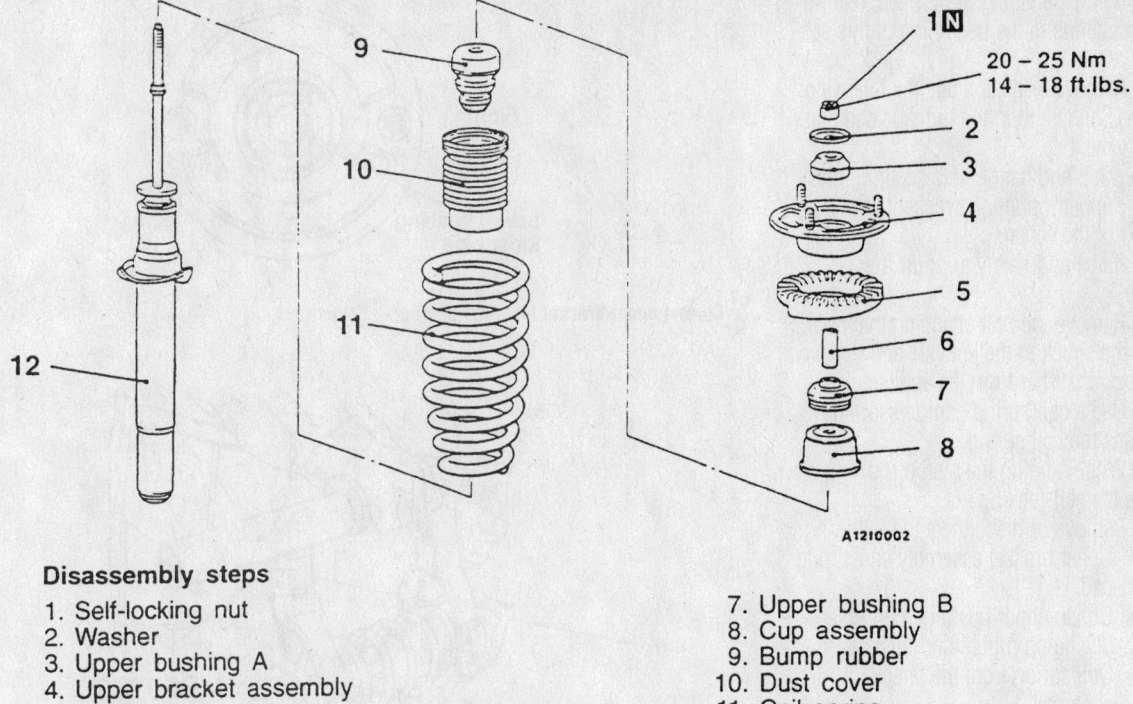

20 – 25 Nm
14 – 18 ft.lbs.

A1210002

Disassembly steps

1. Self-locking nut
2. Washer
3. Upper bushing A
4. Upper bracket assembly
5. Upper spring pad
6. Collar
7. Upper bushing B
8. Cup assembly
9. Bump rubber
10. Dust cover
11. Coil spring
12. Shock absorber assembly

7923PGB3

Exploded view of the coil spring removal procedure—Eclipse

- Shock absorber upper nuts and remove the strut assembly from the vehicle
5. Compress the coil spring with a special compression tool.
  - Self-locking nut and washer
  - Upper bushing, upper bracket assembly, the upper spring pad, and the collar
  - Other upper bushing, cup assembly, bump rubber, dust cover, and the coil spring. Carefully remove the coil spring compression tool

To install:
6. Install or connect the following:
  - Compressed coil spring to the shock absorber assembly. Be sure to align the edge of coil spring to the stepped part of the spring seat. Install the dust cover, bump rubber, cup assembly, upper bushing, collar, and upper spring pad.
  - Upper bracket assembly and position it so that the 3 bolts are in the correct position
  - Upper bushing, washer, and locknut. Torque the locknut to 18 ft. lbs. (24 Nm).
  - Shock absorber and tighten the upper mounting nuts to 32 ft. lbs. (44 Nm)

7. Align the shock to the damper fork and install the damper fork. Tighten the lower through-bolt/nut to 65 ft. lbs. (88 Nm) and the upper pinch bolt to 76 ft. lbs. (103 Nm).

8. Connect the sway bar link to the damper fork and tighten the link nut to 29 ft. lbs. (39 Nm).

9. Install the wheel and tire assembly.

10. Perform a front end alignment.

Rear

DIAMANTE

1. Before servicing the vehicle, refer to the precautions in the beginning of this section.

2. Disconnect the negative battery cable.

3. Raise and properly support vehicle. Remove both rear wheels.

4. Support the lower control arm with a jack.

5. Matchmark the positioning of the upper spring plate to the vehicle for reinstallation purposes.

6. If equipped with Active Electronic Control Suspension (Active-ECS), perform the following:
   a. Loosen the nut that secures the air

line to the to the top of the strut and discard the O-ring.

   b. Remove the bolts that secure the actuator to the top of the strut and remove the component. Disconnect the wiring harness.

7. Remove the shock absorber lower mounting bolt and remove the 2 nuts that secure the shock upper plate to the vehicle.

8. Lower the support jack and remove the shock from the vehicle.

To install

9. Position the upper spring plate and install the strut. Use the support jack to assist with installation.

10. Tighten the upper strut mounting nuts to 33 ft. lbs. (45 Nm).

11. Tighten the lower strut mounting bolt to 71 ft. lbs. (98 Nm).

12. If equipped with Active-ECS perform the following:
   a. Using a new O-ring, tighten the nut that secures the air line to the to the top of the strut to 84 inch lbs. (9 Nm).

   b. Install the actuator to the top of the shock absorber and secure with mounting bolts. Connect the wiring harness.

13. Remove the support jack, install wheels and lower vehicle.

14. Connect the negative battery cable.

ECLIPSE

1. Before servicing the vehicle, refer to the precautions in the beginning of this section.

2. Remove or disconnect the following:
   - Service lid in the luggage compartment
   - Cap and flange nuts securing the upper mounting bracket to the body of the vehicle

3. Raise and safely support the vehicle.

4. Remove the bolt attaching the lower end of the shock to the knuckle and remove the shock absorber from the vehicle.

5. Use a coil spring compressor and compress the coil spring.

6. While holding the piston rod, remove the self-locking nut.

7. Remove or disconnect the following:
   - Upper bracket assembly and spring pad
   - Collar, upper bushing, cup assembly, bump rubber and dust cover
   - Coil spring from the shock absorber

To install:

8. Align the end of the coil spring with the stepped part of the spring seat and install the compressed coil spring on the shock absorber.

9. Install or connect the following:
   - Dust cover, bump rubber, cup assembly, upper bushing, collar, upper spring pad and bracket assembly on the shock absorber
   - Upper bushing and washer on the piston rod
   - New self-locking nut on the piston rod. Temporarily tighten the nut.

10. Remove the spring compressor from the spring. Torque the self-locking nut to 16 ft. lbs. (25 Nm).

11. Install the upper bracket of the shock to the vehicle. Torque the mounting nuts to 32 ft. lbs. (44 Nm).

12. Raise the suspension up with a jack or adjustable stand to align the shock absorber lower mounting holes.

13. Install the lower mounting bolt. Torque the bolt to 71 ft. lbs. (96 Nm).

14. Remove the jack or stand and safely lower the vehicle to the floor.

15. Install the cap and service lid.

GALANT

1. Before servicing the vehicle, refer to the precautions in the beginning of this section.

2. Raise and support the vehicle chassis.

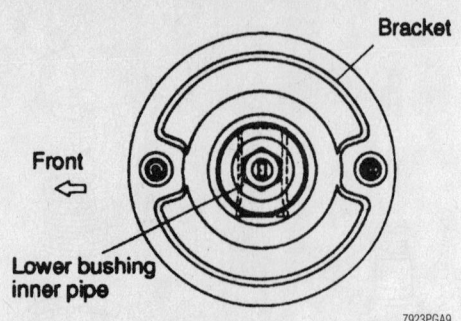

Correct upper bracket installed position—Eclipse

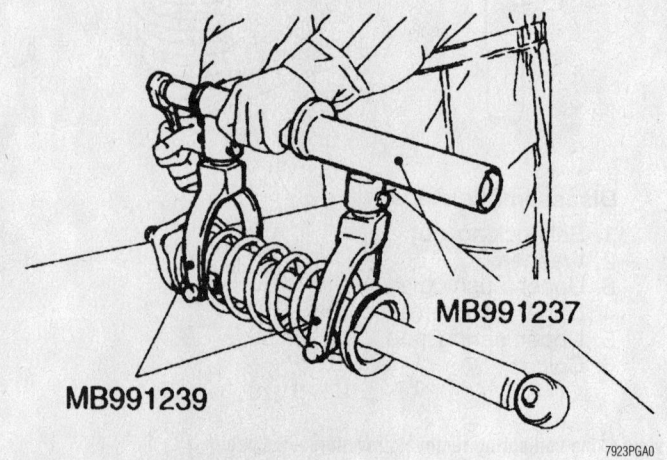

Correct method for compressing the coil spring

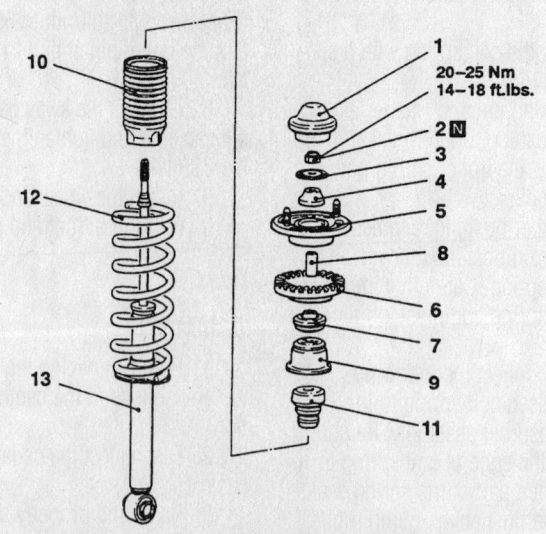

1. Cap	8. Collar
2. Self-locking nut	9. Cup
3. Washer	10. Dust cover
4. Upper bushing A	11. Bump rubber
5. Bracket	12. Coil spring
6. Spring pad	13. Shock absorber assembly
7. Upper bushing B	

Exploded view of the rear shock absorber assembly—Galant and Mirage

3. Raise and support the lower control arm assembly slightly.

4. In order to gain access to the top mounting nuts, remove the rear seat as follows:

a. While pulling the rear seat stopper outward, lift the lower cushion upward. Remove the lower cushion.

b. Remove the seat back mounting bolts.

c. Lift the seat back upward and remove the seat.

5. Remove or disconnect the following:
- Shock upper mounting nuts
- Shock lower mounting bolt and remove the assembly from the vehicle

6. Use a coil spring compressor and compress the coil spring.

7. Remove the shock cap.

8. While holding the piston rod, remove the self-locking nut.

9. Remove or disconnect the following:
- Upper bracket assembly and spring pad
- Collar, upper bushing, cup assembly, bump rubber and dust cover
- Coil spring from the shock

To install

10. Align the end of the coil spring with the stepped part of the spring seat and install the compressed coil spring on the shock.

11. Install or connect the following:
- Dust cover, bump rubber, cup assembly, upper bushing, collar, upper spring pad and bracket assembly on the shock
- Upper bushing and washer on the piston rod
- New self-locking nut on the piston rod. Temporarily tighten the nut.

12. Remove the spring compressor from the spring. Torque the self-locking nut to 16 ft. lbs. (25 Nm).

13. Install the shock cap.

14. Position the shock assembly so that the lower mounting bolt can be installed and lightly tightened.

15. Use a jack to raise or lower the lower control arm, so that the top shock plate studs align through the body. Raise the jack to hold the shock assembly in position.

16. Install the top plate nuts on the studs and tighten the mounting nuts to 32 ft. lbs. (44 Nm).

17. With the vehicle on the ground, tighten the lower mounting bolt to 71 ft. lbs. (98 Nm).

18. Install the rear seat back and cushion.

LANCER

1. Before servicing the vehicle, refer to the precautions in the beginning of this section.

2. Remove or disconnect the following:
- Stabilizer link connection

3. Support the lower control arm with a jack.
- Lower control arm and trailing arm bolt
- Upper shock absorber mounting nut
- Shock absorber-to-lower control arm attaching bolt
- Shock absorber assembly

To install:

4. Position the shock absorber into the vehicle. Install the spring seat stepped section so that it points toward the rear side of the vehicle.

5. Install or connect the following:
- Shock absorber-to-lower control arm bolt and nut. Tighten the nut to 70 ft. lbs. (95 Nm).
- Upper shock mounting nut and tighten to 32 ft. lbs. (44 Nm)
- Lower control arm and trailing arm
- Stabilizer link. Tighten the self-locking nuts so that the end of the stabilizer line bolt protrudes 0.24–0.31 in. (6–8mm).

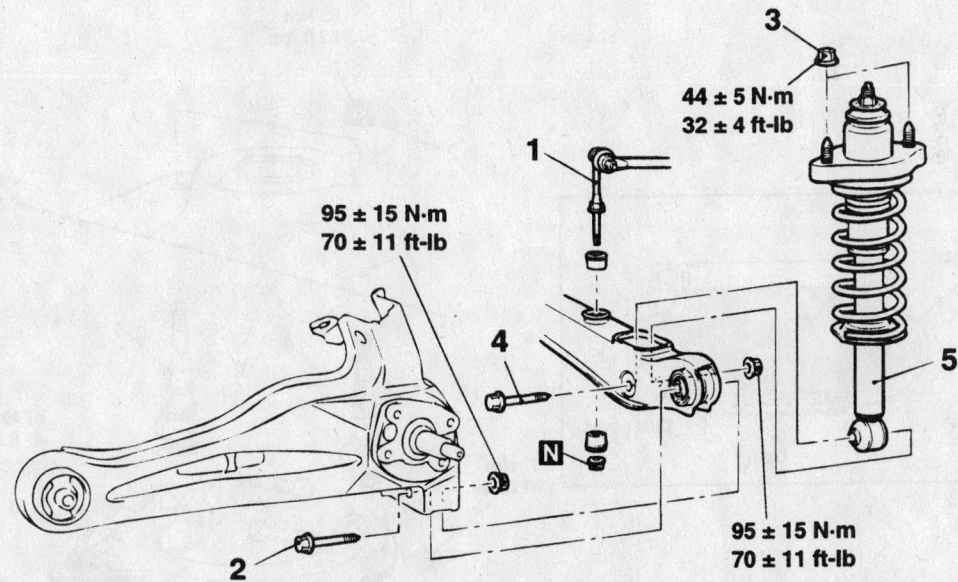

1. STABILIZER LINK CONNECTION
2. LOWER ARM AND TRAILING ARM CONNECTION
3. SHOCK ABSORBER MOUNTING NUT
4. SHOCK ABSORBER AND LOWER ARM CONNECTING BOLT
5. SHOCK ABSORBER ASSEMBLY

9357QG29

Exploded view of the rear shock absorber mounting—Lancer

MIRAGE

1. Before servicing the vehicle, refer to the precautions in the beginning of this section.

2. Remove or disconnect the following:
   - Trunk interior trim to gain access to the top mounting nuts
   - Top cap and upper shock mounting nuts

3. Raise and support vehicle chassis.

4. Support the trailing arm assembly with a jack.

5. Matchmark the upper spring plate to the vehicle chassis for reassembly and remove the upper spring plate mounting nuts.

6. Remove the shock lower mounting bolt and remove the assembly from the vehicle.

7. Compress the coil spring using the proper spring compressor.

8. Hold the piston rod with a wrench and remove the self-locking nut.

9. Remove or disconnect the following:
   - Washer, upper bushing A, bracket, spring pad, upper bushing B, collar, cup, dust cover and bump rubber
   - Coil spring

To install

10. Install or connect the following:
    - Coil spring on the shock
    - Bump rubber, dust cover, cup, collar, upper bushing A, spring pad, bracket, upper bushing B and the washer

11. Temporarily install a new self-locking nut, carefully release the spring from the compressor and tighten the self-locking nut to specifications.

12. Position the shock assembly so that lower mounting bolt can be installed and lightly tightened.

13. Use jack to raise or lower the axle assembly so that top shock plate studs aligns through body. Raise jack to hold the shock assembly in position.

14. Install the top plate nuts and tighten them to 20 ft. lbs. (28 Nm).

15. Lower the vehicle and tighten the lower mounting bolt to 65 ft. lbs. (90 Nm).

16. Install top cap and interior trim.

Upper Ball Joint

REMOVAL & INSTALLATION

The upper ball joints are an integral part of the upper control arm. If the ball joint becomes worn or damaged, the control arm must be replaced.

Upper Control Arm

REMOVAL & INSTALLATION

Diamante and Mirage

These vehicles use a strut type front suspension. No upper control arm is used.

Eclipse and Galant

1. Before servicing the vehicle, refer to the precautions in the beginning of this section.

2. Raise and safely support the vehicle.

3. Remove or disconnect the following:
   - Front wheel
   - Upper arm ball joint from the steering knuckle
   - Upper arm shaft mounting nuts from the body
   - Upper arm
   - Through-bolts that attach the upper arm to the shafts

To install:

4. Assembly the upper arm to the shafts at the proper angle. Torque the through-bolts and nuts to 41 ft. lbs. (57 Nm). The proper angle is 84–86°. After the arm and

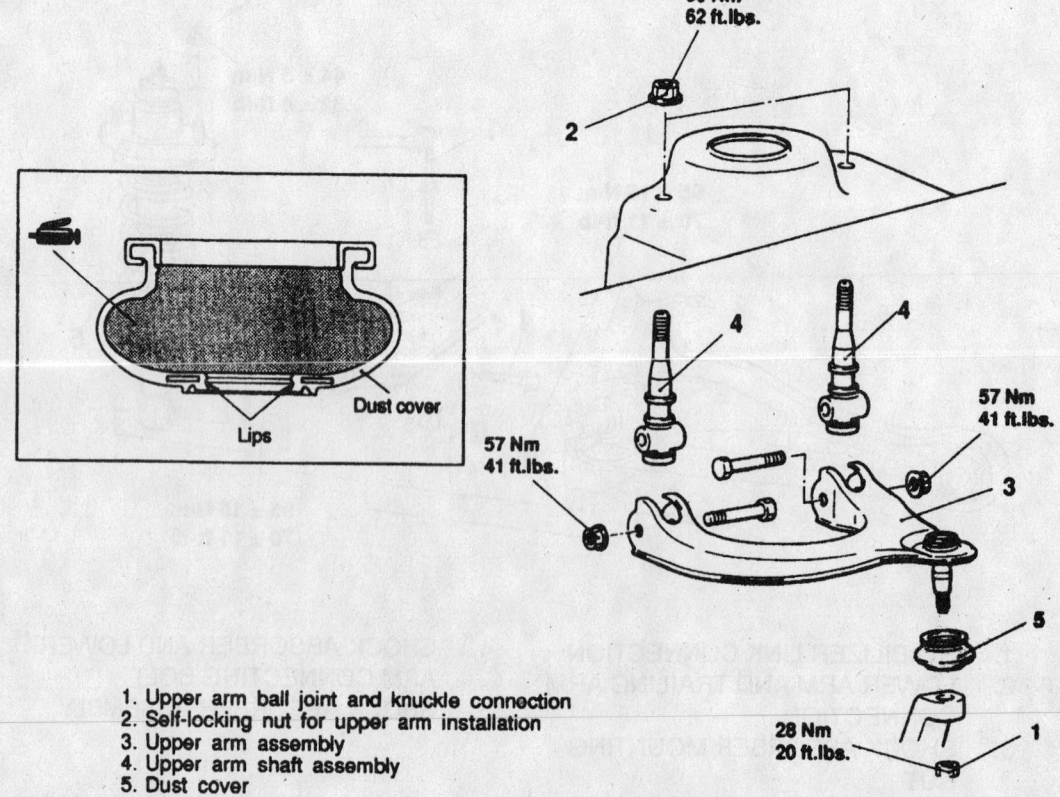

1. Upper arm ball joint and knuckle connection
2. Self-locking nut for upper arm installation
3. Upper arm assembly
4. Upper arm shaft assembly
5. Dust cover

Upper control arm assembly—Eclipse and Galant

7923PGB7

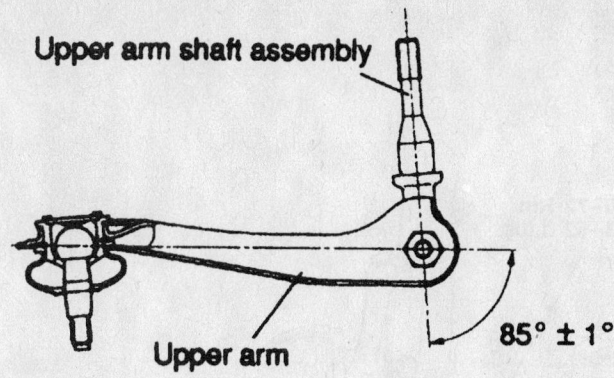

Correct angle of control arm and shafts—Eclipse and Galant

A : 299.9 mm (11.8 in.)
B : 234.0 mm (9.2 in.)

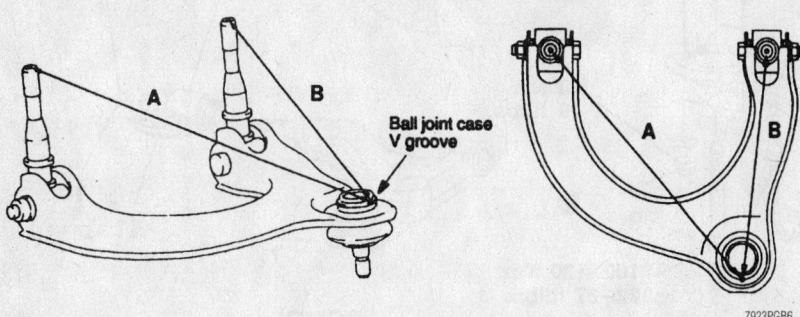

Measure the dimensions A and B as shown—Eclipse and Galant

the shafts are connected at the right angle, measure dimensions A and B to insure correct assembly.

- A O-ring: 11.8 in. (299.9mm)
- B O-ring: 9.2 in. (234.0mm)

5. Install or connect the following:
- Control arm assembly to the body with new self-locking nuts. Torque the self-locking nuts to 62 ft. lbs. (86 Nm).
- Upper arm ball joint to the steering knuckle with a new self-locking nut. Torque the locking nut to 20 ft. lbs. (28 Nm).
- Front wheel

6. Check the front wheel alignment and adjust if necessary.

7. Lower the vehicle.

Lower Ball Joint

REMOVAL & INSTALLATION

The lower ball joint is an integral part of the lower control arm assembly, and can not be serviced separately. A worn or damaged ball joint requires replacement of lower control arm assembly.

Lower Control Arm

REMOVAL & INSTALLATION

Front

DIAMANTE

1. Before servicing the vehicle, refer to the precautions in the beginning of this section.

2. Disconnect the negative battery cable.

3. Raise the vehicle and support safely allowing wheels and suspension to hang freely.

4. Remove or disconnect the following:
- Sway bar links from the lower control arm
- Ball joint stud from the steering knuckle
- Inner mounting frame through-bolt and nut
- Rear mount bolts. Remove the clamp if equipped
- Rear rod bushing if servicing

To install:

5. Assemble the control arm and bushing.

6. Install or connect the following:
- Control arm to the vehicle and install the through-bolt. Replace the nut and snug temporarily.
- Rear mount clamp, bolts and replacement nuts. Torque the bolts to 72–87 ft. lbs. (100–120 Nm). Torque the nuts to 29 ft. lbs. (40 Nm).
- Ball joint stud to the knuckle
- New nut and tighten to 43–52 ft. lbs. (60–72 Nm)
- Sway bar and links

7. Lower the vehicle to the floor for the final tightening of the frame mount through-bolt.

8. Once the full weight of the vehicle is on the floor, tighten the frame mount through-bolt nuts to 75–90 ft. lbs. (102–122 Nm).

9. Connect the negative battery cable.

10. Check the wheel alignment and adjust if necessary.

ECLIPSE AND GALANT

The lower lateral arm ball joint and the compression arm ball joint are integral components of the lateral arm and the compression arm respectively. If the ball joints are to be serviced, the arms must be replaced.

1. Before servicing the vehicle, refer to the precautions in the beginning of this section.

2. Raise and support the vehicle safely.

3. Disconnect both ball joint studs from the steering knuckle.

4. To remove the lower lateral arm, remove the crossmember brackets.

5. Remove or disconnect the following:
- Inner lateral arm mounting bolts and nut
- Arm from the vehicle
- 2 bolts holding the compression arm
- Compression arm

To install:

6. Assemble the control arms and bushings.

7. Install or connect the following:
- Lateral control arm to the vehicle and install the inner mounting bolts. Install a new nut and snug temporarily.
- Compression arm to the vehicle
- Ball joint studs to the knuckle
- New nuts and tighten to 43–51 ft. lbs. (59–71 Nm)

8. Lower the vehicle to the floor for final tightening.

9. Once the full weight of the vehicle is

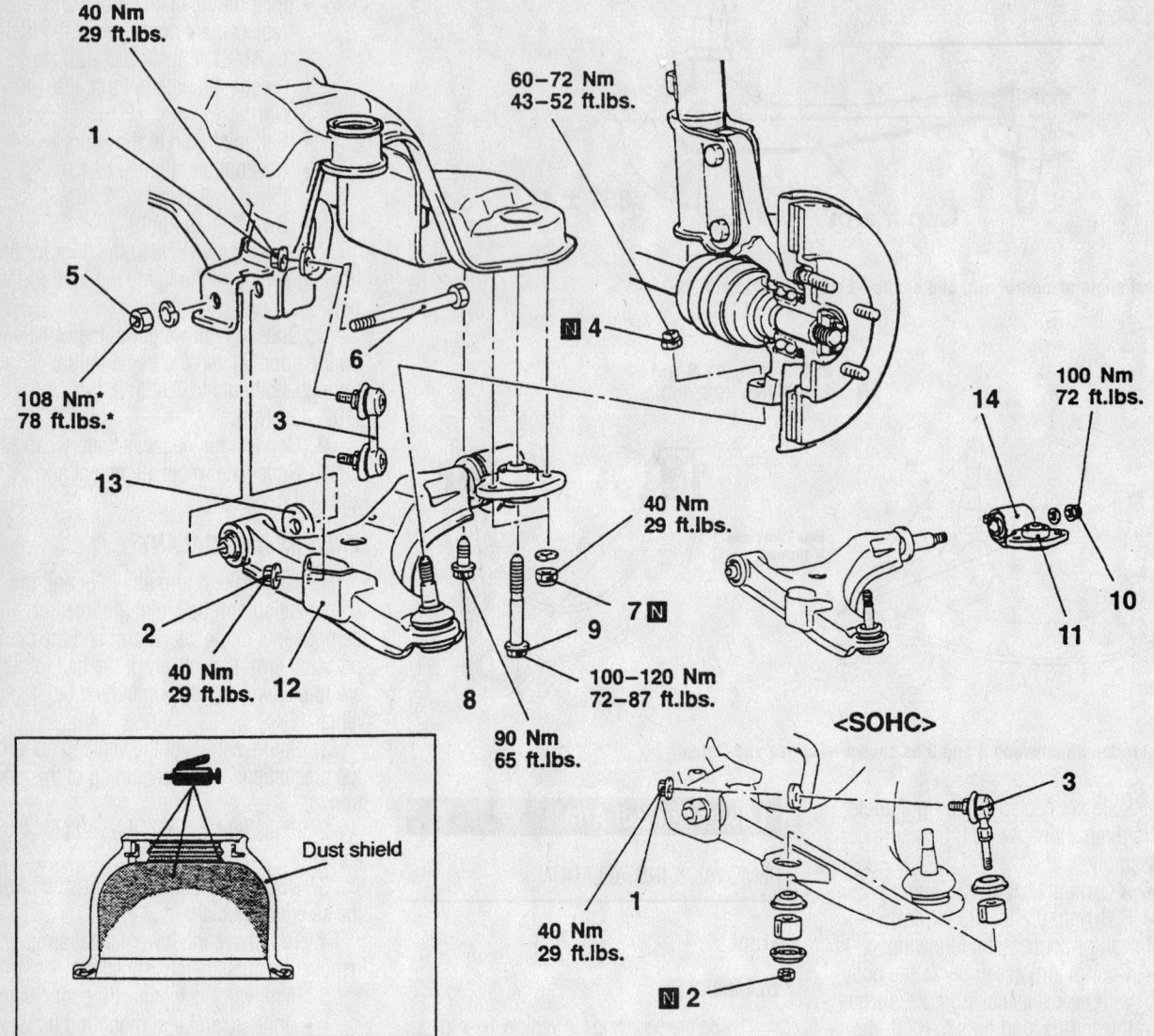

Removal steps

1. Stabilizer link mounting nut (stabilizer bar side)
2. Stabilizer link mounting nut (lower arm side)
3. Stabilizer link
4. Self-locking nut connecting lower arm ball joint to knuckle
5. Lower arm mounting nut
6. Lower arm mounting bolt
7. Clamp mounting self-locking nut
8. Clamp mounting bolt (small)
9. Clamp mounting bolt (large)
10. Lower arm clamp mounting self-locking nut
11. Lower arm mounting clamp
12. Lower arm
13. Stopper
14. Rod bushing

Caution
*: Indicates parts which should be temporarily tightened, and then fully tightened with the vehicle on the ground in an unladen condition.

Exploded view of the lower control arm removal procedure—Diamante is similar

7923PGB9

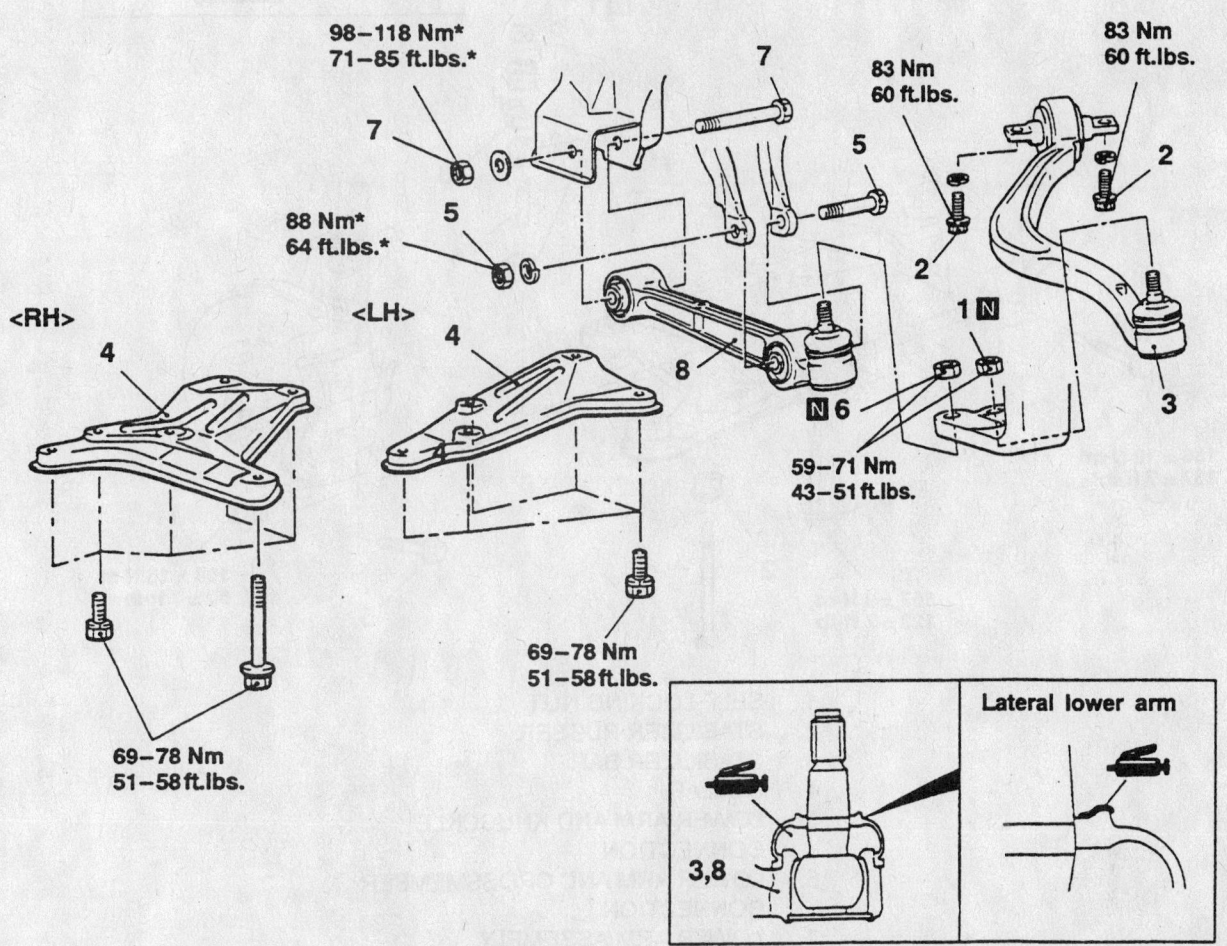

Exploded view of the lower control arms—Eclipse and Galant

Compression lower arm assembly removal steps

1. Connection for compression lower arm ball joint and knuckle
2. Compression lower arm mounting bolt
3. Compression lower arm assembly

Lateral lower arm assembly removal steps

4. Stay
5. Shock absorber lower mounting bolt and nut
6. Connection for lateral lower arm ball joint and knuckle
7. Lateral lower arm mounting bolt and nut
8. Lateral lower arm assembly

Caution
*: Indicates parts which should be temporarily tightened, and then fully tightened with the vehicle on the ground in the unladen condition.

7923PGB8

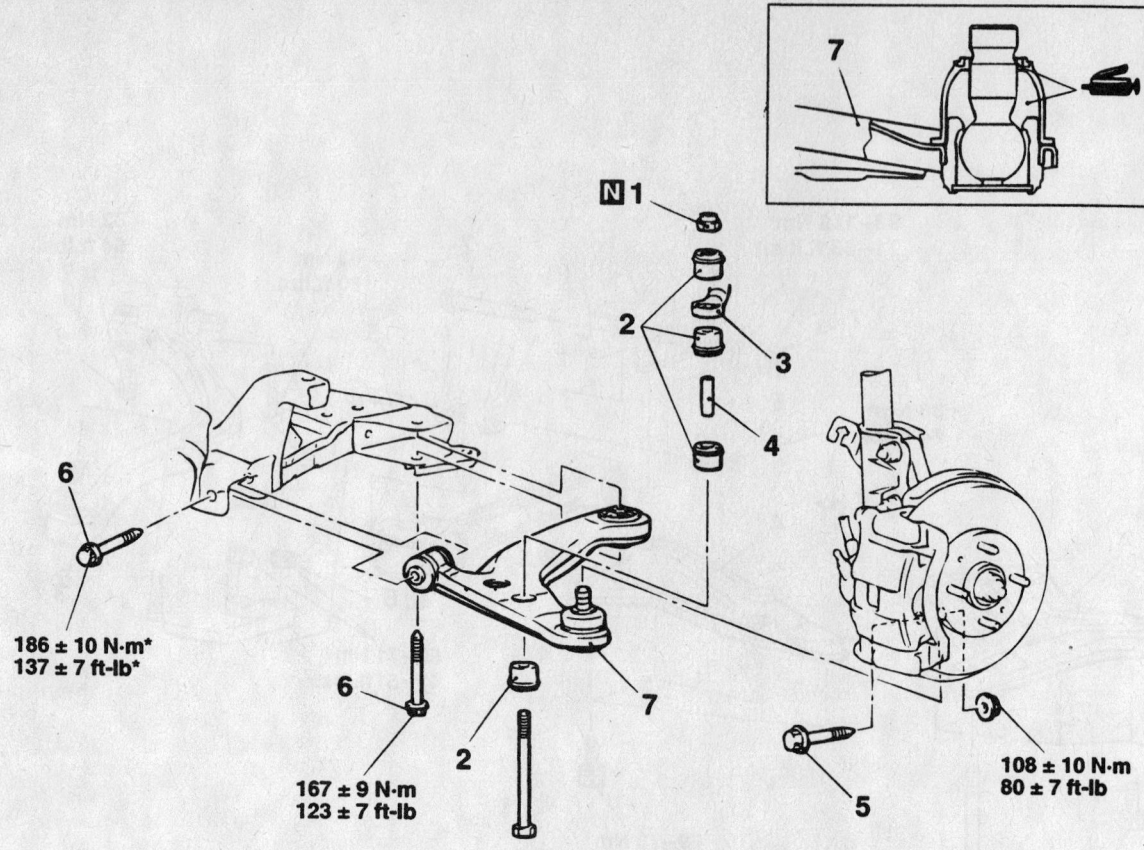

186 ± 10 N·m*
137 ± 7 ft-lb*

167 ± 9 N·m
123 ± 7 ft-lb

108 ± 10 N·m
80 ± 7 ft-lb

1. SELF-LOCKING NUT
2. STABILIZER RUBBER
3. STABILIZER BAR
4. COLLAR
5. LOWER ARM AND KNUCCKLE CONNECTION
6. LOWER ARM AND CROSSMEMBER CONNECTION
7. LOWER ARM ASSEMBLY

93570G30

Exploded view of the lower control arm mounting—Lancer

on the suspension, tighten the lateral arm rear bolt to 71–85 ft. lbs. (98–118 Nm) and the front bolt to the damper fork to 64 ft. lbs. (88 Nm).

10. Torque the bolts for the compression arm to 60 ft. lbs. (83 Nm).

11. Reinstall the crossmember brackets with their mounting bolts. Torque the mounting bolts to 51–58 ft. lbs. (69–78 Nm).

12. Perform an alignment on the vehicle.

LANCER

➡The suspension components should not be tightened until the vehicle's weight is resting on its wheels.

1. Before servicing the vehicle, refer to the precautions in the beginning of this section.

2. Raise the vehicle and support safely.

3. Remove or disconnect the following:
- Wheel and tire assembly
- Stabilizer bar self-locking nut, rubber bushings, stabilizer bar and collar. Discard the nut.
- Lower control arm-to-knuckle bolt and nut

4. Lift the transaxle with a jack, then remove the front control arm-to-crossmember bolt.
- Lower control arm

To install:

5. Install or connect the following:
- Lower control arm into the vehicle
- Lower control arm-to-crossmember

bolts. Torque the bottom bolt to 123 ft. lbs. (167 Nm) and the side bolt snug until the vehicle is lowered.
- Lower control arm-to-steering knuckle bolt. Torque to 80 ft. lbs. (108 Nm).
- Stabilizer bar collar, stabilizer bar, bushings and new self-locking nut.

6. Lower the vehicle, install the wheels, then with the weight of the vehicle on the wheels, torque the side control arm-to-crossmember bolt to 137 ft. lbs. (186 Nm).

MIRAGE

➡The suspension components should not be tightened until the vehicle's weight is resting on its wheels.

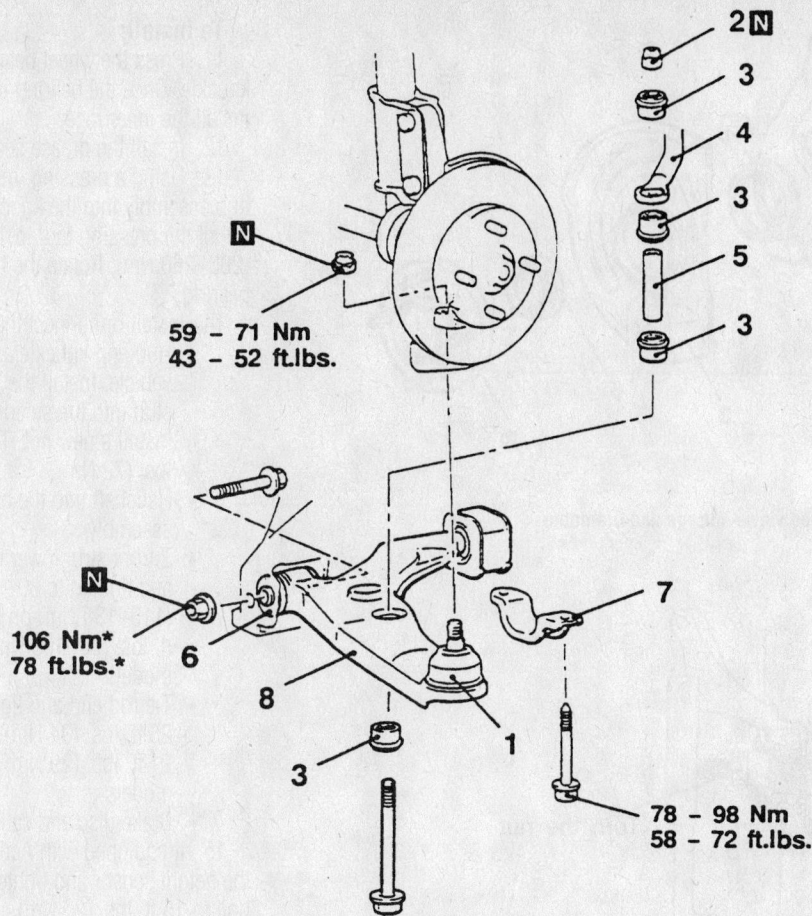

59 – 71 Nm
43 – 52 ft.lbs.

106 Nm*
78 ft.lbs.*

78 – 98 Nm
58 – 72 ft.lbs.

7923PGB0

Removal steps

1. Lower arm ball joint connection
2. Self-locking nut
3. Stabilizer rubber
4. Stabilizer bar
5. Collar
6. Lower arm front bushing connection
7. Support bracket
8. Lower arm assembly

Caution

*: **Indicates parts which should be temporarily tightened, and then fully tightened with the vehicle on the ground in the unladen condition.**

Lower control arm assembly and related components—Mirage

1. Before servicing the vehicle, refer to the precautions in the beginning of this section.
2. Raise the vehicle and support safely.
3. Remove or disconnect the following:
   - Wheel and tire assembly
   - Stabilizer bar links or mounting nuts and bolts from lower control arm. Remove the joint cups and bushings.
   - Ball joint stud from the steering knuckle
   - Inner lower arm mounting bolt and nut
   - Rear mount bolts from the retaining clamp. Remove the rear retainer clamp if equipped.
   - Arm from the vehicle

To install:

4. Install or connect the following:
   - Control arm to the vehicle and install the inner mounting bolt. Install new nut and tighten to 78 ft. lbs. (108 Nm).
   - Rear mount clamp and bolts. Torque the clamp mounting bolts to 65 ft. lbs. (90 Nm).
   - Ball joint stud to the knuckle. Install a new nut and tighten to 43–52 ft. lbs. (60–72 Nm).
   - Sway bar and links
5. Lower the vehicle to the floor for the final tightening of the inner frame mount bolt.
6. Install the wheel and tire assembly.

Wheel Bearings

ADJUSTMENT

The front and rear wheel bearings on these vehicles are not adjustable. If the bearings are noisy or become loose, they must be replaced.

REMOVAL & INSTALLATION

Front

DIAMANTE AND MIRAGE

1. Before servicing the vehicle, refer to the precautions in the beginning of this section.

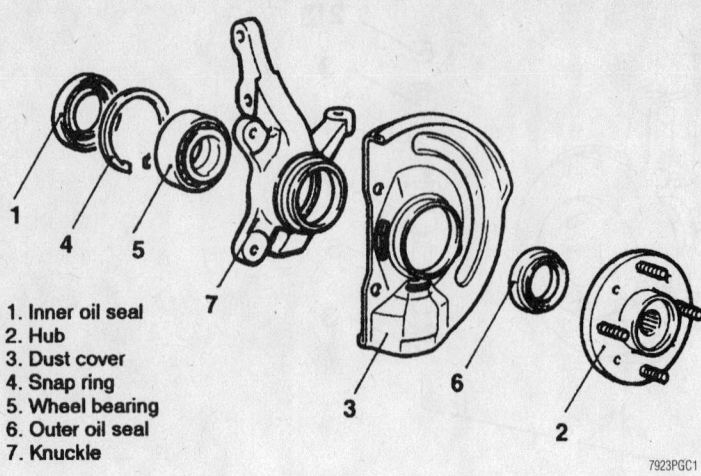

1. Inner oil seal
2. Hub
3. Dust cover
4. Snap ring
5. Wheel bearing
6. Outer oil seal
7. Knuckle

7923PGC1

Front wheel bearing assembly exploded view—Mirage and Diamante

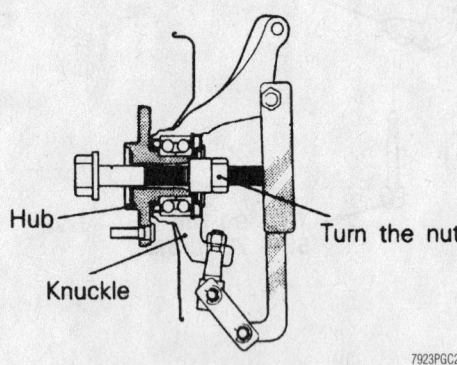

7923PGC2

Use of press tool for hub removal—Mirage and Diamante

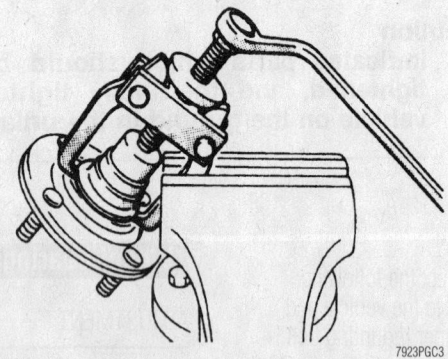

7923PGC3

Removing inner race from hub—Mirage and Diamante

2. Disconnect the negative battery cable.

3. Raise the vehicle and support safely. Remove the halfshaft nut.

4. If equipped with Anti-Lock Brake (ABS), remove the front wheel speed sensor.

5. If equipped with Active Electronic Control Suspension (Active-ECS), disconnect the height sensor from the lower control arm.

6. Remove the caliper assembly and brake pads. Suspend the caliper with a wire.

7. Ball joint and tie rod end from the steering knuckle.

8. Remove the halfshaft from the hub.

9. Unbolt the lower end of the strut and remove the hub and steering knuckle assembly from the vehicle.

10. Press the hub from the bearing and remove the bearing races from the knuckle.

To install:

11. Press the wheel bearing into the knuckle. Once the bearing is installed, install the inner race.

12. Install the grease seal.

13. Using a pressing, mount the front hub assembly into the knuckle. Tighten the nut of the pressing tool to 144–188 ft. lbs. (200–260 Nm). Rotate the hub to seat the bearing.

14. Install or connect the following:
- Hub and knuckle assembly onto the vehicle. Install the lower ball joint stud into the steering knuckle and install a new nut. Tighten to 52 ft. lbs. (72 Nm).
- Halfshaft into the hub/knuckle assembly
- 2 front strut lower mounting bolts and tighten to 80–94 ft. lbs. (110–130 Nm) on Mirage or 65–76 ft. lbs. (90–105 Nm) on Diamante models.
- Tie rod end and tighten the nut to 25 ft. lbs. (34 Nm) for Mirage and 21 ft. lbs. (29 Nm) on Diamante models
- Brake disc and caliper assembly

15. If equipped with Active-ECS, connect the height sensor and tighten the mounting bolt to 15 ft. lbs. (20 Nm).

16. Install or connect the following:
- Front speed sensor, if removed
- Washer and new locknut to the end of the halfshaft. Tighten the locknut snugly to 144–188 ft. lbs. (200–260 Nm).
- Tire and wheel assembly onto the vehicle. Lower the vehicle to the ground.

ECLIPSE

1. Before servicing the vehicle, refer to the precautions in the beginning of this section.

2. Remove or disconnect the following:
- Front wheel
- Axle nut
- Wheel speed sensor, vehicles with Anti-Lock Brake (ABS)
- Caliper and suspend it out of the way with wire or string
- Brake rotor
- Steering knuckle from the upper arm

3. Pull the knuckle away from the vehicle to access the hub mounting bolts on the inboard side of the hub. Be careful not to damage the ball joint boot or the ABS rotor if equipped.

4. Remove the mounting bolts and the front hub assembly.

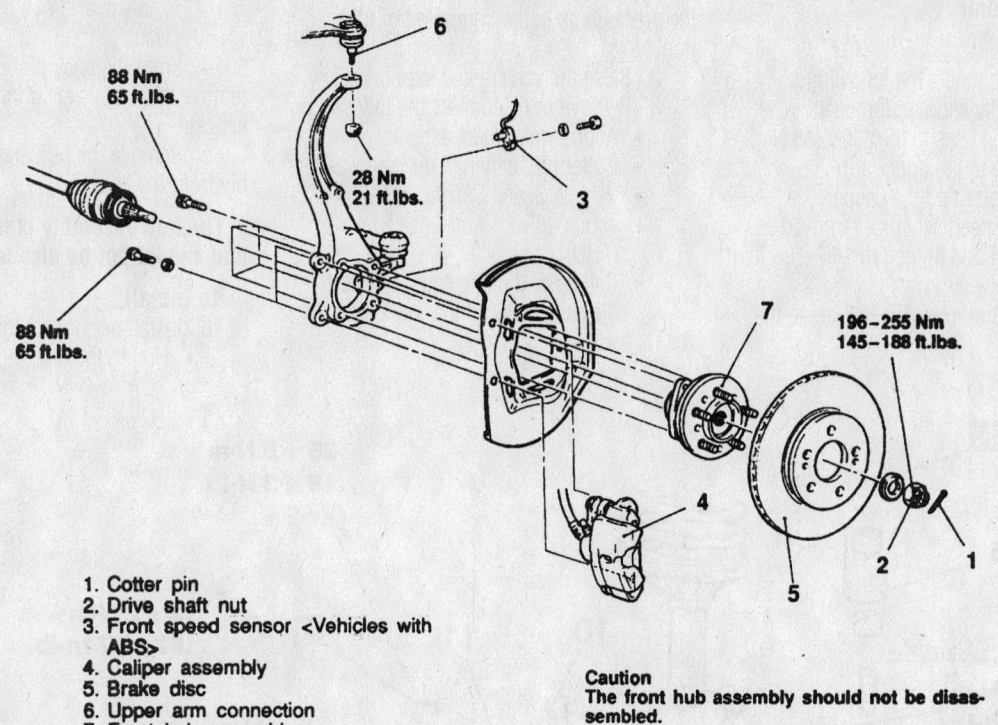

88 Nm
65 ft.lbs.

28 Nm
21 ft.lbs.

88 Nm
65 ft.lbs.

196–255 Nm
145–188 ft.lbs.

1. Cotter pin
2. Drive shaft nut
3. Front speed sensor <Vehicles with ABS>
4. Caliper assembly
5. Brake disc
6. Upper arm connection
7. Front hub assembly

Caution
The front hub assembly should not be disassembled.

7923PGC4

Front hub and related components—Eclipse

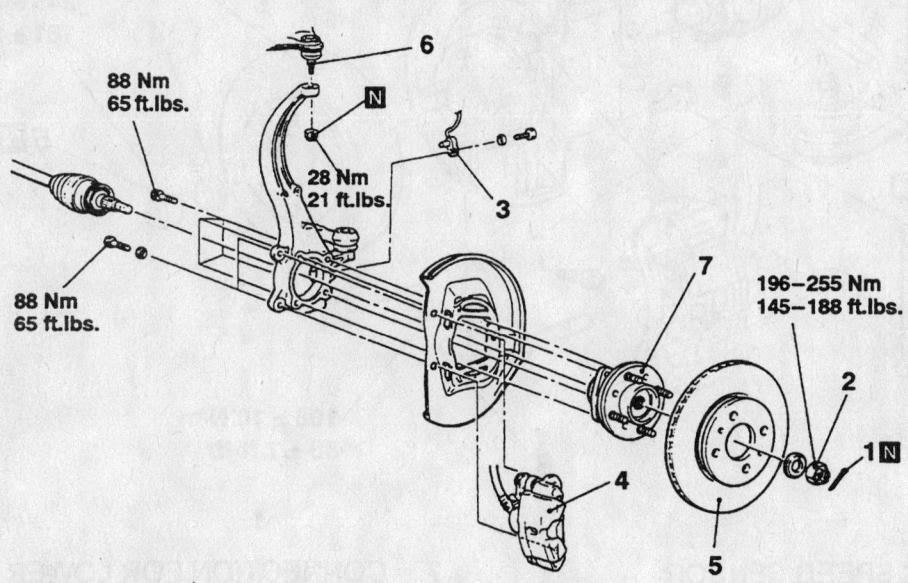

88 Nm
65 ft.lbs.

28 Nm
21 ft.lbs.

88 Nm
65 ft.lbs.

196–255 Nm
145–188 ft.lbs.

Removal steps

1. Cotter pin
2. Drive shaft nut
3. Front speed sensor <Vehicles with ABS>
4. Caliper assembly
5. Brake disc
6. Connection for upper arm
7. Front hub assembly

Caution
The front hub assembly should not be disassembled.

7923PGC5

Exploded view of the front hub removal—Galant

➡ Do not disassemble the hub assembly. If binding or damaged, it must be replaced as a unit.

To install:

5. Install or connect the following:
  - Hub to the knuckle. Torque the mounting bolts to 65 ft. lbs. (88 Nm).
  - Knuckle to the upper arm
  - Brake rotor and the caliper
  - Wheel speed sensor if removed
  - Axle nut and tighten to 145–188 ft. lbs. (196–255 Nm)
  - Wheel and lower the vehicle to the floor

GALANT

1. Before servicing the vehicle, refer to the precautions in the beginning of this section.
2. Raise the vehicle and support safely.
3. Remove or disconnect the following:
  - Appropriate wheel assembly
  - Cotter pin, halfshaft nut and washer
  - Vehicle Speed Sensor (VSS()), if equipped with Anti-Lock Brake (ABS)
  - Caliper and brake pads. Support the caliper out of the way using wire.

- Brake rotor from the hub assembly
- Upper ball joint from the steering knuckle and pull the knuckle outward

4. From the back of the knuckle, remove the 4 bolts securing the hub to the knuckle.

5. Remove the hub and bearing assembly from the knuckle.

➡ The hub assembly is not serviceable and should not be disassembled.

To install

6. Install or connect the following:
  - Hub to the steering knuckle and

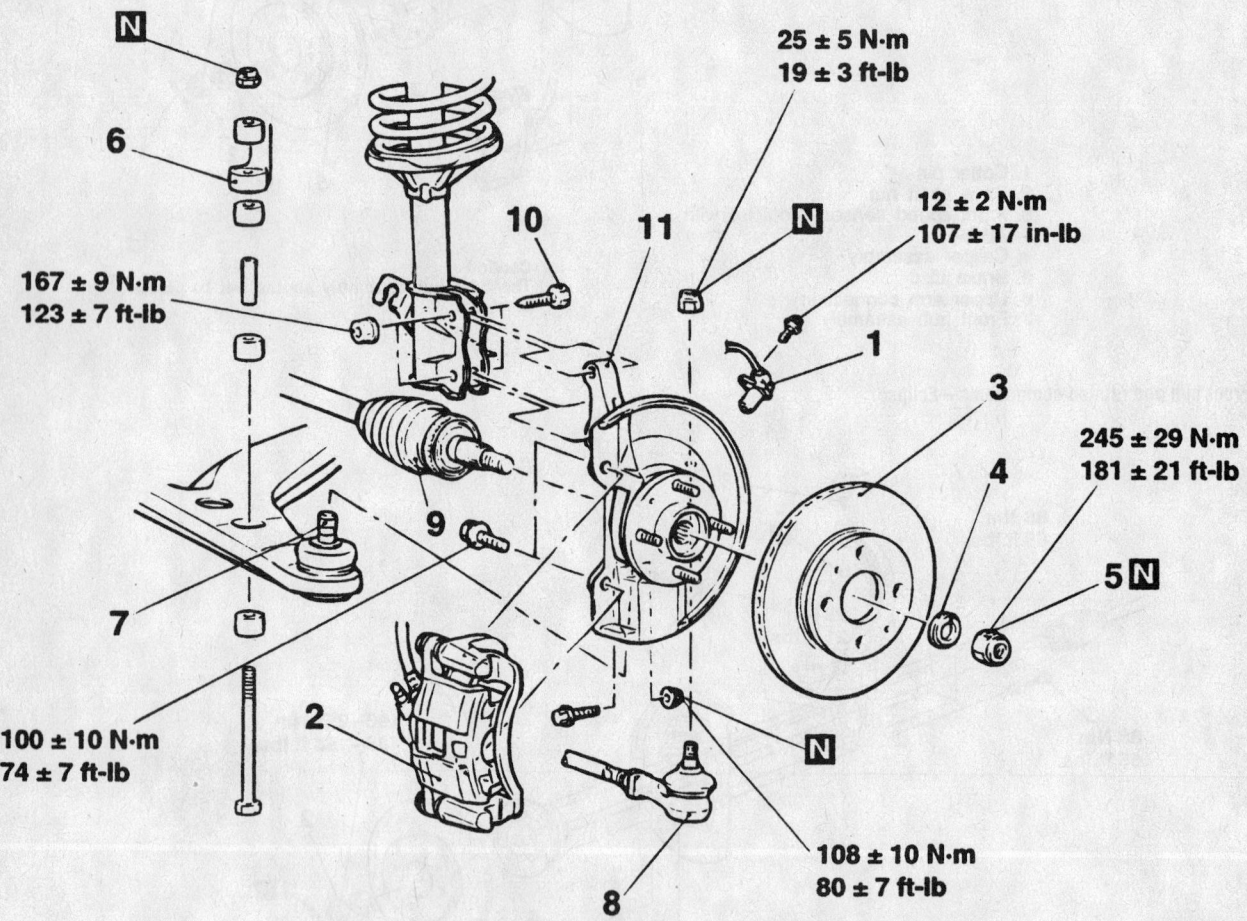

1. FRONT ABS SPEED SENSOR <VEHICLES WITH ABS>
2. CALIPER ASSEMBLY
3. BRAKE DISC
4. WASHER
5. DRIVESHAFT NUT
6. CONNECTION FOR STABILIZER BAR
7. CONNECTION FOR LOWER ARM BALL JOINT
8. CONNECTION FOR TIE ROD END
9. DRIVESHAFT
10. FRONT STRUT TO HUB AND KNUCKLE MOUNTING BOLT AND NUT
11. HUB AND KNUCKLE

93570G31

Exploded view of the front wheel bearing and related components

tighten the mounting bolts to 65 ft. lbs. (88 Nm)

- Upper ball joint to the steering knuckle and tighten the self-locking nut to 21 ft. lbs. (28 Nm)
- Axle washer and nut. Tighten the nut to 145–188 ft. lbs. (200–260 Nm).

7. Position the rotor on the hub.

8. Install the caliper holder and the brake caliper.

9. If equipped with ABS, install the VSS.

10. Install the wheel assembly and lower the vehicle.

LANCER

1. Before servicing the vehicle, refer to the precautions in the beginning of this section.

2. Raise and safely support the vehicle.

3. Remove or disconnect the following:
- Front wheel
- Wheel Speed Sensor (WSS), vehicles with Anti-Lock Brake (ABS)
- Caliper and suspend it out of the way with wire or string
- Brake rotor
- Axle nut, using special tool MB990767 to hold the hub secure while removing the nut. Discard the nut.
- Stabilizer bar from the lower control arm
- Lower ball joint from the steering knuckle
- Tie rod end from the steering knuckle. Do not remove the nut from the tie rod end. Loosen the nut and use special tool MB991113 or MB990635 to avoid damaging the threads.
- Halfshaft from the hub and knuckle using the proper puller
- Front strut-to-hub and knuckle mounting bolt and nut
- Hub and knuckle from the vehicle

➡**Do not disassemble the hub assembly. If binding or damaged, it must be replaced as a unit.**

To install:

4. Install or connect the following:
- Hub and knuckle assembly
- Front strut-to-hub bolt and nut. Torque to 123 ft. lbs. (167 Nm).
- Halfshaft
- Tie rod end. Torque the nut to 19 ft. lbs. (25 Nm).
- Lower control arm ball joint. Tighten the nut to 19 ft. lbs. (25 Nm).

- Stabilizer bar
- New axle nut and washer and tighten to 181 ft. lbs. (245 Nm)
- Brake rotor and caliper. Torque the caliper bolts to 74 ft. lbs. (100 Nm).
- Front WSS, if equipped
- Front wheel

5. Lower the vehicle

Rear

ECLIPSE

➡**The hub and bearing assembly is serviced as a unit.**

1. Before servicing the vehicle, refer to the precautions in the beginning of this section.

2. Disconnect the negative battery cable.

3. Raise and safely support the vehicle.

4. Remove or disconnect the following:
- Wheel and tire assembly
- Rear wheel speed sensor if equipped with Anti-Lock Brake (ABS)
- Brake drum. Or, if equipped with disc brakes, remove the caliper assembly and rotor. Suspend the caliper out of the way with wire.

5. On vehicles with rear disc brakes, remove the parking brake shoes.

6. On vehicles equipped with All Wheel Drive (AWD), remove the axle shaft locking nut, and using a suitable tool, separate the hub from the axle shaft.

7. Remove the hub mounting bolts from behind the backing plate and remove the hub.

➡**The rotor for the ABS must be removed and installed using a press.**

To install:

8. Press the rotor (ABS) to the hub.

9. On vehicles with AWD, engage the splines of the axle shaft with the hub

assembly and tighten the axle shaft locking nut to 145–188 ft. lbs. (196–255 Nm).

10. Install or connect the following:
- Hub and tighten the mounting bolts to 54–65 ft. lbs. (74–88 Nm)
- Parking brake shoes if equipped
- Rotor and caliper or drum
- Speed sensor if equipped
- Wheel and tire assembly

11. Lower the vehicle to the floor.

12. Connect the negative battery cable.

GALANT AND DIAMANTE

1. Before servicing the vehicle, refer to the precautions in the beginning of this section.

2. Raise the vehicle and support safely.

3. Remove the appropriate wheel assembly.

4. If equipped with Anti-Lock Brake (ABS), remove the Vehicle Speed Sensor (VSS).

5. Remove the brake drum from the hub assembly.

6. From the back of the knuckle, remove the 4 bolts securing the hub to the knuckle.

7. Remove the hub and bearing assembly from the knuckle.

➡**The hub assembly is not serviceable and should not be disassembled.**

8. If replacing the hub, use special socket MB991248 and a press, to remove the wheel sensor rotor from the hub.

To install

9. Press the wheel sensor rotor onto the hub.

10. Install or connect the following:
- Hub to the knuckle and tighten the mounting bolts to 54–65 ft. lbs. (74–88 Nm)
- Brake drum on the hub
- VSS, if equipped with ABS
- Wheel assembly and lower the vehicle

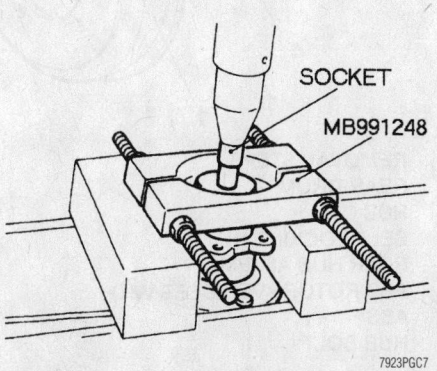

SOCKET

MB991248

7923PGC7

Use a press to remove the speed sensor rotor from the hub—Galant

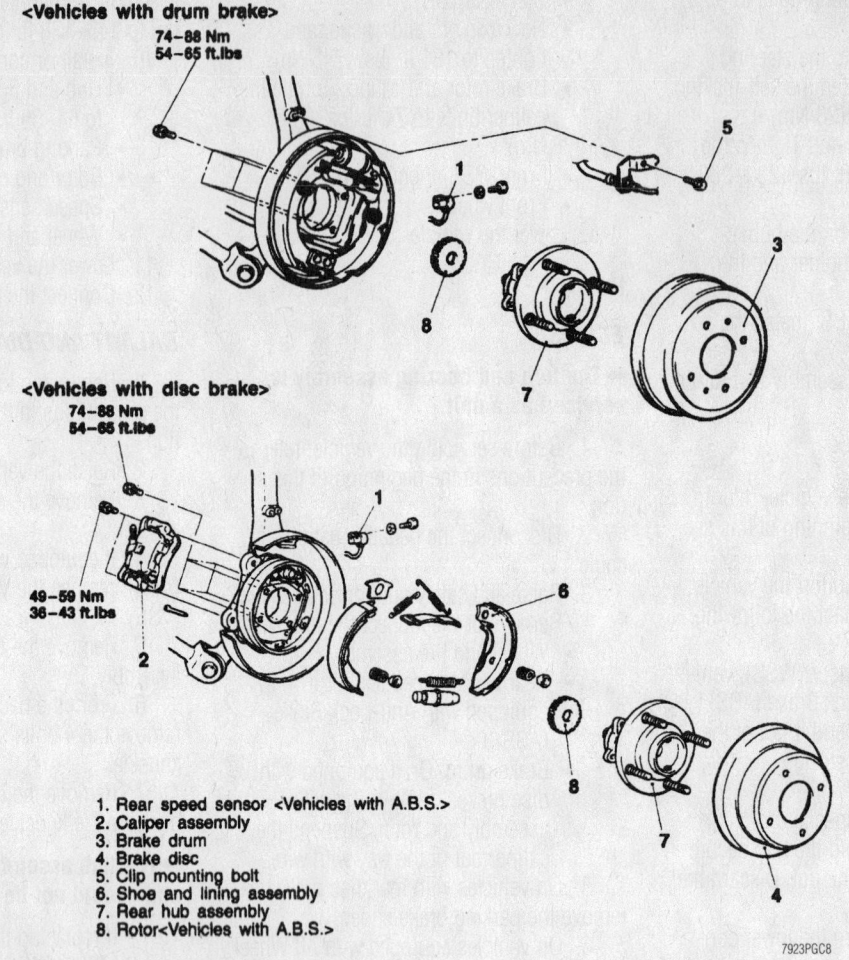

\<Vehicles with drum brake\>
74–88 Nm
54–65 ft.lbs

\<Vehicles with disc brake\>
74–88 Nm
54–65 ft.lbs

49–59 Nm
36–43 ft.lbs

1. Rear speed sensor \<Vehicles with A.B.S.\>
2. Caliper assembly
3. Brake drum
4. Brake disc
5. Clip mounting bolt
6. Shoe and lining assembly
7. Rear hub assembly
8. Rotor\<Vehicles with A.B.S.\>

7923PGC8

Exploded view of the rear hub/bearing assembly and related components—Galant

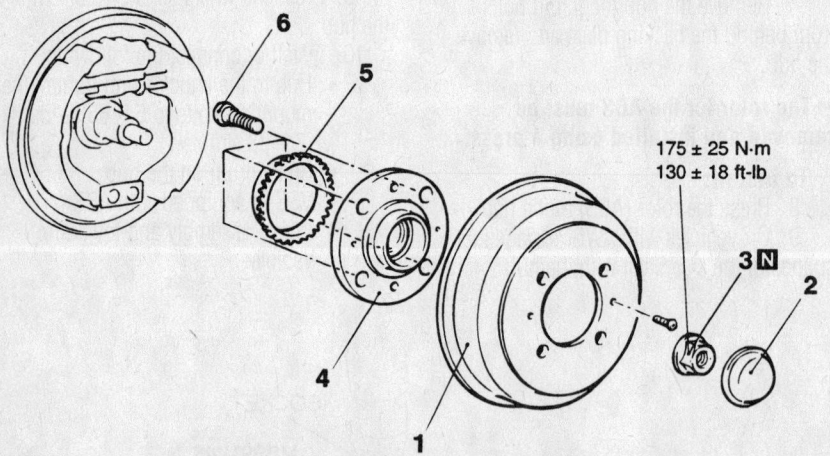

175 ± 25 N·m
130 ± 18 ft-lb

REMOVAL STEPS
1. REAR DRUM
2. HUB CAP
3. SELF-LOCKING NUT
4. REAR HUB ASSEMBLY
5. ABS ROTOR \<VEHICLES WITH ABS\>
6. HUB BOLT

9357QG32

Exploded of the rear wheel hub and bearing—Lancer shown, Mirage similar

LANCER & MIRAGE

➡The wheel bearing is serviced by replacement of the hub.

1. Before servicing the vehicle, refer to the precautions in the beginning of this section.
2. If equipped with Anti-Lock Brake (ABS), remove the wheel speed sensor.
3. Raise and safely support the vehicle.
4. Remove or disconnect the following:
   - Rear wheel
   - Caliper and brake disc or brake drum
   - Dust cap and flange nut
   - Rear hub assembly

To install:
5. Install or connect the following:
   - Rear hub assembly using a new flange nut. Torque the flange nut to 130 ft. lbs. (180 Nm).
   - Dust cap
   - Wheel speed sensor if removed. The air gap should be 0.012–0.035 in. (0.3–0.9mm).
   - Brake disc and caliper, or brake drum
   - Rear wheel assembly and lower the vehicle to the floor

BRAKES

Brake Caliper

REMOVAL & INSTALLATION

Mirage and Lancer

FRONT

1. Before servicing the vehicle, refer to the precautions in the beginning of this section.
2. Remove or disconnect the following:
   - Wheels
   - Brake hose
   - Caliper guide and lock pins
   - Caliper assembly from the caliper support

To install

3. Install or connect the following:
   - Brake caliper into position on the caliper support
   - Guide and lock pins
   - Brake hose. Bleed the brake system.
   - Wheels

Eclipse

FRONT

1. Before servicing the vehicle, refer to the precautions in the beginning of this section.

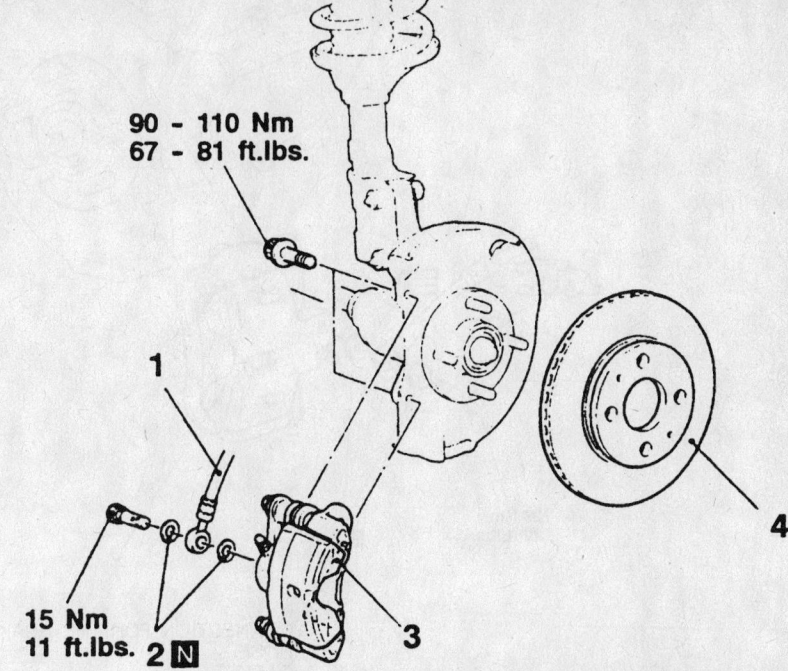

90 – 110 Nm
67 – 81 ft.lbs.

15 Nm
11 ft.lbs. 2 N

1. Brake hose connection
2. Gasket
3. Disc brake assembly
4. Brake disc

93016G36

Front disc brakes—Mirage

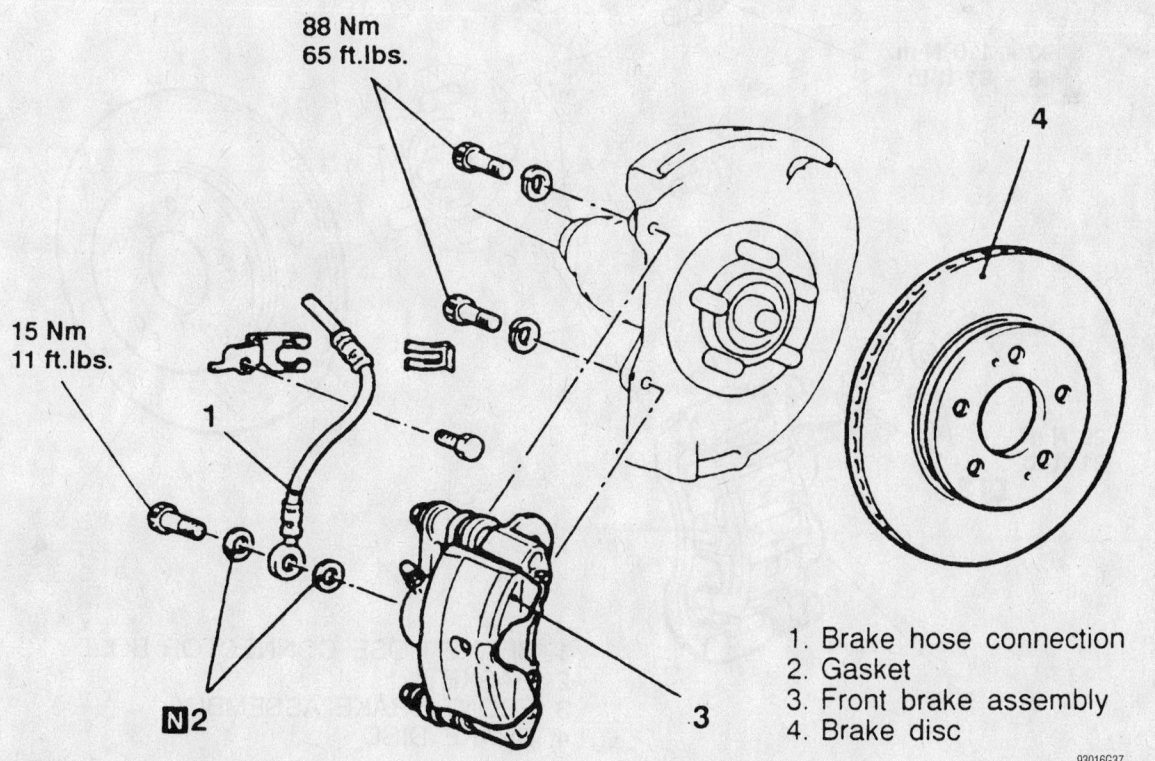

88 Nm
65 ft.lbs.

15 Nm
11 ft.lbs.

N2

1. Brake hose connection
2. Gasket
3. Front brake assembly
4. Brake disc

93016G37

Front and rear disc brakes—Eclipse

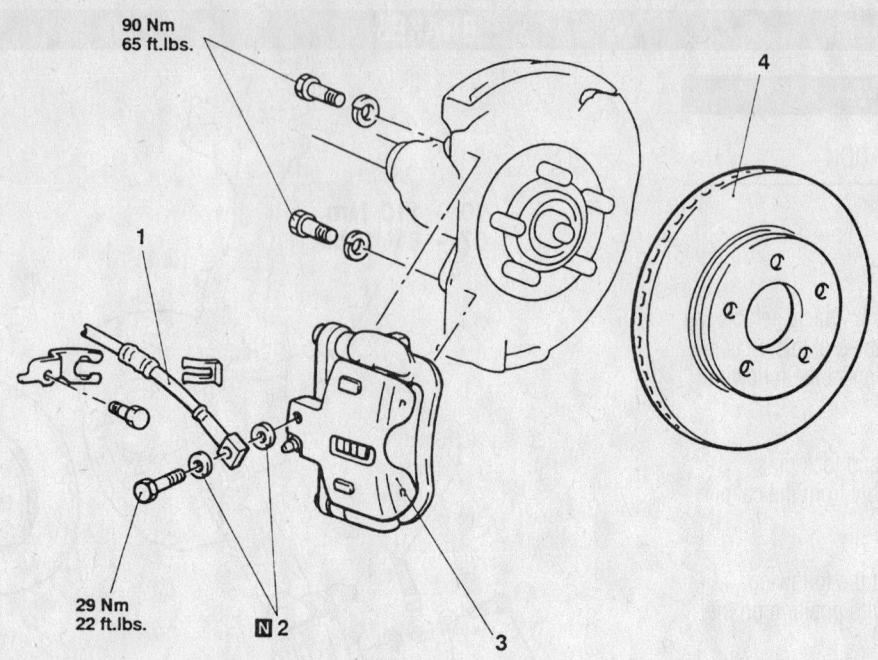

90 Nm
65 ft.lbs.

4

1

29 Nm
22 ft.lbs.

N 2

3

1. CONNECTION FOR THE BRAKE
   HOSE
2. GASKET
3. FRONT BRAKE ASSEMBLY
4. BRAKE DISC

93016G39

Front disc brakes—Diamante

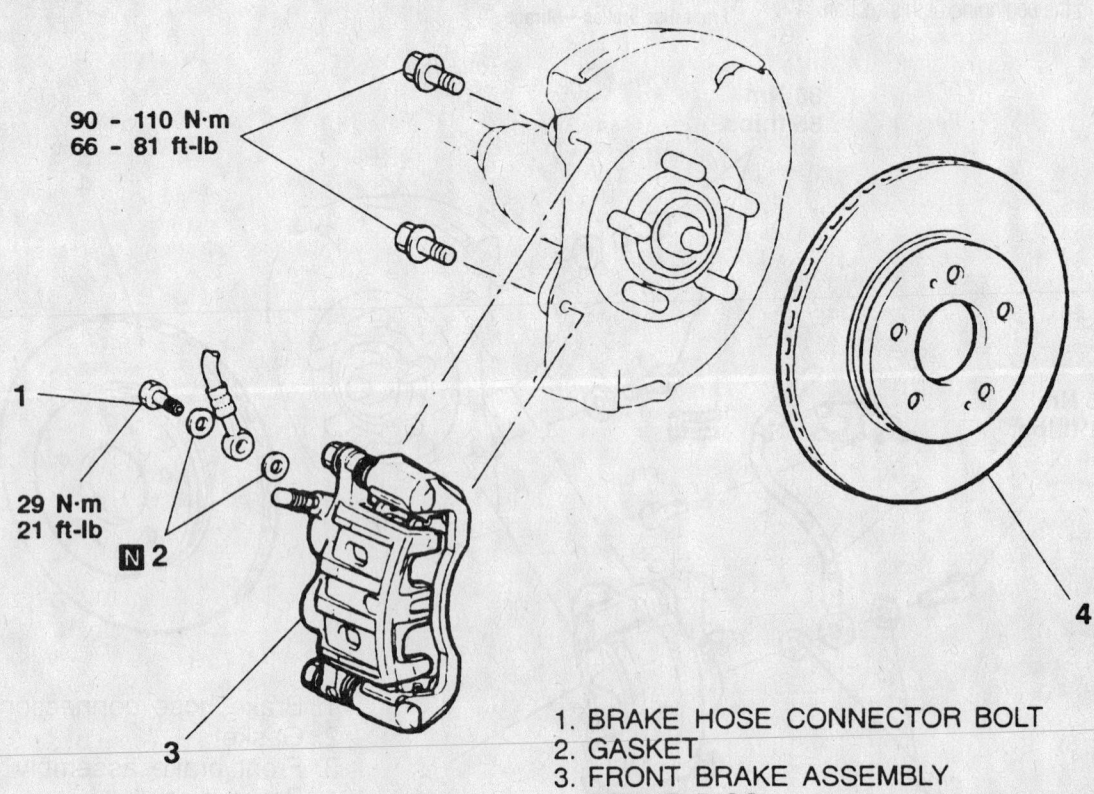

90 – 110 N·m
66 – 81 ft-lb

1

29 N·m
21 ft-lb

N 2

4

3

1. BRAKE HOSE CONNECTOR BOLT
2. GASKET
3. FRONT BRAKE ASSEMBLY
4. BRAKE DISC

93016G40

Front disc brakes—Galant

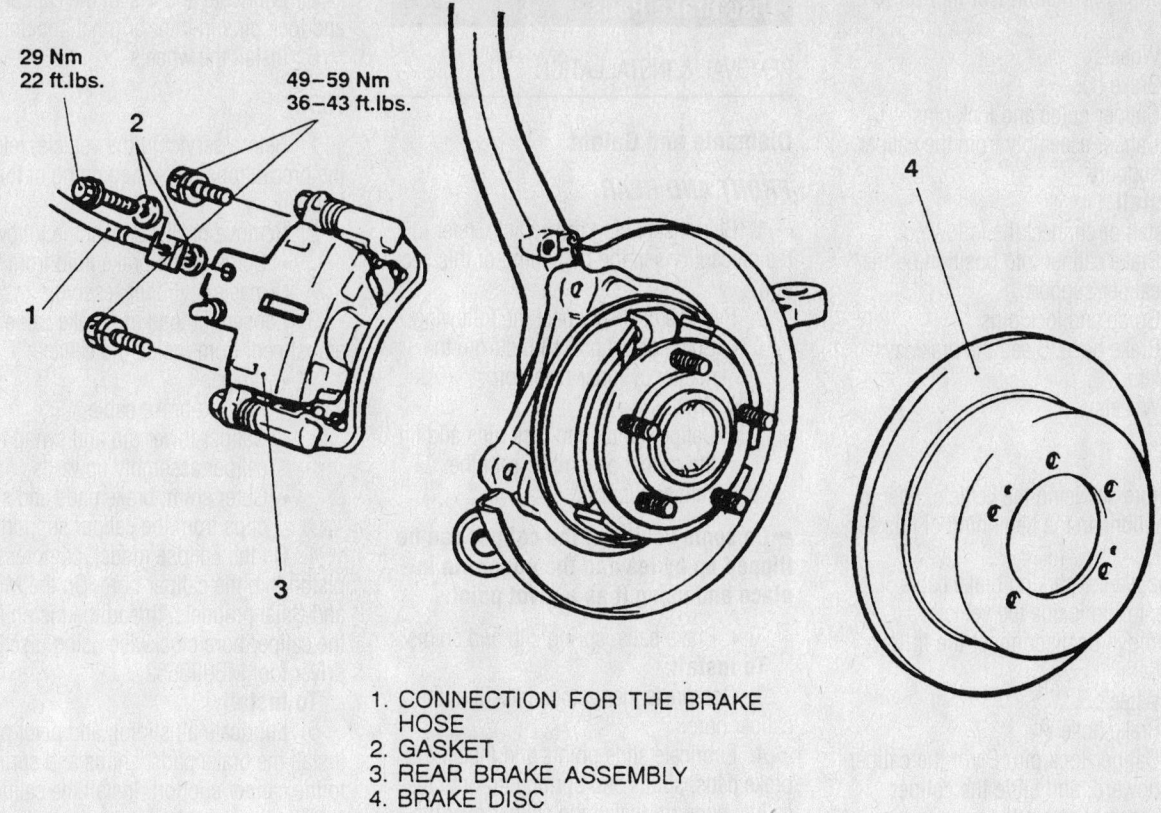

29 Nm
22 ft.lbs.

49–59 Nm
36–43 ft.lbs.

2

1

3

1. CONNECTION FOR THE BRAKE
   HOSE
2. GASKET
3. REAR BRAKE ASSEMBLY
4. BRAKE DISC

4

93016G41

Rear disc brakes—Diamante

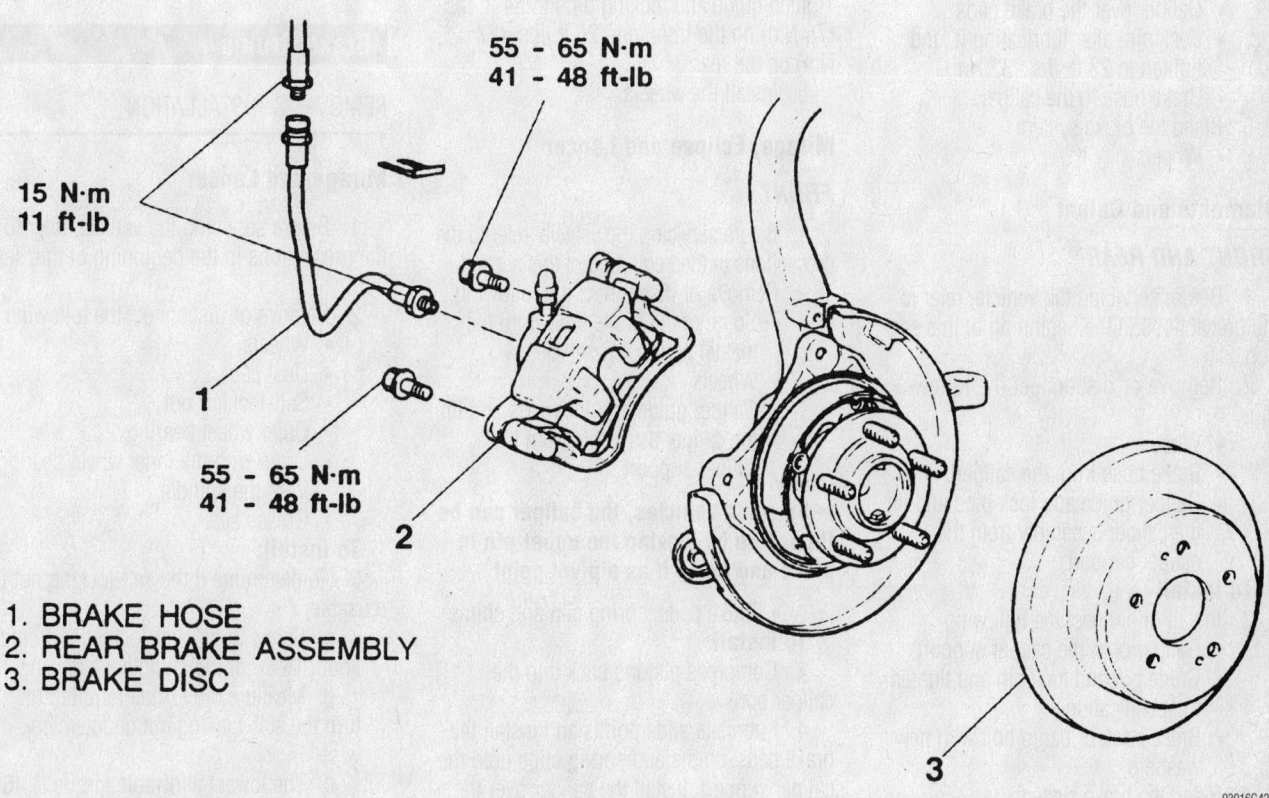

55 - 65 N·m
41 - 48 ft-lb

15 N·m
11 ft-lb

1

55 - 65 N·m
41 - 48 ft-lb

2

3

1. BRAKE HOSE
2. REAR BRAKE ASSEMBLY
3. BRAKE DISC

93016G42

Rear disc brakes—Galant

2. Remove or disconnect the following:
- Wheels
- Brake hose
- Caliper guide and lock pins
- Caliper assembly from the caliper support

To install

3. Install or connect the following:
- Brake caliper into position on the caliper support
- Guide and lock pins
- Brake hose. Bleed the brake system.
- Wheels

REAR

1. Before servicing the vehicle, refer to the precautions in the beginning of this section.
2. Loosen the parking brake cable adjustment from inside the vehicle.
3. Remove or disconnect the following:
- Wheels
- Brake hose
- Caliper lock pin. Pivot the caliper upward, and slide the caliper assembly from the caliper support.

To install:

4. Install or connect the following:
- Caliper over the brake pads
- Lock pin, after lubricating it, and tighten to 23 ft. lbs. (32 Nm)
- Brake hose to the caliper
5. Bleed the brake system.
- Wheels

Diamante and Galant
FRONT AND REAR

1. Before servicing the vehicle, refer to the precautions in the beginning of this section.
2. Remove or disconnect the following:
- Wheels
- Brake hose from the caliper
- Caliper guide and lock pins and lift the caliper assembly from the caliper support

To install

3. Install or connect the following:
- Caliper onto the caliper support
- Guide pin and lock pin and tighten to specification
- Brake hose or banjo bolt with new washers
4. Bleed the brake system.
- Wheels

Disc Brake Pads

REMOVAL & INSTALLATION

Diamante and Galant
FRONT AND REAR

1. Before servicing the vehicle, refer to the precautions in the beginning of this section.
2. Remove or disconnect the following:
- Some of the brake fluid from the master cylinder reservoir
- Wheels
- Caliper guide and lock pins and lift the caliper assembly from the caliper support.

➡ **On some vehicles, the caliper can be flipped up by leaving the upper pin in place and using it as a pivot point.**

- Brake pads, spring clip and shims

To install:

3. Compress the pistons back into the caliper bore.
4. Lubricate slide points and install the brake pads, shims and spring clip onto the caliper support. Install the caliper over the brake pads.
5. Lubricate and install the caliper guide and lock pins in their original positions. Tighten guide and locking pins to 54 ft. lbs. (75 Nm) on the front, and 20 ft. lbs. (27 Nm) on the rear.
6. Install the wheels.

Mirage, Eclipse and Lancer
FRONT

1. Before servicing the vehicle, refer to the precautions in the beginning of this section.
2. Remove or disconnect the following:
- Some of the brake fluid from the master cylinder reservoir
- Wheels
- Caliper guide and lock pins and lift the caliper assembly from the caliper support.

➡ **On some vehicles, the caliper can be flipped up by leaving the upper pin in place and using it as a pivot point.**

- Brake pads, spring clip and shims

To install:

3. Compress pistons back into the caliper bore.
4. Lubricate slide points and install the brake pads, shims and spring clips onto the caliper support. Install the caliper over the brake pads.

5. Lubricate and install the caliper guide and lock pins in their original positions.
6. Install the wheels.

REAR

1. Before servicing the vehicle, refer to the precautions in the beginning of this section.
2. Remove or disconnect the following:
- Some of the brake fluid from the master cylinder reservoir
3. Loosen the parking brake cable adjustment from inside the vehicle.
- Wheels
- Parking brake cable
- Caliper lower pin and swing the caliper assembly upwards
- Outer shim, brake pads and spring clips from the caliper support.
4. On the Eclipse model, compress the piston into the caliper bore. On the Mirage and Galant models, thread the piston into the caliper bore clockwise using disc brake driver tool MB990652.

To install:

5. Lubricate all sliding and pivot points. Install the brake pads, shims and spring clip to the caliper support. Install the caliper over the brake pads.
6. Lubricate, install and tighten the lower pin.
7. Install the wheels.

Brake Drums

REMOVAL & INSTALLATION

Mirage and Lancer

1. Before servicing the vehicle, refer to the precautions in the beginning of this section.
2. Remove or disconnect the following:
- Wheels
- Dust cap
- Self-locking nut
- Outer wheel bearing
- Drum with the inner wheel bearing from the spindle
- Grease seal

To install:

3. To determine if the self-locking nut is reusable:
   a. Screw in the self-locking nut until about 1/8 in. of the spindle is showing.
   b. Measure the torque required to turn the self-locking nut counterclockwise.
   c. The lowest allowable torque is 48 inch lbs. (5.5 Nm). If the measured

torque is less than the specification, replace the nut.

4. Install or connect the following:
   - Inner wheel bearing, after lubricating it
   - New grease seal
   - Drum to the spindle
   - Outer wheel bearing, making sure to lubricate it first
   - Self-locking nut. Torque the nut to 108–145 ft. lbs. (150–200 Nm).
   - Grease cap
   - Wheels

Galant and Eclipse

1. Before servicing the vehicle, refer to the precautions in the beginning of this section.
2. Remove or disconnect the following:
   - Rear wheels
   - Drum from the rear hub assembly

To install:

3. Install or connect the following:
   - Drum on the rear hub assembly
   - Wheels

Brake Shoes

REMOVAL & INSTALLATION

Mirage and Lancer

1. Before servicing the vehicle, refer to the precautions in the beginning of this section.
2. Remove or disconnect the following:
   - Wheels
   - Brake drum
   - Shoe-to-shoe spring
   - Shoe-to-lever spring and adjuster assembly
   - Shoe hold-down clips and the brake shoes

- Parking brake cable from the rear shoes by spreading the horseshoe clip apart.

To install

3. Lubricate the backing plate bosses, anchor pin, and parking brake actuating mechanism with lithium-based grease.
4. Install or connect the following:
   - Parking brake arm to the appropriate brake shoe
   - Brake shoes and the shoe hold-down clips
   - Adjuster assembly and the shoe-to-lever spring
   - Shoe-to-shoe spring
5. Pre-adjust the shoes so the drum slides on with a light drag and install brake drum.
   - New wheel bearing self-locking nut and torque to 130 ft. lbs. (180 Nm)
   - Wheel bearing dust cap and adjust the rear brake shoes

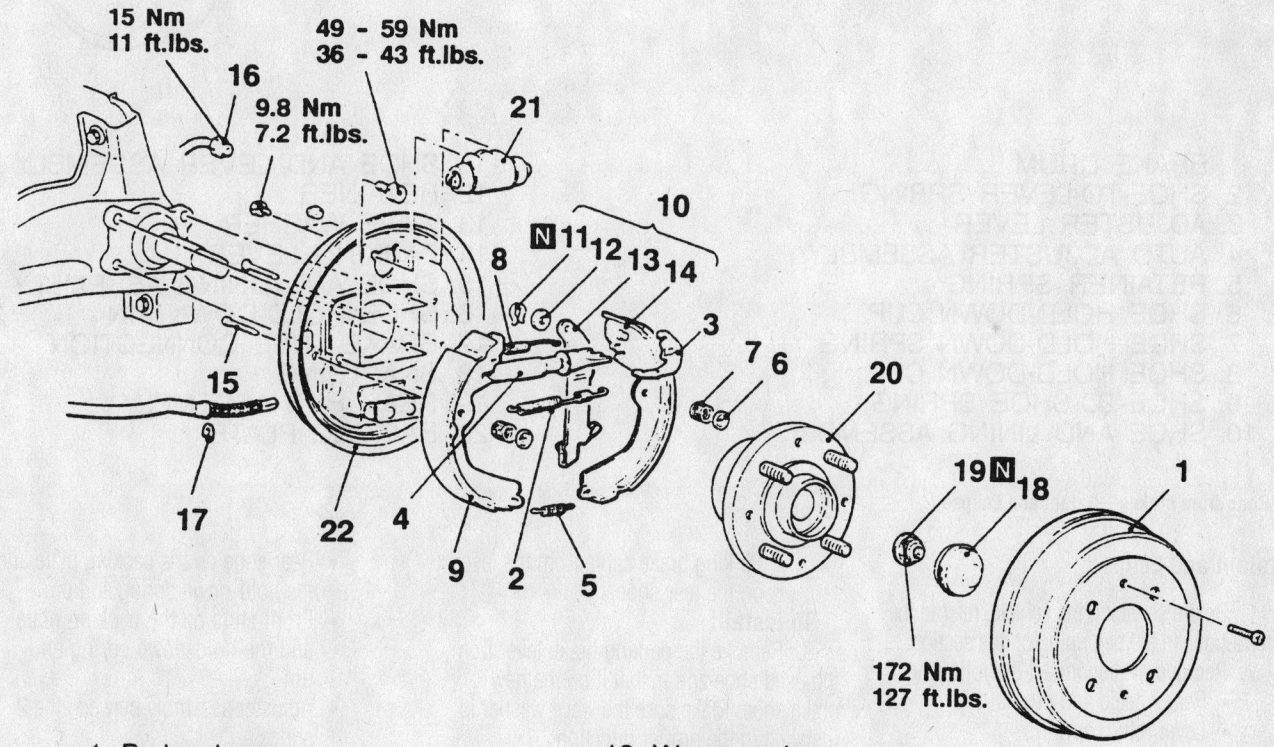

1. Brake drum
2. Shoe-to-lever spring
3. Adjuster lever
4. Auto adjuster assembly
5. Retainer spring
6. Shoe hold-down cup
7. Shoe hold-down spring
8. Shoe-to-shoe spring
9. Shoe and lining assembly
10. Shoe, lining and lever assembly
11. Retainer
12. Wave washer
13. Parking lever
14. Shoe and lining assembly
15. Shoe hold-down pin
16. Brake pipe connection
17. Snap ring
18. Hub cap
19. Flange nut
20. Rear hub assembly
21. Wheel cylinder
22. Backing plate

Rear drum brakes—Mirage

93016G46

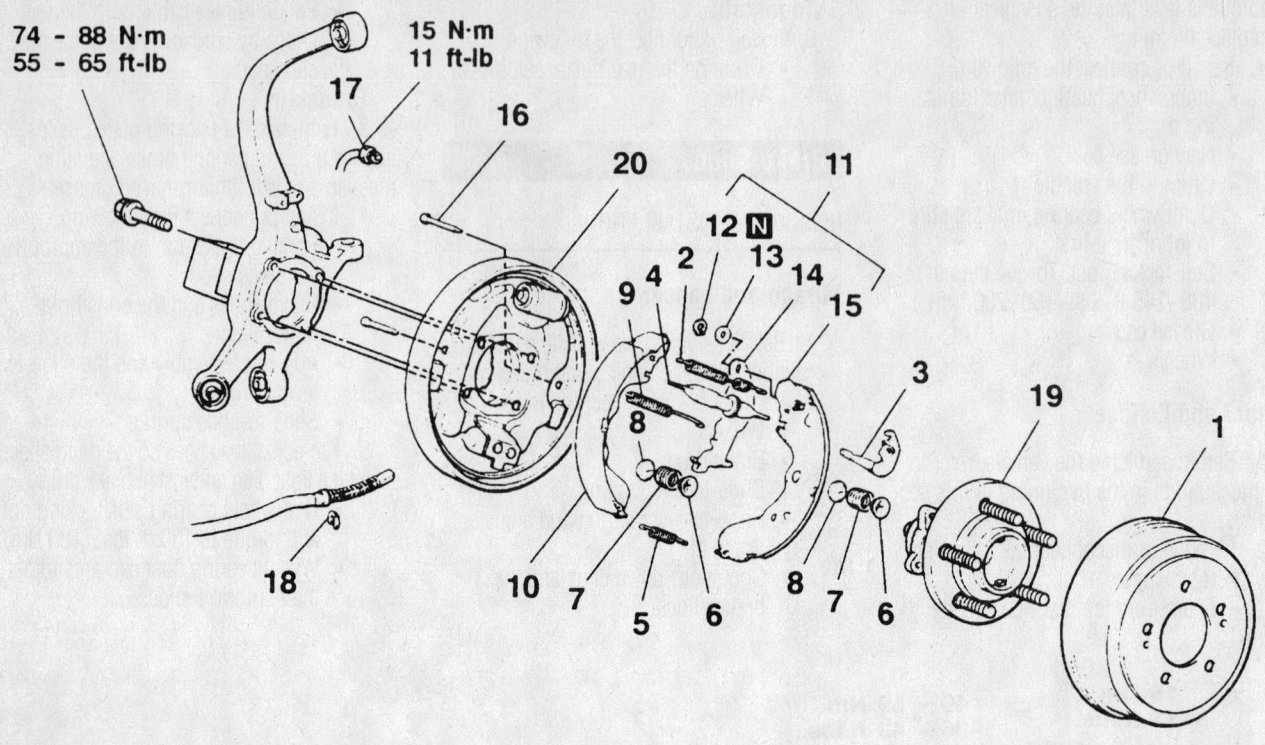

74 – 88 N·m
55 – 65 ft-lb

15 N·m
11 ft-lb

1. BRAKE DRUM
2. SHOE-TO-LEVER SPRING
3. ADJUSTER LEVER
4. AUTO ADJUSTER ASSEMBLY
5. RETAINER SPRING
6. SHOE HOLD-DOWN CUP
7. SHOE HOLD-DOWN SPRING
8. SHOE HOLD-DOWN CUP
9. SHOE-TO-SHOE SPRING
10. SHOE AND LINING ASSEMBLY
11. SHOE AND LEVER ASSEMBLY
12. RETAINER
13. WAVE WASHER
14. PARKING LEVER
15. SHOE AND LINING ASSEMBLY
16. SHOE HOLD-DOWN PIN
17. BRAKE TUBE CONNECTION
18. SNAP RING
19. REAR HUB ASSEMBLY
20. BACKING PLATE

93016G47

Rear drum brakes—Galant and Eclipse

Galant and Eclipse

1. Before servicing the vehicle, refer to the precautions in the beginning of this section.
2. Remove or disconnect the following:
- Rear wheels and drums
- Lever return spring
- Shoe-to-lever spring
- Adjuster lever
- Auto-adjuster assembly
- Retainer spring
- Brake shoe hold-down springs and spring cups
- Shoe-to-shoe spring
- Brake shoes
- Parking brake cable from the lever on the rear shoe

To install:
3. Remove the parking brake lever from the used shoe and install it on the new brake shoe. Make sure the wave washer is installed in the proper direction.
4. Clean the backing plate and lightly apply brake grease to the 6 shoe support pads.
5. Clean the adjuster assembly and apply brake grease to the threads.
6. Install or connect the following:
- Parking brake cable to the lever on the rear shoe
- Rear shoe on the backing plate and the hold-down spring and pin
- Front shoe on the backing plate and the hold-down spring and pin
- Adjuster assembly between the 2 shoes
- Shoe-to-shoe spring
- Retainer spring
- Adjuster lever
- Shoe-to-lever spring
- Lever return spring
7. Adjust the brake shoes
- Drum
- Wheels

MITSUBISHI

Montero • Montero Sport

SPECIFICATION CHARTS

ENGINE AND VEHICLE IDENTIFICATION CHART

			Engine						Model Year	
Code	Liters (cc)	Cu. In.	Cyl.	Fuel Sys.	Engine Type	Eng. Mfg.		Code	Year	
M	3.5 (3479)	213.4	6	MFI	SOHC	Mitsubishi		Y	2000	
P	3.0 (2972)	181.4	6	MFI	SOHC	Mitsubishi		1	2001	
S	3.8 (3828)	233.6	6	MFI	SOHC	Mitsubishi		2	2002	
								3	2003	
								4	2004	

MFI: Multi-port Fuel Injection

42356-MONT-C01

GENERAL ENGINE SPECIFICATIONS

Year	Engine Displacement Liters (cc)	Engine VIN	Fuel System Type	Net Horsepower @ rpm	Net Torque @ rpm (ft. lbs.)	Bore x Stroke (in.)	Compression Ratio	Oil Pressure @ rpm
2000	3.5 (3497)	M	MFI	200@5000	228@3500	3.66x3.38	9.0:1	30-80@2000
	3.0 (2972)	P	MFI	173@5500	188@4500	3.59x2.99	9.0:1	30-80@2000
2001	3.5 (3497)	M	MFI	200@5000	228@3500	3.66x3.38	9.0:1	30-80@2000
	3.0 (2972)	P	MFI	173@5500	188@4500	3.59x2.99	9.0:1	30-80@2000
2002	3.5 (3497)	M	MFI	200@5000	228@3500	3.66x3.38	9.0:1	30-80@2000
	3.0 (2972)	P	MFI	173@5500	188@4500	3.59x2.99	9.0:1	30-80@2000
2003-04	3.5 (3497)	M	MFI	200@5000	228@3500	3.66x3.38	9.0:1	30-80@2000
	3.0 (2972)	P	MFI	173@5500	188@4500	3.59x2.99	9.0:1	30-80@2000
	3.8 (3828)	S	MFI	215@5500	248@3250	3.74x3.54	10.0:1	①

① 11.6 psi @ idle

42356-MONT-C02

ENGINE TUNE-UP SPECIFICATIONS

Year	Engine Displacement Liters (cc)	Engine VIN	Spark Plugs Gap (in.)	Ignition Timing (deg.) MT	Ignition Timing (deg.) AT	Fuel Pump (psi)	Idle Speed (rpm) MT	Idle Speed (rpm) AT	Valve Clearance In.	Valve Clearance Ex.
2000	3.5 (3479)	M	0.039-0.043	—	5B	38 ①	—	700	HYD	HYD
	3.0 (2972)	P	0.039-0.043	—	5B	38 ①	—	750	HYD	HYD
2001	3.5 (3479)	M	0.039-0.043	—	5B	38 ①	—	700	HYD	HYD
	3.0 (2972)	P	0.039-0.043	—	5B	38 ①	—	750	HYD	HYD
2002	3.5 (3479)	M	0.039-0.043	—	5B	38 ①	—	700	HYD	HYD
	3.0 (2972)	P	0.039-0.043	—	5B	38 ①	—	750	HYD	HYD
2003-04	3.5 (3479)	M	0.039-0.043	—	5B	38 ①	—	700	HYD	HYD
	3.0 (2972)	P	0.039-0.043	—	5B	38 ①	—	750	HYD	HYD
	3.8 (3828)	S	0.028-0.031	—	5B	38 ①	—	750	HYD	HYD

B: Before top dead center

HYD: Hydraulic

① With vacuum hose connected

42356-MONT-C03

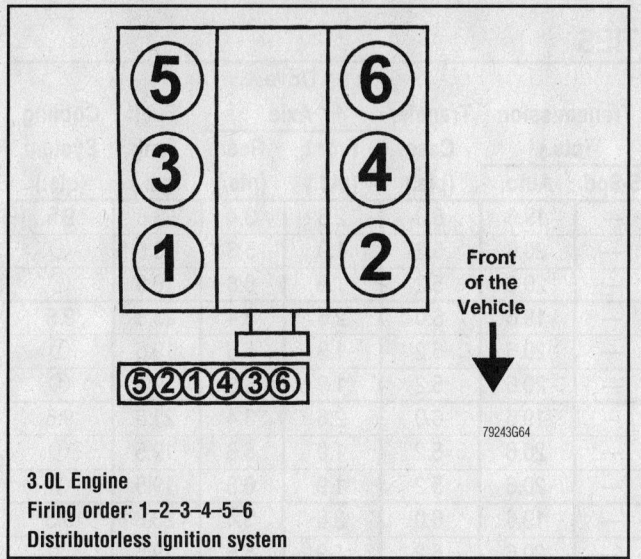

3.0L Engine
Firing order: 1–2–3–4–5–6
Distributorless ignition system

79243G64

Front
of the
Vehicle

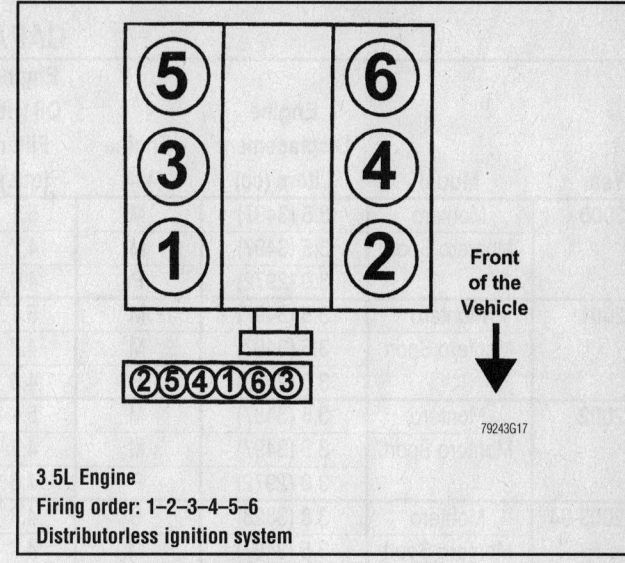

3.5L Engine
Firing order: 1–2–3–4–5–6
Distributorless ignition system

79243G17

Front
of the
Vehicle

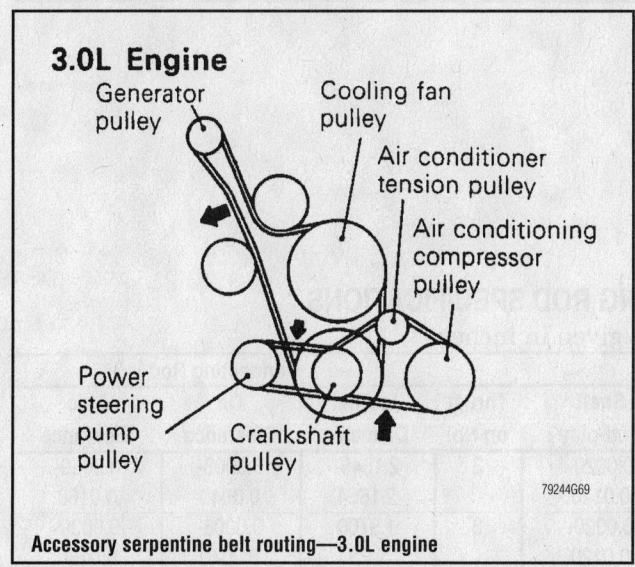

Accessory serpentine belt routing—3.0L engine

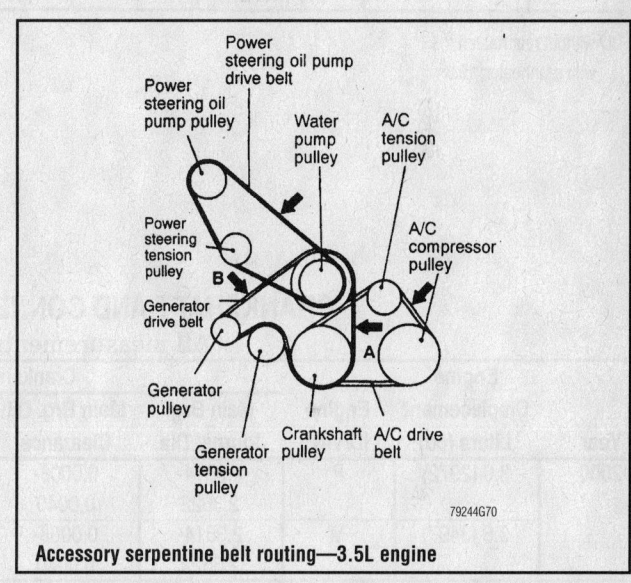

Accessory serpentine belt routing—3.5L engine

CAPACITIES

Year	Model	Engine Displacement Liters (cc)	Engine VIN	Engine Oil with Filter (qts.)	Transmission (pts.) 5-Spd	Transmission (pts.) Auto.	Transfer Case (pts.)	Drive Axle Front (pts.)	Drive Axle Rear (pts.)	Fuel Tank (gal.)	Cooling System (qts.)
2000	Montero	3.5 (3497)	M	5.1	—	19.6	6.0	2.6	3.4	23.8	9.5
	Montero Sport	3.5 (3497)	M	4.7	—	20.6	5.2	1.9	5.6	19.5	①
		3.0 (2972)	P	4.6	—	20.6	5.2	1.9	6.8	19.5	①
2001	Montero	3.5 (3497)	M	5.1	—	19.6	6.0	2.6	3.4	23.8	9.5
	Montero Sport	3.5 (3497)	M	4.7	—	20.6	5.2	1.9	5.6	19.5	①
		3.0 (2972)	P	4.6	—	20.6	5.2	1.9	6.8	19.5	①
2002	Montero	3.5 (3497)	M	5.1	—	19.6	6.0	2.6	3.4	23.8	9.5
	Montero Sport	3.5 (3497)	M	4.7	—	20.6	5.2	1.9	5.6	19.5	①
		3.0 (2972)	P	4.6	—	20.6	5.2	1.9	6.8	19.5	①
2003-04	Montero	3.8 (3828)	S	4.8	—	19.6	6.0	2.6	3.4	23.8	9.5
	Montero Sport	3.5 (3497)	M	4.7	—	20.6	5.2	1.9	5.6	19.5	①
		3.0 (2972)	P	4.6	—	20.6	5.2	1.9	6.8	19.5	①

① without rear heater: 9.5
with rear heater: 10.6

42356-MONT-C04

CRANKSHAFT AND CONNECTING ROD SPECIFICATIONS
All measurements are given in inches.

Year	Engine Displacement Liters (cc)	Engine ID/VIN	Crankshaft Main Brg. Journal Dia.	Crankshaft Main Brg. Oil Clearance	Crankshaft Shaft End-play	Crankshaft Thrust on No.	Connecting Rod Journal Diameter	Connecting Rod Oil Clearance	Connecting Rod Side Clearance
2000	3.0 (2972)	P	2.3614-2.3622	0.0008-0.0040	0.0020-0.0120	3	2.1646-2.1654	0.0008-0.0040	0.0039-0.0160
	3.5 (3497)	M	2.3614-2.3622	0.0008-0.0040	0.0020-0.0120	3	1.9700	0.0008-0.0040	0.0039-0.0160
2001	3.0 (2972)	P	2.3614-2.3622	0.0008-0.0040	0.0020-0.0120	3	2.1646-2.1654	0.0008-0.0040	0.0039-0.0160
	3.5 (3497)	M	2.3614-2.3622	0.0008-0.0040	0.0020-0.0120	3	1.9700	0.0008-0.0040	0.0039-0.0160
2002	3.0 (2972)	P	2.3614-2.3622	0.0008-0.0040	0.0020-0.0120	3	2.1646-2.1654	0.0008-0.0040	0.0039-0.0160
	3.5 (3497)	M	2.3614-2.3622	0.0008-0.0040	0.0020-0.0120	3	1.9700	0.0008-0.0040	0.0039-0.0160
2003-04	3.0 (2972)	P	2.3614-2.3622	0.0008-0.0040	0.0020-0.0120	3	2.1646-2.1654	0.0008-0.0040	0.0039-0.0160
	3.5 (3497)	M	2.3614-2.3622	0.0008-0.0040	0.0020-0.0120	3	1.9700	0.0008-0.0040	0.0039-0.0160
	3.8 (3828)	S	2.3614-2.3622	0.0008-0.0040	0.0020-0.0120	3	1.9700	0.0008-0.0040	0.0039-0.0160

42356-MONT-C05

PISTON AND RING SPECIFICATIONS
All measurements are given in inches.

Year	Engine Displacement Liters (cc)	Engine ID/VIN	Piston Clearance	Ring Gap			Ring Side Clearance		
				Top Compression	Bottom Compression	Oil Control	Top Compression	Bottom Compression	Oil Control
2000	3.0 (2972)	P	0.0008-0.0020	0.0118-0.0310	0.0177-0.0310	0.0079-0.0390	0.0012-0.0040	0.0008-0.0040	Snug
	3.5 (3497)	M	0.0008-0.0020	0.0118-0.0310	0.0177-0.0310	0.0079-0.0390	0.0012-0.0040	0.0008-0.0040	Snug
2001	3.0 (2972)	P	0.0008-0.0020	0.0118-0.0310	0.0177-0.0310	0.0079-0.0390	0.0012-0.0040	0.0008-0.0040	Snug
	3.5 (3497)	M	0.0008-0.0020	0.0118-0.0310	0.0177-0.0310	0.0079-0.0390	0.0012-0.0040	0.0008-0.0040	Snug
2002	3.0 (2972)	P	0.0008-0.0020	0.0118-0.0310	0.0177-0.0310	0.0079-0.0390	0.0012-0.0040	0.0008-0.0040	Snug
	3.5 (3497)	M	0.0008-0.0020	0.0118-0.0310	0.0177-0.0310	0.0079-0.0390	0.0012-0.0040	0.0008-0.0040	Snug
2003-04	3.0 (2972)	P	0.0008-0.0020	0.0118-0.0310	0.0177-0.0310	0.0079-0.0390	0.0012-0.0040	0.0008-0.0040	Snug
	3.5 (3497)	M	0.0008-0.0020	0.0118-0.0310	0.0177-0.0310	0.0079-0.0390	0.0012-0.0040	0.0008-0.0040	Snug
	3.8 (3828)	S	0.0008-0.0019	0.0012-0.0170	0.0180-0.0230	0.0080-0.0230	0.0012-0.0027	0.0008-0.0023	Snug

42356-MONT-C06

VALVE SPECIFICATIONS

Year	Engine Displacement Liters (cc)	Engine VIN	Seat Angle (deg.)	Face Angle (deg.)	Spring Test Pressure (lbs. @ in.)	Spring Installed Height (in.)	Stem-to-Guide Clearance (in.)		Stem Diameter (in.)	
							Intake	Exhaust	Intake	Exhaust
2000	3.5 (3497)	M	44-44.5	45-45.5	60@1.74	1.740	0.0008-0.0020	0.0016-0.0028	0.236	0.236
	3.0 (2972)	P	44-44.5	45-45.5	60@1.74	1.740	0.0008-0.0020	0.0016-0.0028	0.240	0.240
2001	3.5 (3497)	M	44-44.5	45-45.5	60@1.74	1.740	0.0008-0.0020	0.0016-0.0028	0.236	0.236
	3.0 (2972)	P	44-44.5	45-45.5	60@1.74	1.740	0.0008-0.0020	0.0016-0.0028	0.240	0.240
2002	3.5 (3497)	M	44-44.5	45-45.5	60@1.74	1.740	0.0008-0.0020	0.0016-0.0028	0.236	0.236
	3.0 (2972)	P	44-44.5	45-45.5	60@1.74	1.740	0.0008-0.0020	0.0016-0.0028	0.240	0.240
2003-04	3.5 (3497)	M	44-44.5	45-45.5	60@1.74	1.740	0.0008-0.0020	0.0016-0.0028	0.236	0.236
	3.0 (2972)	P	44-44.5	45-45.5	60@1.74	1.740	0.0008-0.0020	0.0016-0.0028	0.240	0.240
	3.8 (3828)	S	44-44.5	45-45.5	60@1.74	1.740	0.0008-0.0019	0.0016-0.0027	0.240	0.240

42356-MONT-C07

TORQUE SPECIFICATIONS
All readings in ft. lbs.

Year	Engine ID/VIN	Engine Displacement Liters (cc)	Cylinder Head Bolts	Main Bearing Bolts	Rod Bearing Bolts	Crankshaft Damper Bolts	Flywheel Bolts	Manifold Intake *	Manifold Exhaust	Spark Plugs	Lug Nut
2000	M	3.5 (3497)	80	54	38	134	54	16	22	18	100
	P	3.0 (2972)	80	69	37	134	55	16	33	18	100
2001	M	3.5 (3497)	80	54	38	134	54	16	22	18	100
	P	3.0 (2972)	80	69	37	134	55	16	33	18	100
2002	M	3.5 (3497)	80	54	38	134	54	16	22	18	100
	P	3.0 (2972)	80	69	37	134	55	16	33	18	100
2003-04	M	3.5 (3497)	80	54	38	134	54	16	22	18	100
	P	3.0 (2972)	80	69	37	134	55	16	33	18	100
	S	3.8 (3828)	80	74	①	137	54	②	33	18	100

① 20 ft. lbs., plus an additional 90 degrees

② 1st step (right bank nuts): 58 inch lbs.
   2nd step (left bank nuts): 16 ft. lbs.
   3rd step (right bank nuts): 16 ft. lbs.
   4th step (left bank nuts): 16 ft. lbs.
   5th step (right bank nuts): 16 ft. lbs.

42356-MONT-C08

BRAKE SPECIFICATIONS
All measurements in inches unless noted

Year	Model		Brake Disc Original Thickness	Brake Disc Minimum Thickness	Brake Disc Maximum Runout	Brake Drum Diameter Original Inside Diameter	Brake Drum Diameter Max. Wear Limit	Brake Drum Diameter Maximum Machine Diameter	Minimum Lining Thickness Front	Minimum Lining Thickness Rear	Brake Caliper Bracket Bolts (ft. lbs.)	Brake Caliper Mounting Bolts (ft. lbs.)
2000	Montero	F	1.060	1.000	0.002	—	—	—	0.079	—	65	54
		R	0.710	0.646	0.003	—	—	—	—	0.040	65	32
	Montero Sport	F	0.940	0.880	0.002	—	—	—	0.079	—	65	55
		R	0.700	0.650	0.003	10.63	—	10.71	—	①	94	32
2001	Montero	F	1.023	0.960	0.002	—	—	—	0.079	—	65	54
		R	0.866	0.803	0.003	—	—	—	—	0.079	65	32
	Montero Sport	F	0.940	0.880	0.001	—	—	—	0.079	—	65	55
		R	0.710	0.650	0.003	10.63	—	10.71	—	①	94	32
2002	Montero	F	1.023	0.960	0.002	—	—	—	0.079	—	65	54
		R	0.866	0.803	0.003	—	—	—	—	0.079	65	32
	Montero Sport	F	0.940	0.880	0.001	—	—	—	0.079	—	65	55
		R	0.710	0.650	0.003	10.63	—	10.71	—	①	94	32
2003-04	Montero	F	1.023	0.960	0.002	—	—	—	0.080	—	83	66
		R	0.870	0.800	0.002	—	—	—	—	0.080	74	33
	Montero Sport	F	0.940	0.880	0.001	—	—	—	0.079	—	65	55
		R	0.710	0.650	0.003	10.63	—	10.71	—	①	94	32

① disc pad: 0.79
   brake shoe: 0.04

42356-MONT-C09

WHEEL ALIGNMENT

Year	Model		Caster Range (+/-Deg.)	Caster Preferred Setting (Deg.)	Camber Range (+/-Deg.)	Camber Preferred Setting (Deg.)	Toe-in (in.)	Axis Inclination (Deg.)
2000	Montero	F	1.00	+3.00	0.50	+0.66	0.14+/-0.14	—
		R	—	—	1.00	0	0	—
	Montero Sport	F	1.00	+2.66	0.50	+4.00	0.14+/-0.14	—
		R	—	—	—	0	0	—
2001	Montero	F	0.50	+3.50	0.50	0	0.10+/-0.10	—
		R	—	—	0.50	0	0.12+/-0.12	—
	Montero Sport	F	1.00	+2.66	0.50	①	0.14+/-0.14	—
		R	—	—	—	0	0	—
2002	Montero	F	0.50	+3.50	0.50	0	0.10+/-0.10	—
		R	—	—	0.50	0	0.12+/-0.12	—
	Montero Sport	F	1.00	+2.66	0.50	①	0.14+/-0.14	—
		R	—	—	—	0	0	—
2003-04	Montero	F	0.30	+3.50	0.30	0	0.10+/-0.10	—
		R	—	—	0.30	0	0.12+/-0.12	—
	Montero Sport	F	1.00	+2.66	0.50	①	0.14+/-0.14	—
		R	—	—	—	0	0	—

① Right: 0.42
   Left: 0.92

42356-MONT-C10

TIRE, WHEEL AND BALL JOINT SPECIFICATIONS

Year	Model	OEM Tires Standard	OEM Tires Optional	Tire Pressures (psi) Front	Tire Pressures (psi) Rear	Wheel Size	Ball Joint Inspection
2000	Montero	P265/70HR15	None	26	26	7-JJ	U: 7-30 in. ① L: 0.010 in.
	Montero Sport	P235/75R15	P265/70R15	26	26	Std: 6-JJ Opt: 7-JJ	U: 7-30 in. ① L: 0.010 in.
2001	Montero	P265/70HR16	None	29	29	7-JJ	U: 7-30 in. ① L: 0.010 in.
	Montero Sport	P235/75R15	P255/70R16	26	26	Std: 6-JJ Opt: 7-JJ	U: 7-30 in. ① L: 0.010 in.
2002	Montero	P265/70HR16	None	29	29	7-JJ	U: 7-30 in. ① L: 0.010 in.
	Montero Sport	P235/75R15	P255/70R16	26	26	Std: 6-JJ Opt: 7-JJ	U: 7-30 in. ① L: 0.010 in.
2003-04	Montero	P265/70R16	None	29	29	7-JJ	U: 7-30 in. ① L: 0.010 in.
	Montero Sport	P235/75R15	P255/70R16	26	26	Std: 6-JJ Opt: 7-JJ	U: 7-30 in. ① L: 0.010 in.

OEM: Original Equipment Manufacturer

PSI: Pounds Per Square Inch

STD: Standard

OPT: Optional

① Torque required in inch lbs. to rotate ball joint when removed from the knuckle

42356-MONT-C11

SCHEDULED MAINTENANCE INTERVALS
Mitsubishi—Montero, Montero Sport

TO BE SERVICED	TYPE OF SERVICE	VEHICLE MILEAGE INTERVAL (x1000)												
		7.5	15	22.5	30	37.5	45	52.5	60	67.5	75	82.5	90	97.5
Engine oil & filter	R	✓	✓	✓	✓	✓	✓	✓	✓	✓	✓	✓	✓	✓
Automatic transmission & transfer oil	S/I		✓		✓		✓		✓		✓		✓	
Brake hoses	S/I		✓		✓		✓		✓		✓		✓	
Disc brake pads & rotors	S/I		✓		✓		✓		✓		✓		✓	
Drive shaft boots	S/I		✓		✓		✓		✓		✓		✓	
Air cleaner filter	R				✓				✓				✓	
Automatic transmission & transfer oil (4WD)	R				✓				✓				✓	
Engine coolant	R				✓				✓				✓	
Ball joints & steering linkage seals	S/I				✓				✓				✓	
Drive belt(s)	S/I				✓				✓				✓	
Drum brake linings & wheel cylinders	S/I				✓				✓				✓	
Exhaust system	S/I				✓				✓				✓	
Front & rear axle	S/I				✓				✓				✓	
Fuel hoses	S/I				✓				✓				✓	
Manual transmission & transfer oil (4WD)	S/I				✓				✓				✓	
Propeller shaft joint	S/I				✓				✓				✓	
Ignition cables	R								✓					
Timing belt	R								✓					
Distributor cap & rotor	S/I								✓					
EVAP system (except EVAP canister)	S/I								✓					
EGR valve ①	S/I													
EVAP canister ①	S/I													
PCV system ②	S/I													
Spark plugs ③	R													

R: Replace   S/I: Service or Inspect

① Replace at 100,000 miles.

② PCV system (except EVAP canister): service or inspect at 100,000 miles.

③ Iron tips: 30,000 miles
   Platinum tips: 60,000 miles
   Iridium tips: 100,000 miles

FREQUENT OPERATION MAINTENANCE (SEVERE SERVICE)

If a vehicle is operated under any of the following conditions it is considered severe service:

- Extremely dusty areas.

- 50% or more of the vehicle operation is in 32°C (90°F) or higher temperatures, or constant operation in temperatures below 0°C (32°F).

- Prolonged idling (vehicle operation in stop and go traffic).

- Frequent short running periods (engine does not warm to normal operating temperatures).

- Police, taxi, delivery usage or trailer towing usage.

Oil & oil filter: replace every 3000 miles.

Front disc brake pads (dusty or salty conditions): service or inspect every 6000 miles.

Front disc brake pads: service or inspect every 7500 miles.

Air cleaner filter: service or inspect every 15,000 miles.

Rear drum brake linings & rear wheel cylinders: service or inspect every 15,000 miles.

Spark plugs (iron tip): replace every 15,000 miles.

PCV system: service or inspect every 60,000 miles.

PRECAUTIONS

Before servicing any vehicle, please be sure to read all of the following precautions, which deal with personal safety, prevention of component damage, and important points to take into consideration when servicing a motor vehicle:

• Never open, service or drain the radiator or cooling system when the engine is hot; serious burns can occur from the steam and hot coolant.

• Observe all applicable safety precautions when working around fuel. Whenever servicing the fuel system, always work in a well-ventilated area. Do not allow fuel spray or vapors to come in contact with a spark, open flame, or excessive heat (a hot drop light, for example). Keep a dry chemical fire extinguisher near the work area. Always keep fuel in a container specifically designed for fuel storage; also, always properly seal fuel containers to avoid the possibility of fire or explosion. Refer to the additional fuel system precautions later in this section.

• Fuel injection systems often remain pressurized, even after the engine has been turned **OFF**. The fuel system pressure must be relieved before disconnecting any fuel lines. Failure to do so may result in fire and/or personal injury.

• Brake fluid often contains polyglycol ethers and polyglycols. Avoid contact with the eyes and wash your hands thoroughly after handling brake fluid. If you do get brake fluid in your eyes, flush your eyes with clean, running water for 15 minutes. If eye irritation persists, or if you have taken brake fluid internally, IMMEDIATELY seek medical assistance.

• The EPA warns that prolonged contact with used engine oil may cause a number of skin disorders, including cancer! You should make every effort to minimize your exposure to used engine oil. Protective gloves should be worn when changing oil. Wash your hands and any other exposed skin areas as soon as possible after exposure to used engine oil. Soap and water, or waterless hand cleaner should be used.

• All new vehicles are now equipped with an air bag system. The system must be disabled before performing service on or around system components, steering column, instrument panel components, wiring and sensors. Failure to follow safety and disabling procedures could result in accidental air bag deployment, possible personal injury and unnecessary system repairs.

• Always wear safety goggles when working with, or around, the air bag system. When carrying a non-deployed air bag, be sure the bag and trim cover are pointed away from your body. When placing a non-deployed air bag on a work surface, always face the bag and trim cover upward, away from the surface. This will reduce the motion of the module if it is accidentally deployed. Refer to the additional air bag system precautions later in this section.

• Clean, high quality brake fluid from a sealed container is essential to the safe and proper operation of the brake system. You should always buy the correct type of brake fluid for your vehicle. If the brake fluid becomes contaminated, completely flush the system with new fluid. Never reuse any brake fluid. Any brake fluid that is removed from the system should be discarded. Also, do not allow any brake fluid to come in contact with a painted surface; it will damage the paint.

• Never operate the engine without the proper amount and type of engine oil; doing so WILL result in severe engine damage.

• Timing belt maintenance is extremely important! Many models utilize an interference-type, non-freewheeling engine. If the timing belt breaks, the valves in the cylinder head may strike the pistons, causing potentially serious (also time-consuming and expensive) engine damage. Refer to the maintenance interval charts in the front of this section for the recommended replacement interval for the timing belt, and to the timing belt procedure for belt replacement and inspection.

• Disconnecting the negative battery cable on some vehicles may interfere with the functions of the on-board computer system(s) and may require the computer to undergo a relearning process once the negative battery cable is reconnected.

• When servicing drum brakes, only disassemble and assemble one side at a time, leaving the remaining side intact for reference.

ENGINE REPAIR

➡**Disconnecting the negative battery cable on some vehicles may interfere with the functions of the on board computer systems and may require the computer to undergo a relearning process, once the negative battery cable is reconnected.**

Distributor

All of the engines covered in this section are distributorless.

Alternator

REMOVAL

1. Before servicing the vehicle, refer to the precautions in the beginning of this section.
2. Remove or disconnect the following:

• Negative battery cable
• Under cover
• Air cleaner assembly, ducts and air intake hose
• Drive belt(s)
• Wires
• Mounting bracket, if equipped
• Alternator

To install:
3. Install or connect the following:

• Alternator. Torque the through-bolt to 38 ft. lbs. (52 Nm) and the mounting bolt to 16 ft. lbs. (22 Nm).
• Mounting bracket, if equipped. Torque the bolt to 14–18 ft. lbs. (20–25 Nm).
• Wires. Torque the nut to 124 inch lbs. (14 Nm).
• Drive belt(s)
• Air cleaner assembly, ducts and air intake hose

• Under cover
• Negative battery cable

Ignition Timing

ADJUSTMENT

The ignition timing is controlled by the ECM and is not adjustable. The ECM determines the timing based on input from the crankshaft position sensor.

TIMING CHECK

1. Before servicing the vehicle, refer to the precautions in the beginning of this section.

Before attempting to adjust the ignition timing, be sure of the following:

• The engine should be at normal operating temperature.

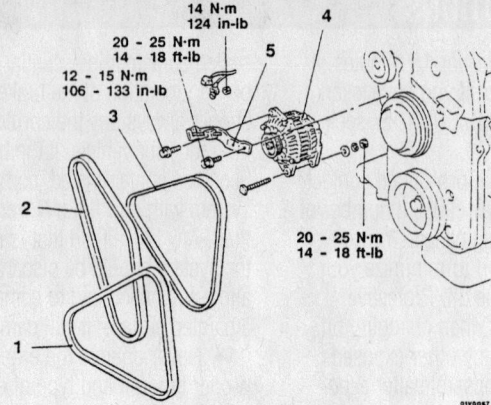

<2.4L ENGINE>

14 N·m
124 in-lb

20 - 25 N·m
14 - 18 ft-lb

12 - 15 N·m
106 - 133 in-lb

20 - 25 N·m
14 - 18 ft-lb

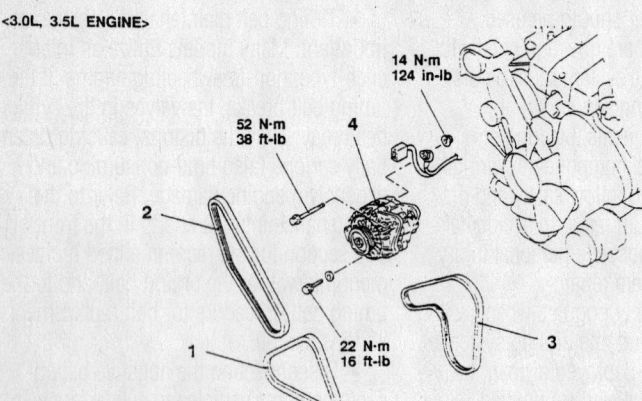

<3.0L, 3.5L ENGINE>

14 N·m
124 in-lb

52 N·m
38 ft-lb

22 N·m
16 ft-lb

1. DRIVE BELT (FOR A/C)
2. DRIVE BELT
   (FOR POWER STEERING)
3. DRIVE BELT (FOR GENERATOR)

4. GENERATOR
5. GENERATOR BRACE ASSEMBLY
   <2.4L ENGINE>

Alternator mounting and related components

• The lights and all accessories should be OFF.

• If equipped with an automatic transmission, the transmission should be in **P** or **N**.

• Connect scan tool MB991502 to the data link connector

• Set up the timing light.

• Start the engine and run at idle.

• Verify that the idle speed is 600–800 rpm.

• Select scan tool MB991502 actuator test "item number 17".

• Check that basic timing is with in standard, it should be 3–7° BTDC.

• Press the clear key on the scan tool, select forced drive stop mode and cancel the actuator test.

❄❄ CAUTION

If the actuator test is not canceled, the forced drive will continue for 27 minutes. Driving in this state could lead to engine failure.

2. If the base timing is out of specification:

3. Check to see if the distributor is aligned properly

4. Check to see if the timing belt cover and Crankshaft Position (CKP) sensor installation is conditions.

5. Crankshaft sensing blade conditions.

Engine Assembly

REMOVAL & INSTALLATION

1. Before servicing the vehicle, refer to the precautions in the beginning of this section.

2. Relieve the fuel system pressure.

❄❄ CAUTION

The fuel injection system remains under pressure after the engine has beenOFF. Properly relieve fuel pressure before disconnecting any fuel

lines. Failure to do so may result in fire or personal injury.

3. Drain the engine oil.

4. Drain the cooling system.

5. Remove or disconnect the following:

• Battery
• Hood, matchmark for reassembly
• Oil dipstick
• Engine undercover
• Starter
• Exhaust pipe from the exhaust manifolds
• Transfer case, if equipped with 4WD
• Transmission, if equipped with a manual transmission

6. If equipped with an automatic transmission and 2WD:

a. Remove the inspection plate.

b. Matchmark the flexplate to the converter; remove the torque converter bolts and move the torque converter back as far as it will go.

c. Remove the lower bell housing bolts.

7. Remove or disconnect the following:

• Air cleaner assembly, ducts and air intake hose
• Linkages and cables from the throttle body
• Fuel lines and plug the lines
• Air conditioning compressor, if equipped and position it aside. It is not necessary to remove the lines from the compressor.
• Radiator, shroud
• Cooling fan
• Heater hoses
• Accessory belts
• Power steering pump and wires from its brackets and position it to the side. Do not remove the hoses from the pump.
• Alternator and wires
• Ignition coil and power transistor assembly, if equipped
• MDP sensor connector
• EGR connector
• TP sensor connector
• IAC motor connector
• Magnetic clutch and refrigerant temperature switch connector
• EVAP Purge Solenoid
• ECT sensor and gauge connectors
• Front and injector wiring harness
• CMP sensor
• CKP sensor
• Distributor signal Generator
• Compactor connector
• Left and right heated O2sensor
• Oil pressure switch connector

- Vacuum hoses

8. Attach an engine removal device to the engine support eyes on the engine.

9. If equipped with an automatic transmission, support the transmission with a floor jack. Remove the remaining bell housing bolts.

10. Remove the engine mount nuts and remove the engine from the vehicle.

To install:

11. Lower the engine into position and install the engine mount nuts and bolts. Tighten the nuts to 18–20 ft. lbs. (25–27 Nm), and the bolts to tighten to 33 ft. lbs. (44 Nm).

12. Install or connect the following:
- Bell housing bolts

13. Remove the engine removal device and the transmission support.

14. Install or connect the following:
- Transfer case, if equipped
- Manual transmission, if equipped
- Automatic transmission, if equipped align the torque converter and flexplate and the bolts.
- Inspection plate
- Starter motor
- Exhaust pipe to the exhaust manifolds using new gaskets
- Lower radiator hose
- Heater hoses
- Alternator and wires
- Power steering pump and brackets
- Air conditioning compressor
- Linkages and cables to the carburetor or throttle body
- Ignition coil and power transistor assembly, if equipped
- MDP sensor connector
- EGR connector
- TP sensor connector
- IAC motor connector
- Magnetic clutch and refrigerant temperature switch connector
- EVAP Purge Solenoid
- ECT sensor and gauge connectors
- Front and injector wiring harness
- CMP sensor
- CKP sensor
- Distributor signal Generator
- Compactor connector
- Left and right heated O$_2$ sensor
- Oil pressure switch connector
- Vacuum hoses
- Air cleaner assembly, ducts and air intake hose
- Accessory belts
- Radiator, shroud and upper hose
- Cooling fan
- Battery
- Oil dipstick
- Hood

15. Refill the engine with the specified amount of oil.

16. Refill the radiator with coolant.

17. Check fuel system for leaks.

18. Check the automatic transmission fluid level, if equipped.

19. Recheck all engine adjustments.

Water Pump

REMOVAL & INSTALLATION

3.0L and 3.5L Engines

1. Before servicing the vehicle, refer to the precautions in the beginning of this section.

2. If necessary, properly release the fuel pressure.

3. Drain the cooling system.

4. Remove or disconnect the following:
- Negative battery cable

✳✳ CAUTION

Wait at least 90 seconds after the negative battery cable is disconnected to prevent possible deployment of the air bag.

- Upper radiator shroud
- Accessory belts
- Air conditioning compressor tensioner pulley, if equipped
- Cooling fan and clutch assembly and the water pump pulley
- Thermostat and housing on 3.0L, 3.5L engines
- Water outlet, gasket and houses
- Radiator hoses from the water pump
- Crankshaft pulley(s)
- Timing belt covers. If the same timing belt will be reused, mark the direction of the timing belt's rotation, for installation in the same direction. Be sure the engine is positioned so the No. 1 cylinder is at the TDC of its compression stroke and the sprockets timing marks are aligned with the engine's timing mark indicators.
- Timing belt
- Water pump bolts are different lengths, note their positions before removing.
- Water pump from the block
- Water pipe connection and O-ring

To install:

5. Clean and dry the mating surfaces of the block and water pump

6. Install or connect the following:

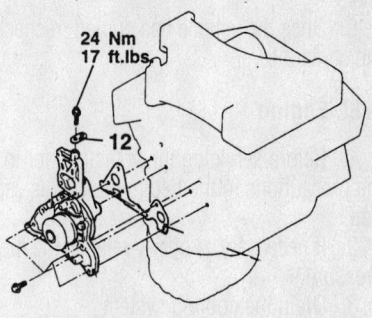

Water pump mounting—3.0L engine

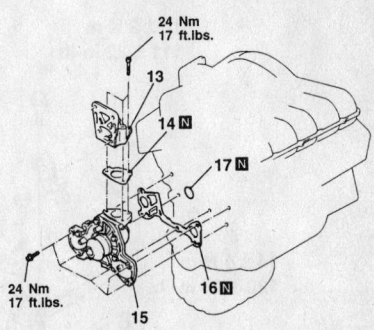

Water pump and related components—3.5L engine

- New O-ring on the water pipe connection, wet the new O-ring with water to aid in installation
- Water pump, with a new gasket, Torque the bolts to 17 ft. lbs. (23 Nm) on 3.0L and 3.5L engines
- Alternator bracket bolt to 17 ft. lbs. (23 Nm)
- Timing belt(s) and covers
- Crankshaft pulley(s)
- Thermostat and housing on 3.0L, 3.5L engines. Torque the bolts to 12–14 ft. lbs. (17–20 Nm).
- Radiator hose to the water pump
- Water outlet, new gasket and houses. Torque the bolts to 12–14 ft. lbs. (17–20 Nm).
- Water pump pulley
- Cooling fan and clutch assembly
- Air conditioning compressor tensioner pulley, if equipped
- Accessory belts
- Upper radiator shroud
- Thermostat and housing on 3.0L, 3.5L engines
- Negative battery cable

7. Refill the radiator with coolant. This cooling system has a self-bleeding thermostat, so system bleeding is not required.

8. Run the vehicle until the thermostat

opens and fill the overflow tank. Check for leaks.

9. Once the vehicle has cooled, recheck the coolant level.

3.8L Engine

1. Before servicing the vehicle, refer to the precautions in the beginning of this section.

2. If necessary, properly release the fuel pressure.

3. Drain the cooling system.

4. Remove or disconnect the following:
- Negative battery cable

✳✳ CAUTION

Wait at least 90 seconds after the negative battery cable is disconnected to prevent possible deployment of the air bag.

- Timing belt
- Camshaft sprocket
- Engine Coolant Temperature (ECT) sensor connector
- ECT gauge unit connector
- Spark plug cable support
- Upper radiator hose

- Water hoses
- Water outlet fitting bracket
- Water outlet fitting, O-ring and gasket
- Water pump assembly, gasket and O-ring
- Fitting, gaskets and thermostat case

To install:

5. Clean and dry the mating surfaces of the block and water pump

6. Install or connect the following:
- Thermostat case with new gaskets
- Fitting
- Water pump assembly, using a new gasket and O-ring. Torque the water

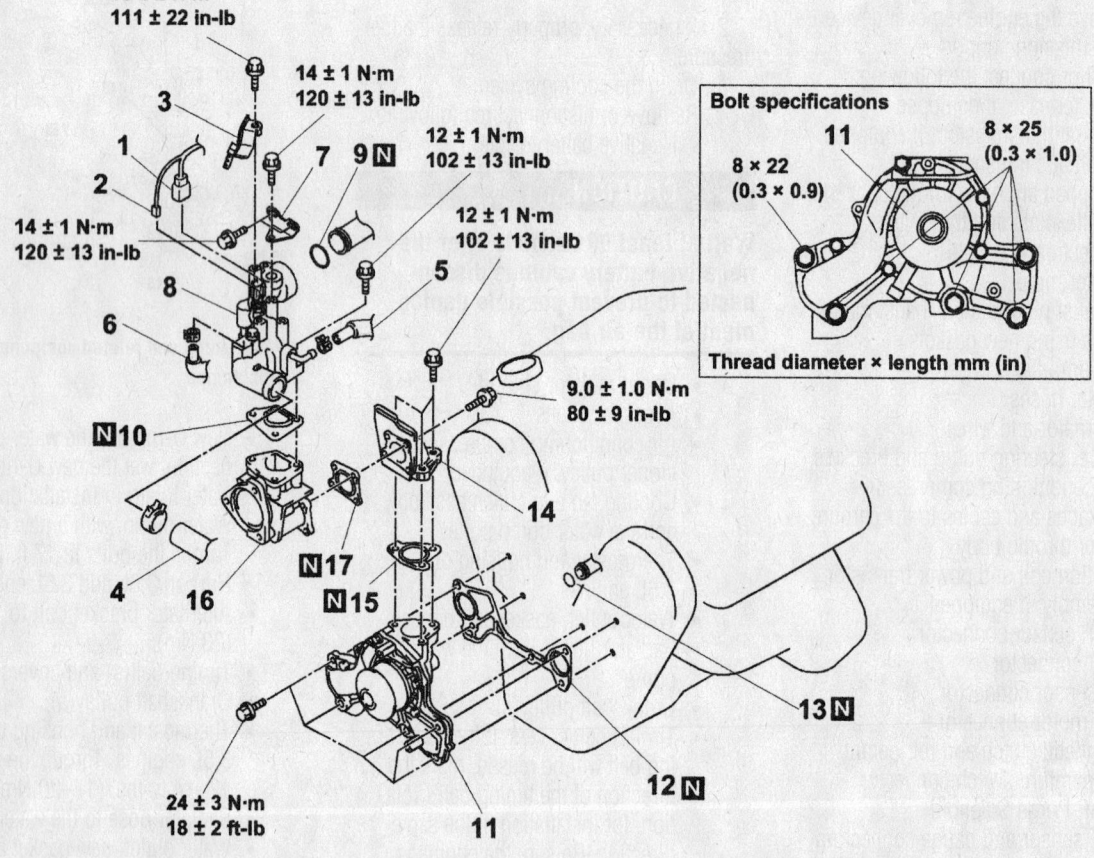

Exploded view of the water pump and related components—3.8L engine

1. ENGINE COOLANT TEMPERATURE SENSOR CONNECTOR	7. WATER OUTLET FITTING BRACKET	
2. ENGINE COOLANT TEMPERATURE GAUGE UNIT CONNECTOR	8. WATER OUTLET FITTING	
3. SPARK PLUG CABLE SUPPORT	9. O-RING	
4. RADIATOR UPPER HOSE CONNECTION	10. GASKET	
5. WATER HOSE	11. WATER PUMP ASSEMBLY	
6. WATER HOSE	12. GASKET	
	13. O-RING	
	14. FITTING	
	15. GASKET	
	16. GASKET	
	17. THERMOSTAT CASE	

42356-MONT-G01

pump bolts to 16–20 ft. lbs. (21–27 Nm).
- Water outlet fitting, O-ring and gasket
- Water outlet fitting bracket
- Water hoses
- Upper radiator hose
- Spark plug cable support
- ECT gauge unit connector
- ECT sensor connector
- Camshaft sprocket
- Timing belt

7. Refill the radiator with coolant. This cooling system has a self-bleeding thermostat, so system bleeding is not required.

8. Run the vehicle until the thermostat opens and fill the overflow tank. Check for leaks.

9. Once the vehicle has cooled, recheck the coolant level.

Heater Core

REMOVAL & INSTALLATION

Montero

1. Before servicing the vehicle, refer to the precautions in the beginning of this section.

2. Place the wheels in the straight-ahead position.

3. Drain the cooling system into a clean container for reuse.

4. Discharge and recover the air conditioning system refrigerant.

5. Remove or disconnect the following:
- Negative battery cable

✳✳ CAUTION

Wait at least 60 seconds after disconnecting the battery cable before performing any work on the air bag or instrument panel.

- Floor console assembly
- Steering column-to-instrument panel cover screws and the cover
- Air bag module, carefully, from the steering wheel
- Electrical connectors from the air bag module

✳✳ CAUTION

Store the air bag module facing up.

- Steering wheel nut
- Press the steering wheel from the steering column, using a steering wheel puller

- Passenger's side foot shower duct
- Glove box stoppers and the glove box
- Passenger's air bag module nut and the air bag module
- Electrical connector from the air bag module
- Hood lock release handle
- Fuel filler door lock release handle
- Knee protector assembly and bracket
- Meter bezel assembly
- Under cover, the corner cover and the stopper
- Glove box assembly
- Center under cover assembly
- Radio and tape/CD player
- Heater/air conditioning control assembly
- Combination meter and the speaker
- Glove box striker and the upper glove box frame
- Multi-meter panel and the multi-meter assembly
- Side defroster grille
- Instrument panel assembly, carefully, with the help of an assistant
- Blower motor assembly
- Both foot ducts

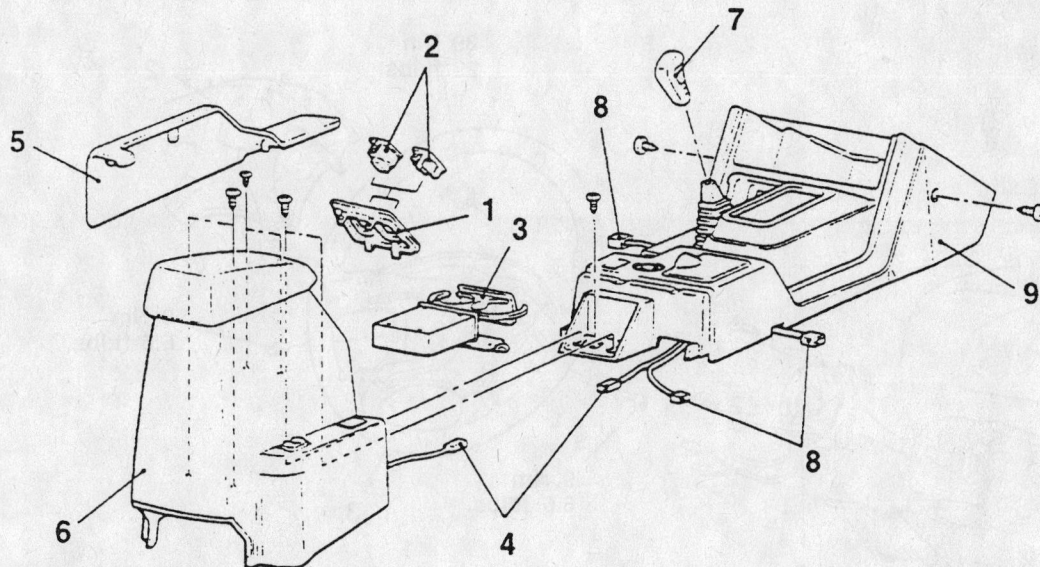

1. Switch panel
2. Suspension control switch or hole cover
3. Cup holder assembly
4. Rear console harness connector
5. Side panel A
6. Rear console assembly
7. Transfer shift lever knob
8. Floor console harness connector
9. Front console assembly

93113GE2

Exploded view of the floor console and related components—Montero

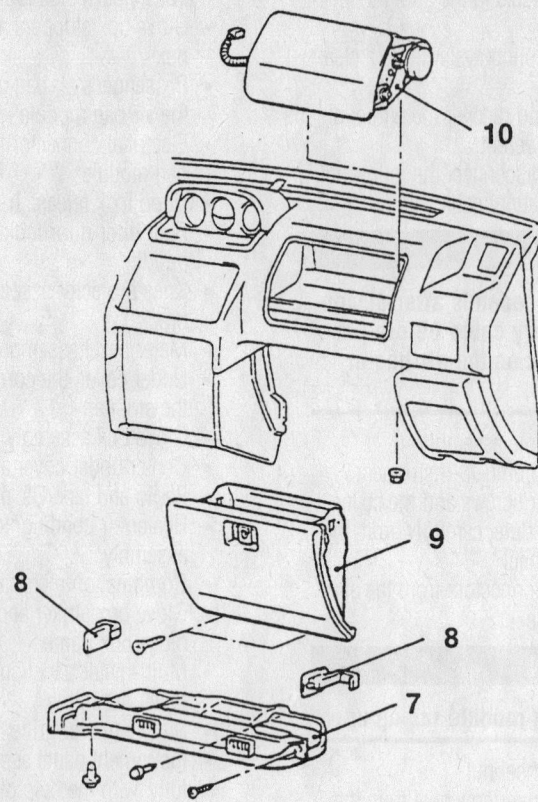

7. Foot shower duct (R.H.)
8. Stopper
9. Glove box
10. Air bag module (Passenger's side)

93113GE3

Exploded view of the passenger's side air bag module—Montero

39 Nm
29 ft.lbs.

9 Nm
6.6 ft.lbs.

9 Nm
6.6 ft.lbs.

2. Air bag module (Driver's side)
3. Steering wheel
4. Column cover lower
5. Clock spring and body wiring harness connection
6. Clock spring
• Pre-installation inspection

93113GE4

Exploded view of the steering wheel and air bag module—Montero

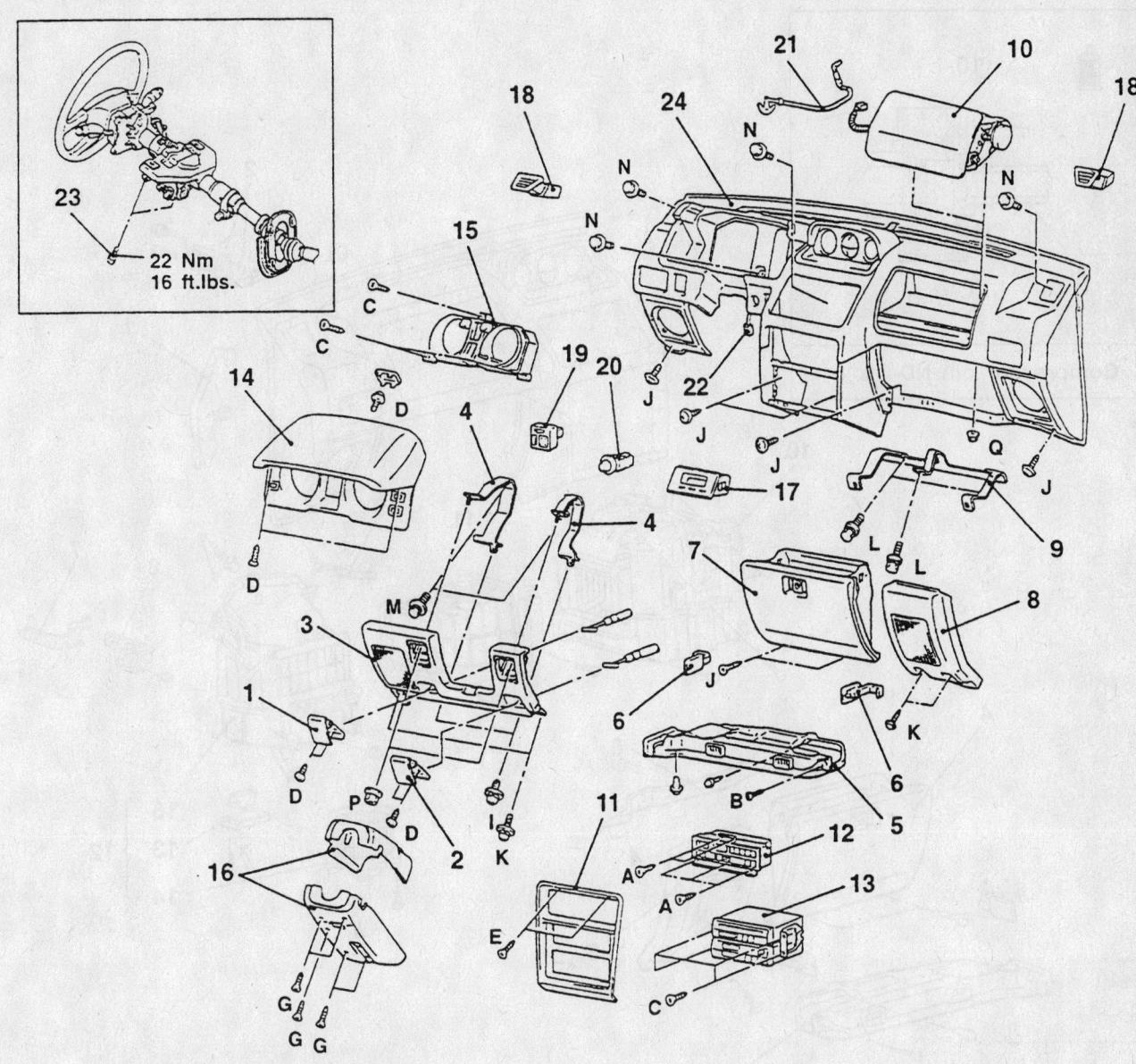

22 Nm
16 ft.lbs.

1. Hood lock release handle
2. Fuel filler door lock release handle
3. Knee protector
4. Stay
5. Foot shower duct (R.H.)
6. Glove box stopper
7. Glove box assembly
8. Corner cover
9. Stay
10. Passenger-side air bag module assembly
11. Center panel
12. Heater control assembly
13. Radio and tape player
14. Meter bezel assembly
15. Combination meter
16. Column cover
17. Clock
18. Side defroster garnish
19. Door mirror control switch
20. Rheostat
21. Ventilation control wire
22. Harness connector
23. Steering column installation bolts
24. Instrument panel assembly

93113GE5

Exploded view of the instrument panel and related components—Montero

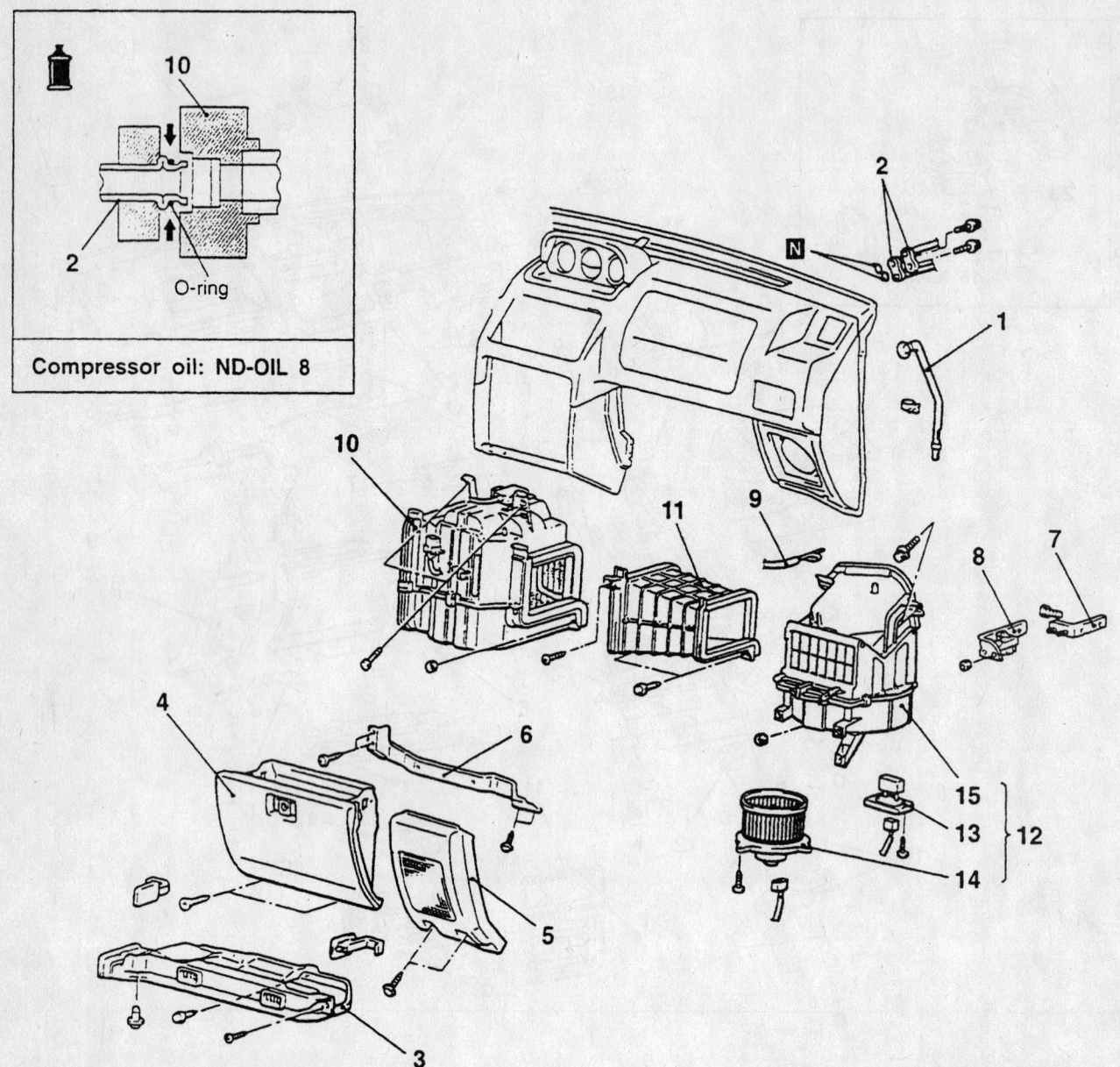

Compressor oil: ND-OIL 8

O-ring

1. Drain hose
2. Liquid pipe and suction hose connection
3. Foot shower duct (R.H.)
4. Glove box
5. Corner cover
6. Lower frame
7. Engine control relay assembly
8. Bracket
9. Air selection control wire connection
10. Evaporator
11. Duct joint
12. Blower assembly
13. Resistor
14. Blower motor assembly
15. Blower case assembly

93113GE6

Exploded view of the air conditioning evaporator housing, blower motor assembly and related components—Montero

- Joint duct from the air conditioning evaporator housing assembly
- Foot distribution duct
- Center reinforcement
- Center ventilation duct
- Drain hose from the air conditioning evaporator housing assembly
- Heater hoses from the heater housing assembly
- Refrigerant lines from the air conditioning evaporator housing assembly and discard the O-rings
- Heater housing assembly
- Center duct assembly
- Heater core from the heater housing

To install:

6. Install or connect the following:
   - Heater core to the heater housing
   - Center duct assembly

- Heater housing assembly
- Refrigerant lines to the air conditioning evaporator housing assembly, using new O-rings
- Heater hoses to the heater housing assembly
- Drain hose to the air conditioning evaporator housing assembly
- Center ventilation duct
- Center reinforcement
- Foot distribution duct
- Joint duct to the air conditioning evaporator housing assembly
- Both foot shower ducts
- Blower motor assembly
- Instrument panel, carefully with the help of an assistant
- Side defroster grille
- Multi-meter assembly and the multi-meter panel

- Upper glove box frame and the glove box striker
- Speaker and the combination meter
- Heater/air conditioning control assembly
- Radio and tape/CD player
- Center under cover assembly
- Glove box assembly
- Under cover, the corner cover and the stopper
- Meter bezel assembly
- Knee protector assembly and bracket
- Filler door lock release handle
- Hood lock release handle
- Electrical connector to the passenger's side air bag module
- Passenger's air bag module and the air bag module nut

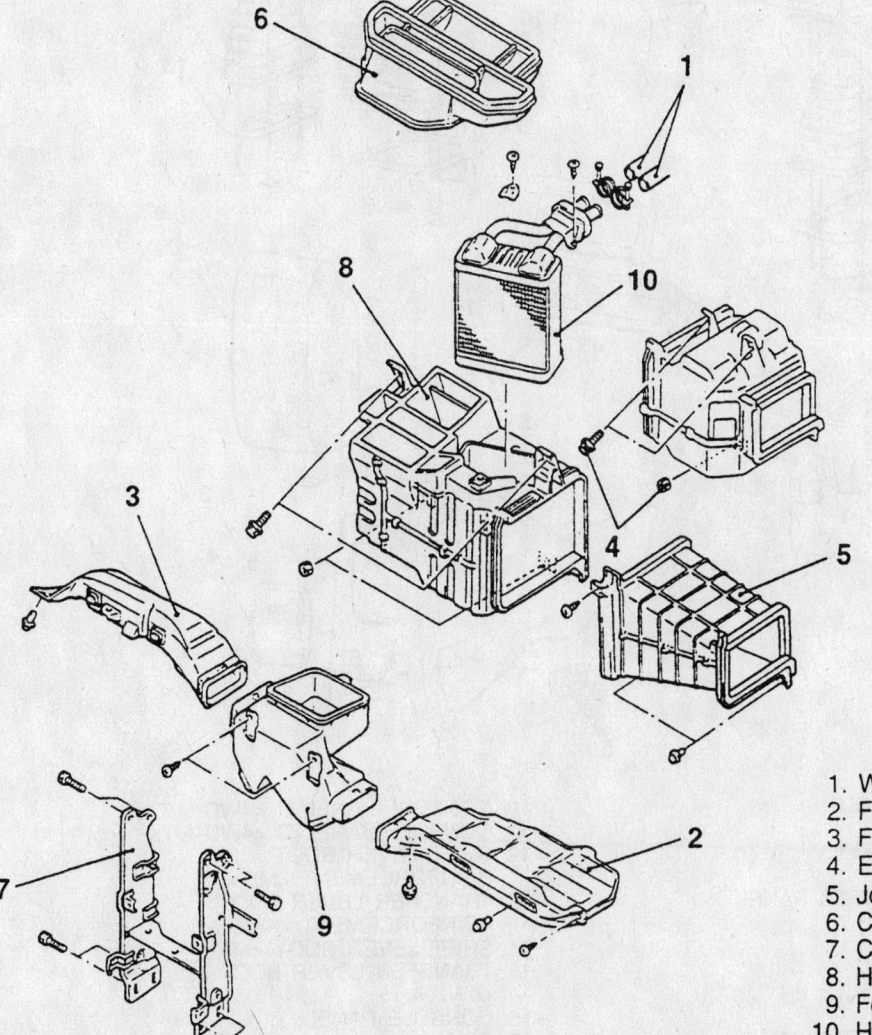

1. Water hoses connection
2. Foot shower duct (RH)
3. Foot shower duct (LH)
4. Evaporator mounting bolt and nut
5. Joint duct
6. Center duct assembly
7. Center reinforcement
8. Heater unit
9. Foot distribution duct
10. Heater core

93113GE7

Exploded view of the heater housing, air conditioning evaporator housing and related components—Montero

- Glove box and the glove box stoppers
- Foot shower duct
- Steering wheel to the steering column
- Steering wheel nut and torque the nut to 29 ft. lbs. (39 Nm)
- Electrical connectors to the air bag module
- Air bag module to the steering wheel

- Steering column-to-instrument panel cover and the cover screws
- Floor console assembly

7. Refill the cooling system.
8. Connect the negative battery.
9. Evacuate and charge the air conditioning system refrigerant.
10. Run the engine to normal operating temperatures; then, check the climate control operation and check for leaks.

Montero Sport

FRONT HEATER SYSTEM

1. Before servicing the vehicle, refer to the precautions in the beginning of this section.
2. Place the wheels in the straight-ahead position.
3. Disconnect the negative battery.

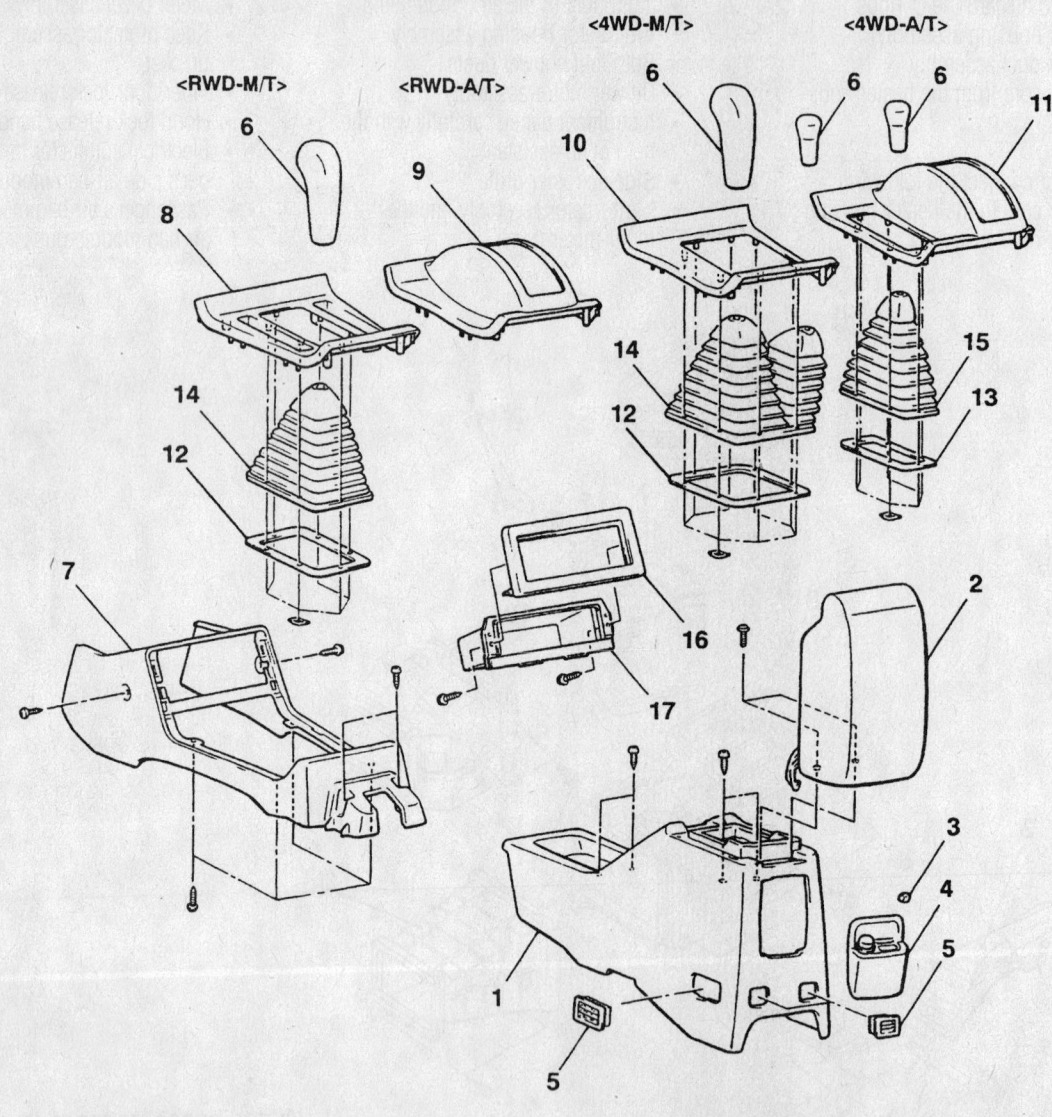

1. REAR FLOOR CONSOLE ASSEMBLY
2. CONSOLE LID ASSEMBLY
3. KNOB
4. REAR HEATER CONTROL PANEL ASSEMBLY
5. FOOT GRILL
6. SHIFT LEVER KNOB
7. FRONT FLOOR CONSOLE ASSEMBLY
8. CONSOLE PANEL A <RWD-M/T>
9. CONSOLE PANEL B <RWD-A/T>
10. CONSOLE PANEL C <4WD-M/T>
11. CONSOLE PANEL D <4WD-A/T>
12. SHIFT LEVER BOOT REINFORCEMENT <M/T>
13. TRANSFER LEVER BOOT REINFORCEMENT <4WD-A/T>
14. SHIFT LEVER BOOT <M/T>
15. TRANSFER LEVER BOOT <4WD-A/T>
16. CONSOLE PANEL
17. BOX

93113GD6

Exploded view of the floor console and related components—Montero Sport

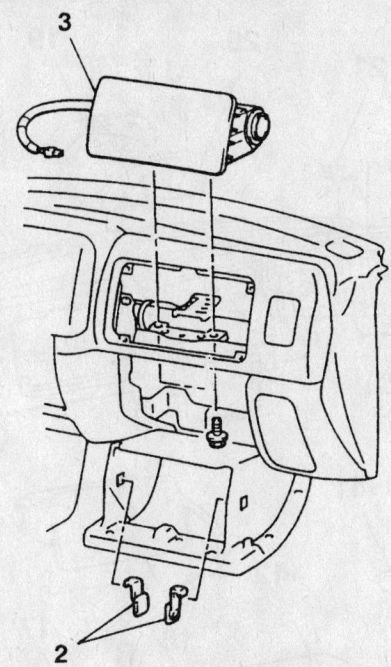

1. NEGATIVE (-) BATTERY CABLE
   CONNECTION
2. STOPPER
3. AIR BAG MODULE
• PRE-INSTALLATION INSPECTION

93113GD7

Exploded view of the passenger's side air bag module—Montero Sport

Wait at least 60 seconds after disconnecting the battery cable before performing any work on the air bag or instrument panel.

 4. Drain the cooling system into a clean container for reuse.

 5. Discharge and recover the air conditioning system refrigerant.

 6. Remove or disconnect the following:

• Floor console assembly
• Steering column-to-instrument panel cover screws and the cover
• Air bag module, carefully, from the steering wheel
• Electrical connectors from the air bag module

Store the air bag module facing up.

• Steering wheel nut
• Press the steering wheel from the steering column, using a steering wheel puller
• Glove box stoppers and lower the glove box

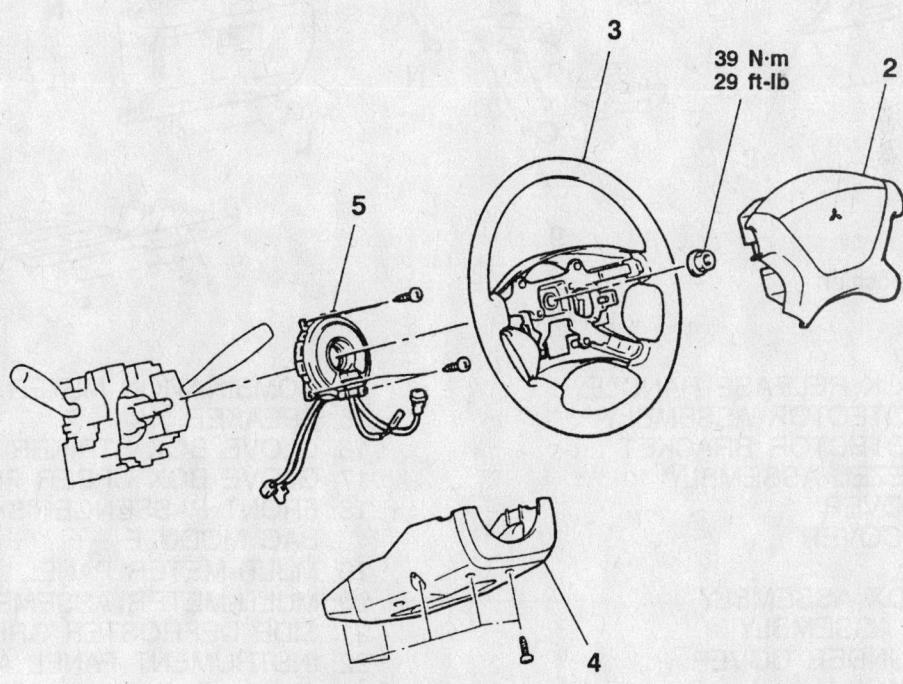

39 N·m
29 ft-lb

2. AIR BAG MODULE
3. STEERING WHEEL
4. COLUMN COVER LOWER
5. CLOCK SPRING

93113GD8

Exploded view of the steering wheel and air bag module—Montero Sport

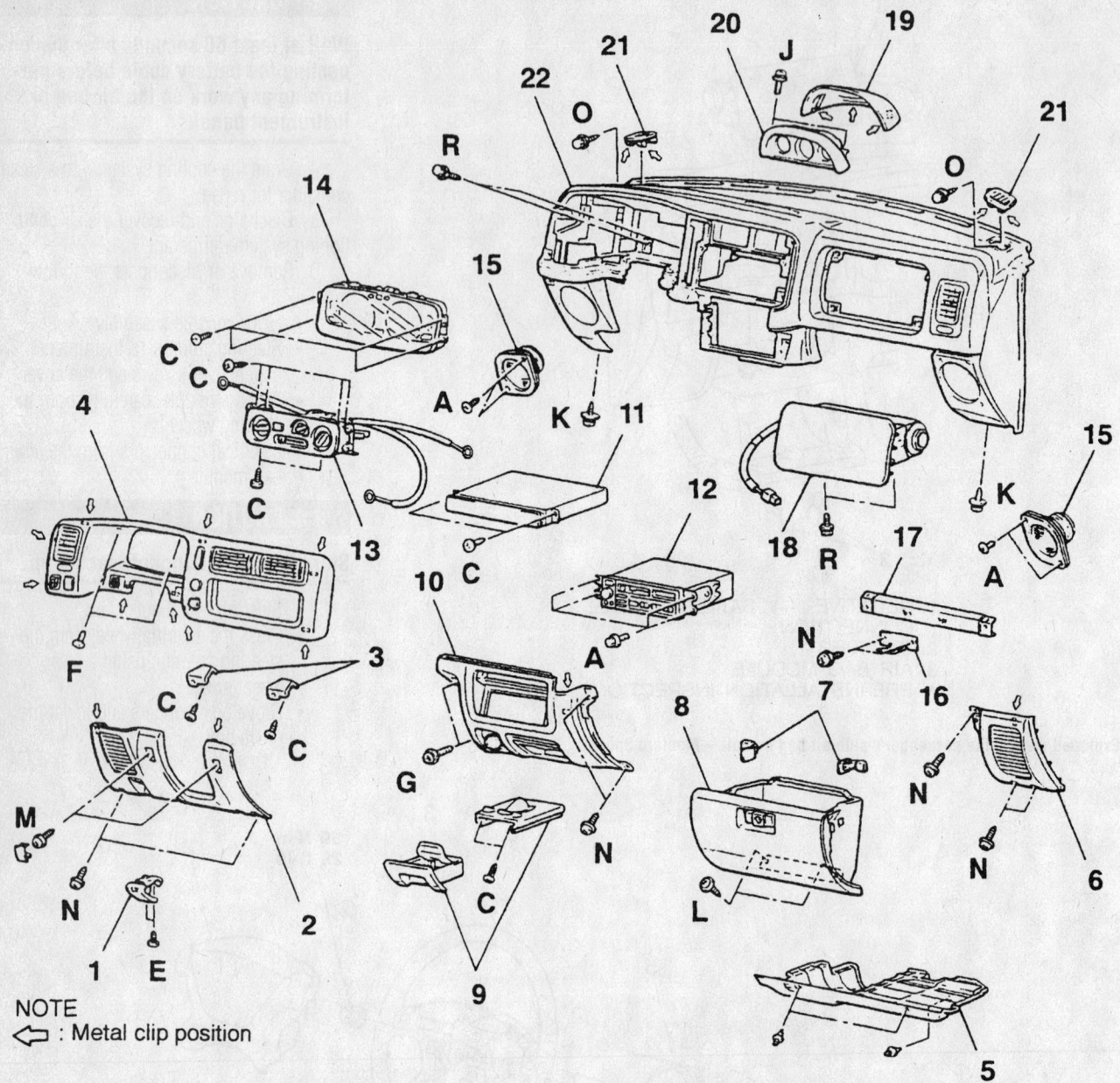

NOTE
⟸ : Metal clip position

1. HOOD LOCK RELEASE HANDLE
2. KNEE PROTECTOR ASSEMBLY
3. KNEE PROTECTOR BRACKET
4. METER BEZEL ASSEMBLY
5. UNDER COVER
6. CORNER COVER
7. STOPPER
8. GLOVE BOX ASSEMBLY
9. ASHTRAY ASSEMBLY
10. CENTER UNDER COVER ASSEMBLY
11. CUP HOLDER ASSEMBLY
12. RADIO AND TAPE PLAYER
13. HEATER CONTROL ASSEMBLY

14. COMBINATION METER
15. SPEAKER
16. GLOVE BOX STRIKER
17. GLOVE BOX UPPER FRAME
18. FRONT PASSENGER'S SIDE AIR BAG MODULE
19. MULTI-METER PANEL
20. MULTI-METER ASSEMBLY
21. SIDE DEFROSTER GRILL
22. INSTRUMENT PANEL ASSEMBLY

93113GD9

Exploded view of the instrument panel and related components—Montero Sport

- Passenger's air bag module bolts and air bag module
- Electrical connector from the air bag module
- Hood lock release handle
- Knee protector assembly and bracket
- Meter bezel assembly
- Under cover, the corner cover and the stopper
- Glove box assembly and the ashtray
- Center under cover assembly and the cup holder assembly

- Radio and tape/CD player
- Heater/air conditioning control assembly
- Combination meter and the speaker
- Glove box striker and the upper glove box frame
- Multi-meter panel and the multimeter assembly
- Side defroster grille
- Instrument panel assembly, carefully, with the help of an assistant
- Blower motor assembly
- Joint duct from the air conditioning evaporator housing assembly

- Center reinforcement
- Center ventilation duct
- Drain hose from the air conditioning evaporator housing assembly
- Heater hoses from the heater housing assembly
- Refrigerant lines from the air conditioning evaporator housing assembly and discard the O-rings
- Heater housing assembly
- Heater core from the heater housing

To install:
7. Install or connect the following:

PIPING CONNECTION

5

COMPRESSOR OIL: SUN PAG56

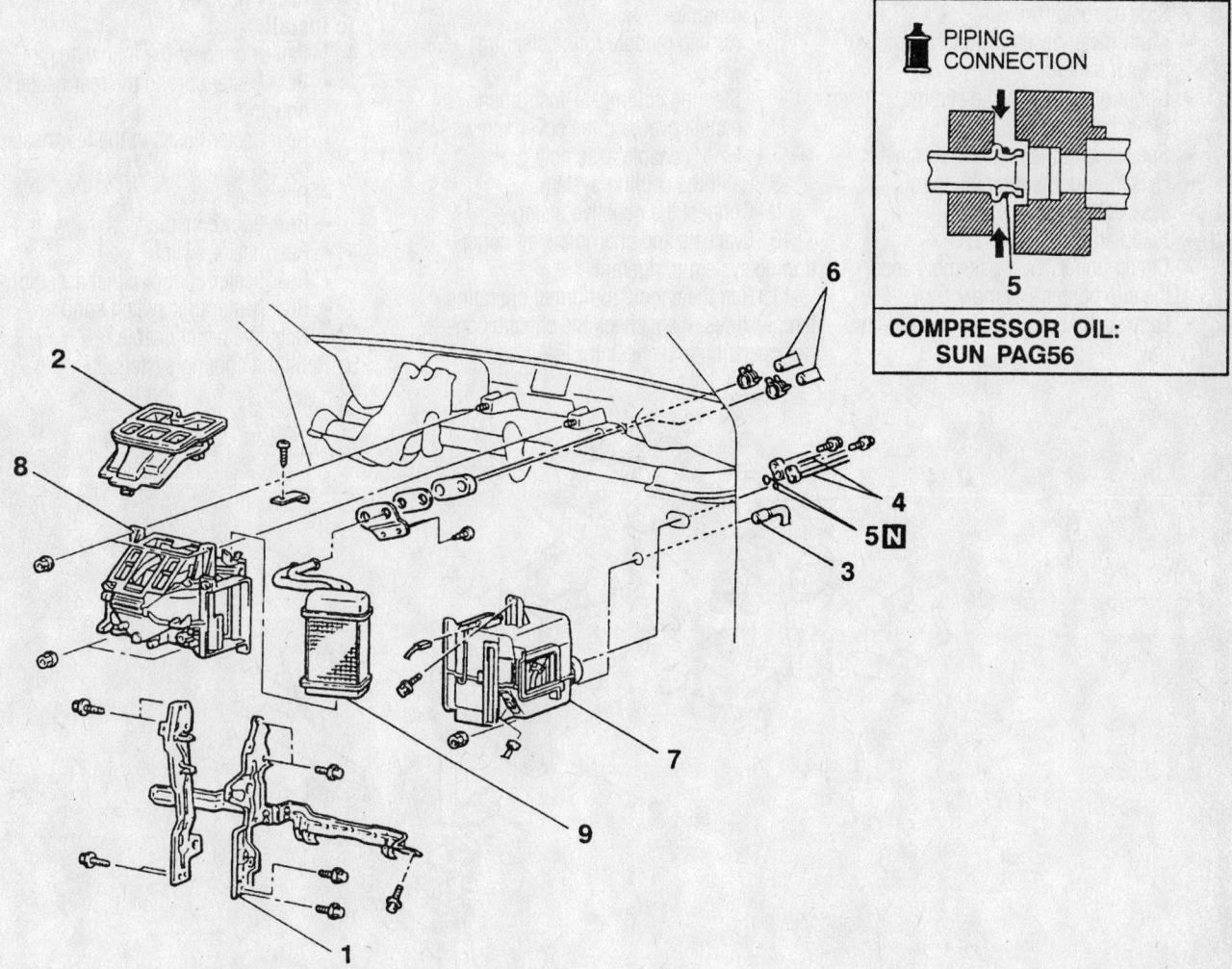

1. CENTER REINFORCEMENT
2. CENTER VENTILATION DUCT
3. DRAIN HOSE <VEHICLES WITH A/C>
4. SUCTION PIPE OR HOSE AND DISCHARGE PIPE CONNECTION <VEHICLES WITH A/C>
5. O-RING
6. HEATER HOSE CONNECTION
7. EVAPORATOR <VEHICLES WITH A/C>
8. HEATER UNIT
9. HEATER CORE

93113GD0

Exploded view of the heater housing, air conditioning evaporator housing and related components—Montero Sport

- Heater core to the heater housing
- Heater housing assembly
- Refrigerant lines to the air conditioning evaporator housing assembly, using new O-rings
- Heater hoses to the heater housing assembly
- Drain hose to the air conditioning evaporator housing assembly
- Center ventilation duct
- Center reinforcement
- Joint duct to the air conditioning evaporator housing assembly
- Blower motor assembly
- Instrument panel assembly, carefully, with the help of an assistant
- Side defroster grille
- Multi-meter assembly and the multimeter panel
- Upper glove box frame and the glove box striker
- Speaker and the combination meter
- Heater/air conditioning control assembly
- Radio and tape/CD player
- Center under cover assembly and the cup holder assembly
- Glove box assembly and the ashtray

- Under cover, the corner cover and the stopper
- Meter bezel assembly
- Knee protector assembly and bracket
- Hood lock release handle
- Electrical connector to the passenger air bag module
- Passenger air bag module and the module bolts
- Glove box and the glove box stoppers
- Steering wheel to the steering column
- Steering wheel nut and torque the nut to 29 ft. lbs. (39 Nm)
- Electrical connectors to the air bag module
- Air bag module to the steering wheel
- Steering column-to-instrument panel cover and the cover screws
- Floor console assembly

8. Refill the cooling system.
9. Connect the negative battery.
10. Evacuate and charge the air conditioning system refrigerant.
11. Run the engine to normal operating temperatures; then, check the climate control operation and check for leaks.

REAR AUXILIARY SYSTEM

1. Before servicing the vehicle, refer to the precautions in the beginning of this section.
2. Drain the cooling system into a clean container for reuse.
3. Remove or disconnect the following:
- Negative battery cable
- Rear heater unit switch knob
- Rear heater control panel assembly
- Rear heater switch
- Rear floor console
- Resistor
- Rear heater hoses from the rear heater core
- Rear heater core from the rear heater housing

To install:
4. Install or connect the following:
- Rear heater core to the rear heater housing
- Rear heater hoses to the rear heater core
- Resistor
- Rear floor console
- Rear heater switch
- Rear heater control panel assembly
- Rear heater unit switch knob
- Negative battery cable

5. Refill the cooling system.

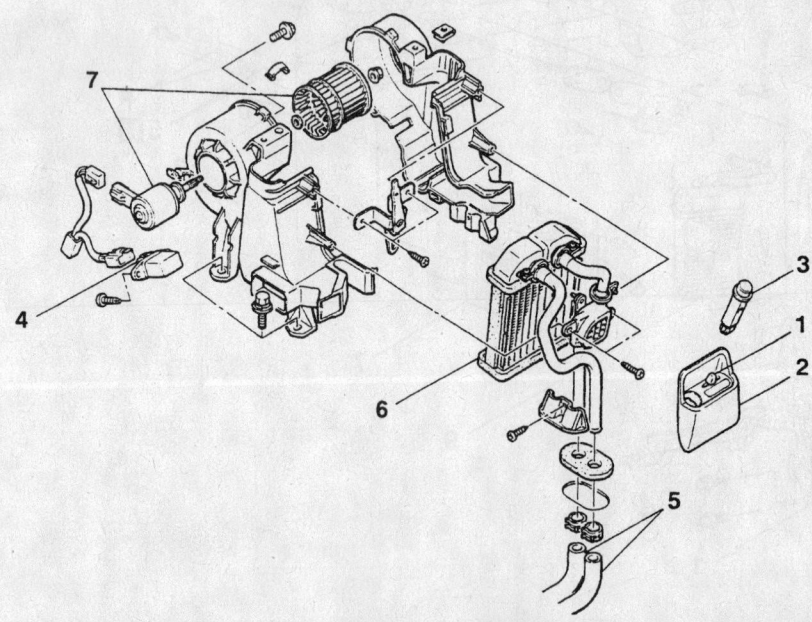

1. KNOB
2. REAR HEATER CONTROL PANEL ASSEMBLY
3. REAR HEATER SWITCH
4. RESISTOR
• DRAINING AND SUPPLYING OF COOLANT
5. REAR HEATER HOSE CONNECTION
6. REAR HEATER CORE ASSEMBLY
7. REAR BLOWER MOTOR ASSEMBLY

93113GE1

Exploded view of the rear heater core and related components—Montero Sport

6. Run the engine to normal operating temperatures; then, check the climate control operation and check for leaks.

Cylinder Head

REMOVAL & INSTALLATION

3.0L Engines

1. Before servicing the vehicle, refer to the precautions in the beginning of this section.
2. Relieve the fuel system pressure.
3. Drain the cooling system.
4. Remove or disconnect the following:
   - Negative battery cable
   - Air cleaner assembly, ducts and air intake hose
   - Upper radiator hose
   - Accessory drive belts
   - Cooling fan and pulleys
   - Air conditioning compressor, if equipped
   - Power steering pump and mounting brackets and position them to the side, without disconnecting the lines.
   - Timing belt covers
5. Remove the timing belt as follows:
   a. Rotate the crankshaft and bring the No. 1 piston to Top Dead Center (TDC) on the compression stroke. Align the camshaft and crankshaft sprocket timing marks.
   b. Mark the timing belt in the direction of rotation for reinstallation purposes.
   c. Loosen the timing belt tensioner bolt and turn the tensioner counterclockwise.

❊❊ WARNING

Do not rotate the crankshaft or camshaft sprockets after the timing belt has been removed.

6. Remove or disconnect the following:
   - Timing belt
   - Fuel lines and plug
   - Wiring connectors, vacuum lines and hoses from the air intake plenum, intake manifold and cylinder head.
   - Air intake plenum
   - Intake manifold
   - Exhaust manifold
   - Camshaft sprocket bolt and camshaft sprocket, if necessary
   - Alternator bracket and/or timing belt rear cover

- Oil dipstick, on left side only
- Crankshaft position (CKP) sensor on left side only
- Spark plug wires from the spark plugs
- Valve cover
- Cylinder head bolts starting from the outside and working inward
- Cylinder head from the engine

To install:
7. Clean the gasket mounting surfaces.
8. Install or connect the following:
   - New cylinder head gasket
   - Cylinder head on the engine. Torque the cylinder head bolts in sequence using 3 even steps, to 80 ft. lbs. (108 Nm).
   - Exhaust manifold
   - Intake manifold and air intake plenum
   - Fuel lines
   - Wiring connectors, vacuum lines and hoses to the air intake plenum, intake manifold and cylinder head
   - Valve cover
   - Spark plug wires
   - CKP sensor, if removed
   - Oil dipstick, if removed
   - Alternator bracket and/or timing belt rear cover
   - Camshaft sprocket bolt and camshaft sprocket, if necessary
9. Be sure the camshaft and crankshaft sprocket timing marks are aligned.
10. Turn the timing belt tensioner to the extreme counter-clockwise position and temporarily tighten the bolt.
11. Install the timing belt in the original rotation direction. Loosen the timing belt tensioner bolt and allow the spring force of the tensioner to tension the belt.
12. Turn the crankshaft 2 turns in the normal direction of rotation and check the timing mark alignment.
13. If the timing is correct, tighten the tensioner bolt to 21 ft. lbs. (30 Nm). If the

timing is incorrect, repeat the belt installation procedure.
14. Install or connect the following:
   - Timing belt covers
   - Alternator, alternator cover and alternator stay, if removed
   - Air conditioning compressor
   - Power steering pump with the brackets
   - Pulleys. Torque the crankshaft pulley bolt to 134 ft. lbs. (181 Nm).
   - Cooling fan
   - Accessory drive belts
   - Air cleaner assembly, ducts and air intake hose
   - Upper radiator hose
   - Negative battery
15. Refill the cooling system.
16. Start the engine and check for leaks. Check the ignition timing.

3.5L Engine

1. Before servicing the vehicle, refer to the precautions in the beginning of this section.

❊❊ CAUTION

The fuel injection system remains under pressure after the engine has beenOFF. Properly relieve fuel pressure before disconnecting any fuel lines. Failure to do so may result in fire or personal injury.

2. Relieve fuel system pressure.
3. Drain the cooling system.
4. Remove or disconnect the following:
   - Negative battery cable

❊❊ CAUTION

Work must be started after 90 seconds from the time the ignition switch is turned to theLOCK position and the negative battery cable is disconnected.

- Air intake hoses
- Air intake plenum and intake manifold
- Exhaust manifold
- Engine under cover
- Radiator and shroud
- Alternator
- Cooling fan
- Timing belt
- Breather hose
- Oil dipstick
- Camshaft Position (CMP) sensor
- Spark plug cable center cover and remove the spark plug cables
- Valve cover
- Intake camshaft sprocket

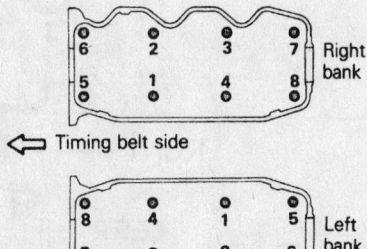

7924UG12

Cylinder head bolt tightening sequence—3.0L engines

- Rear timing belt cover
- Ignition coil
- Water hoses from the thermostat housing and the housing
- Water inlet from the front head and discard O-ring
- Water passage

5. Loosen the cylinder head mounting bolts in 3 steps, starting from the outside and working inward. Lift off the cylinder head assembly and remove the head gasket.

To install:

6. Thoroughly clean and dry the mating surfaces of the head and block. Check the cylinder head for cracks, damage or engine coolant leakage. Remove scale, sealing compound and carbon. Clean oil passages thoroughly. Check the head for flatness. End to end, the head should be within 0.0012 in. (0.030mm), normally with 0.008 in. (0.203mm) the maximum allowed out of true. The total thickness allowed to be removed from the head and block is 0.008 in. (0.203mm) maximum.

7. Place a new head gasket on the cylinder block with the identification marks in the front top (upward) position. Do not use sealer on the gasket.

8. Install or connect the following:

- Cylinder head on the block. Be sure the head bolt washers are installed with the chamfered edge upward. Using 3 even steps, torque the head bolts in sequence, to 76–83 ft. lbs. (105–115 Nm).
- New O-ring and the water inlet to the front head
- New gaskets, thermostat housing and connect the hoses
- Water passage and new gaskets
- Ignition coil and center rear timing belt cover
- Intake camshaft sprocket. Use hex flange on camshaft to secure and tighten the retaining bolt to 65 ft. lbs. (90 Nm).

- New gasket and the valve cover. Torque the bolts to 84 inch lbs. (10 Nm).
- Spark plug cables and the center cover
- Oil dipstick
- CMP sensor
- Breather hose
- Radiator and shroud
- Timing belt
- Cooling fan
- Alternator
- Engine under cover
- Intake manifold and new gasket. Torque the nuts to 16 ft. lbs. (21 Nm).
- Air intake plenum and new gaskets. Torque the bolts to 13 ft. lbs. (18 Nm).
- Exhaust manifold and new gaskets. Torque the nuts to 22 ft. lbs. (29 Nm).
- Air intake hoses
- Negative battery cable

9. Change the engine oil and oil filter.
10. Refill the system with coolant.
11. Run the vehicle until the thermostat opens.
12. Once the vehicle has cooled, recheck the coolant level.

3.8L Engine

1. Before servicing the vehicle, refer to the precautions in the beginning of this section.

✳✳ CAUTION

The fuel injection system remains under pressure after the engine has beenOFF. Properly relieve fuel pressure before disconnecting any fuel lines. Failure to do so may result in fire or personal injury.

2. Relieve fuel system pressure.
3. Drain the cooling system.
4. Remove or disconnect the following:

- Negative battery cable

✳✳ CAUTION

Work must be started after 90 seconds from the time the ignition switch is turned to theLOCK position and the negative battery cable is disconnected.

- Intake manifold
- Timing belt
- Front exhaust pipe

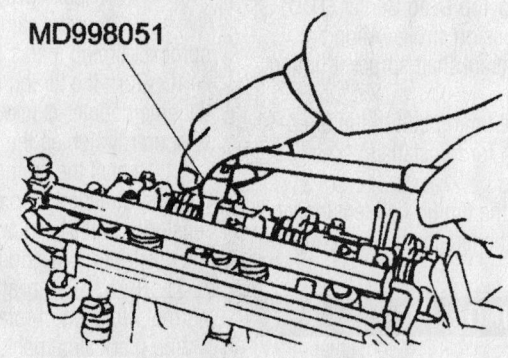

MD998051

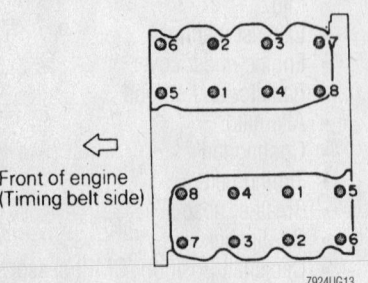

Cylinder head bolt tightening sequence—3.5L engines

Front of engine (Timing belt side)

7924UG13

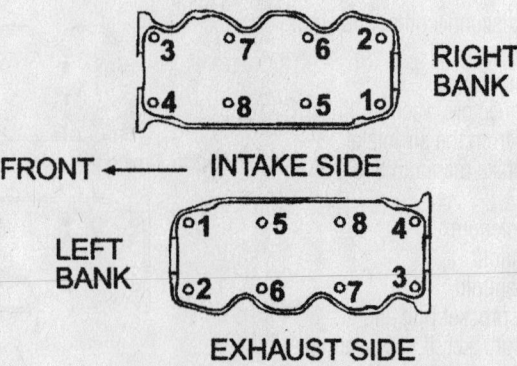

EXHAUST SIDE

RIGHT BANK

FRONT ← INTAKE SIDE

LEFT BANK

EXHAUST SIDE

Loosen the cylinder head bolts in the proper sequence—3.8L engine

42356-MONT-G02

- Water outlet pipe assembly and O-ring
- Heater hose connections
- Water passage assembly and gasket
- Water pipe and O-ring
- Breather hose
- Spark plug wires; tag before disconnecting
- Ignition coil
- Oxygen (O_2) sensor connector
- Engine oil dipstick assembly and O-ring
- Intake manifold plenum stay
- Rocker cover
- Camshaft Position (CMP) sensor connector

5. Loosen the cylinder head mounting bolts in 3 steps, in the sequence shown. Lift off the cylinder head assembly and remove the head gasket.

To install:

6. Thoroughly clean and dry the mating surfaces of the head and block. Check the cylinder head for cracks, damage or engine coolant leakage. Remove scale, sealing compound and carbon. Clean oil passages thoroughly. Check the head for flatness. End to end, the head should be within 0.0012 in. (0.030mm), normally with 0.008 in.

(0.203mm) the maximum allowed out of true. The total thickness allowed to be removed from the head and block is 0.008 in. (0.203mm) maximum.

7. Place a new head gasket on the cylinder block with the identification marks in the front top (upward) position. Do not use sealer on the gasket.

8. Install or connect the following:
- Cylinder head on the block. Be sure the head bolt washers are installed with the chamfered edge upward. Using 3 even steps, torque the head bolts in sequence, to 77–83 ft. lbs. (105–113 Nm) with a torque wrench and Special Tool No. MD998501.
- Rocker cover
- Intake manifold plenum stay
- Engine oil dipstick assembly and O-ring
- O_2 sensor connector
- Ignition coil
- Spark plug wires
- Breather hose
- Water pipe and O-ring
- Water passage assembly and gasket
- Heater hose connections
- Water outlet pipe assembly and O-ring

- Front exhaust pipe
- Timing belt
- Intake manifold
- Negative battery cable

9. Change the engine oil and oil filter.
10. Refill the system with coolant.
11. Run the vehicle until the thermostat opens.
12. Once the vehicle has cooled, recheck the coolant level.

Rocker Arms/Shafts

REMOVAL & INSTALLATION

3.0L Engines

1. Before servicing the vehicle, refer to the precautions in the beginning of this section.
2. Remove or disconnect the following:
- Negative battery cable

✶✶ CAUTION

Work must be started after 90 seconds from the time the ignition switch is turned to theLOCK position and the negative battery cable is disconnected.

- Valve cover
- Auto lash adjuster retainers SST MD998443 on the rocker arms
- Rocker arms, rocker shafts and bearing caps, as an assembly

To install:

3. Inspect the bearing journals on the camshaft and the cylinder head.
4. Lubricate the camshaft journals and camshaft with clean engine oil.
5. Install the rocker arms, rocker arm shaft and the rocker shaft spring as follows:

   a. Temporarily tighten the rocker shaft with the bolts so that the intake valve rocker arms do not push on the valves.

   b. Insert the rocker shaft spring from above and mount it at right angles to the plug guide.

   c. Before installing the exhaust rocker arms and the rocker arm shaft, mount the rocker shaft spring.

   d. Remove tool SST MD998443 used to hold the lash adjuster in position.

   e. Check to ensure that the flat side of the rocker shaft is perpendicular to the cylinder head, and facing the valves.

   f. Gradually tighten the bearing caps in 2 or 3 steps. In the final step, tighten to 23 ft. lbs. (31 Nm).

6. Install or connect the following:

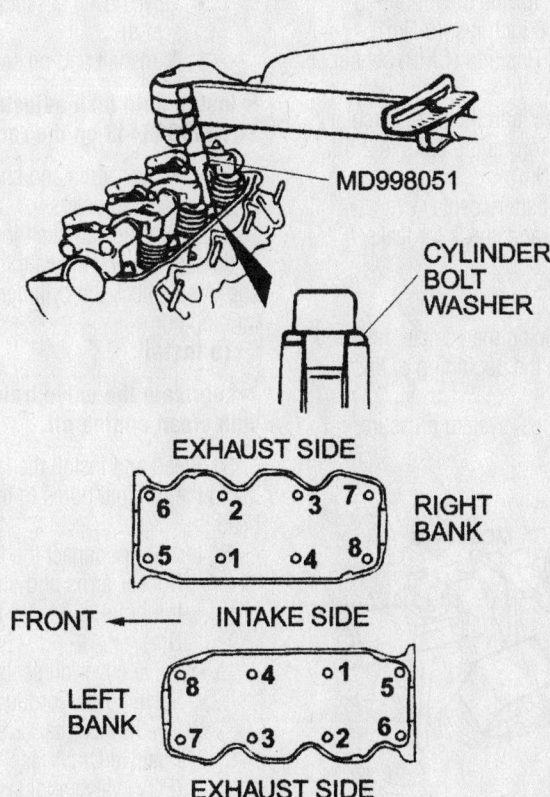

42356-MONT-G03

The cylinder head bolts must be installed with the chamfered edge up and tighten in the proper sequence—3.8L engine

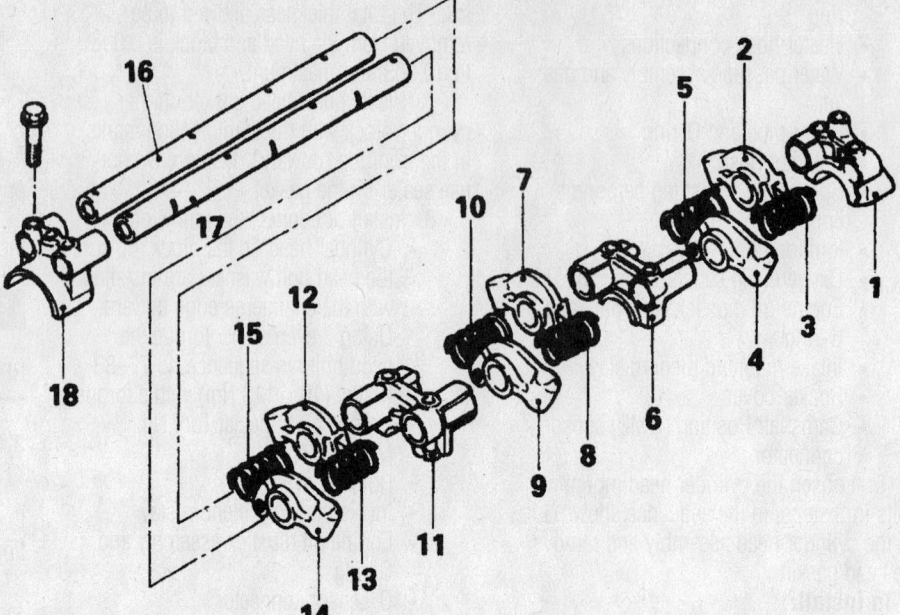

1. Bearing cap No. 4
2. Rocker arm (B)
3. Spring
4. Rocker arm (A)
5. Spring
6. Bearing cap No. 3
7. Rocker arm (B)
8. Spring
9. Rocker arm (A)
10. Spring
11. Bearing cap No. 2
12. Rocker arm (B)
13. Spring
14. Rocker arm (A)
15. Spring
16. Rocker arm shaft (B)
17. Rocker arm shaft (A)
18. Bearing cap No. 1

7924UG15

Exploded view of the rocker arms and shafts—3.0L engine

- Valve cover and new gasket. Torque the bolt to 2–3 ft. lbs. (3–4 Nm).
- Negative battery cable

7. Start the engine and check for leaks and proper operation.

3.5L Engine

1. Before servicing the vehicle, refer to the precautions in the beginning of this section.
2. Relieve the fuel system pressure.
3. Remove or disconnect the following:
   - Negative battery cable
   - Valve cover and the semi-circular packing.
   - Crankshaft Position (CKP) sensor, matchmark for reassembly
   - Camshaft Position (CMP) sensor, if equipped

➡Install auto lash adjuster retainers SST MD998443 on the rocker arms

- Rocker arms and shafts
- Lash adjusters

4. Check the camshaft journals for wear or damage. Check the cam lobes for damage. Also, check the cylinder head oil holes for clogging.

To install:

➡Lubricate the valve train components with clean engine oil.

5. Bleed and install the lash adjusters to the to the original bores in the cylinder head.

6. Install or connect the following:
   - Rocker arms and shafts. Torque the bolts to 23 ft. lbs. (31 Nm).
   - Camshaft position sensor, if removed. Torque the mounting bolts to 78 inch lbs. (9 Nm).
   - Camshaft Position (CMP) sensor, if equipped
   - Valve cover and the semi-circular packing. Torque the bolts to 2.5 ft. lbs. (3.5 Nm).
   - Negative battery cable
7. Run vehicle and check for leaks.

3.8L Engine

1. Before servicing the vehicle, refer to the precautions in the beginning of this section.
2. Relieve the fuel system pressure.

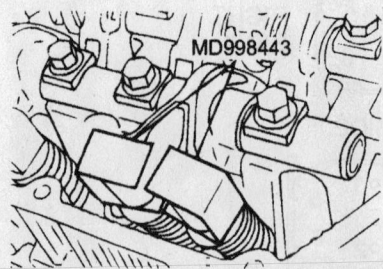

MD998443

42356-MONT-G04

Install the special tool to prevent the lash adjusters from falling to the floor during rocker arm removal—3.8L engine

3. Remove or disconnect the following:
   - Negative battery cable
   - Oil filler cap
   - Positive Crankcase Ventilation (PCV) valve and gasket
   - Valve cover, gasket and oil seal(s)
   - Camshaft and oil seal

➡Install auto lash adjuster retainers SST MD998443 on the rocker arms

- Rocker arms and shafts
- Lash adjusters

4. Check the camshaft journals for wear or damage. Check the cam lobes for damage. Also, check the cylinder head oil holes for clogging.

To install:

➡Lubricate the valve train components with clean engine oil.

5. Bleed and install the lash adjusters to the to the original bores in the cylinder head.

6. Install or connect the following:
   - Rocker arms and shafts. Torque the bolts to 21–25 ft. lbs. (28–34 Nm).
   - Valve cover oil seals and gasket
   - Valve cover. Torque the bolts to 22–30 inch lbs. (2.9–3.4 Nm).
   - Positive Crankcase Ventilation (PCV) valve and gasket
   - Oil filler cap
   - Negative battery cable
7. Run vehicle and check for leaks.

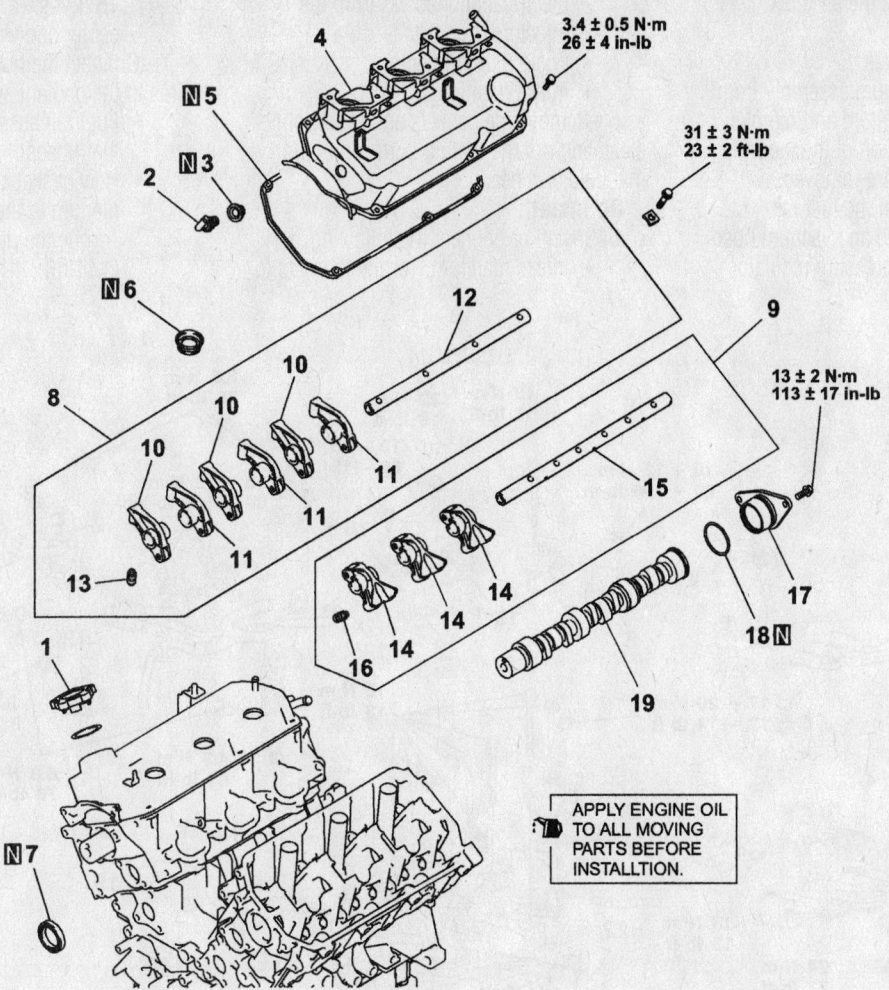

3.4 ± 0.5 N·m
26 ± 4 in-lb

31 ± 3 N·m
23 ± 2 ft-lb

13 ± 2 N·m
113 ± 17 in-lb

APPLY ENGINE OIL
TO ALL MOVING
PARTS BEFORE
INSTALLTION.

1.	OIL FILLER CAP	11.	ROCKER ARM B
2.	PCV VALVE	12.	ROCKER ARM SHAFT
3.	PCV VALVE GASKET	13.	LASH ADJUSTER
4.	ROCKER COVER	14.	ROCKER ARM C
5.	ROCKER COVER GASKET	15.	ROCKER ARM SHAFT
6.	OIL SEAL	16.	LASH ADJUSTER
7.	CAMSHAFT OIL SEAL	17.	THRUST CASE
8.	ROCKER ARMS AND SHAFT		(RIGHT BANK ONLY)
0.	ROCKER ARMS AND SHAFT	18.	O-RING (RIGHT BANK ONLY)
10.	ROCKER ARM A	19.	CAMSHAFT

42356-MONT-G05

Exploded view of the rocker arms, camshafts and related components—3.8L engine

Intake Manifold

REMOVAL & INSTALLATION

3.0L Engine

1. Before servicing the vehicle, refer to the precautions in the beginning of this section.
2. Relieve the fuel pressure.

❊❊ CAUTION

The fuel injection system remains under pressure after the engine has beenOFF. Properly relieve fuel pressure before disconnecting any fuel lines. Failure to do so may result in fire or personal injury.

3. Drain the engine coolant.
4. Remove or disconnect the following:
   • Negative battery cable

❊❊ CAUTION

Work must be started after 90 seconds from the time the ignition switch is turned to theLOCK position and the negative battery cable is disconnected.

• Air intake hose from the throttle body
• Positive Crankcase Ventilation (PCV) hose
• Exhaust Gas Recirculation (EGR) valve
• Manifold Differential Pressure (MDP) sensor
• Vacuum hoses from the throttle body and air intake plenum
• Accelerator cable and the throttle control cable
• Coolant hoses
• Engine oil filler neck bracket from the air intake plenum

- EGR tube from the air intake plenum
- Plenum brackets
- Air intake plenum assembly from the intake manifold and remove. Note the position of the mounting bolts as they are removed.
- Fuel hose from the fuel rail
- Fuel return line and vacuum hose from the fuel pressure regulator

- Electrical connectors from the injectors
- Fuel rail and injectors
- Intake manifold

5. Remove the gaskets and thoroughly clean and dry the mating surfaces of the manifold and heads.

To install:

6. Install or connect the following:
- Intake manifold. Torque the nuts to

16 ft. lbs. (21 Nm) start from the center and working outward.

7. Connect the hoses and connect the wires to the coolant switches.
- Fuel rail assembly and connect the fuel hoses
- New gasket and the air intake plenum to the intake manifold. Torque the nuts/bolts to 13 ft. lbs. (17 Nm).

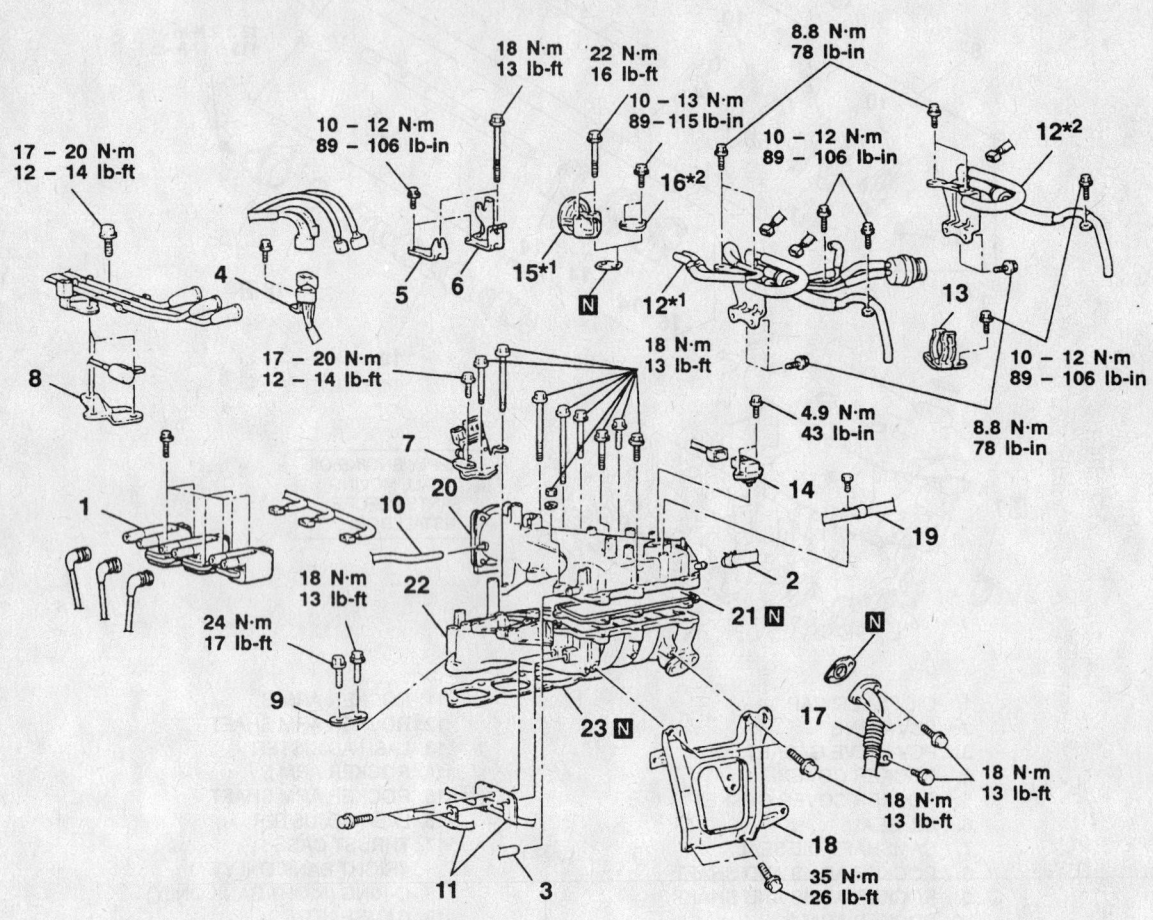

1. IGNITION COILS
2. BRAKE BOOSTER VACUUM HOSE CONNECTION
3. PCV HOSE CONNECTION
4. CRANKSHAFT POSITION SENSOR AND CAM POSITION SENSOR CONNECTOR
5. ACCELERATOR CABLE BRACKET <M/T>
6. THROTTLE CABLE BRACKET <A/T>
7. IGNITION POWER TRANSISTOR
8. WATER OUTLET FITTING BRACKET
9. WATER PUMP STAY
10. VACUUM HOSE CONNECTION
11. FUEL PIPE CONNECTION
12. SOLENOID VALVE AND VACUUM HOSE ASSEMBLY

13. VCV BRACKET
14. MDP SENSOR
15. EGR VALVE
16. COVER
17. EGR PIPE CONNECTION
18. INTAKE MANIFOLD PLENUM STAY
19. THROTTLE CABLE CONNECTION
20. AIR INTAKE FITTING
21. AIR INTAKE FITTING GASKET
22. UPPER INTAKE MANIFOLD
23. INTAKE MANIFOLD PLENUM GASKET

NOTE
*1: Vehicles for Federal
*2: Vehicles for California

Exploded view of the upper intake manifold and related components—3.0L 24-valve engine shown

7924UG41

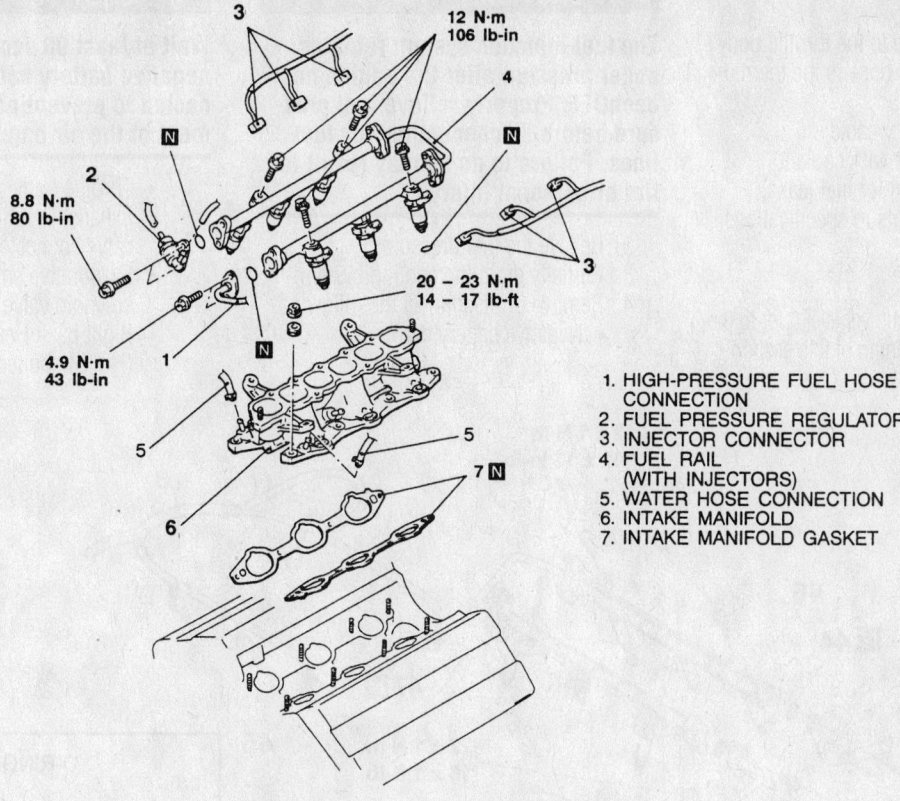

12 N·m
106 lb-in

8.8 N·m
80 lb-in

20 – 23 N·m
14 – 17 lb-ft

4.9 N·m
43 lb-in

1. HIGH-PRESSURE FUEL HOSE
   CONNECTION
2. FUEL PRESSURE REGULATOR
3. INJECTOR CONNECTOR
4. FUEL RAIL
   (WITH INJECTORS)
5. WATER HOSE CONNECTION
6. INTAKE MANIFOLD
7. INTAKE MANIFOLD GASKET

7924UG42

Exploded view of the lower intake manifold and related components—3.0L 24-valve engine shown

- Plenum brackets
- PCV hose and vacuum hose cluster to the plenum
- EGR tube
- EGR temperature sensor wire
- Wires, hoses and linkages to the throttle body
- Air intake hose to the throttle body
- Upper radiator hose to the thermostat housing
- Negative battery cable

8. Refill the radiator with coolant.
9. Check fuel system for leaks.

3.5L Engines

1. Before servicing the vehicle, refer to the precautions in the beginning of this section.

✳✳ CAUTION

The fuel injection system remains under pressure after the engine has beenOFF. Properly relieve fuel pressure before disconnecting any fuel lines. Failure to do so may result in fire or personal injury.

2. Relieve the fuel pressure.
3. Partially drain the cooling system.
4. Remove or disconnect the following:
   - Negative battery cable

✳✳ CAUTION

Wait at least 90 seconds after the negative battery cable is disconnected to prevent possible deployment of the air bag.

- Air intake hose from the throttle body
- Electrical connectors and vacuum hoses from the throttle body and air intake plenum
- Accelerator cable and the throttle control cable
- Coolant hoses
- Positive Crankcase Ventilation (PCV) hose
- Exhaust Gas Recirculation (EGR) temperature sensor connector
- EGR tube from the air intake plenum
- Intake manifold plenum cover
- Intake manifold plenum stay brackets
- Air intake plenum assembly from the intake manifold and remove. Note the position of the mounting bolts as they are removed
- Induction control valve assembly
- Fuel hose from the fuel rail

- Fuel return line and vacuum hose from the fuel pressure regulator
- Electrical connectors from the injectors
- Fuel rail and injectors
- Intake manifold

5. Remove the gaskets and thoroughly clean and dry the mating surfaces of the manifold and heads.

To install:

6. Install or connect the following:
   - Intake manifold. Tighten the nuts to 16 ft. lbs. (21 Nm). Start from the center and work outward.
   - Fuel rail assembly and connect the fuel hoses
   - Induction control valve assembly and tighten to 72 inch lbs. (9 Nm).
   - New gasket and air intake plenum to the intake manifold. Torque the nuts/bolts to 13 ft. lbs. (18 Nm).
   - Plenum to engine brackets
   - Hoses and wires to the coolant switches
   - PCV hose and vacuum hose cluster to the plenum
   - EGR tube. Torque the bolts to 13 ft. lbs. (18 Nm).
   - EGR temperature sensor wire

- Wires, hoses and linkages to the throttle body
- Air intake hose to the throttle body
- Upper radiator hose to the thermostat housing
- Negative battery cable

7. Refill the radiator with coolant.
8. Check the system for fuel leaks.
9. Set all adjustments to specifications.

3.8L Engines

1. Before servicing the vehicle, refer to the precautions in the beginning of this section.

✳✳ CAUTION

The fuel injection system remains under pressure after the engine has beenOFF. Properly relieve fuel pressure before disconnecting any fuel lines. Failure to do so may result in fire or personal injury.

2. Relieve the fuel pressure.
3. Partially drain the cooling system.
4. Remove or disconnect the following:
   - Negative battery cable

✳✳ CAUTION

Wait at least 90 seconds after the negative battery cable is disconnected to prevent possible deployment of the air bag.

- Throttle body
- Exhaust Gas Recirculation (EGR) valve connector
- Evaporative emission (EVAP) purge solenoid valve connector
- Right bank Heated Oxygen Sensor (HO2S) connector connection

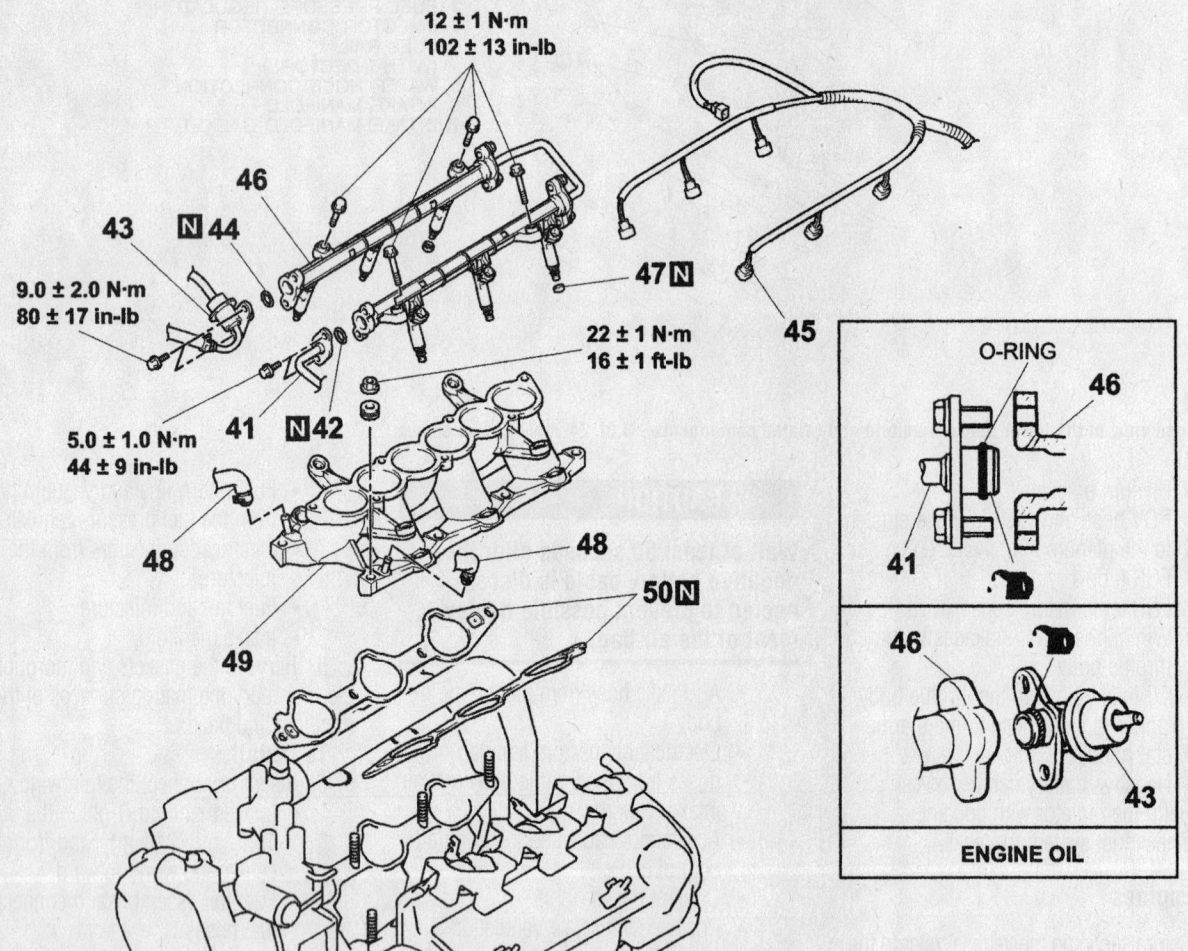

41. FUEL HIGH-PRESSURE HOSE CONNECTION
42. O-RING
43. FUEL PRESSURE REGULATOR
44. O-RING
45. INJECTOR CONNECTOR
46. FUEL RAIL (WITH INJECTORS)
47. INSULATORS
48. WATER HOSE CONNECTION
49. INTAKE MANIFOLD
50. INTAKE MANIFOLD GASKET

42356-MONT-G06

Exploded view of the intake manifold—3.8L engine

- Manifold Differential Pressure (MDP) sensor connector
- Capacitor connector
- Knock Sensor (KS) connector
- Control wiring harness and Camshaft Position (CMP) sensor wiring harness combination connector
- Ground cable
- Fuel injector connector
- Control wiring harness and injector wiring harness combination connector
- Intake manifold tuning solenoid connector
- Engine Coolant Temperature (ECT) sensor connector
- ECT gauge unit connector
- Crankshaft Position (CKP) sensor connector
- Ground cable
- Knock Sensor (KS) and Camshaft Position (CMP) sensor combination connector
- Control wiring harness and injector harness combination connector
- Connector bracket
- Positive Crankcase Ventilation (PCV) hose connection
- Fuel pipe
- Vacuum hose connection
- Water outlet fitting bracket
- EGR pipe
- EGR pipe and gasket
- Intake manifold plenum stay
- Right bank HO$_2$S connector
- Fuel pipe clip
- Intake manifold plenum
- Intake manifold plenum gasket
- Manifold Differential Pressure (MDP) sensor and O-ring
- Solenoid valve and vacuum hose assembly
- Capacitor
- EGR valve and gasket
- Purge hose

- EVAP purge solenoid valve
- Intake manifold tuning valve assembly
- Intake manifold tuning valve gaskets P and S
- High pressure fuel hose connection and O-ring
- Fuel pressure regulator and O-ring
- Fuel injector connectors
- Fuel rail (with injectors attached)
- Insulators
- Water hose connection
- Intake manifold retainers, manifold and gasket. Thoroughly clean and dry the mating surfaces of the manifold and heads.

To install:

5. Install or connect the following:
   - New intake manifold gasket. Make sure the gaskets are installed with the protrusions as shown in the illustration.
   - Intake manifold

6. Install the intake manifold bolts and tighten as follows:

   a. 1st step: Right bank nuts to 58 inch lbs. (6.5 Nm).

   b. 2nd step: Left bank nuts to 16 ft. lbs. (22 Nm).

   c. 3rd step: Right bank nuts to 16 ft. lbs. (22 Nm).

   d. 4th step: Left bank nuts to 16 ft. lbs. (22 Nm).

   e. 5th step: Right bank nuts to 16 ft. lbs. (22 Nm).
   - Water hose connection
   - Insulators
   - Fuel rail and injector assembly
   - Fuel injector connectors
   - Fuel pressure regulator and O-ring
   - High pressure fuel hose connection and O-ring
   - Intake manifold tuning valve gaskets P and S
   - Intake manifold tuning valve assembly

- EVAP purge solenoid valve
- Purge hose
- EGR valve and gasket
- Capacitor
- Solenoid valve and vacuum hose assembly
- Manifold Differential Pressure (MDP) sensor and O-ring
- Intake manifold plenum gasket
- Intake manifold plenum
- Fuel pipe clip
- Right bank HO$_2$S connector
- Intake manifold plenum stay
- EGR pipe and gasket
- EGR pipe
- Water outlet fitting bracket
- Vacuum hose connection
- Fuel pipe
- PCV hose connection
- Connector bracket
- Control wiring harness and injector harness combination connector
- KS and CMP sensor combination connector
- Ground cable
- CKP sensor connector
- ECT gauge unit connector
- ECT sensor connector
- Intake manifold tuning solenoid connector
- Control wiring harness and injector wiring harness combination connector
- Fuel injector connector
- Ground cable
- Control wiring harness and CMP sensor wiring harness combination connector
- KS connector
- Capacitor connector
- MDP sensor connector
- Right bank HO$_2$S connector
- EVAP purge solenoid valve connector
- EGR valve connector

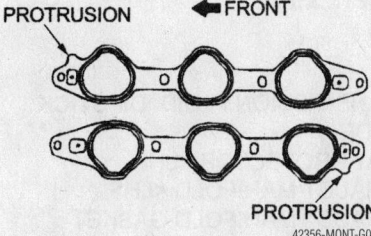

PROTRUSION   ◄FRONT

PROTRUSION

42356-MONT-G07

The intake manifold gaskets must be installed properly—3.8L engine

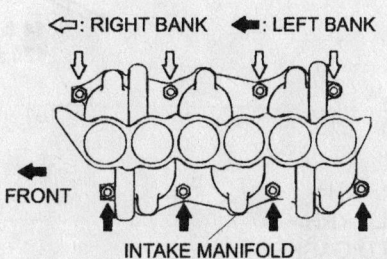

◁: RIGHT BANK ◀: LEFT BANK

FRONT

INTAKE MANIFOLD

42356-MONT-G08

Left and right bank nut locations—3.8L engine

◄ FRONT

PROTRUSION

42356-MONT-G09

Proper installation of the intake manifold plenum gasket—3.8L engine

- Throttle body
- Negative battery cable
7. Refill the radiator with coolant.
8. Check the system for fuel leaks.
9. Set all adjustments to specifications.

Exhaust Manifold

REMOVAL & INSTALLATION

3.0L and 3.5L Engines

1. Remove or disconnect the following:
   - Negative battery cable
   - Exhaust pipe from the exhaust manifolds
   - Oil dipstick, guide and O-ring
   - Heat shields
   - Exhaust manifolds
2. Clean the gasket mounting surfaces. Inspect the manifolds for cracks, flatness and/or damage.

To install:

3. Install or connect the following:
   - New gasket and exhaust manifold. Torque the nuts to 33 ft. lbs. (44 Nm) on 3.0L engines and 22 ft. lbs. (29 Nm) on 3.5L engines.
   - Heat shield, Torque the bolts to 10 ft. lbs. (14 Nm).
   - Exhaust pipe to the exhaust manifolds. Torque the nuts to 35ft. (49 Nm).
   - Oil dipstick, guide and new O-ring
   - Negative battery cable
4. Start the engine and check for exhaust leaks.

3.8L Engines

1. Remove or disconnect the following:
   - Negative battery cable
   - Front exhaust pipe from the exhaust manifolds
   - Air cleaner assembly

- Battery and battery tray
- Exhaust Gas Recirculation (EGR) pipe and gasket
- Right side heat shield
- Right side exhaust manifold and gasket
- Oil dipstick, guide and O-ring
- Transmission fluid dipstick guide
- Left side heat shield
- Left side exhaust manifold and gasket
2. Clean the gasket mounting surfaces. Inspect the manifolds for cracks, flatness and/or damage.

To install:

3. Install or connect the following:
   - New gasket and left side exhaust manifold. Torque the nuts to 30–36 ft. lbs. (39–49 Nm).
   - Left side heat shield
   - Transmission fluid dipstick guide
   - Oil dipstick, guide and O-ring

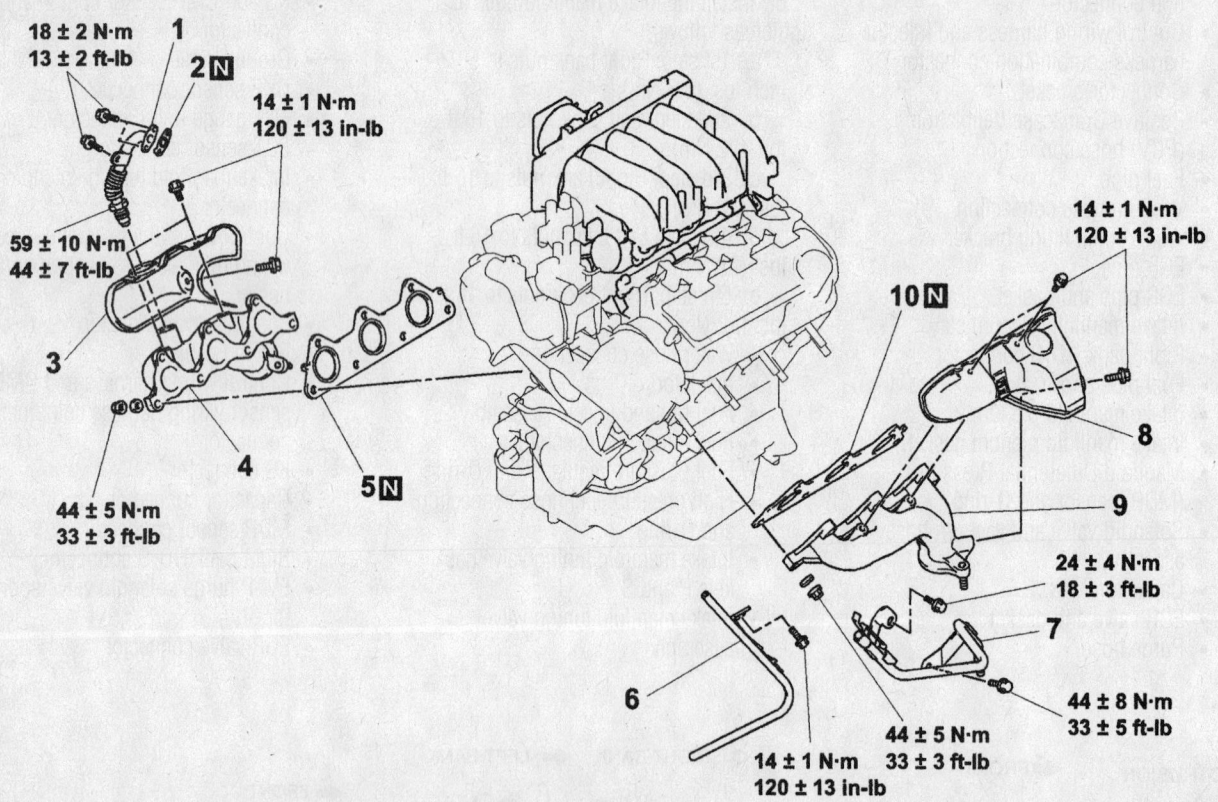

1. EGR PIPE
2. EGR PIPE GASKET
3. HEAT PROTECTOR <RH>
4. EXHAUST MANIFOLD <RH>
5. EXHAUST MANIFOLD GASKET <RH>
6. ENGINE OIL DIPSTICK GUIDE
7. TRANSMISSION FLUID DIPSTICK GUIDE
8. HEAT PROTECTOR <LH>
9. EXHAUST MANIFOLD <LH>
10. EXHAUST MANIFOLD GASKET <LH>

42356-MONT-G10

Exploded view of the exhaust manifolds—3.8L engine

- New gasket and right side exhaust manifold. Torque the nuts to 30–36 ft. lbs. (39–49 Nm).
- Right side heat shield
- EGR gasket and pipe
- Battery tray and battery
- Air cleaner assembly
- Front exhaust pipe to the exhaust manifolds. Torque the nuts to 35ft. (49 Nm).
- Oil dipstick, guide and new O-ring
- Negative battery cable

4. Start the engine and check for exhaust leaks.

Front Crankshaft Seal

REMOVAL & INSTALLATION

3.0L and 3.5L Engines

1. Before servicing the vehicle, refer to the precautions in the beginning of this section.
2. Drain the crankcase.
3. Drain and recycle the engine coolant.
4. Remove or disconnect the following:
- Negative battery cable

✳✳ CAUTION

Wait at least 90 seconds after the negative battery cable is disconnected to prevent possible deployment of the air bag.

- Cooling fan
- Accessory drive belts
- Alternator
- Engine undercover, if equipped
- Power steering oil pump assembly
- Air conditioner compressor and bracket, if equipped
- Timing indicator bracket
- Accessory mount assembly
- Crankshaft pulley
- Timing belt covers and the timing belt
- Crankshaft sprocket

5. Cut out a portion in the crankshaft oil seal lip and pry out the oil seal with a flat prying tool, being careful not to damage the crankshaft.

To install:

6. Coat the lip of the new seal with oil and install the seal using the proper seal driver.

7. Install or connect the following:
- Crankshaft sprocket and the timing belt
- Timing belt covers
- Crankshaft pulley. Torque the bolt to 134 ft. lbs. (181Nm).
- Accessory mount Assembly. Torque the bolts to 33 ft. lbs. (44 Nm).
- Timing indicator bracket. Torque the bolts to 97 inch lbs. (11 Nm).
- Air conditioner compressor and bracket, if equipped
- Power steering oil pump assembly
- Engine undercover, if equipped
- Alternator

- Accessory drive belts
- Cooling fan
- Negative battery cable

8. Refill the crankcase.
9. Refill the cooling system.
10. Start the engine and check for proper operation.

3.8L Engine

1. Before servicing the vehicle, refer to the precautions in the beginning of this section.
2. Drain the crankcase.
3. Drain and recycle the engine coolant.
4. Remove or disconnect the following:
- Negative battery cable

✳✳ CAUTION

Wait at least 90 seconds after the negative battery cable is disconnected to prevent possible deployment of the air bag.

- Timing belt
- Crankshaft sprocket
- Crankshaft Position (CKP) sensor
- Crankshaft sensing blade
- Crankshaft spacer and key
- Front oil seal

To install:

- Front oil seal. Apply oil to the seal, and install using Crankshaft Front Oil Seal Installer tool no. MD998717.
- Crankshaft key and spacer.
- Crankshaft sensing blade
- CKP sensor

➡**To be sure the crankshaft pulley bolt does not loosen, make sure the clean the mating areas of the crankshaft, spacer, sensing blade and sprocket.**

- Crankshaft sprocket
- Timing belt

Camshaft and Valve Lifters

REMOVAL & INSTALLATION

3.0L and 3.5L Engines

1. Before servicing the vehicle, refer to the precautions in the beginning of this section.
2. Relieve the fuel system pressure.
3. Drain the engine oil and coolant.
4. Remove or disconnect the following:
- Negative battery cable

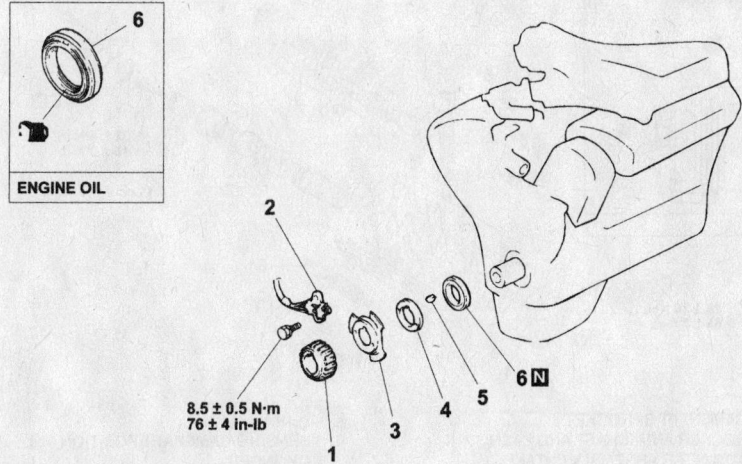

ENGINE OIL

8.5 ± 0.5 N·m
76 ± 4 in-lb

1. CRANKSHAFT SPROCKET
2. CRANKSHAFT POSITION SENSOR
3. CRANKSHAFT SENSING BLADE
4. CRANKSHAFT SPACER
5. KEY
6. CRANKSHAFT FRONT OIL SEAL

42356-MONT-G11

Exploded view of the crankshaft front oil seal—3.8L engine

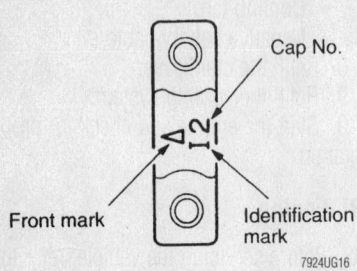

Cap No.

Front mark

Identification mark

7924UG16

The camshaft bearing caps have identification marks on them—3.5L engine

✳ CAUTION

Work must be started after 90 seconds from the time the ignition switch is turned to the LOCK position and the negative battery cable is disconnected.

- Intake manifold plenum
- Valve cover
- Timing belt
- Sprocket from the camshaft

5. Install auto lash adjuster retainers SST MD998443 on the rocker arms.

6. Remove or disconnect the following:
- Distributor and the distributor extension, if equipped
- Rocker arms, rocker shafts and bearing caps, as an assembly
- Thrust cage and O-ring
- Camshaft from the cylinder head

7. Inspect the bearing journals on the camshaft and the cylinder head.

To install:

8. Lubricate the camshaft journals and camshaft with clean engine oil

9. Install or connect the following:
- Camshaft in the cylinder head.
- Thrust cage and O-ring. Torque the bolts to 109 inch lbs. (12 Nm).
- Rocker arms, rocker arm shaft and the rocker shaft spring.

10. Temporarily tighten the rocker shaft with the bolts positioned so that the intake valve rocker arms do not push the valves.

11. Install or connect the following:
- Rocker shaft spring from above and mount it at right angles to the plug guide.

12. Before installing the exhaust rocker arms and the rocker arm shaft, mount the rocker shaft spring.

13. Remove the SST used to hold the lash adjuster in position.

14. Check to ensure that the flat side of

the rocker shaft is perpendicular to the cylinder head, and facing the valves.

15. Gradually tighten the bearing caps in 2 or 3 steps. In the final step tighten to 23 ft. lbs. (31 Nm).

16. Install or connect the following:
- Distributor, if removed
- Sprockets. Torque the bolts to 65 ft. lbs. (88 Nm).
- Timing belt and timing belt cover
- Valve cover. Torque the bolts to 26 inch lbs. (3.4 Nm).
- Intake manifold plenum
- Negative battery cable

17. Start the engine and check for leaks and proper operation.

18. Refill the coolant and crankcase.

3.8L Engine

1. Before servicing the vehicle, refer to the precautions in the beginning of this section.

2. Relieve the fuel system pressure.

3. Remove or disconnect the following:
- Negative battery cable
- Cylinder head assembly
- Camshaft sprocket, using special tools M998715 and MB990767

➡Install auto lash adjuster retainers SST MD998443 on the rocker arms

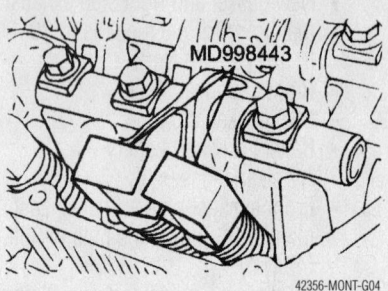

Install the special tool to prevent the lash adjusters from falling to the floor during rocker arm removal—3.8L engine

- Intake rocker arm, shaft and lash adjuster assembly. Loosen the rocker arm assembly mounting bolt and remove the assembly with the bolt still attached.
- Exhaust rocker arm, shaft and lash adjuster assembly. Loosen the rocker arm assembly mounting bolt and remove the assembly with the bolt still attached.
- Camshaft Position (CMP) sensor support and O-ring
- Sensing CMP cylinder
- Camshaft

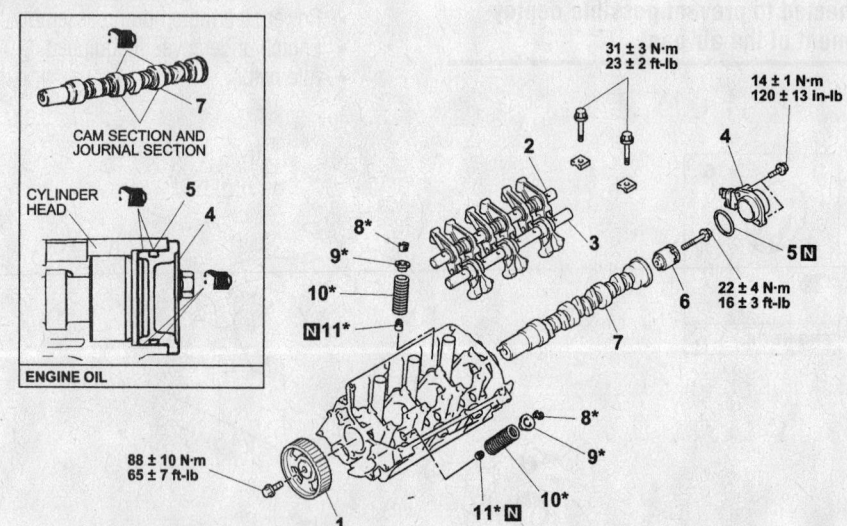

1. CAMSHAFT SPROCKET
2. ROCKER ARM, SHAFT AND LASH ADJUSTER ASSEMBLY (INTAKE SIDE)
3. ROCKER ARM, SHAFT AND LASH ADJUSTER ASSEMBLY (EXHAUST SIDE)
4. CAMSHAFT POSITION SENSOR SUPPORT
5. O-RING
6. SENSING CAMSHAFT POSITION CYLINDER
7. CAMSHAFT
8. VALVE SPRING RETAINER LOCKS
9. VALVE SPRING RETAINERS
10. VALVE SPRINGS
11. VALVE STEM SEALS

42356-MONT-G12

Exploded view of the camshaft and related components—3.8L engine

To install:
- Camshaft
- Sensing CMP cylinder
- CMP sensor support and O-ring
- Exhaust rocker arm, shaft and lash adjuster assembly
- Intake rocker arm, shaft and lash adjuster assembly
- Camshaft sprocket
- Cylinder head assembly
- Negative battery cable

4. Run vehicle and check for leaks.

Starter Motor

REMOVAL & INSTALLATION

1. Before servicing the vehicle, refer to the precautions in the beginning of this section.
2. Remove or disconnect the following:
   - Negative battery cable
   - Engine under cover, if equipped
   - Front engine mount heat protector
   - Starter cover
   - Wires
   - Starter motor

To install:

3. Install or connect the following:

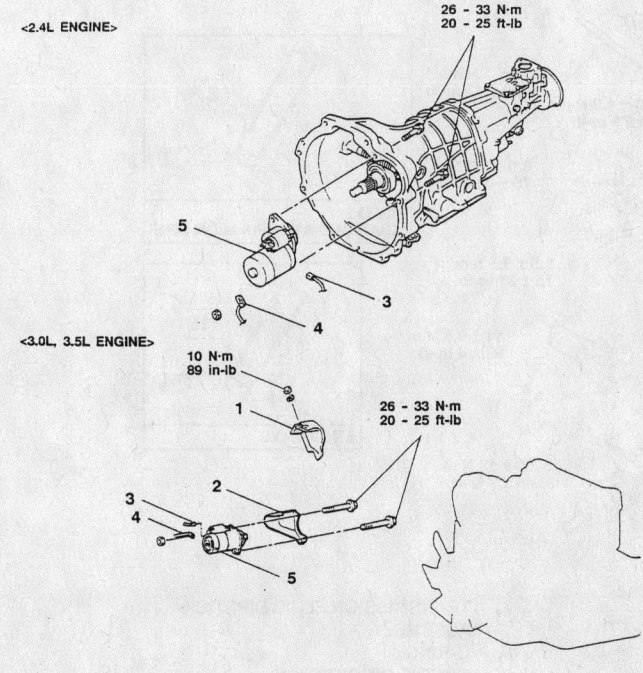

<2.4L ENGINE>

26 - 33 N·m
20 - 25 ft-lb

<3.0L, 3.5L ENGINE>

10 N·m
89 in-lb

26 - 33 N·m
20 - 25 ft-lb

1. FRONT ENGINE MOUNT HEAT PROTECTOR <RH>
2. STARTER COVER
3. STARTER CONNECTOR
4. BATTERY CABLE
5. STARTER ASSEMBLY

9308UG03

Starter motor mounting

- Starter motor and cover. Torque the bolts to 20–25 ft. lbs. (26–33 Nm).
- Wires
- Front engine mount heat protector. Torque the nut to 89 inch lbs. (10 Nm).
- Engine under cover, if equipped
- Negative battery cable

Oil Pan

REMOVAL & INSTALLATION

3.0L Engines

1. Before servicing the vehicle, refer to the precautions in the beginning of this section.
2. Drain the engine oil.
3. Remove or disconnect the following:
   - Negative battery
   - Engine under cover
   - Alternator and belt
   - Stabilizer bar
   - Front exhaust pipe
   - Actuator assembly and heat protector
   - Oil dipstick
   - Crossmember assembly
   - Automatic transmission oil dipstick assembly

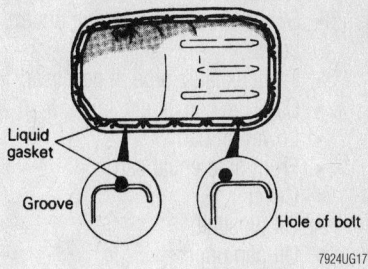

Liquid gasket

Groove

Hole of bolt

7924UG17

Apply a bead of sealant around the oil pan flange as shown—all engines are similar

- Exhaust pipe support bracket
- Transmission stay
- Oil pan, lower
- Oil screen and baffle plate
- Oil pan upper

To install:

4. Before installing, thoroughly clean the oil pan and cylinder block mating surfaces.

5. Apply liquid gasket around the surface of the oil pan.

➡ **Assemble the oil pan to the cylinder block within 15 minutes after applying the liquid gasket.**

6. Install or connect the following:
   - Oil pan upper. Torque the bolts to 53 inch lbs. (6.0 Nm).
   - Oil screen and baffle plate. Torque the bolts to 14 ft. lbs. (19 Nm).
   - Oil pan, lower. Torque the bolts to 53 inch lbs. (6.0 Nm).
   - Transmission stay. Torque the bolts to 26 ft. lbs. (35 Nm).
   - Exhaust pipe support bracket. Torque the bolts to 35 ft. lbs. (49 Nm).
   - Automatic transmission oil dipstick assembly. Torque the bolts to 33 ft. lbs. (44 Nm).
   - Crossmember assembly. Torque the bolts to 80 ft. lbs. (108 Nm).
   - Oil dipstick. Torque the bolts to 35 ft. lbs. (48 Nm).
   - Actuator assembly and heat protector
   - Front exhaust pipe
   - Stabilizer bar
   - Alternator and belt
   - Engine under cover
   - Negative battery

3.5L Engines

1. Before servicing the vehicle, refer to the precautions in the beginning of this section.
2. Drain the engine oil.

3. Remove or disconnect the following:
- Negative battery cable
- Skid plate and the engine under-cover
- Front exhaust pipe, if necessary
- Catalytic converter
- Lower oil pan
- Front differential carrier
- Cover
- Oil dipstick
- Oil pan upper
- Oil screen

To install:

4. Before installing, thoroughly clean the oil pan and cylinder block mating surfaces.

5. Apply liquid gasket around the surface of the oil pan.

➡**Assemble the oil pan to the cylinder block within 15 minutes after applying the liquid gasket.**

6. Install or connect the following:
- Oil screen. Torque the bolts to 13 ft. lbs. (19 Nm).

- Oil pan upper. Torque the bolts to 48 inch lbs. (6 Nm).
- Oil dipstick. Torque the bolts to 39 inch lbs. (4.8 Nm).
- Cover. Torque the bolts to 84–108 inch lbs. (10–12 Nm).
- Front differential carrier
- Lower oil pan. Torque the bolts to 84–108 inch lbs. (10–12 Nm).
- Catalytic converter
- Front exhaust pipe, if necessary. Torque the bolts to 35 ft. lbs. (49 Nm).
- Skid plate and the engine under-cover
- Negative battery cable

3.8L Engines

1. Before servicing the vehicle, refer to the precautions in the beginning of this section.

2. Drain the engine oil.

3. Remove or disconnect the following:

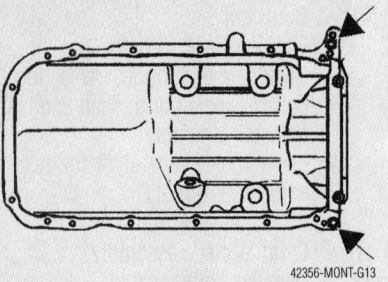

42356-MONT-G13

Screw the M10 bolts that hold the oil pan to the transmission assembly into the bolt holes shown by arrows—3.8L engine

- Negative battery cable
- Skid plate and the engine under-cover
- Starter motor
- Right and left halfshaft connections
- Front differential No. 2 crossmember
- Drain plug, gasket and cover

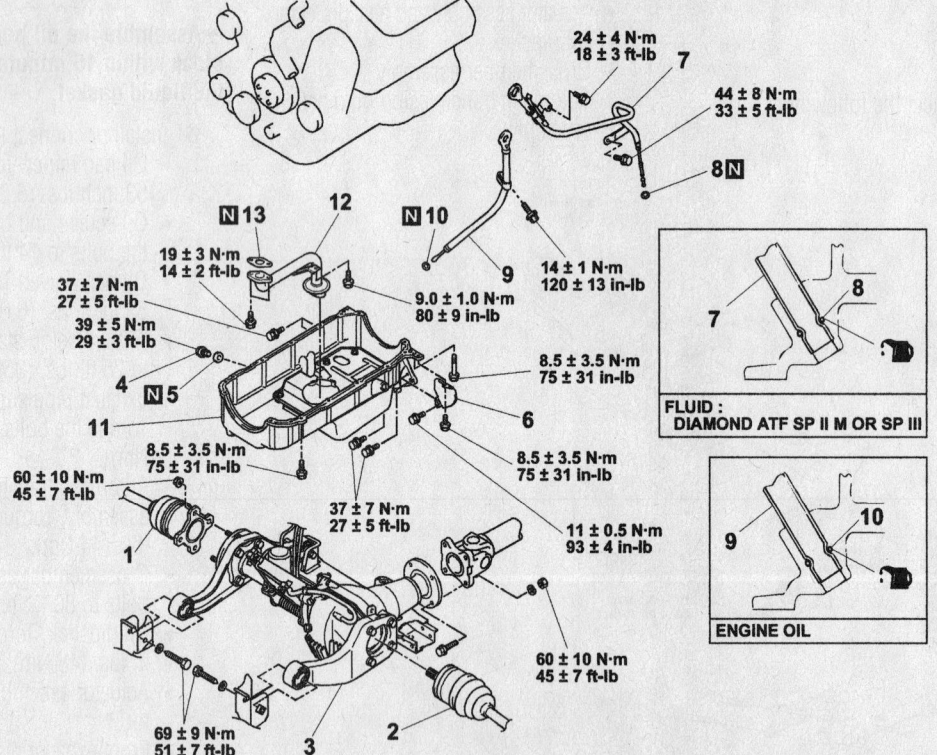

1. DRIVE SHAFT (RH) CONNECTION
2. DRIVE SHAFT (LH) CONNECTION
3. FRONT DIFFERENTIAL NUMBER 2 CROSSMEMBER ASSEMBLY
4. DRAIN PLUG
5. DRAIN PLUG GASKET
6. COVER
7. TRANSMISSION FLUID DIPSTICK ASSEMBLY
8. O-RING
9. ENGINE OIL DIPSTICK ASSEMBLY
10. O-RING
11. OIL PAN
12. OIL SCREEN
13. GASKET

42356-MONT-G14

Exploded view of the oil pan and related components—3.8L engine

- Transmission fluid dipstick assembly and O-ring
- Engine oil dipstick and O-ring
- Oil pan retainers and oil pan. Screw the M10 bolts holding the oil pan to the transmission assembly into the bolt holes shown in the illustration to remove the pan.
- Oil screen
- Oil pan gasket

To install:

4. Before installing, thoroughly clean the oil pan and cylinder block mating surfaces.

5. Apply liquid gasket around the surface of the oil pan.

➥**Assemble the oil pan to the cylinder block within 30 minutes after applying the liquid gasket.**

6. Install or connect the following:
- Oil screen. Torque the bolts to 12–16 ft. lbs. (16–22 Nm).
- Oil pan. Torque the bolts, in sequence, to specifications shown the illustration. The bolt holes for bolts 13 and 14 are cut away on the transmission side. Be sure you do not insert the bolts at an angle.
- Engine oil dipstick and O-ring

- Transmission fluid dipstick assembly and O-ring
- Drain plug, gasket and cover
- Front differential No. 2 crossmember
- Right and left halfshaft connections
- Skid plate and the engine undercover
- Negative battery cable

Oil Pump

REMOVAL & INSTALLATION

1. Before servicing the vehicle, refer to the precautions in the beginning of this section.
2. Drain the engine oil.
3. Remove or disconnect the following:
- Negative battery cable

- Timing belt
- Oil pressure switch
- Oil dipstick
- Oil pans from the engine
- Oil baffle and screen
- Oil pump mounting bolts and the pump from the front of the engine

➥**Note the position of each oil pump case retaining bolts to facilitate installation. The bolts are of different length.**

To install:

4. Clean the gasket mounting surfaces of the pump and engine block.

5. Prime the pump by pouring fresh oil into the inlet and turning the rotors or by packing pump with petroleum jelly. Using a new gasket, install the oil pump on the engine and tighten all bolts to 10 ft. lbs. (14 Nm).

6. Clean out the oil pick-up or replace

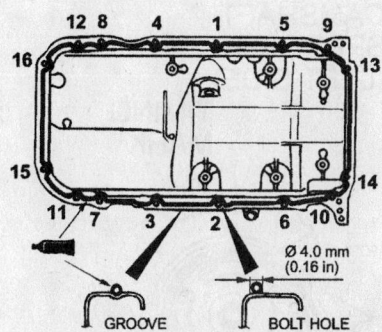

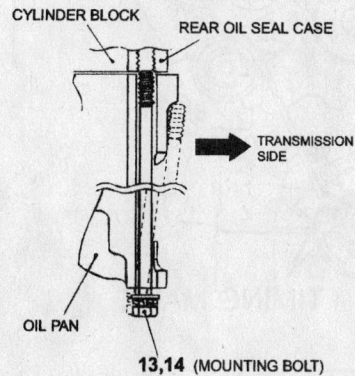

Oil pan bolt tightening sequence—3.8L engine

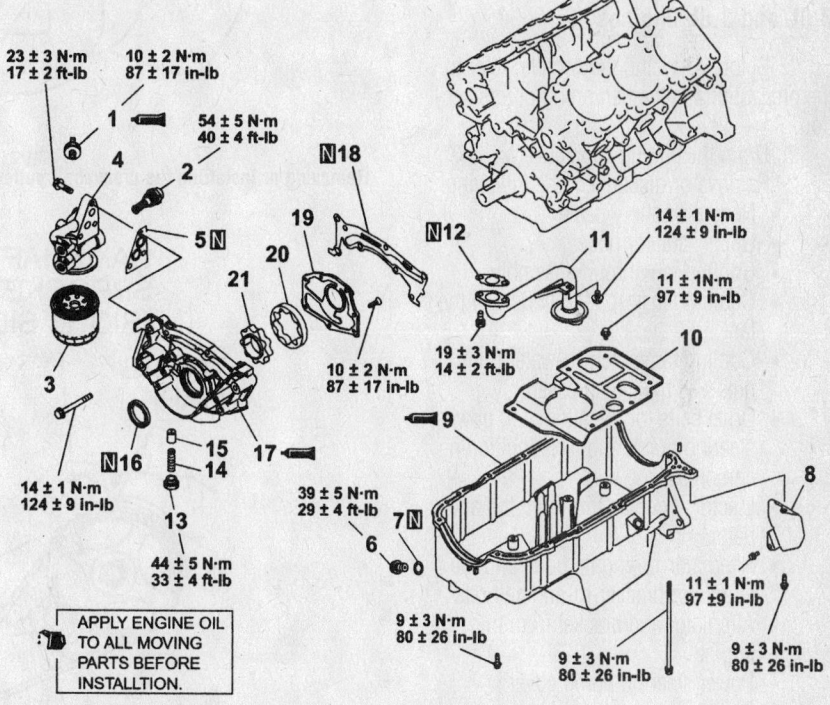

1. ENGINE OIL PRESSURE SWITCH
2. OIL COOLER BY-PASS VALVE
3. OIL FILTER
4. OIL FILTER BRACKET
5. OIL FILTER BRACKET GASKET
6. DRAIN PLUG
7. DRAIN PLUG GASKET
8. COVER
9. OIL PAN
10. BAFFLE PLATE
11. OIL SCREEN
12. OIL SCREEN GASKET
13. RELIEF PLUG
14. RELIEF SPRING
15. RELIEF PLUNGER
16. CRANKSHAFT OIL SEAL
17. OIL PUMP CASE
18. OIL PUMP CASE GASKET
19. OIL PUMP COVER
20. OIL PUMP OUTER ROTOR
21. OIL PUMP INNER ROTOR

Exploded view of the oil pump, oil pan and related components—3.8L engine shown, others similar

as required. Replace the oil pick-up gasket ring and install the pick-up to the pump.

7. Install or connect the following:
- Oil filter and the bracket. Torque the bolts to 17 ft. lbs. (23 Nm).
- Oil baffle and screen. Torque the bolts to 13 ft. lbs. (18 Nm).
- Oil pans. Torque the engines to 52 inch lbs. (5.9 Nm).
- Oil pressure switch. Torque the switch to 87 inch lbs. (9.8 Nm).
- Timing belt
- Dipstick
- Negative battery cable

8. Refill the engine with the proper amount of oil.

9. Start the engine and check for proper oil pressure. Check for leaks.

Timing Belt

REMOVAL & INSTALLATION

3.0L and 3.5L Engines

1. Before servicing the vehicle, refer to the precautions in the beginning of this section.

2. Drain the engine coolant.

3. Remove or disconnect the following:
- Negative battery cable
- Upper radiator hose
- Cooling fan shroud assembly
- Cooling fan-to-clutch bolts and the fan
- Cooling fan clutch-to-water pump nuts and the clutch assembly
- Drive belts for the alternator, power steering pump and air conditioning compressor
- Electrical connectors from the alternator
- Alternator-to-engine bolts and the alternator bracket-to-engine bolts
- Alternator and bracket from the engine
- Power steering pump cover
- Power steering pump-to-engine bolts and move the pump aside with the hoses and electrical connector attached
- Air conditioning compressor-to-bracket bolts and move the compressor aside with the lines and electrical connector attached
- Air conditioning compressor bracket-to-engine bolts and the bracket
- Timing indicator bracket (near crankshaft pulley) bolts and the bracket

- Accessory mount assembly-to-engine bolts and the mount assembly
- Upper timing belt cover assembly

4. Using the End Yoke Holder tool MD990767 and 2 Crankshaft Pulley Holder Pin tools MD998715, or equivalent to hold the crankshaft pulley, and a socket wrench, remove the crankshaft pulley bolt and the pulley.
- Lower timing belt cover

5. Rotate the crankshaft clockwise to align the timing marks to position the No. 1 cylinder at the Top Dead Center (TDC) of its compression stroke.

6. Use chalk to mark the rotating (clockwise) direction of the timing belt for reinstallation purposes.

7. Loosen the auto-tensioner pulley center bolt and remove the timing bolt.

8. Remove the auto-tensioner pulley and the auto-tensioner arm assembly.

To install:

9. Press the end of the auto-tensioner inward with 72–145 ft. lbs. (98–196 Nm) of force and measure the distance that the pushrod is pushed in. If the standard distance is not 0.04 in. (1mm), replace the auto-tensioner.

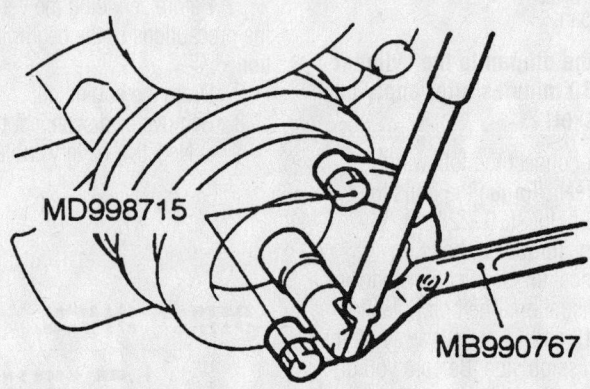

93025G14

Removing or installing the crankshaft pulley bolt—3.0L and 3.5L engines

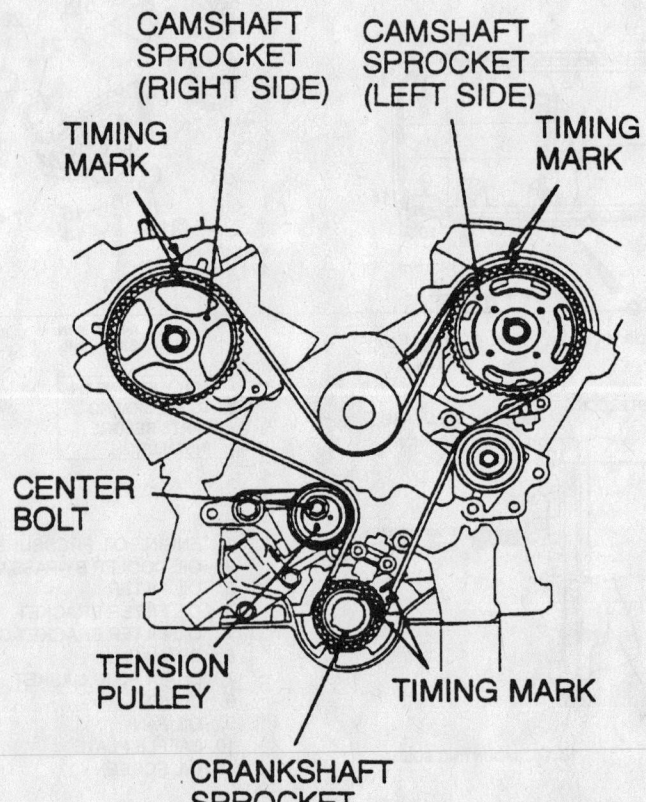

93025G15

View of the timing belt alignment marks—3.0L and 3.5L engines

10. Position the auto-tensioner in a soft-jawed vise and SLOWLY compress the pushrod until the pushrod and housing holes align; then, install a setting pin to secure the auto-tensioner in the retracted position.

11. Align the camshaft and crankshaft TDC timing marks.

12. Install the timing belt (noting its rotational direction) so that there is no deflection between the sprockets and pulleys in the following manner:
- Crankshaft sprocket
- Idler pulley
- Left camshaft sprocket
- Water pump pulley
- Right camshaft sprocket
- Tension pulley

13. Turn the camshaft sprocket counterclockwise until the tension side of the timing belt is firmly stretched, then, recheck the timing marks.

14. Using the Tension Pulley Socket Wrench tool MD998767, or equivalent, push the tensioner pulley into the timing belt and secure the center bolt.

15. Using the Crankshaft Pulley Spacer tool MD998769, or equivalent, rotate the crankshaft 1/4 turn counterclockwise, then, turn it again clockwise to align the timing marks.

16. Loosen the timing belt tensioner center bolt. Using the Tension Pulley Socket Wrench tool MD998767, or equivalent, and a torque wrench, apply 39 inch lbs. (4.4 Nm) pressure on the timing belt. Torque the tensioner pulley center bolt to 35 ft. lbs. (48 Nm).

17. Remove the setting pin from the auto-tensioner.

18. Rotate the crankshaft 2 complete revolutions and realign the timing marks. Then, wait for 5 minutes until the auto-tensioner pushrod extends to its standard value. If the standard value is not 0.15–0.20 in. (3.8–5.0mm), repeat the adjustment procedure. If the standard value is still not achieved, replace the auto-tensioner.

19. Install the lower timing belt cover and crankshaft pulley.

20. Using the End Yoke Holder tool MD990767 and 2 Crankshaft Pulley Holder Pin tools MD998715, or equivalent to hold the crankshaft pulley, and a socket torque wrench, torque the crankshaft pulley bolt to 134 ft. lbs. (181 Nm).
- Install or connect the following:
- Upper timing belt cover assembly
- Remaining items by reversing the removal procedures
- Negative battery cable

21. Refill the cooling system.

3.8L ENGINE

1. Before servicing the vehicle, refer to the precautions in the beginning of this section.

2. Drain the engine coolant.

3. Remove or disconnect the following:
- Negative battery cable
- Skid plate and under cover
- Battery and battery tray
- Air cleaner
- Radiator shroud cover
- Drive belt
- Cooling fan and pulley
- Drive belt auto tensioner
- Accessory mount stay
- Power steering pump. Unbolt and position aside; do not disconnect the fluid lines.
- A/C compressor. Unbolt and position aside; do not disconnect the refrigerant lines.
- Compressor bracket
- Cooling fan bracket assembly
- Accessory mount assembly
- Timing belt upper cover
- Crankshaft pulley, using special tools MB991800 and MB991802

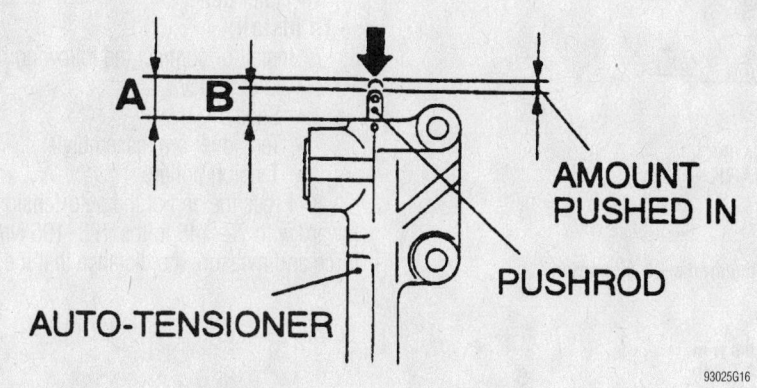

Inspecting the auto-tensioner movement—3.0L and 3.5L engines

93025G16

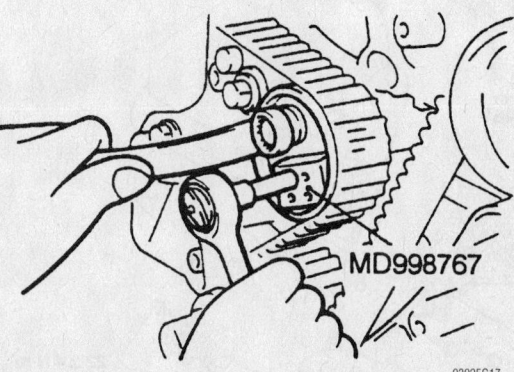

Adjusting the timing belt tensioner pulley—3.0L and 3.5L engines

93025G17

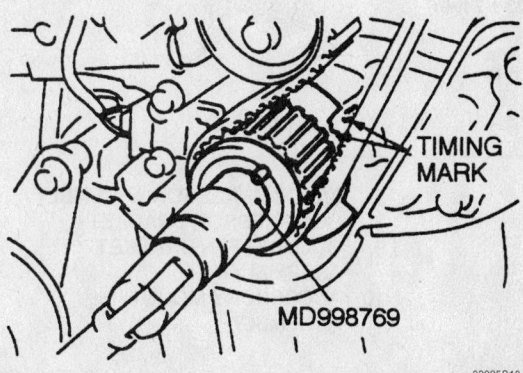

Using crankshaft spacer tool to rotate the crankshaft—3.0L and 3.5L engines

93025G18

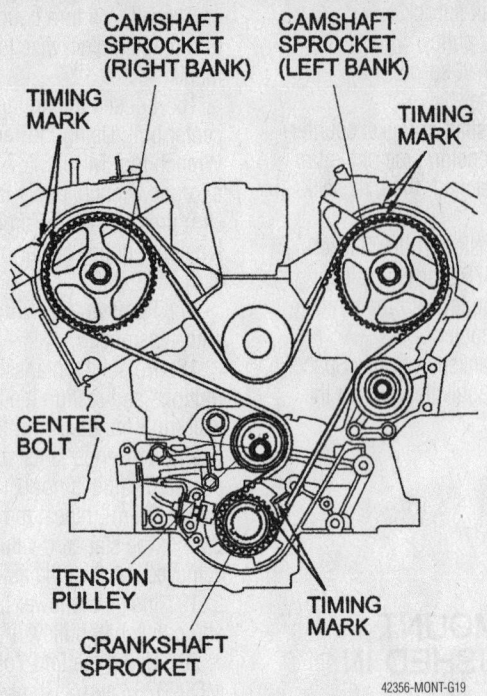

CAMSHAFT SPROCKET (RIGHT BANK)
CAMSHAFT SPROCKET (LEFT BANK)
TIMING MARK
TIMING MARK
CENTER BOLT
TENSION PULLEY
CRANKSHAFT SPROCKET
TIMING MARK

42356-MONT-G19

Installed view of the timing belt and alignment of the timing marks—3.8L engine

- Timing belt indicator bracket
- Timing belt lower cover
- Auto tensioner

4. Turn the crankshaft clockwise to align the timing marks and set the No. 1 cylinder at Top Dead Center (TDC). If you are reusing the timing belt, mark the flat side of the belt with an arrow showing the clockwise direction.

5. Loosen the center bolt of the tension pulley, and then remove the timing belt.

6. Remove or disconnect the following:
- Tension pulley
- Tensioner arm assembly
- Shaft
- Idler pulley

To install:

7. Install or connect the following:
- Idler pulley
- Shaft
- Tensioner arm assembly
- Tension pulley

8. Press the end of the auto-tensioner inward with 72–145 ft. lbs. (98–196 Nm) of force and measure the distance that the

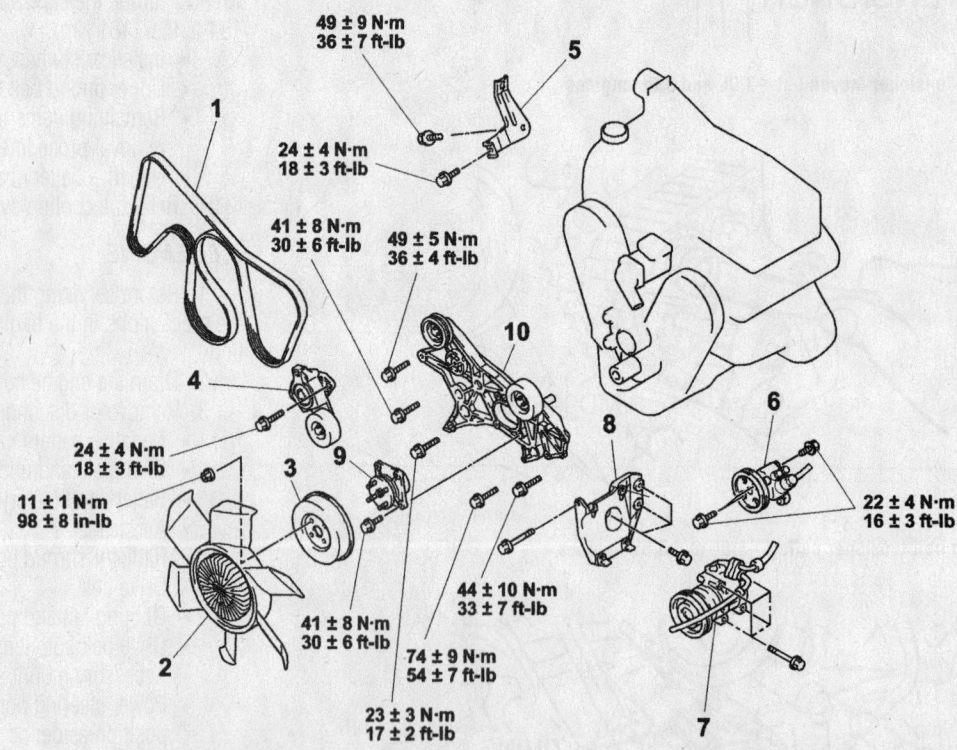

49 ± 9 N·m
36 ± 7 ft-lb

24 ± 4 N·m
18 ± 3 ft-lb

41 ± 8 N·m
30 ± 6 ft-lb

49 ± 5 N·m
36 ± 4 ft-lb

24 ± 4 N·m
18 ± 3 ft-lb

11 ± 1 N·m
98 ± 8 in-lb

44 ± 10 N·m
33 ± 7 ft-lb

41 ± 8 N·m
30 ± 6 ft-lb

74 ± 9 N·m
54 ± 7 ft-lb

23 ± 3 N·m
17 ± 2 ft-lb

22 ± 4 N·m
16 ± 3 ft-lb

1. DRIVE BELT
2. COOLING FAN
3. COOLING FAN PULLEY
4. DRIVE BELT AUTO TENSIONER
5. ACCESSORY MOUNT STAY
6. POWER STEERING OIL PUMP ASSEMBLY
7. A/C COMPRESSOR ASSEMBLY
8. COMPRESSOR BRACKET
9. COOLING FAN BRACKET ASSEMBLY
10. ACCESSORY MOUNT ASSEMBLY

42356-MONT-G17

Exploded view of the components you need to remove for timing belt removal—3.8L engine

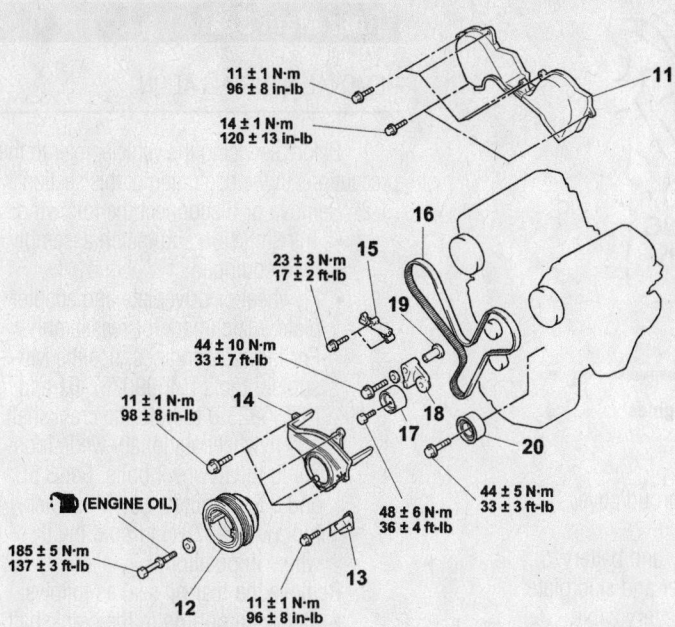

11 ± 1 N·m
96 ± 8 in-lb
— 11

14 ± 1 N·m
120 ± 13 in-lb

16

23 ± 3 N·m
17 ± 2 ft-lb   15

19

44 ± 10 N·m
33 ± 7 ft-lb

11 ± 1 N·m
98 ± 8 in-lb   14   18

17

20

(ENGINE OIL)

185 ± 5 N·m
137 ± 3 ft-lb

12

13

11 ± 1 N·m
96 ± 8 in-lb

48 ± 6 N·m
36 ± 4 ft-lb

44 ± 5 N·m
33 ± 3 ft-lb

11. TIMING BELT UPPER COVER
    ASSEMBLY
12. CRANKSHAFT PULLEY
13. TIMING BELT INDICATOR
    BRACKET
14. TIMING BELT LOWER COVER
    ASSEMBLY

15. AUTO-TENSIONER
16. TIMING BELT
17. TENSION PULLEY
18. TENSIONER ARM ASSEMBLY
19. SHAFT
20. IDLER PULLEY

42356-MONT-G18

Exploded view of the timing belt and related components—3.8L engine

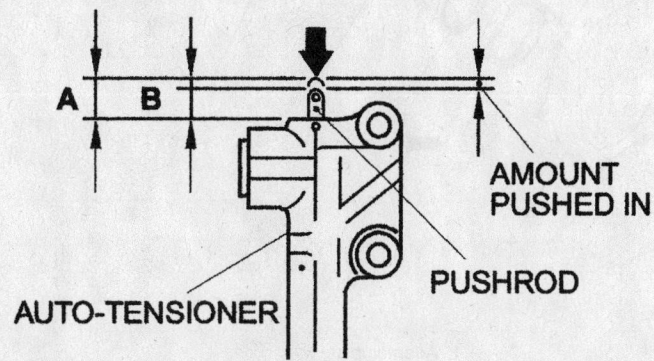

A   B

AMOUNT
PUSHED IN

PUSHROD

AUTO-TENSIONER

42356-MONT-G20

Inspecting the auto-tensioner movement—3.8L engines

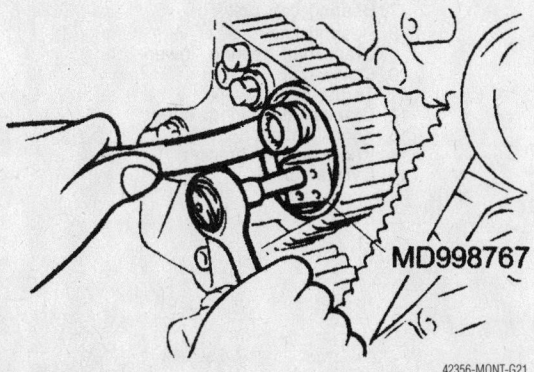

MD998767

42356-MONT-G21

Adjusting the timing belt tensioner pulley—3.8L engines

pushrod is pushed in. If the standard distance is not 0.04 in. (1mm), replace the auto-tensioner.

9. Position the auto-tensioner in a soft-jawed vise and SLOWLY compress the pushrod until the pushrod and housing holes align; then, install a setting pin to secure the auto-tensioner in the retracted position.

10. Align the camshaft and crankshaft TDC timing marks.

11. Install the timing belt (noting its rotational direction) so that there is no deflection between the sprockets and pulleys in the following manner:
   • Crankshaft sprocket
   • Idler pulley
   • Left camshaft sprocket
   • Water pump pulley
   • Right camshaft sprocket
   • Tension pulley

12. Turn the camshaft sprocket counterclockwise until the tension side of the timing belt is firmly stretched, then, recheck the timing marks.

13. Using the Tension Pulley Socket Wrench tool MD998767, or equivalent, push the tensioner pulley into the timing belt and secure the center bolt.

14. Using the Crankshaft Pulley Spacer tool MD998769, or equivalent, rotate the crankshaft 1/4 turn counterclockwise, then, turn it again clockwise to align the timing marks.

15. Loosen the timing belt tensioner center bolt. Using the Tension Pulley Socket Wrench tool MD998767, or equivalent, and a torque wrench, apply 39 inch lbs. (4.4 Nm) pressure on the timing belt. Torque the tensioner pulley center bolt to 35 ft. lbs. (48 Nm).

16. Remove the setting pin from the auto-tensioner.

17. Rotate the crankshaft 2 complete revolutions and realign the timing marks. Then, wait for 5 minutes until the auto-tensioner pushrod extends to its standard value. If the standard value is not 0.15–0.20 in. (3.8–5.0mm), repeat the adjustment procedure. If the standard value is still not achieved, replace the auto-tensioner.

18. Install the lower timing belt cover and crankshaft pulley.

19. Using the End Yoke Holder tool MD990767 and 2 Crankshaft Pulley Holder Pin tools MD998715, or equivalent to hold the crankshaft pulley, and a socket torque wrench, torque the crankshaft pulley bolt to 134 ft. lbs. (181 Nm).
   • Install or connect the following:
   • Timing belt upper cover
   • Accessory mount assembly

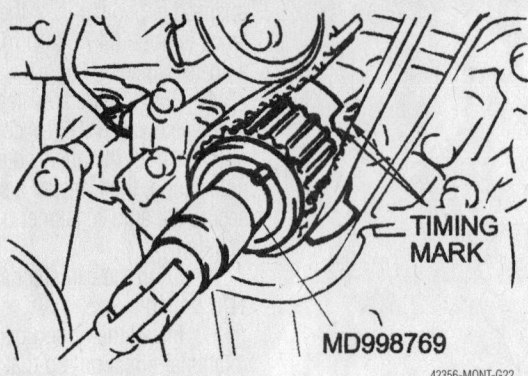

MD998769

42356-MONT-G22

Using crankshaft spacer tool to rotate the crankshaft—3.0L and 3.5L engines

- Cooling fan bracket assembly
- Compressor bracket
- A/C compressor
- Power steering pump
- Accessory mount stay
- Drive belt auto tensioner
- Cooling fan pulley and fan

- Drive belt
- Radiator shroud cover
- Air cleaner
- Battery tray and battery
- Under cover and skid plate
- Negative battery cable

20. Refill the cooling system.

Rear Main Seal

REMOVAL & INSTALLATION

1. Before servicing the vehicle, refer to the precautions in the beginning of this section.
2. Remove or disconnect the following:
   - Transmission and clutch assembly, if so equipped.
   - Flywheel or driveplate and adapter plate, matchmark for reassembly. For the 3.0L engines, use the Mitsubishi tools (MB990767-01 and MIT308239) to hold the crankshaft and flywheel stationary while loosening the flywheel bolts. For 3.5L and 3.8L engines, use Mitsubishi tool (MD998781) to hold the flywheel in position.
3. Remove the rear oil seal as follows:
   a. Cut out a portion in the crankshaft oil seal lip.

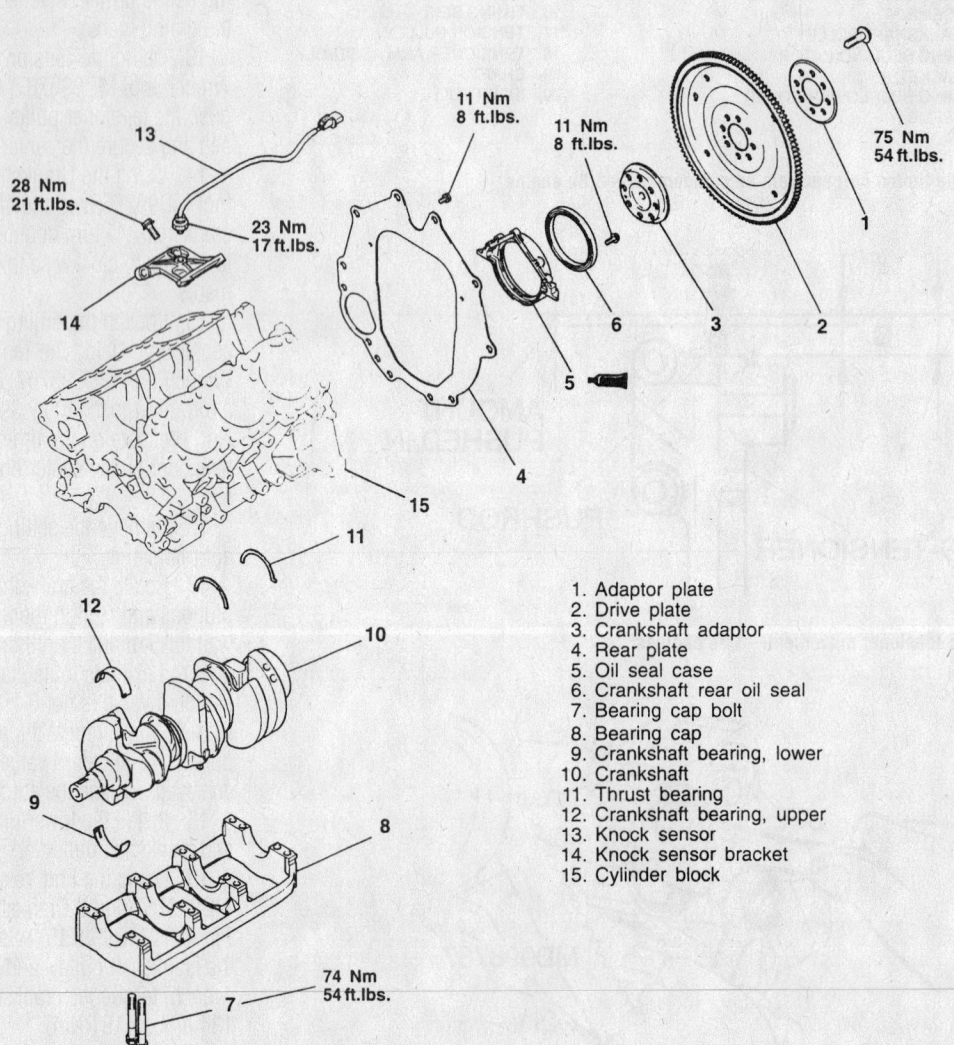

1. Adaptor plate
2. Drive plate
3. Crankshaft adaptor
4. Rear plate
5. Oil seal case
6. Crankshaft rear oil seal
7. Bearing cap bolt
8. Bearing cap
9. Crankshaft bearing, lower
10. Crankshaft
11. Thrust bearing
12. Crankshaft bearing, upper
13. Knock sensor
14. Knock sensor bracket
15. Cylinder block

7924UG22

Exploded view of the crankshaft, rear main seal and related components—3.5L engine shown, others similar

b. Cover the tip of a small prytool with a cloth and apply it to the cutout in the oil seal to pry the oil seal out.

❈❈ CAUTION

Take care not to damage the crankshaft and oil seal case.

To install:

4. Inspect the sealing surface at the rear of the crankshaft. If a deep groove is worn into the surface, the crankshaft will have to be replaced. Coat the sealing lip of the seal with fresh, clean engine oil. Press the new seal into the case with a seal installing tool. The seal must be pressed in squarely until it bottoms in the case. It is necessary to use the proper tool (MD998718-01) to fit the seal into place.

5. Install or connect the following:
- Rear plate
- Transmission mounting plate
- Flywheel or drive plate and adapter
- Transmission and related components as necessary

Piston and Ring

POSITION

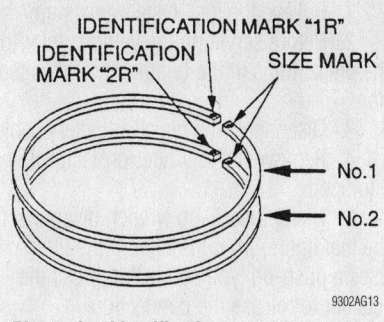

Piston ring identification

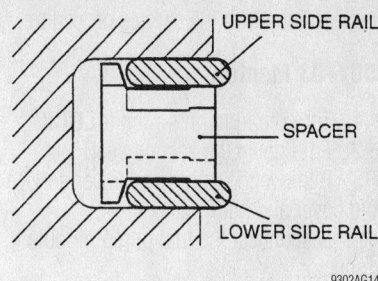

Oil ring identification

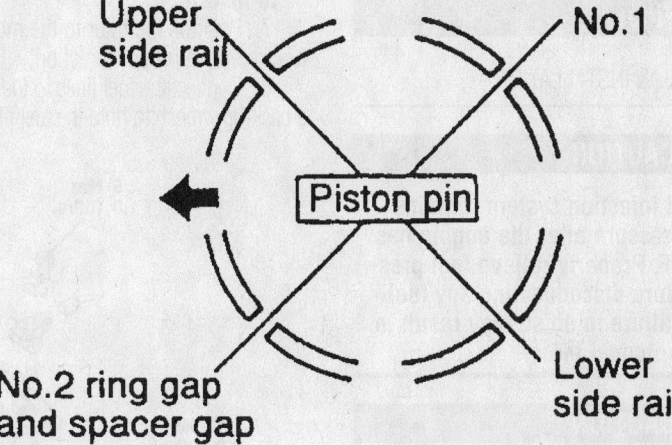

Piston ring end-gap spacing

FUEL SYSTEM

Fuel System Service Precautions

Safety is the most important factor when performing not only fuel system maintenance but any type of maintenance. Failure to conduct maintenance and repairs in a safe manner may result in serious personal injury or death. Maintenance and testing of the vehicle's fuel system components can be accomplished safely and effectively by adhering to the following rules and guidelines.

- To avoid the possibility of fire and personal injury, always disconnect the negative battery cable unless the repair or test procedure requires that battery voltage be applied.
- Always relieve the fuel system pressure prior to disconnecting any fuel system component (injector, fuel rail, pressure regulator, etc.), fitting or fuel line connection. Exercise extreme caution when relieving fuel system pressure, to avoid exposing your skin, face and eyes to fuel spray. Please be advised that fuel under pressure may penetrate the skin or any part of the body that it contacts.
- Always place a shop towel or cloth around the fitting or connection prior to loosening to absorb any excess fuel due to spillage. Ensure that all fuel spillage (should it occur) is quickly removed from engine surfaces. Ensure that all fuel soaked cloths or towels are deposited into a suitable waste container.
- Always keep a dry chemical (Class B) fire extinguisher near the work area.
- Do not allow fuel spray or fuel vapors to come into contact with a spark or open flame.
- Always use a back-up wrench when loosening and tightening fuel line connection fittings. This will prevent unnecessary stress and torsion to fuel line piping. Always follow the proper torque specifications.
- Always replace worn fuel fitting O-rings with new. Do not substitute fuel hose where fuel pipe is installed.

Fuel System Pressure

RELIEVING

❈❈ CAUTION

The fuel system is under constant pressure, even with the engine off. This pressure must be relieved before disconnecting any fuel system component, fitting or fuel line connection. Failure to do so may result in personal injury.

2000–03 Montero Sport and 2000 Montero

1. Disconnect the fuel pump electrical connector, located at the rear side of the fuel tank.
2. Start the engine.
3. After the engine stalls, turn the ignition switch **OFF** and reconnect the fuel pump connector.

4. Disconnect the negative battery cable, then continue with the service procedure.

2001–03 Montero

1. Turn the ignition switch to **LOCK**.
2. Fold down the second seat.
3. Remove the upper and lower service hole cover and packing.
4. Disconnect the fuel pump module connector.
5. Start the engine and let it run out of fuel.

Fuel Filter

REMOVAL & INSTALLATION

✳✳ CAUTION

The fuel injection system remains under pressure after the engine has been **OFF**. Properly relieve fuel pressure before disconnecting any fuel lines. Failure to do so may result in fire or personal injury.

✳✳ CAUTION

Do not allow fuel spray or fuel vapors to come in contact with a spark or open flame. Keep a dry chemical fire extinguisher nearby. Never store fuel in an open container due to risk of fire or explosion.

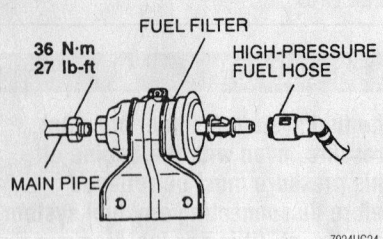

36 N·m 27 lb-ft
FUEL FILTER
HIGH-PRESSURE FUEL HOSE
MAIN PIPE

7924UG24

Fuel filter removal—Montero Sport shown

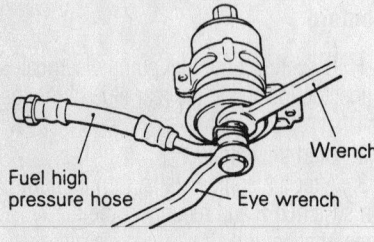

Fuel high pressure hose — Eye wrench — Wrench

7924UG25

Always use a back-up wrench when removing or installing fuel lines to the filter

1. Relieve the fuel system pressure.
2. Before servicing the vehicle, refer to the precautions in the beginning of this section.
3. Disconnect the negative battery cable.
4. Remove the fuel filter protector if equipped.
5. Using a back-up wrench disconnect the fuel line(s) from the filter. If the filter uses a push-on type connector, press the retainer to release the connection.
6. Remove the filter from the mounting bracket.

To install:

7. Position the filter to the mounting bracket in the proper direction.
8. Connect the fuel lines to the filter. Use a back-up wrench to hold the fuel filter. Torque the banjo bolt(s) to 18–25 ft. lbs. (25–35 Nm) or the line fitting to 27 ft. lbs. (36 Nm).
9. Install the fuel filter protector if equipped.
10. Connect the negative battery cable.
11. Start the engine and check for leaks.

Fuel Pump

REMOVAL & INSTALLATION

Montero

➡ **The manufacturer recommends draining of the fuel tank.**

1. Before servicing the vehicle, refer to the precautions in the beginning of this section.

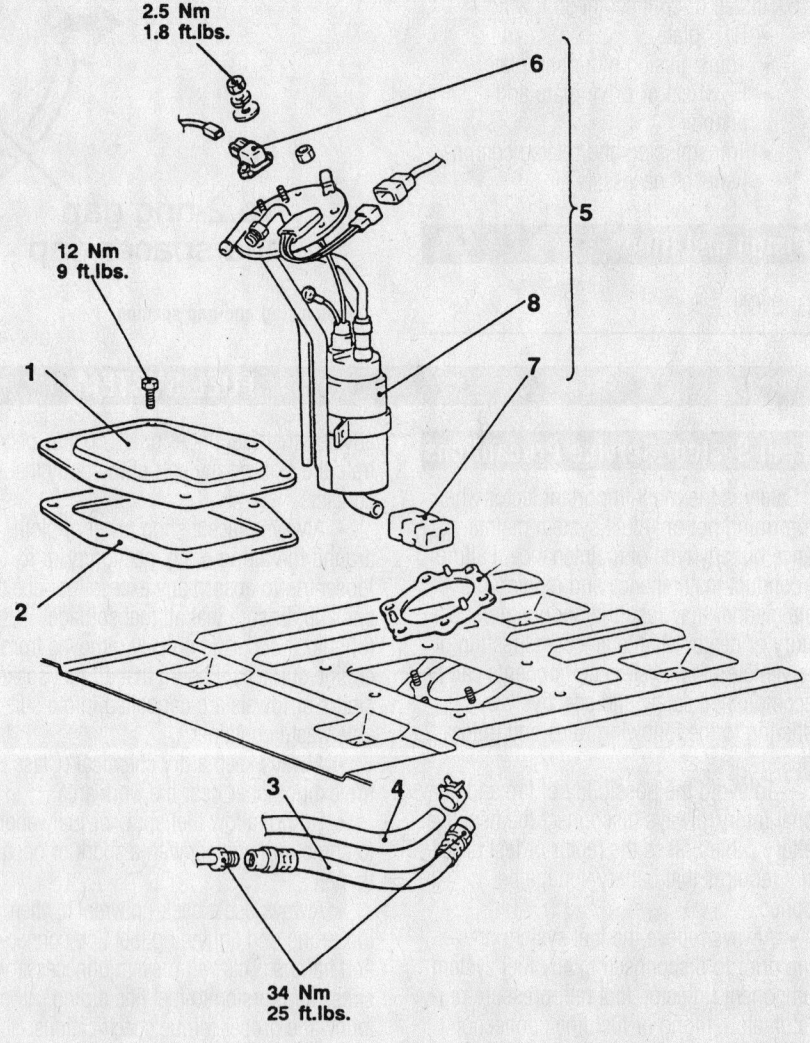

2.5 Nm 1.8 ft.lbs.

12 Nm 9 ft.lbs.

34 Nm 25 ft.lbs.

1. Floor cover
2. Packing
3. High-pressure fuel hose
4. Fuel return hose connection
5. Fuel pump and filter assembly
6. Fuel tank differential pressure sensor
7. Filter
8. Fuel pump assembly

7924UG26

The fuel pump on the Montero is removed through the rear floor pan

2. Relieve the fuel system pressure.

3. Remove or disconnect the following:
- Negative battery cable

✳✳ CAUTION

The fuel injection system remains under pressure after the engine has beenOFF. Properly relieve fuel pressure before disconnecting any fuel lines. Failure to do so may result in fire or personal injury. Do not allow fuel spray or fuel vapors to come in contact with a spark or open flame. Keep a dry chemical fire extinguisher nearby. Never store fuel in an open container due to risk of fire or explosion.

- Rear floor carpeting
- Fuel pump cover
- Fuel pump connector and the fuel hoses
- Fuel pump assembly

To install:

4. Install or connect the following:
- Fuel pump assembly into the fuel tank. Torque the nuts to 24 inch lbs. (2.5 Nm).
- Fuel lines and the fuel pump connector.
- Fuel pump cover. Torque the bolts to 108 inch lbs. (12 Nm).
- Rear floor carpeting.
- Negative battery cable

5. Refill the fuel tank, if drained

6. Start the vehicle; check for leaks and proper operation.

Montero Sport

1. Before servicing the vehicle, refer to the precautions in the beginning of this section.

2. Properly relieve the fuel system pressure.

3. Remove the fuel tank drain plug and drain the fuel from the tank.

✳✳ CAUTION

The fuel injection system remains under pressure after the engine has beenOFF. Properly relieve fuel pressure before disconnecting any fuel lines. Failure to do so may result in fire or personal injury.

4. Remove or disconnect the following:
- Negative battery cable

✳✳ CAUTION

Wait at least 90 seconds after the negative battery cable is disconnected to prevent possible deployment of the air bag.

- Fuel tank protector, if equipped
- Fuel tank from the vehicle
- Fuel pump retaining screws and the pump from the tank

To install:

5. Clean the seal area of the tank.

6. Install or connect the following:
- New gasket
- Fuel pump in the same position as originally installed.
- Fuel pump retaining screws, Torque the nuts to 22 inch lbs. (2.5 Nm).
- Fuel tank. Torque the bolts to 20 ft. lbs. (27 Nm).
- Fuel tank drain plug and the fuel tank protector, if equipped
- Negative battery cable

7. Refill the fuel tank and install the cap.

8. Check fuel system for leaks.

Fuel Injector

REMOVAL & INSTALLATION

3.0L and 3.5L Engines

1. Before servicing the vehicle, refer to the precautions in the beginning of this section.

2. Properly relieve the fuel system pressure.

3. Remove or disconnect the following:
- Air cleaner assembly, ducts and air intake hose
- Intake manifold plenum
- Fuel return line
- Pressure regulator, vacuum line and O-ring
- High pressure fuel hose and O-ring
- Injector connector
- Fuel pipe and O-rings
- Fuel rails and insulators
- Fuel injectors and insulators
- O-rings and grommets

To install:

4. Install or connect the following:
- O-rings and grommets
- Fuel injectors and insulators
- Fuel rail and insulators. Torque the bolts to 106 inch lbs. (12 Nm).
- Injector connector
- High pressure fuel hose and O-ring. Torque the bolts to 43 inch lbs. (4.9 Nm).
- Pressure regulator, vacuum line and O-ring. Torque the bolts to 78 inch lbs. (8.8 Nm).
- Fuel return line
- Intake manifold plenum
- Air cleaner assembly, ducts and air intake hose

DRIVE TRAIN

Transmission Assembly

REMOVAL & INSTALLATION

1. Before servicing the vehicle, refer to the precautions in the beginning of this section.

2. Drain the transmission fluid.

3. Remove or disconnect the following:
- Negative battery cable
- Transmission and transfer case shift lever assembly. On manual transmissions
- Transfer case protector, if equipped
- Front exhaust pipe from the 2 exhaust manifolds, then disconnect it from the intermediate pipe/cat-

alytic converter (make certain to retain the bolts and nuts for reassembly).
- Rear driveshaft at both the rear axle and the transfer case flanges. Matchmark for reassembly.
- Front driveshaft from the front axle, by sliding it forward, plug the transfer case to prevent residual fluid leakage, on 4-wheel drive vehicles
- Dust seal from the rear extension housing.
- Ground cables
- 4WD indicator light switch connector
- Pulse generator connector

- Speed Sensor Connector
- Oxygen (O2S) sensor connector
- Back-up light switch connector
- HI/LO detection switch connector
- Center differential lock detection switch connection.

There may be others depending on the particular year, model and engine with which the vehicle came equipped.

4. Remove or disconnect the following:
- Speedometer cable out of the transmission

5. On manual transmissions:

6. Remove or disconnect the following:
- Clutch cylinder heat protector
- Clutch release cylinder (with the clutch hose connected to it) from

the transmission. Suspend it from the body by using a piece of wire or a similarly safe method.
- Starter motor and the heat shield

7. On automatic transmissions:

8. Remove or disconnect the following:
- Bolts attaching the torque converter to the flexplate
- Dipstick
- Fluid cooling lines
- Shift linkage at the transmission

9. Place a floor jack and a block of wood below the engine oil pan.

10. Lift the floor jack under the engine just until the weight of the engine is taken onto the jack—the engine should only barely be lifted by the jack.

11. Use a transmission jack or second floor jack to place under the transmission. Don't support the transmission yet, only lift the jack until it is slightly below the transmission.

12. Remove or disconnect the following:
- Left-hand and right-hand side transmission stays from the front of the transmission, if equipped
- Bell housing lower cover
- Transfer case mounting bracket, if equipped

13. Lift the floor jack up until the transmission is being slightly supported by it.

14. Remove or disconnect the following:
- Transmission-to-crossmember bolts. Lift the jack about ¼ in. (6mm) off of the crossmember support.
- Crossmember from the vehicle
- Transmission mounting blots. Pull the transmission away from the engine and lower it from the vehicle.

To install:

15. Lift the transmission and transfer assembly into position with the floor jack.

16. On the engine side, there are 2 centering locations. Be sure that the transmission mounting bolt holes are aligned with them before mounting the transmission and transfer assembly to the engine. Lowering the rear of the engine SLIGHTLY may help align the 2 assemblies.

17. Install or connect the following:
- Transmission assembly onto the engine making sure the aligning areas stay aligned. Torque the bolts to 54 ft. lbs. (75 Nm).

18. Lift the transmission/transfer assembly with the floor jack. Since the engine is now attached to the transmission, it also will rise slightly. Adjust its jack to keep only slight support.

19. Install or connect the following:
- Crossmember in place and secure

with the mounting bolts. Torque the bolts to 47 ft. lbs. (65 Nm).
- Transmission and transfer case assembly onto the crossmember
- Crossmember-to-transmission bolts. Torque the bolts to 15–18 ft. lbs. (20–24 Nm) on Montero sport and to 36 ft. lbs. (49 Nm) on Montero.
- Mounting bracket back onto the transfer case, if equipped
- Bell housing lower cover
- Left-hand and right-hand side transmission stays
- Starter motor and heat shield
- Clutch release cylinder
- Speedometer cable into the transmission and secure it there with the retaining ring
- Center differential lock detection switch connection
- HI/LO detection switch connector
- Back-up light switch connector
- O₂ sensor connector
- Speed Sensor Connector
- Pulse generator connector
- 4WD indicator light switch connector
- Ground cables
- Flexplate-to-torque converter bolts. Torque the bolts to 25–30 ft. lbs. (35–42 Nm). On automatic transmissions
- Dipstick tube
- Shift linkage

- Fluid cooler lines. Torque the line fittings to 32 ft. lbs. (44 Nm), on automatic transmissions

20. Tap the dust seal guard back onto the rear extension housing with a rubber or plastic mallet.

21. Install or connect the following:
- Front driveshaft into the transfer case, then attach it to the front differential
- Rear driveshaft, make certain that the matchmarks line up.
- Front exhaust pipe to the catalytic converter and the exhaust manifolds
- Transfer case protector, if equipped
- Transmission and transfer case shift lever assembly, on manual transmissions
- Negative battery cable to the battery

22. Refill the transmission and transfer case with oil.

23. Start the vehicle and check for any leaks.

Clutch

REMOVAL & INSTALLATION

1. Before servicing the vehicle, refer to the precautions in the beginning of this section.

2. Remove or disconnect the following:

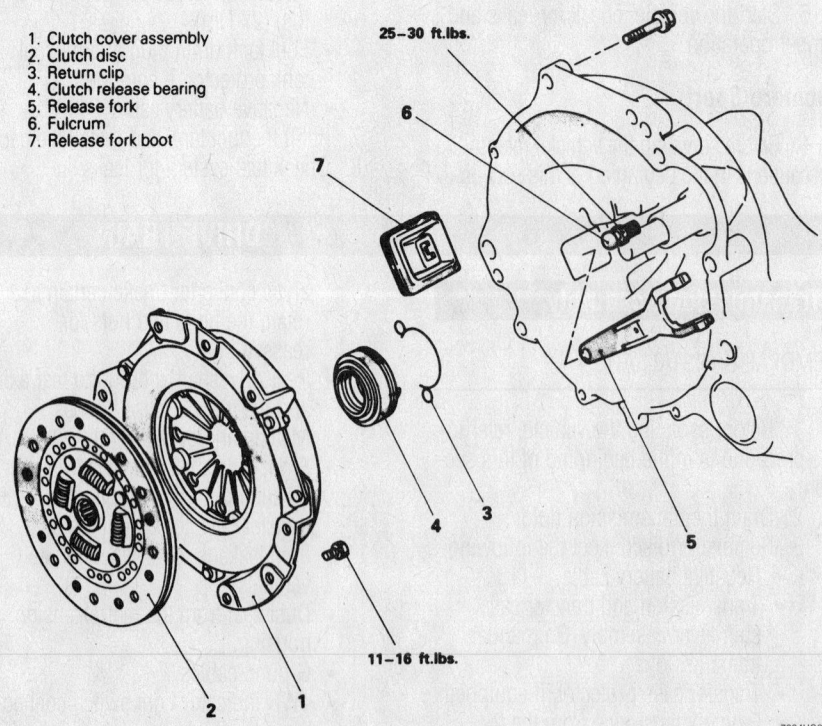

1. Clutch cover assembly
2. Clutch disc
3. Return clip
4. Clutch release bearing
5. Release fork
6. Fulcrum
7. Release fork boot

25–30 ft.lbs.

11–16 ft.lbs.

Exploded view of the typical clutch assembly components

7924UG29

- Negative battery cable
- Transmission assembly

3. Insert a suitable tool in the flywheel pilot bearing hole to keep the clutch disc from falling off. Loosen the clutch cover retainer bolts gradually in a crisscross fashion.

4. Remove or disconnect the following:
  - Clutch cover and disc

5. Check the release bearing for scorching, damage or strange noise. Replace, if necessary.

6. Inspect the flywheel surface for heat cracks or scoring. Reface or replace the flywheel as required.

To install:

7. Apply high temperature grease to the clutch disc splines, input shaft, contact points of the release fork and inside diameter of the release bearing.

➡**Do not allow oil or grease to contact the clutch facing and pressure plate.**

8. Install or connect the following:
  - Flywheel, align using a suitable tool

➡**When installing the clutch disc, be sure that the surface having the manufacturer's stamped mark is on the pressure plate side.**

- Clutch cover with the dowel pin holes in alignment with the dowel pins in the flywheel and tighten the bolt gradually in a crisscross fashion. Torque the bolts to 14 ft. lbs. (19 Nm).
- Transmission assembly
- Negative battery cable

9. Road test the vehicle for proper operation.

Hydraulic Clutch System

BLEEDING

✳✳ WARNING

When bleeding, keep the facial area well away from the slave cylinder and protect all painted surfaces from fluid contact. Brake fluid will damage painted surfaces and could cause physical injury.

1. Fill the clutch master cylinder with fresh DOT 3 brake fluid.
2. Have a helper sit in the vehicle.
3. Remove the bleeder screw cap.
4. If the system is empty, the most efficient way to get fluid down to the cylinder is:

a. Loosen the bleeder about ½–¾ turn

b. Place a finger firmly over the bleeder

c. Have a helper pump the brakes slowly until fluid pressure is felt at the bleeder

d. Once fluid is at the bleeder, close it before the pedal is released.

➡**If the pedal is pumped rapidly, the fluid will churn and create small air bubbles, which are difficult and time consuming to remove from the system. These air bubbles will eventually congregate and will result in a spongy pedal.**

5. Once fluid has been pumped to the slave cylinder:
  - Open the bleeder screw
  - Have a helper depress the clutch pedal
  - Lock the bleeder and have the helper release the pedal
  - Wait 15 seconds and repeat the procedure (including the 15 second wait) until no air bubbles flow from the bleeder.

Remember to close the bleeder before the pedal is released. If the bleeder is left open when the pedal is released, air will be induced into the system.

6. If a helper is not available, connect a small hose to the bleeder, submerge the other end in a clean container of fresh brake fluid placed in a position that is visible from the driver's seat. Pump the pedal until no air comes out of the tube.

Transfer Case Assembly

REMOVAL & INSTALLATION

The transfer case is removed from the vehicle along with the transmission. Refer to the Transmission Removal and Installation procedure for information.

Halfshaft

REMOVAL & INSTALLATION

Outer Axle Shafts

1. Before servicing the vehicle, refer to the precautions in the beginning of this section.

2. Remove or disconnect the following:
  - Negative battery cable
  - Undercover
  - Wheels

- Hub cover dust cap
- Snapring from the inside of the hub and the shim
- Front brake caliper assembly and support with mechanics wire
- Speed sensor, if equipped with ABS
- Tie rod from the steering knuckle assembly
- Upper and lower ball joints from the steering knuckle assembly
- Front hub/knuckle assembly with the inner and outer bearings intact

3. On the left side, pull the halfshaft from the differential carrier.

4. For the right side, remove the fasteners and the halfshaft from the vehicle.

To install:

5. Install or connect the following:
  - New circlip, on the left side halfshaft
  - Inner shaft. Torque the nuts to 36–43 ft. lbs. (49–59 Nm), on the right side halfshaft.
  - Front hub/knuckle and bearing assembly
  - Upper ball joint to the knuckle. Torque the nut to 54 ft. lbs. (74 Nm).
  - Lower ball joint to knuckle. Torque the nut to 108 ft. lbs. (147 Nm).
  - New cotter pins
  - Tie rod end to the steering knuckle. Torque the nut to 33 ft. lbs. (44 Nm) and a new cotter pin.
  - Speed sensor, if removed
  - Front brake assembly
  - Shim and snapring to the axle shaft. Install the front hub dust cover
  - Wheels and the undercover
  - Negative battery cable

Inner Axle Shafts

1. Before servicing the vehicle, refer to the precautions in the beginning of this section.

2. Remove or disconnect the following:
  - Negative battery cable
  - Undercover
  - Right side wheel
  - Right outer halfshaft
  - Lower shock absorber mounting bolts
  - Inner shaft from housing, using a slide hammer with tool MB990241

To install:

3. Install or connect the following:
  - New circlip on the inner halfshaft
  - Inner shaft into the housing, drive the axle into position.
  - Lower shock absorber mounting

bolts. Torque the bolts to 65–76 ft. lbs. (88–103 Nm).
- Right halfshaft assembly
- Undercover
- Wheel
- Negative battery cable

CV-Joints

OVERHAUL

1. Before servicing the vehicle, refer to the precautions in the beginning of this section.
2. Remove or disconnect the following:

- Front wheel
- Driveshaft from the car
- Small and larger band
- Circlip
- Double Offset Joint (DOJ) outer race
- Dust cover
- Circlip
- Balls from the cage
- Cage from the inner race. Turn the cage so that the projections of the inner race align with the recesses of the cage.
- Snapring from the shaft

- DOJ inner race
- Slide the boot off
- Birfield Joint (BJ) small and larger bands
- BJ boot
- BJ assembly

To install:

3. Check the shaft and splines for damage or wear. Inspect the cage, race and balls for any sign of corrosion, wear, cracking or damage. Clean all the parts thoroughly and air dry them completely before installation. Any remaining cleaning solvent can dissolve the lubricating grease.

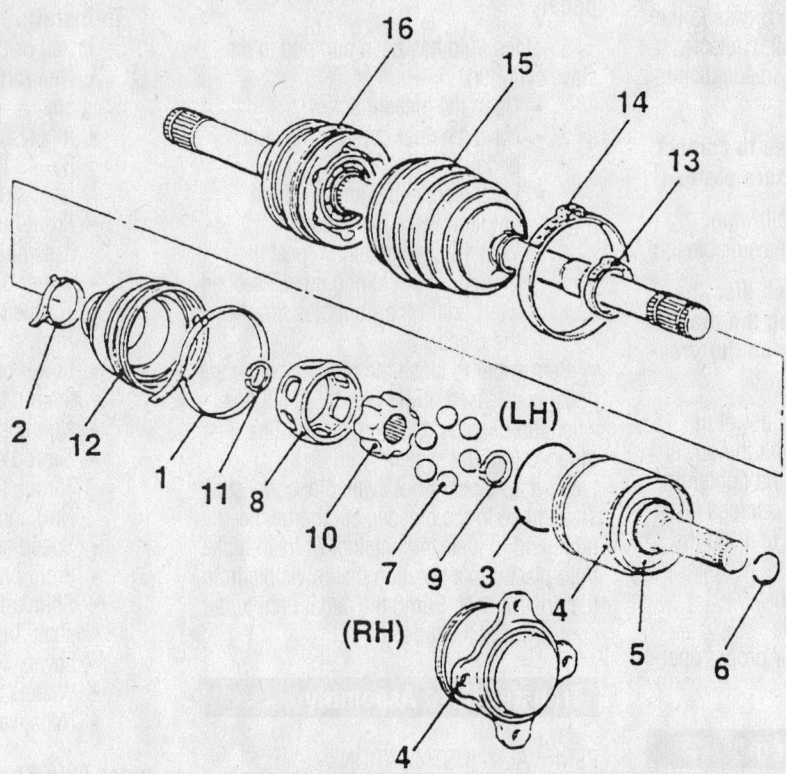

1. D.O.J. boot band (large)
2. D.O.J. boot band (small)
3. Circlip
4. D.O.J. outer race
5. Dust cover
6. Circlip
7. Ball
8. D.O.J. cage
9. Snap ring
10. D.O.J. inner race
11. Circlip
12. D.O.J. boot
13. B.J. boot band (small)
14. B.J. boot band (large)
15. B.J. boot
16. B.J. assembly

9308UG06

CV-Joint, exploded view

4. Tool MB991561 can be used to crimp the bands in place.

5. Install or connect the following:
   • BJ assembly
   • BJ boot, slid the small end of the boot until only one shaft groove cone be seen.
   • BJ small band, crimp the band. Fill the BJ boot with 4.6 oz (130 g) of grease.
   • BJ larger band, crimp the band
   • DOJ small band and boot, fill the boot with grease
   • DOJ cage onto the driveshaft so that the smaller diameter side is installed first.
   • Circlip
   • DOJ inner race and new snap ring, apply grease to the inner race
   • Balls into the cage, grease to the ball areas of the cage and race
   • Outer race, fill the outer race about⅓ full of grease.
   • Dust cover
   • Circlip
   • Large boot band, release the air from the boot then crimp
   • Driveshaft into the car

Automatic Locking Hubs

REMOVAL & INSTALLATION

1. Place the locking hub in the free position. To do this, shift the transfer shift lever to the 2H position, then move the vehicle 4–7 ft. (1–2 m) backwards.

2. Before servicing the vehicle, refer to the precautions in the beginning of this section.

3. Remove or disconnect the following:
   • Front wheels of the vehicle
   • Hub cover
   • Snapring from the axle shaft
   • Shim
   • Drive flange
   • Front brake assembly
   • Speed sensor, if equipped
   • Lock washer
   • Lock nut
   • Front hub assembly

To install:
4. Install or connect the following:
   • Front hub assembly
   • Lock nut. Torque the nut as follows:
   a. Step 1: Torque the nut to 119 ft. lbs. (162 Nm).
   b. Step 2: Loosen to 0 ft. lbs. (0 Nm).
   c. Step 3: Torque the nut to 18 ft lbs. (25 Nm).
   d. Step 4: Loosen the nut 30–40 degrees.

5. Install or connect the following:
   • Lock washer
   • Speed sensor, if equipped
   • Front brake assembly. Torque the bolts to 65 ft. lbs. (88 Nm).
   • Drive flange. Torque the bolts to 36–43 ft. lbs. (49–59 Nm).
   • Shim
   • Snapring to the axle shaft
   • Hub cover
   • Front wheels of the vehicle

Axle Shaft, Bearing and Seal

REMOVAL & INSTALLATION

Rear

1. Before servicing the vehicle, refer to the precautions in the beginning of this section.

2. Remove or disconnect the following:
   • Wheel assembly
   • Brake line
   • Rear brake assembly
   • Parking brake cable and assembly
   • Axle shaft
   • Snapring
   • Retainer
   • Bearing inner race, inner and outer
   • Oil seal
   • Bearing case
   • O-ring
   • Oil seal

To install:
3. Install or connect the following:
   • Bearing case
   • Bearing inner race, outer. Press the bearing into the bearing case.
   • Oil seal. Press the seal using tools MB990932 and MB990938.
   • Bearing inner race, inner. Press the bearing into the bearing case.
   • Axle shaft, Place into bearing case
   • Retainer, press onto shaft

Removal steps

1. Cover
Adjustment of drive shaft end play
2. Snap ring
3. Shim
4. Front brake assembly

5. Bolts
6. Automatic free-wheeling hub assembly
7. Shim
8. Lock washer
9. Lock nut
10. Front hub assembly

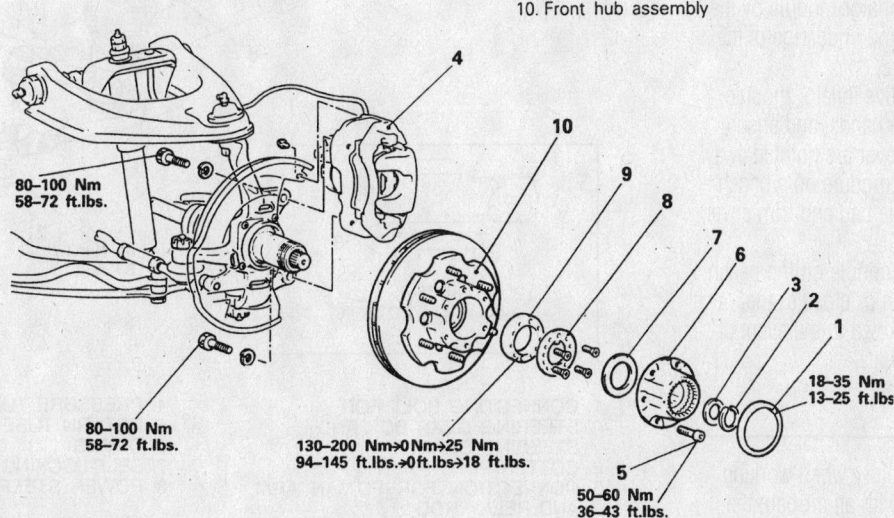

80–100 Nm
58–72 ft.lbs.

80–100 Nm
58–72 ft.lbs.

130–200 Nm→0 Nm→25 Nm
94–145 ft.lbs.→0 ft.lbs→18 ft.lbs.

50–60 Nm
36–43 ft.lbs.

18–35 Nm
13–25 ft.lbs

7924UG45B

Axle hub and locking hub removal and installation—automatic hubs

- Snapring
- Oil seal and O-ring into axle shaft
4. Axle shaft assembly into axle.
- Parking brake cable end
- Parking brake cable attaching bolt
- Rear brake assembly
- Brake line
- Wheel assembly

Pinion Seal

REMOVAL & INSTALLATION

1. Before servicing the vehicle, refer to the precautions in the beginning of this section.

2. Remove or disconnect the following:
- Driveshaft, matchmark for reassembly
3. Check the turning torque of the pinion before proceeding. It should be 2.6–4.5 inch lbs. (0.4–0.5 Nm). This is the torque that must be reached during installation of the pinion nut.
- Pinion nut and washer using a suitable pinion flange holding tool
- Companion flange from the drive pinion
4. Pry the pinion seal out of the differential carrier.

To install:
5. Clean and inspect the sealing surface of the housing.

6. Install or connect the following:
- New seal into the housing until the flange on the seal is flush with the carrier. Using a seal driver.
7. With the seal installed, the pinion bearing preload must be set.
- Pinion nut (a new self-locking pinion nut must be used) while holding the flange, until the turning torque is the same as before removal. The final pinion nut torque must be between 137–181 ft. lbs. (190–250 Nm).
- Driveshaft, align the matchmarks
8. Check the level of the differential lubricant when finished.

STEERING AND SUSPENSION

Air Bag

✳✳ CAUTION

Some vehicles are equipped with an air bag system. The system must be disabled before performing service on or around system components, steering column, instrument panel components, wiring and sensors. Failure to follow safety and disabling procedures could result in accidental air bag deployment, possible personal injury and unnecessary system repairs.

PRECAUTIONS

Several precautions must be observed when handling the inflator module to avoid accidental deployment and possible personal injury.
- Never carry the inflator module by the wires or connector on the underside of the module.
- When carrying a live inflator module, hold securely with both hands, and ensure that the bag and trim cover are pointed away.
- Place the inflator module on a bench or other surface with the bag and trim cover facing up.
- With the inflator module on the bench, never place anything on or close to the module that may be thrown in the event of an accidental deployment.

DISARMING

To avoid personal injury when working on vehicles equipped with an air bag, the negative battery cable must be disconnected and at least 60 seconds must elapse before

working on the system. Failure to do so may result in deployment of the air bag.

Recirculating Ball Power Steering Gear

REMOVAL & INSTALLATION

1. On vehicles equipped with a supplemental restraint system (SRS), turn the front wheel to the straight ahead position and

remove the ignition key to prevent the steering wheel from turning.
2. Drain the power steering fluid.
3. Remove or disconnect the following:
- Negative battery cable
- Pinch bolt securing the steering shaft to the steering gear
- Pitman arm from the relay rod
- Fluid lines from the steering gear
- Mounting bolts securing the gear to the frame rail and steering gear

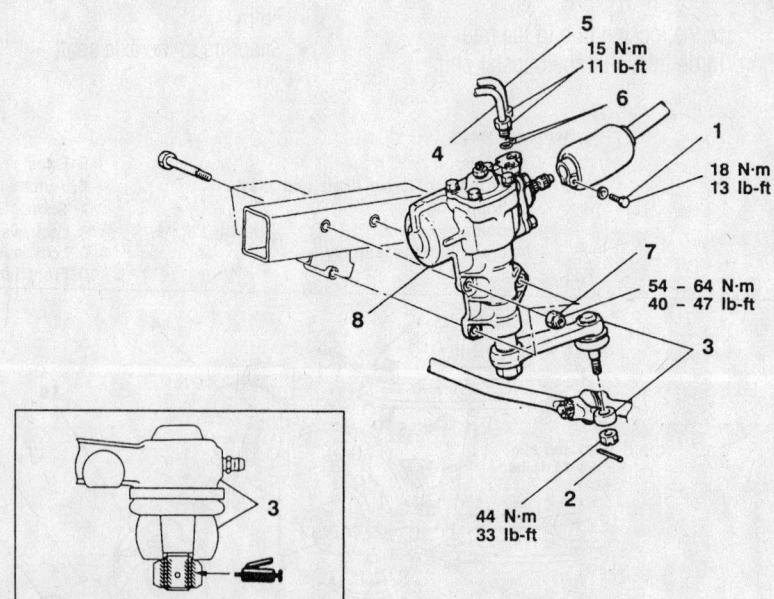

1. CONNECTING BOLT FOR STEERING GEAR BOX AND STEERING SHAFT
2. COTTER PIN
3. CONNECTION FOR PITMAN ARM AND RELAY ROD
4. PRESSURE TUBE
5. RETURN TUBE
6. O-RING
7. SELF-LOCKING NUT
8. POWER STEERING GEAR BOX

7924UG33

Exploded view of a typical power steering gear mounting

To install:

4. Install or connect the following:

- Steering gear on the frame rail. Torque the nuts to 40–47 ft. lbs. (54–64 Nm).
- Fluid lines to the steering gear use a new O-rings. Torque the fittings to 11 ft. lbs. (15 Nm).
- Relay rod on the Pitman arm. Torque the nut to 33 ft. lbs. (44 Nm).
- Steering shaft on the steering gear. Torque the bolt to 13 ft. lbs. (18 Nm).
- Negative battery cable

5. Refill and bleed the power steering system.

Rack and Pinion Steering Gear

REMOVAL & INSTALLATION

1. On vehicles equipped with a supplemental restraint system (SRS), turn the front wheel to the straight ahead position and remove the ignition key to prevent the steering wheel from turning.

2. Drain the power steering fluid.

3. Remove or disconnect the following:

- Negative battery cable
- Engine under cover
- Tie rod ends
- Steering hoses
- Left side differential mount bracket
- Intermediate shaft-to-gear box bolt
- Gear mounting clamps and gear

4. Installation is the reverse of removal. Observe the following torques:

- Mounting clamp bolts: 51 ft. lbs. (69Nm)
- Tie rod end ball stud nuts: 29 ft. lbs. (39Nm)
- Intermediate shaft pinch bolt: 13 ft. lbs. (18Nm)

Shock Absorber

REMOVAL & INSTALLATION

Front

1. Before servicing the vehicle, refer to the precautions in the beginning of this section.

2. Remove the upper shock mounting nut, washer and bushing.

3. Remove the lower mounting bolts.

4. Remove the shock absorber.

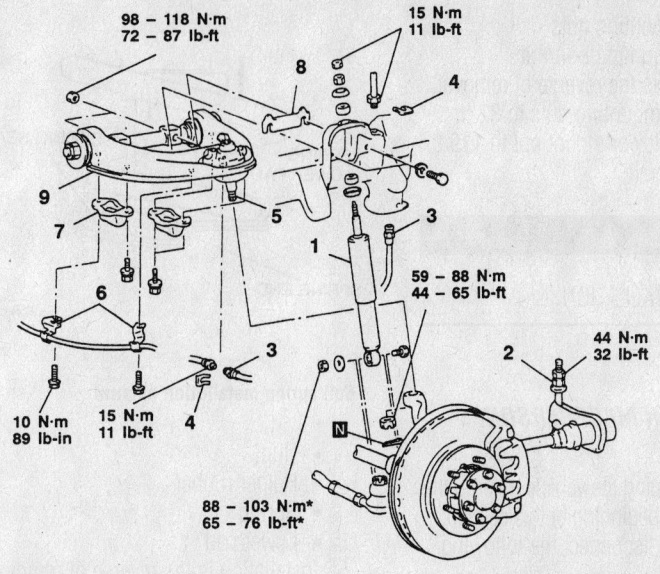

98 – 118 N·m
72 – 87 lb-ft

15 N·m
11 lb-ft

10 N·m
89 lb-in

15 N·m
11 lb-ft

59 – 88 N·m
44 – 65 lb-ft

44 N·m
32 lb-ft

88 – 103 N·m*
65 – 76 lb-ft*

1. SHOCK ABSORBER
- BUMP STOPPER AND BUMP STOPPER BRACKET CLEARANCE ADJUSTMENT
2. REAR ANCHOR ARM ADJUSTING NUT
3. BRAKE HOSE CONNECTION
4. HOSE CLIP
5. UPPER ARM BALL JOINT CONNECTION
6. SPEED SENSOR BRACKET <VEHICLES WITH ABS>
7. REBOUND STOPPER
8. SHIMS
9. UPPER ARM
10. UPPER ARM BALL JOINT ASSEMBLY

Caution
*: Indicates parts which should be temporarily tightened, and then fully tightened with the vehicle on the ground in an unladen condition.

7924UG34

Common shock absorber and upper control arm components

To install:

➡ **If the shock absorber has a white paint mark on the lower end, be sure the mark faces the outside of the vehicle when installed.**

5. Install the shock absorber. Torque the lower nut to 65–76 ft. lbs. (88–103 Nm) and the upper nut to 11 ft. lbs. (15 Nm).

6. Test drive the vehicle and check the alignment.

Rear

1. Before servicing the vehicle, refer to the precautions in the beginning of this section.

2. Support the rear axle assembly with a hydraulic floor jack, so that the shock absorber may be removed.

3. Remove the upper and lower mounting nuts and bolts that attach the shock to the frame and bracket.

4. Remove the shock absorber from the vehicle.

To install:

5. Install the shock absorber

- 2000 lower bolt: 159–181 ft. lbs. (216–245 Nm) on the Montero and 16 ft. lbs. (22 Nm) on the Montero Sport

- 2001–03 lower bolt: 159–181 ft. lbs. (216–245 Nm) on the Montero Sport and 113 ft. lbs. (152Nm) on the Montero
- 2000 upper mounting nut: 11 ft. lbs. (15 Nm) on the Montero and 16 ft. lbs. (22 Nm) on the Montero Sport
- 2001–03 upper mounting nut: 16 ft. lbs. (22Nm) on the Montero Sport and 33 ft. lbs. (44Nm) on the Montero

6. Remove the floor jack from under the axle assembly.

Strut

REMOVAL & INSTALLATION

1. On vehicles equipped with a supplemental restraint system (SRS), turn the front wheel to the straight ahead position and remove the ignition key to prevent the steering wheel from turning.

2. Remove or disconnect the following:

- Negative battery cable
- Upper control arm
- Battery and battery tray
- A/C condenser
- Air cleaner

- Wheel
- Upper mounting nuts
- Lower mounting bolt/nut

3. Installation is the reverse of removal. Torque the upper mounting nuts to 32 ft. lbs. (44 Nm); the lower mount nut to 119 ft. lbs. (162 Nm).

Coil Spring

REMOVAL & INSTALLATION

Front

MONTERO WITH MACPHERSON STRUTS

1. Before servicing the vehicle, refer to the precautions in the beginning of this section.
2. Remove or disconnect the following:
   - Strut
3. Compress the coil spring until there is a clearance on both ends
4. Remove or disconnect the following:
   - Strut center nut
   - Seat
   - Collar
   - Bushing
   - Bracket
   - Upper pad

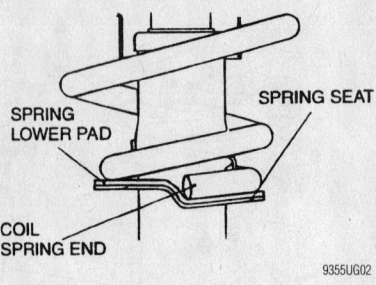

Coil spring installation on strut

- Cup
- Helper rubber
- Spring
- Lower pad

5. Installation is the reverse of removal. Torque the center nut to 17 ft. lbs. (22 Nm).

Rear

MONTERO

1. Before servicing the vehicle, refer to the precautions in the beginning of this section.
2. Support the weight of the axle.
3. Remove or disconnect the following:

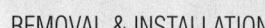

N 1 — 22 ± 2 N·m / 17 ± 1 ft-lb

1. SELF-LOCKING NUT
2. SEAT
3. COLLAR
4. UPPER BUSHING
5. SPRING BRACKET ASSEMBLY
6. SPRING UPPER PAD
7. CUP ASSEMBLY
8. HELPER RUBBER
9. COIL SPRING
10. SPRING LOWER PAD
11. SHOCK ABSORBER ASSEMBLY

Strut exploded view

- Breather hose
- Parking brake cable attaching bolt
- ABS speed sensor attaching bolt
- Brake hose connection
- Lower shock mounting bolt
- Bolt that attaches the lateral rod to the body
- Stabilizer bar

4. Lower the axle and remove the coil spring and seat
5. Installation is the reverse of removal. Torque the lateral rod-to-body bolt to 159–181 ft. lbs. (216–245 Nm).

MONTERO SPORT

1. Before servicing the vehicle, refer to the precautions in the beginning of this section.
2. Remove or disconnect the following:
   - Shock absorber lower bolt
   - Lower arm mounting bolt
   - Coil spring
3. Installation is the reverse of removal. Torque the lower arm bolt to 113 ft. lbs. (152 Nm).

Torsion Bars

REMOVAL & INSTALLATION

1. Before servicing the vehicle, refer to the precautions in the beginning of this section.
2. Support the lower arm with a jack.
3. Remove or disconnect the following:
   - Heat protector, right side only
   - Bump stopper
   - Anchor adjustment nut and arm assembly
   - Anchor collar
   - Torsion bar
   - Dust covers
   - Heat covers, right side only

To install:

4. Install or connect the following:
   - Heat covers, right side only
   - Dust covers
   - Torsion bar
   - Anchor collar
   - Anchor adjustment nut and arm assembly. Torque the nut to 32 ft. lbs. (44 Nm).
   - Heat protector, right side only

Upper Ball Joint

REMOVAL & INSTALLATION

1. Before servicing the vehicle, refer to the precautions in the beginning of this section.

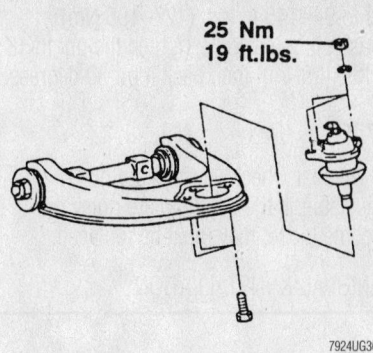

25 Nm
19 ft.lbs.

7924UG36

Exploded view of the upper ball joint and related components—2000 models

2. Remove the front wheel.
3. Support the lower control arm.
4. Remove the upper ball joint from the steering knuckle.
5. Remove the ball joint from the upper control arm.

To install:

6. Install the ball joint in the upper control arm. Tighten the bolts to 18 ft. lbs. (25 Nm) for 2000 models; 22 ft. lbs. (30 Nm) for 2001–03 models.
7. Install the ball joint stud to the steering knuckle. Torque the nut to 54 ft. lbs. (74 Nm).

8. Install a new cotter pin.
9. Install the front wheel.
10. Grease the upper ball joint and all other suspension components with a grease fitting.

Lower Ball Joint

REMOVAL & INSTALLATION

1. Before servicing the vehicle, refer to the precautions in the beginning of this section.
2. Apply upward pressure to the lower control arm with a jack or an adjustable stand.

✳✳ CAUTION

Do not disconnect the lower ball joint stud from the steering knuckle unless the lower control arm has a stand or a jack under it.

3. Remove the ball joint stud nut/stud from the steering knuckle.
4. Remove the ball joint retaining nuts/bolts and the ball joint from the arm

To install:

5. Install the lower ball joint on the control arm. Torque the ball joint retaining

nuts/bolts to 60 ft. lbs. (81 Nm) on 2000 models and 2001–03 Montero Sport; 70 ft. lbs. (95Nm) on 2001–03 Montero models.

6. Install the ball stud to the knuckle. Torque the nut to 108 ft. lbs. (147 Nm) and a new cotter pin.
7. Lubricate the ball joint with a grease gun.
8. Check and adjust the alignment if necessary.

Upper Control Arm

REMOVAL & INSTALLATION

1. Before servicing the vehicle, refer to the precautions in the beginning of this section.
2. Remove or disconnect the following:
   - Wheel assembly
   - Bumper stop
   - Anchor arm assembly adjustment nut
   - Brake hose clip and connection
   - Upper ball joint from knuckle
   - Brake hose clip
   - Rebound stopper(s)
   - Speed sensor bracket
   - Shim
   - Upper control arm

To install:

3. Install or connect the following:
   - Upper control arm and shim. Torque the nuts to 80 ft. lbs. (108 Nm).
   - Speed sensor bracket
   - Rebound stopper(s)
   - Brake hose clip
   - Upper ball joint from knuckle. Torque the nut to 54 ft. lbs. (74 Nm).
   - New cotter pin
   - Brake hose clip and connection
   - Anchor arm assembly adjustment nut. Torque the nuts to 33 ft. lbs. (44 Nm).
   - Bumper stop
   - Wheel assembly

Lower Control Arm

REMOVAL & INSTALLATION

1. Before servicing the vehicle, refer to the precautions in the beginning of this section.
2. Remove or disconnect the following:
   - Skid plate and undercover
   - Bumper stop
   - Rear anchor assembly

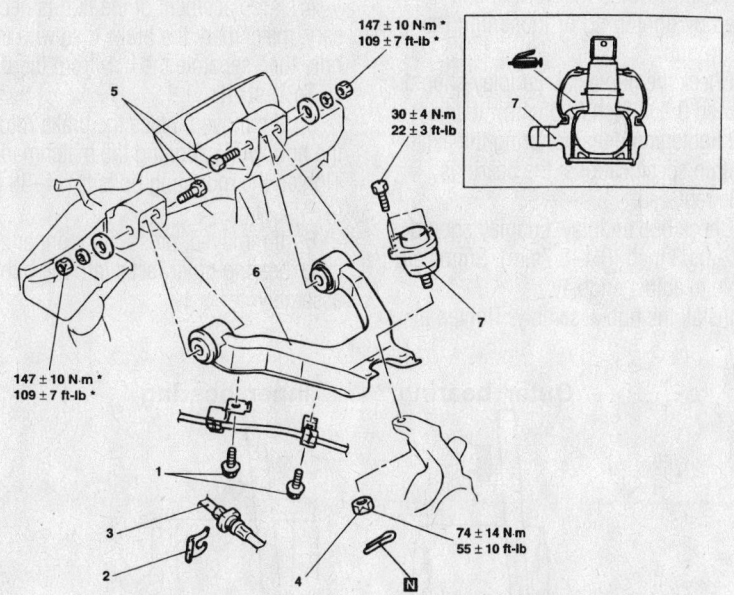

⚠ CAUTION
*: Indicates parts which should be temporarily tightened, and then fully tightened with the vehicle on the ground in an unladen condition.

147 ± 10 N·m *
109 ± 7 ft-lb *

30 ± 4 N·m
22 ± 3 ft-lb

147 ± 10 N·m *
109 ± 7 ft-lb *

74 ± 14 N·m
55 ± 10 ft-lb

1. FRONT WHEEL SPEED SENSOR BRACKET MOUNTING BOLT
2. CLIP
3. BRAKE HOSE
4. UPPER ARM ASSEMBLY AND KNUCKLE CONNECTION
5. UPPER ARM ASSEMBLY AND FRONT FRAME CONNECTION
6. UPPER ARM ASSEMBLY
7. UPPER ARM BALL JOINT ASSEMBLY

9355UG03

Exploded view of the upper ball joint and related components—2001–03 models

- Torsion bar
- Lower ball joint from knuckle
- Stabilizer link assembly
- Shock absorber mounting bolts
- Lower arm shaft
- Anchor arm
- Lower control arm

To install:

3. Install or connect the following:
- Lower control arm. Torque the mounting bolts to 108 ft. lbs. (147 Nm).
- Anchor arm. Torque the bolt to 33 ft. Lbs. (44 Nm).
- Lower arm shaft. Torque the bolt to 108 ft. lbs. (147 Nm).
- Shock absorber mounting bolts. Torque the bolt to 65–76 ft. lbs. (88–103 Nm).
- Stabilizer link assembly
- Lower ball joint from knuckle. Torque the nut to 108 ft. lbs. (147 Nm).
- Torsion bar
- Rear anchor assembly. Torque the nuts to 33ft. lbs. (44 Nm).
- Bumper stop. Torque the nut to 18 ft. lbs. (25 Nm).
- Skid plate and undercover

CONTROL ARM BUSHING REPLACEMENT

Rear

1. Before servicing the vehicle, refer to the precautions in the beginning of this section.
2. Remove the wheel.
3. Remove the lower control arm.
4. Using tool MB991522 press out the bushing

To install:

5. Position the bushing with the larger end facing the front of the vehicle.
6. Using tool MB991522 press the bushing into the bracket.
7. Install the lower control arm.
8. Install the wheel.

Front

1. Before servicing the vehicle, refer to the precautions in the beginning of this section.
2. Remove the wheel.
3. Remove the lower control arm and place in a vise.
4. Using tool MB990883 remove the bushing

To install:

5. Position the bushing with the larger end facing the front of the vehicle.

➡**Coat the bushing with a soap solution and take care not to twist.**

6. Using tool MB991522, press the bushing into the bracket.
7. Install the lower control arm.
8. Install the wheel.

Wheel Bearings

ADJUSTMENT

Front

2000 MODELS

1. Tighten the wheel bearing nut to 119 ft. lbs. (162 Nm) while turning the rotor.
2. Loosen the wheel bearing adjusting nut completely.
3. Tighten the nut to 18 ft. lbs. (25 Nm), then loosen the nut approximately 30°.
4. Using a dial indicator, check the wheel bearing end-play. The specification is 0.002 in. (0.05mm).
5. Install the locknut.

2001–03 MONTERO

The bearings are integral with the hub. No adjustment is possible.

2001–03 MONTERO SPORT

1. With the caliper removed, check the rotational starting torque. Rotational torque should be 2.7–11.5 inch lbs. (0.3–1.3 Nm). Rotational torque can be adjusted by tightening or loosening the adjusting nut.
2. Check the hub axial. Endplay should not exceed 0.002 inch (0.05mm). If adjusting nut tightening does not bring the axial play within specifications, the bearings must be replaced.
3. Check hub endplay. Endplay should be 0.02–0.03 inch (0.4–0.7mm). Shims are available to adjust endplay.
4. Install the hub assembly. Tighten the

nut to 94–145 ft. lbs. (127–196 Nm). Loosen it completely. Tighten the nut to 18 ft. lbs. (25Nm), then back it off 30 degrees.

Rear

The rear wheel bearings are not adjustable. If the bearings are noisy or become loose, they must be replaced.

REMOVAL & INSTALLATION

Front

2000 MONTERO WITH 2WD AND 2000–01 MONTERO SPORT WITH 2WD

1. Before servicing the vehicle, refer to the precautions in the beginning of this section.
2. Remove or disconnect the following:
- Tire and wheel assembly
- Caliper assembly and suspend it from the upper arm
- Dust cap
- Cotter pin, castellated nut lock, wheel bearing nut and washer from the spindle
- Outer wheel bearing
- Hub and rotor as an assembly
- Grease seal and inner wheel bearing

3. If required, press the inner and outer bearing outer races from the hub assembly.
4. If replacement of the hub is necessary, matchmark the brake disc with the hub, then separate the hub from the disc.

To install:

5. If removed, place the brake rotor on the hub, while aligning the matchmarks. Tighten the mounting bolts to 34–38 ft. lbs. (47–52 Nm).
6. If removed, press-fit the inner and outer bearing outer races into the hub assembly.

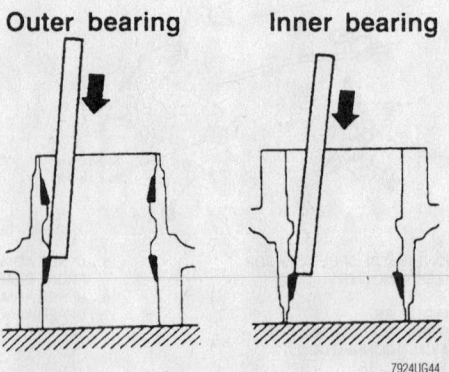

Outer bearing   **Inner bearing**

7924UG44

The bearing races can be removed from the hub using a drift and hammer

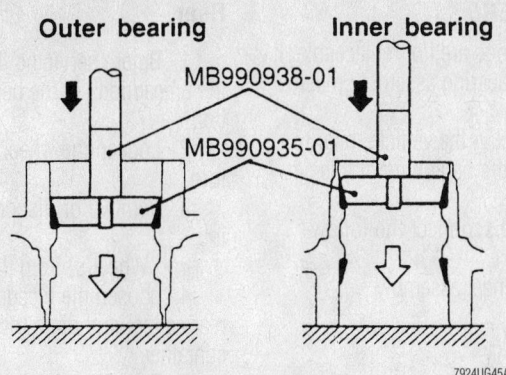

Outer bearing   Inner bearing

MB990938-01

MB990935-01

7924UG45A

Install the new races into the hub using the proper size driver

7. Lubricate the seal lip and inside surface of the front hub with MP grease.

8. Install or connect the following:
   - Inner wheel bearing and repack
   - New grease seal
   - Hub assembly on the spindle
   - Outer wheel bearing, washer and nut, lubricate. When the bearing preload is properly set, install the nut lock and a new cotter pin.
   - Grease cap
   - Caliper assembly
   - Tire and wheel assembly

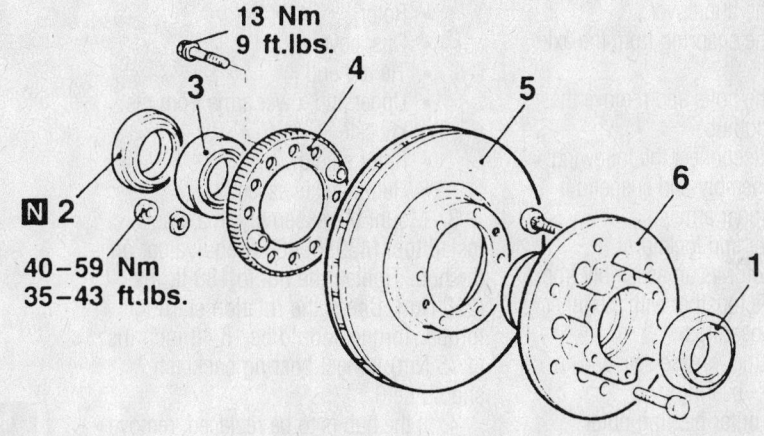

13 Nm
9 ft.lbs.

3

4

5

6

1

N 2

40–59 Nm
35–43 ft.lbs.

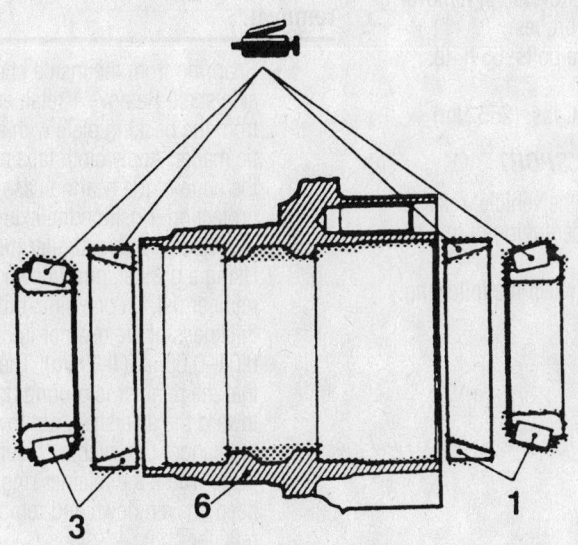

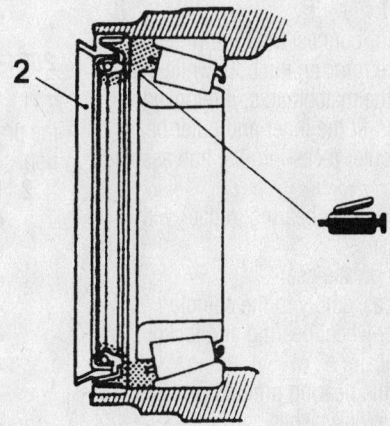

2

1. Outer bearing
2. Oil seal
3. Inner bearing

4. Rotor
5. Brake disc
6. Front hub

7924UG43

Exploded view of the hub and wheel bearing assembly—Montero

2000 MONTERO AND MONTERO SPORT WITH 4WD

1. Before servicing the vehicle, refer to the precautions in the beginning of this section.

2. Remove or disconnect the following:
- Tire and wheel assembly

3. If equipped with locking hub:

a. Place the locking hub in the free position.

➡ A free position can be obtained by shifting the transfer shift lever to the 2H position, then moving the vehicle in reverse for approximately 3–6 ft. (1–2 m).

b. Remove the hub cover.

c. Remove the snapring from the axle shaft.

d. Remove the bolts and remove the automatic locking hub.

4. Remove or disconnect the following:
- Caliper assembly and suspend it from the upper arm.
- Lockwasher and locknut
- Hub and rotor as an assembly from the knuckle together with the inner and outer bearings
- Outer bearing, grease seal and inner wheel bearing
- Inner and outer bearing outer races from the hub assembly, if required

5. If replacement of the hub is necessary, matchmark the brake disc with the hub, then separate the hub from the disc.

To install:

6. Install or connect the following:
- Brake rotor on the hub, while aligning the matchmarks, if removed
- Press-fit the inner and outer bearing outer races into the hub assembly
- Inner wheel bearing, repack with grease
- New grease seal
- Hub assembly to the spindle
- Outer wheel bearing and locknut, lubricate

7. When the bearing preload is properly set, install the lockwasher.

8. If equipped with locking hubs:

a. Apply a coating of semi-drying sealant to the locking hub body and front hub contact surfaces.

b. Align the key of the brake**B** and the keyway of the knuckle spindle and loosely install the automatic locking hub assembly. Tighten the mounting bolts to 36–43 ft. lbs. (50–60 Nm).

9. Install the wheel and tire assembly

2001–03 MONTERO

The wheel bearings are not replaceable. If defective, the hub/bearing assembly must be replaced.

1. Before servicing the vehicle, refer to the precautions in the beginning of this section.

2. Remove or disconnect the following:

- Tire and wheel assembly
- Hub cover
- Nut
- Washer
- Brake hose
- Speed sensor
- Caliper
- Rotor
- Dust cover
- Tie rod end
- Upper and lower arms from the knuckle
- Rotor shield
- Hub/knuckle assembly

3. Mount the assembly in a vise. Install tool MB990998, or equivalent on the hub. Tighten the nut to 188 ft. lbs. (255 Nm). Check the rotation starting torque. Torque should be 15.48 inch lbs. (1.75 Nm). Wheel bearing backlash should be 0.

4. If the hub is to be replaced, remove the hub-to-knuckle bolts.

5. Installation is the reverse of removal. Observe the following torques:
- Hub-to-knuckle bolts: 65 ft. lbs. (88 Nm)
- Hub nut: 188 ft. lbs. (255 Nm)

2001–03 MONTERO SPORT

1. Before servicing the vehicle, refer to the precautions in the beginning of this section.

2. Remove or disconnect the following:
- Wheel
- Caliper
- Hub cover
- Snapring
- Shim
- Drive flange
- Spring washer
- Nut
- Hub and outer bearing
- Oil seal
- Inner bearing
- Races

3. Installation is the reverse of removal.

4. Install the hub assembly. Tighten the nut to 94–145 ft. lbs. (127–196 Nm). Loosen it completely. Tighten the nut to 18 ft. lbs. (25 Nm), then back it off 30 degrees.

Rear

1. Before servicing the vehicle, refer to the precautions in the beginning of this section.

2. Loosen the wheel lug nuts only½ a turn.

3. Remove or disconnect the following:
- Wheel(s) from the vehicle

4. Loosen the bleeder valve on the right rear caliper and drain the brake fluid into a container.

5. Remove or disconnect the following:
- Rear brake hose from the hard line on the frame
- Rear brake caliper
- Rear disc off of the rear axle
- Parking cable attaching bolt and cable end from the brake assembly
- Parking brake assembly from the end of the axle
- Speed sensor, on vehicles with Anti-lock Brakes (ABS)
- Rear axle shaft out of the axle housing. If the rear axle shaft is difficult to remove, use a slide hammer (impact puller) to remove it.

✱✱ WARNING

Do not damage the oil seal during removal.

- Snapring from the inside end of the axle shaft. Remove 1 retainer bolt from the backing plate with a plastic mallet. Apply cloth tape around the edge of the bearing case for protection. Position the axle shaft in a vise or with a similar method. Using a grinder, grind down the retainer flat, on one side, until the thickness of the retainer is only 0.04–0.08 in. (1–2mm). That is that the retainer is ground down toward the axle shaft, not toward the flange. Cut, with a chisel, the place where the retainer ring has been shaven down and remove the retainer.

✱✱ CAUTION

Be careful not to damage the bearing case and the axle shaft.

➡ Only the retainer ring is to be ground down, NOT the axle shaft, the axle flange, the bearing or any other component.

6. Grind the plate of special tool MB990861 with a grinder (see illustration) so that there will be no interference between the plate and the bearing case. While adjusting the height of the hanger, secure the washers, plate and nuts in order so that the processed plate is as shown in the illustration.

➡ **The washers are used to eliminate the difference in height of the bearing case so that the plate and the bearing case are parallel.**

Place the end of the bolt against the center of the axle shaft, then tighten the nuts to

remove the axle shaft from the bearing case assembly.

➡ **The hanger and plate must be placed so that they are parallel.**

7. Remove the bearing inner race and the bearing outer race. To remove the races, install the tool MB990560 and use a press to remove the bearing race from the axle shaft.

8. Remove the oil seal and the dust cover on vehicles without ABS.

9. On vehicles without ABS, insert an iron plate of approximately 0.04 in. (1mm)

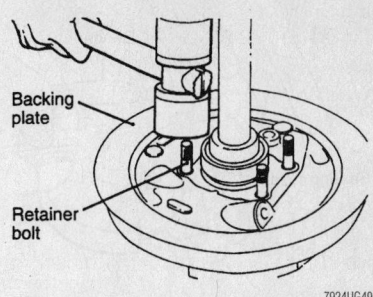

Remove one of the rear axle studs before attempting to grind down the retainer

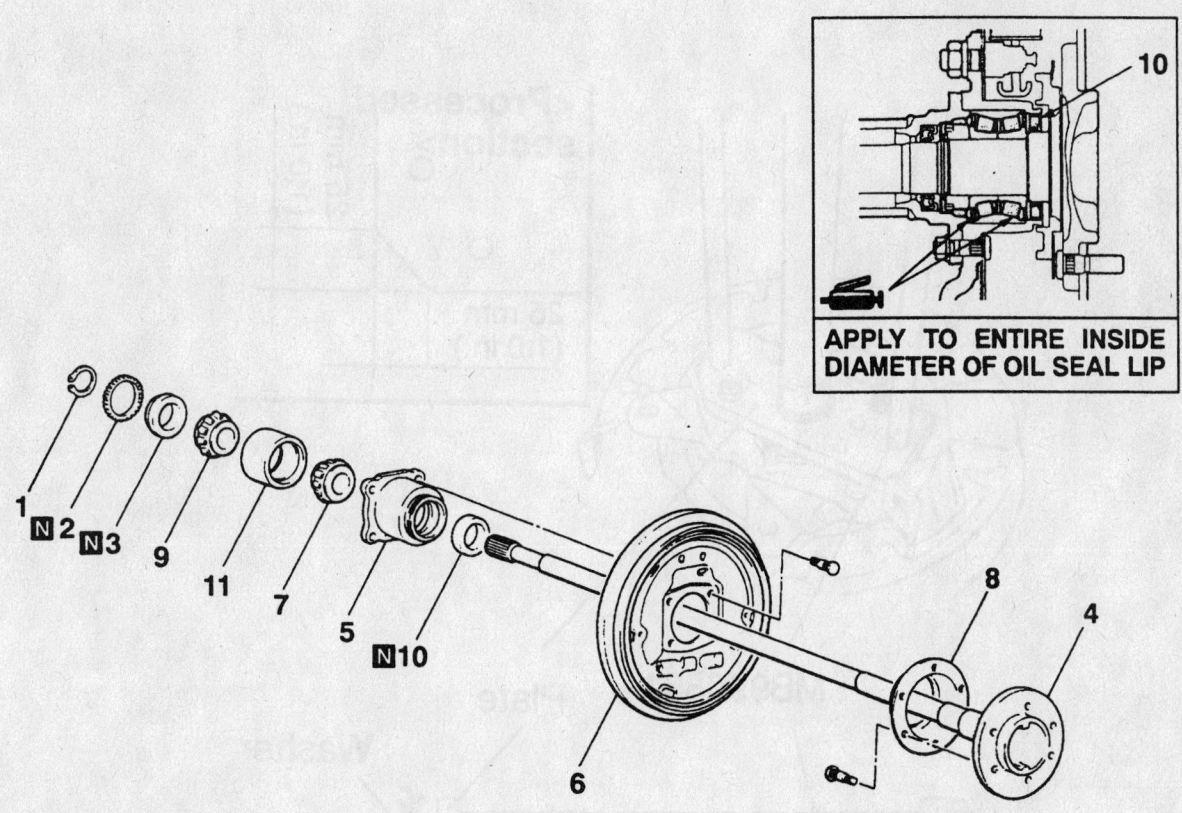

APPLY TO ENTIRE INSIDE DIAMETER OF OIL SEAL LIP

DISASSEMBLY STEPS
1. SNAP RING
2. ABS ROTOR
3. RETAINER
4. AXLE SHAFT
5. BEARING CASE
6. BACKING PLATE
7. OUTER BEARING INNER RACE
8. DUST COVER
9. INNER BEARING INNER RACE
10. OIL SEAL
11. BEARING OUTER RACE

ASSEMBLY STEPS
11. BEARING OUTER RACE
9. INNER BEARING INNER RACE
7. OUTER BEARING INNER RACE
10. OIL SEAL
8. DUST COVER
6. BACKING PLATE
5. BEARING CASE
4. AXLE SHAFT
3. RETAINER
2. ABS ROTOR
1. SNAP RING

Exploded view of the typical rear axle shaft, bearings and races

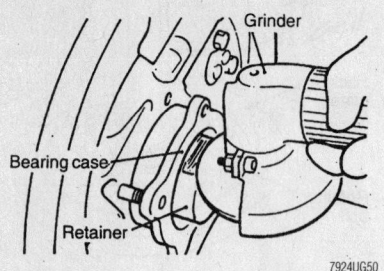

Using a grinder, grind the retainer, on one side, down to 1–2mm (0.04–0.08 in.) thickness

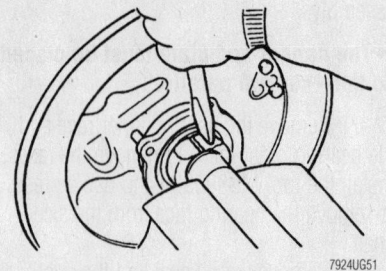

Use a chisel on the ground-down spot on the rear axle bearing retainer to split the retainer, then remove it

thickness between the rotor assembly and the axle shaft, then use a press to remove the rotor assembly.

✳✳ WARNING

In order not to bend the rotor assembly plate, place the support in contact with the axle shaft when using the press.

10. Remove or disconnect the following:
   • Axle shaft from the remaining bearings and components
   • Backing plate

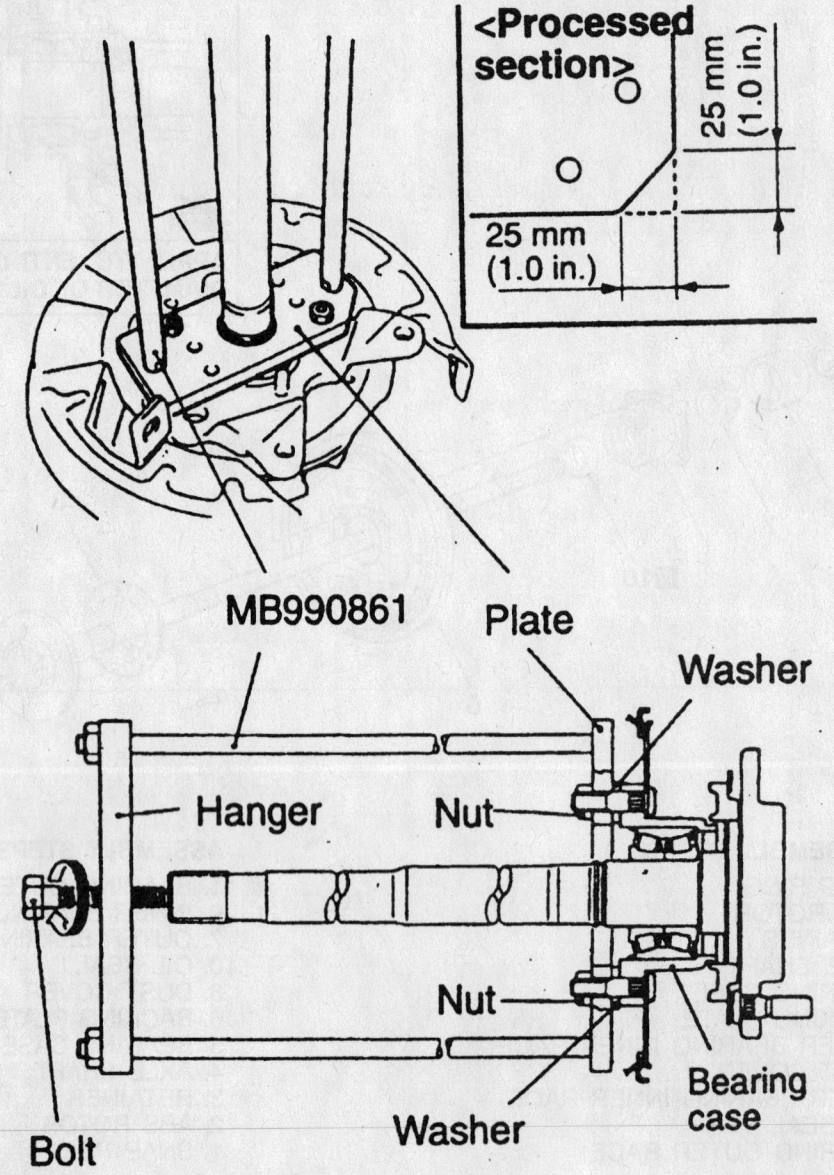

Use the special tool MB990861 to remove the rear axle shaft from the bearing case

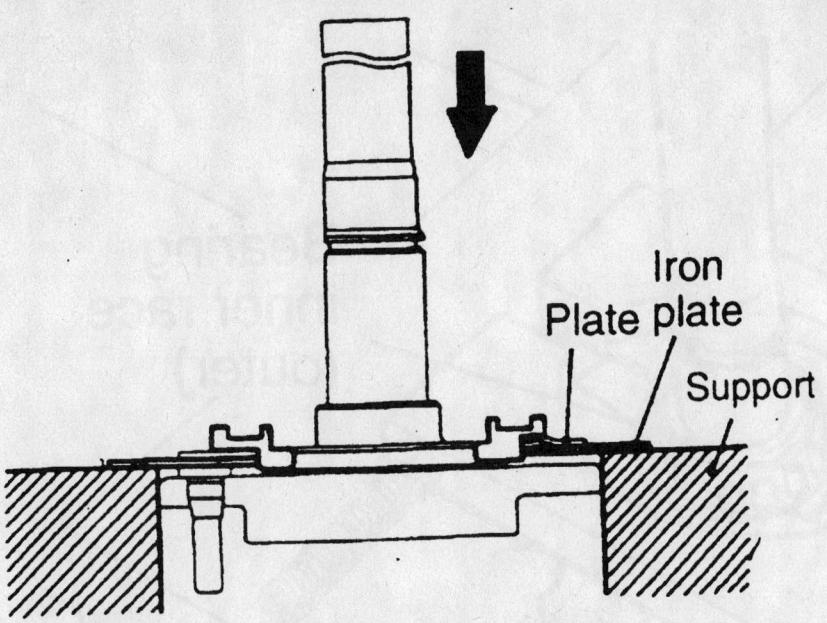

Use an iron plate and supports to remove the rotor assembly

11. Reinstall the bearing inner race that was removed previously, then use the tool MB990799-01 and press to remove the bearing outer race.

12. Remove or disconnect the following:
- Bearing case
- O-ring from the end of the axle housing tube
- Oil seal from the end of the rear axle housing using the tool

MB990211-01 (slide hammer with a hooked end), if necessary

13. Check the dust cover for deformation or damage. Check the oil seal for damage. Check the inner and outer bearings for seizure, discoloration and rough raceway surface. Check the axle shaft for cracks, wear and damage. For there are any of these indications, replace the part with a new one. The retainer, the bearing inner (inner and

outer) and outer races and the oil seal need to be replaced with new components upon reassembly. After all of this work, it is probably a good idea to replace the bearings and the axle housing tube oil seals.

To install:

14. Install or connect the following:
- New oil seal into the end of the rear axle housing using the tools MB990932-01 and MB990938-01, if necessary.
- New O-ring into the axle tube

15. Apply multi-purpose grease to the external surface of the bearing out race. Press-fit the bearing outer race into the bearing case by using the tool MB990890-01.

16. Install or connect the following:
- Speed sensor bracket to the back of the backing plate
- Rotor assembly to the axle shaft by press-fitting (plastic mallet will also work) it on using the special tool MB991388
- Backing plate onto the axle shaft
- Dust cover to the backing plate if the vehicle is equipped with ABS.
- Bearing inner race (outer) to the bearing case
- Oil seal to the front end of the bearing case. To do this, apply multi-purpose grease to the outside of the oil seal. Use the special tools MB990936-01 and MB990938-01 to press-fit the oil seal until it is flush with the end of the bearing case. Apply multi-purpose grease to the lip of the oil seal.
- Axle shaft through the bearing inner race, the bearing case and the second bearing inner race in that order. Use the special tool MB990799 to press-fit the bearing inner race to the axle shaft.

✳✳ WARNING

Both bearing inner race sets should be press-fitted together. The left and right lengths of the axle shaft are different in vehicles with rear differential locks. The right side is longer; be careful when installing it.

17. Use the tool MB990799-01 to press-fit the retainer onto the axle shaft, while checking that the press-fitting force is at the following values:
- Initial press-fitting force: 11,016 lbs. (5000 kg) or more.
- Final press-fitting force: 22,031–24,280 lbs. (98,000–108,000 N).

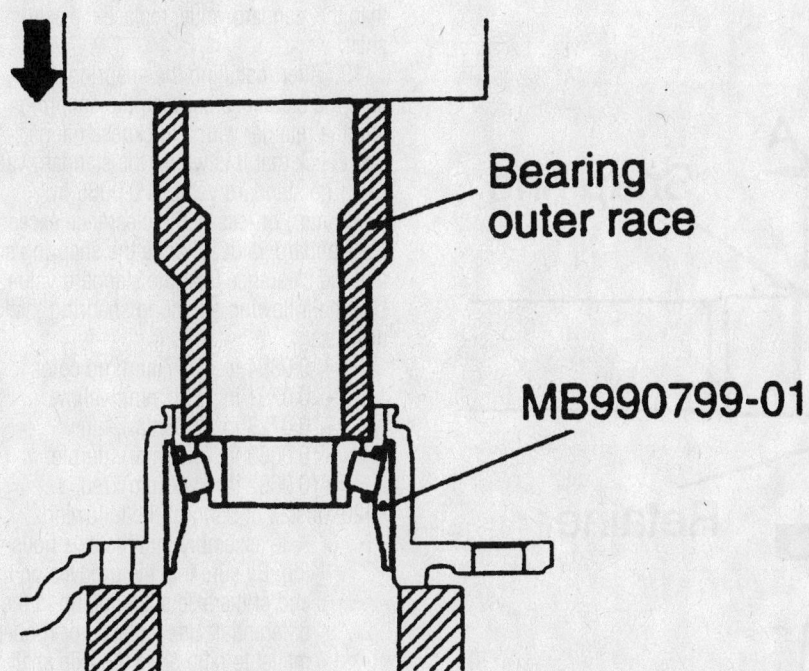

Use the tool MB990799-01 to install and remove the rear axle bearing races

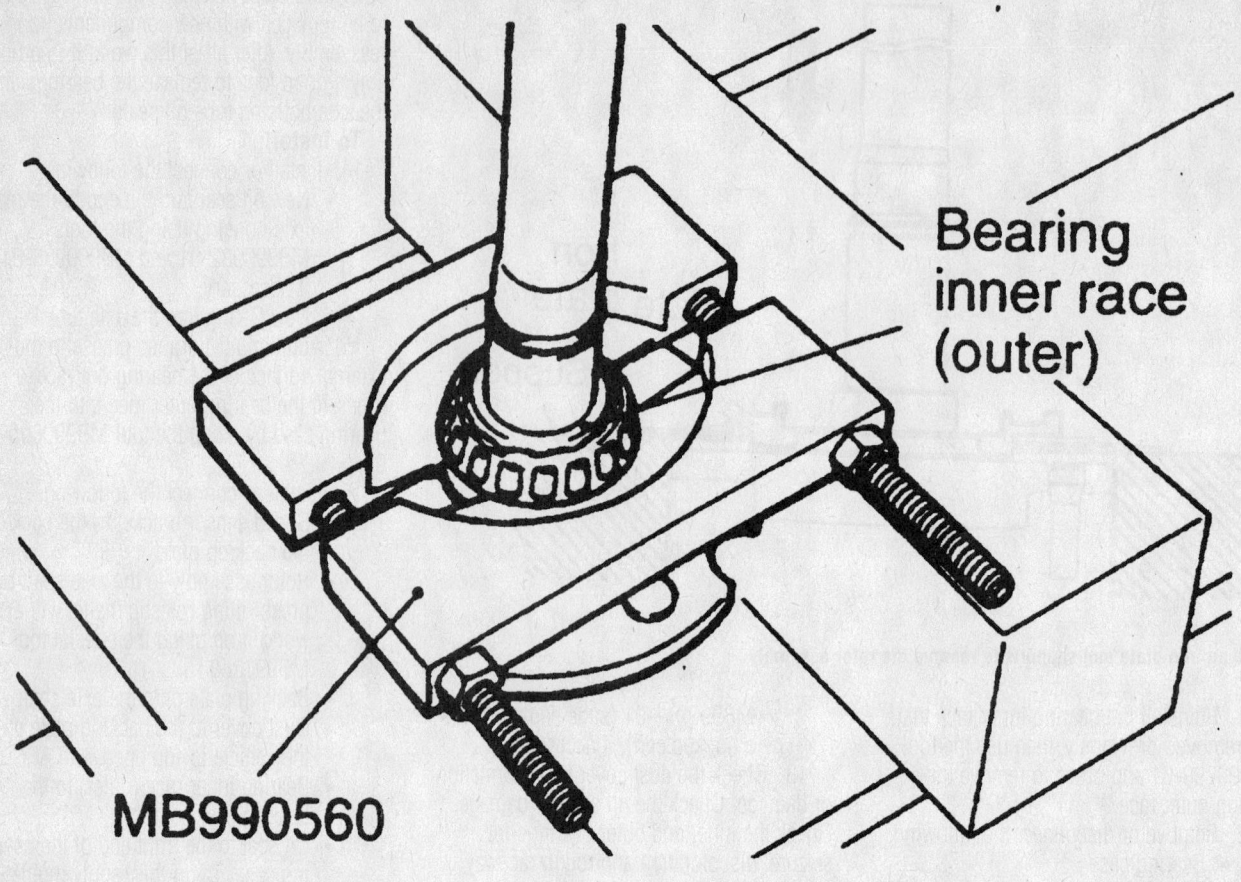

MB990560

Bearing inner race (outer)

7924UG55

Use the MB990560 tool to hold the bearing inner race (outer), then use a plastic hammer to drive the axle out of the race—do not let the axle fall onto a hard floor

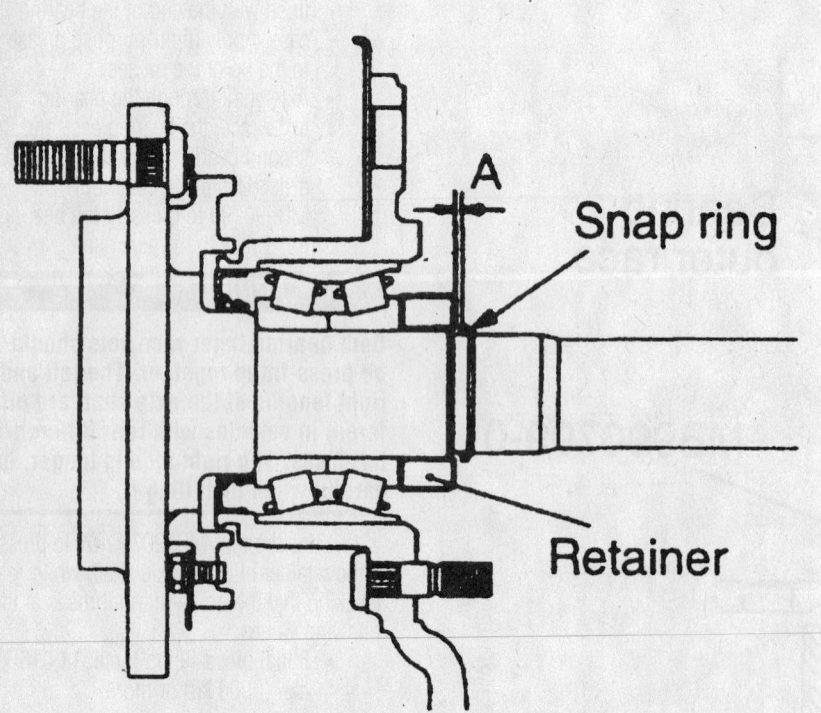

Snap ring

Retainer

7924UG56

Measure the clearance (A) between the snapring spond the retainer edge

18. If the initial press-fitting force is less than the standard value, replace the axle shaft.

19. After installing the snapring, measure the clearance between the snapring and the retainer with a thickness gauge, and check that it is within the standard values. The standard value is 0.0065 in. (0.166mm) or less. If the clearance exceeds the standard value, change the snapring so that the clearance is at the standard value. Use the following adjusting snapring thicknesses:

- 0.0854 in. (2.17mm): no color.
- 0.0791 in. (2.01mm): yellow.
- 0.0728 in. (1.85mm): blue.
- 0.0665 in. (1.69mm): purple.
- 0.0602 in. (1.53mm): red.

20. Install or connect the following:
- Axle assembly into the axle housing. Be sure that the grooves on the end of the axle shaft line up in the differential. Use a plastic or rubber mallet to help drive the axle shaft into the differential unit. Tighten the 4 retaining bolts for the axle shafts to 36–43 ft. lbs. (49–59 Nm).

- Speed sensor
- Parking brake assembly components to the axle flange.
- Parking brake cable to the parking brake assembly, then secure it in place with the cable bracket.
- Brake rotor onto the axle shaft, and

the brake caliper. Torque the caliper bolts to 65 ft. lbs. (88 Nm).
- Brake hose to the frame brake line. Torque the flare nut to 11 ft. lbs. (15 Nm).
- Wheels. Torque the lug nuts as tight as possible with the vehicle not on the ground.

21. Bleed the brake system.
22. Lower the vehicle until the wheels are touching the ground, then finish tightening the lug nuts. Lower the vehicle the rest of the way to the ground.
23. Road test the vehicle and check for leaks.

BRAKES

Brake Caliper

REMOVAL & INSTALLATION

Front

1. Before servicing the vehicle, refer to the precautions in the beginning of this section.
2. Raise and safely support the vehicle.
3. Remove or disconnect the following:
   - Wheel and tire assembly
   - Brake hose from the caliper brake line and remove the retaining clip
   - Caliper guide pin bolts
   - Caliper, by lifting it from the caliper support

To install:
4. Make sure the disc brake pad shims and clips are properly positioned.
5. Position the caliper over the rotor so the caliper engages the adapter correctly
6. Install or connect the following:
   - Mounting pins and tighten to 54 ft. lbs. (74 Nm)
   - Brake hose to the caliper brake line
   - Retaining clip
7. Bleed the brake system.
   - Wheel and tire assembly

Rear

1. Before servicing the vehicle, refer to the precautions in the beginning of this section.

2. Raise and safely support the vehicle.
3. Remove or disconnect the following:
   - Wheel and tire assembly
   - Brake hose from the caliper brake line and remove the retaining clip
   - Caliper guide pin bolts
   - Caliper by lifting it from the caliper support

To install:
4. Make sure the disc brake pad shims and clips are properly positioned.
5. Position the caliper over the rotor so the caliper engages the adapter correctly.
6. Install or connect the following:
   - Mounting pins and tighten them to 32 ft. lbs. (44 Nm)

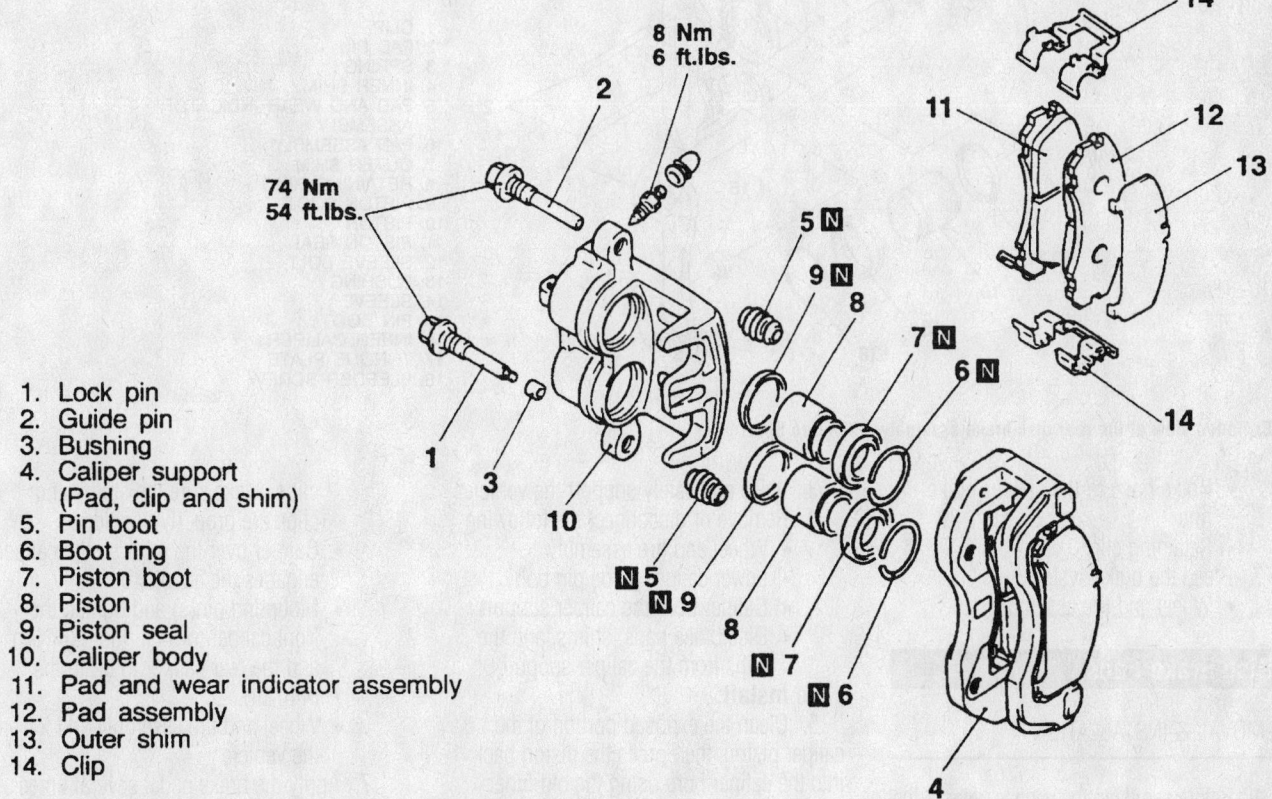

1. Lock pin
2. Guide pin
3. Bushing
4. Caliper support (Pad, clip and shim)
5. Pin boot
6. Boot ring
7. Piston boot
8. Piston
9. Piston seal
10. Caliper body
11. Pad and wear indicator assembly
12. Pad assembly
13. Outer shim
14. Clip

Exploded view of the front disc brake assembly—Montero and Montero Sport

93026G99

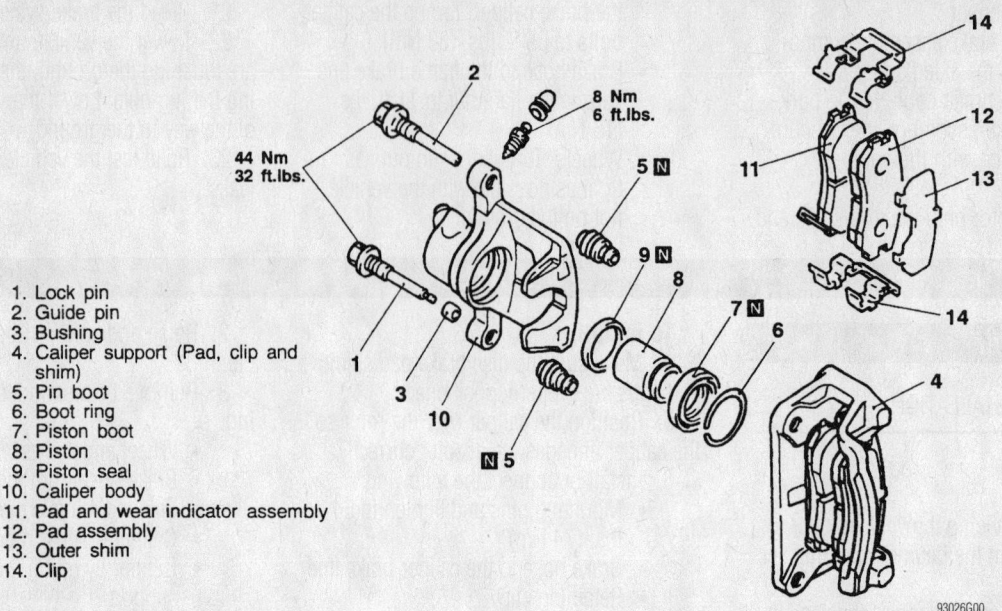

1. Lock pin
2. Guide pin
3. Bushing
4. Caliper support (Pad, clip and shim)
5. Pin boot
6. Boot ring
7. Piston boot
8. Piston
9. Piston seal
10. Caliper body
11. Pad and wear indicator assembly
12. Pad assembly
13. Outer shim
14. Clip

Exploded view of the rear disc brake assembly—Montero

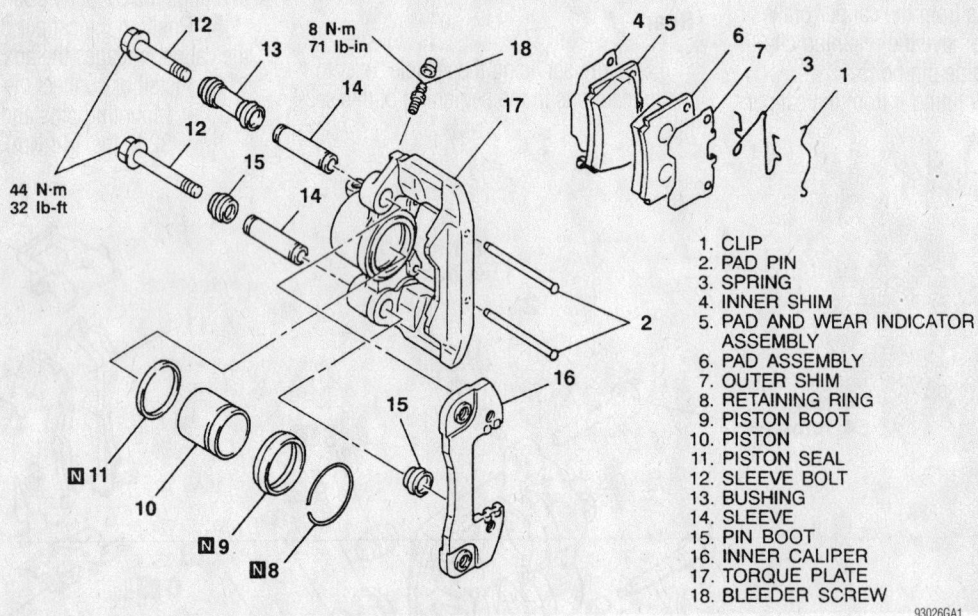

1. CLIP
2. PAD PIN
3. SPRING
4. INNER SHIM
5. PAD AND WEAR INDICATOR ASSEMBLY
6. PAD ASSEMBLY
7. OUTER SHIM
8. RETAINING RING
9. PISTON BOOT
10. PISTON
11. PISTON SEAL
12. SLEEVE BOLT
13. BUSHING
14. SLEEVE
15. PIN BOOT
16. INNER CALIPER
17. TORQUE PLATE
18. BLEEDER SCREW

Exploded view of the rear disc brake assembly—Montero Sport

- Brake hose to the caliper brake line
- Retaining clip
7. Bleed the brake system.
- Wheel and tire assembly

Disc Brake Pads

REMOVAL & INSTALLATION

1. Before servicing the vehicle, refer to the precautions in the beginning of this section.

2. Remove ½ of the brake fluid from the master cylinder.

3. Raise and safely support the vehicle.
4. Remove or disconnect the following:
- Wheel and tire assembly
- Lower caliper guide pin bolt
- Caliper from the caliper support
- Disc brake pads, shims, and the clips from the caliper support

To install:
5. Clean the exposed portion of the caliper piston, then press the piston back into the caliper bore using the old inner brake pad and a C-clamp.
6. Install or connect the following:
- Disc brake pads, shims, and the

clips. Make sure the shims and clips are properly positioned.
- Caliper over the rotor so the caliper engages the adapter correctly
- Mounting pin(s) and tighten the front caliper to 54 ft. lbs. (74 Nm) and the rear caliper to 32 ft. lbs. (44 Nm).
- Wheel and tire assembly and lower the vehicle

7. Apply the brake pedal several times until a firm pedal is obtained. Check the fluid level in the master cylinder and add fluid, as necessary.

NISSAN

Altima • Maxima • Sentra

23

SPECIFICATION CHARTS

ENGINE AND VEHICLE IDENTIFICATION

Code ①	Liters (cc)	Cu. In.	Cyl.	Fuel Sys.	Engine Type	Eng. Mfg.
QG18DE	1.8 (1769)	108	4	MFI	DOHC	Nissan
SR20DE	2.0 (1998)	122	4	MFI	DOHC	Nissan
KA24DE	2.4 (2389)	146	4	MFI	DOHC	Nissan
QR25DE	2.5 (2488)	152	4	MFI	DOHC	Nissan
VQ30DE	3.0 (2988)	182	6	MFI	DOHC	Nissan
VQ35DE	3.5 (3498)	213	6	MFI	DOHC	Nissan

Model Year	
Code ②	Year
Y	2000
1	2001
2	2002
3	2003
4	2004

MFI: Multi-port Fuel Injection

DOHC: Double Overhead Camshaft

① The Engine Code is stamped on the engine block near the starter.

② 10th position of the Vehicle Identification Number (VIN)

42356-MAXI-C01

GENERAL ENGINE SPECIFICATIONS

Year	Model	Engine Displacement Liters (cc)	Engine Series (ID/VIN)	Net Horsepower @ rpm	Net Torque @ rpm (ft. lbs.)	Bore x Stroke (in.)	Com-pression Ratio	Oil Pressure @ rpm
2000	Altima	2.4 (2389)	KA24DE	150@5600	154@5600	3.50x3.78	9.2:1	60@3000
	Maxima	3.0 (2988)	VQ30DE	190@5600	205@4000	3.66x2.89	10.0:1	63@3000
	Sentra	1.8 (1769)	QG18DE	126@6000	129@2400	3.15x3.46	9.5:1	50@3000
	Sentra	2.0 (1998)	SR20DE	145@6400	136@4800	3.39x3.39	9.5:1	46@3200
2001	Altima	2.4 (2389)	KA24DE	150@5600	154@5600	3.50x3.78	9.2:1	60@3000
	Maxima	3.0 (2988)	VQ30DE	190@5600	205@4000	3.66x2.89	10.0:1	63@3000
	Sentra	1.8 (1769)	QG18DE	126@6000	129@2400	3.15x3.46	9.5:1	50@3000
	Sentra	2.0 (1998)	SR20DE	145@6400	136@4800	3.39x3.39	9.5:1	46@3200
2002	Altima	2.5 (2488)	QR25DE	150@5600	154@5600	3.50X3.94	9.5:1	60@3000
	Altima	3.5 (3498)	VQ35DE	260@6000	260@4800	3.76X3.20	10.3:1	43@2000
	Maxima	3.5 (3498)	VQ35DE	260@6000	260@4800	3.76X3.20	10.3:1	43@2000
	Sentra	2.0 (1998)	SR20DE	145@6400	136@4800	3.39x3.39	9.5:1	46@3200
2003-04	Altima	2.5 (2488)	QR25DE	150@5600	154@5600	3.50X3.94	9.5:1	60@3000
	Altima	3.5 (3498)	VQ35DE	260@6000	260@4800	3.76X3.20	10.3:1	43@2000
	Maxima	3.5 (3498)	VQ35DE	260@6000	260@4800	3.76X3.20	10.3:1	43@2000
	Sentra	2.0 (1998)	SR20DE	145@6400	136@4800	3.39x3.39	9.5:1	46@3200

42356-MAXI-C02

ENGINE TUNE-UP SPECIFICATIONS

Year	Engine Displacement Liters (cc)	Engine ID/VIN	Spark Plug Gap (in.)	Ignition Timing (deg.) MT	Ignition Timing (deg.) AT	Fuel Pump (psi) ①	Idle Speed (rpm) MT	Idle Speed (rpm) AT ②	Valve Clearance Intake ③	Valve Clearance Exhaust ③
2000	1.8 (1769)	QG18DE	0.041	9B	9B	36	625 ④	725	0.015	0.016
	2.0 (1998)	SR20DE	0.033	15B	15B	36	800	800	HYD	HYD
	2.4 (2389)	KA24DE	0.041	20B	20B	33	650	650	0.015	0.016
	3.0 (2988)	VQ30DE	0.041	15B	15B	34	650	700	0.014	0.015
2001	1.8 (1769)	QG18DE	0.041	9B	9B	36	625 ④	725	0.015	0.016
	2.0 (1998)	SR20DE	0.033	15B	15B	36	800	800	HYD	HYD
	2.4 (2389)	KA24DE	0.041	20B	20B	33	650	650	0.015	0.016
	3.0 (2988)	VQ30DE	0.041	15B	15B	34	650	700	0.014	0.015
2002	2.0 (1998)	SR20DE	0.033	15B	15B	36	800	800	HYD	HYD
	2.5 (2488))	QR25DE	0.043	15B	20B	33	650	650	HYD	HYD
	3.5 (3498)	VQ35DE	0.043	—	15B	34	—	600-700	HYD	HYD
2003-04	2.0 (1998)	SR20DE	0.033	15B	15B	36	800	800	HYD	HYD
	2.5 (2488)	QR25DE	0.043	15B	20B	33	650	650	HYD	HYD
	3.5 (3498)	VQ35DE	0.043	—	15B	34	—	600-700	HYD	HYD

NOTE: The Vehicle Emission Control Information label often reflects specification changes made during production. The label figures must be used if they differ from those in this chart.

B: Before top dead center

HYD: Hydraulic

① System pressure at idle with vacuum hose connected; should increase to 43 psi when disconnected

② Automatic transmission in neutral

③ Engine warm

④ Canada: 750

42356-MAXI-C03

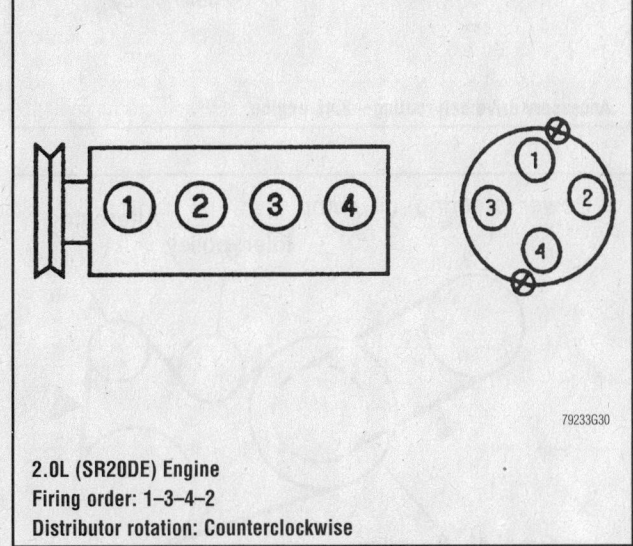

2.0L (SR20DE) Engine
Firing order: 1–3–4–2
Distributor rotation: Counterclockwise

79233G30

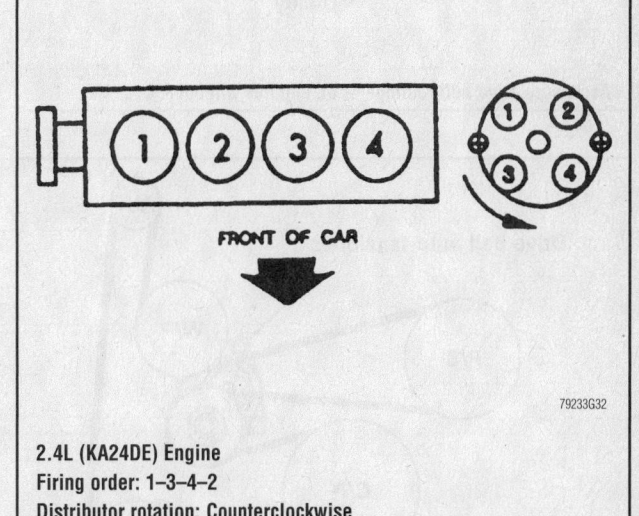

FRONT OF CAR

2.4L (KA24DE) Engine
Firing order: 1–3–4–2
Distributor rotation: Counterclockwise

79233G32

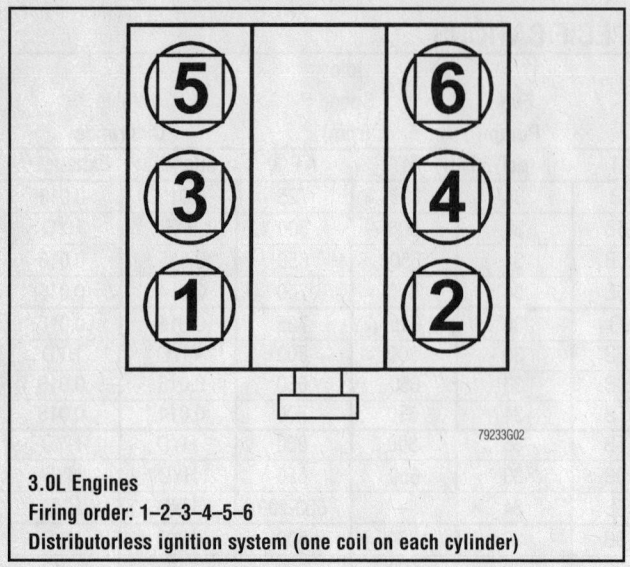

3.0L Engines
Firing order: 1–2–3–4–5–6
Distributorless ignition system (one coil on each cylinder)

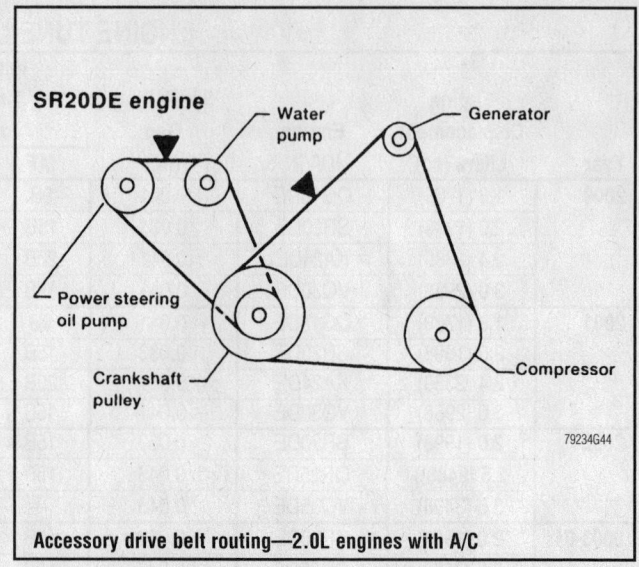

Accessory drive belt routing—2.0L engines with A/C

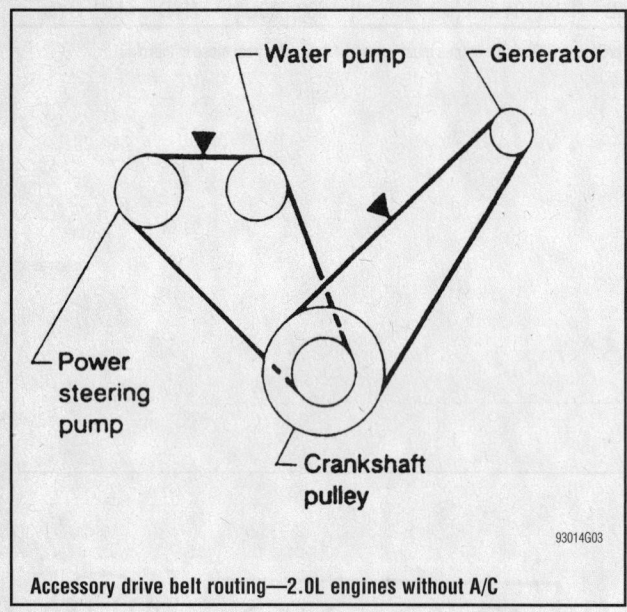

Accessory drive belt routing—2.0L engines without A/C

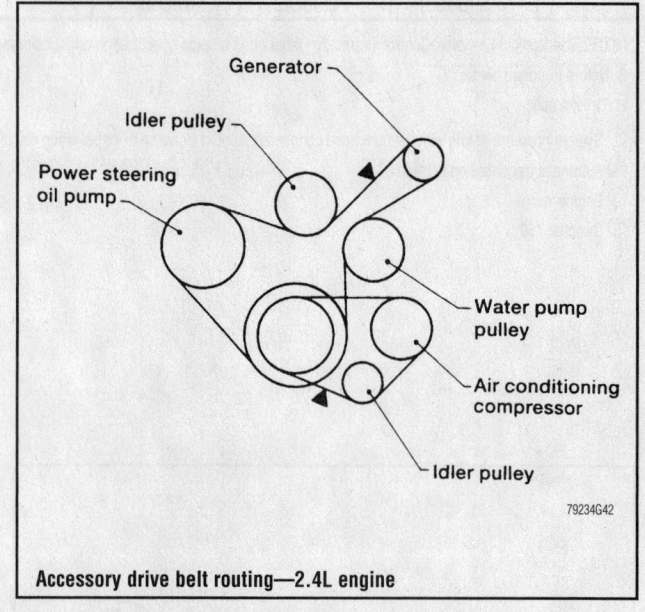

Accessory drive belt routing—2.4L engine

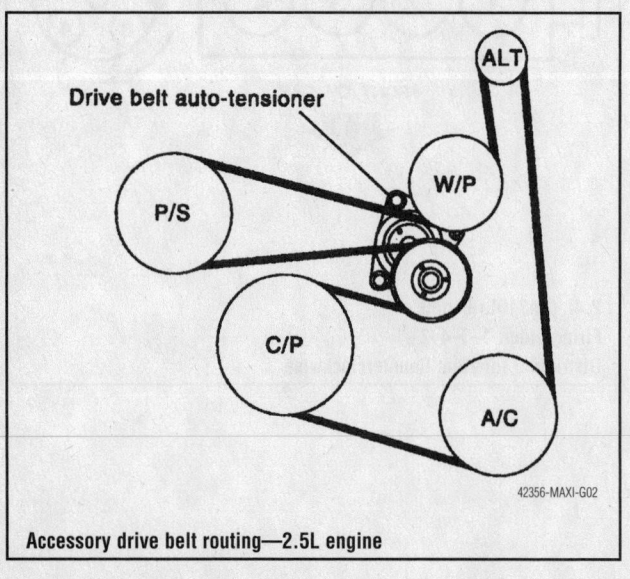

Accessory drive belt routing—2.5L engine

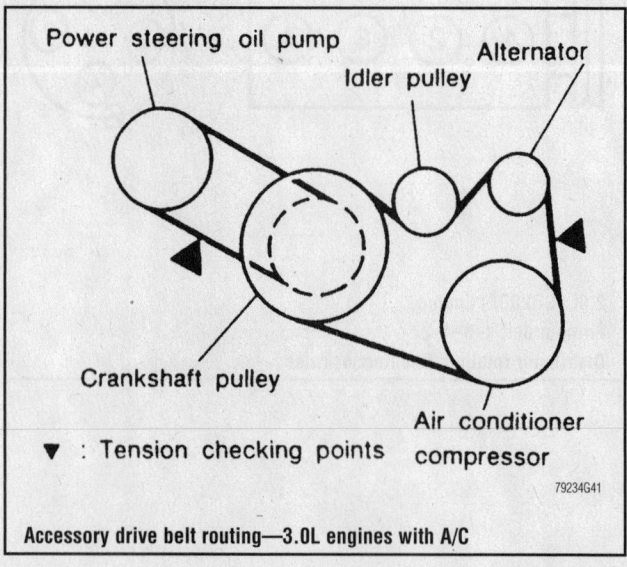

▼ : Tension checking points

Accessory drive belt routing—3.0L engines with A/C

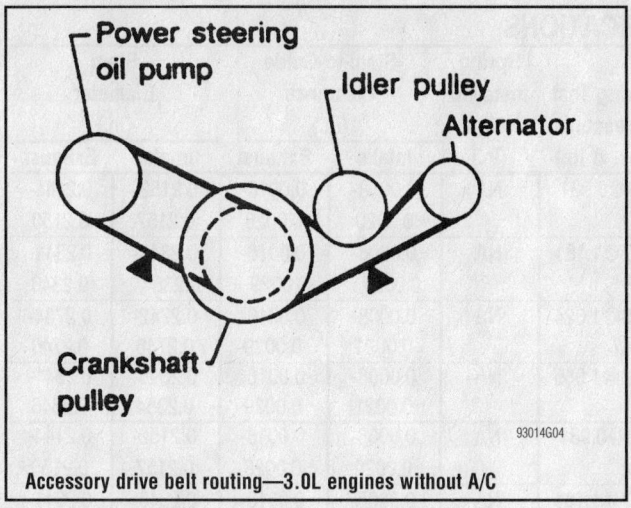

Accessory drive belt routing—3.0L engines without A/C

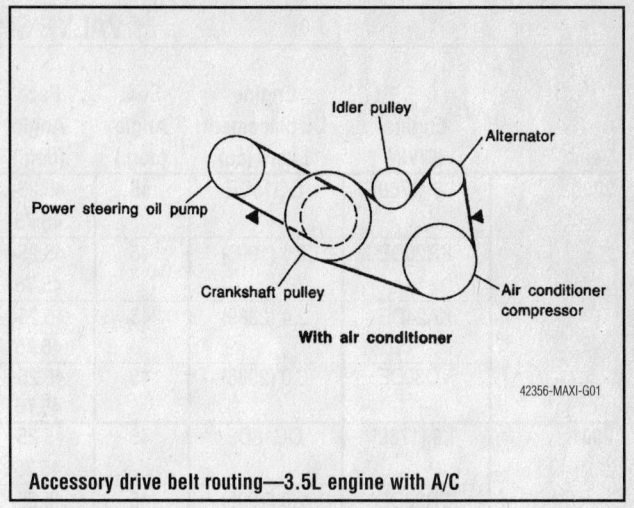

Accessory drive belt routing—3.5L engine with A/C

CAPACITIES

Year	Model	Engine ID/VIN	Engine Displacement Liters (cc)	Engine Oil with Filter (qts.)	Transmission (pts.) 5-Spd	Transmission (pts.) Auto.	Drive Axle Rear (pts.)	Fuel Tank (gal.)	Cooling System (qts.)
2000	Altima	KA24DE	2.4 (2389)	4.1	10.0	20.0	—	15.9	7.8
	Maxima	VQ30DE	3.0 (2988)	4.3	9.5	20.0	—	18.5	9.0
	Sentra	QG18DE	1.8 (1769)	3.5	①	14.8	—	13.2	②
	Sentra	SR20DE	2.0 (1998)	3.4	7.5	14.8	—	13.0	6.0
2001	Altima	KA24DE	2.4 (2389)	4.1	10.0	20.0	—	15.9	7.8
	Maxima	VQ30DE	3.0 (2988)	4.3	9.5	20.0	—	18.5	9.0
	Sentra	QG18DE	1.8 (1769)	3.5	①	14.8	—	13.2	②
	Sentra	SR20DE	2.0 (1998)	3.4	7.5	14.8	—	13.0	6.0
2002	Altima	QR25DE	2.5 (2488)	4.5	6.8	19.5	—	15.9	8.2
	Altima	VQ35DE	3.5 (3498)	4.3	6.8	19.0	—	15.9	9.3
	Maxima	VQ35DE	3.5 (3498)	5.0	—	21.8	—	18.5	9.0
	Sentra	SR20DE	2.0 (1998)	3.4	7.5	14.8	—	13.0	6.0
2003-04	Altima	QR25DE	2.5 (2488)	4.5	6.8	19.5	—	15.9	8.2
	Altima	VQ35DE	3.5 (3498)	4.3	6.8	19.0	—	15.9	9.3
	Maxima	VQ35DE	3.5 (3498)	5.0	—	21.8	—	18.5	9.0
	Sentra	SR20DE	2.0 (1998)	3.4	7.5	14.8	—	13.0	6.0

NOTE: All capacities are approximate. Add fluid gradually and check to be sure a proper fluid level is obtained.

① RS5F31A: 6.5 pts.
RS5F32V: 8.0 pts.

② GA16DE with MT: 5.5 qts.
GA16DE with AT: 6.0 qts.

42356-MAXI-C04

VALVE SPECIFICATIONS

Year	Engine ID/VIN	Engine Displacement Liters (cc)	Seat Angle (deg.)	Face Angle (deg.)	Spring Test Pressure (lbs. @ in.)	Spring Installed Height (in.)	Stem-to-Guide Clearance (in.)		Stem Diameter (in.)	
							Intake	Exhaust	Intake	Exhaust
2000	1.8 (1769)	QG18DE	45	45.25-45.75	83@0.931	NA	0.0008-0.0020	0.0016-0.0028	0.2152-0.2157	0.2144-0.2150
	SR20DE	2.0 (1998)	45	45.25-45.75	137@1.181	NA	0.0008-0.0021	0.0016-0.0029	0.2348-0.2354	0.2341-0.2346
	KA24DE	2.4 (2389)	45	45.25-45.75	123@1.024	NA	0.0008-0.0021	0.0016-0.0029	0.2742-0.2748	0.2734-0.2740
	VQ30DE	3.0 (2988)	45	45.25-45.75	102@1.085	NA	0.0008-0.0021	0.0016-0.0029	0.2348-0.2354	0.2341-0.2346
2001	1.8 (1769)	QG18DE	45	45.25-45.75	83@0.931	NA	0.0008-0.0020	0.0016-0.0028	0.2152-0.2157	0.2144-0.2150
	SR20DE	2.0 (1998)	45	45.25-45.75	137@1.181	NA	0.0008-0.0021	0.0016-0.0029	0.2348-0.2354	0.2341-0.2346
	KA24DE	2.4 (2389)	45	45.25-45.75	123@1.024	NA	0.0008-0.0021	0.0016-0.0029	0.2742-0.2748	0.2734-0.2740
	VQ30DE	3.0 (2988)	45	45.25-45.75	102@1.085	NA	0.0008-0.0021	0.0016-0.0029	0.2348-0.2354	0.2341-0.2346
2002	SR20DE	2.0 (1998)	45	45.25-45.75	137@1.181	NA	0.0008-0.0021	0.0016-0.0029	0.2348-0.2354	0.2341-0.2346
	QR25DE	2.5 (2488)	45.15-45.45	NA	34-39@1.39	NA	0.0009-0.0013	0.010-0.013	0.2348-0.2354	0.2344-0.2350
	VQ35DE	3.5 (3498)	45.15-45.45	NA	91.5-103.2@1.094	1.796	0.010-0.013	0.011-0.015	0.2348-0.2354	0.2344-0.2350
2003-04	SR20DE	2.0 (1998)	45	45.25-45.75	137@1.181	NA	0.0008-0.0021	0.0016-0.0029	0.2348-0.2354	0.2341-0.2346
	QR25DE	2.5 (2488)	45.15-45.45	NA	34-39@1.39	NA	0.0009-0.0013	0.010-0.013	0.2348-0.2354	0.2344-0.2350
	VQ35DE	3.5 (3498)	45.15-45.45	NA	91.5-103.2@1.094	1.796	0.010-0.013	0.011-0.015	0.2348-0.2354	0.2344-0.2350

NA: Not Available

42356-MAXI-C05

TORQUE SPECIFICATIONS
All readings in ft. lbs.

Year	Engine Displacement Liters (cc)	Engine ID/VIN	Cylinder Head Bolts	Main Bearing Bolts	Rod Bearing Bolts	Crankshaft Damper Bolts	Flywheel Bolts	Manifold		Spark Plugs	Lug Nuts
								Intake	Exhaust		
2000	1.8 (1769)	QG18DE	①	34-38	②	98-112	③	14	19	18	79
	2.0 (1998)	SR20DE	④	⑤	⑥	105-112	61-69	14	30	18	79
	2.4 (2389)	KA24DE	⑦	34-41	⑥	105-112	105-112	14	32	18	80
	3.0 (2988)	VQ30DE	⑧	⑨	⑩	⑪	61-69	⑫	23	18	80
2001	1.8 (1769)	QG18DE	①	34-38	②	98-112	③	14	19	18	79
	2.0 (1998)	SR20DE	④	⑤	⑥	105-112	61-69	14	30	18	79
	2.4 (2389)	KA24DE	⑦	34-41	⑥	105-112	105-112	14	32	18	80
	3.0 (2988)	VQ30DE	⑧	⑨	⑩	⑪	61-69	⑫	23	18	80
2002	2.0 (1998)	SR20DE	④	⑤	⑥	105-112	61-69	14	30	18	79
	2.5 (2488)	QR25DE	⑬	⑭	⑮	⑪	105-112	13-15	29-32	18	80
	3.5 (3498)	VQ35DE	⑯	⑰	⑱	⑪	61-69	⑲	21-23	15-21	72-87
2003-04	2.0 (1998)	SR20DE	④	⑤	⑥	105-112	61-69	14	30	18	79
	2.5 (2488)	QR25DE	⑬	⑭	⑮	⑪	105-112	13-15	29-32	18	80
	3.5 (3498)	VQ35DE	⑯	⑰	⑱	⑪	61-69	⑲	21-23	15-21	72-87

① Bolt Nos. 1-10:
Step 1: 22 ft. lbs.
Step 2: 43 ft. lbs.
Step 3: Loosen completely then retorque to 22 ft. lbs.
Step 4: 43 ft. lbs. or an additional 50-55 degrees
Bolt Nos. 11-14: Torque last, to 72 inch lbs.

② Step 1: 12 ft. lbs.
Step 2: 19 ft. lbs. or an additional 35-40 degrees

③ Manual transmission: 61-69 ft. lbs.
Automatic transmission: 69-76 ft. lbs.

④ Step 1: 29 ft. lbs.
Step 2: 58 ft. lbs.
Step 3: Loosen completely then retorque to 30 ft. lbs.
Step 4: Turn each bolt, in sequence,
an additional 90-100 degrees
Step 5: Repeat Step 4

⑤ Step 1: 20-24 ft. lbs.
Step 2: 75-80 degrees
Step 3: Loosen completely and retorque to 24-28 ft. lbs.
Step 4: 45-50 degree turn

⑥ 12 ft. lbs. plus an additional 60-65 degrees

⑦ Step 1: 22 ft. lbs.
Step 2: 58 ft. lbs.
Step 3: Loosen completely then retorque to 22 ft. lbs.
Step 4: 58 ft. lbs. or an additional 80-85 degrees

⑧ Step 1: 29-36 ft. lbs.
Step 2: Plus 60-65 degrees

⑨ Step 1: 3.6-7.2 ft. lbs.
Step 2: 20-23 ft. lbs.

⑩ Step 1: 29 ft. lbs.
Step 2: 90 ft. lbs.
Step 3: Loosen completely and retorque to 25-33 ft. lbs.
Step 4: Plus 90 ft. lbs. or 70 degrees
Step 5: Tighten two bolts marked with an "X" to 7-9 ft. lbs.

⑪ Step 1: 29-36 ft. lbs.
Step 2: 60-66 degrees

⑫ Step 1: 10-12 ft. lbs.
Step 2: 43-48 ft. lbs. or an additional 60-65 degrees

⑬ Step 1: 72 ft. lbs.
Step 2: Loosen completely, then retorque to 26-32 ft. lbs.
Step 3: Turn each bolt, in sequence, an additional 75-80 degrees
Step 4: Turn each bolt, in sequence, an additional 75-80 degrees

⑭ Bolt Nos. 1-10:
Step 1: 27-31 ft. lbs.
Step 2: Torque an additional 60-65 degrees
Bolt Nos. 11-14: Torque last, to 15-18 ft. lbs.

⑮ Step 1: 14-15 ft. lbs.
Step 2: 85-95 degrees

⑯ Step 1: 72 ft. lbs.
Step 2: Loosen bolts completely
Step 3: 25-33 ft. lbs.
Step 4: Tighten an additional 90-95 degrees
Step 5: Repeat Step 4

⑰ Step 1: Shift crankshaft to align the bearing beam
Step 2: Tighten all bolts to 24-28 ft. lbs.
Step 3: Tighten an additional 90-95 degrees

⑱ Step 1: Tighten to 15 ft. lbs.
Step 2: Tighten an additional 90-95 degrees

⑲ Step 1: Tighten to 4-7 ft. lbs.
Step 2: Tighten to 20-23 ft. lbs.
Step 3: Tighten, again, to 20-23 ft. lbs.

PISTON AND RING SPECIFICATIONS
All measurements are given in inches.

Year	Engine Displacement Liters (cc)	Engine ID/VIN	Piston Clearance	Ring Gap			Ring Side Clearance		
				Top Compression	Bottom Compression	Oil Control	Top Compression	Bottom Compression	Oil Control
2000	1.8 (1769)	QG18DE	0.0010-0.0018	0.0079-0.0154	0.0126-0.0220	0.0079-0.0272	0.0018-0.0031	0.0012-0.0028	0.0026-0.0053
	2.0 (1998)	SR20DE	0.0006-0.0014	0.0079-0.0118	0.0138-0.0197	0.0079-0.0236	0.0018-0.0031	0.0012-0.0026	SNUG
	2.4 (2389)	KA24DE	0.0006-0.0014	0.0110-0.0205	0.0079-0.0272	0.0100-1.0000	0.0016-0.0031	0.0012-0.0028	SNUG
	3.0 (2988)	VQ30DE	0.0006-0.0014	0.0087-0.0126	0.0126-0.0185	0.0079-0.0236	0.0016-0.0031	0.0012-0.0028	SNUG
2001	1.8 (1769)	QG18DE	0.0010-0.0018	0.0079-0.0154	0.0126-0.0220	0.0079-0.0272	0.0018-0.0031	0.0012-0.0028	0.0026-0.0053
	2.0 (1998)	SR20DE	0.0006-0.0014	0.0079-0.0118	0.0138-0.0197	0.0079-0.0236	0.0018-0.0031	0.0012-0.0026	SNUG
	2.4 (2389)	KA24DE	0.0006-0.0014	0.0110-0.0205	0.0079-0.0272	0.0100-1.0000	0.0016-0.0031	0.0012-0.0028	SNUG
	3.0 (2988)	VQ30DE	0.0006-0.0014	0.0087-0.0126	0.0126-0.0185	0.0079-0.0236	0.0016-0.0031	0.0012-0.0028	SNUG
2002	2.0 (1998)	SR20DE	0.0006-0.0014	0.0079-0.0118	0.0138-0.0197	0.0079-0.0236	0.0018-0.0031	0.0012-0.0026	SNUG
	2.5 (2488)	QR25DE	0.0004-0.0012	0.0083-0.0122	0.0126-0.0185	0.0079-0.0236	0.0018-0.0031	0.0012-0.0028	0.0026-0.0053
	3.5 (3498)	VQ35DE	0.0004-0.0012	0.0091-0.0130	0.0130-0.0189	0.0079-0.0197	0.0018-0.0031	0.0012-0.0028	0.0026-0.0053
2003-04	2.0 (1998)	SR20DE	0.0006-0.0014	0.0079-0.0118	0.0138-0.0197	0.0079-0.0236	0.0018-0.0031	0.0012-0.0026	SNUG
	2.5 (2488)	QR25DE	0.0004-0.0012	0.0083-0.0122	0.0126-0.0185	0.0079-0.0236	0.0018-0.0031	0.0012-0.0028	0.0026-0.0053
	3.5 (3498)	VQ35DE	0.0004-0.0012	0.0091-0.0130	0.0130-0.0189	0.0079-0.0197	0.0018-0.0031	0.0012-0.0028	0.0026-0.0053

42356-MAXI-C07

CRANKSHAFT AND CONNECTING ROD SPECIFICATIONS
All measurements are given in inches.

Year	Engine Displacement Liters (cc)	Engine ID/VIN	Crankshaft				Connecting Rod		
			Main Brg. Journal Dia.	Main Brg. Oil Clearance	Shaft End-play	Thrust on No.	Journal Diameter	Oil Clearance	Side Clearance
2000	1.8 (1769)	QG18DE	1.9668-1.9671	0.0007-0.0017	0.0024-0.0071	3	1.6929-1.6934	0.0006-0.0015	0.0079-0.0185
	2.0 (1998)	SR20DE	2.1643-2.1646	0.0002-0.0009	0.0039-0.0102	3	1.8885-1.8887	0.0008-0.0018	0.0079-0.0138
	2.4 (2389)	KA24DE	2.3609-2.3612	0.0008-0.0019	0.0020-0.0070	3	1.9672-1.9675	0.0004-0.0014	0.0080-0.0160
	3.0 (2988)	VQ30DE	2.3610-2.3612	0.0014-0.0021	0.0039-0.0098	3	1.7704-1.7706	0.0013-0.0023	0.0079-0.0138
2001	1.8 (1769)	QG18DE	1.9668-1.9671	0.0007-0.0017	0.0024-0.0071	3	1.6929-1.6934	0.0006-0.0015	0.0079-0.0185
	2.0 (1998)	SR20DE	2.1643-2.1646	0.0002-0.0009	0.0039-0.0102	3	1.8885-1.8887	0.0008-0.0018	0.0079-0.0138
	2.4 (2389)	KA24DE	2.3609-2.3612	0.0008-0.0019	0.0020-0.0070	3	1.9672-1.9675	0.0004-0.0014	0.0080-0.0160
	3.0 (2988)	VQ30DE	2.3610-2.3612	0.0014-0.0021	0.0039-0.0098	3	1.7704-1.7706	0.0013-0.0023	0.0079-0.0138
2002	2.0 (1998)	SR20DE	2.1643-2.1646	0.0002-0.0009	0.0039-0.0102	3	1.8885-1.8887	0.0008-0.0018	0.0079-0.0138
	2.5 (2488)	QR25DE	2.1636-2.1645	①	0.0039-0.0102	3	1.8898-1.8903	0.0004-0.0014	0.0079-0.0138
	3.5 (3498)	VQ35DE	2.3603-2.3612	0.0014-0.0021	0.0039-0.0098	3	1.7704-1.7706	0.0013-0.0023	0.0079-0.0138
2003-04	2.0 (1998)	SR20DE	2.1643-2.1646	0.0002-0.0009	0.0039-0.0102	3	1.8885-1.8887	0.0008-0.0018	0.0079-0.0138
	2.5 (2488)	QR25DE	2.1636-2.1645	①	0.0039-0.0102	3	1.8898-1.8903	0.0004-0.0014	0.0079-0.0138
	3.5 (3498)	VQ35DE	2.3603-2.3612	0.0014-0.0021	0.0039-0.0098	3	1.7704-1.7706	0.0013-0.0023	0.0079-0.0138

① Nos. 1, 3 and 5: 0.0005-0.0009 in.
Nos. 2 and 4: 0.0007-0.0011

42356-MAXI-C08

WHEEL ALIGNMENT

Year	Model		Caster Range (+/-Deg.)	Caster Preferred Setting (Deg.)	Camber Range (+/-Deg.)	Camber Preferred Setting (Deg.)	Toe-in (in.)	Steering Axis Inclination (Deg.)
2000	Altima	F	0.75	+2.66	0.75	-0.10	0.04 +/- 0.04	14.09
		R	—		0.75	-1.25	0.08 +/- 0.04	—
	Maxima ①	F	0.75	+2.75	0.75	-0.25	0.04 +/- 0.04	—
		R	—	—	0.75	-1.00	0.04 +/- 0.16	—
	Maxima ②	F	0.75	+2.75	0.75	-0.33	0.04 +/- 0.04	—
		R	—	—	0.75	-1.00	0.04 +/- 0.16	—
	Sentra	F	0.75	+1.42	0.75	-0.58	0.08 +/- 0.08	—
		R	—	—	0.75	+1.00	0.04 +/- 0.15	—
2001	Altima	F	0.75	+2.66	0.75	-0.10	0.04 +/- 0.04	14.09
		R	—	—	0.75	-1.25	0.08 +/- 0.04	—
	Maxima ①	F	0.75	+2.75	0.75	-0.25	0.04 +/- 0.04	—
		R	—	—	0.75	-1.00	0.04 +/- 0.16	—
	Maxima ②	F	0.75	+2.75	0.75	-0.33	0.04 +/- 0.04	—
		R	—	—	0.75	-1.00	0.04 +/- 0.16	—
	Sentra	F	0.75	+1.42	0.75	-0.58	0.08 +/- 0.08	—
		R	—	—	0.75	+1.00	0.04 +/- 0.15	—
2002	Altima	F	0.75	+2.66	0.75	-0.10	0.04 +/- 0.04	14.09
		R	—	—	0.75	-1.25	0.08 +/- 0.04	—
	Maxima ①	F	0.75	+2.75	0.75	-0.25	0.04 +/- 0.04	—
		R	—	—	0.75	-1.00	0.04 +/- 0.16	—
	Maxima ②	F	0.75	+2.75	0.75	-0.33	0.04 +/- 0.04	—
		R	—	—	0.75	-1.00	0.04 +/- 0.16	—
	Sentra	F	0.75	+1.42	0.75	-0.58	0.08 +/- 0.08	—
		R	—	—	0.75	+1.00	0.04 +/- 0.15	—
2003-04	Altima	F	0.75	+2.66	0.75	-0.10	0.04 +/- 0.04	14.09
		R	—	—	0.75	-1.25	0.08 +/- 0.04	—
	Maxima ①	F	0.75	+2.75	0.75	-0.25	0.04 +/- 0.04	—
		R	—	—	0.75	-1.00	0.04 +/- 0.16	—
	Maxima ②	F	0.75	+2.75	0.75	-0.33	0.04 +/- 0.04	—
		R	—	—	0.75	-1.00	0.04 +/- 0.16	—
	Sentra	F	0.75	+1.42	0.75	-0.58	0.08 +/- 0.08	—
		R	—	—	0.75	+1.00	0.04 +/- 0.15	—

① With P225/55R16, P215/55R16 tires
② With P205/65R15 tires

42356-MAXI-C09

TIRE, WHEEL AND BALL JOINT SPECIFICATIONS

Year	Model	OEM Tires		Tire Pressures (psi)		Wheel Size	Ball Joint Inspection
		Standard	Optional	Front	Rear		
2000	Altima	P195/65R15	P205/60R15	30	30	6-JJ	①
	Maxima GLE	P215/55R16	None	29	29	6.5-JJ	①
	Maxima GXE	P205/65SR15	None	29	29	6JJ	①
	Maxima SE	P215/55R16	P225/50R17	29	29	6.5J/7J	①
	Sentra, base	P155/80R13	None	26	26	5-J	①
	Sentra XE	P175/70R13	None	26	26	5-J	①
	Sentra GXE	P175/65R14	None	26	26	5.5-JJ	①
	Sentra GLE	P175/65R14	None	26	26	5.5-JJ	①
	Sentra SE	P195/55R15	None	30	30	6-JJ	①
2001	Altima	P195/65R15	P205/60R15	30	30	6-JJ	①
	Maxima GLE	P215/55R16	None	29	29	6.5-JJ	①
	Maxima GXE	P205/65SR15	None	29	29	6JJ	①
	Maxima SE	P215/55R16	P225/50R17	29	29	6.5J/7J	①
	Sentra, base	P155/80R13	None	26	26	5-J	①
	Sentra XE	P175/70R13	None	26	26	5-J	①
	Sentra GXE	P175/65R14	None	26	26	5.5-JJ	①
	Sentra GLE	P175/65R14	None	26	26	5.5-JJ	①
	Sentra SE	P195/55R15	None	30	30	6-JJ	①
2002	Altima	P195/65R15	P205/60R15	30	30	6-JJ	①
	Maxima GLE	P215/55R16	None	29	29	6.5-JJ	①
	Maxima GXE	P205/65SR15	None	29	29	6JJ	①
	Maxima SE	P215/55R16	P225/50R17	29	29	6.5J/7J	①
	Sentra, base	P155/80R13	None	26	26	5-J	①
	Sentra XE	P175/70R13	None	26	26	5-J	①
	Sentra GXE	P175/65R14	None	26	26	5.5-JJ	①
	Sentra GLE	P175/65R14	None	26	26	5.5-JJ	①
	Sentra SE	P195/55R15	None	30	30	6-JJ	①
2003-04	Altima	P195/65R15	P205/60R15	30	30	6-JJ	①
	Maxima GLE	P215/55R16	None	29	29	6.5-JJ	①
	Maxima GXE	P205/65SR15	None	29	29	6JJ	①
	Maxima SE	P215/55R16	P225/50R17	29	29	6.5J/7J	①
	Sentra, base	P155/80R13	None	26	26	5-J	①
	Sentra XE	P175/70R13	None	26	26	5-J	①
	Sentra GXE	P175/65R14	None	26	26	5.5-JJ	①
	Sentra GLE	P175/65R14	None	26	26	5.5-JJ	①
	Sentra SE	P195/55R15	None	30	30	6-JJ	①

OEM: Original Equipment Manufacturer

PSI: Pounds Per Square Inch

STD: Standard

OPT: Optional

① Replace if any measurable movement is found.

42356-MAXI-C10

BRAKE SPECIFICATIONS
All measurements in inches unless noted

| Year | Model | | Brake Disc | | | Brake Drum Diameter | | | Minimum Lining Thickness | | Brake Caliper | |
			Original Thickness	Minimum Thickness	Maximum Run-out	Original Inside Diameter	Max. Wear Limit	Maximum Machine Diameter	Front	Rear	Bracket Bolts (ft. lbs.)	Mounting Bolts (ft. lbs.)
2000	Altima	F	0.870	0.787	0.003	—	—	—	0.079	—	53-72	16-23
		R	0.390	0.315	0.003	9.00	NA	9.06	—	0.059	—	—
	Maxima	F	0.870	0.787	0.003	—	—	—	0.079	—	53-72	16-23
		R	0.350	0.315	0.003	—	—	—	—	0.059	—	—
	Sentra	F	0.710	0.630	0.003	—	—	—	0.079	—	40-47	12-14
		R	0.280	0.236	0.003	7.09	7.13	7.13	—	0.059	—	—
2001	Altima	F	0.870	0.787	0.003	—	—	—	0.079	—	53-72	16-23
		R	0.390	0.315	0.003	9.00	NA	9.06	—	0.059	—	—
	Maxima	F	0.870	0.787	0.003	—	—	—	0.079	—	53-72	16-23
		R	0.350	0.315	0.003	—	—	—	—	0.059	—	—
	Sentra	F	0.710	0.630	0.003	—	—	—	0.079	—	40-47	12-14
		R	0.280	0.236	0.003	7.09	7.13	7.13	—	0.059	—	—
2002	Altima	F	1.020	0.866	0.003	—	—	—	0.079	—	53-72	16-23
		R	0.350	0.310	0.003	—	—	—	—	0.059	—	—
	Maxima	F	0.940	0.866	0.003	—	—	—	0.079	—	53-72	16-23
		R	0.350	0.315	0.003	—	—	—	—	0.059	—	—
	Sentra	F	0.710	0.630	0.003	—	—	—	0.079	—	40-47	12-14
		R	0.280	0.236	0.003	7.09	7.13	7.13	—	0.059	—	—
2003-04	Altima	F	1.020	0.866	0.003	—	—	—	0.079	—	53-72	16-23
		R	0.350	0.310	0.003	—	—	—	—	0.059	—	—
	Maxima	F	0.940	0.866	0.003	—	—	—	0.079	—	53-72	16-23
		R	0.350	0.315	0.003	—	—	—	—	0.059	—	—
	Sentra	F	0.710	0.630	0.003	—	—	—	0.079	—	40-47	12-14
		R	0.280	0.236	0.003	7.09	7.13	7.13	—	0.059	—	—

NA: Not Available

42356-MAXI-C11

SCHEDULED MAINTENANCE INTERVALS
Nissan—Altima, Sentra & Maxima

TO BE SERVICED	TYPE OF SERVICE	VEHICLE MILEAGE INTERVAL (x1000)												
		7.5	15	22.5	30	37.5	45	52.5	60	67.5	75	82.5	90	97.5
Engine oil & filter	R	✓	✓	✓	✓	✓	✓	✓	✓	✓	✓	✓	✓	✓
Brake lines & cables	S/I		✓		✓		✓		✓		✓		✓	
Brake pads, discs, drums & linings	S/I		✓		✓		✓		✓		✓		✓	
Driveshaft boots	S/I		✓		✓		✓		✓		✓		✓	
Exhaust system	S/I				✓				✓				✓	
Transmission or transaxle fluid	S/I		✓		✓		✓		✓		✓		✓	
Air cleaner filter	R				✓				✓				✓	
Spark plugs (except below)	R				✓				✓				✓	
Spark plugs (platinum tip) (Sentra,	R								✓					
Idle RPM (Sentra)	S/I				✓				✓				✓	
Steering gear & linkage, axle & suspension parts	S/I				✓				✓				✓	
Engine coolant	R								✓					
Timing belt	R								✓					
Drive belts	S/I								✓					
Fuel lines	S/I								✓					
Vapor lines	S/I								✓					

R: Replace   S/I: Service or Inspect

FREQUENT OPERATION MAINTENANCE (SEVERE SERVICE)

If a vehicle is operated under any of the following conditions it is considered severe service:

- Extremely dusty areas.

- 50% or more of the vehicle operation is in 32°C (90°F) or higher temperatures, or constant operation in temperatures below 0°C (32°F).

- Prolonged idling (vehicle operation in stop and go traffic).

- Frequent short running periods (engine does not warm to normal operating temperatures).

- Police, taxi, delivery usage or trailer towing usage.

Oil & oil filter: change every 3750 miles.

Brake pads & discs: service or inspect every 7500 miles.

Driveshaft boots: service or inspect every 7500 miles.

Exhaust system: service or inspect every 7500 miles.

Steering gear & linkage, axle & suspension parts: service or inspect every 7500 miles.

Steering linkage ball joints & front suspension ball joints: service or inspect every 7500 miles.

Air cleaner filter: service or inspect every 15,000 miles.

42356-MAXI-C12

PRECAUTIONS

Before servicing any vehicle, please be sure to read all of the following precautions, which deal with personal safety, prevention of component damage and important points to take into consideration when servicing a motor vehicle:

• Never open, service or drain the radiator or cooling system when the engine is hot; serious burns can occur from the steam and hot coolant.

• Observe all applicable safety precautions when working around fuel. Whenever servicing the fuel system, always work in a well-ventilated area. Do not allow fuel spray or vapors to come in contact with a spark, open flame, or excessive heat (a hot drop light, for example). Keep a dry chemical fire extinguisher near the work area. Always keep fuel in a container specifically designed for fuel storage; also, always properly seal fuel containers to avoid the possibility of fire or explosion. Refer to the additional fuel system precautions later in this section.

• Fuel injection systems often remain pressurized, even after the engine has been turned **OFF**. The fuel system pressure must be relieved before disconnecting any fuel lines. Failure to do so may result in fire and/or personal injury.

• Brake fluid often contains polyglycol ethers and polyglycols. Avoid contact with the eyes and wash your hands thoroughly after handling brake fluid. If you do get brake fluid in your eyes, flush your eyes with clean, running water for 15 minutes. If

eye irritation persists, or if you have taken brake fluid internally, IMMEDIATELY seek medical assistance.

• The EPA warns that prolonged contact with used engine oil may cause a number of skin disorders, including cancer! You should make every effort to minimize your exposure to used engine oil. Protective gloves should be worn when changing oil. Wash your hands and any other exposed skin areas as soon as possible after exposure to used engine oil. Soap and water, or waterless hand cleaner should be used.

• All new vehicles are now equipped with an air bag system. The system must be disabled before performing service on or around system components, steering column, instrument panel components, wiring and sensors. Failure to follow safety and disabling procedures could result in accidental air bag deployment, possible personal injury and unnecessary system repairs.

• Always wear safety goggles when working with, or around, the air bag system. When carrying a non-deployed air bag, be sure the bag and trim cover are pointed away from your body. When placing a non-deployed air bag on a work surface, always face the bag and trim cover upward, away from the surface. This will reduce the motion of the module if it is accidentally deployed. Refer to the additional air bag system precautions later in this section.

• Clean, high quality brake fluid from a

sealed container is essential to the safe and proper operation of the brake system. You should always buy the correct type of brake fluid for your vehicle. If the brake fluid becomes contaminated, completely flush the system with new fluid. Never reuse any brake fluid. Any brake fluid that is removed from the system should be discarded. Also, do not allow any brake fluid to come in contact with a painted surface; it will damage the paint.

• Never operate the engine without the proper amount and type of engine oil; doing so WILL result in severe engine damage.

• Timing belt maintenance is extremely important! Many models utilize an interference-type, non-freewheeling engine. If the timing belt breaks, the valves in the cylinder head may strike the pistons, causing potentially serious (also time-consuming and expensive) engine damage. Refer to the maintenance interval charts in the front of this section for the recommended replacement interval for the timing belt and to the timing belt procedure for belt replacement and inspection.

• Disconnecting the negative battery cable on some vehicles may interfere with the functions of the on-board computer system(s) and may require the computer to undergo a relearning process once the negative battery cable is reconnected.

• When servicing drum brakes, only disassemble and assemble one side at a time, leaving the remaining side intact for reference.

ENGINE REPAIR

Distributor

REMOVAL

The Nissan 3.0L and 3.5L engines are equipped with a Distributorless Ignition System (DIS).

1.8L, 2.0L and 2.4L Engines

1. Before servicing the vehicle, refer to the precautions in the beginning of this section.
2. Set the engine to Top Dead Center (TDC) with the No. 1 piston on compression stroke.
3. Remove or disconnect the following:

• Negative battery cable

• Distributor spark plug wires from the distributor cap
• Distributor cap. Scribe a mark on the engine block to show the rotor and distributor position prior to removal.
• Wiring connections to the distributor
• Bolt(s) holding distributor to engine
• Distributor by pulling it upward from the cylinder block

➡Do not disturb the camshaft or crankshaft position after the distributor is removed from the engine. If any of these components are moved, TDC on cylinder No. 1 will have to be found again before reinstalling the distributor.

INSTALLATION

1.8L, 2.0L and 2.4L Engines

ENGINE NOT DISTURBED

1. Install or connect the following:
• New distributor housing O-ring
• Distributor so the rotor is aligned with the matchmark on the housing and the housing is aligned with the matchmark on the engine

➡Be sure the distributor is fully seated and the distributor gear is fully engaged.

• Snug the hold-down bolt
• Distributor pick-up lead wires
• Distributor cap and tighten the screws

- Splash shield
- Spark plug wires
- Negative battery cable

2. After the ignition timing has been adjusted, tighten the hold-down bolt(s) as follows:

- 1.8L engines: 80–104 inch lbs. (9–11 Nm)
- 2.0L and 2.4L engines: 108–144 inch lbs. (13–16 Nm)

ENGINE DISTURBED

1. Install the a new distributor housing O-ring

2. Position the engine so the No. 1 piston is at Top Dead Center (TDC) of its compression stroke and the mark on the vibration damper is aligned with **0** on the timing indicator.

- Distributor in the engine so the rotor is aligned with the position of the No. 1 ignition wire on the distributor cap. Be sure the distributor is fully seated and that the distributor shaft is fully engaged.
- Snug the hold-down bolt
- Distributor pick-up lead wires
- Distributor cap and tighten the screws. Install the splash shield, if equipped.
- Spark plug wires
- Negative battery cable.

3. After the ignition timing has been adjusted, tighten the hold-down bolt(s) as follows:

- 1.8L engines: 80–104 inch lbs. (9–11 Nm)
- 2.0L and 2.4L engines: 108–144 inch lbs. (13–16 Nm)

Ignition Timing

ADJUSTMENT

1.8L, 2.0L and 2.4L Engines

Visually check the air cleaner, intake hoses, ducts, Exhaust Gas Recirculation (EGR) valve operation and electrical connections prior to the adjustment of the ignition timing. Correct or repair any problem as required. Be sure to inspect the throttle valve and the Throttle Position (TP) sensor for proper operation.

1. Before servicing the vehicle, refer to the precautions in the beginning of this section.

2. Locate the timing marks on the crankshaft pulley and the front of the engine.

3. Clean the timing marks.

4. Using chalk or white paint, color the mark on the crankshaft pulley and the mark on the scale which will indicate the correct timing when aligned with the notch on the crankshaft pulley.

5. Attach a tachometer to the engine.

6. Attach a timing light to the engine, to No. 1 cylinder's ignition wire.

7. Check to be sure all of the wires clear the fan; start the engine and allow it to reach normal operating temperatures.

8. Block the front wheels and set the parking brake. Shift the transmission into **NEUTRAL** for automatic and manual transaxles; do not stand in front of the vehicle when making adjustments.

9. Perform the following procedures:

a. Race the engine at 2000 rpm for about 2 minutes under a no-load condition; be sure all of the accessories are turned off.

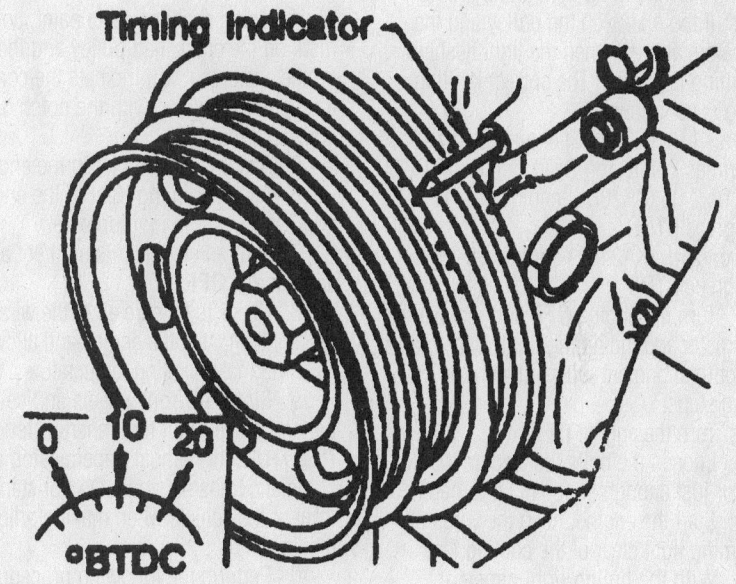

Point the timing light at the crankshaft pulley to see the timing marks—1.8L and 2.0L engines

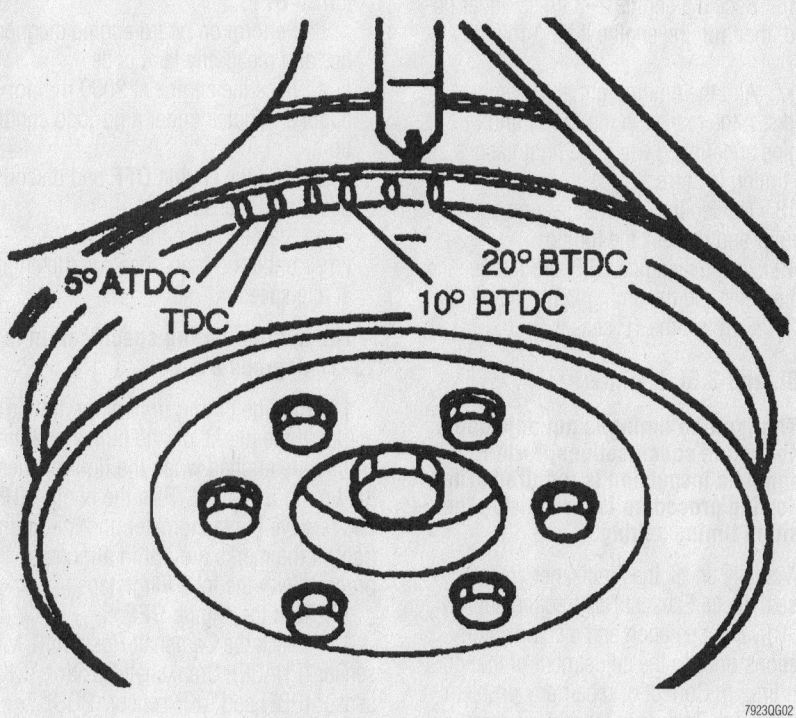

The timing marks are located on the crankshaft pulley—2.4L engine

b. Perform on board engine diagnostics and repair any fault code.

c. Race the engine 2–3 times under no-load, then run the engine it for 1 minute at idle.

d. Stop the engine and disconnect the Throttle Position (TP) sensor.

e. Race the engine at 2000 rpm for about 2 minutes under a no-load condition; be sure all of the accessories are turned OFF.

f. Run the engine at idle speed.

10. Aim the timing light at the timing marks. If the marks on the pulley and the engine are aligned when the light flashes, the timing is correct. The correct ignition timing is as follows:

a. 1.8L (QG18DE) engine: 6–10 degrees Before Top Dead Center (BTDC)

b. 2.0L (SR20DE) engine: 13–17 degrees BTDC

c. 2.4L (KA24DE) engine: 18–22 degrees BTDC

11. Turn the engine OFF and remove the tachometer and the timing light. If the marks are not in alignment, proceed with the following steps.

12. Turn the engine OFF.

13. Loosen the bolts that secure the distributor just enough so it can be turned.

14. Start the engine. Keep the wires of the timing light clear of the cooling fan.

15. With the timing light aimed at the pulley and the marks on the engine, turn the distributor for the proper adjustment.

16. Race the engine 2–3 times under no-load, then run the engine it for 1 minute at idle.

17. Aim the timing light at the timing marks. If the marks on the pulley and the engine are aligned when the light flashes, the timing is correct.

18. Tighten the bolt that secures the distributor and recheck the timing.

19. Turn the engine OFF and remove the tachometer and the timing light.

20. Connect the TP sensor.

3.0L and 3.5L Engines

➡ The ignition timing is not adjustable. If not within specifications, further diagnostic inspection is required. The following procedure is for viewing the ignition timing setting.

Visually check the air cleaner, intake hoses, ducts, Exhaust Gas Recirculation (EGR) valve operation and electrical connections prior to the adjustment of the ignition timing. Correct or repair any problem as required. Be sure to inspect the throttle valve and Throttle Position (TP) sensor for proper operation.

1. Before servicing the vehicle, refer to the precautions in the beginning of this section.

2. Locate the timing marks on the crankshaft pulley and the front of the engine.

3. Clean the timing marks.

➡ The ignition timing specification is 13–17 degrees Before Top Dead Center (BTDC).

4. Using chalk or white paint, color the mark on the crankshaft pulley and the mark on the scale, that will indicate the correct timing when aligned with the notch on the crankshaft pulley.

5. Attach a tachometer to the engine.

6. Attach a timing light to the engine to number 1 cylinder ignition wire.

7. Turn all electrical equipment and accessories OFF.

8. Check to be sure all of the wires clear the fan, then, start the engine and allow it to reach normal operating temperatures.

9. Block the front wheels and set the parking brake. Shift the transmission into NEUTRAL for manual transmission and automatic transmissions. Do not stand in front of the vehicle when making adjustments.

10. Perform the following procedures:

a. Race the engine at 2000 rpm for about 2 minutes under a no-load condition; be sure all of the accessories are turned OFF.

b. Perform on board engine diagnostics and repair any fault code.

c. Race the engine at 2000 rpm for about 2 minutes under a no-load condition.

d. Turn the engine OFF and disconnect the TP sensor.

e. Start and race the engine 2–3 times under no-load, then run the engine at idle speed.

➡ The ignition timing specification is 13–17 degrees BTDC.

11. Aim the timing light at the timing marks. If the marks on the pulley and the engine are aligned when the light flashes, the timing is correct. Turn the engine OFF and remove the tachometer and the timing light. If the marks are not in alignment, proceed with the following steps.

12. Turn the engine OFF.

13. Check the Camshaft Position (CMP) sensor (PHASE), Crankshaft Position (CKP) sensor (REF) and CKP sensor (POS). Replace if necessary.

14. If the ignition timing is still not correct, substitute a known good Electronic Control Module (ECM).

➡ The ECM may be the cause of the problem but this is rarely the case.

15. Turn the engine OFF and remove the tachometer and the timing light.

Alternator

REMOVAL

1.8L and 2.0L Engines

1. Before servicing the vehicle, refer to the precautions in the beginning of this section.

2. Remove or disconnect the following:

- Negative battery cable
- 2 lead wires and connector from the alternator
- Drive belt adjusting bolt, loosen only
- Drive belt
- Alternator

2.4L Engine

1. Before servicing the vehicle, refer to the precautions in the beginning of this section.

2. Drain the cooling system below the upper radiator hose level.

3. Remove or disconnect the following:

- Negative battery cable
- Upper radiator hose
- Alternator electrical harness, harness stay and the harness-to-A/C compressor
- Throttle cable
- Loosen the adjusting bolt
- Accessory drive belt
- Alternator mounting bolts
- Alternator from the engine

3.0L Engines

1. Before servicing the vehicle, refer to the precautions in the beginning of this section.

2. Remove or disconnect the following:
- Negative battery cable
- Splash guard on the right side of the vehicle
- Drive belt
- Four A/C mounting bolts
- Radiator fan and shroud
- A/C compressor forward
- Alternator harness connector

- Alternator mounting bolts and lower the alternator from the vehicle

3.5L Engines

1. Before servicing the vehicle, refer to the precautions in the beginning of this section.
2. Remove or disconnect the following:
   - Negative battery cable
   - Right side engine undercover and side inspection cover
   - Radiator
   - Drive belt
   - Alternator and A/C compressor harness connectors
   - Upper and lower alternator bolts
   - Alternator

INSTALLATION

1.8L and 2.0L Engines

1. Install or connect the following:
   - Alternator and retaining bolts loosely
   - Belt and connect the wiring
2. Adjust the drive belt.
3. Torque the retaining bolts to 25 ft. lbs. (34 Nm).
4. Connect the negative battery cable.

2.4L Engines

1. Install or connect the following:
   - Alternator and torque the bolts to 11–15 ft. lbs. (16–20 Nm)
   - Harness-to-A/C compressor, harness stay and the alternator electrical harness
   - Throttle cable
   - Drive belt. Properly tension the belt
   - Upper radiator hose
   - Negative battery cable
2. Top off the cooling system.
3. Start the vehicle, check for leaks and repair if necessary.

3.0L Engines

Install or connect the following:
   - Alternator and torque the bolts to 38 ft. lbs. (52 Nm)
   - Alternator harness connector
   - A/C compressor back into location
   - Radiator cooling fan and shroud
   - A/C compressor mounting bolts
   - Drive belt and tension the belt
   - Splash guard
   - Negative battery cable

➥**Proper belt tension is important. A belt that is too tight may cause alterna-** tor bearing failure; one that is too loose will cause a gradual battery discharge and/or belt slippage, resulting in belt breakage from overheating.

3.5L Engine

Install or connect the following:
   - Alternator
   - Upper and lower alternator bolts. Tighten the upper bolt to 12–15 ft. lbs. (16 –20 Nm) and the lower bolt to 32–38 ft. lbs. (44–52 Nm).
   - Alternator and A/C compressor harness connectors
   - Drive belt
   - Radiator
   - Right side engine undercover and side inspection cover
   - Negative battery cable

Engine Assembly

REMOVAL & INSTALLATION

Sentra

➥**The engine and transaxle are removed as one unit from the underside of the vehicle.**

1. Before servicing the vehicle, refer to the precautions in the beginning of this section.
2. Relieve the fuel system pressure.
3. Drain the coolant from the radiator and the engine block.
4. Drain the engine oil.
5. Remove or disconnect the following:
   - Negative and positive battery cables
   - Battery and tray from the vehicle
   - Both front wheels
   - Engine undercovers and the engine side covers
   - Air cleaner assembly and air duct
   - Vacuum hoses. Make sure to note the locations prior to disconnection them.
   - Heater hoses from the engine
   - Automatic transmission cooler hoses from the transaxle, if equipped
   - Power steering hoses
   - Fuel hoses from the engine
   - Harness and wiring connections. Make sure to note the locations prior to disconnecting them.
   - Throttle cable and the cruise control cable
   - Control cable, if equipped with an automatic transmission

   - Cooling fans, radiator and the recovery tank
   - Front halfshafts from the vehicle
   - Front exhaust pipe
   - Starter motor and intake manifold support brackets
   - Engine drive belts
   - Alternator and adjusting brackets
   - Power steering pump and A/C compressor. It is not necessary to disconnect the lines.
6. Position a transmission jack under the transaxle and support the engine with engine slinger.
   - Center crossmember
   - Front stabilizer bar, if necessary
   - Engine mounting bolts from both sides of the engine
7. Slowly lower the jacking devices and remove the engine and transaxle from the vehicle.

   To install:
8. Install or connect the following:
   - Engine and transaxle assembly
   - Mounting bolts to both sides of the engine and torque the bolts to 44 ft. lbs. (60 Nm)
9. For vehicles with manual transaxles, adjust the height of the mounting bracket (buffer rod). The distance between the 2 through-bolts should be 2.13–2.20 in. (54–56mm).
   - Center crossmember and torque the bolts to 40 ft. lbs. (54 Nm)
10. Remove the engine support jacks and engine slinger.
11. Install or connect the following:
   - A/C compressor and power steering pump
   - Alternator and brackets
   - Starter motor and intake manifold support bracket
   - Front exhaust pipe
   - Drive belts
   - Both front halfshafts
   - Radiator, cooling fans and recovery tank
   - Control cable, automatic transmissions only
   - Throttle and cruise control cables, if equipped
   - Wiring harness and electrical connections
   - Power steering hoses and fuel line
   - Transmission cooler lines, if equipped
   - Vacuum hoses
   - Air cleaner assembly
   - Engine side and under covers
   - Both front wheels
   - Battery tray and battery

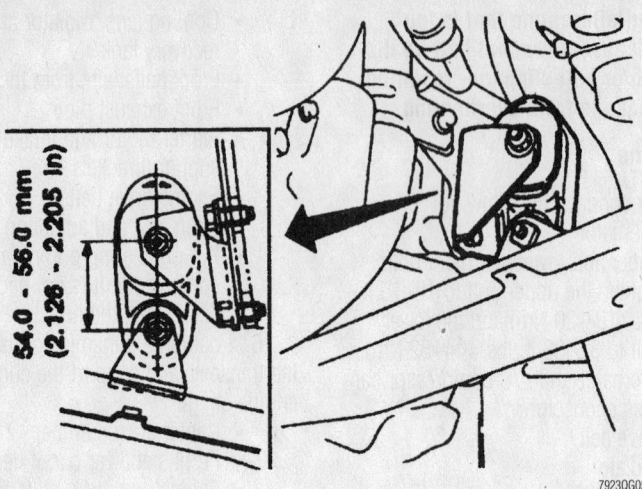

Be sure to adjust the height of the engine mount for manual transmission vehicles—2.0L engine

54.0 - 56.0 mm
(2.126 - 2.205 In)

79230G05

- Both battery cables
12. Fill the engine with clean oil.
13. Fill the cooling system.
14. Start the engine and check for leaks. Make all the necessary adjustments.

Altima

➡**The engine and transaxle must be removed as a single unit. The engine and transaxle are removed from under the vehicle.**

1. Before servicing the vehicle, refer to the precautions in the beginning of this section.
2. Release fuel system pressure.
3. Drain the cooling system.
4. Drain the engine oil.
5. Remove or disconnect the following:
- Battery cables and the battery tray
- Air cleaner assembly
- Both front wheels
- Engine under cover and engine hood
- Cooler lines from the radiator, if equipped with an automatic transaxle
- Upper and lower hoses from the radiator
- Radiator assembly
- Heater hoses from the engine
- Throttle cable and cruise control cable, if equipped
- Fuel feed and return hoses
- All the necessary vacuum hoses and electrical connectors. Label all wires and hoses before disconnecting them.
- Wiring from starter motor
- Slave cylinder from the transaxle, if equipped. It is not necessary to disconnect the hydraulic hose.

- Engine drive belts. Be sure to mark belts for reinstallation.
- Alternator, A/C compressor and the power steering pump
- Both halfshafts from the transaxle and support the engine with slinger and support the transaxle with proper jack
- Left and right engine mounting through-bolts
- Crossmember
- Front and rear engine mounts
6. Lower the transaxle and engine assembly from the vehicle.

➡**The engine and transaxle assembly should be removed through the bottom of the vehicle. Do not attempt to remove the assembly from above.**

To install:

7. Raise the transaxle and engine assembly to the vehicle.
8. Install or connect the following:
- Front and rear engine mounts. Torque the mounting bolts to 55 ft. lbs. (75 Nm).
- Crossmember and torque the bolts to 57–72 ft. lbs. (77–98 Nm)
- Left and right engine mounting through bolts. Torque the bolts to 72 ft. lbs. (98 Nm).
9. Remove the engine and transaxle support jacks.
- Both halfshafts
- Power steering pump, A/C compressor and alternator
- Slave cylinder, if equipped
- Starter motor
- Drive belts
- Vacuum hoses and electrical connectors

- Fuel feed and return lines
- Throttle and cruise control cables, if equipped
- Radiator
- Heater and radiator hoses
- Cooler lines, if equipped
- Engine side and under covers
- Both front wheels
- Air cleaner assembly
- Battery tray and battery
- Both battery cables
- Hood
10. Fill the cooling system.
11. Fill the engine with clean oil.
12. Start the vehicle, check for leaks and repair if necessary.

Maxima

3.0L ENGINE

It is recommended the engine and transaxle be removed as a single unit. If need be, the units may be separated after removal.

➡**The engine and transaxle assembly must be removed from the underside of the vehicle.**

1. Before servicing the vehicle, refer to the precautions in the beginning of this section.
2. Release the fuel system pressure.
3. Drain the cooling system.
4. Drain the engine oil.
5. Drain the automatic transaxle, if equipped.
6. Remove or disconnect the following:
- Negative battery cable
- Hood
- Engine under cover
- Air cleaner, the air intake tube, the air flow meter and the throttle linkage
- Drive belts
- Engine ground cable
- Electrical connector from the crank angle sensor
- Engine electrical harness connectors
- Fuel feed and fuel return hoses
- Upper and lower radiator hoses
- Heater inlet and outlet hoses
- Engine vacuum hoses
- Power steering pump, A/C compressor and alternator
- Carbon canister
- Auxiliary fan, washer tank and the radiator (with the fan assembly)
- Clutch release cylinder from the clutch housing, if equipped with a manual transaxle
- Shift control rod and the shift sup-

port rod, on some models with a manual transaxle
- Control cable from the transaxle, on models with an automatic transaxle

7. Install engine slingers to the block and connect a suitable lifting device to the slingers. Do not tension the lifting device at this point.
- Exhaust pipe at both the manifold connections
- Front exhaust pipe from the vehicle and support the engine and transaxle assembly with proper jack
- Right and left side halfshafts from their side flanges
- Bolt holding the radius link support

8. Lower the shifter and selector rods and remove the bolts from the motor mount brackets. Remove the nuts holding the front and rear motor mounts to the frame.

9. On some models it will be necessary to remove the center crossmember assembly from the vehicle.

10. Lower the engine/transaxle assembly onto an engine stand.

To install:

11. Raise the engine/transaxle assembly into the vehicle. When raising the engine onto the mounts, be sure to keep it as level as possible.

12. After installing the motor mounts, adjust and install the buffer rods; the front should be 3.50–5.58 in. (89–91mm) and the rear should be 3.90–3.98 in. (99–101mm).

13. Check the clearance between the frame and clutch housing and be sure the engine mount bolts are seated in the groove of the mounting bracket.

14. Remove the transaxle and engine jack assembly.

15. Install or connect the following:
- Center crossmember, if removed. Torque the bolts to 72 ft. lbs. (98 Nm).
- Halfshafts
- Radius link support
- Front exhaust pipe and remove the engine slingers and supports
- Control cable, if equipped
- Shift control and support rods, if equipped
- Clutch release cylinder, if equipped
- Radiator, auxiliary fan and washer tank
- Carbon canister
- Power steering pump, A/C compressor and alternator
- All engine vacuum hoses
- Heater and radiator hoses
- Fuel feed and return lines

- Engine electrical connectors and ground cables
- Drive belts
- Air cleaner, air intake tube and air flow meter
- Throttle linkage
- Engine under cover
- Negative battery cable
- Hood

16. Fill the transmission fluid.
17. Fill the cooling system.
18. Fill the engine with clean oil.
19. Start the vehicle, check for leaks and repair if necessary.

3.5L ENGINE

It is recommended the engine and transaxle be removed as a single unit. If need be, the units may be separated after removal.

➡**The engine and transaxle assembly must be removed from the underside of the vehicle.**

1. Before servicing the vehicle, refer to the precautions in the beginning of this section.
2. Release the fuel system pressure.
3. Drain the cooling system.
4. Drain the engine oil.
5. Drain the automatic transaxle, if equipped.

6. Remove or disconnect the following:
- Negative battery cable
- Hood
- Engine under cover
- All vacuum hoses, fuel lines, wires and connectors; tag before disconnecting
- Front exhaust pipe from the manifold
- Ball joints from the steering knuckle
- Halfshafts
- Radiator and fans
- Drive belts
- Alternator
- A/C compressor. Position it aside with the lines attached. Do NOT disconnect the refrigerant lines.
- Power steering pump and position aside with the lines attached. Do NOT disconnect the fluid lines.

7. Place a suitable jack under the transaxle. Install engine slingers and a suitable engine hoist. Raise the engine for access to the left side engine mount.
- Left side engine mount
- Control and support rods from the transaxle, manual transaxle only
- Control cable from the transaxle, automatic transaxle only
- Right side engine mount

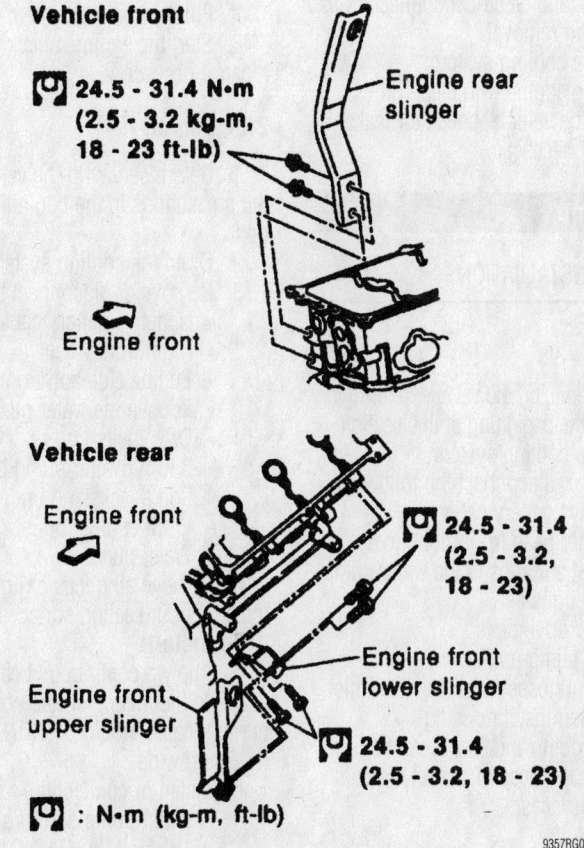

Vehicle front

⊡ 24.5 - 31.4 N•m
(2.5 - 3.2 kg-m,
18 - 23 ft-lb)

Engine rear slinger

Engine front

Vehicle rear

Engine front

⊡ 24.5 - 31.4
(2.5 - 3.2,
18 - 23)

Engine front lower slinger

Engine front upper slinger

⊡ 24.5 - 31.4
(2.5 - 3.2, 18 - 23)

⊡ : N•m (kg-m, ft-lb)

9357RG01

Installation of engine slingers to lift the engine

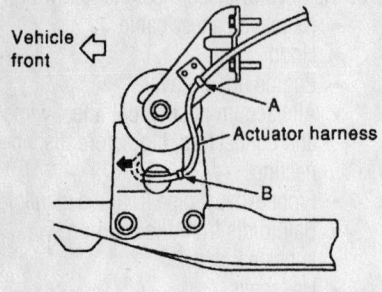

Vehicle front ←

A — Actuator harness

B

9357RG02

For electronically controlled engine mounts, the proper length from A to B is 6.69 in. (170mm)

- Center member, then carefully and slowly lower the transmission jack
8. Lower the engine/transaxle assembly onto an engine stand.

➡ **When lowering the engine out, guide it carefully to avoid hitting any other components.**

To install:

9. Installation is the reverse of the removal procedure, noting the following points:

a. If equipped with electronically controlled engine mounts, install them to the specifications shown in the accompanying figure

b. Make sure to connect all vacuum hoses, lines, and electrical connectors as tagged during removal.

c. Fill the cooling system.

d. Fill the engine with clean oil.

e. Start the vehicle, check for leaks and repair if necessary.

Water Pump

REMOVAL & INSTALLATION

1.8L Engine

1. Before servicing the vehicle, refer to the precautions in the beginning of this section.
2. Drain the cooling system.
3. Remove or disconnect the following:
- Negative battery cable
- Cylinder head front mounting bracket and loosen the water pump pulley bolts
- Engine drive belts
- Water pump pulley
- Coolant hoses from the water inlet and thermostat housing
- Water pump and thermostat housing
4. Remove all traces of gasket material from sealing surfaces.

To install:

5. Apply a continuous bead of liquid

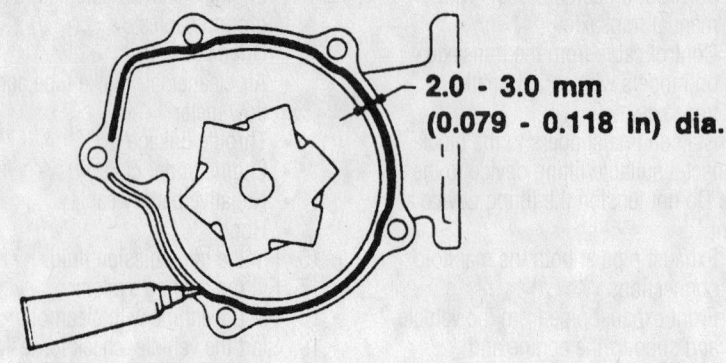

2.0 - 3.0 mm
(0.079 - 0.118 in) dia.

7923QG06

Apply RTV sealant to the water pump sealing surface as shown— 1.8L engines

sealer to the sealing surface of the thermostat housing. The sealant should be 0.079–0.118 in. (2–3mm) diameter.
6. Install or connect the following:
- Water pump. Torque the bolts to 56–73 inch lbs. (7–8 Nm).
- Pulley to the water pump and tighten the mounting bolts to 56–73 inch lbs. (7–8 Nm).
- Coolant hoses to the thermostat housing
- Drive belts and adjust as needed
- Cylinder head front mounting bracket
- Negative battery cable
7. Fill the cooling system
8. Start the engine, check for leaks and repair if necessary.

2.0L Engine

1. Before servicing the vehicle, refer to the precautions in the beginning of this section.
2. Drain the cooling system.
3. Remove or disconnect the following:
- Negative battery cable
- Right front wheel
- Engine side and front covers
- Loosen the water pump pulley bolts
- Drive belts
- 3 lower water pump bolts and position a jackstand under the engine
- Front engine mount
- Water pump
4. Remove all traces of liquid gasket material from sealing surfaces.

To install:

5. Apply a continuous bead of liquid sealer to the mating surface of the water pump. Sealer should be 0.079–0.118 in. (2–3mm) wide.
6. Install or connect the following:
- Water pump and torque the bolts to 12–15 ft. lbs. (16–21 Nm)

- Front engine mount and remove the engine support
- Water pump pulley. Torque the mounting bolts to 55–73 inch lbs. (6–8 Nm).
- Drive belts and adjust as needed
- Engine front and side cover
- Right front wheel
- Negative battery cable
7. Fill the cooling system.
8. Start the vehicle, check for leaks and repair if necessary.

2.4L Engine

1. Before servicing the vehicle, refer to the precautions in the beginning of this section.
2. Drain the cooling system.
3. Remove or disconnect the following:
- Negative battery cable
- Right lower splash cover
- Alternator and A/C compressor
- Coolant tube
- Water pump

➡ **Do not disconnect the air conditioning compressor lines. Unbolt the compressor and lay it off to the side.**

➡ **The mounting bolts are different sizes and must be reinstalled in the correct location; therefore it is a good idea to arrange the bolts so that they can be easily identified during installation.**

To install:

4. Be sure all gasket surfaces are clean and properly apply a continuous bead of silicone sealer to the pump.
5. Install or connect the following:
- Water pump and torque the 6mm bolts to 57–66 inch lbs. (6–8 Nm) and the 8mm bolts to 12–14 ft. lbs. (16–19 Nm)
- Coolant tube
- Alternator and A/C compressor

- Right side lower splash shield
- Negative battery cable

6. Fill the cooling system.

7. Start the engine, check for leaks and repair if necessary.

3.0L and 3.5L Engines

1. Before servicing the vehicle, refer to the precautions in the beginning of this section.

2. Drain the cooling system.

3. Position a jack under the oil pan for support. Be sure to place a block of wood on the jack for protection to the engine parts.

4. Remove or disconnect the following:
- Negative battery cable
- Right side engine mount and bracket
- Drive belts and the idler pulley bracket
- Chain tensioner cover and the water pump cover

5. Push the timing chain tensioner sleeve and apply a stopper pin so it does not return.
- Timing chain tensioner assembly
- 3 bolts that secure the water pump

6. Rotate the crankshaft 20 degrees counterclockwise to provide timing chain slack.

7. Put M8 bolts in 2 M8 threaded holes of the water pump.

Exploded view of the water pump assembly–2.4L engine

9347UG01

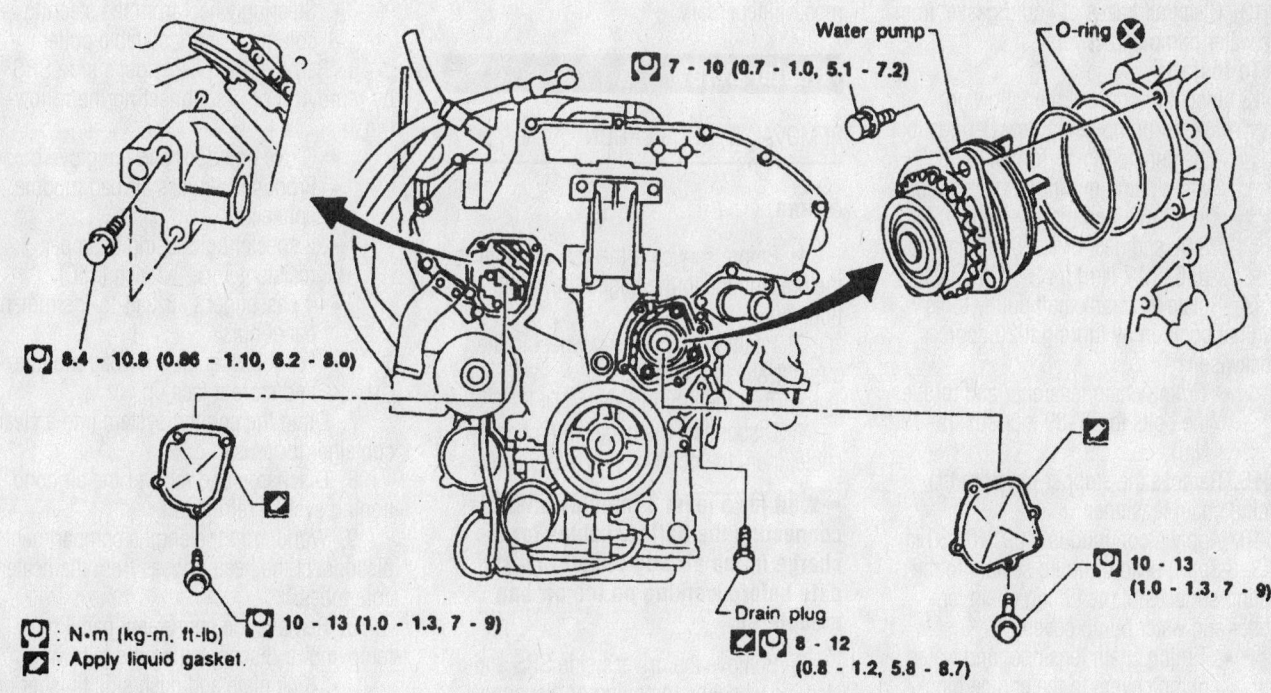

Water pump and timing cover assembly—3.0L engine

7923QG07

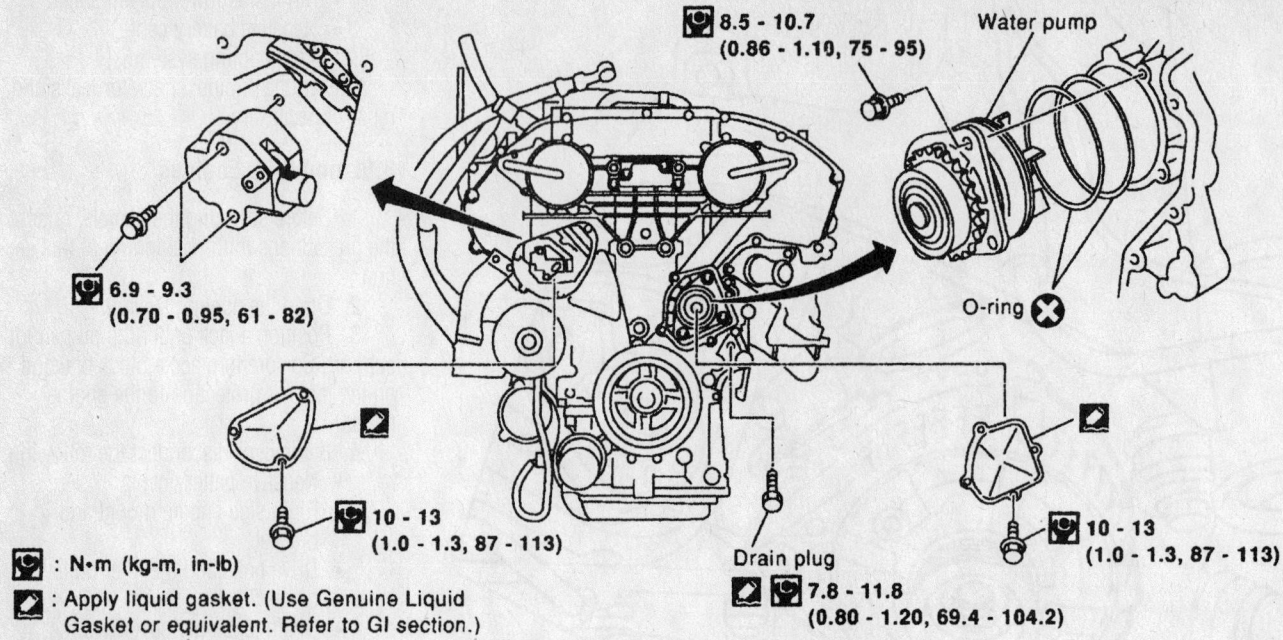

8.5 - 10.7
(0.86 - 1.10, 75 - 95)

Water pump

6.9 - 9.3
(0.70 - 0.95, 61 - 82)

O-ring

10 - 13
(1.0 - 1.3, 87 - 113)

10 - 13
(1.0 - 1.3, 87 - 113)

: N•m (kg-m, in-lb)

: Apply liquid gasket. (Use Genuine Liquid Gasket or equivalent. Refer to GI section.)

Drain plug

7.8 - 11.8
(0.80 - 1.20, 69.4 - 104.2)

9357RG03

Water pump and timing cover assembly—3.5L engine

8. Tighten each bolt by turning alternately ½ turn until they reach the timing chain rear case. Be sure to turn each bolt ½ turn at a time to prevent damage.

9. Lift up the water pump and remove it.

10. When removing the water pump, do not allow the water pump gear to hit the timing chain.

11. Remove and discard the O-rings from the water pump.

12. Clean all traces of liquid gasket from the water pump and covers.

To install:

13. Install or connect the following:
- Water pump using new O-rings to the engine block. Torque the 3 water pump mounting bolts evenly to 62–89 inch lbs. (7–10 Nm) for 3.0L engines, or to 75–95 inch lbs. (8.5–10.7 Nm) for 3.5L engines.

14. Rotate the crankshaft pulley to its original position by turning it 20 degrees clockwise.
- Timing chain tensioner and torque the bolts to 75–89 inch lbs. (9–10 Nm)

15. Remove the stopper pin from the timing chain tensioner.

16. Apply a continuous 0.091–0.130 in. (2.3–3.3mm) bead of liquid sealant to the mating surfaces of the timing chain tensioner and water pump covers.
- Timing chain tensioner and water pump covers to the engine block. Torque the bolts to 89–108 inch lbs. (10–13 Nm).

- Drive belts and the idler pulley bracket
- Right side engine mounting bracket and the engine mount
- Negative battery cable

17. Remove the jack from under the engine and install the drain plugs to the cylinder block.

18. Fill the cooling system.

19. Start the engine, check for leaks and repair if necessary.

Heater Core

REMOVAL & INSTALLATION

Altima

1. Before servicing the vehicle, refer to the precautions in the beginning of this section.

2. Position the steering wheel in the straight-ahead position.

3. Turn the ignition switch OFF.

4. Disconnect the negative (() battery cable; then, the positive (+) battery cable.

➡**Wait for a least 3 minutes after disconnecting the battery cables for the charge in the air bag circuit to dissipate before working on the air bag module(s).**

5. Remove the driver's side SRS and steering wheel by removing or disconnecting the following:
- Lower lid from the steering wheel

and disconnect the driver's air bag module connector
- Left and right side lids from the steering wheel
- Special bolts from both side of the steering wheel using a tamper resistant Torx® wrench (T50)
- Air bag module and store it face up
- Horn's electrical connector and remove the steering wheel nut
- Steering wheel from the steering column using a suitable puller

6. Remove the passenger's side SRS by removing or disconnecting the following:
- Glove box door and the glove box
- Front passenger's air bag module connector
- 2 special bolts using a tamper resistant Torx® wrench (T50)
- 4 passenger's air bag-to-instrument panel nuts
- Front passenger's air bag module and store it face up.

7. Drain the cooling system into a clean container for reuse.

8. Discharge and recover the air conditioning system refrigerant.

9. Working in the engine compartment, disconnect the heater hoses from the heater core tubes.

10. Remove the instrument panel by removing or disconnecting the following:
- Kick plate and dash side finisher on the driver's side
- 2 lower panel-to-instrument panel

screws and the lower panel on the driver's side
- 2 lower reinforcement panel-to-instrument panel screws and the lower reinforcement panel
- 6 steering column cover screws, the covers, the spiral cable and combination switch
- 2 cluster lid "A" screws and the cluster lid "A"
- 3 combination meter screws, disconnect the electrical harness connector and remove the combination meter
- Switch panel
- Instrument panel lower covers
- Snap out the transmission shifter finisher (boot)
- 4 cluster lid "C" screws and the cluster lid "C"
- 4 audio and deck pocket-to-instrument panel screws and the audio and deck pocket

- 5 center console screws and the center console
- 2 center instrument panel screws and the center panel
- front defroster grilles
- Front pillar garnish
- Instrument panel 3 nuts/4 screws and the instrument panel
- 8 instrument stay assembly nuts and the stay
- Steering member assembly 5 nuts/1 bolt and the steering member

11. Remove the air conditioning housing assembly by removing or disconnecting the following:
- Refrigerant lines from the air conditioning housing assembly
- Thermo control amp
- Air conditioning housing assembly

12. Remove or disconnect the following:
- Heater unit
- Heater core from the heater unit

To install:
13. Install or connect the following:
- Heater core to the heater unit
- Heater unit

14. Install the air conditioning housing assembly by installing or connecting the following:
- Air conditioning housing assembly
- Thermo control amp
- Refrigerant lines to the air conditioning housing assembly

15. Install the instrument panel by installing or connecting the following:
- Steering member assembly and the steering member 5 nuts/1 bolt
- Instrument stay assembly and the 8 stay nuts
- Instrument panel and the instrument panel 3 nuts/4 screws
- Front pillar garnish
- Front defroster grilles
- Center instrument panel and the 2 center panel screws

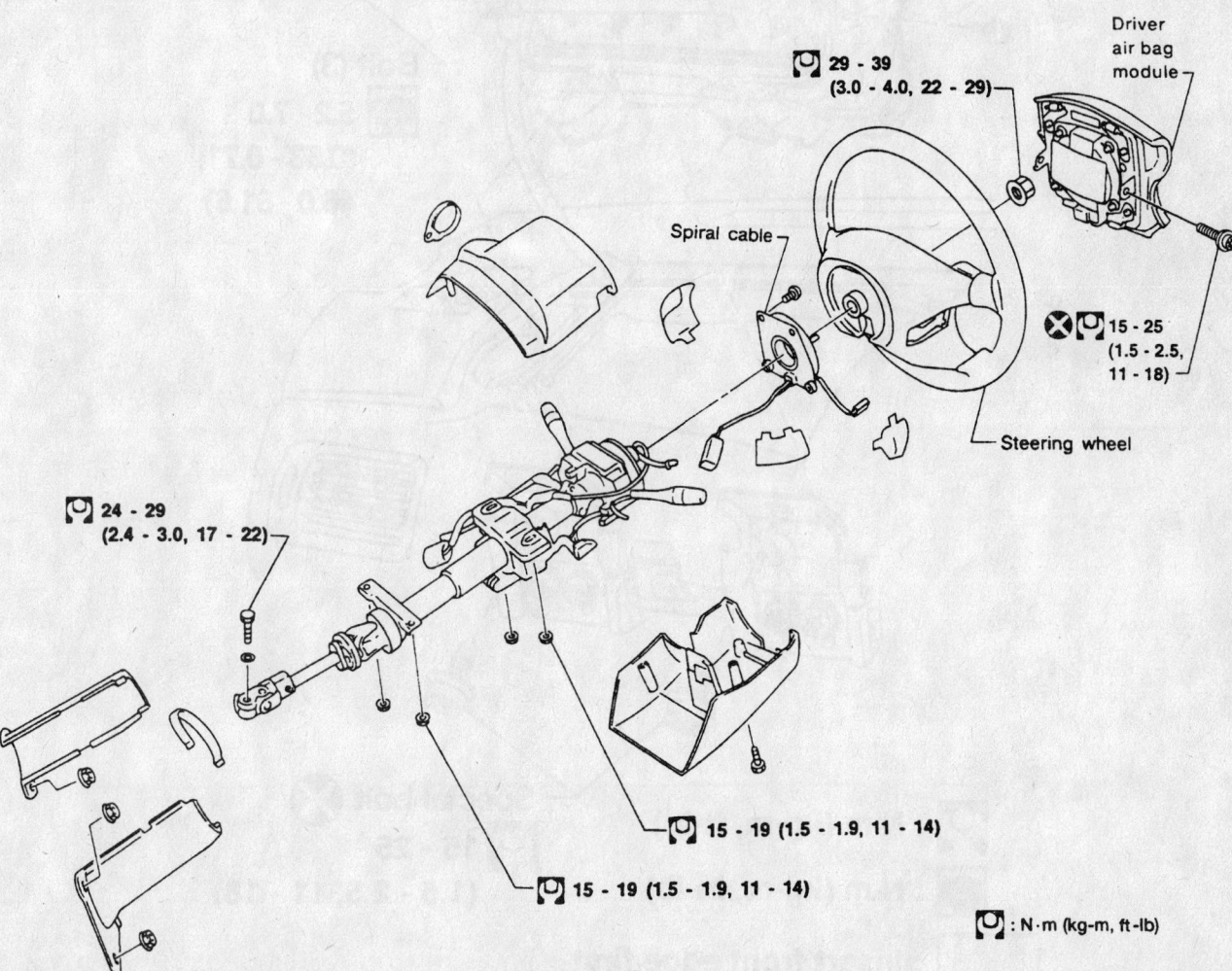

29 - 39 (3.0 - 4.0, 22 - 29)

Driver air bag module

Spiral cable

15 - 25 (1.5 - 2.5, 11 - 18)

Steering wheel

24 - 29 (2.4 - 3.0, 17 - 22)

15 - 19 (1.5 - 1.9, 11 - 14)

15 - 19 (1.5 - 1.9, 11 - 14)

: N·m (kg-m, ft-lb)

93112GD9

Exploded view of the steering wheel and air bag module—Altima

- Center console and the 5 center console screws
- Audio and deck pocket and the 4 audio and deck pocket-to-instrument panel screws
- Cluster lid "C" and the 4 cluster lid "C" screws
- Snap in the transmission shifter finisher (boot)
- Instrument panel lower covers
- Switch panel
- Combination meter, connect the electrical harness connector and install the 3 combination meter screws
- Cluster lid "A" and the 2 cluster lid "A" screw
- Combination switch, the spiral cable, the covers and the 6 steering column cover screws
- Lower reinforcement panel and the 2 lower reinforcement panel-to-instrument panel screws
- Lower panel and the 2 lower panel-to-instrument panel screws, on the driver's side
- Kick plate and dash side finisher, on the driver's side

16. Working in the engine compartment, connect the heater hoses to the heater core tubes.

17. Install the passenger's side SRS by installing or connecting the following:
- Front passenger's air bag module
- 4 passenger's air bag-to-instrument panel nuts

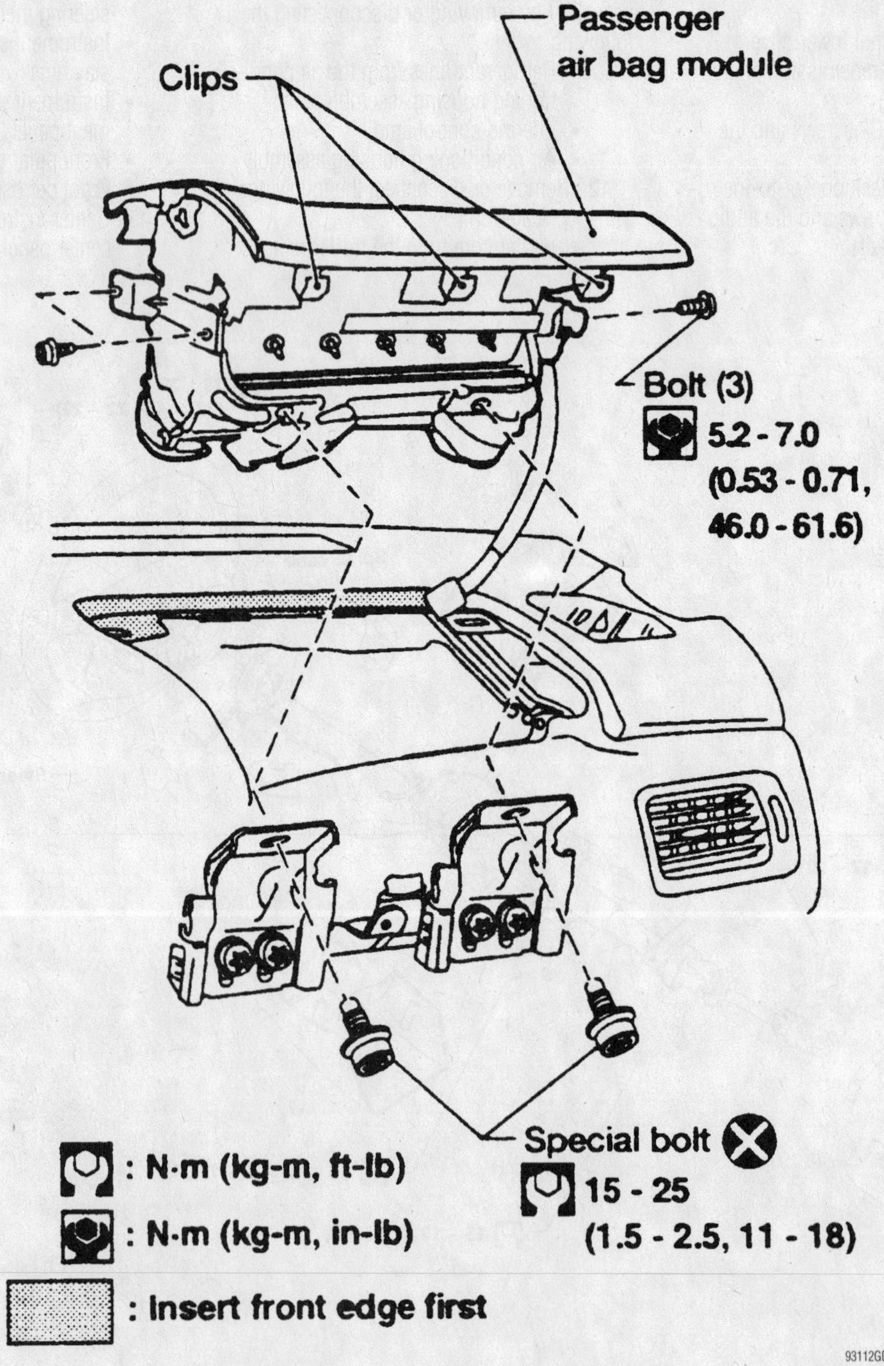

Clips

Passenger air bag module

Bolt (3)
5.2 - 7.0
(0.53 - 0.71, 46.0 - 61.6)

 : N·m (kg-m, ft-lb)

: N·m (kg-m, in-lb)

: Insert front edge first

Special bolt ⊗
15 - 25
(1.5 - 2.5, 11 - 18)

93112GD0

Exploded view of the passenger's side air bag module—Altima

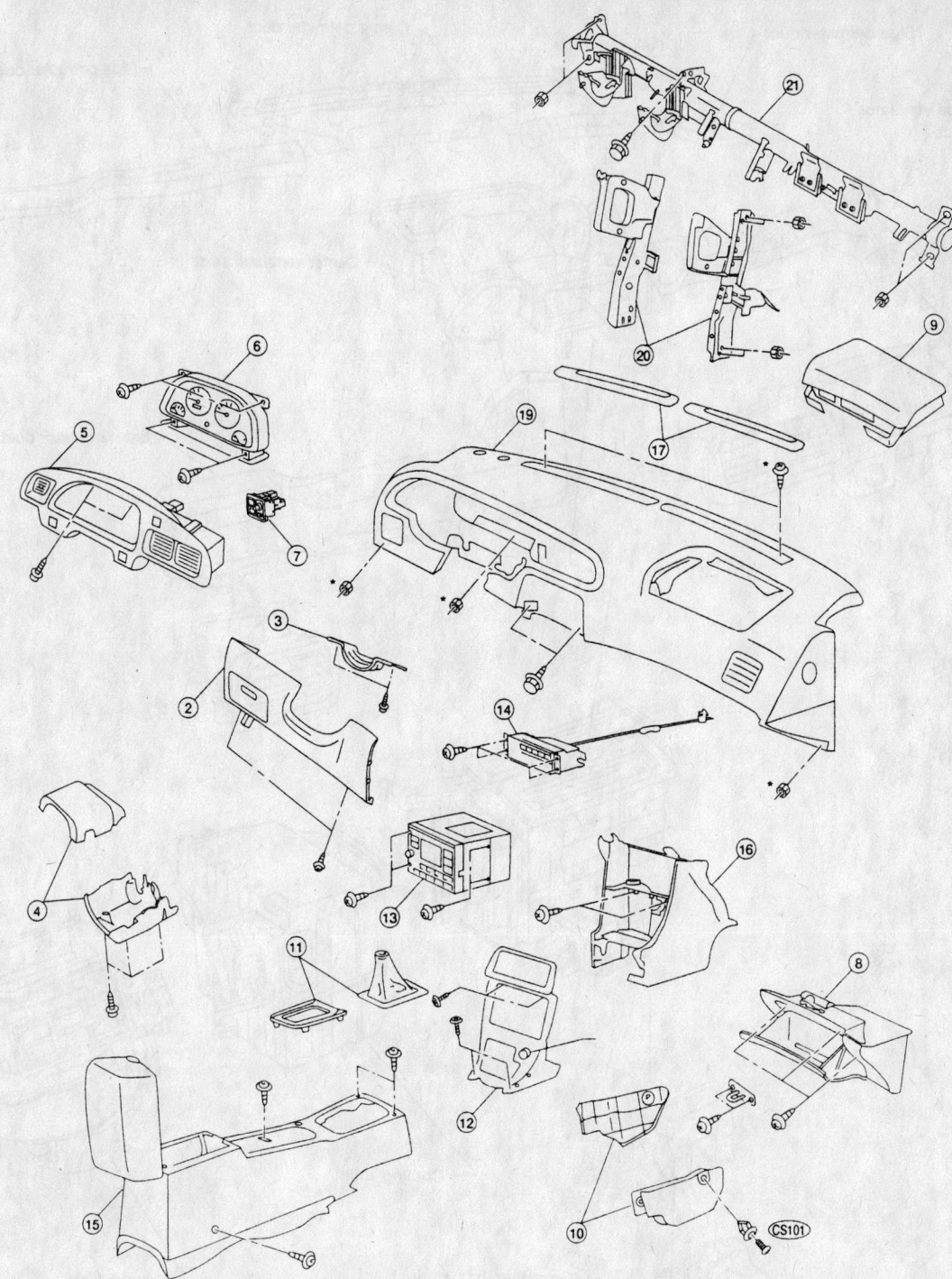

*: Instrument panel assembly mounting bolts, screws and nuts.

1. Remove kick plate and dash side finisher on driver side
2. Instrument lower panel on driver side
3. Dash lower reinforcement panel
4. Steering column covers, spiral cable and combination switch
5. Cluster lid A
6. Combination meter
7. Switch panel
8. Glove box assembly
9. Remove passenger side air bag moldule
10. Instrument lower covers
11. A/T finisher or M/T boot

12. Cluster lid C
13. Audio and deck pocket
14. A/C & heater control
15. Center console assembly
16. Instrument center panel
17. Front defroster grilles
18. Front pillar garnish
19. Instrument panel assembly
20. Instrument stay assemblies, if necesary
21. Steering member assembly, if necessary

93112GE1

Exploded view of the instrument panel assembly—Altima

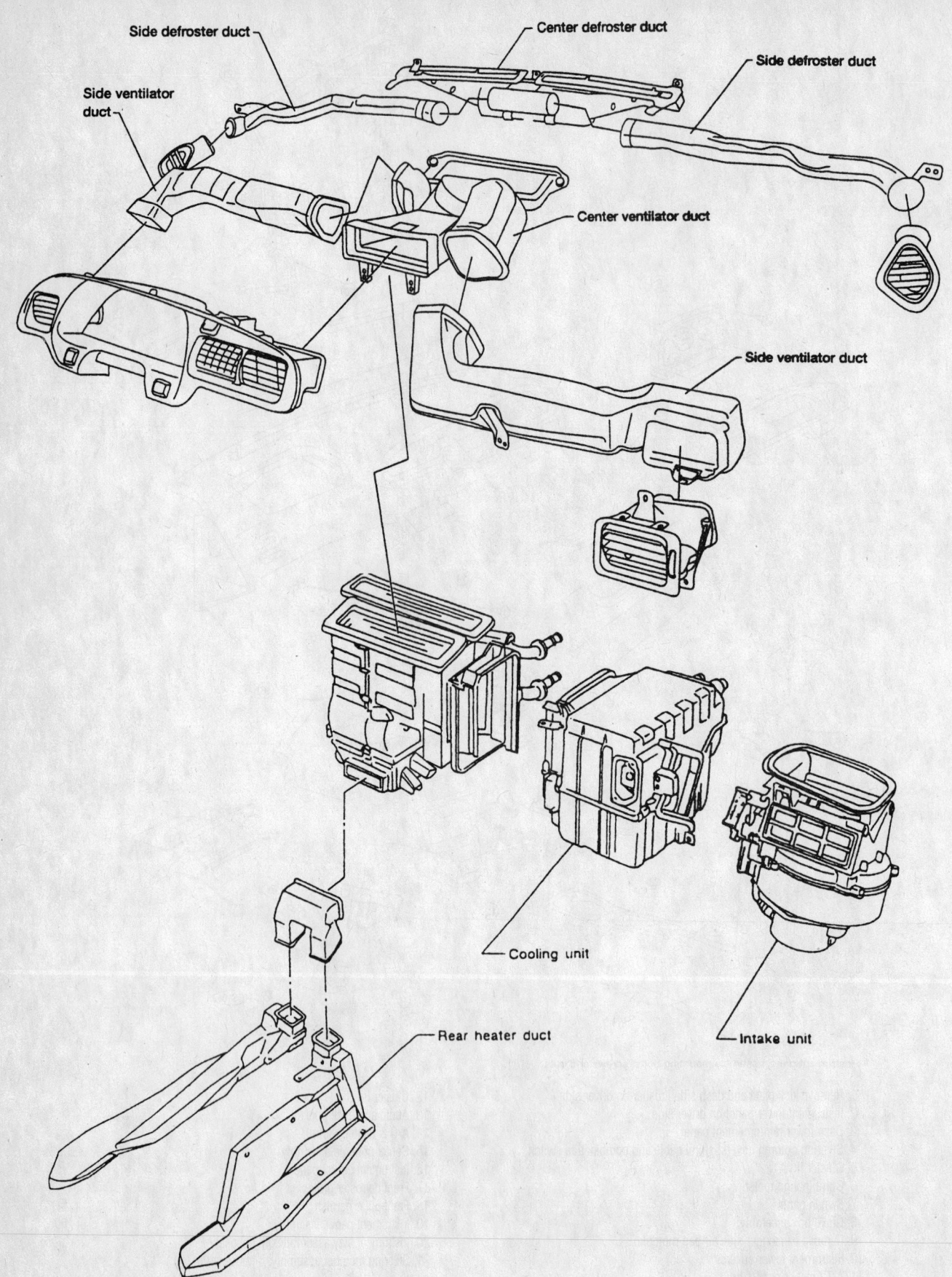

Side defroster duct

Center defroster duct

Side defroster duct

Side ventilator duct

Center ventilator duct

Side ventilator duct

Cooling unit

Intake unit

Rear heater duct

93112GE2

Exploded view of the heater housing assembly and related components—Altima

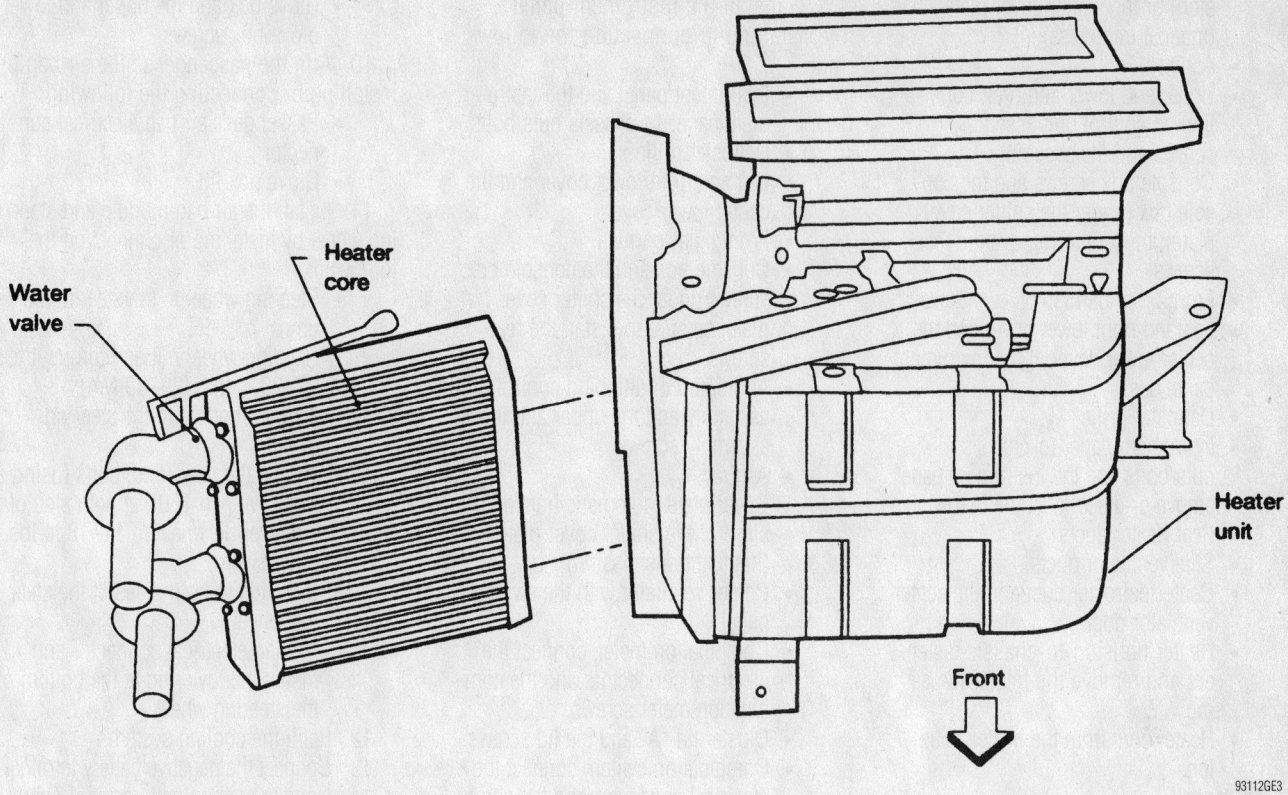

Water valve

Heater core

Heater unit

Front

93112GE3

View of the heater core and heater housing—Altima

- 2 new special bolts and torque using a tamper resistant Torx® wrench (T50) to 11–18 ft. lbs. (15–25 Nm)
- Front passenger's air bag module connector
- Glove box door and the glove box

18. Install the driver's side SRS and steering wheel by installing or connecting the following:

- Steering wheel to the steering column
- Steering wheel nut and torque the nut to 22–29 ft. lbs.
- Horn's electrical connector
- Air bag module
- New special bolts to both sides of the steering wheel and torque the bolts using a tamper resistant Torx® wrench (T50) to 11–18 ft. lbs. (15–25 Nm).
- Both the left and right side lids to the steering wheel
- Lower lid to the steering wheel and connect the driver's air bag module connector

19. Refill the cooling system.

20. Connect the positive (+) battery cable; then, the negative (()) battery cable.

21. Evacuate, charge and leak test the air conditioning system refrigerant.

22. Operate the engine to normal operat-ing temperatures; then, check the climate control operation and check for leaks.

Maxima

1. Before servicing the vehicle, refer to the precautions in the beginning of this section.

2. Disconnect the negative battery terminal.

✳✳ **CAUTION**

After disconnecting the negative battery cable, wait for at least 3 minutes for the SRS modules to deplete its energy.

3. Drain the cooling system into a clean container for reuse.

4. Remove the air bag module and steering wheel by removing or disconnecting the following:

- Place the front wheels in the straight-ahead position
- Lower lid and disconnect the air bag electrical connector at the bottom of the steering wheel
- Side lids from both sides of the steering wheel
- Torx® bolts using a Torx® wrench T50 from both side of the steering wheel; then, discard the bolts

- Air bag module from the steering wheel
- Steering wheel nut
- Steering wheel from the steering column using a suitable puller

5. Disarm the passenger's side air bag by removing or disconnecting the following:

- Glove box lid
- Passenger's air bag electrical connector

6. Remove the instrument panel by removing or disconnecting the following:

- Upper and lower glove box screws and remove the glove box
- Lower instrument panel screws and the panel at the driver's side
- Knee protector screws and the knee protector
- Steering column cover screws and the cover
- Combination switch-to-steering column screws, disconnect the electrical connector and the combination switch
- Cluster lid "A" screws and the lid
- Combination meter screws, disconnect the electrical connectors and the combination meter
- Center ventilator with the switch panel using a suitable prytool
- Cover plate (automatic transmis-

sion) or the shifter cover plate (manual transmission)
- Ashtray
- Upper and lower audio/air conditioning control unit assembly screws and the assembly
- Console box screws and the console box (under the shifter cover plate); be sure to remove the rear screws
- Front pillar garnish
- Left and right lower cover and the center lower cover at the instrument panel dash
- Defroster grille
- Instrument panel-to-chassis nuts/bolts and the instrument panel

7. Remove or disconnect the following:
- Rear heater ducts
- Side ventilator ducts
- Center defroster duct and the center ventilator duct
- Heater housing-to-chassis fasteners and remove the heater housing
- Heater core from the heater housing

To install:

8. Install or connect the following:
- Heater core to the heater housing
- Heater housing and the heater housing-to-chassis fasteners
- Center ventilator duct and the center defroster duct
- Side ventilator ducts
- Rear heater ducts

9. Install the instrument panel by installing or connecting the following:
- Instrument panel and the instrument panel-to-chassis nuts/bolts
- Defroster grille
- Left and right lower cover and the center lower cover
- Front pillar garnish
- Console box and the console box screws under the shifter cover plate; be sure to install the rear screws
- Audio/air conditioning control unit assembly and the upper and lower assembly screws
- Ashtray
- Cover plate (automatic transmission) or the shifter cover plate (manual transmission)
- Center ventilator with the switch panel
- Combination meter, connect the electrical connectors and the combination meter screws
- Cluster lid "A" and the lid screws
- Combination switch, connect the electrical connector and the combination switch-to-steering column screws
- Steering column cover and the cover screws
- Knee protector and the knee protector screws
- Lower instrument panel and the panel screws on the driver's side

- Glove box and the upper and lower glove box screws

10. Arm the passenger's side air bag by installing or connecting the following:
- Passenger's air bag electrical connector
- Glove box lid

11. Install the air bag module and steering wheel by installing or connecting the following:
- Steering wheel to the steering column
- Steering wheel nut and torque it to 22–29 ft. lbs. (29–39 Nm)
- Air bag module to the steering wheel
- Torque the new Torx® bolts (using a Torx® wrench T50), at both side of the steering wheel to 11–18 ft. lbs. (15–25 Nm).
- Side lids to both sides of the steering wheel
- Air bag electrical connector and install the lower lid at the bottom of the steering wheel

12. Refill the cooling system.
13. Connect the negative battery terminal.
14. Operate the engine to normal operating temperatures; then, check the climate control operation and check for leaks.

Sentra

1. Before servicing the vehicle, refer to the precautions in the beginning of this section.

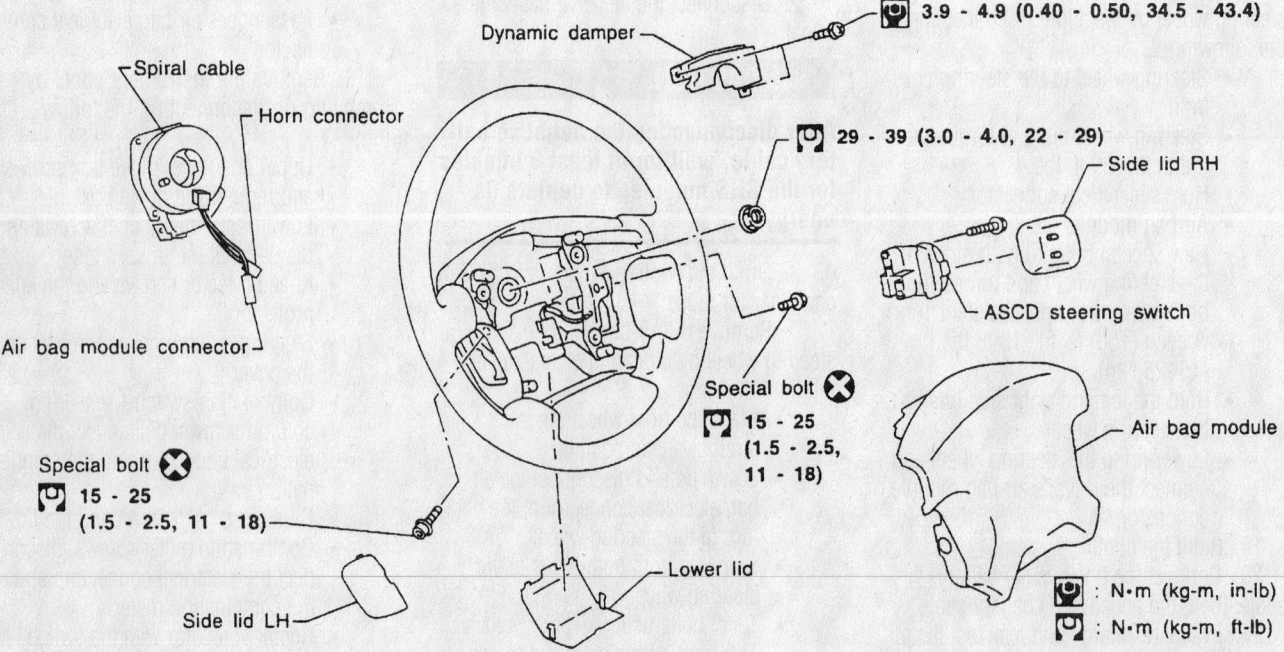

Exploded view of the air bag module and steering wheel—Maxima

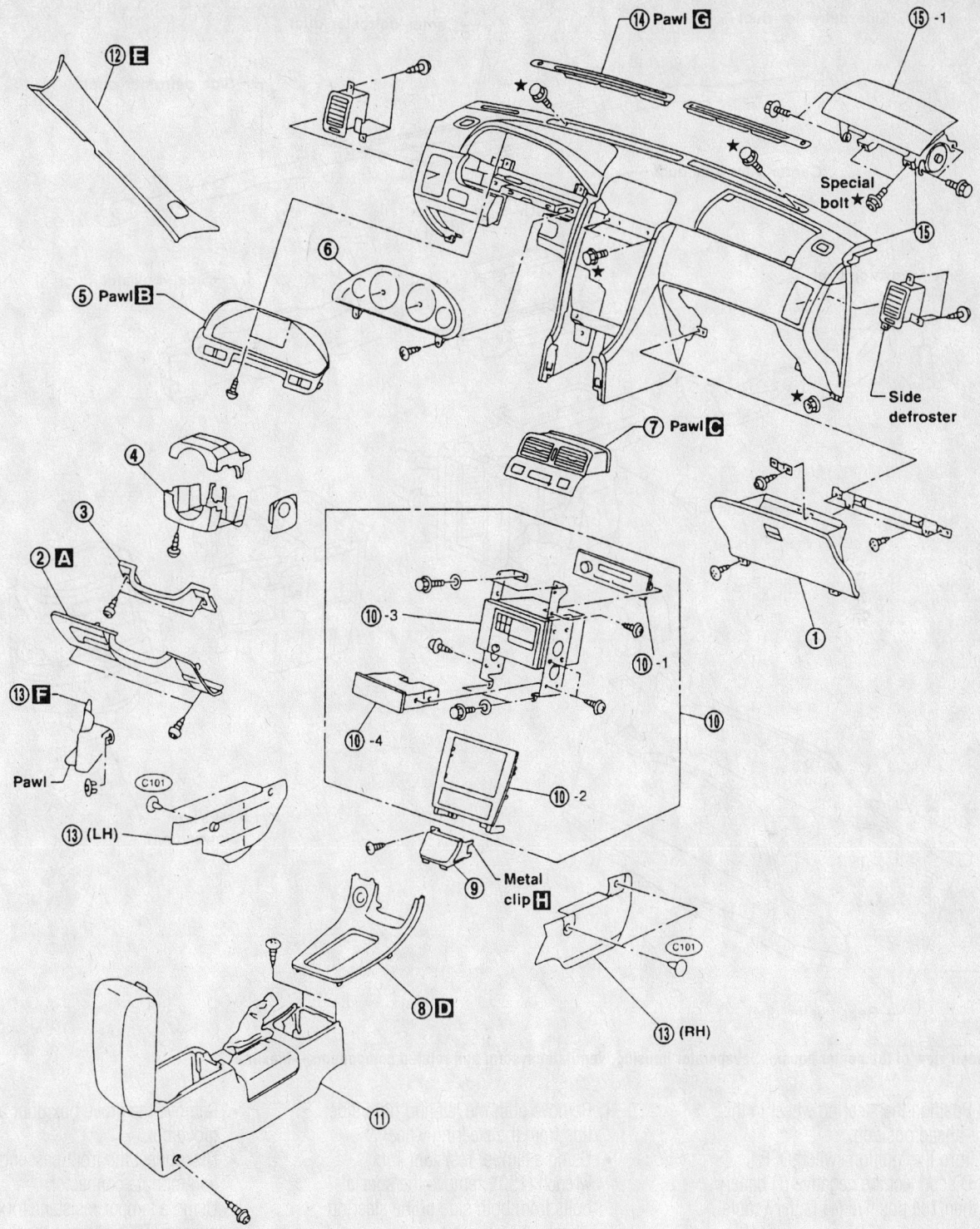

★ : Instrument panel assembly mounting bolts and nuts

1. Glove box assembly
2. Instrument lower panel on driver side
3. Knee protector assembly
4. Steering column cover & combination switch
5. Cluster lid A

6. Combination meter
7. Center ventilator with switch panel
8. A/T shifter cover plate or M/T shifter cover plate
9. Ashtray
10. Audio & A/C control unit assembly

11. Console box
12. Front pillar garnish
13. Instrument dash: lower cover and center lower cover on LH, RH
14. Defroster grille
15. Instrument panel assembly
15. -1 Passenger air bag module

93112GK2

Exploded view of the instrument panel, console and related components—Maxima

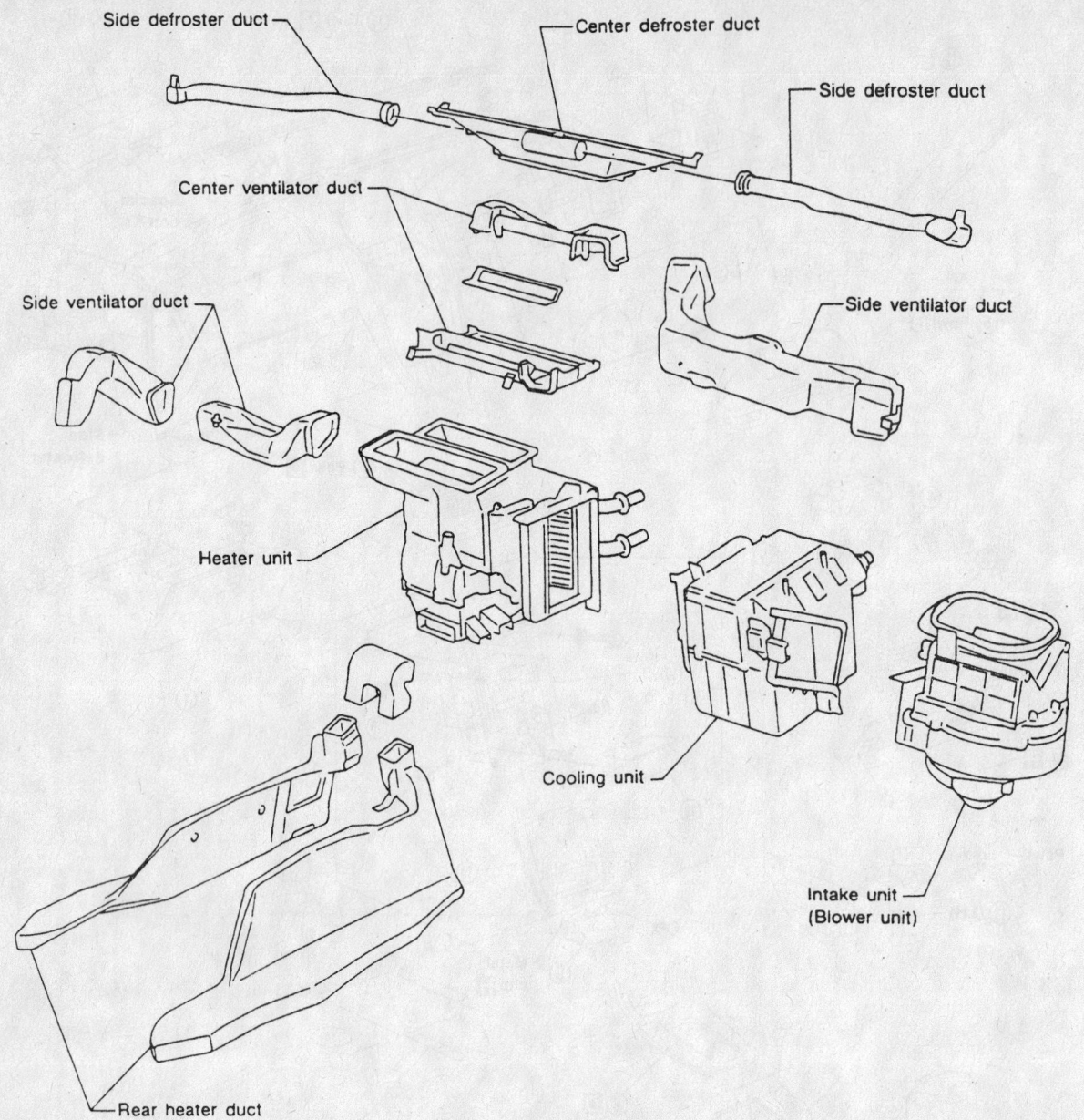

Side defroster duct

Center defroster duct

Side defroster duct

Center ventilator duct

Side ventilator duct

Side ventilator duct

Heater unit

Cooling unit

Intake unit
(Blower unit)

Rear heater duct

93112GK3

Exploded view of the heater housing, evaporator housing, ventilator system and related components—Maxima

2. Position the steering wheel in the straight-ahead position.

3. Turn the ignition switch OFF.

4. Disconnect the negative (()) battery cable; then, the positive (+) battery cable.

➡**Wait for a least 3 minutes after disconnecting the battery cables for the charge in the air bag circuit to dissipate before working on the air bag module(s).**

5. Remove the driver's side SRS and steering wheel by performing the following procedure:

- Remove the lower lid from the steering wheel and disconnect the driver's air bag module connector.

- Remove both the left and right side lids from the steering wheel.
- Using a tamper resistant Torx® wrench (T50), remove the special bolts from both side of the steering wheel.
- Carefully, remove the air bag module and store it face up.
- Disconnect the horn's electrical connector and remove the steering wheel nut.
- Using a steering wheel puller, press the steering wheel from the steering column.

6. Remove the passenger's side SRS by performing the following procedure:

- Remove the glove box door and the glove box.
- Disconnect the front passenger's air bag module connector.
- Using a tamper resistant Torx® wrench (T50), remove the 2 special bolts.
- Remove the 4 passenger's air bag-to-instrument panel nuts.
- Carefully, remove the front passenger's air bag module and store it face up.

7. Drain the cooling system into a clean container for reuse.

8. Discharge and recover the air conditioning system refrigerant.

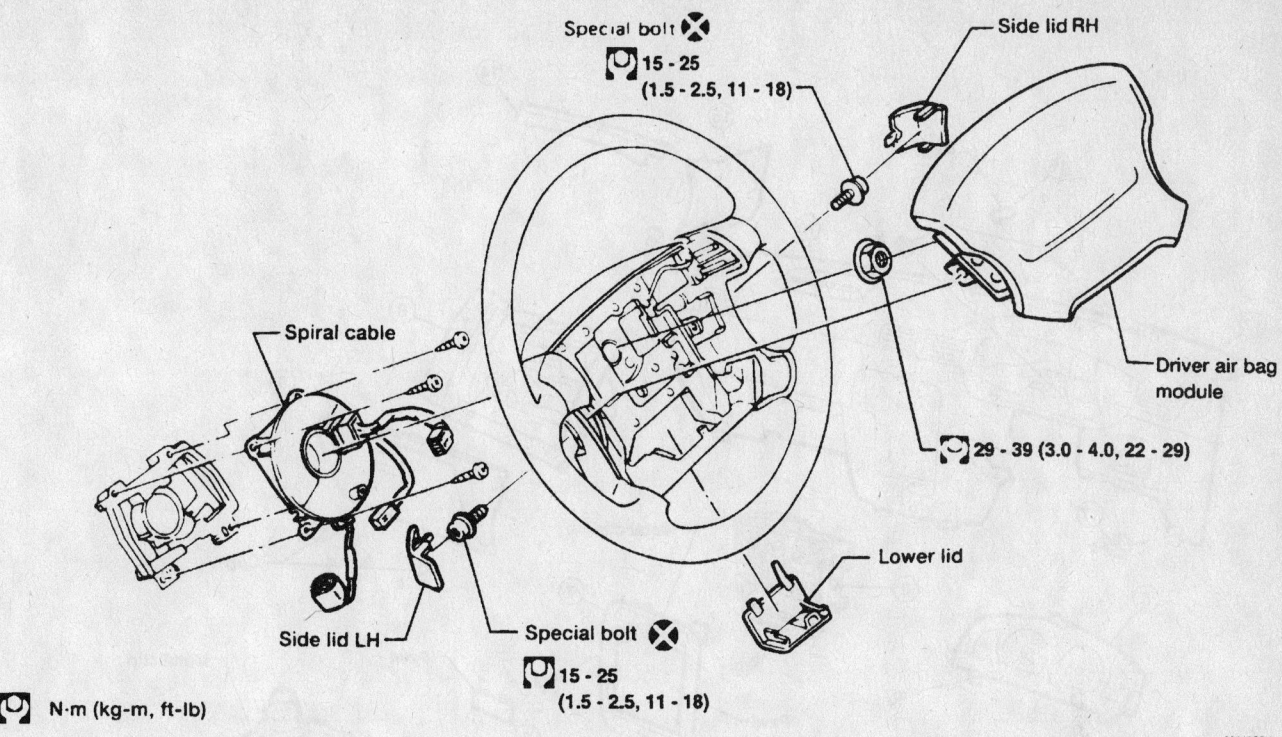

Special bolt ⊗
15 - 25
(1.5 - 2.5, 11 - 18)

Side lid RH

Spiral cable

Driver air bag module

29 - 39 (3.0 - 4.0, 22 - 29)

Lower lid

Side lid LH

Special bolt ⊗
15 - 25
(1.5 - 2.5, 11 - 18)

N·m (kg-m, ft-lb)

93112GD4

Exploded view of the steering wheel and air bag module—Sentra

9. Working in the engine compartment, disconnect the heater hoses from the heater core tubes.

10. Remove the instrument panel by performing the following procedures:
- On the driver's side, remove the 2 lower panel-to-instrument panel screws and the lower panel.
- Remove the 2 lower reinforcement panel-to-instrument panel screws and the lower reinforcement panel.
- Remove the 6 steering column cover screws, the cover and combination switch.
- Remove the 2 cluster lid "A" screws and the cluster lid "A".
- Remove the 3 combination meter screws, disconnect the electrical harness connector and remove the combination meter.
- Remove the ashtray.
- Remove the cluster lid "C" mask, screw and the cluster lid "C".
- Remove the 8 audio and air conditioning control assembly-to-instrument panel screws, the electrical connectors and the audio and air conditioning control assembly.
- Remove the transmission shifter finisher.
- Remove the rear console mask, the 4 screws and the rear console.

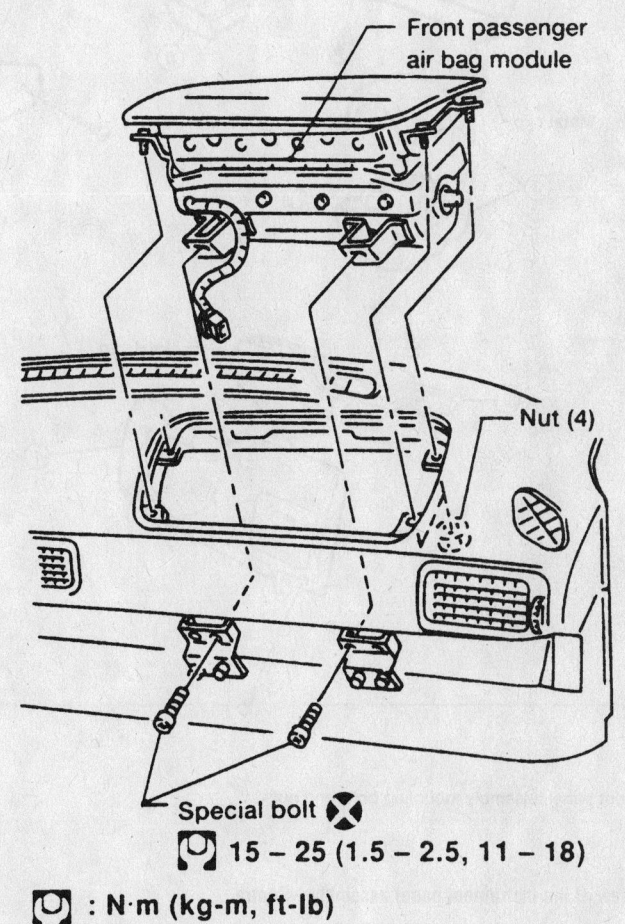

Front passenger air bag module

Nut (4)

Special bolt ⊗
15 – 25 (1.5 – 2.5, 11 – 18)

: N·m (kg-m, ft-lb)

93112GD5

Exploded view of the passenger's side air bag module—Sentra

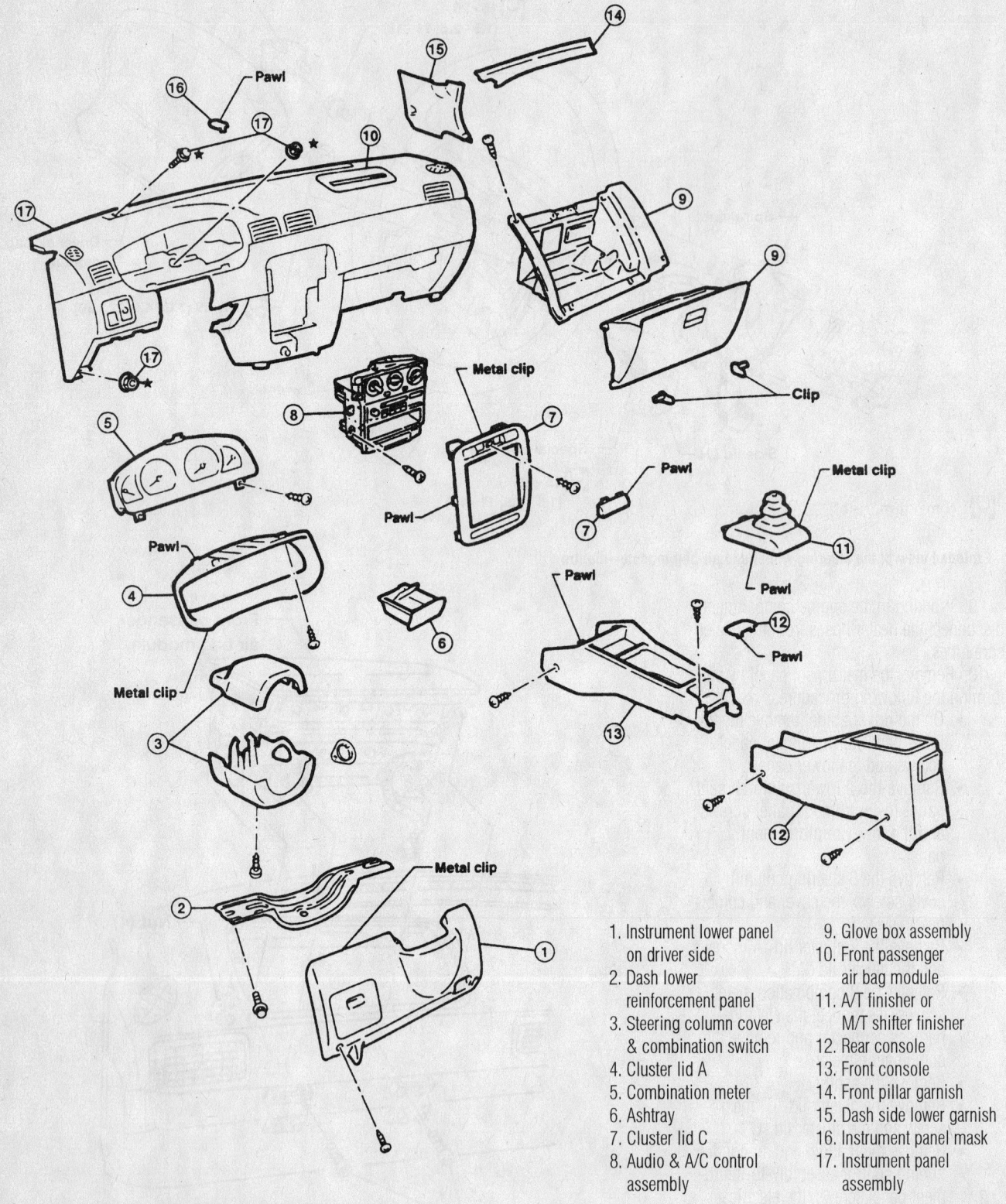

1. Instrument lower panel on driver side
2. Dash lower reinforcement panel
3. Steering column cover & combination switch
4. Cluster lid A
5. Combination meter
6. Ashtray
7. Cluster lid C
8. Audio & A/C control assembly
9. Glove box assembly
10. Front passenger air bag module
11. A/T finisher or M/T shifter finisher
12. Rear console
13. Front console
14. Front pillar garnish
15. Dash side lower garnish
16. Instrument panel mask
17. Instrument panel assembly

★ : Instrument panel assembly mounting bolts and nuts.

93112GD6

Exploded view of the instrument panel assembly—Sentra

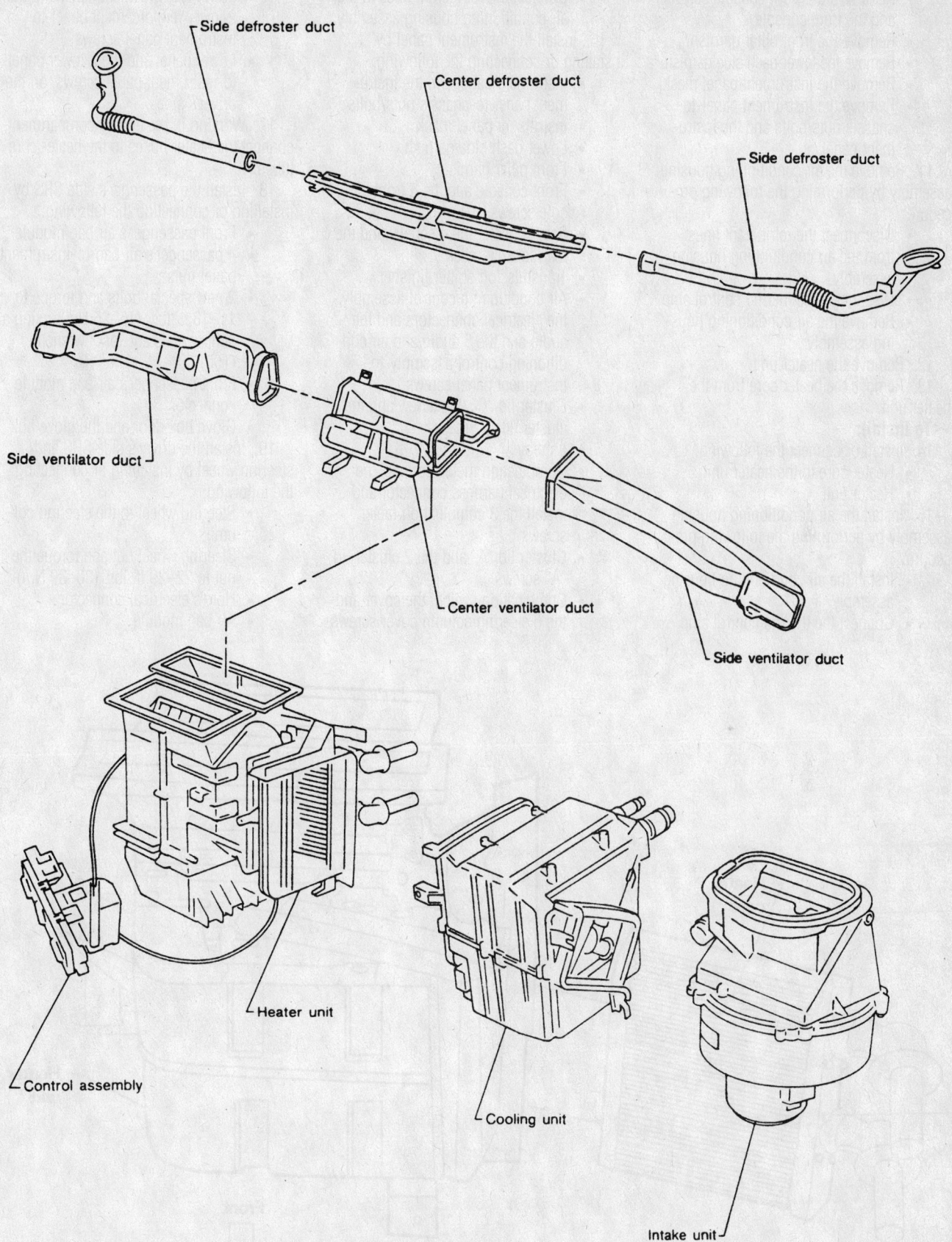

- Side defroster duct
- Center defroster duct
- Side defroster duct
- Side ventilator duct
- Center ventilator duct
- Side ventilator duct
- Control assembly
- Heater unit
- Cooling unit
- Intake unit

93112GD7

Exploded view of the heater housing assembly and related components—Sentra

- Remove the 4 front console screws and the front console.
- Remove the front pillar garnish.
- Remove the lower dash side garnish.
- Remove the instrument panel mask.
- Remove the instrument panel-to-chassis nuts/bolts and the instrument panel.

11. Remove the air conditioning housing assembly by performing the following procedure:

- Disconnect the refrigerant lines from the air conditioning housing assembly.
- Disconnect the thermo control amp.
- Remove the air conditioning housing assembly.

12. Remove the heater unit.

13. Remove the heater core from the heater unit.

To install:

14. Install or connect the following:

- Heater core to the heater unit
- Heater unit

15. Install the air conditioning housing assembly by performing the following procedure:

- Install the air conditioning housing assembly.
- Connect the thermo control amp.

- Connect the refrigerant lines to the air conditioning housing assembly.

16. Install the instrument panel by installing or connecting the following:

- Instrument panel and the instrument panel-to-chassis nuts/bolts
- Instrument panel mask
- Lower dash side garnish
- Front pillar garnish
- Front console and the 4 front console screws
- Rear console, the 4 screws and the rear console mask
- Transmission shifter finisher
- Air conditioning control assembly, the electrical connectors and the audio and the 8 audio and air conditioning control assembly-to-instrument panel screws.
- Cluster lid "C", the screw and the cluster lid "C" mask
- Ashtray
- Combination meter, connect the electrical harness connector and install the 3 combination meter screws
- Cluster lid "A" and the 2 cluster lid "A" screws
- Combination switch, the cover and the 6 steering column cover screws

- Lower reinforcement panel and the 2 lower reinforcement panel-to-instrument panel screws
- Lower panel and the 2 lower panel-to-instrument panel screws, on the driver's side

17. Working in the engine compartment, connect the heater hoses to the heater core tubes.

18. Install the passenger's side SRS by installing or connecting the following:

- Front passenger's air bag module
- 4 passenger's air bag-to-instrument panel nuts
- 2 new special bolts and torque to 11–18 ft. lbs. (15–25 Nm), using a tamper resistant Torx® wrench (T50)
- Front passenger's air bag module connector
- Glove box door and the glove box

19. Install the driver's side SRS and steering wheel by installing or connecting the following:

- Steering wheel to the steering column
- Steering wheel nut and torque the nut to 22–29 ft. lbs. (30–39 Nm)
- Horn's electrical connector
- Air bag module

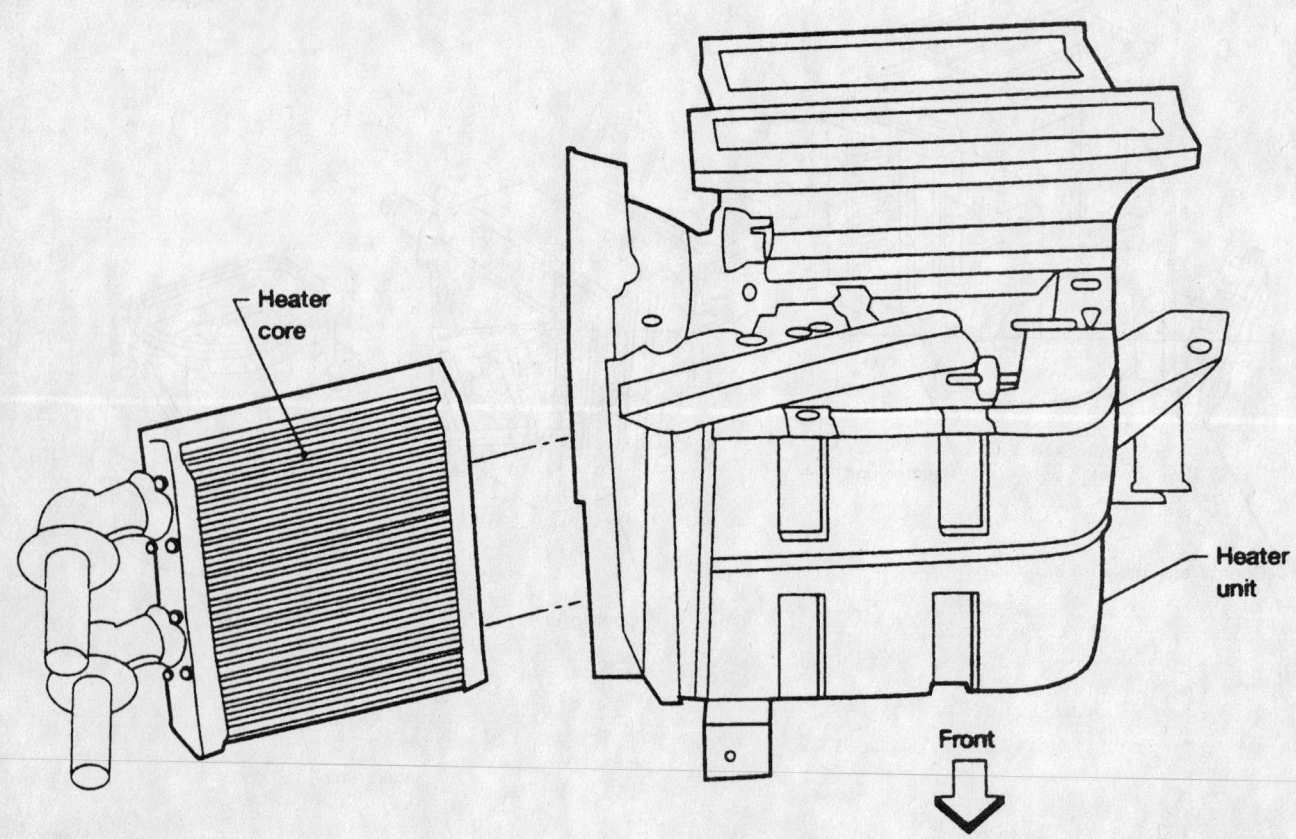

View of the heater core and heater housing—Sentra

93112GD8

- New special bolts to both side of the steering wheel and torque the bolts to 11–18 ft. lbs. (15–25 Nm), using a tamper resistant Torx® wrench (T50)
- Left and right side lids to the steering wheel
- Lower lid to the steering wheel
- Driver's air bag module connector.

20. Refill the cooling system.

21. Connect the positive (+) battery cable; then, the negative (()) battery cable.

22. Evacuate, charge and leak test the air conditioning system refrigerant.

23. Operate the engine to normal operating temperatures; then, check the climate control operation and check for leaks.

Cylinder Head

REMOVAL & INSTALLATION

1.8L Engines

1. Before servicing the vehicle, refer to the precautions in the beginning of this section.

2. Drain the cooling system.

3. Properly relieve the fuel system pressure.

4. Remove or disconnect the following:
- Negative battery cable
- Engine drive belts
- Power steering pulley
- Oil pump and bracket
- Air duct to the intake manifold collector
- Right front wheel
- Engine side and under covers
- Front exhaust tube
- Cylinder head front mounting bracket
- Rocker cover by loosening the bolts in numerical order
- Distributor, plug wires and spark plugs
- Spark plugs
- Intake manifold support and set the No. 1 cylinder at the Top Dead Center (TDC) position
- Idler pulley, camshaft sprockets and timing chains
- Camshafts

5. Loosen the cylinder head bolts in 2–3 steps in the reverse order of the tightening sequence to prevent warpage or cracking of the cylinder head assembly.
- Cylinder head (carefully), from the block, pulling the head up evenly from both ends. If the head seems stuck, do not pry it off. Tap lightly

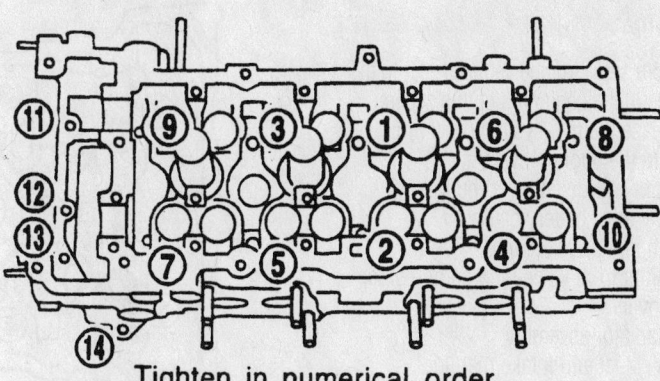

Tighten in numerical order.

9347QG01

Cylinder head torque sequence—1.8L engine

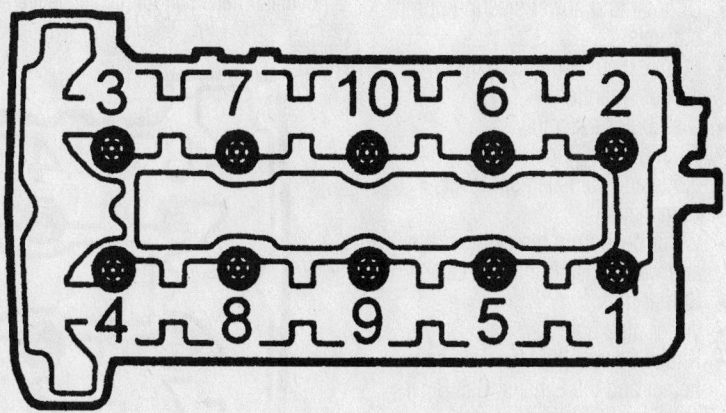

9347UG03

Tighten the rocker cover bolts in sequence—1.8L engines

around the lower perimeter of the head with a rubber mallet to help break the seal. The cylinder head and the intake and exhaust manifolds are removed together.
- Cylinder head gasket(s)

To install:

6. Thoroughly clean both the cylinder block and head mating surfaces. Avoid scratching either surface.

7. Coat the threads and the seating surface of the head bolts with clean engine oil. Install the cylinder head assembly (always replace the head gasket). Install head bolts (with washers) in their proper locations.

8. For 1.8L engines, tighten the bolts in sequence, as follows:
   a. Step 1: Bolts 1–10 to 22 ft. lbs. (29 Nm).
   b. Step 2: Bolts 1–10 to 43 ft. lbs. (59 Nm).
   c. Step 3: Loosen bolts 1–10 completely.
   d. Step 4: Bolts 1–10 to 22 ft. lbs. (29 Nm).
   e. Step 5: Bolts 1–10 plus 50–55 degrees.
   f. Step 6: Bolts 11–14 to 74 inch lbs. (8 Nm).

9. Install or connect the following:
- Camshafts
- Idler pulley, camshaft sprockets and timing chains
- Intake manifold support
- Distributor
- Spark plugs and wires
- Distributor cap
- Rocker arm cover and torque the bolts to 34 inch lbs. (4 Nm)
- Cylinder head front mounting bracket
- Front exhaust tube
- Engine side and under covers
- Right front wheel
- Air duct to the intake manifold collector
- Oil pump and bracket
- Power steering pulley
- Drive belts
- Negative battery cable

10. Fill the cooling system.

11. Start the vehicle, check for leaks and repair if necessary.

2.0L Engine

1. Before servicing the vehicle, refer to the precautions in the beginning of this section.
2. Release the fuel system pressure.
3. Drain the cooling system.
4. Remove or disconnect the following:
- Negative battery cable
- Engine under covers
- Right front wheel and engine side cover
- Radiator assembly
- Air duct and intake manifold
- Drive belts and water pump pulley
- Alternator
- Power steering pump
- Cylinder head cover and oil separator
- Oil filter and power steering pump brackets
- Front exhaust pipe from the exhaust manifold
- Distributor assembly
- Timing chain, tensioner, chain guide and camshaft sprockets
- Camshafts
- Water hose from the cylinder block and water hose from the heater
- Starter motor
- Water pipe bolt
- Knock Sensor (KS) harness connector and the Exhaust Gas Recirculation (EGR) tube
- Cylinder outside bolts. Remove the cylinder head bolts in 2 or 3 steps.
- Cylinder head completely with manifolds attached

To install:

5. Check all components for wear. Replace as necessary. Clean all mating surfaces and replace the cylinder head gasket.

➡ If the length of any cylinder head bolt exceeds 6.22 in. (158.2mm), replace the bolt.

6. Install cylinder head. Torque the cylinder head bolts in the following sequence:
   a. Step 1: 29 ft. lbs. (39 Nm).
   b. Step 2: 58 ft. lbs. (78 Nm).
   c. Step 3: Loosen all bolts in sequence completely.
   d. Step 4: 25–33 ft. lbs. (34–44 Nm).
   e. Step 5: Plus 90–95 degrees clockwise in sequence.
   f. Step 6: Plus additional 90–95 degrees.

➡ Do not turn any bolt 180–200 degrees clockwise all at once.

7. Install or connect the following:

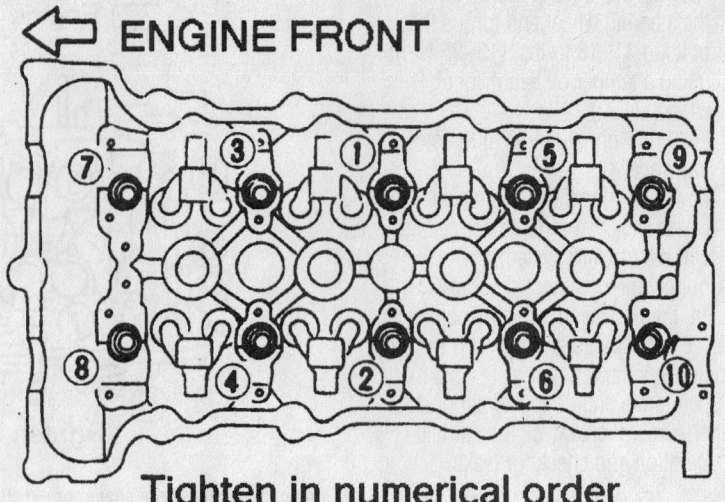

Tighten in numerical order

Cylinder head bolt torque sequence—2.0L engine

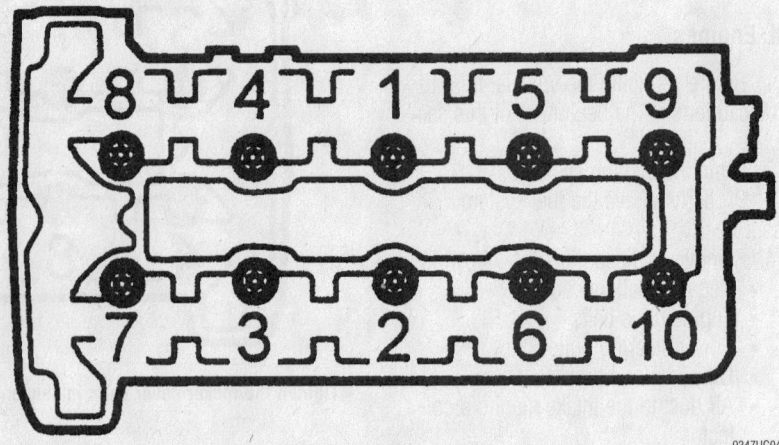

Tighten the rocker cover bolts in sequence—2.0L engines

- KS connector and EGR tube
- Starter motor and the wiring
- Water hoses to the engine block and heater
- Camshafts
- Camshaft sprockets timing chain guide, tensioner and timing chain
- Distributor assembly
- Front exhaust pipe to the exhaust manifold
- Oil filter and power steering pump brackets
- Cylinder head cover and oil separator
- Power steering pump
- Alternator
- Drive belts
- Air duct and intake manifold
- Radiator
- Engine side cover and right front wheel
- Engine under covers
- Negative battery cable

8. Fill the cooling system.
9. Start the vehicle, check for leaks and repair if necessary.

2.4L Engine

1. Before servicing the vehicle, refer to the precautions in the beginning of this section.
2. Drain cooling system.
3. Relieve the fuel system pressure.
4. Remove or disconnect the following:
- Negative battery cable
- Intake manifold collector, exhaust manifold and all related components
- Distributor assembly. Using a block of wood, set a jack under the aluminum oil pan and remove the front engine mount.
- Cylinder head cover
- Timing chain and camshaft sprockets
- Camshafts

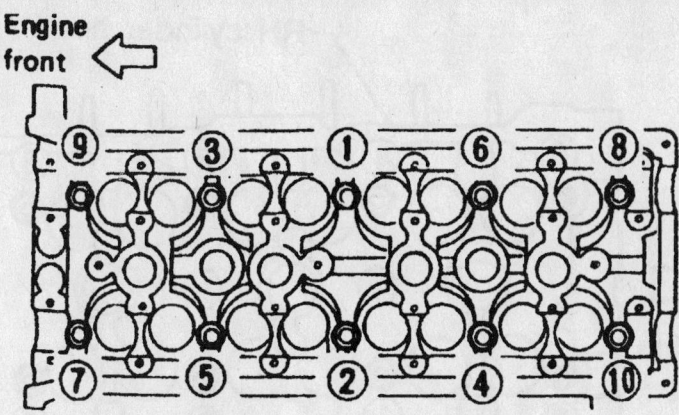

Cylinder head bolt torque sequence—2.4L engine

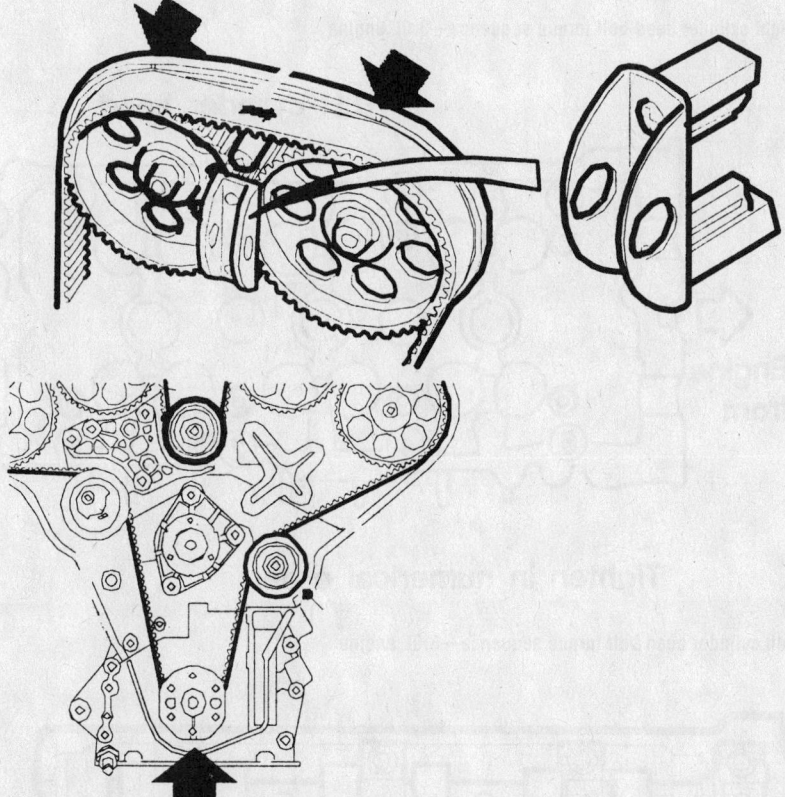

Tighten the rocker arm bolts in sequence—2.4L engines

➡ **The valvetrain components must be reassembled in their original positions.**

5. Loosen the cylinder head bolts in reverse order of tightening.

➡ **A warped or cracked cylinder head could result from loosening in incorrect order. The cylinder head bolts should be loosened in 2 or 3 steps.**

6. Remove the cylinder head and the intake manifold. Remove the cylinder head gasket. The lower timing chain will not be disengaged from crankshaft sprocket.

To install:

7. Clean the gasket surfaces.

8. Install or connect the following:
- New cylinder head gasket
- Cylinder head and temporarily tighten the cylinder head bolts. This is necessary to avoid damaging the cylinder head gasket. Be sure to install washers between the bolts and cylinder head.

- Idler shaft assembly
- Upper timing chain and cover

9. Tighten the cylinder head bolts in sequence as follows:
   a. Step 1: 22 ft. lbs. (29 Nm).
   b. Step 2: 59 ft. lbs. (79 Nm).
   c. Step 3: Loosen all the bolts completely.
   d. Step 4: 18–25 ft. lbs. (25–34 Nm).
   e. Step 5: Plus 86–91 degrees clockwise.

10. Install or connect the following:
- Camshafts
- Timing chains, chain tensioner and camshaft sprockets
- Cylinder head cover
- Distributor assembly
- Intake manifold collector
- Negative battery cable

11. Fill the cooling system.

12. Start the vehicle, check for leaks and repair if necessary.

3.0L Engine

1. Before servicing the vehicle, refer to the precautions in the beginning of this section.

2. Relieve the fuel system pressure.

3. Drain the engine oil.

4. Drain the cooling system.

➡ **Before detaching any hoses or connectors, note the locations for reassembly.**

5. Remove or disconnect the following:
- Negative battery cable
- Intake manifold collector
- Fuel tube
- Intake manifold
- Cylinder head covers
- Ignition coils
- Exhaust Gas Recirculation (EGR) guide tube
- Engine under cover
- Right front wheel and engine side cover
- Drive belts and idler pulley
- Steel (lower) and aluminum (upper) oil pans
- Water pump cover
- Timing chain case cover
- Timing chains, camshaft sprockets and related components
- Crankshaft sprocket

6. Loosen the bolts that secure the rear timing chain case. The bolts must be loosened in the reverse order of installation sequence.
- Rear timing case cover using seal cutter tool

➡ **Remove the O-rings from the front of the engine block.**

- Camshafts
- Cylinder head bolts in the reverse order of the tightening sequence. The bolts should be loosened in the reverse order of installation sequence.

➡ **A warped or cracked cylinder head could result from removing the bolts in incorrect order.**

- Cylinder heads from the vehicle.
- Discard the head gaskets

7. Remove all traces of liquid gasket from the timing chain case and from the water pump covers.

8. Remove all traces of liquid gasket from the engine block.

9. Inspect the timing chain for excessive wear or damage and replace as necessary.

To install:

10. Turn the crankshaft until the No. 1 piston is set 240 degrees before Top Dead Center (TDC) on compression stroke.

11. Using new head gaskets, install the cylinder heads.

➡ **If possible, replacement of the head bolts is suggested.**

12. If replacement of the head bolts is not possible, perform the following bolt measurement:

  a. Measure the diameter of the head bolt 0.43 in. (11mm) from the bottom of the bolt.

  b. Measure the diameter of the head bolt 1.89 in. (48mm) from the bottom of the bolt.

  c. Whenever the size difference between the 2 measurements exceeds 0.0043 in. (0.11mm) the head bolts must be replaced.

13. Install the cylinder head bolts and torque in sequence as follows:

  a. Step 1: 72 ft. lbs. (98 Nm).

  b. Step 2: Completely loosen all bolts.

  c. Step 3: 25–33 ft. lbs. (34–44 Nm).

  d. Step 4: plus 90–95 degrees clockwise.

  e. Step 5: plus 90–95 degrees clockwise.

14. Install or connect the following:

- Camshafts and related components
- New O-rings to the front of the engine block

15. Apply sealant to the hatched portion of the of the rear timing chain case.

16. Align the rear timing chain case with the dowel pins and install onto the cylinder heads and engine block.

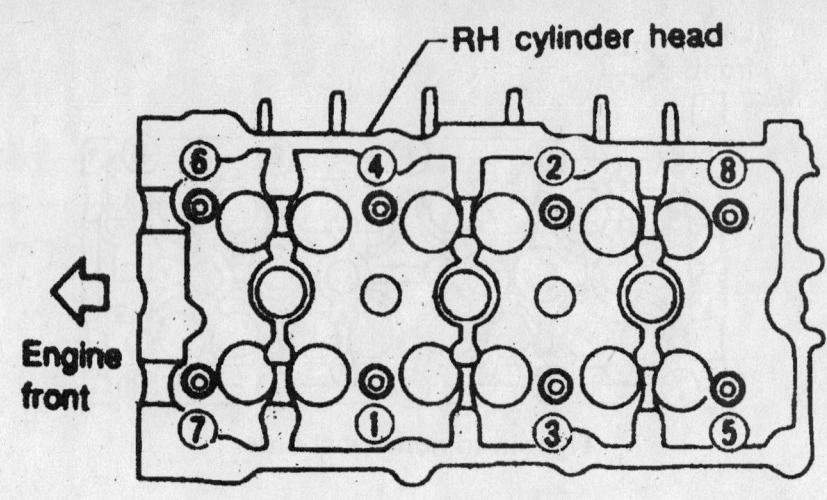

Tighten in numerical order.

Right cylinder head bolt torque sequence—3.0L engine

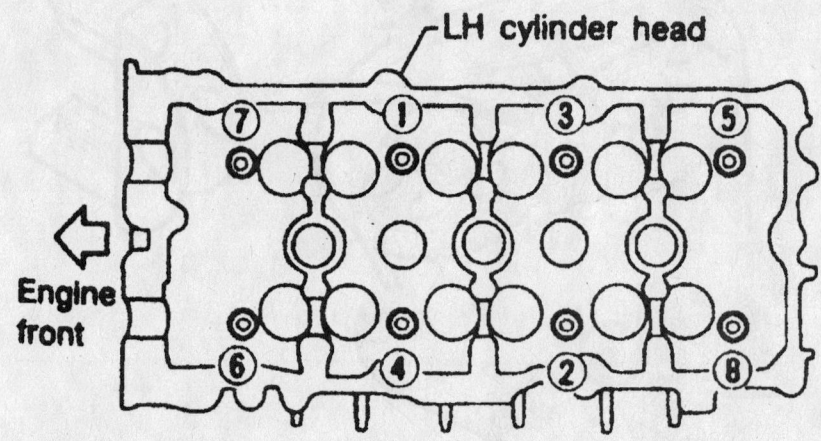

Tighten in numerical order.

Left cylinder head bolt torque sequence—3.0L engine

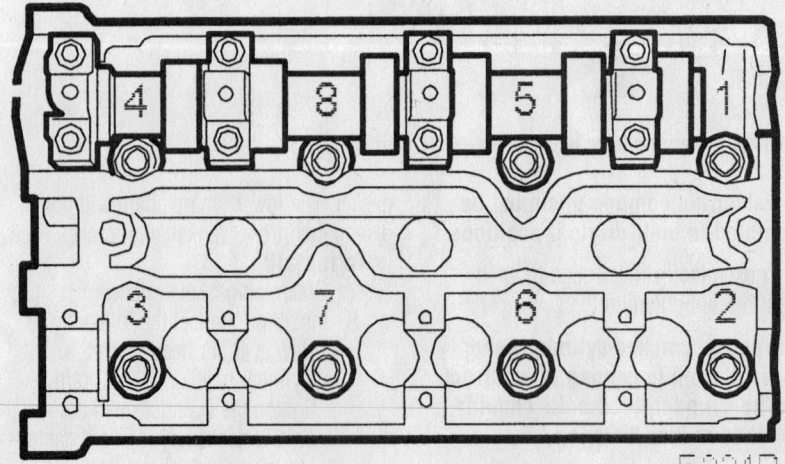

Tighten the rocker arm bolts in sequence—3.0L and 3.5L engines

17. Torque the rear timing chain case mounting bolts in sequence to 105–121 inch lbs. (11.8–13.7 Nm).

18. Install or connect the following:
- Water pump cover
- Upper and lower oil pans
- Idler pulley and drive belts
- Cylinder head covers
- Intake manifold
- Fuel tube
- Intake manifold collector
- Negative battery cable

19. Fill the cooling system.

20. Fill the engine with clean oil.

21. Start the vehicle, check for leaks and repair if necessary.

3.5L Engine

➡ **You must remove the engine from the vehicle in order to remove the cylinder head, for this procedure.**

1. Before servicing the vehicle, refer to the precautions in the beginning of this section.

2. Relieve the fuel system pressure.

3. Drain the engine oil.

4. Drain the cooling system.

➡ **Before detaching any hoses or connectors, note the locations for reassembly.**

5. Remove or disconnect the following:
- Negative battery cable
- Engine assembly
- Exhaust manifold

6. Place the engine on a suitable workstand.
- Oil pan
- Timing chain
- Intake manifold
- Water outlet
- Rear timing chain case bolts, in the sequence shown
- Rear timing chain case
- O-rings from the cylinder head and block
- Intake valve timing control solenoid valves

➡ **For installation purposes, matchmark the camshaft brackets before removing them.**

- Intake and exhaust camshafts and brackets. Loosen the bracket bolts in several steps, in the sequence shown.
- Right and left side cam chain tensioner from the cylinder head
- Cylinder head bolts. Loosen in several steps, in the sequence shown.

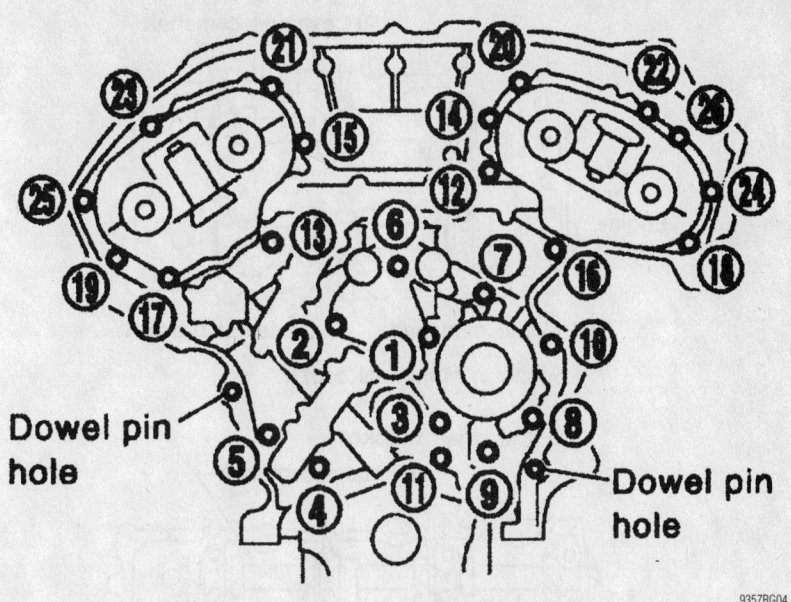

Loosen the rear timing chain case bolts in sequence—3.5L engine

9357RG04

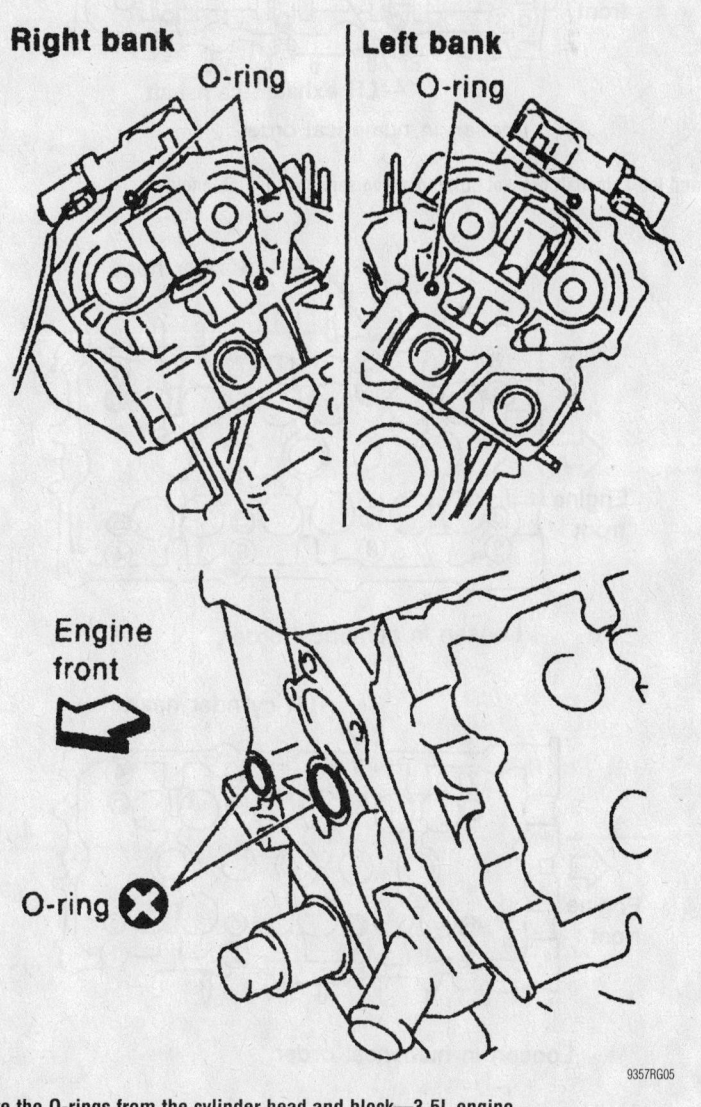

Remove the O-rings from the cylinder head and block—3.5L engine

9357RG05

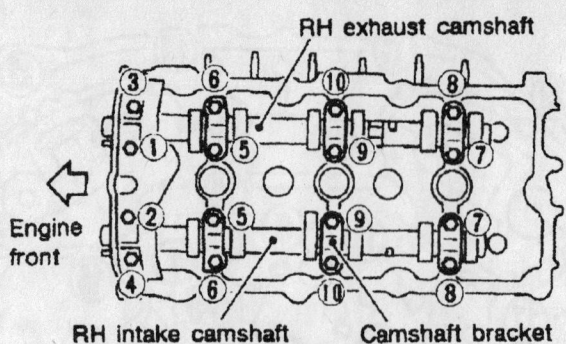

Loosen in numerical order.

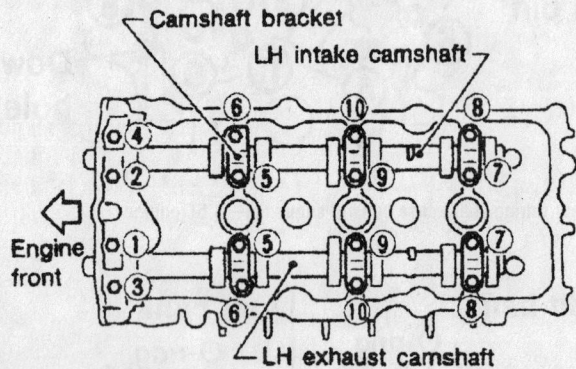

Right and left camshaft bracket bolt loosening sequence—3.5L engine

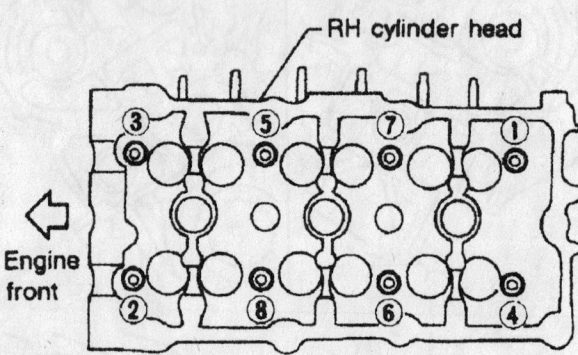

Loosen in numerical order.

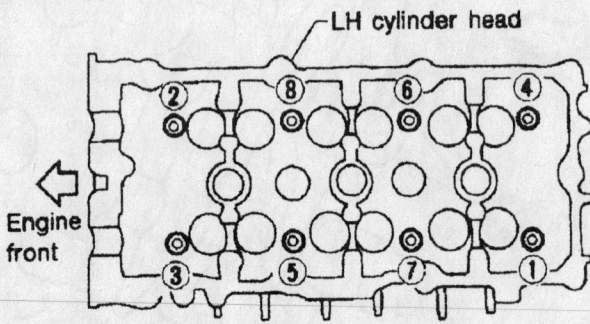

Loosen in numerical order.

Cylinder head bolt loosening sequence—3.5L engine

➡ **A warped or cracked cylinder head could result from removing the bolts in incorrect order.**

- Cylinder heads from the vehicle
- Discard the head gaskets

7. Remove all traces of liquid gasket from the timing chain case and from the water pump covers.

8. Remove all traces of liquid gasket from the engine block.

9. Inspect the timing chain for excessive wear or damage and replace as necessary.

To install:

10. Turn the crankshaft until the No. 1 piston is a Top Dead Center (TDC) on compression stroke. The crankshaft key should face toward the right bank.

11. Using new head gaskets, install the cylinder heads.

➡ **If possible, replacement of the head bolts is suggested.**

12. If replacement of the head bolts is not possible, perform the following bolt measurement:

   a. Measure the diameter of the head bolt 0.43 in. (11mm) from the bottom of the bolt.

   b. Measure the diameter of the head bolt 1.89 in. (48mm) from the bottom of the bolt.

   c. Whenever the size difference between the 2 measurements exceeds 0.0043 in. (0.11mm) the head bolts must be replaced.

13. Install the cylinder head bolts and torque in sequence as follows:

   a. Step 1: 72 ft. lbs. (98 Nm).

   b. Step 2: Completely loosen all bolts.

   c. Step 3: 26–32 ft. lbs. (34–44 Nm).

   d. Step 4: plus 90–95 degrees clockwise.

   e. Step 5: plus 90–95 degrees clockwise.

14. Install or connect the following:

   - Camshafts and related components
   - Intake valve timing control solenoid valves
   - New O-rings to the front of the engine block and cylinder head

15. Apply sealant to the hatched portion of the of the rear timing chain case.

16. Align the rear timing chain case with the dowel pins and install onto the cylinder heads and engine block.

17. Torque the rear timing chain case mounting bolts in sequence to 105–121 inch lbs. (11.8–13.7 Nm).

18. Install or connect the following:

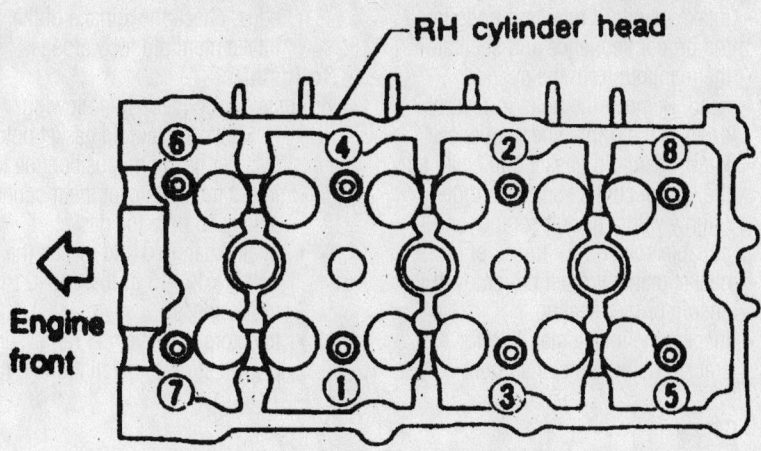

Tighten in numerical order.

79230G11

Right cylinder head bolt torque sequence—3.5L engine

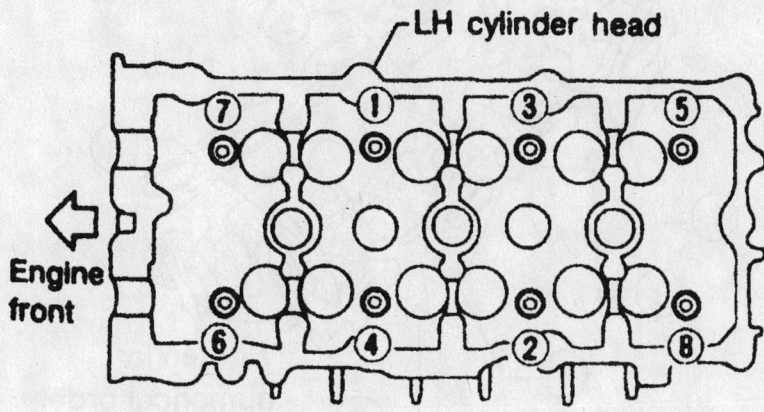

Tighten in numerical order.

79230G12

Left cylinder head bolt torque sequence—3.5L engine

- Water outlet
- Intake manifold
- Timing chain
- Oil pan
- Exhaust manifold
- Engine assembly into the vehicle
- Negative battery cable
19. Fill the cooling system.
20. Fill the engine with clean oil.
21. Start the vehicle, check for leaks and repair if necessary.

Rocker Arms

REMOVAL & INSTALLATION

Except 2.0L Engine

Nissan engines, with the exception of the 2.0L engine, do not utilize rocker arms. The valves are actuated directly by the camshafts.

2.0L Engine

1. Before servicing the vehicle, refer to the precautions in the beginning of this section.
2. Release the fuel pressure following the proper procedure.
3. Remove or disconnect the following:
   - Negative battery cable
   - Rocker arm cover, gasket and the oil separator
   - Intake manifold supports, oil filter bracket and the power steering pump
4. Set the No. 1 cylinder at Top Dead Center (TDC) on the compression stroke.
   - Timing chain tensioner from the side of the head

5. Matchmark the position of the rotor and housing and remove the distributor.
   - Timing chain guide
   - Camshaft sprockets while holding the camshaft stationary with a large wrench. Secure the timing chain with wire so the timing is not lost. The front cover will have to be removed if the chain timing is lost.

➡ **When removing the camshafts, loosen the journal caps in the opposite sequence of tightening. Camshaft damage may result if this step is not followed.**

   - Camshafts, brackets, oil tubes and the baffle plate. Label all components for proper installation.

➡ **It is essential that all parts be kept in the same order and orientation for reinstallation. Be sure to mark and separate parts to keep them from getting mixed. This will aid assembly.**

   - Rocker arms, shims, rocker arm guides and the hydraulic lash adjusters. Label all components for proper installation.

➡ **The valve lifters must be stored in the vertical position or submersed in clean oil to prevent air from entering the lifters.**

6. Inspect the surfaces of the rockers and replace if there are any signs of damage.
To install:
7. Lubricate the rocker arms, shims, rocker arm guides and the hydraulic lash adjusters.
8. Install or connect the following:
   - Rocker arms, shims, rocker arm guides and the hydraulic lash adjusters in their original locations
   - Camshafts, brackets, oil tubes and the baffle plate in the proper location
9. Tighten the bolts in sequence as follows:

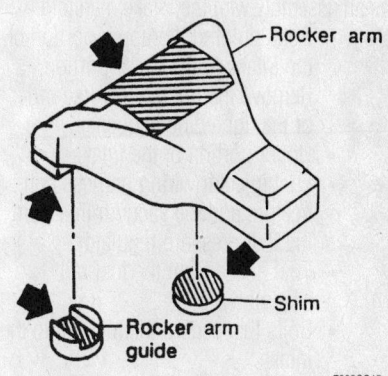

79230G15

Rocker arm, guide and shim—2.0L engine

a. Right camshaft bolts Nos. 9 and No. 10: 17 inch lbs. (2 Nm).

b. Right camshaft bolts No. 1 through 8: 17 inch lbs. (2 Nm).

c. Left camshaft bolts Nos. 11 and 12: 17 inch lbs. (2 Nm).

d. Left camshaft bolts Nos. 1 through 10: 17 inch lbs. (2 Nm).

e. All camshaft bolts in numerical sequence: 52 inch lbs. (6 Nm).

f. All camshaft bolts in numerical sequence: 87–104 inch lbs. (10–11 Nm).

g. Rear 2 bolts of the left-hand camshaft: 13–19 ft. lbs. (18–25 Nm).

10. Install or connect the following:
- Camshaft sprockets while holding the camshaft stationary with a large wrench
- Remaining components in the reverse order from which they were removed
- Negative battery cable

11. Check and adjust the ignition and valve timing. If there is air in the lifters, bleed the air by running the engine at 1000 rpm for 10 minutes.

Intake Manifold

REMOVAL & INSTALLATION

1.8L Engine

1. Before servicing the vehicle, refer to the precautions in the beginning of this section.
2. Relieve the fuel system pressure.
3. Drain the cooling system.
4. Remove or disconnect the following:
- Negative battery cable
- Air cleaner assembly
- Throttle linkage, electrical connections and vacuum lines from the throttle body
- Intake manifold collector support brackets

5. The throttle body can be removed from the manifold at this point or can be removed as an assembly with the intake manifold.
- Bolts holding the upper portion of the intake to the lower portion. Remove the bolts in reverse order of the tightening sequence.
- Upper portion of the intake
- Fuel injector wiring harness connectors and the vacuum line from the fuel pressure regulator
- Fuel hoses from the fuel rail assembly
- Bolts that secure the fuel rail to the intake
- Injectors with the fuel rail assembly

- Intake manifold retaining bolts in the proper sequence and separate the manifold from the cylinder head. Remove the bolts in reverse order of the tightening sequence.
- Intake manifold gasket and clean all the gasket contact surfaces thoroughly with a gasket scraper and suitable solvent. All traces of old gasket material must be removed to ensure proper sealing.
- Inspect the intake manifold for cracks. Using a metal straight-edge, check the surface of the intake manifold for warpage.

To install:

6. Install or connect the following:
- New intake manifold gasket onto the cylinder head and position the lower intake manifold over the mounting studs and onto the gasket.
- Intake manifold and torque the bolts to 13–15 ft. lbs. (18–21 Nm) in sequence
- Injectors with the fuel rail assembly. Be sure to install the fuel rail

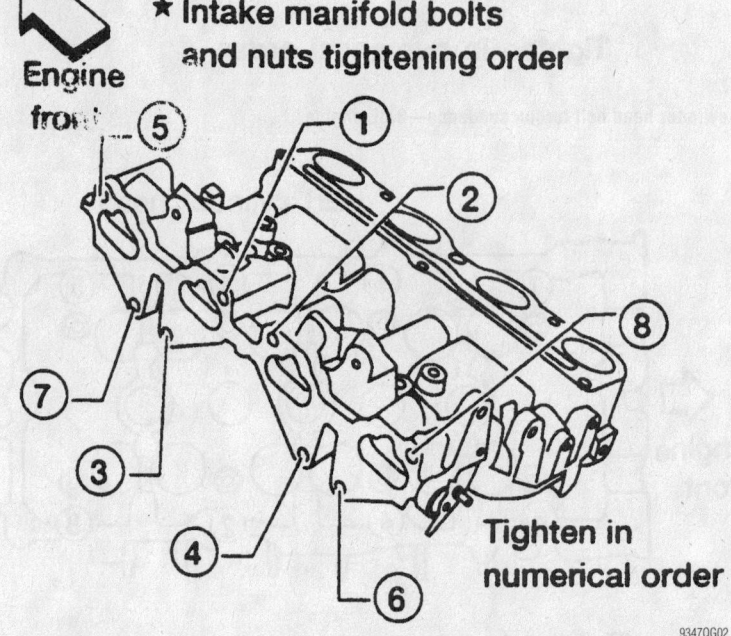

Lower intake manifold torque sequence—1.8L engine

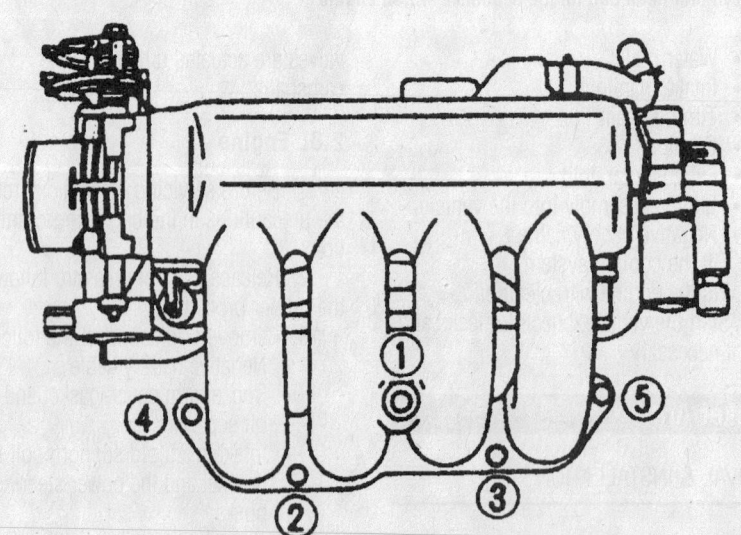

Upper intake manifold torque sequence—1.8L engine

insulators. Torque the bolts in 2 steps to 13–15 ft. lbs. (18–21 Nm).
- Fuel injector wiring harness connectors and the vacuum line to the fuel pressure regulator
- Fuel hoses to the fuel rail assembly using new hose clamps
- Intake manifold collector using a new gasket. Torque the bolts to 13–15 ft. lbs. (18–21 Nm) in sequence.
- Throttle body or throttle chamber, if removed. Torque the bolts in a crisscross pattern to 13–16 ft. lbs. (18–22 Nm).

➡**Be sure to properly position the throttle body gasket with the cut out facing down.**

- Intake manifold collector support brackets
- Throttle linkage, electrical connections and vacuum lines
- Air cleaner
- Negative battery cable

7. Fill the cooling system to the proper level.

8. Start the engine, check for leaks and repair if necessary.

2.0L Engine

1. Before servicing the vehicle, refer to the precautions in the beginning of this section.

2. Relieve the fuel system pressure.

3. Drain the cooling system.

4. Remove or disconnect the follow-ing:
- Negative battery cable
- Air cleaner assembly
- Manifold support brackets
- Throttle linkage, electrical connections and vacuum lines from the throttle body. Be sure to note the locations of all connections.
- Exhaust Gas Recirculation (EGR) tube from the manifold
- Fuel rail assembly
- Drive belts and water pump pulley
- Alternator and power steering pump
- Oil filter bracket and power steering bracket
- Intake manifold collector retaining bolts in the reverse of installation sequence and separate the collector from the manifold
- Intake manifold assembly retaining bolts in the reverse order of the tightening sequence and separate the manifold from the cylinder head
- All gasket material and clean all the gasket contact surfaces thoroughly with a gasket scraper and a suitable

solvent. All traces of old gasket material must be removed to ensure proper sealing.
- Inspect the intake manifold for cracks. Using a metal straightedge, check the surface of the intake manifold for warpage.

To install:

5. Install or connect the following:
- Intake manifold with new gaskets and torque the bolts, in sequence, to 13–15 ft. lbs. (18–21 Nm)
- Intake manifold collector with new gaskets and torque the bolts, in sequence, to 13–15 ft. lbs. (18–21 Nm)
- Oil filter bracket
- Power steering bracket

- Alternator and power steering pump
- Water pump pulley and drive belts
- Fuel rail assembly with new insulators and torque the bolts to 35 inch lbs. (4 Nm)
- EGR tube
- Throttle linkage, vacuum lines and electrical connections to the throttle body
- Manifold support brackets with new gaskets and torque the bolts to 20 ft. lbs. (26 Nm)
- Air cleaner assembly
- Negative battery cable

6. Fill the cooling system to the proper level.

7. Start the vehicle, check for leaks and repair if necessary.

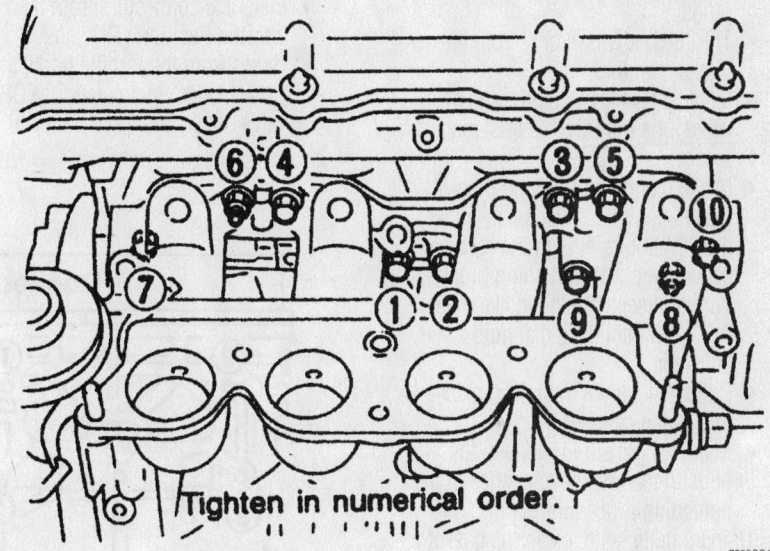

Lower intake manifold torque sequence—2.0L engine

7923QG18

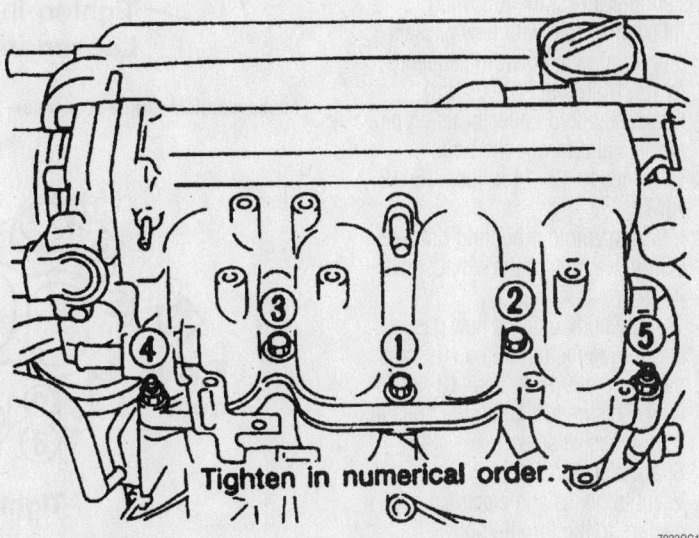

Upper intake manifold torque sequence—2.0L and 1.8L engines

7923QG19

2.4L Engine

1. Before servicing the vehicle, refer to the precautions in the beginning of this section.

2. Relieve the fuel system pressure.

3. Drain the cooling system.

4. Remove or disconnect the following:
- Negative battery cable
- Air duct between the air flow meter and the throttle body
- Throttle cable and the cruise control cable, if equipped
- Fuel supply and return lines from the fuel injector assembly. Plug the lines to prevent leakage.
- Electrical connectors and the vacuum hoses to the throttle body and intake manifold/collector assembly
- Spark plug wires from the spark plugs
- Throttle body assembly from the intake manifold
- Exhaust Gas Recirculation (EGR) valve tube from the exhaust manifold
- Intake manifold mounting brackets
- Intake manifold collector-to-intake manifold bolts/nuts in the reverse sequence of the tightening procedure and separate the intake manifold from the intake manifold collector
- Bolts that secure the intake manifold to the cylinder head
- Manifold. Be sure to loosen the bolts in the reverse sequence of the tightening procedure.

5. Using a putty knife, clean the gasket mounting surfaces. Check the intake manifold/collector for cracks and warpage.

To install:

6. Install or connect the following:
- Intake manifold with new gaskets and torque the bolts, in sequence, to 12–14 ft. lbs. (16–19 Nm)
- Intake manifold collector using new gaskets and torque the bolts in sequence to 12–14 ft. lbs. (16–19 Nm)
- Intake manifold mounting brackets
- EGR valve tube to the exhaust manifold
- Throttle body using a new gasket and torque the bolts in a crisscross pattern to 13–16 ft. lbs. (18–22 Nm). Be sure to tighten the bolts in 2 progressive steps.
- Spark plug wires
- Vacuum hoses and electrical connectors to the throttle body
- EGR valve to the exhaust manifold

- Fuel return and supply lines
- Throttle and cruise control cables, if equipped
- Air duct
- Negative battery cable

7. Fill the cooling system to the proper level.

8. Start the vehicle, check for leaks and repair if necessary.

3.0L and 3.5L Engines

1. Before servicing the vehicle, refer to the precautions in the beginning of this section.

2. Drain the cooling system.

3. Release the fuel system pressure.

4. Remove or disconnect the following:
- Negative battery cable
- Throttle body coolant hoses
- Electrical connectors from the Throttle Position (TP) sensor
- Hoses from the throttle body, the Exhaust Gas Recirculation (EGR) valve, intake manifold collector, Idle Air Control (IAC) valve and the fuel pressure regulator
- Canister purge hose and blow-by hose
- EGR guide tube
- Accelerator cable from the throttle body
- Intake manifold collector support brackets
- Right side electrical connectors from the ignition coils
- Electrical connector from the crank angle sensor and the power transistor, if necessary
- Intake manifold collector-to-intake manifold bolts/nuts and the intake manifold collector

5. Remove the fuel injector assembly by performing the following procedures:

   a. Detach the electrical connectors from the fuel injectors.

   b. Disconnect the fuel lines from the fuel injector assembly.

   c. Remove the fuel rail-to-cylinder head bolts.

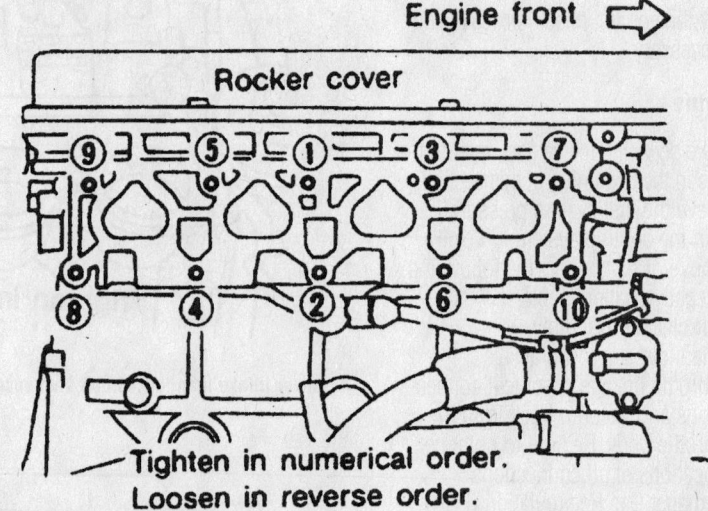

Intake manifold torque sequence—2.4L engine

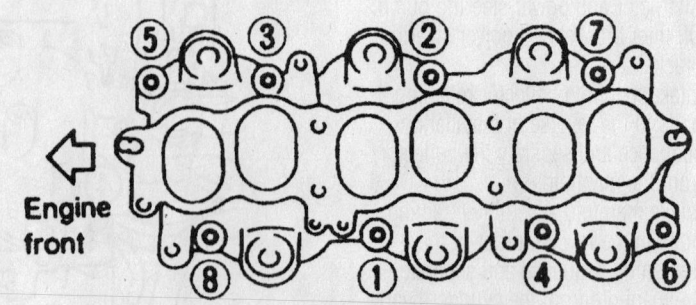

Intake manifold torque sequence—3.0L and 3.5L engines

d. Remove the fuel rail assembly from the engine.

6. Remove or disconnect the following:
- Intake manifold bolts/nuts in the reverse of the installation sequence
- Intake manifold from the engine and discard the gaskets

7. Clean all gasket mounting surfaces.

To install:

8. Using new gaskets, install the intake manifold to the engine.

9. Torque the bolts in sequence as follows:

    a. Step 1: 44–89 inch lbs. (5–10 Nm).

    b. Step 2: 20–23 ft. lbs. (26–31 Nm).

10. Install the fuel injector assembly by performing the following procedures:

    a. Install the fuel rail assembly to the engine.

    b. Install the fuel rail-to-cylinder head bolts and tighten the bolts to 15–20 ft. lbs. (21–26 Nm) in 2 progressive steps.

    c. Connect the fuel lines to the fuel injector assembly.

    d. Connect the electrical connectors to the fuel injectors.

11. Install the intake manifold collector. Torque the bolts to 8–11 ft. lbs. (11–15 Nm).

12. Install or connect the following:
- Crank angle sensor and transmitter electrical connectors
- Right side ignition coil electrical connectors
- Intake manifold collector support brackets
- Accelerator cable to the throttle body
- EGR guide tube
- Canister purge and blow by hoses
- Throttle body, EGR valve and intake manifold collector hoses
- IAC valve and fuel pressure regulator hoses
- TP sensor electrical connector
- Throttle body coolant hose
- Negative battery cable

13. Fill the cooling system.

14. Start the vehicle, check for leaks and repair if necessary.

Exhaust Manifold

REMOVAL & INSTALLATION

1.8L and 2.0L Engines

1. Before servicing the vehicle, refer to the precautions in the beginning of this section.

2. Remove or disconnect the following:
- Negative battery cable
- Engine undercovers
- Air cleaner or collector assembly
- Heat shields from the manifold and front exhaust pipe
- Front exhaust pipe from the exhaust manifold
- Temperature sensors, Oxygen (O_2) sensors and air induction pipes from the manifold
- Manifold support brackets
- Exhaust manifold attaching nuts and the manifold from the block. Discard the exhaust manifold gaskets.

3. Clean the gasket surfaces and check the manifold for cracks and warpage.

To install:

4. Install the exhaust manifold with a new gasket.

5. On 1.8L engines, tighten the mounting nuts with washers in sequence to 19–21 ft. lbs. (26–29 Nm).

6. On 2.0L engines, torque the fasteners from the center outward in several stages to 27–35 ft. lbs. (37–48 Nm).

7. Install or connect the following:
- Temperature sensors, O_2 sensors and air induction pipes
- Manifold support brackets
- Exhaust pipe to the manifold using a new gasket. Torque the nuts to 21–25 ft. lbs. (28–33 Nm) for 1.8L engines, or 32–37 ft. lbs. (43–50 Nm) for 2.0L engine models.
- Heat shields
- Air cleaner or collector assembly
- Engine undercovers
- Negative battery cable

8. Start the engine and check for exhaust leaks.

2.4L Engine

1. Before servicing the vehicle, refer to the precautions in the beginning of this section.

2. Remove or disconnect the following:

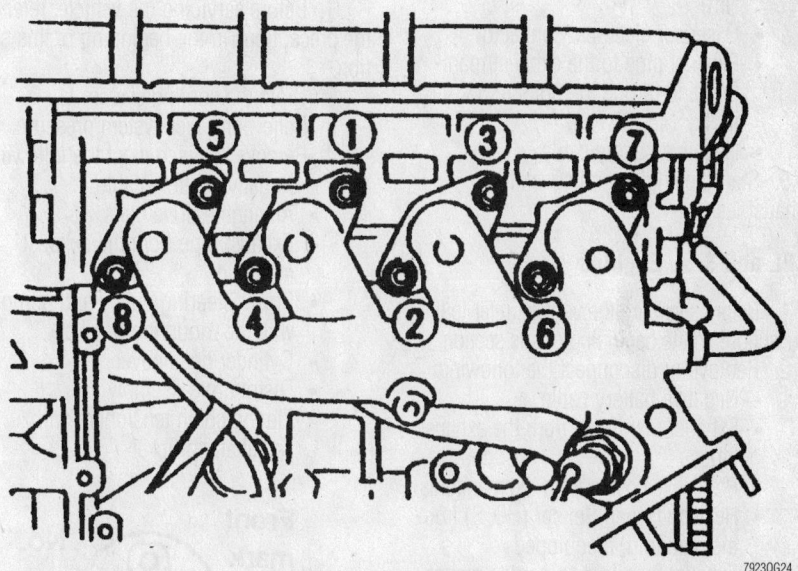

Exhaust manifold bolt tightening sequence—2.4L engine (except California models)

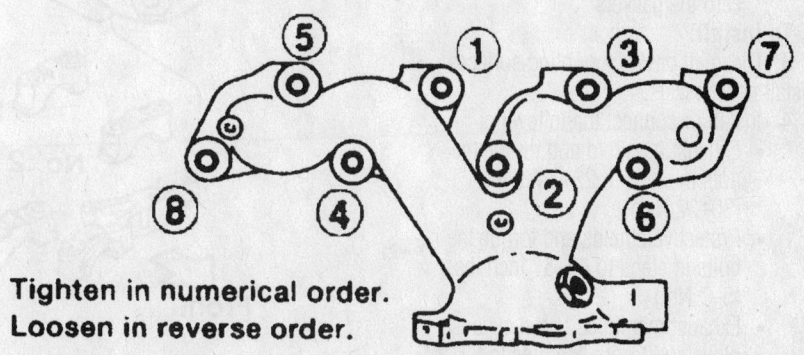

Tighten in numerical order.
Loosen in reverse order.

Exhaust manifold bolt tightening sequence—2.4L engine (California models)

- Negative battery cable
- Exhaust pipe from the exhaust manifold
- Oxygen (O_2) sensor electrical connector
- Exhaust manifold cover
- Exhaust Gas Recirculation (EGR) tube from the exhaust manifold
- Exhaust manifold-to-engine bolts and nuts and discard the gaskets
- Retaining bolts and nuts in reverse of the tightening sequence
- Exhaust manifold from the vehicle

To install:

3. Clean all gasket mounting surfaces and install new gaskets.
4. Install or connect the following:
   - Exhaust manifold to the engine and torque the nuts to 27–35 ft. lbs. (37–48 Nm)
   - EGR tube to the exhaust manifold and torque the nuts to 29–36 ft. lbs. (39–49 Nm)
   - Exhaust manifold cover and torque the bolts to 45–57 inch lbs. (5–7 Nm)
   - O_2 sensor electrical connector
   - Exhaust pipe to the exhaust manifold and torque the bolts to 33–44 ft. lbs. (45–60 Nm)
   - Negative battery cable
5. Start the engine and check for exhaust leaks.

3.0L and 3.5L Engines

1. Before servicing the vehicle, refer to the precautions in the beginning of this section.
2. Remove or disconnect the following:
   - Negative battery cable
   - Exhaust manifolds from the exhaust pipes
   - Protective covers from the manifolds
   - Heated Oxygen Sensor (HO2S) from the manifold, if equipped
   - Exhaust manifold-to-engine mounting nuts
   - Manifolds from the engine and discard the gaskets

To install:

3. Clean all gasket mounting surfaces. Install new gaskets.
4. Install or connect the following:
   - Exhaust manifold and torque the nuts in steps to 22–24 ft. lbs. (30–32 Nm)
   - Protective shields and torque the bolts in steps to 46–57 inch lbs. (5–7 Nm)
   - Exhaust manifolds to the exhaust pipes and torque the nuts to 32–37 ft. lbs. (43–50 Nm)

- HO2S sensor to the manifold and torque the fastener to 30–44 ft. lbs. (40–60 Nm), if equipped
- Negative battery cable

5. Start the engine and check for exhaust leaks.

Front Crankshaft Seal

REMOVAL & INSTALLATION

➡ **The front crankshaft seal procedure is applicable to timing belt-equipped engines only. For the front seal on engines equipped with timing chains, refer to the timing chain, sprockets, front cover and seal procedure later in this section.**

Camshaft and Valve Lifters

REMOVAL & INSTALLATION

1.8L Engines

1. Before servicing the vehicle, refer to the precautions in the beginning of this section.
2. Drain the cooling system.
3. Relieve the fuel system pressure.
4. Remove or disconnect the following:
   - Negative battery cable
   - All engine drive belts
   - Exhaust pipe from the exhaust manifold
   - Power steering pulley and pump with the mounting bracket
   - Cylinder head cover
   - Distributor assembly
   - Timing chain tensioners and camshaft sprocket

➡ **Before the camshafts are removed from the cylinder head, note the positioning of the pins at the end of the camshafts for reassembly purposes.**

- Camshaft bearing caps in sequence
- Camshafts from the cylinder head
- Idler sprocket bolt. These parts should be reassembled in their original position.
- Shims from the tops of the lifters. Be sure to note the position of each shim.
- Valve lifters from the bores in the cylinder head. Note the positioning of the lifters for reassembly.

5. Measure the diameter of the lifters. The diameter should be 1.1795–1.1801 in. (29.960–29.975mm).
6. Measure the diameter of the lifter bores. The diameter should be 1.1811–1.1819 in. (30.000–30.021mm).
7. Clearance between the lifter and bore should be 0.0010–0.0024 in. (0.025–0.061mm).

To install:

8. Install or connect the following:
   - Lifters and shims to the cylinder head in the proper locations as noted during removal

➡ **The exhaust and intake camshafts are marked with identification stamps. (E for exhaust and I for intake).**

- Camshafts to the cylinder head and position the intake camshaft knock pin at the 9 o'clock position and the exhaust camshaft at the 12 o'clock position

9. Install the camshaft bearing caps and tighten the mounting bolts as follows:

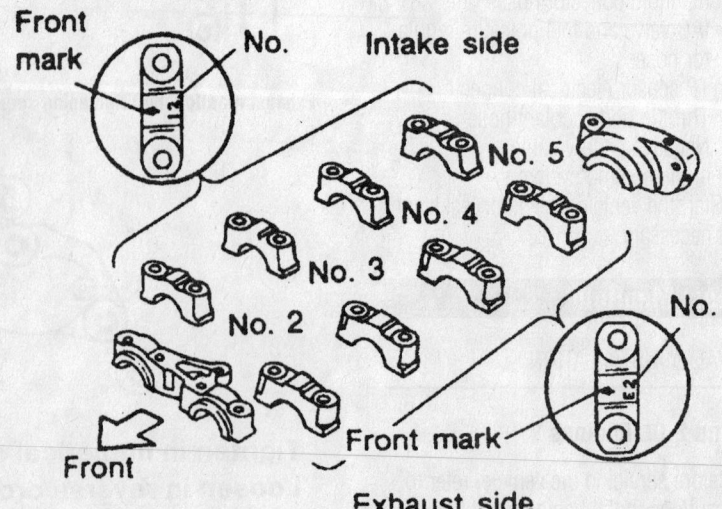

Be sure to install the camshaft bearing caps in their original positions—1.8L engine

79230G29

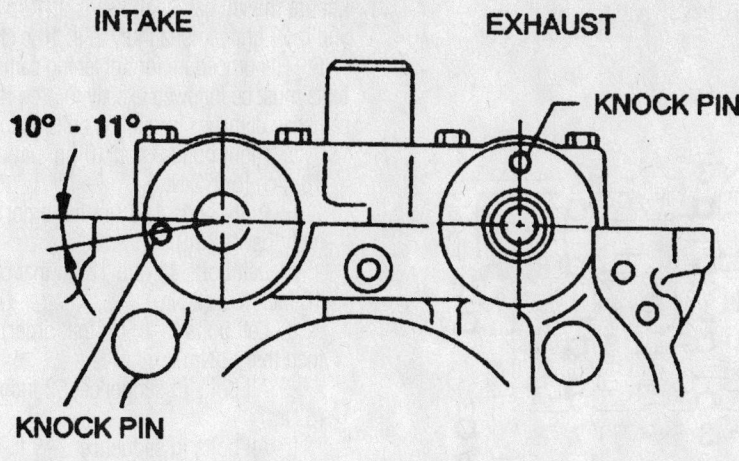

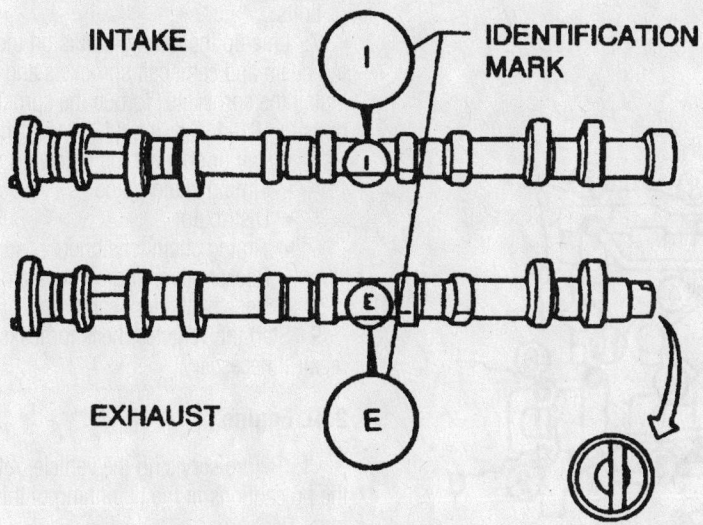

Positioning and identification of the camshafts—1.8L engine

7923QG30

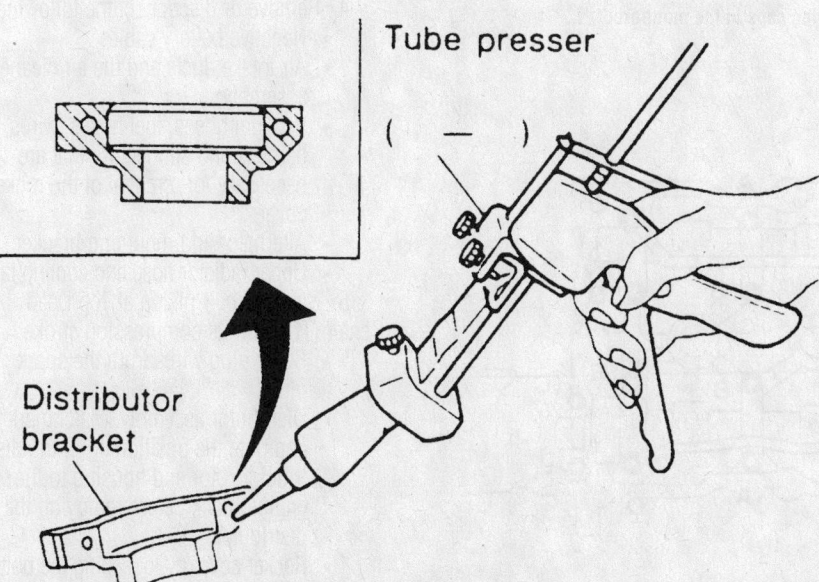

If removed, apply liquid gasket to the distributor bracket as shown—1.8L engine

9307QG08

a. Bolts 11 through 15, then bolts 1 through 10: 18 inch lbs. (2 Nm).

b. Bolts 1 through 15: 53 inch lbs. (6 Nm).

c. Bolts 1 through 14: 87–105 inch lbs. (10–12 Nm).

d. Bolt 15: 56–73 inch lbs. (7–8 Nm).

➡ **If any part of the valvetrain has been has been replaced, the valve adjustment must be checked. DO NOT adjust the valves or rotate the camshafts at this point. Internal engine damage will result.**

10. Install or connect the following:
   • Camshaft sprockets with timing chains
   • Distributor assembly

11. Check and adjust the valve clearance.

   • Cylinder head cover
   • Power steering pulley pump
   • Exhaust pipe from the exhaust manifold
   • All engine drive belts
   • Negative battery cable

12. Fill the cooling system.

13. Start the vehicle, check for leaks and repair if necessary.

2.0L Engine

1. Before servicing the vehicle, refer to the precautions in the beginning of this section.

2. Properly relieve the fuel system pressure.

3. Remove or disconnect the following:

   • Negative battery cable
   • Rocker cover and oil separator

4. Rotate the crankshaft until the No.1 piston is at Top Dead Center (TDC) on the compression stroke. Then, rotate the crankshaft until the mating marks on the camshaft sprockets line up with the mating marks on the timing chain.

   • Timing chain tensioner
   • Distributor
   • Timing chain guide
   • Camshaft sprockets. Use a wrench to hold the camshaft while loosening the sprocket bolt.
   • Camshaft bracket bolts in the opposite order of the tightening sequence
   • Camshaft

To install:

5. Clean the left-hand camshaft end bracket and coat the mating surface with liquid gasket. Install the camshafts, camshaft brackets, oil tubes and baffle plate.

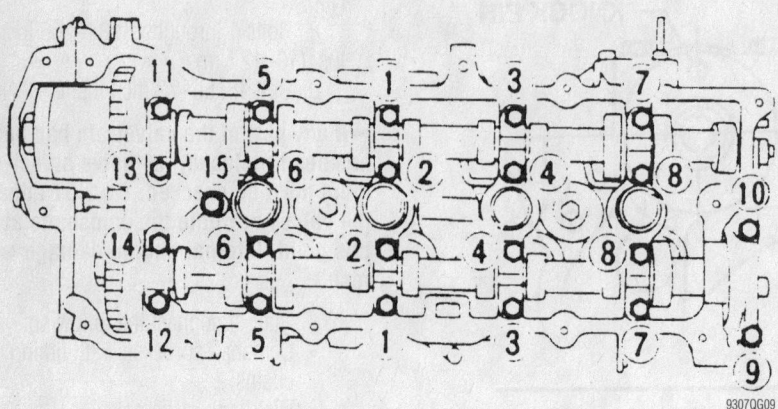

Camshaft bolt torque sequence—1.8L engine

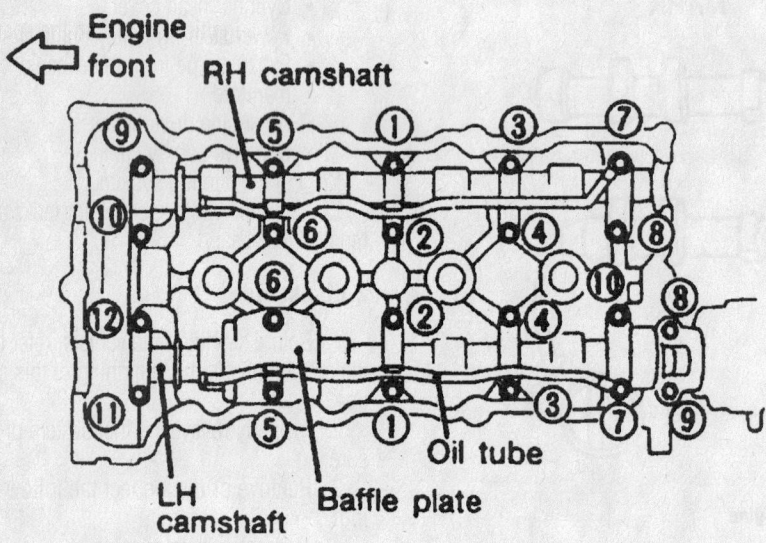

To prevent damage to the camshafts, torque the bearing caps in the numbered . . .

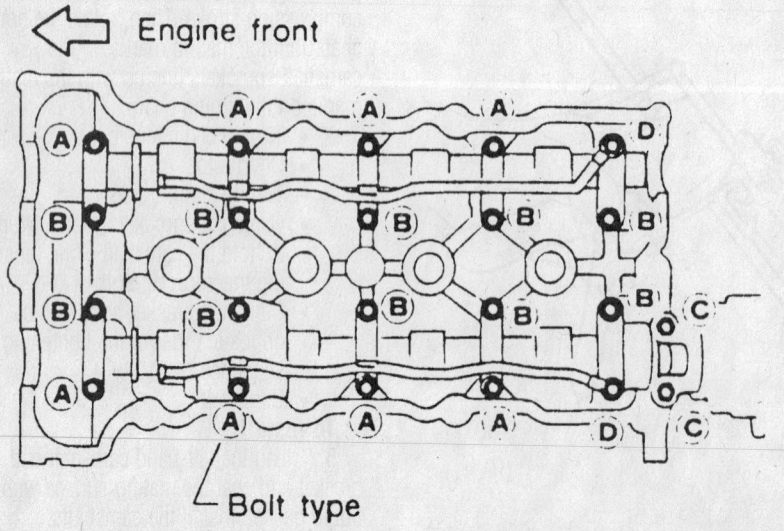

. . . and lettered sequence shown. Refer to the text for the proper torque—2.0L engines

Ensure the left camshaft key is at 12 o'clock and the right camshaft key is at 10 o'clock.

6. The procedure for tightening camshaft bolts must be followed exactly to prevent camshaft damage. Torque the bolts as follows:

  a. Right bolts 9 and 10 (in that order): 18 inch lbs. (2 Nm).

  b. Right bolts 1–8 (in that order): 18 inch lbs. (2 Nm).

  c. Left bolts 11 and 12 (in that order): 18 inch lbs. (2 Nm).

  d. Left bolts 1–10 (in that order): 18 inch lbs. (2 Nm).

  e. All bolts in sequence: 52 inch lbs. (6 Nm).

  f. All bolts in sequence: 7–9 ft. lbs. (9–12 Nm) for type A, B and C bolts and 13–19 ft. lbs. (18–25 Nm) for type D bolts.

7. Line up the mating marks on the timing chain and camshaft sprockets and install the sprockets. Torque the sprocket bolts to 101–116 ft. lbs. (137–157 Nm).

8. Install or connect the following:
• Timing chain guide
• Distributor
• Timing chain tensioner
• Rocker cover and oil separator
• Negative battery cable

9. Start the vehicle, check for leaks and repair if necessary.

2.4L Engine

1. Before servicing the vehicle, refer to the precautions in the beginning of this section.

2. Relieve the fuel system pressure.

3. Drain coolant from the engine and radiator.

4. Remove or disconnect the following:
• Negative battery cable
• Air intake ducts and the air cleaner assembly
• Vacuum hoses, fuel hoses, wires, harness and connectors that are necessary for removal of the rocker cover
• Alternator and mounting bracket
• Upper radiator hose and cooling fan

5. Set the No. 1 piston at Top Dead Center (TDC) on its compression stroke.
• Spark plug wires from the spark plugs
• Distributor assembly. Matchmark and note the positioning of the distributor rotor and housing to the engine block before removing the distributor.
• Rocker cover by loosening the bolts in the reverse order of installation

6. Wire the chain to the sprocket so the

chain does not fall off during sprocket removal. Hold the flats of the camshaft with a wrench just behind the first camshaft bearing cap. Loosen the bolts and remove the sprockets.

- Camshaft sprockets

➡ **The stoppers on camshaft covers prevent the upper timing chain from disengaging from the idle sprocket. Also, after removal of the camshaft sprockets, note the positioning of the pins at the end of the camshafts for reinstallation purposes.**

- Camshaft bearing caps in reverse order of installation
- Camshafts. The camshaft brackets must be loosened in the correct sequence to prevent damage to the camshaft.

➡ **All the valvetrain components must be reassembled in their original positions.**

- Valve lifter adjusting shims from the tops of the of the lifters. Be sure to note the location and positioning of each shim.
- Valve lifters from the bores in the cylinder heads. Be sure to note the location and positioning of each lifter.

7. Check the diameter of the valve lifter and the valve lifter bore and compare to the following specifications.

8. The valve lifter diameter should be 1.3370–1.3376 in. (33.960–33.975mm).

9. The lifter guide bore diameter should be 1.3386–1.3394 in. (33.960–33.975mm).

10. The valve lifter to lifter guide bore clearance should be 0.0010–0.0024 in. (0.025–0.061mm).

To install:

➡ **When installing the valve components, apply a coat of clean engine oil to the component.**

11. Install or connect the following:
- Lifters into the lifter bores from which they were removed
- Valve shims to the lifters from which they came

12. Install the camshafts in the same position as noted during removal and camshaft bearing caps; torque cap bolts in the proper sequence as follows:
   a. Step 1: 17 inch lbs. (2 Nm).
   b. Step 2: 81–104 inch lbs. (9–12 Nm).

➡ **When installing the timing chain and sprockets, align the marks on the sprockets with the colored links of the chain.**

13. Install or connect the following:
- Camshaft sprockets with the timing

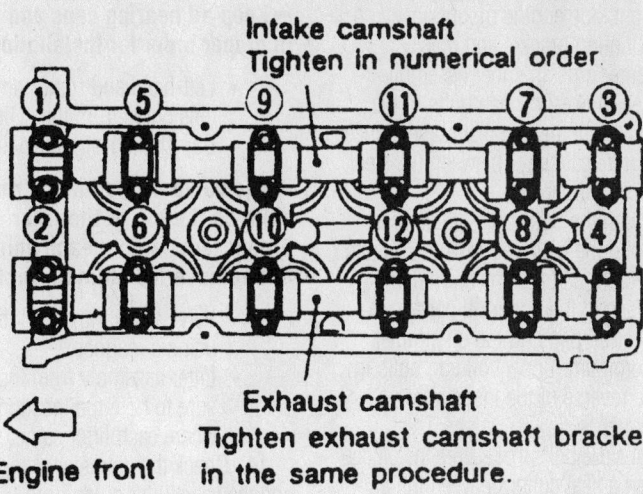

Tighten the camshaft bearing caps in sequence to prevent damage to the camshaft and cylinder head—2.4L engines

chain and torque the bolts to 123–130 ft. lbs. (167–177 Nm). Install the chain guide between both camshaft sprockets. The alignment marks on the upper portion of the timing chain should now be aligned with the marks on the sprockets.
- Rocker cover
- Distributor
- Spark plugs and wires
- Upper radiator hose and cooling fan
- Alternator
- Vacuum hoses, fuel lines and electrical connectors
- Air cleaner assembly
- Negative battery cable

14. Fill the cooling system.

15. Start the vehicle, check for leaks and repair if necessary.

3.0L and 3.5L Engines

1. Before servicing the vehicle, refer to the precautions in the beginning of this section.

2. Relieve the fuel system pressure.

3. Drain the engine oil.

4. Drain the cooling system.

5. Remove or disconnect the following:
- Negative battery cable
- Left side rocker cover ornament

➡ **Before detaching any hoses or connectors, note the locations for reassembly.**

- Air duct to intake manifold hose, collector hose, blow-by hose and vacuum hoses
- Fuel hoses and detach the harness connections
- Canister purge hoses
- Water hoses from the cylinder head and intake manifold
- All 6 ignition coils from the spark plugs
- Spark plugs
- Bolts that secure the Exhaust Gas Recirculation (EGR) tube
- EGR tube
- Intake manifold collector supports and the collector
- Bolts that secure the fuel tube and the fuel tube
- Bolts that secure the intake manifold to the engine block and the manifold. Loosen the bolts in the reverse sequence of the tightening procedure.
- Left-hand and right-hand rocker covers from the cylinder head
- Engine undercovers
- Right front wheel and engine side covers
- Drive belts and idler pulley
- Power steering oil pump belt and the power steering oil pump assembly
- Camshaft Position (CMP) sensor (PHASE) and Crankshaft Position (CKP) sensors (REF)/(POS)

6. Set the No. 1 piston to Top Dead Center (TDC) of compression stroke by rotating the crankshaft.
- Ring gear cover access plate. Loosen the crankshaft pulley bolt while securing the ring gear so the crankshaft cannot rotate
- Crankshaft pulley, using a suitable puller
- Air conditioning compressor and bracket
- Front exhaust pipe and its support

7. Hang the engine at the right and left side engine slingers with a suitable hoist.

8. Support the transaxle with jack.

- Right side engine mounting, mounting bracket and nuts
- Center crossmember assembly
- Steel (lower) oil pan bolts in the reverse of the installation sequence

9. Insert a seal cutter between the steel and aluminum oil pan

10. Tapping the cutter with a hammer, slide it around the entire edge of the oil pan. Be careful not to damage the aluminum mating surface of the upper oil pan.

- Steel oil pan and the oil strainer
- Aluminum (upper) oil pan bolts in the reverse of the installation sequence
- Transaxle bolts that secure the oil pan

11. Insert a seal cutter between the aluminum oil pan and the engine block.

12. Tapping the cutter with a hammer, slide it around the entire edge of the oil pan. Be careful not to damage the mating surfaces of the oil pan or engine block.

- Oil pan from the vehicle
- Water pump cover and the bolts that secure the front timing chain case cover
- Timing chain case cover, using the seal cutter
- Internal timing chain guide and the upper chain guide
- Timing chain tensioner and slack side chain guide
- Left and right intake camshaft sprockets first. Be sure to hold the flats of the camshafts while removing the sprocket bolts.
- Lower timing chain assembly. Be sure to note the aligning marks of the chain before removal.

13. Insert a suitable stopper pin for the left and right camshaft tensioners.

- Left and right exhaust camshaft sprocket bolts. Be sure to hold the flats of the camshafts while removing the sprocket bolts.
- Upper timing chain assembly. Be sure to note the aligning marks of the chain before removal.
- Lower timing chain guide
- Crankshaft sprocket
- Bolts that secure the rear timing chain case. The bolts must be loosened in sequence.
- Rear timing case cover, using the seal cutter

➤Remove the O-rings from the front of the engine block.

- Camshaft bearing caps in several steps. The bearing caps MUST be loosened in sequence.

➤Keep all bearing caps and camshafts in proper order for installation.

- Left-hand and right-hand camshaft tensioners from the cylinder head
- Camshafts from the cylinder heads

➤The valve lifters have a replaceable shim on the top of the lifter. Note the proper locations of each shim to lifter and remove the shims from the lifters.

- Valve adjusting shim from the lifter, using a magnet
- Lifter assembly from the bore. Be sure to note the locations from where each lifter came.

14. Check the diameter of the valve lifter and the valve lifter guide bore.

15. The diameter of the lifter should be 1.3764–1.3770 in. (34.960–34.975mm) and the diameter of the bore should be 1.3780–1.3788 in. (35.000–35.021mm).

16. Remove all traces of liquid gasket from the timing chain case and from the water pump covers.

17. Remove all traces of liquid gasket from the engine block.

18. Inspect the camshafts for excessive wear or damage and replace as necessary.

To install:

➤Before installing the camshaft brackets, apply RTV sealant to the mating surface of the No. 1 journal head.

19. Lubricate the valve lifters with clean engine oil and install the lifters into the bore from which they were removed.

20. Lubricate the valve lifter shims with clean engine oil and install the shims into the lifter from which they were removed.

21. Turn the crankshaft clockwise until the No. 1 piston is set 240 degrees before TDC on compression stroke.

22. Install or connect the following:

- Camshaft tensioners on both sides of the cylinder heads and torque the bolts to 75–96 inch lbs. (8.4–10.8 Nm)

➤The camshafts can be identified by the paint marks on the camshaft. The left cylinder head camshafts have a YELLOW paint mark and the right cylinder head camshafts have a WHITE paint mark.

- Exhaust and intake camshafts and install the bearing caps. Before installing the No. 1 bearing cap, apply liquid gasket to the corners of the cap.

➤When installing the camshafts, position the camshaft keys at the 12 o'clock position in respect to the cylinder head angle.

23. Torque the camshaft bearing caps as follows:
   a. Bolts No. 7–10: 17 inch lbs. (2 Nm).
   b. Bolts No. 1–6: 17 inch lbs. (2 Nm).
   c. Bolts No. 1–10: 52 inch lbs. (6 Nm).
   d. Bolts No. 1–10: 81–104 inch lbs. (9–11 Nm).

24. Install new O-rings to the front of the engine block.

25. Apply sealant to the hatched portion of the of the rear timing chain case.

26. Align the rear timing chain case with the dowel pins and install onto the cylinder heads and engine block.

27. Torque the rear timing chain case mounting bolts in sequence to 105–121 inch lbs. (11.8–13.7 Nm).

28. Install the crankshaft sprocket with the mating mark facing out.

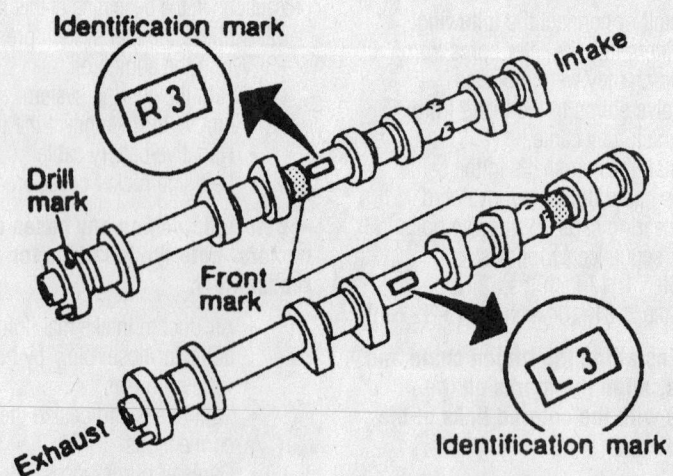

Camshaft identification marks—3.0L and 3.5L engines

79230G33

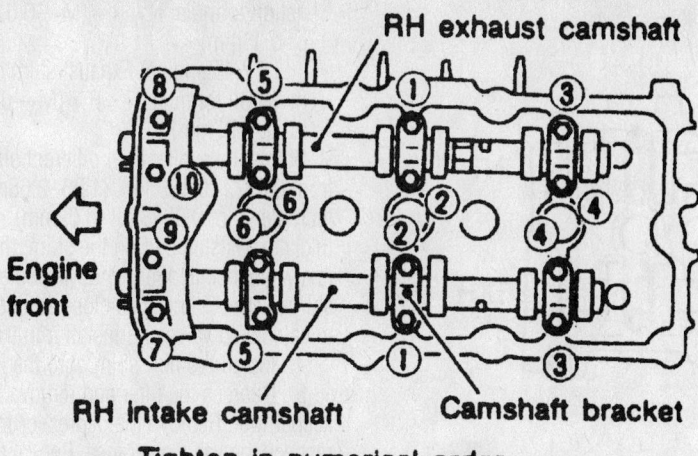

RH exhaust camshaft

RH intake camshaft   Camshaft bracket

Tighten in numerical order.

79230G34

Right cylinder head camshaft bearing cap tightening sequence—3.0L and 3.5L engines

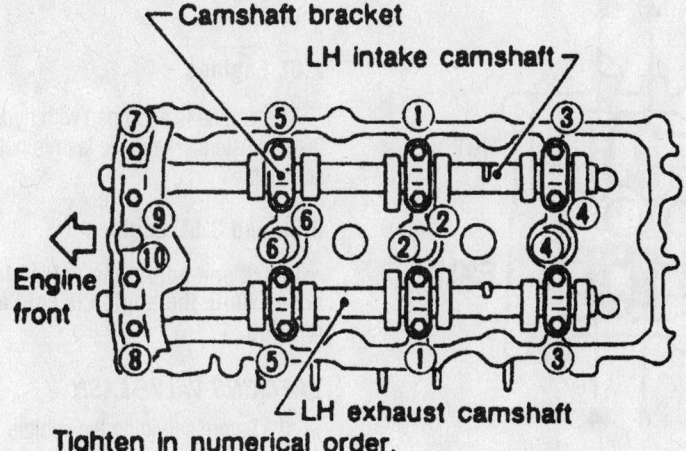

Camshaft bracket

LH intake camshaft

LH exhaust camshaft

Tighten in numerical order.

79230G35

Left cylinder head camshaft bearing cap tightening sequence—3.0L and 3.5L engines

29. Rotate the crankshaft clockwise and position the crankshaft to TDC of compression stroke and align the dowels of the camshaft sprockets to the 12 o'clock position in respect to the cylinder head.

30. Install the lower chain guide on the dowel pin with the front mark on the guide facing upward.

31. On a workbench, align the marks on the intake and exhaust camshaft sprockets with the marks of the chain.

32. Put the exhaust camshaft sprockets onto the dowel pin and torque the bolts to 88–95 ft. lbs. (119–128 Nm). Be sure to secure the camshafts while tightening the bolts.

33. Install or connect the following:
- Timing chains, sprockets and related components
- Transaxle bolts that secure the oil pan
- Oil pan strainer and torque the bolts to 12–14 ft. lbs. (16–19 Nm)

34. Apply a 0.177–0.217 in. (4.5–5.5mm) continuous bead of liquid gasket to the lower oil pan mating surface and install the oil pan. Torque the bolts in sequence to 57–66 inch lbs. (6.4–7.5 Nm).
- Center crossmember assembly
- Right side engine mounting bracket and mount assembly

35. Remove the engine slinger assembly.
- Front exhaust pipe and its support
- Air conditioning compressor and bracket
- Crankshaft pulley to the crankshaft and install the mounting bolt. Torque the bolt to 14–22 ft. lbs. (20–29 Nm). Torque the crankshaft bolt an additional 60–66 degrees clockwise. This is about the angle from one hexagon bolt head corner to another
- CMP sensor, PHASE and CKP sensors
- Power steering pump

- Idler pulley and all belts
- Engine side and under covers
- Right front wheel
- Intake manifold
- Rocker covers
- Fuel tube
- Intake manifold support and collector
- EGR tube
- Spark plugs and ignition coils
- Coolant hoses
- Canister purge hoses
- Fuel feed and return lines
- Vacuum hoses
- Negative battery cable

36. Fill the cooling system.
37. Fill the engine with clean oil.
38. Start the vehicle, check for leaks and repair if necessary.

Valve Lash

ADJUSTMENT

1.8L and 2.4L Engines

CHECKING VALVE LASH

1. Before servicing the vehicle, refer to the precautions in the beginning of this section.

2. Run the engine until it reaches normal operating temperature and shut if off.

3. Remove the cylinder head cover and all the spark plugs.

4. Set the No. 1 cylinder at Top Dead Center (TDC) on its compression stroke. Align the pointer with the TDC mark on the crankshaft pulley. Check that the valve lifters on the No. 1 cylinder are loose and valve lifters on the No. 4 cylinder are tight. If not, turn the crankshaft 1 revolution (360 degrees) and align the pointer with the TDC mark on the crankshaft pulley.

5. Check the following valves:
- Both No. 1 intake valves
- Both No. 1 exhaust valves
- Both No. 2 intake valves
- Both No. 3 exhaust valves

6. Using a feeler gauge, measure the clearance between the valve lifter and the camshaft. Record any valve clearance measurements which are out of specification.

7. Turn the crankshaft 1 revolution (360 degrees) and align the mark on the crankshaft pulley with the pointer. Check the following valves:
- Both No. 2 exhaust valves
- Both No. 3 intake valves
- Both No. 4 intake valves
- Both No. 4 exhaust valves

8. Using a feeler gauge, measure the clearance between the valve lifter and the

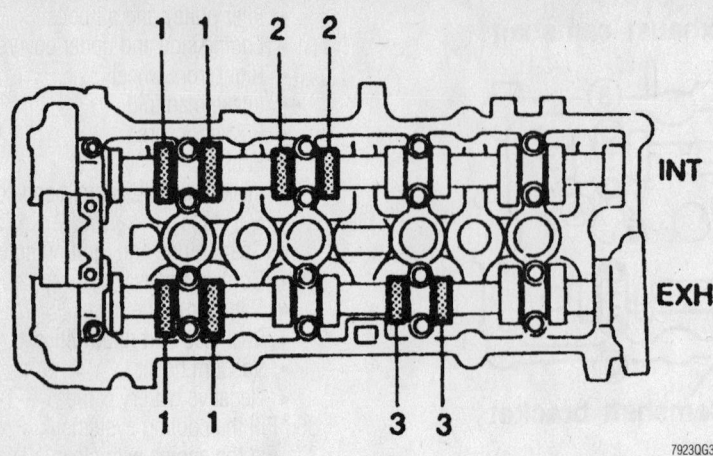

Measure the clearance of the valves indicated when the No. 1 piston is at TDC on compression—1.8L and 2.4L engines

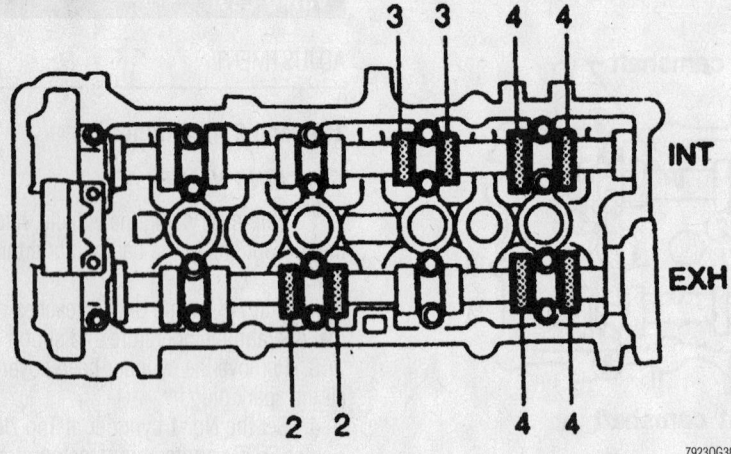

Measure the clearance of the valves indicated when the No. 4 piston is at TDC on compression—1.8L and 2.4L engines

camshaft. Record any valve clearance measurements which are out of specification.

9. If all the valve clearances are within specification, install the cylinder head cover and the spark plugs.

ADJUSTING VALVE LASH

1. Before servicing the vehicle, refer to the precautions in the beginning of this section.

2. If an adjustment is necessary, adjust the valve clearance while engine is cold by removing the adjusting shim. The adjusting shim can be removed by using the following procedures:

   a. Turn the crankshaft so the camshaft lobe of the valve to be adjusted is pointed straight up.

   b. Turn the lifter so the notch is pointed towards the center of the cylinder head; this will facilitate the shim removal process.

   c. Using a depressor tool, push down on the lifter and insert a keeper tool on the edge of the lifter to keep the lifter in the depressed position.

   d. Remove the depressor tool and remove the shim with a magnet.

3. Determine the replacement adjusting shim size by using the following procedures and formula:

   a. Using a micrometer determine thickness of the removed shim.

   b. Calculate the thickness of a new adjusting shim so valve clearance is within the specified values.

   c. R = thickness of the removed shim.

   d. N = thickness of the new shim.

   e. M = measured valve clearance.

   f. 1.8L engine: Intake shim determination formula: $N = R + (M - 0.0146$ in. or 0.37mm)

   g. 1.8L engine: Exhaust shim determination formula: $N = R + (M - 0.0157$ in. or 0.40mm)

   h. 2.4L engine: Intake shim determination formula: $N = R + (M - 0.0138$ in. or 0.35mm)

   i. 2.4L engine: Exhaust shim determination formula: $N = R + (M - 0.0146$ in. or 0.37mm)

Shims are available in different sizes from 0.0772–0.1055 in. (1.96–2.68mm) in increments of 0.0008 in. (0.02mm). The thickness is stamped on the shim; this side is always installed facing down. Select new shims with thickness as close as possible to calculated valve and install it in the lifter.

4. Install the new shim onto the lifter.

5. Depress the lifter and remove the keeper tool. Remove the depressor tool and recheck the valve clearance. Repeat this procedure for any other valves requiring adjustment.

6. Install the cylinder head cover and spark plugs when all valve adjustments are finished.

2.0L Engines

The engine is equipped with hydraulic lash adjusters. The valve lash is not adjustable.

3.0L and 3.5L Engines

➥Check and adjust the valve clearances while the engine is cold and not running.

CHECKING VALVE LASH

1. Before servicing the vehicle, refer to the precautions in the beginning of this section.

2. Remove or disconnect the following:
- Intake manifold collector
- Left and right rocker covers
- Spark plugs

3. Set the No. 1 cylinder at Top Dead Center (TDC) on its compression stroke. Align the pointer with the TDC mark on the crankshaft pulley. Check that the valve lifters on the No. 1 cylinder are loose and valve lifters on the No. 4 cylinder are tight. If not, turn the crankshaft 1 revolution (360 degrees) and align the pointer with the TDC mark on the crankshaft pulley.

4. Check the following valves:
- Both No. 1 intake valves
- Both No. 2 exhaust valves
- Both No. 3 exhaust valves
- Both No. 6 intake valves

5. Using a feeler gauge, measure the clearance between the valve lifter and the camshaft. Record any valve clearance measurements that are out of specification. Intake valve clearance (cold) is 0.010–0.013 in. (0.26–0.34mm) and exhaust valve clear-

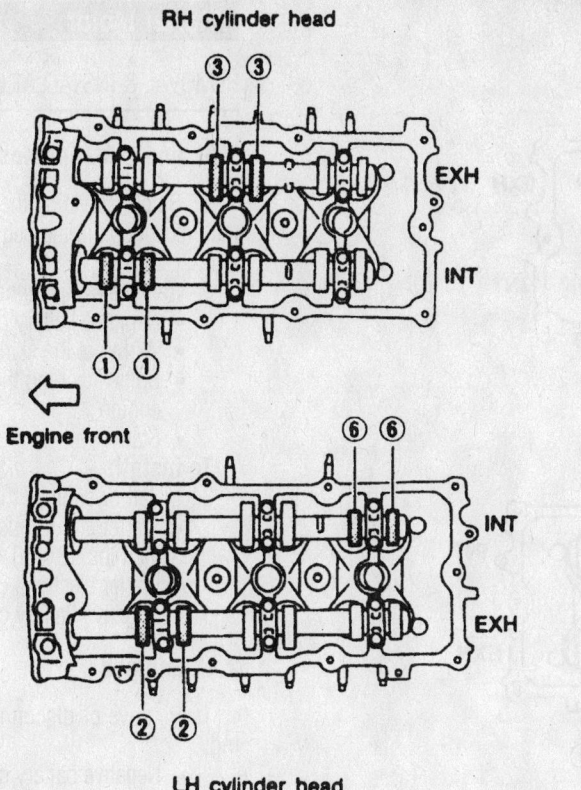

RH cylinder head

Engine front

LH cylinder head

7923QG39

Measure the valves indicated while the No. 1 piston is at TDC on the compression stroke—3.0L and 3.5L engines

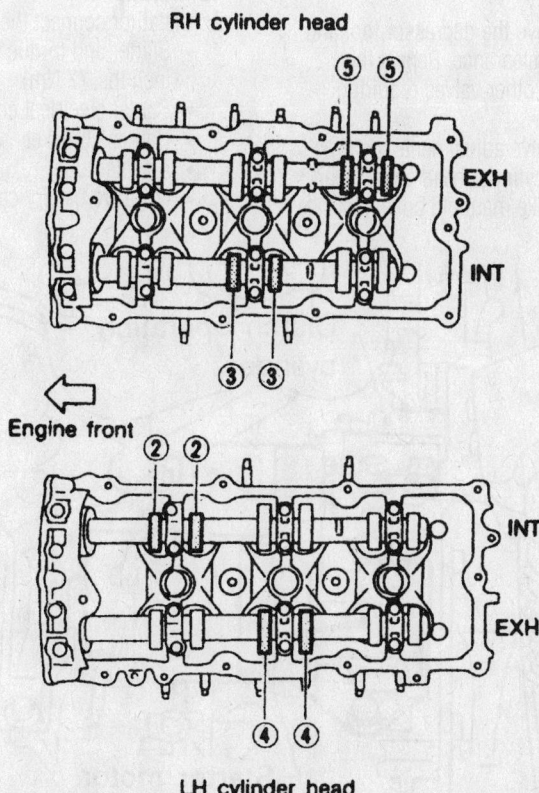

RH cylinder head

Engine front

LH cylinder head

7923QG40

Measure the valves indicated while the No. 3 piston is at TDC on the compression stroke—3.0L and 3.5L engines

ance (cold) is 0.011–0.015 in. (0.29–0.37mm).

6. Turn the crankshaft 240 degrees and set the No. 3 cylinder to TDC of its compression stroke.

7. Check the following valves:
  • Both No. 2 intake valves
  • Both No. 3 intake valves
  • Both No. 4 exhaust valves
  • Both No. 5 exhaust valves

8. Using a feeler gauge, measure the clearance between the valve lifter and the camshaft. Record any valve clearance measurements that are out of specification. Intake valve clearance (cold) is 0.010–0.013 in. (0.26–0.34mm) and exhaust valve clearance (cold) is 0.011–0.015 in. (0.29–0.37mm).

9. Turn the crankshaft 240 degrees and set the No. 5 cylinder to TDC of its compression stroke.

10. Check the following valves:
  • Both No. 1 exhaust valves
  • Both No. 4 intake valves
  • Both No. 5 intake valves
  • Both No. 6 exhaust valves

11. Using a feeler gauge, measure the clearance between the valve lifter and the camshaft. Record any valve clearance measurements that are out of specification. Intake valve clearance (cold) is 0.010–0.013 in. (0.26–0.34mm) and exhaust valve clearance (cold) is 0.011–0.015 in. (0.29–0.37mm).

12. If all the valve clearances are within specification, install the cylinder head cover, spark plugs and the intake manifold collector.

ADJUSTING VALVE LASH

1. Before servicing the vehicle, refer to the precautions in the beginning of this section.

2. If an adjustment is necessary, adjust the valve clearance while engine is cold by removing the adjusting shim. The adjusting shim can be removed by using the following procedures:

  a. Turn the crankshaft so the camshaft lobe of the valve to be adjusted is pointed straight up.

  b. Turn the lifter so the notch is pointed towards the center of the cylinder head; this will facilitate the shim removal process.

  c. Using a depressor tool, push down on the lifter and insert a keeper tool on the edge of the lifter to keep the lifter in the depressed position.

  d. Remove the depressor tool and remove the shim with a magnet.

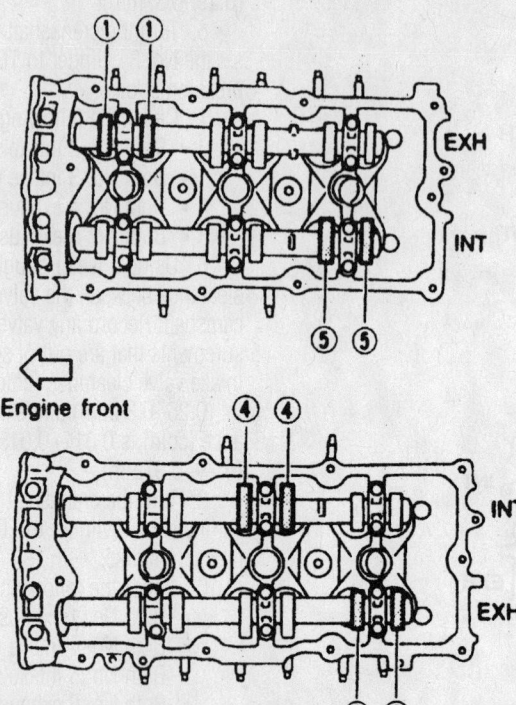

RH cylinder head

EXH

INT

⬅ Engine front

INT

EXH

LH cylinder head

79230G41

Measure the valves indicated while the No. 5 piston is at TDC on compression—3.0L and 3.5L engines

➡**Compressed air can be blown into the hole of the lifter to separate the adjusting shim from the lifter.**

3. Determine the replacement adjusting shim size by using the following procedures and formula:

   a. Using a micrometer determine thickness of the removed shim.

   b. Calculate the thickness of a new adjusting shim so valve clearance is within the specified values.

   c. R = thickness of the removed shim.

   d. N = thickness of the new shim.

   e. M = measured valve clearance.

- Intake shim determination formula: N = R + (M—0.0118 in. or 0.30mm)

- Exhaust shim determination formula: N = R + (M—0.0130 in. or 0.33mm)

4. Shims are available in 64 sizes from 0.0913–0.1161 in. (2.32–2.95mm) in steps of 0.004 in. (0.01mm). The thickness is stamped on the shim; this side is always installed facing down. Select new shims with thickness as close as possible to calculated valve and install it in the lifter.

5. Install the new shim onto the lifter.

6. Depress the lifter and remove the

keeper tool. Remove the depressor tool and recheck the valve clearance. Repeat this procedure for any other valves requiring adjustment.

7. When all valve adjustments are finished, install the cylinder head cover, spark plugs and the intake manifold collector.

REMOVAL & INSTALLATION

1.8L and 2.0L Engines

1. Before servicing the vehicle, refer to the precautions in the beginning of this section.

2. Remove or disconnect the following:
- Negative battery cable
- Wiring at the starter
- Bolts attaching the starter to the engine
- Starter

To install:

3. Install or connect the following:
- Starter and torque the bolts to 70 inch lbs. (8 Nm)
- Starter electrical connections
- Negative battery cable

2.4L Engines

1. Remove or disconnect the following:
- Negative battery cable
- Air inlet tube
- Harness bracket
- Wiring at the starter
- Starter

To install:

2. Install or connect the following:
- Starter and torque the bolts to 60 inch lbs. (7 Nm)
- Starter electrical connectors
- Harness bracket
- Air inlet tube
- Negative battery cable

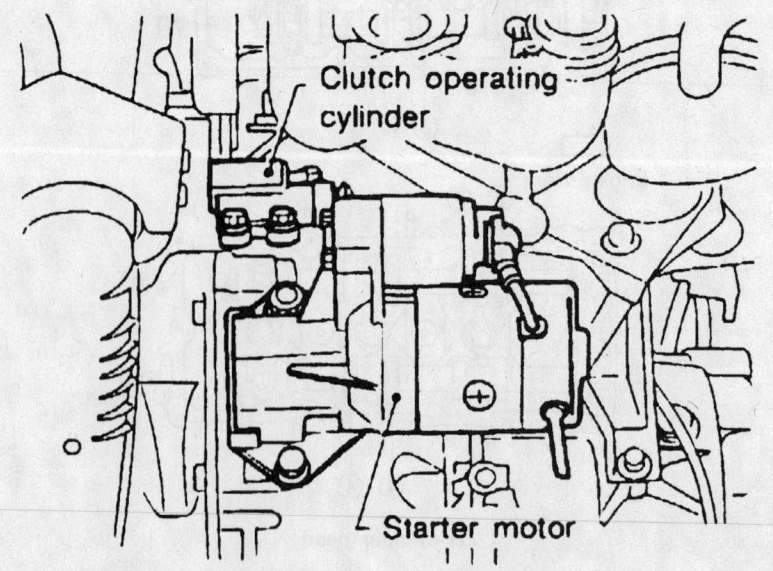

Clutch operating cylinder

Starter motor

89612G12

Starter location and mounting detail—3.0L Maxima with a manual transaxle

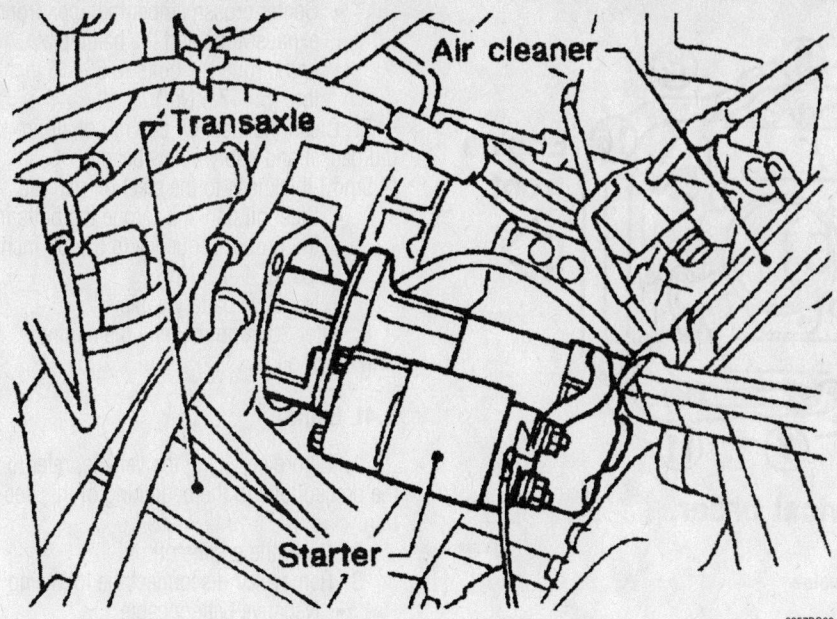

Starter location and mounting detail—3.5L Maxima

9357RG08

3.0L and 3.5L Engines

1. Remove or disconnect the following:
   - Negative battery cable
   - Air duct
   - Harness protector from the harness
   - Starter wiring at the starter
   - Starter-to-engine bolts
   - Starter from the vehicle

   To install:
2. Install or connect the following:
   - Starter and torque the long bolt to 57–72 ft. lbs. (77–98 Nm) and the short bolt to 22–30 ft. lbs. (30–41 Nm)
   - Starter wiring
   - Harness protector
   - Air duct
   - Negative battery cable

Oil Pan

REMOVAL & INSTALLATION

1.8L Engine

1. Before servicing the vehicle, refer to the precautions in the beginning of this section.
2. Drain the engine oil.
3. Remove or disconnect the following:
   - Negative battery cable
   - Engine undercovers
   - Front exhaust tube and properly support the transaxle assembly
   - Center crossmember
   - Support brackets from the sides of the oil pan

   - Rear cover plate, models equipped with a automatic transaxle
   - Oil pan mounting bolts
4. Using an oil pan seal cutter, separate the oil pan from the engine.

> ⁂ **WARNING**
>
> **Do not drive the seal cutter into the oil pump or rear oil seal retainer portion, for the aluminum mating surfaces will be damaged. Do not use a prybar to remove the oil pan; the flange will be deformed.**

5. Clean all the sealing surfaces.
   To install:
6. Apply sealant to the rear oil seal retainer.
7. Apply a 0.128–0.177 in. (3.5–4.5mm) continuous bead of liquid gasket to the oil pan mating surface.
8. Install or connect the following:

- Oil pan and torque bolts, in sequence, to 56–73 inch lbs. (6.3–8.3 Nm)
- Rear cover plate, models equipped with a automatic transaxle
- Oil pan support brackets
- Center crossmember
- Front exhaust tube
- Engine undercovers
- Oil pan plug, using a new gasket and tighten the plug to 21–28 ft. lbs. (7–8 Nm)
- Negative battery cable

9. After 30 minutes of gasket curing time, refill the oil pan with the specified quantity of clean oil.
10. Start the vehicle, check for leaks and repair if necessary.

2.0L Engine

1. Before servicing the vehicle, refer to the precautions in the beginning of this section.
2. Drain the engine oil.
3. Remove or disconnect the following:
   - Negative battery cable
   - Engine undercover
   - Lower steel oil pan bolts in the reverse of installation sequence
   - Steel oil pan. Insert a cutting tool between steel oil pan and aluminum oil pan. Tap the tool around the perimeter of the pan to cut the gasket material.
   - Oil baffle bolts and oil baffle
   - Front exhaust tube and set a suitable jack under the transaxle and raise the engine
   - Center crossmember from the vehicle
   - Transaxle shift control cable, if equipped with an automatic transaxle
   - Compressor gussets and the rear cover plate
   - Aluminum oil pan bolts. Loosen

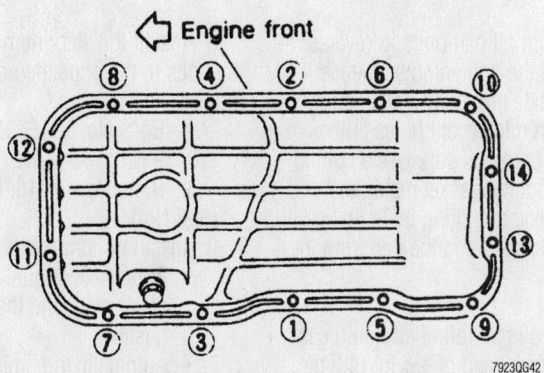

7923QG42

Tighten the oil pan bolts in the correct sequence to prevent oil leakage—1.8L engines

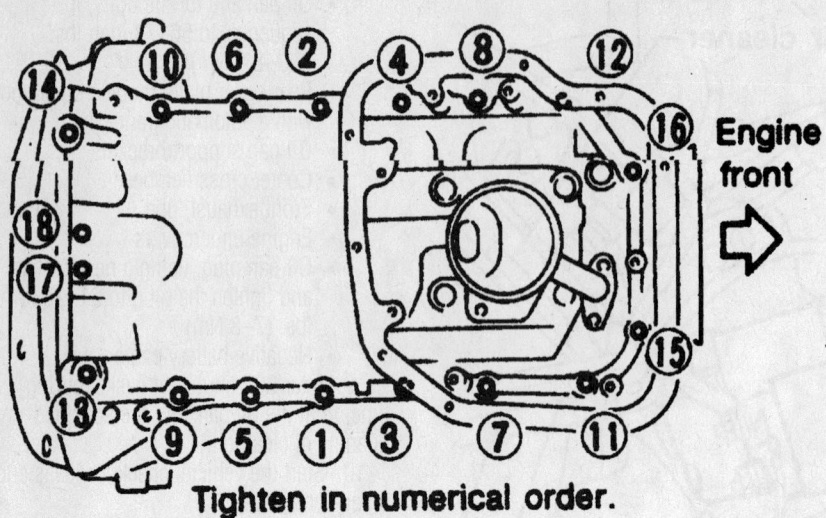

Tighten in numerical order.

7923QG43

Aluminum oil pan bolt tightening sequence—2.0L engine

Engine front

Tighten in numerical order.

7923QG44

Steel oil pan bolt tightening sequence—2.0L engine

aluminum oil pan bolts in reverse order of the tightening sequence.
• 2 transaxle mounting bolts and refit the them into vacant holes at the bottom of the oil pan. Use a cutting tool to cut the gasket material.
• 2 transaxle mounting bolts that were relocated and the pan from the vehicle

To install:

4. Clean the oil pan rail of all liquid gasket and apply a new bead of 5/32 in. (4.5mm) thickness to the aluminum oil pan rail.

5. Install the aluminum oil pan. Torque the bolts in the opposite order of removal as follows:
   a. Bolts No. 1–16: 12–14 ft. lbs. (16–19 Nm).
   b. Bolts No. 17–18: 60–72 inch lbs. (6–8 Nm).

6. Install or connect the following:
• 2 transaxle mounting bolts, rear cover plate and the compressor gussets
• Automatic transmission shift control cable, if equipped

• Center crossmember member, front exhaust tube and the baffle plate and torque the bolts to 56–66 inch lbs. (6.4–7.5 Nm)

7. Clean the steel oil pan rail of all liquid gasket and apply a new bead of 5/32 in. (4.5mm) thickness to the steel oil pan rail.
• Steel oil pan and torque the bolts in the proper sequence to 56–66 inch lbs. (6.4–7.5 Nm)
• Negative battery cable

8. After 30 minutes, refill the engine with clean oil

2.4L Engine

1. Before servicing the vehicle, refer to the precautions in the beginning of this section.

2. Drain the engine oil.

3. Remove or disconnect the following:
• Negative battery cable
• Engine undercover
• Bolts securing the steel oil pan to the aluminum oil pan in reverse order of the tightening sequence

4. Install a seal cutter between the steel oil pan and the aluminum oil pan

5. Tapping the cutter with a hammer, slide it around the entire edge of the oil pan. Take care not to damage the aluminum oil pan.

6. Remove or disconnect the following:
• Steel oil pan
• Baffle plate and oil strainer
• Front exhaust tube
• Front suspension member
• A/C compressor gussets
• Rear cover plate
• Aluminum oil pan retaining bolts in reverse order of the tightening sequence

7. Insert a seal cutter between the oil pan and the cylinder block.

8. Tapping the cutter with a hammer, slide it around the entire edge of the oil pan. Take care not to damage the aluminum oil pan.

9. Lower the oil pan from the cylinder block and remove it from the engine.

To install:

10. Carefully scrape the old gasket material away from the pan and cylinder block mounting surfaces, then apply a continuous 3.5–4.5mm wide bead of liquid gasket around the oil pan. Install the pan within 5 minutes or else this step will have to be repeated.

11. Install or connect the following:
• Aluminum oil pan and torque the bolts, in sequence, to 13 ft. lbs. (17.5 Nm)

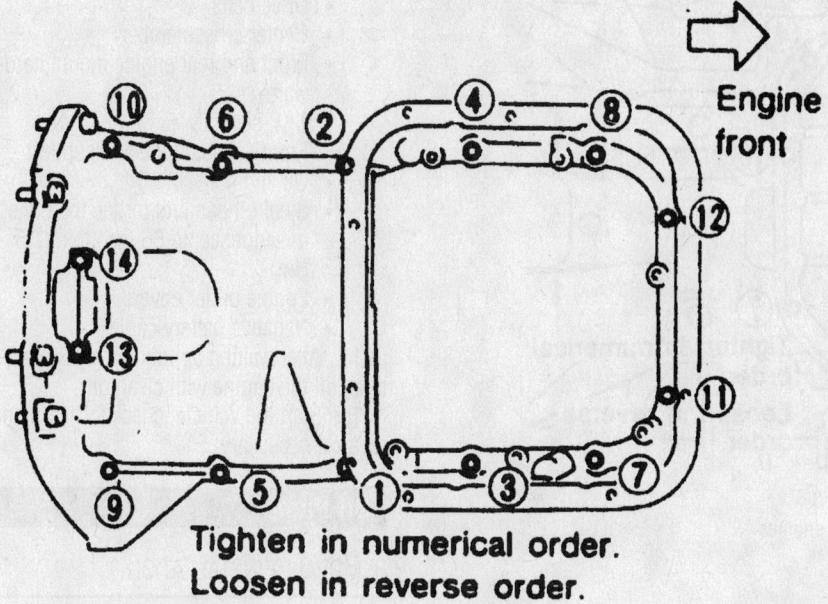

Tighten in numerical order.
Loosen in reverse order.

79230G45

Oil pan bolt loosening and tightening sequence—2.4L engine

- Baffle plate
- Steel oil pan. Torque the bolts in sequence to 61 inch lbs. (7 Nm).
- Rear cover plate
- Front suspension member
- Front exhaust tube
- A/C compressor gussets
- Front suspension member
- Engine undercovers
- Negative battery cable

12. Wait 30 minutes before refilling the crankcase with clean oil.

13. Fill the engine with clean oil.

14. Start the vehicle, checks for leaks and repair if necessary.

3.0L and 3.5L Engines

1. Before servicing the vehicle, refer to the precautions in the beginning of this section.

2. Drain the engine oil.

3. Remove or disconnect the following:
- Negative battery cable
- Engine undercovers
- Steel (lower) oil pan bolts in the reverse of the installation sequence

4. Insert a seal cutter between the steel and aluminum oil pan.

5. Tapping the cutter with a hammer, slide it around the entire edge of the oil pan. Be careful not to damage the aluminum mating surface of the upper oil pan.
- Steel oil pan and the oil strainer
- Front exhaust pipe and its support

6. Hang the engine at the right and left side engine slingers with a suitable hoist.

7. Position a suitable jack under the transaxle.
- Crankshaft Position (CKP) sensors (REFERENCE and POSITION) from the oil pan
- Front and rear engine mounting nuts and bolts
- Center crossmember assembly
- Engine drive belts
- A/C compressor and mounting bracket

- Rear cover plate and the lower transaxle bolts
- Aluminum (upper) oil pan bolts in the reverse of the installation sequence

8. Insert a seal cutter between the aluminum oil pan and the engine block.

9. Tapping the cutter with a hammer, slide it around the entire edge of the oil pan. Be careful not to damage the mating surfaces of the oil pan or engine block.

10. Remove or disconnect the following:
- Oil pan assembly
- Bolts that secure the baffle plate and the baffle plate
- O-rings from the cylinder block and oil pump body

To install:

11. Install or connect the following:
- Baffle plate to the oil pan and torque the bolts to 22–27 inch lbs. (2.5–3.1 Nm) and apply sealant to the front and rear seal of the oil pan
- New O-rings to the cylinder block and the oil pump body

12. Apply a 4.5–5.5mm wide continuous bead of liquid gasket to the upper oil pan mating surface and install the oil pan. Torque the bolts in sequence to 12–14 ft. lbs. (16–19 Nm).

13. Install or connect the following:
- Oil pan strainer and torque the bolts to 12–14 ft. lbs. (16–19 Nm)
- Rear cover plate and lower transaxle bolts

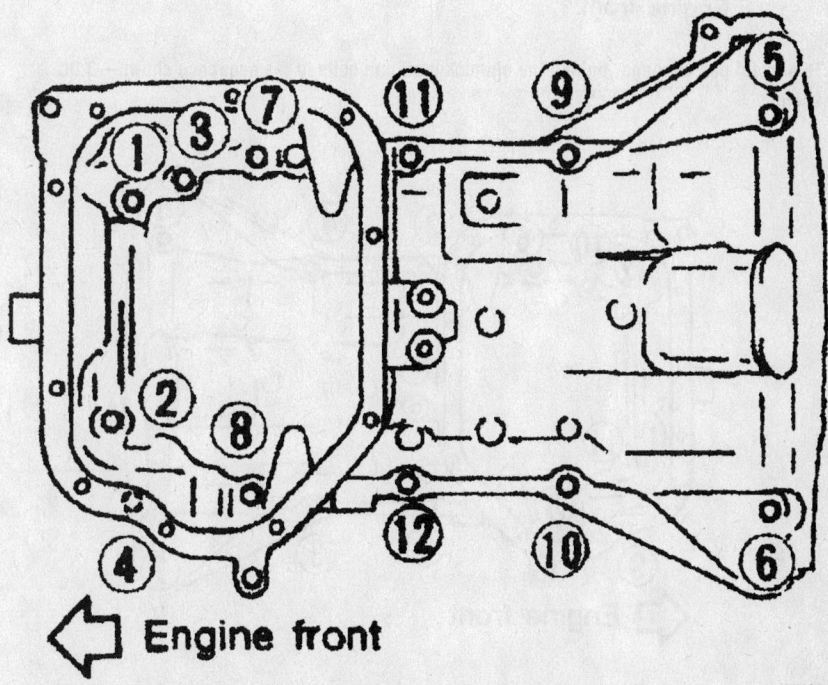

Engine front

9357RG09

Aluminum (upper) oil pan bolt loosening sequence—3.0L and 3.5L engine

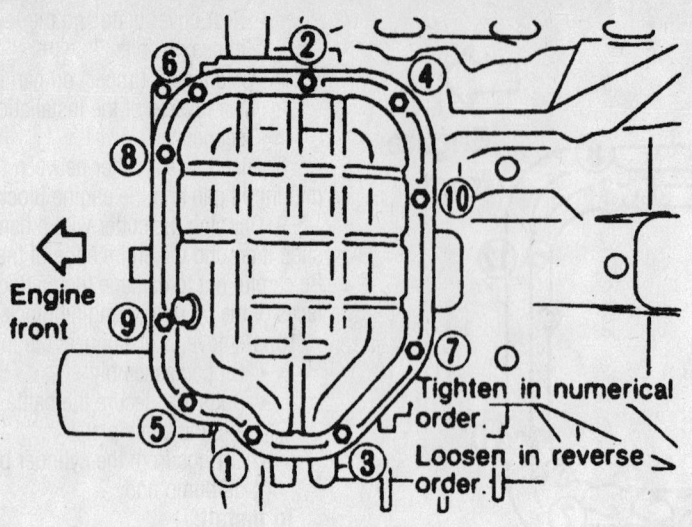

Bolt tightening sequence for the steel oil pan—3.0L engines

7923QG46

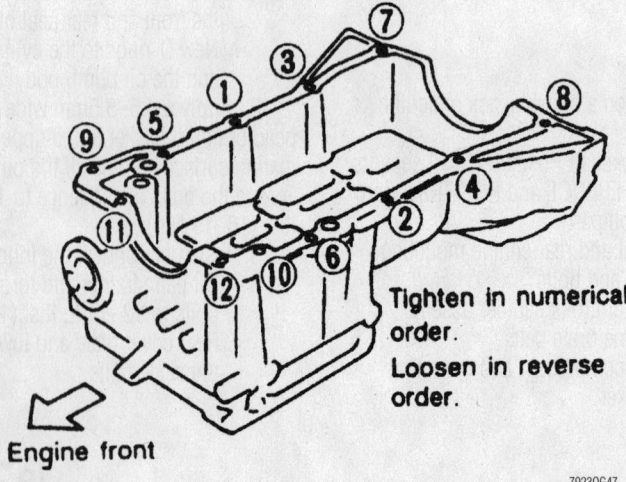

To prevent pan warpage, tighten the aluminum oil pan bolts in the sequence shown—3.0L engines

7923QG47

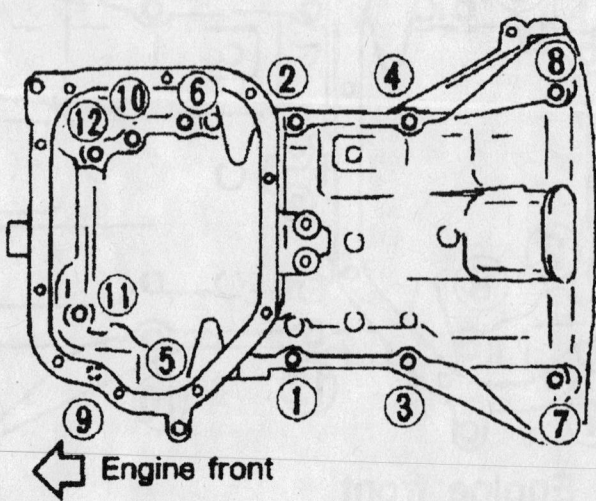

To prevent pan warpage, tighten the aluminum oil pan bolts in the sequence shown—3.5L engines

9357RG10

- A/C compressor and bracket
- Drive belts
- Center crossmember
- Front and rear engine mount hardware
- CKP sensors
- Front exhaust tube and support
- Oil strainer
- Steel oil pan and torque the bolts, in sequence, to 66 inch lbs. (7.5 Nm)
- Engine under covers
- Negative battery cable

14. After waiting approximately 30 minutes, fill the engine with clean oil.

15. Start the vehicle, check for leaks and repair if necessary.

Oil Pump

REMOVAL & INSTALLATION

1.8L and 2.0L Engines

1. Before servicing the vehicle, refer to the precautions in the beginning of this section.

2. Drain the engine oil.

3. Remove or disconnect the following:
- Negative battery cable
- Drive belts
- Cylinder head
- Oil pan and strainer
- Engine front cover
- Oil pump from the front cover

To install:

4. Install or connect the following:
- Oil pump cover and torque the long bolt to 70 inch lbs. (8 Nm) and the short bolt to 44 inch lbs. (5 Nm)
- Front cover and torque the bolts to 43 ft. lbs. (58 Nm)
- Oil strainer and oil pan
- Cylinder head
- Drive belts
- Negative battery cable

5. Fill the engine with clean oil.

6. Start the vehicle, check for leaks and repair if necessary.

2.4L Engine

1. Before servicing the vehicle, refer to the precautions in the beginning of this section.

2. Drain the engine oil.

3. Remove or disconnect the following:
- Negative battery cable
- Engine front cover
- Oil pump cover and pump

To install:

4. Install or connect the following:

- Oil pump
- Oil pump cover and torque the long bolt to 15 ft. lbs. (20 Nm) and shorter bolt to 69 inch lbs. (8 Nm)
- Front cover
- Negative battery cable

5. Fill the engine with clean oil.
6. Start the vehicle, check for leaks and repair if necessary

3.0L and 3.5L Engines

1. Before servicing the vehicle, refer to the precautions in the beginning of this section.
2. Drain the engine oil
3. Remove or disconnect the following:
   - Negative battery cable
   - Drive belts
   - Camshaft Position (CMP) sensor (PHASE) and the Crankshaft Position (CKP) sensor (REF)/(POS)
   - Engine lower covers
   - Crankshaft pulley
   - Front exhaust tube and support
   - Right side mounting insulator and bracket
   - Center member
   - A/C compressor and move it aside
   - Oil pans
   - Water pump cover
   - Front cover
   - Timing chain
   - Oil pump assembly
4. Clean all mating surfaces.
To install:
5. Install or connect the following:
   - Oil pump

- Timing chain
- Front cover and torque the long bolt to 95 inch lbs. (11 Nm) and the short bolt to 71 inch lbs. (8 Nm)
- Water pump cover
- Oil pans
- A/C compressor
- Center member
- Right side mounting insulator and bracket
- Front exhaust tube and support
- Crankshaft pulley
- CMP and CKP sensors
- Engine lower covers
- Drive belts
- Negative battery cable

6. Fill the engine with clean oil.
7. Start the vehicle, check for leaks and repair if necessary.

Rear Main Seal

REMOVAL & INSTALLATION

1.8L and 2.0L Engines

1. Before servicing the vehicle, refer to the precautions in the beginning of this section.
2. Remove or disconnect the following:
   - Transaxle
   - Driveplate/ flywheel
   - Oil seal retainer
3. Carefully pry the seal from the retainer. Be sure not to scratch the sealing surface of the crankshaft or oil seal bore.

To install:
4. Apply clean engine oil to the new seal. Position the seal on the rear of the engine in the proper direction.
5. Using a suitable seal driver, tap the seal into position in the seal retainer.
6. Install or connect the following:
   - Flywheel/flexplate
   - Transaxle assembly

2.4L Engine

1. Before servicing the vehicle, refer to the precautions in the beginning of this section.
2. Remove or disconnect the following:
   - Transaxle
   - Driveplate/flywheel
   - Rear oil seal retainer with the oil seal
3. Tap the oil seal out of the retainer with a hammer and drift.

To install:
4. Apply clean engine oil to the new seal.
5. Install the new seal in the retainer with a suitable seal driver.
6. Apply a continuous bead of RTV silicone sealant, 2–3mm wide, to the seal retainer. Be sure to apply around the inner side of the bolt holes.
7. Install or connect the following:
   - Tap the seal into position in the seal retainer, using a suitable seal driver
   - Oil seal retainer and torque the bolts to 56–66 inch lbs. (6.5–7.5 Nm)

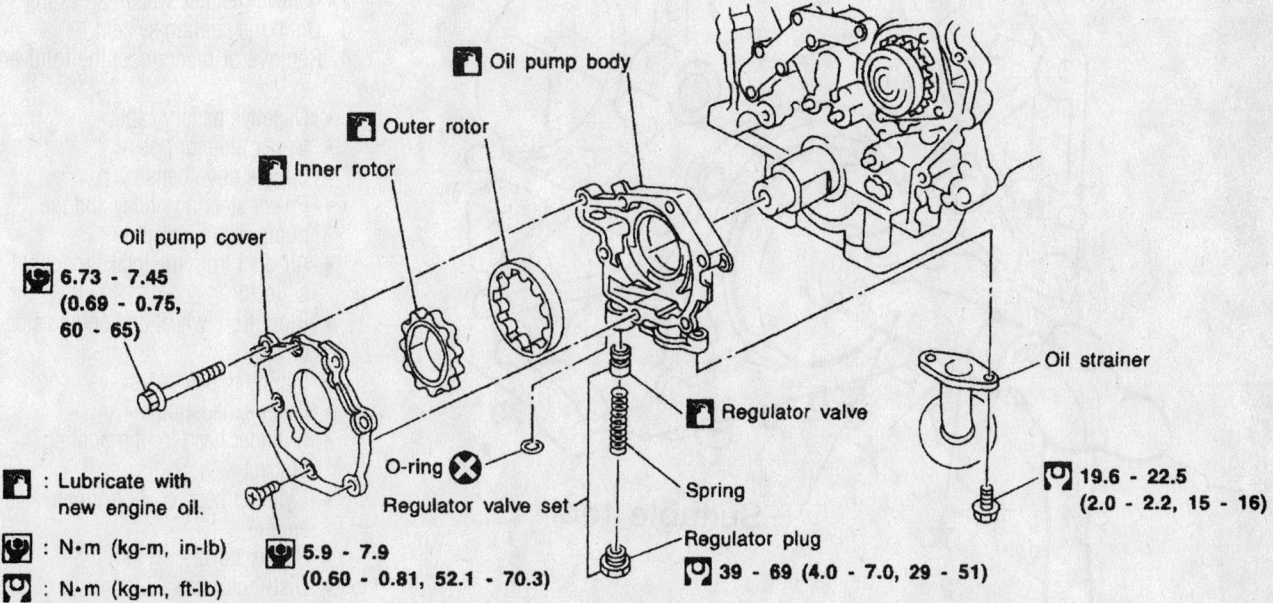

Exploded view of the oil pump assembly—3.0L and 3.5L engines

9357RG11

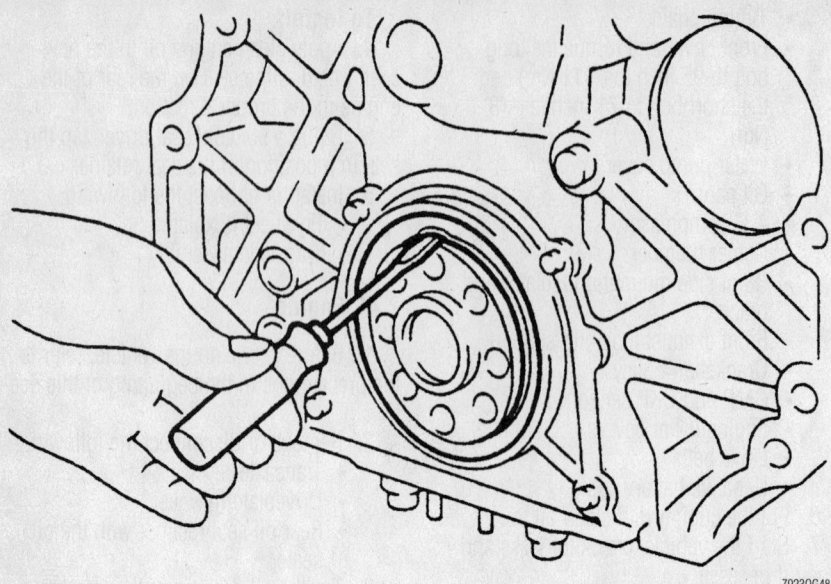

7923QG48

Carefully pry the rear main seal out of the retainer on the rear of the engine—1.8L and 2.0L

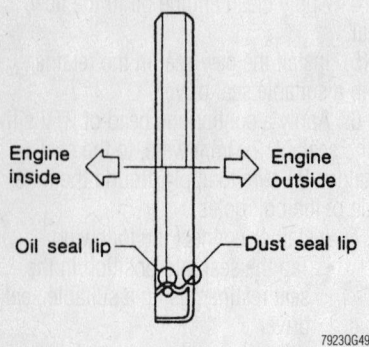

Engine inside ⇦   ⇨ Engine outside

Oil seal lip ——   —— Dust seal lip

7923QG49

Be sure to install the seal in the correct orientation—1.8L and 2.0L engines

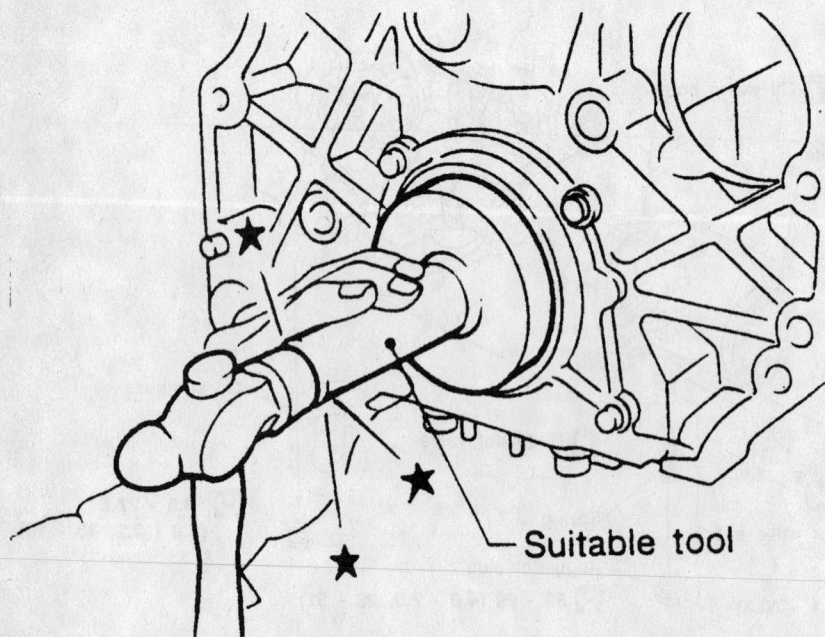

★ Suitable tool

7923QG50

Install the rear main seal using a suitable driver—1.8L and 2.0L engines

• Driveplate/flywheel
• Transaxle

3.0L and 3.5L Engines

1. Before servicing the vehicle, refer to the precautions in the beginning of this section.
2. Drain the engine oil.
3. Remove or disconnect the following:

• Transaxle
• Driveplate/flywheel
• Oil pan
• Oil seal retainer

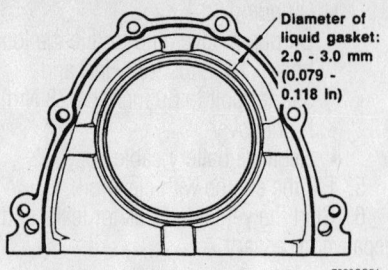

Diameter of liquid gasket: 2.0 - 3.0 mm (0.079 - 0.118 In)

7923QG51

Apply sealant to the seal retainer as shown—2.4L engine

4. Tap the oil seal out of the retainer with a hammer and drift.
5. Clean all mating surfaces of any residual liquid gasket.

To install:

6. Install or connect the following:

• New seal into the retainer
• Oil seal retainer
• Oil pan
• Driveplate/flywheel
• Transaxle

7. Fill the engine with clean oil.
8. Start the vehicle, check for leaks and repair if necessary.

Timing Chain, Sprockets, Front Cover and Seal

REMOVAL & INSTALLATION

1.8L Engine

1. Before servicing the vehicle, refer to the precautions in the beginning of this section.
2. Relieve the fuel system pressure.
3. Drain the cooling system.
4. Remove or disconnect the following:

• Negative battery cable
• Upper radiator hose
• Engine drive belts
• Power steering pulley and the pump with bracket
• Air duct from the intake manifold collector
• Right front wheel and engine side covers
• Engine undercovers
• Front exhaust pipe
• Cylinder head front mounting bracket
• Cylinder head cover from the engine
• Rocker cover
• Distributor cap
• Spark plugs
• Intake manifold support and set the

No. 1 piston at the Top Dead Center (TDC) compression stroke
- Distributor
- Cylinder head front cover
- Water pump pulley
- Thermostat housing
- Lower timing chain tensioner
- Upper timing chain tensioner and slack side timing chain guide
- Idler sprocket bolt
- Camshaft sprocket bolts and the sprockets from the camshafts. Be sure to mark the sprockets for proper reinstallation.
- Camshaft mounting caps by loosening the bolts in 2 or 3 steps
- Camshafts from the engine
- Idler sprocket bolt
- Cylinder head with the manifolds
- Idler sprocket shaft from the rear side
- Upper timing chain and support the engine assembly
- Center crossmember
- Oil pan and strainer assembly
- Crankshaft pulley
- Engine front mount and bracket
- Bolts that secure the front timing cover and the cover from the engine. Once the timing chain cover is removed, drive out the old oil seal.
- Idler sprocket and the lower timing chain
- Oil pump drive spacer and the crankshaft sprocket
- Timing chain guide

To install:

5. Drive a new oil seal into the front cover. Lubricate the oil seal lip with clean engine oil.

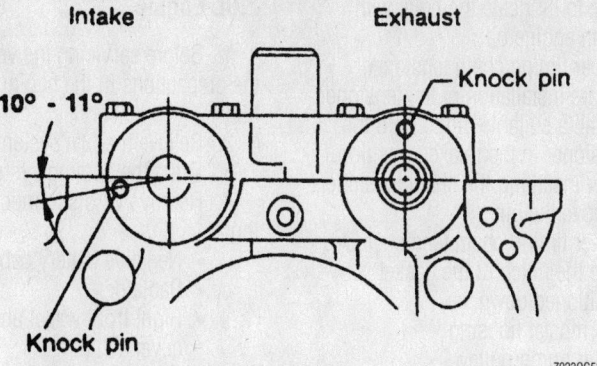

Positioning of camshaft knock pins during assembly—1.8L engine

6. Confirm that No. 1 piston is set at Top Dead Center (TDC) on compression stroke.

7. Install or connect the following:
- Crankshaft sprocket with the marks of the sprocket facing the front of the engine
- Oil pump drive spacer and the chain guide
- Lower timing chain. Set the chain by aligning its mating mark with the one on the crankshaft sprocket. Be sure the sprocket's mating mark faces the front of the engine.

➡**The number of links between the alignment marks are the same for the left and the right side.**

- Crankshaft sprocket and the lower timing chain. Set the timing chain by aligning its mating mark with the one on the crankshaft sprocket. Be sure sprocket's mating mark faces engine front.

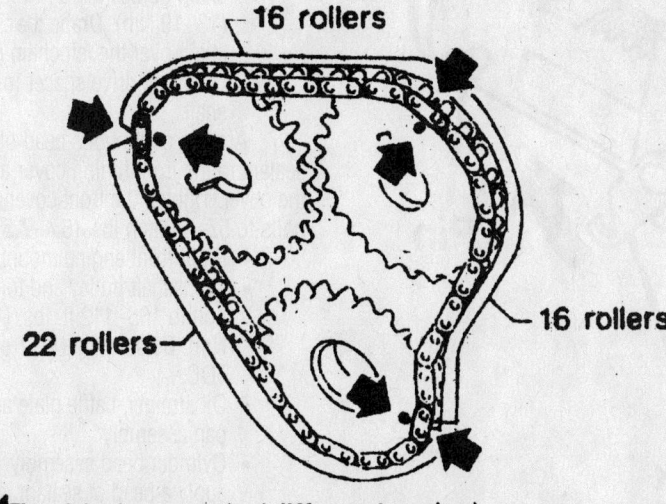

◄ : Mating mark (different color)

Be sure to align the camshaft sprockets with the timing chain—1.8L engine

- Front cover assembly, using liquid gasket
- Engine front mounting bracket and the engine mount
- Oil strainer, oil pan assembly and the crankshaft pulley
- Center crossmember

8. Set the idler sprocket by aligning the mating mark on the larger sprocket with the silver mating mark on the lower timing chain.
- Upper timing chain and set it by aligning the mating mark on the smaller sprocket with the silver mating marks on the upper timing chain. Be sure sprocket marks face engine front.
- Idler sprocket shaft to the rear side
- Cylinder head assembly
- Idler sprocket bolt. Be sure to lubricate the bolt with clean engine oil.
- Exhaust and intake camshafts. The camshafts and marked **I** for intake and **E** for exhaust.

9. Position the intake camshaft knock pin at the 9 o'clock position and the exhaust camshaft knock pin at the 12 o'clock position.

10. Install or connect the following:
- Camshaft bearing caps and distributor bracket. Apply liquid sealant to the distributor bracket.

11. Torque the mounting bolts in sequence as follows:
   a. Bolts 11–15, then bolts 1–10: 18 inch lbs. (2.0 Nm).
   b. Bolts 1–15: 52 inch lbs. (6 Nm).
   c. Bolts 1–14: 104 inch lbs. (12 Nm).
   d. Bolt 15: 73 inch lbs. (8 Nm).

12. Install or connect the following:
- Camshaft sprockets with timing chain. Set the camshaft sprockets by aligning the mating marks of the timing chain with the marks on the camshaft sprockets.
- Camshaft sprocket bolts. Torque the bolts to 86 ft. lbs. (117 Nm). Be

sure to lubricate the bolts with clean engine oil.

- Upper timing chain tensioner. Before installation of the tensioner, install a suitable pin to hold the tensioner in the relaxed position. After installing the chain tensioner, remove the pin.
- Lower timing chain tensioner. Be sure the notch of the gasket is positioned down.
- Thermostat housing
- Water pump pulley
- Cylinder head front cover
- Distributor
- Intake manifold support
- Spark plugs and leads
- Distributor cap
- Rocker cover
- Cylinder head cover
- Front exhaust pipe
- Engine under covers
- Engine side covers and the right front wheel
- Air duct to the intake manifold collector
- Power steering pulley and oil pump
- Drive belts
- Upper radiator hose
- Negative battery cable

13. Fill the cooling system.
14. Start the vehicle, check for leaks and repair if necessary.

2.0L Engine

1. Before servicing the vehicle, refer to the precautions in the beginning of this section.
2. Relieve the fuel system pressure.
3. Drain the cooling system.
4. Remove or disconnect the following:

- Negative battery cable
- Radiator
- Right front wheel and engine side cover
- Spark plugs and rotate the engine and position the No. 1 cylinder to Top Dead Center (TDC)
- Air duct to the intake manifold
- Drive belts and the water pump pulley
- Alternator and the power steering pump from the engine
- Vacuum hoses, fuel hoses and the wiring harness connectors
- Cylinder head cover
- Intake manifold supports
- Oil filter bracket and the power steering pump bracket
- Timing chain tensioner
- Distributor
- Timing chain guide and holding the flats of the camshaft sprockets, remove the bolts that secure the sprockets

- Timing chain sprockets from the camshafts
- Oil tubes, baffle plate and camshaft brackets
- Camshafts from the cylinder head
- Starter motor
- Coolant hoses from the engine block
- Knock sensor (KS) harness connector
- Exhaust Gas Recirculation (EGR) tube
- Cylinder head
- Oil pan, oil strainer and the baffle plate
- Crankshaft pulley using a suitable puller
- Engine front mount
- Front cover and oil pump drive spacer
- Timing chain guides and timing chain. Check the timing chain for excessive wear at the roller links.

To install:

5. Clean all gasket mating surfaces.
6. Install or connect the following:

- Crankshaft sprocket. Position the crankshaft so that No. 1 piston is set at TDC (keyway at 12 o'clock, mating mark at 4 o'clock) fit timing chain to crankshaft sprocket so the mating mark is in line with mating mark on crankshaft sprocket. The mating marks on timing chain for the camshaft sprockets should be silver. The mating mark on the timing chain for the crankshaft sprocket should be gold.
- Timing chain to the crankshaft sprocket and install the timing chain guides. Tighten the timing chain guides to 10–14 ft. lbs. (13–19 Nm). Drape the timing chain over the left chain guide.
- Oil pump drive spacer to the crankshaft

7. Apply a continuous bead of liquid sealant to the front timing cover and install the cover. Tighten the front cover mounting bolts to 57–66 inch lbs. (6.4–7.5 Nm).

- Right front engine mount
- Crankshaft pulley and torque the bolt to 105–112 ft. lbs. (142–152 Nm). Be sure the No. 1 piston is at TDC.
- Oil strainer, baffle plate and the oil pan assembly
- Cylinder head assembly. Be sure to apply a bead of sealant to the joint of the block and front timing cover.
- EGR tube
- KS harness connector

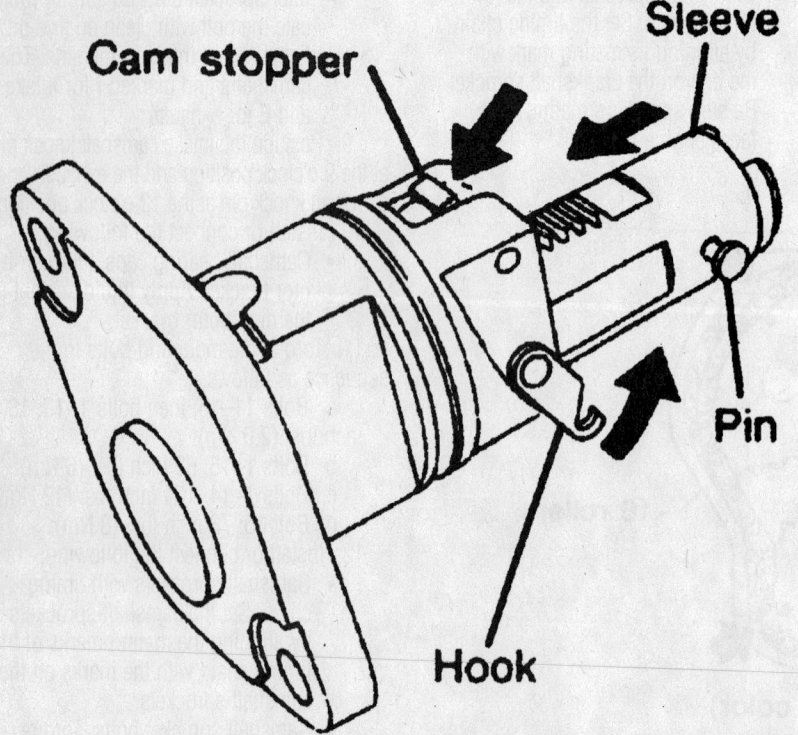

Cam stopper

Sleeve

Pin

Hook

79230G54

Timing chain tensioner—2.0L engines

- Coolant hoses to the engine block, using new hose clamps
- Starter motor
- Camshafts, camshaft bearing caps, oil tubes and the baffle plate

➡**When installing the camshafts, be sure to position the left-hand and right-hand camshaft keys at 12 o'clock. Also be sure the camshaft brackets are facing in the correct direction.**

- Camshaft sprockets by lining up the mating marks on the timing chain with the mating marks on the camshaft sprockets and torque the bolts to 101–116 ft. lbs. (137–157 Nm)

- Timing chain guide and distributor
- Chain tensioner. Press the cam stopper down and the press-in sleeve until the hook can be engaged on the pin. When tensioner is bolted in position the hook will release automatically.

➡**Ensure the arrow on the outside of the tensioner faces the front of the engine.**

- Oil filter bracket and power steering pump bracket
- Intake manifold supports
- Cylinder head cover
- Vacuum hoses and fuel lines
- Alternator and power steering pump

- Water pump pulley and drive belts
- Air duct
- Spark plugs after making certain that the No. 1 piston is at the TDC position
- Engine side cover and right front wheel
- Radiator
- Negative battery cable

8. Fill the cooling system.
9. Start the engine, check for leaks and repair if necessary.

2.4L Engine

1. Before servicing the vehicle, refer to the precautions in the beginning of this section.
2. Drain the cooling system.
3. Drain the engine oil.
4. Remove or disconnect the following:

- Negative battery cable
- Spark plug wires and set the No. 1 piston at the Top Dead Center (TDC) of the compression stroke
- Engine undercover
- Vacuum hoses, fuel hoses, wires, harness and connectors
- Drive belts
- Power steering reservoir
- Alternator and bracket, the upper radiator hose, the air duct and the front exhaust tube
- Intake manifold collector supports, intake manifold collector and the exhaust manifold
- Distributor

5. Using a block of wood, set a transmission jack under the aluminum oil pan and remove the front engine mounting.

- Rocker cover. Remove the rocker cover bolts in the proper sequence.
- Camshaft sprockets

➡**The stoppers on camshaft covers prevent the upper timing chain from disengaging from the idle sprocket.**

- Cam bearing caps in sequence
- Camshafts. The camshaft brackets must be loosened in reverse order of tightening to prevent damage to the camshaft.

➡**These parts must be reassembled in their original positions.**

- Cylinder head bolts in the reverse order of installation

➡**A warped or cracked cylinder head could result from loosening in incorrect order. The cylinder head bolts should be loosened in 2 or 3 steps.**

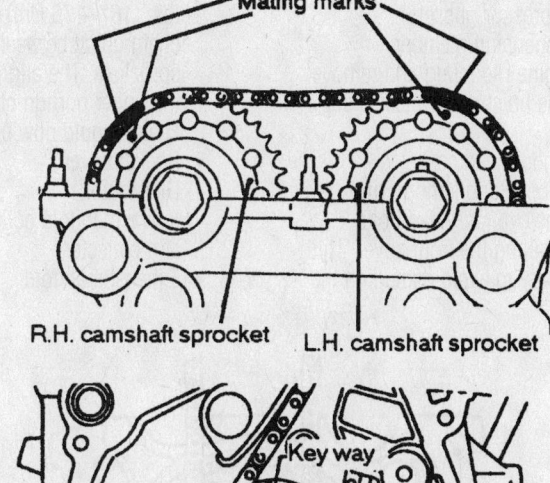

R.H. camshaft sprocket   L.H. camshaft sprocket

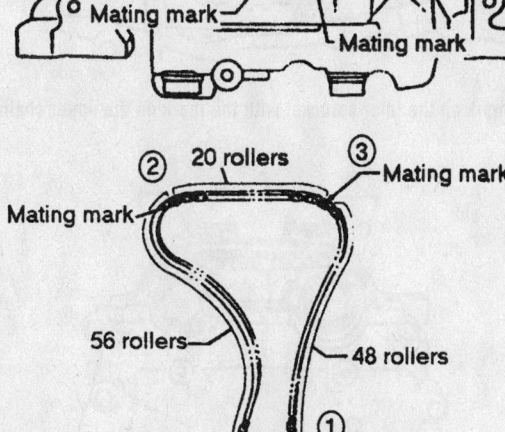

* Mating mark color on timing chain
①:Gold   ②③ :Silver

79230G55

Timing chain sprocket alignment marks—2.0L engines

- Cam sprocket cover
- Upper chain tensioner and upper chain guides
- Upper timing chain
- Idler sprocket bolt
- Cylinder head and the intake manifold
- Cylinder head gasket. The lower timing chain will not be disengaged from crankshaft sprocket.

➡ **The cast portion of the front cover is located on the lower side of the crankshaft sprocket, so the lower timing chain need not be disengaged from idler sprocket.**

- Steel oil pan bolts in the reverse sequence of the tightening procedure

6. Install a seal cutter between the steel oil pan and the aluminum oil pan.

7. Tapping the cutter with a hammer, slide it around the entire edge of the oil pan. Take care not to damage the aluminum oil pan.

8. Remove or disconnect the following:
- Steel oil pan
- Baffle plate, oil strainer and the front tube

9. Support the transaxle with a jack and the engine with a engine hoist.
- Front suspension member
- A/C compressor gussets
- Rear cover plate
- Aluminum oil pan retaining bolts in sequence

10. Insert a seal cutter between the oil pan and the cylinder block.

11. Tapping the cutter with a hammer, slide it around the entire edge of the oil pan. Take care not to damage the aluminum oil pan.
- Oil pan from the engine
- Crankshaft pulley
- Front timing chain cover
- Oil pump drive spacer
- Lower timing chain tensioner, tensioner arm and lower timing chain guide
- Lower timing chain and idler sprocket

To install:

12. Install or connect the following:
- Crankshaft sprocket and oil pump drive spacer
- Idler sprocket and lower timing chain

13. Set the lower timing chain on the sprockets, aligning the mating marks. The mating marks on the timing chain assembly will be silver.
- Chain tension arm and chain guide
- Lower timing chain tensioner

14. Apply a continuous bead of liquid

gasket to the front cover and install the front cover. Install a new oil seal.
- Crankshaft pulley and torque tighten bolt to 105–112 ft. lbs. (142–152 Nm)

15. Carefully scrape the old gasket material away from the pan and cylinder block mounting surfaces, then apply a continuous 3.5–4.5mm wide bead of liquid gasket around the oil pan and the cylinder block.
- Aluminum oil pan and torque the bolts, in sequence, to 13 ft. lbs. (17.5 Nm)
- Baffle plate, oil strainer and the front tube
- Steel oil pan and torque the bolts, in sequence, to 61 inch lbs. (7 Nm)
- Rear cover plate
- A/C compressor gussets
- Front suspension member
- Front engine mounting and remove the engine hoist and transaxle support
- New cylinder head gasket
- Cylinder head and temporarily tighten the cylinder head bolts when installing the front cover. This is necessary to avoid damaging the

cylinder head gasket. Be sure to install washers between the bolts and cylinder head.
- Upper timing chain, chain tensioner and chain guide

16. Set the upper timing chain on idler sprockets, aligning the mating marks.
- Cam sprocket cover. Apply a continuous bead of liquid gasket to front cover. Be careful not to damage the cylinder head gasket. Be careful that the upper timing chain does not slip or jump when installing cam sprocket cover.

17. Tighten cylinder head bolts.
- Camshafts and camshaft bearing caps
- Camshaft sprockets, then torque the sprocket bolts to 123–130 ft. lbs. (167–176 Nm). Install the chain guide between both camshaft sprockets. The alignment marks on the upper portion of the timing chain should now be aligned.
- Rocker cover
- Distributor
- Intake manifold collector supports and collector
- Exhaust manifold

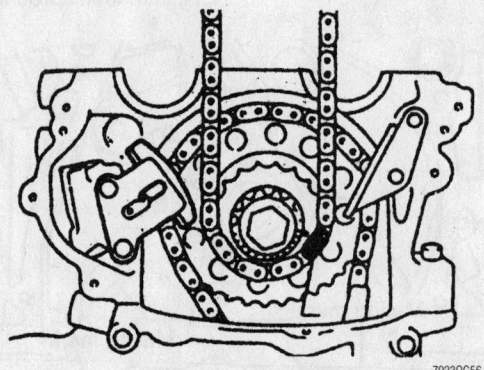

79230G56

Be sure to align the mark on the idler sprocket with the mark on the upper chain—2.4L engines

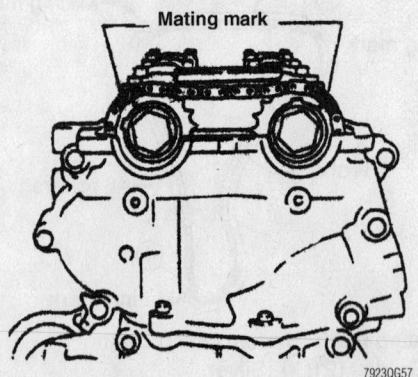

Mating mark

79230G57

Align the marks on the camshaft sprockets with the upper portion of the upper timing chain and mating marks—2.4L engine

- Alternator and bracket
- Upper radiator hose
- Air duct assembly
- Front exhaust tube
- Power steering reservoir
- Drive belts
- Vacuum hoses and fuel lines
- Engine under cover
- Spark plug wires
- Negative battery cable

18. Fill the cooling system.
19. Fill the engine with clean oil.
20. Start the vehicle, check for leaks and repair if necessary.

3.0L and 3.5L Engines

1. Before servicing the vehicle, refer to the precautions in the beginning of this section.
2. Drain the engine oil.
3. Drain the cooling system.
4. Relieve the fuel system pressure.
5. Remove or disconnect the following:
   - Negative battery cable
   - Left side rocker cover ornament

→**Before detaching any hoses or connectors, note the locations for reassembly.**

- Air duct to intake manifold hose, collector hose, blow-by hose and vacuum hoses
- Fuel hoses and detach the harness connections
- Canister purge hoses
- Water hoses from the cylinder head and intake manifold
- All 6 ignition coils from the spark plugs
- Spark plugs
- Bolts that secure the Exhaust Gas Recirculation (EGR) tube and remove the tube
- Intake manifold collector supports and the collector
- Fuel tube assembly
- Intake manifold. Loosen the bolts in the reverse sequence of the tightening procedure.
- Left-hand and right-hand intake valve timing control solenoid valves, 3.5L engine only
- Left-hand and right-hand rocker covers from the cylinder head
- Engine undercovers
- Right front wheel and the engine side covers
- Drive belts and the idler pulley
- Power steering oil pump belt and the power steering oil pump assembly
- Camshaft Position (CMP) sensor (PHASE) and Crankshaft Position (CKP) sensors (REF)/(POS)

6. Set the No. 1 piston to Top Dead Center (TDC) of compression stroke by rotating the crankshaft.
7. Loosen the crankshaft pulley bolt while securing the ring gear so the crankshaft cannot rotate.
   - Ring gear cover access plate

→**Use care not to damage the ring gear teeth.**

- Crankshaft pulley using a suitable puller
- Intake valve timing control valve cover, 3.5L engine only. Loosen the bolts in the reverse order shown in the accompanying figure. In the cover, the shaft is engaged with the center hole of the intake camshaft sprocket.

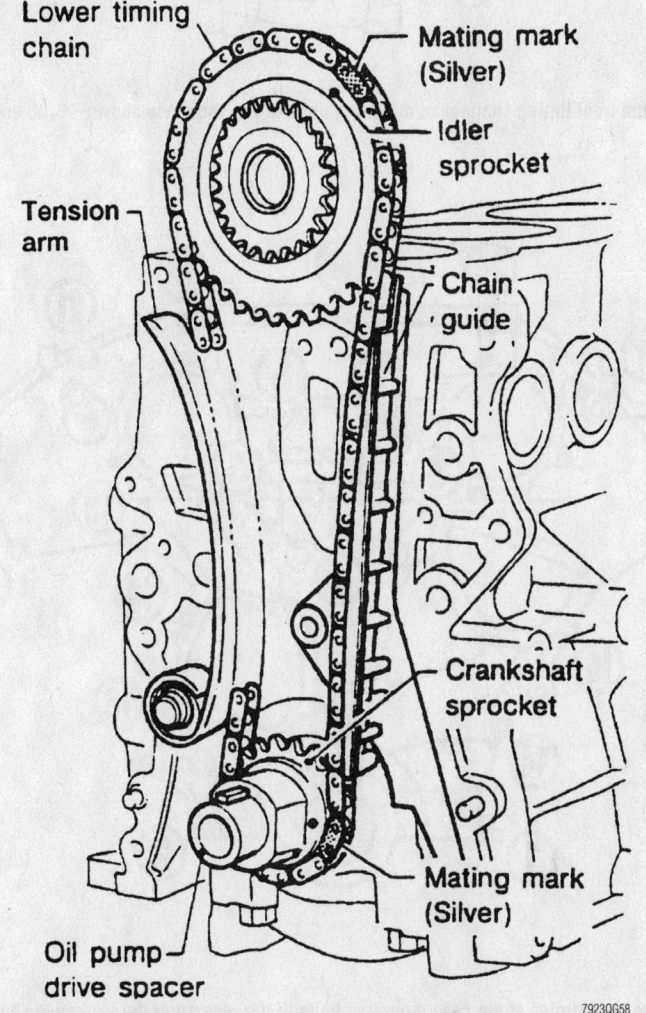

Lower timing chain alignment marks—2.4L engine

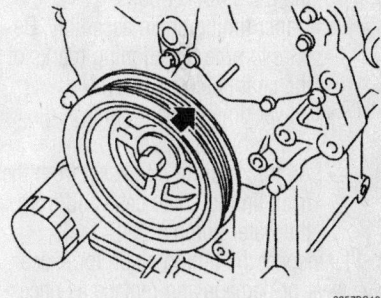

Set the No. 1 piston to Top Dead Center (TDC)—3.0L and 3.5L engines

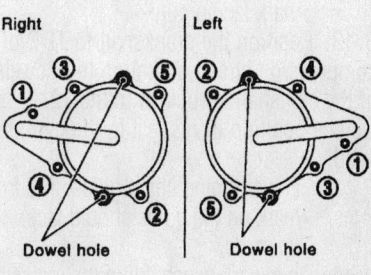

Loosen the intake valve timing control valve cover bolts in the reverse of the order shown—3.5L engine

Remove it straight out until the engagement comes off.
- A/C compressor and bracket
- Front exhaust pipe and its support

8. Hang the engine at the right and left side engine slingers with a suitable hoist.

9. Support the transaxle with jack.
- Right side engine mounting and bracket
- Center crossmember assembly
- Upper and lower oil pans
- Water pump cover
- Bolts that secure the front timing chain case, in sequence
- Timing chain case cover using a seal cutter
- Internal timing chain guide and the upper chain guide
- Timing chain tensioner and slack side chain guide
- Left and right intake camshaft sprockets first. Be sure to hold the flats of the camshafts while removing the sprocket bolts.
- Lower timing chain assembly. Be sure to note the aligning marks of the chain before removal.

10. Insert a suitable stopper pin for the left and right camshaft tensioners.
- Left and right exhaust camshaft sprocket bolts. Be sure to hold the flats of the camshafts while removing the sprocket bolts.
- Upper timing chain assembly. Be sure to note the aligning marks of the chain before removal.
- Lower timing chain guide
- Crankshaft sprocket
- All traces of liquid gasket from the front timing chain case and from the water pump

11. Inspect the timing chain for excessive wear or damage and replace as necessary.

To install:

12. Install or connect the following:
- Crankshaft sprocket with the mating mark facing out

13. Position the crankshaft to TDC of compression stroke and align the dowels of the camshaft sprockets to the 12 o'clock position in respect to the cylinder head.
- Lower timing chain guide. The front mark on the guide should face upwards.

14. On a work bench, align the marks on the intake and exhaust camshaft sprockets with the marks of the chain.

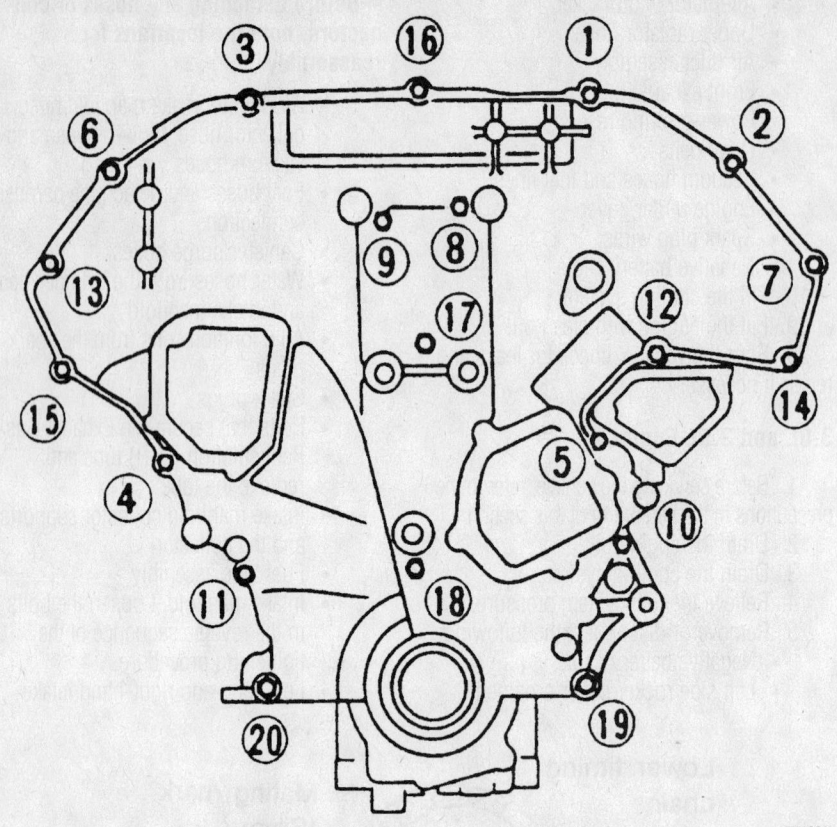

Remove the front timing chain case mounting bolts in the sequence shown—3.0L engines

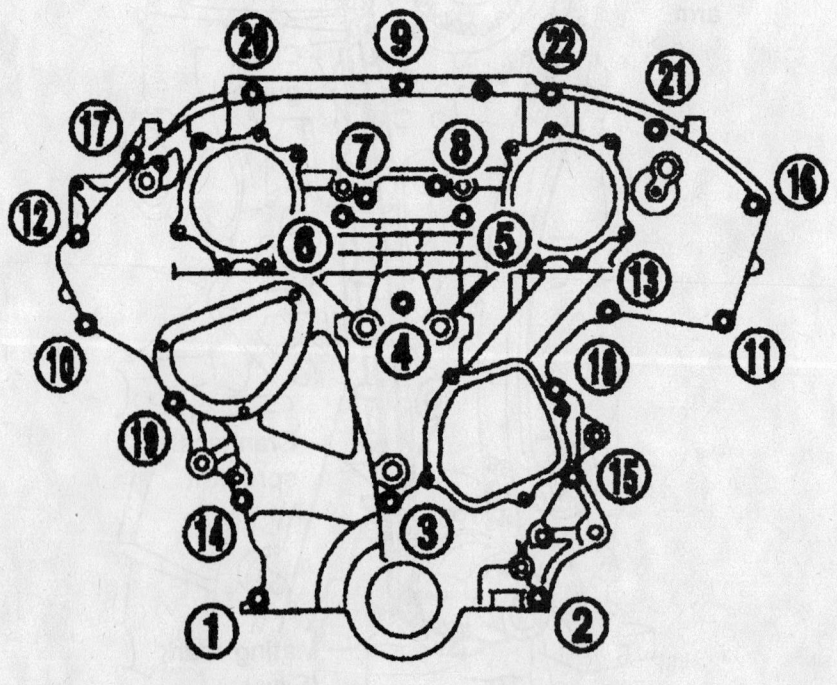

Remove the front timing chain case mounting bolts in the reverse of the sequence shown—3.5L engines

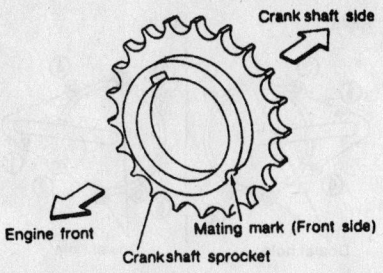

Crankshaft sprocket with mating marks—3.0L engine

- Exhaust camshaft sprockets onto the dowel pin and torque the mounting bolts to 88–95 ft. lbs. (119–128 Nm). Be sure to secure the camshafts while tightening the bolts.
- Timing chains and sprockets to the intake camshafts. Be sure to align the timing chain and sprocket mating marks.
- Left and right camshaft tensioner stopper pins

15. Align the mating mark on the crankshaft with the matchmark (gold link) on the lower timing chain.

16. Install the lower timing chain to the water pump sprocket.

17. Working counterclockwise, install the lower timing chain camshaft sprockets. Be sure to align the sprocket marks with the blue links of the timing chain during installation.

18. Install or connect the following:
- Intake sprocket and torque the bolts to 88–95 ft. lbs. (119–128 Nm). Be sure to secure the camshafts while tightening the bolts.
- Internal timing chain guide, upper timing chain guide, lower timing chain tensioner and slack side timing chain guide

19. Torque the tensioner mounting bolt to 75–96 inch lbs. (8.4–10.8 Nm) and the guide bolts to 108–168 inch lbs. (13–19 Nm).

20. Apply a 0.102–0.142 in. (2.6–3.6mm) continuous bead of liquid gasket to all necessary areas as shown on the front timing cover.
- Timing cover evenly and gently. Be sure to align the dowel pin holes.

21. Torque the mounting bolts in sequence as follows:
  a. Bolts No. 1 and 2: 19–23 ft. lbs. (26–31 Nm).
  b. Bolts No. 3–20: 105–121 inch lbs. (11.8–13.7 Nm).

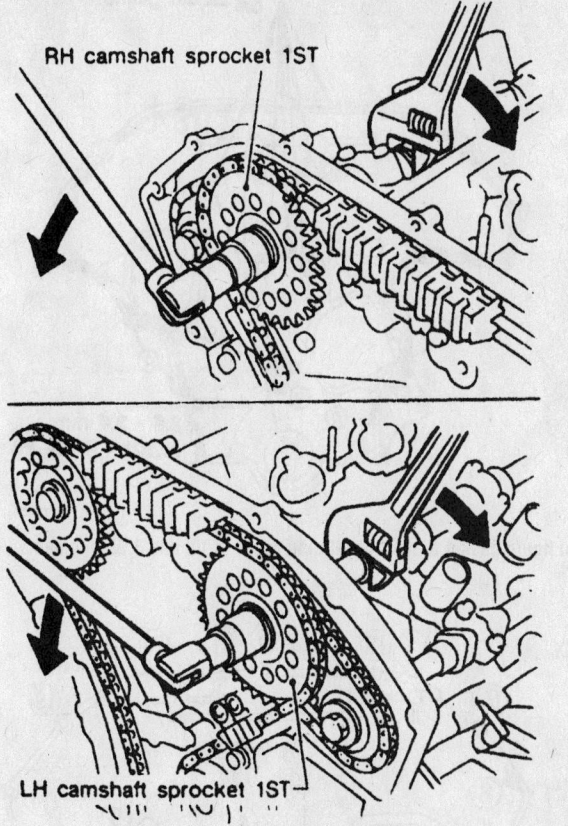

Hold the camshaft with a wrench while removing the sprocket bolts—3.0L and 3.5L engines

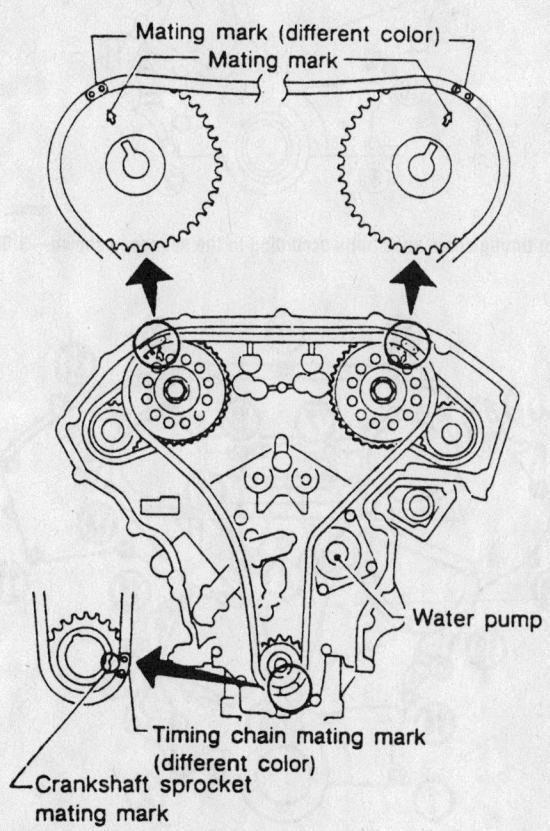

Timing chain alignment marks—3.0L and 3.5L engines

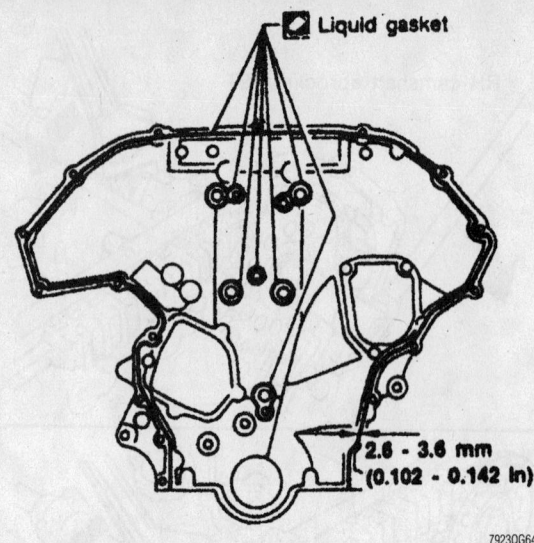

Application of liquid gasket to the front timing case—3.0L and 3.5L engines

2.6 - 3.6 mm
(0.102 - 0.142 in)

7923QG64

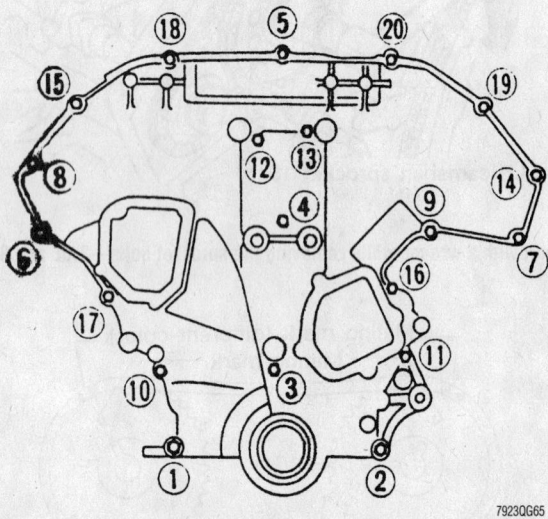

Tighten the front timing chain case bolts according to the sequence shown—3.0L engines

7923QG65

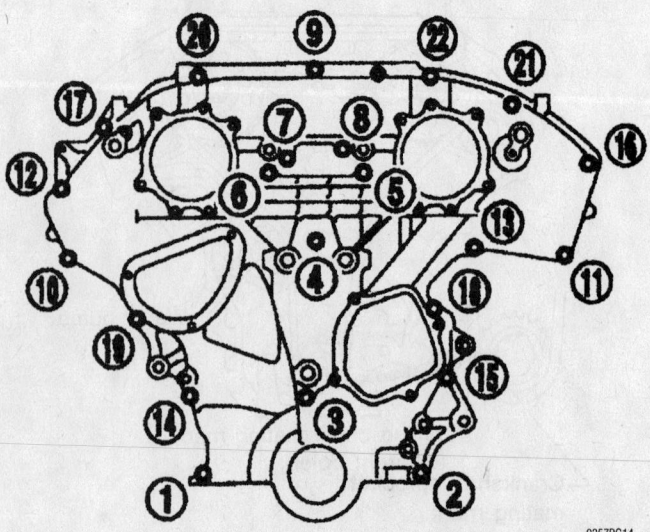

Tighten the front timing chain case bolts according to the sequence shown—3.5L engines

9357RG14

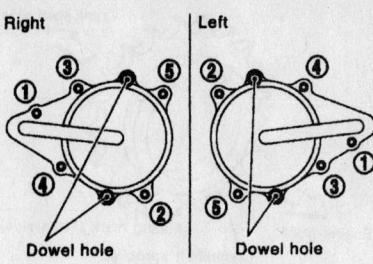

Tighten the intake valve timing control valve cover bolts in sequence—3.5L engine

9357RG13

➡ **Leave the bolts unattended for 30 minutes or more after tightening. This will allow the liquid gasket to cure sufficiently.**

22. Apply a 0.091–0.130 in. (2.3–3.3mm) continuous bead of liquid gasket to the water pump cover and install the cover. Torque the bolts to 84–108 inch lbs. (10–13 Nm).

23. Install or connect the following:
   - Oil pans
   - Center crossmember
   - Right side engine mount and bracket
   - Front exhaust pipe and remove the transaxle support
   - A/C compressor and bracket
   - Crankshaft pulley
   - Ring gear access cover plate
   - CMP sensor and CKP sensors
   - Power steering pump
   - Idler pulley and drive belts
   - Engine side cover and right front wheel
   - Engine under covers
   - Intake valve timing control solenoid valves with new covers, 3.5L engine only. Tighten the cover bolts in the proper sequence.
   - Rocker covers
   - Intake manifold
   - Fuel tube assembly
   - Intake manifold collector and support
   - EGR tube
   - Spark plugs and ignition coils
   - Coolant hoses
   - Fuel hoses
   - Air duct assembly and hoses
   - Left side rocker cover ornament
   - Negative battery cable

24. Fill the cooling system.
25. Fill the engine with clean oil.
26. Start the vehicle, check for leaks and repair if necessary.

Piston and Ring

POSITIONING

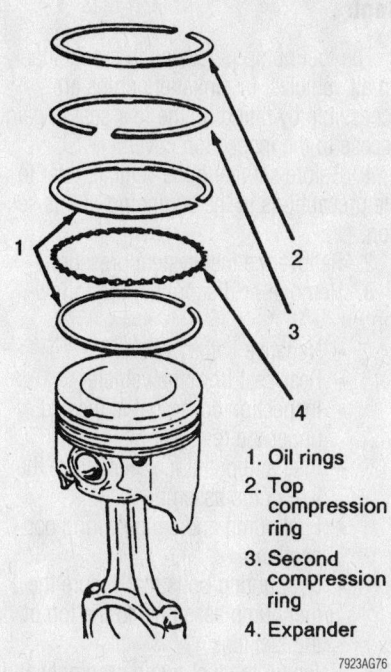

1. Oil rings
2. Top compression ring
3. Second compression ring
4. Expander

7923AG76

Exploded view of common piston ring mounting

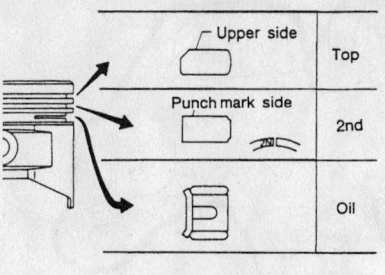

7923AG72

Piston ring positioning—2.0L and 2.4L engines

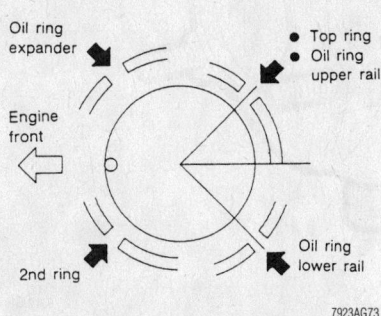

7923AG73

Piston ring end-gap spacing

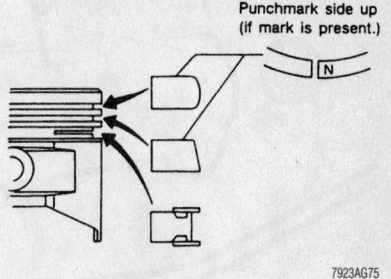

7923AG75

Piston ring positioning—3.0L and 3.5L engines

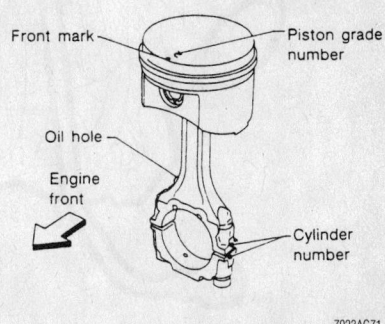

7923AG71

Piston and connecting rod assembly positioning

FUEL SYSTEM

Fuel System Service Precautions

Safety is the most important factor when performing not only fuel system maintenance but any type of maintenance. Failure to conduct maintenance and repairs in a safe manner may result in serious personal injury or death. Maintenance and testing of the vehicle's fuel system components can be accomplished safely and effectively by adhering to the following rules and guidelines.

• To avoid the possibility of fire and personal injury, always disconnect the negative battery cable unless the repair or test procedure requires that battery voltage be applied.

• Always relieve the fuel system pressure prior to disconnecting any fuel system component (injector, fuel rail, pressure regulator, etc.), fitting or fuel line connection. Exercise extreme caution whenever relieving fuel system pressure, to avoid exposing skin, face and eyes to fuel spray. Please be advised that fuel under pressure may penetrate the skin or any part of the body that it contacts.

• Always place a shop towel or cloth

around the fitting or connection prior to loosening to absorb any excess fuel due to spillage. Ensure that all fuel spillage (should it occur) is quickly removed from engine surfaces. Ensure that all fuel soaked cloths or towels are deposited into a suitable waste container.

• Always keep a dry chemical (Class B) fire extinguisher near the work area.

• Do not allow fuel spray or fuel vapors to come into contact with a spark or open flame.

• Always use a back-up wrench when loosening and tightening fuel line connection fittings. This will prevent unnecessary stress and torsion to fuel line piping. Always follow the proper torque specifications.

• Always replace worn fuel fitting O-rings with new. Do not substitute fuel hose where fuel pipe is installed.

Fuel System Pressure

RELIEVING

The fuel pump fuse is located in the dash fuse box or in the engine compartment fuse

box. Check the lid of the fuse box for exact location.

1. Before servicing the vehicle, refer to the precautions in the beginning of this section.

2. Remove the fuel pump fuse.

3. Start the engine.

4. Start the engine and run until the engine stalls.

5. After the engine stalls, try to restart the engine; if the engine will not start, the fuel pressure has been released.

6. Turn the ignition switch **OFF**. Reinstall the fuel pump fuse into the fuse block.

➡**Do not crank the engine or turn the ignition switch ON after the fuel pump fuse has been reinstalled, or the fuel pressure will be re-established.**

Fuel Filter

REMOVAL & INSTALLATION

Except Maxima

1. Before servicing the vehicle, refer to the precautions in the beginning of this section.

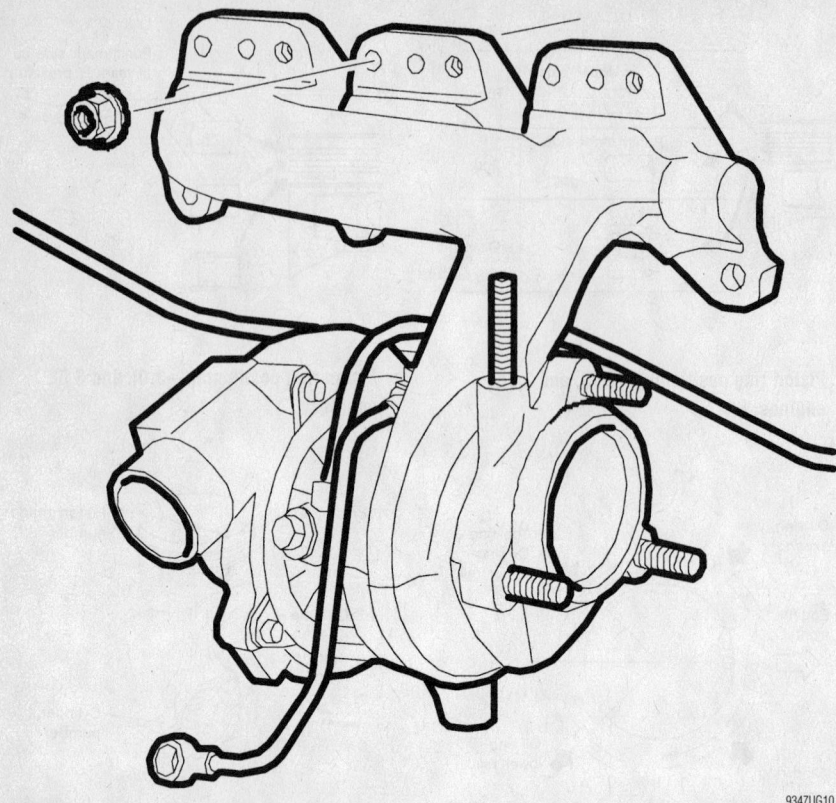

Remove the fuel filter from the fuel chamber—Maxima

2. Properly relieve fuel system pressure.
3. Remove or disconnect the following:
   - Negative battery cable
   - Fuel hose clamps
   - Hoses from the fuel filter
   - Bolt securing the filter to the bracket or the filter from the bracket clips
   - Filter

To install:
4. Install or connect the following:
   - New filter and secure the filter in the bracket
5. If necessary, replace the fuel line hoses and hose clamps
   - Fuel hoses and tighten the clamps
   - Negative battery cable
   - Fuel pump fuse
6. Start the engine and check for leaks.

Maxima

1. Before servicing the vehicle, refer to the precautions in the beginning of this section.
2. Properly relieve fuel system pressure.
3. Remove or disconnect the following:
   - Negative battery cable
   - Rear seat bottom
   - Inspection hole cover
   - Electrical and quick connectors
   - Six screws
   - Fuel level sensor unit and fuel pump assembly
   - Flange and snap fit portion of the fuel pump
   - Fuel tank temperature sensor harness
   - Fuel level sensor flange
   - Fuel pump connector
   - Quick connectors from the fuel level sensor
   - Fuel level sensor from the chamber
   - Fuel filter from the chamber

To install:
4. Install or connect the following:
   - Fuel filter to the chamber
   - Fuel level sensor to the chamber
   - Quick connectors to the fuel level sensor
   - Fuel pump connector
   - Fuel level sensor flange
   - Fuel tank temperature sensor harness
   - Fuel pump assembly to the fuel tank
   - Screws and electrical connectors
   - Quick connectors
   - Negative battery cable
5. Start the vehicle, check for leaks and repair if necessary.
   - Inspection hole cover
   - Rear seat bottom

Fuel Pump

REMOVAL & INSTALLATION

Sentra

The fuel pump is located in the fuel tank on all vehicles. In-tank fuel pumps are accessible by lifting up the rear seat to gain access to the inspection cover.
1. Before servicing the vehicle, refer to the precautions in the beginning of this section.
2. Relieve the fuel system pressure.
3. Remove or disconnect the following:
   - Negative battery cable
   - Rear seat from the vehicle
   - Inspection cover that is located under the rear seat
   - Inlet and outlet fuel lines from the fuel pump assembly
   - Fuel pump and gauge wiring connections
   - 6 mounting bolts that secure the fuel pump assembly to the top of the fuel tank
4. Raise up the fuel pump assembly and detach the fuel tubes and connector.
   - Fuel gauge assembly
   - Fuel pump with the fuel chamber
5. Pull up the front of the fuel pump chamber and slide the chamber forward.
   - Fuel pump from the chamber
   - O-ring seal or gasket

To install:
6. Install or connect the following:
   - Fuel pump to the fuel pump chamber and slide chamber rearward
   - Fuel pump with the fuel pump chamber
   - Fuel gauge assembly using a new O-ring
   - Fuel tubes and connector. Use new hoses and clamps.
   - 6 mounting bolts to the top of the fuel gauge unit and torque the bolts to 17–22 inch lbs. (2–3 Nm)
   - Fuel pump and gauge wiring connections
   - New inlet and outlet fuel lines to the fuel pump assembly
   - Negative battery cable
7. Start the vehicle, check for leaks and repair if necessary.
   - Inspection cover and the rear seat

Altima

1. Before servicing the vehicle, refer to the precautions in the beginning of this section.

2. Relieve the pressure from the fuel system.

3. Remove or disconnect the following:

- Negative battery cable
- Rear seat and the access cover
- Fuel pump electrical connector
- Fuel lines from the fuel pump assembly
- Locking ring
- Fuel gauge assembly
- Fuel tube and connector from the fuel gauge

➡ **When the fuel sending unit needs to be removed, pull the tab upwards. The tab is located on the sending unit, opposite the end of the float. After the tab is pulled, the sending unit will lift straight out of the tank bracket.**

- Fuel pump by pinching the 2 locking tabs together. Lift the fuel pump assembly straight upward and out of fuel tank.
- O-ring and discard

4. Place a clean rag in the hole to keep out dirt.

To install:

5. Remove the rag

6. Install or connect the following:

- New O-ring and fuel pump
- Electrical connection and fuel tube to the fuel gauge sending unit
- Fuel sending unit into the tank

➡ **Verify that the mark on the fuel tank and the components are aligned when installing the pump and fuel gauge sending unit.**

- Locking ring and torque the ring to 22–26 ft. lbs. (30–35 Nm)
- Fuel lines and fuel pump electrical connector. Always install new clamps on the fuel lines.
- Negative battery cable

7. Start the engine, check for leaks and repair if necessary.

- Fuel pump access cover
- Rear seat

Maxima

1. Before servicing the vehicle, refer to the precautions in the beginning of this section.

2. Relieve the fuel system pressure

3. Remove or disconnect the following:

- Negative battery cable
- Rear seat or open the access panel in the trunk
- Fuel gauge electrical connector and pump electrical connector

- Fuel outlet and the return hoses
- Fuel tank, if necessary

4. On some models you need to remove the fuel pump assembly-to-fuel tank bolts and lift the fuel pump assembly from the fuel tank.

5. On other models you need to remove the locking ring and raise the fuel pump from the tank. Disconnect the feed tube while raising the pump.

6. Discard the O-ring. Plug the fuel tank opening with a clean rag to prevent dirt from entering the system.

➡ **When removing or installing the fuel pump assembly, be careful not to damage or deform it and always install a new O-ring.**

To install:

7. Remove the rag

8. Install or connect the following:

- Fuel pump assembly into the fuel tank using a new O-ring
- Fuel pump assembly-to-fuel tank bolts and torque the bolts to 17–22 inch lbs. (2.0–2.5 Nm)
- Locking ring assembly and tighten
- Fuel tank assembly, if removed
- Fuel lines and the electrical connectors. Always use new clamps when reconnecting fuel line hoses.

➡ **When installing the upper plate, be sure to align the mark with the center marks on the fuel tank.**

- Negative battery cable

9. Start the engine, check for fuel leaks and repair if necessary.

10. Install the fuel pump access cover.

➡ **On some models, the Check Engine Light will stay ON after installation is completed. The memory code in the control unit must be erased. This code is stored for an open fuel pump circuit, this is caused when the fuel pressure is released. To erase the code, disconnect the battery cable for 10 seconds, then reconnect after installation of fuel pump.**

Fuel Injector

REMOVAL & INSTALLATION

Except Maxima

1. Before servicing the vehicle, refer to the precautions in the beginning of this section.

2. Relieve the fuel system pressure

3. Remove or disconnect the following:

- Negative battery cable
- Intake manifold collector
- Vacuum hose from the pressure regulator
- Fuel hoses from the rail
- Injector electrical connectors
- Fuel rail bolts
- Injector rail assembly with injectors from the intake manifold
- Injector from the rail by pushing on the injector tail piece
- Discard injector O-rings

To install:

4. Clean the injector tail piece and lubricate new O-rings with a smear of clean engine oil.

5. Install or connect the following:

- New O-rings
- Injector to the fuel rail
- Fuel rail and the injectors as an assembly to the intake manifold

6. Install the fuel rail bolts and tighten in 2 steps as follows:

a. Step 1: Bolts to 84–96 inch lbs. (9–10 Nm).

b. Step 2: Bolts to 15–20 ft. lbs. (21–26 Nm).

7. Install or connect the following:

- Injector electrical connectors
- Vacuum hose to the pressure regulator
- Fuel hoses to the rail
- All remaining components in the reverse order of removal

Maxima

1. Remove or disconnect the following:

- Negative battery cable
- Intake manifold collector
- Vacuum hose from the pressure regulator
- Fuel hoses from the rail
- Injector electrical connectors
- Fuel rail bolts

2. To remove the fuel injector from the fuel rail, expand and remove the clips securing the injectors and press the fuel injector out from the fuel rail. Discard the O-rings.

To install:

3. Apply a thin coat of engine oil to the new O-rings, install them on the injectors, then press the injector into the fuel rail.

4. Install or connect the following:

- New injector retaining clips

- New injector gaskets onto the manifold
- Fuel rail assembly to the engine
- Fuel rail-to-cylinder head bolts and torque the bolts to 84–96 inch lbs.

(9.3–10.8 Nm). Then tighten them again to 16–19 ft. lbs. (21–26 Nm).
- Fuel lines to the rail assembly
- Vacuum hose to the fuel pressure regulator

- Electrical connectors to the fuel injectors
- Intake manifold collector
- Negative battery cable.
5. Start the engine and check for leaks.

DRIVE TRAIN

Transaxle Assembly

REMOVAL & INSTALLATION

Manual

SENTRA

1. Before servicing the vehicle, refer to the precautions in the beginning of this section.
2. Drain the fluid from the transaxle.
3. Remove or disconnect the following:
   - Both battery cables
   - Battery and bracket from the vehicle
   - Crankshaft Position (CKP) sensor from the transaxle
   - Air cleaner assembly
   - All electrical connectors from the transaxles
   - Control cable from the transaxle

- Speed sensor, OD position switch and Back-up lamp switch
- Neutral position switch
- Starter
- Shift control rod
- Halfshafts and properly support the transmission
- Left side and rear engine to transmission mounts

4. Slide the transmission away from the engine and lower the transmission assembly.

To install:

5. Install the transaxle mounting bolts in the proper location as noted during removal.
   a. On 1.8L engines: torque the 2 bottom bolts to 12–15 ft. lbs. (16–21 Nm) and all other bolts to 22–30 ft. lbs. (30–40 Nm).
   b. On 2.0L engines: torque the 2 bottom bolts to 23–31 ft. lbs. (31–42 Nm) and all other bolts to 51–59 ft. lbs. (70–79 Nm).
6. Install or connect the following:
   - Left side and rear engine to transmission mounts
   - Halfshafts
   - Shift control rod
   - Starter
   - Neutral position switch
   - Speed sensor, OD position switch and back-up lamp switch
   - Clutch control cable
   - CKP sensor
   - Air cleaner assembly
   - Battery and both cables
7. Fill the transmission with clean oil to the proper level.
8. Start the vehicle, check for leaks and repair if necessary.

GA engine models
⊙ M/T to engine
⊗ Engine (gusset) to M/T

Bolt No.	Tightening torque N·m (kg-m, ft-lb)	"ℓ" mm (in)
①	30 - 40 (3.1 - 4.1, 22 - 30)	70 (2.76)
②	30 - 40 (3.1 - 4.1, 22 - 30)	85 (3.35)
③	30 - 40 (3.1 - 4.1, 22 - 30)	30 (1.18)
④	16 - 21 (1.6 - 2.1, 12 - 15)	25 (0.98)
Front gusset to engine	30 - 40 (3.1 - 4.1, 22 - 30)	20 (0.79)
Rear gusset to engine	16 - 21 (1.6 - 2.1, 12 - 15)	16 (0.63)

9307QG13

Bolt locations and torque specifications—1.8L engine with manual transaxle

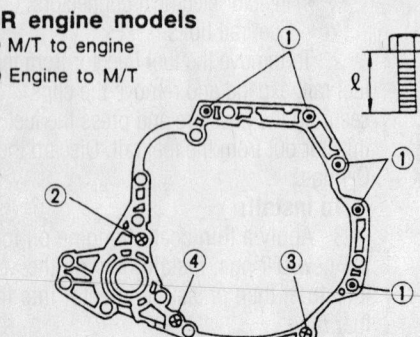

SR engine models
⊙ M/T to engine
⊗ Engine to M/T

Bolt No.	Tightening torque N·m (kg-m, ft-lb)	"ℓ" mm (in)
①	70 - 79 (7.1 - 8.1, 51 - 59)	55 (2.17)
②	70 - 79 (7.1 - 8.1, 51 - 59)	65 (2.56)
③	31 - 42 (3.2 - 4.3, 23 - 31)	35 (1.38)
④	31 - 42 (3.2 - 4.3, 23 - 31)	45 (1.77)

9307QG14

Bolt locations and torque specifications—2.0L engine with manual transaxle

ALTIMA

1. Before servicing the vehicle, refer to the precautions in the beginning of this section.
2. Drain the transmission fluid.
3. Remove or disconnect the following:
   - Battery cables
   - Battery and tray
   - Air cleaner box with the Mass Air Flow (MAF) sensor
   - Air duct
   - Clutch operating cylinder
   - Speedometer pinion electrical connectors
   - Park/Neutral Position (PNP) switch electrical connectors
   - Starter
   - Crankshaft Position (CKP) sensor
   - Shift control rod
   - Front wheels
   - Halfshafts and properly support the engine
   - Rear and left side engine mounts
   - Transaxle assembly

To install:

4. Install the transaxle assembly into the vehicle.
5. Torque the 4 lower mounting bolts to 22–30 ft. lbs. (30–40 Nm) and all remaining bolts to 29–36 ft. lbs. (39–49 Nm).
6. Install or connect the following:
   - Rear and left side engine mounts
   - Halfshafts and remove the engine support

- Both front wheels
- Shift control rod
- CKP sensor
- Starter and torque the bolts to 30 ft. lbs. (41 Nm)
- PNP switch electrical connectors
- Speedometer pinion connectors
- Clutch operating cylinder
- Air duct and air cleaner assembly
- Battery tray and battery
- Both battery cables

7. Fill the transmission with clean fluid.
8. Start the vehicle, check for leaks and repair if necessary.

MAXIMA

1. Before servicing the vehicle, refer to the precautions in the beginning of this section.
2. Drain the fluid from the transaxle.
3. Remove or disconnect the following:

   - Battery cables
   - Battery and the battery tray
   - Air cleaner and Mass Air Flow (MAF) sensor
   - Air breather hose
   - Control cable and cable mounting bracket
   - Clutch operating cylinder and hose clamps
   - Speedometer pinion, Park/Neutral Position (PNP) and ground harness switch connectors
   - Starter

- Crankshaft Position (CKP) sensor (POS) from the transaxle
- Shift control rod and support rod bracket
- Front wheels
- Halfshafts and properly support the engine with a jack under the oil pan

✴✴ WARNING

Do not place the jack under the oil pan drain plug.

- Center member
- Left-hand mounting bracket from the transaxle and body
- Transaxle

To install:

➡**The transaxle mounting bolts are different lengths and require special torque specifications. Use care when installing and tightening these bolts.**

4. Install or connect the following:
   - Transaxle assembly into the vehicle
   - Transaxle mounting bolts in the proper location as noted during removal
   - Left hand mounting bracket
   - Center member
   - Halfshafts and front wheels
   - Shift control rod and support rod bracket
   - Starter
   - CKP sensor

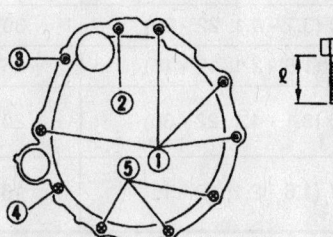

◉ M/T to engine
⊗ Engine to M/T

Bolt No.	Tightening torque N·m (kg-m, ft-lb)	"ℓ" mm (in)
①	70 - 79 (7.1 - 8.1, 51 - 59)	52 (2.05)
②	70 - 79 (7.1 - 8.1, 51 - 59)	65 (2.56)
③	70 - 79 (7.1 - 8.1, 51 - 59)	124 (4.88)
④	35.1 - 47.1 (3.58 - 4.80, 25.89 - 34.74)	40 (1.57)
⑤	35.1 - 47.1 (3.58 - 4.80, 25.89 - 34.74)	40 (1.57)

③ with starter
④ with support rod bracket

79230G73

Manual transaxle bolt torque specifications and locations—Maxima with 3.0L engine

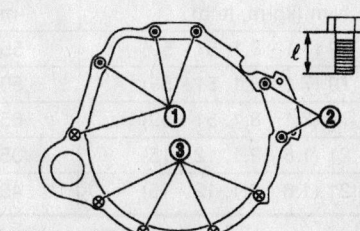

◉ M/T to engine
⊗ Engine or oil pan to M/T

Bolt No.	Tightening torque N·m (kg-m, ft-lb)	"ℓ" mm (in)
1	69.6 - 79.4 (7.1 - 8.0, 52 - 58)	52 (2.05)
2	69.6 - 79.4 (7.1 - 8.0, 52 - 58)	113 (4.45)
3	36 - 47 (3.7 - 4.7, 27 - 34)	40 (1.57)

9357RG15

Manual transaxle bolt torque specifications and locations—Maxima with 3.5L engine

- Speedometer pinion, PNP switch and ground harness switch connectors
- Clutch operating cylinder and hose clamp
- Control cable and cable mounting bracket
- Air breather hose
- Air cleaner and MAF sensor
- Battery and tray
- Battery cables

5. Fill the transaxle with clean fluid.

6. Start the vehicle, check for leaks and repair if necessary.

Automatic

SENTRA

1. Before servicing the vehicle, refer to the precautions in the beginning of this section.

2. Drain the fluid from the transaxle.

3. Remove or disconnect the following:
- Both battery cables
- Battery and bracket from the vehicle
- Crankshaft Position (CKP) sensor from the transaxle
- Air cleaner assembly
- Torque converter clutch solenoid valve electrical connector
- Inhibitor switch and Vehicle Speed Sensor (VSS) electrical connectors

- Throttle wire from the engine side
- Control cable
- Oil cooler hoses
- Halfshafts
- Intake manifold support bracket
- Starter
- Upper engine to transmission bolts and properly support the transmission
- Center member
- Front and rear gussets and engine rear plate
- Rear transaxle to engine bracket
- Rear transaxle mount
- Transaxle assembly

To install:

When connecting the torque converter to the transaxle, be sure to measure the distance between the mounting lug of the converter and the front edge of the transaxle.

4. The measured distance between the converter and the front of the transaxle should be:
   a. 1.8L engines: 0.831 in. (21.1mm) or more.
   b. 2.0L engines: 0.626 in. (15.9mm) or more.

5. Raise the transaxle and install to engine drive plate.

6. Install the transaxle mounting bolts

in the proper location as noted during removal.

7. On 1.8L engines torque the 2 bottom bolts to 12–15 ft. lbs. (16–21 Nm) and all other bolts to 22–30 ft. lbs. (30–40 Nm).

8. On 2.0L engines torque the 2 bottom bolts to 12–15 ft. lbs. (16–21 Nm) and all other bolts to 51–59 ft. lbs. (70–79 Nm).

9. Install or connect the following:
- Torque converter to the drive plate and torque the bolts to 33–43 ft. lbs. (44–59 Nm)
- Rear transmission mount
- Rear transmission bracket
- Front and rear gussets and the rear engine plate
- Center member
- Starter and torque the bolts to 31 ft. lbs. (42 Nm)
- Intake manifold support bracket
- Both half shafts
- Control cable
- Throttle wire
- CKP sensor
- Torque converter clutch solenoid valve, VSS and inhibitor switch electrical connectors
- Air duct
- Battery and both cables

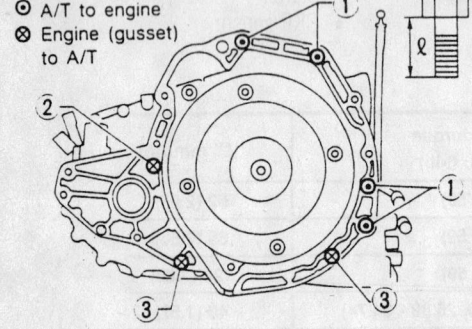

Bolt No.	Tightening torque N·m (kg-m, ft-lb)	Bolt length "ℓ" mm (in)
①	30 - 40 (3.1 - 4.1, 22 - 30)	50 (1.97)
②	30 - 40 (3.1 - 4.1, 22 - 30)	30 (1.18)
③	16 - 21 (1.6 - 2.1, 12 - 15)	25 (0.98)
Front gusset to engine	30 - 40 (3.1 - 4.1, 22 - 30)	20 (0.79)
Rear gusset to engine	16 - 21 (1.6 - 2.1, 12 - 15)	16 (0.63)

9307QG11

Bolt locations and torque specifications—1.8L engine with automatic transaxle

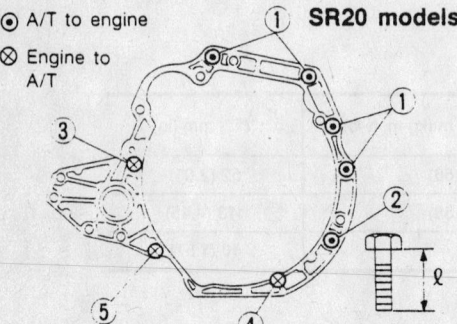

Bolt No.	Tightening torque N·m (kg-m, ft-lb)	Bolt length "ℓ" mm (in)
①	70 - 79 (7.1 - 8.1, 51 - 59)	55 (2.17)
②	70 - 79 (7.1 - 8.1, 51 - 59)	50 (1.97)
③	70 - 79 (7.1 - 8.1, 51 - 59)	65 (2.56)
④	16 - 21 (1.6 - 2.1, 12 - 15)	35 (1.38)
⑤	16 - 21 (1.6 - 2.1, 12 - 15)	45 (1.77)

9307QG12

Bolt locations and torque specifications—2.0L engine with automatic transaxle

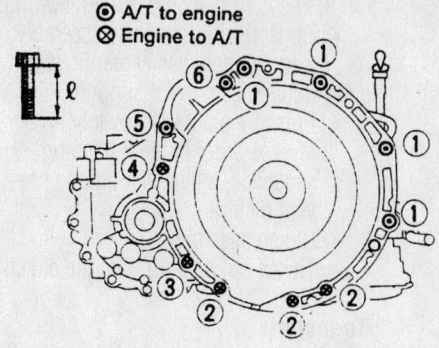

⊙ A/T to engine
⊗ Engine to A/T

Bolt No.	Tightening torque N·m (kg-m, ft-lb)	ℓ mm (in)
①	39 - 49 (4.0 - 5.0, 29 - 36)	45 (1.77)
②	30 - 36 (3.1 - 3.7, 22 - 27)	30 (1.18)
③	30 - 36 (3.1 - 3.7, 22 - 27)	40 (1.57)
④	74 - 83 (7.5 - 8.5, 54 - 61)	45 (1.77)
⑤	30 - 36 (3.1 - 3.7, 22 - 27)	80 (3.15)
⑥	30 - 36 (3.1 - 3.7, 22 - 27)	65 (2.56)

7923QG72

Be sure to install the bolts in the correct location and tighten them to specification—Altima with automatic transaxle

10. Fill the transmission to the proper level.

11. Start the vehicle, check for leaks and repair if necessary.

ALTIMA

1. Before servicing the vehicle, refer to the precautions in the beginning of this section.

2. Drain the transmission fluid.

3. Remove or disconnect the following:
- Battery cables
- Battery and tray
- Air cleaner and resonator
- Park/Neutral Position (PNP) switch
- Revolution sensor and Vehicle Speed Sensor (VSS) electrical connectors
- Crankshaft Position (CKP) sensor
- Left hand mounting bracket from the transaxle and body
- Control cable
- Both front wheels
- Halfshafts
- Oil cooler pipes
- Starter and properly support the engine
- Center member
- Rear cover plate
- Torque converter
- Transaxle assembly

➡When removing the torque converter, turn the crankshaft for access to the bolts. Place alignment marks on the converter and drive plate, so the converter can be installed in its original position.

➡The transaxle mounting bolts are different lengths. Tagging the bolts upon removal will facilitate proper tightening during installation.

To install:

➡When installing the torque converter to the transaxle, measure the depth of the converter to ensure proper installation.

4. Using a straight edge across the mounting flange, measure the depth of the converter. The measurement is to the bolt mounting flange of the converter.

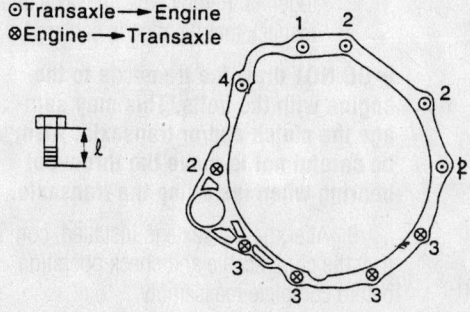

⊙ Transaxle → Engine
⊗ Engine → Transaxle

Bolt No.	Tightening torque N·m (kg-m, ft-lb)	ℓ mm (in)
1	70 - 79 (7.1 - 8.1, 52 - 58)	65 (2.56)
2	70 - 79 (7.1 - 8.1, 52 - 58)	52 (2.05)
3	70 - 79 (7.1 - 8.1, 52 - 58)	40 (1.57)
4	78 - 98 (7.9 - 10.0, 58 - 72)	124 (4.88)

9307QG20

Automatic transaxle bolt torque specifications and locations—Maxima with 3.0L engine

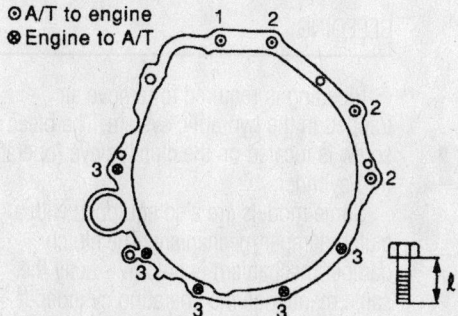

⊙ A/T to engine
⊗ Engine to A/T

Bolt No.	Tightening torque N·m (kg-m, ft-lb)	ℓ mm (in)
1	69.6 - 79.4 (7.1 - 8.0, 52 - 58)	65 (2.56)
2	69.6 - 79.4 (7.1 - 8.0, 52 - 58)	52 (2.05)
3	69.6 - 79.4 (7.1 - 8.0, 52 - 58)	40 (1.57)

9357RG16

Automatic transaxle bolt torque specifications and locations—Maxima with 3.5L engine

5. The depth measurement of the converter should be 0.75 in. (19mm) or more.

➡ **The transaxle mounting bolts are different lengths and require special torque specifications. Use care when installing and tightening these bolts.**

6. Install the transaxle assembly into the vehicle.

7. Refer to the diagram for the automatic transaxle mounting bolt torque specifications.

8. Torque the bolts holding the converter to the flexplate to 33–43 ft. lbs. (44–59 Nm).

9. Install or connect the following:
- Rear cover plate
- Center member
- Starter and remove the engine support
- Oil cooler pipes
- Halfshafts and both front wheels
- Control cable
- Left hand mounting bracket
- CKP sensor
- Revolution and VSS sensor electrical connectors
- PNP switch
- Air cleaner and resonator
- Battery, tray and both cables

10. Fill the transaxle with the proper type and amount of fluid.

11. Start the vehicle, check for leaks and repair if necessary.

MAXIMA

1. Before servicing the vehicle, refer to the precautions in the beginning of this section.

2. Drain the transaxle fluid.

3. Remove or disconnect the following:
- Battery cables
- Battery and tray
- Air cleaner and resonator
- Park/Neutral Position (PNP) switch
- Revolution sensor and Vehicle Speed Sensor (VSS) electrical connectors
- Crankshaft Position (CKP) sensor
- Left hand mounting bracket from the transaxle and body
- Control cable
- Both front wheels
- Halfshafts
- Oil cooler pipes
- Starter and properly support the engine
- Center member
- Rear cover plate
- Torque converter
- Transaxle assembly

➡ **When removing the torque converter, turn the crankshaft for access to the bolts. Place alignment marks on the converter and drive plate, so the converter can be installed in its original position.**

➡ **The transaxle mounting bolts are different lengths. Tagging the bolts upon removal will facilitate proper tightening during installation.**

To install:

➡ **When installing the torque converter to the transaxle, measure the depth of the converter to ensure proper installation.**

4. Using a straight edge across the mounting flange, measure the depth of the converter. The measurement is to the bolt mounting flange of the converter.

5. The depth measurement of the converter should be 0.75 in. (19mm) or more.

➡ **The transaxle mounting bolts are different lengths and require special torque specifications. Use care when installing and tightening these bolts.**

6. Install the transaxle assembly into the vehicle.

7. Refer to the diagram for the automatic transaxle mounting bolt torque specifications.

8. Torque the bolts holding the converter to the flexplate to 33–43 ft. lbs. (44–59 Nm).

9. Install or connect the following:
- Rear cover plate
- Center member
- Starter and remove the engine support
- Oil cooler pipes
- Halfshafts and both front wheels
- Control cable
- Left hand mounting bracket
- CKP sensor
- Revolution and VSS sensor electrical connectors
- PNP switch
- Air cleaner and resonator
- Battery, tray and both cables

10. Fill the transaxle with the proper type and amount of fluid.

11. Start the vehicle, check for leaks and repair if necessary.

Clutch

REMOVAL & INSTALLATION

1. Before servicing the vehicle, refer to the precautions in the beginning of this section.

2. Remove or disconnect the following:
- Transmission/transaxle assembly

3. Insert a clutch disc centering tool into the clutch disc hub for support.
- Pressure plate bolts evenly in reverse order of the tightening sequence, a little at a time to prevent distortion
- Clutch assembly
- Throw-out bearing from the clutch lever

To install:

4. Apply a light coating of chassis lube to the clutch disc spleens, input shaft and pilot bearing. Use a disc centering tool to aid installation.

5. Install or connect the following:
- Disc and pressure plate

6. On all except Maxima, torque the pressure plate bolts in a crisscross pattern and in several steps to 16–22 ft. lbs. (20–26 Nm).

7. On Maxima, torque the pressure plate bolts in a crisscross pattern in the following 2 steps:
   a. Step 1: 7–14 ft. lbs. (10–20 Nm).
   b. Step 2: 25–33 ft. lbs. (34–44 Nm).

8. Install or connect the following:
- New throw-out bearing in the clutch release lever. Remove the clutch disc centering tool.
- Transaxle into the vehicle. If the mating surfaces will not come together, do not force the units together. Remove the transaxle and recheck that the disc is centered.

➡ **DO NOT draw the transaxle to the engine with the bolts. This may damage the clutch and/or transaxle. Also, be careful not to move the throw-out bearing when installing the transaxle.**

9. After the transaxle is installed, connect the clutch cable and check operation before complete reassembly.

10. Adjust the clutch pedal as necessary.

Hydraulic Clutch System

BLEEDING

Bleeding is required to remove air trapped in the hydraulic system. The bleed screw is located on the clutch slave (operating) cylinder.

Some models are also equipped with a clutch damper mechanism. The clutch damper mechanism is bled in exactly the same manner as the operating cylinder. It should be bled along with the operating cylinder.

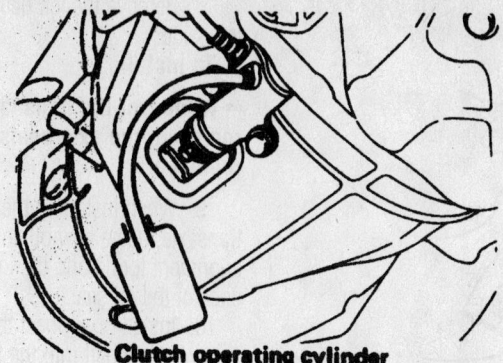

Clutch operating cylinder

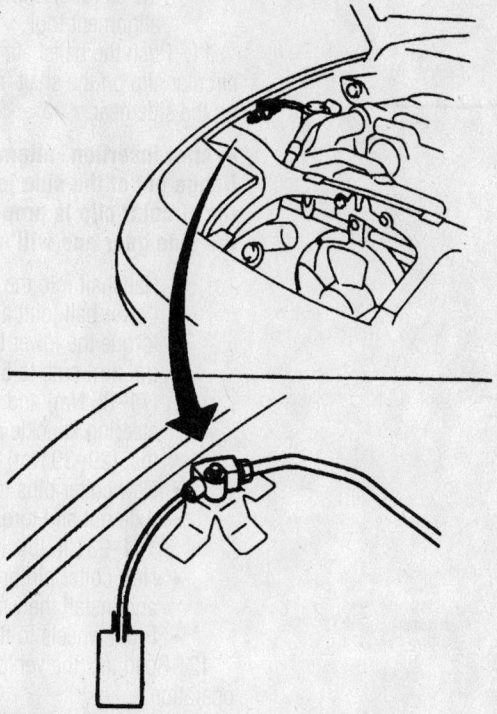

79230G74

Clutch system bleeding points—Altima and Maxima

1. Before servicing the vehicle, refer to the precautions in the beginning of this section.

2. Remove the bleed screw dust cap.

3. Attach a transparent vinyl tube to the bleed screw, immersing the free end in a clean container of clean brake fluid.

4. Fill the master cylinder with the proper fluid.

5. Open the bleed screw about ¾ turn.

6. Depress the clutch pedal quickly. Hold it down. Have an assistant tighten the bleed screw. Allow the pedal to return slowly.

7. Repeat the above procedure until no more air bubbles are seen in the fluid container.

8. Remove the bleed tube.

9. Replace the dust cap and refill the master cylinder.

10. Bleed the clutch damper, if equipped.

Halfshaft

REMOVAL & INSTALLATION

Sentra

➡The halfshafts will require a special tool for the spline alignment of the halfshaft end into the transaxle case. Do not perform this procedure without access to this tool. The Kent Moore tool Number is J-34296 and J-34297

1. Before servicing the vehicle, refer to the precautions in the beginning of this section.

2. Raise the front of the vehicle and support it on jackstands, then remove the wheel and the tire assembly.

3. Remove or disconnect the following:
• Wheel

• Hub nut using a bar to hold the wheel from turning
• Clip and separate the brake hose from the strut
• Caliper assembly and support it with a wire. Do not allow the caliper to hang from the brake hose.
• Bolts that secure the strut to the steering knuckle

➡Cover the halfshaft boots with shop towels to protect them during removal of the shaft.

• Halfshaft from the knuckle by lightly tapping it with a hammer. If it is hard to remove, use a puller.

4. Remove the halfshaft from the transaxle as follows:
  a. Models without support bearing: Pry the halfshaft from the transaxle.
  b. Models with support bearing: Remove the support bearing bolts and pull the halfshaft from transaxle.

➡When removing the halfshaft from the transaxle, do not pull on the halfshaft. The halfshaft will separate at the sliding joint (damaging the boot). Use a small prybar to remove it from the transaxle. Be sure to replace the oil seal in the transaxle.

5. Remove the halfshaft from the vehicle.

To install:

6. Use a new circlip on the halfshaft and install a new oil seal to the transaxle.

➡When installing the halfshaft into the transaxle, use a oil seal protector tool to protect the oil seal from damage.

7. Install or connect the following:
• Halfshaft assembly into the transaxle

➡After installation of the halfshaft, try to pull the flange out by hand. If it pulls out, the circular clip is not locked into the transaxle.

• Support bearing bracket and torque the bolts to 19–26 ft. lbs. (25–35 Nm)

8. Lubricate the splines of the halfshaft and insert the shaft through the steering knuckle.

9. Align the steering knuckle with the lower strut mount. Torque the bolts to 68–82 ft. lbs. (92–111 Nm).

• Disc brake caliper and the brake hose to the strut with the clip
• Washer and hub nut to the halfshaft and torque the nut to 145–202 ft. lbs. (197–274 Nm)

RH

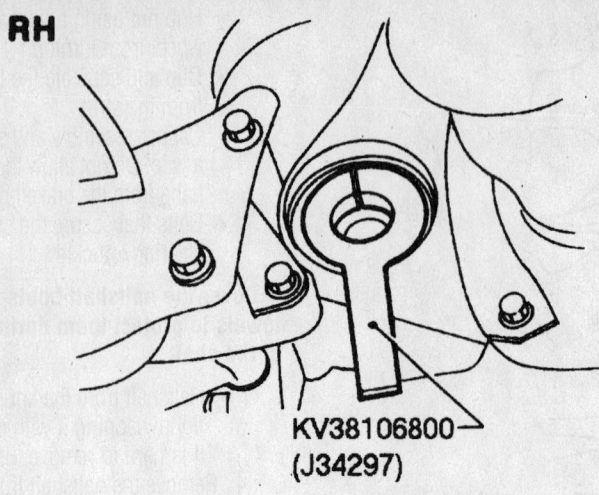

KV38106800 —
(J34297)

LH

KV38106700 —
(J34296)

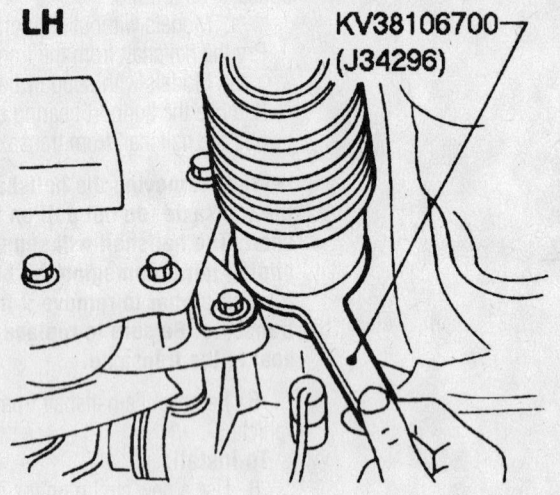

7923QG75

Halfshaft installation tools—Sentra

- Adjusting cap and a new cotter pin in drive axle
- Wheel and tire assembly and lower the vehicle

10. Road test the vehicle for proper operation.

Altima

1. Before servicing the vehicle, refer to the precautions in the beginning of this section.

2. Raise and safely support the vehicle with the front wheels hanging freely.

3. Remove or disconnect the following:
- Front wheels from the vehicle

➡️**The brake caliper does not need to be disconnected from the knuckle.**

- Cotter pin from the castellated nut on the wheel hub
- Wheel bearing locknut

➡️**Cover the CV-joint boots with a shop towel or waste cloth so not to damage them when removing the halfshaft.**

- Cotter pin and castle nut from the lower ball joint

4. Strike the knuckle with a hammer and pull down on the transverse link to separate the lower ball joint from the knuckle.

- Tie rod end from the steering knuckle
- Halfshaft from the steering knuckle by tapping it with a block of wood and a mallet

5. Using a prybar, reach through the engine crossmember and carefully pry the right inner CV-joint from the transaxle.

6. If equipped with manual transaxle, carefully pry the left inner CV-joint from the transaxle.

7. If equipped with automatic transaxle, insert a long tool into the opening for the right halfshaft and strike the tool to with a hammer.

8. Remove the left halfshaft from the transaxle.

To install:

➡️**Whenever the halfshafts are removed, the axle seals should be replaced.**

9. When installing the shafts into the transaxle, use a new oil seal and install an alignment tool along the inner circumference of the oil seal.

10. Install or connect the following:
- Halfshaft into the transaxle, align the serration's and remove the alignment tool

11. Push the halfshaft, then press-fit the circular clip on the shaft into the clip groove on the side gear.

➡️**After insertion, attempt to pull the flange out of the side joint to be sure the circular clip is properly seated in the side gear and will not come out.**

- Halfshaft into the steering knuckle
- Lower ball joint and tie rod end and torque the lower ball joint-to-control arm nuts to 52–64 ft. lbs. (71–86 Nm) and the tie rod end-to-steering knuckle nut to 22–29 ft. lbs. (29–39 Nm)
- New cotter pins to the castle nuts
- Axle nut and torque the locknut to 174–231 ft. lbs. (235–314 Nm)
- New cotter pin on the wheel hub and install the wheel
- Front wheels to the vehicle

12. Road test the vehicle for proper operation.

13. Check the transaxle fluid level and top off as necessary.

Maxima

1. Before servicing the vehicle, refer to the precautions in the beginning of this section.

2. Raise and support the front of the vehicle safely and remove the wheels.

3. Remove or disconnect the following:
- Anti-Lock Brake (ABS) wheel sensor and move it out of the way
- Brake hose from the strut
- Wheel bearing locknut
- Bolts attaching the steering knuckle to the strut. Matchmark the bolts before removal.

➡️**Cover axle boots with waste cloth so as not to damage them when removing halfshaft.**

- Halfshaft from the knuckle by slightly tapping it

- Bolts attaching the support bearing to the support bearing bracket
- Halfshaft from the transaxle with a flat-bladed tool, if equipped with a manual transaxle

4. If equipped with a automatic transaxle perform the following:

  a. Remove the right halfshaft from the vehicle.

  b. Insert a flat-bladed tool into the transaxle where the right halfshaft was, place the end of the tool on the halfshaft, then, drive the left shaft from the pinion side gear.

5. Remove or disconnect the following:
- Support bearing bolts and the halfshaft from the vehicle
- Discard the circlip on the end of the halfshaft
- Seal from the transaxle

To install:

6. Install or connect the following:
- New seal into the transaxle and install a halfshaft alignment tool into the transaxle seal
- New circlip to the halfshaft, then insert the halfshaft into the transaxle

7. With the serration's aligned remove the alignment tool.

8. Push the halfshaft fully into the transaxle to seat the circlip. Try to pull the halfshaft from the transaxle by hand to verify that the circlip is properly seated.
- Support bearing and torque the bolts to 10–14 ft. lbs. (13–19 Nm)

- Halfshaft into the steering knuckle and install the hub locknut, do not tighten the hub nut
- Steering knuckle to the strut
- Strut mounting bolts to the matchmarks and torque the bolts to 103–117 ft. lbs. (140–159 Nm)
- Brake hose to the strut
- ABS wheel sensor and torque the bolt to 13–17 ft. lbs. (18–24 Nm)
- Front wheels and torque hub locknut to 174–231 ft. lbs. (235–314 Nm)

9. Check and/or adjust the wheel alignment as necessary.

CV-Joints

OVERHAUL

Sentra

TRANSAXLE SIDE

1. Before servicing the vehicle, refer to the precautions in the beginning of this section.

2. Disassemble the joint as follows:

  a. Remove the boot bands.

  b. Matchmark the slide joint housing and inner race before separating the assembly.

  c. Matchmark the spider assembly and drive shaft.

  d. Remove the snapring.

  e. Remove the spider assembly.

  f. Remove the boot.

➡ Cover the halfshaft serrations with tape, so as not to damage the boot.

To install:

3. Assemble the joint as follows:

  a. Thoroughly clean all parts in solvent and dry with compressed air. Check parts for evidence of damage and replace as necessary.

  b. Install the boot and new small boot band on the halfshaft.

4. Install the spider assembly. Confirm that the matchmarks are aligned.

5. Install a new outer snapring.

6. Pack the CV joint with 5.5–5.8 ounces (155–165g) of grease.

  a. Install the slide joint housing.

7. Set the boot so that it does not swell or deform when its length is 4–4.07 in. (101.5–103.5mm).

8. Lock the new boot bands securely.

WHEEL SIDE

1. Before servicing the vehicle, refer to the precautions in the beginning of this section.

  The joint on the wheel side cannot be disassembled.

2. Prior to separating the joint assembly, matchmark the halfshaft and joint assembly.

3. Separate the joint using a slide hammer.

4. Remove the boot bands.

To assemble:

5. Thoroughly clean all parts in solvent and dry with compressed air. Check parts for evidence of damage and replace as necessary.

➡ Cover the halfshaft serrations with tape, so as not to damage the boot.

6. Install the boot and small boot band on the halfshaft.

7. Set the joint assembly onto the halfshaft and align the matchmarks.

8. Attach the joint assembly to the halfshaft by lightly tapping the serrated end with a plastic hammer.

➡ Using a metal hammer may damage the threads on the end of the joint.

9. Pack the CV joint with 4.6–4.41 ounces (115–125g) of grease.

10. Ensure that the boot is properly installed on the halfshaft groove.

11. Set the boot so that it does not swell or deform when its length is 3.78–3.86 in. (96–98mm).

12. Lock the new boot bands securely.

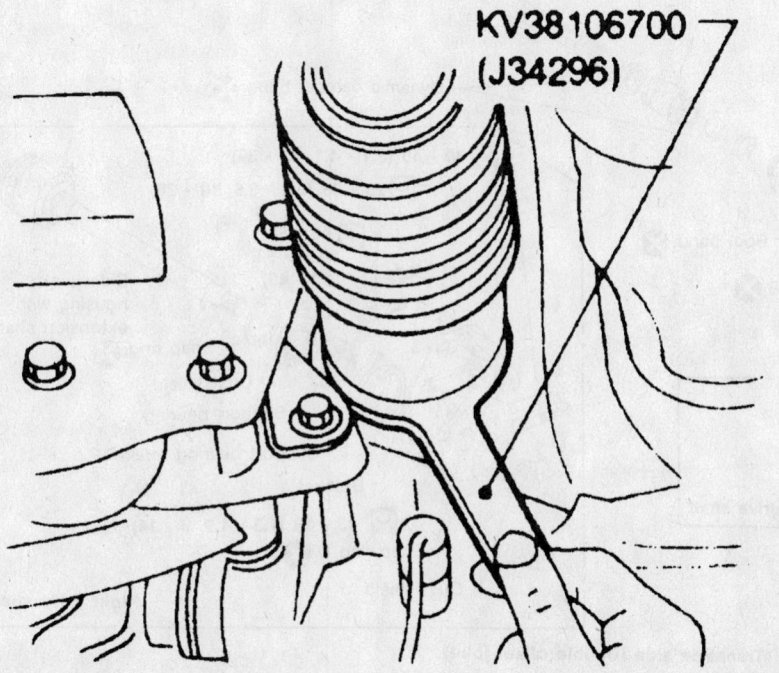

KV38106700 (J34296)

Left halfshaft alignment tool—Maxima

7923QG76

Altima

TRANSAXLE SIDE

1. Remove the boot bands.
2. Matchmark the slide joint housing and inner race, prior to separating the joint assembly.
3. Pry off the snapring and remove the ball cage, inner race and balls as a unit.
4. Remove the snapring and withdraw the boot.

➡**Cover the halfshaft serrations with tape, so as not to damage the boot.**

To install:

5. Thoroughly clean all parts in solvent and dry with compressed air. Check parts for evidence of damage and replace as necessary.
6. Install or connect the following:
   - Boot and new boot band on the halfshaft
   - New inner snapring
   - Ball cage, inner race and balls as a unit. Ensure that the matchmarks are aligned.
   - New outer snapring

7. Pack the halfshaft with 5.0–6.0 ounces (165–175 g) of grease.
8. Ensure that the boot is properly installed on the halfshaft groove.
9. Set the boot so that it does not swell or deform when its length is 3.82–3.90 in. (97–99mm).
10. Lock the new boot bands securely.

WHEEL SIDE

The joint on the wheel side cannot be disassembled.

1. Prior to separating the joint assembly, matchmark the halfshaft and joint assembly.
2. Separate the joint using a slide hammer.
3. Remove the boot bands.

To assemble:

4. Thoroughly clean all parts in solvent and dry with compressed air. Check parts for evidence of damage and replace as necessary.

➡**Cover the halfshaft serrations with tape, so as not to damage the boot.**

5. Install the boot and small boot band on the halfshaft.
6. Set the joint assembly onto the halfshaft and align the matchmarks.
7. Attach the joint assembly to the halfshaft by lightly tapping the serrated end with a plastic hammer.

➡**Using a metal hammer may damage the threads on the end of the joint.**

8. Pack the halfshaft with 3.5–4.0 ounces (100–115 g) of grease.
9. Ensure that the boot is properly installed on the halfshaft groove.
10. Set the boot so that it does not swell or deform when its length is 3.327–3.406 in. (84.5–86.5mm).
11. Lock the new boot bands securely.

Maxima

TRANSAXLE SIDE

1. Remove the boot bands.
2. Matchmark the slide joint housing and inner race, prior to separating the joint assembly.

Circular clip:
Make sure circular clip is properly meshed with side gear (transaxle side) and joint assembly (wheel side), and will not come out.
Be careful not to damage boots. Use suitable protector or cloth during removal and installation.

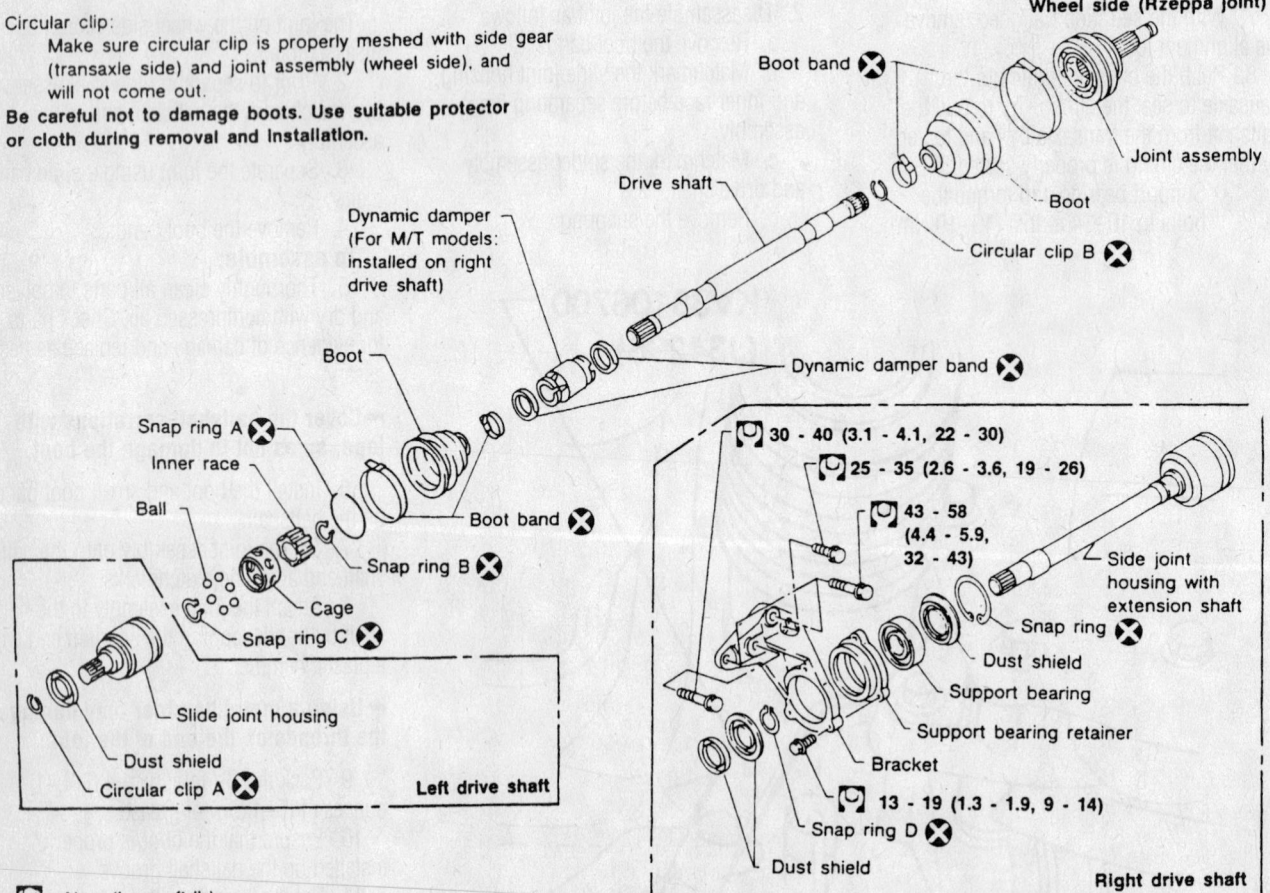

□ : N·m (kg-m, ft-lb)

Exploded view of the halfshafts and related components

89617G09

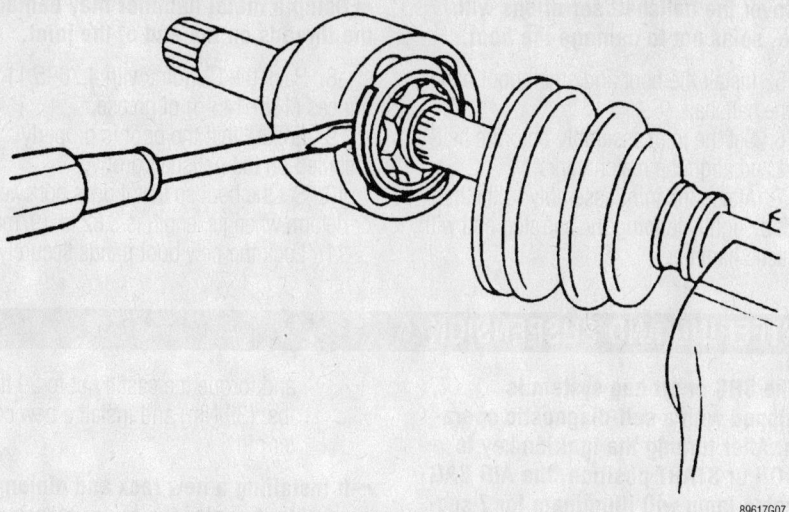

The inner CV joint uses a large C-clip to retain the ball and cage assembly in the outer housing

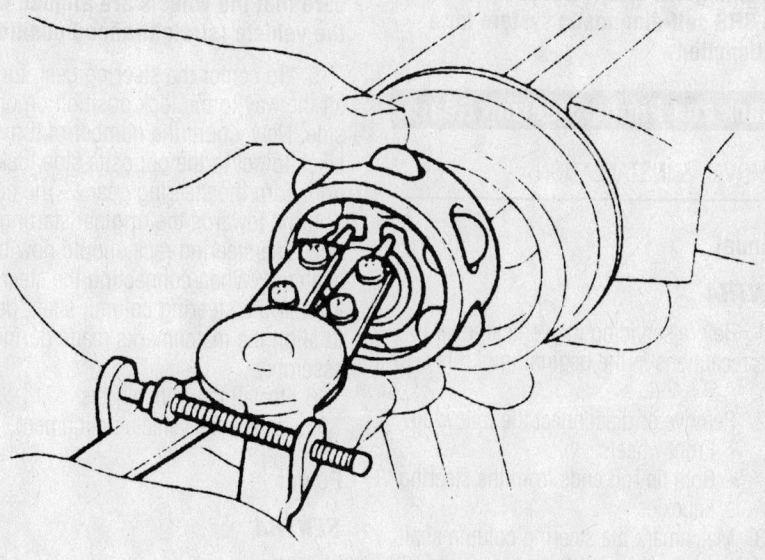

After the outer housing is removed, the ball and cage assembly can slide from the shaft by removing the C-clip

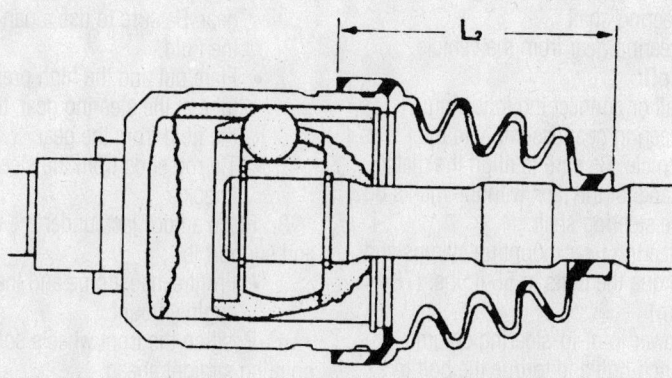

Make sure to properly position the boot before tightening the boot clamps

3. Pry off the snapring and remove the ball cage, inner race and balls as a unit.

4. Remove the snapring and withdraw the boot.

To install:

➡ **Cover the halfshaft serrations with tape, so as not to damage the boot.**

5. Thoroughly clean all parts in solvent and dry with compressed air. Check parts for evidence of damage and replace as necessary.

6. Install or connect the following:
- Boot and new boot band on the halfshaft
- New inner snapring
- Ball cage, inner race and balls as a unit. Confirm that the matchmarks are aligned.
- New outer snapring

7. Pack the CV joint with 5.0–6.0 ounces (165–175 g) of grease.

8. Ensure that the boot is properly installed on the halfshaft groove.

9. Set the boot so that it does not swell or deform when its length is 3.86 in. (98mm).

10. Lock the new boot bands securely.

WHEEL SIDE

The joint on the wheel side cannot be disassembled.

1. Prior to separating the joint assembly, matchmark the halfshaft and joint assembly.

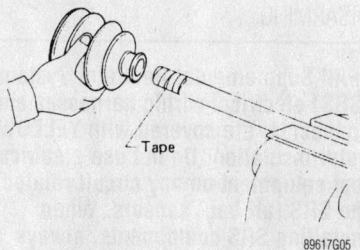

Use vinyl tape and wrap the end of the shaft to protect the boot during installation

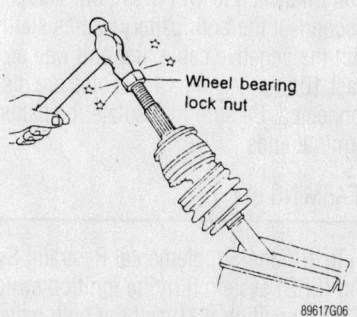

Use an old nut to protect the threads when tapping the outer CV joint onto the shaft

2. Separate the joint using a slide hammer.

3. Remove the boot bands and the boot.

To install:

4. Thoroughly clean all parts in solvent and dry with compressed air. Check parts for evidence of damage and replace as necessary.

➡**Cover the halfshaft serrations with tape, so as not to damage the boot.**

5. Install the boot and small boot band on the halfshaft.

6. Set the joint assembly onto the halfshaft and align the matchmarks.

7. Attach the joint assembly to the halfshaft by lightly tapping the serrated end with a plastic hammer.

➡**Using a metal hammer may damage the threads on the end of the joint.**

8. Pack the CV joint with 4.76–5.11 ounces (135–145 g) of grease.

9. Ensure that the boot is properly installed on the halfshaft groove.

10. Set the boot so that it does not swell or deform when its length is 3.82 in. (97mm).

11. Lock the new boot bands securely.

STEERING AND SUSPENSION

Air Bag

PRECAUTIONS

Several precautions must be observed when handling the inflator module to avoid accidental deployment and possible personal injury.

1. Never carry the inflator module by the wires or connector on the underside of the module.

2. When carrying a live inflator module, hold securely with both hands and ensure that the bag and trim cover are pointed away.

3. Place the inflator module on a bench or other surface with the bag and trim cover facing up.

4. With the inflator module on the bench, never place anything on or close to the module that may be thrown in the event of an accidental deployment.

DISARMING

➡**All Supplemental Restraint System (SRS) electrical wiring harnesses and connectors are covered with YELLOW outer insulation. Do not use electrical test equipment on any circuit related to the SRS (air bag) sensors. When installing SRS components, always install with the arrow marks facing the front of the vehicle.**

To disarm the SRS system turn the ignition switch to **OFF** position. Then, disconnect the both battery cables starting with the negative cable first and wait at least 10 minutes after the cables are disconnected. Be sure to insulate the battery terminal ends.

REARMING

To arm the Supplemental Restraint System (SRS) system turn the ignition switch to **OFF** position. Connect the both battery cables starting with the positive cable first.

➡**The SRS or air bag system is equipped with a self-diagnostic operation. After turning the ignition key to the ON or START position, the AIR BAG warning lamp will illuminate for 7 seconds. After 7 seconds, the AIR BAG lamp will extinguish if no malfunction is detected. If the AIR BAG lamp does not extinguish after 7 seconds, check the SRS self-diagnostic system for a malfunction.**

Rack and Pinion Steering Gear

REMOVAL & INSTALLATION

Manual

SENTRA

1. Before servicing the vehicle, refer to the precautions in the beginning of this section.

2. Remove or disconnect the following:
   • Front wheels
   • Both tie rod ends from the steering knuckles

3. Matchmark the steering column shaft to the lower joint and remove the pinch bolt from the joint.
   • Steering gear mounting bolts
   • Mounting clamps from the steering gear
   • Steering gear by sliding it off the steering shaft
   • Steering gear from the vehicle

To install:

4. Install or connect the following:
   • Steering gear assembly to the vehicle. Be sure to align the matchmarks of the rack with the marks on the steering shaft.
   • Steering gear mounting clamps and torque the bolts to 58 ft. lbs. (78 Nm)
   • Lower joint-to-steering column pinch bolt and torque the bolt to 22 ft. lbs. (29 Nm)
   • Tie rod end to the steering knuckle

and torque the castle nut to 29 ft. lbs. (39 Nm) and install a new cotter pin

➡**If installing a new rack and pinion assembly, transfer the lower steering joint to the new rack and pinion prior to installation. When installing the lower steering joint to the steering gear, be sure that the wheels are aligned with the vehicle (straight-ahead position).**

5. To center the steering gear, turn it all the way to the lock position on one side. Now, count the number of turns it takes to get to the opposite side lock position. Turn the steering gear ½ the number of turns towards the original starting position. The steering rack should now be centered. When connecting the steering joint to the steering column shaft, be sure to align the matchmarks made during disassembly.

6. Install the front wheels

7. Check the vehicle's alignment.

Power

SENTRA

1. Before servicing the vehicle, refer to the precautions in the beginning of this section.

2. Remove or disconnect the following:
   • Low pressure hose clamp
   • Low pressure hose at the steering gear. Be sure to use a pan to catch the fluid.
   • Flare nut and the high pressure tube at the steering gear, then drain the fluid from the gear
   • Tie rod ends from the steering knuckle

3. Place a floor jack under the transaxle and support it.
   • Front exhaust pipe and the rear engine mount

4. Position the front wheels so they are pointing straight ahead.

5. Matchmark the steering column lower joint to the steering gear.

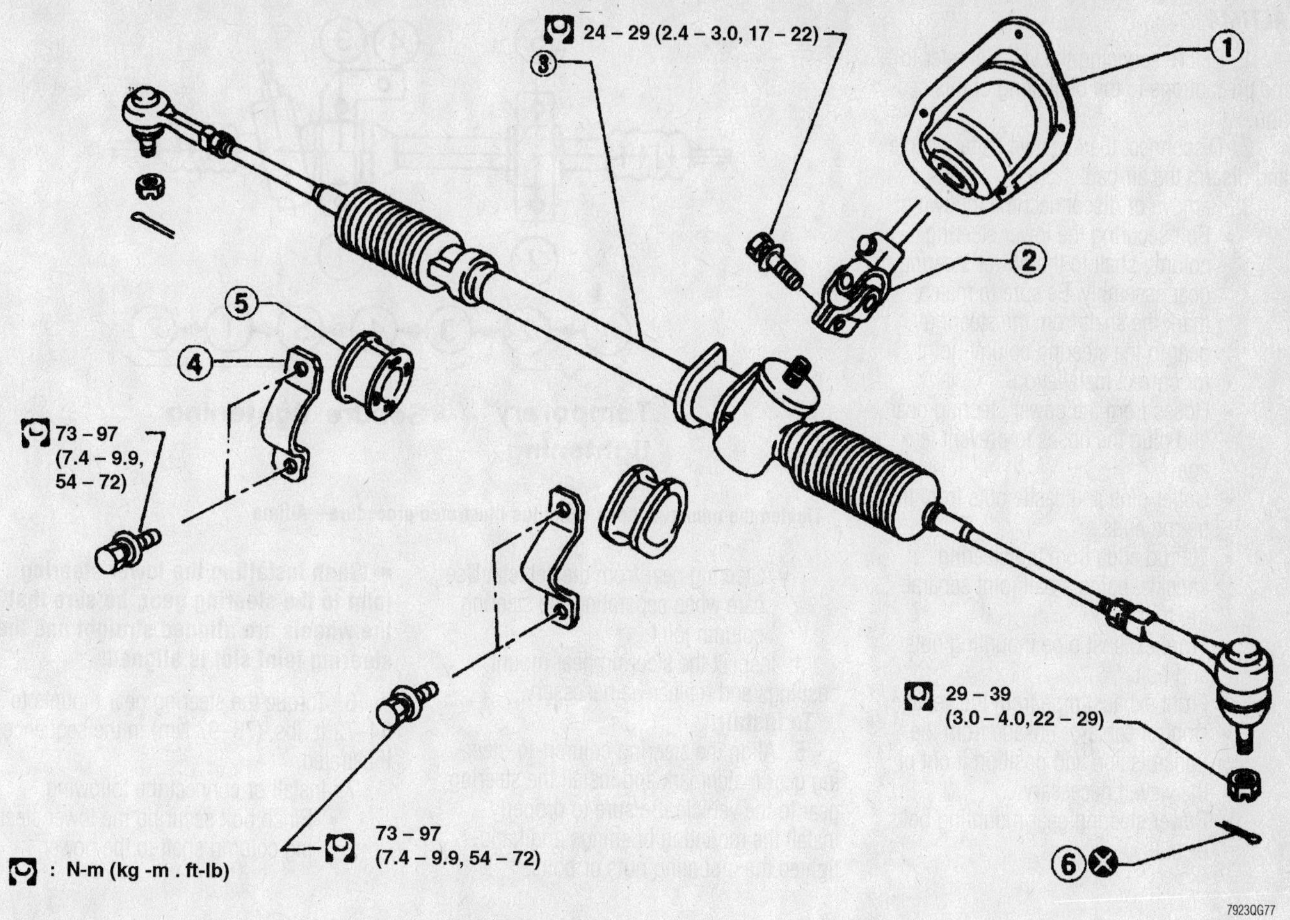

24 – 29 (2.4 – 3.0, 17 – 22)

73 – 97
(7.4 – 9.9,
54 – 72)

73 – 97
(7.4 – 9.9, 54 – 72)

29 – 39
(3.0 – 4.0, 22 – 29)

: N-m (kg -m . ft-lb)

7923QG77

Exploded view of the manual rack and pinion steering gear mounting—Sentra

➡ **The steering gear splines have a flat spot or keyway. Be sure to note this during removal.**

6. Remove or disconnect the following:
   • Bolt that secures steering column lower joint
   • Bolts, steering gear unit and the linkage

To install:

7. Install or connect the following:
   • Power steering gear assembly to the vehicle. Align the steering column to the steering gear.

➡ **Be sure to align the flat spot or keyway during installation.**

   • Steering gear mounts and torque the bolts in sequence to 54–72 ft. lbs. (73–97 Nm)
   • Pinch bolt for the steering column-to-gear connection and torque the bolt to 17–22 ft. lbs. (24–29 Nm)
   • Tie rod ends to the steering knuckle and torque the nut to 22–29 ft. lbs. (29–39 Nm). Tighten the tie rod mounting nut further so the groves

in the nut align with first cotter pin hole. Install a new cotter pin.
   • Power steering low pressure hose and torque the fitting to 20–29 ft. lbs. (27–39 Nm)
   • Power steering high pressure and torque the fitting to 11–18 ft. lbs. (15–25 Nm)

   • Rear engine mount and remove the floor jack
   • Front exhaust pipe assembly using new gaskets
8. Fill the power steering system and start the engine.
9. Check the wheel alignment.

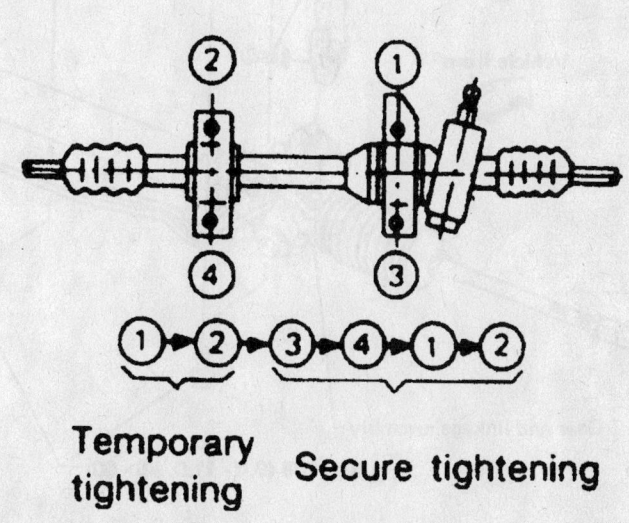

Temporary tightening   **Secure tightening**

7923QG78

Tighten the power steering gear mounting bolts according to the sequence shown—Sentra

ALTIMA

1. Before servicing the vehicle, refer to the precautions in the beginning of this section.

2. Disconnect the negative battery cable and disarm the air bag.

3. Remove or disconnect the following:
- Bolt securing the lower steering column shaft to the power steering gear assembly. Be sure to matchmark the shaft from the steering gear to the steering column joint for correct installation.
- Hoses from the power steering gear and plug the hoses to prevent leakage
- Cotter pins and castle nuts from the tie rod ends
- Tie rod ends from the steering knuckle, using a ball joint separating tool
- Front exhaust pipe mounting nuts and bolts
- Front exhaust pipe from the vehicle
- Control cable or linkage from the transmission and position it out of the way, if necessary
- Power steering gear mounting bolts or nuts

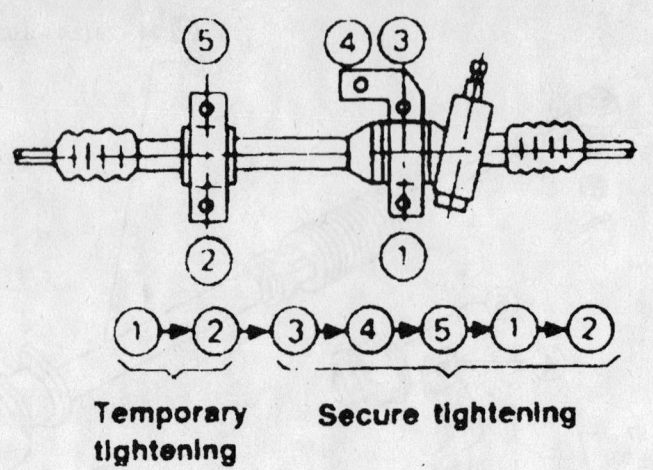

Tighten the mounting bolts using the illustrated procedure—Altima

- Steering gear from the vehicle. Use care when separating the steering column joint.

4. Inspect the steering gear mount bushings and replace as necessary.

To install:

5. Align the steering column-to-steering gear matchmark and install the steering gear to the vehicle. Be sure to properly install the mounting bushings and hand-tighten the mounting nuts or bolts.

➡ When installing the lower steering joint to the steering gear, be sure that the wheels are aligned straight and the steering joint slot is aligned.

6. Torque the steering gear mounts to 54–72 ft. lbs. (73–97 Nm) in the sequence illustrated.

7. Install or connect the following:
- Pinch bolt securing the lower steering column shaft to the power

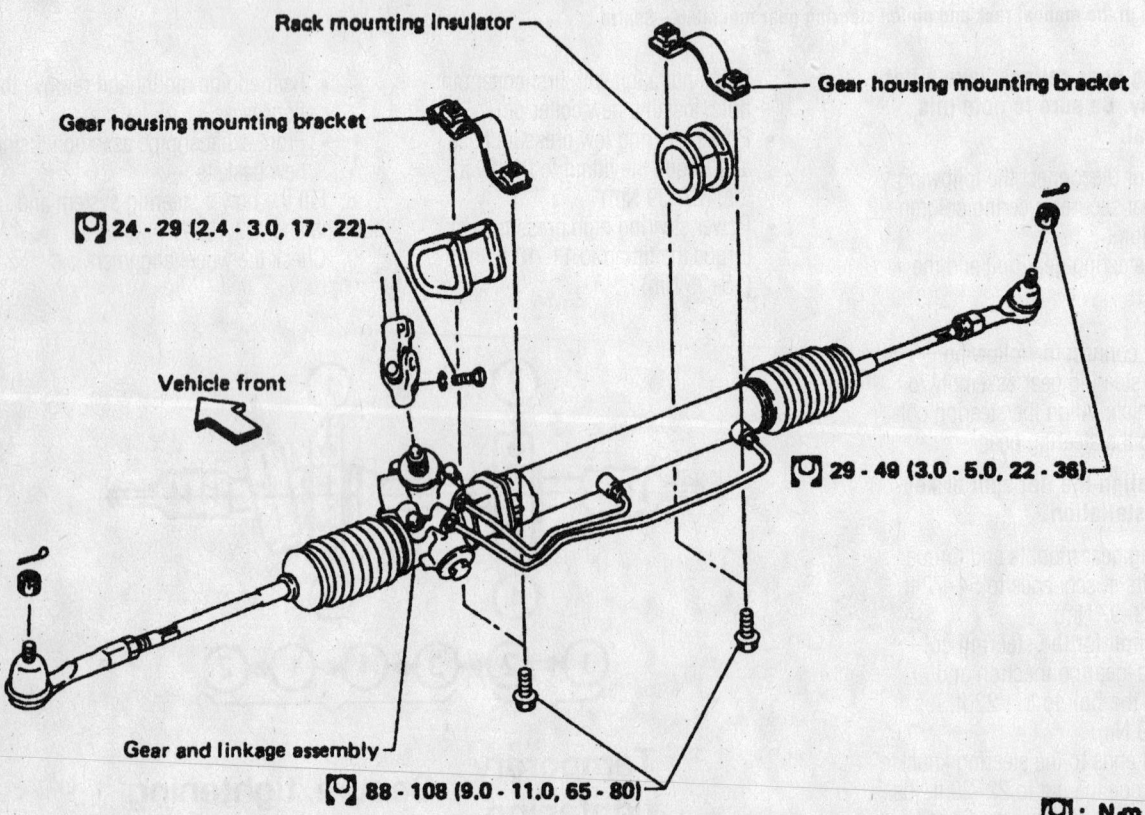

Exploded view of the power steering gear mounting—Altima

steering gear assembly and torque the pinch bolt to 17–22 ft. lbs. (24–29 Nm).
- Tie rod end to steering knuckle and torque the castle nut to 22–29 ft. lbs. (29–39 Nm). Tighten the castle nut further to align the slot in the castle nut with the cotter pin hole and install a new cotter pin.
- Control cable or linkage to the transmission, if removed
- Front exhaust pipe assembly, using new gaskets
- Power steering hoses to the steering gear

8. Start the engine and fill the power steering reservoir.

9. Perform a front end alignment.

10. Connect the negative battery cable.

11. If equipped, enable the air bag system.

MAXIMA

1. Before servicing the vehicle, refer to the precautions in the beginning of this section.

2. Disconnect both battery cables and wait at least 10 minutes after the battery cables are disconnected. This will disarm the air bag system so the steering wheel can be removed.

3. Point the front tires straight ahead and lock the steering in this position.

✳✳ WARNING

Do not turn the steering wheel or column with the lower joint removed

from the steering column or the spiral cable may be damaged.

4. Remove the steering wheel.

➡**The steering wheel must be removed before disconnecting the steering column lower joint to avoid damaging the Supplemental Restraint System (SRS) spiral cable.**

5. Raise and support the vehicle safely and remove the front wheels.

6. Remove or disconnect the following:
- Tie rod ends from the steering knuckles
- Front exhaust tube and properly support the engine
- Bolts attaching the engine mounts to the engine mounting center member
- Engine mounting center member and rear engine mount
- Front stabilizer bar from the vehicle
- Nuts attaching the hole cover to the bulkhead

7. Move the hole cover aside and disconnect the lower joint from the rack and pinion. Matchmark the pinion shaft and the pinion housing to record the steering neutral position.
- Power steering fluid pipes from the rack and pinion
- Bolts attaching the mounting brackets
- Rack and pinion from the vehicle

To install:

8. Position the rack and pinion in the vehicle and install the mounting brackets.

Torque the mounting nuts and bolts in the proper sequence to 54–72 ft. lbs. (73–97 Nm).

9. Install or connect the following:
- New O-rings to the power steering fluid pipes and connect them to the rack and pinion. Torque the low pressure line 20–29 ft. lbs. (27–39 Nm) and the high pressure line to 11–18 ft. lbs. (15–25 Nm).

10. Align the lower steering joint to the pinion shaft and install the joint onto the pinion shaft. Torque the bolt to 17–22 ft. lbs. (24–29 Nm).
- Hole cover and torque the nuts to 36–43 inch lbs. (4–5 Nm)
- Front stabilizer
- Engine mounting center member and torque the bolts to 57–72 ft. lbs. (77–98 Nm)
- Engine mounts to the center member. Torque the bolts to 57–72 ft. lbs. (77–98 Nm). Remove the support from the engine.
- Remaining components in the reverse order of removal

11. Torque the tie rod end nuts to 22–29 ft. lbs. (29–39 Nm), then install a new cotter pin.

12. Fill the power steering reservoir with fluid and bleed the air from the power steering system.

13. Check the vehicle front end alignment and adjust as necessary.

Strut and Coil Spring

REMOVAL & INSTALLATION

Front

SENTRA

1. Before servicing the vehicle, refer to the precautions in the beginning of this section.

2. Raise and support the vehicle on jackstands.

3. Remove or disconnect the following:

- Wheel
- Brake tube from the strut
- Anti-Lock Brake (ABS) wiring from the strut, if equipped

4. Support the transverse link with a jackstand.
- Steering knuckle from the strut

➡**Note the positioning of the strut alignment mark for reassembly purposes.**

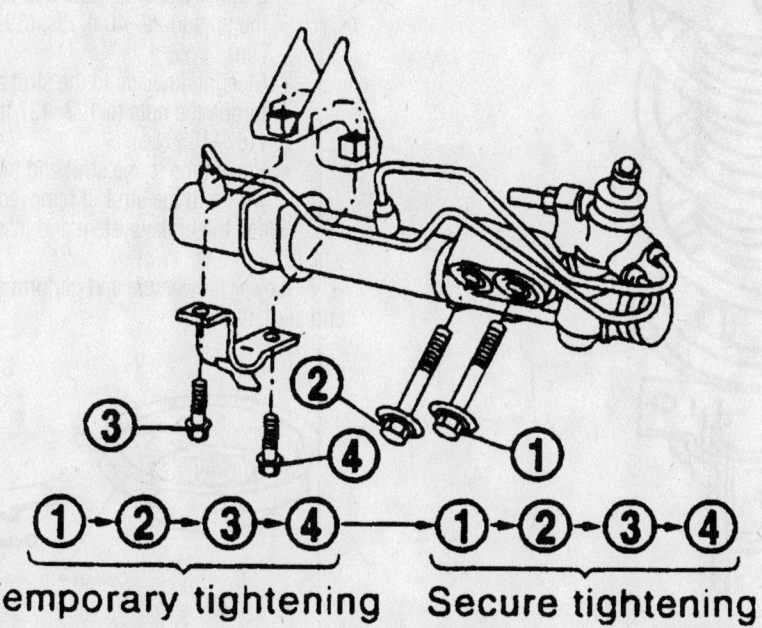

Temporary tightening   **Secure tightening**

9307QG21

Tighten the mounting bolts using the illustrated procedure—Maxima

5. Support the strut and remove the 3 upper attaching nuts. Remove the strut from the vehicle.

❋❋ CAUTION

Never loosen the center spring retaining nut until the coil spring is compressed, or serious injury or vehicle damage may occur.

6. Place the strut assembly in a vise with a holding tool or in a spring compressor.

7. Loosen the piston rod locknut.

8. Compress the spring with the spring compressor, then remove the piston rod locknut.

➡️**Before removing the strut from the coil spring, note the positioning of the strut in relationship to the coil spring for reassembly.**

9. Remove or disconnect the following:
- Strut mounting insulator bracket, strut mounting bearing, upper spring seat and the upper spring rubber seat
- Strut, leaving the coil spring compressed

- Piston boot and rebound bumper from the strut

To install:

10. Install or connect the following:
- Rebound bumper and the boot to the strut piston
- Strut into the coil spring, be sure the strut and spring are properly positioned
- Upper spring rubber seat, upper spring seat, strut mounting bearing and the strut mounting insulator bracket. Be sure that the cutout on the upper spring seat is facing the outside of the vehicle.
- Piston rod locknut. Remove the tool and torque the piston rod locknut to 43–54 ft. lbs. (59–74 Nm).

➡️**When installing the strut, be sure to position the alignment mark toward the outside of the vehicle.**

- Strut to the vehicle
- 3 upper attaching nuts and torque the nuts to 18–22 ft. lbs. (25–29 Nm)
- Steering knuckle to the strut and torque the bolts to 68–82 ft. lbs. (92–111 Nm)

- Brake tube to the strut and the ABS wiring to the strut, if it was removed

11. Bleed the brake system and install the wheel.

12. Perform a front end alignment.

ALTIMA

1. Before servicing the vehicle, refer to the precautions in the beginning of this section.

2. Raise and support the vehicle on jackstands.

3. Remove or disconnect the following:
- Wheel
- Brake tube from the strut
- Anti-Lock Brake (ABS) wiring from the strut, if equipped with ABS

4. Support the transverse link with a jackstand.
- Steering knuckle from the strut and properly support the strut
- 3 upper attaching nuts
- Strut from the vehicle

❋❋ WARNING

Never loosen the center spring retaining nut until the coil spring is compressed, or serious injury or vehicle damage may occur.

To install:

➡️**When installing the strut, be sure to position the alignment mark toward the outside of the vehicle.**

5. Install or connect the following:
- Strut to the vehicle
- 3 upper attaching nuts and torque the nuts to 29–40 ft. lbs. (39–54 Nm)
- Steering knuckle to the strut and torque the nuts to 123–137 ft. lbs. (167–186 Nm)
- Brake tube to the strut and the ABS wiring to the strut, if removed

6. Bleed the brake system and install the wheel.

7. Lower the vehicle and perform a front end alignment.

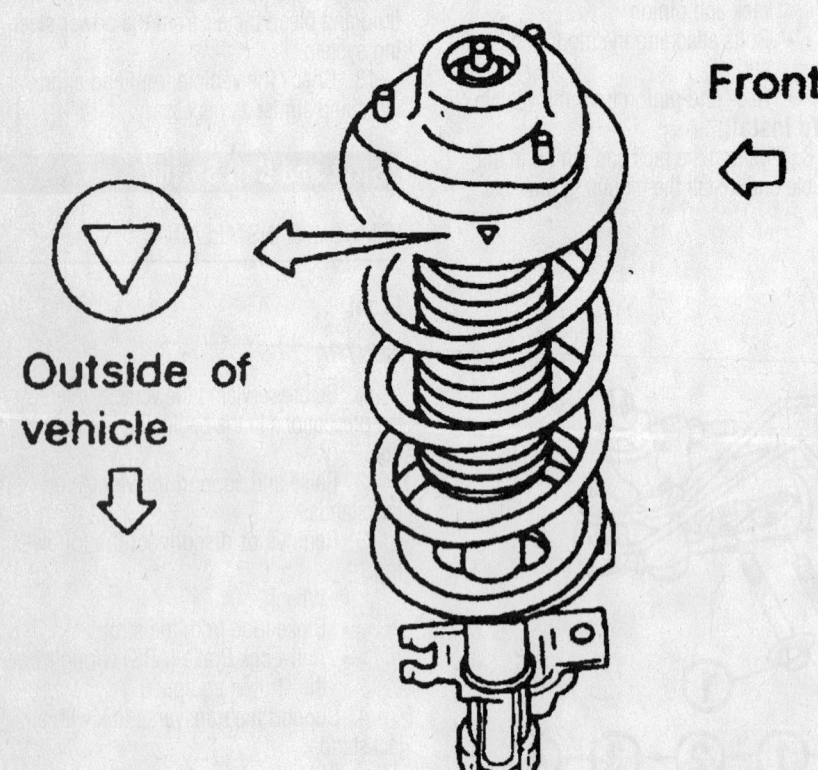

During assembly, be sure to point the alignment mark toward the outside of the vehicle—Sentra

7923QG81

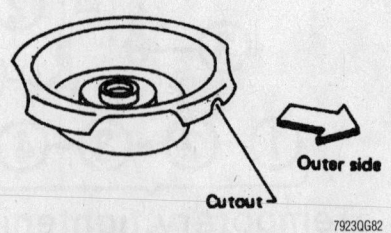

Position the alignment mark toward the outside of the vehicle—Altima

7923QG82

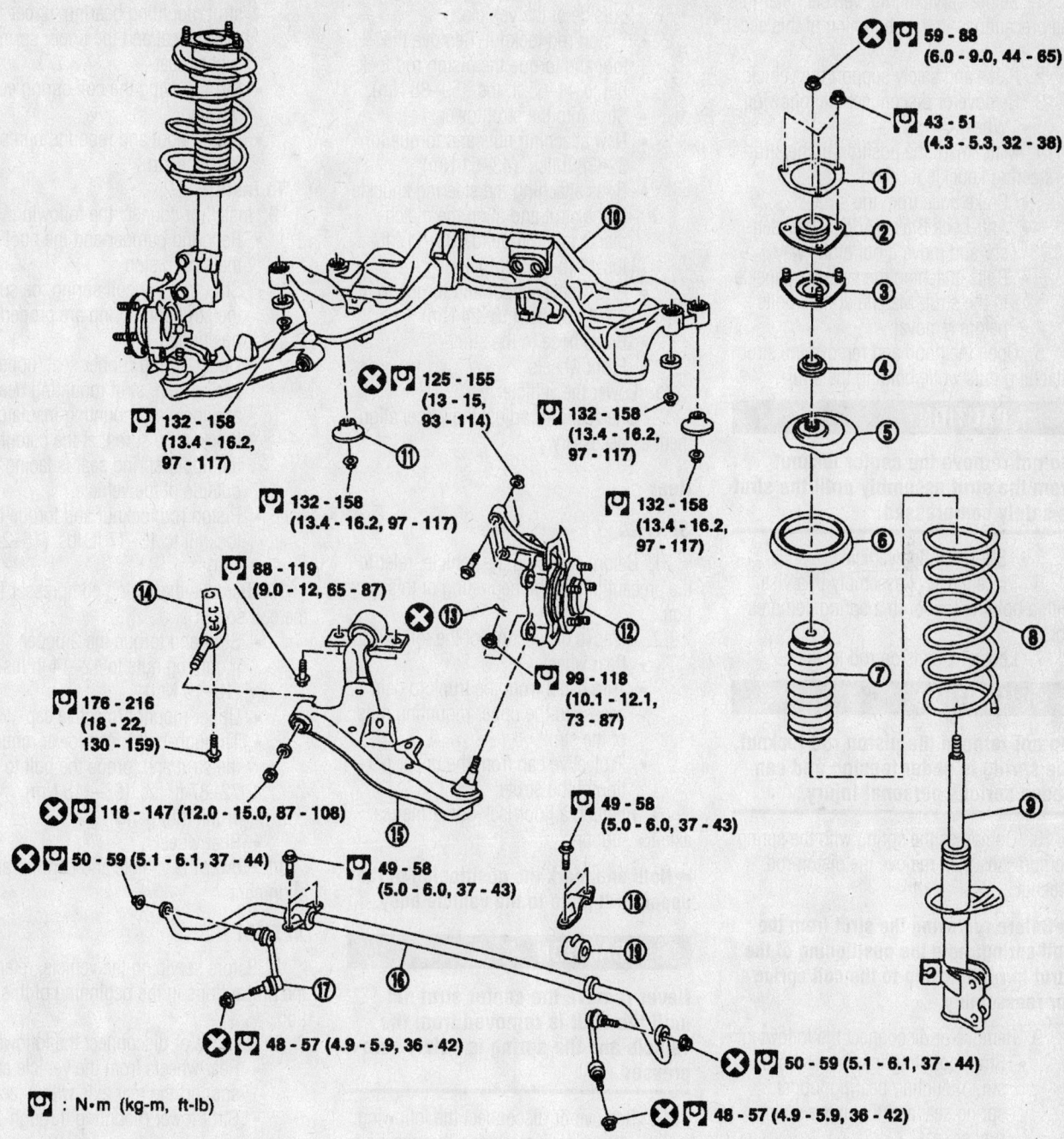

59 - 88
(6.0 - 9.0, 44 - 65)

43 - 51
(4.3 - 5.3, 32 - 38)

125 - 155
(13 - 15,
93 - 114)

132 - 158
(13.4 - 16.2,
97 - 117)

132 - 158
(13.4 - 16.2,
97 - 117)

132 - 158
(13.4 - 16.2, 97 - 117)

132 - 158
(13.4 - 16.2,
97 - 117)

88 - 119
(9.0 - 12, 65 - 87)

176 - 216
(18 - 22,
130 - 159)

99 - 118
(10.1 - 12.1,
73 - 87)

118 - 147 (12.0 - 15.0, 87 - 108)

49 - 58
(5.0 - 6.0, 37 - 43)

50 - 59 (5.1 - 6.1, 37 - 44)

49 - 58
(5.0 - 6.0, 37 - 43)

48 - 57 (4.9 - 5.9, 36 - 42)

50 - 59 (5.1 - 6.1, 37 - 44)

48 - 57 (4.9 - 5.9, 36 - 42)

: N•m (kg-m, ft-lb)

1. Strut spacer
2. Strut mount insulator
3. Strut mount bracket
4. Strut mount bearing
5. Spring upper seat
6. Spring rubber seat
7. Bound bumper rubber
8. Coil spring
9. Shock absorber
10. Suspension member
11. Rebound stopper
12. Wheel hub and steering knuckle
13. Cotter pin
14. Bush link pin
15. Transverse link
16. Stabilizer
17. Connecting rod
18. Stabilizer clamp
19. Bushing

9357RG17

Exploded view of the front suspension—2002 Maxima shown

MAXIMA

1. Before servicing the vehicle, refer to the precautions in the beginning of this section.
2. Raise and safely support the vehicle.
3. Remove or disconnect the following:
   - Wheel
4. Matchmark the position of the strut-to-steering knuckle location.
   - Brake hose from the strut
   - Anti-Lock Brake (ABS) wheel sensor and move it out of the way
   - Bolts attaching the steering knuckle to the strut. Matchmark the bolts before removal.
5. Open the hood and remove the strut attaching nuts while holding the strut.

✳✳ CAUTION

Do not remove the center locknut from the strut assembly until the strut is safely compressed.

   - Strut from the vehicle
6. Place the strut assembly in a vise with a holding tool or in a spring compressor.
7. Loosen the piston rod locknut.

✳✳ CAUTION

Do not remove the piston rod locknut, the spring is under tension and can cause serious personal injury.

8. Compress the spring with the spring compressor, then remove the piston rod locknut.

➡**Before removing the strut from the coil spring, note the positioning of the strut in relationship to the coil spring for reassembly.**

9. Remove or disconnect the following:
   - Strut mounting insulator bracket, strut mounting bearing, upper spring seat and the upper spring rubber seat
   - Strut, leaving the coil spring compressed
   - Piston boot and rebound bumper from the strut

To install:
10. Install or connect the following:
   - Rebound bumper and the boot to the strut piston
   - Strut into the coil spring, be sure the strut and spring are properly positioned
   - Upper spring rubber seat, upper spring seat, strut mounting bearing and the strut mounting insulator bracket. Be sure that the cutout on the upper spring seat is facing the outside of the vehicle.
   - Piston rod locknut. Remove the tool and torque the piston rod locknut to 44–65 ft. lbs. (59–88 Nm).
   - Strut into the strut tower
   - New attaching nuts and torque to 32–38 ft. lbs. (43–51 Nm)
   - Bolts attaching the steering knuckle to the strut and align the matchmarks and torque to 103–117 ft. lbs. (140–159 Nm)
   - ABS wheel sensor and torque to 13–17 ft. lbs. (18–24 Nm)
   - Brake hose to the strut
   - Front wheels
11. Lower the vehicle.
12. Check and/or adjust the wheel alignment as necessary.

Rear

SENTRA

1. Before servicing the vehicle, refer to the precautions in the beginning of this section.
2. Remove or disconnect the following:
   - Rear wheel
   - Trim panel from the trunk to gain access to the upper mounting nuts of the strut
   - Protective cap from the upper portion of the strut
3. Position a floor jack under the rear axle for support.

➡**Note and mark the positioning of the upper strut plate to the vehicle body.**

✳✳ CAUTION

Never remove the center strut nut until the strut is removed from the vehicle and the spring is safely compressed.

4. Remove or disconnect the following:
   - Lower strut mounting through-bolt
   - 2 upper strut mounting nuts and the strut from the vehicle
5. Place the strut assembly in a vise with a holding tool or in a spring compressor.
6. Loosen the piston rod locknut.
7. Compress the spring with the spring compressor, then remove the piston rod locknut.

➡**Before removing the strut from the coil spring, note the positioning of the strut in relationship to the coil spring for reassembly.**

8. Remove or disconnect the following:
   - Strut mounting insulator bracket, strut mounting bearing, upper spring seat and the upper spring rubber seat
   - Strut, leaving the coil spring compressed
   - Piston boot and rebound bumper from the strut

To install:
9. Install or connect the following:
   - Rebound bumper and the boot to the strut piston
   - Strut into the coil spring, be sure the strut and spring are properly positioned
   - Upper spring rubber seat, upper spring seat, strut mounting bearing and the strut mounting insulator bracket. Be sure that the cutout on the upper spring seat is facing the outside of the vehicle.
   - Piston rod locknut and torque the locknut to 13–17 ft. lbs. (18–24 Nm)
10. Remove the spring compressor from the coil spring.
   - Strut and torque the 2 upper mounting nuts to 12–14 ft. lbs. (16–19 Nm)
   - Upper mount protective cap
   - Through-bolt to the lower mount of the strut and torque the bolt to 72–87 ft. lbs. (98–118 Nm)
   - Trunk trim panel
   - Rear wheel
11. Lower the vehicle and perform an alignment.

ALTIMA

1. Before servicing the vehicle, refer to the precautions in the beginning of this section.
2. Remove or disconnect the following:
   - Rear wheels from the vehicle and support the rear axle with a jack
   - Strut lower mounting through-bolts

➡**Be sure to note the position the strut upper plate to the vehicle for reinstallation purposes.**

   - 2 nuts from the top of the strut
   - Strut as an assembly

✳✳ CAUTION

Do not remove the center locknut from the strut assembly until the strut is safely compressed.

3. Compress the strut coil spring with a spring compressor.

- Strut assembly center locknut

➡**Before removing the strut from the coil spring, note the positioning of the strut in relationship to the coil spring for reassembly.**

- Strut leaving the coil spring compressed

➡**Mark the coil spring position to the strut assembly for reinstallation purposes.**

4. To remove the spring from the strut assembly, perform the following steps:

a. Compress the coil spring with the proper compressor tool.

b. Remove the center retaining nut holding strut mounting insulator.

c. Slowly decompress the coil spring.

d. Remove the strut mounting insulator.

e. Remove coil spring.

To install:

5. Install or connect the following:

- Coil spring onto the strut assembly. Be sure to align the matchmarks made during the removal procedure.

- Strut mounting insulator and compress the coil spring assembly

➡**It will be necessary to use a new locknut for the center retaining nut of the coil spring.**

- Center retaining nut and torque to 43–58 ft. lbs. (59–78 Nm). Be sure the spring is seated properly on the strut and in the mounting insulator.

6. Slowly remove the spring compressor tool.

- Strut assembly and torque the upper nuts to 31–40 ft. lbs. (42–54 Nm)

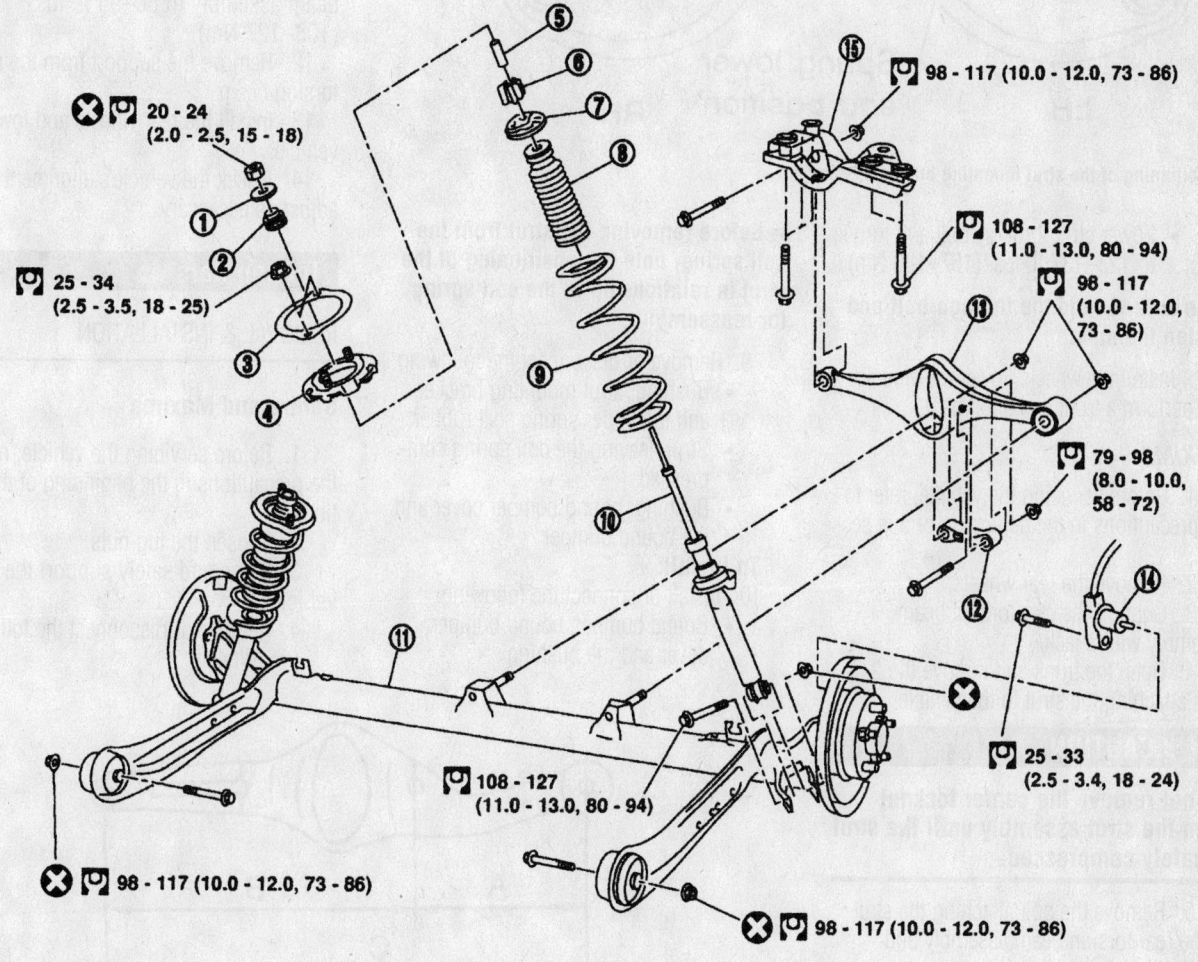

20 - 24 (2.0 - 2.5, 15 - 18)

25 - 34 (2.5 - 3.5, 18 - 25)

98 - 117 (10.0 - 12.0, 73 - 86)

108 - 127 (11.0 - 13.0, 80 - 94)

98 - 117 (10.0 - 12.0, 73 - 86)

79 - 98 (8.0 - 10.0, 58 - 72)

25 - 33 (2.5 - 3.4, 18 - 24)

108 - 127 (11.0 - 13.0, 80 - 94)

98 - 117 (10.0 - 12.0, 73 - 86)

98 - 117 (10.0 - 12.0, 73 - 86)

: N•m (kg-m, ft-lb)

1. Washer
2. Bushing
3. Shock absorber mounting seal
4. Shock absorber mounting bracket
5. Distance tube
6. Bushing
7. Bound bumper cover
8. Bound bumper
9. Coil spring
10. Shock absorber
11. Torsion beam
12. Control rod
13. Lateral link
14. ABS sensor
15. Suspension member

Exploded view of the rear suspension—2002 Maxima shown

9357RG18

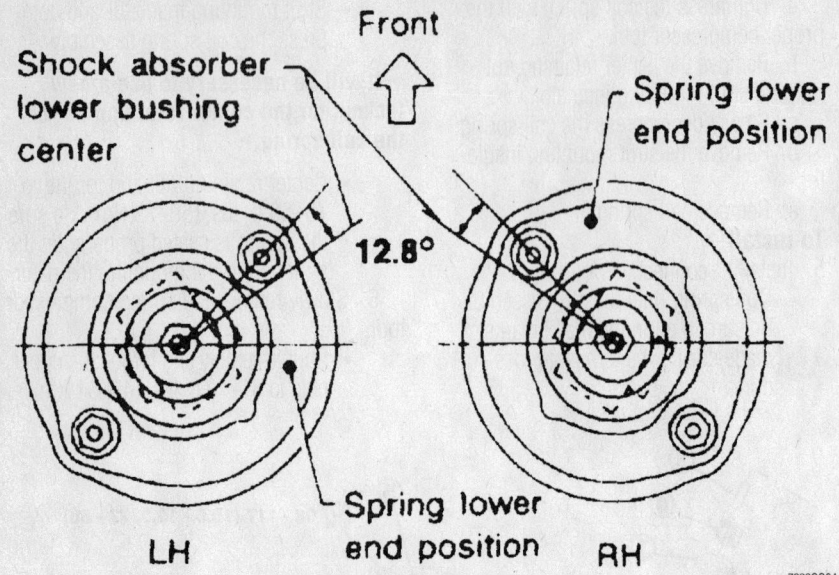

Front

Shock absorber lower bushing center

Spring lower end position

12.8°

Spring lower end position

LH   RH

7923QG84

Positioning of the strut mounting brackets—Maxima

- Lower strut through-bolt and torque to 123–137 ft. lbs. (167–186 Nm)

➡ **Be sure to hold the through-bolt and tighten the nuts.**

7. Install the wheels, lower the vehicle and perform a front end alignment.

MAXIMA

1. Before servicing the vehicle, refer to the precautions in the beginning of this section.
2. Remove the rear wheels.
3. Support the rear torsion beam assembly with a jack.
4. Open the trunk and remove the 2 nuts attaching the strut to the vehicle.

❋❋ CAUTION

Do not remove the center locknut from the strut assembly until the strut is safely compressed.

5. Remove the bolt attaching the strut to the rear torsion beam assembly and remove the strut.
6. Place the strut assembly in a vise with a holding tool or in a spring compressor.
7. Loosen the piston rod locknut.

❋❋ CAUTION

Do not remove the piston rod locknut, the spring is under tension and can cause serious personal injury.

8. Compress the spring with the spring compressor, then remove the piston rod locknut.

➡ **Before removing the strut from the coil spring, note the positioning of the strut in relationship to the coil spring for reassembly.**

9. Remove or disconnect the following:
- Bushing, strut mounting bracket and the upper spring seat rubber
- Strut, leaving the coil spring compressed
- Bushing, bound bumper cover and the bound bumper

To install:

10. Install or connect the following:
- Bound bumper, bound bumper cover and the bushing

- Strut into the coil spring, be sure the strut and spring are properly positioned
- Upper spring seat rubber, strut mounting bracket and the bushing. Be sure that the mounting bracket is properly positioned.
- Piston rod locknut. Remove the tool and torque the piston rod locknut to 15–18 ft. lbs. (20–24 Nm).
- Strut and torque the new nuts to 18–25 ft. lbs. (25–34 Nm)

11. Position the strut on the rear torsion beam and install the bolt. Torque the bolt attaching the strut to the torsion beam assembly to 80–94 ft. lbs. (108–127 Nm).

12. Remove the support from the rear torsion beam.

13. Install the rear wheels and lower the vehicle.

14. Check the vehicle's alignment and adjust as necessary.

Torsion Bars

REMOVAL & INSTALLATION

Sentra and Maxima

1. Before servicing the vehicle, refer to the precautions in the beginning of this section.
2. Loosen the lug nuts.
3. Raise and safely support the vehicle
4. Remove or disconnect the following:
- Wheels

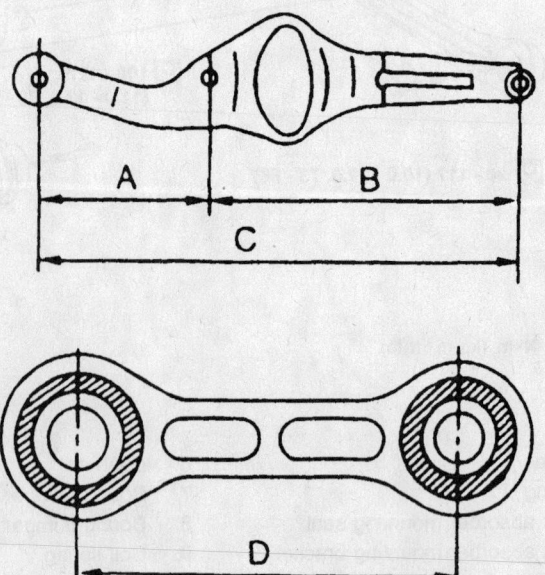

89618G33

Measure the control rod and lateral links at these points—Sentra and Maxima

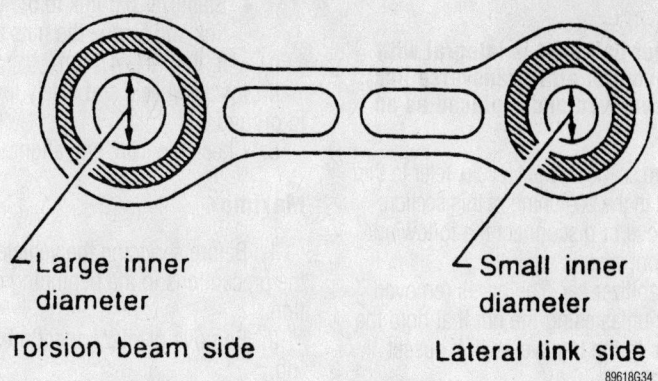

Large inner diameter

Torsion beam side

Small inner diameter

Lateral link side

89618G34

Be sure to install the control rod correctly—Sentra and Maxima

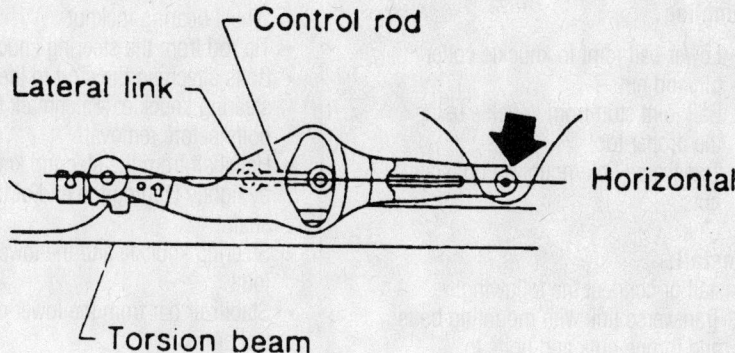

Control rod

Lateral link

Horizontal

Torsion beam

89618G35

The lateral link must be in the horizontal position when tightening the bolts—Sentra and Maxima

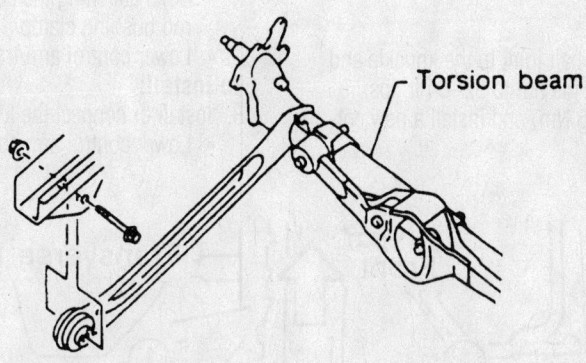

Torsion beam

89618G36

Tighten the torsion beam-to-chassis bolts with the suspension unloaded—Sentra and Maxima

---

✳✳ WARNING

Be sure to disconnect the Anti-lock Brake System (ABS) wheel sensor from the assembly. Failure to do so may result in damage to the sensor wire and the sensor becoming inoperative.

- Brake calipers and suspend them with a piece of wire. Do not let them hang by the hose.

5. Using a transmission jack, raise the torsion beam a little, then remove the suspension mounting bolts.

6. Lower the jack and remove the suspension assembly.

7. Remove the lateral link and control rod.

8. Inspect the torsion beam and control rod for cracks, wear and deformation. The length of the lateral link and control rod is as follows:

- A—8.15–8.19 in. (207–208mm)
- B—15.51–15.55 in. (394–395mm)

- C—23.66–23.74 in. (601–603mm)
- D—4.17–4.25 in. (106–108mm)

To install:

9. When installing the control rod, connect the bushing with the smaller inner diameter to the lateral link. Install the lateral link and the control rod on the torsion beam. Place the lateral link with the arrow topside.

10. Place the lateral link and control rod horizontally against the beam and tighten the bolts. Refer to the illustration.

11. Secure the torsion beam to the vehicle. Make sure the lateral link is horizontal, then tighten the link to the chassis.

12. Attach the struts to the torsion beam and tighten the fasteners.

13. Tighten the torsion beam-to-chassis bolts.

14. Install the calipers, ABS sensor and wheels. Lower the vehicle to the ground.

Lower Ball Joint

REMOVAL & INSTALLATION

The ball joint is an integral part of the lower control arm. If the ball joint is defective the control arm must be replaced.

Lower Control Arm (Transverse Link)

REMOVAL & INSTALLATION

Sentra

1. Before servicing the vehicle, refer to the precautions in the beginning of this section.

2. Remove or disconnect the following:

- Front wheels
- Disc brake caliper from the steering knuckle

✳✳ WARNING

DO NOT allow the disc brake caliper to hang from the brake hose. Support the disc caliper with safety wire.

- Cotter pin and loosen the wheel bearing locknut
- Cotter pin and the castle nut from the tie rod ball joint. Separate the tie rod with a suitable puller.
- 2 bolts that secure the lower portion of the strut to the steering knuckle

3. Using a plastic or rubber mallet, tap on the loosened wheel bearing locknut to loosen the halfshaft in the knuckle. Remove the locknut and remove the halfshaft from the steering knuckle. Be sure to cover the CV-joints with a shop rag.

➡ **Support the halfshaft assembly with wire. Do not allow the halfshaft to hang by the inner joint.**

- Nut that secures the stabilizer link to the lower control arm
- Link from the control arm. Note the positioning of the washers and spacers for reassembly.
- Cotter pin and castle nut from the lower ball joint
- Lower ball joint from the knuckle
- Knuckle from the vehicle
- Mounting nuts/bolts that secure the lower control arm to the frame
- Control arm from the vehicle

To install:

➡ **Final tightening of all suspension components should take place with the weight of the vehicle on the wheels.**

4. Install the lower control arm assembly and torque mounting bolts/nuts as follows:

a. Through bolt and nut: 76–90 ft. lbs. (103–123 Nm).

b. 2 saddle bracket mounting bolts: 58–72 ft. lbs. (78–98 Nm).

5. Install or connect the following:
- Steering knuckle to the lower ball joint and torque the castle nut to 43–54 ft. lbs. (59–74 Nm). Install a new cotter pin.
- Stabilizer link to the lower control arm and torque the nut to 12–16 ft. lbs. (16–22 Nm).
- Halfshaft through the wheel bearing
- Wheel bearing locknut. Do not torque the locknut at this time.
- Steering knuckle to the strut and torque the bolts to 68–82 ft. lbs. (92–111 Nm).
- Tie rod end and torque the castle nut to 22–29 ft. lbs. (29–39 Nm). Install a new cotter pin.
- Disc brake caliper to the steering knuckle

6. Tighten the halfshaft mounting nut (hub nut) and torque the nut to 145–202 ft. lbs. (196–274 Nm). It may be necessary to have an assistant hold the brake pedal while tightening the locknut. Install the adjusting cap and a new cotter pin.

7. Install the front wheels, lower the vehicle and perform a front end alignment.

Altima

➡ **The lower ball joint is integral with the lower control arm (transverse link). They are removed and replaced as an assembly.**

1. Before servicing the vehicle, refer to the precautions in the beginning of this section.

2. Remove or disconnect the following:
- Front wheels
- Stabilizer bar. The bar is removed by unfastening the nut that hold the bar to the transverse link gusset plate.

➡ **Take note of position of marks on clamp face and stabilizer bar for reassembling.**

- Lower ball joint to knuckle cotter pin and nut
- Ball joint stud from knuckle using the proper tool
- Transverse link mounting bolts and nuts
- Link

To install:

3. Install or connect the following:
- Transverse link with mounting bolts and torque nuts and bolts to 87–108 ft. lbs. (118–147 Nm)

➡ **The final tightening of suspension components must be done with wheels on the ground and vehicle at curb weight.**

- Lower ball joint to the knuckle and torque the nut to 52–64 ft. lbs. (71–86 Nm) and install a new cotter pin

- Stabilizer bar link to the transverse link and torque the nuts to 30–35 ft. lbs. (41–47 Nm)

4. Install wheels and safely lower vehicle to ground.

5. Check the front end alignment.

Maxima

1. Before servicing the vehicle, refer to the precautions in the beginning of this section.

2. Remove or disconnect the following:

- Front wheels
- Anti-Lock Brake (ABS) wheel sensor and move it out of the way
- Wheel bearing locknut
- Tie rod from the steering knuckle
- Bolts attaching the strut to the steering knuckle. Matchmark the bolts before removal.
- Halfshaft from the steering knuckle by lightly tapping the end of the shaft
- Steering knuckle and the lower ball joint
- Stabilizer bar from the lower control arm
- Bolts attaching the link bushing pin to the chassis
- Nut attaching the link to the control arm and the link, if necessary
- Bolts attaching the compression rod bushing clamp
- Lower control arm/traverse link

To install:

3. Install or connect the following:
- Lower control arm and the com-

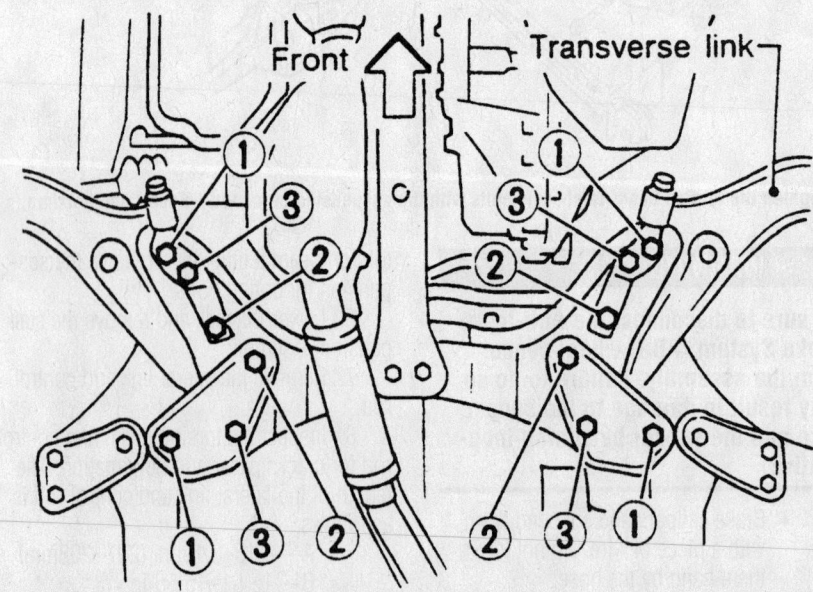

Bolt tightening sequence for the lower control arms—2000 Maxima

9301QG01

pression rod bushing clamp into the vehicle
- Link bushing pin, if removed from the control arm

4. Tighten all bolts and nuts until they are snug enough to support the weight of the vehicle but not fully tight, the bolts should be torqued to specification with the vehicle on the floor.

➡**Always use a new nut when installing the ball joint to the control arm.**

- Steering knuckle to the strut and to the halfshaft
- Strut mounting bolts and torque the bolts to 103–117 ft. lbs. (140–159 Nm)
- Tie rod ball joint and torque the nut to 22–29 ft. lbs. (29–39 Nm)
- Wheel bearing locknut
- ABS wheel sensor and torque the bolt to 13–17 ft. lbs. (18–24 Nm)
- Front wheels, lower the vehicle and torque hub locknut to 174–231 ft. lbs. (235–314 Nm)

5. Torque the bolts attaching the compression rod bushing clamp and the link bushing pin, in the proper sequence to 87–108 ft. lbs. (118–147 Nm).

6. If the link bushing pin was removed from the control arm torque the attaching nut to 87–108 ft. lbs. (118–147 Nm).

7. Torque the sway bar attaching nut to 30–35 ft. lbs. (41–47 Nm).

8. Check the vehicle alignment.

CONTROL ARM BUSHING REPLACEMENT

The bushing is an integral part of the lower control arm. If the bushing is defective the control arm must be replaced.

Wheel Bearings

ADJUSTMENT

Front

➡**Whenever the hub or bearing assemblies are removed, the wheel bearing must be replaced. Never reuse the old bearing assembly.**

The wheel bearings are sealed and are not adjustable. If defective, replacement is the only option.

Rear

If the wheel hub bearing assembly is removed, it must be replaced.

➡**The wheel hub bearing assembly is not repairable; it must be replaced when defective.**

1. Before servicing the vehicle, refer to the precautions in the beginning of this section.

2. Torque the wheel bearing locknut to 138–188 ft. lbs. (187–255 Nm).

3. Verify that the wheel bearings operate smoothly.

4. Install a new cotter pin into the spindle to hold the wheel bearing locknut.

5. Install a dial indicator to the rear wheel hub bearing assembly and check the axial end-play; it should be less than 0.0020 in. (0.05mm).

6. Install the grease cap.

7. If the axial end-play exceeds specifications, the wheel bearing must be replaced.

REMOVAL & INSTALLATION

Front

SENTRA

➡**Whenever the hub or bearing assembly is removed, the wheel bearing assembly must be replaced. Never reuse the old bearing assembly.**

1. Before servicing the vehicle, refer to the precautions in the beginning of this section.

2. Remove or disconnect the following:
- Front wheel
- Wheel bearing/axle shaft locknut while depressing the brake pedal
- Brake caliper and support it with a piece of wire. It is not necessary to disconnect the brake line from the caliper.
- Anti-Lock Brake System (ABS) sensor from the steering knuckle

➡**Do not depress the brake pedal or twist the brake line.**

- Tie rod end
- Halfshaft from the knuckle by slightly tapping with a soft hammer. Position the axle shaft nut on the threads of the shaft to protect them when lightly tapping.
- Lower ball joint nut and separate
- 2 strut-to-knuckle retaining bolts and separate
- Steering knuckle from the vehicle

3. Place the assembly in a vise. Drive the hub with the inner race from the knuckle with a suitable tool. Remove the inner and outer grease seals.
- Bearing inner race and outer grease seal from the hub
- Snapring and press the bearing outer race to remove the bearing from the steering knuckle

To install:

4. Press a new wheel bearing into the knuckle assembly not exceeding 3.3 tons (2994 kg) pressure.

5. Install or connect the following:
- Snapring and pack the grease seal lips with chassis grease
- Inner and outer grease seals

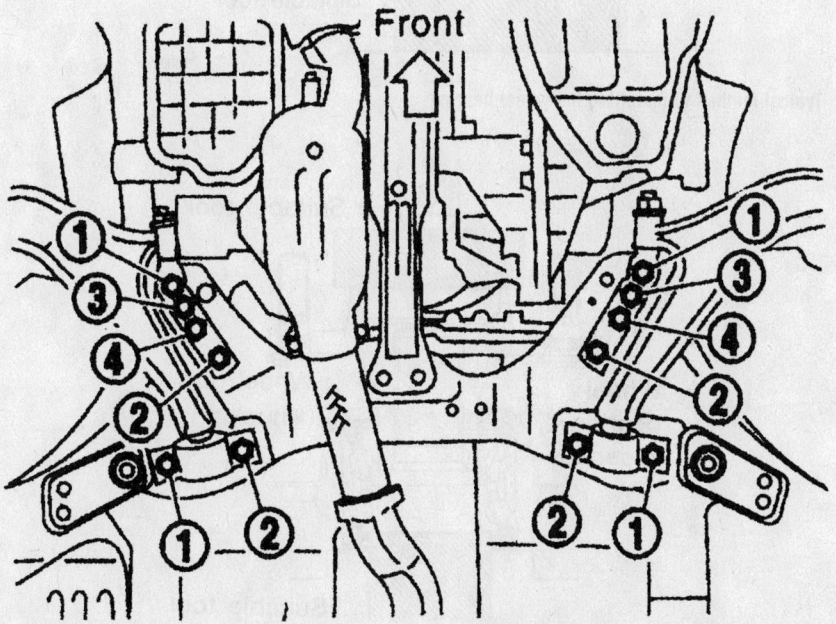

Bolt tightening sequence for the lower control arms—2001–03 Maxima

9357RG19

6. Press the wheel hub into the knuckle not exceeding 3.3 tons (2994 kg) pressure.

7. Check bearing operation and by applying 3.9–5.5 tons (3538–4990 kg) pressure to the hub assembly. Spin the hub several times in both directions.

8. Be sure the bearings rotate freely. If the bearings do not rotate freely, replace the bearings.

9. Install or connect the following:
- Knuckle and wheel hub assembly
- Lower ball joint and torque the nut to 43–54 ft. lbs. (59–74 Nm). Install a new cotter pin.
- Strut and torque the bolts to 68–82 ft. lbs. (92–118 Nm)
- Tie rod end and tighten the nut to 22–29 ft. lbs. (29–39 Nm). Install a new cotter pin.
- Disc brake caliper
- Torque the wheel bearing locknut to 145–203 ft. lbs. (196–275 Nm). Install a new cotter pin.
- Front wheels and lower the vehicle

10. Check the vehicle's alignment.

11. Road test the vehicle and verify proper operation.

MAXIMA AND ALTIMA

→Whenever the hub or bearing assembly is removed, the wheel bearing assembly must be replaced. Never reuse the old bearing assembly.

1. Before servicing the vehicle, refer to the precautions in the beginning of this section.

2. Remove or disconnect the following:
- Knuckle assembly from the vehicle
- Hub with the inner race from the steering knuckle, using a shop press and a suitable tool
- Bearing inner race from the hub, using a shop press and a suitable tool
- Outer grease seal
- Inner grease seal from the steering knuckle, using a prybar
- Inner and outer snaprings from the steering knuckle, using snapring pliers
- Sealed bearing assembly from the steering knuckle, using a shop press and a suitable tool

3. Inspect the hub, steering knuckle and snaprings for cracks and/or wear; if necessary, replace the damaged part(s).

To install:

4. Install or connect the following:
- Inner snapring in the steering knuckle groove
- New wheel bearing assembly into the steering knuckle, using a shop press and a suitable tool, until it seats, using a maximum pressure of 3 tons (2722 kg)
- Outer snapring

5. Pack the new grease seal lips with multi-purpose grease.
- New outer grease seal into the steering knuckle, using a shop press and a suitable tool
- Hub into the steering knuckle, using a shop press and a suitable tool, until it seats, using a maximum pressure of 5.5 tons (4990 kg); be careful not to damage the grease seal

6. To check the bearing operation, perform the following procedures:
a. Increase the press pressure to 3.5–5.0 tons (3175–4536 kg).
b. Spin the steering knuckle, several turns, in both directions.
c. Be sure the wheel bearings operate smoothly.

7. If the wheel bearings do not operate smoothly, replace the wheel bearing assembly.

8. Install the knuckle assembly.

9. Install the halfshaft into the hub. Torque the locknut to 174–231 ft. lbs. (235–314 Nm).

10. Install the wheel assembly and lower the vehicle.

11. Road test the vehicle and verify proper operation.

Rear

If the wheel hub bearing assembly is removed, it must be replaced.

→If the vehicle is equipped with Anti-Lock Brake (ABS), the sensor must be removed to protect the sensor and its wiring.

1. Before servicing the vehicle, refer to the precautions in the beginning of this section.

2. Raise and safely support the vehicle. Remove the rear wheel(s).

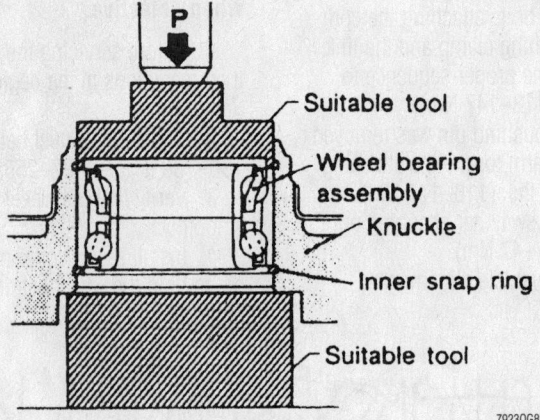

Typical method of installing the wheel bearing

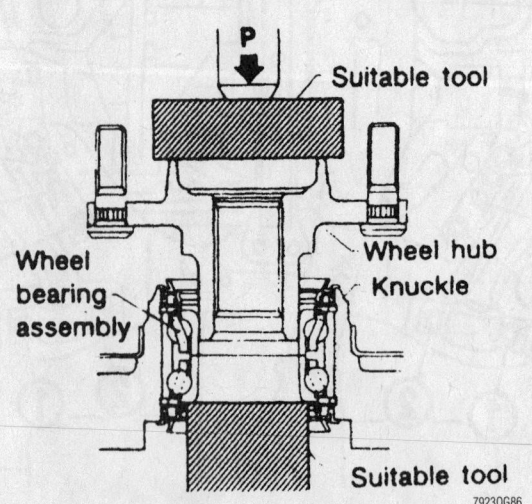

Use a press to install the hub into the knuckle assembly

3. If equipped with disc brakes, remove or disconnect the following:
- Brake caliper and hang it by a piece of wire
- Brake caliper support
- Disc brake pads
- Brake rotor

4. If equipped with drum brakes, remove or disconnect the following:
- Brake drum
- Brake shoe assembly, if necessary
- Grease cap

5. Remove the cotter pin, wheel bearing locknut, washer and the wheel hub bearing assembly. A slide hammer may be needed to remove the hub bearing assembly.

➡The wheel hub bearing assembly is not repairable; it must be replaced when defective.

To install:

➡If the vehicle is equipped with ABS, the sensor ring must be removed and installed on the new hub.

6. Apply oil to the threaded portion of the spindle and both sides of the plain washer.

7. Install the wheel hub bearing assembly, the washer and the wheel bearing locknut. Torque the wheel bearing locknut to 138–188 ft. lbs. (187–255 Nm).

8. Verify that the wheel bearings operate smoothly.

9. Install or connect the following:
- New cotter pin into the spindle to hold the wheel bearing locknut

10. Install a dial micrometer to the rear wheel hub bearing assembly and check the axial end-play. It should be less than 0.0020 in. (0.05mm).
- Grease cap
- ABS sensor and its wiring, if removed
- Brake assembly and the wheels

BRAKES

Brake Caliper

REMOVAL & INSTALLATION

Altima, Maxima and Sentra

FRONT

1. Before servicing the vehicle, refer to the precautions in the beginning of this section.

2. Remove or disconnect the following:
- Front wheels
- Brake fluid hose
- Pin bolts
- Caliper assembly from the vehicle

To install:

3. Use a large C-clamp to press the caliper piston back into the caliper.

4. Install or connect the following:

- New pads, new shims and pad retainers
- Brake caliper and torque the pin bolts to 23 ft. lbs. (31 Nm)
- Brake line to the caliper, using new copper washers, and torque the connecting bolt to 12–14 ft. lbs. (17–20 Nm).
- Wheels

5. Bleed the brake system and top off the master cylinder as necessary.

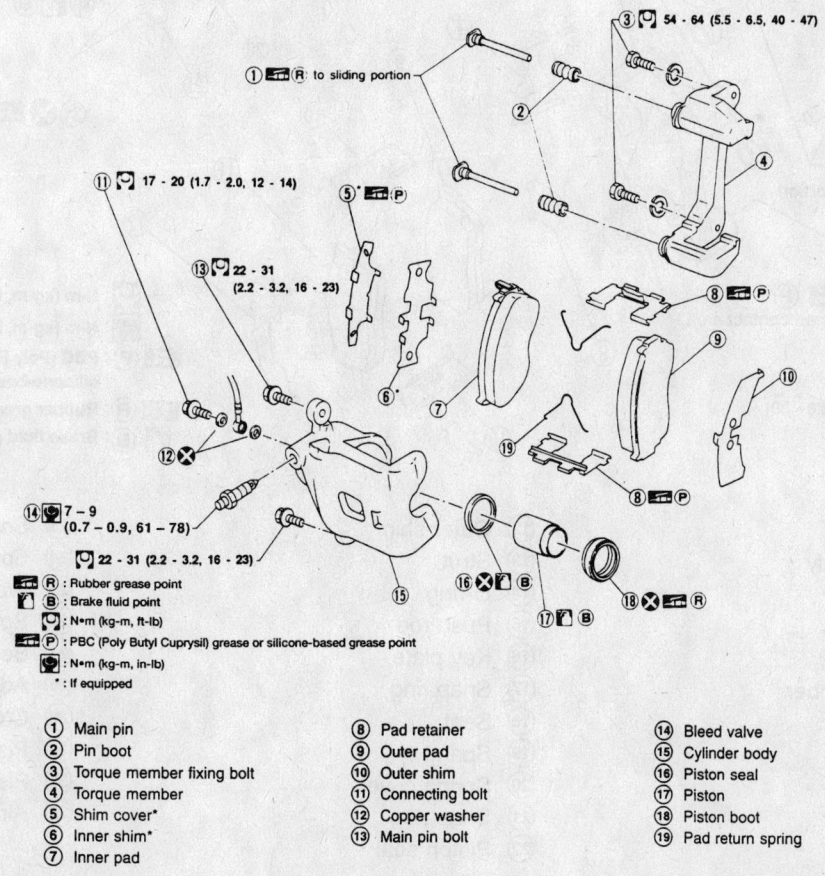

③ 54 - 64 (5.5 - 6.5, 40 - 47)
① ⊞ ℝ : to sliding portion
⑪ 17 - 20 (1.7 - 2.0, 12 - 14)
⑤ ⊞ ℗
⑬ 22 - 31 (2.2 - 3.2, 16 - 23)
⑭ 7 - 9 (0.7 - 0.9, 61 - 78)
22 - 31 (2.2 - 3.2, 16 - 23)

⊞ ℝ : Rubber grease point
⊞ ℬ : Brake fluid point
N•m (kg-m, ft-lb)
⊞ ℗ : PBC (Poly Butyl Cuprysil) grease or silicone-based grease point
N•m (kg-m, in-lb)
* : If equipped

① Main pin
② Pin boot
③ Torque member fixing bolt
④ Torque member
⑤ Shim cover*
⑥ Inner shim*
⑦ Inner pad
⑧ Pad retainer
⑨ Outer pad
⑩ Outer shim
⑪ Connecting bolt
⑫ Copper washer
⑬ Main pin bolt
⑭ Bleed valve
⑮ Cylinder body
⑯ Piston seal
⑰ Piston
⑱ Piston boot
⑲ Pad return spring

93016G51

Front brake caliper—Sentra

REAR

1. Before servicing the vehicle, refer to the precautions in the beginning of this section.
2. Remove or disconnect the following:
   - Rear wheels
   - Parking brake cable and the lock spring
   - Brake fluid hose from the caliper
   - Caliper pin bolts and remove the caliper

To install:

3. Turn the piston clockwise back into the caliper body. Remove some brake fluid from the master cylinder, if necessary. Take care not to damage the piston boot.
4. Coat the pad contact area on the mounting support with a silicone based grease.
5. Install or connect the following:
   - New pads, shims and the pad springs
   - Caliper body into position and torque the caliper pin bolts to 16–23 ft. lbs. (22–31 Nm)
   - Brake fluid hose, using new copper washers, and tighten the flare nut to 12–14 ft. lbs. (17–20 Nm)
   - Lock spring and the parking brake cable
6. Bleed the brake system and top off the master cylinder as necessary.
7. Replace the wheels.

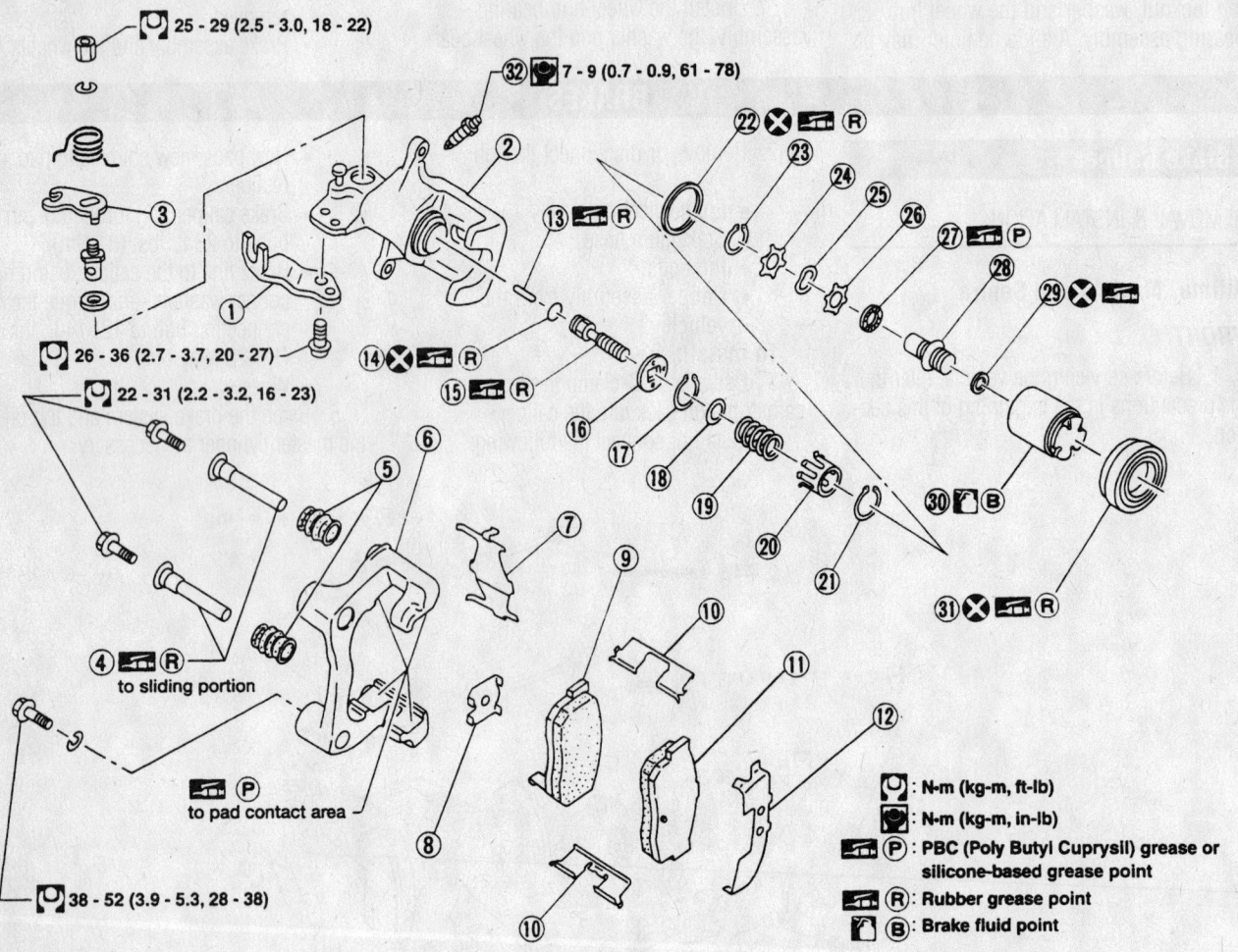

① Cable guide
② Cylinder body
③ Toggle lever
④ Pin
⑤ Pin boot
⑥ Torque member
⑦ Retainer
⑧ Inner shim
⑨ Inner pad
⑩ Pad retainer
⑪ Outer pad
⑫ Outer shim
⑬ Strut
⑭ O-ring
⑮ Push rod
⑯ Key plate
⑰ Snap ring
⑱ Seat
⑲ Spring
⑳ Spring cover
㉑ Snap ring
㉒ Piston seal
㉓ Snap ring
㉔ Spacer
㉕ Wave washer
㉖ Spacer
㉗ Bearing
㉘ Adjuster
㉙ Cup
㉚ Piston
㉛ Piston boot
㉜ Air bleeder

N•m (kg-m, ft-lb)
N•m (kg-m, in-lb)
P : PBC (Poly Butyl Cuprysil) grease or silicone-based grease point
R : Rubber grease point
B : Brake fluid point

93016G52

Rear disc brakes—Sentra shown

Disc Brake Pads

REMOVAL & INSTALLATION

Sentra

FRONT

1. Before servicing the vehicle, refer to the precautions in the beginning of this section.
2. Remove or disconnect the following:
   - Wheels
   - Bottom guide pin from the caliper and swing the caliper cylinder body up
   - Brake pad retainers and the pads

To install:

3. Compress the piston of the disc brake caliper.
4. Install or connect the following:
   - Brake pads, shims, and retainers
   - Caliper assembly. Torque the guide pin to 23–30 ft. lbs. (31–41 Nm).
   - Wheels
5. Check the master cylinder and add fluid if necessary.

REAR

1. Before servicing the vehicle, refer to the precautions in the beginning of this section.
2. Remove or disconnect the following:
   - Wheels
   - Parking brake cable bracket bolt
   - Pin bolts and lift off the caliper body
   - Pad springs and the pads and shims

To install:

3. Turn the piston clockwise back into the caliper body. Take care not to damage the piston boot.
4. Coat the pad contact area on the mounting support with a silicone based grease.
5. Install or connect the following:
   - Pads, shims and retainer springs
   - Caliper body into position in the mounting support and tighten the pin bolts to 28–38 ft. lbs. (38–52 Nm)
   - Wheels and bleed the system if necessary.

Altima and Maxima

FRONT

1. Before servicing the vehicle, refer to the precautions in the beginning of this section.
2. Remove or disconnect the following:

- Wheels
- Bottom guide pin from the caliper and swing the caliper cylinder body up
- Brake pad retainers and the pads

To install:

3. Compress the piston of the disc brake caliper.
4. Install or connect the following:
   - Brake pads, retainers, and caliper assembly. Torque the guide pin to 16–23 ft. lbs. (22–31 Nm).
   - Wheels
5. Check the master cylinder and add fluid if necessary.

REAR

1. Before servicing the vehicle, refer to the precautions in the beginning of this section.
2. Remove or disconnect the following:
   - Rear wheels
   - Parking brake cable bracket bolt
   - Pin bolts and lift off the caliper body
   - Pad springs by pulling them out
   Pads and shims

To install:

3. Turn the piston clockwise back into the caliper body. Take care not to damage the piston boot.
4. Coat the pad contact area on the mounting support with a silicone based grease.
5. Install or connect the following:

- Pads, shims, and the pad springs
- Caliper body into position in the mounting support and tighten the pin bolts to 16–23 ft. lbs. (22–31 Nm)
- Wheels
6. Check the master cylinder and add fluid if necessary.

Brake Drums

REMOVAL & INSTALLATION

1. Before servicing the vehicle, refer to the precautions in the beginning of this section.
2. Remove the wheels.
3. Remove the brake drum from the brake shoes. Two 8mm x 1.25 bolts can be used to press the drum from the hub.

To install:

4. Install the drum assembly to the vehicle.
5. Install the wheels.
6. Adjust the rear brakes.

Brake Shoes

REMOVAL & INSTALLATION

1. Before servicing the vehicle, refer to the precautions in the beginning of this section.

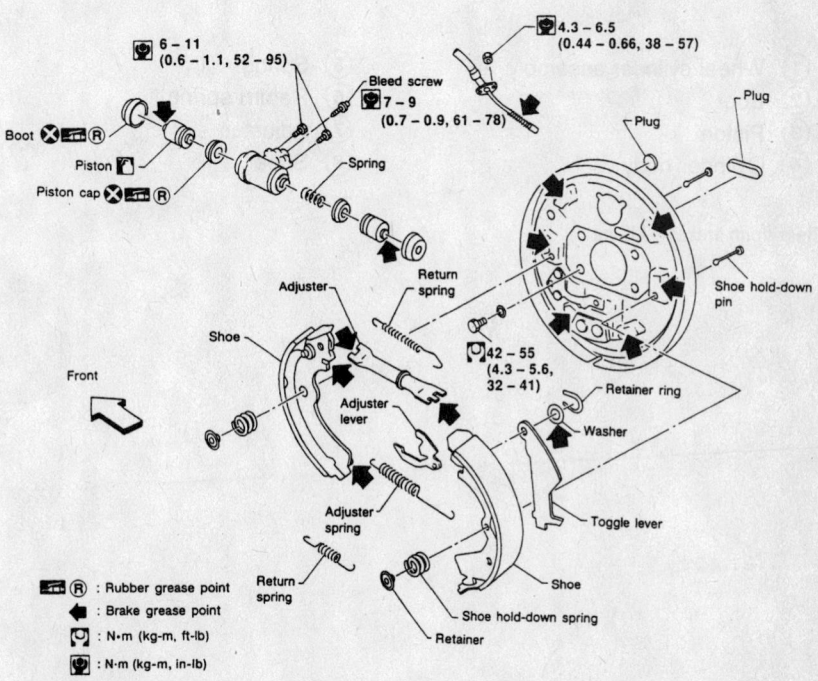

6 – 11
(0.6 – 1.1, 52 – 95)

4.3 – 6.5
(0.44 – 0.66, 38 – 57)

Bleed screw
7 – 9
(0.7 – 0.9, 61 – 78)

Boot
Piston
Piston cap
Spring
Plug
Plug
Plug

Adjuster
Return spring
Shoe hold-down pin

Shoe
42 – 55
(4.3 – 5.6, 32 – 41)
Retainer ring

Front
Adjuster lever
Washer

Adjuster spring
Toggle lever

Return spring
Shoe

Retainer
Shoe hold-down spring

R : Rubber grease point
: Brake grease point
N·m (kg-m, ft-lb)
N·m (kg-m, in-lb)

93016G53

Exploded view of the rear drum brakes—Sentra

2. Remove or disconnect the following:
   - Brake drum
   - Return springs, adjuster assembly, hold-down springs, and brake shoes
   - Parking brake cable from the toggle lever

To install:

3. Install or connect the following:
   - Parking brake cable
   - Shoes with the hold-down springs
   - Return springs, by hooking them into the new shoes

   - Adjuster assembly
   - Drums and wheels. Adjust the brakes and bleed the hydraulic system, if necessary.

4. Check the parking brake adjustment.

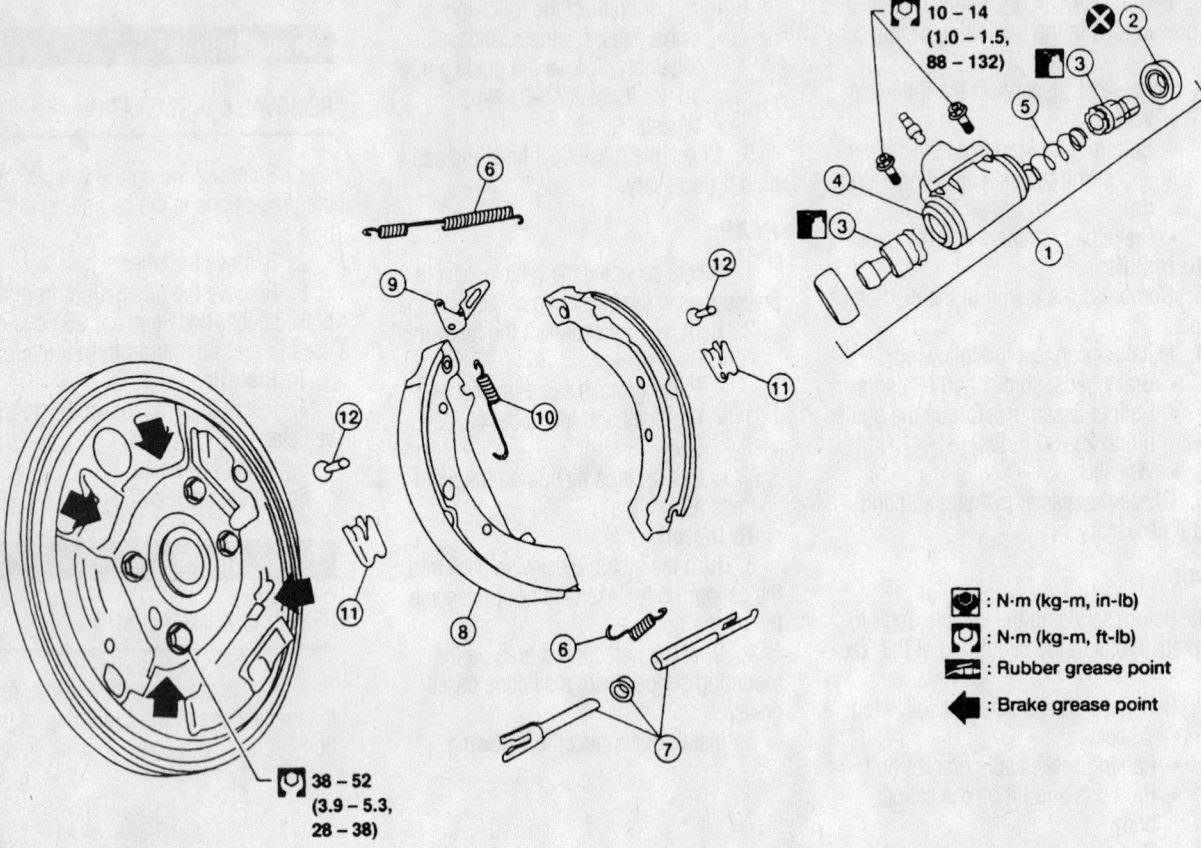

10 – 14
(1.0 – 1.5,
88 – 132)

38 – 52
(3.9 – 5.3,
28 – 38)

: N·m (kg-m, in-lb)

: N·m (kg-m, ft-lb)

: Rubber grease point

: Brake grease point

① Wheel cylinder assembly	⑤ Spring	⑨ Adjuster lever
② Boot	⑥ Return spring	⑩ Adjuster spring
③ Piston	⑦ Adjuster	⑪ Retainer
④ Cylinder body	⑧ Shoe	⑫ Shoe hold-down pin

Rear drum brakes—Altima

93016G55

NISSAN

Frontier • Xterra

24

SPECIFICATION CHARTS

ENGINE AND VEHICLE IDENTIFICATION

		Engine						Model Year	
Code ①	Liters (cc)	Cu. In.	Cyl.	Fuel Sys.	Engine Typ	Eng. Mfg.		Code ②	Year
KA24DE	2.4 (2389)	146	4	MFI	DOHC	Nissan		Y	2000
VG33E	3.3 (3277)	199	6	MFI	SOHC	Nissan		1	2001
VG33ER	3.3 (3277)	199	6	MFI	SOHC	Nissan		2	2002
								3	2003
								4	2004

MFI: Multi-port Fuel Injection

SOHC: Single Overhead Camshaft

DOHC: Double Overhead Camshafts

① Located on the timing belt cover

② 10th digit of the Vehicle Identification Number (VIN)

42356-FRON-C01

GENERAL ENGINE SPECIFICATIONS

Year	Model	Engine Displacement Liters (cc)	Engine ID	Fuel System Type	Net Horsepower @ rpm	Net Torque @ rpm (ft. lbs.)	Bore x Stroke (in.)	Compression Ratio	Oil Pressure @ rpm
2000	Frontier	2.4 (2389)	KA24DE	MFI	143@5200	154@4000	3.50x3.78	9.2:1	60-70@3000
	Frontier	3.3 (3277)	VG33E	MFI	170@4800	200@2800	3.60X3.27	8.9:1	60-65@2000
	Xterra	3.3 (3277)	VG33E	MFI	170@4800	200@2800	3.60X3.27	8.9:1	60-65@2000
2001	Frontier	2.4 (2389)	KA24DE	MFI	143@5200	154@4000	3.50x3.78	9.2:1	60-70@3000
	Frontier	3.3 (3277)	VG33E	MFI	170@4800	200@2800	3.60X3.27	8.9:1	60-65@2000
	Xterra	2.4 (2398)	KA24DE	MFI	143@5200	154@4000	3.50X3.78	9.2:1	60-70@3000
	Xterra	3.3 (3277)	VG33E	MFI	170@4800	200@2800	3.60X3.27	8.9:1	60-65@2000
2002	Frontier	2.4 (2389)	KA24DE	MFI	143@5200	154@4000	3.50x3.78	9.2:1	60-70@3000
	Frontier	3.3 (3277)	VG33E	MFI	170@4800	200@2800	3.60X3.27	8.9:1	60-65@2000
	Frontier	3.3 (3277)	VG33ER	MFI	210@4800	231@2800	3.60X3.27	8.3:1	60-65@2000
	Xterra	2.4 (2398)	KA24DE	MFI	143@5200	154@4000	3.50X3.78	9.2:1	60-70@3000
	Xterra	3.3 (3277)	VG33E	MFI	170@4800	200@2800	3.60X3.27	8.9:1	60-65@2000
	Xterra	3.3 (3277)	VG33ER	MFI	210@4800	231@2800	3.60X3.27	8.3:1	60-65@2000
2003	Frontier	2.4 (2389)	KA24DE	MFI	143@5200	154@4000	3.50x3.78	9.2:1	60-70@3000
	Frontier	3.3 (3277)	VG33E	MFI	170@4800	200@2800	3.60X3.27	8.9:1	60-65@2000
	Frontier	3.3 (3277)	VG33ER	MFI	210@4800	246@2800	3.60X3.27	8.3:1	60-65@2000
	Xterra	2.4 (2398)	KA24DE	MFI	143@5200	154@4000	3.50X3.78	9.2:1	60-70@3000
	Xterra	3.3 (3277)	VG33E	MFI	170@4800	200@2800	3.60X3.27	8.9:1	60-65@2000
	Xterra	3.3 (3277)	VG33ER	MFI	210@4800	231@2800	3.60X3.27	8.3:1	60-65@2000

MFI: Multi-port Fuel Injection

42356-FRON-C02

ENGINE TUNE-UP SPECIFICATIONS

Year	Engine Displacement Liters (cc)	Engine ID	Spark Plug Gap (in.)	Ignition Timing (deg.)		Fuel Pump (psi) ①	Idle Speed (rpm)		Valve Clearance (in.)	
				MT	AT		MT	AT ②	In.	Ex.
2000	2.4 (2389)	KA24DE	0.039-0.043	18-22B	18-22B	34	750-850	750-850	0.012-0.015	0.013-0.016
	3.3 (3277)	VG33E	0.039-0.043	13-17B	13-17B	34	700-800	700-800	HYD	HYD
2001	2.4 (2389)	KA24DE	0.039-0.043	18-22B	18-22B	34	750-850	750-850	0.012-0.015	0.013-0.016
	3.3 (3277)	VG33E	0.039-0.043	13-17B	13-17B	34	700-800	700-800	HYD	HYD
2002	2.4 (2389)	KA24DE	0.039-0.043	18-22B	18-22B	34	750-850	750-850	0.012-0.015	0.013-0.016
	3.3 (3277)	VG33E	0.039-0.043	13-17B	13-17B	34	700-800	700-800	HYD	HYD
	3.3 (3277)	VG33ER	0.043	10B	10B	35	700-800	700-800	HYD	HYD
2003	2.4 (2389)	KA24DE	0.039-0.043	18-22B	18-22B	34	750-850	750-850	0.012-0.015	0.013-0.016
	3.3 (3277)	VG33E	0.039-0.043	13-17B	13-17B	34	700-800	700-800	HYD	HYD
	3.3 (3277)	VG33ER	0.043	10B	10B	35	700-800	700-800	HYD	HYD

NOTE: The Vehicle Emission Control Information label often reflects specification changes made during production. The label figures must be used if they differ from those in this chart.

B: Before top dead center

HYD: Hydraulic

① System pressure at idle with vacuum hose connected

    Should increase to 43 psi when disconnected

② Automatic transmission in Neutral

42356-FRON-C03

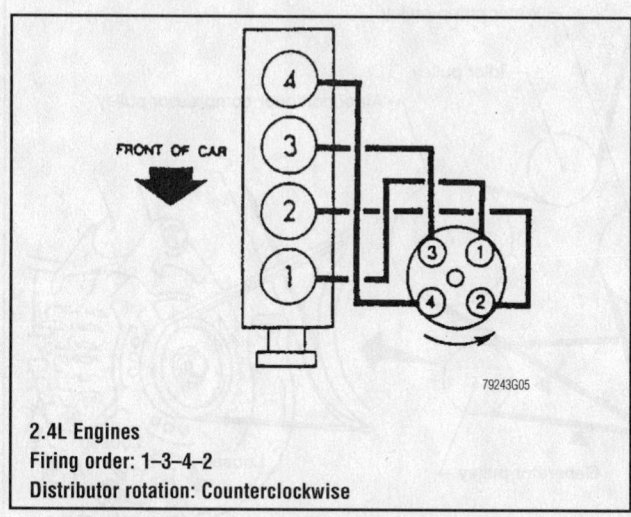

2.4L Engines
Firing order: 1–3–4–2
Distributor rotation: Counterclockwise

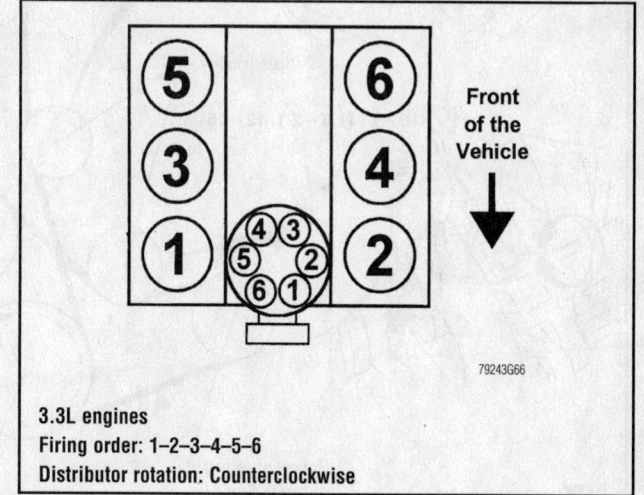

3.3L engines
Firing order: 1–2–3–4–5–6
Distributor rotation: Counterclockwise

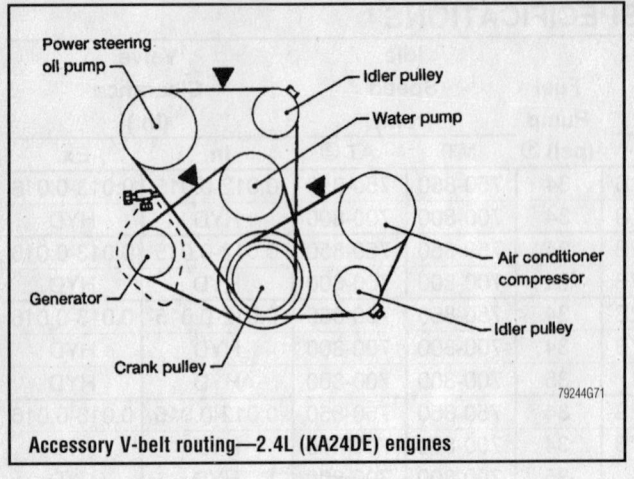

Accessory V-belt routing—2.4L (KA24DE) engines

Labels: Power steering oil pump, Idler pulley, Water pump, Air conditioner compressor, Idler pulley, Crank pulley, Generator

79244G71

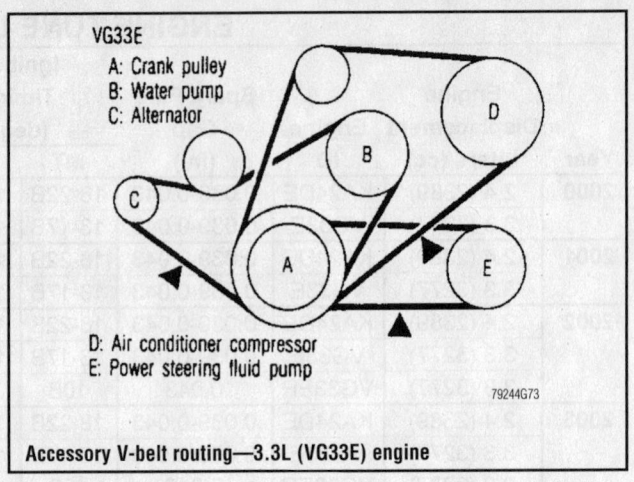

VG33E
A: Crank pulley
B: Water pump
C: Alternator
D: Air conditioner compressor
E: Power steering fluid pump

Accessory V-belt routing—3.3L (VG33E) engine

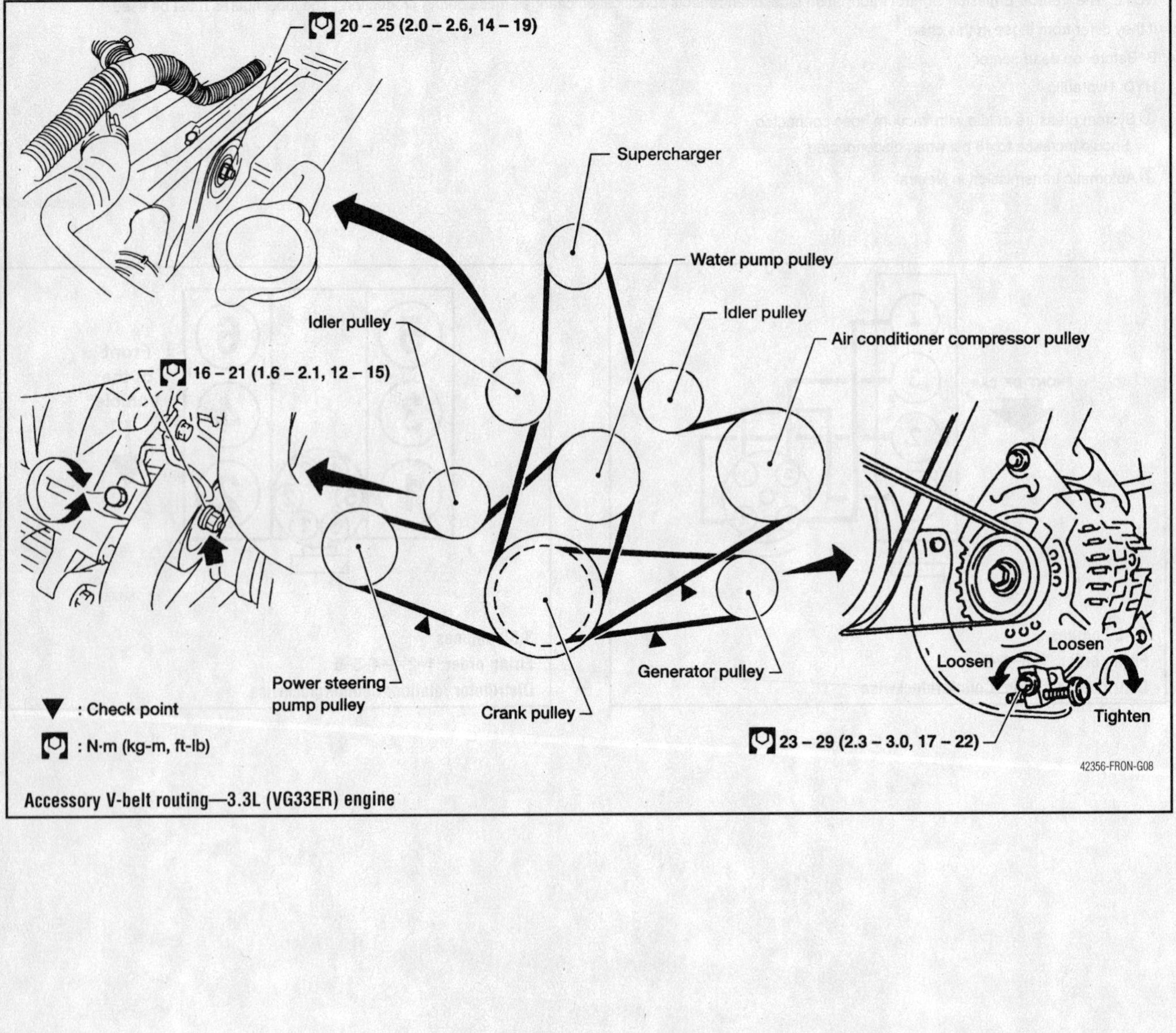

79244G73

20 – 25 (2.0 – 2.6, 14 – 19)

16 – 21 (1.6 – 2.1, 12 – 15)

Supercharger

Water pump pulley

Idler pulley

Idler pulley

Air conditioner compressor pulley

Loosen

Loosen

Tighten

Power steering pump pulley

Crank pulley

Generator pulley

▼ : Check point

: N·m (kg-m, ft-lb)

23 – 29 (2.3 – 3.0, 17 – 22)

42356-FRON-G08

Accessory V-belt routing—3.3L (VG33ER) engine

CAPACITIES

Year	Model	Engine Displacement Liters (cc)	Engine ID	Engine Oil with Filter (qts.)	Transmission (pts.) 5-Spd	Transmission (pts.) Auto.	Transfer Case (pts.)	Drive Axle Front (pts.)	Drive Axle Rear (pts.)	Fuel Tank (gal.)	Cooling System (qts.)
2000	Frontier	2.4 (2389)	KA24DE	①	4.25	—	—	—	②	15.9	7.75
	Frontier	3.3 (3277)	VG33E	①	②	16.8	4.8	2.8	②	19.4	④
	Xterra	3.3 (3277)	VG33E	①	③	16.8	4.8	2.8	③	19.4	④
2001	Frontier	2.4 (2389)	KA24DE	①	4.25	—	—	—	②	15.9	7.75
	Frontier	3.3 (3277)	VG33E	①	③	16.8	4.8	2.8	②	19.4	④
	Xterra	2.4 (2398)	KA24DE	3.75	8.5	—	—	—	3.1	19.4	7.75
	Xterra	3.3 (3277)	VG33E	3.8	⑤	⑥	4.8	4.4	4.9	19.4	11.25
2002	Frontier	2.4 (2389)	KA24DE	3.5	4.25	—	—	—	2.4	15.9	7.75
	Frontier	3.3 (3277)	VG33E	3.5	⑤	⑥	4.8	5.8	3.12	19.4	11.6
	Xterra	2.4 (2398)	KA24DE	3.75	8.5	—	—	—	3.1	19.4	7.75
	Xterra	3.3 (3277)	VG33E	3.5	⑤	⑥	4.8	3.1	5.9	19.4	11.6
	Xterra	3.3 (3277)	VG33ER	3.5	⑤	⑥	4.8	3.1	5.9	19.4	11.6
2003	Frontier	2.4 (2389)	KA24DE	3.5	4.25	—	—	—	2.4	15.9	7.75
	Frontier	3.3 (3277)	VG33E	3.5	⑤	⑥	4.8	5.8	3.12	19.4	11.6
	Frontier	3.3 (3277)	VG33ER	3.5	⑤	⑥	4.8	5.8	3.12	19.4	11.6
	Xterra	2.4 (2398)	KA24DE	3.75	8.5	—	—	—	3.1	19.4	7.75
	Xterra	3.3 (3277)	VG33E	3.5	⑤	⑥	4.8	3.1	5.9	19.4	11.6
	Xterra	3.3 (3277)	VG33ER	3.5	⑤	⑥	4.8	3.1	5.9	19.4	11.6

NOTE: All capacities are approximate. Add fluid gradually and check to be sure a proper fluid level is obtained.

① 2WD: 3.75 qts.
   4WD: 4.125 qts.

② H190A: 3.1 pts.
   C200: 2.75 pts.
   H233B: 5.9 pts.

③ 2WD: 4.25 pts.
   4WD: 10.4 pts.

④ 2WD: 8.6 qts.; 4WD: 9.5 qts.

⑤ MT: 5.8; AT: 10.75

⑥ MT: 17.5; AT: 18 pts.

42356-FRON-C04

CRANKSHAFT AND CONNECTING ROD SPECIFICATIONS

All measurements are given in inches.

Year	Engine Displacement Liters (cc)	Engine ID/VIN	Crankshaft				Connecting Rod		
			Main Brg. Journal Dia.	Main Brg. Oil Clearance	Shaft End-play	Thrust on No.	Journal Diameter	Oil Clearance	Side Clearance
2000	2.4 (2389)	KA24DE	2.3609-2.3612	0.0008-0.0019	0.0020-0.0071	3	1.9672-1.9675	0.0004-0.0014	0.0080-0.0160
	3.3 (3277)	VG33E	2.4790-2.4793	0.0011-0.0022	0.0020-0.0067	4	1.9967-1.9675	0.0006-0.0021	0.0079-0.0138
2001	2.4 (2389)	KA24DE	2.3609-2.3612	0.0008-0.0019	0.0020-0.0071	3	1.9672-1.9675	0.0004-0.0014	0.0080-0.0160
	3.3 (3277)	VG33E	2.4790-2.4793	0.0011-0.0022	0.0020-0.0067	4	1.9967-1.9675	0.0006-0.0021	0.0079-0.0138
2002	2.4 (2389)	KA24DE	2.3609-2.3612	0.0008-0.0019	0.0020-0.0071	3	1.9672-1.9675	0.0004-0.0014	0.0080-0.0160
	3.3 (3277)	VG33E	2.4790-2.4793	0.0011-0.0022	0.0020-0.0067	4	1.9967-1.9675	0.0006-0.0021	0.0079-0.0138
	3.3 (3277)	VG33ER	①	②	0.0020-0.0067	4	1.9667-1.9675	0.0009-0.0025	0.0079-0.0138
2003	2.4 (2389)	KA24DE	2.3609-2.3612	0.0008-0.0019	0.0020-0.0071	3	1.9672-1.9675	0.0004-0.0014	0.0080-0.0160
	3.3 (3277)	VG33E	2.4790-2.4793	0.0011-0.0022	0.0020-0.0067	4	1.9967-1.9675	0.0006-0.0021	0.0079-0.0138
	3.3 (3277)	VG33ER	①	②	0.0020-0.0067	4	1.9667-1.9675	0.0009-0.0025	0.0079-0.0138

① Except No. 1
  Grade 0: 2.4790-2.4793
  Grade 1: 2.4787-2,4790
  Grade 2: 2.4784-2.4787
  No. 1
  Grade 3: 2.4683-2.4793
  Grade 4: 2.4789-2.4791
  Grade 5: 2.4786-2.4789
  Grade 6: 2.4784-2.4786

② No. 1: 0.0012-0.0019
  Nos. 2, 3, 4: 0.0015-0.0026

42356-FRON-C05

VALVE SPECIFICATIONS

Year	Engine Displacement Liters (cc)	Engine ID	Seat Angle (deg.)	Face Angle (deg.)	Spring Test Pressure (lbs. @ in.)	Spring Installed Height (in.)	Stem-to-Guide Clearance (in.)		Stem Diameter (in.)	
							Intake	Exhaust	Intake	Exhaust
2000	2.4 (2389)	KA24DE	45	45.5	93.9@1.15	NA	0.0008-0.0021	0.0016-0.0029	0.2742-0.2748	0.2734-0.2740
	3.3 (3277)	VG33E	45	45.25-46.75	①	NA	0.0008-0.0021	0.0016-0.0029	0.2742-0.2748	0.3135-0.3138
2001	2.4 (2389)	KA24DE	45	45.5	93.9@1.15	NA	0.0008-0.0021	0.0016-0.0029	0.2742-0.2748	0.2734-0.2740
	3.3 (3277)	VG33E	45	45.25-46.75	①	NA	0.0008-0.0021	0.0016-0.0029	0.2742-0.2748	0.3135-0.3138
2002	2.4 (2389)	KA24DE	45	45.5	93.9@1.15	NA	0.0008-0.0021	0.0016-0.0029	0.2742-0.2748	0.2734-0.2740
	3.3 (3277)	VG33E	45	45.25-46.75	①	NA	0.0008-0.0021	0.0016-0.0029	0.2742-0.2748	0.3135-0.3138
	3.3 (3277)	VG33ER	45	45.25-46.75	①	NA	0.0008-0.0021	0.0012-0.0019	0.2742-0.2748	0.3135-0.3138
2003	2.4 (2389)	KA24DE	45	45.5	93.9@1.15	NA	0.0008-0.0021	0.0016-0.0029	0.2742-0.2748	0.2734-0.2740
	3.3 (3277)	VG33E	45	45.25-46.75	①	NA	0.0008-0.0021	0.0016-0.0029	0.2742-0.2748	0.3135-0.3138
	3.3 (3277)	VG33ER	45	45.25-46.75	①	NA	0.0008-0.0021	0.0012-0.0019	0.2742-0.2748	0.3135-0.3138

NA: Not Available

① Inner: 57.3 @ 0.984
  Outer: 117.7 @ 1.181

42356-FRON-C06

PISTON AND RING SPECIFICATIONS

All measurements are given in inches.

Year	Engine Displacement Liters (cc)	Engine ID	Piston Clearance	Ring Gap			Ring Side Clearance		
				Top Comp.	Bottom Comp.	Oil Control	Top Comp.	Bottom Comp.	Oil Control
2000	2.4 (2389)	KA24E	0.0008-0.0016	0.011-0.021	0.018-0.027	0.008-0.027	0.0016-0.0031	0.0012-0.0028	0.0026-0.0053
	3.3 (3277)	VG33E	①	0.0083-0.0157	0.0197-0.0272	0.0079-0.0272	0.0009-0.0030	0.0012-0.0028	0.0006-0.0073
2001	2.4 (2389)	KA24E	0.0008-0.0016	0.011-0.021	0.018-0.027	0.008-0.027	0.0016-0.0031	0.0012-0.0028	0.0026-0.0053
	3.3 (3277)	VG33E	①	0.0083-0.0157	0.0197-0.0272	0.0079-0.0272	0.0009-0.0030	0.0012-0.0028	0.0006-0.0073
2002	2.4 (2389)	KA24E	0.0008-0.0016	0.011-0.021	0.018-0.027	0.008-0.027	0.0016-0.0031	0.0012-0.0028	0.0026-0.0053
	3.3 (3277)	VG33E	①	0.0083-0.0157	0.0197-0.0272	0.0079-0.0272	0.0009-0.0030	0.0012-0.0028	0.0006-0.0073
	3.3 (3277)	VG33ER	②	0.0083-0.0122	0.0197-0.0236	0.0079-0.0236	0.0016-0.0031	0.0012-0.0028	0.0006-0.0073
2003	2.4 (2389)	KA24E	0.0008-0.0016	0.011-0.021	0.018-0.027	0.008-0.027	0.0016-0.0031	0.0012-0.0028	0.0026-0.0053
	3.3 (3277)	VG33E	①	0.0083-0.0157	0.0197-0.0272	0.0079-0.0272	0.0009-0.0030	0.0012-0.0028	0.0006-0.0073
	3.3 (3277)	VG33ER	②	0.0083-0.0122	0.0197-0.0236	0.0079-0.0236	0.0016-0.0031	0.0012-0.0028	0.0006-0.0073

① Except cylinders 3 and 4: 0.0010 - 0.0018 in.
Cylinders 3 and 4: 0.0006 - 0.0010 in.

② Cylinders 3, 4: 0.0006-0.0010 in.
Cylinders 1, 2, 5, 6: 0.0010-0.0018 in.

42356-FRON-C07

TORQUE SPECIFICATIONS
All readings in ft. lbs.

Year	Engine Displacement Liters (cc)	Engine ID	Cylinder Head Bolts	Main Bearing Bolts	Rod Bearing Bolts	Crankshaft Damper Bolts	Flywheel Bolts	Manifold Intake	Manifold Exhaust	Spark Plugs	Lug Nuts
2000	2.4 (2389)	KA24DE	①	34-41	②	105-112	105-112	12-14	27-35	14-22	87-108
	3.3 (3277)	VG33E	③	67-74	②	141-156	61-69	③	21-25	14-22	87-108
2001	2.4 (2389)	KA24DE	①	34-41	②	105-112	105-112	12-14	27-35	14-22	87-108
	3.3 (3277)	VG33E	③	67-74	②	141-156	61-69	③	21-25	14-22	87-108
2002	2.4 (2389)	KA24DE	①	34-41	②	105-112	105-112	12-14	27-35	14-22	87-108
	3.3 (3277)	VG33E	③	67-74	②	141-156	61-69	③	21-25	14-22	87-108
	3.3 (3277)	VG33ER	③	67-74	④	141-156	61-69	③	21-25	14-22	87-108
2003	2.4 (2389)	KA24DE	①	34-41	②	105-112	105-112	12-14	27-35	14-22	87-108
	3.3 (3277)	VG33E	③	67-74	②	141-156	61-69	③	21-25	14-22	87-108
	3.3 (3277)	VG33ER	③	67-74	④	141-156	61-69	③	21-25	14-22	87-108

① Step 1: 22 ft. lbs.
Step 2: 59 ft. lbs.
Step 3: Loosen completely then retorque to 22 ft. lbs.
Step 4: 18-25 ft. lbs.
Step 5: Plus 86-91 degrees

② 10-12 ft. lbs. plus 60-65 degrees or 28-33 ft. lbs.

③ The cylinder heads and the lower intake manifold are installed together
Step 1: Tighten the cylinder head bolts to 22 ft. lbs.
Step 2: Tighten the cylinder head bolts to 43 ft. lbs.
Step 3: Loosen the cylinder head bolts completely
Step 4: Tighten the cylinder head bolts to 84 inch lbs.
Step 5: Tighten the intake manifold fasteners to 35 inch lbs.
Step 6: Tighten the intake manifold fasteners to 13 ft. lbs.
Step 7: Tighten the intake manifold fasteners to 12-14 ft. lbs.
Step 8: Loosen all intake manifold fasteners completely
Step 9: Tighten the cylinder head bolts to 22 ft. lbs.
Step 10: Tighten the cylinder head bolts 60-65 degrees
Step 11: Tighten the cylinder head sub-bolts to 80-105 inch lbs.
Step 12: Tighten the intake manifold fasteners to 35 inch lbs.
Step 13: Tighten the intake manifold fasteners to 78 inch lbs.
Step 14: Tighten the intake manifold fasteners to 70-84 inch lbs.

④ 10-12 ft. lbs. +60-65 degrees

42356-FRON-C08

WHEEL ALIGNMENT

Year	Model		Caster Range (+/-Deg.)	Caster Preferred Setting (Deg.)	Camber Range (+/-Deg.)	Camber Preferred Setting (Deg.)	Toe-in (in.)	Axis Inclination (Deg.)
2000	Frontier	2.4L	0.50	+0.60	0.50	+0.42	0.12+/-0.04	—
		3.3L	0.50	+2.17	0.50	+0.60	0.16+/-0.04	—
	Xterra	ALL	0.75	+0.60	0.50	+0.42	0.12+/-0.04	—
2001	Frontier	2.4L	0.50	+0.60	0.50	+0.42	0.12+/-0.04	—
		3.3L	0.50	+2.17	0.50	+0.60	0.16+/-0.04	—
	Xterra	ALL	0.75	+0.60	0.50	+0.42	0.12+/-0.04	—
2002	Frontier	2.4L	0.50	+0.60	0.50	+0.42	0.12+/-0.04	—
		3.3L	0.50	+2.17	0.50	+0.60	0.16+/-0.04	—
	Xterra	2WD	0.50	+2.57	0.50	+0.33	0.16+/-0.04	—
	Xterra	4WD	0.50	+2.10	0.50	+0.60	0.16+/-0.04	—
2003	Frontier	2.4L	0.50	+0.60	0.50	+0.42	0.12+/-0.04	—
		3.3L	0.50	+2.17	0.50	+0.60	0.16+/-0.04	—
	Xterra	2WD	0.50	+2.57	0.50	+0.33	0.16+/-0.04	—
	Xterra	4WD	0.50	+2.10	0.50	+0.60	0.16+/-0.04	—

42356-FRON-C09

TIRE, WHEEL AND BALL JOINT SPECIFICATIONS

| Year | Model | OEM Tires | | Tire Pressures (psi) | | Wheel Size | Ball Joint Inspection |
		Standard	Optional	Front	Rear		
2000	Frontier 2wd 4-Cyl.	P215/65R15	None	26	29	6JJ	U: 0.020 in. L: ①
	Frontier 2wd 6-Cyl.	P235/70R15	P265/70R15	26	26	6.5JJ/7JJ	U: 0.020 in. L: ①
	Frontier 4wd XE	P215/75R15	P235/70R15	30	30	7-JJ	U: 0.020 in. L: ①
	Frontier 4wd SE	P255/65R16	None	26	26	7JJ	U: 0.020 in. L: ①
	Xterra	P235/70R15	P265/70R15	26	26	7JJ	U: 0.020 in. L: ①
2001	Frontier 2wd 4-Cyl.	P215/65R15	None	26	29	6JJ	U: 0.020 in. L: ①
	Frontier 2wd 6-Cyl.	P235/70R15	P265/70R15	26	26	6.5JJ/7JJ	U: 0.020 in. L: ①
	Frontier 4wd XE	P215/75R15	P235/70R15	30	30	7-JJ	U: 0.020 in. L: ①
	Frontier 4wd SE	P255/65R16	None	26	26	7JJ	U: 0.020 in. L: ①
	Xterra	P235/70R15	P265/70R15	26	26	7JJ	U: 0.020 in. L: ①
2002	Frontier 2wd 4-Cyl.	P225/70R15	None	②	②	NA	U: 0.020 in. L: ①
	Frontier 4wd 6-Cyl.	P255/65R16	None	②	②	NA	U: 0.020 in. L: ①
	Frontier 2wd SC 6-Cyl.	P265/55R17	None	②	②	NA	U: 0.020 in. L: ①
	Frontier 4wd SC 6-Cyl	P265/65R17	None	②	②	NA	U: 0.020 in. L: ①
	Frontier 4wd XE 6-cyl.	P265/70R15	None	②	②	NA	U: 0.020 in. L: ①
	Frontier 6-Cyl. XE Desert Runner	P265/70R15	None	②	②	NA	U: 0.020 in. L: ①
	Frontier 6-Cyl. SE Desert Runner	P255/65R16	None	②	②	NA	U: 0.020 in. L: ①
	Frontier SE V6	P265/70R16	None	②	②	NA	U: 0.020 in. L: ①
	Frontier 2wd SC V6 Crew Cab	P265/55R17	None	②	②	NA	U: 0.020 in. L: ①
	Frontier 4wd SC V6 Crew Cab	P265/65R17	None	②	②	NA	U: 0.020 in. L: ①
	Xterra SE, SE S/C, and XE S/C	P265/70R16	None	②	②	7JJ	③
	Xterra XE, XE V6	P265/70R15	None	②	②	7JJ	③

42356-FRON-C10

TIRE, WHEEL AND BALL JOINT SPECIFICATIONS

Year	Model	OEM Tires		Tire Pressures (psi)		Wheel Size	Ball Joint Inspection
		Standard	Optional	Front	Rear		
2003	Frontier 2wd 4-Cyl.	P225/70R15	None	②	②	NA	U: 0.020 in. L: ①
	Frontier 4wd 6-Cyl.	P255/65R16	None	②	②	NA	U: 0.020 in. L: ①
	Frontier 2wd SC 6-Cyl.	P265/55R17	None	②	②	NA	U: 0.020 in. L: ①
	Frontier 4wd SC 6-Cyl	P265/65R17	None	②	②	NA	U: 0.020 in. L: ①
	Frontier 4wd XE 6-cyl.	P265/70R15	None	②	②	NA	U: 0.020 in. L: ①
	Frontier 6-Cyl. XE Desert Runner	P265/70R15	None	②	②	NA	U: 0.020 in. L: ①
	Frontier 6-Cyl. SE Desert Runner	P255/65R16	None	②	②	NA	U: 0.020 in. L: ①
	Frontier SE V6	P265/70R16	None	②	②	NA	U: 0.020 in. L: ①
	Frontier 2wd SC V6 Crew Cab	P265/55R17	None	②	②	NA	U: 0.020 in. L: ①
	Frontier 4wd SC V6 Crew Cab	P265/65R17	None	②	②	NA	U: 0.020 in. L: ①
	Xterra SE, SE S/C, and XE S/C	P265/70R16	None	②	②	7JJ	③
	Xterra XE, XE V6	P265/70R15	None	②	②	7JJ	③

OEM: Original Equipment Manufacturer

PSI: Pounds Per Square Inch

STD: Standard

OPT: Optional

L: Lower

U: Upper

① Replace if any measurable movement is found.

② See placard on vehicle

③ Axial play

Upper: 0

Lower: 0.008 in.

42356-FRON-C11

BRAKE SPECIFICATIONS

All measurements in inches unless noted

| Year | Model | Brake Disc | | | Brake Drum Diameter | | | Minimum Lining Thickness | | Brake Caliper | |
		Original Thickness	Minimum Thickness	Maximum Runout	Original Inside Diameter	Max. Wear Limit	Maximum Machine Diameter	Front	Rear	Bracket Bolts (ft. lbs.)	Mounting Bolts (ft. lbs.)
2000	Frontier	①	②	0.003	③	NA	④	0.079	0.059	53-72	24-31
	Xterra	①	②	0.003	③	NA	④	0.079	0.059	53-72	24-31
2001	Frontier	①	②	0.003	③	NA	④	0.079	0.059	53-72	24-31
	Xterra	①	②	0.003	③	NA	④	0.079	0.059	53-72	24-31
2002	Frontier	①	②	0.003	③	NA	④	0.079	0.059	53-72	17-22
	Xterra	1.100	1.024	0.003	11.61	NA	11.67	0.079	0.059	⑤	⑤
2003	Frontier	NA	⑥	0.003	③	NA	⑦	0.079	0.059	⑧	17-22
	Xterra	1.100	1.024	0.003	11.61	NA	11.67	0.079	0.059	⑤	⑤

NA: Not Available

① 2WD: 0.870
   4WD: 1.020

② 2WD: 0.787
   4WD: 0.945

③ 2WD: 10.20
   4WD: 11.60

④ 2WD: 10.30
   4WD: 11.67

⑤ Torque member mounting bolt: 101-130
   Main pin bolt: 17-22

⑥ 4-cyl.: 0.945
   6-cyl.: 1.024

⑦ 4-cyl.: 10.30
   6-cyl.: 11.67

⑧ 4-cyl.: 53-72 ft. lbs.
   6-cyl.: 101-130 ft. lbs.

42356-FRON-C12

SCHEDULED MAINTENANCE INTERVALS
Nissan Frontier and Xterra

TO BE SERVICED	TYPE OF SERVICE	VEHICLE MILEAGE INTERVAL (x1000)												
		7.5	15	22.5	30	37.5	45	52.5	60	67.5	75	82.5	90	97.5
Engine oil & filter	R	✓	✓	✓	✓	✓	✓	✓	✓	✓	✓	✓	✓	✓
Brake lines & cables	S/I		✓		✓		✓		✓		✓		✓	
Brake pads, discs, drums & linings	S/I		✓		✓		✓		✓		✓		✓	
Driveshaft boots & propeller shaft	S/I				✓				✓				✓	
Front wheel bearings (4x2)	S/I				✓				✓				✓	
Front wheel bearings (4x4)	S/I				✓				✓				✓	
Automatic & manual transmission, transfer & differential gear oil ①	S/I		✓		✓		✓		✓		✓		✓	
Air cleaner filter	R				✓				✓				✓	
Engine coolant	R				✓				✓				✓	
PCV filter (KA24E)	R				✓				✓				✓	
Spark plugs	R				✓				✓				✓	
Drive belt(s)	S/I				✓				✓				✓	
Exhaust system	S/I				✓				✓				✓	
Fuel lines	S/I				✓				✓				✓	
Steering gear (box) & linkage, axle & suspension parts	S/I				✓				✓				✓	
Vapor lines	S/I				✓				✓				✓	
Timing belt ②	R													

R: Replace   S/I: Service or Inspect

① Differential (w/limited-slip differential) oil: replace oil every 30,000 miles.

② Timing belt: replace at 105,000 miles.

FREQUENT OPERATION MAINTENANCE (SEVERE SERVICE)

If a vehicle is operated under any of the following conditions it is considered severe service:

- Extremely dusty areas.

- 50% or more of the vehicle operation is in 32°C (90°F) or higher temperatures, or constant operation in temperatures below 0°C (32°F).

- Prolonged idling (vehicle operation in stop and go traffic).

- Frequent short running periods (engine does not warm to normal operating temperatures).

- Police, taxi, delivery usage or trailer towing usage.

Oil & oil filter: replace every 3750 miles.

Brake pads, discs, drums & linings: service or inspect every 7500 miles.

Driveshaft boots & propeller shaft: service or inspect every 7500 miles.

Exhaust system: service or inspect every 7500 miles.

Steering gear (box) & linkage, (steering damper-4x4), axle & suspension parts: service or inspect every 7500 miles.

Steering linkage ball joints & front suspension ball joints: service or inspect every 7500 miles.

42356-FRON-C13

PRECAUTIONS

Before servicing any vehicle, please be sure to read all of the following precautions, which deal with personal safety, prevention of component damage, and important points to take into consideration when servicing a motor vehicle:

• Never open, service or drain the radiator or cooling system when the engine is hot; serious burns can occur from the steam and hot coolant.

• Observe all applicable safety precautions when working around fuel. Whenever servicing the fuel system, always work in a well-ventilated area. Do not allow fuel spray or vapors to come in contact with a spark, open flame, or excessive heat (a hot drop light, for example). Keep a dry chemical fire extinguisher near the work area. Always keep fuel in a container specifically designed for fuel storage; also, always properly seal fuel containers to avoid the possibility of fire or explosion. Refer to the additional fuel system precautions later in this section.

• Fuel injection systems often remain pressurized, even after the engine has been turned **OFF**. The fuel system pressure must be relieved before disconnecting any fuel lines. Failure to do so may result in fire and/or personal injury.

• Brake fluid often contains polyglycol ethers and polyglycols. Avoid contact with the eyes and wash your hands thoroughly after handling brake fluid. If you do get brake fluid in your eyes, flush your eyes with clean, running water for 15 minutes. If

eye irritation persists, or if you have taken brake fluid internally, IMMEDIATELY seek medical assistance.

• The EPA warns that prolonged contact with used engine oil may cause a number of skin disorders, including cancer! You should make every effort to minimize your exposure to used engine oil. Protective gloves should be worn when changing oil. Wash your hands and any other exposed skin areas as soon as possible after exposure to used engine oil. Soap and water, or waterless hand cleaner should be used.

• All new vehicles are now equipped with an air bag system. The system must be disabled before performing service on or around system components, steering column, instrument panel components, wiring and sensors. Failure to follow safety and disabling procedures could result in accidental air bag deployment, possible personal injury and unnecessary system repairs.

• Always wear safety goggles when working with, or around, the air bag system. When carrying a non-deployed air bag, be sure the bag and trim cover are pointed away from your body. When placing a non-deployed air bag on a work surface, always face the bag and trim cover upward, away from the surface. This will reduce the motion of the module if it is accidentally deployed. Refer to the additional air bag system precautions later in this section.

• Clean, high quality brake fluid from a sealed container is essential to the safe and

proper operation of the brake system. You should always buy the correct type of brake fluid for your vehicle. If the brake fluid becomes contaminated, completely flush the system with new fluid. Never reuse any brake fluid. Any brake fluid that is removed from the system should be discarded. Also, do not allow any brake fluid to come in contact with a painted surface; it will damage the paint.

• Never operate the engine without the proper amount and type of engine oil; doing so WILL result in severe engine damage.

• Timing belt maintenance is extremely important! Many models utilize an interference-type, non-freewheeling engine. If the timing belt breaks, the valves in the cylinder head may strike the pistons, causing potentially serious (also time-consuming and expensive) engine damage. Refer to the maintenance interval charts in the front of this manual for the recommended replacement interval for the timing belt, and to the timing belt section for belt replacement and inspection.

• Disconnecting the negative battery cable on some vehicles may interfere with the functions of the on-board computer system(s) and may require the computer to undergo a relearning process once the negative battery cable is reconnected.

• When servicing drum brakes, only disassemble and assemble one side at a time, leaving the remaining side intact for reference.

ENGINE REPAIR

➡**Disconnecting the negative battery cable on some vehicles may interfere with the functions of the on board computer system. The computer may undergo a relearning process once the negative battery cable is reconnected.**

Distributor

REMOVAL

1. Before servicing the vehicle, refer to the precautions in the beginning of this section.
2. Remove or disconnect the following:
   • Negative battery cable
   • Distributor cap
   • Distributor wiring harness connector
3. Matchmark the rotor to the distributor housing and the distributor housing to the cylinder head.
4. Remove the distributor.

INSTALLATION

Timing Not Disturbed

1. Install or connect the following:
   • Distributor by aligning the matchmarks made during removal
   • Distributor wiring harness connector
   • Distributor cap
   • Negative battery cable
2. Check the ignition timing and adjust, as necessary.

Timing Disturbed

2.4L ENGINE

1. Set the engine to Top Dead Center (TDC) of the compression stroke for the No. 1 cylinder.
2. Install the distributor so that the distributor shaft engages the oil pump driveshaft.

3. Check that the distributor rotor is aligned, as shown.
4. Install or connect the following:
   • Distributor cap
   • Distributor harness connector
5. Check the ignition timing and adjust, as necessary.

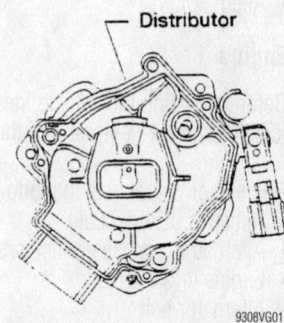

9308VG01

Distributor rotor alignment with the engine at Top Dead Center (TDC)—2.4L engine

3.3L ENGINE

1. Set the engine to Top Dead Center (TDC) of the compression stroke for the No. 1 cylinder.

2. Align the index mark on the distributor shaft with the protrusion on the distributor housing.

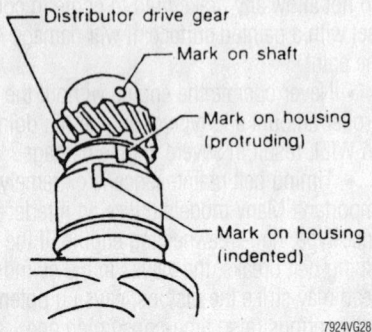

- Distributor drive gear
- Mark on shaft
- Mark on housing (protruding)
- Mark on housing (indented)

7924VG28

Distributor shaft alignment—3.3L engine

3. Install the distributor and check that the distributor rotor is aligned.

4. Install or connect the following:
- Distributor cap
- Distributor harness connector

5. Check the ignition timing and adjust, as necessary.

Alternator

REMOVAL

2.4L Engine

1. Before servicing the vehicle, refer to the precautions in the beginning of this section.

2. Remove or disconnect the following:
- Negative battery cable
- Engine under cover
- Right splash shield
- Alternator harness connectors
- Alternator belt
- Alternator

3.3L Engine

1. Before servicing the vehicle, refer to the precautions in the beginning of this section.

2. Remove or disconnect the following:
- Negative battery cable
- Alternator harness connectors
- Engine under cover
- Alternator belt
- Alternator

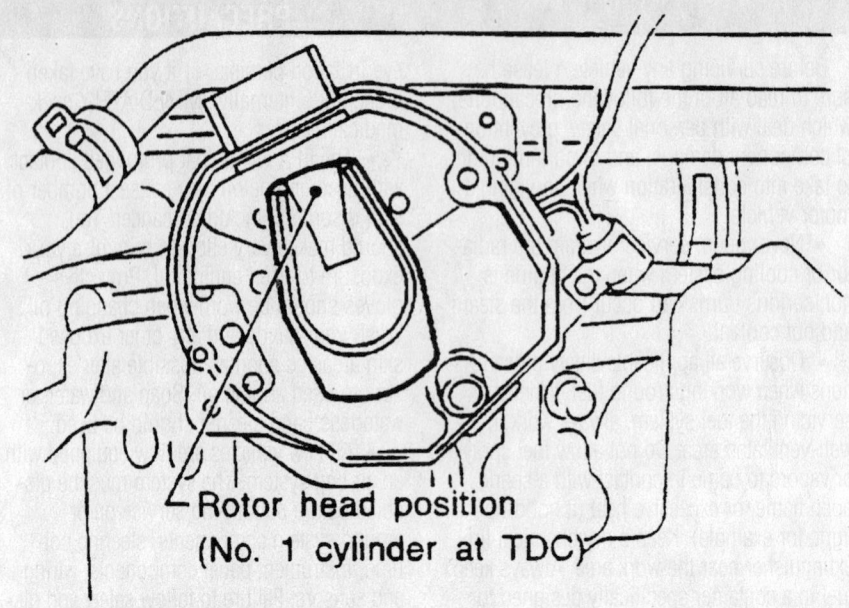

Rotor head position
(No. 1 cylinder at TDC)

9308VG03

Distributor rotor alignment—3.3L engine

INSTALLATION

2.4L Engine

1. Install or connect the following:
- Alternator
- Alternator belt. Tighten the adjustment bolt to 12–14 ft. lbs. (16–19 Nm) and the pivot bolt to 32–38 ft. lbs. (44–52 Nm).
- Alternator harness connectors
- Right splash shield
- Engine under cover
- Negative battery cable

3.3L Engine

1. Install or connect the following:
- Alternator
- Alternator belt. Tighten the adjust-

ment bolt to 12–14 ft. lbs. (16–19 Nm) and the pivot bolts to 16–22 ft. lbs. (22–30 Nm).
- Engine under cover
- Alternator harness connectors
- Negative battery cable

Ignition Timing

ADJUSTMENT

➡Ignition timing is set with the engine at operating temperature, transmission in Neutral and all electrical accessories OFF.

1. Before servicing the vehicle, refer to the precautions in the beginning of this section.

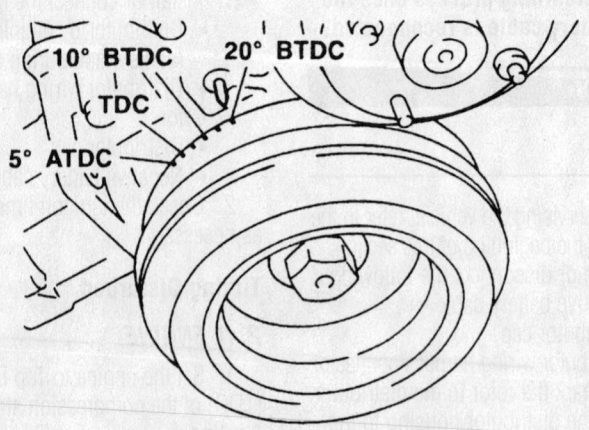

10° BTDC   20° BTDC
TDC
5° ATDC

7924VG04

Timing indicator—3.3L engine shown

2. Attach a timing light to the No. 1 spark plug wire.

3. Start the engine and allow it to reach normal operating temperature.

4. Check that the idle speed is less than 1000 rpm.

5. Run the engine at 2000 rpm for 2 minutes.

6. Rev the engine to 3000 rpm 2–3 times and allow it to idle for 1 minute.

7. Check for the presence of Diagnostic Trouble Codes (DTC) and service as necessary.

8. Run the engine at 2000 rpm for 2 minutes.

9. Stop the engine and disconnect the Throttle Position (TP) sensor.

10. Start the engine and rev it to 3000 rpm 2–3 times and allow it to idle.

11. Set the base timing to 8–12 degrees Before Top Dead Center (BTDC).

12. Tighten the distributor lockbolt to 83–113 inch lbs. (9–13 Nm).

13. Set the base idle speed to 700–800 rpm.

14. Stop the engine and connect the TP sensor.

Engine Assembly

REMOVAL & INSTALLATION

Frontier

2.4L ENGINE

1. Before servicing the vehicle, refer to the precautions in the beginning of this section.

2. Drain the cooling system.

3. Relieve the fuel system pressure.

4. Remove or disconnect the following:
- Negative battery cable
- Hood
- Air cleaner assembly
- Idle Air Control (IAC) valve and solenoid connectors
- Throttle Position (TP) sensor and switch connectors
- Engine Coolant Temperature (ECT) sensor connector
- Manifold Absolute Pressure (MAP) sensor connector and vacuum line
- Evaporative Emissions (EVAP) canister purge valve connector and vacuum line

- Mass Air Flow (MAF) sensor connector
- Brake booster vacuum line
- Fuel lines
- Exhaust Gas Recirculation (EGR) temperature sensor connector
- Throttle cable
- Accessory drive belts
- Radiator and hoses
- Heater hoses
- Exhaust manifold heat shield
- Heated Oxygen (HO$_2$S) sensor connectors
- Exhaust front pipe
- A/C compressor, if equipped
- Power steering pump, if equipped
- Crankshaft Position (CKP) sensor
- Starter motor
- Transmission
- Left and right engine mounts
- Engine

To install:

5. Install or connect the following:
- Engine. Tighten the engine mount nuts to 30–38 ft. lbs. (41–52 Nm).
- Transmission
- Starter motor
- CKP sensor

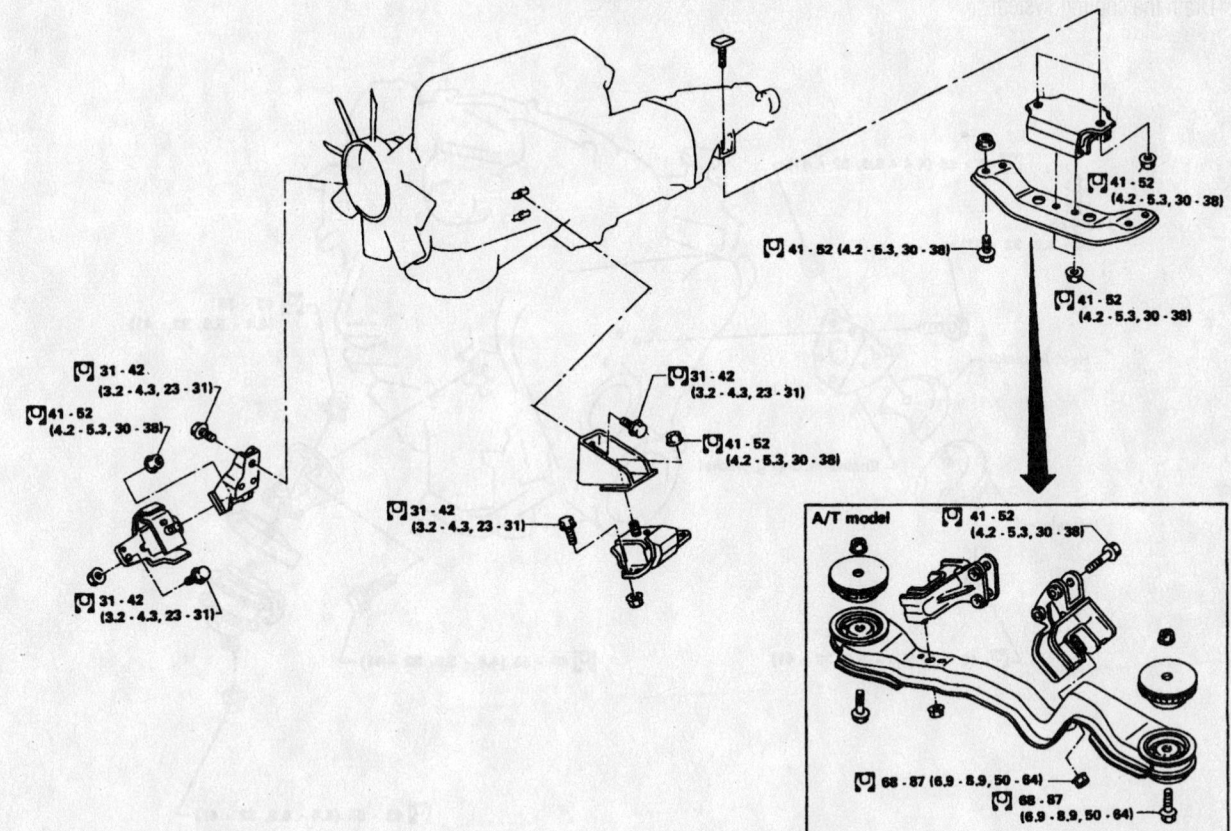

Engine and transmission mounts—2.4L engine

- Power steering pump, if equipped
- A/C compressor, if equipped
- Exhaust front pipe
- HO2S sensor connectors
- Exhaust manifold heat shield
- Heater hoses
- Radiator and hoses
- Accessory drive belts
- Throttle cable
- EGR temperature sensor connector
- Fuel lines
- Brake booster vacuum line
- MAF sensor connector
- EVAP canister purge valve connector and vacuum line
- MAP sensor connector and vacuum line
- ECT sensor connector
- TP sensor and switch connectors
- IAC valve and solenoid connectors
- Air cleaner assembly
- Hood
- Negative battery cable

6. Fill the cooling system.
7. Start the engine and check for leaks.

3.3L Engine

1. Before servicing the vehicle, refer to the precautions in the beginning of this section.
2. Drain the cooling system.

3. Relieve the fuel system pressure.
4. Recover the A/C refrigerant, if equipped.
5. Remove or disconnect the following:
- Negative battery cable
- Hood
- Air cleaner assembly
- Idle Air Control (IAC) valve and solenoid connectors
- Throttle Position (TP) sensor and switch connectors
- Engine Coolant Temperature (ECT) sensor connector
- Manifold Absolute Pressure (MAP) sensor connector and vacuum line
- Evaporative Emissions (EVAP) canister purge valve connector and vacuum line
- Mass Air Flow (MAF) sensor connector
- Brake booster vacuum line
- Fuel lines
- Exhaust Gas Recirculation (EGR) temperature sensor connector
- Throttle cable
- Accessory drive belts
- Cooling fan and shroud
- Radiator and hoses
- Engine under cover
- A/C compressor manifold

- Power steering pump
- Heated Oxygen (HO2S) sensor connectors
- Exhaust front pipes
- Crankshaft Position (CKP) sensor
- Starter motor
- Transmission
- Left and right engine mounts
- Engine

➡ When removing the engine mounts, do not loosen the 4 mount cover nuts. The mount is fluid filled and will not function if the fluid leaks out.

To install:
6. Install or connect the following:
- Engine. Tighten the engine mount nuts to 43–58 ft. lbs. (59–78 Nm).
- Transmission
- Starter motor
- CKP sensor
- Exhaust front pipes
- HO2S sensor connectors
- Power steering pump
- A/C compressor manifold
- Engine under cover
- Radiator and hoses
- Cooling fan and shroud
- Accessory drive belts
- Throttle cable

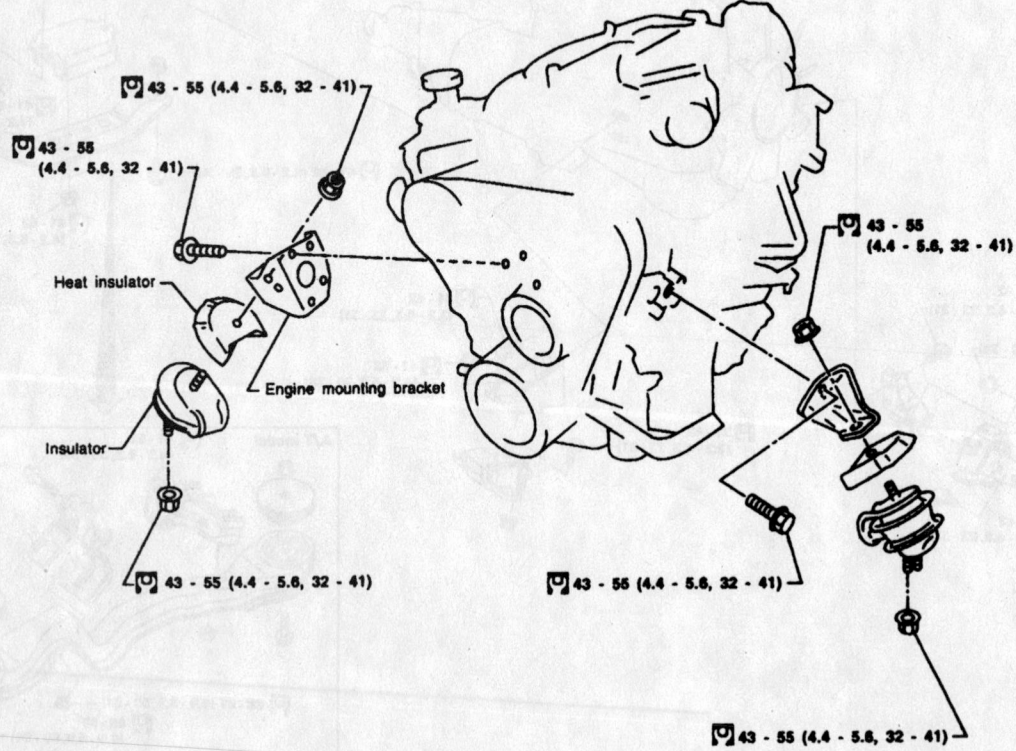

43 - 55 (4.4 - 5.6, 32 - 41)

43 - 55 (4.4 - 5.6, 32 - 41)

Heat insulator

Engine mounting bracket

Insulator

43 - 55 (4.4 - 5.6, 32 - 41)

43 - 55 (4.4 - 5.6, 32 - 41)

43 - 55 (4.4 - 5.6, 32 - 41)

43 - 55 (4.4 - 5.6, 32 - 41)

: N·m (kg-m, ft-lb)

7924VG11

Engine mounts and related components—3.3L engine

- EGR temperature sensor connector
- Fuel lines
- Brake booster vacuum line
- MAF sensor connector
- EVAP canister purge valve connector and vacuum line
- MAP sensor connector and vacuum line
- ECT sensor connector
- TP sensor and switch connectors
- IAC valve and solenoid connectors
- Air cleaner assembly
- Hood
- Negative battery cable

7. Fill the cooling system.
8. Recharge the A/C system, if equipped.
9. Start the engine and check for leaks.

Xterra

➥**Do not loosen front engine mounting insulator cover securing bolts. When cover is removed, damper oil flows out and mounting insulator will not function.**

1. Remove engine undercover and hood.
2. Drain coolant from cylinder block and radiator.
3. Remove vacuum hoses, fuel tubes, wires, harnesses and connectors.

4. Before disconnecting fuel hose, release fuel pressure from fuel line.
5. Remove radiator with shroud and cooling fan.
6. Remove drive belts.
7. Discharge refrigerant.
8. Remove A/C compressor manifold.
9. Remove power steering oil pump from engine.
10. Remove front exhaust tubes.
11. Remove transmission from vehicle.
12. Install engine slingers. Tighten the slinger bolts to 15–20 ft. lbs. (20–26 Nm).
13. Hoist engine with engine slingers and remove engine mounting nuts from both sides.
14. Lift and remove engine from vehicle.
15. Installation is the reverse of removal. See the accompanying illustration for installation torques.

Heater Core

REMOVAL & INSTALLATION

2000–02 Frontier and Xterra

1. Disconnect both the negative (1st) and positive (2nd) battery cables.

2. Remove the steering wheel by performing the following procedure:
   a. Turn the ignition switch to the OFF position.

❄❄ **CAUTION**

Wait 3 minutes after disconnecting the battery cables and turning the ignition switch to the OFF position before servicing the air bag system.

   b. Remove the lower lid and disconnect the driver's air bag module connector.
   c. Remove both side lids.
   d. Using the Tamper Resistant Torx® Wrench size T50, remove the special bolts from both sides of the steering wheel and discard them.
   e. Remove the SRS module from the steering wheel.

❄❄ **CAUTION**

Always store the SRS module face up.

   f. Position the steering wheel in the straight-ahead position.
   g. Disconnect the horn connector and remove the steering wheel nut.

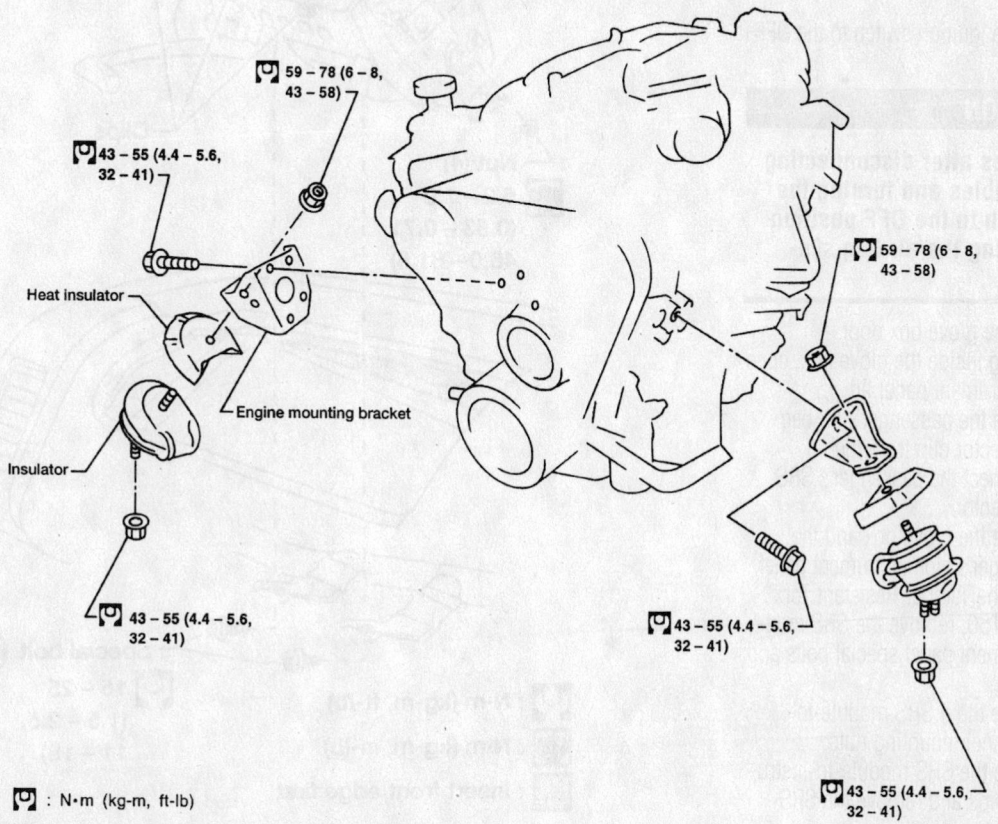

Heat insulator
Engine mounting bracket
Insulator

59 – 78 (6 – 8, 43 – 58)

43 – 55 (4.4 – 5.6, 32 – 41)

59 – 78 (6 – 8, 43 – 58)

43 – 55 (4.4 – 5.6, 32 – 41)

43 – 55 (4.4 – 5.6, 32 – 41)

43 – 55 (4.4 – 5.6, 32 – 41)

: N·m (kg-m, ft-lb)

Engine mounting—Xterra 3.3L engine

9359VG34

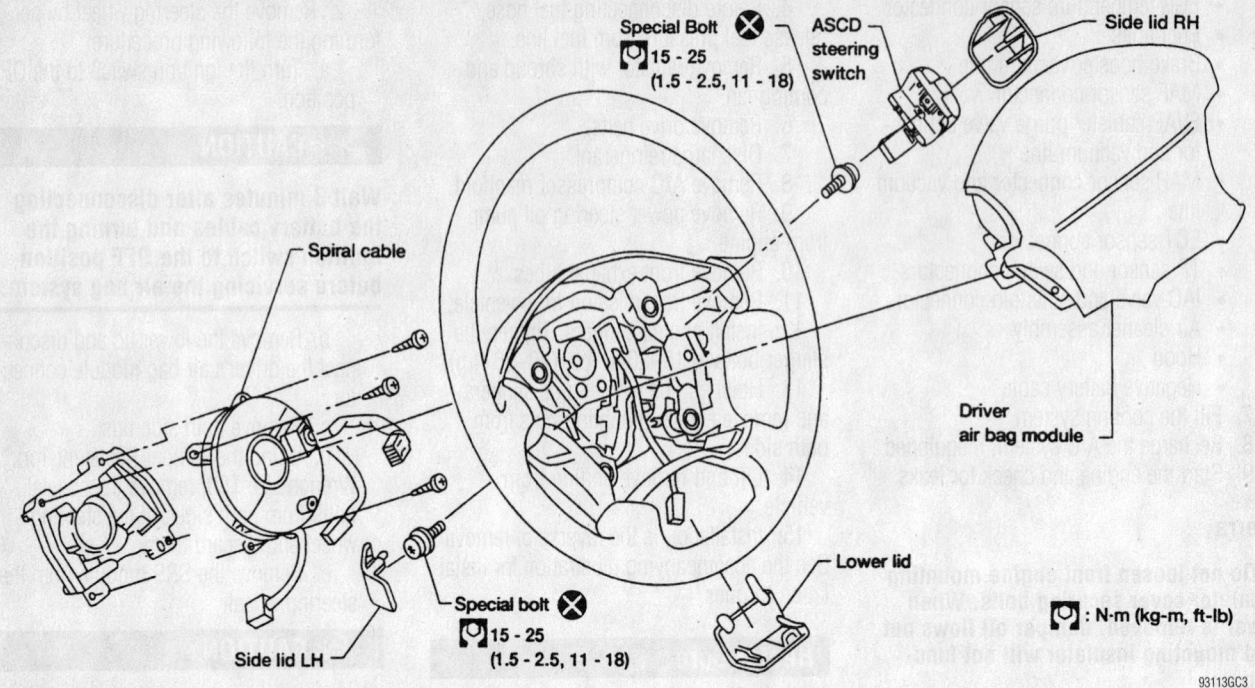

Special Bolt ⊗
🔧 15 - 25
(1.5 - 2.5, 11 - 18)

ASCD steering switch

Side lid RH

Spiral cable

Driver air bag module

Side lid LH

Special bolt ⊗
🔧 15 - 25
(1.5 - 2.5, 11 - 18)

Lower lid

🔧 : N·m (kg-m, ft-lb)

93113GC3

Exploded view of the steering wheel and SRS module and related components—2000–02 Frontier

h. Using a steering wheel puller, press the steering wheel from the steering column.

3. Remove the passenger's side air bag by disconnecting or removing the following items:

a. Turn the ignition switch to the OFF position.

✳✳ CAUTION

Wait 3 minutes after disconnecting the battery cables and turning the ignition switch to the OFF position before servicing the air bag system.

b. Open the glove box door.

c. Working inside the glove box, open the lower instrument panel lid.

d. Remove the passenger's air bag module connector clip from the lid.

e. Disconnect the passenger's SRS module connector.

f. Remove the glove box and the lower passenger's side instrument panel.

g. Using the Tamper Resistant Torx® Wrench size T50, remove the SRS module-to-instrument panel special bolts and discard them.

h. Remove the 4 SRS module-to-instrument panel mounting nuts.

i. Release the SRS module-to-instrument panel clips and remove the SRS module.

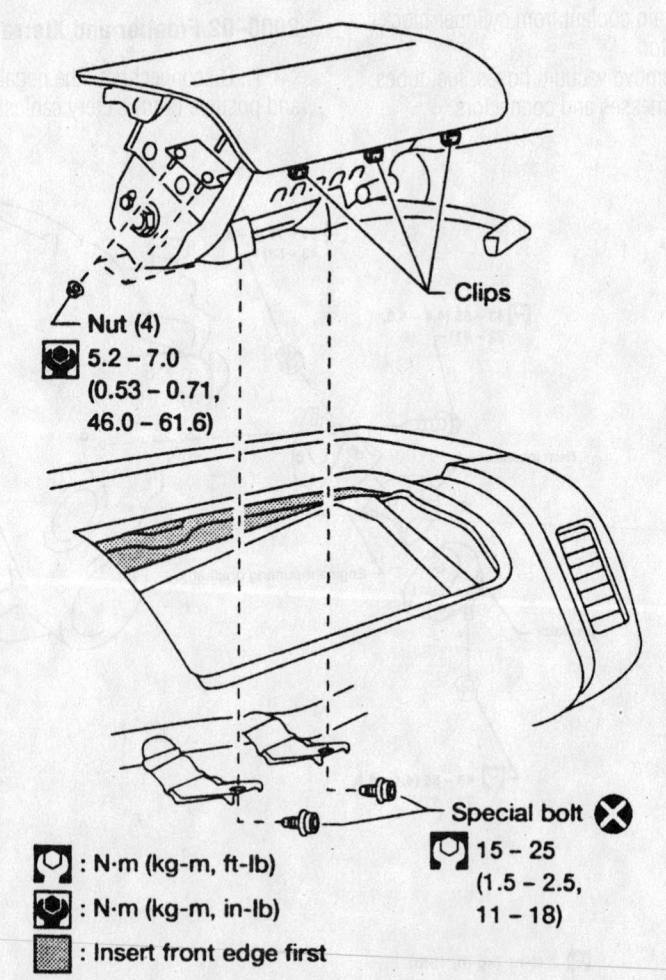

Clips

Nut (4)
⚙ 5.2 – 7.0
(0.53 – 0.71, 46.0 – 61.6)

Special bolt ⊗
🔧 15 – 25
(1.5 – 2.5, 11 – 18)

🔧 : N·m (kg-m, ft-lb)

⚙ : N·m (kg-m, in-lb)

▦ : Insert front edge first

93113GC4

View of the passenger's side SRS module and related components—2000–02 Frontier

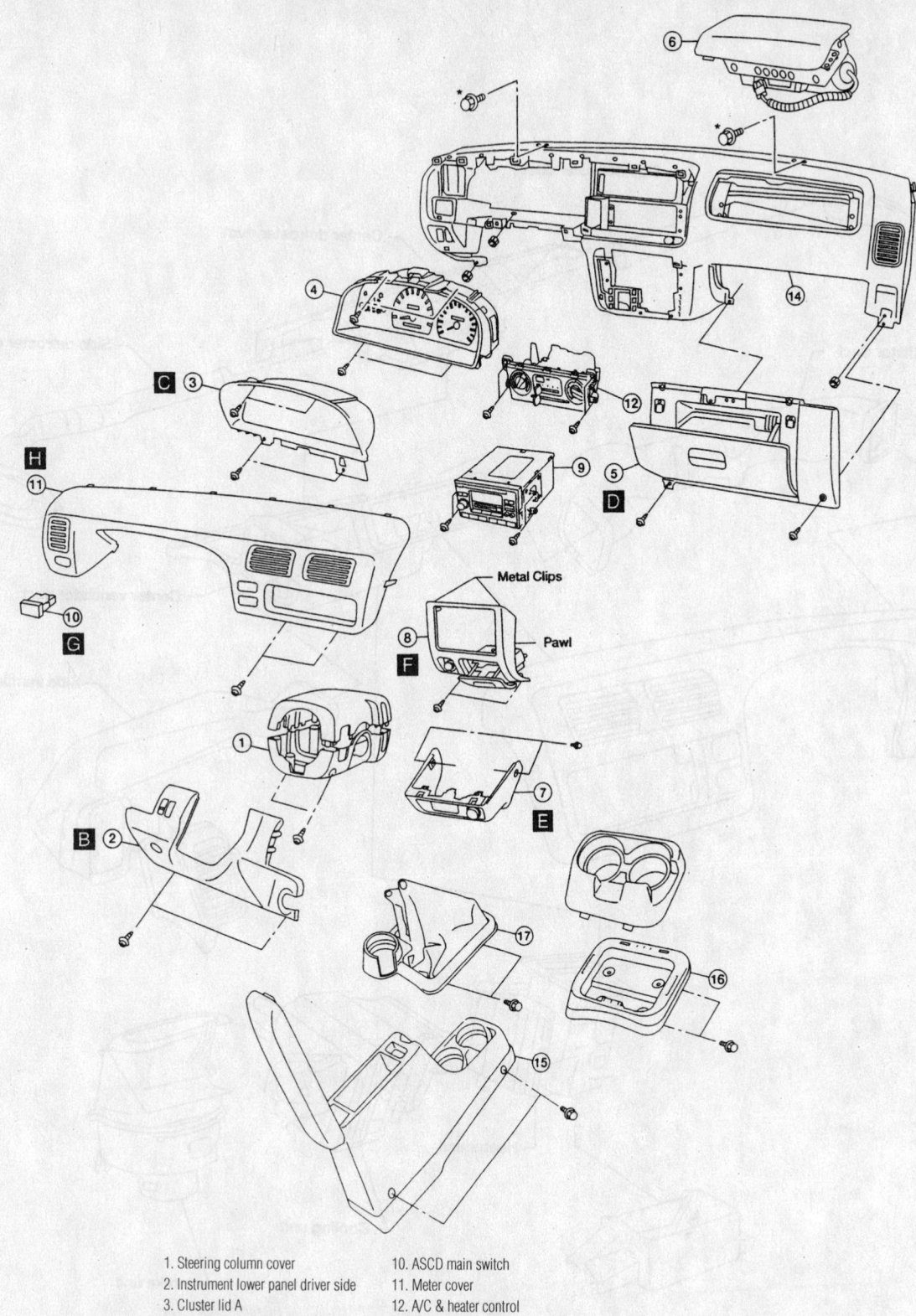

Metal Clips

Pawl

1. Steering column cover
2. Instrument lower panel driver side
3. Cluster lid A
4. Combination meter
5. Glove box assembly
6. Passenger side air bag module
7. Instrument stay cover
8. Cluster lid C
9. Audio and deck pocket
10. ASCD main switch
11. Meter cover
12. A/C & heater control
13. Front pillar garnish
14. Instrument panel assembly
15. Center console assembly
16. Cup holder assembly
17. M/T boot assembly

93113GC5

Exploded view of the instrument panel and related components—2000–02 Frontier

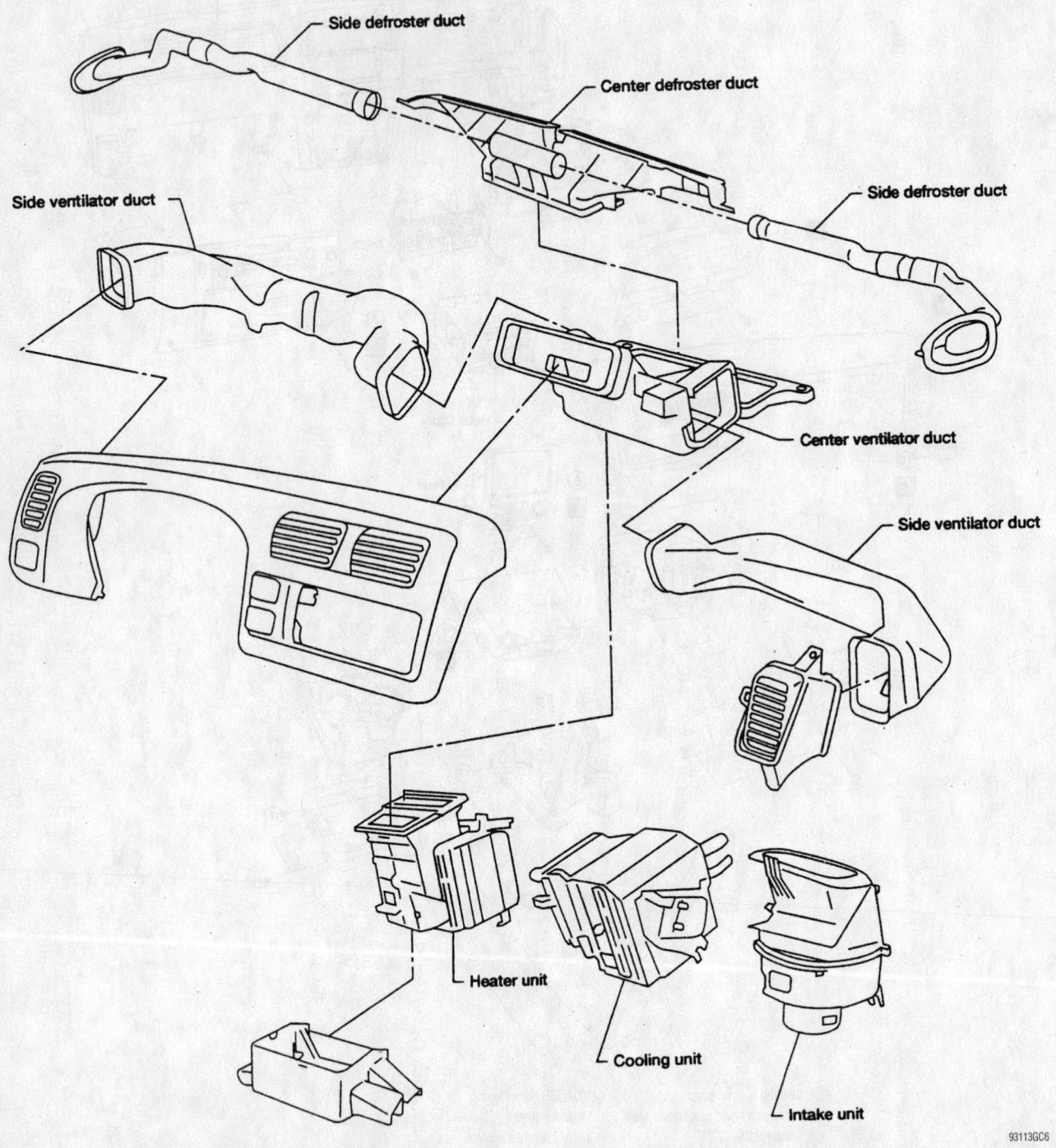

Side defroster duct

Center defroster duct

Side defroster duct

Side ventilator duct

Center ventilator duct

Side ventilator duct

Heater unit

Cooling unit

Intake unit

93113GC6

Exploded view of the heater housing, air conditioning housing and related components—2000–02 Frontier

Always store the SRS module face up.

4. Drain the cooling system into a clean container for reuse.

5. Working inside the engine compartment, disconnect the 2 heater hoses from the heater core.

6. Discharge and recover the air conditioning system refrigerant.

7. Disconnect both refrigerant lines from the evaporator core. Plug the lines to prevent moisture from entering the system.

8. Remove the glove box and the mating trim.

9. Disconnect the thermal amp connector.

10. Remove the air conditioning housing assembly from the vehicle.

11. Remove the instrument panel assembly by performing the following procedure:

   a. Remove the 4 steering column cover screws; then, separate and remove the steering column covers.

   b. Remove the 2 driver's side lower instrument panel screws and the lower instrument panel.

   c. Remove the 4 cluster cover screws and the cluster cover.

   d. Remove the 6 combination meter screws; then, disconnect the combination meter electrical connector and remove the meter.

   e. Remove the 2 glove box screws and the glove box.

   f. Remove the 2 instrument stay cover screws; then, disconnect the electrical harness connectors and remove the stay cover.

   g. Remove the 2 cluster lid "C" screws; then, disconnect the electrical harness connectors and remove the cluster lid "C".

   h. Remove the 4 audio and deck pocket screws; then, disconnect the electrical harness connectors and remove the audio and deck pocket.

   i. Disconnect the ASCD main switch connector.

   j. Remove the 2 meter cover screws; then, disconnect the electrical harness connectors and remove the meter cover.

   k. Remove the 4 air conditioning-heater control screws; then, disconnect the control cables and remove the air conditioning-heater control.

   l. Remove the front pillar garnish.

   m. Remove the 3 instrument panel assembly nuts and 2 bolts; then, remove the instrument panel.

12. Remove the heater housing assembly.

13. Remove the heater core from the heater housing assembly.

To install:

14. Install the heater core to the heater housing assembly.

15. Install the heater housing assembly.

16. Install the instrument panel assembly by performing the following procedure:

   a. Install the instrument panel and the 3 instrument panel assembly nuts and 2 bolts.

   b. Install the front pillar garnish.

   c. Install the air conditioning-heater control, connect the control cables and install the 4 air conditioning/heater control screws.

   d. Install the meter cover, connect the electrical harness connectors and install the 2 meter cover screws.

   e. Connect the ASCD main switch connector.

   f. Install the audio and deck pocket, connect the electrical harness connectors and install the 4 audio and deck pocket screws.

   g. Install the cluster lid "C", connect the electrical harness connectors and install the 2 cluster lid "C" screws.

   h. Install the stay cover, connect the electrical harness connectors and the 2 instrument stay cover screws.

   i. Install the glove box and the 2 glove box screws.

   j. Install the meter, connect the combination meter electrical connector and install the 6 combination meter screws.

   k. Install the cluster cover and the 4 cluster cover screws.

   l. Install the lower instrument panel and the 2 driver's side lower instrument panel screws.

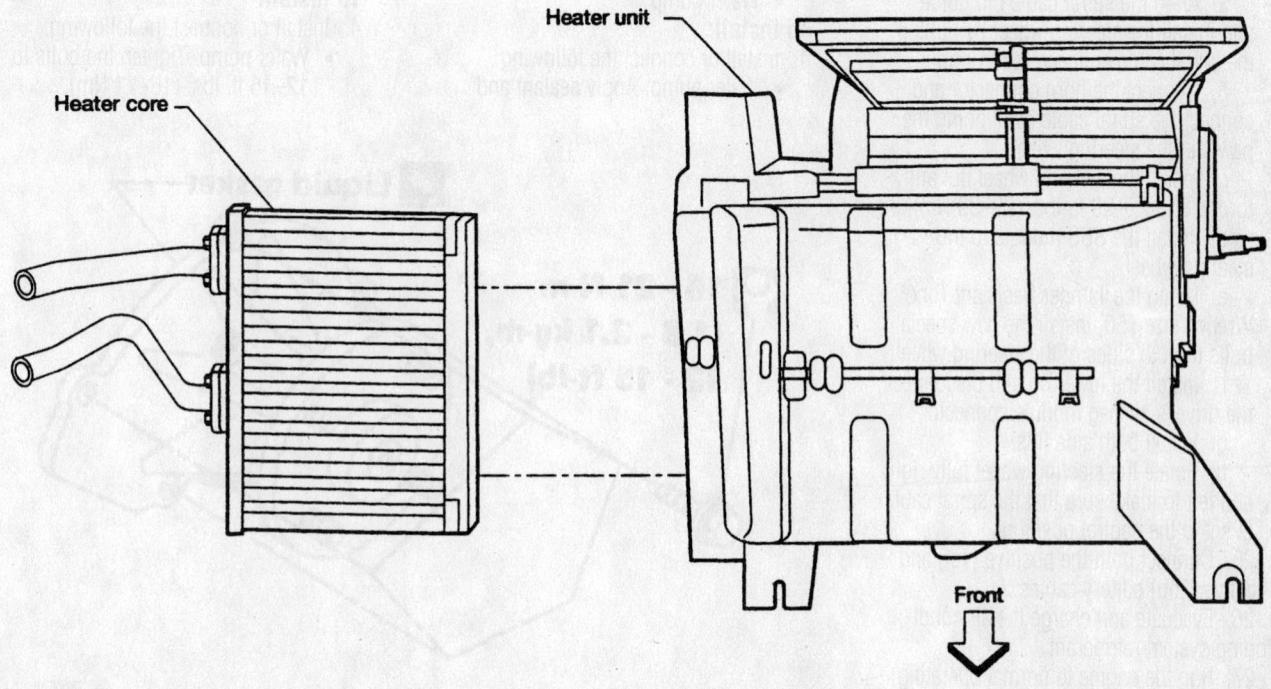

Heater core

Heater unit

Front

93113GC7

View of the heater core and heater housing—2000–02 Frontier

m. Install the steering column covers and the 4 steering column cover screws.

17. Install the air conditioning housing assembly to the vehicle.

18. Connect the thermal amp connector.

19. Install the glove box and the mating trim.

20. Connect both refrigerant lines to the evaporator core.

21. Inside the engine compartment, connect the heater hoses to the heater core.

22. Refill the cooling system.

23. Install the passenger's side air bag by performing the following procedure:

   a. Install the SRS module and secure the SRS module-to-instrument panel clips.

   b. Install the 4 SRS module-to-instrument panel mounting nuts.

   c. Using the Tamper Resistant Torx® Wrench size T50, install the new special SRS module-to-instrument panel bolts.

   d. Install the lower passenger's side instrument panel and the glove box.

   e. Connect the passenger's SRS module connector.

   f. Install the passenger's air bag module connector clip to the lid.

   g. Inside the glove box, close the lower instrument panel lid.

   h. Close the glove box door.

24. Install the steering wheel by performing the following procedure:

   a. Align the spiral cable pin guide and install the steering wheel by pulling the spiral cable connectors through it.

   b. Connect the horn connector and connect the spiral cable by aligning the pawls in the steering wheel.

   c. Install the steering wheel nut and torque it to 22–29 ft. lbs. (29–39 Nm).

   d. Install the SRS module to the steering wheel.

   e. Using the Tamper Resistant Torx® Wrench size T50, install the new special bolts to both sides of the steering wheel.

   f. Install the lower lid and disconnect the driver's air bag module connector.

   g. Install both side lids.

   h. Rotate the steering wheel fully right and left to make sure that the spiral cable is set in the neutral position.

25. Connect both the positive (1st) and negative (2nd) battery cables.

26. Evacuate and charge the air conditioning system refrigerant.

27. Run the engine to normal operating temperatures; then, check the climate control operation and check for leaks.

2003 Frontier and Xterra

1. Drain cooling system.

2. Disconnect the two heater hoses from the engine compartment side.

3. Discharge the A/C system.

4. Disconnect the two evaporator core refrigerant lines from the engine compartment side. Cap the refrigerant lines to prevent moisture from entering the system.

5. Remove the glove box and mating trim.

6. Disconnect the thermal amp. connector.

7. Remove the cooling unit.

8. Separate the cooling unit case, and remove the evaporator.

9. Remove the steering column.

10. Remove the heater unit.

11. Remove the heater core.

12. Installation is the reverse of removal.

Water Pump

REMOVAL & INSTALLATION

2.4L Engine

1. Before servicing the vehicle, refer to the precautions in the beginning of this section.

2. Drain the cooling system.

3. Remove or disconnect the following:
   • Negative battery cable
   • Accessory drive belts
   • Cooling fan
   • Water pump

To install:

4. Install or connect the following:
   • Water pump. Apply sealant and

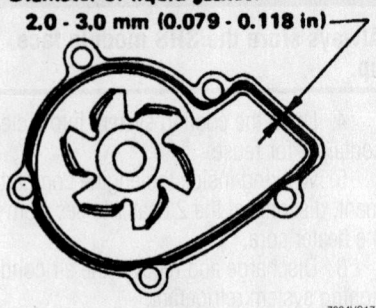

Diameter of liquid gasket:
2.0 - 3.0 mm (0.079 - 0.118 in)

7924VG17

Liquid gasket application—2.4L engine

tighten the bolts to 12–15 ft. lbs. (16–21 Nm).
   • Cooling fan
   • Accessory drive belts
   • Negative battery cable

5. Fill the cooling system.

6. Start the engine and check for leaks.

3.3L Engine

1. Before servicing the vehicle, refer to the precautions in the beginning of this section.

2. Drain the cooling system.

3. Remove or disconnect the following:
   • Negative battery cable
   • Accessory drive belts
   • Radiator hoses
   • Cooling fan and shroud
   • Water pump pulley
   • Front cover
   • Timing belt
   • Water pump

To install:

4. Install or connect the following:
   • Water pump. Tighten the bolts to 12–15 ft. lbs. (16–21 Nm).

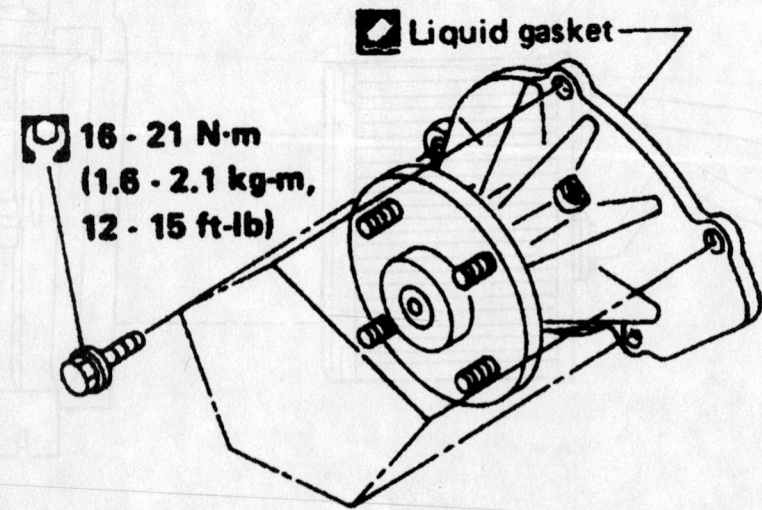

🔲 Liquid gasket

🔲 16 - 21 N·m
(1.6 - 2.1 kg-m,
12 - 15 ft-lb)

7924VG16

Water pump assembly—2.4L engine

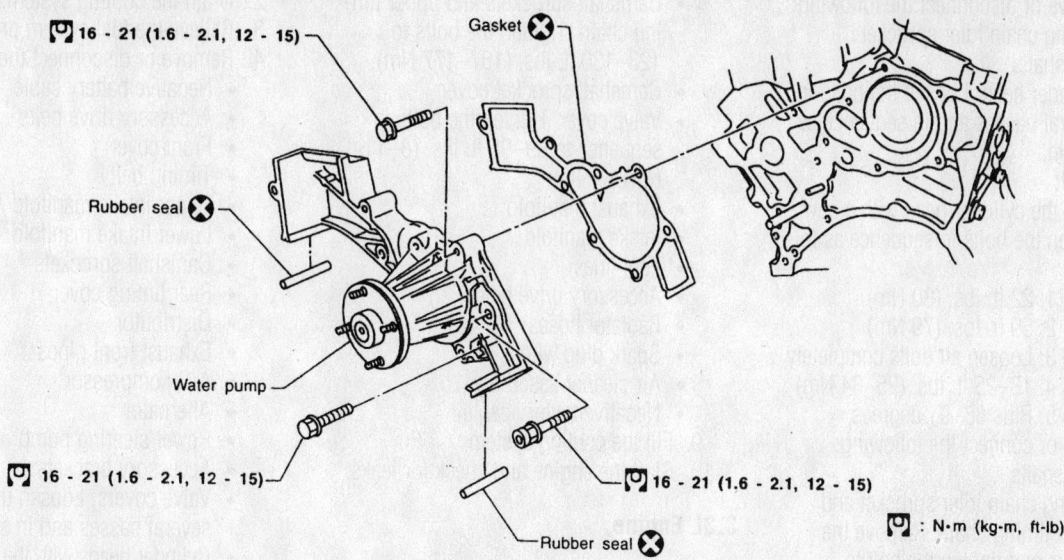

Exploded view of the water pump assembly—3.3L engine

- Timing belt
- Front cover
- Water pump pulley
- Cooling fan and shroud
- Radiator hoses
- Accessory drive belts
- Negative battery cable
5. Fill the cooling system.
6. Start the engine and check for leaks.

Cylinder Head

REMOVAL & INSTALLATION

2.4L Engine

1. Before servicing the vehicle, refer to the precautions in the beginning of this section.
2. Drain the cooling system.
3. Relieve the fuel system pressure.
4. Remove or disconnect the following:

- Negative battery cable
- Air cleaner assembly
- Spark plug wires
- Radiator hoses
- Accessory drive belts
- Fuel lines
- Intake manifold
- Exhaust manifold
- Valve cover. Remove the bolts in the sequence shown.
- Camshaft sprocket cover
- Camshaft sprockets and upper timing chain

5. Wedge the lower timing chain in place to prevent the chain tensioner from expanding.

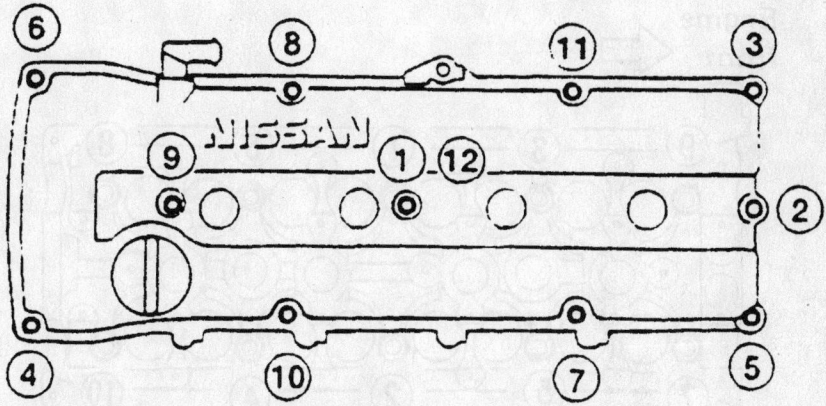

Loosen in numerical order.

Valve cover loosening sequence—2.4L engine

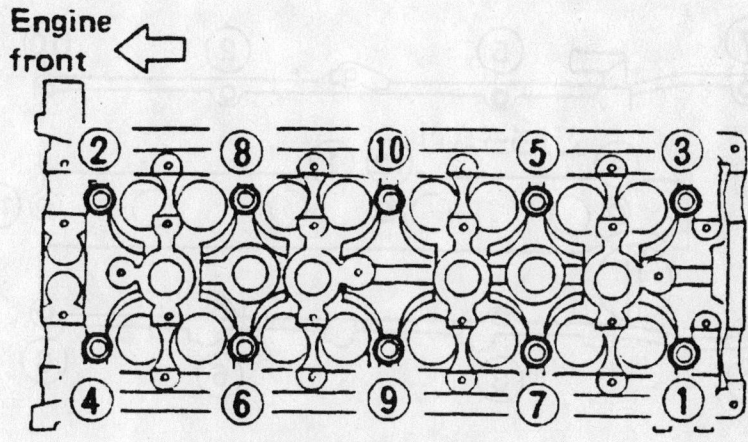

Loosen in numerical order.

Cylinder head loosening sequence—2.4L engine

6. Remove or disconnect the following:
   - Timing chain idler sprocket
   - Camshafts
   - Cylinder head. Loosen the bolts in several passes and in sequence as shown.

To install:

7. Install the cylinder head with a new gasket. Tighten the bolts in sequence as follows:
   a. Step 1: 22 ft. lbs. (30 Nm)
   b. Step 2: 59 ft. lbs. (79 Nm)
   c. Step 3: Loosen all bolts completely
   d. Step 4: 18–25 ft. lbs. (25–34 Nm)
   e. Step 5: Plus 86–91 degrees

8. Install or connect the following:
   - Camshafts
   - Timing chain idler sprocket and lower timing chain. Remove the wedge and tighten the bolt to 48–61 ft. lbs. (66–83 Nm).

- Camshaft sprockets and upper timing chain. Tighten the bolts to 123–130 ft. lbs. (167–177 Nm).
- Camshaft sprocket cover
- Valve cover. Tighten the bolts in sequence to 69–95 ft. lbs. (8–11 Nm).
- Exhaust manifold
- Intake manifold
- Fuel lines
- Accessory drive belts
- Radiator hoses
- Spark plug wires
- Air cleaner assembly
- Negative battery cable

9. Fill the cooling system.
10. Start the engine and check for leaks.

3.3L Engine

1. Before servicing the vehicle, refer to the precautions in the beginning of this section.

2. Drain the cooling system.
3. Relieve the fuel system pressure.
4. Remove or disconnect the following:
   - Negative battery cable
   - Accessory drive belts
   - Front cover
   - Timing belt
   - Upper intake manifold
   - Lower intake manifold
   - Camshaft sprockets
   - Rear timing cover
   - Distributor
   - Exhaust front pipes
   - A/C compressor
   - Alternator
   - Power steering pump
   - Accessory brackets
   - Valve covers. Loosen the bolts in several passes and in sequence.
   - Cylinder heads with the exhaust manifolds attached. Loosen the bolts in several passes and in sequence.

➡**The cylinder head bolts vary in length. Note the bolt locations for assembly.**

To install:

5. Install the cylinder heads and the lower intake manifold at the same time. Tighten the bolts in sequence as follows:
   a. Step 1: Tighten the cylinder head bolts to 22 ft. lbs. (29 Nm)
   b. Step 2: Tighten the cylinder head bolts to 43 ft. lbs. (59 Nm)
   c. Step 3: Loosen all cylinder head bolts completely
   d. Step 4: Tighten the cylinder head bolts to 84 inch lbs. (10 Nm)
   e. Step 5: Tighten the intake manifold fasteners to 35 inch lbs. (4 Nm)

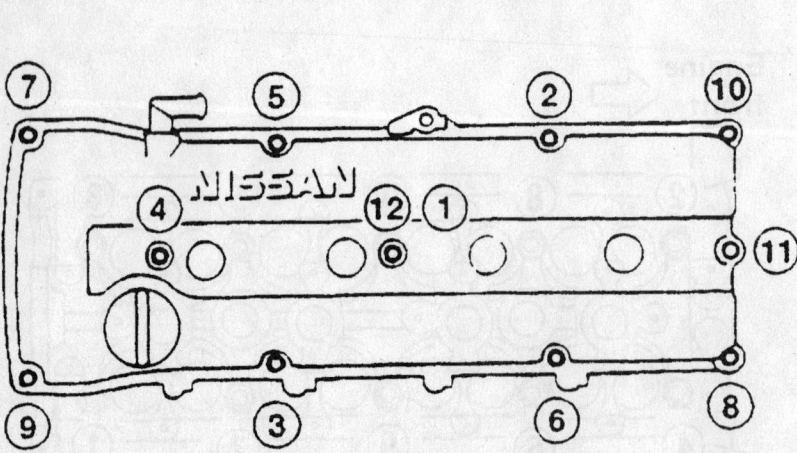

Cylinder head torque sequence—2.4L DOHC engine

Tighten in numerical order.

9308VG05

Tighten in numerical order.

9308VG07

Valve cover torque sequence—2.4L engine

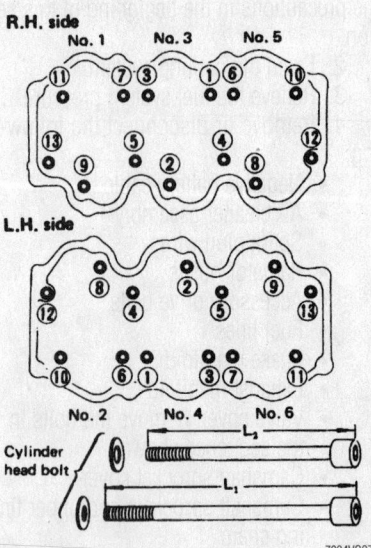

7924VG27

Cylinder head torque sequence—3.3L engine

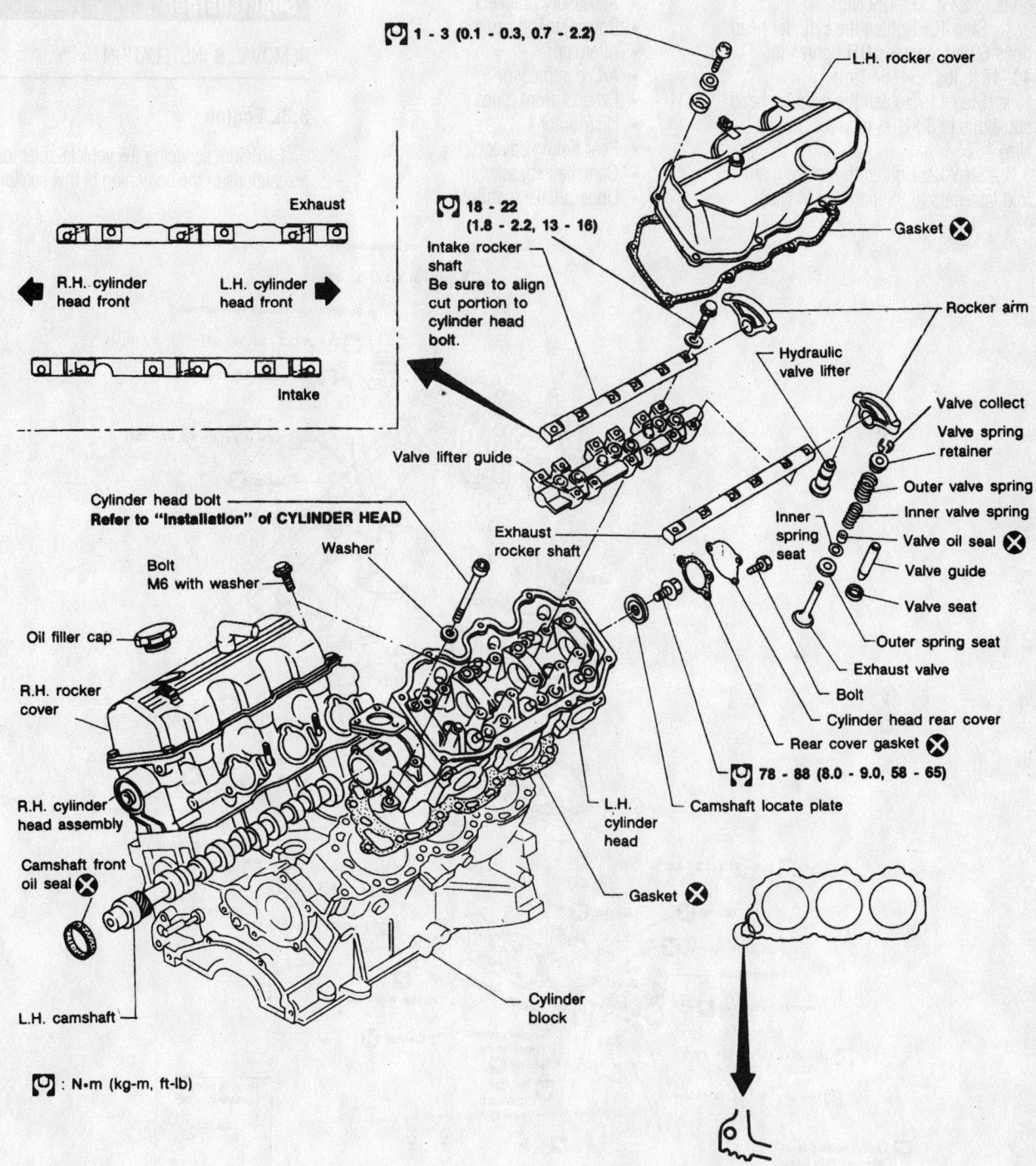

1 - 3 (0.1 - 0.3, 0.7 - 2.2)

— L.H. rocker cover

Gasket ⊗

18 - 22
(1.8 - 2.2, 13 - 16)

Intake rocker
shaft
Be sure to align
cut portion to
cylinder head
bolt.

Rocker arm

Hydraulic
valve lifter

Valve collect

Valve spring
retainer

Outer valve spring

Inner valve spring

Valve oil seal ⊗

Valve guide

Valve seat

Inner
spring
seat

Exhaust
rocker shaft

Valve lifter guide

Exhaust
Outer spring seat

Exhaust valve

Bolt

Cylinder head rear cover

Rear cover gasket ⊗

78 - 88 (8.0 - 9.0, 58 - 65)

Camshaft locate plate

L.H.
cylinder
head

Gasket ⊗

Cylinder head bolt
Refer to "Installation" of CYLINDER HEAD

Washer

Bolt
M6 with washer

Oil filler cap

R.H. rocker
cover

R.H. cylinder
head assembly

Camshaft front
oil seal ⊗

L.H. camshaft

Cylinder
block

: N•m (kg-m, ft-lb)

Exhaust

R.H. cylinder
head front

L.H. cylinder
head front

Intake

Exploded view of the cylinder head assembly—3.3L engine

7924VG25

f. Step 6: Tighten the intake manifold fasteners to 13 ft. lbs. (18 Nm)

g. Step 7: Tighten the intake manifold fasteners to 12–14 ft. lbs. (16–20 Nm)

h. Step 8: Loosen all intake fasteners completely

i. Step 9: Tighten the cylinder head bolts to 22 ft. lbs. (29 Nm)

j. Step 10: Tighten the cylinder head bolts 60–65 degrees **OR** tighten to 40–47 ft. lbs. (54–64 Nm)

k. Step 11: Tighten the cylinder head sub-bolts to 80–105 inch lbs. (9–12 Nm)

l. Step 12: Tighten the intake manifold fasteners to 35 inch lbs. (4 Nm)

m. Step 13: Tighten the intake manifold fasteners to 78 inch lbs. (9 Nm)

n. Step 14: Tighten the intake manifold fasteners to 70–84 inch lbs. (6–7 Nm)

6. Install or connect the following:
- Valve covers
- Accessory brackets
- Power steering pump
- Alternator
- A/C compressor
- Exhaust front pipes
- Distributor
- Rear timing cover
- Camshaft sprockets
- Upper intake manifold
- Timing belt
- Front cover
- Accessory drive belts
- Negative battery cable

7. Fill the cooling system.

8. Start the engine and check for leaks.

Supercharger

REMOVAL & INSTLLATION

3.3L Engine

1. Before servicing the vehicle, refer to the precautions in the beginning of this section.

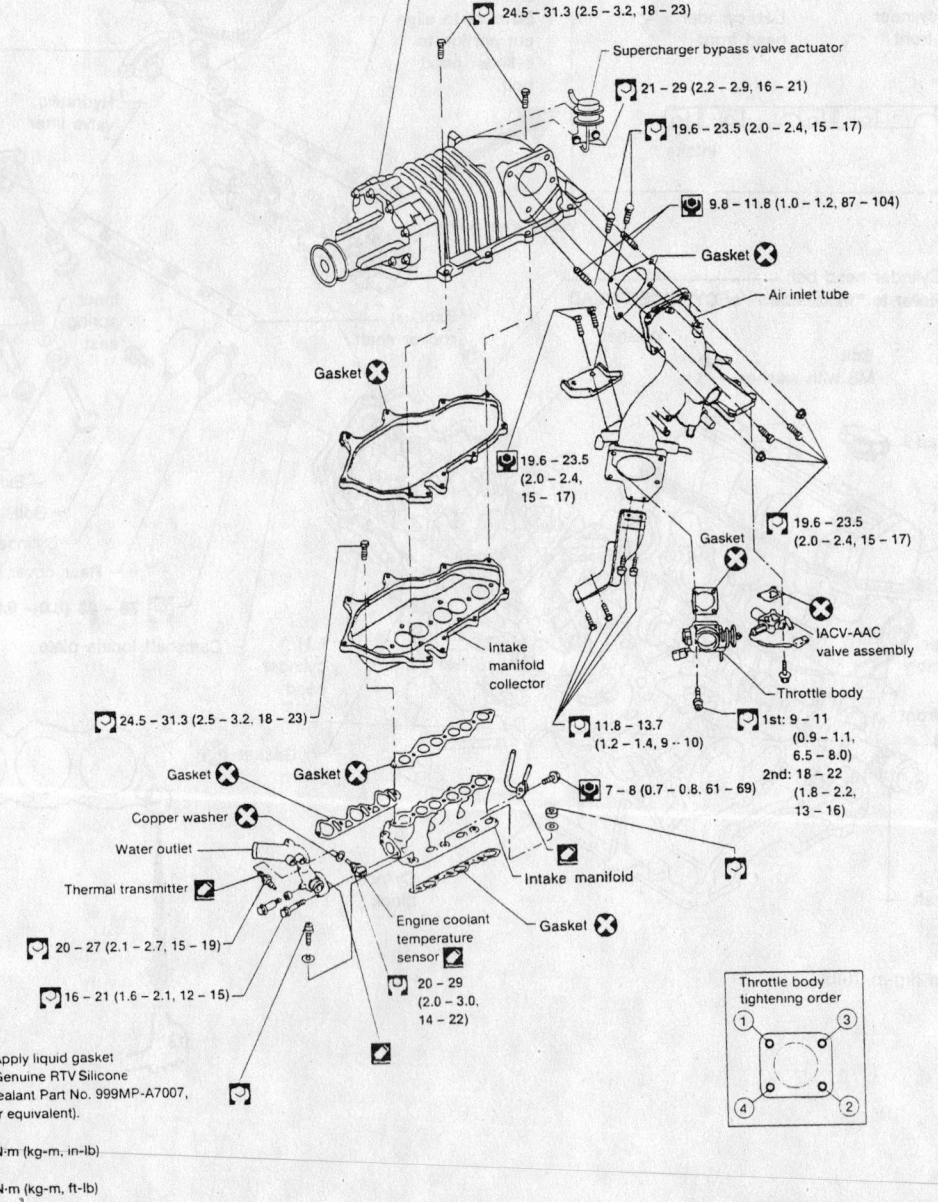

- Supercharger (do not disassemble)
- 24.5 – 31.3 (2.5 – 3.2, 18 – 23)
- Supercharger bypass valve actuator
- 21 – 29 (2.2 – 2.9, 16 – 21)
- 19.6 – 23.5 (2.0 – 2.4, 15 – 17)
- 9.8 – 11.8 (1.0 – 1.2, 87 – 104)
- Gasket
- Air inlet tube
- Gasket
- 19.6 – 23.5 (2.0 – 2.4, 15 – 17)
- 19.6 – 23.5 (2.0 – 2.4, 15 – 17)
- Gasket
- IACV-AAC valve assembly
- Throttle body
- Intake manifold collector
- 24.5 – 31.3 (2.5 – 3.2, 18 – 23)
- 11.8 – 13.7 (1.2 – 1.4, 9 – 10)
- 1st: 9 – 11 (0.9 – 1.1, 6.5 – 8.0)
- 2nd: 18 – 22 (1.8 – 2.2, 13 – 16)
- Gasket
- Gasket
- 7 – 8 (0.7 – 0.8, 61 – 69)
- Copper washer
- Water outlet
- Intake manifold
- Thermal transmitter
- Gasket
- 20 – 27 (2.1 – 2.7, 15 – 19)
- Engine coolant temperature sensor
- 16 – 21 (1.6 – 2.1, 12 – 15)
- 20 – 29 (2.0 – 3.0, 14 – 22)
- : Apply liquid gasket (Genuine RTV Silicone Sealant Part No. 999MP-A7007, or equivalent).
- : N·m (kg-m, in-lb)
- : N·m (kg-m, ft-lb)

Throttle body tightening order

9348VG92

Supercharger and related parts—3.3L engine

2. Drain the coolant.
3. Remove or disconnect the following:
   • Negative battery cable
   • Accelerator cable
   • ASCD cable at the throttle body
   • Air inlet duct
   • PCV hoses
   • Resonator hose
   • Supercharger pulley cover
   • Supercharger drive belt
   • Air inlet tube supports
   • Air inlet tube
   • EVAP vacuum hose
   • Brake booster hose
   • All remaining hoses and wires in the way of removal
   • Intake manifold collector
   • Heater hoses
   • Supercharger
4. Installation is the reverse of removal. Observe the following torques:

• Supercharger mounting bolts: 18–23 ft. lbs. (24–31 Nm).
• Air inlet tube-to-supercharger: 15–17 ft. lbs. (20–24 Nm).

Rocker Arms/Shafts

REMOVAL & INSTALLATION

2.4L Engine

This engine is not equipped with rocker arms. The camshafts act directly on the valve lifters.

3.3L Engine

1. Before servicing the vehicle, refer to the precautions in the beginning of this section.
2. Remove or disconnect the following:
   • Negative battery cable

• Supercharger or upper intake manifold
• Valve covers
• Rocker arm and shaft assemblies
• Rocker arms from the shafts

➡ **Keep all valvetrain components in order for assembly.**

To install:

3. Lubricate all contact points with clean engine oil and assemble the rocker arms to the shafts in their original positions.
4. Install or connect the following:
   • Rocker arm and shaft assemblies. Tighten the bolts to 13–16 ft. lbs. (18–22 Nm).
   • Valve covers
   • Upper intake manifold
   • Negative battery cable
5. Start the engine and check for leaks.

Intake Manifold

REMOVAL & INSTALLATION

2.4L Engine

1. Before servicing the vehicle, refer to the precautions in the beginning of this section.
2. Drain the cooling system.
3. Relieve the fuel system pressure.
4. Remove or disconnect the following:
   • Negative battery cable
   • Air cleaner assembly
   • Coolant hoses
   • Fuel lines
   • Accelerator cable
   • Cruise control cable, if equipped
   • Positive Crankcase Ventilation (PCV) valve and hose
   • Exhaust Gas Recirculation (EGR) tube
   • EGR temperature sensor connector
   • Idle Air Control (IAC) valve and solenoid connectors
   • Throttle Position (TP) sensor and switch connectors
   • Engine Coolant Temperature (ECT) sensor connector
   • Manifold Absolute Pressure (MAP) sensor connector and vacuum line
   • Evaporative Emissions (EVAP) canister purge valve vacuum line
   • Brake booster vacuum line
   • Fuel injector connectors
   • Intake manifold bracket
   • Intake manifold. Loosen the fasteners in reverse of the torque sequence.

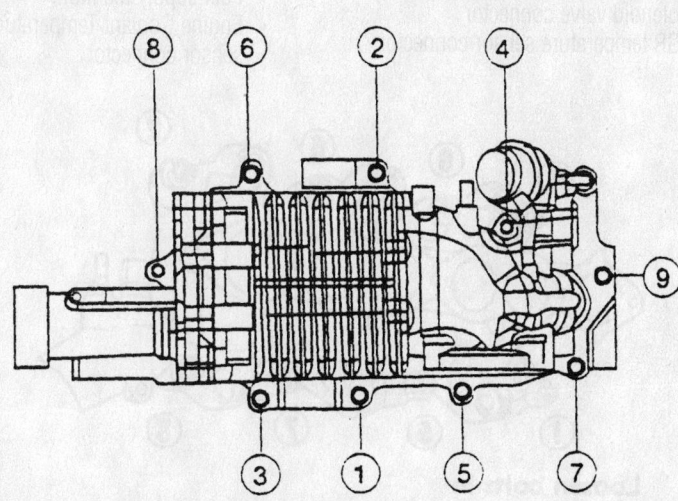

9348VG90

Supercharger torque sequence. Loosen in reverse order—3.3L engine

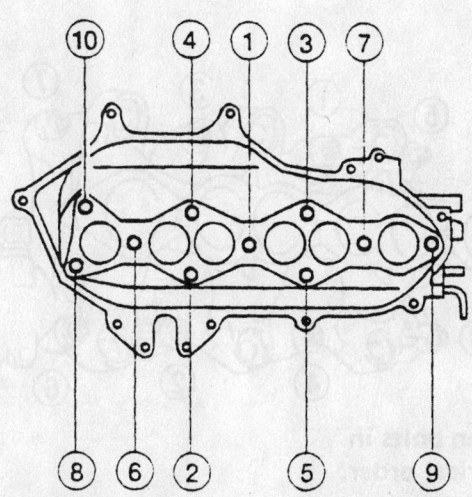

9348VG91

Intake manifold collector torque sequence. Loosen in reverse order—3.3L supercharged engine

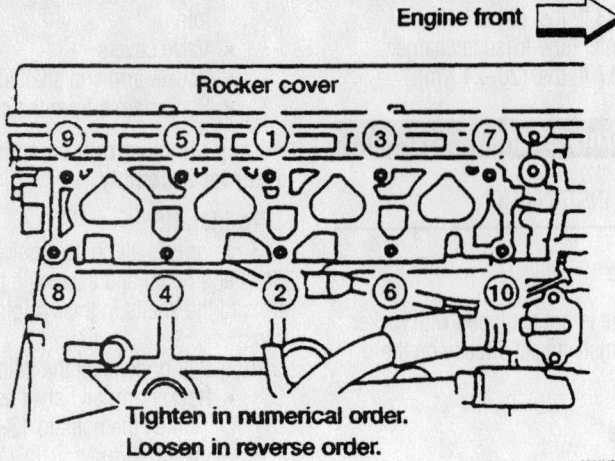

Intake manifold torque sequence—2.4L engines

To install:

5. Install or connect the following:
    *   Intake manifold. Tighten the bolts to 12–14 ft. lbs. (16–19 Nm).
    *   Intake manifold bracket. Tighten the bolts to 24–28 ft. lbs. (32–38 Nm).
    *   Fuel injector connectors
    *   Brake booster vacuum line
    *   EVAP canister purge valve vacuum line
    *   MAP sensor connector and vacuum line
    *   ECT sensor connector
    *   TP sensor and switch connectors
    *   IAC valve and solenoid connectors
    *   EGR temperature sensor connector
    *   EGR tube
    *   PCV valve and hose
    *   Cruise control cable, if equipped
    *   Accelerator cable
    *   Fuel lines
    *   Coolant hoses
    *   Air cleaner assembly
    *   Negative battery cable
6. Fill the cooling system.
7. Start the engine and check for leaks.

3.3L Engine

1. Before servicing the vehicle, refer to the precautions in the beginning of this section.
2. Drain the cooling system.
3. Relieve the fuel system pressure.
4. Remove or disconnect the following:
    *   Negative battery cable
    *   Air intake duct
    *   Accelerator cable
    *   Cruise control cable
    *   Idle Air Control (IAC) valve connector
    *   Throttle Position (TP) sensor and switch connectors

*   Ignition coil and power transistor connectors
*   Exhaust Gas Recirculation (EGR) Solenoid valve connector
*   EGR temperature sensor connector

*   Radiator hoses
*   Heater hoses
*   Positive Crankcase Ventilation (PCV) valve and hose
*   Evaporative Emissions (EVAP) canister vacuum and purge hoses
*   Brake booster vacuum hose
*   Fuel pressure regulator vacuum hose
*   EGR tube
*   Spark plug wires
*   Distributor
*   Left bank injector connectors
*   Thermal transmitter
*   Upper intake manifold ground cable (VG33)
*   Supercharger (VG33ER)
*   Breather pipe
*   Upper intake manifold (VG33)
*   Intake manifold collector (VG33ER)
*   Fuel lines
*   Right bank injector connectors
*   Fuel supply manifold
*   Engine Coolant Temperature (ECT) sensor connector

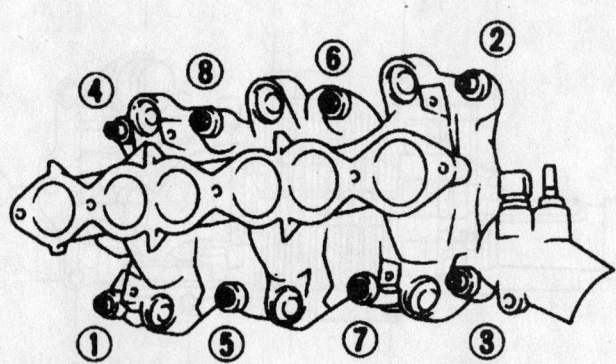

Intake manifold loosening sequence—3.3L engine

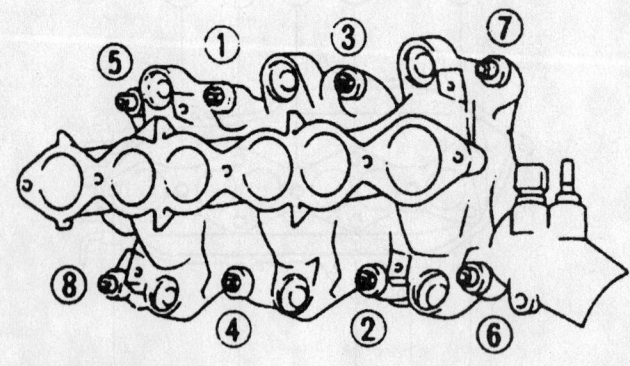

Intake manifold torque sequence—3.3L engine

- Lower intake manifold. Loosen the fasteners in the sequence shown.

To install:

5. Install the lower intake manifold with a new gasket.

6. Tighten the fasteners in sequence as follows:

  a. Step 1: 35 inch lbs. (4 Nm)
  b. Step 2: 78 inch lbs. (9 Nm)
  c. Step 3: 70–84 inch lbs. (8–10 Nm)

7. Install or connect the following:
- ECT sensor connector
- Fuel supply manifold
- Right bank injector connectors
- Fuel lines
- Upper intake manifold
- Breather pipe
- Upper intake manifold ground cable
- Thermal transmitter
- Left bank injector connectors
- Distributor
- Spark plug wires
- EGR tube
- Fuel pressure regulator vacuum hose
- Brake booster vacuum hose
- EVAP canister vacuum and purge hoses
- PCV valve and hose
- Heater hoses
- Radiator hoses
- EGR temperature sensor connector
- EGR Solenoid valve connector
- Ignition coil and power transistor connectors
- TP sensor and switch connectors
- IAC valve connector
- Cruise control cable
- Accelerator cable
- Air intake duct
- Negative battery cable

8. Fill the cooling system.

9. Start the engine and check for leaks.

Exhaust Manifold

REMOVAL & INSTALLATION

2.4L Engine

1. Before servicing the vehicle, refer to the precautions in the beginning of this section.

2. Remove or disconnect the following:
- Negative battery cable
- Heated Oxygen (HO$_2$S) sensor connector
- Exhaust manifold heat shield
- Exhaust Gas Recirculation (EGR) tube
- Exhaust front pipe

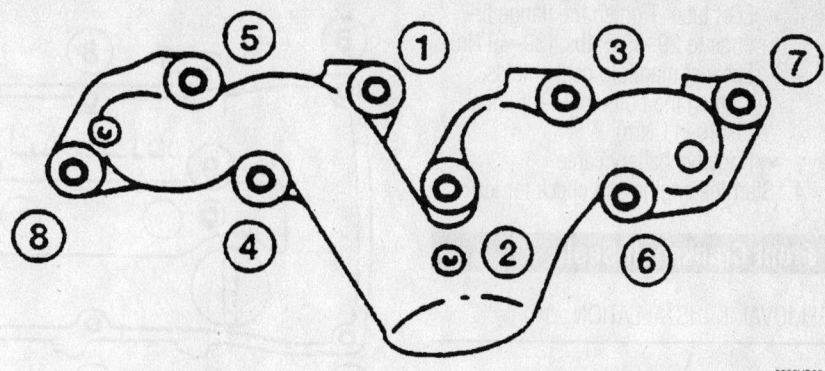

Exhaust manifold torque sequence—2.4L engine

9308VG09

- Exhaust manifold. Loosen the nuts in reverse of the torque sequence.

To install:

3. Install or connect the following:
- Exhaust manifold. Tighten the nuts in sequence to 28–35 ft. lbs. (37–48 Nm).
- Exhaust front pipe. Tighten the fasteners to 32–37 ft. lbs. (43–50 Nm).
- EGR tube. Tighten the flange fittings to 29–36 ft. lbs. (39–49 Nm).
- Exhaust manifold heat shield. Tighten the bolts to 45–57 inch lbs. (5–7 Nm).
- HO$_2$S sensor connector
- Negative battery cable

4. Start the engine and check for leaks.

3.3L Engine

1. Before servicing the vehicle, refer to the precautions in the beginning of this section.

2. Remove or disconnect the following:
- Negative battery cable
- Exhaust manifold heat shields
- Exhaust Gas Recirculation (EGR) tube
- Heated Oxygen (HO$_2$S) sensor connectors
- Exhaust front pipes
- Exhaust manifolds with catalytic converters attached. Loosen the nuts in the reverse of the torque sequence.

To install:

3. Install or connect the following:
- Exhaust manifolds with catalytic converters attached. Tighten the nuts in sequence to 21–25 ft. lbs. (28–33 Nm).
- Exhaust front pipes. Tighten the bolts to 21–25 ft. lbs. (28–33 Nm).
- Heated Oxygen (HO$_2$S) sensor connectors

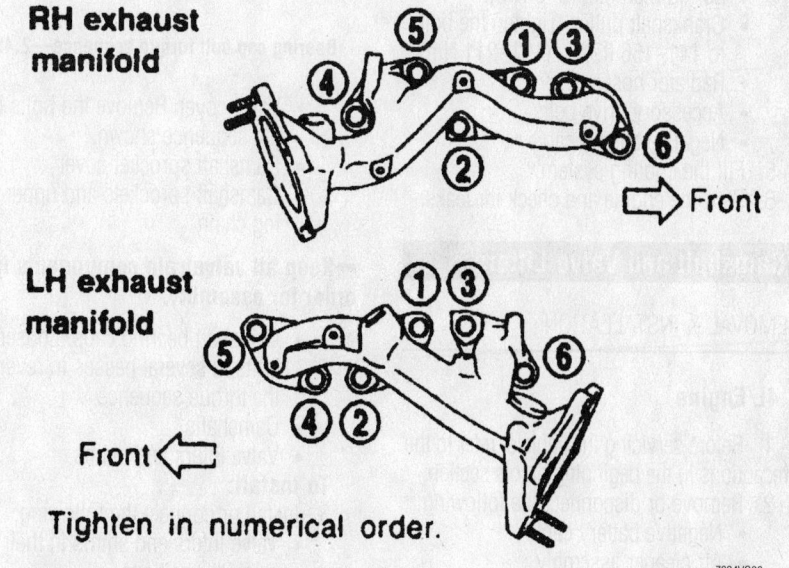

RH exhaust manifold

LH exhaust manifold

Front

Tighten in numerical order.

Exhaust manifold torque sequence—3.3L engine

7924VG36

- EGR tube. Tighten the flange fittings to 29–36 ft. lbs. (39–49 Nm).
- Exhaust manifold heat shields. Tighten the bolts to 84–96 inch lbs. (9–11 Nm)
- Negative battery cable
4. Start the engine and check for leaks.

Front Crankshaft Seal

REMOVAL & INSTALLATION

2.4L Engine

Refer to the Timing Chain, Sprockets, Front Cover and Seal procedure in this section.

3.3L Engine

1. Before servicing the vehicle, refer to the precautions in the beginning of this section.
2. Drain the cooling system.
3. Remove or disconnect the following:
   - Negative battery cable
   - Accessory drive belts
   - Radiator hoses
   - Crankshaft pulley
   - Front cover
   - Timing belt
   - Crankshaft timing sprocket
   - Front crankshaft seal

To install:
4. Install or connect the following:
   - Front crankshaft seal flush with the oil pump housing
   - Crankshaft timing sprocket
   - Timing belt
   - Front cover. Tighten the bolts to 26–43 inch lbs. (3–5 Nm).
   - Crankshaft pulley. Tighten the bolt to 141–156 ft. lbs. (191–211 Nm).
   - Radiator hoses
   - Accessory drive belts
   - Negative battery cable
5. Fill the cooling system.
6. Start the engine and check for leaks.

Camshaft and Valve Lifters

REMOVAL & INSTALLATION

2.4L Engine

1. Before servicing the vehicle, refer to the precautions in the beginning of this section.
2. Remove or disconnect the following:
   - Negative battery cable
   - Air cleaner assembly
   - Spark plug wires

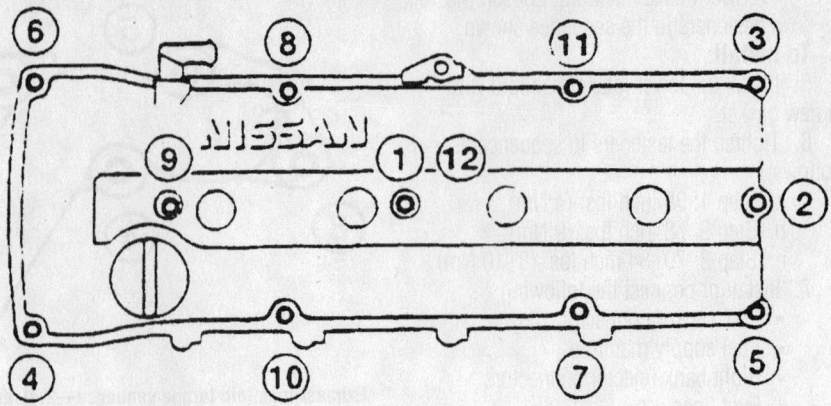

Loosen in numerical order.

9308VG06

Valve cover loosening sequence—2.4L engine

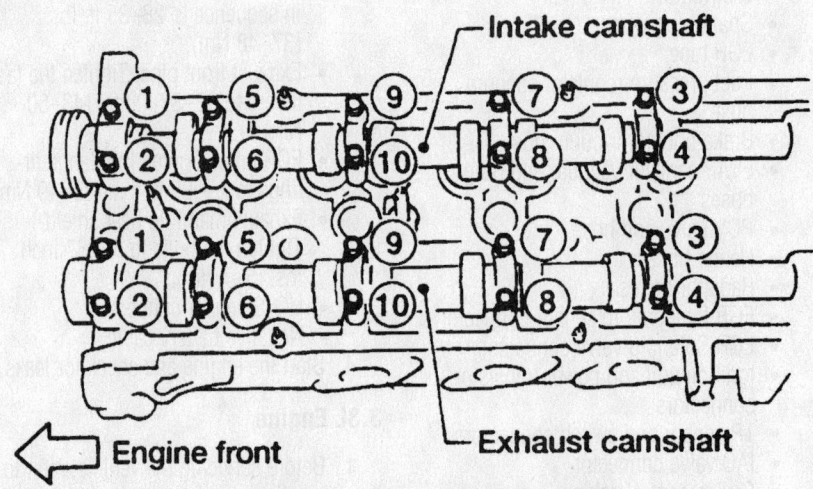

Tighten in numerical order.
Loosen in reverse order.

7924VG51

Bearing cap bolt torque sequence—2.4L engine

- Valve cover. Remove the bolts in the sequence shown.
- Camshaft sprocket cover
- Camshaft sprockets and upper timing chain

➡ **Keep all valvetrain components in order for assembly.**

- Camshaft bearing caps. Loosen the bolts in several passes in reverse of the torque sequence.
- Camshafts
- Valve lifters and shims

To install:
3. Install or connect the following:
   - Valve lifters and shims in their original positions
   - Camshafts

4. Install the bearing caps. Tighten the bolts in sequence as follows:
   a. Step 1: 17 inch lbs. (2 Nm)
   b. Step 2: 80–104 inch lbs. (9–12 Nm)
5. Install or connect the following:
   - Camshaft sprockets and upper timing chain. Tighten the sprocket bolts to 123–130 ft. lbs. (167–177 Nm).
   - Camshaft sprocket cover
   - Valve cover. Tighten the bolts in sequence to 69–95 inch lbs. (8–11 Nm).
   - Spark plug wires
   - Air cleaner assembly
   - Negative battery cable

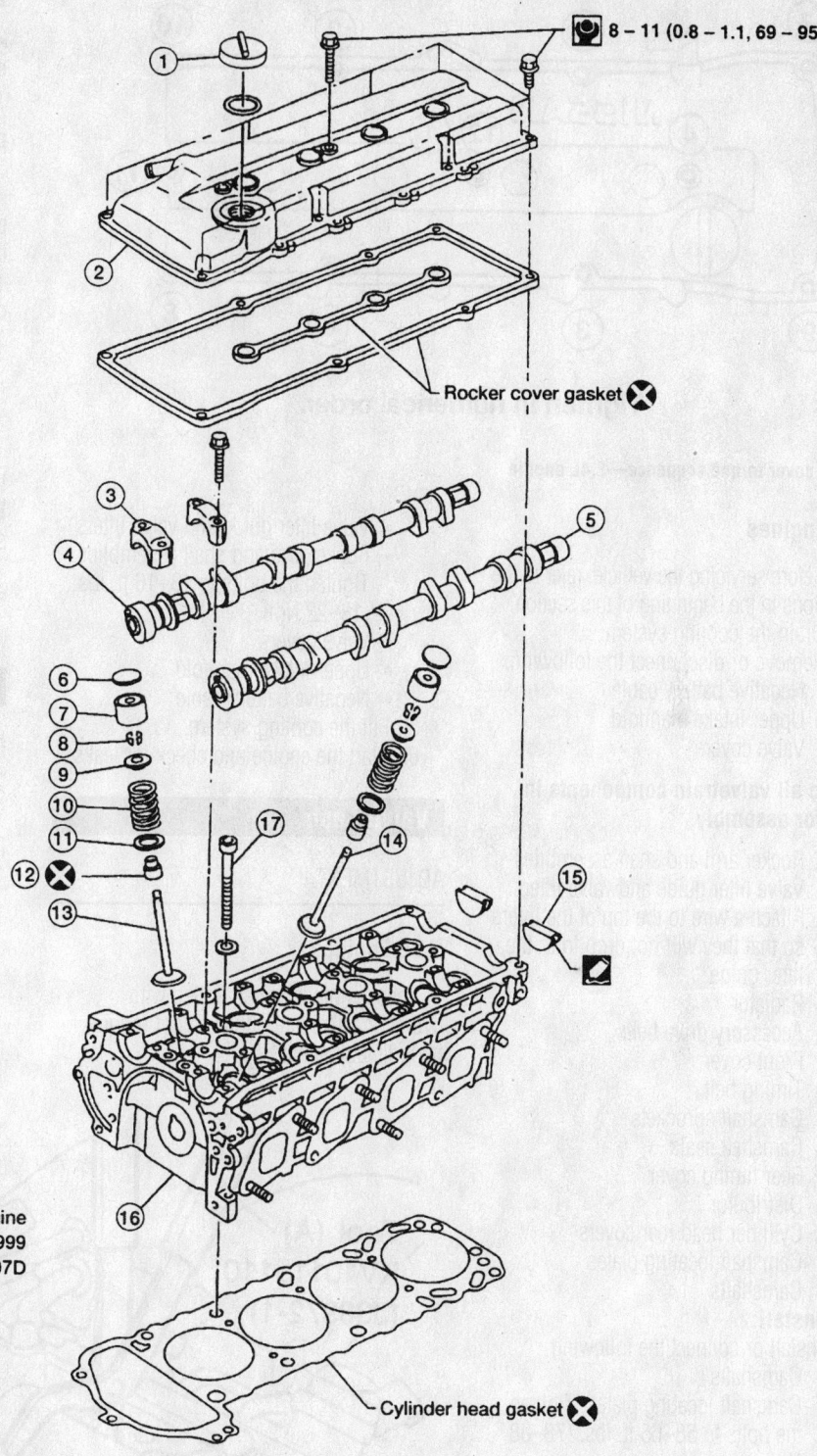

8 – 11 (0.8 – 1.1, 69 – 95)

Rocker cover gasket ⊗

: Apply liquid gasket. Use Genuine RTV silicone sealant, Part No. 999 MP-A7007, Three Bond TB 1207D or equivalent.

: Lubricate with new engine oil.

: N·m (kg-m, in-lb)

: N·m (kg-m, ft-lb)

Cylinder head gasket ⊗

① Oil filler cap	⑦ Valve lifter	⑬ Intake valve
② Rocker cover	⑧ Valve cotter	⑭ Exhaust valve
③ Camshaft bracket	⑨ Spring retainer	⑮ Rubber plug
④ Intake camshaft	⑩ Valve spring	⑯ Cylinder head
⑤ Exhaust camshaft	⑪ Spring seat	⑰ Cylinder head bolt
⑥ Shim	⑫ Valve oil seal	

7924VG53

Exploded view of the camshafts and related components—2.4L engine

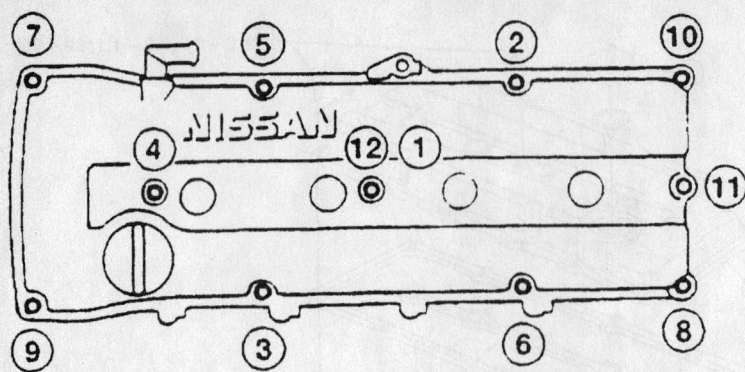

Tighten in numerical order.

9308VG07

Valve cover torque sequence—2.4L engine

3.3L Engines

1. Before servicing the vehicle, refer to the precautions in the beginning of this section.
2. Drain the cooling system.
3. Remove or disconnect the following:
   - Negative battery cable
   - Upper intake manifold
   - Valve covers

➡ **Keep all valvetrain components in order for assembly.**

- Rocker arm and shaft assemblies
- Valve lifter guide and valve lifters. Attach a wire to the top of the lifters so that they will not drop from the lifter guide.
- Radiator
- Accessory drive belts
- Front cover
- Timing belt
- Camshaft sprockets
- Camshaft seals
- Rear timing cover
- Distributor
- Cylinder head rear covers
- Camshaft locating plates
- Camshafts

To install:
4. Install or connect the following:
   - Camshafts
   - Camshaft locating plates. Tighten the bolts to 58–65 ft. lbs. (78–88 Nm).
   - Cylinder head rear covers
   - Distributor
   - Rear timing cover
   - Camshaft seals
   - Camshaft sprockets. Tighten the bolts to 58–65 ft. lbs. (78–88 Nm).
   - Timing belt
   - Front cover
   - Accessory drive belts
   - Radiator

- Valve lifter guide and valve lifters
- Rocker arm and shaft assemblies. Tighten the bolts to 13–16 ft. lbs. (18–22 Nm).
- Valve covers
- Upper intake manifold
- Negative battery cable
5. Fill the cooling system.
6. Start the engine and check for leaks.

Valve Lash

ADJUSTMENT

3.3L Engines

These engines are equipped with hydraulic valve lifters that do not require periodic adjustment.

2.4L Engine

➡ **Measure valve clearance with the engine warm.**

1. Before servicing the vehicle, refer to the precautions in the beginning of this section.
2. Remove the valve cover.
3. Set the engine to the top of the compression stroke with the valves closed for the cylinder to be measured.
4. Check the valve clearance. The valve clearance specifications are as follows:
   - Intake: 0.012–0.015 in. (0.31–0.39mm)
   - Exhaust: 0.013–0.016 in. (0.33–0.41mm)
5. If adjustment is necessary, compress the valve spring with Tool **A** and insert Tool **B** to hold the valve in the open position as shown.
6. Replace the shims as necessary to achieve the correct valve clearance.
7. Repeat for each valve to be adjusted.

Starter Motor

REMOVAL & INSTALLATION

1. Before servicing the vehicle, refer to the precautions in the beginning of this section.
2. Remove or disconnect the following:
   - Negative battery cable
   - Engine under cover
   - Starter harness connectors
   - Starter motor

To install:
3. Install or connect the following:

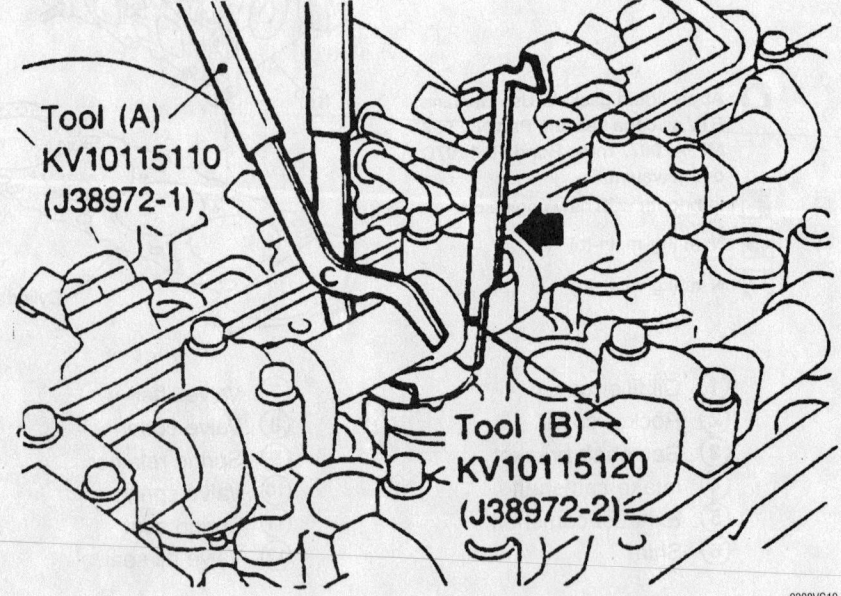

Valve adjustment tools (A) and (B)—2.4L engine

9308VG10

- Starter motor. Tighten the bolts to 22–27 ft. lbs. (30–36 Nm).
- Starter harness connectors
- Engine under cover
- Negative battery cable

Oil Pan

REMOVAL & INSTALLATION

2.4L Engine

1. Before servicing the vehicle, refer to the precautions in the beginning of this section.
2. Drain the engine oil.
3. Remove or disconnect the following:
   - Negative battery cable
   - Engine under cover
   - Stabilizer bar
   - Oil pan. Loosen the bolts in the sequence shown.

To install:

4. Apply a continuous bead of sealant 0.138–0.177 in. (3.5–4.5mm) to the oil pan mating surface.
5. Install or connect the following:
   - Oil pan. Tighten the bolts in sequence to 60–72 inch lbs. (7–8 Nm).
   - Stabilizer bar. Tighten the bracket bolts to 38–45 ft. lbs. (51–61 Nm) and the link nuts to 12–16 ft. lbs. (16–22 Nm).
   - Engine under cover
   - Negative battery cable

➡ **Wait 30 minutes after installation of the oil pan to allow the sealant to cure before adding oil.**

6. Fill the crankcase to the correct level.
7. Start the engine and check for leaks.

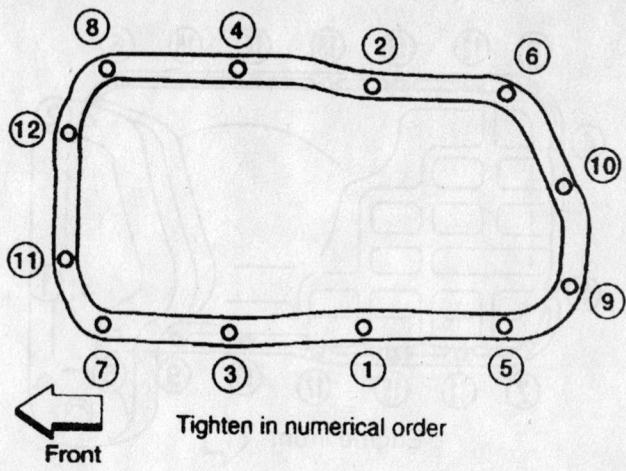

Tighten in numerical order

Front

Oil pan bolt installation sequence—2.4L engine

7924VG40

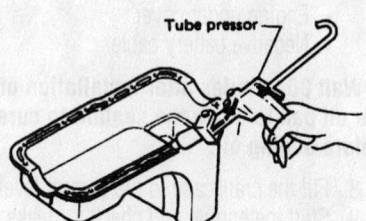

Tube pressor

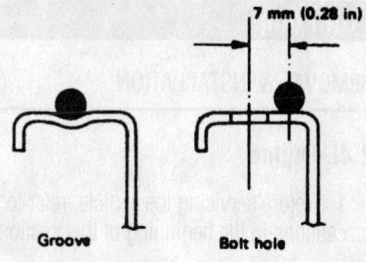

7 mm (0.28 in)

Groove   Bolt hole

7924VG39

Oil pan sealant application—2.4L engine shown

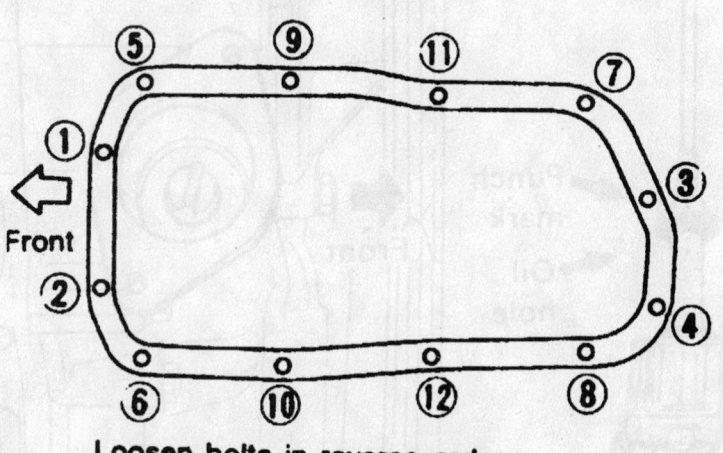

Front

Loosen bolts in reverse order.

7924VG38

Oil pan bolt removal sequence—2.4L engine

3.3L Engine

2WD MODELS

1. Before servicing the vehicle, refer to the precautions in the beginning of this section.
2. Drain the engine oil.
3. Remove or disconnect the following:
   - Negative battery cable
   - Engine under cover
   - Stabilizer bar
   - Front crossmember
   - Starter motor
   - Transmission mount
   - Left and right motor mounts
   - Power steering gear
4. Raise and support the engine for clearance.
5. Remove or disconnect the following:
   - Oil pan bolts in the sequence
   - Oil pan

To install:

6. Apply a continuous bead of sealant 0.138–0.177 in. (3.5–4.5mm) to the oil pan mating surface.
7. Install or connect the following:
   - Oil pan. Tighten the bolts in reverse of the removal sequence to 62 inch lbs. (7 Nm).
   - Power steering gear
   - Left and right motor mounts
   - Transmission mount
   - Starter motor
   - Front crossmember
   - Stabilizer bar
   - Engine under cover
   - Negative battery cable

➡ **Wait 30 minutes after installation of the oil pan to allow the sealant to cure before adding oil.**

8. Fill the crankcase to the correct level.
9. Start the engine and check for leaks.

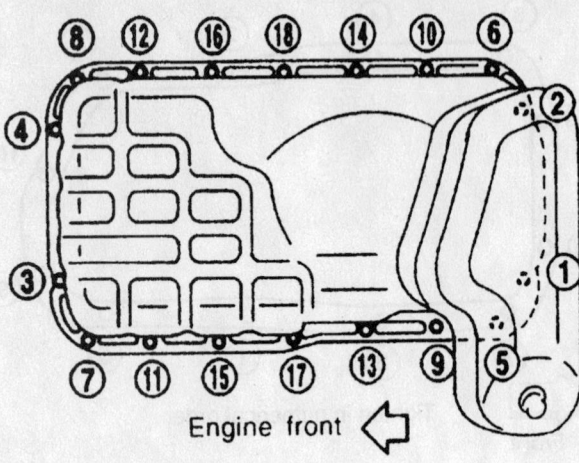

Oil pan bolt removal sequence—3.3L engine

4WD MODELS

1. Before servicing the vehicle, refer to the precautions in the beginning of this section.
2. Drain the engine oil.
3. Remove or disconnect the following:
   - Negative battery cable
   - Engine under cover
   - Stabilizer bar brackets
   - Front driveshaft
   - Axle halfshafts
   - Front suspension crossmember
   - Front differential and mounting bracket
   - Starter motor
   - Transmission mount
   - Left and right motor mounts
   - Power steering gear
   - Relay rod
4. Raise and support the engine for clearance.
5. Remove or disconnect the following:
   - Oil pan bolts in the sequence
   - Oil pan

To install:

6. Apply a continuous bead of sealant 0.138–0.177 in. (3.5–4.5mm) to the oil pan mating surface.
7. Install or connect the following:
   - Oil pan. Tighten the bolts in reverse of the removal sequence to 62 inch lbs. (7 Nm).
   - Relay rod
   - Power steering gear
   - Left and right motor mounts
   - Transmission mount
   - Starter motor
   - Front differential and mounting bracket
   - Front suspension crossmember
   - Axle halfshafts
   - Front driveshaft

- Stabilizer bar brackets
- Engine under cover
- Negative battery cable

➡ **Wait 30 minutes after installation of the oil pan to allow the sealant to cure before adding oil.**

8. Fill the crankcase to the correct level.
9. Start the engine and check for leaks.

Oil Pump

REMOVAL & INSTALLATION

2.4L Engine

1. Before servicing the vehicle, refer to the precautions in the beginning of this section.

2. Set the engine to Top Dead Center (TDC) of the compression stroke for the No. 1 cylinder.
3. Remove or disconnect the following:
   - Distributor cap
   - Distributor
   - Engine under cover
   - Stabilizer bar
   - Oil pump and drive spindle

To install:

4. Fill the pump housing with engine oil, then align the punch mark on the spindle with the hole in the oil pump as shown.
5. Install or connect the following:
   - Oil pump and drive spindle. Tighten the mounting bolts to 96–132 inch lbs. (11–15 Nm).
   - Stabilizer bar
   - Engine under cover
   - Distributor
   - Distributor cap
6. Start the engine and check for leaks.
7. Check the ignition timing and adjust, as necessary.

3.3L Engine

1. Before servicing the vehicle, refer to the precautions in the beginning of this section.
2. Drain the engine oil.
3. Drain the cooling system.
4. Remove or disconnect the following:
   - Negative battery cable
   - Accessory drive belts
   - Radiator hoses
   - Crankshaft pulley
   - Front cover
   - Timing belt

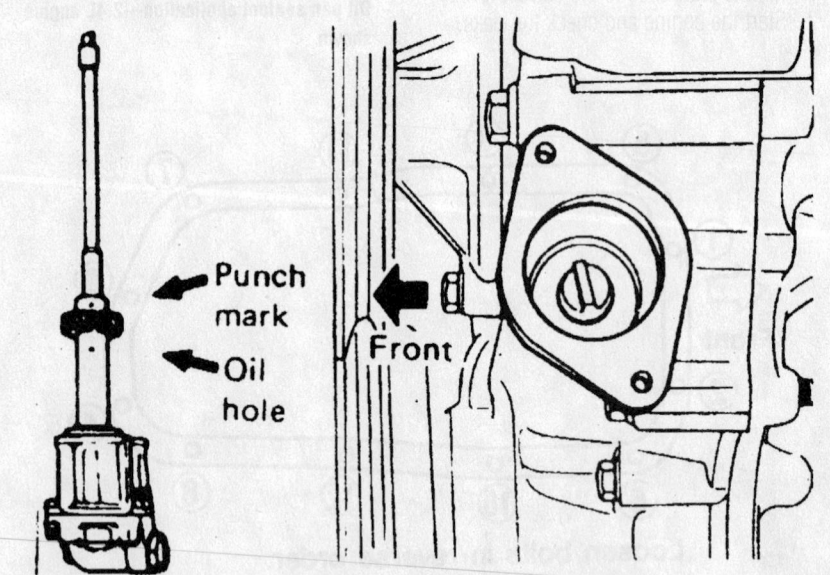

Align the punch mark with the oil hole before oil pump installation—2.4L engine

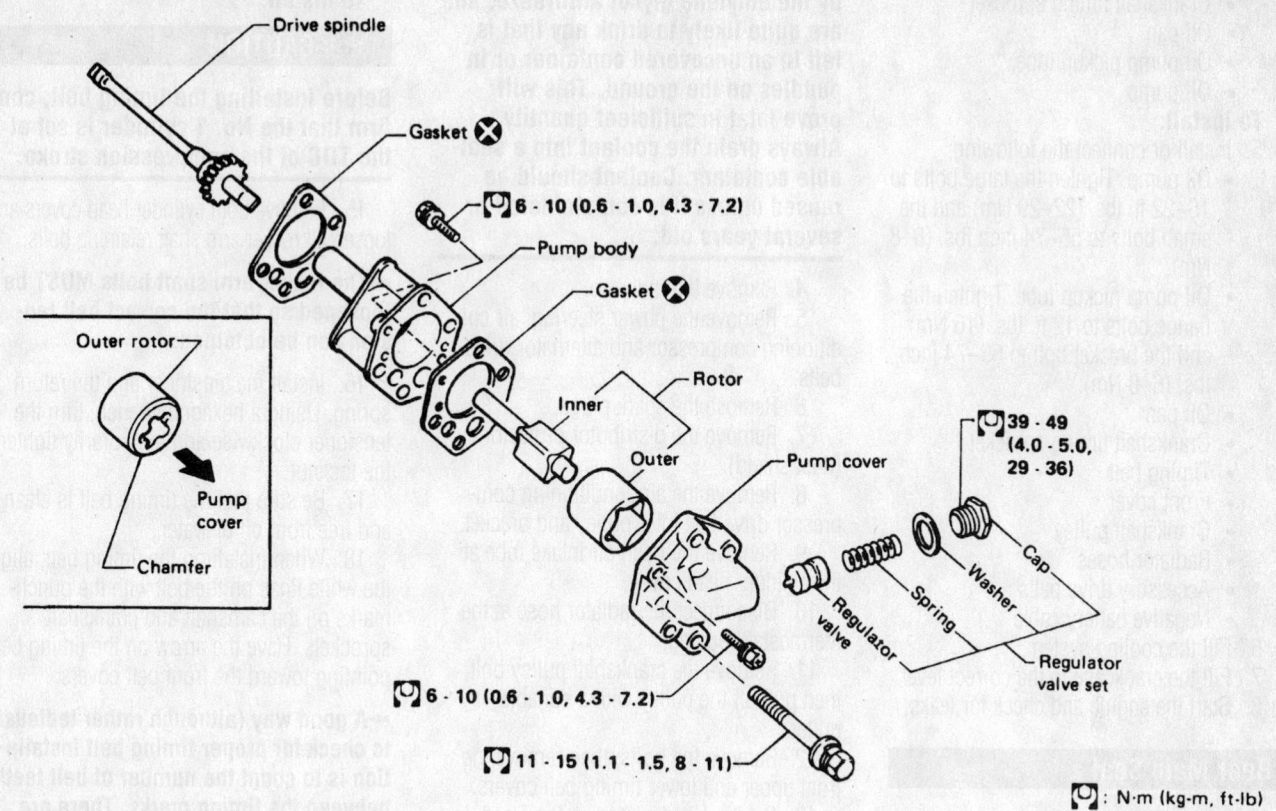

Exploded view of oil pump assembly—2.4L engine

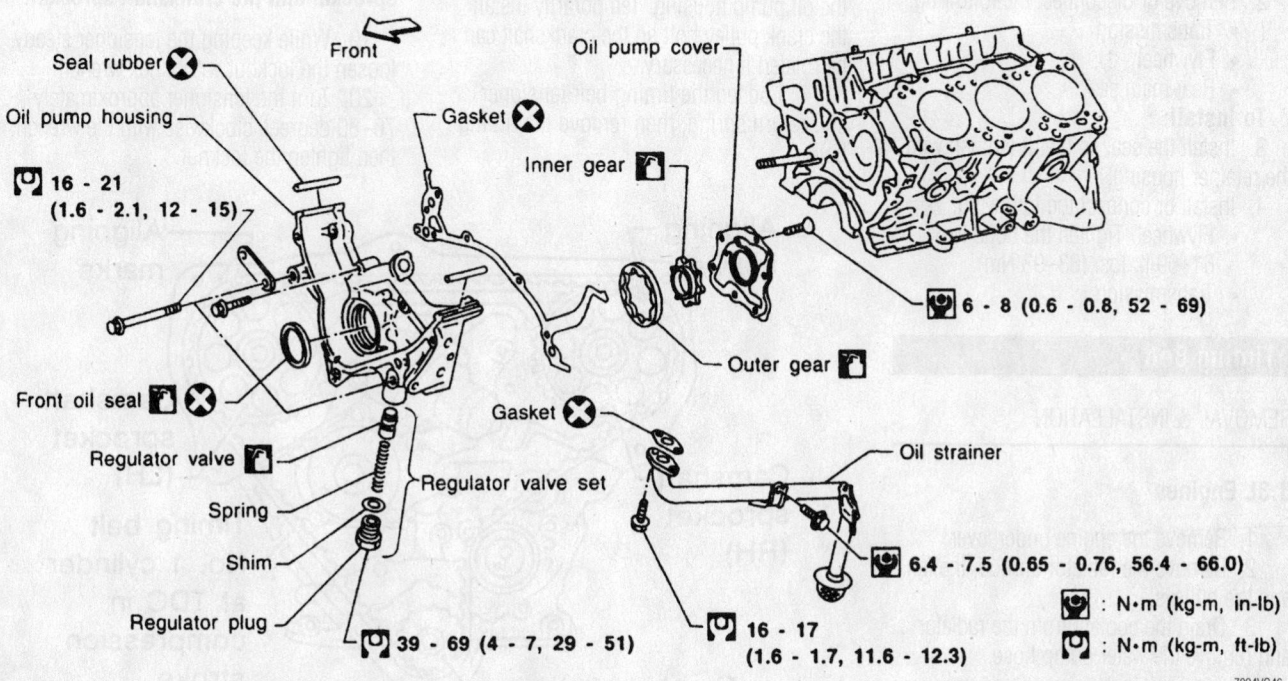

Oil pump assembly exploded view—3.3L engine

- Crankshaft timing sprocket
- Oil pan
- Oil pump pickup tube
- Oil pump

To install:

5. Install or connect the following:
- Oil pump. Tighten the large bolts to 16–22 ft. lbs. (22–29 Nm) and the small bolts to 55–74 inch lbs. (6–8 Nm).
- Oil pump pickup tube. Tighten the flange bolts to 12 ft. lbs. (16 Nm) and the bracket bolt to 55–74 inch lbs. (6–8 Nm).
- Oil pan
- Crankshaft timing sprocket
- Timing belt
- Front cover
- Crankshaft pulley
- Radiator hoses
- Accessory drive belts
- Negative battery cable

6. Fill the cooling system.
7. Fill the crankcase to the correct level.
8. Start the engine and check for leaks.

Rear Main Seal

REMOVAL & INSTALLATION

1. Before servicing the vehicle, refer to the precautions in the beginning of this section.
2. Remove or disconnect the following:
- Transmission
- Flywheel
- Rear main seal

To install:

3. Install the seal so that it is flush with the retainer housing.
4. Install or connect the following:
- Flywheel. Tighten the bolts to 61–69 ft. lbs. (83–93 Nm).
- Transmission

Timing Belt

REMOVAL & INSTALLATION

3.3L Engines

1. Remove the engine undercover.
2. Remove the radiator shroud, the fan and the pulleys.
3. Drain the coolant from the radiator and remove the water pump hose.

✷✷ CAUTION

When draining the coolant, keep in mind that cats and dogs are attracted

by the ethylene glycol antifreeze, and are quite likely to drink any that is left in an uncovered container or in puddles on the ground. This will prove fatal in sufficient quantity. Always drain the coolant into a sealable container. Coolant should be reused unless it is contaminated or several years old.

4. Remove the radiator.
5. Remove the power steering, air conditioning compressor and alternator drive belts.
6. Remove the spark plugs.
7. Remove the distributor protector (dust shield).
8. Remove the air conditioning compressor drive belt idler pulley and bracket.
9. Remove the fresh air intake tube at the cylinder head cover.
10. Disconnect the radiator hose at the thermostat housing.
11. Remove the crankshaft pulley bolt, then pull off the pulley with a suitable puller.
12. Remove the bolts, then remove the front upper and lower timing belt covers.
13. Set the No. 1 piston at Top Dead Center (TDC) of its compression stroke. Align the punchmark on the left camshaft sprocket with the punchmark on the timing belt upper rear cover. Align the punchmark on the crankshaft sprocket with the notch on the oil pump housing. Temporarily install the crank pulley bolt so the crankshaft can be rotated if necessary.
14. Loosen the timing belt tensioner and return spring, then remove the timing belt.

To install:

✷✷ CAUTION

Before installing the timing belt, confirm that the No. 1 cylinder is set at the TDC of the compression stroke.

15. Remove both cylinder head covers and loosen all rocker arm shaft retaining bolts.

➡ The rocker arm shaft bolts MUST be loosened so that the correct belt tension can be obtained.

16. Install the tensioner and the return spring. Using a hexagon wrench, turn the tensioner clockwise and temporarily tighten the locknut.
17. Be sure that the timing belt is clean and free from oil or water.
18. When installing the timing belt, align the white lines on the belt with the punchmarks on the camshaft and crankshaft sprockets. Have the arrow on the timing belt pointing toward the front belt covers.

➡ A good way (although rather tedious!) to check for proper timing belt installation is to count the number of belt teeth between the timing marks. There are 133 teeth on the belt; there should be 40 teeth between the timing marks on the left and right side camshaft sprockets, and 43 teeth between the timing marks on the left side camshaft sprocket and the crankshaft sprocket.

19. While keeping the tensioner steady, loosen the locknut with a hex wrench.
20. Turn the tensioner approximately 70–80 degrees clockwise with the wrench, then tighten the locknut.

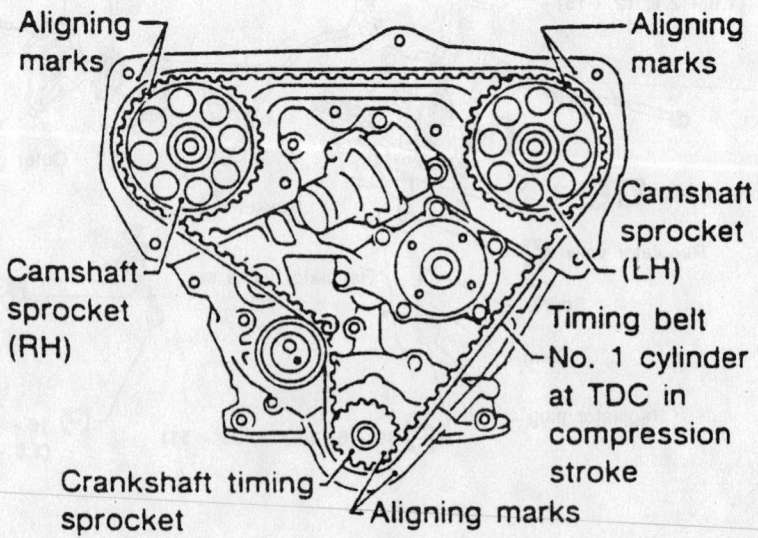

Timing belt alignment mark locations—3.3L engines

79245G35

✳✳ WARNING

If any binding is felt when adjusting the timing belt tension by turning the crankshaft, STOP turning the engine, because the pistons may be hitting the valves.

21. Turn the crankshaft in a clockwise direction several times, then **slowly** set the No. 1 piston to TDC of the compression stroke.

22. Apply 22 lbs. (10 kg) of pressure (push it in!) to the center span of the timing belt between the right side camshaft sprocket and the tensioner pulley, then loosen the tensioner locknut.

23. Using a 0.0138 in. (0.35mm) thick feeler gauge (the actual width of the blade **must** be ½ in. or 13mm!), turn the crankshaft clockwise (**slowly!**). The timing belt should move approximately 2½ teeth. Tighten the tensioner locknut, turn the crankshaft slightly and remove the feeler gauge.

24. Slowly rotate the crankshaft clockwise several more times, then set the No. 1 piston to TDC of the compression stroke.

25. Position the 2 timing covers on the block, then tighten the mounting bolts to 24 ft. lbs. (35 Nm).

26. Press the crankshaft pulley onto the shaft, then tighten the bolt to 90–98 ft. lbs. (123–132 Nm).

27. Connect the radiator hose to the thermostat housing.

28. Reconnect the fresh air intake tube at the cylinder head cover.

29. Install the air conditioning compressor drive belt idler pulley and bracket.

30. Install the distributor protector (dust shield).

31. Install the spark plugs.

32. Install the power steering, air conditioning compressor and alternator drive belts.

33. Install the radiator.

34. Reconnect the water pump hose and fill the engine with coolant. Install the fan shroud and pulleys.

35. Install the engine undercover.

36. Start the engine and check for any leaks.

Timing Chain, Sprockets, Front Cover and Seal

REMOVAL & INSTALLATION

2.4L Engine

1. Before servicing the vehicle, refer to the precautions in the beginning of this section.

2. Drain the cooling system.

3. Drain the engine oil.

4. Set the engine to Top Dead Center (TDC) of the compression stroke for the No. 1 cylinder.

5. Remove or disconnect the following:

- Negative battery cable
- Air cleaner assembly
- Spark plug wires
- Cooling fan and shroud
- Distributor
- Valve cover
- Accessory drive belts
- Power steering pump and brackets
- A/C compressor and bracket
- Idler pulleys
- Water pump pulley
- Crankshaft pulley
- Front crankshaft seal
- Oil pump and drive spindle
- Oil pan
- Upper timing cover
- Lower timing cover
- Upper timing chain tensioner

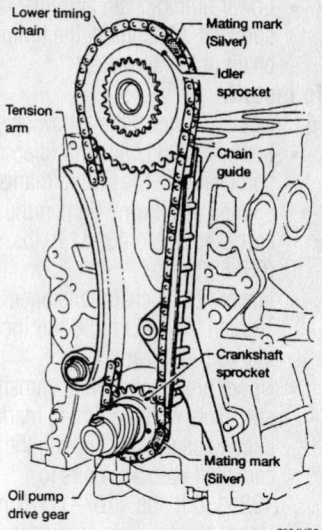

Lower timing chain alignment—2.4L engine

- Upper timing chain and camshaft sprockets. Matchmark the timing chain to the sprockets.
- Lower timing chain tensioner

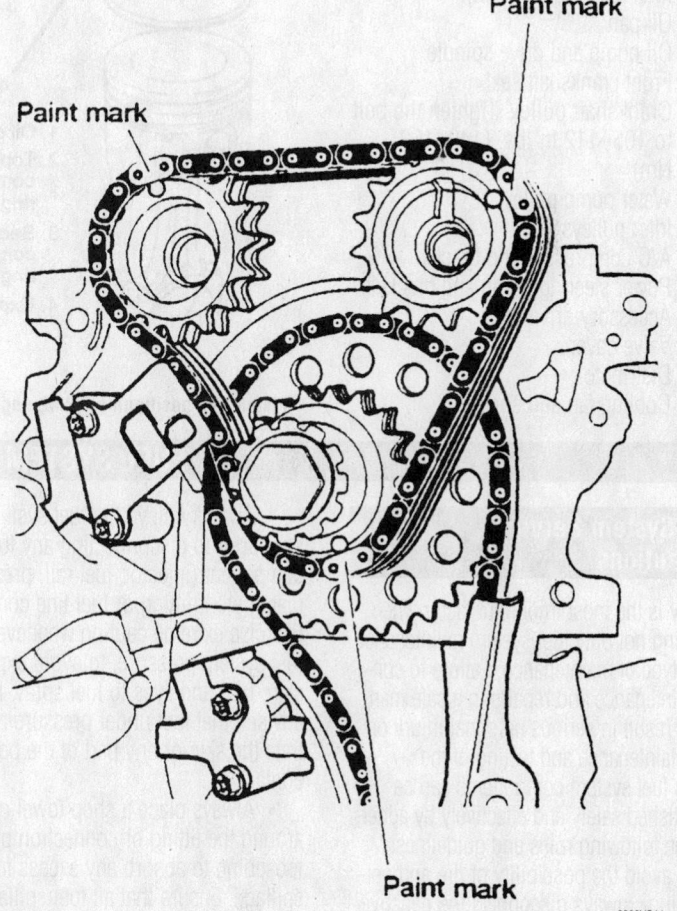

Upper timing chain alignment—2.4L engine

- Lower timing chain and idler sprocket. Matchmark the timing chain to the sprockets.

To install:

6. Install or connect the following:

- Lower timing chain and idler sprocket with the timing marks aligned as shown. Tighten the idler sprocket bolt to 48–61 ft. lbs. (66–83 Nm).
- Lower timing chain tensioner. Tighten the bolts to 56–66 inch lbs. (6.5–7.5 Nm).
- Upper timing chain and camshaft sprockets with the timing marks aligned as shown. Tighten the camshaft sprocket bolts to 123–130 ft. lbs. (167–177 Nm).
- Upper timing chain tensioner. Tighten the bolts to 56–66 inch lbs. (6.5–7.5 Nm).
- Lower timing cover. Tighten the large bolts to 12–14 ft. lbs. (16–19 Nm) and the small bolts to 56–66 inch lbs. (6.5–7.5 Nm).
- Upper timing cover. Tighten the large bolts to 12–14 ft. lbs. (16–19 Nm) and the small bolts to 56–66 inch lbs. (6.5–7.5 Nm).
- Oil pan
- Oil pump and drive spindle
- Front crankshaft seal
- Crankshaft pulley. Tighten the bolt to 105–112 ft. lbs. (142–152 Nm).
- Water pump pulley
- Idler pulleys
- A/C compressor and bracket
- Power steering pump and brackets
- Accessory drive belts
- Valve cover
- Distributor
- Cooling fan and shroud

- Spark plug wires
- Air cleaner assembly
- Negative battery cable

7. Fill the cooling system.
8. Fill the crankcase to the correct level.
9. Start the engine and check for leaks.
10. Check the ignition timing and adjust, as necessary.

Piston and Ring

POSITIONING

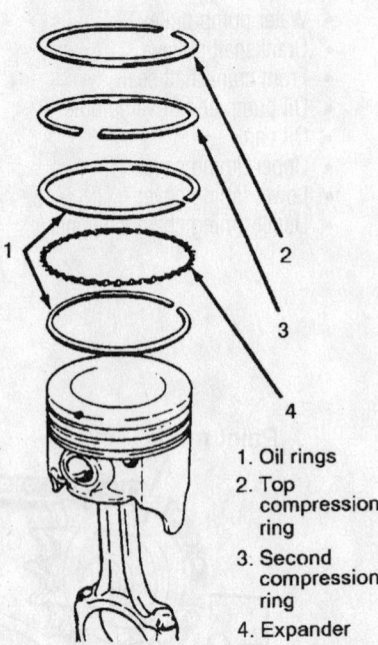

1. Oil rings
2. Top compression ring
3. Second compression ring
4. Expander

7924AG82

Piston ring positioning—2.4L engine

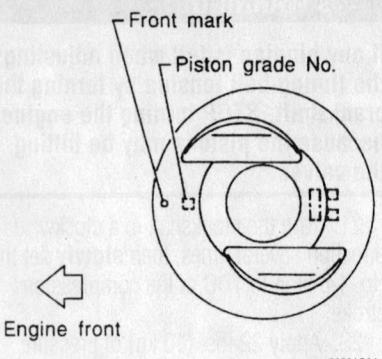

9302AG04

Piston ring positioning—3.3L engine

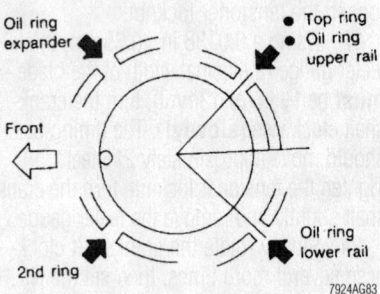

7924AG83

Piston ring end-gap spacing

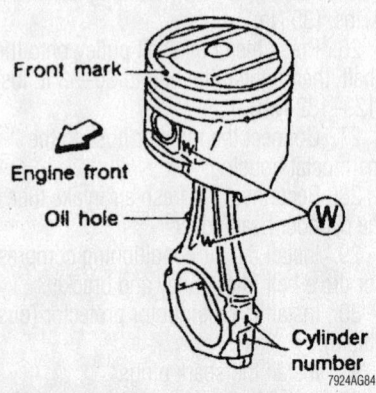

7924AG84

Piston and connecting rod positioning

FUEL SYSTEM

Fuel System Service Precautions

Safety is the most important factor when performing not only fuel system maintenance but any type of maintenance. Failure to conduct maintenance and repairs in a safe manner may result in serious personal injury or death. Maintenance and testing of the vehicle's fuel system components can be accomplished safely and effectively by adhering to the following rules and guidelines.

- To avoid the possibility of fire and personal injury, always disconnect the negative battery cable unless the repair or test procedure requires that battery voltage be applied.

- Always relieve the fuel system pressure prior to disconnecting any fuel system component (injector, fuel rail, pressure regulator, etc.), fitting or fuel line connection. Exercise extreme caution whenever relieving fuel system pressure, to avoid exposing skin, face and eyes to fuel spray. Please be advised that fuel under pressure may penetrate the skin or any part of the body that it contacts.

- Always place a shop towel or cloth around the fitting or connection prior to loosening to absorb any excess fuel due to spillage. Ensure that all fuel spillage (should it occur) is quickly removed from engine surfaces. Ensure that all fuel soaked

cloths or towels are deposited into a suitable waste container.

- Always keep a dry chemical (Class B) fire extinguisher near the work area.

- Do not allow fuel spray or fuel vapors to come into contact with a spark or open flame.

- Always use a back-up wrench when loosening and tightening fuel line connection fittings. This will prevent unnecessary stress and torsion to fuel line piping. Always follow the proper torque specifications.

- Always replace worn fuel fitting O-rings with new. Do not substitute fuel hose or equivalent, where fuel pipe is installed.

Fuel System Pressure

RELIEVING

1. Before servicing the vehicle, refer to the precautions in the beginning of this section.
2. Remove the fuel pump fuse from the panel.
3. Start the engine and allow it to run until it stalls. Crank the engine for a few seconds to relieve additional fuel pressure.
4. Disconnect the negative battery cable.
5. When repairs are complete, replace the fuel pump fuse and connect the negative battery cable.

Fuel Filter

REMOVAL & INSTALLATION

➡ **The fuel filter is located under the vehicle near the fuel tank.**

1. Before servicing the vehicle, refer to the precautions in the beginning of this section.
2. Relieve the fuel system pressure.
3. Remove or disconnect the following:
   - Fuel filter shield, if equipped

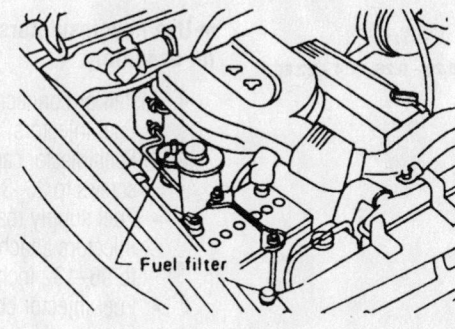

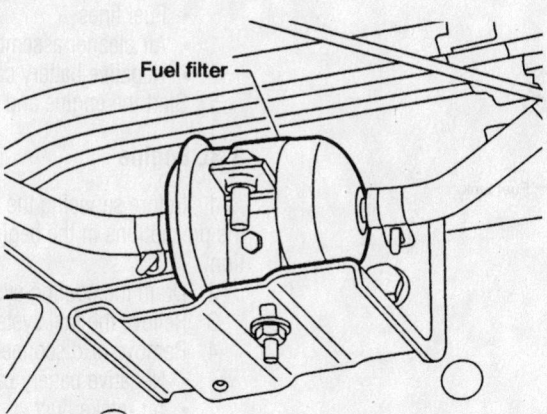

Typical fuel filter locations

   - Fuel lines
   - Fuel filter from the bracket

To install:

4. Install or connect the following:
   - Fuel filter to the bracket
   - Fuel lines
   - Fuel filter shield, if equipped
5. Start the engine and check for leaks.

Fuel Pump

REMOVAL & INSTALLATION

Frontier

1. Before servicing the vehicle, refer to the precautions in the beginning of this section.
2. Relieve the fuel system pressure.
3. Drain the fuel tank.
4. Remove or disconnect the following:
   - Negative battery cable
   - Fuel pump module harness connectors
   - Filler hose shield
   - Fuel pressure and return lines
   - Filler hose
   - Vent hose
   - Evaporative Emissions (EVAP) hose
   - Fuel tank skid plate
   - Fuel tank

   - Fuel level sender
   - Fuel pump

To install:

5. Install or connect the following:
   - Fuel pump
   - Fuel level sender. Tighten the screws to 17–23 inch lbs. (2.0–2.5 Nm).
   - Fuel tank. Tighten the bolts to 27–36 ft. lbs. (37–49 Nm).
   - Fuel tank skid plate. Tighten the bolts to 27–36 ft. lbs. (37–49 Nm).
   - EVAP hose
   - Vent hose
   - Filler hose
   - Fuel pressure and return lines
   - Filler hose shield
   - Fuel pump module harness connectors
   - Negative battery cable
6. Fill the fuel tank.
7. Start the engine and check for leaks.

Xterra

2000 VEHICLES

1. Before servicing the vehicle, refer to the precautions in the beginning of this section.
2. Relieve the fuel system pressure.
3. Remove the rear seat and the access panel.
4. Drain the fuel tank.
5. Remove or disconnect the following:
   - Negative battery cable
   - Fuel pump module harness connectors
   - Filler hose shield
   - Fuel pressure and return lines
   - Filler hose
   - Vent hose
   - Evaporative Emissions (EVAP) hose
   - Fuel tank skid plate
   - Fuel tank
   - Fuel level sender
   - Fuel pump

To install:

6. Install or connect the following:
   - Fuel pump
   - Fuel level sender. Tighten the screws to 17–23 inch lbs. (2.0–2.5 Nm).
   - Fuel tank. Tighten the bolts to 27–36 ft. lbs. (37–49 Nm).
   - Fuel tank skid plate. Tighten the bolts to 27–36 ft. lbs. (37–49 Nm).
   - EVAP hose
   - Vent hose
   - Filler hose
   - Fuel pressure and return lines
   - Filler hose shield
   - Fuel pump module harness connectors
   - Negative battery cable

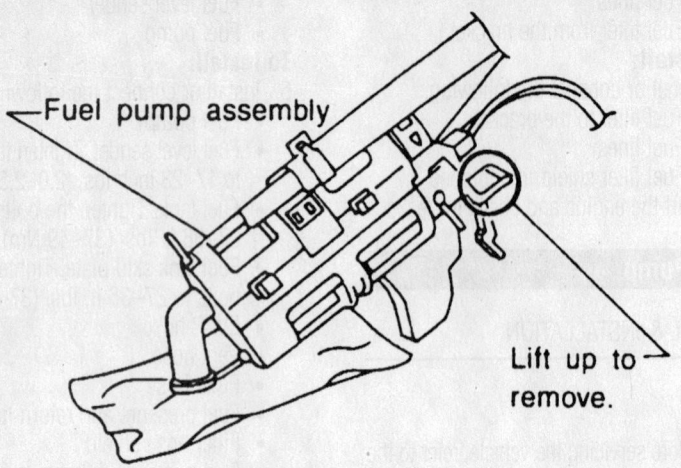

Fuel pump assembly

Lift up to remove.

7924VG58

Remove the fuel pump with bracket while lifting the pawl of the pump bracket upward

7. Install the access panel and the rear seat.

8. Fill the fuel tank.

9. Start the engine and check for leaks.

2001–03 VEHICLES

1. Before servicing the vehicle, refer to the precautions in the beginning of this section.

2. Relieve the fuel system pressure.

3. Remove the rear seat cushion and the access panel.

4. Remove or disconnect the following:
- Negative battery cable
- Fuel pump electrical connectors

5. Matchmark the installed position of the fuel line quick connect fittings, then disconnect the fittings by holding the sides of the connector, push in the tabs and pull out the tube inserted in the retainer.

➡**The tube can be removed when the tabs are completely pushed in. Do NOT use any tools to remove the quick connector.**

2.0 – 2.5 (0.20 – 0.26, 17.4 – 22.6)

Quick connectors

Fuel level sensor

Fuel pump

Front

Fuel tank

O-ring

: N·m (kg-m, in-lb)

9355WG05

Exploded view of the fuel pump—2001–03 Xterra shown

- Six screws
- Fuel level sensor retainer and fuel level sensor
- Fuel pump with the bracket, while lifting the pawl of the fuel bracket upward
- Fuel level sensor

To install:

6. Installation is the reverse of the removal procedure.

7. Start the engine and check for leaks.

Fuel Injectors

REMOVAL & INSTALLATION

2.4L Engine

1. Before servicing the vehicle, refer to the precautions in the beginning of this section.

2. Relieve the fuel system pressure.

3. Remove or disconnect the following:
- Negative battery cable
- Air cleaner assembly
- Fuel lines
- Fuel pressure regulator vacuum line
- Fuel injector connectors
- Fuel supply manifold with the injectors attached
- Fuel injector caps
- Fuel injectors

To install:

➡**Use new insulators and O-ring seals for assembly.**

4. Install or connect the following:
- Fuel injectors
- Fuel injector caps. Tighten the screws to 26–34 inch lbs. (3–4 Nm).
- Fuel supply manifold with the injectors attached. Tighten the bolts to 96–132 inch lbs. (11–15 Nm).
- Fuel injector connectors
- Fuel pressure regulator vacuum line
- Fuel lines
- Air cleaner assembly
- Negative battery cable

5. Start the engine and check for leaks.

3.3L Engine

1. Before servicing the vehicle, refer to the precautions in the beginning of this section.

2. Drain the cooling system.

3. Relieve the fuel system pressure.

4. Remove or disconnect the following:
- Negative battery cable
- Air intake duct
- Accelerator cable
- Cruise control cable

- Idle Air Control (IAC) valve connector
- Throttle Position (TP) sensor and switch connectors
- Ignition coil and power transistor connectors
- Exhaust Gas Recirculation (EGR) Solenoid valve connector
- EGR temperature sensor connector
- Radiator hoses
- Heater hoses
- Positive Crankcase Ventilation (PCV) valve and hose
- Evaporative Emissions (EVAP) canister vacuum and purge hoses
- Brake booster vacuum hose
- Fuel pressure regulator vacuum hose
- EGR tube
- Left bank injector connectors
- Thermal transmitter
- Upper intake manifold ground cable
- Breather pipe

- Supercharger or upper intake manifold
- Fuel lines
- Right bank injector connectors
- Fuel supply manifold with the injectors attached
- Fuel injector caps
- Fuel injectors

To install:

➡ **Use new insulators and O-ring seals for assembly.**

5. Install or connect the following:
- Fuel injectors
- Fuel injector caps. Tighten the screws to 26–34 inch lbs. (3–4 Nm).
- Fuel supply manifold with the injectors attached. Tighten the bolts to 96–132 inch lbs. (11–15 Nm).
- Right bank injector connectors
- Fuel lines
- Upper intake manifold
- Breather pipe

- Upper intake manifold ground cable
- Thermal transmitter
- Left bank injector connectors
- EGR tube
- Fuel pressure regulator vacuum hose
- Brake booster vacuum hose
- EVAP canister vacuum and purge hoses
- PCV valve and hose
- Heater hoses
- Radiator hoses
- EGR temperature sensor connector
- EGR Solenoid valve connector
- Ignition coil and power transistor connectors
- TP sensor and switch connectors
- IAC valve connector
- Cruise control cable
- Accelerator cable
- Air intake duct
- Negative battery cable
6. Fill the cooling system.
7. Start the engine and check for leaks.

DRIVE TRAIN

Manual Transmission

REMOVAL & INSTALLATION

2 Wheel Drive

1. Before servicing the vehicle, refer to the precautions in the beginning of this section.

2. Remove or disconnect the following:
- Negative battery cable
- Shift lever
- Crankshaft Position (CKP) sensor
- Clutch slave cylinder
- Vehicle Speed (VSS) sensor connector
- Back-up lamp switch connector
- Park/Neutral Position (PNP) switch connector

- Rear Heated Oxygen (HO2S) sensor connector
- Starter motor
- Driveshaft
- Exhaust mounting bracket
- Transmission mount and crossmember. Support the transmission.
- Transmission flange bolts
- Transmission

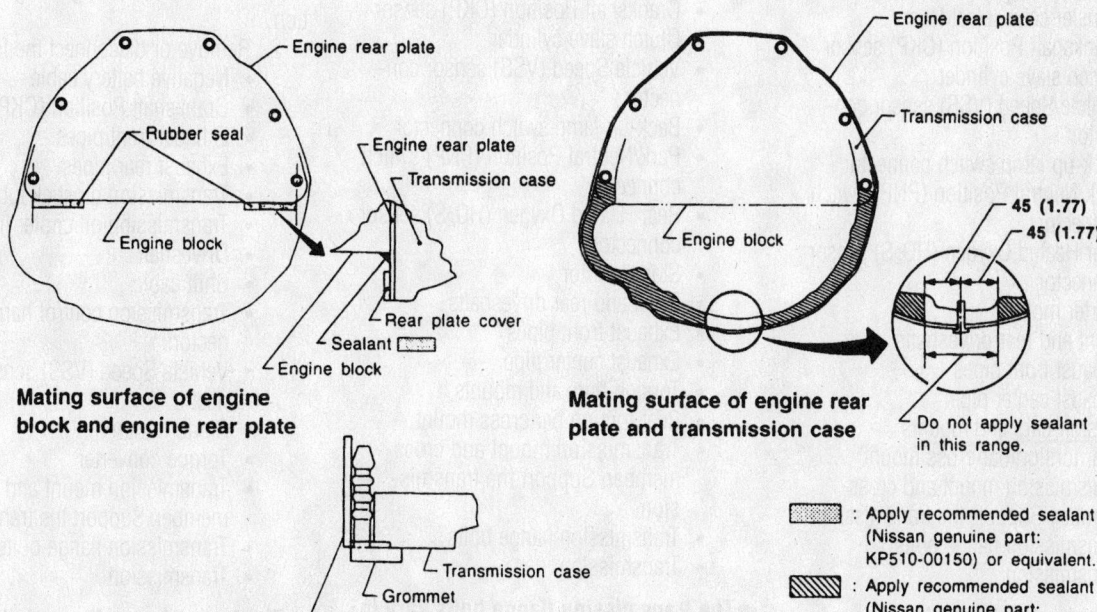

Apply sealant to the indicated areas between the engine block, transmission and engine rear plate—4 Wheel Drive shown

7924VG61

➡ **The transmission flange bolts vary in length. Note their positions for assembly.**

To install:

3. Apply sealant to the transmission flange, engine block and engine rear plate as shown.

4. Install or connect the following:
- Transmission. Tighten the large bolts to 29–36 ft. lbs. (39–49 Nm) and the small bolts to 12–16 ft. lbs. (16–22 Nm).
- Transmission mount and crossmember. Tighten the mount and crossmember fasteners to 30–38 ft. lbs. (41–52 Nm).
- Exhaust mounting bracket
- Driveshaft
- Starter motor
- HO2S sensor connector
- PNP switch connector
- Back-up lamp switch connector
- VSS sensor connector
- Clutch slave cylinder
- CKP sensor
- Shift lever
- Negative battery cable

4 Wheel Drive

FRONTIER

1. Before servicing the vehicle, refer to the precautions in the beginning of this section.

2. Remove or disconnect the following:
- Negative battery cable
- Shift lever
- Transfer case select lever
- Crankshaft Position (CKP) sensor
- Clutch slave cylinder
- Vehicle Speed (VSS) sensor connector
- Back-up lamp switch connector
- Park/Neutral Position (PNP) switch connector
- Rear Heated Oxygen (HO2S) sensor connector
- Starter motor
- Front and rear driveshafts
- Exhaust front pipes
- Exhaust center pipe
- Torsion bars and mounts
- Rear torsion bar cross mount
- Transmission mount and crossmember. Support the transmission.
- Transmission flange bolts
- Transmission

➡ **The transmission flange bolts vary in length. Note their positions for assembly.**

To install:

3. Apply sealant to the transmission flange, engine block, and engine rear plate as shown.

4. Install or connect the following:
- Transmission. Tighten the large bolts to 29–36 ft. lbs. (39–49 Nm) and the small bolts to 22–29 ft. lbs. (29–39 Nm).
- Transmission mount and crossmember. Tighten the mount and crossmember fasteners to 30–38 ft. lbs. (41–52 Nm).
- Rear torsion bar cross mount
- Torsion bars and mounts
- Exhaust center pipe
- Exhaust front pipes
- Front and rear driveshafts
- Starter motor
- HO2S sensor connector
- PNP switch connector
- Back-up lamp switch connector
- VSS sensor connector
- Clutch slave cylinder
- CKP sensor
- Transfer case select lever
- Shift lever
- Negative battery cable

XTERRA

1. Before servicing the vehicle, refer to the precautions in the beginning of this section.

2. Remove or disconnect the following:
- Negative battery cable
- Shift lever
- Transfer case select lever
- Crankshaft Position (CKP) sensor
- Clutch slave cylinder
- Vehicle Speed (VSS) sensor connector
- Back-up lamp switch connector
- Park/Neutral Position (PNP) switch connector
- Rear Heated Oxygen (HO2S) sensor connector
- Starter motor
- Front and rear driveshafts
- Exhaust front pipes
- Exhaust center pipe
- Torsion bars and mounts
- Rear torsion bar cross mount
- Transmission mount and crossmember. Support the transmission.
- Transmission flange bolts
- Transmission

➡ **The transmission flange bolts vary in length. Note their positions for assembly.**

To install:

3. Apply sealant to the transmission flange, engine block, and engine rear plate as shown.

4. Install or connect the following:
- Transmission. Tighten the large bolts to 29–36 ft. lbs. (39–49 Nm) and the small bolts to 22–29 ft. lbs. (29–39 Nm).
- Transmission mount and crossmember. Tighten the mount and crossmember fasteners to 30–38 ft. lbs. (41–52 Nm).
- Rear torsion bar cross mount
- Torsion bars and mounts
- Exhaust center pipe
- Exhaust front pipes
- Front and rear driveshafts
- Starter motor
- HO2S sensor connector
- PNP switch connector
- Back-up lamp switch connector
- VSS sensor connector
- Clutch slave cylinder
- CKP sensor
- Transfer case select lever
- Shift lever
- Negative battery cable

Automatic Transmission

REMOVAL & INSTALLATION

2 Wheel Drive

1. Before servicing the vehicle, refer to the precautions in the beginning of this section.

2. Remove or disconnect the following:
- Negative battery cable
- Crankshaft Position (CKP) sensor
- Exhaust front pipes
- Exhaust rear pipes
- Transmission dipstick tube
- Transmission oil cooler lines
- Driveshaft
- Shift cable
- Transmission control harness connectors
- Vehicle Speed (VSS) sensor connector
- Starter motor
- Torque converter
- Transmission mount and crossmember. Support the transmission.
- Transmission flange bolts
- Transmission

➡ **The transmission flange bolts vary in length. Note their positions for assembly.**

To install:

3. Install or connect the following:
- Transmission. Tighten the large bolts to 29–36 ft. lbs. (39–49 Nm) and the small bolts to 22–29 ft. lbs. (29–39 Nm).
- Transmission mount and crossmember. Tighten the mount and crossmember fasteners to 30–38 ft. lbs. (41–52 Nm).
- Torque converter. Tighten the bolts to 33–43 ft. lbs. (44–59 Nm).
- Starter motor
- VSS sensor connector
- Transmission control harness connectors
- Shift cable
- Driveshaft
- Transmission oil cooler lines
- Transmission dipstick tube
- Exhaust rear pipes
- Exhaust front pipes
- CKP sensor
- Negative battery cable

4 Wheel Drive

1. Before servicing the vehicle, refer to the precautions in the beginning of this section.
2. Remove or disconnect the following:
- Negative battery cable
- Crankshaft Position (CKP) sensor

- Exhaust front pipes
- Exhaust rear pipes
- Transmission dipstick tube
- Transmission oil cooler lines
- Front and rear driveshafts
- Transfer case linkage
- Shift cable
- Transmission control harness connectors
- Vehicle Speed (VSS) sensor connector
- Starter motor
- Torque converter
- Transmission mount and crossmember. Support the transmission.
- Transmission flange bolts
- Transmission

➡ **The transmission flange bolts vary in length. Note their positions for assembly.**

To install:

3. Install or connect the following:
- Transmission. Tighten the large bolts to 29–36 ft. lbs. (39–49 Nm) and the small bolts to 22–29 ft. lbs. (29–39 Nm).
- Transmission mount and crossmember. Tighten the mount and crossmember fasteners to 30–38 ft. lbs. (41–52 Nm).
- Torque converter. Tighten the bolts to 33–43 ft. lbs. (44–59 Nm).
- Starter motor

- VSS sensor connector
- Transmission control harness connectors
- Shift cable
- Transfer case linkage
- Front and rear driveshafts
- Transmission oil cooler lines
- Transmission dipstick tube
- Exhaust rear pipes
- Exhaust front pipes
- CKP sensor
- Negative battery cable

Clutch

REMOVAL & INSTALLATION

1. Before servicing the vehicle, refer to the precautions in the beginning of this section.
2. Remove or disconnect the following:
- Negative battery cable
- Transmission
- Pressure plate. Loosen the bolts evenly in ½ turn steps.
- Clutch disc

To install:

3. Install or connect the following:
- Clutch disc and pressure plate. Tighten the pressure plate bolts evenly in ½ turns to 16–22 ft. lbs. (22–29 Nm).
- Transmission
- Negative battery cable

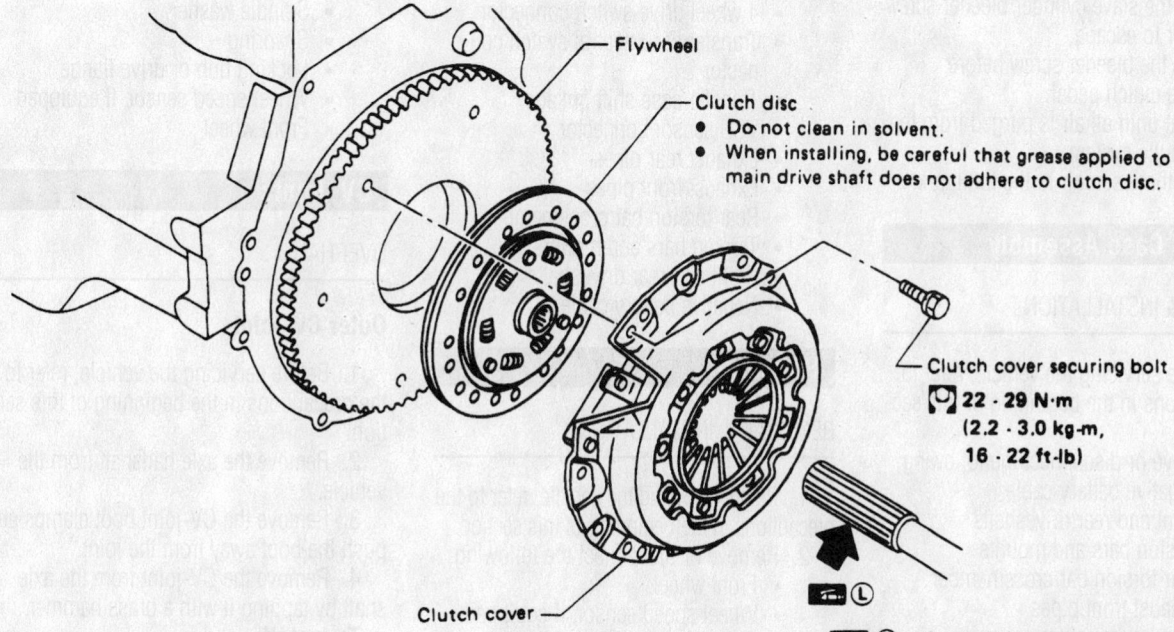

Exploded view of the pressure plate and clutch disc and related components—all models

- Flywheel
- Clutch disc
  - Do not clean in solvent.
  - When installing, be careful that grease applied to main drive shaft does not adhere to clutch disc.
- Clutch cover securing bolt
  22 - 29 N·m
  (2.2 - 3.0 kg-m, 16 - 22 ft-lb)
- Clutch cover
- Ⓛ : Apply lithium-based grease including molybdenum disulphide.

7924VG63

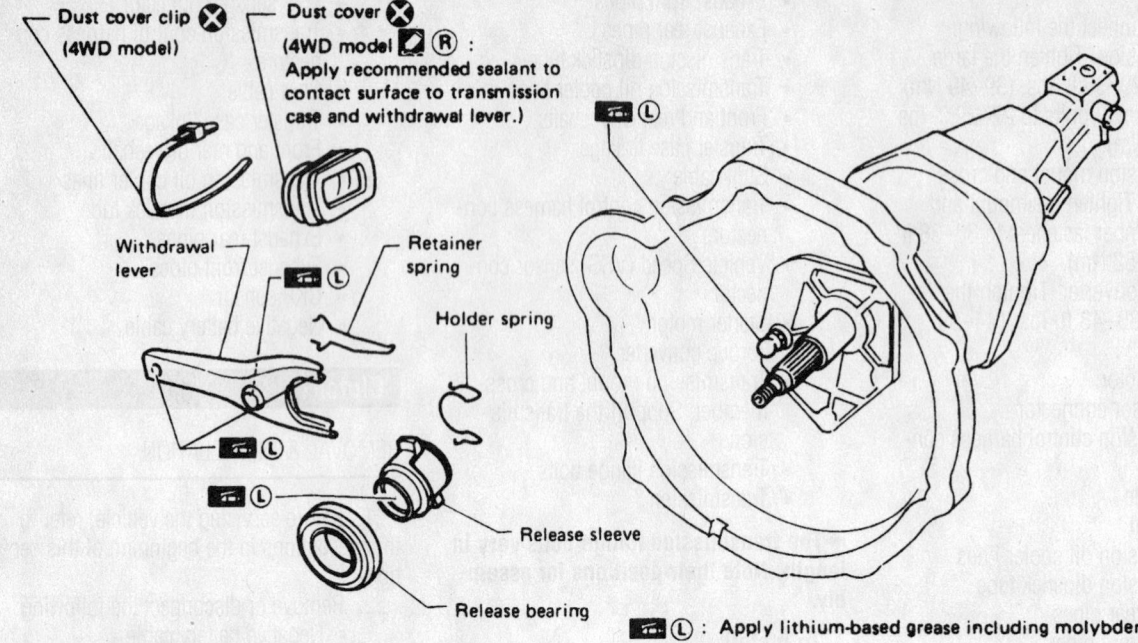

Dust cover clip ⊗
(4WD model)

Dust cover ⊗
(4WD model 🖊️ ℝ) :
Apply recommended sealant to
contact surface to transmission
case and withdrawal lever.)

Withdrawal
lever

Retainer
spring

Holder spring

Release sleeve

Release bearing

🔧 ℓ : Apply lithium-based grease including molybdenum disulphide

7924VG64

Clutch release mechanism exploded view—all models

Hydraulic Clutch System

BLEEDING

1. Before servicing the vehicle, refer to the precautions in the beginning of this section.

2. Have an assistant pump the clutch pedal slowly several times and hold it depressed.

3. Open the slave cylinder bleeder screw and allow air to escape.

4. Close the bleeder screw before releasing the clutch pedal.

5. Repeat until all air is purged from the clutch hydraulic system.

6. Refill the reservoir to the full mark.

Transfer Case Assembly

REMOVAL & INSTALLATION

1. Before servicing the vehicle, refer to the precautions in the beginning of this section.

2. Remove or disconnect the following:
- Negative battery cable
- Front and rear driveshafts
- Torsion bars and mounts
- Rear torsion bar crossmember
- Exhaust front pipes
- Exhaust rear pipes
- Vehicle Speed (VSS) sensor connector
- Transfer case shift linkage

- Transfer case neutral switch connector
- 4 wheel drive switch connector
- Vent hose
- Transfer case flange bolts
- Transfer case

To install:

3. Install or connect the following:
- Transfer case. Tighten the flange bolts to 23–30 ft. lbs. (31–41 Nm).
- Vent hose
- 4 wheel drive switch connector
- Transfer case neutral switch connector
- Transfer case shift linkage
- VSS sensor connector
- Exhaust rear pipes
- Exhaust front pipes
- Rear torsion bar crossmember
- Torsion bars and mounts
- Front and rear driveshafts
- Negative battery cable

Halfshaft

REMOVAL & INSTALLATION

1. Before servicing the vehicle, refer to the precautions in the beginning of this section.

2. Remove or disconnect the following:
- Front wheel
- Wheel speed sensor, if equipped
- Locking hub or drive flange
- Snapring
- Spindle washer
- Thrust washer

- Inner CV-joint bolts
- Axle halfshaft. Separate the stub shaft from the spindle by tapping with a plastic hammer.

To install:

3. Install or connect the following:
- Axle halfshaft. Guide the stub shaft into the spindle and tighten the inner CV-joint bolts to 25–33 ft. lbs. (34–44 Nm).
- Thrust washer
- Spindle washer
- Snapring
- Locking hub or drive flange
- Wheel speed sensor, if equipped
- Front wheel

CV-Joints

OVERHAUL

Outer CV-Joint

1. Before servicing the vehicle, refer to the precautions in the beginning of this section.

2. Remove the axle halfshaft from the vehicle.

3. Remove the CV-joint boot clamps and push the boot away from the joint.

4. Remove the CV-joint from the axle shaft by tapping it with a brass hammer.

To install:

➡Use new circlips and boot clamps for assembly.

5. Install the CV-joint to the axle shaft by tapping it with a brass hammer.

6. Pack the joint with grease.

7. Install the boot clamps.

8. Install the axle halfshaft to the vehicle.

Inner Tri-Pot Joint

1. Before servicing the vehicle, refer to the precautions in the beginning of this section.

2. Remove the axle halfshaft from the vehicle.

3. Remove the plug seal by tapping around the joint housing flange with a brass hammer.

4. Remove or disconnect the following:
   - CV-joint boot clamps
   - Snapring
   - Spider assembly
   - CV-joint housing
   - CV-joint boot

To install:

➡ **Use new snaprings and plug seals for assembly.**

5. Install or connect the following:
   - CV-joint boot
   - CV-joint housing
   - Spider assembly
   - Snapring. Pack the joint with grease.
   - CV-joint boot clamps
   - Plug seal

6. Install the axle halfshaft to the vehicle.

Spindle Bearings

REMOVAL, PACKING AND INSTALLATION

1. Before servicing the vehicle, refer to the precautions in the beginning of this section.

2. Remove or disconnect the following:
   - Front wheel
   - Locking hub or drive flange
   - Brake caliper and support
   - Wheel speed sensor, if equipped
   - Axle halfshaft
   - Outer tie rod ends
   - Upper ball joint or steering knuckle bracket bolts
   - Lower ball joint
   - Steering knuckle
   - Inner seal
   - Thrust washer
   - Spindle bearing

To install:

3. Install or connect the following:
   - Spindle bearing. Coat the bearing with multi-purpose grease.

- Thrust washer
- Inner seal
- Steering knuckle
- Lower ball joint
- Upper ball joint or steering knuckle bracket bolts
- Outer tie rod ends
- Axle halfshaft
- Wheel speed sensor, if equipped
- Brake caliper and support
- Locking hub or drive flange
- Front wheel

Axle Shaft, Bearing and Seal

REMOVAL & INSTALLATION

1. Before servicing the vehicle, refer to the precautions in the beginning of this section.

2. Remove or disconnect the following:
   - Rear wheel
   - Wheel speed sensor, if equipped
   - Brake drum
   - Brake shoes
   - Parking brake cable
   - Brake fluid line
   - Bearing cage and backing plate bolts
   - Axle shaft assembly
   - Axle seal
   - Wheel speed sensor rotor, if equipped
   - Lockwasher
   - Bearing locknut
   - Flat washer
   - Wheel bearing
   - Wheel bearing cage grease seal

To install:

➡ **Use new lockwashers, seals and bearings for assembly.**

3. Install or connect the following:
   - Wheel bearing cage grease seal
   - Wheel bearing
   - Flat washer
   - Bearing locknut
   - Lockwasher
   - Wheel speed sensor rotor, if equipped
   - Axle seal
   - Axle shaft assembly
   - Bearing cage and backing plate bolts
   - Brake fluid line
   - Parking brake cable
   - Brake shoes
   - Brake drum
   - Wheel speed sensor, if equipped
   - Rear wheel

4. Bleed the rear brakes and check the rear axle lubricant level.

Pinion Seal

Front

1. Before servicing the vehicle, refer to the precautions in the beginning of this section.

2. Remove or disconnect the following:
   - Driveshaft
   - Front wheels
   - Front brake calipers

➡ **The front brake calipers must be removed so that there is no additional drag when measuring pinion bearing preload.**

3. Use an inch lb. torque wrench and measure the amount of torque required to maintain pinion rotation through several revolutions.

4. Remove or disconnect the following:
   - Pinion flange
   - Oil seal

To install:

5. Install or connect the following:
   - Pinion seal
   - Pinion flange

6. Rotate the pinion flange occasionally while tightening the flange nut to make sure the pinion bearings seat correctly.

7. Take frequent bearing preload torque readings. Tighten the flange nut to achieve the preload torque readings originally recorded. Do not exceed 137–180 ft. lbs. (186–245 Nm) torque when tightening the pinion flange nut.

✱✱ CAUTION

If the bearing preload can not be achieved at the specified torque, remove the pinion bearing and install a new adjustment spacer.

8. Install or connect the following:
   - Front brake calipers
   - Front wheels
   - Driveshaft. Tighten the fasteners to 29–33 ft. lbs. (39–44 Nm).

9. Fill the differential with gear lubricant and check for leaks.

Rear

2 WHEEL DRIVE

1. Before servicing the vehicle, refer to the precautions in the beginning of this section.

2. Remove or disconnect the following:
   - Driveshaft
   - Rear wheels
   - Brake drums

➡ **The rear brake drums must be removed so that there is no additional drag when measuring pinion bearing preload.**

3. Use an inch lb. torque wrench and measure the amount of torque required to maintain pinion rotation through several revolutions.

4. Remove or disconnect the following:
- Pinion flange
- Wheel speed sensor and rotor, if equipped
- Oil seal
- Pinion bearing
- Collapsible spacer

To install:

➡ **Use a new collapsible spacer and wheel speed sensor rotor for assembly.**

5. Install or connect the following:
- Collapsible spacer
- Pinion bearing
- Pinion seal
- Pinion flange

6. Rotate the pinion flange occasionally while tightening the flange nut to make sure the pinion bearings seat correctly.

7. Take frequent bearing preload torque readings. Tighten the flange nut to achieve the preload torque readings originally recorded. Do not exceed 137–180 ft. lbs. (186–245 Nm) torque when tightening the pinion flange nut.

※ CAUTION

Never loosen the pinion nut to reduce bearing preload. If it is necessary to reduce bearing preload, install a new collapsible spacer.

8. Install or connect the following:
- Brake drums
- Rear wheels
- Driveshaft. Tighten the fasteners to 58–65 ft. lbs. (78–88 Nm).

9. Fill the differential with gear lubricant and check for leaks.

4 WHEEL DRIVE

1. Before servicing the vehicle, refer to the precautions in the beginning of this section.

2. Remove or disconnect the following:
- Driveshaft
- Rear wheels
- Brake drums

➡ **The rear brake drums must be removed so that there is no additional drag when measuring pinion bearing preload.**

3. Use an inch lb. torque wrench and measure the amount of torque required to maintain pinion rotation through several revolutions.

4. Remove or disconnect the following:
- Pinion flange
- Oil seal

To install:

5. Install or connect the following:
- Pinion seal
- Pinion flange

6. Rotate the pinion flange occasionally while tightening the flange nut to make sure the pinion bearings seat correctly.

7. Take frequent bearing preload torque readings. Tighten the flange nut to achieve the preload torque readings originally recorded. Do not exceed 137–180 ft. lbs. (186–245 Nm) torque when tightening the pinion flange nut.

※ CAUTION

If the bearing preload can not be achieved at the specified torque, remove the pinion bearing and install a new adjustment spacer.

8. Install or connect the following:
- Brake drums
- Rear wheels
- Driveshaft. Tighten the fasteners to 58–65 ft. lbs. (78–88 Nm).

9. Fill the differential with gear lubricant and check for leaks.

STEERING AND SUSPENSION

Air Bag

※ CAUTION

Some vehicles are equipped with an air bag system. The system must be disarmed before performing service on, or around, system components, the steering column, instrument panel components, wiring and sensors. Failure to follow the safety precautions and the disarming procedure could result in accidental air bag deployment, possible injury and unnecessary system repairs.

PRECAUTIONS

Several precautions must be observed when handling the inflator module to avoid accidental deployment and possible personal injury.

- Never carry the inflator module by the wires or connector on the underside of the module.
- When carrying a live inflator module, hold securely with both hands, and ensure that the bag and trim cover are pointed away.
- Place the inflator module on a bench or other surface with the bag and trim cover facing up.
- With the inflator module on the bench, never place anything on or close to the module which may be thrown in the event of an accidental deployment.

DISARMING

To disarm the **SRS** system turn the ignition switch to the **OFF** position. Then, disconnect both battery cables starting with the negative cable first and wait at least 3 minutes after the cables are disconnected.

To rearm the **SRS** system, turn the ignition switch to the **OFF** position. Connect both battery cables starting with the positive cable first.

Recirculating Ball Power Steering Gear

REMOVAL & INSTALLATION

1. Before servicing the vehicle, refer to the precautions in the beginning of this section.

2. Remove or disconnect the following:
- Pitman arm
- Steering column intermediate shaft
- Power steering hoses
- Steering gear

To install:

3. Install or connect the following:
- Steering gear. Tighten the bolts to 62–71 ft. lbs. (84–96 Nm).
- Power steering hoses. Tighten the banjo fittings to 29–38 ft. lbs. (39–51 Nm).
- Steering column intermediate shaft. Tighten the pinch bolt to 17–22 ft. lbs. (24–29 Nm).
- Pitman arm. Tighten the nut to

102-130 ft. lbs. (138-176 Nm) (4-cyl.) or 174–195 ft. lbs. (235–265 Nm) (6-cyl.).

4. Check the wheel alignment and adjust, as necessary.

Shock Absorber

REMOVAL & INSTALLATION

Front

FRONTIER

1. Before servicing the vehicle, refer to the precautions in the beginning of this section.
2. Support the lower control arm.
3. Remove or disconnect the following:
   - Front wheel
   - Lower shock absorber mounting bolt
   - Upper shock absorber mounting nut
   - Shock absorber

To install:

4. Install or connect the following:
   - Shock absorber
   - Upper shock absorber mounting nut. Tighten the nut to 12–16 ft. lbs. (16–22 Nm).
   - Lower shock absorber mounting bolt. Tighten the bolt to 87–106 ft. lbs. (118–147 Nm).
   - Front wheel

XTERRA

1. Before servicing the vehicle, refer to the precautions in the beginning of this section.
2. Support the lower control arm.
3. Remove or disconnect the following:
   - Front wheel
   - Lower shock absorber mounting bolt
   - Upper shock absorber mounting nut
   - Shock absorber

To install:

4. Install or connect the following:
   - Shock absorber
   - Upper shock absorber mounting nut. Tighten the nut to 12–16 ft. lbs. (16–22 Nm).
   - Lower shock absorber mounting bolt. Tighten the bolt to 87–108 ft. lbs. (118–147 Nm).
   - Front wheel

Rear

FRONTIER

1. Before servicing the vehicle, refer to the precautions in the beginning of this section.
2. Support the rear axle.
3. Remove or disconnect the following:
   - Lower shock absorber bolt
   - Upper shock absorber bolt
   - Shock absorber

To install:

➡ **Use new fasteners for assembly.**

4. Install the shock absorber and tighten the bolts to 49–65 ft. lbs. (67–88 Nm).
5. Before servicing the vehicle, refer to the precautions in the beginning of this section.
6. Remove or disconnect the following:
   - Upper and lower shock absorber nuts
   - Shock absorber

To install:

➡ **Use new nuts for assembly.**

7. Install the shock absorber and tighten the nuts to 30–37 ft. lbs. (40–50 Nm).

XTERRA

1. Before servicing the vehicle, refer to the precautions in the beginning of this section.
2. Remove or disconnect the following:
   - Upper and lower shock absorber nuts
   - Shock absorber

To install:

➡ **Use new nuts for assembly.**

3. Install the shock absorber and tighten the nuts to 30–37 ft. lbs. (40–50 Nm).

Leaf Springs

REMOVAL & INSTALLATION

1. Before servicing the vehicle, refer to the precautions in the beginning of this section.
2. Support the vehicle at the frame.
3. Support the axle with a floor jack.
4. Remove or disconnect the following:
   - Rear wheels
   - Shock absorbers
   - Axle U-bolts and spring pad
   - Spring shackle
   - Front mount bolt
   - Leaf spring

To install:

➡ **Use new fasteners for assembly.**

5. Install or connect the following:
   - Leaf spring. Tighten the front mount bolt to 86–108 ft. lbs. (117–147 Nm).
   - Spring shackle. Tighten the nuts to 58–72 ft. lbs. (78–98 Nm).
   - Axle U-bolts and spring pad. Tighten the nuts to 72–80 ft. lbs. (98–108 Nm).
   - Shock absorbers
   - Rear wheels

Torsion Bar

1. Before servicing the vehicle, refer to the precautions in the beginning of this section.

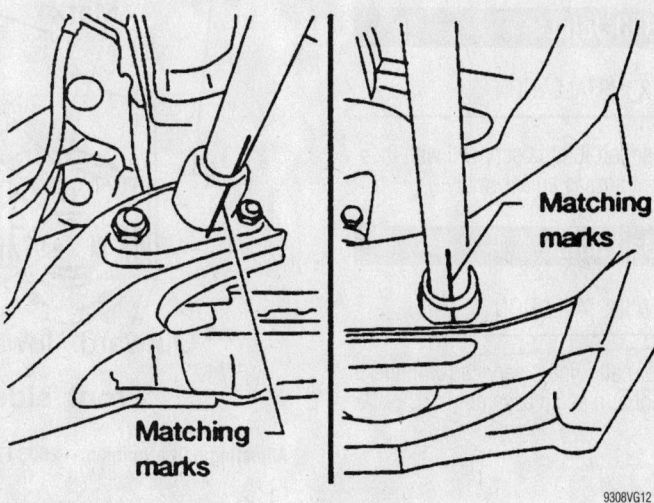

Torsion bar matchmarks

9308VG12

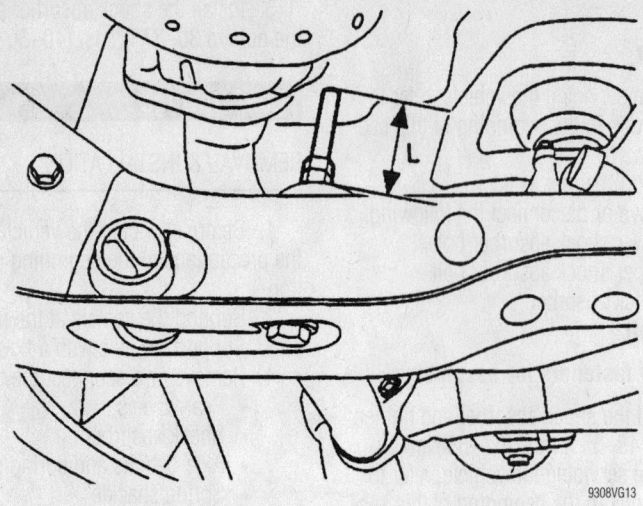

Adjustment bolt measurement (L)

2. Matchmark the torsion bar to the control arm mount and the anchor arm.

3. Measure the adjustment bolt protrusion as shown and note the length (L) for assembly.

4. Loosen the adjustment bolt so that all tension is released.

5. Remove the torsion bar mount from the control arm and remove the torsion bar.

To install:

6. Align the matchmarks and install the torsion bar. Tighten the large mount nut to 66–87 ft. lbs. (89–118 Nm) and the small nut to 33–44 ft. lbs. (45–60 Nm).

7. Tighten the adjustment bolt to achieve the measurement (L) noted earlier. Tighten the locknut to 22–30 ft. lbs. (30–40 Nm).

8. If a new torsion bar is being installed, set length (L) to 2.68 inches.

Upper Ball Joint

REMOVAL & INSTALLATION

The upper ball joint is serviced with the upper control arm as an assembly.

Lower Ball Joint

REMOVAL & INSTALLATION

The lower ball joint is serviced with the lower control arm as an assembly.

Upper Control Arm

REMOVAL & INSTALLATION

2000–02 Frontier and Xterra

1. Before servicing the vehicle, refer to the precautions in the beginning of this section.

2. Support the lower control arm.

3. Remove or disconnect the following:
   - Front wheel
   - Shock absorber
   - Upper ball joint
   - Control arm mounting bolts
   - Upper control arm

To install:

4. Install or connect the following:
   - Upper control arm. Tighten the

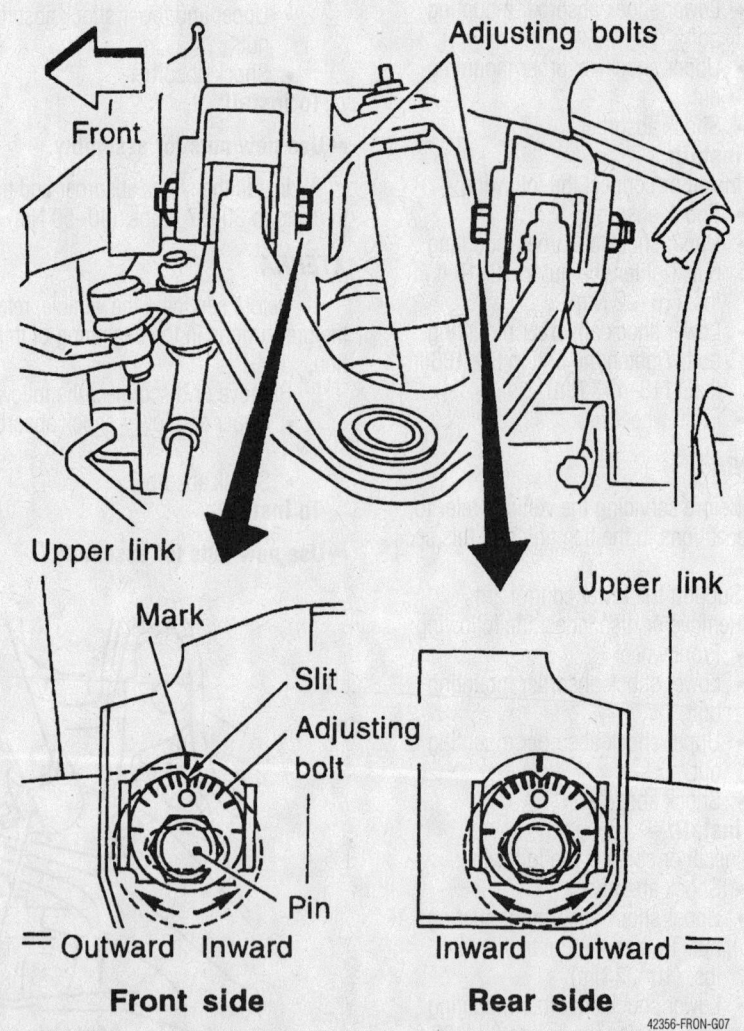

Adjusting bolt alignment—2003 Frontier and Xterra

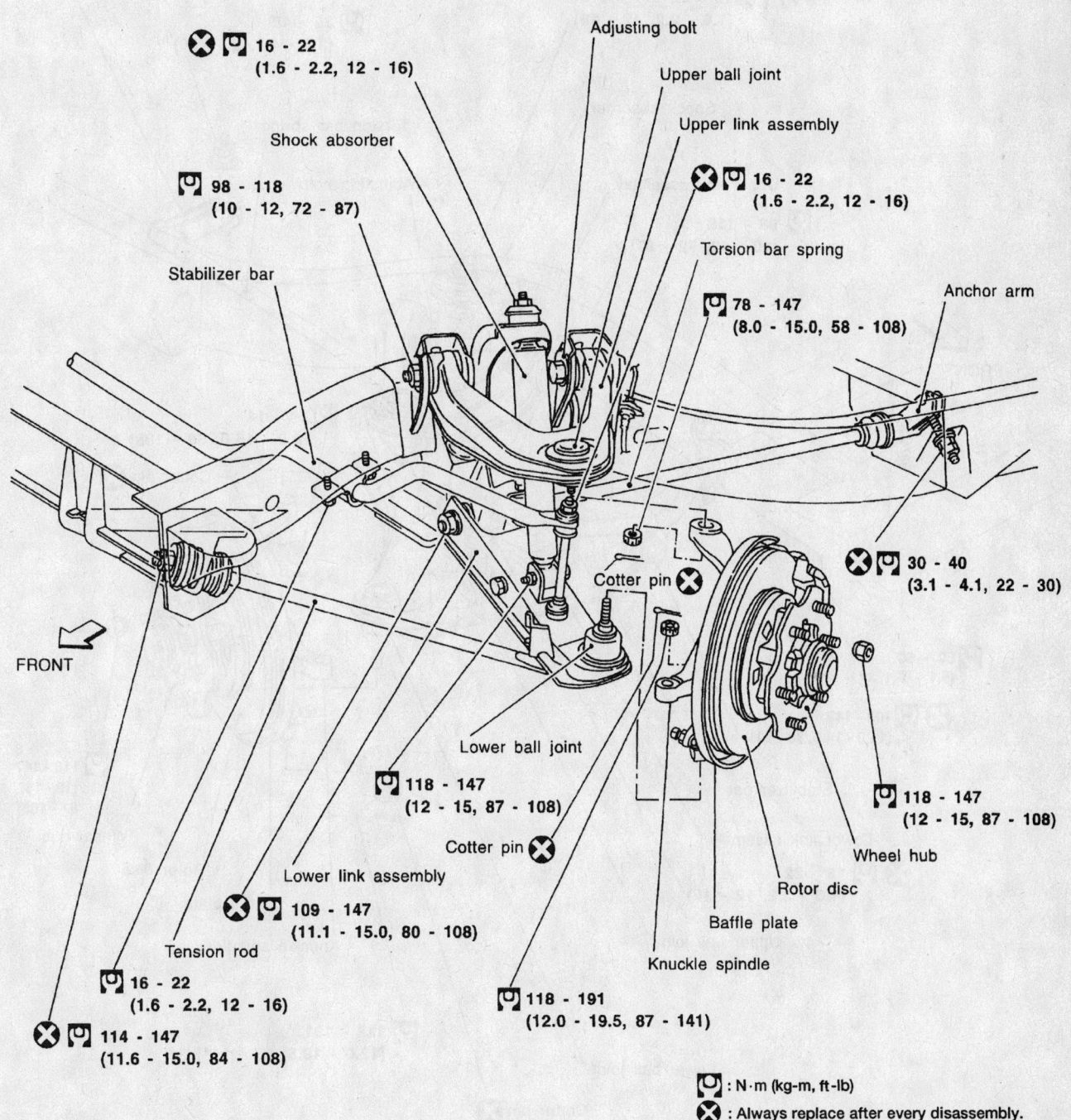

16 - 22
(1.6 - 2.2, 12 - 16)

Shock absorber

98 - 118
(10 - 12, 72 - 87)

Stabilizer bar

Adjusting bolt

Upper ball joint

Upper link assembly

16 - 22
(1.6 - 2.2, 12 - 16)

Torsion bar spring

78 - 147
(8.0 - 15.0, 58 - 108)

Anchor arm

30 - 40
(3.1 - 4.1, 22 - 30)

Cotter pin

FRONT

Lower ball joint

118 - 147
(12 - 15, 87 - 108)

118 - 147
(12 - 15, 87 - 108)

Cotter pin

Lower link assembly

109 - 147
(11.1 - 15.0, 80 - 108)

Tension rod

16 - 22
(1.6 - 2.2, 12 - 16)

114 - 147
(11.6 - 15.0, 84 - 108)

Wheel hub

Rotor disc

Baffle plate

Knuckle spindle

118 - 191
(12.0 - 19.5, 87 - 141)

: N·m (kg-m, ft-lb)

: Always replace after every disassembly.

42356-FRON-G04

Front suspension components—2wd 4-cyl. 2003 Frontier

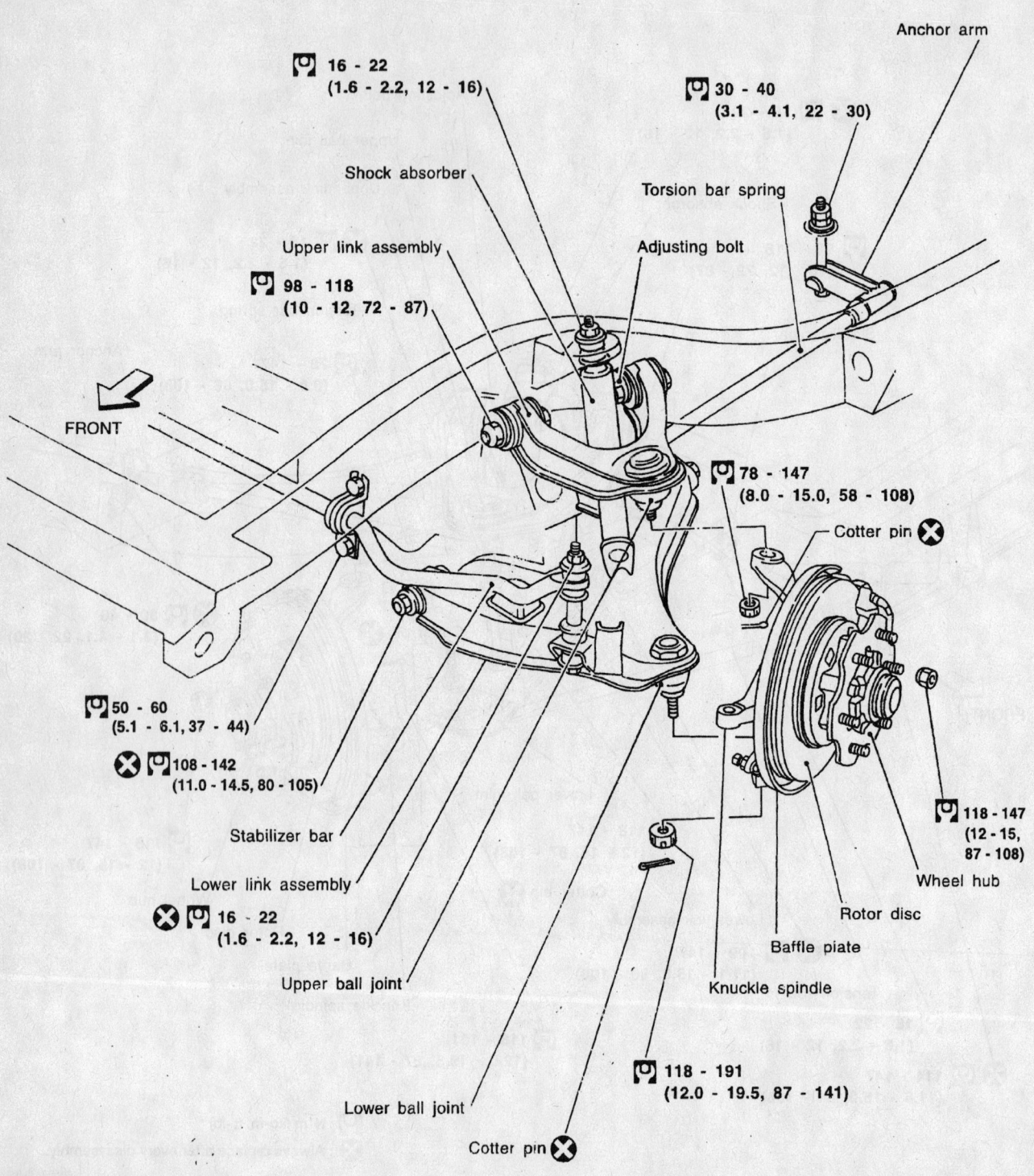

Anchor arm

16 - 22
(1.6 - 2.2, 12 - 16)

30 - 40
(3.1 - 4.1, 22 - 30)

Shock absorber

Torsion bar spring

Upper link assembly

Adjusting bolt

98 - 118
(10 - 12, 72 - 87)

FRONT

78 - 147
(8.0 - 15.0, 58 - 108)

Cotter pin ⊗

50 - 60
(5.1 - 6.1, 37 - 44)

⊗ 108 - 142
(11.0 - 14.5, 80 - 105)

Stabilizer bar

Lower link assembly

⊗ 16 - 22
(1.6 - 2.2, 12 - 16)

Upper ball joint

118 - 147
(12 - 15,
87 - 108)

Wheel hub

Rotor disc

Baffle piate

Knuckle spindle

118 - 191
(12.0 - 19.5, 87 - 141)

Lower ball joint

Cotter pin ⊗

: N•m (kg-m, ft-lb)

42356-FRON-G05

Front suspension components—2wd 6-cyl. 2003 Frontier and Xterra

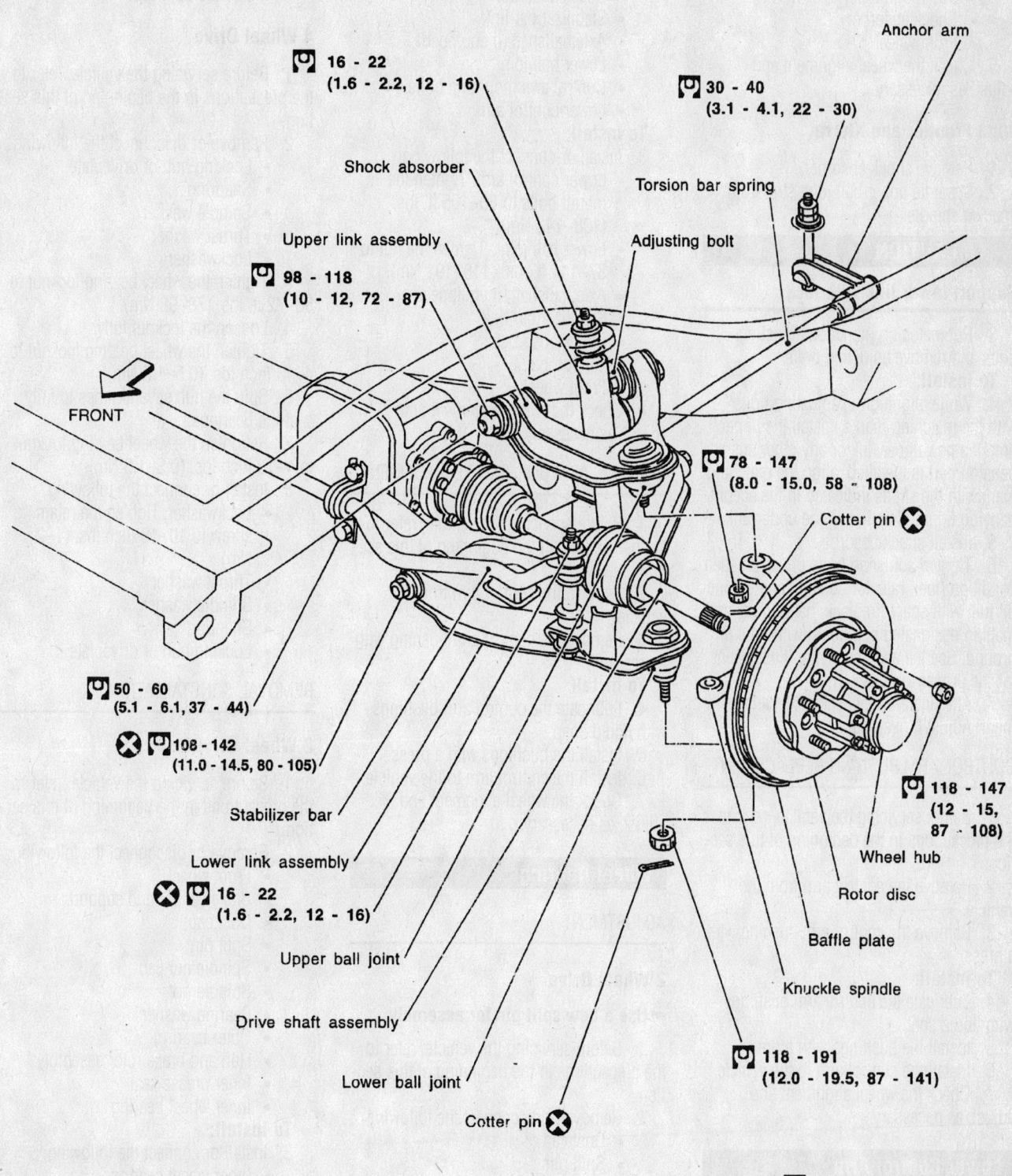

16 - 22
(1.6 - 2.2, 12 - 16)

30 - 40
(3.1 - 4.1, 22 - 30)

Anchor arm

Shock absorber

Torsion bar spring

Upper link assembly

Adjusting bolt

98 - 118
(10 - 12, 72 - 87)

FRONT

78 - 147
(8.0 - 15.0, 58 - 108)

Cotter pin

50 - 60
(5.1 - 6.1, 37 - 44)

108 - 142
(11.0 - 14.5, 80 - 105)

Stabilizer bar

Lower link assembly

16 - 22
(1.6 - 2.2, 12 - 16)

Upper ball joint

Drive shaft assembly

Lower ball joint

Cotter pin

118 - 147
(12 - 15,
87 - 108)

Wheel hub

Rotor disc

Baffle plate

Knuckle spindle

118 - 191
(12.0 - 19.5, 87 - 141)

: N•m (kg-m, ft-lb)

42356-FRON-G06

Front suspension components—4wd 2003 Frontier and Xterra

mounting bolts to 72–87 ft. lbs. (98–118 Nm).

- Upper ball joint. Tighten the nut to 58–108 ft. lbs. (78–147 Nm).
- Shock absorber
- Front wheel

5. Check the wheel alignment and adjust, as necessary.

2003 Frontier and Xterra

1. Remove shock absorber.
2. Separate upper ball joint stud from knuckle spindle.

✳✳ CAUTION

Support lower link with jack.

3. Put matching marks on adjusting bolts and remove adjusting bolts.

To install:

4. While aligning the adjusting bolts with the matching marks, install the upper link. If a new upper link or any other suspension part is installed, align the matching mark with the slit as indicated in the accompanying figure, then install the upper link.

5. Install shock absorber.

6. Tighten adjusting bolts under unladen condition (fuel, radiator coolant, and engine oil full; with spare tire, jack, hand tools, and mats in designated positions) with tires on ground. See the accompanying illustrations for the proper torques.

7. After installing, check wheel alignment. Adjust if necessary.

CONTROL ARM BUSHING REPLACEMENT

1. Before servicing the vehicle, refer to the precautions in the beginning of this section.

2. Remove the control arm from the vehicle.

3. Remove the control arm bushing with a press.

To install:

4. Lubricate the control arm bushings with liquid soap.

5. Install the bushings with a press.

6. Install the control arm to the vehicle.

7. Check the wheel alignment and adjust, as necessary.

Lower Control Arm

REMOVAL & INSTALLATION

1. Before servicing the vehicle, refer to the precautions in the beginning of this section.

2. Remove or disconnect the following:

- Front wheel
- Torsion bar
- Shock absorber
- Stabilizer bar link
- Axle halfshaft, if equipped
- Lower ball joint
- Control arm mounting bolts
- Lower control arm

To install:

3. Install or connect the following:

- Lower control arm. Tighten the mount bolts to 80–105 ft. lbs. (108–142 Nm).
- Lower ball joint. Tighten the nut to 87–141 ft. lbs. (118–191 Nm).
- Axle halfshaft, if equipped
- Stabilizer bar link
- Shock absorber
- Torsion bar
- Front wheel

4. Check the wheel alignment and adjust, as necessary.

CONTROL ARM BUSHING REPLACEMENT

1. Before servicing the vehicle, refer to the precautions in the beginning of this section.

2. Remove the control arm from the vehicle.

3. Remove the control arm bushing with a press.

To install:

4. Lubricate the control arm bushings with liquid soap.

5. Install the bushings with a press.

6. Install the control arm to the vehicle.

7. Check the wheel alignment and adjust, as necessary.

Wheel Bearings

ADJUSTMENT

2 Wheel Drive

➡**Use a new split pin for assembly.**

1. Before servicing the vehicle, refer to the precautions in the beginning of this section.

2. Remove or disconnect the following:

- Dust cap
- Split pin
- Spindle nut cap

3. Tighten the spindle nut to 25–29 ft. lbs. (34–39 Nm).

4. Spin the hub several times to fully seat the bearings.

5. Retighten the spindle nut to 25–29 ft. lbs. (34–39 Nm).

6. Loosen the spindle nut 45–60 degrees and install the spindle nut cap and split pin.

7. Install the dust cap.

4 Wheel Drive

1. Before servicing the vehicle, refer to the precautions in the beginning of this section.

2. Remove or disconnect the following:

- Locking hub or driveplate
- Snapring
- Spindle washer
- Thrust washer
- Lockwasher

3. Tighten the wheel bearing locknut to 58–72 ft. lbs. (78–98 Nm).

4. Loosen the locknut fully.

5. Tighten the wheel bearing locknut to 4–13 inch lbs. (0.5–1.5 Nm).

6. Spin the hub several times to fully seat the bearings.

7. Retighten the wheel bearing locknut to 4–13 inch lbs. (0.5–1.5 Nm).

8. Install or connect the following:

- Lockwasher. Tighten the retaining screw to 10–16 inch lbs. (1–2 Nm).
- Thrust washer
- Spindle washer
- Snapring
- Locking hub or driveplate

REMOVAL & INSTALLATION

2 Wheel Drive

1. Before servicing the vehicle, refer to the precautions in the beginning of this section.

2. Remove or disconnect the following:

- Front wheel
- Brake caliper and support
- Dust cap
- Split pin
- Spindle nut cap
- Spindle nut
- Bearing washer
- Outer bearing
- Hub and brake rotor assembly
- Inner grease seal
- Inner wheel bearing

To install:

3. Install or connect the following:

- Inner wheel bearing
- Inner grease seal
- Hub and brake rotor assembly
- Outer bearing
- Bearing washer
- Spindle nut. Adjust the wheel bearings.

- Spindle nut cap
- Split pin
- Dust cap
- Brake caliper and support
- Front wheel

4 Wheel Drive

1. Before servicing the vehicle, refer to the precautions in the beginning of this section.
2. Remove or disconnect the following:
   - Front wheel
   - Brake caliper and support

- Locking hub or driveplate
- Snapring
- Spindle washer
- Thrust washer
- Lockwasher
- Wheel bearing locknut
- Outer bearing
- Hub and brake rotor assembly
- Inner grease seal
- Inner wheel bearing

To install:

3. Install or connect the following:
   - Inner wheel bearing

- Inner wheel bearing
- Inner grease seal
- Hub and brake rotor assembly
- Outer bearing
- Wheel bearing locknut. Adjust the wheel bearings.
- Lockwasher
- Thrust washer
- Spindle washer
- Snapring
- Locking hub or driveplate
- Brake caliper and support
- Front wheel

BRAKES

Brake Caliper

REMOVAL & INSTALLATION

1. Raise the vehicle and support safely.
2. Remove the appropriate tire and wheel assembly.
3. Remove the bolt attaching the brake hose to the caliper. Plug the brake hose to prevent brake fluid loss.
4. Remove the caliper support mounting bolts and lift the caliper assembly from the knuckle.

To install

5. Position the caliper assembly onto

the knuckle and install the bolts. Make sure the rotor fits between the brake pads. Torque the bolts to 53–72 ft. lbs. (72–92 Nm) for 2003 4-cyl. Frontier; 101–130 ft. lbs. (137–177 Nm) for all others.

6. Using new copper washers, connect the brake hose to the caliper. Torque the brake hose attaching bolt to 12–14 ft. lbs. (17–20 Nm).
7. Bleed the brake system.
8. Apply the brake pedal and inspect the system. Ensure proper operation and no leakage.
9. Install tire and wheel assembly. Lower the vehicle and road-test.

Disc Brake Pads

REMOVAL & INSTALLATION

1. Raise and support the front of the vehicle, then remove the wheels.
2. Remove the bottom pin from the caliper and swing the caliper cylinder body upward; support the caliper with a wire.
3. Remove the brake pad retainers, shims and the pads.

To install:

4. Compress the piston of the disc brake caliper.
5. Install the brake pads and caliper

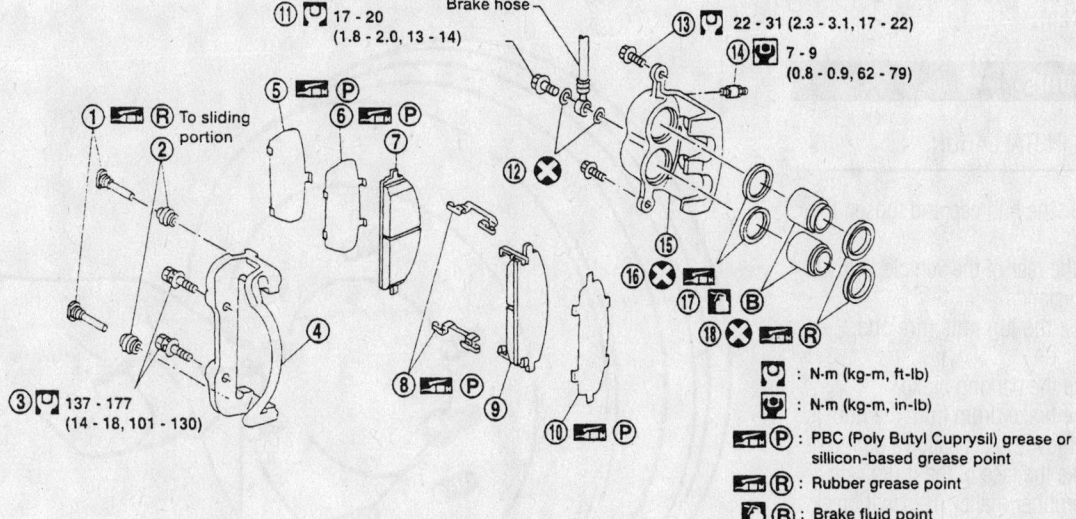

① 🔧 Ⓡ To sliding portion
②
③ 🔧 137 - 177 (14 - 18, 101 - 130)
⑤ 🔧 Ⓟ
⑥ 🔧 Ⓟ
⑦
④
Brake hose
⑪ 🔧 17 - 20 (1.8 - 2.0, 13 - 14)
⑬ 🔧 22 - 31 (2.3 - 3.1, 17 - 22)
⑭ 🔧 7 - 9 (0.8 - 0.9, 62 - 79)
⑫ ✕
⑯ ✕ 🔧
⑮
⑰ 🔧 Ⓑ
⑱ ✕ 🔧 Ⓡ
⑧ 🔧 Ⓟ
⑨
⑩ 🔧 Ⓟ

🔧 : N-m (kg-m, ft-lb)
🔧 : N-m (kg-m, in-lb)
🔧 Ⓟ : PBC (Poly Butyl Cuprysil) grease or sillicon-based grease point
🔧 Ⓡ : Rubber grease point
🔧 Ⓑ : Brake fluid point

1.	Main pin	7.	Inner pad	13.	Main pin bolt
2.	Pin boot	8.	Pad retainer	14.	Bleed valve
3.	Torque member fixing bolt	9.	Outer pad	15.	Cylinder body
4.	Torque member	10.	Outer shim	16.	Piston seal
5.	Shim cover (if so equipped)	11.	Connecting bolt	17.	Piston
6.	Inner shim	12.	Copper washer	18.	Piston boot

9348VG95

Front disc brake components—2000–02

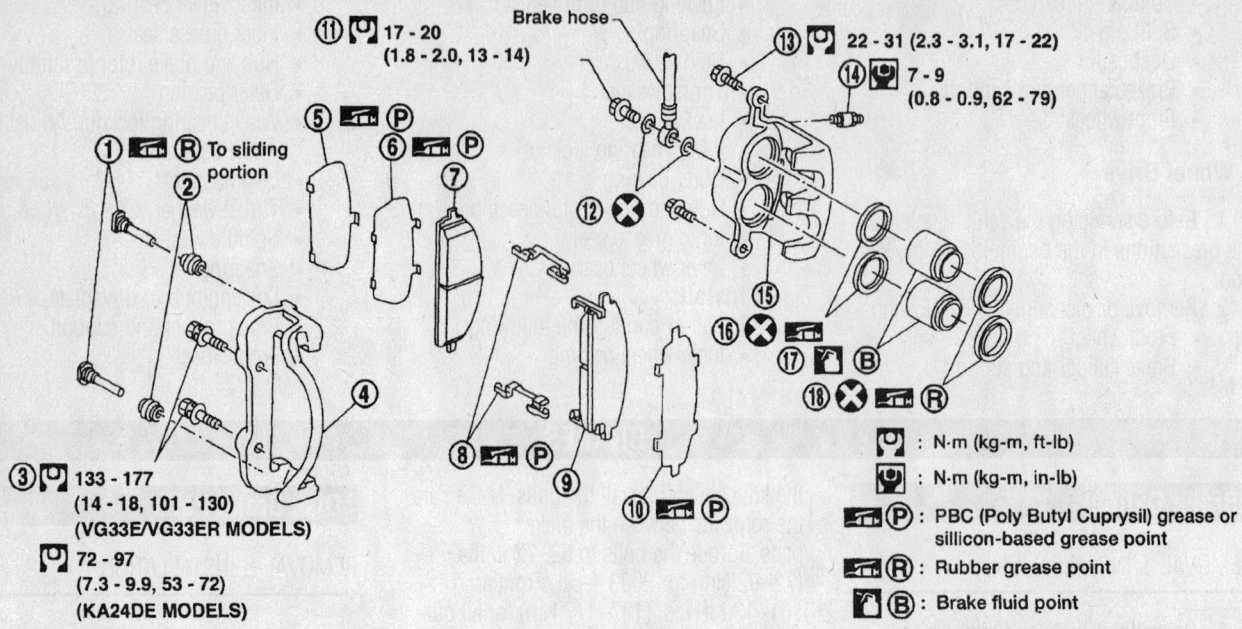

Brake hose

⑪ 🔧 17 - 20
(1.8 - 2.0, 13 - 14)

⑬ 🔧 22 - 31 (2.3 - 3.1, 17 - 22)

⑭ 🔧 7 - 9
(0.8 - 0.9, 62 - 79)

① 🔧 Ⓡ To sliding portion

③ 🔧 133 - 177
(14 - 18, 101 - 130)
(VG33E/VG33ER MODELS)

🔧 72 - 97
(7.3 - 9.9, 53 - 72)
(KA24DE MODELS)

🔧 : N·m (kg-m, ft-lb)

🔧 : N·m (kg-m, in-lb)

🔧Ⓟ : PBC (Poly Butyl Cuprysil) grease or sillicon-based grease point

🔧Ⓡ : Rubber grease point

🔧Ⓑ : Brake fluid point

1.	Main pin	2.	Pin boot	3.	Torque member fixing bolt
4.	Torque member	5.	Shim cover (if equipped)	6.	Inner shim
7.	Inner pad	8.	Pad retainer	9.	Outer pad
10.	Outer shim	11.	Connecting bolt	12.	Copper washer
13.	Main pin bolt	14.	Bleed valve	15.	Cylinder body
16.	Piston seal	17.	Piston	18.	Piston boot

42356-FRON-G01

Front disc brake components—2003

assembly. Torque the guide pin to 17–22 ft. lbs. (22–31 Nm).

Brake Drums

REMOVAL & INSTALLATION

1. Remove the hub cap and loosen the lug nuts.
2. Raise the rear of the vehicle and support it on jackstands.
3. Remove the lug nuts, tire and wheel.
4. Release the parking brake.
5. Pull the brake drum from the hub. If difficult to remove try the following:
   a. Strike the face of the drum with a plastic or rubber mallet. This will break free any rust that may develop between the drum and the hub.
   b. Install 2, M8x1.25mm bolts into the holes in the drum and gradually tighten them to pull the drum off the hub.

To install:
6. Install the brake drum to the hub.
7. Install the wheel.

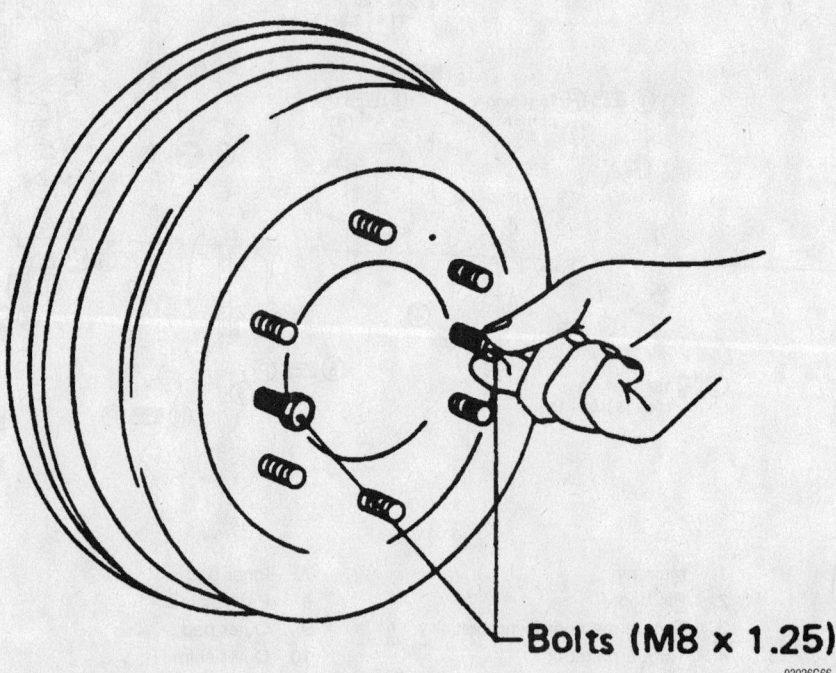

Bolts (M8 x 1.25)

93026G66

Install and tighten 2 bolts to remove a stubborn brake drum

8. Remove the jackstands and lower the vehicle.

9. Road-test the vehicle to ensure that the brakes are working properly.

Brake Shoes

REMOVAL & INSTALLATION

4-Cyl. Frontier and 2000–02 6-Cyl. Frontier and Xterra

1. Release the parking brake.
2. Safely raise and support the vehicle.
3. Remove the rear wheel and drum.
4. Remove the hold-down pin retainers.
5. Remove the leading shoe and then the trailing shoe.
6. Remove the adjuster.
7. Disconnect the parking brake cable from the toggle lever on the rear shoe.

To install:

8. Transfer the toggle lever to the new rear shoe.
9. Apply a small amount of brake grease to the tips of the shoes and the 6 pads on the backing plate that contact the brake shoe.
10. Shorten the adjuster by turning it.
11. Connect the parking brake cable to the toggle lever on the rear shoe.
12. Install the lower return spring to both shoes and install the shoes on the backing plate with the hold down pins and retainers.
13. Install the adjuster and the remaining springs. Pay attention to the direction of the adjuster assembly.
14. Inspect the complete assembly and install the brake drum.
15. Adjust the shoe to drum clearance.
16. Install the wheel assembly and lower the vehicle to the floor.

2003 6-Cyl. Frontier and Xterra

1. Remove the drum.
2. After removing shoe hold pin by rotating retainer, remove leading shoe then remove trailing shoe. Remove spring by rotating shoes in direction arrow.

❋❋ WARNING

Be careful not to damage wheel cylinder piston boots. Be careful not to damage parking brake cable when separating it.

3. Remove the adjuster.
4. Disconnect the parking brake cable from toggle lever.
5. Remove retainer clip with a suitable tool. Then separate toggle lever and brake shoe (trailing side).
6. Installation is the reverse of removal.

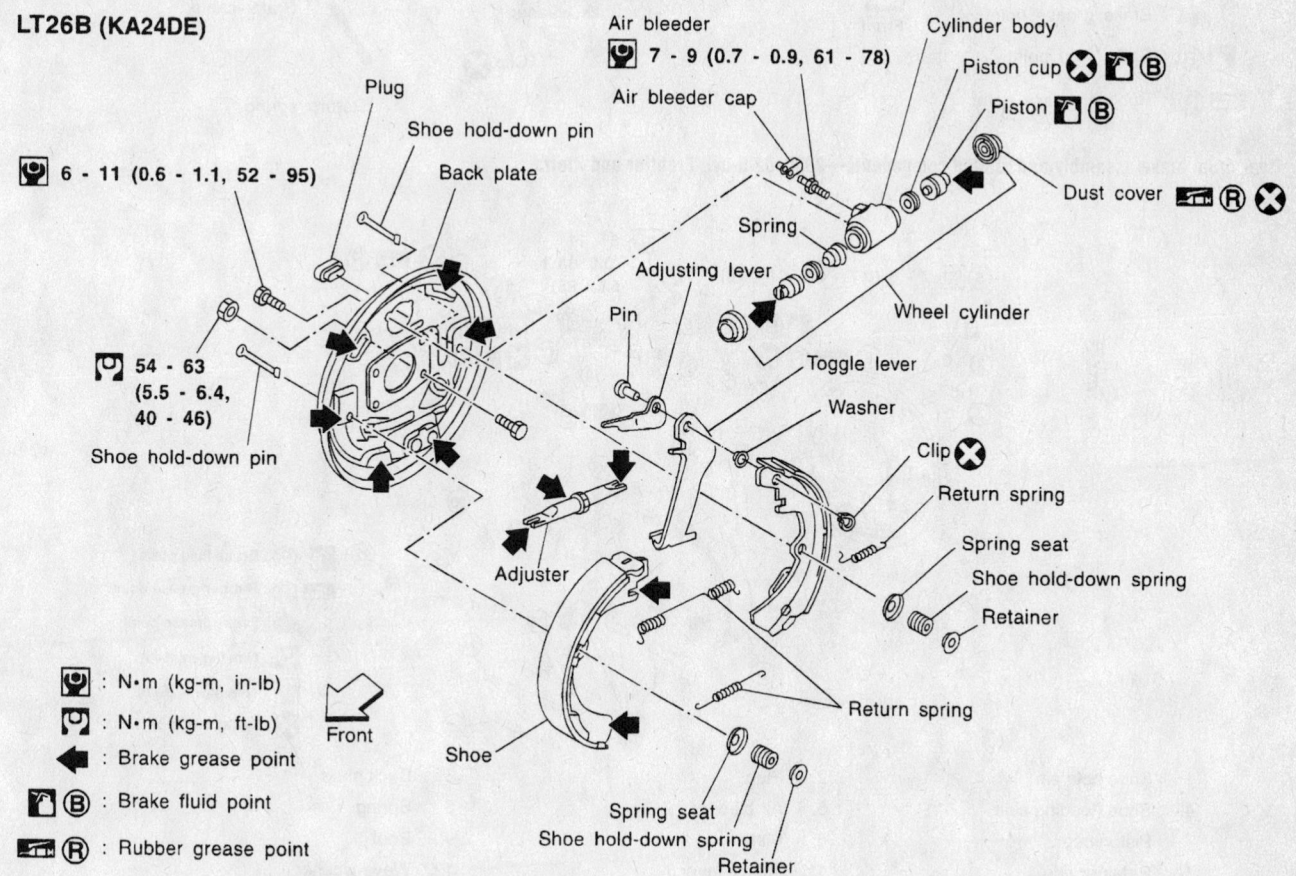

LT26B (KA24DE)

Rear drum brake assembly and related components—4-cyl. Frontier

9348VG96

LT30A (VG33E and VG33ER)

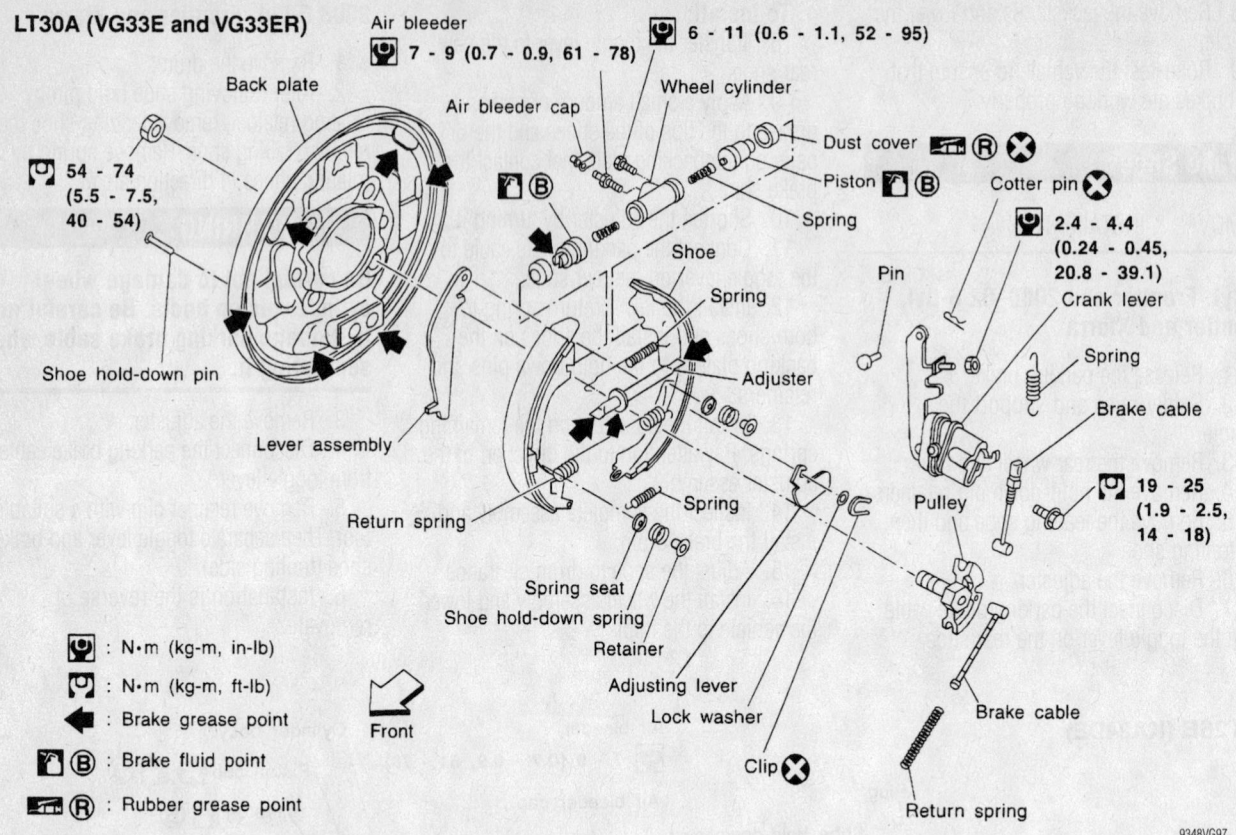

Rear drum brake assembly and related components—2000–02 6-cyl. Frontier and Xterra

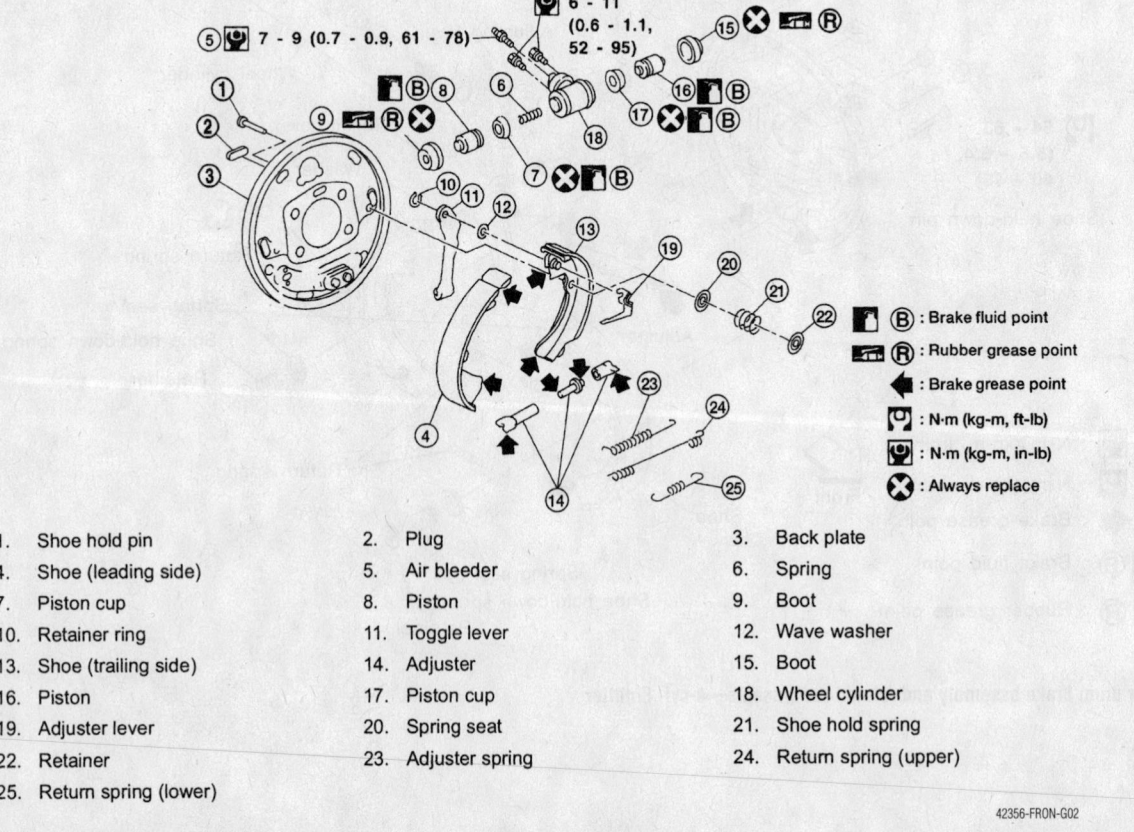

1.	Shoe hold pin	2.	Plug	3.	Back plate
4.	Shoe (leading side)	5.	Air bleeder	6.	Spring
7.	Piston cup	8.	Piston	9.	Boot
10.	Retainer ring	11.	Toggle lever	12.	Wave washer
13.	Shoe (trailing side)	14.	Adjuster	15.	Boot
16.	Piston	17.	Piston cup	18.	Wheel cylinder
19.	Adjuster lever	20.	Spring seat	21.	Shoe hold spring
22.	Retainer	23.	Adjuster spring	24.	Return spring (upper)
25.	Return spring (lower)				

42356-FRON-G02

Rear drum brake assembly and related components—2003 6-cyl. Frontier and Xterra

NISSAN

Murano

25

SPECIFICATION CHARTS

ENGINE AND VEHICLE IDENTIFICATION

	Engine						Model Year	
Code ①	Liters (cc)	Cu. In.	Cyl.	Fuel Sys.	Engine	Eng. Mfg.	Code	Year
VQ35DE	3.5 (3498)	213	6	MFI	DOHC	Nissan	3	2003
							4	2004

MFI: Multi-port Fuel Injection

DOHC: Double Overhead Camshafts

42356-MURA-C01

GENERAL ENGINE SPECIFICATIONS

Year	Model	Engine Displacement Liters (cc)	Engine ID/VIN	Fuel System Type	Net Horsepower @ rpm	Net Torque @ rpm (ft. lbs.)	Bore x Stroke (in.)	Com- pression Ratio	Oil Pressure @ rpm
2003	Murano	3.5 (3498)	VQ35DE	MFI	240@6000	265@3200	3.76X3.20	10.0:1	43@2000

MFI: Multi-port Fuel Injection

42356-MURA-C02

ENGINE TUNE-UP SPECIFICATIONS

Year	Engine Displacement Liters (cc)	Engine ID/VIN	Spark Plug Gap (in.)	Ignition Timing	Fuel Pump (psi)	Idle Speed RPM	Valve Clearance (in.) In.	Valve Clearance (in.) Ex.
2003	3.5 (3498)	VQ35DE	0.043	15B	51 ①	600-700	HYD	HYD

NOTE: The Vehicle Emission Control Information label often reflects specification changes made during production. The label figures must be used

if they differ from those in this chart.

B: Before top dead center

HYD: Hydraulic

① At idle

42356-MURA-C03

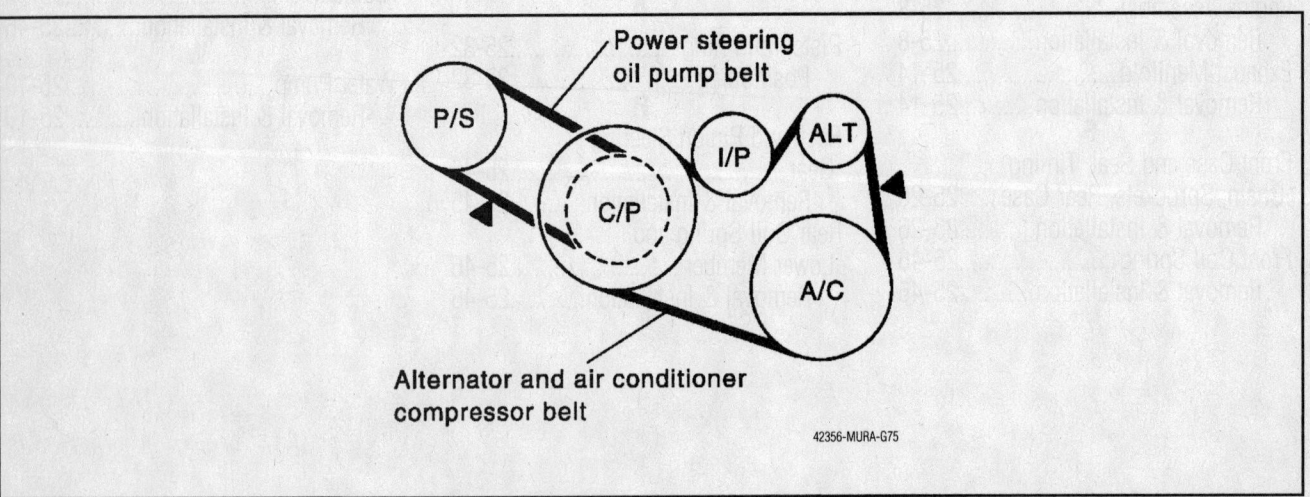

Power steering oil pump belt

Alternator and air conditioner compressor belt

42356-MURA-G75

CAPACITIES

Year	Model	Engine Displacement Liters (cc)	Engine ID/VIN	Engine Oil with Filter (qts.)	Transmission (pts.)	Transfer Case (pts.)	Drive Axle Front (pts.)	Rear (pts.)	Fuel Tank (gal.)	Cooling System (qts.)
2003	Murano	3.5 (3498)	VQ35DE	4.25	21.5	0.32	—	1.5	21.6	9.75

NOTE: All capacities are approximate. Add fluid gradually and check to be sure a proper fluid level is obtained.

42356-MURA-C04

CRANKSHAFT AND CONNECTING ROD SPECIFICATIONS
All measurements are given in inches.

Year	Engine Displacement Liters (cc)	Engine ID/VIN	Crankshaft Main Brg. Journal Dia.	Main Brg. Oil Clearance	Shaft End-play	Thrust on No.	Connecting Rod Journal Diameter	Oil Clearance	Side Clearance
2003	3.5 (3498)	VQ35DE	①	0.0014-0.0018	0.0118	4	②	0.0013-0.0023	0.0079-0 0138

① There are 24 different grades, ranging from A (2.3612) to 7 (2.3603)
② Grade 0: 2.0460-2.0462
  Grade 1: 2.0457-2.0460
  Grade 2: 2.0445-2.0457

42356-MURA-C05

VALVE SPECIFICATIONS

Year	Engine Displacement Liters (cc)	Engine ID/VIN	Seat Angle (deg.)	Face Angle (deg.)	Spring Test Pressure (lbs. @ in.)	Spring Installed Height (in.)	Stem-to-Guide Clearance (in.) Intake	Exhaust	Stem Diameter (in.) Intake	Exhaust
2003	3.5 (3498)	VQ35DE	45.15-45.45	45	45.4@1.457	1.457	0.0008-0.0021	0.0016-0.0029	0.2348-0.2354	0.2341-0.2346

42356-MURA-C06

PISTON AND RING SPECIFICATIONS

All measurements are given in inches.

Year	Engine Displacement Liters (cc)	Engine ID/VIN	Piston Clearance	Ring Gap Top Comp.	Ring Gap Bottom Comp.	Ring Gap Oil Control	Ring Side Clearance Top Comp.	Ring Side Clearance Bottom Comp.	Ring Side Clearance Oil Control
2003	3.5 (3498)	VQ35DE	0.0004-0.0012	0.0091-0.0130	0.0130-0.0189	0.0079-0.0236	0.0016-0.0031	0.0012-0.0028	0.0006-0.0020

42356-MURA-C07

TORQUE SPECIFICATIONS

All readings in ft. lbs.

Year	Engine Displacement Liters (cc)	Engine ID/VIN	Cylinder Head Bolts	Main Bearing Bolts	Rod Bearing Bolts	Crankshaft Damper Bolts	Driveplate Bolts	Manifold Intake	Manifold Exhaust	Spark Plugs	Lug Nuts
2003	3.5 (3498)	VQ35DE	①	②	③	④	61-69	⑤	21-24	14-22	90

① Step 1: 72 ft. lbs.

   Step 2: Loosen all bolts completely

   Step 3: 25-33 ft. lbs.

   Step 4: +90 degrees

   Step 5: +90 degrees

② Step 1: 24-28 ft. lbs.

   Step 2: +90 degrees

③ Step 1: 15 ft. lbs.

   Step 2: +90 degrees

④ 29-36 ft. lbs. +60-66 degrees

⑤ Step 1: 44-86 inch lbs.

   Step 2: 20-23 ft. lbs.

42356-MURA-C08

WHEEL ALIGNMENT

Year	Model	Caster Range (+/-Deg.)	Caster Preferred Setting (Deg.)	Camber Range (+/-Deg.)	Camber Preferred Setting (Deg.)	Toe-in (in.)	Kingpin Inclination (Deg.)
2003	Murano	0.75	+2.58	0.25	-0.33	0.02+/-0.04	14.33

42356-MURA-C09

TIRE, WHEEL AND BALL JOINT SPECIFICATIONS

| Year | Model | OEM Tires | | Tire Pressures (psi) | | Wheel Size | Ball Joint Inspection |
		Standard	Optional	Front	Rear		
2003	Murano	P235/65SR18	none	33	33	7.5	①

OEM: Original Equipment Manufacturer

PSI: Pounds Per Square Inch

NA: Not available

① Rotating torque: 5-30 inch lbs.

42356-MURA-C10

BRAKE SPECIFICATIONS

All measurements in inches unless noted

| Year | Model | Front Brake Disc | | | Rear Brake Disc | | | Minimum Lining Thickness | | Brake Caliper | |
		Original Thickness	Minimum Thickness	Maximum Runout	Original Thickness	Minimum Thickness	Maximum Runout	Front	Rear	Bracket Bolts (ft. lbs.)	Mounting Bolts (ft. lbs.)
2003	Murano	1.102	1.024	0.0016	0.630	0.551	0.002	0.079	0.08	101-129	17-22

42356-MURA-C11

SCHEDULED MAINTENANCE INTERVALS
2003 Nissan—Murano

TO BE SERVICED	TYPE OF SERVICE	VEHICLE MILEAGE INTERVAL (x1000)															
		3.8	7.5	11	15	19	22.5	26	30	34	37.5	41	45	49	52.5	56	60
Engine oil & filter	R		✓		✓		✓		✓		✓		✓		✓		✓
Brake lines & cables	S/I				✓				✓				✓				✓
Brake pads, discs, drums & linings	I				✓				✓				✓				✓
Driveshaft boots & propeller shaft	L/I				✓				✓				✓				✓
CVT, transfer case and differential fluid	I				✓				✓				✓				✓
Air cleaner filter	R								✓								✓
Drive belt(s) ①	S/I																
Engine coolant ②	R																✓
Spark plugs	R	platinum tipped plugs every 105,000 miles															
Cabin air filter	R				✓				✓				✓				✓
Exhaust system	I						✓										✓
Fuel lines	S/I								✓								✓
Steering gear (box) & linkage, axle & suspension parts	I				✓				✓				✓				✓
Vapor lines	S/I								✓								✓

R: Replace   S/I: Service or Inspect   L: Lubricate

① First a 60,000, then every 15,000 miles

② After 60,000, replace every 30,000

FREQUENT OPERATION MAINTENANCE (SEVERE SERVICE)

If a vehicle is operated under any of the following conditions it is considered severe service:

- Extremely dusty areas.

- 50% or more of the vehicle operation is in 32°C (90°F) or higher temperatures, or constant operation in temperatures below 0°C (32°F).

- Prolonged idling (vehicle operation in stop and go traffic).

- Frequent short running periods (engine does not warm to normal operating temperatures).

- Police, taxi, delivery usage or trailer towing usage.

Oil & oil filter: replace every 3750 miles.

Brake pads, discs, drums & linings: service or inspect every 7500 miles.

Driveshaft boots & propeller shaft: service or inspect every 7500 miles.

Exhaust system: service or inspect every 7500 miles.

Steering gear (box) & linkage, (steering damper-4x4), axle & suspension parts: service or inspect every 7500 miles.

Steering linkage ball joints & front suspension ball joints: service or inspect every 7500 miles.

42356-MURA-C12

PRECAUTIONS

Before servicing any vehicle, please be sure to read all of the following precautions, which deal with personal safety, prevention of component damage, and important points to take into consideration when servicing a motor vehicle:

• Never open, service or drain the radiator or cooling system when the engine is hot; serious burns can occur from the steam and hot coolant.

• Observe all applicable safety precautions when working around fuel. Whenever servicing the fuel system, always work in a well-ventilated area. Do not allow fuel spray or vapors to come in contact with a spark, open flame, or excessive heat (a hot drop light, for example). Keep a dry chemical fire extinguisher near the work area. Always keep fuel in a container specifically designed for fuel storage; also, always properly seal fuel containers to avoid the possibility of fire or explosion. Refer to the additional fuel system precautions later in this section.

• Fuel injection systems often remain pressurized, even after the engine has been turned **OFF**. The fuel system pressure must be relieved before disconnecting any fuel lines. Failure to do so may result in fire and/or personal injury.

• Brake fluid often contains polyglycol ethers and polyglycols. Avoid contact with the eyes and wash your hands thoroughly after handling brake fluid. If you do get brake fluid in your eyes, flush your eyes with clean, running water for 15 minutes. If

eye irritation persists, or if you have taken brake fluid internally, IMMEDIATELY seek medical assistance.

• The EPA warns that prolonged contact with used engine oil may cause a number of skin disorders, including cancer! You should make every effort to minimize your exposure to used engine oil. Protective gloves should be worn when changing oil. Wash your hands and any other exposed skin areas as soon as possible after exposure to used engine oil. Soap and water, or waterless hand cleaner should be used.

• All new vehicles are now equipped with an air bag system. The system must be disabled before performing service on or around system components, steering column, instrument panel components, wiring and sensors. Failure to follow safety and disabling procedures could result in accidental air bag deployment, possible personal injury and unnecessary system repairs.

• Always wear safety goggles when working with, or around, the air bag system. When carrying a non-deployed air bag, be sure the bag and trim cover are pointed away from your body. When placing a non-deployed air bag on a work surface, always face the bag and trim cover upward, away from the surface. This will reduce the motion of the module if it is accidentally deployed. Refer to the additional air bag system precautions later in this section.

• Clean, high quality brake fluid from a sealed container is essential to the safe and

proper operation of the brake system. You should always buy the correct type of brake fluid for your vehicle. If the brake fluid becomes contaminated, completely flush the system with new fluid. Never reuse any brake fluid. Any brake fluid that is removed from the system should be discarded. Also, do not allow any brake fluid to come in contact with a painted surface; it will damage the paint.

• Never operate the engine without the proper amount and type of engine oil; doing so WILL result in severe engine damage.

• Timing belt maintenance is extremely important! Many models utilize an interference-type, non-freewheeling engine. If the timing belt breaks, the valves in the cylinder head may strike the pistons, causing potentially serious (also time-consuming and expensive) engine damage. Refer to the maintenance interval charts in the front of this manual for the recommended replacement interval for the timing belt, and to the timing belt section for belt replacement and inspection.

• Disconnecting the negative battery cable on some vehicles may interfere with the functions of the on-board computer system(s) and may require the computer to undergo a relearning process once the negative battery cable is reconnected.

• When servicing drum brakes, only disassemble and assemble one side at a time, leaving the remaining side intact for reference.

ENGINE REPAIR

➡️ **Disconnecting the negative battery cable on some vehicles may interfere with the functions of the on board computer system. The computer may undergo a relearning process once the negative battery cable is reconnected.**

Alternator

REMOVAL & INSTALLATOIN

1. Before servicing the vehicle, refer to the precautions in the beginning of this section.
2. Remove or disconnect the following:
   • Negative battery cable
   • Alternator harness connectors
   • Engine right side under cover
   • Radiator
   • Remove alternator and air conditioner compressor belt.

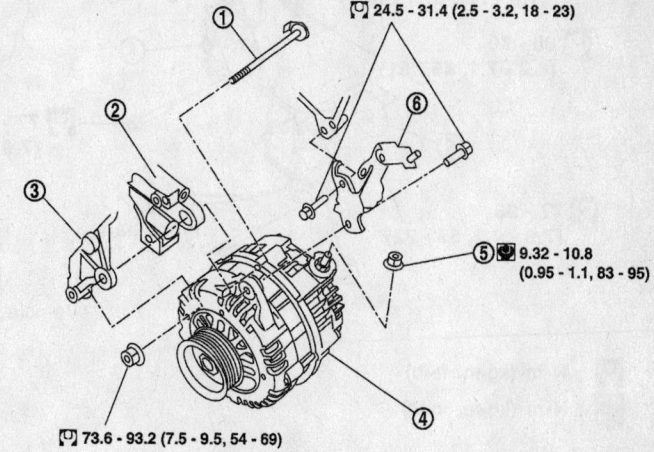

24.5 - 31.4 (2.5 - 3.2, 18 - 23)

5 9.32 - 10.8 (0.95 - 1.1, 83 - 95)

 : N·m (kg-m, in-lb)
 : N·m (kg-m, ft-lb)

73.6 - 93.2 (7.5 - 9.5, 54 - 69)

1. Through bolt	2. Cylinder block	3. Timing chain case
4. Alternator	5. B terminal nut	6. Alternator bracket

42356-MURA-G01

Alternator exploded view

- Idler pulley
- Alternator

To install:

3. Install or connect the following:
- Alternator
- Idler pulley
- Alternator belt. Tighten the through-bolts to 18–23 ft. lbs. (25–31 Nm).
- Engine under cover
- Radiator
- Alternator harness connectors
- Negative battery cable

Ignition Timing

ADJUSTMENT

Timing is not adjustable.

Engine Assembly

REMOVAL & INSTALLATION

1. Release fuel pressure.
2. Remove the engine cover, and the splash guards.
3. Drain engine coolant.
4. Remove or disconnect the following:
- Battery and tray
- Air inlet duct
- Air duct and air cleaner case (upper) assembly with mass air flow sensor
- Power brake booster vacuum hose
- Drive belts
- Radiator assembly, coolant reservoir, and system hoses
- RH windshield wiper arm and right font cowl top cover

- Engine room harness from the ECM side
- Heater hoses
- Wheel and tires
- A/C compressor with piping connected, and temporarily secure it aside
- Fuel hose quick connector at vehicle piping side
- Transaxle shift control cable.
- Starter motor
- Front exhaust tube
- Reservoir tank for the power steering from engine compartment bracket and position it aside.
- Power steering gear from steering lower joint
- Steering outer socket from steering knuckle
- Stabilizer connecting rod
- Propeller shaft (AWD models)

2WD models

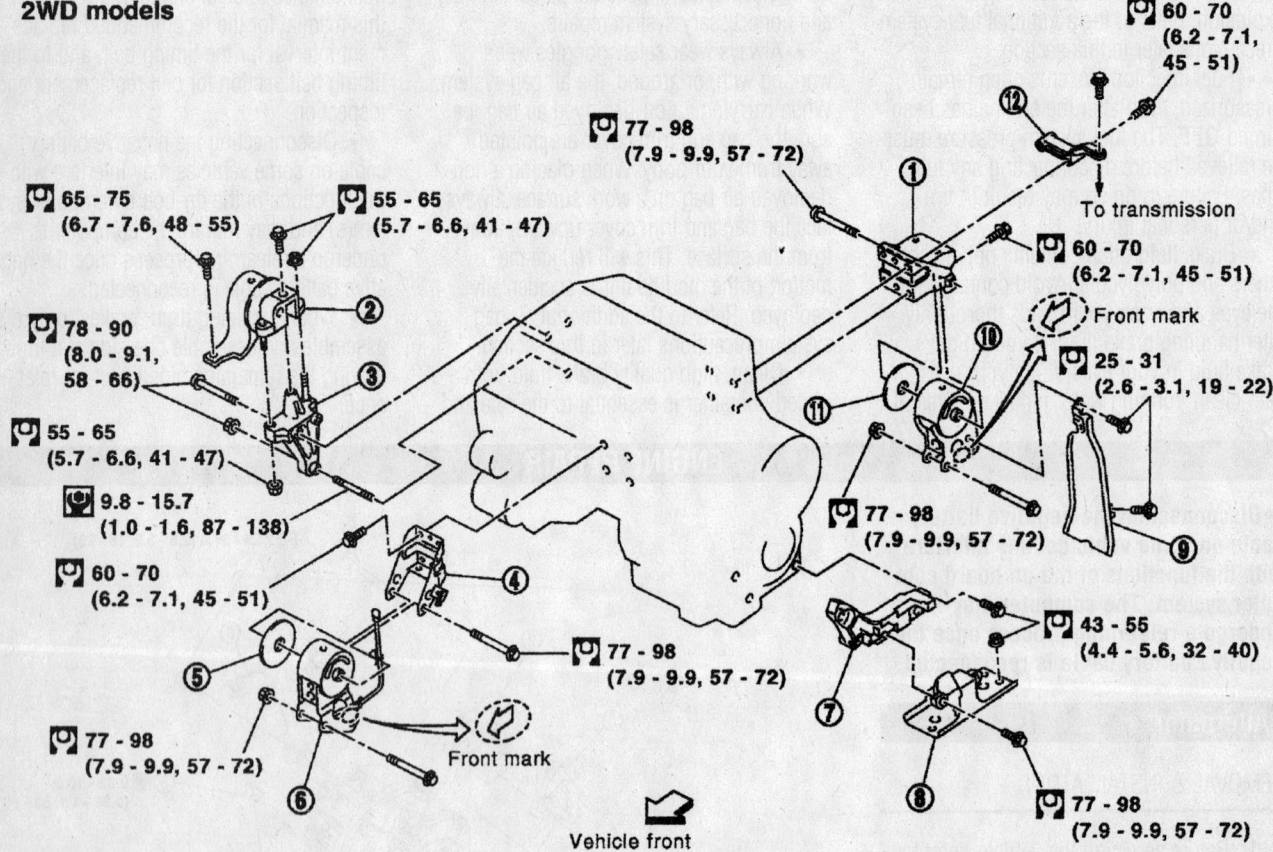

1. Rear engine mounting bracket
2. RH engine mounting insulator
3. RH engine mounting bracket
4. Front engine mounting bracket
5. Stopper
6. Front engine mounting insulator

42356-MURA-G02

Engine mounting—2wd

AWD models

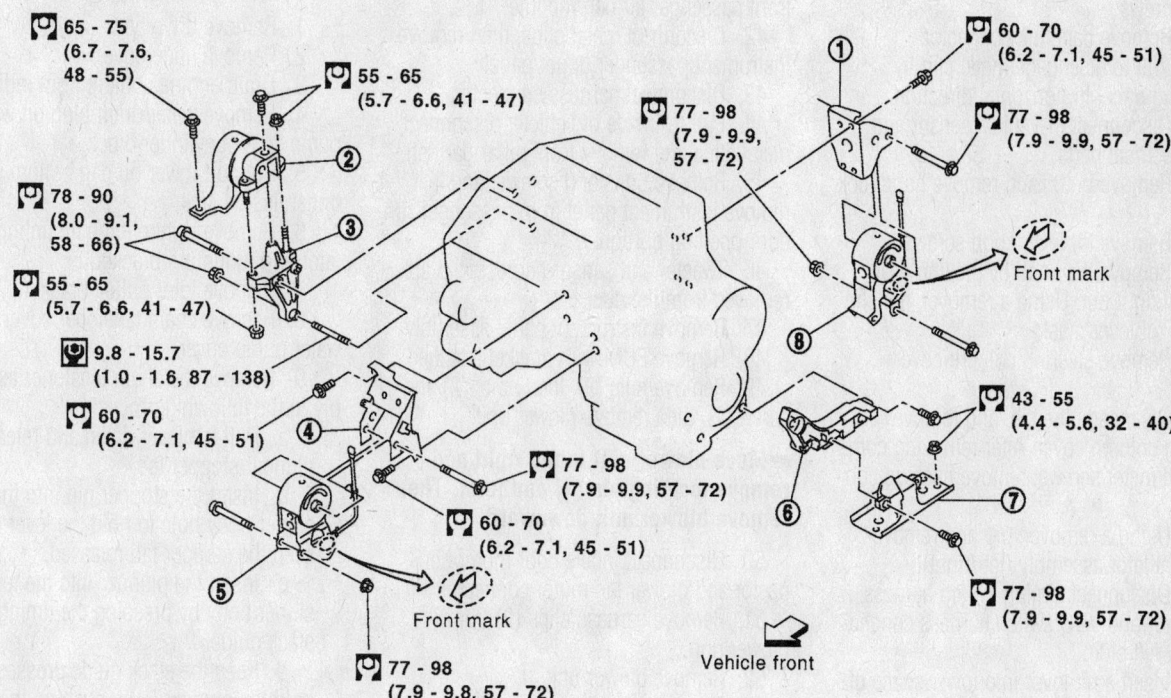

65 - 75
(6.7 - 7.6, 48 - 55)

55 - 65
(5.7 - 6.6, 41 - 47)

78 - 90
(8.0 - 9.1, 58 - 66)

55 - 65
(5.7 - 6.6, 41 - 47)

9.8 - 15.7
(1.0 - 1.6, 87 - 138)

60 - 70
(6.2 - 7.1, 45 - 51)

77 - 98
(7.9 - 9.8, 57 - 72)

Front mark

60 - 70
(6.2 - 7.1, 45 - 51)

77 - 98
(7.9 - 9.9, 57 - 72)

60 - 70
(6.2 - 7.1, 45 - 51)

77 - 98
(7.9 - 9.9, 57 - 72)

77 - 98
(7.9 - 9.9, 57 - 72)

Front mark

43 - 55
(4.4 - 5.6, 32 - 40)

77 - 98
(7.9 - 9.9, 57 - 72)

Vehicle front

: N·m (kg-m, ft-lb)

: N·m (kg-m, in-lb)

1. Rear engine mounting bracket
2. RH engine mounting insulator
3. RH engine mounting bracket
4. Front engine mounting bracket
5. Front engine mounting insulator
6. LH engine mounting bracket
7. LH engine mounting insulator
8. Rear engine mounting insulator

42356-MURA-G03

Engine mounting—AWD

- Left and right front halfshafts
- Power steering piping from power steering oil cooler

5. Position a manual lift table caddy under the engine and transaxle assembly.

6. Remove the right engine mounting insulator

7. Remove mounting bolt between transverse link and front suspension member

8. Carefully lower the engine, transaxle, transfer case (AWD models) and front suspension member assembly with the manual lift table caddy, avoiding interference with the vehicle body.

✳✳ WARNING

Before and during this procedure, always check if any harnesses are left connected. Avoid any damage to, or any oil or grease smearing or spills onto the engine mounting insulators.

9. Remove the crankshaft position sensor (POS).

10. Disconnect front suspension mounting nuts and bolts to remove engine, transaxle, transfer case (AWD models) and front suspension member assembly as a unit.

11. Separate the engine, transaxle and transfer case (AWD models) assembly and front suspension member.

12. Installation is the reverse order of removal. See the accompanying illustrations for the proper torque values.

Heater Core

REMOVAL & INSTALLATION

1. Use a refrigerant collecting equipment (for HFC-134a) to discharge refrigerant.

2. Drain coolant from cooling system.

3. Remove both right/left wiper arms.

4. Remove cowl top seal rubber.

5. Remove clips from cowl top cover (right) and remove cowl top cover (right).

6. Remove clips from cowl top cover (left) and remove cowl top cover (left)

7. Remove washer nozzles and hose from cowl top cover.

8. Remove cowl top cover.

9. Disconnect evaporator-side one-touch joints.

   a. Install a disconnector tool (High-pressure side: 92530-89908, Low pressure side: 92530-89916) on A/C piping.

   b. Slide a disconnector toward vehicle front until it clicks.

   c. Slide A/C piping toward vehicle front and disconnect it.

✳✳ WARNING

Seal connection opening of piping with a cap or vinyl tape to avoid exposure to atmosphere.

10. Disconnect two heater hoses from heater core.

11. Remove fuse lid.

12. Remove instrument driver lower panel screws.

13. Remove data link connector.

14. Pull to disengage metal clip by removing panel in horizontal direction.

15. Disconnect in-vehicle sensor and each electrical parts.

16. Remove bolts, and remove hood lock opener.

17. Remove tilt lever knob screws.

18. Remove the knob by picking it up and pulling it out. Using a remover, ply and remove tilt lever mask.

19. Remove steering column cover screws.

20. Disengage the tab, and remove steering column cover. After removing combination meter screws, remove harness connector.

21. Using a remover, pry and remove side ventilator assembly (left/right).

22. Disconnect aiming switch harness connector and VDC switch harness connector only left side.

23. Insert a remover into lower space of instrument side finisher (left/right) and remove by lifting.

24. Remove screws and glove box striker, disconnect connectors, and remove instrument passenger lower panel assembly.

25. Detach the damper from glove box right side.

26. Remove glove box pins, and remove glove box.

27. Insert a remover into lower space of center ventilator and remove by lifting.

28. Insert a remover into upper space of center ventilator and the upper clip is removed.

29. Remove screws. Disconnect harness connector, and remove tweeter with right and left part.

30. Insert a remover into front space of instrument stay cover (left/right) and detach.

31. Disconnect the left side harness connector only.

32. Remove cluster lid screws.

33. Disconnect A/C and AV harness connectors, and remove cluster lid C.

34. Remove display unit screws.

35. Disconnect harness connector and remove display.

36. Using a remover and disengage the ignition key finisher metal clips

37. Disconnect harness connector.

38. Using a remover, pry and remove instrument mask.

39. Disconnect harness connector.

40. Remove screw. Disengage the metal clip and remove instrument driver upper panel.

41. Remove bolt and screws Remove front passenger air bag module.

42. Disconnect metal clips, then remove instrument passenger upper panel.

43. Disconnect harness connector.

44. Pull to inside of vehicle, disconnect metal clips and remove front pillar garnish.

45. Remove bolts and screws, and remove instrument panel from passenger door opening portion.

46. Tweeter and sensor harness clip are removed from the duct.

47. Remove instrument panel assembly.

48. Remove ECM with bracket attached.

49. Remove nuts (2), then bolts (2) and screw (1), then remove blower unit.

➡**Move blower unit to the right and remove locating pin (1) and joint. Then remove blower unit downward.**

50. Disconnect intake door motor connector and blower fan motor connector.

51. Remove harness clips (2) from blower unit.

52. Remove blower unit.

53. Remove clips from vehicle harness from steering member.

54. Remove instrument stays (driver side and passenger side).

55. Remove rear ventilator duct1 and front floor duct.

56. Remove mounting screws from heater & cooling unit.

57. Remove the steering member, and then remove heater & cooling unit.

58. Remove foot duct (right).

59. Remove heater core cover.

60. Remove heater pipe support and heater pipe grommet.

61. Slide heater core to passenger side.

62. Installation is the reverse of removal. Note the following points:

- Replace O-rings for A/C piping with new ones, coated with compressor oil.
- Connection point for female-side piping is thin. So, when inserting male-side piping, take care not to deform female-side piping. Slowly insert in axial direction.
- Insert one-touch joint connection point securely until it clicks.
- After piping has been connected, pull male-side piping by hand to check that piping does not come off.
- When recharging refrigerant, check for leaks.

Water Pump

REMOVAL & INSTALLATION

1. Remove drive belts.

2. Remove undercover.

3. Drain engine coolant from radiator.

4. Remove water drain plug on water pump side of cylinder block.

5. Support lower oil pan bottom with a transmission jack.

6. Remove right engine mounting insulator and mounting bracket.

7. Remove idler pulley bracket.

8. Remove chain tensioner cover and water pump cover.

9. Remove the chain tensioner assembly in the following procedure.

   a. Pull the lever down and release the plunger stopper tab.

   b. Insert the stopper pin into the tensioner body hole to hold the lever and keep the stopper tab released.

   c. Insert the plunger into the tensioner body by pressing the timing chain slack guide.

   d. Keep the slack guide pressed and hold the plunger in by pushing the stopper pin deeper through the lever and into the tensioner body hole.

   e. Turn crankshaft pulley approximately 20 degrees clockwise so that the timing chain on the chain tensioner side is loose.

10. Remove chain tensioner.

11. Remove the 3 water pump bolts. Secure a gap between water pump gear and timing chain, by turning crankshaft pulley approximately 20 degrees counterclockwise.

12. Screw M8 bolts, pitch: 1.25mm (0.049in) length: Approx. 50 mm (1.97in) into water pump's upper and lower mounting-bolt holes until they reach timing chain case. Then, alternately tighten each bolt for a half turn, and pull out water pump.

13. Pull straight out while preventing vane from contacting socket in installation area. Remove water pump without causing sprocket to contact timing chain.

14. Remove M8 bolts and O-rings from water pump.

To install:

15. Install new O-rings to water pump.

16. Apply engine oil and engine coolant to the O-rings as shown. Locate the O-ring with white paint mark to engine front side.

17. Install the water pump.

➡**Do not allow cylinder block to nip the O-rings when installing the water pump.**

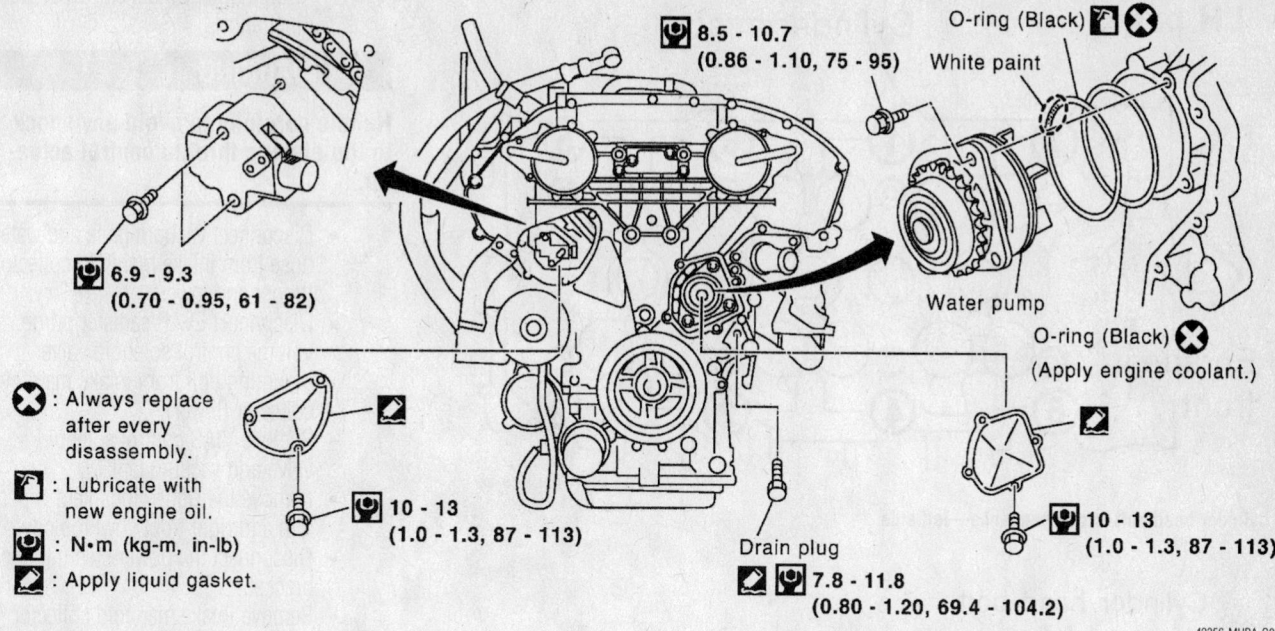

**8.5 - 10.7
(0.86 - 1.10, 75 - 95)**

O-ring (Black)
White paint

Water pump

O-ring (Black) ✗
(Apply engine coolant.)

**6.9 - 9.3
(0.70 - 0.95, 61 - 82)**

✗ : Always replace after every disassembly.

🔧 : Lubricate with new engine oil.

🔧 : N•m (kg-m, in-lb)

📐 : Apply liquid gasket.

**10 - 13
(1.0 - 1.3, 87 - 113)**

Drain plug

**7.8 - 11.8
(0.80 - 1.20, 69.4 - 104.2)**

**10 - 13
(1.0 - 1.3, 87 - 113)**

42356-MURA-G04

Water pump mounting

18. Check that timing chain and water pump sprocket are engaged.

19. Insert water pump by tightening mounting bolts alternately and evenly.

20. Remove dust and foreign material completely from backside of chain tensioner and from installation area of rear timing chain case.

21. Turn the crankshaft pulley clockwise so that the timing chain on the timing chain tensioner side is loose.

➡**When installing the timing chain tensioner, engine oil should be applied to the oil hole and tensioner.**

22. Install the timing chain tensioner.

23. Remove the stopper pin.

24. Install chain tensioner cover and water pump cover.

a. Before installing, remove all traces of liquid gasket from mating surface of water pump cover and chain tensioner cover using a scraper. Also remove traces of liquid gasket from the mating surface of the front cover.

b. Apply a continuous bead of liquid gasket, to mating surface of chain tensioner cover and water pump cover. Use RTV silicon sealant or equivalent

25. Install water drain plug on water pump side of cylinder block.

26. Installation is in the reverse order of removal for remaining parts.

27. After starting engine, let idle for three minutes, then rev engine up to 3,000 rpm under no load to purge air from the high-pressure chamber of the chain tensioner.

The engine may produce a rattling noise. This indicates that air still remains in the chamber and is not a matter of concern.

Cylinder Head

REMOVAL & INSTALLATION

1. Before servicing the vehicle, refer to the precautions in the beginning of this section.

2. Remove or disconnect the following:
- Negative battery cable
- Fuel tube and fuel injector assembly
- Intake manifold
- Exhaust manifold

- Water inlet and thermostat assembly
- Water outlet and water piping
- Camshaft
- Cylinder head bolts in reverse of the tightening sequence

3. Inspect the bolts. Cylinder head bolts are tightened by plastic zone tightening method. Whenever the size difference between d1 and d2 exceeds the limit, replace them with new one. If reduction of outer diameter appears in a position other than d2, use it as d2 point.

To install:

4. Install cylinder head gasket.

5. Turn the crankshaft until No. 1 piston is set at TDC on the compression stroke.

RH bank

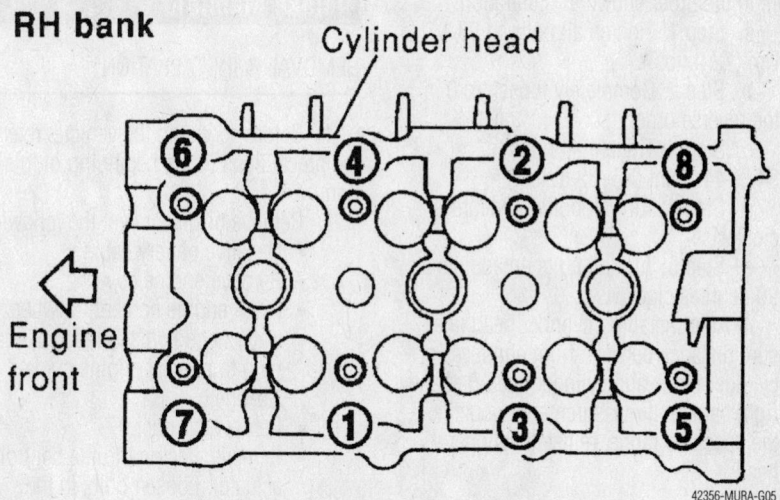

Cylinder head

Engine front

42356-MURA-G05

Cylinder head bolt torque sequence—right side

LH bank

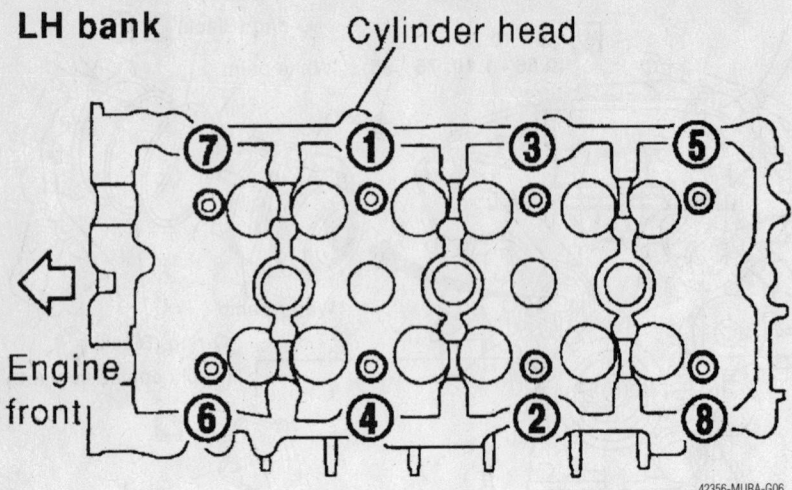

Cylinder head

Engine front

Cylinder head bolt torque sequence—left side

42356-MURA-G06

Cylinder head bolt

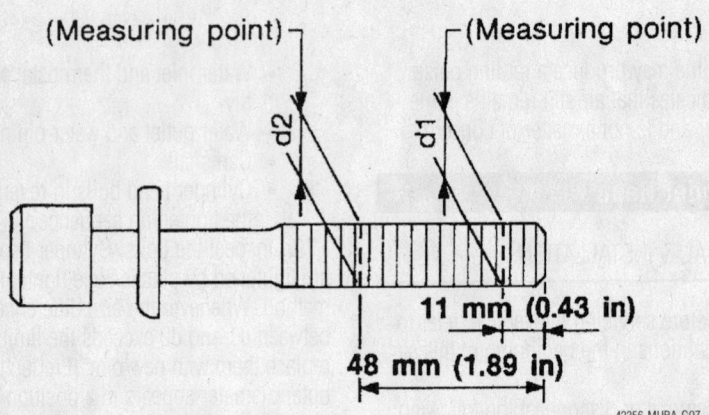

(Measuring point)   (Measuring point)

11 mm (0.43 in)

48 mm (1.89 in)

42356-MURA-G07

Cylinder head bolt measurement

The crankshaft key should line up with the right bank cylinder center line as shown.

6. Install cylinder head. Tighten the head bolts in the order shown in illustration.

  a. Step 1: Tighten all bolts to 98.1 Nm (72 ft. lbs.).

  b. Step 2: Completely loosen to 0 in the reverse order.

  c. Step 3: Tighten all bolts to 34.3–44.1 Nm (26–32 ft. lbs.).

  d. Step 4: Turn all bolts 90 degrees clockwise.

  e. Step 5: Turn all bolts an additional 90 degrees clockwise.

4. After installing cylinder head, measure distance between front end faces of cylinder block and cylinder head (left and right banks). If measurement is outside the specified range, reinstall cylinder head.

5. The remainder of installation is the reverse of removal.

Intake Manifold

REMOVAL & INSTALLATION

1. Before servicing the vehicle, refer to the precautions in the beginning of this section.

2. Remove or disconnect the following:
  • Negative battery cable
  • Remove engine cover.
  • Drain engine coolant, or when water hose is disconnected, attach plug to prevent engine coolant leakage.
  • Remove air duct.
  • Remove electric throttle control actuator. Loosen bolts in the

reverse order of that shown in the illustration.

✳✳ WARNING

Handle carefully to avoid any shock to the electric throttle control actuator.

  • Disconnect vacuum hose and water hose from intake manifold collector (upper and lower).
  • Disconnect EVAP canister purge volume control solenoid valve mounting bolt from intake manifold collector (lower).
  • Remove VIAS control solenoid valve and vacuum tank.
  • Remove the right windshield wiper arm and right front cowl top cover.
  • Disconnect the power steering hose bracket.
  • Remove intake manifold collector support bracket.
  • Remove PCV hose
  • Loosen bolts in reverse order of illustration, and remove intake manifold collector (upper and lower) assembly.
  • Loosen bolts in reverse order of illustration to remove intake manifold collector (upper)
  • Remove power valve in reverse order of illustration.
  • Remove fuel tube and fuel injector assembly.
  • Loosen bolts and nuts in reverse order of illustration to remove intake manifold assembly

3. Using straightedge and feeler gauge, inspect the surface distortion of each surface on intake manifold. If it exceeds the limit, replace the intake manifold.

4. Using straightedge and feeler gauge, inspect the surface distortion of intake manifold collector (lower). If it exceeds the limit, replace the intake manifold collector.

5. Installation is the reverse of removal paying attention to the following.

  • If the intake manifold stud bolts were removed, install them and tighten to the torque specified below.
  • Tighten all mounting bolts and nuts to specified torque in two or more steps in numerical order shown in illustration.
  • Tighten the power valve bolts in the order shown in the illustration.
  • Tighten the upper and lower intake manifold collector bolts in the order shown in the illustration.

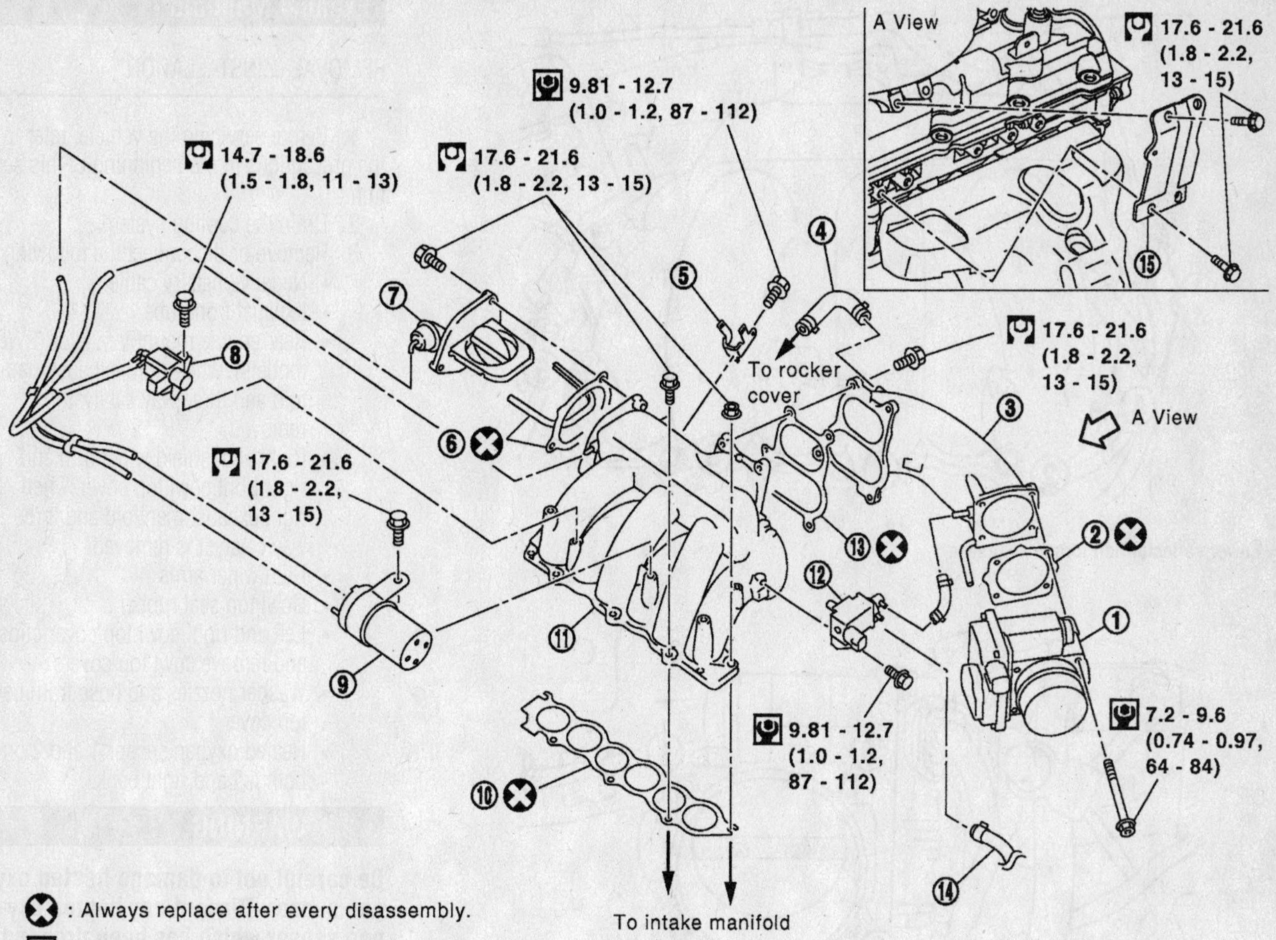

A View

🔧 17.6 - 21.6
(1.8 - 2.2, 13 - 15)

🔧 9.81 - 12.7
(1.0 - 1.2, 87 - 112)

🔧 14.7 - 18.6
(1.5 - 1.8, 11 - 13)

🔧 17.6 - 21.6
(1.8 - 2.2, 13 - 15)

④

⑤

To rocker cover

🔧 17.6 - 21.6
(1.8 - 2.2, 13 - 15)

⑦

⑧

③

A View

⑥ ✕

🔧 17.6 - 21.6
(1.8 - 2.2, 13 - 15)

⑬ ✕

② ✕

⑫

⑨

⑪

①

🔧 7.2 - 9.6
(0.74 - 0.97, 64 - 84)

⑩ ✕

🔧 9.81 - 12.7
(1.0 - 1.2, 87 - 112)

⑭

To intake manifold

✕ : Always replace after every disassembly.

🔧 : N•m (kg-m, in-lb)

🔧 : N•m (kg-m, ft-lb)

1. Electric throttle control actuator
2. Gasket
3. Intake manifold collector (upper)
4. PCV hose
5. Harness bracket
6. Gasket
7. Power valve
8. VIAS control solenoid valve
9. Vacuum tank
10. Gasket
11. Intake manifold collector (lower)
12. EVAP canister purge volume control solenoid valve
13. Gasket
14. EVAP hose
15. Support bracket

42356-MURA-G08

Upper and lower intake manifold collectors

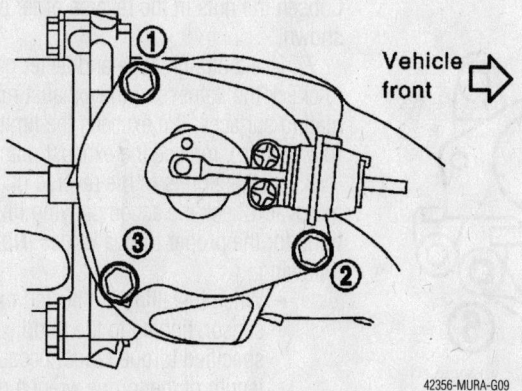

Power valve installation

42356-MURA-G09

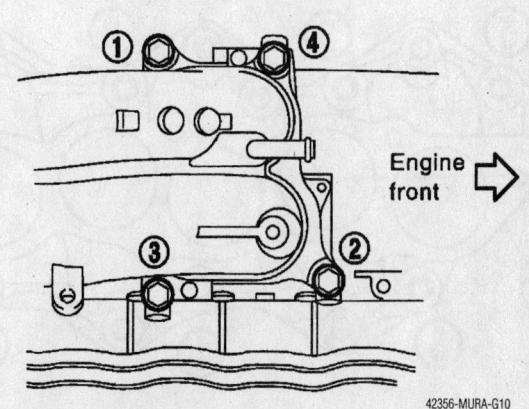

Upper collector bolt torque sequence

42356-MURA-G10

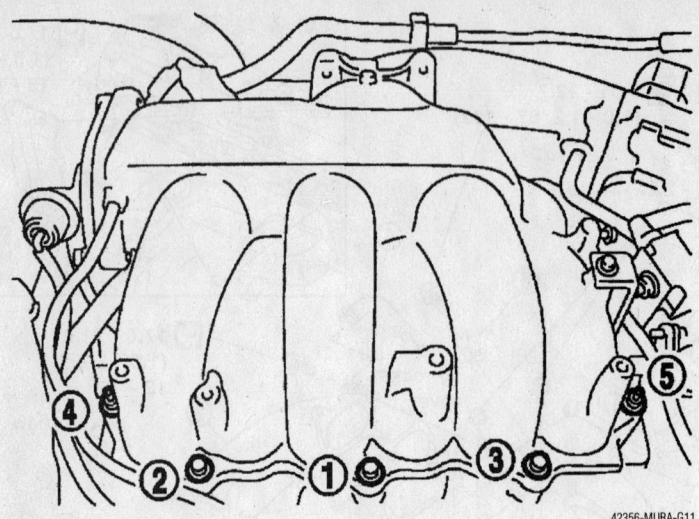

Lower collector bolt torque sequence

42356-MURA-G11

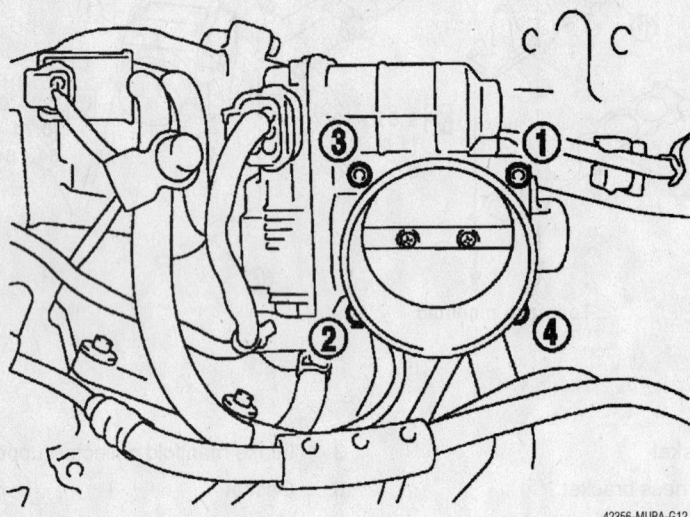

Throttle control actuator bolt torque sequence

42356-MURA-G12

⬅ Engine front

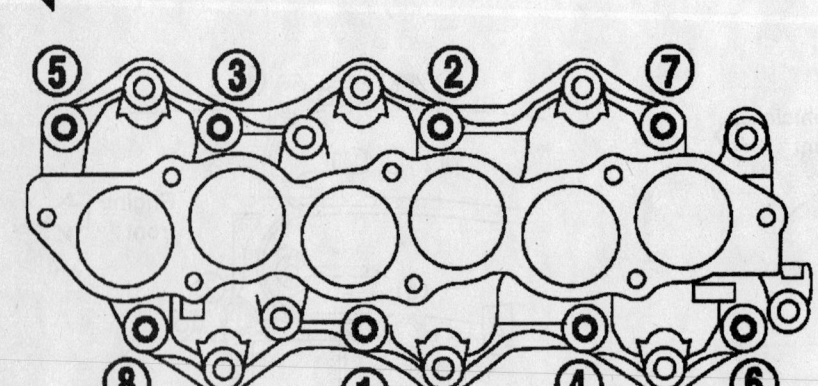

Intake manifold bolt torque sequence

42356-MURA-G13

Exhaust Manifold

REMOVAL & INSTALLATION

1. Before servicing the vehicle, refer to the precautions in the beginning of this section.

2. Drain the cooling system.

3. Remove or disconnect the following:
   - Negative battery cable
   - Exhaust front tube
   - Rear engine mount insulator (2WD models), when right exhaust manifold and three way catalyst is removed
   - Right windshield wiper arm and right front cowl top cover, when right exhaust manifold and three way catalyst is removed
   - Both wiper arms
   - Cowl top seal rubber
   - Left and right cowl top cover clips and remove cowl top covers
   - Washer nozzles and hose from cowl top cover
   - Heated oxygen sensor 1 and 2 on both left and right bank.

✳✳ WARNING

Be careful not to damage heated oxygen sensor. Discard any heated oxygen sensor which has been dropped from a height of more than 0.5 m (19.7 inches) onto a hard surface such as a concrete floor; replace with a new sensor.

   - Exhaust manifold covers and the three way catalyst heat shields

4. Remove bolts in the reverse order of illustration to remove three way catalyst supports.

5. Remove the left and right three way catalysts by loosening the bolts first and then removing the nuts.

6. Remove the exhaust manifolds. Loosen the nuts in the reverse order as shown.

7. Use a straightedge and feeler gauge to check the flatness of the exhaust manifold mating surfaces. If it exceeds the limit (0.012 inch), replace the exhaust manifold.

8. Installation is in the reverse of removal. Check the accompanying illustrations for the proper torque values. Note the following:
   - When installing the heated oxygen sensor, tighten to the middle of specified torque range, because the length of the torque wrench may increase the actual tightness. Do

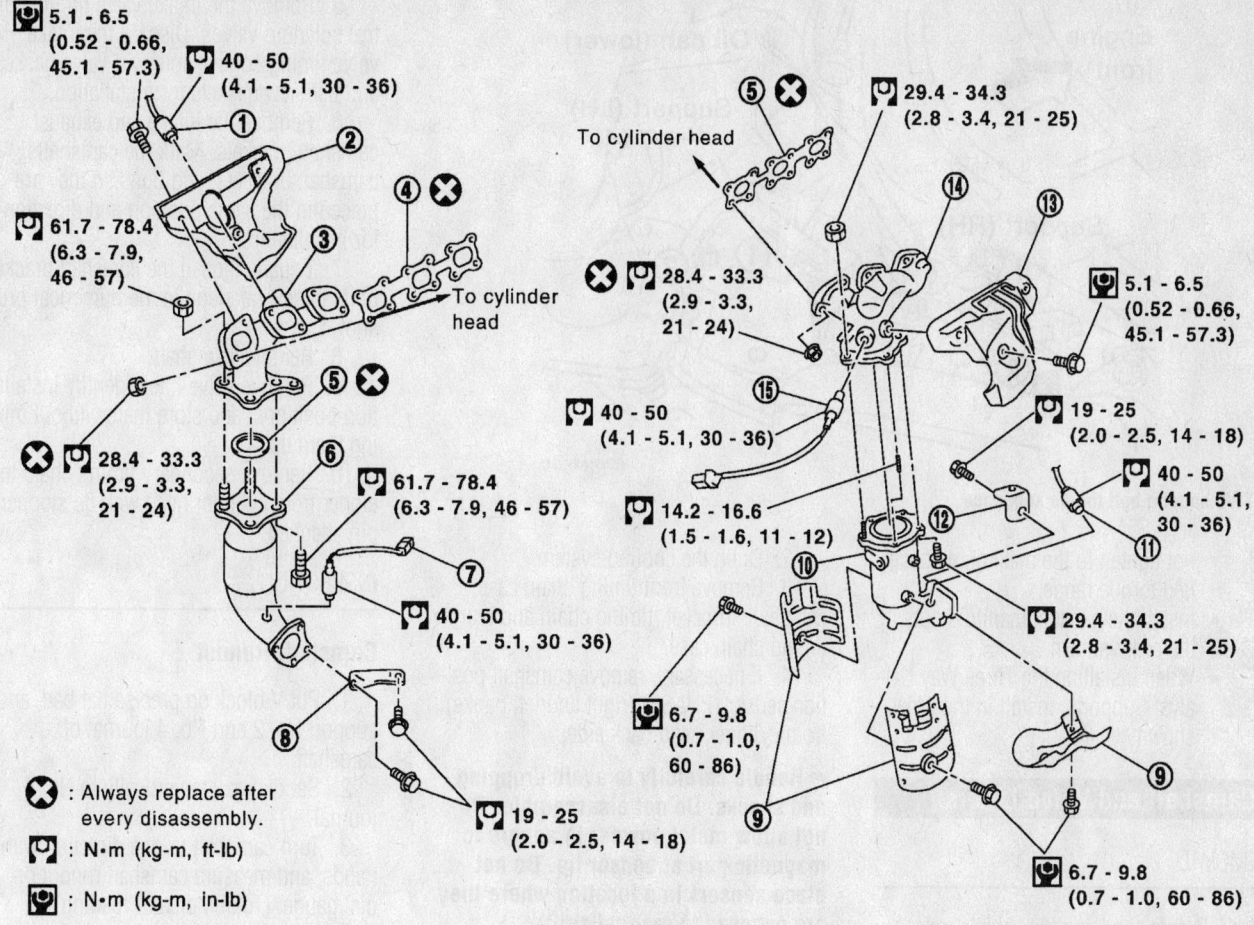

5.1 - 6.5 (0.52 - 0.66, 45.1 - 57.3)

40 - 50 (4.1 - 5.1, 30 - 36)

61.7 - 78.4 (6.3 - 7.9, 46 - 57)

28.4 - 33.3 (2.9 - 3.3, 21 - 24)

61.7 - 78.4 (6.3 - 7.9, 46 - 57)

40 - 50 (4.1 - 5.1, 30 - 36)

To cylinder head

To cylinder head

29.4 - 34.3 (2.8 - 3.4, 21 - 25)

5.1 - 6.5 (0.52 - 0.66, 45.1 - 57.3)

28.4 - 33.3 (2.9 - 3.3, 21 - 24)

40 - 50 (4.1 - 5.1, 30 - 36)

14.2 - 16.6 (1.5 - 1.6, 11 - 12)

19 - 25 (2.0 - 2.5, 14 - 18)

40 - 50 (4.1 - 5.1, 30 - 36)

29.4 - 34.3 (2.8 - 3.4, 21 - 25)

6.7 - 9.8 (0.7 - 1.0, 60 - 86)

19 - 25 (2.0 - 2.5, 14 - 18)

6.7 - 9.8 (0.7 - 1.0, 60 - 86)

❌ : Always replace after every disassembly.

: N·m (kg-m, ft-lb)

: N·m (kg-m, in-lb)

1. Heated oxygen sensor 1 (bank 1)
2. Exhaust manifold cover
3. Exhaust manifold (RH bank)
4. Gasket
5. Gasket
6. Three way catalyst (manifold) (RH bank)
7. Heated oxygen sensor 2 (bank 1)
8. Support (RH)
9. Three way catalyst heat shield
10. Three way catalyst (manifold) (LH bank)
11. Heated oxygen sensor 2 (bank 2)
12. Support (LH)
13. Exhaust manifold cover
14. Exhaust manifold (LH bank)
15. Heated oxygen sensor 1 (bank 2)

42356-MURA-G14

Exhaust manifold removal and installation

RH bank

LH bank

Engine front ⇨

⇦ Engine front

42356-MURA-G15

42356-MURA-G16

Exhaust manifold bolt torque sequence—right side

Exhaust manifold bolt torque sequence—left side

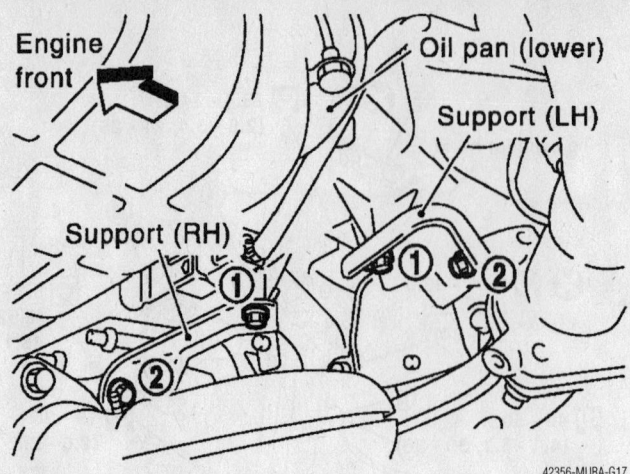

Engine front

Oil pan (lower)

Support (LH)

Support (RH)

42356-MURA-G17

TWC support bolt torque sequence

not tighten to the maximum specified torque range.
- Install the exhaust manifold nuts in the order shown.
- When installing the Three Way Catalyst Supports, install in the order shown.

Camshaft and Valve Lifters

REMOVAL

1. Before servicing the vehicle, refer to the precautions in the beginning of this section.

2. Drain the cooling system.
3. Remove front timing chain case, camshaft sprocket, timing chain and rear timing chain case.
4. If necessary, remove camshaft position sensor (PHASE) (right and left banks) from cylinder head back side.

➡ **Handle carefully to avoid dropping and shocks. Do not disassemble. Do not allow metal powder to adhere to magnetic part at sensor tip. Do not place sensors in a location where they are exposed to magnetism.**

5. Remove the intake valve timing control solenoid valves. Discard the intake valve timing control solenoid valve gaskets and use new gaskets for installation.
6. Remove the intake and exhaust camshaft brackets. Mark the camshafts, camshaft brackets, and bolts so they are placed in the same position and direction for installation.
7. Equally loosen the camshaft bracket bolts in several steps in the numerical order shown.
8. Remove camshaft.
9. Remove valve lifter. Identify installation positions, and store them without mixing them up.
10. Remove secondary timing chain tensioner from cylinder head with its stopper pin attached.

INSPECTION

Camshaft Runout

1. Put V block on precise flat bed, and support No. 2 and No. 4 journal of camshaft.
2. Set dial gauge vertically to No. 3 journal.
3. Turn camshaft to one direction with hands, and measure camshaft runout on dial gauge. (Total indicator reading)
4. If it exceeds the limit, replace camshaft.

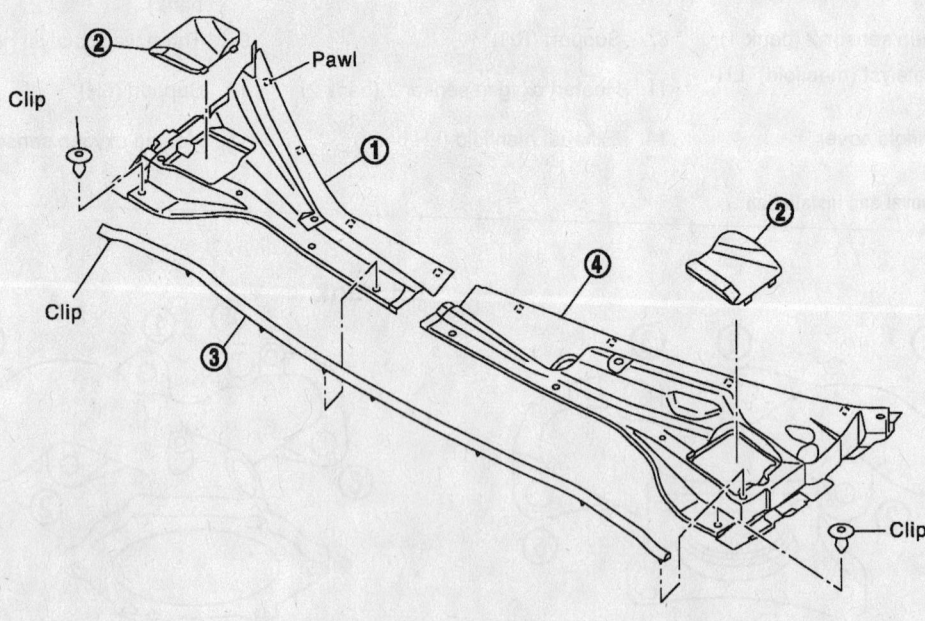

Pawl

Clip

Clip

Clip

1. Cowl top cover (right)
2. Cap (left / right)
3. Cowl top seal rubber
4. Cowl top cover (left)

42356-MURA-G18

Cowl top removal

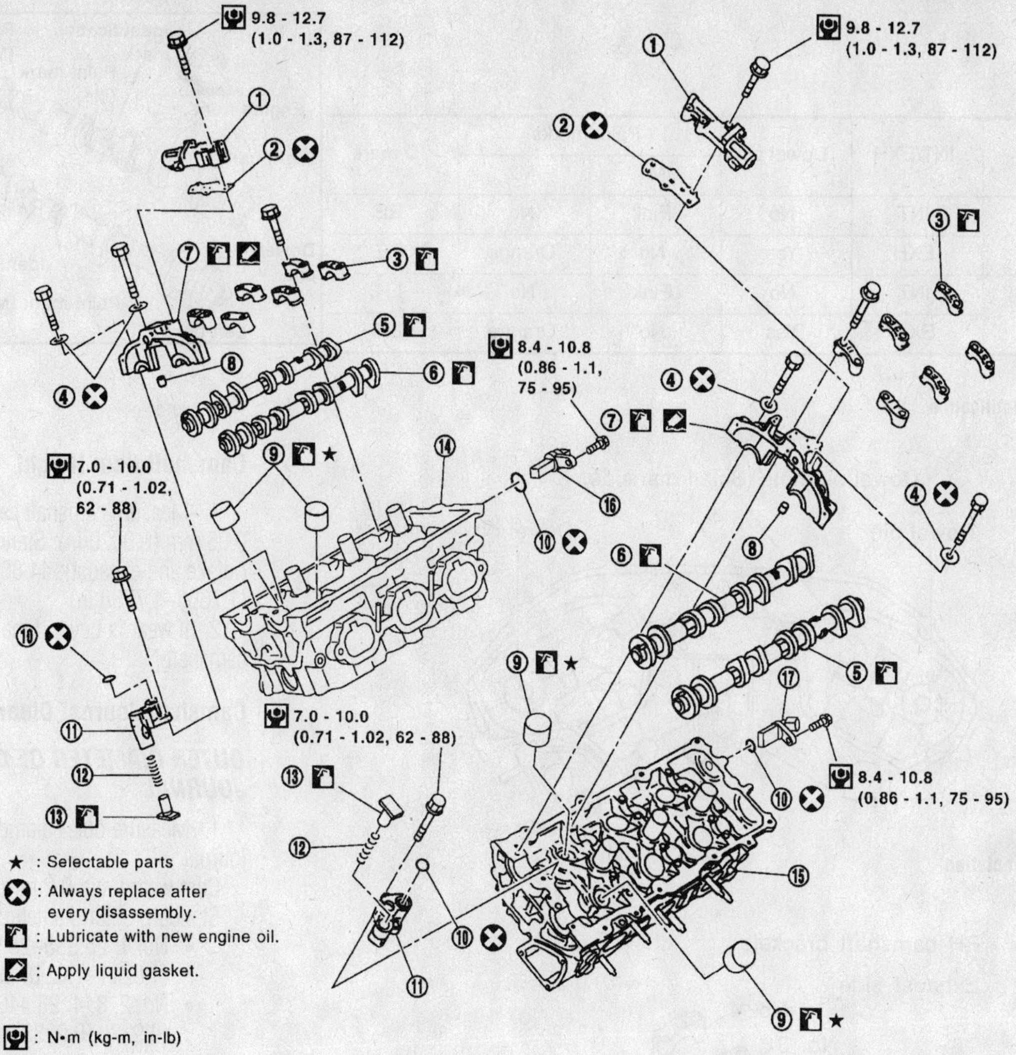

9.8 - 12.7
(1.0 - 1.3, 87 - 112)

9.8 - 12.7
(1.0 - 1.3, 87 - 112)

8.4 - 10.8
(0.86 - 1.1, 75 - 95)

7.0 - 10.0
(0.71 - 1.02, 62 - 88)

7.0 - 10.0
(0.71 - 1.02, 62 - 88)

8.4 - 10.8
(0.86 - 1.1, 75 - 95)

★ : Selectable parts

⊗ : Always replace after every disassembly.

▦ : Lubricate with new engine oil.

▨ : Apply liquid gasket.

▦ : N•m (kg-m, in-lb)

▨ : N•m (kg-m, ft-lb)

1. Intake valve timing control solenoid valve
2. Gasket
3. Camshaft bracket (No.2 to No.4)
4. Seal washer
5. Camshaft (EXH)
6. Camshaft (INT)
7. Camshaft bracket (No.1)
8. Dowel pin
9. Valve lifter
10. O-ring
11. Chain tensioner
12. Spring
13. Plunger
14. Cylinder head (RH bank)
15. Cylinder head (LH bank)
16. Camshaft position sensor (PHASE) (RH bank)
17. Camshaft position sensor (PHASE) (LH bank)

42356-MURA-G19

Camshaft removal and installation

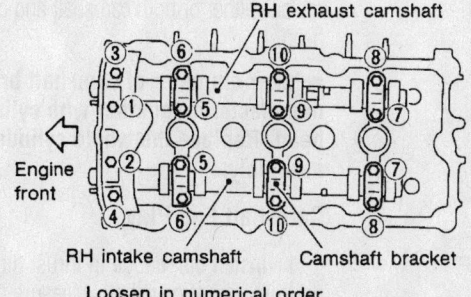

RH exhaust camshaft

Engine front

RH intake camshaft   Camshaft bracket

Loosen in numerical order.

42356-MURA-G20

Camshaft bracket bolt loosening sequence—right side

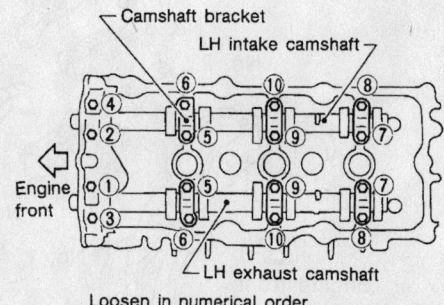

Camshaft bracket
LH intake camshaft

Engine front

LH exhaust camshaft

Loosen in numerical order.

42356-MURA-G21

Camshaft bracket bolt loosening sequence—left side

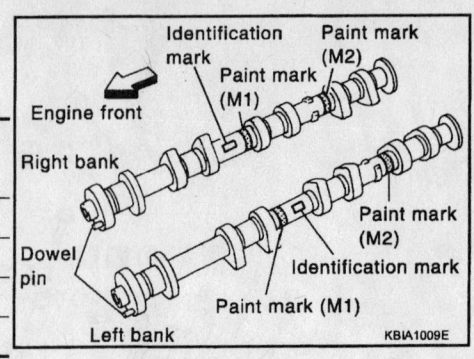

Bank	INT/EXH	Dowel pin	Paint marks		ID mark
			M1	M2	
RH	INT	No	Pink	No	RE
	EXH	Yes	No	Orange	RE
LH	INT	No	Pink	No	LH
	EXH	Yes	No	Orange	LH

Camshaft identification

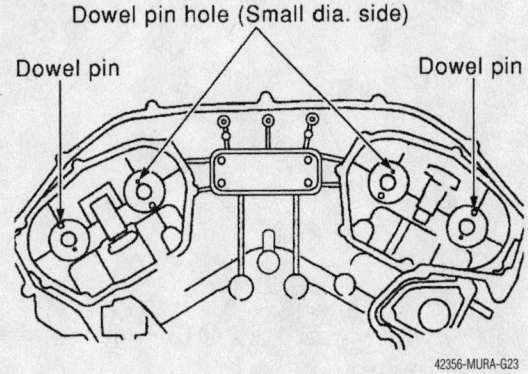

Dowel pin orientation

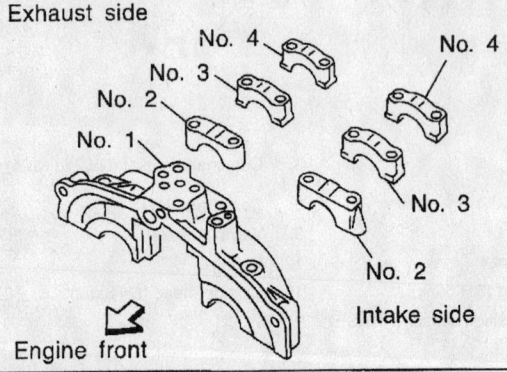

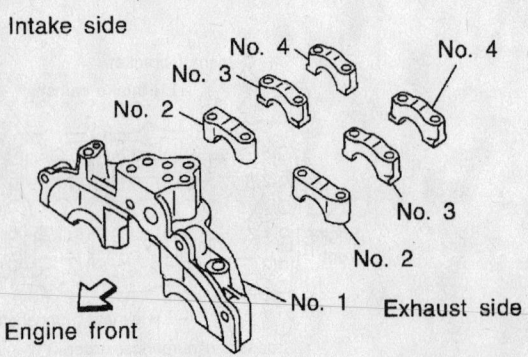

Camshaft bracket identification

Camshaft Cam Height

1. Measure camshaft cam height. Limit: 0.05 mm (0.0020 in). Standard cam height (intake and exhaust): 44.865–45.055 mm (1.7663–1.7738 in)

2. If wear is beyond the limit, replace camshaft.

Camshaft Journal Clearance

OUTER DIAMETER OF CAMSHAFT JOURNAL

1. Measure outer diameter of camshaft journal.
   Cam wear limit: 0.2 mm (0.008 inch)
   Standard outer diameter:
   - No. 1: 25.935–25.955 mm (1.0211–1.0218 in)
   - No. 2, 3, 4: 23.445–23.465 mm (0.9230–0.9238 in)

INNER DIAMETER OF CAMSHAFT BRACKET

1. Tighten camshaft bracket bolt with specified torque.
2. Using inside micrometer, measure inner diameter "A" of camshaft bracket.

CALCULATION OF CAMSHAFT JOURNAL CLEARANCE

Journal clearance = inner diameter of camshaft bracket—outer diameter of camshaft journal. When outside the limit, replace either or both camshaft and cylinder head.

➡ **Inner diameter of camshaft bracket is manufactured together with cylinder head. Replace the whole cylinder head assembly.**

Camshaft End Play

1. Install dial gauge in thrust direction on front end of camshaft. Measure end play of dial gauge when camshaft is moved forward and backward.

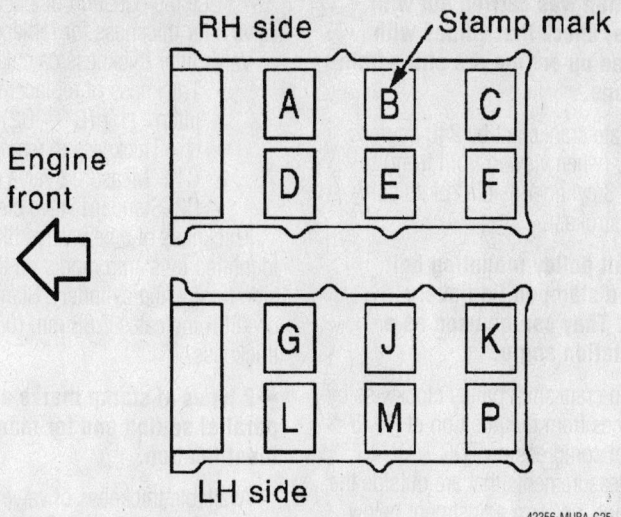

RH side — Stamp mark

Engine front ←

| A | B | C |
| D | E | F |

| G | J | K |
| L | M | P |

LH side

42356-MURA-G25

Align the stamp marks as shown

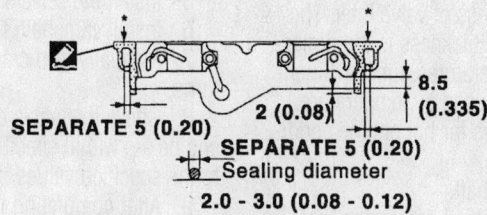

SEPARATE 5 (0.20)

2 (0.08)

8.5 (0.335)

SEPARATE 5 (0.20)
Sealing diameter
2.0 - 3.0 (0.08 - 0.12)

* : Remove the protruding sealant from front face. (Remove the hardended sealant from surface only.)

◢ : Apply liquid gasket

Unit: mm (in)

42356-MURA-G26

RTV sealer application

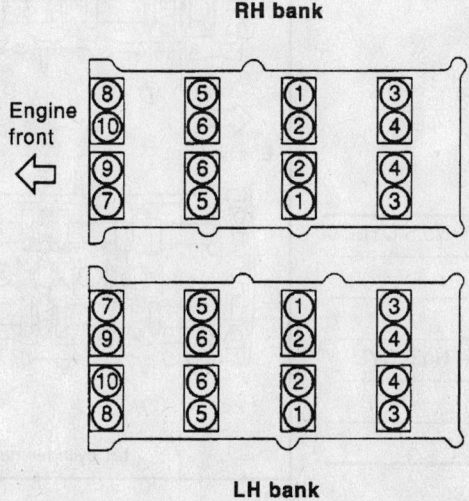

RH bank

Engine front ←

LH bank

42356-MURA-G27

Camshaft bracket bolt torque sequence

2. When out of the limit, replace with new camshaft and measure again.

3. When out of the limit again, replace with new cylinder head.

INSTALLATION

1. Install secondary chain tensioners on both sides of cylinder head.

   a. Install chain tensioner with its stopper pin attached.

   b. Install tensioner with sliding part facing downward on right side cylinder head, and with sliding part facing upward on left side cylinder head.

   c. Install O-ring as shown.

2. Install valve lifters in their original position.

   Valve lifter outer diameter (Intake and exhaust)

   • 33.977–33.987mm (1.3377–1.3381 inch)
   Standard (Intake and exhaust)

   • 34.000–34.016 mm (1.3386–1.3392 inch)
   Standard (Intake and exhaust)

   • 0.013–0.039 mm (0.0005–0.0015 in)

3. Install camshafts.

   a. Install camshaft with dowel pin attached to its front end face on the exhaust side.

   b. Follow your identification marks made during removal, or follow the identification marks that are present on the new camshafts for proper placement and direction.

   c. Install camshaft so that dowel pin hole and dowel pin on front end face are positioned as shown in illustration. (No. 1 cylinder TDC on its compression stroke)

➡**Large- and small-pin holes are located on front end face of intake camshaft, at intervals of 180 degrees. Face small dia. side pin hole upward (in cylinder head upper face direction).**

4. Install camshaft brackets.

   a. Remove foreign material completely from camshaft bracket backside and from cylinder head installation face.

   b. Install camshaft bracket in original position and direction as shown in illustration.

   c. Install No.2 to 4 camshaft brackets aligning the stamp marks as shown.

➡**There are no identification marks indicating left and right for No. 1 camshaft bracket.**

   d. Apply sealant to mating surface of No.1 camshaft bracket as shown on right and left banks.

➡**Use RTV silicone sealant or equivalent.**

5. Tighten the camshaft brackets in the following steps, in numerical order as shown.

a. Tighten No. 7 to 10, then tighten No.1 to 6 in order as shown.

b. Tighten No.1 to 10 in numerical order as shown.

c. Tighten No. 1 to 6 in the numerical order as shown.

d. Tighten No. 7 to 10 in the numerical order as shown.

- 1.96 Nm (17 inch lbs.)
- 5.88 Nm (52 inch lbs.)
- 9.02–11.8 Nm (80–104 inch lbs.)
- 8.3–10.3 Nm (74–91 inch lbs.)

6. Measure difference in levels between front end faces of No. 1 camshaft bracket and cylinder head. If measurement is outside the specified range, re-install camshaft and camshaft bracket.

7. Inspect and adjust valve clearance.

8. Install in the reverse order of removal after this step.

Valve Clearance

Perform inspection as follows after removal, installation or replacement of camshaft or valve-related parts, or if there is unusual engine conditions regarding valve clearance.

1. Remove right and left rocker covers.

2. Measure valve clearance as below:

a. Set No.1 cylinder at TDC of its compression stroke. Align crankshaft pulley timing mark (grooved line without color) with timing indicator. Check that No. 1 cylinder intake and exhaust cam nose is facing in direction shown in illustration. If not, rotate crankshaft pulley 360 degrees clockwise (when viewed from front).

b. Using a feeler gauge, measure valve clearance. Standard: -0.14 to 0.14 mm (-0.0055 to 0.0055 in)

➡**If inspection was carried out with cold engine, check that values with fully warmed up engine are still within specifications.**

c. Rotate crankshaft by 240 degrees clockwise (when viewed from front) to align No. 3 cylinder at TDC of its compression stroke.

➡**Crankshaft pulley mounting bolt flange has a stamped line every 60degrees. They can be used as a guide to rotation angle.**

d. Turn crankshaft pulley clockwise by 240 degrees from the position of No. 5 cylinder at compression TDC.

3. For measurements that are outside the specified range, perform adjustment below.

ADJUSTMENT

Perform adjustment depending on selected head thickness of valve lifter. The specified valve lifter thickness is the dimension at normal temperatures. Ignore dimensional differences caused by temperature. Use the specifications for hot engine condition to adjust.

1. Remove camshaft.

2. Remove the valve lifters at the locations that are outside the standard.

3. Measure the center thickness of the removed valve lifters with a micrometer.

4. Use the equation below to calculate valve lifter thickness for replacement. Valve lifter thickness calculation:

- Thickness of replacement valve lifter = t1 + (C1—C2)
- t1 = Thickness of removed valve lifter
- C1 = Measured valve clearance
- C2=Standard valve clearance:

Thickness of a new valve lifter can be identified by stamp marks on the reverse side (inside the cylinder). Stamp mark 788U or 788R indicates 7.88 mm (0.3102 in) in thickness.

➡**2 types of stamp marks are used for parallel setting and for manufacturer identification.**

Available thickness of valve lifter: 27 sizes with range 7.88 to 8.40 mm (0.3102 to 0.3307 in) in steps of 0.02 mm (0.0008 in) (when manufactured at factory).

5. Install the selected valve lifter.

6. Install camshaft.

7. Manually turn crankshaft pulley a few turns.

8. Check that valve clearances for cold engine are within specifications by referring to the specified values.

9. After completing the repair, check valve clearances again with the specifications for warmed engine. Make sure the values are within specifications.

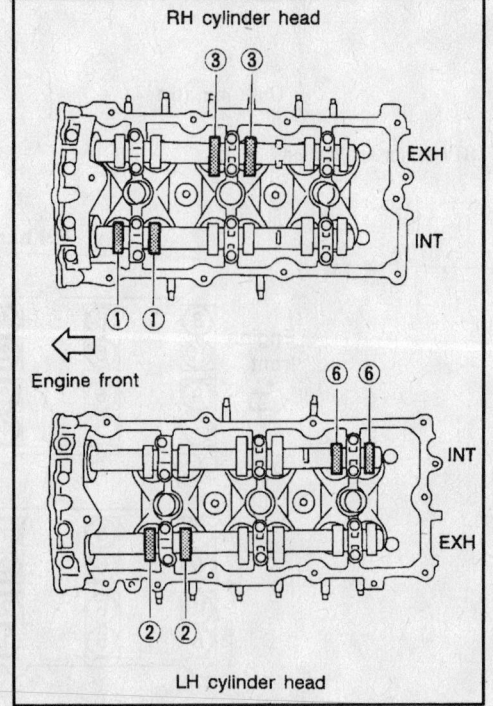

Measuring position (RH bank)		No.1 CYL.	No.3 CYL.	No.5 CYL.
No.1 cylinder at TDC	EXH		×	
	INT	×		
Measuring position (LH bank)		No.2 CYL.	No.4 CYL.	No.6 CYL.
No.1 cylinder at TDC	INT			×
	EXH	×		

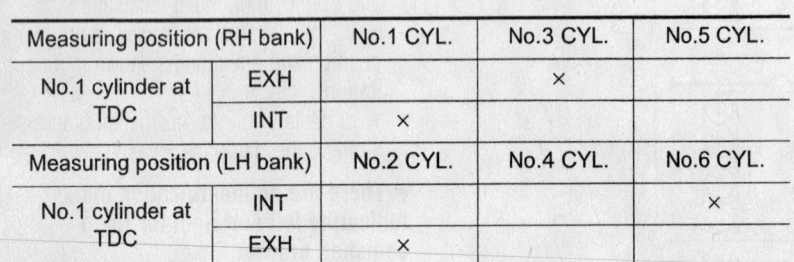

Valve clearance inspection—No.1 TDC

42356-MURA-G28

Measuring position (RH bank)		No.1 CYL.	No.3 CYL.	No.5 CYL.
No.3 cylinder at TDC	EXH			×
	INT		×	
Measuring position (LH bank)		No.2 CYL.	No.4 CYL.	No.6 CYL.
No.3 cylinder at TDC	INT	×		
	EXH		×	

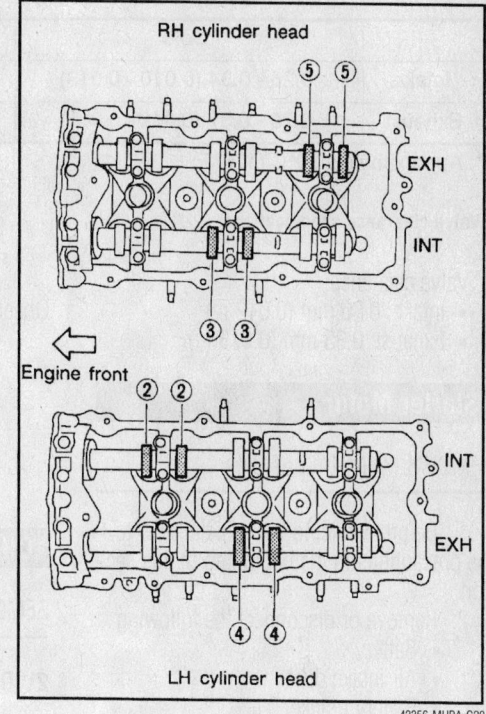

42356-MURA-G29

Valve clearance inspection—No.3 TDC

Measuring position (RH bank)		No.1 CYL.	No.3 CYL.	No.5 CYL.
No.5 cylinder at TDC	EXH	×		
	INT			×
Measuring position (LH bank)		No.2 CYL.	No.4 CYL.	No.6 CYL.
No.5 cylinder at TDC	INT		×	
	EXH			×

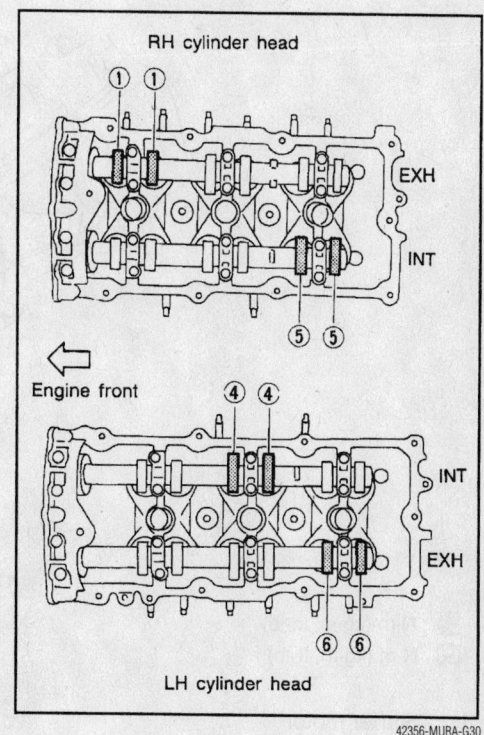

42356-MURA-G30

Valve clearance inspection—No.5 TDC

Valve clearance:

Unit: mm (in)

	Cold	Hot * (reference data)
Intake	0.26 - 0.34 (0.010 - 0.013)	0.304 - 0.416 (0.012 - 0.016)
Exhaust	0.29 - 0.37 (0.011 - 0.015)	0.308 - 0.432 (0.012 - 0.016)

*: Approximately 80°C (176°F)

42356-MURA-G31

Valve clearance specifications

Valve clearance:
• Intake: 0.30 mm (0.012 in)
• Exhaust: 0.33 mm (0.013 in)

Starter Motor

REMOVAL & INSTALLATION

1. Before servicing the vehicle, refer to the precautions in the beginning of this section.
2. Remove or disconnect the following:
   • Battery
   • Air intake duct
   • Battery bracket
   • Fluid charging pipe
   • S connector
   • B terminal nut
   • Starter motor mounting bolts
   • Starter motor to the direction of upper side the vehicle

3. Install in the reverse order of removal. Observe the following toques:
   • B terminal nut: 9.8–11.8 Nm (87–104 inch lbs.)
   • Starter motor mounting bolt: 47–63 Nm (35–46 ft. lbs.)
   • Battery bracket mounting bolt: 14–20 Nm (10–15 ft. lbs.)

Oil Pan

REMOVAL & INSTALLATION

2WD Models

1. Before servicing the vehicle, refer to the precautions in the beginning of this section.

➡**When removing the upper oil pan from the engine, first remove the** crankshaft position sensor (POS). Be careful not to damage sensor edges or signal plate teeth.

2. Remove engine cover.
3. Remove right splash guard.
4. Remove the front right road wheel and tire.
5. Drain engine oil.
6. Drain engine coolant.
7. Remove oil filter.
8. Remove oil cooler and water pipes.
9. Remove all drive belts.
10. Remove A/C compressor with piping connected, and temporarily secure it aside.
11. Remove exhaust front tube.
12. Remove the heated oxygen sensor 2 (bank 2) and remove the three way catalyst (manifold) (bank 2) from the exhaust manifold.
13. Loosen lower oil pan bolts in reverse order of illustration.
14. Insert a seal cutter (special service tool) between the lower oil pan and the upper oil pan.

➡**Be careful not to damage the mating surface. Do not insert a screwdriver; this will damage the mating surfaces.**

15. Slide seal cutter (special service tool) by tapping on the side of the tool with a hammer.
16. Remove lower oil pan.

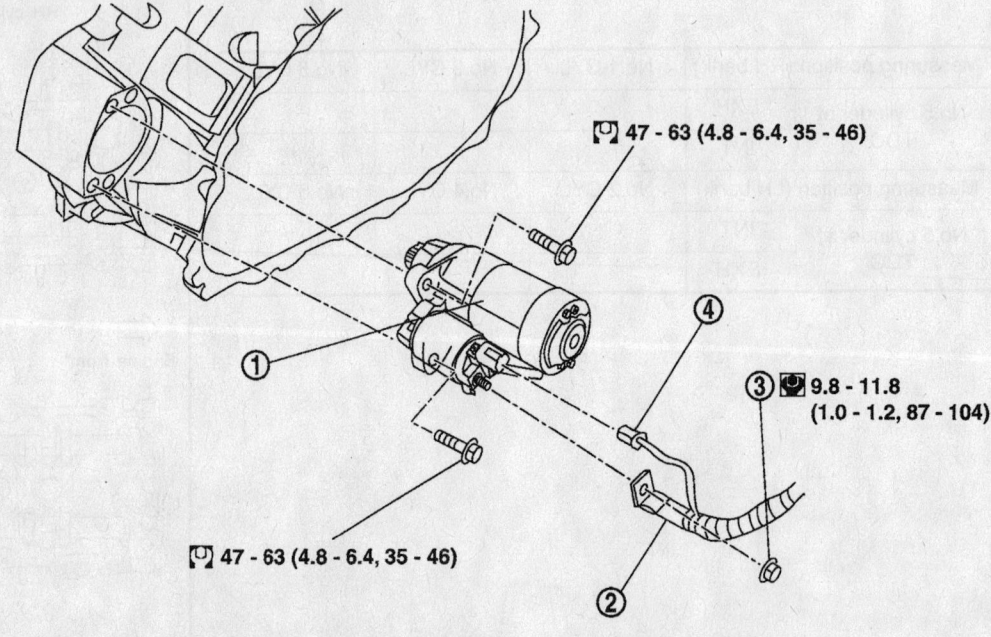

🔧 : N·m (kg-m, in-lb)
🔧 : N·m (kg-m, ft-lb)

1. Starter motor
2. B terminal harness
3. B terminal nut
4. S connector

Starter removal

42356-MURA-G32

17. Remove oil strainer.
18. Remove the oil pressure switch.
19. Remove crankshaft position sensor (POS).

➡**Handle carefully to avoid dropping and shocks. Do not disassemble. Do not allow metal powder to adhere to magnetic part at sensor tip. Do not place sensors in a location where they are exposed to magnetism.**

20. Remove the four engine-to-transaxle bolts.

21. Remove upper oil pan. Loosen bolts in reverse order shown.

22. Insert an appropriate size tool into the notch of the upper oil pan shown (1). Pry off the upper oil pan by moving the tool up and down shown (2).

23. Remove O-rings from the bottom of the cylinder block and oil pump body.

24. Remove oil pan gasket.

To install:
Installation is the reverse of removal. Note the following:

• Use a scraper to remove old liquid gasket from mating surfaces.

• Also remove the old liquid gasket from mating surface of the cylinder block.

• Remove the old liquid gasket from the bolt holes and threads.

➡**Do not scratch or damage the mating surfaces when cleaning off the old liquid gasket.**

• Apply Genuine RTV Silicone Sealant or equivalent, to the front timing chain case gasket and the rear oil seal retainer gasket shown.

• To install, align protrusion of oil pan gasket with notches of front timing chain case and rear oil seal retainer.

• Install oil pan gasket with smaller arc to front timing chain case side.

• Install new O-rings on the cylinder block and oil pump side.

• Apply a continuous bead of sealant to the cylinder block mating surface of the upper oil pan to a limited portion shown. Use RTV silicone sealant or equivalent.

• For bolt holes with marks (5 locations), apply liquid gasket outside the holes.

• Apply a bead of 4.5 to 5.5 mm (0.177 to 0.217 in) diameter to area "A".

• Attaching within 5 minutes after coating.

• Install the upper oil pan. Tighten bolts in numerical order shown. There are two

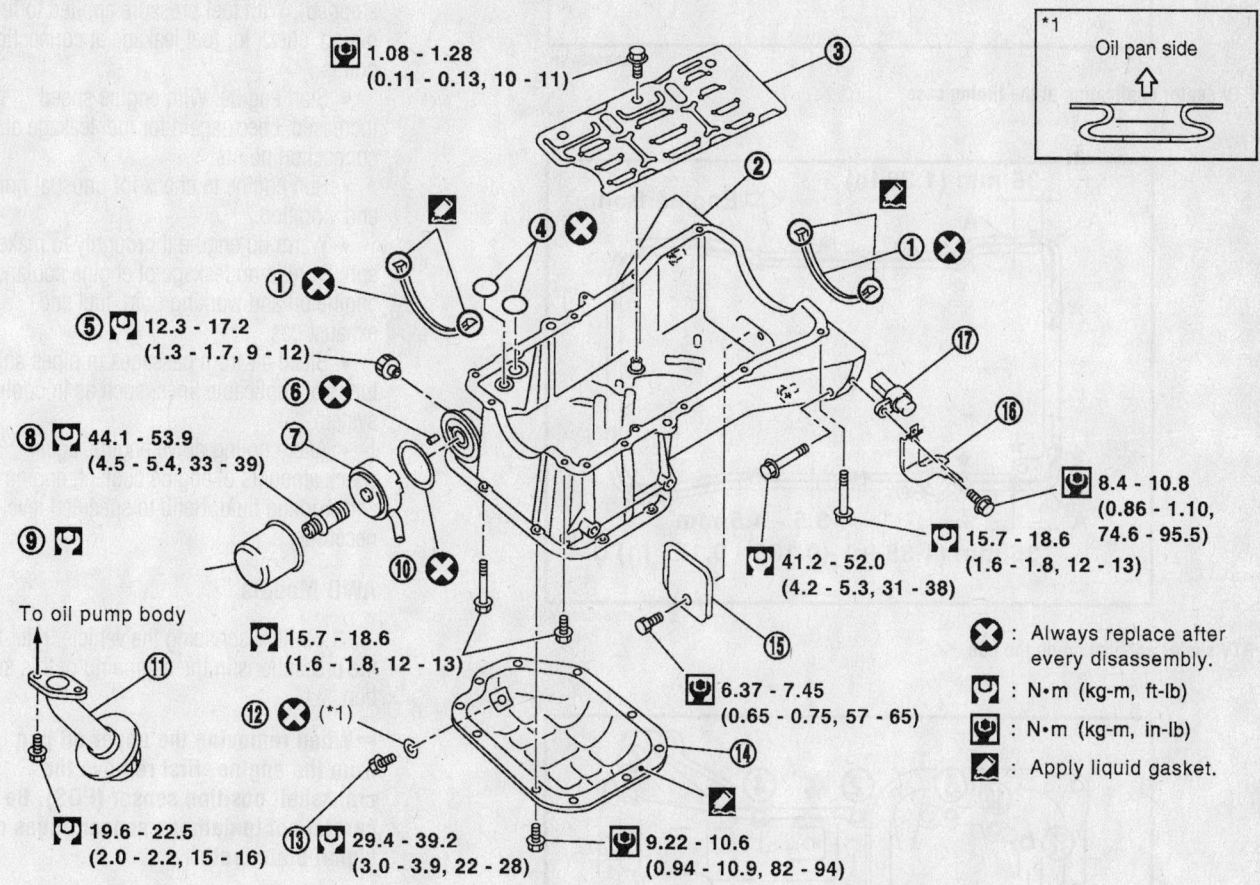

*1 Oil pan side

1.08 - 1.28
(0.11 - 0.13, 10 - 11)

⑤ 12.3 - 17.2
(1.3 - 1.7, 9 - 12)

⑧ 44.1 - 53.9
(4.5 - 5.4, 33 - 39)

⑨

To oil pump body

15.7 - 18.6
(1.6 - 1.8, 12 - 13)

⑫ ✕ (*1)

19.6 - 22.5
(2.0 - 2.2, 15 - 16)

⑬ 29.4 - 39.2
(3.0 - 3.9, 22 - 28)

9.22 - 10.6
(0.94 - 10.9, 82 - 94)

6.37 - 7.45
(0.65 - 0.75, 57 - 65)

41.2 - 52.0
(4.2 - 5.3, 31 - 38)

8.4 - 10.8
(0.86 - 1.10, 74.6 - 95.5)

15.7 - 18.6
(1.6 - 1.8, 12 - 13)

✕ : Always replace after every disassembly.
🔧 : N•m (kg-m, ft-lb)
🔧 : N•m (kg-m, in-lb)
📝 : Apply liquid gasket.

1.	Gasket	2.	Upper oil pan	3.	Baffle plate
4.	O-ring	5.	Oil pressure switch	6.	Relief valve
7.	Oil cooler	8.	Oil cooler connector	9.	Oil filter
10.	Gasket	11.	Oil strainer	12.	Gasket
13.	Drain plug	14.	Lower oil pan	15.	Rear cover plate
16.	Heated oxygen sensor (bank 2) harness clamp (2WD models)	17.	Crankshaft position sensor (POS)		

Oil pan exploded view

42356-MURA-G33

types of mounting bolts. Refer to the accompanying illustration.

• Install the four engine-to-transaxle bolts.

• Install oil strainer to oil pump.

• Use a scraper to remove old liquid gasket from mating surfaces. Also remove old liquid gasket from mating surface of upper oil pan.

• Apply a continuous bead of sealant to the lower oil pan. Use RTV silicone sealant. Be sure the sealant is 4.5–5.5 mm (0.177–0.217 inch) wide. Attach within 5 minutes after coating.

• Install lower oil pan. Tighten the bolts in the numerical order shown.

• Install oil pan drain plug.

• Refer to illustration for installation of washer.

• Install in the reverse order of removal after this step.

• Wait at least 30 minutes after oil pan is installed, before adding engine oil.

• Before starting engine, check the levels of engine coolant, engine oil and working fluid. If less than required quantity, fill to the specified level.

• Use procedure below to check for fuel leakage.

• Turn ignition switch ON (with engine stopped). With fuel pressure applied to fuel piping, check for fuel leakage at connection points.

• Start engine. With engine speed increased, check again for fuel leakage at connection points.

• Run engine to check for unusual noise and vibration.

• Warm up engine thoroughly to make sure there is no leakage of engine coolant, engine oil and working fluid, fuel and exhaust gas.

• Bleed air from passages in pipes and tubes of applicable lines, such as in cooling system.

• After cooling down engine, again check amounts of engine coolant, engine oil and working fluid. Refill to specified level, if necessary.

AWD Models

1. Before servicing the vehicle, refer to the precautions in the beginning of this section.

➡ **When removing the upper oil pan from the engine, first remove the crankshaft position sensor (POS). Be careful not to damage sensor edges or signal plate teeth.**

2. Remove engine assembly from vehicle, and separate front suspension member, transaxle and transfer case assembly from engine.

3. Install an engine crane.

4. Remove the engine and mount it onto the engine stand.

5. Drain engine oil.

6. Remove oil filter.

7. Remove oil cooler and water pipes.

8. Remove the heated oxygen sensor 2

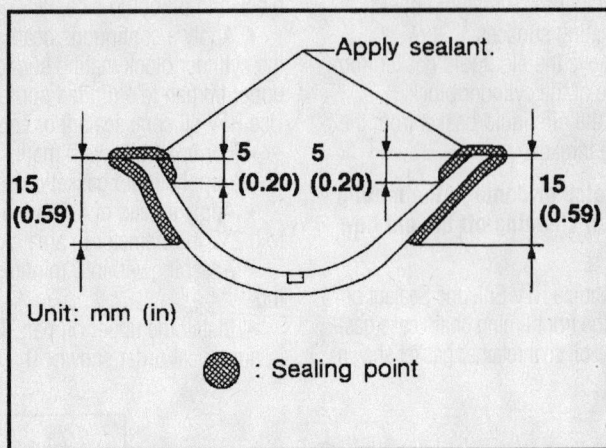

RTV sealer application at the timing case

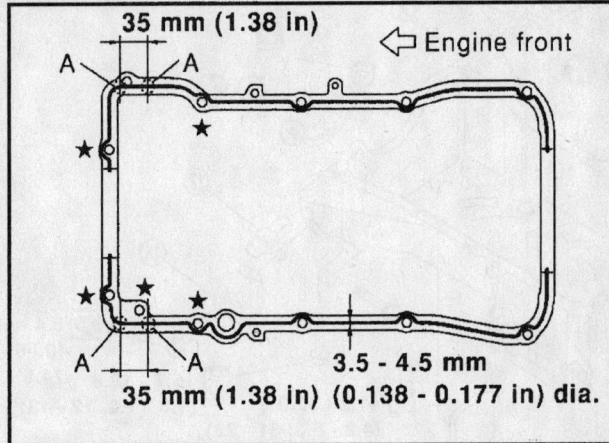

RTV sealer application on the pan

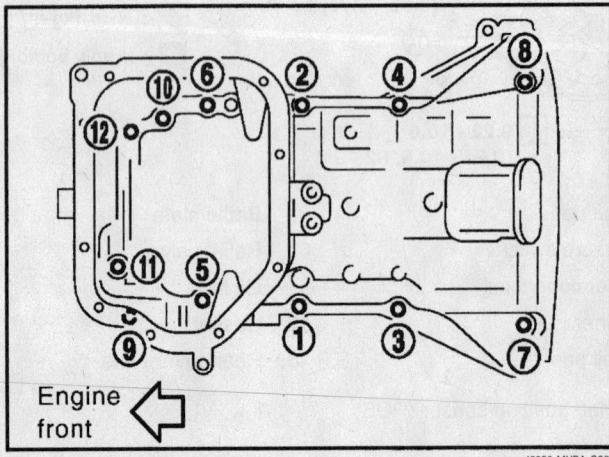

Oil pan bolt torque sequence

(bank 2) and remove the three way catalyst (manifold) (bank 2) from the exhaust manifold.

9. Loosen lower oil pan bolts in reverse order of illustration.

10. Insert a seal cutter (special service tool) between the lower oil pan and the upper oil pan.

→**Be careful not to damage the mating surface. Do not insert a screwdriver, this will damage the mating surfaces.**

11. Slide seal cutter (special service tool) by tapping on the side of the tool with a hammer. Remove lower oil pan.

12. Remove oil strainer.

13. Remove the oil pressure switch.

14. Remove upper oil pan. Loosen bolts in reverse order shown.

15. Insert an appropriate size tool into the notch of the upper oil pan shown (1). Pry off the upper oil pan by moving the tool up and down shown (2).

16. Remove O-rings from the bottom of the cylinder block and oil pump body.

17. Remove oil pan gasket.

To install:
Installation is the reverse of removal. Note the following:

• Use a scraper to remove old liquid gasket from mating surfaces.

• Also remove the old liquid gasket from mating surface of the cylinder block.

• Remove the old liquid gasket from the bolt holes and threads.

→**Do not scratch or damage the mating surfaces when cleaning off the old liquid gasket.**

• Apply Genuine RTV Silicone Sealant or equivalent, to the front timing chain case gasket and the rear oil seal retainer gasket shown.

• To install, align protrusion of oil pan gasket with notches of front timing chain case and rear oil seal retainer.

• Install oil pan gasket with smaller arc to front timing chain case side.

• Install new O-rings on the cylinder block and oil pump side.

• Apply a continuous bead of sealant to the cylinder block mating surface of the upper oil pan to a limited portion shown. Use RTV silicone sealant or equivalent.

• For bolt holes with marks (5 locations), apply liquid gasket outside the holes.

• Apply a bead of 4.5 to 5.5 mm (0.177 to 0.217 in) diameter to area "A".

• Attaching within 5 minutes after coating.

• Install the upper oil pan. Tighten bolts in numerical order shown. There are two types of mounting bolts. Refer to the accompanying illustration.

• Install oil strainer to oil pump.

• Use a scraper to remove old liquid gasket from mating surfaces. Also remove old liquid gasket from mating surface of upper oil pan.

• Apply a continuous bead of sealant to the lower oil pan. Use RTV silicone sealant. Be sure the sealant is 4.5–5.5 mm (0.177–0.217 inch) wide. Attach within 5 minutes after coating.

• Install lower oil pan. Tighten the bolts in the numerical order shown.

• Install oil pan drain plug.

• Refer to illustration for installation of washer.

• Install in the reverse order of removal.

• Wait at least 30 minutes after oil pan is installed, before adding oil.

• Before starting engine, check the levels of engine coolant, engine oil and working fluid. If less than required quantity, fill to the specified level.

• Use procedure below to check for fuel leakage.

• Turn ignition switch ON (with engine stopped). With fuel pressure applied to fuel piping, check for fuel leakage at connection points.

• Start engine. With engine speed increased, check again for fuel leakage at connection points.

• Run engine to check for unusual noise and vibration.

• Warm up engine thoroughly to make sure there is no leakage of engine coolant, engine oil and working fluid, fuel and exhaust gas.

• Bleed air from passages in pipes and tubes of applicable lines, such as in cooling system.

• After cooling down engine, again check amounts of engine coolant, engine oil and working fluid. Refill to specified level, if necessary.

Oil Pump

REMOVAL & INSTALLATION

1. Before servicing the vehicle, refer to the precautions in the beginning of this section.

2. Remove oil pan and oil strainer.

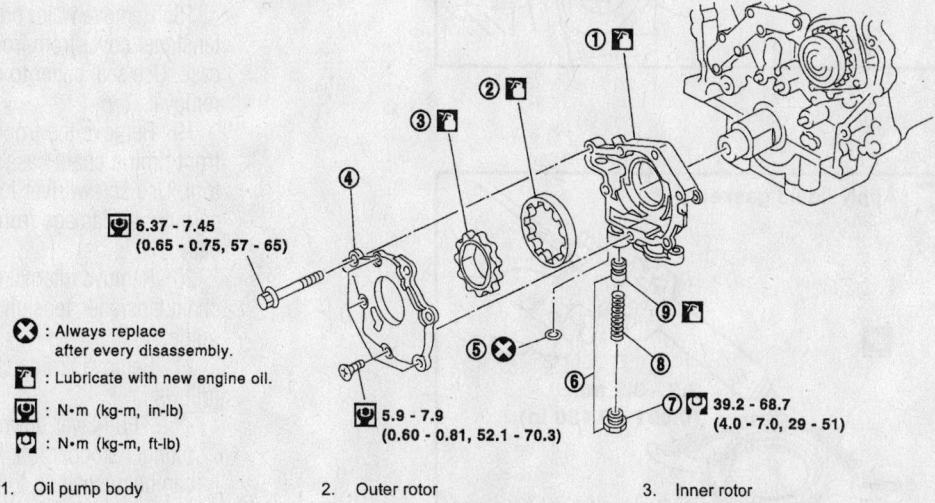

6.37 - 7.45
(0.65 - 0.75, 57 - 65)

⊗ : Always replace after every disassembly.

🔖 : Lubricate with new engine oil.

🔧 : N•m (kg-m, in-lb)

🔧 : N•m (kg-m, ft-lb)

5.9 - 7.9
(0.60 - 0.81, 52.1 - 70.3)

39.2 - 68.7
(4.0 - 7.0, 29 - 51)

1.	Oil pump body	2.	Outer rotor	3.	Inner rotor
4.	Oil pump cover	5.	O-ring	6.	Regulator valve set
7.	Regulator valve plug	8.	Spring	9.	Regulator valve

42356-MURA-G37

Oil pump exploded view

3. Remove front timing chain case and timing chain (primary).

4. Remove oil pump assembly.

5. Installation is the reverse of removal. When installing, align crankshaft flat faces with inner rotor flat faces.

Rear Main Seal

REMOVAL & INSTALLATION

1. Before servicing the vehicle, refer to the precautions in the beginning of this section.

2. Remove engine from vehicle, and separate front suspension member, transaxle and transfer case (AWD models) assembly from engine.

3. Remove drive plate.

4. Remove upper oil pan.

5. Use a seal cutter (special service tool) to cut away liquid gasket and remove rear oil seal retainer.

To install:

6. Remove old liquid gasket from mating surface of cylinder block and oil pan using scraper.

7. Apply liquid gasket to rear oil seal retainer as shown in the illustration. Use

RTV silicone sealant. Assembly should be done within 5 minutes after coating.

8. Install rear oil seal retainer to cylinder block.

9. Perform the remaining steps in the reverse order of removal.

Front Case and Seal, Timing Chain, Sprockets, Rear Case

REMOVAL & INSTALLATION

1. Before servicing the vehicle, refer to the precautions in the beginning of this section.

2. Remove engine assembly from vehicle, and separate front suspension member, transaxle and transfer case (AWD models) assembly from engine.

3. Drain engine oil.

4. Remove engine harnesses.

5. Remove water hoses.

6. Remove EVAP canister purge volume control solenoid valve.

7. Remove drive belts and idler pulley bracket.

8. Remove power steering oil pump assembly.

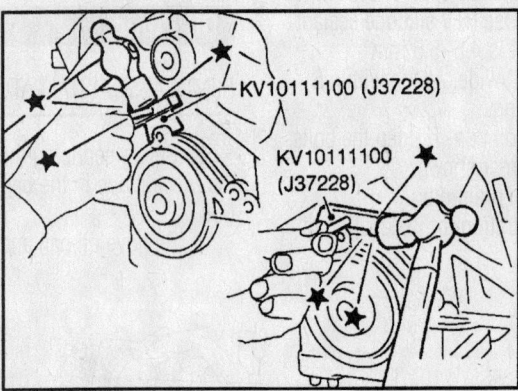

9. Remove alternator.

10. Remove right and left rocker covers.

11. Remove crankshaft position sensor (POS).

➡ **Handle carefully to avoid dropping and shocks. Do not disassemble. Do not allow metal powder to adhere to magnetic part at sensor tip. Do not place sensors in a location where they are exposed to magnetism.**

12. Obtain compression TDC of No.1 cylinder as follows. Rotate crankshaft pulley clockwise to align timing mark (grooved line without color) with timing indicator.

13. Remove lower and upper oil pans.

14. Remove crankshaft pulley as follows:

   a. Lock crankshaft with a hammer handle or similar tool to loosen bolts.

   b. Remove crankshaft pulley with a suitable puller.

15. Remove the right and left intake valve timing control covers. Loosen bolts in reverse order as shown. Use seal cutter to cut liquid gasket for removal.

➡ **Shaft is internally jointed with intake camshaft sprocket center hole. When removing, keep it horizontal until it is completely disconnected.**

16. Remove right engine mounting bracket.

17. Remove front timing chain case.

   a. Loosen mounting bolts in reverse order as shown.

   b. Insert the appropriate size tool into the notch at the top of the front timing chain case as shown (1).

   c. Pry off the case by moving the tool as shown (2). Use seal cutter to cut liquid gasket for removal.

18. Remove water pump cover and chain tensioner cover from front timing chain case. Use seal cutter to cut liquid gasket for removal.

19. Remove the front oil seal from the front timing chain case using a suitable tool. Use screwdriver for removal. Exercise care not to damage front timing chain case.

20. Remove internal chain guide, timing chain tensioner, tension guide and slack guide.

21. Remove timing chain tensioner as follows:

   a. Pull lever down and release plunger stopper tab. Plunger stopper tab can be pushed up to release (coaxial structure with lever).

   b. Insert stopper pin into tensioner body hole to hold lever, and keep the tab released.

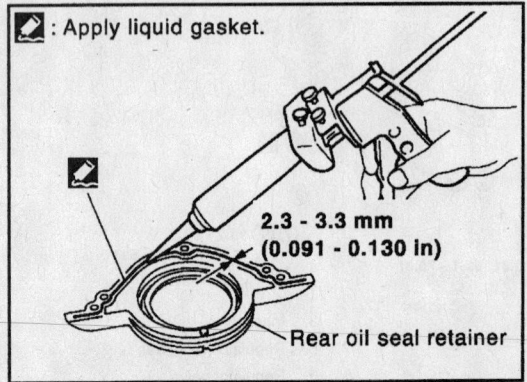

: Apply liquid gasket.

2.3 - 3.3 mm
(0.091 - 0.130 in)

Rear oil seal retainer

42356-MURA-G38

Applying sealer to the rear main seal

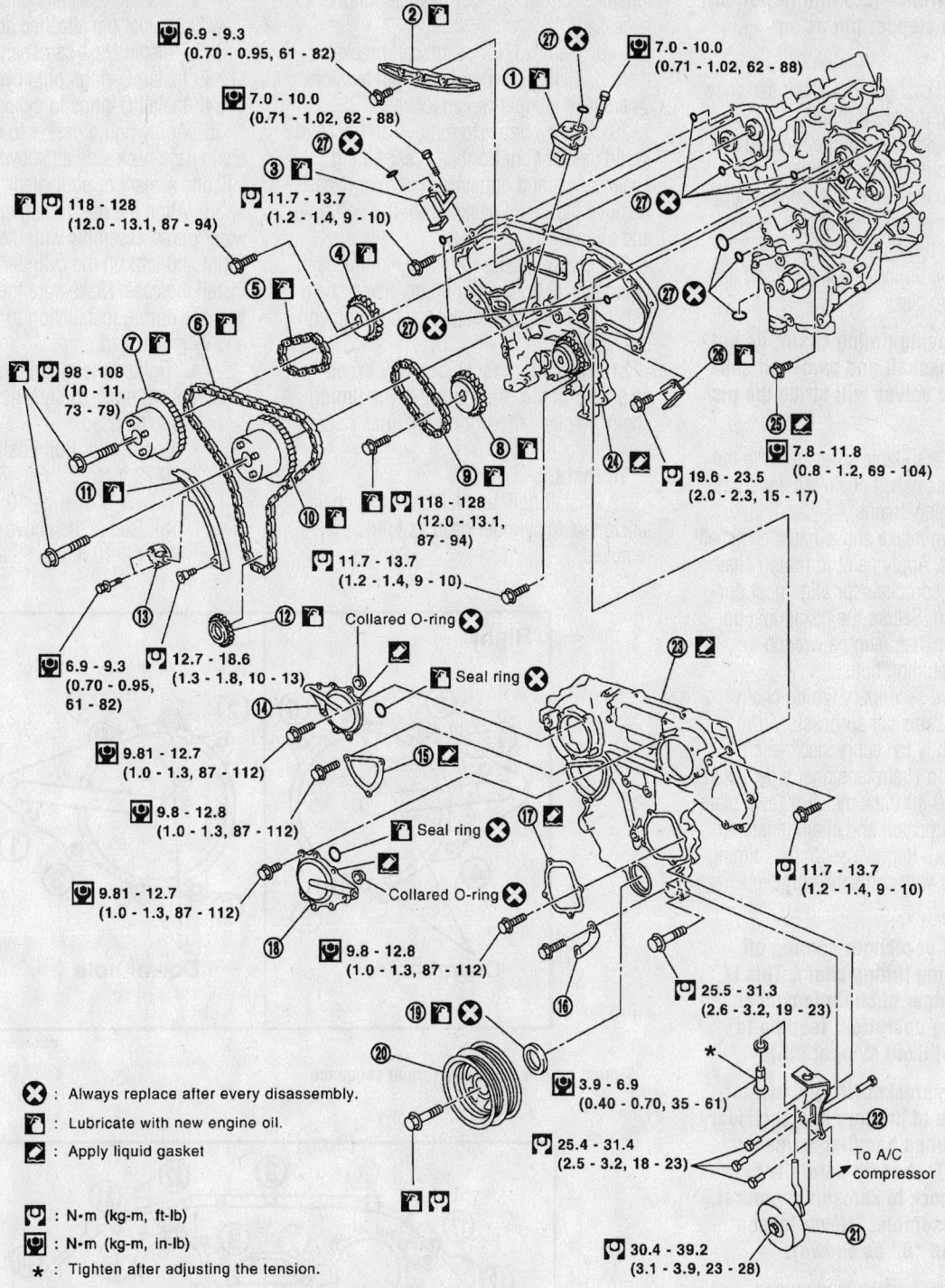

6.9 - 9.3
(0.70 - 0.95, 61 - 82)

7.0 - 10.0
(0.71 - 1.02, 62 - 88)

7.0 - 10.0
(0.71 - 1.02, 62 - 88)

118 - 128
(12.0 - 13.1, 87 - 94)

11.7 - 13.7
(1.2 - 1.4, 9 - 10)

98 - 108
(10 - 11, 73 - 79)

118 - 128
(12.0 - 13.1, 87 - 94)

11.7 - 13.7
(1.2 - 1.4, 9 - 10)

19.6 - 23.5
(2.0 - 2.3, 15 - 17)

7.8 - 11.8
(0.8 - 1.2, 69 - 104)

6.9 - 9.3
(0.70 - 0.95, 61 - 82)

12.7 - 18.6
(1.3 - 1.8, 10 - 13)

Collared O-ring

Seal ring

9.81 - 12.7
(1.0 - 1.3, 87 - 112)

9.8 - 12.8
(1.0 - 1.3, 87 - 112)

Seal ring

11.7 - 13.7
(1.2 - 1.4, 9 - 10)

9.81 - 12.7
(1.0 - 1.3, 87 - 112)

Collared O-ring

9.8 - 12.8
(1.0 - 1.3, 87 - 112)

25.5 - 31.3
(2.6 - 3.2, 19 - 23)

3.9 - 6.9
(0.40 - 0.70, 35 - 61)

*

25.4 - 31.4
(2.5 - 3.2, 18 - 23)

To A/C
compressor

30.4 - 39.2
(3.1 - 3.9, 23 - 28)

❌ : Always replace after every disassembly.

🛢 : Lubricate with new engine oil.

🔧 : Apply liquid gasket

🔩 : N·m (kg-m, ft-lb)

🔩 : N·m (kg-m, in-lb)

★ : Tighten after adjusting the tension.

1. Timing chain tensioner (secondary)	2. Internal chain guide	3. Timing chain tensioner (secondary)
4. Camshaft sprocket (EXH)	5. Timing chain (secondary)	6. Timing chain (primary)
7. Camshaft sprocket (INT)	8. Camshaft sprocket (EXH)	9. Timing chain (secondary)
10. Camshaft sprocket (INT)	11. Slack guide	12. Crankshaft sprocket
13. Timing chain tensioner (primary)	14. Intake valve timing control cover	15. Chain tensioner cover
16. Water hose clamp	17. Water pump cover	18. Intake valve timing control cover
19. Front oil seal	20. Crankshaft pulley	21. Idler pulley
22. Idler pulley bracket	23. Front timing chain case	24. Rear timing chain case
25. Water drain plug	26. Tension guide	27. O-ring

42356-MURA-G39

Timing chain and related components

➡**An Allen wrench [2.5 mm (0.098 in)] is used for a stopper pin as an example.**

c. Insert plunger into tensioner body by pressing the slack guide.

d. Keep the slack guide pressed and hold it by pushing the stopper pin through the lever hole and body hole.

e. Remove the mounting bolts and remove the timing chain tensioner.

22. Remove timing chain (primary) and crankshaft sprocket.

➡**After removing timing chain, do not turn the crankshaft and camshaft separately, or the valves will strike the piston heads.**

23. Attach a suitable stopper pin to the right and left camshaft chain tensioners (for secondary timing chains).

24. Remove intake and exhaust camshaft sprocket bolts. Apply paint to timing chain and camshaft sprockets for alignment during installation. Secure the hexagonal portion of the camshaft using a wrench to loosen the mounting bolts.

25. Remove secondary timing chain together with camshaft sprockets. Turn camshaft slightly to secure slackness of timing chain on chain tensioner side. Insert 0.5 mm (0.020 in) thick metal or resin plate between timing chain and chain tensioner plunger (guide). Remove secondary timing chain together with camshaft sprockets from guide groove.

➡**Be careful of plunger coming off when removing timing chain. This is because plunger of chain tensioner moves during operation, leading to coming off of fixed stopper pin.**

➡**Camshaft sprocket (INT) is two-for-one structure of primary and secondary sprockets. When handling camshaft sprocket (INT), handle carefully to avoid any shock to camshaft sprocket. Do not disassemble. (Never loosen bolts "A" and "B" as shown).**

26. Remove chain tension guide.

27. Remove rear timing chain case as follows:

a. Loosen and remove mounting bolts in reverse order as shown.

b. Cut the sealant using a seal cutter and remove rear timing chain case.

➡**Do not remove plate metal cover of oil passage. After removing chain case, do not apply any load which affects flatness.**

28. Remove right and left camshaft chain

tensioners from cylinder head as follows if necessary.

a. Remove No.1 camshaft brackets.

b. Remove secondary chain tensioners with stopper pin attached.

29. Use a scraper to remove all traces of liquid gasket from front and rear timing chain cases, and opposite mating surfaces. Remove old liquid gasket from the bolt hole and thread.

30. Use a scraper to remove all traces of liquid gasket from water pump cover, chain tensioner cover and intake valve timing control covers.

31. Check for cracks and any excessive wear at the roller links of the timing chain. Replace the timing chain as necessary.

To install:

32. Install right and left camshaft chain tensioners to cylinder head as follows if removed.

a. Install secondary chain tensioners with stopper pin attached and new O-ring.

b. Install No.1 camshaft brackets.

33. Install O-rings onto cylinder block.

34. Install O-rings to cylinder head.

35. Apply liquid gasket to rear timing chain case back side as shown. Use RTV silicone sealant or equivalent.

36. Align the rear timing chain case and water pump assembly with the dowel pins (right and left) on the cylinder block and install the case. Make sure the O-rings stay in place during installation to cylinder block and cylinder head.

a. Tighten the mounting bolts in the numerical order as shown. There are two bolt lengths used.

- Bolt length: Bolt position 20 mm (0.79 inch)
- 1, 2, 3, 6, 7, 8, 9, 10 16 mm (0.63 in): Except the above 11.7–13.7 Nm (9–10 ft. lbs.) Standard Rear

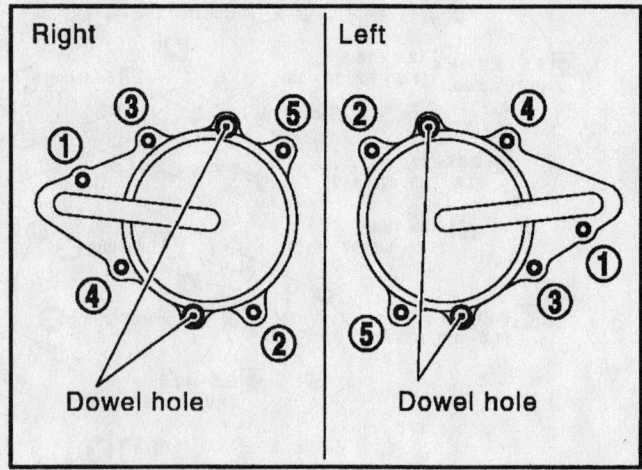

Timing control cover bolt torque sequence

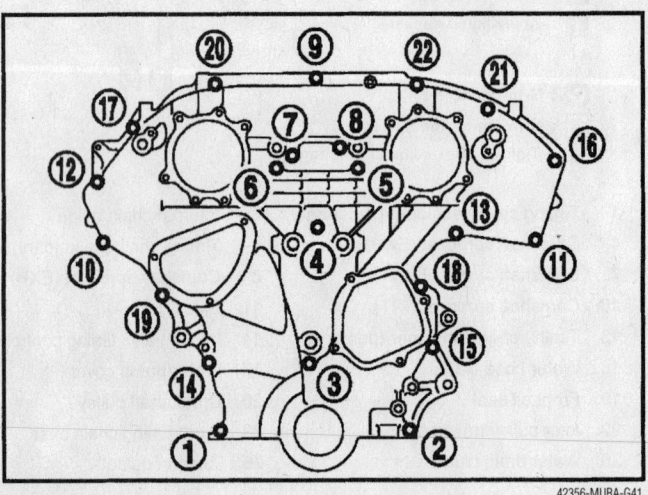

Front timing chain case bolt torque sequence

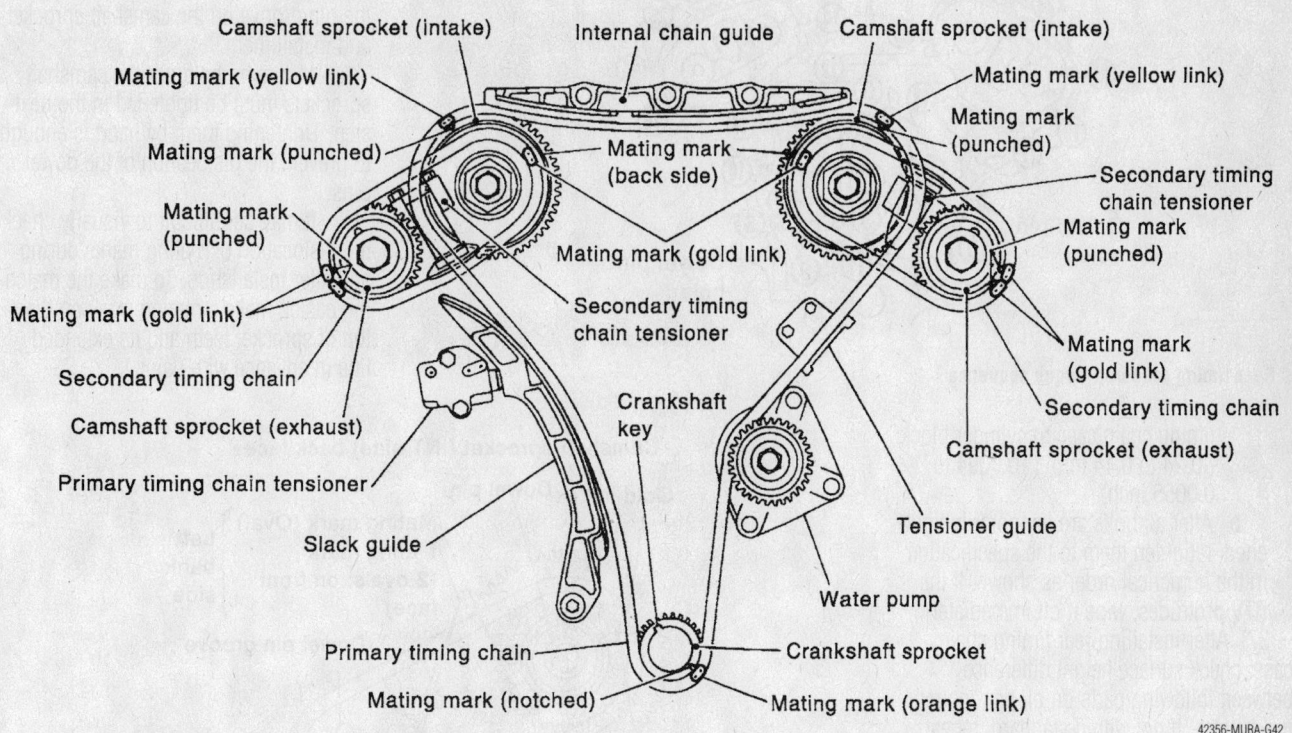

Camshaft sprocket (intake)
Internal chain guide
Camshaft sprocket (intake)
Mating mark (yellow link)
Mating mark (yellow link)
Mating mark (punched)
Mating mark (punched)
Mating mark (back side)
Mating mark (punched)
Secondary timing chain tensioner
Mating mark (gold link)
Mating mark (punched)
Mating mark (gold link)
Mating mark (gold link)
Secondary timing chain
Secondary timing chain tensioner
Secondary timing chain
Camshaft sprocket (exhaust)
Crankshaft key
Camshaft sprocket (exhaust)
Primary timing chain tensioner
Tensioner guide
Slack guide
Water pump
Primary timing chain
Crankshaft sprocket
Mating mark (notched)
Mating mark (orange link)

42356-MURA-G42

Timing chain mating marks

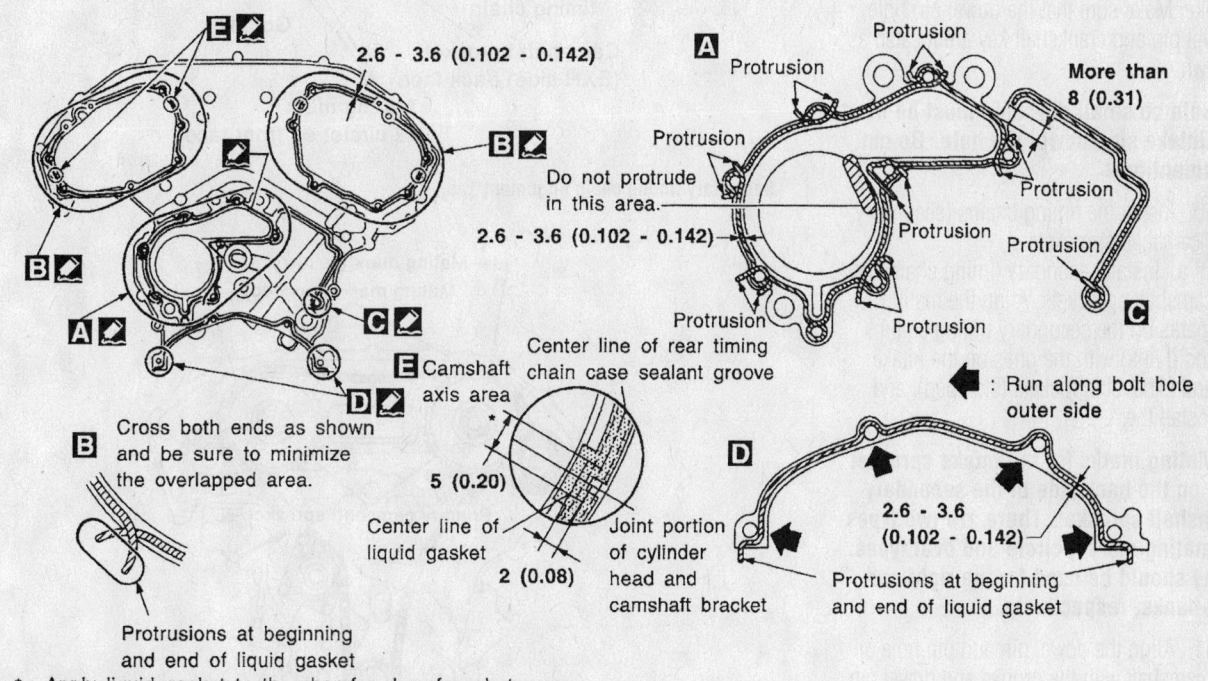

2.6 - 3.6 (0.102 - 0.142)

A Protrusion
Protrusion
Protrusion
Do not protrude in this area.
2.6 - 3.6 (0.102 - 0.142)
Protrusion
Protrusion
Protrusion
Protrusion
More than 8 (0.31)
Protrusion
C

E Camshaft axis area
Center line of rear timing chain case sealant groove
5 (0.20)
Center line of liquid gasket
2 (0.08)
Joint portion of cylinder head and camshaft bracket

B Cross both ends as shown and be sure to minimize the overlapped area.
Protrusions at beginning and end of liquid gasket

: Run along bolt hole outer side

D 2.6 - 3.6 (0.102 - 0.142)
Protrusions at beginning and end of liquid gasket

* : Apply liquid gasket to the chamfered surface between camshaft bracket and cylinder head.

: Apply liquid gasket.

Unit: mm (in)

42356-MURA-G43

Rear timing case sealer application

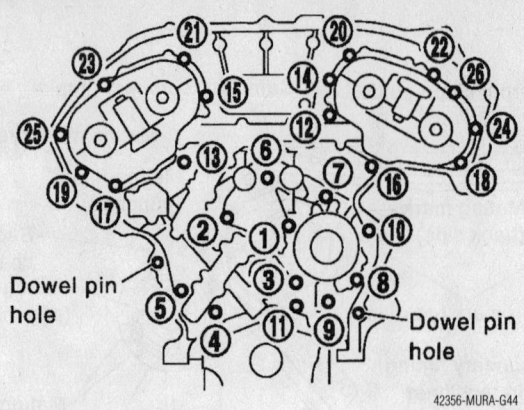

Rear timing case bolt torque sequence

42356-MURA-G44

b. On the exhaust side, align the dowel pin on the camshaft front end with the pin groove on the camshaft sprocket, and install them.

c. Mounting bolts for the camshaft sprockets must be tightened in the next step. Tightening them by hand is enough to prevent the dislocation of the dowel pins.

d. It may be difficult to visually check the dislocation of mating marks during and after installation. To make the matching easier, make a mating mark on the top of sprocket teeth and its extended line in advance with paint.

timing chain case to cylinder block: −0.24 to 0.14 mm (−0.0094 to 0.0055 inch)

b. After all bolts are temporarily tightened, retighten them to the specification in the numerical order as shown. If the RTV protrudes, wipe it off immediately.

37. After installing rear timing chain case, check surface height difference between following parts on oil pan mounting surface. If not within standard, repeat above installation procedure.

38. Install chain tension guide.

39. Position the crankshaft so No. 1 piston is set at TDC on the compression stroke. Make sure that the dowel pin hole, dowel pin and crankshaft key are located as shown.

➡**Hole on small dia. side must be used for intake side dowel pin hole. Do not misidentify.**

40. Install the timing chains (secondary) and camshaft sprockets.

a. Install secondary timing chains and camshaft sprockets. Align the mating marks on the secondary timing chain (gold link) with the ones on the intake and exhaust sprockets (stamped), and install them.

➡**Mating marks for the intake sprocket are on the back side of the secondary camshaft sprocket. There are two types of mating marks, circle and oval types. They should be used for the right and left banks, respectively.**

41. Align the dowel pin and pin hole on the camshaft with the groove and dowel pin on the sprocket, and install them.

a. On the intake side, align the pin hole on the small diameter side of the camshaft front end with the dowel pin on the back side of the camshaft sprocket, and install them.

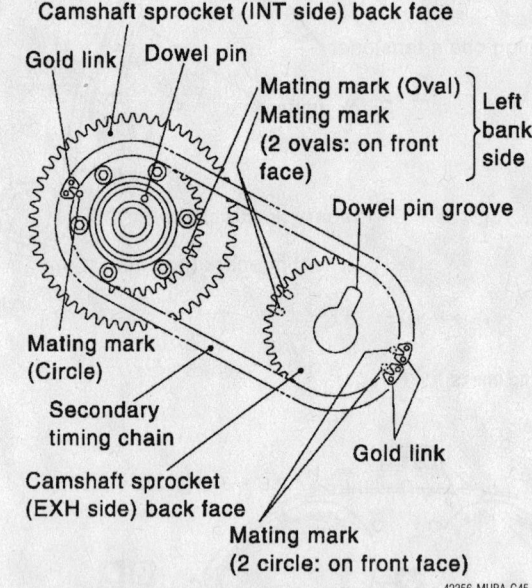

Secondary timing chain alignment

42356-MURA-G45

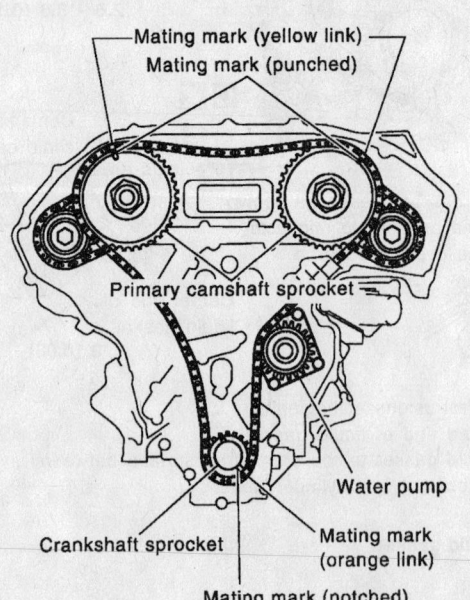

Primary timing chain alignment

42356-MURA-G46

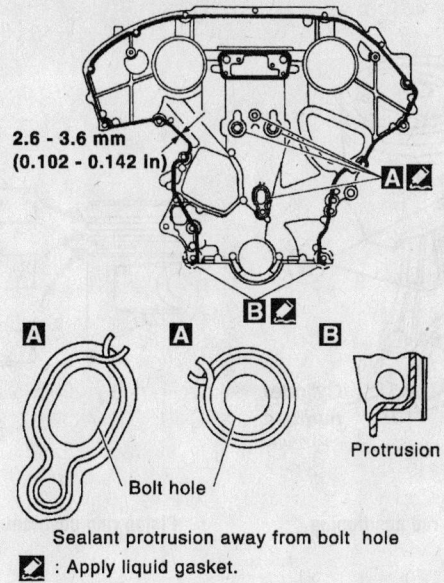

2.6 - 3.6 mm
(0.102 - 0.142 in)

Bolt hole

Protrusion

Sealant protrusion away from bolt hole

✎ : Apply liquid gasket.

42356-MURA-G47

Sealer application on front case

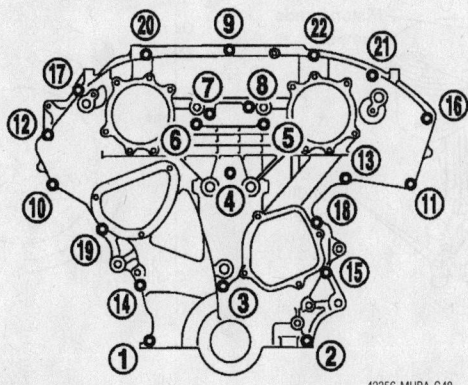

Front case bolt torque sequence

42356-MURA-G48

42. After confirming the mating marks are aligned, tighten the camshaft sprocket mounting bolts. Secure the camshaft using a wrench at the hexagonal portion to tighten the mounting bolts.

43. Pull the stopper pins out from the secondary timing chain tensioners.

44. Install the primary timing chain as follows:

a. Install the crankshaft sprocket. Make sure the mating marks on the crankshaft sprocket face the front of the engine.

b. Install the primary timing chain. Install primary timing chain so the mating mark (punched) on camshaft sprocket is aligned with the yellow link on the timing chain, while the mating mark (notched) on the crankshaft sprocket is aligned with the orange one on the timing chain, as shown.

c. When it is difficult to align mating marks of the primary timing chain with each sprocket, gradually turn the camshaft using a wrench on the hexagonal portion to align it with the mating marks.

d. During alignment, be careful to prevent dislocation of mating mark alignments of the secondary timing chains.

45. Install the internal chain guide and tension guide.

46. Install slack guide. Do not over tighten the slack guide mounting bolts. It is normal for a gap to exist under the bolt seats when the mounting bolts are tightened to specification.

47. Install chain tensioner for slack guide. When installing the chain tensioner, push in the sleeve and keep it pressed in with the stopper pin. Remove any dirt and foreign materials completely from the back and the mounting surfaces of the chain tensioner. After installation, pull out the stopper pin by pressing the slack guide.

48. Reconfirm that the mating marks on the sprockets and the timing chain have not slipped out of alignment.

49. Install new O-rings on the rear timing chain case.

50. Install the front oil seal on the front timing chain case. Apply new engine oil to the oil seal edges. Install it so that each seal lip is oriented as shown in illustration. Using a suitable drift, press-fit oil seal until it becomes flush with timing chain case end face. Make sure the garter spring is in position and seal lip is not inverted.

51. Install the water pump cover and the chain tensioner cover to front cover. Apply RTV silicone sealant.

52. Install front timing chain case as follows:

a. Apply liquid gasket to front timing chain case back side as shown.

b. Install dowel pin on the rear timing chain case into dowel pin hole on front timing chain case.

c. Tighten bolts to the specified torque in order shown in the illustration.

- 8 mm (0.31 in) dia. bolts 1, 2: 25.5–31.3 Nm (19–23 ft. lbs.)
- 6 mm (0.24 in) dia. bolts Except the above: 11.7–13.7 Nm (9–10 ft. lbs.)

d. After tightening, retighten them to specified torque in numerical order shown in illustration.

e. After tightening, retighten them to specified torque in numerical order shown in illustration.

53. After installing the front timing chain case, check the surface height difference between the following parts on the oil pan mounting surface. If not within specification, repeat the installation procedure.

54. Install right and left intake valve timing control covers as follows:

a. Install seal rings in shaft grooves.

b. Apply liquid gasket to the intake valve timing control covers. Use RTV Silicone Sealant.

c. Install collared O-ring in front cover oil hole (left and right sides).

d. Being careful not to move the seal ring from the installation groove, align the dowel pins on the chain case with the holes to install the intake valve timing control covers.

e. Tighten bolts in the numerical order as shown.

55. Install right and left rocker covers.

56. Install crankshaft pulley as follows:

a. Fix crankshaft using a hammer shaft or an equivalent tool.

b. Install crankshaft pulley, taking care not to damage front oil seal. When press-fitting crankshaft pulley with a plastic hammer, tap on its center portion (not circumference).

c. Tighten bolt to 39.2 to 49.0 Nm (29 to 36 ft. lbs.).

d. Put a paint mark on crankshaft pulley aligning with angle mark on crankshaft pulley bolt. Then, further retighten bolt by 60 to 65 degrees.

57. Rotate crankshaft pulley in normal direction (clockwise when viewed from front) to confirm it turns smoothly.

58. For the following operations, perform steps in the reverse order of removal.

➡**If hydraulic pressure inside chain tensioner drops after removal/installation, slack in the guide may generate a pounding noise during and just after engine start. However, this is not unusual. Noise will stop after hydraulic pressure rises.**

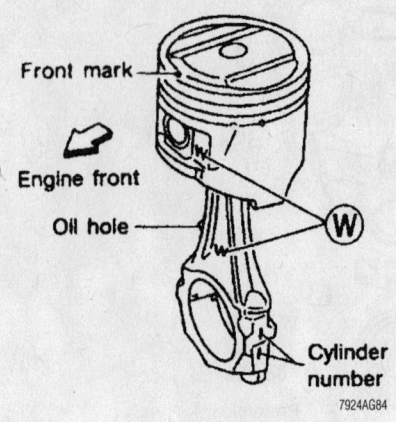

Piston and connecting rod positioning

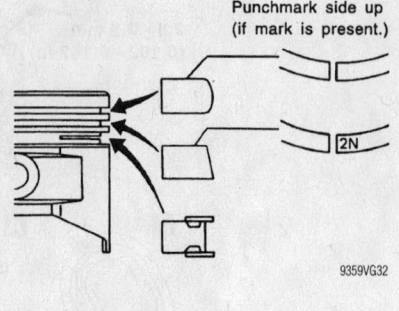

Piston ring positioning

Piston and Ring

POSITIONING

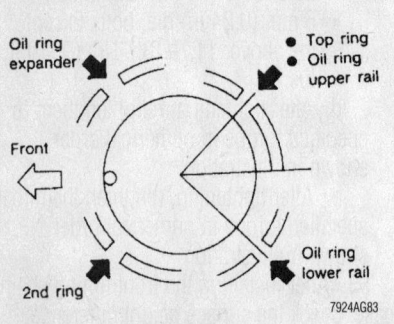

Piston ring end-gap spacing

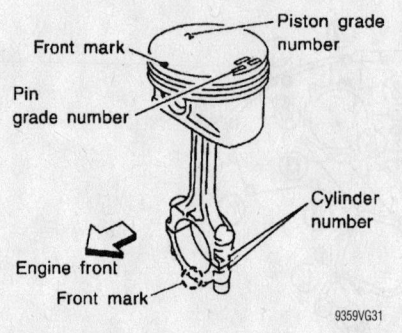

Piston and connecting rod positioning

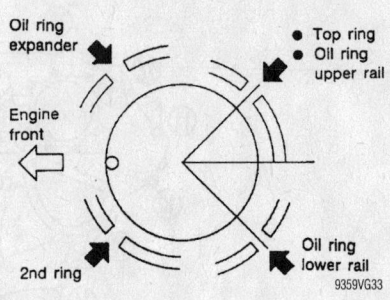

Piston ring positioning

FUEL SYSTEM

Fuel System Service Precautions

Safety is the most important factor when performing not only fuel system maintenance but any type of maintenance. Failure to conduct maintenance and repairs in a safe manner may result in serious personal injury or death. Maintenance and testing of the vehicle's fuel system components can be accomplished safely and effectively by adhering to the following rules and guidelines.

• To avoid the possibility of fire and personal injury, always disconnect the negative battery cable unless the repair or test procedure requires that battery voltage be applied.

• Always relieve the fuel system pressure prior to disconnecting any fuel system

component (injector, fuel rail, pressure regulator, etc.), fitting or fuel line connection. Exercise extreme caution whenever relieving fuel system pressure, to avoid exposing skin, face and eyes to fuel spray. Please be advised that fuel under pressure may penetrate the skin or any part of the body that it contacts.

• Always place a shop towel or cloth around the fitting or connection prior to loosening to absorb any excess fuel due to spillage. Ensure that all fuel spillage (should it occur) is quickly removed from engine surfaces. Ensure that all fuel soaked cloths or towels are deposited into a suitable waste container.

• Always keep a dry chemical (Class B) fire extinguisher near the work area.

• Do not allow fuel spray or fuel vapors

to come into contact with a spark or open flame.

• Always use a back-up wrench when loosening and tightening fuel line connection fittings. This will prevent unnecessary stress and torsion to fuel line piping. Always follow the proper torque specifications.

• Always replace worn fuel fitting O-rings with new. Do not substitute fuel hose or equivalent, where fuel pipe is installed.

Fuel System Pressure

RELIEVING

1. Before servicing the vehicle, refer to the precautions in the beginning of this section.

2. Remove the fuel pump fuse from the panel.

3. Start the engine and allow it to run until it stalls. Crank the engine for a few seconds to relieve additional fuel pressure.

4. Disconnect the negative battery cable.

5. When repairs are complete, replace the fuel pump fuse and connect the negative battery cable.

Fuel Pump

REMOVAL & INSTALLATION

1. Before servicing the vehicle, refer to the precautions in the beginning of this section.

2. Relieve the fuel system pressure.

3. Open the fuel filler lid.

4. Open the filler cap and release the pressure inside the fuel tank.

5. Remove or disconnect the following:

- Rear seat cushion trim and pad bolts, then lift up rear seat cushion.
- Inspection hole cover for main and sub fuel level sensor unit by turning clips counterclockwise by 90°.
- Harness connector and quick connectors for EVAP/Vent line hose and fuel feed tube. Disconnect EVAP/Vent line hose connector (push in tubs and pull out).

- Remove the retainer for main and sub fuel level sensor unit with fuel tank lock ring wrench (SST) by turning counterclockwise.
- Remove main fuel level sensor unit, fuel filter and fuel pump assembly, and sub fuel level sensor unit. Raise the main fuel level sensor unit, fuel filter and fuel pump assembly, and disconnect the fuel hose connector (push in tabs and pull out) and sub fuel level sensor unit harness connector. Raise and release the sub fuel level sensor unit to remove.

6. Installation is the reverse of removal. Note the following:

- Connect fuel hose connector (push in until it stops) and sub fuel level sensor unit harness connector.
- Align the direction mark on main and sub fuel level sensor unit with that on fuel tank as shown in the illustration.
- Install the inspection hole cover with front mark (arrow) facing front of the vehicle (Both for right and left). Lock the clips by turning clockwise.

7. Connect the quick connector as follows.

- Check the connection for damage or any foreign materials.
- Align the connector with the tube, then insert the connector straight into the tube until a click is heard.
- After connecting, make sure that the connection is secure by following method.
- Pull the tube and the connector to make sure they are securely connected. Visually confirm that the two retainer tabs are connected to the connector.

8. Turn ignition switch "ON" (with engine stopped), then check connections for leaks by applying fuel pressure to fuel piping.

9. Start the engine and let it idle and make sure there are no fuel leaks at the fuel system connections.

Fuel Injectors

REMOVAL & INSTALLATION

1. Before servicing the vehicle, refer to the precautions in the beginning of this section.

2. Remove the engine cover.

3. Release the fuel pressure.

4. Remove the right windshield wiper arm and the right cowl top cover.

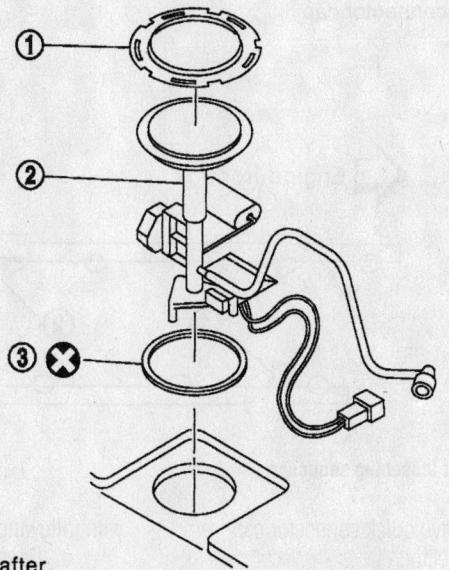

Left side

❌ : Always replace after every disassembly.

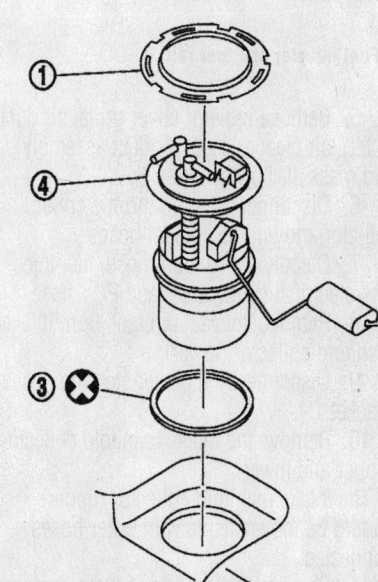

1. Retainer

4. Main fuel level sensor unit, fuel filter and fuel pump assembly

2. Sub fuel level sensor unit

3. O-ring

Fuel level sensor unit, filter and pump

42356-MURA-G49

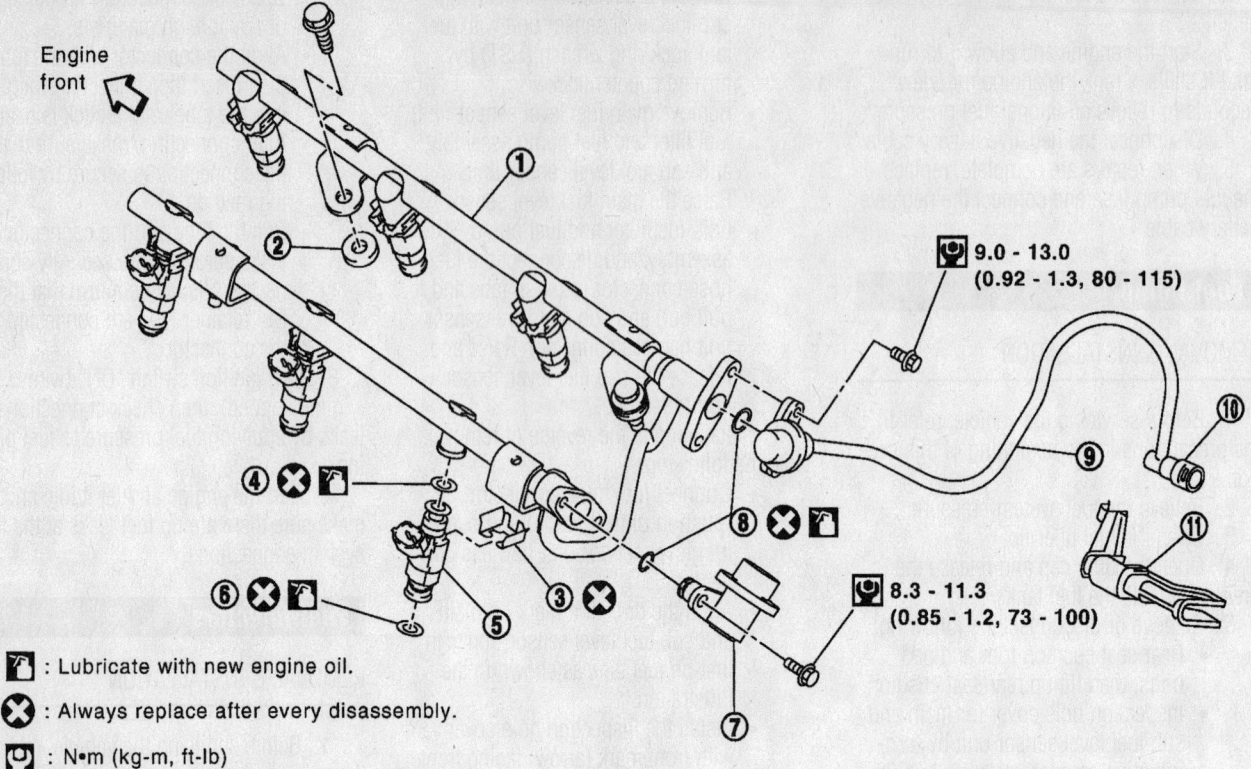

Engine front

9.0 - 13.0
(0.92 - 1.3, 80 - 115)

8.3 - 11.3
(0.85 - 1.2, 73 - 100)

🛢 : Lubricate with new engine oil.

✖ : Always replace after every disassembly.

: N•m (kg-m, ft-lb)

: N•m (kg-m, in-lb)

1. Fuel tube	2. Insulator	3. Clip
4. O-ring (black)	5. Fuel injector	6. O-ring (green)
7. Fuel damper	8. O-ring	9. Fuel feed hose (with damper)
10. Quick connector	11. Quick connector cap	

42356-MURA-G50

Fuel injector and fuel rail

5. Remove radiator cover grille, air duct (inlet), air cleaner case, air duct assembly and mass air flow sensor.

6. Disconnect electric throttle control actuator and engine coolant hoses.

7. Disconnect vacuum hose, fuel injectors electrical connectors, and PCV hose.

8. Remove the vacuum tank from intake manifold collector (lower).

9. Disconnect the power steering hose bracket.

10. Remove the intake manifold collector (upper and lower).

The intake manifold collector (upper) should be moved aside with water hoses connected.

11. Remove fuel feed hose (with damper) from fuel tube.

12. Disconnect fuel feed hose (with damper) quick connector at vehicle piping side. When separating fuel feed hose and centralized under-floor piping connection, disconnect quick connector with the following procedure.

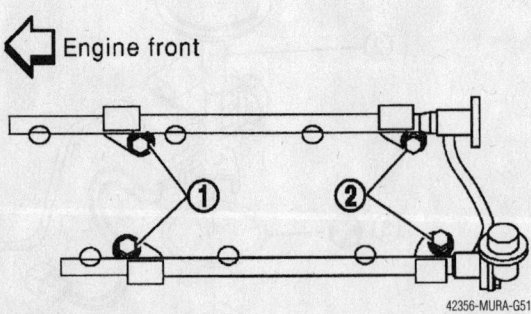

Engine front

Fuel rail bolt loosening sequence

42356-MURA-G51

a. Remove quick connector cap from quick connector.

b. Disconnect quick connector from centralized under-floor piping.

13. Remove harness connector from fuel injector.

14. Loosen mounting bolts in numerical order in the illustration, and remove fuel tube and fuel injector assembly.

15. Remove fuel injector from fuel tube

with following procedure.

a. Open and remove clip.

b. Remove fuel injector from the fuel tube by pulling straight.

16. Remove fuel damper from fuel tube.

To install:

17. Install fuel damper. Insert fuel damper straight into fuel tube. Tighten mounting bolts evenly in turn. After tightening mounting bolts, make sure that there is no gap

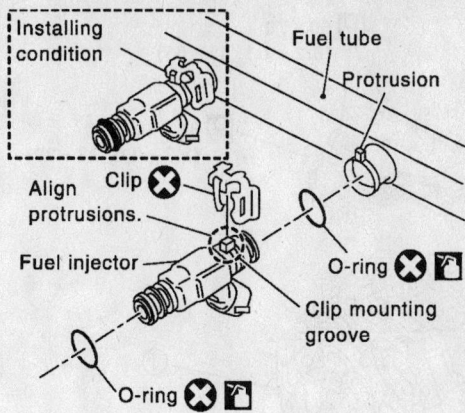

Installing condition
Fuel tube
Protrusion
Align protrusions.
Clip ⊗
Fuel injector
O-ring ⊗ 🛢
Clip mounting groove
O-ring ⊗ 🛢

⊗ : Always replace after every disassembly.

🛢 : Lubricate with new engine oil.

42356-MURA-G52

Fuel injector installation

between flange and fuel tube. When handling O-rings, be careful of the following:

- Handle O-ring with bare hands. Never wear gloves.
- Lubricate O-ring with new engine oil.
- Do not clean O-ring with solvent.
- Make sure that O-ring and its mating part are free of foreign material.
- When installing O-ring, be careful not to scratch it with tool or fingernails. Also be careful not to twist or stretch O-ring. If O-ring was stretched while it was being attached, do not insert it quickly into fuel tube.

18. Install O-rings on the fuel injector. Upper and lower O-ring are different.

19. Install fuel injector to fuel tube with the following procedure.

a. Insert clip into clip mounting groove on fuel injector. Insert clip so that lug "A" of fuel injector matches notch "A" of the clip.

➡**Do not reuse clip. Replace it with a new one. Be careful to keep clip from interfering with O-ring. If interference occurs, replace O-ring.**

b. Insert fuel injector into fuel tube with clip attached. Insert it while matching it to the axial center. Insert fuel injector so that lug "B" of fuel tube matches notch "B" of the clip. Make sure that fuel tube flange is securely fixed in flange groove on clip.

c. Make sure that installation is complete by checking that fuel injector does not rotate or come off.

20. Tighten mounting bolts in two steps in numerical order shown in illustration.

21. Connect fuel injector harness.

22. Install intake manifold collector (upper and lower).

23. Connect quick connector between fuel feed hose (with damper) and centralized under-floor piping connection with the following procedure:

a. Check the connection for damage and foreign materials.

b. Align the quick connector with the tube, then insert the connector straight into the tube until a click is heard.

c. After connecting the quick connector, use the following method to make sure it is full connected. Visually confirm that the two retainer tabs are connected to the connector. Pull the tube and the connector to make sure they are securely connected.

d. Install quick connector cap to quick connector connection. Install quick connector cap with arrow on surface facing in direction of quick connector.

➡**If cap cannot be installed smoothly, quick connector may have not been installed correctly. Check connection again.**

24. The remainder of installation is the reverse of removal.

25. Turn ignition switch ON (with engine stopped). With fuel pressure applied to fuel piping, check for fuel leakage at connection points.

26. Start engine. With engine speed increased, check again for fuel leakage at connection points.

DRIVE TRAIN

Transaxle

REMOVAL & INSTALLATION

Remove the transaxle assembly and engine assembly together from the vehicle.

1. Remove exhaust front tube.
2. Remove dust cover from converter housing part.
3. Turn crankshaft clockwise and remove the four tightening nuts for drive plate and torque converter.
4. Remove the four bolts in the illustration.
5. Remove transaxle assembly and engine assembly together from the vehicle.

6. Remove halfshaft.

➡**Be sure to replace the new differential side oil seal every removal of halfshaft.**

7. Remove transfer case gusset. (AWD models)
8. Remove transfer case assembly.

➡**Be sure to replace the new differential side oil seal (converter housing side only) whenever the transfer case is removed.**

9. Remove filler pipe.
10. Disconnect harness connector and wire harness.
11. Remove POS sensor, from engine assembly.

12. Remove starter motor.
13. Remove CVT fluid cooler valve assembly. (With CVT fluid cooler tube assembly and heater hose).
14. Install slinger to transaxle assembly.
15. Remove rear gusset.
16. Remove left engine mounting bracket and left engine mounting insulator.
17. Remove front suspension member from transaxle assembly and engine assembly.
18. Remove transaxle assembly bolts.
19. Remove transaxle assembly from engine assembly with a hoist. Secure torque converter to prevent it from dropping.

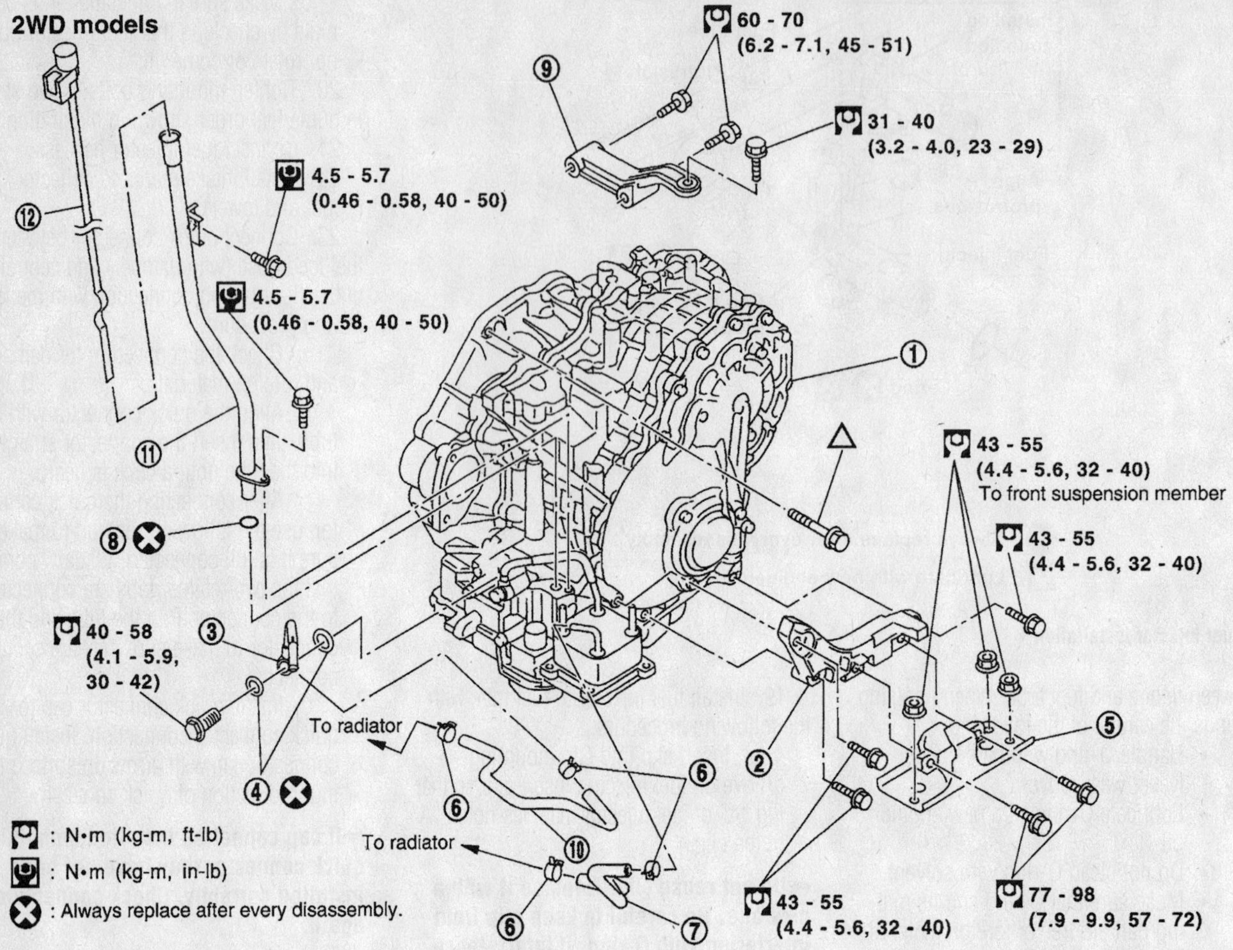

2WD models

60 - 70
(6.2 - 7.1, 45 - 51)

31 - 40
(3.2 - 4.0, 23 - 29)

4.5 - 5.7
(0.46 - 0.58, 40 - 50)

4.5 - 5.7
(0.46 - 0.58, 40 - 50)

43 - 55
(4.4 - 5.6, 32 - 40)
To front suspension member

43 - 55
(4.4 - 5.6, 32 - 40)

40 - 58
(4.1 - 5.9,
30 - 42)

To radiator

43 - 55
(4.4 - 5.6, 32 - 40)

77 - 98
(7.9 - 9.9, 57 - 72)

To radiator

: N•m (kg-m, ft-lb)

: N•m (kg-m, in-lb)

: Always replace after every disassembly.

1. Transaxle assembly
2. LH engine mounting bracket
3. Fluid cooler tube
4. Copper washer
5. LH engine mounting insulator
6. Hose clamp
7. CVT fluid cooler hose
8. O-ring
9. Rear gusset
10. CVT fluid cooler hose
11. CVT fluid charging pipe
12. CVT fluid level gauge

42356-MURA-G53

Transaxle and related parts—2wd

Bolt No.	1	2	3	4
Number of bolts	1	2	2	4
Bolt length "ℓ"mm (in)	52 (2.05)	36 (1.42)	105 (4.13)	35 (1.38)
Tightening torque N-m (kg-m, ft-lb)		70 - 79 (7.1 - 8.0, 51 - 58)		42-52 (4.3 - 5.3, 31 - 38)

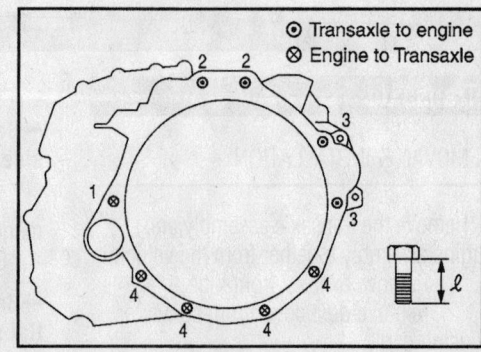

⊙ Transaxle to engine
⊗ Engine to Transaxle

42356-MURA-G55

Transaxle bolt location and torque sequence

➡ **After installing a torque converter to a transaxle, be sure to check dimension A to ensure it is within the reference value limit.**

20. Installation is the reverse of the removal. Note the following:
• Screw and set the locator into the stud bolts for the torque converter locate.

• Rotate the torque converter to allow the locator to go down.
• Rotate the drive plate so that the hole of the drive plate locator faces down.

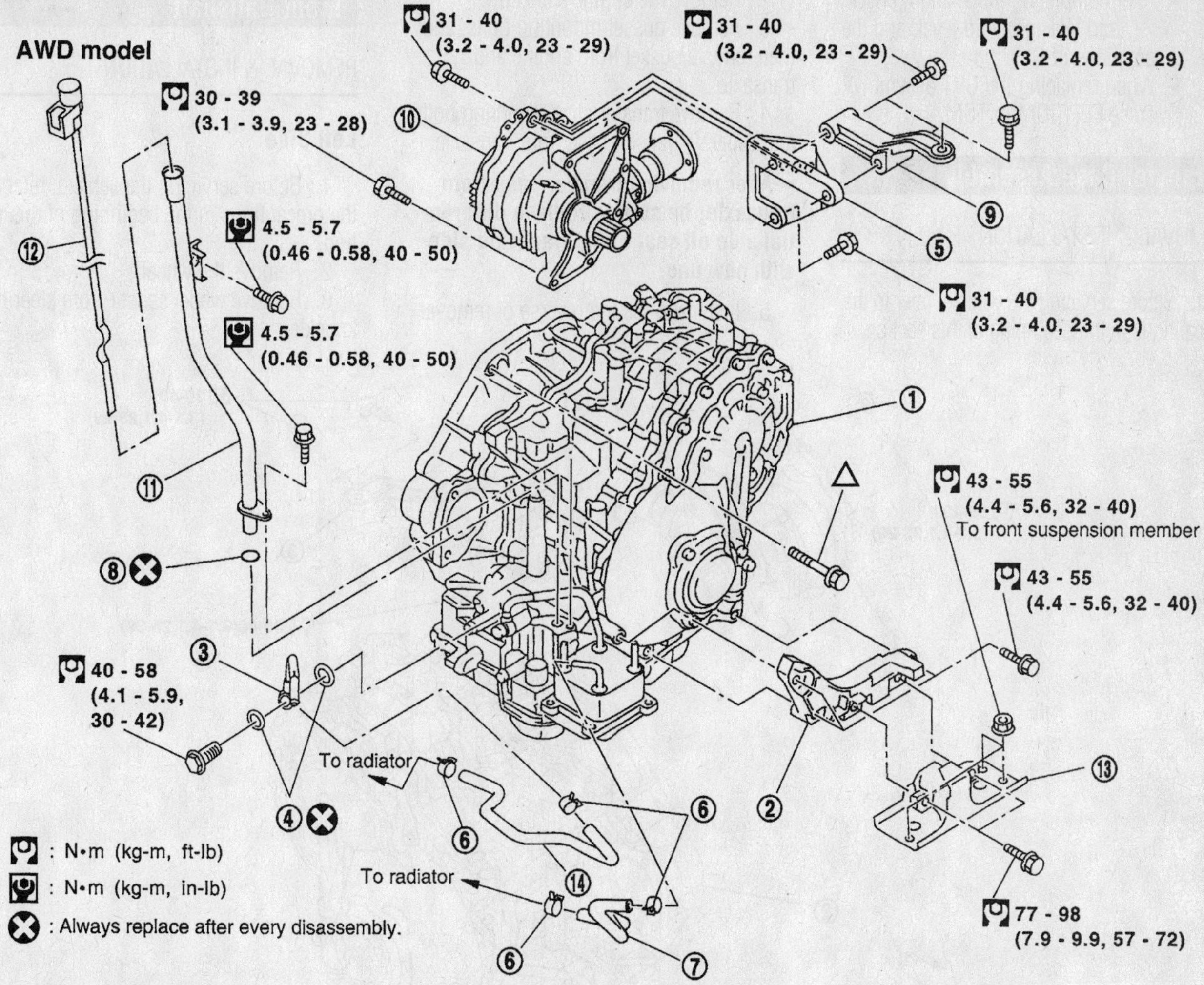

AWD model

31 - 40
(3.2 - 4.0, 23 - 29)

31 - 40
(3.2 - 4.0, 23 - 29)

31 - 40
(3.2 - 4.0, 23 - 29)

30 - 39
(3.1 - 3.9, 23 - 28)

⑫

⑩

⑨

⑤

31 - 40
(3.2 - 4.0, 23 - 29)

4.5 - 5.7
(0.46 - 0.58, 40 - 50)

4.5 - 5.7
(0.46 - 0.58, 40 - 50)

⑪

①

⚠

43 - 55
(4.4 - 5.6, 32 - 40)
To front suspension member

43 - 55
(4.4 - 5.6, 32 - 40)

⑧✖

40 - 58
(4.1 - 5.9,
30 - 42)

③

To radiator

②

⑬

④✖

⑥

⑥

To radiator

⑭

⑥

⑦

77 - 98
(7.9 - 9.9, 57 - 72)

: N•m (kg-m, ft-lb)

: N•m (kg-m, in-lb)

✖ : Always replace after every disassembly.

1. Transaxle assembly	2. LH engine mounting bracket	3. Fluid cooler tube
4. Copper washer	5. Transfer gusset	6. Hose clamp
7. CVT fluid cooler hose	8. O-ring	9. Rear gusset
10. Transfer assembly	11. CVT fluid charging pipe	12. CVT fluid level gauge
13. LH engine mounting insulator	14. CVT fluid cooler hose	

42356-MURA-G54

Transaxle and related parts—AWD

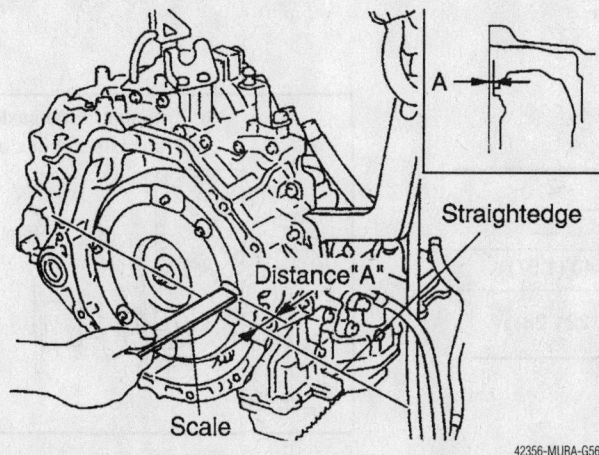

Straightedge

Distance "A"

A

Scale

42356-MURA-G56

Torque converter installation inspection

- Installing transaxle assembly from engine assembly with a hoist.
- When installing fluid cooler tube to transaxle assembly, transaxle assembly the part with the tube aligned with the rib.
- When installing CVT fluid cooler valve assembly to the engine, torque the bolts to: 28–32 Nm (21–23 ft. lbs.)
- Align the positions of tightening nuts for drive plate with those of the torque converter, and temporarily tighten the nuts. Then, tighten the nuts to the specified torque.
- Install POS sensor.

- After completing installation, check for fluid leakage, fluid level, and the positions of CVT.
- When replacing the CVT assembly, erase EEP ROM in TCM.

Transfer Case Assembly

REMOVAL & INSTALLATION

1. Before servicing the vehicle, refer to the precautions in the beginning of this section.

2. Remove the engine assembly.
3. Remove gusset mounting bolts, and then remove gusset from engine and transaxle.
4. Remove transfer case mounting bolts and separate transfer case from transaxle.

➡**After removing transfer case from transaxle, be sure to replace differential side oil seal of the transaxle side with new one.**

5. Installation is the reverse of removal.

Front Halfshaft

REMOVAL & INSTALLATION

Left Side

1. Before servicing the vehicle, refer to the precautions in the beginning of this section.
2. Remove the wheel.
3. Remove wheel sensor from steering knuckle.

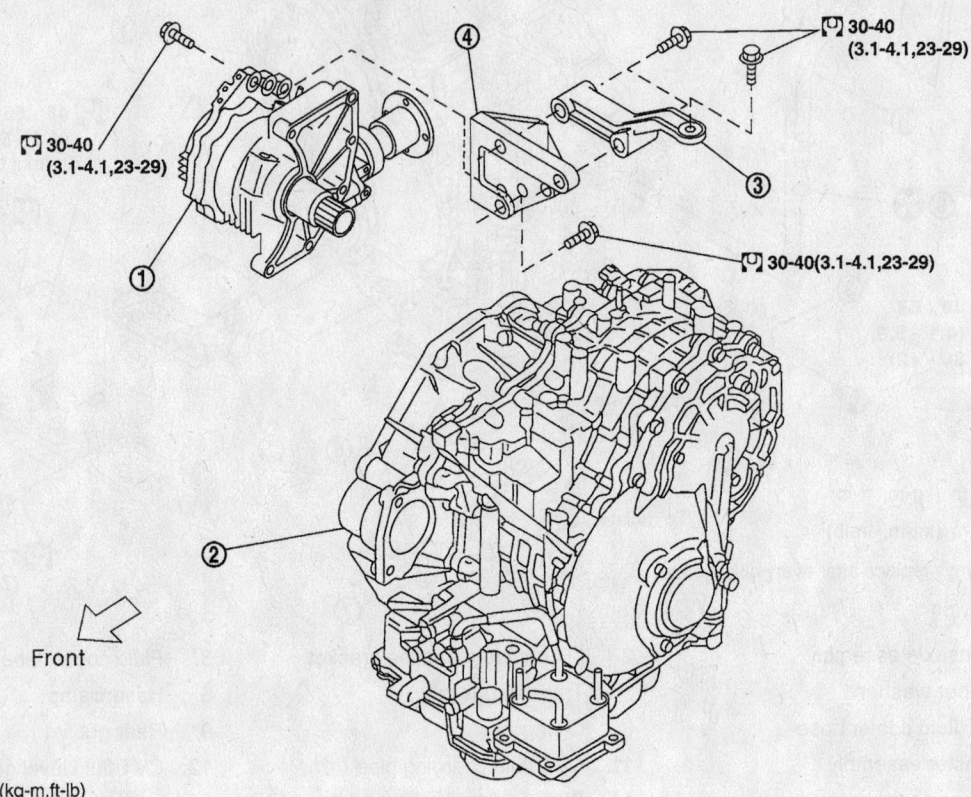

Front

⌼ : N·m(kg-m,ft-lb)

1.	Transfer assembly	2.	Transaxle assembly	3.	Rear gusset
4.	Transfer gusset		

Transfer case and related parts

Bolt No.	1	2
Quantity	4	2
Nominal length mm (in)	65 (2.56)	40 (1.57)
Tightening torque [N·m (kg·m, ft.-lb.)]	29.4 - 39.2 (3.0 - 3.9, 22 - 28)	

42356-MURA-G57

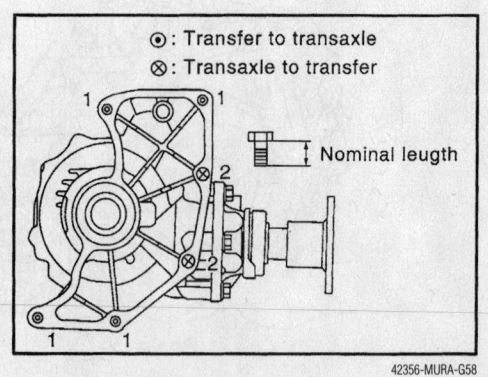

42356-MURA-G58

Transfer case bolt location and torque sequence

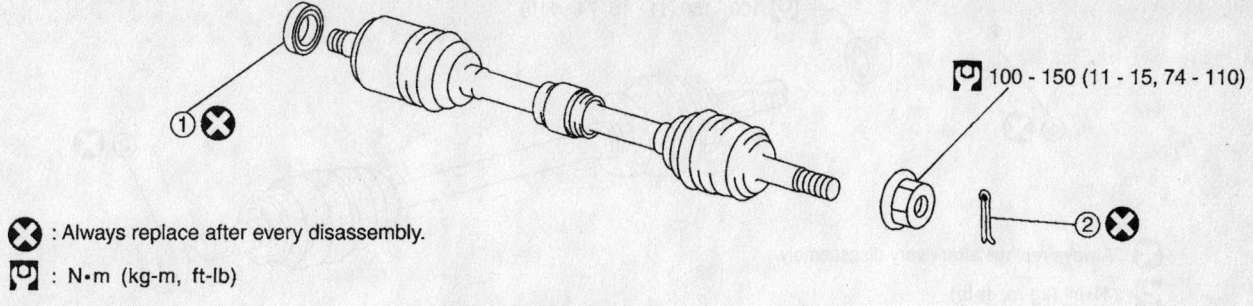

⊗ : Always replace after every disassembly.

🔧 : N•m (kg-m, ft-lb)

1.   Dust shield   2.   Cotter pin

Left side front halfshaft

42356-MURA-G59

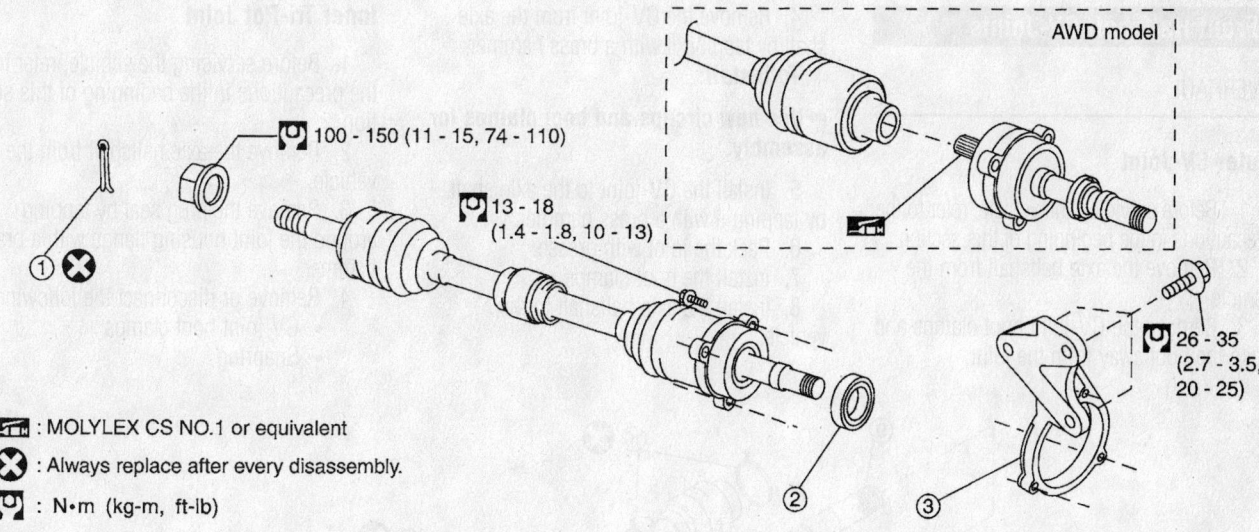

🔧 100 - 150 (11 - 15, 74 - 110)

🔧 13 - 18 (1.4 - 1.8, 10 - 13)

AWD model

🔧 26 - 35 (2.7 - 3.5, 20 - 25)

🔩 : MOLYLEX CS NO.1 or equivalent

⊗ : Always replace after every disassembly.

🔧 : N•m (kg-m, ft-lb)

1.   Cotter pin   2.   Dust shield   3.   Support bearing bracket

Right side front halfshaft

42356-MURA-G60

4.   Remove cotter pin. Then remove lock nut from halfshaft.

5.   Remove brake hose lock plate. Then remove brake hose from strut assembly.

6.   Remove strut assembly and steering knuckle bolt and nut.

7.   Using a puller, remove halfshaft from steering knuckle.

8.   Remove halfshaft from transaxle.

9.   Installation is the reverse of removal.

Right Side

2WD MODELS

1.   Before servicing the vehicle, refer to the precautions in the beginning of this section.

2.   Remove the wheel.

3.   Remove wheel sensor from steering knuckle.

4.   Remove cotter pin. Then remove lock nut from halfshaft.

5.   Remove brake hose lock plate. Then remove brake hose from strut assembly.

6.   Remove strut assembly and steering knuckle bolt and nut.

7.   Using a puller, remove halfshaft from axle.

8.   Remove support bearing bolts, and pull halfshaft from transaxle. Pry off halfshaft from transaxle.

9.   Installation is the reverse of removal.

AWD MODELS

1.   Before servicing the vehicle, refer to the precautions in the beginning of this section.

2.   Remove the wheel.

3.   Remove wheel sensor from steering knuckle.

4.   Remove brake hose lock plate. Then remove brake hose from strut assembly.

5.   Remove cotter pin. Then remove lock nut from halfshaft.

6.   Remove strut assembly and steering knuckle bolt and nut.

7.   Using a puller, remove halfshaft from knuckle.

8.   Remove halfshaft from transaxle.

9.   Installation is the reverse of removal.

Rear Halfshaft

REMOVAL & INSTALLATION

1.   Remove the wheel.

2.   Remove wheel sensor from axle.

3.   Remove cotter pin. Then remove lock nut from halfshaft.

4.   Remove parking cable and parking brake shoe from back plate.

5.   5.   Remove wheel hub and bearing assembly bolts, then remove wheel hub and bearing assembly from axle.

6.   Use a wheel wrench or other tool to remove halfshaft from final drive.

7.   Installation is the reverse of removal.

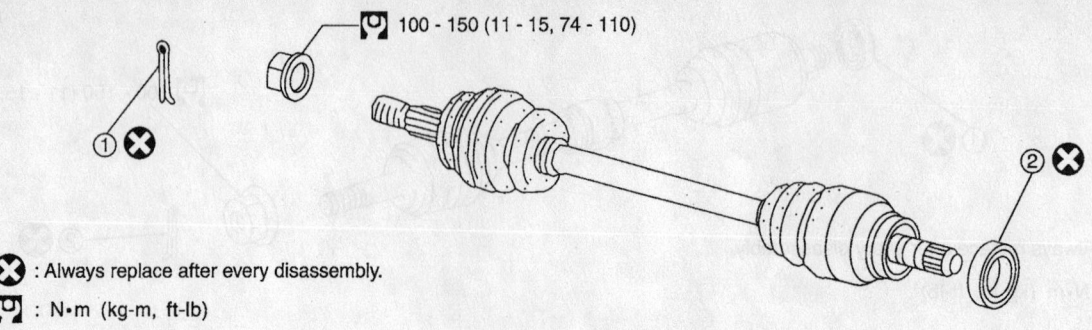

100 - 150 (11 - 15, 74 - 110)

❌ : Always replace after every disassembly.

🔧 : N•m (kg-m, ft-lb)

1. Cotter pin

2. Dust shield

42356-MURA-G63

Rear halfshaft

Front Halfshaft CV-Joints

OVERHAUL

Outer CV-Joint

1. Before servicing the vehicle, refer to the precautions in the beginning of this section.
2. Remove the axle halfshaft from the vehicle.
3. Remove the CV-joint boot clamps and push the boot away from the joint.

4. Remove the CV-joint from the axle shaft by tapping it with a brass hammer.

To install:

➡ **Use new circlips and boot clamps for assembly.**

5. Install the CV-joint to the axle shaft by tapping it with a brass hammer.
6. Pack the joint with grease.
7. Install the boot clamps.
8. Install the axle halfshaft to the vehicle.

Inner Tri-Pot Joint

1. Before servicing the vehicle, refer to the precautions in the beginning of this section.
2. Remove the axle halfshaft from the vehicle.
3. Remove the plug seal by tapping around the joint housing flange with a brass hammer.
4. Remove or disconnect the following:
   - CV-joint boot clamps
   - Snapring

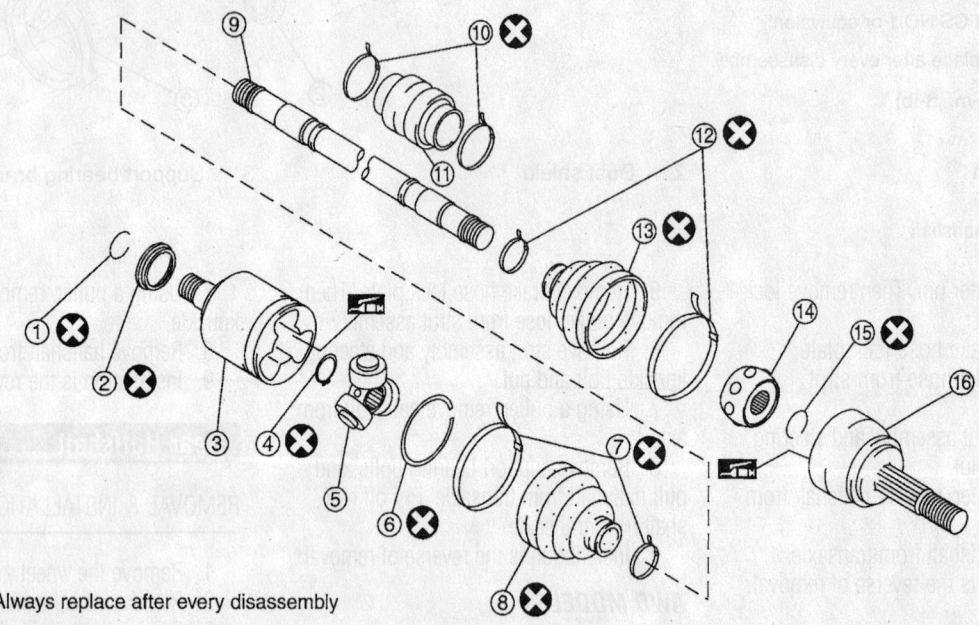

❌ : Always replace after every disassembly

1.	Circular clip	2.	Dust cover	3.	Slide joint assembly
4.	Snap ring	5.	Spider assembly	6.	Stopper ring
7.	Boot band	8.	Boot	9.	Shaft
10.	Damper band	11.	Damper	12.	Boot band
13.	Boot	14.	Ball cage / Steel ball / Inner race assembly	15.	Circular clip
16.	Joint sub-assembly				

42356-MURA-G61

Left side front halfshaft exploded view

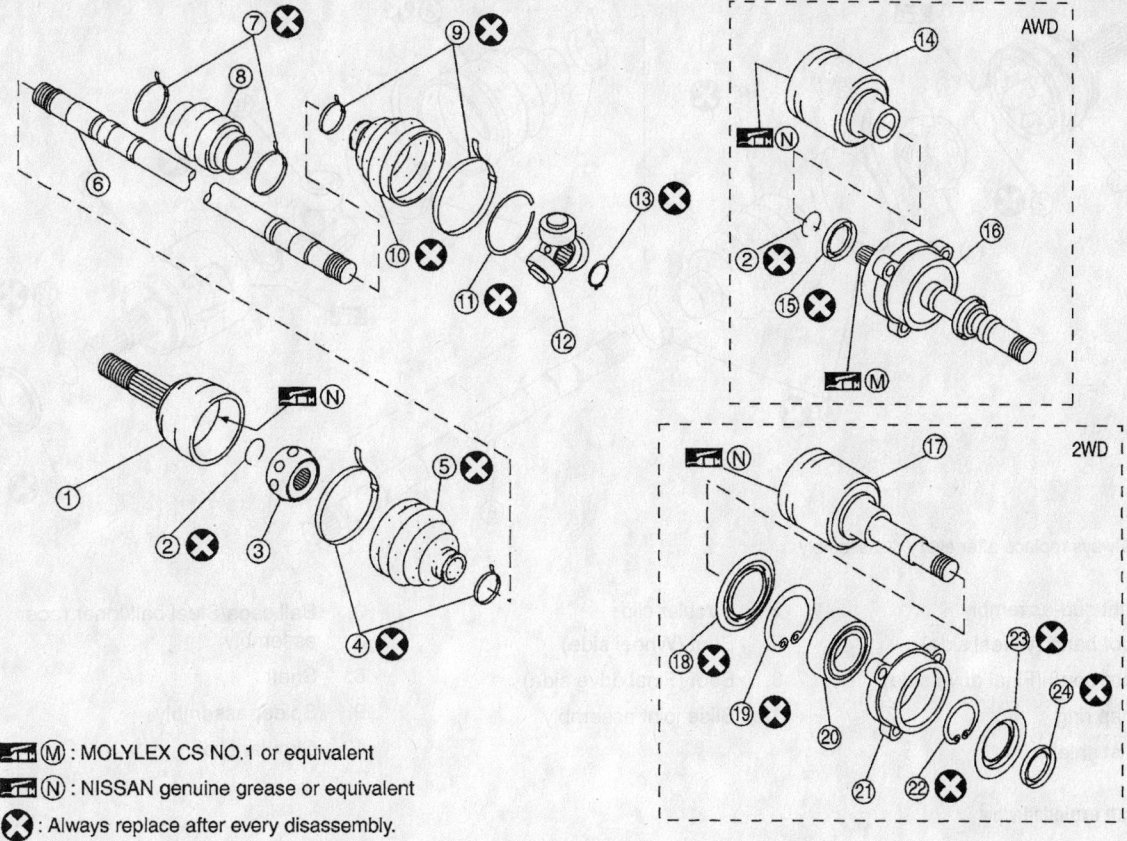

(M) : MOLYLEX CS NO.1 or equivalent

(N) : NISSAN genuine grease or equivalent

(X) : Always replace after every disassembly.

1.	Joint sub-assembly	2.	Circular clip	3.	Ball cage / Steel ball / Inner race assembly
4.	Boot band	5.	Boot	6.	Shaft
7.	Damper band	8.	Damper	9.	Boot band
10.	Boot	11.	Stopper ring	12.	Spider assembly
13.	Circular clip	14.	Slide joint assembly	15.	Dust cover
16.	Support bearing	17.	Slide joint assembly	18.	Dust cover
19.	Snap ring	20.	Bearing	21.	Bracket
22.	Snap ring	23.	Dust cover	24.	Dust cover

42356-MURA-G62

Right side front halfshaft exploded view

- Spider assembly
- CV-joint housing
- CV-joint boot

To install:

➡**Use new snaprings and plug seals for assembly.**

5. Install or connect the following:
- CV-joint boot
- CV-joint housing
- Spider assembly
- Snapring. Pack the joint with grease.
- CV-joint boot clamps
- Plug seal
6. Install the axle halfshaft to the vehicle.

Rear Halfshaft CV-Joints

DISASSEMBLY

Final Drive Side

1. Press shaft in a vise.
2. Remove boot band.

➡**When retaining shaft in a vice, always use copper or aluminum plates between vise and shaft.**

3. Put matching marks on slide joint assembly and shaft before separating slide joint assembly.
4. Put matching marks on spider assembly and shaft.

5. Remove snap ring, then remove spider assembly from shaft.
6. Remove boot from shaft.
7. Remove old grease on slide joint assembly with paper towels.
8. Remove circular clip and dust shield from slide joint assembly.

Wheel Side

1. Place shaft in a vise.

➡**When retaining shaft in a vise, always use copper or aluminum plates between vise and shaft.**

2. Remove boot bands. Then remove boot from joint sub-assembly.
3. Screw a halfshaft puller (suitable tool)

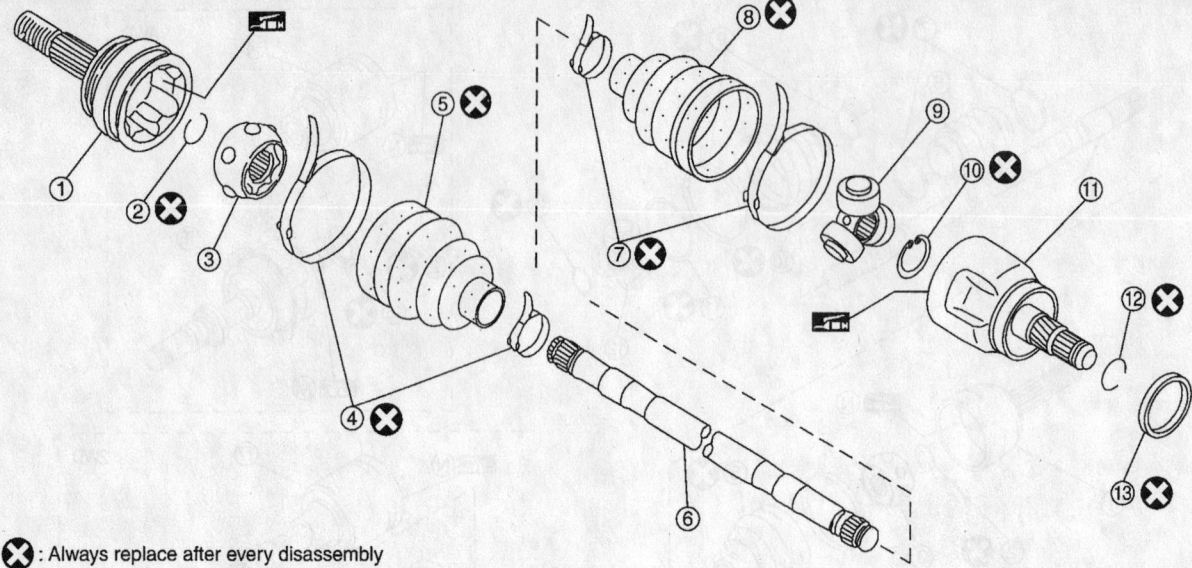

⊗: Always replace after every disassembly

1. Joint sub-assembly
4. Boot band (Wheel side)
7. Boot band (Final drive side)
10. Snap ring
13. Dust shield

2. Circular clip
5. Boot (Wheel side)
8. Boot (Final drive side)
11. Slide joint assembly

3. Ball cage/Steel ball/Inner race assembly
6. Shaft
9. Spider assembly
12. Circular clip

42356-MURA-G64

Rear halfshaft exploded view

30 mm (1.18 in) or more into threaded part of joint sub-assembly. Pull joint sub-assembly out of shaft.

➡**If joint sub-assembly cannot be removed after five or more unsuccessful attempts, replace the entire half-shaft assembly. Align sliding hammer and halfshaft and remove them by pulling directly.**

4. Remove boot from shaft.
5. Remove circular clip from shaft.
6. While rotating ball cage, remove old grease on joint sub-assembly with paper towels.
7. Replace halfshaft if there is any runout, cracking, or other damage.

➡**If there are any irregular conditions of joint sub-assembly components, replace the entire joint sub-assembly.**

ASSEMBLY

Final Drive Side

1. Install new boot and new small boot band on shaft.

➡**Cover shaft serration with tape to prevent damage to boot during installation.**

2. Remove protective tape wound around serrated part of shaft.
3. Install spider assembly securely, making sure the matching marks which were made during disassembly are properly aligned.
4. Install new snap ring.
5. Insert the amount of new grease listed below into housing from large end of boot. Grease amount: 85–95 g (3.0–3.35 oz)
6. Install slide joint assembly.
7. Install boot securely into grooves (indicated by * marks) shown in the illustration.

➡**If there is grease on boot mounting surfaces (indicated by * marks) of shaft and housing, boot may come off. Remove all grease from surfaces.**

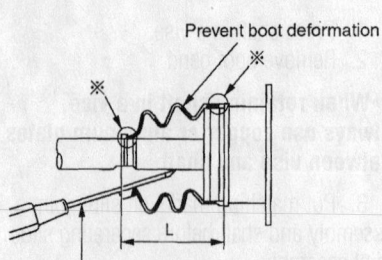

Prevent boot deformation

Flat-bladed screwdriver or similar tool

42356-MURA-G65

Checking boot length

8. Make sure boot installation length "L" is the length indicated below. Insert a flat-bladed screwdriver or similar tool into smaller side of boot. Bleed air from boot to prevent boot deformation. Boot installation length "L": 79.6 mm (3.13 in)

➡**Boot may break if boot installation length is less than standard value. Take care not to touch the tip of screwdriver to inside of boot.**

9. Secure big and small ends of boot with new boot bands as shown in the illustration.

➡**Discard old boot bands; replace with new ones.**

10. After installing housing and shaft, rotate boot to check whether or not the actual position is correct. If boot position is not correct, secure boot with new boot band again.

Wheel Side

1. Insert the amount of new grease into joint sub-assembly serration hole until grease begins to ooze from ball groove and serration hole. After inserting grease, use a shop cloth to wipe off old grease that has oozed out.
2. Wind serrated part of shaft with tape.

Install new boot band and boot to shaft. Be careful not to damage boot.

➡**Discard old boot band and boot; replace with new ones.**

3. Remove protective tape wound around serrated part of shaft.

4. Install new circular clip to shaft. At this time, circular clip must fit securely into shaft groove. Attach nut to joint sub-assembly. Use a wooden hammer to press-fit.

➡**Discard old circular clip; replace with new one.**

5. Insert the amount of new grease listed below into housing from large end of boot. Grease amount: 75–85 g (2.65–3.0 oz)

6. Install boot securely into grooves (indicated by * marks) shown in the illustration.

➡**If there is grease on boot mounting surfaces (indicated by * marks) of shaft**

and housing, boot may come off. **Remove all grease from surfaces.**

7. Make sure boot installation length "L" is the length indicated below. Insert a flat-bladed screwdriver or similar tool into smaller side of boot. Bleed air from boot to prevent boot deformation. Boot installation length "L": 67.7mm (2.67 in)

➡**Boot may break if boot installation length is less than standard value. Be careful that screwdriver tip does not contact inside surface of boot.**

8. Secure big and small ends of boot with new boot bands as shown in the illustration.

➡**Discard old boot bands; replace with new ones.**

9. Check installation status of boot. Rotate joint to make sure boot is securely in place. If not, reinstall using a new boot band.

Rear Final Drive

REMOVAL & INSTALLATION

1. Before servicing the vehicle, refer to the precautions in the beginning of this section.
2. Remove rear propeller shaft.
3. Remove rear stabilizer mounting bracket.
4. Remove wheel sensor.
5. Remove rear halfshaft.
6. Remove electric controlled coupling connector.
7. Remove electric controlled coupling breather hose and rear final drive breather hose.
8. Remove canister.
9. Set Transmission Jack to rear final drive assembly, and then remove nuts from rear suspension member.

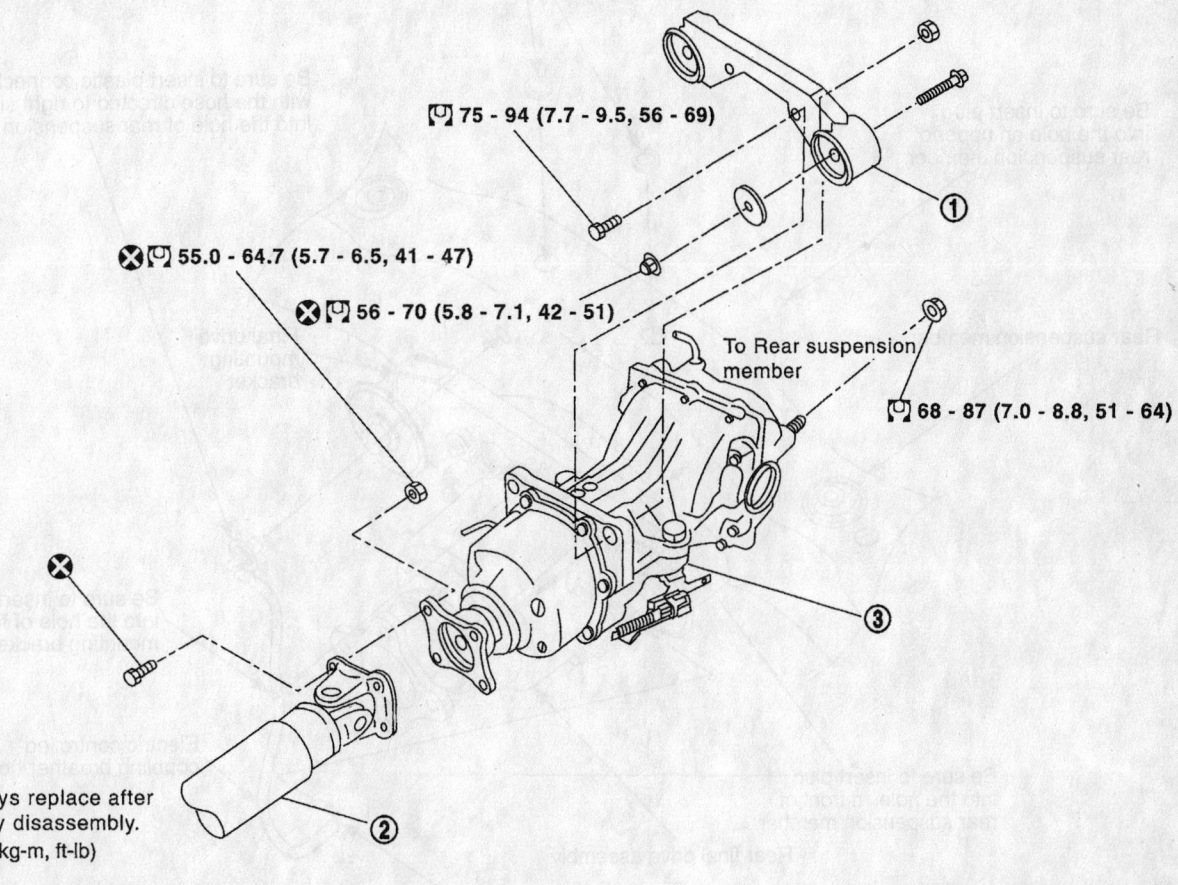

🔩 75 - 94 (7.7 - 9.5, 56 - 69)

✖🔩 55.0 - 64.7 (5.7 - 6.5, 41 - 47)

✖🔩 56 - 70 (5.8 - 7.1, 42 - 51)

To Rear suspension member

🔩 68 - 87 (7.0 - 8.8, 51 - 64)

①

③

②

✖ : Always replace after every disassembly.
🔩 : N·m (kg-m, ft-lb)

1. Final drive mount bracket

2. Rear propeller shaft

3. Rear final drive assembly

42356-MURA-G66

Rear final drive

➡**Do not place a transmission jack on the rear cover (aluminum case).**

10. Remove bolt and nut from final drive mount bracket, and then remove rear final drive assembly from vehicle.

11. Installation is the reverse of removal. Supporting rear final drive assembly securely with transmission jack, install it to final drive mount bracket and rear suspension member with bolt and nut.

Rear Final Drive Seals

REMOVAL & INSTALLATION

Pinion Seal

1. Remove the propeller shaft.
2. Put a mark on the end of the drive pinion corresponding to the position mark on the final drive companion flange.
3. Using the drive pinion flange wrench, Remove companion flange nut.
4. Using the puller, remove the companion flange.
5. Using the side bearing outer race puller, remove front oil seal.

To install:

6. Apply multi-purpose grease to sealing lips of oil seal. Press front oil seal into carrier with tool.
7. Align the matching mark of drive pinion with the matching mark of companion flange, then install the companion flange.
8. Apply oil or grease on the screw part of drive pinion and the seating surface of companion flange nut.
9. Install companion flange nut with tool.

➡**Never reuse companion flange nut.**

10. Install propeller shaft.

Side Seal

1. Remove rear wheel sensor.
2. Remove rear axle assembly.
3. Remove rear halfshaft.
4. Using flat tip screwdriver, remove side oil seal.

To install:

5. Apply multi-purpose grease to sealing lips of oil seal.
6. Using the drift, press-fit oil seal so that its surface comes face to face with the end surface of the case.
7. Install rear halfshaft.
8. Install rear axle assembly.

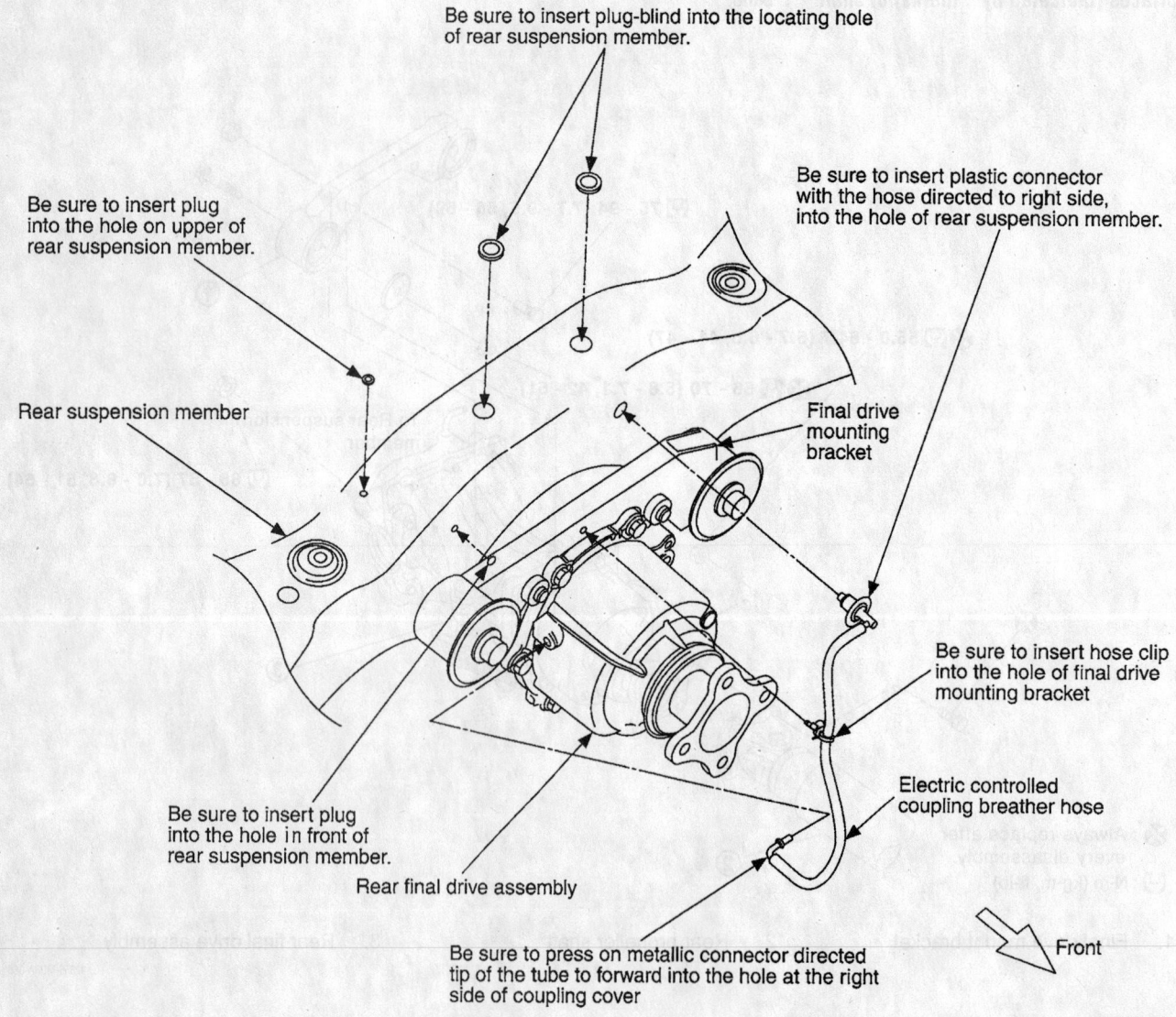

Be sure to insert plug-blind into the locating hole of rear suspension member.

Be sure to insert plastic connector with the hose directed to right side, into the hole of rear suspension member.

Be sure to insert plug into the hole on upper of rear suspension member.

Rear suspension member

Final drive mounting bracket

Be sure to insert hose clip into the hole of final drive mounting bracket

Be sure to insert plug into the hole in front of rear suspension member.

Rear final drive assembly

Electric controlled coupling breather hose

Be sure to press on metallic connector directed tip of the tube to forward into the hole at the right side of coupling cover

Front

42356-MURA-G67

Electronically controlled coupling breather hose

STEERING AND SUSPENSION

Air Bag

✳✳ CAUTION

Some vehicles are equipped with an air bag system. The system must be disarmed before performing service on, or around, system components, the steering column, instrument panel components, wiring and sensors. Failure to follow the safety precautions and the disarming procedure could result in accidental air bag deployment, possible injury and unnecessary system repairs.

PRECAUTIONS

Several precautions must be observed when handling the inflator module to avoid accidental deployment and possible personal injury.

- Never carry the inflator module by the wires or connector on the underside of the module.
- When carrying a live inflator module, hold securely with both hands, and ensure that the bag and trim cover are pointed away.
- Place the inflator module on a bench or other surface with the bag and trim cover facing up.

- With the inflator module on the bench, never place anything on or close to the module which may be thrown in the event of an accidental deployment.

DISARMING

To disarm the **SRS** system turn the ignition switch to the **OFF** position. Then, disconnect both battery cables starting with the negative cable first and wait at least 3 minutes after the cables are disconnected.

To rearm the **SRS** system, turn the ignition switch to the **OFF** position. Connect both battery cables starting with the positive cable first.

Rack and Pinion Steering Gear

REMOVAL & INSTALLATION

2WD

1. Set wheels in the straight ahead position.
2. Remove lock nut and bolt, then separate lower joint from upper joint.
3. Remove wheels.
4. Confirm that the slit on lower joint fits with the projection on rear cover cap, also the matchmark on steering gear assembly fits with the projection on rear cover cap.

5. Remove cotter pin at steering knuckle, then loosen mounting nut.
6. Use a ball joint remover to remove steering outer socket from steering knuckle. Be careful not to damage ball joint boot.

➡ **To prevent damage to threads and to prevent ball joint remover from coming off, temporarily tighten mounting nut.**

7. Remove oil pipes (high pressure side and low pressure side) from steering gear assembly, then drain fluid from pipes.
8. Remove mounting bolt (lower side) from lower joint.
9. Remove mounting bolts and nut from steering gear assembly, and then remove steering gear assembly, rack mounting bracket, rack mounting insulator and sleeve from vehicle.
10. Installation is the reverse of removal.

➡ **When steering wheel is set in the straight ahead direction, confirm slit of lower joint fits with the projection on rear cover cap, also the matchmarks on steering gear assembly fit with the projection on rear cover cap.**

11. After installation, bleed air from piping.
12. Check if steering wheel turns smoothly when it is turned several times fully to the end of the left and right.

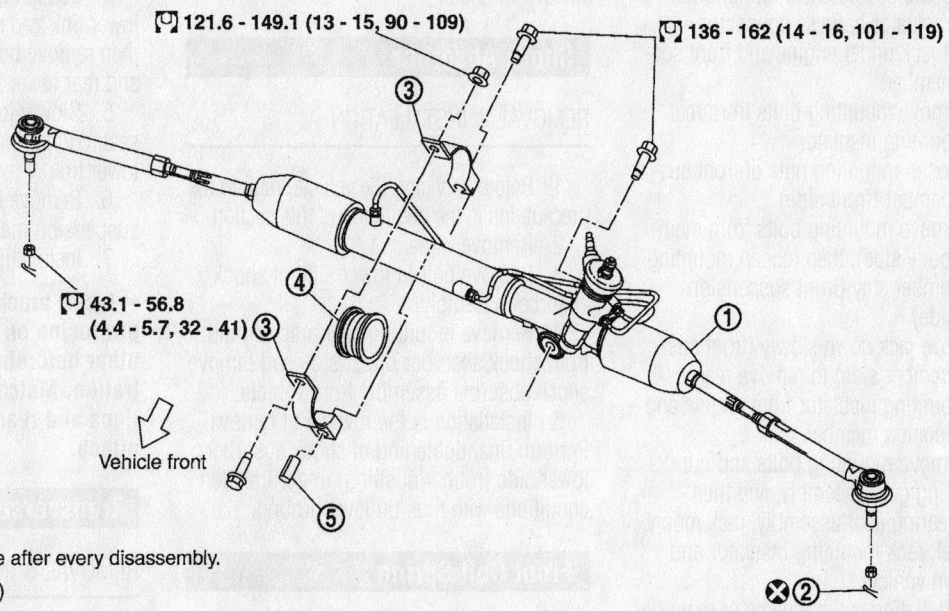

121.6 - 149.1 (13 - 15, 90 - 109)

136 - 162 (14 - 16, 101 - 119)

43.1 - 56.8 (4.4 - 5.7, 32 - 41)

Vehicle front

❌ : Always replace after every disassembly.

⟦ ⟧ : N·m(kg-m,ft-lb)

| 1. Steering gear assembly | 2. Cotter pin | 3. Rack mounting bracket |
| 4. Rack mounting insulator | 5. Sleeve | |

Steering gear and linkage

42356-MURA-G68

AWD

1. Set wheels in the straight-ahead position.

2. Remove lock nut and bolt, then separate lower joint from upper joint.

3. Remove tires from vehicle.

4. Remove undercover from vehicle.

5. Confirm slit of lower joint fits with the projection on rear cover cap, furthermore marking position on steering gear assembly nearly fits with the projection on rear cover cap.

6. Remove oil pipes (high pressure side and low pressure side) from steering gear assembly, then drain fluid from pipes.

7. Remove cotter pin at steering knuckle, then loosen mounting nut.

8. Use a ball joint remover to remove steering outer socket from steering knuckle. Be careful not to damage ball joint boot.

➡ **To prevent damage to threads and to prevent ball joint remover from coming off, and temporarily tighten mounting nut.**

9. Remove mounting bolt (lower side) from lower joint.

10. Remove front exhaust tube.

11. Remove rear propeller shaft.

12. Remove mounting nuts on lower position from stabilizer connecting rod.

13. Remove mounting bolts from stabilizer clamp and hang stabilizer on vehicle.

14. Remove steering hydraulic piping bracket from front suspension member.

15. Disconnect electrical rear engine mounting actuator harness connector.

16. Set jack under engine and front suspension member.

17. Remove mounting bolts from rear engine mounting insulator.

18. Loosen mounting nuts of front suspension member (front side).

19. Remove mounting bolts from member stay (body side), then loosen mounting nuts of member stay (front suspension member side).

20. Move jack down slowly (front suspension member side) to remove rear engine mounting insulator from engine and front suspension member.

21. Remove mounting bolts and nut from steering gear assembly, and then remove steering gear assembly, rack mounting bracket, rack mounting insulator and sleeve from vehicle.

22. Installation is the reverse of removal.

23. When steering wheel is set in the straight ahead direction, confirm slit of lower joint fits with the projection on rear cover cap, also the matchmarks on steering

gear assembly nearly fits with the projection on rear cover cap.

24. After installation, bleed air from piping.

25. Check if steering wheel turns smoothly when it is turned several times fully to the end of the left and right.

Strut

REMOVAL & INSTALLATION

1. Before servicing the vehicle, refer to the precautions in the beginning of this section.

2. Remove wheel.

3. Remove cowl top grille.

4. Remove brake caliper. Hang it in a place where it will not interfere with work.

5. Remove lock plate from brake hose from strut assembly.

6. Remove harness from wheel sensor from strut assembly. Do not pull on wheel sensor harness.

7. Remove mounting nut between strut assembly and connecting rod.

8. Remove mounting bolt and nut between strut assembly and steering knuckle.

9. Remove mounting nuts on mounting insulator bracket, then remove strut assembly from vehicle.

10. Installation is the reverse of removal. Perform final tightening of strut assembly lower side (rubber bushing) under unladen conditions with tires on level ground. Check wheel alignment.

Shock Absorber

REMOVAL & INSTALLATION

1. Before servicing the vehicle, refer to the precautions in the beginning of this section.

2. Remove wheel.

3. Remove bolt in lower side of shock absorber assembly.

4. Remove mounting seal bracket nuts from shock absorber upper side and remove shock absorber assembly from vehicle.

5. Installation is the reverse of removal. Perform final tightening of shock absorber lower side (rubber bushing) under unladen conditions with tires on level ground.

Front Coil Spring

REMOVAL & INSTALLATION

1. Before servicing the vehicle, refer to the precautions in the beginning of this section.

2. Remove the strut assembly.

3. Compress the coil spring and remove the piston rod nut.

4. Remove or disconnect the following:
   - Upper strut mount
   - Strut mount bracket
   - Upper strut bearing
   - Spring upper seat
   - Coil spring

To install:

➡ **Use new fasteners for assembly.**

5. Install or connect the following:
   - Coil spring
   - Spring upper seat
   - Upper strut bearing
   - Strut mount bracket
   - Upper strut mount. Tighten the piston rod nut to 43–58 ft. lbs. (59–78 Nm).

6. Remove the spring compressor and install the strut assembly to the vehicle.

7. Check the wheel alignment and adjust, as necessary.

Rear Coil Spring and Lower Member

REMOVAL & INSTALLATION

1. Before servicing the vehicle, refer to the precautions in the beginning of this section.

2. Remove wheel.

3. Set jack under rear lower link.

4. Loosen bolt and nut between rear lower link and suspension member, and then remove bolt and nut between rear axle and rear lower link.

5. Slowly lower jack, then remove upper seat, coil spring and rubber seat from rear lower link.

6. Remove bolt and nut between rear suspension member and rear lower link.

7. Installation is the reverse of removal.

➡ **Insert bracket tabs (3) and the inside protrusion on upper seat into each other beforehand as shown in the illustration. Match up rubber seat indentions and rear lower link grooves and attach.**

Transverse Link and Ball Joint

REMOVAL & INSTALLATION

1. Before servicing the vehicle, refer to the precautions in the beginning of this section.

2. Remove wheel.

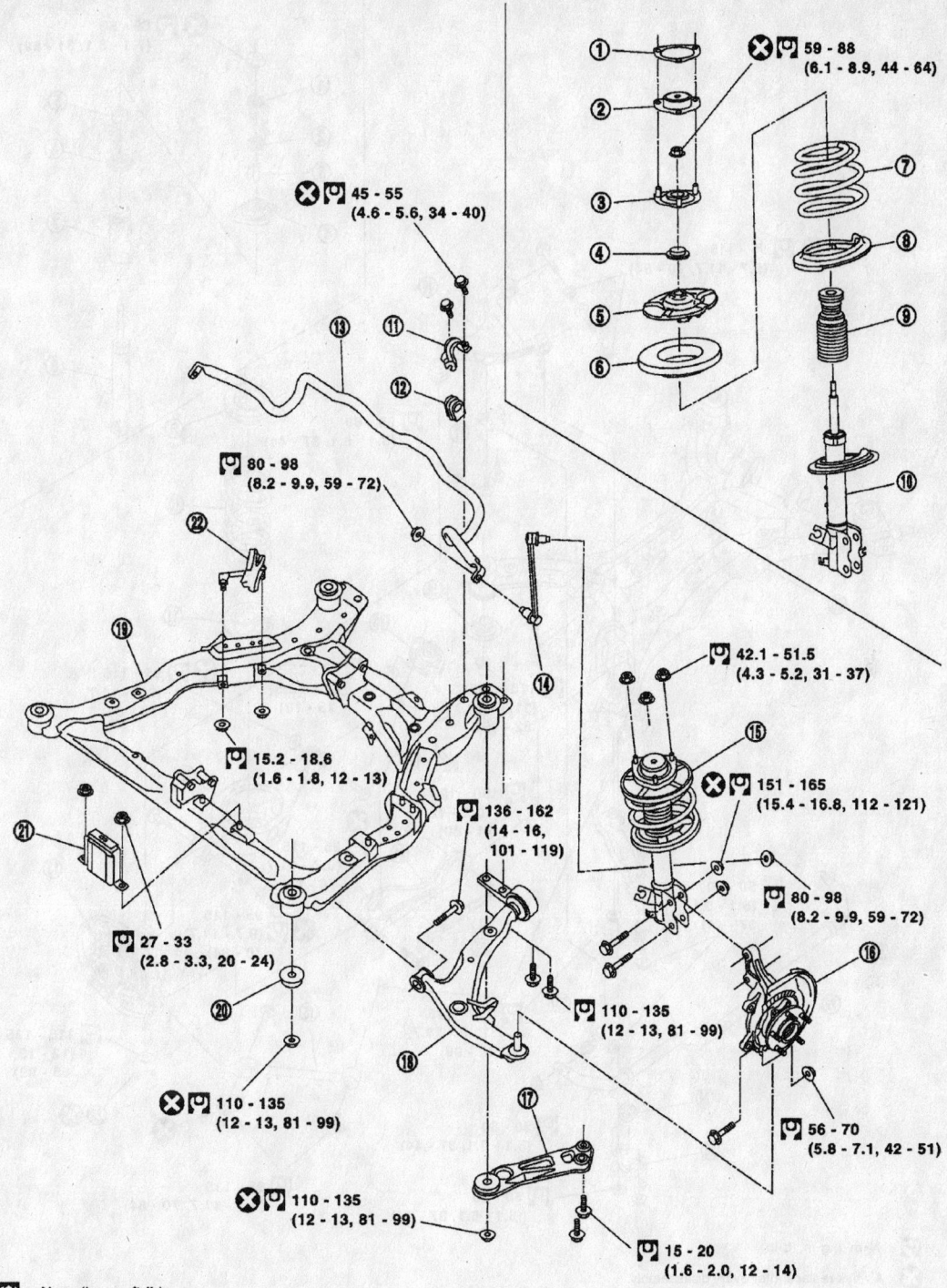

❌ 🔧 59 - 88
(6.1 - 8.9, 44 - 64)

❌ 🔧 45 - 55
(4.6 - 5.6, 34 - 40)

🔧 **80 - 98**
(8.2 - 9.9, 59 - 72)

🔧 **42.1 - 51.5**
(4.3 - 5.2, 31 - 37)

🔧 **15.2 - 18.6**
(1.6 - 1.8, 12 - 13)

❌ 🔧 151 - 165
(15.4 - 16.8, 112 - 121)

🔧 **136 - 162**
(14 - 16,
101 - 119)

🔧 **80 - 98**
(8.2 - 9.9, 59 - 72)

🔧 **27 - 33**
(2.8 - 3.3, 20 - 24)

🔧 **110 - 135**
(12 - 13, 81 - 99)

❌ 🔧 110 - 135
(12 - 13, 81 - 99)

🔧 **56 - 70**
(5.8 - 7.1, 42 - 51)

❌ 🔧 110 - 135
(12 - 13, 81 - 99)

🔧 **15 - 20**
(1.6 - 2.0, 12 - 14)

🔧 : N•m (kg-m, ft-lb)

❌ : Always replace after every disassembly.

1.	Upper mounting plate	2.	Mounting insulator	3.	Mounting insulator bracket
4.	Mounting bearing	5.	Spring upper seat	6.	Spring upper rubber seat
7.	Coil spring	8.	Spring lower rubber seat	9.	Bound bumper
10.	Strut	11.	Stabilizer clamp	12.	Stabilizer bushing
13.	Stabilizer	14.	Connecting rod	15.	Strut assembly
16.	Front axle	17.	Member stay	18.	Transverse link
19.	Front suspension member	20.	Rebound stopper	21.	Damper assembly
22.	Air guide				

42356-MURA-G69

Front suspension

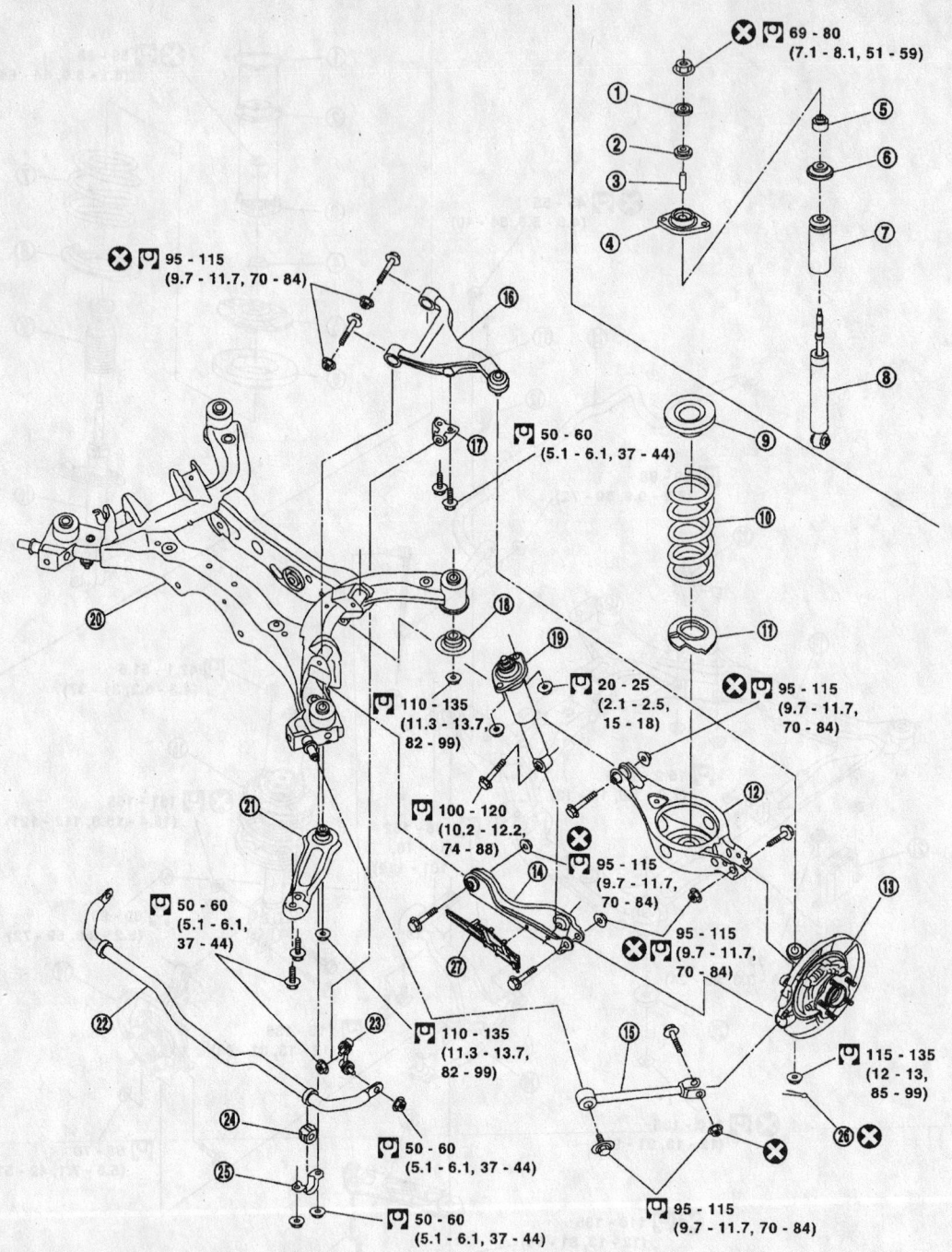

❌ 🔧 69 - 80
(7.1 - 8.1, 51 - 59)

❌ 🔧 95 - 115
(9.7 - 11.7, 70 - 84)

🔧 50 - 60
(5.1 - 6.1, 37 - 44)

🔧 110 - 135
(11.3 - 13.7, 82 - 99)

🔧 20 - 25
(2.1 - 2.5, 15 - 18)

❌ 🔧 95 - 115
(9.7 - 11.7, 70 - 84)

🔧 100 - 120
(10.2 - 12.2, 74 - 88)

❌ 🔧 95 - 115
(9.7 - 11.7, 70 - 84)

95 - 115
(9.7 - 11.7, 70 - 84)

🔧 50 - 60
(5.1 - 6.1, 37 - 44)

🔧 110 - 135
(11.3 - 13.7, 82 - 99)

🔧 115 - 135
(12 - 13, 85 - 99)

🔧 50 - 60
(5.1 - 6.1, 37 - 44)

🔧 50 - 60
(5.1 - 6.1, 37 - 44)

❌ 🔧 95 - 115
(9.7 - 11.7, 70 - 84)

🔧 : N•m (kg-m, ft-lb)

❌ : Always replace after every disassembly.

1. Outer washer	2. Bushing A	3. Distance tube
4. Mounting seal bracket	5. Bushing B	6. Bound bumper cover
7. Bound bumper	8. Shock absorber	9. Upper seat
10. Coil spring	11. Rubber seat	12. Rear lower link
13. Axle	14. Front lower link	15. Radius rod
16. Suspension arm	17. Stabilizer connecting rod mount bracket	18. Rebound stopper
19. Shock absorber assembly	20. Rear suspension member	21. Member stay
22. Stabilizer bar	23. Stabilizer connecting rod	24. Stabilizer bushing
25. Stabilizer clamp	26. Cotter pin	27. Front lower link protector

42356-MURA-G70

Rear suspension

3. Remove mounting bolt between transverse link and front suspension member.

4. Remove transverse link from steering knuckle.

5. Remove transverse link from vehicle.

6. Check transverse link and bushing for deformation, cracks, or damage. If any non-standard condition is found, replace it.

7. Check boot of ball joint for cracks or other damage, and also for grease leakage. If any non-standard condition is found, replace it.

8. Manually move ball stud to confirm it moves smoothly with no binding.

➡**Before measurement, move ball joint at least ten times by hand to check for smooth movement. Hook spring scale at ball stud. Confirm spring scale measurement value is within specifications when ball stud begins moving. If it is outside the specified range, replace suspension arm assembly.**

9. Attach mounting nut to ball stud. Check that sliding torque is within specifications with a preload gauge (SST). If it is outside the specified range, replace suspension arm assembly.

10. Move tip of ball joint in axial direction to check for looseness. If it is outside the specified range, replace suspension arm assembly.

11. Installation is the reverse of removal.

➡**Perform final tightening of front suspension member installation position and strut assembly lower side (rubber bushing) under unladen conditions with tires on level ground.**

Front Wheel Bearings

ADJUSTMENT

The front wheel bearings are part of a unitized hub and are not adjustable.

REMOVAL & INSTALLATION

1. Before servicing the vehicle, refer to the precautions in the beginning of this section.

2. Remove tire from vehicle.

3. Remove brake caliper. Hang it in a place where it will not interfere with work. Avoid depressing brake pedal while brake caliper is removed.

4. Put alignment marks on disc rotor and wheel hub and bearing assembly, then remove disc rotor.

5. Remove wheel sensor from steering knuckle.

6. Remove cotter pin, then remove lock nut from halfshaft.

7. Remove steering outer socket and cotter pin at steering knuckle, then loosen mounting nut.

8. Use a ball joint remover (SST) to remove steering outer socket from steering knuckle. Be careful not to damage ball joint boot.

➡**To prevent damage to threads and to prevent ball joint remover (SST) from coming off suddenly, temporarily tighten mounting nut.**

9. Using a puller (suitable tool), remove wheel hub and bearing assembly from halfshaft.

10. Remove wheel hub and bearing assembly bolt.

11. Remove splash guard and wheel hub and bearing assembly from steering knuckle.

12. Remove strut assembly and steering knuckle bolts and nuts.

13. Remove transverse link and steering knuckle bolt and nut.

14. Remove steering knuckle from vehicle.

15. Check for deformity, cracks and damage on each parts, replace if necessary.

16. Check for boot breakage, axial looseness, and torque of transverse link ball joint.

17. Installation is the reverse of removal.

Rear Wheel Bearings

ADJUSTMENT

The front wheel bearings are part of a unitized hub and are not adjustable.

REMOVAL & INSTALLATION

1. Before servicing the vehicle, refer to the precautions in the beginning of this section.

2. Remove the wheel.

3. Remove brake caliper. Hang it in a place where it will not interfere with work.

4. Put alignment marks on disc rotor and wheel hub and bearing assembly, then remove disc rotor.

5. Remove wheel sensor from axle.

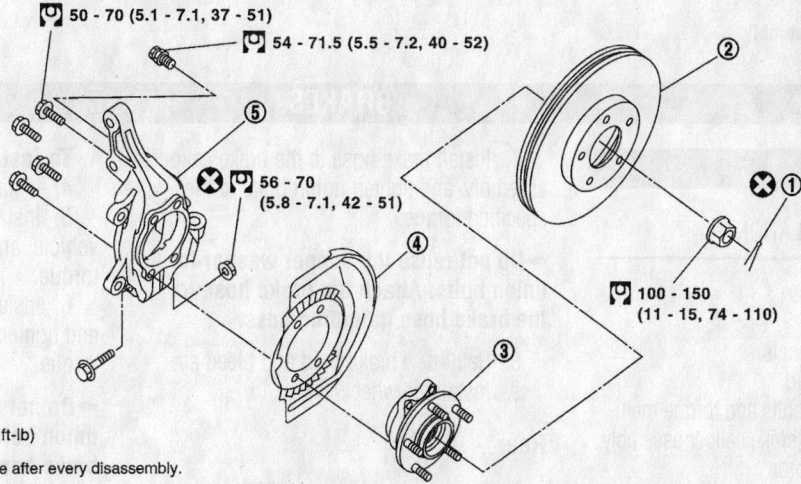

50 - 70 (5.1 - 7.1, 37 - 51)

54 - 71.5 (5.5 - 7.2, 40 - 52)

56 - 70 (5.8 - 7.1, 42 - 51)

100 - 150 (11 - 15, 74 - 110)

🔧 : N•m (kg-m, ft-lb)

✖ : Always replace after every disassembly.

1. Cotter pin
2. Disc rotor
3. Wheel hub and bearing assembly
4. Splash guard
5. Steering knuckle

42356-MURA-G71

Front wheel hub/bearing assembly

6. Remove parking cable and parking brake shoe from back plate.

7. Remove cotter pin. Then remove lock nut from halfshaft. (AWD models)

8. Using a puller (suitable tool), remove wheel hub and bearing assembly from halfshaft. (AWD models)

9. Remove wheel hub and bearing assembly from axle.

10. Loosen bolts and nuts of front lower link, radius rod and rear lower link in side of suspension member.

11. Remove shock absorber bolt (lower), front lower link bolt and nut (axle-side) while supporting rear lower link with jack.

12. Remove bolt and nut in axle side of rear lower link. Then remove coil spring.

13. Remove bolt and nut in axle side of radius rod.

14. Remove suspension arm and cotter pin at axle, then loosen mounting nut.

15. Use a ball joint remover (suitable tool) to remove suspension arm from axle. Be careful not to damage ball joint boot.

➡ To prevent damage to threads and to prevent ball joint remover (suitable tool) from coming off suddenly, and temporarily tighten mounting nut.

16. Remove axle from vehicle.

17. Remove nuts from anchor block, then remove anchor block and back plate from axle.

18. Remove axle cap (2WD) or dust shield (AWD) from axle.

19. Check for deformity, cracks and damage on each parts, replace if necessary.

20. Check for boot breakage, axial looseness, and torque of suspension arm ball joint.

21. Installation is the reverse of removal.

22. Perform final tightening of installation position of suspension links (rubber bushing) under unladen conditions with tires on level ground.

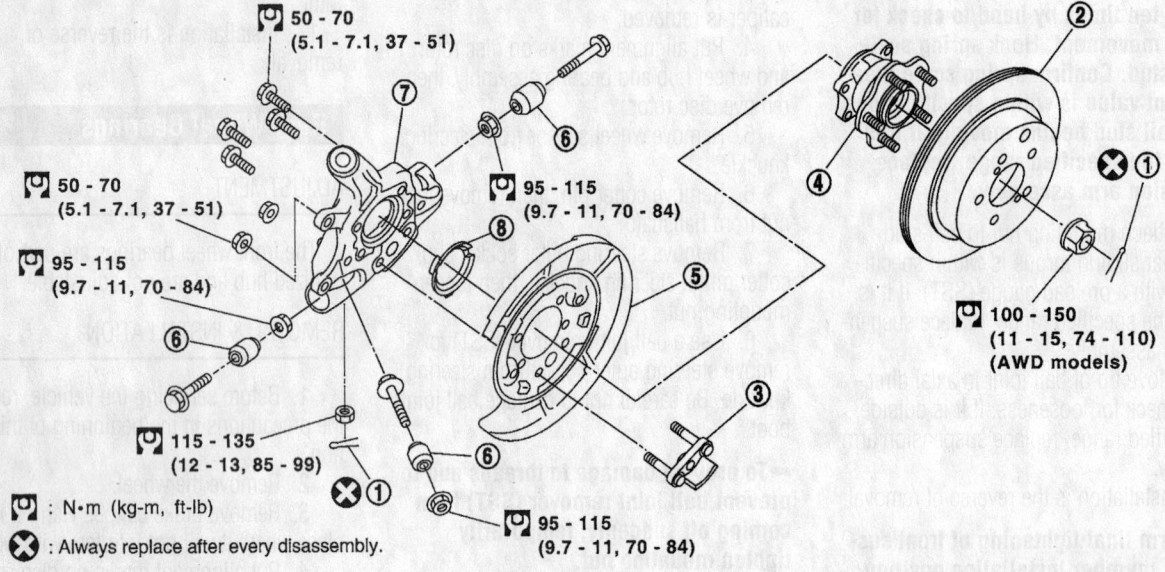

50 - 70
(5.1 - 7.1, 37 - 51)

50 - 70
(5.1 - 7.1, 37 - 51)

95 - 115
(9.7 - 11, 70 - 84)

95 - 115
(9.7 - 11, 70 - 84)

115 - 135
(12 - 13, 85 - 99)

95 - 115
(9.7 - 11, 70 - 84)

100 - 150
(11 - 15, 74 - 110)
(AWD models)

⬚ : N·m (kg-m, ft-lb)

✕ : Always replace after every disassembly.

1. Cotter pin (AWD models)
2. Disc rotor
3. Anchor block
4. Wheel hub and bearing assembly
5. Back plate
6. Bushing
7. Axle
8. Axle cap (2WD models)
   Dust shield (AWD models)

42356-MURA-G72

Rear hub/bearing assembly

BRAKES

Brake Caliper

REMOVAL & INSTALLATION

Front

1. Remove the wheels.
2. Drain brake fluid.
3. Remove union bolts and torque member bolts, and remove brake caliper assembly.
4. Remove disc rotor.

To install:
5. Install disc rotor.
6. Install caliper assembly to the vehicle, and tighten bolts to the specified torque.

7. Install brake hose to the brake caliper assembly, and tighten union bolts to the specified torque.

➡ Do not reuse the copper washer for union bolts. Attach the brake hose to the brake hose mounting boss.

8. Refill new brake fluid and bleed air.
9. Install the wheels.

Rear

1. Remove wheels from vehicle.
2. Drain brake fluid.
3. Remove union bolts and torque member bolts, and remove brake caliper assembly.

To install:
4. Install disc rotor.
5. Install caliper assembly to the vehicle, and tighten bolts to the specified torque.
6. Install brake hose to caliper assembly and tighten union bolts to the specified torque.

➡ Do not reuse the copper washer for union bolts. Attach brake hose to the brake hose mounting boss.

7. Refill new brake fluid and bleed air.
8. Install the tires to the vehicle.

Disc Brake Pads

REMOVAL & INSTALLATION

Front

1. Remove the wheels.
2. Remove sliding pin bolt (top).
3. Suspend cylinder body with a wire,

and remove pads, pad retainers, shims from torque member.
To install:
4. Apply PBC (Poly Butyl Cuprysil) grease or silicon-based grease to the rear of the pad and to both sides of the shim, and attach the inner shim and shim cover to the inner pad, and the outer shim and outer shim cover to the outer pad.

5. Attach the pad retainer and pad to the torque member.
6. Push the piston in so that the pad is attached and attach the cylinder body to the torque member.
7. Install the sliding pin bolt (top) and tighten to the specified torque.
8. Check brake for drag.
9. Install the tires to the vehicle.

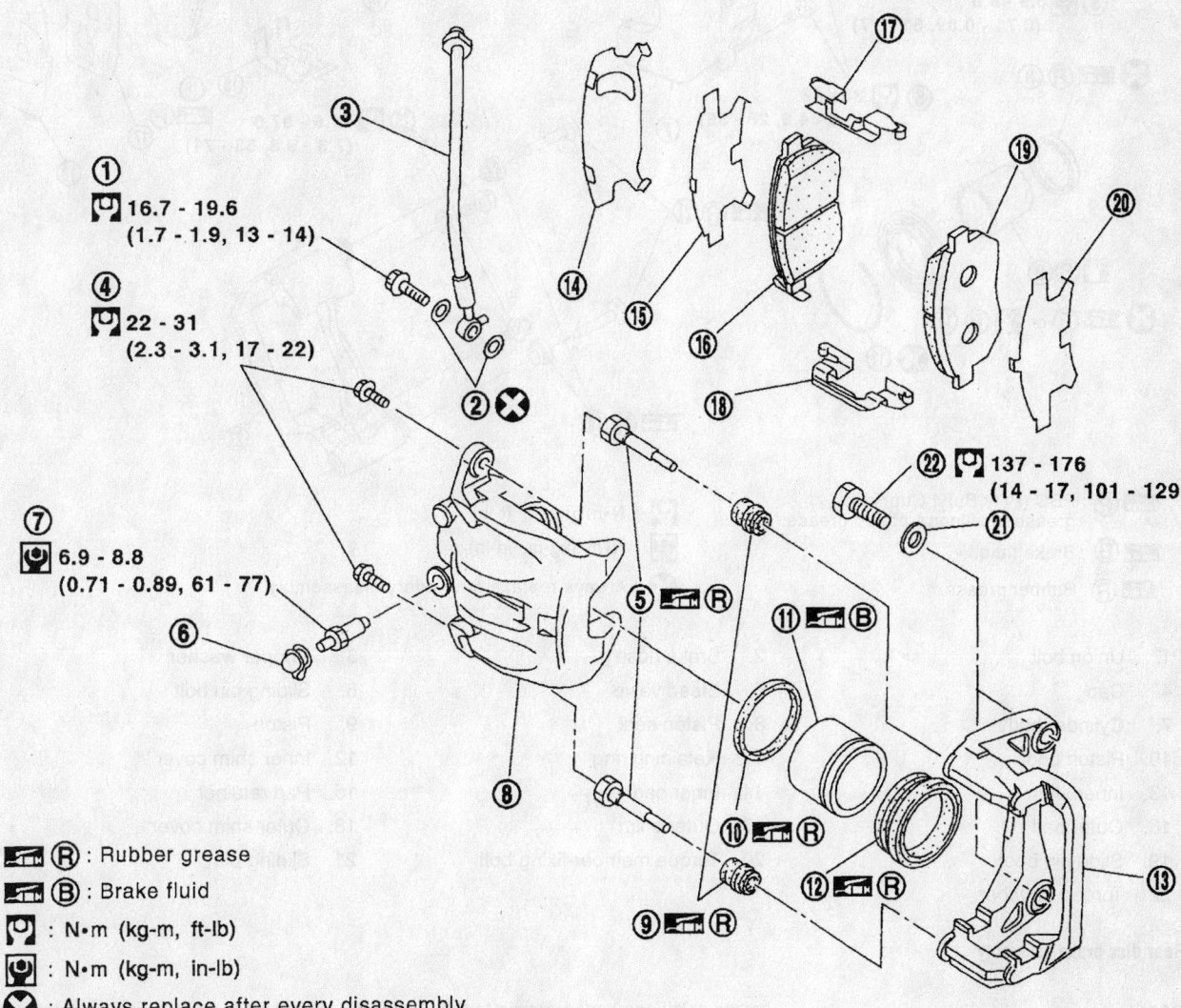

① 16.7 - 19.6 (1.7 - 1.9, 13 - 14)
④ 22 - 31 (2.3 - 3.1, 17 - 22)
⑦ 6.9 - 8.8 (0.71 - 0.89, 61 - 77)
㉒ 137 - 176 (14 - 17, 101 - 129)

: Rubber grease
: Brake fluid
N•m (kg-m, ft-lb)
N•m (kg-m, in-lb)
: Always replace after every disassembly.

1. Union bolt	2. Copper washer	3. Brake hose
4. Sliding pin bolt	5. Sliding pin	6. Cap
7. Bleed valve	8. Cylinder body	9. Sliding pin boot
10. Piston seal	11. Piston	12. Piston boot
13. Torque member	14. Inner shim cover	15. Inner shim
16. Inner pad	17. Pad retainer (Upper)	18. Pad retainer (Lower)
19. Outer pad	20. Outer shim	21. Washer
22. Torque member fixing bolt		

Front disc brake assembly

42356-MURA-G73

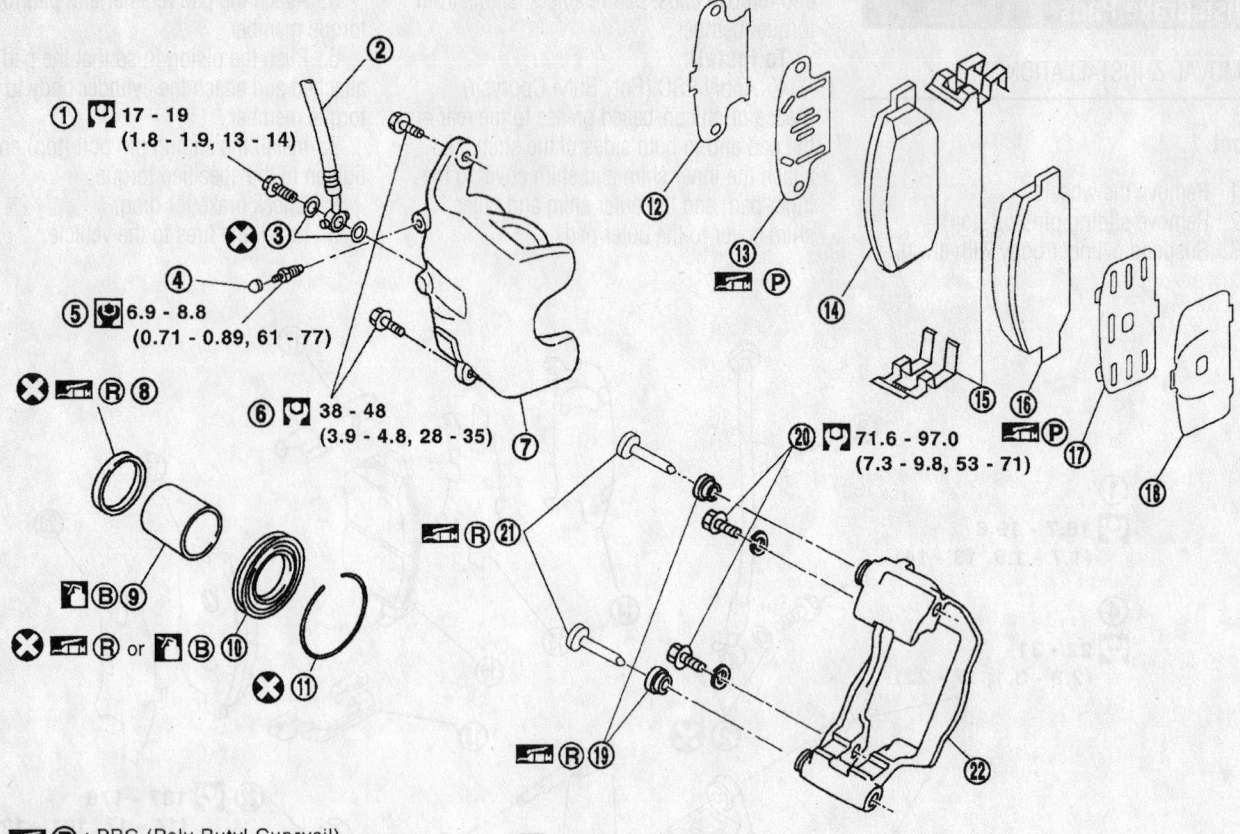

① 🔧 17 - 19
(1.8 - 1.9, 13 - 14)

✕ ③

④

⑤ 🔧 6.9 - 8.8
(0.71 - 0.89, 61 - 77)

✕ 🔧 ⓡ ⑧

🔧 ⓑ ⑨

✕ 🔧 ⓡ or 🔧 ⓑ ⑩

✕ ⑪

⑥ 🔧 38 - 48
(3.9 - 4.8, 28 - 35) ⑦

②

⑫

⑬

🔧 ⓟ ⑭

⑮ ⑯

🔧 ⓟ ⑰

⑱

⑳ 🔧 71.6 - 97.0
(7.3 - 9.8, 53 - 71)

🔧 ⓡ ㉑

🔧 ⓡ ⑲

㉒

🔧 ⓟ : PBC (Poly Butyl Cuprysil)
grease or silicone-based grease.

🔧 ⓑ : Brake fluid.

🔧 ⓡ : Rubber grease.

🔧 : N•m (kg-m, ft-lb)

🔧 : N•m (kg-m, in-lb)

✕ : Always replace after every disassembly.

1. Union bolt	2. Brake hose	3. Copper washer
4. Cap	5. Bleed valve	6. Sliding pin bolt
7. Cylinder body	8. Piston seal	9. Piston
10. Piston boot	11. Retaining ring	12. Inner shim cover
13. Inner shim	14. Inner pad	15. Pad retainer
16. Outer pad	17. Outer shim	18. Outer shim cover
19. Slide pin boot	20. Torque member fixing bolt	21. Sliding pin
22. Torque member		

42356-MURA-G74

Rear disc brake assembly

Rear

1. Remove tires from vehicle.
2. Remove sliding pin bolt (top).
3. Suspend cylinder body with a wire, and remove pads, pad retainers, shims from torque member.

To install:

4. Apply PBC (Poly Butyl Cuprysil) grease or silicon based grease to the rear of the pad and to both sides of the shim, and attach the inner shim and shim cover to the inner pad, and the outer shim and outer shim cover to the outer pad.

5. Attach the pad retainer and pad to the torque member.

6. Push the piston in so that the pad is attached and attach the cylinder body to the torque member.

7. Install the sliding pin bolt (one on top) and tighten to the specified torque.

8. Check brake for drag.
9. Install the tires to the vehicle.

NISSAN

Quest

26

SPECIFICATION CHARTS

VEHICLE AND ENGINE IDENTIFICATION CHART

Engine Code							Model Year	
Code	Liters (cc)	Cu. In.	Cyl.	Fuel Sys.	Engine Type	Eng. Mfg.	Code	Year
T	3.3 (3275)	200	6	SEFI	SOHC	Nissan	Y	2000
							1	2001
							2	2002
							3	2003
							4	2004

MFI: Multi-port Fuel Injection

SEFI: Sequential Multi-port Fuel Injection

42356-QUES-C01

GENERAL ENGINE SPECIFICATIONS

Year	Engine Displacement Liters (cc)	Engine VIN	Fuel System Type	Net Horsepower @ rpm	Net Torque @ rpm (ft. lbs.)	Bore x Stroke (in.)	Compression Ratio	Oil Pressure @ rpm
2000	3.3 (3275)	T	SEFI	195@4500	190@3800	3.60x3.27	8.9:1	60-65@2000
2001	3.3 (3275)	T	SEFI	195@4500	190@3800	3.60x3.27	8.9:1	60-65@2000
2002	3.3 (3275)	T	SEFI	195@4500	190@3800	3.60x3.27	8.9:1	60-65@2000

MFI: Multiport fuel injection

SEFI: Sequential Multi-port Fuel Injection

42356-QUES-C02

ENGINE TUNE-UP SPECIFICATIONS

Year	Engine Displacement Liters (cc)	Engine ID/VIN	Spark Plug Gap (in.)	Ignition Timing (deg.) MT	AT	Fuel Pump (psi) ①	Idle Speed (rpm) MT	AT ②	Valve Clearance In.	Ex.
2000	3.3 (3277)	T	0.043	—	13-17B	34	—	650-750	HYD	HYD
2001	3.3 (3277)	T	0.043	—	13-17B	34	—	650-750	HYD	HYD
2002	3.3 (3277)	T	0.043	—	13-17B	34	—	650-750	HYD	HYD

NOTE: The Vehicle Emission Control Information label must be used if they differ from those in this chart.

B: Before top dead center

HYD: Hydraulic

① System pressure at idle with vacuum hose connected should increase to 43 psi when disconnected

② Transmission in Neutral

42356-QUES-C03

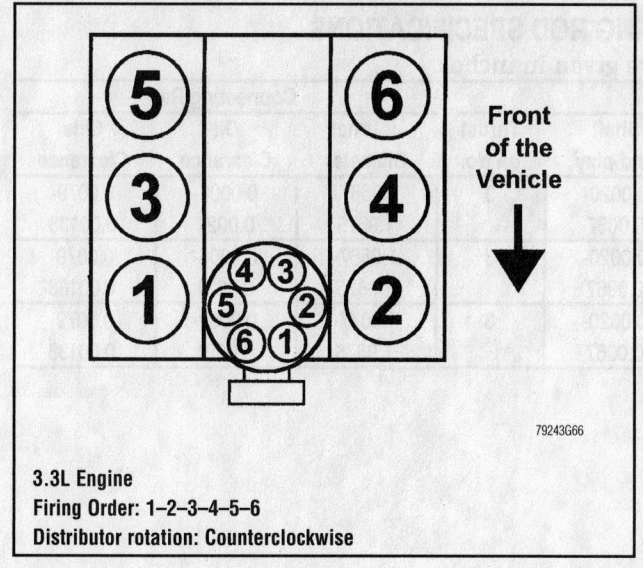

3.3L Engine
Firing Order: 1–2–3–4–5–6
Distributor rotation: Counterclockwise

79243G66

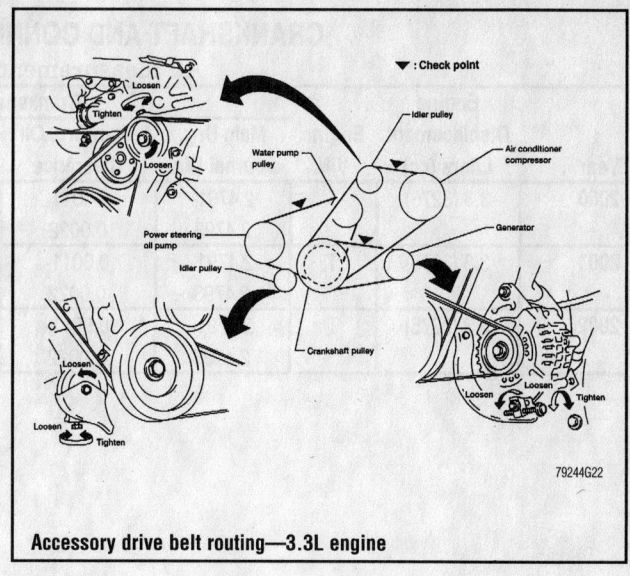

Accessory drive belt routing—3.3L engine

79244G22

CAPACITIES

Year	Model	Engine Displacement Liters (cc)	Engine VIN	Engine Oil with Filter (qts.)	Transmission (pts.) 4-Spd	Transmission (pts.) 5-Spd	Transmission (pts.) Auto.	Drive Axle Front (pts.)	Drive Axle Rear (pts.)	Fuel Tank (gal.)	Cooling System (qts.) ②
2000	Quest	3.3 (3275)	T	4.0	—	—	20.0	①	—	20	11.25
2001	Quest	3.3 (3275)	T	4.0	—	—	20.0	①	—	20	11.25
2002	Quest	3.3 (3275)	T	4.0	—	—	20.0	①	—	20	11.25

NOTE: All capacities are approximate. Add fluid gradually and check to be sure a proper fluid level is obtained.

① Included in transaxle capacity
② Includes reservoir tank.

42356-QUES-C04

VALVE SPECIFICATIONS

Year	Engine VIN	Engine Displacement Liters (cc)	Seat Angle (deg.)	Face Angle (deg.)	Spring Test Pressure (lbs. @ in.)	Spring Installed Height (in.)	Stem-to-Guide Clearance (in.) Intake	Stem-to-Guide Clearance (in.) Exhaust	Stem Diameter (in.) Intake	Stem Diameter (in.) Exhaust
2000	T	3.3 (3275)	45	45	①	②	0.0008-0.0021	0.0012-0.0019	0.2742-0.2748	0.3136-0.3138
2001	T	3.3 (3275)	45	45	①	②	0.0008-0.0021	0.0012-0.0019	0.2742-0.2748	0.3136-0.3138
2002	T	3.3 (3275)	45	45	①	②	0.0008-0.0021	0.0012-0.0019	0.2742-0.2748	0.3136-0.3138

① Outer spring: 118@1.81
   Inner spring: 57.3@0.984
② Spring height measured unloaded
   Minimum length. outer spring: 2.016
   Minimum length. inner spring: 1.736

42356-QUES-C05

CRANKSHAFT AND CONNECTING ROD SPECIFICATIONS
All measurements are given in inches.

Year	Engine Displacement Liters (cc)	Engine VIN	Crankshaft				Connecting Rod		
			Main Brg. Journal Dia.	Main Brg. Oil Clearance	Shaft End-play	Thrust on No.	Journal Diameter	Oil Clearance	Side Clearance
2000	3.3 (3275)	T	2.4791-2.4793	0.0011-0.0022	0.0020-0.0067	3	1.9667-1.9675	0.0006-0.0021	0.0079-0.00138
2001	3.3 (3275)	T	2.4791-2.4793	0.0011-0.0022	0.0020-0.0067	3	1.9667-1.9675	0.0006-0.0021	0.0079-0.00138
2002	3.3 (3275)	T	2.4791-2.4793	0.0011-0.0022	0.0020-0.0067	3	1.9667-1.9675	0.0006-0.0021	0.0079-0.00138

42356-QUES-C06

PISTON AND RING SPECIFICATIONS
All measurements are given in inches.

Year	Engine Displacement Liters (cc)	Engine VIN	Piston Clearance	Ring Gap			Ring Side Clearance		
				Top Compression	Bottom Compression	Oil Control	Top Compression	Bottom Compression	Oil Control
2000	3.3 (3275)	T	①	0.0083-0.0122	0.0197-0.0236	0.0079-0.0236	0.0016-0.0031	0.0012-0.0028	0.0006-0.0073
2001	3.3 (3275)	T	①	0.0083-0.0122	0.0197-0.0236	0.0079-0.0236	0.0016-0.0031	0.0012-0.0028	0.0006-0.0073
2002	3.3 (3275)	T	①	0.0083-0.0122	0.0197-0.0236	0.0079-0.0236	0.0016-0.0031	0.0012-0.0028	0.0006-0.0073

① Journals 1, 2 and 6: 0.0010 - 0.0018 in.
  Journals 3 and 4: 0.0006 - 0.0010 in.
  Journal 5: 0.0012 - 0.0016 in.

42356-QUES-C07

TORQUE SPECIFICATIONS
All readings in ft. lbs.

Year	Engine VIN	Engine Displacement Liters (cc)	Cylinder Head Bolts	Main Bearing Bolts	Rod Bearing Bolts	Crankshaft Damper Bolts	Flywheel Bolts	Manifold Intake	Manifold Exhaust	Spark Plugs	Lug Nut
2000	T	3.3 (3275)	①	②	②	141-156	61-69	①	13-16	14-22	80
2001	T	3.3 (3275)	①	②	②	141-156	61-69	①	13-16	14-22	80
2002	T	3.3 (3275)	①	②	②	141-156	61-69	①	13-16	14-22	80

① Intake manifold and cylinder heads are installed at the same time.

   Step 1: cylinder head bolts to 22 ft. lbs.

   Step 2: cylinder head bolts to 43 ft. lbs.

   Step 3: Loosen all bolts completely

   Step 4: cylinder head bolts to 7 ft. lbs.

   Step 5: Intake manifold bolts to 2.9 (4Nm) ft. lbs.

   Step 6: Intake manifold bolts to 13 ft. lbs.

   Step 7: Intake manifold bolts to 14 ft. lbs.

   Step 8: Loosen all maifold bolts completely

   Step 9: Cylinder head bolts to 26 inch lbs.

   Step 10: cylinder head bolts to 47 ft. lbs. Or, an additional 65 degrees

   Step 11: cylinder head sub-bolts to 104 inch lbs.

   Step 12: Intake manifold bolts to 36 inch lbs.

   Step 13: Intake manifold bolts to 78 inch lbs.

   Step 14: Intake manifold bolts to 84 inch lbs.

② Step 1: 34-37 ft. lbs.

   Step 2: 67-74 ft. lbs.

42356-QUES-C08

WHEEL ALIGNMENT

Year	Model		Caster Range (Deg.)	Caster Preferred Setting (Deg.)	Camber Range (Deg.)	Camber Preferred Setting (Deg.)	Toe-in (in.)	Steering Axis Inclination (Deg.)
2000	Quest	F	0.75	2.75	0.75	-0.25	0.04 +/- 0.04	—
		R	—	—	0.75	-1.00	0.04 +/- 0.16	—
2001	Quest	F	0.75	2.75	0.75	-0.25	0.04 +/- 0.04	—
		R	—	—	0.75	-1.00	0.04 +/- 0.16	—
2002	Quest	F	0.75	2.75	0.75	-0.25	0.04 +/- 0.04	—
		R	—	—	0.75	-1.00	0.04 +/- 0.16	—

42356-QUES-C09

TIRE, WHEEL AND BALL JOINT SPECIFICATIONS

Year	Model	OEM Tires		Tire Pressures (psi)		Wheel Size	Ball Joint Inspection
		Standard	Optional	Front	Rear		
2000	Quest	P215/70R15	P225/60R16	30	30	Std: 5.5-JJ Opt: 6.5-JJ	①
2001	Quest	P215/70R15	P225/60R16	30	30	Std: 5.5-JJ Opt: 6.5-JJ	①
2002	Quest	P215/70R15	P225/60R16	30	30	Std: 5.5-JJ Opt: 6.5-JJ	①

OEM: Original Equipment Manufacturer

PSI: Pounds Per Square Inch

① Replace if any measurable movement is found.

42356-QUES-C10

BRAKE SPECIFICATIONS
All measurements in inches unless noted

Year	Model		Brake Disc			Brake Drum Diameter			Minimum Lining Thickness	Brake Caliper	
			Original Thickness	Minimum Thickness	Maximum Runout	Original Inside Diameter	Max, Wear Limit	Maximum Machine Diameter		Bracket-to-Hub Bolt (ft. lbs.)	Mounting Pin or Bolt (ft. lbs.)
2000	Quest	F	1.002	0.940	0.0028	—	—	—	0.079	—	18-25
		R	—	—	—	9.84	9.90	9.86	0.079	—	—
2001	Quest	F	1.002	0.940	0.0028	—	—	—	0.079	—	18-25
		R	—	—	—	9.84	9.90	9.86	0.079	—	—
2002	Quest	F	1.002	0.940	0.0028	—	—	—	0.079	—	18-25
		R	—	—	—	9.84	9.90	9.86	0.079	—	—

NOTE: Due to changes made during production, refer to the manufacturer's specifications if they differ from those in this chart

F: Front

R: Rear

42356-QUES-C11

SCHEDULED MAINTENANCE INTERVALS
NISSAN QUEST

TO BE SERVICED	TYPE OF SERVICE	VEHICLE MILEAGE INTERVAL (x1000)												
		5	10	15	20	25	30	35	40	45	50	55	60	65
Engine oil & filter	R	✓	✓	✓	✓	✓	✓	✓	✓	✓	✓	✓	✓	✓
Rotate tires	S/I	✓		✓		✓		✓		✓		✓		✓
Engine coolant strength hoses & clamps	S/I			✓			✓			✓			✓	
Air cleaner filter	R						✓						✓	
Automatic transmission fluid & filter	R						✓						✓	
Engine coolant ①	R						✓						✓	
PCV valve	R												✓	
Spark plugs ②	R													
Drive belts ③	S/I			✓			✓			✓			✓	
Timing belt ④	S/I													
Exhaust system & heat shields	S/I			✓			✓			✓			✓	
Drive shaft boots	S/I			✓			✓			✓		✓		
Front & rear brake components	S/I	✓	✓	✓	✓	✓	✓	✓	✓	✓	✓	✓	✓	✓

R: Replace   S/I: Service or Inspect

① Engine coolant: change every 30,000 miles or 36 months.

② Replace every 105,000 miles

③ Inspect every 15, 000 miles or 12 months and replace every 60,000 miles or 48 months.

④ Replace every 105,000 miles

FREQUENT OPERATION MAINTENANCE (SEVERE SERVICE)

If a vehicle is operated under any of the following conditions it is considered severe service:

- Extremely dusty areas.
- 50% or more of the vehicle operation is in 32°C (90°F) or higher temperatures, or constant operation in temperatures below 0°C (32°F).
- Prolonged idling (vehicle operation in stop and go traffic.
- Frequent short running periods (engine does not warm to normal operating temperatures).
- Police, taxi, delivery usage or trailer towing usage.

Engine oil & filter: replace every 3000 miles.

Rotate tires initially at 6000 miles and every 9000 miles thereafter.

Air cleaner filter: change every 15,000 miles.

Engine coolant strength, hoses & clamps: check every 15,000 miles.

Exhaust system: check every 15,000 miles.

Automatic transmission fluid & filter: change every 21,000 miles.

42356-QUES-C12

PRECAUTIONS

Before servicing any vehicle, please be sure to read all of the following precautions, which deal with personal safety, prevention of component damage, and important points to take into consideration when servicing a motor vehicle:

• Never open, service or drain the radiator or cooling system when the engine is hot; serious burns can occur from the steam and hot coolant.

• Observe all applicable safety precautions when working around fuel. Whenever servicing the fuel system, always work in a well-ventilated area. Do not allow fuel spray or vapors to come in contact with a spark, open flame or excessive heat (a hot drop light, for example). Keep a dry chemical fire extinguisher near the work area. Always keep fuel in a container specifically designed for fuel storage; also, always properly seal fuel containers to avoid the possibility of fire or explosion. Refer to the additional fuel system precautions later in this section.

• Fuel injection systems often remain pressurized, even after the engine has been turned **OFF**. The fuel system pressure must be relieved before disconnecting any fuel lines. Failure to do so may result in fire and/or personal injury.

• Brake fluid often contains polyglycol ethers and polyglycols. Avoid contact with the eyes and wash your hands thoroughly after handling brake fluid. If you do get brake fluid in your eyes, flush your eyes with clean, running water for 15 minutes. If

eye irritation persists, or if you have taken brake fluid internally, IMMEDIATELY seek medical assistance.

• The EPA warns that prolonged contact with used engine oil may cause a number of skin disorders, including cancer! You should make every effort to minimize your exposure to used engine oil. Protective gloves should be worn when changing oil. Wash your hands and any other exposed skin areas as soon as possible after exposure to used engine oil. Soap and water, or waterless hand cleaner should be used.

• All new vehicles are now equipped with an air bag system. The system must be disabled before performing service on or around system components, steering column, instrument panel components, wiring and sensors. Failure to follow safety and disabling procedures could result in accidental air bag deployment, possible personal injury and unnecessary system repairs.

• Always wear safety goggles when working with, or around, the air bag system. When carrying a non-deployed air bag, be sure the bag and trim cover are pointed away from your body. When placing a non-deployed air bag on a work surface, always face the bag and trim cover upward, away from the surface. This will reduce the motion of the module if it is accidentally deployed. Refer to the additional air bag system precautions later in this section.

• Clean, high quality brake fluid from a sealed container is essential to the safe and proper operation of the brake system. You should always buy the correct type of brake fluid for your vehicle. If the brake fluid becomes contaminated, completely flush the system with new fluid. Never reuse any brake fluid. Any brake fluid that is removed from the system should be discarded. Also, do not allow any brake fluid to come in contact with a painted surface; it will damage the paint.

• Never operate the engine without the proper amount and type of engine oil; doing so WILL result in severe engine damage.

• Timing belt maintenance is extremely important! Many models utilize an interference-type, non-freewheeling engine. If the timing belt breaks, the valves in the cylinder head may strike the pistons, causing potentially serious (also time-consuming and expensive) engine damage.

• Disconnecting the negative battery cable on some vehicles may interfere with the functions of the on-board computer system(s) and may require the computer to undergo a relearning process once the negative battery cable is reconnected.

• When servicing drum brakes, only disassemble and assemble one side at a time, leaving the remaining side intact for reference.

• Only an MVAC-trained, EPA-certified automotive technician should service the air conditioning system or its components.

ENGINE REPAIR

Distributor

REMOVAL

1. Before servicing the vehicle, refer to the precautions in the beginning of this section.
2. Remove or disconnect the following:
   • Negative battery cable
   • Distributor cap
   • Distributor wiring harness connector
3. Matchmark the rotor to the distributor housing and the distributor housing to the cylinder head.
4. Remove the distributor hold-down bolt and the distributor.

INSTALLATION

Timing Not Disturbed

1. Install or connect the following:
   • Distributor and align the matchmarks made during removal. Tighten the hold-down bolt to 10–12 ft. lbs. (14–17 Nm).
   • Distributor wiring harness connector
   • Distributor cap
   • Negative battery cable
2. Check the ignition timing and adjust, as necessary.

Timing Disturbed

1. Set the engine to Top Dead Center (TDC) of the compression stroke for the No. 1 cylinder.

2. Align the index mark on the distributor shaft with the protrusion on the distributor housing.
3. Install the distributor and check that the distributor rotor is aligned.

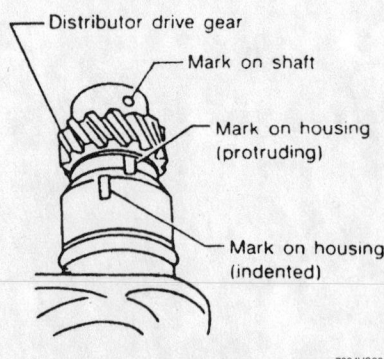

Distributor drive gear
Mark on shaft
Mark on housing (protruding)
Mark on housing (indented)

7924VG28

Distributor shaft alignment

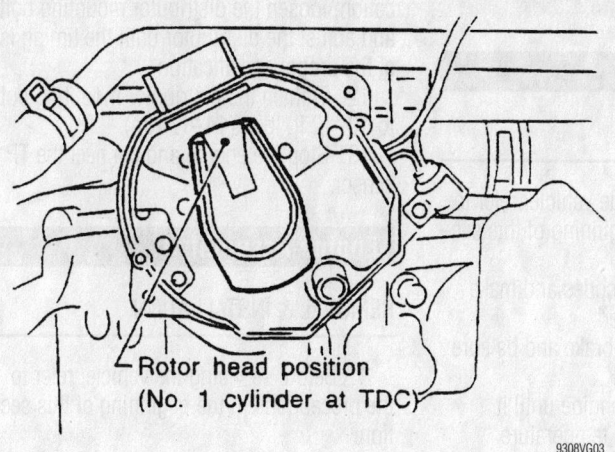

Rotor head position
(No. 1 cylinder at TDC)

9308VG03

Distributor rotor alignment

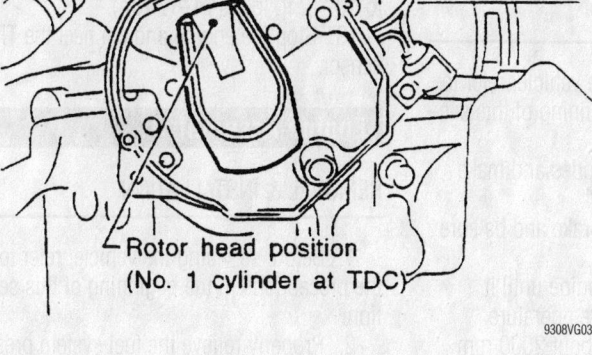

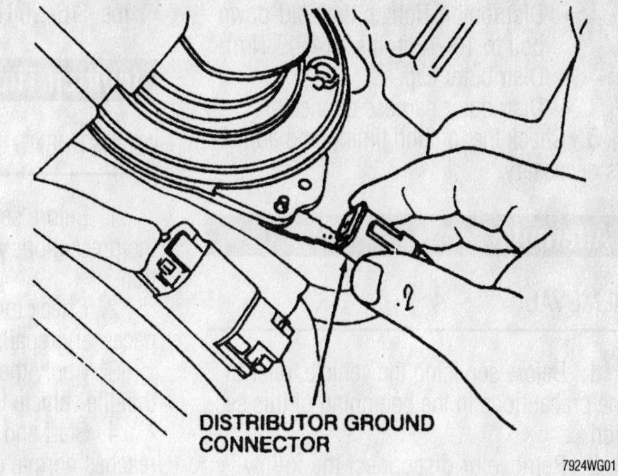

DISTRIBUTOR GROUND
CONNECTOR

7924WG01

Disengage the distributor ground connector when removing the distributor

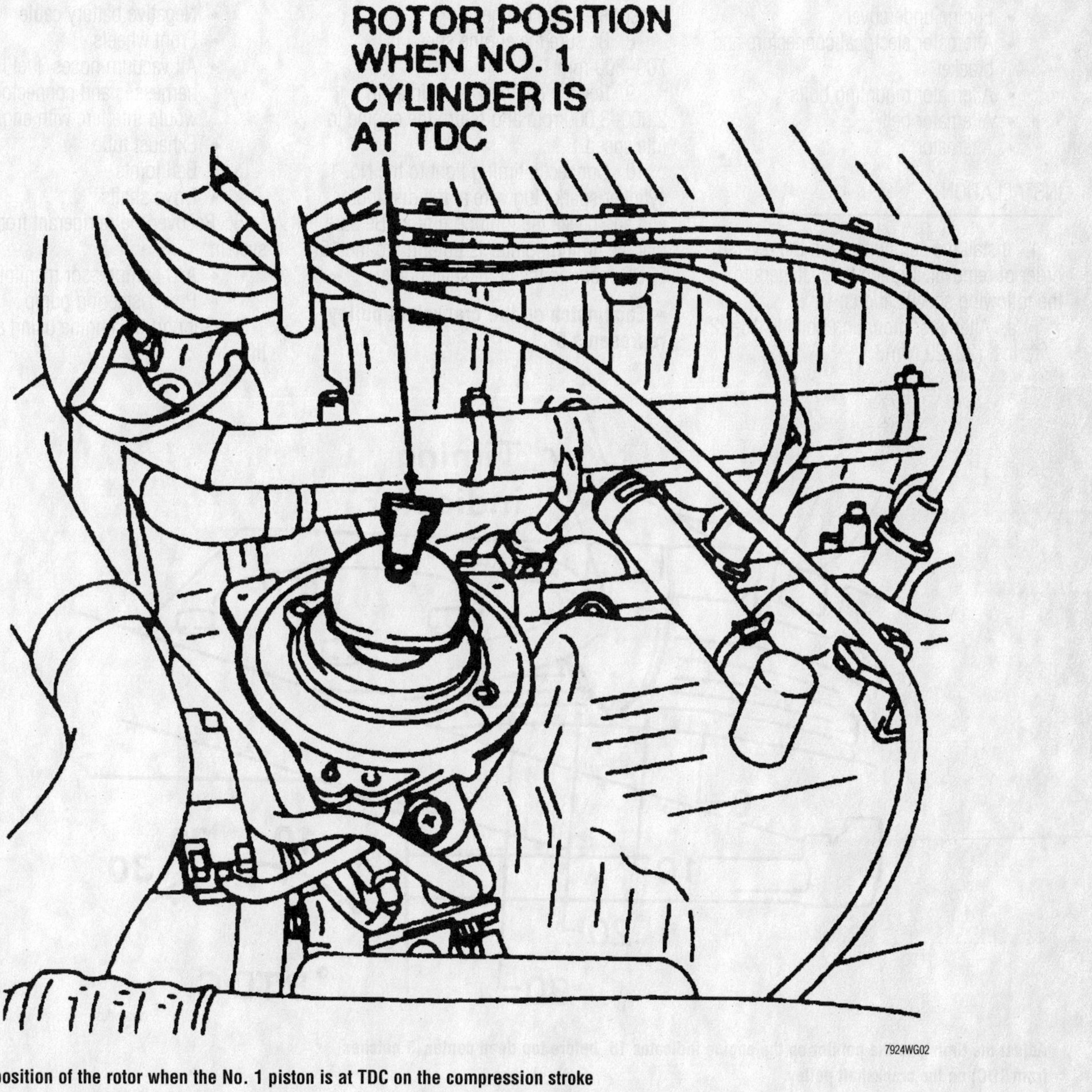

ROTOR POSITION
WHEN NO. 1
CYLINDER IS
AT TDC

7924WG02

Note the position of the rotor when the No. 1 piston is at TDC on the compression stroke

4. Install or connect the following:
- Distributor. Tighten the hold-down bolt to 10–12 ft. lbs. (14–17 Nm).
- Distributor cap
- Distributor harness connector

5. Check the ignition timing and adjust, as necessary.

Alternator

REMOVAL

1. Before servicing the vehicle, refer to the precautions in the beginning of this section.

2. Remove or disconnect the following:
- Negative battery cable
- Idler adjusting bolt, loosen
- A/C belt
- Engine undercover
- Alternator electrical connectors and bracket
- Alternator mounting bolts
- Alternator belt
- Alternator

INSTALLATION

1. Install the components in the reverse order of removal. Tighten the fasteners to the following specifications:
   a. Alternator mounting bolts to 16–22 ft. lbs. (22–29 Nm).

   b. Alternator bracket bolt to 12–15 ft. lbs. (16–20 Nm).

Ignition Timing

ADJUSTMENT

1. Before servicing the vehicle, refer to the precautions in the beginning of this section.

2. Check for trouble codes and make necessary repairs if needed.

3. Apply the parking brake and be sure that the vehicle is in PARK.

4. Start and run the engine until it reaches normal operating temperature.

5. Run the engine at about 2000 rpm for 2 minutes under no-load.

6. Turn off all electrical loads.

7. Disconnect the Throttle Position (TP) sensor electrical connector.

8. Be sure the engine speed is 700–800 rpm.

9. Rev the engine 2 or 3 times to 2,000–3,000 rpm and return the engine to idle speed.

10. Connect a timing light to the No. 1 cylinder spark plug wire at the distributor end and check the ignition timing. Be sure that the timing pointer is pointing to the 15° BTDC mark on the crankshaft pulley.

➡ Each notch on the crankshaft pulley represents 5°.

11. If the timing is not within the specification, loosen the distributor mounting bolt and adjust the distributor until the timing is at the proper specification.

12. Tighten the distributor mounting bolt to 10–12 ft. lbs., (14–17 Nm).

13. Stop the engine and connect the TP sensor.

Engine Assembly

REMOVAL & INSTALLATION

1. Before servicing the vehicle, refer to the precautions in the beginning of this section.

2. Properly relieve the fuel system pressure.

3. Drain the coolant and crankcase.

4. Remove or disconnect the following:
- Negative battery cable
- Front wheels
- All vacuum hoses, fuel lines, wires, harnesses and connectors that would interfere with engine removal
- Exhaust tube
- Ball joints
- Drive shafts

5. Recover the refrigerant from the A/C system
- A/C compressor manifold
- Power steering pump

6. Support the engine using a suitable lift.

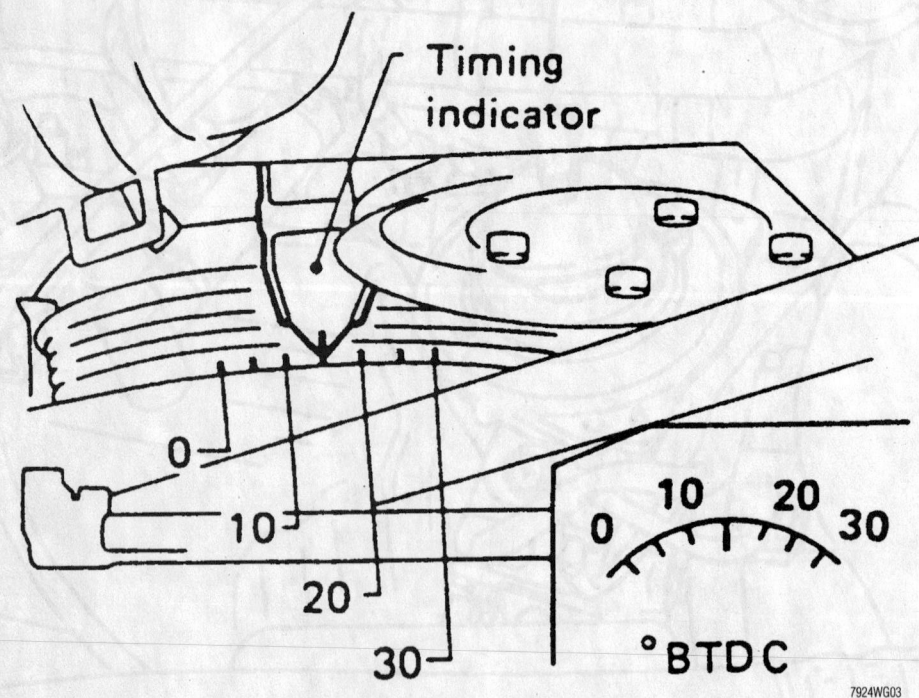

Adjust the timing so the pointer on the engine indicates 15° before top dead center (3 notches from TDC) on the crankshaft pulley.

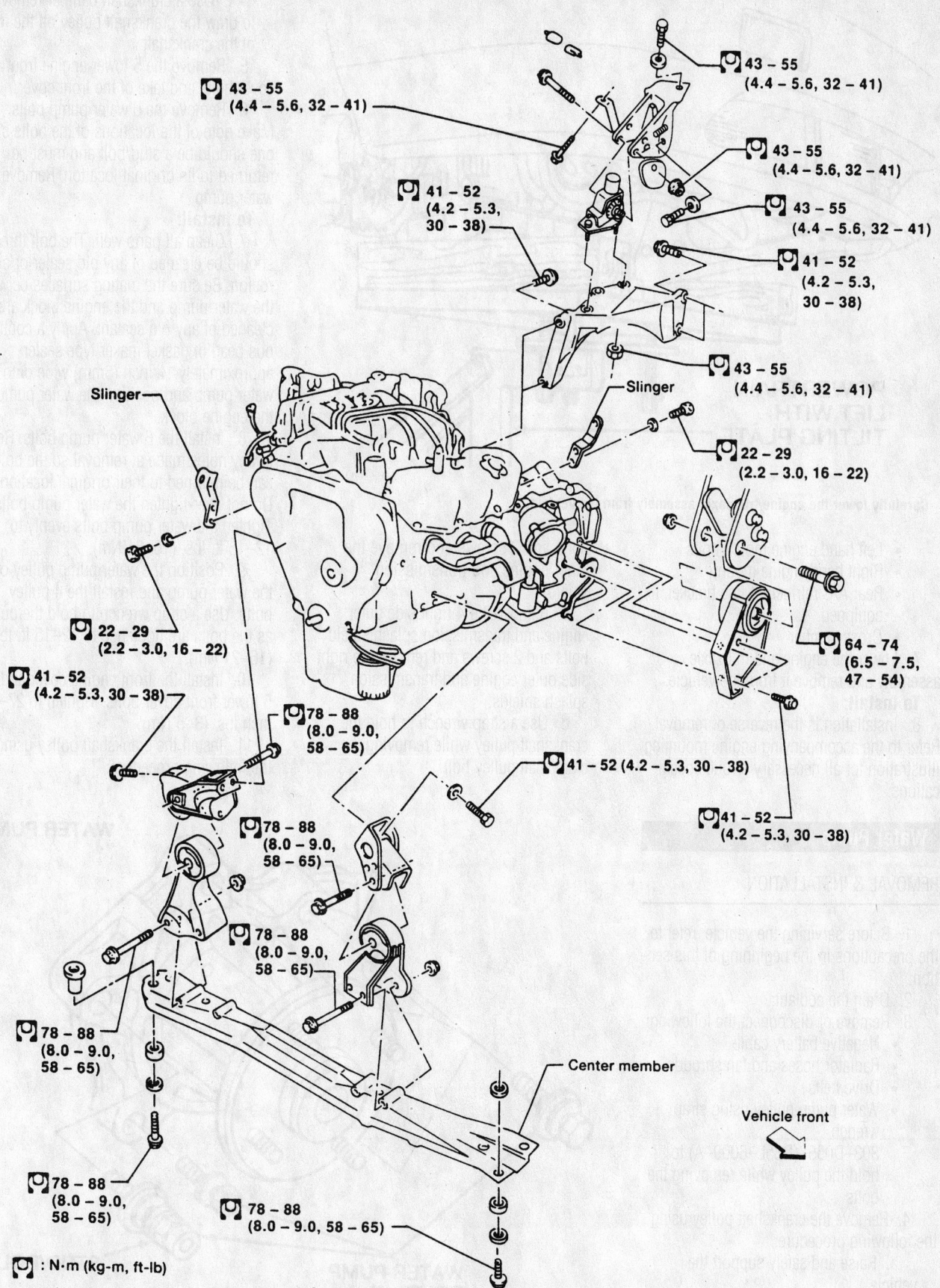

43 − 55
(4.4 − 5.6, 32 − 41)

43 − 55
(4.4 − 5.6, 32 − 41)

43 − 55
(4.4 − 5.6, 32 − 41)

41 − 52
(4.2 − 5.3,
30 − 38)

43 − 55
(4.4 − 5.6, 32 − 41)

41 − 52
(4.2 − 5.3,
30 − 38)

43 − 55
(4.4 − 5.6, 32 − 41)

22 − 29
(2.2 − 3.0, 16 − 22)

Slinger

Slinger

22 − 29
(2.2 − 3.0, 16 − 22)

64 − 74
(6.5 − 7.5,
47 − 54)

41 − 52
(4.2 − 5.3, 30 − 38)

78 − 88
(8.0 − 9.0,
58 − 65)

41 − 52 (4.2 − 5.3, 30 − 38)

41 − 52
(4.2 − 5.3, 30 − 38)

78 − 88
(8.0 − 9.0,
58 − 65)

78 − 88
(8.0 − 9.0,
58 − 65)

Center member

78 − 88
(8.0 − 9.0,
58 − 65)

Vehicle front

78 − 88
(8.0 − 9.0,
58 − 65)

78 − 88
(8.0 − 9.0, 58 − 65)

: N·m (kg-m, ft-lb)

Engine mounting components and specifications—3.3L engines

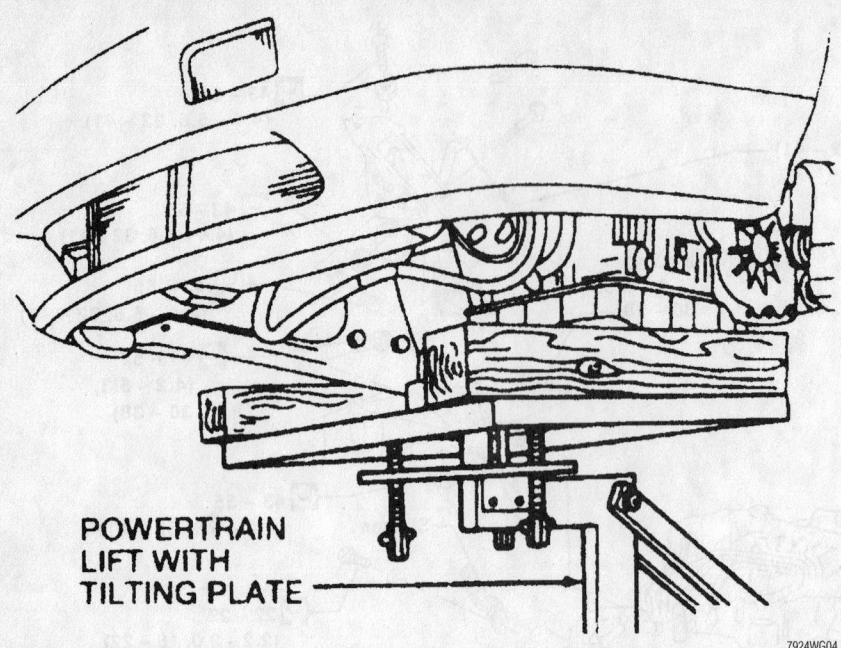

POWERTRAIN
LIFT WITH
TILTING PLATE

7924WG04

Carefully lower the engine/transaxle assembly from the vehicle.

- Left hand engine mount bolts
- Right hand engine mount
- Rear A/C refrigerant line bracket, if equipped
- Crossmember

7. Lower the engine and transaxle assembly and remove it from the vehicle.

To install:

8. Installation is the reverse of removal. Refer to the accompanying engine mounting illustration for all necessary torque specifications.

Water Pump

REMOVAL & INSTALLATION

1. Before servicing the vehicle, refer to the precautions in the beginning of this section.
2. Drain the coolant.
3. Remove or disconnect the following:
- Negative battery cable
- Radiator hoses and fan shroud
- Drive belts
- Water pump pulley using strap wrench 303–D055–(D85L–6000–A) to hold the pulley while removing the bolts

4. Remove the crankshaft pulley using the following procedure:

   a. Raise and safely support the vehicle.

   b. Remove the 5 right side inner engine and transmission splash shield bolts and 2 screws and remove the inner engine and transmission shield.

   c. Remove the 4 right side outer engine and transmission splash shield bolts and 2 screws and remove the right side outer engine and transmission splash shields.

   d. Use a strap wrench to hold the crankshaft pulley while removing the crankshaft pulley bolt.

   e. Use a crankshaft damper remover to draw the crankshaft pulley off the front of the crankshaft.

5. Remove the 5 lower engine front cover bolts and take of the front cover.

6. Remove the 6 water pump bolts. Make note of the locations of the bolts since one should be a stud/bolt and must be returned to its original location. Remove the water pump.

To install:

7. Clean all parts well. The bolt threads should be cleaned of any old sealer or corrosion. Be sure the mating surfaces between the water pump and the engine block are cleaned of any old sealant. Apply a continuous bead of gasket maker type sealer approximately ⅛ inch (3mm) wide onto the water pump and position the water pump on the engine block.

8. Install the 6 water pump bolts. Refer to any notes made at removal so the bolts can be returned to their original locations. Do not over-tighten the water pump bolts. Tighten the water pump bolts evenly to 12–15 ft. lbs. (16–21 Nm).

9. Position the water pump pulley on the water pump and install the 4 pulley bolts. Use a strap wrench to hold the pulley as the bolts are tightened to 12–15 ft. lbs. (16–21 Nm).

10. Install the front engine cover and the 5 lower front cover bolts. Tighten to 27–44 inch lbs. (3–5 Nm).

11. Install the crankshaft pulley using the following procedure:

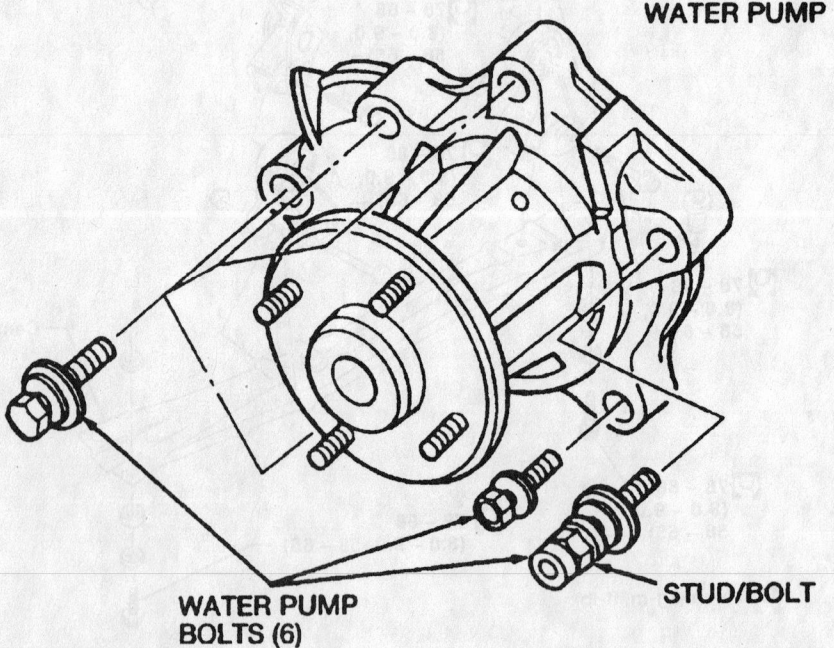

WATER PUMP

WATER PUMP BOLTS (6)

STUD/BOLT

7924WG05

Water pump mounting. Note the location of the stud/bolt

a. Install the crankshaft pulley and pulley bolt.

b. Hold the pulley with a strap wrench. Tighten the crankshaft pulley bolt to 90–98 ft. lbs. (123–132 Nm).

c. Install the inner and outer engine and transmission splash shields.

12. Install the drive belts.
13. Connect the negative battery cable.
14. Refill the cooling system.
15. Start the engine and check for leaks.

Heater Core

REMOVAL & INSTALLATION

Front System

1. Disconnect the negative battery cable.
2. Drain the cooling system into a clean container for reuse.

3. Remove or disconnect the following:

- Heater hoses at the bulkhead and plug
- Storage bin, then both side covers by the bin and the footlamp, if equipped
- Control console bezel (1 screw in the center), then the ashtray assembly
- Climate control console screws, pull the console rearward and detach the electrical connectors
- 4 radio assembly screws and take the radio out of the vehicle
- Floor duct and the right and left knee reinforcement plates
- ABS control module.

4. The speed control module, keyless entry module (if equipped) and the passive restraint (air bag) module are all located behind the center console and can be removed after detaching the respective connectors and removing the retaining nuts or screws.

The control modules are very sensitive to static electricity and can be damaged if exposed to static or stray electrical impulses.

5. Remove or disconnect the following:
- Center air duct
- 2 ground wire bolts, the U-bracket and the 2 console brackets
- Glove box and lamp
- Accelerator pedal and pedal stop
- Floor air duct
- Temperature blend sir door actuator and mode door actuator by unfastening the attaching bracket bolts

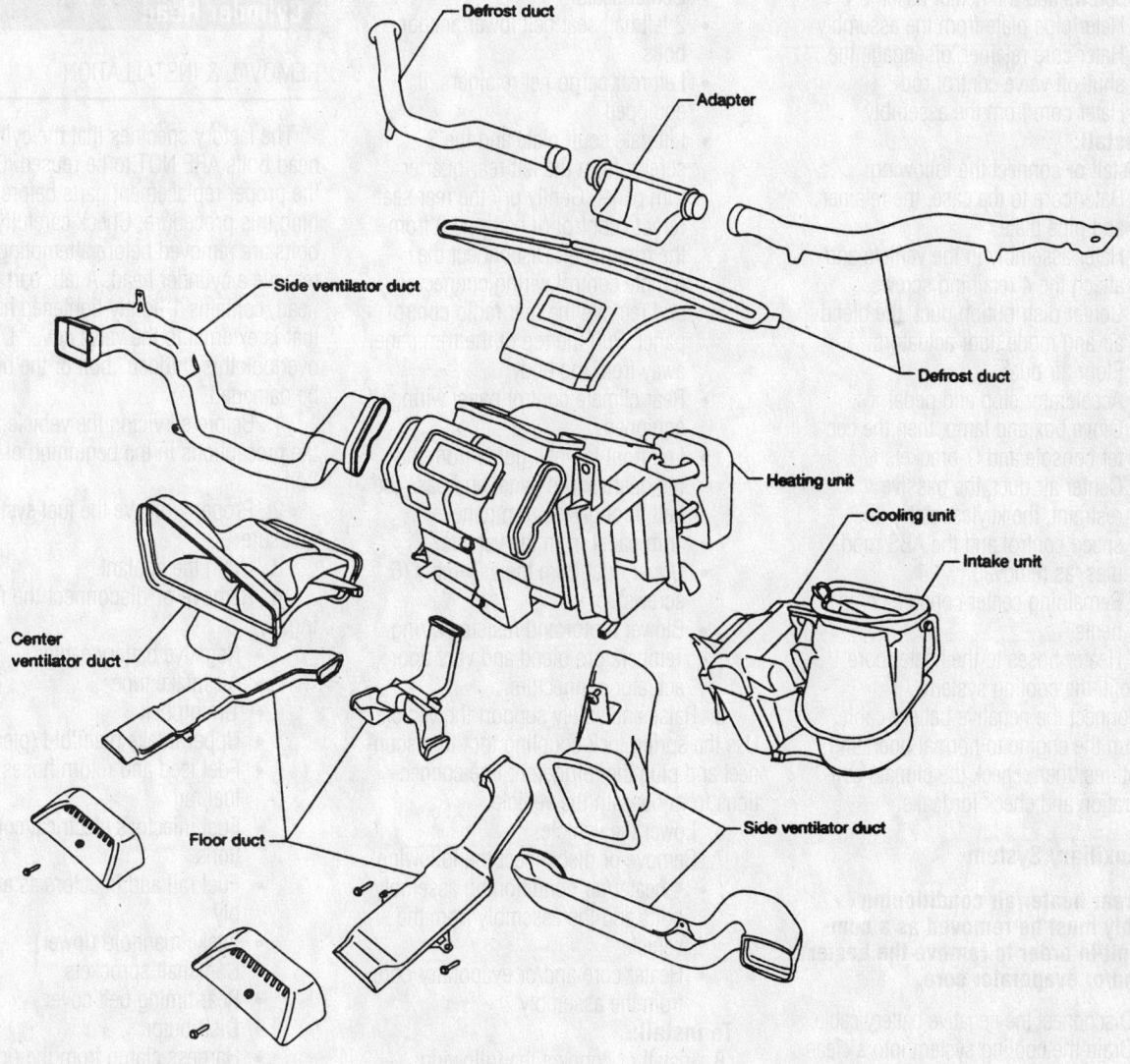

Exploded view the front heater/air conditioning assembly

93113GC2

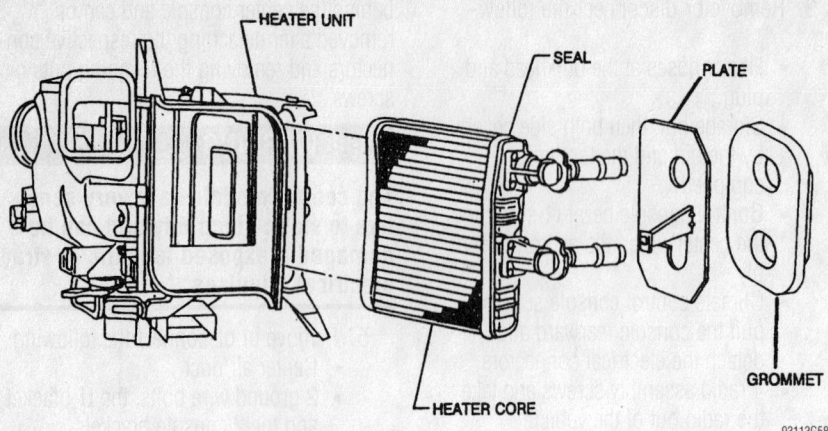

View the front heater core and heater housing assembly

and detaching the electrical connections
- Center distribution duct
- 4 evaporator/blower assembly screws, the 4 heater assembly screws and the heater assembly
- Hater pipe plate from the assembly
- Hater core retainer, disengage the shut-off valve control rod
- Hater core from the assembly

To install:

6. Install or connect the following:
- Hater core to the case, the retainer and pipe plate
- Hater assembly in the vehicle and attach the 4 retaining screws
- Center distribution duct, the blend air and mode door actuators
- Floor air duct
- Accelerator stop and pedal
- Glove box and lamp, then the center console and U-brackets
- Center air duct, the passive restraint, the keyless entry, the speed control and the ABS modules, as removed
- Remaining center console components
- Heater hoses to the heater core

7. Refill the cooling system.
8. Connect the negative battery cable.
9. Run the engine to normal operating temperatures; then, check the climate control operation and check for leaks.

Rear Auxiliary System

➡**The rear heater/air conditioning assembly must be removed as a complete unit in order to remove the heater core and/or evaporator core.**

1. Disconnect the negative battery cable.
2. Drain the cooling system into a clean container for reuse.

3. Discharge and recover the air conditioning system refrigerant.
4. Remove or disconnect the following:
- Heater hoses at the bulkhead and plug
- Center seats
- 2 left half seat belt lower anchor bolts
- Left rear cargo net retainers, if equipped
- Lift gate scuff plate and the 3 screws from the left rear quarter trim panel. Gently pry the rear seat remote control (if equipped) from the trim panel. Disconnect the remote control wiring connector and remove the rear radio control panel. Pull the top of the trim panel away from the body.
- Rear climate control panel wiring, if equipped
- Left front lap belt guide from the left quarter trim panel and pass the belt through the trim panel
- Trim panel from the vehicle
- Upper duct from the assembly (6 screws)
- Blower motor and resistor wiring
- Temperature blend and vent door actuator connectors

5. Raise and safely support the vehicle. Use the spring lock coupling tool to disconnect and plug the refrigerant line connections from beneath the vehicle.
6. Lower the vehicle.
7. Remove or disconnect the following:
- 4 heater/air conditioning assembly bolts and the assembly from the vehicle
- Heater core and/or evaporator core from the assembly

To install:

8. Install or connect the following:
- Heater core and/or evaporator core

into the assembly and the 4 retaining bolts
9. Raise and safely support the vehicle.
10. Using new O-rings, reconnect the refrigerant lines to the evaporator.
11. Lower the vehicle.
12. Install or connect the following:
- All wiring connectors
- Upper air duct with the 6 screws
- Trim panel and pass the lap seat belt through the panel slot
- Rear climate control panel
- Rear radio and rear remote control
- Remaining trim panel and components

13. Refill the cooling system.
14. Connect the negative battery cable.
15. Evacuate and charge the air conditioning system.
16. Run the engine to normal operating temperatures; then, check the climate control operation and check for leaks.

Cylinder Head

REMOVAL & INSTALLATION

The factory specifies that the cylinder head bolts ARE NOT to be reused. Obtain the proper replacement parts before beginning this procedure. Check carefully that all bolts are removed before attempting to remove a cylinder head. A tab, part of the head, contains 1 lightly tightened head bolt that is external to the valve cover. Do not overlook this "hidden" bolt or the head will be damaged.

1. Before servicing the vehicle, refer to the precautions in the beginning of this section.
2. Properly relieve the fuel system pressure.
3. Drain the coolant.
4. Remove or disconnect the following:
- Negative battery cable
- Air intake tube
- Timing belt
- Upper intake manifold (plenum)
- Fuel feed and return hoses from the fuel rail
- Fuel injector's electrical connections
- Fuel rail and injectors as an assembly
- Intake manifold (lower)
- Camshaft sprockets
- Rear timing belt cover
- Distributor
- Harness clamp from the right hand rocker cover

- Exhaust tube from the left hand manifold
- Left hand exhaust manifold from the right hand exhaust manifold
- Left hand manifold-to-bracket bolt
- A/C compressor, alternator and their brackets
- Rocker covers
- Cylinder head bolts in the sequence illustrated using several passes
- Cylinder head with the exhaust manifold and gasket. Discard the gasket.
- Exhaust manifold from the head

To install:

5. Clean all parts well.

6. Inspect the cylinder head for damage, cracks and leakage of water and oil. If necessary, replace the head. Check the head gasket surface for burrs and nicks. If the head is cracked, it must be replaced.

7. Install the exhaust manifold on the cylinder head.

8. Position a new head gasket and the cylinder head on the block. Examine the head bolt washers. Note that the washers have a chamfer or bevel on one side. The beveled side should face "up" when installed. Examine the new replacement head bolts. There are different lengths. The head bolts in positions 4, 7, 9 and 12 are 5.00 inches (127mm) long and the rest are 4.17 inches (106mm) long. Be sure the new cylinder head bolts are installed in the correct positions.

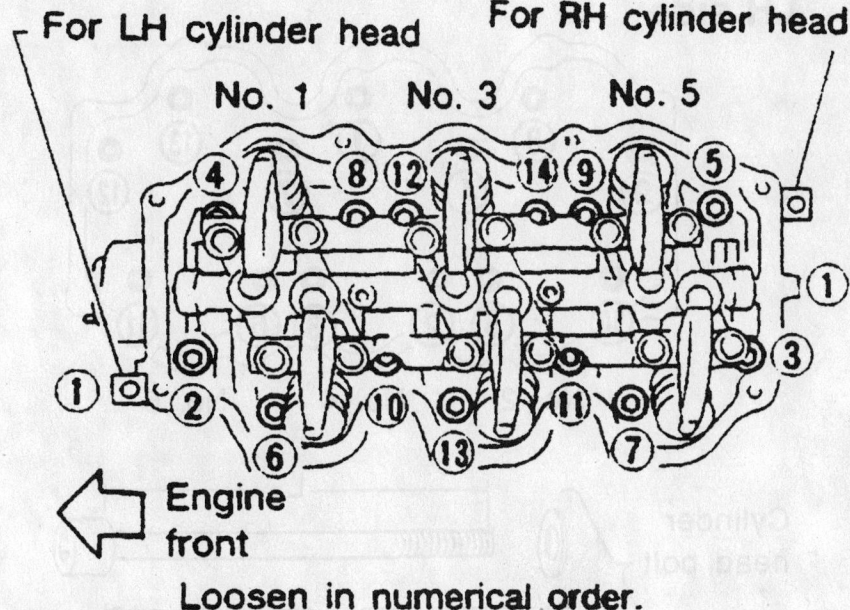

Remove the cylinder head bolts in the sequence shown—3.3L engines

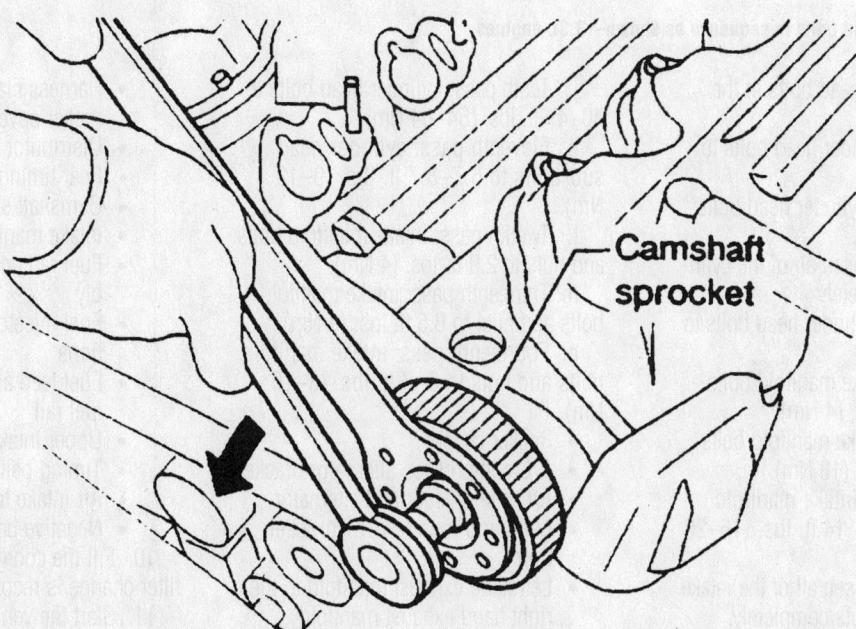

Hold the camshaft sprocket while removing the sprocket retaining bolt—3.3L engines

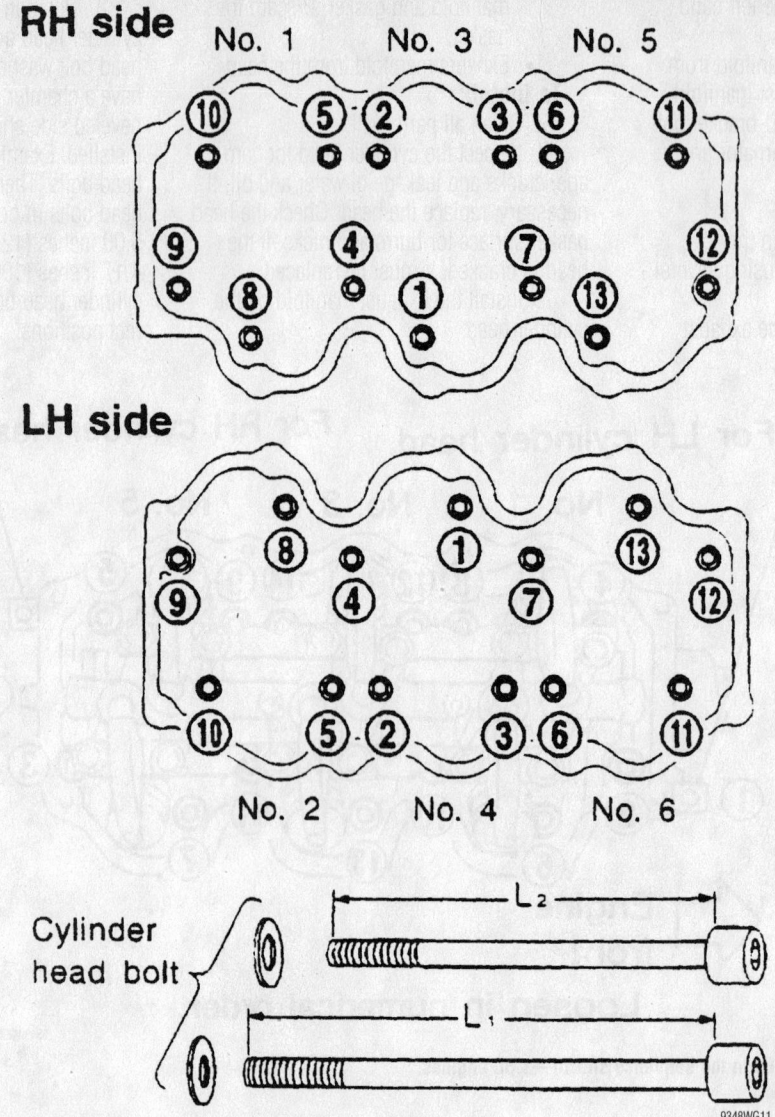

RH side

No. 1   No. 3   No. 5

LH side

No. 2   No. 4   No. 6

Cylinder head bolt

9348WG11

Tighten the cylinder head bolts in sequence as shown—3.3L engines

9. Tighten the new head bolts in the following sequence:

a. First pass: cylinder head bolts to 22 ft. lbs. (29 Nm).

b. Second pass: cylinder head bolts to 43 ft. lbs. (59 Nm).

c. Third pass: Loosen all of the cylinder head bolts completely.

d. Fourth pass: cylinder head bolts to 7 ft. lbs. (10 Nm).

e. Fifth pass: intake manifold bolts and nuts to 2.9 ft. lbs. (4 Nm).

f. Sixth pass: intake manifold bolts and nuts to 13 ft. lbs. (18 Nm).

g. Seventh pass: intake manifold bolts and nuts to 12–14 ft. lbs. (16–20 Nm).

h. Eight pass: Loosen all of the intake manifold bolts and nuts completely.

i. Ninth pass: cylinder head bolts to 22 ft. lbs. (29 Nm).

j. Tenth pass: cylinder head bolts to 40–47 ft. lbs. (54–64 Nm).

k. Eleventh pass: cylinder head sub-bolts to 6.7–8.7 ft. lbs. (9–12 Nm).

l. Twelfth pass: intake manifold bolts and nuts to 2.9 ft. lbs. (4 Nm).

m. Thirteenth pass: intake manifold bolts and nuts to 6.5 ft. lbs. (9 Nm).

n. Fourteenth pass: intake manifold bolts and nuts to 6–7 ft. lbs. (8–10 Nm).

- Rocker covers
- A/C compressor, alternator brackets
- A/C compressor and alternator
- Left hand manifold-to-bracket bolt
- Left hand exhaust manifold to the right hand exhaust manifold
- Exhaust tube to the left hand manifold

- Harness clamp to the right hand rocker cover
- Distributor
- Rear timing belt cover
- Camshaft sprockets
- Intake manifold (lower)
- Fuel rail and injectors as an assembly
- Fuel injector's electrical connections
- Fuel feed and return hoses to the fuel rail
- Upper intake manifold (plenum)
- Timing belt
- Air intake tube
- Negative battery cable

10. Fill the cooling system. An oil and filter change is recommended.

11. Start the vehicle and check for leaks. Check the ignition timing and adjust as required.

Rocker Arms/Shafts

REMOVAL & INSTALLATION

1. Before servicing the vehicle, refer to the precautions in the beginning of this section.
2. Remove or disconnect the following:
   - Negative battery cable
   - Upper intake manifold
   - Valve covers
   - Rocker arm and shaft assemblies
   - Rocker arms from the shafts

➡ **Keep all valvetrain components in order for assembly.**

To install:

3. Lubricate all contact points with clean engine oil and assemble the rocker arms to the shafts in their original positions.
4. Install or connect the following:
   - Rocker arm and shaft assemblies. Tighten the bolts to 13–16 ft. lbs. (18–22 Nm).
   - Valve covers
   - Upper intake manifold
   - Negative battery cable
5. Start the engine and check for leaks.

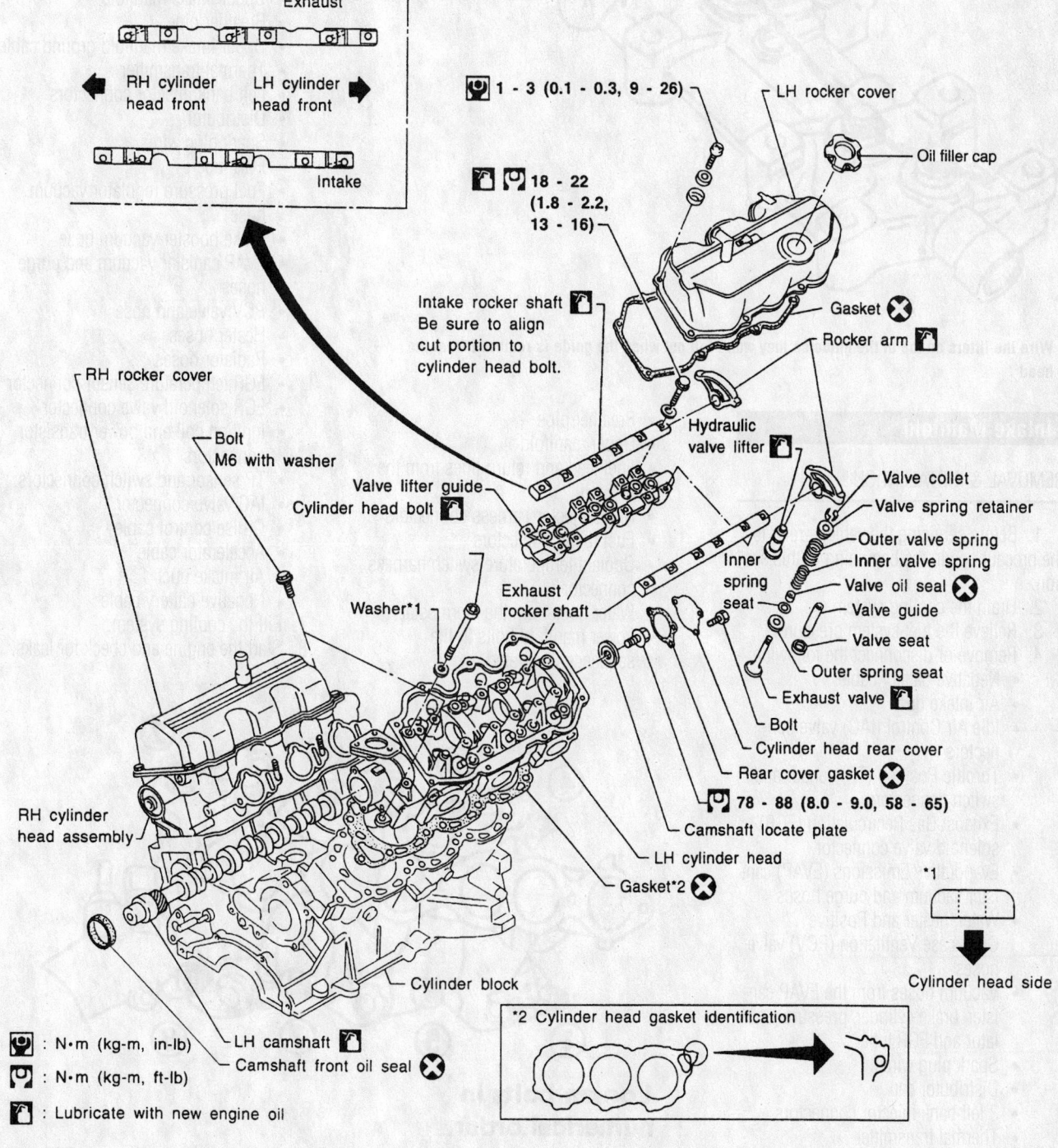

Rocker arm and shaft components

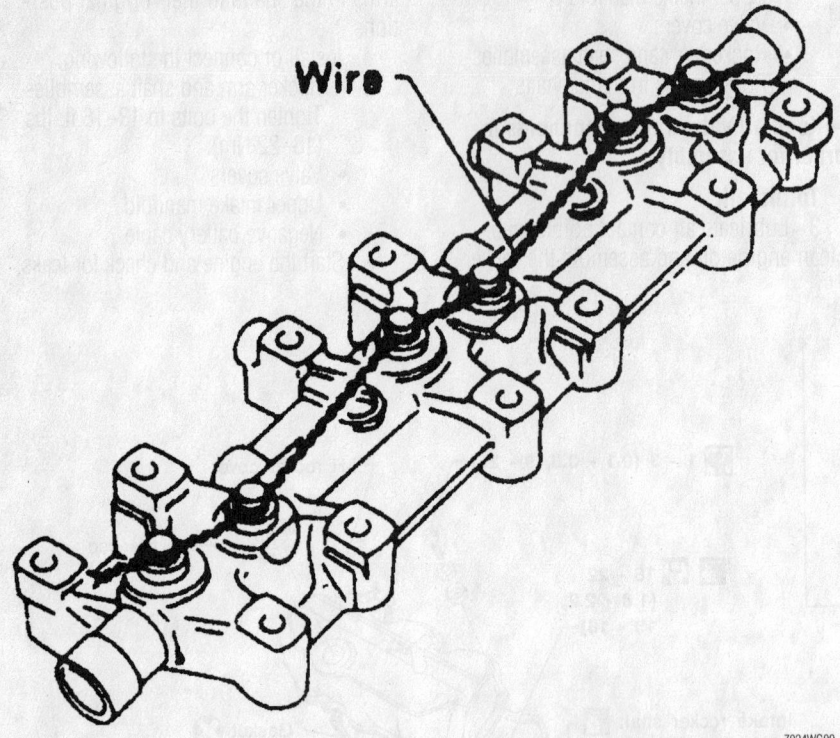

Wire

Wire the lifters on top of the guide so they won't fall out when the guide is removed from the head

7924WG09

Intake Manifold

REMOVAL & INSTALLATION

1. Before servicing the vehicle, refer to the precautions in the beginning of this section.
2. Drain the cooling system.
3. Relieve the fuel system pressure.
4. Remove or disconnect the following:
   - Negative battery cable
   - Air intake duct
   - Idle Air Control (IAC) valve connectors
   - Throttle Position (TP) sensor and switch connectors
   - Exhaust Gas Recirculation (EGR) solenoid valve connector
   - Evaporative Emissions (EVAP) canister vacuum and purge hoses
   - Water, heater and Positive Crankcase Ventilation (PCV) valve hoses
   - Vacuum hoses from the EVAP canister, brake cylinder, pressure regulator and EGR tube
   - Spark plug wires
   - Distributor cap
   - 3 left bank injector connectors
   - Thermal transmitter
   - Ground harness
   - Breather pipe
   - Upper manifold
   - Fuel feed and return lines from the fuel rail
   - Right injector harness connectors
   - Fuel rail and injectors
   - Coolant temperature switch harness connector
   - Water hose from the thermostat
   - Lower manifold bolts in the sequence illustrated.

   - Manifold gasket and discard

To install:
5. Install the lower intake manifold with a new gasket.
   a. Step 1: 35 inch lbs. (4 Nm)
   b. Step 2: 78 inch lbs. (9 Nm)
   c. Step 3: 70–84 inch lbs. (8–10 Nm)
6. Install or connect the following:
   - ECT sensor connector
   - Fuel supply manifold
   - Right bank injector connectors
   - Fuel lines
   - Upper intake manifold
   - Breather pipe
   - Upper intake manifold ground cable
   - Thermal transmitter
   - Left bank injector connectors
   - Distributor
   - Spark plug wires
   - EGR tube
   - Fuel pressure regulator vacuum hose
   - Brake booster vacuum hose
   - EVAP canister vacuum and purge hoses
   - PCV valve and hose
   - Heater hoses
   - Radiator hoses
   - EGR temperature sensor connector
   - EGR solenoid valve connector
   - Ignition coil and power transistor connectors
   - TP sensor and switch connectors
   - IAC valve connector
   - Cruise control cable
   - Accelerator cable
   - Air intake duct
   - Negative battery cable
7. Fill the cooling system.
8. Start the engine and check for leaks.

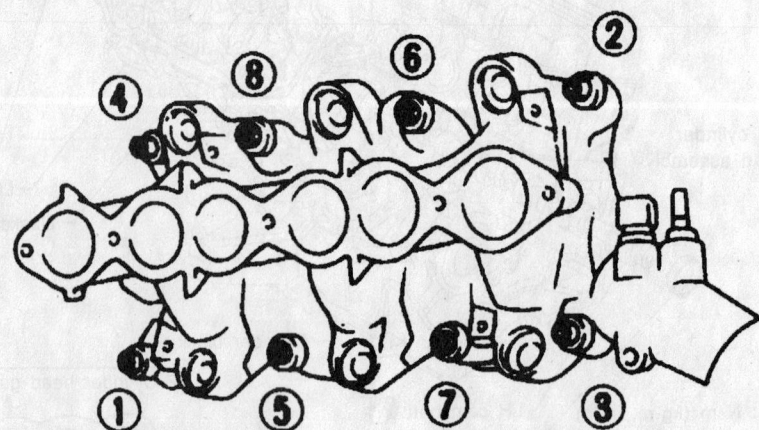

Loosen bolts in numerical order.

7924VG32

Intake manifold loosening sequence—3.3L engine

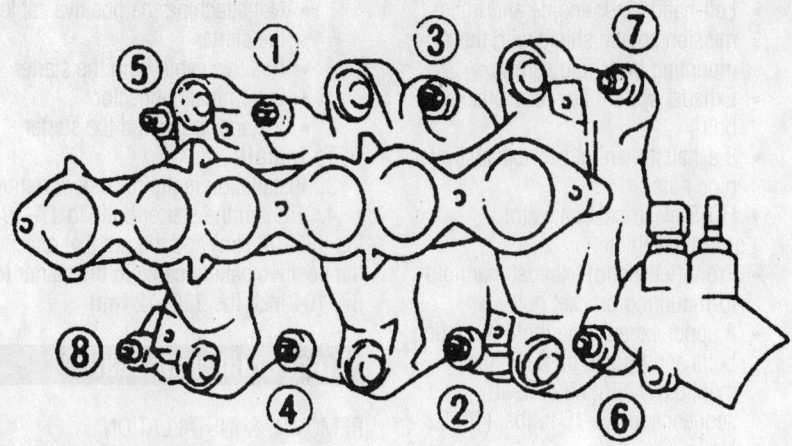

Tighten bolts in numerical order.

Intake manifold tightening sequence—3.3L engine

7924VG33

Exhaust Manifold

REMOVAL & INSTALLATION

Rear (Right-Hand) Exhaust Manifold

1. Before servicing the vehicle, refer to the precautions in the beginning of this section.

2. Remove or disconnect the following:

- Negative battery cable
- Radiator overflow hose from the radiator
- Radiator coolant-recovery reservoir off of the bracket
- Reservoir
- Air cleaner intake tube and the engine air intake resonator
- 6 rear (right-hand) exhaust manifold crossover tube heat-shield bolts and the heat shields
- 2 nuts and the 1 bolt securing the rear (right-hand) exhaust manifold tube to the front (left-hand) exhaust manifold. Discard the gasket.
- Transmission fluid level indicator tube heat shield

3. Disengage the following electrical connectors:

- Idle switch
- Throttle Position (TP) sensor
- Exhaust Gas Recirculation (EGR) control solenoid

4. Raise and safely support the vehicle.

5. Remove or disconnect the following:

- EGR valve-to-back-pressure transducer valve tube nut and position it out of the way

- 2 EGR valve-to-exhaust manifold tube nuts and tube
- 6 rear exhaust manifold nuts in the reverse order of the tightening sequence

6. Safely lower the vehicle, remove the exhaust manifold and discard the exhaust manifold gasket.

To install:

7. Raise and safely support the vehicle.

8. Be sure that both the exhaust manifold and the cylinder head mating surfaces are clean of any old gasket material.

9. Install or connect the following:

RH exhaust

LH exhaust

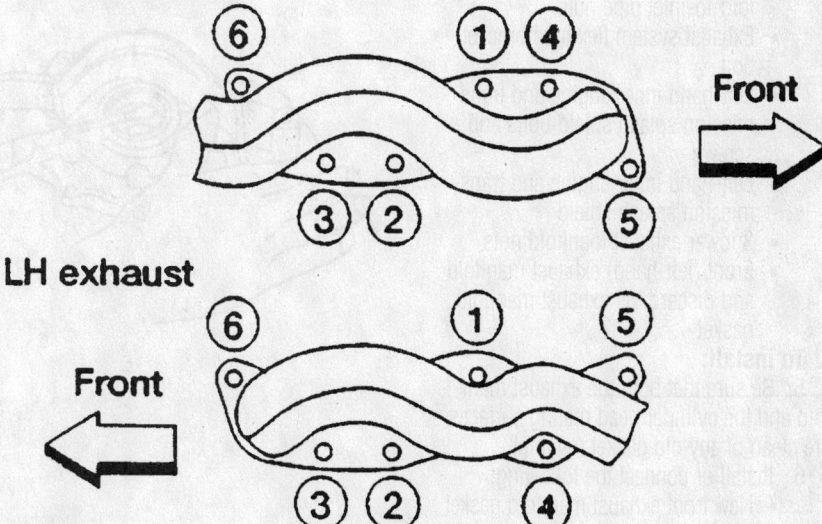

Tighten in numerical order.

9348WG12

To avoid warping the exhaust manifolds, use this sequence when loosening the bolts—3.3L engine

- Rear (right-hand) exhaust manifold gasket onto the exhaust manifold mounting studs

10. Lower the vehicle safely.

- Rear (right-hand) exhaust manifold onto the studs
- 6 rear (right-hand) exhaust manifold nuts. Tighten the nuts in sequence to 13–16 ft. lbs. (18–22 Nm).
- EGR valve-to-exhaust manifold tube and tube nuts
- EGR valve-to-back-pressure transducer valve tube nut

11. Lower the vehicle carefully.

12. Reconnect the following electrical connectors:

- EGR solenoid
- TP sensor
- Idle switch
- Transmission fluid level indicator tube heat shield
- New gasket between the front (left-hand) exhaust manifold and the rear exhaust manifold crossover tube
- 2 nuts and the 1 bolt securing the rear (right-hand) exhaust manifold crossover tube to the front (left-hand) exhaust manifold. Tighten the rear exhaust manifold crossover tube-to-front (left-hand) exhaust manifold nuts and bolt.
- Rear (right-hand) exhaust manifold crossover tube heat shield with the 6 mounting bolts

- Rear (right-hand) exhaust manifold crossover tube bolts
- Air cleaner intake tube and the engine air intake resonator
- Radiator coolant recovery reservoir
- Radiator overflow hose to the radiator
- Negative battery cable

13. Start the engine and check for leaks and proper operation.

Front (Left-Hand) Exhaust Manifold

1. Before servicing the vehicle, refer to the precautions in the beginning of this section.

2. Remove or disconnect the following:
- Negative battery cable and wait at least 90 seconds before performing any work. This allows time for the SRS or air bag system to deplete its back up energy supply.
- 2 nuts and the 1 bolt securing the front (left-hand) exhaust manifold to the rear (right-hand) exhaust manifold crossover tube. Discard the gasket.

3. Remove the transmission fluid level indicator tube heat shield.
- 6 front (left-hand) exhaust manifold nuts in 2 steps in the reverse order of the tightening sequence. Do not remove the 3 lower front (left-hand) exhaust manifold nuts.
- Front (left-hand) exhaust manifold-to-mounting bracket bolt

4. Raise and safely support the vehicle.
- Heated Oxygen Sensor (HO2S) electrical connector
- 3 front (left-hand) exhaust manifold-to-inlet pipe nuts
- Exhaust system flex tube bracket bolt
- Left-hand inner engine and transmission splash shield bolts and screws
- Left-hand inner engine and transmission splash shield
- 3 lower exhaust manifold nuts
- Front (left-hand) exhaust manifold and discard the exhaust manifold gasket

To install:

5. Be sure that both the exhaust manifold and the cylinder head mating surfaces are clean of any old gasket material.

6. Install or connect the following:
- New front exhaust manifold gasket in place
- Front (left-hand) exhaust manifold
- 3 lower exhaust manifold mounting nuts. Do not tighten the nuts at this time.

- Left-hand inner engine and transmission splash shield with their mounting bolts and screws
- Exhaust system flex tube bracket bolt
- 3 exhaust manifold-to-exhaust inlet pipe nuts
- HO2S electrical connector

7. Lower the vehicle.
- Front (left-hand) exhaust manifold-to-mounting bracket bolt
- 3 upper exhaust manifold mounting bolts and tighten all 6 exhaust manifold mounting bolts in sequence to 13–16 ft. lbs. (18–22 Nm)
- Transmission fluid level indicator tube heat shield
- 2 nuts and the 1 bolt securing the front (left-hand) exhaust manifold to the rear (right-hand) exhaust manifold crossover tube
- Negative battery cable

8. Start the engine, check for leaks and road test for proper operation.

Starter

REMOVAL & INSTALLATION

1. Before servicing the vehicle, refer to the precautions in the beginning of this section.

2. Remove or disconnect the following:
- Battery negative cable
- Air cleaner

- Nut attaching the positive cable to the starter
- Positive cable from the starter
- S-terminal connector
- 2 starter bolts and the starter

To install:

3. Installation is the reverse of removal.

4. Tighten the starter bolts to 17–19 ft. lbs. (23–26 Nm) and the nut that attaches the positive battery cable to the starter to 87–104 inch lbs. (10–12 Nm).

Front Crankshaft Seal

REMOVAL & INSTALLATION

1. Before servicing the vehicle, refer to the precautions in the beginning of this section.

2. Remove or disconnect the following:
- Negative battery cable
- Drive belts
- Radiator hoses
- Crankshaft pulley
- Front cover
- Timing belt
- Crankshaft timing sprocket
- Crankshaft seal using a suitable prytool

To install:

3. Install or connect the following:
- Crankshaft seal using a driver and a hammer until its flush with the housing
- Crankshaft timing sprocket
- Timing belt

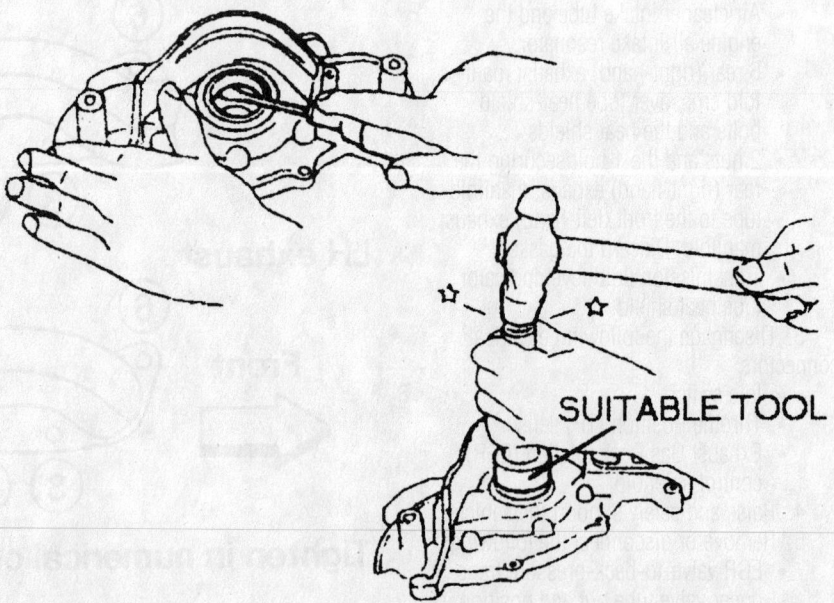

SUITABLE TOOL

Removal and installation of the front crankshaft seal—3.3L engines

7924WG13

- Front cover and tighten the bolts to 26–43 inch lbs. (3–5 Nm)
- Crankshaft pulley and tighten the bolt to 141–156 ft. lbs. (191–211 Nm)
- Radiator hoses
- Drive belts
- Negative battery cable

4. Fill the cooling system, start the vehicle and check for leaks.

Camshaft And Valve Lifters

REMOVAL & INSTALLATION

1. Before servicing the vehicle, refer to the precautions in the beginning of this section.
2. Drain the cooling system.
3. Remove or disconnect the following:
   - Negative battery cable
   - Upper intake manifold
   - Valve covers

➡**Keep all valvetrain components in order for assembly.**

- Rocker arm and shaft assemblies
- Valve lifter guide and valve lifters. Attach a wire to the top of the lifters so that they will not drop from the lifter guide.
- Radiator
- Accessory drive belts
- Front cover
- Timing belt
- Camshaft sprockets
- Camshaft seals
- Rear timing cover
- Distributor
- Cylinder head rear covers
- Camshaft locating plates
- Camshafts

To install:
4. Install or connect the following:
   - Camshafts
   - Camshaft locating plates. Tighten the bolts to 58–65 ft. lbs. (78–88 Nm).
   - Cylinder head rear covers
   - Distributor
   - Rear timing cover
   - Camshaft seals
   - Camshaft sprockets. Tighten the bolts to 58–65 ft. lbs. (78–88 Nm).
   - Timing belt
   - Front cover
   - Accessory drive belts
   - Radiator
   - Valve lifter guide and valve lifters
   - Rocker arm and shaft assemblies. Tighten the bolts to 13–16 ft. lbs. (18–22 Nm).

- Valve covers
- Upper intake manifold
- Negative battery cable

5. Fill the cooling system.
6. Start the engine and check for leaks.

Valve Lash

ADJUSTMENT

The engines covered in this section use hydraulic valve lifters that automatically adjust the valve lash. No periodic adjustment is needed.

Oil Pan

REMOVAL & INSTALLATION

1. Before servicing the vehicle, refer to the precautions in the beginning of this section.
2. Drain the engine oil.
3. Remove or disconnect the following:
   - Negative battery cable
   - Front engine mount (support) insulator through-bolt
   - Rear engine mount (support) through-bolt

- 2 rear refrigerant/heater pipe hold down bracket bolts
- 4 crossmember (also called a transverse member) bolts, and remove the crossmember.
- Exhaust inlet pipe
- 4 rear transaxle-to-engine brace bolts and the 5 front transaxle-to-engine brace bolts
- Front transaxle-to-engine brace
- Low oil level sensor electrical connector
- 18 oil pan bolts in the reverse order of the tightening sequence, working from the outside, towards the center bolts.
- Oil pan and discard the seals

To install:
4. Clean all parts well. Be sure that all old sealing material is removed from the oil pan and engine mating surfaces.
5. Position new oil pan seals. Apply Loctite® Ultra Gray 599 Silicone Sealer, or equivalent, to the ends of the oil pan seals.
6. Apply a bead of Loctite® Ultra Gray 599 Silicone Sealer or equivalent to the oil pan gasket rail inboard of the bolt holes.
7. Install or connect the following:
   - Oil pan on the engine block. Tighten the 18 oil pan bolts in

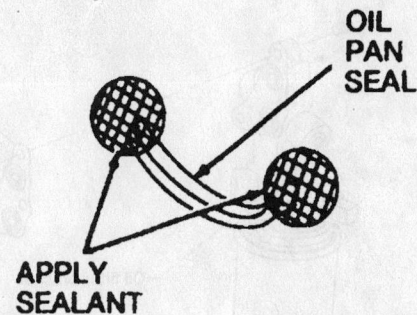

OIL PAN SEAL

APPLY SEALANT

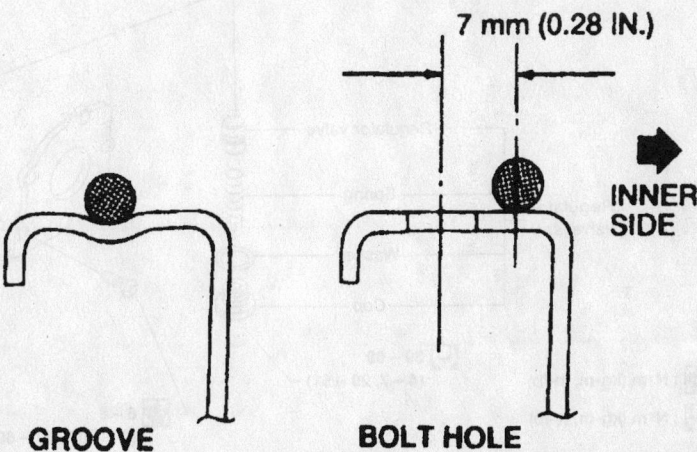

7 mm (0.28 IN.)

INNER SIDE

GROOVE   BOLT HOLE

7924WG14

Apply RTV silicone sealer to the seal ends and to the oil pan gasket rail

TIGHTENING SEQUENCE

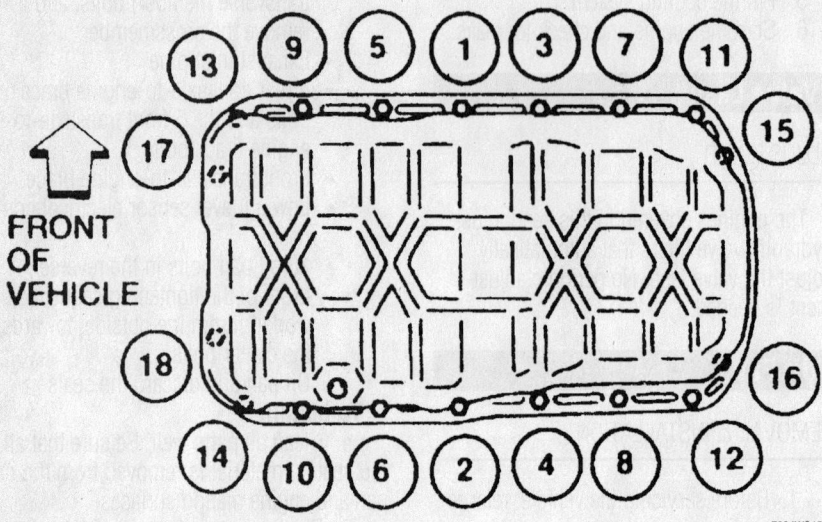

FRONT OF VEHICLE

7924WG15

Tighten the 18 oil pan bolts in sequence, working from the inside, towards the outer bolts

sequence, working from the inside, towards the outer bolts. Do not over-tighten. Tighten to 62–70 inch lbs. (7–8 Nm).
- Low oil level sensor electrical connector.

- Front and rear transaxle braces. Tighten all bolts to 22–30 ft. lbs. (30–40 Nm).
- Exhaust inlet pipe
- Crossmember and tighten the bolts to 58–65 ft. lbs. (78–88 Nm).

- Both engine support through-bolts and tighten to 58–65 ft. lbs. (78–88 Nm).
8. Remove the support jack from under the crankshaft pulley.
9. Lower the vehicle.
10. Fill the engine with the specified engine oil to the required level.
11. Connect the negative battery cable. Start the engine and check for leaks.

Oil Pump

REMOVAL & INSTALLATION

1. Before servicing the vehicle, refer to the precautions in the beginning of this section.
2. Drain the engine oil.
3. Drain the cooling system.
4. Remove or disconnect the following:
   - Negative battery cable
   - Oil pan
5. After removing the oil pan, reinstall the crossmember and the mount bolts.
   - Timing belt
   - Crankshaft sprocket and timing belt plate
   - Oil pump

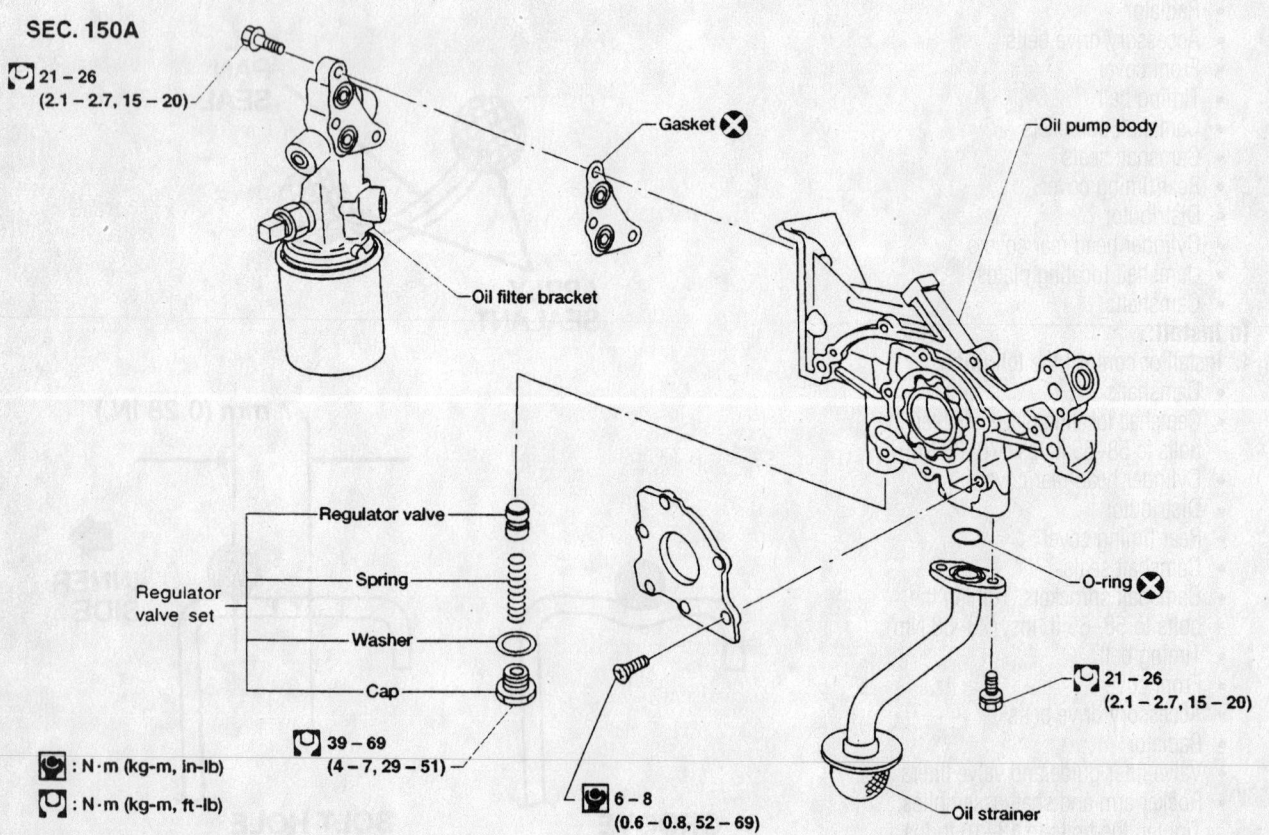

SEC. 150A

21 – 26
(2.1 – 2.7, 15 – 20)

Gasket

Oil pump body

Oil filter bracket

Regulator valve
Spring
Regulator valve set
Washer
Cap

O-ring

Oil strainer

21 – 26
(2.1 – 2.7, 15 – 20)

: N·m (kg-m, in-lb)
: N·m (kg-m, ft-lb)

39 – 69
(4 – 7, 29 – 51)

6 – 8
(0.6 – 0.8, 52 – 69)

9348WG14

Exploded view of the oil pump assembly—3.3L engine

To install:

6. Install or connect the following:
- Oil pump-to-body bolts to 52–69 inch lbs. (6–8 Nm)
- Timing belt plate and tighten the bolts to 52–69 inch lbs. (6–8 Nm)
- Crankshaft sprocket
- Timing belt
- Oil pan
- Negative battery cable

7. Fill the cooling system.
8. Fill the crankcase to the correct level.
9. Start the engine and check for leaks.

Rear Main Seal

REMOVAL & INSTALLATION

3.3L Engine

1. Before servicing the vehicle, refer to the precautions in the beginning of this section.
2. Disconnect the negative battery cable.
3. Remove the transaxle from the vehicle.
4. Remove the flexplate from the crankshaft.
5. Remove the rear oil seal retainer.

❋❋ WARNING

Do not scratch the seal bore of the oil seal retainer when removing the oil seal.

6. Remove the oil seal from the seal retainer.

To install:

7. Apply clean engine oil to the lip and outer surface of the new seal to aid during installation.
8. Install the seal in the retainer using a suitable seal driver.

9. Using a new gasket install the retainer on the engine. Tighten the bolts to 52–61 inch lbs. (6–7 Nm).
10. Install the flexplate. Tighten the bolts to 61–69 ft. lbs. (83–93 Nm).
11. Install the transaxle and remaining components.

Timing Belt

REMOVAL & INSTALLATION

3.3L engines

On this vehicle, right side refers to the "rear" components (near the firewall) and left side refers to the "front" components (near the radiator).

1. If the timing belt is to be removed, it is good practice to turn the crankshaft until the engine is at Top Dead Center (TDC) of the No. 1 cylinder, compression stroke (firing position), before beginning work. This should align all timing marks and serve as a reference for all work that follows. After verifying that the engine is at TDC for the No. 1 cylinder, do not crank the engine or allow the crankshaft or camshaft sprockets to be turned otherwise engine timing will be lost.
2. Before servicing the vehicle, refer to the precautions in the beginning of this section.
3. Drain the cooling system.
4. Remove or disconnect the following:
- Negative battery cable
- Alternator drive belt, water pump and power steering pump belt and the air conditioning compressor belt, if equipped
- 3 air conditioning compressor drive belt idler pulley bolts and the idler pulley, if equipped with air conditioning
- Upper radiator hose bracket bolt

- Upper hose with the bracket from the vehicle
- Water bypass hose from between the thermostat housing and the lower water hose connection
- Main wiring harness from the upper engine front cover
- 8 upper engine front cover bolts and the upper cover
- Right side front wheel and tire assembly
- 4 right side engine and transmission splash shield bolts and 2 screws, and right side outer engine and transaxle splash shield

5. Use a strap wrench to hold the water pump pulley. Remove the 4 pulley bolts, and the water pump pulley.
6. Use a strap wrench to hold the crankshaft pulley. Remove the center pulley bolt, and the crankshaft pulley using a harmonic balancer (damper) puller to draw the pulley from the front of the crankshaft.
- 5 lower engine front cover bolts, then remove the lower engine front cover

7. Be sure that the timing marks between the crankshaft sprocket and the oil pump housing align.
8. If the timing belt is to be reused, mark an arrow on the belt indicating the direction of rotation. The directional arrow is necessary to ensure that the timing belt, if it to be reused, can reinstalled in the same direction.
9. Loosen the timing belt tensioner nut and slip the timing belt off of the sprockets.
10. If necessary, the camshaft sprockets can be removed. A special spanner tool is designed to hold the sprocket to keep it from turning while the center bolt is being loosened. Use care if using substitutes.

➡**The sprockets are not interchangeable.**

11. If necessary, the crankshaft sprocket can be removed. The outer timing belt guide (looks like a large washer) and the crankshaft sprocket simply pull off the front of the crankshaft.

➡**Be careful, there are 2 crankshaft keys. Use care not to loose them.**

To install:

12. Clean all parts well. If removed, inspect the crankshaft sprocket for warping or abnormal wear. Check the sprocket teeth for wear, deformation, chipping or other damage. Replace as necessary. Clean the sprocket mounting surface to ease installation. Install the key. Slip the sprocket onto

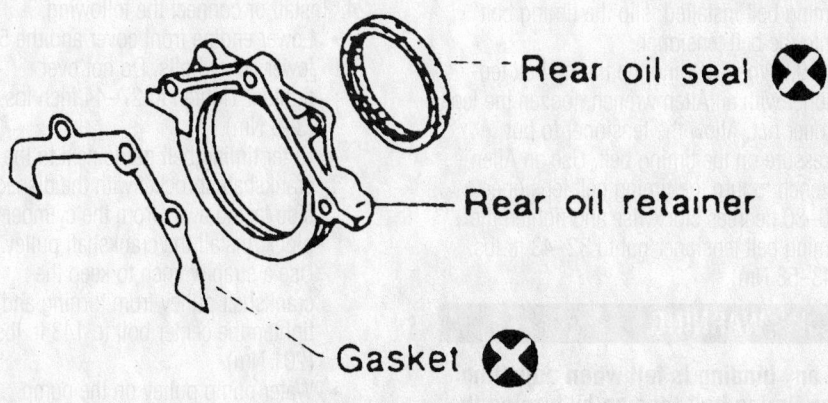

Exploded view of the oil seal, retainer and gasket—3.3L engine

7924WG17

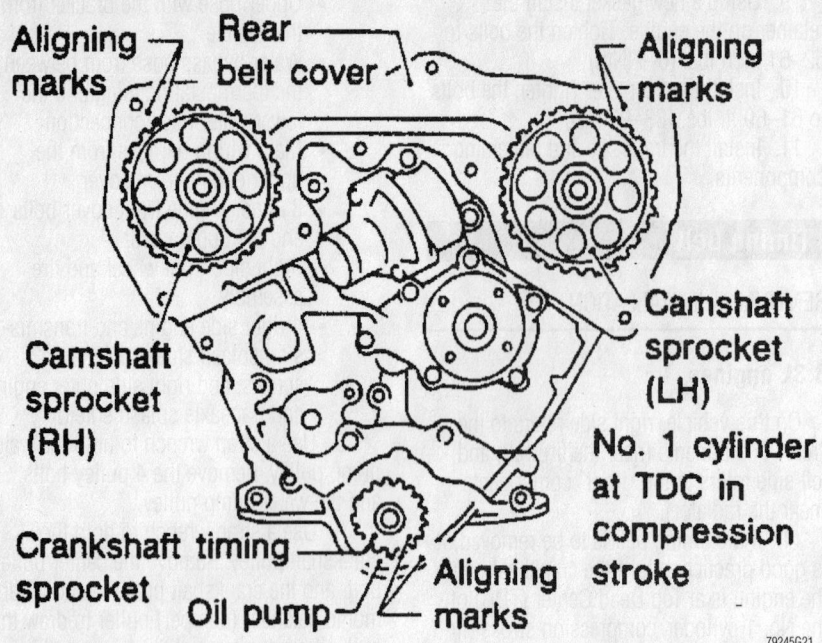

Aligning marks — **Rear belt cover** — **Aligning marks** — **Camshaft sprocket (LH)** — **No. 1 cylinder at TDC in compression stroke** — **Aligning marks** — **Oil pump** — **Crankshaft timing sprocket** — **Camshaft sprocket (RH)**

79245G21

Use a shop rag to clean the alignment marks for the timing belt— 3.3L engines

the crankshaft. Tap it in place with a suitably-sized socket.

13. If removed, inspect the camshaft sprockets for damage and wear. Replace as required. The sprockets should be marked **L3** to designate the front, or left side camshaft and **R3** to designate the rear, or right side camshaft. Use care to install the sprockets properly. A special spanner tool is designed to hold the sprocket to keep it from turning while the center bolt is being tightened. Use care if using a substitute. Tighten the camshaft sprocket center bolts to 61 ft. lbs. (83 Nm). Verify that the timing marks on the camshaft sprockets and the timing marks on the rear cover (called the seal plate) are aligned.

14. Use an Allen wrench to turn the timing belt tensioner clockwise until the belt tensioner spring is fully extended. Temporarily tighten the tensioner nut to 32–43 ft. lbs. (43–58 Nm).

15. If a new timing belt is to be installed, look for a printed arrow on the belt. Be sure the arrow is pointing away from the engine. If the original timing belt is to be reused, be sure that the directional arrow that was marked at disassembly is facing the correct direction.

16. A new Original Equipment Manufacture (OEM) timing belt should have 3 white timing marks on it that indicate the correct timing positions of the camshafts and the crankshaft. These marks are to help ensure that the engine is properly timed. When the engine is properly timed, each white timing mark on the timing belt will be aligned with

the corresponding camshaft and crankshaft timing mark on the sprocket. Because the white timing marks are not evenly spaced, the technician needs to use care in installing the belt. There should be 40 timing belt teeth between the timing marks on the front and rear camshaft sprockets and 43 teeth between the timing mark on the front camshaft sprocket and the timing mark on the crankshaft sprocket.

17. Verify that the camshaft timing marks are aligned with the timing marks on the rear cover (seal plate) and that the crankshaft sprocket timing mark is aligned with the timing mark on the oil pump housing.

18. Install the timing belt starting at the crankshaft sprocket and moving around the camshaft sprockets following a counterclockwise path. Do not allow any slack in the timing belt between the sprockets. After all of the timing marks are aligned with the timing belt installed, slip the timing belt onto the belt tensioner.

19. While holding the timing belt tensioner with an Allen wrench, loosen the tensioner nut. Allow the tensioner to put pressure on the timing belt. Use an Allen wrench to turn the timing belt tensioner 70–80 degrees clockwise and tighten the timing belt tensioner nut to 32–43 ft. lbs. (43–58 Nm).

❄❄ WARNING

If any binding is felt when adjusting the timing belt tension by turning the crankshaft, STOP turning the engine,

because the pistons may be hitting the valves.

20. Rotate the crankshaft clockwise twice and align the No. 1 piston to TDC on the compression stroke (firing position).

21. Apply 22 lbs. (10kg) of force on the timing belt between the rear camshaft sprocket and the timing belt tensioner. An assistant may be needed. While holding the timing belt tensioner steady with an Allen wrench, loosen the timing belt tensioner nut. Remove the Allen wrench and adjust the timing belt tensioner using the following procedure:

a. Install a 0.0138 in. (0.35mm) thick and 0.500 in. (12.7mm) wide feeler gauge where the timing belt just starts to go around the tensioner (approximately the 4 o'clock position, looking at the tensioner).

b. Turn the crankshaft sprocket clockwise, which should force the feeler gauge between the timing belt and the tensioner, up to a position on the tensioner of about 1 o'clock.

c. Tighten the timing belt tensioner nut to 61 ft. lbs. (83 Nm).

d. Turn the crankshaft clockwise to rotate the feeler gauge out from between the timing belt tensioner and the timing belt.

22. Rotate the crankshaft clockwise twice, and once again align the No. 1 piston to TDC on the compression stroke (firing position).

23. Apply 22 lbs. (10kg) of force on the timing belt between the front and rear camshaft sprockets. Measure the amount of belt deflection. Belt deflection should be between 0.51–0.59 in. (13–15mm). If belt deflection is out of specification, repeat Steps 29 through 33. If the timing belt deflection cannot be adjusted into specification, the timing belt will have to be replaced.

24. Install or connect the following:
- Lower engine front cover and the 5 lower cover bolts. Do not over tighten. Tighten to 27–44 inch lbs. (3–5 Nm).
- Outer timing belt guide next to the crankshaft sprocket with the dished side facing away from the cylinder block. Install the crankshaft pulley. Use a strap wrench to keep the crankshaft pulley from turning and tighten the center bolt to 148 ft. lbs. (201 Nm).
- Water pump pulley on the pump. Install the 4 bolts. Use a strap wrench to keep the water pump

- pulley from turning and tighten the 4 water pump pulley bolts to 89 inch lbs. (10 Nm).
- Right side outer engine and transaxle splash shield, and secure with the 4 bolts and 2 screws
- Right side front wheel. Tighten the lug nuts to 72–87 ft. lbs. (98–118 Nm).
- Upper engine timing belt front cover, and tighten the 8 bolts to 27–44 inch lbs. (3–5 Nm)
- Main wiring harness on the upper engine front cover
- Water bypass hose between the thermostat housing and water connection
- Upper radiator hose between the radiator and the water hose connection. Secure the hoses with clamps. Install the upper radiator hose bracket. Tighten the bracket bolt to 34–58 ft. lbs. (46–65 Nm).
- Air conditioning compressor drive belt idler pulley and install the 3 bolts. Tighten to 15 ft. lbs. (21 Nm), if equipped

- Alternator drive belt, the water pump and power steering pump drive belt and the air conditioning compressor drive belt, if equipped.
- Negative battery cable

25. Fill the cooling system.

26. Start the engine and allow it to warm to operating temperature. Check and adjust the ignition timing. Road test to verify correct engine operation.

Piston and Ring Positioning

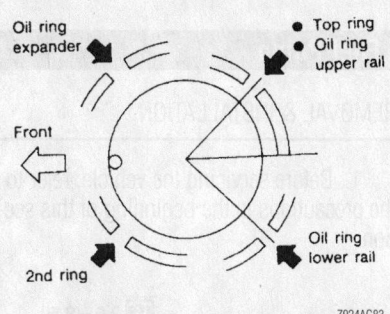

3.3L engines piston ring end-gap spacing

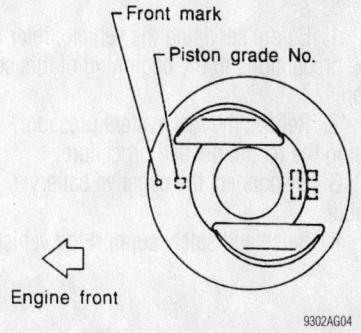

3.3L engines piston and connecting rod assembly positioning

3.3L engine piston positioning

FUEL SYSTEM

Fuel System Service Precautions

Safety is the most important factor when performing not only fuel system maintenance but any type of maintenance. Failure to conduct maintenance and repairs in a safe manner may result in serious personal injury or death. Maintenance and testing of the vehicle's fuel system components can be accomplished safely and effectively by adhering to the following rules and guidelines.

- To avoid the possibility of fire and personal injury, always disconnect the negative battery cable unless the repair or test procedure requires that battery voltage be applied.
- Always relieve the fuel system pressure prior to disconnecting any fuel system component (injector, fuel rail, pressure regulator, etc.), fitting or fuel line connection. Exercise extreme caution whenever relieving fuel system pressure, to avoid exposing skin, face and eyes to fuel spray. Please be advised that fuel under pressure may penetrate the skin or any part of the body that it contacts.
- Always place a shop towel or cloth around the fitting or connection prior to loosening to absorb any excess fuel due to spillage. Ensure that all fuel spillage (should it occur) is quickly removed from

engine surfaces. Ensure that all fuel soaked cloths or towels are deposited into a suitable waste container.

- Always keep a dry chemical (Class B) fire extinguisher near the work area.
- Do not allow fuel spray or fuel vapors to come into contact with a spark or open flame.
- Always use a back-up wrench when loosening and tightening fuel line connection fittings. This will prevent unnecessary stress and torsion to fuel line piping. Always follow the proper torque specifications.
- Always replace worn fuel fitting O-rings with new. Do not substitute fuel hose or equivalent, where fuel pipe is installed.

Fuel System Pressure

RELIEVING

1. Before servicing the vehicle, refer to the precautions in the beginning of this section.

2. Remove the left side engine compartment relay panel cover.

3. Locate and remove the fuel pump relay from the relay panel.

4. Start the engine.

5. Allow the engine to run until it stalls

from fuel starvation. After the engine stalls, crank the engine over 2 more times to ensure all pressure has been released.

6. Turn the ignition switch to the **OFF** position and install the fuel pump relay.

7. Most service work that follows fuel pressure relief also requires that the negative battery cable (ground) be disconnected before service work begins. This also prevents accidental fuel pump energizing that could pressurize the system.

Fuel Filter

REMOVAL & INSTALLATION

In-Line—Except California

1. Before servicing the vehicle, refer to the precautions in the beginning of this section.

2. Relieve the fuel system pressure using the recommended procedure.

3. Disconnect the negative battery cable.

4. Raise and safely support the vehicle.

5. Remove the fuel hose clamps.

6. Disconnect and plug the hoses to prevent leakage.

7. Remove the fuel filter from the bracket.

To install:

8. Install the fuel filter into the bracket with the arrow facing up, in the direction of the fuel travel to the engine.

9. Reconnect the fuel hoses.

10. Install and tighten the hose clamps. Verify that the clamps are properly tightened. System operating pressure is approximately 36 psi (248 kPa) and fuel will leak is connections are not properly made.

11. Lower the vehicle.

12. Reconnect the negative battery cable.

13. Check for leaks.

In-Line—California

1. Before servicing the vehicle, refer to the precautions in the beginning of this section.

2. Relieve the fuel system pressure using the recommended procedure.

3. Disconnect the negative battery cable.

4. Raise and safely support the vehicle.

5. Remove the filter splash shield bolts.

6. Disconnect the lines from each end of the filter and plug the hoses to prevent leakage.

7. Loosen the filter bracket nuts.

8. Remove the fuel filter from the bracket.

To install:

9. Install the fuel filter into the bracket with the arrow facing forward. Tighten the bracket bolts to 44 inch lbs. (5 Nm).

10. Reconnect the fuel hoses.

11. Lower the vehicle.

12. Reconnect the negative battery cable.

13. Check for leaks.

Fuel Pump

REMOVAL & INSTALLATION

1. Before servicing the vehicle, refer to the precautions in the beginning of this section.

2. Properly relieve the fuel system pressure.

3. Disconnect the negative battery cable.

4. Raise and safely support the vehicle.

5. Remove the fuel tank as follows:

   a. Drain the fuel from the tank.

   b. Remove the filler protector.

   c. Disconnect the filler tube.

   d. Detach any electrical connectors related to the fuel pump and fuel level sending unit.

   e. Detach the fuel line quick connectors.

   f. Safely support the fuel tank.

   g. Remove the tank mounting straps, then lower the tank out of the vehicle.

6. Remove the 6 fuel pump bolts.

7. Lift the fuel pump out of the fuel tank. Use care. The fuel level sensor and fuel pump and bracket must be tipped to remove it from the fuel tank. Do not lift the fuel sensor and pump assembly straight out

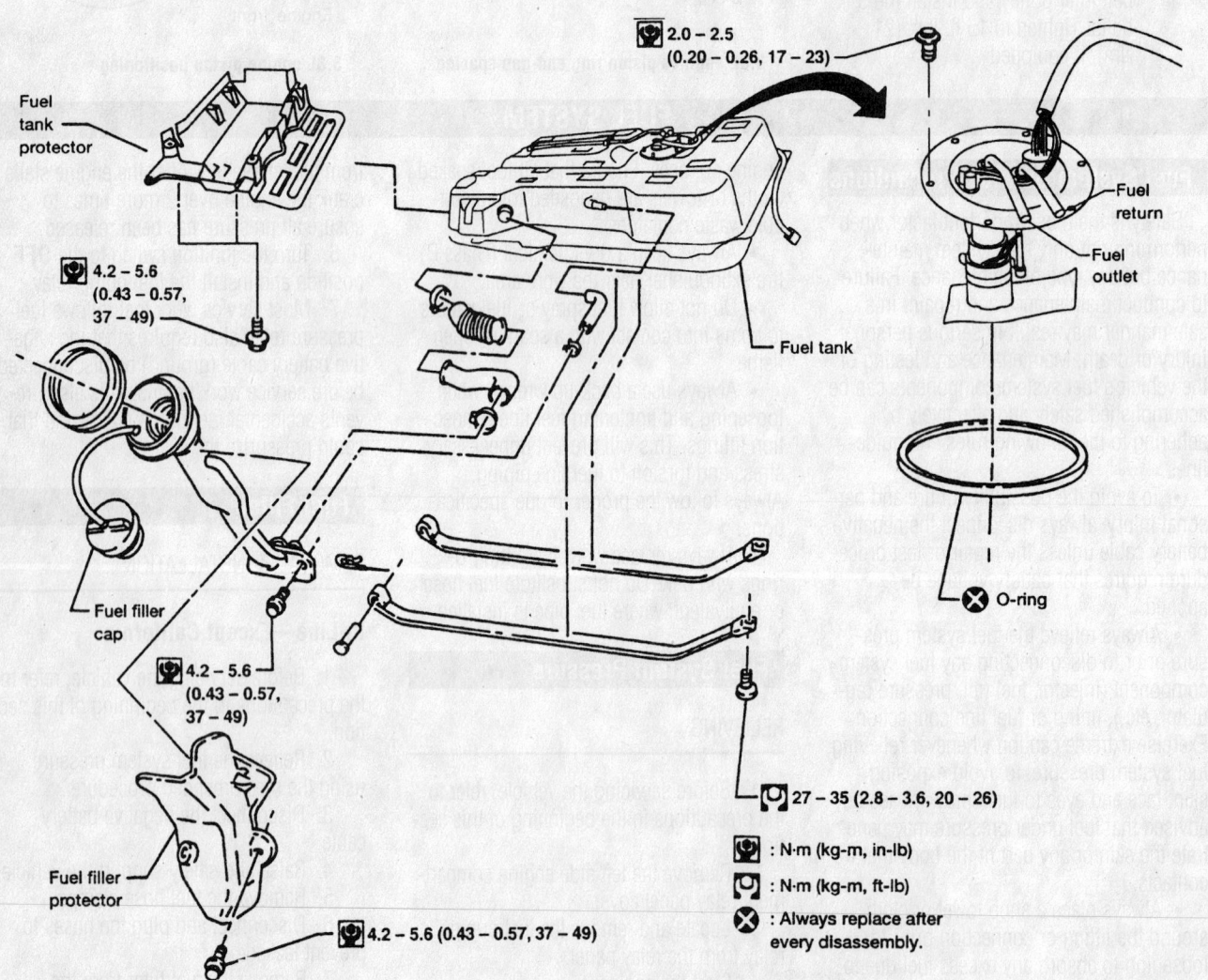

Fuel tank and related components

9302WG03

of the fuel tank or damage to the level sensor may occur.

8. Remove the 2 bolts attaching the level sensor to the fuel pump.

9. Remove the fuel pump level sensor and the gasket.

10. Discard the gasket.

11. Remove the fuel pump from the bracket.

To install:

12. Position the fuel level sensor on the fuel pump and bracket and install the 2 bolts.

13. Install a new level sensor gasket. Carefully install the level sensor and pump assembly.

14. Install the 6 fuel pump bolts. Do not over-tighten the bolts. Tighten the bolts to just 17–23 inch lbs. (2–3 Nm).

15. Install the fuel tank in the reverse order of removal, be sure to tighten the tank mounting straps to 20–26 ft. lbs. (27–35 Nm).

16. Lower the vehicle. Refill the fuel tank as required.

17. Connect the negative battery cable. Verify that the fuel pump relay has been properly installed. Start the engine and check for proper operation.

Fuel Injector

REMOVAL & INSTALLATION

1. Before servicing the vehicle, refer to the precautions in the beginning of this section.

2. Disconnect the negative battery cable.

3. If removing a rear injector, remove the upper intake manifold.

4. Disengage the injector electrical connection.

➡ **When removing the fuel injectors, use a screwdriver head socket to remove the injector cap screws.**

5. Remove the injector cap screws and the cap.

6. Pull the injector from the fuel rail.

7. Remove and discard the injector O-rings.

To install:

➡ **Use new insulators and O-ring seals for assembly.**

8. Install or connect the following:
   - Fuel injectors with the rail and tighten the fasteners to 8–11 ft. lbs. (11–15 Nm)
   - Fuel injector caps. Tighten the screws to 26–33 inch lbs. (3–4 Nm).
   - Injector electrical connections.
   - Intake manifold, if removed.
   - Negative battery cable.

9. Start the vehicle and check for leaks.

DRIVE TRAIN

Automatic Transaxle Assembly

REMOVAL & INSTALLATION

1. Before servicing the vehicle, refer to the precautions in the beginning of this section.

2. Remove or disconnect the following:
   - Negative battery cable
   - Battery and tray
   - Resonator
   - Terminal cord assembly harness connector
   - Vacuum lines
   - Starter motor
   - Transaxle fluid from the unit
   - Halfshafts
   - Transaxle cooler hose and control cable
   - Front exhaust manifold

   - Crankshaft Position (CKP) sensor
   - Engine gusset and torque converter undercover
   - Bolts from the drive plate from the torque converter. Rotate the crankshaft to access all the bolts.

3. Support the transaxle with a suitable jack.

   - Front mount
   - Rear mount
   - Bolts attaching the transaxle to the engine

4. Carefully separate the transaxle assembly from the engine assembly. Lower the assembly from the vehicle.

To install:

5. Be sure that the transaxle is secured firmly to the transaxle jack.

6. Carefully raise the transaxle into the vehicle and align the transaxle to the engine assembly, making sure that the

alignment dowels are positioned properly.

7. Install or connect the remaining components in the reverse order of removal. Refer to the accompanying transaxle torque specification illustration for bolt locations and their specifications.

8. Connect the negative battery cable.

9. Fill the transaxle with the correct amount and type of fluid.

10. Start the engine.

11. Check for leaks and proper operation.

Halfshaft

REMOVAL & INSTALLATION

1. Before servicing the vehicle, refer to the precautions in the beginning of this section.

2. Raise and safely support the vehicle.

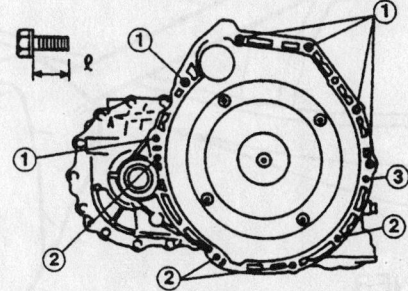

Bolt No.	Tightening torque N·m (kg-m, ft-lb)	ℓ mm (in)
1	39 - 49 (4.0 - 5.0, 29 - 36)	60 (2.36)
2	30 - 40 (3.1 - 4.1, 22 - 30)	25 (0.98)
3*	30 - 40 (3.1 - 4.1, 22 - 30)	25 (0.98)

*: TORX bolt

9348WG17

Transaxle torque specification and locations—3.3L engine

3. Remove or disconnect the following:
- Wheel
- Fender splash shield
- Cotter pin, nut retainer, and the hub retainer washers from the front hub assembly
- Lower ball joint from the knuckle
- Sway bar from the lower control arm at the sway bar link nut
- Halfshaft and CV-joint from the wheel hub

4. Position a drain pan under the transaxle since some fluid may run out when the inner joint is disengaged from the transaxle.

5. A prybar is used to separate the inner CV-joint from the transaxle. Use great care that the prybar does not damage the transaxle case, differential oil seal, outer race or boot. If removing the left side half-shaft, position prybars on both sides of the outer race, between the outer race and the transaxle case. Gently pry outward to unseat the circlip.

6. When removing the right side half-shaft, it is not be necessary to remove the halfshaft bearing retainer bracket from the cylinder block. Remove the 3 bearing retainer bolts and pull the right side half-shaft CV-joint with the bearing retainer from the differential side gear.

7. Support the halfshafts and remove them from the vehicle. Use care not to damage the boots. Place the halfshafts on a flat, protected work area.

To install:

✳✳ CAUTION

Do not reuse the circlip used on the left side halfshaft.

8. To prevent over-expanding the cir-clip, install the circlip carefully, starting one end in the shaft groove, then working the circlip over the CV-joint splined end. Always use a new circlip. No circlip is used on the right side halfshaft.

9. Inspect the CV-joint boots. If service is required, replace the CV-joint boots.

10. Inspect the differential oil seals. If dam-aged, the factory recommends using a hook-type puller and slide hammer arrangement to remove the seals. A seal driver is used to install the replacement differential oil seals.

11. If installing the left side halfshaft and CV-joint assembly, position the CV-joint so the splines are aligned with the differential side gear splines, then push the halfshaft joint into the differential case. As the circlip locks into the differential side gear groove, a click will be felt.

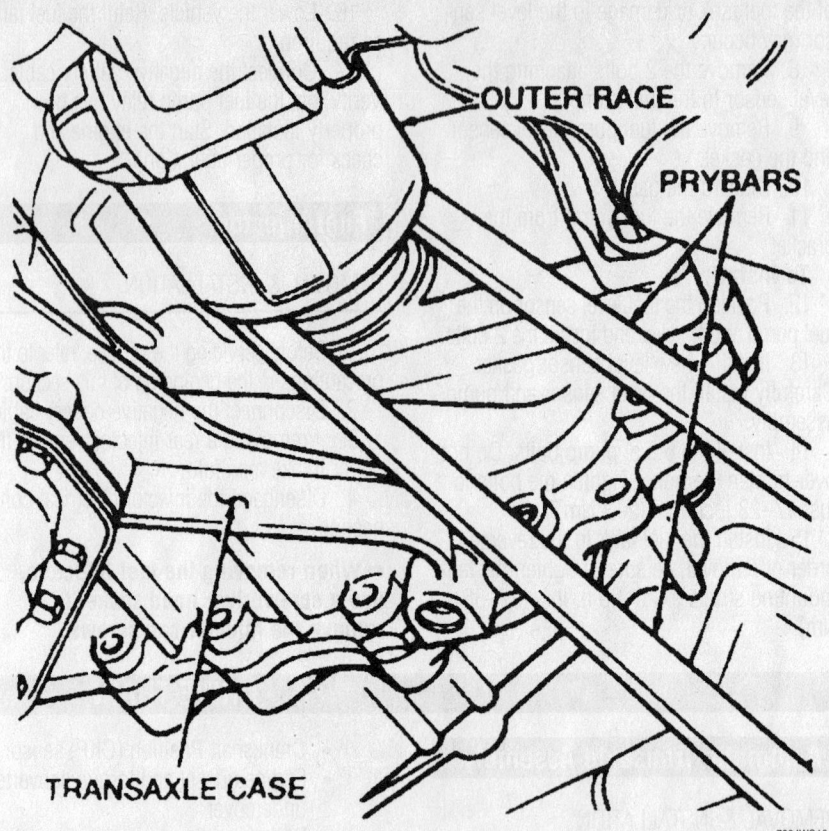

OUTER RACE

PRYBARS

TRANSAXLE CASE

7924WG18

Removing the left side halfshaft by gently prying with 2 prybars to unseat the circlip

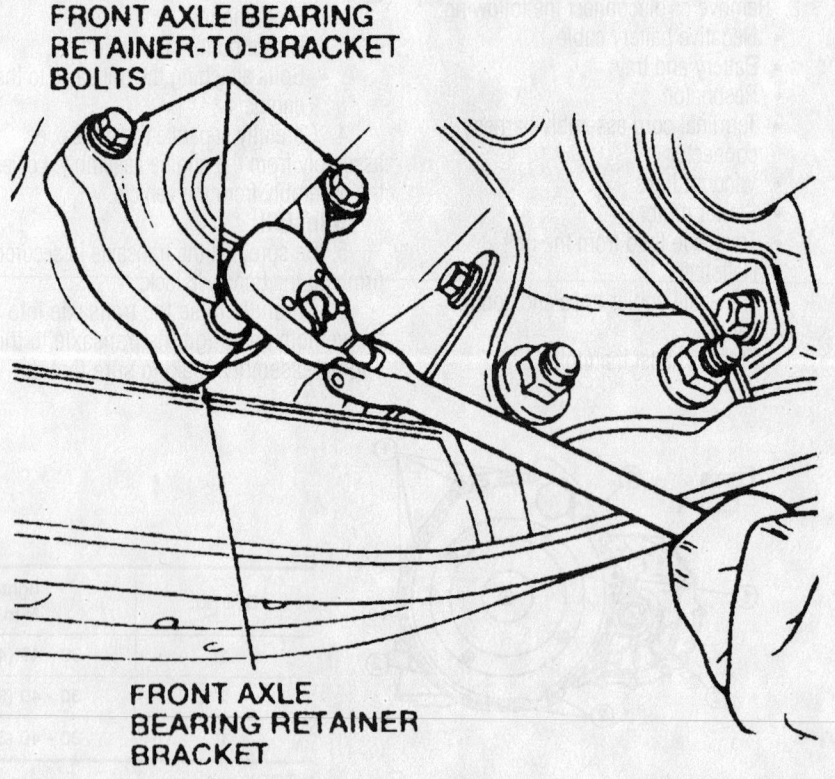

FRONT AXLE BEARING RETAINER-TO-BRACKET BOLTS

FRONT AXLE BEARING RETAINER BRACKET

7924WG19

Right side halfshaft bearing retainer bracket

12. If installing the right side halfshaft and CV-joint assembly, simply push the CV-joint into the differential side gear. Position the bearing retainer onto the bearing retainer bracket that should still be on the cylinder block. Install the 3 bolts and tighten to 8–14 ft. lbs. (13–19 Nm).

13. Install or connect the following:
- Halfshaft
- Lower ball joint and tighten the lower ball joint stud nut to 52–63 ft. lbs. (71–86 Nm). Secure the nut with a new cotter pin.
- Sway bar link to the lower control arm and tighten the link nut to 12–16 ft. lbs. (16–22 Nm).
- Wheel outer bearing retainer, washer and axle nut. Tighten the hub nut to 174–231 ft. lbs. (235–314 Nm). Install the nut retainer and secure with a new cotter pin.
- Splash shield
- Wheel. Tighten the lug nuts to 72–87 ft. lbs. (98–118 Nm).

14. Lower the vehicle.
15. Check the transaxle fluid level.
16. Road test the vehicle to verify correct operation and no noise or vibration.

CV-Joint

OVERHAUL

Inner

1. Remove the boot bands.
2. Matchmark the slide joint housing and inner race, prior to separating the joint assembly.
3. Pry off the snapring and remove the ball cage, inner race and balls as a unit.
4. Remove the snapring and withdraw the boot.

To install:

➡ **Cover the halfshaft serrations with tape, so as not to damage the boot.**

5. Thoroughly clean all parts in solvent and dry with compressed air. Check parts for evidence of damage and replace as necessary.
6. Install the boot and new boot band on the halfshaft.
7. Install a new inner snapring.
8. Install the ball cage, inner race and balls as a unit. Confirm that the matchmarks are aligned.
9. Install a new outer snapring.
10. Pack the CV-joint with 5.0–6.0 ounces (165–175 g) of grease.
11. Ensure that the boot is properly installed on the halfshaft groove.
12. Set the boot so that it does not swell or deform when its length is 3.86 in. (98mm).
13. Lock the new boot bands securely.

Outer

The joint on the wheel side cannot be disassembled.
1. Prior to separating the joint assem-

Exploded view of the halfshafts and related components

The inner CV-joint uses a large C-clip to retain the ball and cage assembly in the outer housing

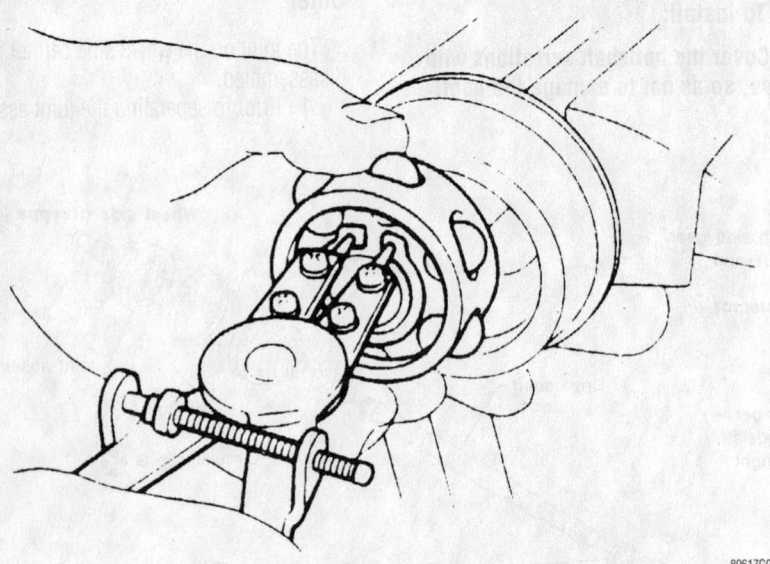

After the outer housing is removed, the ball and cage assembly can slide from the shaft by removing the C-clip

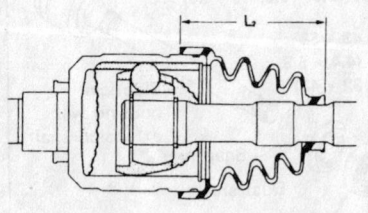

Make sure to properly position the boot before tightening the boot clamps

bly, matchmark the halfshaft and joint assembly.

2. Separate the joint using a slide hammer.

3. Remove the boot bands and the boot.

To install:

4. Thoroughly clean all parts in solvent and dry with compressed air. Check parts for evidence of damage and replace as necessary.

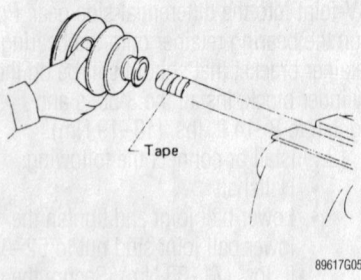

Use vinyl tape and wrap the end of the shaft to protect the boot during installation

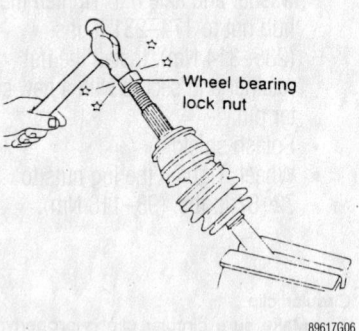

Use an old nut to protect the threads when tapping the outer CV-joint onto the shaft

➡️Cover the halfshaft serrations with tape, so as not to damage the boot.

5. Install the boot and small boot band on the halfshaft.

6. Set the joint assembly onto the half-shaft and align the matchmarks.

7. Attach the joint assembly to the half-shaft by lightly tapping the serrated end with a plastic hammer.

➡️Using a metal hammer may damage the threads on the end of the joint.

8. Pack the CV-joint with 4.76–5.11 ounces (135–145 g) of grease.

9. Ensure that the boot is properly installed on the halfshaft groove.

10. Set the boot so that it does not swell or deform when its length is 3.82 in. (97mm).

11. Lock the new boot bands securely.

STEERING AND SUSPENSION

Air Bag

PRECAUTIONS

Several precautions must be observed when handling the inflator module to avoid accidental deployment and possible personal injury.

• Never carry the inflator module by the wires or connector on the underside of the module.

• When carrying a live inflator module, hold securely with both hands, and ensure that the bag and trim cover are pointed away.

• Place the inflator module on a bench or other surface with the bag and trim cover facing up.

• With the inflator module on the bench, never place anything on or close to the module which may be thrown in the event of an accidental deployment.

DISARMING

✳✳ CAUTION

To avoid rendering the Supplemental Restraint System (SRS) inoperative, which could lead to personal injury or death in the event of a severe frontal collision, extreme caution must be taken when servicing the electrical related systems.

➡**All SRS electrical wiring harnesses and connectors are covered with YELLOW outer insulation. Do not use electrical test equipment on any circuit related to the SRS (air bag) sensors. When installing SRS components, always install with the arrow marks facing the front of the vehicle.**

Disarming

To disarm the Supplemental Restraint System (SRS) system turn the ignition switch to the **OFF** position. Then, disconnect the both battery cables starting with the negative cable first and wait at least 10 minutes after the cables are disconnected. Be sure to insulate the battery terminal ends.

Arming

To arm the Supplemental Restraint System (SRS) system turn the ignition switch to **OFF** position. Connect the both battery cables starting with the positive cable first.

➡The SRS or air bag system is equipped with a self-diagnostic operation. After turning the ignition key to the ON or START position, the AIR BAG warning lamp will illuminate for 7 seconds. After 7 seconds, the AIR BAG lamp will extinguish if no malfunction is detected. If the AIR BAG lamp does not extinguish after 7 seconds, check the SRS self-diagnostic system for a malfunction.

Power Rack and Pinion

REMOVAL & INSTALLATION

The power steering gear is held in position by 2 steering gear brackets and insulators. Note that the housing may move slightly when the steering wheel is turned. If the housing moves more than 0.080 inch (2mm), replace the steering gear insulators. If one or both of the brackets move, check the torque of the bracket bolts. The correct torque for these bolts is 54–72 ft. lbs. (73–97 Nm).

1. Before servicing the vehicle, refer to the precautions in the beginning of this section.

2. Place a drain pan under the steering rack.

3. Remove or disconnect the following:

• Brake master cylinder remote reservoir bracket screws. Position the reservoir out of the way and secure with wire.

• Junction block/high pressure line from the steering rack. Position the junction block and line out of the way.

• Both front wheels

• Front sway bar

• Tie rod ends from the steering knuckles

• Lower steering column shaft clamp bolt

• Power steering fluid return hose and position out of the way.

• The steering rack clamp bracket bolts

4. Lower the steering rack from the vehicle.

To install:

5. Carefully slide the steering gear rack and pinion assembly in place from the left side of the vehicle. Position the input shaft so it is just below the lower steering column shaft clamp.

6. Raise the steering gear until the plastic aligning tab on the input shaft enters the clamp bolt gap on the lower column shaft. Do not install the clamp bolt yet.

7. Examine the steering gear brackets. They should be marked UP with arrows pointing to one end of the bracket. Be sure the brackets are installed correctly. Tighten the steering gear bracket bolts to 54–72 ft. lbs. (73–97Nm) in sequence, working counterclockwise from the number 1 bolt (upper right side).

8. Install or connect the following:

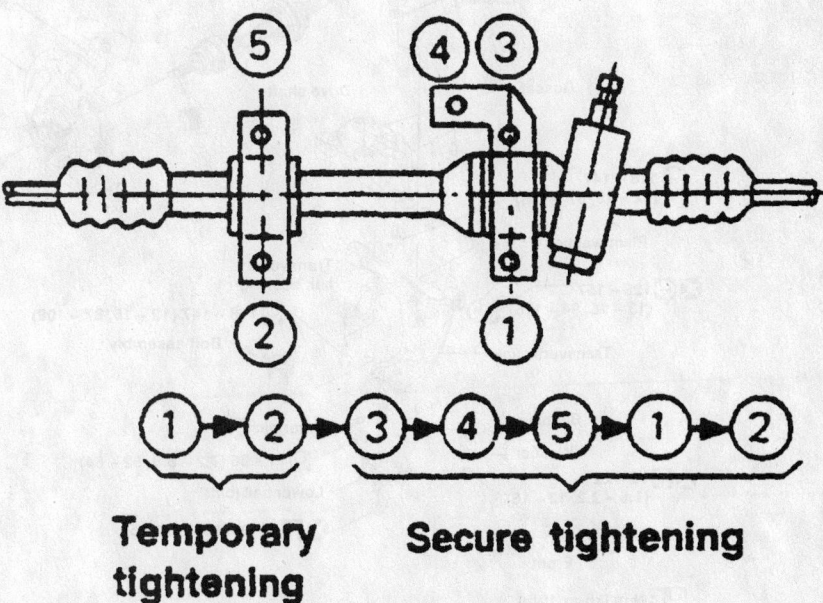

Temporary tightening **Secure tightening**

9348WG18

Tighten the power steering rack mounting bolts in the sequence shown

- Fluid return line to the steering gear
- Steering column shaft clamp bolt. Tighten the bolt to 17–22 ft lbs. (24–29 Nm). Install the dust cover.
- Tie rod ends
- Stabilizer bar
- Wheel. Tighten the lug nuts to 72–87 ft. lbs. (98–118 Nm).
- Junction block. Tighten the high-pressure line to 11–18 ft. lbs. (15–25 Nm).

- Brake master cylinder reservoir
9. Check for leaks and proper operation.

MacPherson Strut

REMOVAL & INSTALLATION

1. Before servicing the vehicle, refer to the precautions in the beginning of this section.

2. Disconnect the negative battery cable.

3. Matchmark the front strut upper mounting bracket and the chassis strut tower.

4. Raise and safely support the vehicle.

5. Remove the front wheel.

6. If equipped, remove the 2 front brake anti-lock sensor cable bracket bolts and position the anti-lock sensor cable out of the way.

7. Detach the brake tube from the strut.

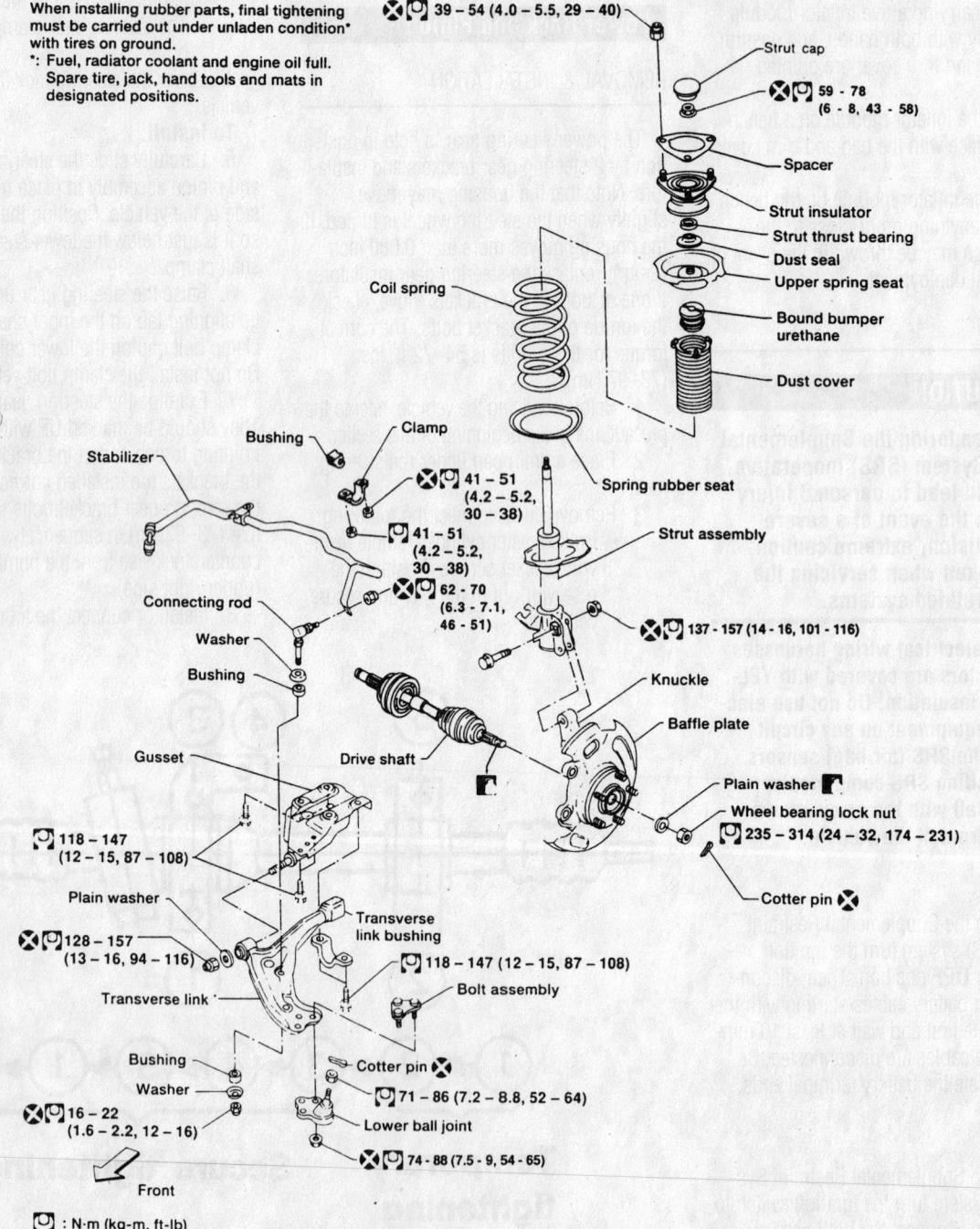

When installing rubber parts, final tightening must be carried out under unladen condition* with tires on ground.
*: Fuel, radiator coolant and engine oil full. Spare tire, jack, hand tools and mats in designated positions.

: N·m (kg-m, ft-lb)

Coil spring and strut assembly

9302WG04

8. Support the control arm.

9. Matchmark the knuckle to the strut so it can installed in the same position. This is important for the camber angle of the front wheel.

10. Remove the strut-to-steering knuckle bolts.

11. Support the strut and remove the 3 upper strut-to-chassis nuts. Remove the strut from the vehicle.

✳✳ WARNING

Never loosen the strut center nut until the spring is compressed or serious injury or vehicle damage may occur.

12. Place the strut and coil spring assembly in a suitable vise and remove the strut nut cover.

13. Slightly loosen, but **do not** remove the front strut nut.

If desired, use the following steps to remove the coil spring from the strut.

14. Using an approved coil spring compressor, compress the coil spring.

15. Remove the strut assembly top nut.

16. Remove the following components from the strut assembly:
  * Upper mounting bracket
  * Strut bearing
  * The bearing seat.
  * Upper coil spring seat and dust boot
  * Coil spring

17. Slowly release the tension of the coil spring compressor and remove the coil spring from the compressor tool.

18. Remove the coil spring insulator and slide the jounce bumper off of the strut assembly.

To install:

19. Slide the jounce bumper onto the strut assembly and install the coil spring insulator.

20. Carefully compress the coil spring with an approved coil spring compressor.

21. Reinstall the following components to the strut assembly:
  * Coil spring

➡**Install the coil spring to the strut assembly with the end of the spring in the lower coil spring seat indentation.**

  * Upper coil spring seat and dust boot
  * Bearing seat and the bearing
  * Upper mounting bracket

22. Install and tighten the strut assembly nut and tighten the nut to 43–58 ft. lbs. (59–78 Nm).

23. Install the strut assembly onto the vehicle and tighten the following:

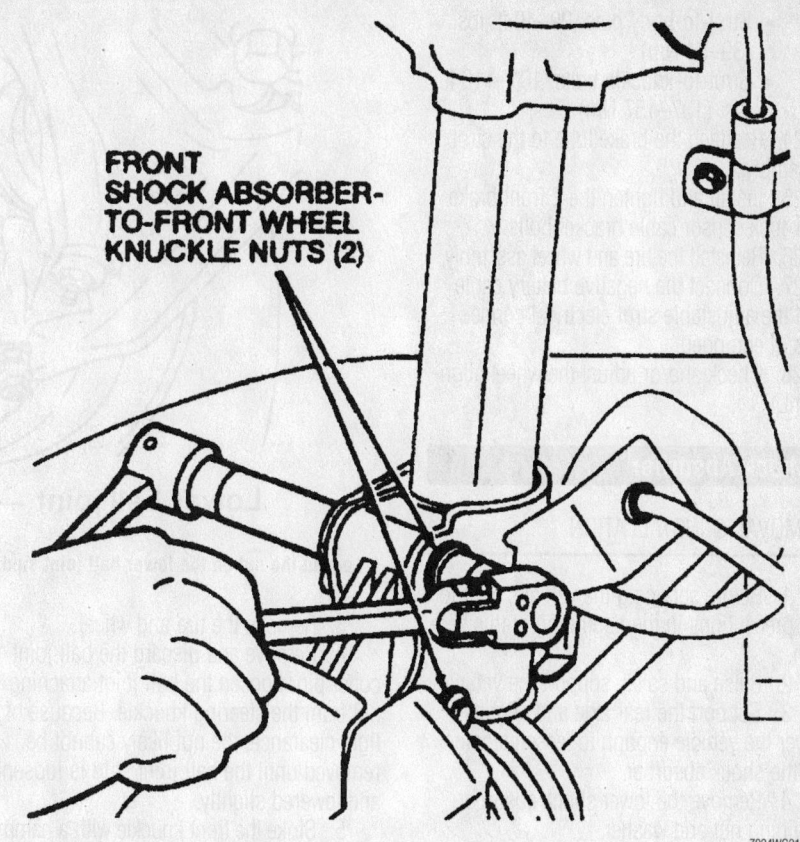

FRONT SHOCK ABSORBER-TO-FRONT WHEEL KNUCKLE NUTS (2)

7924WG21

The strut is attached to the knuckle with 2 large bolts

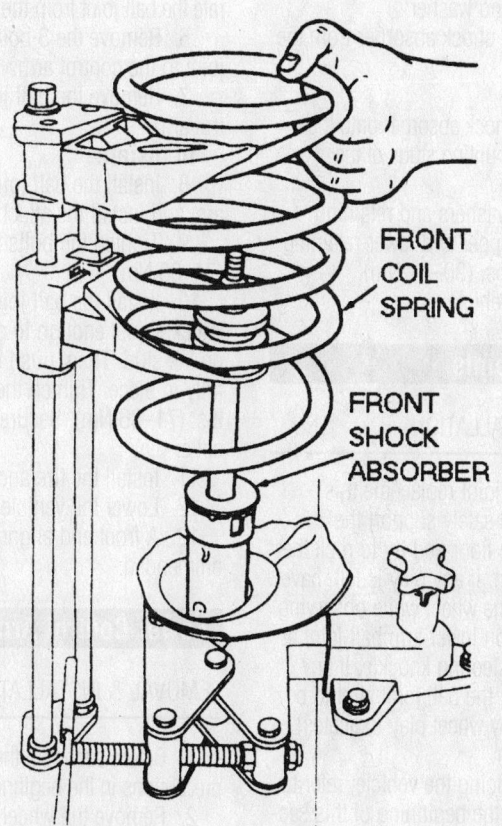

FRONT COIL SPRING

FRONT SHOCK ABSORBER

7924WG22

Compress the coil spring in an approved spring compressor

- Strut-to-body nuts: 29–40 ft. lbs. (39–54 Nm)
- Strut-to-knuckle bolts: 101–116 ft. lbs. (137–157 Nm)

24. Reattach the brake tube to the strut assembly.

25. Install and tighten the 2 front brake anti-lock sensor cable bracket bolts.

26. Reinstall the tire and wheel assembly.

27. Connect the negative battery cable and the adjustable strut electrical connectors, if equipped.

28. Check and/or adjust the wheel alignment.

Shock Absorber

REMOVAL & INSTALLATION

1. Before servicing the vehicle, refer to the precautions in the beginning of this section.

2. Raise and safely support the vehicle.

3. Support the rear axle and slightly lower the vehicle enough to lessen tension on the shock absorber.

4. Remove the lower shock absorber retaining nut and washer.

5. Disconnect the lower end of the shock absorber from the mounting stud.

6. Remove the shock absorber upper end retaining nut and washer.

7. Remove the shock absorber from the vehicle.

To install:

8. Install the shock absorber onto the upper and lower mounting studs of the vehicle.

9. Install the washers and retaining nuts. Tighten the upper and lower retaining nuts to 22–30 ft. lbs. (30–41 Nm).

10. Lower the vehicle.

Lower Ball Joints

REMOVAL & INSTALLATION

To check if ball joint replacement is required, raise and safely support the vehicle clear of the floor and try to rock the wheel up and down. If any play is felt, have an assistant rock the wheel while observing the front suspension lower arm ball joint at the bottom of the steering knuckle. If any movement is seen, the ball joint should be replaced. If not, any wheel play indicates wheel bearing wear.

1. Before servicing the vehicle, refer to the precautions in the beginning of this section.

2. Raise and safely support the vehicle.

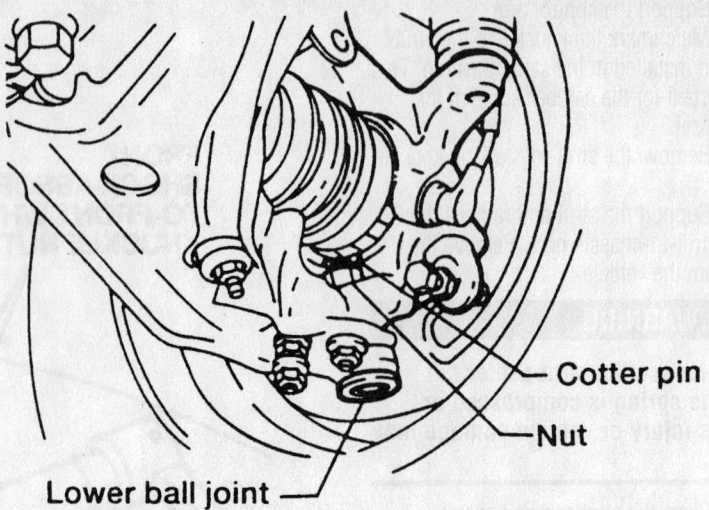

Loosen the nut on the lower ball joint stud

3. Remove the tire and wheel.

4. Remove and discard the ball joint cotter pin. Loosen the ball joint attaching nut from the steering knuckle. Because of tight clearance, the nut likely cannot be removed until the ball joint stud is loosened and lowered slightly.

5. Strike the front knuckle with a hammer while pulling down on the lower control arm. There should now be enough clearance to allow removal of the ball joint stud nut. Separate the ball joint from the steering knuckle.

6. Remove the 3 bolts attaching the ball joint to the control arm.

7. Remove the ball joint from the control arm.

To install:

8. Install the ball joint to the control arm and install the attaching bolts.

9. Tighten the bolts to 54–65 ft. lbs. (74–88 Nm).

10. Install the ball joint into the steering knuckle, just enough to get the nut started on the stud. Then, push the ball joint stud fully in place. Tighten the nut to 52–63 ft. lbs. (71–86 Nm). Secure the nut with a new cotter pin.

11. Install the tire and wheel.

12. Lower the vehicle.

13. A front end alignment check is recommended.

Lower Control Arm

REMOVAL & INSTALLATION

1. Before servicing the vehicle, refer to the precautions in the beginning of this section.

2. Remove the wheel.

3. Disconnect the ball joint.

4. Disconnect the stabilizer bar from the control arm.

5. Remove the 2 rear arm bolts and the mounting bracket.

6. Remove the lower arm nut.

7. Pull the rear of the arm down and gently pry the arm forward and off the gusset.

8. Installation is the reverse of removal. Observe the following torques:

- Stabilizer bar-to-lower arm: 12–16 ft. lbs. (16–22 Nm)
- Lower arm rear bolts: 87–108 ft. lbs. (118–147 Nm)
- Lower arm nuts: 94–115 ft. lbs. (128–156 Nm)
- Ball stud nut: 56–80 ft. lbs. (76–109 Nm)

BUSHING REPLACEMENT

The bushings are press-fit types. Support the arm in a press, using the proper adapters. Ford tool numbers are: T93P-5493-A, T75L-1165-B and -DA.

Wheel Bearings

ADJUSTMENT

The wheel bearings are not adjustable. If the bearings become loose or make noise, they must be replaced using the following procedure.

REMOVAL & INSTALLATION

Front

1. Before servicing the vehicle, refer to the precautions in the beginning of this section.

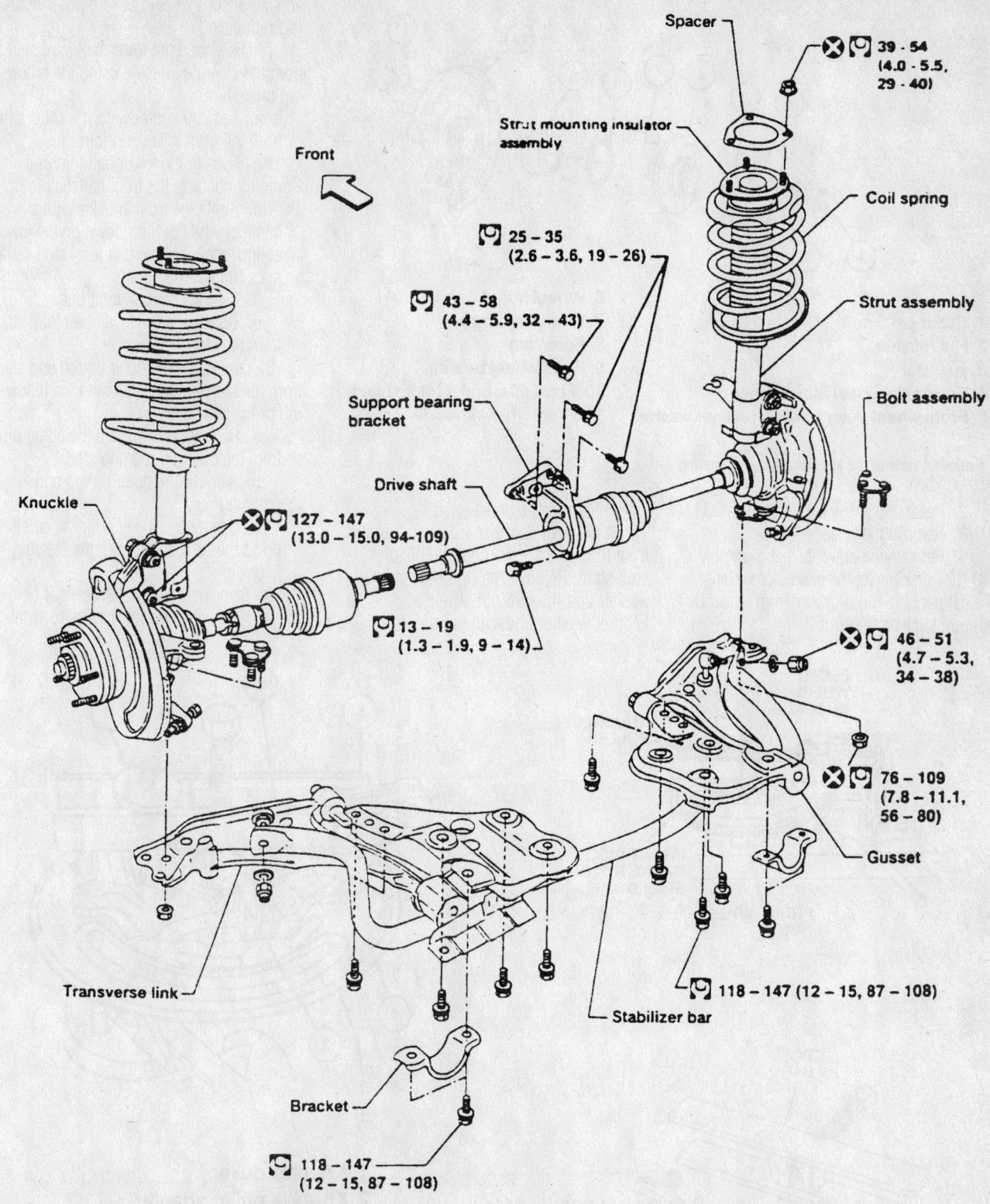

Spacer

⊗ 🔧 **39 - 54**
**(4.0 - 5.5,
29 - 40)**

Strut mounting insulator assembly

Coil spring

🔧 **25 - 35**
(2.6 - 3.6, 19 - 26)

Strut assembly

🔧 **43 - 58**
(4.4 - 5.9, 32 - 43)

Front

Bolt assembly

Support bearing bracket

Drive shaft

Knuckle

⊗ 🔧 **127 - 147**
(13.0 - 15.0, 94-109)

🔧 **13 - 19**
(1.3 - 1.9, 9 - 14)

⊗ 🔧 **46 - 51**
**(4.7 - 5.3,
34 - 38)**

⊗ 🔧 **76 - 109**
**(7.8 - 11.1,
56 - 80)**

Gusset

Transverse link

🔧 **118 - 147 (12 - 15, 87 - 108)**

Stabilizer bar

Bracket

🔧 **118 - 147**
(12 - 15, 87 - 108)

When installing rubber parts, final tightening must be carried out under unladen condition*
with tires on ground.
*: Fuel, radiator coolant and engine oil full.
Spare tire, jack, hand tools and mats in designated positions.

🔧 : N·m (kg-m, ft-lb)

Exploded view of the front suspension and drive axles

7924WG25

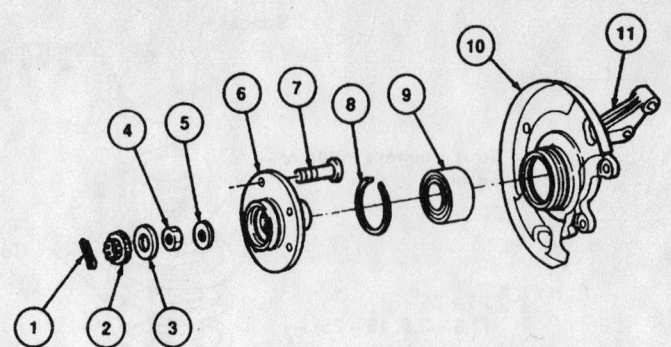

1. Cotter pin
2. Nut retainer
3. Insulator
4. Front axle wheel hub retainer
5. Front wheel outer bearing retainer washer
6. Wheel hub
7. Wheel hub bolt
8. Snap ring
9. Front wheel bearing
10. Front disc brake rotor shield
11. Front wheel knuckle

7924WG23

Exploded view of the knuckle, hub and bearing

2. Raise and safely support the vehicle.
3. Remove the wheel and tire.
4. Remove the brake caliper assembly. DO NOT disconnect the brake hose. Hang the caliper on a piece of wire from a near by support such as the strut.

5. Remove the brake rotor.
6. Remove and discard the cotter pin from the end of the outboard CV-joint stub shaft. Remove the hub nut retainer, washer and the hub nut. There should be another washer under the hub nut that

acts as a front wheel bearing outer bearing retainer.
7. Disengage the lower ball joint stud from the steering knuckle using the following procedure.
   a. Remove and discard the cotter pin from the front lower ball joint.
   b. Loosen the lower ball joint nut until it contacts the front halfshaft joint.
   c. Strike the front knuckle with a hammer while pulling down on the lower control arm until the ball joint stud separates from the knuckle.
   d. Remove the ball joint nut.
   e. Disengage the lower ball joint stud from the steering knuckle.
8. Disengage the outer tie rod end stud from the steering knuckle using the following procedure.
   a. Remove and discard the cotter pin from the outer tie rod end stud.
   b. Remove the outer tie rod end retaining nut.
   c. Use a tie rod end puller to carefully press the tie rod end from the steering knuckle.
9. Remove the front ABS sensor bolt.
10. Remove the 2 front strut-to-front

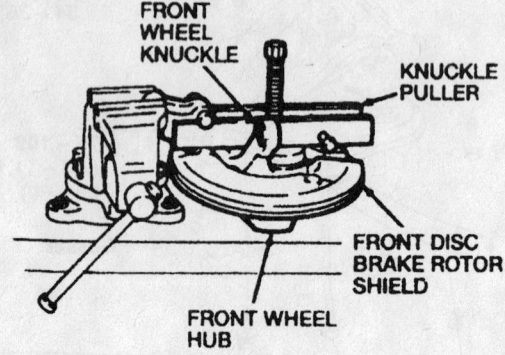

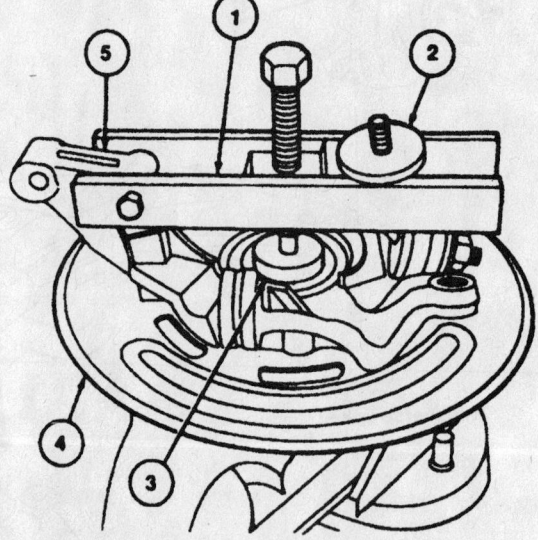

1. Knuckle puller
2. Knuckle puller adapter
3. Step plate adapter
4. Front disc brake rotor shield
5. Front wheel knuckle

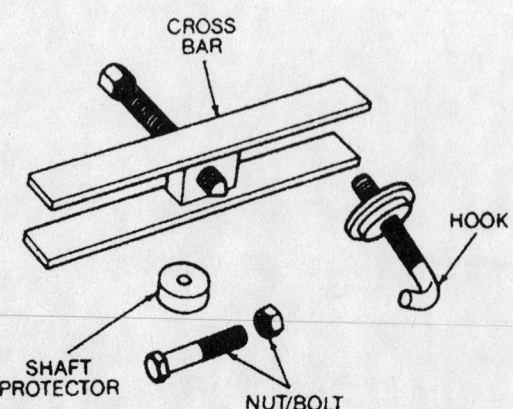

Example of a puller set up to bear against the front wheel bearing inner race

7924WG27

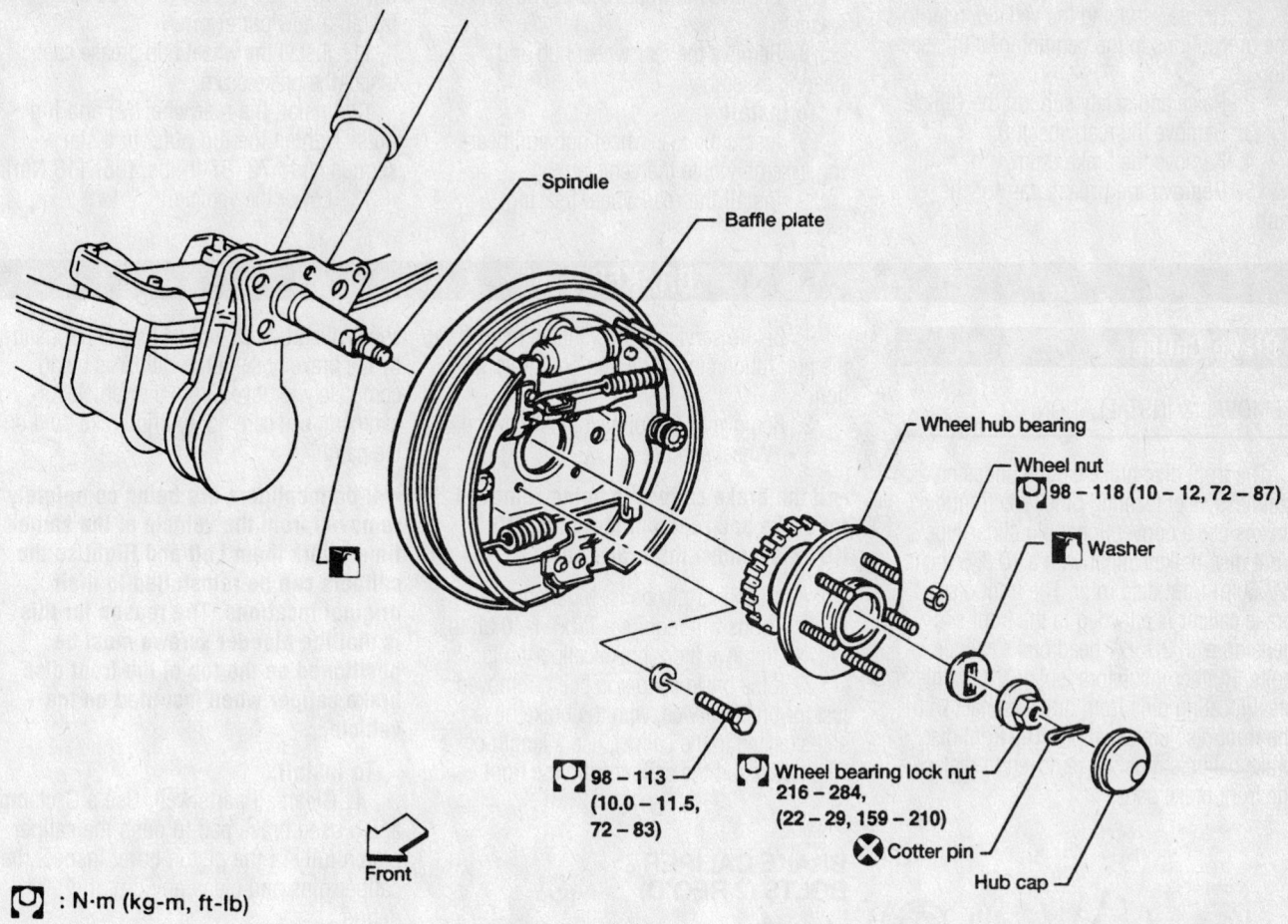

Spindle

Baffle plate

Wheel hub bearing

Wheel nut
⊙ 98 – 118 (10 – 12, 72 – 87)

Washer

98 – 113
(10.0 – 11.5,
72 – 83)

Wheel bearing lock nut
216 – 284,
(22 – 29, 159 – 210)

⊗ Cotter pin

Hub cap

Front

⊙ : N·m (kg-m, ft-lb)

9302WG05

Rear hub assembly

knuckle nuts and remove the 2 bolts. Disengage the strut from the steering knuckle.

11. Use a 2-jaw puller to separate the front halfshaft outboard CV-joint stub shaft from the knuckle/bearing assembly.

12. Remove the front wheel hub, knuckle and wheel bearing assembly from the vehicle.

13. If the knuckle is being replaced with a service part, change over the steering stop bolt and jam nut from the old knuckle to the replacement part.

14. To remove the front wheel bearing, jig up a puller to bear against the front wheel bearing inner race and pull the race from the hub/knuckle assembly.

15. Use a shop press to press out damaged wheel studs and also to press out the outer bearing race.

16. Use a shop press to press out the inner bearing race.

To install:

17. If the front wheel bearings were removed, assemble the ABS sensing ring, if

removed and the disc brake dust shield under the steering knuckle. Use a shop press to push in new front wheel bearing inner and outer races. Support the knuckle and press the front wheel bearing into the knuckle and install the snapring retainer. Support the bearing assemblies and press the hub onto the knuckle and wheel bearing assembly.

18. Install the hub, knuckle and bearings as an assembly. Position the assembly on the halfshaft outer CV-joint stub axle end. Guide the knuckle into the front strut and install the 2 knuckle-to-strut bolts and nuts. Tighten the nuts to 83–91 ft. lbs. (113–123 Nm).

19. Install the ABS sensor bolt. Do not over-tighten. Tighten to just 16–21 inch lbs. (1.8–2.4 Nm).

20. Install the outer tie rod end to the steering knuckle. Tighten the nut to 22–29 ft. lbs. (29–39 Nm). If the cotter pin holes do not align, tighten the nut slightly until they do. Never loosen the nut to align the holes. Secure the nut with a new cotter pin.

21. Start the lower ball joint stud to the steering knuckle and partially install the nut, then push the ball joint stud fully in place. Tighten the ball joint stud nut to 52–63 ft. lbs. (71–86 Nm). Secure the nut with a new cotter pin.

22. Install the front wheel outer bearing retaining washer and the hub retainer nut. Tighten to 174–231 ft. lbs. (235–314 Nm). Install the nut retainer, insulator and a new cotter pin.

23. Install the front brake rotor and install the disc brake caliper.

24. If removed, install the steering stop bolt.

25. Install the tire and wheel assembly. Tighten the lug nuts to 72–87 ft. lbs. (98 to 118 Nm).

26. Lower the vehicle. Pump the brake pedal slowly to seat the front brake pads. Do not move the vehicle until a firm pedal is obtained.

27. A front end alignment is recommended.

Rear

1. Before servicing the vehicle, refer to the precautions in the beginning of this section.
2. Raise and safely support the vehicle.
3. Remove the rear wheel(s).
4. Remove the brake drum.
5. Remove the grease cap for the hub.

6. Remove and discard the cotter pin.
7. Remove the wheel bearing nut and washer.
8. Remove the rear wheel hub and bearing assembly.

To install:

9. Install the rear wheel hub and bearing assembly onto the vehicle.
10. Install the rear wheel bearing

washer and nut and tighten the bearing nut to 159–210 ft. lbs. (216–284 Nm). Install a new cotter pin.
11. Install the wheel hub grease cap. Install the brake drum.
12. Install the rear wheel(s) and lug nuts. Tighten the lug nuts, in a star sequence, to 72–87 ft. lbs. (98–118 Nm).
13. Lower the vehicle.

BRAKES

Brake Caliper

REMOVAL & INSTALLATION

The front disc brake caliper slides on 2 stainless steel locating pins. The front disc brakes use a conventional pin slider-type front disc brake caliper with a 10.875 inch (27.6cm) front disc rotor. The front disc brake caliper is attached to the front suspension with 2 Torx® head brake caliper bolts. Rubber insulators isolate the stainless steel locating pins from direct contact with the front disc brake caliper. The front disc brake calipers must be removed to replace the front brake pads.

1. Before servicing the vehicle, refer to the precautions in the beginning of this section.
2. Remove or disconnect the following:
   • Wheel and tire

➡**If the brake caliper is being removed for brake pad replacement only, DO NOT disconnect the brake hose.**

   • 2 caliper pin bolts. Most applications will require a Torx® T-40 bit to remove the 2 brake caliper bolts.
3. If the brake caliper is being removed just for brake service, with the brake hose still attached to the caliper, use a length of wire to support the caliper from the front

shock absorber. Do not let the caliper hang by the brake hose. If the caliper is being completely removed from the vehicle for overhaul, use care not to drip brake fluid on the paint.

➡**If both calipers are being completely removed from the vehicle at the same time, mark them Left and Right so the calipers can be reinstalled to their original locations. The reason for this is that the bleeder screws must be positioned on the top of the front disc brake caliper when installed on the vehicle.**

To install:

4. Clean all parts well. Use a C-clamp and a used brake pad to push the caliper piston fully in the piston bore. Inspect the caliper pins and clean any dirt and debris.
5. Install the caliper onto the rotor. Make sure the inboard and outboard brake pads are properly positioned.
6. Lubricate the stainless steel locating pins with a Silicone Dielectric Compound such as Ford DZAZ-19A331-A or equivalent silicone grease. Install the 2 caliper pin bolts and torque to 18–25 ft. lbs. (24–34 Nm).
7. If disconnected, install the brake hose using a new replacement copper washer, install the banjo bolt and torque to 12–14 ft. lbs. (17–20 Nm).
8. If the brake hose had been disconnected, bleed the brake system.
9. Install the wheel and tire.
10. Torque the lug nuts to 72–87 ft. lbs. (98–118 Nm).
11. Check the master cylinder reservoir and add fresh DOT 3 brake fluid as required.
12. Lower the vehicle. Pump the brake pedal slowly until a firm brake pedal is obtained, indicating that the brake pads are properly seated, before attempting to move the vehicle. Road-test and check for proper brake operation.

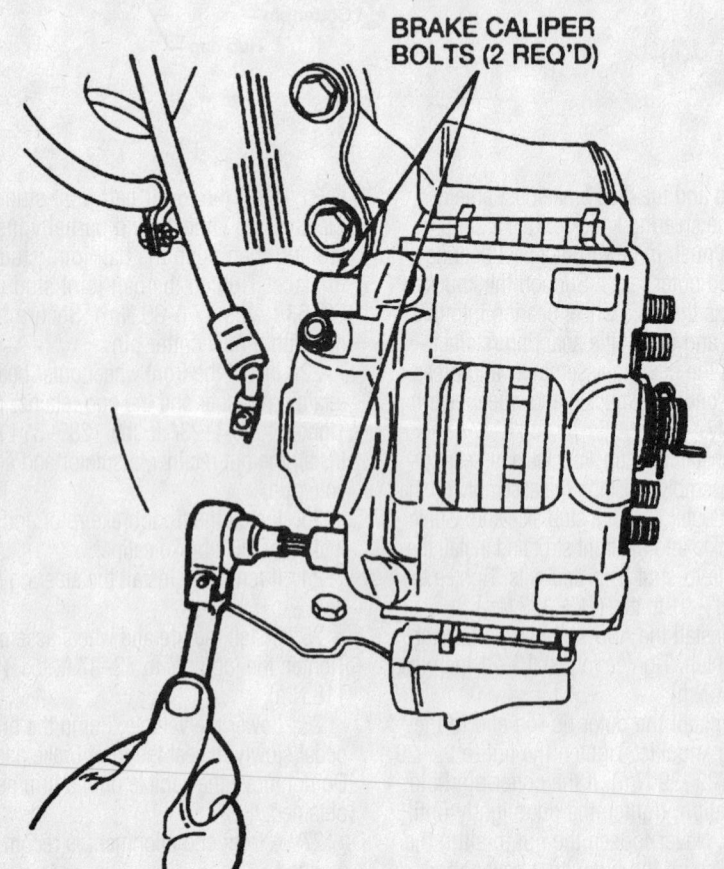

BRAKE CALIPER BOLTS (2 REQ'D)

93026G10

Caliper pin bolt removal

Disc Brake Pads

REMOVAL & INSTALLATION

Front

1. Before servicing the vehicle, refer to the precautions in the beginning of this section.
2. Remove or disconnect the following:
   • Wheels
   • Bottom guide pin from the caliper and swing the caliper cylinder body upward; support the caliper with a wire
   • Brake pad retainers and the pads

To install:

3. Compress the piston of the disc brake caliper.
4. Install or connect the following:
   • Brake pads and caliper assembly. Torque the guide pin to 23–30 ft. lbs. (31–41 Nm).
   • Wheels
5. Apply the brakes a few times to seat the pads. Check the master cylinder and add fluid if necessary. Bleed the brakes, if necessary.

Rear

1. Before servicing the vehicle, refer to the precautions in the beginning of this section.

➡**Do not press the piston into the bore as performed on the front disc brakes. Due to the parking brake mechanism, the caliper piston must be turned into the bore using a special tool.**

2. Remove or disconnect the following:
   • Rear wheels
3. Release the parking brake.
   • Parking brake cable bracket bolt
   • Pin bolts and lift off the caliper body
   • Pad springs
   • Pads and shims

To install:

4. Clean the piston end of the caliper body and the area around the pin holes. Be careful not to get oil on the rotor.
5. Using the proper tool, carefully turn the piston clockwise back into the caliper body. Take care not to damage the piston boot.
6. Coat the pad contact area on the mounting support with a silicone based grease.
7. Install or connect the following:
   • Pads, shims, and the pad springs. Always use new shims.
   • Caliper body in the mounting support and tighten the pin bolts to 28–38 ft. lbs. (38–52 Nm)
   • Wheels
8. Bleed the system if necessary.

Brake Drums

REMOVAL & INSTALLATION

1. Before servicing the vehicle, refer to the precautions in the beginning of this section.

The rear drum brakes used on these vehicles are conventional expanding shoe-type with the brake shoe lining applied to the inside of the rotating drum. An incremental brake adjuster screw is designed to actuate whenever sufficient wear occurs.

2. Remove or disconnect the following:
   • Wheel and tire
   • Brake drum by pulling it from the wheel studs
3. If necessary for brake drum removal, pry off the access hole plug from the access hole. Insert a screwdriver and a brake adjustment tool. Press the screwdriver against the adjusting lever to disengage it from the adjuster. Loosen the adjuster using the brake adjusting tool.

To install:

4. Clean all parts well. It is good practice to inspect the wheel cylinder for leaks anytime the brake drum is removed. If a new replacement brake drum is being installed, inspect it for a protective coating on the machined inside braking surface. Remove any coating with suitable solvent.
5. Install or connect the following:

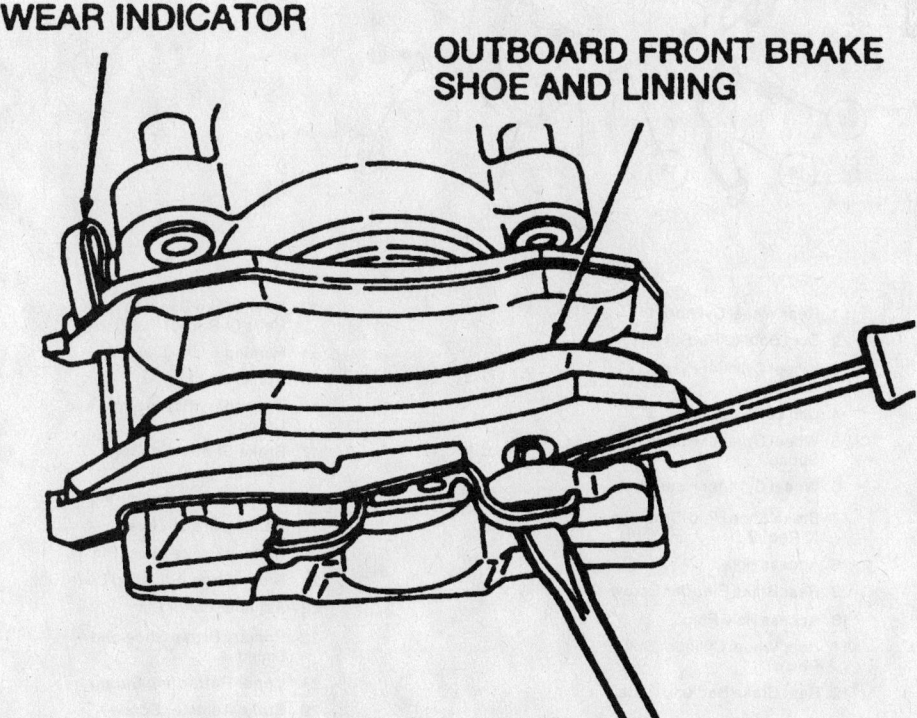

WEAR INDICATOR

OUTBOARD FRONT BRAKE SHOE AND LINING

93026G33

Replacing the disc brake pads

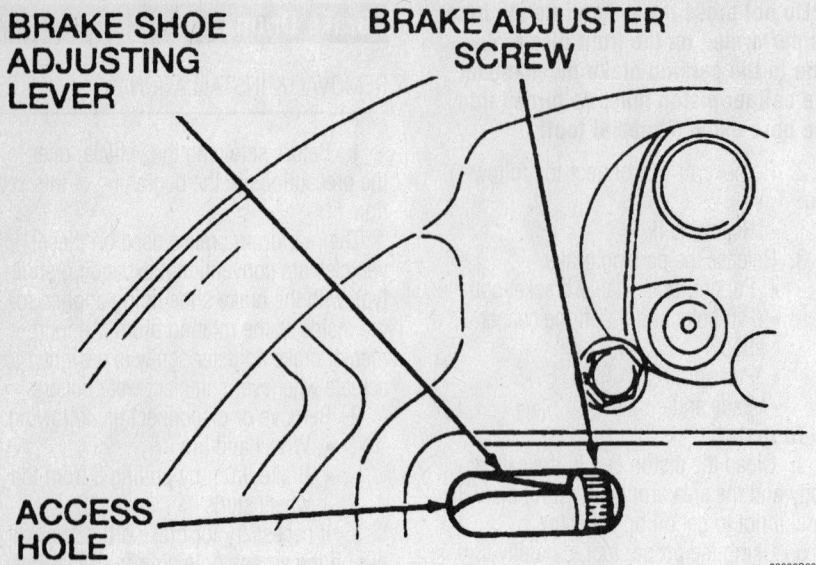

BRAKE SHOE ADJUSTING LEVER

BRAKE ADJUSTER SCREW

ACCESS HOLE

93026G28

Brake shoe adjustment may need to be loosened to remove the brake drum

- Brake drum onto the wheel studs
6. In most all cases, manual brake adjustment IS NOT recommended. Adjustment is performed by driving the vehicle and applying the brakes.
- Tire and wheel and torque the fasteners to 72–87 ft. lbs. (98–118 Nm)
7. Adjust the rear brake shoes by sharply applying the brakes several times while driving the vehicle alternately forwards and backwards. Check the brake operation by making several stops while driving forward.

Brake Shoes

REMOVAL & INSTALLATION

1. Before servicing the vehicle, refer to the precautions in the beginning of this section.

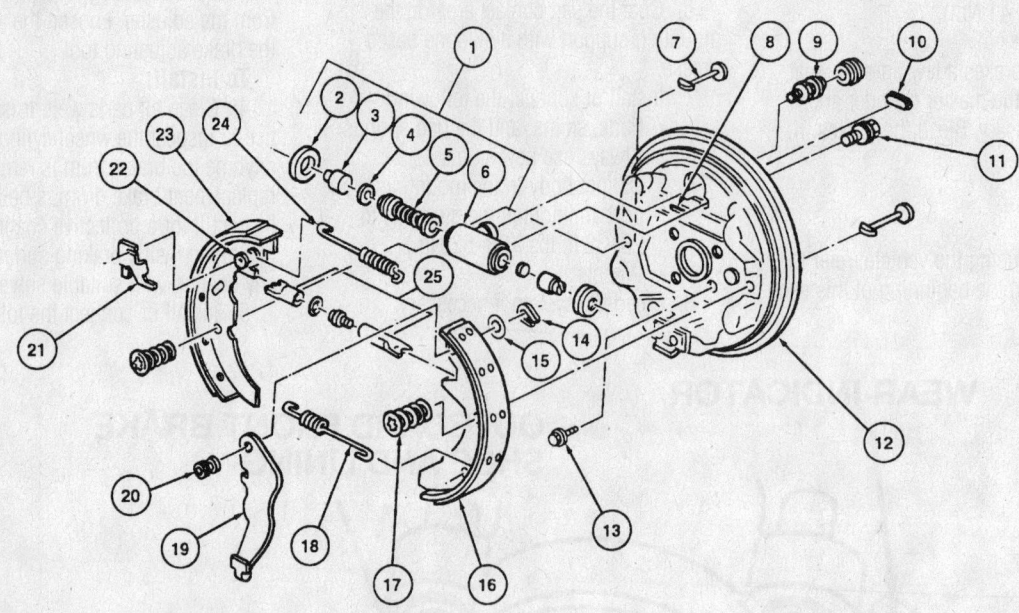

1 Rear Wheel Cylinder	13 Rear Brake Backing Plate Bolts (4 Req'd)
2 Dust Boot (2 Req'd)	14 Parking Brake Lever Clip
3 Wheel Cylinder Piston (2 Req'd)	15 Spring Washer
4 Cup (2 Req'd)	16 Secondary Brake Shoe and Lining
5 Wheel Cylinder Piston Cup Spring	17 Brake Shoe Hold-Down Spring
6 Wheel Cylinder Housing	18 Lower Retracting Spring
7 Brake Shoe Hold-Down Pin (2 Req'd)	19 Parking Brake Lever
8 Access Hole	20 Parking Brake Lever Pin
9 Rear Brake Bleeder Screw	21 Brake Shoe Adjusting Lever
10 Access Hole Plug	22 Adjuster Lever Pin
11 Rear Wheel Cylinder Bolt (2 Req'd)	23 Primary Brake Shoe and Lining
12 Rear Brake Backing Plate	24 Upper Retracting Spring
	29 Brake Adjuster Screw

93026G29

Rear drum brake assembly and related components

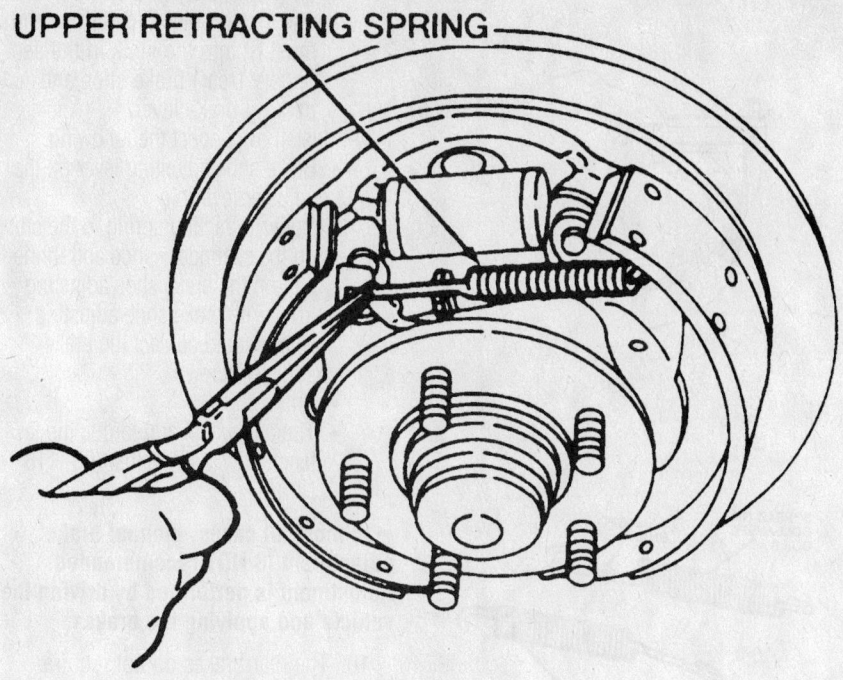

UPPER RETRACTING SPRING

93026G30

Remove the upper retracting spring

The rear drum brakes use an internal rear wheel cylinder with expanding shoes and lining that are applied against a rotating brake drum. An incremental brake adjuster screw is actuated whenever sufficient wear occurs. Brake adjustment takes place in forward or reverse braking but not with parking brake application.

2. Remove or disconnect the following:

- Wheels
- Brake drum

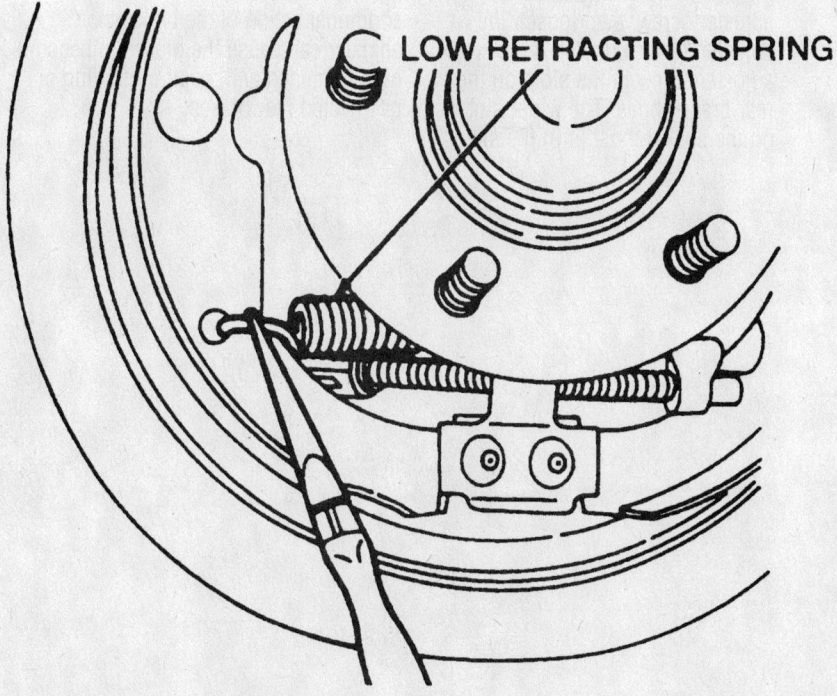

LOW RETRACTING SPRING

93026G31

Remove the lower retracting spring

- Parking brake rear cable and conduit from the parking brake lever
- 2 brake shoe hold-down springs and the 2 brake shoe hold-down pins
- Upper retracting spring
- Lower retracting spring
- Brake adjuster screw
- Rear brake shoes and linings from the brake backing plate
- Parking brake lever clip and washer
- Parking brake lever from the secondary brake shoe and lining

To install:

3. Clean all parts well.

4. Inspect the wheel cylinder for signs of leaking. Service as required.

5. Inspect the retracting springs for heat damage, bends or damage to the coils or shank or loss of tension. A good retracting spring will make a full thud when dropped on a concrete floor. A heat-damaged retracting spring that has lost tension will make a distinctive ringing sound when dropped on a concrete floor.

6. Check the brake backing plate for signs of scoring. The shoe contact points must be smooth and have a light coating of lithium grease. Verify that the brake lining thickness is between 0.059–0.232 in. (1.5–5.9mm). Failure to replace worn rear brake shoes will result in a scored drum.

7. Inspect the brake drum for scratches, scoring, bell mouth and out-of-round conditions. Remove minor scores on a brake drum with sandpaper. Do not refinish brake drums to remove scoring marks. A brake drum surface that is highly polished can cause the brakes to lock up. Remove polished surfaces with sandpaper or refinish the brake drum. Refinish a brake drum that is out-of-round enough to cause vehicle vibration or noise when braking. Remove only enough surface metal to true-up the brake drum. Brake drum maximum inside diameter is shown on each drum. If the maximum inside diameter shown on the brake drum is exceeded through wear or refinishing, replace the brake drum. After a brake drum is refinished, wipe the refinished surface with a cloth soaked in clean denatured alcohol. If one brake drum is refinished, the brake drum on the opposite side of the vehicle should also be refinished to the same diameter. The standard inner brake drum diameter is 9.840 inches (250.0mm). Replace the brake drum if worn beyond 9.900 inches (251.5mm).

8. Install or connect the following:

- Parking brake lever to the secondary brake shoe and lining with a new parking brake lever clip

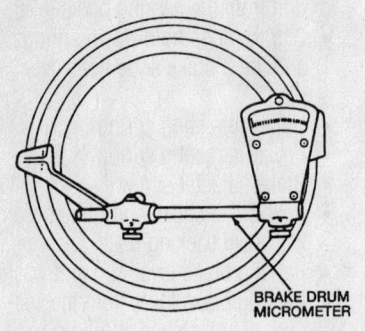

BRAKE DRUM MICROMETER

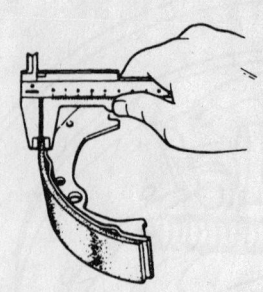

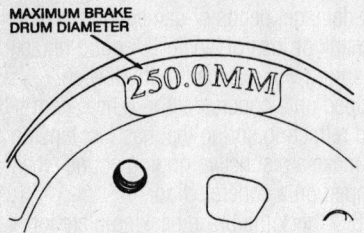

MAXIMUM BRAKE DRUM DIAMETER

250.0MM

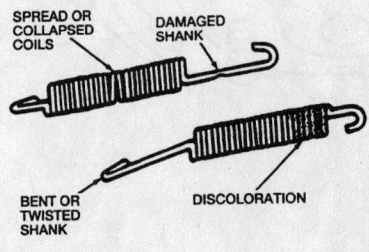

SPREAD OR COLLAPSED COILS

DAMAGED SHANK

BENT OR TWISTED SHANK

DISCOLORATION

93026G32

These size checks should be made and the retracting springs' condition checked. Replace questionable parts

- Secondary (rear) shoe on the backing plate and the brake shoe hold-down spring and pin
- Primary (front) shoe on the backing plate and the brake shoe hold-down spring and pin
- Parking brake rear cable and conduit to the parking brake lever
- Lower retracting spring to the rear brake shoes

- Apply a light coat of high-quality grease to the threaded areas of the adjuster nut and adjuster socket. Turn the adjuster nut all the way down on the brake adjuster screw, then loosen the adjuster ½ turn. Install the adjuster screw in the slots on the rear brake shoes. The wider slot on the socket must fit in the slot

on the primary (front) brake shoe. The slot on the adjuster nut end must fit into the slots in the secondary (rear) brake shoe and parking brake lever.

9. Install or connect the following:
- Brake shoe adjusting lever on the adjuster lever pin
- Upper retracting spring in the slot on the secondary shoe and in the slot on the brake shoe adjusting lever. The brake shoe adjusting lever should contact the brake adjuster screw.
- Brake drum
- Tire and wheel and torque the fasteners to 72–87 ft. lbs. (98–118 Nm).

➡ **In most all cases, manual brake adjustment IS NOT recommended. Adjustment is performed by driving the vehicle and applying the brakes.**

10. The rear brakes do not require adjustment when being serviced to obtain a firm brake pedal feel. To achieve a firm brake pedal after servicing the rear brakes, sharply apply the brake pedal several times while driving the vehicle alternately forwards and backwards. Check the brake operation by making several stops while driving forward. The self-adjusting mechanism will sufficiently adjust the rear brake shoes without any manual tightening at the brake shoe adjuster. If the rear brake shoes are manually adjusted, the additional action of the brake shoe adjuster can cause the brakes to become over-tightened and result in binding or overheated rear brakes.

SPECIFICATION CHARTS

ENGINE AND VEHICLE IDENTIFICATION

Engine							Model Year	
Code ①	Liters (cc)	Cu. In.	Cyl.	Fuel Sys.	Engine	Eng. Mfg.	Code ②	Year
VG33E	3.3 (3277)	199.8	6	MFI	SOHC	Nissan	Y	2000
VQ35DE	3.5 (3498)	213	6	MFI	DOHC	Nissan	1	2001
							2	2002
							3	2003
							4	2004

MFI: Multi-port Fuel Injection

SOHC: Single Overhead Camshaft

DOHC: Double Overhead Camshafts

① Located on the timing belt cover

② 10th digit of the Vehicle Identification Number (VIN)

42356-PATH-C01

GENERAL ENGINE SPECIFICATIONS

Year	Model	Engine Displacement Liters (cc)	Engine ID/VIN	Fuel System Type	Net Horsepower @ rpm	Net Torque @ rpm (ft. lbs.)	Bore x Stroke (in.)	Compression Ratio	Oil Pressure @ rpm
2000	QX4	3.3 (3277)	VG33E	MFI	170@4800	200@2800	3.60x3.27	8.9:1	60-65@2000
	Pathfinder	3.3 (3277)	VG33E	MFI	170@4800	200@2800	3.60X3.27	8.9:1	60-65@2000
2001	QX4	3.3 (3277)	VG33E	MFI	170@4800	200@2800	3.60x3.27	8.9:1	60-65@2000
	Pathfinder	3.3 (3277)	VG33E	MFI	170@4800	200@2800	3.60X3.27	8.9:1	60-65@2000
2002	QX4	3.5 (3498)	VQ35DE	MFI	240@6000	265@3200	3.76X3.20	10.0:1	43@2000
	Pathfinder	3.5 (3498)	VQ35DE	MFI	240@6000	265@3200	3.76X3.20	10.0:1	43@2000
2003	QX4	3.5 (3498)	VQ35DE	MFI	240@6000	265@3200	3.76X3.20	10.0:1	43@2000
	Pathfinder	3.5 (3498)	VQ35DE	MFI	240@6000	265@3200	3.76X3.20	10.0:1	43@2000

MFI: Multi-port Fuel Injection

42356-PATH-C02

ENGINE TUNE-UP SPECIFICATIONS

Year	Engine Displacement Liters (cc)	Engine ID/VIN	Spark Plug Gap (in.)	Ignition Timing (deg.) MT	Ignition Timing (deg.) AT	Fuel Pump (psi) ①	Idle Speed (rpm) MT	Idle Speed (rpm) AT ②	Valve Clearance (in.) In.	Valve Clearance (in.) Ex.
2000	3.3 (3277)	VG33E	0.039-0.043	13-17B	13-17B	34	700-800	700-800	HYD	HYD
2001	3.3 (3277)	VG33E	0.039-0.043	13-17B	13-17B	34	700-800	700-800	HYD	HYD
2002	3.5 (3498)	VQ35DE	0.044	15B	15B	35	700-800	700-800	HYD	HYD
2003	3.5 (3498)	VQ35DE	0.044	15B	15B	35	700-800	700-800	HYD	HYD

NOTE: The Vehicle Emission Control Information label often reflects specification changes made during production. The label figures must be used if they differ from those in this chart.

B: Before top dead center

HYD: Hydraulic

① System pressure at idle with vacuum hose connected

    Should increase to 43 psi when disconnected

② Automatic transmission in Neutral

42356-PATH-C03

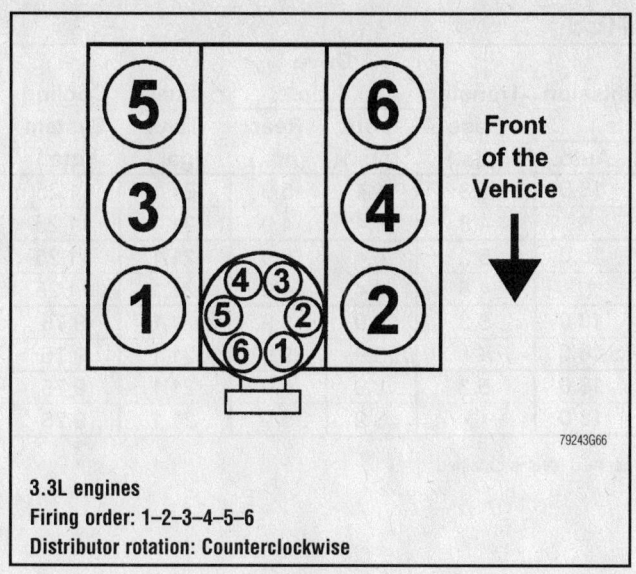

Front of the Vehicle

3.3L engines
Firing order: 1–2–3–4–5–6
Distributor rotation: Counterclockwise

79243G66

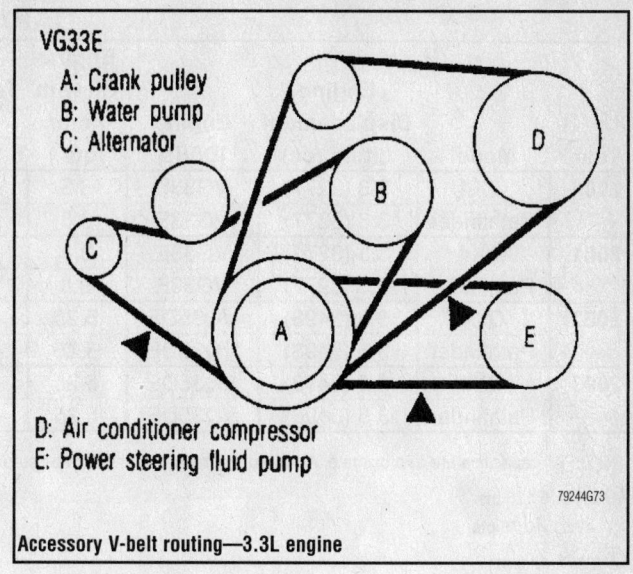

VG33E
A: Crank pulley
B: Water pump
C: Alternator

D: Air conditioner compressor
E: Power steering fluid pump

79244G73

Accessory V-belt routing—3.3L engine

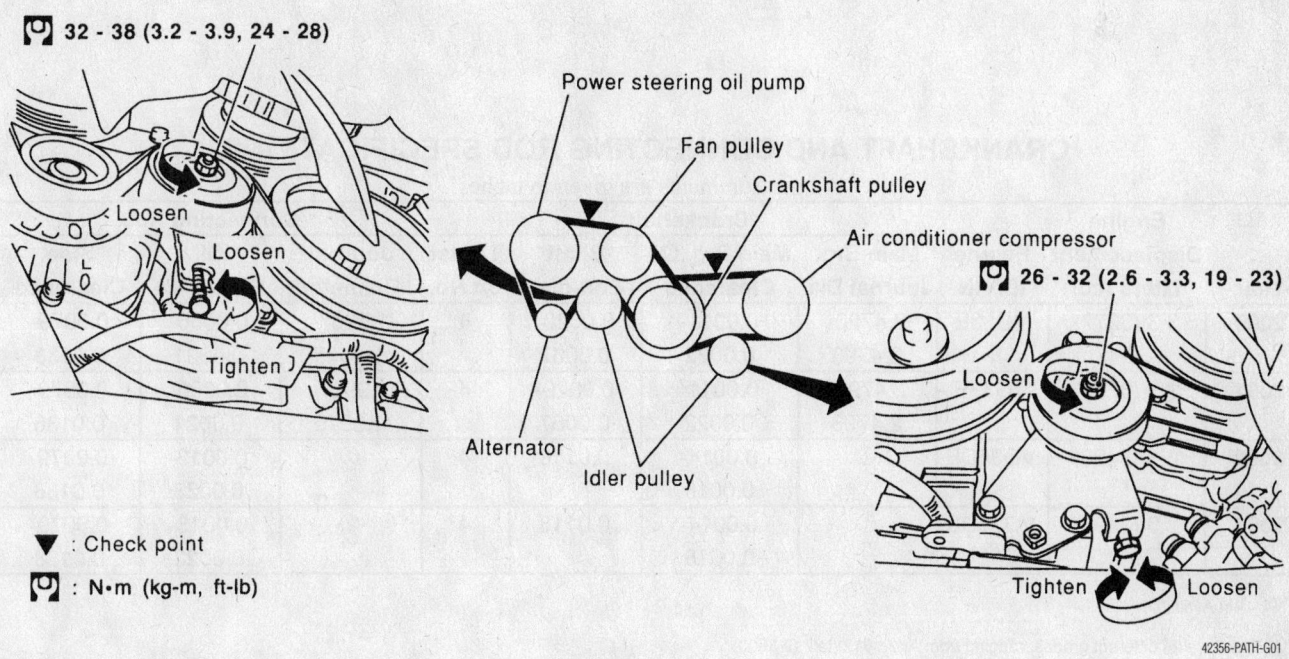

32 - 38 (3.2 - 3.9, 24 - 28)

Loosen
Loosen
Tighten

Power steering oil pump
Fan pulley
Crankshaft pulley
Air conditioner compressor

26 - 32 (2.6 - 3.3, 19 - 23)

Loosen

Alternator
Idler pulley

Tighten   Loosen

▼ : Check point

: N•m (kg-m, ft-lb)

42356-PATH-G01

Accessory V-belt routing—3.5L engine

CAPACITIES

Year	Model	Engine Displacement Liters (cc)	Engine ID/VIN	Engine Oil with Filter (qts.)	Transmission (pts.) 5-Spd	Transmission (pts.) Auto.	Transfer Case (pts.)	Drive Axle Front (pts.)	Drive Axle Rear (pts.)	Fuel Tank (gal.)	Cooling System (qts.)
2000	QX4	3.3 (3277)	VG33E	3.8	—	18.0	5.3	4.4	5.9	21.1	11.25
	Pathfinder	3.3 (3277)	VG33E	3.8	①	②	4.8	4.4	4.9	21.1	11.25
2001	QX4	3.3 (3277)	VG33E	3.8	—	18.0	5.3	4.4	5.9	21.1	11.25
	Pathfinder	3.3 (3277)	VG33E	3.8	①	②	4.8	4.4	4.9	21.1	11.25
2002	QX4	3.5 (3498)	VQ35DE	5.25	—	18.0	5.3	3.9	5.9	21.1	9.75
	Pathfinder	3.5 (3498)	VQ35DE	5.25	—	18.0	③	3.9	5.9	21.1	9.75
2003	QX4	3.5 (3498)	VQ35DE	5.25	—	18.0	5.3	3.9	5.9	21.1	9.75
	Pathfinder	3.5 (3498)	VQ35DE	5.25	—	18.0	③	3.9	5.9	21.1	9.75

NOTE: All capacities are approximate. Add fluid gradually and check to be sure a proper fluid level is obtained.

① 2WD: 5.125 pts.
   4WD: 10.75 pts.

② 2WD: 17.5 pts.
   4WD: 18 pts.

③ Part time: 2.375; full time: 2.625 pts.

42356-PATH-C04

CRANKSHAFT AND CONNECTING ROD SPECIFICATIONS
All measurements are given in inches.

Year	Engine Displacement Liters (cc)	Engine ID/VIN	Crankshaft Main Brg. Journal Dia.	Crankshaft Main Brg. Oil Clearance	Crankshaft Shaft End-play	Crankshaft Thrust on No.	Connecting Rod Journal Diameter	Connecting Rod Oil Clearance	Connecting Rod Side Clearance
2000	3.3 (3277)	VG33E	2.4790-2.4793	0.0011-0.0022	0.0020-0.0067	4	1.9967-1.9675	0.0006-0.0021	0.0079-0.0138
2001	3.3 (3277)	VG33E	2.4790-2.4793	0.0011-0.0022	0.0020-0.0067	4	1.9967-1.9675	0.0006-0.0021	0.0079-0.0138
2002	3.5 (3498)	VQ35DE	①	0.0014-0.0018	0.0118	4	②	0.0013-0.0023	0.0079-0.0138
2003	3.5 (3498)	VQ35DE	①	0.0014-0.0018	0.0118	4	②	0.0013-0.0023	0.0079-0.0138

NA - Not Available

① There are 24 different grades, ranging from A (2.3612) to 7 (2.3603)

② Grade 0: 2.0460-2.0462
   Grade 1: 2.0457-2.0460
   Grade 2: 2.0445-2.0457

42356-PATH-C05

VALVE SPECIFICATIONS

Year	Engine Displacement Liters (cc)	Engine ID/VIN	Seat Angle (deg.)	Face Angle (deg.)	Spring Test Pressure (lbs. @ in.)	Spring Installed Height (in.)	Stem-to-Guide Clearance (in.)		Stem Diameter (in.)	
							Intake	Exhaust	Intake	Exhaust
2000	3.3 (3277)	VG33E	45	45.25-46.75	①	NA	0.0008-0.0021	0.0016-0.0029	0.2742-0.2748	0.3135-0.3138
2001	3.3 (3277)	VG33E	45	45.25-46.75	①	NA	0.0008-0.0021	0.0016-0.0029	0.2742-0.2748	0.3135-0.3138
2002	3.5 (3498)	VQ35DE	45.15-45.45	45	45.4@1.457	1.457	0.0008-0.0021	0.0016-0.0029	0.2348-0.2354	0.2341-0.2346
2003	3.5 (3498)	VQ35DE	45.15-45.45	45	45.4@1.457	1.457	0.0008-0.0021	0.0016-0.0029	0.2348-0.2354	0.2341-0.2346

NA: Not Available

① Inner: 57.3 @ 0.984
  Outer: 117.7 @ 1.181

42356-PATH-C06

PISTON AND RING SPECIFICATIONS
All measurements are given in inches.

Year	Engine Displacement Liters (cc)	Engine ID/VIN	Piston Clearance	Ring Gap			Ring Side Clearance		
				Top Comp.	Bottom Comp.	Oil Control	Top Comp.	Bottom Comp.	Oil Control
2000	3.3 (3277)	VG33E	①	0.0083-0.0157	0.0197-0.0272	0.0079-0.0272	0.0009-0.0030	0.0012-0.0028	0.0006-0.0073
2001	3.3 (3277)	VG33E	①	0.0083-0.0157	0.0197-0.0272	0.0079-0.0272	0.0009-0.0030	0.0012-0.0028	0.0006-0.0073
2002	3.5 (3498)	VQ35DE	0.0004-0.0012	0.0091-0.0130	0.0130-0.0189	0.0079-0.0236	0.0016-0.0031	0.0012-0.0028	0.0006-0.0020
2003	3.5 (3498)	VQ35DE	0.0004-0.0012	0.0091-0.0130	0.0130-0.0189	0.0079-0.0236	0.0016-0.0031	0.0012-0.0028	0.0006-0.0020

① Cylinders 1, 2, 6: 0.0010 - 0.0018 in.
  Cylinders 3 and 4: 0.0006 - 0.0010 in.
  Cylinder 5: 0.0012-0.0016 in.

42356-PATH-C07

TORQUE SPECIFICATIONS
All readings in ft. lbs.

Year	Engine Displacement Liters (cc)	Engine ID/VIN	Cylinder Head Bolts	Main Bearing Bolts	Rod Bearing Bolts	Crankshaft Damper Bolts	Flywheel Bolts	Manifold Intake	Manifold Exhaust	Spark Plugs	Lug Nuts
2000	3.3 (3277)	VG33E	①	67-74	②	141-156	61-69	①	21-25	14-22	87-108
2001	3.3 (3277)	VG33E	①	67-74	②	141-156	61-69	①	21-25	14-22	87-108
2002	3.5 (3498)	VQ35DE	③	④	⑤	⑥	61-69	⑦	21-24	14-22	87-108
2003	3.5 (3498)	VQ35DE	③	④	⑤	⑥	61-69	⑦	21-24	14-22	87-108

① The cylinder heads and the lower intake manifold are installed together

Step 1: Tighten the cylinder head bolts to 22 ft. lbs.

Step 2: Tighten the cylinder head bolts to 43 ft. lbs.

Step 3: Loosen the cylinder head bolts completely

Step 4: Tighten the cylinder head bolts to 84 inch lbs.

Step 5: Tighten the intake manifold fasteners to 35 inch lbs.

Step 6: Tighten the intake manifold fasteners to 13 ft. lbs.

Step 7: Tighten the intake manifold fasteners to 12-14 ft. lbs.

Step 8: Loosen all intake manifold fasteners completely

Step 9: Tighten the cylinder head bolts to 22 ft. lbs.

Step 10: Tighten the cylinder head bolts 60-65 degrees

Step 11: Tighten the cylinder head sub-bolts to 80-105 inch lbs.

Step 12: Tighten the intake manifold fasteners to 35 inch lbs.

Step 13: Tighten the intake manifold fasteners to 78 inch lbs.

Step 14: Tighten the intake manifold fasteners to 70-84 inch lbs.

② 10-12 ft. lbs. plus 60-65 degrees or 28-33 ft. lbs.

③ Step 1: 72 ft. lbs.

Step 2: Loosen all bolts completely

Step 3: 25-33 ft. lbs.

Step 4: +90 degrees

Step 5: +90 degrees

④ Step 1: 24-28 ft. lbs.

Step 2: +90 degrees

⑤ Step 1: 15 ft. lbs.

Step 2: +90 degrees

⑥ 29-36 ft. lbs. +60-66 degrees

⑦ Step 1: 44-86 inch lbs.

Step 2: 20-23 ft. lbs.

42356-PATH-C08

WHEEL ALIGNMENT

Year	Model	Caster Range (+/-Deg.)	Caster Preferred Setting (Deg.)	Camber Range (+/-Deg.)	Camber Preferred Setting (Deg.)	Toe-in (in.)	Axis Inclination (Deg.)
2000	Pathfinder	0.75	+3.00	0.75	+0.17	0.08+/-0.04	—
	QX4	0.75	+3.00	0.75	+0.10	0.08+/-0.04	—
2001	Pathfinder	0.75	+3.00	0.75	+0.17	0.08+/-0.04	—
	QX4	0.75	+3.00	0.75	+0.10	0.08+/-0.04	—
2002	Pathfinder	0.75	+3.00	0.75	+0.17	0.08+/-0.04	—
	QX4 ①	0.75	+3.00	0.75	+0.17	0.08+/-0.04	—
2003	Pathfinder	0.75	+3.00	0.75	+0.17	0.08+/-0.04	—
	QX4 ①	0.75	+3.00	0.75	+0.17	0.08+/-0.04	—

① Assumes P245/65R17 tire

42356-PATH-C09

TIRE, WHEEL AND BALL JOINT SPECIFICATIONS

Year	Model	OEM Tires Standard	OEM Tires Optional	Tire Pressures (psi) Front	Tire Pressures (psi) Rear	Wheel Size	Ball Joint Inspection
2000	Pathfinder	P235/70R15	P265/70R15	30	30	Std: 6.5-JJ Opt: 7-JJ	U: 0.020 in. L: ①
	QX4	P245/70R16	None	35	35	7-JJ	①
2001	Pathfinder	P235/70R15	P265/70R15	30	30	Std: 6.5-JJ Opt: 7-JJ	U: 0.020 in. L: ①
	QX4	P245/70R16	None	35	35	7-JJ	①
2002	Pathfinder LE	P245/65SR17	none	②	②	8J	③
	Pathfinder SE	P255/65SR16	None	②	②	7JJ	③
	QX4	P245/70R16	P245/65R17	②	②	Std: 7J/Opt: 8J	③
2003	Pathfinder LE	P245/65SR17	none	②	②	8J	③
	Pathfinder SE	P255/65SR16	None	②	②	7JJ	③
	QX4	P245/65R17	None	②	②	Std: 7J/Opt: 8J	③

OEM: Original Equipment Manufacturer

PSI: Pounds Per Square Inch

STD: Standard

OPT: Optional

L: Lower

U: Upper

① Replace if any measurable movement is found.

② See placard on vehicle

③ Axial play
   Upper: 0
   Lower: 0.008 in.

④ Turning torque: 4.3-43 inch lbs.

42356-PATH-C10

BRAKE SPECIFICATIONS
All measurements in inches unless noted

Year	Model	Brake Disc Original Thickness	Brake Disc Minimum Thickness	Brake Disc Maximum Runout	Brake Drum Diameter Original Inside Diameter	Brake Drum Diameter Max. Wear Limit	Brake Drum Diameter Maximum Machine Diameter	Minimum Lining Thickness Front	Minimum Lining Thickness Rear	Brake Caliper Bracket Bolts (ft. lbs.)	Brake Caliper Mounting Bolts (ft. lbs.)
2000	QX4	1.100	1.024	0.004	11.60	NA	11.67	0.079	0.059	53-72	24-31
	Pathfinder	1.100	1.024	0.004	11.60	NA	11.67	0.079	0.059	53-72	16-23
2001	QX4	1.100	1.024	0.004	11.60	NA	11.67	0.079	0.059	53-72	24-31
	Pathfinder	1.100	1.024	0.004	11.60	NA	11.67	0.079	0.059	53-72	16-23
2002	Pathfinder	1.100	1.024	0.003	11.61	NA	11.67	0.079	0.059	①	①
	QX4	1.100	1.024	0.003	11.61	NA	11.67	0.079	0.059	②	②
2003	Pathfinder	1.100	1.024	0.003	11.61	NA	11.67	0.079	0.059	①	①
	QX4	1.100	1.024	0.003	11.61	NA	11.67	0.079	0.059	②	②

NA: Not Available

① Torque member mounting bolt: 127-134
   Main pin bolt: 24-31

② 2WD: 0.870
   4WD: 1.020

42356-PATH-C11

SCHEDULED MAINTENANCE INTERVALS
2000 Nissan—Pathfinder; Infiniti—QX4

TO BE SERVICED	TYPE OF SERVICE	VEHICLE MILEAGE INTERVAL (x1000)												
		7.5	15	22.5	30	37.5	45	52.5	60	67.5	75	82.5	90	97.5
Engine oil & filter	R	✓	✓	✓	✓	✓	✓	✓	✓	✓	✓	✓	✓	✓
Brake lines & cables	S/I		✓		✓		✓		✓		✓		✓	
Brake pads, discs, drums & linings	S/I		✓		✓		✓		✓		✓		✓	
Driveshaft boots & propeller shaft	S/I				✓				✓				✓	
Front wheel bearings (4x2)	S/I				✓				✓				✓	
Automatic & manual transmission, transfer & differential gear oil ①	S/I		✓		✓		✓		✓		✓		✓	
Front wheel bearings (4x4)	S/I				✓				✓				✓	
Air cleaner filter	R				✓				✓				✓	
Engine coolant	R				✓				✓				✓	
Spark plugs	R				✓				✓				✓	
Drive belt(s)	S/I				✓				✓				✓	
Exhaust system	S/I				✓				✓				✓	
Fuel lines	S/I				✓				✓				✓	
Steering gear (box) & linkage, axle & suspension parts	S/I				✓				✓				✓	
Vapor lines	S/I				✓				✓				✓	
Timing belt ②	R													

R: Replace   S/I: Service or Inspect

① Differential (w/limited-slip differential) oil: replace oil every 30,000 miles.

② Timing belt: replace at 105,000 miles.

FREQUENT OPERATION MAINTENANCE (SEVERE SERVICE)

If a vehicle is operated under any of the following conditions it is considered severe service:

- Extremely dusty areas.

- 50% or more of the vehicle operation is in 32°C (90°F) or higher temperatures, or constant operation in temperatures below 0°C (32°F).

- Prolonged idling (vehicle operation in stop and go traffic).

- Frequent short running periods (engine does not warm to normal operating temperatures).

- Police, taxi, delivery usage or trailer towing usage.

Oil & oil filter: replace every 3750 miles.

Brake pads, discs, drums & linings: service or inspect every 7500 miles.

Driveshaft boots & propeller shaft: service or inspect every 7500 miles.

Exhaust system: service or inspect every 7500 miles.

Steering gear (box) & linkage, (steering damper-4x4), axle & suspension parts: service or inspect every 7500 miles.

Steering linkage ball joints & front suspension ball joints: service or inspect every 7500 miles.

42356-PATH-C12

SCHEDULED MAINTENANCE INTERVALS
2001-03 Nissan—Pathfinder; Infiniti—QX4

TO BE SERVICED	TYPE OF SERVICE	VEHICLE MILEAGE INTERVAL (x1000)															
		3.8	7.5	11	15	19	22.5	26	30	34	37.5	41	45	49	52.5	56	60
Engine oil & filter	R	✓	✓	✓	✓	✓	✓	✓	✓	✓	✓	✓	✓	✓	✓	✓	✓
Brake lines & cables	S/I				✓				✓				✓				✓
Brake pads, discs, drums & linings	I		✓		✓		✓		✓		✓		✓		✓		✓
Driveshaft boots & propeller shaft	L/I		✓		✓		✓		✓		✓		✓		✓		✓
Front wheel bearings (4x2)	I								✓								✓
Automatic & manual transmission, transfer & differential gear oil ①	I				✓				✓				✓				✓
LSD gear oil	R								✓								✓
Front wheel bearing grease (4x4)	R								✓								✓
Timing belt ②	R																
Air cleaner filter	R								✓								✓
Engine coolant ③	R																✓
Spark plugs	R					platinum tipped plugs every 105,000 miles											
Drive belt(s)	S/I								✓								✓
Cabin air filter	I/R		I		R		I		R		I		R		I		R
Exhaust system	I		✓		✓		✓		✓		✓		✓		✓		✓
Fuel lines	S/I								✓								✓
Steering gear (box) & linkage, axle & suspension parts	I		✓		✓		✓		✓		✓		✓		✓		✓
Vapor lines	S/I								✓								✓

R: Replace   S/I: Service or Inspect   L: Lubricate

① Differential (w/limited-slip differential) oil: replace oil every 30,000 miles.

② Timing belt: replace at 105,000 miles.

③ After 60,000, replace every 30,000

FREQUENT OPERATION MAINTENANCE (SEVERE SERVICE)

If a vehicle is operated under any of the following conditions it is considered severe service:

- Extremely dusty areas.

- 50% or more of the vehicle operation is in 32°C (90°F) or higher temperatures, or constant operation in temperatures below 0°C (32°F).

- Prolonged idling (vehicle operation in stop and go traffic).

- Frequent short running periods (engine does not warm to normal operating temperatures).

- Police, taxi, delivery usage or trailer towing usage.

Oil & oil filter: replace every 3750 miles.

Brake pads, discs, drums & linings: service or inspect every 7500 miles.

Driveshaft boots & propeller shaft: service or inspect every 7500 miles.

Exhaust system: service or inspect every 7500 miles.

Steering gear (box) & linkage, (steering damper-4x4), axle & suspension parts: service or inspect every 7500 miles.

Steering linkage ball joints & front suspension ball joints: service or inspect every 7500 miles.

42356-PATH-C13

PRECAUTIONS

Before servicing any vehicle, please be sure to read all of the following precautions, which deal with personal safety, prevention of component damage, and important points to take into consideration when servicing a motor vehicle:

• Never open, service or drain the radiator or cooling system when the engine is hot; serious burns can occur from the steam and hot coolant.

• Observe all applicable safety precautions when working around fuel. Whenever servicing the fuel system, always work in a well-ventilated area. Do not allow fuel spray or vapors to come in contact with a spark, open flame, or excessive heat (a hot drop light, for example). Keep a dry chemical fire extinguisher near the work area. Always keep fuel in a container specifically designed for fuel storage; also, always properly seal fuel containers to avoid the possibility of fire or explosion. Refer to the additional fuel system precautions later in this section.

• Fuel injection systems often remain pressurized, even after the engine has been turned **OFF**. The fuel system pressure must be relieved before disconnecting any fuel lines. Failure to do so may result in fire and/or personal injury.

• Brake fluid often contains polyglycol ethers and polyglycols. Avoid contact with the eyes and wash your hands thoroughly after handling brake fluid. If you do get brake fluid in your eyes, flush your eyes with clean, running water for 15 minutes. If eye irritation persists, or if you have taken brake fluid internally, IMMEDIATELY seek medical assistance.

• The EPA warns that prolonged contact with used engine oil may cause a number of skin disorders, including cancer! You should make every effort to minimize your exposure to used engine oil. Protective gloves should be worn when changing oil. Wash your hands and any other exposed skin areas as soon as possible after exposure to used engine oil. Soap and water, or waterless hand cleaner should be used.

• All new vehicles are now equipped with an air bag system. The system must be disabled before performing service on or around system components, steering column, instrument panel components, wiring and sensors. Failure to follow safety and disabling procedures could result in accidental air bag deployment, possible personal injury and unnecessary system repairs.

• Always wear safety goggles when working with, or around, the air bag system. When carrying a non-deployed air bag, be sure the bag and trim cover are pointed away from your body. When placing a non-deployed air bag on a work surface, always face the bag and trim cover upward, away from the surface. This will reduce the motion of the module if it is accidentally deployed. Refer to the additional air bag system precautions later in this section.

• Clean, high quality brake fluid from a sealed container is essential to the safe and proper operation of the brake system. You should always buy the correct type of brake fluid for your vehicle. If the brake fluid becomes contaminated, completely flush the system with new fluid. Never reuse any brake fluid. Any brake fluid that is removed from the system should be discarded. Also, do not allow any brake fluid to come in contact with a painted surface; it will damage the paint.

• Never operate the engine without the proper amount and type of engine oil; doing so WILL result in severe engine damage.

• Timing belt maintenance is extremely important! Many models utilize an interference-type, non-freewheeling engine. If the timing belt breaks, the valves in the cylinder head may strike the pistons, causing potentially serious (also time-consuming and expensive) engine damage. Refer to the maintenance interval charts in the front of this section for the recommended replacement interval for the timing belt, and to the timing belt section for belt replacement and inspection.

• Disconnecting the negative battery cable on some vehicles may interfere with the functions of the on-board computer system(s) and may require the computer to undergo a relearning process once the negative battery cable is reconnected.

• When servicing drum brakes, only disassemble and assemble one side at a time, leaving the remaining side intact for reference.

ENGINE REPAIR

➡ **Disconnecting the negative battery cable on some vehicles may interfere with the functions of the on board computer system. The computer may undergo a relearning process once the negative battery cable is reconnected.**

Distributor

REMOVAL

1. Before servicing the vehicle, refer to the precautions in the beginning of this section.
2. Remove or disconnect the following:
  • Negative battery cable
  • Distributor cap

• Distributor wiring harness connector
3. Matchmark the rotor to the distributor housing and the distributor housing to the cylinder head.
4. Remove the distributor.

INSTALLATION

Timing Not Disturbed

1. Install or connect the following:
  • Distributor by aligning the matchmarks made during removal
  • Distributor wiring harness connector
  • Distributor cap
  • Negative battery cable
2. Check the ignition timing and adjust, as necessary.

Timing Disturbed

3.3L ENGINE

1. Set the engine to Top Dead Center (TDC) of the compression stroke for the No. 1 cylinder.

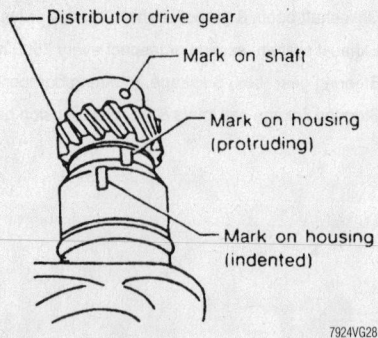

Distributor shaft alignment—3.3L engine

7924VG28

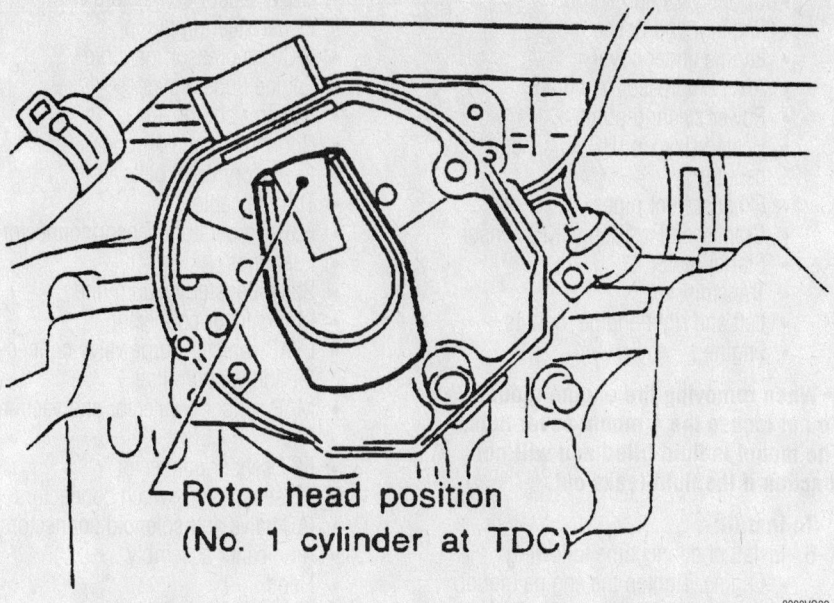

Distributor rotor alignment—3.3L engine

2. Align the index mark on the distributor shaft with the protrusion on the distributor housing.

3. Install the distributor and check that the distributor rotor is aligned.

4. Install or connect the following:
- Distributor cap
- Distributor harness connector

5. Check the ignition timing and adjust, as necessary.

Alternator

REMOVAL & INSTALLATOIN

3.3L and 3.5L Engines

1. Before servicing the vehicle, refer to the precautions in the beginning of this section.

2. Remove or disconnect the following:
- Negative battery cable
- Alternator harness connectors
- Engine under cover
- Alternator belt
- Alternator

To install:

3. Install or connect the following:
- Alternator
- Alternator belt. Tighten the adjustment bolt to 12–14 ft. lbs. (16–19 Nm) and the pivot bolts to 16–22 ft. lbs. (22–30 Nm).
- Engine under cover
- Alternator harness connectors
- Negative battery cable

Ignition Timing

ADJUSTMENT

➡**Ignition timing is set with the engine at operating temperature, transmission in Neutral and all electrical accessories OFF.**

1. Before servicing the vehicle, refer to the precautions in the beginning of this section.

2. Attach a timing light to the No. 1 spark plug wire.

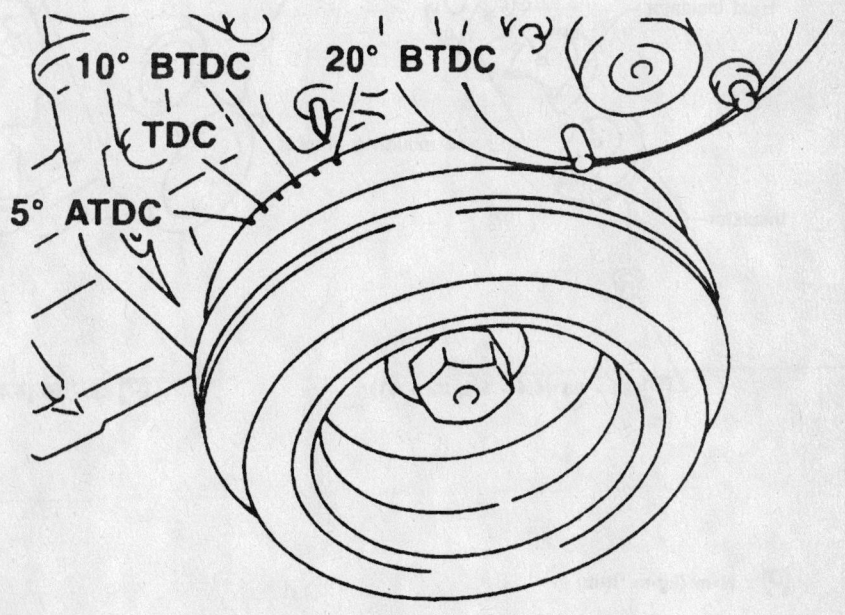

Timing indicator—3.3L engines

3. Start the engine and allow it to reach normal operating temperature.

4. Check that the idle speed is less than 1000 rpm.

5. Run the engine at 2000 rpm for 2 minutes.

6. Rev the engine to 3000 rpm 2–3 times and allow it to idle for 1 minute.

7. Check for the presence of Diagnostic Trouble Codes (DTC) and service as necessary.

8. Run the engine at 2000 rpm for 2 minutes.

9. Stop the engine and disconnect the Throttle Position (TP) sensor.

10. Start the engine and rev it to 3000 rpm 2–3 times and allow it to idle.

11. Set the base timing to 8–12 degrees Before Top Dead Center (BTDC) for 3.3L engines.

12. Tighten the distributor lockbolt to 83–113 inch lbs. (9–13 Nm).

13. Set the base idle speed to 700–800 rpm.

14. Stop the engine and connect the TP sensor.

Engine Assembly

REMOVAL & INSTALLATION

3.3L Engine

1. Before servicing the vehicle, refer to the precautions in the beginning of this section.

2. Drain the cooling system.

3. Relieve the fuel system pressure.

4. Recover the A/C refrigerant, if equipped.

5. Remove or disconnect the following:
- Negative battery cable
- Hood
- Air cleaner assembly
- Idle Air Control (IAC) valve and solenoid connectors
- Throttle Position (TP) sensor and switch connectors
- Engine Coolant Temperature (ECT) sensor connector
- Manifold Absolute Pressure (MAP) sensor connector and vacuum line
- Evaporative Emissions (EVAP) canister purge valve connector and vacuum line
- Mass Air Flow (MAF) sensor connector
- Brake booster vacuum line
- Fuel lines
- Exhaust Gas Recirculation (EGR) temperature sensor connector
- Throttle cable
- Accessory drive belts

- Cooling fan and shroud
- Radiator and hoses
- Engine under cover
- A/C compressor manifold
- Power steering pump
- Heated Oxygen (HO2S) sensor connectors
- Exhaust front pipes
- Crankshaft Position (CKP) sensor
- Starter motor
- Transmission
- Left and right engine mounts
- Engine

➡ When removing the engine mounts, do not loosen the 4 mount cover nuts. The mount is fluid filled and will not function if the fluid leaks out.

To install:

6. Install or connect the following:
- Engine. Tighten the engine mount nuts to 43–58 ft. lbs. (59–78 Nm).
- Transmission
- Starter motor
- CKP sensor
- Exhaust front pipes

- HO2S sensor connectors
- Power steering pump
- A/C compressor manifold
- Engine under cover
- Radiator and hoses
- Cooling fan and shroud
- Accessory drive belts
- Throttle cable
- EGR temperature sensor connector
- Fuel lines
- Brake booster vacuum line
- MAF sensor connector
- EVAP canister purge valve connector and vacuum line
- MAP sensor connector and vacuum line
- ECT sensor connector
- TP sensor and switch connectors
- IAC valve and solenoid connectors
- Air cleaner assembly
- Hood
- Negative battery cable

7. Fill the cooling system.

8. Recharge the A/C system, if equipped.

9. Start the engine and check for leaks.

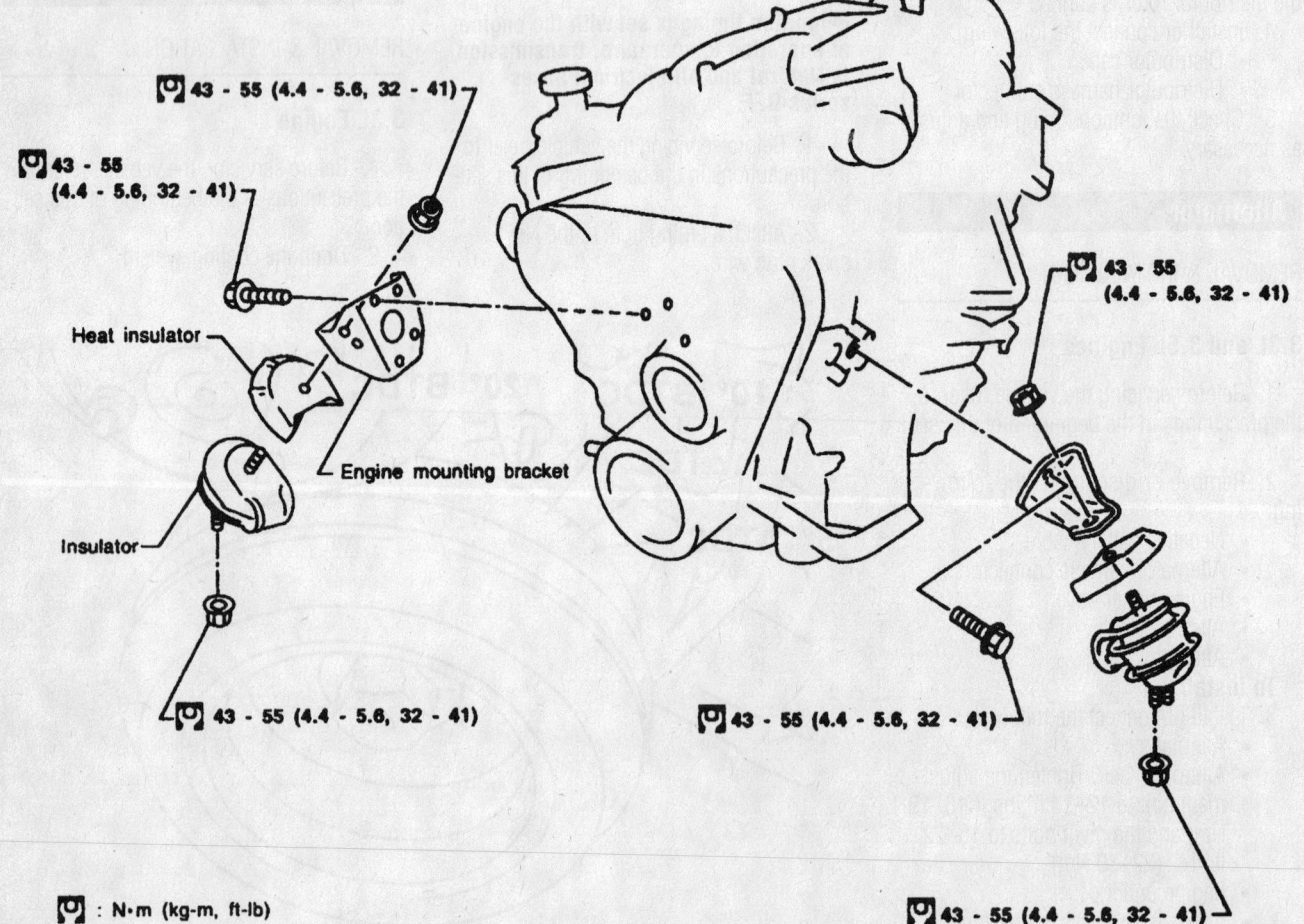

Heat insulator

Engine mounting bracket

Insulator

43 - 55 (4.4 - 5.6, 32 - 41)

43 - 55 (4.4 - 5.6, 32 - 41)

43 - 55 (4.4 - 5.6, 32 - 41)

43 - 55 (4.4 - 5.6, 32 - 41)

43 - 55 (4.4 - 5.6, 32 - 41)

: N•m (kg-m, ft-lb)

7924VG11

Engine mounts and related components—3.3L engine

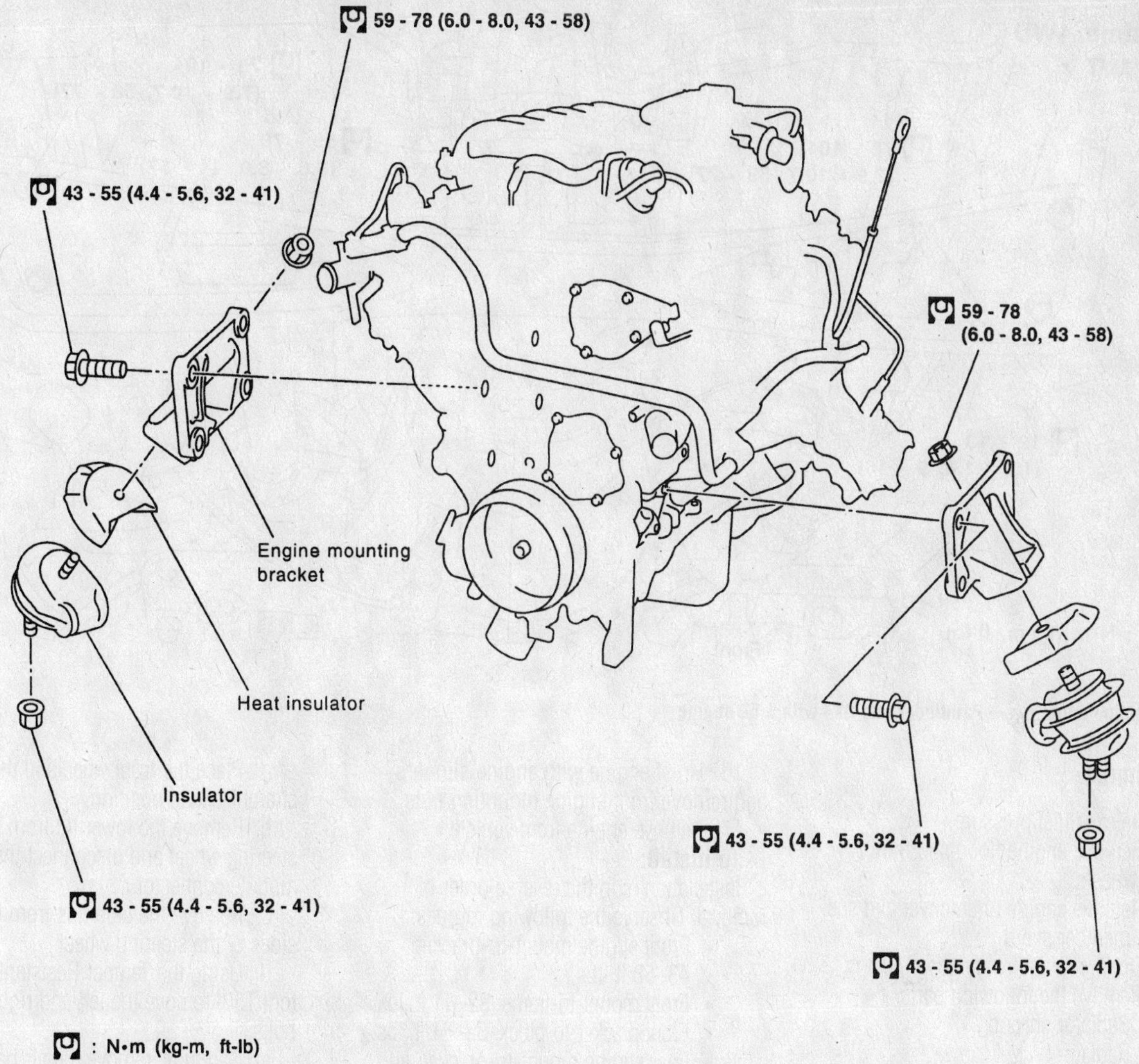

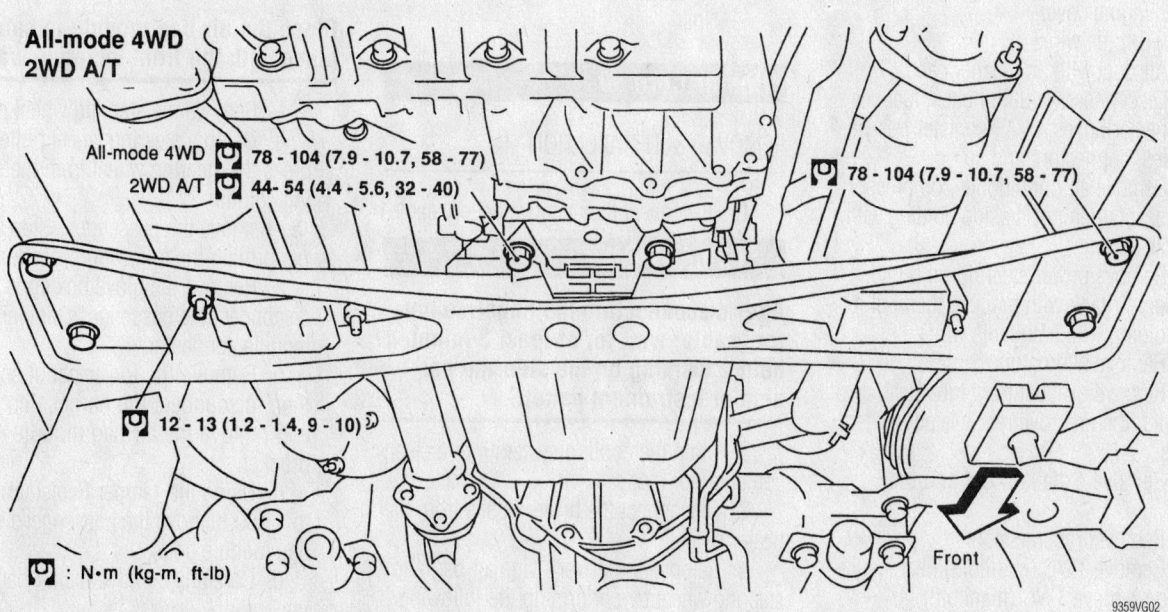

🔧 59 - 78 (6.0 - 8.0, 43 - 58)

🔧 43 - 55 (4.4 - 5.6, 32 - 41)

🔧 59 - 78 (6.0 - 8.0, 43 - 58)

Engine mounting bracket

Heat insulator

Insulator

🔧 43 - 55 (4.4 - 5.6, 32 - 41)

🔧 43 - 55 (4.4 - 5.6, 32 - 41)

🔧 43 - 55 (4.4 - 5.6, 32 - 41)

🔧 : N•m (kg-m, ft-lb)

9359VG01

Front engine mounting—Pathfinder and QX4 with 3.5L engine

All-mode 4WD
2WD A/T

All-mode 4WD 🔧 78 - 104 (7.9 - 10.7, 58 - 77)

2WD A/T 🔧 44 - 54 (4.4 - 5.6, 32 - 40)

🔧 78 - 104 (7.9 - 10.7, 58 - 77)

🔧 12 - 13 (1.2 - 1.4, 9 - 10)

Front

🔧 : N•m (kg-m, ft-lb)

9359VG02

Rear engine mounting— Pathfinder and QX4 with 3.5L engine

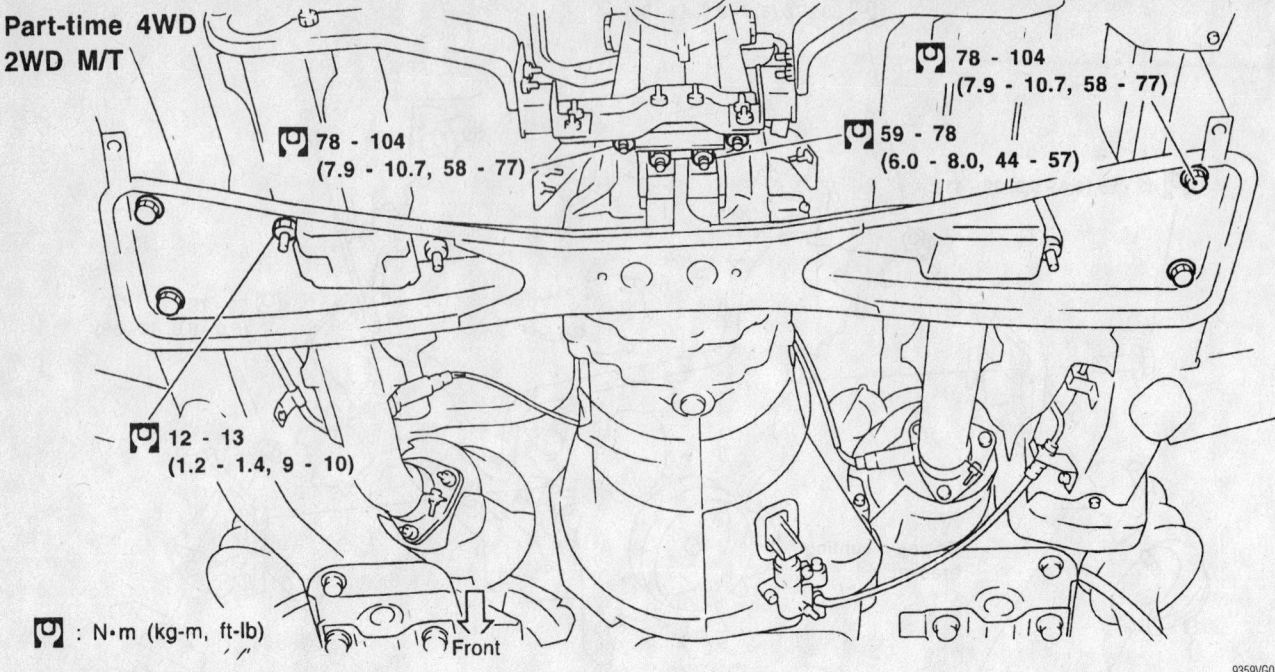

Part-time 4WD
2WD M/T

78 - 104
(7.9 - 10.7, 58 - 77)

78 - 104
(7.9 - 10.7, 58 - 77)

59 - 78
(6.0 - 8.0, 44 - 57)

12 - 13
(1.2 - 1.4, 9 - 10)

: N·m (kg-m, ft-lb)

Front

9359VG03

Rear engine mounting— Pathfinder and QX4 with 3.5L engine

3.5L Engine

1. Release fuel pressure.
2. Remove engine hood and front RH and LH wheels.
3. Remove engine undercover and suspension member stay.
4. Drain coolant from radiator.
5. Remove the following parts.
   - Radiator shroud
   - Radiator
   - Cooling fan
   - Drive belts
   - Battery
   - Engine cover
   - Throttle wires
6. Air duct with air cleaner case.
7. Disconnect vacuum hoses, fuel hoses, heater hoses, EVAP canister hoses, harnesses, connectors and so on.
8. Remove air conditioner compressor from bracket, then put it aside holding with a suitable wire.
9. Remove power steering oil pump and reservoir tank with bracket, then put it aside holding with a suitable wire.
10. Remove alternator.
11. Remove exhaust front tube heat insulators, then remove rear heated oxygen sensors.
12. Remove exhaust front and rear tubes.
13. Remove transmission.
14. Remove TWC (manifold) heat insulators, then remove TWC (manifold).
15. Install engine slingers.

16. Hoist engine with engine slingers and remove front engine mounting nuts.
17. Remove engine from vehicle.

To install:

Installation is in the reverse order of removal. Observe the following torques:
- Front engine mount-to-bracket: 43–58 ft. lbs.
- Front mount-to-frame: 32–41 ft. lbs.
- Front bracket-to-block: 32–41 ft. lbs.
- Rear engine mount-to-bracket: all exc. 2wd with AT: 58–77 ft. lbs.; 2wd with AT: 32–40 ft. lbs.
- Crossmember-to-frame: 58–77 ft. lbs.

Heater Core

REMOVAL & INSTALLATION

1. Disconnect the negative battery cable.

✳✳ CAUTION

After disconnecting the negative battery cable, wait for at least 3 minutes before working on the steering column or instrument panel.

2. Drain the cooling system into a clean container for reuse.
3. Disconnect the heater hoses from the heater core.
4. Remove the driver's side air bag and steering wheel by performing the following procedure:

a. Place the front wheels in the straight-ahead position.
b. Remove the lower lid from the steering wheel and disconnect the air bag module connector.
c. Remove the side lids from both sides of the steering wheel.
d. Using the Tamper Resistant Torx® tool T50, remove the left and right Torx® bolts.
e. Carefully, remove the air bag module.

✳✳ CAUTION

Place the air bag module in safe place with the front facing upward.

f. Remove the steering wheel nut.
g. Using a steering wheel puller, press the steering wheel from the steering column.
5. Remove the passenger's side air bag by performing the following procedure:

a. Remove the glove box clips and disconnect the passenger's side air bag module connector.
b. Remove the lower panel screws; then, disconnect the harness connector and remove the air bag module bracket.
c. Using the Tamper Resistant Torx® tool T50, remove the passenger's side air bag module bolts.
d. Carefully, remove the air bag module.

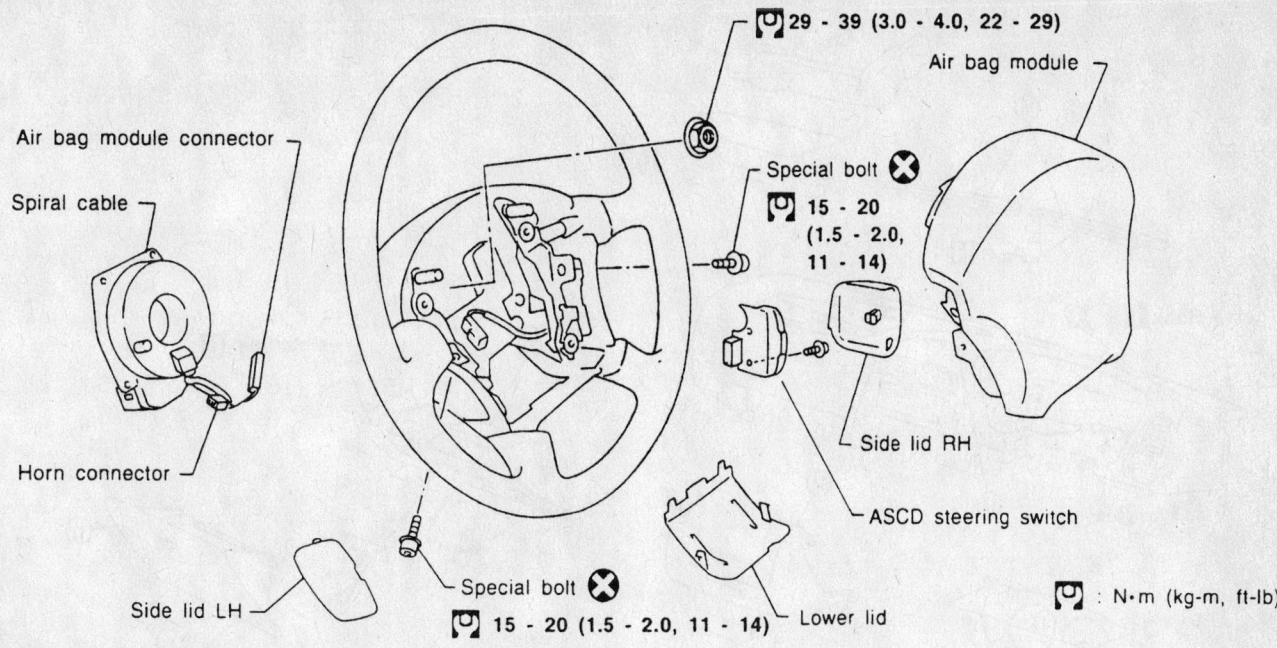

29 - 39 (3.0 - 4.0, 22 - 29)

Air bag module

Air bag module connector

Spiral cable

Special bolt ⊗
15 - 20
(1.5 - 2.0,
11 - 14)

Horn connector

Side lid RH

ASCD steering switch

Side lid LH

Special bolt ⊗
15 - 20 (1.5 - 2.0, 11 - 14)

Lower lid

⊡ : N•m (kg-m, ft-lb)

93113GH8

Exploded view of the driver's side air bag module and steering wheel

❊❊ CAUTION

Place the air bag module in safe place with the front facing upward.

6. Remove the instrument panel by performing the following procedure:

a. Remove the steering column cover and the combination switch.

b. Remove the instrument panel side lower finisher.

c. At the driver's side, remove the lower panel screws, disconnect the electrical harness connectors and remove the panel.

d. Remove the cluster lid "A" screws and the cluster lid "A".

e. Remove the combination meter screws, disconnect the electrical harness connectors and remove the combination meter.

f. Remove the cluster lid "C" screws, disconnect the electrical harness connectors and remove the cluster lid "C".

g. Remove the audio assembly screws and the audio assembly.

h. Remove the air conditioning control unit screws, disconnect the electrical harness connectors and the air conditioning control unit.

i. Remove the ashtray.

j. Remove the shifter (automatic transmission) or shift lever boot (manual transmission); then, remove the screw and disconnect the harness connector.

k. Remove the console box; then, remove the screw and disconnect the harness connector.

l. Remove the lower instrument center panel screws and the lower instrument center panel.

m. Remove the defroster grille.

n. At both sides, remove the pillar garnishes.

o. Remove the instrument panel and pads nuts and bolts.

p. Using an assistant, remove the instrument panel.

7. Remove the defroster nozzle and the heater nozzle from the heater housing.

8. Disconnect the electrical connector and/or control cable from the heater housing.

9. Remove the heater housing-to-chassis fasteners and remove the heater housing.

10. Separate the heater core from the heater housing and remove the heater core.

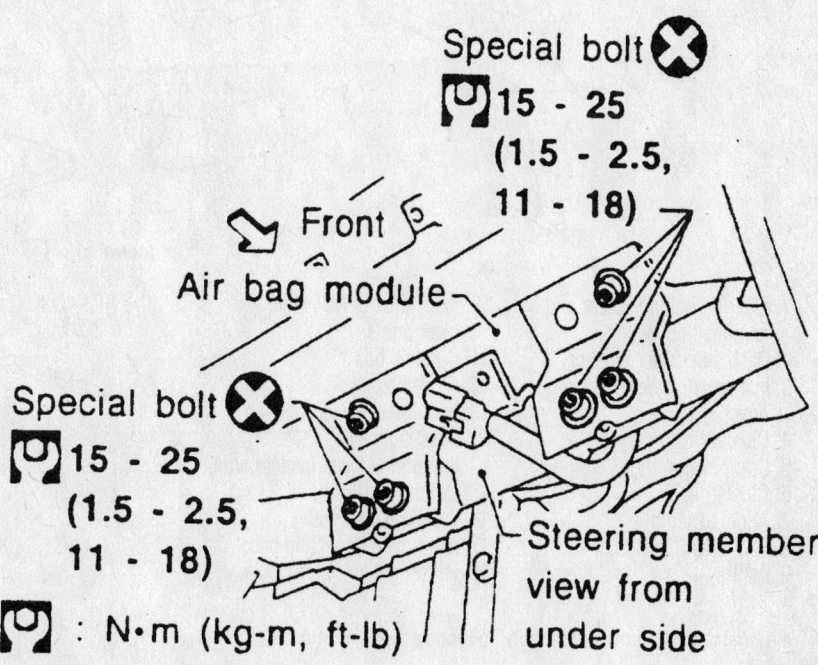

Special bolt ⊗
15 - 25
(1.5 - 2.5,
11 - 18)

Front

Air bag module

Special bolt ⊗
15 - 25
(1.5 - 2.5,
11 - 18)

⊡ : N•m (kg-m, ft-lb)

Steering member view from under side

93113GH9

Exploded view of the passenger's side air bag module

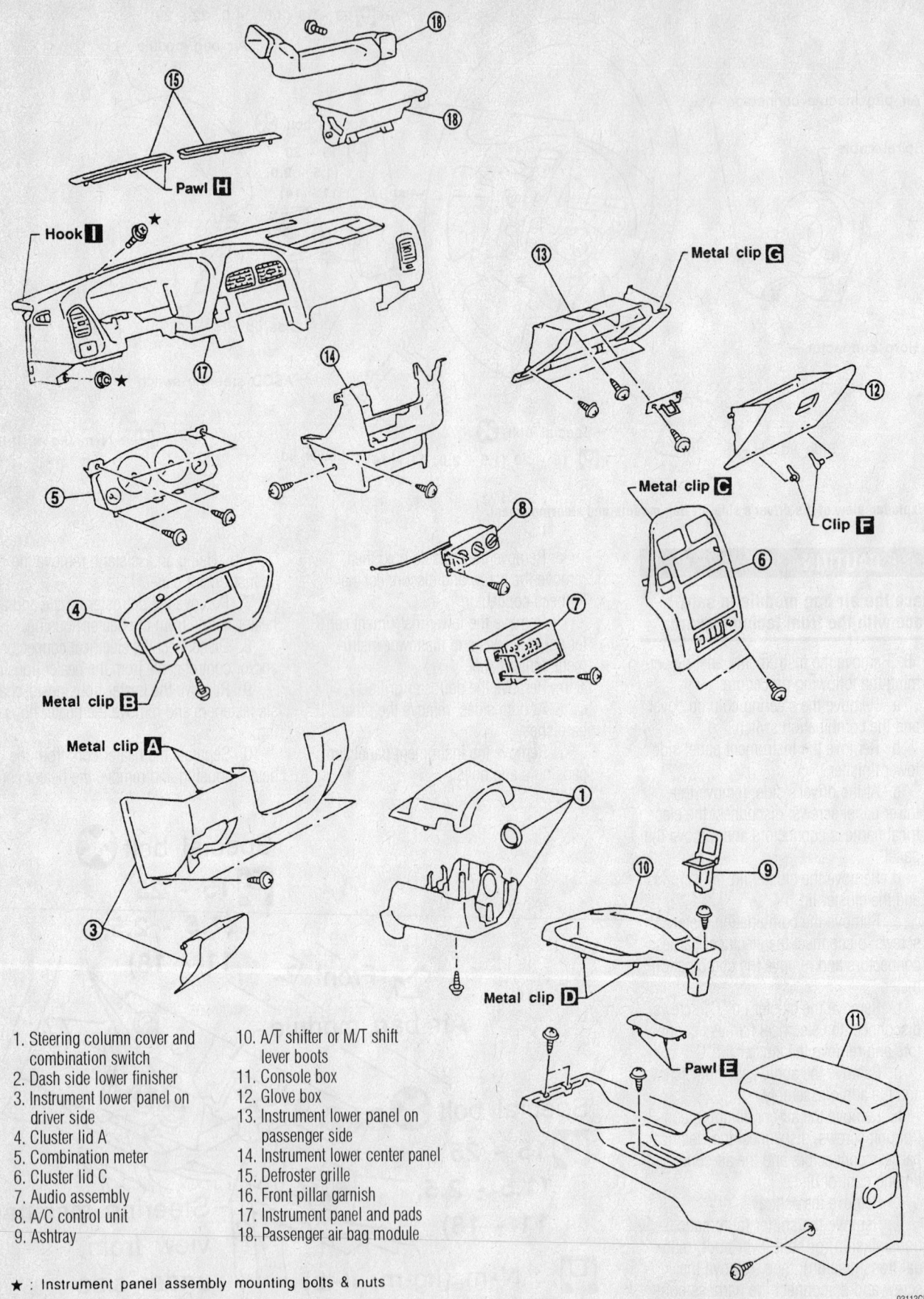

1. Steering column cover and combination switch
2. Dash side lower finisher
3. Instrument lower panel on driver side
4. Cluster lid A
5. Combination meter
6. Cluster lid C
7. Audio assembly
8. A/C control unit
9. Ashtray
10. A/T shifter or M/T shift lever boots
11. Console box
12. Glove box
13. Instrument lower panel on passenger side
14. Instrument lower center panel
15. Defroster grille
16. Front pillar garnish
17. Instrument panel and pads
18. Passenger air bag module

★ : Instrument panel assembly mounting bolts & nuts

Exploded view of the instrument panel and related accessories

93113GH0

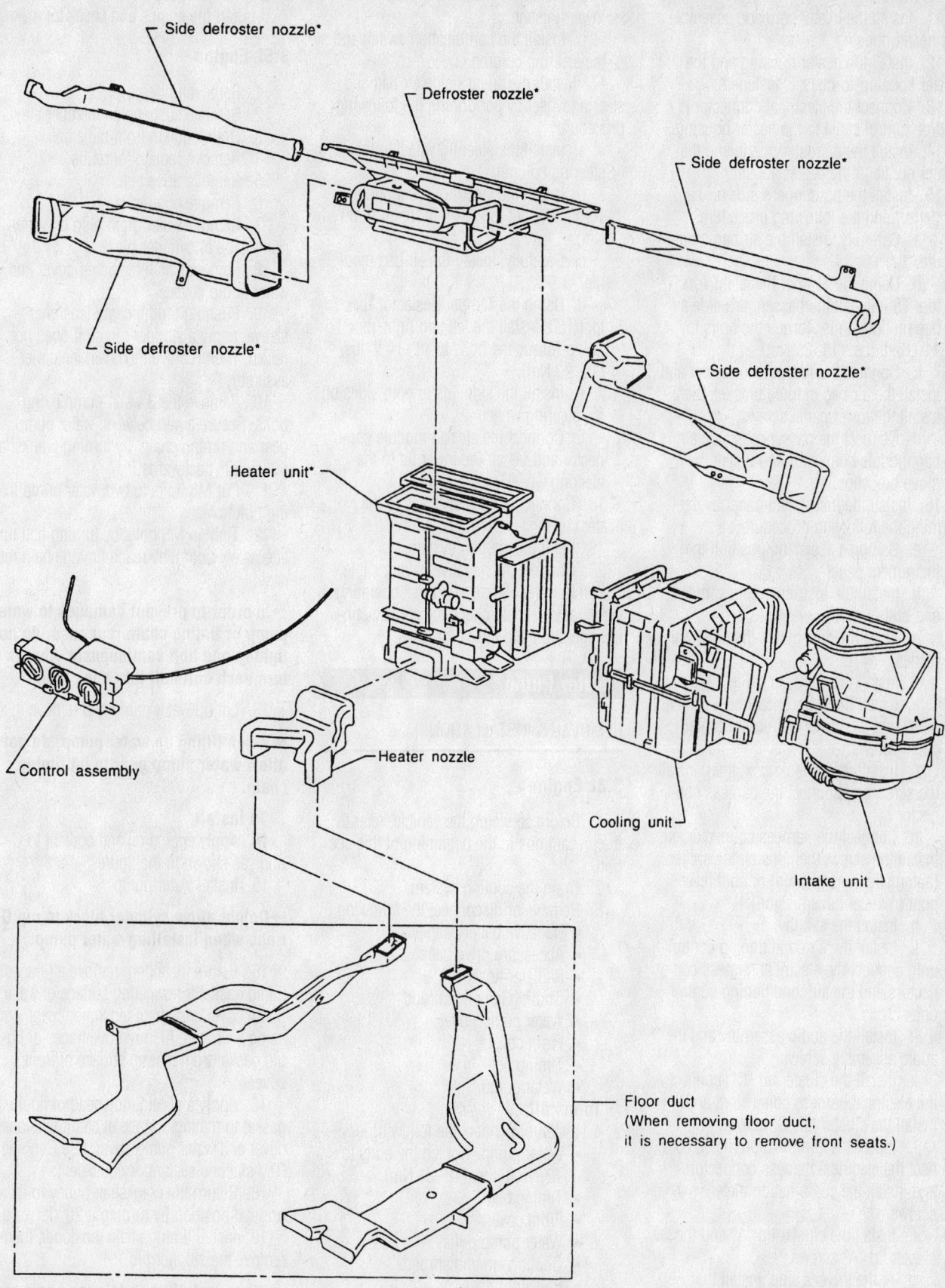

Side defroster nozzle*

Defroster nozzle*

Side defroster nozzle*

Side defroster nozzle*

Side defroster nozzle*

Heater unit*

Control assembly

Heater nozzle

Cooling unit

Intake unit

Floor duct
(When removing floor duct,
it is necessary to remove front seats.)

93113GI1

Exploded view of the heater housing, the evaporator housing, the ventilation dusts and related accessories

To install:

11. Install the heater core and assemble the heater housing.

12. Install the heater housing and the heater housing-to-chassis fasteners.

13. Connect the electrical connector and/or control cable to the heater housing.

14. Install the defroster nozzle and the heater nozzle to the heater housing.

15. Install the passenger's side air bag by performing the following procedure:

   a. Carefully, install the air bag module.

   b. Using the Tamper Resistant Torx® tool T50, install the passenger's side air bag module bolts. Torque the bolts to 11–18 ft. lbs. (15–25 Nm).

   c. Connect the harness connector and install the air bag module bracket; then, install the lower panel screws.

   d. Connect the passenger's side air bag module connector and install the glove box clips.

16. Install the instrument panel by performing the following procedure:

   a. Using an assistant, position the instrument panel.

   b. Install the instrument pads, nuts and bolts.

   c. At both sides, install the pillar garnishes.

   d. Install the defroster grille.

   e. Install the lower instrument center panel and the lower instrument center panel screws.

   f. Install the console box; then, install the screw and connect the harness connector.

   g. Connect the harness connector and install the screw; then, install the shifter (automatic transmission) or shift lever boot (manual transmission).

   h. Install the ashtray.

   i. Install the air conditioning control unit, connect the electrical harness connectors and the air conditioning control unit screws.

   j. Install the audio assembly and the audio assembly screws.

   k. Install the cluster lid "C", connect the electrical harness connectors and install the cluster lid "C" screws.

   l. Install the combination meter, connect the electrical harness connectors and install the combination meter screws.

   m. Install the cluster lid "A" and the cluster lid "A" screws.

   n. At the driver's side, install the lower panel, connect the electrical harness connectors and install the panel screws.

   o. Install the instrument panel side lower finisher.

   p. Install the combination switch and the steering column cover.

17. Install the driver's side air bag and steering wheel by performing the following procedure:

   a. Install the steering wheel to the steering column.

   b. Install the steering wheel nut. Torque the nut to 22–29 ft. lbs. (29–39 Nm).

   c. Carefully, install the air bag module.

   d. Using the Tamper Resistant Torx® tool T50, install the left and right Torx® bolts. Torque the bolts to 11–14 ft. lbs. (15–20 Nm).

   e. Install the side lids to both sides of the steering wheel.

   f. Connect the air bag module connector and install the lower lid to the steering wheel.

18. Connect the heater hoses to the heater core.

19. Refill the cooling system.

20. Connect the negative battery cable.

21. Run the engine to normal operating temperatures; then, check the climate control operation and check for leaks.

Water Pump

REMOVAL & INSTALLATION

3.3L Engine

1. Before servicing the vehicle, refer to the precautions in the beginning of this section.

2. Drain the cooling system.

3. Remove or disconnect the following:
   - Negative battery cable
   - Accessory drive belts
   - Radiator hoses
   - Cooling fan and shroud
   - Water pump pulley
   - Front cover
   - Timing belt
   - Water pump

To install:

4. Install or connect the following:
   - Water pump. Tighten the bolts to 12–15 ft. lbs. (16–21 Nm).
   - Timing belt
   - Front cover
   - Water pump pulley
   - Cooling fan and shroud
   - Radiator hoses
   - Accessory drive belts
   - Negative battery cable

5. Fill the cooling system.

6. Start the engine and check for leaks.

3.5L Engine

1. Remove undercover.

2. Remove suspension member stay.

3. Drain coolant from radiator.

4. Remove radiator shrouds.

5. Remove drive belts.

6. Remove cooling fan.

7. Remove water drain plug on water pump side of cylinder block.

8. Remove chain tensioner cover and water pump cover.

9. Pushing timing chain tensioner sleeve, apply a stopper pin so it does not return. Then remove the chain tensioner assembly.

10. Remove the 3 water pump fixing bolts. Secure a gap between water pump gear and timing chain, by turning crankshaft pulley 20° backwards.

11. Put M8 bolts to two water pump fixing bolt holes.

12. Tighten M8 bolts by turning half turn alternately until they reach timing chain rear case.

➡ **In order to prevent damages to water pump or timing chain rear case, do not tighten one bolt continuously. Always turn each bolt half turn each time.**

13. Lift up water pump and remove it.

➡ **When lifting up water pump, do not allow water pump gear to hit timing chain.**

To install:

14. Apply engine oil and coolant to O-rings as shown in the figure.

15. Install water pump.

➡ **Do not allow cylinder block to nip O-rings when installing water pump.**

16. Before installing, remove all traces of liquid gasket from mating surface of water pump cover and chain tensioner cover using a scraper. Also remove traces of liquid gasket from mating surface of front cover.

17. Apply a continuous bead of liquid gasket to mating surface of chain tensioner cover and water pump cover. Use Genuine RTV silicone sealant or equivalent.

18. Return the crankshaft pulley to its original position by turning it 20° forward.

19. Install timing chain tensioner, then remove the stopper pin.

➡ **When installing the timing chain tensioner, engine oil should be applied to the oil hole and tensioner.**

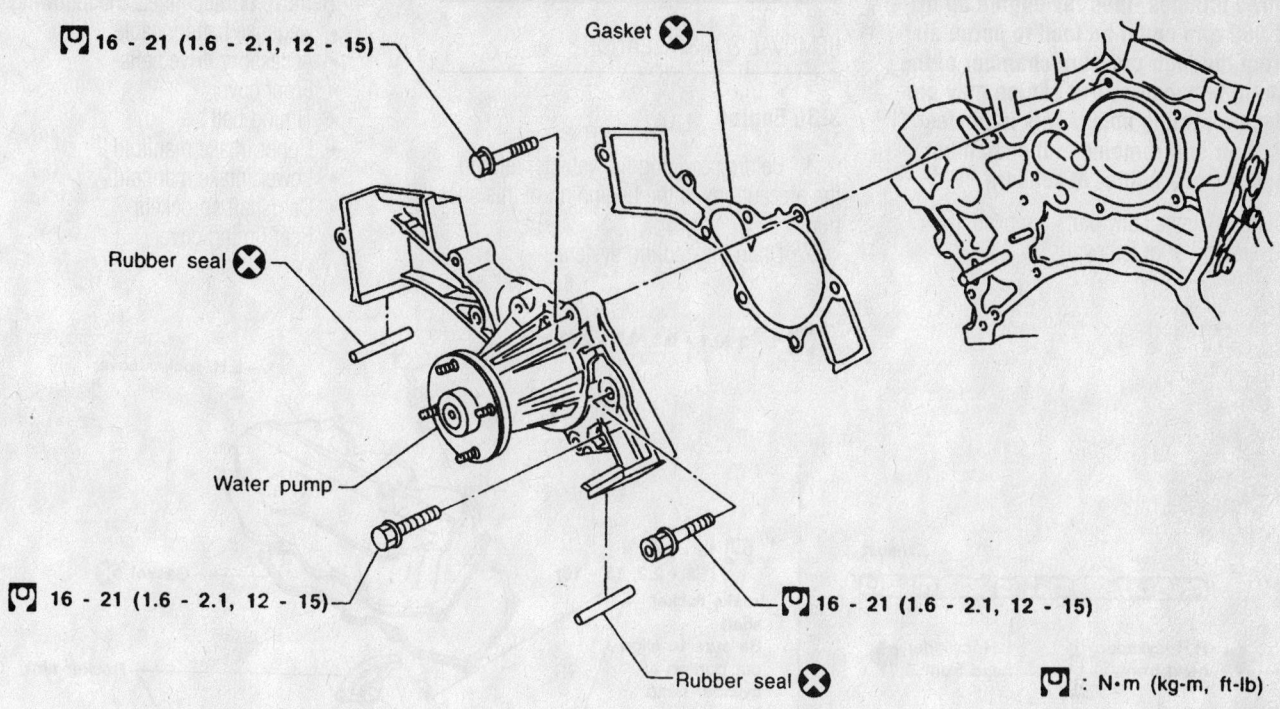

16 - 21 (1.6 - 2.1, 12 - 15)

Gasket ⊗

Rubber seal ⊗

Water pump

16 - 21 (1.6 - 2.1, 12 - 15)

16 - 21 (1.6 - 2.1, 12 - 15)

Rubber seal ⊗

: N•m (kg-m, ft-lb)

7924VG20

Exploded view of the water pump assembly—3.3L engine

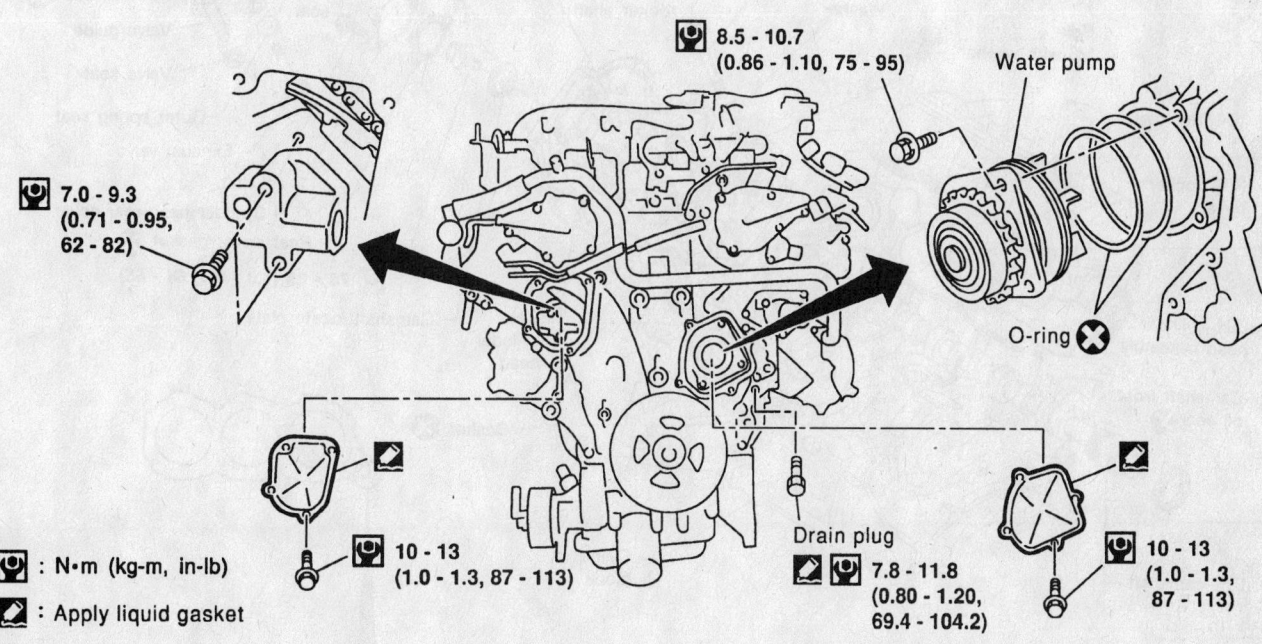

8.5 - 10.7
(0.86 - 1.10, 75 - 95)

Water pump

7.0 - 9.3
(0.71 - 0.95,
62 - 82)

O-ring ⊗

: N•m (kg-m, in-lb)

: Apply liquid gasket

10 - 13
(1.0 - 1.3, 87 - 113)

Drain plug

7.8 - 11.8
(0.80 - 1.20,
69.4 - 104.2)

10 - 13
(1.0 - 1.3,
87 - 113)

9359VG04

Exploded view of the water pump assembly—3.5L engine

➡ **After starting engine, let idle for three minutes, then rev engine up to 3,000 rpm under no load to purge air from the high-pressure chamber of the chain tensioners. The engine may produce a rattling noise. This indicates that air still remains in the chamber and is not a matter of concern.**

20. Reinstall any parts removed in reverse order of removal.

Cylinder Head

REMOVAL & INSTALLATION

3.3L Engine

1. Before servicing the vehicle, refer to the precautions in the beginning of this section.
2. Drain the cooling system.

3. Relieve the fuel system pressure.
4. Remove or disconnect the following:
   • Negative battery cable
   • Accessory drive belts
   • Front cover
   • Timing belt
   • Upper intake manifold
   • Lower intake manifold
   • Camshaft sprockets
   • Rear timing cover

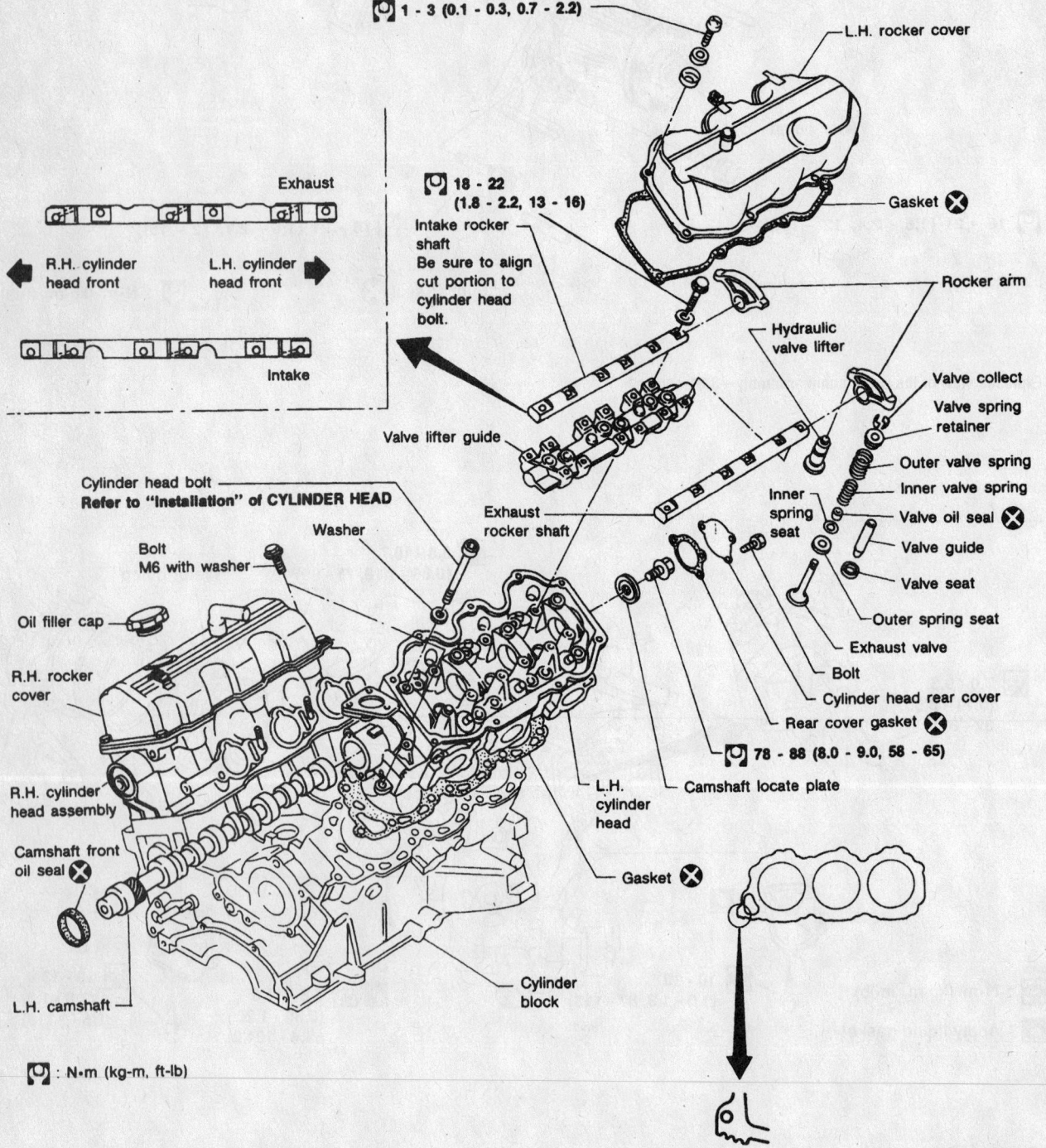

Exploded view of the cylinder head assembly—3.3L engine

7924VG25

- Distributor
- Exhaust front pipes
- A/C compressor
- Alternator
- Power steering pump
- Accessory brackets
- Valve covers. Loosen the bolts in several passes and in sequence.
- Cylinder heads with the exhaust manifolds attached. Loosen the bolts in several passes and in sequence.

➡ **The cylinder head bolts vary in length. Note the bolt locations for assembly.**

To install:

5. Install the cylinder heads and the lower intake manifold at the same time. Tighten the bolts in sequence as follows:

   a. Step 1: Tighten the cylinder head bolts to 22 ft. lbs. (29 Nm)

   b. Step 2: Tighten the cylinder head bolts to 43 ft. lbs. (59 Nm)

   c. Step 3: Loosen all cylinder head bolts completely

   d. Step 4: Tighten the cylinder head bolts to 84 inch lbs. (10 Nm)

   e. Step 5: Tighten the intake manifold fasteners to 35 inch lbs. (4 Nm)

   f. Step 6: Tighten the intake manifold fasteners to 13 ft. lbs. (18 Nm)

   g. Step 7: Tighten the intake manifold fasteners to 12–14 ft. lbs. (16–20 Nm)

   h. Step 8: Loosen all intake fasteners completely

   i. Step 9: Tighten the cylinder head bolts to 22 ft. lbs. (29 Nm)

   j. Step 10: Tighten the cylinder head bolts 60–65 degrees **OR** tighten to 40–47 ft. lbs. (54–64 Nm)

   k. Step 11: Tighten the cylinder head sub-bolts to 80–105 inch lbs. (9–12 Nm)

   l. Step 12: Tighten the intake manifold fasteners to 35 inch lbs. (4 Nm)

   m. Step 13: Tighten the intake manifold fasteners to 78 inch lbs. (9 Nm)

   n. Step 14: Tighten the intake manifold fasteners to 70–84 inch lbs. (6–7 Nm)

6. Install or connect the following:
   - Valve covers
   - Accessory brackets
   - Power steering pump
   - Alternator
   - A/C compressor
   - Exhaust front pipes
   - Distributor
   - Rear timing cover
   - Camshaft sprockets
   - Upper intake manifold

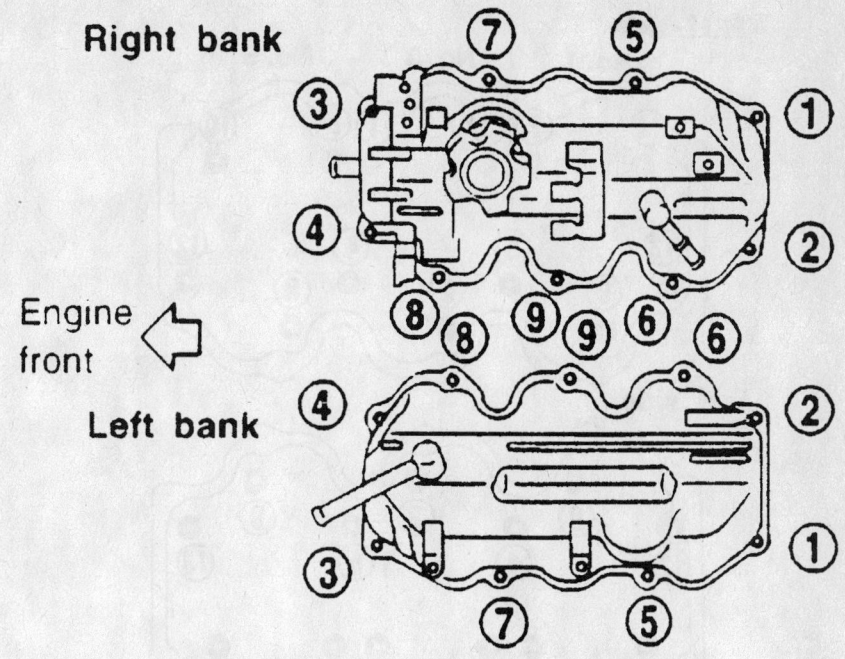

Valve cover bolt loosening sequence—3.3L engine

- Timing belt
- Front cover
- Accessory drive belts
- Negative battery cable

7. Fill the cooling system.
8. Start the engine and check for leaks.

3.5L Engine

1. Remove engine from vehicle.
2. Remove exhaust manifolds in reverse order of installation.
3. Place engine on a work stand.
4. Remove aluminum oil pan
5. Remove timing chain.
6. Remove intake manifold in reverse order of installation.
7. Remove water outlet.
8. Remove rear timing chain case bolts. Loosen in numerical order as shown in the figure.
9. Remove rear timing chain case.
10. Remove O-rings to cylinder head.
11. Remove O-rings to cylinder block.
12. Remove intake valve timing control solenoid valves.
13. Remove intake and exhaust camshafts and camshaft brackets. Equally loosen camshaft bracket bolts in several

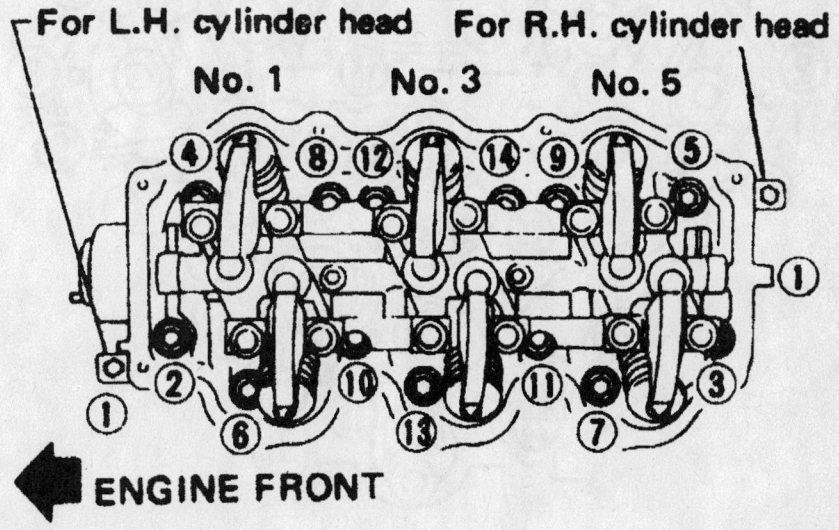

Cylinder head loosening sequence—3.3L engine

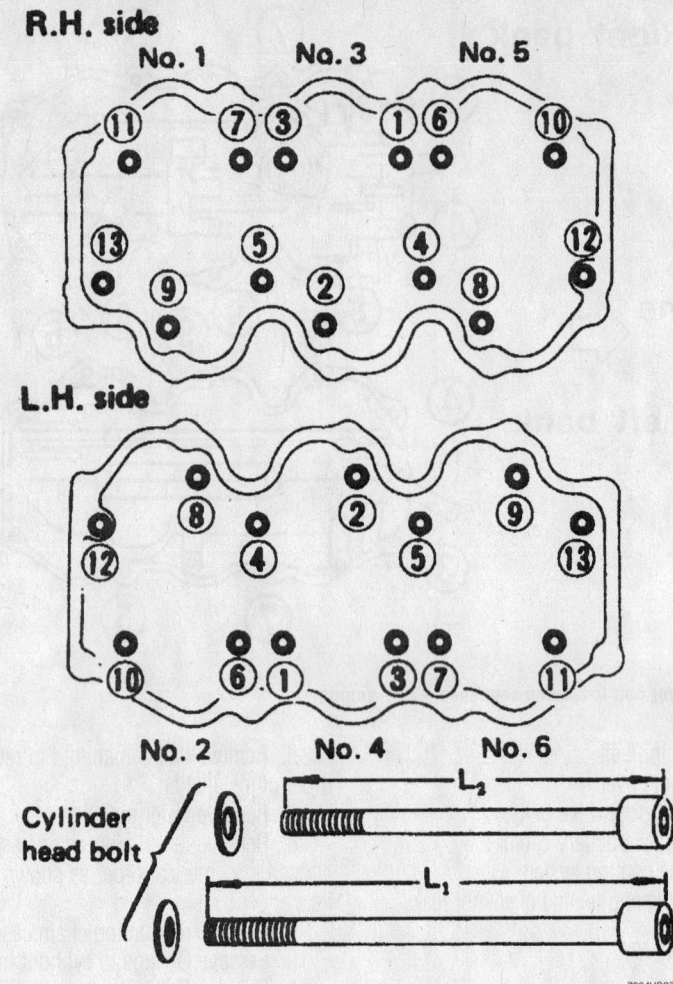

R.H. side

L.H. side

Cylinder head bolt

Cylinder head torque sequence—3.3L engine

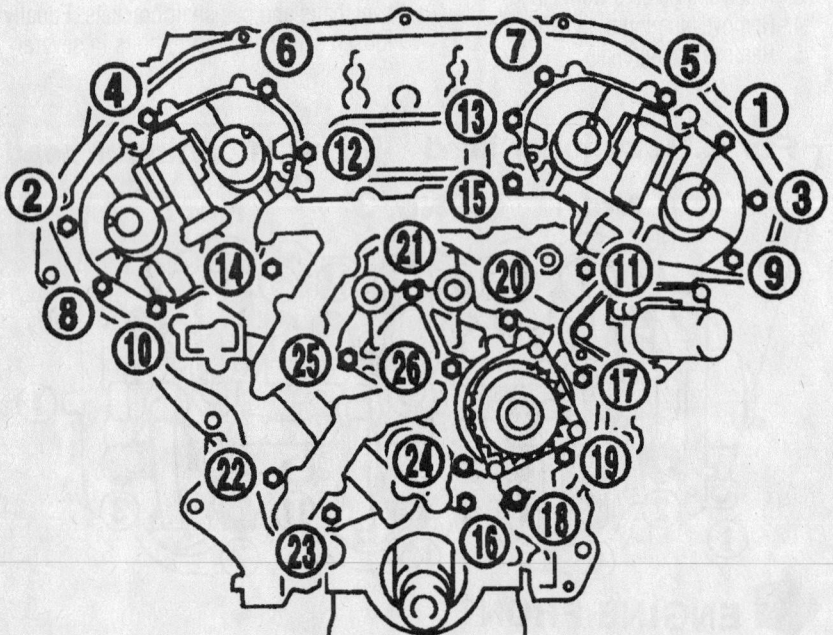

Rear timing case loosening sequence—3.5L engine

9359VG05

steps in the numerical order shown in the figure. For reinstallation, be sure to put marks on camshaft bracket before removal.

14. Remove RH and LH camshaft chain tensioners from cylinder head.

15. Remove cylinder head bolts. Cylinder head bolts should be loosened in two or three steps.

16. Remove cylinder head.

To install:

17. Before installing rear timing chain case, remove old liquid gasket from mating surface using a scraper. Also remove old liquid gasket from mating surface of cylinder block. Remove old liquid gasket from the bolt hole and thread.

18. Before installing cam bracket, remove old liquid gasket from mating surface using a scraper.

19. Before installing the cylinder head gasket, be sure that No. 1 cylinder is at TDC. At this time, the crankshaft key should face toward the right bank.

20. Install cylinder heads with new gaskets.

➡ **Do not rotate crankshaft and camshaft separately, or valves will strike piston heads.**

✳✳ CAUTION

Cylinder head bolts are tightened by plastic zone tightening method. Whenever the size difference between d1 and d2 exceeds the limit, replace them with new ones. Limit (d1—d2): 0.0043 in. Lubricate threads and seat surfaces of the bolts with new engine oil.

21. Install cylinder head outside bolts Tighten in numerical order shown in the figure. Tightening procedure:

   a. Tighten all bolts to 98 N·m (10 kg-m, 72 ft-lb).

   b. Completely loosen all bolts.

   c. Tighten all bolts to 34 to 44 N·m (3.5 to 4.5 kg-m, 25 to 33 ft-lb).

   d. Turn all bolts 90 to 95 degrees clockwise.

   e. Turn all bolts 90 to 95 degrees clockwise.

22. Install camshaft chain tensioners on both sides of cylinder head.

23. Install exhaust and intake camshafts and camshaft brackets.

➡ **Intake camshaft has a drill mark on camshaft sprocket mounting flange. Install it on the intake side. Position camshaft. RH exhaust camshaft dowel**

pin at about 10 o'clock; LH exhaust camshaft dowel pin at about 2 o'clock

24. Before installing camshaft brackets, apply sealant to mating surface of No. 1 journal head. Use Genuine RTV silicone sealant or equivalent. Install camshaft brackets in their original positions. Align stamp mark as shown in the figure. If any part of valve assembly or camshaft is replaced, check valve clearance according to reference data. After completing assembly check valve clearance. Valve clearance (Cold):

- Intake 0.26—0.34 mm (0.010—0.013 in)

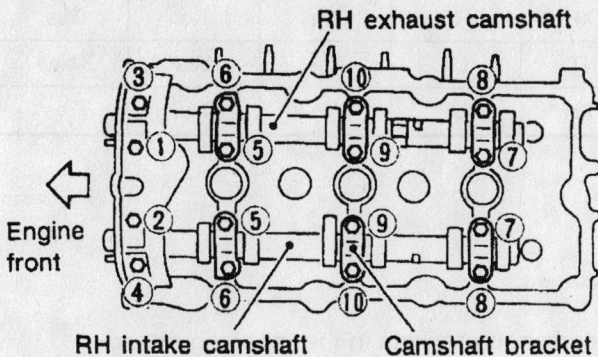

Right camshaft loosening sequence—3.5L engine

9359VG06

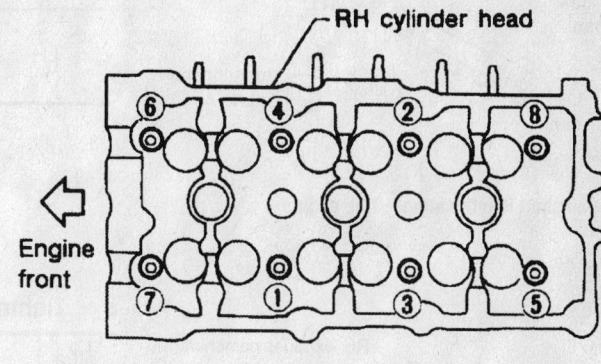

Right cylinder head bolt torque sequence—3.5L engine

9359VG09

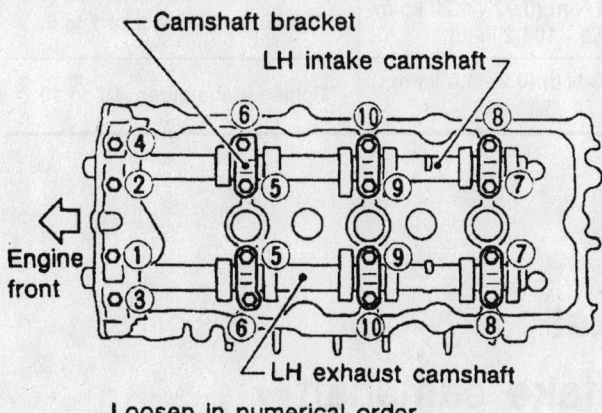

Left camshaft loosening sequence—3.5L engine

9359VG07

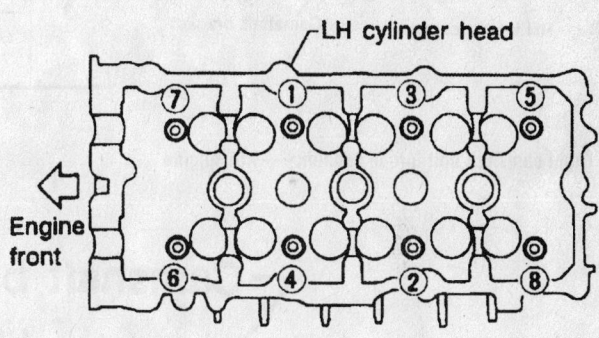

Left cylinder head bolt torque sequence—3.5L engine

9359VG10

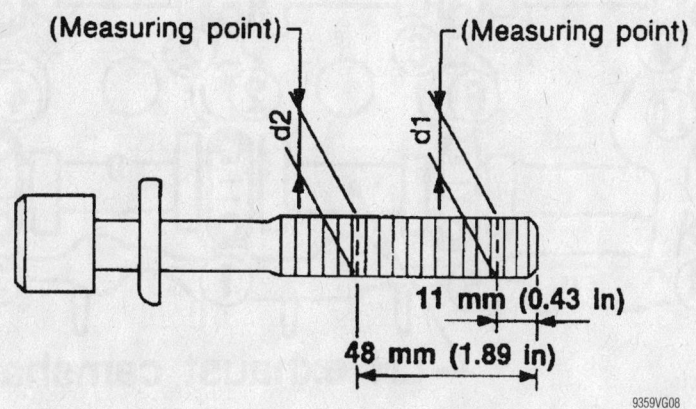

Head bolt checking—3.5L engine

9359VG08

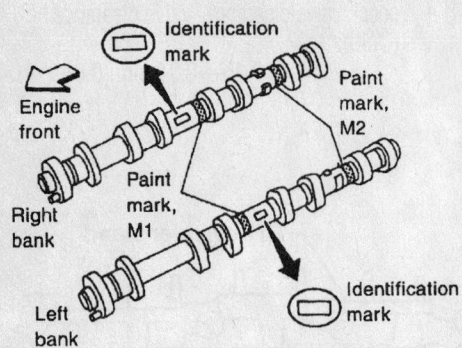

Camshaft identification—3.5L engine

● **Identification marks are present on camshafts.**

Bank	INT/EXH	ID mark	Drill mark	Paint mark	
				M1	M2
RH	INT	R3	Yes	Yes	No
	EXH	R3	No	No	Yes
LH	INT	L3	Yes	Yes	No
	EXH	L3	No	No	Yes

9359VG11

● Tighten the camshaft brackets in the following steps.

Step	Tightening torque	Tightening order
1	1.96 N·m (0.2 kg-m, 17 in-lb)	Tighten in the order of 7 to 10, then tighten 1 to 6.
2	5.88 N·m (0.6 kg-m, 52 in-lb)	Tighten in the numerical order.
3	9.02 - 11.8 N·m (0.92 - 1.20 kg-m, 79.9 - 104.2 in-lb)	Tighten in the order of 1 to 6.
	8.3 - 10.3 N·m (0.9 - 1.0 kg-m, 74 - 91 in-lb)	Tighten in the order of 7 to 10.

9359VG12

Right camshaft bolt torque sequence—3.5L engine

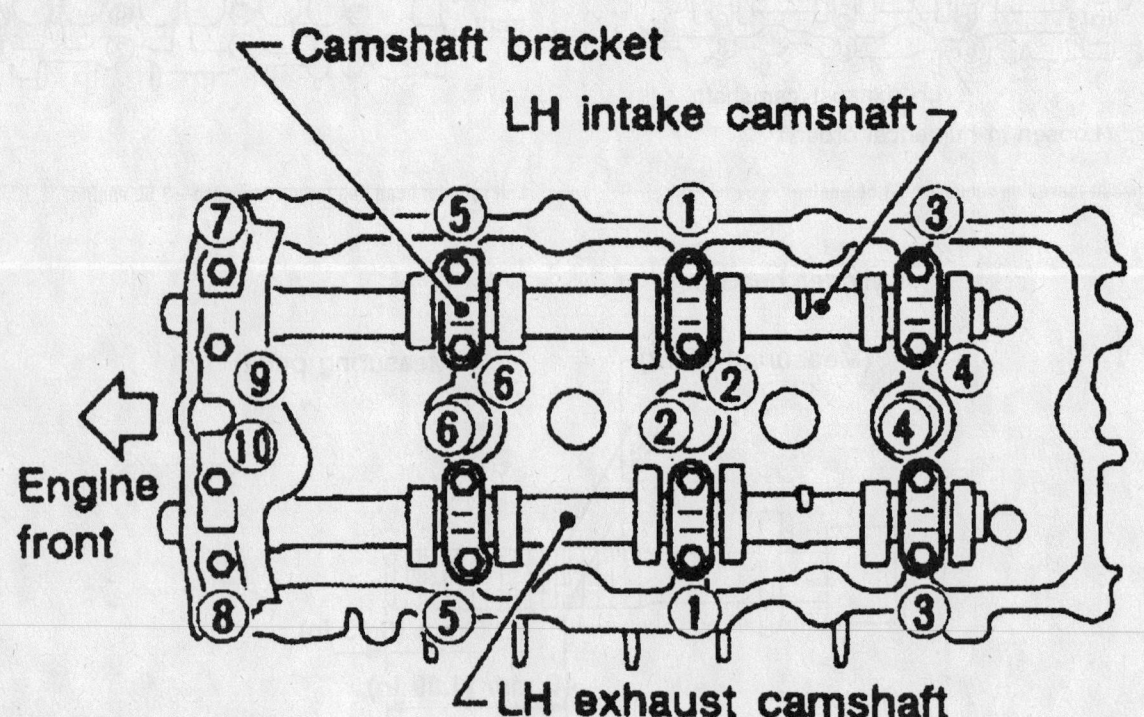

Left camshaft bolt torque sequence—3.5L engine

9359VG13

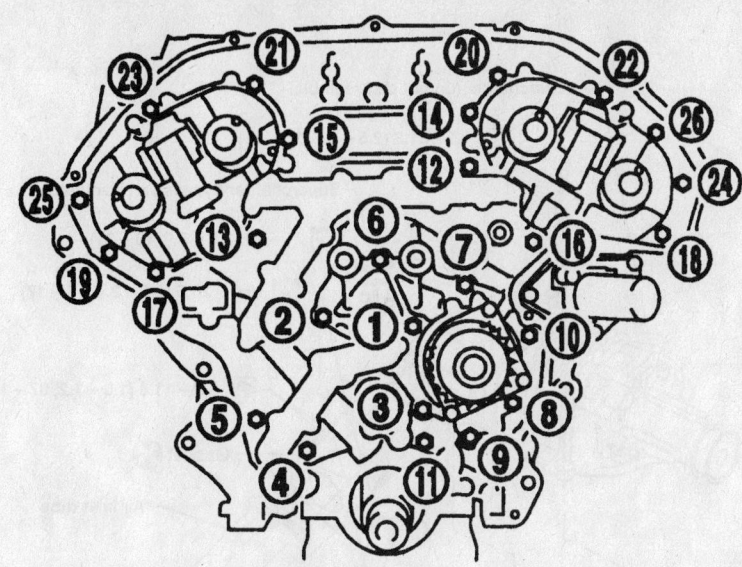

**12 - 13 N•m
(1.2 - 1.4 kg-m, 9 - 10 ft-lb)**

9359VG14

Rear timing case bolt torque sequence—3.5L engine

- Exhaust 0.29—0.37 mm (0.011—0.015 in)

➡️**Lubricate threads and seat surfaces of camshaft bracket bolts with new engine oil before installing them.**

25. Install intake valve timing control solenoid valves.
26. Install O-rings to cylinder block.
27. Install O-rings to cylinder head.
28. Apply sealant to the hatched portion of rear timing chain case. Apply continuous bead of liquid gasket to mating surface of rear timing chain case. Before installation, wipe off the protruding sealant.
29. Align rear timing chain case with dowel pins, then install on cylinder head and block.
30. Tighten rear chain case bolts.
   a. Tighten bolts in numerical order shown in the figure.
   b. Repeat above step a.
31. Reinstall all removed parts in reverse order of removal.

Rocker Arms/Shafts

REMOVAL & INSTALLATION

3.3L Engine

1. Before servicing the vehicle, refer to the precautions in the beginning of this section.

2. Remove or disconnect the following:
   - Negative battery cable
   - Upper intake manifold
   - Valve covers
   - Rocker arm and shaft assemblies
   - Rocker arms from the shafts

➡️**Keep all valvetrain components in order for assembly.**

To install:

3. Lubricate all contact points with clean engine oil and assemble the rocker arms to the shafts in their original positions.
4. Install or connect the following:
   - Rocker arm and shaft assemblies. Tighten the bolts to 13–16 ft. lbs. (18–22 Nm).
   - Valve covers
   - Upper intake manifold
   - Negative battery cable
5. Start the engine and check for leaks.

Supercharger

REMOVAL & INSTALLATION

✳✳ CAUTION

Do not disassemble or adjust the supercharger.

1. Disconnect the negative battery cable.
2. Disconnect the accelerator cable from the throttle body and the air inlet tube bracket.
3. Disconnect the ASCD cable from the throttle body and the air inlet tube bracket, if equipped.
4. Remove the air inlet duct
   a. Disconnect the PCV hoses.
   b. Disconnect the resonator hose.
5. Partially drain the cooling system.
6. Remove the supercharger pulley cover and the supercharger/air conditioning drive belt.
7. Remove the air inlet tube upper and lower supports.
8. Remove the air inlet tube bolts, nuts, and studs. Position the air inlet tube aside.
   a. Disconnect the evaporative emission vacuum hose.
   b. Disconnect the brake booster vacuum hose.
   c. Disconnect the TPS sensor electrical connector.
   d. Disconnect the TPS switch electrical connector.
9. Remove the supercharger bolts and the supercharger assembly.
   a. Disconnect the boost control valve vacuum hose.
   b. Disconnect the PCV hose.

INSPECTION

Supercharger Flange

1. Clean the mating surface of the supercharger flange.
2. Check the flange surface for any deformation and flatness.
   Use a reliable straightedge and feeler gauge, or attach the supercharger flange to the intake collector mating flange, and check that the flatness is within specification. Flange flatness limit: 0.12 mm (0.005 in).

Rotor System

1. Check that the supercharger pulley rotates smoothly when turning it by hand in a clockwise direction. Rotating torque must not exceed specification. Rotating torque: 0.5 N.m (0.05 kg-m, 4 in-lb).
2. Check that both the left and right rotors are free from any cracks or contamination.

Supercharger Bypass Valve Actuator

1. Apply air pressure of less than 12 kPa (90 mmHg, 3.54 inHg) to the supercharger bypass valve actuator's lower side hose port and check for any leakage.
2. Check the supercharger bypass valve

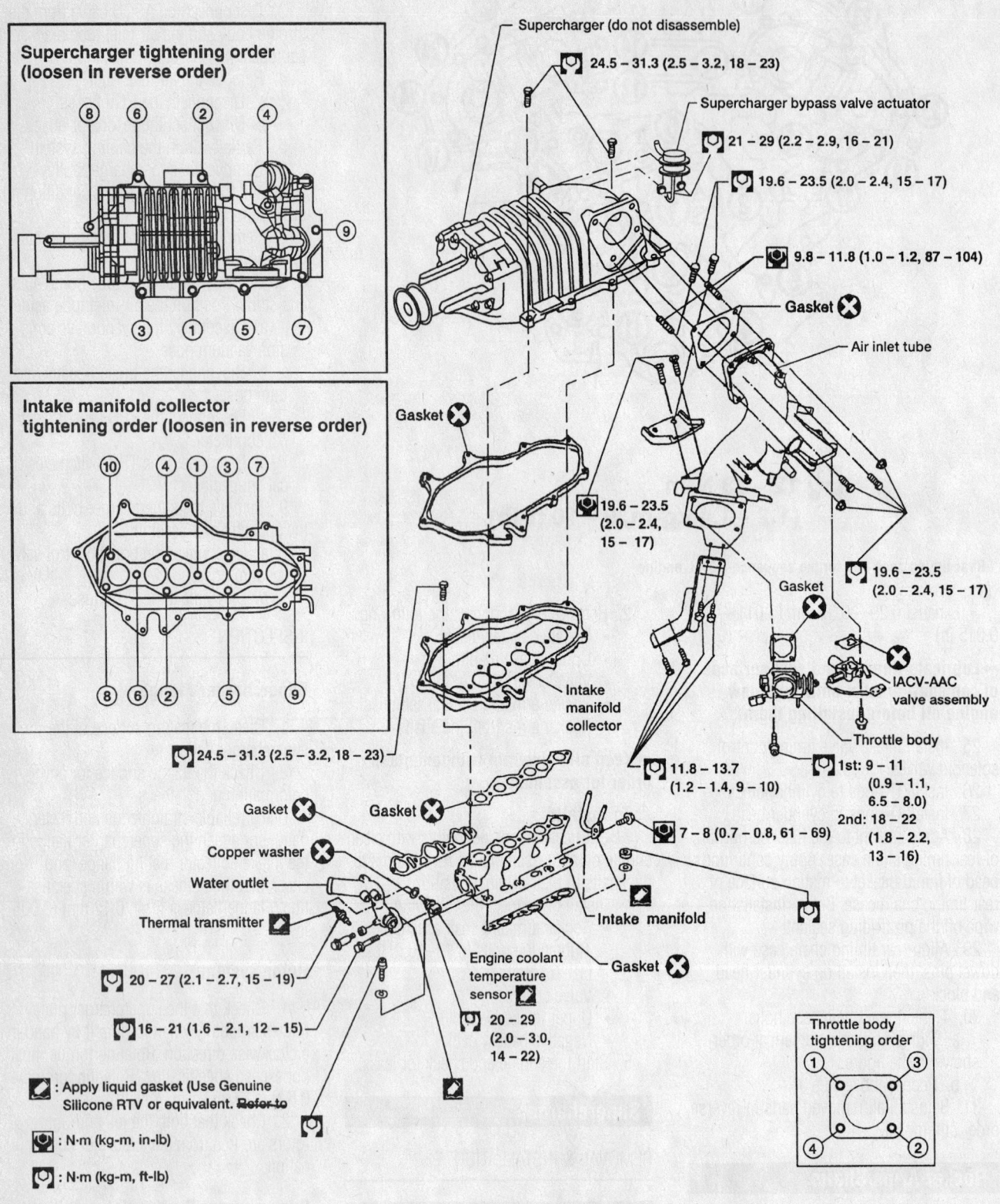

Supercharger tightening order (loosen in reverse order)

Intake manifold collector tightening order (loosen in reverse order)

Supercharger (do not disassemble)

24.5 – 31.3 (2.5 – 3.2, 18 – 23)

Supercharger bypass valve actuator

21 – 29 (2.2 – 2.9, 16 – 21)

19.6 – 23.5 (2.0 – 2.4, 15 – 17)

9.8 – 11.8 (1.0 – 1.2, 87 – 104)

Gasket

Air inlet tube

Gasket

19.6 – 23.5 (2.0 – 2.4, 15 – 17)

19.6 – 23.5 (2.0 – 2.4, 15 – 17)

Gasket

IACV-AAC valve assembly

Throttle body

Intake manifold collector

11.8 – 13.7 (1.2 – 1.4, 9 – 10)

1st: 9 – 11 (0.9 – 1.1, 6.5 – 8.0)
2nd: 18 – 22 (1.8 – 2.2, 13 – 16)

7 – 8 (0.7 – 0.8, 61 – 69)

24.5 – 31.3 (2.5 – 3.2, 18 – 23)

Gasket

Gasket

Copper washer

Water outlet

Intake manifold

Thermal transmitter

Gasket

Engine coolant temperature sensor

20 – 27 (2.1 – 2.7, 15 – 19)

16 – 21 (1.6 – 2.1, 12 – 15)

20 – 29 (2.0 – 3.0, 14 – 22)

: Apply liquid gasket (Use Genuine Silicone RTV or equivalent. ~~Refer to~~

: N·m (kg-m, in-lb)

: N·m (kg-m, ft-lb)

Throttle body tightening order

9359VG35

Supercharger components

actuator rod for smooth movement while maintaining the pressure at the specified levels below:

- Rod starts to extend at approximately: 12 Kpa (90 mmHg, 3.54 inHg)
- Rod is fully extended at approximately: 33.3 kPa (250 mmHg, 9.84 inHg)
- Rod full extended length: 20.83–22.71 mm (0.82–0.89 in)

INSTALLATION

To install the supercharger, follow the removal steps in reverse order. Replace all gaskets; make sure that all gasket surfaces are clean and undamaged. Follow all torque sequences for tightening. Refill the cooling system.

Intake Manifold

REMOVAL & INSTALLATION

3.3L and 3.5L Engines

1. Before servicing the vehicle, refer to the precautions in the beginning of this section.
2. Drain the cooling system.
3. Relieve the fuel system pressure.
4. Remove or disconnect the following:
   - Negative battery cable
   - Air intake duct
   - Accelerator cable
   - Cruise control cable
   - Idle Air Control (IAC) valve connector
   - Throttle Position (TP) sensor and switch connectors
   - Ignition coil and power transistor connectors
   - Exhaust Gas Recirculation (EGR) Solenoid valve connector
   - EGR temperature sensor connector
   - Radiator hoses
   - Heater hoses
   - Positive Crankcase Ventilation (PCV) valve and hose
   - Evaporative Emissions (EVAP) canister vacuum and purge hoses
   - Brake booster vacuum hose
   - Fuel pressure regulator vacuum hose
   - EGR tube
   - Spark plug wires
   - Distributor
   - Left bank injector connectors
   - Thermal transmitter
   - Upper intake manifold ground cable

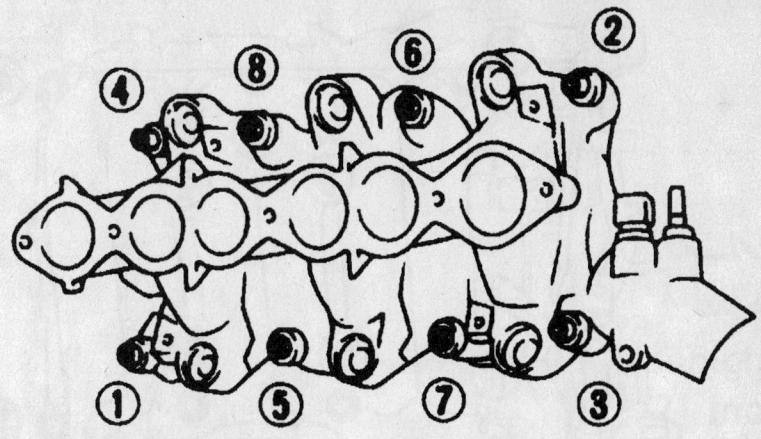

Loosen bolts in numerical order.

7924VG32

Intake manifold loosening sequence—3.3L engine

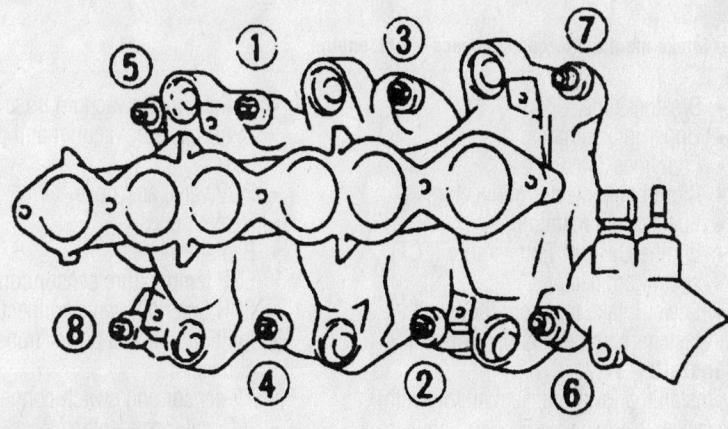

Tighten bolts in numerical order.

7924VG33

Intake manifold torque sequence—3.3L engine

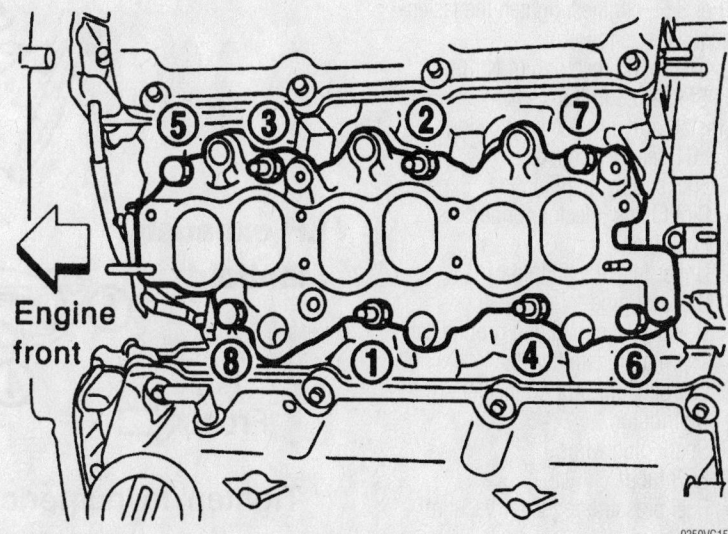

Engine front

9359VG15

Lower intake manifold torque sequence—3.5L engine

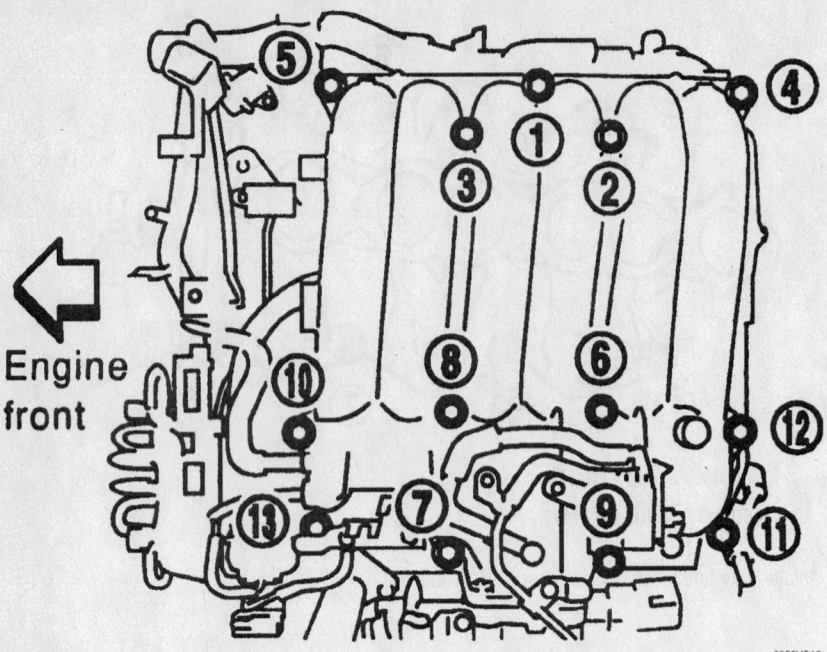

Upper intake manifold torque sequence—3.5L engine

- Cruise control cable
- Accelerator cable
- Air intake duct
- Negative battery cable
9. Fill the cooling system.
10. Start the engine and check for leaks.

Exhaust Manifold

REMOVAL & INSTALLATION

3.3L and 3.5L Engines

1. Before servicing the vehicle, refer to the precautions in the beginning of this section.
2. Remove or disconnect the following:
   - Negative battery cable
   - Exhaust manifold heat shields
   - Exhaust Gas Recirculation (EGR) tube
   - Heated Oxygen (HO₂S) sensor connectors
   - Exhaust front pipes
   - Exhaust manifolds with catalytic converters attached. Loosen the nuts in the reverse of the torque sequence.

To install:
3. Install or connect the following:
   - Exhaust manifolds with catalytic converters attached. Tighten the nuts in sequence to 21–25 ft. lbs. (28–33 Nm).
   - Exhaust front pipes. Tighten the bolts to 21–25 ft. lbs. (28–33 Nm).
   - Heated Oxygen (HO₂S) sensor connectors

- Breather pipe
- Upper intake manifold
- Fuel lines
- Right bank injector connectors
- Fuel supply manifold
- Engine Coolant Temperature (ECT) sensor connector
- Lower intake manifold. Loosen the fasteners in the sequence shown.

To install:
5. Install the lower intake manifold with a new gasket.
6. For 3.3L engines, tighten the fasteners in sequence as follows:
   a. Step 1: 35 inch lbs. (4 Nm)
   b. Step 2: 78 inch lbs. (9 Nm)
   c. Step 3: 70–84 inch lbs. (8–10 Nm)
7. For 3.5L engines, tighten the fasteners in sequence as follows:
   a. Step 1: 86 inch lbs. (4 Nm)
   b. Step 2: 23 ft. lbs. (9 Nm)
8. Install or connect the following:
   - ECT sensor connector
   - Fuel supply manifold
   - Right bank injector connectors
   - Fuel lines
   - Upper intake manifold
   - Breather pipe
   - Upper intake manifold ground cable
   - Thermal transmitter
   - Left bank injector connectors
   - Distributor
   - Spark plug wires
   - EGR tube
   - Fuel pressure regulator vacuum hose

- Brake booster vacuum hose
- EVAP canister vacuum and purge hoses
- PCV valve and hose
- Heater hoses
- Radiator hoses
- EGR temperature sensor connector
- EGR Solenoid valve connector
- Ignition coil and power transistor connectors
- TP sensor and switch connectors
- IAC valve connector

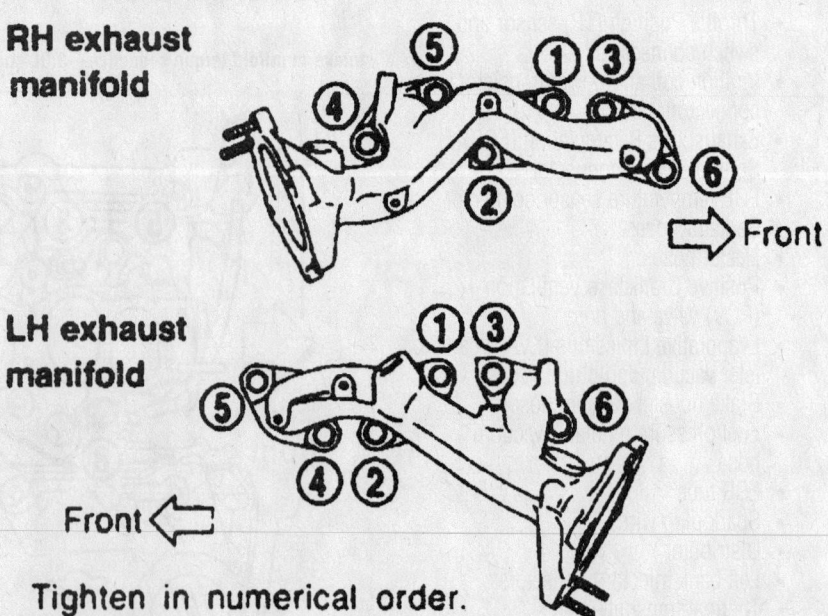

RH exhaust manifold

LH exhaust manifold

Tighten in numerical order.

Exhaust manifold torque sequence—3.3L engine

Right bank

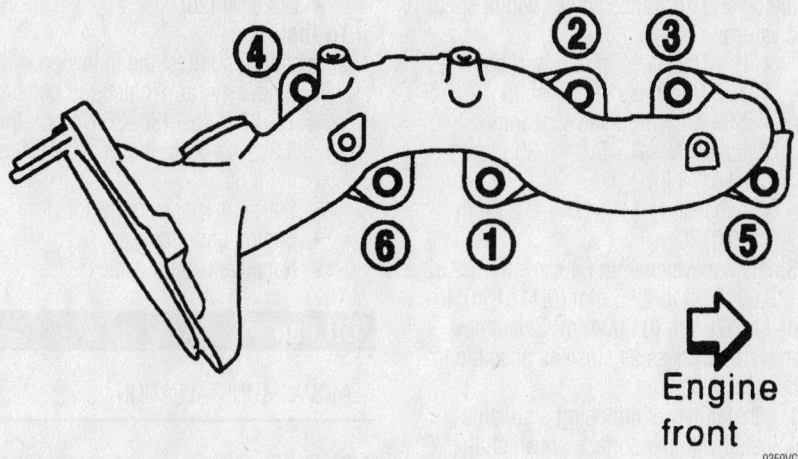

Right exhaust manifold torque sequence—3.5L engine

9359VG17

Left bank

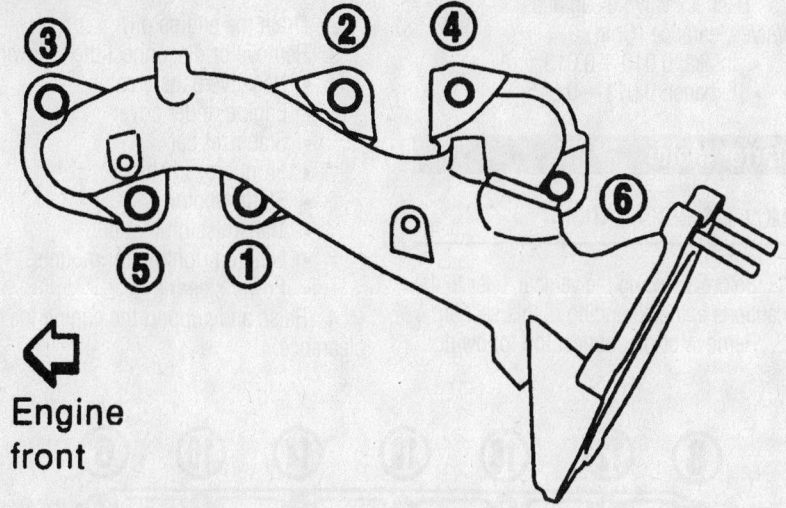

Exhaust manifold torque sequence—3.5L engine

9359VG18

- EGR tube. Tighten the flange fittings to 29–36 ft. lbs. (39–49 Nm).
- Exhaust manifold heat shields. Tighten the bolts to 84–96 inch lbs. (9–11 Nm)
- Negative battery cable
4. Start the engine and check for leaks.

Front Crankshaft Seal

REMOVAL & INSTALLATION

3.3L Engine

1. Before servicing the vehicle, refer to the precautions in the beginning of this section.

2. Drain the cooling system.
3. Remove or disconnect the following:

- Negative battery cable
- Accessory drive belts
- Radiator hoses
- Crankshaft pulley
- Front cover
- Timing belt
- Crankshaft timing sprocket
- Front crankshaft seal

To install:
4. Install or connect the following:
- Front crankshaft seal flush with the oil pump housing
- Crankshaft timing sprocket
- Timing belt

- Front cover. Tighten the bolts to 26–43 inch lbs. (3–5 Nm).
- Crankshaft pulley. Tighten the bolt to 141–156 ft. lbs. (191–211 Nm).
- Radiator hoses
- Accessory drive belts
- Negative battery cable
5. Fill the cooling system.
6. Start the engine and check for leaks.

Camshaft and Valve Lifters

REMOVAL & INSTALLATION

3.3L Engines

1. Before servicing the vehicle, refer to the precautions in the beginning of this section.
2. Drain the cooling system.
3. Remove or disconnect the following:
- Negative battery cable
- Upper intake manifold
- Valve covers

➡ Keep all valvetrain components in order for assembly.

- Rocker arm and shaft assemblies
- Valve lifter guide and valve lifters. Attach a wire to the top of the lifters so that they will not drop from the lifter guide.
- Radiator
- Accessory drive belts
- Front cover
- Timing belt
- Camshaft sprockets
- Camshaft seals
- Rear timing cover
- Distributor
- Cylinder head rear covers
- Camshaft locating plates
- Camshafts

To install:
4. Install or connect the following:
- Camshafts
- Camshaft locating plates. Tighten the bolts to 58–65 ft. lbs. (78–88 Nm).
- Cylinder head rear covers
- Distributor
- Rear timing cover
- Camshaft seals
- Camshaft sprockets. Tighten the bolts to 58–65 ft. lbs. (78–88 Nm).
- Timing belt
- Front cover
- Accessory drive belts
- Radiator
- Valve lifter guide and valve lifters
- Rocker arm and shaft assemblies.

Tighten the bolts to 13–16 ft. lbs. (18–22 Nm).
- Valve covers
- Upper intake manifold
- Negative battery cable

5. Fill the cooling system.
6. Start the engine and check for leaks.

3.5L Engines

See the Cylinder Head Removal and Installation procedure.

Valve Lash

ADJUSTMENT

3.3L Engines

These engines are equipped with hydraulic valve lifters that do not require periodic adjustment.

3.5L Engines

➡ Adjust valve clearance while engine is cold.

1. Turn crankshaft, to position cam lobe on camshaft of valve that must be adjusted upward.
2. Thoroughly wipe off engine oil around adjusting shim using a rag.
3. Using an extra-fine screwdriver, turn the round hole of the adjusting shim in the direction of the arrow.
4. Place Tool (A) around camshaft as shown in figure.

Before placing Tool (A), rotate notch toward center of cylinder head (See figure.), to simplify shim removal later.

✳✳ CAUTION

Be careful not to damage cam surface with Tool (A).

5. Rotate Tool (A) (See figure.) so that valve lifter is pushed down.
6. Place Tool (B) between camshaft and the edge of the valve lifter to retain valve lifter.

✳✳ CAUTION

Tool (B) must be placed as close to camshaft bracket as possible. Be careful not to damage cam surface with Tool (B).

7. Remove Tool (A).
8. Blow air into the hole to separate adjusting shim from valve lifter.
9. Remove adjusting shim using a small screwdriver and a magnetic finger.
10. Determine replacement adjusting

shim size following formula. Using a micrometer determine thickness of removed shim. Calculate thickness of new adjusting shim so valve clearance comes within specified values.

- R = Thickness of removed shim
- N = Thickness of new shim
- M = Measured valve clearance
- Intake: N = R + [M—0.30 mm (0.0118 in)]
- Exhaust: N = R + [M—0.33 mm (0.0130 in)]

Shims are available in 64 sizes from 2.32 mm (0.0913 in) to 2.95 mm (0.1161 in), in steps of 0.01 mm (0.0004 in). Select new shim with thickness as close as possible to calculated value.

11. Install new shim using a suitable tool. Install with the surface on which the thickness is stamped facing down.
12. Place Tool (A) as mentioned in steps 2 and 3.
13. Remove Tool (B).
14. Remove Tool (A).
15. Recheck valve clearance.
Valve clearance (Cold)
- Intake: 0.010—0.013
- Exhaust: 0.011—0.015

Starter Motor

REMOVAL & INSTALLATION

1. Before servicing the vehicle, refer to the precautions in the beginning of this section.
2. Remove or disconnect the following:

- Negative battery cable
- Engine under cover
- Starter harness connectors
- Starter motor

To install:

3. Install or connect the following:
- Starter motor. Tighten the bolts to 22–27 ft. lbs. (30–36 Nm) on the 3.3L; 37–45 ft. lbs. (61-69NM) on the 3.5L.
- Starter harness connectors
- Engine under cover
- Negative battery cable

Oil Pan

REMOVAL & INSTALLATION

3.3L Engine

2WD MODELS

1. Before servicing the vehicle, refer to the precautions in the beginning of this section.
2. Drain the engine oil.
3. Remove or disconnect the following:

- Negative battery cable
- Engine under cover
- Stabilizer bar
- Front crossmember
- Starter motor
- Transmission mount
- Left and right motor mounts
- Power steering gear

4. Raise and support the engine for clearance.

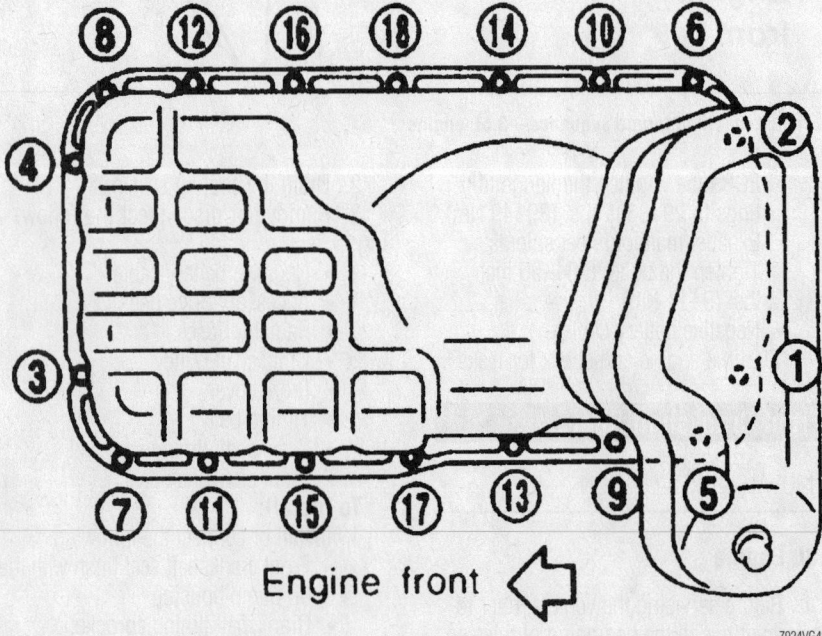

Engine front ⬅

Oil pan bolt removal sequence—3.3L engine

7924VG42

5. Remove or disconnect the following:
  • Oil pan bolts in the sequence
  • Oil pan

To install:

6. Apply a continuous bead of sealant 0.138–0.177 in. (3.5–4.5mm) to the oil pan mating surface.

7. Install or connect the following:
  • Oil pan. Tighten the bolts in reverse of the removal sequence to 62 inch lbs. (7 Nm).
  • Power steering gear
  • Left and right motor mounts
  • Transmission mount
  • Starter motor
  • Front crossmember
  • Stabilizer bar
  • Engine under cover
  • Negative battery cable

➥**Wait 30 minutes after installation of the oil pan to allow the sealant to cure before adding oil.**

8. Fill the crankcase to the correct level.
9. Start the engine and check for leaks.

4WD MODELS

1. Before servicing the vehicle, refer to the precautions in the beginning of this section.
2. Drain the engine oil.
3. Remove or disconnect the following:
  • Negative battery cable
  • Engine under cover
  • Stabilizer bar brackets
  • Front driveshaft
  • Axle halfshafts
  • Front suspension crossmember
  • Front differential and mounting bracket
  • Starter motor
  • Transmission mount
  • Left and right motor mounts
  • Power steering gear
  • Relay rod
4. Raise and support the engine for clearance.
5. Remove or disconnect the following:
  • Oil pan bolts in the sequence
  • Oil pan

To install:

6. Apply a continuous bead of sealant 0.138–0.177 in. (3.5–4.5mm) to the oil pan mating surface.

7. Install or connect the following:
  • Oil pan. Tighten the bolts in reverse of the removal sequence to 62 inch lbs. (7 Nm).
  • Relay rod
  • Power steering gear
  • Left and right motor mounts
  • Transmission mount
  • Starter motor
  • Front differential and mounting bracket
  • Front suspension crossmember
  • Axle halfshafts
  • Front driveshaft
  • Stabilizer bar brackets
  • Engine under cover
  • Negative battery cable

➥**Wait 30 minutes after installation of the oil pan to allow the sealant to cure before adding oil.**

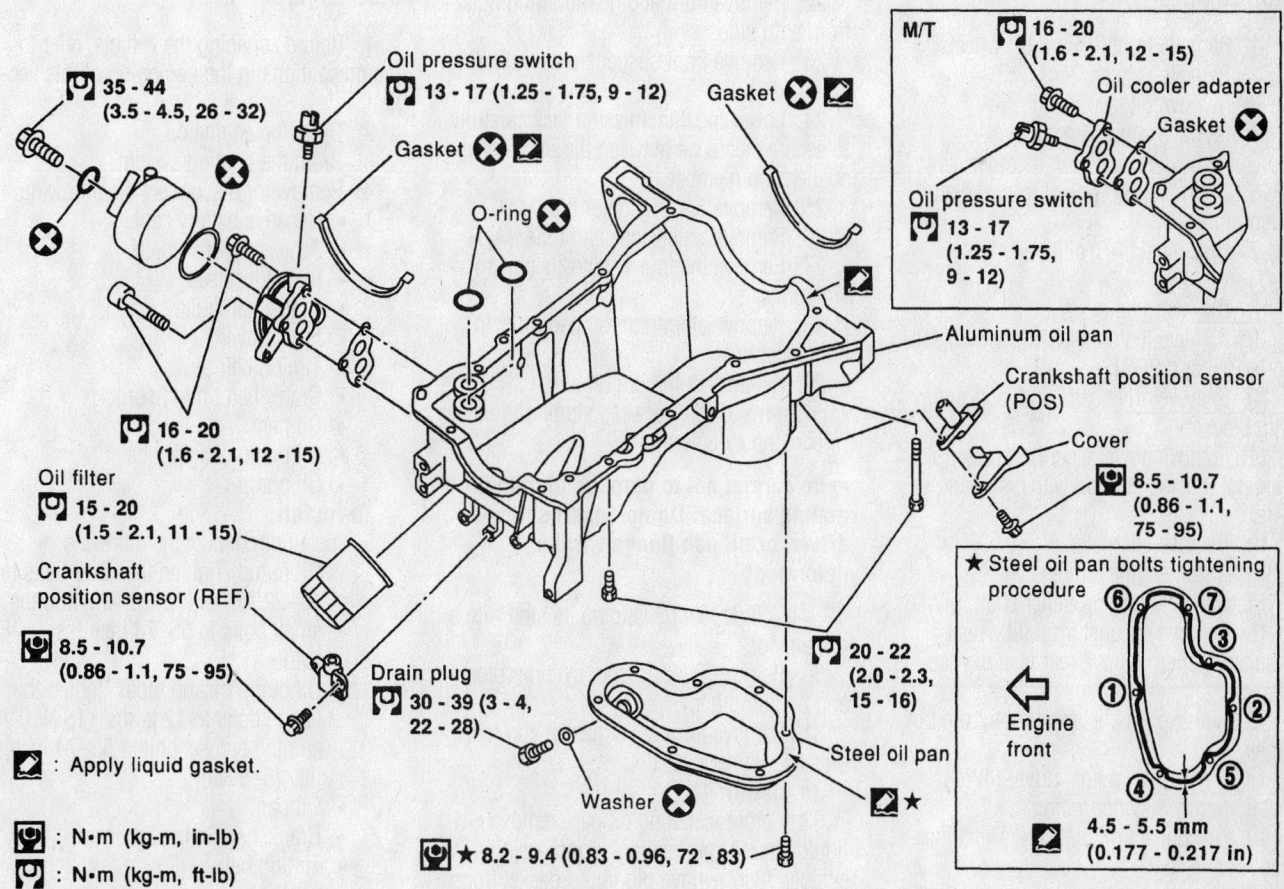

Oil pan exploded view—3.5L engine

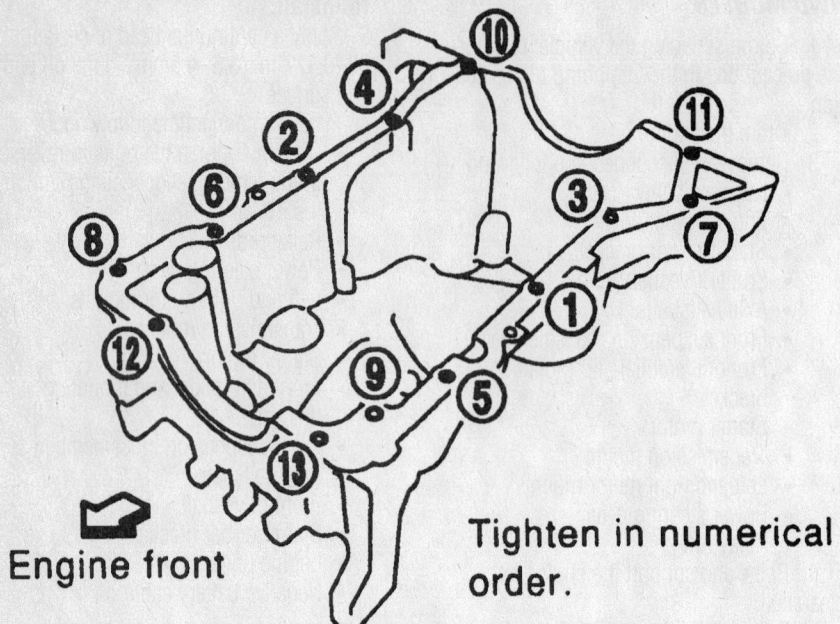

Engine front

Tighten in numerical order.

9359VG20

Oil pan bolt torque sequence—3.5L engine

8. Fill the crankcase to the correct level.
9. Start the engine and check for leaks.

3.5L Engines

1. Remove front RH and LH wheels.
2. Remove battery.
3. Remove oil level gauge.
4. Remove engine undercover.
5. Remove suspension member stay.
6. Drain engine coolant from radiator drain plug.
7. Disconnect A/T oil cooler hoses. (A/T)
8. Drain engine oil.
9. Remove the crankshaft position sensors (REF and POS).
10. Remove drive belts and idler pulley with bracket.
11. Remove power steering oil pump, then put it aside holding with a suitable wire.
12. Remove alternator.
13. Install engine slingers.
14. Remove front propeller shaft. (4WD)
15. Remove exhaust front tube heat insulators, then remove rear heat oxygen sensors.
16. Remove exhaust front tube from both sides.
17. Remove front final drive. (4WD)
18. Remove starter motor.
19. Disconnect oil pressure switch harness connector.
20. Loosen and disconnect the bolts fixing the steering column assembly lower joint and the power steering gear.

21. Set a suitable transmission jack under the front suspension member and hoist engine with engine slingers.
22. Remove front engine mounting nuts from both sides.
23. Remove front suspension member bolts.
24. Lower the transmission jack carefully to secure clearance between the oil pan and suspension member.
25. Remove A/T oil cooler tube. (A/T)
26. Remove water hose and tube. (A/T)
27. Remove the four engine-to-transmission bolts.
28. Remove aluminum oil pan bolts in numerical order.
29. Remove aluminum oil pan.
   a. Insert tool between aluminum oil pan and cylinder block.

➡ **Be careful not to damage aluminum mating surface. Do not insert screwdriver, or oil pan flange will be deformed.**

   b. Slide tool by tapping its side with a hammer.
30. Remove O-rings from cylinder block and oil pump body.
31. Remove front cover gasket and rear oil seal retainer gasket.
   To install:
32. Before installing oil pan, remove old liquid gasket from mating surface using a scraper. Also remove old liquid gasket from mating surface of cylinder block. Remove old liquid gasket from the bolt hole and thread.

33. Apply sealant to front cover gasket and rear oil seal retainer gasket.
34. Install front cover gasket and rear oil seal retainer gasket.
35. Apply a continuous bead of liquid gasket to mating surface of aluminum oil pan. Use RTV silicone sealant or equivalent.
36. Apply liquid gasket to inner sealing surface as shown in figure. Be sure liquid gasket is 4.0 to 5.0 mm (0.157 to 0.197 in) or 4.5 to 5.5 mm (0.177 to 0.217 in) wide. Attaching should be done within 5 minutes after coating.
37. Install O-rings, cylinder block and oil pump body.
38. Install aluminum oil pan. Tighten bolts in numerical order. Wait at least 30 minutes before refilling engine oil.
39. Install the four engine-to-transmission bolts.
40. Reinstall in the reverse order of removal.

Oil Pump

REMOVAL & INSTALLATION

3.3L Engine

1. Before servicing the vehicle, refer to the precautions in the beginning of this section.
2. Drain the engine oil.
3. Drain the cooling system.
4. Remove or disconnect the following:
   • Negative battery cable
   • Accessory drive belts
   • Radiator hoses
   • Crankshaft pulley
   • Front cover
   • Timing belt
   • Crankshaft timing sprocket
   • Oil pan
   • Oil pump pickup tube
   • Oil pump
To install:
5. Install or connect the following:
   • Oil pump. Tighten the large bolts to 16–22 ft. lbs. (22–29 Nm) and the small bolts to 55–74 inch lbs. (6–8 Nm).
   • Oil pump pickup tube. Tighten the flange bolts to 12 ft. lbs. (16 Nm) and the bracket bolt to 55–74 inch lbs. (6–8 Nm).
   • Oil pan
   • Crankshaft timing sprocket
   • Timing belt
   • Front cover
   • Crankshaft pulley
   • Radiator hoses

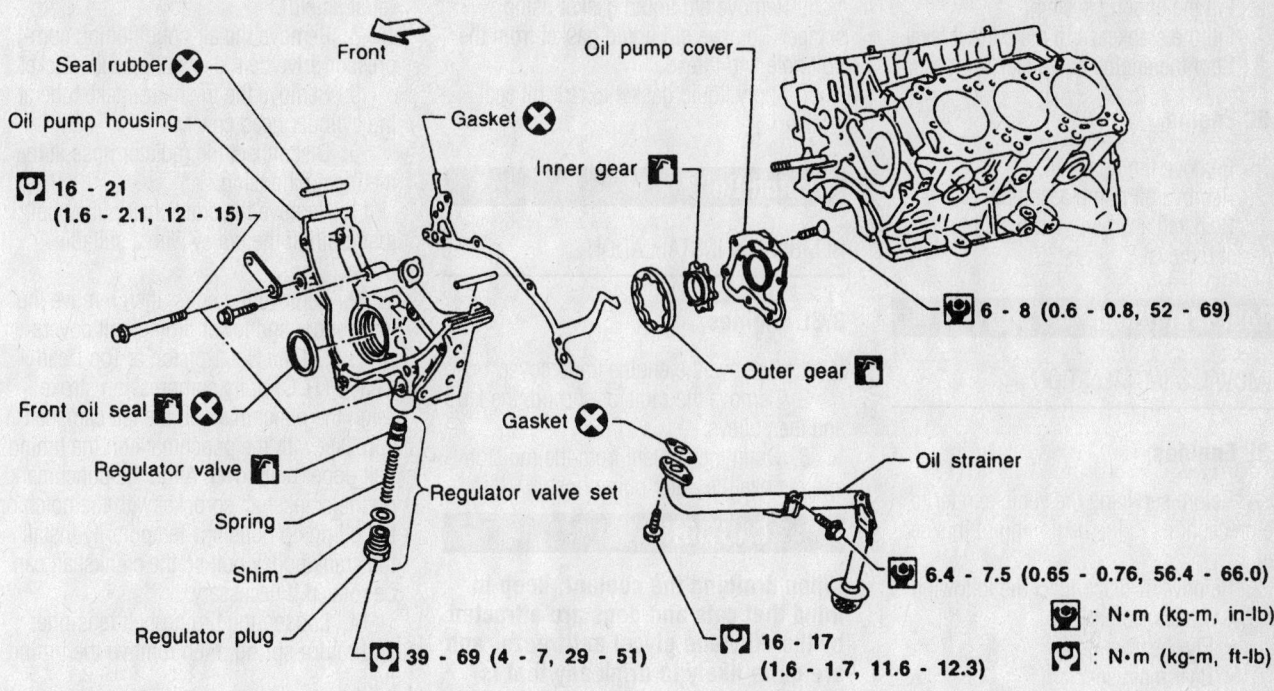

Oil pump assembly exploded view—3.3L engine

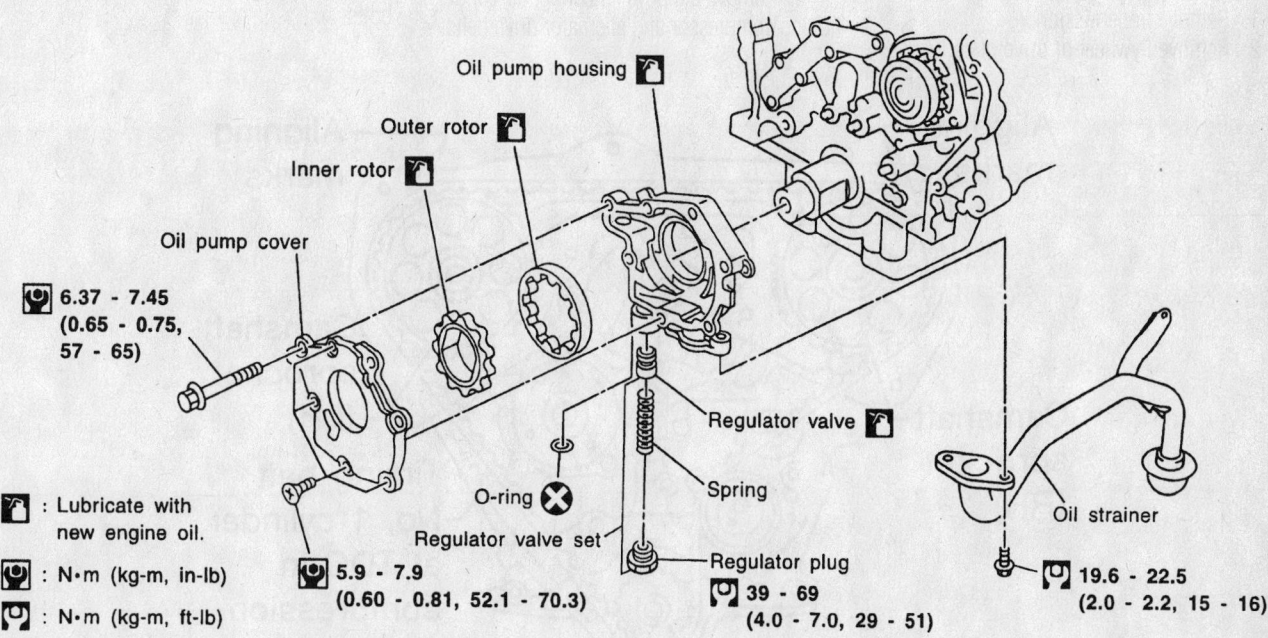

Oil pump assembly exploded view—3.5L engine

- Accessory drive belts
- Negative battery cable
6. Fill the cooling system.
7. Fill the crankcase to the correct level.
8. Start the engine and check for leaks.

3.5L Engine

1. Remove timing chain.
2. Remove oil pump assembly.
3. Reinstall any parts removed in reverse order of removal.

Rear Main Seal

REMOVAL & INSTALLATION

3.3L Engines

1. Before servicing the vehicle, refer to the precautions in the beginning of this section.
2. Remove or disconnect the following:
   - Transmission
   - Flywheel
   - Rear main seal

To install:
3. Install the seal so that it is flush with the retainer housing.
4. Install or connect the following:
   - Flywheel. Tighten the bolts to 61–69 ft. lbs. (83–93 Nm).
   - Transmission

3.5L Engine

1. Remove transmission.
2. Remove flywheel or drive plate.

3. Remove oil pan.
4. Remove rear oil seal retainer.
5. Remove old liquid gasket using scraper. Remove old liquid gasket from the bolt hole and thread.
6. Apply liquid gasket to rear oil seal retainer.

Timing Belt

REMOVAL & INSTALLATION

3.3L Engines

1. Remove the engine undercover.
2. Remove the radiator shroud, the fan and the pulleys.
3. Drain the coolant from the radiator and remove the water pump hose.

❋❋ CAUTION

When draining the coolant, keep in mind that cats and dogs are attracted by the ethylene glycol antifreeze, and are quite likely to drink any that is left in an uncovered container or in puddles on the ground. This will prove fatal in sufficient quantity. Always drain the coolant into a sealable container. Coolant should be reused unless it is contaminated or several years old.

4. Remove the radiator.
5. Remove the power steering, air conditioning compressor and alternator drive belts.

6. Remove the spark plugs.
7. Remove the distributor protector (dust shield).
8. Remove the air conditioning compressor drive belt idler pulley and bracket.
9. Remove the fresh air intake tube at the cylinder head cover.
10. Disconnect the radiator hose at the thermostat housing.
11. Remove the crankshaft pulley bolt, then pull off the pulley with a suitable puller.
12. Remove the bolts, then remove the front upper and lower timing belt covers.
13. Set the No. 1 piston at Top Dead Center (TDC) of its compression stroke. Align the punchmark on the left camshaft sprocket with the punchmark on the timing belt upper rear cover. Align the punchmark on the crankshaft sprocket with the notch on the oil pump housing. Temporarily install the crank pulley bolt so the crankshaft can be rotated if necessary.
14. Loosen the timing belt tensioner and return spring, then remove the timing belt.

To install:

❋❋ CAUTION

Before installing the timing belt, confirm that the No. 1 cylinder is set at the TDC of the compression stroke.

15. Remove both cylinder head covers and loosen all rocker arm shaft retaining bolts.

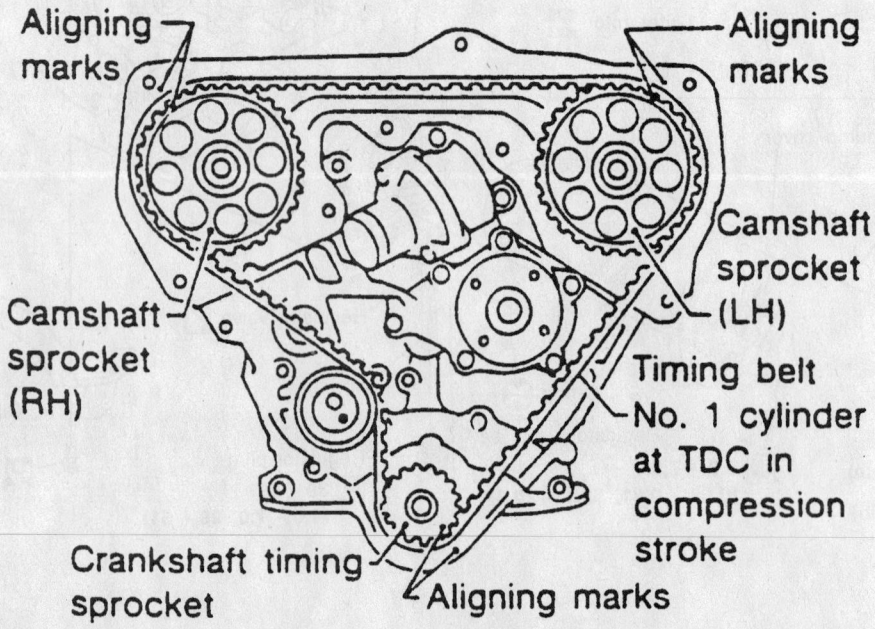

Timing belt alignment mark locations —3.3L engines

Aligning marks

Aligning marks

Camshaft sprocket (RH)

Camshaft sprocket (LH)

Timing belt No. 1 cylinder at TDC in compression stroke

Crankshaft timing sprocket

Aligning marks

79245G35

➥**The rocker arm shaft bolts MUST be loosened so that the correct belt tension can be obtained.**

16. Install the tensioner and the return spring. Using a hexagon wrench, turn the tensioner clockwise and temporarily tighten the locknut.

17. Be sure that the timing belt is clean and free from oil or water.

18. When installing the timing belt, align the white lines on the belt with the punchmarks on the camshaft and crankshaft sprockets. Have the arrow on the timing belt pointing toward the front belt covers.

➥**A good way (although rather tedious!) to check for proper timing belt installation is to count the number of belt teeth between the timing marks. There are 133 teeth on the belt; there should be 40 teeth between the timing marks on the left and right side camshaft sprockets, and 43 teeth between the timing marks on the left side camshaft sprocket and the crankshaft sprocket.**

19. While keeping the tensioner steady, loosen the locknut with a hex wrench.

20. Turn the tensioner approximately 70 – 80 degrees clockwise with the wrench, then tighten the locknut.

❄❄ WARNING

If any binding is felt when adjusting the timing belt tension by turning the crankshaft, STOP turning the engine, because the pistons may be hitting the valves.

21. Turn the crankshaft in a clockwise direction several times, then **slowly** set the No. 1 piston to TDC of the compression stroke.

22. Apply 22 lbs. (10 kg) of pressure (push it in!) to the center span of the timing belt between the right side camshaft sprocket and the tensioner pulley, then loosen the tensioner locknut.

23. Using a 0.0138 in. (0.35mm) thick feeler gauge (the actual width of the blade **must** be $1/2$ in. or 13mm!), turn the crankshaft clockwise (**slowly!**). The timing belt should move approximately 2 $1/2$ teeth. Tighten the tensioner locknut, turn the crankshaft slightly and remove the feeler gauge.

24. Slowly rotate the crankshaft clockwise several more times, then set the No. 1 piston to TDC of the compression stroke.

25. Position the 2 timing covers on the block, then tighten the mounting bolts to 24 ft. lbs. (35 Nm).

26. Press the crankshaft pulley onto the shaft, then tighten the bolt to 90 – 98 ft. lbs. (123 – 132 Nm).

27. Connect the radiator hose to the thermostat housing.

28. Reconnect the fresh air intake tube at the cylinder head cover.

29. Install the air conditioning compressor drive belt idler pulley and bracket.

30. Install the distributor protector (dust shield).

31. Install the spark plugs.

32. Install the power steering, air conditioning compressor and alternator drive belts.

33. Install the radiator.

34. Reconnect the water pump hose and fill the engine with coolant. Install the fan shroud and pulleys.

35. Install the engine undercover.

36. Start the engine and check for any leaks.

Timing Chain, Sprockets, Front Cover and Seal

REMOVAL & INSTALLATION

3.5L Engine

1. Release fuel pressure.
2. Remove battery.
3. Remove radiator.
4. Drain engine oil.
5. Remove drive belts and idler pulley with brackets.
6. Remove cooling fan with bracket.
7. Remove engine cover.
8. Remove air duct with air cleaner case, collector, blow-by hose, vacuum hoses, fuel hoses, water hoses, wires, harnesses, connectors and so on.
9. Remove the air compressor, and tie it down using rope or the like to keep it from interfering.
10. Remove the power steering oil pump and reservoir tank. Tie them down using rope or the like to keep them from interfering.
11. Remove alternator.
12. Remove the following.
   - Vacuum gallery
   - Water bypass pipe
   - Brackets
13. Remove camshaft position sensor (PHASE), intake valve timing control position sensors and crankshaft position sensor.

➥**Avoid impact such as dropping. Do not disassemble the components. Do not place them on areas where iron powder may adhere. Keep away from the objects susceptible to magnetism.**

14. Remove upper intake manifold collector in reverse order of installation.

15. Remove intake manifold collector support bolts.

16. Remove lower intake manifold collector in reverse order of installation.

17. Disconnect injector harness connectors.

18. Remove fuel tube assembly in reverse order of installation.

19. Remove ignition coils.

20. Remove RH and LH rocker covers from cylinder head.

21. Set No. 1 piston at TDC on the compression stroke by rotating crankshaft. Align pointer with TDC mark on crankshaft pulley. Check that intake and exhaust cam nose on No. 1 cylinder are installed as shown left. If not, turn the crankshaft one revolution (360°) and align as above.

22. Remove starter motor, and set ring gear stopper using the mounting bolt hole. Be careful not to damage the signal plate teeth.

23. Loosen the crankshaft pulley bolt.

24. Remove crankshaft pulley with a suitable puller.

25. Remove aluminum oil pan.

26. Temporarily install the suspension member bolts and engine mounting nuts.

27. Remove intake valve timing control valve covers. Loosen bolts in numerical order as shown in the figure. In the cover, the shaft is engaged with the center hole of the intake cam sprocket. Remove it straight out until the engagement comes off.

28. Remove front timing chain case bolts. Loosen bolts in numerical order as shown in the figure.

29. Remove front timing chain case. Do not scratch sealing surfaces.

30. Remove internal chain guide.

31. Remove upper tension guide.

32. Remove timing chain tensioner and slack guide. Remove timing chain tensioner. (Push piston and insert a suitable pin into pinhole.)

33. Attach a suitable stopper pin to RH and LH camshaft chain tensioners.

34. Remove intake and exhaust camshaft sprocket bolts. I Apply paint to timing chain and camshaft sprockets for alignment during installation. Secure the hexagonal head of the camshaft using a spanner to loosen mounting bolts.

35. Remove primary and secondary timing chains along with the camshaft sprock-

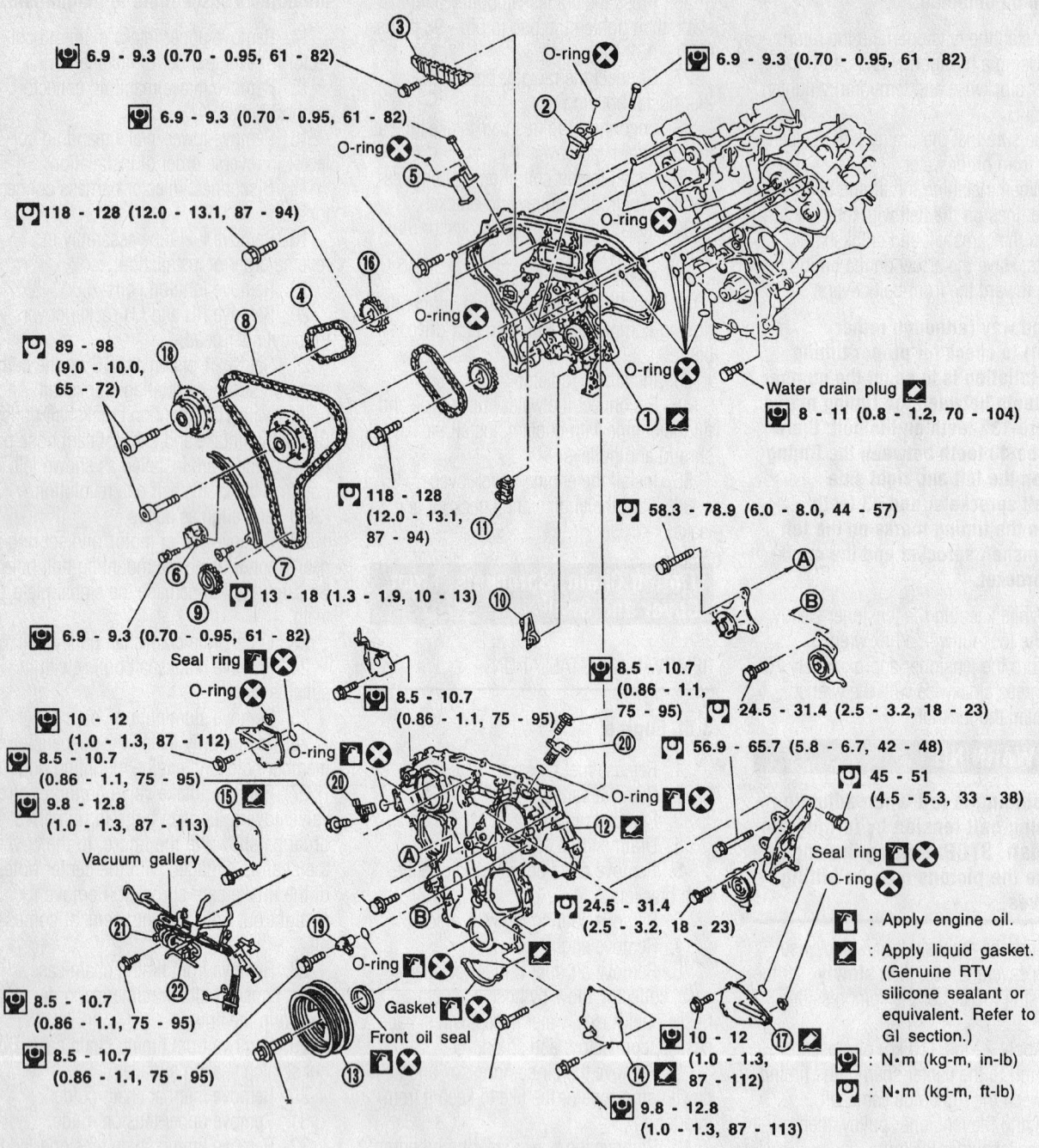

6.9 - 9.3 (0.70 - 0.95, 61 - 82)

6.9 - 9.3 (0.70 - 0.95, 61 - 82)

O-ring

6.9 - 9.3 (0.70 - 0.95, 61 - 82)

O-ring

O-ring

118 - 128 (12.0 - 13.1, 87 - 94)

O-ring

O-ring

Water drain plug

8 - 11 (0.8 - 1.2, 70 - 104)

89 - 98 (9.0 - 10.0, 65 - 72)

118 - 128 (12.0 - 13.1, 87 - 94)

58.3 - 78.9 (6.0 - 8.0, 44 - 57)

13 - 18 (1.3 - 1.9, 10 - 13)

6.9 - 9.3 (0.70 - 0.95, 61 - 82)

Seal ring

O-ring

8.5 - 10.7 (0.86 - 1.1, 75 - 95)

8.5 - 10.7 (0.86 - 1.1, 75 - 95)

24.5 - 31.4 (2.5 - 3.2, 18 - 23)

10 - 12 (1.0 - 1.3, 87 - 112)

O-ring

56.9 - 65.7 (5.8 - 6.7, 42 - 48)

8.5 - 10.7 (0.86 - 1.1, 75 - 95)

45 - 51 (4.5 - 5.3, 33 - 38)

9.8 - 12.8 (1.0 - 1.3, 87 - 113)

Vacuum gallery

Seal ring

O-ring

24.5 - 31.4 (2.5 - 3.2, 18 - 23)

: Apply engine oil.

O-ring

: Apply liquid gasket. (Genuine RTV silicone sealant or equivalent. Refer to GI section.)

8.5 - 10.7 (0.86 - 1.1, 75 - 95)

Gasket

Front oil seal

10 - 12 (1.0 - 1.3, 87 - 112)

: N·m (kg-m, in-lb)

8.5 - 10.7 (0.86 - 1.1, 75 - 95)

: N·m (kg-m, ft-lb)

9.8 - 12.8 (1.0 - 1.3, 87 - 113)

1. Rear timing chain case
2. Left camshaft chain tensioner
3. Internal guide
4. Timing chain (Secondary)
5. Right camshaft chain tensioner
6. Timing chain tensioner
7. Slack guide
8. Timing chain (Primary)
9. Crankshaft sprocket
10. Lower tension guide
11. Upper tension guide
12. Front timing chain case
13. Crankshaft pulley
14. Water pump cover
15. Chain tensioner cover
16. Exhaust camshaft sprocket
17. Intake valve timing control valve cover
18. Intake camshaft sprocket
19. Camshaft position sensor (PHASE)
20. Intake valve timing control position sensor
21. Power valve actuator (A/T)
22. Swirl control valve control solenoid valve

Timing chain components—3.5L engine

9359VG30

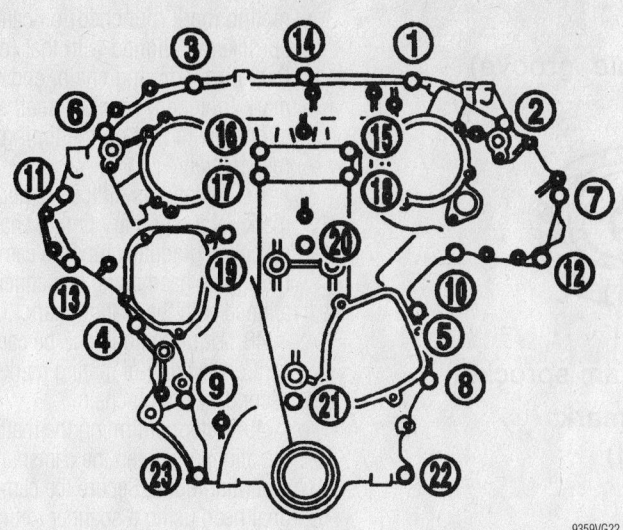

Rear timing case removal sequence—3.5L engine

9359VG22

Back side

Primary sprocket

Secondary sprocket

Front

Trigger teeth section (left bank only)

Primary and secondary sprockets—3.5L engine

9359VG23

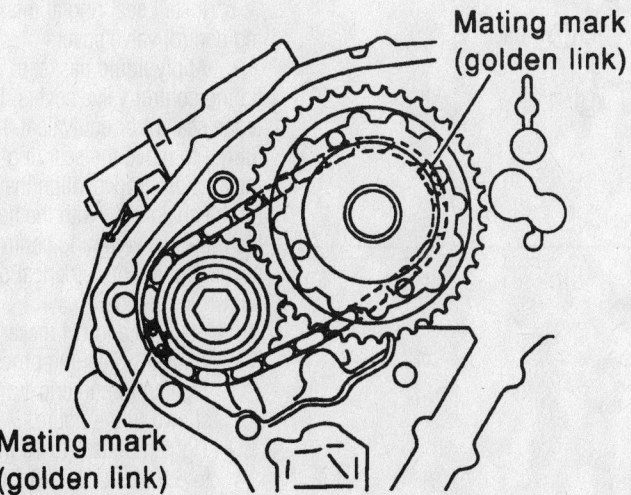

Mating mark (golden link)

Mating mark (golden link)

Secondary timing chain installed—3.5L engine

9359VG24

ets. Do not disassemble the intake camshaft sprocket. Avoid damaging the signal mark protrusion area at the front of the left bank intake camshaft sprocket. Keep it away from magnetized objects.

36. Remove lower chain guide.

37. Remove crankshaft sprocket.

38. Use a scraper to remove all traces of liquid gasket from front timing chain case. Remove old liquid gasket from the bolt hole and thread.

39. Use a scraper to remove all traces of liquid gasket from intake valve timing control valve cover.

To install:

40. Position crankshaft so that No. 1 piston is set at TDC on compression stroke.

41. Install crankshaft sprocket on crankshaft. Make sure that mating marks on crankshaft sprocket face front of engine.

42. Install lower chain guide on dowel pin, with front mark on the guide facing upside.

43. Press and shrink the secondary chain tensioner sleeve, and fix it using stopper pins. Lubricate threads and seat surfaces of camshaft sprocket bolts with new engine oil.

44. Install secondary timing chain and sprocket to one of the banks (Right bank shown in the figure) as described below.

   a. Align mating marks (golden links) on secondary timing chain with those (punched marks) on the intake and exhaust sprockets.

   b. Align camshaft knock pins with the sprocket groove and hole. Because camshaft sprocket mounting bolts are tightened in step 7, perform manual tightening to the extent necessary to keep camshaft knock pin from dislocating. Matching marks of the intake sprocket are on the back side of the secondary sprockets. There are two types of the marks; round and oval types, which should be used for right and left banks respectively.

   • Right bank: Round
   • Left bank: Oval

It may be difficult to visually check the dislocation of mating marks during and after installation. To make the matching easier, make a mating mark on the sprocket teeth in advance using paint.

45. Install secondary timing chain and sprocket to the other bank. Install primary timing chain at the same time. Installation of the secondary timing chain follows the procedure described in step 5.

46. Install primary timing chain so that

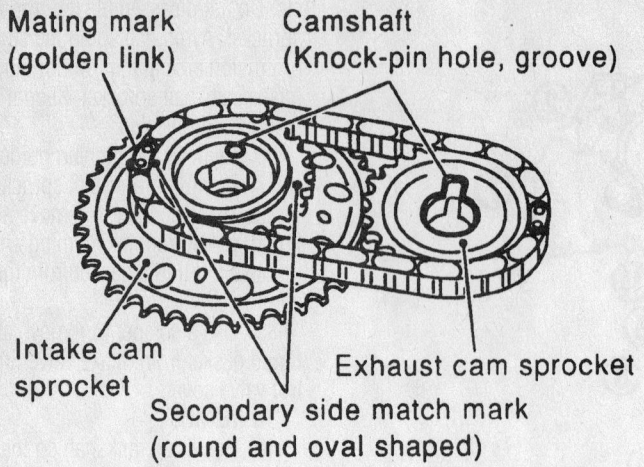

Intake sprocket mating marks—3.5L engine

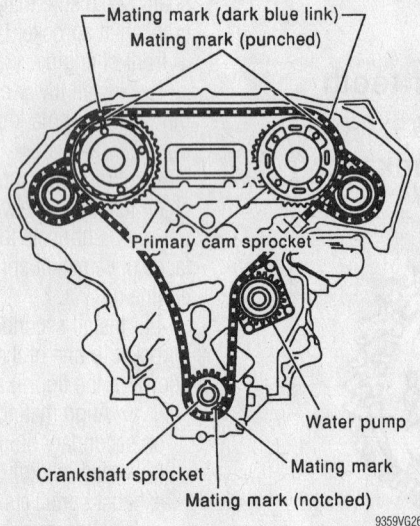

Primary timing chain installation—3.5L engine

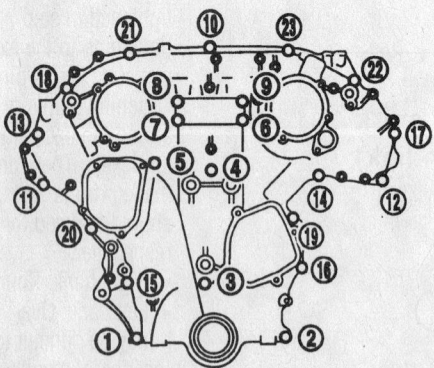

⟨symbol⟩ ① - ② 8 mm dia. bolts
25.5 - 31.4 N•m
(2.6 - 3.2 kg-m, 18.8 - 23.1 ft-lb)
③ - ㉔ 6 mm dia. bolts
11.8 - 13.7 N•m
(1.2 - 1.4 kg-m, 8.7 - 10.1 ft-lb)

9359VG27

Rear timing case installation—3.5L engine

mating mark (punched) on camshaft sprocket is aligned with that (dark blue link) on the timing chain, and mating mark (notched) on crankshaft sprocket is aligned with that on the timing chain, respectively.

47. When it is difficult to align mating marks of the primary timing chain with each sprocket, gradually turn the camshaft hexagonal head using a spanner so it is aligned with the mating mark.

48. During alignment, be careful to prevent dislocation of mating marks on the secondary timing chain.

49. After confirming the mating marks are aligned, tighten the camshaft sprocket mounting bolts. Secure the camshaft hexagonal head using a spanner to tighten mounting bolts.

50. Pull out the stopper pin from the secondary timing chain tensioner.

51. Install internal guide.

52. Install upper tension guide and slack guide.

53. Install timing chain tensioner, then remove the stopper pin. When installing the timing chain tensioner, engine oil should be applied to the oil hole and tensioner.

54. Install O-rings on rear timing chain case.

55. Apply liquid gasket to front timing chain case. Before installation, wipe off the protruding sealant.

56. Install rear case pin into dowel pin hole on front timing chain case.

57. Tighten bolts to the specified torque in order shown in the figure. Leave the bolts unattended for 30 minutes or more after tightening.

58. Install intake valve timing control valve cover.

   a. Install O-rings at front timing chain case.

   b. Install seal ring at intake valve timing control valve covers.

   c. Apply liquid gasket to intake valve timing control valve covers. Use RTV silicone sealant or equivalent. I Being careful not to move the seal ring from the installation groove, align the dowel pins on the chain case with the holes to install the intake valve timing control valve cover. Tighten in numerical order as shown in the figure.

59. Install RH and LH rocker covers. Rocker cover tightening procedure:

- Tighten in numerical order as shown in the figure.
- Tighten bolts 1 to 10 in that order to 6.9 to 8.8 N•m (0.7 to 0.9 kg-m, 61 to 78 in-lb).
- Then tighten bolts 1 to 10 as indi-

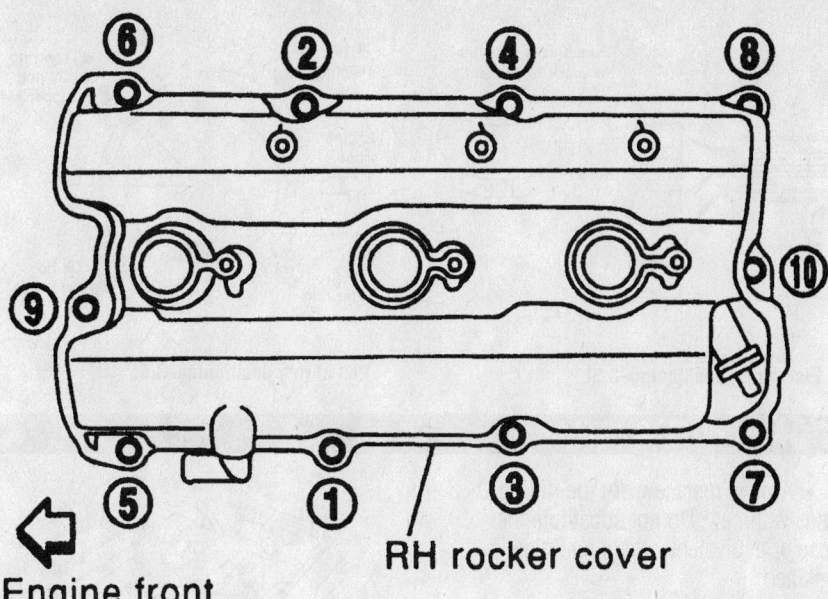

Right rocker cover installation—3.5L engine

9359VG28

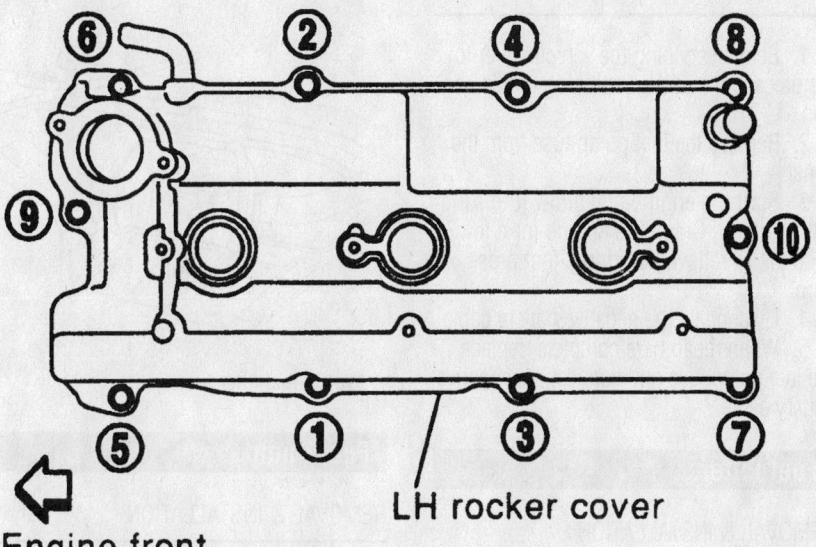

Left rocker cover installation—3.5L engine

9359VG29

cated in figure to 6.9 to 8.8 N·m (0.7 to 0.9 kg-m, 61 to 78 in-lb).

60. Hang engine using the right and left side engine slingers with a suitable hoist.

61. Set a suitable transmission jack under the suspension member.

62. Remove right and left side engine mounting nuts.

63. Remove right and left side suspension member bolts.

64. Install aluminum oil pan.

65. Set ring gear stopper using the mounting bolt hole. Be careful not to damage the signal plate teeth.

66. Install crankshaft pulley to crankshaft. Align pointer with TDC mark on crankshaft pulley.

67. Install crankshaft pulley bolt. Lubricate thread and seat surface of the bolt with new engine oil. Tighten to 39 to 49 N·m (4.0 to 5.0 kg-m, 29 to 36 ft-lb). Put a paint mark on the crankshaft pulley. Again tighten by turning 60° to 66°, about the angle from one hexagon bolt head corner to another.

68. Install camshaft position sensor (PHASE), crankshaft position sensors (REF)/(POS) and intake valve timing control position sensors.

69. Reinstall removed parts in the reverse order of removal. After starting engine, keep idling for three minutes. Then rev engine up to 3,000 rpm under no load to purge air from the high-pressure chamber of the chain tensioners. The engine may produce a rattling noise. This indicates that air still remains in the chamber and is not a matter of concern.

Piston and Ring

POSITIONING

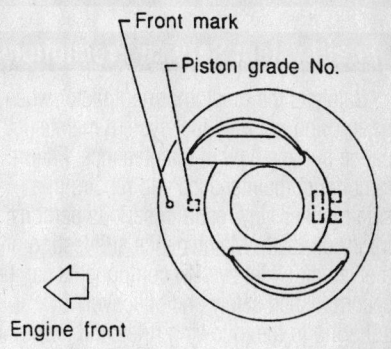

Piston ring positioning—3.3L engine

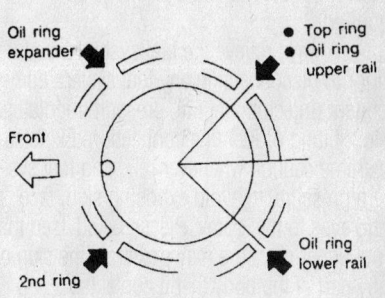

Piston ring end-gap spacing

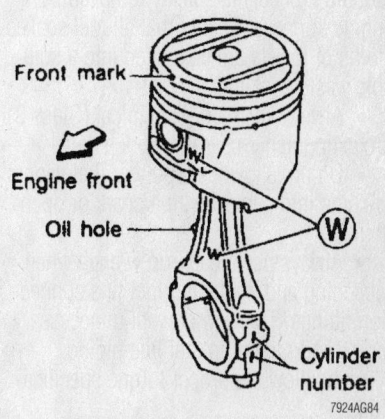

Piston and connecting rod positioning

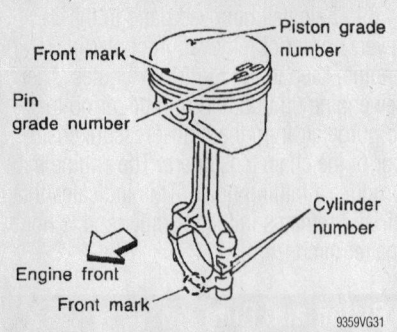

Piston and connecting rod positioning–3.5L

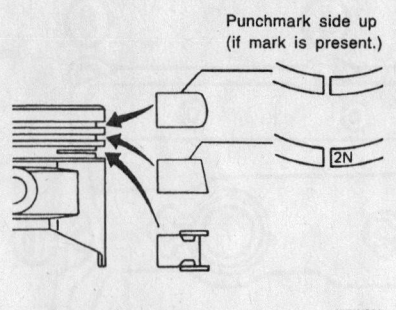

Piston ring positioning–3.5L

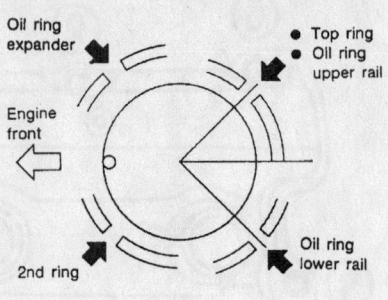

Piston ring positioning–3.5L

FUEL SYSTEM

Fuel System Service Precautions

Safety is the most important factor when performing not only fuel system maintenance but any type of maintenance. Failure to conduct maintenance and repairs in a safe manner may result in serious personal injury or death. Maintenance and testing of the vehicle's fuel system components can be accomplished safely and effectively by adhering to the following rules and guidelines.

• To avoid the possibility of fire and personal injury, always disconnect the negative battery cable unless the repair or test procedure requires that battery voltage be applied.

• Always relieve the fuel system pressure prior to disconnecting any fuel system component (injector, fuel rail, pressure regulator, etc.), fitting or fuel line connection. Exercise extreme caution whenever relieving fuel system pressure, to avoid exposing skin, face and eyes to fuel spray. Please be advised that fuel under pressure may penetrate the skin or any part of the body that it contacts.

• Always place a shop towel or cloth around the fitting or connection prior to loosening to absorb any excess fuel due to spillage. Ensure that all fuel spillage (should it occur) is quickly removed from engine surfaces. Ensure that all fuel soaked cloths or towels are deposited into a suitable waste container.

• Always keep a dry chemical (Class B) fire extinguisher near the work area.

• Do not allow fuel spray or fuel vapors to come into contact with a spark or open flame.

• Always use a back-up wrench when loosening and tightening fuel line connection fittings. This will prevent unnecessary stress and torsion to fuel line piping. Always follow the proper torque specifications.

• Always replace worn fuel fitting O-rings with new. Do not substitute fuel hose or equivalent, where fuel pipe is installed.

Fuel System Pressure

RELIEVING

1. Before servicing the vehicle, refer to the precautions in the beginning of this section.
2. Remove the fuel pump fuse from the panel.
3. Start the engine and allow it to run until it stalls. Crank the engine for a few seconds to relieve additional fuel pressure.
4. Disconnect the negative battery cable.
5. When repairs are complete, replace the fuel pump fuse and connect the negative battery cable.

Fuel Filter

REMOVAL & INSTALLATION

➡ **The fuel filter is located under the vehicle near the fuel tank.**

1. Before servicing the vehicle, refer to the precautions in the beginning of this section.
2. Relieve the fuel system pressure.
3. Remove or disconnect the following:
   • Fuel filter shield, if equipped
   • Fuel lines
   • Fuel filter from the bracket

To install:

4. Install or connect the following:
   • Fuel filter to the bracket
   • Fuel lines
   • Fuel filter shield, if equipped
5. Start the engine and check for leaks.

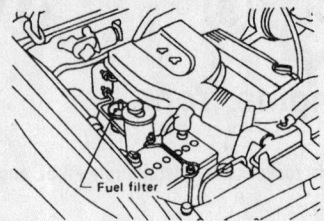

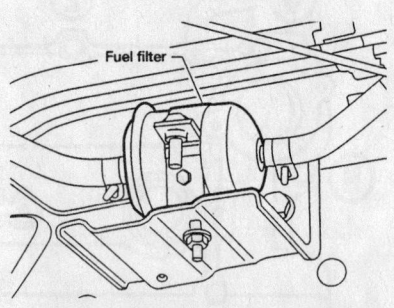

Typical fuel filter locations

Fuel Pump

REMOVAL & INSTALLATION

1. Before servicing the vehicle, refer to the precautions in the beginning of this section.
2. Relieve the fuel system pressure.
3. Remove or disconnect the following:
   • Negative battery cable
   • Access panel behind the rear seat
   • Fuel lines
   • Fuel pump and gauge harness connectors
   • Fuel gauge sender
   • Fuel pump

To install:

4. Install or connect the following:
   • Fuel pump
   • Fuel gauge sender. Tighten the screws to 17–23 inch lbs. (2.0–2.5 Nm).

- Fuel pump and gauge harness connectors
- Fuel lines
- Access panel
- Negative battery cable

5. Start the engine and check for leaks.

Fuel Injectors

REMOVAL & INSTALLATION

3.3L Engine

1. Before servicing the vehicle, refer to the precautions in the beginning of this section.
2. Drain the cooling system.
3. Relieve the fuel system pressure.
4. Remove or disconnect the following:
   - Negative battery cable
   - Air intake duct
   - Accelerator cable
   - Cruise control cable
   - Idle Air Control (IAC) valve connector
   - Throttle Position (TP) sensor and switch connectors
   - Ignition coil and power transistor connectors
   - Exhaust Gas Recirculation (EGR) Solenoid valve connector
   - EGR temperature sensor connector
   - Radiator hoses
   - Heater hoses
   - Positive Crankcase Ventilation (PCV) valve and hose
   - Evaporative Emissions (EVAP) canister vacuum and purge hoses
   - Brake booster vacuum hose
   - Fuel pressure regulator vacuum hose
   - EGR tube
   - Left bank injector connectors
   - Thermal transmitter
   - Upper intake manifold ground cable

- Breather pipe
- Upper intake manifold
- Fuel lines
- Right bank injector connectors
- Fuel supply manifold with the injectors attached
- Fuel injector caps
- Fuel injectors

To install:

➡**Use new insulators and O-ring seals for assembly.**

5. Install or connect the following:
   - Fuel injectors
   - Fuel injector caps. Tighten the screws to 26–34 inch lbs. (3–4 Nm).
   - Fuel supply manifold with the injectors attached. Tighten the bolts to 96–132 inch lbs. (11–15 Nm).
   - Right bank injector connectors
   - Fuel lines
   - Upper intake manifold
   - Breather pipe
   - Upper intake manifold ground cable
   - Thermal transmitter
   - Left bank injector connectors
   - EGR tube
   - Fuel pressure regulator vacuum hose
   - Brake booster vacuum hose
   - EVAP canister vacuum and purge hoses
   - PCV valve and hose
   - Heater hoses
   - Radiator hoses
   - EGR temperature sensor connector
   - EGR Solenoid valve connector
   - Ignition coil and power transistor connectors
   - TP sensor and switch connectors
   - IAC valve connector
   - Cruise control cable
   - Accelerator cable
   - Air intake duct
   - Negative battery cable

6. Fill the cooling system.
7. Start the engine and check for leaks.

3.5L

1. Release fuel pressure to zero.
2. Remove intake manifold collector.
3. Remove fuel tube assemblies in numerical sequence as shown in the figure at left.
4. Expand and remove clips securing fuel injectors.
5. Extract fuel injectors straight from fuel tubes.

➡**Be careful not to damage injector nozzles during removal. Do not bump or drop fuel injectors.**

6. Carefully install O-rings, including the one used with the pressure regulator. Lubricate O-rings with a smear of engine oil.

➡**Be careful not to damage O-rings with service tools, finger nails or clips. Do not expand or twist O-rings. Discard old clips; replace with new ones.**

7. Position clips in grooves on fuel injectors. Make sure that protrusions of fuel injectors are aligned with cutouts of clips after installation.
8. Align protrusions of fuel tubes with those of fuel injectors. Insert fuel injectors straight into fuel tubes.
9. After properly inserting fuel injectors, check to make sure that fuel tube protrusions are engaged with those of fuel injectors, and that flanges of fuel tubes are engaged with clips.
10. Tighten fuel tube assembly mounting nuts in numerical sequence (indicated in the figure at left) and in two stages. Tighten to:
    Step 1: 84–96 inch lbs.
    Step 2: 16–19 ft. lbs.
11. Install all parts removed in reverse order of removal.

DRIVE TRAIN

Manual Transmission

REMOVAL & INSTALLATION

2 Wheel Drive

1. Before servicing the vehicle, refer to the precautions in the beginning of this section.
2. Remove or disconnect the following:
   - Negative battery cable
   - Shift lever
   - Crankshaft Position (CKP) sensor
   - Clutch slave cylinder
   - Vehicle Speed (VSS) sensor connector
   - Back-up lamp switch connector
   - Park/Neutral Position (PNP) switch connector
   - Rear Heated Oxygen (HO2S) sensor connector
   - Starter motor
   - Driveshaft
   - Exhaust mounting bracket
   - Transmission mount and crossmember. Support the transmission.
   - Transmission flange bolts
   - Transmission

→The transmission flange bolts vary in length. Note their positions for assembly.

To install:

3. Apply sealant to the transmission flange, engine block and engine rear plate as shown.
4. Install or connect the following:
   - Transmission. Tighten the large bolts to 29–36 ft. lbs. (39–49 Nm) and the small bolts to 12–16 ft. lbs. (16–22 Nm).
   - Transmission mount and crossmember. Tighten the mount and crossmember fasteners to 30–38 ft. lbs. (41–52 Nm).
   - Exhaust mounting bracket
   - Driveshaft
   - Starter motor
   - HO2S sensor connector
   - PNP switch connector
   - Back-up lamp switch connector
   - VSS sensor connector
   - Clutch slave cylinder
   - CKP sensor
   - Shift lever
   - Negative battery cable

4 Wheel Drive

PATHFINDER

1. Before servicing the vehicle, refer to the precautions in the beginning of this section.
2. Remove or disconnect the following:
   - Negative battery cable
   - Shift lever
   - Transfer case select lever
   - Crankshaft Position (CKP) sensor
   - Clutch slave cylinder
   - Vehicle Speed (VSS) sensor connector
   - Back-up lamp switch connector
   - Park/Neutral Position (PNP) switch connector
   - Rear Heated Oxygen (HO2S) sensor connector
   - Starter motor
   - Front and rear driveshafts
   - Exhaust front pipes
   - Exhaust center pipe
   - Transmission mount and crossmember. Support the transmission.
   - Transmission flange bolts
   - Transmission

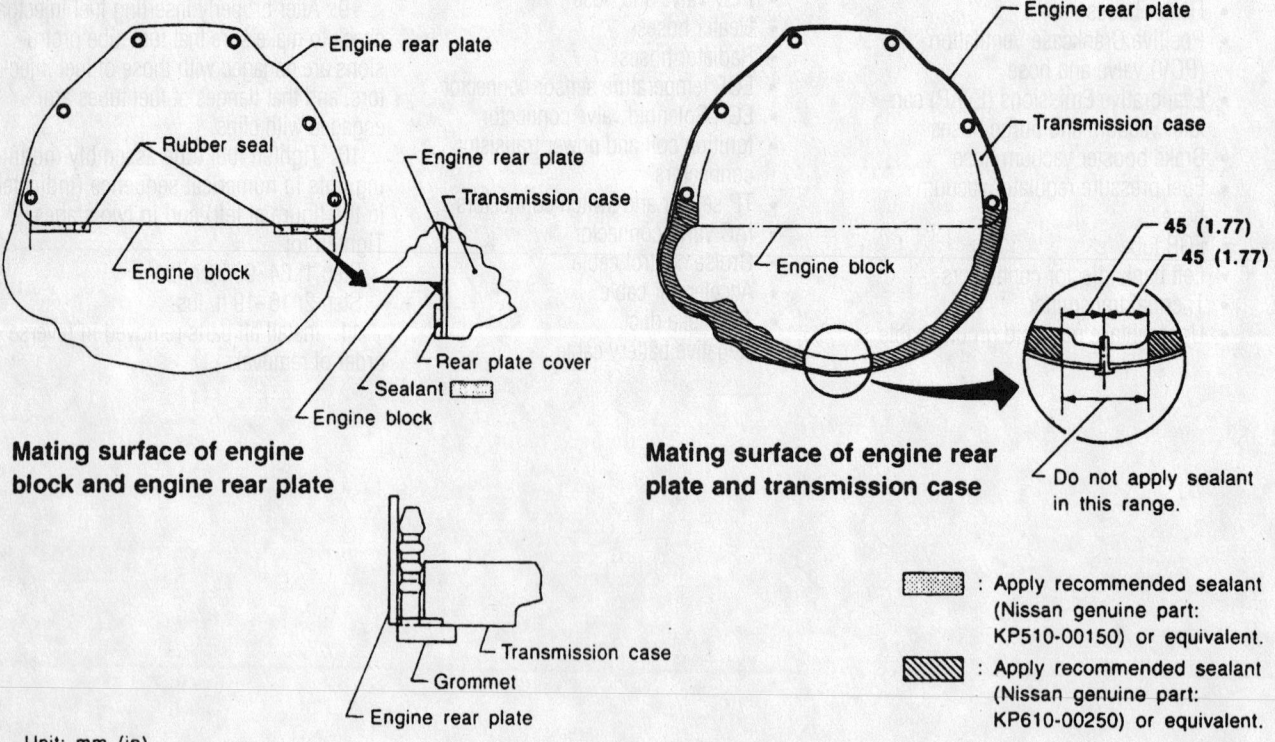

Mating surface of engine block and engine rear plate

Mating surface of engine rear plate and transmission case

45 (1.77)
45 (1.77)

Do not apply sealant in this range.

▭ : Apply recommended sealant (Nissan genuine part: KP510-00150) or equivalent.

▨ : Apply recommended sealant (Nissan genuine part: KP610-00250) or equivalent.

Unit: mm (in)

Apply sealant to the indicated areas between the engine block, transmission and engine rear plate—4 Wheel Drive shown

7924VG61

➡ **The transmission flange bolts vary in length. Note their positions for assembly.**

To install:

3. Apply sealant to the transmission flange, engine block, and engine rear plate as shown.

4. Install or connect the following:
- Transmission. Tighten the large

bolts to 29–36 ft. lbs. (39–49 Nm) and the small bolts to 22–29 ft. lbs. (29–39 Nm).
- Transmission mount and crossmember. Tighten the mount and crossmember fasteners to 30–38 ft. lbs. (41–52 Nm).
- Exhaust center pipe
- Exhaust front pipes
- Front and rear driveshafts

- Starter motor
- HO$_2$S sensor connector
- PNP switch connector
- Back-up lamp switch connector
- VSS sensor connector
- Clutch slave cylinder
- CKP sensor
- Transfer case select lever
- Shift lever
- Negative battery cable

🔧 : N•m (kg-m, in-lb)

🔧 : N•m (kg-m, ft-lb)

🛢 1 : Fill multi-purpose grease up.

🛢 2 : Apply multi-purpose grease.

*1 : Securely bend pawls during assembly. Be careful not to damage boot.

*2 : Do not touch boot with a sharp-pointed or a hard tool as it breaks easily.

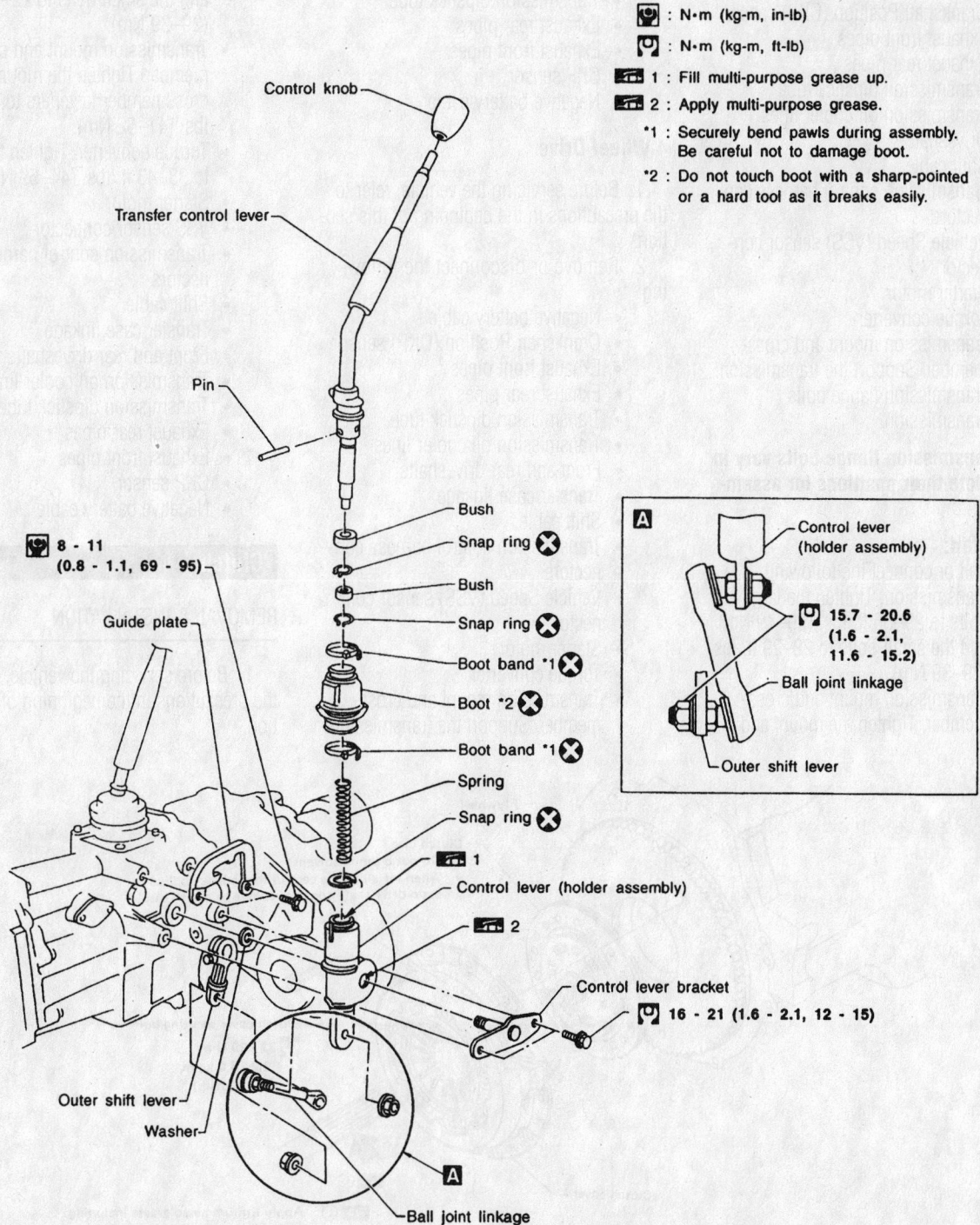

Exploded view of the transfer case shifter lever and related components—Pathfinder 4WD

7924VG60

Automatic Transmission

REMOVAL & INSTALLATION

2 Wheel Drive

1. Before servicing the vehicle, refer to the precautions in the beginning of this section.

2. Remove or disconnect the following:
- Negative battery cable
- Crankshaft Position (CKP) sensor
- Exhaust front pipes
- Exhaust rear pipes
- Transmission dipstick tube
- Transmission oil cooler lines
- Driveshaft
- Shift cable
- Transmission control harness connectors
- Vehicle Speed (VSS) sensor connector
- Starter motor
- Torque converter
- Transmission mount and crossmember. Support the transmission.
- Transmission flange bolts
- Transmission

➡ The transmission flange bolts vary in length. Note their positions for assembly.

To install:

3. Install or connect the following:
- Transmission. Tighten the large bolts to 29–36 ft. lbs. (39–49 Nm) and the small bolts to 22–29 ft. lbs. (29–39 Nm).
- Transmission mount and crossmember. Tighten the mount and

crossmember fasteners to 30–38 ft. lbs. (41–52 Nm).
- Torque converter. Tighten the bolts to 33–43 ft. lbs. (44–59 Nm).
- Starter motor
- VSS sensor connector
- Transmission control harness connectors
- Shift cable
- Driveshaft
- Transmission oil cooler lines
- Transmission dipstick tube
- Exhaust rear pipes
- Exhaust front pipes
- CKP sensor
- Negative battery cable

4 Wheel Drive

1. Before servicing the vehicle, refer to the precautions in the beginning of this section.

2. Remove or disconnect the following:
- Negative battery cable
- Crankshaft Position (CKP) sensor
- Exhaust front pipes
- Exhaust rear pipes
- Transmission dipstick tube
- Transmission oil cooler lines
- Front and rear driveshafts
- Transfer case linkage
- Shift cable
- Transmission control harness connectors
- Vehicle Speed (VSS) sensor connector
- Starter motor
- Torque converter
- Transmission mount and crossmember. Support the transmission.

- Transmission flange bolts
- Transmission

➡ The transmission flange bolts vary in length. Note their positions for assembly.

To install:

3. Install or connect the following:
4. Install or connect the following:
- Transmission. Tighten the large bolts to 29–36 ft. lbs. (39–49 Nm) and the small bolts to 22–29 ft. lbs. (29–39 Nm).
- Transmission mount and crossmember. Tighten the mount and crossmember fasteners to 30–38 ft. lbs. (41–52 Nm).
- Torque converter. Tighten the bolts to 33–43 ft. lbs. (44–59 Nm).
- Starter motor
- VSS sensor connector
- Transmission control harness connectors
- Shift cable
- Transfer case linkage
- Front and rear driveshafts
- Transmission oil cooler lines
- Transmission dipstick tube
- Exhaust rear pipes
- Exhaust front pipes
- CKP sensor
- Negative battery cable

Clutch

REMOVAL & INSTALLATION

1. Before servicing the vehicle, refer to the precautions in the beginning of this section.

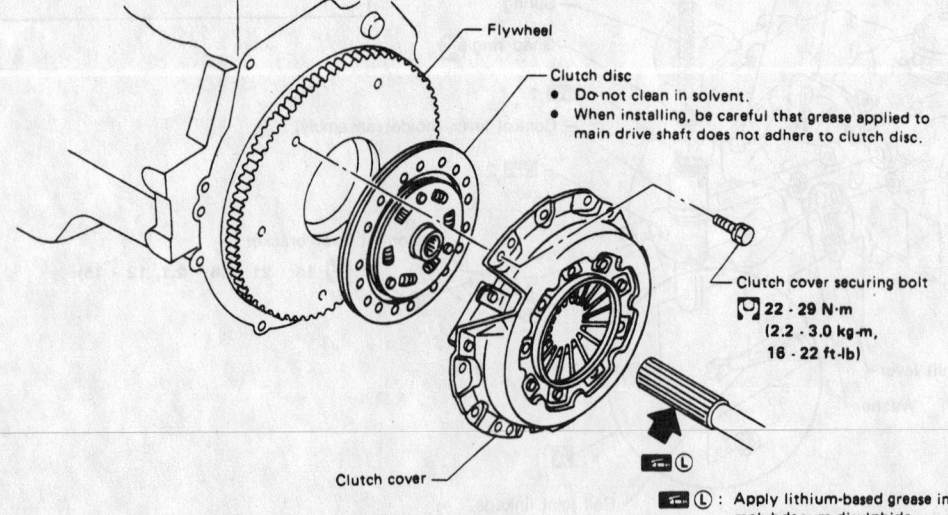

- Flywheel
- Clutch disc
  - Do not clean in solvent.
  - When installing, be careful that grease applied to main drive shaft does not adhere to clutch disc.
- Clutch cover securing bolt
  22 - 29 N·m
  (2.2 - 3.0 kg-m,
  16 - 22 ft-lb)
- Clutch cover
- (L): Apply lithium-based grease including molybdenum disulphide.

7924VG63

Exploded view of the pressure plate and clutch disc and related components—all models

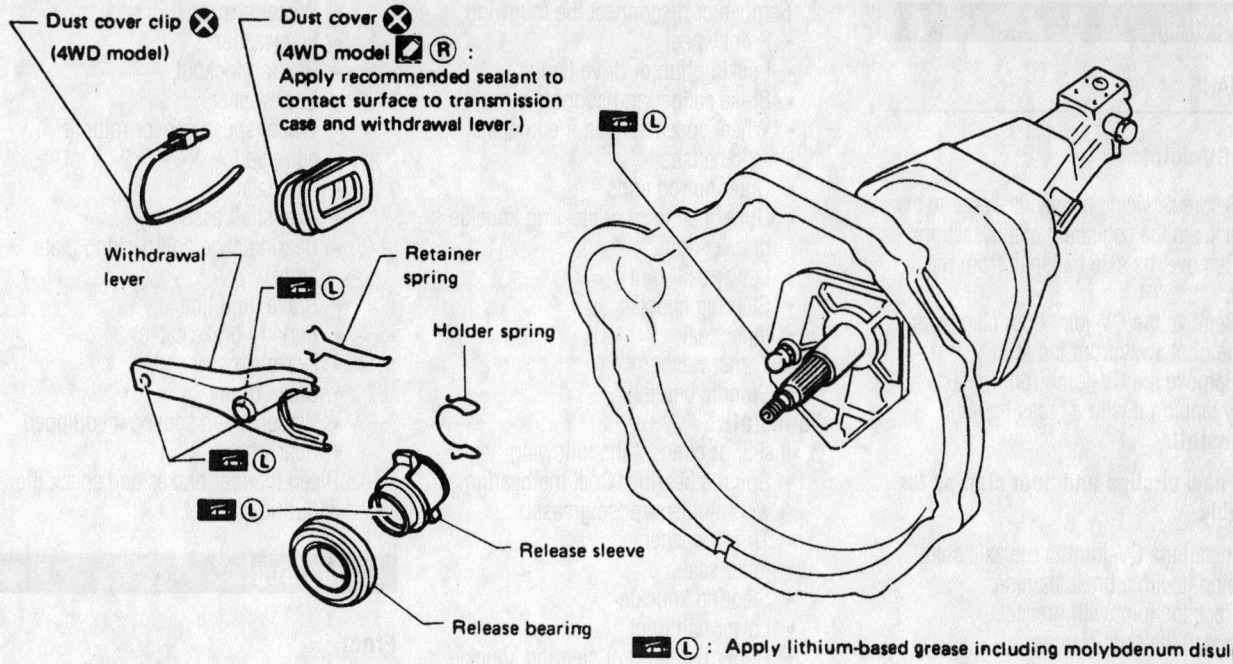

Clutch release mechanism exploded view—all models

2. Remove or disconnect the following:
- Negative battery cable
- Transmission
- Pressure plate. Loosen the bolts evenly in ½ turn steps.
- Clutch disc

To install:
3. Install or connect the following:
- Clutch disc and pressure plate. Tighten the pressure plate bolts evenly in ½ turns to 16–22 ft. lbs. (22–29 Nm).
- Transmission
- Negative battery cable

Hydraulic Clutch System

BLEEDING

1. Before servicing the vehicle, refer to the precautions in the beginning of this section.
2. Fill the clutch master cylinder reservoir with fresh clean brake fluid.
3. Connect a clear plastic hose to the air bleeder.
4. Have an assistant pump the clutch pedal slowly several times and hold it depressed.
5. Open the slave cylinder bleeder screw and allow air to escape.
6. Close the bleeder screw before releasing the clutch pedal.
7. Repeat until all air is purged from the clutch hydraulic system.
8. Refill the reservoir to the full mark.

Transfer Case Assembly

REMOVAL & INSTALLATION

1. Before servicing the vehicle, refer to the precautions in the beginning of this section.
2. Remove or disconnect the following:
- Negative battery cable
- Front and rear driveshafts
- Torsion bars and mounts
- Rear torsion bar crossmember
- Exhaust front pipes
- Exhaust rear pipes
- Vehicle Speed (VSS) sensor connector
- Transfer case shift linkage
- Transfer case neutral switch connector
- 4 wheel drive switch connector
- Vent hose
- Transfer case flange bolts
- Transfer case

To install:
3. Install or connect the following:
- Transfer case. Tighten the flange bolts to 23–30 ft. lbs. (31–41 Nm).
- Vent hose
- 4 wheel drive switch connector
- Transfer case neutral switch connector
- Transfer case shift linkage
- VSS sensor connector
- Exhaust rear pipes
- Exhaust front pipes

- Rear torsion bar crossmember
- Torsion bars and mounts
- Front and rear driveshafts
- Negative battery cable

Halfshaft

REMOVAL & INSTALLATION

1. Before servicing the vehicle, refer to the precautions in the beginning of this section.
2. Remove or disconnect the following:
- Front wheel
- Wheel speed sensor, if equipped
- Locking hub or drive flange
- Snapring
- Spindle washer
- Thrust washer
- Inner CV-joint bolts
- Axle halfshaft. Separate the stub shaft from the spindle by tapping with a plastic hammer.

To install:
3. Install or connect the following:
- Axle halfshaft. Guide the stub shaft into the spindle and tighten the inner CV-joint bolts to 25–33 ft. lbs. (34–44 Nm).
- Thrust washer
- Spindle washer
- Snapring
- Locking hub or drive flange
- Wheel speed sensor, if equipped
- Front wheel

CV-Joints

OVERHAUL

Outer CV-Joint

1. Before servicing the vehicle, refer to the precautions in the beginning of this section.
2. Remove the axle halfshaft from the vehicle.
3. Remove the CV-joint boot clamps and push the boot away from the joint.
4. Remove the CV-joint from the axle shaft by tapping it with a brass hammer.

To install:

→**Use new circlips and boot clamps for assembly.**

5. Install the CV-joint to the axle shaft by tapping it with a brass hammer.
6. Pack the joint with grease.
7. Install the boot clamps.
8. Install the axle halfshaft to the vehicle.

Inner Tri-Pot Joint

1. Before servicing the vehicle, refer to the precautions in the beginning of this section.
2. Remove the axle halfshaft from the vehicle.
3. Remove the plug seal by tapping around the joint housing flange with a brass hammer.
4. Remove or disconnect the following:
   - CV-joint boot clamps
   - Snapring
   - Spider assembly
   - CV-joint housing
   - CV-joint boot

To install:

→**Use new snaprings and plug seals for assembly.**

5. Install or connect the following:
   - CV-joint boot
   - CV-joint housing
   - Spider assembly
   - Snapring. Pack the joint with grease.
   - CV-joint boot clamps
   - Plug seal
6. Install the axle halfshaft to the vehicle.

Spindle Bearings

REMOVAL, PACKING AND INSTALLATION

1. Before servicing the vehicle, refer to the precautions in the beginning of this section.

2. Remove or disconnect the following:
   - Front wheel
   - Locking hub or drive flange
   - Brake caliper and support
   - Wheel speed sensor, if equipped
   - Axle halfshaft
   - Outer tie rod ends
   - Upper ball joint or steering knuckle bracket bolts
   - Lower ball joint
   - Steering knuckle
   - Inner seal
   - Thrust washer
   - Spindle bearing

To install:

3. Install or connect the following:
   - Spindle bearing. Coat the bearing with multi-purpose grease.
   - Thrust washer
   - Inner seal
   - Steering knuckle
   - Lower ball joint
   - Upper ball joint or steering knuckle bracket bolts
   - Outer tie rod ends
   - Axle halfshaft
   - Wheel speed sensor, if equipped
   - Brake caliper and support
   - Locking hub or drive flange
   - Front wheel

Axle Shaft, Bearing and Seal

REMOVAL & INSTALLATION

1. Before servicing the vehicle, refer to the precautions in the beginning of this section.
2. Remove or disconnect the following:
   - Rear wheel
   - Wheel speed sensor, if equipped
   - Brake drum
   - Brake shoes
   - Parking brake cable
   - Brake fluid line
   - Bearing cage and backing plate bolts
   - Axle shaft assembly
   - Axle seal
   - Wheel speed sensor rotor, if equipped
   - Lockwasher
   - Bearing locknut
   - Flat washer
   - Wheel bearing
   - Wheel bearing cage grease seal

To install:

→**Use new lockwashers, seals and bearings for assembly.**

3. Install or connect the following:
   - Wheel bearing cage grease seal
   - Wheel bearing
   - Flat washer
   - Bearing locknut
   - Lockwasher
   - Wheel speed sensor rotor, if equipped
   - Axle seal
   - Axle shaft assembly
   - Bearing cage and backing plate bolts
   - Brake fluid line
   - Parking brake cable
   - Brake shoes
   - Brake drum
   - Wheel speed sensor, if equipped
   - Rear wheel

4. Bleed the rear brakes and check the rear axle lubricant level.

Pinion Seal

Front

1. Before servicing the vehicle, refer to the precautions in the beginning of this section.
2. Remove or disconnect the following:
   - Driveshaft
   - Front wheels
   - Front brake calipers

→**The front brake calipers must be removed so that there is no additional drag when measuring pinion bearing preload.**

3. Use an inch lb. torque wrench and measure the amount of torque required to maintain pinion rotation through several revolutions.
4. Remove or disconnect the following:
   - Pinion flange
   - Oil seal

To install:

5. Install or connect the following:
   - Pinion seal
   - Pinion flange
6. Rotate the pinion flange occasionally while tightening the flange nut to make sure the pinion bearings seat correctly.
7. Take frequent bearing preload torque readings. Tighten the flange nut to achieve the preload torque readings originally recorded. Do not exceed 137–217 ft. lbs. (186–294 Nm) torque when tightening the pinion flange nut.

❈ CAUTION

If the bearing preload can not be achieved at the specified torque, remove the pinion bearing and install a new adjustment spacer.

8. Install or connect the following:
   - Front brake calipers
   - Front wheels
   - Driveshaft. Tighten the fasteners to 29–33 ft. lbs. (39–44 Nm).
9. Fill the differential with gear lubricant and check for leaks.

Rear

2 WHEEL DRIVE

1. Before servicing the vehicle, refer to the precautions in the beginning of this section.
2. Remove or disconnect the following:
   - Driveshaft
   - Rear wheels
   - Brake drums

➡**The rear brake drums must be removed so that there is no additional drag when measuring pinion bearing preload.**

3. Use an inch lb. torque wrench and measure the amount of torque required to maintain pinion rotation through several revolutions.
4. Remove or disconnect the following:
   - Pinion flange
   - Wheel speed sensor and rotor, if equipped
   - Oil seal
   - Pinion bearing
   - Collapsible spacer

To install:

➡**Use a new collapsible spacer and wheel speed sensor rotor for assembly.**

5. Install or connect the following:

   - Collapsible spacer
   - Pinion bearing
   - Pinion seal
   - Pinion flange
6. Rotate the pinion flange occasionally while tightening the flange nut to make sure the pinion bearings seat correctly.
7. Take frequent bearing preload torque readings. Tighten the flange nut to achieve the preload torque readings originally recorded. Do not exceed 137–217 ft. lbs. (186–294 Nm) torque when tightening the pinion flange nut.

❊❊ CAUTION

Never loosen the pinion nut to reduce bearing preload. If it is necessary to reduce bearing preload, install a new collapsible spacer.

8. Install or connect the following:
   - Brake drums
   - Rear wheels
   - Driveshaft. Tighten the fasteners to 58–65 ft. lbs. (78–88 Nm).
9. Fill the differential with gear lubricant and check for leaks.

4 WHEEL DRIVE

1. Before servicing the vehicle, refer to the precautions in the beginning of this section.
2. Remove or disconnect the following:
   - Driveshaft
   - Rear wheels
   - Brake drums

➡**The rear brake drums must be removed so that there is no additional**

drag when measuring pinion bearing preload.

3. Use an inch lb. torque wrench and measure the amount of torque required to maintain pinion rotation through several revolutions.
4. Remove or disconnect the following:
   - Pinion flange
   - Oil seal

To install:

5. Install or connect the following:
   - Pinion seal
   - Pinion flange
6. Rotate the pinion flange occasionally while tightening the flange nut to make sure the pinion bearings seat correctly.
7. Take frequent bearing preload torque readings. Tighten the flange nut to achieve the preload torque readings originally recorded. Do not exceed 137–217 ft. lbs. (186–294 Nm) torque when tightening the pinion flange nut.

❊❊ CAUTION

If the bearing preload can not be achieved at the specified torque, remove the pinion bearing and install a new adjustment spacer.

8. Install or connect the following:
   - Brake drums
   - Rear wheels
   - Driveshaft. Tighten the fasteners to 58–65 ft. lbs. (78–88 Nm).
9. Fill the differential with gear lubricant and check for leaks.

STEERING AND SUSPENSION

Air Bag

❊❊ CAUTION

Some vehicles are equipped with an air bag system. The system must be disarmed before performing service on, or around, system components, the steering column, instrument panel components, wiring and sensors. Failure to follow the safety precautions and the disarming procedure could result in accidental air bag deployment, possible injury and unnecessary system repairs.

PRECAUTIONS

Several precautions must be observed when handling the inflator module to avoid

accidental deployment and possible personal injury.
   • Never carry the inflator module by the wires or connector on the underside of the module.
   • When carrying a live inflator module, hold securely with both hands, and ensure that the bag and trim cover are pointed away.
   • Place the inflator module on a bench or other surface with the bag and trim cover facing up.
   • With the inflator module on the bench, never place anything on or close to the module which may be thrown in the event of an accidental deployment.

DISARMING

To disarm the **SRS** system turn the ignition switch to the **OFF** position. Then,

disconnect both battery cables starting with the negative cable first and wait at least 3 minutes after the cables are disconnected.

To rearm the **SRS** system, turn the ignition switch to the **OFF** position. Connect both battery cables starting with the positive cable first.

Recirculating Ball Power Steering Gear

REMOVAL & INSTALLATION

1. Before servicing the vehicle, refer to the precautions in the beginning of this section.
2. Remove or disconnect the following:
   - Pitman arm

- Steering column intermediate shaft
- Power steering hoses
- Steering gear

To install:

3. Install or connect the following:
- Steering gear. Tighten the bolts to 62–71 ft. lbs. (84–96 Nm).
- Power steering hoses. Tighten the banjo fittings to 29–38 ft. lbs. (39–51 Nm).
- Steering column intermediate shaft. Tighten the pinch bolt to 17–22 ft. lbs. (24–29 Nm).
- Pitman arm. Tighten the nut to 174–195 ft. lbs. (235–265 Nm).

4. Check the wheel alignment and adjust, as necessary.

Rack and Pinion Steering Gear

REMOVAL & INSTALLATION

1. Before servicing the vehicle, refer to the precautions in the beginning of this section.

2. Remove or disconnect the following:
- Front wheels
- Outer tie rod ends
- Steering shaft coupler
- Power steering hoses
- Steering gear

To install:

3. Install or connect the following:
- Steering gear. Tighten the bolts to 101 ft. lbs. (137 Nm).

- Power steering hoses. Tighten the fittings to 25 ft. lbs. (35 Nm).
- Steering shaft coupler. Tighten the bolt to 22 ft. lbs. (29 Nm).
- Outer tie rod ends. Tighten the nuts to 65 ft. lbs. (88 Nm).
- Front wheels

Strut

REMOVAL & INSTALLATION

Front

1. Before servicing the vehicle, refer to the precautions in the beginning of this section.

When installing rubber parts, final tightening must be carried out under unladen condition* with tires on ground.
Fuel, radiator coolant and engine oil full.
Spare tire, jack, hand tools and mats in designated positions.

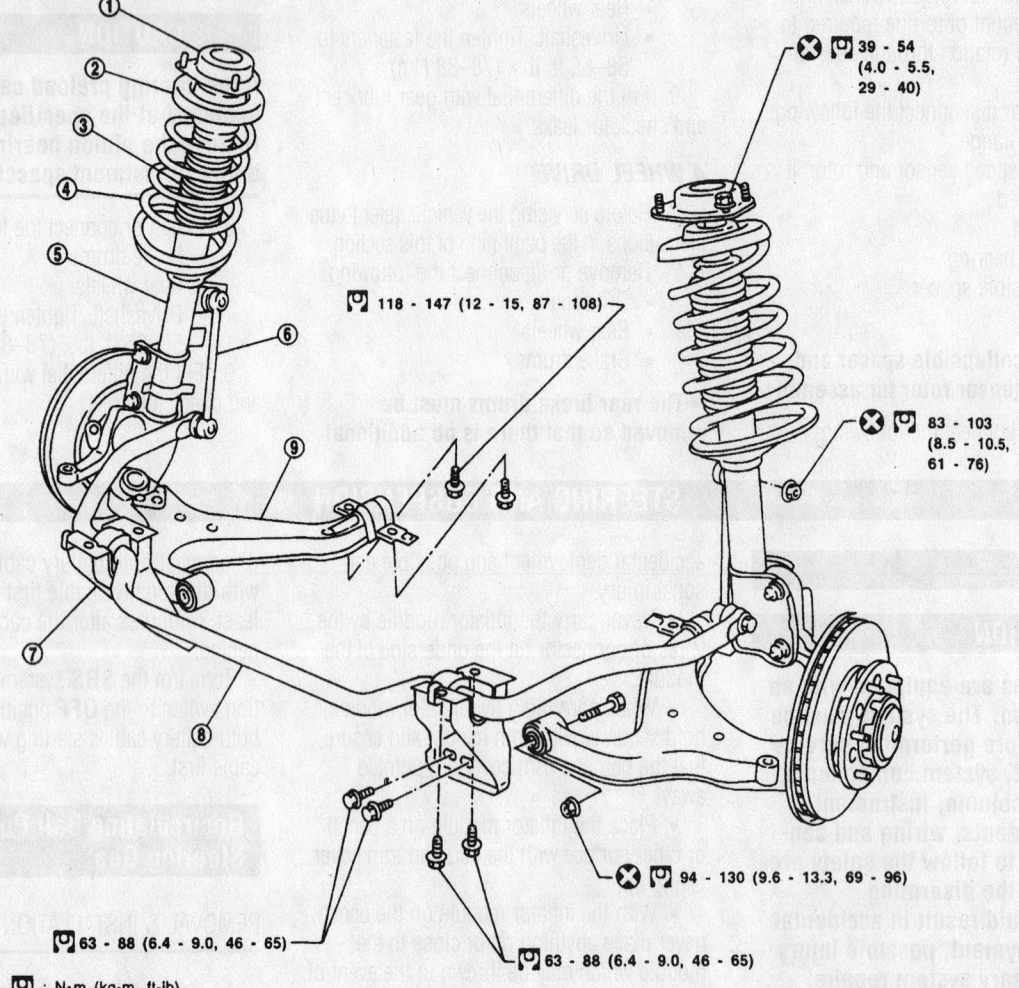

118 - 147 (12 - 15, 87 - 108)

39 - 54 (4.0 - 5.5, 29 - 40)

83 - 103 (8.5 - 10.5, 61 - 76)

94 - 130 (9.6 - 13.3, 69 - 96)

63 - 88 (6.4 - 9.0, 46 - 65)

63 - 88 (6.4 - 9.0, 46 - 65)

: N·m (kg-m, ft-lb)

① Strut mounting insulator	④ Coil spring	⑦ Bracket
② Spring upper seat	⑤ Strut assembly	⑧ Stabilizer bar
③ Bound bumper	⑥ Stabilizer connecting rod	⑨ Transverse link

7924VG66

Exploded view of the front suspension—2WD Pathfinder shown

2. Remove or disconnect the following:
- Front wheel
- Stabilizer bar link
- Steering knuckle bracket bolts
- Upper strut mount nuts
- Strut

To install:

➡ **Use new nuts and bolts for assembly.**

3. Install or connect the following:
- Strut. Tighten the upper strut mount nuts to 29–40 ft. lbs. (39–54 Nm) and the knuckle bracket bolts to 111–122 ft. lbs. (151–165 Nm).
- Stabilizer bar link. Tighten the nut to 61–76 ft. lbs. (83–103 Nm).
- Front wheel

4. Check the wheel alignment and adjust, as necessary.

Shock Absorber

REMOVAL & INSTALLATION

Rear

1. Before servicing the vehicle, refer to the precautions in the beginning of this section.
2. Support the rear axle.
3. Remove or disconnect the following:

- Lower shock absorber bolt
- Upper shock absorber bolt
- Shock absorber

To install:

➡ **Use new fasteners for assembly.**

4. Install the shock absorber and tighten the bolts to 49–65 ft. lbs. (67–88 Nm).

Coil Spring

REMOVAL & INSTALLATION

Front

1. Before servicing the vehicle, refer to the precautions in the beginning of this section.
2. Remove the strut assembly.
3. Compress the coil spring and remove the piston rod nut.
4. Remove or disconnect the following:
- Upper strut mount
- Strut mount bracket
- Upper strut bearing
- Spring upper seat
- Coil spring

To install:

➡ **Use new fasteners for assembly.**

5. Install or connect the following:

- Coil spring
- Spring upper seat
- Upper strut bearing
- Strut mount bracket
- Upper strut mount. Tighten the piston rod nut to 43–58 ft. lbs. (59–78 Nm).

6. Remove the spring compressor and install the strut assembly to the vehicle.

7. Check the wheel alignment and adjust, as necessary.

Rear

1. Before servicing the vehicle, refer to the precautions in the beginning of this section.
2. Support the vehicle at the frame.
3. Support the axle with a floor jack.
4. Remove or disconnect the following:
- Rear wheels
- Shock absorbers
- Stabilizer bar links
- Lateral control rod
- Coil springs

To install:

➡ **Use new fasteners for assembly.**

5. Install or connect the following:
- Coil springs
- Lateral control rod. Tighten the nut to 80–94 ft. lbs. (108–127 Nm).

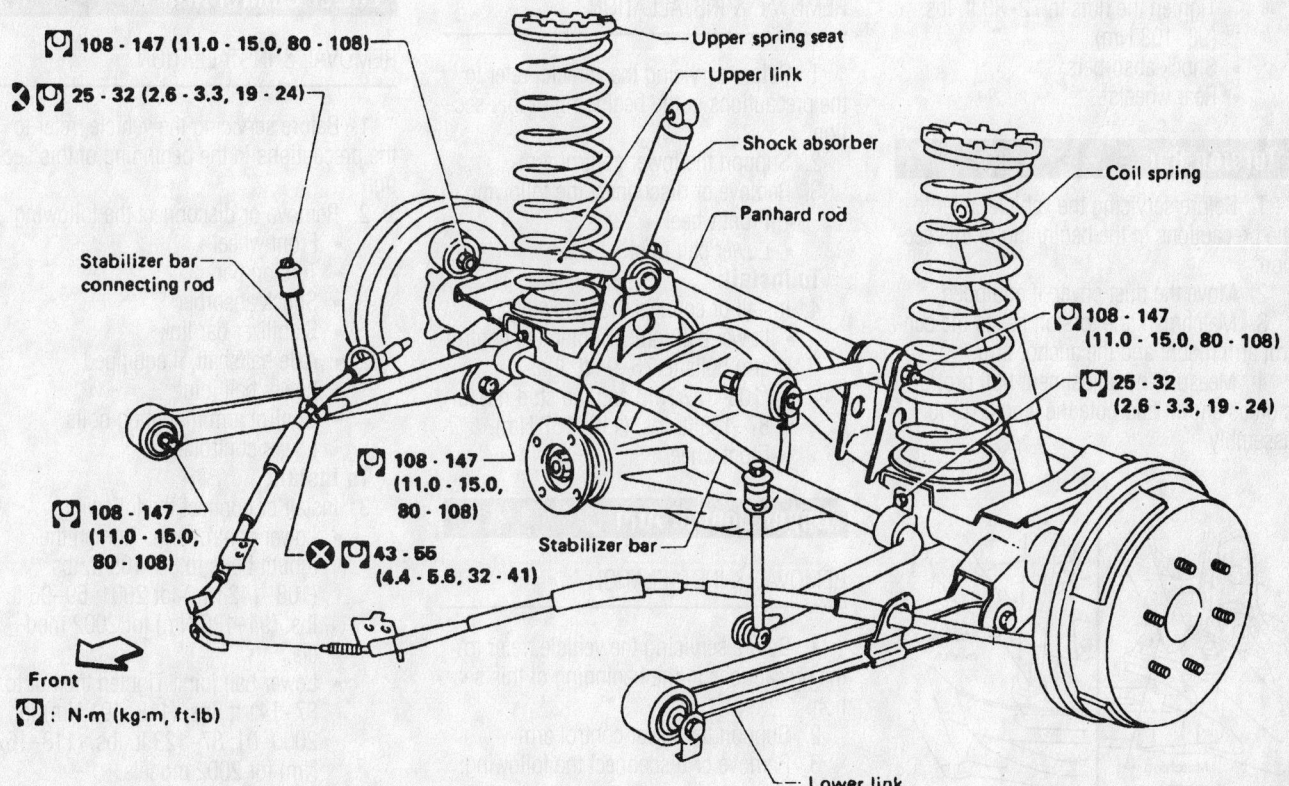

108 - 147 (11.0 - 15.0, 80 - 108)

25 - 32 (2.6 - 3.3, 19 - 24)

Stabilizer bar connecting rod

Upper spring seat

Upper link

Shock absorber

Panhard rod

Coil spring

108 - 147 (11.0 - 15.0, 80 - 108)

25 - 32 (2.6 - 3.3, 19 - 24)

108 - 147 (11.0 - 15.0, 80 - 108)

43 - 55 (4.4 - 5.6, 32 - 41)

Stabilizer bar

108 - 147 (11.0 - 15.0, 80 - 108)

Front

: N·m (kg-m, ft-lb)

Lower link

7924VG69

Rear suspension component identification—Pathfinder

- Stabilizer bar links. Tighten the nuts to 30–35 ft. lbs. (41–47 Nm).
- Shock absorbers
- Rear wheels

Leaf Springs

REMOVAL & INSTALLATION

1. Before servicing the vehicle, refer to the precautions in the beginning of this section.
2. Support the vehicle at the frame.
3. Support the axle with a floor jack.
4. Remove or disconnect the following:

- Rear wheels
- Shock absorbers
- Axle U-bolts and spring pad
- Spring shackle
- Front mount bolt/pin
- Leaf spring

To install:

➡ **Use new fasteners for assembly.**

5. Install or connect the following:

- Leaf spring. Tighten the front mount bolt to 86–108 ft. lbs. (117–147 Nm).
- Spring shackle. Tighten the nuts to 58–72 ft. lbs. (78–98 Nm).
- Axle U-bolts and spring pad. Tighten the nuts to 72–80 ft. lbs. (98–108 Nm).
- Shock absorbers
- Rear wheels

Torsion Bar

1. Before servicing the vehicle, refer to the precautions in the beginning of this section.
2. Move the dust cover, if equipped.
3. Matchmark the torsion bar to the control arm mount and the anchor arm.
4. Measure the adjustment bolt protrusion as shown and note the length (L) for assembly.

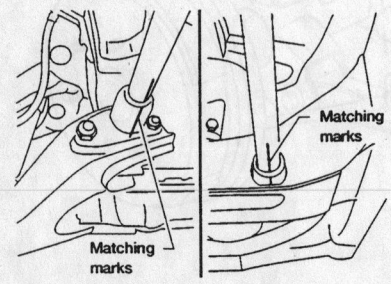

Torsion bar matchmarks

9308VG12

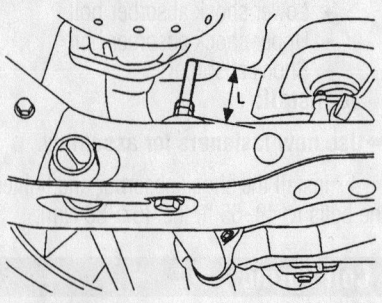

Adjustment bolt measurement (L)

9308VG13

5. Loosen the adjustment bolt so that all tension is released.
6. Remove the torsion bar mount from the control arm and remove the torsion bar.

To install:

7. Align the matchmarks and install the torsion bar. Tighten the large mount nut to 66–87 ft. lbs. (89–118 Nm) and the small nut to 33–44 ft. lbs. (45–60 Nm).
8. Tighten the adjustment bolt to achieve the measurement (L) noted earlier. Tighten the locknut to 22–30 ft. lbs. (30–40 Nm).
9. If a new torsion bar is being installed, set length (L) as follows:

- 2 wheel drive: 2.13 inches (54mm)
- 4 wheel drive: 2.76 inches (70mm)

Lower Ball Joint

REMOVAL & INSTALLATION

1. Before servicing the vehicle, refer to the precautions in the beginning of this section.
2. Support the lower control arm.
3. Remove or disconnect the following:

- Front wheel
- Lower ball joint

To install:

4. Install or connect the following:

- Lower ball joint. Tighten the control arm bolts to 76–94 ft. lbs. (103–127 Nm) and the stud nut to 87–123 ft. lbs. (118–167 Nm).
- Front wheel

Upper Control Arm

REMOVAL & INSTALLATION

1. Before servicing the vehicle, refer to the precautions in the beginning of this section.
2. Support the lower control arm.
3. Remove or disconnect the following:

- Front wheel
- Shock absorber

- Upper ball joint
- Control arm mounting bolts
- Upper control arm

To install:

4. Install or connect the following:

- Upper control arm. Tighten the mounting bolts to 72–87 ft. lbs. (98–118 Nm).
- Upper ball joint. Tighten the nut to 58–108 ft. lbs. (78–147 Nm).
- Shock absorber
- Front wheel

5. Check the wheel alignment and adjust, as necessary.

CONTROL ARM BUSHING REPLACEMENT

1. Before servicing the vehicle, refer to the precautions in the beginning of this section.
2. Remove the control arm from the vehicle.
3. Remove the control arm bushing with a press.

To install:

4. Lubricate the control arm bushings with liquid soap.
5. Install the bushings with a press.
6. Install the control arm to the vehicle.
7. Check the wheel alignment and adjust, as necessary.

Lower Control Arm

REMOVAL & INSTALLATION

1. Before servicing the vehicle, refer to the precautions in the beginning of this section.
2. Remove or disconnect the following:

- Front wheel
- Torsion bar
- Shock absorber
- Stabilizer bar link
- Axle halfshaft, if equipped
- Lower ball joint
- Control arm mounting bolts
- Lower control arm

To install:

3. Install or connect the following:

- Lower control arm. Tighten the mount bolts to 80–105 ft. lbs. (108–142 Nm) for 2001; 69–96 ft. lbs. (94–130 Nm) for 2002 models..
- Lower ball joint. Tighten the nut to 87–141 ft. lbs. (118–191 Nm) for 2000–01; 87–123 ft. lbs. (118–167 Nm) for 2002 models.
- Axle halfshaft, if equipped
- Stabilizer bar link

- Shock absorber
- Torsion bar
- Front wheel

4. Check the wheel alignment and adjust, as necessary.

CONTROL ARM BUSHING REPLACEMENT

1. Before servicing the vehicle, refer to the precautions in the beginning of this section.

2. Remove the control arm from the vehicle.

3. Remove the control arm bushing with a press.

To install:

4. Lubricate the control arm bushings with liquid soap.

5. Install the bushings with a press.

6. Install the control arm to the vehicle.

7. Check the wheel alignment and adjust, as necessary.

Wheel Bearings

ADJUSTMENT

2 Wheel Drive

➡**Use a new split pin for assembly.**

1. Before servicing the vehicle, refer to the precautions in the beginning of this section.

2. Remove or disconnect the following:
- Dust cap
- Split pin
- Spindle nut cap

3. Tighten the spindle nut to 25–29 ft. lbs. (34–39 Nm).

4. Spin the hub several times to fully seat the bearings.

5. Retighten the spindle nut to 25–29 ft. lbs. (34–39 Nm).

6. Loosen the spindle nut 45–60 degrees and install the spindle nut cap and split pin.

7. Install the dust cap.

4 Wheel Drive

1. Before servicing the vehicle, refer to the precautions in the beginning of this section.

2. Remove or disconnect the following:
- Locking hub or driveplate
- Snapring
- Spindle washer
- Thrust washer
- Lockwasher

3. Tighten the wheel bearing locknut to 58–72 ft. lbs. (78–98 Nm).

4. Loosen the locknut fully.

5. Tighten the wheel bearing locknut to 4–13 inch lbs. (0.5–1.5 Nm).

6. Spin the hub several times to fully seat the bearings.

7. Retighten the wheel bearing locknut to 4–13 inch lbs. (0.5–1.5 Nm).

8. Install or connect the following:
- Lockwasher. Tighten the retaining screw to 10–16 inch lbs. (1–2 Nm).
- Thrust washer
- Spindle washer
- Snapring
- Locking hub or driveplate

REMOVAL & INSTALLATION

2 Wheel Drive

1. Before servicing the vehicle, refer to the precautions in the beginning of this section.

2. Remove or disconnect the following:
- Front wheel
- Brake caliper and support
- Dust cap
- Split pin
- Spindle nut cap
- Spindle nut
- Bearing washer
- Outer bearing
- Hub and brake rotor assembly
- Inner grease seal
- Inner wheel bearing

To install:

3. Install or connect the following:
- Inner wheel bearing
- Inner grease seal
- Hub and brake rotor assembly
- Outer bearing
- Bearing washer
- Spindle nut. Adjust the wheel bearings.
- Spindle nut cap
- Split pin
- Dust cap
- Brake caliper and support
- Front wheel

4 Wheel Drive

1. Before servicing the vehicle, refer to the precautions in the beginning of this section.

2. Remove or disconnect the following:
- Front wheel
- Brake caliper and support
- Locking hub or driveplate
- Snapring
- Spindle washer
- Thrust washer
- Lockwasher
- Wheel bearing locknut
- Outer bearing
- Hub and brake rotor assembly
- Inner grease seal
- Inner wheel bearing

To install:

3. Install or connect the following:
- Inner wheel bearing
- Inner wheel bearing
- Inner grease seal
- Hub and brake rotor assembly
- Outer bearing
- Wheel bearing locknut. Adjust the wheel bearings.
- Lockwasher
- Thrust washer
- Spindle washer
- Snapring
- Locking hub or driveplate
- Brake caliper and support
- Front wheel

BRAKES

Brake Caliper

REMOVAL. & INSTALLATION

Front

1. Raise the vehicle and support safely.
2. Remove the appropriate tire and wheel assembly.
3. Remove the bolt attaching the brake hose to the caliper. Plug the brake hose to prevent brake fluid loss.
4. Remove the caliper support mounting bolts and lift the caliper assembly from the knuckle.

To install

5. Position the caliper assembly onto the knuckle and install the bolts. Make sure the rotor fits between the brake pads. Torque the bolts to 53–72 ft. lbs. (72–97 Nm).

6. Using new copper washers, connect the brake hose to the caliper. Torque the brake hose attaching bolt to 12–14 ft. lbs. (17–20 Nm).
7. Bleed the brake system.
8. Apply the brake pedal and inspect the system. Ensure proper operation and no leakage.
9. Install tire and wheel assembly. Lower the vehicle and road-test.

Rear

➡️ **Unlike most rear disc brake designs, this system does not incorporate the parking brake system into the rear brake caliper. The rear brake system is serviced in the same manner as the front system.**

1. Raise the vehicle and support safely.
2. Remove the appropriate tire and wheel assembly.

3. Remove the caliper support mounting bolts and lift the caliper assembly from the baffle plate.
4. Loosen the brake fluid hose with a wrench and turn the caliper to disconnect it from the brake hose. Plug the brake hose to prevent brake fluid loss.

To install:

5. Use a new copper washer and connect the brake hose to the caliper. Torque the hose fitting to 11 ft. lbs. (15 Nm).
6. Position the caliper assembly over the baffle plate and install the bolts. Make sure the rotor fits between the brake pads. Torque the bolts to 28–38 ft. lbs. (38–52 Nm).
7. Bleed the brake system.
8. Apply the brake pedal and inspect the system. Ensure proper operation and no leakage.
9. Install tire and wheel assembly. Lower the vehicle and road-test.

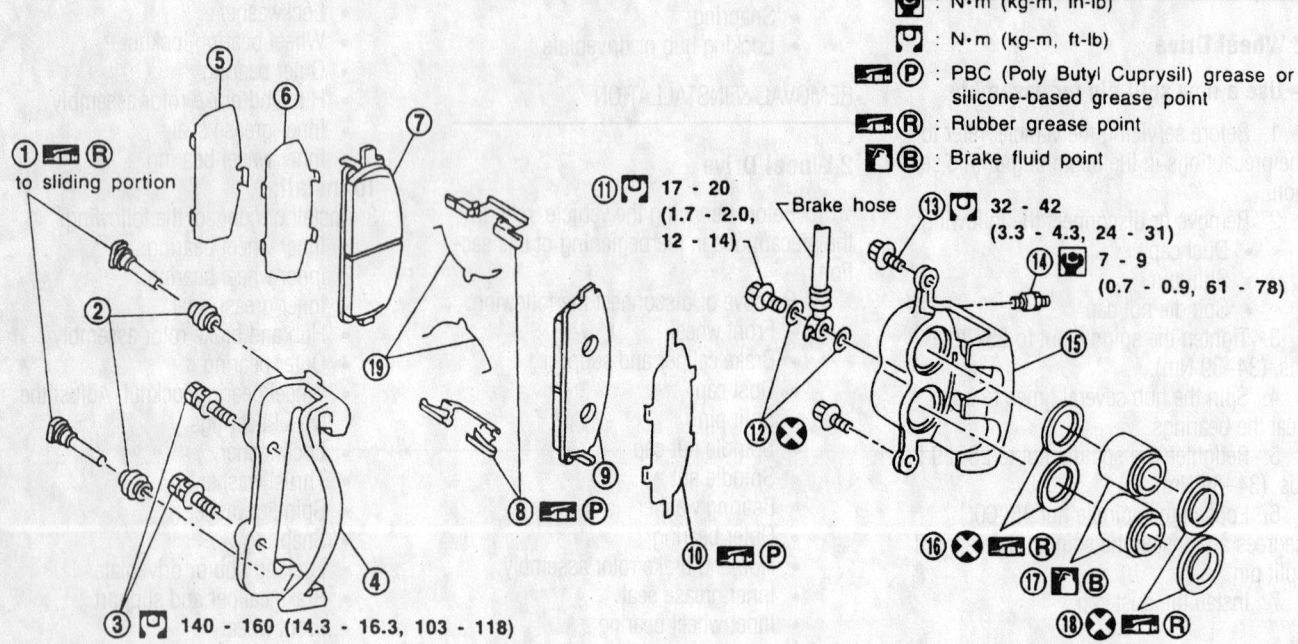

- 🔧 : N·m (kg-m, in-lb)
- 🔧 : N·m (kg-m, ft-lb)
- 🔧(P) : PBC (Poly Butyl Cuprysil) grease or silicone-based grease point
- 🔧(R) : Rubber grease point
- 🔧(B) : Brake fluid point

①	Main pin	⑧	Pad retainer	⑭	Bleed valve
②	Pin boot	⑨	Outer pad	⑮	Cylinder body
③	Torque member fixing bolt	⑩	Outer shim	⑯	Piston seal
④	Torque member	⑪	Connecting bolt	⑰	Piston
⑤	Shim cover	⑫	Copper washer	⑱	Piston boot
⑥	Inner shim	⑬	Main pin bolt	⑲	Pad spring
⑦	Inner pad				

Exploded view of the dual piston caliper front brake components

93026G60

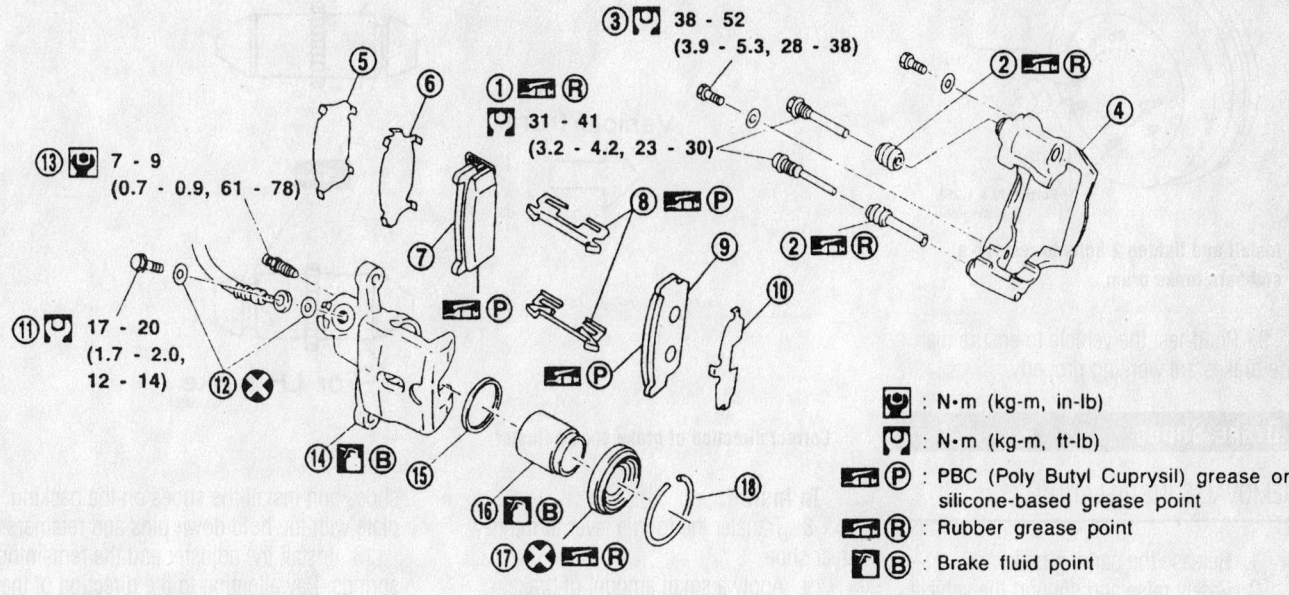

③ 🔧 38 - 52
(3.9 - 5.3, 28 - 38)

① 🔧 R
🔩 31 - 41
(3.2 - 4.2, 23 - 30)

② 🔧 R

④

⑬ 🔩 7 - 9
(0.7 - 0.9, 61 - 78)

⑧ 🔧 P

⑨

② 🔧 R

⑩

⑪ 🔩 17 - 20
(1.7 - 2.0,
12 - 14)

⑫ ⊗

🔧 P

🔧 P

⑭ 🔩 B

⑮

⑯ 🔩 B

⑰ ⊗ 🔧 R

⑱

🔩 : N·m (kg-m, in-lb)

🔧 : N·m (kg-m, ft-lb)

🔧 P : PBC (Poly Butyl Cuprysil) grease or silicone-based grease point

🔧 R : Rubber grease point

🔩 B : Brake fluid point

① Main pin bolt	⑦ Inner pad	⑬ Bleed valve
② Pin boot	⑧ Pad retainer	⑭ Cylinder body
③ Torque member fixing bolt	⑨ Outer pad	⑮ Piston seal
④ Torque member	⑩ Outer shim	⑯ Piston
⑤ Shim cover	⑪ Connecting bolt	⑰ Piston boot
⑥ Inner shim	⑫ Copper washer	⑱ Retainer

93026G61

Exploded view of the rear disc brake components—QX4

Disc Brake Pads

REMOVAL & INSTALLATION

➠**Both the front and rear disc brake pads can be serviced using the same procedure.**

1. Using a syringe, siphon brake fluid from the reservoir, leaving reservoir approximately ½ full.
2. Raise and properly support the vehicle.
3. Remove the wheel assemblies.
4. Remove the lower pin bolt from the brake caliper.
5. Swivel the caliper up and away from the torque member. Tie the caliper to a suspension member so that it is out of the way.
6. Lift the 2 brake pads out of the torque member.
7. Remove the inner and outer shims. Remove the 2 pad retainers if they are not attached to the pads.
8. Check the pad thickness and replace the pads if they are less than 0.079 in. (2mm) thick.

To install:

9. Install the inner and outer shims into the torque member.
10. Install a pad retainer to the bottom of each pad.
11. Install the pads into the torque member.
12. Use a C-clamp or hammer handle and press the caliper piston(s) back into the housing.
13. Untie the caliper and swivel it back into position over the torque plate so that the dust boot is not pinched. Install the pin bolt and torque it to 16–23 ft. lbs. (22–31 Nm).
14. Check the condition of the pin boot. Gently pull on it to expel any trapped air.
15. Install the wheel and lower the vehicle.
16. Pump the brakes until the pedal is firm and check the level of brake fluid. Road-test the vehicle.

Brake Drums

REMOVAL & INSTALLATION

1. Remove the hub cap and loosen the lug nuts.
2. Raise the rear of the vehicle and support it on jackstands.
3. Remove the lug nuts, tire and wheel.
4. Release the parking brake.
5. Pull the brake drum from the hub. If difficult to remove try the following:
   a. Strike the face of the drum with a plastic or rubber mallet. This will break free any rust that may develop between the drum and the hub.
   b. Install 2, M8x1.25mm bolts into the holes in the drum and gradually tighten them to pull the drum off the hub.

To install:

6. Install the brake drum to the hub.
7. Install the wheel.
8. Remove the jackstands and lower the vehicle.

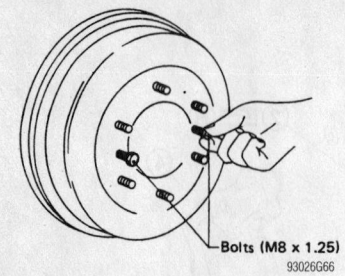

Install and tighten 2 bolts to remove a stubborn brake drum

93026G66

9. Road-test the vehicle to ensure that the brakes are working properly.

Brake Shoes

REMOVAL & INSTALLATION

1. Release the parking brake.
2. Safely raise and support the vehicle.
3. Remove the rear wheel and drum.
4. Remove the hold-down pin retainers.
5. Remove the leading shoe and then the trailing shoe.
6. Remove the adjuster.
7. Disconnect the parking brake cable from the toggle lever on the rear shoe.

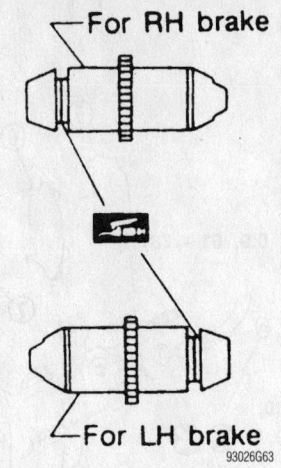

93026G63

Correct direction of brake shoe adjuster

To install:

8. Transfer the toggle lever to the new rear shoe.
9. Apply a small amount of brake grease to the tips of the shoes and the 6 pads on the backing plate that contact the brake shoe.
10. Shorten the adjuster by turning it.
11. Connect the parking brake cable to the toggle lever on the rear shoe.
12. Install the lower return spring to both

shoes and install the shoes on the backing plate with the hold down pins and retainers.
13. Install the adjuster and the remaining springs. Pay attention to the direction of the adjuster assembly.
14. Inspect the complete assembly and install the brake drum.
15. Adjust the shoe to drum clearance.
16. Install the wheel assembly and lower the vehicle to the floor.

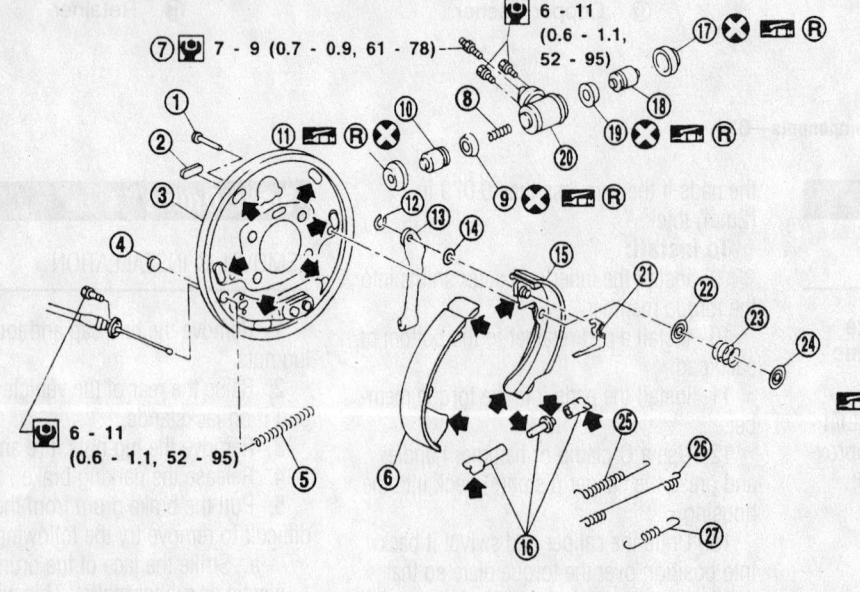

1. Shoe hold pin	10. Piston	19. Piston cup
2. Plug	11. Boot	20. Wheel cylinder
3. Back plate	12. Retainer ring	21. Adjuster lever
4. Check plug	13. Toggle lever	22. Spring seat
5. Spring	14. Wave washer	23. Shoe hold spring
6. Shoe (leading side)	15. Shoe (trailing side)	24. Retainer
7. Air bleeder	16. Adjuster	25. Adjuster spring
8. Spring	17. Boot	26. Return spring (upper)
9. Piston cup	18. Piston	27. Return spring (lower)

🔧Ⓡ : Rubber grease point

◀ : Brake grease point

🔧 : N•m (kg-m, ft-lb)

🔧 : N•m (kg-m, in-lb)

93026G69

Drum brake assembly exploded view

SUBARU

28

Legacy • Outback • SUS • Impreza • Outback • Outback Sport • WRX • Baja

SPECIFICATION CHARTS

ENGINE AND VEHICLE IDENTIFICATION CHART

Code ①	Liters (cc)	Cu. In.	Cyl.	Fuel Sys.	Type	Eng. Mfg.
4	2.2 (2212)	135	4	MFI	SOHC	Subaru
6	2.5 (2457)	150	4	MFI	DOHC	Subaru
6	2.5 (2457)	150	4	MFI	SOHC	Subaru
8	3.0 (3000)	183	6	MFI	DOHC	Subaru
2	2.0 (1994)	121	4	MFI	DOHC	Subaru

Code ②	Year
Y	2000
1	2001
2	2002
3	2003
4	2004

MFI: Multiport Fuel Injection

SOHC: Single Overhead Camshaft

DOHC: Double Overhead Camshaft

① 6th digit of the VIN

② 10th digit of the VIN

42356-SBCR-C01

GENERAL ENGINE SPECIFICATIONS

Year		Engine Displacement Liters (cc)	Engine ID/VIN	Fuel System Type	Net Horsepower @ rpm	Net Torque @ rpm (ft. lbs.)	Bore x Stroke (in.)	Compression Ratio	Oil Pressure @ rpm
2000	Impreza ①	2.2 (2212)	4	MFI	142@5600	149@3600	3.82x2.95	10.0:1	14@600
	Impreza RS	2.5 (2457)	6	MFI	165@5600	166@4000	3.92x3.11	10.0:1	14@600
	Outback	2.5 (2457)	6	MFI	165@5600	166@4000	3.92x3.11	10.0:1	14@600
	Legacy ②	2.5 (2457)	6	MFI	165@5600	166@4000	3.92x3.11	10.0:1	14@600
2001	Impreza ①	2.2 (2212)	4	MFI	142@5600	149@3600	3.82x2.95	10.0:1	14@600
	Impreza RS	2.5 (2457)	6	MFI	165@5600	166@4000	3.92x3.11	10.0:1	14@600
	Outback	2.5 (2457)	6	MFI	165@5600	166@4000	3.92x3.11	10.0:1	14@600
	Outback	3.0 (3000)	8	MFI	212@6000	210@4400	3.51x3.15	10.7:1	14@600
	Legacy	2.5 (2457)	6	MFI	165@5600	166@4000	3.92x3.11	10.0:1	14@600
2002	Impreza RS	2.5 (2457)	6	MFI	165@5600	166@4000	3.92x3.11	10.0:1	14@600
	WRX	2.0 (1994)	2	MFI	227@6000	217@4000	3.62x2.95	8.0:1	14@800
	Outback	2.5 (2457)	6	MFI	165@5600	166@4000	3.92x3.11	10.0:1	14@600
	Outback	3.0 (3000)	8	MFI	212@6000	210@4400	3.51x3.15	10.7:1	14@600
	Legacy	2.5 (2457)	6	MFI	165@5600	166@4000	3.92x3.11	10.0:1	14@600
2003-04	Impreza RS	2.5 (2457)	6	MFI	165@5600	166@4000	3.92x3.11	10.0:1	14@600
	WRX	2.0 (1994)	2	MFI	227@6000	217@4000	3.62x2.95	8.0:1	14@800
	Outback	2.5 (2457)	6	MFI	165@5600	166@4000	3.92x3.11	10.0:1	14@600
	Outback	3.0 (3000)	8	MFI	212@6000	210@4400	3.51x3.15	10.7:1	14@600
	Baja	2.5 (2457)	6	MFI	165@5600	166@4000	3.92x3.11	10.0:1	14@600
	Legacy	2.5 (2457)	6	MFI	165@5600	166@4000	3.92x3.11	10.0:1	14@600

MFI: Multi-port Fuel Injection

Note: All capacities are approximate. Add fluid gradually and check to be sure a proper fluid level is obtained.

① Includes Outback

② Includes Outback and Sport Utility Sedan

42356-SBCR-C02

ENGINE TUNE-UP SPECIFICATIONS

Year	Model	Engine Displacement Liters (cc)	Engine ID/VIN	Spark Plug Gap (in.)	Ignition Timing (deg.) ① MT	Ignition Timing (deg.) ① AT	Fuel Pump (psi)	Idle Speed (rpm) MT	Idle Speed (rpm) AT	Valve Clearance ② In.	Valve Clearance ② Ex.
2000	Impreza/ Outback	2.2 (2212)	4	0.039-0.043	6-22	12-28	34-38	600-800	600-800	0.0071-0.0087	0.0090-0.0106
	Impreza RS	2.5 (2457)	6	0.039-0.043	7-23	7-23	34-38	600-800	600-800	0.0071-0.0087	0.0090-0.0106
	Legacy/ Outback/SUS ③	2.5 (2457)	6	0.039-0.043	7-23	7-23	34-38	600-800	600-800	0.0071-0.0087	0.0090-0.0106
2001	Impreza/ Outback	2.2 (2212)	4	0.039-0.043	6-22	12-28	34-38	600-800	600-800	0.0071-0.0087	0.0090-0.0106
	Impreza RS	2.5 (2457)	6	0.039-0.043	7-23	7-23	34-38	600-800	600-800	0.0071-0.0087	0.0090-0.0106
	Outback/SUS ③	3.0 (3000)	8	0.039-0.043	7-23	7-23	34-38	600-800	600-800	0.0063-0.0095	0.0078-0.0118
	Legacy	2.5 (2457)	6	0.039-0.043	7-23	7-23	34-38	600-800	600-800	0.0071-0.0087	0.0090-0.0106
2002	Impreza RS	2.5 (2457)	6	0.039-0.043	7-23	7-23	34-38	600-800	600-800	0.0071-0.0087	0.0090-0.0106
	Outback/SUS ③	2.5 (2457)	6	0.039-0.043	7-23	7-23	34-38	600-800	600-800	0.0071-0.0087	0.0090-0.0106
	Outback/SUS ③	3.0 (3000)	8	0.039-0.043	2-18 ④	2-18 ④	34-38	600	600	0.0063-0.0095	0.0078-0.0118
	WRX	2.0 (1994)	2	0.028-0.031	2-22	2-22	33-38	650-850	650-850	0.0071-0.0087	0.0090-0.0106
	Legacy	2.5 (2457)	6	0.039-0.043	7-23	7-23	34-38	600-800	600-800	0.0071-0.0087	0.0090-0.0106
2003-04	Impreza RS	2.5 (2457)	6	0.039-0.043	7-23	7-23	34-38	600-800	600-800	0.0071-0.0087	0.0090-0.0106
	Outback/SUS ③	2.5 (2457)	6	0.039-0.043	7-23	7-23	34-38	600-800	600-800	0.0071-0.0087	0.0090-0.0106
	Outback/SUS ③	3.0 (3000)	8	0.039-0.043	7-23	7-23	34-38	600-800	600-800	0.0063-0.0095	0.0078-0.0118
	WRX	2.0 (1994)	2	0.028-0.031	2-22	2-22	33-38	650-850	650-850	0.0071-0.0087	0.0090-0.0106
	Baja	2.5 (2457)	6	0.039-0.043	7-23	7-23	34-38	600-800	600-800	0.0071-0.0087	0.0090-0.0106
	Legacy	2.5 (2457)	6	0.039-0.043	7-23	7-23	34-38	600-800	600-800	0.0071-0.0087	0.0090-0.0106

Note: The Vehicle Emission Control Information label often reflects specification changes made during production. The lable figures mudst be used if they differ from those in this chart.

① Before Top Dead Center
② Measured with engine cold
③ Sport Utility Sedan
④ @ 600 rpm

42356-SBCR-C03

2.2L and 2.5L Engines
Firing order: 1–3–2–4
Distributorless ignition system

79233G37

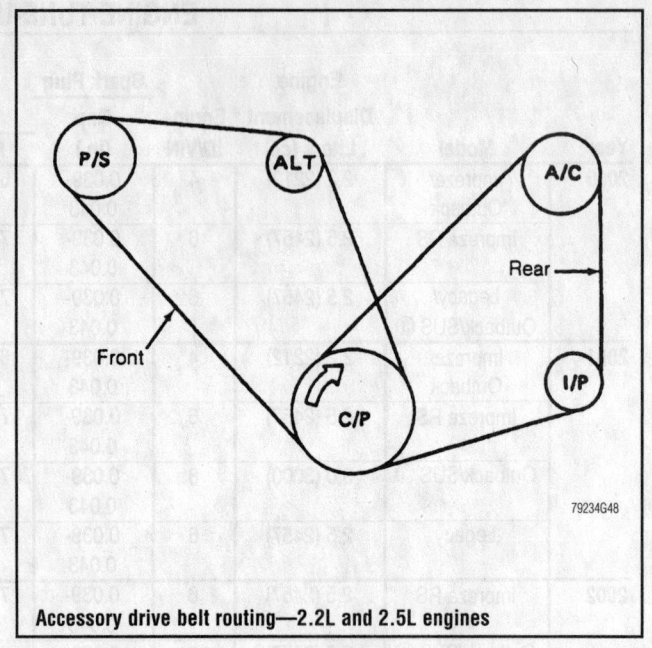

Accessory drive belt routing—2.2L and 2.5L engines

79234G48

CAPACITIES

Year	Model	Engine Displacement Liters (cc)	Engine ID/VIN	Engine Oil with Filter (qts.)	Transmission (pts.)		Transfer Case (pts.)	Drive Axle		Fuel Tank (gal.)	Cooling System (qts.)
					5-Spd	Auto.		Front ① (pts.)	Rear (pts.)		
2000	Impreza ②	2.2 (2212)	4	4.4	7.4	16.8	–	2.6	1.6	13.2	6.6
	Impreza RS	2.5 (2457)	6	4.2	7.4	20.0	–	2.6	1.6	13.2	6.6
	Legacy ③	2.5 (2457)	6	4.2	7.4	20.0	–	2.6	1.6	16.9	7.1
2001	Impreza ②	2.2 (2212)	4	4.4	7.4	16.8	–	2.6	1.6	13.2	6.6
	Impreza RS	2.5 (2457)	6	4.2	7.4	20.0	–	2.6	1.6	13.2	6.6
	Outback	3.0 (3000)	8	5.9	7.4	20.2	–	2.6	1.6	16.9	8.4
	Legacy	2.5 (2457)	6	4.2	7.4	20.0	–	2.6	1.6	16.9	7.1
2002	Impreza RS	2.5 (2457)	6	4.2	7.4	20.0	–	2.6	1.6	13.2	6.6
	WRX	2.0 (1994)	2	4.8	7.4	19.6	–	2.6	1.6	15.9	8.0
	Outback	3.0 (3000)	8	5.9	7.4	20.2	–	2.6	1.6	16.9	8.4
	Legacy	2.5 (2457)	6	4.2	7.4	20.0	–	2.6	1.6	16.9	7.1
2003-04	Impreza RS	2.5 (2457)	6	4.2	7.4	20.0	–	2.6	1.6	13.2	6.6
	WRX	2.0 (1994)	2	4.8	7.4	19.6	–	2.6	1.6	15.9	8.0
	Baja	2.5 (2457)	6	4.2	7.4	20.0	–	2.6	1.6	16.9	7.1
	Outback	3.0 (3000)	8	5.9	7.4	20.2	–	2.6	1.6	16.9	8.4
	Legacy	2.5 (2457)	6	4.2	7.4	20.0	–	2.6	1.6	16.9	7.1

Note: All capacities are approximate. Add fluid gradually and check to be sure a proper fluid level is obtained.

① A/T differential only
② Includes Outback
③ Includes Outback and Sport Utility Sedan

42356-SBCR-C04

VALVE SPECIFICATIONS

Year	Engine Displacement Liters (cc)	Engine ID/VIN	Seat Angle (deg.)	Face Angle (deg.)	Spring Test Pressure (lbs. @ in.)	Spring Installed Height (in.)	Stem-to-Guide Clearance (in.)		Stem Diameter (in.)	
							Intake	Exhaust	Intake	Exhaust
2000	2.2 (2212)	4	45	45	91 - 103@ 1.110	①	0.0014- 0.0059	0.0016- 0.0059	0.2344- 0.2350	0.2341- 0.2346
	2.5 (2457)	6	45	45	102 - 118@ 1.315	②	0.0014- 0.0024	0.0016- 0.0026	0.2343- 0.2348	0.2341- 0.2346
2001	2.2 (2212)	4	45	45	91 - 103@ 1.110	①	0.0014- 0.0059	0.0016- 0.0059	0.2344- 0.2350	0.2341- 0.2346
	2.5 (2457)	6	45	45	102 - 118@ 1.315	②	0.0014- 0.0024	0.0016- 0.0026	0.2343- 0.2348	0.2341- 0.2346
	3.0 (3000)	8	45	45	102 - 118@ 1.315	④	0.0012- 0.0022	0.0016- 0.0026	002148- 0.2154	0.2148- 0.2150
2002	2.0 (2212)	2	45	45	91 - 103@ 1.110	③	0.0014- 0.0059	0.0016- 0.0059	0.2343- 0.2348	0.2341- 0.2346
	2.5 (2457)	6	45	45	102 - 118@ 1.315	②	0.0014- 0.0024	0.0016- 0.0026	0.2343- 0.2348	0.2341- 0.2346
	3.0 (3000)	8	45	45	102 - 118@ 1.315	④	0.0012- 0.0022	0.0016- 0.0026	002148- 0.2154	0.2148- 0.2150
2003-04	2.0 (2212)	2	45	45	91 - 103@ 1.110	③	0.0014- 0.0059	0.0016- 0.0059	0.2343- 0.2348	0.2341- 0.2346
	2.5 (2457)	6	45	45	102 - 118@ 1.315	②	0.0014- 0.0024	0.0016- 0.0026	0.2343- 0.2348	0.2341- 0.2346
	3.0 (3000)	8	45	45	102 - 118@ 1.315	④	0.0012- 0.0022	0.0016- 0.0026	002148- 0.2154	0.2148- 0.2150

① Free length: 1.7342 in.
② Free length: 1.8913 in.
③ Free length: 1.7587 in.
④ Free length: 1.8421 in.

42356-SBCR-C05

CRANKSHAFT AND CONNECTING ROD SPECIFICATIONS
All measurements are given in inches.

Year	Engine Displacement Liters (cc)	Engine ID/VIN	Crankshaft				Connecting Rod		
			Main Brg. Journal Dia.	Main Brg. Oil Clearance	Shaft End-play	Thrust on No.	Journal Diameter	Oil Clearance	Side Clearance
2000	2.2 (2212)	4	2.3619-2.3625	①	0.0012-0.0098	3	2.0466-2.0472	0.0006-0.0020	0.0028-0.0160
	2.5 (2457)	6	2.3619-2.3625	②	0.0012-0.0098	3	1.8891-1.8898	0.0005-0.0015	0.0028-0.0130
2001	2.2 (2212)	4	2.3619-2.3625	①	0.0012-0.0098	3	2.0466-2.0472	0.0006-0.0020	0.0028-0.0160
	2.5 (2457)	6	2.3619-2.3625	②	0.0012-0.0098	3	1.8891-1.8898	0.0005-0.0015	0.0028-0.0130
	3.0 (3000)	8	2.3619-2.3625	③	0.0012-0.0098	3	1.8891-1.8898	0.0009-0.0020	0.0028-0.0130
2002	2.0 (1994)	2	2.3619-2.3625	0.0004-0.0012	0.0012-0.0048	3	1.8891-1.8898	0.0008-0.0018	0.0028-0.0130
	2.5 (2457)	6	2.5194-2.5200	②	0.0012-0.0098	3	1.8891-1.8898	0.0005-0.0015	0.0028-0.0130
	3.0 (3000)	8	2.3619-2.3625	③	0.0012-0.0098	3	1.8891-1.8898	0.0009-0.0020	0.0028-0.0130
2003-04	2.0 (1994)	2	2.3619-2.3625	0.0004-0.0012	0.0012-0.0048	3	1.8891-1.8898	0.0008-0.0018	0.0028-0.0130
	2.5 (2457)	6	2.3619-2.3625	②	0.0012-0.0098	3	1.8891-1.8898	0.0005-0.0015	0.0028-0.0130
	3.0 (3000)	8	2.3619-2.3625	③	0.0012-0.0098	3	1.8891-1.8898	0.0009-0.0020	0.0028-0.0130

① Journals 1 and 5: 0.0001 - 0.0016 in.
  Journals 2 and 4: 0.0004 - 0.0014 in.
  Journal 3: 0.0004 - 0.0014 in.

② Journals 1 and 5: 0.0001 - 0.0016 in.
  Journals 2 and 4: 0.0004 - 0.0018 in.
  Journal 3: 0.0004 - 0.0016 in.

③ 0.0006 - 0.0012 in.

42356-SBCR-C06

PISTON AND RING SPECIFICATIONS

All measurements are given in inches.

Year	Engine Displacement Liters (cc)	Engine ID/VIN	Piston Clearance	Ring Gap			Ring Side Clearance		
				Top Compression	Bottom Compression	Oil Control	Top Compression	Bottom Compression	Oil Control
2000	2.2 (2212)	4	0.0004-0.0020	0.0079-0.0390	0.0079-0.0390	0.0079-0.0590	0.0016-0.0590	0.0012-0.0590	NA
	2.5 (2457)	6	0.0004-0.0020	0.0079-0.0390	0.0146-0.0390	0.0079-0.0590	0.0016-0.0059	0.0012-0.0059	NA
2001	2.2 (2212)	4	0.0004-0.0020	0.0079-0.0390	0.0079-0.0390	0.0079-0.0590	0.0016-0.0590	0.0012-0.0590	NA
	2.5 (2457)	6	0.0004-0.0012	0.0079-0.0138	0.0146-0.0250	0.0079-0.0197	0.0016-0.0031	0.0012-0.0028	NA
	3.0 (3000)	8	0.0004-0.0012	0.0079-0.0138	0.0138-0.0197	0.0079-0.0236	0.0016-0.0031	0.0012-0.0028	NA
2002	2.0 (1994)	2	0.0004-0.0012	0.0079-0.0138	0.0138-0.0197	0.0079-0.0276	0.0016-0.0031	0.0012-0.0028	NA
	2.5 (2457)	6	0.0004-0.0012	0.0079-0.0138	0.0146-0.0250	0.0079-0.0197	0.0016-0.0031	0.0012-0.0028	NA
	3.0 (3000)	8	0.0004-0.0012	0.0079-0.0138	0.0138-0.0197	0.0079-0.0236	0.0016-0.0031	0.0012-0.0028	NA
2003-04	2.0 (1994)	2	0.0004-0.0012	0.0079-0.0138	0.0138-0.0197	0.0079-0.0276	0.0016-0.0031	0.0012-0.0028	NA
	2.5 (2457)	6	0.0004-0.0012	0.0079-0.0138	0.0146-0.0250	0.0079-0.0197	0.0016-0.0031	0.0012-0.0028	NA
	3.0 (3000)	8	0.0004-0.0012	0.0079-0.0138	0.0138-0.0197	0.0079-0.0236	0.0016-0.0031	0.0012-0.0028	NA

NA: Not Available

42356-SBCR-C07

TORQUE SPECIFICATIONS
All readings in ft. lbs.

Year	Engine Displacement Liters (cc)	Engine ID/VIN	Cylinder Head Bolts	Main ① Bearing Bolts	Rod Bearing Bolts	Crankshaft Damper Bolts	Flywheel Bolts	Manifold Intake	Manifold Exhaust	Spark Plugs	Lug Nut
2000	2.2 (2212)	4	②	③	31 - 44	87 - 101	51 - 55	17 - 20	19 - 26	13 - 17	58 - 72
	2.5 (2457)	6	②	③	31 - 44	123 - 137	51 - 55	17 - 20	19 - 26	13 - 17	58 - 72
2001	2.2 (2212)	4	②	③	31 - 44	87 - 101	51 - 55	17 - 20	19 - 26	13 - 17	58 - 72
	2.5 (2457)	6	②	③	31 - 44	123 - 137	51 - 55	17 - 20	19 - 26	13 - 17	58 - 72
	3.0 (3000)	8	④	14	39	131	60	18	22	15	65
2002	2.0 (1994)	2	②	③	31 - 44	94	51 - 55	18	26	13 - 17	58 - 72
	2.5 (2457)	6	②	③	31 - 44	123 - 137	51 - 55	17 - 20	19 - 26	13 - 17	58 - 72
	3.0 (3000)	8	④	14	39	131	60	18	22	15	65
2003-04	2.0 (1994)	2	②	③	31 - 44	94	51 - 55	18	26	13 - 17	58 - 72
	2.5 (2457)	6	②	③	31 - 44	123 - 137	51 - 55	17 - 20	19 - 26	13 - 17	58 - 72
	3.0 (3000)	8	④	14	39	131	60	18	22	15	65

① Engine block connecting bolts

② Step 1: Tighten all bolts to 22 ft. lbs.
Step 2: Tighten all bolts to 51 ft. lbs.
Step 3: Loosen all botls 180 degrees.
Step 4: Repeat Step 3.
Step 5: Tighten bolts 1 and 2 to 25 ft. lbs.
Step 6: Tighten bolts 3, 4, 5 and 6 to 11 ft. lbs.
Step 7: Tighten all bolts 80 to 90 degrees.
Step 8: Repeat Step 7. Do not exceed 180 degrees total tightening.

③ Split engine case connecting bolts:
Short bolts: 17-20 ft. lbs.
Long bolts: 33-37 ft. lbs.
Smaller short bolts (if used) 5 ft. lbs.

④ Step 1: Tighten all bolts to 14 ft. lbs.
Step 2: Tighten all bolts to 37 ft. lbs.
Step 3: Loosen all bolts 180 degrees the an additional 180 degrees
in two steps in the reverse order of tightening sequence.
Step 4: Tighten all bolts to 18 ft. lbs.
Step 5: Tighten all bolts to 18 ft. lbs.
Step 6: Tighten bolts 90 degrees
Step 7: Tighten bolts 1, 2, 3 and 4 90 degrees.
Step 8: Tighten bolts 5, 6, 7 and 8 45 degrees.

42356-SBCR-C08

WHEEL ALIGNMENT

Year	Model		Caster Range (+/-Deg.)	Caster Preferred Setting (Deg.)	Camber Range (+/-Deg.)	Camber Preferred Setting (Deg.)	Toe-in (in.)	Steering Axis Inclination (Deg.)
2000	Impreza 2.2L	F	1.00	+3.00	0.50	0	0+/-0.12	—
		R	—	—	0.75	-0.92	0+/-0.12	—
	Impreza 2.5L	F	1.00	+3.05	0.50	-0.42	0+/-0.12	—
		R	—	—	0.75	-1.17	0+/-0.12	—
	Legacy Sedan	F	0.75	+3.05	0.50	-0.05	0+/-0.12	—
		R	—		0.75	-0.50	0+/-0.12	—
	Legacy Wagon	F	0.75	+2.05	0.50	-0.05	0+/-0.12	—
		R	—		0.75	-0.20	0+/-0.12	—
	Outback	F	0.75	+2.40	0.50	+0.33	0+/-0.12	—
		R	—		0.75	-0.17	0+/-0.12	—
2001	Impreza 2.2L	F	1.00	+3.00	0.50	0	0+/-0.12	—
		R	—	—	0.75	-0.92	0+/-0.12	—
	Impreza 2.5L	F	1.00	+3.05	0.50	-0.42	0+/-0.12	—
		R	—	—	0.75	-1.17	0+/-0.12	—
	Legacy Sedan	F	0.75	+3.05	0.50	-0.05	0+/-0.12	—
		R	—		0.75	-0.50	0+/-0.12	—
	Legacy Wagon	F	0.75	+2.05	0.50	-0.05	0+/-0.12	—
		R	—		0.75	-0.20	0+/-0.12	—
	Outback	F	0.75	+2.40	0.50	+0.33	0+/-0.12	—
		R	—		0.75	-0.17	0+/-0.12	—
2002	Impreza 2.5L	F	1.00	+3.05	0.50	-0.42	0+/-0.12	—
		R	—	—	0.75	-1.17	0+/-0.12	—
	Legacy Sedan	F	0.75	+3.05	0.50	-0.05	0+/-0.12	—
		R	—		0.75	-0.50	0+/-0.12	—
	Legacy Wagon	F	0.75	+2.05	0.50	-0.05	0+/-0.12	—
		R	—		0.75	-0.20	0+/-0.12	—
	Outback	F	0.75	+2.40	0.50	+0.33	0+/-0.12	—
		R	—		0.75	-0.17	0+/-0.12	—
	WRX	F	0.75	+2.40	0.50	+0.33	0+/-0.12	—
		R	—		0.75	-0.17	0+/-0.12	—
2003-04	Impreza 2.5L	F	1.00	+3.05	0.50	-0.42	0+/-0.12	—
		R	—	—	0.75	-1.17	0+/-0.12	—
	Legacy Sedan	F	0.75	+3.05	0.50	-0.05	0+/-0.12	—
		R	—		0.75	-0.50	0+/-0.12	—
	Legacy Wagon	F	0.75	+2.05	0.50	-0.05	0+/-0.12	—
		R	—		0.75	-0.20	0+/-0.12	—
	Baja	F	1.00	+3.05	0.50	-0.42	0+/-0.12	—
		R	—	—	0.75	-1.17	0+/-0.12	—
	Outback	F	0.75	+2.40	0.50	+0.33	0+/-0.12	—
		R	—		0.75	-0.17	0+/-0.12	—
	WRX	F	0.75	+2.40	0.50	+0.33	0+/-0.12	—
		R	—		0.75	-0.17	0+/-0.12	—

42356-SBCR-C09

TIRE, WHEEL AND BALL JOINT SPECIFICATIONS

| Year | Model | OEM Tires | | Tire Pressures (psi) | | Wheel Size | Ball Joint Inspection |
		Standard	Optional	Front	Rear		
2000	Impreza	P195/60R15	None	32	29	6-JJ	0.012 in.
	Impreza RS	P205/55R16	None	32	29	7-JJ	0.012 in.
	Impreza Outback	P225/60R16	T145/80R16	32	30	16x16 1/2JJ	0.012 in.
	Legacy	P185/70R14	P195/60R15 P205/55R16 P205/70R15	32	30	6.5-JJ	0.012 in.
2001	Impreza	P195/60R15	None	32	29	6-JJ	0.012 in.
	Impreza RS	P205/55R16	None	32	29	7-JJ	0.012 in.
	Impreza Outback	P205/60R15	None	32	29	7-JJ	0.012 in.
	Legacy	P185/70R14	P195/60R15 P205/55R16 P205/70R15	32	30	6.5-JJ	0.012 in.
2002	Impreza	P195/60R15	None	32	29	6-JJ	0.012 in.
	Impreza RS	P205/55R16	None	32	29	7-JJ	0.012 in.
	Impreza Outback	P225/60R16	T145/80R16	32	30	16x16 1/2JJ	0.012 in.
	Impreza WRX	P195/60R15	P205/55R16 P215/45R17	32	29	6-JJ 7-JJ	0.012 in.
	Legacy	P185/70R14	P195/60R15 P205/55R16 P205/70R15	32	30	6.5-JJ	0.012 in.
2003-04	Impreza	P195/60R15	None	32	29	6-JJ	0.012 in.
	Impreza RS	P205/55R16	None	32	29	7-JJ	0.012 in.
	Impreza Outback	P225/60R16	T145/80R16	32	30	16x16 1/2JJ	0.012 in.
	Impreza WRX	P195/60R15	P205/55R16 P215/45R17	32	29	6-JJ 7-JJ	0.012 in.
	Baja	P185/70R14	P195/60R15 P205/55R16 P205/70R15	32	30	6.5-JJ	0.012 in.
	Legacy	P185/70R14	P195/60R15 P205/55R16 P205/70R15	32	30	6.5-JJ	0.012 in.

OEM: Original Equipment Manufacturer

PSI: Pounds Per Square Inch

42356-SBCR-C10

BRAKE SPECIFICATIONS
All measurements in inches unless noted

Year	Model		Brake Disc Original Thickness	Brake Disc Minimum Thickness	Brake Disc Maximum Runout	Brake Drum Diameter Original Inside Diameter	Brake Drum Diameter Max. Wear Limit	Brake Drum Diameter Maximum Machine Diameter	Minimum Lining Thickness Front	Minimum Lining Thickness Rear	Brake Caliper Bracket Bolts (ft. lbs.)	Brake Caliper Mounting Bolts (ft. lbs.)
2000	Impreza	F	0.940	0.870	0.003	—	—	—	0.059	—	59	23-30
		R	0.390	0.335	0.004	9.000 ①	9.079 ②	NA	—	0.059	—	23-30
	Outback	F	0.945	0.866	0.003	—	—	—	0.295	—	59	29
		R	0.390	0.335	0.003	9.000 ①	9.079 ②	NA	—	0.256	59	29
	Legacy	F	0.940	0.870	0.003	—	—	—	0.059	—	59	23-30
		R	0.390	0.335	0.004	9.000 ①	9.079 ②	NA	—	0.059	—	23-30
2001	Impreza	F	0.940	0.870	0.003	—	—	—	0.059	—	59	23-30
		R	0.390	0.335	0.004	9.000 ①	9.079 ②	NA	—	0.059	—	23-30
	Outback	F	0.940	0.870	0.003	—	—	—	0.059	—	59	23-30
		R	0.390	0.335	0.004	9.000 ①	9.079 ②	NA	—	0.059	59	23-30
	Legacy	F	0.940	0.870	0.003	—	—	—	0.059	—	59	23-30
		R	0.390	0.335	0.004	9.000 ①	9.079 ②	NA	—	0.059	—	23-30
2002	Impreza	F	0.940	0.870	0.003	—	—	—	0.059	—	59	23-30
		R	0.390	0.335	0.004	9.000 ①	9.079 ②	NA	—	0.059	—	23-30
	WRX	F	0.940	0.870	0.003	—	—	—	0.295	—	59	19.5
		R	0.390	0.335	0.004	9.000 ①	9.080 ②	9.000 ①	—	0.059	59	27.5
	Outback	F	0.945	0.866	0.003	—	—	—	0.295	—	59	29
		R	0.390	0.335	0.003	9.000 ①	9.079 ②	NA	—	0.256	59	29
	Legacy	F	0.940	0.870	0.003	—	—	—	0.059	—	59	23-30
		R	0.390	0.335	0.004	9.000 ①	9.079 ②	NA	—	0.059	—	23-30
2003-04	Impreza	F	0.940	0.870	0.003	—	—	—	0.059	—	59	23-30
		R	0.390	0.335	0.004	9.000 ①	9.079 ②	NA	—	0.059	—	23-30
	WRX	F	0.940	0.870	0.003	—	—	—	0.295	—	59	19.5
		R	0.390	0.335	0.004	9.000 ①	9.080 ②	9.000 ①	—	0.059	59	27.5
	Outback	F	0.945	0.866	0.003	—	—	—	0.295	—	59	29
		R	0.390	0.335	0.003	9.000 ①	9.079 ②	NA	—	0.256	59	29
	Baja	F	0.940	0.870	0.003	—	—	—	0.059	—	59	23-30
		R	0.390	0.335	0.004	9.000 ①	9.079 ②	NA	—	0.059	—	23-30
	Legacy	F	0.940	0.870	0.003	—	—	—	0.059	—	59	23-30
		R	0.390	0.335	0.004	9.000 ①	9.079 ②	NA	—	0.059	—	23-30

NA: Not Available

① Parking brake drum on vehicles with rear disc brakes: 6.69 in.

② Specification is for the parking brake drum.

42356-SBCR-C11

SCHEDULED MAINTENANCE INTERVALS

SUBARU—IMPREZA, OUTBACK, WRX, LEGACY & BAJA

TO BE SERVICED	TYPE OF SERVICE	VEHICLE MILEAGE INTERVAL (x1000)												
		7.5	15	22.5	30	37.5	45	52.5	60	67.5	75	82.5	90	97.5
Engine oil & filter	R	✓	✓	✓	✓	✓	✓	✓	✓	✓	✓	✓	✓	✓
Brake lines	S/I		✓		✓		✓		✓		✓		✓	
Clutch & hill holder system	S/I		✓		✓		✓		✓		✓		✓	
Disc brake pads & discs, front & rear axle boots & axle shaft joint portions	S/I		✓		✓		✓		✓		✓		✓	
Parking brake	S/I		✓		✓		✓		✓		✓		✓	
Steering & suspension	S/I		✓		✓		✓		✓		✓		✓	
Air filter element	R				✓				✓				✓	
Engine coolant	R				✓				✓				✓	
Fuel filter	R				✓				✓				✓	
Spark plugs	R								✓					
Automatic transmission fluid & filter	S/I				✓				✓				✓	
Brake fluid	S/I				✓				✓				✓	
Brake linings & drums	S/I				✓				✓				✓	
Camshaft drive belt ①	S/I				✓				✓				✓	
Coolant level, hoses & clamps	S/I				✓				✓				✓	
Drive belts	S/I				✓				✓				✓	
Fuel system, hoses & connections	S/I				✓				✓				✓	
Transmission and/or differential gear fluid	S/I				✓								✓	
Front & rear wheel bearing repack	S/I								✓					

R: Replace   S/I: Service or Inspect

① Non-California vehicles: replace every 60,000 miles.

FREQUENT OPERATION MAINTENANCE (SEVERE SERVICE)

If a vehicle is operated under any of the following conditions it is considered severe service:

- Extremely dusty areas.

- 50% or more of the vehicle operation is in 32°C (90°F) or higher temperatures, or constant operation in temperatures below 0°C (32°F).

- Prolonged idling (vehicle operation in stop and go traffic).

- Frequent short running periods (engine does not warm to normal operating temperatures).

- Police, taxi, delivery usage or trailer towing usage.

Oil & oil filter change: change every 3750 miles.

Clutch & hill holder system: service or inspect every 7500 miles.

Disc brake pads & discs, front & rear axle boots & axle shaft joint portions: service or inspect every 7500 miles.

Steering & suspension: service or inspect every 7500 miles.

Air filter element: service or inspect every 15,000 miles.

Automatic transmission fluid: service or inspect every 15,000 miles.

Brake linings & drums: service or inspect every 15,000 miles.

Coolant level, hoses & clamps: service or inspect every 15,000 miles.

Drive belts: service or inspect every 15,000 miles.

Transmission/differential gear oil (except SVX): service or inspect every 15,000 miles.

Front & rear wheel bearing repack: service or inspect every 30,000 miles.

PRECAUTIONS

Before servicing any vehicle, please be sure to read all of the following precautions, which deal with personal safety, prevention of component damage, and important points to take into consideration when servicing a motor vehicle:

• Never open, service or drain the radiator or cooling system when the engine is hot; serious burns can occur from the steam and hot coolant.

• Observe all applicable safety precautions when working around fuel. Whenever servicing the fuel system, always work in a well-ventilated area. Do not allow fuel spray or vapors to come in contact with a spark, open flame or excessive heat (a hot drop light, for example). Keep a dry chemical fire extinguisher near the work area. Always keep fuel in a container specifically designed for fuel storage; also, always properly seal fuel containers to avoid the possibility of fire or explosion. Refer to the additional fuel system precautions later in this section.

• Fuel injection systems often remain pressurized, even after the engine has been turned **OFF**. The fuel system pressure must be relieved before disconnecting any fuel lines. Failure to do so may result in fire and/or personal injury.

• Brake fluid often contains polyglycol ethers and polyglycols. Avoid contact with the eyes and wash your hands thoroughly after handling brake fluid. If you do get brake fluid in your eyes, flush your eyes with clean, running water for 15 minutes. If

eye irritation persists, or if you have taken brake fluid internally, IMMEDIATELY seek medical assistance.

• The EPA warns that prolonged contact with used engine oil may cause a number of skin disorders, including cancer! You should make every effort to minimize your exposure to used engine oil. Protective gloves should be worn when changing oil. Wash your hands and any other exposed skin areas as soon as possible after exposure to used engine oil. Soap and water, or waterless hand cleaner should be used.

• All new vehicles are now equipped with an air bag system. The system must be disabled before performing service on or around system components, steering column, instrument panel components, wiring and sensors. Failure to follow safety and disabling procedures could result in accidental air bag deployment, possible personal injury and unnecessary system repairs.

• Always wear safety goggles when working with, or around, the air bag system. When carrying a non-deployed air bag, be sure the bag and trim cover are pointed away from your body. When placing a non-deployed air bag on a work surface, always face the bag and trim cover upward, away from the surface. This will reduce the motion of the module if it is accidentally deployed. Refer to the additional air bag system precautions later in this section.

• Clean, high quality brake fluid from a

sealed container is essential to the safe and proper operation of the brake system. You should always buy the correct type of brake fluid for your vehicle. If the brake fluid becomes contaminated, completely flush the system with new fluid. Never reuse any brake fluid. Any brake fluid that is removed from the system should be discarded. Also, do not allow any brake fluid to come in contact with a painted surface; it will damage the paint.

• Never operate the engine without the proper amount and type of engine oil; doing so WILL result in severe engine damage.

• Timing belt maintenance is extremely important! Many models utilize an interference-type, non-freewheeling engine. If the timing belt breaks, the valves in the cylinder head may strike the pistons, causing potentially serious (also time-consuming and expensive) engine damage.

• Disconnecting the negative battery cable on some vehicles may interfere with the functions of the on-board computer system(s) and may require the computer to undergo a relearning process once the negative battery cable is reconnected.

• When servicing drum brakes, only disassemble and assemble one side at a time, leaving the remaining side intact for reference.

• Only an MVAC-trained, EPA-certified automotive technician should service the air conditioning system or its components.

ENGINE REPAIR

➡**Disconnecting the negative battery cable on some vehicles may interfere with the functions of the on board computer systems and may require the computer to undergo a relearning process, once the negative battery cable is reconnected.**

Alternator

REMOVAL & INSTALLATION

1. Remove or disconnect the following:
   • Negative battery cable
   • Connector and terminal from the alternator
   • V-belt cover, if equipped
   • Front side V-belt
   • Alternator to bracket bolts
   • Alternator from the vehicle

To install:
2. Install or connect the following:
   • Alternator into the vehicle
   • Alternator to bracket bolts
   • Front side V-belt
   • V-belt cover, if equipped
   • Connector and terminal to the alternator
   • Negative battery cable
3. Check and adjust the belt tension.

Ignition Timing

ADJUSTMENT

All Subaru models are equipped with Distributorless Ignition System (DIS). The ignition timing is controlled by the Powertrain Control Module (PCM) and is not adjustable.

Engine Assembly

❉❉ CAUTION

Some models covered by this manual may be equipped with an air bag. Whenever working near any of the Supplemental Restraint System (SRS) components, such as the impact sensors, the air bag module, steering column and instrument panel, properly disable the SRS.

REMOVAL & INSTALLATION

2.0L Engines

1. Before servicing the vehicle, refer to the precautions in the beginning of this section.

2. Relieve the fuel system pressure.

3. Drain the engine oil and coolant into suitable containers.

4. Raise the rear seat and turn the floor mat up.

5. Disconnect the fuel pump relay connector, start the engine and let it stall. Once the engine stalls, crank it for a further 5 seconds to ensure the fuel system is properly relieved.

6. Remove the fuel filler cap.

7. Remove or disconnect the following:
- Air cleaner cover and element
- Battery cables
- Radiator
- Coolant filler tank

8. If equipped with air conditioning, discharge the system using an approved recovery/recycling machine. Disconnect and cap the lines from the compressor.
- Intercooler

9. Disconnect the following electrical connections:
- Engine harness connector
- Engine ground terminal
- Alternator connector, terminal and A/C compressor connections

10. Remove or disconnect the following:
- Accelerator cable
- Clutch release spring
- Brake booster hose
- Heater inlet and outlet hoses

11. Remove the power steering pump from the bracket by performing the following steps:

a. Loosen the lock and slider bolts.

b. Remove the V-belt.

c. Disconnect the power steering switch connection.

d. Remove the pipe with bracket from the intake manifold.

e. Remove the power steering pump from the engine.

f. Remove the power steering tank from the bracket by pulling it upwards.

g. Place the power steering pump on the wheel apron on the right.

12. Remove or disconnect the following:

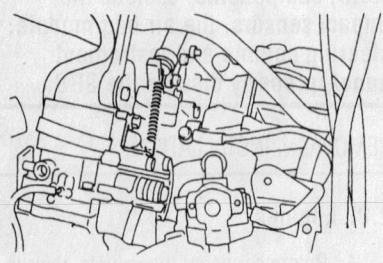

View of the clutch release spring location—2.0L engines

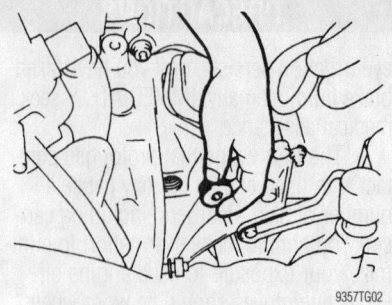

Using a 10mm wrench, remove the plug—2.0L engines

- Center exhaust pipe
- Nuts which attach the lower side of the engine to the transmission
- Nuts which attach the front cushion rubber onto the crossmember

13. Disconnect the clutch release fork from the release bearing as follows:

a. Remove the clutch cylinder from the transmission.

b. Using a 10mm wrench, remove the plug.

c. Screw a 6mm diameter bolt into the release fork and remove it.

d. Raise the release fork and unfasten the release tabs to free the release fork.

14. Disconnect the torque converter clutch from the drive plate on models equipped with an automatic transaxle as follows:

a. Remove the service hole plug.

b. Remove the torque converter clutch-to-drive plate bolts.

c. Remove the remaining bolts while rotating the engine using a crankshaft pulley wrench.

15. Remove or disconnect the following:
- Pitching stopper
- Fuel delivery, return and evaporation hoses
- Fuel filter and bracket

16. Attach a lifting device to the engine.

17. Using a floor jack, support the transmission.

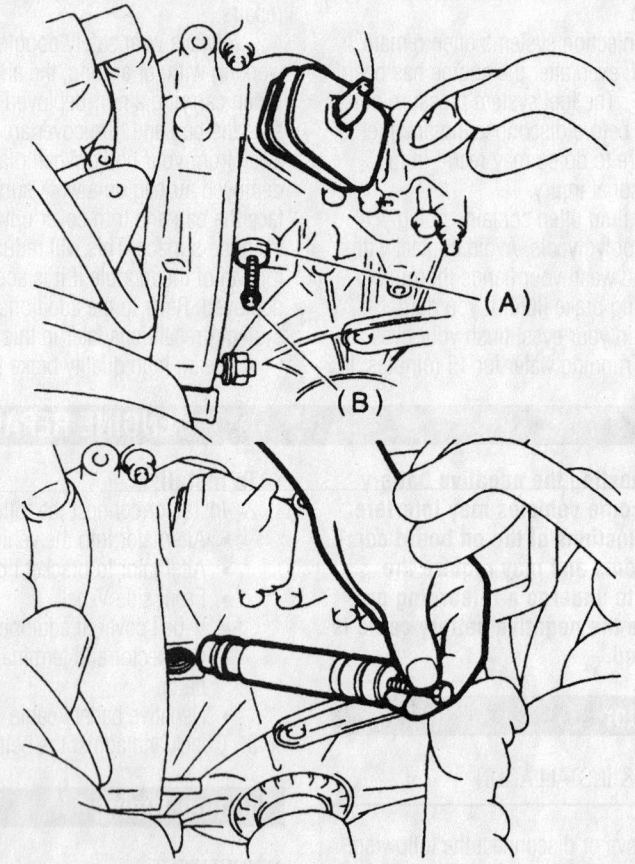

(A) Shaft
(B) Bolt

Screw a 6mm diameter bolt into the release fork and remove it—2.0L engines

18. Remove the starter.

19. Separate the engine from the transmission.

20. Remove the upper right transmission-to-engine bolts.

21. Remove the engine as follows:

    a. Raise the engine slightly.

    b. Using the floor jack, raise the transmission.

    c. Move the engine horizontally until the mainshaft is withdrawn from the clutch cover.

    d. Remove the engine.

To install:

22. Installation is the reverse of removal, please note the following torques:

- Clutch release fork plug: 32 ft. lbs. (44 Nm)
- Front cushion rubbers: 25 ft. lbs. (34 Nm)
- Bolts attaching the upper right side of the transmission to the engine: 37 ft. lbs. (50 Nm)
- Pitching stopper-to-fender bolt: 37 ft. lbs. (50 Nm)
- Pitching stopper-to-engine bolt: 43 ft. lbs. (58 Nm)
- Torque converter clutch-to-drive plate bolts, while rotating the engine: 18 ft. lbs. (25 Nm)
- Power steering pump bolts: 15 ft. lbs. (20 Nm)
- Bolts attaching the lower side of the transmission to the engine: 37 ft. lbs. (50 Nm)
- Front cushion rubber-to-crossmember bolts: 61 ft. lbs. (83 Nm)

23. Fill the engine with the recommended oil.

24. Fill and bleed the cooling system.

25. Charge the air conditioning system using an approved recovery/recycling machine.

26. Adjust the clutch cable.

27. If equipped, check the automatic transaxle fluid level and add Dexron®II if necessary.

28. Start the engine and allow it to reach normal operating temperature. Check for leaks.

2.2L and 2.5L Engines

1. Before servicing the vehicle, refer to the precautions in the beginning of this section.

2. Relieve the fuel system pressure.

3. Drain the engine oil and coolant into suitable containers.

4. Remove or disconnect the following:
- Battery cables
- Battery from the vehicle
- Radiator hoses
- Fan motor harness
- Radiator

5. If equipped with air conditioning, discharge the system using an approved recovery/recycling machine. Disconnect and cap the lines from the compressor.
- Air intake duct
- Air cleaner element and upper cover
- Evaporator canister and bracket
- Oxygen Sensor (O2S) connector
- Engine ground terminal
- Crankshaft Position (CKP) sensor connector
- Camshaft Position (CMP) sensor connector
- Knock Sensor (KS) connector
- Alternator connector and terminal
- Air conditioning compressor connectors, if equipped
- Accelerator cable
- Cruise control cable, if equipped
- Clutch release spring, clutch cable and hill holder cable, if equipped with a manual transaxle
- Brake booster hose(s)
- Heater inlet and outlet hoses
- Alternator drive belt
- Spark plug wires from left side of engine
- Power steering pump line bracket
- Power steering pump, leave the lines connected and position aside
- Exhaust Y-pipe
- Lower starter nuts
- Lower engine-to-transaxle nuts
- Front engine mount-to-crossmember nuts
- Starter

6. If equipped with an automatic transaxle, perform the following:

    a. Remove the torque converter service hole plug.

    b. Rotate the engine. Remove the torque converter-to-drive plate bolts as they become accessible.

7. Remove or disconnect the following:
- Pitching stopper
- Fuel delivery, return and evaporation hoses

8. Support the engine with a suitable lifting device attached to the engine lifting eyes.

9. Slightly raise the engine.

10. Raise the transaxle with a floor jack.

11. Slowly remove the engine from the vehicle.

To install:

12. Apply a small amount of grease to the splines of the mainshaft.

13. Position the engine in the engine compartment and align it with the transaxle.

14. Install or connect the following:
- Engine upper bolts and tighten to 34–40 ft. lbs. (44–54 Nm)

15. Remove the lifting device and floor jack.

16. Install the pitching stopper and tighten the bolts to the following specifications:

    a. Body side: 49 ft. lbs. (67 Nm).

    b. Bracket side: 40 ft. lbs. (54 Nm).

17. If equipped with an automatic transaxle, perform the following:

    a. Install the torque converter-to-drive plate bolts while rotating the engine, and tighten to 20 ft. lbs. (26 Nm).

    b. Install the service hole cover.

18. Install or connect the following:
- Evaporator canister and bracket
- Power steering pump. Torque retainers to 22–36 ft. lbs. (29–47 Nm).
- Drive belt, adjust tension
- Starter. Tighten bolts to 34–40 ft. lbs. (44–52 Nm).
- Lower engine-to-transaxle nuts. Tighten to 34–40 ft. lbs. (44–52 Nm).
- Lower engine mounting nuts. Tighten to 61 ft. lbs. (83 Nm) in the inner most elliptical hole in the front crossmember so the clearance is 0.16–0.24 in. (4–6mm).
- Exhaust Y-pipe with new gaskets and nuts
- Brake booster hose
- Heater inlet and outlet hoses
- Accelerator and the cruise control cables, if equipped

19. If equipped with a manual transaxle, install the following:
- Clutch release spring
- Clutch cable
- Hill holder cable

20. Install or connect the following:
- Engine harness connectors
- O2 sensor connector
- Engine ground terminal
- CKP sensor connector
- CMP sensor connector
- Knock sensor connector
- Alternator connector and terminal
- Air conditioning compressor connectors, if equipped
- Air cleaner element and cover
- Air conditioning lines, if equipped, with new O-rings. Tighten the bolts to 23 ft. lbs. (31 Nm).
- Radiator
- Engine cover
- Battery

21. Fill the engine with the recommended oil.

22. Fill and bleed the cooling system.

23. Charge the air conditioning system using an approved recovery/recycling machine.

24. Adjust the clutch cable.

25. If equipped, check the automatic transaxle fluid level and add Dexron®II if necessary.

26. Start the engine and allow it to reach normal operating temperature. Check for leaks.

3.0L Engines

1. Before servicing the vehicle, refer to the precautions in the beginning of this section.

2. Relieve the fuel system pressure.

3. Drain the engine oil and coolant into suitable containers.

4. Remove or disconnect the following:
- Battery cables
- Battery
- Air intake duct
- Engine undercover
- Radiator
- Drive belt

5. If equipped with air conditioning, discharge the system using an approved recovery/recycling machine. Disconnect and cap the lines from the compressor.
- Engine ground terminal
- Engine harness connectors
- Alternator connector and terminal
- Air conditioning compressor connectors, if equipped
- Accelerator and the cruise control cables, if equipped
- Brake booster hose
- Heater inlet and outlet hoses
- Power steering pump line bracket
- Power steering pump, leave the lines connected and position aside
- Exhaust Y-pipe
- Lower engine-to-transaxle nuts
- Front engine mount-to-crossmember nuts

6. If equipped with an automatic transaxle, perform the following:
    a. Remove the torque converter service hole plug.
    b. Rotate the engine. Remove the torque converter-to-drive plate bolts as they become accessible.

7. Remove or disconnect the following:
- Pitching stopper
- Fuel delivery, return and evaporation hoses

8. Support the engine with a suitable lifting device attached to the engine lifting eyes.

9. Support the transaxle with a floor jack.

10. Remove or disconnect the following:
- Starter
- Upper engine-to-transmission bolts

11. Slightly raise the engine.

12. Raise the transaxle with a floor jack.

13. Slowly remove the engine from the vehicle.

To install:

14. Apply a small amount of grease to the splines of the mainshaft.

15. Position the engine in the engine compartment and align it with the transaxle.

16. Install or connect the following:
- Engine upper bolts and tighten to 36 ft. lbs. (50 Nm)

17. Remove the lifting device and floor jack.

18. Install the pitching stopper and tighten the bolts to the following specifications:
    a. Body side: 42 ft. lbs. (57 Nm).
    b. Bracket side: 42 ft. lbs. (49 Nm).

19. Install the starter. Tighten bolts to 37 ft. lbs. (50 Nm).

20. If equipped with an automatic transaxle, perform the following:
    a. Install the torque converter-to-drive plate bolts while rotating the engine, and tighten to 18 ft. lbs. (25 Nm).
    b. Install the service hole cover.

21. Install or connect the following:
- Power steering pump. Torque retainers to 14 ft. lbs. (20 Nm).
- Lower engine-to-transaxle nuts. Tighten to 36 ft. lbs. (50 Nm).
- Lower engine mounting nuts. Tighten to 54 ft. lbs. (74 Nm).
- Exhaust Y-pipe with new gaskets and nuts
- Fuel delivery, return and evaporation hoses
- Heater inlet and outlet hoses
- Brake booster hose
- Engine ground terminal
- Engine harness connectors
- Alternator connector and terminal
- Air conditioning compressor connectors, if equipped
- Accelerator and the cruise control cables, if equipped
- Air conditioning lines, if equipped, with new O-rings. Tighten the bolts to 10 ft. lbs. (15 Nm).
- Radiator
- Drive belt, adjust tension
- Air cleaner element and cover
- Engine under cover
- Battery
- Battery cables

22. Fill the engine with the recommended oil.

23. Fill and bleed the cooling system.

24. Charge the air conditioning system using an approved recovery/recycling machine.

25. Check the automatic transaxle fluid level and add Dexron®II if necessary.

26. Start the engine and allow it to reach normal operating temperature. Check for leaks.

Water Pump

REMOVAL & INSTALLATION

Except 3.0L Engines

1. Before servicing the vehicle, refer to the precautions in the beginning of this section.

2. Remove or disconnect the following:
- Negative battery cable
- Engine undercover, if equipped

3. Drain the coolant into a suitable container.

4. Remove or disconnect the following:
- Radiator fan connector(s)
- Radiator outlet and heater hoses
- Heater bypass hose or overflow hose, if equipped
- Reservoir tank, on Legacy models
- Radiator fan motor assembly(ies)
- Accessory drive belts
- Timing belt
- Belt tension adjuster
- Belt idler No. 2
- Camshaft Position (CMP) sensor
- Left side camshaft pulley(s)
- Left side rear timing belt cover
- Tensioner bracket
- Radiator and heater hoses from water pump
- Water pump retainer bolts
- Water pump

5. Inspect the radiator hoses for deterioration and replace as necessary.

To install:

6. Clean the gasket mating surfaces thoroughly. Always use new gaskets during installation.

7. Install or connect the following:
- Water pump, tighten the bolts in sequence to 10 ft. lbs. (13 Nm). After tightening the bolts once, retighten to the same specification again.
- Radiator heater hoses to water pump
- Tensioner bracket and tighten to 18 ft. lbs. (25 Nm)

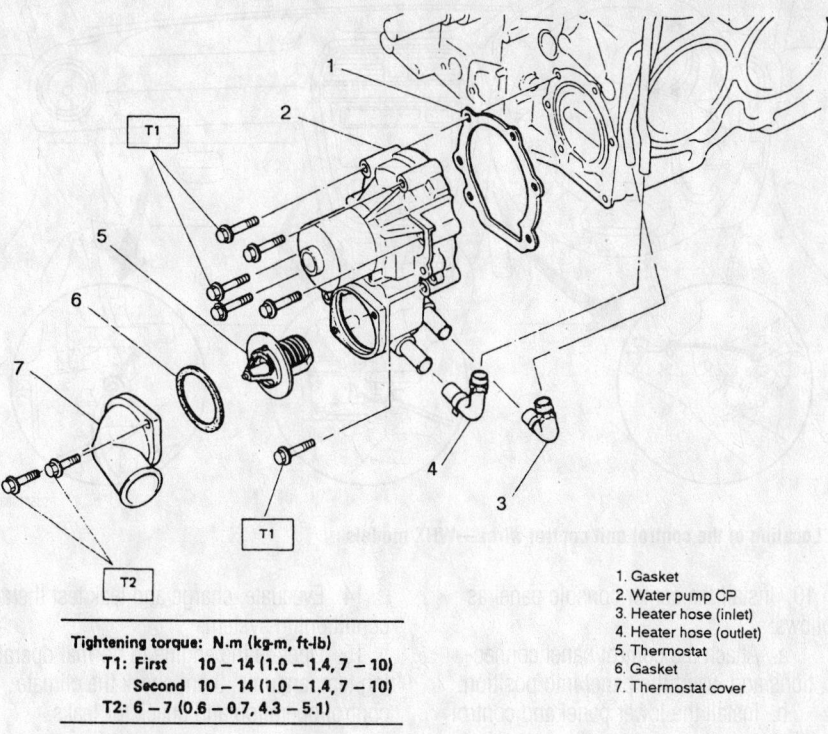

Tightening torque: N.m (kg-m, ft-lb)
  T1: First 10 – 14 (1.0 – 1.4, 7 – 10)
      Second 10 – 14 (1.0 – 1.4, 7 – 10)
  T2: 6 – 7 (0.6 – 0.7, 4.3 – 5.1)

1. Gasket
2. Water pump CP
3. Heater hose (inlet)
4. Heater hose (outlet)
5. Thermostat
6. Gasket
7. Thermostat cover

7923TG01

Exploded view of the water pump assembly—except 3.0L engines

- Left side rear timing belt cover
- Left side camshaft pulley(s). Tighten to 58 ft. lbs. (78 Nm) on non-turbo models and 72 ft. lbs. (98 Nm) on turbo models.
- CMP sensor
- Belt idler No. 2 and tighten to 29 ft. lbs. (39 Nm)
- Belt tension adjuster
- Timing belt
- Accessory drive belts
- Radiator fan assembly(ies)
- Reservoir tank, if removed

- Heater bypass hose or overflow hose, if equipped
- Air intake duct
- Radiator outlet and heater hoses
- Radiator fan connector(s)
- Engine undercover, if removed

8. Fill the system with coolant and connect the negative battery cable.

9. Start the engine and allow it to reach operating temperature.

10. Check for leaks.

3.0L Engines

1. Before servicing the vehicle, refer to the precautions in the beginning of this section.

2. Remove or disconnect the following:
- Negative battery cable
- Engine undercover, if equipped

3. Drain the coolant into a suitable container.

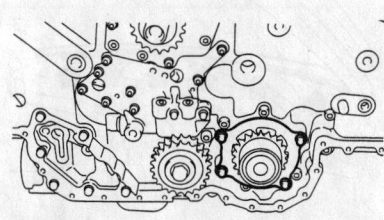

9357TG04

Tighten the water pump bolts is two steps using the following sequence—2.0L engine Turbo model shown, others similar

42356-SBCR-G01

View of the water pump assembly—3.0L engines

4. Remove or disconnect the following:
- Radiator
- Accessory drive belts
- Timing chain
- Water pump retainer bolts
- Water pump

5. Inspect the radiator hoses for deterioration and replace as necessary.

To install:

6. Clean the gasket mating surfaces thoroughly. Always use new gaskets during installation.

7. Apply coolant to the new O-ring before installation

8. Install or connect the following:
- Water pump with a new O-ring, tighten the bolts to 5 ft. lbs. (7 Nm).
- Timing chain
- Front chain cover
- Accessory drive belts
- Radiator
- Engine undercover, if removed

9. Fill the system with coolant and connect the negative battery cable.

10. Start the engine and allow it to reach operating temperature.

11. Check for leaks.

Heater Core

REMOVAL & INSTALLATION

WRX

1. Before servicing the vehicle, refer to the precautions in the beginning of this section.

2. Disconnect the negative battery cable.

❋❋ CAUTION

If equipped with an air bag system, wait 10 minutes after disconnecting the negative battery cable before performing any further work while the system fully de-energizes in order to avoid accidental deployment. All air bag system wiring is yellow. Do not use electrically powered test equipment on these circuits.

3. Drain the cooling system into a clean container for reuse.

4. Remove or disconnect the following:
- Bolts retaining the expansion valve and pipe
- Heater hoses from the heater core. Plug the heater core and heater hoses.
- Lower cover from the instrument panel

- Glove box

5. Remove the center console panel as follows:

a. Remove lower panel and control wires.

b. Pull the control panel out and disconnect the connections.

6. Remove the passenger side air bag module as follows:

a. Disconnect the airbag connector from the support beam bracket and then unplug the connector.

b. Unfasten the 3 airbag module bolts and remove the module.

7. Remove or disconnect the following:
- 4 screws and two nuts then remove the lower console panel
- Hooks that retain the defroster panel
- Nuts and remove the electrical connections as shown in the illustration
- Instrument panel bolts and remove the panel
- Support beam
- Blower motor
- Heater core retainers
- Heater core

To install:

8. Install or connect the following:
- Heater core
- Heater core retainers
- Blower motor
- Support beam
- Instrument panel and tighten the bolts
- Electrical connections and tighten the nuts
- Hooks to the defroster panel
- Lower console panel and tighten the 4 screws and two nuts

9. Install the passenger side air bag module as follows:

a. Install the module and tighten the 3 airbag module bolts.

b. Connect the airbag connector and then attach it to the support beam bracket.

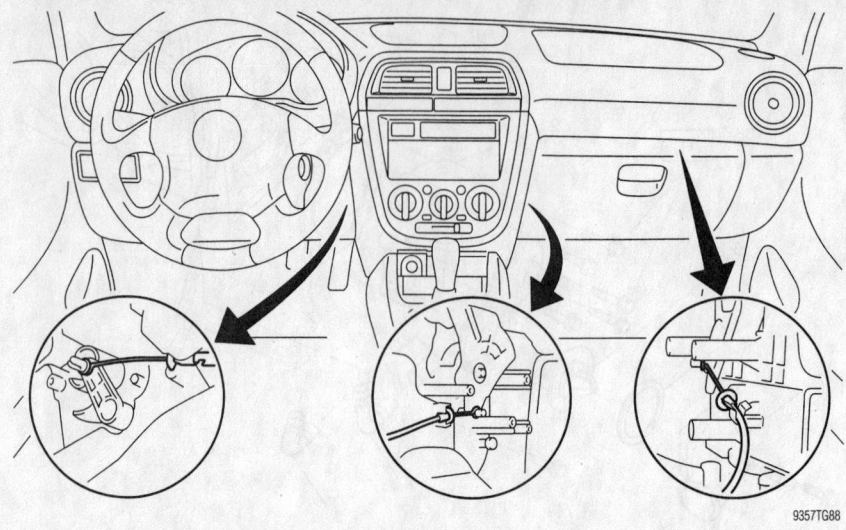

Location of the control unit control wires—WRX models

10. Install the center console panel as follows:

a. Attach the control panel connections and insert the panel into position.

b. Install the lower panel and control wires.

11. Install or connect the following:
- Glove box
- Lower cover to the instrument panel
- Heater hoses to the heater core
- Bolts retaining the expansion valve and pipe

12. Refill the cooling system.

13. Connect the negative battery cable.

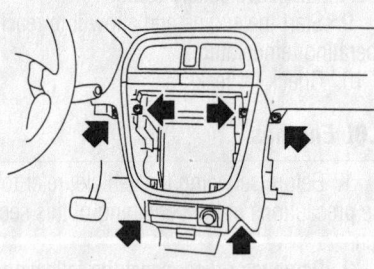

Location of the lower console panel retainers—WRX models

14. Evacuate, charge and leak test the air conditioning system.

15. Operate the engine to normal operating temperatures; then, check the climate control operation and check for leaks.

Impreza

1. Before servicing the vehicle, refer to the precautions in the beginning of this section.

2. Disconnect the negative battery cable.

✳✳ CAUTION

If equipped with an air bag system, wait 10 minutes after disconnecting the negative battery cable before performing any further work while the system fully de-energizes in order to avoid accidental deployment. All air bag system wiring is yellow. Do not use electrically powered test equipment on these circuits.

3. Drain the cooling system into a clean container for reuse.

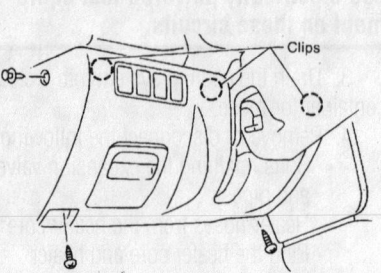

Location of the lower cover retainers— WRX models

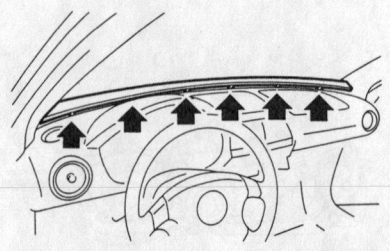

Loosen the hooks that retain the defroster panel—WRX models

Disconnect the two electrical connections after removing the console panel—WRX models

4. Remove or disconnect the following:

  • Heater hoses from the heater core. Plug the heater core and heater hoses.
  • Radio box or console

5. Remove the instrument panel as follows:

  a. Remove the rear console box.
  b. Pull the cup holder.
  c. Turn over the shift lever boot (manual transaxle models) or remove select lever cover (automatic transaxle models).
  d. Remove the console cover.
  e. Remove the audio assembly and disconnect the antenna cable and connectors.
  f. Remove the lower cover and then disconnect the seat belt timer connector.
  g. Remove the glove box.
  h. Remove the instrument panel console.
  i. Remove the 2 bolts and lower the steering column.
  j. Remove the column cover.
  k. Remove the hood opening lever.
  l. Set the temperature control switch to MAX COLD, and mode selector switch to the defroster position.
  m. Disconnect both the temperature

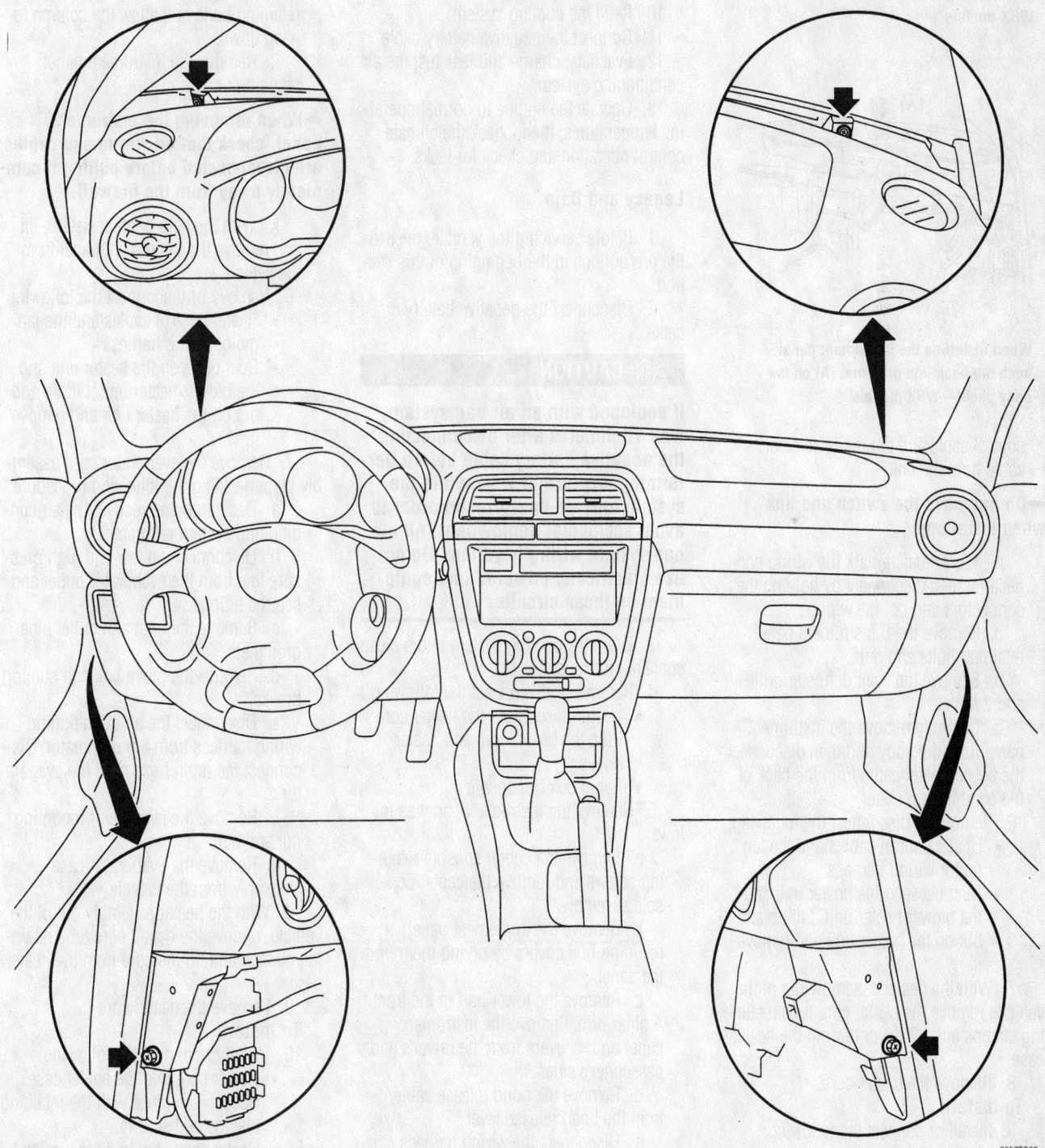

Location of the instrument panel bolts—WRX models

9357TG85

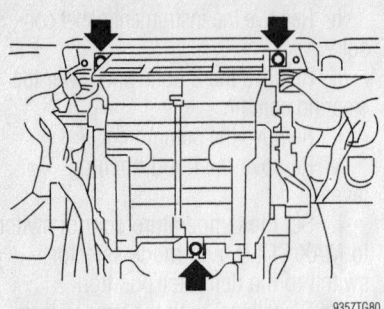

Location of the heater core retainers—WRX models

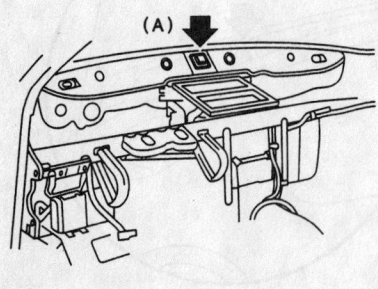

When installing the instrument panel, push the hook into grommet (A) on the body panel—WRX models

control cable and the mode selector cable from the link.

➡ **Do not move the switch and link when installing.**

n. Tag or match mark the wiring connectors, then disconnect by holding the connectors and not the wiring.

o. Remove the 6 instrument panel retaining bolts and nuts.

p. Remove the front defroster grille and 2 bolts.

q. Carefully remove the instrument panel from the body and then disconnect the speedometer cable from the back of the combination meter.

6. Remove or disconnect the following:
- Heater control cables and the fan motor wiring harness
- Duct between the heater unit and the blower heater unit. Lift up and out on the heater unit and remove it.

7. With the heater assembly out of the vehicle, remove the heater core tube retaining clamps and lift the core from the heater case.

8. Remove the heater core.

To install:

9. Install or connect the following:
- Heater core into the heater case.

Secure it in place with the retaining clamps and screws.
- Heater assembly to its mounting position under the dash. Torque the mounting bolts to 48–84 inch lbs. (5–9 Nm).
- Heater control cables and fan motor wiring harness connectors
- Instrument panel
- Radio and console assemblies
- Heater hoses in the engine compartment

10. Refill the cooling system.

11. Connect the negative battery cable.

12. Evacuate, charge and leak test the air conditioning system.

13. Operate the engine to normal operating temperatures; then, check the climate control operation and check for leaks.

Legacy and Baja

1. Before servicing the vehicle, refer to the precautions in the beginning of this section.

2. Disconnect the negative battery cable.

✻✻ CAUTION

If equipped with an air bag system, wait 10 minutes after disconnecting the negative battery cable before performing any further work while the system fully de-energizes in order to avoid accidental deployment. All air bag system wiring is yellow. Do not use electrically powered test equipment on these circuits.

3. Drain the cooling system into a clean container for reuse.

4. Remove or disconnect the following:
- Heater hoses from the heater core. Plug the heater core and heater hoses.
- Radio box or console

5. Remove the instrument panel as follows:

a. Remove the center console retaining screws and remove the center console assembly.

b. Remove the instrument panel retaining bolt covers by prying them from the panel.

c. Remove the lower part of the front A pillar trim. Remove the instrument panel under covers from the driver's and passenger's sides.

d. Remove the hood release cable from the hood release lever.

e. Disconnect the wiring harness connectors under the instrument panel.

f. Remove the instrument cluster assembly. Remove the glove box assembly

g. Disconnect the ventilation control cables and electrical connectors at the heater unit. Disconnect the vacuum line at the blower housing.

h. Disconnect the radio antenna feeder wire. Disconnect the main harness connector at the fuse box.

i. Remove the lower steering column covers. Remove the steering column retaining bolts and allow the column to hang down.

j. Remove the instrument panel retaining bolts.

➡ **When removing the instrument panel, check that all wiring and cables are disconnected before pulling it completely away from the firewall.**

k. With the help of an assistant, lift and remove the instrument panel from the vehicle.

6. Remove or disconnect the following:
- Heater control cables and the fan motor wiring harness
- Duct between the heater unit and the blower heater unit. Lift up and out on the heater unit and remove it.

7. Remove the evaporator case assembly by performing the following procedure:

a. Discharge and recover the air conditioning system refrigerant.

b. Disconnect the low and high pressure line from the evaporator outlet and cap the fittings.

c. Remove the inlet and outlet pipe grommets.

d. Remove the glove box and support bracket.

e. Disconnect the air conditioning wiring harness from the evaporator. Disconnect the drain hose from the evaporator.

f. Remove the evaporator mounting nut and bolt.

g. Remove the evaporator case assembly from the vehicle.

8. With the heater assembly out of the vehicle, remove the heater core tube retaining clamps and lift the core from the heater case.

9. Remove the heater core.

To install:

10. Install or connect the following:
- Heater core into the heater case. Secure it in place with the retaining clamps and screws.
- Heater assembly to its mounting position under the dash Torque the

mounting bolts to 48–84 inch lbs. (5–9 Nm).

11. Install the evaporator by performing the following procedure:

a. Install the evaporator case assembly and the nuts and bolts.

b. Adjust the position of the evaporator assembly so the inlet and outlet connections are aligned with the heater and blower unit connections.

c. Install the drain hose. Connect the air conditioning wiring harness.

d. Install the inlet and outlet pipe grommets.

e. Install the glove box and the lower support bracket.

f. Using new O-rings, lubricate them with clean refrigerant oil and install them on the pipe fittings.

g. Connect the suction hose to the evaporator inlet fitting.

h. Connect the discharge hose to the evaporator outlet fitting.

i. Connect the heater control cables

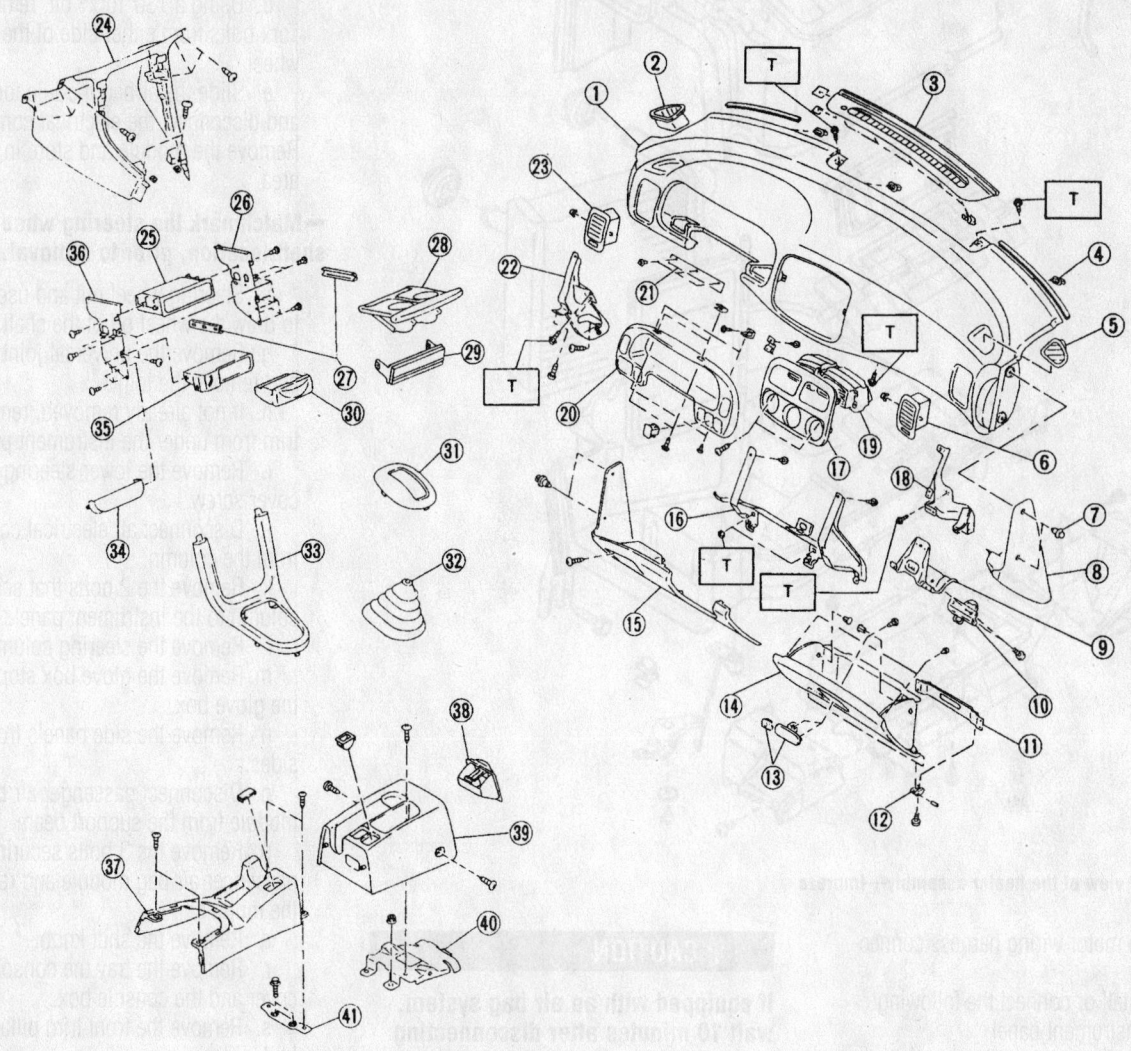

① Pad & frame	⑯ Reinf. CTR	㉛ Panel (AT) ASSY
② Grille SD def. (D)	⑰ Panel CTR (A)	㉜ Shift boot
③ Front def. grille	⑱ Reinf. (P)	㉝ Console cover
④ Grommet	⑲ Grille CTR def.	㉞ Panel (Airbag)
⑤ Grille SD def. (P)	⑳ Meter visor	㉟ Housing (Ash tray)
⑥ Grille vent (P)	㉑ Cover	㊱ BRKT (Radio) LH
⑦ Clip	㉒ Reinf. (D)	㊲ Center console
⑧ SD panel (P)	㉓ Grille vent (D)	㊳ Ash tray
⑨ Reinforcement striker	㉔ Instrument panel console	㊴ Rear console box
⑩ Striker	㉕ Pocket CTR	㊵ Rear console BRKT
⑪ Frame pocket	㉖ BRKT (Radio) RH	㊶ Center console BRKT
⑫ Hinge	㉗ Rail (Cup holder)	
⑬ Lock ASSY	㉘ Cup holder	**Tightening torque: N·m (kg-cm, in-lb)**
⑭ Pocket ASSY	㉙ Panel (Radio)	T: 6.9 ± 1.0 (70 ± 10, 60.8 ± 8.7)
⑮ Lower cover ASSY	㉚ Ash tray	

87970G59

Exploded view of the instrument panel and center console—Impreza

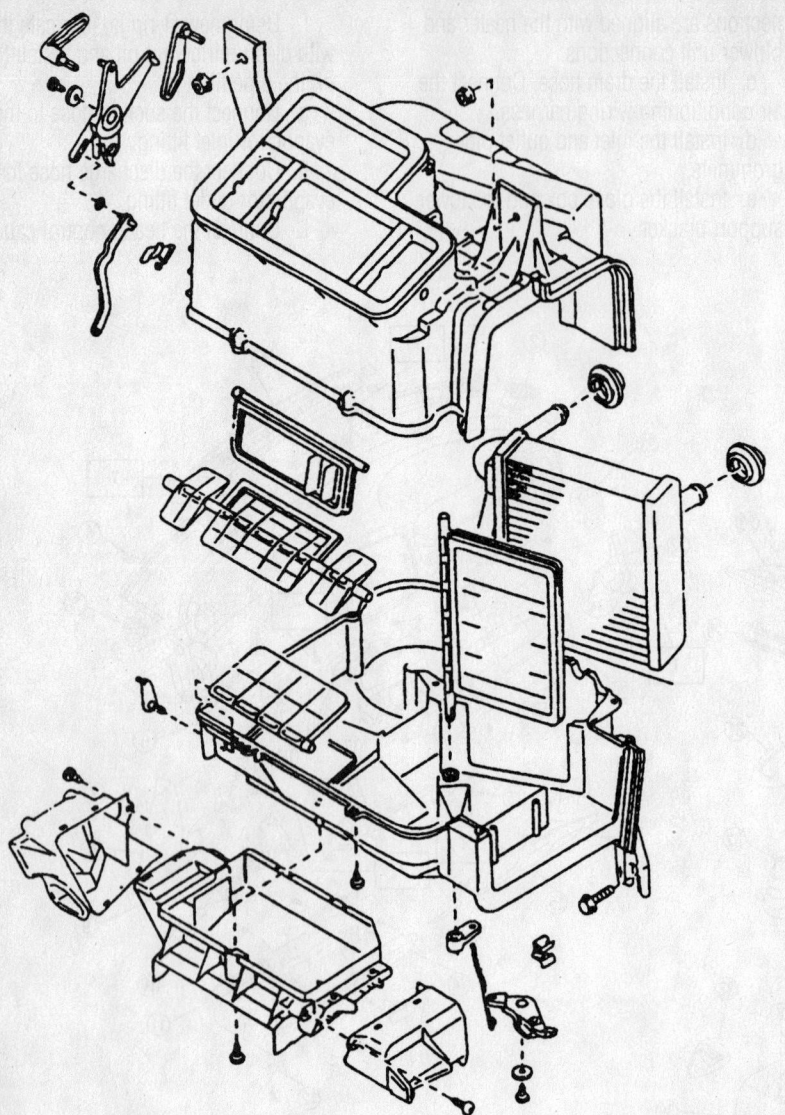

93112G95

Exploded view of the heater assembly—Impreza

and fan motor wiring harness connectors.

12. Install or connect the following:
   - Instrument panel
   - Radio and console assemblies
   - Heater hoses in the engine compartment
13. Refill the cooling system.
14. Connect the negative battery cable.
15. Evacuate, charge and leak test the air conditioning system.
16. Operate the engine to normal operating temperatures; then, check the climate control operation and check for leaks.

Outback

1. Before servicing the vehicle, refer to the precautions in the beginning of this section.
2. Disconnect the negative battery cable.

✼✼ CAUTION

If equipped with an air bag system, wait 10 minutes after disconnecting the negative battery cable before performing any further work while the system fully de-energizes in order to avoid accidental deployment. All air bag system wiring is yellow. Do not use electrically powered test equipment on these circuits.

3. Drain the cooling system into a clean container for reuse.
4. Remove or disconnect the following:
   - Negative battery cable
   - LLC
   - Heater hoses from the heater core. Plug the heater core and heater hoses.
   - Bolts attaching the expansion valve

and pipe in the engine compartment
5. Remove the instrument panel as follows:
   a. Remove the lower cover.
   b. Remove the lower column cover and disconnect the harness.
   c. Set the tires in the straight ahead position.
   d. Using a T30 Torx® bit, remove the Torx bolts from either side of the steering wheel.
   e. Slide the airbag module forward and disconnect the electrical connection. Remove the module and store in a safe area.

➡ **Matchmark the steering wheel to shaft location, prior to removal.**

   f. Steering wheel nut and use a puller to draw the wheel off of the shaft.
   g. Remove the universal joint bolts and remove the joint.
   h. If not already removed, remove the trim from under the instrument panel.
   i. Remove the lower steering column cover screw.
   j. Disconnect all electrical connectors from the column.
   k. Remove the 2 bolts that secure the column to the instrument panel.
   l. Remove the steering column.
   m. Remove the glove box stopper and the glove box.
   n. Remove the side panels from both sides.
   o. Disconnect passenger air bag module from the support beam.
   p. Remove the 3 bolts securing the passenger air bag module and remove the module.
   q. Remove the shift knob.
   r. Remove the tray the console box cover and the console box.
   s. Remove the front trim pillar from both sides.
   t. Remove the front pillar lower trim from the passenger side.
   u. Set the temperature control switch to **FULL HOT** and disconnect the control cable from the bottom of the heater unit. Do not move the switch and link.
   v. Remove the instrument panel bolts, harness connectors and the panel.
6. Remove or disconnect the following:
   - Keyless and CRU units
   - Sunroof connector
   - Servo motor connector
   - Heater blower transistor connector
   - Blower motor and in-vehicle temperature sensor connectors
   - Intake unit bolts and nuts

- Drain hose from the intake unit
- Intake unit
- Heater unit case screws
- Heater core from the case

To install:

7. Install or connect the following:

- Heater core into the heater case. Secure it in place with the retaining screws.
- Drain hose to the intake unit
- Intake unit to its mounting position. Torque the mounting bolts to 48–84 inch lbs. (5–9 Nm).
- Blower motor and in-vehicle temperature sensor connectors
- Heater blower transistor connector
- Servo motor connector
- Sunroof connector

- Keyless and CRU units

8. Install the instrument panel as follows:

a. Install the instrument panel, tighten the bolts and connect the harness connectors

b. Connect the control cable to the bottom of the heater unit. Do not move the switch and link during installation.

c. Install the front pillar lower trim to the passenger side.

d. Install the front trim pillar on both sides.

e. Install the console box, box cover and tray.

f. Install the shift knob.

g. Passenger side airbag module and tighten the 3 bolts securing the module.

h. Connect passenger air bag module to the support beam.

i. Install the side panels on both sides.

j. Install the glove box and stopper.

k. Install the steering column.

l. Install the 2 bolts that secure the column to the instrument panel. Tighten the bolts to 18 ft. lbs. (25 Nm).

m. Connect all electrical connectors to the column.

n. Install the lower steering column cover with the tilt lever held in the lowered position.

o. Install the universal joint. Align the bolt hole on the long yoke side of the joint with the cutout at the serrated section of the shaft end and insert the joint.

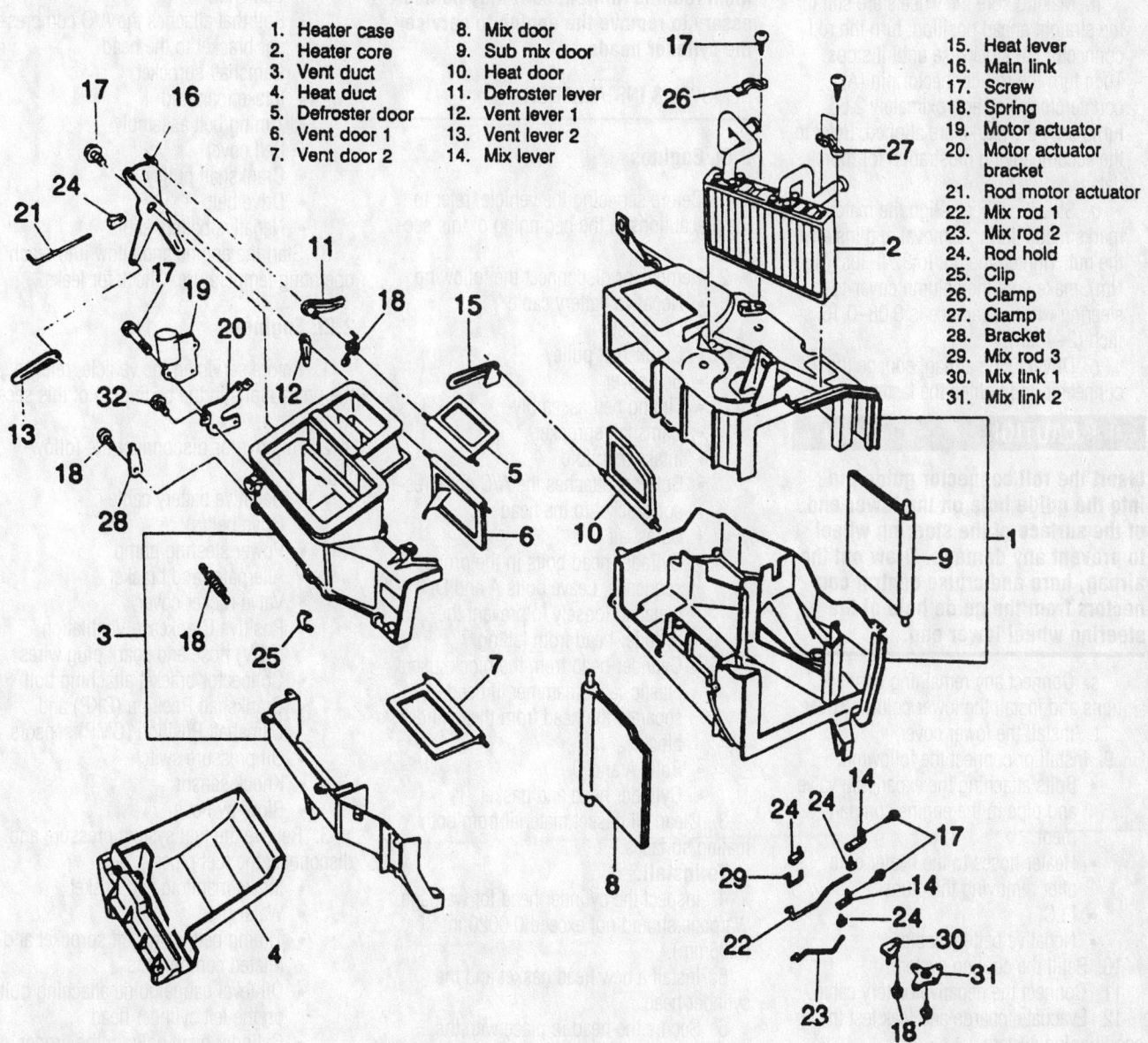

1. Heater case	8. Mix door
2. Heater core	9. Sub mix door
3. Vent duct	10. Heat door
4. Heat duct	11. Defroster lever
5. Defroster door	12. Vent lever 1
6. Vent door 1	13. Vent lever 2
7. Vent door 2	14. Mix lever

15. Heat lever
16. Main link
17. Screw
18. Spring
19. Motor actuator
20. Motor actuator bracket
21. Rod motor actuator
22. Mix rod 1
23. Mix rod 2
24. Rod hold
25. Clip
26. Clamp
27. Clamp
28. Bracket
29. Mix rod 3
30. Mix link 1
31. Mix link 2

Exploded view of the heater assembly—Legacy

93112G93

Align the bolt hole on the short yoke side of the joint at the serrated section of the gearbox assembly and lower the joint completely. Temporarily tighten the bolt on the short yoke side and raise the joint to ensure the bolt is passing properly through the cutout. Tighten the bolt on the long yoke side, then the short yoke side to 17 ft. lbs. (24 Nm).

✳✳ CAUTION

Make sure the joint bolt is tightened through the notch in the shaft serration. Do not overtighten the joint bolts as this can lead to heavy steering wheel operation. Make sure the clearance between the gearbox is over 0.59 inch (15mm).

p. Making sure the wheels are still in the straight ahead position, turn the roll connector (A) clockwise until it stops. Then turn the roll connector pin (A) counterclockwise approximately 2.65 turns until the marks are aligned. Refer to the accompanying illustration for more detail.

q. Steering wheel, align the matchmarks made during removal and install the nut. Tighten the nut to 32 ft. lbs. (45 Nm). make sure the column cover-to-steering wheel clearance is 0.08–0.16 inch (2–4mm).

r. Drivers side airbag, engage the connector and tighten the fasteners.

✳✳ CAUTION

Insert the roll connector guide pin into the guide hole on the lower end of the surface of the steering wheel to prevent any damage. Draw out the airbag, horn and cruise control connectors from the guide hole of the steering wheel lower end.

s. Connect any remaining connections and install the lower column cover.

t. Install the lower cover.

9. Install or connect the following:
- Bolts attaching the expansion valve and pipe in the engine compartment
- Heater hoses to the heater core after removing the plugs.
- LLC
- Negative battery cable

10. Refill the cooling system.

11. Connect the negative battery cable.

12. Evacuate, charge and leak test the air conditioning system.

13. Operate the engine to normal operat-

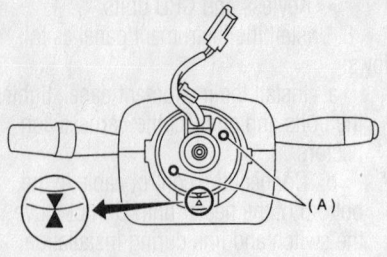

42356-SBCR-G02

Make sure the roll pin is properly aligned—Outback

ing temperatures; then, check the climate control operation and check for leaks.

Cylinder Head

➡**On some models, engine compartment room is limited, so it may be necessary to remove the engine to service the cylinder heads.**

REMOVAL & INSTALLATION

2.0L Engines

1. Before servicing the vehicle, refer to the precautions in the beginning of this section.

2. Remove or disconnect the following:
- Negative battery cable
- Drive belt
- Crankshaft pulley
- Belt cover
- Timing belt assembly
- Camshaft sprocket
- Intake manifold
- Bolt that attaches the A/C compressor bracket to the head
- Camshaft
- Cylinder head bolts in the proper sequence. Leave bolts A and D installed loosely to prevent the cylinder head from falling.
- Cylinder head from the block, use a plastic-faced hammer, if needed, to separate the head from the cylinder block
- Bolts A and D
- Cylinder head and gasket

3. Clean all gasket material from both mating surfaces.

To install:

4. Inspect the cylinder head for warpage. Warpage should not exceed 0.0020 in. (0.05mm).

5. Install a new head gasket and the cylinder head.

6. Secure the head in place with the mounting bolts. Coat each bolts with clean engine oil, and hand-tighten. Tighten the

cylinder head bolts, in sequence, to the following specifications:

a. Step 1: 22 ft. lbs. (29 Nm).

b. Step 2: 51 ft. lbs. (69 Nm).

c. Step 3: loosen all bolts by 180 degrees, then loosen an additional 180 degrees.

d. Step 4: bolts 1 and 2 to 25 ft. lbs. (34 Nm).

e. Step 5: bolts 3, 4, 5 and 6 to 11 ft. lbs. (15 Nm).

f. Step 6: all bolts plus 80–90 degrees.

g. Step 7: all bolts plus 80–90 degrees.

➡**Do not exceed 180 degrees total tightening.**

7. Install or connect the following:
- Camshaft
- Bolt that attaches the A/C compressor bracket to the head
- Camshaft sprocket
- Intake manifold
- Timing belt assembly
- Belt cover
- Crankshaft pulley
- Drive belt
- Negative battery cable

8. Start the engine and allow it to reach operating temperature. Check for leaks.

2.2L Engines

1. Before servicing the vehicle, refer to the precautions in the beginning of this section.

2. Remove or disconnect the following:
- Negative battery cable
- Drive belt(s)
- Power steering pump
- Alternator and bracket
- Valve rocker cover
- Positive Crankcase Ventilation (PCV) hose and spark plug wires
- Connector bracket attaching bolt
- Crankshaft Position (CKP) and Camshaft Position (CMP) sensors
- Oil pressure switch
- Knock sensor
- Blow-by hose

3. Relieve the fuel system pressure and disconnect the fuel pipes.
- Intake manifold and gasket
- Water pipe
- Timing belt, camshaft sprocket and related components
- Oil level gauge guide attaching bolt on the left cylinder head
- Cylinder head bolts in the proper sequence. Leave bolts 1 and 3

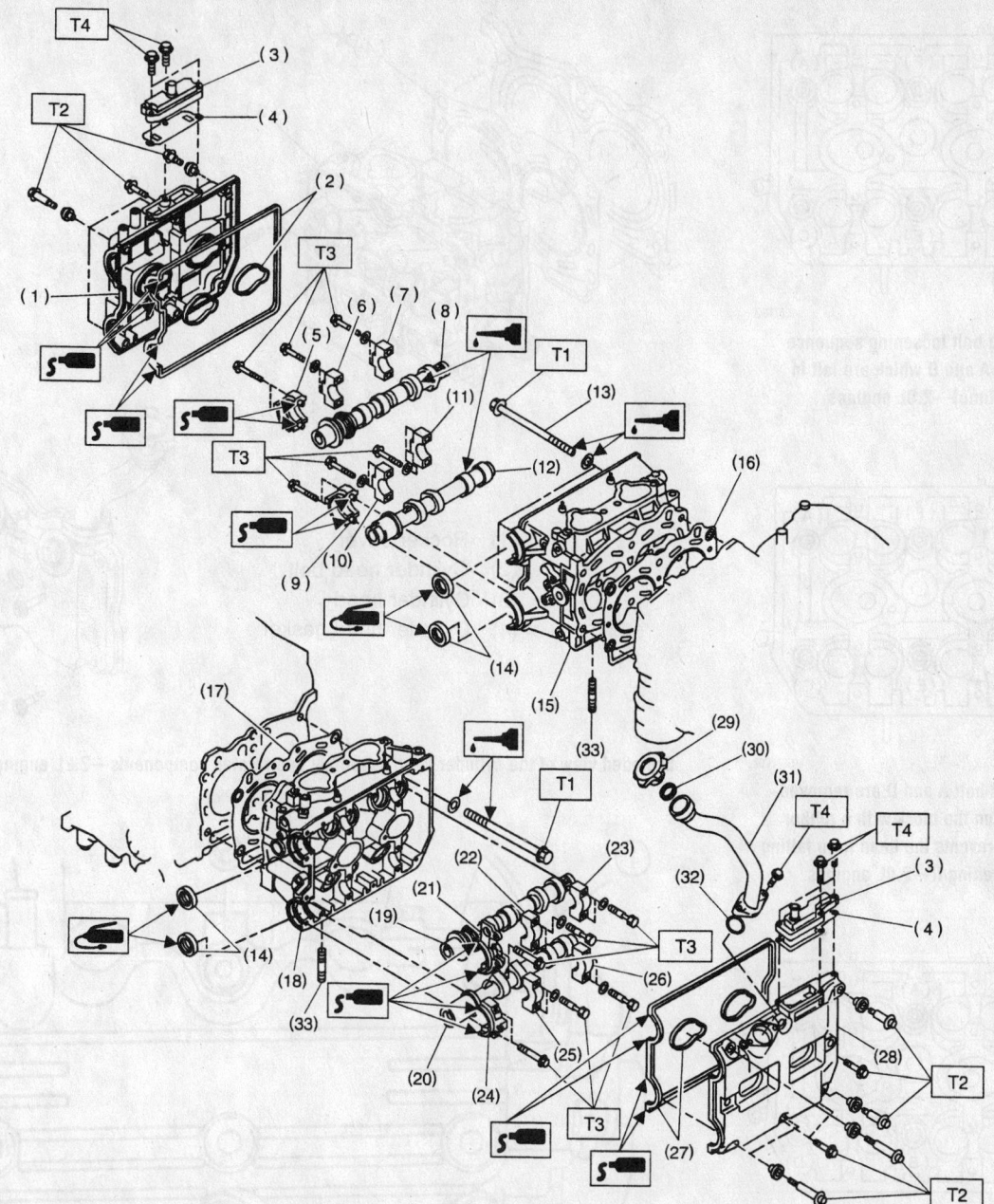

(1)	Rocker cover (RH)	(15)	Cylinder head (RH)	(29)	Oil filler cap
(2)	Rocker cover gasket (RH)	(16)	Cylinder head gasket (RH)	(30)	Gasket
(3)	Oil separator cover	(17)	Cylinder head gasket (LH)	(31)	Oil filler duct
(4)	Gasket	(18)	Cylinder head (LH)	(32)	O-ring
(5)	Intake camshaft cap (Front RH)	(19)	Intake camshaft (LH)	(33)	Stud bolt
(6)	Intake camshaft cap (Center RH)	(20)	Exhaust camshaft (LH)		
(7)	Intake camshaft cap (Rear RH)	(21)	Intake camshaft cap (Front LH)		
(8)	Intake camshaft (RH)	(22)	Intake camshaft cap (Center LH)		
(9)	Exhaust camshaft cap (Front RH)	(23)	Intake camshaft cap (Rear LH)		
(10)	Exhaust camshaft cap (Center RH)	(24)	Exhaust camshaft (Front LH)		
(11)	Exhaust camshaft cap (Rear RH)	(25)	Exhaust camshaft cap (Center LH)		
(12)	Exhaust camshaft (RH)	(26)	Exhaust camshaft cap (Rear LH)		
(13)	Cylinder head bolt	(27)	Rocker cover gasket (LH)		
(14)	Oil seal	(28)	Rocker cover (LH)		

Tightening torque: N·m (kgf-m, ft-lb)

T1: *<Ref. to ME(DOHC TURBO)-64, INSTALLATION, Cylinder Head Assembly.>*

T2: *5 (0.5, 3.6)*

T3: *10 (1.0, 7)*

T4: *6.4 (0.65, 4.7)*

9357TG05

Exploded view of the cylinder head assembly and related components—2.0L engines

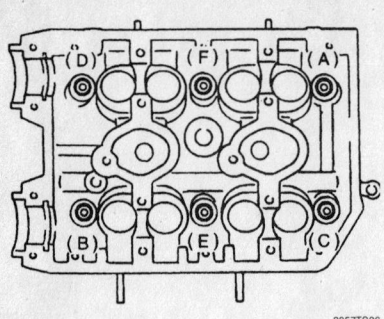

9357TG06

Cylinder head bolt loosening sequence (except bolts A and D which are left in place at this time)—2.0L engines

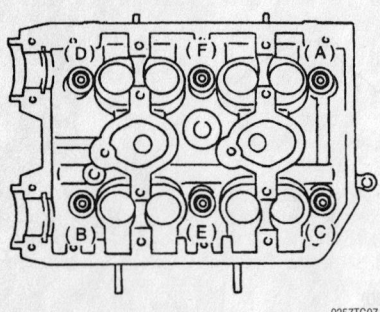

9357TG07

Cylinder head bolt A and D are removed after tapping on the block with a rubber mallet, this prevents the head from falling off while loosening it—2.0L engines

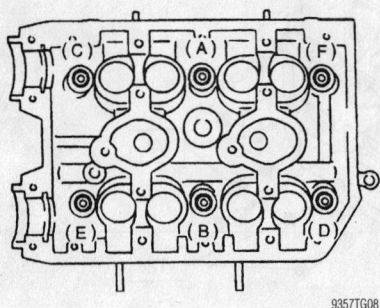

9357TG08

Cylinder head bolt tightening sequence—2.0L engines

installed loosely to prevent the cylinder head from falling.

- Cylinder head from the block, use a plastic-faced hammer, if needed, to separate the head from the cylinder block
- Bolts 1 and 3
- Cylinder head and gasket

4. Clean all gasket material from both mating surfaces.

To install:

5. Inspect the cylinder head for warpage. Warpage should not exceed 0.0020 in. (0.05mm).

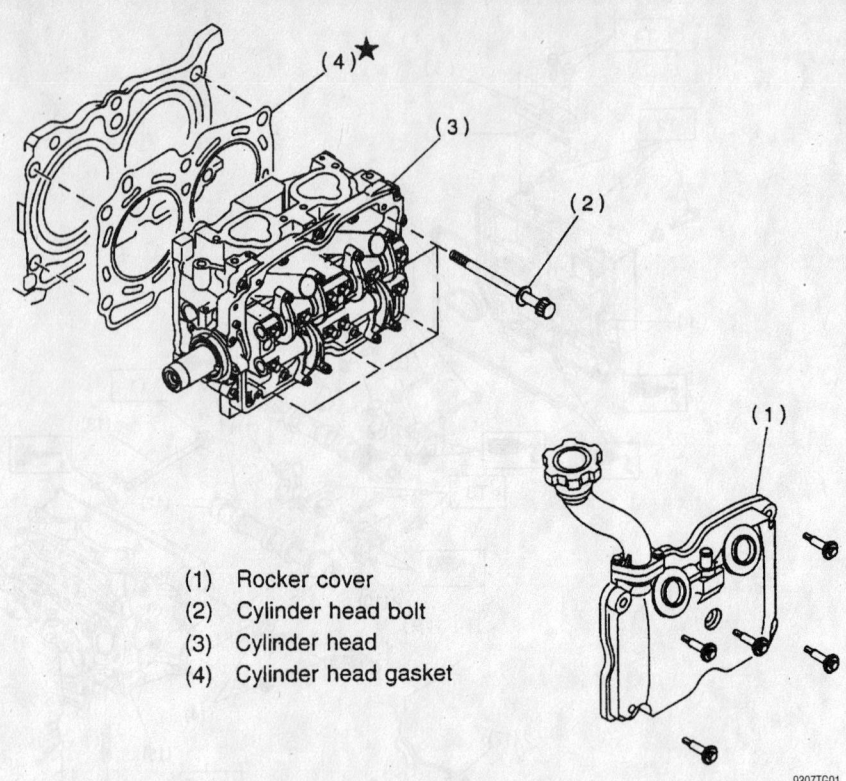

(1) Rocker cover
(2) Cylinder head bolt
(3) Cylinder head
(4) Cylinder head gasket

9307TG01

Exploded view of the cylinder head assembly and related components—2.2L engines

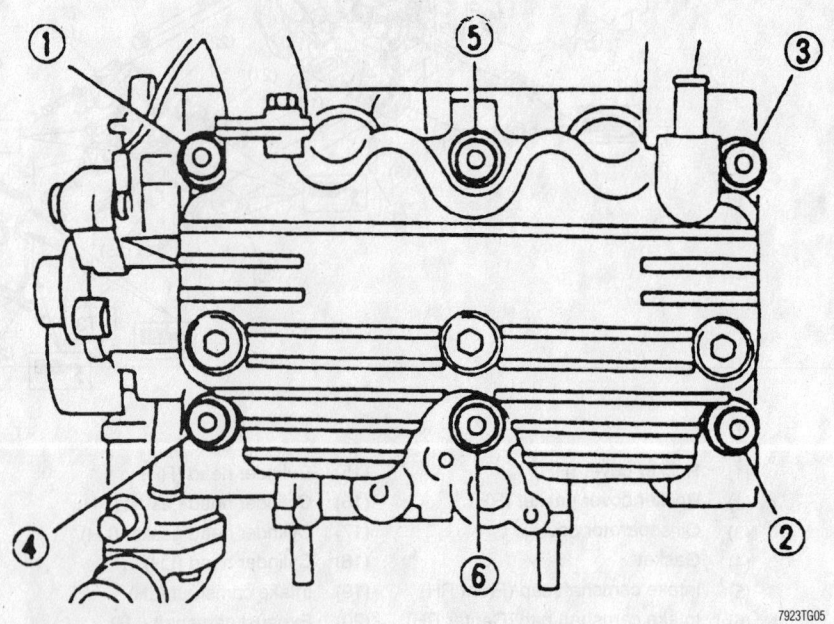

7923TG05

Cylinder head bolt loosening sequence—2.2L engines

6. Install a new head gasket and the cylinder head.

7. Secure the head in place with the mounting bolts. Coat each bolts with clean engine oil, and hand-tighten. Tighten the cylinder head bolts, in sequence, to the following specifications:

a. Step 1: 22 ft. lbs. (29 Nm).
b. Step 2: 51 ft. lbs. (69 Nm).
c. Step 3: loosen all bolts by 180 degrees, then loosen an additional 180 degrees.
d. Step 4: bolts 1 and 2 to 25 ft. lbs. (34 Nm).
e. Step 5: bolts 3, 4, 5 and 6 to 11 ft. lbs. (15 Nm).
f. Step 6: all bolts plus 80–90 degrees.

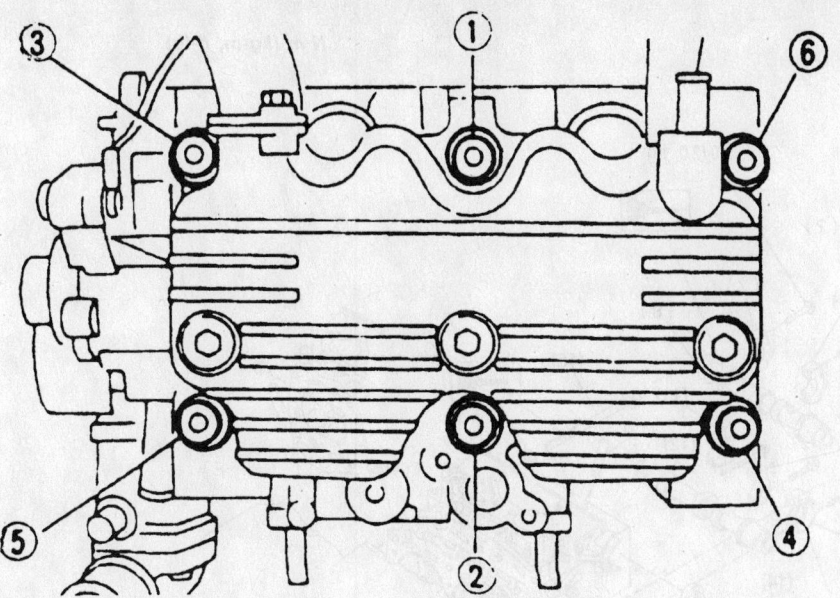

Cylinder head bolt tightening sequence—2.2L engines

a. Step 1: all bolts to 22 ft. lbs. (29 Nm).

b. Step 2: all bolts to 51 ft. lbs. (69 Nm).

c. Step 3: loosen all bolts by 180 degrees, then loosen an additional 180 degrees.

d. Step 4: bolts 1 and 2 to 25 ft. lbs. (24 Nm).

e. Step 5: bolts 3, 4, 5 and 6 to 11 ft. lbs. (15 Nm).

f. Step 6: all bolts plus 80–90 degrees.

g. Step 7: All bolts plus 80–90 degrees.

➡**Do not exceed 180 degrees total tightening.**

8. Install or connect the following:
   • Camshafts, refer to the procedure in this section
   • Valve covers
   • Oil level gauge guide attaching bolt on the left cylinder head
   • Timing belt, camshaft sprockets and related components
   • Water pipe
   • Intake manifold and tighten the bolts to 21–25 ft. lbs. (28–34 Nm)
   • Fuel delivery pipes
   • Blow-by hose
   • Knock sensor
   • CKP and CMP sensors
   • Connector bracket attaching bolt
   • Spark plug wires
   • Valve rocker cover and tighten the bolts to 48 inch lbs. (9 Nm)
   • Alternator
   • Power steering pump
   • Accessory drive belt
   • Negative battery cable

9. Start the engine and allow it to reach operating temperature. Check for leaks.

3.0L Engine

1. Before servicing the vehicle, refer to the precautions in the beginning of this section.

2. Remove or disconnect the following:
   • Negative battery cable
   • Crankshaft pulley cover
   • Crankshaft pulley bolt using tool 499977100 to hold the crankshaft
   • Crankshaft pulley
   • Timing chain cover and the chain. Refer to the procedure in this section.
   • Camshaft and crankshaft sprockets. Refer to the procedure in this section.

g. Step 7: all bolts plus 80–90 degrees.

➡**Do not exceed 180 degrees total tightening.**

8. Install or connect the following:
   • Oil level gauge guide attaching bolt on the left cylinder head
   • Timing belt, camshaft sprocket and related components
   • Water pipe
   • Intake manifold and tighten the bolts to 21–25 ft. lbs. (28–34 Nm)
   • Fuel delivery pipes
   • Blow-by hose
   • Knock sensor
   • Oil pressure switch connector
   • CKP and CMP sensors
   • Connector bracket attaching bolt
   • Spark plug wires
   • PCV hose
   • Valve rocker cover and tighten the bolts to 44 inch lbs. (5 Nm)
   • Alternator
   • Power steering pump
   • Accessory drive belt
   • Negative battery cable

9. Start the engine and allow it to reach operating temperature. Check for leaks.

2.5L Engine

1. Before servicing the vehicle, refer to the precautions in the beginning of this section.

2. Remove or disconnect the following:
   • Negative battery cable
   • Accessory drive belts
   • Power steering pump
   • Alternator and bracket
   • Valve rocker cover
   • Connector bracket attaching bolt
   • Crankshaft Position (CKP) and Camshaft Position (CMP) sensors
   • Coolant filler tank

3. Relieve the fuel system pressure and disconnect the fuel pipes.
   • Intake manifold and gasket
   • Water pipe
   • Timing belt, camshaft sprocket and related components
   • Oil level gauge guide attaching bolt on the left cylinder head
   • Valve covers
   • Camshafts, refer to the camshaft procedure in this section
   • Cylinder head bolts, in the proper sequence. Leave bolts 1 and 3 installed loosely to prevent the cylinder head from falling.
   • Cylinder head from the block using a plastic-faced hammer, if needed
   • Bolts 1 and 3
   • Cylinder head and gasket

4. Clean all gasket material from both mating surfaces.

To install:

5. Inspect the cylinder head for warpage. Warpage should not exceed 0.0020 in. (0.05mm).

6. Install a new head gasket and the cylinder head.

7. Secure the head in place with the mounting bolts. Coat each bolts with clean engine oil, and hand-tighten. Tighten the cylinder head bolts to the following specifications:

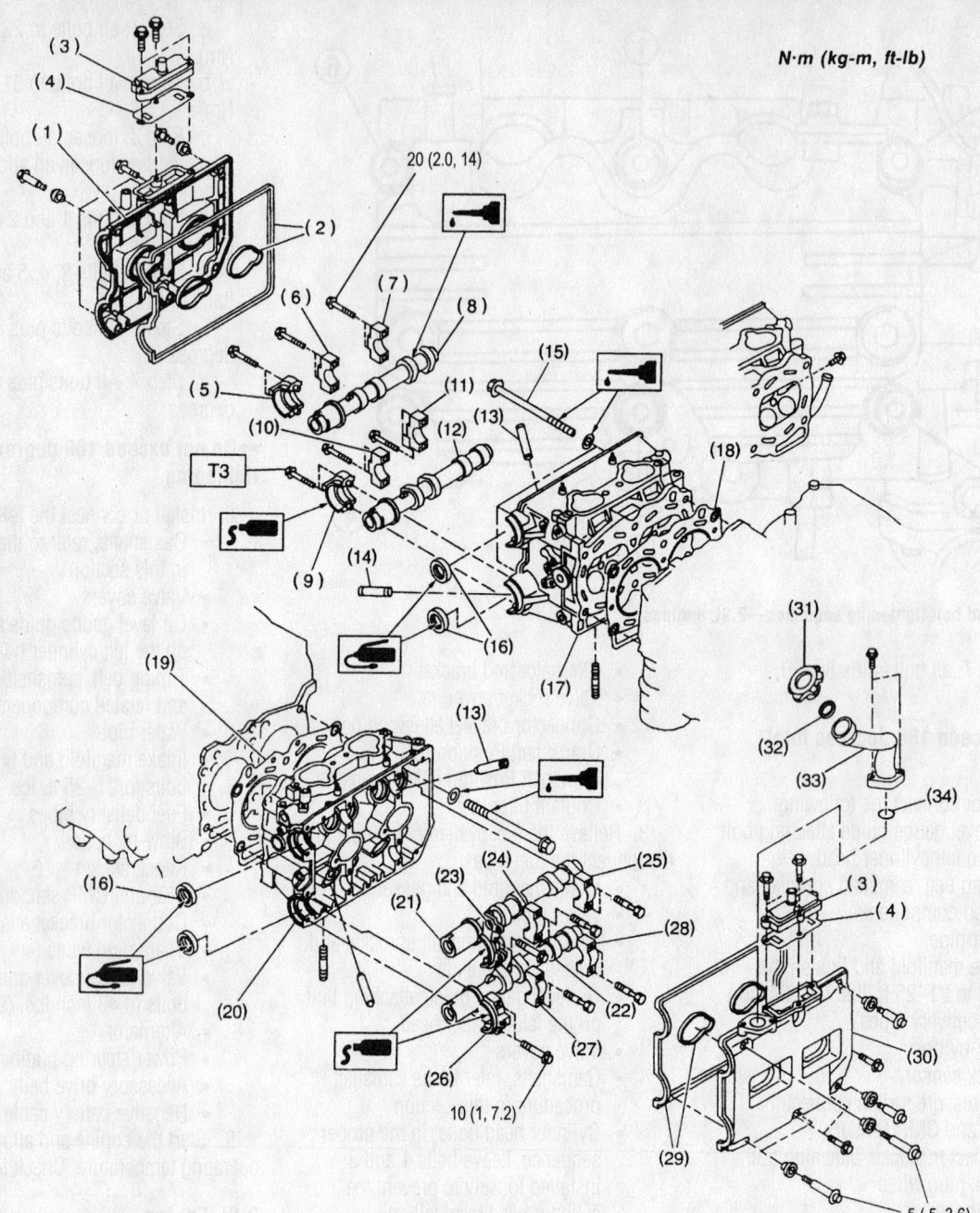

N·m (kg-m, ft-lb)

20 (2.0, 14)

T3

10 (1, 7.2)

5 (.5, 3.6)

(1) Rocker cover (RH)
(2) Rocker cover gasket (RH)
(3) Oil separator cover
(4) Gasket
(5) Intake camshaft cap (Front RH)
(6) Intake camshaft cap (Center RH)
(7) Intake camshaft cap (Rear RH)
(8) Intake camshaft (RH)
(9) Exhaust camshaft cap (Front RH)
(10) Exhaust camshaft cap (Center RH)
(11) Exhaust camshaft cap (Rear RH)
(12) Exhaust camshaft (RH)
(13) Intake valve guide
(14) Exhaust valve guide

(15) Cylinder head bolt
(16) Oil seal
(17) Cylinder head (RH)
(18) Cylinder head gasket (RH)
(19) Cylinder head gasket (LH)
(20) Cylinder head (LH)
(21) Intake camshaft (LH)
(22) Exhaust camshaft (LH)
(23) Intake camshaft cap (Front LH)
(24) Intake camshaft cap (Center LH)
(25) Intake camshaft cap (Rear LH)
(26) Exhaust camshaft (Front LH)
(27) Exhaust camshaft cap (Center LH)
(28) Exhaust camshaft cap (Rear LH)

(29) Rocker cover gasket (LH)
(30) Rocker cover (LH)
(31) Oil filler cap
(32) Gasket
(33) Oil filler duct
(34) O-ring

Exploded view of the cylinder head and related components—2.5L engine

7923TG07

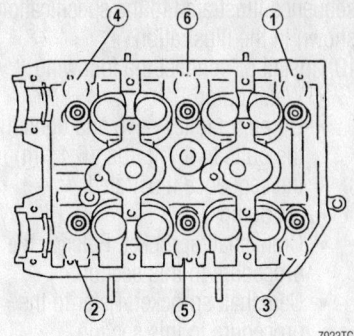

Cylinder head bolt loosening sequence—2.5L engine

7923TG08

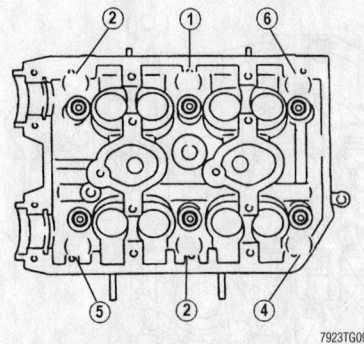

Cylinder head bolt tightening sequence—2.5L engines

7923TG09

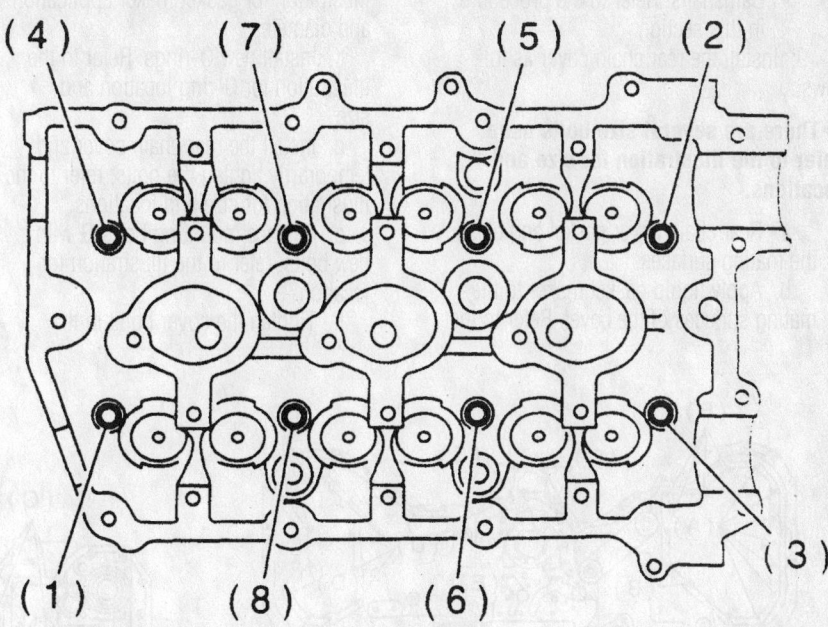

Cylinder head bolt loosening sequence—3.0L engines

42356-SBCR-G03

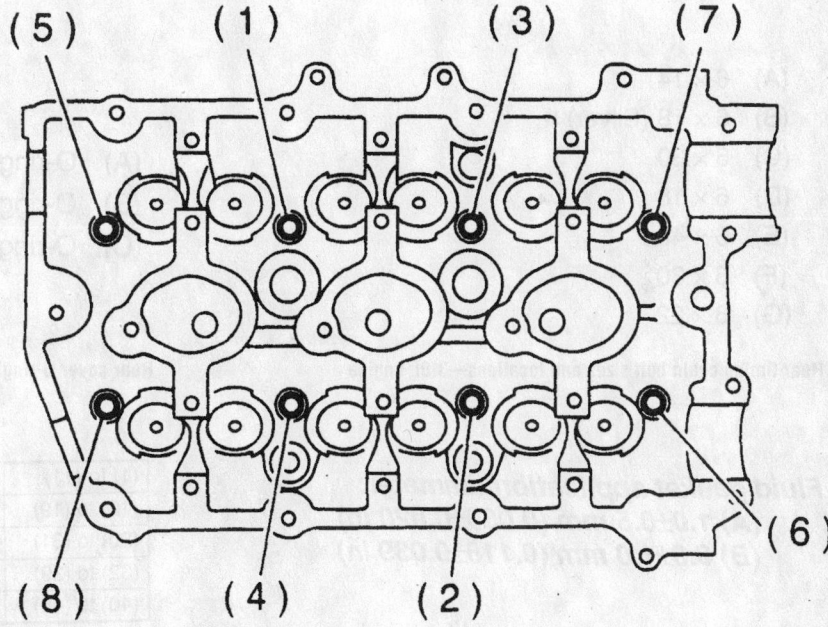

Cylinder head bolt tightening sequence—3.0L engines

42356-SBCR-G04

- Oil pump. Refer to the procedure in this section.
- Oil pump relief valve
- Water pump. Refer to the procedure in this section.
- Rear chain cover
- Camshafts. Refer to the procedure in this section.
- Cylinder head bolts in the sequence illustrated. Leave bolts 2 and 4 connected by a few threads to prevent the head from falling. Tap the head with a plastic mallet to separate it from the block.
- Bolts 2 and 4, the cylinder head and gasket

3. Clean all gasket material from both mating surfaces.

To install:

4. Inspect the cylinder head for warpage. Warpage should not exceed 0.0020 in. (0.05mm).

5. Make sure not to scratch or damage the mating surfaces of the cylinder head, block and oil pump.

6. Install a new head gasket and the cylinder head.

7. Secure the head in place with the mounting bolts. Coat each bolts with clean engine oil, and hand-tighten. Tighten the cylinder head bolts to the following specifications:

a. Step 1: all bolts to 14 ft. lbs. (20 Nm).

b. Step 2: all bolts to 37 ft. lbs. (50 Nm).

c. Step 3: loosen all bolts by 180 degrees, then loosen an additional 180 degrees.

d. Step 4: all bolts to 18 ft. lbs. (25 Nm).

e. Step 5: all bolts to 18 ft. lbs. (25 Nm).

f. Step 6: all bolts plus 90 degrees.

g. Step 7: bolts 1, 2, 3 and 4 plus 80–90 degrees.

h. Step 8: bolts 5, 6, 7 and 8 plus 45 degrees.

➡**Do not exceed 180 degrees total tightening.**

8. Install or connect the following:

- Camshafts. Refer to the procedure in this section.

9. Install the rear chain cover as follows:

➡ **There are several size bolts used, refer to the illustration for size and locations.**

a. Rear chain cover gasket and clean the mating surfaces.

b. Apply liquid gasket maker to the mating surfaces of the cover. Refer to the illustration for gasket maker application and diameter.

c. Install new O-rings. Refer to the illustration for O-ring location and size

d. Install the rear chain cover and temporarily tighten the bolts, refer to the illustration for size and locations.

e. Replace mounting bolts **G** with new bolts, refer to the illustration for location.

f. Tighten the cover bolts in the sequence illustrated to the specifications shown in the illustration.

10. Install or connect the following:
- Water pump
- Oil pump relief valve and tighten the bolts to 4.7 ft. lbs. (6.4 Nm) in the sequence illustrated.
- Oil pump
- Crankshaft sprocket. Refer to the procedure in this section.
- Camshaft sprocket. Refer to the procedure in this section.

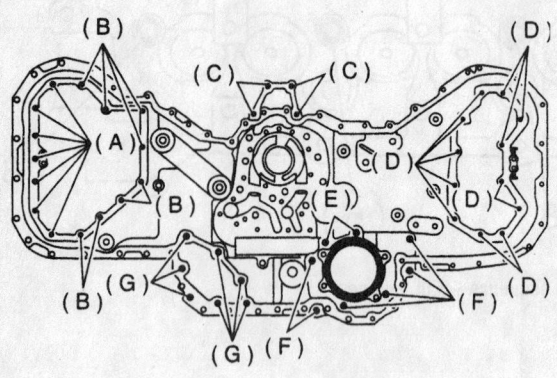

(A)	6 × 14
(B)	6 × 18 (Silver)
(C)	6 × 30
(D)	6 × 18
(E)	6 × 40
(F)	6 × 30
(G)	6 × 22

42356-SBCR-G05

Rear timing chain bolt sizes and locations—3.0L engine

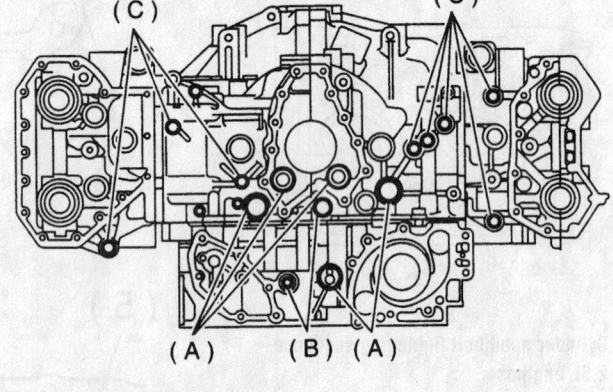

(A)	O-ring (Large)
(B)	O-ring (Medium)
(C)	O-ring (Small)

42356-SBCR-G07

Rear cover O-ring sizes and locations—3.0L engine

Fluid gasket application diameter:
(A) 1.0±0.5 mm (0.039±0.020 in)
(B) 3.0±1.0 mm (0.118±0.039 in)

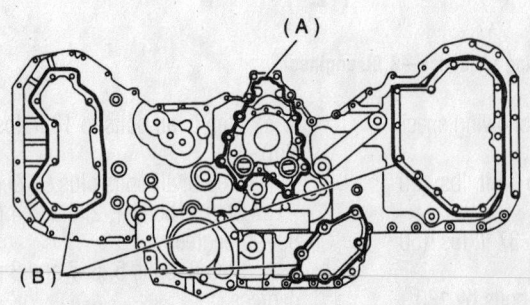

42356-SBCR-G06

Apply liquid gasket maker to the mating surfaces of the cover as shown—3.0L engine

(1) to (11)	9 N·m (0.9 kgf-m, 6.5 ft-lb)
(12) to (19)	20 N·m (2.0 kgf-m, 14 ft-lb)
(20) to (31)	9 N·m (0.9 kgf-m, 6.5 ft-lb)
(32) to (39)	12 N·m (1.2 kgf-m, 8.7 ft-lb)
(40) to (46)	9 N·m (0.9 kgf-m, 6.5 ft-lb)

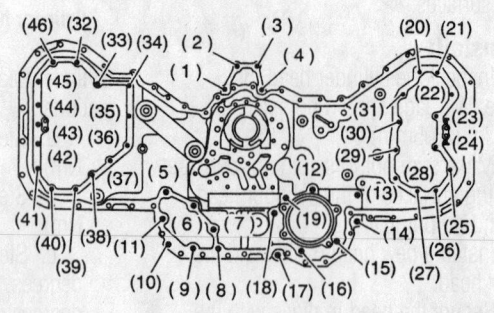

42356-SBCR-G08

Rear cover bolt tightening sequence and torque specifications— 3.0L engine

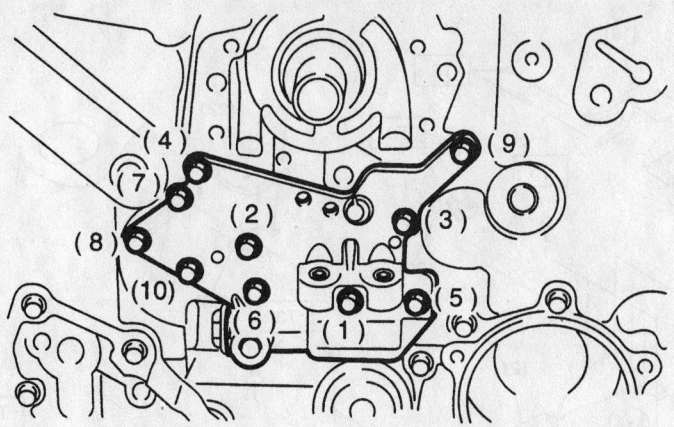

Bolt installation position	Bolt dimension
(1) and (5)	6 x 26
(2), (3), (4) and (9)	6 x 35
(6), (7), (8) and (10)	6 x 16

42356-SBCR-G09

Oil pump relief valve bolt tightening sequence and bolt specifications—3.0L engine

- Timing chain. Refer to the procedure in this section.
- Crankshaft pulley, apply clean oil to the bolt threads and tighten the bolt to 131 ft. lbs. (178 Nm)
- Camshaft pulley cover and tighten the bolts to 5 ft. lbs. (7 Nm)
- Negative battery cable

11. Start the engine and allow it to reach operating temperature. Check for leaks.

Rocker Arms/Shafts

REMOVAL & INSTALLATION

2.2L Engines

1. Before servicing the vehicle, refer to the precautions in the beginning of this section.

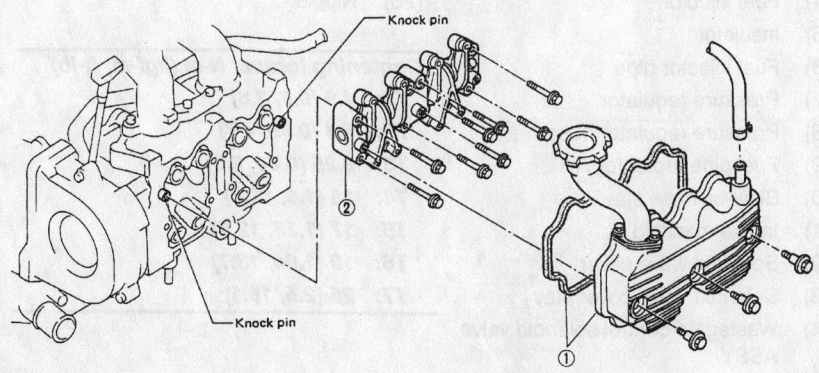

Exploded view of the cylinder head and rocker assembly—2.2L engines

2. Remove or disconnect the following:

- Positive Crankcase Ventilation (PCV) hose
- Rocker cover
- Valve rocker assembly by removing bolts 2 through 4 in numerical sequence

3. Loosen bolt 1, but leave it engaged to retain the valve rocker assembly.

4. Remove or disconnect the following:
- Bolts 5 through 8, taking care not to gouge the dowel pin
- Valve rocker assembly

To install:

5. Install the valve rocker assembly on the cylinder head.

6. Temporarily tighten bolts 1 through 4 equally.

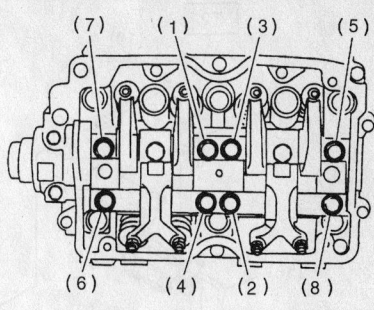

9307TG02

Rocker shaft bolt loosening/tightening sequence—2000–01 2.2L engines

➡ **Do not allow the valve rocker assembly to gouge the dowel pins.**

7. Tighten bolts 5–8 to 108 inch lbs. (12 Nm).

8. Tighten bolts 1–4 to 108 inch lbs. (12 Nm).

9. Install the rocker cover and connect the PCV hose.

Except 2.2L Engines

These engines are not equipped with either rocker shafts or rocker arms. Instead, the camshafts act directly on the individual valves.

Intake Manifold

REMOVAL & INSTALLATION

2.0L Engines

1. Before servicing the vehicle, refer to the precautions in the beginning of this section.

2. Release the fuel system pressure.

3. Remove or disconnect the following:

- Negative battery cable
- Engine cover, if necessary
- Air cleaner upper cover and boot
- Air cleaner filter
- Intercooler
- Accelerator cable
- Coolant filler tank

4. Remove the power steering pump from the bracket by performing the following steps:

a. Loosen the lock and slider bolts.
b. Remove the V-belt.
c. Disconnect the power steering switch connection.
d. Remove the bolts that attach the power steering pump pipe brackets to the intake manifold. Do not disconnect the hose.

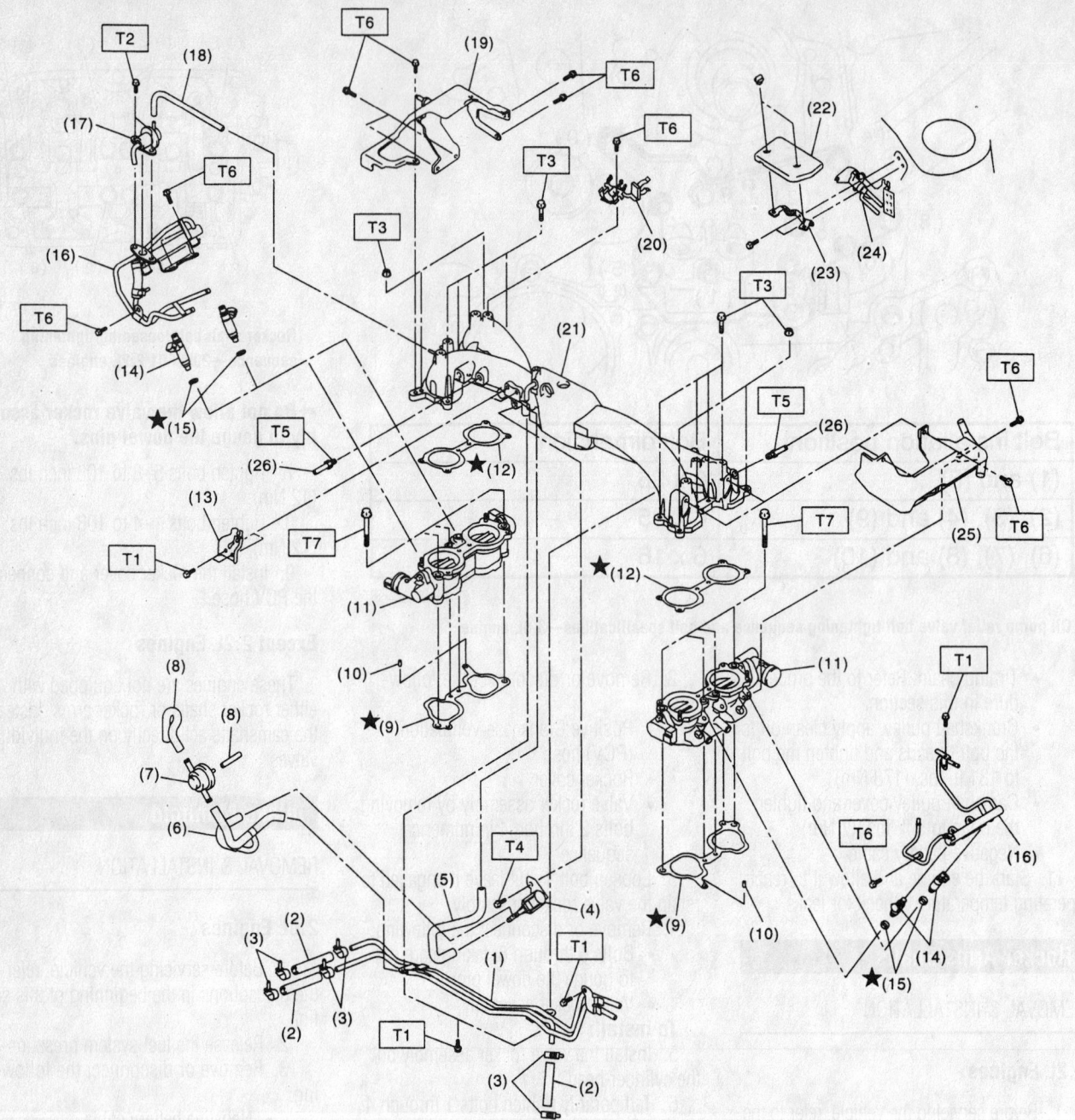

(1)	Fuel pipe ASSY	(13)	Accelerator cable bracket
(2)	Fuel hose	(14)	Fuel injector
(3)	Clip	(15)	Insulator
(4)	Purge control solenoid valve	(16)	Fuel injector pipe
(5)	Vacuum hose	(17)	Pressure regulator
(6)	Vacuum control hose	(18)	Pressure regulator hose
(7)	Purge valve	(19)	Fuel pipe protector RH
(8)	Purge hose	(20)	Blow-by hose stay
(9)	Intake manifold gasket	(21)	Intake manifold
(10)	Guide pin	(22)	Solenoid valve cover
(11)	Tumble generator valve ASSY	(23)	Solenoid valve cover stay
(12)	Tumble generator valve gasket	(24)	Wastegate control solenoid valve ASSY

(25) Fuel pipe protector LH
(26) Nipple

Tightening torque: N·m (kgf-m, ft-lb)

T1:	4.9 (0.5, 3.6)
T2:	6.4 (0.65, 4.7)
T3:	8.25 (0.84, 6.1)
T4:	16 (1.6, 11.8)
T5:	17 (1.73, 12.5)
T6:	19 (1.94, 13.7)
T7:	25 (2.5, 18.1)

Exploded view of the intake manifold assembly—2.0L engine

9357TG09

e. Bolts that attach the power steering pump bracket.

f. Remove the power steering tank from the bracket by pulling it upwards.

g. Place the power steering pump on the wheel apron on the right.

5. Remove or disconnect the following:
- Emission hose from the Positive Crankcase Ventilation (PCV) valve
- Engine coolant temperature hoses from the throttle body
- Brake booster hose
- Pressure hose from the intake duct
- Engine harness connectors from the bulkhead connections

6. Disconnect the following electrical connections:
- Engine Coolant Temperature (ECT) sensor
- Oil pressure switch
- Crankshaft Position (CKP) sensor
- Knock Sensor (KS)
- Camshaft Position (CMP) sensor
- Ignition coil

7. Remove or disconnect the following:
- Engine harness fixed clip from the bracket
- Fuel delivery, return and evaporative hoses
- Intake manifold bolts
- Intake manifold and gasket

To install:

8. Install or connect the following:
- Intake manifold and gasket
- Intake manifold bolts and tighten to 18 ft. lbs. (25 Nm)
- Fuel delivery, return and evaporative hoses
- Engine harness fixed clip from the bracket

9. Connect the following electrical connections:
- Oil pressure switch
- CKP sensor
- ECT sensor
- KS sensor
- CMP sensor
- Ignition coil

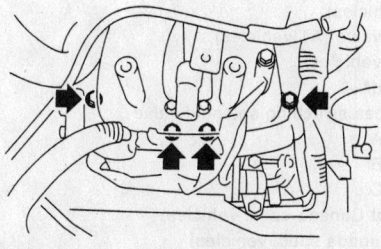

Location of the intake manifold bolts— 2.0L engine

9357TG10

- Engine harness connectors to the bulkhead connections

10. Install or connect the following:
- Brake booster hose
- Engine coolant temperature hoses to the throttle body
- Emission hose to the PCV valve
- Pressure hose to the intake duct

11. Install the power steering pump on the bracket by performing the following steps:

a. Install the power steering tank on the bracket.

b. Attach the power steering switch connection.

c. Install the bolts that attach the power steering pump bracket and tighten to 16 ft. lbs. (22 Nm).

d. Attach the power steering pump pipe brackets to the intake manifold.

e. Install the V-belt.

12. Install or connect the following:
- Coolant filler tank
- Accelerator cable
- Intercooler
- Air cleaner filter
- Air cleaner upper cover and boot
- Engine undercover
- Negative battery cable

13. Fill the cooling system.

14. Start the engine and allow it to reach operating temperature. Check for leaks and test drive the vehicle.

2.2L Engines

1. Before servicing the vehicle, refer to the precautions in the beginning of this section.

2. Release the fuel system pressure.

3. Remove or disconnect the following:
- Negative battery cable
- Engine cover, if necessary

4. Drain the cooling system into a suitable container.
- Mass Air Flow (MAF) sensor electrical connector, if equipped
- Intake Air Temperature (IAT) sensor electrical connector, if equipped
- Clamp that connects the air intake duct to the air intake chamber
- Air cleaner cover clips
- Blow-by hose from the air intake duct
- Air intake duct and air cleaner cover as an assembly
- Air cleaner element

5. Loosen the clamp that connects the air intake chamber to the throttle body
- Air hoses and air intake chamber
- Accelerator cable and cruise control cable, if equipped

- Vacuum hoses from the pressure sources switching solenoid valve
- Resonator chamber
- Power steering pump drive belt cover(s) and belt
- Power steering pipe brackets from the right side of the intake manifold
- Power steering pump from bracket, then position on the right side wheel apron; do not disconnect the fluid lines
- Fuel pipe protector, if equipped
- Spark plug wires and electrical connector from ignition coil
- Positive Crankcase Ventilation (PCV) hose and pressure regulator vacuum hose from intake manifold
- Coolant hoses from the throttle body
- Engine coolant hose and air by-pass hose from the idle air control solenoid valve
- Brake booster hose
- Cruise control vacuum hose, if equipped
- Exhaust Gas Recirculation (EGR) pipe, on 2.2L engines with automatic transaxle
- Canister hoses from the pipes
- Engine harness connectors from bulkhead harness connectors, then remove from bracket
- Engine Coolant Temperature (ECT) sensor connector and thermometer connector, if equipped
- Knock sensor connector
- Crankshaft Position (CKP) and Camshaft Position (CMP) sensor connectors
- Oil pressure switch connector
- Fuel supply lines
- Intake manifold bolts
- Intake manifold and discard the gaskets

6. Clean all gasket material from both mating surfaces.

To install:

7. Use a straightedge and a feeler gauge to inspect the intake manifold for flatness. Distortion should not exceed 0.020 in. (0.5mm).

8. Install or connect the following:
- Intake manifold to the engine. Tighten the retaining bolts to 16.7–19.5 ft. lbs. (23–27 Nm).
- Fuel lines
- Oil pressure switch electrical connector
- CKP and CMP sensors
- Knock sensor
- ECT sensor and thermometer connectors

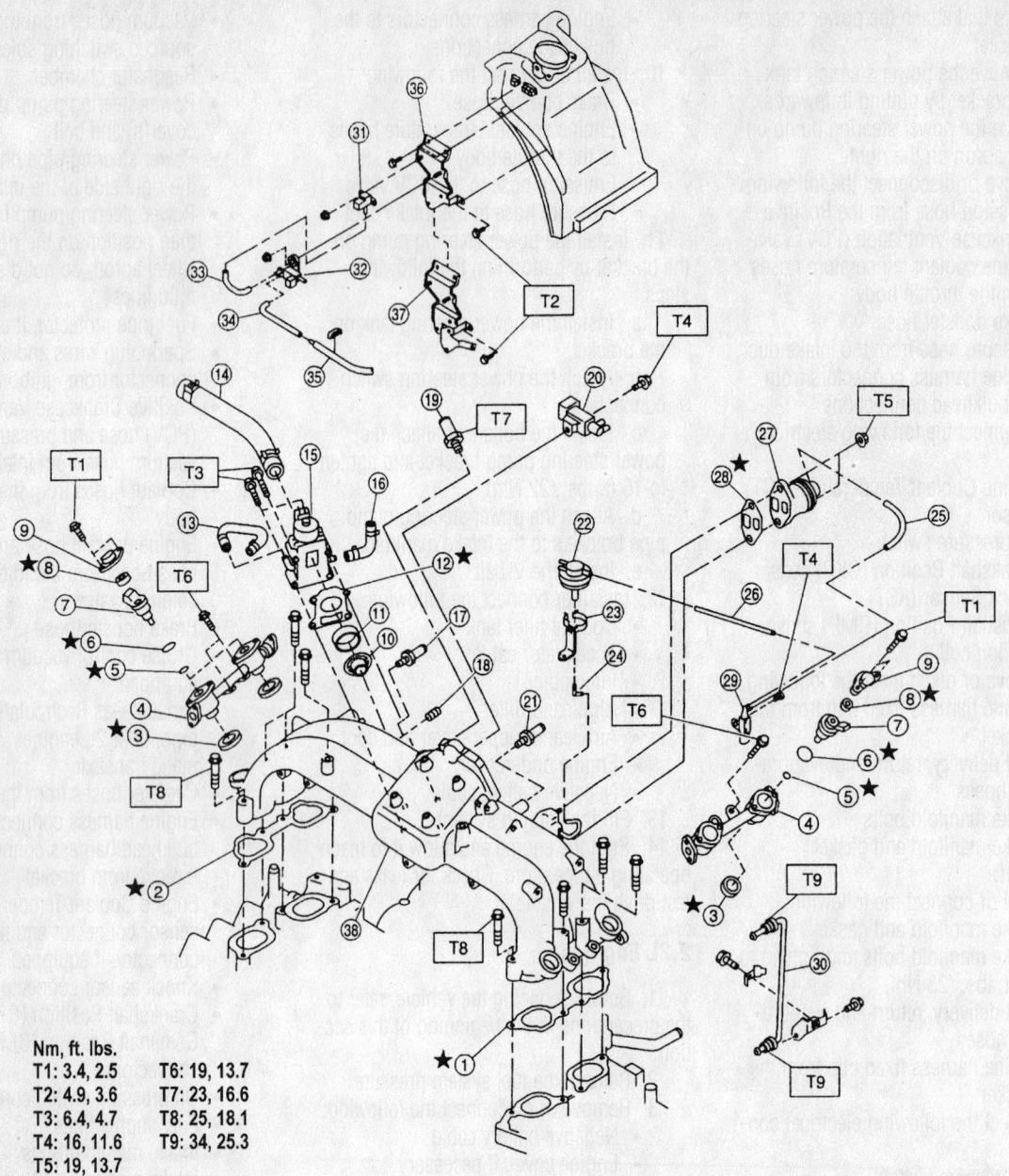

Nm, ft. lbs.

T1: 3.4, 2.5	T6: 19, 13.7
T2: 4.9, 3.6	T7: 23, 16.6
T3: 6.4, 4.7	T8: 25, 18.1
T4: 16, 11.6	T9: 34, 25.3
T5: 19, 13.7	

1. Intake manifold gasket LH
2. Intake manifold gasket RH
3. Fuel injector pipe insulator
4. Fuel injector pipe
5. O-ring A
6. O-ring B
7. Fuel injector
8. Insulator
9. Fuel injector cap
10. Plate
11. Sealing
12. Gasket
13. Engine coolant hose B
14. Air by-pass hose
15. Idle air control solenoid valve
16. Engine coolant hose A
17. Nipple (AT vehicles)
18. Plug
19. PCV valve

20. Purge control solenoid valve
21. Nipple
22. BPT (AT vehicles)
23. BPT holder bracket (AT vehicles)
24. Back pressure hose (AT vehicles)
25. EGR vacuum hose A (AT vehicles)
26. EGR vacuum hose B (AT vehicles)
27. EGR valve (AT vehicles)
28. Gasket (AT vehicles)
29. EGR solenoid valve (AT vehicles)
30. EGR pipe (AT vehicles)
31. Pressure sensor
32. Pressure sources switching solenoid valve
33. Vacuum hose A
34. Vacuum hose B
35. Vacuum hose C
36. Bracket (Except Canada spec. vehicles)
37. Bracket (For Canada spec. vehicles)
38. Intake manifold

7923TG16

Exploded view of the intake manifold assembly—2.2L engine

- Engine harness connector to the bracket and bulkhead
- Canister hoses, if disconnected
- Install the EGR pipe, if removed
- Cruise control vacuum hose, if equipped
- Brake booster vacuum hose
- Air bypass hose to the FICD solenoid valve
- PCV valve hose to the manifold
- Coolant hoses to throttle body
- Electrical connector and spark plug wires to ignition coil
- Fuel pipe protector, if equipped
- Power steering pump to bracket
- Power steering pump bolts and tighten to 13–16.6 ft. lbs. (17.6–22.6 Nm)
- Power steering brackets to the right side of the intake manifold
- Power steering pump belt and adjust the belt
- Power steering pump belt cover
- Vacuum hoses to pressure sources switching solenoid valve
- Accelerator cable and cruise control cable, if equipped
- Air intake chamber and hoses
- Air cleaner element, and the upper cover and duct as a unit
- MAF sensor electrical connector
- Engine cover, if equipped
- Negative battery cable

9. Fill the cooling system.

10. Start the engine and allow it to reach operating temperature. Check for leaks and test drive the vehicle.

2.5L Engine

1. Before servicing the vehicle, refer to the precautions in the beginning of this section.

2. Disconnect the negative battery cable.

3. Drain the cooling system into a suitable container.

4. Remove or disconnect the following:
- Mass Air Flow (MAF) sensor connector, if necessary
- Air intake duct, air cleaner upper cover and the air cleaner element

5. Properly release the fuel pressure.
- Accelerator cable and the cruise control cable, if equipped
- Resonator chamber, if equipped
- V-belt cover(s)

6. Loosen the lock bolt and slider bolt, then remove the power steering belt.

7. Remove or disconnect the following:
- Power steering pipe bracket-to-manifold bolts

- Bolts holding the power steering pump to the bracket
- Connector from the power steering pump switch, if equipped
- Power steering pump, and place on the right side wheel apron. Do NOT disconnect the fluid lines.
- Spark plug wires from the spark plugs
- Positive Crankcase Ventilation (PCV) hose and vacuum hose from the intake manifold
- Engine coolant hoses from the throttle body
- Brake booster hose
- Air cleaner case stay (right side) and engine harness bracket, if necessary
- Engine harness connectors form the bulkhead harness connectors
- Engine Coolant Temperature (ECT) sensor connector
- Knock sensor, Camshaft Position (CMP) sensor and Crankshaft Position (CKP) sensor electrical connectors
- Oil pressure switch
- Fuel hoses from the fuel pipes
- Intake manifold mounting bolts
- Intake manifold and discard the gaskets

➡ **The intake manifold sits on pins that protrude from the cylinder heads. Be sure the pins remain in the cylinder heads.**

To install:

8. Install or connect the following:
- New gaskets
- Intake manifold to the engine. Tighten the mounting bolts to 16.7–19.5 ft. lbs. (23–27 Nm).
- Fuel hoses to the fuel pipes, be sure to secure the hoses with new clamps
- Knock sensor, CMP sensor, CKP sensor, oil pressure switch and ECT sensor wiring
- Engine harness bracket and engine harness connectors to bulkhead connectors
- EGR pipe, if removed
- Brake booster hose
- Engine coolant hose and air bypass hose to idle air control solenoid valve, if removed
- Engine coolant hoses to the throttle body
- PCV hoses to intake manifold
- Spark plug wires to ignition coil or plugs, as applicable
- Power steering pump to bracket.

Tighten the bolts to 13–16.6 ft. lbs. (17.6–22.6 Nm).
- Power steering pipe brackets on right side of the intake manifold
- Bolt, which installs the power steering pump stiffener on the engine block, to 14.4–17.3 ft. lbs. (20–24 Nm)
- Power steering pump belt and adjust the belt as necessary
- V-belt cover(s)
- Resonator chamber, if equipped
- Accelerator cable and the cruise control cable, if equipped
- Air cleaner assembly
- MAF sensor connector, if disconnected

9. Connect the negative battery cable and refill the cooling system. Start the engine, and bleed the cooling system. Check for leaks.

3.0L Engine

1. Before servicing the vehicle, refer to the precautions in the beginning of this section.

2. Disconnect the negative battery cable.

3. Properly release the fuel pressure.

4. Remove or disconnect the following:
- Air intake duct, air cleaner upper cover and the air cleaner element
- Resonator chamber
- Accelerator cable and the cruise control cable, if equipped
- Power steering pipe bracket-to-manifold bolts
- V-belt
- Bolts holding the power steering pump to the bracket
- Connector from the power steering pump switch
- Power steering pump, and place on the right side wheel apron. Do NOT disconnect the fluid lines.
- Washer reservoir bolts, hose from the front washer motor and connector from the rear washer motor.
- Rear washer hose from the motor and plug the hose
- Washer reservoir aside
- Positive Crankcase Ventilation (PCV) hose from the cylinder head cover
- Engine coolant hoses from the throttle body
- Brake booster hose
- Exhaust Gas Recirculation (EGR) pipe from the valve being careful not to drop the gaskets
- Engine harness connectors from the bulkhead connectors

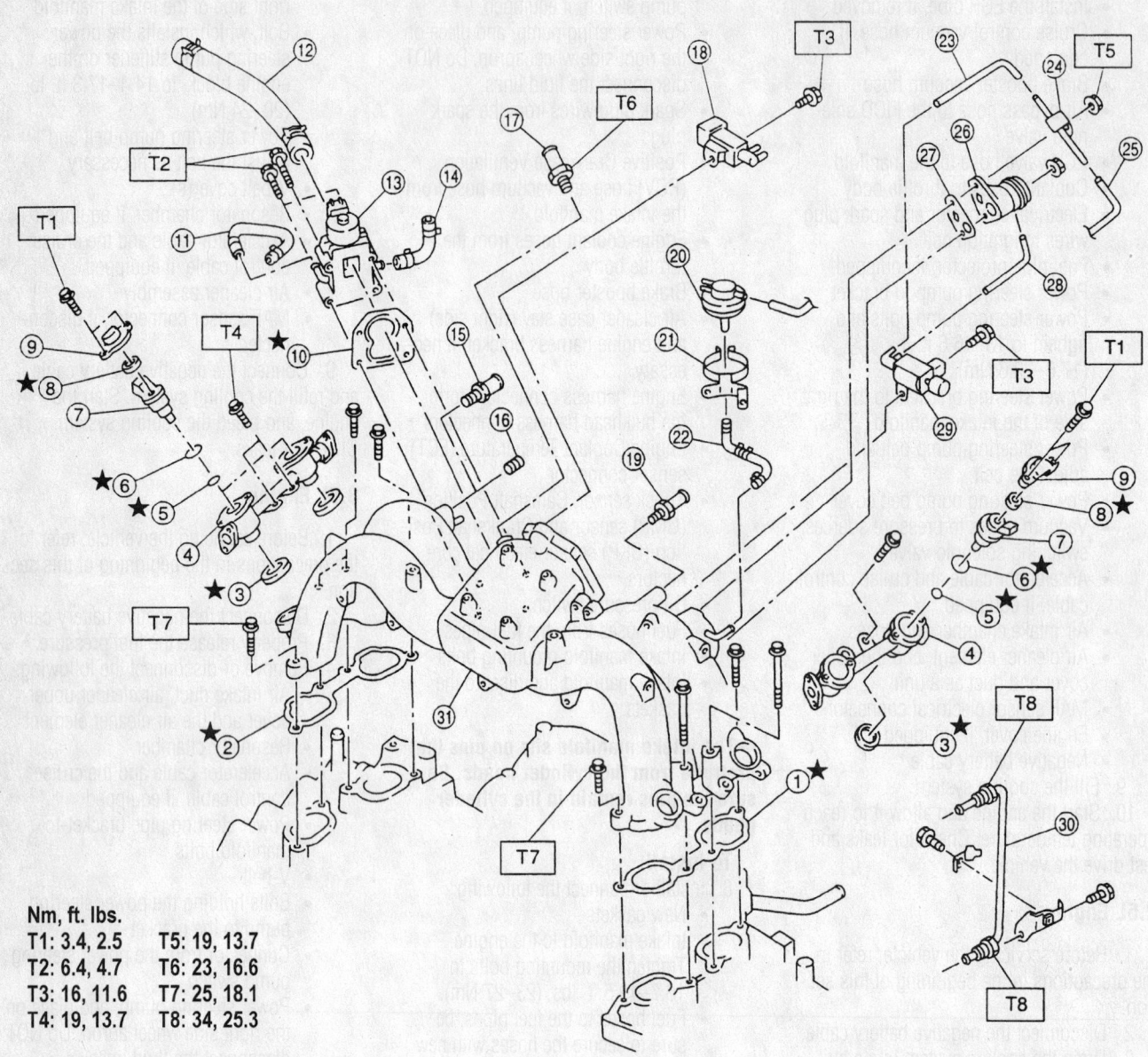

Nm, ft. lbs.

T1: 3.4, 2.5	T5: 19, 13.7
T2: 6.4, 4.7	T6: 23, 16.6
T3: 16, 11.6	T7: 25, 18.1
T4: 19, 13.7	T8: 34, 25.3

① Intake manifold gasket RH
② Intake manifold gasket LH
③ Fuel injector pipe insulator
④ Fuel injector pipe
⑤ O-ring A
⑥ O-ring B
⑦ Fuel injector
⑧ Insulator
⑨ Fuel injector cap
⑩ Gasket
⑪ Engine coolant hose B

⑫ Air by-pass hose
⑬ Idle air control solenoid valve
⑭ Engine coolant hose A
⑮ Nipple (AT model)
⑯ Plug
⑰ PCV valve
⑱ Purge control solenoid valve
⑲ Nipple
⑳ BPT
㉑ BPT holder bracket

㉒ Back pressure hose
㉓ EGR vacuum hose A
㉔ EGR vacuum pipe
㉕ EGR vacuum hose C
㉖ EGR valve
㉗ Gasket
㉘ EGR vacuum hose B
㉙ EGR solenoid valve
㉚ EGR pipe
㉛ Intake manifold

7923TG17

Exploded view of the intake manifold and related components—2.5L engine

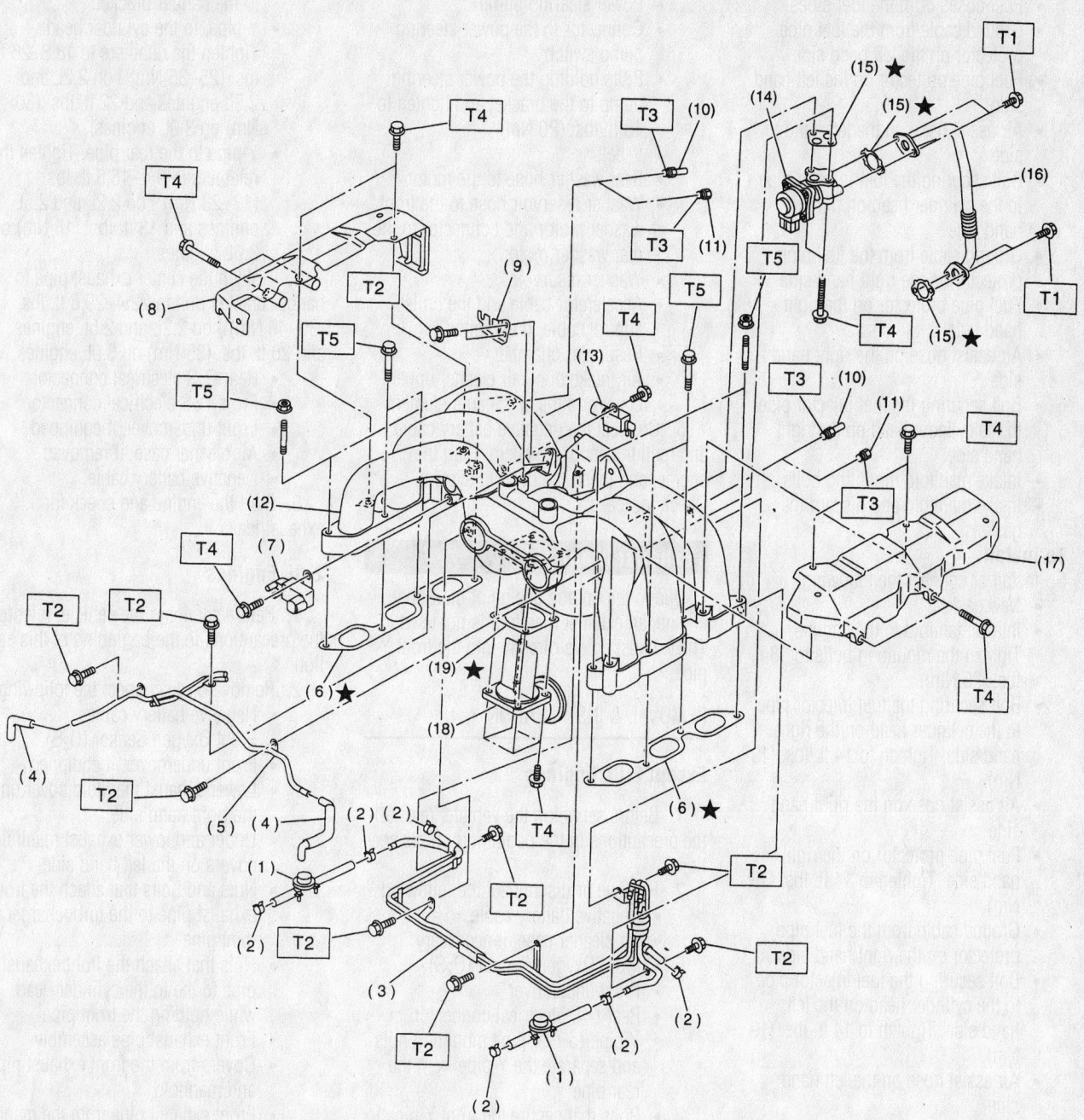

(1)	Fuel damper valve	(10)	Nipple	(19)	Gasket
(2)	Clamp	(11)	Plug		
(3)	Fuel pipe ASSY	(12)	Intake manifold		
(4)	Air assist hose	(13)	Induction valve control solenoid		
(5)	Air assist and purge pipe ASSY	(14)	EGR valve		
(6)	Gasket	(15)	Gasket		
(7)	Purge control solenoid valve	(16)	EGR pipe		
(8)	Fuel pipe protector RH	(17)	Fuel pipe protector LH		
(9)	Accelerator cable bracket	(18)	Induction valve		

Tightening torque: N·m (kgf-m, ft-lb)

T1:	6.4 (0.65, 4.7)
T2:	5.0 (0.51, 3.7)
T3:	17 (1.7, 12)
T4:	19 (1.9, 14)
T5:	25 (2.5, 18)

42356-SBCR-G10

Exploded view of the intake manifold and related components—3.0L engine

- Engine ground terminal from the manifold
- Fuel hoses from the fuel pipes
- Ground cable from the fuel pipe protector on the left hand side
- Fuel pipe protector on the left hand side
- Air assist hose on the left hand side
- Bolt securing the fuel injector pipe to the cylinder head on the left hand side
- Ground cable from the fuel pipe protector on the right hand side
- Fuel pipe protector on the right hand side
- Air assist hose on the right hand side
- Bolt securing the fuel injector pipe to the cylinder head on the right hand side
- Intake manifold mounting bolts
- Intake manifold and discard the gaskets

To install:

5. Install or connect the following:
- New gaskets
- Intake manifold to the engine. Tighten the mounting bolts to 18 ft. lbs. (25 Nm).
- Bolt securing the fuel injector pipe to the cylinder head on the right hand side. Tighten to 14 ft. lbs. (19 Nm).
- Air assist hose on the right hand side
- Fuel pipe protector on the right hand side. Tighten to 14 ft. lbs. (19 Nm).
- Ground cable from the fuel pipe protector on the right hand side
- Bolt securing the fuel injector pipe to the cylinder head on the left hand side. Tighten to 14 ft. lbs. (19 Nm).
- Air assist hose on the left hand side
- Fuel pipe protector on the left hand side. Tighten to 14 ft. lbs. (19 Nm).
- Ground cable to the fuel pipe protector on the left hand side
- EGR pipe to the valve using new gaskets and tighten to 5 ft. lbs. (7 Nm)
- Fuel hoses to the fuel pipes
- Engine ground terminal to the manifold
- Engine harness connectors to the bulkhead connectors
- Brake booster hose
- Engine coolant hoses to the throttle body

- PCV hose to the cylinder head cover
- Power steering pump
- Connector to the power steering pump switch
- Bolts holding the power steering pump to the bracket and tighten to 15 ft. lbs. (20 Nm)
- V-belt
- Rear washer hose to the motor
- Washer reservoir hose to the front washer motor and connector to the rear washer motor.
- Washer reservoir
- Accelerator cable and the cruise control cable, if equipped
- Resonator chamber
- Air intake duct, air cleaner upper cover and the air cleaner element

6. Connect the negative battery cable and refill the cooling system. Start the engine, and bleed the cooling system. Check for leaks.

Exhaust Manifold

Due to the unique design of the Subaru engine, an exhaust manifold is not used. The exhaust enters directly into the front Y-pipe.

REMOVAL & INSTALLATION

Except 2.0L Engines

1. Before servicing the vehicle, refer to the precautions in the beginning of this section.
2. Remove or disconnect the following:
- Negative battery cable
- Air cleaner case, if necessary
- Front Oxygen Sensor (O2S)
- Front undercover
- Rear O2S electrical connector
- Y-pipe-to-rear pipe mounting nuts and separate the Y-pipe from the rear pipe
- Bolts that secure the front Y-pipe to the cylinder head
- Y-pipe from the hanger bracket
- Front exhaust pipe from the catalytic converter and discard the gaskets

To install:

3. Clean all gasket surfaces completely.
4. Install or connect the following:
- New gaskets
- Front catalytic converter to front exhaust pipe. Tighten the bolts to 18.8–26 ft. lbs. (25–35 Nm) on 2.2L and 2.5L engines and 22 ft. lbs. (30 Nm) on 3.0L engines.

- Y-pipe. Temporarily tighten the bolt that holds the center exhaust pipe to the hanger bracket.
- Y-pipe, to the cylinder head. Tighten the retainers to 18.8–26 ft. lbs. (25–35 Nm)) on 2.2L and 2.5L engines and 22 ft. lbs. (30 Nm) on 3.0L engines.
- Y-pipe to the rear pipe. Tighten the retainers to 9.4–16.6 ft. lbs. (13–23 Nm)) on 2.2L and 2.5L engines and 13 ft. lbs. (18 Nm) on 3.0L engines.

5. Tighten the center exhaust pipe to hanger bracket bolt to 22.4–29.6 ft. lbs. (30–40 Nm)) on 2.2L and 2.5L engines and 26 ft. lbs. (35 Nm) on 3.0L engines..
- Rear O2S electrical connector
- Front O2S electrical connector
- Front undercover, if equipped
- Air cleaner case, if removed
- Negative battery cable

6. Start the engine and check for exhaust leaks.

2.0L Engines

1. Before servicing the vehicle, refer to the precautions in the beginning of this section.
2. Remove or disconnect the following:
- Negative battery cable
- Front Oxygen Sensor (O2S)
- Front undercover, if equipped
- Lower exhaust manifold cover on the right hand side
- Upper and lower exhaust manifold covers on the left hand side
- Nuts and bolts that attach the front exhaust pipe to the turbocharger joint pipe
- Nuts that attach the front exhaust pipe to the to the cylinder head while holding the front pipe
- Front exhaust pipe assembly
- Covers from the front exhaust pipe and manifold
- Front exhaust pipe from the manifolds and discard the gaskets

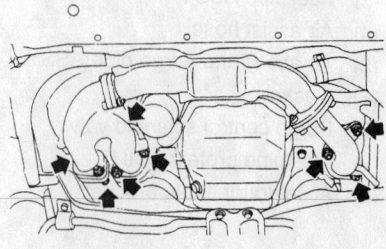

9357TG11

Location of the front exhaust pipe retainers—2.0L engines

To install:

3. Clean all gasket surfaces completely.

4. Install or connect the following:
- New gaskets
- Front exhaust pipe to the manifolds and tighten the retainers to 26 ft. lbs. (35 Nm)
- Covers to the front exhaust pipe and tighten to 18 ft. lbs. (25 Nm)
- Upper exhaust manifold cover on the right hand side and tighten the retainers to 13 ft. lbs. (19 Nm)
- Front exhaust pipe assembly and tighten the retainers to 26 ft. lbs. (35 Nm)
- Right hand side manifold to the turbocharger joint pipe and tighten the retainers to 13 ft. lbs. (19 Nm)
- Upper and lower manifold covers on the left hand side to 13 ft. lbs. (19 Nm)
- Front O_2S
- Front undercover, if equipped
- Negative battery cable

Turbocharger

REMOVAL & INSTALLATION

2.0L Engines

1. Before servicing the vehicle, refer to the precautions in the beginning of this section.

2. Remove or disconnect the following:
- Negative battery cable
- Center exhaust pipe
- Turbocharger joint pipe from the turbocharger
- Engine coolant hose from the filler tank
- Clamp that attaches the turbocharger to the inlet duct
- Bolt that attaches the bracket of the oil pipe to the turbocharger
- Oil pipe from the turbocharger
- Turbocharger bracket
- Oil outlet hose from the pipe
- Turbocharger.

To install:

3. Installation is the reverse of removal.

4. Tighten the front pipe to the turbocharger retainers to 22 ft. lbs. (30 Nm).

Front Crankshaft Seal

REMOVAL & INSTALLATION

2.0L Engines

1. Before servicing the vehicle, refer to the precautions in the beginning of this section.

2. Remove or disconnect the following:
- Negative battery cable
- Drive belt

3. Secure the crankshaft pulley with tool No. 499977300.

4. Remove or disconnect the following:
- Crankshaft pulley bolt and pulley
- Left, right and center timing belt cover(s) mounting bolts
- Belt covers
- Timing belt
- Crankshaft seal from the oil pump housing

To install:

5. Using a suitable seal driver, install a new crankshaft seal.

6. Install or connect the following:
- Timing belt crankshaft sprocket and timing belt
- Belt covers and tighten the bolts to 36–48 inch lbs. (4–5 Nm)
- Crankshaft pulley and tighten the bolt to 33 ft. lbs. (44 Nm)
- Drive belt
- Negative battery cable

2.2L Engines

1. Before servicing the vehicle, refer to the precautions in the beginning of this section.

2. Remove or disconnect the following:
- Negative battery cable
- Accessory drive belts
- Power steering pump and alternator
- Air conditioner compressor brackets

3. Secure the crankshaft pulley with tool No. 499977000.

4. Remove or disconnect the following:
- Crankshaft pulley bolt and pulley
- Timing belt cover mounting bolts
- Belt covers
- Timing belt
- Timing belt crankshaft sprocket
- Crankshaft seal from the oil pump housing

To install:

5. Using a suitable seal driver, install a new crankshaft seal.

6. Install or connect the following:
- Timing belt crankshaft sprocket and timing belt
- Belt covers and tighten the bolts to 36–48 inch lbs. (4–5 Nm)
- Crankshaft pulley and tighten the bolt to 69–76 ft. lbs. (93–103 Nm)
- Power steering pump, alternator, air conditioning compressor and associated brackets
- Accessory drive belts
- Negative battery cable

2.5L Engine

1. Before servicing the vehicle, refer to the precautions in the beginning of this section.

2. Remove or disconnect the following:
- Negative battery cable
- Radiator electric fan motor wiring connectors
- Coolant reservoir tank
- 4 bolts that secure the radiator shroud, then remove the fan assembly

3. Position the No. 1 piston to Top Dead Center (TDC) of its compression stroke.
- Accessory drive belt cover
- Air conditioning compressor drive belt and tensioner

4. Secure the crankshaft pulley with tool No. ST499977000.
- Crankshaft pulley bolt and pulley
- Left timing belt cover mounting bolts and the left cover
- Right timing belt cover mounting bolts and the right cover
- Center timing belt cover mounting bolts and the center cover
- Timing belt
- Timing belt crankshaft sprocket
- Crankshaft seal from the oil pump housing

To install:

5. Install or connect the following:
- New crankshaft seal, using a suitable seal driver
- Timing belt crankshaft sprocket and the timing belt
- Center, right, then the left timing belt covers. Tighten the bolts to 44 inch lbs. (5 Nm).
- Crankshaft pulley and tighten the bolt to 94 ft. lbs. (127 Nm)
- Air conditioning compressor drive belt tensioner and the drive belts
- Fan shroud and fan motor assembly
- Accessory drive belt cover
- Negative battery cable

3.0L Engine

1. Before servicing the vehicle, refer to the precautions in the beginning of this section.

2. Remove or disconnect the follow-ing:
- Negative battery cable
- Drive belt

3. Secure the crankshaft pulley with tool No. 499977100.

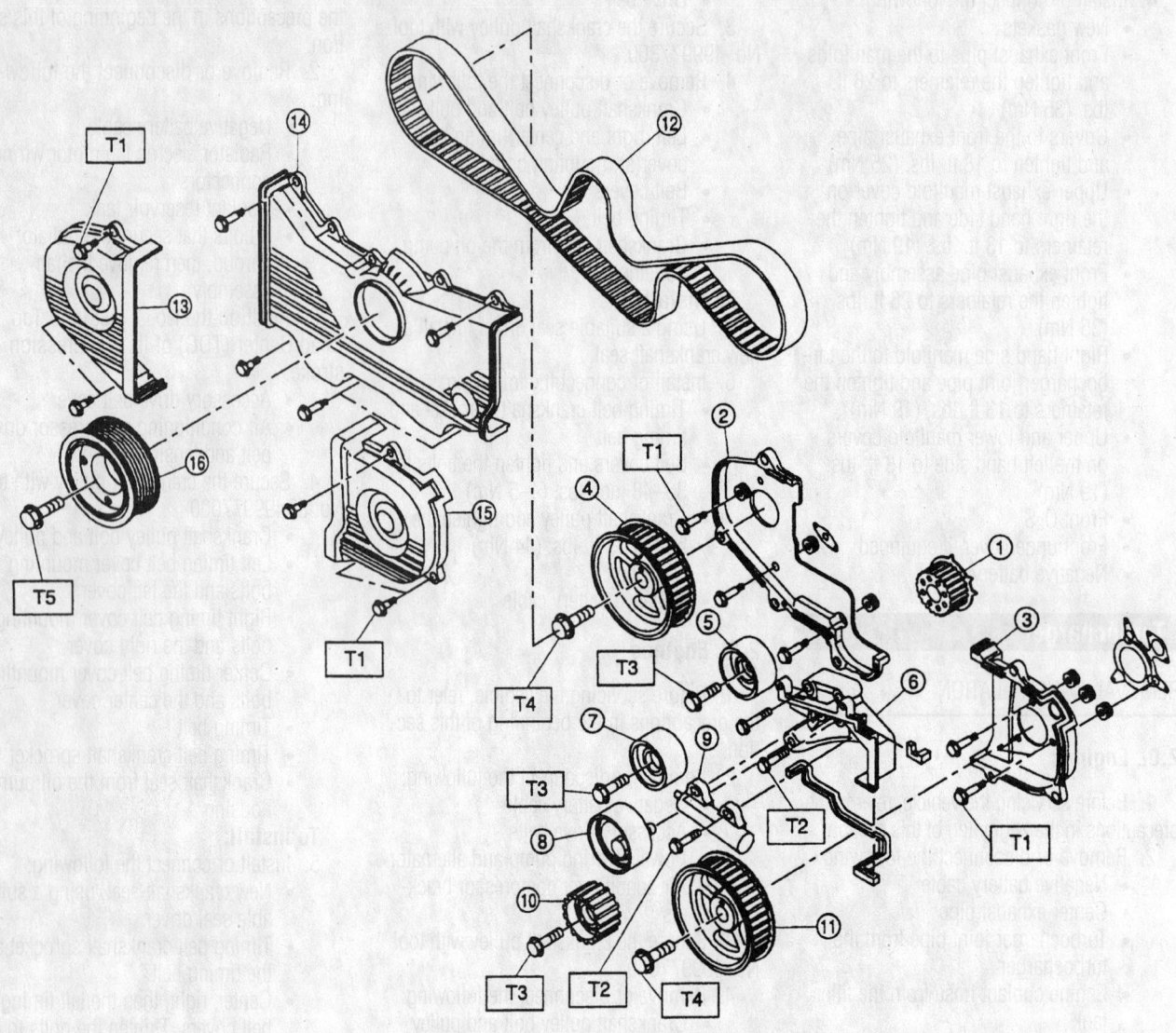

1. Crankshaft sprocket
2. Belt cover No. 2 (RH)
3. Belt cover No. 2 (LH)
4. Camshaft sprocket (RH)
5. Belt idler
6. Tensioner bracket
7. Belt idler
8. Belt tensioner
9. Tensioner adjuster
10. Belt idler No. 2
11. Camshaft sprocket (LH)
12. Timing belt

13. Belt cover (RH)
14. Front belt cover
15. Belt cover (LH)
16. Crankshaft pulley

Tightening torque: N·m (kg-m, ft-lb)
T1: 5 ± 1 (0.5 ± 0.1, 3.6 ± 0.7)
T2: 25 ± 2 (2.5 ± 0.2, 18.1 ± 1.4)
T3: 39 ± 4 (4.0 ± 0.4, 28.9 ± 2.9)
T4: 78 ± 5 (8.0 ± 0.5, 57.9 ± 3.6)
T5: 108^{+10}_{-5} ($11^{+1.0}_{-0.5}$, $79.6^{+7.2}_{-3.6}$)

7923TG20

Exploded view of the timing belt covers and components—2.2L engine

4. Remove or disconnect the following:
- Crankshaft pulley bolt and pulley
- Front chain cover. Refer to the timing chain procedure in this section.

- Timing chain
- Crankshaft seal

To install:

5. Using a suitable seal driver, install a new crankshaft seal.

6. Install or connect the following:
- Timing chain
- Front chain cover. Refer to the timing chain procedure in this section.

Nm, ft. lbs.

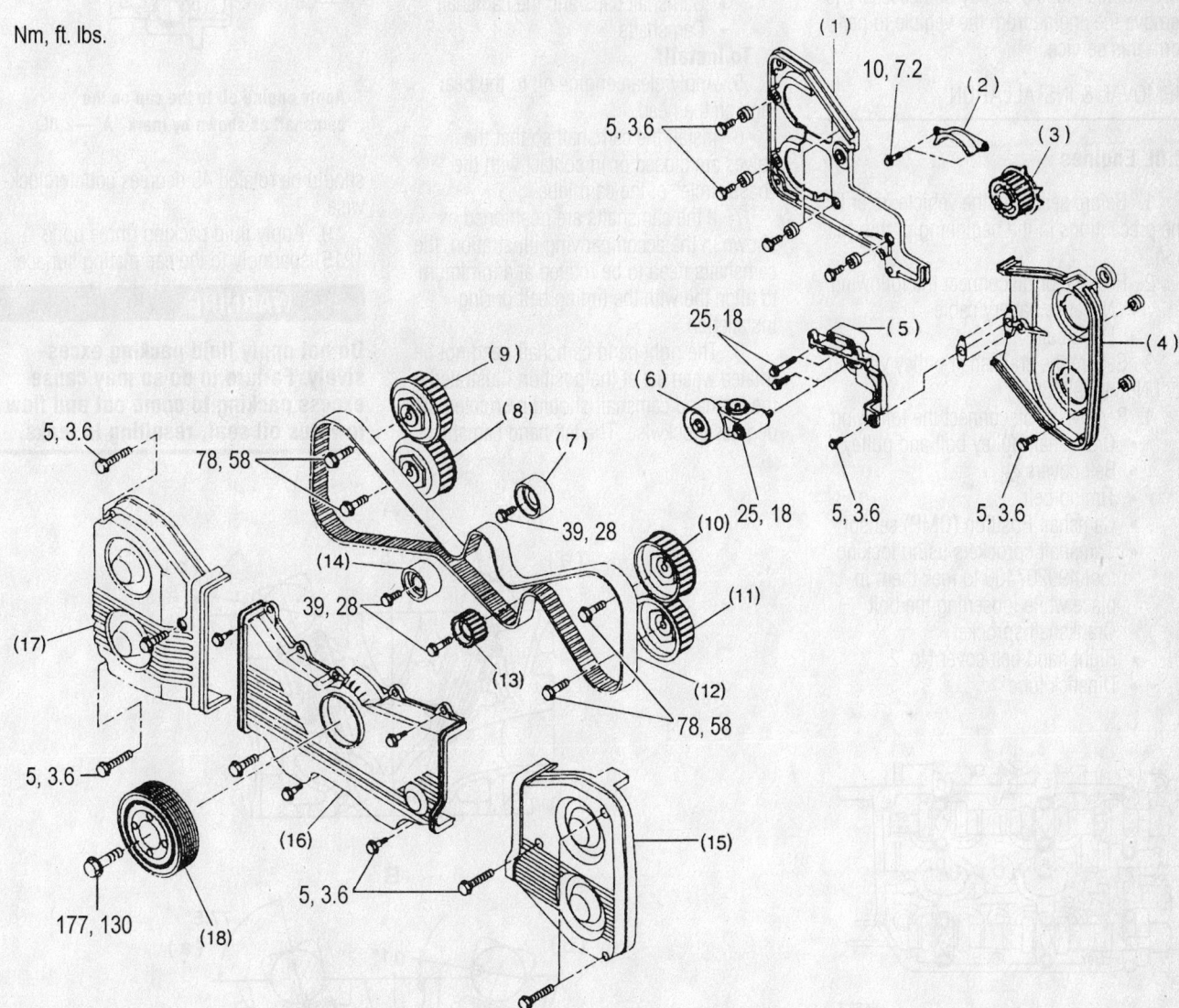

(1) Right-hand belt cover No. 2
(2) Timing belt guide (MT vehicles only)
(3) Crankshaft sprocket
(4) Left-hand belt cover No. 2
(5) Tensioner bracket
(6) Automatic belt tension adjuster ASSY
(7) Belt idler
(8) Right-hand exhaust camshaft sprocket

(9) Right-hand intake camshaft sprocket
(10) Left-hand intake camshaft sprocket
(11) Left-hand exhaust camshaft sprocket
(12) Timing belt
(13) Belt idler No. 2
(14) Belt idler
(15) Left-hand belt cover
(16) Front belt cover

(17) Right-hand belt cover
(18) Crankshaft pulley

7923TG21

Exploded view of the timing belt covers and components—2.5L engine

- Crankshaft pulley and tighten the bolt to 131 ft. lbs. (178 Nm)
- Drive belt
- Negative battery cable

Camshaft

On some models, it may be necessary to remove the engine from the vehicle to perform this service.

REMOVAL & INSTALLATION

2.0L Engines

1. Before servicing the vehicle, refer to the precautions in the beginning of this section.

2. Remove or disconnect the following:
- Negative battery cable
- Drive belt

3. Secure the crankshaft pulley with tool No. 499977300.

4. Remove or disconnect the following:
- Crankshaft pulley bolt and pulley
- Belt covers
- Timing belt
- Camshaft Position (CMP) sensor
- Camshaft sprockets using locking tool 499207400 to lock them in place while loosening the bolt
- Crankshaft sprocket
- Right hand belt cover No. 2
- Dipstick tube

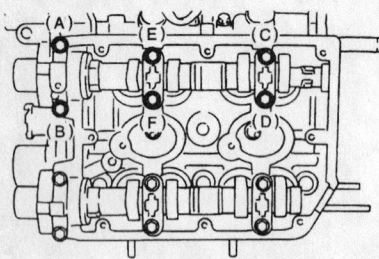

Loosen the intake camshaft caps as shown—2.0L engines

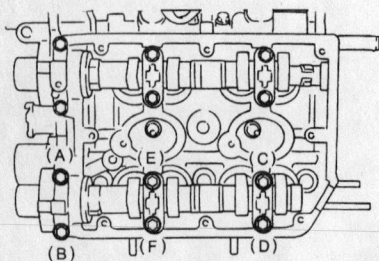

Loosen the exhaust camshaft caps as shown—2.0L engines

- Spark plug cord
- Rocker cover and gasket
- Intake camshaft bolts as illustrated a little at a time
- Camshaft caps and the camshaft
- Exhaust camshaft bolts as illustrated a little at a time
- Camshaft caps and the camshaft
- Camshafts

To install:

5. Apply clean engine oil to the bearings on the head.

6. Install the camshaft so that the valves are closed or in contact with the "base circle" of the cam lobe.

7. If the camshafts are positioned as shown in the accompanying illustration, the camshafts need to be rotated at a minimum to align the with the timing belt during installation.

8. The right hand camshaft need not be rotated when set at the position illustrated, the left hand camshaft should be rotated 80 degrees clockwise. The left hand camshaft

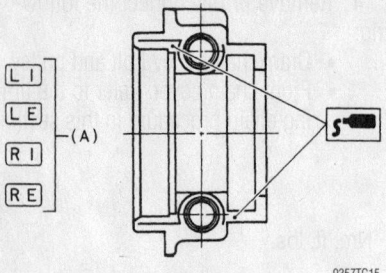

Apply engine oil to the cap on the camshaft as shown by mark "A"—2.0L

should be rotated 45 degrees counterclockwise.

9. Apply fluid packing (three bond 1215) sparingly to the cap mating surface.

❊❊ WARNING

Do not apply fluid packing excessively. Failure to do so may cause excess packing to come out and flow towards oil seal, resulting in leaks.

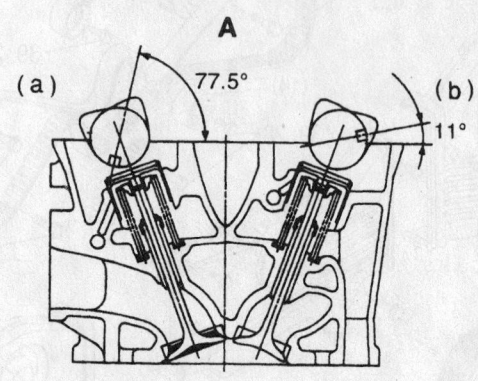

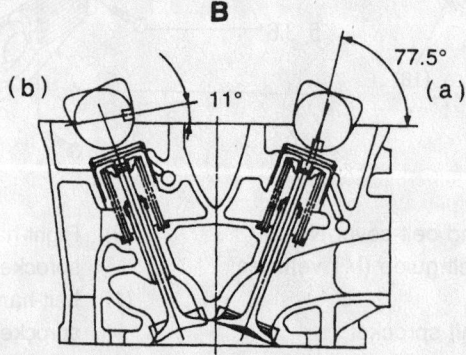

A Left side cylinder head
B Right side cylinder head
(a) Intake camshaft
(b) Exhaust camshaft

The right hand camshaft need not be rotated when set at the position illustrated, the left hand camshaft should be rotated 80 degrees clockwise. The left hand camshaft should be rotated 45 degrees counterclockwise—2.0L engines

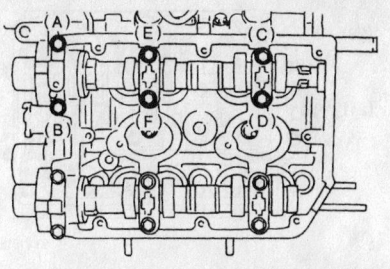

9357TG16

Tighten the cap in two stages in the sequence shown—2.0L engines

10. Apply engine oil to the cap on the camshaft as shown by mark "A" in the accompanying illustration.

11. Tighten the cap in two stages to 14.5 ft. lbs. (20 Nm) in the sequence illustrated.

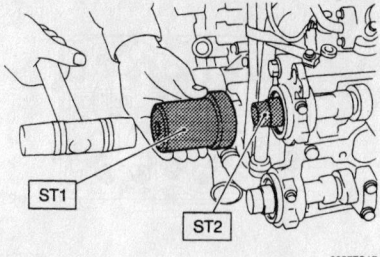

9357TG17

Using guide tool 499597200 and installer 499587600, install the new camshaft seal—2.0L engines

12. Apply oil to the lip of the new camshaft seal and using guide tool 499597200 and installer 499587600, install the new seal.

13. Install or connect the following:
- Rocker cover, making sure to apply fluid packing (three bond 1215) to the four front open edges of the gasket
- Spark plug cord
- Right hand belt cover No. 2 and tighten to 3 ft. lbs. (4 Nm)
- Tensioner bracket and tighten to 18 ft. lbs. (25 Nm)
- Left hand belt cover No. 2 and tighten to 3 ft. lbs. (4 Nm)
- Crankshaft sprocket
- Camshaft sprockets using locking tool 499207400 to lock them in place while tightening the bolt to 72 ft. lbs. (98 Nm)
- Timing belt

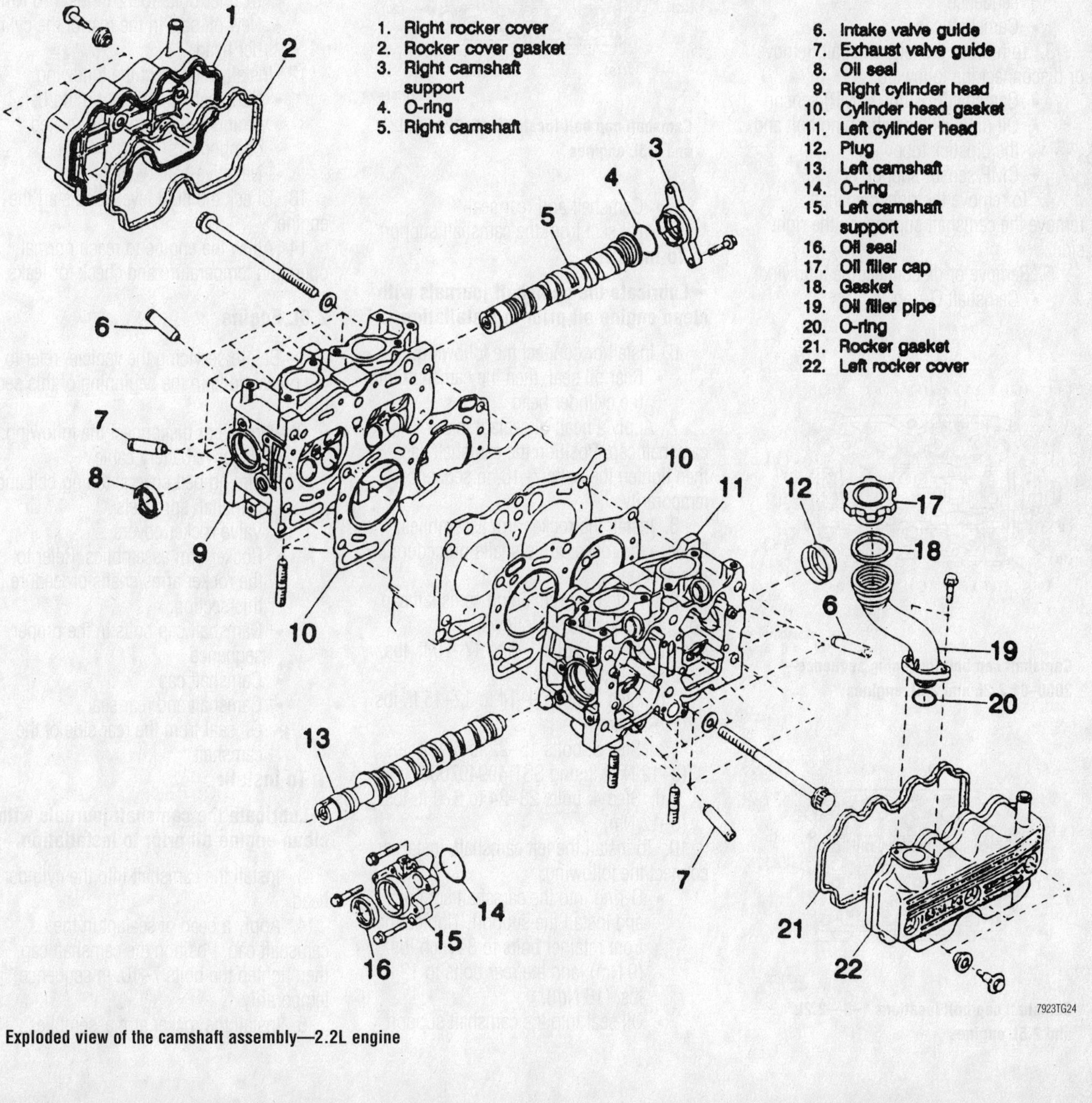

1. Right rocker cover
2. Rocker cover gasket
3. Right camshaft support
4. O-ring
5. Right camshaft
6. Intake valve guide
7. Exhaust valve guide
8. Oil seal
9. Right cylinder head
10. Cylinder head gasket
11. Left cylinder head
12. Plug
13. Left camshaft
14. O-ring
15. Left camshaft support
16. Oil seal
17. Oil filler cap
18. Gasket
19. Oil filler pipe
20. O-ring
21. Rocker gasket
22. Left rocker cover

Exploded view of the camshaft assembly—2.2L engine

7923TG24

- Belt cover(s)
- Crankshaft pulley and tighten the bolt to 33 ft. lbs. (44 Nm)
- Drive belt
- Negative battery cable

2.2L Engines

1. Before servicing the vehicle, refer to the precautions in the beginning of this section.

2. Remove or disconnect the following:
- Negative battery cable
- Timing belt covers, timing belt and camshaft sprockets
- Valve rocker covers
- Rocker arm assemblies. Refer to the rocker arms/shafts procedure in this section.
- Camshaft cap bolts in the proper sequence
- Camshaft cap

3. To remove the left camshaft, remove or disconnect the following:
- Camshaft Position (CMP) sensor
- Oil dipstick tube attaching bolt and the dipstick tube
- CMP sensor support

4. To remove the right camshaft, remove the camshaft support on the right side.

5. Remove or disconnect the following:
- Camshaft O-ring

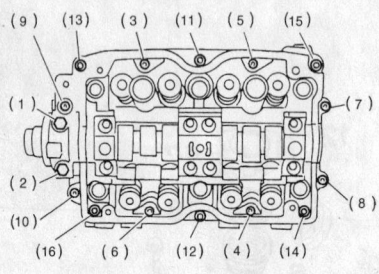

Camshaft cap bolt loosening sequence— 2000–01 2.2L and 2.5L engines

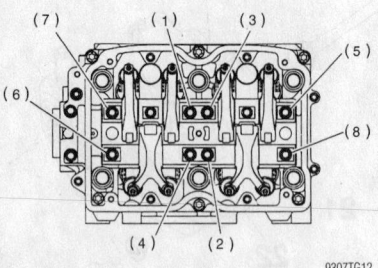

Camshaft cap bolt locations 1–8—2.2L and 2.5L engines

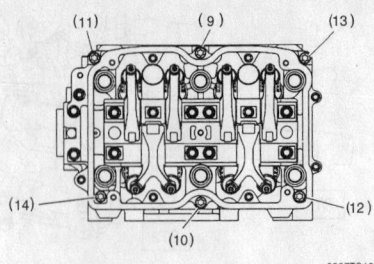

Camshaft cap bolt locations 9–14—2.2L and 2.5L engines

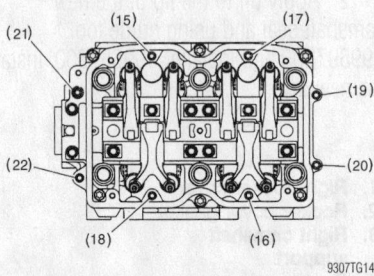

Camshaft cap bolt locations 15–22—2.2L and 2.5L engines

- Camshaft and rear seal
- Oil seal from the camshaft support

To install:

➡ Lubricate the camshaft journals with clean engine oil prior to installation.

6. Install or connect the following:
- Rear oil seal, then the camshaft into the cylinder head

7. Apply a bead of sealant in the camshaft cap. Position the camshaft cap, then tighten the bolts 7–10, in sequence, temporarily

8. Install the rocker arm assemblies. Refer to the rocker arms/shafts procedure in this section.

9. Tighten the remaining camshaft cap bolts, in sequence, as follows:
  a. Step 1: bolts 1–8 to 17–20 ft. lbs. (23–27 Nm).
  b. Step 2: bolts 9–14 to 12–15 ft. lbs. (16–20 Nm).
  c. Step 3: bolts 15–22 to 6–9 ft. lbs. (8–12 Nm), using SST 499497000.
  d. Step 4: bolts 23–24 to 6–9 ft. lbs. (8–12 Nm).

10. To install the left camshaft, install or connect the following:
- O-ring into the camshaft support and install the support. Tighten the front retainer bolts to 84 inch lbs. (9 Nm), and the rear bolts to 12 ft. lbs. (16 Nm).
- Oil seal into the camshaft support

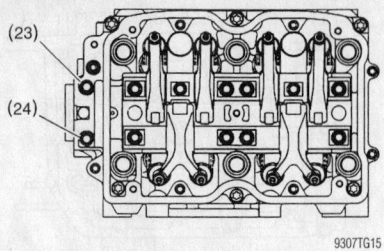

Camshaft cap bolt locations 23–24—2.2L and 2.5L engines

- Dipstick tube. Tighten the retaining bolt to 10 ft. lbs. (13 Nm).
- CMP sensor

11. To install the right camshaft, install or connect the following:
- O-ring into the camshaft support and install the support. Tighten the retainer bolts to 12 ft. lbs. (16 Nm).
- New oil seal in the rear of the cylinder head

12. Install or connect the following:
- Camshaft sprockets, timing belt, timing belt covers, and related components
- Negative battery cable

13. Check the fluid levels and start the engine.

14. Allow the engine to reach normal operating temperature and check for leaks.

2.5L Engine

1. Before servicing the vehicle, refer to the precautions in the beginning of this section.

2. Remove or disconnect the following:
- Negative battery cable
- Timing belt covers, timing belt and camshaft sprockets
- Valve rocker covers
- Rocker arm assemblies. Refer to the rocker arms/shafts procedure in this section.
- Camshaft cap bolts in the proper sequence
- Camshaft cap
- Camshaft and rear seal
- Oil seal from the rear side of the camshaft

To install:

➡ Lubricate the camshaft journals with clean engine oil prior to installation.

3. Install the camshaft into the cylinder head

4. Apply a bead of sealant in the camshaft cap. Position the camshaft cap, then tighten the bolts 7–10, in sequence, temporarily

5. Install the rocker arm assemblies.

Refer to the rocker arms/shafts procedure in this section.

6. Tighten the remaining camshaft cap bolts, in sequence, as follows:

    a. Step 1: bolts 1–8 to 17–20 ft. lbs. (23–27 Nm).

    b. Step 2: bolts 9–14 to 12–15 ft. lbs. (16–20 Nm).

    c. Step 3: bolts 15–22 to 6–9 ft. lbs. (8–12 Nm), using SST 499497000.

    d. Step 4: bolts 23–24 to 6–9 ft. lbs. (8–12 Nm).

7. Install or connect the following:

➥**Lubricate the seals lips with clean engine oil prior to installation**

- Oil seal and plug with suitable tools
- Camshaft sprockets, timing belt, timing belt covers, and related components
- Negative battery cable

8. Check the fluid levels and start the engine.

9. Allow the engine to reach normal operating temperature and check for leaks.

3.0L Engine

1. Before servicing the vehicle, refer to the precautions in the beginning of this section.

2. Remove or disconnect the following:

- Negative battery cable
- Crankshaft pulley cover
- Crankshaft pulley bolt using tool 499977100 to hold the crankshaft
- Crankshaft pulley
- Timing chain cover and the chain. Refer to the procedure in this section.
- Camshaft and crankshaft sprockets. Refer to the procedure in this section.
- Oil pump. Refer to the procedure in this section.
- Oil pump relief valve
- Water pump. Refer to the procedure in this section.
- Rear chain cover
- Right hand valve cover
- Front camshaft cap bolts equally in small increments in the sequence illustrated on the right hand side

3. Install or connect the following:

- Camshaft cap and intake camshaft on the right hand side
- Camshaft cap bolts equally in small increments in the sequence illustrated
- Camshaft cap and exhaust camshaft of the right hand side

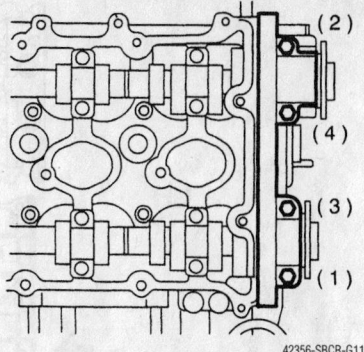

Front camshaft cap bolts equally in small increments in the sequence shown—3.0L engine

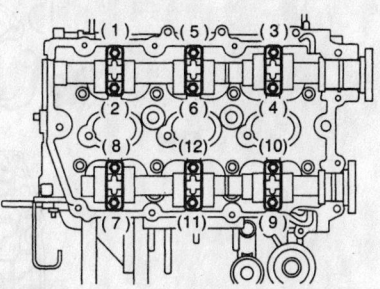

Remove the camshaft cap bolts equally in small increments in the sequence shown—3.0L engine

➥**Mark the camshaft caps so that they can be reinstalled in their original positions.**

- Plug from the left hand side
- Left hand camshaft components in the same manner as the right hand components

To install:

4. Apply a coat of engine oil to the journals on the camshafts and place the camshafts into position.

5. When installing the camshaft, adjust the camshaft front flange knock pin (A) at the 12 O'clock position on the left hand side and the 10 O'clock position on the right

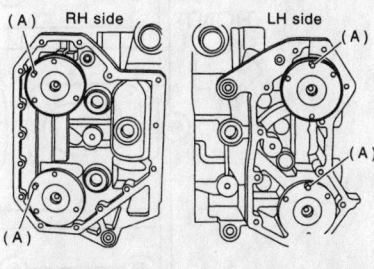

Install the camshafts so the flange knock pin (A) is positioned as shown—3.0L engine

hand side. Refer to the illustration for more detail.

6. Apply 0.059–0.099 inch (1.5–2.5mm) of fluid packing sparingly to the front and back of the camshaft cap as illustrated. Subaru recommends Three Bond 1280B for this purpose.

✲✲ CAUTION

Do not apply too much gasket maker. This may cause excess fluid gasket maker to come out and flow towards the camshaft journal resulting in engine damage.

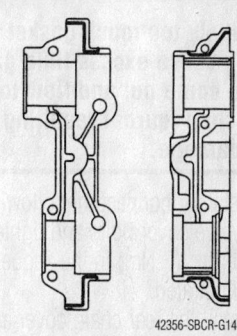

Apply 0.059–0.099 inch (1.5–2.5mm) of fluid gasket sparingly to the front and back of the camshaft cap where indicated—3.0L engine

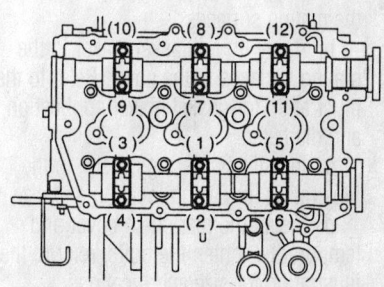

Tighten the camshaft cap bolts in the sequence shown to the proper specification—3.0L engine

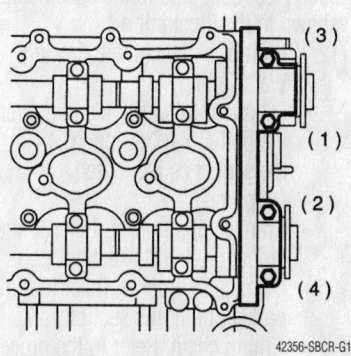

Tighten the front camshaft cap bolts in the sequence shown—3.0L engine

7. Apply oil to the cap bearing surface.

8. Install or connect the following:
   - Camshaft cap in the their original location
   - Camshaft cap bolts and tighten to 11.6 ft. lbs. (16 Nm) in the sequence illustrated
   - Front camshaft cap bolts and tighten to 7 ft. lbs. (10 Nm) in the sequence illustrated

9. Apply fluid gasket maker of the of the cylinder heads and valve covers as illustrated. Subaru recommends Three Bond 1280B for this purpose.

✳✳ CAUTION

Do not apply too much gasket maker. This may cause excess fluid gasket maker to come out and flow towards the camshaft journal resulting in engine damage.

10. Install or connect the following:
    - Valve cover bolts and tighten to 5 ft. lbs. (7 Nm) in the sequence illustrated.

11. Install the rear chain cover as follows:

➡**There are several size bolts used, refer to the illustration for size and locations.**

   a. Rear chain cover gasket and clean the mating surfaces.

   b. Apply liquid gasket maker to the mating surfaces of the cover. Refer to the illustration for gasket maker application and diameter.

   c. Install new O-rings. Refer to the illustration for O-ring location and size.

   d. Install the rear chain cover and temporarily tighten the bolts, refer to the illustration for size and locations.

   e. Replace mounting bolts **G** with new bolts, refer to the illustration for location.

   f. Tighten the cover bolts in the sequence illustrated to the specifications shown in the illustration.

12. Install or connect the following:
    - Water pump
    - Oil pump relief valve and tighten the bolts to 4.7 ft. lbs. (6.4 Nm) in the sequence illustrated.
    - Oil pump
    - Crankshaft sprocket. Refer to the procedure in this section.
    - Camshaft sprocket. Refer to the procedure in this section.
    - Timing chain. Refer to the procedure in this section.

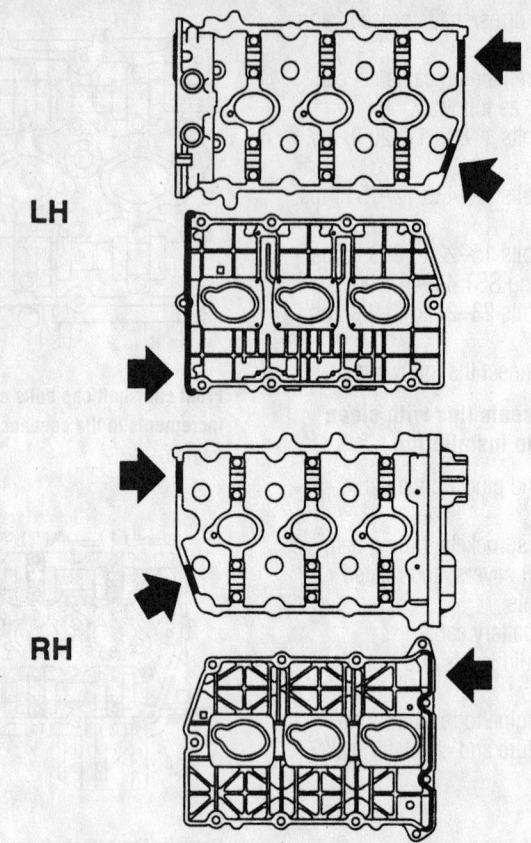

42356-SBCR-G17

Apply fluid gasket maker of the of the cylinder heads and valve covers as shown—3.0L engine

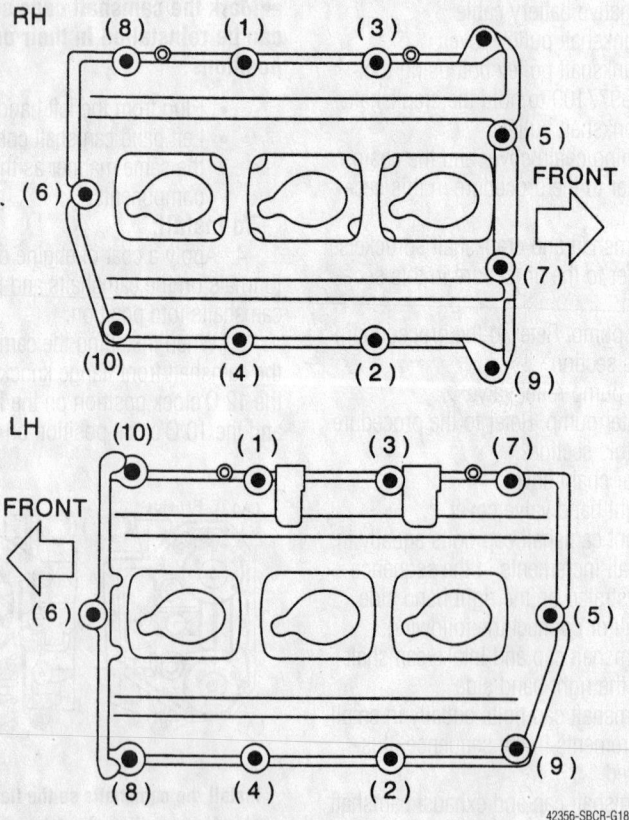

42356-SBCR-G18

Tighten the valve cover bolts in this sequence—3.0L engine

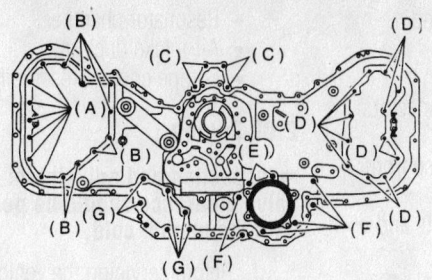

(A) 6 × 14
(B) 6 × 18 (Silver)
(C) 6 × 30
(D) 6 × 18
(E) 6 × 40
(F) 6 × 30
(G) 6 × 22

42356-SBCR-G05

Rear timing chain bolt sizes and locations—3.0L engine

Fluid gasket application diameter:
(A) 1.0±0.5 mm (0.039±0.020 in)
(B) 3.0±1.0 mm (0.118±0.039 in)

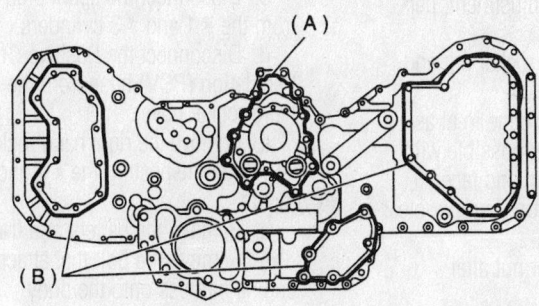

42356-SBCR-G06

Apply liquid gasket maker to the mating surfaces of the cover as shown—3.0L engine

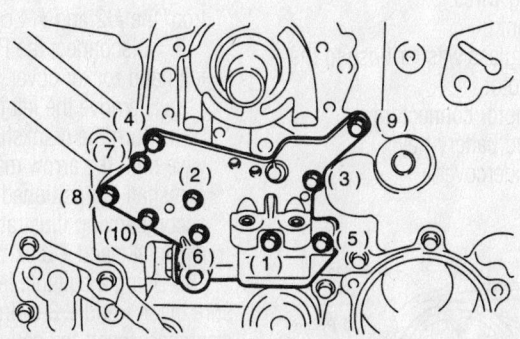

42356-SBCR-G09

Oil pump relief valve bolt tightening sequence and bolt specifications—3.0L engine

Bolt installation position	Bolt dimension
(1) and (5)	6 × 26
(2), (3), (4) and (9)	6 × 35
(6), (7), (8) and (10)	6 × 16

(A) O-ring (Large)
(B) O-ring (Medium)
(C) O-ring (Small)

42356-SBCR-G07

Rear cover O-ring sizes and locations—3.0L engine

(1) to (11)	9 N·m (0.9 kgf-m, 6.5 ft-lb)
(12) to (19)	20 N·m (2.0 kgf-m, 14 ft-lb)
(20) to (31)	9 N·m (0.9 kgf-m, 6.5 ft-lb)
(32) to (39)	12 N·m (1.2 kgf-m, 8.7 ft-lb)
(40) to (46)	9 N·m (0.9 kgf-m, 6.5 ft-lb)

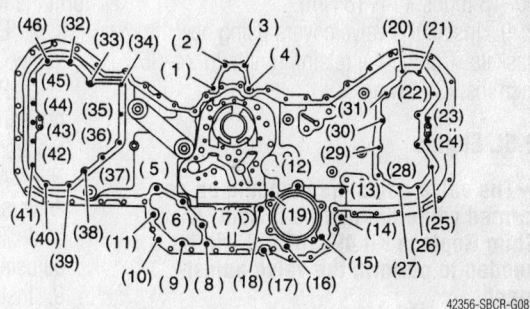

42356-SBCR-G08

Rear cover bolt tightening sequence and torque specifications—3.0L engine

- Crankshaft pulley, apply clean oil to the bolt threads and tighten the bolt to 131 ft. lbs. (178 Nm)
- Camshaft pulley cover and tighten the bolts to 5 ft. lbs. (7 Nm)
- Negative battery cable

13. Start the engine and allow it to reach operating temperature. Check for leaks.

Valve Lash

INSPECTION AND ADJUSTMENT

2.2L Engine

1. Before servicing the vehicle, refer to the precautions in the beginning of this section.
2. With the engine cold, rotate the

engine so that the No. 1 piston is at Top Dead Center (TDC) of its compression stroke.

3. Check the clearance of both the intake and exhaust valves of the No. 1 cylinder by inserting a feeler gauge between each valve stem and rocker arm.

4. If the clearance is not within specifications, loosen the locknut with the proper size wrench and turn the adjusting stud either in or out until the valve clearance is correct.

➡**Proper valve clearance is obtained when the feeler gauge slides between the valve stem and the rocker arm with a minimum amount of resistance.**

5. Tighten the locknut and recheck the valve stem-to-rocker clearance.

6. The rest of the valves are adjusted in the same way. Bring each piston to TDC of its compression stroke, then check and adjust the valves for that cylinder. The proper valve adjustment sequence is 1–3–2–4.

7. Rotate the crankshaft at least 2 revolutions, then recheck the valve clearance.

8. Tighten the rocker arm locknuts to 10–13 ft. lbs. (14–18 Nm).

9. Install the valve covers using new gaskets. Tighten the retaining nuts to 24–36 inch lbs. (3–4 Nm).

2.5L Engine

➡**The valve adjustment should be performed while the engine is cold. A Shim Replace Kit 498187100 will be needed to perform the valve adjustment.**

1. Before servicing the vehicle, refer to the precautions in the beginning of this section.

2. Adjustment should be performed when engine is cold.

3. Remove or disconnect the following:
- Negative battery cable
- Engine coolant reservoir tank
- Timing belt cover on the left hand side

4. When inspecting the No. 1 and 3 cylinders remove the following:
- Air intake duct as a unit
- Resonator chamber
- Spark plug wires from the No. 1 and 3 cylinders
- Blow-by house from valve cover
- Engine undercover
- Timing belt cover on the right hand side
- Valve cover on the right hand side

5. When inspecting the No. 2 and 4 cylinders remove the following:
- Battery and battery tray
- Window washer motor connectors front and rear
- Rear gate glass washer hose from the washer motor
- Washer tank mounting bolts and secure out of the way
- Spark plug wires from the No. 2 and 4 cylinders
- Blow by house from valve cover
- Timing belt cover on the right hand side
- Valve cover on the left hand side

6. Set No. 1 cylinder to Top Dead Center (TDC).

➡**When arrow mark on the left hand side comes exactly to the top, No. 1 cylinder piston is brought to TDC of the compression stroke.**

7. Check the valve clearance:
- Intake valve: 0.0071–0.0087 in. (0.18–0.22mm)
- Exhaust valve: 0.0090–0.0106 in. (0.23–0.27mm)

8. If any valve needs adjustment, perform the following:
   a. Loosen the valve rocker nut and screw.
   b. Place a thickness gage in at as horizontal a direction as possible with respect to the valve stem and face.
   c. Adjust the screw until proper clearance is obtained.
   d. Tighten the rocker nut after adjusted.

9. Install or connect the following:
- Valve covers left and right
- Timing belt covers
- Blow-by houses to valve covers
- Spark plug wires
- Washer tank
- Rear gate glass washer hose to the washer motor
- Washer motor connectors
- Battery and battery tray
- Engine undercover

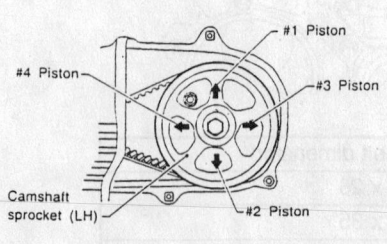

Position the camshaft for adjustment to valves

9308XG05

- Resonator chamber
- Air intake duct unit
- Engine coolant reservoir tank

2.0L Engine

➡**Inspection and adjustment of the valve clearance should be performed with the engine cold.**

1. Before servicing the vehicle, refer to the precautions in the beginning of this section.

2. Remove or disconnect the following:
- Negative battery cable
- Air intake duct
- Bolt which attaches the right hand timing cover
- Engine undercover
- Remaining bolts attaching the right hand timing belt cover and the cover

3. When inspecting the # 1 and # 3 cylinders:
   a. Pull out the engine harness connector with the bracket from the air cleaner upper cover.
   b. Remove the air cleaner case.
   c. Disconnect the spark plug wires from the # 1 and # 3 cylinders.
   d. Disconnect the Positive Crankcase Ventilation (PCV) hose from the right hand rocker cover.
   e. Remove the right hand rocker cover.

4. When inspecting the # 2 and # 4 cylinders:
   a. Remove the battery and tray.
   b. Remove the bolt that attaches the engine harness onto the body.
   c. Disconnect the washer motor connectors.
   d. Remove the washer tank bolts and lift the tank upwards.
   e. Disconnect the spark plug wires from the # 2 and # 4 cylinders.
   f. Disconnect the PCV hose from the left hand rocker cover.
   g. Remove the left hand rocker cover.
   h. Turn the crankshaft pulley clockwise until the arrow mark on the camshaft is positioned as shown in the accompanying illustration to measure the # 1 intake and # 3 exhaust valves.

5. Using a suitable feeler gauge, measure the # 1 and # 3 cylinder exhaust valve clearance. Insert the gauge in as horizontal a direction with respect to the shim. Make sure to measure the exhaust valve clearances while lifting up the vehicle.

6. The intake valve clearance should be 0.0071–0.0087 inch (0.18–0.22mm). The exhaust valve clearance should be 0.0090–0.0106 inch (0.23–0.27mm).

7. If not within specification, adjust the valve as outlined in the adjustment steps.

8. Turn the crankshaft pulley clockwise to measure the valve clearance for the # 2 exhaust and # 3 intake valves as shown in the accompanying illustration.

9. Turn the crankshaft pulley clockwise to measure the valve clearance for the # 2 intake and # 4 exhaust valves as shown in the accompanying illustration.

10. Turn the crankshaft pulley clockwise to measure the valve clearance for the # 1 exhaust and # 4 intake valves as shown in the accompanying illustration.

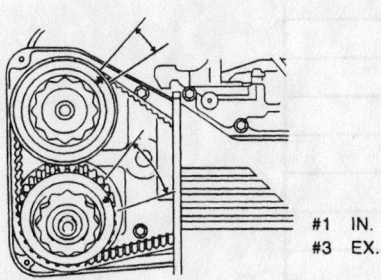

#1 IN.
#3 EX.

9357TG18

Turn the crankshaft pulley clockwise until the arrow mark on the camshaft is positioned as shown to measure the # 1 intake and # 3 exhaust valves—2.0L engines

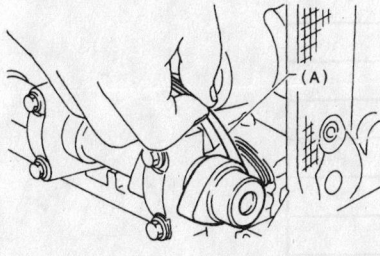

(A)

9357TG19

Use a feeler gauge to inspect the valve clearance—2.0L engines

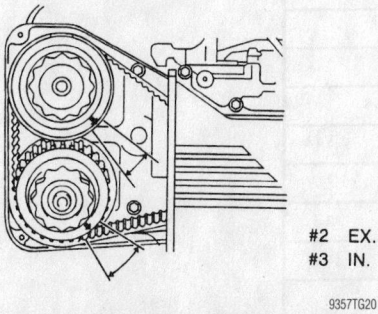

#2 EX.
#3 IN.

9357TG20

Turn the crankshaft pulley clockwise until the arrow mark on the camshaft is positioned as shown to measure the # 2 exhaust and # 3 intake valves—2.0L engines

11. Adjust the valve clearance as follows:

a. Measure and record all valve clearances using the procedures outlined in the inspection steps in this section.

b. Prepare shim replacer tool 498187200.

c. Rotate the notch of the valve lifter outwards 45 degrees.

d. Adjust the shim replacer tool notch to the lifter and set it.

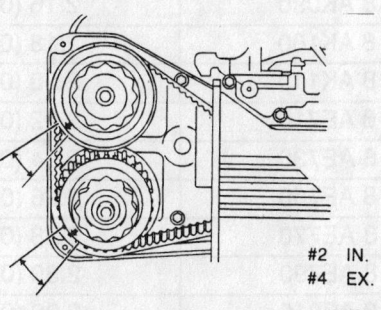

#2 IN.
#4 EX.

9357TG21

Turn the crankshaft pulley clockwise until the arrow mark on the camshaft is positioned as shown to measure the # 2 intake and # 4 exhaust valves—2.0L engines

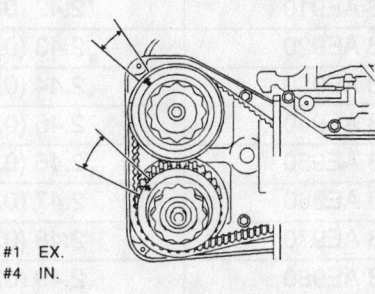

#1 EX.
#4 IN.

9357TG22

Turn the crankshaft pulley clockwise until the arrow mark on the camshaft is positioned as shown to measure the # 1 exhaust and # 4 intake valves—2.0L engines

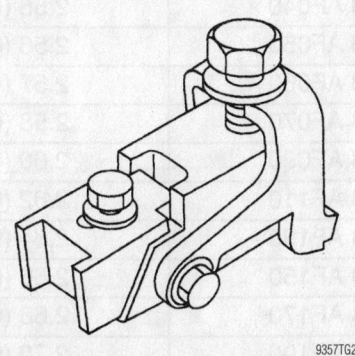

9357TG23

Shim replacer tool 498187200 is required to adjust the valves—2.0L engines

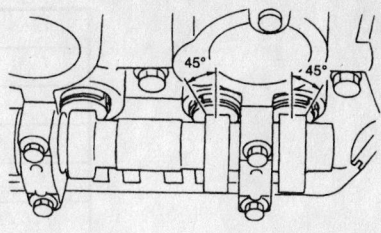

45° 45°

9357TG24

Rotate the notch of the valve lifter outwards 45 degrees—2.0L engines

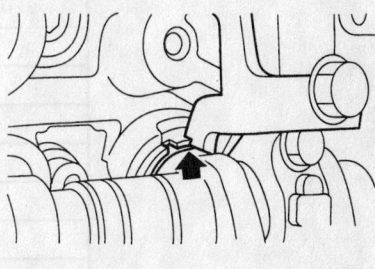

9357TG25

Adjust the shim replacer tool notch to the lifter and set it—2.0L engines

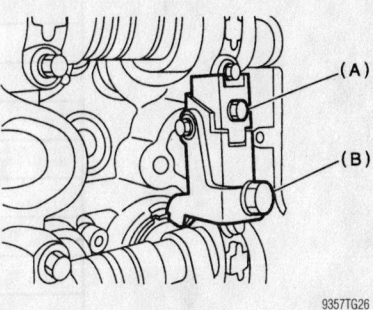

(A)

(B)

9357TG26

Location of bolts "A" and "B" on the shim replacer tool—2.0L engines

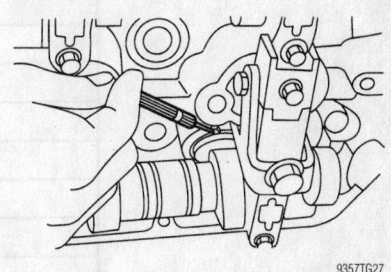

9357TG27

Remove the shim from the lifter—2.0L engines

	Unit: mm
Intake valve:S = (V + T) – 0.20	
Exhaust valve:S = (V + T) – 0.25	
S: Shim thickness to be used	
V: Measured valve clearance	
T: Shim thickness required	

9357TG28

Use this table to help you select a suitable shim—2.0L engines

Part No.	Thickness mm (in)
13218 AK010	2.00 (0.0787)
13218 AK020	2.02 (0.0795)
13218 AK030	2.04 (0.0803)
13218 AK040	2.06 (0.0811)
13218 AK050	2.08 (0.0819)
13218 AK060	2.10 (0.0827)
13218 AK070	2.12 (0.0835)
13218 AK080	2.14 (0.0843)
13218 AK090	2.16 (0.0850)
13218 AK100	2.18 (0.0858)
13218 AK110	2.20 (0.0866)
13218 AE710	2.22 (0.0874)
13218 AE730	2.24 (0.0882)
13218 AE750	2.26 (0.0890)
13218 AE770	2.28 (0.0898)
13218 AE790	2.30 (0.0906)
13218 AE810	2.32 (0.0913)
13218 AE830	2.34 (0.0921)
13218 AE850	2.36 (0.0929)
13218 AE870	2.38 (0.0937)
13218 AE890	2.40 (0.0945)
13218 AE910	2.42 (0.0953)
13218 AE920	2.43 (0.0957)
13218 AE930	2.44 (0.0961)
13218 AE940	2.45 (0.0965)
13218 AE950	2.46 (0.0969)
13218 AE960	2.47 (0.0972)
13218 AE970	2.48 (0.0976)
13218 AE980	2.49 (0.0980)
13218 AE990	2.50 (0.0984)
13218 AF000	2.51 (0.0988)
13218 AF010	2.52 (0.0992)
13218 AF020	2.53 (0.0996)
13218 AF030	2.54 (0.1000)
13218 AF040	2.55 (0.1004)
13218 AF050	2.56 (0.1008)
13218 AF060	2.57 (0.1012)
13218 AF070	2.58 (0.1016)
13218 AF090	2.60 (0.1024)
13218 AF110	2.62 (0.1031)
13218 AF130	2.64 (0.1039)
13218 AF150	2.66 (0.1047)
13218 AF170	2.68 (0.1055)
13218 AF190	2.70 (0.1063)

9357TG30

Valve adjusting shim chart—2.0L engines

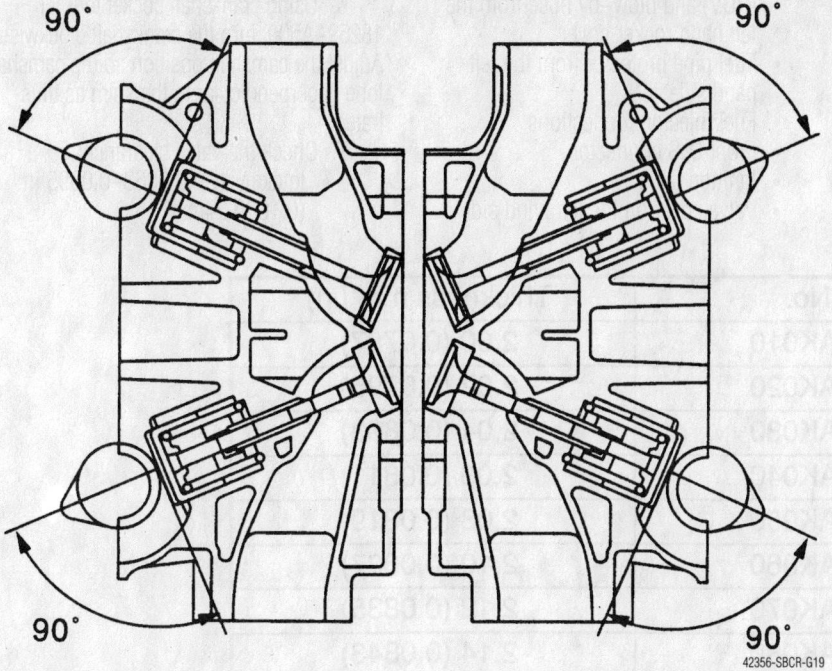

90°   90°

90°   90°

42356-SBCR-G19

Turn the crankshaft clockwise. Adjust the camshaft position so the camshaft lobe is perpendicular to the shim —3.0L engine

➡**Make sure when setting the tool that the edge does not touch the shim.**

e. Tighten bolt "A" and attach it to the cylinder head. Refer to the accompanying illustration for bolt locations.

f. Tighten bolt "B" and insert the lifter. Refer to the accompanying illustration for bolt locations.

g. Use tweezers and remove the shim from the lifter. A magnet can also be used to remove the shim.

h. Measure the shim thickness using a micrometer.

i. Using the table supplied, select a suitable shim using measured valve clearance and shim thickness.

j. Install the replacement shim to the lifter.

k. After all shims have been adjusted, inspect the valve clearances again.

l. After completion, install all removed components.

3.0L Engine

➡**The valve adjustment should be performed while the engine is cold.**

1. Before servicing the vehicle, refer to the precautions in the beginning of this section.

2. Adjustment should be performed when engine is cold.

3. Remove or disconnect the following:
- Negative battery cable
- Engine undercover

4. To inspect the right hand side perform remove the following:
- Drive belt
- Power steering hose from the bracket
- Power steering pump bracket bolts and place the pump assembly on the right side wheel apron with the lines still attached
- Fuel pipe protector from the right hand side

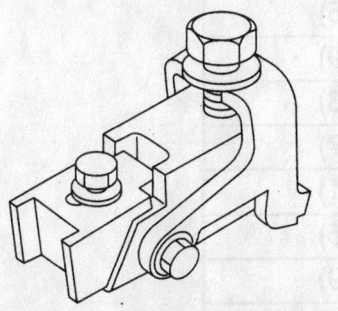

42356-SBCR-G20

Use shim replacer tool ST 18329AA000 to remove the shim from the lifter —3.0L engine

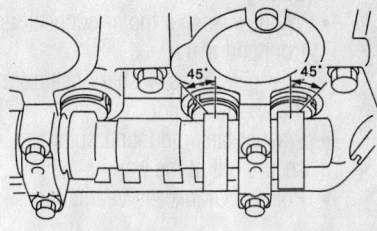

42356-SBCR-G21

Rotate the notch of the valve lifter outwards by 45 degrees—3.0L engine

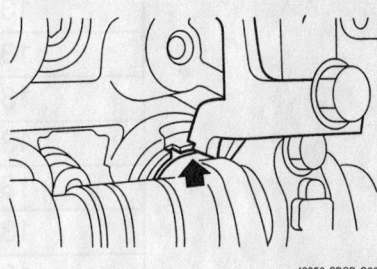

42356-SBCR-G22

Adjust the shim replacer notch to the valve lifter and set it as shown—3.0L engine

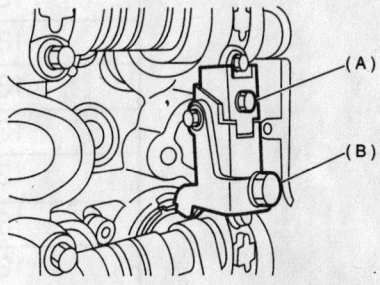

42356-SBCR-G23

Tighten bolt A on the tool and attach it to the cylinder head, then tighten bolt B and insert the valve lifter—3.0L engine

- Fuel injector connections
- Front Oxygen Sensor (O_2S) connector
- Oil pressure switch connector
- Ignition coils
- Valve cover on the right hand side

5. To inspect the left hand side perform remove the following:

	Unit: mm
Intake valve: $S = (V + T) - 0.20$	
Exhaust valve: $S = (V + T) - 0.25$	
S: Shim thickness to be used	
V: Measured valve clearance	
T: Shim thickness required	

9357TG28

Use this table to help you select a suitable shim—3.0L engines

- Battery
- Window washer motor connectors front and rear
- Rear gate glass washer hose from the washer motor
- Washer tank mounting bolts and secure out of the way
- Positive Crankcase Ventilation

(PCV) and blow–by hose from the left hand rocker cover
- Fuel pipe protector from the left hand side
- Fuel injector connections
- Front O_2S connector
- Ignition coils
- Valve cover on the left hand side

6. Using crankshaft socket tool ST 18252AA000, turn the crankshaft clockwise. Adjust the camshaft position so the camshaft lobe is perpendicular to the shim as illustrated.

7. Check the valve clearance:
- Intake valve: 0.0063–0.0095 in. (0.16–0.24mm).

Part No.	Thickness mm (in)
13218 AK010	2.00 (0.0787)
13218 AK020	2.02 (0.0795)
13218 AK030	2.04 (0.0803)
13218 AK040	2.06 (0.0811)
13218 AK050	2.08 (0.0819)
13218 AK060	2.10 (0.0827)
13218 AK070	2.12 (0.0835)
13218 AK080	2.14 (0.0843)
13218 AK090	2.16 (0.0850)
13218 AK100	2.18 (0.0858)
13218 AK110	2.20 (0.0866)
13218 AE710	2.22 (0.0874)
13218 AE720	2.23 (0.0878)
13218 AE730	2.24 (0.0882)
13218 AE740	2.25 (0.0886)
13218 AE750	2.26 (0.0890)
13218 AE760	2.27 (0.0894)
13218 AE770	2.28 (0.0898)
13218 AE780	2.29 (0.0902)
13218 AE790	2.30 (0.0906)
13218 AE800	2.31 (0.0909)
13218 AE810	2.32 (0.0913)
13218 AE820	2.33 (0.0917)
13218 AE830	2.34 (0.0921)
13218 AE840	2.35 (0.0925)
13218 AE850	2.36 (0.0929)
13218 AE860	2.37 (0.0933)
13218 AE870	2.38 (0.0937)
13218 AE880	2.39 (0.0941)
13218 AE890	2.40 (0.0945)
13218 AE900	2.41 (0.0949)
13218 AE910	2.42 (0.0953)

42356-SBCR-G24

Valve adjusting shim chart (1 of 2)—3.0L engines

• Exhaust valve: 0.0078–0.0118 in. (0.20–0.25mm).

8. If any valve needs adjustment, perform the following:

a. Record each valve measurement after it has been measured.

b. Using shim replacer tool ST 18329AA000, remove the shim from the lifter.

c. Rotate the notch of the valve lifter outwards by 45 degrees.

d. Adjust the shim replacer notch to the valve lifter and set it as illustrated. Make sure when setting the replacer, the edge does not touch the shim.

e. Tighten bolt **A** on the tool and attach it to the cylinder head, then tighten bolt **B** and insert the valve

lifter. Refer to the accompanying illustration.

f. Use tweezers to remove the shim. A magnet may be used as well to remove the shim without dropping it.

g. Use a micrometer to measure the shim.

9. Measure the shim thickness using a micrometer.

Part No.	Thickness mm (in)
13218 AE920	2.43 (0.0957)
13218 AE930	2.44 (0.0961)
13218 AE940	2.45 (0.0965)
13218 AE950	2.46 (0.0969)
13218 AE960	2.47 (0.0972)
13218 AE970	2.48 (0.0976)
13218 AE980	2.49 (0.0980)
13218 AE990	2.50 (0.0984)
13218 AF000	2.51 (0.0988)
13218 AF010	2.52 (0.0992)
13218 AF020	2.53 (0.0996)
13218 AF030	2.54 (0.1000)
13218 AF040	2.55 (0.1004)
13218 AF050	2.56 (0.1008)
13218 AF060	2.57 (0.1012)
13218 AF070	2.58 (0.1016)
13218 AF090	2.60 (0.1024)
13218 AF100	2.61 (0.1028)
13218 AF110	2.62 (0.1031)
13218 AF120	2.63 (0.1035)
13218 AF130	2.64 (0.1039)
13218 AF140	2.65 (0.1043)
13218 AF150	2.66 (0.1047)
13218 AF160	2.67 (0.1051)
13218 AF170	2.68 (0.1055)
13218 AF180	2.69 (0.1059)
13218 AF190	2.70 (0.1063)
13218 AF200	2.71 (0.1067)
13218 AF210	2.72 (0.1071)
13218 AF220	2.73 (0.1075)
13218 AF230	2.74 (0.1079)
13218 AF240	2.75 (0.1083)
13218 AF250	2.76 (0.1087)
13218 AF260	2.77 (0.1091)
13218 AF270	2.78 (0.1094)
13218 AF280	2.79 (0.1098)
13218 AF290	2.80 (0.1102)
13218 AF300	2.81 (0.1106)

42356-SBCR-G25

Valve adjusting shim chart (2 of 2)—3.0L engines

10. Using the table supplied, select a suitable shim using measured valve clearance and shim thickness.

11. Install the replacement shim to the lifter.

12. After all shims have been adjusted, inspect the valve clearances again.

13. After completion, install all removed components.

Starter Motor

REMOVAL & INSTALLATION

1. Before servicing the vehicle, refer to the precautions in the beginning of this section.

2. Remove or disconnect the following:
- Negative battery cable
- Intake Air Temperature (IAT) connector, on Legacy models equipped with a manual transaxle
- Air cleaner case and duct
- Air cleaner case stay, on Legacy models
- Connector and terminal from starter
- Retaining bolts and/or nuts
- Starter from transmission

To install:

3. Install or connect the following:
- Starter to the transmission
- Starter retaining bolts and/or nuts and tighten to 34–40 ft. lbs. (46–54 Nm)
- Connector and terminal to starter

- Air cleaner case stay, on Legacy models
- Air cleaner case and duct
- IAT connector, on Legacy models equipped with a manual transaxle
- Negative battery cable

Oil Pan

REMOVAL & INSTALLATION

1. Before servicing the vehicle, refer to the precautions in the beginning of this section.

2. Remove or disconnect the following:
- Negative battery cable
- Air intake duct

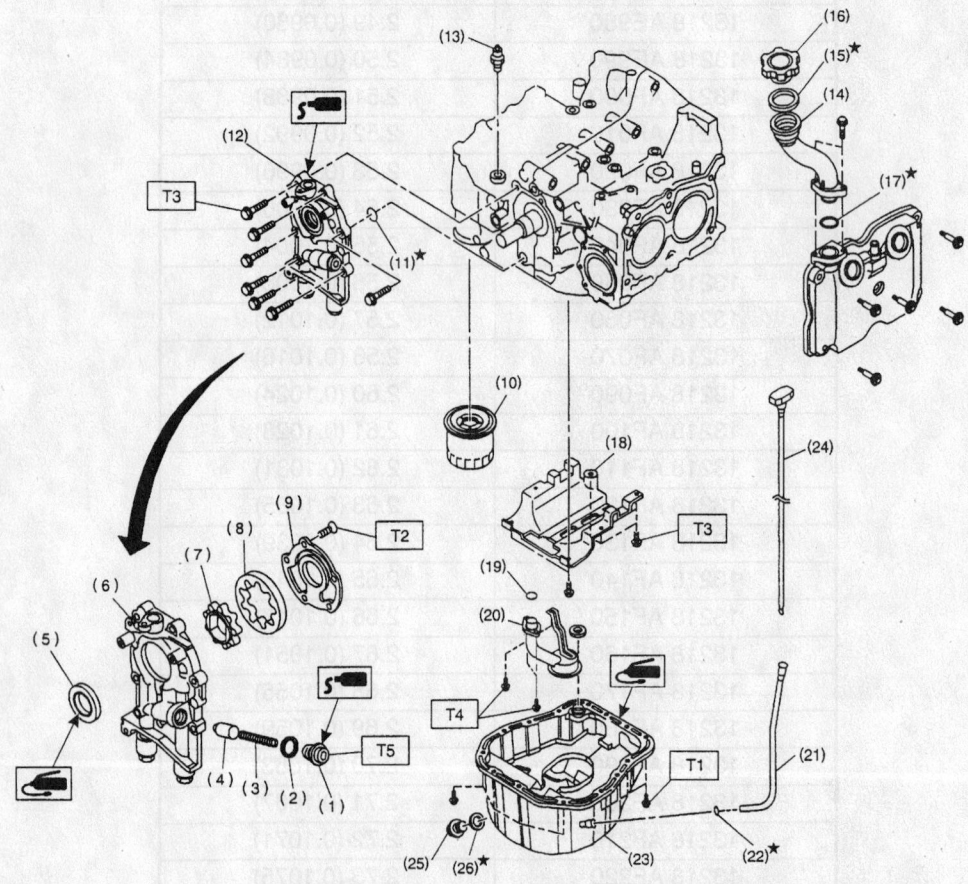

1) Plug	13) Oil pressure switch	25) Drain plug
2) Washer	14) Oil filler duct	26) Metal gasket
3) Relief valve spring	15) O-ring	
4) Relief valve	16) Oil filler cap	
5) Oil seal	17) O-ring	**Tightening torque: N·m (kg-m, ft-lb)**
6) Oil pump case	18) Baffle plate	T1: 5 (0.5, 3.6)
7) Inner rotor	19) O-ring	T2: 5^{+1}_{-0} $(0.5^{+0.1}_{-0}, 3.6^{+0.7}_{-0})$
8) Outer rotor	20) Oil strainer	T3: 6.4 (0.65, 4.7)
9) Oil pump cover	21) Oil level gauge guide	T4: 10 (1.0, 7.2)
10) Oil filter	22) O-ring	T5: 44.1±3.4 (4.5±0.35, 32.5±2.5)
11) O-ring	23) Oil pan	
12) Oil pump ASSY	24) Oil level gauge	

Exploded view of the oil pan and lubrication components—2.2L and 2.5L engines

9307TG03

- Mass Air Flow (MAF) sensor on turbo models
- Air intake boot and air cleaner upper cover on turbo models
- Intercooler on turbo models
- Front Oxygen Sensor (O₂S) electrical connector

- Pitching stopper
- Upper radiator brackets

3. Support the engine with a suitable lifting device.
- Front wheel and tire assemblies

4. Lift up the engine slightly.
- Engine undercover

5. Drain the oil from the engine into a suitable container.

6. Install the drain plug with a new gasket and tighten it to 33–36 ft. lbs. (43–47 Nm).

7. Remove or disconnect the following:
- Rear O₂S electrical connector

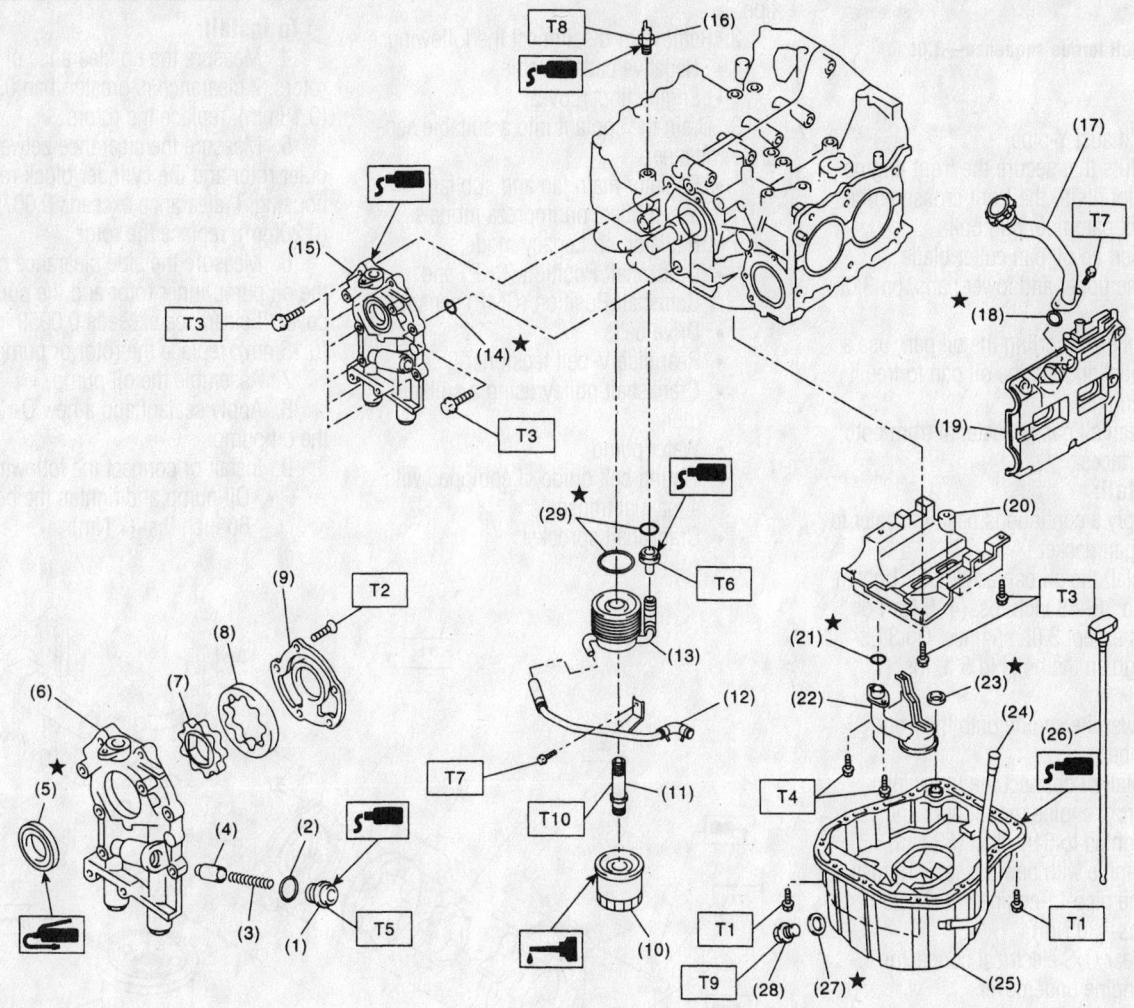

(1)	Plug	(15)	Oil pump ASSY	(29)	O-ring
(2)	Washer	(16)	Oil pressure switch		
(3)	Relief valve spring	(17)	Oil filler duct		
(4)	Relief valve	(18)	O-ring		
(5)	Oil seal	(19)	Cylinder head cover		
(6)	Oil pump case	(20)	Baffle plate		
(7)	Inner rotor	(21)	O-ring		
(8)	Outer rotor	(22)	Oil strainer		
(9)	Oil pump cover	(23)	Gasket		
(10)	Oil filter	(24)	Oil level gauge guide		
(11)	Connector	(25)	Oil pan		
(12)	Water by-pass pipe	(26)	Oil level gauge		
(13)	Oil cooler	(27)	Metal gasket		
(14)	O-ring	(28)	Drain plug		

Tightening torque: N·m (kgf-m, ft-lb)

T1:	5 (0.5, 3.6)
T2:	5 (0.5, 3.6)
T3:	6.4 (0.65, 4.7)
T4:	10 (1.0, 7.0)
T5:	44.1 (4.5, 32.5)
T6:	69 (7.0, 4.7)
T7:	6.4 (0.65, 50.6)
T8:	25 (2.5, 18.1)
T9:	44 (4.5, 33)
T10:	54 (5.5, 40)

Exploded view of the oil pan and lubrication components—2.0L turbocharged engines

9357TG31

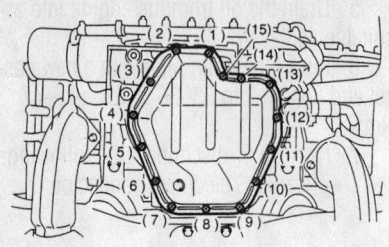

Oil pan bolt torque sequence—3.0L engines

- Exhaust Y-pipe
- Nuts that secure the front engine mounts to the front crossmember
- Oil pan mounting bolts

8. Insert an oil pan cutter blade between the upper and lower pans, on 3.0L engines

9. While supporting the oil pan, use a rubber mallet and tap the oil pan to free it from the engine.

10. Clean all gasket material from both mating surfaces.

To install:

11. Apply a continuous bead of sealer to a new oil pan gasket.

12. Install the oil pan assembly. Tighten the bolts to 36–48 inch lbs. (4–5 Nm) on all models except 3.0L engines. On 3.0L engines tighten the bolts to 5 ft. lbs. (7 Nm).

13. Lower the engine onto the front crossmember.

14. Install or connect the following:
- Front engine mount nuts and tighten to 61 ft. lbs. (83 Nm)
- Y-pipe with new gaskets. Tighten the pipe-to-engine nuts to 23 ft. lbs. (30 Nm)
- Rear O₂S electrical connector
- Engine undercover
- Front wheel and tire assemblies

15. Remove the engine lifting device.
- Front O₂S sensor electrical connector
- Pitching stopper. Tighten the front bolt to 40 ft. lbs. (54 Nm) and the rear bolt to 49 ft. lbs. (67 Nm).
- Upper radiator brackets
- MAF sensor on turbo models
- Air intake boot and air cleaner upper cover on turbo models
- Intercooler on turbo models
- Air intake duct
- Negative battery cable

16. Fill the engine to the proper level with the recommended oil and run the engine. Check for leaks.

Oil Pump

REMOVAL & INSTALLATION

2.2L and 2.5L Engines

1. Before servicing the vehicle, refer to the precautions in the beginning of this section.

2. Remove or disconnect the following:
- Negative battery cable
- Engine undercover

3. Drain the coolant into a suitable separate container.
- Radiator main fan and sub fan assemblies, on Impreza models
- Radiator, on Legacy models
- Crankshaft Position (CKP) and Camshaft Position (CMP) sensors
- Drive belts
- Rear side V-belt tensioner
- Crankshaft pulley using a suitable tool
- Water pump
- Timing belt guide, if equipped with a manual transaxle
- Crankshaft sprocket
- Oil pump mounting bolts
- Oil pump by carefully prying it from the engine block

✳✳ WARNING

Use extreme care not to damage the engine block or the oil pump during removal of the pump.

To install:

4. Measure the tip clearance of the rotors. If clearance is greater than 0.0071 in. (0.18mm), replace the rotors.

5. Measure the clearance between the outer rotor and the cylinder block rotor housing. If clearance exceeds 0.0079 in. (0.20mm), replace the rotor.

6. Measure the side clearance between the oil pump inner rotor and the pump cover. If clearance exceeds 0.0059 in. (0.15mm), replace the rotor or pump body.

7. Assemble the oil pump.

8. Apply sealant and a new O-ring to the oil pump.

9. Install or connect the following:
- Oil pump and tighten the bolts to 60 inch lbs. (7 Nm)

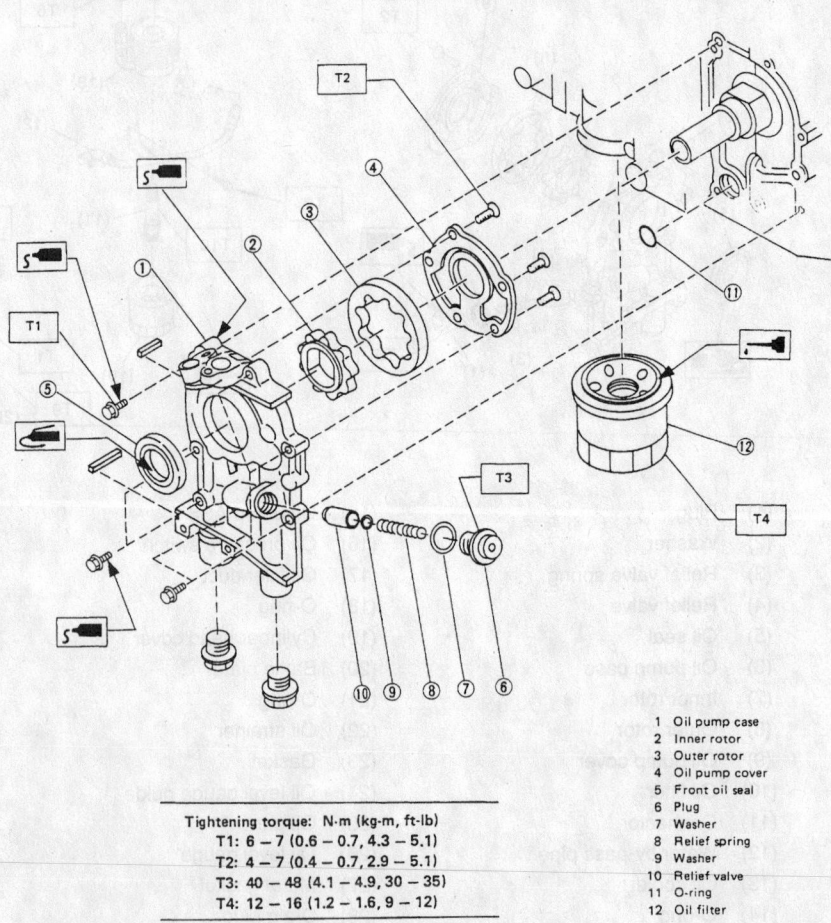

Tightening torque: N·m (kg-m, ft-lb)
T1: 6 – 7 (0.6 – 0.7, 4.3 – 5.1)
T2: 4 – 7 (0.4 – 0.7, 2.9 – 5.1)
T3: 40 – 48 (4.1 – 4.9, 30 – 35)
T4: 12 – 16 (1.2 – 1.6, 9 – 12)

1	Oil pump case
2	Inner rotor
3	Outer rotor
4	Oil pump cover
5	Front oil seal
6	Plug
7	Washer
8	Relief spring
9	Washer
10	Relief valve
11	O-ring
12	Oil filter

Oil pump and components

7923TG37

- Crankshaft sprocket
- Timing belt guide, if equipped with a manual transaxle
- Water pump
- Crankshaft pulley using a suitable tool
- Rear side V-belt tensioner
- Drive belts
- CKP and CMP sensors
- Radiator, on Legacy models
- Radiator main fan and sub fan assemblies, on Impreza models
- Engine undercover
- Negative battery cable

10. Fill and bleed the cooling system.

11. Start the engine and check for leaks.

3.0L Engine

1. Before servicing the vehicle, refer to the precautions in the beginning of this section.

2. Remove or disconnect the following:
   - Negative battery cable
   - Engine under cover

3. Drain the coolant.
   - Radiator
   - Drive belt
   - Front timing chain cover and chain. Refer to the timing chain removal procedure in this section.
   - Oil pump cover and crankshaft sprocket
   - Inner and outer rotor

To install:

4. Apply engine oil to the entire surface area of the inner and outer rotor

5. Install or connect the following:
   - Inner rotor by fitting it into the groove on the crankshaft and then assemble the outer rotor
   - Oil pump cover and tighten the bolts in the sequence illustrated to 5 ft. lbs. (7 Nm)
   - Crankshaft sprocket
   - Timing chain and cover. Refer to

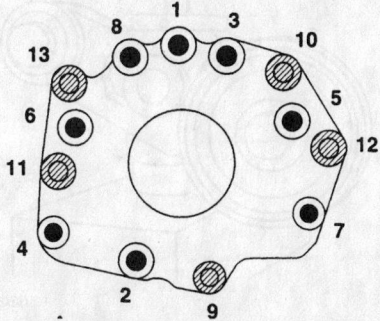

Oil pump cover torque sequence—3.0L engines

42356-SBCR-G27

the timing chain removal procedure in this section.
   - Drive belt
   - Radiator
   - Engine undercover
   - Negative battery cable

6. Fill and bleed the cooling system.

7. Start the engine and check for leaks.

Rear Main Seal

REMOVAL & INSTALLATION

1. Before servicing the vehicle, refer to the precautions in the beginning of this section.

2. Remove or disconnect the following:
   - Engine from the vehicle
   - Clutch assembly/flywheel using the Clutch Disc Guide tool 499747000, if equipped with a manual transmission
   - Torque converter flexplate from the crankshaft, if equipped with an automatic transmission
   - Oil seal from the cylinder block using a small prybar

To install:

3. Install or connect the following:
   - New oil seal by pressing it into the cylinder block using the appropriate driver and hammer

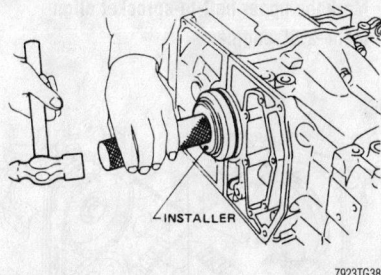

Installing the rear main seal—except 3.0L engines

7923TG38

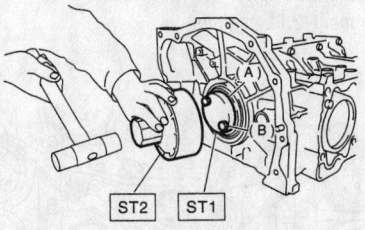

(A) Rear oil seal
(B) Drive plate attaching bolt

42356-SBCR-G28

Installing the rear main seal using oil seal guide ST1 499597100 and ST2 499598200—3.0L engines

- Flywheel housing using new gaskets and sealant where necessary.
- Flywheel and tighten the bolts to 50–54 ft. lbs. (69–75 Nm) on all models except 3.0L engines. On 3.0L engines tighten the bolts to 60 ft. lbs. (81 Nm).
- Engine

Timing Belt

REMOVAL & INSTALLATION

2.0L Engine

1. Before servicing the vehicle, refer to the precautions in the beginning of this section.

2. Disconnect the negative battery cable.

3. Remove the V-belt.

4. Remove the crankshaft pulley.

5. Remove the belt cover as follows:
   a. Crankshaft pulley.
   b. Left hand belt cover.
   c. Right hand belt cover.
   d. Front hand belt cover.

6. Remove the timing belt guides on models equipped with a manual transmission.

7. If the alignment marks that indicate rotation are faded, put new marks on the belt before removal as follows:

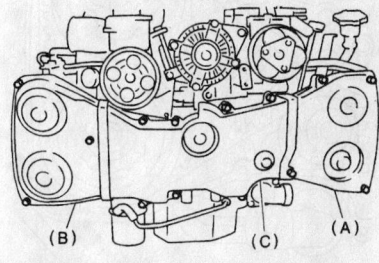

9357TG46

Remove the left hand (A), right hand (B) and front (C) belt covers–2.0L engines

9357TG47

Location of the upper belt guide–2.0L engines

a. Turn the crankshaft using crankshaft sprocket tool 499987500 and a breaker bar to align the crankshaft sprocket, left hand intake camshaft sprocket, left hand exhaust camshaft, right hand intake camshaft sprocket and right hand exhaust camshaft sprocket with the on the cover and cylinder block.

b. Using white paint such as white out, place alignment marks on the belts in relation to the sprockets.

8. Remove the belt idler (A), illustrated in the accompanying illustration.

9. Remove the timing belt.

10. If necessary, remove belt idlers (B) and (C).

11. Remove the belt idler 2.

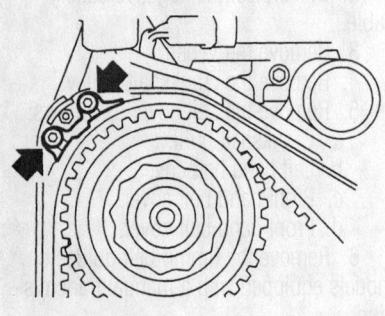

Location of the upper left belt guide—2.0L engines

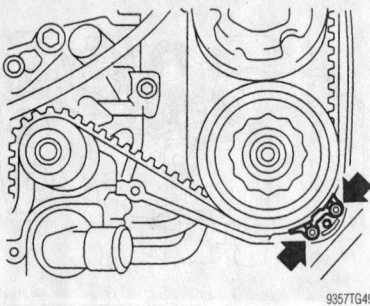

Location of the lower right belt guide—2.0L engines

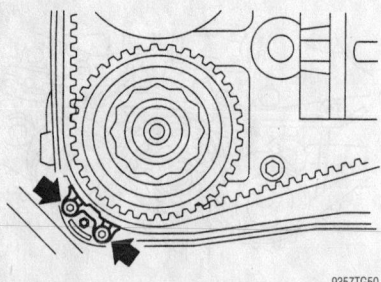

Location of the lower left belt guide—2.0L engines

12. Remove the automatic belt tension adjuster assembly.

To install:

13. To prepare the automatic belt tensioner for assembly, perform the following steps:

a. Always use a vertical type pressing tool to move the adjuster rod down.

b. Do not use a lateral type vise.

c. Always push the adjuster rod vertically.

d. Make sure to slowly move the adjuster rod down applying a pressure of 66 lbs. (294 N).

e. Press in the push adjuster rod gradually taking more than 3 minutes.

f. Never allow the press pressure to exceed 2,205 lbs. (9,807 N).

g. Press the adjuster rod as far as the end surface of the cylinder. Do not press the rod into the cylinder as doing so may damage the cylinder.

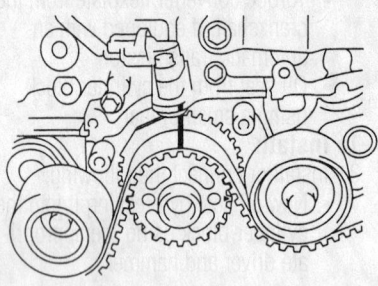

Mark the upper belt-to-sprocket alignment—2.0L engines

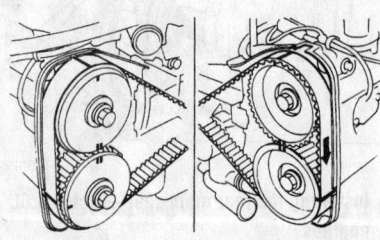

Mark the lower belt-to-sprocket alignment—2.0L engines

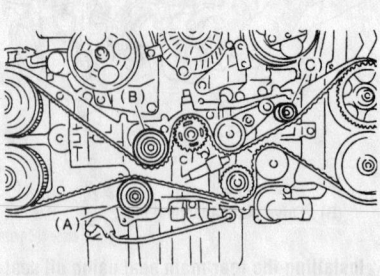

Remove the belt idler (A)—2.0L engines

h. Never release the press pressure until the stopper pin has been fully inserted.

14. Attach the automatic belt tension adjuster assembly to the vertical pressing tool.

15. Move the adjuster rod down slowly using a pressure of 66 lbs. (294 N) until the rod is aligned with the stopper pin hole in the cylinder.

16. Insert a 0.08 inch (2mm) stopper pin or diameter Allen wrench into the stopper pin hole in the cylinder to retain the rod.

17. Install the adjuster assembly and tighten the retainers to 29 ft. lbs. (39 Nm).

18. Install belt idle 2 and tighten the retainers to 29 ft. lbs. (39 Nm).

19. Install the belt idler and tighten to 29 ft. lbs. (39 Nm).

20. Align the mark on the crankshaft sprocket with the mark on the oil pump.

21. Align the single line mark on the right hand exhaust camshaft sprocket with the notch on the belt cover.

22. Align the single line mark on the right hand intake camshaft with the notch on the belt cover. Make sure the double lines on the intake camshaft and exhaust sprockets are aligned as shown in the accompanying illustration.

23. Align the single line mark on the left hand exhaust camshaft sprocket with the notch on the belt cover by turning the sprocket counterclockwise (as viewed from the front of the engine).

24. Align the single line mark on the left hand intake camshaft sprocket with the notch on the belt cover by turning the sprocket counterclockwise (as viewed from the front of the engine). Make sure the double lines on the intake camshaft and

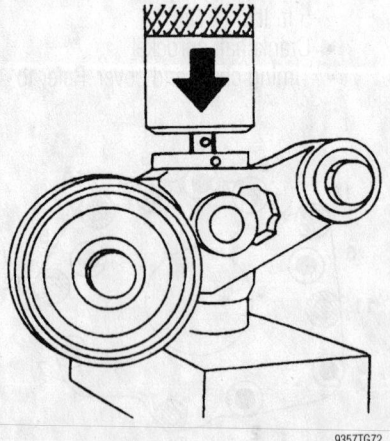

Attach the automatic belt tension adjuster assembly to the vertical pressing tool—2.0L engines

exhaust sprockets are aligned as shown in the accompanying illustration.

➡Make sure the camshaft and crankshaft sprockets are positioned correctly. The intake and exhaust camshafts on this engine can be rotated independently with the timing belt removed. By looking at the accompanying illustration it will show you that if the intake and exhaust valve are lift together the heads will hit each other and bend.

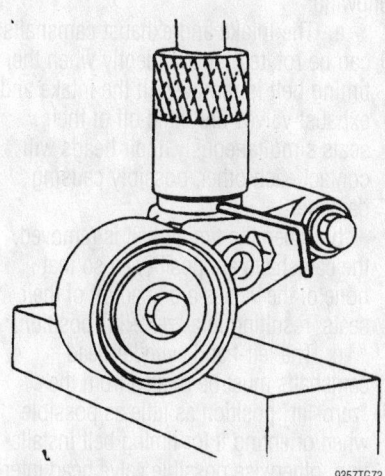

Insert a 0.08 inch (2mm) stopper pin or diameter Allen wrench into the stopper pin hole in the cylinder to retain the rod–2.0L engines

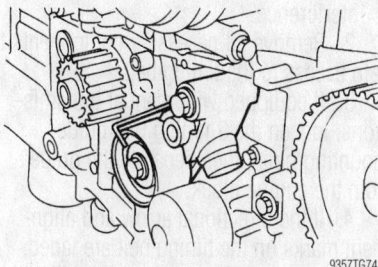

Install the adjuster assembly–2.0L

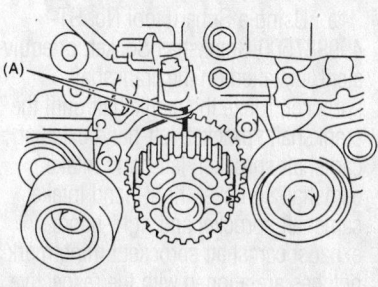

Align the mark on the crankshaft sprocket with the mark on the oil pump–2.0L engines

➡When the timing belts are not installed, 4 camshafts are held at "zero lift" position, where all cams on the camshafts do not push the intake and exhaust valves down (under this condition all valves remain unlifted). When the camshafts are rotated to install the timing belts, # 2 intake and # 4 exhaust cam of the left hand camshafts are held to push their corresponding valves

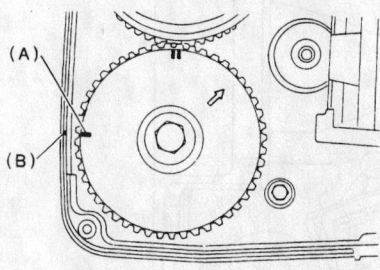

Align the single line mark on the right hand exhaust camshaft sprocket with the notch on the belt cover–2.0L engines

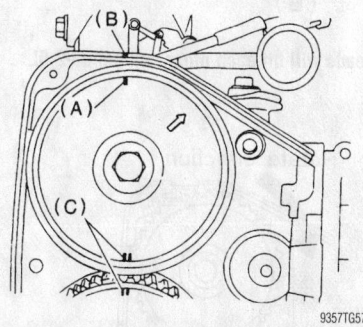

Align the single line mark on the right hand intake camshaft with the notch on the belt cover. Make sure the double lines on the intake camshaft and exhaust sprockets are aligned–2.0L engines

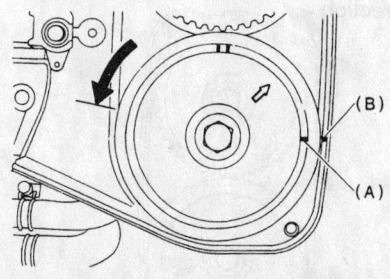

Align the single line mark on the left hand exhaust camshaft sprocket with the notch on the belt cover by turning the sprocket counterclockwise (as viewed from the front of the engine)–2.0L engines

down. Under this condition these valves are held lifted. The right side camshafts are held in so that their cams do not push the valves down. The left hand camshafts must be rotated from the "zero lift" position to the position where the timing belt is to be installed at as small an angle as possible, in order to prevent mutual interference of intake and exhaust valve heads. Do not allow the camshafts to rotate in the direction illustrated as this causes both the intake and exhaust valves to lift off at the same time with will cause valve damage.

25. When installing the belt, make sure to align the marks made during removal or if using a new belt, align the in alphabetical order as shown in the accompanying illustration.

✳✳ WARNING

Disengagement of more than 3 timing belt teeth may result in contact between the valve and piston. Always make sure the belts rotation is correct.

26. Install the belt idlers and tighten to 29 ft. lbs. (39 Nm).

✳✳ WARNING

Make sure the marks on the belt and sprockets are properly aligned.

27. Once the marks on the belt and sprockets are aligned, remove the stopper pin from the tensioner adjuster.
28. Install the timing belt guide on models with manual transmission. measure the clearance between the belt and guide. the clearance should be 0.019–0.059 inch

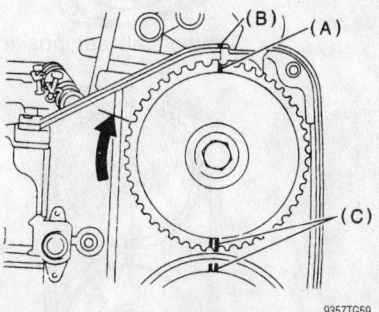

Align the single line mark on the left hand intake camshaft sprocket with the notch on the belt cover by turning the sprocket counterclockwise (as viewed from the front of the engine). Make sure the double lines on the intake camshaft and exhaust sprockets are aligned–2.0L engines

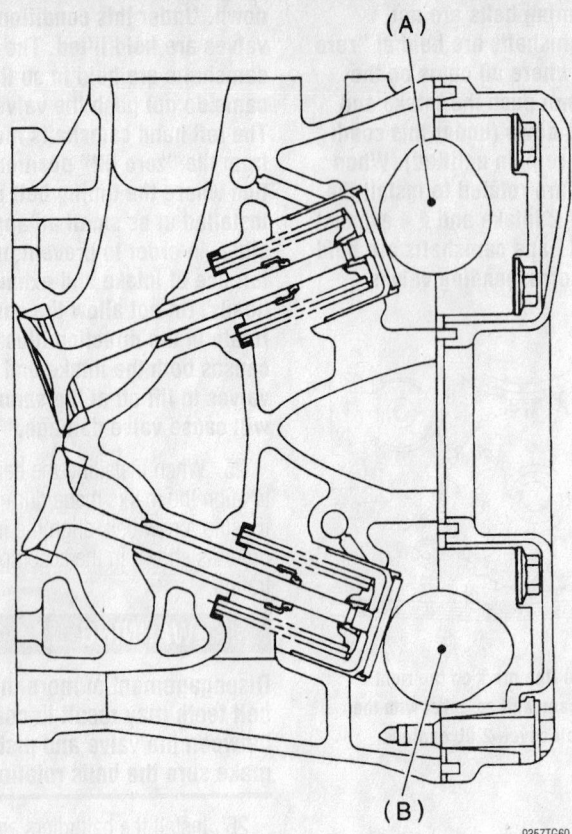

(A)

(B)

9357TG60

If the intake and exhaust valve are lift together the heads will hit each other and bend –2.0L engines

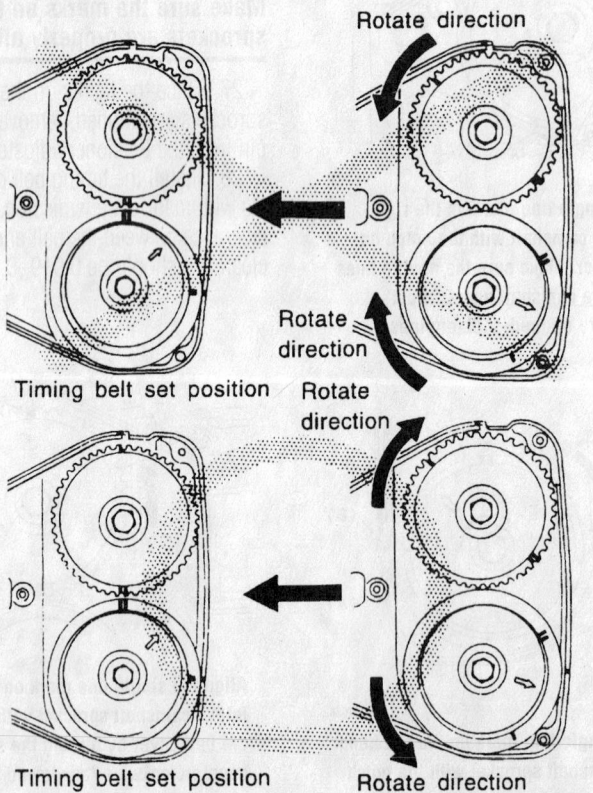

Rotate direction

Timing belt set position

Rotate direction

Rotate direction

Rotate direction

Timing belt set position

Rotate direction

9357TG61

Do not allow the camshafts to rotate in the direction shown as this causes both the intake and exhaust valves to lift off at the same time with will cause valve damage–2.0L engines

(0.5–1.5mm) and tighten the retainers to 7 ft. lbs. (10 Nm).

29. Install the belt covers and tighten to 3.5 ft. lbs. (5 Nm).

30. Install the crankshaft pulley and tighten the bolt to 94 ft. lbs. (127 Nm).

31. Install the V-belt.

2.2L and 2.5L Engine

1. Before servicing the vehicle, refer to the precautions in the beginning of this section.

When servicing the timing belt, note the following:

a. The intake and exhaust camshafts can be rotated independently when the timing belt is removed. If the intake and exhaust valves are lifted off of their seats simultaneously, their heads will contact each other, possibly causing damage.

b. When the timing belt is removed, the camshafts are positioned so that none of the valves are lifted off of their seats, resulting in a "zero-lift" position.

c. The left-hand cylinder head camshafts must be rotated from the "zero-lift" position as little as possible when orienting it for timing belt installation, otherwise possible valve head interference may occur.

d. Never allow the camshafts to rotate in the direction shown in the accompanying illustration, which would cause both the intake and exhaust valves to lift simultaneously, causing interference.

2. Remove all necessary components to gain access to the timing belt.

3. If equipped with manual transmissions, loosen the 2 timing belt guide mounting bolts, then separate the guide from the engine block.

4. If the directional arrow and alignment marks on the timing belt are faded, and the belt is to be reused, remark the belt with white paint or a grease pencil as follows:

a. Using a Subaru tool No. ST-499987500 Crankshaft Socket, or equivalent, installed on the crankshaft sprocket, rotate the crankshaft until the crankshaft sprocket, left-hand exhaust camshaft sprocket, left-hand intake camshaft sprocket, right-hand intake camshaft sprocket and right-hand exhaust camshaft sprocket timing mark notches are aligned with the respective marks on the belt cover and engine block.

b. Make alignment and/or arrow

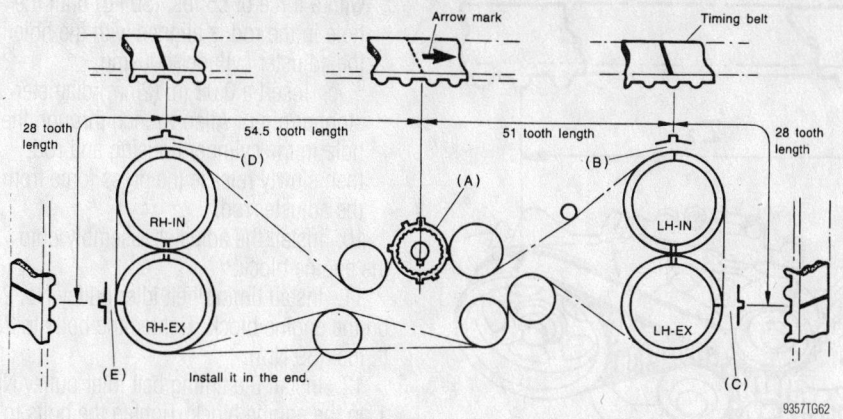

Align the marks in alphabetical order as shown if using a new belt–2.0L engines

Remove the timing belt guide on vehicles equipped with manual transmission—2.5L engine shown, 2.2L similar

marks on the timing belt in relation to the sprockets as indicated in the accompanying illustration.

- Z1: 44 tooth length on all 2.2L engines and 2000 2.5L engines and 46.8 tooth length on 2001–04 2.5L engines
- Z2: 40.5 tooth length on all 2.2L engines and 2000 2.5L engines or 43.7 tooth length on 2001–04 2.5L engines

5. Loosen the center bolt from the timing belt idler pulley, then remove the idler pulley from the engine block.

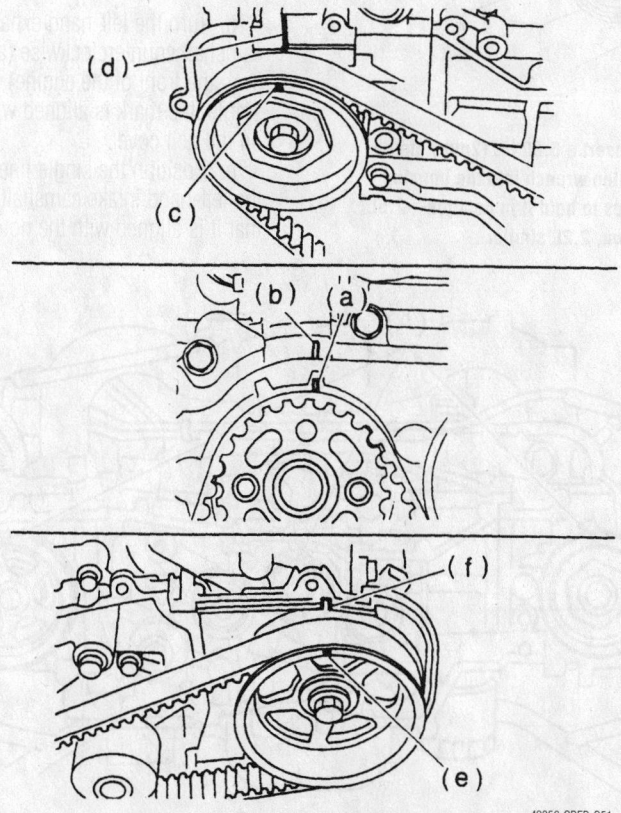

Before removing the timing belt, turn the crankshaft sprocket until all of the alignment marks are aligned as indicated—2.5L engine shown, 2.2L similar

⁂ WARNING

After removing the timing belt, DO NOT rotate the camshafts. Damage to the valves may occur.

6. Carefully remove the timing belt from all of the sprockets.

7. Remove the automatic belt tension adjuster assembly as follows:

  a. Remove the 2 timing belt idler pulleys, as indicated in the accompanying illustration.

  b. Loosen the automatic tension adjuster assembly mounting bolts, then separate the adjuster assembly from the engine block.

To install:

⁂ WARNING

Do not allow oil, grease, or coolant to come in contact with the timing belt. If this occurs, quickly and thoroughly remove all traces of the compound. Also, never bend the timing belt sharply; the minimum bending radius is 2.36 in. (60mm).

8. Inspect the camshaft and crankshaft sprocket teeth for abnormal or excessive wear or scratches. Ensure there is no free-play between the sprocket and the key. Inspect the crankshaft sprocket sensor notch for damage or contamination with debris or dirt.

➡️When preparing the automatic tension adjuster assembly for installation, adhere to the following points:

- Always use a vertical press, rather than a horizontal press or vise, to depress the adjuster assembly rod
- Depress the adjuster rod in a vertical position ONLY
- Depress the adjuster rod slowly (taking more than 3 minutes) with a force of 66 lbs. (30 kg)
- Do not allow the press force to exceed 2205 lbs. (1000 kg)
- Press the adjuster rod in as far as the end surface of the cylinder — do not press the rod into the cylinder, which may cause damage to the assembly
- Do not release the press force from the rod until the stopper pin is completely inserted in the cylinder

9. Prepare the automatic timing belt tension adjuster assembly for installation as follows:

  a. Position the adjuster assembly in a vertical press.

  b. Slowly depress the adjuster rod

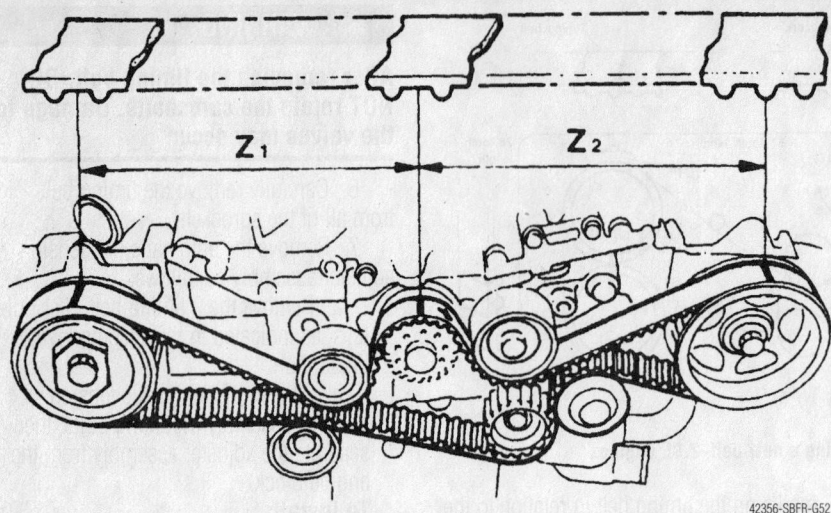

42356-SBFR-G52

If the original marks on the timing belt are worn or faded, make new alignment marks in the positions indicated—2.5L engine shown, 2.2L similar

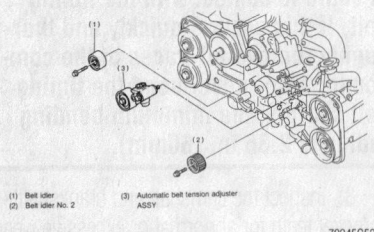

(1) Belt idler
(2) Belt idler No. 2
(3) Automatic belt tension adjuster ASSY

79245G52

It is necessary to remove the automatic adjuster assembly and reset the pushrod for timing belt installation—2.5L engine shown, 2.2L similar

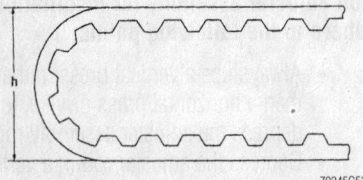

79245G53

Never bend the timing belt into a radius tighter than 2.36 in./60mm (h), otherwise it will be damaged beyond—2.5L engine shown, 2.2L similar

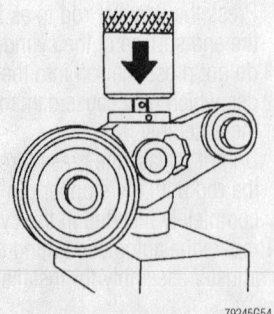

79245G54

Use a vertical press to push the adjuster rod into its housing until it is flush with the assembly's outer surface . . .

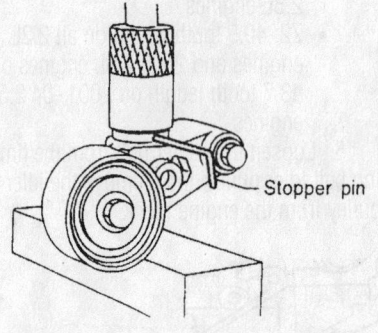

Stopper pin

79245G55

. . . then insert a 0.08 in. (2mm) diameter pin or Allen wrench into the housing and rod holes to hold it in position—2.5L engine shown, 2.2L similar

with a force of 66 lbs. (30 kg) until the hole in the rod is aligned with the hole in the adjuster cylinder housing.

   c. Insert a 0.08 in. (2mm) diameter stopper pin or Allen wrench through the hole in the cylinder housing and rod, then slowly release the press force from the adjuster rod.

10. Install the adjuster assembly onto the engine block.

11. Install timing belt idler pulley No. 2 on the engine block. Tighten the bolts to 28 ft. lbs. (39 Nm).

12. Install the timing belt idler pulley No. 1 on the engine block. Tighten the bolts to 28 ft. lbs. (39 Nm).

13. If the camshaft and crankshaft timing marks are no longer aligned, perform the following:

   a. Position the crankshaft sprocket so that its mark is aligned with the mark on the oil pump cover on the engine block.

   b. Align the single line mark on the right-hand exhaust camshaft sprocket with the notch on the belt cover.

   c. Rotate the right-hand intake camshaft so that the single line mark is aligned with the notch on the belt cover.

➡**At this point, the double line marks on both right-hand camshaft sprockets should be aligned.**

   d. Turn the left-hand exhaust (lower) camshaft counterclockwise (as viewed from the front of the engine) until the single line mark is aligned with the notch on the belt cover.

   e. Position the single line mark on the left-hand intake camshaft sprocket so that it is aligned with the notch on the

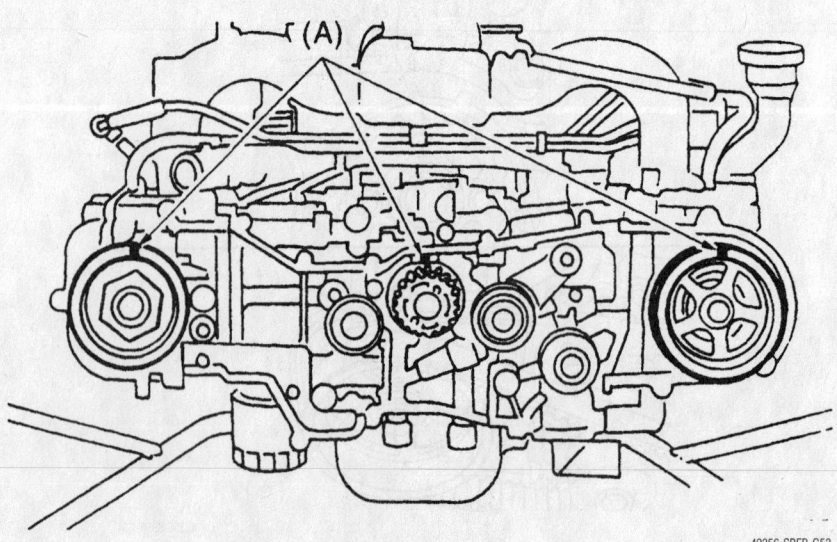

42356-SBFR-G53

Make sure the sprockets are aligned as shown—2.5L engine shown, 2.2L similar

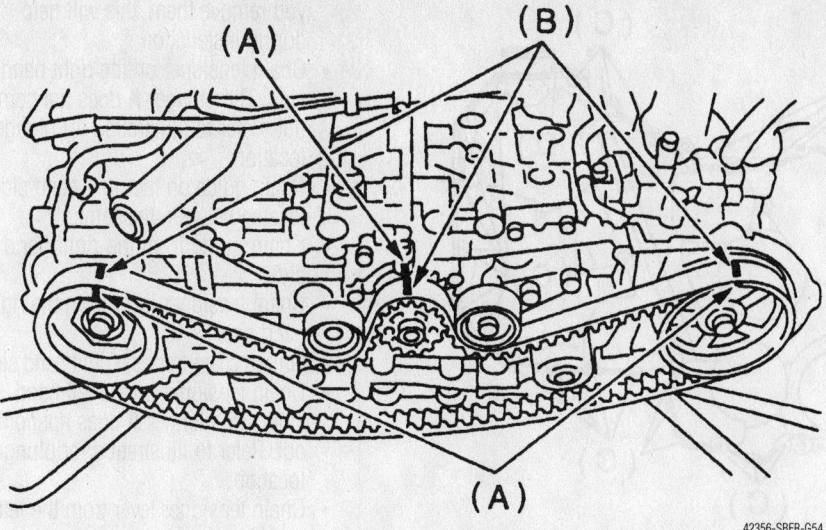

42356-SBFR-G54

After installing the timing belt, the alignment marks should be positioned as shown. If not, remove the belt and align the sprockets and reinstall the belt—2.5L engine shown, 2.2L similar

79245G64

On models equipped with manual transmissions, ensure the timing belt-to-guide clearance (arrows) is correct before tightening the mounting bolts—2.5L engine shown, 2.2L similar

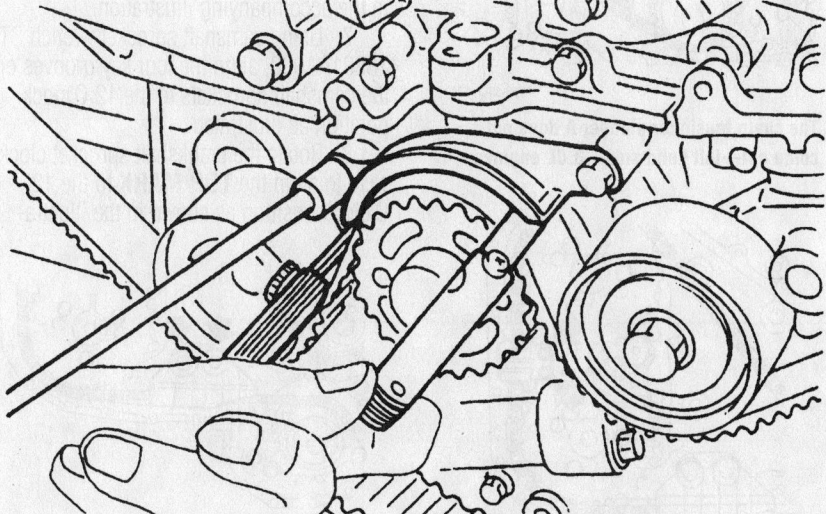

42356-SBFR-G55

On models equipped with manual transmissions, use a feeler gauges to adjust the clearance between the timing belt and the belt guide—2.5L engine shown, 2.2L similar

belt cover. When rotating the camshaft, do so only in a clockwise direction (as viewed from the front of the engine).

➡ **At this point, the double line marks on both left-hand camshaft sprockets should be aligned.**

　f. Ensure the timing marks are aligned as shown in the accompanying illustration. If they are not, repeat Sub-steps 12a through 12e until they are properly aligned.

14. Install the timing belt around the camshaft, crankshaft and idler pulleys so that the positioning marks on the timing belt are aligned with the marks on the sprockets as follows:

　a. Position the timing belt on the crankshaft sprocket so that the marks are aligned.

　b. Route the belt down and under the left-hand, upper idler pulley, then up and around the left-hand intake camshaft sprocket, ensuring the camshaft sprocket mark is aligned with the mark on the belt.

　c. Route the belt down and around the left-hand exhaust camshaft sprocket, making sure the marks are properly aligned, then up and over the first lower idler pulley and down and around the second lower idler pulley.

　d. While holding the timing belt on the inner, left-hand, lower idler pulley, route the other side of the timing belt (from the crankshaft sprocket) down and under the right-hand upper idler pulley.

　e. Route the timing belt up and around the right-hand intake camshaft sprocket so that the belt and sprocket marks are aligned.

　f. Position the belt down and around the right-hand exhaust camshaft sprocket, ensuring the positioning marks are aligned.

15. Install the right-hand lower idler pulley so that the timing belt is routed over the top side of it.

➡ **Once the belt is completely installed on all of the pulleys and sprockets, ensure that the positioning marks are still all aligned.**

16. After ensuring all of the marks are still aligned, use a pair of pliers to withdraw the stopper pin or Allen wrench from the adjuster assembly housing.

17. On models with manual transmissions, perform the following:

　a. Install the timing belt guide by temporarily tightening the mounting bolts.

　b. Position the timing belt guide so that there is 0.019–0.059 in. (0.5–1.5mm) clearance between the timing belt and the belt guide.

　c. Tighten the guide mounting bolts securely, then double check the guide clearance.

18. Install the timing belt covers and all remaining engine components.

Timing Chain

REMOVAL & INSTALLATION

3.0L Engines

1. Before servicing the vehicle, refer to the precautions in the beginning of this section.

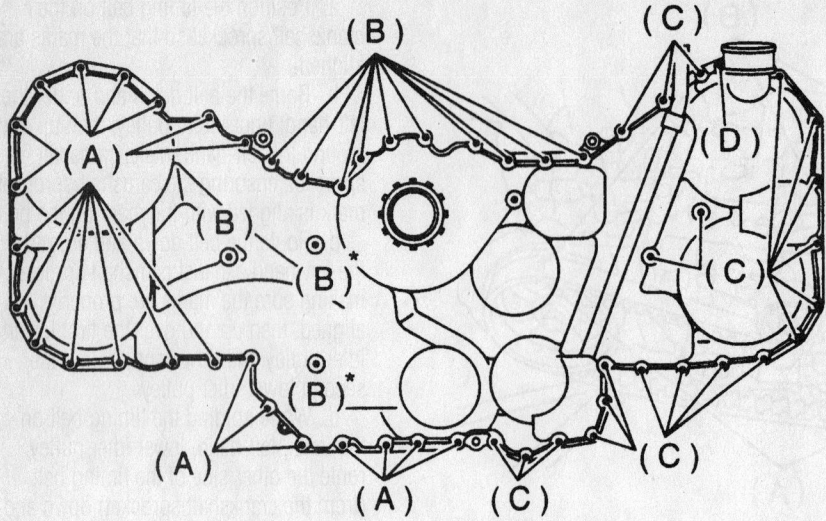

(B)   (C)

(A)   (D)

(B)

(B)*

(C)

(B)*

(A)   (C)

(A)   (C)

42356-SBCR-G29

Front timing cover bolt sizes and locations—3.0L engines

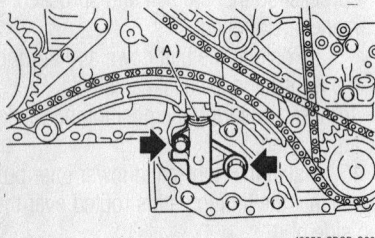

(A)

42356-SBCR-G30

The chain tensioner plunger A does not come out—right hand side—3.0L engines

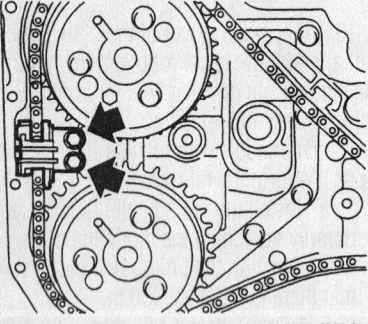

42356-SBCR-G31

Location of the chain guide between the cams—right hand side—3.0L engines

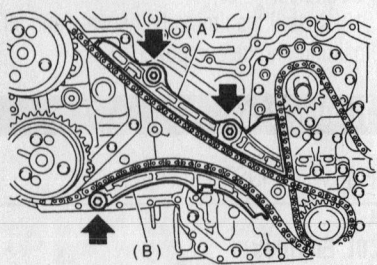

(A)

(B)

42356-SBCR-G32

Location of the chain guide—right hand side—3.0L engines

2. Remove or disconnect the following:
   • Negative battery cable
   • Crankshaft pulley cover
   • Crankshaft pulley bolt using tool 499977100 to hold the crankshaft
   • Crankshaft pulley

➥**There are 4 different types of front cover bolts. Note their sizes and keep them separate to avoid a problem during installation.**

   • Front timing cover bolts. Note the location and the size of the bolts as

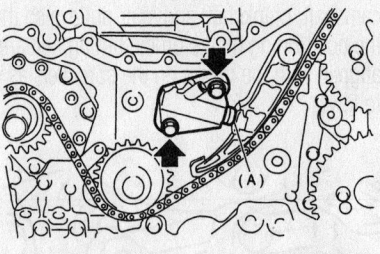

(A)

42356-SBCR-G33

The chain tensioner plunger A does not come out—left hand side—3.0L engines

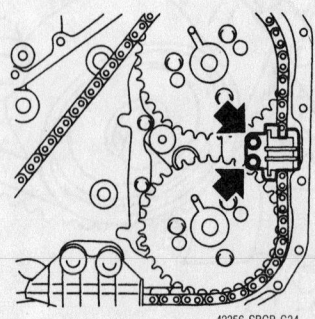

42356-SBCR-G34

Location of the chain guide between the cams—left hand side—3.0L engines

you remove them, this will help during installation
   • Chain tensioner on the right hand side. The plunger **A** does not come out. Refer to illustration for plunger location.
   • Chain guide on the right hand side located between the cams
   • Chain guide from the right hand side
   • Chain tensioner lever from the right hand side
   • Timing chain from the right hand side
   • Chain tensioner on the left hand side. The plunger **A** does not come out. Refer to illustration for plunger location.
   • Chain tensioner lever from the left hand side
   • Chain guide on the left hand side located between the cams
   • Chain guide from the left hand side
   • Center chain guide
   • Upper idler sprocket
   • Timing chain from the left hand side
   • Lower idler sprocket

To install:
   3. Make sure all components are clean. Apply oil to the chain guide, tensioner lever and idler sprockets.
   4. Place the screw, spring, pin and tension rod into the tensioner body.
   5. While pressing the tensioner onto a rubber mat, twist it to the left and right to shorten the rod. Place a thin pin into the holes between the rod and body to hold it in place. Always perform this task on a rubber mat.
   6. Using the crankshaft socket tool, align the **TOP MARK** on the crankshaft sprocket to the 9 O'clock position as shown in the accompanying illustration.
   7. Using camshaft sprocket wrench ST 18231AA000, align the four key grooves on the camshaft sprockets to the 12 O'clock position as illustrated.
   8. Rotate the crankshaft sprocket clockwise to align the **TOP MARK** to the 12 O'clock position as shown in the illustra-

42356-SBCR-G35

Location of the chain guide—left hand side—3.0L engines

tion. Piston 1 is now at Top Dead Center (TDC).

✳✳ CAUTION

Do not rotate the camshaft or crankshaft sprockets until the chain is completely routed or damage will occur.

9. Install the lower idler sprocket and tighten the bolt to 50 ft. lbs. (69 Nm).

10. Install the left hand timing chain, align the mark **B** on the crankshaft sprocket with the matching mark **A** on the timing chain. Refer to the illustration for more detail

11. Route the left hand timing chain onto the lower idler sprocket, water pump, exhaust cam sprocket and the intake cam sprocket in that order.

12. Make sure the mark **A** on the chain and the mark **B** camshaft sprocket are aligned the same way as the one on the crankshaft sprocket or damage will occur.

13. Install or connect the following:
- Upper chain idler and tighten the bolt to 50 ft. lbs. (69 Nm)
- Chain guide on the left hand side between the cams and tighten the bolt to 4.6 ft. lbs. (6 Nm) using a NEW bolt
- Chain guide on the left hand side and tighten the bolts to 11 ft. lbs. (16 Nm)
- Tensioner lever on the left hand side and tighten the bolt to 11 ft. lbs. (16 Nm)
- Chain tensioner on the left hand side and tighten the bolts to 11 ft. lbs. (16 Nm)
- Right hand timing chain. On the lower idler sprocket align the match marks on the timing chain on the left and right hand sides as illustrated. Route the chain onto the intake cam sprocket and the exhaust cam sprocket.

14. Make sure the mark **A** on the chain and the mark **B** camshaft sprocket are aligned the same way as the one on the crankshaft sprocket or damage will occur.

15. Install or connect the following:
- Right hand chain guide
- Right hand chain tensioner lever and tighten the bolts to 11 ft. lbs. (16 Nm)
- Right hand chain guide and tighten the NEW bolt to 4.6 ft. lbs. (6 Nm)
- Right hand chain tensioner and tighten the bolts to 11 ft. lbs. (16 Nm)

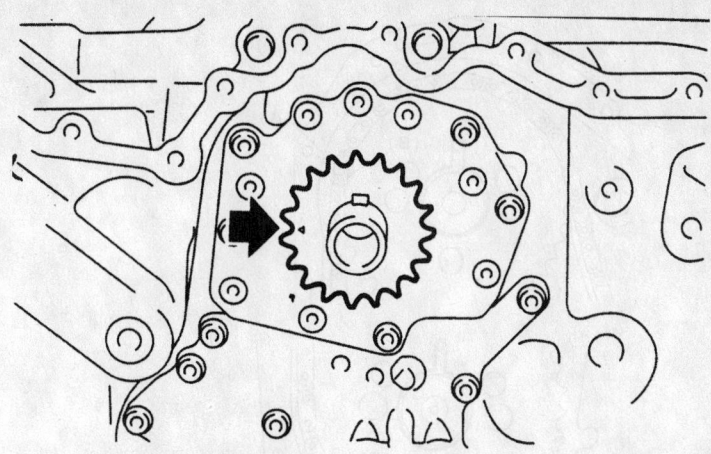

Align the TOP MARK on the crankshaft sprocket to the 9 O'clock position —3.0L engines

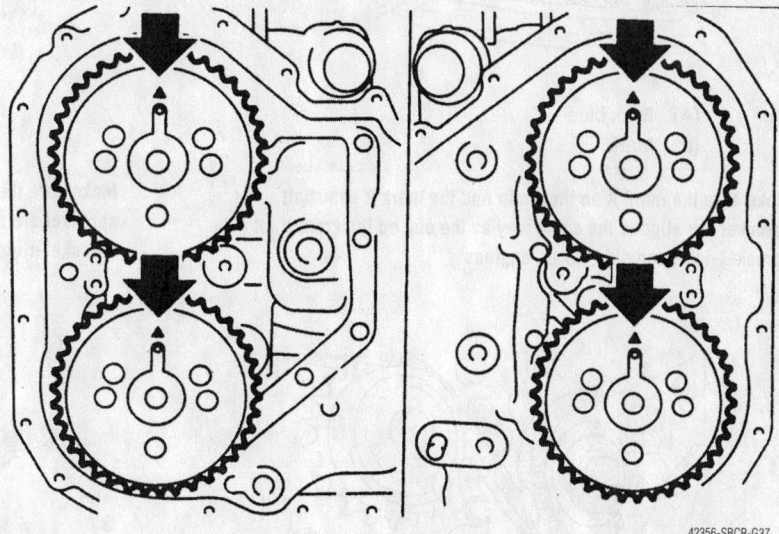

Align the four key grooves on the camshaft sprockets to the 12 O'clock position—3.0L engines

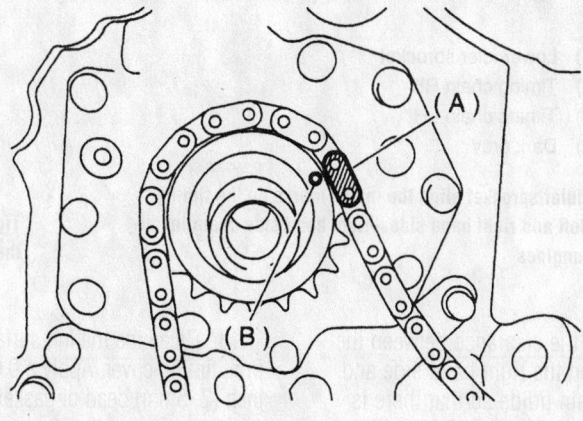

(A) Gold
(B) Mark

Align the mark B on the crankshaft sprocket with the matching mark A on the left hand timing chain—3.0L engines

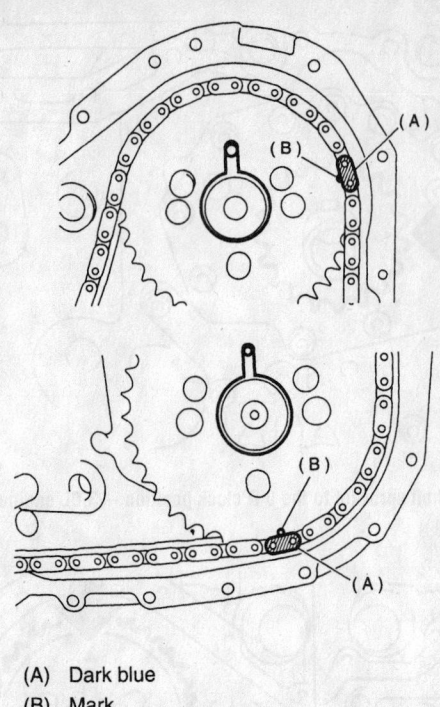

(A) Dark blue
(B) Mark

42356-SBCR-G39

Make sure the mark A on the chain and the mark B camshaft sprocket are aligned the same way as the one on the crankshaft sprocket–left hand side—3.0L engines

Make sure the mark A on the chain and the mark B camshaft sprocket are aligned the same way as the one on the crankshaft sprocket–right hand side—3.0L engines

42356-SBCR-G41

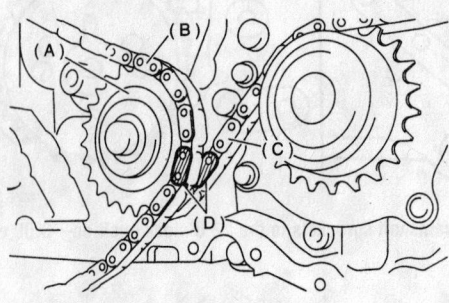

(A) Lower idler sprocket
(B) Timing chain RH
(C) Timing chain LH
(D) Dark gray

42356-SBCR-G40

On the lower idler sprocket align the match marks on the timing chain on the left and right hand sides–right hand side chain installation—3.0L engines

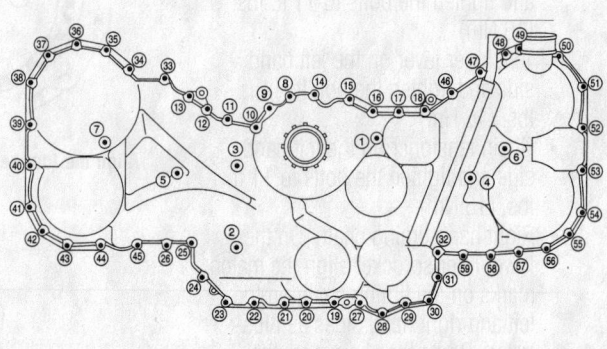

Tighten the front timing cover bolts in the sequence illustrated to the correct specification—3.0L engines

42356-SBCR-G42

16. Adjust the clearance between the chain guide on the right hand side and the center chain guide so that there is range between 0.331–0.339 inch (8.4–8.6mm).
- Center chain guide and tighten the NEW bolt to 6 ft. lbs. (8 Nm)

17. Check the match marks on each sprocket and corresponding timing chain are correct, remove the stopper from the tensioner.

18. Clean the mating surfaces on the front timing cover. Apply a 0.078–0.126 inch (2–5mm) bead of gasket maker to the mating surface of the front cover. Subaru recommends Three Bond 1280B for this procedure.

19. Install the front chain cover and temporarily tighten the bolts. use the illustration showing the bolt sizes and location for proper installation.

20. Tighten the front timing cover bolts in the sequence illustrated to 5 ft. lbs. (7 Nm).

21. Install or connect the following:
- Crankshaft pulley, apply clean oil to the bolt threads and tighten the bolt to 131 ft. lbs. (178 Nm)
- Camshaft pulley cover and tighten the bolts to 5 ft. lbs. (7 Nm)
- Negative battery cable

22. Start the engine and allow it to reach operating temperature. Check for leaks.

Piston and Ring

POSITIONING

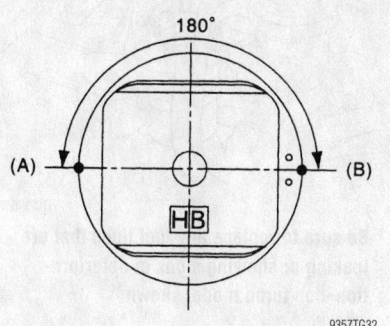

Subaru 2.0L engine—top ring end-gap spacing

9357TG32

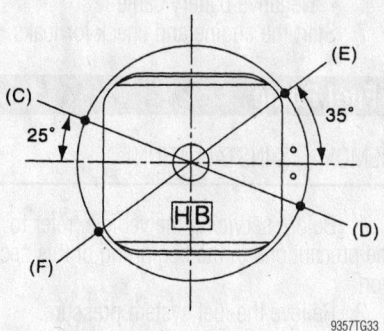

Subaru 2.0L engine—upper rail end-gap spacing

9357TG33

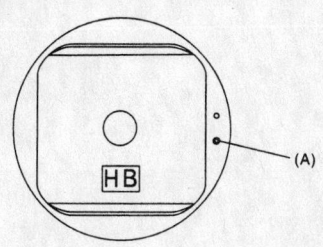

(A) Front mark

9357TG34

Subaru 2.0L engine—piston front mark faces towards the front of the engine

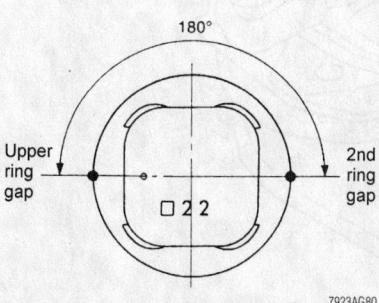

Subaru 2.2L engine—compression ring end-gap spacing

7923AG80

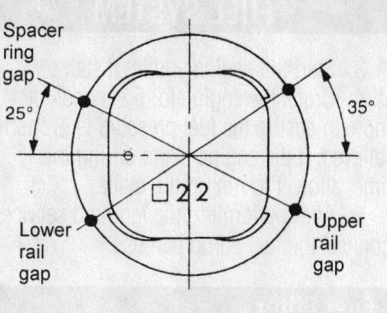

Subaru 2.2L engine—upper, spacer and lower oil ring end-gap spacing

7923AG82

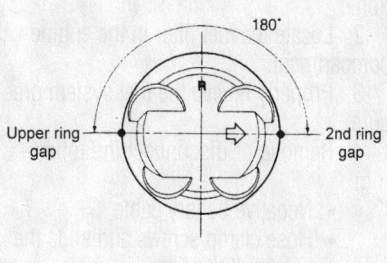

Subaru 2.5L engine—compression ring end-gap spacing

7923AG77

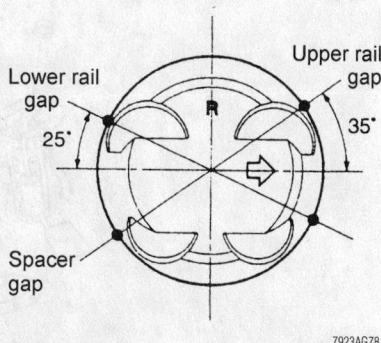
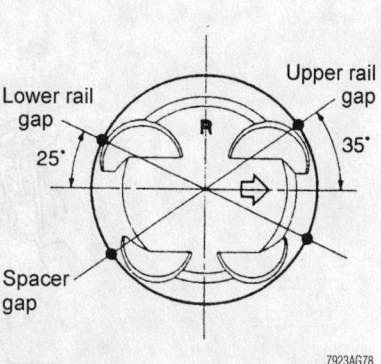

Subaru 2.5L engine—upper, spacer and lower oil ring end-gap spacing

7923AG78

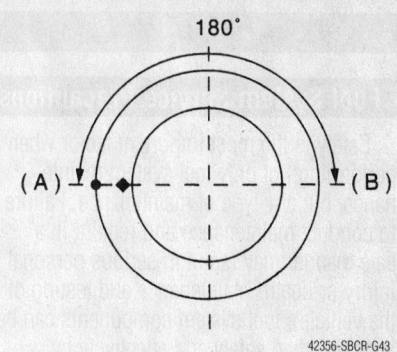

Top ring end-gap (A), second ring gap (B)—3.0L engines

42356-SBCR-G43

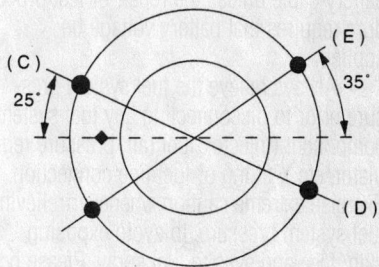

Upper rail end-gap (C), expander gap (D) and lower rail gap (E)—3.0L engines

42356-SBCR-G44

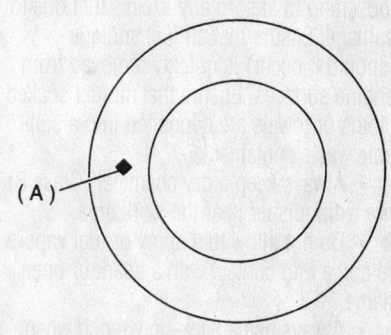

Piston front mark (A) faces towards the front of the engine—3.0L engine

42356-SBCR-G45

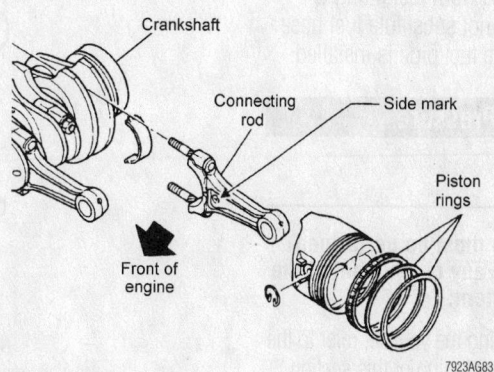

Subaru engines—piston and connecting rod assembly positioning

7923AG83

FUEL SYSTEM

Fuel System Service Precautions

Safety is the most important factor when performing not only fuel system maintenance, but any type of maintenance. Failure to conduct maintenance and repairs in a safe manner may result in serious personal injury or death. Maintenance and testing of the vehicle's fuel system components can be accomplished safely and effectively by adhering to the following rules and guidelines.

• To avoid the possibility of fire and personal injury, always disconnect the negative battery cable unless the repair or test procedure requires that battery voltage be applied.

• Always relieve the fuel system pressure prior to disconnecting any fuel system component (injector, fuel rail, pressure regulator, etc.), fitting or fuel line connection. Exercise extreme caution whenever relieving fuel system pressure, to avoid exposing skin, face and eyes to fuel spray. Please be advised that fuel under pressure may penetrate the skin or any part of the body that it contacts.

• Always place a shop towel or rag around the fitting or connection prior to loosening to absorb any excess fuel due to spillage. Ensure that all fuel spillage (should it occur) is quickly removed from engine surfaces. Ensure that all fuel soaked cloths or towels are deposited into a suitable waste container.

• Always keep a dry chemical (Class B) fire extinguisher near the work area.

• Do not allow fuel spray or fuel vapors to come into contact with a spark or open flame.

• Always use a back-up wrench when loosening and tightening fuel line connection fittings. This will prevent unnecessary stress and torsion to fuel line piping.

• Always replace worn fuel fitting O-rings with new. Do not substitute fuel hose or equivalent, where fuel pipe is installed.

Fuel System Pressure

RELIEVING

➡**This procedure must be performed prior to servicing any component of the fuel injection system.**

1. Before servicing the vehicle, refer to the precautions in the beginning of this section.
2. Disconnect the fuel pump connector from the fuel pump relay.

3. Start the engine and let it stall.
4. Crank the engine for 5 seconds or more to ensure the fuel pressure is properly relieved. If the engine starts during this time, allow it to run until it stalls.
5. After performing the required service, connect the fuel pump harness.

Fuel Filter

REMOVAL & INSTALLATION

1. Before servicing the vehicle, refer to the precautions in the beginning of this section.
2. Locate the fuel filter in the engine compartment.
3. Properly relieve the fuel system pressure.
4. Remove or disconnect the following:

• Negative battery cable
• Hose clamp screws and slide the hoses off the filter
• Filter from the bracket

To install:
5. Inspect the hoses for wear or cracks, and replace if needed.

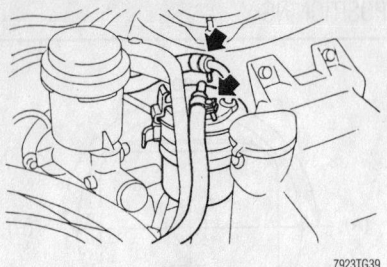

7923TG39

Be sure to replace any fuel lines that are leaking or showing signs of deterioration—non turbo model shown

6. Install or connect the following:
• New filter into the bracket and tighten the hose clamp screws
• Negative battery cable
7. Start the engine and check for leaks.

Fuel Pump

REMOVAL & INSTALLATION

1. Before servicing the vehicle, refer to the precautions in the beginning of this section.
2. Relieve the fuel system pressure.

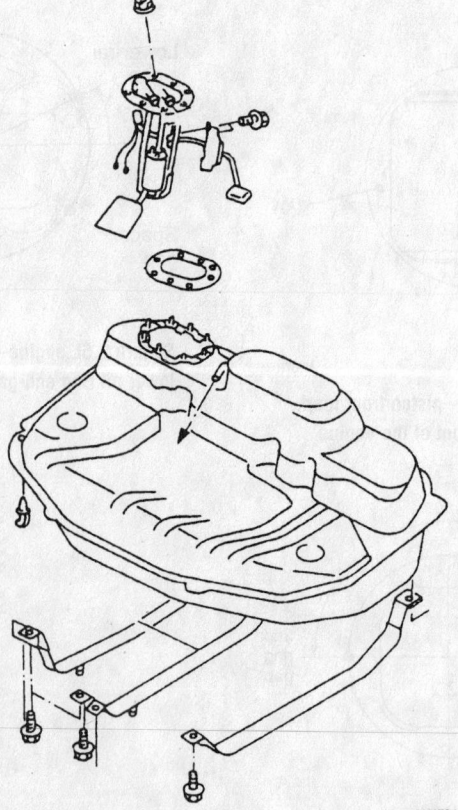

Exploded view of the fuel pump assembly

7923TG41

3. Disconnect the negative battery cable.

4. On the 2000–03 Legacy, Outback and turbocharged models, perform the following:

a. Raise and safely support the vehicle.

b. Remove the front side fuel tank cover.

c. Drain the fuel tank into a suitable container.

d. Tighten the drain plug to 14–24 ft. lbs. (19–33 Nm).

e. Install the front side fuel tank cover Tighten the retainers to 9.4–16.6 ft. lbs. (13–23 Nm).

5. Remove or disconnect the following:
- Rear seat bottom, to reach the fuel pump access cover, if not already done

6. On Legacy models, fold the seat back, then roll the floor mat back.
- Fuel pump cover mounting bolts and the cover
- Electrical harness from the pump assembly

→**Label the fuel lines before disconnecting them from the pump.**

- Fuel lines from the fuel pump
- Fuel pump mounting nuts
- Fuel pump assembly from the tank

To install:

7. Install or connect the following:
- New gasket
- Fuel pump assembly into the fuel tank and secure with the mounting nuts. Tighten the nuts to 24–48 inch lbs. (3–6 Nm) on non turbo models and 3 ft. lbs. (4 Nm) on turbo models.
- Electrical harness to the fuel pump assembly
- Fuel lines to the pump assembly, then tighten the clamps and fittings
- Fuel pump service cover and cover mounting bolts
- Rear seat bottom
- Negative battery cable

8. Start the engine and check for leaks.

Fuel Injector

REMOVAL & INSTALLATION

Except 2.0L Engines

1. Before servicing the vehicle, refer to the precautions in the beginning of this section.

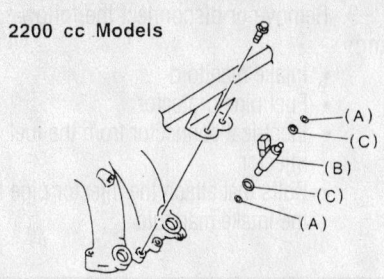

2200 cc Models

(A) O-ring
(C)
(B)
(C)
(A)

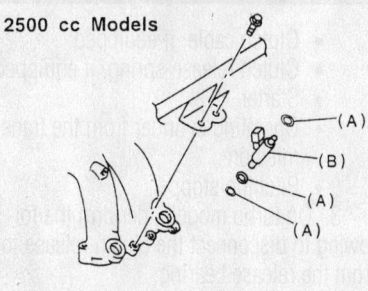

2500 cc Models

(A)
(B)
(A)
(A)

(A) O-ring
(B) Fuel injector
(C) Insulator

9307TG05

Exploded view of the fuel injector

2. Relieve the fuel system pressure.

3. To remove the right side injectors, remove or disconnect the following:
- Air cleaner ducts and resonator chamber, on California vehicles
- Mass Air Flow (MAF) sensor connector and air intake duct and air cleaner upper cover as a unit, on non-California vehicles
- Air cleaner element
- Spark plug wires from the right side spark plugs
- V-belt covers and power steering pump belt
- Power steering pump brackets-to-intake manifold bolts
- Power steering pump-to-bracket bolts, then position the pump on the right side wheel apron

4. To remove the injectors on the left side, remove or disconnect the following:
- Windshield washer motor electrical connector
- Electrical connector from the rear window washer, on station wagon only
- Rear window washer hose from the washer motor and plug or cap the line
- Two bolts that secure the washer tank to the body
- Washer tank and secure it out of the way
- Spark plug wires from the left side spark plugs

- Fuel pipe protector

5. Remove or disconnect the following (for either side):
- Band that secures the engine harness to the fuel injector pipe, if equipped
- Intake manifold protector, if equipped
- Fuel injector electrical connector(s)
- Bolts that hold the fuel injector pipe (fuel rail) to the intake manifold, if applicable

→**Automatic transaxle equipped Legacy's may have a retaining clip that must be removed before the injector can be removed.**

6. Pull up on the injector pipe (fuel rail), then remove the fuel injector(s) from the intake manifold. Remove and discard the injector O-rings.

To install:
- Install or connect the following:
- New injector O-rings
- Fuel injector(s) into the intake manifold
- Retaining clips, if applicable
- Injector pipe (fuel rail) and secure with the retaining bolts. Tighten the bolts to 14 ft. lbs. (19 Nm).
- Fuel injector electrical connector(s)
- Intake manifold protector, if equipped
- Band that secures the engine harness to the fuel injector pipe, if equipped

7. To install the injectors on the left side, install or connect the following:
- Fuel pipe protector
- Spark plug wires to the left side spark plugs
- Washer tank and secure with the two mounting bolts
- Rear window washer hose to the washer motor
- Electrical connector to the rear window washer, on station wagon only
- Windshield washer motor electrical connector

8. To install the right side injectors, install or connect the following:
- Power steering pump into position
- Pump-to-bracket bolts
- Power steering pump brackets-to-intake manifold bolts
- Power steering pump belt and V-belt covers
- Spark plug wires to the right side spark plugs
- Air cleaner element
- Air intake duct and upper cover and the MAF sensor connector

- Air cleaner ducts and resonator chamber, on California vehicles

2.0L Engines

1. Before servicing the vehicle, refer to the precautions in the beginning of this section.
2. Relieve the fuel system pressure.

3. Remove or disconnect the following:

- Intake manifold
- Fuel pipe protector
- Electrical connector from the fuel injector
- Bolts that attach the injector pipe to the intake manifold

- Fuel injector while lifting up the fuel injector pipe

To install:

- Install or connect the following:
- New injector O-rings
- Remaining components in the reverse order of removal. Tighten the injector pipe bolts to 13 ft. lbs. (19 Nm).

DRIVE TRAIN

Transaxle Assembly

REMOVAL & INSTALLATION

Manual

1. Before servicing the vehicle, refer to the precautions in the beginning of this section.
2. Remove or disconnect the following:

- Negative battery cable
- Air intake duct and cleaner case
- Air cleaner stay, if equipped
- Intercooler on turbo models
- Front Oxygen Sensor (O2S) connector
- Neutral position switch connector
- Back-up light switch connector
- Vehicle Speed Sensor (VSS) connector, if equipped
- Transmission ground cable, if necessary

- Clutch cable, if equipped
- Clutch release spring, if equipped
- Starter
- Operating cylinder from the transmission
- Pitching stopper

3. On turbo models, perform the following to disconnect the clutch release fork from the release bearing:

   a. Remove the clutch cylinder from the transmission.

   b. Using a 10mm wrench, remove the plug.

   c. Screw a 6mm diameter bolt into the release fork and remove it.

   d. Raise the release fork and unfasten the release tabs to free the release fork.

4. Remove or disconnect the following:

- Drive belt cover, if necessary
- Slave (operating) cylinder, on 2.5L engines

5. Install engine support assembly 927670000, on 3.0L engines, install support assembly 41099AA00.

- Bolt securing the right upper side of the transaxle to the engine
- Engine undercover, if equipped
- Rear O2S connector
- Front Y-pipe
- Rear exhaust pipe and muffler, on Legacy models
- Heat shield cover
- Hanger bracket from the right side of the transaxle
- Driveshaft
- Spring, and disconnect the shifter stay and rod from the transaxle
- Bolts securing the sway bar clamps to the crossmember
- Ball joints from the steering knuckle
- Halfshafts from the transaxle
- Nuts securing the lower side of the transaxle to the engine

6. Support the transaxle with a jack.

- Rear transaxle crossmember
- Transaxle from the vehicle. Move the jack rearward until the mainshaft is withdrawn from the clutch cover.

To install:

7. Install or connect the following:

- Transaxle assembly and secure it to the engine block
- Crossmember

8. Tighten the crossmember retainers to the following specifications:

   a. Step 1: T1 to 40–62 ft. lbs. (54–84 Nm), on all engines except the 3.0L. On the 3.0L engine, tighten the T1 bolts to 55 ft. lbs. (75 Nm)

   b. Step 2: T2 to 87–115 ft. lbs. (117–157 Nm), on all engines except the 3.0L. On the 3.0L engine, tighten the T2 bolts to 103 ft. lbs. (140 Nm)

9. Remove the transmission jack.
10. Install or connect the following:

- Nuts securing the lower portion of the engine to the transaxle and tighten to 40 ft. lbs. (54 Nm) on all except turbo and 3.0L models. On

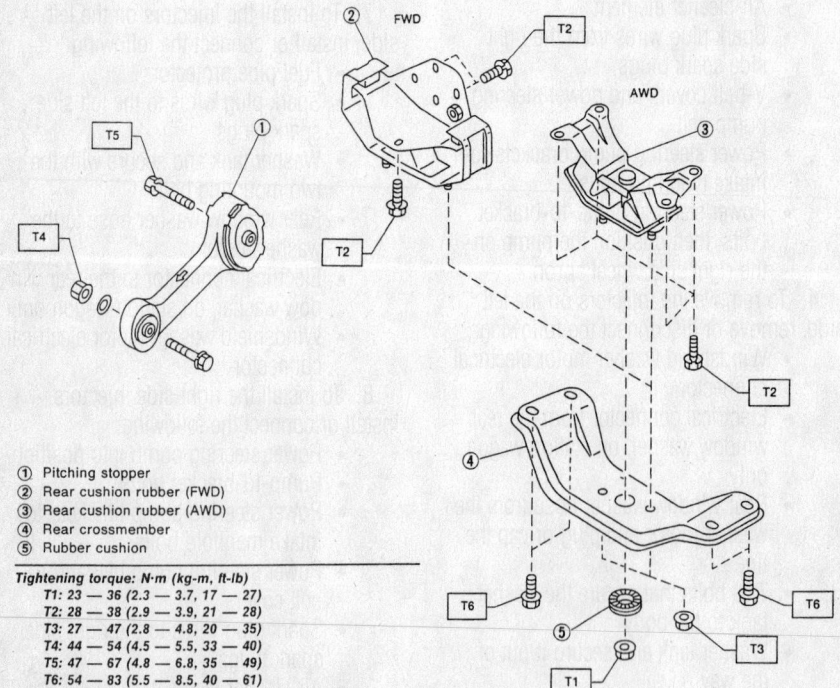

① Pitching stopper
② Rear cushion rubber (FWD)
③ Rear cushion rubber (AWD)
④ Rear crossmember
⑤ Rubber cushion

Tightening torque: N·m (kg-m, ft-lb)
T1: 23 — 36 (2.3 — 3.7, 17 — 27)
T2: 28 — 38 (2.9 — 3.9, 21 — 28)
T3: 27 — 47 (2.8 — 4.8, 20 — 35)
T4: 44 — 54 (4.5 — 5.5, 33 — 40)
T5: 47 — 67 (4.8 — 6.8, 35 — 49)
T6: 54 — 83 (5.5 — 8.5, 40 — 61)

7923TG44

Exploded view of the transaxle mounting—Impreza models

turbo and 3.0L models, tighten to 37 ft. lbs. (50 Nm).

- Bolt securing the right upper side of the transaxle to the engine and tighten it to 40 ft. lbs. (54 Nm) on all except turbo and 3.0L models. On turbo and 3.0L models, tighten to 37 ft. lbs. (50 Nm).

11. Remove the engine support.

- Drive belt cover
- Slave (operating) cylinder, on 2.5L engines

12. Install the pitching stopper and tighten the bolts to the following specifications:

    a. Step 1: T1 to 33–40 ft. lbs. (44–54 Nm) on all except turbo and 3.0L models. On turbo and 3.0L models, tighten to 37 ft. lbs. (50 Nm).

    b. Step 2: T2 to 35–49 ft. lbs. (47–67 Nm). on all except turbo and 3.0L models. On turbo and 3.0L models, tighten to 43 ft. lbs. (58 Nm).

13. Install or connect the following:

- Halfshafts into the transaxle with new roll pins
- Ball joint to the steering knuckle and tighten the bolt to 29–43 ft. lbs. (39–59 Nm) on all except turbo and 3.0L models. On turbo and 3.0L models, tighten to 36 ft. lbs. (49 Nm).
- Sway bar to the crossmember and tighten the clamp bolts to 15–21 ft. lbs. (21–29 Nm) on all except turbo and 3.0L models. On turbo models, tighten to 33 ft. lbs. (45 Nm). On 3.0L models, tighten the clamp bolts to 22 ft. lbs. (30 Nm).
- Shift control rod and stay to the transaxle and install the spring
- Driveshaft
- Heat shield cover, if removed
- Rear exhaust pipe and muffler, if removed
- Y-pipe with new gaskets and nuts
- Hanger bracket on the right side of the transaxle, if removed
- Rear O₂S connector
- Engine undercover, if removed
- Transaxle connectors bracket
- Drive belt cover
- Pitching stopper
- Starter
- Front O₂S connector
- VSS connector, if equipped
- Neutral position switch connector
- Back-up light switch connector
- Clutch cable (if equipped)
- Clutch release spring
- Air cleaner case stay and case

- Air intake duct and attach the air-flow sensor connector
- Negative battery cable

Automatic

EXCEPT WRX MODELS

1. Before servicing the vehicle, refer to the precautions in the beginning of this section.

2. Drain the transaxle fluid.

3. Remove or disconnect the following:

- Negative battery cable
- Air intake duct with air cleaner case
- Air cleaner case stay
- Front Oxygen Sensor (O₂S) connector
- Speedometer cable or electronic wiring connector from the speed sensor
- Transaxle harness connector
- Inhibitor switch connector, if equipped
- Revolution sensor connector, if equipped
- Transaxle ground terminal
- Clip band that secures the air breather hose to the pitching stopper, if equipped
- Starter and air intake boot
- Pitching stopper
- Timing hole inspection plug

- 4 bolts that hold the torque converter to the driveplate
- Automatic Transaxle Fluid (ATF) level gauge
- Engine-to-transaxle mounting nut and bolt on the right side
- Buffer rod

4. Support the engine assembly with special engine support tool.

- Exhaust system
- Exhaust brackets or hangers that attach to the transaxle, as necessary

5. Drain the transmission fluid.

- ATF cooler hoses from the pipes of the transmission side
- ATF level gauge guide

➡**Matchmark the installed position of the driveshaft before removal.**

- Driveshaft. Plug the opening at the rear of extension housing to prevent oil from flowing out.
- Gearshift cable from the transaxle select lever
- Stabilizer from the transverse link
- Parking brake cable bracket from the transverse link
- Transverse link bolts and lower the link
- Spring pins
- Halfshafts from the transaxle

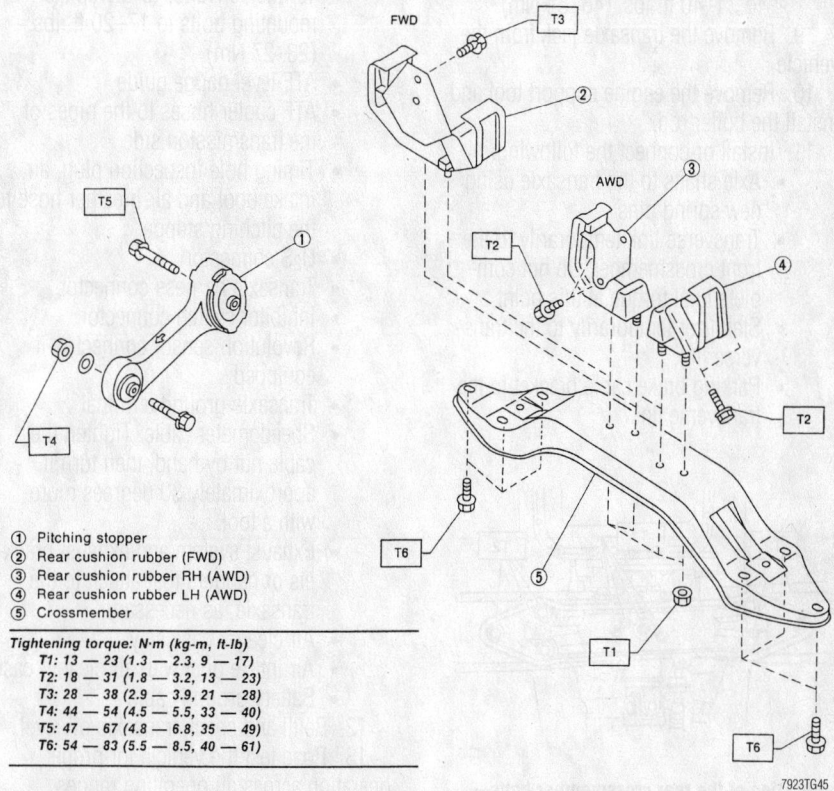

① Pitching stopper
② Rear cushion rubber (FWD)
③ Rear cushion rubber RH (AWD)
④ Rear cushion rubber LH (AWD)
⑤ Crossmember

Tightening torque: N·m (kg-m, ft-lb)
    T1: 13 — 23 (1.3 — 2.3, 9 — 17)
    T2: 18 — 31 (1.8 — 3.2, 13 — 23)
    T3: 28 — 38 (2.9 — 3.9, 21 — 28)
    T4: 44 — 54 (4.5 — 5.5, 33 — 40)
    T5: 47 — 67 (4.8 — 6.8, 35 — 49)
    T6: 54 — 83 (5.5 — 8.5, 40 — 61)

7923TG45

Exploded view of the engine and transaxle mounts—Impreza and Legacy models

➡ **Discard the old spring pin and always install a new pin.**

- Oil cooler hoses

6. Place a transaxle jack under the transaxle.

- Engine to transaxle mounting nuts

➡ **Do not place the jack under the oil pan otherwise the oil pan may be damaged.**

- Rear cushion rubber mounting nuts and the rear crossmember

7. Move the torque converter and transaxle as a unit away from the engine and lower it from the vehicle.

To install:

8. Install or connect the following:

- Transaxle to the engine and temporarily tighten the engine-to-transaxle mounting nuts
- Rear crossmember to the rear cushion rubber mounts. Align the rear cushion guide with the rear crossmember guide hole and tighten nuts.
- Rear crossmember to the chassis. Tighten the rear crossmember bolts to 39–49 ft. lbs. (53–66 Nm) on all models except the 3.0L engine. On the 3.0L engine tighten the T1 bolts to 26 ft. lbs. (35 Nm) and the T2 bolts to 55 ft. lbs. (75 Nm).
- Engine to transaxle retaining nuts to 34–40 ft. lbs. (46–54 Nm)

9. Remove the transaxle jack from the vehicle.

10. Remove the engine support tool and install the buffer rod.

11. Install or connect the following:

- Axle shafts to the transaxle using new spring pins
- Transverse link temporarily to the front crossmember. Do not complete final torque at this point.
- Stabilizer temporarily to the transverse link
- Parking brake cable bracket to the transverse link

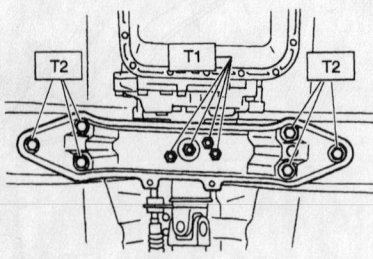

9357TG35

Location of the rear crossmember bolts— 3.0L engine

- Transverse link-to-front crossmember mounting bolts and transverse link-to-stabilizer mounting bolts, with the tires placed on the ground
- Transverse link to front crossmember (self-locking nuts) to 40–62 ft. lbs. (54–84 Nm) and the transverse link to stabilizer to 18–32 ft. lbs. (24–44 Nm).
- Gearshift cable to the select lever. Be sure the lever operates smoothly all across the operating range.
- Driveshaft. Tighten the driveshaft-to-rear differential retaining bolts to 17–24 ft. lbs. (23–33 Nm) and center bearing location retaining bolts to 25–33 ft. lbs. (34–45 Nm).
- Oil cooler hoses
- Engine to transaxle bolts to 34–40 ft. lbs. (46–54 Nm)
- Starter
- Pitching stopper. Be sure to tighten the bolt for the body side first. Tightening torque for the body side bolt is 35–49 ft. lbs. (47–67 Nm) on all except 3.0L; on the 3.0L tighten the bolt to 37 ft. lbs. (50 Nm). The engine or transaxle side bolt is torque to 33–40 ft. lbs. (44–54 Nm)) on all except 3.0L; on the 3.0L tighten the bolt to 43 ft. lbs. (58 Nm).
- Torque converter-to-driveplate mounting bolts to 17–20 ft. lbs. (23–27 Nm)
- ATF level gauge guide
- ATF cooler hoses to the pipes of the transmission side
- Timing hole inspection plug, air intake boot and air breather hose to the pitching stopper
- O_2S connector
- Transaxle harness connector
- Inhibitor switch connector
- Revolution sensor connector, if equipped
- Transaxle ground terminal
- Speedometer cable. Tighten the cable nut by hand, then turn it approximately 30 degrees more with a tool.
- Exhaust system and exhaust brackets or hangers that attach to the transaxle, as necessary
- Air cleaner case stay
- Air intake duct with air cleaner case
- Battery ground cable

12. Refill and check transaxle oil level.

13. Road test the vehicle for proper operation across all operating ranges.

WRX MODELS

1. Before servicing the vehicle, refer to the precautions in the beginning of this section.

2. Drain the transaxle fluid.

3. Remove or disconnect the following:

- Negative battery cable
- Intercooler
- Center and rear exhaust pipes and the muffler
- Transaxle harness connector
- Transaxle ground terminal
- Starter
- Pitching stopper
- Torque converter service hole plug
- Bolts that hold the torque converter to the driveplate
- Automatic Transaxle Fluid (ATF) level gauge

4. Support the engine assembly with special engine support tool.

- Engine-to-transaxle mounting and bolt(s) on the right side
- Undercover
- Heat shield cover
- Buffer rod

5. Drain the transmission fluid.

- ATF cooler hoses from the pipes of the transmission side
- ATF level gauge guide

➡ **Matchmark the installed position of the driveshaft before removal.**

- Driveshaft. Plug the opening at the rear of extension housing to prevent oil from flowing out.
- Gearshift cable from the transaxle select lever
- Stabilizer from the transverse link
- Transverse link bolts and lower the link
- Spring pins
- Halfshafts from the transaxle

➡ **Discard the old spring pin and always install a new pin.**

6. Place a transaxle jack under the transaxle.

- Engine to transaxle mounting nuts

➡ **Do not place the jack under the oil pan otherwise the oil pan may be damaged.**

- Rear cushion rubber mounting nuts and the rear crossmember

7. Move the torque converter and transaxle as a unit away from the engine and lower it from the vehicle.

To install:

8. Install or connect the following:

- Transaxle to the engine and tem-

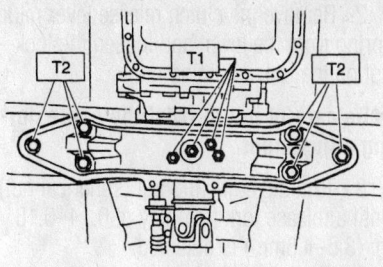

9357TG35

Location of the rear crossmember bolts—WRX models

porarily tighten the engine-to-transaxle mounting nuts

- Rear crossmember to the rear cushion rubber mounts. Align the rear cushion guide with the rear crossmember guide hole and tighten nuts.
- Rear crossmember to the chassis. Tighten the rear crossmember T1 bolts to 26 ft. lbs. (35 Nm) and the T2 bolts to 51 ft. lbs. (70 Nm). Refer to the accompanying illustration for bolt location.
- Engine to transaxle retaining nuts to 36 ft. lbs. (50 Nm)
- Starter
- Torque converter clutch plate bolts to 18 ft. lbs. (25 Nm0
- Torque converter plug
- Pitching stopper. Be sure to tighten the bolt for the body side first. Tightening torque for the body side bolt is 43 ft. lbs. (58 Nm). The engine or transaxle side bolt is torque to 37 ft. lbs. (50 Nm).
- Axle shafts to the transaxle using new spring pins
- Transverse link temporarily to the front crossmember. Do not complete final torque at this point.
- Stabilizer temporarily to the transverse link
- Transverse link-to-front crossmember mounting bolts and transverse link-to-stabilizer mounting bolts, with the tires placed on the ground
- Transverse link to front crossmember (self-locking nuts) to 22 ft. lbs. (30 Nm) and the transverse link to stabilizer to 37 ft. lbs. (50 Nm).
- Gearshift cable to the select lever. Be sure the lever operates smoothly all across the operating range.
- ATF level gauge guide
- Oil cooler hoses
- Driveshaft
- Heat shield cover
- Center, rear exhaust pipes and the muffler

- Undercover
- ATF fluid level gauge
- Transmission harness connectors
- Transmission ground terminal
- Air cleaner case stay
- Intercooler
- Battery ground cable

9. Refill and check transaxle oil level.
10. Road test the vehicle for proper operation across all operating ranges.

Clutch

ADJUSTMENT

Some models are equipped with a mechanical clutch system that is adjustable. Other models are equipped with a hydraulic system that is not adjustable.

Cable

The clutch cable can be adjusted at the cable bracket where the cable is attached to the side of the transaxle housing.

1. Before servicing the vehicle, refer to the precautions in the beginning of this section.
2. Remove the circlip and clamp.
3. Slide the cable end in the direction desired, then replace the circlip and clamp into the nearest gutters on the cable end.

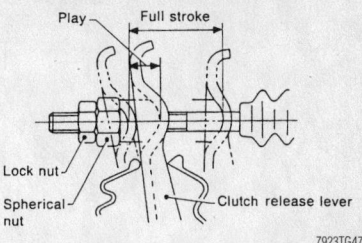

7923TG47

Be sure to tighten the locknut after making the necessary adjustments—mechanical clutch, non-turbo models

➡**The cable should not be stretched out straight nor should it have right angle kinks in it. Any straightening should be gradual.**

4. Check the clutch for proper operation.

Pedal Height

Adjust the pedal with the return stop bolt, so that its pad is on the same level as the brake pedal pad.

Check to be sure that the stroke of the pedal is 5.12–5.31 in. (130–135mm). Check the clutch release fork stroke. It should be 0.67 in. (17mm).

Free-Play

1. Before servicing the vehicle, refer to the precautions in the beginning of this section.

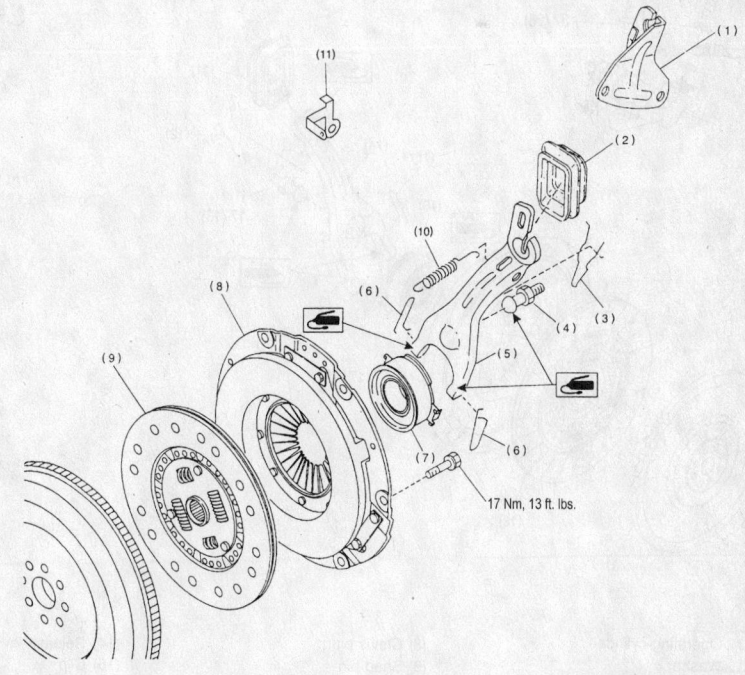

17 Nm, 13 ft. lbs.

(1) Clutch cable bracket	(6) Clip	(10) Return spring (Models without hill holder only)
(2) Clutch release lever sealing	(7) Clutch release bearing	(11) Clutch return spring bracket
(3) Retainer spring	(8) Clutch cover	
(4) Pivot	(9) Clutch disc	
(5) Clutch release lever		

7923TG48

Exploded view of the clutch system components—mechanical clutch, non-turbo models

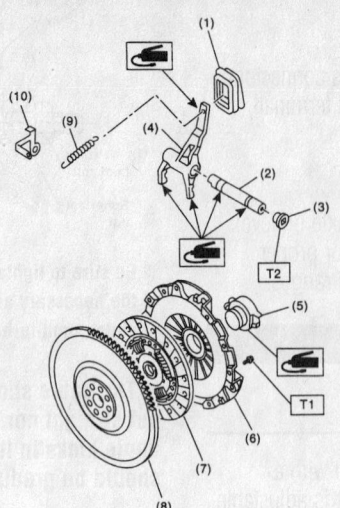

(1)	Clutch release lever sealing	(6)	Clutch cover
(2)	Release lever shaft	(7)	Clutch disc
(3)	Plug	(8)	Flywheel
(4)	Release lever	(9)	Spring
(5)	Release bearing	(10)	Bracket

Tightening torque: N·m (kgf-m, ft-lb)
T1: 15.7 (1.6, 11.6)
T2: 44 (4.5, 32.5)

9357TG36

Exploded view of the clutch system components—mechanical clutch, turbo models

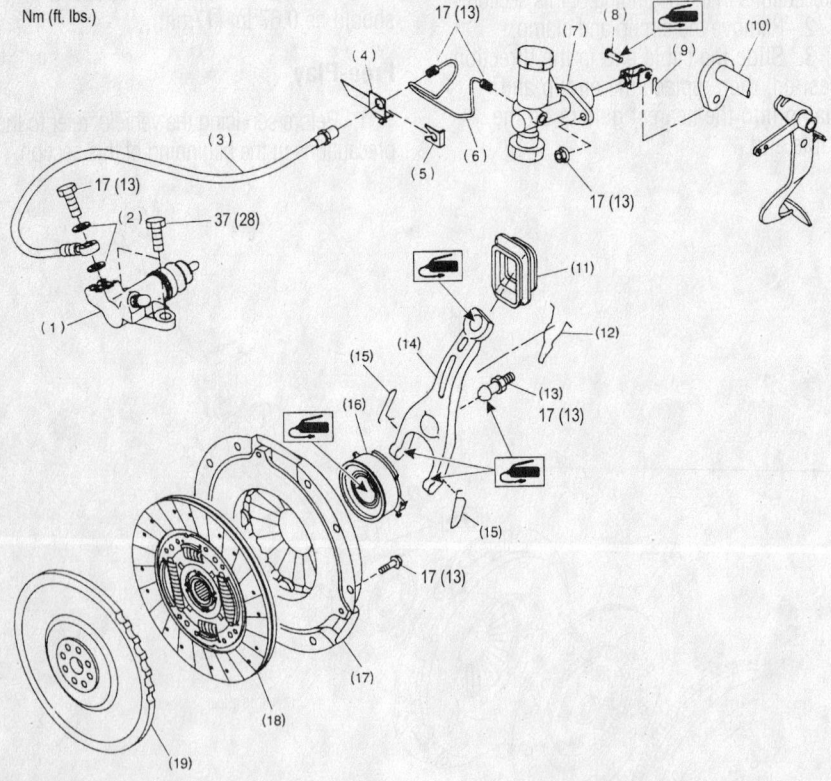

(1)	Operating cylinder	(8)	Clevis pin	(14)	Release lever
(2)	Washer	(9)	Snap pin	(15)	Clip
(3)	Clutch hose	(10)	Lever	(16)	Release bearing
(4)	Bracket	(11)	Clutch release lever sealing	(17)	Clutch cover
(5)	Clamp	(12)	Retainer spring	(18)	Clutch disc
(6)	Pipe	(13)	Pivot	(19)	Flywheel
(7)	Master cylinder ASSY				

7923TG49

Exploded view of the clutch system components—hydraulic clutch

2. Remove the clutch release lever return spring from the lever, and loosen the locknut on the fork adjusting nut.

➡ **Be careful not to twist the cable during adjustment**

3. Turn the adjusting nut (spherical nut) until a release fork free-play of 0.14–0.18 in. (3.5–4.5mm) is obtained.

4. Tighten the locknut.

5. Install the return spring on the lever. Hook the long hook side of the return spring with the lever.

6. Check the pedal free-play. It should be 0.12–0.16 in. (3.0–4.0mm).

7. Adjust the pedal free-play, as necessary, with the pedal adjusting bolt.

REMOVAL & INSTALLATION

✴✴ CAUTION

The clutch driven disc may contain asbestos that has been determined to be a cancer causing agent. Never clean clutch surfaces with compressed air. Avoid inhaling any dust from any clutch surface. When cleaning clutch surfaces, use a commercially available brake cleaning fluid.

1. Before servicing the vehicle, refer to the precautions in the beginning of this section.

2. Remove or disconnect the following:
   • Negative battery cable
   • Transaxle

3. Gradually unscrew the bolts which hold the pressure plate assembly on the flywheel. Loosen the bolts only 1 turn at a time, working around the pressure plate.

4. When all of the bolts have been removed, remove the clutch plate and disc.

✴✴ WARNING

Do not get oil or grease on the clutch facing.

5. Remove the 2 retaining springs and remove the throwout bearing and the release fork.

➡ **Do not disassemble either the clutch cover or disc. Inspect the parts for wear or damage and replace any parts as necessary. Replace the clutch disc if there is any oil or grease on the facing. Do not wash or attempt to lubricate the throwout bearing. If it requires replacement, the bearing may be removed and a new one installed in the holder by means of a press.**

To install:

6. Fit the release fork boot on the front of the transaxle housing.

7. Install or connect the following:
- Release fork
- Throwout bearing assembly and secure it with the 2 springs. Coat the inside diameter of the bearing holder and the fork-to-holder contact points with grease.

8. Insert a pilot shaft through the clutch cover and disc, then insert the end of the pilot into the needle bearing.

9. If equipped, position the **O** marks on the clutch cover and flywheel 120 degrees apart.

10. Tighten the pressure plate bolts gradually, 1 turn at a time, until the proper torque is reached. Tighten to 11 ft. lbs. (15 Nm).

❊❊ WARNING

When installing the clutch pressure plate assembly, be sure that the O marks on the flywheel and the clutch pressure plate assembly are at least 120 degrees apart. These marks indicate the direction of residual unbalance. Also, be sure that the clutch disc is installed properly, noting the FRONT and REAR markings.

11. After installation of the transaxle in the car, perform the adjustments outlined above.

Hydraulic Clutch System

BLEEDING

➡**To properly bleed the system, it must be bled at the slave cylinder and at the damper.**

1. Before servicing the vehicle, refer to the precautions in the beginning of this section.

2. Connect a vinyl tube to the air bleeder

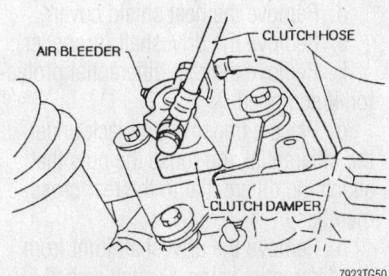

AIR BLEEDER   CLUTCH HOSE

CLUTCH DAMPER

7923TG50

Bleeding the hydraulic clutch at the clutch damper

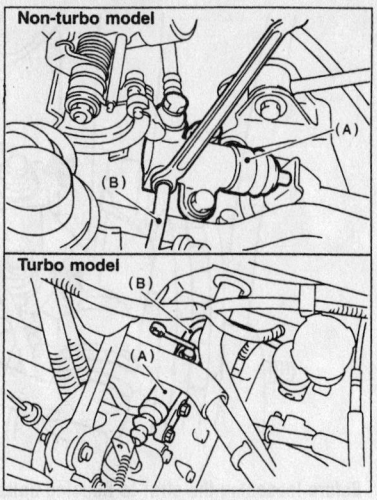

Non-turbo model

(A)

(B)

Turbo model

(B)

(A)

(A) Operating cylinder
(B) Vinyl tube

9357TG37

Bleeding the hydraulic clutch at the slave cylinder

on the clutch operating (slave) cylinder. Put the other end in a jar with clean clutch fluid.

3. With the help of an assistant depressing the clutch pedal, slowly open the bleeder valve. Close the bleeder valve and release the pedal. Repeat this process until no air bubbles appear in the jar.

4. Move the tube to the bleeder on the slave cylinder and repeat the process. Check the operation of the clutch after the bleed procedure is complete.

Transfer Case Assembly

REMOVAL & INSTALLATION

The transfer case must be removed as an assembly with the transaxle.

Halfshaft

REMOVAL & INSTALLATION

Front

1. Before servicing the vehicle, refer to the precautions in the beginning of this section.

2. Remove or disconnect the following:
- Negative battery cable
- Wheel
- Axle nut, unstake the nut before attempting removal
- Stabilizer link from the transverse link
- Transverse link ball joint from the housing

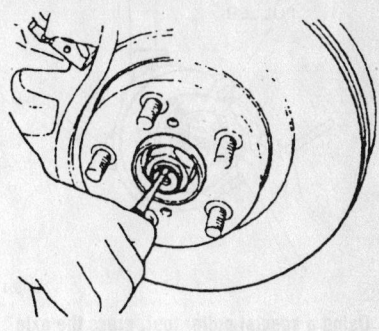

7923TG52

Unstaking the axle nut

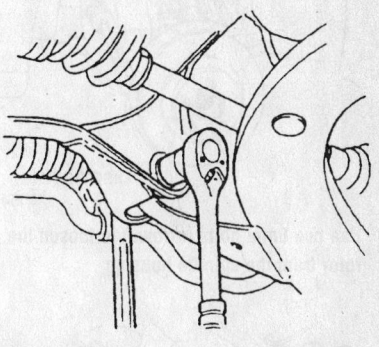

7923TG53

Remove the transverse link arm from the crossmember

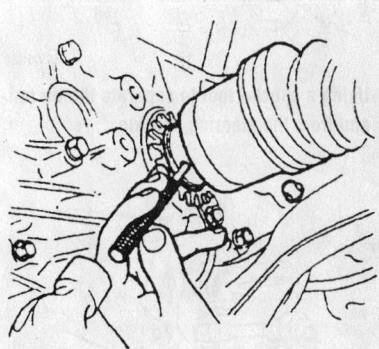

7923TG54

Drive out the halfshaft-to-transaxle roll pin

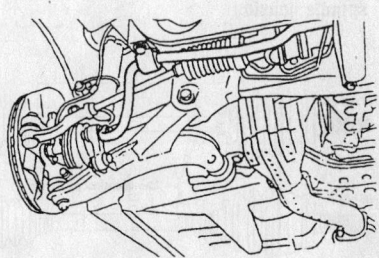

7923TG55

Remove the sway bar bracket

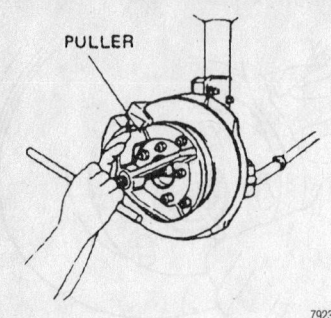

Using a special puller tool, press the axle shaft from the spindle housing

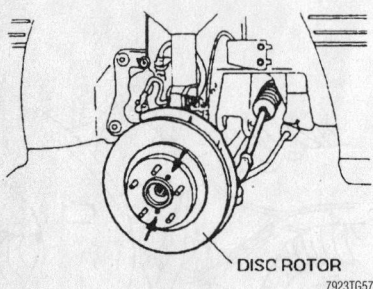

Use two 8mm bolts (arrows) to loosen the rotor from the spindle housing

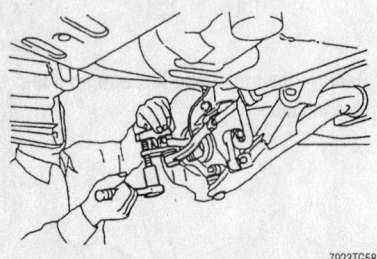

Using a special tool to separate the tie rod end from the steering knuckle

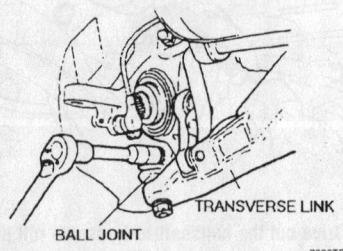

Remove the transverse link arm from the spindle housing

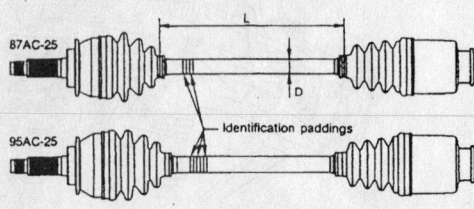

Be sure to identify the correct halfshaft—Except WRX models

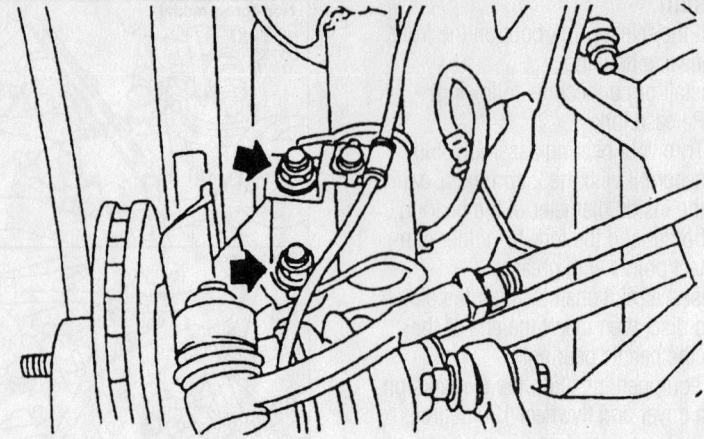

Before loosening the strut-to-housing bolts (arrows), matchmark the camber adjustment bolt and strut

- Halfshaft-to-transaxle roll/spring pin and discard it
- Sway bar bracket
- Halfshaft from the transaxle
- Halfshaft from the hub using puller 92707000

To install:

3. Install or connect the following:
  - Halfshaft into the hub

4. Using installer 922431000 and adapter 927390000, pull the halfshaft through the hub.

5. Install or connect the following:
  - Temporarily tighten a new axle nut
  - Align the halfshaft roll/spring pin hole
  - Halfshaft onto the transaxle
  - New pin.
  - Transverse link to the knuckle and tighten a new self-locking nut to 36 ft. lbs. (50 Nm)
  - Sway bar bracket
  - New axle nut to 152 ft. lbs. (206 Nm) on all models except WRX and models equipped with the 3.0L engines. On WRX models, tighten the nut to 137 ft. lbs. (186 Nm). On models equipped with the 3.0L engines, tighten the nut to 159 ft. lbs. (216 Nm). Stake the nut
  - Wheel
  - Negative battery cable

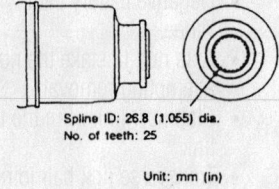

Spline ID: 26.8 (1.055) dia.
No. of teeth: 25

Unit: mm (in)

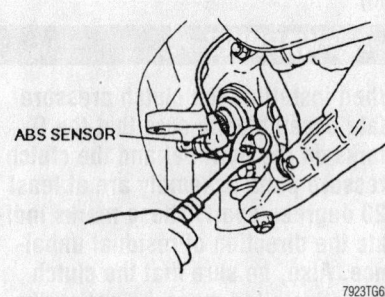

Removing the ABS sensor

Rear

EXCEPT IMPREZA

1. Before servicing the vehicle, refer to the precautions in the beginning of this section.

2. Remove or disconnect the following:
  - Negative battery cable
  - Wheel
  - Axle nut, unstake the nut before attempting removal

3. Remove the rear differential from models equipped with a T–type as follows:
  a. Place the gear shifter into the **Neutral** position.
  b. Release the parking brake.
  c. Remove the rear exhaust pipe and muffler.
  d. Remove the heat shield cover.
  e. Remove the driveshaft (propeller).
  f. Remove the rear differential protector, if equipped.
  g. Place a transmission jack under the differential and loose the nuts that attach the differential to the rear cross-member.
  h. Remove the driveshaft joint from the differential using a suitable shaft removal tool such as ST 28099PA100.
  i. Remove the protector nut.

j. Remove the differential front member and support the differential assembly with the transmission jack making sure to securely attach the differential to the jack with a chain or strap.

k. Remove the nuts attaching the differential to the crossmember.

l. Remove the differential stud bolt from the rear crossmember bushing. You may have to carefully adjust the angle and position of the jack to facilitate stud removal.

m. Once the stud bolt has been removed, lower the jack, making sure the rear drive shaft does not strike the lateral link bolt.

n. Pull the driveshaft out of the differential and remove the differential.

4. Remove the rear differential from models equipped with a VA–type as follows:

a. Place the gear shifter into the **Neutral** position.

b. Release the parking brake.

c. Remove the rear exhaust pipe and muffler.

d. Remove the heat shield cover.

e. Remove the driveshaft (propeller).

f. Place a transmission jack under the differential and loosen the nuts that attach the differential to the rear crossmember.

g. Remove the driveshaft joint from the differential using a suitable shaft removal tool such as ST 28099PA100.

h. Remove the protector nut.

i. Remove the nuts with attach the differential to the front member. Support the differential assembly with the transmission jack.

j. Remove the differential front member and securely attach the differential to the jack with a chain or strap.

k. Remove the nuts attaching the differential to the crossmember.

l. Remove the differential stud bolt from the rear crossmember bushing. You may have to carefully adjust the angle and position of the jack to facilitate stud removal.

m. Once the stud bolt has been removed, lower the jack, making sure the rear drive shaft does not strike the lateral link bolt.

n. Pull the driveshaft out of the differential and remove the differential.

5. Remove the axle nut and using a suitable puller, remove the halfshaft from the being careful not to damage the tone wheel using puller ST 926470000 and plate ST 927140000.

To install:

6. Insert the axle into the hub splines

being careful not to damage the tone wheel using adapter installer ST1 922431000 and adapter ST2 927390000. Tighten the axle nut until snug.

7. Install the rear differential from models equipped with a VA–type as follows:

a. Place the differential on a transmission jack and fasten securely with a chain or band.

b. Place a seal protector tool such as ST28099PA090 on the differential.

c. Insert the splined shaft of the halfshaft until the spline portion is inside the oil seal.

d. Remove the seal protector.

e. Insert the driveshaft into the differential until it is fully seated.

f. Adjust the transmission jack as necessary so the stud bolt is inserted correctly into the crossmember bushing.

g. Once the stud bolt is inserted, raise the transmission jack until the differential is level.

h. Temporarily tighten the rear crossmember nuts.

i. Remove the band or chain securing the differential to the transmission jack and raise the differential enough to move the jack away. Install the differential member and two inner bolts to 81 ft. lbs. (110 Nm) and the outer bolt to 48 ft. lbs. (65 Nm).

j. Tighten the crossmember self locking nut to 51 ft. lbs. (70 Nm) and the protector nut to 47 ft. lbs. (64 Nm).

k. Remove the jack and install the driveshaft (propeller).

l. Install the heat shield cover.

m. Install the rear exhaust pipe and muffler.

n. Apply the parking brake.

o. Place the gear shifter into the **Park** position.

8. Remove the filler plug and fill the differential to the level. Tighten the filler plug to 25 ft. lbs. (34 Nm).

9. Install the rear differential from models equipped with a T–type as follows:

a. Place the differential on a transmission jack and fasten securely with a chain or band.

b. Place seal protector tool ST 28099PA090 over the differential side oil seal and insert the shaft until the spline portion is inside the seal. Remove the seal protector.

c. Insert the axle shaft completely into the differential

d. Adjust the transmission jack as necessary so the stud bolt is inserted correctly into the crossmember bushing.

e. Once the stud bolt is inserted, raise

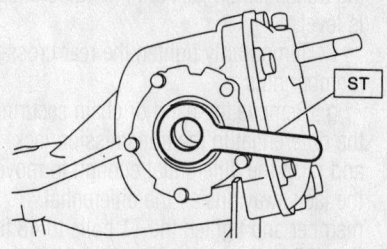

42356-SBCR-G46

Place seal protector tool ST 28099PA090 over the differential side oil seal—Legacy/Outback/Baja

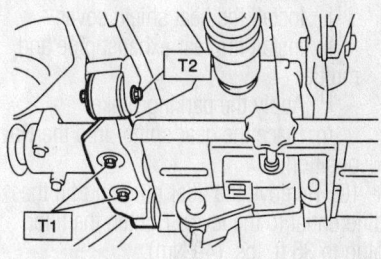

42356-SBCR-G47

Tighten the differential front member and tighten the T1 and T2 bolts to the specifications outlined in the procedure—Legacy/Outback/Baja

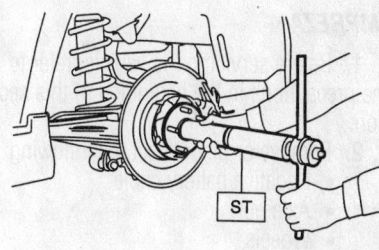

42356-SBCR-G48

Install the axle into the hub splines using adapter installer ST1 922431000 and adapter ST2 927390000—Legacy/Outback/Baja

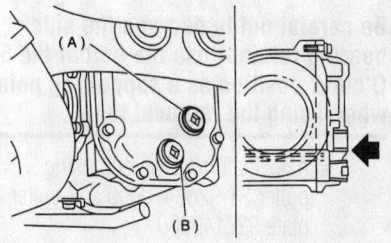

(A) Filler plug
(B) Drain plug

42356-SBCR-G49

Location of the rear differential drain and fill plugs—Legacy/Outback/Baja

the transmission jack until the differential is level.

f. Temporarily tighten the rear cross-member nuts .

g. Remove the band or chain securing the differential to the transmission jack and raise the jack enough to move the jack away. Install the differential member and tighten the T1 bolts to 48 ft. lbs. (65 Nm) and the T2 bolt to 81 ft. lbs. (110 Nm).

h. Tighten the rear crossmember self locking nut to 51 ft. lbs. (70 Nm) and the protector nut to 47 ft. lbs. (64 Nm).

i. Remove the jack and install the driveshaft (propeller).

j. Install the heat shield cover.

k. Install the rear exhaust pipe and muffler.

l. Apply the parking brake.

m. Place the gear shifter into the **Park** position.

10. Remove the filler plug and fill the differential to the level. Tighten the filler plug to 36 ft. lbs. (49 Nm).

11. Install or connect the following:
  • Axle nut and tighten to 174 ft. lbs. (235 Nm). Stake the nut.
  • Wheels
  • Negative battery cable

12. Check all fluids and road test the vehicle.

IMPREZA

1. Before servicing the vehicle, refer to the precautions in the beginning of this section.

2. Remove or disconnect the following:
  • Negative battery cable
  • Axle nut
  • Wheels
  • Halfshaft from the differential using remover tool ST 28099PA100

➡**The side spline shaft circlip comes out with the shaft.**

✳✳ CAUTION

Be careful not to damage the side bearing retainer use the bolt at the 5 O'clock position as a supporting point when using the removal tool.

  • Axle shaft from the hub using puller ST 9266470000 and puller plate 927140000

To install:

3. Install the axle shaft into the hub. Use installer ST 922431000 and adapter ST 927390000 to pull the drive shaft into place and hand tighten the axle nut.

4. Place a seal protector tool such as ST28099PA090 on the differential.

5. Insert the splined shaft of the half-shaft until the spline portion is inside the oil seal.

6. Remove the seal protector.

7. Insert the driveshaft into the differential until it is fully seated.

8. Insert the splined shaft of the half-shaft until the spline portion is inside the oil seal.

9. Torque the new axle nut to 137 ft. lbs. (186 Nm) on 2000–01 models or 174 ft. lbs. (235 Nm) on 2002–03 models and stake the nut.
  • Wheel
  • Negative battery cable

CV-Joints

OVERHAUL

Front

1. Before servicing the vehicle, refer to the precautions in the beginning of this section.

2. Place alignment marks on the shaft and outer race.

3. Remove the inner boot band and boot.

4. Remove the circlip from the inner joint outer race using a suitable prytool.

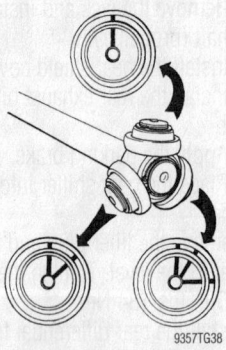
9357TG38

Place alignment marks on the free ring and trunnion as shown

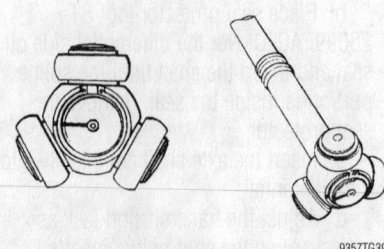
9357TG39

Place alignment marks on the trunnion and shaft as shown

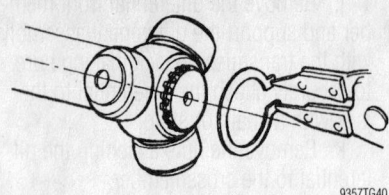
9357TG40

Remove the snapring and trunnion

5. Outer race from the shaft assembly and wipe off the grease.

6. Place alignment marks on the free ring and trunnion as shown in the illustration.

7. Remove the free ring from the trunnion.

8. Place alignment marks on the trunnion and shaft as shown in the illustration.

9. Remove the snapring and trunnion.

10. Place the shaft in a vise between wooden blocks.

11. Using a suitable prytool, raise the outer boot band claws.

12. Cut and remove the boot.

13. Only the boot can be replaced, the joint is not serviceable and must be replaced if damaged.

To install:

14. Place the half shaft in a vise.

15. Place the outer boot and small band on the shaft.

16. Apply 2.12–2.47 oz. (60–70g) of supplied grease to the joint.

17. Apply 0.71–1.06 oz. (20–30g) of supplied grease to the whole inner surface of the boot, and apply some grease to the shaft.

18. Install the boot to the joint groove, and attach the large boot band as shown.

19. Install the boot to the shaft groove, and attach the small boot band as shown.

20. Using boot band plier tool 28099A000 to tighten the large band to 116 ft. lbs. (157 Nm) and the small band to 98 ft. lbs. (133 Nm).

21. Place the inner boot on the center of the shaft.

22. Align the alignment marks from earlier and install the trunnion and snapring. Make sure the snapring is fully engaged.

23. Apply 3.53–3.88 oz. (100–110g) of supplied grease to the joint outer race.

24. Apply a coat of supplied grease to the free ring and trunnion.

25. Align the marks on the free ring and trunnion and install the free ring.

26. Align the marks on the shaft and outer race and install the outer race.

27. Pull on the shaft lightly to ensure the circlip is completely engaged.

28. Apply an even coat 1.06–1.41 oz.

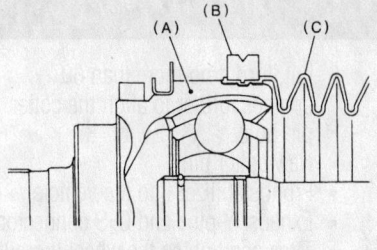

(A) EBJ
(B) Lorge boot band
(C) Boot

9357TG41

Position the boot to the joint groove, and attach the large boot band as shown

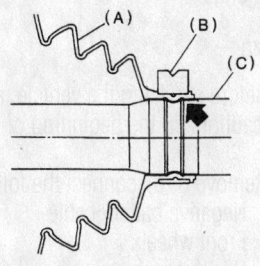

(A) Boot
(B) Small boot band
(C) Shaft

9357TG42

Position the boot to the shaft groove, and attach the small boot band as shown

(30–40g) of the supplied grease to the entire inner surface of the boot.

29. Install the boot and band.

30. Once the band is properly tightened, cut off any excess to leave only 0.39 inch (10mm) and bend it over.

31. Install the shaft.

Rear

The Double Offset Joint (DOJ), is the only part of the assembly that can be replaced, if any of the other components are defective then the shaft should be replaced.

1. Before servicing the vehicle, refer to the precautions in the beginning of this section.

2. Straighten the bent claw of the large clamp at the Double Offset Joint (DOJ) end of the boot.

3. Loosen the band using pliers being careful not to damage the bolt.

4. Remove the small boot band from the DOJ using the same technique.

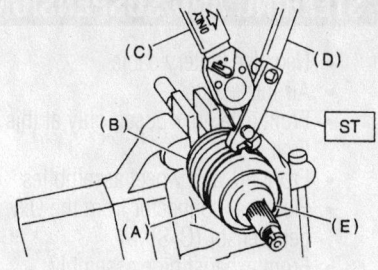

(A) Large boot band
(B) Boot
(C) Torque wrench
(D) Socket flex handle
(E) BJ

9357TG43

Using boot band plier tool 28099A000 tighten the boot bands

5. Remove the boot from the large end of the DOJ outer race.

6. Remove the round circlip using a suitable prytool from the neck of the joint outer race.

7. Remove the joint outer race.

❋❋ CAUTION

The grease used is for CV–Joints, do not replace with another type of grease

8. Clean the grease and remove the balls. Be careful not to loose any of the 6 balls.

9. Turn the cage by a half pitch to the track groove of the inner race and shift the cage.

10. Remove the snap ring that secures the inner race to the shaft and remove the inner race.

11. Take the cage from the shaft and remove the boot.

12. The other boots may be removed in the same manner as the DOJ boot.

13. Wrap the shaft splines with tape to prevent damage.

To install:

14. The following grease must be used during assembly:

   a. BJ side (Non-Turbo): Molylex No. 2 #723223010.

   b. EBJ side (Turbo): NTG2218 # 28093AA000.

15. DOJ side: VU–3A702 (Yellow) # 23223GA050.

16. Install the BJ or EBJ boots and fill it with 2.12–2.47 oz. (60–70g) of grease.

17. Place the DOJ boot on the center of the shaft.

18. Insert the DOJ cage onto the shaft. Make sure to insert the cage with the cut–out portion facing shaft end.

19. Install the inner race on the shaft and fasten with the snap ring. Make sure the snap ring is firmly engaged.

20. Install the cage (previously fitted) to the inner race on the shaft. Fit the cage with the protuded part aligned with the track on the inner race and then turn a half pitch.

21. Fill the DOJ inner race with 2.82–3.17 oz. (80–90g) of grease.

22. Apply a coat of grease to the to the cage pocket and 6 balls.

23. Insert the 6 balls.

24. Align the outer race track and ball positions and place where the shaft, inner race, cage and balls were located prior to removal and then install the outer race.

25. Install the circlip into the groove on the outer race.

➡**Make sure the balls, cage and inner race are fully seated. make sure not to place the matched position of the circlip in the ball groove of the outer race. Pull the shaft lightly to make sure the circlip is fully engaged.**

26. Apply an even coat 0.71–1.06 oz. (20–30g) of grease to the entire inner surface of the boot and to the shaft.

27. Make sure the boot is free from any dirt or foreign materials prior to installation.

28. Place the outer race of the boot at the center of its travel.

29. Put a band through the boot clip and wind twice in alignment with the groove on the boot.

30. Pinch the end of the band using tool ST 9250910000 and tighten securely until it cannot be moved by hand. Make sure there is appropriate air inside the boot.

31. Tap on the clip with a punch to lock it making sure not to damage the damaged while tapping. Cut any excess off the band leaving 0.39 inch (10mm) and bend the remaining portion over the clip. Make sure the end of the band is close to the clip.

32. Install the remaining boot clamps in the same manner.

STEERING AND SUSPENSION

Air Bag

✳✳ CAUTION

Some vehicles are equipped with an air bag system. The system must be disabled before performing service on or around system components, steering column, instrument panel components, wiring and sensors. Failure to follow safety and disabling procedures could result in accidental air bag deployment, possible personal injury and unnecessary system repairs.

PRECAUTIONS

Several precautions must be observed when handling the inflator module to avoid accidental deployment and possible personal injury.

• Never carry the inflator module by the wires or connector on the underside of the module.

• When carrying a live inflator module, hold it securely with both hands, and ensure that the bag and trim cover are pointed away.

• Place the inflator module on a bench or other surface with the bag and trim cover facing up.

• With the inflator module on the bench, never place anything on or close to the module that may be thrown in the event of an accidental deployment.

DISARMING

1. Before servicing the vehicle, refer to the precautions in the beginning of this section.
2. Disconnect the negative battery cable.
3. Disconnect the positive battery cable.
4. Wait more than 20 seconds before starting work.

Power Rack and Pinion Steering Gear

REMOVAL & INSTALLATION

Legacy, Baja and Outback Models

1. Before servicing the vehicle, refer to the precautions in the beginning of this section.
2. Remove or disconnect the following:

• Negative battery cable
• Air intake duct
• Front axle nut, loosen only at this time
• Front tire and wheel assemblies
• Electrical connector from the Oxygen Sensor (O$_2$S)
• Front exhaust pipe assembly
• Tie rod end cotter pin and loosen the castle nut
• Tie rod ends from the steering knuckle arm using a ball joint puller
• Jack up plate and the front stabilizer bar

3. From the power steering rack, remove the center pressure pipe, connect a vinyl hose to the pipe and joint, then turn the steering wheel to discharge the fluid into a container.

➡ **When discharging the power steering fluid (line A and B), turn the steering wheel fully, left and right. Be sure to disconnect the other pipe and drain the fluid in the same manner.**

4. From the control valve of the gearbox assembly, remove the power steering **C** and **D** pressure pipes. Remove pipe **D** first and pipe **C** second.
5. If not disconnected when draining the fluid from the control valve of the gearbox assembly, remove the power steering **A** and **B** pressure pipes. Remove pipe **A** first and pipe **B** second.
6. Remove or disconnect the following:

• Universal joint assembly. Matchmark the assembly before removal.
• Power steering gearbox-to-crossmember assembly bolts
• Gearbox assembly from the vehicle

To install:

7. Install or connect the following:

• Power steering rack and tighten the rack to crossmember bolts to 35–52 ft. lbs. (47–70 Nm)
• Universal joint assembly making sure to align the matchmarks
• Power steering pressure pipes and tighten to 7–12 ft. lbs. (10–16 Nm)
• Universal joint assembly-to-power steering gearbox bolts and tighten to 16–19 ft. lbs. (22–24 Nm) and the universal joint assembly-to-steering shaft bolts 16–19 ft. lbs. (22–24 Nm)
• Tie rod end to steering knuckle nut and tighten to 18–22 ft. lbs. (25–29 Nm). After tightening this

nut, turn it no more than 60 degrees further to align the cotter pin hole.
• New cotter pin
• Front stabilizer into the vehicle
• Exhaust Y-pipe and O$_2$S connector
• Tires and tighten the wheel lug nuts to specification

8. Partially lower the vehicle, then refill and bleed the power steering system.
9. Check for fluid leaks and the fluid level, then install the jack up plate.
10. Check and adjust the toe-in and the steering angle.

Impreza

1. Before servicing the vehicle, refer to the precautions in the beginning of this section.
2. Remove or disconnect the following:

• Negative battery cable
• Front wheels
• Under cover
• Bolt, clip on the sub-frame
• Sub-frame bolts. Leave bolt (1) connected by a few threads and remove the bolts in the sequence illustrated. Once the other bolts are removed, remove bolt (1) and the sub-frame. Refer to the illustration for bolt location.
• Front Y-pipe
• Tie rod end cotter pin and nut
• Tie rod ends from the steering knuckle using a puller
• Jack-up plate and front sway bar
• Fluid lines from the rack and pinion

3. Matchmark the universal joint to the serration in the steering rack for installation reference.

• Lower and upper universal joint bolts and lift the joint upward, disconnecting it from the rack and pinion shaft
• Clamp bolts securing the rack and pinion to the crossmember
• Rack and pinion

To install:

4. Install the rack and pinion. Tighten the clamp bolts to 43 ft. lbs. (59 Nm).

• Fluid pipes and tighten to 9 ft. lbs. (13 Nm)

5. Align the steering rack to the universal joint. Push the long yoke of the joint all the way into the serrated position of the steering shaft, setting the bolt hole in the cut-out. Pull the short yoke all the way out of the serrated portion of the rack and pinion, setting the bolt hole in the cut-out.

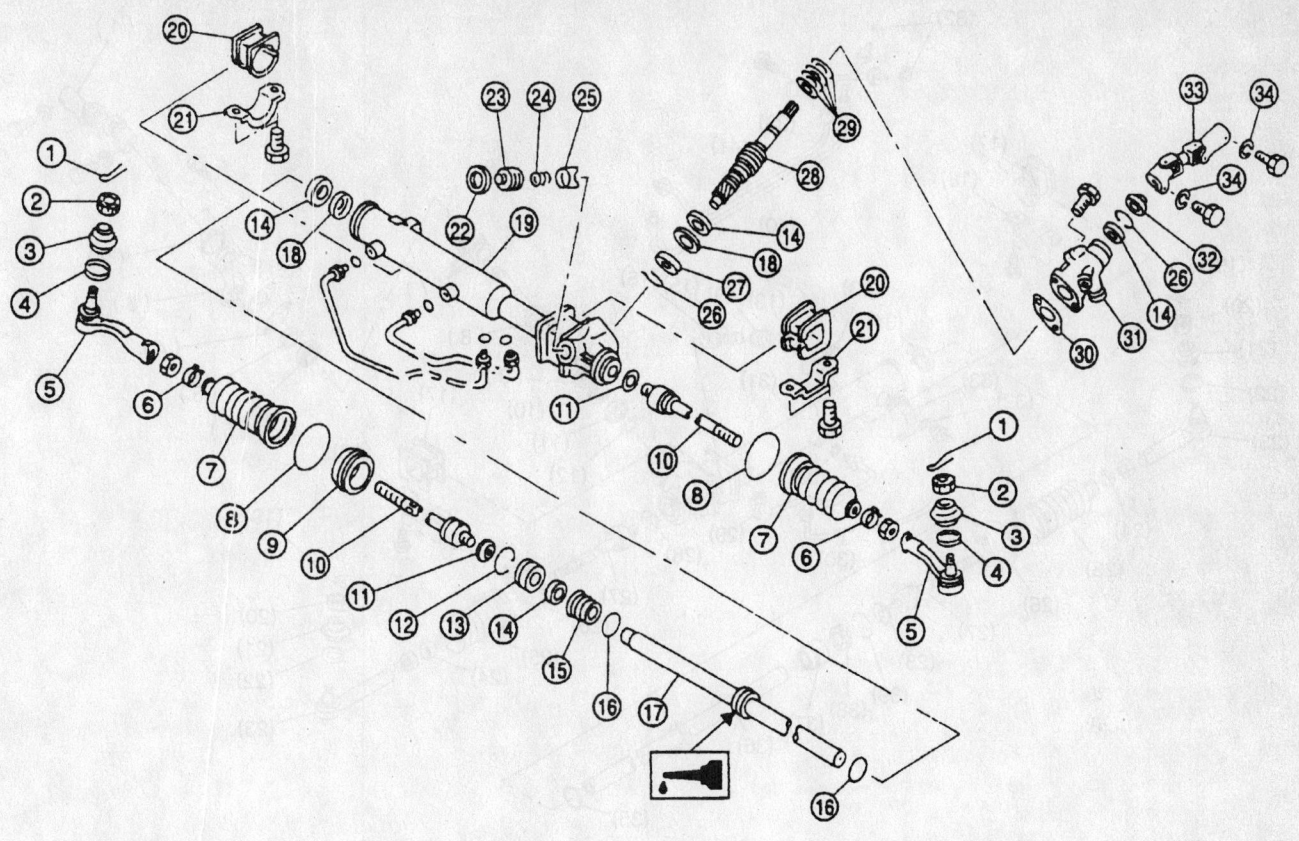

① Cotter pin	⑬ Rack stopper	㉔ Spring
② Castle nut	⑭ Oil seal	㉕ Sleeve
③ Dust cover	⑮ Rack bushing	㉖ C-ring
④ Clip	⑯ O-ring	㉗ Ball bearing
⑤ Tie-rod end	⑰ Rack	㉘ Valve
⑥ Clip	⑱ Back-up washer	㉙ Seal ring
⑦ Boot	⑲ Rack housing	㉚ Packing
⑧ Clip	⑳ Adapter	㉛ Valve housing
⑨ Spacer	㉑ Clamp	㉜ Dust seal
⑩ Tie-rod	㉒ Lock nut	㉝ Universal joint
⑪ Lock washer	㉓ Adjusting screw	㉞ Spring washer
⑫ Circlip		

Exploded view of the steering rack assembly—Legacy models

7923TG63

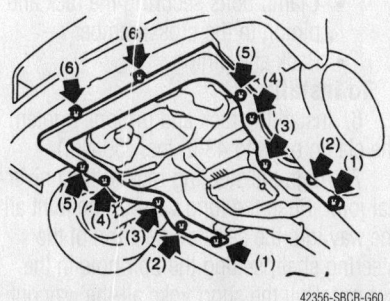

42356-SBCR-G60
Remove the sub-frame bolts as illustrated

Insert the bolt through the short yoke. Pull the yoke and ensure the bolt is properly engaged in the cut-out. Fasten the short yoke side with the spring washer and bolt, then fasten the yoke side. Tighten the bolts to 17 ft. lbs. (24 Nm).

6. Install or connect the following:
  • Tie rod ends to the steering knuckle
  • Sway bar and jack-up plate
  • Y-pipe with new gaskets and nuts
  • Sub-frame. Tighten the T1 bolts to 25 ft. lbs. (34 Nm), the T2 bolts to 41 ft. lbs. (55 Nm) and the T3 bolts to 52 ft. lbs. (71 Nm).
  • Wheels
7. Fill and bleed the steering system.

WRX Models

1. Before servicing the vehicle, refer to the precautions in the beginning of this section.

2. Remove or disconnect the following:
  • Negative battery cable
  • Front wheels

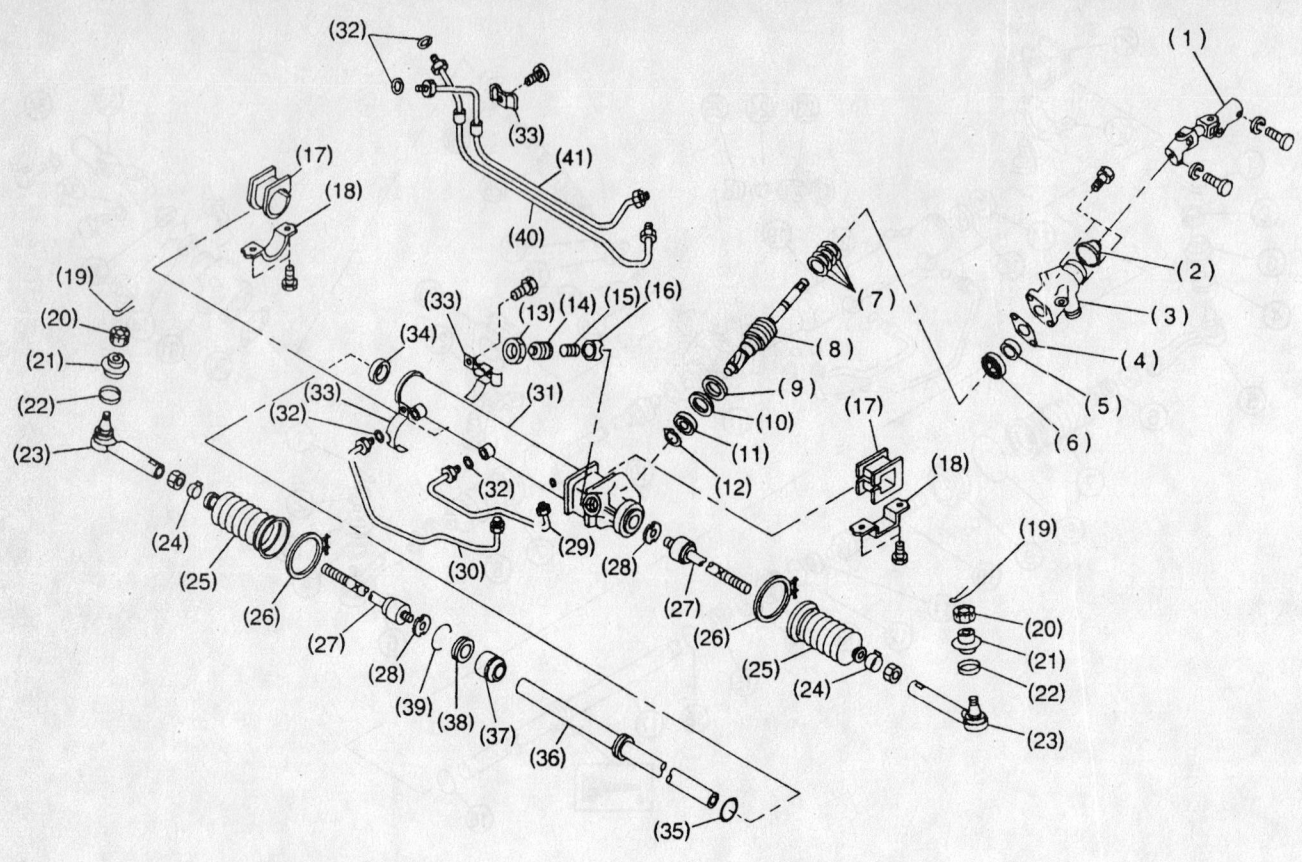

(1) Universal joint	(15) Spring	(29) Pipe B
(2) Dust cover	(16) Sleeve	(30) Pipe A
(3) Valve housing	(17) Adapter	(31) Steering body
(4) Gasket	(18) Clamp	(32) O-ring
(5) Oil seal	(19) Cotter pin	(33) Clamp
(6) Special bearing	(20) Castle nut	(34) Oil seal
(7) Seal ring	(21) Dust cover	(35) Piston ring
(8) Pinion and valve ASSY	(22) Clip	(36) Rack
(9) Oil seal	(23) Tie-rod end	(37) Rack bushing
(10) Back-up washer	(24) Clip	(38) Rack stopper
(11) Ball bearing	(25) Boot	(39) Circlip
(12) Snap ring	(26) Band	(40) Pipe E
(13) Lock nut	(27) Tie-rod	(41) Pipe F
(14) Adjusting screw	(28) Lock washer	

7923TG64

Exploded view of the steering rack assembly—Impreza models

- Engine undercover
3. Remove the sub-frame as follows:
  a. Remove the bolt cover.
  b. Remove the clip.
  c. Loosen the sub-frame bolt (1) and leave it screwed in a few threads. Remove the remaining bolts in the following order: 2, 3, 4, 5 and 6. See illustration for bolt location.
4. Remove or disconnect the following:
  - Front exhaust pipe
  - Tie rod end cotter pin and nut

- Tie rod ends from the steering knuckle using a puller
- Jack-up plate and front sway bar
- Fluid lines from the rack and pinion
5. Matchmark the universal joint to the serration in the steering rack for installation reference.
  - Lower and upper universal joint bolts and lift the joint upward, disconnecting it from the rack and pinion shaft

- Clamp bolts securing the rack and pinion to the crossmember
- Rack and pinion
To install:
6. Install the rack and pinion. Tighten the clamp bolts to 43 ft. lbs. (59 Nm).
7. Align the steering rack to the universal joint. Push the long yoke of the joint all the way into the serrated position of the steering shaft, setting the bolt hole in the cut-out. Pull the short yoke all the way out of the serrated portion of the rack and pin-

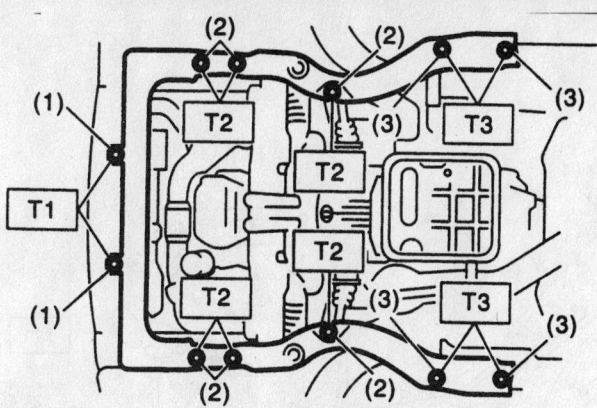

(1) M8 bolt
(2) M12 bolt
(3) M10 bolt

42356-SBCR-G61

Tighten the sub-frame bolts to the sequences outlined in the procedure

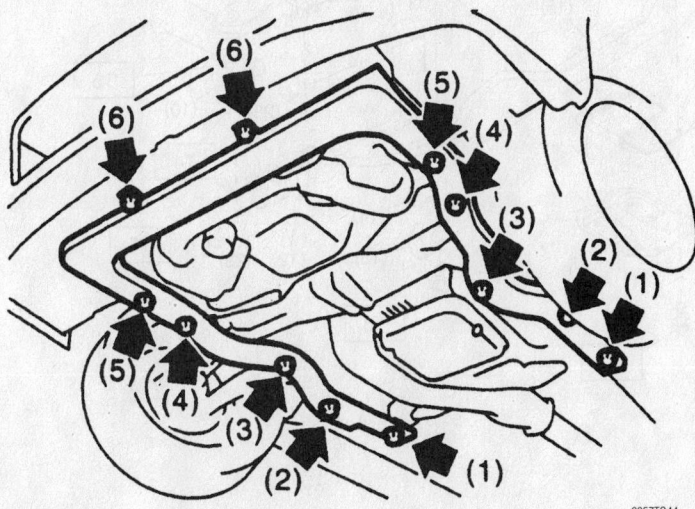

9357TG44

Location of the sub-frame bolts–WRX models

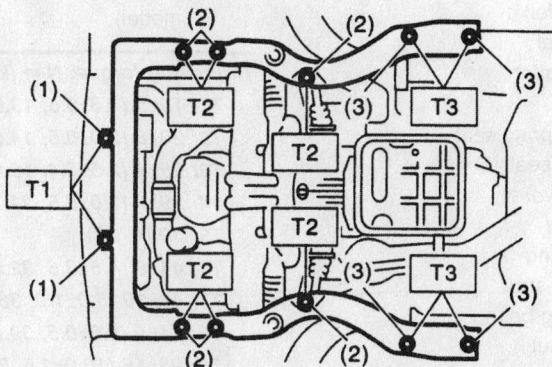

(1) M8 bolt
(2) M12 bolt
(3) M10 bolt

9357TG45

Tighten the sub-frame bolts to the specification outlined in the procedure–WRX models

ion, setting the bolt hole in the cut-out. Insert the bolt through the short yoke. Pull the yoke and ensure the bolt is properly engaged in the cut-out. Fasten the short yoke side with the spring washer and bolt, then fasten the yoke side. Tighten the bolts to 19 ft. lbs. (27 Nm).

8. Install or connect the following:
   • Tie rod ends to the steering knuckle
   • Sway bar and jack-up plate
9. Install the sub-frame and bolts. Tighten the bolts as follows, referring to the illustration for location:
   a. T1: 25 ft. lbs. (34 Nm).
   b. T2: 41 ft. lbs. (55 Nm).
   c. T3: 52 ft. lbs. (71 Nm).
   • Bolt and clip
   • Exhaust pipe with new gaskets and nuts
   • Engine undercover
   • Wheels
10. Fill and bleed the steering system.

Strut

REMOVAL & INSTALLATION

Front

WITH STANDARD STRUTS

1. Before servicing the vehicle, refer to the precautions in the beginning of this section.
2. Remove or disconnect the following:
   • Negative battery cable
   • Front wheel assembly
   • Caliper, if necessary, leaving the line connected, and suspend it out of the way with a piece of wire or string
   • Clip or bolt attaching the brake line to the strut housing
   • Bolt securing the Anti-lock Brake System (ABS) sensor harness, if equipped

➡ Scribe a matchmark on the camber adjusting bolt which secures the strut to the housing.

   • 2 bolts and nuts securing the strut to the steering knuckle. Notice that the shaft of the top bolt is not round. This bolt is used for camber adjustment, and must always be installed in the top hole.
   • 3 nuts securing the strut to the body in the engine compartment
   • Strut and coil spring assembly from the vehicle

To install:

3. Install or connect the following:

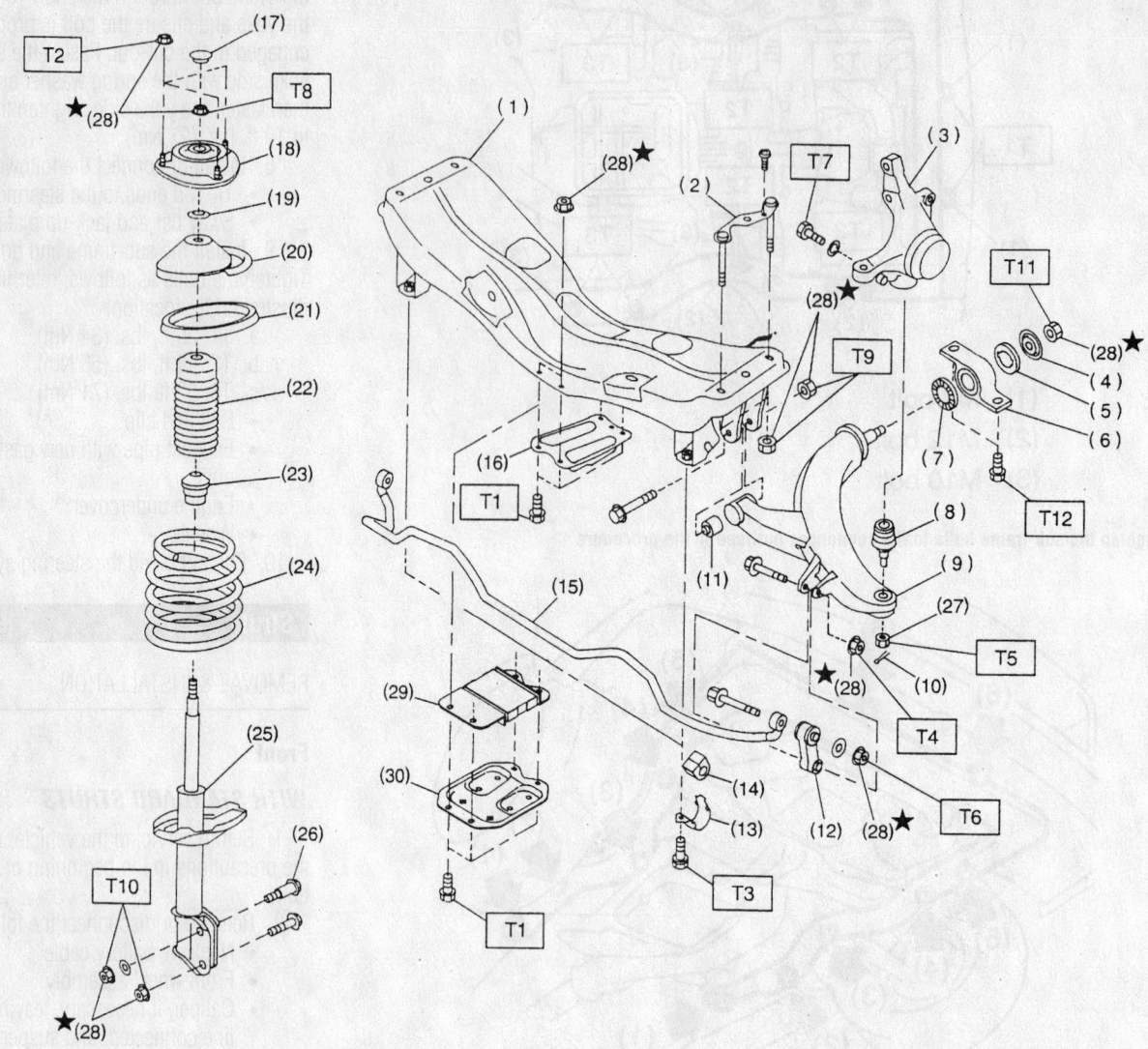

(1) Crossmember
(2) Bolt ASSY
(3) Housing
(4) Washer
(5) Stop rubber (Rear)
(6) Rear bushing
(7) Stop rubber (Front)
(8) Ball joint
(9) Transverse link
(10) Cotter pin
(11) Front bushing
(12) Stabilizer link
(13) Clamp
(14) Bushing
(15) Stabilizer

(16) Jack-up plate (Except 2500 cc MT model)
(17) Dust seal
(18) Strut mount
(19) Spacer
(20) Upper spring seat
(21) Rubber seat
(22) Dust cover
(23) Helper
(24) Coil spring
(25) Damper strut
(26) Adjusting bolt
(27) Castle nut
(28) Self-locking nut
(29) Dynamic damper (2500 cc MT model)

(30) Jack-up plate (2500 cc MT model)

Tightening torque: N·m (kg-m, ft-lb)

T1: 18±5 (1.8±0.5, 13.0±3.6)

T2: 20±6 (2.0±0.6, 14.5±4.3)

T3: 25±4 (2.5±0.4, 18.1±2.9)

T4: 29±5 (3.0±0.5, 21.7±3.6)

T5: 39 (4, 29)

T6: 44±6 (4.5±0.6, 32.5±4.3)

T7: 49±10 (5.0±1.0, 36±7)

T8: 54±5 (5.5±0.5, 39.8±3.6)

T9: 98±15 (10.0±1.5, 72±11)

T10: 152±20 (15.5±2.0, 112±14)

T11: 186±10 (19.0±1.0, 137±7)

T12: 245±49 (25.0±5.0, 181±36)

Exploded view of the front suspension assembly—Legacy and Impreza models

- Strut assembly into the vehicle
- Upper strut retainer nuts and tighten the nuts to 15 ft. lbs. (20 Nm)
- Lower strut nuts and bolts. Be sure the alignment adjustment bolt is installed in the top mounting hole. Tighten the nuts to 112 ft. lbs. (152 Nm) for Impreza and Outback models, or to 130 ft. lbs. (177 Nm) for Legacy and WRX models.
- ABS sensor harness, if equipped and tighten the bolt to 24 ft. lbs. (32 Nm)
- Caliper, if removed
- Brake line to the strut and install the clip or bolt
- Front wheel
- Negative battery cable

4. Check the front end alignment and adjust as necessary.

WITH PNEUMATIC STRUTS

1. Before servicing the vehicle, refer to the precautions in the beginning of this section.

2. Remove or disconnect the following:

- Negative battery cable
- Air line and height sensor harness (from inside the engine compartment) from the strut assembly
- Front wheel assembly
- Anti-lock Brake System (ABS) sensor, if equipped
- Caliper, leaving the line connected, and suspend it out of the way with a piece of wire or string
- Clip attaching the brake line to the strut housing

3. Matchmark the camber adjustment bolt to the strut housing as reference for installation.

- Bolt securing the ABS sensor harness, if equipped
- 2 bolts and nuts securing the strut to the steering knuckle. Notice that the shaft of the top bolt is not round. This bolt is used for camber adjustment, and must always be installed in the top hole.
- 3 nuts securing the strut to the body in the engine compartment
- Strut and coil spring assembly

To install:

4. Install or connect the following:

- Strut assembly
- Upper strut retainer nuts and tighten to 15 ft. lbs. (20 Nm)
- ABS sensor harness (if equipped), and tighten the bolt to 15 ft. lbs. (20 Nm)

- Lower strut nuts and bolts. Be sure the alignment adjustment bolt is installed in the top mounting hole. Tighten the nuts to 112 ft. lbs. (152 Nm).
- Caliper
- Brake line to the strut and install the clip
- Height sensor harness and air line
- Front wheel
- Negative battery cable

5. Start the vehicle and allow enough time for the struts to pressurize before driving.

6. Check the front end alignment and adjust as necessary.

Rear

WITH STANDARD STRUTS

1. Before servicing the vehicle, refer to the precautions in the beginning of this section.

2. Remove or disconnect the following:

- Rear seat assembly, on Sedan only
- Rear speaker grille and service hole cap, on Wagon only
- Strut mount cap
- Wheel and tire assembly
- Brake hose clip
- Union bolt from the brake caliper. Move the brake hose out of the way.
- Lower nuts and bolts securing the strut to the rear wheel housing
- Retainer nuts securing the strut bearing cap to the strut tower, from inside the vehicle
- Strut from the vehicle

To install:

3. Install or connect the following:

- Strut onto the vehicle, making sure to position the strut properly in the upper strut tower mounts. Refer to the illustration.
- Strut retainer nuts and tighten to 14 ft. lbs. (20 Nm) on all models except Legacy/Outback. On Legacy/Outback, tighten the nut to 22 ft. lbs. (30 Nm).
- Strut to the rear wheel knuckle assembly using the retainer nuts and bolts, and tighten the bolts to 162 ft. lbs. (220 Nm) on WRX models and 145 ft. lbs. (196 Nm) on all other models
- Brake union bolt and tighten to 13 ft. lbs. (18 Nm)
- Brake hose clip

4. Bleed the brakes.

- Wheel
- Strut mount cap
- Rear seat, on Sedan
- Speaker grille, on Wagon

WITH PNEUMATIC STRUTS

1. Before servicing the vehicle, refer to the precautions in the beginning of this section.

2. Remove or disconnect the following:

- Negative battery cable
- Rear seat assembly, on the Sedan
- Rear speaker grille and service hole cap, on the Wagon
- Strut mount cap
- Air line from the top of the strut assembly
- Height sensor and solenoid valve wiring harnesses from the strut assembly
- Wheel and tire assembly
- Brake hose clip
- Union bolt from the brake caliper. Move the brake hose out of the way.
- Lower nuts and bolts securing the strut to the rear wheel housing
- Retainer nuts securing the strut bearing cap to the strut tower (from inside the vehicle)
- Strut from the vehicle

To install:

3. Install or connect the following:

- Strut on to the vehicle, making sure to position the strut properly in the upper strut tower mounts. Refer to the illustration if needed. Install the retainer nuts, and tighten to 11 ft. lbs. (15 Nm).
- Strut to the rear wheel knuckle assembly, using the retainer nuts and bolts, and tighten the bolts to 145 ft. lbs. (196 Nm)
- Brake union bolt and tighten to 13 ft. lbs. (18 Nm)
- Brake hose clip

4. Bleed the brakes.

- Wheel
- Height sensor and solenoid valve wiring harnesses to the strut
- Air line to the top of the strut
- Strut mount cap
- Rear seat, on Sedan
- Speaker grille, on Wagon
- Negative battery cable

5. Start the vehicle, and allow enough time for the shock to pressurize before driving the vehicle.

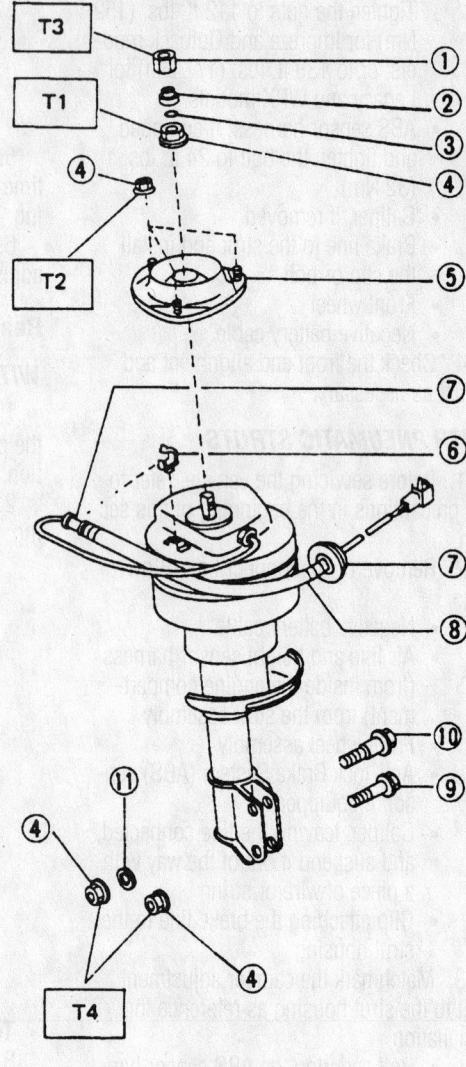

Rear

Front

Tightening torque: N·m (kg-m, ft-lb)

T1: 49 — 69 (5 — 7, 36 — 51)

T2: 14 — 25 (1.4 — 2.6, 10 — 19)

T3: 7 — 17 (0.7 — 1.7, 5.1 —12.3)

T4: 186 — 235 (19 — 24, 137 — 174)

1 Cap
2 Air bushing
3 O-ring
4 Self lock nut
5 Strut mount
6 Clip
7 Grommet
8 Corrugate tube
9 Flange bolt
10 Adjusting bolt
11 Washer
12 Solenoid valve
13 Insulator
14 Air pipe for solenoid valve
15 Air pipe
16 Connector

7923TG69

Exploded view of the front and rear pneumatic suspension assembly—Legacy and Impreza models

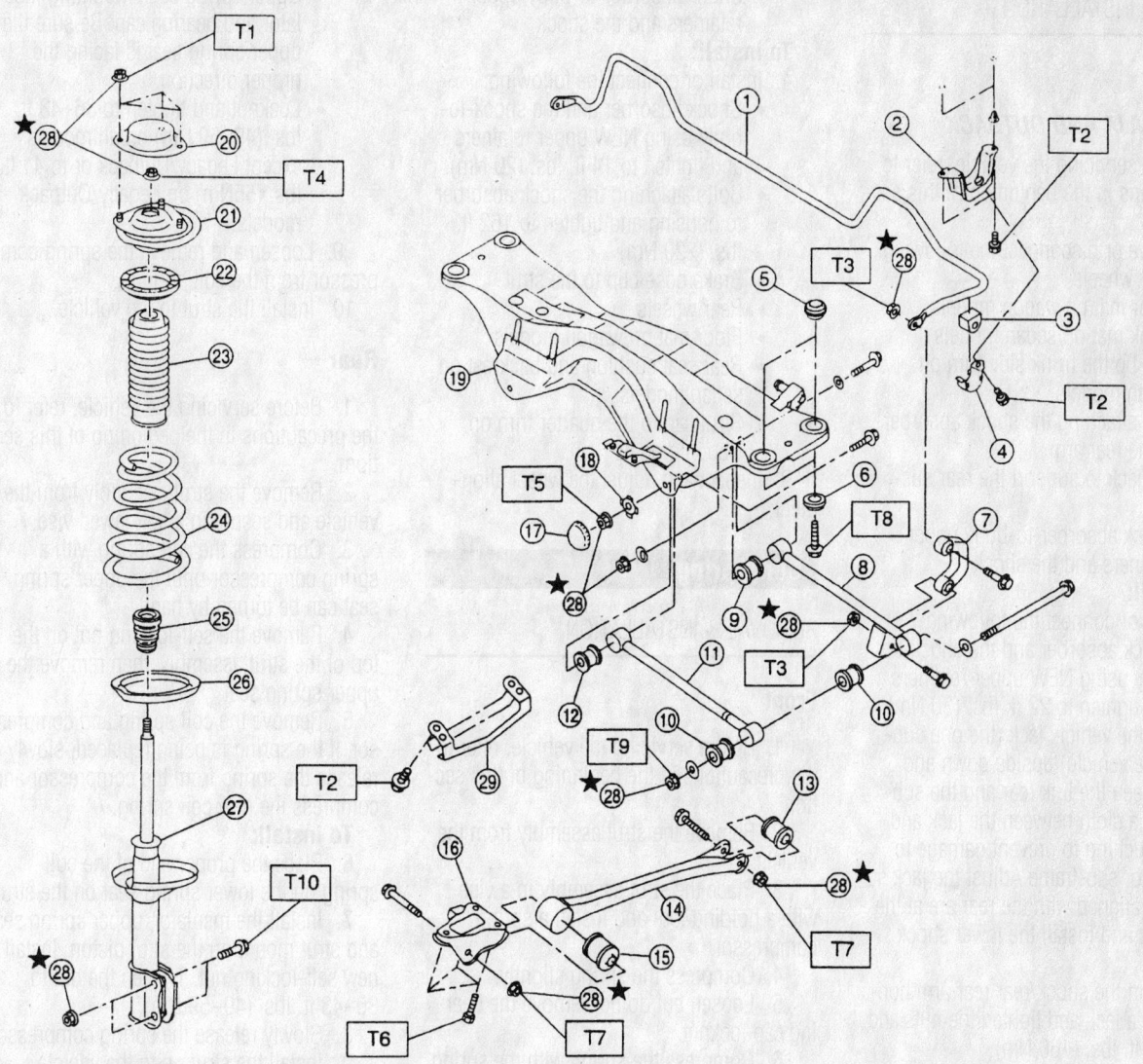

1. Stabilizer
2. Stabilizer bracket
3. Stabilizer bushing
4. Clamp
5. Floating bushing
6. Stopper
7. Stabilizer link
8. Rear lateral link
9. Bushing (C)
10. Bushing (A)
11. Front lateral link
12. Bushing (B)
13. Trailing link rear bushing
14. Trailing link
15. Trailing link front bushing
16. Trailing link bracket
17. Cap (Protection)
18. Washer
19. Crossmember
20. Strut mount cap
21. Strut mount
22. Rubber seat upper
23. Dust cover
24. Coil spring
25. Helper
26. Rubber seat lower
27. Damper strut
28. Self-locking nut
29. Crossmember reinforcement lower (Sedan model only)

Tightening torque: N·m (kg-m, ft-lb)
T1: 20 ± 6 (2.0 ± 0.6, 14.5 ± 4.3)
T2: 25 ± 7 (2.5 ± 0.7, 18.1 ± 5.1)
T3: 44 ± 6 (4.5 ± 0.6, 32.5 ± 4.3)
T4: 59 ± 10 (6.0 ± 1.0, 43 ± 7)
T5: 98 ± 15 (10.0 ± 1.5, 72 ± 11)
T6: 98 ± 20 (10.0 ± 2.0, 72 ± 14)
T7: 113 ± 15 (11.5 ± 1.5, 83 ± 11)
T8: 127 ± 20 (13.0 ± 2.0, 94 ± 14)
T9: 137 ± 20 (14.0 ± 2.0, 101 ± 14)
T10: 196^{+39}_{-10} ($20.0^{+4.0}_{-1.0}$, 145^{+29}_{-7})

Exploded view of the rear standard strut assembly—Legacy and Impreza models

7923TG71

Shock Absorber

REMOVAL & INSTALLATION

Rear

LEGACY, BAJA AND OUTBACK

1. Before servicing the vehicle, refer to the precautions in the beginning of this section.

2. Remove or disconnect the following:
   - Rear wheels
   - Floor mat on wagon models
   - Trunk mat on sedan models
   - Roll up the trunk side trim on sedan models
   - Bolt attaching the shock absorber to the rear arm

3. Use a jack to support the rear suspension.
   - Shock absorber-to-body upper retainers and the shock

To install:

4. Install or connect the following:
   - Shock absorber and the shock-to-body using NEW upper retainers and tighten to 22 ft. lbs. (30 Nm).

5. Place the vehicle jack (the one supplied with the vehicle) upside down and place it between the link rear and the sub-frame. Place a cloth between the jack and areas it is touching to prevent damage to the link rear or sub-frame Adjust the jack so the shock is aligned with the rear are at the correct holes and install the lower shock bolts.

6. Support the shock/rear rear arm horizontally with a jack and tighten the nuts and bolts to 118 ft. lbs. (160 Nm).

7. Install or connect the following:
   - Floor mat on wagon models
   - Trunk side trim on sedan models
   - Trunk mat on sedan models
   - Rear wheels

8. Inspect and adjust the wheel alignment.

Rear

IMPREZA

1. Before servicing the vehicle, refer to the precautions in the beginning of this section.

2. Remove or disconnect the following:
   - Rear seat cushion and backrest on sedan models
   - Strut cap from the quarter trim on wagon models
   - Rear wheels
   - Brake hose clip from the strut
   - Bolts attaching the shock absorber to housing

3. Use a jack to support the rear suspension.
   - Shock absorber-to-body upper retainers and the shock

To install:

4. Install or connect the following:
   - Shock absorber and the shock-to-body using NEW upper retainers and tighten to 14 ft. lbs. (20 Nm).
   - Bolts attaching the shock absorber to housing and tighten to 162 ft. lbs. (220 Nm)
   - Brake hose clip to the strut
   - Rear wheels
   - Floor mat on wagon models
   - Rear seat cushion and backrest on sedan models
   - Strut cap to the quarter trim on wagon models

5. Inspect and adjust the wheel alignment.

Coil Spring

REMOVAL & INSTALLATION

Front

1. Before servicing the vehicle, refer to the precautions in the beginning of this section.

2. Remove the strut assembly from the vehicle.

3. Place the strut assembly in a vise with a holding tool and install a spring compressor.

4. Compress the spring slightly.

5. Loosen but do not remove the bearing cap locknut.

6. Compress the spring with the spring compressor, then remove the locknut.

7. Remove or disconnect the following:
   - Strut bearing cap, mounting insulator bracket and upper spring seat
   - Coil assembly, leaving the spring compressed
   - Strut boot and rebound bumper from the strut. Inspect and replace if worn.
   - Strut retainer nut using a suitable wrench
   - Strut insert from the assembly

To install:

8. Install or connect the following:
   - Strut into the chamber and install the retainer nut. Tighten the nut snugly.
   - Rebound bumper and the boot to the strut piston rod
   - Coil spring on the strut assembly.

Be sure the spring is properly positioned on the lower bracket.
   - Upper spring seat, mounting insulator and bearing cap. Be sure the upper spring seat is facing the proper direction.
   - Locknut and tighten to 36–43 ft. lbs. (49–59 Nm) on all models except Legacy/Outback or to 41 ft. lbs. (55Nm) on Legacy/Outback models.

9. Loosen and remove the spring compressor from the coil spring.

10. Install the strut to the vehicle.

Rear

1. Before servicing the vehicle, refer to the precautions in the beginning of this section.

2. Remove the strut assembly from the vehicle and secure in a soft jawed vise.

3. Compress the coil spring with a spring compressor until the upper spring seat can be turned by hand.

4. Remove the self-locking nut on the top of the strut assembly, then remove the upper spring seat.

5. Remove the coil spring and compressor. If the spring is being replaced, slowly release the spring from the compressor and compress the new coil spring.

To install:

6. Place the proper end of the coil spring on the lower spring seat on the strut.

7. Install the insulator, upper spring seat and strut mount on the strut piston. Install a new self-locking nut. Tighten the nut to 36–43 ft. lbs. (49–59 Nm).

8. Slowly release the spring compressor.

9. Install the strut on to the vehicle.

Lower Ball Joint

REMOVAL & INSTALLATION

1. Before servicing the vehicle, refer to the precautions in the beginning of this section.

2. Remove or disconnect the following:
   - Negative battery cable
   - Front wheel and tire assembly
   - Ball joint castle nut cotter pin and discard the cotter pin
   - Castle nut
   - Ball joint from the lower control arm using a suitable puller or pry-tool
   - Bolt securing the ball joint to the steering knuckle
   - Ball joint using a suitable wedge to

expand the steering knuckle connection point

To install:

3. Install or connect the following:
- Ball joint to the steering knuckle
- Retaining bolt and tighten to 37 ft. lbs. (50 Nm)
- Ball joint to the lower control arm and tighten the castle nut on all models except WRX to 29 ft. lbs. (39 Nm). On WRX sedan models, tighten to 22 ft. lbs. (30 Nm) and 33 ft. lbs. (45 Nm) on all other WRX models. Then, tighten the castle nut an additional 60 degrees until the slot in the castle nut is aligned with the cotter pin hole in the ball joint.
- New cotter pin
- Wheel
- Negative battery cable

Front Lower Control Arm

REMOVAL & INSTALLATION

Except WRX Models

1. Before servicing the vehicle, refer to the precautions in the beginning of this section.
2. Remove or disconnect the following:
- Tire and wheel assembly
3. On Impreza models, remove the sub-frame as follows:
   a. Remove the bolt cover.
   b. Remove the clip.
   c. Loosen the sub-frame bolt (1) and

leave it screwed in a few threads. Remove the remaining bolts in the following order: 2, 3, 4, 5 and 6. See illustration for bolt location.
- Sway link from the lower control arm
- Bolt securing the ball joint to the steering knuckle
- Nuts (NOT the bolts) securing the lower control arm to the crossmember
- 2 bolts holding the bushing bracket of the control arm to the body
- Ball joint from the steering knuckle
- Bolts securing the lower control arm to the crossmember, then the lower control

To install:

4. Install or connect the following:
- Lower control arm; temporarily tighten the 2 bolts used to secure the rear bushing of the lower control arm to the body

➡ **These bolts should be tightened so they can still move back and forth in the oblong shaped hole in the bracket that holds the bushing.**

5. On Impreza models, install the subframe as follows:
6. Install the sub-frame and bolts. Tighten the bolts as follows, referring to the illustration for location:
   a. T1: 25 ft. lbs. (34 Nm).
   b. T2: 41 ft. lbs. (55 Nm).
   c. T3: 52 ft. lbs. (71 Nm).
7. Install or connect the following:
- Bolts used to secure the lower con-

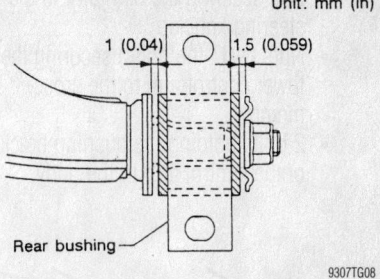

Unit: mm (in)

1 (0.04)   1.5 (0.059)

Rear bushing

9307TG08

Proper control arm-to-rear bushing clearance specifications—all models

trol arm to the crossmember and temporarily tighten the nuts
- Ball joint into the steering knuckle and secure with the retaining bolt
- Sway link to the control arm and temporarily tighten the bolts

✱✱ WARNING

Discard loosened self-locking nut and replace with a new one.

8. Lower the vehicle, then tighten the bolts to the following specifications:
   a. Lower control arm-to-sway bar: 22 ft. lbs. (30 Nm). On Impreza models tighten the retainers on models except sedan turbo to 22 ft. lbs. (30 Nm) turbo and 33 ft. lbs. on sedan turbo models.

➡ **Move the rear bushing back and forth until the control arm-to-rear bushing clearance if established. Refer to the illustration for specifications.**

   b. Lower control arm-to-crossmember: 74 ft. lbs. (100 Nm).
   c. Lower control arm-to-rear link bushing-to-body: 184 ft. lbs. (250 Nm).
9. Check the wheel alignment and adjust if necessary.

WRX Models

1. Before servicing the vehicle, refer to the precautions in the beginning of this section.
2. Remove or disconnect the following:
- Tire and wheel assembly
3. Remove the sub-frame as follows:
   a. Remove the bolt cover.
   b. Remove the clip.
   c. Loosen the sub-frame bolt (1) and leave it screwed in a few threads. Remove the remaining bolts in the following order: 2, 3, 4, 5 and 6. See illustration for bolt location.
- Sway link from the lower control arm

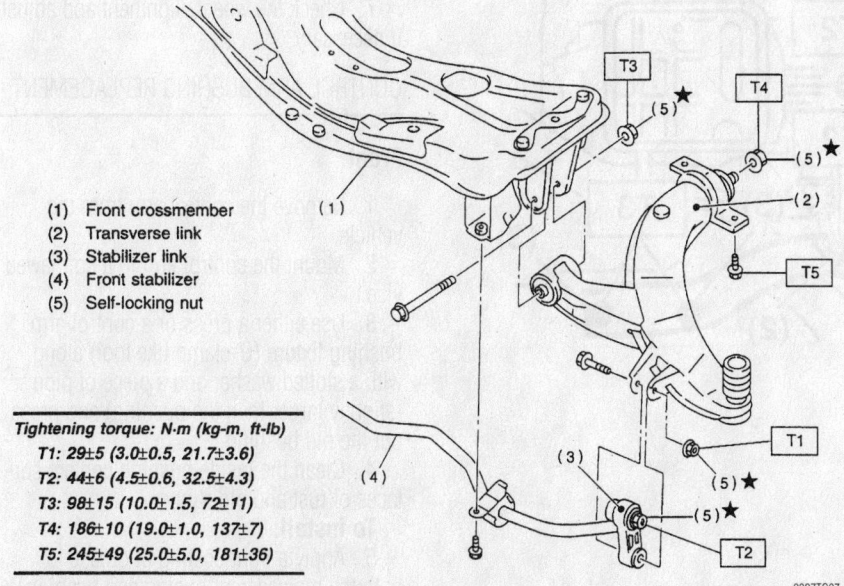

(1) Front crossmember
(2) Transverse link
(3) Stabilizer link
(4) Front stabilizer
(5) Self-locking nut

Tightening torque: N·m (kg-m, ft-lb)
T1: 29±5 (3.0±0.5, 21.7±3.6)
T2: 44±6 (4.5±0.6, 32.5±4.3)
T3: 98±15 (10.0±1.5, 72±11)
T4: 186±10 (19.0±1.0, 137±7)
T5: 245±49 (25.0±5.0, 181±36)

9307TG07

Exploded view of the lower control arm (transverse link)—Impreza and Legacy

- Bolt securing the ball joint to the steering knuckle
- Nuts (NOT the bolts) securing the lower control arm to the cross-member
- 2 bolts holding the bushing bracket of the control arm to the body

- Ball joint from the steering knuckle
- Bolts securing the lower control arm to the crossmember, then the lower control

To install:

4. Install or connect the following:
- Lower control arm; temporarily

tighten the 2 bolts used to secure the rear bushing of the lower control arm to the body

➡ **These bolts should be tightened so they can still move back and forth in the oblong shaped hole in the bracket that holds the bushing.**

- Bolts used to secure the lower control arm to the crossmember and temporarily tighten the nuts
- Ball joint into the steering knuckle and secure with the retaining bolt
- Sway link to the control arm and temporarily tighten the bolts

❄❄ WARNING

Discard loosened self-locking nut and replace with a new one.

5. Lower the vehicle, then tighten the bolts to the following specifications:
   a. Lower control arm-to-sway bar: 33 ft. lbs. (45 Nm) on sedan models or 22 ft. lbs. (30 Nm) on all except sedan models.
   b. Lower control arm and crossmember to 74 ft. lbs. (100 Nm).
   c. Lower control arm rear bushing and body to 184 ft. lbs. (250 Nm).

➡ **Move the rear bushing back and forth until the control arm-to-rear bushing clearance if established. Refer to the illustration for specifications.**

6. Install the sub-frame and bolts. Tighten the bolts as follows referring to the illustration for location:
   a. T1: 25 ft. lbs. (34 Nm).
   b. T2: 41 ft. lbs. (55 Nm).
   c. T3: 52 ft. lbs. (71 Nm).
7. Check the wheel alignment and adjust if necessary.

CONTROL ARM BUSHING REPLACEMENT

Front

1. Remove the control arm from the vehicle.
2. Mount the control arm in a soft jawed vise.
3. Use either a press or a control arm bushing fixture (C-clamp like tool) along with a slotted washer and a piece of pipe (slightly larger than the bushing) and press out the old bushing.
4. Clean the inside bushing contact surfaces of rust and old rubber.

To install:

5. Apply a light coating of grease to both the replacement busing and bushing contact surfaces on the control arm.

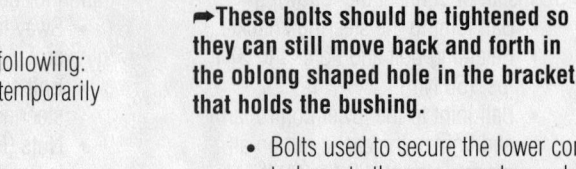

Location of the sub-frame bolts—WRX models

9357TG44

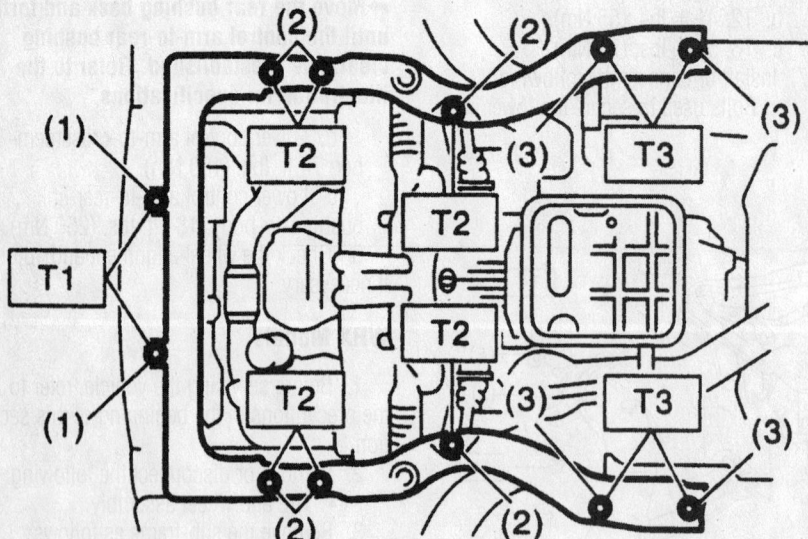

(1) M8 bolt

(2) M12 bolt

(3) M10 bolt

9357TG45

Tighten the sub-frame bolts to the specification outlined in the procedure—WRX models

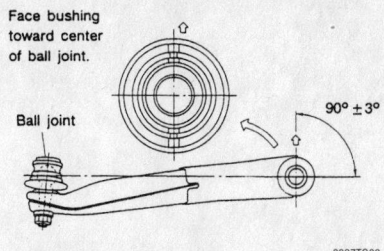

Face bushing toward center of ball joint.

Ball joint

90° ± 3°

9307TG09

The front control arm bushing must be installed in the proper direction

6. Align the bushing according to the illustration.

7. Install the bushing using the press tool. A bushing install clamp can also be used to compress the bushing into the control arm.

8. Install the control arm on the vehicle.

Rear

1. Remove the control arm from the vehicle.

2. Scribe a matchmark on the control arm and rear bushing.

3. Loosen the nut and remove the rear bushing. Discard the nut.

To install:

4. Install the rear bushing to the control arm, making sure to align the marks made during removal.

5. Install a new nut and tighten to 140 ft. lbs. (190 Nm) on WRX models or 137 ft. lbs. (168 Nm) on all models except WRX.

Rear Lower Control Arm

REMOVAL & INSTALLATION

Impreza

TRAILING LINK

1. Before servicing the vehicle, refer to the precautions in the beginning of this section.

2. Remove or disconnect the following:
- Tire and wheel assembly
- Rear parking bracket clamp and Anti-lock Brake System (ABS) sensor harness, if equipped
- Bolts that secure the trailing link to the bracket
- Bolt that secures the trailing link to the rear housing
- Trailing link from the vehicle

To install:

3. Install or connect the following:
- Trailing link and the through-bolts and nuts. DO NOT tighten the nuts and bolts at this time.

- ABS sensor bracket, if equipped, and parking brake cable to the trailing link.
- Tire and wheel assembly

4. Tighten the trailing link-to-bracket bolt to 72 ft. lbs. (98 Nm) and the nut to 83 ft. lbs. (113 Nm).

5. Tighten the trailing link-to-rear housing to 83 ft. lbs. (113 Nm).

6. Check the wheel alignment and adjust if necessary.

LATERAL LINK

1. Before servicing the vehicle, refer to the precautions in the beginning of this section.

2. Remove or disconnect the following:
- Tire and wheel assembly
- Stabilizers on turbo models
- Anti-lock Brake System (ABS) sensor harness from the trailing link, if equipped.
- Bolts that secure the lateral link to the rear housing.

➡**Discard the old self-locking nuts and replace with new ones during installation.**

- Bolts which secure the trailing link to the rear housing
- Halfshaft from the rear differential using a suitable tool.

➡**On all except 2.2L engines, do not remove the circlip attached to the inside of the differential. On 2.2L engines, the side spline circlip comes out together with the shaft. Be careful not to damage the side bearing retainer.**

3. Scribe an alignment mark on the rear lateral link adjusting bolt and crossmember.

4. Remove or disconnect the following:
- Outer lateral link bolt securing the lateral link to the housing
- Bolts securing the front and rear lateral links to the crossmember
- Lateral links from the vehicle

To install:

5. Install or connect the following:
- Bolts securing the front and rear lateral links to the crossmember and hand-tighten
- Outer lateral link bolt securing the lateral link to the housing and hand-tighten
- Halfshaft to the rear differential
- Bolts that secure the lateral link to the rear housing
- Bolts which secure the trailing link to the rear housing
- ABS sensor harness to the trailing link, if equipped

- Tire and wheel assembly

6. Tighten the lateral link bolts as shown in the illustration.

7. Check the wheel alignment and adjust if necessary.

Legacy/Baja

1. Before servicing the vehicle, refer to the precautions in the beginning of this section.

2. Remove or disconnect the following:
- Tire and wheel assembly
- Wheel bearing assembly, refer to the procedure in this section
- Bolt holding the parking brake cable to the control arm
- Bolt securing the brake hose to the rear arm
- Bolt securing the Anti-lock Brake Sensor (ABS) sensor to the rear arm
- Brake line from the wheel cylinder with a flare nut wrench, if equipped with drum brakes. Plug the line to avoid contaminating the system. Suspend the brake backing plate from the sub-frame.
- Nut securing the stabilizer link to the rear arm
- Bolt holding the shock absorber to the rear arm

3. Use a suitable transaxle jack to support the rear arm horizontally.

4. Remove or disconnect the following:
- Bolt securing the rear arm to the body
- Nut securing the front link to the rear arm, loosen
- Nut securing the rear link to the rear arm, loosen
- Bolts holding the rear arm to the links
- Rear arm

To install:

5. Use a transaxle jack to support the rear arm.

6. Install or connect the following:
- Rear arm and temporarily tighten the bolts securing the rear arm to the link
- Wheel bearing unit
- Bolt securing the ABS sensor to the rear arm
- Brake hose-to-rear arm bolt
- Parking brake cable clamp-to-rear arm bolt

➡**Place a rag or cloth between the jack and its mating area to avoid scratching the rear link and sub-frame.**

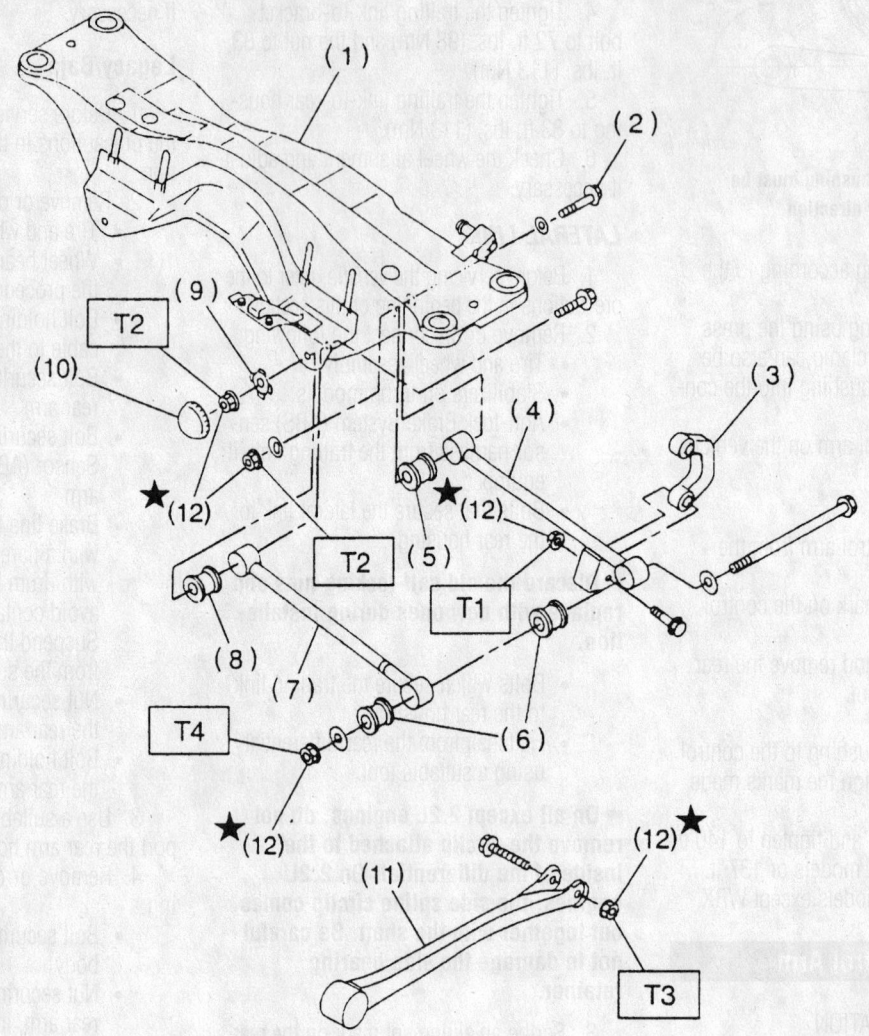

(1) Crossmember
(2) Adjusting bolt
(3) Stabilizer link
(4) Rear lateral link
(5) Bushing (C)
(6) Bushing (A)
(7) Front lateral link

(8) Bushing (B)
(9) Washer
(10) Cap
(11) Trailing link
(12) Self-locking nut

Tightening torque: N·m (kg-m, ft-lb)
T1: 44±6 (4.5±0.6, 32.5±4.3)
T2: 98±15 (10.0±1.5, 72±11)
T3: 113±15 (11.5±1.5, 83±11)
T4: 137±20 (14.0±2.0, 101±14)

9307TG10

Lateral link mounting and tightening specifications—Impreza

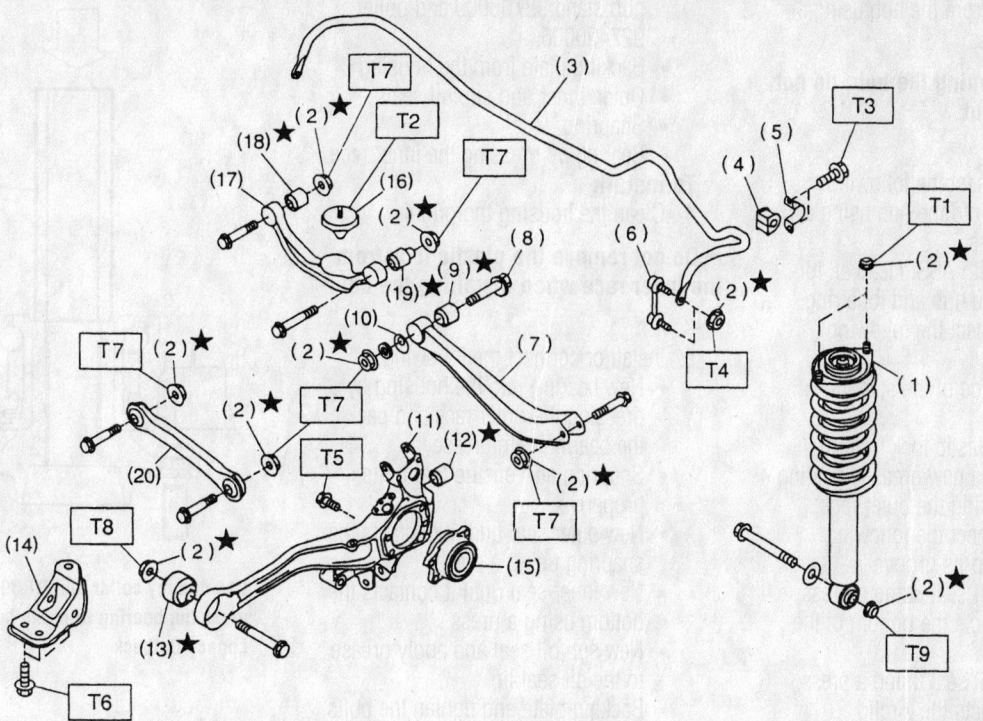

Rear control arm mounting and tightening specifications—Legacy

(1) Shock absorber
(2) Self-locking nut
(3) Stabilizer
(4) Stabilizer bushing
(5) Clamp
(6) Stabilizer link
(7) Link rear
(8) Adjusting bolt
(9) Link rear bushing
(10) Adjusting washer
(11) Rear arm

(12) Rear arm rear bushing
(13) Rear arm front bushing
(14) Rear arm bracket
(15) Hub bearing unit
(16) Helper
(17) Link upper
(18) Link upper bushing (Inside)
(19) Link upper bushing (Outside)
(20) Link front

Tightening torque: N·m (kg-m, ft-lb)
T1: 30±7 (3.1±0.7, 22.4±5.1)
T2: 32±10 (3.3±1.0, 23.9±7.2)
T3: 39±7 (4.0±0.7, 28.9±5.1)
T4: 44±6 (4.5±0.6, 32.5±4.3)
T5: 66±10 (6.7±1.0, 48.5±7.2)
T6: 108±15 (11±1.5, 80±11)
T7: 123±15 (12.5±1.5, 90±11)
T8: 147±20 (15±2, 108±14)
T9: 157±20 (16±2, 116±14)

9307TG16

7. Place the tire changing jack (supplied with the car) upside down and position between the rear link and sub-frame. Adjust the jack position so the rear shock absorber is aligned with the rear arm at their corresponding holes. Install the lower shock absorber bolts.

8. Using the transmission jack, support the rear arm horizontally, then tighten the nuts and bolts holding the rear arm, front and rear links, upper link and shock absorber. Refer to the specifications in the illustration.

9. Install the tire and wheel assembly.

10. Check and adjust the alignment, if necessary.

BUSHING REPLACEMENT

1. Use a suitable press to remove and install the bushings.

Wheel Bearings

ADJUSTMENT

the wheel bearings are not adjustable.

REMOVAL & INSTALLATION

Front

1. Before servicing the vehicle, refer to the precautions in the beginning of this section.

2. Remove or disconnect the following:
   • Steering knuckle assembly from the vehicle

3. Position the steering knuckle in a soft-jawed vise.

4. Press the hub from the steering knuckle. If the inner bearing race remains in the hub, press it out.

5. Remove or disconnect the following:
   • Rotor shield
   • Inner and outer seals
   • Snapring from the steering knuckle

6. Press the inner bearing race to remove the outer bearing.

7. Remove or disconnect the following:
   • Tone ring, if equipped with Anti-lock Brake System (ABS)

- Wheel lugs from the hub using a suitable press

➡ **To prevent deforming the hub, do not hammer the lugs out.**

To install:

8. Install or connect the following:
   - Wheel lugs into the hub using a suitable press
9. If equipped with ABS, clean all foreign material from the hub and tone ring.
10. Install or connect the following:
    - Tone ring
11. Clean the inside of the steering knuckle.
12. Remove the plastic lock from the inner race and press a new greased bearing into the hub by pressing the outer race.
13. Install or connect the following:
    - Snapring into its groove
    - New outer oil seal using a press, until it contacts the bottom of the housing
    - New inner oil seal using a press, until it contacts the circlip
14. Apply grease to the oil seal lips.
15. Install or connect the following:
    - Rotor shield and tighten the bolts to 10 ft. lbs. (14 Nm)
    - Hub to the steering knuckle
16. Press a new bearing into the hub by driving the inner race.
17. Install the steering knuckle on the vehicle.

Rear

EXCEPT LEGACY/OUTBACK/BAJA

1. Before servicing the vehicle, refer to the precautions in the beginning of this section.
2. Loosen the parking brake adjustment.
3. Remove or disconnect the following:
   - Wheel assembly
   - Unstake and remove the axle nut
   - Caliper, leaving the line connected, and suspend it aside
   - Rotor
   - Parking brake cable
   - Sway bar clamp
   - Bolt securing the lateral link to the housing
   - Bolts securing the trailing link to the housing
   - Halfshaft
   - Bolts securing the strut to the housing
   - Speed sensor from the backing plate, if equipped with Anti-lock Brake System (ABS)
   - Housing assembly
   - Hub from the rear housing using

hub stand 92708000 and puller 927420000
   - Backing plate from the housing
   - Outer, inner and sub oil seals
   - Snapring
   - Bearing by pressing the inner race

To install:

4. Clean the housing thoroughly.

➡ **Do not remove the plastic lock from the inner race when installing the bearing.**

5. Install or connect the following:
   - New bearing into the housing by pressing the outer race and pack the bearing with grease
   - Snapring and ensure that it fits properly
   - New outer seal until it contacts the snapring using a press
   - New inner seal until it contacts the bottom using a press
   - New sub oil seal and apply grease to the oil seal lip
   - Backing plate and tighten the bolts to 43 ft. lbs. (58 Nm)
   - Hub into the housing using installer 927450000 to press it into position
   - Housing to the strut and tighten the bolts to 119 ft. lbs. (162 Nm)
   - Speed sensor, if equipped with ABS
   - Halfshaft
   - Trailing link to the housing and tighten the bolt and new nut to 94 ft. lbs. (127 Nm)
   - Lateral link to the housing and tighten the bolt and new nut to 116 ft. lbs. (157 Nm)
   - Sway bar clamp
   - Parking brake cable
   - Rear brake assembly
   - New axle nut and tighten it to 152 ft. lbs. (206 Nm). Stake the nut.
   - Wheel
6. Adjust the parking brake cable.

LEGACY/OUTBACK/BAJA

1. Before servicing the vehicle, refer to the precautions in the beginning of this section.
2. Loosen the parking brake adjustment.
3. Remove or disconnect the following:
   - Wheel assembly
   - Unstake and remove the axle nut
   - Parking brake lever
   - ABS sensor
   - Caliper, leaving the line connected, and suspend it aside
   - Rotor
   - Four bolts from the rear arm and the hub and bearing assembly

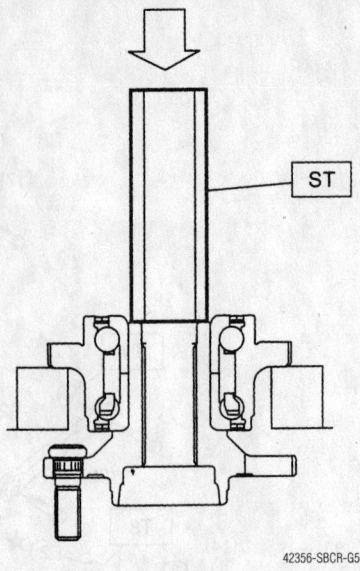

Use dummy collar tool ST 398507703 to press the bearing from the hub—Legacy/Outback

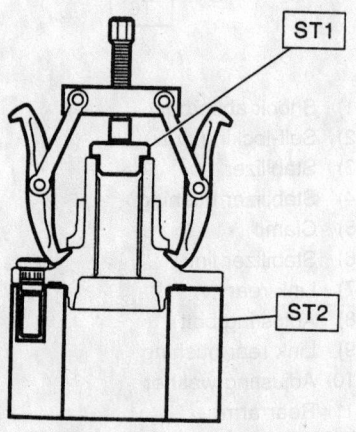

Place the hub assembly on stand ST 927080000 and use a common puller and tool ST 399520105, remove the bearing inner race—Legacy/Outback

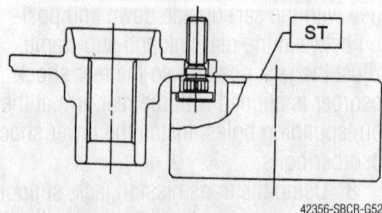

Using the hub stand tool, press out the hub bolt—Legacy/Outback

4. Disaasemble the hub/bearing assembly as follows:
   a. Place the hub/bearing in a press.
   b. Using dummy collar tool ST 398507703 to press the bearing from the hub.

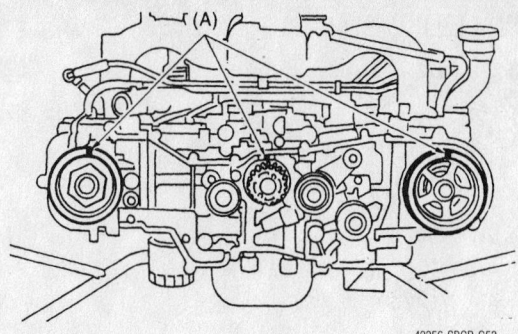

Installing the bearing into the hub—Legacy/Outback

42356-SBCR-G53

c. Place the hub assembly on hub stand ST 927080000 and using a common puller and tool ST 399520105, remove the inner race of the bearing. Discard the bearing as it should not be reused.

d. Using the hub stand tool, press out the hub bolt.

To install:

5. Install the bearing assembly to the hub as follows:

a. Press a new hub bolt into place using the hub stand. Make sure the bolt closely contacts the hub. use a 0.47 inch (12mm) hole in the hub stand to prevent the bolt from tilting while installing.

b. Using the hub stand, hub installer ST 927450000 and spacer ST 28499AE000, press the NEW bearing assembly into the hub. Make sure to always press on the inner race while installing.

6. Install or connect the following:

- Hub assembly with the mounting holes on the backing plate and install the hub assembly and backing plate. Temporarily tighten the axle nuts. be careful not to damage the tone ring.
- Four bolts and tighten to 48 ft. lbs. (66 Nm).

7. Remove the axle nut, use axle shaft installer tool ST 922431000 and adapter 927390000 to pull the axle shaft into position and hand tighten the axle nut

- Rotor
- Caliper
- ABS sensor
- Parking brake lever
- Tighten the axle nut to 174 ft. lbs. (235 Nm) and stake the nut.
- Wheel

8. Adjust the parking brake cable.

BRAKES

Brake Caliper

REMOVAL & INSTALLATION

WRX

FRONT

1. Before servicing the vehicle, refer to the precautions in the beginning of this section.

2. Remove or disconnect the following:
- Front wheels
- Brake hose from the caliper body
- Caliper retainer bolts and the caliper
- Caliper bracket, if necessary

To install:

3. Compress the piston assembly into the cylinder bore.

4. Install or connect the following:
- Caliper bracket to the spindle assembly, and secure in place with the retainer bolts. Tighten the retainer bolts to 59 ft. lbs. (80 Nm).
- Caliper and tighten the retainers to 19 ft. lbs. (26 Nm)
- Brake hose using new sealing washers, and tighten the fitting to 13 ft. lbs. (18 Nm)

5. Bleed the brake system.

6. Install the wheels and check the fluid level in the master cylinder.

REAR

1. Before servicing the vehicle, refer to the precautions in the beginning of this section.

2. Remove or disconnect the following:
- Rear wheels
- Brake hose from the caliper body
- Caliper bracket retainer bolts
- Caliper and bracket assembly off the rotor

To install:

3. Compress the piston assembly into the cylinder bore.

4. Install or connect the following:
- Caliper bracket to the spindle assembly, and secure in place with the retainer bolts. Tighten the retainer bolts to 27 ft. lbs. (37 Nm).
- Brake hose using new sealing washers, and tighten the fitting to 13 ft. lbs. (18 Nm)

5. Bleed the brake system.

6. Install the wheels and check the fluid level in the master cylinder.

Except WRX

FRONT

1. Before servicing the vehicle, refer to the precautions in the beginning of this section.

2. Remove or disconnect the following:
- Front wheels

- Brake hose from the caliper body
- Caliper retainer bolts and the caliper
- Caliper bracket, if necessary

To install:

3. Compress the piston assembly into the cylinder bore.

4. Install or connect the following:
- Caliper bracket to the spindle assembly, and secure in place with the retainer bolts. Tighten the retainer bolts to 58 ft. lbs. (78 Nm).
- Caliper and tighten the retainers to 29 ft. lbs. (40 Nm) on 2000–02 models and Outback models 19 ft. lbs. (26 Nm)
- Brake hose using new sealing washers, and tighten the fitting to 13 ft. lbs. (18 Nm)

5. Bleed the brake system.

6. Install the wheels and check the fluid level in the master cylinder.

7. Pump the brake pedal several times to seat the brakes before attempting to move the vehicle and road test the vehicle.

REAR

1. Before servicing the vehicle, refer to the precautions in the beginning of this section.

2. Install or connect the following:
- Rear wheels
- Brake hose from the caliper body

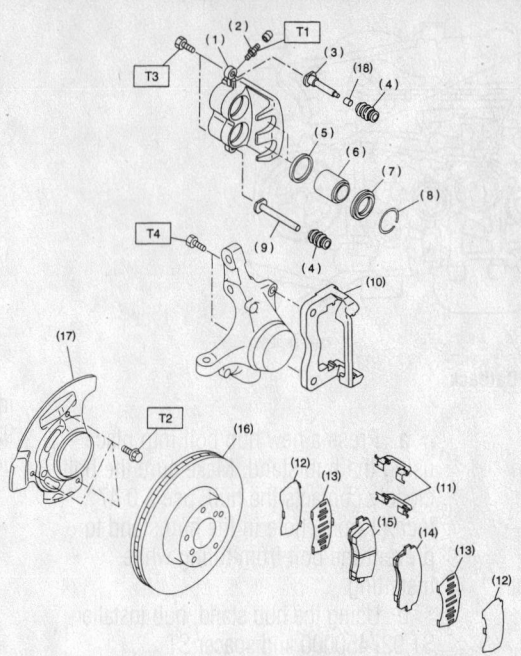

(1)	Caliper body	(9)	Lock pin (Yellow)	(17)	Disc cover
(2)	Air bleeder screw	(10)	Support	(18)	Bush
(3)	Guide pin (Green)	(11)	Pad clip		
(4)	Pin boot	(12)	Outer shim		
(5)	Piston seal	(13)	Inner shim		
(6)	Piston	(14)	Pad (Outside)		
(7)	Piston boot	(15)	Pad (Inside)		
(8)	Boot ring	(16)	Disc rotor		

Tightening torque: N·m (kgf-m, ft-lb)
T1: 8 (0.8, 5.8)
T2: 18 (1.8, 13.0)
T3: 37 (3.8, 27.5)
T4: 80 (8.2, 59)

9357TG65

Exploded view of the front disc brakes—Subaru WRX

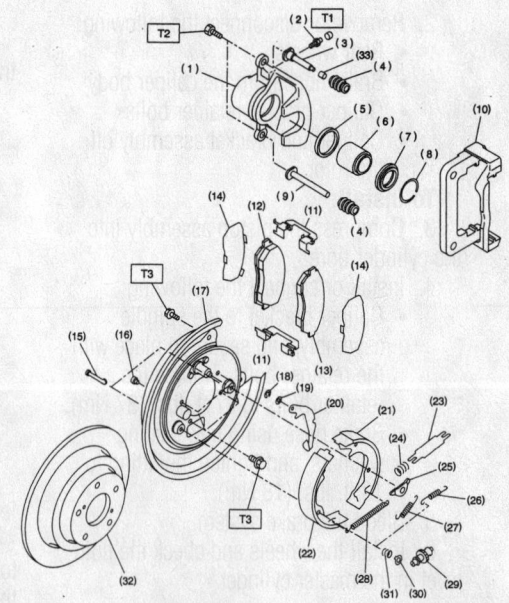

(1)	Caliper body	(14)	Shim	(27)	Primary shoe return spring
(2)	Air bleeder screw	(15)	Shoe hold-down pin	(28)	Adjusting spring
(3)	Guide pin (Green)	(16)	Cover	(29)	Adjuster
(4)	Pin boot	(17)	Back plate	(30)	Shoe hold-down cup
(5)	Piston seal	(18)	Retainer	(31)	Shoe hold-down spring
(6)	Piston	(19)	Spring washer	(32)	Disc rotor
(7)	Piston boot	(20)	Parking brake lever	(33)	Bush
(8)	Boot ring	(21)	Parking brake shoe (Secondary)		
(9)	Lock pin (Yellow)	(22)	Parking brake shoe (Primary)		
(10)	Support	(23)	Strut	**Tightening torque: N·m (kgf-m, ft-lb)**	
(11)	Pad clip	(24)	Strut shoe spring	T1: 8 (0.8, 5.8)	
(12)	Inner pad	(25)	Shoe guide plate	T2: 37 (3.8, 27.5)	
(13)	Outer pad	(26)	Secondary shoe return spring	T3: 53 (5.4, 39.1)	

9357TG66

Exploded view of the rear disc brakes—Subaru WRX

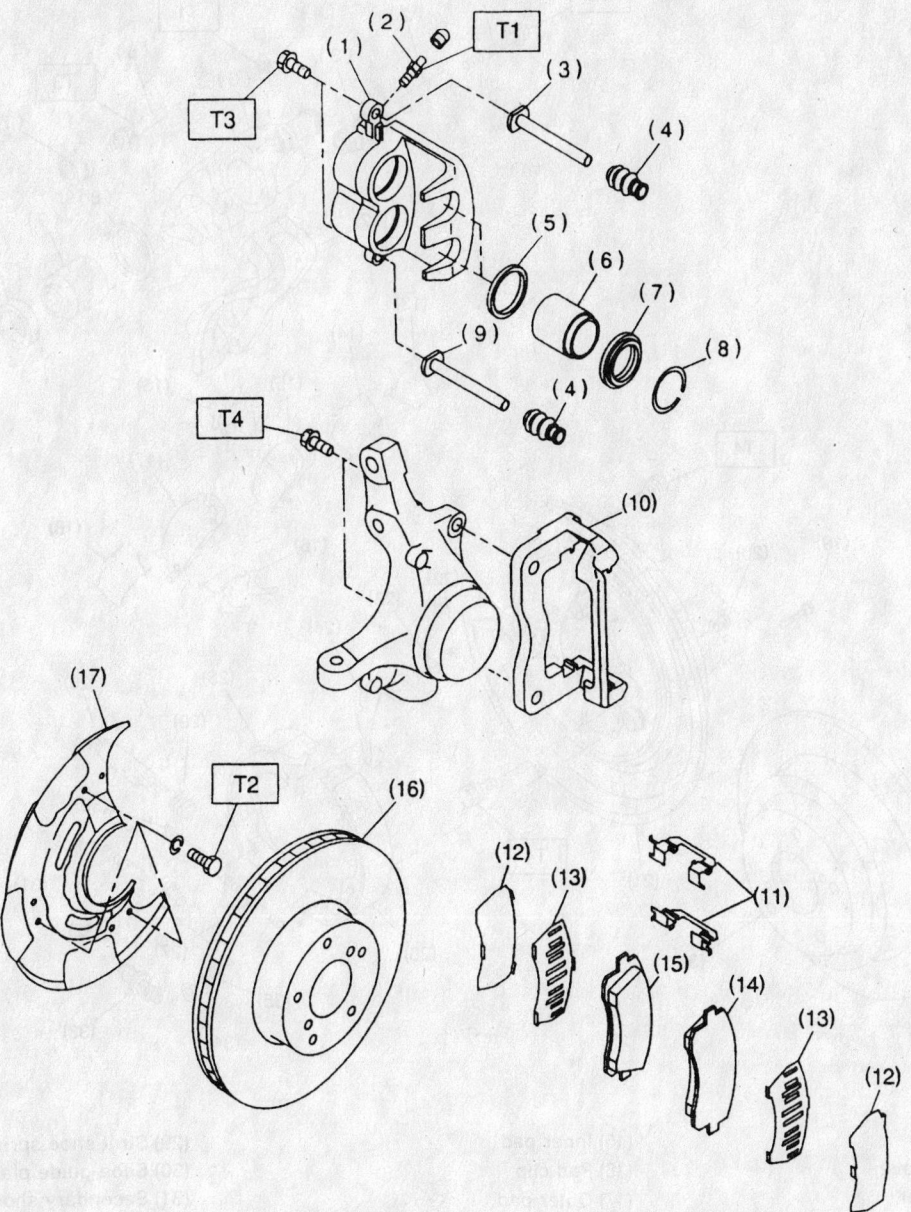

(1) Caliper body	(9) Lock pin (Yellow)	(17) Disc cover
(2) Air bleeder screw	(10) Support	
(3) Guide pin (Green)	(11) Pad clip	
(4) Pin boot	(12) Outer shim	
(5) Piston seal	(13) Inner shim	
(6) Piston	(14) Pad (Outside)	
(7) Piston boot	(15) Pad (Inside)	
(8) Boot ring	(16) Disc rotor	

Tightening torque: N·m (kg-m, ft-lb)

T1: 8±1 (0.8±0.1, 5.8±0.7)

T2: 18±5 (1.8±0.5, 13.0±3.6)

T3: 37±5 (3.8±0.5, 27.5±3.6)

T4: 78±10 (8.0±1.0, 58±7)

93016G60

Exploded view of the front disc brakes—Subaru models, except WRX

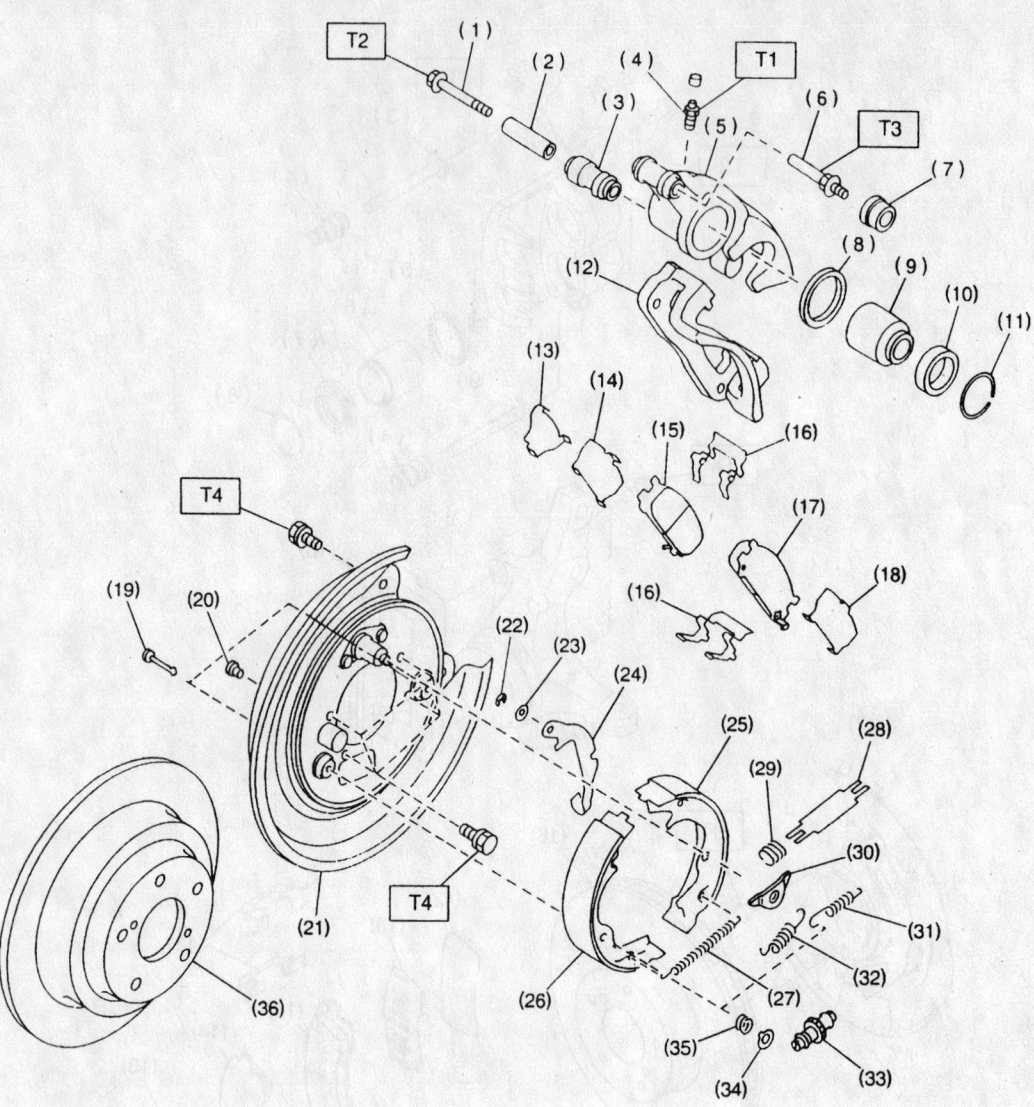

(1) Lock pin
(2) Lock pin sleeve
(3) Lock pin boot
(4) Air bleeder screw
(5) Caliper body
(6) Guide pin
(7) Guide pin boot
(8) Piston seal
(9) Piston
(10) Piston boot
(11) Boot ring
(12) Support
(13) Shim
(14) Inner shim

(15) Inner pad
(16) Pad clip
(17) Outer pad
(18) Outer shim
(19) Shoe hold-down pin
(20) Cover
(21) Back plate
(22) Retainer
(23) Spring washer
(24) Parking brake lever
(25) Parking brake shoe (Secondary)
(26) Parking brake shoe (Primary)
(27) Adjusting spring
(28) Strut

(29) Strut shoe spring
(30) Shoe guide plate
(31) Secondary shoe return spring
(32) Primary shoe return spring
(33) Adjuster
(34) Shoe hold-down cup
(35) Shoe hold-down spring
(36) Disc rotor

Tightening torque: N·m (kg-m, ft-lb)
T1: 8 ± 1 (0.8 ± 0.1, 5.8 ± 0.7)
T2: 20 ± 4 (2.0 ± 0.4, 14.5 ± 2.9)
T3: 26 ± 5 (2.7 ± 0.5, 19.5 ± 3.6)
T4: 52 ± 6 (5.3 ± 0.6, 38.3 ± 4.3)

93016G61

Exploded view of the rear disc brakes—Subaru models, except WRX

- Caliper bracket retainer bolts
- Caliper and bracket assembly off the rotor

To install:

3. Compress the piston assembly into the cylinder bore.

4. Install or connect the following:
- Caliper bracket to the spindle assembly, and secure in place with the retainer bolts, torque the bolts to 58 ft. lbs. (78 Nm)
- Caliper and tighten the retainers to 29 ft. lbs. (40 Nm) on 2000–02 models except Outback and 19 ft. lbs. (26 Nm) on Outback models
- Brake hose using new sealing washers, and tighten the fitting to 13 ft. lbs. (18 Nm)

5. Bleed the brake system. Install the wheels. Check the fluid level in the master cylinder.

6. Pump the brake pedal several times to seat the brakes before attempting to move the vehicle and road test the vehicle.

Brake Pads

REMOVAL & INSTALLATION

Except WRX

FRONT

1. Before servicing the vehicle, refer to the precautions in the beginning of this section.

2. Remove or disconnect the following:
- Wheels
- Lock pin bolts from the lower portion of the caliper
- Caliper by swinging it upward to access the pads
- Disc brake pads

To install:

3. Compress the caliper pistons.

4. Install or connect the following:
- New pads into the caliper brackets, being sure all shims and clips are in their original positions
- Caliper and tighten the retainers to 29 ft. lbs. (40 Nm) on 2000–02 models except Outback and 19 ft. lbs. (26 Nm) on Outback models
- Wheels

5. Check the fluid level in the master cylinder, pump the brake pedal several times to seat the brakes before attempting to move the vehicle and road test the vehicle.

REAR

1. Before servicing the vehicle, refer to the precautions in the beginning of this section.

2. Remove or disconnect the following:
- Wheels
- Small portion of brake fluid from the master cylinder reservoir
- Parking brake cable from the caliper lever, if equipped
- Lock pin bolts from the lower portion of the caliper
- Caliper by swinging it upward to access the pads
- Disc brake pads

To install:

3. Compress the caliper pistons.

4. Install or connect the following:
- New pads into the caliper brackets, being sure all shims and clips are in their original positions
- Caliper down into position and install the lock pin bolts. Tighten the lock pin bolts to 29 ft. lbs. (40 Nm).
- Parking brake cable
- Wheels

5. Check the fluid level in the master cylinder, pump the brake pedal several times to seat the brakes before attempting to move the vehicle and road test the vehicle.

WRX

FRONT

1. Before servicing the vehicle, refer to the precautions in the beginning of this section.

2. Remove or disconnect the following:
- Wheels
- Lock pin bolts from the lower portion of the caliper
- Caliper upward to access the pads
- Disc brake pads

To install:

3. Compress the caliper piston.

4. Install or connect the following:
- New pads into the caliper brackets, being sure all shims and clips are in their original positions
- Calipers down into position and install the lock pin bolts. Tighten the lock pin bolts to 19 ft. lbs. (26 Nm).
- Wheels

5. Check the fluid level in the master cylinder, pump the brake pedal several times to seat the brakes before attempting to move the vehicle and road test the vehicle.

REAR

1. Before servicing the vehicle, refer to the precautions in the beginning of this section.

2. Remove or disconnect the following:
- Portion of brake fluid from the master cylinder reservoir
- Wheels
- Lock pin bolts from the lower portion of the caliper
- Caliper upward to access the pads
- Disc brake pads

To install:

3. Compress the caliper piston.

4. Install or connect the following:
- New pads into the caliper bracket, being sure all shims and clips are in their original positions
- Caliper down into position and install the lock pin bolts. Tighten the lock pin bolt to 27 ft. lbs. (37 Nm).
- Wheels

5. Check the fluid level in the master cylinder, pump the brake pedal several times to seat the brakes before attempting to move the vehicle and road test the vehicle.

Brake Drums

REMOVAL & INSTALLATION

Except WRX

1. Before servicing the vehicle, refer to the precautions in the beginning of this section.

2. Remove or disconnect the following:
- Rear wheels
- Center cap by prying it off
- Cotter pin, castle nut, and center retainer washer
- Drum

To install:

3. Install or connect the following:
- Drum
- Center retainer washer and castle nut. Tighten the castle nut to 108 ft. lbs. (147 Nm). Install a new cotter pin.
- Center cap

4. Adjust the brake shoes.

5. Install the rear wheels.

WRX

1. Before servicing the vehicle, refer to the precautions in the beginning of this section.

2. Remove the rear wheels.

3. Release the parking brake.

4. If necessary, remove the adjusting hole cover from the backing plate and using a suitable tool, back off the shoe adjuster.

5. If the drum is difficult to remove, insert an 8mm bolt into the hole on the drum to push it off.

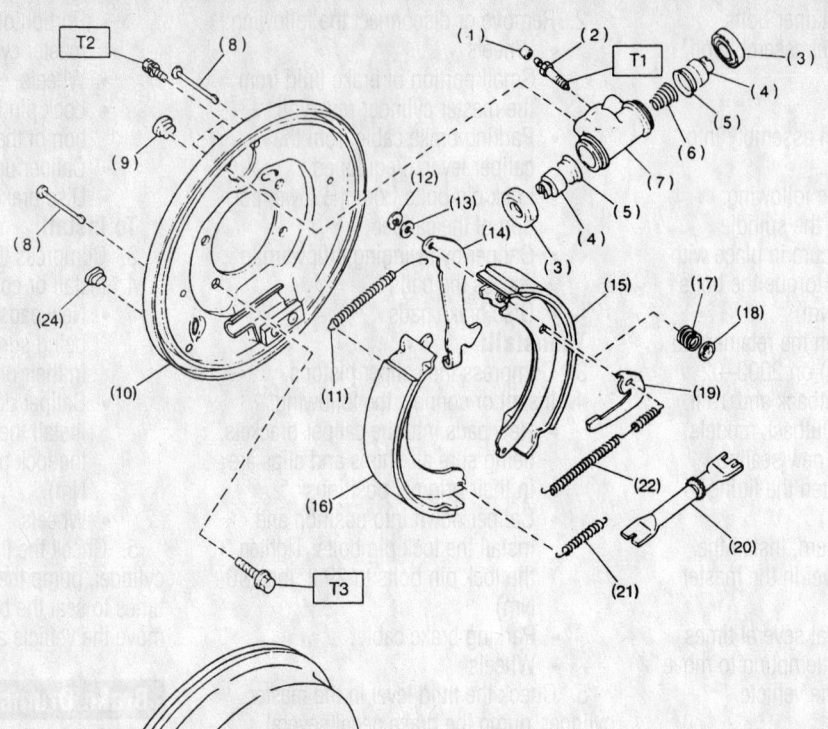

(1)	Air bleeder cap	(11)	Upper shoe return spring	(21) Lower shoe return spring
(2)	Air bleeder screw	(12)	Retainer	(22) Adjusting spring
(3)	Boot	(13)	Washer	(23) Drum
(4)	Piston	(14)	Parking brake lever	(24) Plug
(5)	Cup	(15)	Brake shoe (Trailing)	
(6)	Spring	(16)	Brake shoe (Leading)	
(7)	Wheel cylinder body	(17)	Shoe hold-down spring	
(8)	Pin	(18)	Cup	
(9)	Plug	(19)	Adjusting lever	
(10)	Back plate	(20)	Adjuster	

Tightening torque: N·m (kg-m, ft-lb)
T1: 8±1 (0.8±0.1, 5.8±0.7)
T2: 10±2 (1.0±0.2, 7.2±1.4)
T3: 52±6 (5.3±0.6, 38.3±4.3)

93016G62

Exploded view of a typical rear drum brake assembly

6. Remove the drum.
To install:
7. Install the drum.
8. Adjust the brake shoes.
9. Install the rear wheels.

Brake Shoes

REMOVAL & INSTALLATION

1. Before servicing the vehicle, refer to the precautions in the beginning of this section.
2. Remove or disconnect the following:
   • Wheels
   • Brake drum
   • Both return springs
   • Both retaining clips
   • Brake shoes from the adjuster side first, then the wheel cylinder side, and pull them off the backing plate
   • Parking brake cable from the parking lever on the trailing brake shoe, if equipped with rear drum parking brakes

To install:
3. Apply brake grease to the backing plate where the brake shoes contact it.
4. Install or connect the following:
   • Parking brake cable to the parking lever on the trailing brake shoe, if equipped with rear drum parking brakes
   • Brake shoes to the wheel cylinder, then to the adjuster. Secure in place with the 2 pins and retaining clips.
   • Return springs. The upper spring is thinner.
   • Drum and adjust the brake shoes
   • Wheels
5. Adjust the parking brake.
6. Check the fluid level in the master cylinder, pump the brake pedal several times to seat the brakes before attempting to move the vehicle and road test the vehicle.

SUBARU

Forrester

29

SPECIFICATION CHARTS

ENGINE AND VEHICLE IDENTIFICATION CHART

Code ①	Liters (cc)	Cu. In.	Cyl.	Fuel Sys.	Type	Eng. Mfg.	Code ②	Year
6	2.5 (2457)	150	4	MFI	DOHC	Subaru	Y	2000
							1	2001
							2	2002
							3	2003
							4	2004

MFI: Multiport Fuel Injection

DOHC: Double Overhead Camshafts

① 6th digit of the VIN.

② 10th digit of the VIN.

42356-SBFR-C01

GENERAL ENGINE SPECIFICATIONS

Year	Model	Engine Displacement Liters (cc)	Engine ID/VIN	Fuel System Type	Net Horsepower @ rpm	Net Torque @ rpm (ft. lbs.)	Bore x Stroke (in.)	Compression Ratio	Oil Pressure psi @ rpm
2000	Forester	2.5 (2457)	6	MFI	165@5600	162@4000	3.92x3.11	9.7:1	14@800
2001	Forester	2.5 (2457)	6	MFI	165@5600	162@4000	3.92x3.11	9.7:1	14@800
2002	Forester	2.5 (2457)	6	MFI	165@5600	162@4000	3.92x3.11	9.7:1	14@800
2003-04	Forester	2.5 (2457)	6	MFI	165@5600	162@4000	3.92x3.11	9.7:1	14@800

MFI: Multi-port Fuel Injection

42356-SBFR-C02

ENGINE TUNE-UP SPECIFICATIONS

Year	Engine Displacement Liters (cc)	Engine ID/VIN	Spark Plugs Gap (in.)	Ignition Timing (deg.) ① MT	AT	Fuel Pump (psi)	Idle Speed (rpm) ② MT	AT	Valve Clearance ③ In.	Ex.
2000	2.5 (2457)	6	0.039-0.043	7-23 BTDC	7-23 BTDC	34-38	600-800	600-800	0.0071-0.0087	0.0090-0.0106
2001	2.5 (2457)	6	0.039-0.043	7-23 BTDC	7-23 BTDC	34-38	600-800	600-800	0.0071-0.0087	0.0090-0.0106
2002	2.5 (2457)	6	0.039-0.043	7-23 BTDC	7-23 BTDC	34-38	600-800	600-800	0.0071-0.0087	0.0090-0.0106
2003-04	2.5 (2457)	6	0.039-0.043	7-23 BTDC	7-23 BTDC	34-38	600-800	600-800	0.0071-0.0087	0.0090-0.0106

BTDC: Before Top Dead Center

① At idle speed.

② With engine under no load.

③ With engine cold.

42356-SBFR-C03

2.5L engine
Firing order: 1–3–2–4
Distributorless ignition system

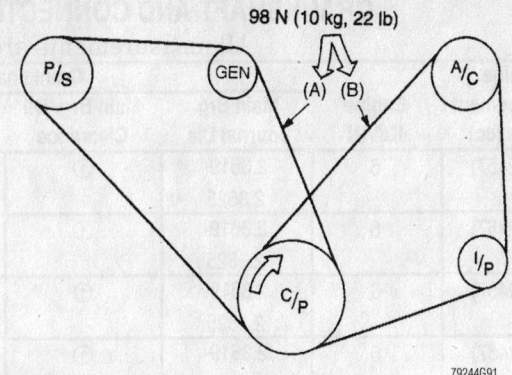

98 N (10 kg, 22 lb)

Accessory V-belt routing—2.5L engine

CAPACITIES

Year	Model	Engine Displacement Liters (cc)	Engine ID/VIN	Engine Oil with Filter (qts.)	Transmission (pts.)			Transfer Case (pts.)	Drive Axle		Fuel Tank (gal.)	Cooling System (qts.)
					4-Spd	5-Spd	Auto.		Front (pts.)	Rear (pts.)		
2000	Forester	2.5 (2457)	6	4.2	—	7.4	20	—	2.6 ①	1.6	15.9	6.3
2001	Forester	2.5 (2457)	6	4.2	—	7.4	20	—	2.6 ①	1.6	15.9	6.3
2002	Forester	2.5 (2457)	6	4.2	—	7.4	20	—	2.6 ①	1.6	15.9	6.3
2003-04	Forester	2.5 (2457)	6	4.2	—	7.4	20	—	2.6 ①	1.6	15.9	6.3

① A/T differential only.

42356-SBFR-C04

VALVE SPECIFICATIONS

Year	Engine Displacement Liters (cc)	Engine ID/VIN	Seat Angle (deg.)	Face Angle (deg.)	Spring Test Pressure (lbs. @ in.)	Spring Installed Height (in.)	Stem-to-Guide Clearance (in.)		Stem Diameter (in.)	
							Intake	Exhaust	Intake	Exhaust
2000	2.5 (2457)	6	①	①	33-38 @ 1.654 ②	③	0.0014- ③ 0.0024	0.0016- ④ 0.0026	0.2343- 0.2348	0.2343- 0.2348
2001	2.5 (2457)	6	①	①	33-38 @ 1.654 ②	③	0.0014- ③ 0.0024	0.0016- ④ 0.0026	0.2343- 0.2348	0.2343- 0.2348
2002	2.5 (2457)	6	①	①	33-38 @ 1.654 ②	③	0.0014- ③ 0.0024	0.0016- ④ 0.0026	0.2343- 0.2348	0.2343- 0.2348
2003-04	2.5 (2457)	6	①	①	33-38 @ 1.654 ②	③	0.0014- ③ 0.0024	0.0016- ④ 0.0026	0.2343- 0.2348	0.2343- 0.2348

① Refacing angle: 90 degrees
② 102-118 lbs. @ 1.315 in.
③ Free length: 1.8913 in.
④ Wear limit: 0.0059 in.

42356-SBFR-C05

CRANKSHAFT AND CONNECTING ROD SPECIFICATIONS
All measurements are given in inches.

Year	Engine Displacement Liters (cc)	Engine ID/VIN	Crankshaft				Connecting Rod		
			Main Brg. Journal Dia.	Main Brg. Oil Clearance	Shaft End-play	Thrust on No.	Journal Diameter	Oil Clearance	Side Clearance
2000	2.5 (2457)	6	2.3619-2.3625	①	0.0012-0.0098	3	1.8891-1.8898	0.0004-0.0020	0.0028-0.0160
2001	2.5 (2457)	6	2.3619-2.3625	①	0.0012-0.0098	3	1.8891-1.8898	0.0004-0.0020	0.0028-0.0160
2002	2.5 (2457)	6	2.3619-2.3625	①	0.0012-0.0098	3	1.8891-1.8898	0.0004-0.0020	0.0028-0.0160
2003-04	2.5 (2457)	6	2.3619-2.3625	①	0.0012-0.0098	3	1.8891-1.8898	0.0004-0.0020	0.0028-0.0160

① Journals 1 and 5: 0.0001-0.0016
Journals 2 and 4: 0.0004-0.0018
Journal 3: 0.0004-0.0016

42356-SBFR-C06

PISTON AND RING SPECIFICATIONS
All measurements are given in inches.

Year	Engine Displacement Liters (cc)	Engine ID/VIN	Piston Clearance	Ring Gap			Ring Side Clearance		
				Top Compression	Bottom Compression	Oil Control	Top Compression	Bottom Compression	Oil Control
2000	2.5 (2457)	6	0.0004-0.0020	0.0079-0.0390	0.0146-0.0390	0.0079-0.0590	0.0016-0.0059	0.0012-0.0059	NA
2001	2.5 (2457)	6	0.0004-0.0020	0.0079-0.0390	0.0146-0.0390	0.0079-0.0590	0.0016-0.0059	0.0012-0.0059	NA
2002	2.5 (2457)	6	0.0004-0.0020	0.0079-0.0390	0.0146-0.0390	0.0079-0.0590	0.0016-0.0059	0.0012-0.0059	NA
2003-04	2.5 (2457)	6	0.0004-0.0020	0.0079-0.0390	0.0146-0.0390	0.0079-0.0590	0.0016-0.0059	0.0012-0.0059	NA

NA: Not Available

42356-SBFR-C07

TORQUE SPECIFICATIONS
All readings in ft. lbs.

Year	Engine Displacement Liters (cc)	Engine ID/VIN	Cylinder Head Bolts	Main Bearing Bolts	Rod Bearing Bolts	Crankshaft Damper Bolts	Flywheel Bolts	Manifold		Spark Plugs	Lug Nut
								Intake	Exhaust		
2000	2.5 (2457)	6	①	②	31-34	123-137	51-55	14-17	19-26 ③	13-17	58-72
2001	2.5 (2457)	6	①	②	31-34	123-137	51-55	14-17	19-26 ③	13-17	58-72
2002	2.5 (2457)	6	①	②	31-34	123-137	51-55	14-17	19-26 ③	13-17	58-72
2003-04	2.5 (2457)	6	①	②	31-34	123-137	51-55	14-17	19-26 ③	13-17	58-72

① Step 1: Tighten all bolts, in sequence, to 22 ft. lbs.
Step 2: Tighten all bolts, in sequence, to 51 ft. lbs.
Step 3: Loosen all bolts 180 degrees (one-half turn)
Step 4: Loosen all bolts another 180 degrees (one-half turn)
Step 5: Tighetn bolts A and B, in sequence, to 25 ft. lbs.
Step 6: Tighten bolts C, D, E and F, in sequence, to 11 ft. lbs.
Step 7: Tighten all bolts, in sequence, 80-90 degrees
Step 8: Tighten all bolts, in sequence, another 80-90 degrees

② Split engine case bolts:
10mm bolts: 33-37 ft. lbs.
8mm bolts: A thru G to 17-20 ft. lbs. and H to 5 ft. lbs.

③ No separate exhaust manifold is used, the front pipe bolts directly to the cylinder heads

42356-SBFR-C08

WHEEL ALIGNMENT

Year	Model		Caster Range (+/-Deg.)	Caster Preferred Setting (Deg.)	Camber Range (+/-Deg.)	Camber Preferred Setting (Deg.)	Toe-in (in.)	Steering Axis Inclination (Deg.)
2000	Forester	F	0.75	+2.58	0.50	-0.25	0+/-0.12	—
		R	—	—	0.75	-0.58	0.08+/-0.04	—
2001	Forester	F	0.75	+2.58	0.50	-0.25	0+/-0.12	—
		R	—	—	0.75	-0.58	0.08+/-0.04	—
2002	Forester	F	0.75	+2.58	0.50	-0.25	0+/-0.12	—
		R	—	—	0.75	-0.58	0.08+/-0.04	—
2003-04	Forester	F	0.75	+2.58	0.50	-0.25	0+/-0.12	—
		R	—	—	0.75	-0.58	0.08+/-0.04	—

42356-SBFR-C09

TIRE, WHEEL AND BALL JOINT SPECIFICATIONS

Year	Model	OEM Tires Standard	OEM Tires Optional	Tire Pressures (psi) Front	Tire Pressures (psi) Rear	Wheel Size	Ball Joint Inspection
2000	Forester	P205/70R15 95S	P215/60R16 94H	29	26 ①	②	0.012 in. ③
2001	Forester	P205/70R15 95S	P215/60R16 94H	29	26 ①	②	0.012 in. ③
2002	Forester	P205/70R15 95S	P215/60R16 94H	29	26 ①	②	0.012 in. ③
2003-04	Forester	P205/70R15 95S	P215/60R16 94H	29	26 ①	②	0.012 in. ③

OEM: Original Equipment Manufacturer

PSI: Pounds Per Square Inch

STD: Standard

OPT: Optional

① With full load: 36 psi.

② With standard tires: 6-JJ

With optional tires: 6.5-JJ

③ Apply 154 lbs. vertical force

42356-SBFR-C10

BRAKE SPECIFICATIONS
All measurements in inches unless noted

Year	Model		Brake Disc Original Thickness	Brake Disc Minimum Thickness	Brake Disc Maximum Runout	Brake Drum Diameter Original Inside Diameter	Brake Drum Diameter Max. Wear Limit	Maximum Machine Diameter	Minimum Lining Thickness Front	Minimum Lining Thickness Rear	Brake Caliper Bracket Bolts (ft. lbs.)	Brake Caliper Mounting Bolts (ft. lbs.)
2000	Forester	F	0.945	0.660	0.003	—	—	—	0.295	0.256	59	28
		R	0.390	0.335	0.0028	9.00 ①	9.08 ②	NA	—	0.059	38	28
2001	Forester	F	0.945	0.660	0.003	—	—	—	0.295	0.256	59	28
		R	0.390	0.335	0.0028	9.00 ①	9.08 ②	NA	—	0.059	38	28
2002	Forester	F	0.945	0.660	0.003	—	—	—	0.295	0.256	59	28
		R	0.390	0.335	0.0028	9.00 ①	9.08 ②	NA	—	0.059	38	28
2003-04	Forester	F	0.940	0.870	0.003	—	—	—	0.06	0.059	59	28
		R	0.390	0.335	0.0027	9.00 ①	9.08 ②	NA	—	0.059	38	28

NA: Not Available

① Parking brake drum on vehicles with rear disc brakes: 6.69 in.

② Parking brake drum on vehicles with rear disc brakes: 6.73 in.

42356-SBFR-C11

SCHEDULED MAINTENANCE INTERVALS
SUBARU—FORESTER

TO BE SERVICED	TYPE OF SERVICE	3	7.5	15	22.5	30	37.5	45	52.5	60	67.5	75	82.5	90	97.5	105	112.5	120
		VEHICLE MILEAGE INTERVAL (x1000)																
Accessory drive belts	R									✓								✓
Accessory drive belts	S/I					✓								✓				
Air cleaner filter	R					✓				✓				✓				✓
Automatic transmission fluid	S/I					✓				✓				✓				✓
Axle shaft joints	S/I			✓				✓		✓		✓		✓		✓		✓
Brake fluid	R					✓				✓				✓				✓
Brake system lines	S/I			✓				✓		✓		✓		✓		✓		✓
Clutch operation	S/I			✓		✓		✓		✓		✓		✓		✓		✓
Disc brake pads & rotors	S/I			✓				✓		✓		✓		✓		✓		✓
Drums brake linings & drums	S/I					✓				✓				✓				✓
Engine coolant	R					✓				✓				✓				✓
Engine cooling system, hoses & connections	S/I					✓				✓				✓				✓
Engine oil & filter	R	✓	✓	✓	✓	✓	✓	✓	✓	✓	✓	✓	✓	✓	✓	✓	✓	✓
Front & rear axle boots	S/I			✓				✓		✓		✓		✓		✓		✓
Front & rear wheel bearings	S/I & L									✓								✓
Fuel filter	R					✓				✓				✓				✓
Parking & service brake systems' operation	S/I			✓		✓		✓		✓		✓		✓		✓		✓
Spark plugs	R									✓								✓
Steering & suspension	S/I			✓		✓		✓		✓		✓		✓		✓		✓
Timing belt	R															✓		
Timing belt	S/I					✓				✓				✓				
Transmission & differential fluid levels	S/I					✓				✓				✓				✓
Valve clearance	S/I															✓		

R: Replace   S/I: Inspect and service, if needed   L: Lubricate

FREQUENT OPERATION MAINTENANCE (SEVERE SERVICE)

If a vehicle is operated under any of the following conditions it is considered severe service:

- Towing a trailer or using a camper or car-top carrier.
- Repeated short trips of less than 5 miles in temperatures below freezing, or trips of less than 10 miles in any temperature.
- Extensive idling or low-speed driving for long distances as in heavy commercial use, such as delivery, taxi or police cars.
- Operating on rough, muddy or salt-covered roads, or extensive mountain driving.
- Operating on unpaved or dusty roads.
- Driving in extremely hot (over 90°) conditions.

Engine oil and filter: replace every 3000 miles or 3 months, whichever occurs first.

Fuel filter: replace every 7500 miles or 7.5 months, whichever occurs first.

Fuel system, hoses & connections: inspect every 7500 miles or 7.5 months, whichever occurs first.

Transmission & differential fluid: replace every 15,000 miles.

Automatic transmission fluid: replace every 15,000 miles.

Brake fluid: replace every 15,000 miles.

Disc brake pads & rotors: inspect every 7500 miles or 7.5 months, whichever occurs first.

Front & rear axle boots: inspect every 7500 miles or 7.5 months, whichever occurs first.

Axle shaft boots: inspect every 7500 miles or 7.5 months, whichever occurs first.

Drum brake linings & drums: inspect every 7500 miles or 7.5 months, whichever occurs first.

Brake lines: inspect every 7500 miles or 7.5 months, whichever occurs first.

Parking & service brake system operation: inspect every 7500 miles or 7.5 months, whichever occurs first.

Clutch operation: inspect every 7500 miles or 7.5 months, whichever occurs first.

42356-SBFR-C12

PRECAUTIONS

Before servicing any vehicle, please be sure to read all of the following precautions, which deal with personal safety, prevention of component damage, and important points to take into consideration when servicing a motor vehicle:

• Never open, service or drain the radiator or cooling system when the engine is hot; serious burns can occur from the steam and hot coolant.

• Observe all applicable safety precautions when working around fuel. Whenever servicing the fuel system, always work in a well-ventilated area. Do not allow fuel spray or vapors to come in contact with a spark, open flame or excessive heat (a hot drop light, for example). Keep a dry chemical fire extinguisher near the work area. Always keep fuel in a container specifically designed for fuel storage; also, always properly seal fuel containers to avoid the possibility of fire or explosion. Refer to the additional fuel system precautions later in this section.

• Fuel injection systems often remain pressurized, even after the engine has been turned **OFF**. The fuel system pressure must be relieved before disconnecting any fuel lines. Failure to do so may result in fire and/or personal injury.

• Brake fluid often contains polyglycol ethers and polyglycols. Avoid contact with the eyes and wash your hands thoroughly after handling brake fluid. If you do get brake fluid in your eyes, flush your eyes with clean, running water for 15 minutes. If eye irritation persists, or if you have taken brake fluid internally, IMMEDIATELY seek medical assistance.

• The EPA warns that prolonged contact with used engine oil may cause a number of skin disorders, including cancer! You should make every effort to minimize your exposure to used engine oil. Protective gloves should be worn when changing oil. Wash your hands and any other exposed skin areas as soon as possible after exposure to used engine oil. Soap and water, or waterless hand cleaner should be used.

• All new vehicles are now equipped with an air bag system. The system must be disabled before performing service on or around system components, steering column, instrument panel components, wiring and sensors. Failure to follow safety and disabling procedures could result in accidental air bag deployment, possible personal injury and unnecessary system repairs.

• Always wear safety goggles when working with, or around, the air bag system. When carrying a non-deployed air bag, be sure the bag and trim cover are pointed away from your body. When placing a non-deployed air bag on a work surface, always face the bag and trim cover upward, away from the surface. This will reduce the motion of the module if it is accidentally deployed. Refer to the additional air bag system precautions later in this section.

• Clean, high quality brake fluid from a sealed container is essential to the safe and proper operation of the brake system. You should always buy the correct type of brake fluid for your vehicle. If the brake fluid becomes contaminated, completely flush the system with new fluid. Never reuse any brake fluid. Any brake fluid that is removed from the system should be discarded. Also, do not allow any brake fluid to come in contact with a painted surface; it will damage the paint.

• Never operate the engine without the proper amount and type of engine oil; doing so WILL result in severe engine damage.

• Timing belt maintenance is extremely important! Many models utilize an interference-type, non-freewheeling engine. If the timing belt breaks, the valves in the cylinder head may strike the pistons, causing potentially serious (also time-consuming and expensive) engine damage.

• Disconnecting the negative battery cable on some vehicles may interfere with the functions of the on-board computer system(s) and may require the computer to undergo a relearning process once the negative battery cable is reconnected.

• When servicing drum brakes, only disassemble and assemble one side at a time, leaving the remaining side intact for reference.

• Only an MVAC-trained, EPA-certified automotive technician should service the air conditioning system or its components.

ENGINE REPAIR

➡**Disconnecting the negative battery cable on some vehicles may interfere with the functions of the on board computer systems and may require the computer to undergo a relearning process, once the negative battery cable is reconnected.**

Distributor

The Forester is equipped with a distributorless ignition system.

Alternator

REMOVAL & INSTALLATION

1. Before servicing the vehicle, refer to the precautions in the beginning of this section.

2. Remove or disconnect the following:
• Negative battery cable
• Wires
• Belt cover
• Drive belt
• Mounting bolts
• Alternator
To install:
Install or connect the following:
• Alternator
• Mounting bolts
• Drive belt. Adjust the tension to 0.276–0.354 in. (7–9mm) for new or 0.354–0.433 in. (9–11mm) for used. Torque the slider bolt to 4–6 ft. lbs. (6–10 Nm) and the lockbolt to 16.5 ft. lbs. (19.5 Nm).
• Belt cover
• Wires
• Negative battery cable

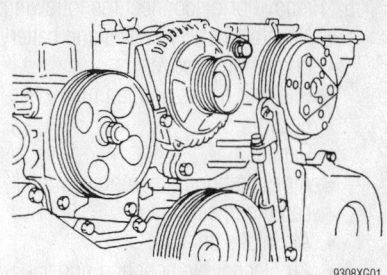

View of alternator mounting

Ignition Timing

ADJUSTMENT

The ignition timing is controlled by the engine control computer and is not adjustable. To check the ignition timing proceed as follows:

1. Before servicing the vehicle, refer to the precautions in the beginning of this section.

2. Warm up the engine, then turn the ignition **OFF**.

3. Connect a timing light to the No. 1 spark plug wire according to the manufactures directions.

4. Start the engine. With the vehicle at idle check the timing.

5. The timing should be 7–23 degrees Before Top Dead Center (BTDC) at 700 RPM on vehicles with an automatic transmission and 2–18 degrees at 650 RPM on models with a manual transmission.

6. If the timing is not correct, there could be a problem in the ignition control system.

Engine Assembly

REMOVAL & INSTALLATION

2000–01 Models

※ CAUTION

The fuel injection system remains under pressure after the engine has been turned OFF. Properly relieve fuel pressure before disconnecting any fuel lines. Failure to do so may result in fire or personal injury.

1. Before servicing the vehicle, refer to the precautions in the beginning of this section.

2. Relieve the fuel system pressure.

3. Drain the engine oil and coolant.

4. Discharge and recover the air conditioning system.

5. Remove or disconnect the following:
- Negative battery cables and battery
- Engine undercover
- Radiator hoses and fan motor harness
- Radiator
- Air conditioning compressor and cap the lines
- Air intake duct
- Air cleaner element and upper cover
- Evaporative Emissions (EVAP) canister and bracket
- Front Oxygen (O$_2$S) sensor

6. If equipped with California emissions specifications, disconnect the rear O$_2$S sensor.

7. Remove or disconnect the following:
- Engine ground terminal
- Crankshaft Position (CKP) sensor connector

- Camshaft Position (CMP) sensor connector
- Knock Sensor (KS) connector
- Alternator connector and terminal
- Air conditioning compressor connectors, if equipped
- Accelerator cable
- Cruise control cable, if equipped
- Brake booster hose
- Heater inlet and outlet hoses
- Alternator drive belt
- Wires from the spark plugs on the left side of the engine
- Power steering pump line bracket
- Power steering pump, leaving the lines connected and position it aside
- Exhaust Y-pipe
- Lower starter nuts
- Lower engine-to-transmission nuts
- Front engine mount-to-crossmember nuts
- Starter

8. If equipped with an automatic transmission, perform the following:

a. Remove the torque converter service hole plug.

b. Matchmark the torque converter-to-driveplate.

c. Rotate the engine to remove the torque converter-to-driveplate bolts as they become accessible.

9. Remove or disconnect the following:
- Flywheel cover, if equipped with a manual transmission
- Pitching stopper
- Fuel delivery, return and evaporation hoses

10. Support the engine with a suitable lifting device attached to the engine lifting eyes.

11. Slightly raise the engine.

12. Raise the transmission with a floor jack.

13. If equipped with a manual transmission, pull the engine forward then up and out of the vehicle to clear the transmission mainshaft.

14. If equipped with an automatic transmission, pull the engine forward then up and out of the vehicle.

To install:

15. If equipped with a manual transmission, apply a small amount of grease to the splines of the mainshaft.

16. Position the engine in the engine compartment and align it with the transmission.

17. Install the engine. Torque the upper bolts to 37 ft. lbs. (50 Nm).

18. Remove the lifting device and floor jack.

19. Install or connect the following:
- Pitching stopper. Torque the bolts to 43 ft. lbs. (58 Nm) on the body side and 37 ft. lbs. (50 Nm) on the bracket side.
- Flywheel cover, if equipped with a manual transmission

20. If equipped with an automatic transmission, perform the following:

a. Align the matchmarks, install the torque converter-to-driveplate bolts while rotating the engine and tighten to 18 ft. lbs. (25 Nm).

b. Install the service hole cover.

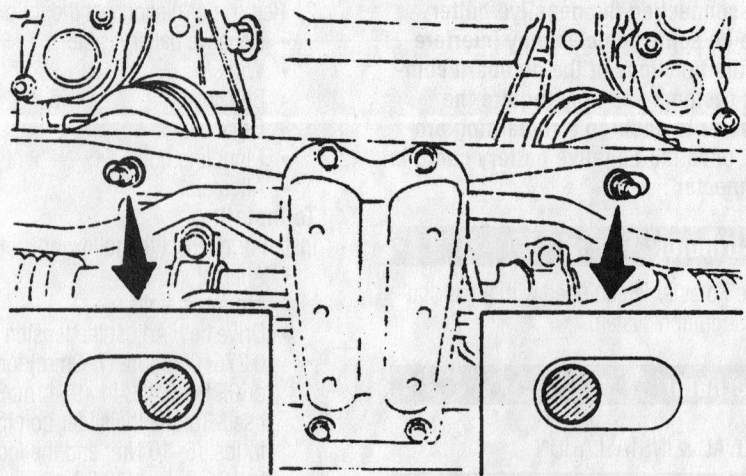

7924XG25

Be sure to tighten the front cushion rubber mounting bolts in the innermost elliptical hole in the front crossmember—2000–01 Models

21. Install or connect the following:
- EVAP canister and bracket
- Power steering pump. Tighten the retainer bolts to 14 ft. lbs. (20 Nm).
- Accessory drive belt
- Starter. Torque the bolts to 34–40 ft. lbs. (44–52 Nm).
- Lower engine-to-transmission nuts. Tighten them to 36 ft. lbs. (50 Nm).
- Lower engine mounting nuts. Tighten them to 61 ft. lbs. (83 Nm) in the inner most elliptical hole in the front crossmember so the clearance is 0.16–0.24 in. (4–6mm).
- Exhaust Y-pipe with new gaskets and nuts
- Brake booster hose
- Heater inlet and outlet hoses
- Accelerator cable
- Cruise control cable, if equipped
- Engine harness connectors
- Engine ground terminal
- CKP sensor connector
- CMP sensor connector
- Knock sensor connector
- Alternator connector and terminal
- Air conditioning compressor connectors, if equipped
- Front O_2S sensor, and if removed, the rear O_2S sensor.
- Air cleaner element and cover
- Air conditioning lines with new O-rings, if equipped. Torque the bolts to 23 ft. lbs. (31 Nm).
- Radiator
- Engine undercover
- Negative battery cable

22. Fill the crankcase to the proper level with clean engine oil.

23. Fill and bleed the cooling system.

24. Charge the air conditioning system using an approved recovery/recycling machine.

25. If equipped, check the automatic transmission fluid level and add Dexron®II if necessary.

26. Start the engine and allow it to reach normal operating temperature. Check for leaks.

2002–04 Models

1. Before servicing the vehicle, refer to the precautions in the beginning of this section.

2. Relieve the fuel system pressure.

3. Drain the engine oil and coolant.

4. Discharge and recover the air conditioning system.

5. Remove or disconnect the following:
- Negative battery cables and battery
- Air cleaner assembly
- Engine undercover
- Radiator hoses and fan motor harness
- Radiator
- Air conditioning compressor and cap the lines
- Air cleaner case stay
- Front Oxygen (O_2S) sensor

6. If equipped with California emissions specifications, disconnect the rear O_2S sensor.

7. Remove or disconnect the following:
- Engine ground terminal
- Crankshaft Position (CKP) sensor connector
- Camshaft Position (CMP) sensor connector
- Knock Sensor (KS) connector
- Alternator connector and terminal
- Air conditioning compressor connectors, if equipped
- Accelerator cable
- Cruise control cable, if equipped
- Pressure Switch
- Brake booster hose
- Heater inlet and outlet hoses
- Resonator chamber
- Front side V–belt
- Pipe with bracket from the intake manifold
- Power steering pump, leaving the lines connected and position it aside
- Exhaust Y-pipe
- Lower starter nuts
- Lower engine-to-transmission nuts
- Front engine mount-to-crossmember nuts

8. If equipped with an automatic transmission, perform the following:
  a. Remove the torque converter service hole plug.
  b. Matchmark the torque converter-to-driveplate.
  c. Rotate the engine to remove the torque converter-to-driveplate bolts as they become accessible.

9. Remove or disconnect the following:
- Flywheel cover, if equipped with a manual transmission
- Pitching stopper
- Fuel delivery, return and evaporation hoses

10. Support the engine with a suitable lifting device attached to the engine lifting eyes.

11. Slightly raise the engine.

12. Raise the transmission with a floor jack.
- Starter
- Upper engine-to-transmission bolts

13. If equipped with a manual transmission, pull the engine forward then up and out of the vehicle to clear the transmission mainshaft.

14. If equipped with an automatic transmission, pull the engine forward then up and out of the vehicle.

To install:

15. If equipped with a manual transmission, apply a small amount of grease to the splines of the mainshaft.

16. Position the engine in the engine compartment and align it with the transmission.

17. Install the engine. Torque the upper bolts to 37 ft. lbs. (50 Nm).

18. Remove the lifting device and floor jack.

19. Install or connect the following:
- Pitching stopper. Torque the bolts to 43 ft. lbs. (58 Nm) on the body side and 37 ft. lbs. (50 Nm) on the bracket side.
- Starter
- Flywheel cover, if equipped with a manual transmission

20. If equipped with an automatic transmission, perform the following:
  a. Align the matchmarks, install the torque converter-to-driveplate bolts while rotating the engine and tighten to 18 ft. lbs. (25 Nm).
  b. Install the service hole cover.

21. Install or connect the following:
- Power steering pump. Tighten the retainer bolts to 14 ft. lbs. (20 Nm).
- Power steering switch connector
- Accessory drive belt
- Resonataor chamber and tighten the bolts to 24 ft. lbs. (33 Nm)
- Lower engine-to-transmission

42356-SBFR-G01

Attach a suitable lift device to the engine

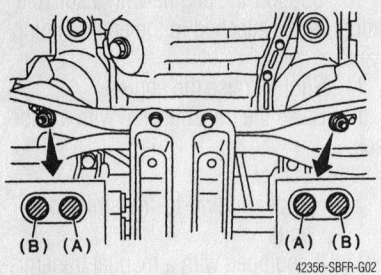

42356-SBFR-G02

Be sure to tighten the front cushion rubber mounting bolts in the innermost elliptical hole in the front crossmember—2002–04 Models

nuts. Tighten them to 36 ft. lbs. (50 Nm).

- Lower engine mounting nuts. Tighten them to 63 ft. lbs. (85 Nm) in the inner most elliptical hole in the front crossmember so the clearance is 0.16–0.24 in. (4–6mm).

- Exhaust Y-pipe with new gaskets and nuts
- Brake booster hose
- Heater inlet and outlet hoses
- Accelerator cable
- Cruise control cable, if equipped
- Engine harness connectors
- Engine ground terminal
- CKP sensor connector
- CMP sensor connector
- Knock sensor connector
- Alternator connector and terminal
- Air conditioning compressor connectors, if equipped
- Front O_2S sensor, and if removed, the rear O_2S sensor.
- Air cleaner element and cover
- Air conditioning lines with new O-rings, if equipped. Torque the bolts to 18 ft. lbs. (25 Nm).
- Radiator
- Engine undercover
- Negative battery cable

22. Fill the crankcase to the proper level with clean engine oil.

23. Fill and bleed the cooling system.

24. Charge the air conditioning system using an approved recovery/recycling machine.

25. If equipped, check the automatic transmission fluid level and add Dexron®II if necessary.

26. Start the engine and allow it to reach normal operating temperature. Check for leaks.

Water Pump

REMOVAL & INSTALLATION

1. Before servicing the vehicle, refer to the precautions in the beginning of this section.

2. Drain the coolant into a suitable container.

3. Remove or disconnect the following:

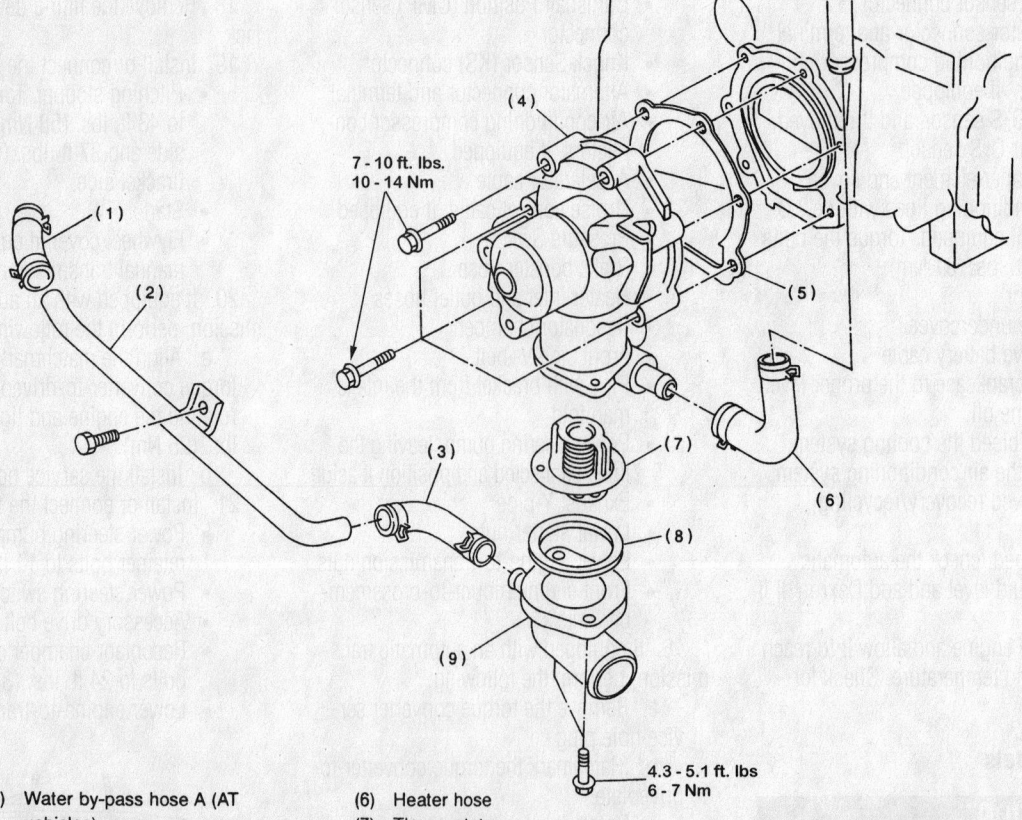

7 - 10 ft. lbs.
10 - 14 Nm

4.3 - 5.1 ft. lbs
6 - 7 Nm

(1) Water by-pass hose A (AT vehicles)
(2) Water by-pass pipe (AT vehicles)
(3) Water by-pass hose B (AT vehicles)
(4) Water pump ASSY
(5) Gasket

(6) Heater hose
(7) Thermostat
(8) Gasket
(9) Thermostat case

Exploded view of the water pump mounting and related components

7924XG01

- Negative battery cable
- Engine undercover
- Electrical connections from the radiator fan motor and sub motor(s)
- Bolt that retains the water by-pass pipe of the oil cooler onto the oil pump on vehicles equipped with an automatic transmission
- Radiator outlet hose
- Radiator fan motor assembly
- Accessory drive belts
- Timing belt and tensioner
- Belt idler number 2
- Camshaft Position (CMP) sensor
- Left side camshaft pulleys and left side rear timing belt cover
- Tensioner bracket
- Radiator hose and heater hose from the water pump
- Water pump retainer bolts
- Water pump

To install:

4. Clean the gasket mating surfaces

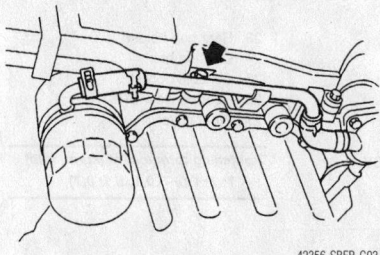

42356-SBFR-G03

Remove the bolt retaining the water by-pass pipe of the oil cooler onto the oil pump on vehicles equipped with A/T

42356-SBFR-G04

Remove the automatic belt tensioner

42356-SBFR-G05

Remove belt idler number 2

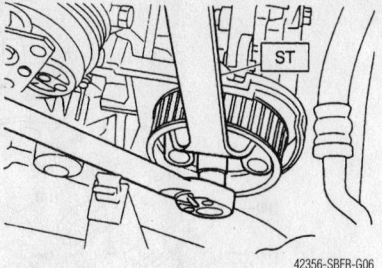

42356-SBFR-G06

Remove the left side camshaft pulleys

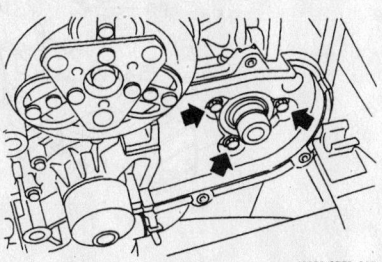

42356-SBFR-G07

Remove the left hand belt cover number 2

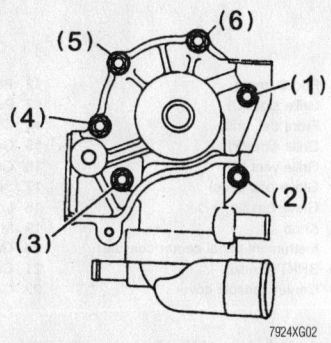

7924XG02

Water pump bolt tightening sequence

thoroughly. Always use new gaskets during installation.

5. Install or connect the following:
- Water pump. Torque the bolts, in sequence, to 9 ft. lbs. 12 Nm). After tightening the bolts once, retighten to the same specification again.
- Radiator hose and heater hose to the water pump
- Left side rear timing belt cover, left side camshaft pulleys and tensioner bracket
- CMP sensor
- Timing belt and tensioner
- Accessory drive belts
- Water pipe bypass pipe retaining bolt
- Radiator fan motor assembly
- Radiator outlet hose
- Engine undercover
- Negative battery cable

6. Fill the system with coolant.
7. Start the engine and allow it to reach operating temperature.
8. Check for leaks.

Heater Core

REMOVAL & INSTALLATION

2000–02 Models

1. Before servicing the vehicle, refer to the precautions in the beginning of this section.
2. Disconnect the negative battery cable.

✻✻ CAUTION

After disconnecting the negative battery cable, wait for at least 20 seconds for the air bag module to deplete its energy.

3. Drain the engine coolant into a clean container for reuse.
4. Disconnect the heater hoses from the heater core.
5. Remove the instrument panel by performing the following procedure:
   a. If equipped with a manual transmission, remove the shift knob.
   b. Remove both the front and rear console covers.
   c. Remove the console box-to-chassis screws and the console box.
   d. Remove the 3 lower left side cover assembly screws, disengage the 3 upper clips, and remove the cover assembly.
   e. Using a screwdriver, disconnect the data link connector from the lower cover.
   f. Remove the knee panel.
   g. At the glove box, remove the right side cover screw, the clip and the side cover.
   h. Remove the glove box screws and remove the glove box.
   i. Remove the center panel bezel.
   j. Remove the audio assembly screws, disconnect the electrical connectors and remove the audio assembly.
   k. Remove the 2 steering column-to-instrument panel bolts and lower the steering column.
   l. Move the temperature control switch to FULL HOT, the mode selector switch to DEF and the recirculation switch to FRESH positions.
   m. Disconnect the temperature control cable and the mode control cable from the heater housing, then the recirculation control cable from the intake housing.
   n. Disconnect the electrical harness connectors.
   o. Remove the instrument panel-to-chassis bolts.

p. Remove the 2 front defroster grille bolts.

q. Carefully, remove the instrument panel.

6. Remove the steering support beam bracket nuts and the steering support beam.

7. Remove the evaporator housing by performing the following procedure:

a. Discharge and recover the air conditioning system refrigerant.

b. Remove the refrigerant line-to-cowl connector bolt, separate the lines, discard the O-rings and plug the openings to prevent contamination.

c. Disconnect the electrical harness connector from the evaporator housing.

d. Disconnect the drain hose.

e. Remove the evaporator housing nut/bolts and the evaporator housing.

8. Remove the heater housing-to-chassis bolts and the heater housing.

9. Remove the heater core from the heater housing.

To install:

10. Install the heater core to the heater housing.

11. Install the heater housing and the heater housing-to-chassis bolts.

12. Install the steering support beam and the steering support beam bracket nuts.

13. Install the evaporator housing by performing the following procedure:

a. Install the evaporator housing and the evaporator housing nut/bolts.

b. Connect the drain hose.

c. Connect the electrical harness connector to the evaporator housing.

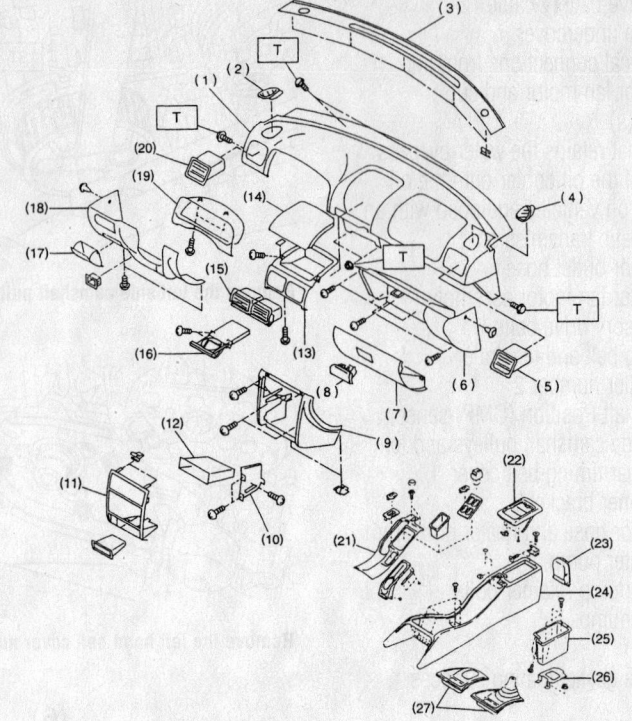

1 Pad & frame	12 Pocket	23 Rear cup holder
2 Grille side (D)	13 Panel center	24 Console box
3 Front def. grille	14 Center pocket lid	25 Console pocket
4 Grille side (P)	15 Grille center	26 Rear console BRKT
5 Grille vent (P)	16 Cup holder	27 Front cover
6 Glove box panel	17 Side pocket	
7 Glove box lid	18 Lower cover ASSY	
8 Knob	19 Meter visor	
9 Instrument panel center console	20 Grille vent (D)	
10 BRKT (Radio)	21 Console cover	
11 Center console cover	22 Console lid	

Tightening torque: N·m (kg-m, ft-lb)
T: 7±1 (0.7±0.1, 5.1±0.7)

93113GI8

Exploded view of the instrument panel assembly

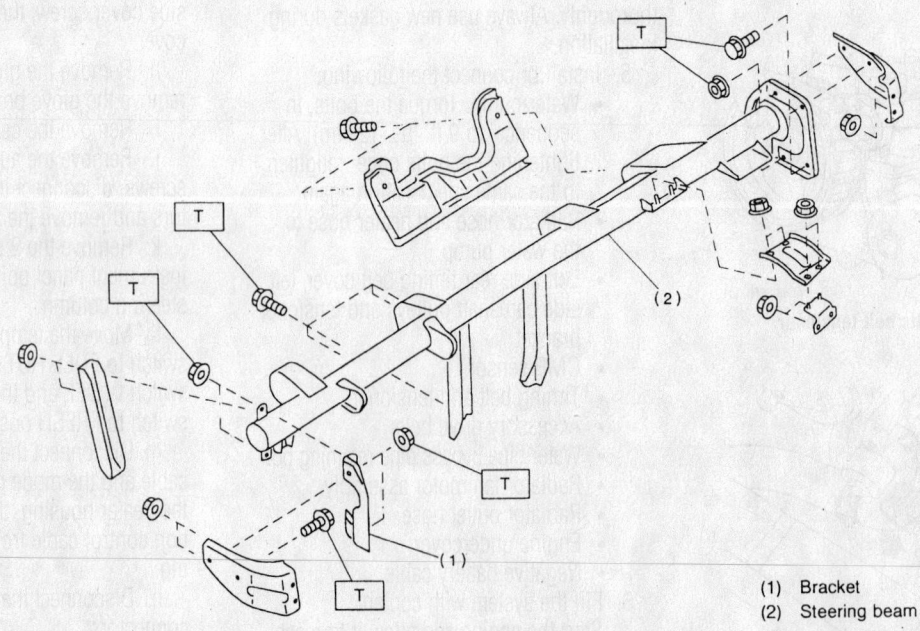

(1) Bracket
(2) Steering beam

93113GK8

Exploded view of the steering support beam assembly

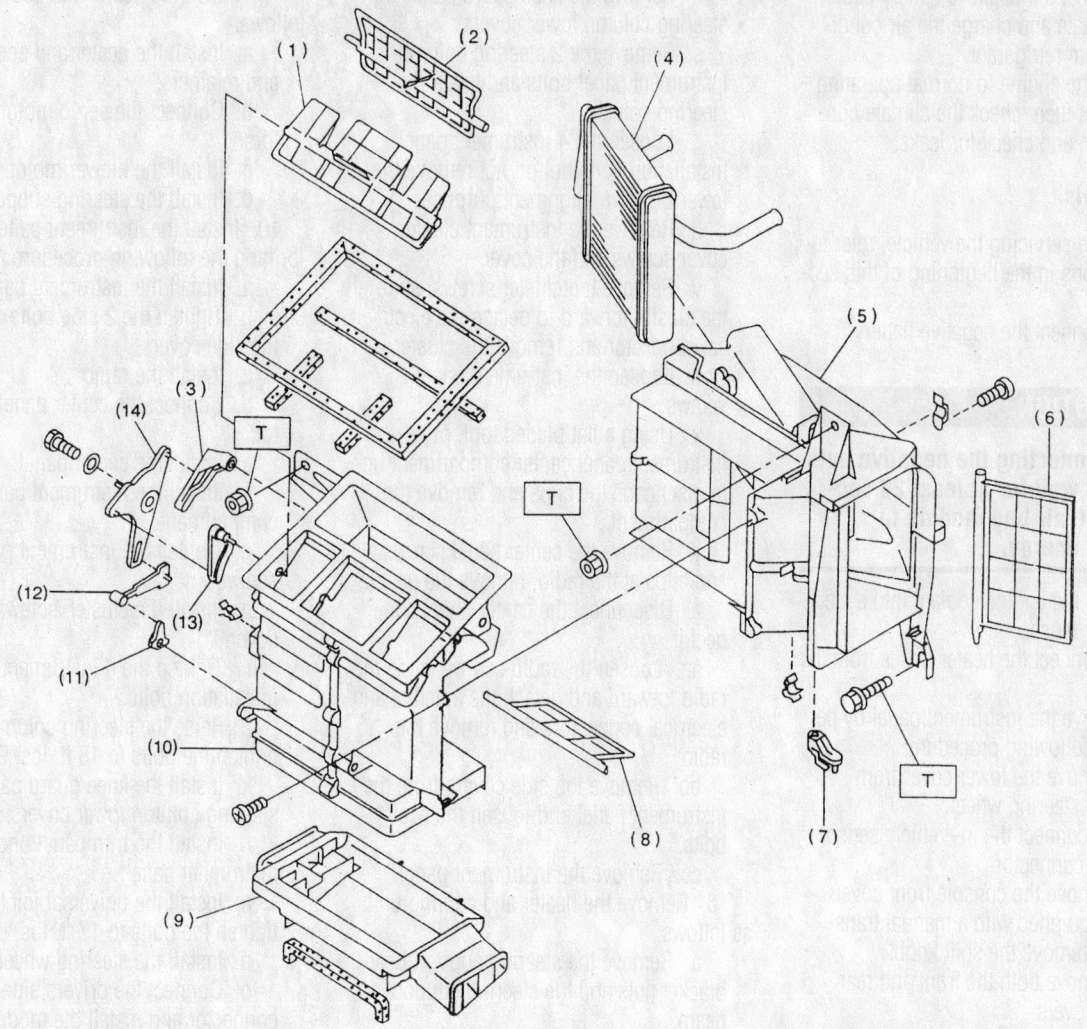

1	Vent door	7	**Mix** lever	13	Vent lever
2	DEF door	8	Foot door	14	Side link
3	DEF lever	9	Foot duct		
4	Heater core	10	Heater case REAR		
5	Heater case FRONT	11	Foot lever lower		
6	Mix door	12	Foot lever upper		

Tightening torque: N·m (kg-m, ft-lb)
T: 7.35±1.96
(0.750±0.200, 5.421±1.446)

93113GI7

Exploded view of the heater core, heater housing and related components

d. Using new O-rings, assemble the refrigerant lines and install the refrigerant line-to-cowl connector bolt.

14. Install the instrument panel by performing the following procedure:

a. Carefully, install the instrument panel.

b. Install the 2 front defroster grille bolts.

c. Install the instrument panel-to-chassis bolts.

d. Connect the electrical harness connectors.

e. Connect the temperature control cable and the mode control cable to the heater housing. Then, the recirculation control cable to the intake housing.

f. Install the steering column and lower the 2 steering column-to-instrument panel bolts and torque to 14–21 ft. lbs. (20–30 Nm).

g. Install the audio assembly, connect the electrical connectors and install the audio assembly screws.

h. Install the center panel bezel.

i. Install the glove box and the glove box screws.

j. At the glove box, install the right side cover, the clip and the side cover screw.

k. Install the knee panel.

l. Connect the data link connector to the lower cover.

m. Install the lower left side cover assembly, engage the 3 upper clips and install the cover assembly screws.

n. Install the console box and the console box-to-chassis screws.

o. Install both the front and rear console covers.

p. If equipped with a manual transmission, install the shift knob.

15. Connect the heater hoses to the heater core.

16. Refill the cooling system.

17. Connect the negative battery cable.

18. Evacuate and charge the air conditioning system refrigerant.

19. Run the engine to normal operating temperatures; then, check the climate control operation and check for leaks.

2003 Models

1. Before servicing the vehicle, refer to the precautions in the beginning of this section.

2. Disconnect the negative battery cable.

✳✳ CAUTION

After disconnecting the negative battery cable, wait for at least 20 seconds for the air bag module to deplete its energy.

3. Drain the engine coolant into a clean container for reuse.

4. Disconnect the heater hoses from the heater core.

5. Remove the instrument panel by performing the following procedure:

   a. Remove the lower cover from below the steering wheel.

   b. Disconnect the in–vehicle sensor hose and connector.

   c. Remove the console front cover.

   d. If equipped with a manual transmission, remove the shift knob.

   e. Remove both the front and rear console covers.

   f. Remove the console box-to-chassis screws and the console box.

   g. Remove the 3 lower left side cover assembly screws, disengage the 3 upper clips, and remove the cover assembly.

   h. Disconnect the A/C and hazard switch connectors.

   i. Remove the side console panel.

   j. Remove the glove box screws and remove the glove box.

   k. Disconnect the passenger side air bag module connector from the support beam and the module.

   l. Unfasten the air bag module screws and remove the module.

   m. Remove the drivers side air bag module Torx® bolts from each side of the module.

   n. Slide the module forward and disconnect the air bag connector and remove the module.

   o. Remove the steering wheel.

   p. Matchmark the universal joint and remove the bolts and the joint.

   q. Remove the trim panel from under the instrument panel.

   r. Remove the knee guard panel and steering column lower covers.

   s. Remove the 2 steering column-to-instrument panel bolts and lower the steering column.

   t. Loosen the 4 instrument panel installation bolts but do not remove the lower bolts for alignment purposes.

   u. Remove the instrument cluster cover screws and the cover.

   v. Remove the cluster screws, slide the cluster forward to detach the electrical connector and remove the cluster.

   w. Loosen the instrument panel screws.

   x. Using a flat bladed tool, pry the instrument panel center compartment up to disengage the clips and remove the compartment.

   y. Remove the center panel screws and clips at the radio, remove the panel.

   z. Disconnect the center panel connector.

   aa. Loosen the radio screws, slide the radio forward and detach the antenna and electrical connectors and remove the radio.

   bb. Remove the side covers from the instrument panel and loosen the two bolts.

   cc. Remove the instrument panel.

6. Remove the heater and cooling unit as follows:

   a. Remove the steering support beam bracket nuts and the steering support beam.

   b. Remove the blower motor.

   c. Disconnect the servo motor connectors.

   d. Remove the heater and cooling unit retainers and the unit.

7. Remove the heater core as follows:

   a. Open the heater core pipe cover.

   b. Loosen the screws and remove the mode actuator.

   c. Loosen the foot duct screws and remove the duct.

   d. Remove the evaporator cover screws and the cover.

   e. Remove the lower case cover screws and the cover

   f. Remove the heater core.

To install:

8. Install the heater core as follows:

   a. Install the heater core.

   b. Install the lower case cover and screws

   c. Install the evaporator cover and screws.

   d. Install the foot duct and screws.

   e. Install the mode actuator.

   f. Close the heater core pipe cover.

9. Install the heater and cooling unit as follows:

   a. Install the heater and cooling unit and retainers.

   b. Connect the servo motor connectors.

   c. Install the blower motor.

   d. Install the steering support beam.

10. Install the instrument panel by performing the following procedure:

   a. Install the instrument panel.

   b. Tighten the 2 side bolts and install the side covers.

   c. Install the radio.

   d. Connect the center panel connector.

   e. Install the center panel.

   f. Install the instrument panel center compartment.

   g. Tighten the instrument panel screws.

   h. Install the cluster, screws and the cover.

   i. Tighten the 4 instrument panel installation bolts.

   j. Raise the steering column and tighten the bolts to 18 ft. lbs. 925 Nm).

   k. Install the knee guard panel and steering column lower covers.

   l. Install the trim panel under the instrument panel.

   m. Install the universal joint and tighten the bolts to 17 ft. lbs. (24 Nm).

   n. Install the steering wheel.

   o. Connect the drivers side air bag connector and install the module.

   p. Install the drivers side air bag module Torx® bolts on each side of the module.

   q. Install the air bag module and tighten the screws.

   r. Connnect the passenger side air bag module connector to the module and the support beam.

   s. Install the glove box.

   t. Install the side console panel.

   u. Connect the A/C and hazard switch connectors.

   v. Install the left side cover assembly.

   w. Install the console box.

   x. Install both the front and rear console covers.

   y. If equipped with a manual transmission, install the shift knob.

   z. Install the console front cover.

   aa. Connect the in–vehicle sensor hose and connector.

   bb. Install the lower cover below the steering wheel.

11. Connect the heater hoses to the heater core.

12. Refill the cooling system.

13. Connect the negative battery cable.

14. Evacuate and charge the air conditioning system refrigerant.

15. Run the engine to normal operating temperatures; then, check the climate control operation and check for leaks.

Cylinder Head

REMOVAL & INSTALLATION

1. Before servicing the vehicle, refer to the precautions in the beginning of this section.

2. Properly relieve the fuel system pressure.

3. Remove or disconnect the following:
- Negative battery cable
- Oxygen (O_2S) sensor. If equipped with California emissions, disconnect the rear O_2S sensor.
- Engine undercover
- Exhaust Y-pipe and lower it just enough to clear the studs in the heads. Do not allow the Y-pipe to hang without support.
- Accessory drive belts
- Engine accessories and brackets from the side of the engine the cylinder head is being removed
- Connector bracket attaching bolt, if necessary

4. On the left cylinder head, remove the CMP sensor.

5. Remove or disconnect the following:

- Fuel pipes
- Timing belt, camshaft sprockets, and related components
- Intake manifold and gasket
- Valve covers, camshafts and related components
- Oil dipstick tube attaching bolt on the left cylinder head

- Bolt attaching the A/C compressor to the head
- Cylinder head bolts in the proper sequence. Leave bolts 1 and 3 installed loosely to prevent the cylinder head from falling.

6. Separate the cylinder head from the block. Use a plastic-faced hammer, if needed.

7. Remove bolts 1 and 3. Remove the cylinder head and gasket.

8. Clean all gasket material from both mating surfaces.

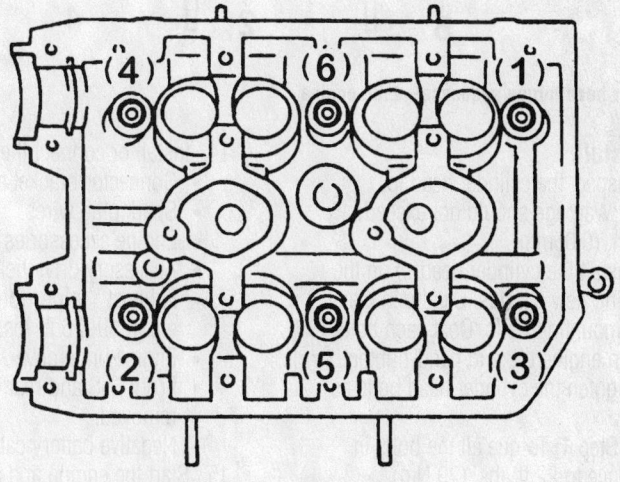

7924XG05

Cylinder head bolt loosening sequence

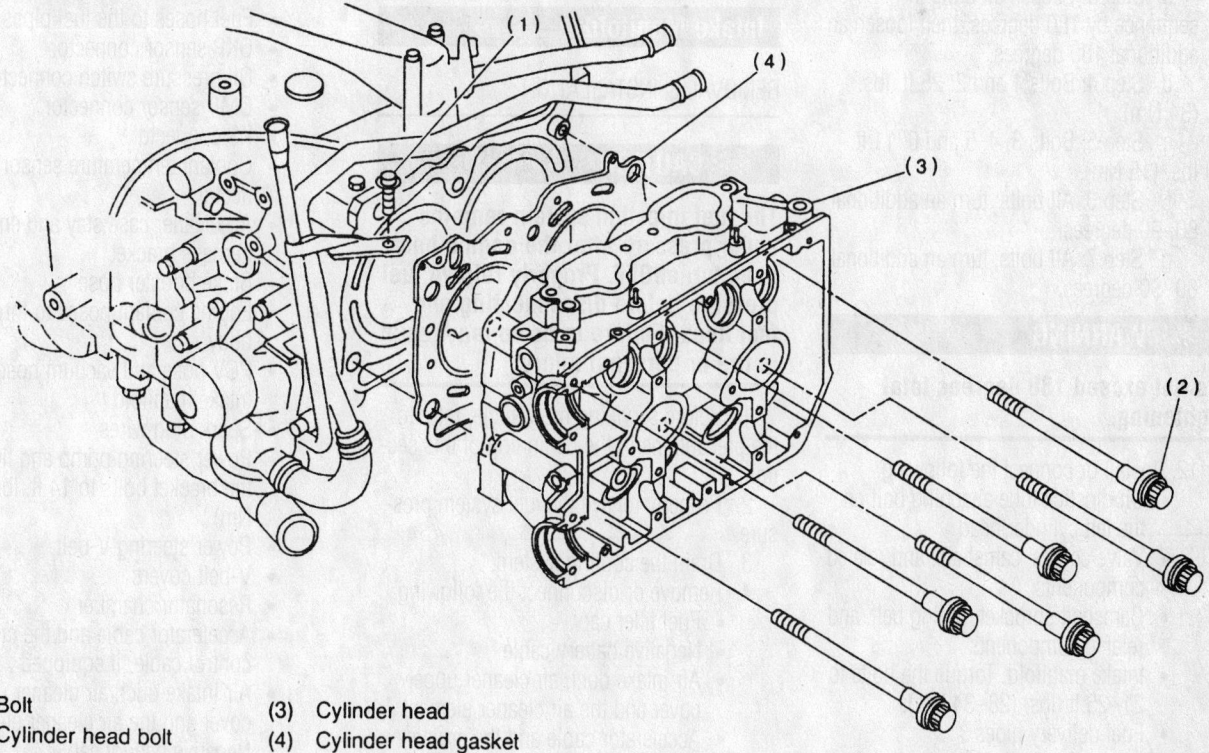

(1) Bolt
(2) Cylinder head bolt
(3) Cylinder head
(4) Cylinder head gasket

7924XG03

Exploded view of the cylinder head mounting

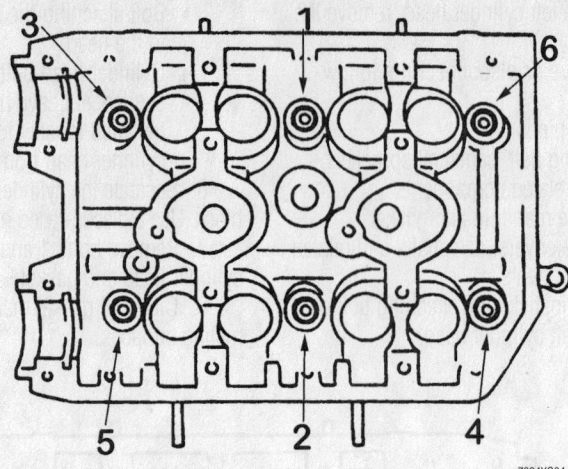

7924XG04

Cylinder head torque sequence—2.5L engine

To install:

9. Inspect the cylinder head for warpage. Warpage should not exceed 0.0020 in. (0.05mm).

10. Install the cylinder head(s) on the block using new gaskets. Secure in place with the mounting bolts. Coat each bolts with clean engine oil, and hand-tighten.

11. Tighten the cylinder head bolts as follows:

    a. Step 1: Torque all the bolts in sequence to 22 ft. lbs. (29 Nm).

    b. Step 2: Torque all the bolts in sequence to 51 ft. lbs. (69 Nm).

    c. Step 3: Loosen all bolts in sequence by 180 degrees, then loosen an additional 180 degrees.

    d. Step 4: Bolts 1 and 2: 25 ft. lbs. (34 Nm).

    e. Step 5: Bolts 3, 4, 5 and 6: 11 ft. lbs. (15 Nm).

    f. Step 6: All bolts: turn an additional 80–90 degrees.

    g. Step 7: All bolts: turn an additional 80–90 degrees.

✹✹ WARNING

Do not exceed 180 degrees total tightening.

12. Install or connect the following:
- Oil dipstick tube attaching bolt on the left cylinder head
- Valve covers, camshafts and related components
- Camshaft sprocket, timing belt, and related components
- Intake manifold. Torque the bolts to 21–25 ft. lbs. (28–34 Nm).
- Fuel delivery pipes

13. On the left cylinder head, install the CMP sensor.

14. Install or connect the following:
- Connector bracket attaching bolt
- Spark plug wires
- Engine accessories and brackets
- Accessory drive belts
- Exhaust Y-pipe. Torque the fasteners to 19–26 ft. lbs. (25–35 Nm).
- Engine undercover
- Front O$_2$S and rear O$_2$S, if removed.
- Negative battery cable

15. Start the engine and allow it to reach operating temperature.

16. Check for leaks.

Intake Manifold

REMOVAL & INSTALLATION

✹✹ CAUTION

The fuel injection system remains under pressure after the engine has been turnedOFF. Properly relieve fuel pressure before disconnecting any fuel lines. Failure to do so may result in fire or personal injury.

1. Before servicing the vehicle, refer to the precautions in the beginning of this section.

2. Properly relieve the fuel system pressure.

3. Drain the cooling system.

4. Remove or disconnect the following:
- Fuel filler cap
- Negative battery cable
- Air intake duct, air cleaner upper cover and the air cleaner element
- Accelerator cable and the cruise control cable, if equipped
- Resonator chamber
- V-belt covers
- Power steering V-belt
- Power steering hose brackets from intake manifold
- Power steering pump and set aside, do not disconnect hoses
- Spark plug wires
- Positive Crankcase Ventilation (PCV) hose and vacuum hose from intake manifold
- Engine coolant hoses from throttle body
- Brake booster hose
- Air cleaner case stay and engine harness bracket
- Coolant temperature sensor connector
- Knock Sensor (KS) connector
- Crankshaft Position (CMP) sensor connector
- Oil pressure switch connector
- Camshaft Position (CKP) sensor connector
- Fuel hoses from fuel pipes
- Intake manifold

➡ **The intake manifold sits on pins that protrude from the cylinder heads. Be sure the pins remain in the cylinder heads.**

To install:

5. Install or connect the following:
- Intake manifold and tighten the bolts to 18 ft. lbs. (25 Nm)
- Fuel hoses to the fuel pipes
- CKP sensor connector
- Oil pressure switch connector
- CMP sensor connector
- KS connector
- Coolant temperature sensor connector
- Air cleaner case stay and engine harness bracket
- Brake booster hose
- Engine coolant hoses to throttle body
- PCV hose and vacuum hose to the intake manifold
- Spark plug wires
- Power steering pump and tighten the bracket bolts to 14 ft. lbs. (20 Nm)
- Power steering V-belt
- V-belt covers
- Resonator chamber
- Accelerator cable and the cruise control cable, if equipped
- Air intake duct, air cleaner upper cover and the air cleaner element
- Negative battery cable
- Fuel filler cap

6. Refill the cooling system.

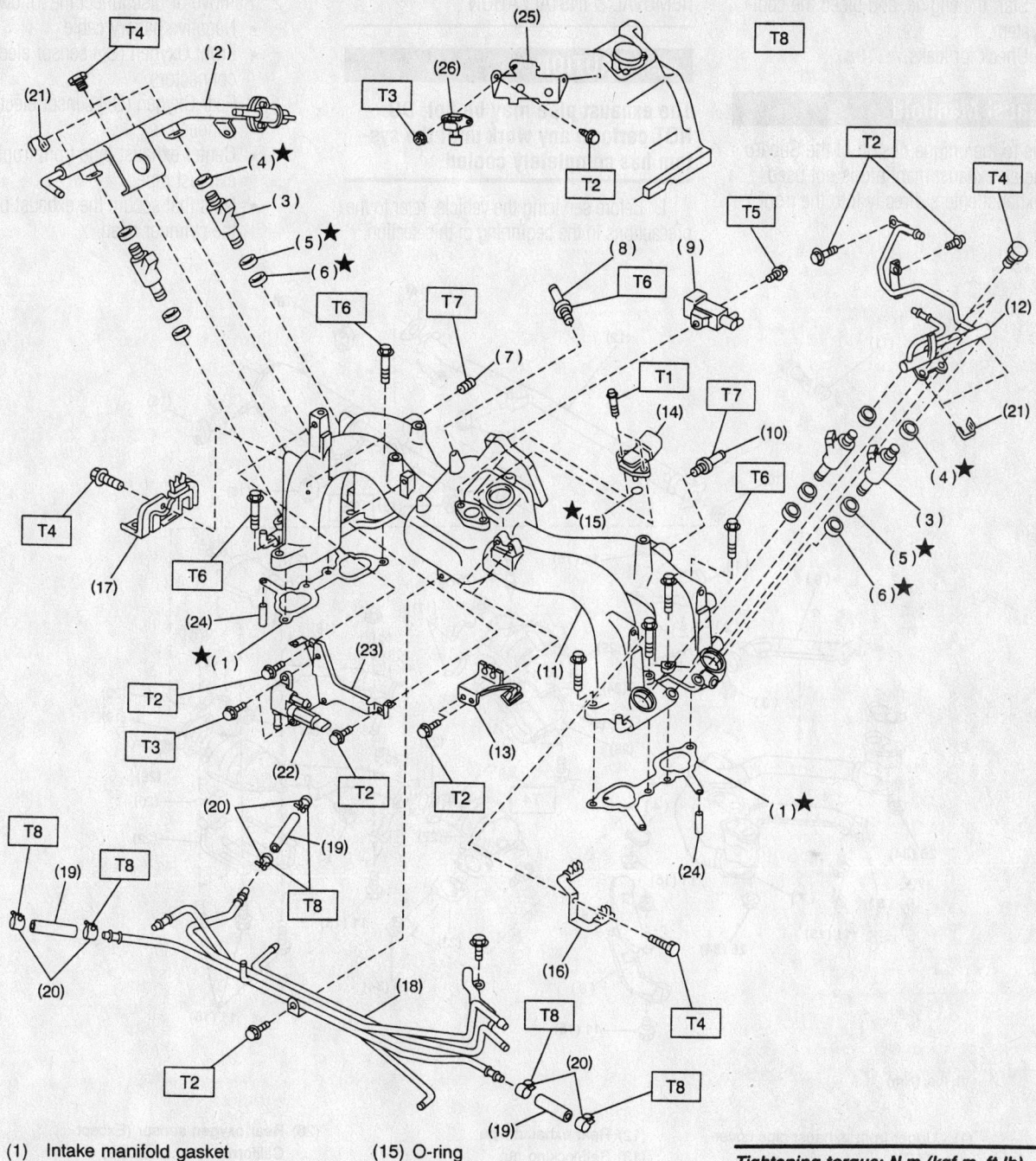

(1)	Intake manifold gasket	(15) O-ring
(2)	Fuel injector pipe RH	(16) Plug cord holder LH
(3)	Fuel injector	(17) Plug cord holder RH
(4)	O-ring	(18) Fuel pipe ASSY
(5)	O-ring	(19) Fuel hose
(6)	O-ring	(20) Clip
(7)	Plug	(21) Clip
(8)	PCV valve	(22) Air assist injector solenoid valve
(9)	Purge control solenoid valve	(23) Air assist injector solenoid valve
(10)	Nipple	bracket
(11)	Intake manifold	(24) Guide pin
(12)	Fuel injector pipe LH	(25) Atmospheric pressure sensor
(13)	Accelerator cable bracket	bracket
(14)	Intake air temperature and pres-	(26) Atmospheric pressure sensor
	sure sensor	

Tightening torque: N·m (kgf-m, ft-lb)
T1: 3.4 (0.35, 2.5)
T2: 5.0 (0.51, 3.7)
T3: 6.4 (0.65, 4.7)
T4: 19 (1.9, 13.7)
T5: 16 (1.6, 12)
T6: 25 (2.6, 18.8)
T7: 17 (1.7, 12)
T8: 1.5 (0.15, 1.1)

42356-SBFR-G08

Exploded view of the intake manifold—2002 model shown, other models similar

7. Start the engine, and bleed the cooling system.

8. Check for leaks.

Exhaust Manifold

Due to the unique design of the Subaru engine an exhaust manifold is not used. The exhaust enters directly into the front Y-pipe.

REMOVAL & INSTALLATION

✳✳ CAUTION

The exhaust pipe may be hot; DO NOT perform any work until the system has completely cooled.

1. Before servicing the vehicle, refer to the precautions in the beginning of this section.

2. Remove or disconnect the following:
- Negative battery cable
- Front Oxygen (O₂) sensor electrical connectors
- Rear Oxygen (O₂) sensor electrical connectors
- Center exhaust pipe from front exhaust pipe
- Nuts that secure the exhaust pipe to the cylinder head

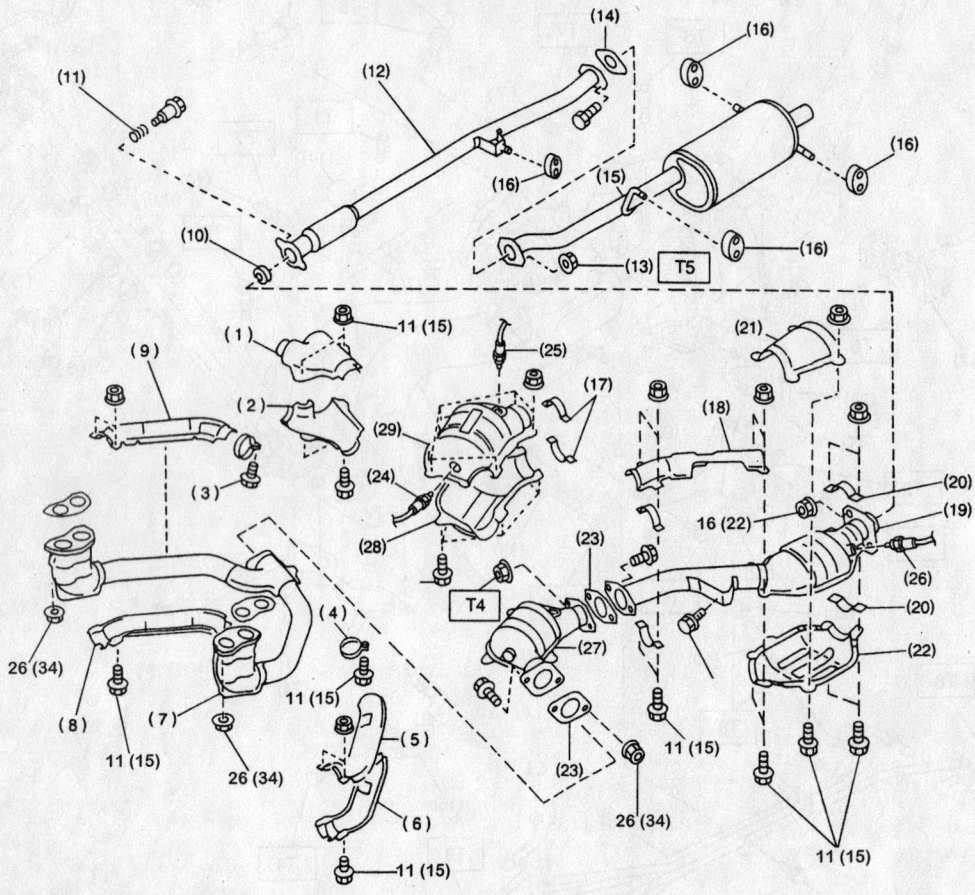

ft. lbs. (Nm)

(1) Upper front exhaust pipe cover CTR
(2) Lower front exhaust pipe cover CTR
(3) Band RH
(4) Band LH
(5) Upper front exhaust pipe cover LH
(6) Lower front exhaust pipe cover LH
(7) Front exhaust pipe
(8) Lower front exhaust pipe cover RH
(9) Upper front exhaust pipe cover RH
(10) Gasket
(11) Spring

(12) Rear exhaust pipe
(13) Self-locking nut
(14) Gasket
(15) Muffler
(16) Cushion rubber
(17) Clamp
(18) Upper center exhaust pipe cover
(19) Center exhaust pipe
(20) Clamp B
(21) Upper rear catalytic converter cover
(22) Lower rear catalytic converter cover
(23) Gasket
(24) Front oxygen sensor
(25) Rear oxygen sensor (California spec. vehicles)

(26) Rear oxygen sensor (Except California spec. vehicles)
(27) Front catalytic converter
(28) Lower front catalytic converter cover
(29) Upper front catalytic converter cover

Exploded view of the exhaust system and related components

7924XG07

- Front pipe-to-front catalytic converter mounting nuts
3. Discard the gaskets.
To install:
4. Clean all gasket surfaces completely.
5. Install or connect the following:
   - Catalytic converter to front exhaust pipe using new gasket. Torque the bolts to 22 ft. lbs. (30 Nm).
   - Exhaust pipe to the cylinder head using new gaskets. Torque the mounting nuts to 22 ft. lbs. (30 Nm).
   - Exhaust pipe to the center pipe using new gaskets. Torque the mounting nuts to 26 ft. lbs. (35 Nm).
   - Rear O$_2$ sensors electrical connectors
   - Front O$_2$ sensors electrical connectors
   - Negative battery cable
6. Start the engine and check for exhaust leaks.

Front Crankshaft Seal

REMOVAL & INSTALLATION

The front crankshaft seal is mounted in the oil pump. The removal and installation is covered in the oil pump procedure.

Camshaft and Valve Lifters

REMOVAL & INSTALLATION

1. Before servicing the vehicle, refer to the precautions in the beginning of this section.
2. Remove or disconnect the following:
   - Negative battery cable
   - Timing belt covers
   - Timing belt
   - Camshaft sprockets
   - Spark plug wires
   - Oil level gauge guide and Camshaft Position (CMP) sensor support
   - Positive Crankcase Ventilation (PCV) hose
   - Valve cover
   - Rocker arm assembly
3. Remove the camshaft cap as follows:
   a. Bolts "A" through "B" in alphabetical sequence
   b. Loosen bolts "C" through "J" equally all the way in alphabetical sequence
   c. Bolts "K" through "P" in alphabetical sequence using tool 499497000
4. Remove or disconnect the following:
   - Camshaft

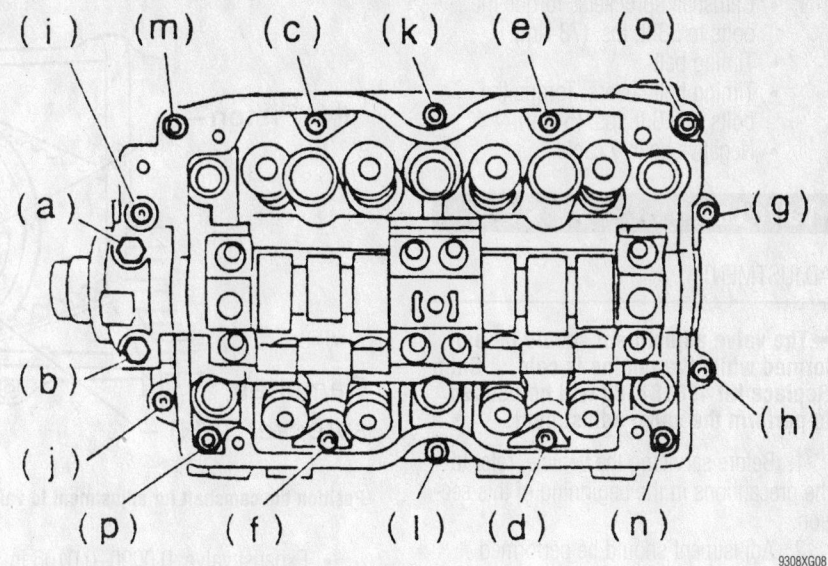

Camshaft cap removal sequence

9308XG08

- Oil seal, if necessary
- Plug from the rear side of the camshaft

To install:

➡**Lubricate the camshaft bearings prior to camshaft installation.**

5. Install or connect the following:
   - Camshaft
   - Camshaft cap. Apply liquid gasket on the edge of the cam cap mating surface 0.12 inch (3mm) thick
6. Temporarily tighten bolts "G" through "J" in alphabetical sequence
7. Install the valve rocker assembly. Torque the bolts "A" through "H" to 18 ft. lbs. (25 Nm).

8. Torque bolts "I" through "N" to 13 ft. lbs. (25 Nm) in alphabetical sequence using tool 499497000
9. Torque bolts "O" through "X" to 7.2 ft. lbs. (10 Nm) in alphabetical sequence
10. Install or connect the following:
    - Oil seal to the camshaft using tool 499597000 oil seal guide and 49958700 oil seal installer
    - Plug to the rear side of the camshaft using tool 499587700 oil seal installer
    - Valve cover
    - Oil level gauge guide and CMP sensor support
    - PCV house
    - Spark plug wires

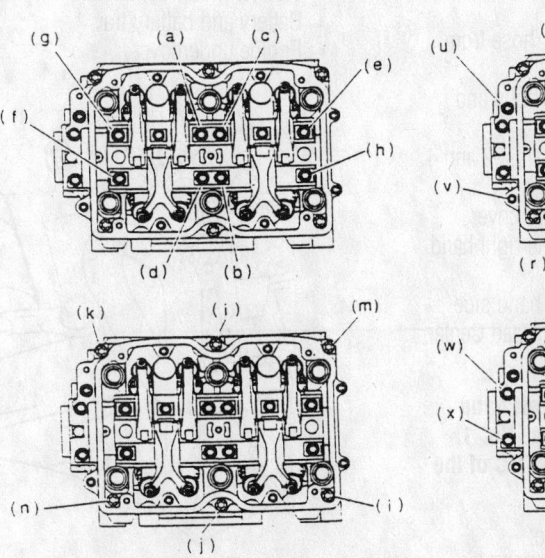
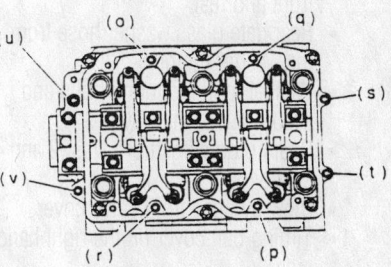
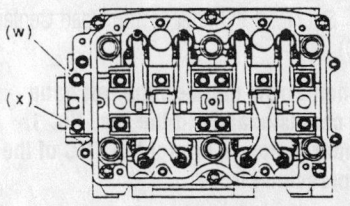

Camshaft cap tightening sequence

9308XG09

- Camshaft sprockets. Torque the bolts to 58 ft. lbs. (78 Nm).
- Timing belt
- Timing belt covers. Torque the bolts to 3.6 ft. lbs. (5 Nm).
- Negative battery cable

Valve Lash

ADJUSTMENT

➡ **The valve adjustment should be performed while the engine is cold. A Shim Replace Kit 498187100 will be needed to perform the valve adjustment.**

1. Before servicing the vehicle, refer to the precautions in the beginning of this section.
2. Adjustment should be performed when engine is cold.
3. Remove or disconnect the following:
   - Negative battery cable
   - Engine coolant reservoir tank
   - Timing belt cover on the left hand side
4. When inspecting Nos. 1 and 3 cylinders remove the following:
   - Air intake duct as a unit
   - Resonator chamber
   - Spark plug wires from Nos. 1 and 3 cylinders
   - Blow-by house from valve cover
   - Engine undercover
   - Timing belt cover on the right hand side
   - Valve cover on the right hand side
5. When inspecting Nos. 2 and 4 cylinders remove the following:
   - Battery and battery tray
   - Window washer motor connectors front and rear
   - Rear gate glass washer hose from the washer motor
   - Washer tank mounting bolts and secure out of the way
   - Spark plug wires from Nos. 2 and 4 cylinders
   - Blow by hose from valve cover
   - Timing belt cover on the right hand side
   - Valve cover on the left hand side
6. Set No. 1 cylinder to Top Dead Center (TDC).

➡ **When arrow mark on the left hand side comes exactly to the top, No. 1 cylinder piston is brought to TDC of the compression stroke.**

7. Check the valve clearance:
   - Intake valve: 0.0071–0.0087 in. (0.18–0.22mm).

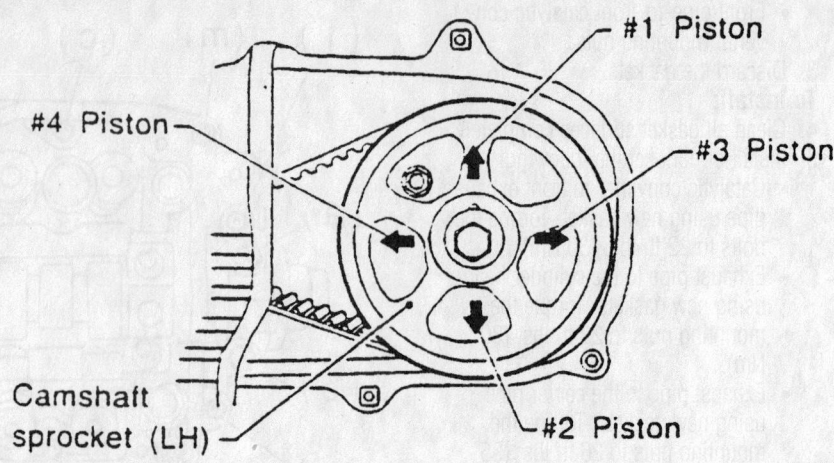

Position the camshaft for adjustment to valves

   - Exhaust valve: 0.0090–0.0106 in. (0.23–0.27mm).
8. If any valve needs adjustment, perform the following:
   a. Loosen the valve rocker nut and screw.
   b. Place a thickness gage in at as horizontal a direction as possible with respect to the vale stem and face.
   c. Adjust the screw until proper clearance is obtained.
   d. Tighten the rocker nut after adjusted to 7 ft. lbs. (10 Nm).
9. Install or connect the following:
   - Valve covers left and right
   - Timing belt covers
   - Blow-by hoses to valve covers
   - Spark plug wires
   - Washer tank
   - Rear gate glass washer hose to the washer motor
   - Washer motor connectors
   - Battery and battery tray
   - Engine undercover

   - Resonator chamber
   - Air intake duct unit
   - Engine coolant reservoir tank

Starter

REMOVAL & INSTALLATION

1. Before servicing the vehicle, refer to the precautions in the beginning of this section.
2. Remove or disconnect the following:
   - Negative battery
   - Air intake duct and assembly
   - Wires
   - Starter

To install:

3. Install or connect the following:
   - Starter. Torque the mounting bolts to 37 ft. lbs. (50 Nm).
   - Wires
   - Air intake duct and assembly
   - Negative battery

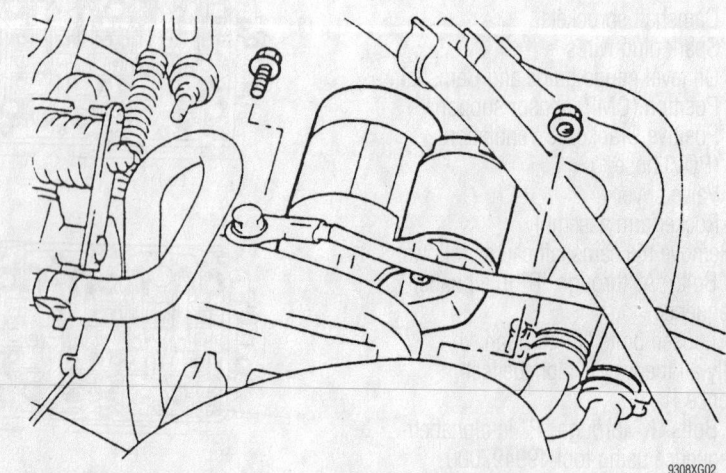

View of the starter mounting

Oil Pan

REMOVAL & INSTALLATION

1. Before servicing the vehicle, refer to the precautions in the beginning of this section.

2. Drain the oil from the engine.
3. Remove or disconnect the following:
   - Air intake duct
   - Oxygen (O2S) sensor connectors, if necessary
   - Pitching stopper
   - Upper radiator brackets

4. Support the engine with a suitable lifting device and lift the engine slightly.
   - Engine undercover
   - Rear O2S sensor connector, if equipped
   - Exhaust front Y-pipe and center pipe

T1: 3.6 ft. lbs. (5Nm)
T2: 3.6 ft. lbs. (5Nm)
T3: 4.7 ft. lbs. (6.4Nm)
T4: 7 ft. lbs. (10Nm)
T5: 32.5 ft. lbs. (44Nm)

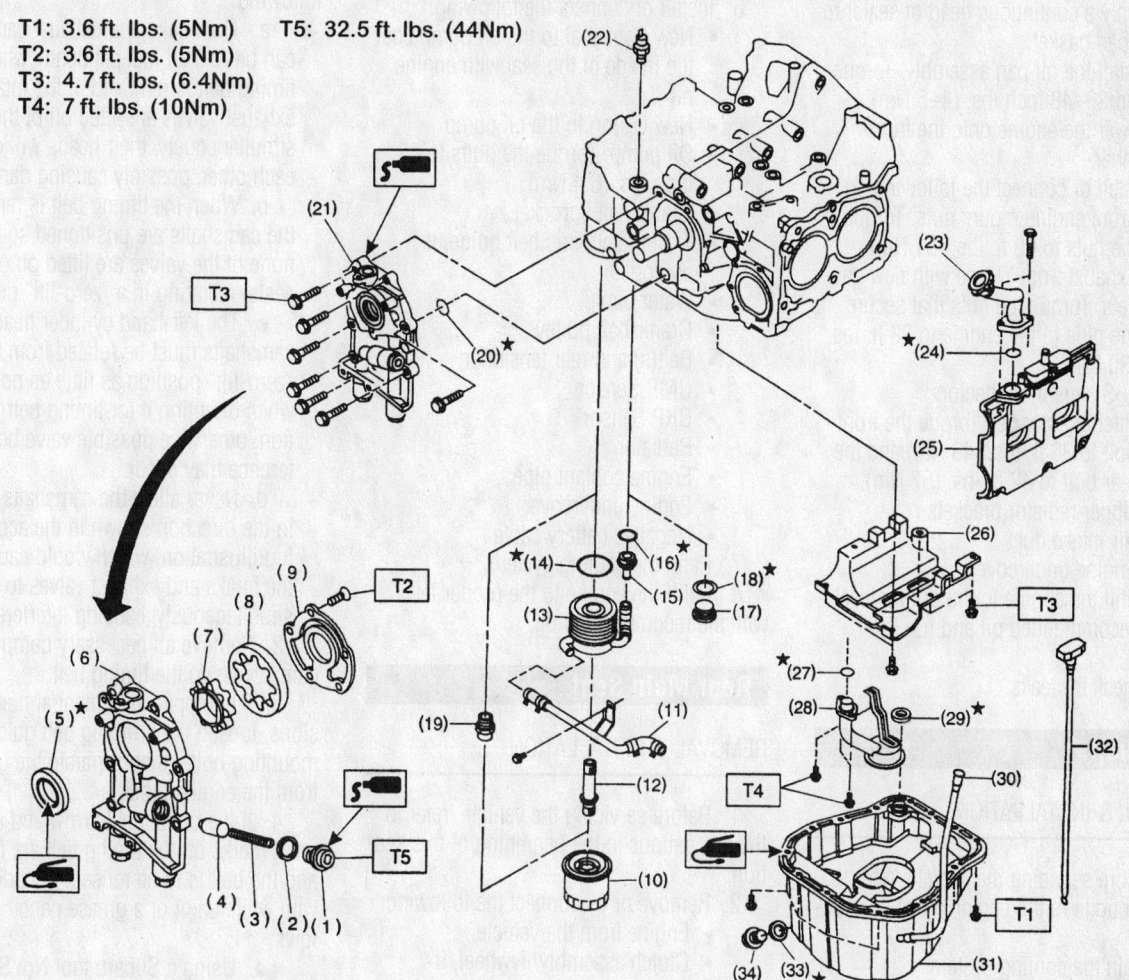

(1) Plug
(2) Washer
(3) Relief valve spring
(4) Relief valve
(5) Oil seal
(6) Oil pump case
(7) Inner rotor
(8) Outer rotor
(9) Oil pump cover
(10) Oil filter
(11) Oil cooler pipe and hose ASSY (AT vehicles)
(12) Connector (AT vehicles)
(13) Oil cooler (AT vehicles)
(14) O-ring (AT vehicles)
(15) Nipple (AT vehicles)
(16) Gasket (AT vehicles)
(17) Oil cooler connector (MT vehicles)
(18) Gasket (MT vehicles)
(19) Oil filter connector (MT vehicles)
(20) O-ring
(21) Oil pump ASSY
(22) Oil pressure switch
(23) Oil filler duct
(24) O-ring
(25) Cylinder head cover
(26) Baffle plate
(27) O-ring
(28) Oil strainer
(29) Gasket
(30) Oil level gauge guide
(31) Oil pan
(32) Oil level gauge
(33) Metal gasket
(34) Drain plug

Oil pan and lubrication components

7924XG28

- Nuts which secure the front engine mounts to the front crossmember

5. Remove the oil pan.

6. Clean all gasket material from both mating surfaces.

To install:

7. Install the drain plug with a new gasket. Torque the bolt to 33–36 ft. lbs. (43–47 Nm).

8. Apply a continuous bead of sealer to a new oil pan gasket.

9. Install the oil pan assembly. Torque the bolts to 36–48 inch lbs. (4–5 Nm).

10. Lower the engine onto the front crossmember.

11. Install or connect the following:

- Front engine mount nuts. Torque the nuts to 69 ft. lbs. (51 Nm).
- Exhaust front Y-pipe with new gaskets. Torque the nuts that secure the pipe to the engine to 23 ft. lbs. (30 Nm).
- O₂S sensor connectors
- Pitching stopper. Torque the front bolt to 36 ft. lbs. (49 Nm) and the rear bolt to 42 ft. lbs. (57 Nm).
- Upper radiator brackets
- Air intake duct
- Engine undercover

12. Refill the engine to the proper level with the recommended oil and run the engine.

13. Check for leaks.

Oil Pump

REMOVAL & INSTALLATION

1. Before servicing the vehicle, refer to the precautions in the beginning of this section.

2. Drain the cooling system.

3. Drain the engine oil into a separate container.

4. Remove or disconnect the following:

- Negative battery cable
- Engine undercover
- Water pipe and hose between oil cooler and water pump
- Radiator
- Crankcase Position (CKP) Sensor
- Camshaft Position (CMP) Sensor
- Belt(s) and rear tensioner
- Crankshaft pulley
- Water pump
- Timing belt, and belt guide, if equipped
- Crankshaft sprocket
- Oil pump mounting bolts and carefully pry the pump from the engine block

✳✳ WARNING

Use extreme care not to damage the engine block or the oil pump during removal of the pump.

To install:

5. Apply a continuous bead sealant to the mating surfaces of the oil pump.

6. Install or connect the following:

- New front seal to the oil pump coat the inside of the seal with engine oil
- New O-ring to the oil pump
- Oil pump. Torque the bolts to 56 inch lbs. (6.4 Nm).
- Crankshaft sprocket
- Timing belt, and belt guide, if equipped
- Water pump
- Crankshaft pulley
- Belt(s) and rear tensioner
- CMP Sensor
- CKP Sensor
- Radiator
- Engine coolant pipe
- Engine undercover
- Negative battery cable

7. Refill the cooling system.

8. Refill the engine to the proper level with the recommended oil.

Rear Main Seal

REMOVAL & INSTALLATION

1. Before servicing the vehicle, refer to the precautions in the beginning of this section.

2. Remove or disconnect the following:

- Engine from the vehicle
- Clutch assembly/flywheel, if equipped with manual transmission
- Torque converter flexplate from the crankshaft, if equipped with an automatic transmission

3. Using a seal removal tool, pry the oil seal from the housing.

To install:

4. Utilizing the appropriate seal installer

5. Install or connect the following:

- New oil seal and press it into the housing using the appropriate driver
- Clutch assembly/flywheel, if equipped
- Flywheel/flexplate and tighten the bolts to 53 ft. lbs. (72 Nm), if equipped
- Engine into the vehicle

Timing Belt

REMOVAL & INSTALLATION

1. Before servicing the vehicle, refer to the precautions in the beginning of this section.

When servicing the timing belt, note the following:

a. The intake and exhaust camshafts can be rotated independently when the timing belt is removed. If the intake and exhaust valves are lifted off of their seats simultaneously, their heads will contact each other, possibly causing damage.

b. When the timing belt is removed, the camshafts are positioned so that none of the valves are lifted off of their seats, resulting in a "zero-lift" position.

c. The left-hand cylinder head camshafts must be rotated from the "zero-lift" position as little as possible when orienting it for timing belt installation, otherwise possible valve head interference may occur.

d. Never allow the camshafts to rotate in the direction shown in the accompanying illustration, which would cause both the intake and exhaust valves to lift simultaneously, causing interference.

2. Remove all necessary components to gain access to the timing belt.

3. If equipped with manual transmissions, loosen the 2 timing belt guide mounting bolts, then separate the guide from the engine block.

4. If the directional arrow and alignment marks on the timing belt are faded, and the belt is to be reused, remark the belt with white paint or a grease pencil as follows:

a. Using a Subaru tool No. ST-499987500 Crankshaft Socket, or equivalent, installed on the crankshaft sprocket, rotate the crankshaft until the crankshaft sprocket, left-hand exhaust camshaft sprocket, left-hand intake camshaft sprocket, right-hand intake

42356-SBFR-G50

Remove the timing belt guide on vehicles equipped with manual transmission

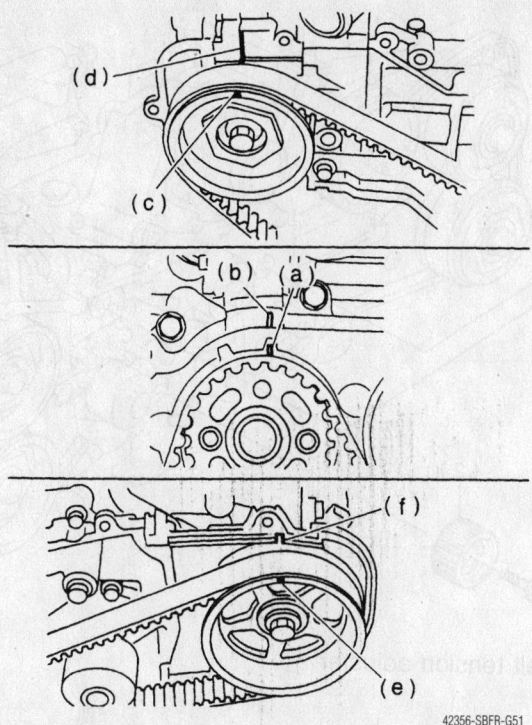

42356-SBFR-G51

Before removing the timing belt, turn the crankshaft sprocket until all of the alignment marks are aligned as indicated

camshaft sprocket and right-hand exhaust camshaft sprocket timing mark notches are aligned with the respective marks on the belt cover and engine block.

b. Make alignment and/or arrow marks on the timing belt in relation to the sprockets as indicated in the accompanying illustration.

- Z1: 44 tooth length on 2000 models and 46.8 tooth length on 2001–04 models

- Z2: 40.5 tooth length on 2000 models or 43.7 tooth length on 2001–04 models

5. Loosen the center bolt from the timing belt idler pulley, then remove the idler pulley from the engine block.

✸✸ WARNING

After removing the timing belt, DO NOT rotate the camshafts. Damage to the valves may occur.

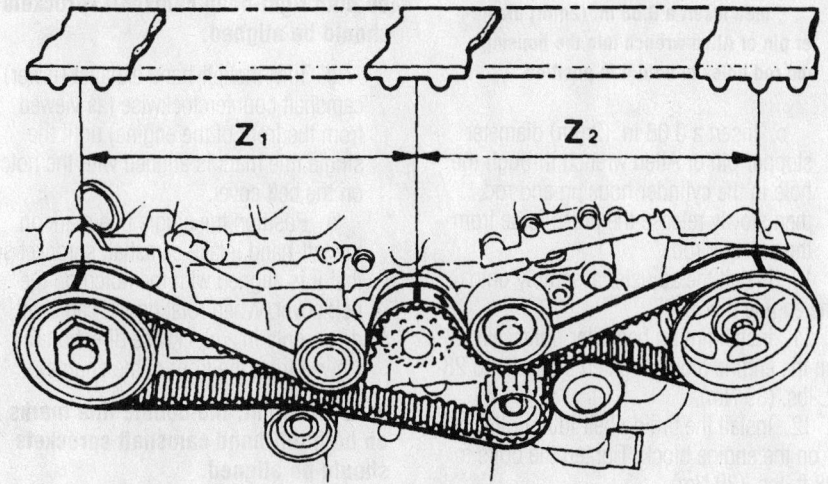

42356-SBFR-G52

If the original marks on the timing belt are worn or faded, make new alignment marks in the positions indicated

6. Carefully remove the timing belt from all of the sprockets.

7. Remove the automatic belt tension adjuster assembly as follows:

a. Remove the 2 timing belt idler pulleys, as indicated in the accompanying illustration.

b. Loosen the automatic tension adjuster assembly mounting bolts, then separate the adjuster assembly from the engine block.

To install:

✸✸ WARNING

Do not allow oil, grease, or coolant to come in contact with the timing belt. If this occurs, quickly and thoroughly remove all traces of the compound. Also, never bend the timing belt sharply; the minimum bending radius is 2.36 in. (60mm).

8. Inspect the camshaft and crankshaft sprocket teeth for abnormal or excessive wear or scratches. Ensure there is no free-play between the sprocket and the key. Inspect the crankshaft sprocket sensor notch for damage or contamination with debris or dirt.

➡**When preparing the automatic tension adjuster assembly for installation, adhere to the following points:**

- Always use a vertical press, rather than a horizontal press or vise, to depress the adjuster assembly rod
- Depress the adjuster rod in a vertical position ONLY
- Depress the adjuster rod slowly (taking more than 3 minutes) with a force of 66 lbs. (30 kg)
- Do not allow the press force to exceed 2205 lbs. (1000 kg)
- Press the adjuster rod in as far as the end surface of the cylinder — do not press the rod into the cylinder, which may cause damage to the assembly
- Do not release the press force from the rod until the stopper pin is completely inserted in the cylinder

9. Prepare the automatic timing belt tension adjuster assembly for installation as follows:

a. Position the adjuster assembly in a vertical press.

b. Slowly depress the adjuster rod with a force of 66 lbs. (30 kg) until the hole in the rod is aligned with the hole in the adjuster cylinder housing.

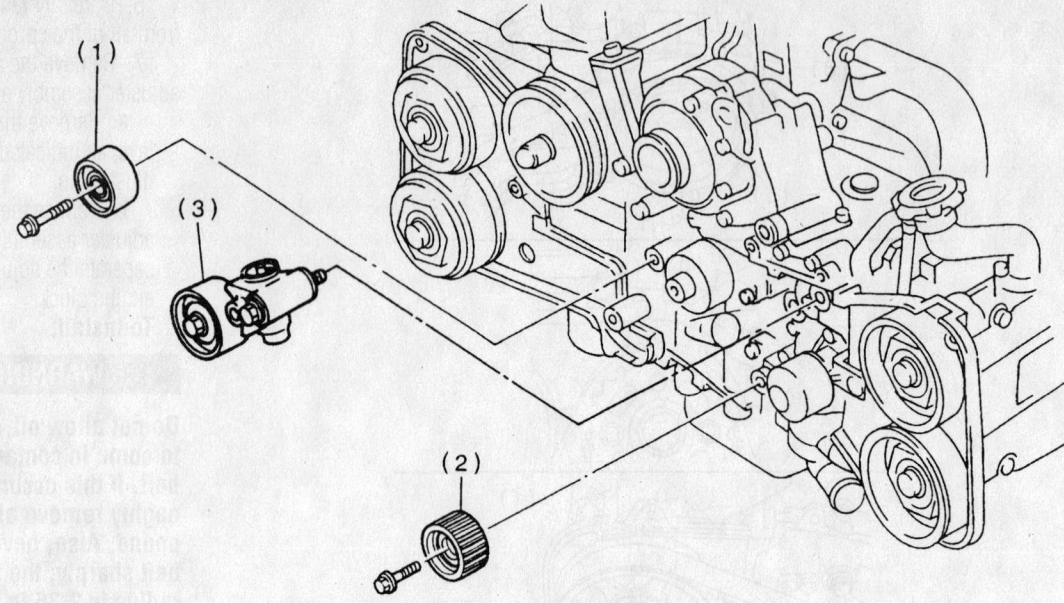

79245G52

(1) Belt idler
(2) Belt idler No. 2
(3) Automatic belt tension adjuster
ASSY

It is necessary to remove the automatic adjuster assembly and reset the pushrod for timing belt installation

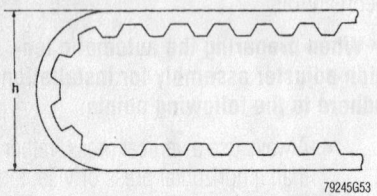

79245G53

Never bend the timing belt into a radius tighter than 2.36 in./60mm (h), otherwise it will be damaged beyond

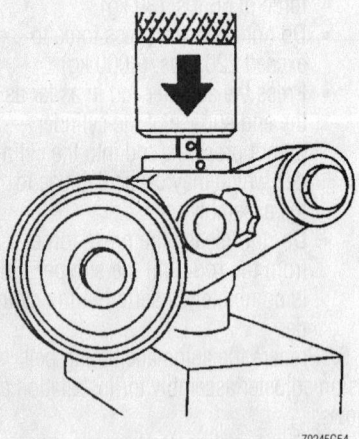

79245G54

Use a vertical press to push the adjuster rod into its housing until it is flush with the assembly's outer surface . . .

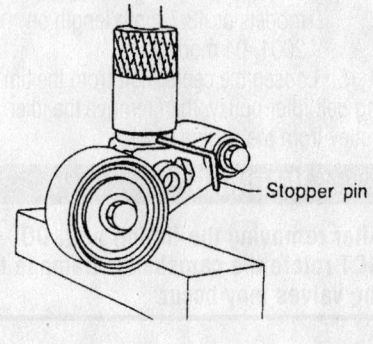

— Stopper pin

79245G55

. . . then insert a 0.08 in. (2mm) diameter pin or Allen wrench into the housing and rod holes to hold it in position

c. Insert a 0.08 in. (2mm) diameter stopper pin or Allen wrench through the hole in the cylinder housing and rod, then slowly release the press force from the adjuster rod.

10. Install the adjuster assembly onto the engine block.

11. Install timing belt idler pulley No. 2 on the engine block. Tighten the bolts to 28 ft. lbs. (39 Nm).

12. Install the timing belt idler pulley No. 1 on the engine block. Tighten the bolts to 28 ft. lbs. (39 Nm).

13. If the camshaft and crankshaft timing marks are no longer aligned, perform the following:

a. Position the crankshaft sprocket so that its mark is aligned with the mark on the oil pump cover on the engine block.

b. Align the single line mark on the right-hand exhaust camshaft sprocket with the notch on the belt cover.

c. Rotate the right-hand intake camshaft so that the single line mark is aligned with the notch on the belt cover.

➡**At this point, the double line marks on both right-hand camshaft sprockets should be aligned.**

d. Turn the left-hand exhaust (lower) camshaft counterclockwise (as viewed from the front of the engine) until the single line mark is aligned with the notch on the belt cover.

e. Position the single line mark on the left-hand intake camshaft sprocket so that it is aligned with the notch on the belt cover. When rotating the camshaft, do so only in a clockwise direction (as viewed from the front of the engine).

➡**At this point, the double line marks on both left-hand camshaft sprockets should be aligned.**

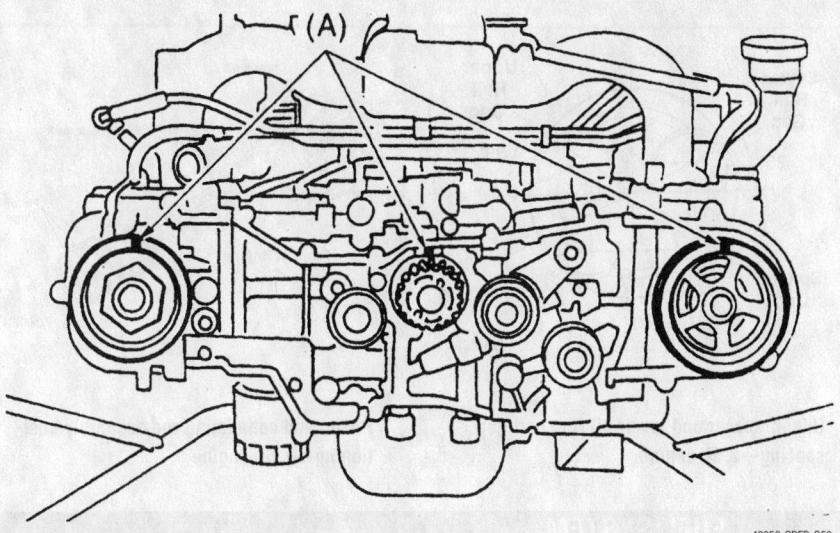

Make sure the sprockets are aligned as shown

42356-SBFR-G53

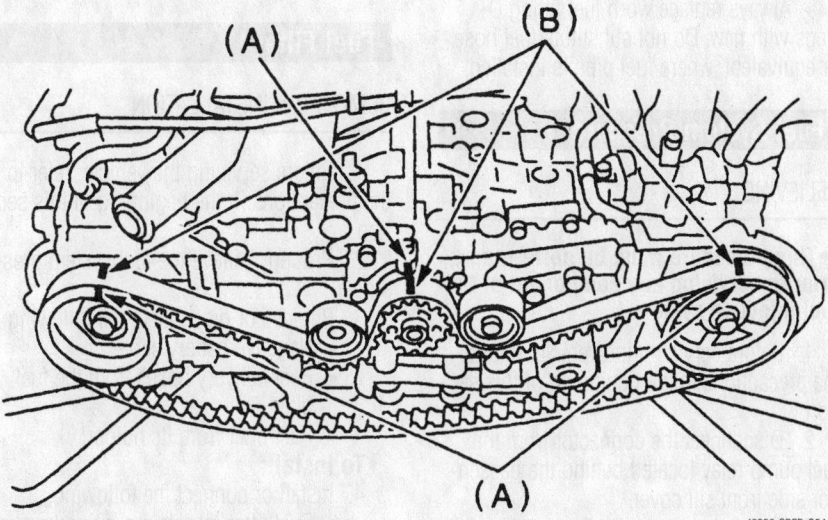

After installing the timing belt, the alignment marks should be positioned as shown. If not, remove the belt and align the sprockets and reinstall the belt

42356-SBFR-G54

f. Ensure the timing marks are aligned as shown in the accompanying illustration. If they are not, repeat Substeps 12a through 12e until they are properly aligned.

14. Install the timing belt around the camshaft, crankshaft and idler pulleys so that the positioning marks on the timing belt are aligned with the marks on the sprockets as follows:

a. Position the timing belt on the crankshaft sprocket so that the marks are aligned.

b. Route the belt down and under the left-hand, upper idler pulley, then up and around the left-hand intake camshaft

sprocket, ensuring the camshaft sprocket mark is aligned with the mark on the belt.

c. Route the belt down and around the left-hand exhaust camshaft sprocket, making sure the marks are properly aligned, then up and over the first lower idler pulley and down and around the second lower idler pulley.

d. While holding the timing belt on the inner, left-hand, lower idler pulley, route the other side of the timing belt (from the crankshaft sprocket) down and under the right-hand upper idler pulley.

e. Route the timing belt up and around the right-hand intake camshaft

sprocket so that the belt and sprocket marks are aligned.

f. Position the belt down and around the right-hand exhaust camshaft sprocket, ensuring the positioning marks are aligned.

15. Install the right-hand lower idler pulley so that the timing belt is routed over the top side of it.

➡ **Once the belt is completely installed on all of the pulleys and sprockets, ensure that the positioning marks are still all aligned.**

16. After ensuring all of the marks are still aligned, use a pair of pliers to withdraw the stopper pin or Allen wrench from the adjuster assembly housing.

17. On models with manual transmissions, perform the following:

a. Install the timing belt guide by temporarily tightening the mounting bolts.

b. Position the timing belt guide so that there is 0.019–0.059 in. (0.5–1.5mm) clearance between the timing belt and the belt guide.

c. Tighten the guide mounting bolts securely, then double check the guide clearance.

18. Install the timing belt covers and all remaining engine components.

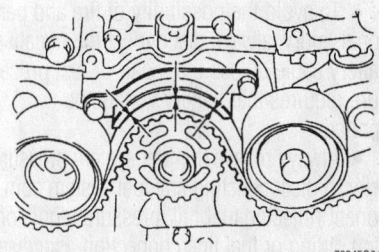

79245G64

On models equipped with manual transmissions, ensure the timing belt-to-guide clearance (arrows) is correct before tightening the mounting bolts

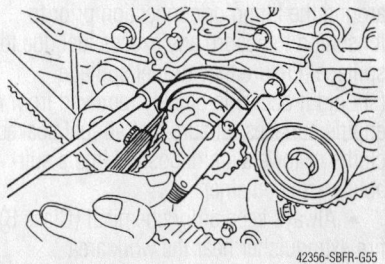

42356-SBFR-G55

On models equipped with manual transmissions, use a feeler gauges to adjust the clearance between the timing belt and the belt guide

Piston and Ring

POSITIONING

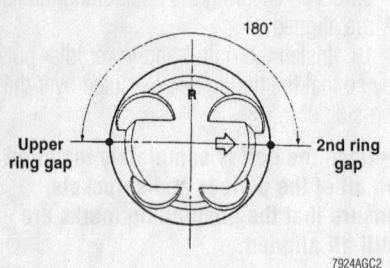

Compression ring end-gap spacing—2.5L engine

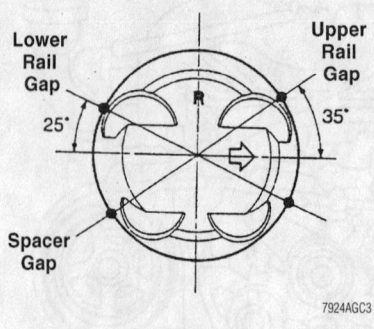

Upper, spacer and lower oil ring end-gap spacing—2.5L engine

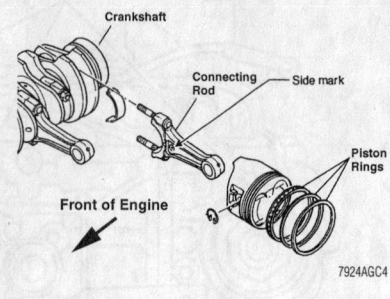

Piston and connecting rod assembly positioning—2.5L engine

FUEL SYSTEM

Fuel System Service Precautions

Safety is the most important factor when performing not only fuel system maintenance but any type of maintenance. Failure to conduct maintenance and repairs in a safe manner may result in serious personal injury or death. Maintenance and testing of the vehicle's fuel system components can be accomplished safely and effectively by adhering to the following rules and guidelines.

• To avoid the possibility of fire and personal injury, always disconnect the negative battery cable unless the repair or test procedure requires that battery voltage be applied.

• Always relieve the fuel system pressure prior to disconnecting any fuel system component (injector, fuel rail, pressure regulator, etc.), fitting or fuel line connection. Exercise extreme caution whenever relieving fuel system pressure, to avoid exposing skin, face and eyes to fuel spray. Please be advised that fuel under pressure may penetrate the skin or any part of the body that it contacts.

• Always place a shop towel or cloth around the fitting or connection prior to loosening to absorb any excess fuel due to spillage. Ensure that all fuel spillage (should it occur) is quickly removed from engine surfaces. Ensure that all fuel soaked cloths or towels are deposited into a suitable waste container.

• Always keep a dry chemical (Class B) fire extinguisher near the work area.

• Do not allow fuel spray or fuel vapors to come into contact with a spark or open flame.

• Always use a backup wrench when loosening and tightening fuel line connection fittings. This will prevent unnecessary

stress and torsion to fuel line piping. Always follow the proper torque specifications.

• Always replace worn fuel fitting O-rings with new. Do not substitute fuel hose or equivalent, where fuel pipe is installed.

Fuel System Pressure

RELIEVING

➡**This procedure must be performed prior to servicing any component of the fuel injection system.**

1. Before servicing the vehicle, refer to the precautions in the beginning of this section.

2. Disconnect the connector from the fuel pump relay located behind the passenger side front sill cover.

3. Crank the engine for 5 seconds or more to relieve the fuel pressure. If the engine starts during this time, allow it to run until it stalls.

4. Connect the fuel pump relay connector after repairs are completed.

Fuel Filter

REMOVAL & INSTALLATION

1. Before servicing the vehicle, refer to the precautions in the beginning of this section.

2. Properly relieve the fuel system pressure.

3. Remove or disconnect the following:
• Negative battery cable
• Fuel delivery hoses from the fuel filter
• Fuel filter from its holder

To install:

4. Install or connect the following:
• Fuel filter into its mounting bracket
• Fuel delivery hoses and tighten the hose clamps
• Negative battery cable

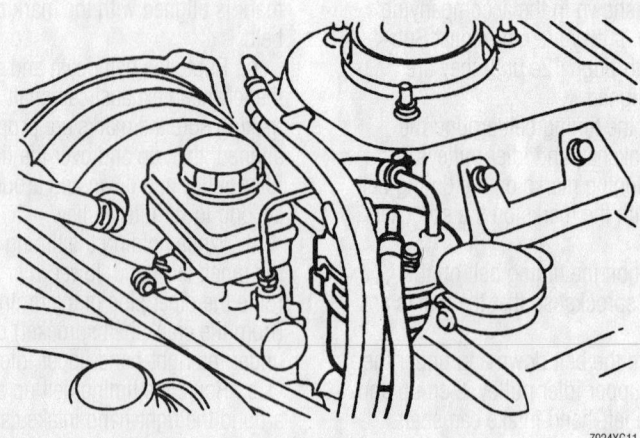

View of fuel filter mounting

Fuel Pump

REMOVAL & INSTALLATION

1. Before servicing the vehicle, refer to the precautions in the beginning of this section.

2. Properly relieve the fuel system pressure.

3. Drain the fuel tank by removing the drain plugs from the tank and draining into an approved container. Once the fuel has been drained, replace the plugs and tighten to 19 ft. lbs. (26 Nm).

4. Remove the rear seat cushion and access panel.

5. Disconnect the negative battery cable.

6. Clean any debris away from the fuel pump mounting to prevent it from entering the tank.

- Fuel pump electrical connector
- Fuel delivery and return hoses
- Fuel pump mounting nuts
- Fuel pump out of the fuel tank

To install:

7. Replace the sealing gaskets for the fuel pump.

8. Install or connect the following:

- Fuel pump into the tank. Torque the mounting nuts in sequence to 39 inch lbs. (4.4 Nm).
- Fuel delivery and return hoses
- Fuel pump electrical connector

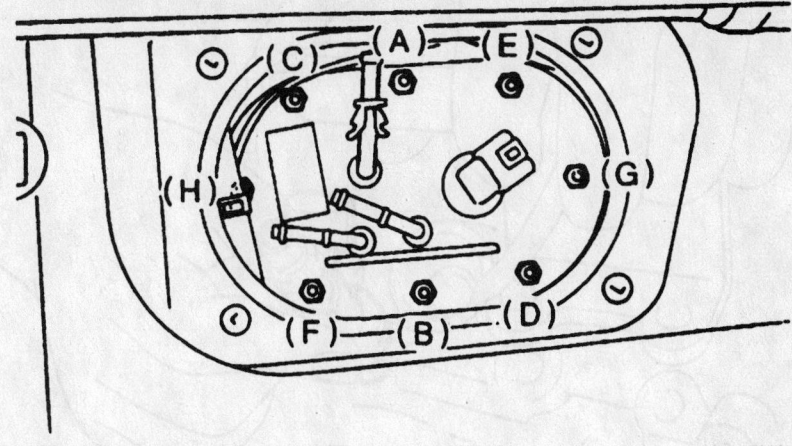

Fuel pump mounting nut tightening sequence

- Fuel filler cap
- Negative battery cable

9. Start the vehicle and check for leaks.

10. Install the fuel pump access cover and rear seat cushion.

Fuel Injector

REMOVAL & INSTALLATION

Right hand Side

1. Before servicing the vehicle, refer to the precautions in the beginning of this section.

2. Properly relieve the fuel system pressure.

3. Remove or disconnect the following:

- Negative battery cable
- Air duct and cleaner assembly
- Resonator chamber
- Spark plug wires No. 1 and 3
- V-belt covers
- Power steering pump belt
- Power steering pipe bracket from manifold
- Power steering pump and set aside
- Wires from fuel injector
- Injector pipe from intake manifold
- Fuel injector

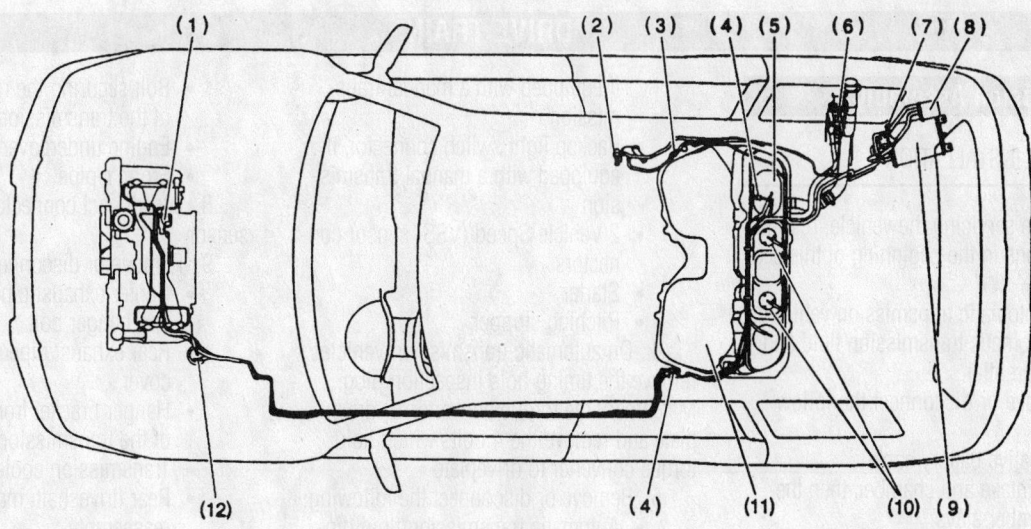

(1) Purge control solenoid valve	(5) Fuel pump	(9) Canister
(2) Roll over valve	(6) Fuel tank pressure sensor	(10) Fuel cut valve
(3) Pressure control solenoid	(7) Vent control solenoid valve	(11) Fuel tank
(4) Quick connector	(8) Air filter	(12) Fuel filter

Fuel system component locations

9308XG03

View of fuel injector removal

To install:
4. Install or connect the following:
- Fuel injector
- Injector pipe to intake manifold
- Wires to fuel injector
- Power steering pump
- Power steering pipe bracket to manifold
- Power steering pump belt
- V-belt covers

- Spark plug wires No. 1 and 3
- Resonator camber
- Air duct and cleaner assembly
- Negative battery cable

Left hand Side

1. Before servicing the vehicle, refer to the precautions in the beginning of this section.

2. Properly relieve the fuel system pressure.
3. Remove or disconnect the following:
- Negative battery cable
- Air duct and cleaner assembly
- Resonator chamber
- Two bolts that attach the washer reservoir to the body
- Connector from the front window washer motor
- Rear window washer hose from the motor and plug the line
- Washer reservoir and secure it to one side
- Spark plug wires No. 2 and 4
- Left hand fuel pipe connector
- Wires from fuel injector
- Injector pipe from intake manifold
- Fuel injector

To install:
4. Install or connect the following:
- Fuel injector
- Injector pipe to intake manifold
- Wires to fuel injector
- Spark plug wires No. 2 and 4
- Washer reservoir
- Rear window washer hose to the motor
- Connector to the front window washer motor
- Two bolts that attach the washer reservoir to the body
- Resonator camber
- Air duct and cleaner assembly
- Negative battery cable

DRIVE TRAIN

Transmission Assembly

REMOVAL & INSTALLATION

1. Before servicing the vehicle, refer to the precautions in the beginning of this section.
2. On automatic transmission vehicles, drain the automatic transmission fluid and the front differential.
3. Remove or disconnect the following:
- Negative battery cable
- Air intake and chamber, then the chamber stays
- Air cleaner case stay
- Front and rear Oxygen (O2S) sensor connectors, if equipped
- Transmission harness connector, if equipped with an automatic transmission
- Transmission ground terminal
- Neutral position switch connector,

if equipped with a manual transmission
- Backup light switch connector, if equipped with a manual transmission
- 2 Vehicle Speed (VSS) sensor connectors
- Starter
- Pitching stopper

4. On automatic transmission vehicles, remove the timing hole inspection plug. Matchmark the torque converter-to-driveplate and remove the 4 bolts which hold torque converter to driveplate.
5. Remove or disconnect the following:
- Automatic transmission fluid dipstick and tube
- Clutch slave cylinder, on manual transmission vehicles

6. Install engine support assembly ST 41099AA020. (Also available as part no. 927670000).
7. Remove or disconnect the following:

- Bolt securing the right upper side of the transmission to the engine
- Engine undercover
- Front Y-pipe
8. Disconnect connector from rear O2S sensor
9. Remove or disconnect the following:
- Center exhaust pipe from rear pipe and hanger bolt
- Rear exhaust pipe and heat shield cover
- Hanger bracket from the right side of the transmission
- Transmission cooler lines
- Rear driveshaft, matchmark for reassembly
- Center bearing bracket

➡ **Plug the opening at the rear of the extension housing to prevent oil from flowing out.**

- Shifter stay and rod from the transmission, on manual transmission

- Gear shift cable from the transmission select lever, on automatic transmission
- Sway bar from the transverse link
- Parking brake cable bracket from the transverse link and bolt holding the transverse link to the crossmember on each side, lower the transverse link
- Lower ball joint from knuckle
- Spring pin and separate the halfshaft from the transmission on each side

➡**Use a small punch to remove the spring pin. Discard old spring pin and always install a new pin.**

10. Disconnect the halfshaft from transmission on each side. Be sure to remove the axle shaft from the transmission by pushing the rear of the tire outward.

11. Remove the engine-to-transmission mounting nuts
12. Support the transmission with a jack.

➡**Do not place jack under the transmission oil pan, otherwise the oil pan may be damaged.**

13. Remove or disconnect the following:
- Rear transmission crossmember
- Transmission

To install:
14. Install or connect the following:
- Special tool 498277200 to torque converter clutch case
- Transmission to the engine
- Transmission crossmember. Torque the front nuts/bolts to 51 ft. lbs. (69 Nm) and the rear nuts to 101 ft. lbs. (137 Nm), on manual transmissions
- Transmission crossmember. Torque the inner nuts/bolts to 26 ft. lbs. (35 Nm) and the outer nuts to 51 ft. lbs. (69 Nm), on automatic transmissions

15. Remove the transmission jack.
16. Install or connect the following:
- Transmission-to-engine mounting nuts. Torque the nuts/bolts to 37 ft. lbs. (50 Nm).
- Torque converter-to-driveplate bolts on automatic transmission vehicles. Torque the bolts to 18 ft. lbs. (25 Nm).
- Clutch slave cylinder on manual transmission vehicles. Torque the mounting bolts to 27 ft. lbs. (37 Nm).

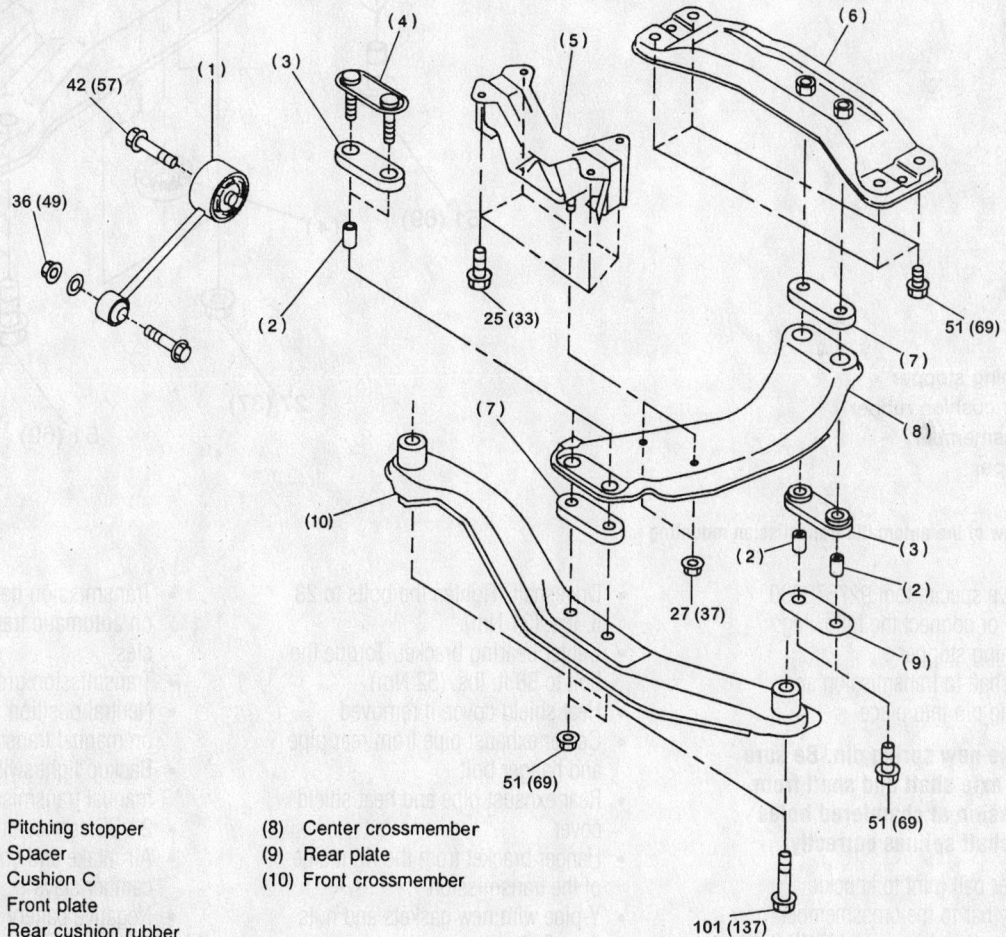

42 (57)
(1)
(3)
(4)
(5)
(6)
36 (49)
(2)
25 (33)
51 (69)
(7)
(8)
(7)
(10)
(3)
(2)
(2)
27 (37)
(9)
51 (69)
51 (69)
101 (137)

(1) Pitching stopper
(2) Spacer
(3) Cushion C
(4) Front plate
(5) Rear cushion rubber
(6) Rear crossmember
(7) Cushion D
(8) Center crossmember
(9) Rear plate
(10) Front crossmember

ft. lbs. (Nm)

Exploded view of the manual transmission mounting

7924XG14

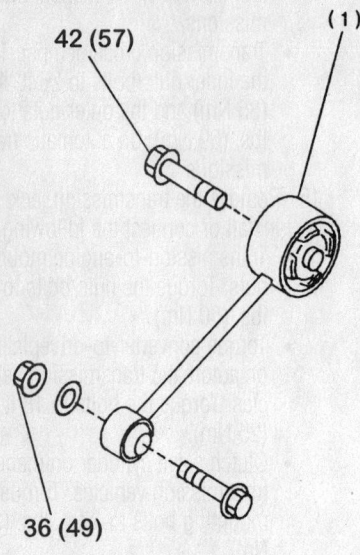

42 (57)

(1)

36 (49)

(2)

28 (38)

28 (38)

(3)

51 (69)

(4)

27 (37)

51 (69)

7924XG15

(1) Pitching stopper
(2) Rear cushion rubber
(3) Crossmember
(4) Stopper

Exploded view of the automatic transmission mounting

17. Remove special tool 927670000.
18. Install or connect the following:
   - Pitching stopper
   - Halfshaft to transmission and spring pin into place

➡**Always use new spring pin. Be sure to align the axle shaft and shaft from the transmission at chamfered holes and install shaft splines correctly.**

   - Lower ball joint to knuckle
   - Sway bar to the crossmember. Torque the clamp bolts to 18 ft. lbs. (25 Nm).
   - Shift control rod, shifter stay to the transmission and the spring, on manual transmission vehicles
   - Gear shift cable to the select lever, on automatic transmission vehicles
   - Fluid cooler lines

   - Driveshaft. Tighten the bolts to 23 ft. lbs. (31 Nm).
   - Center bearing bracket. Torque the bolt to 38 ft. lbs. (52 Nm).
   - Heat shield cover, if removed
   - Center exhaust pipe from rear pipe and hanger bolt
   - Rear exhaust pipe and heat shield cover
   - Hanger bracket from the right side of the transmission
   - Y-pipe with new gaskets and nuts
   - Rear O_2S sensor connector
   - Automatic transmission fluid dipstick tube
   - Transmission connector bracket
   - Starter. Torque the mounting bolts to 37 ft. lbs. (50 Nm).
   - Front and rear O_2S sensor connectors

   - Transmission harness connector, on automatic transmission vehicles
   - Transmission ground terminal
   - Neutral position switch connector, on manual transmission vehicles
   - Backup light switch connector, on manual transmission vehicles
   - 2 VSS connectors
   - Air intake and chamber, and the camber stays
   - Negative battery cable

19. On automatic transmission vehicles, fill the automatic transmission fluid with Dexron®II or III or equivalent.
20. On manual transmission vehicles, check and fill the transmission with 75W-90 gear oil.
21. Road test the vehicle.

Clutch

ADJUSTMENT

This vehicle is equipped with a hydraulic clutch that is self-adjusting, therefore no adjustment is possible or necessary.

REMOVAL & INSTALLATION

✳✳ CAUTION

The clutch driven disc may contain asbestos, which has been deter- mined to be a cancer-causing agent. **Never clean clutch surfaces with compressed air. Avoid inhaling any dust from any clutch surface. When cleaning clutch surfaces, use a commercially available brake cleaning fluid.**

1. Before servicing the vehicle, refer to the precautions in the beginning of this section.
2. Remove or disconnect the following:
   - Negative battery cable
   - Transmission

✳✳ WARNING

Removing the bolts on one side of the pressure plate will warp the pressure plate, rendering it useless.

3. Gradually unscrew the six 6mm bolts that hold the pressure plate assembly on the flywheel. Loosen the bolts only 1 turn at a time, working around the pressure plate. Do not unscrew all the bolts on one side at one time.
4. Remove or disconnect the following:
   - Clutch plate and disc

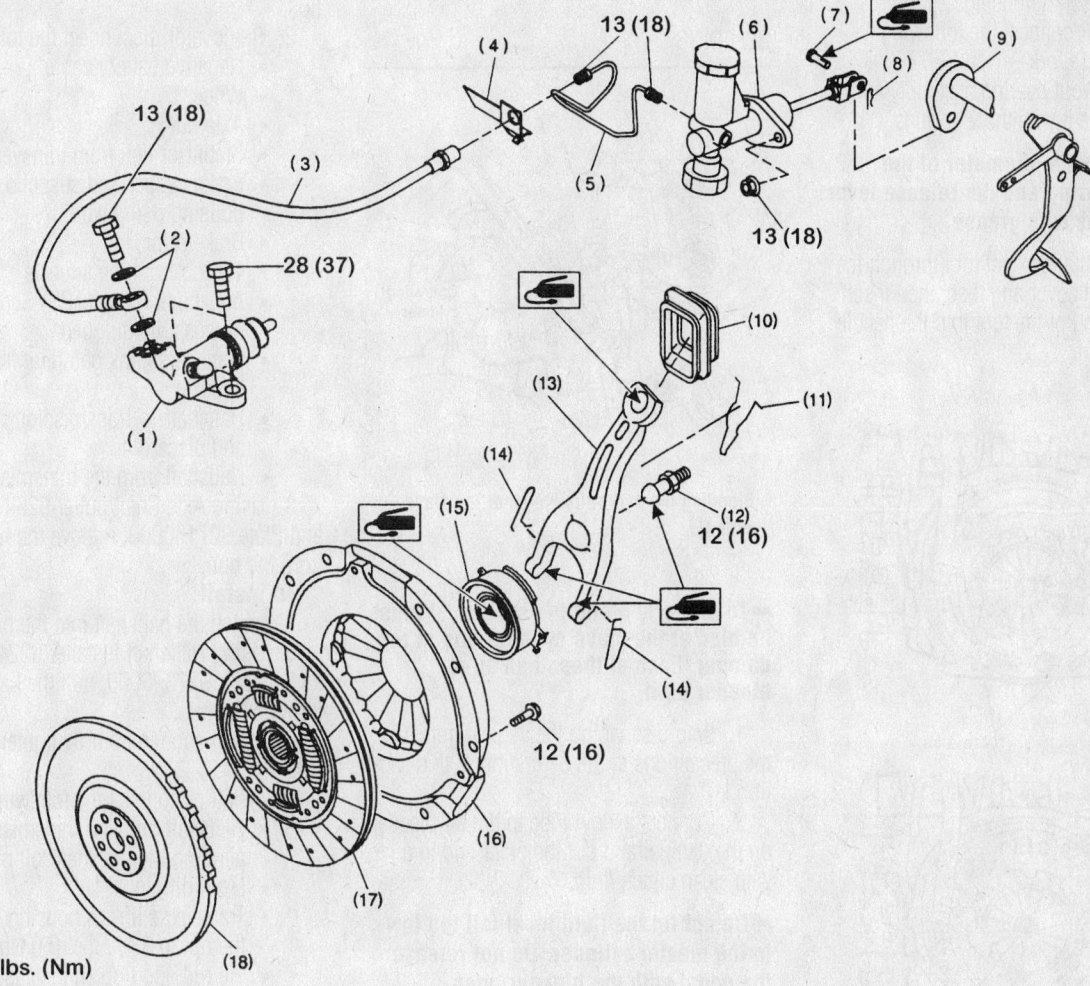

13 (18)
(3)
13 (18)
(2)
28 (37)
(1)
13 (18)
(4)
(5)
(6)
(7)
(8)
(9)
13 (18)
(10)
(13)
(11)
(14)
(12)
12 (16)
(15)
(14)
12 (16)
(16)
(17)
(18)
ft. lbs. (Nm)

(1) Operating cylinder	(9) Lever	(17) Clutch disc
(2) Washer	(10) Clutch release lever sealing	(18) Flywheel
(3) Clutch hose	(11) Retainer spring	
(4) Bracket	(12) Pivot	
(5) Pipe	(13) Release lever	
(6) Master cylinder ASSY	(14) Clip	
(7) Clevis pin	(15) Release bearing	
(8) Snap pin	(16) Clutch cover	

Exploded view of clutch system

- 2 retaining springs, the throwout bearing and the release fork

➡Do not disassemble either the clutch cover or disc. Inspect the parts for wear or damage and replace any parts as necessary. Replace the clutch disc if there is any oil or grease on the facing. Do not wash or attempt to lubricate the throwout bearing, because it is sealed and permanently lubricated. If it requires replacement, the bearing may be removed and a new one installed in the holder by means of a press.

To install:

5. Fit the release fork boot on the front of the transmission housing.
6. Install or connect the following:
   - Release fork
   - Throwout bearing assembly and secure it with the 2 springs

➡Coat the inside diameter of the throwout bearing and the release lever contact points with grease.

   - Clutch alignment tool through the clutch cover and disc, then insert the end of the tool into the needle bearing

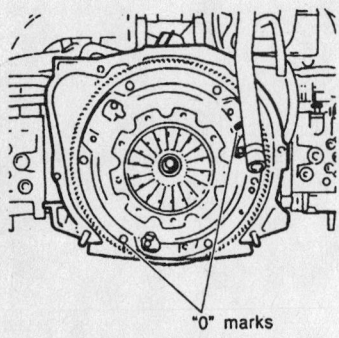

"O" marks

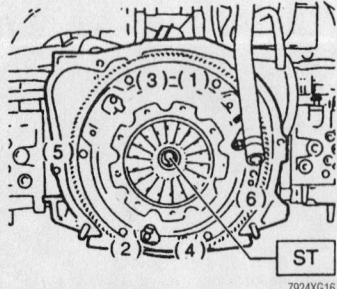

Clutch cover alignment and tightening sequence

7924XG16

7. Tighten the pressure plate bolts following the illustrated sequence, 1 turn at a time, until the proper torque is reached. Tighten to 12 ft. lbs. (16 Nm).

❋❋ WARNING

When installing the clutch pressure plate assembly, be sure that the O marks on the flywheel and the clutch pressure plate assembly are at least 120 degrees apart. These marks indicate the direction of residual unbalance. Also, be sure that the clutch disc is installed properly, noting the FRONT and REAR markings.

8. Install the transmission.

Hydraulic Clutch System

BLEEDING

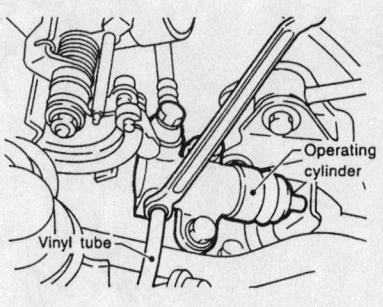

7924XG18

Bleeding the hydraulic clutch at the slave cylinder

➡To properly bleed the system, it must be bled at the slave cylinder and at the damper. Each of these has an air bleeder on it.

1. Before servicing the vehicle, refer to the precautions in the beginning of this section.
2. Connect a vinyl tube to the air bleeder on the damper and put the other end in a jar with clean clutch fluid.

➡Do not let the fluid level fall too low in the master cylinder. Do not release the pedal with the bleeder open.

3. With the help of an assistant depressing the clutch pedal, slowly open the bleeder valve. Close the bleeder valve and release the pedal. Repeat this process until no air bubbles appear in the jar.
4. Move the tube to the bleeder on the slave cylinder and repeat the process. Check the operation of the clutch after the bleed procedure is complete.

Transfer Case Assembly

REMOVAL & INSTALLATION

The transfer case is an integral part of the transmission.

Halfshafts

REMOVAL & INSTALLATION

Front

1. Before servicing the vehicle, refer to the precautions in the beginning of this section.
2. Remove or disconnect the following:
   - Negative battery cable
   - Wheel
   - Axle nut
   - Stabilizer link from transverse link
   - Brake caliper and suspend from the housing using wire
   - Brake rotor
   - Tie rod from the knuckle
   - Anti-Lock Brake (ABS) Sensor and harness, if equipped
   - Transverse link ball joint from the knuckle
   - Halfshaft-to-transmission roll pin and discard it
   - Halfshaft from the transmission
3. Using Axle Shaft puller 926470000 and Plate 927140000, remove the halfshaft from the hub.

To install:

4. Install the halfshaft into the hub.
5. Using Halfshaft installer 922431000 and adapter 927390000, pull the halfshaft through the hub.
6. Install and temporarily tighten a new axle nut.
7. Install or connect the following:
   - Halfshaft onto the transmission, by aligning the halfshaft roll pin hole
   - New roll pin
   - Transverse link to housing. Torque the nut to 37 ft. lbs. (50 Nm).
   - Tie rod and tighten the castellated nut to 20 ft. lbs. (27 Nm) and then up to an additional 60 degrees to align the cotter pin hole with the slot on the nut, then install a new cotter pin
   - Stabilizer link
   - Brake rotor
   - Brake caliper
   - New axle nut. Torque the nut to 140 ft. lbs. (190 Nm) and stake the nut.

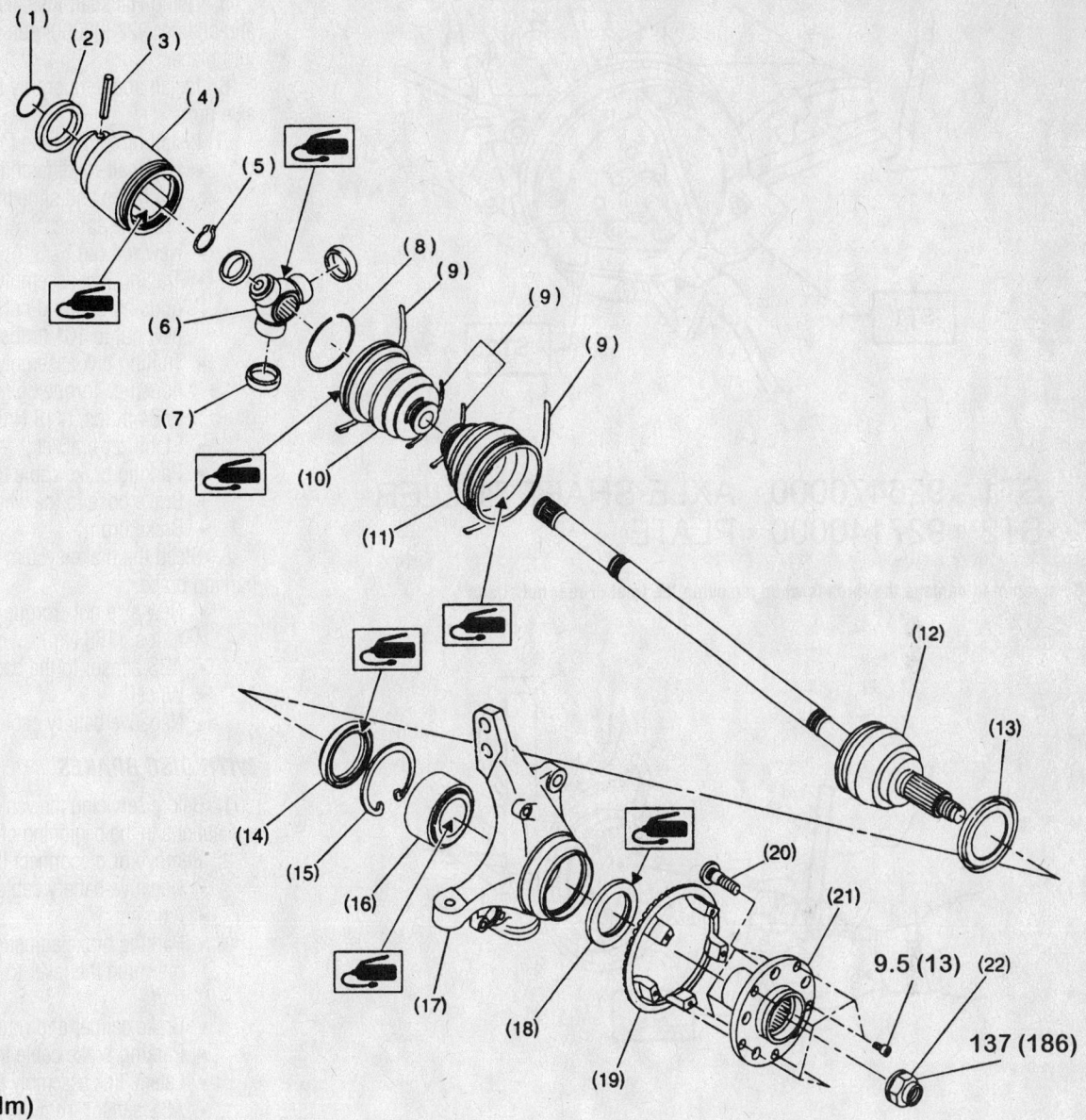

ft. lbs. (Nm)

(1) O-ring	(10) Boot band	(19) Tone wheel
(2) Baffle plate (FTJ)	(11) Boot (BJ)	(20) Hub bolt
(3) Spring pin	(12) BJ ASSY	(21) Hub
(4) Outer race (FTJ)	(13) Baffle plate	(22) Axle nut
(5) Snap ring	(14) Oil seal (IN)	
(6) Trunnion	(15) Snap ring	
(7) Free ring	(16) Bearing	
(8) Circlip	(17) Housing	
(9) Boot band	(18) Oil seal (OUT)	

Exploded view of the front halfshaft

7924XG20

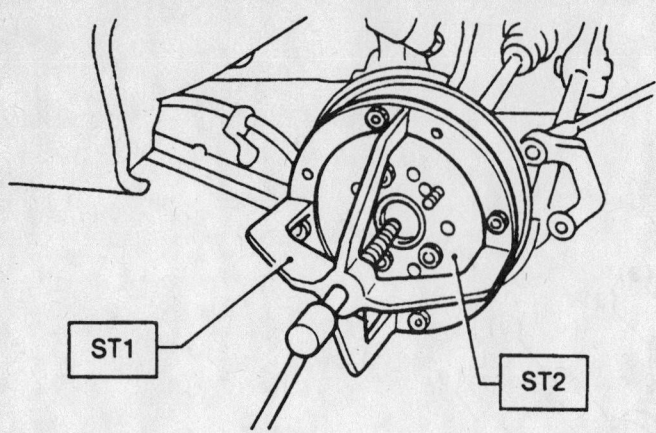

ST1 926470000 AXLE SHAFT PULLER
ST2 927140000 PLATE

7924XG29

Be sure not to damage the threads when removing the front or rear halfshafts

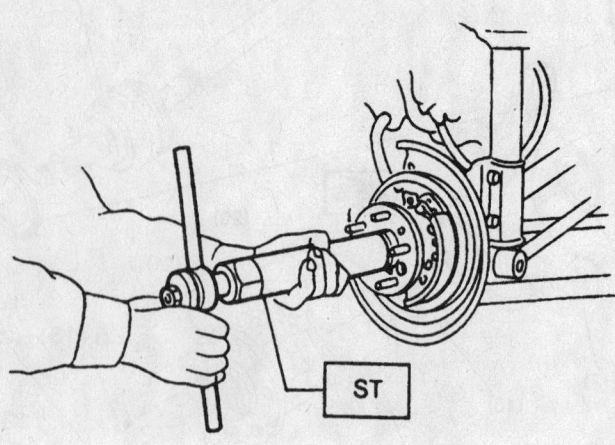

ST1 922431000 AXLE SHAFT INSTALLER
ST2 927390000 ADAPTER

7924XG30

To avoid using a hammer when installing the halfshafts, use the proper tools as shown

- Wheel
- Negative battery cable

Rear

WITH DRUM BRAKES

1. Before servicing the vehicle, refer to the precautions in the beginning of this section.
2. Remove or disconnect the following:
   - Negative battery cable
   - Axle nut
   - Parking brake adjusting nut after returning the lever to the off position
   - Brake drum
   - Brake hose from the wheel cylinder and plug the line and wheel cylinder
   - Parking brake cable from the lever
   - Lateral link assembly to rear housing
   - ABS sensor from the backing plate
   - Halfshaft from the differential
3. Using Axle Shaft puller 926470000 and Plate 927140000, remove the halfshaft from the hub.

   To install:

4. Install the halfshaft into the rear housing.

5. Using Halfshaft Installer 922431000 and adapter 927390000, pull the halfshaft into place.
6. Install and temporarily tighten a new axle nut.
7. Install or connect the following:
   - Halfshaft-to-differential align roll pin holes and slide the halfshaft onto the splines
   - New roll pin
   - Trailing link assembly-to-rear housing bolt and nut. Torque the new nut to 101 ft. lbs. (137 Nm).
   - Trailing link assembly to the rear housing. Torque bolt and new nut to 84 ft. lbs. (113 Nm).
   - Stabilizer bracket
   - Parking brake cable to the lever
   - Brake hose to the wheel cylinder
   - Brake drum
8. Bleed the brake system and adjust the parking brake.
   - New axle nut. Torque the nut to 137 ft. lbs. (186 Nm).
   - ABS sensor to the backing plate
   - Wheel
   - Negative battery cable

WITH DISC BRAKES

1. Before servicing the vehicle, refer to the precautions in the beginning of this section.
2. Remove or disconnect the following:
   - Negative battery cable
   - Axle nut
   - Parking brake adjusting nut after returning the lever to the off position
   - Brake caliper and rotor
   - Parking brake cable from the lever
   - Lateral link assembly to rear housing
   - ABS sensor from the backing plate
   - Halfshaft from the differential
3. Using Axle Shaft puller 926470000 and Plate 927140000, remove the halfshaft from the hub.

   To install:

4. Install the halfshaft into the rear housing.
5. Using Halfshaft Installer 922431000 and adapter 927390000, pull the halfshaft into place.
6. Install and temporarily tighten a new axle nut.
7. Install or connect the following:
   - Halfshaft-to-differential align roll pin holes and slide the halfshaft onto the splines
   - New roll pin
   - Trailing link assembly-to-rear housing bolt and nut. Torque the new nut to 101 ft. lbs. (137 Nm).
   - Trailing link assembly to the rear

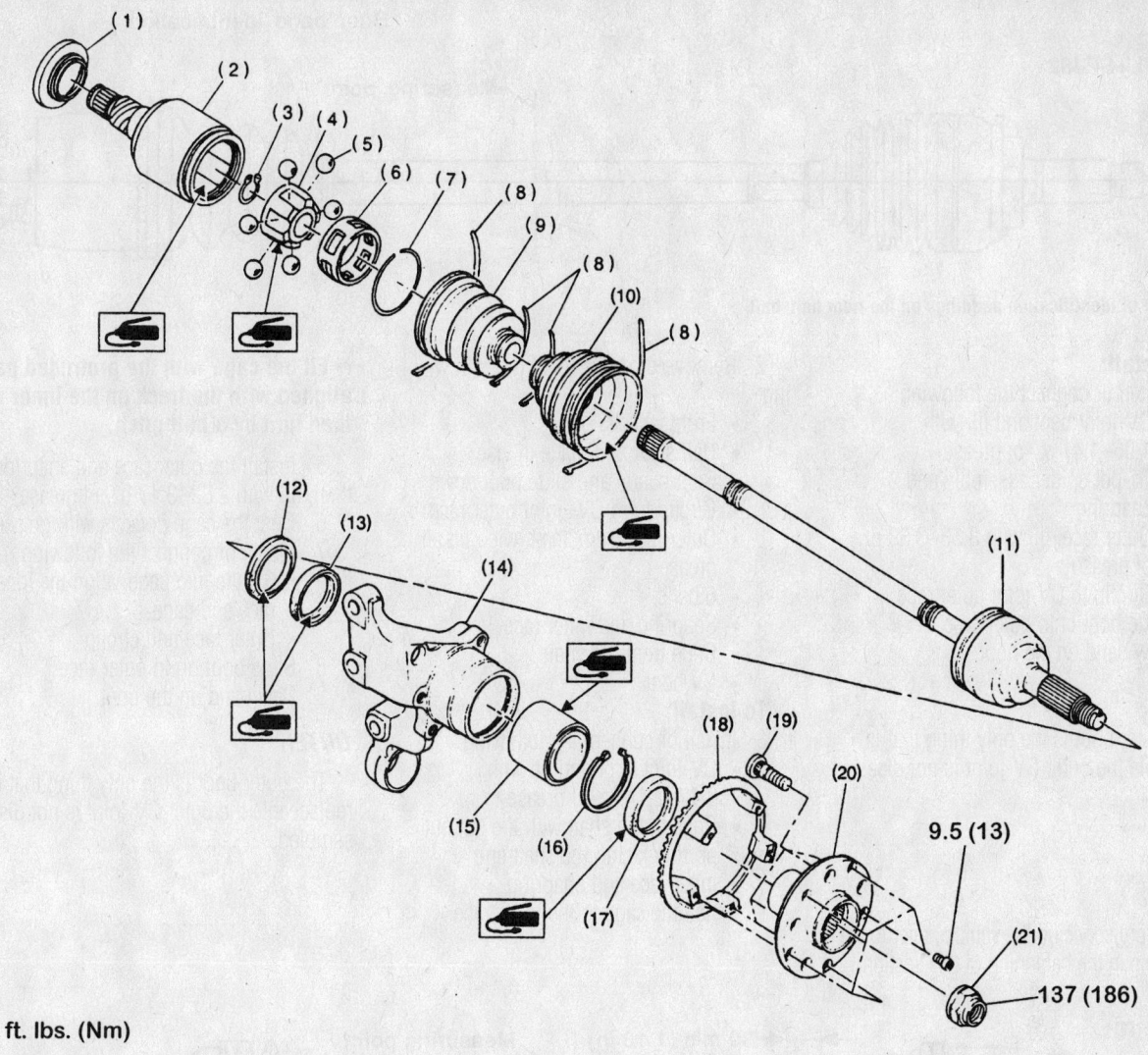

ft. lbs. (Nm)

(1) Baffle plate (DOJ)
(2) Outer race (DOJ)
(3) Snap ring
(4) Inner race
(5) Ball
(6) Cage
(7) Circlip
(8) Boot band
(9) Boot (DOJ)

(10) Boot (BJ)
(11) BJ ASSY
(12) Oil seal (IN. No. 2)
(13) Oil seal (IN. No. 3)
(14) Housing
(15) Bearing
(16) Snap ring
(17) Oil seal (OUT)
(18) Tone wheel

(19) Hub bolt
(20) Hub
(21) Axle nut

7924XG21

Exploded view of the rear halfshaft

housing. Torque bolt and new nut to 84 ft. lbs. (113 Nm).
• Stabilizer bracket
• Parking brake cable to the lever
• Brake rotor and caliper and adjust the parking brake
• New axle nut. Torque the nut to 137 ft. lbs. (186 Nm).
• ABS sensor to the backing plate
• Wheel
• Negative battery cable

CV-Joints

REMOVAL & INSTALLATION

Front

INNER

1. Before servicing the vehicle, refer to the precautions in the beginning of this section.

2. Remove or disconnect the following:
• Front wheel(s)
• Halfshaft and place in vise
• Boot bands and slide boot down
• Circlip from CV-joint outer race
• Outer race from shaft, wipe off all grease
3. Matchmark Tri-pot spider assembly for reassembly.
• Snapring and Tri-pot
• CV-joint boot

• BJ87L+SFJ82

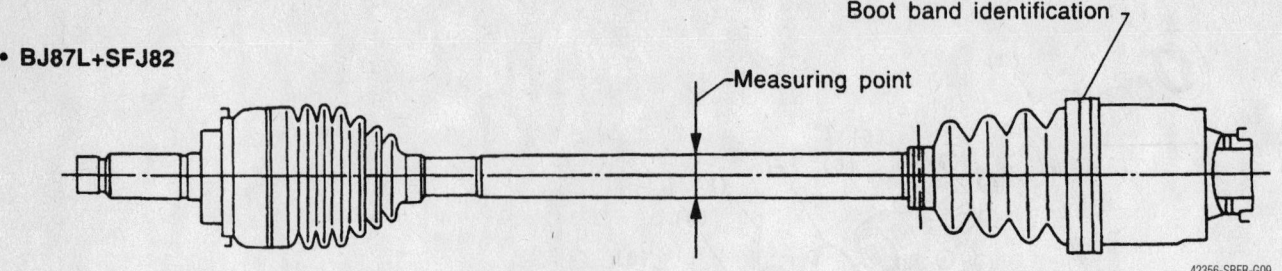

Boot band identification

Measuring point

42356-SBFR-G09

Location of identification paddings on the front halfshaft

To install:
4. Install or connect the following:
   - CV-joint boot and fill with 1.06–1.41 oz. of grease
   - Tri-pot spider assembly and snapring
   - Outer race, fill with 3.53–3.88 oz. of grease
   - Circlip to CV-joint outer race
5. Slide boot onto outer race
6. New band on the boot

OUTER

The outer boot is the only thing that is replaceable the outer CV-joint is not disassembled.

Rear

INNER

1. Before servicing the vehicle, refer to the precautions in the beginning of this section.

2. Remove or disconnect the following:
   - Front wheel(s)
   - Half shaft and place in vise
   - Boot bands and slide boot down
   - Circlip from CV- joint outer race
   - Outer race from shaft, wipe off all grease
   - 6 balls
   - Snapring and inner race
   - Cage from the shaft
   - CV boot

To install:
3. Install or connect the following:
   - CV-joint boot and fill with 2.12–2.47 oz. of grease
   - Cage to the shaft with the cut out portion facing the shaft end
   - Inner race and snapring
4. Install the cage to the inner race.

➡Fit the cage with the protruded part aligned with the track on the inner race then turn by a half pitch.

5. Install the outer race and snapring; then, fill with 2.82–3.17 oz. of grease
6. Coat the cage pockets with grease.
7. Install or connect the following:
   - 6 balls into cage, align the inner race and cage
   - Outer race and circlip
8. Slide boot on to outer race
9. New band on the boot

OUTER

The outer boot is the only thing that is replaceable the outer CV-joint is not disassembled.

• 79AC-RH

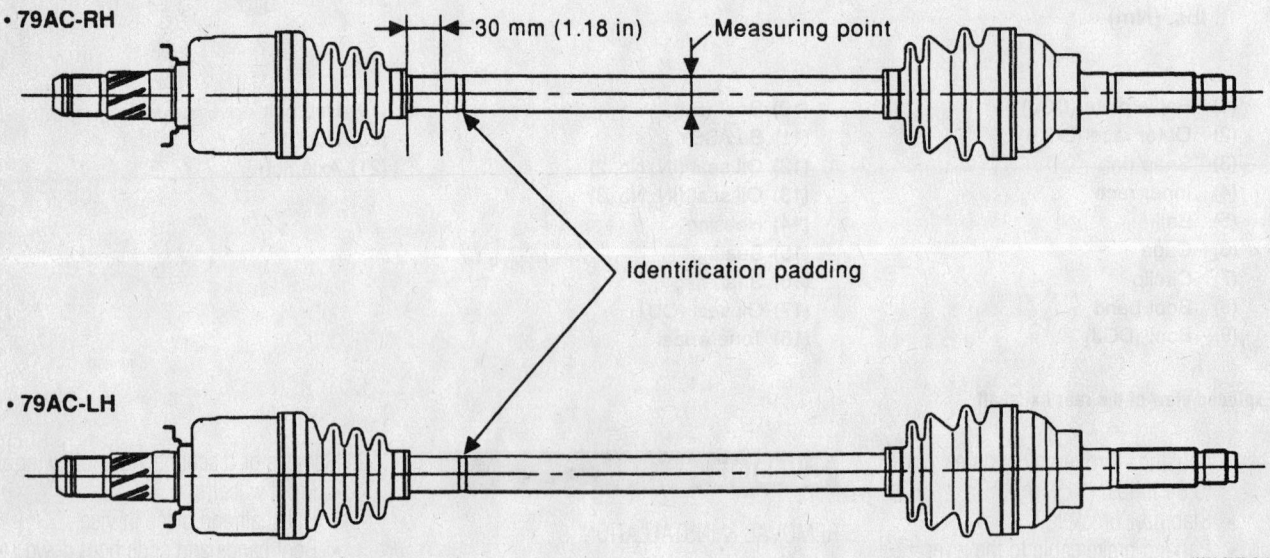

←30 mm (1.18 in)

Measuring point

Identification padding

• 79AC-LH

42356-SBFR-G10

Location of identification paddings on the rear halfshaft

STEERING AND SUSPENSION

Air Bag

❈❈ CAUTION

All vehicles are equipped with an air bag system. The system must be disabled before performing service on or around system components, steering column, instrument panel components, wiring and sensors. Failure to follow safety and disabling procedures could result in accidental air bag deployment, possible personal injury and unnecessary system repairs.

PRECAUTIONS

Several precautions must be observed when handling the inflator module to avoid accidental deployment and possible personal injury.

• Never carry the inflator module by the wires or connector on the underside of the module.

• When carrying a live inflator module, hold securely with both hands, and ensure that the bag and trim cover are pointed away.

• Place the inflator module on a bench or other surface with the bag and trim cover facing up.

• With the inflator module on the bench, never place anything on or close to the module, which may be thrown in the event of an accidental deployment.

DISARMING

1. Before servicing the vehicle, refer to the precautions in the beginning of this section.

2. Disconnect the negative battery cable.

3. Disconnect the positive battery cable.

4. Wait more than 20 seconds to allow the air bag system to deplete its backup power before starting work.

5. To rearm the air bag system, reconnect the positive, then the negative battery cables.

Rack and Pinion Steering Gear

REMOVAL & INSTALLATION

1. Before servicing the vehicle, refer to the precautions in the beginning of this section.

2. Remove or disconnect the following:

• Negative battery cable

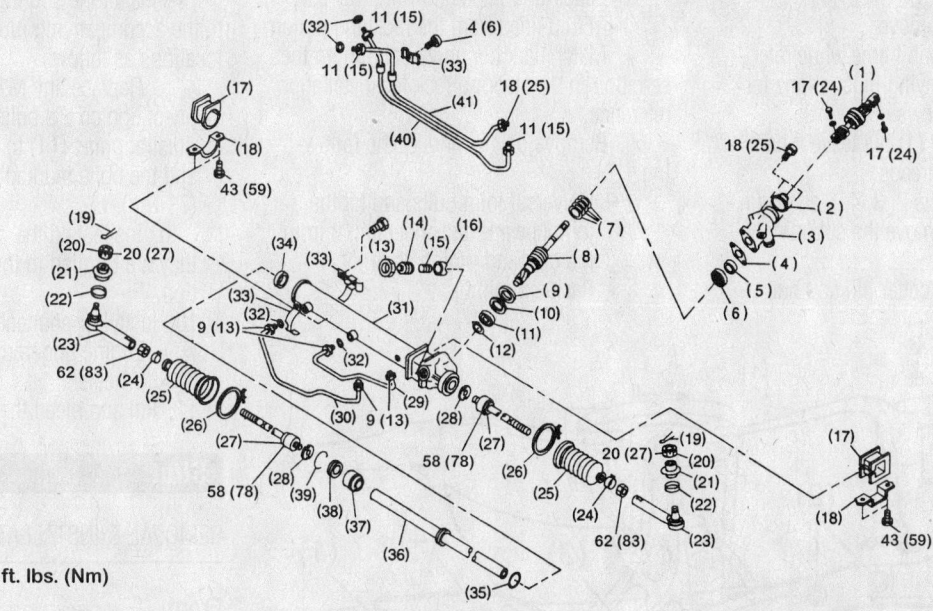

ft. lbs. (Nm)

(1) Universal joint	(19) Cotter pin	(37) Rack bushing
(2) Dust cover	(20) Castle nut	(38) Rack stopper
(3) Valve housing	(21) Dust cover	(39) Circlip
(4) Gasket	(22) Clip	(40) Pipe E
(5) Oil seal	(23) Tie-rod end	(41) Pipe F
(6) Special bearing	(24) Clip	
(7) Seal ring	(25) Boot	
(8) Pinion and valve ASSY	(26) Band	
(9) Oil seal	(27) Tie-rod	
(10) Back-up washer	(28) Lock washer	
(11) Ball bearing	(29) Pipe B	
(12) Snap ring	(30) Pipe A	
(13) Lock nut	(31) Housing ASSY	
(14) Adjusting screw	(32) O-ring	
(15) Spring	(33) Clamp	
(16) Sleeve	(34) Oil seal	
(17) Adapter	(35) Piston ring	
(18) Clamp	(36) Rack	

Exploded view of the rack and pinion steering gear

7924XG22

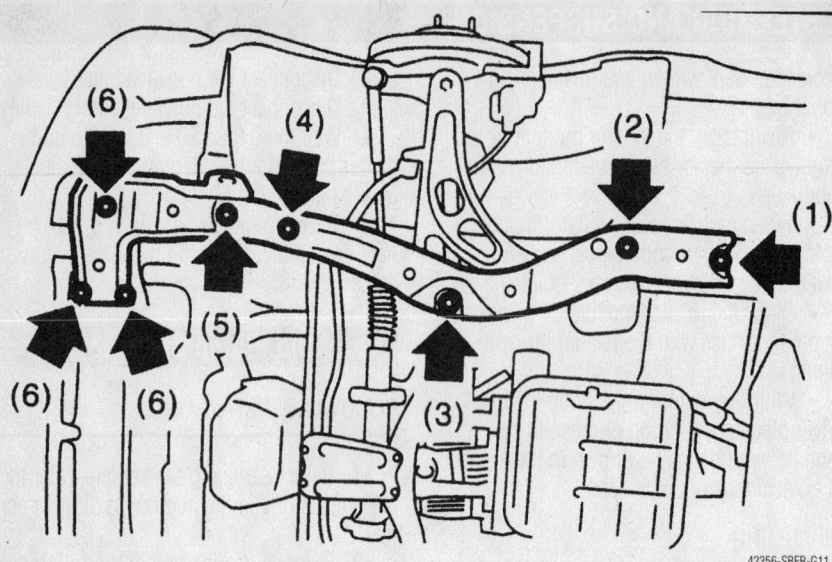

Remove the sub frame bolts in the sequence illustrated following the procedure specified in the text

- Front wheels
- Engine undercover

3. Remove the sub frame while referring to the accompanying illustrations for bolt locations as follows:

  a. Loosen bolt (1) but leave it connected by a few threads.

  b. Remove bolts 2, 3, 4, 5 and 6 (in that order) and remove the subframe.
  - Y-pipe
  - Tie rod end cotter pin and nut

- Jack-up plate and front sway bar
- Fluid lines from the rack and pinion

4. Matchmark the universal joint to the serration in the steering rack for installation reference.

5. Remove or disconnect the following:

- Universal joint bolts and lift the joint upward disconnecting it from the rack and pinion shaft.
- Rack and pinion

To install:

6. Install the rack and pinion. Torque the clamp bolts to 43 ft. lbs. (59 Nm).

7. Align the steering rack to the universal joint. Push the long yoke of the joint all the way into the serrated position of the steering shaft, setting the bolt hole in the cut-out. Pull the short yoke all the way out of the serrated portion of the rack and pinion, setting the bolt hole in the cut-out. Insert the bolt through the short yoke. Pull the yoke and ensure the bolt is properly engaged in the cut-out. Fasten the short yoke side with the spring washer and bolt, then fasten the yoke side. Tighten the bolts to 17 ft. lbs. (24 Nm).

8. Install or connect the following:
- Tie rod ends to the steering knuckle
- Sway bar and jack-up plate
- Y-pipe with new gaskets and nuts

9. Install the sub frame while referring to the accompanying illustrations for bolt locations as follows:

  a. Replace any M12 bolts with new ones. tighten the bolts marked in the illustration as (T1) to 41 ft. lbs. (55 Nm) and the bolts marked (T2) to 52 ft. lbs. (71 Nm).

  b. Inspect all the bolts and make sure they are torqued to the proper specification.

10. Install or connect the following:
- Engine undercover
- Wheels

11. Fill and bleed the steering system.

Strut

REMOVAL & INSTALLATION

Front

✴✴ CAUTION

Do not remove the large nut on top of the strut assembly unless the coil spring is properly compressed with a suitable spring compressor.

1. Before servicing the vehicle, refer to the precautions in the beginning of this section.

2. Remove or disconnect the following:
- Negative battery cable

Install the sub frame bolts and tighten to the specifications listed in the procedure

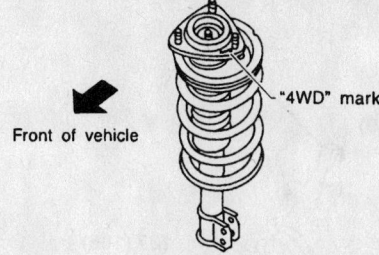

Front of vehicle

"4WD" mark

7924XG31

Position the upper strut bearing as shown—rear strut assembly

- Front wheel assembly
- Brake line to the strut housing bolt
- Caliper, leaving the line connected and suspend it out of the way

3. Matchmark the camber adjustment bolt to the strut housing as reference for installation.

4. Remove or disconnect the following:

- Anti-locking Brakes System (ABS) sensor and harness, if equipped
- Strut from the steering knuckle. Note that the shaft of the top bolt is not round.
- Strut from the body in the engine compartment
- Strut and coil spring assembly

To install:

5. Install the strut and coil assembly. Torque the upper strut retainer nuts to 15 ft. lbs. (20 Nm).

6. Align matchmark on camber adjustment bolt and strut housing.

7. Install or connect the following:

- Lower strut nuts and bolts. Tighten the nuts, while securing the bolts to 112 ft. lbs. (152 Nm) on 2000–01 models and 129 ft. lbs. (175 Nm) on 2002–04 models.
- ABS sensor and harness. Torque the bolt to 24 ft. lbs. (32 Nm), if equipped.
- Brake line to the strut bolt. Torque the bolt to 24 ft. lbs. (32 Nm).
- Caliper
- Front wheel
- Negative battery cable

8. Check and adjust the front end alignment.

Rear

✳✳ CAUTION

Do not remove the large nut on top of the strut assembly unless the coil spring is properly retained with a spring compressor.

1. Before servicing the vehicle, refer to the precautions in the beginning of this section.

2. Remove or disconnect the following:

- Strut mount cap located at the rear interior quarter trim
- Wheel
- Brake hose clip
- Union bolt from the brake caliper, if equipped with disc brakes and move the brake hose out of the way
- Brake hose and pipe from strut and drum, if equipped with drum brakes

3. Support rear with jack.

4. Remove or disconnect the following:

- Retainer nuts securing the strut bearing cap to the strut tower, from inside the vehicle
- Lower nuts and bolts securing the strut to the rear wheel housing
- Strut

To install:

5. Install or connect the following:

- Strut on to the vehicle, making sure to position the strut with the "4WD" mark on the strut mount facing the outside of the vehicle as shown in the illustration. Torque the retaining nuts to 15 ft. lbs. (20 Nm).
- Strut and mount cap. Torque the strut mount cap bolts to 14.5 ft. lbs. (20 Nm).
- Strut to the rear wheel knuckle assembly. Torque the retainer nuts/bolts to 145 ft. lbs. (196 Nm) on 2000—2001 models and 148 ft. lbs. (200 Nm) on 2002–04 models.
- Union bolt, if equipped with disc brakes. Torque the bolt and to 13 ft. lbs. (18 Nm).
- Brake hose to brake pipe, if equipped with drum brakes. Torque to 10 ft. lbs. (15 Nm).
- Brake hose clip
- Wheel
- Strut mount cap

6. Bleed the brakes.

Coil Spring

REMOVAL & INSTALLATION

Front and Rear

✳✳ CAUTION

Do not remove the large nut on top of the strut assembly unless the coil spring is properly compressed with a spring compressor.

1. Before servicing the vehicle, refer to the precautions in the beginning of this section.

2. Remove the strut assembly.

3. Place the strut assembly in a vise with a holding tool and install a spring compressor.

4. Compress the spring slightly.

5. Loosen but do not remove the bearing cap locknut.

6. Unload the spring seat using the spring compressor, then remove the locknut.

7. Remove or disconnect the following:

- Strut bearing cap, mounting insulator bracket and upper spring seat
- Coil spring and compressor. If the spring is being replaced, slowly release the spring from the compressor and compress the new coil spring.
- Strut boot and rebound bumper from the strut, inspect and replace if worn
- Strut retainer nut and the strut insert from the assembly

To install:

8. Install or connect the following:

- Strut into the chamber
- Retainer nut. Tighten the nut until snug.
- Rebound bumper and the boot to the strut piston rod
- Coil spring on the strut assembly
- Upper spring seat, mounting insulator and bearing cap
- Locknut. Tighten it to 41 ft. lbs. (55 Nm).
- Spring compressor from the coil spring
- Strut

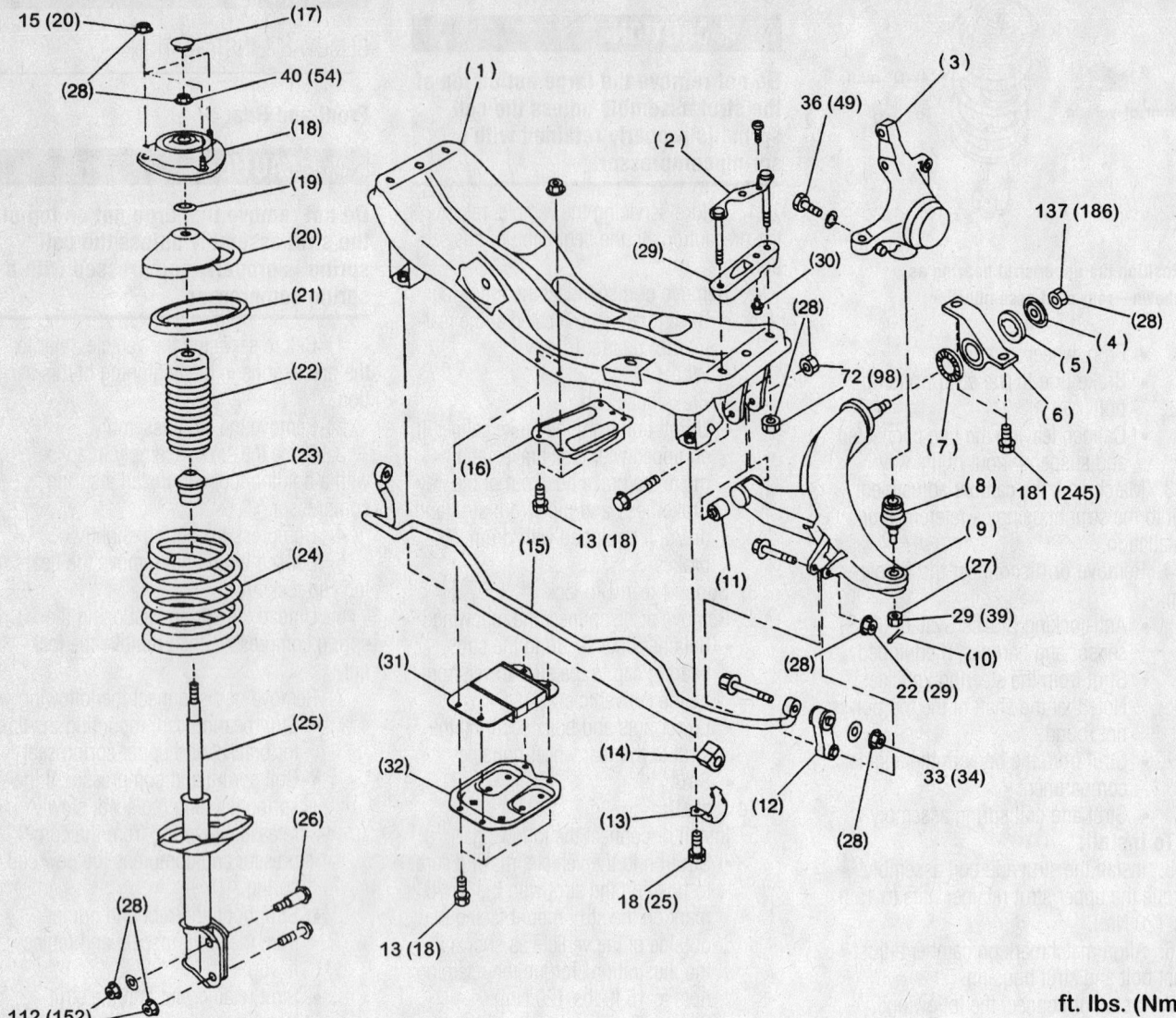

ft. lbs. (Nm)

(1) Front crossmember
(2) Bolt ASSY
(3) Housing
(4) Washer
(5) Stopper rubber (Rear)
(6) Rear bushing
(7) Stopper rubber (Front)
(8) Ball joint
(9) Transverse link
(10) Cotter pin
(11) Front bushing
(12) Stabilizer link
(13) Clamp
(14) Bushing
(15) Stabilizer
(16) Jack-up plate (Except MT model)

(17) Dust seal
(18) Strut mount
(19) Spacer
(20) Upper spring seat
(21) Rubber seat
(22) Dust cover
(23) Helper
(24) Coil spring
(25) Damper strut
(26) Adjusting bolt
(27) Castle nut
(28) Self-locking nut
(29) Adapter front crossmember
(30) Clip
(31) Dynamic damper (MT model)
(32) Jack-up plate (MT model)

7924XG23

Exploded view of the front suspension

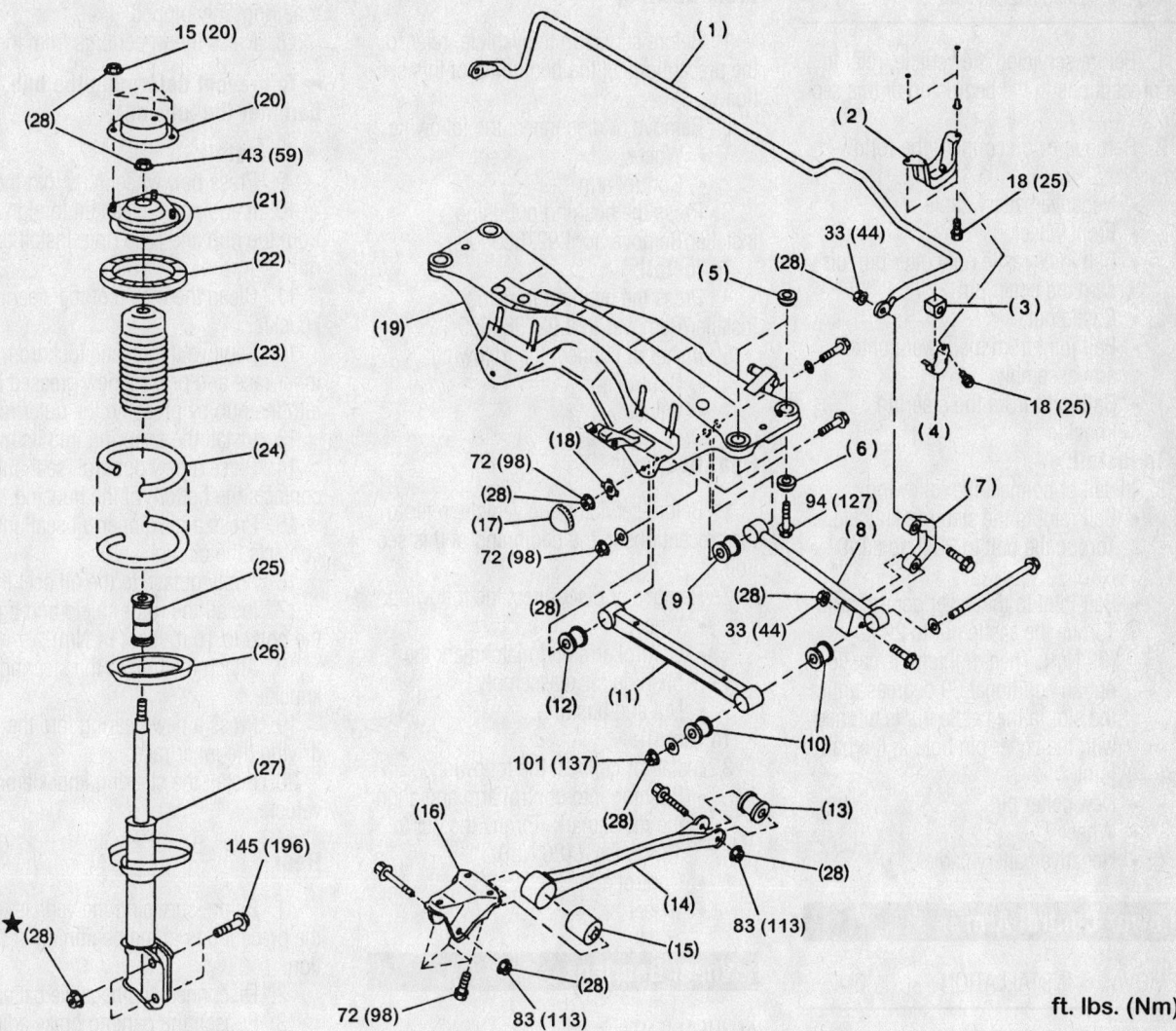

ft. lbs. (Nm)

(1) Stabilizer
(2) Stabilizer bracket
(3) Stabilizer bushing
(4) Clamp
(5) Floating bushing
(6) Stopper
(7) Stabilizer link
(8) Rear lateral link
(9) Bushing (C)
(10) Bushing (A)
(11) Front lateral link
(12) Bushing (B)
(13) Trailing link rear bushing
(14) Trailing link

(15) Trailing link front bushing
(16) Trailing link bracket
(17) Cap (Protection)
(18) Washer
(19) Rear crossmember
(20) Strut mount cap
(21) Strut mount
(22) Rubber seat upper
(23) Dust cover
(24) Coil spring
(25) Helper
(26) Rubber seat lower
(27) Damper strut
(28) Self-locking nut

Exploded view of the rear suspension

7924XG24

Lower Ball Joint

REMOVAL & INSTALLATION

1. Before servicing the vehicle, refer to the precautions in the beginning of this section.
2. Remove or disconnect the following:

- Negative battery cable
- Front wheel
- Ball joint castle nut cotter pin, discard the cotter pin
- Castle nut
- Ball joint from the lower control arm assembly
- Ball joint from the steering knuckle

To install:

3. Install or connect the following:

- Ball joint to the steering knuckle. Torque the bolt to 36 ft. lbs. (49 Nm).
- Ball joint to the lower control arm. Torque the castle nut to 29 ft. lbs. (39 Nm). Then, tighten the castle nut an additional 60 degrees until the slot in the castle nut is aligned with the cotter pin hole in the ball joint.
- New cotter pin
- Wheel
- Negative battery cable

Lower Control Arm

REMOVAL & INSTALLATION

1. Before servicing the vehicle, refer to the precautions in the beginning of this section.
2. Remove or disconnect the following:

- Wheel assembly
- Stabilizer link
- Ball joint from housing
- Mounting bolts
- Control arm

To install:

3. Install or connect the following:

- Control arm to stabilizer. Torque the nut/bolt to 22 ft. lbs. (29 Nm).
- Control arm to crossmember. Torque the nut/bolt to 72 ft. lbs. (98 Nm).
- Control arm to rear mount. Torque the bolts to 181 ft. lbs. (245 Nm).
- Ball joint to housing, Torque the nut to 36 ft. lbs. (49 Nm).
- Wheel assembly

CONTROL ARM BUSHING REPLACEMENT

Front Bushing

1. Before servicing the vehicle, refer to the precautions in the beginning of this section.
2. Remove or disconnect the following:

- Wheel
- Control arm

3. Press the bushing out using Installer/Remover tool 927680000

To install:

4. Press the bushing in using Installer/Remover tool 927680000
5. Install or connect the following:

- Control arm
- Wheel

Rear Bushing

1. Before servicing the vehicle, refer to the precautions in the beginning of this section.
2. Remove or disconnect the following:

- Wheel
- Control arm and matchmark the bushing for reassembly
- Nut and bushing

To install:

3. Install or connect the following:

- Bushing into control arm and align the matchmark. Torque the nut to 137 ft. lbs. (186 Nm).
- Control arm
- Wheel

Wheel Bearings

ADJUSTMENT

The wheel bearings are not adjustable.

REMOVAL & INSTALLATION

Front

1. Before servicing the vehicle, refer to the precautions in the beginning of this section.
2. Remove the steering knuckle assembly.
3. Position the steering knuckle in a soft-jawed vise.
4. Press the hub from the steering knuckle. If the inner bearing race remains in the hub, press it out.
5. Remove or disconnect the following:

- Rotor shield
- Inner and outer seals
- Snapring from the steering knuckle

6. Press the inner bearing race to remove the outer bearing.
7. Remove the Anti-lock Brakes (ABS) tone ring, if equipped
8. Press the wheel lugs from the hub.

➡ **To prevent deforming the hub, do not hammer the lugs out.**

To install:

9. Press new wheel lugs into the hub.
10. If equipped, clean all foreign material from the hub and tone ring. Install the tone ring.
11. Clean the inside of the steering knuckle.
12. Remove the plastic lock from the inner race and press a new greased bearing into the hub by pressing the outer race.
13. Install the snapring into its groove.
14. Press a new outer oil seal until it contacts the bottom of the housing.
15. Press a new inner oil seal until it contacts the circlip.
16. Apply grease to the oil seal lips.
17. Install the rotor shield and tighten the bolts to 10 ft. lbs. (14 Nm).
18. Attach the hub to the steering knuckle.
19. Press a new bearing into the hub by driving the inner race.
20. Install the steering knuckle on the vehicle.

Rear

1. Before servicing the vehicle, refer to the precautions in the beginning of this section.
2. Disconnect the negative battery cable.
3. Loosen the parking brake adjustment.
4. Remove or disconnect the following:

- Wheel
- Axle nut
- Caliper, leaving the line connected if equipped with disc brakes and suspend it aside, then remove the rotor
- Drum and brake line, if equipped with drum brakes
- Parking brake cable
- Rear stabilizer from lateral link
- Trailing link to the housing
- Lateral link to the housing
- Halfshaft
- Anti-lock Brakes (ABS), speed sensor from the backing plate, if equipped
- Strut from the housing
- Housing assembly

5. Using hub stand 92708000 and Hub Remover 927420000

ST1   927080000   HUB STAND
ST2   927420000   HUB REMOVER

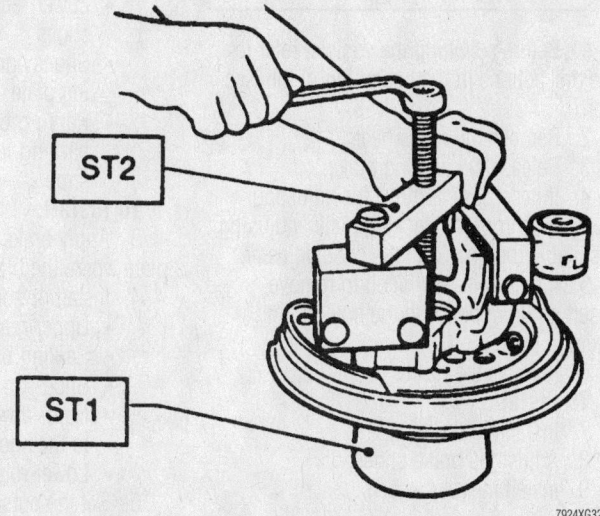

7924XG32

Use the proper tools to separate the hub from the housing to prevent damage

6. Remove or disconnect the following:
- Hub from the rear housing
- Backing plate from the housing.
- Outer, inner and sub oil seals.
- Snapring

7. Remove the bearing by pressing the inner race.

To install:

8. Clean the housing thoroughly.

➡ **Do not remove the plastic lock from the inner race when installing the bearing.**

9. Install the new bearing into the housing by pressing the outer race.

10. Pack the bearing with grease.

11. Install the snapring.

12. Using installer 927460000 seal driver, press in a new outer seal until it comes in contact with the snapring.

13. Using installer 927450000 seal driver, press in a new inner seal until it contacts the bottom.

14. Install or connect the following:
- New sub oil seal, apply grease to the oil seal lip
- Backing plate. Torque the bolts to 38 ft. lbs. (52 Nm).

15. Using installer 927450000 bearing driver, press in the hub into the housing.

16. Install or connect the following:
- Housing to the strut. Torque the bolts to 108 ft. lbs. (147 Nm).
- Halfshaft

- Lateral link to the housing. Torque the bolt and new nut to 101 ft. lbs. (137 Nm).
- Trailing link to the housing. Torque the bolt and new nut to 94 ft. lbs. (127 Nm).
- Stabilizer to rear lateral link
- Parking brake cable and brake
- Brake line, if equipped with drum brakes
- Rotor and caliper, if equipped with disc brakes
- ABS speed sensor, if equipped
- New axle nut and tighten it to 137 ft. lbs. (186 Nm). Stake the nut.
- Wheel
- Negative battery cable

17. Adjust the parking brake cable.

Brake Caliper

REMOVAL & INSTALLATION

Front

1. Before servicing the vehicle, refer to the precautions in the beginning of this section.

2. Remove or disconnect the following:
- Front wheels
- Brake hose from the caliper body
- Caliper retainer bolts and the caliper
- Cliper bracket, if necessary

To install:

3. Compress the piston assembly into the cylinder bore.

4. Install or connect the following:
- Caliper bracket to the spindle assembly, if removed and secure in place with the retainer bolts. Tighten the retainer bolts to 59 ft. lbs. (80 Nm).
- Caliper and tighten the retainers to 28 ft. lbs. (37 Nm).
- Brake hose using new sealing washers, and tighten the fitting to 13 ft. lbs. (18 Nm)

5. Bleed the brake system.

6. Install the wheels and check the fluid level in the master cylinder.

Rear

1. Before servicing the vehicle, refer to the precautions in the beginning of this section.

2. Remove or disconnect the following:
- Rear wheels
- Brake hose from the caliper body
- Caliper retainer bolts and the caliper
- Caliper bracket, if necessary

To install:

3. Compress the piston assembly into the cylinder bore.

4. Install or connect the following:
- Caliper bracket, if removed and secure in place with the retainer bolts. Tighten the retainer bolts to 38 ft. lbs. (52 Nm).
- Caliper and tighten the retainers to 28 ft. lbs. (37 Nm).
- Brake hose using new sealing washers, and tighten the fitting to 13 ft. lbs. (18 Nm)

5. Bleed the brake system.

6. Install the wheels and check the fluid level in the master cylinder.

Disc Brake Pads

REMOVAL & INSTALLATION

Front

1. Before servicing the vehicle, refer to the precautions in the beginning of this section.

2. Remove or disconnect the following:
- Wheels
- Lower caliper bolt
- Swing the caliper upward to access the pads
- Disc brake pads

To install:

3. Compress the caliper piston.
4. Install or connect the following:
   - New pads into the caliper brackets, being sure all shims and clips are in their original positions.

5. Swing the calipers down into position and install the caliper bolt. Tighten the bolt to 28 ft. lbs. (37 Nm).
6. Fill the master cylinder reservoir.
7. Install the wheels.

Rear

1. Before servicing the vehicle, refer to the precautions in the beginning of this section.
2. Remove or disconnect the following:
   - Wheels
   - Lower caliper bolt
   - Swing the caliper upward to access the pads
   - Disc brake pads

To install:

3. Compress the caliper piston.
4. Install or connect the following:
   - New pads into the caliper brackets, being sure all shims and clips are in their original positions.

5. Swing the calipers down into position and install the caliper bolt. Tighten the bolt to 28 ft. lbs. (37 Nm).

6. Fill the master cylinder reservoir.
7. Install the wheels.

Brake Drums

REMOVAL & INSTALLATION

1. Before servicing the vehicle, refer to the precautions in the beginning of this section.
2. Remove the rear wheels.
3. Release the parking brake.
4. If necessary, remove the adjusting hole cover from the backing plate and using a suitable tool, back off the shoe adjuster.
5. If the drum is difficult to remove, insert an 8mm bolt into the hole on the drum to push it off.
6. Remove the drum.

To install:

7. Install the drum.
8. Adjust the brake shoes.
9. Install the rear wheels.

Brake Shoes

REMOVAL & INSTALLATION

1. Before servicing the vehicle, refer to the precautions in the beginning of this section.

2. Remove or disconnect the following:
   - Wheels
   - Brake drum
   - Hold–down pins, springs and cups from the shoes
   - Lower return spring from both shoes
   - Shoes and adjuster from the backing plate
   - Parking brake cable from the parking lever on the trailing brake shoe

To install:

3. Apply brake grease to the backing plate where the brake shoes contact it.
4. Install or connect the following:
   - Upper return to the shoes
   - Parking brake cable to the lever
   - Shoes on the backing plate
   - Hold–down pins, springs and cups to the shoes
   - Lower return spring to both shoes

5. Set the outside diameter of the shoes less than 0.020–0.031 in. (0.5–0.8mm) compared to the inside diameter of the drum.

6. Install the drum and adjust the brake shoes.

7. Install the wheels.

SUZUKI

Esteem • Swift

30

SPECIFICATION CHARTS

VEHICLE AND ENGINE IDENTIFICATION

Engine							Model Year	
Code	Liters (cc)	Cu. in.	Cyl.	Fuel Sys.	Engine Type	Eng. Mfg.	Code	Year
2	1.3 (1298)	79.3	4	TFI	SOHC	Suzuki	Y	2000
3	1.6 (1590)	97.7	4	MFI	SOHC	Suzuki	1	2001
							2	2002
							3	2003
							4	2004

MFI: Multi-port Fuel Injection

TFI: Throttle body Fuel Injection

SOHC: Single Overhead Camshaft

42356-SUZC-C01

GENERAL ENGINE SPECIFICATIONS

Year	Engine ID/VIN	Engine Displacement Liters (cc)	Fuel System Type	Net Horsepower @ rpm	Net Torque @ rpm (ft. lbs.)	Bore x Stroke (in.)	Compression Ratio	Oil Pressure @ rpm
2000	2	1.3 (1298)	TFI	70@5500	74@3000	2.91x2.97	9.5:1	47-61@3000
	3	1.6 (1590)	MFI	98@6000	94@3200	2.95x3.54	9.5:1	47-61@4000
2001	2	1.3 (1298)	TFI	70@5500	74@3000	2.91x2.97	9.5:1	47-61@3000
	3	1.6 (1590)	MFI	98@6000	94@3200	2.95x3.54	9.5:1	47-61@4000
2002	2	1.3 (1298)	TFI	70@5500	74@3000	2.91x2.97	9.5:1	47-61@3000
	3	1.6 (1590)	MFI	98@6000	94@3200	2.95x3.54	9.5:1	47-61@4000
2003	2	1.3 (1298)	TFI	70@5500	74@3000	2.91x2.97	9.5:1	47-61@3000
	3	1.6 (1590)	MFI	98@6000	94@3200	2.95x3.54	9.5:1	47-61@4000

MFI: Multiport Fuel Injection

TFI: Throttle body Fuel Injection

42356-SUZC-C02

GASOLINE ENGINE TUNE-UP SPECIFICATIONS

Year	Engine Displacement Liters (cc)	Engine ID/VIN	Spark Plugs Gap (in.)	Ignition Timing (deg.) MT	Ignition Timing (deg.) AT	Fuel Pump (psi)	Idle Speed (rpm) MT	Idle Speed (rpm) AT	Valve Clearance In.	Valve Clearance Ex.
2000	1.3 (1298)	2	0.029	5B	5B	13-20 ①	750	850	HYD	HYD
	1.6 (1590)	3	0.029	5B	5B	28-34 ①	750-800	750-800	②	②
2001	1.3 (1298)	2	0.029	5B	5B	13-20 ①	750	850	HYD	HYD
	1.6 (1590)	3	0.029	5B	5B	28-34 ①	750-800	750-800	②	②
2002	1.6 (1590)	3	0.029	5B	5B	28-34 ①	750-800	750-800	②	②
2003	1.6 (1590)	3	0.029	5B	5B	28-34 ①	750-800	750-800	②	②

Note: The Vehicle Emission Control Information label often reflects specification changes made during production. The label figures must be used if they differ from those in this chart.

HYD: Hydraulic

B: Before top dead center

① At idle

② When cold: 0.005-0.007

When hot: 0.007-0.008

42356-SUZC-C03

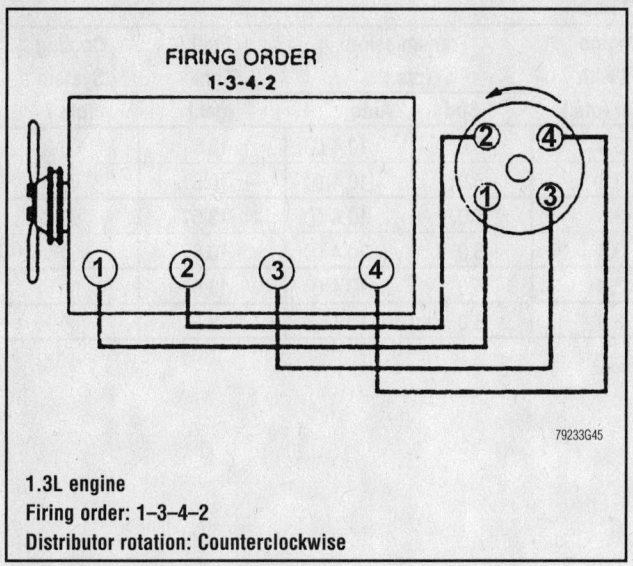

1.3L engine
Firing order: 1–3–4–2
Distributor rotation: Counterclockwise

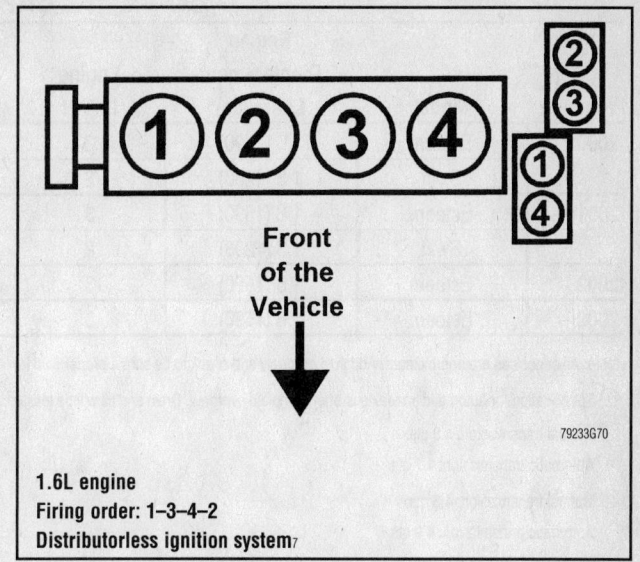

1.6L engine
Firing order: 1–3–4–2
Distributorless ignition system[7]

For vehicle equipped with A/C

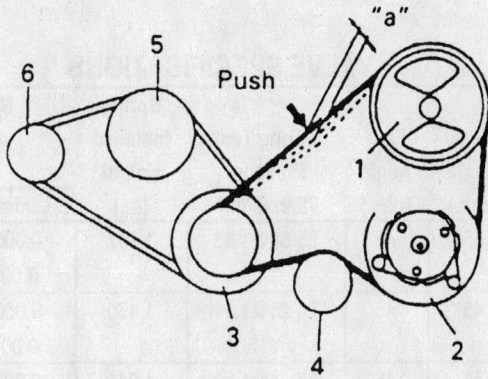

For vehicle not equipped with A/C

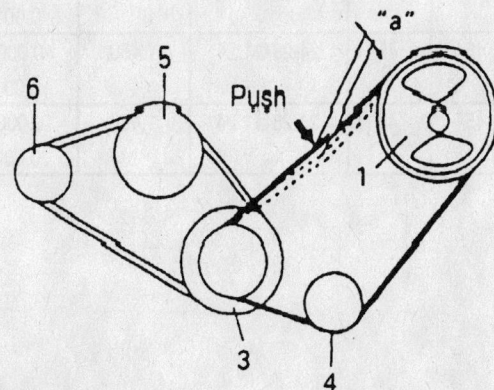

1. P/S pump pulley
2. A/C compressor pulley
3. Crankshaft pulley
4. Tension pulley
5. Water pump pulley
6. Generator

Accessory drive belt routing—1.3L and 1.6L engines

CAPACITIES

Year	Model	Engine Displacement Liters (cc)	Engine ID/VIN	Engine Oil with Filter (qts.)	Transmission (pts.) 5-Spd	Auto.	Fuel Tank (gal.)	Cooling System (qts.)
2000	Esteem	1.6 (1590)	3	3.3	5.0	10.4 ①	13.5	②
	Swift	1.3 (1298)	2	3.3	5.0	10.4 ①	10.6	③
2001	Esteem	1.6 (1590)	3	3.3	5.0	10.4 ①	13.5	②
	Swift	1.3 (1298)	2	3.3	5.0	10.4 ①	10.6	③
2002	Esteem	1.6 (1590)	3	3.3	5.0	10.4 ①	13.5	②
2003	Esteem	1.6 (1590)	3	3.3	5.0	10.4 ①	13.5	②

Note: All capacities are approximate. Add fluid gradualy and check to be sure a proper fluid level is obtained.

① Specification for automatic transaxle is after complete overhaul. Drain and fill will be less

② Manual transmission: 4.8 qts.
Automatic transmission: 4.7 qts.

③ Manual transmission: 4.8 qts.
Automatic transmission: 4.9 qts.

42356-SUZC-C04

VALVE SPECIFICATIONS

Year	Engine ID/VIN	Engine Displacement Liters (cc)	Seat Angle (deg.)	Face Angle (deg.)	Spring Test Pressure (lbs. @ in.)	Spring Installed Height (in.)	Stem-to-Guide Clearance (in.) Intake	Exhaust	Stem Diameter (in.) Intake	Exhaust
2000	2	1.3 (1298)	45	45	55-64@1.63	1.941	0.0008-0.0019	0.0014-0.0025	0.2742-0.2748	0.2737-0.2742
	3	1.6 (1590)	45	45	24-28@1.24	1.450	0.0008-0.0018	0.0018-0.0028	0.2152-0.2157	0.2142-0.2148
2001	2	1.3 (1298)	45	45	55-64@1.63	1.941	0.0008-0.0019	0.0014-0.0025	0.2742-0.2748	0.2737-0.2742
	3	1.6 (1590)	45	45	24-28@1.24	1.450	0.0008-0.0018	0.0018-0.0028	0.2152-0.2157	0.2142-0.2148
2002	3	1.6 (1590)	45	45	24-28@1.24	1.450	0.0008-0.0018	0.0018-0.0028	0.2152-0.2157	0.2142-0.2148
2003	3	1.6 (1590)	45	45	24-28@1.24	1.450	0.0008-0.0018	0.0018-0.0028	0.2152-0.2157	0.2142-0.2148

42356-SUZC-C05

PISTON AND RING SPECIFICATIONS

All measurements are given in inches.

Year	Engine ID/VIN	Engine Displacement Liters (cc)	Piston Clearance	Ring Gap			Ring Side Clearance		
				Top Compression	Bottom Compression	Oil Control	Top Compression	Bottom Compression	Oil Control
2000	2	1.3 (1298)	0.0008-0.0015	0.0079-0.0118	0.0079-0.0118	0.0079-0.0275	0.0012-0.0027	0.0008-0.0023	snug
	3	1.6 (1590)	0.0008-0.0015	0.0079-0.0137	0.0079-0.0137	0.0079-0.0275	0.0012-0.0027	0.0008-0.0023	snug
2001	2	1.3 (1298)	0.0008-0.0015	0.0079-0.0118	0.0079-0.0118	0.0079-0.0275	0.0012-0.0027	0.0008-0.0023	snug
	3	1.6 (1590)	0.0008-0.0015	0.0079-0.0137	0.0079-0.0137	0.0079-0.0275	0.0012-0.0027	0.0008-0.0023	snug
2002	3	1.6 (1590)	0.0008-0.0015	0.0079-0.0137	0.0079-0.0137	0.0079-0.0275	0.0012-0.0027	0.0008-0.0023	snug
2003	3	1.6 (1590)	0.0008-0.0015	0.0079-0.0137	0.0079-0.0137	0.0079-0.0275	0.0012-0.0027	0.0008-0.0023	snug

42356-SUZC-C06

CRANKSHAFT AND CONNECTING ROD SPECIFICATIONS

All measurements are given in inches.

Year	Engine ID/VIN	Engine Displacement Liters (cc)	Crankshaft				Connecting Rod		
			Main Brg. Journal Dia.	Main Brg. Oil Clearance	Shaft End-play	Thrust on No.	Journal Diameter	Oil Clearance	Side Clearance
2000	3	1.3 (1298)	①	0.0008-0.0023	0.0044-0.0149	3	1.6529-1.6535	0.0008-0.0031	0.0039-0.0137
	6	1.6 (1590)	②	0.0008-0.0016	0.0044-0.0122	3	1.7316-1.7322	0.0008-0.0019	0.0039-0.0078
2001	3	1.3 (1298)	①	0.0008-0.0023	0.0044-0.0149	3	1.6529-1.6535	0.0008-0.0031	0.0039-0.0137
	6	1.6 (1590)	②	0.0008-0.0016	0.0044-0.0122	3	1.7316-1.7322	0.0008-0.0019	0.0039-0.0078
2002	6	1.6 (1590)	②	0.0008-0.0016	0.0044-0.0122	3	1.7316-1.7322	0.0008-0.0019	0.0039-0.0078
2003	6	1.6 (1590)	②	0.0008-0.0016	0.0044-0.0122	3	1.7316-1.7322	0.0008-0.0019	0.0039-0.0078

① No. 1: 1.7714-1.7716
No. 2: 1.7712-1.7714
No. 3: 1.7710-1.7712

② No. 1: 2.0470-2.0472
No. 2: 2.0468-2.0470
No. 3: 2.0465-2.0468

42356-SUZC-C07

TORQUE SPECIFICATIONS
All readings in ft. lbs.

Year	Engine ID/VIN	Engine Displacement Liters (cc)	Cylinder Head Bolts	Main Bearing Bolts	Rod Bearing Bolts	Crankshaft Damper Bolts	Flywheel Bolts	Manifold Intake	Manifold Exhaust	Spark Plugs	Lug Nut
2000	2	1.3 (1298)	①	36-41	24-26	76-83 ②	41-47	13-20	13-20	14-21	36-57
	3	1.6 (1590)	①	36-41	24-26	76-83 ②	57	13-20	13-20	14-21	58-80
2001	2	1.3 (1298)	①	36-41	24-26	76-83 ②	41-47	13-20	13-20	14-21	36-57
	3	1.6 (1590)	①	36-41	24-26	76-83 ②	57	13-20	13-20	14-21	58-80
2002	3	1.6 (1590)	①	36-41	24-26	76-83 ②	57	13-20	13-20	14-21	58-80
2003	3	1.6 (1590)	①	36-41	24-26	76-83 ②	57	13-20	13-20	14-21	58-80

① Step 1: 26 ft. lbs. (35 Nm)

  Step 2: 41 ft. lbs. (55 Nm)

  Step 3: 49 ft. lbs. (68 Nm)

② Specification shown is for crankshaft timing sprocket nut

42356-SUZC-C08

WHEEL ALIGNMENT

Year	Model		Caster Range (+/-Deg.)	Caster Preferred Setting (Deg.)	Camber Range (+/-Deg.)	Camber Preferred Setting (Deg.)	Toe-in (in.)	Steering Axis Inclination (Deg.)
2000	Swift	F	2.00	+3.00	1.00	+0.50	0.08+/-0.08	—
		R	—	—	—	—	0.18+/-0.06	—
	Esteem	F	2.00	+2.70	1.00	0	0+/-0.08	—
		R	—	—	—	0	0.08+/-0.08	—
2001	Swift	F	2.00	+3.00	1.00	+0.50	0.08+/-0.08	—
		R	—	—	—	—	0.18+/-0.06	—
	Esteem	F	2.00	+2.70	1.00	0	0+/-0.08	—
		R	—	—	—	0	0.08+/-0.08	—
2002	Esteem	F	2.00	+2.70	1.00	0	0+/-0.08	—
		R	—	—	—	0	0.08+/-0.08	—
2003	Esteem	F	2.00	+2.70	1.00	0	0+/-0.08	—
		R	—	—	—	0	0.08+/-0.08	—

42356-SUZC-C09

TIRE, WHEEL AND BALL JOINT SPECIFICATIONS

Year	Model	OEM Tires		Tire Pressures (psi)		Wheel Size	Ball Joint Inspection
		Standard	Optional	Front	Rear		
2000	Swift	P155/80R13	None	32	32	4.5J	①
	Esteem GL	P175/70R13	None	30	30	5J	①
	Esteem GL Wagon	P185/60R14	None	30	30	5.5JJ	①
	Esteem GLX	P185/60R14	None	30	30	5.5JJ	①
	Esteem GLX Wagon	P195/55R15	None	29	29	5.5JJ	①
2001	Swift	P155/80R13	None	32	32	4.5J	①
	Esteem GL	P175/70R13	None	30	30	5J	①
	Esteem GL Wagon	P185/60R14	None	30	30	5.5JJ	①
	Esteem GLX	P185/60R14	None	30	30	5.5JJ	①
	Esteem GLX Wagon	P195/55R15	None	29	29	5.5JJ	①
2002	Esteem GL	P175/70R13	None	30	30	5J	①
	Esteem GL Wagon	P185/60R14	None	30	30	5.5JJ	①
	Esteem GLX	P185/60R14	None	30	30	5.5JJ	①
	Esteem GLX Wagon	P195/55R15	None	29	29	5.5JJ	①
2003	Esteem GL	P175/70R13	None	30	30	5J	①
	Esteem GL Wagon	P185/60R14	None	30	30	5.5JJ	①
	Esteem GLX	P185/60R14	None	30	30	5.5JJ	①
	Esteem GLX Wagon	P195/55R15	None	29	29	5.5JJ	①

OEM: Original Equipment Manufacturer

PSI: Pounds Per Square Inch

STD: Standard

OPT: Optional

① Replace if any measurable movement is found.

42356-SUZC-C10

BRAKE SPECIFICATIONS
All measurements in inches unless noted

Year	Model	Brake Disc			Brake Drum Diameter			Minimum Lining Thickness		Brake Caliper	
		Original Thickness	Minimum Thickness	Maximum Runout	Original Inside Diameter	Max. Wear Limit	Maximum Machine Diameter	Front	Rear	Bracket bolts ft lbs.	Mounting bolts ft lbs.
2000	Swift ①	0.670	0.620	0.004	7.09	7.87	7.87	0.236 ②	0.110 ②	36	27
	Swift	0.670	0.620	0.004	7.16	7.95	7.95	0.236 ②	0.110 ②	36	27
	Esteem	0.790	0.710	0.004	7.87	7.95	7.95	0.240 ②	0.110 ②	62	16
2001	Swift ①	0.670	0.620	0.004	7.09	7.87	7.87	0.236 ②	0.110 ②	36	27
	Swift	0.670	0.620	0.004	7.16	7.95	7.95	0.236 ②	0.110 ②	36	27
	Esteem	0.790	0.710	0.004	7.87	7.95	7.95	0.240 ②	0.110 ②	62	16
2002	Esteem	0.790	0.710	0.004	7.87	7.95	7.95	0.240 ②	0.110 ②	62	16
2003	Esteem	0.790	0.710	0.004	7.87	7.95	7.95	0.240 ②	0.110 ②	62	16

① Hatchback

② Measurement is for lining and backing together.

42356-SUZC-C11

SCHEDULED MAINTENANCE INTERVALS
SUZUKI—ESTEEM & SWIFT

TO BE SERVICED	TYPE OF SERVICE	VEHICLE MILEAGE INTERVAL (x1000)												
		7.5	15	22.5	30	37.5	45	52.5	60	67.5	75	82.5	90	97.5
Engine oil & filter	R	✓	✓	✓	✓	✓	✓	✓	✓	✓	✓	✓	✓	✓
Automatic transmission fluid & filter ①	S/I	✓	✓	✓	✓	✓	✓	✓	✓	✓	✓	✓	✓	✓
Clutch pedal free travel	S/I	✓	✓	✓	✓	✓	✓	✓	✓	✓	✓	✓	✓	✓
Drive axle boots	S/I	✓	✓	✓	✓	✓	✓	✓	✓	✓	✓	✓	✓	✓
Gear shift control lever/shift operation	S/I	✓	✓	✓	✓	✓	✓	✓	✓	✓	✓	✓	✓	✓
Inspect & rotate tires	S/I	✓	✓	✓	✓	✓	✓	✓	✓	✓	✓	✓	✓	✓
Manual transmission oil ②	S/I	✓	✓	✓	✓	✓	✓	✓	✓	✓	✓	✓	✓	✓
Power steering system	S/I	✓	✓	✓	✓	✓	✓	✓	✓	✓	✓	✓	✓	✓
Suspension system	S/I	✓	✓	✓	✓	✓	✓	✓	✓	✓	✓	✓	✓	✓
Brake discs, pads, drums & shoes	S/I	✓		✓		✓		✓		✓		✓		✓
Brake hoses, pipes, brake lever & cable	S/I	✓		✓		✓		✓		✓		✓		✓
Brake fluid ③	S/I		✓		✓		✓		✓		✓		✓	
Brake pedal	S/I		✓		✓		✓		✓		✓		✓	
Cooling system, hoses & connections	S/I		✓		✓		✓		✓		✓		✓	
Fuel tank, cap & lines	S/I		✓		✓		✓		✓		✓		✓	
Valve lash (clearance)	S/I		✓		✓		✓		✓		✓		✓	
Air cleaner filter element	R				✓				✓				✓	
Engine coolant	R				✓				✓				✓	
Spark plugs	R				✓				✓				✓	
Drive belts	S/I				✓				✓				✓	
Exhaust system	S/I				✓				✓				✓	

42356-SUZC-C12

SCHEDULED MAINTENANCE INTERVALS
SUZUKI—ESTEEM & SWIFT

TO BE SERVICED	TYPE OF SERVICE	VEHICLE MILEAGE INTERVAL (x1000)												
		7.5	15	22.5	30	37.5	45	52.5	60	67.5	75	82.5	90	97.5
Automatic transmission fluid hose	R						✓							
Camshaft timing belt	R								✓					
Ignition wiring	S/I								✓					

R: Replace S/I: Service or Inspect

① Replace every 100,000 miles.

② Replace every 15,000 miles.

③ Replace every 60,000 miles.

FREQUENT OPERATION MAINTENANCE (SEVERE SERVICE)

If a vehicle is operated under any of the following conditions it is considered severe service:

- Extremely dusty areas.

- 50% or more of the vehicle operation is in 32°C (90°F) or higher temperatures, or constant operation in temperatures below 0°C (32°F).

- Prolonged idling (vehicle operation in stop and go traffic).

- Frequent short running periods (engine does not warm to normal operating temperatures).

- Police, taxi, delivery usage or trailer towing usage.

Oil & oil filter: change every 3000 miles.

Brake discs, pads, drums & shoes: service or inspect initially at 3000 miles, 6000 miles, & every 12,000 miles thereafter.

Brake hoses & pipes: service or inspect initially at 3000 miles, 6000 miles & every 12,000 miles thereafter.

Air cleaner filter element: service or inspect ever 3000 miles & replace every 30,000 miles (if not replaced previously).

Automatic transmission fluid & filter: service or inspect every 6000 miles & replace every 15,000 miles (if not replaced previously).

Clutch pedal free travel: service or inspect every 6000 miles.

Inspect & rotate tires: service or inspect every 6000 miles.

Manual transmission oil: service or inspect every 6000 miles & replace every 12,000 miles (if not replaced previously).

Power steering system: service or inspect every 6000 miles.

Steering system: service or inspect every 6000 miles.

Suspension system: service or inspect every 6000 miles.

Drive belts: service or inspect every 15,000 miles.

Exhaust system: service or inspect every 15,000 miles.

42356-SUZC-C13

PRECAUTIONS

Before servicing any vehicle, please be sure to read all of the following precautions, which deal with personal safety, prevention of component damage, and important points to take into consideration when servicing a motor vehicle:

• Never open, service or drain the radiator or cooling system when the engine is hot; serious burns can occur from the steam and hot coolant.

• Observe all applicable safety precautions when working around fuel. Whenever servicing the fuel system, always work in a well-ventilated area. Do not allow fuel spray or vapors to come in contact with a spark, open flame, or excessive heat (a hot drop light, for example). Keep a dry chemical fire extinguisher near the work area. Always keep fuel in a container specifically designed for fuel storage; also, always properly seal fuel containers to avoid the possibility of fire or explosion. Refer to the additional fuel system precautions later in this section.

• Fuel injection systems often remain pressurized, even after the engine has been turned **OFF**. The fuel system pressure must be relieved before disconnecting any fuel lines. Failure to do so may result in fire and/or personal injury.

• Brake fluid often contains polyglycol ethers and polyglycols. Avoid contact with the eyes and wash your hands thoroughly after handling brake fluid. If you do get brake fluid in your eyes, flush your eyes with clean, running water for 15 minutes. If eye irritation persists, or if you have taken brake fluid internally, IMMEDIATELY seek medical assistance.

• The EPA warns that prolonged contact with used engine oil may cause a number of skin disorders, including cancer! You should make every effort to minimize your exposure to used engine oil. Protective gloves should be worn when changing oil. Wash your hands and any other exposed skin areas as soon as possible after exposure to used engine oil. Soap and water, or waterless hand cleaner should be used.

• All new vehicles are now equipped with an air bag system. The system must be disabled before performing service on or around system components, steering column, instrument panel components, wiring and sensors. Failure to follow safety and disabling procedures could result in accidental air bag deployment, possible personal injury and unnecessary system repairs.

• Always wear safety goggles when working with, or around, the air bag system. When carrying a non-deployed air bag, be sure the bag and trim cover are pointed away from your body. When placing a non-deployed air bag on a work surface, always face the bag and trim cover upward, away from the surface. This will reduce the motion of the module if it is accidentally deployed. Refer to the additional air bag system precautions later in this section.

• Clean, high quality brake fluid from a sealed container is essential to the safe and proper operation of the brake system. You should always buy the correct type of brake fluid for your vehicle. If the brake fluid becomes contaminated, completely flush the system with new fluid. Never reuse any brake fluid. Any brake fluid that is removed from the system should be discarded. Also, do not allow any brake fluid to come in contact with a painted surface; it will damage the paint.

• Never operate the engine without the proper amount and type of engine oil; doing so WILL result in severe engine damage.

• Timing belt maintenance is extremely important! Many models utilize an interference-type, non-freewheeling engine. If the timing belt breaks, the valves in the cylinder head may strike the pistons, causing potentially serious (also time-consuming and expensive) engine damage. Refer to the maintenance interval charts in the front of this section for the recommended replacement interval for the timing belt.

• Disconnecting the negative battery cable on some vehicles may interfere with the functions of the on-board computer system(s) and may require the computer to undergo a relearning process once the negative battery cable is reconnected.

• When servicing drum brakes, only dissemble and assemble one side at a time, leaving the remaining side intact for reference.

• Only an MVAC-trained, EPA-certified automotive technician should service the air conditioning system or its components.

ENGINE REPAIR

→**Disconnecting the negative battery cable on some vehicles may interfere with the functions of the on board computer systems and may require the computer to undergo a relearning process, once the negative battery cable is reconnected.**

Distributor

REMOVAL & INSTALLATION

These models utilize a Distributorless Ignition System (DIS). With this system, the Electronic Control Module (ECM) determines proper ignition timing and time for the primary ignition coil circuit to turn **ON** and **OFF**.

Alternator

REMOVAL

Swift

1. Before servicing the vehicle, refer to the precautions in the beginning of this section.
2. Remove or disconnect the following:
   • Negative battery cable
   • **B** terminal wire and coupler from the alternator
   • Alternator drive belt adjusting bolt and loosen the adjuster arm bolt
   • Alternator cover
   • Alternator mounting bolts and nut
   • Alternator

Esteem

1. Before servicing the vehicle, refer to the precautions in the beginning of this section.
2. Remove or disconnect the following:
   • Negative battery cable
   • **B** terminal wire and coupler from the alternator
   • Alternator mounting bolts and loosen the drive belt adjusting bolt
   • Alternator cover
   • Alternator bracket bolts and the bracket
   • Alternator

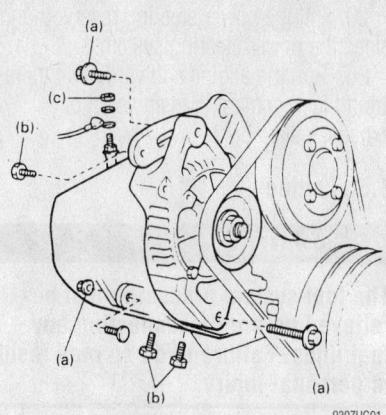

Alternator mounting—Swift models

9307UG01

INSTALLATION

Swift

1. Install or connect the following:
   • Alternator
   • Alternator mounting bolts and nut. Tighten to 13–20 ft. lbs. (18–28 Nm).

• Alternator cover. Tighten the cover retainers to 36–60 inch lbs. (4–7 Nm).
• Alternator drive belt adjusting bolt and tighten the adjuster arm bolt. Tighten to 13–20 ft. lbs. (18–28 Nm).
• **B** terminal wire and coupler to the alternator
• Negative battery cable

Esteem

1. Install or connect the following:
   • Alternator
   • Alternator bracket and bolts. Tighten the bolts to 16 ft. lbs. (23 Nm).
   • Alternator cover
   • Alternator drive belt adjusting bolt and mounting bolts. Tighten to 16 ft. lbs. (23 Nm).
   • **B** terminal wire and coupler to the alternator
   • Negative battery cable

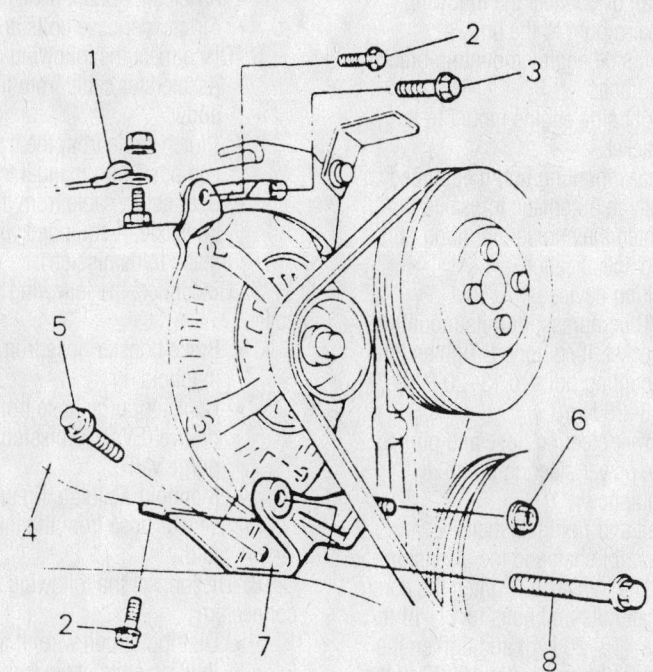

1. "B" terminal wire
2. Cover bolts
3. Drive belt adjusting bolt
4. Cover
5. Bracket bolts
6. Bracket nut
7. Bracket
8. Mounting bolt

9307UG12

Alternator mounting—Esteem models

Ignition Timing

ADJUSTMENT

Ignition timing is controlled by the Electronic Control Module (ECM). The ECM receives signals from various sensors mounted on the engine. No ignition timing adjustment is possible.

Engine Assembly

REMOVAL & INSTALLATION

1.3L Engine

✳✳ CAUTION

The fuel injection system remains under pressure after the engine has been turned OFF. Properly relieve fuel pressure before disconnecting any fuel lines. Failure to do so may result in fire or personal injury.

1. Before servicing the vehicle, refer to the precautions in the beginning of this section.
2. Properly relieve the fuel system pressure.
3. Remove or disconnect the following:
   • Battery and tray
   • Windshield washer hose from the hood
4. Using a grease pencil or marker, mark the hood hinge to hood outline. With the aid of an assistant, remove the hood.
5. Drain the cooling system.
6. Remove or disconnect the following:
   • Air cleaner assembly with the Mass Air Flow (MAF) sensor outlet hose
   • Radiator and cooling fan
7. Disconnect the following electrical wires and release the wiring harness from the clamps:
   • Ignition coil wire from the distributor cap, if equipped
   • Distributor electrical wires, if equipped
   • Ignition coil assembly, if equipped
   • Exhaust Gas Recirculation (EGR) solenoid vacuum valve
   • Radiator fan temperature switch
   • Engine coolant temperature gauge sensor
   • Engine Coolant Temperature (ECT) sensor
   • Idle Air Control (IAC) actuator
   • Ground wires from the intake manifold
   • Throttle Position (TP) sensor

- Fuel injector
- Camshaft Position (CMP) sensor, if equipped
- Oxygen (O₂S) sensor
- Oil pressure gauge sensor
- Alternator
- Starter
- Back-up light switch
- Negative battery cable from the transaxle
- Vehicle Speed Sensor (VSS)
- Noise filter ground wire
- Evaporative (EVAP) emission canister purge valve, if equipped

8. Disconnect the following vacuum hoses:

- Brake booster hose from the intake manifold
- Canister purge hose
- Air conditioning valve hose

9. Remove or disconnect the following:

- Fuel return hose and the fuel feed hose from the throttle body
- Heater inlet and outlet hoses

10. Disconnect the following cables:

- Accelerator cable from the throttle body
- Clutch cable from the transaxle, if equipped
- Speedometer cable from the transaxle, if equipped
- Shift switch, if equipped with an automatic transaxle
- VSS, if equipped

11. Remove or disconnect the following:

- EVAP canister from the vehicle
- Fender apron extensions
- Exhaust pipe from the exhaust manifold
- Control shaft and extension rod from the transaxle

12. Drain the engine and transaxle oil.

13. Remove or disconnect the following:

- Left and right halfshafts

➡ **For engine and transaxle removal, it is not necessary to remove the half-shafts from the steering knuckles.**

14. Remove or disconnect the following:

- Air conditioning compressor from its mounting bracket with the hoses still attached, if equipped with air conditioning

➡ **Suspend the compressor where no damage will occur during engine removal and installation.**

15. Remove or disconnect the following:

- Power steering hoses from the power steering pump, if equipped

➡ **Plug the power steering hose, pipe and pump ports to minimize fluid loss.**

- Rear torque rod bracket from the transaxle, if equipped with an automatic transaxle
- Rear mount from the body, if equipped with a manual transaxle

16. If equipped with an automatic transaxle, remove the rear mounting nut.

17. Install an engine lifting device.

18. Remove or disconnect the following:

- Rear mount from the body
- Left side engine mounting bracket bolts and bracket
- Right side engine mount from its bracket

19. Before lifting the engine and assembly check to be sure that all the hoses, electric wires and cables are disconnected.

20. Remove the engine with the transaxle from the vehicle.

To install:

21. Lower the engine and transaxle into the engine compartment but do not remove the lifting device.

22. Install or connect the following:

- Rear mount to the body
- Left side engine mounting bracket and bolts
- Right side engine mount to its bracket
- Rear mounting nut, if equipped with an automatic transaxle

23. Tighten the engine mounting nuts and bolts to specification.

- Lifting device
- A/C compressor on its mounting bracket, if equipped. Tighten the mounting bolts to 13–20 ft. lbs. (18–28 Nm).
- Power steering hose and pipe to the power steering pump, if equipped
- Left and right halfshafts
- Control shaft and the extension rod to the transaxle. Tighten the control shaft nuts and bolts to 11–14 ft. lbs. (15–20 Nm) and tighten the extension rod nut to 19–29 ft. lbs. (25–40 Nm).
- Exhaust pipe to the exhaust manifold. Tighten the bolts to 29–36 ft. lbs. (40–50 Nm)

24. Fill the transaxle with gear oil.

25. Install the remaining components in the reverse order of removal.

26. Adjust the clutch pedal free-play.

27. Adjust the accelerator cable free-play.

28. Fill the engine with engine oil and the cooling system with coolant.

29. Fill the power steering reservoir and bleed the power steering system.

30. Run the engine and verify that there are no fuel, coolant, transmission or exhaust leaks.

1.6L Engine

✳✳ CAUTION

The fuel system pressure must be relieved before disconnecting any fuel lines. Failure to do so may result in personal injury.

1. Before servicing the vehicle, refer to the precautions in the beginning of this section.

2. Relieve the fuel system pressure.

3. Mark the position of the hood on the hinges for installation reference, then remove the hood with the aid of an assistant.

4. Drain the cooling system.

5. Remove or disconnect the following:

- Radiator and cooling fan
- Air cleaner outlet hose
- Air cleaner case bolts and case

6. Disconnect the following cables:

- Accelerator cable from the throttle body
- Clutch cable from the transaxle, if equipped with manual transmission
- Gear select cable from the transaxle, if equipped with automatic transmission

7. Disconnect the following vacuum hoses:

- Brake booster hose from the intake manifold
- Canister purge hose from the Evaporative (EVAP) emission canister purge valve
- Manifold Absolute Pressure (MAP) sensor hose from the intake manifold

8. Disconnect the following electrical connectors:

- Distributor coil wire, if equipped
- Ignition coils, if equipped
- Camshaft Position (CMP) sensor
- Engine oil pressure switch
- Exhaust Gas Recirculation (EGR) solenoid vacuum valve
- EVAP canister purge valve
- Engine Coolant Temperature (ECT) sensor
- Fuel injectors
- Power steering pressure switch
- Heated Oxygen Sensor (HO₂S)
- Back-up light switch, manual transmission

- Shift switch, automatic transmission
- Forward clutch revolution sensor, automatic transmission
- Automatic transmission Vehicle Speed Sensor (VSS)
- Alternator
- Starter
- Battery negative cable from the transaxle
- Throttle Position (TP) sensor
- Idle Air Control (IAC) valve
- Manifold Absolute Pressure (MAP) sensor
- Crankshaft Position (CKP) sensor, if equipped
- All engine wires from the engine

9. Remove or disconnect the following:
- Fuel feed hose from the feed pipe
- Return hose from the fuel pressure regulator
- Heater inlet and outlet hoses
- Right and left engine undercovers
- Front exhaust pipe from the exhaust manifold and center exhaust pipe
- Gear shift control shaft from the transaxle and the extension rod, if equipped with a manual transaxle

10. Drain the engine and transaxle oil.
11. Remove or disconnect the following:
- Left and right halfshafts
- A/C compressor from the compressor bracket with the hoses still attached, if equipped. Position the air conditioning out of the way from the engine.

12. If equipped with power steering, drain the power steering pump of fluid.
13. Remove or disconnect the following:
- Power steering hose from the power steering pump, if equipped

14. Install a lifting device on the engine.
15. Remove or disconnect the following:
- Center member from the vehicle by unfastening the 7 nuts and 4 bolts
- Left engine mount
- Right engine mount and bracket

16. Check to be sure all cooling hoses, vacuum hoses and electrical wires are disconnected from the engine.
17. Lower the engine with the transaxle from the vehicle.

To install:
18. Raise the engine and transaxle into the engine compartment.
19. Install or connect the following:
- Right engine mount with the bolts and nuts. Tighten the bolts and nuts to 40 ft. lbs. (55 Nm).
- Left engine mount and install the bolts and nuts. Tighten the bolts and nuts to 40 ft. lbs. (55 Nm).

20. Install the center member using the 7 nuts and 4 bolts. Tighten the bolts and nuts as follows:
   a. Center member to the radiator support: 33 ft. lbs. (45 Nm)
   b. Center member to crossmember: 33 ft. lbs. (45 Nm)
   c. Engine mounts to center member nuts: 40 ft. lbs. (55 Nm)
21. Remove the lifting device.
22. Install or connect the following:
- Power steering hose to the power steering pump, if equipped
- A/C compressor on the air conditioning bracket on the engine, if equipped
- Left and right halfshafts
- Gear shift control shaft on the transaxle and the extension rod, if equipped with a manual transaxle
- Front exhaust pipe on the center exhaust pipe and exhaust manifold
- Remaining components in the reverse order of removal

23. Fill the cooling system, engine, transaxle and power steering pump.
24. Adjust all cables and check all connections.
25. Connect the negative battery cable to the battery.
26. Start the vehicle and check for leaks.

Water Pump

REMOVAL & INSTALLATION

1.3L Engines

1. Before servicing the vehicle, refer to the precautions in the beginning of this section.

2. Disconnect the negative battery cable.
3. Drain the cooling system into a suitable container and tighten the drain plug.
4. Remove or disconnect the following:
- Air cleaner assembly and the Mass Air Flow (MAF) sensor and outlet hose
- Air cleaner bracket
- Right side fender apron clips by pushing the center pin

➡ **Do not push the center pin too far in, or it will fall off into the fender.**

- Power steering and air conditioning belt, if equipped
- Loosen the water pump pulley bolts
- Alternator drive belt
- Water pump pulley

5. To remove the crankshaft pulley perform the following:
   a. If equipped with a manual transaxle, insert a suitable flat-bladed tool into the hole in the bell housing next to the exhaust pipe. This will lock the crankshaft in place.
   b. If equipped with an automatic transaxle, hold a suitable flat-bladed tool in line with the oil pan and insert into the teeth of the drive plate. This will lock the crankshaft in place.
   c. Loosen the crankshaft pulley bolts.
   d. Remove the crankshaft timing belt pulley bolt with a 17mm socket.
   e. Remove the pulley from the crankshaft.
   f. Install the crankshaft bolt.
   g. Remove the flat-bladed tool that was used to lock the crankshaft in place.

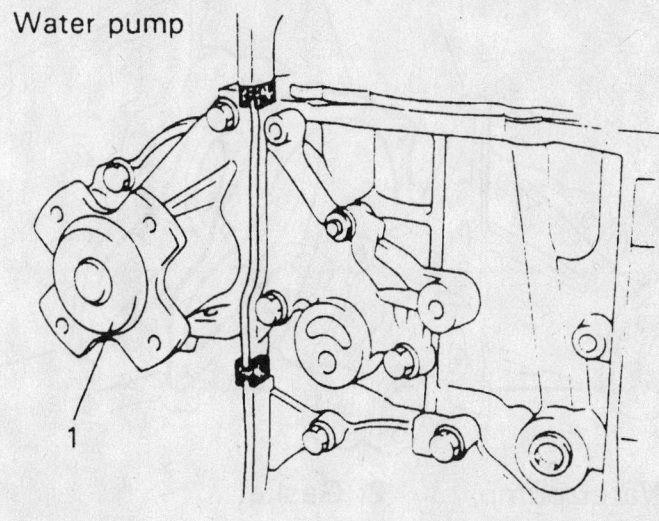

1. Water pump

1

Water pump location—1.3L engine

7923UG05

➡**To remove the crankshaft pulley with the engine assembly mounted on the body, it is necessary to remove the crankshaft timing belt pulley bolt. If the engine assembly is dismounted, the bolt does not need to be removed.**

6. Remove or disconnect the following:
   - Resonator and the timing belt outside cover
   - Loosen the right engine mounting bolt
   - Timing belt

✳✳ WARNING

After the timing belt is removed never turn the camshafts or the crankshaft. Interference may occur between the pistons and the valves causing component damage.

   - Timing belt inside cover
   - Water pump belt adjusting arm

7. Carefully remove the rubber seal between the water and oil pumps, and remove the seal between the water pump and the cylinder head.

8. Remove or disconnect the following:
   - Water pump bolts and the water pump

To install:

9. Clean the water pump mounting surface of old gasket material.

10. Install or connect the following:
   - New water pump gasket on the cylinder block
   - Water pump on the cylinder block and tighten the bolts to 84–108 inch lbs. (10–13 Nm)

   - Rubber seal between the water pump and the oil pump
   - Seal between the water pump and the cylinder head
   - Water pump belt adjusting arm

11. Install the timing belt inside cover.

12. With the crankshaft locked in position, remove the crankshaft bolt and install the crankshaft pulley. Tighten the crankshaft pulley bolts to 10–13 ft. lbs. (14–18 Nm). Using a 17mm socket, tighten the crankshaft timing belt pulley bolt to 76–83 ft. lbs. (105–115 Nm).

13. Install or connect the following:
   - Timing belt
   - Water pump pulley and drive belt. Tighten the water pump pulley bolts to 84–96 inch lbs. (9–12 Nm).
   - Remaining components in the reverse order of removal

14. Fill the cooling system.

15. Connect the negative battery cable.

16. Start the engine and top off the coolant as necessary.

17. Check the cooling system for leaks.

18. Check the ignition timing.

1.6L Engine

1. Before servicing the vehicle, refer to the precautions in the beginning of this section.

2. Disconnect the negative battery cable.

3. Drain the cooling system and tighten the drain plug.

4. Remove or disconnect the following:
   - Timing belt
   - Alternator adjusting arm

   - Oil dipstick guide and dipstick
   - Water pump bolts, gasket, water pump and rubber seal

To install:

5. Clean the water pump mounting surface of old gasket material.

6. Install or connect the following:
   - New water pump gasket on the cylinder block
   - Water pump on the cylinder block and tighten the bolts to 96 inch lbs. (11 Nm)
   - Rubber seal between the water pump and the oil pump
   - Seal between the water pump and the cylinder head
   - Timing belt
   - Alternator adjusting arm
   - Oil dipstick guide and dipstick using a new O-ring

7. Adjust drive belt tension.

8. Fill the cooling system with engine coolant.

9. Connect the negative battery cable.

10. Start the engine and top off the coolant as necessary.

11. Check the ignition timing.

Heater Core

REMOVAL & INSTALLATION

Esteem

1. Before servicing the vehicle, refer to the precautions in the beginning of this section.

2. Disconnect the negative battery cable.

3. Wait for a least 1 minute for the air bag system to deplete its energy before working on any part the instrument panel.

4. Drain the cooling system into a clean container for reuse.

5. Remove the instrument panel.

6. Remove the mode actuator by performing the following procedure:
   - Disconnect the mode actuator coupler.
   - At the heater housing, remove the mode actuator rod.
   - Remove the mode actuator from the heater housing.

7. Remove the rear air duct.

8. Remove the heater housing.

9. Remove the heater housing clips and screws; then, separate the heater housings.

10. Remove the heater core from the heater housing.

To install:

11. Install the heater core to the heater housing.

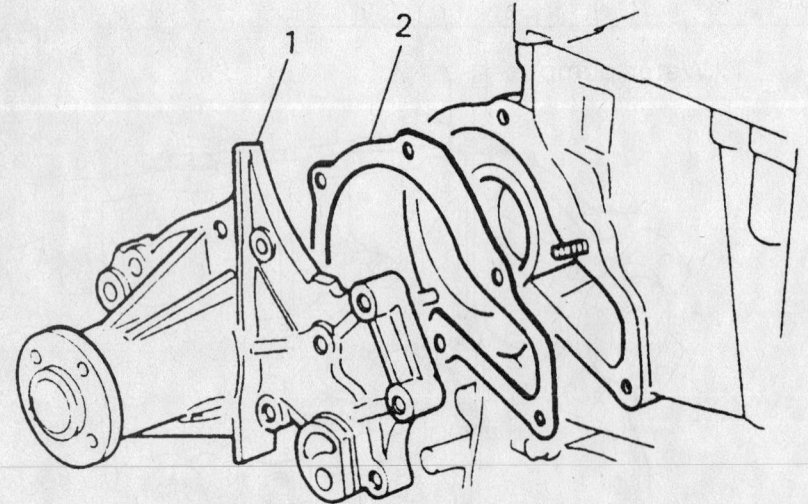

1. Water pump   2. Gasket

7923UG06

Exploded view of the water pump mounting—1.6L engine

12. Assemble the heater housings and install the heater housing clips and screws.

13. Install the heater housing.

14. Install the rear air duct.

15. Install the mode actuator by performing the following procedure:
- Install the mode actuator to the heater housing.
- At the heater housing, install the mode actuator rod.
- Connect the mode actuator coupler.

16. Install the instrument panel.

17. Refill the cooling system.

18. Connect the negative battery cable.

19. Operate the engine to normal operating temperatures; then, check the climate control operation and check for leaks.

Swift

1. Before servicing the vehicle, refer to the precautions in the beginning of this section.

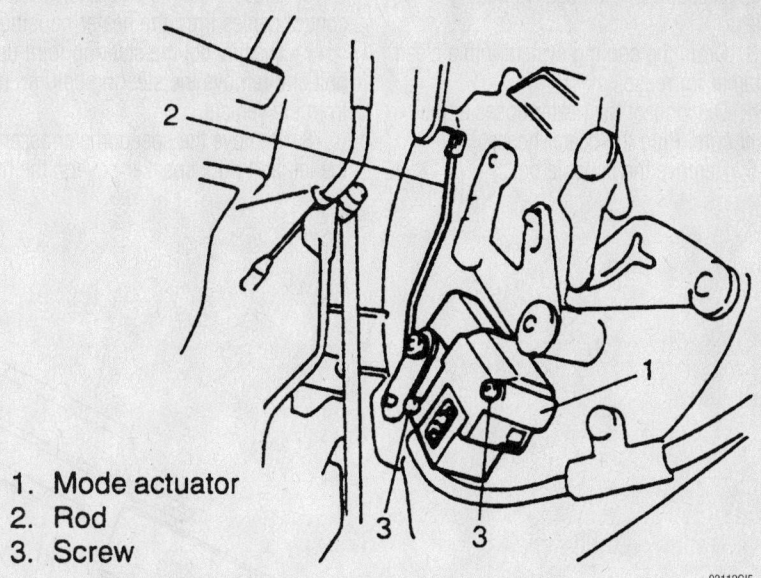

1. Mode actuator
2. Rod
3. Screw

93112GI5

View of the heater housing mode actuator—Esteem

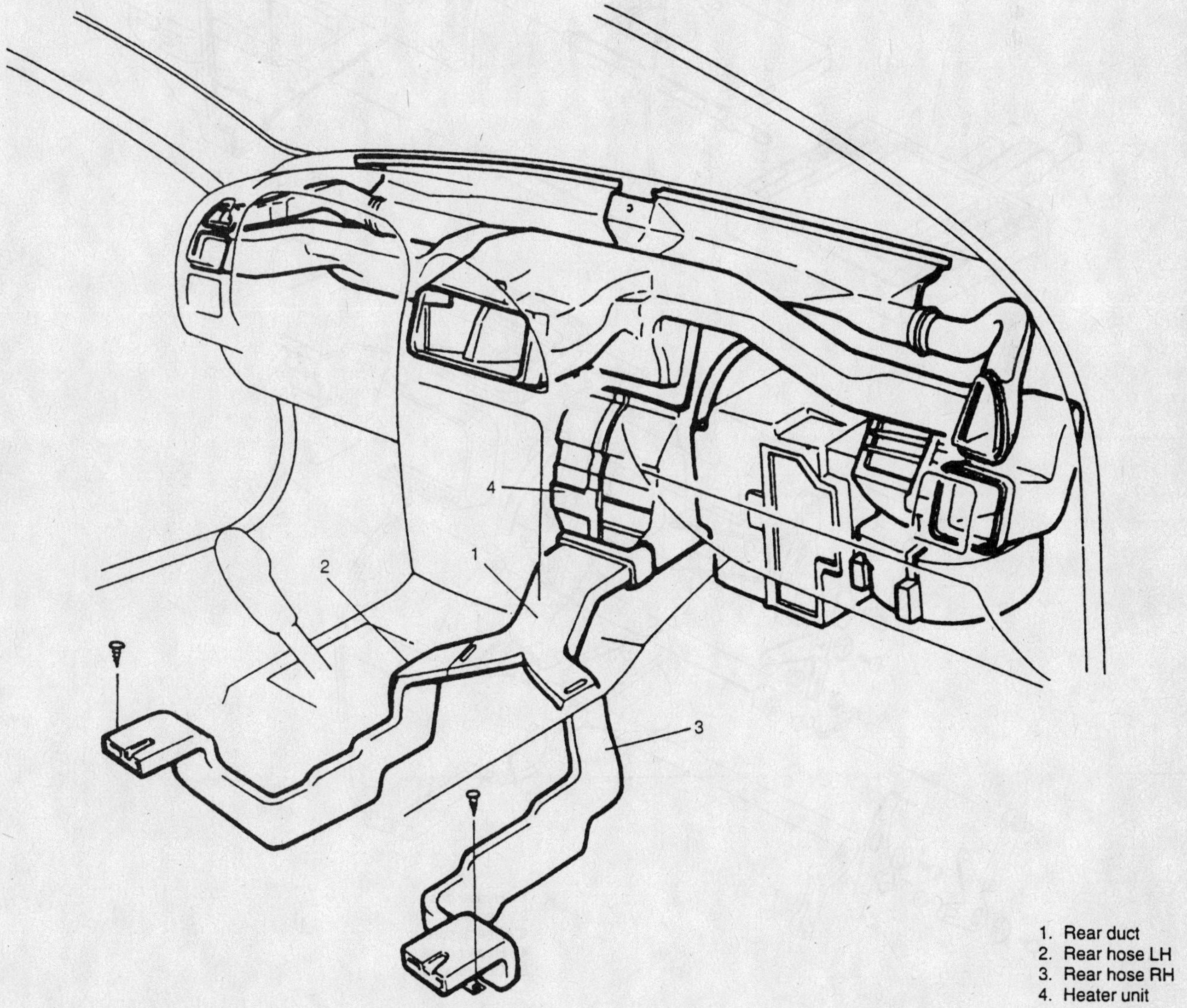

1. Rear duct
2. Rear hose LH
3. Rear hose RH
4. Heater unit

93112GI4

Assembled view of the heater housing and related components—Esteem

2. Disconnect the negative battery cable.

3. Drain the cooling system into a clean container for reuse.

4. Disconnect the heater hoses at the heater core. Plug the heater hoses.

5. Remove the console box.

6. Disconnect the electrical wires and control cables from the heater housing.

7. Disconnect the steering joint upper bolt and remove the steering column unit from the vehicle.

8. Remove the speedometer assembly, the left and right speaker covers, the hood

opening cable and the center garnish from the dash.

9. Remove the dashboard mounting bolts and the dashboard being careful to protect it from damage.

10. Remove the heater housing mounting bolts and remove it from the vehicle.

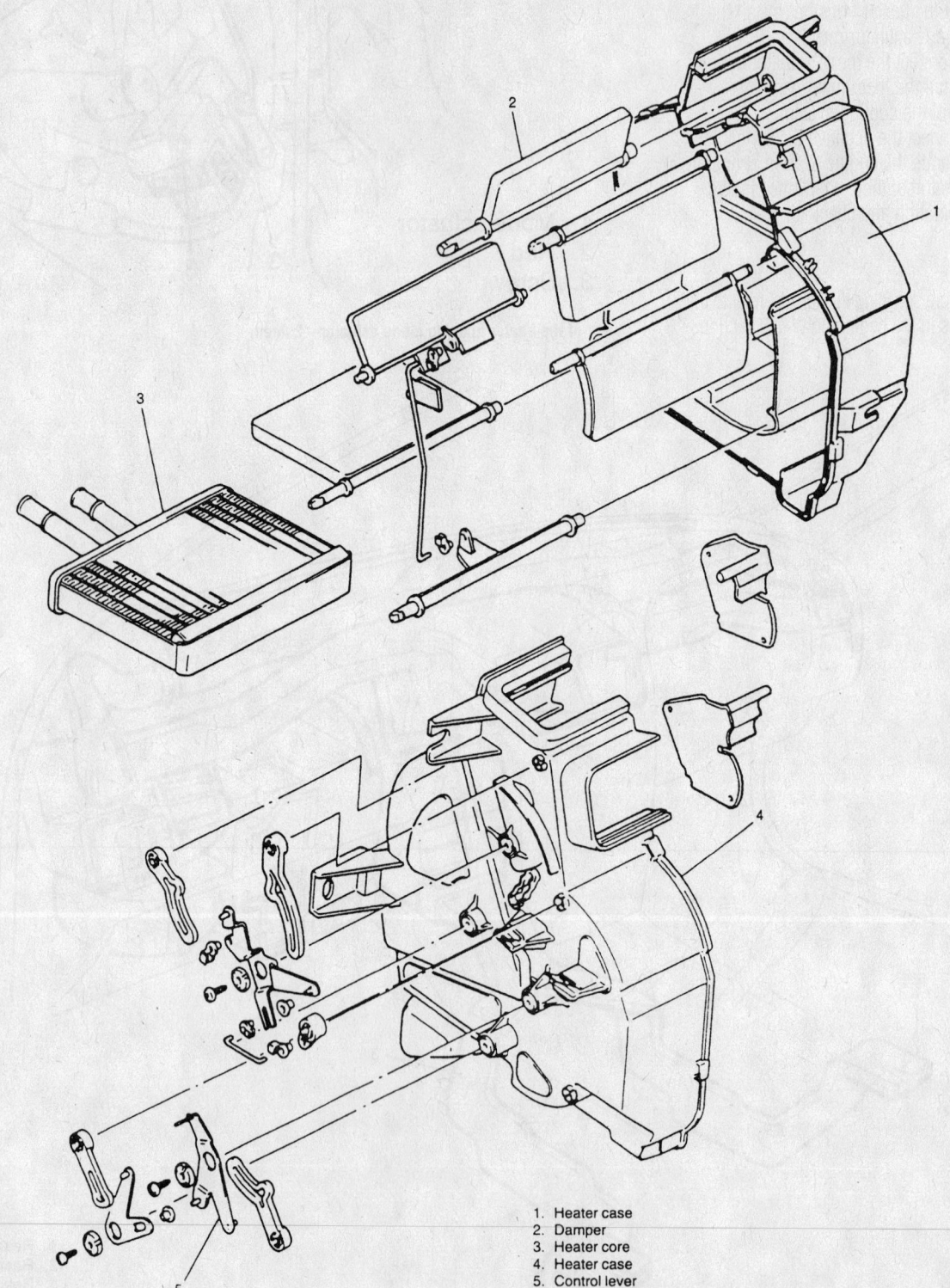

1. Heater case
2. Damper
3. Heater core
4. Heater case
5. Control lever

93112GI6

Exploded view of the heater core, heater housing and related components—Esteem

11. Remove the attaching screws and clips which hold the heater housing together and separate the housing to gain access to the heater core.

12. Remove the heater core from the case.

To install:

13. Install the heater core to the case.

14. Install the heater housing and the attaching screws and clips.

15. Install the heater housing and mounting bolts.

16. Install the dashboard and mounting bolts.

17. Install the speedometer assembly, the left and right speaker covers, the hood opening cable and the center garnish to the dash.

18. Install the steering column unit and connect the steering joint upper bolt.

19. Connect the electrical wires and control cables to the heater housing.

20. Install the console box.

21. Connect the heater hoses at the heater core.

22. Refill the cooling system.

23. Connect the negative battery cable.

24. Operate the engine to normal operating temperatures; then, check the climate control operation and check for leaks.

Cylinder Head

REMOVAL & INSTALLATION

1.3L Engines

❊❊ CAUTION

The fuel injection system remains under pressure after the engine has been turned OFF. Properly relieve fuel pressure before disconnecting any fuel lines. Failure to do so may result in fire or personal injury.

1. Before servicing the vehicle, refer to the precautions in the beginning of this section.

2. Relieve the fuel system pressure.

3. Drain the cooling system.

4. Remove or disconnect the following:
- Air cleaner assembly
- Ignition coil wire from the distributor cap, if equipped
- Distributor electrical wires, if equipped
- Ignition coil assembly, if equipped
- Exhaust Gas Recirculation (EGR) solenoid vacuum valve
- Radiator fan thermo switch
- Engine Coolant Temperature (ECT) sensor
- ECT gauge sensor
- Idle Air Control (IAC) valve
- Throttle Position (TP) sensor
- Fuel injector
- Ground wires from the intake manifold.
- Oxygen (O_2S) sensor
- Radiator hose from the thermostat housing
- Heater hose from the intake manifold
- Throttle body coolant outlet hose from the throttle body
- Manifold Absolute Pressure (MAP) sensor hose from the intake manifold, if equipped
- Canister hose from its pipe
- Canister purge hose from the intake manifold
- Brake booster from the intake manifold
- Fuel return hose and the fuel feed hose from the throttle body
- Throttle cable from the throttle body
- Water pump and crankshaft pulleys
- Timing belt
- Rubber seal between the cylinder head and the water pump
- Exhaust pipe from the exhaust manifold
- Spark plug wire clamps from the cylinder head cover
- Positive Crankcase Ventilation (PCV) hose
- Cylinder head cover

5. Loosen all valve adjusting screws and allow the valves to close.

6. Remove or disconnect the following:
- Cylinder head bolts in the reverse order of the tightening sequence
- Cylinder head with the distributor, intake manifold and exhaust manifold
- Distributor, intake manifold and exhaust manifold from the cylinder head

7. Clean the cylinder block mating surface of any old gasket material and clean any engine coolant from the cylinders.

To install:

8. Install or connect the following:
- Intake and exhaust manifolds
- Distributor on the cylinder head
- New cylinder head gasket with the top mark facing up and toward the crankshaft pulley
- Cylinder head on the engine block

9. Coat the cylinder head mounting bolts threads with clean engine oil.

10. Tighten the bolts as follows:
a. Step 1: 26 ft. lbs. (35 Nm)
b. Step 2: 41 ft. lbs. (55 Nm)
c. Step 3: 49 ft. lbs. (68 Nm)

11. Install or connect the following:
- Rubber seal between the cylinder head and the water pump
- Timing belt
- Exhaust pipe to the exhaust manifold, tighten the attaching bolts to 26–36 ft. lbs. (35–50 Nm)
- Crankshaft and water pump pulleys
- Throttle cable to the throttle body
- Fuel return hose and the fuel feed hose to the throttle body
- Remaining components in the reverse order of removal

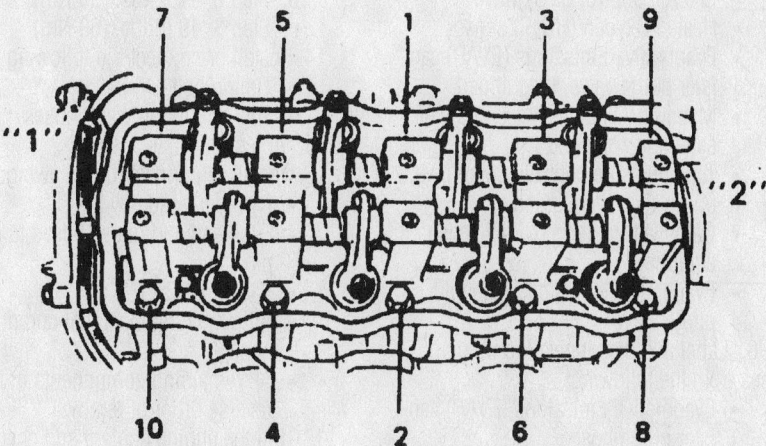

"1": Camshaft pulley side
"2": Distributor side

7923UG07

Cylinder head bolt torque sequence—1.3L engine

12. Refill the cooling system with coolant.
13. Connect the negative battery cable.
14. Adjust the ignition timing.
15. With the engine running, be sure that there are no fuel, coolant or exhaust leaks.

1.6L Engine

✳✳ CAUTION

The fuel system pressure must be relieved before disconnecting any fuel lines. Failure to do so may result in personal injury.

1. Before servicing the vehicle, refer to the precautions in the beginning of this section.
2. Drain the cooling system.
3. Relieve the fuel system pressure.
4. Remove or disconnect the following:
   • Air cleaner outlet hose
   • Intake manifold rear stiffener bolt, alternator adjustment arm reinforcement bolt and right mounting bracket stiffener from the intake manifold
   • Heated Oxygen (HO2S) sensor coupler and release its clamps
   • Exhaust from the manifold
5. Disconnect the electrical connectors from the following components:
   • Distributor, if equipped
   • Ignition coils, if equipped
   • Engine Coolant Temperature (ECT) sensor and gauge
   • Engine ground wire from intake manifold
   • Exhaust Gas Recirculation (EGR) solenoid vacuum valve
   • Fuel injectors
   • Throttle Position (TP) sensor
   • Idle Air Control (IAC) valve
   • Heated Oxygen (HO2S) sensor
   • Evaporative Emissions (EVAP) canister purge valve, if equipped
   • Manifold Absolute Pressure (MAP) sensor, if equipped
   • Tank pressure control solenoid vacuum valve, if equipped
   • Camshaft Position (CMP) sensor, if equipped
   • Evaporative emissions solenoid purge valve
6. Label and disconnect the vacuum hoses from the following:
   • Evaporative Emissions (EVAP) canister purge hose
   • Brake booster supply hose
7. Remove or disconnect the following:
   • Fuel feed and return hoses from each pipe

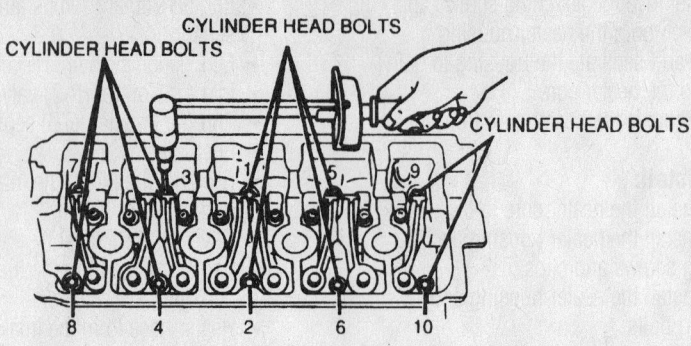

Cylinder head bolt torque sequence—1.6L engine

   • Cylinder head cover
   • Valve lash adjusting screws
8. Disconnect the following hoses:
   • Radiator inlet hose
   • Heater inlet hose
   • IAC valve outlet
9. Remove or disconnect the following:
   • Timing belt
   • Cylinder head bolts in reverse order of tightening. Once each bolt is loose, remove the bolts from the cylinder head.
10. Check to be sure all components are removed or disconnected before removing the cylinder head.
11. Remove the cylinder head with the intake manifold, exhaust manifold and distributor as an assembly.
    To install:
12. Install a new cylinder head gasket and the cylinder head, with the distributor case, onto the cylinder block.
13. Tighten the cylinder head bolts, in sequence, using the following 3 Steps:
    a. Step 1: 26 ft. lbs. (35 Nm)
    b. Step 2: 41 ft. lbs. (55 Nm)
    c. Step 3: 49 ft. lbs. (68 Nm)
14. Install or connect the following:
    • Timing belt
    • Engine cooling water hoses
15. Adjust the valve lash.
16. Install or connect the following:
    • Cylinder head cover
    • Fuel feed and return hoses to each pipe
    • Vacuum hoses
    • All remaining electrical components
    • All remaining components in the reverse order of removal
17. Fill the engine coolant and check all fluids.
18. Connect the negative battery cable to the battery.
19. Start the engine and check for leaks.

Rocker Arms/Shafts

REMOVAL & INSTALLATION

Only the 1.3L engine uses rocker arms/shafts. On the 1.6L engine, the camshaft directly actuates the valves.

1.3L Engine

1. Before servicing the vehicle, refer to the precautions in the beginning of this section.
2. Drain the cooling system into a resealable container.
3. Remove or disconnect the following:
   • Negative battery cable
   • Timing belt
   • Cylinder head cover from the cylinder head
   • Camshaft Position (CMP) sensor coupler from the CMP sensor
   • High tension cords and couplers from the coil assemblies
   • Ignition coil from the exhaust manifold side
   • CMP case from the cylinder head
4. Loosen all valve adjusting screw lock nuts, turn the adjusting screws back all the way to allow the rocker arms to move freely.
5. Remove or disconnect the following:
   • Camshaft housing and the camshaft
   • Timing belt inside cover
   • Intake rocker arm with the clip from the rocker arm shaft. Make sure not to bend the clip when removing the intake rocker arm
   • Rocker arm shaft bolts
   • Exhaust rocker arms and spring by pulling out the shaft on the battery side. If necessary, remove the battery for clearance.
6. Inspect the rocker arms and shafts for wear and/or damage and replace parts as necessary.

a. If the tip of the valve lash adjuster is badly worn, replace the adjuster.

To install:

7. Apply engine oil to the rocker arms and the rocker arm shafts.

8. Install or connect the following:
- O-ring on the shaft
- Rocker arm shaft with its holes facing up
- Rocker arm (exhaust side) and the rocker arm spring
- Rocker arm shaft bolts and tighten the bolts to 96 inch lbs. (11 Nm)

9. Pour a small amount of engine oil into the arm pivot holding part of the shaft.

10. Install or connect the following:
- Rocker arm (intake side), with the clips to the rocker arm shaft
- Camshaft and housing
- Camshaft oil seal
- Timing belt inside cover
- Camshaft timing belt pulley, while fitting the pin on the camshaft into the slot. Tighten the pulley bolt to 43 ft. lbs. (60 Nm).
- Timing belt
- Remaining components in the reverse order of removal.

11. Check the ignition timing.

1.6L Engine

1. Before servicing the vehicle, refer to the precautions in the beginning of this section.

2. Remove or disconnect the following:
- Negative battery cable
- Timing belt

❉❉ WARNING

After the timing belt is removed, never turn the camshaft and crankshaft independently more than 90° in either direction. If turned, interference may occur between the piston and valves causing possible damage to the effected parts.

3. Using a camshaft sprocket holding tool, hold the sprocket stationary and remove the camshaft sprocket bolt.

4. Remove or disconnect the following:
- Cylinder head cover
- Distributor cap and distributor assembly from the engine, if equipped
- Ignition coils and the Camshaft Position (CMP) sensor case from the cylinder head with the CMP sensor coupler disconnected
- Loosen all of the valve adjusting

screw locknuts until the rocker arms move freely
- Camshaft housing and camshaft
- Rocker arm shaft plug from the cylinder head
- Timing belt inner cover-to-cylinder head bolts and the cover
- All intake rocker arms and clips from the rocker shaft. Keep all parts in order so they can be reinstalled in their original locations.
- 6 rocker arm shaft-to-cylinder head bolts. Push the rocker arm shaft through the rear of the cylinder head until the end of the rocker shaft appears.
- O-ring from the rear of the rocker arm shaft
- Exhaust rocker arms, rocker arm springs and rocker shaft by pulling the rocker arm shaft through the front of the cylinder head. Be sure to keep the parts in order for installation purposes.

5. Clean and inspect all parts for wear and/or damage. Replace parts as necessary.

To install:

6. Lubricate the rocker arms and shafts with clean engine oil before installation.

7. Install or connect the following:
- Push the rocker shaft into the front of the cylinder head
- Exhaust rocker arms and springs as the rocker arm shaft is being installed into the cylinder head
- Push the rocker arm shaft through the rear of the cylinder head
- New O-ring onto the rocker shaft

8. Rotate the rocker arm shaft so the flat machined surface is horizontal and facing downward, parallel with the cylinder head mating surface and slide the shaft back into the cylinder head.

9. Install or connect the following:
- 6 rocker arm shaft bolts and tighten the rocker arm-to-cylinder head bolts to 96 inch lbs. (11 Nm). Fill the rocker arm shaft bolt holes with clean engine oil.
- Intake rocker arms and clips onto the rocker arm shaft

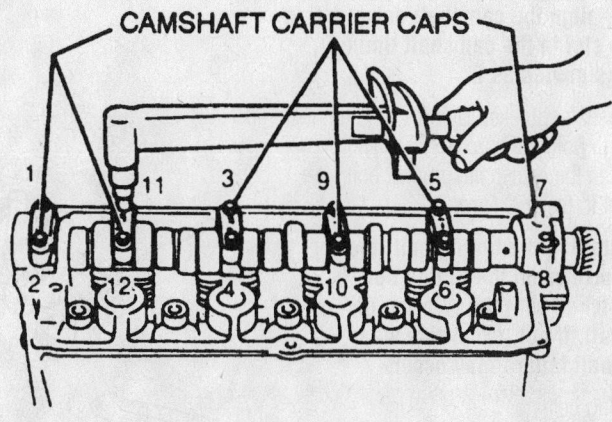

Camshaft carrier cap bolt torque sequence—1.6L engine

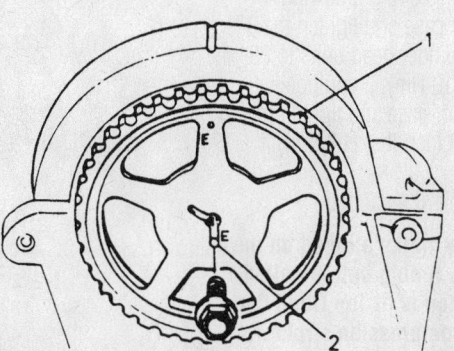

1. Camshaft timing belt pulley   2. Dowel pin

Align the dowel pin with the E-slot in the sprocket—1.6L engine

➡The camshaft carrier caps are embossed with numbers and arrows to ensure correct assembly. The No. 1 camshaft carrier cap must be installed at the front of the cylinder head with the remaining carrier caps following in numerical order. The directional arrows must always point toward the front of the cylinder head.

- Camshaft

※※ WARNING

If the camshaft carrier cap bolts are tightened at random, damage to the camshaft may occur.

10. Lubricate the new camshaft seal lip with clean engine oil
- New camshaft seal into the cylinder head until it is flush with the camshaft carrier surface
- Timing belt inner cover and tighten the cover-to-cylinder head bolts to 89 inch lbs. (10 Nm)
- Rocker arm shaft plug into the cylinder head and tighten to 24 ft. lbs. (33 Nm)

➡During camshaft timing belt sprocket installation, align the camshaft dowel pin with the slot in the camshaft timing belt gear designated as E.

- Camshaft sprocket. Using a holding tool to hold the sprocket in place, tighten the camshaft sprocket bolt to 44 ft. lbs. (60 Nm).

➡When installing the timing belt, the directional arrows on the timing belt must be matched with the rotation of the crankshaft. If not, excessive wear and timing belt failure may occur.

- Timing belt
11. Apply RTV silicone rubber sealant to the surface of the distributor case that mates with the rear of the rocker arm shaft.
12. Install or connect the following:
- Distributor case and tighten the 3 case-to-cylinder head bolts to 89 inch lbs. (10 Nm), if equipped
- CMP sensor case and tighten the bolts to 96 inch lbs. (11 Nm), if equipped
- Ignition coil, if equipped

➡With the timing marks aligned on the sprockets and the timing belt installed, the number 4 piston is at Top Dead Center (TDC) of the compression stroke.

- Distributor into the distributor case, if equipped. Be sure the rotor is

aligned with the No. 4 tower on the distributor cap.
- Distributor cap, if equipped
13. Adjust the valve lash.
14. Install or connect the following:
- Cylinder head cover onto the cylinder head, in the reverse order of removal. Clean all sealing surfaces and use a new gasket and O-rings. Tighten the cylinder head cover bolts to 89 inch lbs. (10 Nm).
- Negative battery cable
15. Start the engine, allow it to reach normal operating temperature and check for leaks.
16. Check and adjust the ignition timing as necessary.

Intake Manifold

REMOVAL & INSTALLATION

※※ CAUTION

The fuel system pressure must be relieved before disconnecting any

fuel lines. Failure to do so may result in personal injury.

1. Before servicing the vehicle, refer to the precautions in the beginning of this section.
2. Properly relieve the fuel system pressure.
3. Drain the coolant from the vehicle.

※※ CAUTION

To help avoid the danger of being burned, do not remove the drain plug and the radiator cap while the engine is still hot. Scalding fluid and steam can be blown out under pressure if the plug and cap are taken off too soon.

4. Remove or disconnect the following:
- Negative battery cable
- Air cleaner assembly
5. Disconnect the following electrical wires:
- Exhaust Gas Recirculation (EGR) solenoid vacuum valve
- Idle Speed Control (ISC) actuator

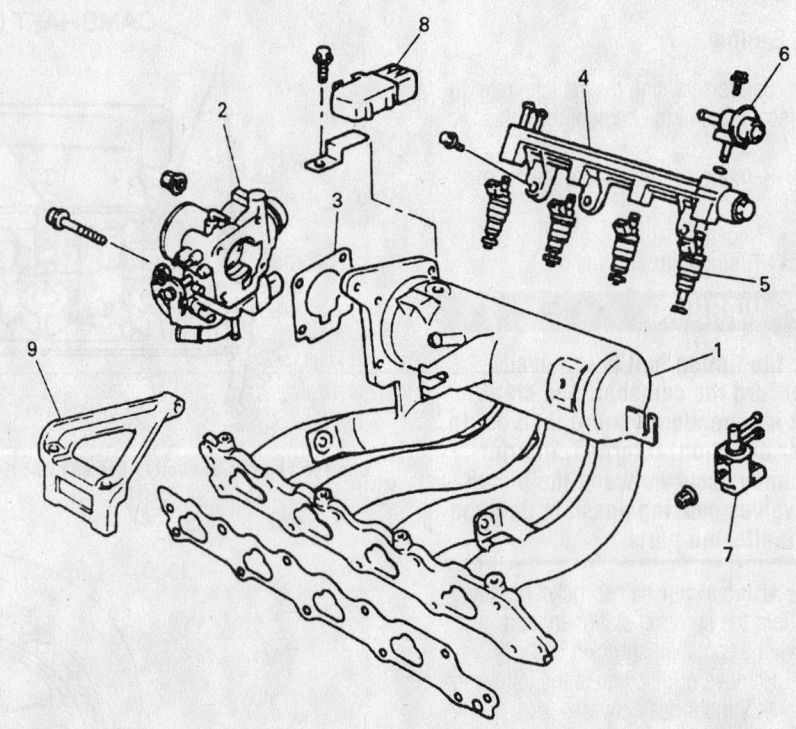

1. Intake manifold
2. Throttle body
3. Gasket
4. Fuel delivery pipe
5. Fuel injector
6. Fuel pressure regulator
7. EVAP canister purge valve
8. MAP sensor
9. Intake manifold upper stiffener

Intake manifold and related components—1.3L engine

7923UG14

- Ground wires from the intake manifold
- Fuel injectors
- Throttle Position (TP) sensor
- Idle Air Control (IAC) valve
- Tank pressure control solenoid vacuum valve, if equipped
- Manifold Absolute Pressure (MAP) sensor

6. Remove or disconnect the following:
- Fuel return and feed hoses from the fuel pipes
- Coolant hoses from the throttle body and the intake manifold

7. Disconnect the following vacuum hoses:
- Canister purge hose from the intake manifold
- Manifold Absolute Pressure (MAP) sensor hose from the intake manifold
- Brake booster hose from the intake manifold
- Positive Crankcase Ventilation (PCV) hose from the PCV valve
- Accelerator cable from the throttle body

8. Remove or disconnect the following:
- Intake manifold attaching nuts and bolts
- Intake manifold and throttle body

To install:
9. Before installing the gasket, make sure that the mating surfaces of the intake manifold and the cylinder head are clean and undamaged.

10. Install or connect the following:
- New intake manifold gasket on the cylinder head
- Intake manifold and throttle body on the cylinder head
- Intake manifold mounting nuts and bolts. Tighten the nuts and bolts to 13–20 ft. lbs. (18–28 Nm). Be sure that the clamps are properly installed on the lower intake manifold bolts.
- Remaining components in the reverse order of removal

11. Refill the cooling system.

12. Connect the negative battery cable.

13. Start the engine and check for fuel and cooling system leaks.

Exhaust Manifold

REMOVAL & INSTALLATION

1.3L Engine

✳✳ CAUTION

To avoid the danger of being burned, do not service the exhaust system while it is hot. Service should be performed only after the system cools down.

1. Before servicing the vehicle, refer to the precautions in the beginning of this section.

2. Remove or disconnect the following:
- Negative battery cable
- Heated Oxygen (HO$_2$S) sensor electrical coupler and release the wire from its clamps
- Exhaust manifold cover
- Exhaust manifold stiffener bolt
- 2 bolts attaching the exhaust pipe to the exhaust manifold
- Exhaust manifold mounting nuts and bolts
- Exhaust manifold and the gasket

To install:
3. Before installing any components check the exhaust manifold and the engine for deterioration or damage and replace as necessary.

4. Install or connect the following:
- Manifold gasket and the exhaust manifold on the engine. Tighten the nuts and bolts to 13–20 ft. lbs. (18–28 Nm).
- Exhaust pipe gasket on the exhaust pipe and position the exhaust pipe
- 2 bolts that attach the exhaust pipe on the exhaust manifold and

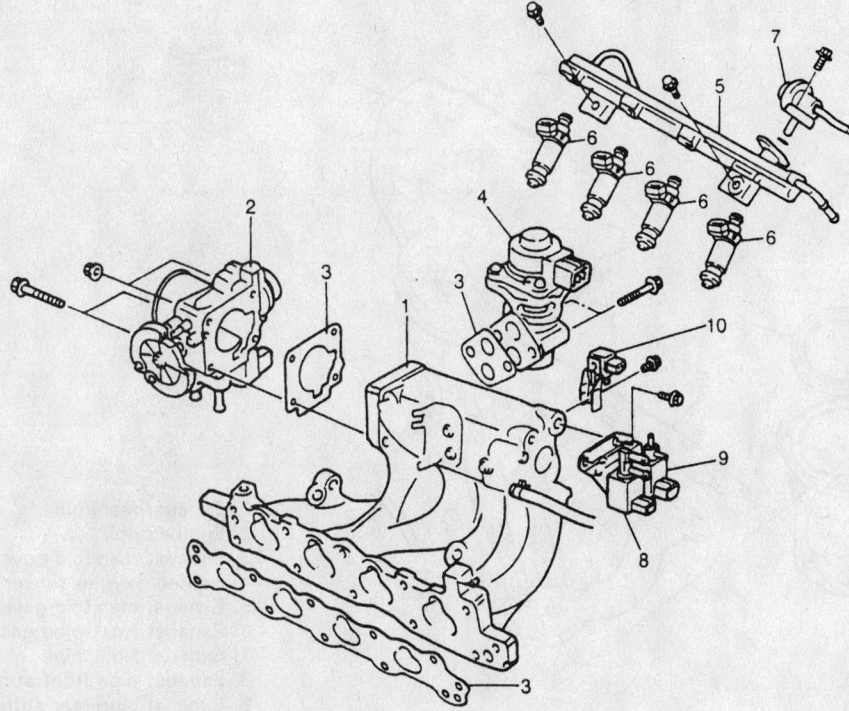

1. Intake manifold
2. Throttle body
3. Gasket
4. EGR valve
5. Fuel delivery pipe
6. Fuel injector
7. Fuel pressure regulator
8. EVAP canister purge valve
9. Tank pressure control solenoid valve
10. MAP sensor

Intake manifold and related components—1.6L engine

7923UG15

tighten the bolts to 25–36 ft. lbs. (35–50 Nm)
- Exhaust manifold stiffener and tighten the bolt to 29–43 ft. lbs. (40–60 Nm)
- Remaining components in the reverse order of removal
- Negative battery cable

5. Run the engine and check for exhaust leaks.

1.6L Engine

1. Before servicing the vehicle, refer to the precautions in the beginning of this section.
2. Remove or disconnect the following:
- Negative cable
- 2 exhaust pipe bolts connecting the exhaust pipe to the exhaust manifold
- Oxygen (O2S) sensor lead wire at the coupler
- Exhaust manifold heat shield from the manifold by unfastening the nut and bolt
- Exhaust manifold mounting bolts and nuts
- Exhaust manifold

To install:
3. Clean and inspect the sealing surfaces of the exhaust manifold and the cylinder head.
4. Install or connect the following:
- New gaskets
- Exhaust manifold and tighten the

mounting bolts and nuts to 17 ft. lbs. (23 Nm)
- O2S lead wire at the coupler
- 3 exhaust pipe bolts connecting the exhaust pipe to the exhaust manifold and tighten to 36 ft. lbs. (50 Nm)
- Manifold heat shield to the exhaust manifold using the nuts and bolts
- Negative battery cable

5. Check for exhaust leaks when finished.

Front Crankshaft Seal

REMOVAL & INSTALLATION

1.3L Engines

1. Remove or disconnect the following:
- Negative battery cable
- Timing belt
- Crankshaft sprocket bolt using a suitable gear stopper to hold the flywheel
- Sprocket bolt, sprocket and key
- Seal from the oil pump housing, using a suitable tool

➡**Be careful not to damage the crankshaft or the oil pump sealing surfaces when removing or installing the seal.**

2. Clean and inspect the surfaces of the crankshaft and the oil pump assembly.

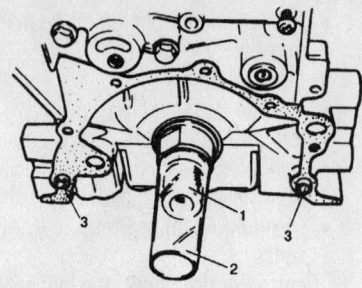

1 Crankshaft
2 Oil seal guide (Vinyl resin) (special tool 09926-18210)
3 Oil pump pin

7923UG17

Oil seal guide tool—1.3L engine

To install:
3. Lubricate the new seal with clean engine oil.
4. Install or connect the following:
- New seal over the crankshaft and into the oil pump, making sure the oil seal lip is not turned up. Use an oil seal guide tool.
- Crankshaft sprocket and timing belt
- Negative battery cable

1.6L Engine

➡**The front oil seal can be removed from the engine without removing the oil pump.**

1. Before servicing the vehicle, refer to the precautions in the beginning of this section.
2. Remove or disconnect the following:

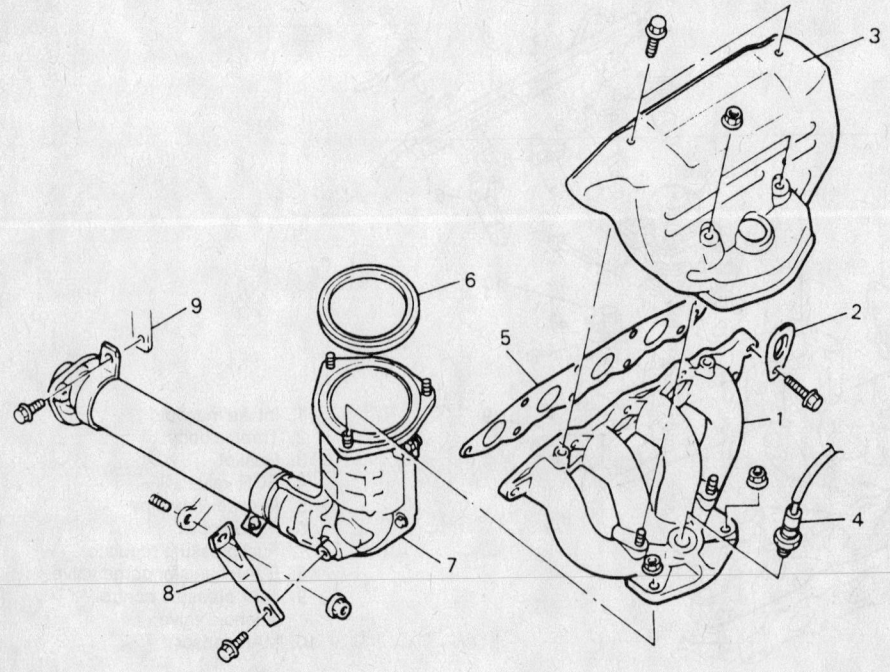

1. Exhaust manifold
2. Engine hook
3. Exhaust manifold cover
4. Heated oxygen sensor
5. Exhaust manifold gasket
6. Exhaust No.1 pipe gasket
7. Exhaust No.1 pipe
8. Exhaust pipe front stiffener
9. Exhaust pipe rear stiffener

7923UG16

Exhaust manifold and related components—1.6L engine

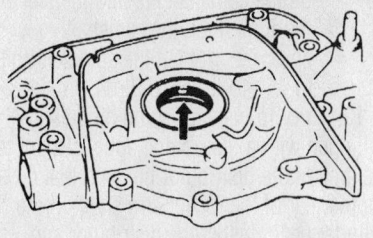

7923UG18

Front crankshaft oil seal location—1.6L engine

- Negative battery cable
- Timing belt
- Crankshaft timing belt sprocket

✳✳ WARNING

When removing the front seal, be extremely careful not to damage the crankshaft.

3. Using a knife, cut off the oil seal lip.

4. Tape the end of a flat-bladed tool to avoid damaging the crankshaft. Pry out the oil seal using the taped end of the tool.

5. Inspect the oil seal contact surface on the crankshaft for signs of wear or damage.

To install:

6. Wipe the seal bore with a clean rag.

7. Apply multipurpose grease to the lip of a new oil seal.

8. Install or connect the following:
- Oil seal into place using a seal installer tool. Be sure the seal surface is flush with the edge of the oil pump case. Work from the front of the cover. Be extremely careful not to damage the seal.
- Crankshaft sprocket
- Timing belt
- Negative battery cable to the battery

9. Start the engine and check for leaks.

Camshaft and Valve Lifters

REMOVAL & INSTALLATION

1.3L Engine

1. Before servicing the vehicle, refer to the precautions in the beginning of this section.

2. Drain the cooling system into a re-sealable container.

3. Relieve the fuel system pressure.

4. Remove or disconnect the following:
- Timing belt
- Cylinder head cover from the cylinder head

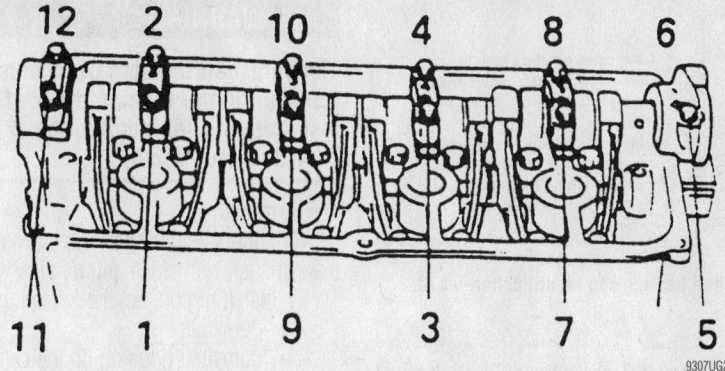

9307UG20

Loosen the camshaft bearing caps using several steps in the sequence shown—1.3L engines

- Camshaft Position (CMP) sensor coupler from the CMP sensor
- High tension cords and couplers from the coil assemblies
- Ignition coil from the exhaust manifold side
- CMP case from the cylinder head

5. Loosen all valve adjusting screw lock nuts, turn the adjusting screws back all the way to allow the rocker arms to move freely.

6. Remove or disconnect the following:
- Camshaft bearing caps in the sequence illustrated in several steps
- Camshaft
- Timing belt inside cover
- Intake rocker arm with the clip from the rocker arm shaft. Make sure not to bend the clip when removing the intake rocker arm
- Rocker arm shaft bolts
- Exhaust rocker arms and spring by pulling out the shaft on the battery side. If necessary, remove the battery for clearance.

7. Inspect the rocker arms and shafts for wear and/or damage and replace parts as necessary.

a. If the tip of the valve lash adjuster is badly worn, replace the adjuster.

To install:

8. Apply engine oil to the rocker arms and the rocker arm shafts.

9. Install or connect the following:
- O-ring on the shaft
- Rocker arm shaft with its holes facing up
- Rocker arm (exhaust side) and the rocker arm spring
- Rocker arm shaft bolts and tighten the bolts to 96 inch lbs. (11 Nm)

10. Pour a small amount of engine oil into the arm pivot holding part of the shaft.

11. Install or connect the following:
- Rocker arm (intake side), with the clips to the rocker arm shaft
- Camshaft and the bearing caps. Tighten the bearing caps in several passes to reach the final torque of 96 inch lbs. (11 Nm). Refer to the accompanying illustration for the proper camshaft bearing cap torque sequence.
- Camshaft oil seal
- Timing belt inside cover
- Camshaft timing belt pulley, while fitting the pin on the camshaft into

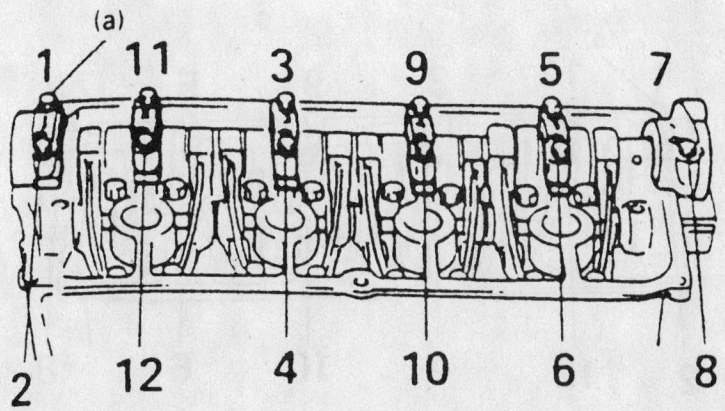

9307UG21

Tighten the camshaft bearing caps using several steps in the sequence shown—1.3L engines

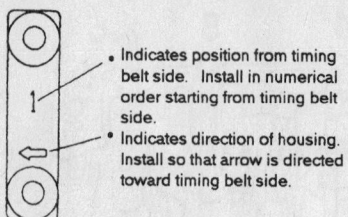

• Indicates position from timing belt side. Install in numerical order starting from timing belt side.
• Indicates direction of housing. Install so that arrow is directed toward timing belt side.

7923UG21

Camshaft bearing cap identification—1.6L engine

the slot. Tighten the pulley bolt to 43 ft. lbs. (60 Nm).
• Timing belt
• Remaining components in the reverse order of removal.
12. Check the ignition timing.

1.6L Engine

1. Before servicing the vehicle, refer to the precautions in the beginning of this section.
2. Relieve the fuel system pressure.
3. Remove or disconnect the following:
• Water pump belt and pulley
• Crankshaft pulley
• Timing belt cover and the timing belt
• Camshaft sprocket
• Cylinder head cover
• Distributor and distributor case, if equipped
• Ignition coils and Camshaft Position (CMP) sensor case, with the CMP sensor housing coupler disconnected, if equipped
4. Loosen the valve adjusting screw locknuts and screws to allow all the valves to close.
5. Remove or disconnect the following:
• Camshaft housing bolts
• Housings
• Camshaft

⁂ WARNING

The camshaft housing bolts must be removed in the reverse order of installation or damage to the camshaft may occur.

To install:

6. Lubricate the lobes and journals of the camshaft with clean engine oil.
7. Install or connect the following:
• Camshaft
• Camshaft housing on the camshaft and cylinder head, starting with the number 1 housing

➡**Embossed marks are provided on each camshaft housing, indicating position and direction for installation.**

8. Apply engine oil to the camshaft journal sliding surface of each housing. Apply sealant to the mating surface of the number 6 housing which will mate with the cylinder head.
9. Apply engine oil to the housing bolts, and hand-tighten the bolts into the housing. Follow the tightening sequence in 3 to 4 even stages, finishing with a final torque of 96 inch lbs. (11 Nm).

⁂ WARNING

The camshaft housing bolts must be tightened in the correct order or damage to the camshaft may occur.

10. Apply engine oil to the camshaft oil seal lip. Install the camshaft oil seal until the surface becomes flush.
11. Install or connect the following:
• Camshaft sprocket
• Timing belt
• Timing belt cover
• Crankshaft pulley
• Water pump pulley
• Water pump belt. Be sure the pin

on the camshaft fits into the slot at the **E** mark on the camshaft sprocket. Tighten the sprocket bolt to 41–46 ft. lbs. (56–64 Nm).
12. Prior to installation on models equipped with a distributor, apply sealant to the area of the distributor housing that covers the rear of the rocker arm shaft on the cylinder head. Install the distributor and distributor housing to the cylinder head. Be sure the distributor is facing the correct firing position.
13. Prior to installation on models equipped with a Distributorless Ignition System (DIS), apply sealant to the area of the housing that covers the rear of the rocker arm shaft on the cylinder head. Install the CMP sensor and housing to the cylinder head and tighten the bolts to 96 inch lbs. (11 Nm).
• Ignition coil assembly, if equipped
14. Adjust the valve lash.
15. Install or connect the following:
• New cylinder head cover gasket and the cover
• Negative battery cable
16. Start the engine and check for any water or oil leaks when finished.
17. Check and/or adjust the ignition timing as necessary.

Valve Lash

ADJUSTMENT

1.3L Engines

Hydraulic valve lash adjusters are used to adjust the valve clearance to **0** lash automatically at all times. Adjustment is not required.

1.6L engine

1. Before servicing the vehicle, refer to the precautions in the beginning of this section.
2. Disconnect the negative battery cable.
3. Remove the cylinder head cover.
4. Remove the right front wheel and fender apron.
5. Turn the crankshaft pulley clockwise until the **V** mark on the pulley is aligned with the **0** calibration mark on the timing belt cover.
6. Remove the Camshaft Position (CMP) sensor and look in the hole for the notch on the camshaft rotor gear. If the notch is not visible in the hole, rotate the crankshaft 1 complete revolution until the notch is visible.

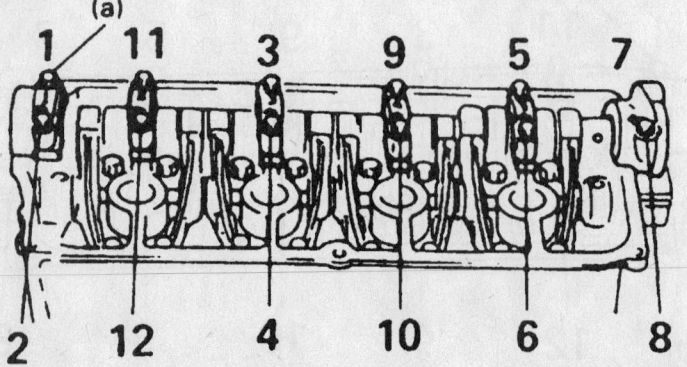

7923UG22

Camshaft housing bolt tightening sequence—1.6L engine

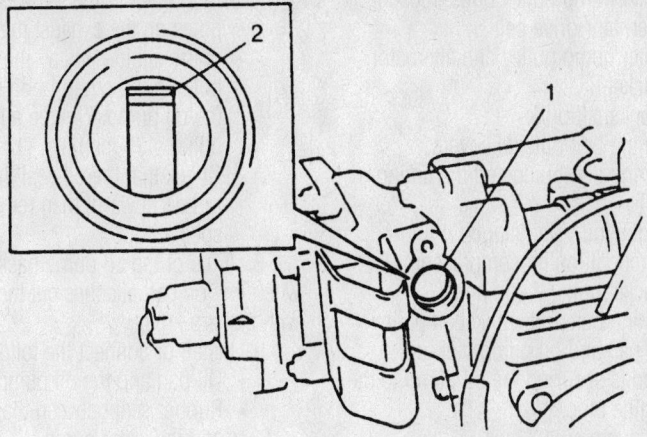

1. Hole for CMP sensor
2. Camshaft rotor gear

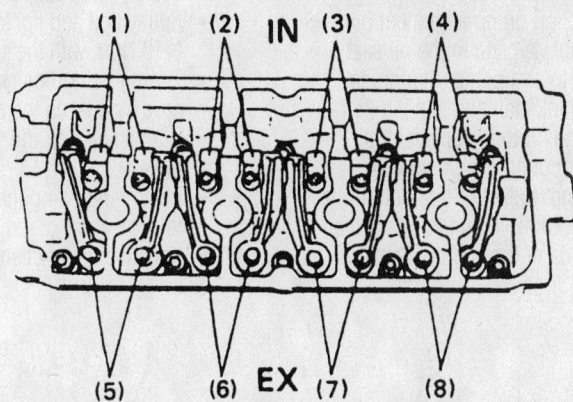

Valve numbered locations and camshaft rotor gear mark—1.6L engines

7923UG23

7. The valve lash is measured between the rocker arm adjusting screw and the valve stem. Use a thickness gauge to measure the gap.

8. Check the valve lash for the following valves:
- Intake valve of cylinder number 1 (ID 1)
- Intake valve of cylinder number 2 (ID 2)
- Exhaust valve of cylinder number 1 (ID 5)
- Exhaust valve of cylinder number 3 (ID 7)

9. If the valve lash is out of specification, adjust the specification after loosening the locknut and turning the adjusting screw. Hold the screw stationary while tightening the locknut. Recheck the specification after tightening the locknut.

10. Rotate the crankshaft 1 rotation clockwise and realign the timing marks.

11. Check the valve lash for the following valves:
- Intake valve of cylinder number 3 (ID 3)
- Intake valve of cylinder number 4 (ID 4)
- Exhaust valve of cylinder number 2 (ID 6)
- Exhaust valve of cylinder number 4 (ID 8)

12. Adjust the valves that are out of specification and recheck after tightening the locknut.

13. Install the remaining components.

14. Connect the negative battery cable.

Starter Motor

REMOVAL & INSTALLATION

1. Remove or disconnect the following:
- Negative battery cable
- Starter electrical connections
- 2 starter mounting bolts
- Starter motor

To install:

2. Install or connect the following:
- Starter
- 2 starter mounting bolts and tighten to 96 inch lbs. (10 Nm)
- Starter electrical connections
- Negative battery cable

Oil Pan

REMOVAL & INSTALLATION

1. Before servicing the vehicle, refer to the precautions in the beginning of this section.

2. Remove or disconnect the following:
- Negative battery cable
- Engine undercovers on 1.6L engines

3. Drain the engine oil into a suitable container.
- Lower plate on 1.3L models, from the clutch housing if equipped with a manual transaxle, or from the torque converter housing if equipped with a automatic transaxle.

4. On 1.6L models, remove or disconnect the following:
- Front exhaust pipe from the vehicle
- Transaxle stiffener plate from the engine and transaxle

5. Support the transmission and engine.

6. Remove or disconnect the following:
- Vehicle center member by removing the 7 nuts and 4 bolts from the center member, on 1.6L models
- Crankshaft Position (CKP) sensor from the oil pan, if equipped
- Oil pan retainer bolts
- Oil pan from the cylinder block
- Oil pump pick-up

To install:

7. Clean the mating surfaces of the oil pan and the engine block.

8. Install or connect the following:
- Oil pump strainer. Tighten the strainer bolt first at the bracket. Tighten the bolts to 96 inch lbs. (11 Nm).

9. Apply silicon sealant to the oil pan mating surface in one continuous bead.

10. Install or connect the following:
- Oil pan to the engine block and install the bolts. Start tightening the bolts at the center and move outward. Tighten the bolts to 96 inch lbs. (11 Nm).
- CKP sensor on the oil pan
- Drain plug and drain plug gasket on the oil pan
- Lower plate on the clutch housing, on 1.3L models; if equipped with a

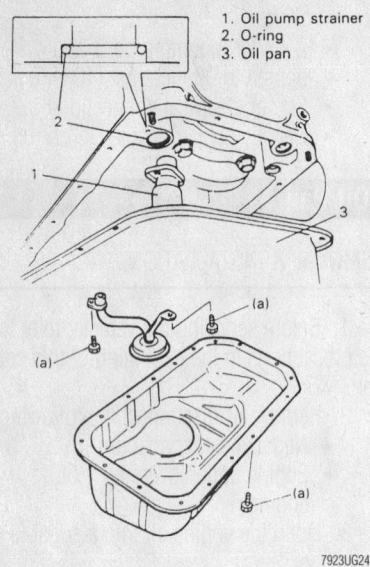

1. Oil pump strainer
2. O-ring
3. Oil pan

Oil pan mounting—1.3L and 1.6L engines

7923UG24

manual transaxle, or on the torque converter housing if equipped with an automatic transaxle

11. On 1.6L models, install or connect the following:

- Center member on the vehicle. Tighten the bolts and nuts at the engine mounts to 40 ft. lbs. (55 Nm) and all other nuts to 33 ft. lbs. (45 Nm).
- Transaxle stiffener plate and tighten the bolts to 37 ft. lbs. (50 Nm)
- Exhaust pipe and tighten the nuts and bolts to 37 ft. lbs. (50 Nm)
- Engine undercovers

12. Refill the engine with oil.

13. Connect the negative battery cable.

14. Start the engine and check for leaks.

Oil Pump

REMOVAL & INSTALLATION

1. Before servicing the vehicle, refer to the precautions in the beginning of this section.

2. Disconnect the negative battery cable.

3. Drain the engine oil.

4. Remove or disconnect the following:
- Right side fender apron clips by pushing the center pin

❋❋ WARNING

Do not push the center pin too far in, or it will fall off into the fender.

- Power steering and air conditioning belt, if equipped

- Water pump pulley bolts, loosen
- Alternator drive belt
- Water pump pulley and alternator bracket
- Crankshaft pulley
- Timing belt outside covers
- Timing belt guide and the timing belt
- Engine oil level gauge
- Air conditioning compressor bracket bolts, if equipped
- Timing belt and crankshaft pulley
- Oil pan and oil pump pick-up
- 7 bolts securing the oil pump to the engine block
- Oil pump

To install:

5. Clean the engine block where the oil pump mounts, then install the oil pump gasket and the 2 oil pump alignment pins.

6. To prevent damage to the oil seal when installing the oil pump, fit a guide sleeve on the crankshaft and apply a thin coating of engine oil to the special tool.

7. Install or connect the following:
- Oil pump. Install a long bolt in the lowest bolt hole on the intake manifold side of the engine. Install 2

long bolts in the 2 lowest bolt holes on the exhaust manifold side of the engine. Install the 4 short bolts in the other 4 bolt holes in the oil pump. Tighten all of the bolts to 96 inch lbs. (11 Nm). Check that the oil seal lip is not turned upward, then remove the special tool.

8. If the of the oil pump gasket bulges where the oil pan attaches cut the excess off with a sharp knife.

9. Install or connect the following:
- Oil pan and the oil pump pick-up
- Rubber seal between the oil pump and the water pump
- Crankshaft key
- Timing belt pulley
- Crankshaft pulley pin
- Pulley bolt and tighten to 80 ft. lbs. (110 Nm), with the crankshaft locked
- Timing belt guide so that the concave side faces the oil pump
- Crankshaft key and the timing belt pulley
- Remaining components in the reverse order of removal
- Negative battery cable

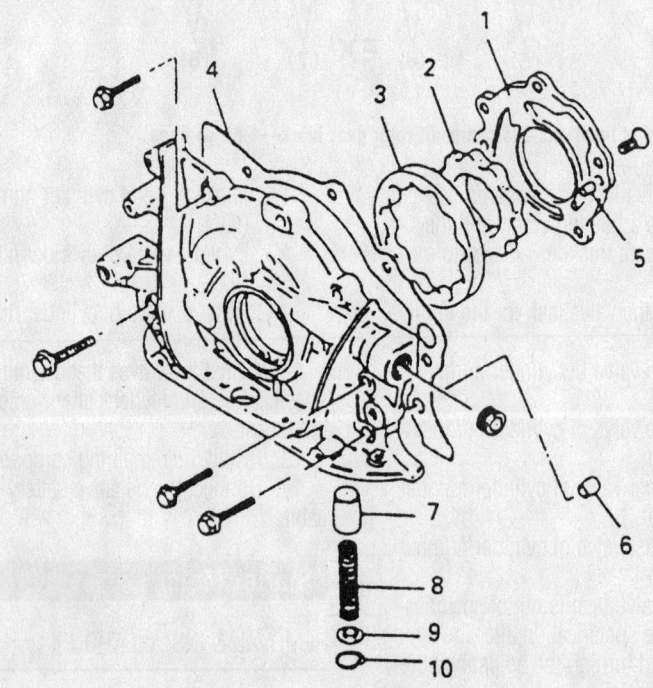

1. Rotor plate	6. Pin
2. Inner rotor	7. Relief valve
3. Outer rotor	8. Spring
4. Gasket	9. Retainer
5. Pin	10. Retainer ring

7923UG25

Exploded view of the oil pump—1.3L and 1.6L engine

10. Start the engine and check the engine oil pressure.

11. Check that no leaks are present.

Rear Main Seal

REMOVAL & INSTALLATION

1. Before servicing the vehicle, refer to the precautions in the beginning of this section.

2. Remove or disconnect the following:

- Transaxle assembly
- Flexplate/flywheel from the crankshaft

3. Carefully pry the oil seal out of the retainer without scratching the sealing surface of the crankshaft.

To install:

4. Apply engine oil the lip of the new seal.

5. Install or connect the following:

- Seal in the retainer using a suitable seal driver
- Flexplate/flywheel
- Transaxle assembly

Timing Belt

REMOVAL & INSTALLATION

1.3L Engines

1. Before servicing the vehicle, refer to the precautions in the beginning of this section.

2. Remove all necessary components for access to the upper and lower timing belt outside covers, then remove the covers.

3. Align the camshaft timing belt pulley with its timing marks. The crankshaft and camshaft marks are straight up.

4. Remove the resonator and the timing belt outside cover.

5. Remove the tensioner stud and loosen the tensioner bolt.

6. Remove the tensioner spring and damper, then remove the timing belt.

❋❋ WARNING

After the timing belt is removed never turn the camshaft or the crankshaft. Interference may occur between the pistons and the valves causing component damage.

7. Remove the tensioner and the tensioner plate.

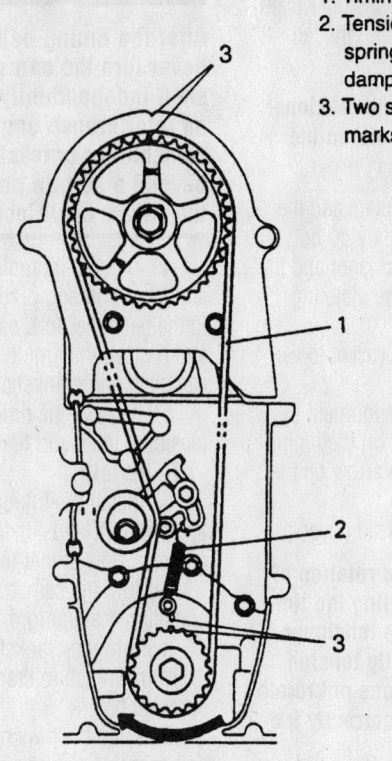

1. Timing belt
2. Tensioner spring & damper
3. Two sets of marks

Direction of crankshaft

93015G30

View of the timing belt and timing marks—Suzuki 1.0L engine

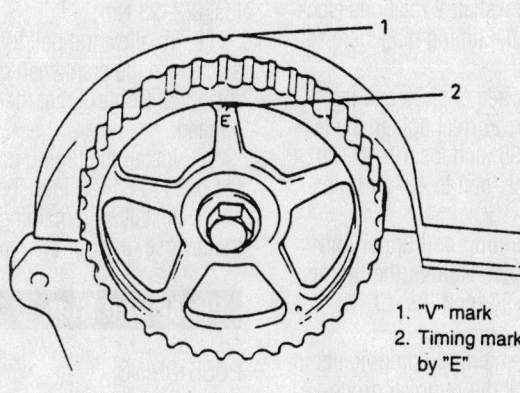

1. "V" mark
2. Timing mark by "E"

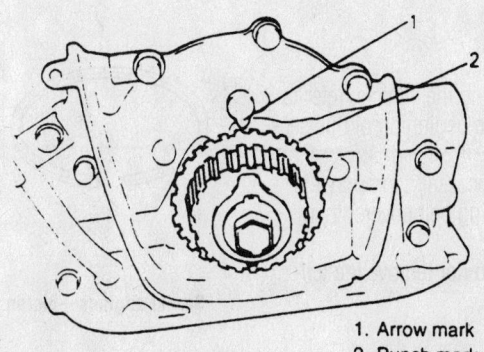

1. Arrow mark
2. Punch mark

79235G69

Match the "V" notch to the "E" mark on the camshaft, and the punch and arrow on the crankshaft to properly position the engine for belt service—Suzuki 1.3L and 1.6L engines

To install:

8. Install the timing belt tensioner plate and tensioner. Only hand-tighten the tensioner bolt.

➥**Be sure that the lug on the tensioner plate is inserted into the hole on the tensioner.**

9. Be sure the tensioner plate and the tensioner move uniformly. If they do not move together remove the tensioner and the tensioner plate and reinsert the plate lug into the tensioner hole.

10. Check the camshaft sprocket to verify that it has not moved.

11. Check the crankshaft alignment by verifying that the punch mark on the timing belt pulley is aligned with the arrow on the oil pump case.

12. Remove the cylinder head cover.

➥**This is to permit the free rotation of the camshaft. When installing the timing belt on the pulleys, the tensioner spring force should correctly tension the belt. If the camshaft does not rotate freely the belt will not be correctly tensioned.**

13. With the timing marks aligned, hold the tensioner plate up by hand and install the timing belt on the pulleys so there is no slack on the drive side of the belt.

14. Turn the crankshaft 2 rotations clockwise. Confirm that the timing marks are still properly aligned.

15. If the belt is free of slack and the alignment marks are correct tighten the tensioner stud to 84–96 inch lbs. (9–12 Nm). Tighten the tensioner bolt to 17–21 ft. lbs. (24–30 Nm).

16. Install the timing belt upper and lower outside covers. Tighten the timing cover bolts to 84–96 inch lbs. (9–12 Nm).

17. Install all remaining components in the reverse order of the removal procedure.

1.6L Engine

1. Before servicing the vehicle, refer to the precautions in the beginning of this section.

2. Remove all necessary components for access to the timing belt covers, then remove the covers.

3. Loosen but do not remove the tensioner bolt.

✳✳ CAUTION

After the timing belt is removed, never turn the camshaft and crankshaft independently. This engine is an interference engine and if the camshaft or crankshaft is turned beyond a certain point, damage to the valves could occur.

4. Loosen the timing belt tensioner adjusting bolt and pivot nut. Apply pressure to the tensioner to loosen the timing belt, and remove the timing belt from the camshaft and crankshaft sprockets.

5. Remove the timing belt tensioner, tensioner plate and tensioner spring.

To install:

6. Install the timing belt tensioner, plate and spring. Hand-tighten the tensioner bolt and stud only at this time.

7. Turn the camshaft sprocket clockwise and align the timing marks.

8. Turn the crankshaft clockwise, using a 17mm wrench to crank the timing belt sprocket bolt.

9. Align the punch mark on the timing belt sprocket with the arrow mark on the oil pump.

10. With the timing marks aligned, remove any slack from the drive side of the belt. Tighten the tensioner bolt to 16–20 ft. lbs. (22–28 Nm).

11. To allow the belt to be free of any slack, turn the crankshaft clockwise 2 full rotations. Confirm that the timing marks are aligned.

12. Install the timing cover and tighten the bolts to 84–96 inch lbs. (9–12 Nm).

13. Install all remaining components in the reverse order of the removal procedure.

Piston and Ring

POSITIONING

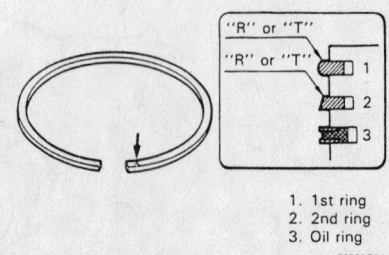

1. 1st ring
2. 2nd ring
3. Oil ring

7923AG84

Suzuki engines—piston ring positioning

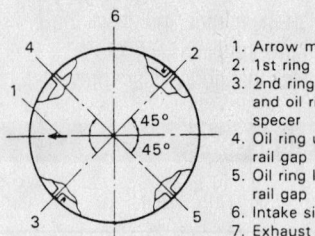

1. Arrow mark
2. 1st ring end gap
3. 2nd ring end gap and oil ring spacer
4. Oil ring upper rail gap
5. Oil ring lower rail gap
6. Intake side
7. Exhaust side

7923AG85

Suzuki engines—piston ring end-gap spacing

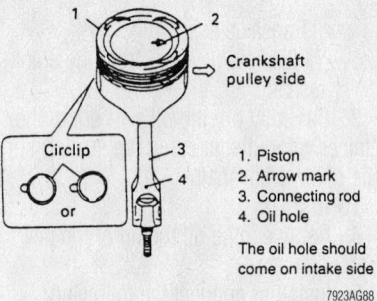

1. Piston
2. Arrow mark
3. Connecting rod
4. Oil hole

The oil hole should come on intake side

7923AG88

Suzuki engines—piston/connecting rod assembly-to-engine positioning

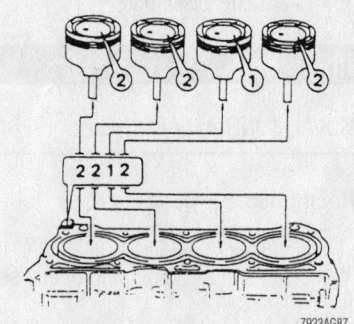

7923AG87

Suzuki engines—the piston ID number must match the number stamped in the engine block

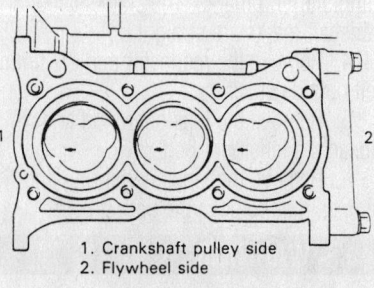

1. Crankshaft pulley side
2. Flywheel side

7923AG86

Suzuki engines—the directional arrow on the piston face must face the crankshaft pulley end of the engine

FUEL SYSTEM

Fuel System Service Precautions

Safety is the most important factor when performing not only fuel system maintenance, but any type of maintenance. Failure to conduct maintenance and repairs in a safe manner may result in serious personal injury or death. Maintenance and testing of the vehicle's fuel system components can be accomplished safely and effectively by adhering to the following rules and guidelines.

• To avoid the possibility of fire and personal injury, always disconnect the negative battery cable unless the repair or test procedure requires that battery voltage be applied.

• Always relieve the fuel system pressure prior to disconnecting any fuel system component (injector, fuel rail, pressure regulator, etc.), fitting or fuel line connection. Exercise extreme caution whenever relieving the fuel system pressure, to avoid exposing your skin, face and eyes to fuel spray. Please be advised that fuel under pressure may penetrate the skin or any part of the body that it contacts.

• Always place a shop towel or cloth around the fitting or connection prior to loosening to absorb any excess fuel due to spillage. Ensure that all fuel spillage (should it occur) is quickly removed from the engine surfaces. Ensure that all fuel soaked cloths or towels are deposited into a suitable waste container.

• Always keep a dry chemical (Class B) fire extinguisher near the work area.

• Do not allow fuel spray or fuel vapors to come into contact with a spark or open flame.

• Always use a back-up wrench when loosening and tightening fuel line connection fittings. This will prevent unnecessary stress and torsion on fuel line piping. Always follow the proper torque specifications.

• Always replace worn fuel fitting O-rings with new. Do not substitute fuel hose or equivalent, where fuel pipe is installed.

Fuel System Pressure

RELIEVING

✳✳ CAUTION

Care should be used when working around the fuel system. DO NOT smoke or expose the fuel system to any open flames. Keep a fire extinguisher handy.

1. Before servicing the vehicle, refer to the precautions in the beginning of this section.
2. Disconnect the negative battery cable from the battery.
3. Place the vehicle in **PARK** for automatic transmission or **NEUTRAL** for manual transmission.
4. Remove the relay box cover.
5. Disconnect the fuel pump relay from the relay box.
6. Remove the fuel filler cap from the filler neck to release the fuel vapor pressure in the fuel tank.
7. Start the vehicle and allow the engine to run until it stalls.
8. Crank the engine 3 more revolutions to eliminate any remaining pressure in the fuel lines.
9. Disconnect the negative battery cable.
10. Connect the fuel pump relay to the relay box.
11. After servicing the fuel system, connect the negative battery cable.
12. Start the engine and check for leaks in the system.

Fuel Filter

REMOVAL & INSTALLATION

✳✳ CAUTION

The fuel system pressure must be relieved before disconnecting any fuel lines. Failure to do so may result in personal injury.

1. Before servicing the vehicle, refer to the precautions in the beginning of this section.
2. Properly relieve the fuel system pressure.
3. Place a container under the fuel filter.
4. Remove or disconnect the following:
   • Fuel inlet hose from the fuel filter

✳✳ CAUTION

A small amount of fuel may be released after the fuel hose is disconnected. Cover the hose and pipe with a shop towel.

   • Outlet hose from the fuel feed pipe
   • 2 fuel filter mounting bracket bolts

and remove the fuel filter from the frame with the outlet hose attached
   • Fuel filter from the bracket by unfastening the mounting bolt
   • Outlet hose from the fuel filter

To install:
5. Install or connect the following:
   • Fuel filter on the bracket
   • Mounting bolt
   • Remaining components in the reverse order of removal
   • Inlet and outlet hoses to the filter
   • Negative battery cable
6. With the ignition **ON** and the engine **OFF** check for leaks.

Fuel Pump

REMOVAL & INSTALLATION

1. Before servicing the vehicle, refer to the precautions in the beginning of this section.
2. Relieve the pressure from the fuel system.
3. Remove the rear seat cushion by removing or disconnecting the following:
   • Spare tire
   • Seat back by unfastening the 2 center mounting nuts and the 4 mounting screws
   • Fitting screws from the rear of the seat cushion
   • Lift the front of the seat cushion and remove the cushion
4. Remove or disconnect the following:
   • Fuel level gauge and the fuel pump lead wire couplers and detach the wire tape
   • Fuel filler hose from the fuel tank
   • Breather hose from the filler neck
5. Drain the fuel from the tank by pumping the fuel out through the fuel tank filler.

✳✳ CAUTION

Use a gasoline safe hand operated pump device to drain the fuel tank.

6. Remove or disconnect the following:
   • Fuel hoses from the fuel pipes, located near the fuel filter

✳✳ CAUTION

A small amount of fuel may be released after the fuel hose is disconnected. Cover the hose and pipe to be disconnected with a shop cloth.

7. Install a support (for example, a transmission jack) under the fuel tank.

8. Remove or disconnect the following:
- Fuel tank mounting hardware
- Tank from the vehicle
- Fuel lines from the fuel pump and sender assembly
- 12 screws that secure the fuel pump and fuel gauge assembly to the tank
- Pump and sender assembly
- Fuel pump electrical connectors
- Fuel strainer
- Fuel pump

To install:

9. Install or connect the following:
- Fuel pump on the fuel gauge assembly
- Fuel strainer on the fuel pump

➠**Always install a new fuel pump strainer when replacing the fuel pump.**

- Electrical connectors to the fuel pump
- Remaining components in the reverse order of removal
- Negative battery cable

10. Turn the ignition switch to the **ON** position, but leave the engine **OFF** and check for fuel leaks.

Fuel Injector

REMOVAL & INSTALLATION

Swift

1. Before servicing the vehicle, refer to the precautions in the beginning of this section.
2. Relieve the pressure from the fuel system.

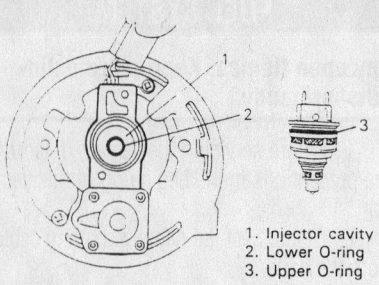

1. Injector cavity
2. Lower O-ring
3. Upper O-ring

9307UG02

Install the lower injector O-ring (2) to the injector cavity (1) and the upper O-ring (3) to the injector

3. Remove or disconnect the following:
- Air cleaner assembly
- Air cleaner mounting stay from the throttle body
- Injector cover
- Injector from the throttle body
- Injector O-rings and discard

To install:

4. Inspect the injector filter for dirt or contamination. If present, clean and check for dirt in the fuel lines and tank.

5. Apply a tin coat of spindle oil or gasoline to the new O-rings.

6. Install or connect the following:
- Lower injector O-ring to the injector cavity and the upper O-ring to the injector
- Injector by pushing straight into the injector cavity. Do not turn the injector while pushing.

7. Make sure the injector cover O-ring is free of contamination and has not deteriorated. Apply a coat of spindle oil or gasoline to the O-ring.

8. Install or connect the following:

- Injector cover and tighten the retainers to 24–36 inch lbs. (3–4 Nm)
- Negative battery cable

9. Turn the ignition switch on and check for fuel leaks
- Air cleaner mounting stay to the throttle body
- Air cleaner assembly

Esteem

1. Before servicing the vehicle, refer to the precautions in the beginning of this section.
2. Relieve the pressure from the fuel system.
3. Remove or disconnect the following:
- Vacuum hose from the fuel pressure regulator
- Fuel injector electrical connections
- Fuel rail bolts
- Injector from the fuel rail
- Injector O-ring and discard

To install:

4. Coat the injector O-rings with gasoline.

5. Install or connect the following:
- Injector O-ring
- Injector to the fuel rail and intake manifold.

6. Make sure the injectors rotate smoothly. If not, the O-ring has been improperly installed and must be replaced.

- Fuel rail bolts. Once the bolts are tightened make sure the injectors rotate smoothly.
- Fuel injector electrical connections
- Vacuum hose to the fuel pressure regulator
- Negative battery cable

7. Start the vehicle and check for leaks.

DRIVE TRAIN

Transaxle Assembly

REMOVAL & INSTALLATION

Manual

1. Before servicing the vehicle, refer to the precautions in the beginning of this section.
2. Drain the transaxle oil.
3. Remove or disconnect the following:
- Battery cables
- Battery and tray
- Clutch cable adjusting nut, joint pin from the cable and cable from the bracket

- All the wiring harness clamps and connectors involved with the transaxle removal, tag if necessary for location to aid during installation
- Speedometer cable boot, case clip and cable from the case
- Radiator outlet pipe from the transmission side cover, on Swift models
- Transaxle retaining bolts
- Starter
- Fender apron extension on the left side
- Bolts connecting the exhaust pipe to the exhaust manifold and disconnect the joint

4. On Esteem models, with the engine supported, remove the vehicle center mounting member by removing the 7 nuts and 4 bolts.
- Gearshift control shaft nut and bolt
- Control shaft from the gearshift control shaft
- Extension rod nut and washers
- Clutch housing lower plate
- Sway bar
- Left and right front wheels
- Left and right ball joints
- Left and right halfshafts
- Transaxle stiffener
- Transmission to engine bolt and nut
- Engine rear mounting bracket bolts

5. Install an engine support.

6. Support the transaxle with a suitable jack.

7. Remove the left engine mounting bracket and stiffener.

8. Lower the transaxle with the engine attached. Pull the transaxle straight out toward the left side.

9. Lower and remove the transaxle.

To install:

10. Install or connect the following:
- Transaxle from the left side of the vehicle. Use care when inserting the pilot shaft into the clutch assembly. If the spline on the input shaft does not align with the clutch assembly spline, turn the crankshaft slightly to aid in spline alignment.

11. Raise the transaxle and engine.
- Left engine mounting bracket and stiffener. Tighten the bolts to 29–43 ft. lbs. (40–60 Nm).
- Rear engine mounting bracket bolts and tighten them to 29–43 ft. lbs. (40–60 Nm)

➡**Before installing the bolts into the rear mounting bracket, apply sealant to the bolt threads.**

- Transmission-to-engine bolt and nut. Tighten the nut and bolt to 29–43 ft. lbs. (40–60 Nm)

12. On Esteem models, install the center member on the vehicle and install the 7 nuts and 4 bolts. Tighten the bolts and nuts as follows:

   a. Center member-to-radiator support: 33 ft. lbs. (45 Nm).

   b. Center member-to-crossmember: 33 ft. lbs. (45 Nm).

   c. Engine mounts-to-center member nuts: 40 ft. lbs. (55 Nm).

13. Install or connect the following:
- Transaxle stiffener

14. Lower the transaxle supporting jack.
- Left and right halfshafts
- Ball joints
- Sway bar
- Clutch housing lower plate
- Extension rod nut and washers. Tighten the rod nut to 18–28 ft. lbs. (25–40 Nm).
- Control shaft on gearshift
- Gearshift control shaft bolt and nut. Tighten the gearshift control shaft bolt and nut to 11–14 ft. lbs. (15–20 Nm).
- Exhaust pipe to the manifold and install the bolts. Tighten the bolts to 29–36 ft. lbs. (40–50 Nm).
- Left fender apron extension

15. Refill the transaxle with the recommended lubricant.

16. Remove the engine support fixture.

17. Install or connect the following:
- Starter
- Transaxle retaining bolts. Tighten the retaining bolts to 29–43 ft. lbs. (40–60 Nm).
- Remaining components in the reverse order of removal
- Negative battery cable and the ground strap on the transaxle

Automatic

1. Before servicing the vehicle, refer to the precautions in the beginning of this section.

2. Drain the cooling system.

3. Drain the transaxle fluid from the transaxle.

4. Remove or disconnect the following:
- Negative battery cable from the battery and transaxle
- Speedometer cable

5. Disconnect the following electrical connectors:
- Solenoids
- Vehicle Speed Sensor (VSS), if equipped
- Shift lever switch
- Forward clutch cylinder revolution sensor

6. Remove or disconnect the following:
- Wiring harness from the clamps on the transaxle
- Select cable from the transaxle
- Cooling system pipe from the transaxle
- Top transaxle-to-engine bolts
- Starter
- Exhaust manifold cover
- Front exhaust pipe from the exhaust manifold

7. Support the engine.
- Engine undercovers, if equipped.
- Transaxle cooler hoses

8. On Esteem models, with the engine supported, remove the vehicle center mounting member by removing the 7 nuts and 4 bolts.
- Front exhaust pipe from the vehicle
- Transaxle stiffener plate
- Transaxle housing lower plate
- Torque converter bolts. To lock the drive plate, engage a flat-bladed tool in the flywheel.
- Sway bar from the control arms
- Left and right front wheels
- Left and right ball joints from the steering knuckles
- Left and right halfshafts
- Engine rear mount and bracket

9. After removing the rear mount, remove the engine-to-transaxle bolt and nut located behind the rear bracket. Remove all bolts holding the engine to the transaxle.

10. Support the transaxle with a transaxle jack.

11. Remove or disconnect the following:
- Bolts from the engine left-hand mount
- Transaxle with the torque converter from the engine compartment

➡**When removing the transaxle from the engine, move it parallel with the crankshaft and use care so not to apply excessive force to the drive plate and torque converter.**

※※ CAUTION

Be sure to keep the transaxle with the torque converter horizontal or facing up throughout the work. Should it be tilted with converter down, the converter may fall off and cause personal injury.

To install:

12. Install or connect the following:
- Transaxle to the engine assembly
- Transaxle attaching nuts and bolts
- Left-hand mounting bolts and tighten to 40 ft. lbs. (55 Nm)
- Engine-to-transaxle bolt and nut before installing the rear transaxle mount. Tighten the nut to 65 ft. lbs. (90 Nm).
- All bolts for the transaxle. Tighten the bolts to 65 ft. lbs. (90 Nm).
- Left halfshafts
- Ball joints
- Sway bar to the control arms
- Torque converter bolts and tighten the bolts to 14 ft. lbs. (19 Nm)
- Transaxle housing lower plate
- Stiffener plate with the 4 bolts. Tighten the bolts to 40 ft. lbs. (55 Nm).
- Exhaust pipe on the center pipe and tighten the bolts to 37 ft. lbs. (50 Nm)

13. On Esteem models, install the center member on the vehicle and install the 7 nuts and 4 bolts. Tighten the bolts and nuts as follows:

   a. Center member-to-radiator support: 33 ft. lbs. (45 Nm).

   b. Center member-to-crossmember: 33 ft. lbs. (45 Nm).

   c. Engine mounts-to-center member nuts: 40 ft. lbs. (55 Nm).

14. Install or connect the following:
   • Oil hoses for the transaxle
   • Engine undercovers
15. Remove the engine support.
   • Exhaust pipe to the exhaust manifold and install the nuts
   • Exhaust manifold cover
   • Starter motor
   • Remaining components in the reverse order of removal
16. Fill the cooling system and the transaxle. Check all fluids.
17. Connect the negative battery cable to the transaxle and battery.

Clutch

ADJUSTMENT

1. Before servicing the vehicle, refer to the precautions in the beginning of this section.
2. Depress the clutch pedal lightly until tension on the clutch cable can be felt.
3. Measure the clutch pedal free-play. It should be 0.6–0.8 in. (15–20mm).
4. Adjust the clutch pedal free-play by tightening or loosening the clutch cable adjustment nut.
5. Measure the clutch lever free-play. It should be 0–0.08 in. (0–2mm). If the clutch release lever free-play exceeds specification, inspect the release shaft return spring for cracks or weakness.

➡ **Be sure the marks on the clutch release lever and release shaft are aligned. If they are not, remove the lever from the shaft, align the marks and repeat the free-play adjustment procedure.**

REMOVAL & INSTALLATION

1. Before servicing the vehicle, refer to the precautions in the beginning of this section.
2. Remove the transaxle.

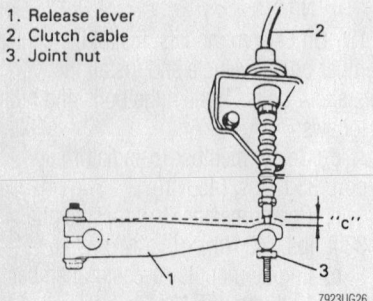

1. Release lever
2. Clutch cable
3. Joint nut

Clutch cable free-play adjustment

7923UG26

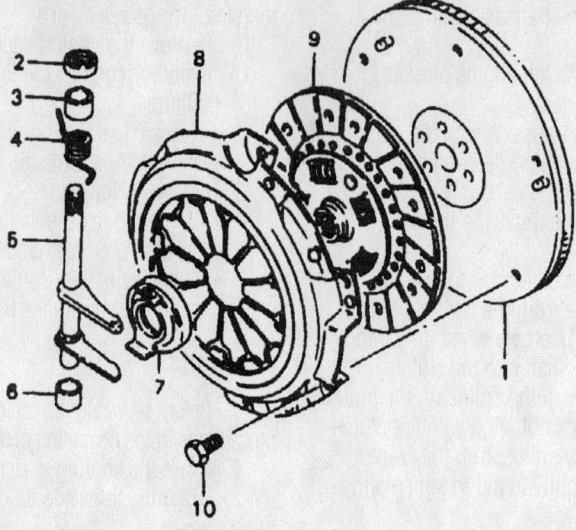

1. Flywheel	5. Release shaft	9. Clutch disc
2. Release shaft seal	6. No. 1 bush	10. Clutch cover bolt
3. No. 2 bush	7. Release bearing	
4. Return spring	8. Clutch cover	

7923UG27

Clutch component identification

3. Hold the flywheel stationary.
4. Matchmark the pressure plate and flywheel for installation reference.
5. Loosen the pressure plate attaching bolts 1 turn at a time (evenly) until the spring pressure is released.
6. Remove the clutch disc and pressure plate.

To install:
7. Clean the flywheel mating surfaces of all oil, grease and metal deposits. Inspect flywheel for cracks, heat checking or other defects and replace or resurface as necessary.
8. Check the wear on the facings of the clutch disc by measuring the depth of each rivet head depression. Replace clutch disc when rivet heads are 0.02 in. (0.5mm) below the surface of clutch surface.
9. Check the diaphragm spring and pressure plate for wear or damage. If the spring or plate is excessively worn, replace the pressure plate assembly.
10. Check the pilot bearing for smooth operation. If the bearing does not spin freely, replace it.
11. Position the clutch disc and pressure plate with the matchmarks aligned and install a clutch alignment tool.
12. Install the pressure plate bolts. Tighten the mounting bolts evenly and in a crisscross pattern to 13–20 ft. lbs. (18–28 Nm). Remove the alignment tool and the flywheel holding tool.

13. Lightly lubricate the transaxle input shaft splines, pilot bearing surface of the input shaft, and the release bearing with grease.
14. Install the transaxle.
15. Adjust the clutch cable.

Halfshaft

REMOVAL & INSTALLATION

1. Before servicing the vehicle, refer to the precautions in the beginning of this section.
2. Remove or disconnect the following:
   • Negative battery cable
   • Undo the sealer on the halfshaft nut
   • Halfshaft nut and washer
3. Drain the oil from the transmission.
4. On the left side shaft, use 2 large prybars to release the snapring fitting on the halfshaft inner joint from the differential.

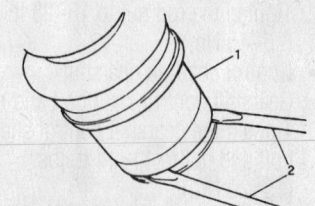

**1 Drive shaft joint (LH)
2 Pry tool**

7923UG28

Disconnecting the left inboard joint

5. On the right side shaft, use a plastic hammer to drive the shaft joint to release the snapring fitting on the halfshaft inner joint from the differential.

6. Remove or disconnect the following:
- Sway bar attaching nut, washer and bushing from the suspension arm
- Ball joint stud bolt and nut
- Inboard joint from the differential by pulling it
- Outer joint from the steering knuckle
- Halfshaft from the vehicle

✳✳ WARNING

To prevent breakage of the boots, be careful not to bring the boots in contact with other components when removing the shaft assembly.

7. If the center shaft requires service, drain the transmission oil and remove the support bolts. Remove the center shaft from the differential, then from the vehicle.

To install:

8. If the center shaft was removed, install the shaft into the differential, then install the support bolts. Tighten the support bolts to 29–43 ft. lbs. (40–60 Nm).

9. Clean the grease seal on the steering knuckle and apply a small amount of fresh grease.

10. Install or connect the following:
- Wheel side joint on the steering knuckle, then the differential side joint on the differential. Seat the differential joint by hand, making sure that the snapring is seated.
- Halfshaft nut on the outer joint loosely, to hold it in position.

✳✳ WARNING

Do not hit the joints with a hammer to seat them. Use hand pressure only or component damage may occur.

- Suspension arm to the steering knuckle
- Ball joint stud
- Sway bar on the suspension arm
- Sway bar bushing, washer and nut. Tighten the nut to 17–20 ft. lbs. (23–28 Nm).

11. If equipped with a manual transaxle fill the transaxle with the specified gear oil.

12. Tighten the halfshaft nut to 109–145 ft. lbs. (150–200 Nm).

13. Connect the negative battery cable.

14. If equipped with an automatic transaxle, fill the transaxle with the specified transmission oil.

CV-Joints

OVERHAUL

Double Offset Joint (DOJ) Type

The Double Offset Joint (DOJ) type is identified by the outside shape of the differential side joint which has no dent.

1. Before servicing the vehicle, refer to the precautions in the beginning of this section.

2. Remove the driveshaft.

3. Remove the boot band from the differential side joint.

4. Slide the boot towards the center of the shaft and remove the snapring from the outer race.

5. Remove the shaft from the outer race.

6. Use a rag to clean off the grease, then remove the circlip that retains the cage using snapring pliers.

7. Using a suitable puller, draw the cage away and remove the boot from the shaft.

8. Inspect all components for wear and/or damage and replace as necessary.

To install:

9. Clean all components and allow them to completely dry.

10. Install the boot onto the shaft until its small diameter side fits to the shaft groove and attach it there with a boot band.

11. Using a pipe whose inner diameter is 0.906 in. (23mm) or more and outer diameter is 1.260 in. (32mm) or less, drive the cage into position. Install the cage using the smaller outside diameter side to the shaft end.

12. Install the circlip.

13. Apply grease to the entire surface of the cage.

14. Insert the cage into the outer race and fit the snapring into the groove of the outer race.

➡**Position the opening of the snapring so that it will not be lined up with a ball.**

15. Apply grease to the inside of the outer race and fit the boot to the outer race.

16. After fitting the boot, insert a screwdriver into the boot on the outer race side and allow air to enter the boot so that the air pressure in the boot equals atmospheric pressure

17. Fix the boot to the outer race with a boot band. Before tightening the band adjust the boot so that measurement D which is 8.10 in. (205.8mm) and measure-

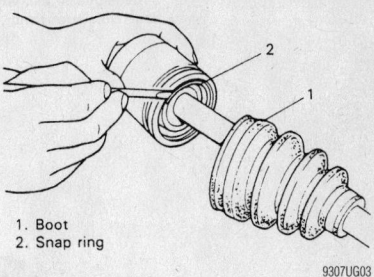

1. Boot
2. Snap ring

9307UG03

Remove the snapring from the outer race—DOJ type

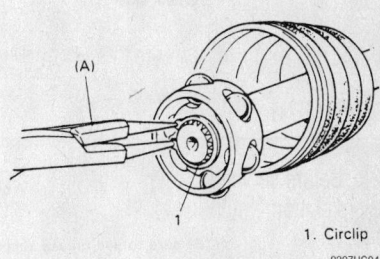

(A)

1. Circlip

9307UG04

Remove the circlip that retains the cage—DOJ type

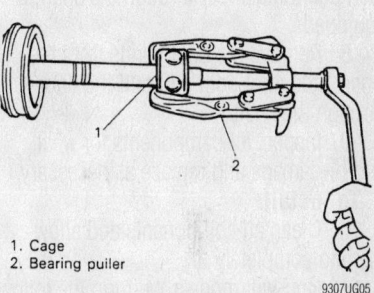

1. Cage
2. Bearing puller

9307UG05

Draw the cage away using a suitable puller—DOJ type

ment E which is 7.34 in. (186.4mm) are as shown in the accompanying illustration.

18. Install the driveshaft in the vehicle.

Tripod Joint Type

The Tripod joint type can be identified by the 3 dent lines on the outside of the differential side joint.

1. Before servicing the vehicle, refer to the precautions in the beginning of this section.

2. Remove the driveshaft.

3. Remove the boot band and the Tripod joint housing.

4. Use a rag to clean off the grease, then remove the circlip using snapring pliers.

5. Remove the spider by using a suitable puller.

6. Remove the boot band and pull the differential side boot from the shaft.

"d" } Dimensions to use when fixing
"e" } boot with boot band.

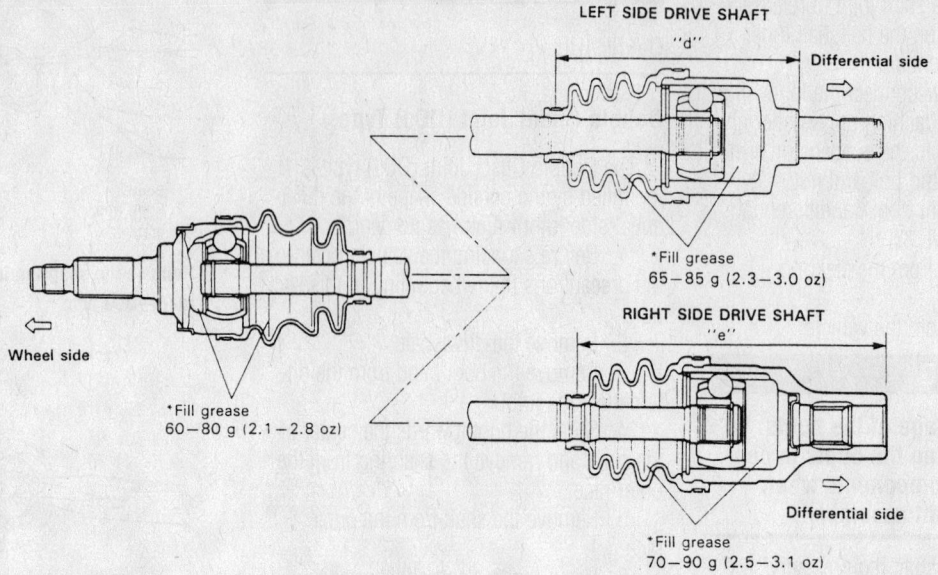

LEFT SIDE DRIVE SHAFT

"d"

Differential side ⇒

*Fill grease
65 – 85 g (2.3 – 3.0 oz)

Wheel side ⇐

*Fill grease
60 – 80 g (2.1 – 2.8 oz)

RIGHT SIDE DRIVE SHAFT

"e"

Differential side ⇒

*Fill grease
70 – 90 g (2.5 – 3.1 oz)

*Be sure to use grease supplied with spare parts.

9307UG06

Adjust the boot so that measurements D and E are as illustrated—DOJ type

7. Remove the dynamic damper band, then pull the damper through the shaft, if equipped.

8. Remove the boot bands from the wheel side joint boot and pull the boot through the shaft.

9. Inspect all components for wear and/or damage and replace as necessary.

To install:

10. Clean all components and allow them to completely dry.

11. On Swift models, perform the following:

a. Apply grease to the wheel side joint. Use black grease in the tube included with the wheel side boot kit.

b. Install the wheel side boot on the shaft.

c. Fill the inside of the boot with grease and fasten the boot with a band.

d. Install the dynamic damper on the shaft, if equipped.

e. Install the differential side boot on the shaft.

f. Apply grease to the Tripod joint. Use the yellow grease in the tube included in the differential side joint kit.

g. Using a pipe whose inner diameter is 0.906 in. (23mm) or more and outer diameter is 1.260 in. (32mm) or less, drive the spider into position. Face the chamfered side inward and fasten it in place with the circlip.

h. Fill the boot with grease, then install the housing and joint it with the boot.

i. Fasten the boot bands.

12. On Esteem models, perform the following:

a. Apply grease to the wheel side joint. Use black grease in the tube included with the wheel side boot kit.

b. Install the wheel side boot on the shaft.

c. Fill the inside of the boot with grease.

d. Fit the boot band into the groove in the boot.

e. Tighten the boot band until its outer diameter is 3.05 in. (77.5mm).

f. Fold the boot band over the metal fixture.

g. Using a punch, caulk the center of the band folded over the fixture.

h. Cut the band about 0.28 in. (7mm) from the clip.

i. Clamp the band with the clip.

j. Fix the small diameter side of the boot band into the groove in the boot.

k. Tighten the boot band until its outer diameter is 1.14 in. (29mm).

l. Fold the boot band over the metal fixture.

m. Using a punch, caulk the center of the band folded over the fixture.

n. Install the dynamic damper on the right side driveshaft.

o. Install the differential side boot on the shaft.

p. Apply grease to the Tripod joint. Use the yellow grease in the tube included in the differential side joint kit.

q. Install the spider into position. Face the chamfered side inward and fasten it in place with the circlip.

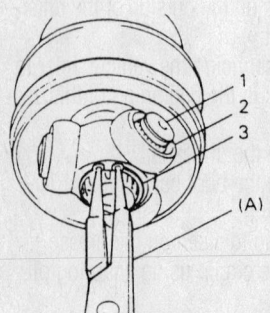

1. Spider
2. Bearing
3. Circlip

9307UG07

Remove the circlip that retains the spider assembly—Tripod type

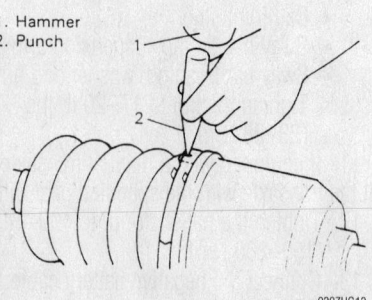

1. Hammer
2. Punch

9307UG13

Using a punch, caulk the center of the band folded over the fixture—Esteem

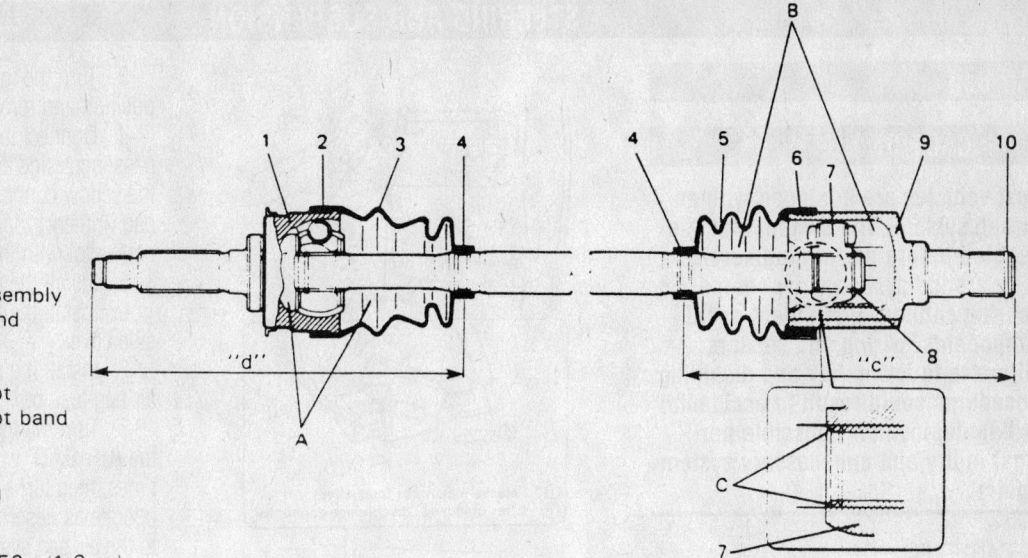

1. Wheel side joint assembly
2. Wheel side boot band
3. Wheel side boot
4. Boot band
5. Differential side boot
6. Differential side boot band
7. Tripod joint spider
8. Circlip
9. Tripod joint housing
10. Snap ring

A: Black grease (about 52 g/1.8 oz)
B: Yellow grease (about 100 g/3.5 oz)
C: Chamfered spline

9307UG14

Adjust the boot so that measurement D is correct—Esteem models equipped with a Tripod type joint

r. Fill the boot with grease, then install the housing and joint it with the boot.

s. Adjust the boot so that the when measured from the tip of the driveshaft (wheels side), to the inner (small) band; the measurement is 8 in. (204mm).

13. Insert a screwdriver into the boot on the outer race side and allow air to enter the boot so that the air pressure in the boot equals atmospheric pressure

a. Fasten the boot band so that diameter E which measures 3.11 in. (79mm) and diameter F which measures 1.14 in. (29mm) are correct. Refer to the accompanying illustration for the diameter locations.

14. Install the driveshaft.

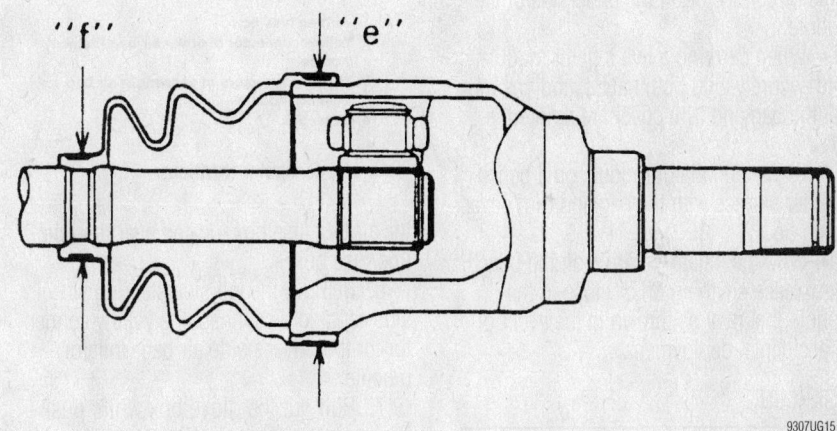

9307UG15

When tightening the bands, make sure that diameters E and F are correct—Esteem models

STEERING AND SUSPENSION

Air Bag

PRECAUTIONS

Several precautions must be observed when handling the inflator module to avoid accidental deployment and possible personal injury.

• Never carry the inflator module by the wires or connector on the underside of the module.

• When carrying a live inflator module, hold securely with both hands, and ensure that the bag and trim cover are pointed away.

• Place the inflator module on a bench or other surface with the bag and trim cover facing up.

• With the inflator module on the bench, never place anything on or close to the module that may be thrown in the event of an accidental deployment.

DISARMING

1. Before servicing the vehicle, refer to the precautions in the beginning of this section.
2. Disconnect the negative battery cable.
3. Turn the steering wheel so the wheels are pointing straight ahead.
4. Turn the ignition switch to the **LOCK** position and remove the key.
5. Remove the **AIR BAG-IG** fuse from

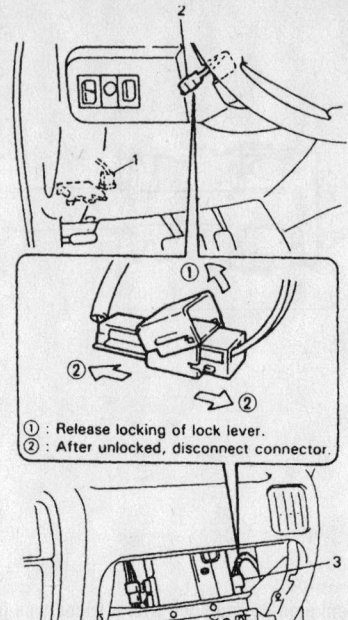

① : Release locking of lock lever.
② : After unlocked, disconnect connector.

1 Air bag fuse box
2 Yellow connector of driver air bag (inflator) module
3 Yellow connectors of passenger air bag (inflator) module
4 Glove box

7923UG29

Air bag connector locations

the air bag fuse box located near the junction/fuse box.

6. Remove the left side steering wheel side cap and disconnect the yellow connector for the driver's side air bag (inflator) module.

7. Pull out the glove box while pushing in on the stoppers from the left and the right sides. Disconnect the yellow connector for the passenger air bag (inflator) module.

REARMING

1. Before servicing the vehicle, refer to the precautions in the beginning of this section.
2. Connect the negative battery cable.

3. Turn the ignition switch to the **LOCK** position and remove the key.
4. Connect the yellow connector for the passenger side air bag (inflator) module and the yellow connector for the driver's side air bag (inflator) module. Be sure to lock each connector with the lock lever.
5. Install the glove box assembly.
6. Install the left side steering wheel side cover.
7. Install the **AIR BAG-IG** fuse in the air bag fuse box.
8. Turn the ignition **ON** and verify that the **AIR BAG** warning lamp flashes 7 times, then turns off. If the system does not operate as described, diagnosis and repairs to the air bag system are necessary.

Rack and Pinion Steering Gear

REMOVAL & INSTALLATION

Manual

1. Before servicing the vehicle, refer to the precautions in the beginning of this section.
2. Disconnect the negative battery cable.
3. Slide the driver's seat back as far as possible.
4. Remove or disconnect the following:

• Pull back the front part of the floor mat on the driver's side
• Steering shaft joint cover
• Steering shaft upper joint bolt, loosen but do not remove it
• Steering shaft lower joint bolt
• Lower joint from the pinion
• Front wheels
• Tie rod ends from the steering knuckles
• Steering gear mounting bolts and the brackets
• Steering gear case from the vehicle

To install:

5. Install or connect the following:
• Steering gear, brackets and mounting bolts. Tighten bolts to 14–21 ft. lbs. (20–30 Nm).

- Tie rod ends to the steering knuckles

6. Be sure the steering wheel is in the straight-ahead position and the front wheels are pointing straight ahead.

7. Install or connect the following:
- Steering shaft to the steering gear
- Lower steering shaft-to-steering gear clinch bolt and tighten both steering joint bolts (upper and lower) to 14–21 ft. lbs. (20–30 Nm)
- Remaining components in the reverse order of removal
- Negative battery cable

8. Check and adjust the front wheel alignment.

Power

> ※※ **WARNING**
>
> **Be sure to set the front wheels straight ahead and remove the ignition key from the cylinder before starting repairs. If equipped with an air bag the contact coil of the air bag system may be damaged if the key is not removed and the wheels are not straight ahead.**

1. Before servicing the vehicle, refer to the precautions in the beginning of this section.

2. Remove or disconnect the following:
- Negative battery cable
- Steering column joint covers
- Steering shaft upper joint bolt, loosen but do not remove
- Steering shaft lower joint bolt
- Lower joint from the pinion
- Front wheels
- Tie rod ends from the steering knuckles
- Front exhaust pipe, on Swift models
- Engine rear torque rod with the torque rod and bracket, Swift models equipped with an automatic transaxle
- Gearshift control shaft and extension rod from the transaxle, if equipped with a manual transaxle
- Rear engine mount together with the bracket from the engine and suspension member, Esteem models
- Mounting member from the suspension frame by removing the 2 bolts, Esteem models
- High and low pressure lines from the rack and pinion

➡**When the lines are disconnected plug the lines or place an oil pan under the vehicle.**

- Cylinder lines from the rack and pinion
- Rack and pinion mounting bolts and brackets
- Rack and pinion case from the vehicle

To install:

3. Install or connect the following:
- Rack and pinion, brackets and mounting bolts. Tighten the bolts to 40 ft. lbs. (55 Nm) on Esteem models or 22–28 ft. lbs. (30–40 Nm) on Swift models.
- Cylinder lines on the rack and pinion and tighten their fittings to 14–21 ft. lbs. (20–30 Nm)
- High and low pressure lines to the rack and pinion. Tighten the fittings to 22–28 ft. lbs. (30–40 Nm).
- Gearshift control shaft and extension rod to the transaxle, if equipped with a manual transaxle. Tighten the extension rod nut to 18–28 ft. lbs. (25–40 Nm) and tighten the control shaft nut and bolt to 11–14 ft. lbs. (15–20 Nm).
- Engine rear torque rod with the torque rod and bracket, Swift mod-

els equipped with an automatic transaxle
- Tie rod ends to the steering knuckles

4. Be sure the steering wheel is straight and the front wheels are pointing straight ahead.
- Steering shaft to the rack and pinion
- Lower steering shaft-to-rack and pinion clinch bolt and tighten both steering joint bolts (upper and lower) to 14–21 ft. lbs. (20–30 Nm)
- Front wheels
- Negative battery cable

5. Bleed the power steering system.
6. Lower the vehicle.
7. Check and adjust the front wheel alignment.

Strut

REMOVAL & INSTALLATION

Front

1. Before servicing the vehicle, refer to the precautions in the beginning of this section.

2. Remove or disconnect the following:

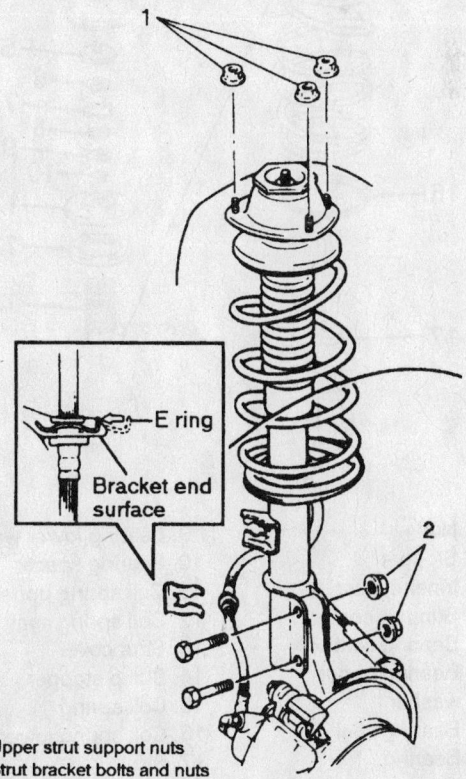

E ring

Bracket end surface

1. Upper strut support nuts
2. Strut bracket bolts and nuts

7923UG30

Strut assembly mounting

- Wheels
- Anti-Lock Brake (ABS) system wheel speed sensor, if equipped
- Brake hose clip, then the hose from the strut

3. Support the lower control arm with a floor jack.

4. Remove the strut bracket bolts.

5. Hold the strut by hand so it will not fall, and remove the upper strut support nuts from the engine compartment.

※※ WARNING

Do not loosen the center nut at this time or serious injury or vehicle damage may result.

6. Remove the strut assembly from the vehicle.

7. Install a pair of coil spring compressors on the coil spring on the strut assembly. Turn the spring compressors alternately until the spring tension is released from the strut assembly. If the spring can be turned slightly, then it has been collapsed enough.

※※ CAUTION

This procedure requires the use of a spring compressor. It cannot be performed without one. If you do not have access to this special tool, do not attempt to disassemble the strut. The coil spring is retained under considerable pressure. It can exert enough force to cause serious injury. Exercise extreme caution.

8. Keeping the spring collapsed, remove the strut center nut and remove the other components from the top of the strut assembly.

9. Remove the spring from the strut.

To install:

10. If you're installing a new spring, compress the spring with a pair of spring compressors. Be sure that the spring compresses to 9 inches (230mm) for installation.

11. Position the coil spring on the strut making sure that the end of the spring is mated to the stepped part of the lower seat.

12. Install or connect the following:
- Bump stop on the strut rod
- Strut cover
- Spring seat
- Upper spring seat
- Bearing spacer. Align the strut bracket with the mark on the upper spring seat.

13. Clean the bearing lower washer and install it on the strut rod.

14. Clean the strut bearing and apply fresh grease to the bearing. Install the bearing on the lower washer.

15. Clean the bearing upper washer and install it.

16. Install these components in the following order:
- Bearing upper seal
- Bearing seat
- Strut support
- Inner spacer
- Washer
- Strut nut. Tighten the nut to 29–43 ft. lbs. (40–60 Nm). Apply a waterproof coating (paint or lacquer) to the nut and strut rod threads.

17. Loosen the spring compressors alternately, checking that the stepped part of the spring seat and spring are properly positioned.

18. Install or connect the following:
- Strut assembly onto the vehicle
- Upper support nuts loosely. Tighten the upper strut support nuts to 16–23 ft. lbs. (22–33 Nm) on Swift models or 20 ft. lbs. (28 Nm) on Esteem models.
- Strut bracket nuts and bolts and tighten to 51–65 ft. lbs. (70–90 Nm)
- ABS wheel speed sensor, if equipped
- Brake hose clip
- Wheels

Rear

ESTEEM

1. Before servicing the vehicle, refer to the precautions in the beginning of this section.

2. Remove wheel and tire assembly.

3. Place a jack under the lower control arm to support the suspension.

4. Remove or disconnect the following:
- Brake line from the brake hose at the strut
- E-ring securing the brake hose to the strut
- Strut upper support nuts and push the strut down

1. Nut
2. Stopper
3. Inner spacer
4. Support comp.
5. Bearing seat
6. Bearing upper washer
7. Bearing seal
8. Bearing
9. Bearing lower washer
10. Bearing spacer
11. Coil spring upper seat
12. Coil spring seat
13. Strut cover
14. Bump stopper
15. Coil spring
16. Coil spring lower seat
17. Strut

7923UG31

Exploded view of the strut assembly

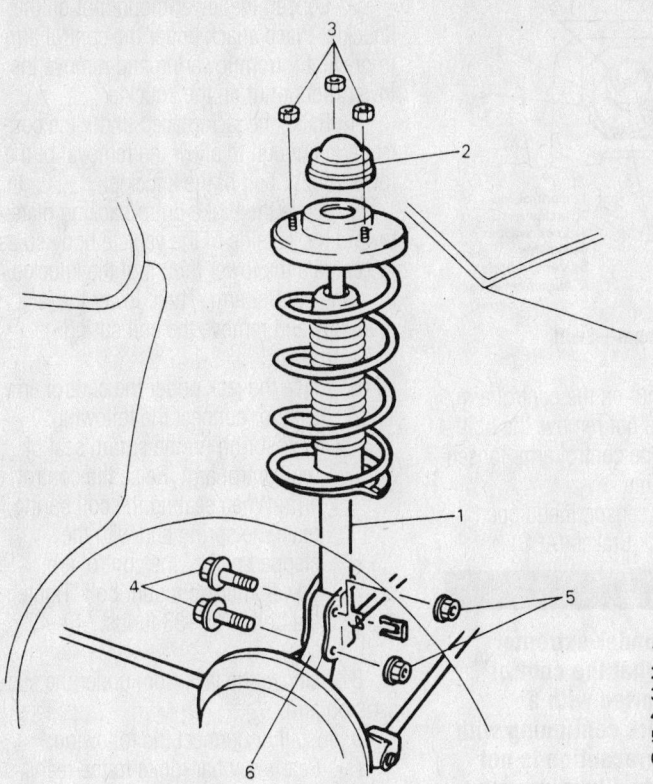

1. Strut assembly
2. Strut upper cap
3. Strut support nut
4. Strut bracket bolt
5. Strut bracket nut
6. Rear knuckle

7923UG32

Rear strut assembly mounting—Esteem

- 2 bolts and nuts holding the strut to the rear knuckle
- Strut from the knuckle
- Strut from the vehicle

5. Install a pair of coil spring compressors on the coil spring on the strut assembly. Turn the spring compressors alternately until the spring tension is released from the strut assembly. If the spring can be turned slightly, then it has been collapsed enough.

❋❋ CAUTION

This procedure requires the use of a spring compressor. It cannot be performed without one. If you do not access to this special tool, do not attempt to disassemble the strut. The coil spring is retained under considerable pressure. It can exert enough force to cause serious injury. Exercise extreme caution.

6. Keeping the spring collapsed, remove the strut center nut and remove the other components from the top of the strut assembly.

7. Remove the spring from the strut.

To install:

8. If installing a new spring, compress the spring with a pair of spring compres-sors. Make sure that the spring compresses to 11½ inches (290mm) for installation.

9. Position the coil spring on the strut making sure that the end of the spring is mated to the stepped part of the lower seat.

10. Install the remaining components.

11. Install the strut support with the center nut. Tighten the nut to 40 ft. lbs. (55 Nm).

12. Loosen the spring compressors alternately, checking that the stepped part of the spring seat and spring are properly positioned.

13. Install or connect the following:
- Strut in the vehicle
- 2 bolts and nuts to hold the strut to the rear knuckle. Tighten the nuts 65 ft. lbs. (90 Nm).

14. Fully extend the strut and position the upper part of the strut into the vehicle's body. If the upper part of the strut does not reach the vehicle body, raise the jack under the control arm a little.

15. Install or connect the following:
- Upper support nuts and tighten them to 20 ft. lbs. (28 Nm)
- Brake hose to the strut and install the E-clip
- Brake hose to the brake line
- Wheels

16. Fill the master cylinder with brake fluid and bleed the brake system.

SWIFT

1. Before servicing the vehicle, refer to the precautions in the beginning of this section.

2. Remove the wheels.

3. Place a jack under the lower control arm to support the suspension.

❋❋ CAUTION

The coil spring is under extreme pressure. Be sure the lower control arm is firmly supported with a hydraulic jack before continuing with procedure. If this caution is not observed, serious bodily injury may result.

4. Remove or disconnect the following:
- Strut support nuts and push the strut down

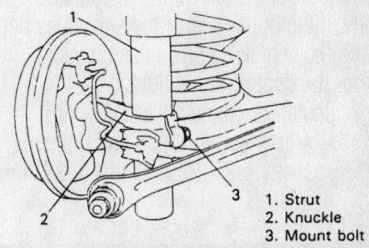

1. Strut
2. Knuckle
3. Mount bolt

7923UG33

Rear strut upper mounting nuts—Swift

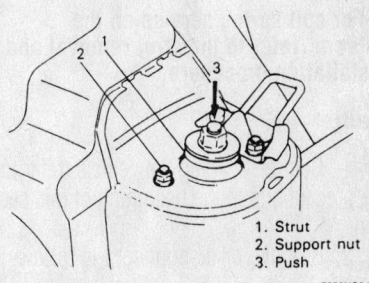

1. Strut
2. Support nut
3. Push

7923UG34

Rear strut lower mounting—Swift

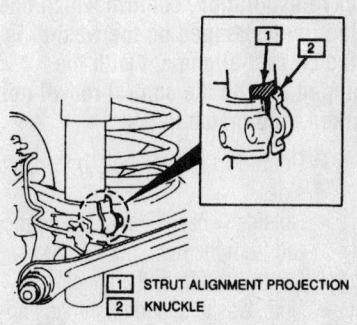

1 STRUT ALIGNMENT PROJECTION
2 KNUCKLE

7923UG35

Shock-to-steering knuckle alignment—Swift

- Strut lower mounting bolt
- Strut from the knuckle. Compress the strut as short as possible for removal. If the strut is hard to remove, open the slit of the knuckle by inserting a wedge.

➡ **Do not open the knuckle slit wider than necessary. Do not lower the jack more than necessary during the strut removal to prevent the coil spring from coming off, or the brake flexible hose from stretching.**

To install:

5. Install or connect the following:
   - Strut in the vehicle. Position the bottom of the alignment projection inside the knuckle opening.
   - Strut lower mounting bolt and tighten it to 36–50 ft. lbs. (50–70 Nm)

6. Fully extend the strut and position the upper part of the strut into the vehicle's body. If the upper part of the strut does not reach the vehicle's body, raise the jack under the control arm a little.

7. Install or connect the following:
   - Upper support nuts and tighten them to 20–27 ft. lbs. (28–38 Nm)
   - Wheels

Coil Spring

REMOVAL & INSTALLATION

➡ **For coil spring service on the esteem, refer to the strut removal and installation procedure.**

Swift

1. Before servicing the vehicle, refer to the precautions in the beginning of this section.

2. Remove or disconnect the following:
   - Rear wheels

➡ **To facilitate the toe-in adjustment after reinstallation, confirm which one of the lines stamped on the washer is in the closest alignment with the stamped line on the control rod. If not marked, add matchmarks.**

- Control rod inside bolt, body center side
- Outside (wheel side) of the control rod from the rear knuckle stud
- Control rod from the knuckle
- Nuts, washers, and bushings connecting the rear sway bar to the rear lower control arms

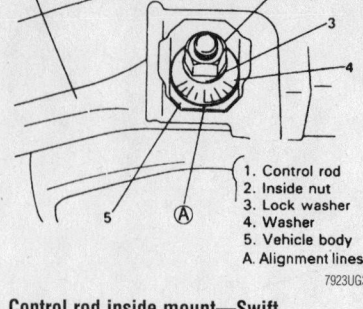

1. Control rod
2. Inside nut
3. Lock washer
4. Washer
5. Vehicle body
A. Alignment lines

7923UG36

Control rod inside mount—Swift

- Rear mount nut on the control arm, loosen but do not remove the bolt
- Front nut of the control arm, loosen only at this time
- Wheel speed sensor, if equipped with Anti-Lock Brakes (ABS)

❈❈ CAUTION

The coil spring is under extreme pressure. Be sure that the control arm is firmly supported with a hydraulic jack before continuing with procedure. If this precaution is not observed, serious bodily injury may result.

3. Loosen the lower mount nut on the knuckle. Place a jack under the control arm to prevent it from lowering and remove the lower mount nut on the knuckle.

4. Raise the jack placed under the control arm enough to allow the removal of the lower mount bolt of the knuckle.

5. Move the brake drum/backing plate toward the outside of the vehicle body so as to separate the lower mount of the knuckle from the control arm. Then, lower the jack gradually and remove the coil spring.

To install:

6. Place the jack under the control arm.

7. Install or connect the following:
   - Coil spring on the spring seat of the control arm. Raise the control arm. When seating the coil spring, mate the spring end with the stepped part of the control arm.
   - Lower knuckle mount bolt. Tighten the bolt to 29–33 ft. lbs. (40–45 Nm).

8. Remove the jack from under the suspension arm.

9. Install or connect the following:
   - Rear sway bar joints to the rear control arms and install the bushings, washers, and nuts. Tighten

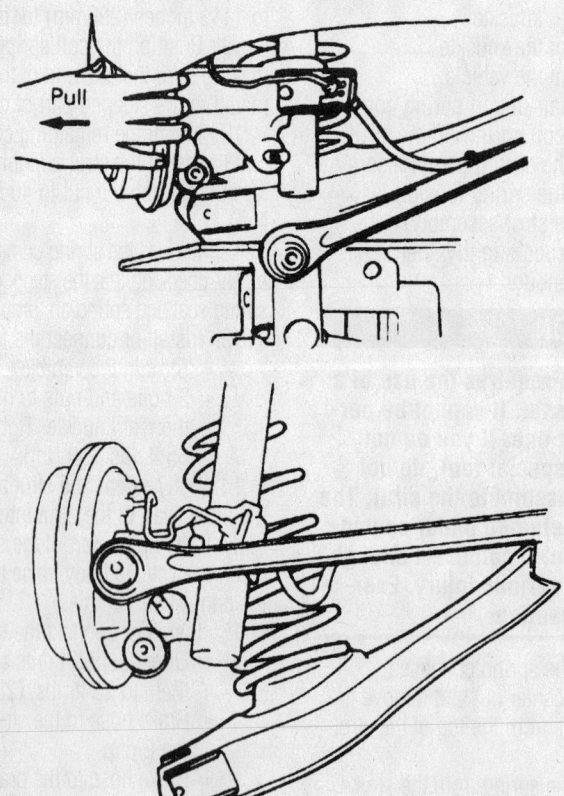

7923UG37

Disconnecting the knuckle and removing the coil spring

the nuts to 16–20 ft. lbs. (22–28 Nm).
- Control rod
- Inside control rod bolt and the outside control rod nut, but do not tighten them at this time
- Wheel speed sensor, if equipped
- Wheels and lower the vehicle

10. Tighten the control rod inside and outside nuts to 51–65 ft. lbs. (70–90 Nm).

➡ **When tightening the nuts, the vehicle should be off the hoist and in a non-loaded state. Also when tightening the inside nut, align the line stamped on the body with the line on the washer as confirmed before removal or align the matchmarks if marked.**

11. Tighten the suspension arm front nuts to 36–50 ft. lbs. (50–70 Nm) and the rear nuts to 29–33 ft. lbs. (40–45 Nm). After tightening the suspension arm front nut, be sure that the washer is not tilted.

12. Check the rear wheel alignment.

Lower Ball Joint

REMOVAL & INSTALLATION

The lower ball joint is an integral part of the lower control arm assembly. If the ball joint is found to be defective the whole lower control arm assembly must be replaced.

Lower Control Arm

REMOVAL & INSTALLATION

The lower control arm and ball joint are a complete unit that will not separate.

1. Before servicing the vehicle, refer to the precautions in the beginning of this section.

2. Remove or disconnect the following:
- Front wheels
- Sway bar link nut, washer, and cushion
- Ball joint stud bolt and nut
- Lower control arm front bushing bolt
- 2 bolts holding the lower control arm bracket to the vehicle
- Lower control arm from the steering knuckle
- Lower control arm from the vehicle

To install:

3. Install or connect the following:
- Ball joint on the knuckle and secure it with the nut and bolt
- Lower control arm rear bracket and bolts

- Lower control arm front mounting bolt
- Sway bar link to the control arm and install the cushion, washer and nut

4. Tighten the retainers to their proper torque specifications as follows:

a. Control arm rear mounting bracket bolts: 27 ft. lbs. (37 Nm).

b. Front mounting bolt: 65 ft. lbs. (90 Nm).

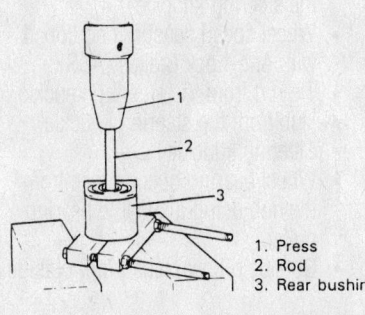

1. Press
2. Rod
3. Rear bushing

9307UG08

Use a suitable hydraulic press to remove the rear lower control arm bushing

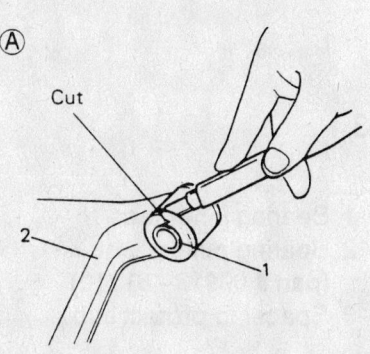

Ⓐ Cut

Ⓑ

1. Front bushing
2. Suspension arm
3. Press
4. Front bushing
5. Suspension arm

9307UG09

Cut the flange from the front bushing, then using a suitable hydraulic press; remove the rear lower control arm bushing

c. Ball joint nut and bolt: 44 ft. lbs. (60 Nm).

d. Sway bar link nut: 18 ft. lbs. (26 Nm).

5. Install the wheels.

6. Check the front wheel alignment.

CONTROL ARM BUSHING REPLACEMENT

1. Before servicing the vehicle, refer to the precautions in the beginning of this section.

2. Remove the lower control arm.

3. Use a hydraulic press to remove the rear bushing.

4. Cut the flange from the front bushing.

5. Use a hydraulic press to remove the front bushing.

To install:

6. Apply a solution of soapy water to the outer diameter of the front bushing, this will aid in installation.

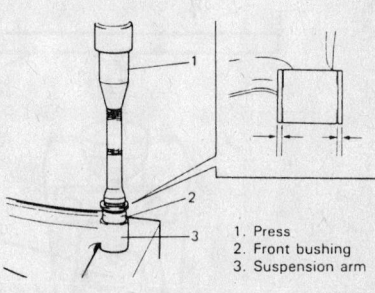

1. Press
2. Front bushing
3. Suspension arm

9307UG10

The front bushing should be positioned equally as shown after being pressed into position

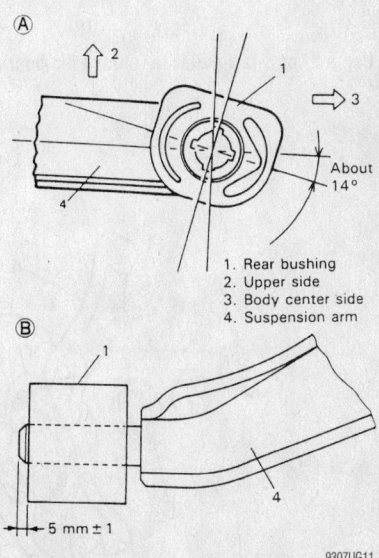

1. Rear bushing
2. Upper side
3. Body center side
4. Suspension arm

9307UG11

Install the rear bushing should be positioned as shown (A) before driving it into position (B)

7. Press the front bushing into its bore using a hydraulic press until the bushing is equal on the right and left of the arm as shown in the accompanying illustration.

8. Apply a solution of soapy water to the outer diameter of the rear bushing, this will aid in installation.

9. Install the rear bushing into the lower control arm in the direction and angle shown in figure A of the accompanying illustration.

10. Drive the bushing into position as shown in figure B of the accompanying illustration.

Wheel Bearings

ADJUSTMENT

The front and rear wheel bearings are a cartridge type design and cannot be adjusted.

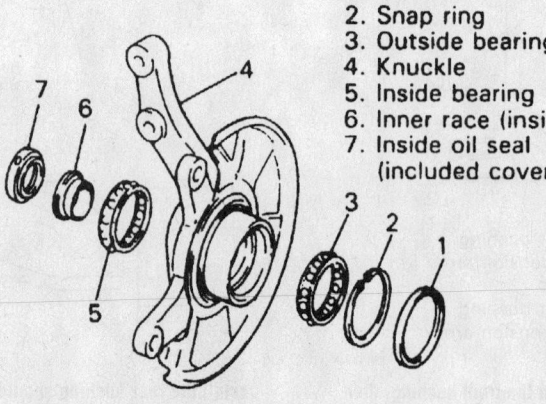

1. Bearing inner race
2. Bearing puller (part # 09913 - 61110)
3. Spacer to protect hub

7923UG38

Use a 2- or 3-jaw puller to remove the bearing race from the hub

1. Outside oil seal
2. Snap ring
3. Outside bearing
4. Knuckle
5. Inside bearing
6. Inner race (inside)
7. Inside oil seal (included cover)

7923UG39

Steering knuckle bearing components—Swift

REMOVAL & INSTALLATION

Front

➡**Always replace bearing races as a complete set.**

1. Before servicing the vehicle, refer to the precautions in the beginning of this section.

2. Remove or disconnect the following:
- Front wheel
- Brake caliper, carrier and disc from the steering knuckle
- Wheel speed sensor, if equipped with Anti-Lock Brakes (ABS)
- Tie rod from the steering knuckle
- Hub from the steering knuckle
- Steering knuckle
- Wheel bearing outside race from the hub using a suitable bearing puller
- Outside oil seal, snapring, outside bearing, inside oil seal and the inside bearing in that order

➡**Once the bearing outer race is removed, the bearing set (outer race, bearings and inner races) should be replaced.**

- Bearing outer race from the knuckle by pressing the race out of the knuckle

To install

➡**When installing the oil seals, be careful not to deform or tilt. Damage to the rubber part of the seal may occur.**

3. Install or connect the following:
- New bearing outer race into the knuckle using a press and the following tools: Bearing Installer Handle 09924–74510, Bearing And Oil Seal Installer 09944–68210 and a Bearing Installer Support 0994–78210

4. Apply lithium grease to the bearing races, bearings and oil seal lips.

5. Install or connect the following:
- Outside bearing on the steering knuckle
- Snapring to hold the outside bearing in place, then the outside oil seal
- Inside bearing to the steering knuckle
- Inside race and oil seal to the steering knuckle. When installing the inside oil seal, drive the oil seal in until it contacts the steering knuckle.

✳✳ WARNING

If equipped with ABS use caution when installing the oil seal, because the seal has a hole that must align with the speed sensor position.

- Outside race to the wheel hub using a bearing installer

6. Using a press and the proper tools, press the hub into the steering knuckle. After installation, be sure the hub is installed straight and turns freely.

7. Install or connect the following:
- Steering knuckle in the vehicle
- Tie rod end
- Brake caliper, carrier and disc on the steering knuckle
- Front wheel

Rear

WITH WHEEL HUBS

1. Before servicing the vehicle, refer to the precautions in the beginning of this section.

2. Set the parking brake.
3. Remove or disconnect the following:
- Rear wheels
- Rear brake drums
4. Use a brass drift and knock the wheel bearings from the drum assembly.

To install:

5. Position the inner wheel bearing on the drum with the sealed side facing out. Using a rear wheel Bearing Installer 09913–76010, install the rear wheel bearing. Install the wheel bearing spacer into the drum.

6. Install or connect the following:
- Outer wheel bearing with the sealed side facing out, using a wheel bearing installer

7. Fill the space in the brake drum in between the wheel bearings to about 40% capacity with wheel bearing grease.

8. Install or connect the following:
- Brake drum
- Spindle washer and a new spindle nut. Tighten the spindle nut to 58–86 ft. lbs. (80–120 Nm).

9. Coat the spindle nut and the spindle dust cap with sealer.

➡**When installing the spindle cap, hammer lightly several times on the collar of the cap until the collar comes closely into contact with the brake drum. If the fitting part of the cap is deformed or damaged or if it fits loose, replace the cap with a new one.**

10. Depress the brake pedal with about 66 lbs. (30 kg) of force 3 to 5 times to obtain proper drum to shoe clearance.

11. Install the wheels.

12. Check to ensure that the brake drum is free from dragging and proper braking is obtained.

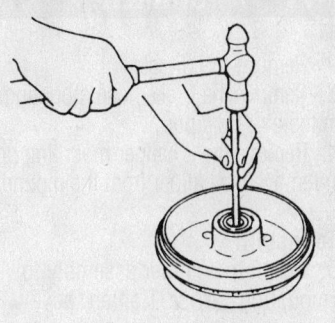

7923UG40

Use a hammer and drift to remove the wheel bearings from the hub—Swift

WITHOUT WHEEL HUBS

1. Before servicing the vehicle, refer to the precautions in the beginning of this section.

2. Set the parking brake.
3. Remove or disconnect the following:
- Rear wheels
- Brake drum, if equipped with drum brakes
- Caliper, carrier and disc, if equipped with rear disc brakes
4. Release the parking brake.
5. Remove or disconnect the following:
- Spindle cap without deforming it
- Sealer from the spindle nut
- Spindle nut and washer
6. Using a brake hub removal tool and a slide hammer remove the hub from the spindle.

➡**The wheel bearing and hub are a solid unit. When the wheel bearing is found defective and it is necessary to replace it, replace the hub assembly.**

To install:

7. Install or connect the following:

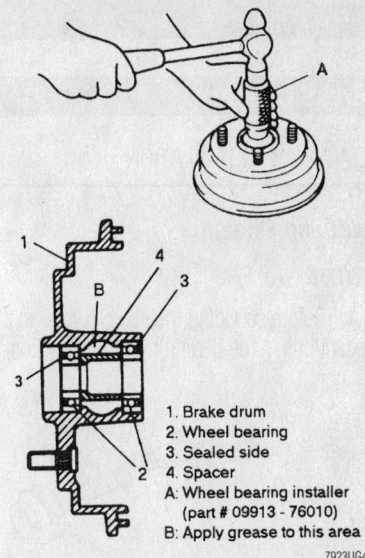

1. Brake drum
2. Wheel bearing
3. Sealed side
4. Spacer
A: Wheel bearing installer (part # 09913 - 76010)
B: Apply grease to this area

7923UG41

Wheel bearing installation—Swift

- Wheel hub, washer and a new spindle nut. Tighten the spindle nut to 108–144 ft. lbs. (150–200 Nm).

8. Coat the spindle nut with sealer
9. Install or connect the following:
- Spindle cap
- Brake drums, if equipped with rear drum brakes
- Brake caliper carrier and disc, if equipped with rear disc brakes

10. Depress the brake pedal with about 66 lbs. (30 kg) of force 3 to 5 times to obtain proper drum/rotor to shoe/pad clearance.

11. Install the wheel and tighten the lug nuts to 36–58 ft. lbs. (50–80 Nm).

12. Check to ensure that the brakes are free from dragging and that proper braking is obtained.

BRAKES

Brake Caliper

REMOVAL & INSTALLATION

Swift and Esteem

FRONT

1. Before servicing the vehicle, refer to the precautions in the beginning of this section.

2. Remove the wheels.

3. Remove the brake hose mounting bolt from the brake caliper.

4. Remove the 2 caliper mounting bolts, then remove the caliper from the mounting carrier.

To install:

5. Install the caliper assembly to the mounting carrier. Tighten the mounting bolts to 16–23 ft. lbs. (22–32 Nm).

6. Install the brake hose to the caliper. Always use new washers and tighten the mounting bolt to 14–18 ft. lbs. (20–25 Nm).

7. Bleed the brake system and install the front wheels.

8. Check the level of the brake fluid in the master cylinder reservoir.

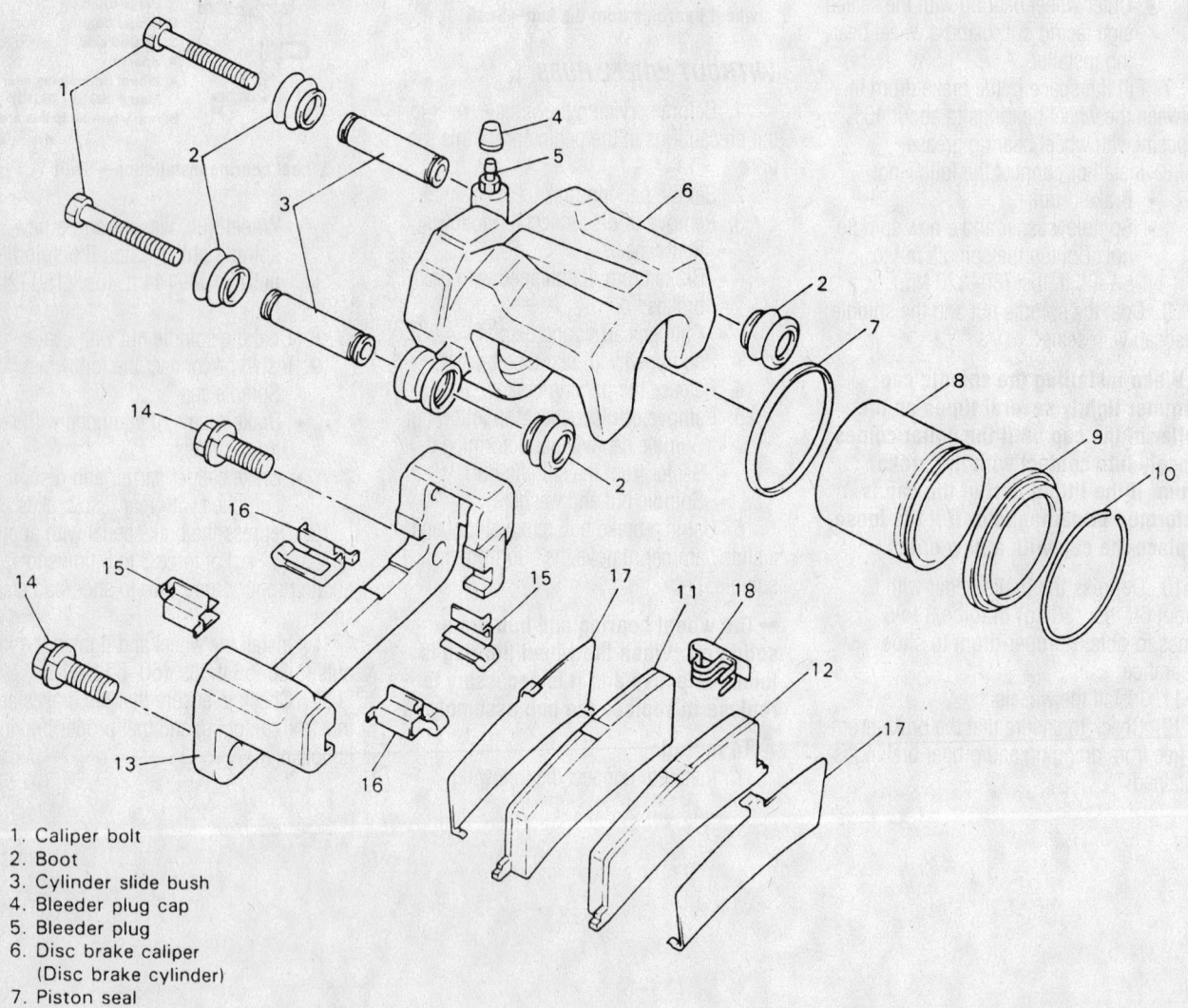

1. Caliper bolt
2. Boot
3. Cylinder slide bush
4. Bleeder plug cap
5. Bleeder plug
6. Disc brake caliper
   (Disc brake cylinder)
7. Piston seal
8. Disc brake piston
9. Cylinder boot
10. Set ring (boot ring)
11. Disc brake inner pad
12. Disc brake outer pad
13. Brake caliper carrier
14. Carrier bolt
15. Pad support plate No.1
16. Pad support plate No.2
17. Anti-noise shim
18. Pad wear plate (wear indicator)

93016G63

Front caliper—Esteem

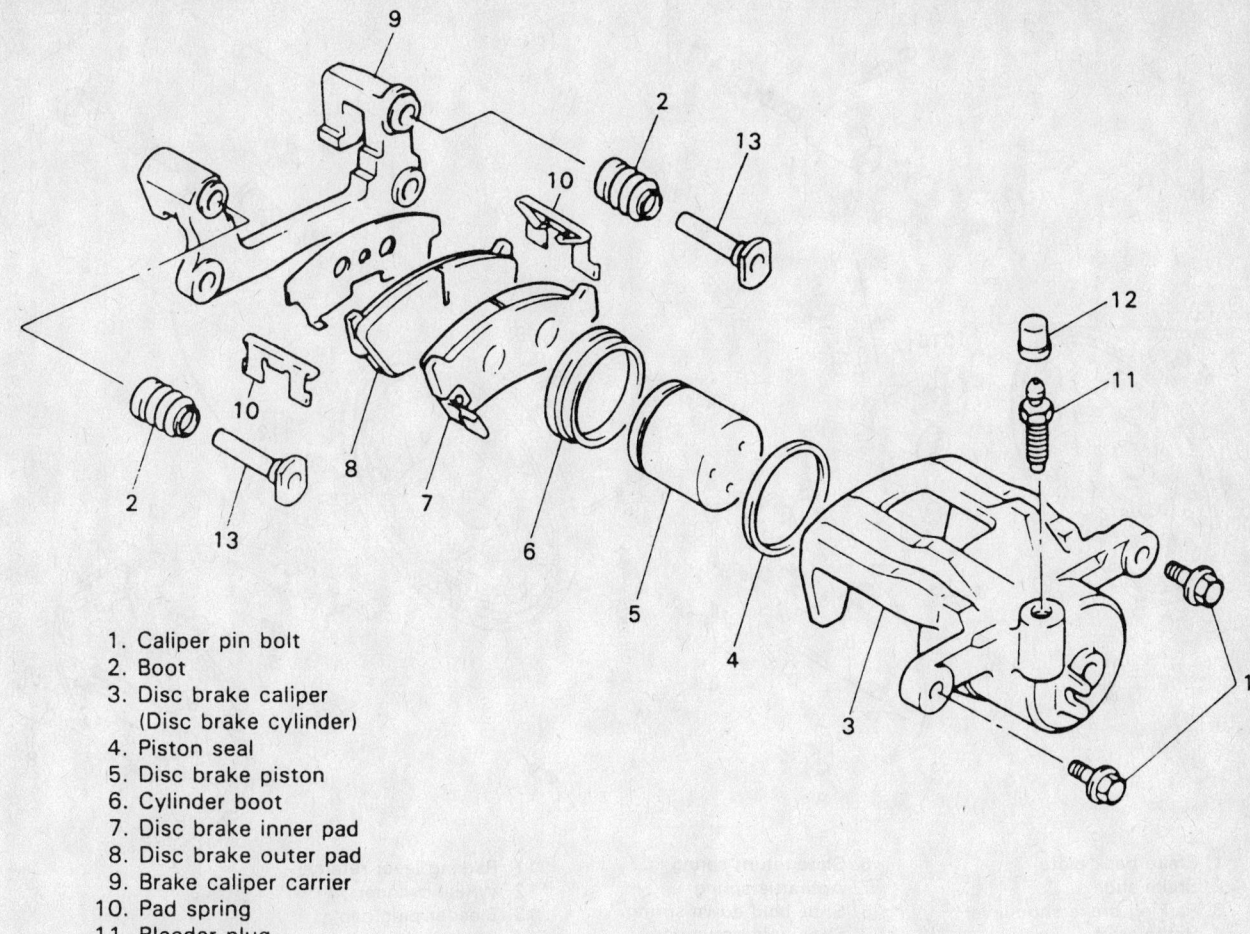

1. Caliper pin bolt
2. Boot
3. Disc brake caliper
   (Disc brake cylinder)
4. Piston seal
5. Disc brake piston
6. Cylinder boot
7. Disc brake inner pad
8. Disc brake outer pad
9. Brake caliper carrier
10. Pad spring
11. Bleeder plug
12. Bleeder plug cap
13. Caliper pin

93016G64

Front caliper—Swift

Disc Brake Pads

REMOVAL & INSTALLATION

Swift and Esteem

FRONT

1. Before servicing the vehicle, refer to the precautions in the beginning of this section.
2. Remove the wheels.
3. Remove the 2 caliper mounting bolts and then remove the caliper from the caliper carrier.
4. Remove the brake pads.
To install:
5. Install the pads into the caliper carrier.
6. On front calipers, press the caliper piston back into the caliper.
7. On rear calipers, use a piston installer 09945-16030, to rotate the caliper piston clockwise into the caliper bore. Lubricate the piston boot with silicone grease to avoid twisting the piston boot.
8. Install the caliper assembly to the caliper carrier. Tighten the bolts to 16–23 ft. lbs. (22–32 Nm).
9. Install the wheels.

Brake Drums

REMOVAL & INSTALLATION

Swift and Esteem

WITHOUT WHEEL HUBS

1. Before servicing the vehicle, refer to the precautions in the beginning of this section.
2. Remove the rear wheels.
3. Remove the spindle dust cap.
4. Remove the sealer from the spindle nut and remove the spindle nut and washer.
5. Install the brake drum removal tool 09943-17911, to the brake drum, then attach a slide hammer to the tool and remove the brake drum.
To install:
6. Install the brake drum.
7. Install the spindle washer and a new spindle nut. Torque the spindle nut to 58–86 ft. lbs. (80–120 Nm).
8. Apply sealer to the spindle nut.
9. Install the spindle dust cap.
10. Install the wheels.

WITH WHEEL HUBS

1. Before servicing the vehicle, refer to the precautions in the beginning of this section.
2. Remove the rear wheels.
3. Remove the 2 brake drum screws.
4. Pull the brake drum off using two 8mm bolts.
To install:
5. Install the brake drum.

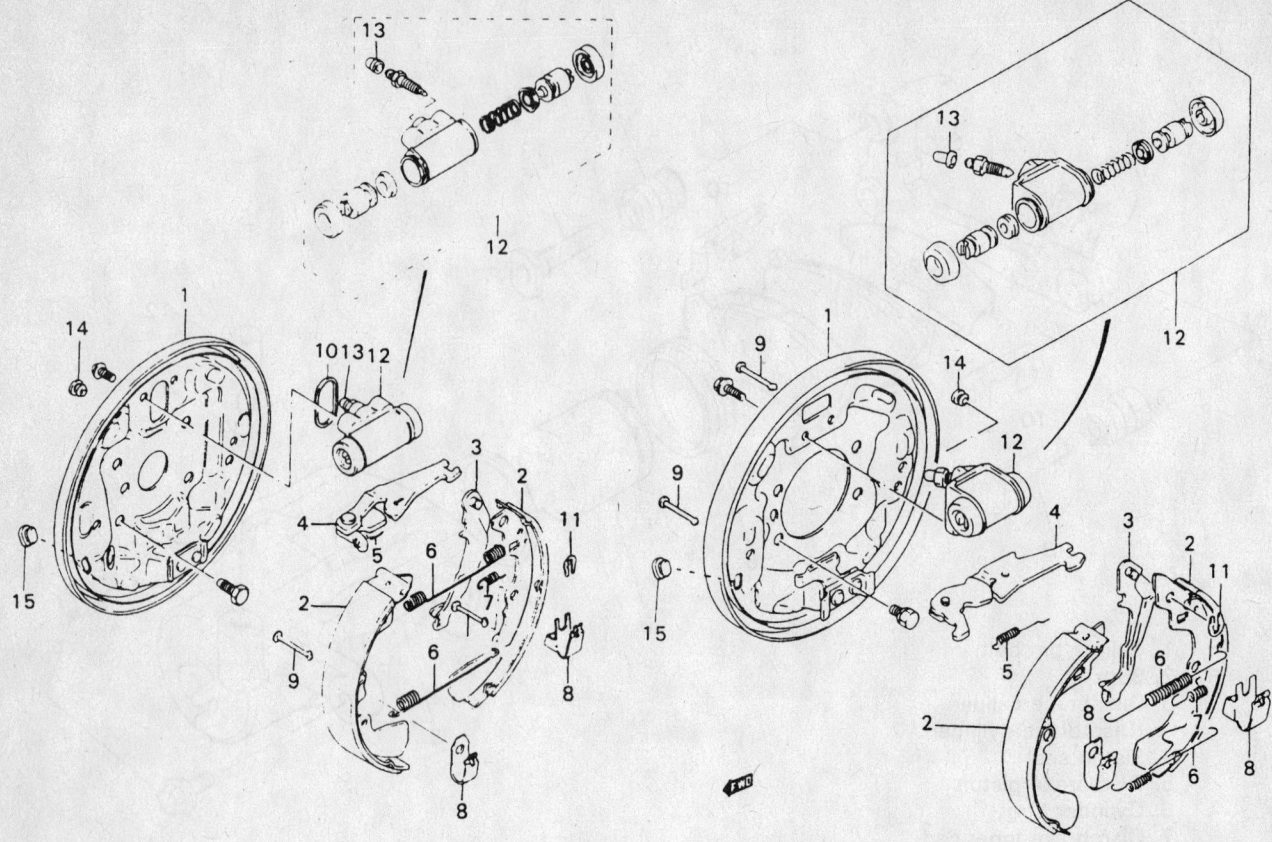

1. Brake back plate
2. Brake shoe
3. Parking brake shoe lever
4. Brake strut
5. Quadrant spring
6. Shoe return spring
7. Antirattle spring
8. Shoe hold down spring
9. Shoe hold down pin
10. Packing
11. Parking lever retainer
12. Wheel cylinder
13. Bleeder plug cap
14. Rubber plug
15. Rubber plug

93016G65

Rear drum brakes—Swift and Esteem

6. Tighten the brake drum screws.
7. Install the wheels.

Brake Shoes

REMOVAL & INSTALLATION

Swift and Esteem

1. Before servicing the vehicle, refer to the precautions in the beginning of this section.
2. Remove the rear wheels.

3. Remove the rear brake drums,
4. Remove the upper and lower springs from the brake shoes.
5. Remove the anti-rattle spring and brake shoe adjustment strut.
6. Remove the primary and secondary brake shoe hold-down springs and remove the shoes from vehicle.
7. Remove the clip securing the parking brake shoe lever to the secondary shoe.

To install:

8. Install the clip securing the parking brake shoe lever to the secondary shoe.

9. Install the primary and secondary brake shoes to the vehicle and secure with the hold-down springs.
10. Install the brake adjustment strut and anti-rattle spring.
11. Install the upper and lower return springs to the primary and secondary brake shoes.
12. Install the brake drum
13. Install the wheels.
14. Press the brake pedal 3–5 times to adjust the brake shoe clearance.
15. Adjust the parking brake cable.

SUZUKI

31

Vitara • Grand Vitara

SPECIFICATION CHARTS

ENGINE AND VEHICLE IDENTIFICATION CHART

		Engine Code						Model Year	
Code	Liters (cc)	Cu. In.	Cyl.	Fuel Sys.	Engine Type	Eng. Mfg.		Code	Year
0	1.6 (1590)	97	4	MFI	SOHC	Suzuki		Y	2000
5	2.0 (1997)	121.8	4	MFI	DOHC	Suzuki		1	2001
6	2.5 (2494)	152	6	MFI	DOHC	Suzuki		2	2002
								3	2003
								4	2004

MFI: Multiport Fuel Injection

DOHC: Dual Overhead Cam

SOHC: Single Overhead Cam

42356-SUZT-C01

GENERAL ENGINE SPECIFICATIONS

Year	Model	Engine Displacement Liters (cc)	Engine ID/VIN	Fuel System Type	Net Horsepower @ rpm	Net Torque @ rpm (ft. lbs.)	Bore x Stroke (in.)	Com- pression Ratio	Oil Pressure @ rpm
2000	Vitara	1.6 (1590)	0	MFI	95@5600	98@4000	2.95x3.54	9.5:1	47-61@3000
		2.0 (1997)	5	MFI	127@6000	134@3000	3.31x3.54	NA	55-67@4000
	Grand Vitara	2.5 (2494)	6	MFI	140@6500	151@4000	3.31x2.95	9.5:1	55-67@4000
2001	Vitara	2.0 (1997)	5	MFI	127@6000	134@3000	3.31x3.54	NA	55-67@4000
	Grand Vitara	2.5 (2494)	6	MFI	140@6500	151@4000	3.31x2.95	9.5:1	55-67@4000
2002	Vitara	2.0 (1997)	5	MFI	127@6000	134@3000	3.31x3.54	NA	55-67@4000
	Grand Vitara	2.5 (2494)	6	MFI	140@6500	151@4000	3.31x2.95	9.5:1	55-67@4000
2003	Vitara	2.0 (1997)	5	MFI	127@6000	134@3000	3.31x3.54	NA	55-67@4000
	Grand Vitara	2.5 (2494)	6	MFI	140@6500	151@4000	3.31x2.95	9.5:1	55-67@4000

MFI: Multi-port Fuel Injection

NA: Not available

42356-SUZT-C02

ENGINE TUNE-UP SPECIFICATIONS

Year	Engine Displacement Liters (cc)	Engine ID/VIN	Spark Plugs Gap (in.)	Ignition Timing (deg.)		Fuel Pump (psi)	Idle Speed (rpm)		Valve Clearance	
				MT	AT		MT	AT	In.	Ex.
2000	1.6 (1590)	0	0.040	5B	5B	30-37	700-800	700-800	0.0050-0.0070	0.0050-0.0070
	2.0 (1997)	5	0.040	5B	5B	30-37	700-800	700-800	HYD	HYD
	2.5 (2494)	6	0.040	5B	5B	30-45	700-800	700-800	HYD	HYD
2001	2.0 (1997)	5	0.040	5B	5B	30-37	700-800	700-800	HYD	HYD
	2.5 (2494)	6	0.040	5B	5B	30-45	700-800	700-800	HYD	HYD
2002	2.0 (1997)	5	0.040	5B	5B	30-37	700-800	700-800	HYD	HYD
	2.5 (2494)	6	0.040	5B	5B	30-45	700-800	700-800	HYD	HYD
2003	2.0 (1997)	5	0.040	5B	5B	30-37	700-800	700-800	HYD	HYD
	2.5 (2494)	6	0.040	5B	5B	30-45	700-800	700-800	HYD	HYD

HYD: Hydraulic

42356-SUZT-C03

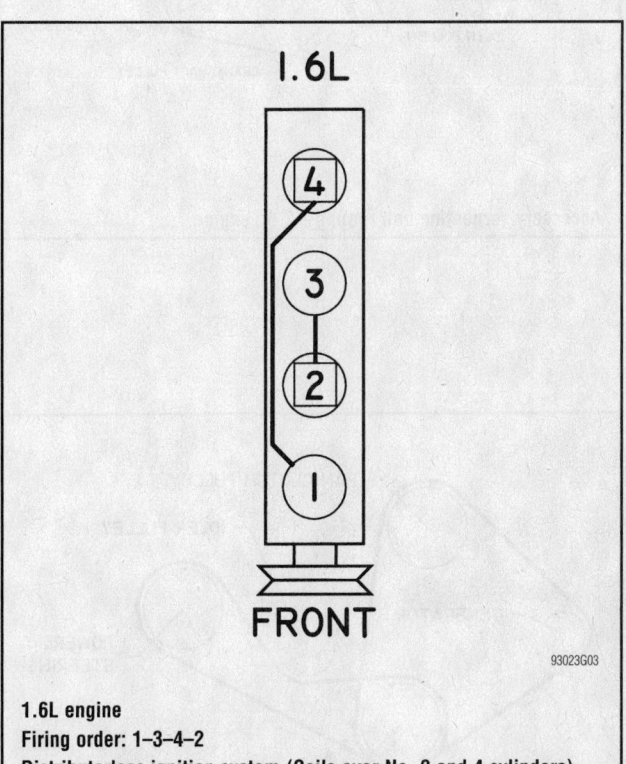

1.6L engine
Firing order: 1–3–4–2
Distributorless ignition system (Coils over No. 2 and 4 cylinders)

93023G03

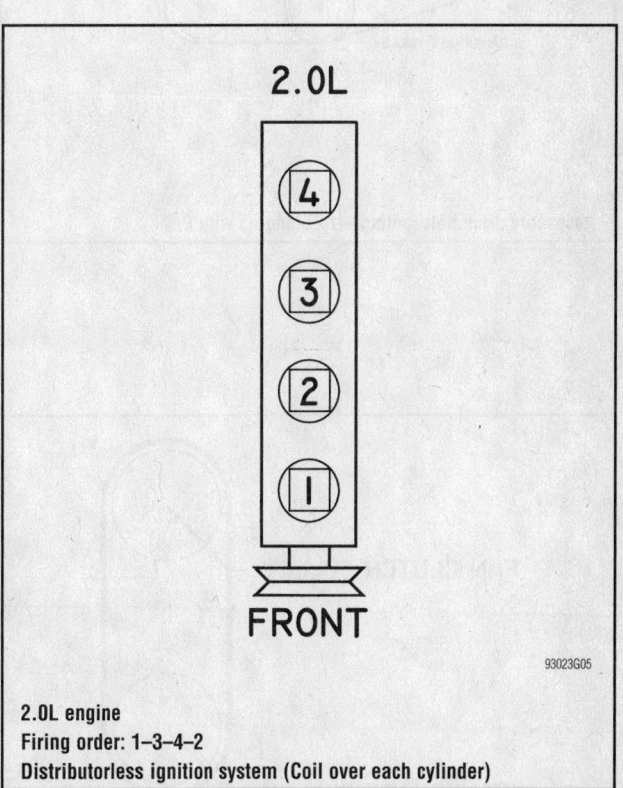

2.0L engine
Firing order: 1–3–4–2
Distributorless ignition system (Coil over each cylinder)

93023G05

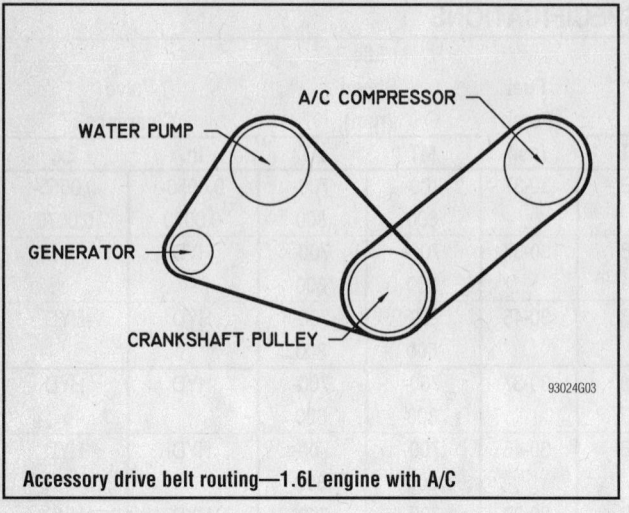

Accessory drive belt routing—1.6L engine with A/C

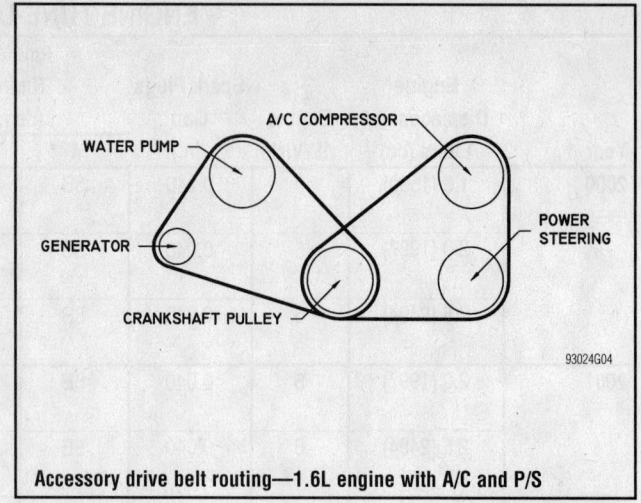

Accessory drive belt routing—1.6L engine with A/C and P/S

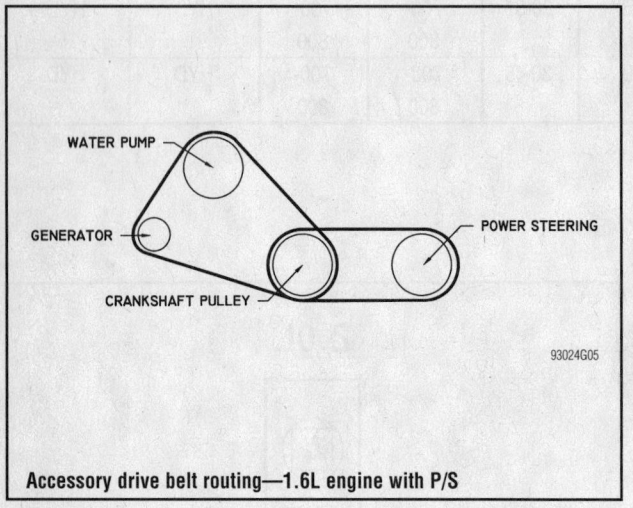

Accessory drive belt routing—1.6L engine with P/S

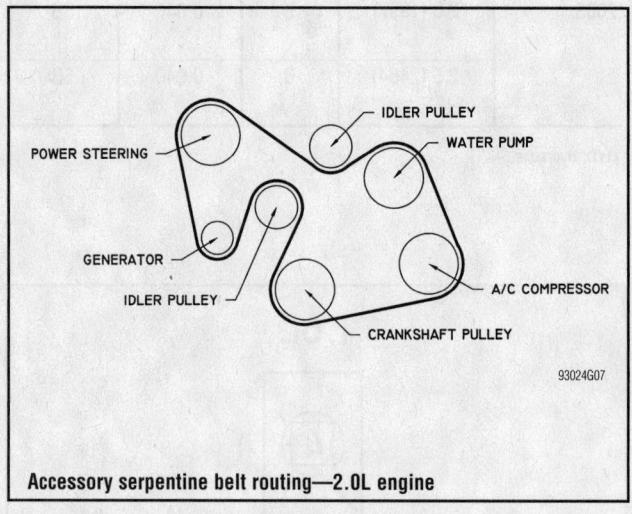

Accessory serpentine belt routing—2.0L engine

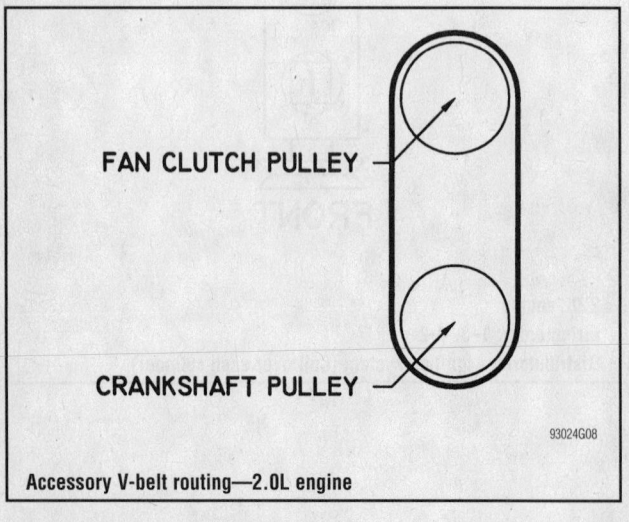

Accessory V-belt routing—2.0L engine

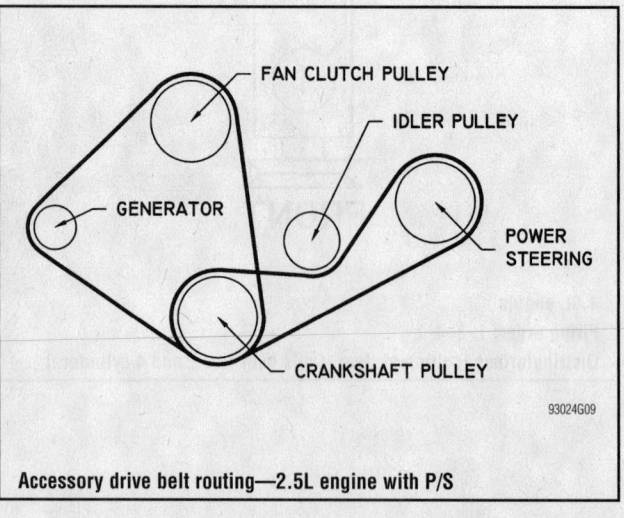

Accessory drive belt routing—2.5L engine with P/S

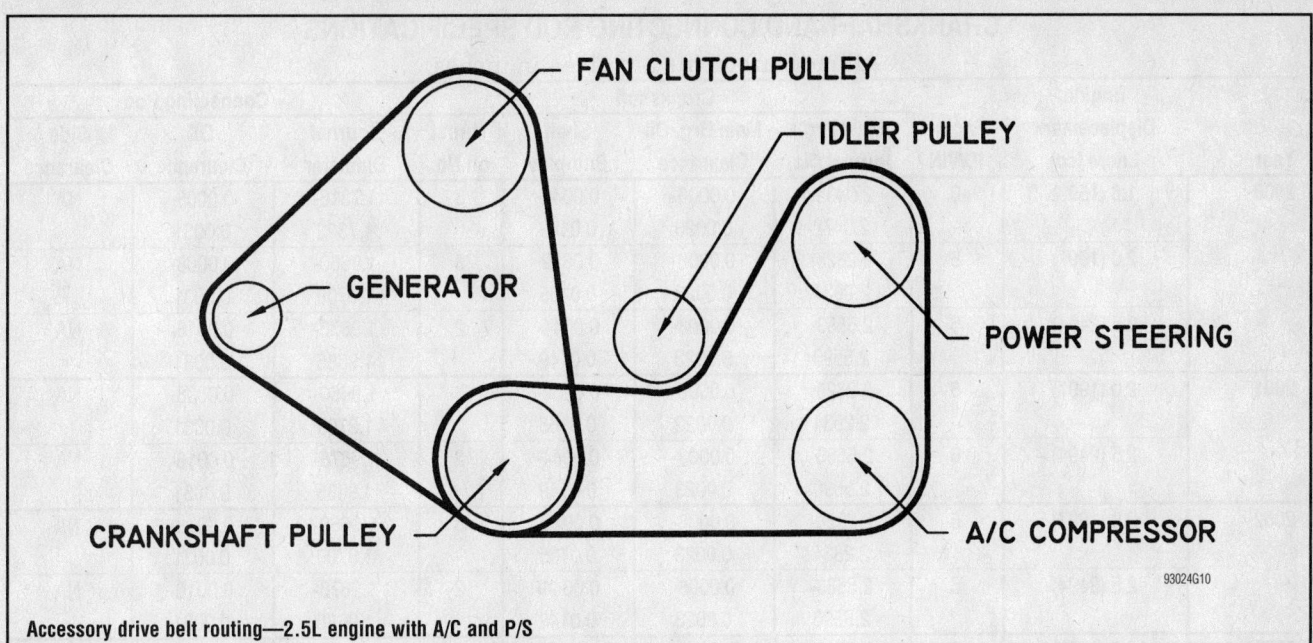

Accessory drive belt routing—2.5L engine with A/C and P/S

CAPACITIES

Year	Model	Engine Displacement Liters (cc)	Engine ID/VIN	Engine Oil with Filter (qts.)	Transmission (pts.)		Transfer Case (pts.)	Drive Axle		Fuel Tank (gal.)	Cooling System (qts.)
					5-Spd	Auto.		Front (pts.)	Rear (pts.)		
2000	Vitara	1.6 (1590)	0	4.75	①	②	3.6	2.1	4.6	11	5.5
		2.0 (1997)	5	5.9	①	②	3.6	2.1	4.6	③	5.5
	Grand Vitara	2.5 (2494)	6	6	5.5	②	3.6	2.1	4.6	18.5	5.5
2001	Vitara	2.0 (1997)	5	5.9	①	④	3.6	2.1	4.6	③	6.9
	Grand Vitara	2.5 (2494)	6	6	5.5	④	3.6	2.1	4.6	18.5	8.5
2002	Vitara	2.0 (1997)	5	5.9	①	④	3.6	2.1	4.6	③	6.9
	Grand Vitara	2.5 (2494)	6	6	5.5	④	3.6	2.1	4.6	18.5	8.5
2003	Vitara	2.0 (1997)	5	5.9	①	④	3.6	2.1	4.6	③	6.9
	Grand Vitara	2.5 (2494)	6	6	5.5	④	3.6	2.1	4.6	18.5	8.5

Note: All capacities are approximate. Add fluid gradually and check to be sure a proper fluid level is obtained.

① 2-wheel drive model: 4.0 pts.
4-wheel drive model: 3.2 pts.

② 3-speed transmission:
Fluid drain, and filter and pan removal only: 5.9 pts.
After complete transmission overhaul: 10.8 pts.
4-speed overdrive transmission:
Fluid drain, and filter and pan removal only: 5.3 pts.
After complete transmission overhaul: 14.6 pts.

③ 2-door model: 11 gals.
4-door model: 14.5 gals.

④ 2WD: 6.0 pts
4WD: 5.2 pts.

42356-SUZT-C04

CRANKSHAFT AND CONNECTING ROD SPECIFICATIONS
All measurements are given in inches.

Year	Engine Displacement Liters (cc)	Engine ID/VIN	Crankshaft Main Brg. Journal Dia.	Crankshaft Main Brg. Oil Clearance	Crankshaft Shaft End-play	Crankshaft Thrust on No.	Connecting Rod Journal Diameter	Connecting Rod Oil Clearance	Connecting Rod Side Clearance
2000	1.6 (1590)	0	2.0465-2.0472	0.0006-0.0023	0.0044-0.0149	3	1.7316-1.7322	0.0008-0.0031	NA
	2.0 (1997)	5	2.2828-2.2834	0.0008-0.0023	0.0039-0.0165	3	1.9660-1.9709	0.0008-0.0031	NA
	2.5 (2494)	6	2.5583-2.5590	0.0008-0.0023	0.0044-0.0149	2	1.9678-1.9685	0.0016-0.0031	NA
2001	2.0 (1997)	5	2.2828-2.2834	0.0008-0.0023	0.0039-0.0165	3	1.9660-1.9709	0.0008-0.0031	NA
	2.5 (2494)	6	2.5583-2.5590	0.0008-0.0023	0.0044-0.0149	2	1.9678-1.9685	0.0016-0.0031	NA
2002	2.0 (1997)	5	2.2828-2.2834	0.0008-0.0023	0.0039-0.0165	3	1.9660-1.9709	0.0008-0.0031	NA
	2.5 (2494)	6	2.5583-2.5590	0.0008-0.0023	0.0044-0.0149	2	1.9678-1.9685	0.0016-0.0031	NA
2003	2.0 (1997)	5	2.2828-2.2834	0.0008-0.0023	0.0039-0.0165	3	1.9660-1.9709	0.0008-0.0031	NA
	2.5 (2494)	6	2.5583-2.5590	0.0008-0.0023	0.0044-0.0149	2	1.9678-1.9685	0.0016-0.0031	NA

NA: Not Available

42356-SUZT-C05

VALVE SPECIFICATIONS

Year	Engine Displacement Liters (cc)	Engine ID/VIN	Seat Angle (deg.)	Face Angle (deg.)	Spring Test Pressure (lbs. @ in.)	Spring Installed Height (in.)	Stem-to-Guide Clearance (in.) Intake	Stem-to-Guide Clearance (in.) Exhaust	Stem Diameter (in.) Intake	Stem Diameter (in.) Exhaust
2000	1.6 (1590)	0	45	45	24-28@1.24	1.245	0.0008-0.0018	0.0018-0.0028	0.2152-0.2157	0.2142-0.2148
	2.0 (1997)	5	45	45	①	1.250	0.0008-0.0027	0.0018-0.0035	0.2348-0.2354	0.2339-0.2344
	2.5 (2494)	6	45	45	①	1.250	0.0008-0.0027	0.0018-0.0035	0.2348-0.2354	0.2339-0.2344
2001	2.0 (1997)	5	45	45	①	1.250	0.0008-0.0027	0.0018-0.0035	0.2348-0.2354	0.2339-0.2344
	2.5 (2494)	6	45	45	①	1.250	0.0008-0.0027	0.0018-0.0035	0.2348-0.2354	0.2339-0.2344
2002	2.0 (1997)	5	45	45	①	1.250	0.0008-0.0027	0.0018-0.0035	0.2348-0.2354	0.2339-0.2344
	2.5 (2494)	6	45	45	①	1.250	0.0008-0.0027	0.0018-0.0035	0.2348-0.2354	0.2339-0.2344
2003	2.0 (1997)	5	45	45	①	1.250	0.0008-0.0027	0.0018-0.0035	0.2348-0.2354	0.2339-0.2344
	2.5 (2494)	6	45	45	①	1.250	0.0008-0.0027	0.0018-0.0035	0.2348-0.2354	0.2339-0.2344

① Inner: 13.6-17.4@1.08
  Outer: 30.4-39.2@1.25

PISTON AND RING SPECIFICATIONS

All measurements are given in inches.

Year	Engine Displacement Liters (cc)	Engine ID/VIN	Piston Clearance	Ring Gap			Ring Side Clearance		
				Top Compression	Bottom Compression	Oil Control	Top Compression	Bottom Compression	Oil Control
2000	1.6 (1590)	0	0.0008-0.0015	0.0079-0.0275	0.0138-0.0275	0.0039-0.0669	0.0012-0.0027	0.0008-0.0023	NA
	2.0 (1997)	5	0.0008-0.0015	0.0079-0.0276	0.0138-0.0276	0.0079-0.0709	0.0012-0.0027	0.0008-0.0023	NA
	2.5 (2494)	6	0.0008-0.0015	0.0079-0.0276	0.0138-0.0276	0.0079-0.0709	0.0012-0.0027	0.0008-0.0023	NA
2001	2.0 (1997)	5	0.0008-0.0015	0.0079-0.0276	0.0138-0.0276	0.0079-0.0709	0.0012-0.0027	0.0008-0.0023	NA
	2.5 (2494)	6	0.0008-0.0015	0.0079-0.0276	0.0138-0.0276	0.0079-0.0709	0.0012-0.0027	0.0008-0.0023	NA
2002	2.0 (1997)	5	0.0008-0.0015	0.0079-0.0276	0.0138-0.0276	0.0079-0.0709	0.0012-0.0027	0.0008-0.0023	NA
	2.5 (2494)	6	0.0008-0.0015	0.0079-0.0276	0.0138-0.0276	0.0079-0.0709	0.0012-0.0027	0.0008-0.0023	NA
2003	2.0 (1997)	5	0.0008-0.0015	0.0079-0.0276	0.0138-0.0276	0.0079-0.0709	0.0012-0.0027	0.0008-0.0023	NA
	2.5 (2494)	6	0.0008-0.0015	0.0079-0.0276	0.0138-0.0276	0.0079-0.0709	0.0012-0.0027	0.0008-0.0023	NA

NA: Not Available

42356-SUZT-C07

TORQUE SPECIFICATIONS
All readings in ft. lbs.

Year	Engine Displacement Liters (cc)	Engine ID/VIN	Cylinder Head Bolts	Main Bearing Bolts	Rod Bearing Bolts	Crankshaft Damper Bolts	Flywheel Bolts	Manifold Intake	Manifold Exhaust	Spark Plugs	Lug Nut
2000	1.6 (1590)	0	①	36-41	24-26	94 ②	57	13-20	13-20	14-21	58-80
	2.0 (1997)	5	③	④	33	109	51	13-20	13-20	14-21	58-80
	2.5 (2494)	6	③	④	33	109	51	16.5	22.0	18	69
2001	2.0 (1997)	5	③	④	33	109	51	13-20	13-20	14-21	70
	2.5 (2494)	6	③	④	33	109	51	17	22	18	70
	2.0 (1997)	C	③	④	33	109	51	17	17	18	70
2002	2.0 (1997)	5	③	④	33	109	51	13-20	13-20	14-21	70
	2.5 (2494)	6	③	④	33	109	51	17	22	18	70
2003	2.0 (1997)	5	③	④	33	109	51	13-20	13-20	14-21	70
	2.5 (2494)	6	③	④	33	109	51	17	22	18	70

① Step 1: 26 ft. lbs.

Step 2: 41 ft. lbs.

Step 3: Loosen in reverse order to 0 ft. lbs.

Step 4: 26 ft. lbs.

Step 5: 52 ft. lbs.

② Value shown is for crankshaft timing belt sprocket

③ Step 1: 38 ft. lbs.

Step 2: 61 ft. lbs.

Step 3: Loosen in reverse order to 0 ft. lbs.

Step 4: 38 ft. lbs.

Step 5: 76 ft. lbs.

Step 6: Tighten 6mm bolt to 8 ft. lbs.

④ 10mm: 43.5 ft. lbs.

8mm: 19.5 ft. lbs.

⑤ Use multiple passes to arrive at final torque.

⑥ Value shown is for crankshaft timing belt sprocket

42356-SUZT-C08

WHEEL ALIGNMENT

Year	Model	Caster Range (+/-Deg.)	Caster Preferred Setting (Deg.)	Camber Range (+/-Deg.)	Camber Preferred Setting (Deg.)	Toe-in (in.)	Steering Axis Inclination (Deg.)
2000	Vitara	1.00	+2.66	1.00	0	0+/-0.08	—
	Grand Vitara	1.00	+2.66	1.00	0	0+/-0.08	—
2001	Vitara	1.00	+2.66	1.00	0	0+/-0.08	—
	Grand Vitara	1.00	+2.66	1.00	0	0+/-0.08	—
2002	Vitara	1.00	+2.66	1.00	0	0+/-0.08	—
	Grand Vitara	1.00	+2.66	1.00	0	0+/-0.08	—
2003	Vitara	1.00	+2.66	1.00	0	0+/-0.08	—
	Grand Vitara	1.00	+2.66	1.00	0	0+/-0.08	—

42356-SUZT-C09

TIRE, WHEEL AND BALL JOINT SPECIFICATIONS

Year	Model	OEM Tires Standard	Optional	Tire Pressures (psi) Front	Rear	Wheel Size	Ball Joint Inspection
2000	Vitara 1.6L 2wd	P195/75SR15	none	26	26	5.5JJ	①
	Vitara 1.6L 4wd	P205/75R15	none	26	26	6.5JJ	①
	Vitara 2.0L	P215/65R16	none	26	26	6.5JJ	①
	Grand Vitara	P235/65SR16	none	26	26	7-JJ	①
2001	Vitara 2.0L	P215/65R16	none	26	26	6.5JJ	①
	Grand Vitara	P235/65SR16	none	26	26	7-JJ	①
2002	Vitara 2.0L	P215/65R16	none	26	26	6.5JJ	①
	Grand Vitara	P235/65SR16	none	26	26	7-JJ	①
2003	Vitara 2.0L	P215/65R16	none	26	26	6.5JJ	①
	Grand Vitara	P235/65SR16	none	26	26	7-JJ	①

OEM: Original Equipment Manufacturer

PSI: Pounds Per Square Inch

STD: Standard

OPT: Optional

① Replace if any measurable movement is found.

42356-SUZT-C10

BRAKE SPECIFICATIONS
All measurements in inches unless noted

Year	Model	Brake Disc Original Thickness	Minimum Thickness	Maximum Runout	Brake Drum Diameter Original Inside Diameter	Max. Wear Limit	Maximum Machine Diameter	Minimum Lining Thickness Front	Rear	Brake Caliper Bracket Bolts (ft. lbs.)	Mounting Bolts (ft. lbs.)
2000	Vitara	0.670	0.590	0.006	8.66	8.74	8.74	0.08	0.04	61.5	19-21
	Grand Vitara	0.866	0.787	0.006	8.66	8.74	8.74	0.08	0.04	61.5	①
2001	Vitara	0.670	0.590	0.006	8.66	8.74	8.74	0.08	0.04	61.5	19-21
	Grand Vitara	0.866	0.787	0.006	8.66	8.74	8.74	0.08	0.04	61.5	①
2002	Vitara	0.670	0.590	0.006	8.66	8.74	8.74	0.08	0.04	61.5	19-21
	Grand Vitara	0.866	0.787	0.006	8.66	8.74	8.74	0.08	0.04	61.5	①
2003	Vitara	0.670	0.590	0.006	8.66	8.74	8.74	0.08	0.04	61.5	19-21
	Grand Vitara	0.866	0.787	0.006	8.66	8.74	8.74	0.08	0.04	61.5	①

① 10mm: 37 ft. lbs.

12mm: 62 ft. lbs.

42356-SUZT-C11

SCHEDULED MAINTENANCE INTERVALS
SUZUKI—VITARA, GRAND VITARA

TO BE SERVICED	TYPE OF SERVICE	VEHICLE MILEAGE INTERVAL (x1000)												
		7.5	15	22.5	30	37.5	45	52.5	60	67.5	75	82.5	90	97.5
Engine oil & filter	R	✓	✓	✓	✓	✓	✓	✓	✓	✓	✓	✓	✓	✓
Automatic transmission fluid ①	S/I	✓	✓	✓	✓	✓	✓	✓	✓	✓	✓	✓	✓	✓
Manual transmission oil ②	S/I	✓	✓	✓	✓	✓	✓	✓	✓	✓	✓	✓	✓	✓
Steering system	S/I	✓	✓	✓	✓	✓	✓	✓	✓	✓	✓	✓	✓	✓
Transfer & differential oil ②	S/I	✓	✓	✓	✓	✓	✓	✓	✓	✓	✓	✓	✓	✓
Wheel discs & free wheeling hubs	S/I	✓	✓	✓	✓	✓	✓	✓	✓	✓	✓	✓	✓	✓
Suspension system	S/I	✓	✓	✓	✓	✓	✓	✓	✓	✓	✓	✓	✓	✓
Brake discs & pads (front)	S/I		✓		✓		✓		✓		✓		✓	
Brake drums & shoes (rear)	S/I		✓		✓		✓		✓		✓		✓	
Brake fluid ③	S/I		✓		✓		✓		✓		✓		✓	
Brake hoses & pipes	S/I		✓		✓		✓		✓		✓		✓	
Brake pedal	S/I		✓		✓		✓		✓		✓		✓	
Brake lever & cable	S/I		✓		✓		✓		✓		✓		✓	
Clutch	S/I		✓		✓		✓		✓		✓		✓	
Idle speed	S/I		✓		✓		✓		✓		✓		✓	
Propeller shafts	S/I		✓		✓		✓		✓		✓		✓	
Valve lash (clearance)	S/I		✓		✓		✓		✓		✓		✓	
Wheel bearings	S/I		✓		✓		✓		✓		✓		✓	
Air cleaner filter element	R				✓				✓				✓	
Engine coolant	R				✓				✓				✓	
Fuel filter	R				✓				✓				✓	
Spark plugs	R				✓				✓				✓	
Cooling system hoses	S/I				✓				✓				✓	
Drive belt(s)	S/I				✓				✓				✓	
Exhaust pipes & mountings	S/I				✓				✓				✓	
Fuel lines & connections	S/I				✓				✓				✓	
Camshaft timing belt	R								✓				✓	
Distributor cap & rotor	S/I								✓					
Emission-related hoses & tubes	S/I								✓					
Oxygen sensor	S/I											✓		
EVAP canister	R	every 100,000 miles												
PCV valve	R							✓						
EGR system	S/I							✓						

42356-SUZT-C12

SCHEDULED MAINTENANCE INTERVALS
SUZUKI—VITARA, GRAND VITARA

TO BE SERVICED	TYPE OF SERVICE	VEHICLE MILEAGE INTERVAL (x1000)												
		7.5	15	22.5	30	37.5	45	52.5	60	67.5	75	82.5	90	97.5
Fuel Injectors	S/I	every 100,000 miles												
TWC converter	S/I	every 100,000 miles												

R: Replace   S/I: Service or Inspect

① Replace at 100,000 miles.

② Replace oil every 30,000 miles.

③ Replace every 60,000 miles.

FREQUENT OPERATION MAINTENANCE (SEVERE SERVICE)

If a vehicle is operated under any of the following conditions it is considered severe service:

- Extremely dusty areas.

- 50% or more of the vehicle operation is in 32°C (90°F) or higher temperatures, or constant operation in temperatures below 0°C (32°F).

- Prolonged idling (vehicle operation in stop and go traffic).

- Frequent short running periods (engine does not warm to normal operating temperatures).

- Police, taxi, delivery usage or trailer towing usage.

Oil & oil filter: replace every 3000 miles.

Air cleaner filter element: service or inspect every 3000 miles & replace every 15,000 miles.

Steering wheel free play, gear box oil & linkage: service or inspect every 3000 miles.

Brake & nuts on chassis: tighten every 6000 miles.

Brake discs & pads (front): service or inspect every 6000 miles.

Brake drums & shoes (rear): service or inspect every 6000 miles.

Exhaust pipes & mountings: tighten every 6000 miles.

Propeller shafts: service or inspect every 6000 miles.

Automatic transmission fluid & filter: replace every 15,000 miles.

Distributor cap & ignition wires: service or inspect every 15,000 miles.

Drive belt(s): service or inspect every 15,000 miles.

Manual transmission oil: replace every 15,000 miles.

Transfer & differential oil: replace every 15,000 miles.

42356-SUZT-C13

PRECAUTIONS

Before servicing any vehicle, please be sure to read all of the following precautions, which deal with personal safety, prevention of component damage and important points to take into consideration when servicing a motor vehicle:

• Never open, service or drain the radiator or cooling system when the engine is hot; serious burns can occur from the steam and hot coolant.

• Observe all applicable safety precautions when working around fuel. Whenever servicing the fuel system, always work in a well-ventilated area. Do not allow fuel spray or vapors to come in contact with a spark, open flame, or excessive heat (a hot drop light, for example). Keep a dry chemical fire extinguisher near the work area. Always keep fuel in a container specifically designed for fuel storage; also, always properly seal fuel containers to avoid the possibility of fire or explosion. Refer to the additional fuel system precautions later in this section.

• Fuel injection systems often remain pressurized, even after the engine has been turned **OFF**. The fuel system pressure must be relieved before disconnecting any fuel lines. Failure to do so may result in fire and/or personal injury.

• Brake fluid often contains polyglycol ethers and polyglycols. Avoid contact with the eyes and wash your hands thoroughly after handling brake fluid. If you do get brake fluid in your eyes, flush your eyes with clean, running water for 15 minutes. If

eye irritation persists, or if you have taken brake fluid internally, IMMEDIATELY seek medical assistance.

• The EPA warns that prolonged contact with used engine oil may cause a number of skin disorders, including cancer. You should make every effort to minimize your exposure to used engine oil. Protective gloves should be worn when changing oil. Wash your hands and any other exposed skin areas as soon as possible after exposure to used engine oil. Soap and water, or waterless hand cleaner should be used.

• All new vehicles are now equipped with an air bag system, often referred to as a Supplemental Restraint System (SRS) or Supplemental Inflatable Restraint (SIR) system. The system must be disabled before performing service on or around system components, steering column, instrument panel components, wiring and sensors. Failure to follow safety and disabling procedures could result in accidental air bag deployment, possible personal injury, and unnecessary system repairs.

• Always wear safety goggles when working with, or around, the air bag system. When carrying a non-deployed air bag, be sure the bag and trim cover are pointed away from your body. When placing a non-deployed air bag on a work surface, always face the bag and trim cover upward, away from the surface. This will reduce the motion of the module if it is accidentally deployed. Refer to the addi-

tional air bag system precautions later in this section.

• Clean, high quality brake fluid from a sealed container is essential to the safe and proper operation of the brake system. You should always buy the correct type of brake fluid for your vehicle. If the brake fluid becomes contaminated, completely flush the system with new fluid. Never reuse any brake fluid. Any brake fluid that is removed from the system should be discarded. Also, do not allow any brake fluid to come in contact with a painted surface; it will damage the paint.

• Never operate the engine without the proper amount and type of engine oil; doing so WILL result in severe engine damage.

• Timing belt maintenance is extremely important. Many models utilize an interference-type, non-freewheeling engine. If the timing belt breaks, the valves in the cylinder head may strike the pistons, causing potentially serious (also time-consuming and expensive) engine damage. Refer to the maintenance interval charts in the front of this section for the recommended replacement interval for the timing belt.

• Disconnecting the negative battery cable on some vehicles may interfere with the functions of the on-board computer system(s) and may require the computer to undergo a relearning process once the negative battery cable is reconnected.

• When servicing drum brakes, only disassemble and assemble one side at a time, leaving the remaining side intact for reference.

ENGINE REPAIR

➡**Disconnecting the negative battery cable on some vehicles may interfere with the functions of the on board computer system. The computer may undergo a relearning process once the negative battery cable is reconnected.**

Distributor

REMOVAL

All engines are equipped with a Distributorless Ignition System (DIS).

Alternator

REMOVAL

1. Before servicing the vehicle, refer to the precautions in the beginning of this section.

2. Remove or disconnect the following:

• Negative battery cable
• Evaporative Emission (EVAP) canister
• Accessory drive belt
• Alternator harness connectors
• Alternator mounting bracket
• Alternator

INSTALLATION

Install or connect the following:

• Alternator
• Alternator mounting bracket. Tighten the bolts to 20 ft. lbs. (27 Nm).
• Alternator harness connectors
• Accessory drive belt. Tighten the alternator bolts to 24 ft. lbs. (33 Nm).
• EVAP canister
• Negative battery cable

Ignition Timing

ADJUSTMENT

1.6L Engine

This engine is equipped with a Distributorless Ignition System (DIS). All timing functions are controlled by the Powertrain Control Module (PCM). No adjustment is possible.

2.0L and 2.5L Engines

➡**The 2.0L and 2.5L engines use a Camshaft Position (CMP) sensor that is rotated to set base timing.**

➡**Check and adjust the ignition timing with the engine at normal operating temperature, all electrical accessories OFF and transmission in P, N for auto-**

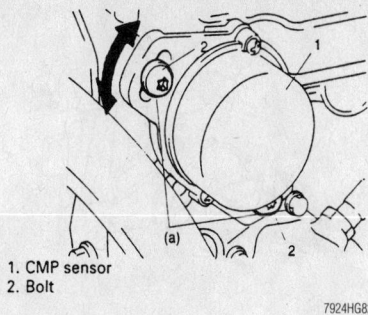

1. CMP sensor
2. Bolt

7924HG82

Camshaft position sensor

matic transmission or neutral for manual transmission.

1. Before servicing the vehicle, refer to the precautions in the beginning of this section.

2. With the engine **OFF**, connect a jumper wire between terminals **D** and **E** of the Data Link Connector (DLC).

3. Connect a timing light to the No. 1 spark plug wire and start the engine.

4. Ignition timing at idle should be 4–6 degrees Before Top Dead Center (BTDC).

5. Adjust the timing as necessary, then turn the engine **OFF**.

6. Remove the jumper wire from the DLC and remove the timing light.

Engine Assembly

REMOVAL & INSTALLATION

1.6L Engine

1. Before servicing the vehicle, refer to the precautions in the beginning of this section.

2. Relieve the fuel system pressure.

3. Drain the cooling system and engine oil.

4. Remove or disconnect the following:
- Negative battery cable
- Hood
- Strut tower bar, if equipped
- Cooling fan and shroud
- Heater hoses
- Radiator hoses
- Bypass hose
- Radiator
- Air intake tube
- Accelerator cable
- Transmission cable, if equipped
- Positive Crankcase Ventilation (PCV) valve and hose
- Exhaust Gas Recirculation (EGR) valve and temperature sensor
- EGR vacuum valve connector
- EGR bypass valve connector

- Idle Air Control (IAC) valve hoses and connector
- Fuel lines
- Main engine control wiring harness connectors at the firewall
- Throttle Position (TP) sensor connector
- Heated Oxygen (HO2S) sensor connectors
- Engine Coolant Temperature (ECT) sensor connector
- Temperature gauge sender connector
- A/C temperature switch, if equipped
- Injector harness connectors
- Alternator wiring connectors
- Manifold Absolute Pressure (MAP) sensor connector and vacuum line
- Brake booster vacuum line
- Evaporative Emission (EVAP) canister and hoses
- Distributor, if equipped
- Front skidplate, if equipped
- Power steering hoses
- A/C compressor, if equipped
- Flywheel access cover
- Torque converter, if equipped
- Clutch cable, if equipped
- Transmission cooler lines, if equipped
- Exhaust front pipe
- Starter motor
- Transmission flange fasteners and support the transmission
- Left and right engine mounts
- Engine

To install:

5. Install or connect the following:
- Engine. Tighten the mount bolts to 40 ft. lbs. (54 Nm).
- Transmission flange fasteners. Tighten them to 62 ft. lbs. (85 Nm).
- Starter motor
- Exhaust front pipe
- Transmission cooler lines, if equipped
- Clutch cable, if equipped
- Torque converter, if equipped. Tighten the bolts to 40 ft. lbs. (54 Nm).
- Flywheel access cover
- A/C compressor, if equipped
- Power steering hoses
- Front skidplate, if equipped
- Distributor, if equipped
- EVAP canister and hoses
- Brake booster vacuum line
- MAP sensor connector and vacuum line
- Alternator wiring connectors
- Injector harness connectors
- A/C temperature switch, if equipped

- Temperature gauge sender connector
- ECT sensor connector
- HO2S sensor connectors
- TP sensor connector
- Main engine control wiring harness connectors at the firewall
- Fuel lines
- IAC valve hoses and connector
- EGR bypass valve connector
- EGR vacuum valve connector
- EGR valve and temperature sensor
- PCV valve and hose
- Transmission cable, if equipped
- Accelerator cable
- Air intake tube
- Radiator
- Bypass hose
- Radiator hoses
- Heater hoses
- Cooling fan and shroud
- Strut tower bar, if equipped
- Hood
- Negative battery cable

6. Fill the crankcase to the correct level.

7. Fill the cooling system.

8. Start the engine and check for leaks.

2.0L and 2.5L Engines

1. Before servicing the vehicle, refer to the precautions in the beginning of this section.

2. Relieve the fuel system pressure.

3. Drain the cooling system.

4. Drain the engine oil.

5. Remove or disconnect the following:
- Negative battery cable
- Hood
- Heater hoses
- Radiator hoses
- Cooling fan and shroud
- Radiator overflow tank
- Radiator
- Accelerator cable
- Transmission cable, if equipped
- Strut tower bar, if equipped
- Air intake assembly
- Engine oil dipstick tube
- Transmission oil dipstick tube, if equipped
- Ignition coil covers
- Ignition coil connectors
- Injector connectors
- Camshaft Position (CMP) sensor connector
- Crankshaft Position (CKP) sensor connector
- Throttle Position (TP) sensor connector
- Mass Air Flow (MAF) sensor connector

- Idle Air Control (IAC) valve
- Intake manifold ground cable
- Evaporative Emission (EVAP) canister purge valve
- Exhaust Gas Recirculation (EGR) valve connector
- Heated Oxygen (HO$_2$S) sensor connectors
- Engine Coolant Temperature (ECT) sensor connector
- Alternator wiring connectors
- Oil pressure gauge sender connector
- Power Steering Pressure (PSP) switch connector
- Alternator bracket ground cable
- Brake booster vacuum line
- Tank pressure control vacuum valve hose
- Fuel lines
- EVAP canister
- Power steering pump
- A/C compressor
- Steering shaft lower assembly
- Front differential housing, if equipped
- Exhaust front pipe and bracket
- Transmission oil cooler lines, if equipped
- Transmission stiffener brackets, if equipped
- Flywheel access cover
- Torque converter, if equipped
- Starter motor
- Transmission flange fasteners and support the transmission
- Left and right engine mounts
- Engine

To install:

6. Install or connect the following:
- Engine
- Left and right engine mounts. Tighten the nuts to 36 ft. lbs. (50 Nm).
- Transmission flange fasteners. Tighten them to 58 ft. lbs. (80 Nm).
- Starter motor
- Torque converter. Tighten the bolts to 47 ft. lbs. (65 Nm).
- Flywheel access cover
- Transmission stiffener brackets. Tighten the bolts to 36 ft. lbs. (50 Nm).
- Transmission oil cooler lines, if equipped
- Exhaust front pipe and bracket
- Front differential housing, if equipped
- Steering shaft lower assembly
- A/C compressor
- Power steering pump
- EVAP canister

- Fuel lines
- Tank pressure control vacuum valve hose
- Brake booster vacuum line
- Alternator bracket ground cable
- PSP switch connector
- Oil pressure gauge sender connector
- Alternator wiring connectors
- ECT sensor connector
- HO$_2$S sensor connectors
- EGR valve connector
- EVAP canister purge valve
- Intake manifold ground cable
- IAC valve
- MAF sensor connector
- TP sensor connector
- CKP sensor connector
- CMP sensor connector
- Injector connectors
- Ignition coil connectors
- Ignition coil covers
- Transmission oil dipstick tube, if equipped
- Engine oil dipstick tube
- Air intake assembly
- Strut tower bar, if equipped
- Transmission cable, if equipped
- Accelerator cable
- Radiator
- Radiator overflow tank
- Cooling fan and shroud
- Radiator hoses
- Heater hoses
- Hood
- Negative battery cable

7. Fill the crankcase to the correct level.
8. Fill the cooling system.
9. Start the engine and check for leaks.

Water Pump

REMOVAL & INSTALLATION

1.6L Engines

1. Before servicing the vehicle, refer to the precautions in the beginning of this section.
2. Drain the cooling system.
3. Remove or disconnect the following:
- Negative battery cable
- Accessory drive belts
- Cooling fan and shroud
- Front cover
- Timing belt
- Oil dipstick tube
- Alternator bracket
- Timing belt tensioner
- Water pump

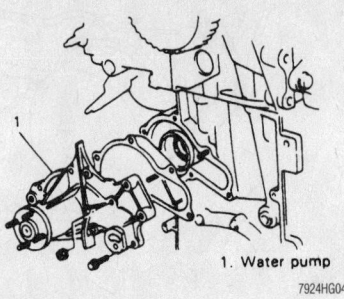

1. Water pump

7924HG04

Exploded view of the water pump mounting—1.6L engines

To install:

4. Install or connect the following:
- Water pump with a new gasket. Tighten the bolts to 106 inch lbs. (12 Nm).
- Timing belt tensioner
- Alternator bracket
- Oil dipstick tube. Tighten the bolt to 97 inch lbs. (11 Nm).
- Timing belt
- Front cover
- Cooling fan and shroud
- Accessory drive belts
- Negative battery cable

5. Fill the cooling system.
6. Start the engine and check for leaks.

2.0L Engines

1. Before servicing the vehicle, refer to the precautions in the beginning of this section.
2. Drain the cooling system.
3. Remove or disconnect the following:
- Negative battery cable
- Radiator hose at the thermostat housing
- Heater outlet pipe bolt
- Alternator belt
- Water pump

To install:

➡ **Use new water pump bolts for assembly.**

4. Install or connect the following:
- Water pump with a new O-ring seal. Tighten the bolts to 19 ft. lbs. (27 Nm).
- Alternator belt
- Heater outlet pipe bolt
- Radiator hose at the thermostat housing
- Negative battery cable

5. Fill the cooling system.
6. Start the engine and check for leaks.

2.5L Engine

1. Before servicing the vehicle, refer to the precautions in the beginning of this section.

2. Drain the cooling system.
3. Remove or disconnect the following:
   - Negative battery cable
   - Accessory drive belts
   - Front cover
   - Water pump

To install:

4. Install or connect the following:
   - Water pump with a new O-ring seal. Tighten the bolts to 19 ft. lbs. (27 Nm).
   - Front cover
   - Accessory drive belts
   - Negative battery cable
5. Fill the cooling system.
6. Start the engine and check for leaks.

Heater Core

REMOVAL & INSTALLATION

1. Disconnect the negative battery cable.
2. To disable the air bag system, perform the following procedure:
   a. Position the front wheels so that they are pointing straight ahead.
   b. Turn the ignition switch to the LOCK position.
   c. In the fuse box, remove the AIR BAG fuse.
   d. Under the steering column, locate the contact coil/combination switch assembly's yellow connector; then, unlock and disconnect the connector.
   e. Pull outward on the glove box while pushing the stopper located at both sides and locate the passenger's side air bag module yellow connector; then, unlock and disconnect the connector.

→**With the AIR BAG fuse removed and the ignition switch turned ON; the air bag warning light may be ON; this is normal operation and does not indicate an air bag malfunction.**

3. Drain the cooling system into a clean container for reuse.
4. Disconnect the heater hoses from the heater core.
5. Remove the instrument panel by performing the following procedure:
   a. Remove the console.
   b. Remove the glove box and the column hole cover.
   c. Disconnect the electrical connector and cables from the heater housing and blower motor assembly.
   d. Remove the steering column.
   e. Disconnect the speedometer connector and the speedometer assembly.
   f. Remove the hood opener.
   g. Disconnect the instrument panel electrical connectors.
   h. Remove the instrument panel-to-chassis screws and bolts.
   i. Using an assistant, remove the instrument panel.
6. If equipped with air conditioning, perform the following procedure:
   a. Discharge and recover the air conditioning refrigerant.
   b. If equipped with a G16 or a J20 engine, disconnect the suction pipe and liquid pipe from the air conditioning housing. Plug the openings to prevent contamination.
   c. If equipped with an H25 engine, disconnect the compressor suction pipe and receiver/drier outlet pipe from the air conditioning housing. Plug the openings to prevent contamination.
   d. Remove the blower motor assembly.
   e. Disconnect the thermistor wire coupler.
   f. Remove the air conditioning housing.
7. Disconnect the rear duct from the heater housing.
8. Disconnect the mode actuator electrical connectors.
9. If equipped, remove the air conditioning controller.
10. If equipped, disconnect and remove the Sensing and Diagnostic Module (SDM) or air bag controller module.
11. Remove the heater housing.
12. Remove the heater core pipe clamps and grommet.
13. Remove the heater core from the heater housing.

To install:

14. Install the heater core to the heater housing.
15. Install the heater core pipe clamps and grommet.
16. Install the heater housing.

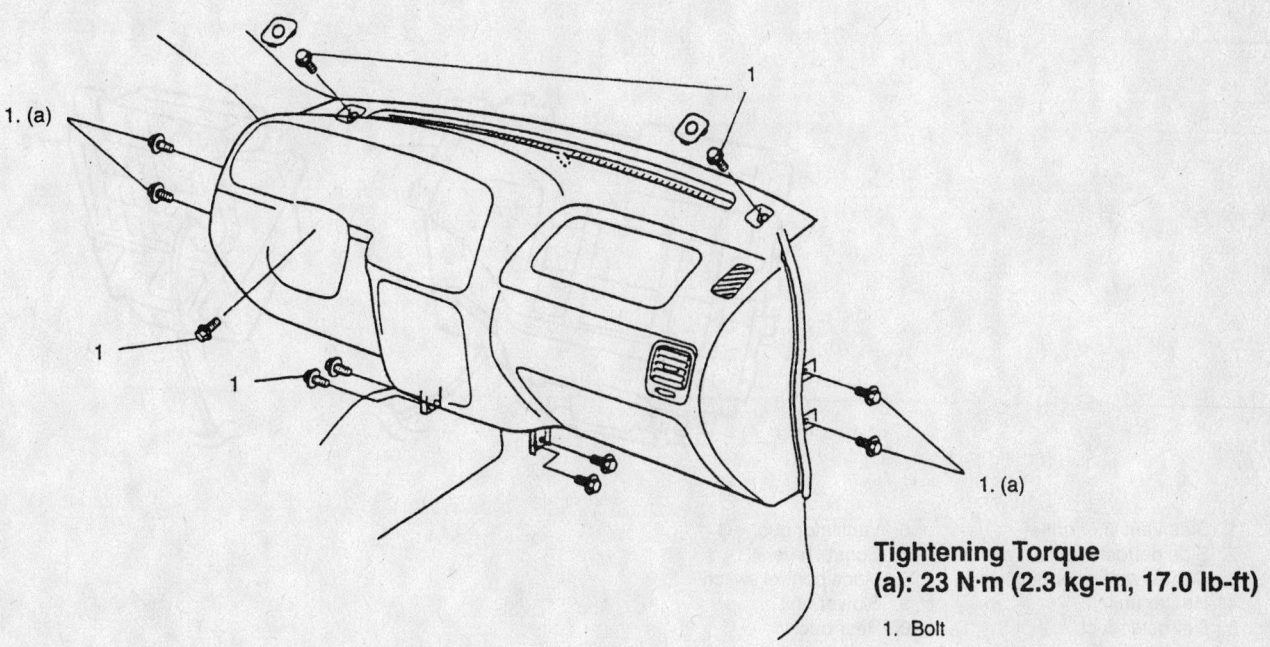

Tightening Torque
(a): 23 N·m (2.3 kg-m, 17.0 lb-ft)

1. Bolt

93113GE8

View of the instrument panel and fasteners—Suzuki Vitara

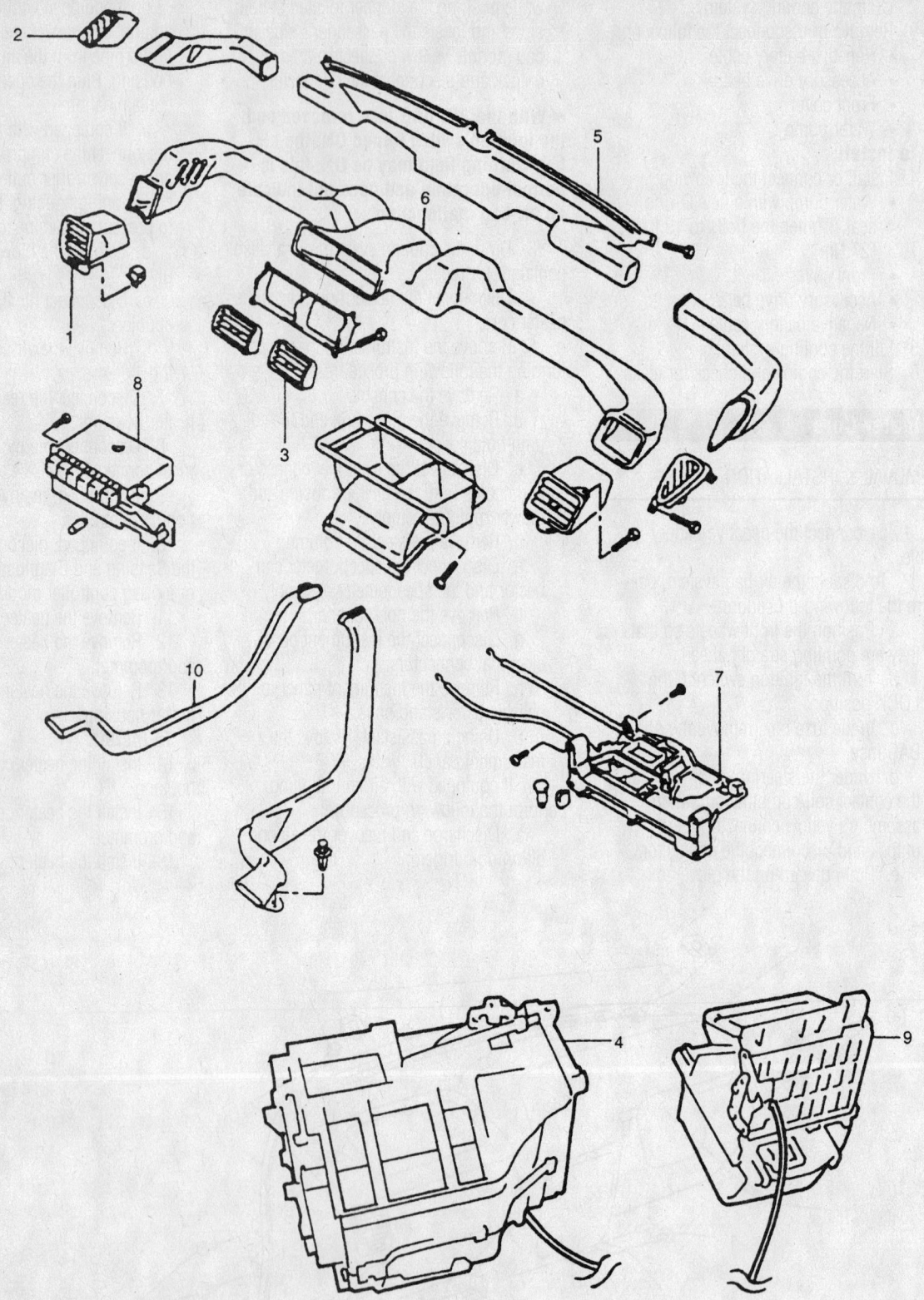

1. Side ventilator outlet
2. Side defroster outlet
3. Center ventilatior outlet
4. Heater unit
5. Defroster duct
6. Ventilator duct
7. Control lever
8. Mode control switch
9. Blower unit
10. Rear duct

93113GE9

Exploded view of the heater housing and ventilation ducts—Suzuki Vitara

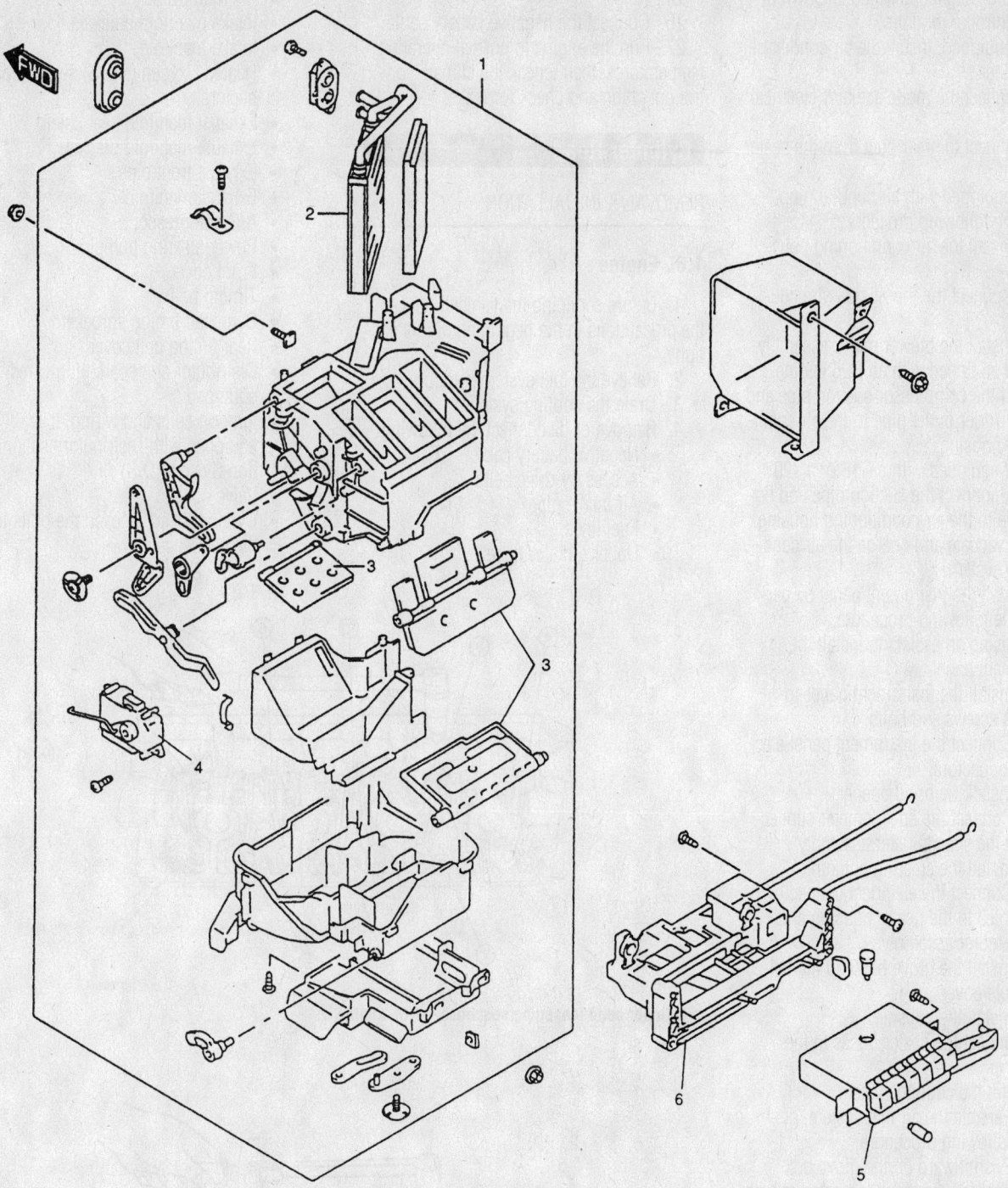

1. Heater assembly
2. Heater core
3. Damper
4. Mode actuator
5. Mode control switch
6. Control lever assembly

93113GE0

Exploded view of the heater core, heater housing and related components—Suzuki Vitara

17. If equipped, connect and install the Sensing and Diagnostic Module (SDM) or air bag controller module.

18. If equipped, install the air conditioning controller.

19. Connect the mode actuator electrical connectors.

20. Connect the rear duct from the heater housing.

21. If equipped with air conditioning, perform the following procedure:

   a. Install the air conditioning housing.

   b. Connect the thermistor wire coupler.

   c. Install the blower motor assembly.

   d. If equipped with an H25 engine, connect the compressor suction pipe and receiver/drier outlet pipe to the air conditioning housing.

   e. If equipped with a G16 or a J20 engine, connect the suction pipe and liquid pipe to the air conditioning housing.

   f. Evacuate and charge the air conditioning system.

22. Install the instrument panel by performing the following procedure:

   a. Using an assistant, install the instrument panel.

   b. Install the instrument panel-to-chassis screws and bolts.

   c. Connect the instrument panel electrical connectors.

   d. Install the hood opener.

   e. Connect the speedometer connector and the speedometer assembly.

   f. Install the steering column.

   g. Connect the electrical connector and cables to the heater housing and blower motor assembly.

   h. Install the glove box and the column hole cover.

   i. Install the console.

23. Connect the heater hoses to the heater core.

24. Refill the cooling system.

25. To enable the air bag system, perform the following procedure:

   a. Push inward on the glove box while pushing the stopper located at both sides and connect the passenger's side air bag module yellow connector and lock it.

   b. Under the steering column, connect the contact coil/combination switch assembly's yellow connector.

   c. In the fuse box, install the AIR BAG fuse.

   d. Turn the ignition switch ON and verify that the AIR BAG warning light flashes 7 times and turns OFF; if the system does not operate as described, per-

form the Air Bad Diagnostic System Check.

26. Connect the negative battery cable.

27. Run the engine to normal operating temperatures; then, check the climate control operation and check for leaks.

Cylinder Head

REMOVAL & INSTALLATION

1.6L Engine

1. Before servicing the vehicle, refer to the precautions in the beginning of this section.

2. Relieve the fuel system pressure.

3. Drain the cooling system.

4. Remove or disconnect the following:

   - Negative battery cable
   - Accessory drive belts
   - Air intake pipe
   - Fuel lines
   - Upper radiator hose
   - Coolant bypass hose
   - Alternator bracket
   - Intake manifold brackets
   - Intake manifold
   - Heated Oxygen (HO$_2$S) sensor connectors
   - Exhaust manifold heat shield
   - Exhaust manifold bracket
   - Exhaust front pipe
   - Exhaust manifold
   - A/C compressor
   - Power steering pump
   - Front cover
   - Timing belt
   - Camshaft timing sprocket
   - Rear timing belt cover
   - Distributor and spark plug wires, if equipped
   - Ignition coils and wiring, if equipped with Distributorless Ignition System (DIS)
   - Valve cover
   - Cylinder head. Loosen the bolts in the sequence shown.

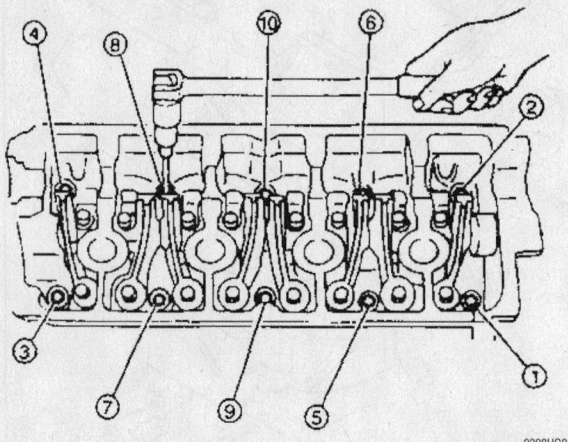

Cylinder head loosening sequence—1.6L engine

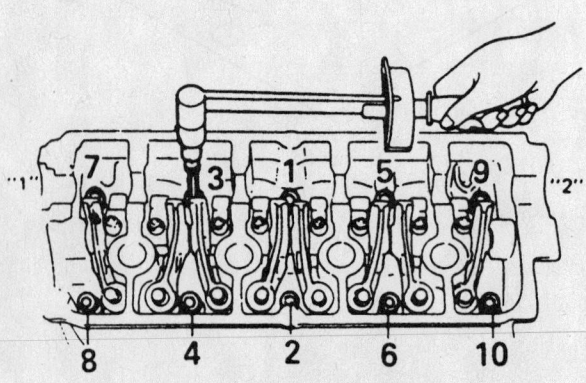

"1": Camshaft pulley side
"2": Distributor side

Cylinder head torque sequence—1.6L engine

To install:

5. Install the cylinder head with a new gasket.

6. Tighten the bolts in sequence as follows:

    a. Step 1: 26 ft. lbs. (35 Nm)

    b. Step 2: 41 ft. lbs. (55 Nm)

    c. Step 3: Loosen all bolts to 0 ft. lbs. (0 Nm)

    d. Step 4: 26 ft. lbs. (35 Nm)

    e. Step 5: 52 ft. lbs. (70 Nm)

7. Install or connect the following:

- Valve cover
- Ignition coils and wiring, if equipped with DIS
- Distributor and spark plug wires, if equipped
- Rear timing belt cover
- Camshaft timing sprocket
- Timing belt
- Front cover
- Power steering pump
- A/C compressor
- Exhaust manifold
- Exhaust front pipe
- Exhaust manifold bracket
- Exhaust manifold heat shield
- HO_2S sensor connectors
- Intake manifold
- Intake manifold brackets
- Alternator bracket
- Coolant bypass hose
- Upper radiator hose
- Fuel lines
- Air intake pipe
- Accessory drive belts
- Negative battery cable

8. Fill the cooling system.

9. Start the engine and check for leaks.

2.0L Engines

1. Before servicing the vehicle, refer to the precautions in the beginning of this section.

2. Relieve the fuel system pressure.

3. Drain the cooling system.

4. Drain the engine oil.

5. Remove or disconnect the following:

- Negative battery cable
- Strut tower brace
- Air intake tube
- Exhaust Gas Recirculation (EGR) valve connector
- Idle Air Control (IAC) valve connector
- Throttle Position (TP) sensor connector
- Evaporative Emission (EVAP) canister purge valve connector and hose
- Intake manifold ground cable

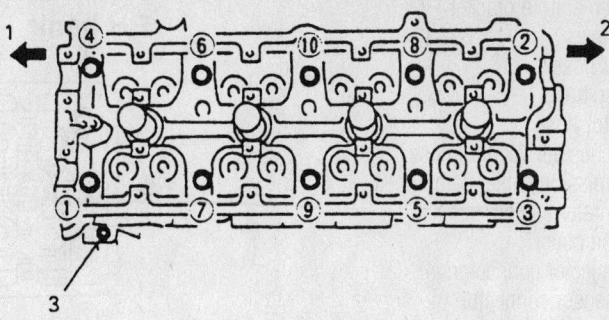

1. Crankshaft pulley side
2. Flywheel side
3. Bolt (M6)

7924HG09

Cylinder head loosening sequence—2.0L engines

- Heated Oxygen (HO_2S) sensor connectors
- Camshaft Position (CMP) sensor connector
- Engine Coolant Temperature (ECT) sensor connector
- Fuel injector connectors
- Ignition coils
- Accelerator cable
- Transmission cable, if equipped
- Brake booster vacuum hose
- Radiator hose
- Bypass hose
- Heater hose
- Fuel lines
- Intake manifold bracket
- Water pipe
- Valve cover
- Accessory drive belts
- Oil pan
- Front cover
- Timing chains
- Camshafts
- Exhaust front pipe
- Exhaust manifold bracket

- Cylinder head. Loosen the bolts in the sequence shown.

To install:

6. Install the cylinder head with a new gasket.

7. Tighten the bolts in sequence as follows:

    a. Step 1: 38 ft. lbs. (52 Nm)

    b. Step 2: 61 ft. lbs. (84 Nm)

    c. Step 3: Loosen all bolts to 0 ft. lbs. (0 Nm)

    d. Step 4: 38 ft. lbs. (52 Nm)

    e. Step 5: 76 ft. lbs. (105 Nm)

    f. Step 6: 6mm bolt to 96 inch lbs. (8 Nm)

8. Install or connect the following:

- Exhaust manifold bracket
- Exhaust front pipe
- Camshafts
- Timing chains
- Front cover
- Oil pan
- Accessory drive belts
- Valve cover
- Water pipe

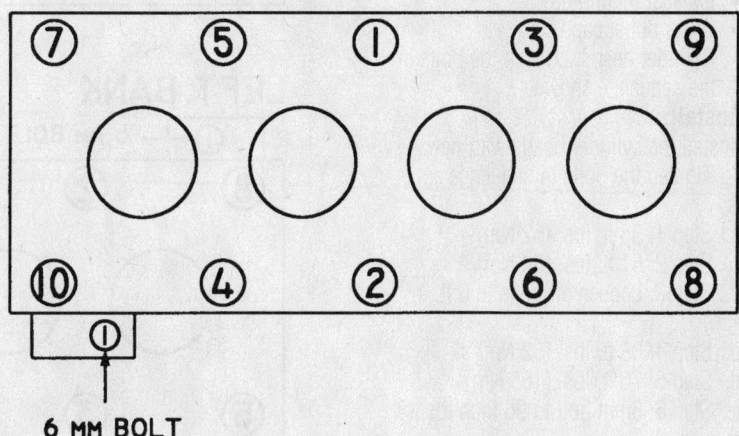

6 MM BOLT

9308HG07

Cylinder head torque sequence—2.0L engines

- Intake manifold bracket
- Fuel lines
- Heater hose
- Bypass hose
- Radiator hose
- Brake booster vacuum hose
- Transmission cable, if equipped
- Accelerator cable
- Ignition coils
- Fuel injector connectors
- ECT sensor connector
- CMP sensor connector
- HO$_2$S sensor connectors
- Intake manifold ground cable
- EVAP canister purge valve connector and hose
- TP sensor connector
- IAC valve connector
- EGR valve connector
- Air intake tube
- Strut tower brace
- Negative battery cable

9. Fill the crankcase to the correct level.
10. Fill the cooling system.
11. Start the engine and check for leaks.

2.5L Engine

1. Before servicing the vehicle, refer to the precautions in the beginning of this section.

2. Relieve the fuel system pressure.

3. Drain the cooling system and engine oil.

4. Remove or disconnect the following:
- Negative battery cable
- Intake manifold
- Ignition coil covers and ignition coils
- Valve covers
- Oil pan
- Timing chain cover and timing chains
- Camshaft Position (CMP) sensor
- Camshafts
- Exhaust manifolds
- Water outlet caps
- Cylinder heads. Loosen the bolts in the sequence shown.

To install:

5. Install the cylinder heads with new gaskets. Tighten the bolts in sequence as follows:
   a. Step 1: 38 ft. lbs. (52 Nm)
   b. Step 2: 61 ft. lbs. (84 Nm)
   c. Step 3: Loosen all bolts to 0 ft. lbs. (0 Nm)
   d. Step 4: 38 ft. lbs. (52 Nm)
   e. Step 5: 76 ft. lbs. (105 Nm)
   f. Step 6: 6mm bolt to 96 inch lbs. (8 Nm)

6. Install or connect the following:

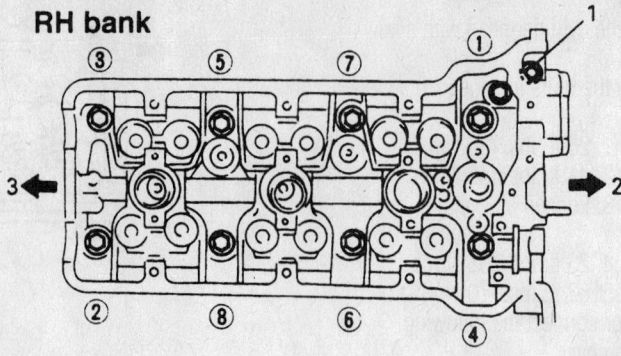
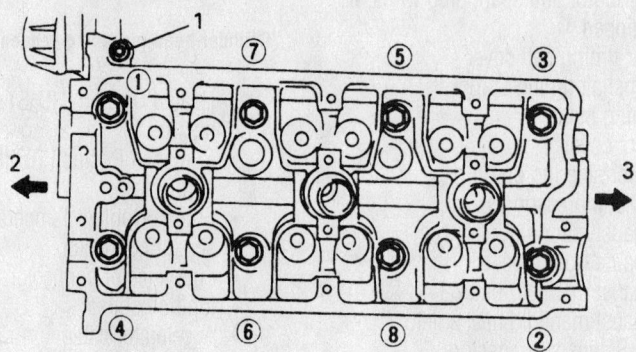

1. Hex hole bolt
2. Timing chain side
3. Flywheel side

9302HG01

Cylinder head loosening sequence—2.5L engine

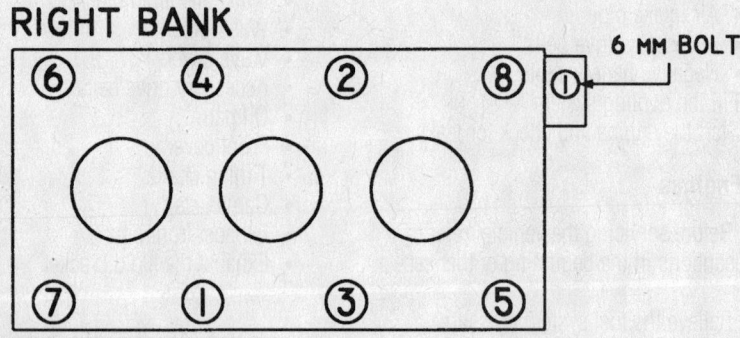
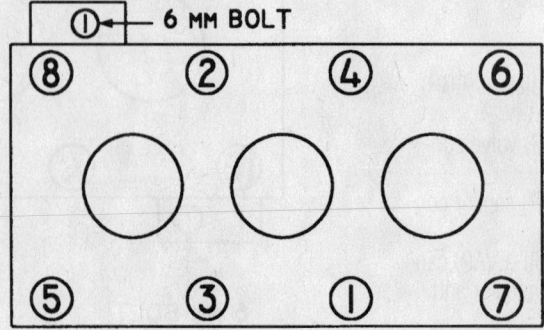

9308HG08

Cylinder head torque sequence—2.5L engine

- Water outlet caps
- Exhaust manifolds
- Camshafts
- CMP sensor
- Timing chain cover and timing chains
- Oil pan
- Valve covers
- Ignition coil covers and ignition coils
- Intake manifold
- Negative battery cable

7. Fill the crankcase to the correct level
8. Fill the cooling system.
9. Start the engine and check for leaks.

Rocker Arms/Shafts

REMOVAL & INSTALLATION

1.6L Engine

1. Before servicing the vehicle, refer to the precautions in the beginning of this section.
2. Drain the cooling system.
3. Remove or disconnect the following:
   - Negative battery cable
   - Accessory drive belts
   - Cooling fan and shroud
   - Radiator and hoses
   - Distributor, if equipped
   - Camshaft Position (CMP) sensor, if equipped
   - Valve cover
   - Front cover
   - Timing belt
   - Camshaft
   - Rocker arm shaft plug
   - Rear timing belt cover
4. Loosen the rocker arm locknuts and back the valve adjusters off until all rocker arms move freely with no tension.

➡**Keep all valvetrain components in order for assembly.**

5. Remove or disconnect the following:
   - Intake rocker arms and clips

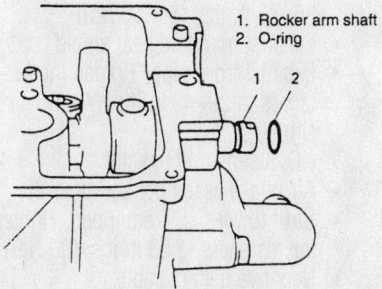

1. Rocker arm shaft
2. O-ring

9308HG02

Rocker arm shaft and O-ring—1.6L engine

- Rocker arm shaft bolts
6. Push the rocker arm shaft towards the rear of the cylinder head and remove the rocker arm shaft O-ring.
7. Remove the exhaust rocker arms and springs by pulling the rocker arm shaft out of the front of the cylinder head.

To install:

8. Insert the rocker arm shaft into the front of the cylinder head, while installing the exhaust rocker arms and springs in their original positions.
9. Push the end of the rocker arm shaft out of the rear of the cylinder head and install a new O-ring.
10. Install or connect the following:
    - Rocker arm shaft bolts. Tighten them to 96 inch lbs. (8 Nm).
    - Intake rocker arms and clips in their original positions
    - Rear timing belt cover
    - Rocker arm shaft plug. Tighten it to 24 ft. lbs. (33 Nm).
    - Camshaft
    - Timing belt and adjust the valve clearance
    - Valve cover
    - Front cover
    - Distributor, if equipped
    - CMP sensor, if equipped
    - Radiator and hoses
    - Cooling fan and shroud
    - Accessory drive belts
    - Negative battery cable
11. Fill the cooling system.
12. Start the engine and check for leaks.

2.0L and 2.5L Engines

The 2.0L and 2.5L engines do not utilize rocker arms or rocker arm shafts.

Intake Manifold

REMOVAL & INSTALLATION

1.6L Engine

1. Before servicing the vehicle, refer to the precautions in the beginning of this section.
2. Relieve the fuel system pressure.
3. Drain the cooling system.
4. Remove or disconnect the following:
   - Negative battery cable
   - Air intake pipe
   - Accelerator cable and bracket
   - Transmission cable, if equipped
   - Throttle Position (TP) sensor connector
   - Idle Air Control (IAC) valve connector and hoses

- Engine Coolant Temperature (ECT) sensor connector
- Coolant temperature gauge sender connector
- A/C coolant temperature switch connector, if equipped
- Evaporative Emission (EVAP) canister purge valve connector and vacuum line
- Exhaust Gas Recirculation (EGR) temperature sensor connector
- EGR vacuum valve connector
- EGR bypass valve connector
- EGR vacuum lines
- Fuel injector connectors
- Intake manifold ground cable
- Transmission vacuum line, if equipped
- Brake booster vacuum line
- Manifold Absolute Pressure (MAP) sensor vacuum line
- Fuel lines
- Upper radiator hose
- Coolant bypass hose
- Alternator bracket
- Intake manifold brackets
- Intake manifold

To install:

5. Install or connect the following:
   - Intake manifold with a new gasket. Tighten the nuts to 17 ft. lbs. (23 Nm).
   - Intake manifold brackets. Tighten the fasteners to 36 ft. lbs. (50 Nm).
   - Alternator bracket. Tighten the fasteners to 36 ft. lbs. (50 Nm).
   - Coolant bypass hose
   - Upper radiator hose
   - Fuel lines
   - MAP sensor vacuum line
   - Brake booster vacuum line
   - Transmission vacuum line, if equipped
   - Intake manifold ground cable
   - Fuel injector connectors
   - EGR vacuum lines
   - EGR bypass valve connector
   - EGR vacuum valve connector
   - EGR temperature sensor connector
   - EVAP canister purge valve connector and vacuum line
   - A/C coolant temperature switch connector, if equipped
   - Coolant temperature gauge sender connector
   - ECT sensor connector
   - IAC valve connector and hoses
   - TP sensor connector
   - Transmission cable, if equipped
   - Accelerator cable and bracket
   - Air intake pipe
   - Negative battery cable

6. Fill the cooling system.
7. Start the engine and check for leaks.

2.0L Engines

1. Before servicing the vehicle, refer to the precautions in the beginning of this section.
2. Relieve the fuel system pressure.
3. Drain the cooling system.
4. Remove or disconnect the following:
   - Negative battery cable
   - Air intake tube
   - Exhaust Gas Recirculation (EGR) valve connector
   - Idle Air Control (IAC) valve connector
   - Throttle Position (TP) sensor connector
   - Evaporative Emissions (EVAP) canister purge valve connector and hose
   - Intake manifold ground cable
   - Manifold Absolute Pressure (MAP) sensor connector
   - Accelerator cable
   - Transmission cable, if equipped
   - Brake booster vacuum hose
   - Positive Crankcase Ventilation (PCV) valve and hose
   - Fuel pressure regulator vacuum hose
   - Intake manifold vacuum hose
   - Throttle body coolant hoses
   - Water bypass pipe
   - Fuel lines
   - Fuel supply manifold with injectors attached
   - Intake manifold support brackets
   - Intake manifold water pipe
   - Intake manifold

To install:

5. Install or connect the following:
   - Intake manifold with a new gasket. Tighten the fasteners to 17 ft. lbs. (23 Nm).
   - Intake manifold water pipe
   - Intake manifold front support bracket. Tighten the bolts to 36 ft. lbs. (50 Nm).
   - Intake manifold rear support bracket. Tighten the bolts to 18 ft. lbs. (25 Nm).
   - Fuel supply manifold with injectors attached
   - Fuel lines
   - Water bypass pipe
   - Throttle body coolant hoses
   - Intake manifold vacuum hose
   - Fuel pressure regulator vacuum hose
   - PCV valve and hose

- Brake booster vacuum hose
- Transmission cable, if equipped
- Accelerator cable
- MAP sensor connector
- Intake manifold ground cable
- EVAP canister purge valve connector and hose
- TP sensor connector
- IAC valve connector
- EGR valve connector
- Air intake tube
- Negative battery cable
6. Fill the cooling system.
7. Start the engine and check for leaks.

2.5L Engine

1. Before servicing the vehicle, refer to the precautions in the beginning of this section.
2. Relieve the fuel system pressure.
3. Drain the cooling system.
4. Remove or disconnect the following:
   - Negative battery cable
   - Strut tower bar
   - Intake Air Temperature (IAT) sensor connector
   - Surge tank cover
   - Air intake assembly
   - Accelerator cable
   - Transmission cable, if equipped
   - Throttle body coolant hoses
   - Fuel injector connectors
   - Throttle Position (TP) sensor connector
   - Mass Air Flow (MAF) sensor connector
   - Idle Air Control (IAC) valve connector
   - Intake manifold ground cables
   - Brake booster vacuum hose
   - Evaporative Emissions (EVAP) canister purge valve connector and hoses
   - Exhaust Gas Recirculation (EGR) valve connector
   - Positive Crankcase Ventilation (PCV) valve and hose
   - Heater hoses
   - EGR pipe
   - Fuel lines
   - Throttle body and intake collector
   - Intake manifold

To install:

5. Install or connect the following:
   - Intake manifold with new gaskets. Tighten the fasteners to 16 ft. lbs. (23 Nm).
   - Throttle body and intake collector with new gaskets. Tighten the fasteners to 102 inch lbs. (12 Nm).
   - Fuel lines

- EGR pipe
- Heater hoses
- PCV valve and hose
- EGR valve connector
- EVAP canister purge valve connector and hoses
- Brake booster vacuum hose
- Intake manifold ground cables
- IAC valve connector
- MAF sensor connector
- TP sensor connector
- Fuel injector connectors
- Throttle body coolant hoses
- Transmission cable, if equipped
- Accelerator cable
- Air intake assembly
- Surge tank cover
- IAT sensor connector
- Strut tower bar
- Negative battery cable
6. Fill the cooling system.
7. Start the engine and check for leaks.

Exhaust Manifold

REMOVAL & INSTALLATION

1.6L and 2.0L Engines

1. Before servicing the vehicle, refer to the precautions in the beginning of this section.
2. Remove or disconnect the following:
   - Negative battery cable
   - Strut tower bar, if equipped
   - Air intake assembly and bracket
   - Heated Oxygen (HO2S) sensor connector
   - Exhaust front pipe
   - Exhaust manifold heat shield
   - Exhaust manifold bracket, if equipped
   - Exhaust manifold

To install:

3. Install or connect the following:
   - Exhaust manifold with a new gasket. Tighten the fasteners to 13–20 ft. lbs. (18–28 Nm).
   - Exhaust manifold bracket, if equipped. Tighten the bolts to 36–43 ft. lbs. (50–60 Nm).
   - Exhaust manifold heat shield
   - Exhaust front pipe. Tighten the fasteners to 29–43 ft. lbs. (40–60 Nm).
   - HO2S sensor connector
   - Air intake assembly and bracket
   - Strut tower bar, if equipped. Tighten the fasteners to 66 ft. lbs. (90 Nm).
   - Negative battery cable
4. Start the engine and check for leaks.

2.5L Engine

1. Before servicing the vehicle, refer to the precautions in the beginning of this section.
2. Remove or disconnect the following:
   - Negative battery cable
   - Strut tower bar
   - Air intake assembly
   - Heated Oxygen (HO2S) sensor connectors
   - Oil dipstick tube
   - Exhaust Gas Recirculation (EGR) pipe
   - Exhaust manifold heat shields
   - Evaporative Emissions (EVAP) canister
   - Front driveshaft, if equipped
   - Exhaust front pipe
   - Exhaust manifold brace
   - Exhaust manifolds

To install:

3. Install or connect the following:
   - Exhaust manifolds with new gaskets. Tighten the nuts to 21 ft. lbs. (30 Nm).
   - Exhaust manifold brace
   - Exhaust front pipe. Tighten the fasteners to 37 ft. lbs. (50 Nm).
   - Front driveshaft, if equipped
   - EVAP canister
   - Exhaust manifold heat shields
   - EGR pipe
   - Oil dipstick tube
   - HO2S sensor connectors
   - Air intake assembly
   - Strut tower bar
   - Negative battery cable
4. Start the engine and check for leaks.

Front Crankshaft Seal

REMOVAL & INSTALLATION

1.6L Engine

1. Before servicing the vehicle, refer to the precautions in the beginning of this section.
2. Drain the cooling system.
3. Remove or disconnect the following:
   - Negative battery cable
   - Accessory drive belts
   - Cooling fan and shroud
   - Water pump pulley
   - Crankshaft pulley
   - Front cover
   - Timing belt
   - Crankshaft timing sprocket
   - Front crankshaft seal

To install:

4. Install or connect the following:

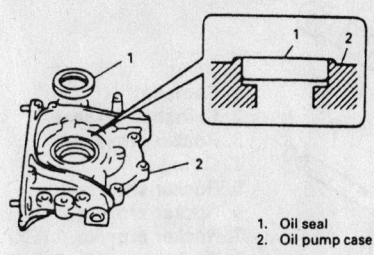

1. Oil seal
2. Oil pump case

7924HG13

Install the new oil pump seal flush with the oil pump housing—1.6L engine

   - Front crankshaft seal flush with the oil pump housing
   - Crankshaft timing sprocket. Tighten the bolt to 94 ft. lbs. (128 Nm).
   - Timing belt
   - Front cover
   - Crankshaft pulley. Tighten the bolts to 12 ft. lbs. (16 Nm).
   - Water pump pulley
   - Cooling fan and shroud
   - Accessory drive belts
   - Negative battery cable
5. Start the engine and check for leaks.

Camshaft and Valve Lifters

REMOVAL & INSTALLATION

1.6L Engine

1. Before servicing the vehicle, refer to the precautions in the beginning of this section.
2. Drain the cooling system.
3. Remove or disconnect the following:
   - Negative battery cable
   - Radiator
   - Accessory drive belts
   - Crankshaft pulley
   - Front cover
   - Timing belt

   - Camshaft sprocket
   - Valve cover
   - Distributor and case, if equipped
   - Camshaft Position (CMP) sensor and case, if equipped
4. Loosen the rocker arm locknuts and back the valve adjusters off until all rocker arms move freely with no tension.
5. Remove or disconnect the following:
   - Camshaft bearing caps. Loosen the bolts in reverse of the tightening sequence.
   - Camshaft

To install:

6. Install or connect the following:
   - Camshaft
   - Camshaft bearing caps. Tighten the bolts in sequence to 96 inch lbs. (11 Nm).
   - CMP sensor and case, if equipped
   - Distributor and case, if equipped
   - Camshaft sprocket. Tighten the bolt to 44 ft. lbs. (60 Nm).
   - Timing belt and adjust the valve clearance
   - Valve cover
   - Front cover
   - Crankshaft pulley. Tighten the bolts to 12 ft. lbs. (16 Nm).
   - Accessory drive belts
   - Radiator
   - Negative battery cable
7. Fill the cooling system.
8. Start the engine and check for leaks.

2.0L Engines

1. Before servicing the vehicle, refer to the precautions in the beginning of this section.
2. Drain the engine oil.
3. Drain the cooling system.
4. Remove or disconnect the following:
   - Negative battery cable
   - Oil pan

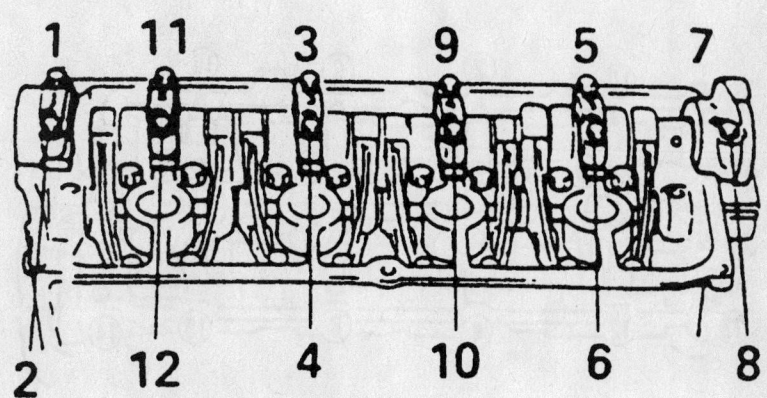

Camshaft housing torque sequence—1.6L engine

7924HG25

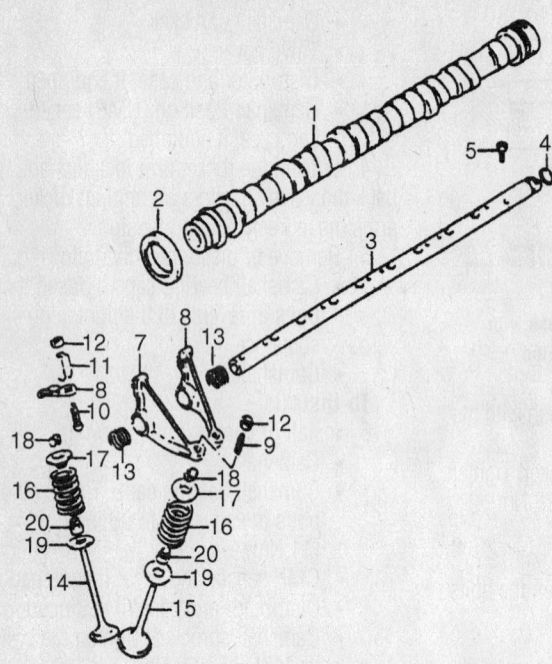

1. Camshaft
2. Camshaft oil seal
3. Rocker arm shaft
4. O ring
5. Rocker shaft bolt
6. Rocker arm (IN)
7. Rocker arm No. 1 (EX)
8. Rocker arm No. 2 (EX)
9. Valve adjusting screw
10. Valve adjusting screw
11. Clip
12. Lock nut
13. Rocker arm spring
14. Intake valve
15. Exhaust valve
16. Valve spring
17. Valve spring retainer
18. Valve cotter
19. Valve spring seat
20. Valve stem seal

7924HG24

Exploded view of the valve train components—1.6L engine

- Valve cover
- Accessory drive belts
- Crankshaft pulley
- Front cover
- Secondary timing chain
- Camshaft Position (CMP) sensor

➡**Keep all valvetrain components in order for installation.**

- Camshaft bearing caps. Loosen the bolts in several steps in reverse of the tightening sequence.
- Camshafts
- Hydraulic lash adjusters

To install:

5. Install or connect the following:
- Hydraulic lash adjusters in their original positions
- Camshafts
- Camshaft bearing caps in their

original positions. Tighten the bolts in several steps in sequence to 96 inch lbs. (11 Nm).
- CMP sensor
- Secondary timing chain
- Front cover
- Crankshaft pulley. Tighten the bolt to 109 ft. lbs. (148 Nm).
- Accessory drive belts
- Valve cover
- Oil pan
- Negative battery cable

6. Fill the crankcase to the correct level.
7. Fill the cooling system.
8. Start the engine and check for leaks.

❊❊ WARNING

Wait ½ hour after installing the lash adjusters and camshafts before cranking or starting the engine to

allow the lash adjusters to bleed down. Operating the engine before this time period may result in interference between the valves and pistons.

2.5L Engine

1. Before servicing the vehicle, refer to the precautions in the beginning of this section.
2. Drain the engine oil.
3. Drain the cooling system.
4. Remove or disconnect the following:
- Negative battery cable
- Intake manifold
- Oil pan
- Accessory drive belts
- Water pump pulley
- Crankshaft pulley
- Timing chain cover
5. Align the timing marks as shown.

❊❊ WARNING

Do not allow the crankshaft or camshafts to rotate once the timing chains have been removed. Valve or piston damage could result.

6. Remove or disconnect the following:
- Left bank secondary timing chain
- Primary timing chain
- Valve covers

➡**Keep all valvetrain components in order for assembly.**

- Right bank camshaft bearing caps. Loosen the bolts in several steps and in the sequence shown.

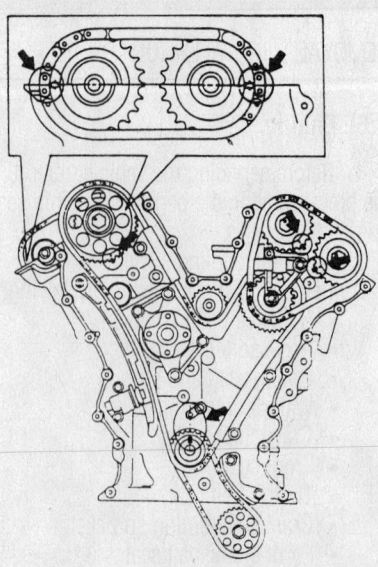

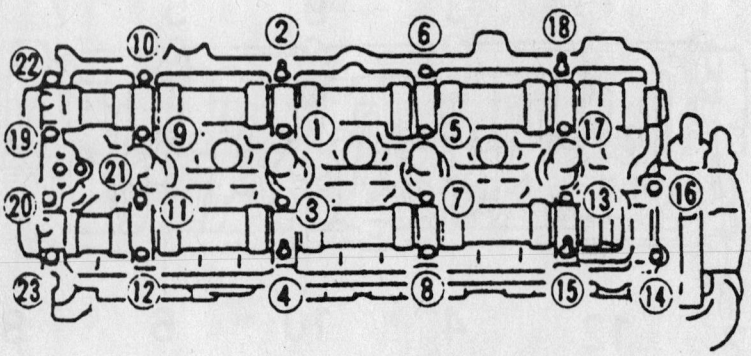

Camshaft housing torque sequence—2.0L engines

7924HG26

Timing mark alignment—2.5L engine

9302HG02

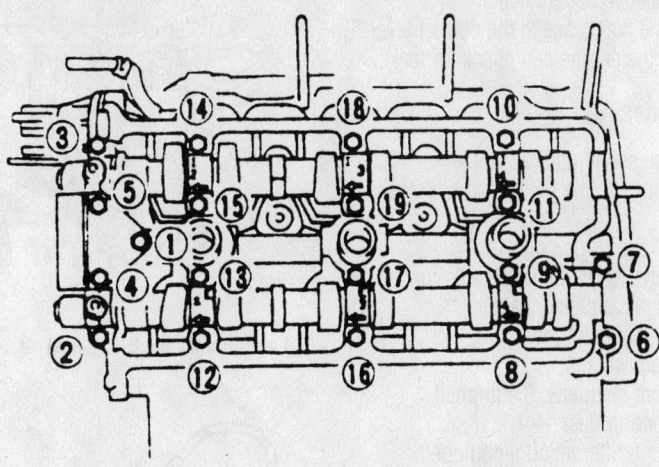

Left bank camshaft housing loosening sequence—2.5L engine

9302HG03

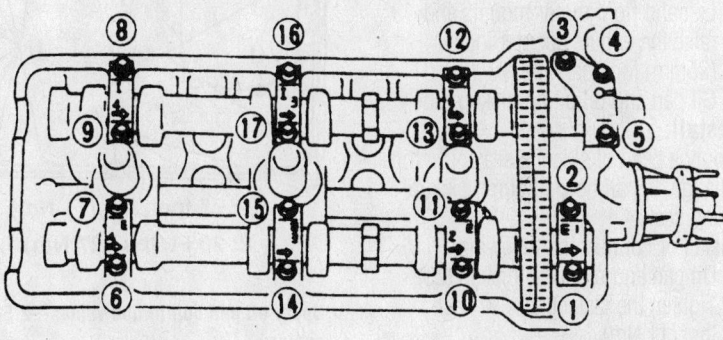

Right bank camshaft housing loosening sequence—2.5L engine

9302HG04

- Right bank secondary timing chain, exhaust and intake camshafts as an assembly
- Camshaft Position (CMP) sensor
- Left bank camshaft bearing caps. Loosen the bolts in several steps and in the sequence shown.
- Left bank camshafts
- Hydraulic lash adjusters

To install:

7. Install or connect the following:
- Hydraulic lash adjusters in their original positions
- Left bank camshafts
- Left bank camshaft bearing caps. Tighten the bolts in several steps and in reverse of the loosening sequence to 102 inch lbs. (12 Nm).
- CMP sensor
- Right bank secondary timing chain, exhaust and intake camshafts as an assembly
- Right bank camshaft bearing caps. Tighten the bolts in several steps and in reverse of the loosening sequence to 102 inch lbs. (12 Nm).

❈❈ WARNING

Wait ½ hour after installing the lash adjusters and camshafts before cranking or starting the engine to allow the lash adjusters to bleed down. Operating the engine before this time period may result in interference between the valves and pistons.

- Valve covers. Tighten the bolts to 90 inch lbs. (10.5 Nm).
- Primary timing chain
- Left bank secondary timing chain
- Timing chain cover
- Crankshaft pulley. Tighten the bolt to 109 ft. lbs. (148 Nm).
- Water pump pulley
- Accessory drive belts
- Oil pan
- Intake manifold
- Negative battery cable

8. Fill the crankcase to the correct level.
9. Fill the cooling system.
10. Start the engine and check for leaks.

Valve Lash

ADJUSTMENT

1.6L Engines

➡**Measure valve clearance with the engine cold.**

1. Before servicing the vehicle, refer to the precautions in the beginning of this section.
2. Remove the valve cover.
3. Set the engine to Top Dead Center (TDC) of the compression stroke for the cylinder to be adjusted.
4. Check the valve clearance. The valve clearance specifications are as follows:
- Intake valves: 0.005–0.007 inches (0.13–0.17mm)
- Exhaust valves: 0.005–0.007 inches (0.13–0.17mm)
5. After adjustment, tighten the locknuts to 11–14 ft. lbs. (15–19 Nm).
6. Repeat for each valve to be adjusted.

2.0L and 2.5L Engines

2.0L and 2.5L engines utilize automatic hydraulic lash adjusters to maintain proper valve lash at all times. Periodic valve lash inspection and adjustment is not necessary or possible.

Starter Motor

REMOVAL & INSTALLATION

1. Before servicing the vehicle, refer to the precautions in the beginning of this section.
2. Remove or disconnect the following:
- Negative battery cable
- Starter motor wiring connectors
- Starter motor

To install:

3. Install or connect the following:
- Starter motor. Tighten the bolts to 22 ft. lbs. (30 Nm).
- Starter motor wiring connectors. Tighten the solenoid nut to 11 ft. lbs. (15 Nm).
- Negative battery cable

Oil Pan

REMOVAL & INSTALLATION

1.6L Engines

1. Before servicing the vehicle, refer to precautions in the beginning of this section.

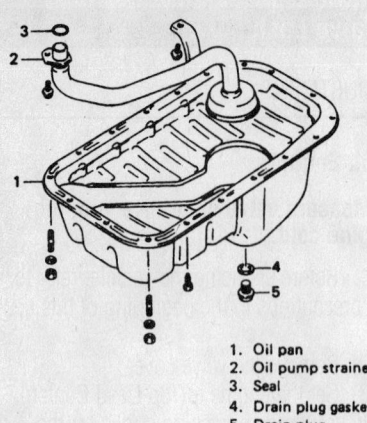

1. Oil pan
2. Oil pump strainer
3. Seal
4. Drain plug gasket
5. Drain plug

7924HG11

Exploded view of the oil pan and pump pickup mounting—1.6L engine

2. Drain the engine oil.
3. Remove or disconnect the following:
   - Negative battery cable
   - Front skidplate, if equipped
   - Front differential, if equipped
   - Crankshaft Position (CKP) sensor
   - Left transmission stiffener bracket, if equipped
   - Flywheel access panel
   - Oil pan and oil pump pickup tube

To install:
4. Apply a bead of silicone sealant to the oil pan flange.
5. Install or connect the following:
   - New oil pump pickup tube O-ring seal
   - Oil pan and oil pump pickup tube. Tighten the fasteners to 97 inch lbs. (11 Nm).
   - Flywheel access panel
   - Left transmission stiffener bracket, if equipped
   - CKP sensor
   - Front differential, if equipped
   - Front skidplate, if equipped. Tighten the bolts to 40 ft. lbs. (55 Nm).

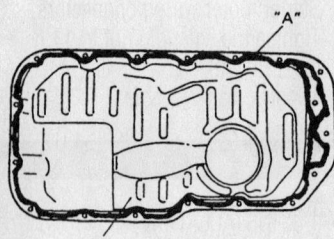

1. Oil pan
A. Sealant

7924HG79

Before installing the oil pan, apply a continuous bead of silicone sealant to the oil pan mating flange—all engines

---

   - Negative battery cable
6. Fill the crankcase to the correct level.
7. Start the engine and check for leaks.

2.0L Engines

1. Before servicing the vehicle, refer to the precautions in the beginning of this section.
2. Drain the engine oil.
3. Remove or disconnect the following:
   - Negative battery cable
   - Oil dipstick tube
   - Front wheels
   - Front skidplate, if equipped
   - Steering gear
   - Front differential, if equipped
   - Left transmission stiffener bracket, if equipped
   - Flywheel access panel
   - Exhaust front pipe
   - Left and right motor mounts and raise the engine about 1 inch (25mm) for clearance
   - Oil pan and oil pump pickup tube

To install:
4. Apply a bead of silicone sealant to the oil pan flange. Install new oil pump pickup tube O-ring seals.
5. Install or connect the following:
   - Oil pan and oil pump pickup tube. Tighten the fasteners to 97 inch lbs. (11 Nm).
   - Left and right engine mounts. Tighten the nuts to 36 ft. lbs. (50 Nm).
   - Exhaust front pipe
   - Flywheel access panel
   - Left transmission stiffener bracket, if equipped
   - Front differential, if equipped
   - Steering gear
   - Front skidplate, if equipped
   - Front wheels
   - Oil dipstick tube
   - Negative battery cable
6. Fill the crankcase to the correct level.
7. Start the engine and check for leaks.

2.5L Engine

1. Before servicing the vehicle, refer to the precautions in the beginning of this section.
2. Drain the engine oil.
3. Remove or disconnect the following:
   - Negative battery cable
   - Oil dipstick tube
   - Front wheels
   - Front skidplate, if equipped
   - Steering gear
   - Front differential, if equipped
   - Lower oil pan

1. O-ring

9302HG05

Lower crankcase O-ring seal

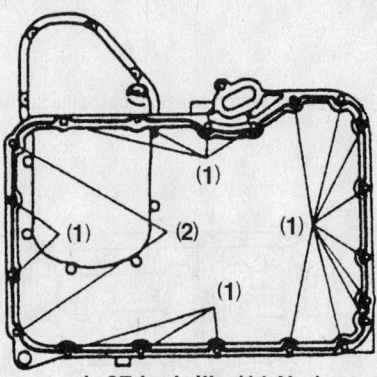

1: 97 Inch ilb. (11 Nm)
2: 20 Ft. lbs. (27 Nm)

9308HG03

Upper oil pan bolt torque values—2.5L engine

   - Oil pickup tube bracket
   - Radiator outlet pipe
   - Upper oil pan and oil pickup tube

To install:
4. Install a new O-ring to the lower crankcase.
5. Apply a bead of silicone sealant to the upper oil pan flange.
6. Install or connect the following:
   - New oil pump pickup tube O-ring seals
   - Upper oil pan and oil pump pickup tube and tighten the fasteners as shown
   - Radiator outlet pipe
   - Oil pickup tube bracket
   - Lower oil pan. Tighten the bolts to 97 inch lbs. (11 Nm).
   - Front differential, if equipped
   - Steering gear
   - Front skidplate, if equipped
   - Front wheels
   - Oil dipstick tube
   - Negative battery cable
7. Fill the crankcase to the correct level.
8. Start the engine and check for leaks.

Oil Pump

REMOVAL & INSTALLATION

1.6L Engines

1. Before servicing the vehicle, refer to the precautions in the beginning of this section.
2. Drain the engine oil.
3. Drain the cooling system.
4. Remove or disconnect the following:
   - Negative battery cable
   - Accessory drive belts
   - Crankshaft pulley
   - Front cover
   - Timing belt
   - Alternator and bracket
   - A/C compressor and bracket, if equipped
   - Oil pan and oil pump pickup tube
   - Crankshaft timing sprocket
   - Oil pump

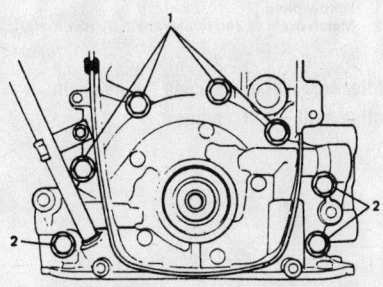

1. No. 1 bolts (short)
2. No. 2 bolts (long)

7924HG14

Oil pump housing short (1) and long bolt (2) locations—1.6L engines

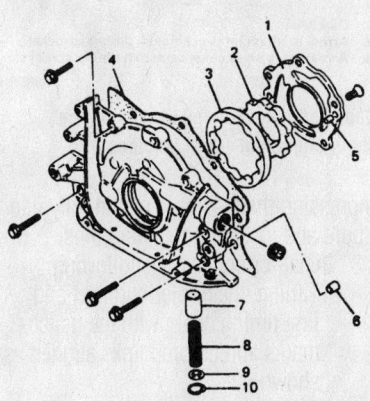

1. Rotor plate
2. Inner rotor
3. Outer rotor
4. Gasket
5. Pin
6. Pin
7. Relief valve
8. Spring
9. Retainer
10. Retainer ring

7924HG12

Exploded view of the oil pump housing—1.6L engines

→The oil pump bolts are different lengths. Note their location for assembly.

To install:
5. Install or connect the following:
   - Oil pump with a new gasket. Tighten the bolts to 97 inch lbs. (11 Nm).
   - Crankshaft timing sprocket. Tighten the bolt to 94 ft. lbs. (130 Nm).
   - Oil pan and oil pump pickup tube
   - A/C compressor and bracket, if equipped
   - Alternator and bracket
   - Timing belt
   - Front cover
   - Crankshaft pulley. Tighten the bolts to 12 ft. lbs. (16 Nm).
   - Accessory drive belts
   - Negative battery cable
6. Fill the crankcase to the correct level.
7. Start the engine and check for leaks.

2.0L Engines

1. Before servicing the vehicle, refer to the precautions in the beginning of this section.
2. Drain the engine oil.
3. Remove or disconnect the following:
   - Negative battery cable
   - Oil pan and pickup tube
   - Oil pump sprocket cover
   - Oil pump

> ⁂ **WARNING**
>
> **Do not remove the sprocket from the oil pump. Damage to the oil pump center shaft and abnormal pump operation may result.**

To install:
4. Install or connect the following:
   - Oil pump. Tighten the bolts to 20 ft. lbs. (27 Nm).
   - Oil pump sprocket cover. Tighten the bolts to 108 inch lbs. (12 Nm).
   - Oil pan and pickup tube
   - Negative battery cable
5. Fill the crankcase to the correct level.
6. Start the engine and check for leaks.

2.5L Engine

1. Before servicing the vehicle, refer to the precautions in the beginning of this section.
2. Drain the cooling system.
3. Drain the engine oil.
4. Remove or disconnect the following:
   - Negative battery cable
   - Accessory drive belts

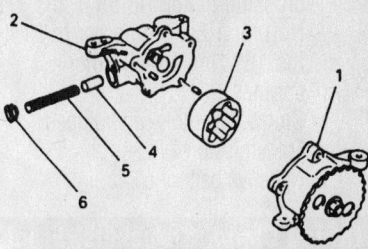

1. Oil pump case No.1
2. Oil pump case No.2
3. Outer rotor
4. Relief valve
5. Relief spring
6. Retainer

7924HG15

Exploded view of oil pump—2.0L and 2.5L engines

   - Intake manifold
   - Oil pan and oil pickup tube
   - Front cover
   - Oil pump chain guide
   - Oil pump

> ⁂ **WARNING**
>
> **Do not remove the sprocket from the oil pump. Damage to the oil pump center shaft and abnormal pump operation may result.**

To install:
5. Install or connect the following:
   - Oil pump. Tighten the bolts to 20 ft. lbs. (27 Nm).
   - Oil pump chain guide. Tighten the bolts to 97 inch lbs. (11 Nm).
   - Front cover
   - Oil pan and oil pickup tube
   - Intake manifold
   - Accessory drive belts
   - Negative battery cable
6. Fill the crankcase to the correct level.
7. Fill the cooling system.
8. Start the engine and check for leaks.

Rear Main Seal

REMOVAL & INSTALLATION

1. Before servicing the vehicle, refer to the precautions in the beginning of this section.
2. Remove or disconnect the following:
   - Negative battery cable
   - Transmission
   - Clutch assembly, if equipped
   - Flywheel
   - Rear main seal

To install:
3. Install or connect the following:
   - Rear main seal flush with the cylinder block
   - Flywheel. Tighten the bolts in a

crossing pattern to 58 ft. lbs. (79 Nm) for 1.6L engines or to 51 ft. lbs. (69 Nm) for all other engines.
- Clutch assembly, if equipped
- Transmission
- Negative battery cable

Timing Chain, Sprockets, Front Cover and Seal

REMOVAL & INSTALLATION

2.0L Engines

1. Before servicing the vehicle, refer to the precautions in the beginning of this section.
2. Drain the cooling system.
3. Drain the engine oil.
4. Remove or disconnect the following:

- Negative battery cable
- Oil pan and pickup tube
- Valve cover
- Bypass pipe and hose
- Accessory drive belts
- Cooling fan and shroud
- Water pump pulley
- Alternator belt tensioner and idler pulleys
- Upper radiator hose
- A/C compressor and bracket, if equipped
- Crankshaft pulley
- Front crankshaft seal
- Front cover

5. Rotate the crankshaft to align the timing marks as shown.

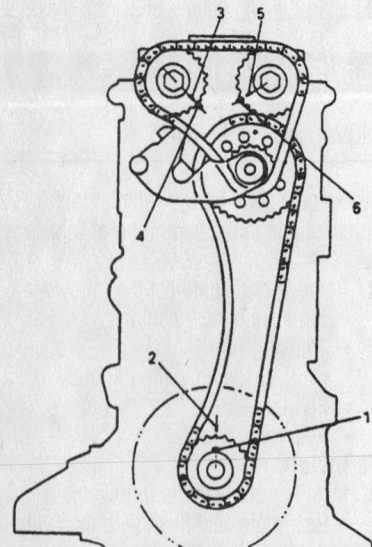

Timing mark alignment—2.0L engines

7924HG22

❄❄ WARNING

Do not allow the crankshaft or camshafts to rotate once the timing chains have been removed. Valve or piston damage could result.

6. Remove or disconnect the following:
- Second timing chain tensioner
- Camshaft sprockets and second timing chain
- First timing chain tensioner
- Timing chain idler sprocket and first timing chain

To install:

7. Prepare the timing chain tensioners for installation by releasing the latches,

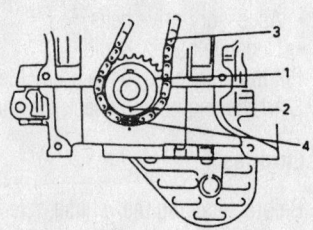

1. Crankshaft timing sprocket   3. 1st timing chain
2. Match mark   4. Yellow plate

7924HG17

Crankshaft and first timing chain alignment—2.0L engines

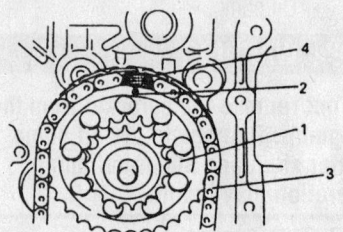

1. Idler sprocket   3. 1st timing chain
2. Match mark on idler sprocket 4. Dark blue plate

7924HG16

Idler sprocket and first timing chain alignment—2.0L engines

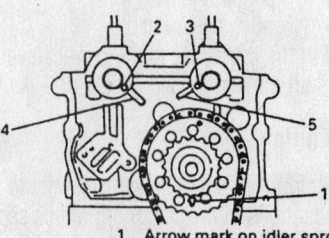

1. Arrow mark on idler sprocket
2. Knock pin of intake camshaft
3. Knock pin of exhaust camsaft
4. Timing mark of intake side
5. Timing mark of exhaust side

7924HG19

Idler sprocket and camshaft alignment— 2.0L engines

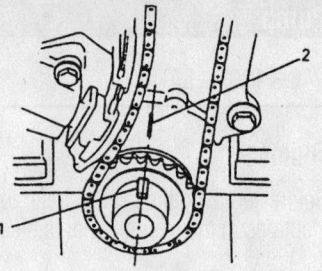

1. Crank timing sprocket key
2. Timing mark

7924HG18

Crankshaft sprocket alignment—2.0L engines

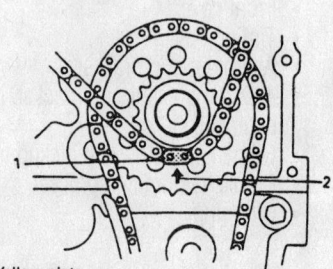

1. Yellow plate
2. Match mark of 2nd timing chain (Arrow mark)

7924HG20

Idler sprocket and second timing chain alignment—2.0L engines

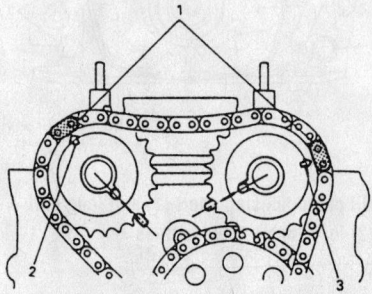

1. Dark blue
2. Arrow mark on intake camshaft timing sprocket
3. Arrow mark on exhaust camshaft timing sprocket

7924HG21

Camshaft sprocket and second timing chain alignment—2.0L engines

compressing the tensioner piston fully into the bore and installing retaining pins.

8. Install or connect the following:
- Timing chain idler sprocket and first timing chain with the matchmarks and colored links aligned as shown
- First timing chain tensioner. Tighten the bolts to 97 inch lbs. (11 Nm).
- Camshaft sprockets and second timing chain with the matchmarks and colored links aligned as shown
- Second timing chain tensioner.

Tighten the bolts to 97 inch lbs. (11 Nm) and the nut to 33 ft. lbs. (45 Nm).

9. Tighten the camshaft sprocket bolts to 59 ft. lbs. (80 Nm).

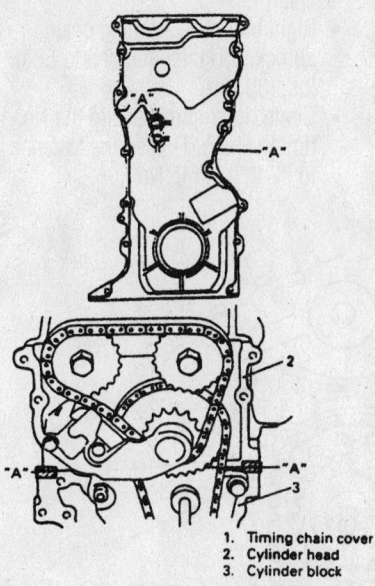

1. Timing chain cover
2. Cylinder head
3. Cylinder block

7924HG10

Prior to installing the timing chain cover on the engine block and cylinder head, apply silicone sealant to the cover as indicated (areas marked A)—2.0L and 2.5L engines

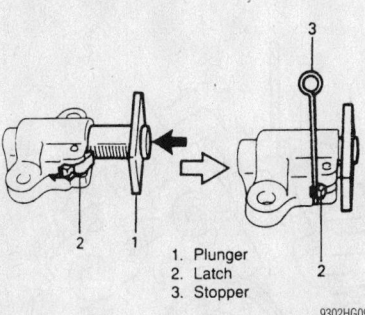

1. Plunger
2. Latch
3. Stopper

9302HG09

Preparing the No. 1 timing chain tensioner adjuster for installation—2.0L and 2.5L engines

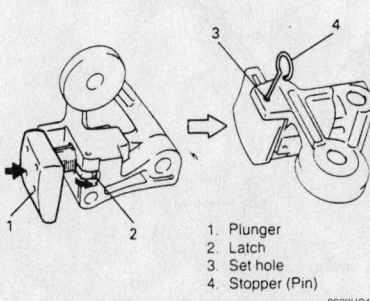

1. Plunger
2. Latch
3. Set hole
4. Stopper (Pin)

9302HG13

No. 2 timing chain tensioner—left bank tensioner shown—2.0L and 2.5L engines

10. Remove the timing chain tensioner retaining pins.

11. Rotate the crankshaft two complete turns and check that the timing marks align.

12. Install or connect the following:
- Front cover. Apply sealant as shown.
- Front crankshaft seal
- Crankshaft pulley. Tighten the bolt to 109 ft. lbs. (130 Nm).
- A/C compressor and bracket, if equipped
- Upper radiator hose
- Alternator belt tensioner and idler pulleys
- Water pump pulley
- Cooling fan and shroud
- Accessory drive belts
- Bypass pipe and hose
- Valve cover
- Oil pan and pickup tube
- Negative battery cable

13. Fill the crankcase to the correct level.

14. Fill the cooling system.

15. Start the engine and check for leaks.

2.5L Engine

1. Before servicing the vehicle, refer to the precautions in the beginning of this section.

2. Drain the cooling system.

3. Drain the engine oil.

4. Remove or disconnect the following:
- Negative battery cable
- Intake manifold
- Ignition coils
- Valve covers
- Accessory drive belts
- Cooling fan and shroud
- Water pump pulley
- Radiator
- Power steering pump and brackets
- Oil pan and pickup tube
- Crankshaft pulley
- Front crankshaft seal
- Crankshaft Position (CKP) sensor
- Front cover

5. Rotate the crankshaft so that the timing marks are aligned as shown.

✳✳ WARNING

Do not allow the crankshaft or camshafts to rotate once the timing chains have been removed. Valve or piston damage could result.

6. Remove or disconnect the following:
- Left bank No. 2 timing chain tensioner
- Left bank intake and exhaust camshaft sprockets with the No. 2 timing chain
- No. 1 timing chain guides

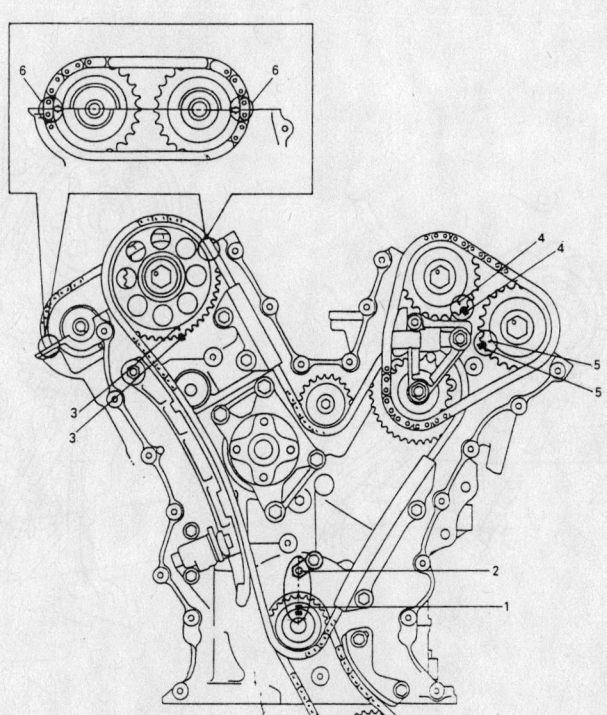

1. Crankshaft Keyway
2. Oil Jet
3. Right Bank No. 1 Chain Marks
4. Left Bank Intake Cam Marks
5. Left Bank Exhaust Cam Marks
6. Right Bank Intake and Exhaust Cam Marks

9302HG07

Timing chain alignment marks—2.5L engine

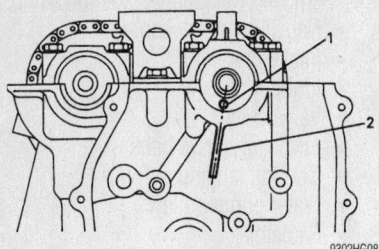

1. Knock pin of intake camshaft
2. Match mark

9302HG08

Right bank camshaft timing marks—2.5L engine

- No. 1 timing chain tensioner
- Center idler sprocket and the No. 1 timing chain
- Right bank No. 1 timing chain sprocket

7. The right bank No. 2 timing chain is removed with the intake and exhaust camshafts.

To install:

8. Prepare the timing chain tensioners for installation by releasing the latches, compressing the tensioner piston fully into the bore and installing retaining pins.

9. Align the timing chain sprocket matchmarks and colored chain links as shown during assembly.

10. Install or connect the following:

- Right bank intake and exhaust camshafts with the No. 2 timing chain
- Right bank No. 1 timing chain sprocket. Tighten the bolt to 58 ft. lbs. (80 Nm).
- Center idler sprocket and the No. 1 timing chain. Tighten the fastener to 32 ft. lbs. (45 Nm).

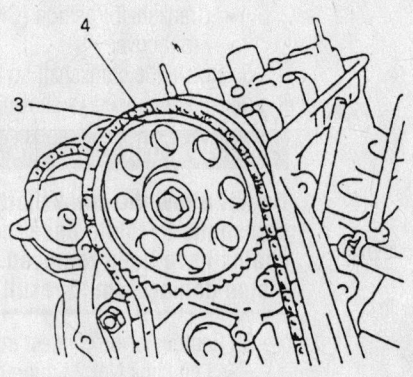

3. Match mark of RH bank 1st timing chain sprocket
4. Silver plate (LH) of 1st timing chain

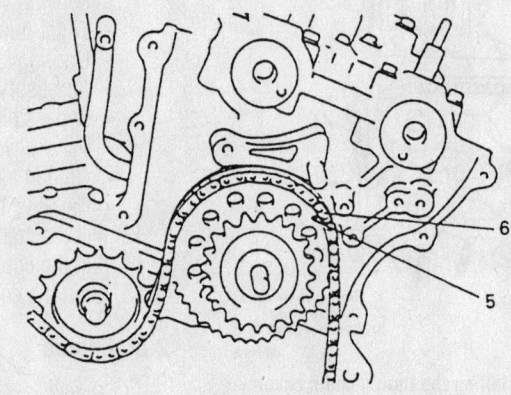

5. Match mark of idler sprocket No.2
6. Silver plate (RH) of 1st timing chain

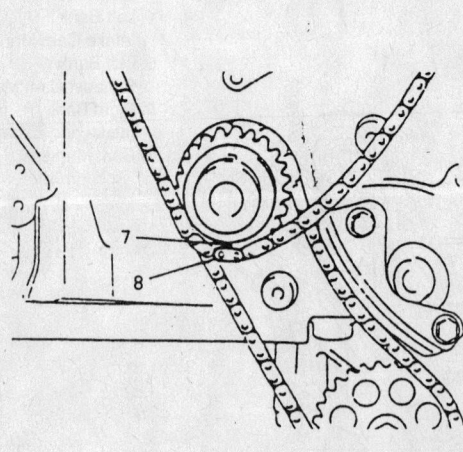

7. Match mark of crankshaft timing sprocket
8. Gold or Yellow plate of 1st timing chain

1. Crank timing pulley key
2. Oil jet

9302HG10

No. 1 timing chain alignment—2.5L engine

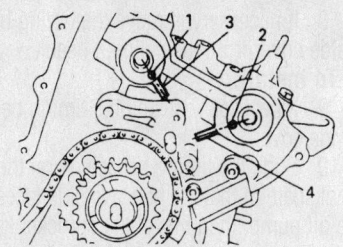

1. Knock pin of LH bank intake camshaft
2. Knock pin of LH bank exhaust camshaft
3. Match mark of intake side
4. Match mark of exhaust side

9302HG11

Left bank camshaft alignment—2.5L engine

- No. 1 timing chain tensioner and guides. Tighten the bolts to 97 inch lbs. (11 Nm).
- Left bank intake and exhaust camshaft sprockets with the No. 2 timing chain. Tighten the bolts to 57 ft. lbs. (80 Nm).
- Left bank No. 2 timing chain ten-

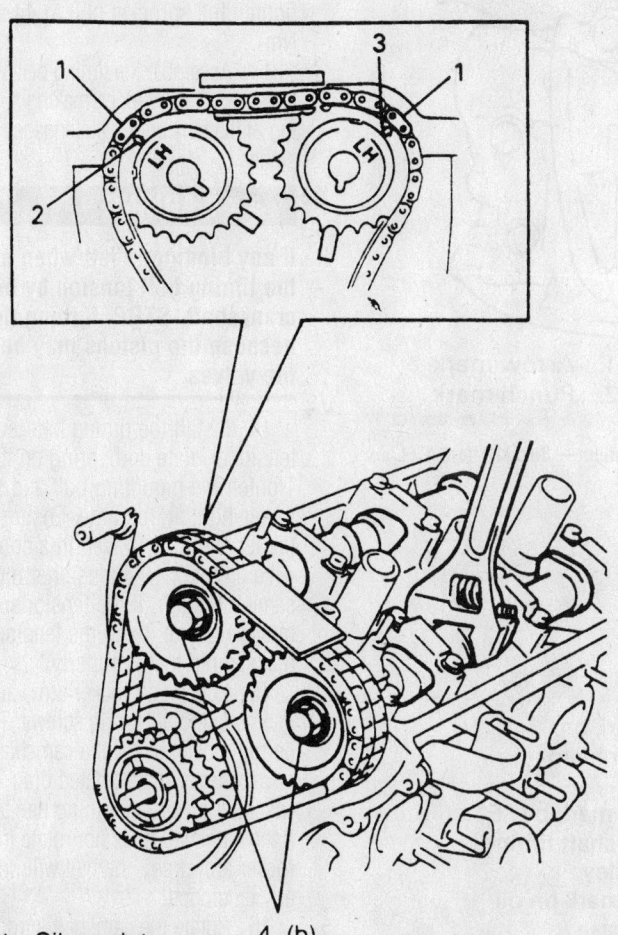

1. Silver plate
2. Arrow mark on intake camshaft timing sprocket
3. Arrow mark on exhaust camshaft timing sprocket
4. Sprocket bolt

9302HG12

Align the left bank No. 2 chain silver links—2.5L engine

sioner. Tighten the bolts to 97 inch lbs. (11 Nm).

11. Remove the retaining pins from the timing chain tensioners.

12. Rotate the crankshaft two complete turns and check that the timing marks align.

13. Install or connect the following:

- Front cover. Tighten the bolts to 97 inch lbs. (11 Nm).
- CKP sensor
- Front crankshaft seal
- Crankshaft pulley. Tighten the bolt to 109 ft. lbs. (148 Nm).
- Oil pan and pickup tube
- Power steering pump and brackets
- Radiator
- Water pump pulley
- Cooling fan and shroud
- Accessory drive belts
- Valve covers
- Ignition coils
- Intake manifold
- Negative battery cable

14. Fill the crankcase to the correct level.
15. Fill the cooling system.
16. Start the engine and check for leaks.

Timing Belt

REMOVAL & INSTALLATION

1.6L Engine

❋❋ WARNING

Do not rotate the crankshaft counterclockwise or attempt to rotate the crankshaft by turning the camshaft sprocket.

1. Remove the timing belt cover.

2. If the timing belt is not already marked with a directional arrow, use white paint, a grease pencil or correction fluid to do so.

3. Rotate the crankshaft clockwise until the timing mark on the camshaft sprocket and the "V" mark on the timing belt inside cover are aligned, and the punch mark on the crankshaft sprocket is aligned with the mark on the engine.

❋❋ WARNING

Do not rotate the crankshaft or camshaft once the timing belt is removed, because the valves and pistons can come into contact, which may cause internal engine damage.

4. Disconnect one end of the tensioner spring. Loosen the timing belt tensioner bolt and stud, then, using your finger, press the tensioner plate up and remove the timing belt from the crankshaft and camshaft sprockets.

5. Remove the timing belt tensioner, tensioner plate and spring from the engine.

6. Install Suzuki tool 09917-68220, or equivalent, onto the camshaft sprocket to hold the camshaft from rotating. Loosen the camshaft sprocket retaining bolt, then pull the camshaft sprocket off of the end of the camshaft.

7. Remove the crankshaft timing belt sprocket by loosening the center bolt, while preventing the crankshaft from rotating. To hold the crankshaft from turning, use Suzuki tool 09927-56010, or equivalent, or a large prybar inserted in the transmission housing slot and the flywheel teeth. Pull the sprocket off of the end of the crankshaft. Be sure to retain the crankshaft sprocket key and belt guide for assembly.

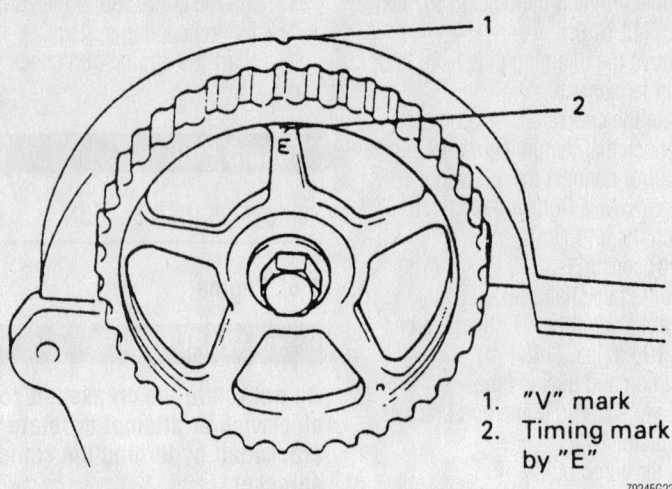

1. "V" mark
2. Timing mark by "E"

79245G22

Camshaft timing marks — Suzuki Vitara 1.6L 16-valve engine

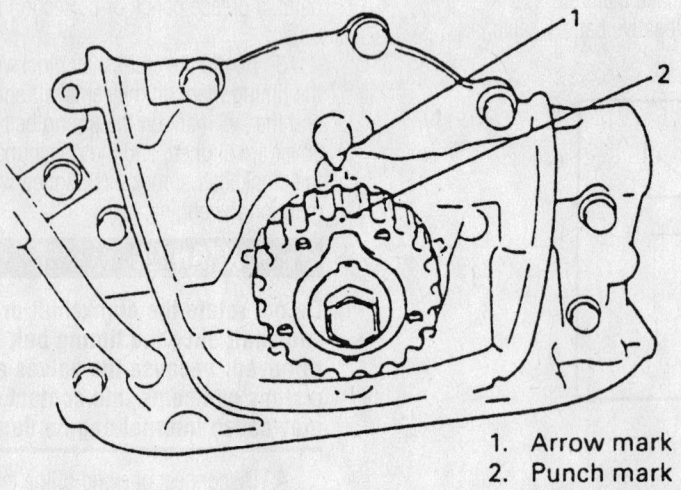

1. Arrow mark
2. Punch mark

79245G23

Align the punch mark with the arrow for proper timing belt installation — Suzuki Vitara 1.6L 16-valve engine

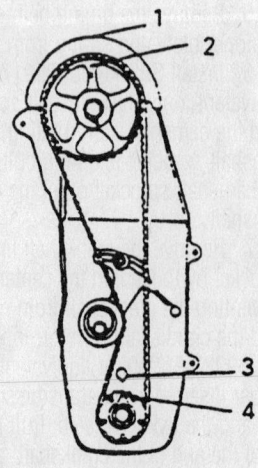

1. "V" mark on cylinder head cover
2. Timing mark by "E" on camshaft timing belt pulley
3. Arrow mark on oil pump case
4. Punch mark on crankshaft timing belt pulley

79245G47

Rotate the crankshaft clockwise until the camshaft and crankshaft timing marks are aligned — Suzuki Vitara 1.6L 16-valve engine

8. If necessary, remove the timing belt inside cover from the cylinder head.

To install:

9. If necessary, install the timing belt inside cover.

10. Slide the timing belt guide on the crankshaft so that the concave side faces the oil pump, then install the sprocket key in the groove in the crankshaft.

11. Slide the pulley onto the crankshaft, and install the center retaining bolt. Tighten the center bolt to 80 ft. lbs. (110 Nm). To hold the crankshaft from turning, use Suzuki tool 09927-56010, or equivalent, or a large prybar inserted in the transmission housing slot and the flywheel teeth.

12. Install the timing belt camshaft sprocket, ensuring that the slot in the sprocket engages the camshaft (pulley) pin; this ensures that the sprocket is properly positioned on the end of the camshaft. Secure the camshaft with the holding tool used during removal, then tighten the sprocket bolt to 44 ft. lbs. (60 Nm).

13. Assemble the timing belt tensioner plate and the tensioner, making sure that the lug of the tensioner plate engages the tensioner.

❋❋ WARNING

If any binding is felt when adjusting the timing belt tension by turning the crankshaft, STOP turning the engine, because the pistons may be hitting the valves.

14. Install the timing belt tensioner, tensioner plate and spring on the engine. Tighten the mounting bolt and stud only finger-tight at this time. Ensure that when the tensioner is moved in a counterclockwise direction, the tensioner moves in the same direction. If the tensioner does not move, remove it and the tensioner plate to reassemble them properly.

15. Loosen all rocker arm valve lash locknuts and adjusting screws. This will permit movement of the camshaft without any rocker arm associated drag, which is essential for proper timing belt tensioning. If the camshaft does not rotate freely (free of rocker arm drag), the belt will not be properly tensioned.

16. Rotate the camshaft sprocket clockwise until the timing mark on the sprocket and the "V" mark on the timing belt inside cover are aligned.

17. Using a wrench, or socket and

breaker bar, on the crankshaft sprocket center bolt, turn the crankshaft clockwise until the punch mark on the sprocket is aligned with the arrow mark on the oil pump.

18. With the camshaft and crankshaft marks properly aligned, push the tensioner up with your finger and install the timing belt on the 2 sprockets, ensuring that the drive side of the belt is free of all slack. Release your finger from the tensioner. Be sure to install the timing belt so that the directional arrow is pointing in the appropriate direction.

➡ **In this position, the No. 4 cylinder is at Top Dead Center (TDC) on the compression stroke.**

19. Rotate the crankshaft clockwise 2 full revolutions, then tighten the tensioner stud to 97 inch lbs. (11 Nm). Then, tighten the tensioner bolt to 18 ft. lbs. (24 Nm).

20. Ensure that all 4 timing marks are still aligned as before; if they are not, remove the timing belt, and install and tension it again.

21. Install the timing belt cover and all related components.

Piston and Ring

POSITIONING

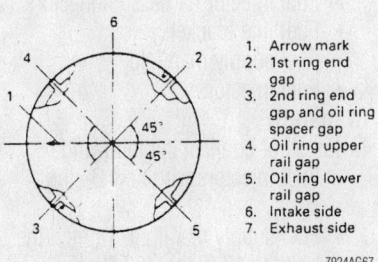

1. Arrow mark
2. 1st ring end gap
3. 2nd ring end gap and oil ring spacer gap
4. Oil ring upper rail gap
5. Oil ring lower rail gap
6. Intake side
7. Exhaust side

7924AG67

Piston ring end-gap spacing—All engines

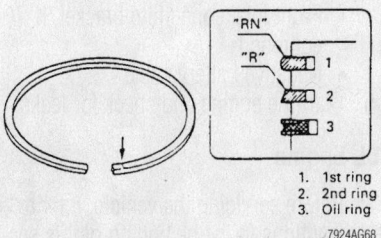

1. 1st ring
2. 2nd ring
3. Oil ring

7924AG68

Compression ring identification marks— All engines

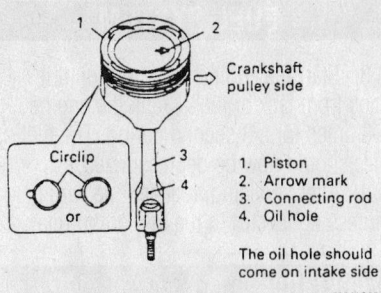

1. Piston
2. Arrow mark
3. Connecting rod
4. Oil hole

The oil hole should come on intake side

7924AG69

Piston and connecting rod positioning— 1.6L engine

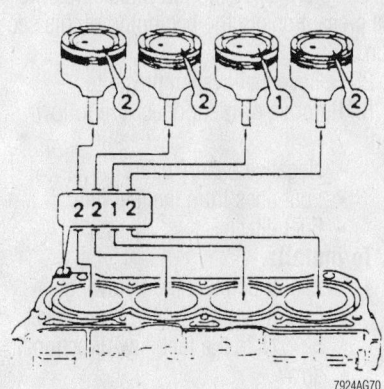

2 2 1 2

7924AG70

Piston installation—1.6L engine

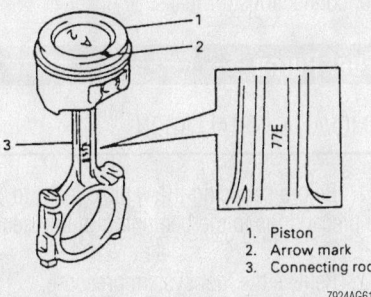

1. Piston
2. Arrow mark
3. Connecting rod

7924AG61

Piston and connecting rod positioning— 2.0L and 2.5L engines

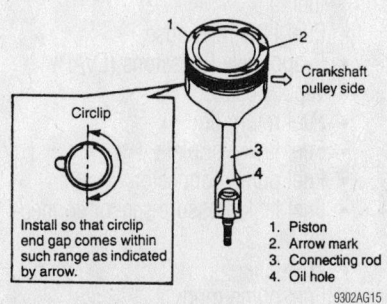

Install so that circlip end gap comes within such range as indicated by arrow.

1. Piston
2. Arrow mark
3. Connecting rod
4. Oil hole

9302AG15

Piston pin circlip installation—2.5L engine

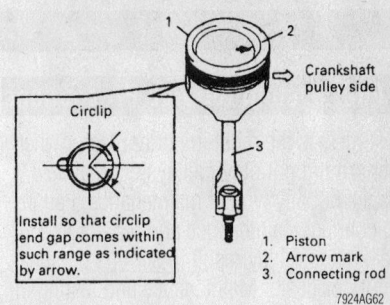

Install so that circlip end gap comes within such range as indicated by arrow.

1. Piston
2. Arrow mark
3. Connecting rod

7924AG62

Piston pin circlip installation—2.0L engines

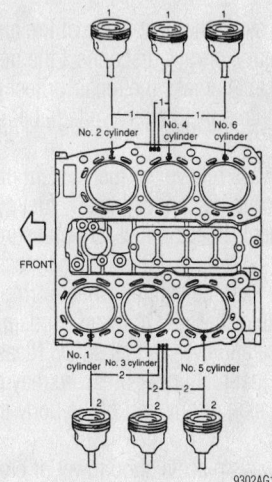

No. 2 cylinder No. 4 cylinder No. 6 cylinder

FRONT

No. 1 cylinder No. 3 cylinder No. 5 cylinder

9302AG16

Piston identification—2.5L engine

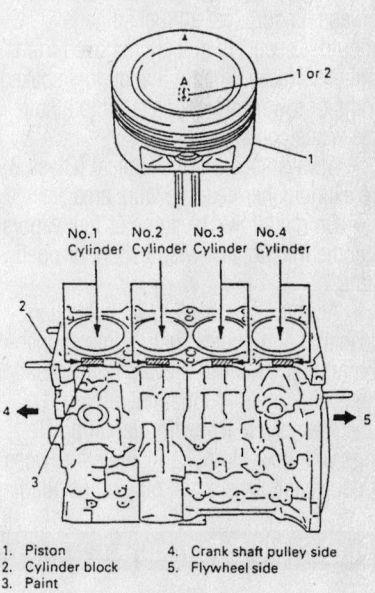

1 or 2

No.1 Cylinder No.2 Cylinder No.3 Cylinder No.4 Cylinder

1. Piston
2. Cylinder block
3. Paint
4. Crank shaft pulley side
5. Flywheel side

7924AG65

Piston identification—2.0L engines
Match pistons with "1" indicators to red cylinder paint marks
Match pistons with "2" indicators to blue cylinder paint marks

FUEL SYSTEM

Fuel System Service Precautions

Safety is the most important factor when performing not only fuel system maintenance but any type of maintenance. Failure to conduct maintenance and repairs in a safe manner may result in serious personal injury or death. Maintenance and testing of the vehicle fuel system components can be accomplished safely and effectively by adhering to the following rules and guidelines.

• To avoid the possibility of fire and personal injury, always disconnect the negative battery cable unless the repair or test procedure requires that battery voltage be applied.

• Always relieve the fuel system pressure prior to disconnecting any fuel system component (injector, fuel rail, pressure regulator, etc.), fitting or fuel line connection. Exercise extreme caution whenever relieving fuel system pressure to avoid exposing skin, face and eyes to fuel spray. Please be advised that fuel under pressure may penetrate the skin or any part of the body that it contacts.

• Always place a shop towel or cloth around the fitting or connection prior to loosening to absorb any excess fuel due to spillage. Ensure that all fuel spillage (should it occur) is quickly removed from engine surfaces. Ensure that all fuel soaked cloths or towels are deposited into a suitable waste container.

• Always keep a dry chemical (Class B) fire extinguisher near the work area.

• Do not allow fuel spray or fuel vapors to come into contact with a spark or open flame.

• Always use a backup wrench when loosening and tightening fuel line connection fittings. This will prevent unnecessary stress and torsion to fuel line piping.

• Always replace worn fuel fitting O-rings with new. Do not substitute fuel hose or equivalent, where fuel pipe is installed.

Fuel System Pressure

RELIEVING

1. Before servicing the vehicle, refer to the precautions in the beginning of this section.

2. Detach the wiring harness connector from the fuel pump relay, located under the left-hand side of the instrument panel near the ECM.

3. Start the engine and run it until it stops from lack of fuel. Crank the engine 2–3 times for a 3 second period. The fuel lines should now be depressurized.

4. After servicing, reattach the wiring harness connector to the fuel pump relay.

Fuel Filter

REMOVAL & INSTALLATION

1. Before servicing the vehicle, refer to the precautions in the beginning of this section.

2. Relieve fuel system pressure.

3. Remove or disconnect the following:

• Negative battery cable
• Fuel lines from the fuel filter
• Fuel filter

To install:

4. Install or connect the following:
• Fuel filter and tighten the bracket bolt. Note the fuel flow directional arrow.
• Fuel lines to the fuel filter.
• Negative battery cable

5. Start the engine and inspect the fuel filter connections for leaks.

Fuel Pump

REMOVAL & INSTALLATION

1. Before servicing the vehicle, refer to the precautions in the beginning of this section.

2. Relieve the fuel system pressure.

3. Remove or disconnect the following:
• Negative battery cable
• Fuel filler hose and vent hose
• Fuel tank inlet valve and drain the fuel tank
• Fuel filter inlet hose
• Evaporative Emissions (EVAP) vapor hose
• Fuel return line
• Fuel tank skidplate
• Fuel pump connector
• Fuel tank pressure sensor connector
• Fuel tank
• Fuel pump module

To install:

4. Install or connect the following:
• Fuel pump module with a new seal. Tighten the bolts to 44 inch lbs. (5 Nm).

• Fuel tank. Tighten the strap bolts to 37 ft. lbs. (50 Nm).
• Fuel tank pressure sensor connector
• Fuel pump connector
• Fuel tank skidplate
• Fuel return line
• EVAP vapor hose
• Fuel filter inlet hose
• Fuel tank inlet valve and drain the fuel tank
• Fuel filler hose and vent hose
• Negative battery cable

5. Fill the fuel tank.

6. Start the engine and check for leaks.

Fuel Injector

REMOVAL & INSTALLATION

1.6L and 2.0L Engines

1. Before servicing the vehicle, refer to the precautions in the beginning of this section.

2. Relieve the fuel system pressure.

3. Remove or disconnect the following:
• Negative battery cable
• Front intake manifold bracket, if equipped
• Positive Crankcase Ventilation (PCV) valve and hose
• Fuel injector harness connectors
• Fuel line bracket
• Fuel supply manifold
• Fuel injectors

To install:

4. Install or connect the following:
• Fuel injectors with new O-ring seals
• Fuel supply manifold. Tighten the bolts to 17 ft. lbs. (23 Nm).
• Fuel line bracket
• Fuel injector harness connectors
• PCV valve and hose
• Front intake manifold bracket, if equipped
• Negative battery cable

5. Start the engine and check for leaks.

2.5L Engine

1. Before servicing the vehicle, refer to the precautions in the beginning of this section.

2. Relieve the fuel system pressure.

3. Remove or disconnect the following:
• Negative battery cable
• Air intake tube

- Throttle body intake collector
- Fuel lines
- Fuel pressure regulator vacuum line
- Fuel injector harness connectors
- Fuel supply manifold connect pipe
- Fuel supply manifolds
- Fuel injectors

To install:
4. Install or connect the following:
- Fuel injectors with new O-ring seals
- Fuel supply manifolds. Tighten the bolts to 17 ft. lbs. (23 Nm).
- Fuel supply manifold connect pipe. Tighten the bolts to 22 ft. lbs. (30 Nm).

- Fuel injector harness connectors
- Fuel pressure regulator vacuum line
- Fuel lines
- Throttle body intake collector
- Air intake tube
- Negative battery cable
5. Start the engine and check for leaks.

DRIVE TRAIN

Transmission Assembly

REMOVAL & INSTALLATION

Manual Transmission

1. Before servicing the vehicle, refer to the precautions in the beginning of this section.
2. Drain the transmission fluid.
3. Drain the transfer case fluid, if equipped.
4. Remove or disconnect the following:
- Negative battery cable
- Shift lever boots
- Gear shift lever
- Transfer case shift lever, if equipped
- 4WD switch connector, if equipped
- Reverse light switch connector
- Starter motor
- Front driveshaft, if equipped
- Rear driveshaft
- Speedometer cable, if equipped
- Vehicle Speed (VSS) sensor, if equipped
- Clutch slave cylinder or cable
- Flywheel access cover
- Transmission flange bolts and nuts
- Transmission braces, if equipped
5. Support the transmission with a jack and remove the transmission mount and crossmember.

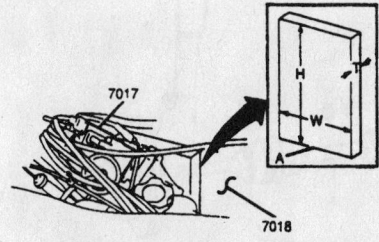

A	WOOD BLOCK
H	200 mm (8.0")
T	45 mm (1.8")
W	100–150 mm (4.0–6.0")
7017	DISTRIBUTOR CAP
7018	BULKHEAD

7924HG37

Support the engine with a wooden block between the cylinder head and the firewall—All models

6. Place a wooden block at the rear of the cylinder head as shown to support the engine when the transmission is removed.
7. Lower the transmission away from the vehicle.

To install:
8. Install or connect the following:
- Transmission. Tighten the flange fasteners to 62–72 ft. lbs. (85–98 Nm).
- Transmission mount and crossmember. Tighten the fasteners to 29–43 ft. lbs. (40–60 Nm).
- Transmission braces, if equipped. Tighten the bolts to 62–72 ft. lbs. (85–98 Nm).
- Flywheel access cover
- Clutch slave cylinder or cable
- VSS sensor, if equipped
- Speedometer cable, if equipped
- Rear driveshaft
- Front driveshaft, if equipped
- Starter motor
- Reverse light switch connector
- 4WD switch connector, if equipped
- Transfer case shift lever, if equipped
- Gear shift lever
- Shift lever boots
- Negative battery cable
9. Fill the transmission to the correct level.
10. Fill the transfer case, if equipped.

Automatic Transmission

1. Before servicing the vehicle, refer to the precautions in the beginning of this section.
2. Drain the transfer case oil, if equipped.
3. Remove or disconnect the following:

- Negative battery cable
- Center console and transfer case shift lever, if equipped
- Transmission dipstick tube
- Transmission wiring harness connectors
- Starter motor
- Front driveshaft, if equipped
- Rear driveshaft

- Gear select cable and bracket
- Throttle Valve (TV) cable, if equipped
- Exhaust front pipe
- Transmission oil cooler lines
- Transmission brace
- Flywheel access cover
- Torque converter
- Speedometer cable, if equipped
- Vehicle Speed (VSS) sensor connector, if equipped
- Transmission flange bolts and nuts
- Transmission braces, if equipped
4. Support the transmission with a jack and remove the transmission mount and crossmember.
5. Place a wooden block at the rear of the cylinder head as shown to support the engine when the transmission is removed.
6. Lower the transmission away from the vehicle.

To install:
7. Install or connect the following:
- Transmission. Tighten the flange fasteners to 62–72 ft. lbs. (85–98 Nm).
- Transmission mount and crossmember. Tighten the fasteners to 29–43 ft. lbs. (40–60 Nm).
- Transmission braces, if equipped. Tighten the bolts to 62–72 ft. lbs. (85–98 Nm).
- VSS sensor connector, if equipped
- Speedometer cable, if equipped
- Torque converter. Tighten the bolts to 47 ft. lbs. (65 Nm).
- Flywheel access cover
- Transmission brace
- Transmission oil cooler lines
- Exhaust front pipe
- TV cable, if equipped
- Gear select cable and bracket
- Rear driveshaft
- Front driveshaft, if equipped
- Starter motor
- Transmission wiring harness connectors
- Transmission dipstick tube

- Center console and transfer case shift lever, if equipped
- Negative battery cable

8. Fill the transmission to the correct level.

9. Fill the transfer case, if equipped.

Clutch

ADJUSTMENTS

These vehicles are equipped with a hydraulic clutch system. No adjustment is necessary.

REMOVAL & INSTALLATION

1. Before servicing the vehicle, refer to the precautions in the beginning of this section.

2. Remove the transmission.

3. Loosen the pressure plate mounting bolts in a 2-step crisscross sequence until the spring tension is relieved.

4. Remove the pressure plate and the clutch disc.

To install:

5. Using a clutch alignment tool, assemble the clutch disc and pressure plate onto the flywheel.

6. Tighten the pressure plate bolts in multiple passes to 17 ft. lbs. (23 Nm).

7. Install the transmission.

8. Check for proper clutch operation.

Hydraulic Clutch System

BLEEDING

1. Before servicing the vehicle, refer to the precautions in the beginning of this section.

2. Fill the master cylinder reservoir to the MAX line with clean brake fluid and keep it at least half full throughout the bleeding procedure.

3. From beneath the vehicle, remove the bleeder plug cap, then attach a clear vinyl tube to the slave cylinder bleeder plug. Insert the open end of the hose into a container.

4. Have an assistant depress the clutch pedal. Open the bleeder after the pedal is depressed.

5. Close the bleeder before releasing the clutch pedal.

6. Repeat until all air bubbles are gone from the hydraulic fluid.

7. Install the bleeder plug cap.

8. Fill the clutch master cylinder fluid reservoir to the specified full level.

Transfer Case Assembly

REMOVAL & INSTALLATION

1. Before servicing the vehicle, refer to the precautions in the beginning of this section.

2. Drain the transfer case oil.

3. Remove or disconnect the following:
- Negative battery cable
- Distributor or Camshaft Position (CMP) sensor, if equipped
- Center console
- Transmission shift lever and case, if equipped with a manual transmission
- Transfer case shift lever
- Front and rear driveshafts
- Exhaust center pipe
- Speedometer cable or Vehicle Speed (VSS) sensor, as equipped
- Vent hose
- 4WD switch connector

4. Support the transmission with a jack and remove the transmission mount and crossmember.

5. Place a wooden block at the rear of the cylinder head as shown to support the engine when the transfer case is removed.

6. Lower the transfer case away from the vehicle.

To install:

7. Install or connect the following:
- Transfer case. Tighten the bolts to 30 ft. lbs. (41 Nm).
- 4WD switch connector
- Vent hose
- Speedometer cable or VSS sensor, as equipped

- Exhaust center pipe
- Front and rear driveshafts. Tighten the bolts to 36 ft. lbs. (50 Nm).
- Transfer case shift lever
- Transmission shift lever and case, if equipped with a manual transmission
- Center console
- Distributor or CMP sensor, if equipped
- Negative battery cable

8. Fill the transfer case.

Halfshaft

REMOVAL & INSTALLATION

Left

1. Before servicing the vehicle, refer to the precautions in the beginning of this section.

2. Remove or disconnect the following:
- Front wheel
- Hub drive flange or locking hub, as equipped
- Snapring
- Thrust washer
- Halfshaft flange fasteners
- Halfshaft

To install:

3. Install or connect the following:
- Halfshaft. Tighten the flange bolts to 37 ft. lbs. (50 Nm).
- Thrust washer
- Snapring
- Locking hub, if equipped. Tighten the bolts to 24 ft. lbs. (33 Nm).

1. Drive shaft oil seal	6. Ball joint assembly (RH side)
2. Double off-set joint (DOJ)	7. Drive shaft assembly (LH side)
3. Joint circlip	8. Left drive shaft
4. DOJ boot	9. Drive shaft bearing circlip
5. Ball joint boot	10. Drive shaft bearing

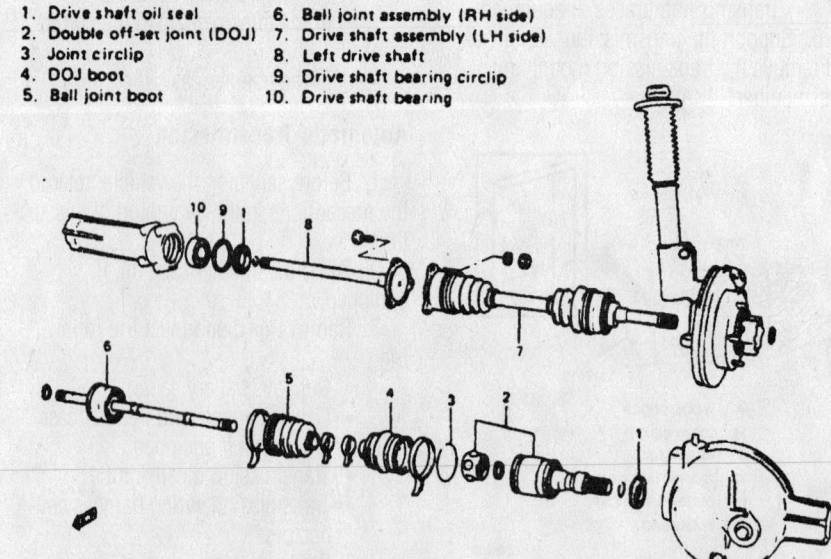

Exploded view of the left- and right-hand halfshaft assemblies

7924HG31

- Hub drive flange, if equipped. Tighten the bolts to 35 ft. lbs. (48 Nm).
- Front wheel

Right

1. Before servicing the vehicle, refer to the precautions in the beginning of this section.
2. Remove or disconnect the following:
   - Front wheel
   - Hub drive flange or locking hub, as equipped
   - Snapring
   - Thrust washer
   - Brake caliper
   - Wheel speed sensor, if equipped
   - Brake rotor
   - Stabilizer bar link
   - Outer tie rod end
   - Lower ball joint
   - Strut bracket bolts
   - Steering knuckle and wheel hub
3. Pry the inboard joint out of the differential and remove the halfshaft.

To install:
4. Insert the inboard joint into the differential until the circlip is felt to seat.
5. Install or connect the following:
   - Steering knuckle and wheel hub
   - Strut bracket bolts. Tighten them to 70 ft. lbs. (95 Nm).
   - Lower ball joint. Tighten the nut to 40 ft. lbs. (55 Nm).
   - Outer tie rod end. Tighten the nut to 35 ft. lbs. (48 Nm).
   - Stabilizer bar link. Tighten the nut to 21 ft. lbs. (29 Nm).
   - Brake rotor
   - Wheel speed sensor, if equipped
   - Brake caliper
   - Thrust washer
   - Snapring
   - Locking hub, if equipped. Tighten the bolts to 24 ft. lbs. (33 Nm).
   - Hub drive flange, if equipped. Tighten the bolts to 35 ft. lbs. (48 Nm).
   - Front wheel
6. Check the wheel alignment and adjust as necessary.

CV-Joints

OVERHAUL

Outer CV-Joint

The outer CV-joint is serviced with the axle shaft as an assembly. The outer CV-joint boot can be serviced by removing the inner CV-joint.

Inner CV-Joint

1. Before servicing the vehicle, refer to the precautions in the beginning of this section.
2. Remove or disconnect the following:
   - Halfshaft from the vehicle
   - Grease boot clamps
   - Outer race snapring
   - Outer race
   - Shaft snapring
   - Inner race, cage and balls

To install:
3. Install or connect the following:
   - Inner race, cage and balls
   - Shaft snapring
   - Outer race
   - Outer race snapring
4. Fill the outer race and the grease boot with CV-joint grease and tighten the boot clamps.
5. Install the axle halfshaft.

Manual Locking Hubs

REMOVAL & INSTALLATION

1. Before servicing the vehicle, refer to the precautions in the beginning of this section.
2. Set the selector knob to the **FREE** position.
3. Remove or disconnect the following:
   - Hub cover assembly
   - Hub body assembly

To install:
4. Install the hub body. Tighten the bolts to 18 ft. lbs. (25 Nm).

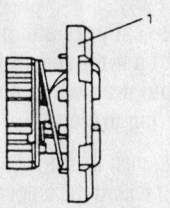

1. Cover

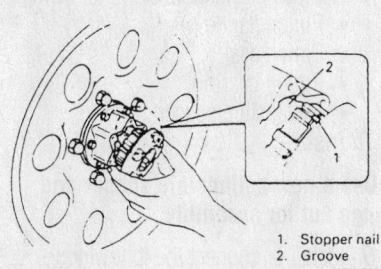

1. Stopper nail
2. Groove

9308HG06

Manual hub alignment—All models

5. Align the hub cover stopper nail with the groove in the hub body and install the hub cover. Tighten the bolts to 90 inch lbs. (10 Nm).
6. Check for proper hub operation.

Automatic Locking Hubs

REMOVAL & INSTALLATION

1. Before servicing the vehicle, refer to the precautions in the beginning of this section.
2. Unlock the hub by setting the transfer case in the 2H position and driving backwards at least 6.5 feet (2 meters).
3. Remove or disconnect the following:
   - Hub sub assembly
   - Hub brake assembly

To install:
4. Align the brake assembly key with the slot in the spindle and install the brake assembly.
5. Align the matchmark on the sub assembly with the mark on the brake assembly and install the sub assembly. Tighten the hub bolts to 24 ft. lbs. (33 Nm).
6. Check for proper hub operation.

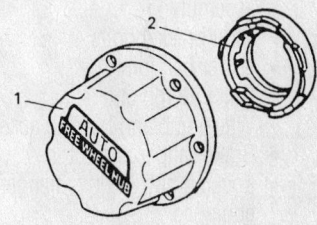

1. Free wheeling hub sub assembly
2. Free wheeling hub brake assembly

9308HG04

Automatic hub—All models

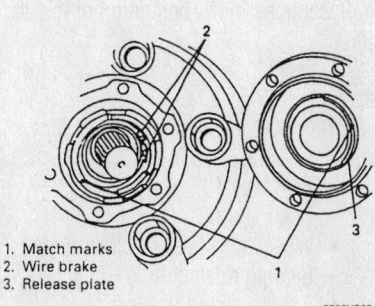

1. Match marks
2. Wire brake
3. Release plate

9308HG05

Automatic hub matchmarks—All models

Spindle Bearings

REMOVAL, PACKING & INSTALLATION

1. Before servicing the vehicle, refer to the precautions in the beginning of this section.
2. Support the control arm with a stand or floor jack.
3. Remove or disconnect the following:

- Front wheel
- Locking hub or drive flange, as equipped
- Brake caliper and rotor
- Wheel hub and bearing assembly
- Outer tie rod end
- Lower ball joint
- Strut bracket bolts
- Wheel spindle and steering knuckle assembly
- Inner oil seal
- Spindle bearing

To install:
4. Fill the recess in the wheel spindle with lithium grease.
5. Coat the spindle bearing and wheel spindle mating surfaces with sealant.
6. Press or drive the spindle bearing into the wheel spindle.
7. Install or connect the following:

- Inner oil seal
- Wheel spindle and steering knuckle assembly
- Strut bracket bolts
- Lower ball joint
- Outer tie rod end
- Wheel hub and bearing assembly
- Brake caliper and rotor
- Locking hub or drive flange, as equipped
- Front wheel

Axle Shaft, Bearing and Seal

REMOVAL & INSTALLATION

1. Before servicing the vehicle, refer to the precautions in the beginning of this section.
2. Loosen the parking brake cable for clearance.
3. Remove or disconnect the following:

- Rear wheel
- Brake drum
- Wheel speed sensor, if equipped
- Bearing retainer nuts
- Axle shaft and bearing
- Axle shaft inner oil seal

4. If equipped with ABS, grind a flat spot on the wheel speed sensor tone ring, then split the ring with a chisel.
5. Grind flat spots on the bearing retainer and split it with a chisel.
6. Press the wheel bearing off the axle shaft.
7. Remove the bearing retainer and the outer oil seal.

To install:
8. Install or connect the following:

- Outer oil seal to the bearing retainer
- Bearing retainer to the axle shaft
- Bearing and retainer ring pressed onto the axle shaft
- Wheel speed sensor tone ring pressed onto the axle shaft, if equipped
- Axle shaft inner oil seal
- Axle shaft and bearing
- Bearing retainer nuts. Tighten them to 17 ft. lbs. (23 Nm).
- Wheel speed sensor, if equipped
- Brake drum
- Rear wheel

9. Fill the rear differential to the correct level.

Pinion Seal

REMOVAL & INSTALLATION

1. Before servicing the vehicle, refer to the precautions in the beginning of this section.
2. Remove or disconnect the following:

- Driveshaft
- Wheels
- Brake calipers and pads or brake drum

➡ **The brake calipers and pads or brake drum must be removed so that there is no additional drag when measuring pinion bearing preload.**

3. Use an inch lb. torque wrench and measure and record the amount of torque required to maintain pinion rotation through several revolutions.
4. Remove or disconnect the following:

- Pinion flange
- Pinion seal
- Pinion bearing
- Collapsible spacer

To install:

➡ **Use a new collapsible spacer and flange nut for assembly.**

5. Install or connect the following:

- Collapsible spacer

- Pinion bearing
- Pinion seal
- Pinion flange

6. Rotate the pinion flange occasionally while tightening the flange nut to make sure the pinion bearings seat correctly.
7. Take frequent bearing preload torque readings. Tighten the flange nut to achieve the preload torque readings originally recorded.

✳✳ CAUTION

Never loosen the pinion nut to reduce bearing preload. If it is necessary to reduce bearing preload, install a new collapsible spacer and pinion nut.

8. Install or connect the following:
- Driveshaft
- Brake calipers and pads or brake drum
- Wheels

9. Fill the differential with gear lubricant and check for leaks.

Axle Housing Assembly

REMOVAL & INSTALLATION

1. Before servicing the vehicle, refer to the precautions in the beginning of this section.
2. Drain the gear oil.
3. Support the vehicle at the frame with a hoist or jackstands.
4. Support the rear axle with a floor jack.
5. Remove or disconnect the following:

- Rear wheels
- Rear brake drums
- Rear axle shafts
- Load sensing proportioning valve linkage, if equipped
- Brake fluid hose
- Brake backing plates
- Wheel speed sensor connector, if equipped
- Axle vent tube
- Rear driveshaft
- Differential carrier assembly
- Shock absorber lower bolts
- Coil springs
- Upper rods
- Lower rods
- Lateral rod
- Axle housing

To install:
6. Install or connect the following:
- Axle housing
- Upper rods
- Lower rods

- Coil springs
- Lateral rod
- Shock absorber lower bolts
- Differential carrier assembly. Tighten the nuts to 40 ft. lbs. (55 Nm).
- Rear driveshaft
- Axle vent tube

- Wheel speed sensor connector, if equipped
- Brake backing plates
- Brake fluid hose
- Load sensing proportioning valve linkage, if equipped
- Rear axle shafts
- Rear brake drums

- Rear wheels
7. Fill the rear axle to the correct level.
8. Lower the vehicle so that the rear suspension is at curb height.
9. Tighten the upper, lower and lateral rod fasteners to 65 ft. lbs. (90 Nm).
10. Tighten the lower shock absorber fasteners to 62 ft. lbs. (85 Nm).

STEERING AND SUSPENSION

Air Bag

✲✲ CAUTION

Some vehicles are equipped with an air bag system. The system must be disarmed before performing service on, or around, system components, the steering column, instrument panel components, wiring and sensors. Failure to follow the safety precautions and the disarming procedure could result in accidental air bag deployment, possible injury and unnecessary system repairs.

PRECAUTIONS

Several precautions must be observed when handling the inflator module to avoid accidental deployment and possible personal injury.

- Never carry the inflator module by the wires or connector on the underside of the module.
- When carrying a live inflator module, hold securely with both hands and ensure that the bag/trim cover are pointed away.
- Place the inflator module on a bench or other surface with the bag and trim cover facing up.
- With the inflator module on the bench, never place anything on or close to the module which may be thrown in the event of an accidental deployment.
- Never use air bag component parts from another vehicle.
- If there is a chance of electrical shock to any of the air bag components, remove the air bag module before servicing the vehicle.

DISARMING

1. Before servicing the vehicle, refer to the precautions in the beginning of this section.
2. Remove or disconnect the following:
   - Negative battery cable
   - AIR BAG fuse

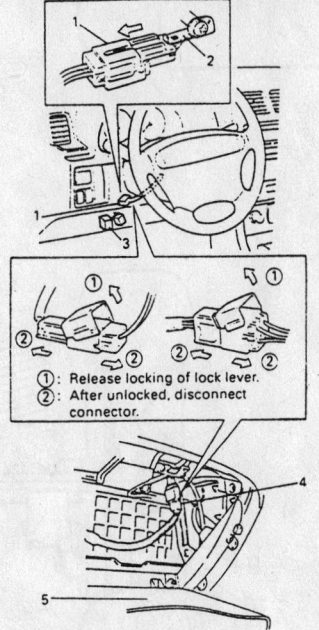

①: Release locking of lock lever.
②: After unlocked, disconnect connector.

1. Yellow connector of driver air bag (inflator) module
2. Connector stay
3. Air bag fuse box
4. Yellow connector of passenger air bag (inflator) module
5. Glove box

7924HG36

Air bag component location and identification—All models

- Driver air bag connector
- Glove box
- Passenger air bag connector

ARMING

When repairs are complete, install or connect the following:
- Passenger air bag connector
- Glove box
- Driver air bag connector
- AIR BAG fuse
- Negative battery cable

Recirculating Ball Steering Gear

REMOVAL & INSTALLATION

1. Before servicing the vehicle, refer to the precautions in the beginning of this section.

2. Remove or disconnect the following:
   - Skidplate, if equipped
   - Coolant overflow tank
   - Intermediate shaft pinch bolt
   - Power steering hoses
   - Pitman arm center link joint
   - Steering gearbox
   - Pitman arm

To install:

3. Install or connect the following:
   - Pitman arm. Tighten the nut to 102 ft. lbs. (140 Nm).
   - Steering gearbox. Tighten the bolts to 62 ft. lbs. (85 Nm).
   - Pitman arm center link joint. Tighten the nut to 37 ft. lbs. (50 Nm).
   - Power steering hoses
   - Intermediate shaft pinch bolt. Tighten it to 18 ft. lbs. (25 Nm).
   - Coolant overflow tank
   - Skidplate, if equipped
4. Fill the power steering system.
5. Start the engine and check for leaks.
6. Check the wheel alignment and adjust as necessary.

Power Rack and Pinion Steering Gear

REMOVAL & INSTALLATION

1. Before servicing the vehicle, refer to the precautions in the beginning of this section.
2. Remove or disconnect the following:
   - Power steering hoses
   - Intermediate shaft pinch bolt
   - Front wheels
   - Outer tie rod ends
   - Steering gear

To install:

3. Install or connect the following:
   - Steering gear. Tighten the bolts to 40 ft. lbs. (55 Nm).
   - Outer tie rod ends. Tighten the nuts to 32 ft. lbs. (43 Nm).
   - Front wheels
   - Intermediate shaft pinch bolt. Tighten it to 18 ft. lbs. (25 Nm).

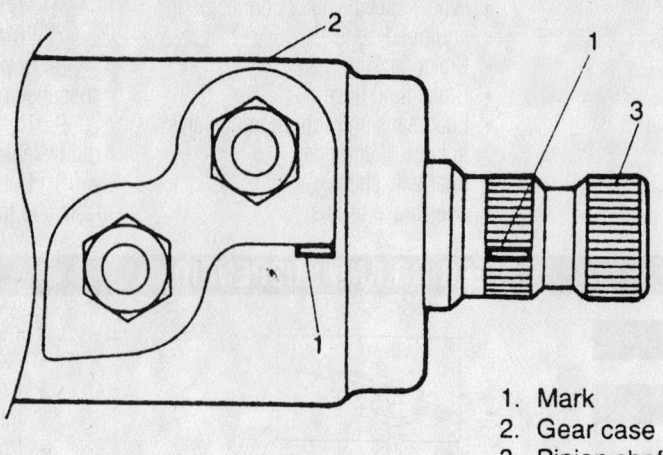

1. Mark
2. Gear case
3. Pinion shaft

9302HG14

Steering gear centering marks

- Power steering hoses
4. Fill the power steering system.
5. Start the engine and check for leaks.
6. Check the wheel alignment and adjust as necessary.

Strut

REMOVAL & INSTALLATION

1. Before servicing the vehicle, refer to the precautions in the beginning of this section.

2. Support the control arm with a stand or floor jack.

3. Remove or disconnect the following:
- Front wheel
- Brake hose bracket
- Strut bracket bolts
- Upper strut mount nuts
- Strut

To install:

4. Install or connect the following:
- Strut. Tighten the upper mount nuts to 40 ft. lbs. (55 Nm) and the bracket bolts to 70 ft. lbs. (95 Nm).
- Brake hose bracket
- Front wheel

5. Check the wheel alignment and adjust as necessary.

Shock Absorber

REMOVAL & INSTALLATION

1. Before servicing the vehicle, refer to the precautions in the beginning of this section.

2. Support the rear axle housing with a hydraulic jack or stand.

3. Remove or disconnect the following:

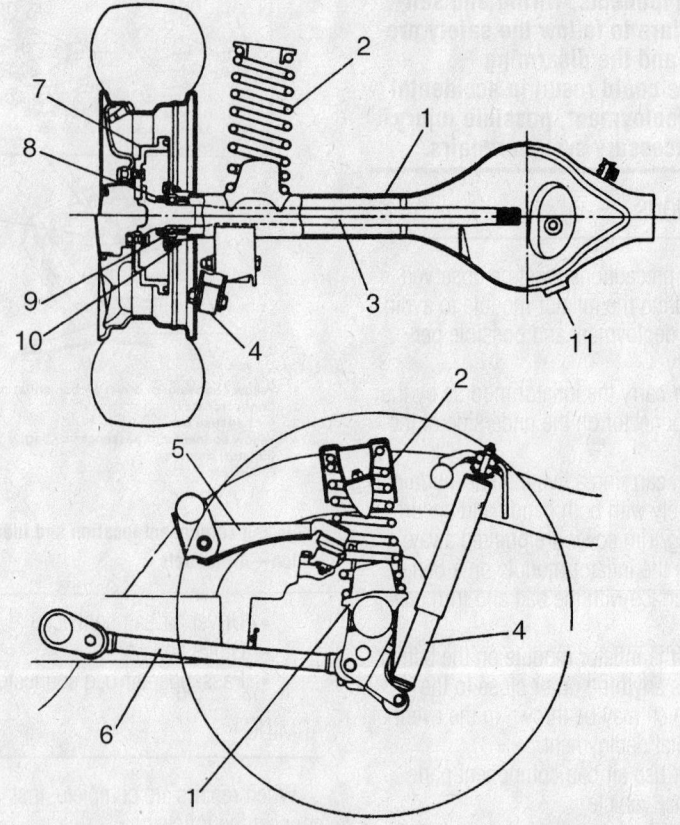

1. Rear axle housing
2. Coil spring
3. Axle shaft
4. Shock absorber
5. Upper arm
6. Trailing rod
7. Brake drum
8. Wheel bearing retainer
9. Rear wheel bearing
10. Brake back plate
11. Oil drain plug

7924HG32

Rear suspension component identification

- Shock absorber upper locknut and retaining nut
- Lower shock absorber mounting nut and bolt
- Rear shock absorber

To install:

4. Install or connect the following:
- Rear shock absorber
- Lower mounting nut and bolt
- Upper retaining nut and locknut

5. Torque the upper mounting nuts to 21 ft. lbs. (22–35 Nm) and the lower mounting nut/bolt to 62 ft. lbs. (85 Nm) for 2000–01 models; 74 ft. lbs. (100 Nm) for 2002–03 models.

6. Remove the jack or stand from the rear axle assembly.

Coil Spring

REMOVAL & INSTALLATION

Front

1. Before servicing the vehicle, refer to the precautions in the beginning of this section.

2. Support the vehicle at the frame with a hoist or jackstand.

3. Support the control arm with a floor jack.

4. Remove or disconnect the following:
- Front wheel
- Brake caliper and rotor
- Locking hub or drive flange, if equipped
- Axle shaft snapring and thrust washer, if equipped
- Wheel speed sensor, if equipped
- Stabilizer bar link
- Lower ball joint
- Strut bracket bolts

5. Lower the floor jack and remove the coil spring.

To install:

➡**The bottom of the spring has a larger diameter than the top.**

6. Install the coil spring onto the control arm and raise the floor jack.

7. Install or connect the following:
- Strut bracket bolts. Tighten them to 70 ft. lbs. (95 Nm).
- Lower ball joint. Tighten the nut to 40 ft. lbs. (55 Nm).
- Stabilizer bar link. Tighten the nut to 21 ft. lbs. (29 Nm).
- Wheel speed sensor, if equipped
- Axle shaft snapring and thrust washer, if equipped

- Locking hub or drive flange, if equipped
- Brake caliper and rotor
- Front wheel

8. Check the wheel alignment and adjust as necessary.

Rear

1. Before servicing the vehicle, refer to the precautions in the beginning of this section.

2. Support the vehicle at the frame with a hoist or jackstand.

3. Support the rear axle housing with a floor jack.

4. Remove or disconnect the following:
- Rear wheels
- Parking brake cable hanger
- Shock absorber lower mounting bolts
- Wheel speed sensor harness clamps, if equipped
- Brake pipe E-ring
- Axle vent hose

5. Lower the floor jack and remove the coil springs.

To install:

6. Install the coil springs onto the axle spring seats and raise the floor jack.

7. Install or connect the following:
- Axle vent hose
- Brake pipe E-ring
- Wheel speed sensor harness clamps, if equipped
- Shock absorber lower mounting bolts. Tighten them to 62 ft. lbs. (85 Nm) for 2000–01 models; 74 ft. lbs. (100 Nm) for 2002–03 models.
- Parking brake cable hanger
- Rear wheels

Lower Ball Joint

REMOVAL & INSTALLATION

The lower ball joint is serviced with the lower control arm as an assembly.

Lower Control Arm

REMOVAL & INSTALLATION

1. Before servicing the vehicle, refer to the precautions in the beginning of this section.

2. Support the vehicle at the frame with a hoist or jackstand.

3. Support the control arm with a floor jack.

4. Remove or disconnect the following:
- Front wheel
- Brake caliper and rotor
- Locking hub or drive flange, if equipped
- Axle shaft snapring and thrust washer, if equipped
- Wheel speed sensor, if equipped
- Stabilizer bar link
- Lower ball joint
- Strut bracket bolts

5. Lower the floor jack and remove the coil spring.

6. Remove the inner control arm bolts and remove the control arm.

To install:

7. Install the inner control arm bolts.

8. Install the coil spring onto the control arm and raise the floor jack.

9. Install or connect the following:
- Strut bracket bolts. Tighten them to 70 ft. lbs. (95 Nm).
- Lower ball joint. Tighten the nut to 40 ft. lbs. (55 Nm).
- Stabilizer bar link. Tighten the nut to 21 ft. lbs. (29 Nm).
- Wheel speed sensor, if equipped
- Axle shaft snapring and thrust washer, if equipped
- Locking hub or drive flange, if equipped
- Brake caliper and rotor
- Front wheel

10. Lower the vehicle so that the front suspension is at curb height.

11. Tighten the front inner bolt to 62 ft. lbs. (85 Nm) and the rear inner bolt to 92 ft. lbs. (127 Nm).

12. Check the wheel alignment and adjust as necessary.

CONTROL ARM BUSHING REPLACEMENT

1. Before servicing the vehicle, refer to the precautions in the beginning of this section.

2. Remove the control arm from the vehicle.

3. Remove the control arm bushings with a hydraulic press.

To install:

4. Lubricate the control arm bushings with liquid soap.

5. Press the bushings into the control arm until the bushing flange contacts the housing edge of the control arm.

6. Install the control arm to the vehicle.

7. Check the wheel alignment and adjust as necessary.

Wheel Bearings

ADJUSTMENT

The wheel bearings are not adjustable.

REMOVAL & INSTALLATION

1. Before servicing the vehicle, refer to the precautions in the beginning of this section.
2. Remove or disconnect the following:
   - Front wheel
   - Brake caliper and rotor
   - Locking hub or hub drive flange, if equipped
   - Hub grease cap, if equipped
   - Wheel speed sensor, if equipped
   - Wheel bearing lockwasher
   - Wheel bearing locknut and inner washer
   - Wheel hub and bearing assembly
   - Wheel hub oil seal
   - Wheel bearing oil seal
   - Snapring
3. Press the wheel bearing and race out of the hub.

To install:

4. Press the wheel bearing and race into the hub so that the race is fully seated in the hub bore.
5. Install or connect the following:
   - Snapring
   - Wheel bearing oil seal
   - Wheel hub oil seal
   - Wheel hub and bearing assembly
   - Wheel bearing locknut and inner washer. Tighten the nut to 157 ft. lbs. (216 Nm).
   - Wheel bearing lockwasher. Tighten the retaining screws to 13 inch lbs. (1.5 Nm).
   - Wheel speed sensor, if equipped
   - Hub grease cap, if equipped
   - Locking hub or hub drive flange, if equipped
   - Brake caliper and rotor
   - Front wheel

BRAKES

Brake Caliper

REMOVAL & INSTALLATION

1. Raise and safely support the vehicle.
2. Remove the wheels.
3. Disconnect and plug the brake line.
4. Remove the caliper mounting bolts (guide pins) and remove the caliper from the vehicle.

To install:

5. Install the caliper on the vehicle. Tighten the mounting bolts as follows:
   - Vitara: 20 ft. lbs. (27 Nm).
   - Grand Vitara: 10mm bolt to 37 ft. lbs. (50 Nm) and 12mm bolt to 62 ft. lbs. (84Nm)
6. Connect the hydraulic brake line, using 2 new washers. Torque the union bolt to 17 ft. lbs. (23 Nm).
7. Replace the front wheels.
8. Lower the vehicle.
9. Fill the brake reservoir and bleed the hydraulic brake system.

Disc Brake Pads

REMOVAL & INSTALLATION

1. Siphon about ⅔ of the fluid out of the master cylinder.
2. Raise and safely support the vehicle.
3. Remove the wheels.
4. Remove the brake caliper mounting bolts and remove the caliper from the mounting bracket.
5. Support the caliper with a wire.

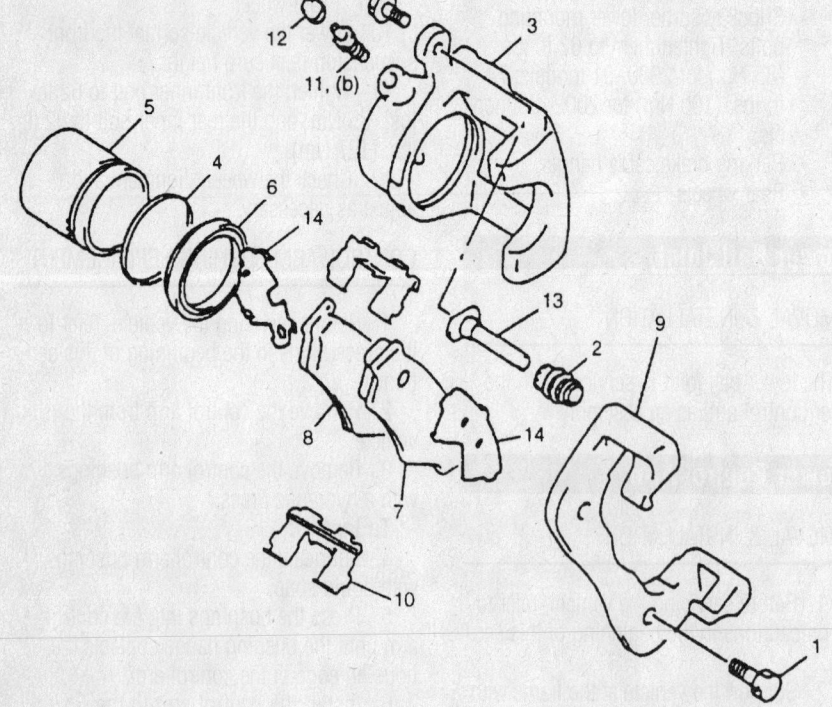

1. Caliper (slide) pin bolt
2. Boot
3. Disc brake caliper (disc brake cylinder)
4. Piston seal
5. Disc brake piston
6. Cylinder boot
7. Disc brake inner pad
8. Disc brake outer pad
9. Brake caliper carrier
10. Pad spring
11. Bleeder plug
12. Bleeder plug cap
13. Caliper pin
14. Anti noise shim
15. Inner shim

Tightening torque
(a): 8.0 N·m (0.80 kg-m, 6.0 lb-ft)
(b): 8.5 N·m (0.85 kg-m, 6.5 lb-ft)

93026G41

Front disc brake components—Vitara

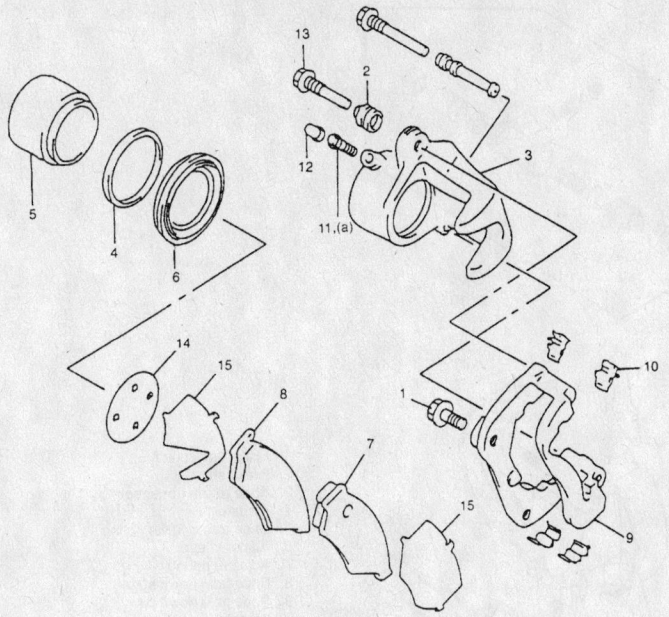

1. Caliper (slide) pin bolt
2. Boot
3. Disc brake caliper (disc brake cylinder)
4. Piston seal
5. Disc brake piston
6. Cylinder boot
7. Disc brake inner pad
8. Disc brake outer pad
9. Brake caliper carrier
10. Pad spring
11. Bleeder plug
12. Bleeder plug cap
13. Caliper pin
14. Anti noise shim
15. Inner shim

Tightening torque
(a): 8.0 N·m (0.80 kg-m, 6.0 lb-ft)

93026G42

Front disc brake components—Grand Vitara

6. Using a large pair of plies or a C-clamp compress the caliper piston back into the bore.

7. Remove the disc brake pads and any shims from the caliper mounting bracket.

To install:

8. Install the brake pads and any shims removed from the caliper mounting bracket.

9. Install the caliper on the mounting bracket and install the mounting bolts.

10. Install the front wheels and lower the vehicle.

❄❄ CAUTION

Do not attempt to drive the vehicle until after the following step is performed.

11. Depress the brake pedal repeatedly until a firm pedal is obtained. Do not attempt to drive the vehicle unless a firm pedal is obtained.

12. Check the fluid level in the master cylinder. Add fresh brake fluid, as necessary.

13. Road-test the vehicle.

Brake Drums

REMOVAL & INSTALLATION

1. Raise and safely support the vehicle.
2. Remove the rear wheel(s).
3. Release the parking brake.
4. Remove the parking brake lever

cover screws and loosen the brake cable locking nut.

5. Install 2, 8mm bolts into the brake drum holes and uniformly tighten each bolt. Tighten each bolt until the brake drum is removed from the vehicle. If there is difficulty in removing the drum, insert a small tool through the hole in the rear of the backing plate, and hold the automatic adjusting lever away from the adjuster. Using another narrow, flat tool at the same time, reduce the brake shoe adjuster by turning the adjusting wheel.

To install:

6. Install the brake drum and pull the parking brake lever all the way up until a clicking sound can no longer be heard.

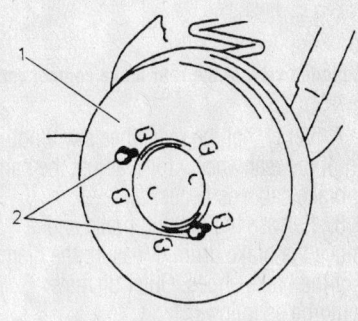

1 DRUM
2 TWO 8mm BOLTS

93026G39

Removing the brake drum with the two 8mm bolts

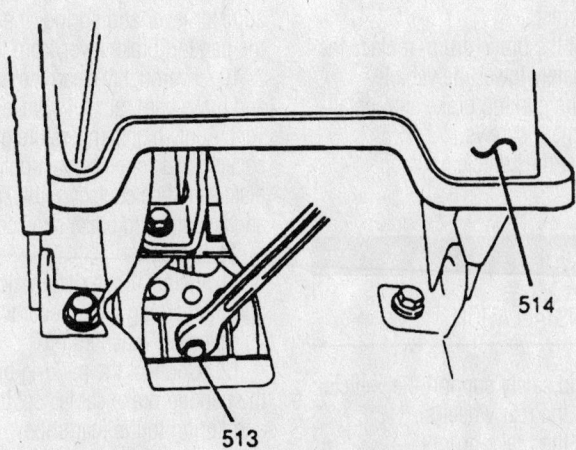

513 PARKING BRAKE CABLE LOCKNUT

514 PARKING BRAKE LEVER COVER

93026G38

Reducing the adjuster to remove the brake drum

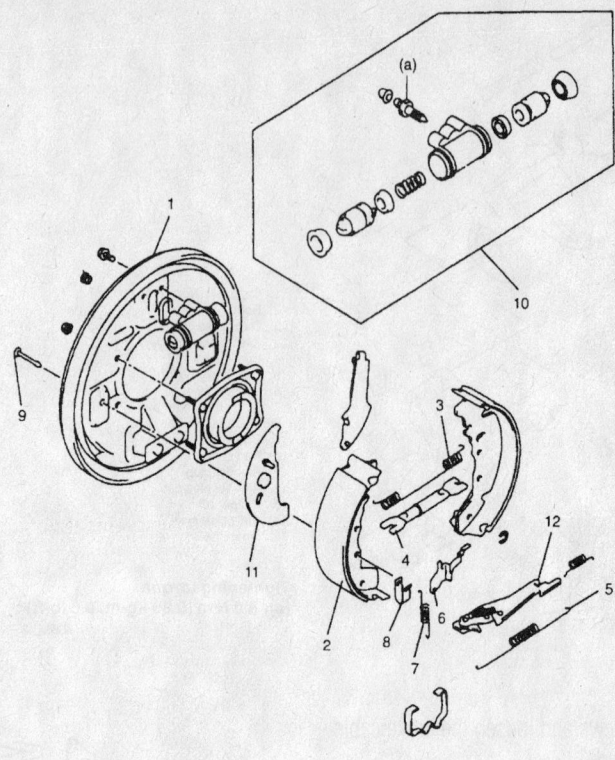

1. Brake back plate
2. Brake shoe
3. Shoe return upper spring
4. Adjuster
5. Shoe return lower spring
6. Adjuster lever
7. Adjuster spring
8. Shoe hold down spring
9. Shoe hold down pin
10. Wheel cylinder
11. Link
12. Brake strut

Tightening torque
(a): 7.5 N·m (0.75 kg-m, 5.5 lb-ft)

93026G43

Exploded view of the rear brake components—Vitara and Grand Vitara

7. Verify that the rear wheels will not turn. If the rear wheels turn, adjust the parking brake cable as necessary.

8. Release the parking brake and remove the brake drum. Measure the diameter of the brake shoes. Outer diameter should be as follows:
- For 2 door models: 8.638 (0.0012 inches (219 (0.3mm)
- For 4 door models: 9.980 (0.0079 inches (253.5 (0.2mm)

9. If the brake shoe clearance is not correct, adjust the brake shoes until the clearance is correct.

10. Reinstall the brake drum, replace the wheel(s), and safely lower the vehicle.

11. Adjust the parking brake and install the cover with the 2 screws.

12. Road-test the vehicle for proper brake operation.

Brake Shoes

REMOVAL & INSTALLATION

1. Raise and safely support the vehicle.
2. Remove the rear wheel(s).
3. Remove the brake drum.
4. Using a suitable tool, remove the brake shoe return spring.
5. Using a brake spring hold-down tool, disengage the hold-down spring and retainers from the front shoe. Remove the hold-down retainer pinch

6. Disconnect the anchor spring from the front shoe and remove the front shoe.

7. Remove the anchor spring from the rear shoe. Using a brake spring hold-down tool, disengage the hold-down spring and retainers from the rear shoe. Remove the hold-down pinch

8. Disengage the parking brake lever from the parking brake cable and remove the rear shoe.

9. Remove the C-washer, the automatic adjuster lever and spring, the C-washer, and the parking brake lever from the rear shoe.

10. Thoroughly clean the backing plate and brake hardware with brake cleaning solvent. Apply high temperature grease to the backing plate shoe contact points, anchor plate and shoe contact points, adjusting bolt, and adjuster and brake shoe contact points.

To install:

11. Reinstall the automatic adjuster lever and the parking brake lever to the rear shoe using new C-washers.

12. Connect the parking brake lever to the parking brake cable. Set the adjuster and spring to the rear shoe.

13. Set the rear brake shoe in place, install the hold-down pin and install the hold-down spring and retainers. Make sure that the shoe is inserted in the wheel cylinder and that the other end is in the anchor plate.

14. Install the anchor spring to the rear shoe.

15. Install the front shoe to the other end of the anchor spring and set the front shoe in place. Make sure that the front shoe engages the wheel cylinder, adjuster mechanism and spring, and the anchor plate.

16. Reinstall the front brake shoe hold-down pin and secure with the hold-down spring and retainers using a suitable tool.

17. Install the return spring.

18. Install the brake drum and pull the parking brake lever all the way up until a clicking sound can no longer be heard.

19. Verify that the rear wheels will not turn. If the rear wheels turn, adjust the parking brake cable as necessary.

20. Release the parking brake and remove the brake drum. Measure the diameter of the brake shoes. Brake diameter should be as follows:
- For 2 door models: 8.638 (0.0012 inches (219 (0.3mm)
- For 4 door models: 9.980 (0.0079 inches (253.5 (0.2mm)

21. If the brake shoe clearance is not correct, adjust the brake shoes until the clearance is correct.

22. Reinstall the brake drum, replace the wheel(s), and safely lower the vehicle.

23. Road-test the vehicle for proper brake operation.

TOYOTA

32

Avalon • Camry • Camry Solara • Celica • Corolla • Echo • GT-S • Supra • MR2

SPECIFICATION CHARTS

ENGINE AND VEHICLE IDENTIFICATION

Engine							Model Year	
Code ①	Liters (cc)	Cu. In.	Cyl.	Fuel Sys.	Engine Type	Eng. Mfg.	Code ②	Year
1MZ-FE	3.0 (2952)	180	6	EFI	DOHC	Toyota	Y	2000
2JZ-GE	3.0 (2997)	183	6	EFI	DOHC	Toyota	1	2001
5S-FE	2.2 (2264)	138	4	EFI	DOHC	Toyota	2	2002
7A-FE	1.8 (1762)	107	4	EFI	DOHC	Toyota	3	2003
2JZ-GTE	3.0 (2997)	183	6	EFI	DOHC	Toyota	4	2004
3S-GTE	2.0 (1998)	122	4	EFI	DOHC	Toyota		
4A-FE	1.6 (1587)	96	4	EFI	DOHC	Toyota		
1ZZ-FE	1.8 (1794)	109	4	EFI	DOHC	Toyota		
2ZZ-GE	1.8 (1796)	109.5	4	EFI	DOHC	Toyota		
2AZ-FE	2.4 (2398)	146	4	EFI	DOHC	Toyota		
1NZ-FE	1.5 (1496)	91	4	EFI	DOHC	Toyota		
1MZ-FE	3.0 (2995)	183	6	EFI	DOHC	Toyota		

EFI: Electronic Fuel Injection

DOHC: Double Overhead Camshaft

① 8th digit of VIN

② 10th digit of VIN

42356-TOYC-C01

GENERAL ENGINE SPECIFICATIONS

Year	Model	Engine Displacement Liters (cc)	Engine Series (ID/VIN)	Fuel System	Net Horsepower @ rpm	Net Torque @ rpm (ft. lbs.)	Bore x Stroke (in.)	Com- pression Ratio	Oil Pressure psi @ idle
2000	Avalon	3.0 (2995)	1MZ-FE	EFI	200@5200	214@5200	3.44x3.27	10.5:1	4.3
	Camry	2.2 (2164)	5S-FE	EFI	133@5200	147@4400	3.43x3.58	9.5:1	4.3
	Camry	3.0 (2995)	1MZ-FE	EFI	194@5200	209@4400	3.44x3.27	10.5:1	4.3
	Camry Solara	2.2 (2164)	5S-FE	EFI	135@5200	147@4400	3.43x3.58	9.5:1	4.3
	Camry Solara	3.0 (2995)	1MZ-FE	EFI	200@5200	214@4400	3.44x3.27	10.5:1	4.3
	MR 2	1.8 (1794)	1ZZ-FE	EFI	140@6400	125@4200	3.11x3.60	10.0:1	4.3
	Corolla	1.8 (1794)	1ZZ-FE	EFI	120@5600	122@4400	3.11x3.60	10.0:1	4.3
	Celica	1.8 (1794)	1ZZ-FE	EFI	140@6400	125@4200	3.11x3.60	10.0:1	4.3
	Celica GT-S	1.8 (1796)	2ZZ-GE	EFI	180@7600	130@6800	3.23x3.35	11.5:1	4.3
	Echo	1.5 (1496)	1NZ-FE	EFI	108@6000	105@4000	2.95x3.33	10.5:1	4.3
2001	Avalon	3.0 (2995)	1MZ-FE	EFI	200@5200	214@5200	3.44x3.27	10.5:1	4.3
	Camry	2.2 (2164)	5S-FE	EFI	133@5200	147@4400	3.43x3.58	9.5:1	4.3
	Camry	3.0 (2995)	1MZ-FE	EFI	194@5200	209@4400	3.44x3.27	10.5:1	4.3
	Camry Solara	2.2 (2164)	5S-FE	EFI	135@5200	147@4400	3.43x3.58	9.5:1	4.3
	Camry Solara	3.0 (2995)	1MZ-FE	EFI	200@5200	214@4400	3.44x3.27	10.5:1	4.3
	MR 2	1.8 (1794)	1ZZ-FE	EFI	140@6400	125@4200	3.11x3.60	10.0:1	4.3
	Corolla	1.8 (1794)	1ZZ-FE	EFI	120@5600	122@4400	3.11x3.60	10.0:1	4.3
	Celica	1.8 (1794)	1ZZ-FE	EFI	140@6400	125@4200	3.11x3.60	10.0:1	4.3
	Celica GT-S	1.8 (1796)	2ZZ-GE	EFI	180@7600	130@6800	3.23x3.35	11.5:1	4.3
	Echo	1.5 (1496)	1NZ-FE	EFI	108@6000	105@4000	2.95x3.33	10.5:1	4.3
2002	Avalon	3.0 (2995)	1MZ-FE	EFI	200@5200	214@5200	3.44x3.27	10.5:1	4.3
	Camry	2.4 (2398)	2AZ-FE	EFI	157@5600	162@4000	3.48x3.84	9.5:1	4.3
	Camry	3.0 (2995)	1MZ-FE	EFI	194@5200	209@4400	3.44x3.27	10.5:1	4.3
	Camry Solara	2.4 (2398)	2AZ-FE	EFI	157@5600	162@4000	3.48x3.84	9.5:1	4.3
	Camry Solara	3.0 (2995)	1MZ-FE	EFI	200@5200	214@4400	3.44x3.27	10.5:1	4.3
	MR 2	1.8 (1794)	1ZZ-FE	EFI	140@6400	125@4200	3.11x3.60	10.0:1	4.3
	Corolla	1.8 (1794)	1ZZ-FE	EFI	120@5600	122@4400	3.11x3.60	10.0:1	4.3
	Celica	1.8 (1794)	1ZZ-FE	EFI	140@6400	125@4200	3.11x3.60	10.0:1	4.3
	Celica GT-S	1.8 (1796)	2ZZ-GE	EFI	180@7600	130@6800	3.23x3.35	11.5:1	4.3
	Echo	1.5 (1496)	1NZ-FE	EFI	108@6000	105@4000	2.95x3.33	10.5:1	4.3
2003	Avalon	3.0 (2995)	1MZ-FE	EFI	200@5200	214@5200	3.44x3.27	10.5:1	4.3
	Camry	2.4 (2398)	2AZ-FE	EFI	157@5600	162@4000	3.48x3.84	9.5:1	4.3
	Camry	3.0 (2995)	1MZ-FE	EFI	194@5200	209@4400	3.44x3.27	10.5:1	4.3
	Camry Solara	2.4 (2398)	2AZ-FE	EFI	157@5600	162@4000	3.48x3.84	9.5:1	4.3
	Camry Solara	3.0 (2995)	1MZ-FE	EFI	200@5200	214@4400	3.44x3.27	10.5:1	4.3
	MR 2	1.8 (1794)	1ZZ-FE	EFI	140@6400	125@4200	3.11x3.60	10.0:1	4.3
	Corolla	1.8 (1794)	1ZZ-FE	EFI	120@5600	122@4400	3.11x3.60	10.0:1	4.3
	Celica	1.8 (1794)	1ZZ-FE	EFI	140@6400	125@4200	3.11x3.60	10.0:1	4.3
	Celica GT-S	1.8 (1796)	2ZZ-GE	EFI	180@7600	130@6800	3.23x3.35	11.5:1	4.3
	Echo	1.5 (1496)	1NZ-FE	EFI	108@6000	105@4000	2.95x3.33	10.5:1	4.3

EFI: Electronic Fuel Injection

① Twin Turbo Charged

ENGINE TUNE-UP SPECIFICATIONS

Year	Engine Displacement Liters (cc)	Engine ID/VIN	Spark Plug Gap (in.)	Ignition Timing (deg.) ①	Fuel Pump (psi)	Idle Speed (rpm)		Valve Clearance	
						MT	AT	In.	Ex.
2000	1.8 (1794)	1ZZ-FE	0.043	10-18 BTDC	44-50	650-750	700-800	0.006-0.010	0.010-0.014
	1.8 (1796)	2ZZ-GE	0.043	8-12 BTDC	44-50	750-850	700-800	0.006-0.010	0.014-0.018
	2.2 (2164)	5S-FE	0.043	8-12 BTDC	44-50	700-800	700-800	0.007-0.011	0.011-0.015
	1.5 (1496)	1NZ-FE	0.043	8-12 BTDC	44-50	700-800	700-800	0.006-0.010	0.011-0.014
	3.0 (2952)	1MZ-FE	0.043	8-12 BTDC	44-50	650-750	650-750	0.006-0.010	0.010-0.014
2001	1.8 (1794)	1ZZ-FE	0.043	10-18 BTDC	44-50	650-750	700-800	0.006-0.010	0.010-0.014
	1.8 (1796)	2ZZ-GE	0.043	8-12 BTDC	44-50	750-850	700-800	0.006-0.010	0.014-0.018
	2.2 (2164)	5S-FE	0.043	8-12 BTDC	44-50	700-800	700-800	0.007-0.011	0.011-0.015
	1.5 (1496)	1NZ-FE	0.043	8-12 BTDC	44-50	700-800	700-800	0.006-0.010	0.011-0.014
	3.0 (2952)	1MZ-FE	0.043	8-12 BTDC	44-50	650-750	650-750	0.006-0.010	0.010-0.014
2002	1.8 (1794)	1ZZ-FE	0.043	8-12 BTDC	44-50	650-750	700-800	0.006-0.010	0.010-0.014
	1.8 (1796)	2ZZ-GE	0.043	8-12 BTDC	44-50	750-850	700-800	0.006-0.010	.018
	2.4 (2398)	2AZ-FE	0.043	8-12 BTDC	44-50	700-800	700-800	0.007-0.011	0.011-0.015
	2.2 (2164)	5S-FE	0.043	8-12 BTDC	44-50	700-800	700-800	0.007-0.011	0.011-0.015
	1.5 (1496)	1NZ-FE	0.043	8-12 BTDC	44-50	700-800	700-800	0.006-0.010	0.011-0.014
	3.0 (2952)	1MZ-FE	0.043	8-12 BTDC	44-50	650-750	650-750	0.006-0.010	0.010-0.014
2003	1.8 (1794)	1ZZ-FE	0.043	8-12 BTDC	44-50	650-750	700-800	0.006-0.010	0.010-0.014
	1.8 (1796)	2ZZ-GE	0.043	8-12 BTDC	44-50	750-850	700-800	0.006-0.010	0.014-0.018
	2.4 (2398)	2AZ-FE	0.043	8-12 BTDC	44-50	700-800	700-800	0.007-0.011	0.011-0.015
	2.2 (2164)	5S-FE	0.043	8-12 BTDC	44-50	700-800	700-800	0.007-0.011	0.011-0.015
	1.5 (1496)	1NZ-FE	0.043	8-12 BTDC	44-50	700-800	700-800	0.006-0.010	0.011-0.014
	3.0 (2952)	1MZ-FE	0.043	8-12 BTDC	44-50	650-750	650-750	0.006-0.010	0.010-0.014

Note: The Vehicle Emission Control Information label often reflects specification changes made during production. The label figures must be used if they differ from those in this chart.

① With terminal TE1 and E1 connected of DLC1

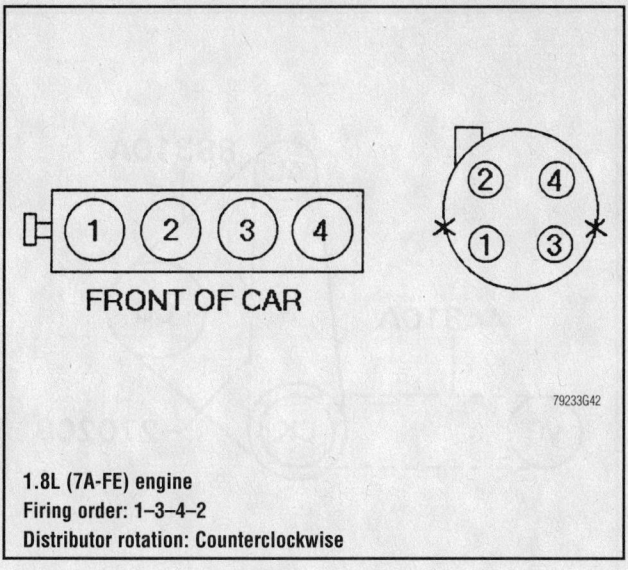

1.8L (7A-FE) engine
Firing order: 1–3–4–2
Distributor rotation: Counterclockwise

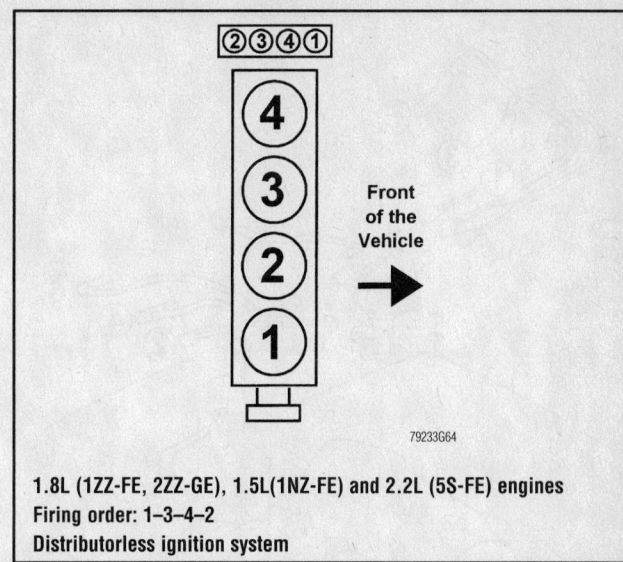

1.8L (1ZZ-FE, 2ZZ-GE), 1.5L(1NZ-FE) and 2.2L (5S-FE) engines
Firing order: 1–3–4–2
Distributorless ignition system

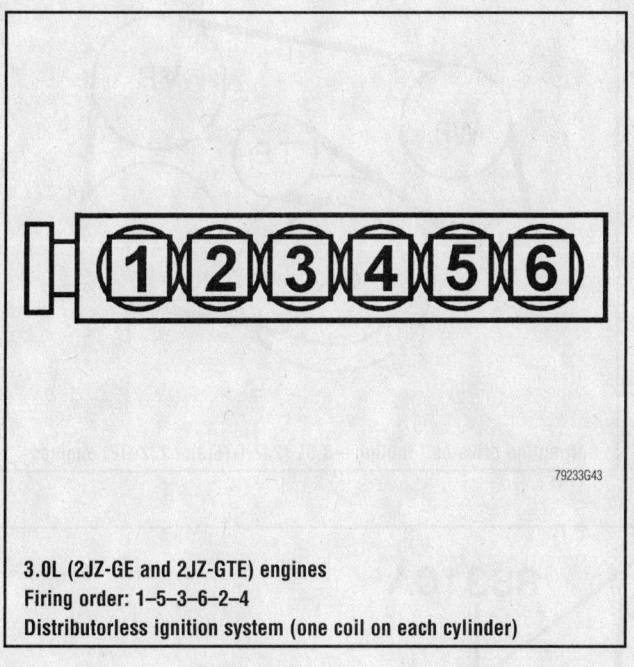

3.0L (2JZ-GE and 2JZ-GTE) engines
Firing order: 1–5–3–6–2–4
Distributorless ignition system (one coil on each cylinder)

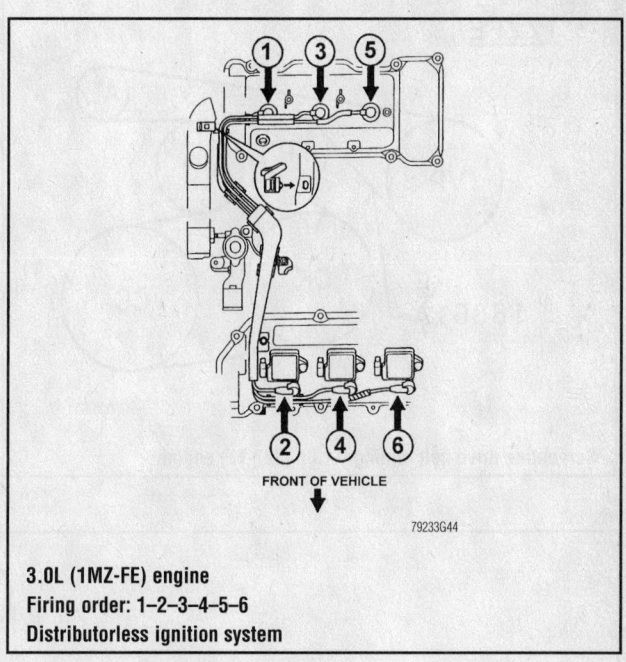

3.0L (1MZ-FE) engine
Firing order: 1–2–3–4–5–6
Distributorless ignition system

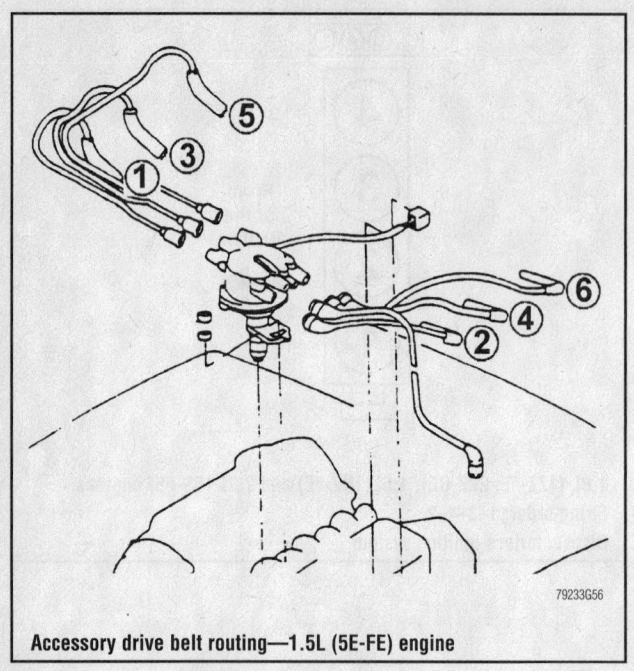

Accessory drive belt routing—1.5L (5E-FE) engine

79233G56

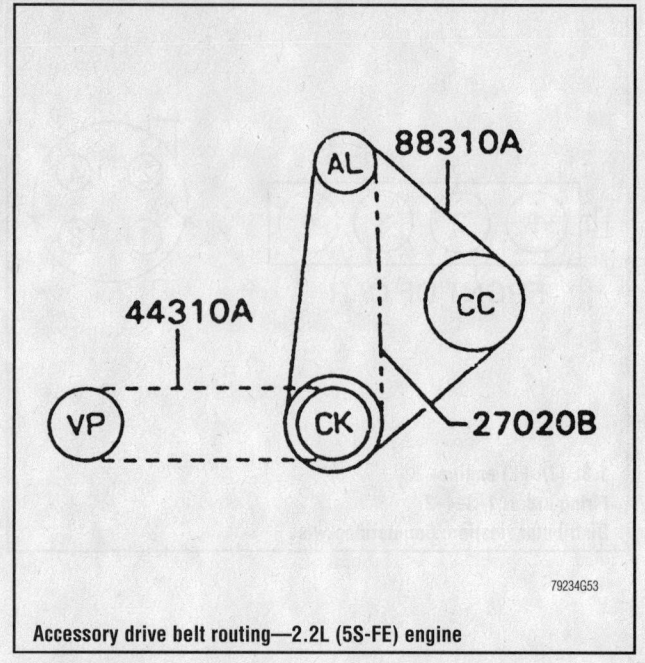

Accessory drive belt routing—2.2L (5S-FE) engine

79234G53

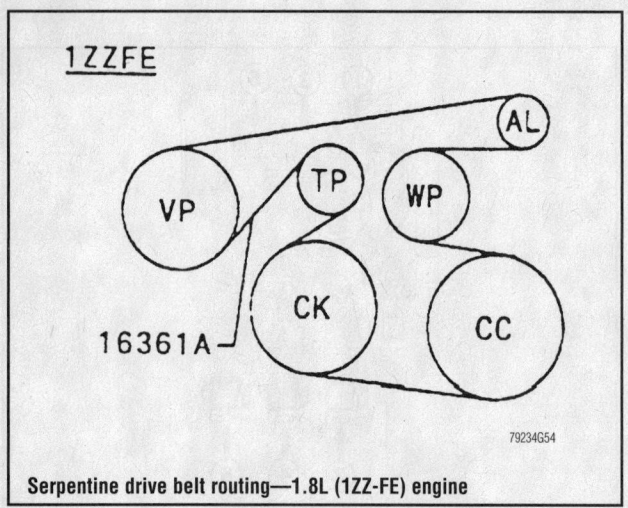

Serpentine drive belt routing—1.8L (1ZZ-FE) engine

79234G54

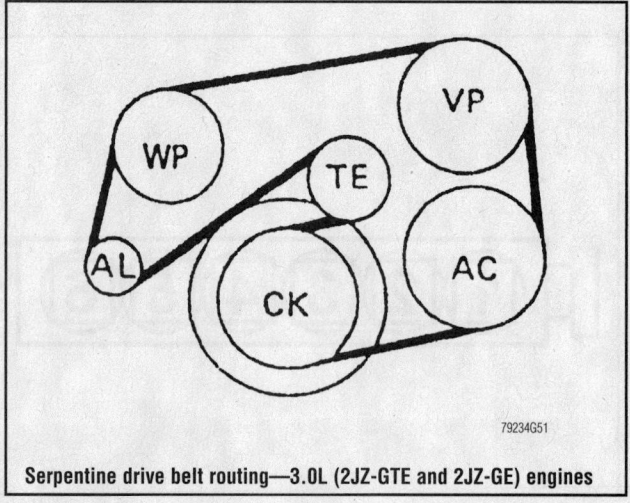

Serpentine drive belt routing—3.0L (2JZ-GTE and 2JZ-GE) engines

79234G51

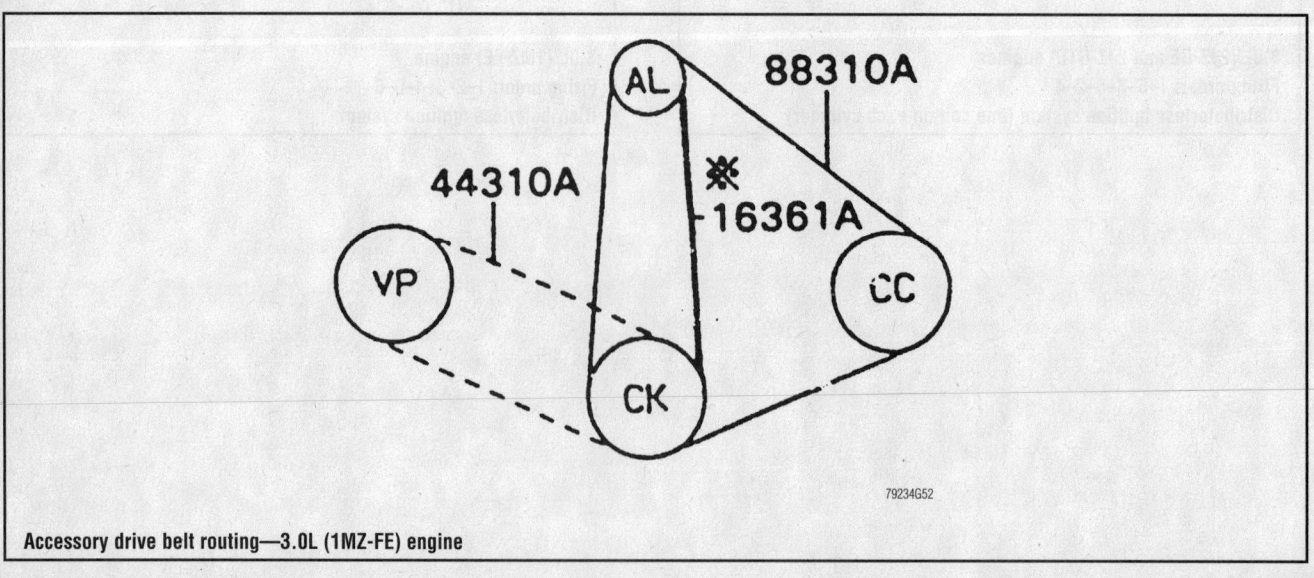

Accessory drive belt routing—3.0L (1MZ-FE) engine

79234G52

CAPACITIES

Year	Model	Engine Displacement Liters (cc)	Engine ID/VIN	Engine Oil with Filter	Transmission (pts.) 4-Spd	5-Spd	Auto.	Drive Axle Front (pts.)	Rear (pts.)	Fuel Tank (gal.)	Cooling System (qts.)
2000	Avalon	3.0 (2995)	1MZ-FE	5.0	—	—	7.4	1.8	—	18.5	9.8
	Camry	2.2 (2164)	5S-FE	3.8	—	4.6	5.2	3.4	—	18.5	7.3
	Camry	3.0 (2995)	1MZ-FE	5.0	—	9.8	7.4	1.8	—	18.5	9.6
	Camry Solara	2.2 (2164)	5S-FE	3.8	—	4.6	5.2	3.4	—	18.5	7.3
	Camry Solara	3.0 (2995)	1MZ-FE	5.0	—	9.8	7.4	1.8	—	18.5	9.6
	MR 2	1.8 (1794)	1ZZ-FE	3.9	—	2.0	—	—	—	12.7	11.0
	Celica	1.8 (1794)	1ZZ-FE	③	—	4.0	8.6	3.0	—	13.2	6.0
	Celica GT-S	1.8 (1796)	2ZZ-GE	④	—	4.8	6.1	⑤	—	14.5	6.0
	Corolla	1.8 (1794)	1ZZ-FE	3.2	—	4.0	5.2	3.0	—	13.2	②
	Echo	1.5 (1496)	1NZ-FE	4.0	—	4.0	6.2	⑤	—	11.9	⑥
2001	Avalon	3.0 (2995)	1MZ-FE	5.0	—	—	7.4	1.8	—	18.5	9.8
	Camry	2.2 (2164)	5S-FE	3.8	—	4.6	5.2	3.4	—	18.5	7.3
	Camry	3.0 (2995)	1MZ-FE	5.0	—	9.8	7.4	1.8	—	18.5	9.6
	Camry Solara	2.2 (2164)	5S-FE	3.8	—	4.6	5.2	3.4	—	18.5	7.3
	Camry Solara	3.0 (2995)	1MZ-FE	5.0	—	9.8	7.4	1.8	—	18.5	9.6
	MR 2	1.8 (1794)	1ZZ-FE	3.9	—	2.0	—	—	—	12.7	11.0
	Celica	1.8 (1794)	1ZZ-FE	③	—	4.0	8.6	3.0	—	13.2	6.0
	Celica GT-S	1.8 (1796)	2ZZ-GE	④	—	4.8	6.1	⑤	—	14.5	6.0
	Corolla	1.8 (1794)	1ZZ-FE	3.2	—	4.0	5.2	3.0	—	13.2	②
	Echo	1.5 (1496)	1NZ-FE	4.0	—	4.0	6.2	⑤	—	11.9	⑥
2002	Avalon	3.0 (2995)	1MZ-FE	5.0	—	—	7.4	1.8	—	18.5	9.8
	Camry	2.4 (2398)	2AZ-FE	3.8	—	4.6	5.2	3.4	—	18.5	7.3
	Camry	3.0 (2995)	1MZ-FE	5.0	—	9.8	7.4	1.8	—	18.5	9.6
	Camry Solara	2.4 (2398)	2AZ-FE	3.8	—	4.6	5.2	3.4	—	18.5	7.3
	Camry Solara	3.0 (2995)	1MZ-FE	5.0	—	9.8	7.4	1.8	—	18.5	9.6
	MR 2	1.8 (1794)	1ZZ-FE	3.9	—	2.0	—	—	—	12.7	11.0
	Celica	1.8 (1794)	1ZZ-FE	③	—	4.0	8.6	3.0	—	13.2	6.0
	Celica GT-S	1.8 (1796	2ZZ-GE	④	—	4.8	6.1	⑤	—	14.5	6.0
	Corolla	1.8 (1794)	1ZZ-FE	3.2	—	4.0	5.2	3.0	—	13.2	②
	Echo	1.5 (1496)	1NZ-FE	4.0	—	4.0	6.2	⑤	—	11.9	⑥
2003	Avalon	3.0 (2995)	1MZ-FE	5.0	—	—	7.4	1.8	—	18.5	9.8
	Camry	2.4 (2398)	2AZ-FE	3.8	—	4.6	5.2	3.4	—	18.5	7.3
	Camry	3.0 (2995)	1MZ-FE	5.0	—	9.8	7.4	1.8	—	18.5	9.6
	Camry Solara	2.4 (2398)	2AZ-FE	3.8	—	4.6	5.2	3.4	—	18.5	7.3
	Camry Solara	3.0 (2995)	1MZ-FE	5.0	—	9.8	7.4	1.8	—	18.5	9.6
	MR 2	1.8 (1794)	1ZZ-FE	3.9	—	2.0	—	—	—	12.7	11.0
	Celica	1.8 (1794)	1ZZ-FE	③	—	4.0	8.6	3.0	—	13.2	6.0
	Celica GT-S	1.8 (1796	2ZZ-GE	④	—	4.8	6.1	⑤	—	14.5	6.0
	Corolla	1.8 (1794)	1ZZ-FE	3.2	—	4.0	5.2	3.0	—	13.2	②
	Echo	1.5 (1496)	1NZ-FE	4.0	—	4.0	6.2	⑤	—	11.9	⑥

Note: All capacities are approximate. Add fluid gradually and check to be sure a proper fluid level is obtained.

① Manual trans.: 6.4
  Automatic trans.: 7.0
② M/T with Nippodenso radiator: 5.6
  A/T with Nippodenso radiator: 6.2
  M/T with Harrison radiator: 6.3
  A/T with Harrison radiator: 6.2
③ w/oil cooler: 4.0
  wo/oil cooler: 3.7

④ w/oil cooler: 4.8
  wo/oil cooler: 4.6
⑤ Included in transaxle capacity
⑥ w/MT: 4.7
  w/AT: 4.5

VALVE SPECIFICATIONS

Year	Engine Displacement Liters (cc)	Engine ID/VIN	Seat Angle (deg.)	Face Angle (deg.)	Spring Test Pressure (lbs. @ in.)	Spring Installed Height (in.)	Stem-to-Guide Clearance (in.)		Stem Diameter (in.)	
							Intake	Exhaust	Intake	Exhaust
2000	1.8 (1794)	1ZZ-FE	45	44.5	31.3-34.8@ 1.252	1.323	0.0010-0.0024	0.0012-0.0025	0.2154-0.2352	0.2152-0.2350
	1.8 (1796)	2ZZ-GE	45	44.5	①	1.516	0.0010-0.0023	0.0012-0.0025	0.2150-0.2156	0.2144-0.2154
	2.2 (2164)	5S-FE	45	44.5	36.8-42.5@ 1.366	1.366	0.0010-0.0024	0.0012-0.0026	0.2350-0.2356	0.2348-0.2354
	1.5 (1496)	1NZ-FE	45	44.5	33.5-37@ 1.280	1.280	0.0010-0.0024	0.0012-0.0026	0.1957-0.1963	0.1955-0.1961
	3.0 (2995)	1MZ-FE	45	44.5	41.9-46.3@ 1.331	1.331	0.0010-0.0024	0.0012-0.0026	0.2154-0.2159	0.2152-0.2157
2001	1.8 (1794)	1ZZ-FE	45	44.5	31.3-34.8@ 1.252	1.323	0.0010-0.0024	0.0012-0.0025	0.2154-0.2352	0.2152-0.2350
	1.8 (1796)	2ZZ-GE	45	44.5	①	1.516	0.0010-0.0023	0.0012-0.0025	0.2150-0.2156	0.2144-0.2154
	2.2 (2164)	5S-FE	45	44.5	36.8-42.5@ 1.366	1.366	0.0010-0.0024	0.0012-0.0026	0.2350-0.2356	0.2348-0.2354
	1.5 (1496)	1NZ-FE	45	44.5	33.5-37@ 1.280	1.280	0.0010-0.0024	0.0012-0.0026	0.1957-0.1963	0.1955-0.1961
	3.0 (2995)	1MZ-FE	45	44.5	41.9-46.3@ 1.331	1.331	0.0010-0.0024	0.0012-0.0026	0.2154-0.2159	0.2152-0.2157
2002	1.8 (1794)	1ZZ-FE	45	44.5	31.3-34.8@ 1.252	1.323	0.0010-0.0024	0.0012-0.0025	0.2154-0.2352	0.2152-0.2350
	1.8 (1796)	2ZZ-GE	45	44.5	①	1.516	0.0010-0.0023	0.0012-0.0025	0.2150-0.2156	0.2144-0.2154
	2.4 (2398)	2AZ-FE	45	45	②	—	0.0010-0.0024	0.0012-0.0026	0.2154-0.2159	0.2152-0.2157
	2.2 (2164)	5S-FE	45	44.5	36.8-42.5@ 1.366	1.366	0.0010-0.0024	0.0012-0.0026	0.2350-0.2356	0.2348-0.2354
	1.5 (1496)	1NZ-FE	45	44.5	33.5-37@ 1.280	1.280	0.0010-0.0024	0.0012-0.0026	0.1957-0.1963	0.1955-0.1961
	3.0 (2995)	1MZ-FE	45	44.5	41.9-46.3@ 1.331	1.331	0.0010-0.0024	0.0012-0.0026	0.2154-0.2159	0.2152-0.2157
2003	1.8 (1794)	1ZZ-FE	45	44.5	31.3-34.8@ 1.252	1.323	0.0010-0.0024	0.0012-0.0025	0.2154-0.2352	0.2152-0.2350
	1.8 (1796)	2ZZ-GE	45	44.5	①	1.516	0.0010-0.0023	0.0012-0.0025	0.2150-0.2156	0.2144-0.2154
	2.4 (2398)	2AZ-FE	45	45	②	—	0.0010-0.0024	0.0012-0.0026	0.2154-0.2159	0.2152-0.2157
	2.2 (2164)	5S-FE	45	44.5	36.8-42.5@ 1.366	1.366	0.0010-0.0024	0.0012-0.0026	0.2350-0.2356	0.2348-0.2354
	1.5 (1496)	1NZ-FE	45	44.5	33.5-37@ 1.280	1.280	0.0010-0.0024	0.0012-0.0026	0.1957-0.1963	0.1955-0.1961
	3.0 (2995)	1MZ-FE	45	44.5	41.9-46.3@ 1.331	1.331	0.0010-0.0024	0.0012-0.0026	0.2154-0.2159	0.2152-0.2157

① Intake: 49.6-55.5@1.516

   Exhaust: 47.6-52.6@1.516

② Inner spring free length: 1.799 in. (45.7 mm)

TORQUE SPECIFICATIONS
All readings in ft. lbs.

Year	Engine Displacement Liters (cc)	Engine ID/VIN	Cylinder Head Bolts	Main Bearing Bolts	Rod Bearing Bolts	Crankshaft Damper Bolts	Flywheel Bolts	Manifold Intake	Manifold Exhaust	Spark Plugs	Lug Nuts
2000	1.8 (1794)	1ZZ-FE	①	⑦	③	102	①	14	27	13	76
	1.8 (1796)	2ZZ-GE	⑧	⑦	⑨	87	①	⑩	37	13	76
	2.2 (2164)	5S-FE	①	43	④	80	③	14	36	13	76
	1.5 (1496)	1NZ-FE	⑪	⑫	⑬	94	①	22	20	13	76
	3.0 (2995)	1MZ-FE	⑤	⑥	④	159	61	11	36	13	76
2001	1.8 (1794)	1ZZ-FE	①	⑦	③	102	①	14	27	13	76
	1.8 (1796)	2ZZ-GE	⑧	⑦	⑨	87	①	⑩	37	13	76
	2.2 (2164)	5S-FE	①	43	④	80	③	14	36	13	76
	1.5 (1496)	1NZ-FE	⑪	⑫	⑬	94	①	22	20	13	76
	3.0 (2995)	1MZ-FE	⑤	⑥	④	159	61	11	36	13	76
2002	1.8 (1794)	1ZZ-FE	①	⑦	③	102	①	14	27	13	76
	1.8 (1796)	2ZZ-GE	⑧	⑦	⑨	87	①	⑩	37	13	76
	2.4 (2398)	2AZ-FE	⑭	⑮	④	125	⑯	22	27	13	76
	1.5 (1496)	1NZ-FE	⑪	⑫	⑬	94	①	22	20	13	76
	3.0 (2995)	1MZ-FE	⑤	⑥	④	159	61	11	36	13	76
2003	1.8 (1794)	1ZZ-FE	①	⑦	③	102	①	14	27	13	76
	1.8 (1796)	2ZZ-GE	⑧	⑦	⑨	87	①	⑩	37	13	76
	2.4 (2398)	2AZ-FE	⑭	⑮	④	125	⑯	22	27	13	76
	1.5 (1496)	1NZ-FE	⑪	⑫	⑬	94	①	22	20	13	76
	3.0 (2995)	1MZ-FE	⑤	⑥	④	159	61	11	36	13	76

① Step 1: 36 ft. lbs.
Step 2: 90 degree turn

② Step 1: 16 ft. lbs.
Step 2: 32 ft. lbs.
Steps 3 and 4: 45 degree turn

③ Step 1: 15 ft. lbs.
Step 2: 90 degree turn

④ Step 1: 18 ft. lbs.
Step 2: 90 degree turn

⑤ Step 1: 40 ft. lbs.
Step 2: 90 degree turn

⑥ 12 pt.: 16 ft. lbs. + 90 degrees
6 pt.: 20 ft. lbs.

⑦ 12 pointed bolts:
Step 1: 16 ft. lbs.
Step 2: 32 ft. lbs.
Step 3: 45 degree turn
Step 4: 45 degree turn
Hex head bolts: 14 ft. lbs.

⑧ Step 1: 26 ft. lbs.
Step 2: 180 degree turn

⑨ Step 1: 22 ft. lbs.
Step 2: 90 degree turn
Recessed head bolt: 13 ft. lbs.

⑩ 4 upper bolts and 1 nut: 20 ft. lbs.
1 lower bolt: 34 ft. lbs.

⑪ Step 1: 22 ft. lbs.
Step 2: 90 degree turn
Step 3: 90 degree turn

⑫ Step 1: 16 ft. lbs.
Step 2: 90 degree turn

⑬ Step 1: 11 ft. lbs.
Step 2: 90 degree turn

⑭ Step 1: Several passes in sequence to 58 ft. lbs.
Step 2: Plus 90 degrees

⑮ Step 1: 15 ft. lbs.
Step 2: 29 ft. lbs.
Step 3: Plus 90 degrees

⑯ Manual Transmission: 96 ft. lbs.
Automatic Transmission: 72 ft. lbs.

42356-TOYC-C06

PISTON AND RING SPECIFICATIONS
All measurements are given in inches.

Year	Engine Displacement Liters (cc)	Engine ID/VIN	Piston Clearance	Ring Gap			Ring Side Clearance		
				Top Compression	Bottom Compression	Oil Control	Top Compression	Bottom Compression	Oil Control
2000	1.8 (1762)	1ZZ-FE	0.0033-0.0041	0.0098-0.0138	0.0138-0.0197	0.0039-0.0157	0.0012-0.0028	0.0012-0.0028	SNUG
	1.8 (1796)	2ZZ-GE	0.0003-0.0015	0.0098-0.0138	0.0138-0.0197	0.0059-0.0157	0.0012-0.0028	0.0012-0.0028	SNUG
	2.2 (2164)	5S-FE	0.0055-0.0063	0.0106-0.0197	0.0138-0.0236	0.0079-0.0217	0.0016-0.0031	0.0012-0.0028	SNUG
	1.5 (1496)	1NZ-FE	0.0022-0.0023	0.0098-0.0138	0.0138-0.0197	0.0039-0.0138	0.0012-0.0028	0.0012-0.0028	SNUG
	3.0 (2995)	1MZ-FE	0.0033-0.0042	0.0098-0.0138	0.0138-0.0177	0.0059-0.0157	0.0008-0.0028	0.0008-0.0024	SNUG
2001	1.8 (1762)	1ZZ-FE	0.0033-0.0041	0.0098-0.0138	0.0138-0.0197	0.0039-0.0157	0.0012-0.0028	0.0012-0.0028	SNUG
	1.8 (1796)	2ZZ-GE	0.0003-0.0015	0.0098-0.0138	0.0138-0.0197	0.0059-0.0157	0.0012-0.0028	0.0012-0.0028	SNUG
	2.2 (2164)	5S-FE	0.0055-0.0063	0.0106-0.0197	0.0138-0.0236	0.0079-0.0217	0.0016-0.0031	0.0012-0.0028	SNUG
	1.5 (1496)	1NZ-FE	0.0022-0.0023	0.0098-0.0138	0.0138-0.0197	0.0039-0.0138	0.0012-0.0028	0.0012-0.0028	SNUG
	3.0 (2995)	1MZ-FE	0.0033-0.0042	0.0098-0.0138	0.0138-0.0177	0.0059-0.0157	0.0008-0.0028	0.0008-0.0024	SNUG
2002	1.8 (1762)	1ZZ-FE	0.0033-0.0041	0.0098-0.0138	0.0138-0.0197	0.0039-0.0157	0.0012-0.0028	0.0012-0.0028	SNUG
	1.8 (1796)	2ZZ-GE	0.0003-0.0015	0.0098-0.0138	0.0138-0.0197	0.0059-0.0157	0.0012-0.0028	0.0012-0.0028	SNUG
	2.4 (2398)	2AZ-FE	0.0020-0.0029	0.0087-0.0126	0.0197-0.0236	0.0039-0.0138	0.0012-0.0028	0.0012-0.0028	SNUG
	1.5 (1496)	1NZ-FE	0.0022-0.0023	0.0098-0.0138	0.0138-0.0197	0.0039-0.0138	0.0012-0.0028	0.0012-0.0028	SNUG
	3.0 (2995)	1MZ-FE	0.0033-0.0042	0.0098-0.0138	0.0138-0.0177	0.0059-0.0157	0.0008-0.0028	0.0008-0.0024	SNUG
2003	1.8 (1762)	1ZZ-FE	0.0033-0.0041	0.0098-0.0138	0.0138-0.0197	0.0039-0.0157	0.0012-0.0028	0.0012-0.0028	SNUG
	1.8 (1796)	2ZZ-GE	0.0003-0.0015	0.0098-0.0138	0.0138-0.0197	0.0059-0.0157	0.0012-0.0028	0.0012-0.0028	SNUG
	2.4 (2398)	2AZ-FE	0.0020-0.0029	0.0087-0.0126	0.0197-0.0236	0.0039-0.0138	0.0012-0.0028	0.0012-0.0028	SNUG
	1.5 (1496)	1NZ-FE	0.0022-0.0023	0.0098-0.0138	0.0138-0.0197	0.0039-0.0138	0.0012-0.0028	0.0012-0.0028	SNUG
	3.0 (2995)	1MZ-FE	0.0033-0.0042	0.0098-0.0138	0.0138-0.0177	0.0059-0.0157	0.0008-0.0028	0.0008-0.0024	SNUG

42356-TOYC-C07

CRANKSHAFT AND CONNECTING ROD SPECIFICATIONS

All measurements are given in inches.

Year	Engine Displacement Liters (cc)	Engine ID/VIN	Crankshaft Main Brg. Journal Dia.	Crankshaft Main Brg. Oil Clearance	Crankshaft Shaft End-play	Thrust on No.	Connecting Rod Journal Diameter	Connecting Rod Oil Clearance	Connecting Rod Side Clearance
2000	1.5 (1496)	1NZ-FE	①	0.0004-0.0009	0.0035-0.0075	3	1.5745-1.5748	0.0006-0.0016	0.0063-0.0142
	1.8 (1762)	1ZZ-FE	②	0.0006-0.0013	0.0008-0.0087	3	1.7320-1.7323	0.0011-0.0024	0.0063-0.0135
	1.8 (1796)	2ZZ-GE	②	0.0006-0.0013	0.0016-0.0094	3	1.7713-1.7717	0.0011-0.0020	0.0063-0.0135
	2.2 (2164)	5S-FE	2.1653-2.6550	0.0010-0.0017	0.0008-0.0087	3	2.0466-2.0472	0.0009-0.0022	0.0063-0.0123
	3.0 (2995)	1MZ-FE	2.4011-2.4016	0.0010-0.0018	0.0016-0.0095	2	2.0863-2.8660	0.0015-0.0025	0.0059-0.0118
2001	1.5 (1496)	1NZ-FE	①	0.0004-0.0009	0.0035-0.0075	3	1.5745-1.5748	0.0006-0.0016	0.0063-0.0142
	1.8 (1762)	1ZZ-FE	②	0.0006-0.0013	0.0008-0.0087	3	1.7320-1.7323	0.0011-0.0024	0.0063-0.0135
	1.8 (1796)	2ZZ-GE	②	0.0006-0.0013	0.0016-0.0094	3	1.7713-1.7717	0.0011-0.0020	0.0063-0.0135
	2.2 (2164)	5S-FE	2.1653-2.6550	0.0010-0.0017	0.0008-0.0087	3	2.0466-2.0472	0.0009-0.0022	0.0063-0.0123
	3.0 (2995)	1MZ-FE	2.4011-2.4016	0.0010-0.0018	0.0016-0.0095	2	2.0863-2.8660	0.0015-0.0025	0.0059-0.0118
2002	1.5 (1496)	1NZ-FE	①	0.0004-0.0009	0.0035-0.0075	3	1.5745-1.5748	0.0006-0.0016	0.0063-0.0142
	1.8 (1762)	1ZZ-FE	②	0.0006-0.0013	0.0008-0.0087	3	1.7320-1.7323	0.0011-0.0024	0.0063-0.0135
	1.8 (1796)	2ZZ-GE	②	0.0006-0.0013	0.0016-0.0094	3	1.7713-1.7717	0.0011-0.0020	0.0063-0.0135
	2.4 (2398)	2AZ-FE	2.1648-2.1654	0.0007-0.0016	0.0063-0.0143	3	1.8894-1.8898	0.0009-0.0022	0.0063-0.0123
	3.0 (2995)	1MZ-FE	2.4011-2.4016	0.0010-0.0018	0.0016-0.0095	2	2.0863-2.8660	0.0015-0.0025	0.0059-0.0118
2002	1.5 (1496)	1NZ-FE	①	0.0004-0.0009	0.0035-0.0075	3	1.5745-1.5748	0.0006-0.0016	0.0063-0.0142
	1.8 (1762)	1ZZ-FE	②	0.0006-0.0013	0.0008-0.0087	3	1.7320-1.7323	0.0011-0.0024	0.0063-0.0135
	1.8 (1796)	2ZZ-GE	②	0.0006-0.0013	0.0016-0.0094	3	1.7713-1.7717	0.0011-0.0020	0.0063-0.0135
	2.4 (2398)	2AZ-FE	2.1648-2.1654	0.0007-0.0016	0.0063-0.0143	3	1.8894-1.8898	0.0009-0.0022	0.0063-0.0123
	3.0 (2995)	1MZ-FE	2.4011-2.4016	0.0010-0.0018	0.0016-0.0095	2	2.0863-2.8660	0.0015-0.0025	0.0059-0.0118

① Reference mark:
0: 1.81102-1.81110
1: 1.81110-1.81118
2: 1.81118-1.81126
3: 1.81126-1.81133
4: 1.81133-1.81141
5: 1.81141-1.81149

② Reference mark:
0: 1.8897-1.8898
1: 1.8896-1.8897
2: 1.8895-1.8896
3: 1.8894-1.8895
4: 1.8893-1.8894
5: 1.8892-1.8893

WHEEL ALIGNMENT

Year	Model		Caster Range (+/-Deg.)	Caster Preferred Setting (Deg.)	Camber Range (+/-Deg.)	Camber Preferred Setting (Deg.)	Toe-in (in.)	Steering Axis Inclination (Deg.)
2000	Avalon	F	0.75	+2.17	0.75	-0.62	0+/-0.08	13.07+/-0.75
		R	—	—	0.75	-0.72	0.16+/-0.08	—
	Camry 4-cyl.	F	0.75	+2.18	0.75	-0.60	0+/-0.08	13.08+/-0.75
		R	—	—	0.75	-0.70	0.16+/-0.08	—
	Camry 6-cyl.	F	0.75	+2.09	0.75	-0.60	0+/-0.08	13.08+/-0.75
		R	—	—	0.75	-0.75	0.16+/-0.08	—
	Celica	F	0.75	+2.05	0.75	-0.77	0+/-0.08	14.97+/-0.75
		R	—	—	0.75	-1.17	0.14+/-0.08	—
	Corolla	F	0.75	+1.19	0.75	-0.18	0.04+/-0.08	12.38+/-0.75
		R	—	—	0.75	-0.92	0.16+/-0.08	—
	Echo	F	0.75	+1.60	0.75	-0.58	0+/-0.08	10.08+/-0.75
		R	—	—	0.75	-0.93	0.11+/-0.12	—
	MR 2	F	0.75	3.08	0.75	-0.47	0.06+/-0.08	14.52+/-0.75
		R	—	—	0.75	-1.05	0.12+/-0.08	—
	Solara	F	0.75	+2.08	0.75	-0.52	0+/-0.08	12.09+/-0.75
		R	—	—	0.75	-0.65	0.16+/-0.08	—
2001	Avalon	F	0.75	+2.17	0.75	-0.62	0+/-0.08	13.07+/-0.75
		R	—	—	0.75	-0.72	0.16+/-0.08	—
	Camry 4-cyl.	F	0.75	+2.18	0.75	-0.60	0+/-0.08	13.08+/-0.75
		R	—	—	0.75	-0.70	0.16+/-0.08	—
	Camry 6-cyl.	F	0.75	+2.09	0.75	-0.60	0+/-0.08	13.08+/-0.75
		R	—	—	0.75	-0.75	0.16+/-0.08	—
	Celica	F	0.75	+2.05	0.75	-0.77	0+/-0.08	14.97+/-0.75
		R	—	—	0.75	-1.17	0.14+/-0.08	—
	Corolla	F	0.75	+1.19	0.75	-0.18	0.04+/-0.08	12.38+/-0.75
		R	—	—	0.75	-0.92	0.16+/-0.08	—
	Echo	F	0.75	+1.60	0.75	-0.58	0+/-0.08	10.08+/-0.75
		R	—	—	0.75	-0.93	0.11+/-0.12	—
	MR 2	F	0.75	3.08	0.75	-0.47	0.06+/-0.08	14.52+/-0.75
		R	—	—	0.75	-1.05	0.12+/-0.08	—
	Solara	F	0.75	+2.08	0.75	-0.52	0+/-0.08	12.09+/-0.75
		R	—	—	0.75	-0.65	0.16+/-0.08	—
2002	Avalon	F	0.75	+2.17	0.75	-0.62	0+/-0.08	13.07+/-0.75
		R	—	—	0.75	-0.72	0.16+/-0.08	—
	Camry 4-cyl.	F	0.75	+2.18	0.75	-0.60	0+/-0.08	13.08+/-0.75
		R	—	—	0.75	-0.70	0.16+/-0.08	—
	Camry 6-cyl.	F	0.75	+2.09	0.75	-0.60	0+/-0.08	13.08+/-0.75
		R	—	—	0.75	-0.75	0.16+/-0.08	—
	Celica	F	0.75	+2.05	0.75	-0.77	0+/-0.08	14.97+/-0.75
		R	—	—	0.75	-1.17	0.14+/-0.08	—
	Corolla	F	0.75	+1.19	0.75	-0.18	0.04+/-0.08	12.38+/-0.75
		R	—	—	0.75	-0.92	0.16+/-0.08	—
	Echo	F	0.75	+1.60	0.75	-0.58	0+/-0.08	10.08+/-0.75
		R	—	—	0.75	-0.93	0.11+/-0.12	—
	MR 2	F	0.75	3.08	0.75	-0.47	0.06+/-0.08	14.52+/-0.75
		R	—	—	0.75	-1.05	0.12+/-0.08	—
	Solara	F	0.75	+2.08	0.75	-0.52	0+/-0.08	12.09+/-0.75
		R	—	—	0.75	-0.65	0.16+/-0.08	—

WHEEL ALIGNMENT

Year	Model		Caster Range (+/-Deg.)	Caster Preferred Setting (Deg.)	Camber Range (+/-Deg.)	Camber Preferred Setting (Deg.)	Toe-in (in.)	Steering Axis Inclination (Deg.)
2003	Avalon	F	0.75	+2.17	0.75	-0.62	0+/-0.08	13.07+/-0.75
		R	—	—	0.75	-0.72	0.16+/-0.08	—
	Camry 4-cyl.	F	0.75	+2.18	0.75	-0.60	0+/-0.08	13.08+/-0.75
		R	—	—	0.75	-0.70	0.16+/-0.08	—
	Camry 6-cyl.	F	0.75	+2.09	0.75	-0.60	0+/-0.08	13.08+/-0.75
		R	—	—	0.75	-0.75	0.16+/-0.08	—
	Celica	F	0.75	+2.05	0.75	-0.77	0+/-0.08	14.97+/-0.75
		R	—	—	0.75	-1.17	0.14+/-0.08	—
	Corolla	F	0.75	+1.19	0.75	-0.18	0.04+/-0.08	12.38+/-0.75
		R	—	—	0.75	-0.92	0.16+/-0.08	—
	Echo	F	0.75	+1.60	0.75	-0.58	0+/-0.08	10.08+/-0.75
		R	—	—	0.75	-0.93	0.11+/-0.12	—
	MR 2	F	0.75	3.08	0.75	-0.47	0.06+/-0.08	14.52+/-0.75
		R	—	—	0.75	-1.05	0.12+/-0.08	—
	Solara	F	0.75	+2.08	0.75	-0.52	0+/-0.08	12.09+/-0.75
		R	—	—	0.75	-0.65	0.16+/-0.08	—

42356-TOYC-C10

TIRE, WHEEL AND BALL JOINT SPECIFICATIONS

Year	Model	OEM Tires		Tire Pressures (psi)		Wheel Size	Ball Joint Inspection
		Standard	Optional	Front	Rear		
2000	Avalon	P205/65HR15	None	32	32	6-JJ	9-30 in. ①
	Camry 4-cyl.	P195/70SR14	None	30	30	5.5-JJ	9-26 in. ①
	Camry 6-cyl.	P205/65HR15	None	32	35	6-JJ	9-26 in. ①
	Celica	205/55VR15	P205/55VR15	33	33	6.5-JJ	9-26 in. ①
	Corolla VE, CE	P175/65R14	None	30	30	5.5-JJ	9-26 in. ①
	Corolla LE	P185/65R14	None	30	30	5.5-JJ	9-26 in. ①
	Echo	P175/65R14	None	32	32	5.5-JJ	9-26 in. ①
	MR 2 Spider	Fr: 205/50R16 Rr: 215/50R16	None	33	33	6-JJ	9-26 in. ①
2001	Avalon	P205/65HR15	None	32	32	6-JJ	9-30 in. ①
	Camry 4-cyl.	P195/70SR14	None	30	30	5.5-JJ	9-26 in. ①
	Camry 6-cyl.	P205/65HR15	None	32	35	6-JJ	9-26 in. ①
	Celica	205/55VR15	P205/55VR15	33	33	6.5-JJ	9-26 in. ①
	Corolla VE, CE	P175/65R14	None	30	30	5.5-JJ	9-26 in. ①
	Corolla LE	P185/65R14	None	30	30	5.5-JJ	9-26 in. ①
	Echo	P175/65R14	None	32	32	5.5-JJ	9-26 in. ①
	MR 2 Spider	Fr: 205/50R16 Rr: 215/50R16	None	33	33	6-JJ	9-26 in. ①
2002	Avalon	P205/65HR15	None	32	32	6-JJ	9-30 in. ①
	Camry 4-cyl.	P195/70SR14	None	30	30	5.5-JJ	9-26 in. ①
	Camry 6-cyl.	P205/65HR15	None	32	35	6-JJ	9-26 in. ①
	Celica	205/55VR15	P205/55VR15	33	33	6.5-JJ	9-26 in. ①
	Corolla VE, CE	P175/65R14	None	30	30	5.5-JJ	9-26 in. ①
	Corolla LE	P185/65R14	None	30	30	5.5-JJ	9-26 in. ①
	Echo	P175/65R14	None	32	32	5.5-JJ	9-26 in. ①
	MR 2 Spider	Fr: 205/50R16 Rr: 215/50R16	None	33	33	6-JJ	9-26 in. ①
2003	Avalon	P205/65HR15	None	32	32	6-JJ	9-30 in. ①
	Camry 4-cyl.	P195/70SR14	None	30	30	5.5-JJ	9-26 in. ①
	Camry 6-cyl.	P205/65HR15	None	32	35	6-JJ	9-26 in. ①
	Celica	205/55VR15	P205/55VR15	33	33	6.5-JJ	9-26 in. ①
	Corolla VE, CE	P175/65R14	None	30	30	5.5-JJ	9-26 in. ①
	Corolla LE	P185/65R14	None	30	30	5.5-JJ	9-26 in. ①
	Echo	P175/65R14	None	32	32	5.5-JJ	9-26 in. ①
	MR 2 Spider	Fr: 205/50R16 Rr: 215/50R16	None	33	33	6-JJ	9-26 in. ①

OEM: Original Equipment Manufacturer

PSI: Pounds Per Square Inch

STD: Standard

OPT: Optional

① Torque required in inch lbs. to rotate ball joint when removed from the knuckle

42356-TOYC-C11

BRAKE SPECIFICATIONS
All measurements in inches unless noted

Year	Model		Brake Disc Original Thickness	Brake Disc Minimum Thickness	Brake Disc Maximum Runout	Brake Drum Diameter Original Inside Diameter	Brake Drum Diameter Max. Wear Limit	Brake Drum Diameter Maximum Machine Diameter	Minimum Lining Thickness	Brake Caliper Bracket Bolts (ft. lbs.)	Brake Caliper Mounting Bolts (ft. lbs.)
2000	Avalon	F	1.102	1.024	0.0020	—	—	—	0.039	25	79
		R	0.354	0.315	0.0059	—	—	—	0.039	25	34
	Camry	F	1.102	1.024	0.0020	—	—	—	0.039	25	79
		R	0.394	0.354	0.0059	9.00	—	9.08	0.039	14	20
	Camry Solara	F	1.102	1.024	0.0020	—	—	—	0.039	25	79
		R	0.394	0.354	0.0059	9.00	—	9.08	0.039	14	20
	Celica GT-S	F	0.984	0.906	0.0020	—	—	—	0.039	25	79
		R	0.354	0.295	0.0059	—	—	—	0.039	—	34
	MR 2	F	0.787	0.709	0.0020	—	—	—	0.039	80	25
		R	0.630	0.591	0.0039	—	—	—	0.039	44	15
	Celica	F	1.102	1.024	0.0020	—	—	—	0.039	25	79
		R	0.354	0.314	0.0059	7.87	—	7.91	0.039	—	34
	Corolla		0.866	0.787	0.0020	7.87	—	7.91	0.039	25	65
	Echo		0.709	0.630	0.0020	7.09	—	7.13	0.039	25	65
2001	Avalon	F	1.102	1.024	0.0020	—	—	—	0.039	25	79
		R	0.354	0.315	0.0059	—	—	—	0.039	25	34
	Camry	F	1.102	1.024	0.0020	—	—	—	0.039	25	79
		R	0.394	0.354	0.0059	9.00	—	9.08	0.039	14	20
	Camry Solara	F	1.102	1.024	0.0020	—	—	—	0.039	25	79
		R	0.394	0.354	0.0059	9.00	—	9.08	0.039	14	20
	Celica GT-S	F	0.984	0.906	0.002	—	—	—	0.039	25	79
		R	0.354	0.295	0.006	—	—	—	0.039	—	34
	MR 2	F	0.787	0.709	0.0020	—	—	—	0.039	80	25
		R	0.630	0.591	0.0039	—	—	—	0.039	44	15
	Celica	F	1.102	1.024	0.0020	—	—	—	0.039	25	79
		R	0.354	0.314	0.0059	7.87	—	7.91	0.039	—	34
	Corolla		0.866	0.787	0.0020	7.87	—	7.91	0.039	25	65
	Echo		0.709	0.630	0.0020	7.09	—	7.13	0.039	25	65

42356-TOYC-C12

BRAKE SPECIFICATIONS
All measurements in inches unless noted

Year	Model		Brake Disc Original Thickness	Brake Disc Minimum Thickness	Brake Disc Maximum Runout	Brake Drum Diameter Original Inside Diameter	Max. Wear Limit	Maximum Machine Diameter	Minimum Lining Thickness	Brake Caliper Bracket Bolts (ft. lbs.)	Brake Caliper Mounting Bolts (ft. lbs.)
2002	Avalon	F	1.102	1.024	0.0020	—	—	—	0.039	25	79
		R	0.354	0.315	0.0059	—	—	—	0.039	25	34
	Camry	F	1.102	1.024	0.0020	—	—	—	0.039	25	79
		R	0.394	0.354	0.0059	9.00	—	9.08	0.039	14	20
	Camry Solara	F	1.102	1.024	0.0020	—	—	—	0.039	25	79
		R	0.394	0.354	0.0059	9.00	—	9.08	0.039	14	20
	Celica GT-S	F	0.984	0.906	0.002	—	—	—	0.039	25	79
		R	0.354	0.295	0.006	—	—	—	0.039	—	34
	MR 2	F	0.787	0.709	0.0020	—	—	—	0.039	80	25
		R	0.630	0.591	0.0039	—	—	—	0.039	44	15
	Celica	F	1.102	1.024	0.0020	—	—	—	0.039	25	79
		R	0.354	0.314	0.0059	7.87	—	7.91	0.039	—	34
	Corolla		0.866	0.787	0.0020	7.87	—	7.91	0.039	25	65
	Echo		0.709	0.630	0.0020	7.09	—	7.13	0.039	25	65
2003	Avalon	F	1.102	1.024	0.0020	—	—	—	0.039	25	79
		R	0.354	0.315	0.0059	—	—	—	0.039	25	34
	Camry	F	1.102	1.024	0.0020	—	—	—	0.039	25	79
		R	0.394	0.354	0.0059	9.00	—	9.08	0.039	14	20
	Camry Solara	F	1.102	1.024	0.0020	—	—	—	0.039	25	79
		R	0.394	0.354	0.0059	9.00	—	9.08	0.039	14	20
	Celica GT-S	F	0.984	0.906	0.002	—	—	—	0.039	25	79
		R	0.354	0.295	0.006	—	—	—	0.039	—	34
	MR 2	F	0.787	0.709	0.0020	—	—	—	0.039	80	25
		R	0.630	0.591	0.0039	—	—	—	0.039	44	15
	Celica	F	1.102	1.024	0.0020	—	—	—	0.039	25	79
		R	0.354	0.314	0.0059	7.87	—	7.91	0.039	—	34
	Corolla		0.866	0.787	0.0020	7.87	—	7.91	0.039	25	65
	Echo		0.709	0.630	0.0020	7.09	—	7.13	0.039	25	65

F: Front
R: Rear

① 7A-FE engine: 0.906
   5S-FE engine: 1.024

② 3S-GTE engine: 1.181
   5S-FE engine: 0.984

③ 2JZ-GTE engine: 1.102
   2JZ-GE engine: 1.181

④ 7A-FE engine: 0.984
   5S-FE engine: 1.102

42356-TOYC-C13

SCHEDULED MAINTENANCE INTERVALS
TOYOTA—AVALON, CAMRY, CELICA, COROLLA, MR 2, ECHO, SOLARA & SUPRA

TO BE SERVICED	TYPE OF SERVICE	VEHICLE MILEAGE INTERVAL (x1000)												
		7.5	15	22.5	30	37.5	45	52.5	60	67.5	75	82.5	90	97.5
Engine oil & filter ①	R	✓	✓	✓	✓	✓	✓	✓	✓	✓	✓	✓	✓	✓
Drive belts	S/I								✓	✓	✓	✓	✓	✓
Automatic transaxle fluid & filter	S/I		✓		✓		✓		✓		✓		✓	
Ball joints & dust covers	S/I		✓		✓		✓		✓		✓		✓	
Bolts & nuts on body & chassis	S/I		✓		✓		✓		✓		✓		✓	
Brake line pipes & hoses	S/I		✓		✓		✓		✓		✓		✓	
Brake linings & drums (except MR2)	S/I		✓		✓		✓		✓		✓		✓	
Brake pads & discs (front & rear if equipped)	S/I		✓		✓		✓		✓		✓		✓	
Differential oil (Camry, Celica, Corolla & Supra) ②	S/I		✓		✓		✓		✓		✓		✓	
Drive shaft boots (except Supra)	S/I		✓		✓		✓		✓		✓		✓	
Manual transaxle oil	S/I		✓		✓		✓		✓		✓		✓	
Steering gear housing oil	S/I		✓		✓		✓		✓		✓		✓	
Steering linkage	S/I		✓		✓		✓		✓		✓		✓	
Air filter	R				✓				✓				✓	
Rear wheel bearings (Tercel)	R				✓				✓				✓	
Spark plugs (Corolla & Tercel)	R				✓				✓				✓	
Spark plugs (platinum tip) (Avalon, Camry, Celica & Supra)	R								✓					
Exhaust system	S/I				✓				✓				✓	
Fuel lines & connections	S/I				✓				✓				✓	
Valve clearance	S/I				✓				✓				✓	
Engine coolant	R						✓				✓			
Fuel tank cap gasket	R								✓					
Charcoal canister	S/I								✓					

R: Replace   S/I: Service or Inspect

① Supra 2JZ-GTE: change every 5000 miles.

② Supra w/LSD: replace every 30,000 miles.

FREQUENT OPERATION MAINTENANCE (SEVERE SERVICE)

If a vehicle is operated under any of the following conditions it is considered severe service:

- Extremely dusty areas.

- 50% or more of the vehicle operation is in 32°C (90°F) or higher temperatures, or constant operation in temperatures below 0°C (32°F).

- Prolonged idling (vehicle operation in stop and go traffic).

- Frequent short running periods (engine does not warm to normal operating temperatures).

- Police, taxi, delivery usage or trailer towing usage.

Oil & oil filter: change every 6000 miles.

Oil & oil filter change (Supra 2JZ-GTE): change every 2500 miles.

Bolts & nuts on chassis & body: tighten every 7500 miles.

Ball joints & dust covers: service or inspect every 12,000 miles.

Brake linings & drums: service or inspect ever 12,000 miles.

Brake pads & discs (front & rear if equipped): service or inspect every 12,000 miles.

Drive shaft boots & except Supra): service or inspect every 12,000 miles.

Steering linkage: service or inspect every 12,000 miles.

Air filter: service or inspect every 15,000 miles.

Exhaust system: service or inspect every 15,000 miles.

Timing belt: replace every 60,000 miles.

PRECAUTIONS

Before servicing any vehicle, please be sure to read all of the following precautions, which deal with personal safety, prevention of component damage, and important points to take into consideration when servicing a motor vehicle:

• Never open, service or drain the radiator or cooling system when the engine is hot; serious burns can occur from the steam and hot coolant.

• Observe all applicable safety precautions when working around fuel. Whenever servicing the fuel system, always work in a well-ventilated area. Do not allow fuel spray or vapors to come in contact with a spark, open flame, or excessive heat (a hot drop light, for example). Keep a dry chemical fire extinguisher near the work area. Always keep fuel in a container specifically designed for fuel storage; also, always properly seal fuel containers to avoid the possibility of fire or explosion. Refer to the additional fuel system precautions later in this section.

• Fuel injection systems often remain pressurized, even after the engine has been turned **OFF**. The fuel system pressure must be relieved before disconnecting any fuel lines. Failure to do so may result in fire and/or personal injury.

• Brake fluid often contains polyglycol ethers and polyglycols. Avoid contact with the eyes and wash your hands thoroughly after handling brake fluid. If you do get brake fluid in your eyes, flush your eyes with clean, running water for 15 minutes. If eye irritation persists, or if you have taken brake fluid internally, IMMEDIATELY seek medical assistance.

• The EPA warns that prolonged contact with used engine oil may cause a number of skin disorders, including cancer! You should make every effort to minimize your exposure to used engine oil. Protective gloves should be worn when changing oil. Wash your hands and any other exposed skin areas as soon as possible after exposure to used engine oil. Soap and water, or waterless hand cleaner should be used.

• All new vehicles are now equipped with an air bag system. The system must be disabled before performing service on or around system components, steering column, instrument panel components, wiring and sensors. Failure to follow safety and disabling procedures could result in accidental air bag deployment, possible personal injury and unnecessary system repairs.

• Always wear safety goggles when working with, or around, the air bag system. When carrying a non-deployed air bag, be sure the bag and trim cover are pointed away from your body. When placing a non-deployed air bag on a work surface, always face the bag and trim cover upward, away from the surface. This will reduce the motion of the module if it is accidentally deployed. Refer to the additional air bag system precautions later in this section.

• Clean, high quality brake fluid from a sealed container is essential to the safe and proper operation of the brake system. You should always buy the correct type of brake fluid for your vehicle. If the brake fluid becomes contaminated, completely flush the system with new fluid. Never reuse any brake fluid. Any brake fluid that is removed from the system should be discarded. Also, do not allow any brake fluid to come in contact with a painted surface; it will damage the paint.

• Never operate the engine without the proper amount and type of engine oil; doing so WILL result in severe engine damage.

• Timing belt maintenance is extremely important! Many models utilize an interference-type, non-freewheeling engine. If the timing belt breaks, the valves in the cylinder head may strike the pistons, causing potentially serious (also time-consuming and expensive) engine damage. Refer to the maintenance interval charts in the front of this section for the recommended replacement interval for the timing belt.

• Disconnecting the negative battery cable on some vehicles may interfere with the functions of the on-board computer system(s) and may require the computer to undergo a relearning process once the negative battery cable is reconnected.

• When servicing drum brakes, only disassemble and assemble one side at a time, leaving the remaining side intact for reference.

ENGINE REPAIR

Distributor

REMOVAL

1. Before servicing the vehicle, refer to the precautions in the beginning of this section.
2. Remove or disconnect the following:
   • Negative battery cable. On vehicles equipped with an air bag, wait at least 90 seconds before proceeding.
   • Electrical connector from the air flow meter. Disconnect the air cleaner hose from the throttle body and remove the air cleaner cover, air flow meter and air duct as 1 unit.
   • Air intake and intercooler to provide clearance. Disconnect the air

temperature sensor, cruise control actuator cable, and the air cleaner hose, if more clearance is necessary.
   • Spark plug wires from the distributor cap.
   • Distributor connector.
   • Distributor mounting bolt
   • O-ring from the distributor housing

➡ **The marks on the distributor drive gear and distributor housing should be aligned. If the marks are not aligned, mark the distributor housing and rotor position.**

INSTALLATION

Engine Disturbed

1. On the 2JZ-GE engine, remove the No. 3 timing belt cover.

2. Bring No. 1 cylinder to TDC
3. Install or connect the following:
   • O-ring
   • Distributor. Tighten the mounting bolts
   • Intercooler or the air intake
   • Spark plug wires
   • On the 2JZ-GE engine, reinstall the No. 3 timing belt cover
   • Air flow meter, air hose and the air cleaner cover if removed
   • Distributor and air flow meter connectors.
   • Negative battery cable
   • Check the ignition timing

Engine Not Disturbed

1. Install or connect the following:
   • O-ring
   • Distributor. Tighten the mounting bolts

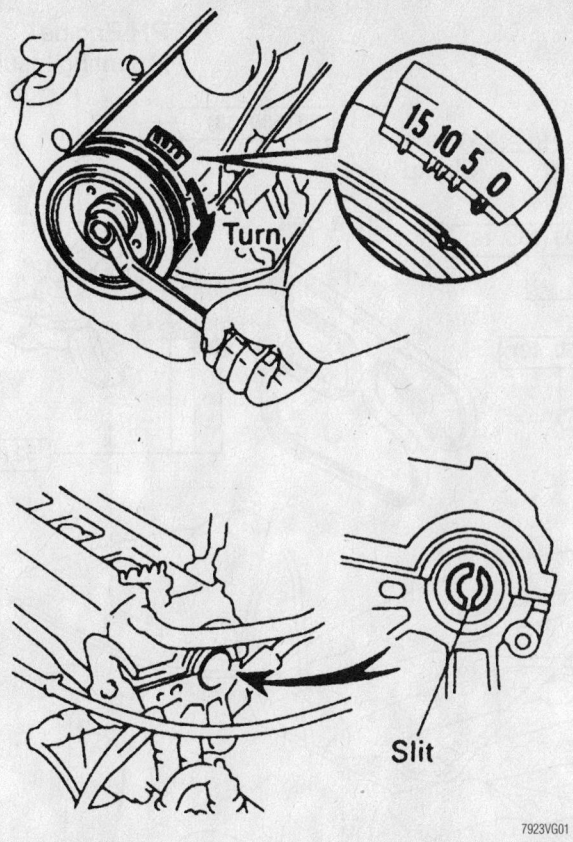

Common method of properly positioning the distributor drive mechanism to TDC

➡**Be sure the mark on the engine aligns with the mark on the distributor made during removal. Also be sure the rotor is in the same position as removal**

- Intercooler or the air intake
- Distributor cap
- Spark plug wires
- Air flow meter, air hose and the air cleaner cover
- Distributor and air flow meter connectors
- Negative battery cable
- Check the ignition timing.

Alternator

REMOVAL & INSTALLATION

Avalon, Camry

1. Before servicing the vehicle, refer to the precautions in the beginning of this section.
2. Remove or disconnect the following:
   - Negative battery cable. On models with an airbag, wait at least 90 seconds from the time that the ignition switch is turned to the LOCK position and the battery is disconnected

before performing any further work.
- On the 5S-FE, the two bolts and the No. 3 right hand engine mounting stay
- On 1MZ-FE, the bolt and nut and then remove the No. 2 right hand engine mounting stay
- Harness and wire (and nut) from the alternator
- Air cleaner, if necessary
- Drive belt from the pulley
- Alternator

To install:
3. Install or connect the following:
   - Alternator
   - Drive belt
   - Wiring
   - Negative and starter battery cables

Celica, GT-S, Supra

1. Before servicing the vehicle, refer to the precautions in the beginning of this section.
2. Remove or disconnect the following:
   - Negative battery cable

➡**On some models it is necessary to remove the fuse/relay block located near the alternator.**

- Drive belt
- Wiring

- Alternator

To install:
3. Install or connect the following:
   - Alternator
   - Drive belt. Torque the pivot bolt and adjusting bolts: 5S-FE: Pivot bolt—40 ft. lbs. (54 Nm); Adjusting bolt—14 ft. lbs. (19 Nm).
   - Wiring
   - Negative battery cable

Corolla

1. Before servicing the vehicle, refer to the precautions in the beginning of this section.

➡**It may be necessary to remove the gravel shield and work from underneath the car in order to gain access to the alternator retaining bolts.**

2. Remove or disconnect the following:
   - Negative battery cable
   - Wiring from the alternator
   - Drive belt
   - Alternator

To install:
3. Install or connect the following:
   - Alternator. Torque the 14mm bolt to 18 ft. lbs. (24 Nm) and the 17 mm bolt to 40 ft. lbs. (54 Nm).
   - Alternator connector and wiring
   - Drive belt
   - Negative battery cable

Echo

1. Before servicing the vehicle, refer to the precautions in the beginning of this section.
2. Remove or disconnect the following:
   - Negative battery cable
   - Wire clamp from the rectifier
   - Alternator harness
   - 2 bolts
   - Alternator

To install:
3. Install or connect the following:
   - Alternator and hand tighten the bolts
   - Drive belt and adjust if necessary. Torque the 14 mm bolt to 14 ft. lbs. (18 Nm) and the 17 mm bolt to 40 ft. lbs. (54 Nm).
   - Alternator connector and clamp
   - Alternator electrical wiring
   - Negative battery cable

MR2

1. Before servicing the vehicle, refer to the precautions in the beginning of this section.

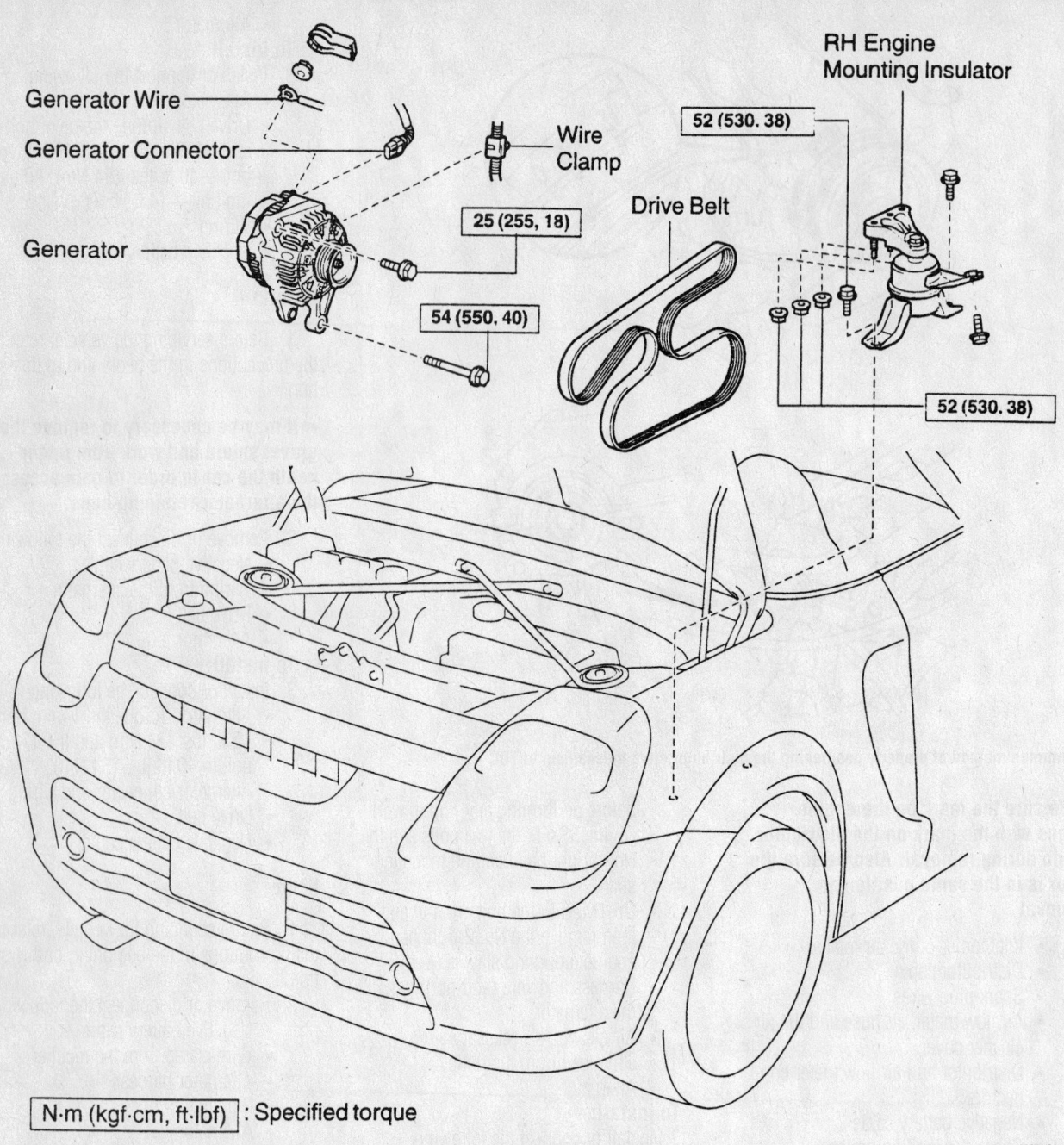

Generator Wire

Generator Connector

Generator

Wire Clamp

25 (255, 18)

54 (550, 40)

Drive Belt

RH Engine Mounting Insulator

52 (530, 38)

52 (530, 38)

N·m (kgf·cm, ft·lbf) : Specified torque

9357WG01

MR2 alternator mounting exploded view

2. Remove or disconnect the following:
   • Negative battery cable
   • Accessory drive belt
   • Right motor mount
   • Wire harness clamp
   • Alternator wiring harness
   • Alternator mounting bolts
   • Alternator

To install:

3. Install or connect the following:
   • Alternator. Tighten the upper mounting bolt to 18 ft. lbs. (25 Nm) and the lower mounting bolt to 40 ft. lbs. (54 Nm).

   • Alternator wiring harness
   • Wire harness clamp
   • Right motor mount. Tighten the bolts to 38 ft. lbs. (52 Nm).
   • Accessory drive belt
   • Negative battery cable

Ignition Timing

ADJUSTMENT

➡The timing on engines equipped with DIS is not adjustable.

1. Before servicing the vehicle, refer to the precautions in the beginning of this section.

2. Perform the following:
   • Start the engine and let it reach operating temperature.
   • Connect the tachometer tester probe to terminal IG of the DLC1.

➡**Never allow the tachometer test probe to touch ground as it could result in damage to the igniter and or ignition coil.**

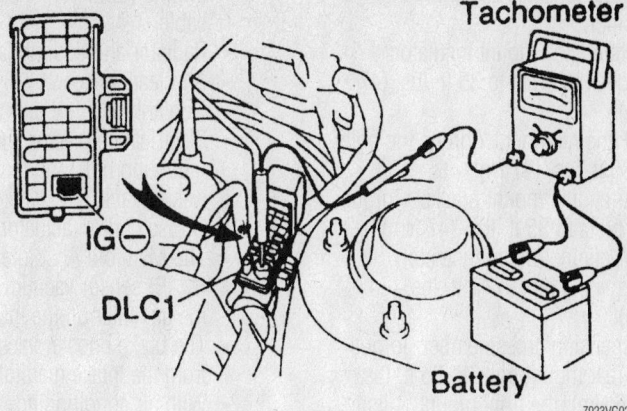

Attach the tachometer test probe to the IG terminal of the data link connector

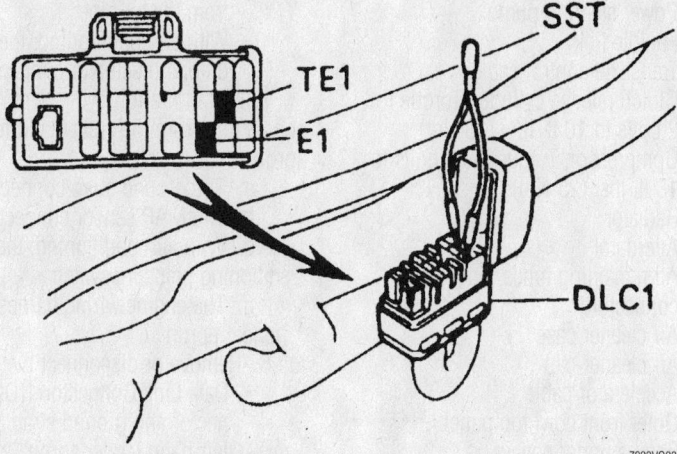

A jumper wire is used to connect terminals TE1 and E1 of the DLC

➡ **Not all tachometers are compatible with this system. Be sure to confirm compatibility of your unit before use.**

- Using a jumper connector, connect terminals TE_1 and E_1 of the DLC1.
- Check the timing. Timing should be 8–12 degrees BTDC. Set the timing as needed.
- Tighten the hold-down bolt.
- Check that the ignition timing advances.

Engine Assembly

REMOVAL & INSTALLATION

5E-FE Engine

1. Before servicing the vehicle, refer to the precautions in the beginning of this section.
2. Drain the cooling system. Drain the engine oil.
3. Drain the transaxle fluid.
4. Remove or disconnect the following:

- Negative battery cable. On vehicles equipped with an air bag, wait at least 90 seconds before proceeding.
- Hood
- Undercovers
- Radiator
- Accelerator cable
- Throttle cable
- Air cleaner assembly and bracket.
- Charcoal canister
- Fuel return and inlet hoses
- Speedometer cable
- Idle up air hoses from the power steering air control valve
- Oxygen sensor wire, the oil pressure switch wire, the coolant fan switch wire, the water temperature gauge wire, the back-up light switch and neutral safety switch wires
- All remaining wiring harnesses connected to the engine
- All necessary vacuum hoses
- Starter wires
- If equipped with cruise control, actuator assembly

- Heater hoses
- On manual transaxles, clutch release cylinder
- Transaxle control cables
- Power steering pump. Do not disconnect the power steering hoses.
- Air conditioning compressor. Leave the refrigerant lines connected.
- Front exhaust pipe
- Halfshafts
- Connect a suitable lifting device to the engine lifting hooks.
- On manual transaxles, rear mounting through-bolt and the rear mounting assembly. On automatic transaxle, front mounting through-bolt and front mounting assembly.
- Right and left side mounting bolts and brackets.
- Lift the engine/transaxle assembly out of the vehicle.
- Starter
- For automatic transaxles, torque converter clutch mounting bolts.
- Engine from the transaxle by removing the bolts.

To install:
5. Install or connect the following:

- Transaxle to the engine. On automatic transaxles, install the torque converter clutch and mounting bolts. Install the gray bolt first. Bolts: 20 ft. lbs. (27 Nm).
- Starter. Bolts: 29 ft. lbs. (39 Nm).
- Lower the engine and transaxle assembly in the vehicle.
- Right side mounting insulator and temporarily install the through-bolt
- Left side mounting bracket. Bolts marked (NT): 47 ft. lbs. (64 Nm), Bolts marked (7T): 35 ft. lbs. (48 Nm).
- Ground strap. Tighten to 35 ft. lbs. (49 Nm).
- Rear mounting bracket. Bolts: 35 ft. lbs. (48 Nm).
- Rear insulator through-bolt. Bolt: 47 ft. lbs. (64 Nm).
- Halfshafts
- 2 bolts, nut and through-bolt of the right side mounting insulator. Bolts and nut: 47 ft. lbs. (64 Nm). Through-bolt: 54 ft. lbs. (73 Nm).
- Front exhaust pipe
- Air conditioning compressor
- Power steering pump
- Transaxle control cables
- On manual transaxles, clutch release cylinder. Bolts: 108 inch lbs. (12 Nm).
- Idle up air hoses
- Speedometer cable

- Heater hoses.
- If equipped with cruise control, actuator assembly.
- Starter wires and nut.
- Vacuum hoses
- Oxygen sensor wire, oil pressure switch, coolant fan switch wire, water temperature gauge wire, back-up light switch, and the neutral safety switch wires.
- All wiring
- Fuel line
- Charcoal canister
- Air cleaner assembly and bracket
- Radiator, coolant hoses and on vehicles with automatic transmission, cooling lines
- All fluids
- Undercovers
- Hood
- Negative battery cable.

1NZ-FE

1. Before servicing the vehicle, refer to the precautions in the beginning of this section.
2. Drain the cooling system.
3. Drain the engine oil.
4. Drain the transaxle fluid.
5. Remove or disconnect the following:

- Hood
- Battery and tray
- Outer front cowl top panel
- Engine under covers
- Accelerator cable
- Air cleaner cap and related parts
- Air cleaner case and related parts
- All tubes, hoses and connectors attached to the engine
- Accessory drive belts
- Alternator
- Radiator
- With A/C, position the compressor out of the way
- With MT, the clutch release cylinder
- Transaxle control cables
- Fusible link
- Power steering pump
- Center exhaust pipe
- Halfshafts
- Suspension crossmember
- Rear engine mount

6. Attach a shop crane to the engine hangers.
7. Remove all remaining engine mount bolts/nuts.
8. Remove the engine/transaxle assembly.

To install:
9. Install or connect the following:

- Engine/transaxle assembly into position
- Right engine mount insulator. Torque the bolts to 35 ft. lbs. (47 Nm).
- Left engine mount. Torque the bolts to 35 ft. lbs. (47 Nm).
- Rear engine mount bracket. Torque the bolts to 35 ft. lbs. (47 Nm).
- Rear engine mount insulator. Torque the bolt to 47 ft. lbs. (64 Nm).
- Suspension crossmember. Torque the rear mount bolts to 86 ft. lbs. (116 Nm); the front mount bolts to 52 ft. lbs. (70 Nm).
- Halfshafts
- Center exhaust pipe
- Power steering pump
- Fusible link
- Transaxle control cables
- Clutch release cylinder. Torque the 2 bolts to 10 ft. lbs. (13 Nm).
- Compressor. Torque the 4 bolts to 18 ft. lbs. (25 Nm).
- Radiator
- Alternator
- All remaining tubes, hoses and connectors
- Air cleaner case
- Air cleaner cap
- Accelerator cable
- Outer front cowl top panel
- Engine under covers
- Battery and tray
- Coolant
- Engine oil
- Transaxle oil
- Hood

10. Start the vehicle, check for leaks and repair if necessary.

1ZZ-FE Engines

COROLLA

1. Before servicing the vehicle, refer to the precautions in the beginning of this section.
2. Relieve the fuel system pressure.
3. Drain the cooling system.
4. Drain the engine oil.
5. Drain the transaxle fluid.
6. Remove or disconnect the following:

- Negative battery cable. On vehicles equipped with an air bag, wait at least 90 seconds before proceeding.
- Battery
- Hood
- Undercover
- Accelerator cable
- With automatic transmission, throttle cable from the accelerator cable.
- Radiator and cooling fan
- Air cleaner assembly
- Coolant reservoir tank stay
- Electrical connector, the hose, the mounting bolt, and remove the washer tank
- Cruise control actuator
- The Manifold Absolute Pressure (MAP) sensor vacuum hose from the gas filter on the intake manifold
- The brake booster vacuum hose from the intake manifold
- With air conditioning: the air conditioning vacuum hose from the actuator
- With power steering: the air hose from the air pipe
- With air conditioning: the air conditioning actuator connector

7. Install the following wires and connectors from the right-hand fender apron:

a. The ground strap connector
b. The MAP sensor connector
c. With air conditioning: the air conditioning pressure switch
d. The engine wiring harness from the fender apron

8. Remove or disconnect the following:

- Data Link Connector 1(DLC1) connector and ground strap from the left-hand fender apron.
- Engine relay box and 4 connectors.
- Charcoal canister
- Heater hoses from water inlet housing
- Fuel inlet and return hoses
- With manual transmission, clutch release cylinder without disconnecting the pipe
- Transaxle control cable(s)

9. To disconnect the engine wiring harness, disconnect or remove the following components:

- Left-hand and right-hand front door scuff plate
- Lower finish panel
- Lower panel with the glove compartment
- Radio and center cluster finish panel
- Rear console box
- On manual transmission, shift lever knob
- On automatic transmission, shifting hole bezel
- Lower center finish panel
- Floor carpet bracket
- The 3 ECM connectors and cowl wire connector

10. Remove or disconnect the remaining components:
- Air conditioning compressor
- Front exhaust pipe
- Halfshafts
- Power steering pump
- Engine mounting center member
- Through-bolt and nut holding the mounting insulator to the mounting bracket
- Engine and transaxle assembly
- Front and rear engine mounting bracket
- Starter
- Separate the transaxle from the engine

To install:

11. Install or connect the following:
- Engine to the transaxle
- Starter
- Rear engine mounting bracket bolts: 57 ft. lbs. (77 Nm).
- Front engine mounting bracket. bolts: 57 ft. lbs. (77 Nm).
- Engine and transaxle assembly into the vehicle
- Engine mounting center member.
- Front engine mounting insulator through-bolt and nut. Torque the bolt to 64 ft. lbs. (87 Nm).
- Halfshafts
- Front exhaust pipe
- Power steering pump. Torque the bolts to 29 ft. lbs. (39 Nm).
- Drive belt
- Air conditioner compressor. Torque the bolts to 18 ft. lbs. (25 Nm).
- Drive belt and reconnect the connector.

12. To install and connect the engine wiring harness, perform the following:
- Push the wire through the cowl
- Connect the 3 ECM connectors
- Attach the cowl wire connector
- Floor carpet bracket
- Center lower finish panel
- With automatic transmission, install the shifting hole bezel, with manual transmission, install the shift lever knob
- Rear console box
- Center cluster finish panel and the radio
- Lower panel with the glove compartment door
- Right and left-hand door scuff plates
- Lower finish panel

13. Install or connect the following:
- With manual transmission, clutch release cylinder
- Transaxle control cable(s)

- Fuel return and inlet hose. Torque the bolt to 22 ft. lbs. (29 Nm).
- Heater hoses to the water inlet housing
- Charcoal canister

14. Connect the following wires and connectors on the left-hand fender apron:
- The 4 connectors to the engine relay box
- Engine relay box
- The DLC1 connector
- The connector on the fender apron
- The ground strap on the fender apron

15. Install or connect on the right-hand fender apron:
- The ground strap connector
- The MAP sensor connector
- With air conditioning, the air conditioning pressure switch
- The engine wire from the fender apron

16. Install or connect the following:
- With A/C, the actuator connector
- With power steering, the air hoses to the air pipe
- The vacuum hose from the MAP sensor to the gas filter to the intake chamber
- The brake booster vacuum hose to the air intake chamber
- With A/C, the vacuum hose from the actuator
- With cruise control, actuator, actuator cable and cover
- Electrical connector and vinyl hose
- Washer tank with the bolt
- Coolant reservoir tank stay
- Air cleaner
- Radiator and cooling fan
- With automatic transmission, connect the throttle cable.
- Accelerator cable
- All fluids
- Negative battery cable
- Undercovers and hood

17. Start the vehicle, check for leaks and repair if necessary.

CELICA

1. Before servicing the vehicle, refer to the precautions in the beginning of this section.
2. Release the fuel system pressure.
3. Drain the engine oil.
4. Drain the cooling system.
5. Drain the transaxle fluid.
6. Remove or disconnect the following:
- Negative battery cable. On vehicles equipped with an air bag, wait at least 90 seconds before proceeding.

- Battery
- Hood
- Undercover
- Accelerator cable, cable bracket, and clamps.
- Air cleaner
- Cruise control actuator cable
- Radiator
- MAP sensor vacuum hose from the gas filter on the intake manifold
- Power steering air hose from the intake manifold
- Power steering hose from the air the air pipe
- Brake booster vacuum hose from the intake manifold
- Air conditioning idle-up valve
- Air conditioning idle-up valve hose from the intake manifold
- Air conditioning idle-up valve hose from the air pipe
- DLC1 from the bracket
- Engine wiring harness from the bracket
- Ground cable from the body and the ground strap from the body
- Heater hoses from the water outlet
- Heater hose from the water bypass pipe
- Fuel inlet hose from the fuel filter and the fuel return hose from the return pipe
- EVAP hose from the charcoal canister
- Engine wiring harness from the engine compartment relay box

7. To remove the engine wiring harness from the passenger's compartment, remove or disconnect the following:
- The scuff plate
- The cowl side trim
- The finish panel from the lower instrument panel
- Front side of the floor carpet
- The wiring harness from the clamp of the ECM bracket
- The 3 ECM connectors
- The circuit opening relay connector
- The 3 connectors from the connectors on the bracket
- The A/C amplifier connector
- The MAP sensor connector
- The MAP sensor wire from the clamp on the bracket
- The wire clamp from the bracket
- The 2 nuts holding the engine wiring harness to the cowl

8. Remove or disconnect the following:
- Front exhaust pipe
- Halfshafts.
- Alternator drive belt
- Air conditioning drive belt, com-

pressor connector, and compressor. Do not disconnect the air conditioning lines
- Remove the drive belt and remove the 4 bolts that secure the power steering pump. Without disconnecting the lines, securely hang the pump out of the way
- A/C relay box
- On manual transmission, clutch release cylinder from the transaxle
- Transaxle control cable(s)
- On automatic transmission, transaxle control cable from the engine mounting center member.
- Exhaust pipe support bracket.

9. To remove the engine mounting center member, remove the following components:
- The 2 dust covers from the rear side of the member
- The A/C pipe from the bracket
- The bolt and nut holding the front engine mounting bracket to the mounting insulator
- The bolt holding the rear engine mounting bracket to the insulator
- The bolt and 2 nuts holding the rear engine mounting insulator to the front suspension member
- The 2 bolts and the rear engine mounting bracket, and the center member with the rear mounting insulator

10. Remove or disconnect the remaining components:
- Attach an engine chain hoist to the engine hangers
- Left-hand engine mounting bracket from the mounting insulator
- Through-bolt and the left-hand mounting insulator
- Ground strap connector
- Right-hand engine mounting bracket from the mounting insulator
- Lift the engine and transaxle assembly from the vehicle
- Transaxle from the engine assembly

To install:
11. Install or connect the following:
- Transaxle to the engine assembly.
- Engine into the engine compartment
- Right-hand engine mounting bracket to the mounting insulator. Temporarily install the 3 nuts.
- Left-hand engine mounting insulator to the body with the through-bolt
- Left-hand engine mounting bracket to the mounting insulator and

install the 2 bolts and nut. Bolts and nut: 47 ft. lbs. (64 Nm).
- Left-hand engine mounting through-bolt to the body. Bolt: 54 ft. lbs. (73 Nm).
- Right-hand mounting bracket to the insulator. 12mm nut: 21 ft. lbs. (28 Nm), 14mm nut: 38 ft. lbs. (52 Nm).
- Engine ground strap connector and remove the engine hoist.

12. To install the engine mounting center member, perform the following:
- Attach the center member together with the rear engine mounting insulator to the front suspension member
- Temporarily install the 2 bolts and nut holding the center member to the body
- Install the rear engine mounting bracket. Bolts: 58 ft. lbs. (78 Nm)
- Temporarily install the bolt and 2 nuts holding the rear engine mounting insulator to the front suspension member
- Temporarily install the bolt holding the rear engine mounting bracket to the insulator
- Temporarily install the bolt and nut holding the front engine mounting bracket to the insulator
- Tighten the 2 bolts holding the center member to the body to 26 ft. lbs. (35 Nm)
- Tighten the bolt and 2 nuts holding the rear mounting insulator to the front suspension member to 59 ft. lbs. (80 Nm)
- Tighten the bolt holding the rear engine mounting bracket to the insulator to 65 ft. lbs. (88 Nm)
- Tighten the bolt and nut holding the front engine mounting bracket to the insulator to 65 ft. lbs. (88 Nm
- Install the air conditioning pipe to the bracket and install the 2 dust covers to the center member

13. Install or connect the following:
- Exhaust pipe support bracket. Bolts to 14 ft. lbs. (19 Nm).
- Transaxle control cable(s)
- On automatic transmission, transaxle control cable to the engine mounting center member.
- On manual transmission, clutch release cylinder. Bolts: 108 inch lbs. (12 Nm), then attach the bracket with the bolt.
- A/C relay box to the body.
- Power steering pump. 12mm bolts:

14 ft. lbs. (19 Nm). 14mm bolts: 29 ft. lbs. (39 Nm). Install the drive belt. Adjusting bolt: 29 ft. lbs. (39 Nm).
- A/C compressor. Bolts: 18 ft. lbs. (25 Nm).
- A/C drive belt with the adjusting bolt. Torque the locknut to 29 ft. lbs. (39 Nm). Connect the connector.
- Alternator drive belt
- Halfshafts
- Front exhaust pipe

14. To install the engine wiring harness in the passenger compartment, perform the following:
- Push the harness through the cowl panel, install the retainer to the cowl with the 2 nuts and install the wire clamp to the bracket
- Connect the harness to the clamp on the ECM
- Connect the 3 ECM connectors and the circuit opening relay connector
- Connect the 3 connectors to the connectors on the bracket
- Connect the A/C amplifier connector
- Install the floor carpet, the lower instrument panel finish panel, the cowl side trim panel, and the scuff plate

15. Install or connect the following:
- Engine wiring harness with the 2 connectors to the engine compartment relay box and install the relay box covers
- MAP sensor connector
- MAP sensor wire to the clamp on the bracket
- MAP sensor vacuum hose to the gas filter on the intake manifold
- Brake booster vacuum hose to the intake manifold
- A/C idle-up valve connector
- A/C idle-up valve hose to the intake manifold
- A/C idle-up valve hose to the air pipe
- DLC1 to the bracket
- Engine harness protector to the bracket
- Ground cable and the ground strap
- Heater hose to the water outlet and the heater hose to the water bypass pipe
- Fuel inlet hose to the fuel filter
- Fuel inlet hose with 2 new gaskets and the union bolt. Bolt: 22 ft. lbs. (30 Nm).
- Fuel return hose to the return pipe

and connect the EVAP hose to the charcoal canister
- Power steering air hoses to the intake manifold and the air pipe.
- Radiator
- On models equipped with cruise control, install the actuator cable to the clamps.
- Accelerator cable to the throttle body, cable bracket, and the clamps.
- Air cleaner

- Battery tray and battery
- Hood
- Refill the transaxle assembly, engine oil and coolant.
- Negative battery cable
- Engine undercover

16. Start the vehicle, check for leaks and repair if necessary.

MR2

1. Before servicing the vehicle, refer to the precautions in the beginning of this section.

2. Drain the cooling system.
3. Relieve the fuel system pressure.
4. Drain the engine oil
5. Drain the transaxle fluid
6. Remove or disconnect the following:
   - Engine hood
   - Rear suspension upper brace
   - Engine undercovers
   - Battery and tray
   - Air cleaner assembly and intake air hose
   - Accelerator cable

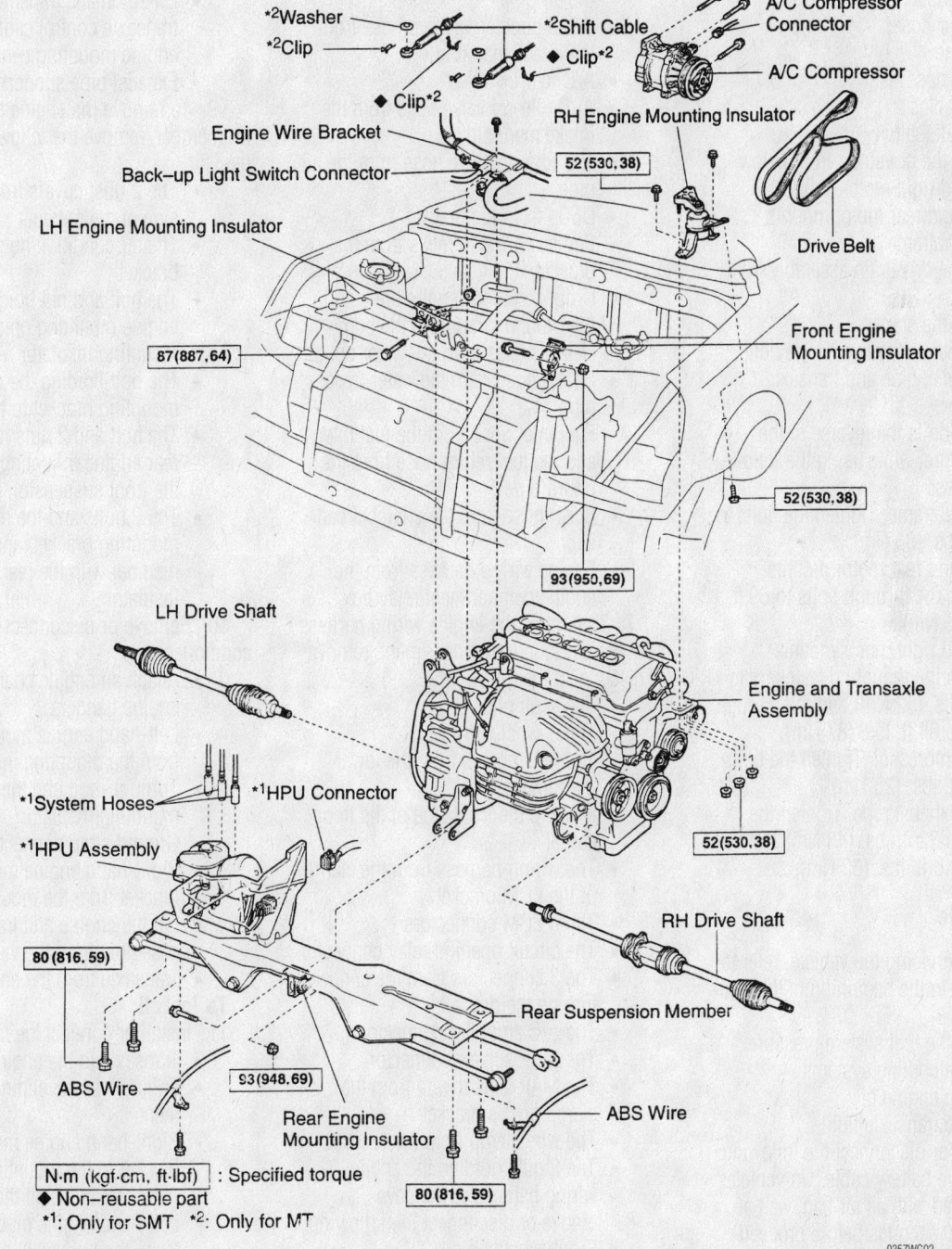

MR2 engine mounting exploded view

9357WG02

- Rear bumper cover
- Heated Oxygen (HO2S) sensor connectors
- Front exhaust pipe
- Accessory drive belt
- A/C compressor
- Engine wiring harness connectors in the luggage compartment. Pull the harness through the firewall.
- Transaxle control ECU harness connector
- Engine wire at Junction Box No. 1
- Engine ground wires
- Wire brackets
- Heater hoses
- Radiator hoses
- Fuel line
- Clutch hose
- Shift cables
- Axle halfshafts

7. Support the drivetrain from below.
- Left and right motor mounts
- Front and rear motor mounts
- Rear subframe

8. Lower the drivetrain assembly from the vehicle.

9. Remove the starter.

10. Remove the transaxle flange bolts and separate the engine and transaxle.

To install:

11. Installation is the reverse of the removal procedure, while using the following torque values:
- Rear subframe. Tighten the bolts to 59 ft. lbs. (80 Nm).
- Front and rear motor mounts. Tighten the through bolts to 69 ft. lbs. (93 Nm).
- Left and right motor mounts. Tighten the mounting fasteners to 38 ft. lbs. (52 Nm) and the through bolts to 64 ft. lbs. (87 Nm).
- A/C compressor. Tighten the bolts to 18 ft. lbs. (25 Nm).
- Front exhaust pipe. Tighten the bolts to 32 ft. lbs. (43 Nm) and the nut to 46 ft. lbs. (62 Nm).

2ZZ-GE Engine

1. Before servicing the vehicle, refer to the precautions in the beginning of this section.

2. Release the fuel system pressure.

3. Drain the cooling system.

4. Drain the engine oil.

5. Drain the transaxle fluid.

6. Remove or disconnect the following:
- Negative battery cable. On vehicles equipped with an air bag, wait at least 90 seconds before proceeding.
- Battery
- Hood
- Undercover
- Accelerator cable, cable bracket, and clamps.
- Air cleaner
- ECM box
- Cruise control actuator cable
- Radiator
- MAP sensor vacuum hose from the gas filter on the intake manifold
- Power steering air hose from the intake manifold
- Power steering hose from the air the air pipe
- Brake booster vacuum hose from the intake manifold
- A/C idle-up valve
- Ai/c idle-up valve hose from the intake manifold
- A/C idle-up valve hose from the air pipe
- DLC1 from the bracket
- Engine wiring harness from the bracket
- Ground cable from the body and the ground strap from the body
- Heater hoses from the water outlet
- Heater hose from the water bypass pipe
- Fuel inlet hose from the fuel filter and the fuel return hose from the return pipe
- EVAP hose from the charcoal canister
- Engine wiring harness from the engine compartment relay box

7. To remove the engine wiring harness from the passenger's compartment, remove or disconnect the following:
- The scuff plate
- The cowl side trim
- The finish panel from the lower instrument panel
- Remove the front side of the floor carpet
- The wiring harness from the clamp of the ECM bracket
- The 3 ECM connectors
- The circuit opening relay connector
- The 3 connectors from the connectors on the bracket
- The A/C amplifier connector
- The MAP sensor connector
- The MAP sensor wire from the clamp on the bracket
- The wire clamp from the bracket
- The 2 nuts holding the engine wiring harness to the cowl

8. Remove or disconnect the following:
- Front exhaust pipe
- Exhaust manifold
- Halfshafts
- Alternator drive belt
- A/C drive belt, compressor connector, and compressor. Do not disconnect the A/C lines
- Remove the drive belt and remove the 4 bolts that secure the power steering pump. Without disconnecting the lines, securely hang the pump out of the way.
- A/C relay box
- On manual transmission, clutch release cylinder from the transaxle
- Transaxle control cable(s)
- On automatic transmission, transaxle control cable from the engine mounting center member.
- Exhaust pipe support bracket

9. To remove the engine mounting center member, remove the following components:
- The 2 dust covers from the rear side of the member
- The air conditioning pipe from the bracket
- The bolt and nut holding the front engine mounting bracket to the mounting insulator
- The bolt holding the rear engine mounting bracket to the insulator
- The bolt and 2 nuts holding the rear engine mounting insulator to the front suspension member
- The 2 bolts and the rear engine mounting bracket, and the center member with the rear mounting insulator

10. Remove or disconnect the remaining components:
- Attach an engine chain hoist to the engine hangers
- Left-hand engine mounting bracket from the mounting insulator
- Through-bolt and the left-hand mounting insulator
- Ground strap connector
- Right-hand engine mounting bracket from the mounting insulator
- Lift the engine and transaxle assembly from the vehicle
- Transaxle from the engine assembly

To install:

11. Install or connect the following:
- Transaxle to the engine assembly
- Engine into the engine compartment
- Right-hand engine mounting bracket to the mounting insulator. Temporarily install the 3 nuts.
- Left-hand engine mounting insulator to the body with the through-bolt

- Left-hand engine mounting bracket to the mounting insulator and install the 2 bolts and nut. Bolts and nut: 47 ft. lbs. (64 Nm).
- Left-hand engine mounting through-bolt to the body. Bolt: 54 ft. lbs. (73 Nm).
- Right-hand mounting bracket to the insulator. 12mm nut: 21 ft. lbs. (28 Nm), 14mm nut: 38 ft. lbs. (52 Nm).
- Engine ground strap connector and remove the engine hoist

12. To install the engine mounting center member, perform the following:

- Attach the center member together with the rear engine mounting insulator to the front suspension member
- Temporarily install the 2 bolts and nut holding the center member to the body
- Install the rear engine mounting bracket. Bolts: 58 ft. lbs. (78 Nm)
- Temporarily install the bolt and 2 nuts holding the rear engine mounting insulator to the front suspension member
- Temporarily install the bolt holding the rear engine mounting bracket to the insulator
- Temporarily install the bolt and nut holding the front engine mounting bracket to the insulator
- Tighten the 2 bolts holding the center member to the body to 26 ft. lbs. (35 Nm)
- Tighten the bolt and 2 nuts holding the rear mounting insulator to the front suspension member to 59 ft. lbs. (80 Nm)
- Tighten the bolt holding the rear engine mounting bracket to the insulator to 65 ft. lbs. (88 Nm)
- Tighten the bolt and nut holding the front engine mounting bracket to the insulator to 65 ft. lbs. (88 Nm
- Install the A/C pipe to the bracket and install the 2 dust covers to the center member
- Exhaust manifold

13. Install or connect the following:

- Exhaust pipe support bracket. Bolts to 14 ft. lbs. (19 Nm).
- Transaxle control cable(s)
- On automatic transmission, transaxle control cable to the engine mounting center member.
- On manual transmission, clutch release cylinder. Bolts: 108 inch lbs. (12 Nm), then attach the bracket with the bolt

- A/C relay box to the body
- Power steering pump. 12mm bolts: 14 ft. lbs. (19 Nm). 14mm bolts: 29 ft. lbs. (39 Nm). Install the drive belt. Adjusting bolt: 29 ft. lbs. (39 Nm).
- A/C compressor. Bolts: 18 ft. lbs. (25 Nm).
- A/C drive belt with the adjusting bolt and tighten the idler pulley locknut to 29 ft. lbs. (39 Nm). Connect the connector.
- Alternator drive belt
- Halfshafts
- Front exhaust pipe

14. To install the engine wiring harness in the passenger compartment, perform the following:

- Push the harness through the cowl panel, install the retainer to the cowl with the 2 nuts and install the wire clamp to the bracket
- Connect the harness to the clamp on the ECM
- Connect the 3 ECM connectors and the circuit opening relay connector
- Connect the 3 connectors to the connectors on the bracket
- A/C amplifier connector
- Install the floor carpet, the lower instrument panel finish panel, the cowl side trim panel, and the scuff plate

15. Install or connect the following:

- Engine wiring harness with the 2 connectors to the engine compartment relay box and install the relay box covers.
- MAP sensor connector
- MAP sensor wire to the clamp on the bracket
- MAP sensor vacuum hose to the gas filter on the intake manifold
- Brake booster vacuum hose to the intake manifold
- A/C idle-up valve connector
- A/C idle-up valve hose to the intake manifold
- A/C idle-up valve hose to the air pipe
- DLC1 to the bracket
- Engine harness protector to the bracket
- Ground cable and the ground strap
- Heater hose to the water outlet and the heater hose to the water bypass pipe
- Fuel inlet hose to the fuel filter
- Fuel inlet hose with 2 new gaskets and the union bolt. Bolt: 22 ft. lbs. (30 Nm).
- Fuel return hose to the return pipe

and connect the EVAP hose to the charcoal canister
- Power steering air hoses to the intake manifold and the air pipe.
- Radiator
- On models equipped with cruise control, install the actuator cable to the clamps.
- Accelerator cable to the throttle body, cable bracket, and the clamps.
- Air cleaner
- Battery tray and battery
- Hood
- Refill the transaxle assembly, engine oil and coolant.
- Negative battery cable
- Engine undercover

16. Start the vehicle, check for leaks and repair if necessary.

5S-FE Engine

CAMRY AND CAMRY SOLARA

1. Before servicing the vehicle, refer to the precautions in the beginning of this section.
2. Drain the cooling system.
3. Drain the engine oil.
4. Drain the transaxle fluid.
5. Remove or disconnect the following:

- Negative battery cable. On vehicles equipped with an air bag, wait at least 90 seconds before proceeding.
- On the Camry Solara, strut tower brace
- Battery and battery tray
- Hood
- Engine undercover
- Accelerator cable from the throttle body. With automatic transmission, throttle cable
- Air cleaner, resonator, and air intake hose
- On models with cruise control, actuator cover, actuator with the bracket.
- Ground strap at the battery carrier
- Radiator and coolant reservoir hose
- Washer tank, electrical lead and hose

6. To disconnect the wiring harness, tag and disconnect the following:

- The 5 connectors to the engine relay box
- The igniter connector
- The noise filter connector
- The connector at the left-hand fender apron
- The 2 ground straps from the left-hand and right-hand fender aprons

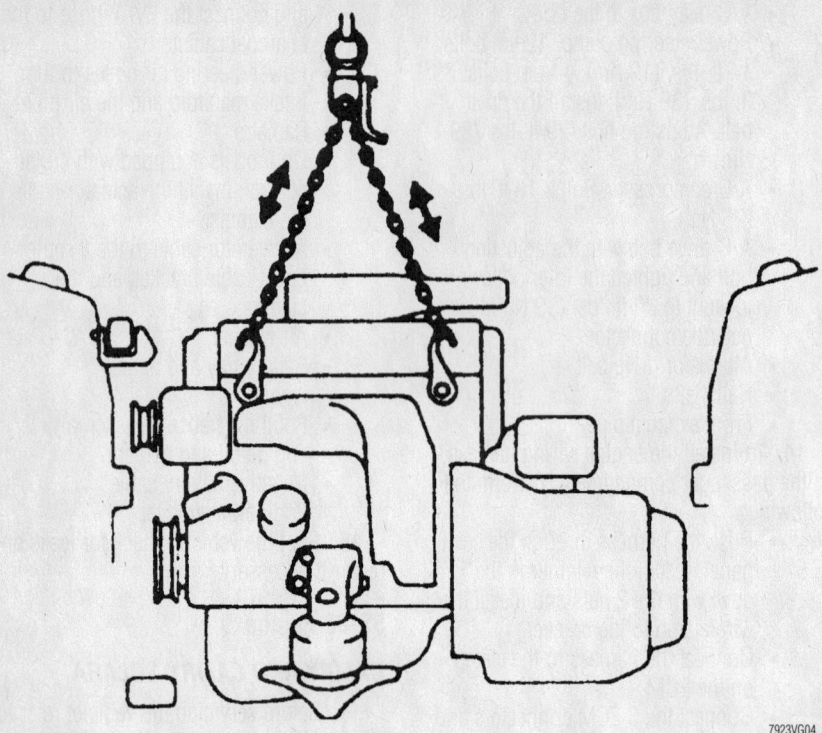

7923VG04

Use a hoist to remove the engine assembly—Camry 5S-FE engine

- The DLC1
- Disconnect the MAP sensor connector
7. Remove or disconnect the following:
- Dash panel undercover, glove compartment door, glove compartment, cowl harness connectors and the 2 ECM connectors
- Heater hoses, fuel return hose, and fuel inlet hose
- With manual transmission, starter and the clutch release cylinder. Don't disconnect the hydraulic line, simply hang the cylinder out of the way
- Transaxle control cables at the transaxle
- Tag and disconnect all remaining vacuum hoses and connectors
- Engine wire from the cowl panel
- Without disconnecting the refrigerant lines, A/C compressor
- Front exhaust pipe bracket and the front pipe from the exhaust manifold
- Halfshafts
- Without disconnecting the hydraulic lines, power steering pump
- Left engine mounting insulator
- Right rear engine mounting insulator
- Front right engine mounting insulator

8. Attach an engine lifting device to the lift hooks. Remove the 3 bolts and disconnect the control rod. Slowly and carefully, lift the engine/transaxle assembly out of the engine compartment.

9. If equipped with automatic transmission, remove the starter. Separate the engine assembly from the transaxle.

To install:

10. Install or connect the following:
- Engine assembly to the transaxle
- With automatic transmission, starter
- Engine control rod. Bolts: 47 ft. lbs. (64 Nm).
- Front right engine mount. Bolts: 59 ft. lbs. (80 Nm).
- Rear engine mount. Nuts: 48 ft. lbs. (66 Nm).
- Left mount. Bolts (3 or 4): 47 ft. lbs. (64 Nm).
- Power steering pump. Bolts: 31 ft. lbs. (43 Nm). Install the drive belt and connect the 2 air hoses to the air pipe.
- Halfshafts
- Front pipe to the manifold. Nuts: 46 ft. lbs. (62 Nm).
- A/C compressor. Bolts: 20 ft. lbs. (27 Nm).

11. Feed the engine harness through the cowl and reattach the clamp to the cowl. Make the following connections:
- The 2 ECM connectors

- The 2 cowl wire connectors
- Install the glove compartment and door
- Install the lower instrument panel and the undercover
12. Install or connect the following:
- Vacuum hoses and the transaxle control cables
- On manual transmission vehicles, release cylinder and the starter
- Fuel inlet hose and tighten it to 22 ft. lbs. (29 Nm). Connect the return hose and the 2 heater hoses.
- Attach the 5 connectors to the relay box
- The connectors from the left-hand fender apron
- Install the engine relay box
- The igniter connector
- On California models, the ignition coil connector
- The noise filter connector
- The 2 ground straps from the left-hand and right-hand fender apron
- The DLC1
- The MAP sensor connector.
- Washer tank and connect the electrical lead and hose
- Coolant reservoir hose and the radiator
- With cruise control, actuator and bracket. Connect the actuator connector and install the cover.
- Ground strap to the battery carrier.
- Air cleaner assembly
- On California models, air hose to the air cleaner assembly and connect the air intake temperature sensor connector
- With automatic transmission, throttle cable
- Accelerator cable
- Battery tray and battery
- On Camry Solara models, strut tower brace
- Hood
- Oil and coolant.
- Negative battery cable
- Undercover
13. Start the vehicle, check for leaks and repair if necessary.

CELICA

1. Before servicing the vehicle, refer to the precautions in the beginning of this section.
2. Properly relieve the fuel system pressure.
3. Drain the cooling system.
4. Drain the engine oil.
5. Drain the transaxle fluid.
6. Remove or disconnect the following:

- Negative battery cable. On vehicles equipped with an air bag, wait at least 90 seconds before proceeding.
- Battery
- Hood
- Undercover
- Accelerator cable from the throttle body, the cable bracket and clamps
- Air cleaner assembly
- With cruise control, actuator from the bracket
- Radiator
- The MAP sensor
- The MAP sensor wire from the clamp
- The MAP sensor vacuum hose from the gas filter on the intake manifold
- The brake booster vacuum hose from the intake manifold
- The DLC1 from the bracket
- The igniter connector
- On California models, the ignition coil connector
- On California models, the ignition coil high tension wire, the noise filter, the wire clamp from bracket, the ignition coil and igniter assembly, and disconnect the wire from the bracket
- The ground cable from the body
- The ground strap from the body
- The heater hose from the water outlet and bypass pipe
- The fuel inlet hose from the fuel filter and the fuel return hose from the return pipe
- The EVAP hose from the charcoal canister
- Engine wiring harness from the engine compartment relay box.

7. To remove the engine wiring harness from the passenger's compartment, remove or disconnect the following:
- The scuff plate
- The cowl side trim
- The finish panel from the lower instrument panel
- Remove the front side of the floor carpet
- The wiring harness from the clamp of the ECM bracket
- The 3 ECM connectors
- The circuit opening relay connector
- The 3 connectors from the connectors on the bracket
- The air conditioning amplifier connector
- The wire clamp from the bracket
- The 2 nuts holding the engine wiring harness to the cowl

8. Remove or disconnect the following:
- Front exhaust pipe
- Halfshafts
- Alternator drive belt
- A/C compressor and suspend it securely out of the way
- Power steering pump. Securely hang the pump out of the way
- On manual transmission, the starter
- On manual transmission, the clutch release cylinder and associated components
- Transaxle control cable(s) from the transaxle
- On automatic transmission model, transaxle control cable from the engine mounting center member.
- Exhaust pipe support bracket

9. To remove the engine mounting center member, remove the following components:
- The 2 dust covers from the rear side of the member
- The air conditioning pipe from the bracket
- The bolt and nut holding the front engine mounting bracket to the mounting insulator
- The bolt holding the rear engine mounting bracket to the insulator
- The bolt and 2 nuts holding the rear engine mounting insulator to the front suspension member
- The 2 bolts (automatic transmission) or 3 bolts (manual transmission) and rear engine mounting bracket.
- Center member with the rear mounting insulator

10. Attach an engine chain hoist to the engine hangers

11. Remove or disconnect the following:
- Left-hand engine mount bracket from the mounting insulator
- Left-hand mounting insulator
- Ground strap connector
- Right-hand engine mount bracket from the mounting insulator
- Halfshaft bearing bracket
- Transaxle

To install:

12. Install or connect the following:
- Transaxle to the engine assembly and reattach the halfshaft bearing bracket. Bolts: 47 ft. lbs. (64 Nm).
- Transaxle/engine assembly into engine compartment.
- Right-hand engine mount bracket to the mounting insulator, and temporarily install the bolt and 2 nuts.
- Left-hand engine mounting insula-

tor to the body with the through-bolt
- Left-hand engine mount bracket to the mounting insulator and install the 2 nuts and bolt (manual transmission) or the 2 bolts (automatic transmission). Bolts and nuts: 47 ft. lbs. (64 Nm).

13. Tighten the left-hand engine mounting through-bolt to the body: 54 ft. lbs. (73 Nm). Tighten the bolt and 2 nuts holding the right-hand mounting bracket to the insulator: 27 ft. lbs. (37 Nm) and nut: 38 ft. lbs. (52 Nm).

14. Install the engine ground strap connector

15. To install the engine mounting center member, perform the following:
- Attach the center member together with the rear engine mounting insulator to the front suspension member
- Temporarily install the 2 bolts holding the center member to the body
- Install the rear engine mounting bracket with the 2 bolts (automatic transmission) or the 3 bolts (manual transmission). Bolt: 58 ft. lbs. (79 Nm).
- Temporarily install the bolt and 2 nuts holding the rear engine mounting insulator to the front suspension member
- Temporarily install the bolt holding the rear engine mounting bracket to the insulator
- Temporarily install the bolt and nut holding the front engine mounting bracket to the insulator
- Tighten the 2 bolts holding the center member to the body to 26 ft. lbs. (35 Nm).
- Tighten the bolt and 2 nuts holding the rear mounting insulator to the front suspension member to 59 ft. lbs. (80 Nm).
- Tighten the bolt holding the rear engine mounting bracket to the insulator to 65 ft. lbs. (88 Nm).
- Tighten the bolt and nut holding the front engine mounting bracket to the insulator to 65 ft. lbs. (88 Nm).
- Install the A/C pipe to the bracket and install the 2 dust covers to the center member

16. Install or connect the following:
- Exhaust pipe support bracket. Bolts: 14 ft. lbs. (19 Nm).
- Transaxle control cable(s) to the transaxle.

17. On the automatic transmission vehi-

cles, transaxle control cable to the engine mounting center member.

18. On manual transmission model only, install and/or connect the following:
- The clutch release cylinder. Bolts: 108 inch lbs. (12 Nm).
- The bracket with the bolt and the tube to the bracket with the clamp
- The tube with the clamp and bolt
- The back-up switch connector

19. Install or connect the following:
- On manual transmission model only, starter, starter wire, and starter connector.
- Power steering pump. 3 mounting bolts: 32 ft. lbs. (44 Nm); adjusting bolt: 29 ft. lbs. (39 Nm); pivot bolt: 32 ft. lbs. (44 Nm)
- Hoses to the air tube
- A/C compressor. Bolts: 18 ft. lbs. (24 Nm).
- Alternator drive belt
- Halfshafts
- Front exhaust pipe

20. To install the engine wiring harness to the passenger's compartment, perform the following:
- Push the harness through the cowl panel, install the retainer to the cowl with the 2 nuts and install the wire clamp to the bracket
- Connect the harness to the clamp on the ECM
- Connect the 3 ECM connectors and the circuit opening relay connector
- Connect the 3 connectors to the connectors on the bracket
- Connect the A/C amplifier connector
- Install the floor carpet, the lower instrument panel finish panel, the cowl side trim panel, and the scuff plate

21. Install or connect the following:
- Engine wiring harness with the 2 connectors to the engine compartment relay box and install the relay box covers
- The MAP sensor connector
- The MAP sensor wire to the clamp on the bracket
- The MAP sensor vacuum hose to the gas filter on the intake manifold
- The brake booster vacuum hose to the intake manifold
- The DLC1 to the bracket
- The engine harness protector to the bracket
- On California models, the engine harness clamp to the bracket and the ignition coil and igniter assembly with the 3 bolts, and install the

harness to the bracket
- The igniter connector
- On California model, the ignition coil connector, high tension wire to the coil, and the noise filter
- The ground cable and the ground strap to the body
- The heater hose to the water outlet and the heater hose to the water bypass pipe
- Fuel inlet hose to the fuel filter
- Fuel inlet hose with 2 new gaskets and the union bolt. Bolt: 22 ft. lbs. (30 Nm).
- Fuel return hose to the return pipe and connect the EVAP hose to the charcoal canister
- Radiator
- With cruise control, actuator and connect the connector
- Accelerator cable to the throttle body, the cable bracket, and the clamps.
- Air cleaner
- Battery tray and battery
- Hood
- Oil and coolant
- Transaxle fluid to the proper level
- Negative battery cable
- Undercover

22. Start the vehicle, check for leaks and repair if necessary.

1MZ-FE Engine

AVALON, CAMRY AND CAMRY SOLARA

1. Before servicing the vehicle, refer to the precautions in the beginning of this section.

2. Properly relieve the fuel system pressure.

3. Drain the cooling system.

4. Drain the engine oil.

5. Drain the transaxle fluid.

6. Remove or disconnect the following:
- Negative battery cable. On vehicles equipped with an air bag, wait at least 90 seconds before proceeding.
- Hood
- Battery and battery tray
- Accelerator and throttle cables
- Cruise control actuator, if equipped
- Air cleaner assembly, mass air flow meter and air cleaner hose
- Radiator
- Engine relay box
- 2 igniter connectors
- Noise filter connector
- Connector from the left-hand fender apron

- 2 ground straps and any other electrical connections keeping them from being removed.
- Vacuum hoses from the engine.
- Fuel inlet and return hoses
- Heater hoses
- Transaxle control cable from the transaxle
- Instrument panel undercover, the lower instrument panel and glove box assembly
- 3 ECM connectors, the 5 cowl wire connectors, and the cooling fan ECM connector. Push the engine wire through the cowl panel
- Front exhaust pipe
- Halfshafts
- Power steering pressure tube
- Power steering pump
- A/C compressor without disconnecting the hoses
- Left-hand engine mounting insulator
- Right-hand engine mounting insulator
- Engine mounting shock absorber
- Front right engine mounting insulator

7. Attach a hoist chain to the engine hangers.

8. Remove or disconnect the following:
- Coolant reservoir hose and reservoir tank
- Right-side engine mounting stay bracket
- Engine control rod and bracket assembly

➡ **Make certain all wires, connectors and hoses are cleared from the engine.**

- Engine/transaxle assembly from the vehicle

To install:

9. Carefully lower the engine position. Keep the engine level while aligning the engine mounts.

10. Install or connect the following:
- Engine control rod and bracket. Tighten to 47 ft. lbs. (64 Nm).
- Right engine mount stay bracket. Tighten to 23 ft. lbs. (31 Nm).
- Engine ground straps
- Coolant reservoir tank
- Front engine insulator. Tighten to 48 ft. lbs. (66 Nm).
- Engine mounting shock absorber. Tighten to 35 ft. lbs. (48 Nm).
- Left and right engine mounts. Tighten to 48 ft. lbs. (66 Nm).
- Power steering pump and A/C compressor
- Power steering pressure tube

- Halfshafts and front exhaust pipe
- Engine wires and connectors
- Transaxle control cable to the transaxle
- Fuel hoses and heater hoses
- All vacuum hoses, wiring and connectors
- Radiator
- Cruise control actuator
- Throttle cable and accelerator cable
- MAF meter, the air cleaner assembly, and air cleaner hose
- Coolant and engine oil
- Battery tray and battery.
- Transaxle fluid to the proper level
- Hood
- Negative battery cable

11. Start the vehicle, check for leaks and repair if necessary.

2JZ-GE and 2JZ-GTE Engines

1. Before servicing the vehicle, refer to the precautions in the beginning of this section.

2. Properly relieve the fuel system pressure.

3. Drain the cooling system.

4. Drain the engine oil.

5. Remove or disconnect the following:

- Negative battery cable. Do not start any work for at least 90 seconds to prevent accidental deployment of the air bag.
- Hood
- Radiator

✳✳ CAUTION

The fuel injection system remains under pressure even after the engine has been turned OFF. The fuel system pressure must be relieved before disconnecting any fuel lines. Failure to do so may result in fire and/or personal injury.

- Accelerator cable and cruise control actuator
- Air cleaner assembly, volume air flow meter, and the air intake hose
- Drive belt by turning the tensioner clockwise
- Fan, fan clutch, and the water pump pulley
- Charcoal canister
- Heater water hoses
- Brake booster vacuum hose
- EVAP hose
- Noise filter connector
- Ignition coil connector
- Engine wire from the wire clamp

- Rubber cap, nut and wire from the alternator
- Engine room main wire
- Igniter connector
- Theft deterrent horn connector
- Engine wire from the 2 wire clamps
- Wire clamp and power steering solenoid valve connector
- Ground strap from the cylinder block by removing the bolt
- Rubber cap, nut, and the wire from the starter
- Fuel inlet hose from the engine. Suspend the hose union upward
- Fuel return hose from the fuel return guide
- Fuel return hose from the fuel return hose. Plug the hose end
- Engine wire from the intake manifold stay
- Power steering pump
- Power steering pressure tube
- A/C compressor without disconnecting the hoses
- Engine wire from the cowl panel
- Remove the scuff plate from the right door
- Take out the front side of the floor carpet
- Remove the 2 nuts and ECM protector
- Remove the nut and disconnect the ECM from the floor panel
- Disconnect the 2 connectors from the ECM
- Disconnect the connector from the instrument panel wire
- Disconnect the connector from the connector cassette
- Pull out the engine wire from the cabin
- With manual transmission, upper console panel, shift lever boots, and holding bolts
- With manual transmission, clutch release cylinder and the ground strap from the transmission.
- No. 2 front exhaust pipe
- Exhaust pipe heat insulator
- Driveshaft
- With automatic transmissions, control rod from the shift lever by removing the nut.

6. Support the transmission with a jack.

7. Remove the rear support member by removing the 8 bolts.

8. Attach the engine hoist chain to the engine and raise the engine slightly.

9. Remove or disconnect the following:

- Nuts holding the engine front mounting insulators to the front suspension crossmember.

- Engine from the vehicle
- Oil dipstick guide from the transmission
- Engine wire from the transmission
- Starter connector, 2 bolts, engine wire bracket, and the starter
- With automatic transmissions, oil cooler tubes from the transmission
- With automatic transmissions, torque converter clutch mounting bolts
- Transmission

To install:

10. Assemble the engine and transmission. Tighten the bolts as follows:

- 14mm: 29 ft. lbs. (39 Nm)
- 17mm: 43 ft. lbs. (72 Nm)

11. Install or connect the following:

- With automatic transmission, torque converter clutch mounting bolts by first installing the gray bolt, then install the other 5 bolts. Bolts: 25 ft. lbs. (33 Nm).
- With automatic transmission, oil cooler tubes to the transmission. Union nuts: 25 ft. lbs. (33 Nm).
- Starter
- Engine wire to the transmission
- With automatic transmission, oil dipstick guide and dipstick
- Engine and transmission as an assembly to the vehicle. Keep slight tension on the engine until the mounting bolts and nuts are installed
- Engine front mounting insulators to the front suspension crossmember. Nuts: 43 ft. lbs. (59 Nm).
- Bolts holding the support member to the body. Bolts: 19 ft. lbs. (25 Nm).
- Nuts holding the support member to the engine rear mounting insulator. Nuts: 10 ft. lbs. (13 Nm).
- Driveshaft

12. With automatic transmissions, transmission control rod as follows:

- Shift the shift lever to the **N** position.
- Fully turn the control shaft lever back and return 2 notches. The control shaft is now in the neutral position
- Connect the control rod to the shift lever with the nut. Tighten the nut to 108 inch lbs. (13 Nm).

13. Install or connect the following:

- Exhaust pipe heat insulator
- No. 2 front exhaust pipe

- With manual transmission, clutch release cylinder and ground strap. Tighten the clutch release cylinder bolts to 108 inch lbs. (13 Nm) and the ground strap bolt to 27 ft. lbs. (37 Nm).
- With manual transmission, upper console panel, shift lever boots, and holding bolts.
- Engine wire through the cowl panel.
- Connect the connector to the connector cassette
- Connect the connector to the instrument panel wire connector
- Connect the 2 connectors to the ECM
- Insert the ECM bracket into the stay on the floor panel
- Install the ECM with the nut
- Install the ECM protector with the 2 nuts
- Install the floor carpet
- Install the scuff plate
- Engine wire to the cowl panel
- Air conditioning compressor to the engine
- Power steering tube with the 2 clamp bolts
- Power steering pump. Lower bolt: 43 ft. lbs. (58 Nm); upper bolt: 29 ft. lbs. (39 Nm); rear stay bolts: 29 ft. lbs. (39 Nm); front bracket bolts: 43 ft. lbs. (58 Nm).
- Engine wire bracket
- Fuel return hose to the fuel return pipe
- Fuel return hose to the clamp of the oil dipstick guide
- Fuel inlet hose with the 2 new gaskets and the union bolt. Bolt: 22 ft. lbs. (29 Nm).
- Wires and connectors
- EVAP hose
- Brake booster vacuum hose
- Heater hoses
- Charcoal canister
- Water pump pulley, fan, and the fan clutch
- Drive belt to the engine
- Air cleaner, VAF meter, and the intake air connector pipe.
- Control cables to the throttle body
- Oil
- Radiator
- Cooling system
- Negative battery cable
- Hood

14. Start the vehicle, check for leaks and repair if necessary.

Water Pump

REMOVAL & INSTALLATION

5E-FE Engine

1. Before servicing the vehicle, refer to the precautions in the beginning of this section.
2. Drain the cooling system.
3. Remove or disconnect the following:

- Negative battery cable. On vehicles equipped with an air bag, wait at least 90 seconds before proceeding.
- Alternator
- With distributorless ignition, intake manifold stay bracket by disconnecting the wire clamps and removing the 2 nuts
- With distributor ignition, remove the intake manifold stay bracket by removing the 2 nuts and 2 bolts.
- Water inlet pipe
- Oil dipstick guide
- Alternator adjusting bar
- Water pump attaching bolt and nuts
- Water pump

To install:

4. Install or connect the following:

- 0.08–0.12 in. (2–3mm) bead of sealant to the groove in the pump.
- O-ring, lubricated with a little soap and water, on the water inlet pipe.

- Water pump. Bolts: 13 ft. lbs. (17 Nm).
- Oil dipstick guide
- Alternator adjusting bar
- Dipstick guide clamp bolt
- Water inlet pipe. Bolt: 65 inch lbs. (7.5 Nm).
- Water bypass, heater inlet and water inlet hoses
- For distributorless ignition engines, install the intake manifold bracket by installing the 2 bolts and the wire clamp. Bolts: 15 ft. lbs. (20 Nm).
- For distributor ignition, install the intake manifold bracket by installing the 2 bolts. Bolts: 15 ft. lbs. (20 Nm).
- Alternator and belt
- Negative battery cable

5. Fill the coolant system to the proper level.
6. Start the vehicle, check for leaks and repair if necessary.

1NZ-FE Engine

1. Before servicing the vehicle, refer to the precautions in the beginning of this section.
2. Drain the cooling system.
3. Remove or disconnect the following:
- Negative battery cable
- Accessory drive belt
- Water pump pulley, using a holding tool
- Water pump and gasket (3 bolts; 2 nuts)

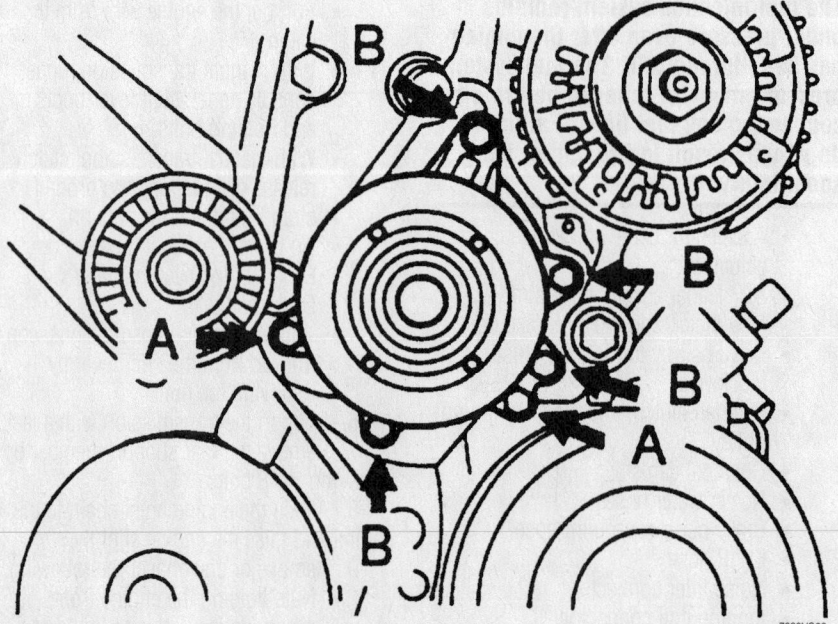

Water pump bolt identification—1.8L (1ZZ-FE) engine

To install:

4. Install or connect:
- Water pump, with a new gasket. Torque the nuts and bolts to 96 inch lbs. (11 Nm).
- Pulley. Torque the bolts to 11 ft. lbs. (15 Nm).
- Drive belt
- Negative battery cable

5. Fill the cooling system to the proper level.

6. Start the vehicle, check for leaks and repair if necessary.

1ZZ-FE Engine

1. Before servicing the vehicle, refer to the precautions in the beginning of this section.
2. Drain the cooling system.
3. Remove or disconnect the following:
- Negative battery cable
- Right-hand engine under cover
- Drive belt
- Water pump

To install:

4. Install or connect the following:
- Water pump. Bolts marked **A** (short): 80 inch lbs. (9 Nm). Bolts marked **B** (long): 96 inch lbs. (11 Nm).
- Drive belt
- Right engine under cover
- Negative battery cable

5. Fill the cooling system to the proper level.

6. Start the vehicle, check for leaks and repair if necessary.

2ZZ-GE Engine

1. Before servicing the vehicle, refer to the precautions in the beginning of this section.
2. Drain the cooling system.
3. Remove or disconnect the following:
- Negative battery cable
- Right-hand engine under cover
- Drive belt
- Water pump pulley
- Water pump and O-ring

To install:

4. Install or connect the following:
- Water pump with new O-ring. Bolts: 80 inch lbs. (9 Nm).
- Water pump pulley. Bolts: 11 ft. lbs. (15 Nm).
- Drive belt
- Right engine under cover
- Negative battery cable

5. Fill the cooling system to the proper level.

6. Start the vehicle, check for leaks and repair if necessary.

5S-FE Engine

1. Before servicing the vehicle, refer to the precautions in the beginning of this section.
2. Drain the cooling system.
3. Remove or disconnect the following:
- Negative battery cable. On vehicles equipped with an air bag, wait at least 90 seconds before proceeding.
- Right engine undercover
- Lower radiator hose from the water outlet
- Timing belt, timing belt tension spring, and the No. 2 idler pulley
- Alternator, drive belt and the adjusting bar if necessary
- 2 nuts holding the water pump to the water bypass pipe and remove the 3 bolts in sequence.
- Water pump cover assembly
- Gasket and 2 O-rings from the water pump and the bypass pipe.
- Water pump from the water pump cover by removing the 3 bolts in sequence.

To install:

4. Install or connect the following:
- Water pump to the water pump cover. Bolts: 78 inch lbs. (9 Nm) in proper sequence.
- Water pump cover to the water bypass pipe, but do not install the nuts yet.
- Water pump and tighten the 3 bolts in sequence. Bolts: 78 inch lbs. (9 Nm). Nuts: 82 inch lbs. (9 Nm).

- Alternator drive belt adjusting bar. Bolt: 13 ft. lbs. (18 Nm).
- No. 2 idler pulley and the timing belt tension spring
- Lower radiator hose
- Timing belt
- Right engine undercover
- Negative battery cable

5. Fill the cooling system to the proper level.

6. Start the vehicle, check for leaks and repair if necessary.

1MZ-FE Engine

1. Before servicing the vehicle, refer to the precautions in the beginning of this section.
2. Drain the cooling system.
3. Remove or disconnect the following:
- Negative battery cable. On vehicles equipped with an air bag, wait at least 90 seconds before proceeding
- Timing belt
- No. 2 idler pulley
- 3 clamps and engine wire from the rear timing belt cover
- Rear timing belt cover
- Water pump

To install:

4. Install or connect the following:
- Liquid sealer to the gasket, water pump and engine block.
- Water pump. Bolts and nuts: 53 inch lbs. (6 Nm).
- Rear timing belt cover. Bolts: 74 inch lbs. (9 Nm).
- Engine wire with the 3 clamps to the rear timing belt cover.
- No. 2 idler pulley. Bolt: 32 ft. lbs. (43 Nm).

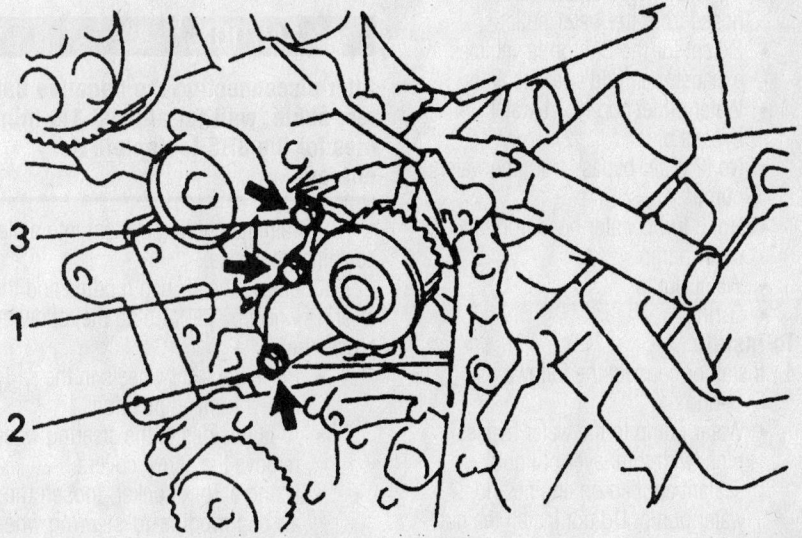

Install the 3 water pump bolts in this sequence—2.2L (5S-FE) engine

7923VG05

- With the flange side **outward**, right-hand camshaft pulley. Align the knock pin hole on the camshaft pulley with the knock pin on the camshaft. Bolt: 65 ft. lbs. (88 Nm).
- With the flange side **inward**, left-hand camshaft pulley. Align the knock pin hole on the camshaft pulley with the knock pin on the camshaft. Bolt: 94 ft. lbs. (125 Nm).
- Timing belt
- Negative battery cable

5. Fill the cooling system to the proper level.

6. Start the vehicle, check for leaks and repair if necessary.

2JZ-GE and 2JZ-GTE Engines

1. Before servicing the vehicle, refer to the precautions in the beginning of this section.

2. Drain the cooling system.

3. Remove or disconnect the following:
- Negative battery cable. On vehicles equipped with an air bag, wait at least 90 seconds before proceeding.
- Air cleaner and MAF meter assembly
- Radiator
- With manual transmission, drive belt tensioner damper.
- Loosen the 4 nuts holding the fan clutch to the water pump.
- Drive belt from the engine
- Fan, fan clutch, and the water pump pulley
- Water inlet, lower radiator hose assembly, and the thermostat
- Timing belt
- Alternator
- On turbo models, turbo water hoses from the water outlet
- Except for the California vehicles, exhaust manifold heat insulator
- Water outlet and No. 1 water bypass pipe
- No. 2 water bypass from the water pump
- No. 3 turbo water hose from the water pump
- Water pump
- O-ring

To install:

4. Install or connect the following:
- O-ring
- Water pump to the water bypass pipe, with thin layer of liquid sealant applied on engine and water pump. Do not install the nut at this time.

- Water pump. Be sure to replace the bolts to their original positions. Bolts: 15 ft. lbs. (21 Nm).
- 2 nuts holding the No. 2 water bypass pipe to the water pump. Nuts: 15 ft. lbs. (21 Nm).
- No. 3 turbo water hose to the water pump.
- Water bypass outlet and No. 1 water bypass pipe.
- Turbo water hoses to the water outlet.
- Alternator
- Except for California vehicle, exhaust manifold heat insulator
- Engine wire bracket
- Timing belt
- Thermostat, water inlet, and the lower radiator hose assembly.
- Water pump pulley, fan, fluid clutch assembly, and the drive belt. Fan nuts: 12 ft. lbs. (16 Nm).
- With manual transmission, drive belt tensioner damper.
- Radiator
- Air cleaner and MAF meter assembly.
- No. 1 air hose
- Negative battery cable

5. Fill the cooling system to the proper level.

6. Start the vehicle, check for leaks and repair if necessary.

Heater Core

REMOVAL & INSTALLATION

Avalon

1. Disconnect the negative battery cable.

✳✳ CAUTION

After disconnecting the negative battery cable, wait for at least 1½ minutes for the SRS to deplete its energy.

2. Drain the cooling system into a clean container for reuse.

3. Remove the air bag module and the steering wheel by performing the following procedure:
- Place the front wheels in the straight-ahead position.
- At both sides of the steering wheel, remove the screw covers.
- Using a Torx socket, loosen the 2 air bag module-to-steering wheel Torx screws until the circumfer-

ence ring catches on the screw case.
- Carefully, remove the air bag module and disconnect the electrical connector.
- Remove the steering wheel nut.
- Using a steering wheel puller, press the steering wheel from the steering column.

4. Remove the instrument panel by performing the following procedure:
- Remove the front pillar garnishes and the door scuff plates.
- Remove the hood lock release lever and the cowl side trims.
- Remove the steering column covers and the combination switch.
- Remove the lower finish panel assembly and the instrument panel finish lower left side panel.
- Remove the fuse box bolt and the No. 2 heater-to-register duct.
- Remove the parking brake release lever and the No. 2 undercover.
- Remove the lower No. 2 finish panel.
- If equipped with a column shifter, disconnect the shift control cable from the shift lever housing; then, disconnect the shift control cable from the steering column cable bracket.
- Matchmark the steering column shaft and the control valve shaft.
- Remove the steering column shaft-to-intermediate shaft bolt.
- Remove the steering column-to-instrument panel nuts and remove the steering column assembly.
- Inside the glove compartment, pry out the glove compartment door finish plate.
- Pull out the air bag electrical connector and disconnect it.
- Remove the 3 glove compartment door-to-instrument panel nuts and the door.
- Remove the 4 glove compartment-to-instrument panel screws and the glove compartment; then, disconnect the glove box light connector.
- Remove the passenger's side air bag module-to-instrument panel 2 bolts and 4 nuts. Carefully, remove the air bag module from the instrument panel.
- Remove the center cluster finish panel and the radio.
- Remove the heater control assembly.
- If equipped with a floor shifter, remove the upper console panel,

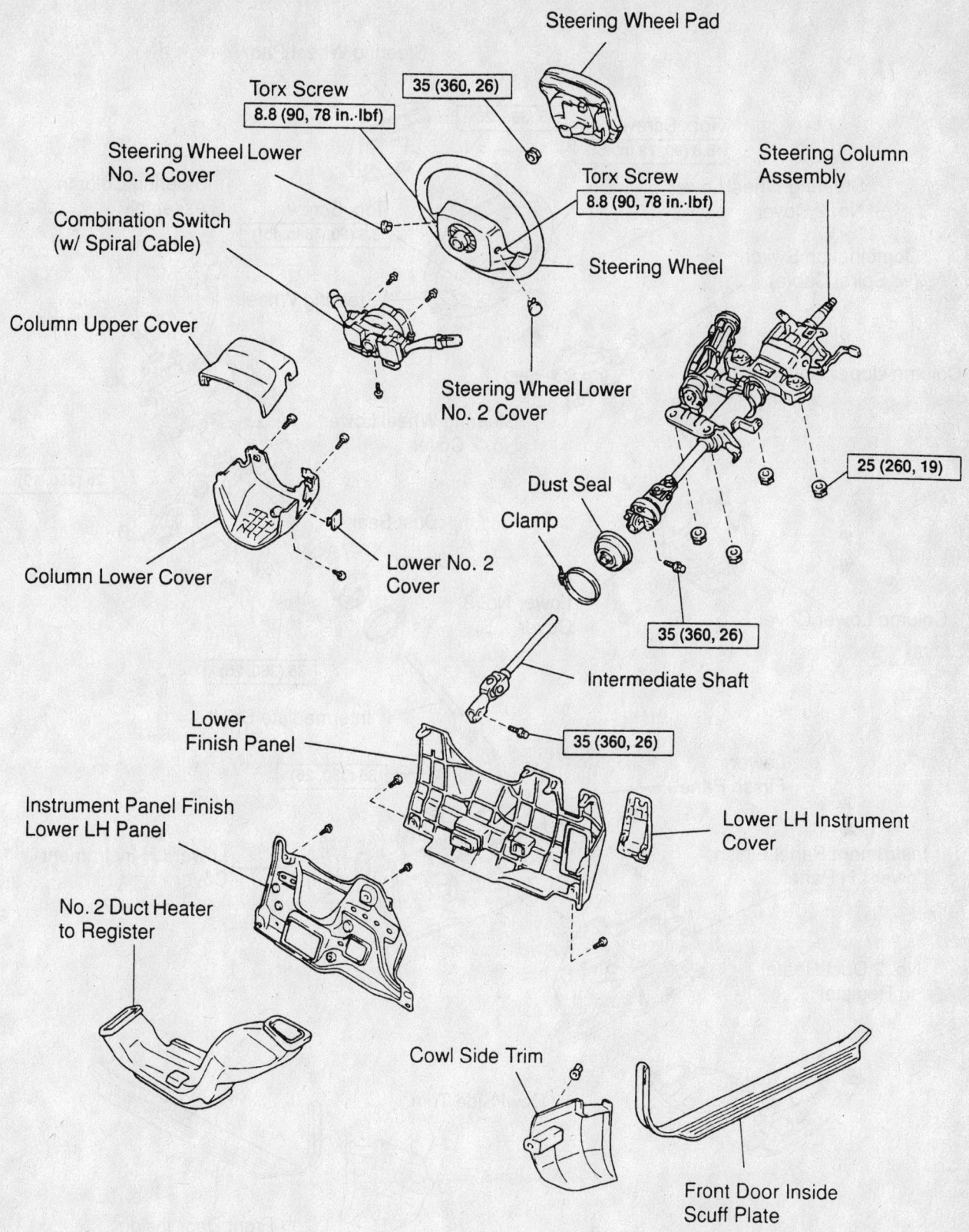

Steering Wheel Pad

Torx Screw
8.8 (90, 78 in.·lbf)

35 (360, 26)

Steering Wheel Lower
No. 2 Cover

Torx Screw
8.8 (90, 78 in.·lbf)

Steering Column
Assembly

Combination Switch
(w/ Spiral Cable)

Steering Wheel

Column Upper Cover

Steering Wheel Lower
No. 2 Cover

25 (260, 19)

Dust Seal

Clamp

Column Lower Cover

Lower No. 2
Cover

35 (360, 26)

Intermediate Shaft

Lower
Finish Panel

35 (360, 26)

Instrument Panel Finish
Lower LH Panel

Lower LH Instrument
Cover

No. 2 Duct Heater
to Register

Cowl Side Trim

Front Door Inside
Scuff Plate

N·m (kgf·cm, ft·lbf) : Specified torque

93112GL2

Exploded view of the air bag module, the steering wheel, the floor shift steering column and related components—Avalon

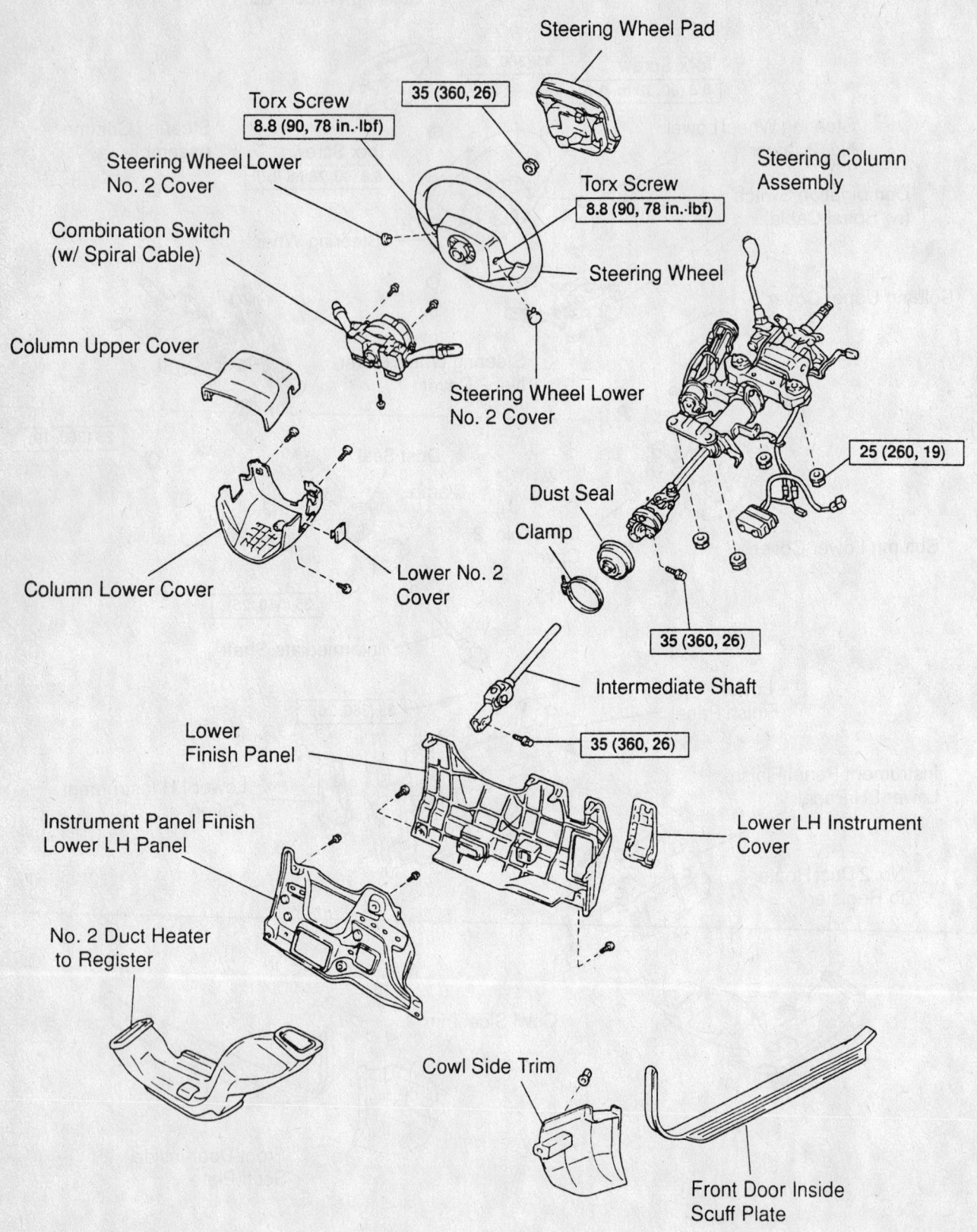

Steering Wheel Pad

Torx Screw
8.8 (90, 78 in.·lbf)

35 (360, 26)

Steering Wheel Lower
No. 2 Cover

Torx Screw
8.8 (90, 78 in.·lbf)

Steering Column
Assembly

Combination Switch
(w/ Spiral Cable)

Steering Wheel

Column Upper Cover

Steering Wheel Lower
No. 2 Cover

Column Lower Cover

Dust Seal

Clamp

25 (260, 19)

Lower No. 2
Cover

35 (360, 26)

Intermediate Shaft

Lower
Finish Panel

35 (360, 26)

Instrument Panel Finish
Lower LH Panel

Lower LH Instrument
Cover

No. 2 Duct Heater
to Register

Cowl Side Trim

Front Door Inside
Scuff Plate

N·m (kgf·cm, ft·lbf) : Specified torque

93112GL3

Exploded view of the air bag module, the steering wheel, the column shift steering column and related components—Avalon

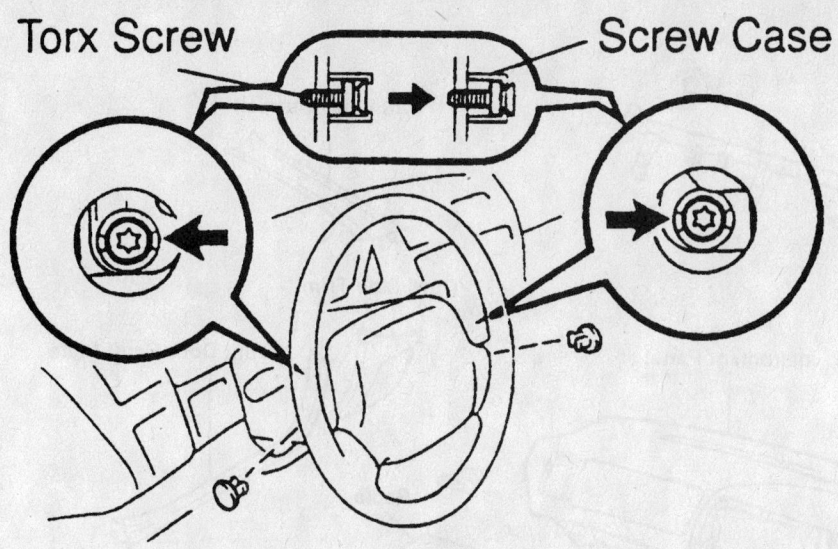

Releasing the air bag module-to-steering wheel screws—Avalon

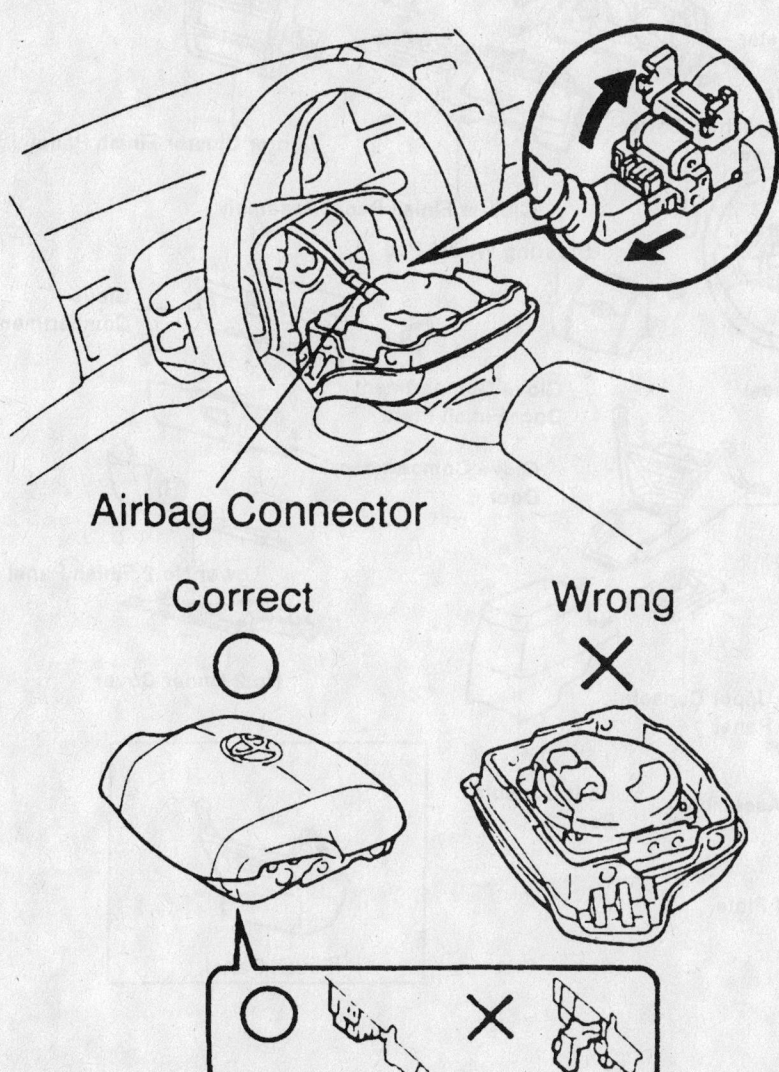

Disconnecting and positioning the air bag module—Avalon

the rear console box and the front console box.

- If equipped with a column shifter, remove the finish panel.
- Pry out the cluster finish panel.
- Remove the 6 cluster finish panel screws, the cluster finish panel assembly.
- Remove the 4 combination meter screws and the combination meter.
- Disconnect instrument panel electrical connectors.
- Remove the instrument panel-to-chassis nuts/bolts and the remove the instrument panel.

5. Remove the instrument panels No. 2 brace.

6. Disconnect the heater hoses from the heater core.

7. Remove the 2 heater pipes-to-heater core screws and clips; then, disconnect the heater pipes from the heater core.

8. Remove the heater core O-rings and discard them.

9. Remove the heater core from the heater housing.

To install:

10. Install the heater core to the heater housing.

11. Install new heater core O-rings.

12. Install the heater pipes to the heater core; then, the 2 heater pipes-to-heater core screws and clips.

13. Install the heater hoses to the heater core.

14. Install the instrument panels No. 2 brace.

15. Install the instrument panel by performing the following procedure:

- Install the instrument panel and the instrument panel-to-chassis nuts/bolts.
- Install instrument panel electrical connectors.
- Install the combination meter and the 4 combination meter screws.
- Install the cluster finish panel, the 6 cluster finish panel assembly screws.
- Install the cluster finish panel.
- If equipped with a column shifter, install the finish panel.
- If equipped with a floor shifter, install the upper console panel, the rear console box and the front console box.
- Install the heater control assembly.
- Install the center cluster finish panel and the radio.
- Carefully, install the air bag module to the instrument panel; then,

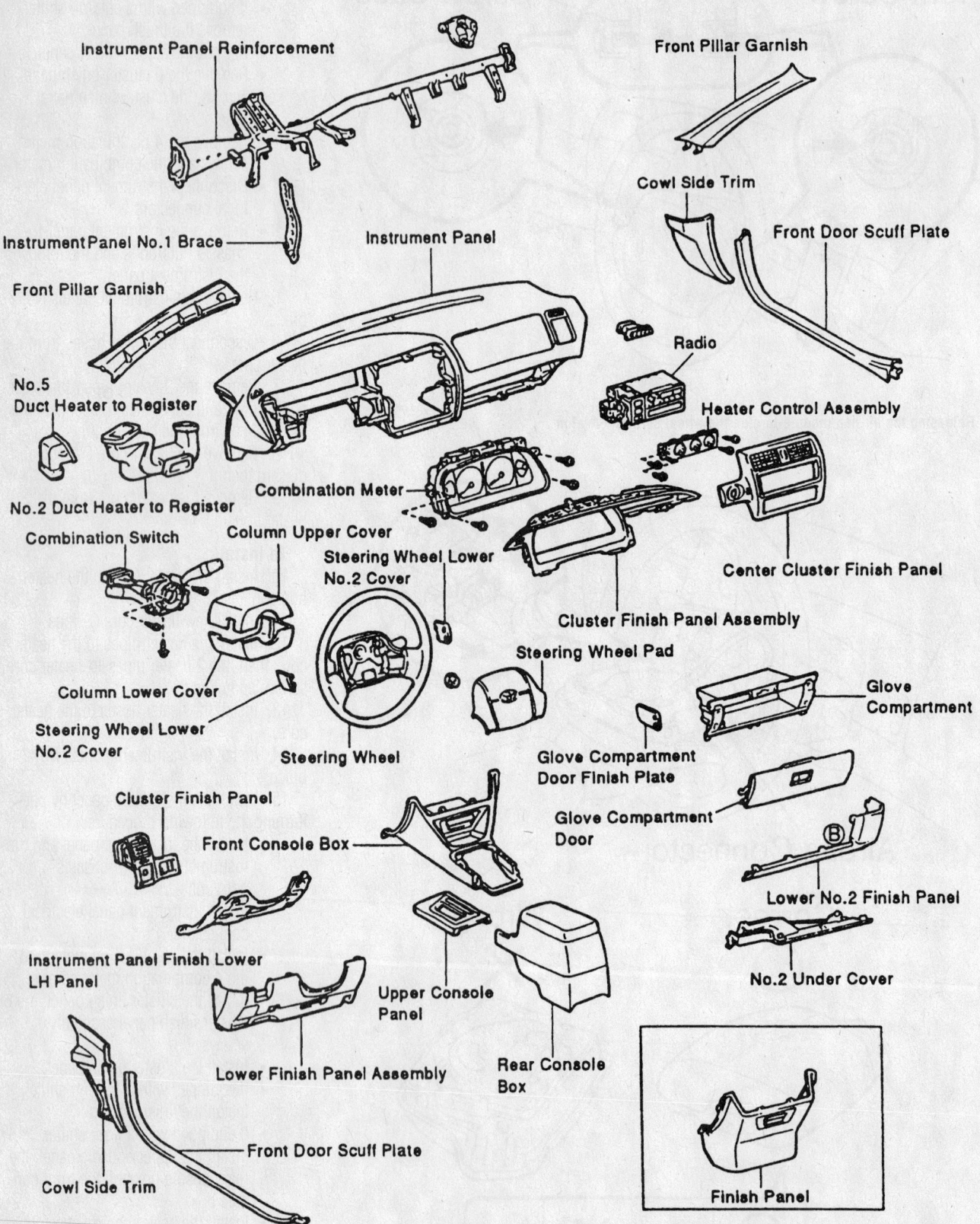

Instrument Panel Reinforcement

Front Pillar Garnish

Cowl Side Trim

Front Door Scuff Plate

Instrument Panel No.1 Brace

Instrument Panel

Front Pillar Garnish

Radio

No.5
Duct Heater to Register

Heater Control Assembly

Combination Meter

No.2 Duct Heater to Register

Center Cluster Finish Panel

Combination Switch

Column Upper Cover

Steering Wheel Lower
No.2 Cover

Cluster Finish Panel Assembly

Steering Wheel Pad

Column Lower Cover

Glove
Compartment

Steering Wheel Lower
No.2 Cover

Steering Wheel

Glove Compartment
Door Finish Plate

Cluster Finish Panel

Glove Compartment
Door

Front Console Box

Lower No.2 Finish Panel

Instrument Panel Finish Lower
LH Panel

No.2 Under Cover

Upper Console
Panel

Lower Finish Panel Assembly

Rear Console
Box

Front Door Scuff Plate

Cowl Side Trim

Finish Panel

93112GA5

Exploded view of the instrument panel and related components—Avalon

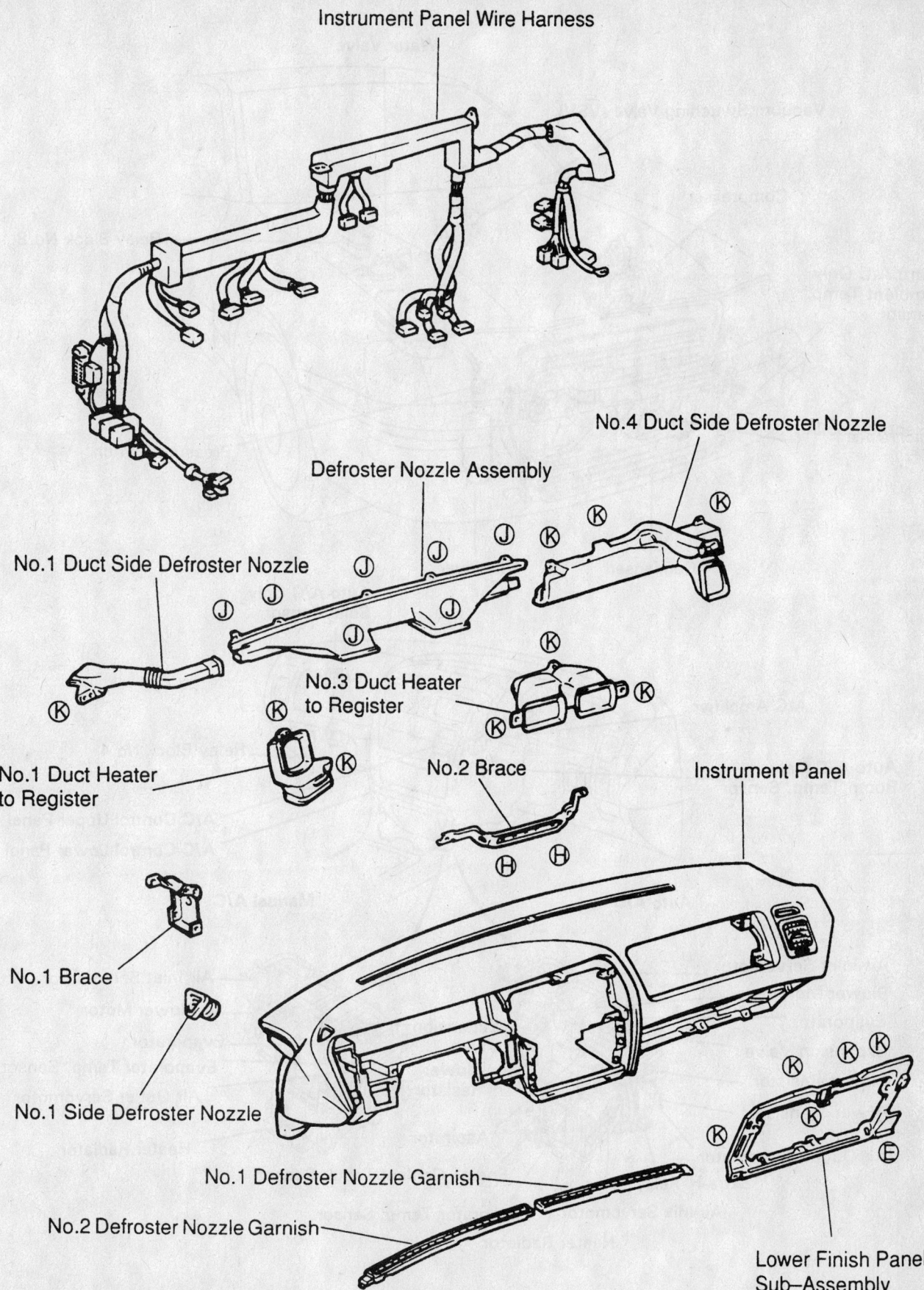

Instrument Panel Wire Harness

No.4 Duct Side Defroster Nozzle

Defroster Nozzle Assembly

No.1 Duct Side Defroster Nozzle

No.3 Duct Heater to Register

No.1 Duct Heater to Register

No.2 Brace

Instrument Panel

No.1 Brace

No.1 Side Defroster Nozzle

No.1 Defroster Nozzle Garnish

No.2 Defroster Nozzle Garnish

Lower Finish Panel Sub–Assembly

93112GL4

Exploded view of the wiring harness, ventilation system and related components—Avalon

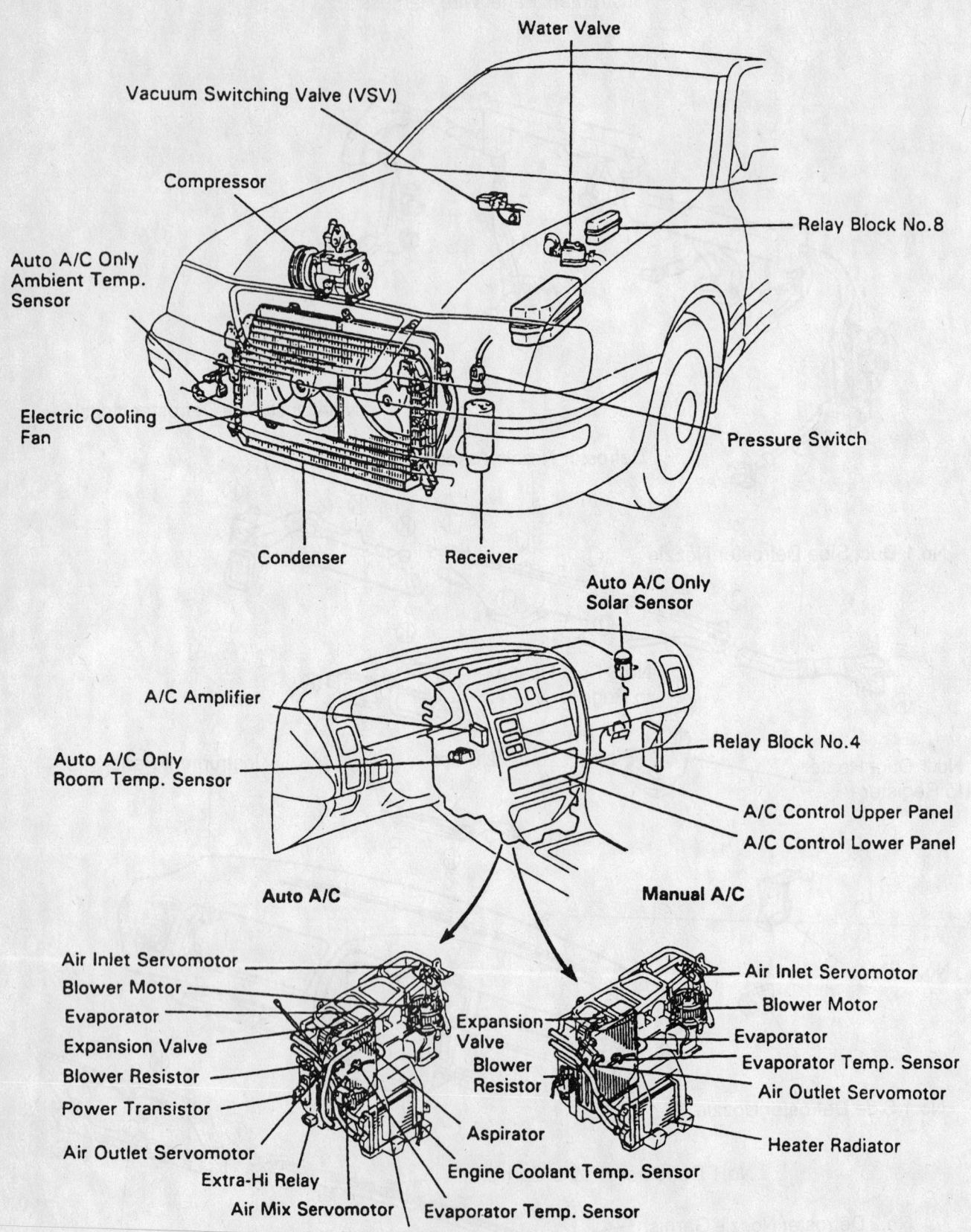

Water Valve

Vacuum Switching Valve (VSV)

Compressor

Auto A/C Only Ambient Temp. Sensor

Relay Block No.8

Electric Cooling Fan

Pressure Switch

Condenser

Receiver

Auto A/C Only Solar Sensor

A/C Amplifier

Auto A/C Only Room Temp. Sensor

Relay Block No.4

A/C Control Upper Panel

A/C Control Lower Panel

Auto A/C

Manual A/C

Air Inlet Servomotor
Blower Motor
Evaporator
Expansion Valve
Blower Resistor
Power Transistor
Air Outlet Servomotor
Extra-Hi Relay
Air Mix Servomotor
Heater Radiator

Expansion Valve
Blower Resistor
Aspirator
Engine Coolant Temp. Sensor
Evaporator Temp. Sensor

Air Inlet Servomotor
Blower Motor
Evaporator
Evaporator Temp. Sensor
Air Outlet Servomotor
Heater Radiator

93112GA6

View of the heater/air conditioning assembly and related components—Avalon

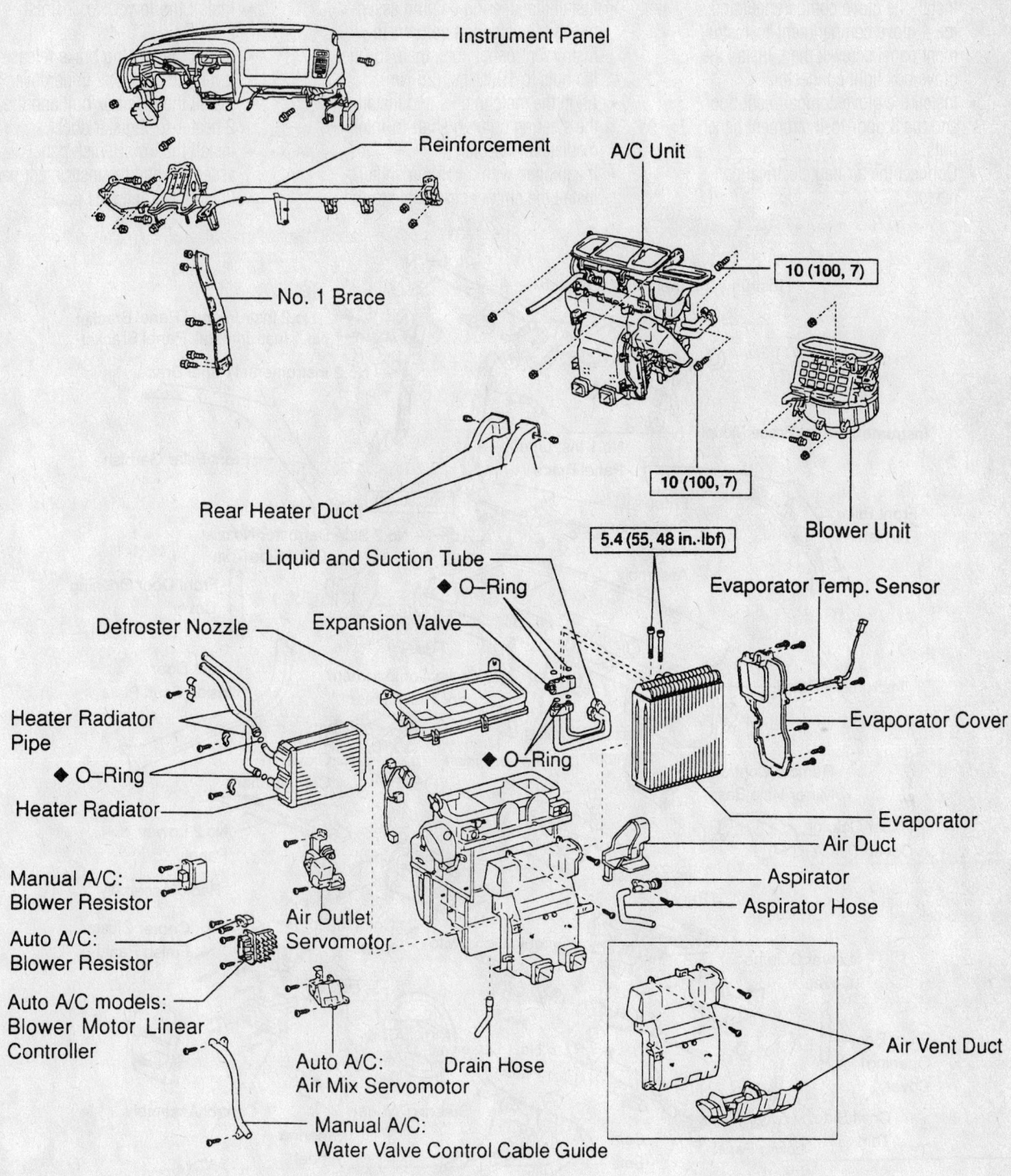

Instrument Panel

Reinforcement

A/C Unit

No. 1 Brace

10 (100, 7)

10 (100, 7)

Blower Unit

Rear Heater Duct

5.4 (55, 48 in.·lbf)

Liquid and Suction Tube

◆ O–Ring

Evaporator Temp. Sensor

Defroster Nozzle

Expansion Valve

Heater Radiator Pipe

◆ O–Ring

Evaporator Cover

◆ O–Ring

Heater Radiator

Evaporator

Air Duct

Manual A/C: Blower Resistor

Aspirator

Auto A/C: Blower Resistor

Aspirator Hose

Air Outlet Servomotor

Auto A/C models: Blower Motor Linear Controller

Air Vent Duct

Auto A/C: Air Mix Servomotor

Drain Hose

Manual A/C: Water Valve Control Cable Guide

N·m (kgf·cm, ft·lbf) : Specified torque

◆ Non–reusable part

93112GL5

Exploded view of the evaporator housing, heater housing, heater core and related components—Avalon

install the passenger's side air bag module-to-instrument panel 2 bolts and 4 nuts.

- Install the glove compartment and the 4 glove compartment-to-instrument panel screws; then, install the glove box light connector.
- Install the glove compartment door and the 3 door-to-instrument panel nuts.
- Connect the air bag electrical connector.

- Inside the glove compartment, install the glove compartment door finish plate.
- Install the steering column assembly and the steering column-to-instrument panel nuts; then, torque the nuts to 19 ft. lbs. (25 Nm).
- Align the matchmarks and install the steering column shaft-to-intermediate shaft bolt.
- If equipped with a column shifter, install the shift control cable to the

shift lever housing; then, install the shift control cable to the steering column cable bracket.

- Install the lower No. 2 finish panel.
- Install the parking brake release lever and the No. 2 undercover.
- Install the fuse box bolt and the No. 2 heater-to-register duct.
- Install the lower finish panel assembly and the instrument panel finish lower left side panel.

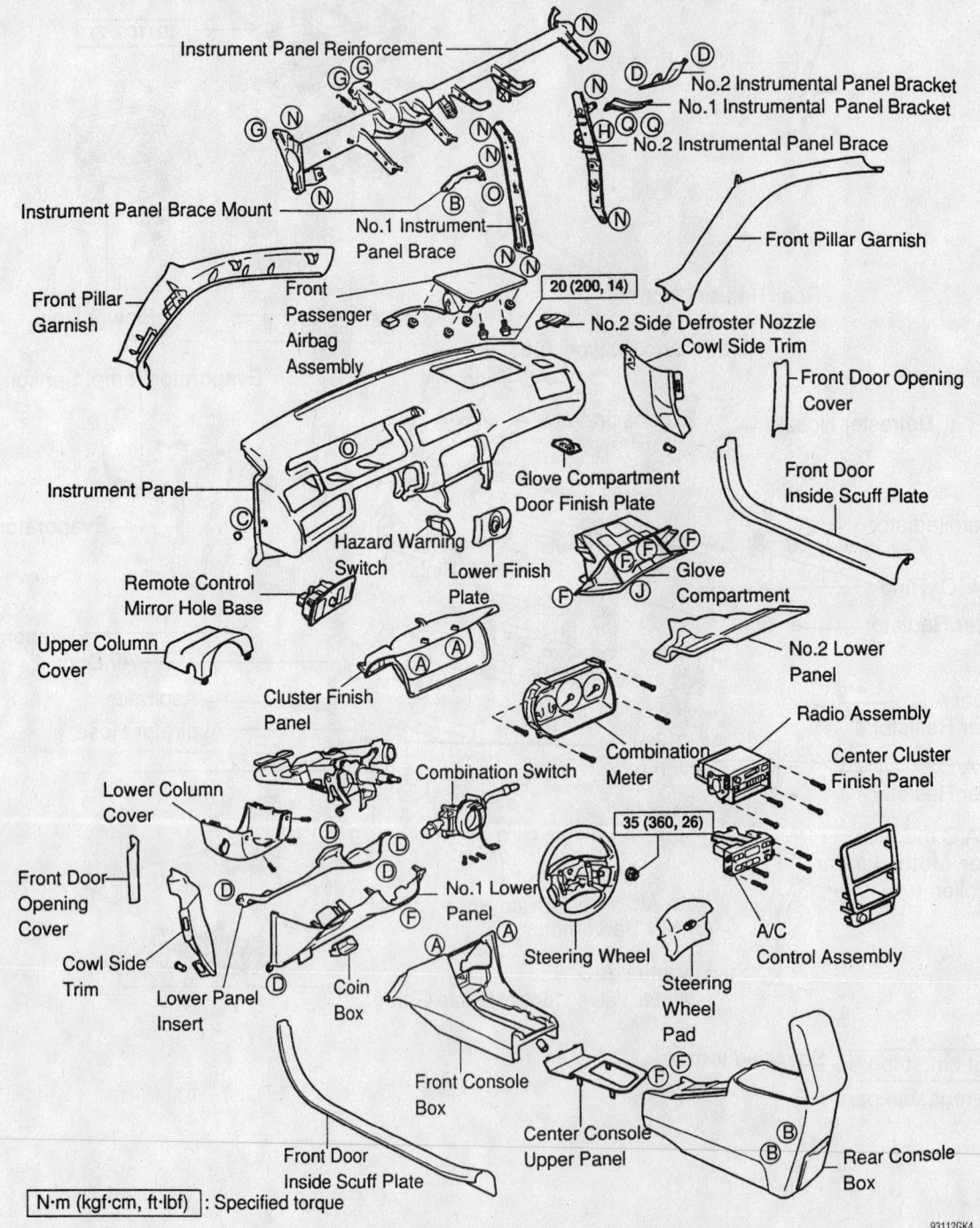

Exploded view of the instrument panel, heater/air conditioning housing and related components—Camry

93112GK4

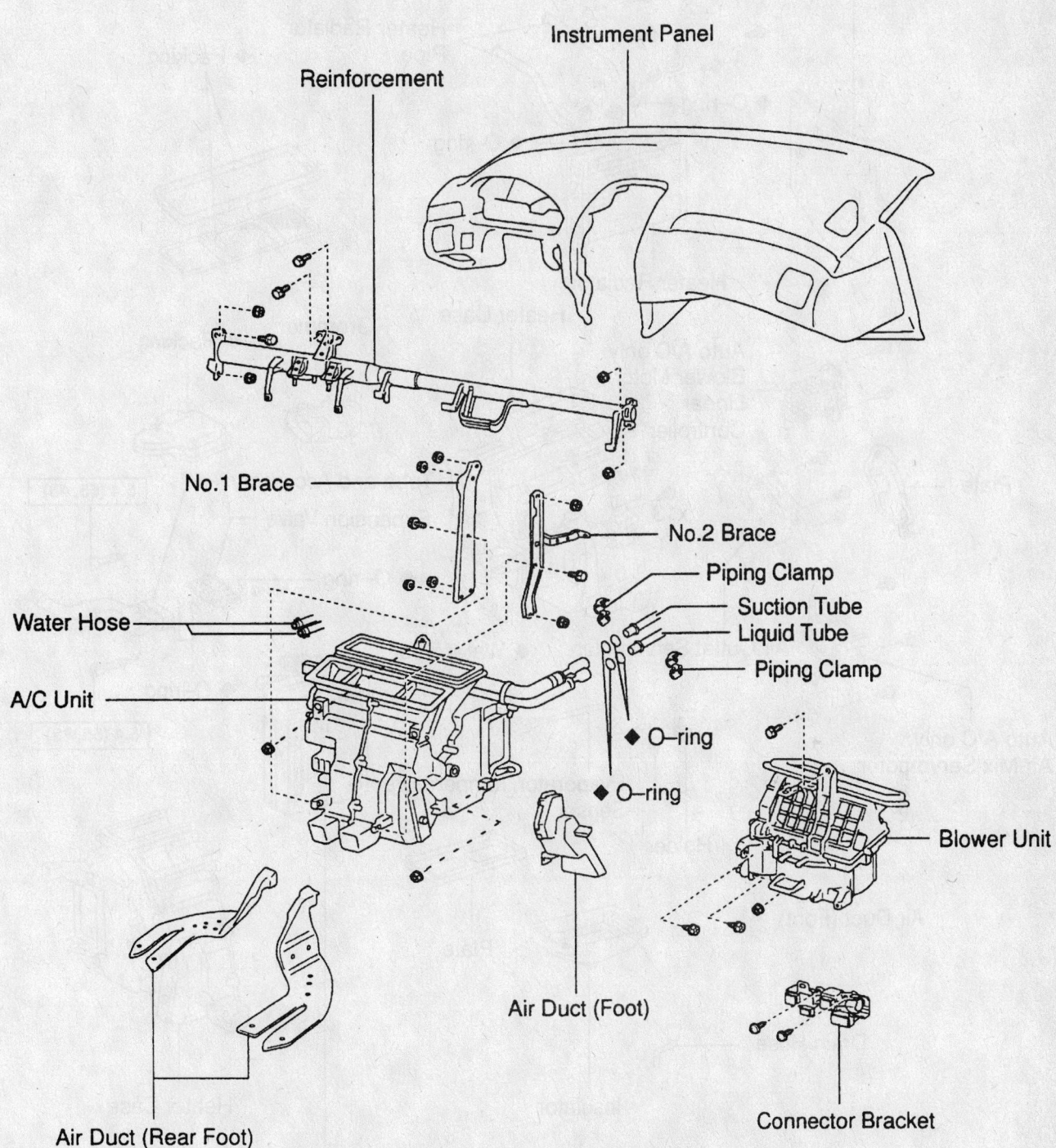

Instrument Panel

Reinforcement

No.1 Brace

No.2 Brace

Piping Clamp

Suction Tube

Liquid Tube

Piping Clamp

Water Hose

A/C Unit

◆ O–ring

◆ O–ring

Blower Unit

Air Duct (Foot)

Connector Bracket

Air Duct (Rear Foot)

◆ Non–reusable part

93112GK5

Exploded view of the instrument panel, heater/air conditioning housing and related components—Camry

93112GK6

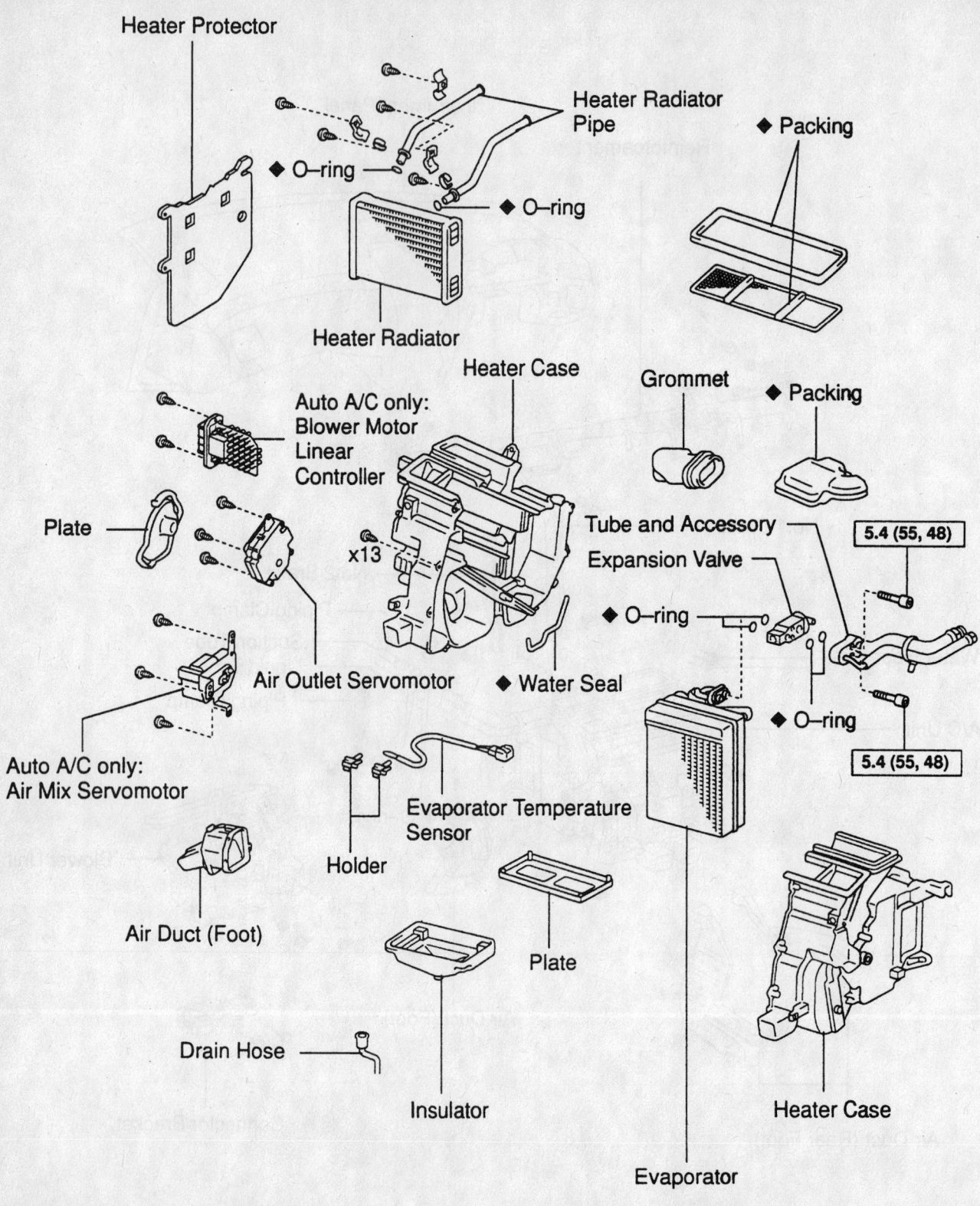

Heater Protector

Heater Radiator Pipe

◆ Packing

◆ O−ring

◆ O−ring

Heater Radiator

Heater Case

Grommet

◆ Packing

Auto A/C only:
Blower Motor
Linear
Controller

Plate

x13

Tube and Accessory

Expansion Valve

5.4 (55, 48)

◆ O−ring

Air Outlet Servomotor

◆ Water Seal

◆ O−ring

Auto A/C only:
Air Mix Servomotor

Evaporator Temperature
Sensor

Holder

5.4 (55, 48)

Air Duct (Foot)

Plate

Drain Hose

Insulator

Heater Case

Evaporator

N·m (kgf·cm, in.·lbf) : Specified torque

◆ Non−reusable part

Exploded view of the heater/air conditioning housing, heater core, evaporator core and related components—Camry

- Install the steering column covers and the combination switch.
- Install the hood lock release lever and the cowl side trims.
- Install the front pillar garnishes and the door scuff plates.

16. Install the air bag module and the steering wheel by performing the following procedure:

- Align the matchmarks and install the steering wheel to the steering column.
- Install the steering wheel nut and torque to 26 ft. lbs. (35 Nm).
- Carefully, connect the electrical connector and install the air bag module.
- Using a Torx socket, tighten the 2 air bag module-to-steering wheel Torx screws to 78 inch lbs. (8.8 Nm).
- At both sides of the steering wheel, install the screw covers.

17. Refill the cooling system.
18. Install the negative battery cable.
19. Operate the engine to normal operating temperatures; then, check the climate control operation and check for leaks.

Camry

1. Disconnect the negative battery cable.
2. Drain the cooling system into a clean container for reuse.
3. Disconnect the heater hoses from the heater core.
4. At the driver's side, remove the lower instrument panel.
5. Remove the left hand instrument lower panel.

6. At the heater/air conditioning housing, disengage the 3 heater protector-to-heater/air conditioning housing clips and remove the heater protector.
7. Remove the 3 heater core pipe clamp screws and the clamps.
8. Remove the 2 heater core pipe clamp screws and the clamps; then, disconnect the pipes from the heater core.
9. Remove the heater core.

To install:
10. Install the heater core.
11. Connect the pipes to the heater core and install the heater core pipe clamp and the 2 clamp screws.
12. Install the heater core pipe clamps and the 3 clamp screws.
13. At the heater/air conditioning housing, install the heater protector and engage the 3 heater protector-to-heater/air conditioning housing clips.
14. Install the left hand instrument lower panel.
15. At the driver's side, install the lower instrument panel.
16. Connect the heater hoses to the heater core.
17. Refill the cooling system.
18. Connect the negative battery cable.

Celica

1. Disconnect the negative battery cable.
2. Drain the cooling system into a clean container for reuse.
3. Discharge and recover the air conditioning system refrigerant.
4. Remove the evaporator housing by performing the following procedure:

- Disconnect the refrigerant lines from the evaporator core. Discard

the O-rings. Plug the openings to prevent contamination.
- Remove the 2 grommets and the drain pipe grommet.
- At the passenger's side, remove the lower finish panel.
- Disconnect the evaporator housing's electrical connector.
- Remove the evaporator housing-to-chassis 3 nuts and 4 bolts.
- Remove the evaporator housing.

5. Remove the heater valve-to-cowl bolt.
6. Remove the heater hoses from the heater core pipes.
7. Remove the heater pipe grommets.
8. Remove the air bag module and the steering wheel by performing the following procedure:

- Place the front wheels in the straight-ahead position.
- At both sides of the steering wheel, remove the screw covers.
- Using a Torx socket, loosen the 2 air bag module-to-steering wheel Torx screws until the circumference ring catches on the screw case.
- Carefully, remove the air bag module and disconnect the electrical connector.
- Remove the steering wheel nut.
- Using a steering wheel puller, press the steering wheel from the steering column.

9. Remove the instrument panel and reinforcement by performing the following procedure:

- Remove the front pillar lower garnishes and the pillar garnishes.
- Remove the front door scuff plates.
- Remove the cowl side trim boards.
- Remove the steering column covers.
- Pry out the upper console panel by disengaging the 4 clips.
- Remove the console box.
- Remove the No. 1 finish panel.
- Remove the finish panel and the combination switch.
- Remove the lower cluster finish panel and the cluster finish panel.
- Remove the No. 1 register and the combination meter.
- Remove the 2 center cluster finish panel screws; then, pry out the finish panel.
- Remove the air conditioning control assembly.
- Remove the lower finish panel.
- Remove the glove compartment door finish plate located inside the lower finish panel.

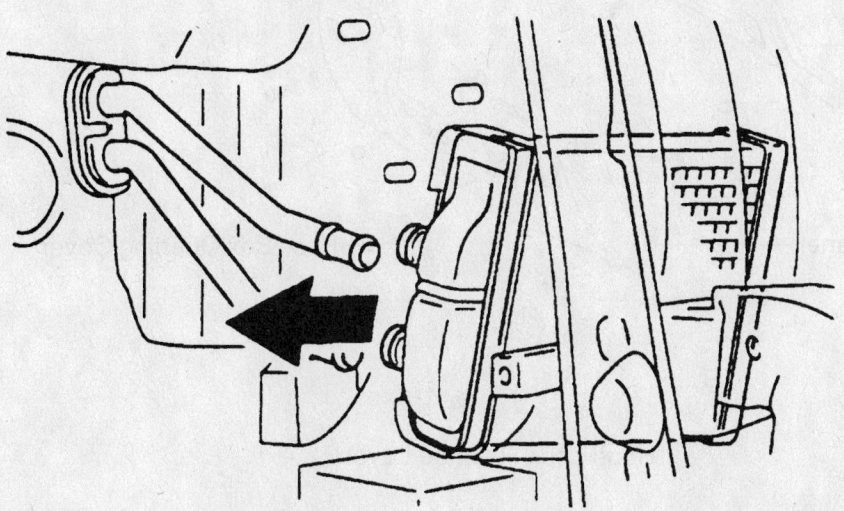

93112GK7

View of the heater core—Camry

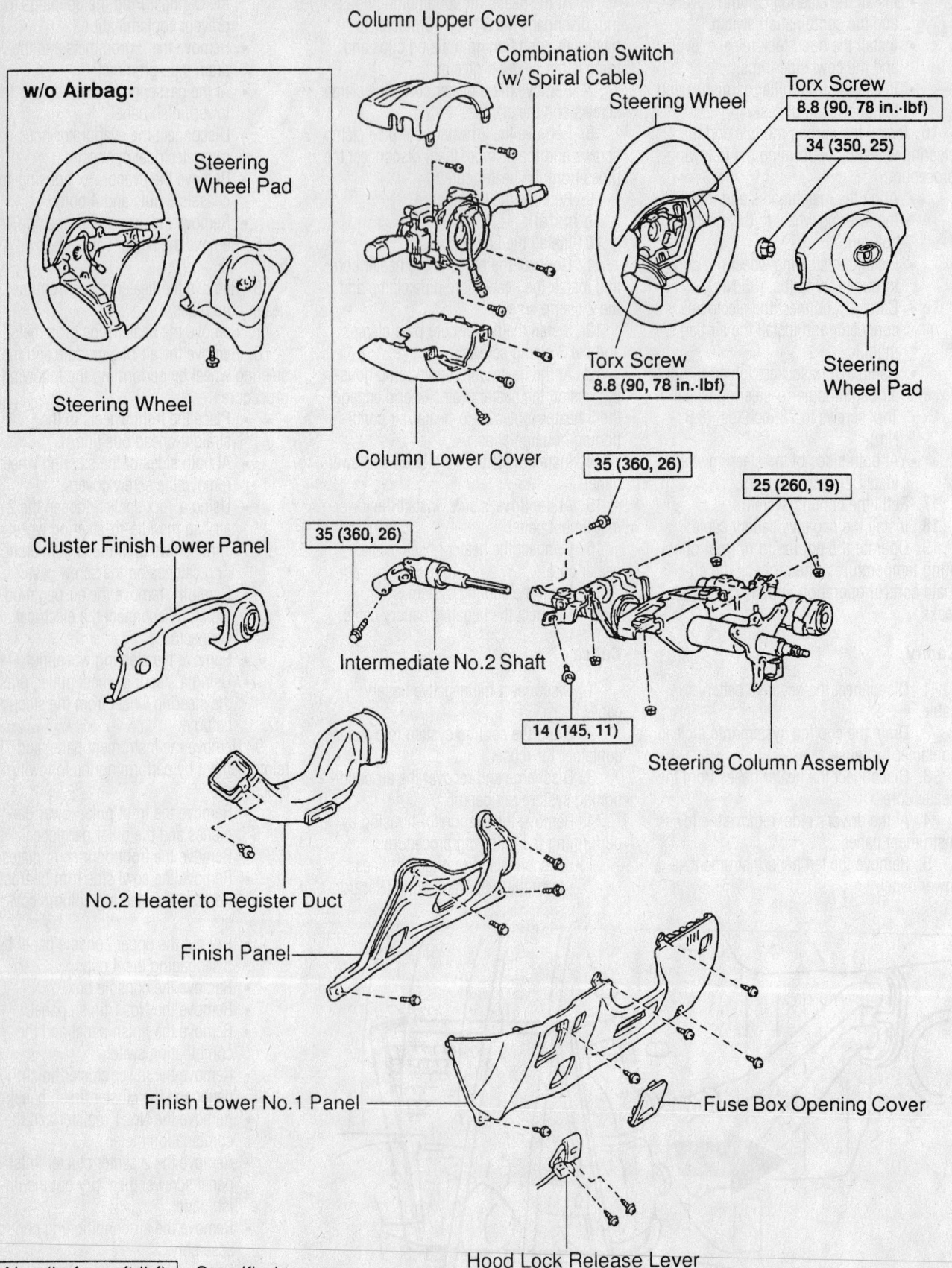

Column Upper Cover

Combination Switch
(w/ Spiral Cable)

Steering Wheel

Torx Screw
8.8 (90, 78 in.·lbf)

34 (350, 25)

w/o Airbag:

Steering
Wheel Pad

Steering Wheel

Column Lower Cover

Torx Screw
8.8 (90, 78 in.·lbf)

Steering
Wheel Pad

Cluster Finish Lower Panel

35 (360, 26)

35 (360, 26)

25 (260, 19)

Intermediate No.2 Shaft

14 (145, 11)

Steering Column Assembly

No.2 Heater to Register Duct

Finish Panel

Finish Lower No.1 Panel

Fuse Box Opening Cover

Hood Lock Release Lever

N·m (kgf·cm, ft·lbf) : Specified torque

93112GK8

Exploded view of the air bag module, steering wheel and related components (typical)—Celica

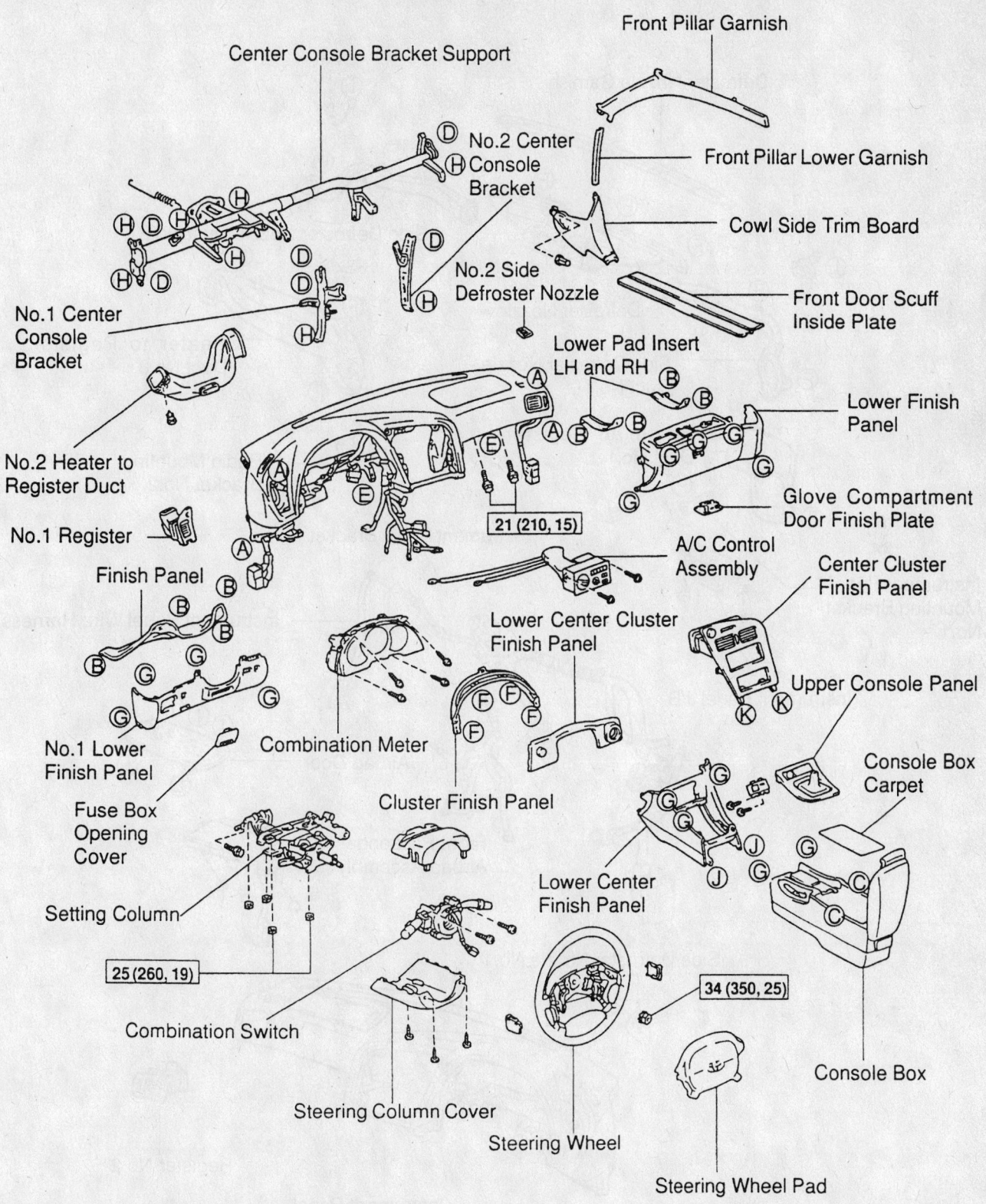

Center Console Bracket Support

Front Pillar Garnish

No.2 Center Console Bracket

Front Pillar Lower Garnish

Cowl Side Trim Board

No.2 Side Defroster Nozzle

No.1 Center Console Bracket

Lower Pad Insert LH and RH

Front Door Scuff Inside Plate

Lower Finish Panel

No.2 Heater to Register Duct

21 (210, 15)

Glove Compartment Door Finish Plate

No.1 Register

A/C Control Assembly

Center Cluster Finish Panel

Finish Panel

Lower Center Cluster Finish Panel

Upper Console Panel

No.1 Lower Finish Panel

Combination Meter

Console Box Carpet

Fuse Box Opening Cover

Cluster Finish Panel

Lower Center Finish Panel

Setting Column

25 (260, 19)

34 (350, 25)

Combination Switch

Steering Column Cover

Console Box

Steering Wheel

Steering Wheel Pad

N·m (kgf·cm, ft·lbf) : Specified torque

93112GK9

Exploded view of the instrument panel and related components—Celica

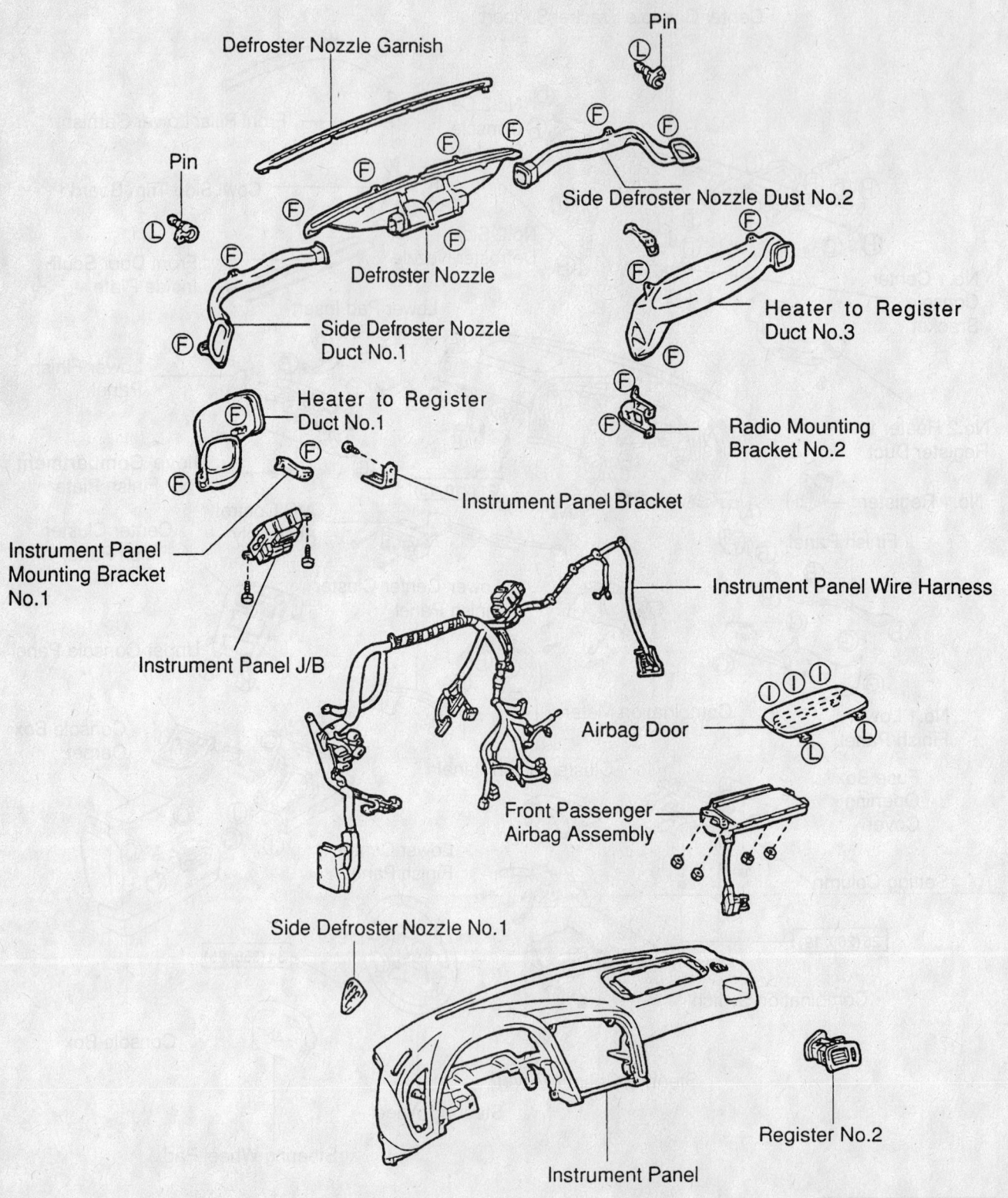

Defroster Nozzle Garnish

Pin

Pin

Side Defroster Nozzle Dust No.2

Defroster Nozzle

Side Defroster Nozzle Duct No.1

Heater to Register Duct No.3

Heater to Register Duct No.1

Radio Mounting Bracket No.2

Instrument Panel Bracket

Instrument Panel Mounting Bracket No.1

Instrument Panel Wire Harness

Instrument Panel J/B

Airbag Door

Front Passenger Airbag Assembly

Side Defroster Nozzle No.1

Register No.2

Instrument Panel

93112GK0

Exploded view of the ventilation system, wiring harness and related components—Celica

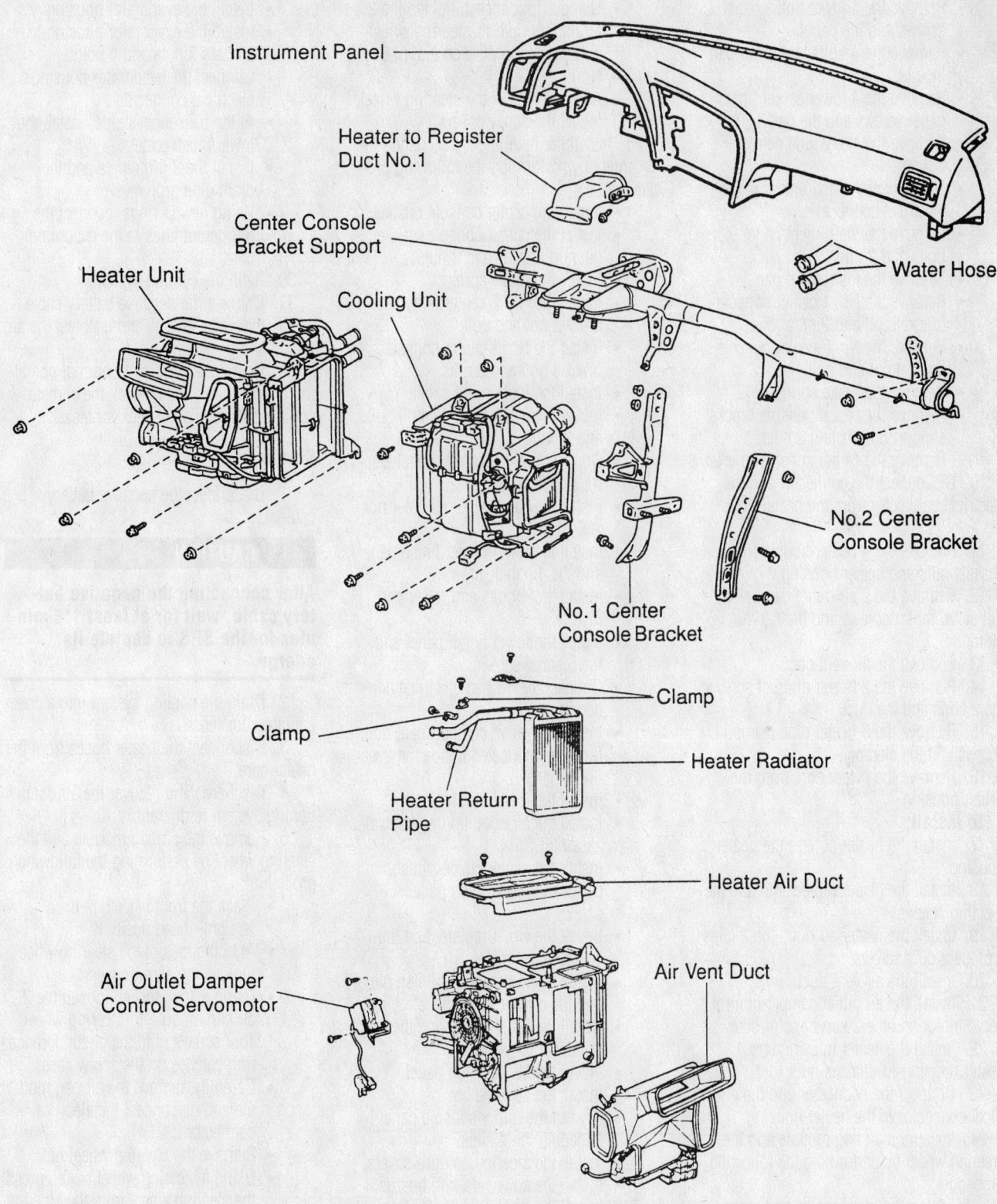

Instrument Panel

Heater to Register
Duct No.1

Center Console
Bracket Support

Heater Unit

Cooling Unit

Water Hose

No.2 Center
Console Bracket

No.1 Center
Console Bracket

Clamp

Clamp

Heater Radiator

Heater Return
Pipe

Heater Air Duct

Air Outlet Damper
Control Servomotor

Air Vent Duct

93112GL1

Exploded view of the heater core, heater housing, evaporator housing and related components—Celica

- Pull up the air bag electrical connector and disconnect it.
- Remove the 5 lower finish panel screws and the panel.
- Remove the 4 bolts and lower pad inserts.
- Remove the 4 lower center finish panel screws and the panel.
- Remove the No. 2 side defroster nozzle.
- Disconnect the instrument panel electrical connectors.
- Remove the instrument panel 9 bolts and 2 nuts.
- Remove the instrument panel.
- Remove the No. 1 center console bracket bolt and 2 nuts.
- Remove the No. 2 center console bracket nut and bolt.
- Remove the brake spring.
- Remove the center console bracket support 6 bolts and 2 nuts.
- Remove the center console bracket.

10. Disconnect the connector and the electrical connector from the heater housing.

11. Remove the 4 heater housing-to-chassis nuts and heater housing.

12. Remove the 2 air outlet damper control servo motor screws and the servo motor.

13. Remove the air vent duct.

14. Remove the 2 heater air duct screws, the 2 clips and the duct.

15. Remove the 3 heater pipe clamp screws and the clamps.

16. Remove the heater core from the heater housing.

To install:

17. Install the heater core to the heater housing.

18. Install the heater pipe clamp and the 3 clamp screws.

19. Install the heater air duct, the 2 clips and the 2 duct screws.

20. Install the air vent duct.

21. Install the air outlet damper control servo motor and the 2 servo motor screws.

22. Install the heater housing and 4 heater housing-to-chassis nuts.

23. Connect the connector and the electrical connector to the heater housing.

24. Install the air bag module and the steering wheel by performing the following procedure:

- Align the matchmarks and install the steering wheel to the steering column.
- Install the steering wheel nut and torque the nut to 26 ft. lbs. (35 Nm).
- Carefully, connect the electrical

connector and install the air bag module.

- Using a Torx socket, tighten the 2 air bag module-to-steering wheel Torx screws to 78 inch lbs. (8.8 Nm).
- At both sides of the steering wheel, install the screw covers.

25. Install the instrument panel and reinforcement by performing the following procedure:

- Install the center console bracket.
- Install the center console bracket support 6 bolts and 2 nuts.
- Install the brake spring.
- Install the No. 2 center console bracket nut and bolt.
- Install the No. 1 center console bracket bolt and 2 nuts.
- Install the instrument panel.
- Install the instrument panel 9 bolts and 2 nuts.
- Connect the instrument panel electrical connectors.
- Install the No. 2 side defroster nozzle.
- Install the lower center finish panel and the 4 panel screws.
- Install the 4 bolts and lower pad inserts.
- Install the lower finish panel and the 5 panel screws.
- Connect the air bag electrical connector.
- Install the glove compartment door finish plate located inside the lower finish panel.
- Install the lower finish panel.
- Install the air conditioning control assembly.
- Install the center cluster finish panel and the 2 finish panel screws.
- Install the No. 1 register and the combination meter.
- Install the lower cluster finish panel and the cluster finish panel.
- Install the finish panel and the combination switch.
- Install the No. 1 finish panel.
- Install the console box.
- Pry out the upper console panel by engaging the 4 clips.
- Install the steering column covers.
- Install the cowl side trim boards.
- Install the front door scuff plates.
- Install the front pillar lower garnishes and the pillar garnishes.

26. Install the heater pipe grommets.

27. Install the heater hoses to the heater core pipes.

28. Install the heater valve-to-cowl bolt.

29. Install the evaporator housing by performing the following procedure:

- Install the evaporator housing.
- Install the evaporator housing-to-chassis 3 nuts and 4 bolts.
- Connect the evaporator housing's electrical connector.
- At the passenger's side, install the lower finish panel.
- Install the 2 grommets and the drain pipe grommet.
- Using new O-rings, connect the refrigerant lines to the evaporator core.

30. Refill the cooling system.

31. Connect the negative battery cable.

32. Evacuate, charge and leak test the air conditioning system.

33. Operate the engine to normal operating temperatures; then, check the climate control operation and check for leaks.

Corolla

1. Disconnect the negative battery cable.

✳✳ CAUTION

After connecting the negative battery cable, wait for at least 1½ minutes for the SRS to deplete its energy.

2. Drain the cooling system into a clean container for reuse.

3. Disconnect the heater hoses from the heater core.

4. Discharge and recover the air conditioning system refrigerant.

5. Remove the air bag module and the steering wheel by performing the following procedure:

- Place the front wheels in the straight-ahead position.
- At both sides of the steering wheel, remove the screw covers.
- Using a Torx socket, loosen the 2 air bag module-to-steering wheel Torx screws until the circumference ring catches on the screw case.
- Carefully, remove the air bag module and disconnect the electrical connector.
- Remove the steering wheel nut.
- Using a steering wheel puller, press the steering wheel from the steering column.

6. Remove the passenger's side air bag module by performing the following procedure:

- Remove the glove box-to-instrument panel 2 bolts and 3 screws;

Cooling Unit
- Expansion Valve
- Evaporator
- Blower Resistor
- Thermistor

Blower Unit

Heater Unit

Relays

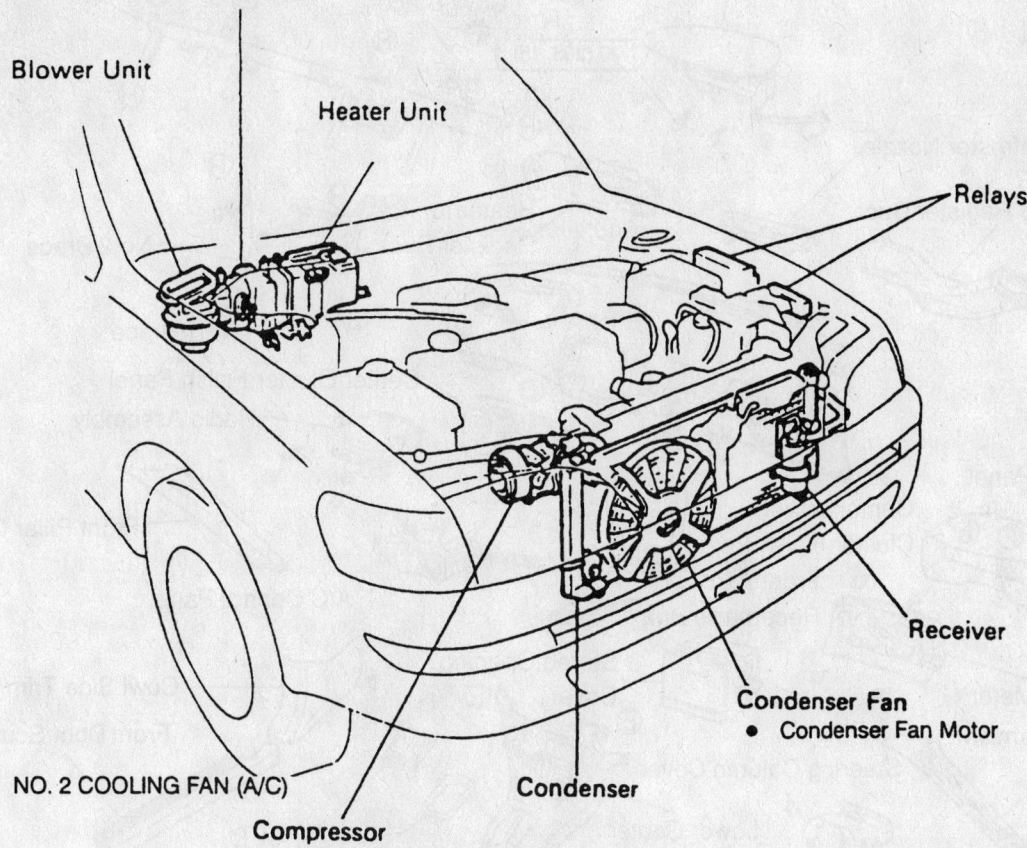

NO. 2 COOLING FAN (A/C)

Receiver

Condenser Fan
- Condenser Fan Motor

Condenser

Compressor
- Magnetic Clutch
- Refrigerant Temperature Switch

Heater Unit

Blower Unit

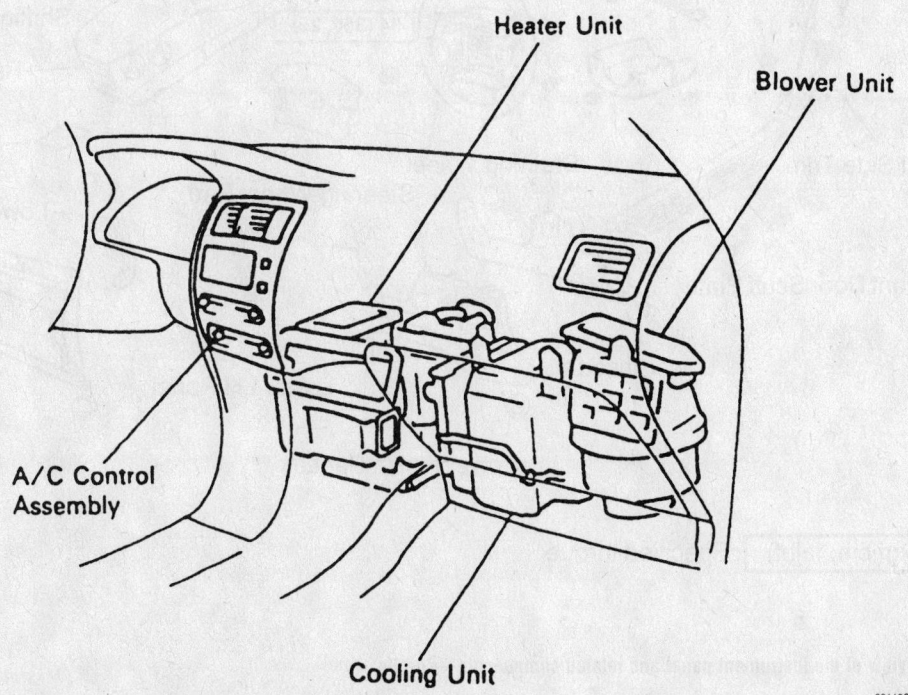

A/C Control Assembly

Cooling Unit

93112GA4

View of the heater/air conditioning assembly and related components—Corolla

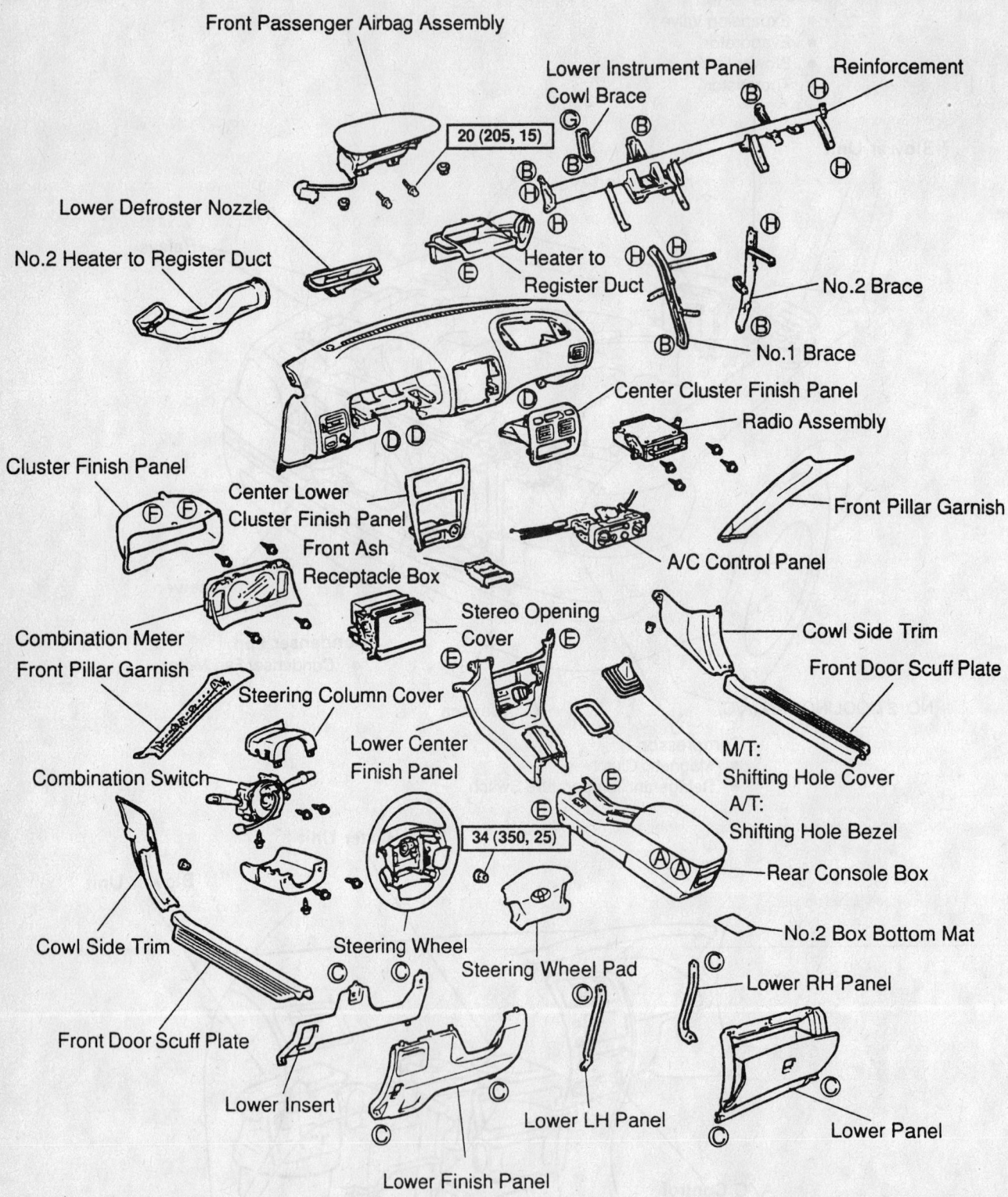

Front Passenger Airbag Assembly

Lower Instrument Panel Cowl Brace

Reinforcement

20 (205, 15)

Lower Defroster Nozzle

No.2 Heater to Register Duct

Heater to Register Duct

No.2 Brace

No.1 Brace

Center Cluster Finish Panel

Radio Assembly

Cluster Finish Panel

Center Lower Cluster Finish Panel

Front Ash Receptacle Box

A/C Control Panel

Front Pillar Garnish

Combination Meter

Stereo Opening Cover

Cowl Side Trim

Front Door Scuff Plate

Front Pillar Garnish

Combination Switch

Steering Column Cover

Lower Center Finish Panel

M/T: Shifting Hole Cover

A/T: Shifting Hole Bezel

Rear Console Box

34 (350, 25)

Cowl Side Trim

Front Door Scuff Plate

Steering Wheel

Steering Wheel Pad

No.2 Box Bottom Mat

Lower RH Panel

Lower Insert

Lower LH Panel

Lower Panel

Lower Finish Panel

N·m (kgf·cm, ft·lbf) : Specified torque

93112GM4

Exploded view of the instrument panel and related components—Corolla

then, pull out the glove box and remove it.
- Disconnect the passenger's side air bag module electrical connector.
- Remove the 2 passenger's air bag module-to-instrument panel bolts and nuts. Carefully, remove the air bag module.

7. Remove the following items in the following order:
- The front door scuff plates
- The cowl side trims
- The front pillar garnishes
- Remove the 2 lower finish panel bolts and the panel.

- Disconnect the hood lock control cable.
- Remove the 2 lower insert bolts and the lower insert.
- Remove the 2 cluster finish panel screws; then, pry out the panel.
- Remove the steering column cover.
- Remove the steering column's combination switch.
- Remove the No. 2 heater-to-register duct.
- Disconnect and remove the combination meter.
- Remove the steering column-to-instrument panel fasteners, the

steering column shaft pinch bolt and the steering column.

8. Remove the following items in the following order:
- The lower left side panel
- The lower right side panel
- The shifting hole knob (manual transmission)
- The shifting hole bezel (automatic transmission)
- The rear console box
- Remove the center lower cluster finish panel by prying it out, disconnect the electrical connector and remove it.

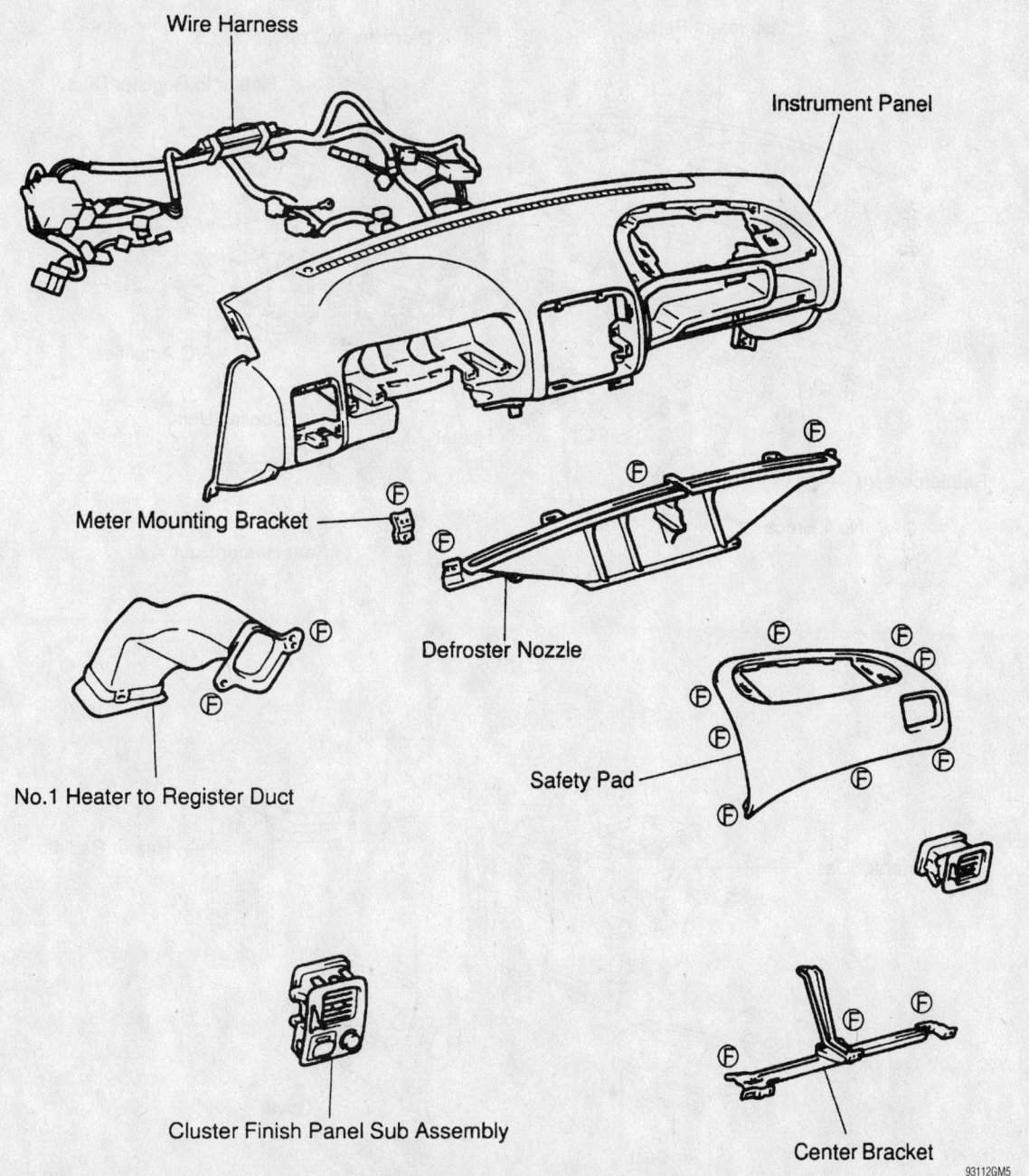

Wire Harness

Instrument Panel

Meter Mounting Bracket

Defroster Nozzle

No.1 Heater to Register Duct

Safety Pad

Cluster Finish Panel Sub Assembly

Center Bracket

93112GM5

Exploded view of the ventilation system, wiring harness and related components—Corolla

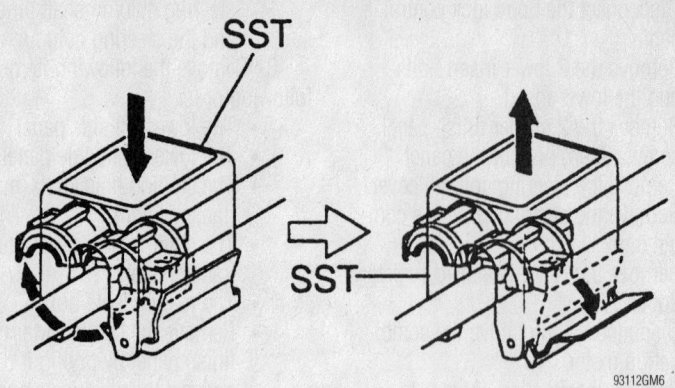

SST

93112GM6

Using the special tool to remove the air conditioning refrigerant line clamps—Corolla

9. Remove the following items in the following order:
  - The stereo opening cover
  - The lower center finish panel
  - The air conditioning control panel
  - Remove the center cluster finish panel by prying it out.
  - Remove the radio.
  - Disconnect the instrument panel electrical connectors.
  - Remove the 4 instrument panel screws and the instrument panel.

10. Remove the following items in the following order:
  - The heater-to-register duct

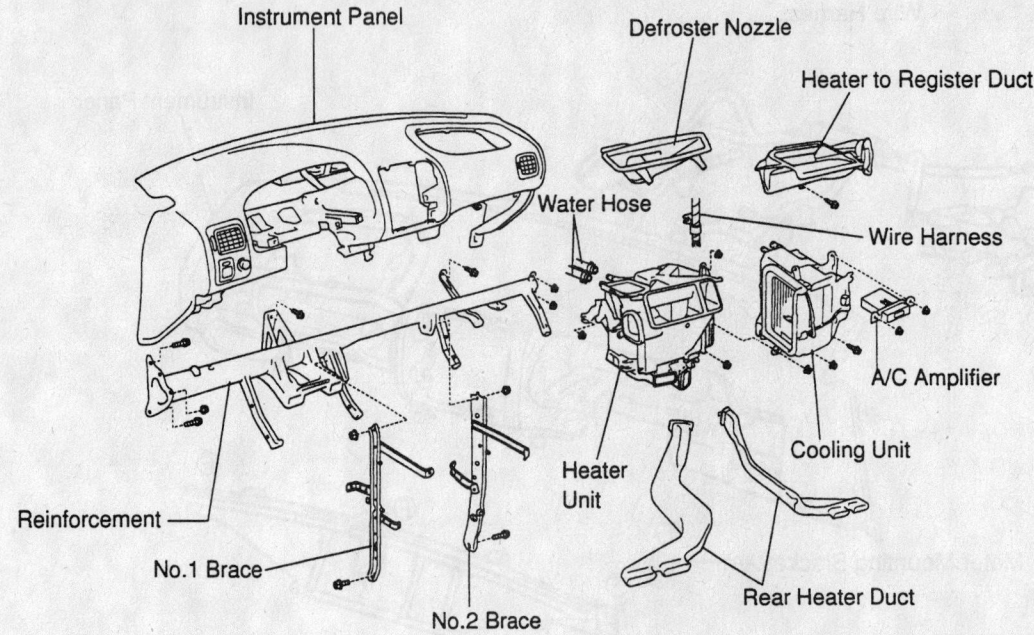

Instrument Panel
Defroster Nozzle
Heater to Register Duct
Water Hose
Wire Harness
Reinforcement
No.1 Brace
Heater Unit
A/C Amplifier
Cooling Unit
No.2 Brace
Rear Heater Duct

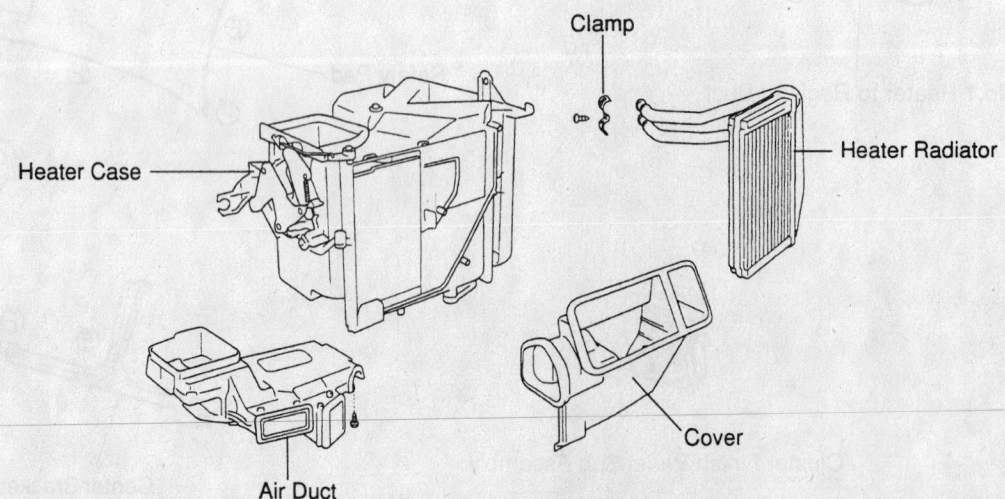

Clamp
Heater Case
Heater Radiator
Air Duct
Cover

93112GM7

Exploded view of the heater core, heater housing, evaporator housing and related components—Corolla

- The lower defroster nozzle
- The No. 1 and No. 2 braces
- The reinforcement

11. Remove the evaporator housing by performing the following procedure:

- Disconnect the cruise control actuator connector.
- Remove the 3 cruise control actuator set bolts.
- Using tool No. 09870-00025 (liquid line) and/or 09870-00015 (suction line), disconnect the refrigerant line clamps. Discard the O-rings and plug the openings to prevent contamination.

- Remove the tube grommet and the drain hose grommet.
- Disconnect the air conditioning amplifier electrical connector.
- Remove the 2 air conditioning amplifier nuts and the amplifier.
- Disconnect the evaporator housing's electrical connectors.
- Remove the evaporator housing-to-chassis 3 screws, nut and the housing.

12. Pull back the carpet and remove the rear heater ducts.

13. Disconnect the wiring harness from the heater housing.

14. Remove the 3 heater housing-to-chassis nuts and the heater housing.

15. Remove the air duct-to-heater housing screw, the 5 clips and the duct.

16. Release the 2 heater core cover clips and the cover.

17. Remove the heater core-to-heater housing screw, clamp and the heater core.

To install:

18. Install the heater core, clamp and the heater core-to-heater housing screw.

19. Install the heater core cover and the 2 cover clips.

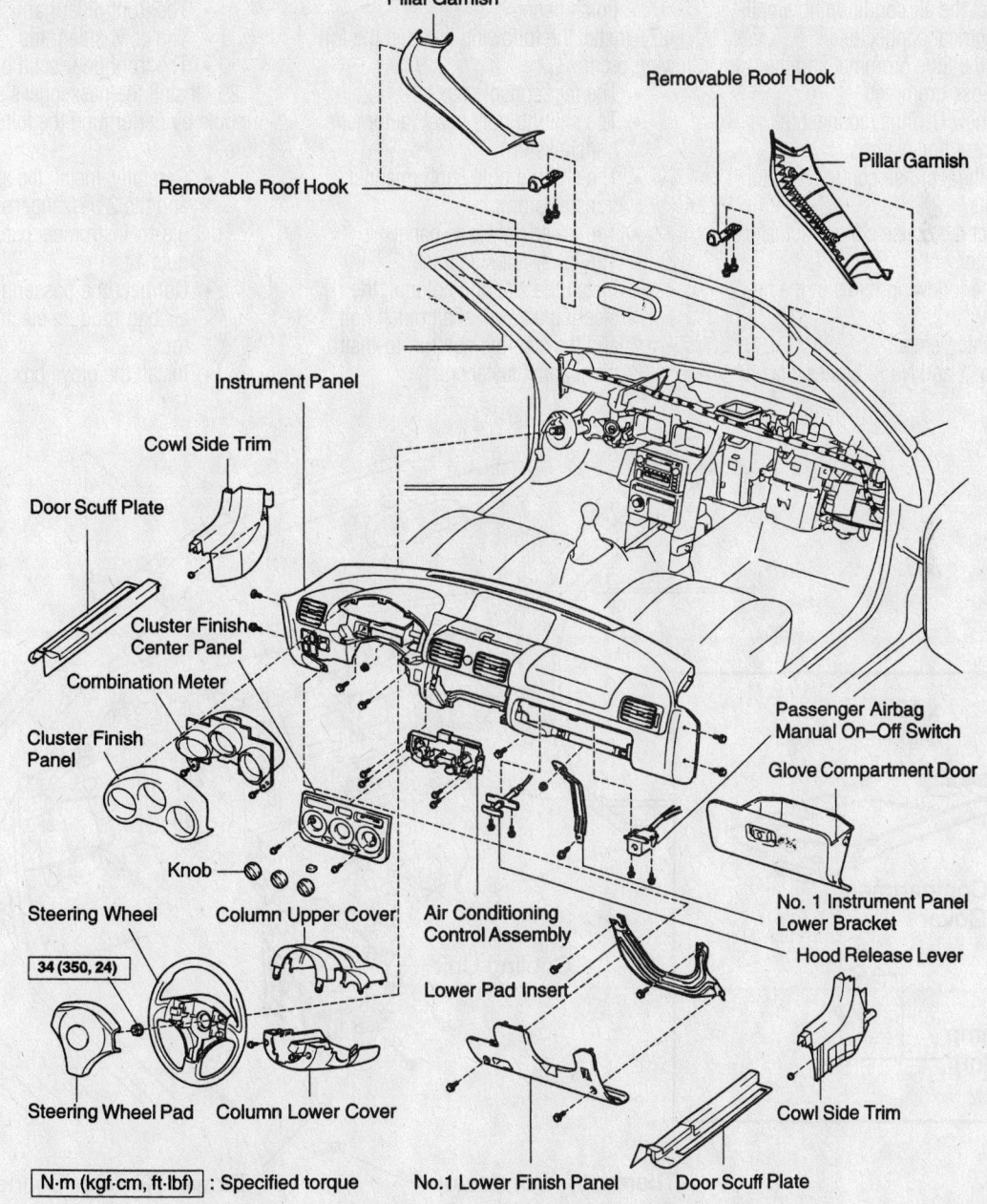

N·m (kgf·cm, ft·lbf) : Specified torque

Instrument panel removal exploded view—MR2

9357WG15

20. Install the air duct, the 5 clips and the duct-to-heater housing screw.

21. Install the heater housing and the 3 heater housing-to-chassis nuts.

22. Connect the wiring harness to the heater housing.

23. Install the rear heater ducts and the carpet.

24. Install the evaporator housing by performing the following procedure:
   - Install the evaporator housing-to-chassis 3 screws, nut and the housing.
   - Connect the evaporator housing's electrical connectors.
   - Install the 2 air conditioning amplifier nuts and the amplifier.
   - Connect the air conditioning amplifier electrical connector.
   - Install the tube grommet and the drain hose grommet.
   - Using new O-rings, connect the refrigerant line clamps.
   - Install the 3 cruise control actuator set bolts.
   - Connect the cruise control actuator connector.

25. Install the following items in the following order:
   - The reinforcement
   - The No. 1 and No. 2 braces

- The lower defroster nozzle
- The heater-to-register duct
- Install the instrument panel and the 4 instrument panel screws.
- Connect the instrument panel electrical connectors.
- Install the radio.
- Install the center cluster finish panel.

26. Install the following items in the following order:
   - The air conditioning control panel
   - The lower center finish panel
   - The stereo opening cover
   - Connect the electrical connector and install the center lower cluster finish panel.

27. Install the following items in the following order:
   - The rear console box
   - The shifting hole bezel (automatic transmission)
   - The shifting hole knob (manual transmission)
   - The lower right side panel
   - The lower left side panel
   - Install the steering column, the steering column shaft pinch bolt and the steering column-to-instrument panel fasteners.

- Connect and install the combination meter.
- Install the No. 2 heater-to-register duct.
- Install the steering column's combination switch.
- Install the steering column cover.
- Install the cluster finish panel and the 2 panel screws.
- Install the lower insert and the 2 lower insert bolts.
- Connect the hood lock control cable.
- Install the lower finish panel and the 2 panel bolts.

28. Install the following items in the following order:
   - The front pillar garnishes
   - The cowl side trims
   - The front door scuff plates

29. Install the passenger's side air bag module by performing the following procedure:
   - Carefully, Install the air bag module and the 2 passenger's air bag module-to-instrument panel bolts and nuts.
   - Connect the passenger's side air bag module electrical connector.
   - Install the glove box and the glove

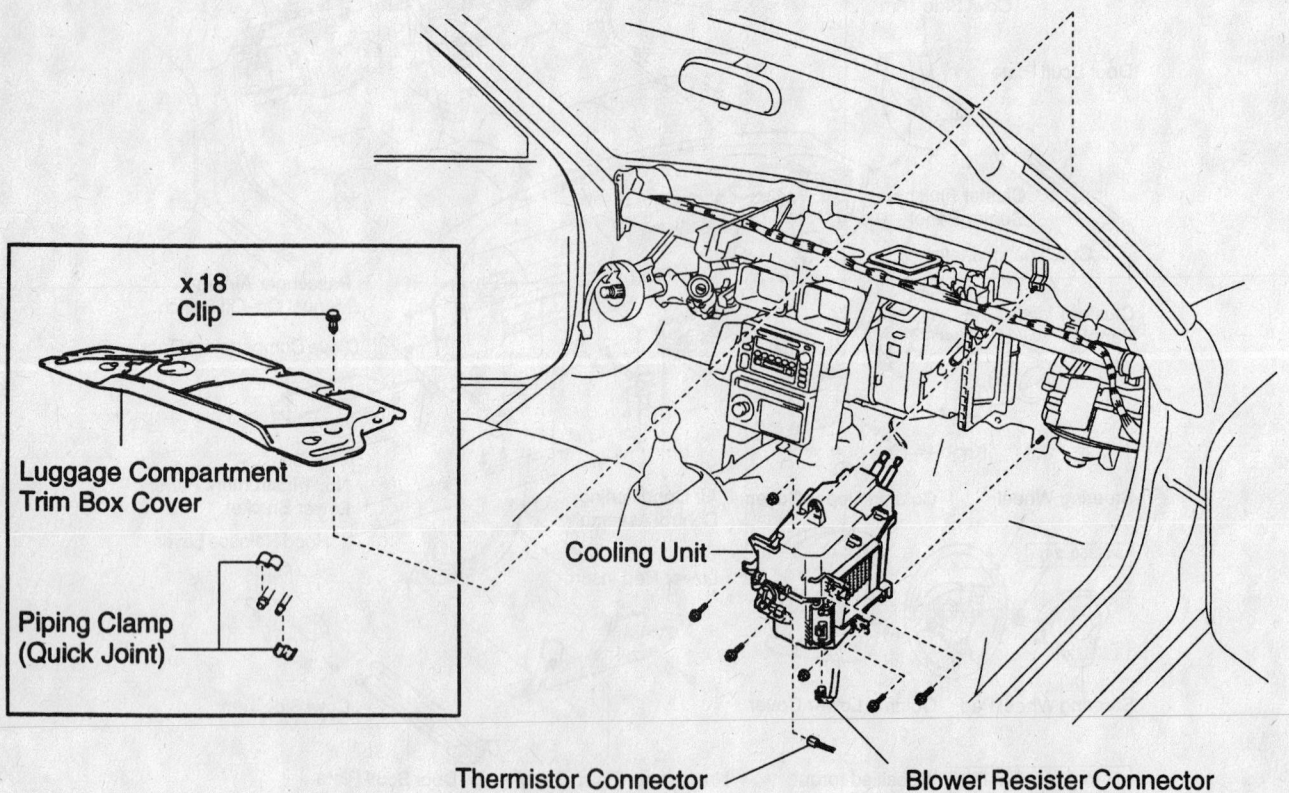

x 18
Clip

Luggage Compartment
Trim Box Cover

Piping Clamp
(Quick Joint)

Cooling Unit

Thermistor Connector

Blower Resister Connector

9357WG16

Evaporator case removal—MR2

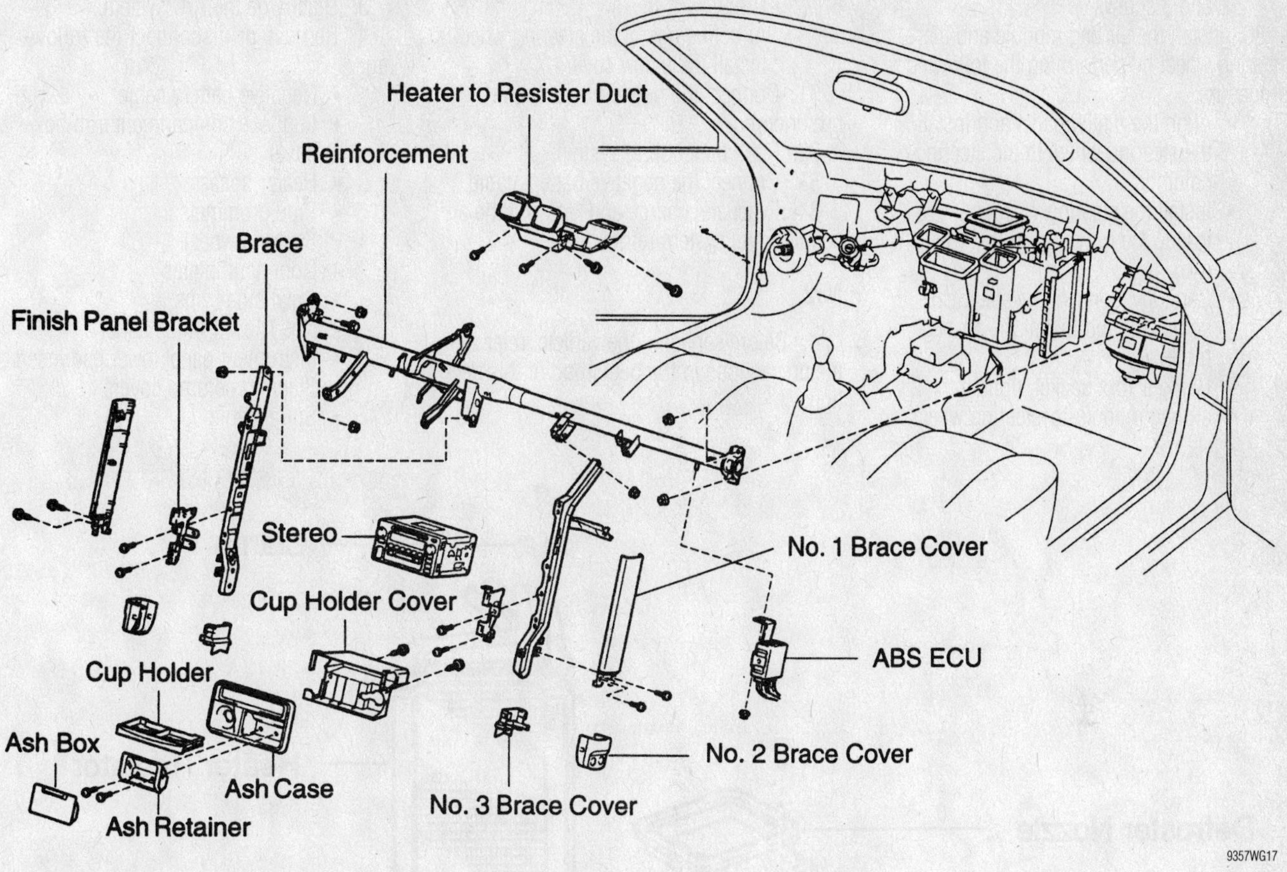

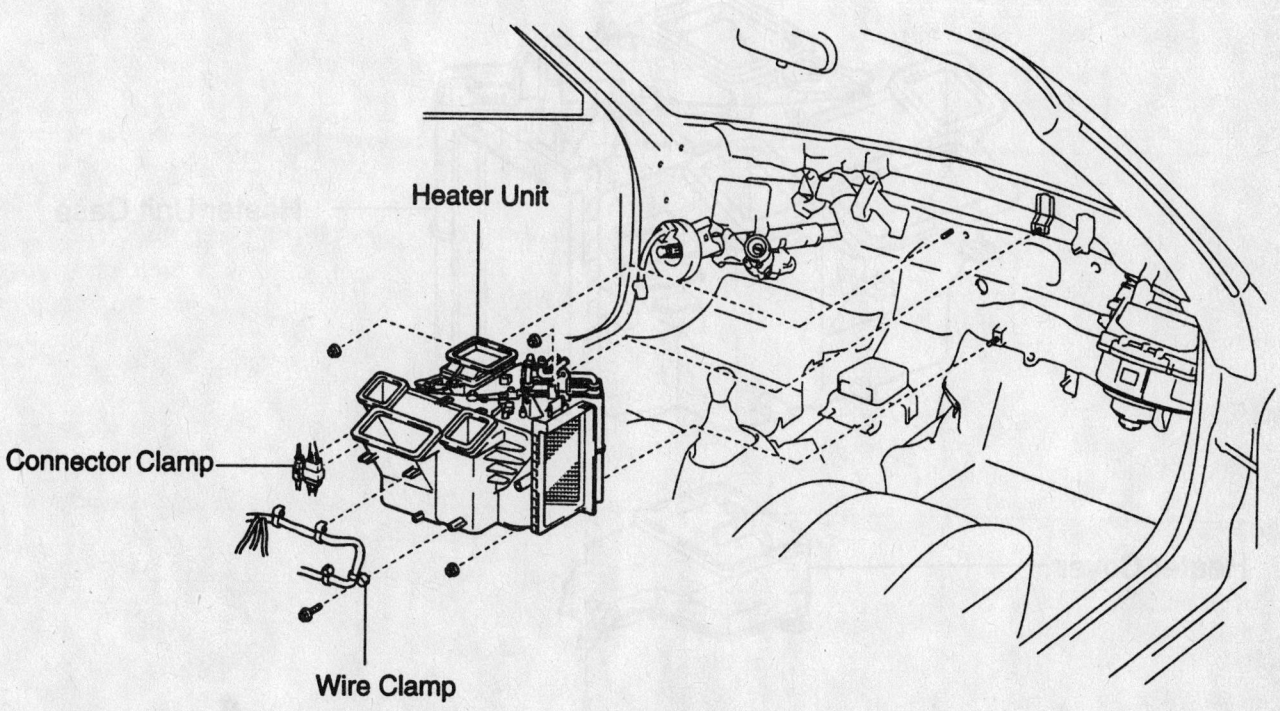

Heater to Resister Duct

Reinforcement

Brace

Finish Panel Bracket

Stereo

Cup Holder Cover

No. 1 Brace Cover

Cup Holder

ABS ECU

Ash Box

No. 2 Brace Cover

Ash Case

No. 3 Brace Cover

Ash Retainer

9357WG17

Instrument panel cross brace removal—MR2

Heater Unit

Connector Clamp

Wire Clamp

9357WG18

Heater unit removal—MR2

box-to-instrument panel 2 bolts and 3 screws.

30. Install the air bag module and the steering wheel by performing the following procedure:

- Align the matchmarks and install the steering wheel to the steering column.
- Install the steering wheel nut and torque the nut to 26 ft. lbs. (35 Nm).
- Carefully, connect the electrical connector and install the air bag module.
- Using a Torx socket, tighten the 2 air bag module-to-steering wheel

Torx screws to 78 inch lbs. (8.8 Nm).

- At both sides of the steering wheel, install the screw covers.

31. Connect the heater hoses to the heater core.
32. Refill the cooling system.
33. Connect the negative battery cable.
34. Evacuate, charge and leak test the air conditioning system refrigerant.

Mr2

1. Before servicing the vehicle, refer to the precautions in the beginning of this section.

2. Drain the cooling system.
3. Discharge the A/C system.
4. Remove or disconnect the following:

- Negative battery cable
- Luggage compartment trim box cover
- Heater hoses
- Pipe grommets
- Steering wheel
- Door scuff plates
- Cowl side trims
- No. 1 lower finish panel
- Instrument panel lower pad insert
- Steering column covers
- Spiral cable

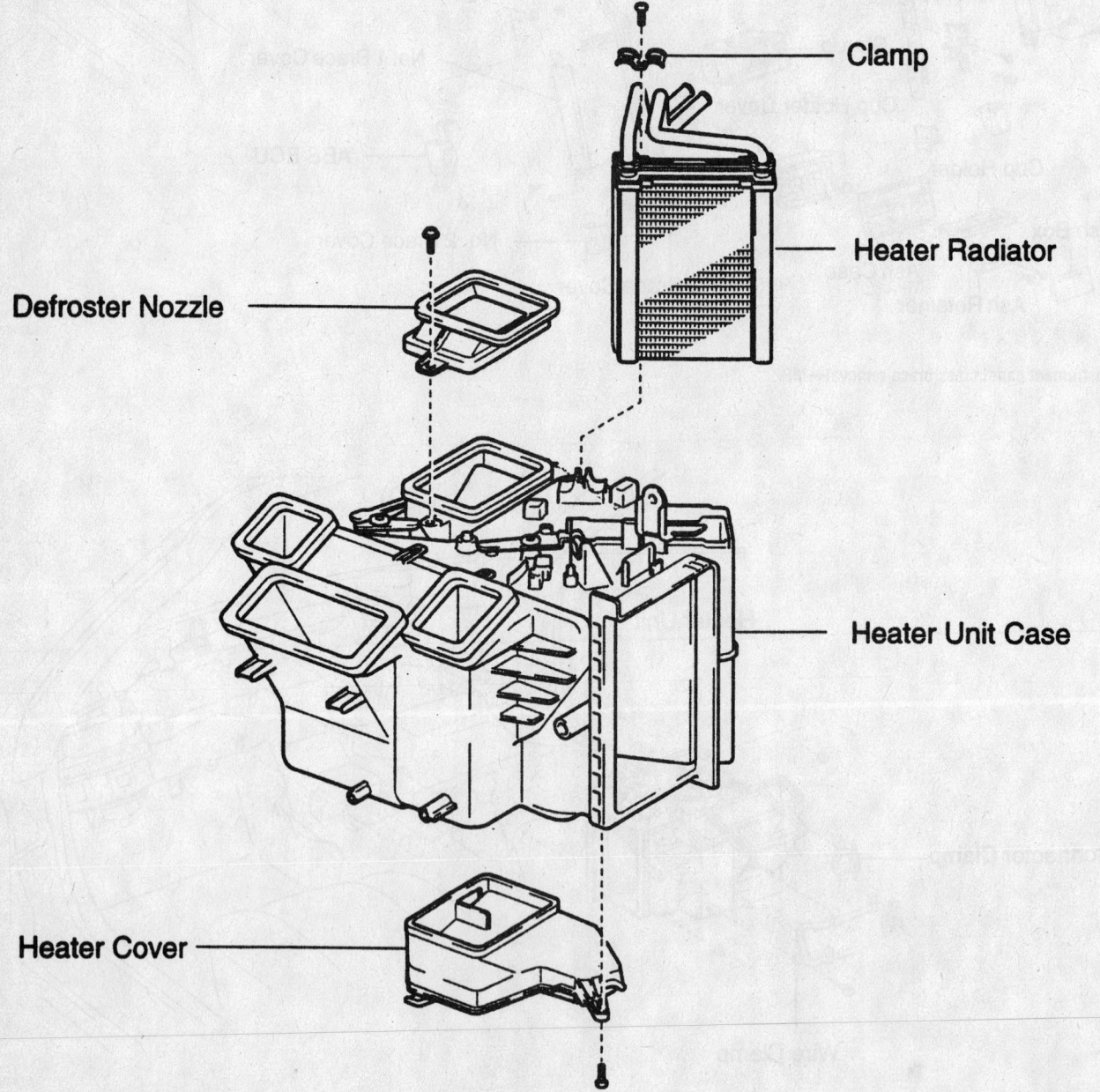

Clamp

Heater Radiator

Defroster Nozzle

Heater Unit Case

Heater Cover

9357WG19

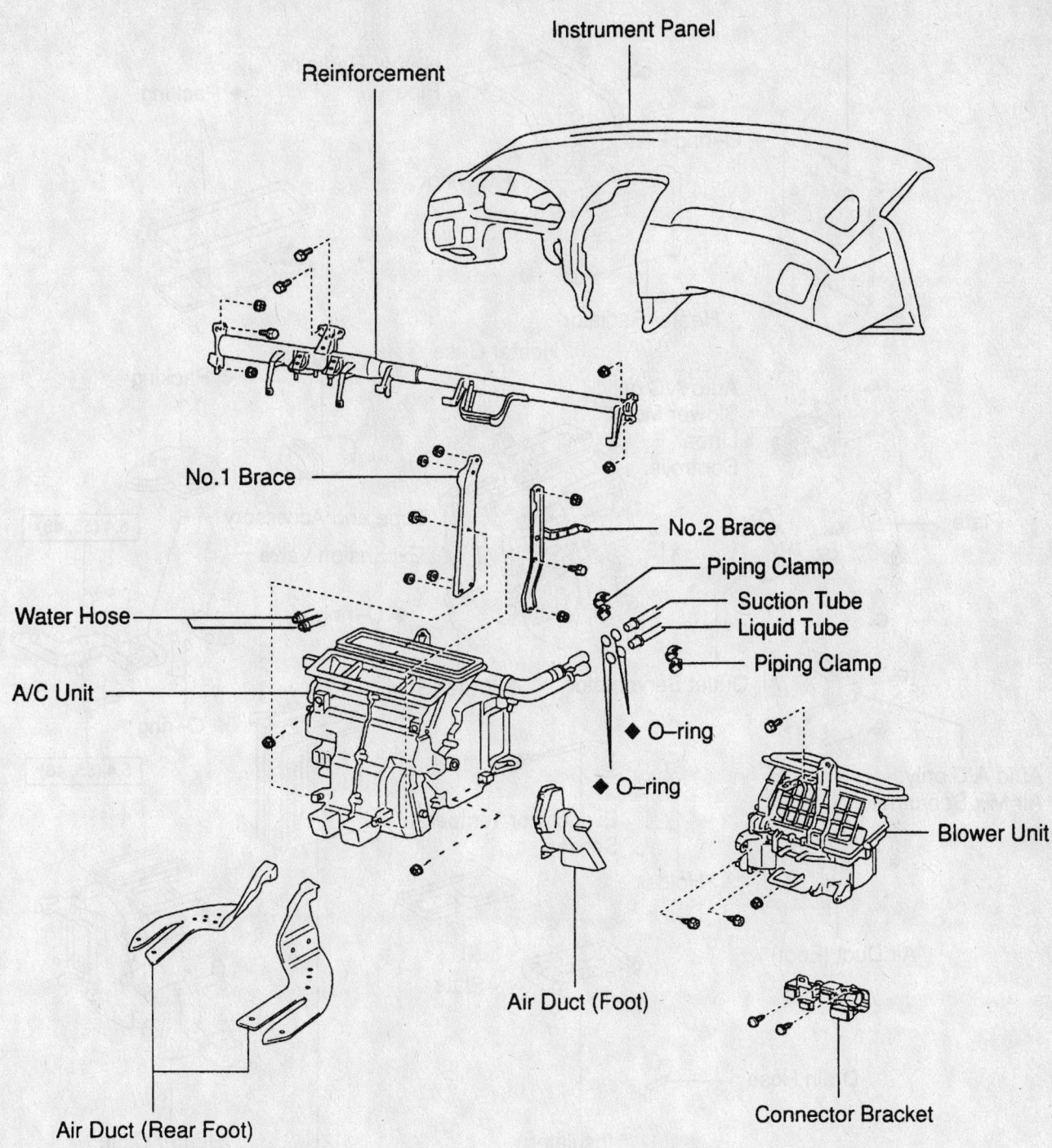

Instrument Panel

Reinforcement

No.1 Brace

No.2 Brace

Piping Clamp

Suction Tube

Liquid Tube

Piping Clamp

Water Hose

A/C Unit

◆ O-ring

◆ O-ring

Blower Unit

Air Duct (Foot)

Air Duct (Rear Foot)

Connector Bracket

◆ Non-reusable part

93112GK5

Exploded view of the instrument panel, heater/air conditioning housing and related components—Solara

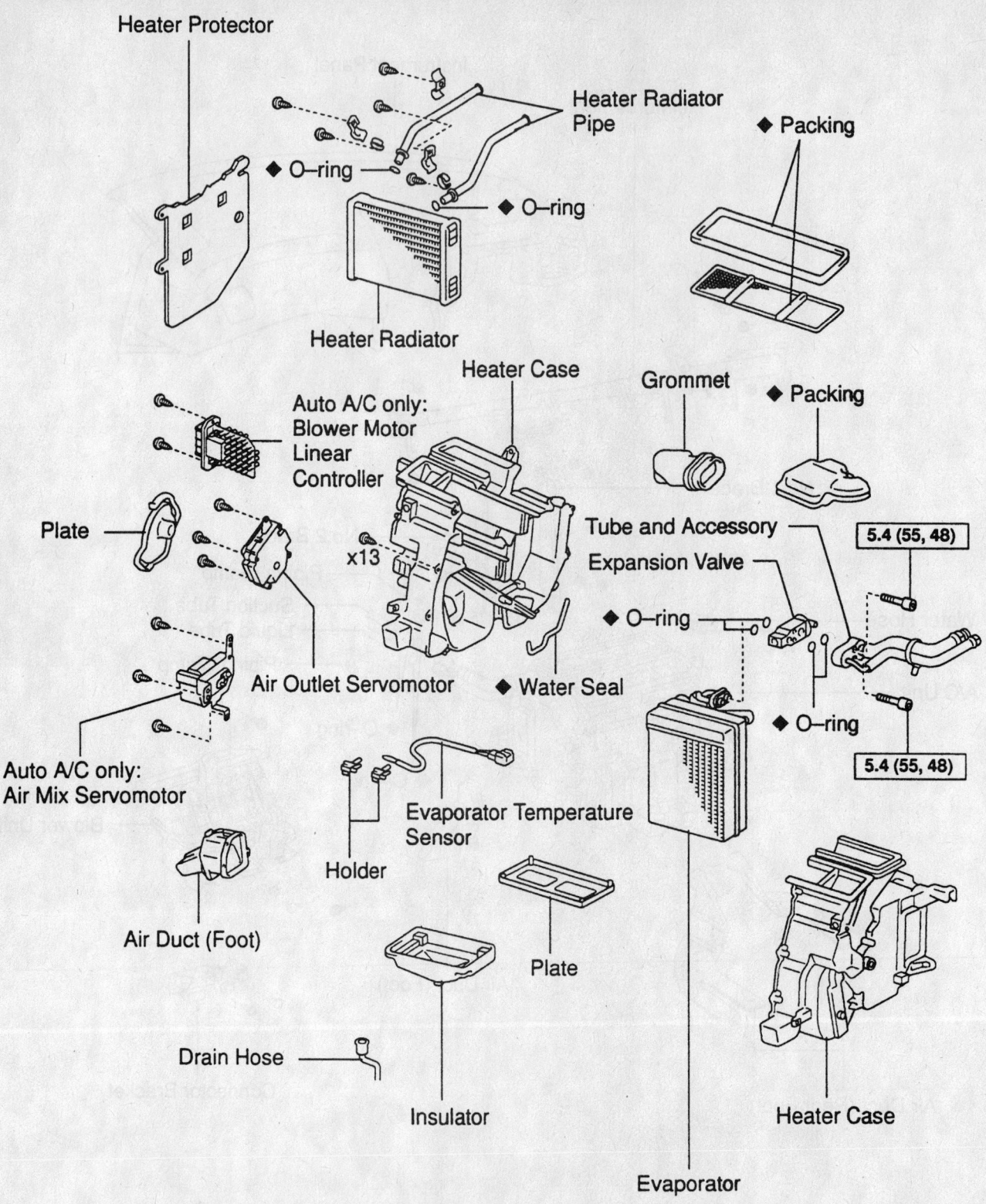

Heater Protector

Heater Radiator Pipe

◆ Packing

◆ O–ring

◆ O–ring

Heater Radiator

Heater Case

Grommet

◆ Packing

Auto A/C only: Blower Motor Linear Controller

Plate

x13

Tube and Accessory

Expansion Valve

5.4 (55, 48)

◆ O–ring

◆ O–ring

Air Outlet Servomotor

◆ Water Seal

5.4 (55, 48)

Auto A/C only: Air Mix Servomotor

Evaporator Temperature Sensor

Holder

Plate

Air Duct (Foot)

Drain Hose

Insulator

Heater Case

Evaporator

N·m (kgf·cm, in.·lbf) : Specified torque

◆ Non–reusable part

93112GK6

Exploded view of the heater/air conditioning housing, heater core, evaporator core and related components—Solara

- Wiper/washer switch
- Light control /headlight dimmer switch
- Instrument cluster finish panel
- Combination meter
- Glove compartment door
- Passenger airbag assembly
- Passenger airbag manual switch
- Hood release lever
- Center instrument cluster finish panel
- Ash tray outer case
- Stereo finish panel
- Instrument panel brace inner and outer covers
- Heater control cables
- Heater control assembly
- Steering column
- Front pillar garnishes
- Instrument panel lower mount bracket
- Center instrument panel bracket
- Instrument panel braces
- Instrument panel
- Instrument panel reinforcement
- A/C system suction tube and liquid line
- Blower resistor connector
- Thermistor connector
- Cooling unit
- Heater unit
- Heater core

To install:

5. Install or connect the following:
- Heater core

- Heater unit
- Cooling unit
- Thermistor connector
- Blower resistor connector
- A/C system suction tube and liquid line
- Instrument panel reinforcement
- Instrument panel
- Instrument panel braces
- Center instrument panel bracket
- Instrument panel lower mount bracket
- Front pillar garnishes
- Steering column
- Heater control assembly
- Heater control cables
- Instrument panel brace inner and outer covers
- Stereo finish panel
- Ash tray outer case
- Center instrument cluster finish panel
- Hood release lever
- Passenger airbag manual switch
- Passenger airbag assembly
- Glove compartment door
- Combination meter
- Instrument cluster finish panel
- Light control /headlight dimmer switch
- Wiper/washer switch
- Spiral cable
- Steering column covers
- Instrument panel lower pad insert
- No. 1 lower finish panel

- Cowl side trims
- Door scuff plates
- Steering wheel
- Pipe grommets
- Heater hoses
- Luggage compartment trim box cover
- Negative battery cable

6. Fill the cooling system.
7. Recharge the A/C system.
8. Start the engine and check for leaks.

Solara

1. Disconnect the negative battery cable.

2. Drain the cooling system into a clean container for reuse.

3. Disconnect the heater hoses from the heater core.

4. At the driver's side, remove the lower instrument panel.

5. Remove the left hand instrument lower panel.

6. At the heater/air conditioning housing, disengage the 3 heater protector-to-heater/air conditioning housing clips and remove the heater protector.

7. Remove the 3 heater core pipe clamp screws and the clamps.

8. Remove the 2 heater core pipe clamp screws and the clamps; then, disconnect the pipes from the heater core.

9. Remove the heater core.

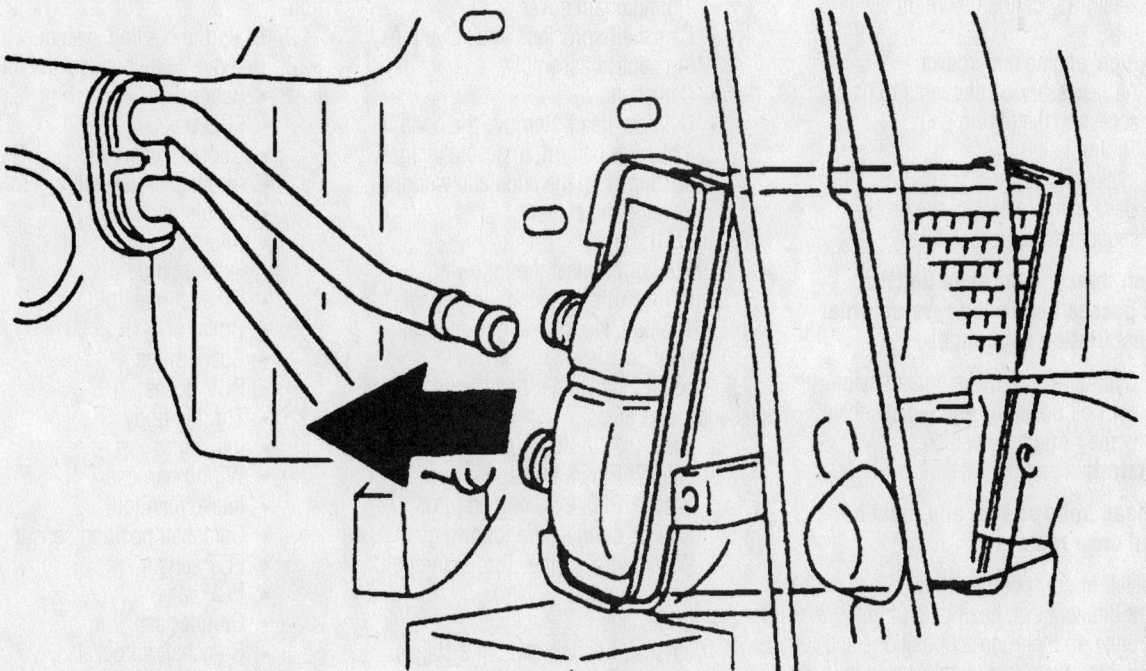

View of the heater core—Solara

93112GK7

To install:

10. Install the heater core.

11. Connect the pipes to the heater core and install the heater core pipe clamp and the 2 clamp screws.

12. Install the heater core pipe clamps and the 3 clamp screws.

13. At the heater/air conditioning housing, install the heater protector and engage the 3 heater protector-to-heater/air conditioning housing clips.

14. Install the left hand instrument lower panel.

15. At the driver's side, install the lower instrument panel.

16. Connect the heater hoses to the heater core.

17. Refill the cooling system.

18. Connect the negative battery cable.

Cylinder Head

REMOVAL & INSTALLATION

5E-FE Engine

1. Before servicing the vehicle, refer to the precautions in the beginning of this section.

2. Properly relieve the fuel system pressure.

3. Drain the engine oil.

4. Drain the cooling system.

5. Remove or disconnect the following:
   - Negative battery cable. On vehicles equipped with an air bag, wait at least 90 seconds before proceeding.
   - Right engine undercover
   - All necessary components to gain access to the timing belt
   - Timing belt
   - Intake and Exhaust manifolds
   - Camshafts, following the proper sequences and procedures.

➡**Loosen the cylinder head bolts in several passes and in the reverse order of the installation sequence.**

   - Cylinder head. Make note of cylinder bolt positions and replace them in their original position.

To install:

➡**The head bolts stretch and must be replaced once removed.**

6. Install or connect the following:
   - 2 different size head bolts, lightly oiled, in their correct positions and tighten them in several passes in the proper sequence, evenly, to 33

ft. lbs. (44 Nm). Mark each bolt with a reference mark and tighten each bolt in sequence an additional 90 degrees.
   - Camshafts and all other components following the proper sequences and procedures.

1NZ-FE Engine

1. Before servicing the vehicle, refer to the precautions in the beginning of this section.

2. Drain the cooling system.

3. Remove or disconnect the following:
   - Negative battery cable
   - Water filler
   - Outer front cowl top panel
   - Alternator
   - Air cleaner
   - Accelerator cable
   - Center exhaust pipe
   - Exhaust manifold support
   - Exhaust manifold
   - Ignition coil
   - Spark plugs
   - PCV hoses
   - Throttle body
   - Engine wiring harness at the head
   - Intake manifold
   - Camshaft position sensor
   - ECT sensor
   - Oil control valve
   - PCV valve
   - Oil filer cap
   - Cylinder head cover
   - Fuel injectors
   - Timing chain cover
   - Camshaft sprockets and valve timing control assembly
   - Camshafts
   - Cylinder head. Remove the bolts in a circular pattern, in several stages, starting from the ends and working towards the center

To install:

4. Install or connect the following:
   - Cylinder head, using a new gasket. The Lod. No. on the gasket faces UP

5. Torque the cylinder head bolts, in sequence, in 3 steps:
   a. Step 1: 22 ft. lbs. (29 Nm).
   b. Step 2: Plus a 90 degree turn.
   c. Step 3: Plus a 90 degree turn.

6. Install or connect the following:
   - Water bypass pipe. Torque the bolt to 80 inch lbs. (9 Nm).
   - Camshafts

7. Camshaft bearing caps in 2 stages:
   a. Step: 10 ft. lbs. (13 Nm).
   b. Step 2: 17 ft. lbs. (23 Nm).

8. Install or connect the following:
   - Sprockets and valve timing controller assembly, aligning the knock pin and hole. Torque the bolts to 47 ft. lbs. (64 Nm).
   - Check and adjust the valves
   - Cylinder head cover
   - Oil filler cap
   - PCV valve
   - ECT sensor
   - Camshaft position sensor
   - Timing chain cover
   - Intake manifold
   - Engine wiring harness
   - Throttle body
   - PCV hoses
   - Spark plugs
   - Ignition coils
   - Exhaust manifold
   - Exhaust manifold support. Torque the bolts to 27 ft. lbs. (37 Nm).
   - Front exhaust pipe. Torque the nuts to 46 ft. lbs. (62 Nm).
   - Accelerator cable
   - Air cleaner
   - Alternator
   - Water filler
   - Negative battery cable

9. Fill the cooling system to the proper level.

10. Start the vehicle, check for leaks and repair if necessary.

1ZZ-FE Engine

1. Before servicing the vehicle, refer to the precautions in the beginning of this section.

2. Drain the cooling system.

3. Remove or disconnect the following:
   - Battery
   - ECU box
   - Coolant reservoir
   - Air cleaner assembly
   - Accelerator cable
   - Alternator
   - Exhaust pipe
   - Exhaust manifold
   - Coils
   - Spark plugs
   - PCV hoses
   - Throttle body
   - Injectors
   - Wiring harness
   - Intake manifold
   - Camshaft position sensor
   - ECT sensor
   - PCV valve
   - Oil filler cap
   - Camshaft sprockets
   - Camshafts
   - Hoses

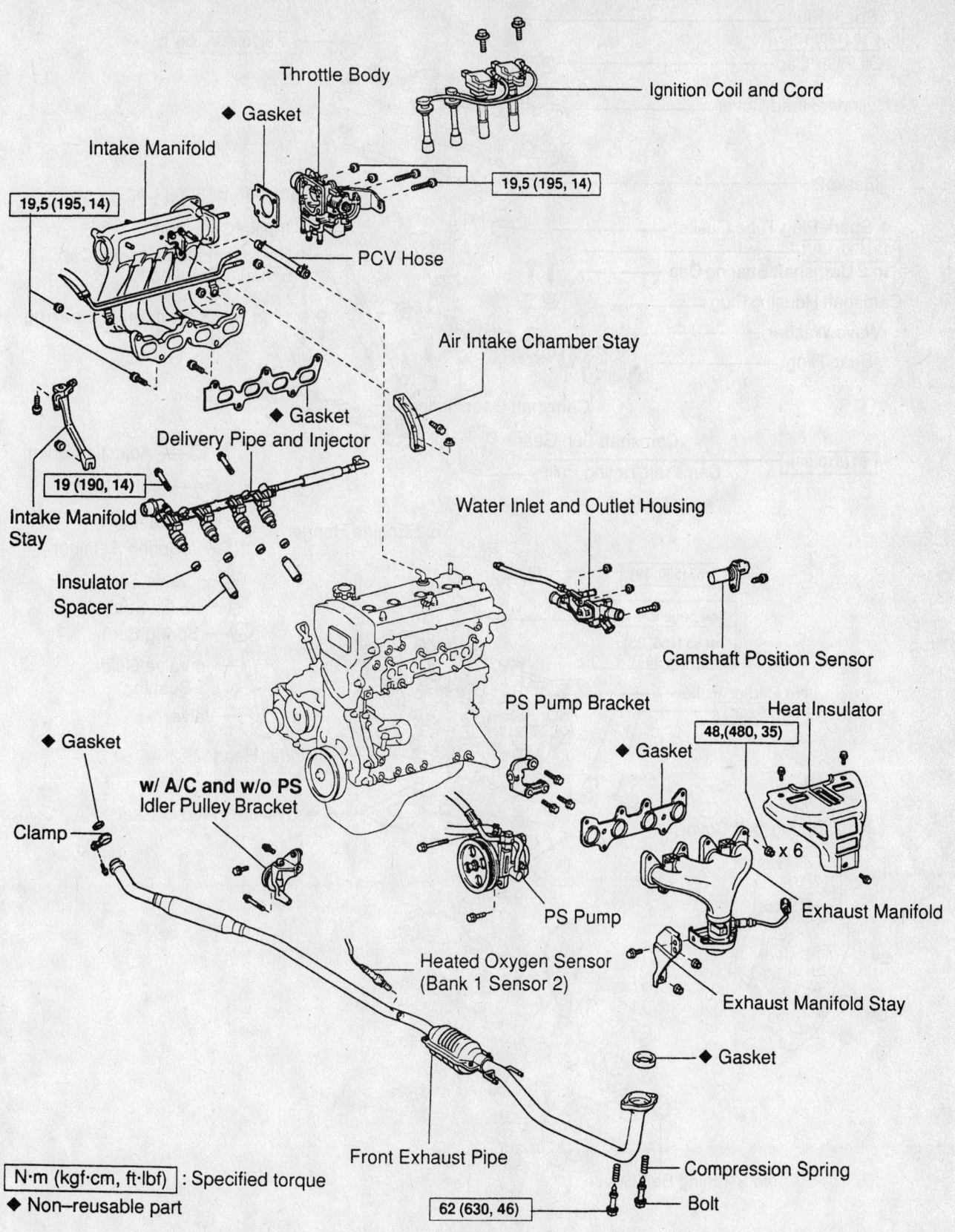

Throttle Body

◆ Gasket

Ignition Coil and Cord

Intake Manifold

19,5 (195, 14)

19,5 (195, 14)

PCV Hose

Air Intake Chamber Stay

◆ Gasket

Delivery Pipe and Injector

19 (190, 14)

Intake Manifold Stay

Water Inlet and Outlet Housing

Insulator

Spacer

Camshaft Position Sensor

PS Pump Bracket

Heat Insulator

◆ Gasket

◆ Gasket

48,(480, 35)

Clamp

w/ A/C and w/o PS
Idler Pulley Bracket

x 6

Exhaust Manifold

PS Pump

Heated Oxygen Sensor
(Bank 1 Sensor 2)

Exhaust Manifold Stay

◆ Gasket

Front Exhaust Pipe

Compression Spring

62 (630, 46)

Bolt

N·m (kgf·cm, ft·lbf) : Specified torque
◆ Non-reusable part

9301WG01

Removal of intake manifold, throttle body, exhaust front pipe and manifold—1.5L (5E-FE) engine

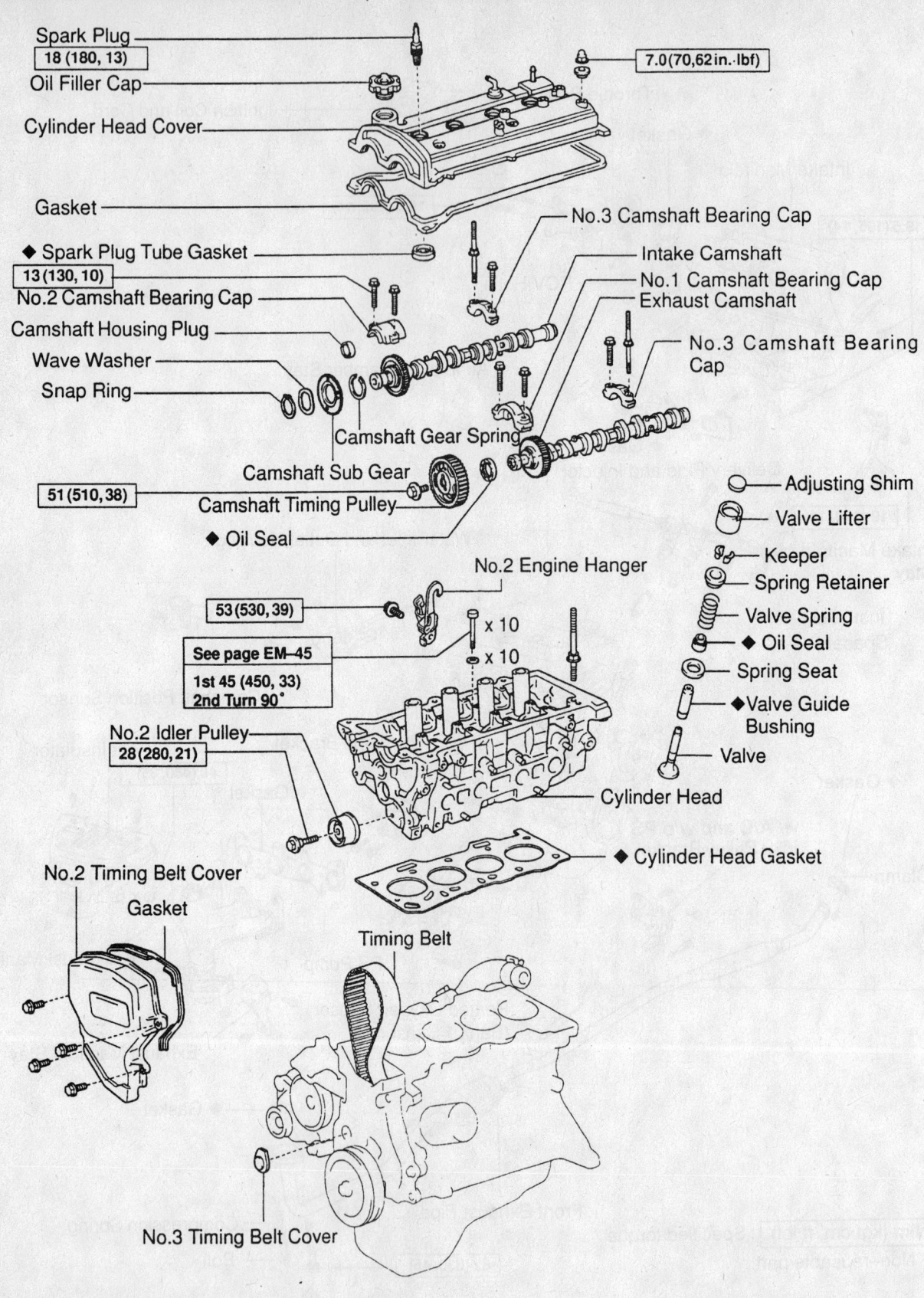

Spark Plug
18 (180, 13)
Oil Filler Cap
Cylinder Head Cover
Gasket

7.0 (70, 62 in.·lbf)

No.3 Camshaft Bearing Cap
Intake Camshaft
No.1 Camshaft Bearing Cap
Exhaust Camshaft

◆ Spark Plug Tube Gasket
13 (130, 10)
No.2 Camshaft Bearing Cap
Camshaft Housing Plug
Wave Washer
Snap Ring

No.3 Camshaft Bearing Cap

Camshaft Gear Spring
Camshaft Sub Gear
51 (510, 38)
Camshaft Timing Pulley
◆ Oil Seal

Adjusting Shim
Valve Lifter
Keeper
Spring Retainer
Valve Spring
◆ Oil Seal
Spring Seat
◆ Valve Guide Bushing
Valve

No.2 Engine Hanger
53 (530, 39)
See page EM-45
1st 45 (450, 33)
2nd Turn 90°

x 10
x 10

No.2 Idler Pulley
28 (280, 21)

Cylinder Head

◆ Cylinder Head Gasket

No.2 Timing Belt Cover
Gasket

Timing Belt

No.3 Timing Belt Cover

N·m (kgf·cm, ft·lbf) : Specified torque

◆ Non-reusable part

9301WG02

Exploded view of cylinder head disassembly—1.5L (5E-FE) engine

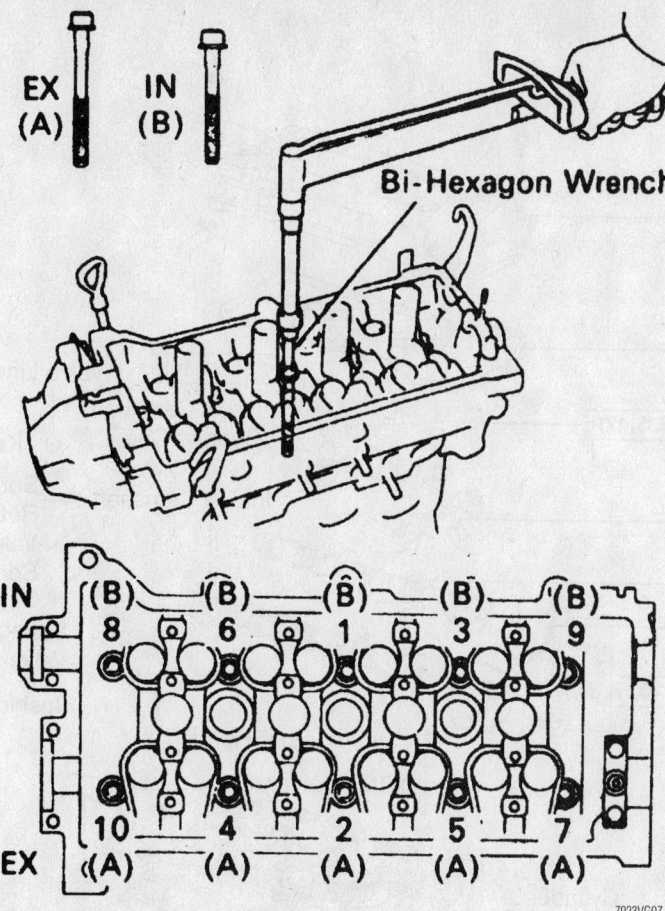

Cylinder head bolt tightening sequence—5E-FE engine

7923VG07

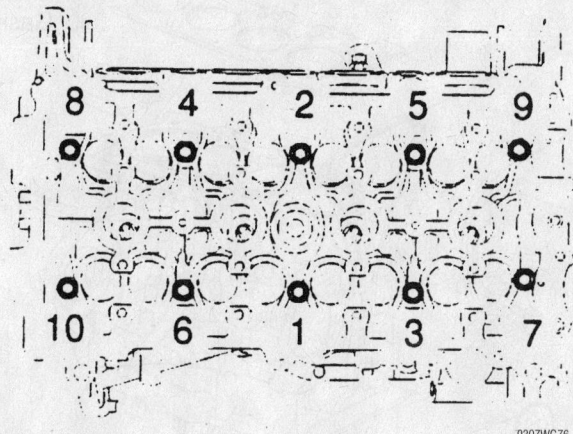

9307WG76

Cylinder head bolt tightening sequence—1NZ-FE engine

- Cylinder head bolts in sequence. To prevent damage to the cylinder head, loosen each bolt about ¼ of a turn during each pass until the bolts are loose.
- Cylinder head

To install:

4. Clean and degrease the surface of the cylinder head and engine block.

5. Install or connect the following:

- New gasket on the engine block with the Lod No. stamp facing up.
- Cylinder head
- Apply a light coat of oil to cylinder head bolt threads and tighten in sequence. Replace any bolt that appears deformed. Bolts: 36 ft. lbs. (49 Nm).
- Tighten each bolt in sequence an additional 90 degree turn.

- Camshafts
- Sprockets
- Oil filler cap
- PCV valve
- ECT sensor
- Intake manifold
- Wiring harness
- Exhaust manifold
- Exhaust pipe
- Alternator
- accelerator cable
- Air cleaner
- ECM box
- Battery

6. Fill the cooling system to the proper level.

7. Start the vehicle, check for leaks and repair if necessary.

2ZZ-GE Engine

1. Before servicing the vehicle, refer to the precautions in the beginning of this section.

2. Drain the cooling system.

3. Remove or disconnect the following:

- Battery
- ECU box
- Coolant reservoir
- Air cleaner assembly
- Accelerator cable
- Alternator
- Exhaust pipe
- Exhaust manifold
- Coils
- Spark plugs
- PCV hoses
- Throttle body
- Injectors
- Wiring harness
- Intake manifold
- Camshaft position sensor
- ECT sensor
- PCV valve
- Oil filler cap
- Camshaft sprockets
- Camshafts
- Hoses
- Cylinder head bolts in sequence. To prevent damage to the cylinder head, loosen each bolt about ¼ of a turn during each pass until the bolts are loose.
- Cylinder head

To install:

4. Clean and degrease the surface of the cylinder head and engine block.

5. Install or connect the following:

- New gasket on the engine block with the Lod No. stamp facing up.
- Cylinder head
- Apply a light coat of oil to cylinder

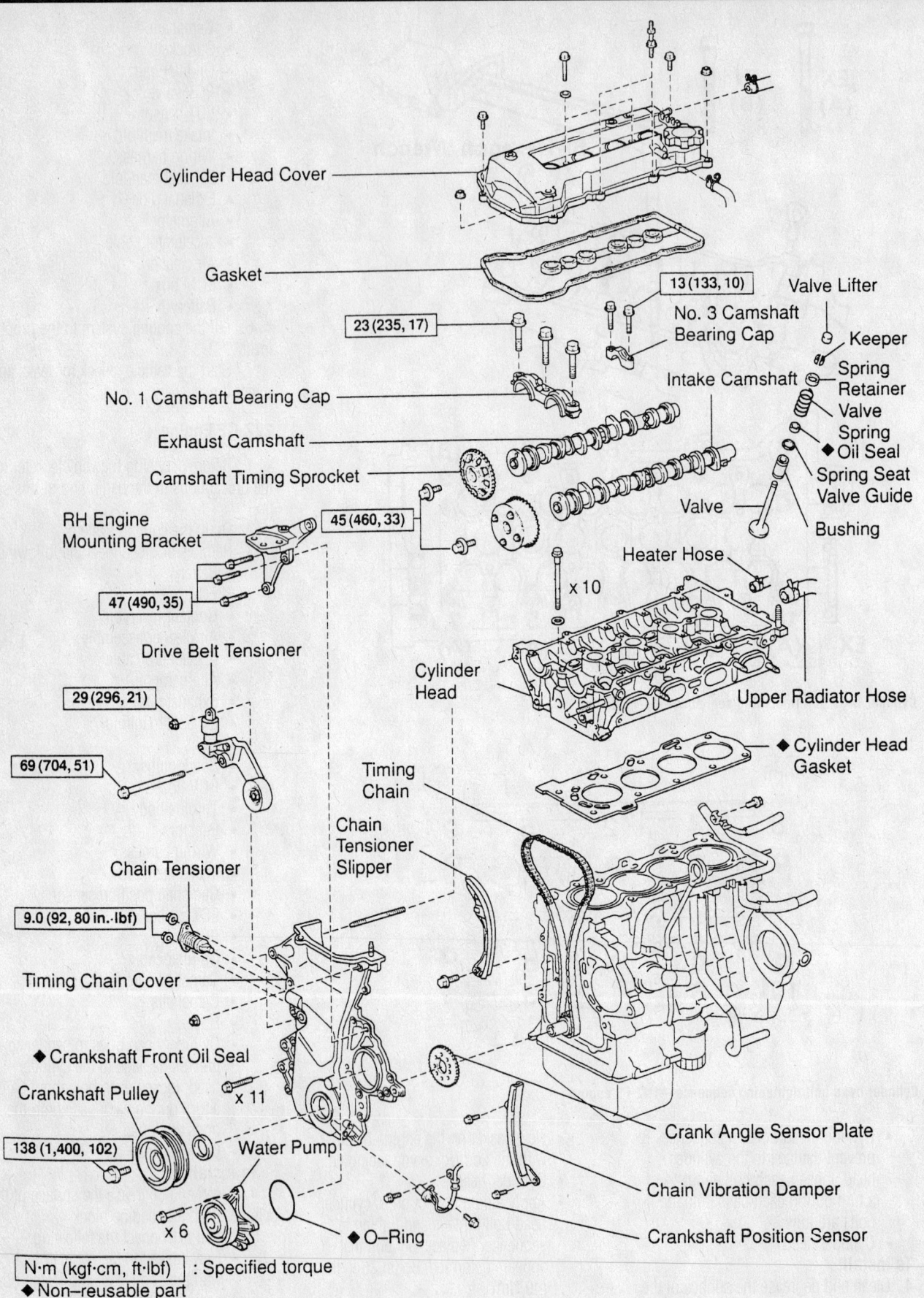

Cylinder Head Cover

Gasket

13 (133, 10)

23 (235, 17)

No. 1 Camshaft Bearing Cap

Exhaust Camshaft

Camshaft Timing Sprocket

RH Engine Mounting Bracket

45 (460, 33)

47 (490, 35)

Drive Belt Tensioner

29 (296, 21)

69 (704, 51)

Chain Tensioner

9.0 (92, 80 in.·lbf)

Timing Chain Cover

◆ Crankshaft Front Oil Seal

Crankshaft Pulley

138 (1,400, 102)

Water Pump

x 6

◆ O–Ring

Valve Lifter

No. 3 Camshaft Bearing Cap

Keeper

Spring Retainer

Intake Camshaft

Valve Spring

◆ Oil Seal

Spring Seat

Valve Guide

Valve

Bushing

Heater Hose

x 10

Cylinder Head

Upper Radiator Hose

◆ Cylinder Head Gasket

Timing Chain

Chain Tensioner Slipper

Crank Angle Sensor Plate

Chain Vibration Damper

Crankshaft Position Sensor

x 11

N·m (kgf·cm, ft·lbf) : Specified torque
◆ Non–reusable part

9307WG92

Cylinder head component removal—1ZZ-FE engine

head bolt threads and tighten in sequence. Replace any bolt that appears deformed. Bolts: 26 ft. lbs. (49 Nm).

- Torque each bolt in sequence an additional 180 degree turn.
- Camshafts
- Sprockets
- Oil filler cap
- PCV valve
- ECT sensor
- Intake manifold
- Wiring harness
- Exhaust manifold
- Exhaust pipe

- Alternator
- accelerator cable
- Air cleaner
- ECM box
- Battery

6. Fill the cooling system to the proper level.

7. Start the vehicle, check for leaks and repair if necessary.

5S-FE Engine

1. Before servicing the vehicle, refer to the precautions in the beginning of this section.

2. Drain the cooling system.

3. Remove or disconnect the following:
- Negative battery cable. On vehicles equipped with an air bag, wait at least 90 seconds before proceeding.
- All necessary components to gain access to the cylinder head using the diagram above.

➡**On California vehicles, remove the Vacuum Switching Valve (VSV) for fuel pressure control and EGR. On all vehicles (except California), remove the VSV for EGR.**

- Timing belt

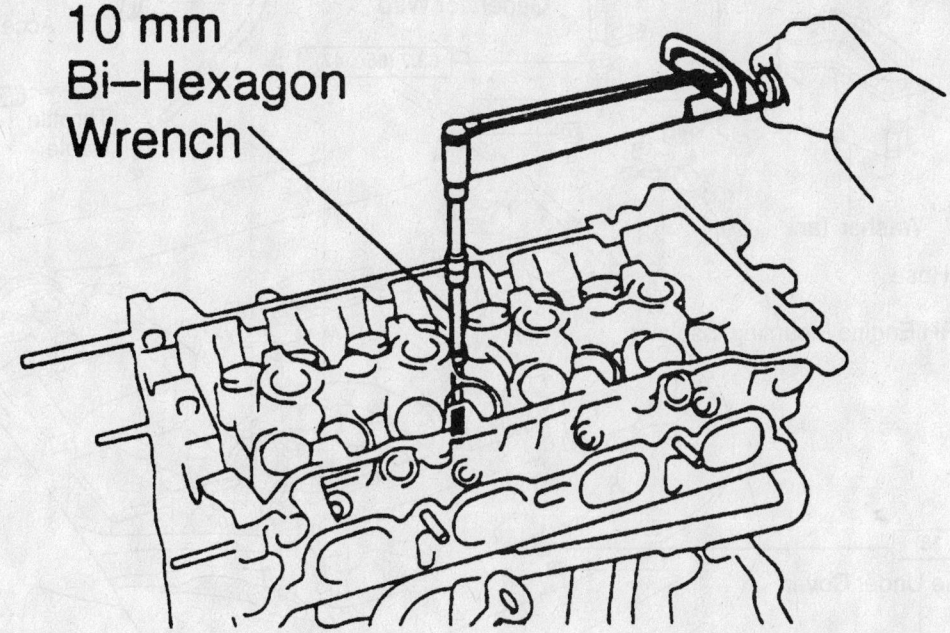

10 mm Bi-Hexagon Wrench

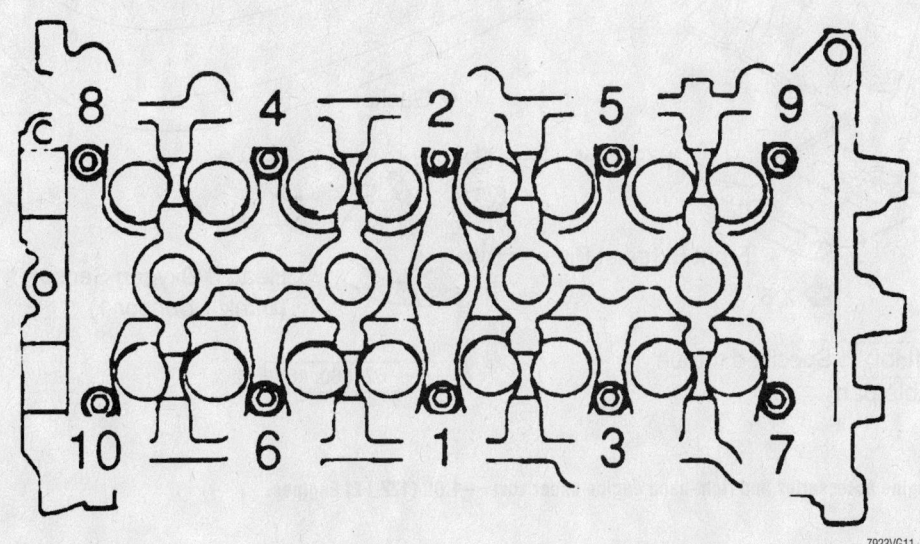

Cylinder head bolt tightening sequence—1ZZ-FE and 2ZZ-GE engines

7923VG11

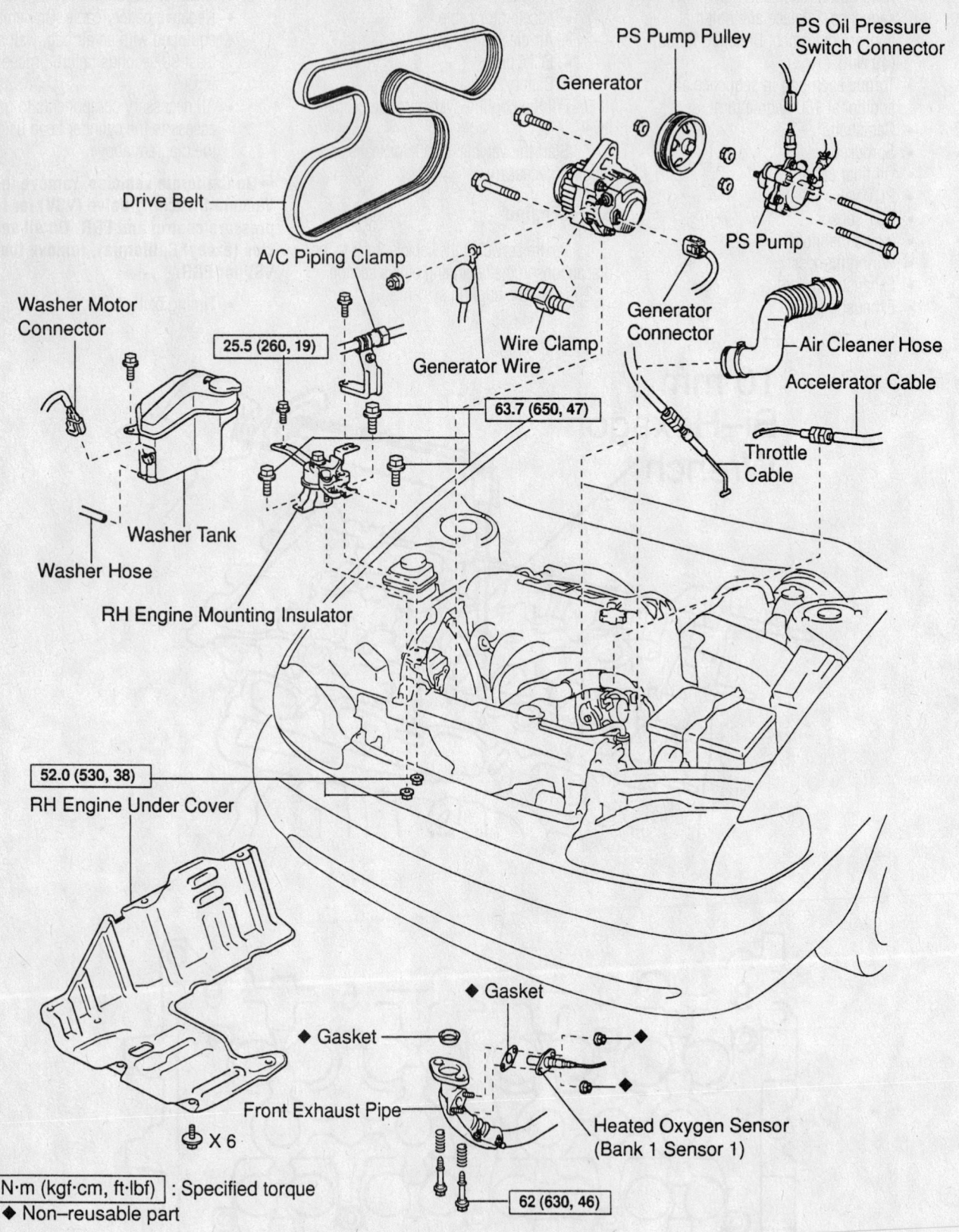

PS Pump Pulley

PS Oil Pressure Switch Connector

Generator

Drive Belt

A/C Piping Clamp

PS Pump

Washer Motor Connector

Wire Clamp

Generator Connector

Generator Wire

Air Cleaner Hose

25.5 (260, 19)

Accelerator Cable

63.7 (650, 47)

Throttle Cable

Washer Tank

Washer Hose

RH Engine Mounting Insulator

52.0 (530, 38)

RH Engine Under Cover

◆ Gasket

◆ Gasket

Front Exhaust Pipe

× 6

Heated Oxygen Sensor (Bank 1 Sensor 1)

N·m (kgf·cm, ft·lbf) : Specified torque

62 (630, 46)

◆ Non–reusable part

9301WG03

Exploded view of engine accessories and right-hand engine under cover—1.8L (1ZZ-FE) Engines.

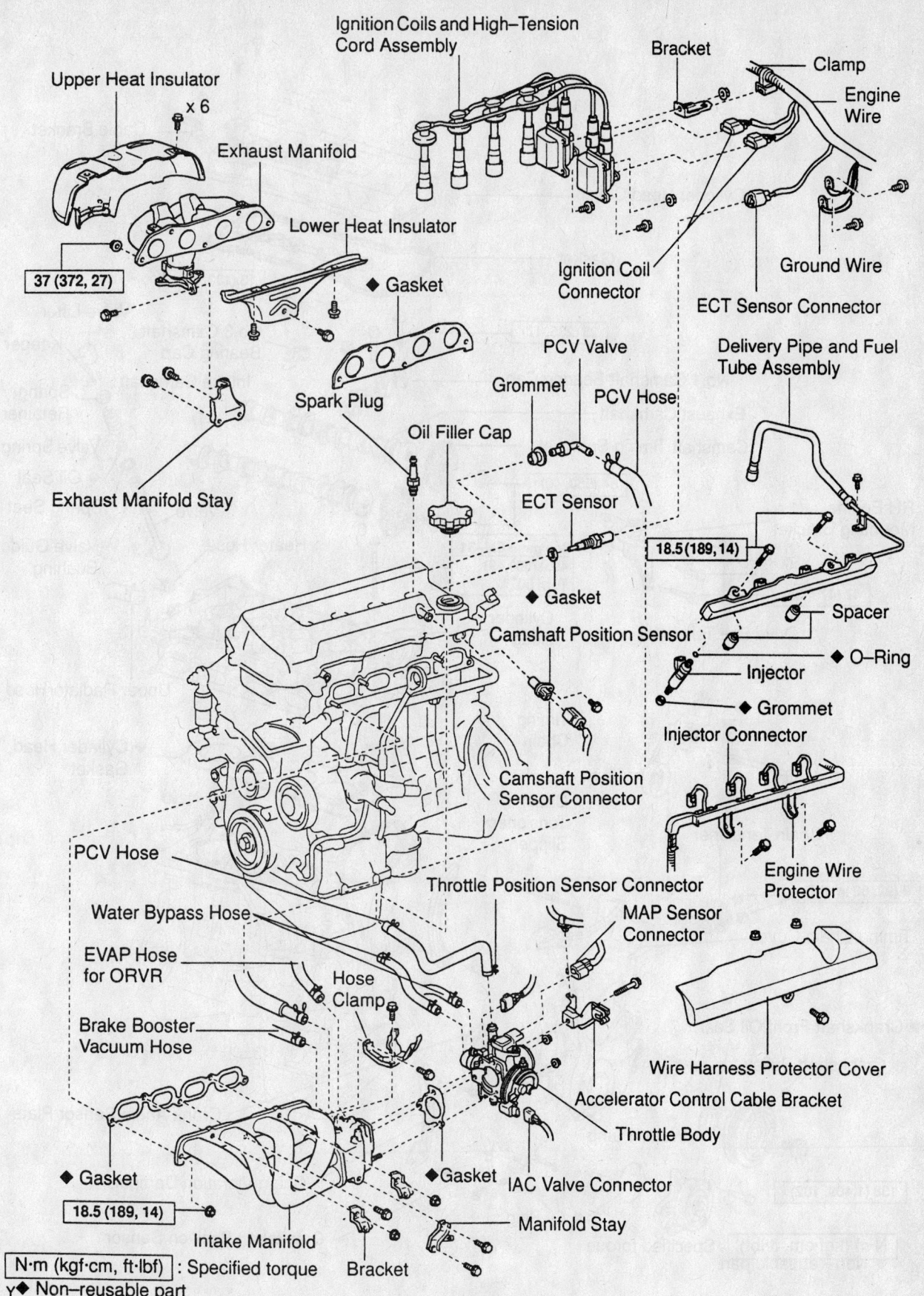

Upper Heat Insulator

x 6

Exhaust Manifold

Lower Heat Insulator

◆ Gasket

37 (372, 27)

Spark Plug

Exhaust Manifold Stay

Oil Filler Cap

Ignition Coils and High–Tension Cord Assembly

Bracket

Clamp

Engine Wire

Ignition Coil Connector

ECT Sensor Connector

Ground Wire

PCV Valve

Grommet

PCV Hose

Delivery Pipe and Fuel Tube Assembly

ECT Sensor

18.5 (189, 14)

◆ Gasket

Camshaft Position Sensor

Camshaft Position Sensor Connector

Spacer

◆ O–Ring

Injector

◆ Grommet

Injector Connector

Engine Wire Protector

PCV Hose

Water Bypass Hose

EVAP Hose for ORVR

Brake Booster Vacuum Hose

Hose Clamp

Throttle Position Sensor Connector

MAP Sensor Connector

Wire Harness Protector Cover

Accelerator Control Cable Bracket

Throttle Body

IAC Valve Connector

◆ Gasket

◆ Gasket

18.5 (189, 14)

Intake Manifold

Bracket

Manifold Stay

N·m (kgf·cm, ft·lbf) : Specified torque
γ◆ Non–reusable part

View of engine intake, exhaust, ignition, and fuel system location—1.8L (1ZZ-FE) engines.

9301WG04

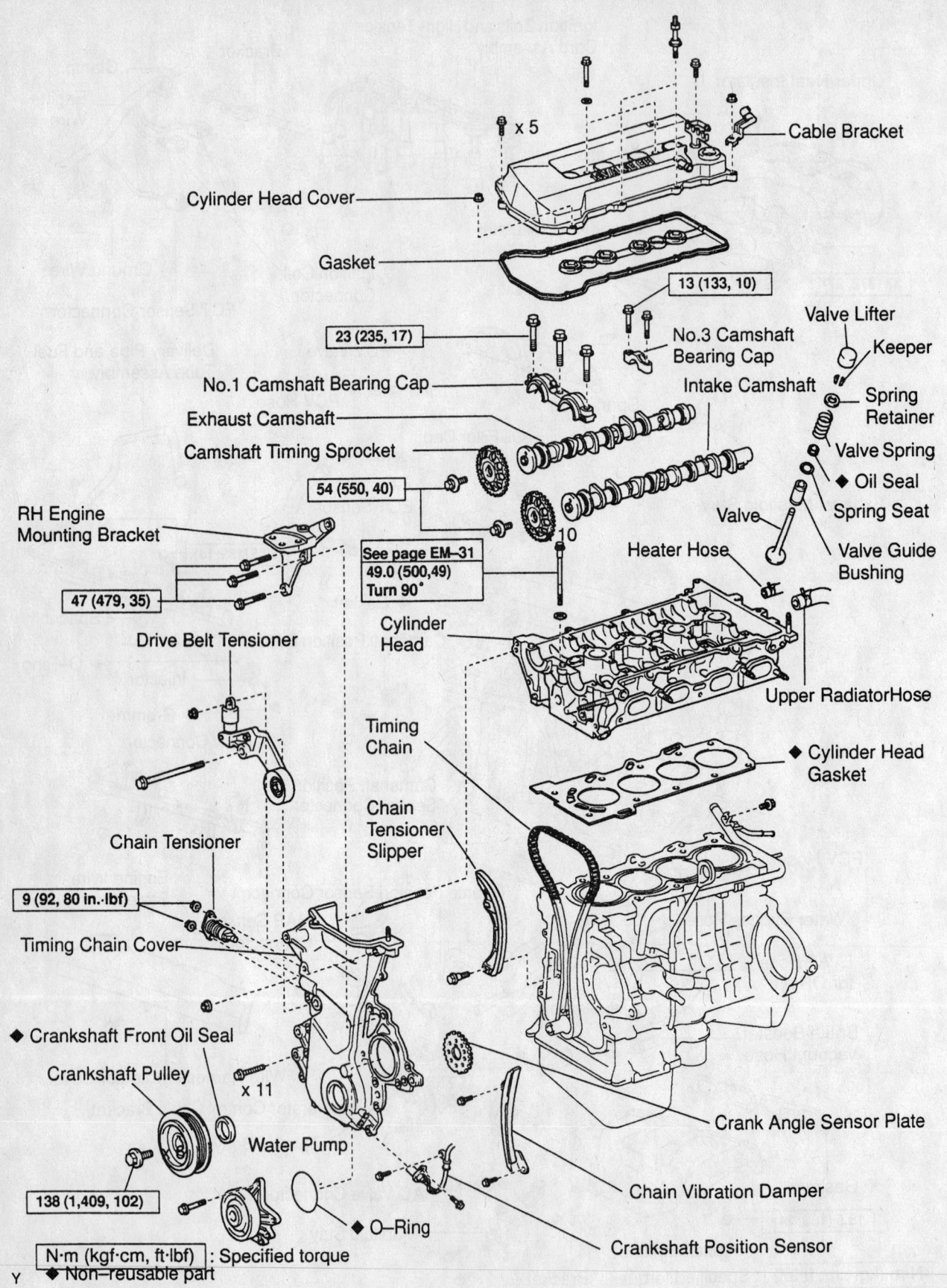

Cable Bracket

Cylinder Head Cover

Gasket

13 (133, 10)

Valve Lifter

Keeper

No.3 Camshaft
Bearing Cap

23 (235, 17)

Intake Camshaft

Spring
Retainer

No.1 Camshaft Bearing Cap

Valve Spring

Exhaust Camshaft

◆ Oil Seal

Camshaft Timing Sprocket

Spring Seat

54 (550, 40)

Valve

10

RH Engine
Mounting Bracket

Heater Hose

Valve Guide
Bushing

See page EM-31
49.0 (500, 49)
Turn 90°

47 (479, 35)

Cylinder
Head

Drive Belt Tensioner

Upper RadiatorHose

Timing
Chain

◆ Cylinder Head
Gasket

Chain
Tensioner
Slipper

Chain Tensioner

9 (92, 80 in.·lbf)

Timing Chain Cover

◆ Crankshaft Front Oil Seal

Crankshaft Pulley

x 11

Crank Angle Sensor Plate

Water Pump

Chain Vibration Damper

138 (1,409, 102)

◆ O–Ring

Crankshaft Position Sensor

N·m (kgf·cm, ft·lbf) : Specified torque

Y ◆ Non–reusable part

9301WG05

Illustration of disassembled cylinder head assembly—1.8L (1ZZ-FE) Engines.

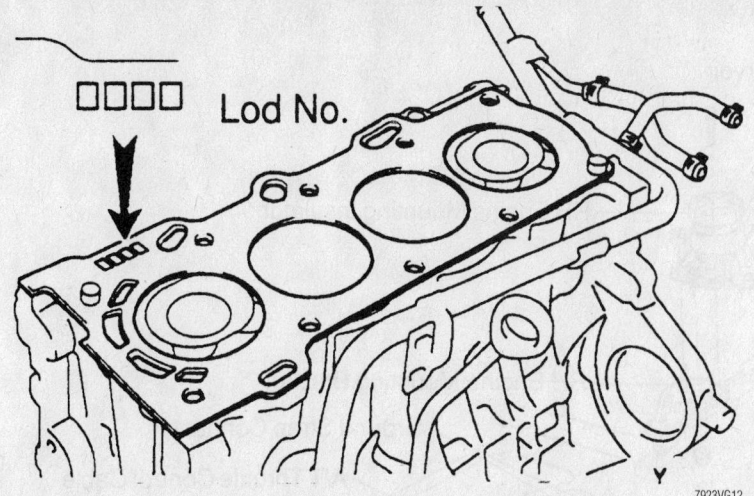

Position the head gasket correctly on the cylinder head—1.8L (1ZZ-FE) engine

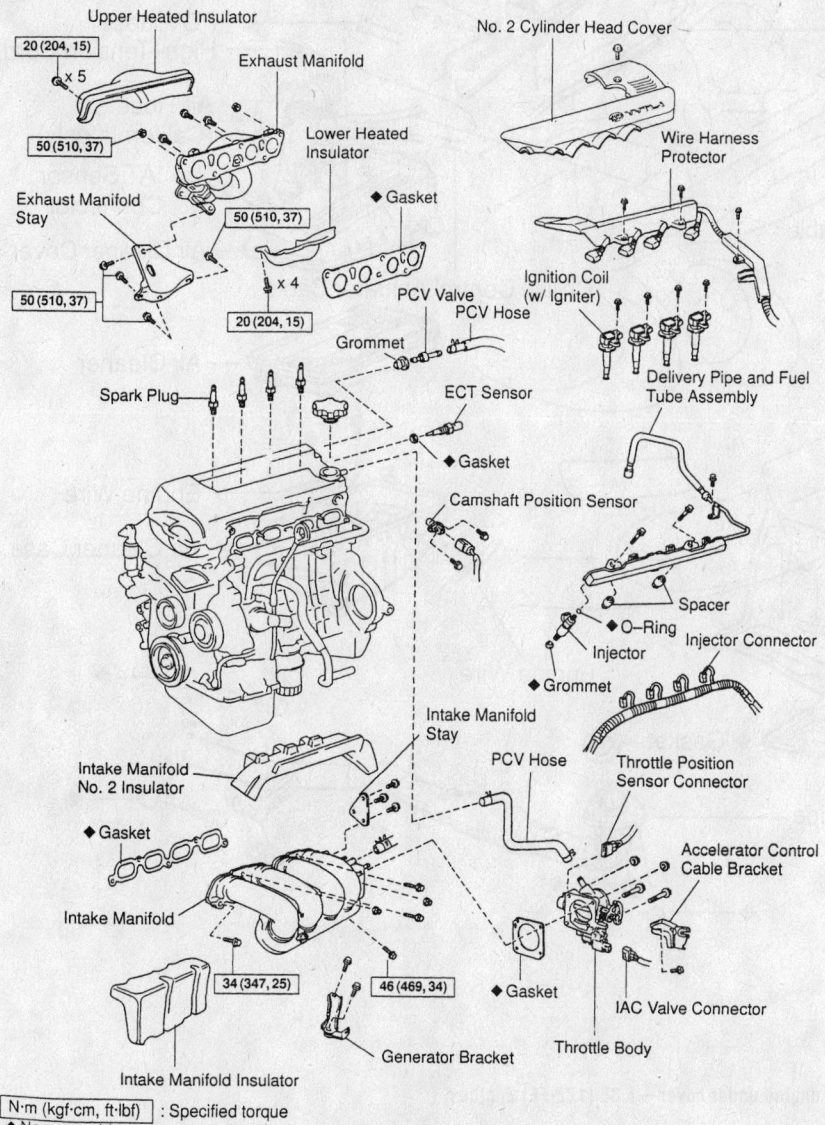

Cylinder head component exploded view—2ZZ-GE engine

➡ Support the timing belt, so that the meshing of the crankshaft timing pulley and the timing belt does not shift. Be careful not to drop anything inside the timing belt cover.

- Engine hangers and the alternator bracket.
- Camshafts following the proper sequences and procedures.
- Cylinder head bolts in several passes and in the reverse order of the installation sequence.
- Cylinder head from the cylinder block, disengaging the cylinder head from the block dowel pins.

To install:

4. Install or connect the following:
- Gasket and cylinder head
- Cylinder head bolts, lightly oiled. Tighten, in several passes and in sequence, to 36 ft. lbs. (49 Nm). Tighten the cylinder head bolts an additional 90 degrees in sequence. Then, tighten an additional 90 degrees.
- Camshafts and all other components following the proper sequences and procedures.
- Check and adjust valve clearance.
- Sealant to the 2 new semi-circular seals and install the seals to the cylinder head.
- Cylinder head cover with new gasket. Nuts: 17 ft. lbs. (23 Nm).

2JZ-GTE Engine

1. Before servicing the vehicle, refer to the precautions in the beginning of this section.
2. Drain the cooling system.
3. Drain the engine oil.
4. Remove or disconnect the following:
- Negative battery cable from the battery. On vehicles equipped with an air bag, wait at least 90 seconds before proceeding.
- Turbocharger
- Exhaust manifold
- With manual transmission, drive belt tensioner damper.
- Drive belt
- Water outlet and No. 1 water bypass pipe.
- Power steering pump without disconnecting the hoses as follows:
- Intake manifold and delivery pipe assembly by removing the 4 bolts and 2 nuts.
- Upper 2 timing belt covers (Nos. 2 and 3).
- Drive belt tensioner

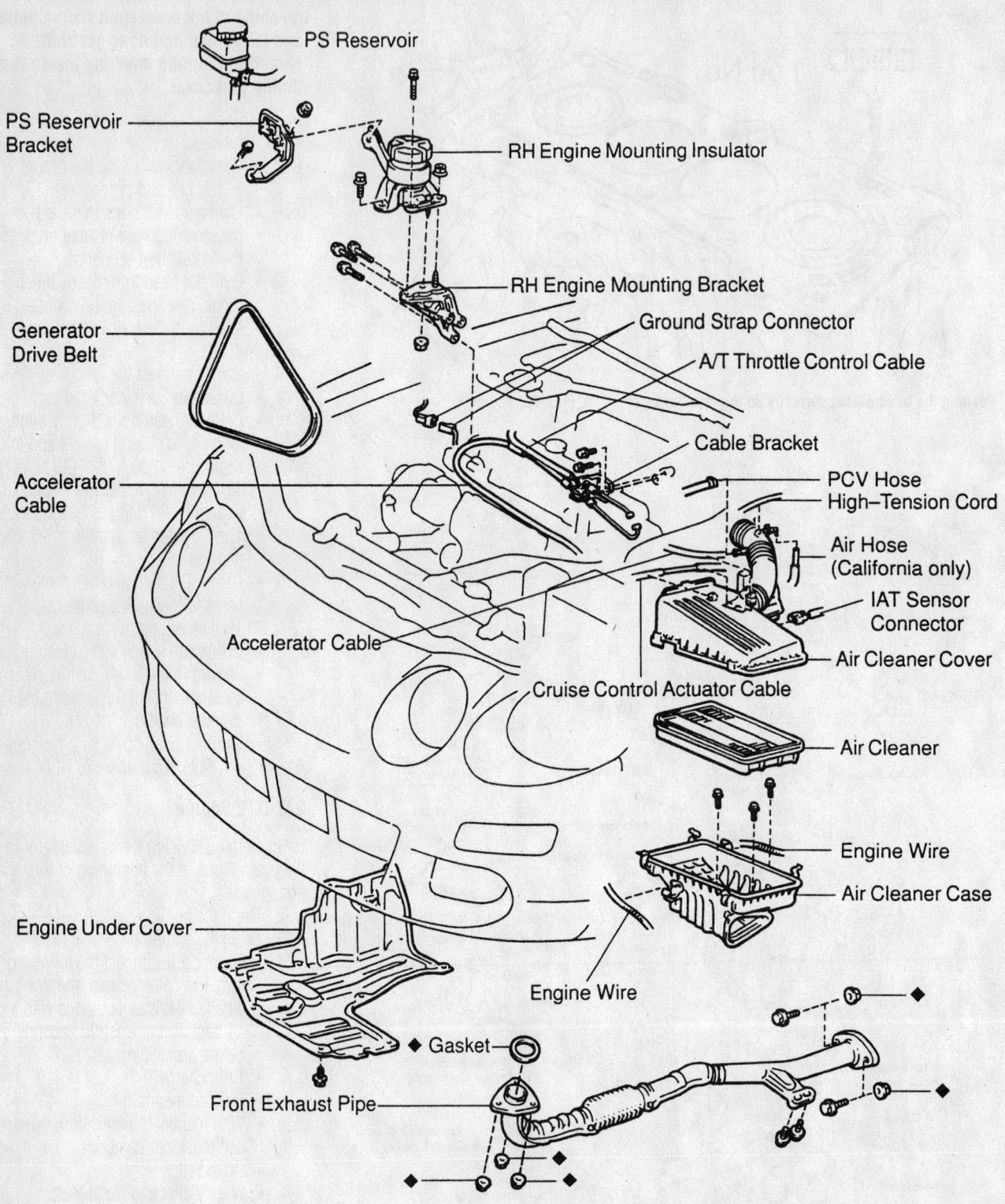

PS Reservoir

PS Reservoir Bracket

RH Engine Mounting Insulator

RH Engine Mounting Bracket

Ground Strap Connector

A/T Throttle Control Cable

Generator Drive Belt

Cable Bracket

Accelerator Cable

PCV Hose

High–Tension Cord

Air Hose (California only)

IAT Sensor Connector

Accelerator Cable

Air Cleaner Cover

Cruise Control Actuator Cable

Air Cleaner

Engine Wire

Air Cleaner Case

Engine Under Cover

Engine Wire

◆ Gasket

Front Exhaust Pipe

◆ Non–reusable part

9301WG09

Exploded view of engine accessories and right-hand engine under cover—1.8L (1ZZ-FE) Engines.

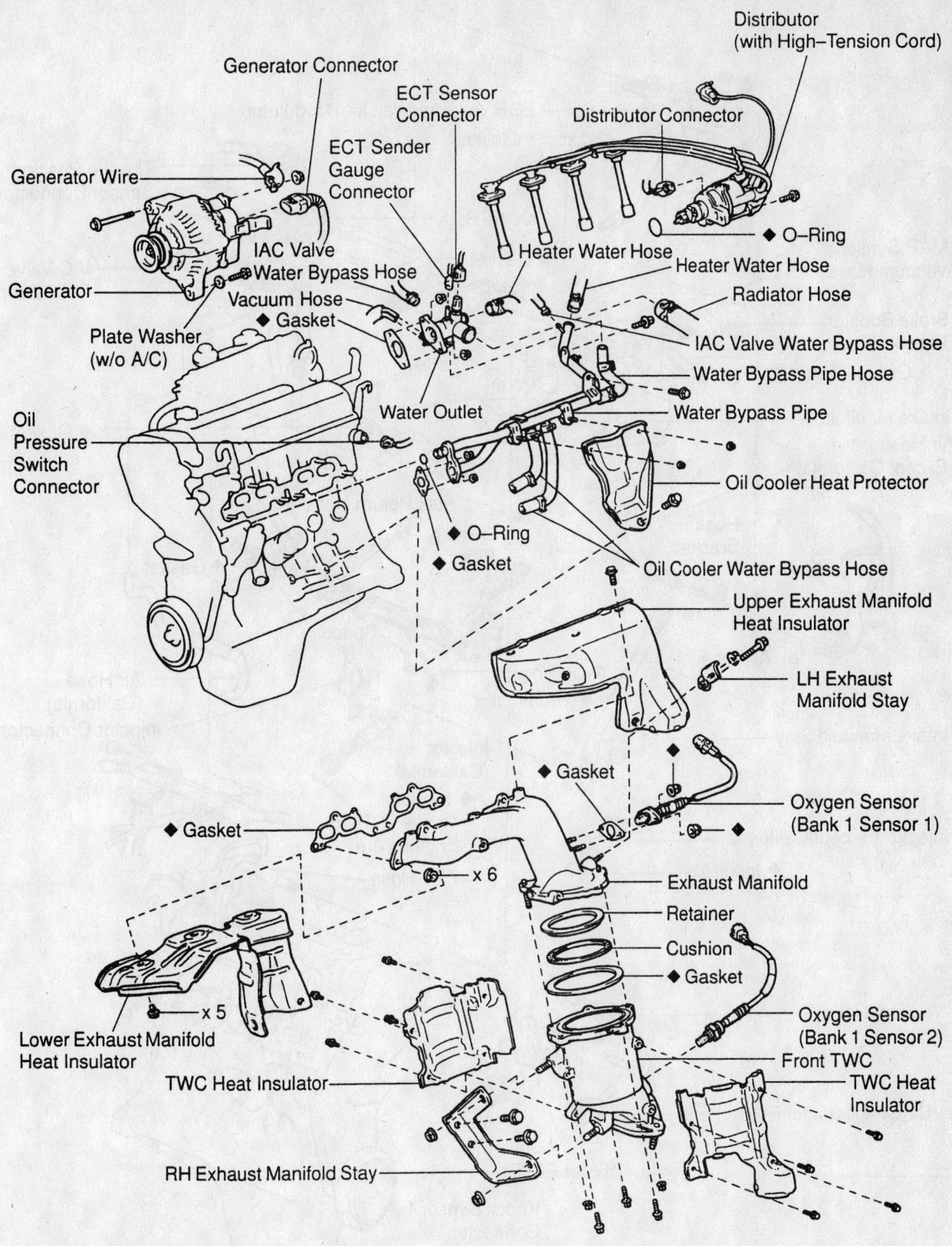

Generator Connector

Generator Wire

Generator

Plate Washer (w/o A/C)

Oil Pressure Switch Connector

ECT Sensor Connector

ECT Sender Gauge Connector

IAC Valve Water Bypass Hose

Vacuum Hose

◆ Gasket

Water Outlet

◆ O-Ring

◆ Gasket

Distributor (with High-Tension Cord)

Distributor Connector

◆ O-Ring

Heater Water Hose

Heater Water Hose

Radiator Hose

IAC Valve Water Bypass Hose

Water Bypass Pipe Hose

Water Bypass Pipe

Oil Cooler Heat Protector

Oil Cooler Water Bypass Hose

Upper Exhaust Manifold Heat Insulator

LH Exhaust Manifold Stay

◆ Gasket

Oxygen Sensor (Bank 1 Sensor 1)

◆ Gasket

x 6

Exhaust Manifold

Retainer

Cushion

◆ Gasket

Oxygen Sensor (Bank 1 Sensor 2)

Front TWC

TWC Heat Insulator

x 5

Lower Exhaust Manifold Heat Insulator

TWC Heat Insulator

RH Exhaust Manifold Stay

◆ Non-reusable part

9301WG10

View of engine, exhaust, ignition and fuel system location—1.8L (1ZZ-FE) Engines.

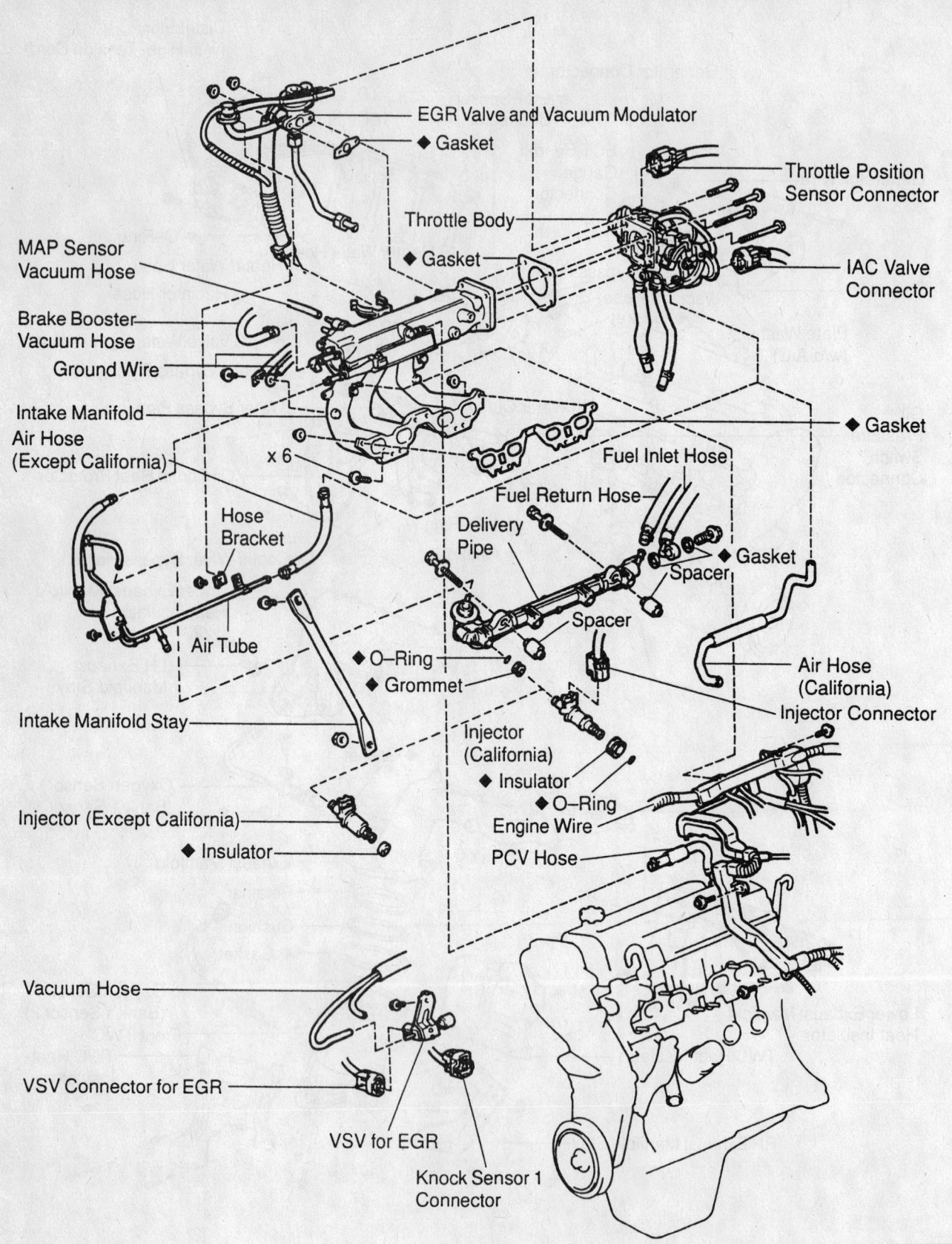

EGR Valve and Vacuum Modulator

◆ Gasket

Throttle Body

◆ Gasket

Throttle Position Sensor Connector

IAC Valve Connector

MAP Sensor Vacuum Hose

Brake Booster Vacuum Hose

Ground Wire

Intake Manifold

Air Hose (Except California)

x 6

◆ Gasket

Fuel Inlet Hose

Fuel Return Hose

Delivery Pipe

◆ Gasket

Spacer

Spacer

Hose Bracket

Air Tube

Intake Manifold Stay

◆ O-Ring

◆ Grommet

Air Hose (California)

Injector Connector

Injector (California)

◆ Insulator

◆ O-Ring

Engine Wire

PCV Hose

Injector (Except California)

◆ Insulator

Vacuum Hose

VSV Connector for EGR

VSV for EGR

Knock Sensor 1 Connector

◆ Non-reusable part

9301WG11

View of engine, intake, ignition and fuel system location—1.8L (1ZZ-FE) Engines.

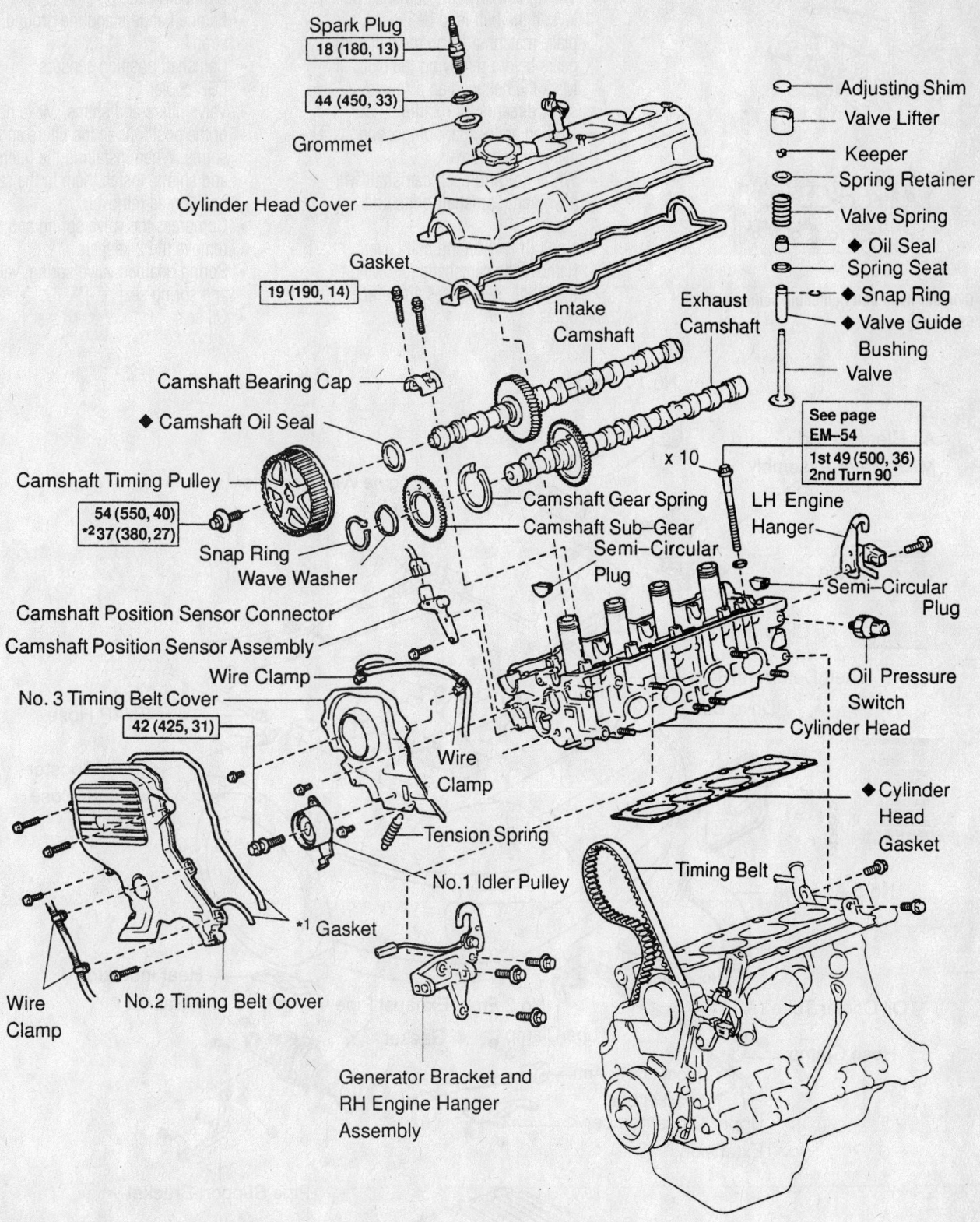

N·m (kgf·cm, ft·lbf) : Specified torque
*¹ Replace only if damaged
*² For use with SST
◆ Non–reusable part

Illustration of disassembled cylinder head assembly—1.8L (1ZZ-FE) Engines.

9301WG12

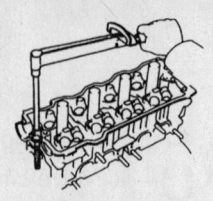

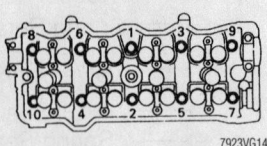

7923VG14

Cylinder head bolt tightening sequence— 5S-FE engine

- Timing belt from the camshaft pulleys. If the belt is to be reused, place matchmarks on the belt and gears before removing the belt. Mark the belt with an arrow to show direction of rotation
- Ignition coils, spark plugs, and cylinder head covers
- While holding each camshaft with a wrench, camshaft bolts and gears
- No. 4 (inner) timing belt cover
- Remove the camshafts following the proper sequences and procedures

- Cylinder head
- Engine hangers and the ground strap
- Camshaft position sensors
- EGR cooler
- Valve lifters and shims. Make note of the positions of the lifters and shims. When installing the lifters and shims, install them in the same position as removal.
- Compress the valve spring and remove the 2 keepers
- Spring retainer, valve spring, valve and spring seat
- Oil seal

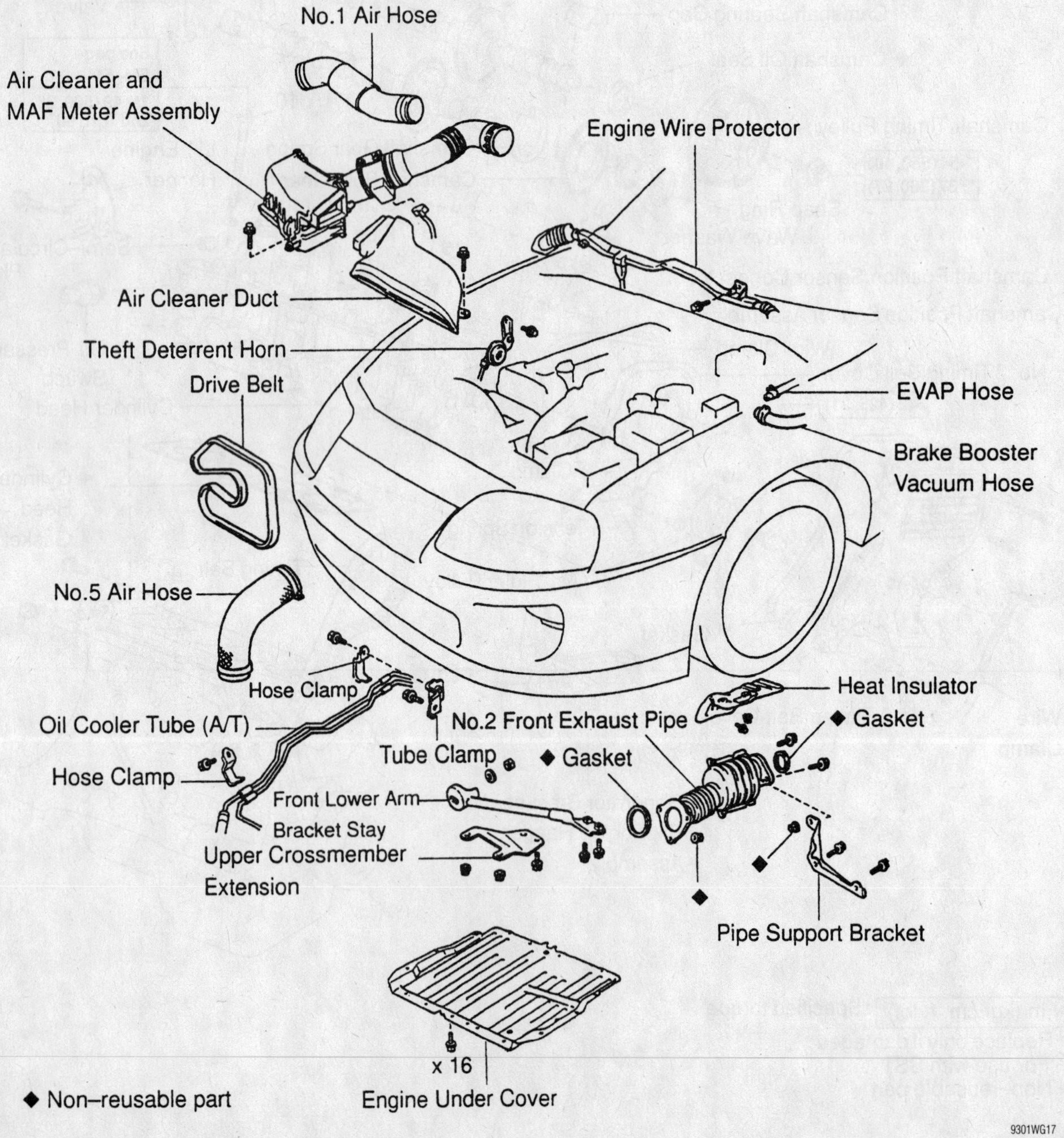

No.1 Air Hose

Air Cleaner and MAF Meter Assembly

Engine Wire Protector

Air Cleaner Duct

Theft Deterrent Horn

Drive Belt

EVAP Hose

Brake Booster Vacuum Hose

No.5 Air Hose

Hose Clamp

Heat Insulator

◆ Gasket

Oil Cooler Tube (A/T)

No.2 Front Exhaust Pipe

Tube Clamp

◆ Gasket

Hose Clamp

Front Lower Arm

Bracket Stay

Upper Crossmember Extension

Pipe Support Bracket

x 16

◆ Non–reusable part

Engine Under Cover

9301WG17

Removal of air box, engine under cover and front exhaust pipe—3.0L (2JZ-GTE) engine

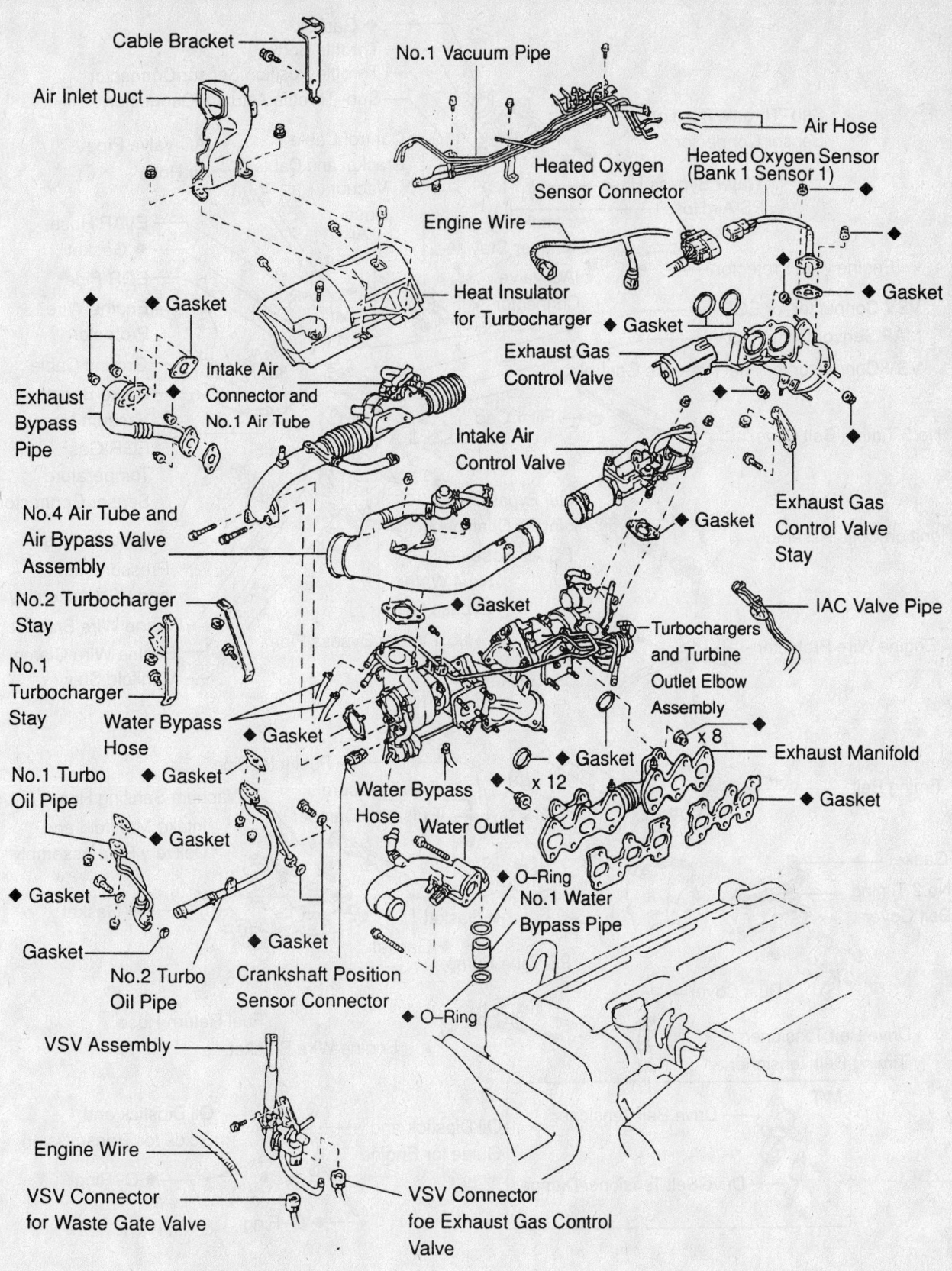

◆ Non-reusable part

9301WG18

Diagram of intake and exhaust manifolds—3.0L (2JZ-GTE) engine

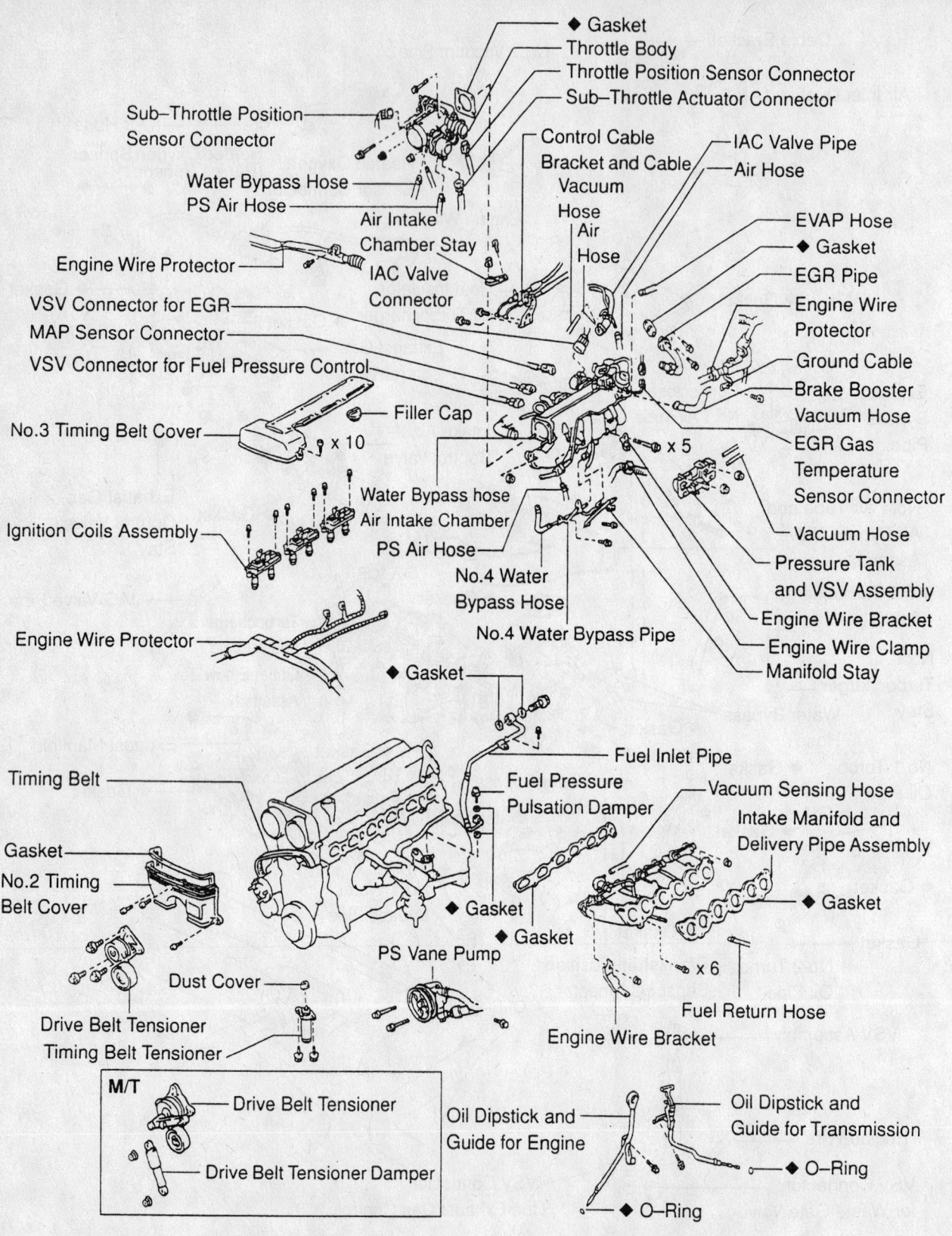

◆ Non-reusable part

Exploded view of intake manifold and throttle body—3.0L (2JZ-GTE) engine

9301WG19

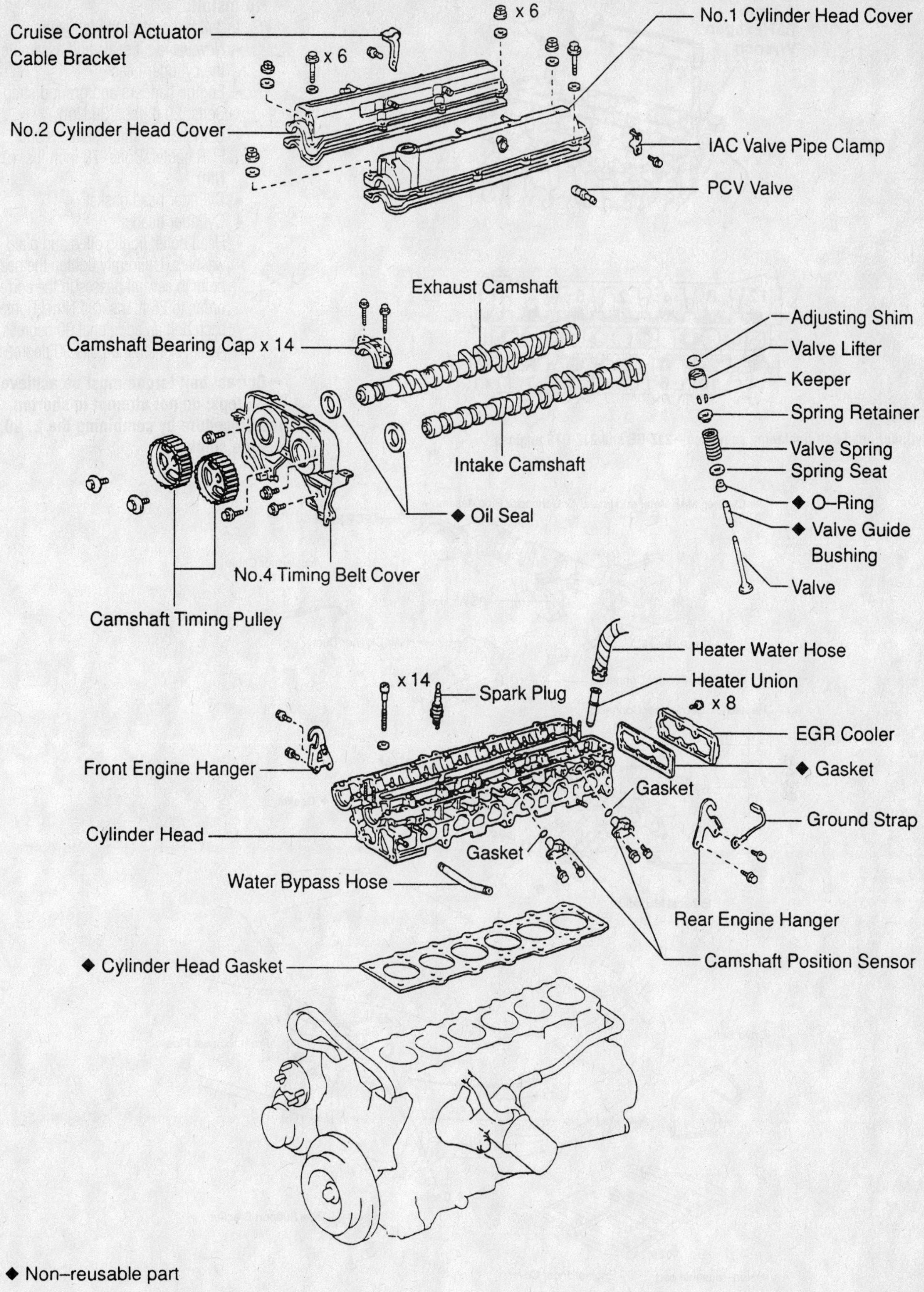

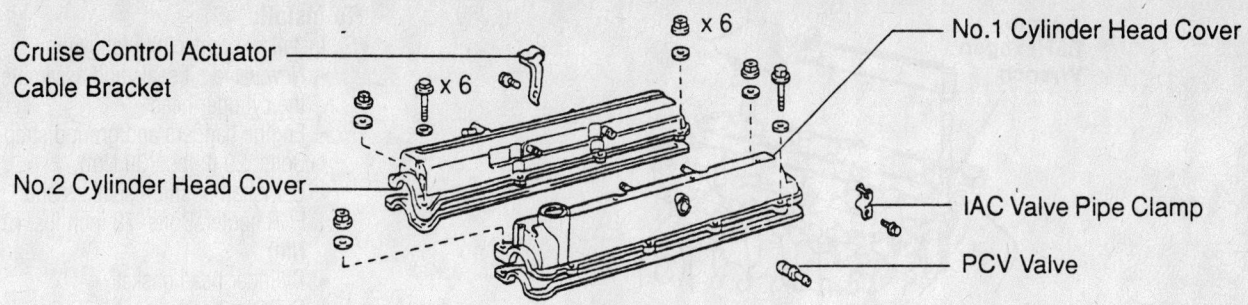

Cruise Control Actuator Cable Bracket

x 6

x 6

x 6

No.1 Cylinder Head Cover

No.2 Cylinder Head Cover

IAC Valve Pipe Clamp

PCV Valve

Exhaust Camshaft

Camshaft Bearing Cap x 14

Adjusting Shim

Valve Lifter

Keeper

Spring Retainer

Valve Spring

Spring Seat

◆ O–Ring

◆ Valve Guide Bushing

Valve

Intake Camshaft

◆ Oil Seal

No.4 Timing Belt Cover

Camshaft Timing Pulley

Heater Water Hose

Heater Union

x 8

x 14   Spark Plug

EGR Cooler

◆ Gasket

Front Engine Hanger

Ground Strap

Cylinder Head

Gasket

Gasket

Water Bypass Hose

Rear Engine Hanger

Camshaft Position Sensor

◆ Cylinder Head Gasket

◆ Non–reusable part

9301WG20

Exploded view of cylinder head—3.0L (2JZ-GTE) engine

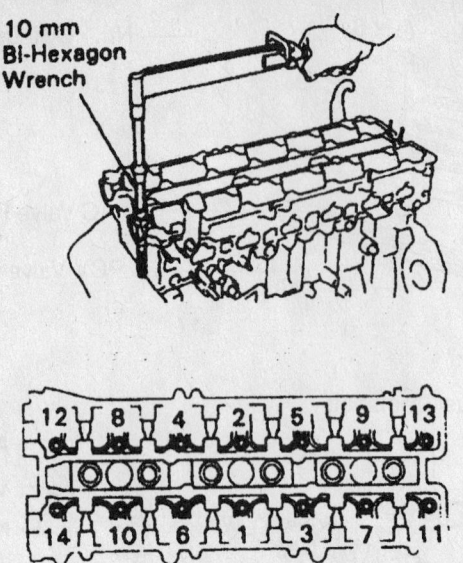

Cylinder head bolt tightening sequence—2JZ-GE and 2JZ-GTE engines

To install:

5. Install or connect the following:
   - New valve oil seals and assemble the cylinder head
   - Engine hangers and ground strap. Bolts: 29 ft. lbs. (39 Nm).
   - Camshaft position sensors and EGR cooler. Bolts: 78 inch lbs. (9 Nm).
   - Cylinder head gasket
   - Cylinder head
   - Head bolts, lightly oiled and plate washers. Uniformly tighten the head bolts in several passes in the correct order, to 25 ft. lbs. (34 Nm). Tighten each bolt an additional 90 degrees. Again, tighten the bolts 90 degrees.

➡ **Correct bolt torque must be achieved in 3 steps; do not attempt to shorten the procedure by combining the 2, 90 degree steps.**

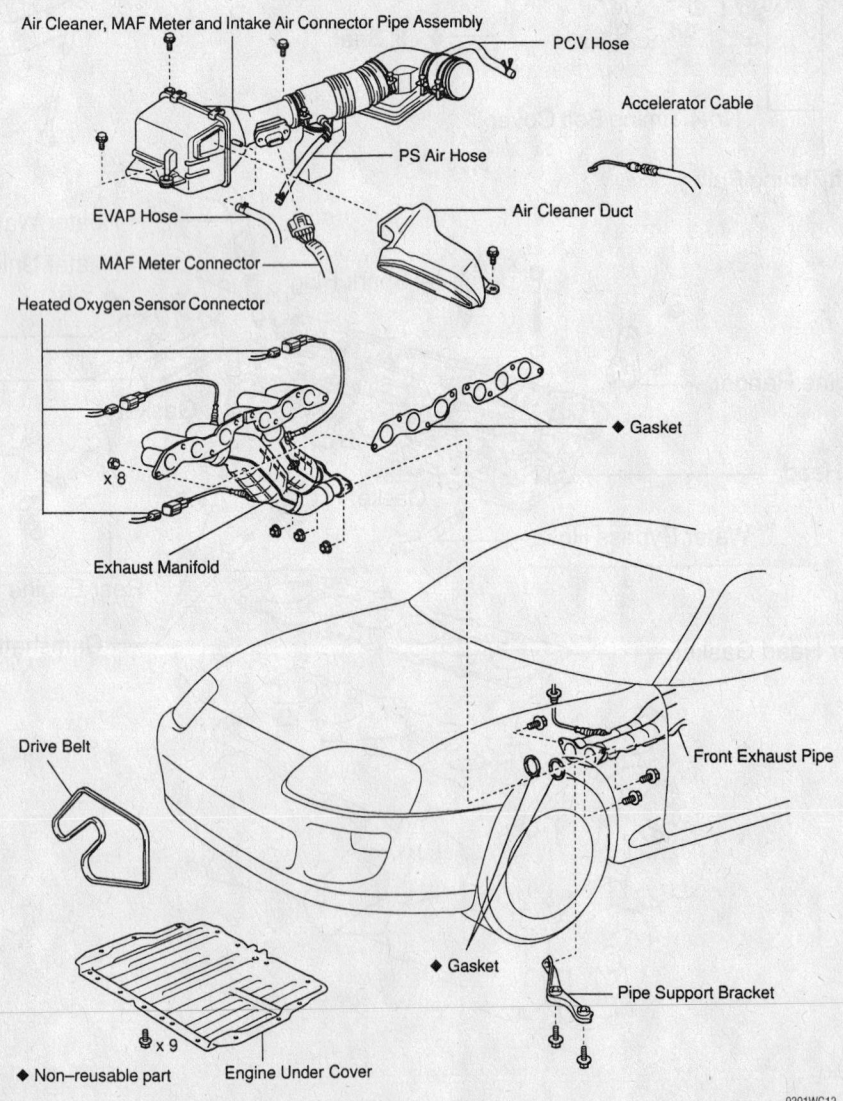

Removal of the air duct assembly, exhaust manifold, front exhaust pipe and engine under cover—3.0L (2JZ-GE) engine

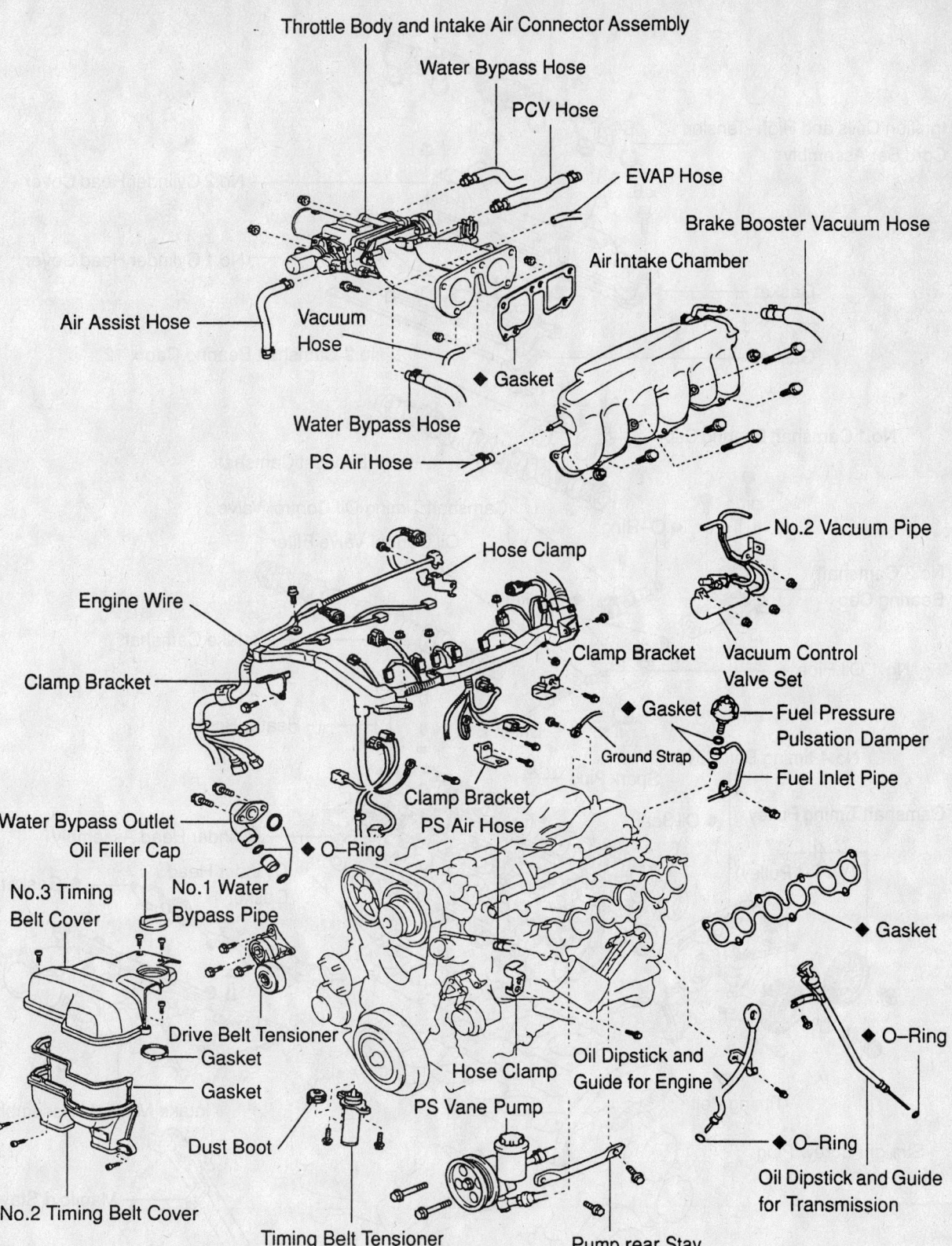

Throttle Body and Intake Air Connector Assembly
Water Bypass Hose
PCV Hose
EVAP Hose
Brake Booster Vacuum Hose
Air Intake Chamber
Air Assist Hose
Vacuum Hose
◆ Gasket
Water Bypass Hose
PS Air Hose

No.2 Vacuum Pipe
Hose Clamp
Engine Wire
Clamp Bracket
Vacuum Control Valve Set
◆ Gasket
Fuel Pressure Pulsation Damper
Ground Strap
Fuel Inlet Pipe
Clamp Bracket
Clamp Bracket
PS Air Hose
Water Bypass Outlet
Oil Filler Cap
◆ O–Ring
No.1 Water Bypass Pipe
No.3 Timing Belt Cover
◆ Gasket
Drive Belt Tensioner
Gasket
Gasket
◆ O–Ring
Hose Clamp
Oil Dipstick and Guide for Engine
PS Vane Pump
Dust Boot
◆ O–Ring
Oil Dipstick and Guide for Transmission
No.2 Timing Belt Cover
Timing Belt Tensioner
Pump rear Stay

◆ Non–reusable part

Removal of the intake manifold and throttle body—3.0L (2JZ-GE) engine

9301WG14

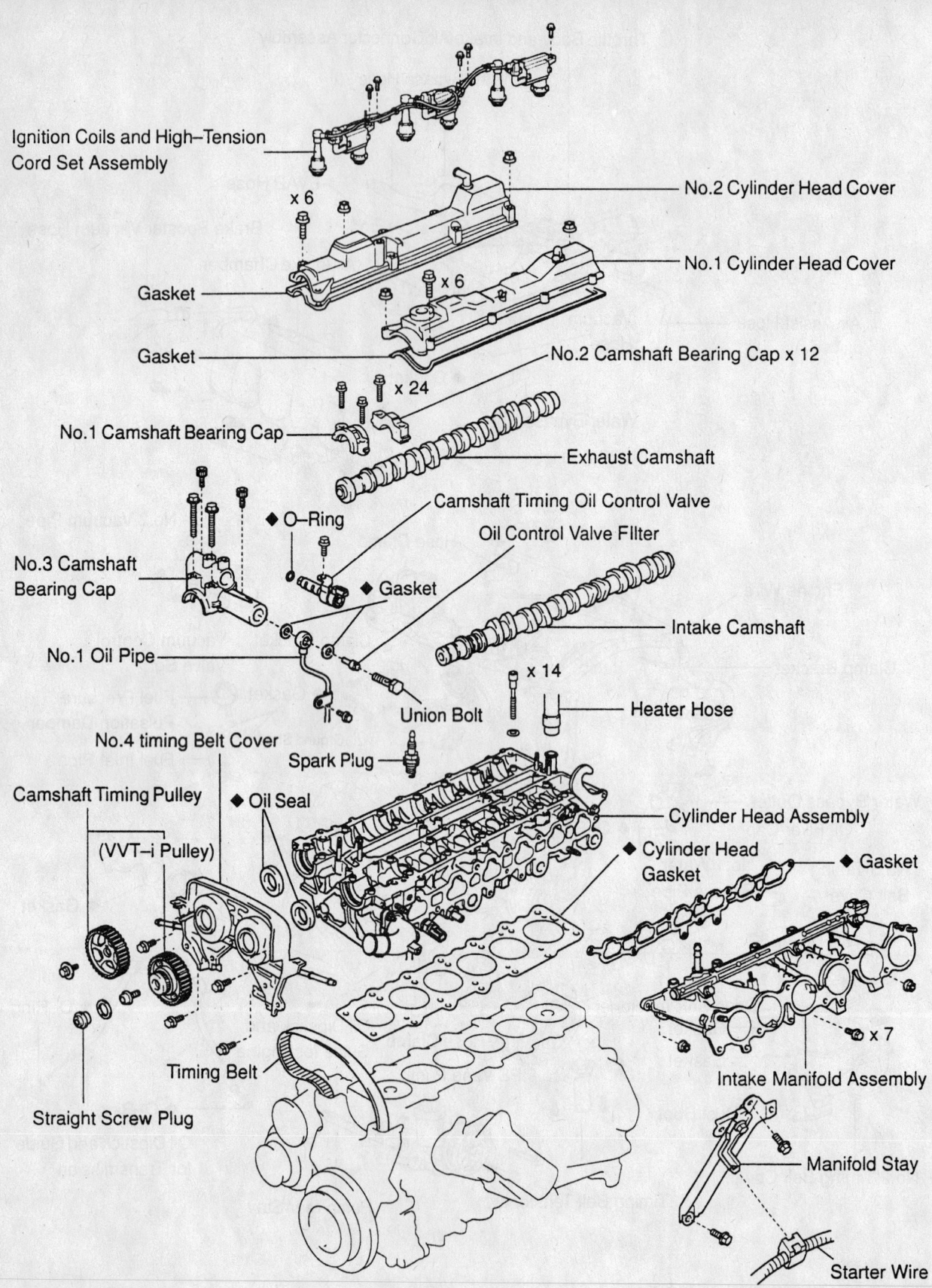

Ignition Coils and High–Tension Cord Set Assembly

No.2 Cylinder Head Cover

No.1 Cylinder Head Cover

x 6

Gasket

Gasket

x 6

No.2 Camshaft Bearing Cap x 12

x 24

No.1 Camshaft Bearing Cap

Exhaust Camshaft

◆ O–Ring

Camshaft Timing Oil Control Valve

Oil Control Valve Filter

No.3 Camshaft Bearing Cap

◆ Gasket

No.1 Oil Pipe

Intake Camshaft

Union Bolt

x 14

Heater Hose

No.4 timing Belt Cover

Spark Plug

Camshaft Timing Pulley

◆ Oil Seal

Cylinder Head Assembly

(VVT–i Pulley)

◆ Cylinder Head Gasket

◆ Gasket

Timing Belt

Straight Screw Plug

Intake Manifold Assembly

x 7

Manifold Stay

Starter Wire

◆ Non–reusable part

View of cylinder head covers, camshafts, and the cylinder head—3.0L (2JZ-GE) engine

9301WG15

- Cylinder head covers, spark plugs, ignition coils, and timing belt
- Intake manifold and delivery pipe assembly by installing a new gasket and engine wire. Bolts and nuts: 20 ft. lbs. (27 Nm)
- Fuel inlet pipe by installing a new gasket and the union bolt. Bolt: 30 ft. lbs. (42 Nm).
- Fuel inlet pipe clamp bolt to the intake manifold
- All other components in the reverse of installation
- Power steering pump. Bolts: 43 ft. lbs. (58 Nm).
- Water outlet and No. 1 water bypass pipe. Bolts: 15 ft. lbs. (21 Nm).
- ECT sensor and sender gauge connectors
- Upper radiator hose to the water outlet
- Drive belt
- With manual transmission torque the belt tensioner damper nuts to 14 ft. lbs. (20 Nm).
- 2 new gaskets to the cylinder head
- Exhaust manifold. Bolts: 29 ft. lbs. (39 Nm).
- Turbocharger
- Negative battery cable

6. Fill the cooling system to the proper level

7. Fill the engine with clean oil.

8. Start the vehicle, check for leaks and repair if necessary.

2JZ-GE Engine

1. Before servicing the vehicle, refer to the precautions in the beginning of this section.

2. Relieve the fuel pressure in the fuel lines.

3. Drain the engine oil.

4. Drain the cooling system.

5. Remove or disconnect the following:
- Negative battery cable. Wait at least 90 seconds before performing any other work
- Undercovers
- Accelerator, throttle control (automatic transmission only) and cruise control cables from the throttle body
- Air cleaner duct
- Air cleaner, airflow meter, and the intake air pipe
- Drive belt, the fan and fluid coupling, and the water pump pulley
- Front exhaust pipe

- Except for California vehicles, the manifold heat insulator
- Oxygen sensor connector(s)
- Exhaust manifolds and gaskets
- Power steering pump without disconnecting the hoses
- Brake booster vacuum hose
- EVAP hose
- Throttle body and intake air connector assembly
- Air intake chamber stays
- No. 2 vacuum pipe and VSV assembly
- No. 3 timing belt cover by removing the oil filler cap
- Cylinder head rear cover
- Spark plug wires from the cylinder head covers
- Distributor and wires
- Spark plugs
- Timing belt
- Water bypass outlet and No. 1 bypass pipe
- Fuel return hose
- Engine wire bracket from the intake manifold
- Oil dipstick and guide
- Starter
- Air intake chamber
- Vacuum control valve set
- Intake manifold and gaskets
- Cylinder head covers (valve covers)
- Camshaft timing pulleys. Hold the hexagon portion of the camshaft with a wrench and remove the pulley mounting bolt and camshaft pulley
- No. 4 timing belt cover
- Camshafts following the proper sequences and procedures
- Cylinder head bolts in the reverse order of the installation sequence
- Cylinder head

To install:

6. Install or connect the following:
- New gasket on the cylinder block
- Plate washers and cylinder head bolts, lightly coated with engine oil. Tighten the bolts to 25 ft. lbs. (34 Nm). Tighten each bolt an additional 90 degrees. Again, tighten the bolts another 90 degrees of rotation.
- Position camshafts in the cylinder head with the cam lobes and the knock pins in the correct position.
- Position the No. 3 and No. 7 bearing caps in place, then uniformly and alternately tighten them temporarily
- New oil seals over the camshafts

7. Clean the surfaces of the No. 1 bear-

ing cap and cylinder head with cleaner. Apply seal packing to the No. 1 bearing cap.

8. Install camshafts following the proper sequences and procedures.

9. Press the 2 oil seals in as far as they will go.

10. Rotate each camshaft until the forward straight (knock) pin is straight up. Loosen exhaust No. 1, 2, and 6 bearing cap bolts until they can be turned by hand; tighten, in several passes, to 14 ft. lbs. (20 Nm). Loosen intake No. 1, 2, and 5 and tighten, in several passes, to 14 ft. lbs. (20 Nm).

11. Turn each camshaft ⅓ of a revolution (120 degrees). Loosen exhaust Nos. 4 and 7 bearing cap bolts; tighten, in several passes, to 14 ft. lbs. (20 Nm). Loosen intake No. 4 and 6 bearing cap bolts; tighten, in several passes, to 14 ft. lbs. (20 Nm).

12. Turn each camshaft an additional ⅓ of a revolution, loosen exhaust bearing cap bolts Nos. 3 and 5, then tighten them, in several passes, to 14 ft. lbs. (20 Nm). Loosen intake bearing cap bolts No. 3 and 7, then tighten them, in several passes to 14 ft. lbs. (20 Nm).

13. Check and adjust the valve clearance.

14. Install or connect the following:
- No. 4 timing belt cover. Bolts: 78 inch lbs. (9 Nm).
- Camshaft timing pulleys. Align the shaft pin with the pulley groove and slide the pulley on. Install the bolt temporarily. Hold the hex portion of the camshaft with a wrench and tighten the pulley bolt to 59 ft. lbs. (79 Nm).
- Cylinder head covers
- Intake manifold with a new gasket. Bolts: 20 ft. lbs. (27 Nm).
- Injectors
- Delivery pipe. Bolts: 20 ft. lbs. (27 Nm).
- Fuel inlet pipe. Bolt: 30 ft. lbs. (42 Nm).
- Clamp bolt to the intake manifold
- Fuel pressure pulsation damper
- Intake manifold stay. Bolts: 29 ft. lbs. (39 Nm).
- Water outlet and No. 1 bypass hose assembly
- Engine wire protector to the intake manifold with the 3 nuts
- 6 injector connectors
- ECT sensor connector and sender gauge
- 2 wire clamps and the 2 ground straps to the intake manifold with the bolts

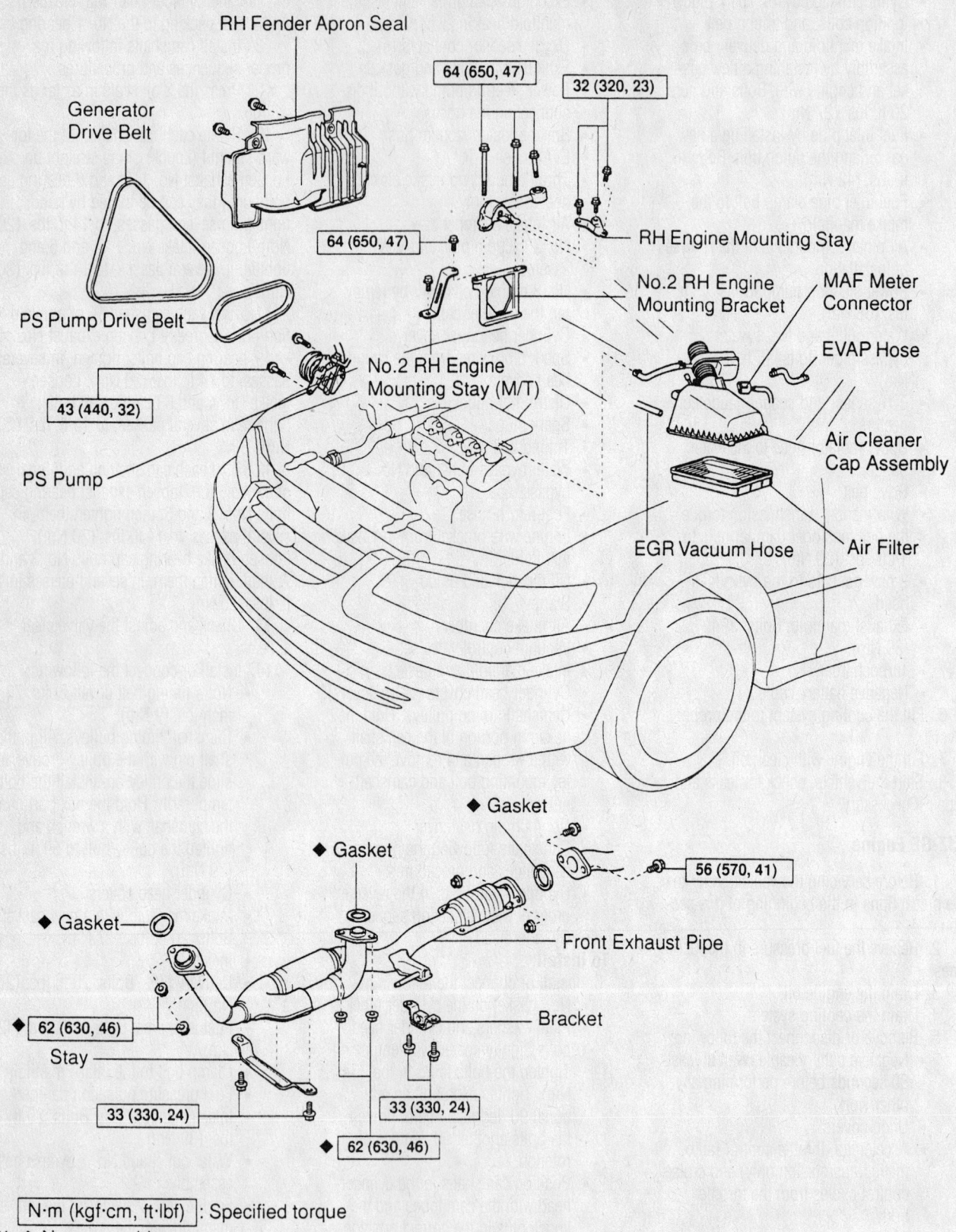

RH Fender Apron Seal

64 (650, 47)

32 (320, 23)

Generator Drive Belt

64 (650, 47)

RH Engine Mounting Stay

No.2 RH Engine Mounting Bracket

MAF Meter Connector

EVAP Hose

PS Pump Drive Belt

43 (440, 32)

No.2 RH Engine Mounting Stay (M/T)

Air Cleaner Cap Assembly

PS Pump

EGR Vacuum Hose

Air Filter

◆ Gasket

◆ Gasket

56 (570, 41)

◆ Gasket

Front Exhaust Pipe

◆ 62 (630, 46)

Bracket

Stay

33 (330, 24)

33 (330, 24)

◆ 62 (630, 46)

N·m (kgf·cm, ft·lbf) : Specified torque

Y ◆ Non–reusable part

Removal of the right-hand fender apron, front exhaust pipe, and air cleaner assembly—3.0L (1MZ-FE) engine

9301WG21

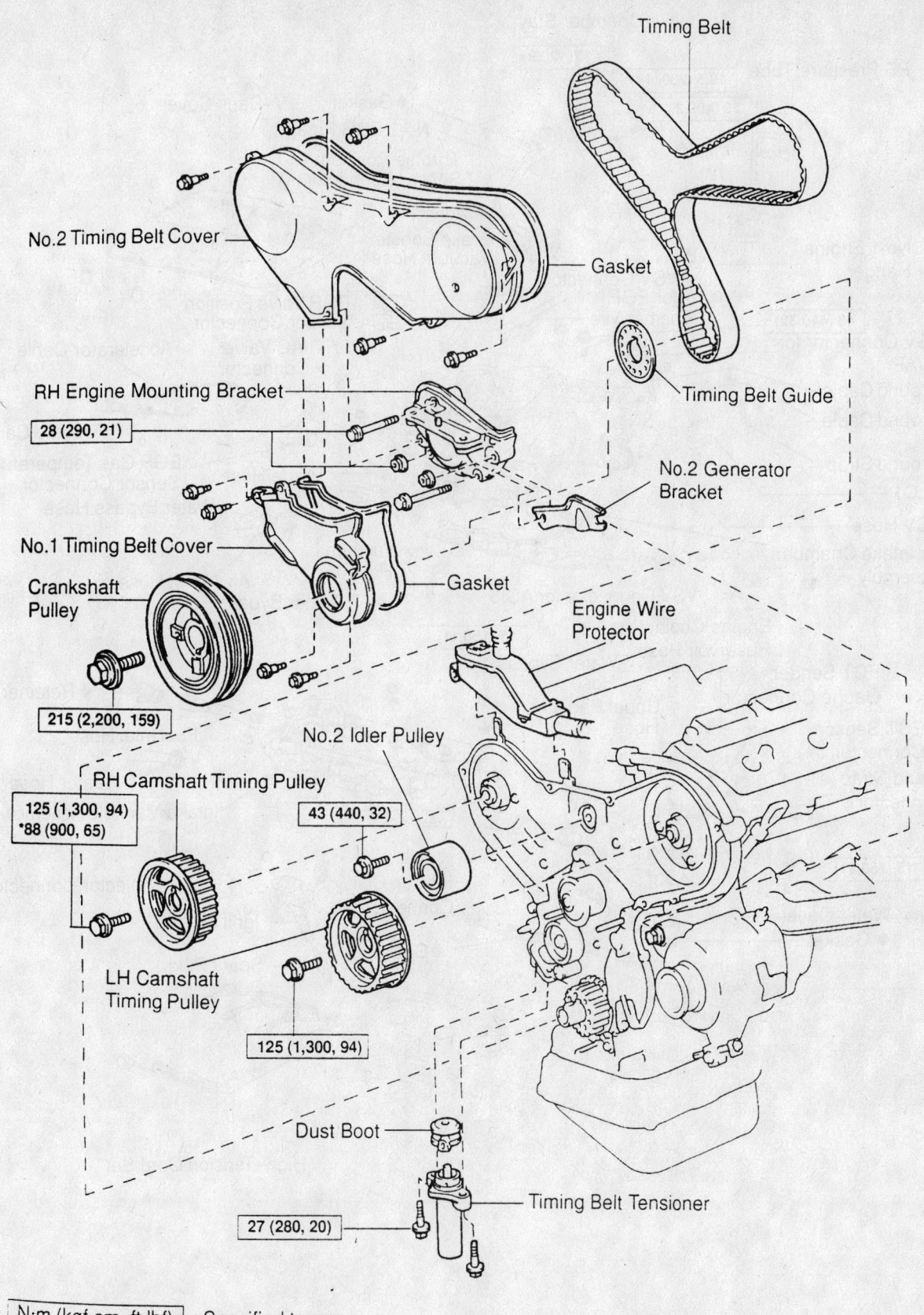

Timing Belt

No.2 Timing Belt Cover

Gasket

Timing Belt Guide

RH Engine Mounting Bracket

28 (290, 21)

No.2 Generator Bracket

No.1 Timing Belt Cover

Gasket

Crankshaft Pulley

Engine Wire Protector

215 (2,200, 159)

No.2 Idler Pulley

RH Camshaft Timing Pulley

125 (1,300, 94)
*88 (900, 65)

43 (440, 32)

LH Camshaft Timing Pulley

125 (1,300, 94)

Dust Boot

Timing Belt Tensioner

27 (280, 20)

N·m (kgf·cm, ft·lbf) : Specified torque
◆ Non–reusable part
* For use with SST

9301WG23

Timing belt and pulleys—3.0L (1MZ-FE) engine

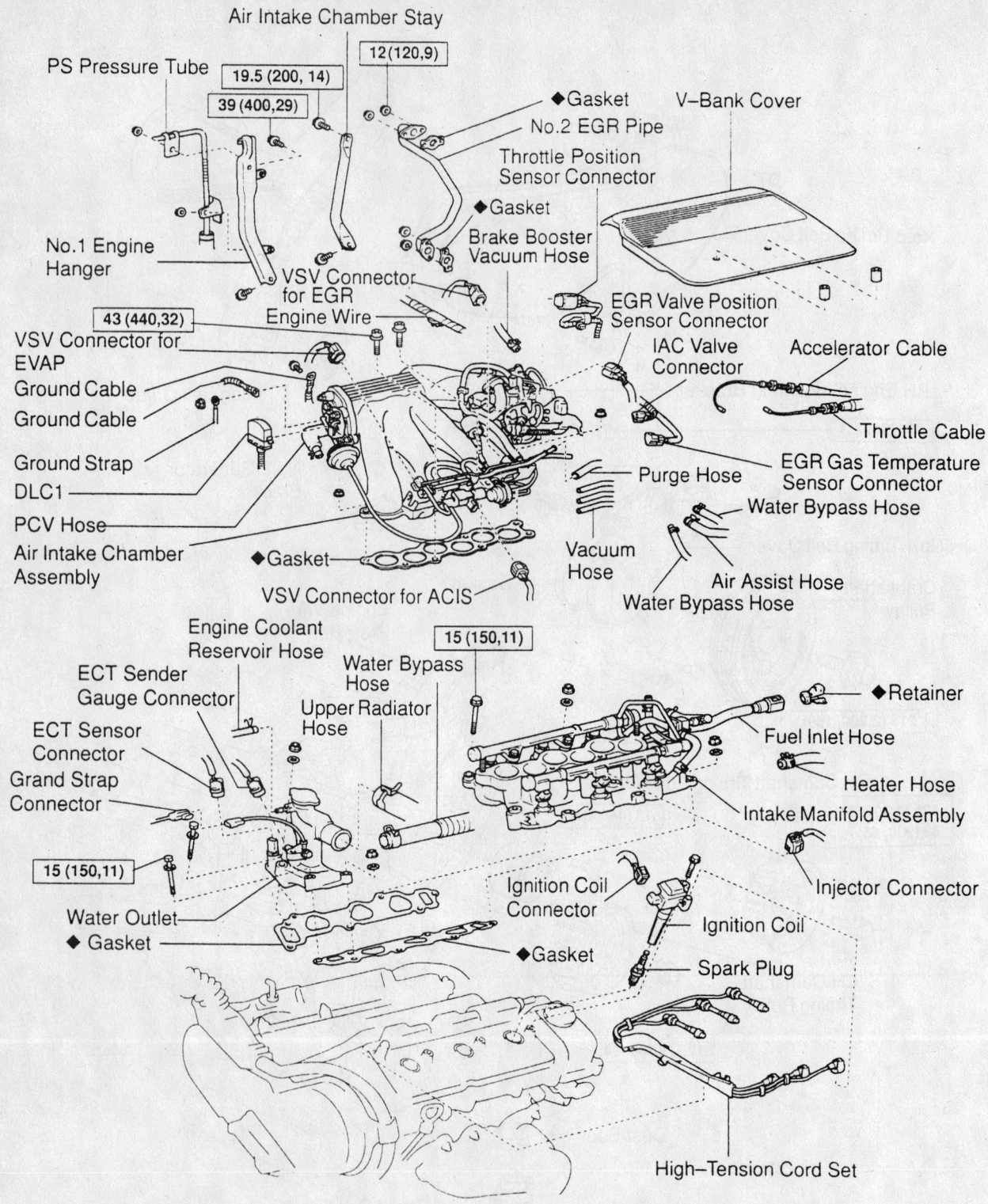

Air Intake Chamber Stay

PS Pressure Tube

12(120,9)

19.5 (200, 14)

39 (400,29)

◆Gasket

No.2 EGR Pipe

V–Bank Cover

Throttle Position
Sensor Connector

◆Gasket

Brake Booster
Vacuum Hose

No.1 Engine
Hanger

VSV Connector
for EGR

Engine Wire

EGR Valve Position
Sensor Connector

43 (440,32)

VSV Connector for
EVAP

IAC Valve
Connector

Accelerator Cable

Ground Cable

Throttle Cable

Ground Cable

EGR Gas Temperature
Sensor Connector

Ground Strap

Purge Hose

Water Bypass Hose

DLC1

PCV Hose

Vacuum
Hose

Water Bypass Hose

Air Assist Hose

Air Intake Chamber
Assembly

◆Gasket

VSV Connector for ACIS

Engine Coolant
Reservoir Hose

Water Bypass
Hose

15 (150,11)

ECT Sender
Gauge Connector

Upper Radiator
Hose

ECT Sensor
Connector

Grand Strap
Connector

◆Retainer

Fuel Inlet Hose

Heater Hose

Intake Manifold Assembly

15 (150,11)

Injector Connector

Water Outlet

◆ Gasket

Ignition Coil
Connector

Ignition Coil

◆Gasket

Spark Plug

High–Tension Cord Set

N·m (kgf·cm, ft·lbf) : Specified torque

◆ Non–reusable part

Intake manifold removal—3.0L (1MZ-FE) engine

9301WG22

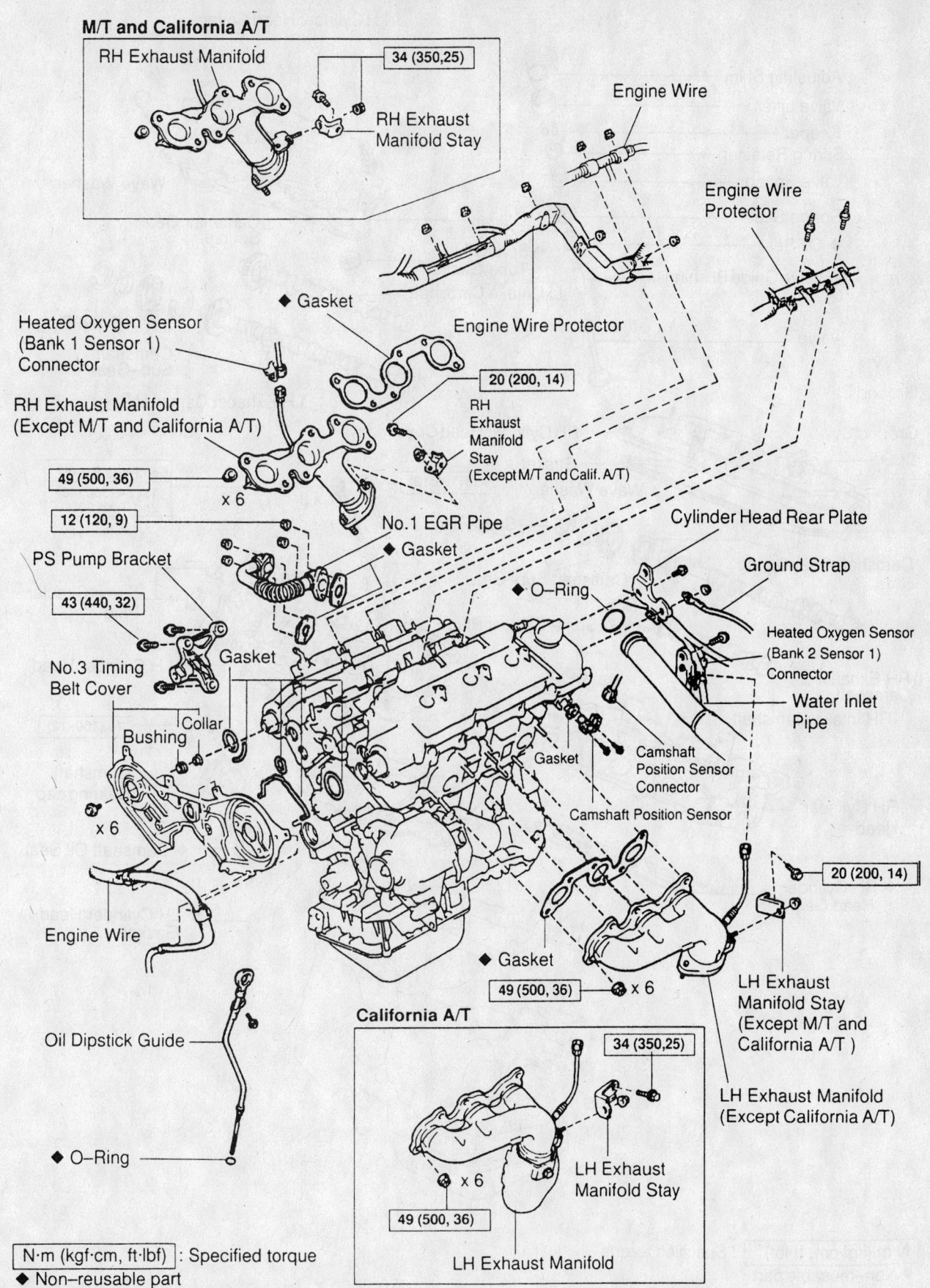

M/T and California A/T

RH Exhaust Manifold

34 (350,25)

RH Exhaust Manifold Stay

Engine Wire

Engine Wire Protector

◆ Gasket

Engine Wire Protector

Heated Oxygen Sensor (Bank 1 Sensor 1) Connector

RH Exhaust Manifold (Except M/T and California A/T)

49 (500, 36) × 6

12 (120, 9)

20 (200, 14)

RH Exhaust Manifold Stay (Except M/T and Calif. A/T)

No.1 EGR Pipe

◆ Gasket

PS Pump Bracket

43 (440, 32)

No.3 Timing Belt Cover

Gasket

Bushing   Collar

◆ O–Ring

Cylinder Head Rear Plate

Ground Strap

Heated Oxygen Sensor (Bank 2 Sensor 1) Connector

Water Inlet Pipe

Gasket

Camshaft Position Sensor Connector

Camshaft Position Sensor

20 (200, 14)

× 6

Engine Wire

Oil Dipstick Guide

◆ O–Ring

◆ Gasket

49 (500, 36) × 6

LH Exhaust Manifold Stay (Except M/T and California A/T)

LH Exhaust Manifold (Except California A/T)

California A/T

34 (350,25)

× 6

49 (500, 36)

LH Exhaust Manifold Stay

LH Exhaust Manifold

N·m (kgf·cm, ft·lbf) : Specified torque

◆ Non–reusable part

9301WG24

Exhaust manifold removal—3.0L (1MZ-FE) engine

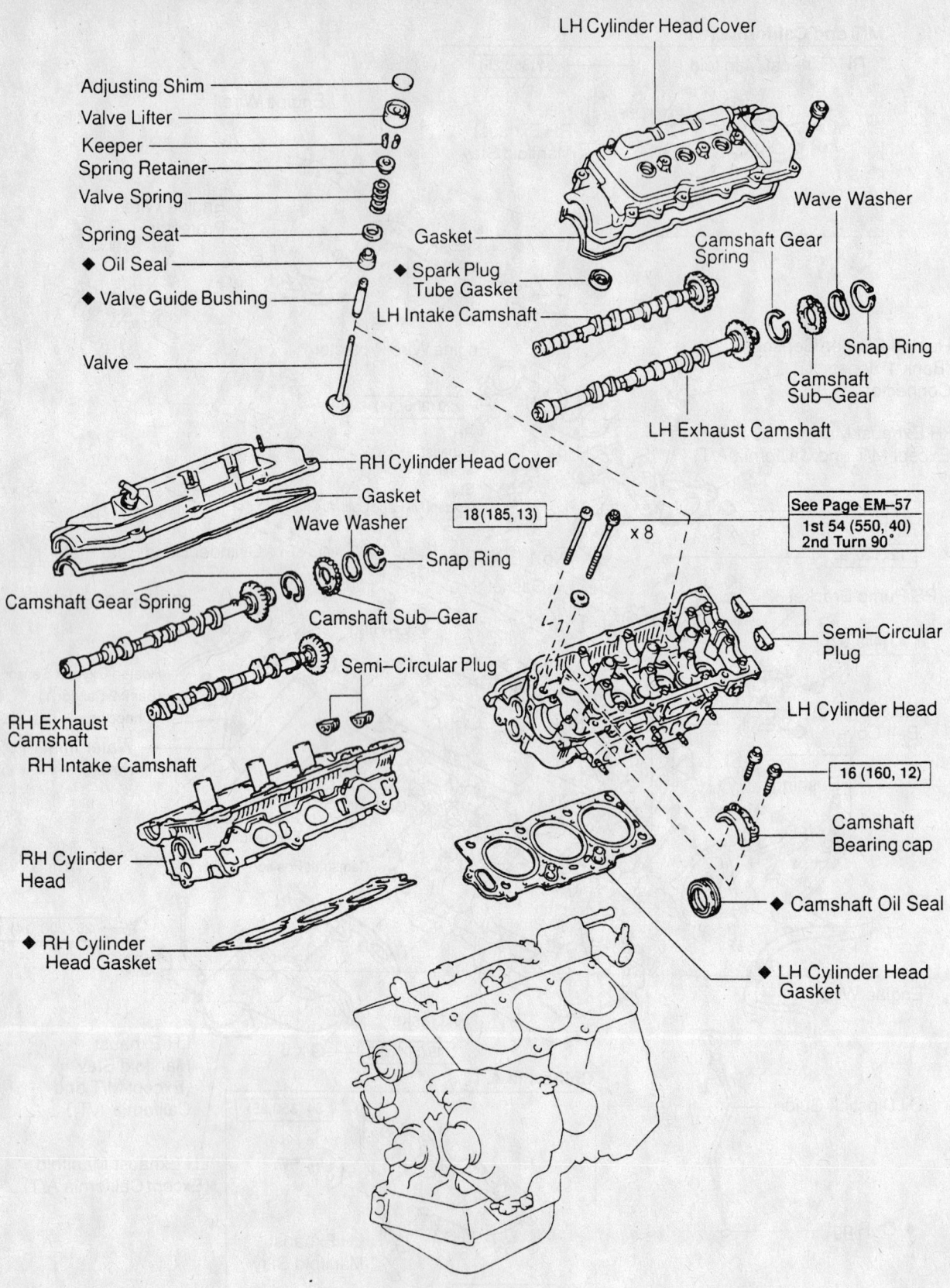

Adjusting Shim

Valve Lifter

Keeper

Spring Retainer

Valve Spring

Spring Seat

◆ Oil Seal

◆ Valve Guide Bushing

Valve

LH Cylinder Head Cover

Gasket

◆ Spark Plug Tube Gasket

LH Intake Camshaft

Camshaft Gear Spring

Wave Washer

Snap Ring

Camshaft Sub–Gear

LH Exhaust Camshaft

RH Cylinder Head Cover

Gasket

Wave Washer

Snap Ring

Camshaft Gear Spring

Camshaft Sub–Gear

RH Exhaust Camshaft

RH Intake Camshaft

Semi–Circular Plug

RH Cylinder Head

◆ RH Cylinder Head Gasket

18(185, 13)

x 8

See Page EM–57
1st 54 (550, 40)
2nd Turn 90°

Semi–Circular Plug

LH Cylinder Head

16 (160, 12)

Camshaft Bearing cap

◆ Camshaft Oil Seal

◆ LH Cylinder Head Gasket

N·m (kgf·cm, ft·lbf) : Specified torque

◆ Non–reusable part

9301WG25

Cylinder head gasket position and torque specification—3.0L (1MZ-FE) engine

- Engine wire bracket to the water pump with the bolt
- Vacuum control valve set. Bolts: 15 ft. lbs. (21 Nm)
- VSV connector
- Air intake chamber as follows. Bolts and nut: 20 ft. lbs. (27 Nm).
- Bolt holding the engine wire protector to the air intake chamber
- Vacuum sensing hose to the fuel pressure regulator (except for California vehicles)
- Starter
- Oil and transmission dipstick tubes
- Engine wire bracket
- Fuel return hose
- Water bypass outlet and No. 1 water bypass pipe

15. Install the timing belt as follows:

a. Turn the crankshaft pulley and align its groove with the timing mark, **0**, on the No. 1 timing belt cover.

b. Align the timing marks on the camshaft timing gears and the No. 4 timing belt cover.

c. Install the timing belt.

d. Double check that all the timing marks for the crankshaft pulley and the camshaft gears are aligned as they were during disassembly.

e. Set the timing belt tensioner:

f. Use a press to slowly push in the pushrod on the tensioner. This will require between 220–2200 pounds of pressure.

g. Align the holes of the pushrod and housing. Place a 1.5mm hex wrench through the holes to keep the pushrod retracted.

h. Release the press and install the dust boot onto the tensioner.

i. Install the tensioner; alternately tighten the bolts: 20 ft. lbs. (26 Nm).

j. Remove the hex wrench from the tensioner with a pair of pliers.

k. Turn the crankshaft pulley 2 full turns clockwise. Check that each pulley's timing marks align correctly after the 2 turns. If any mark does not align, remove the timing belt and reinstall it.

16. Install or connect the following:
- Drive belt tensioner. Bolts: 15 ft. lbs. (21 Nm).
- No. 2 timing belt cover
- Spark plugs
- Distributor and spark plug wires
- No. 3 timing belt cover
- Cylinder head rear cover
- No. 2 vacuum pipe and VSV assembly
- Air intake chamber stays. Bolt and nut: 13 ft. lbs. (18 Nm).

- Throttle body and intake air connector assembly
- EVAP hose
- Brake booster vacuum hose
- Power steering pump
- Exhaust manifolds
- No. 2 front exhaust pipe
- Water pump pulley, fan, fluid coupling assembly, and the drive belt. Pulley bolts: 12 ft. lbs. (16 Nm).
- Air cleaner, VAF meter, and the intake air connector pipe assembly
- Air cleaner duct
- Cruise control cable, throttle control, and accelerator cables
- Negative battery cable
- Engine undercover

17. Fill the cooling system to the proper level.

18. Fill the engine with clean oil.

19. Start the vehicle, check for leaks and repair if necessary.

1MZ-FE Engine

1. Before servicing the vehicle, refer to the precautions in the beginning of this section.

2. Drain the engine oil.

3. Drain the cooling system.

4. Remove or disconnect the following:

- Negative battery cable. On vehicles equipped with an air bag, wait at least 90 seconds before proceeding.
- Accelerator cable and the throttle cable on vehicles equipped with an automatic transaxle.
- Air cleaner cover, air flow meter, air duct, cruise control actuator and bracket (if equipped), the 2 engine ground straps, right engine mounting support, radiator hoses, and 2 heater hoses
- Fuel feed and return lines and plug the lines
- Pressure hose, plug, and remove from V-bank cover
- Vacuum hoses from: Fuel pressure VSV; Fuel pressure regulator; Cylinder head rear plate; Intake air control valve VSV; EGR vacuum modulator; EGR valve
- All necessary connectors and hoses
- Ground straps and the hydraulic motor pressure hose
- 2 nuts and the power steering pressure tube
- Engine hanger and the intake chamber support

- Ignition coils and the spark plugs, timing belt, camshaft pulleys and the timing belt rear cover, cylinder head rear plate, water inlet pipe, and water outlet
- Intake manifold and fuel rail assembly
- EGR pipe
- Exhaust manifolds
- Dipstick assembly and the power steering pump bracket
- Valve covers and the camshaft position sensor
- Camshafts following the proper sequences and procedures
- 2 (1 on each head) 8mm recessed hex bolts. Loosen and remove the 8 head bolts evenly, in 3 passes, in the reverse order of the installation sequence
- Cylinder head gasket

To install:

5. Install or connect the following:
- New cylinder head gasket on cylinder block. Place the cylinder head onto the gasket.
- Cylinder head bolts into the cylinder head. Bolts: 40 ft. lbs. (54 Nm), then an additional 90 degrees
- 2 remaining 8mm bolts. Bolts: 13 ft. lbs. (18 Nm).
- Camshafts following the proper sequences and procedures
- Check and adjust the valves
- Sealant to the cylinder heads where the camshaft supports meet the cylinder heads

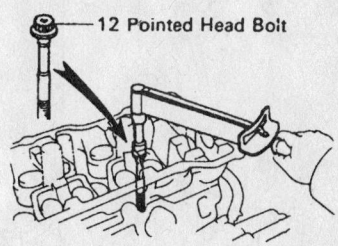

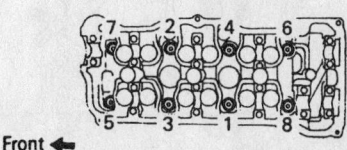

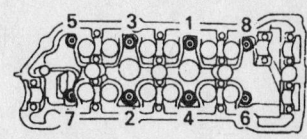

Cylinder head bolt tightening sequence—1MZ-FE engine

7923VG16

- All remaining components in reverse of removal using the diagrams for the proper torque figure.

Turbocharger

REMOVAL & INSTALLATION

3.0L (2JZ-GTE) Engine

1. Before servicing the vehicle, refer to the precautions in the beginning of this section.
2. Drain the cooling system.
3. Remove or disconnect the following:
   - Negative battery cable
   - Cruise control actuator cable
   - No. 1 air hose
   - Air cleaner and MAF meter assembly as follows:
   - Theft deterrent horn
   - Front lower arm bracket stay
   - Front upper crossmember extension
   - No. 2 front exhaust pipe as follows:
   - Heat insulator for the No. 2 front exhaust pipe
   - With automatic transmission, automatic transmission oil cooler tubes
   - Engine wire protector
   - Heater hose from the No. 3 water bypass pipe
   - EVAP hose from the No. 1 vacuum pipe
   - IAC valve pipe from the No. 2 air tube
   - No. 1 vacuum pipe from the air tubes
   - VSV assembly
   - Crankshaft position sensor connector from the clamp
   - Water bypass hose (from the water pump) from the No.1 turbo water pipe
   - Water bypass hose (from the water outlet) from the No. 1 turbo water pipe
   - Water bypass hose (from the water outlet) from the No. 2 turbo water pipe
   - No. 2 turbo water pipe from the No. 4 air tube
   - No. 1 air tube from the No. 1 turbocharger by removing the 2 bolts
   - 2 bolts holding the No. 4 air tube to the No. 1 turbocharger
   - Air hose from the No. 4 air tube
   - Air hose from the intake air connector
   - No. 4 air tube and air bypass valve assembly
   - Intake air control valve and gasket by removing the 2 nuts
   - Air hose from the No. 2 air tube
   - PCV hose from the No. 2 cylinder head cover

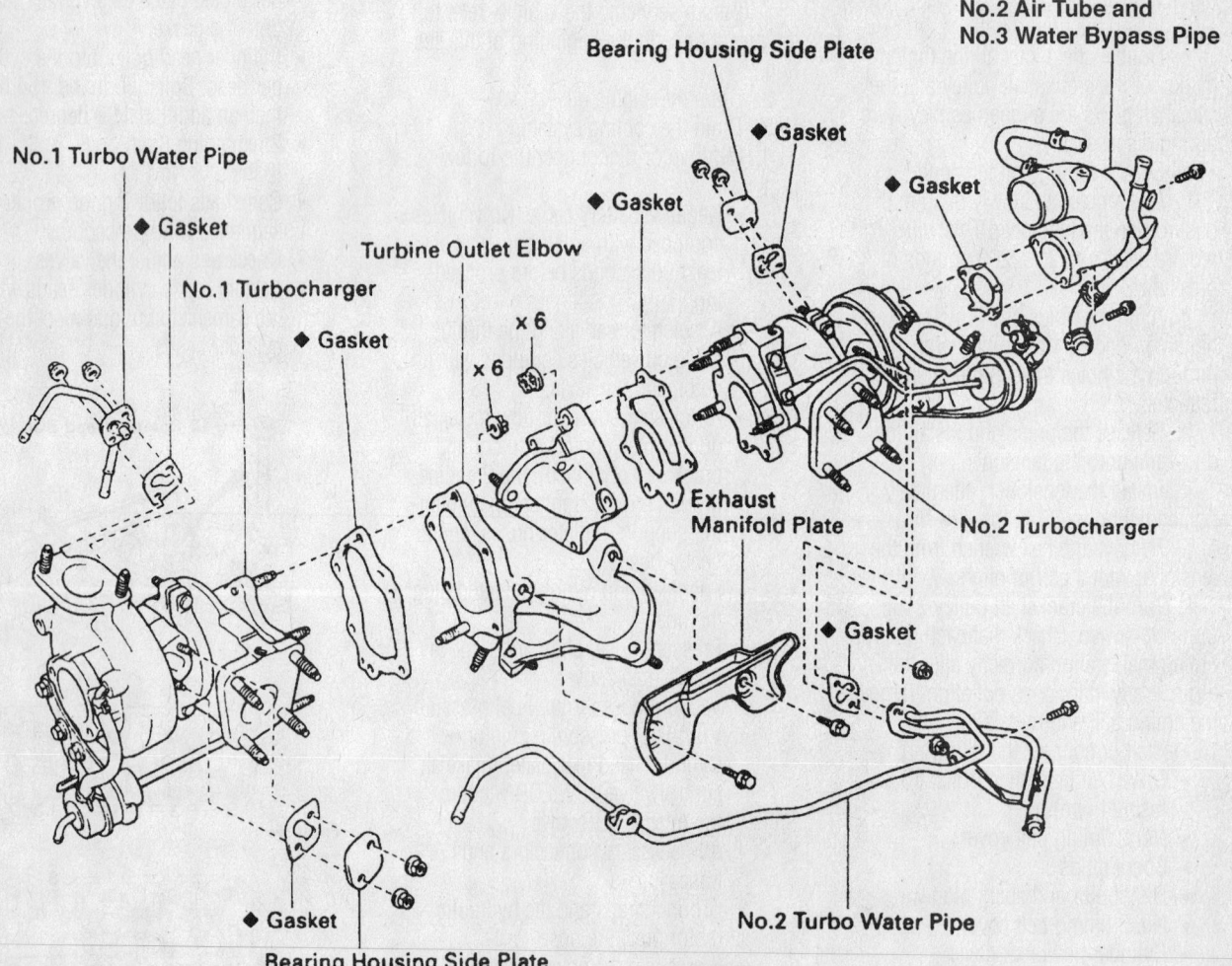

◆ Non-reusable part

Exploded view of the turbocharger component assembly—Supra with 3.0L (2JZ-GTE) engine

7923VG17

- Intake air connector and No. 1 air tube assembly
- Air inlet duct and cable bracket
- Heat insulator for the turbocharger
- Exhaust bypass pipe and gasket
- Exhaust gas control valve stay
- Main heated oxygen sensor
- Exhaust gas control valve
- No. 1 turbocharger stay
- No. 2 turbocharger stay
- Union bolt holding the No. 1 turbo oil pipe to the cylinder block. Remove the 2 gaskets
- 2 nuts and disconnect the turbo oil pipe from the turbocharger. Remove the gaskets
- Turbo oil hose from the turbo oil outlet on the No. 1 oil pan. Remove the No. 1 turbo oil pipe
- Union bolt holding the No. 2 turbo oil pipe to the cylinder block. Remove the 2 gaskets
- Turbo oil pipe from the turbocharger. Remove the 2 gaskets
- Turbo oil hose from the turbo oil outlet on the No. 1 oil pan and remove the turbo oil pipe
- Turbochargers and turbine outlet elbow assembly
- No. 1 vacuum pipe from the No. 2 turbocharger
- No. 2 air tube and No. 3 water bypass pipe assembly from the No. 2 turbocharger
- Exhaust manifold plate from the turbine outlet elbow
- No. 2 turbo water pipe from the No. 2 turbocharger
- Bearing housing side plate from the No. 1 turbocharger
- No. 1 turbo water pipe from the No. 1 turbocharger
- Bearing housing side plate from the No. 2 turbocharger
- No. 1 turbocharger from the turbine outlet elbow
- No. 2 turbocharger from the turbine outlet elbow

To install:
4. Install or connect the following:
- No. 2 turbocharger to the turbine outlet elbow by installing a new gasket and 6 new nuts. Nuts: 18 ft. lbs. (25 Nm).
- No. 1 turbocharger to the turbine outlet elbow by installing a new gasket and 6 new nuts. Nuts: 18 ft. lbs. (25 Nm).
- Bearing housing side plate to the No. 2 turbocharger by installing a new gasket and 2 nuts. Nuts: 78 inch lbs. (9 Nm).

- No. 1 turbo water pipe to the No. 1 turbocharger by installing a new gasket and 2 nuts. Nuts: 78 inch lbs. (9 Nm).
- Bearing housing side plate to the No. 1 turbocharger by installing a new gasket and 2 nuts. Nuts: 78 inch lbs. (9 Nm).
- No. 2 turbo water pipe to the No. 2 turbocharger by installing a new gasket and 2 nuts. Nuts: 78 inch lbs. (9 Nm).
- Exhaust manifold plate to the turbine outlet elbow
- No. 2 air tube and No. 3 water bypass pipe assembly to the No. 2 turbocharger. Bolts: 15 ft. lbs. (21 Nm).
- No. 1 vacuum pipe to the No. 2 turbocharger.
- Turbochargers and turbine outlet elbow assembly. Install 8 **new** nuts and uniformly tighten the nuts in several passes. Nuts: 40 ft. lbs. (54 Nm).
- Water bypass hose (from the No. 2 turbo water pipe) to the No. 2 water bypass pipe
- Heater hose (from the No. 3 water bypass pipe) to the No. 2 water bypass pipe
- Install the No. 2 turbo oil pipe. Nuts: 15 ft. lbs. (21 Nm).
- Union bolt to hold the turbo oil pipe to the cylinder block. Bolt: 29 ft. lbs. (39 Nm).
- No. 1 turbo oil pipe. Nuts: 15 ft. lbs. (21 Nm).
- Union bolt to hold the turbo oil pipe to the cylinder block. Bolts: 29 ft. lbs. (39 Nm).
- No. 2 turbocharger stay. Nut and bolt: 32 ft. lbs. (43 Nm).
- No. 1 turbocharger stay. Nut and bolt: 32 ft. lbs. (43 Nm).
- Exhaust gas control valve by installing 2 new gaskets and the 3 nuts. Nuts: 51 ft. lbs. (69 Nm).
- Main heated oxygen sensor. Nuts: 14 ft. lbs. (20 Nm).
- Exhaust gas control valve stay. Bolt and nut: 32 ft. lbs. (43 Nm).
- Exhaust bypass pipe by installing 2 new gaskets and 4 new nuts. Nuts: 18 ft. lbs. (25 Nm).
- Heat insulator for the turbocharger
- Air inlet duct by installing the cable bracket, bolt, and 2 nuts
- Intake air connector and No. 1 air tube assembly
- PCV hose to the No. 2 cylinder head cover

- Air hose to the No. 2 air tube
- Intake air control valve and gasket. Nuts: 15 ft. lbs. (21 Nm)
- No. 4 air tube and air bypass valve assembly
- Air hose to the intake air connector
- Air hose to the No. 4 air tube
- 2 bolts holding the No. 4 air tube to the No. 1 turbocharger. Bolts: 15 ft. lbs. (21 Nm).
- No. 1 air tube to the No. 1 turbocharger. Bolts: 15 ft. lbs. (21 Nm).
- No. 2 turbo water pipe to the No. 4 air tube by installing the bolt
- Water bypass hose (from the water outlet) to the No. 2 turbo water pipe
- Water hose (from the water outlet) to the No. 1 turbo water pipe
- Water bypass hose (from the water pump) to the No. 1 turbo water pipe
- Crankshaft position sensor connector to the clamp
- VSV assembly and connect the 2 VSV connectors
- Engine wire to the wire clamp
- Air hose to the hose clamp
- Air hose to the actuator for the exhaust gas control valve
- Air hose to the actuator for the waste gate valve
- Install the No. 1 vacuum pipe to the air tubes and install the 3 bolts
- 2 air hoses (from the vacuum pressure tank) to the vacuum pipe
- Air hose to the VSV for the intake air control valve
- Air hose (from the No. 2 air tube) from the vacuum pipe
- 2 air hoses (from the VSV for exhaust bypass valve) to the vacuum pipe
- Vacuum hose (from the air bypass valve) to the No. 1 air tube
- Air hose (from the VSV for exhaust gas control valve) to the vacuum pipe
- Air hose (from the VSV for waste gate valve) to the vacuum pipe
- Air hose to the No. 1 air tube
- Air hose to the No. 4 air tube
- Engine wire to the 3 clamps
- VSV connector for the exhaust bypass valve
- VSV connector for the intake air control valve
- IAC valve pipe to the clamp
- Air hose to the No. 2 air tube
- Air hose (from the No. 1 vacuum pipe) to the IAC valve pipe
- Engine wire to the clamp

- EVAP hose to the No. 1 vacuum pipe
- Heater hose to the No. 3 water bypass pipe
- Engine wire protector
- Automatic transmission oil cooler tubes
- Heat insulator for the No. 2 front exhaust pipe
- No. 2 front exhaust pipe. Nuts: 46 ft. lbs. (62 Nm).
- Pipe support bracket. Bolts: 32 ft. lbs. (43 Nm).
- 2 bolts and nuts to hold the front exhaust pipe to the No. 2 front exhaust pipe. Bolts and nuts: 43 ft. lbs. (58 Nm).
- Upper front crossmember extension. Bolts: 22 ft. lbs. (29 Nm). Nuts: 25 ft. lbs. (33 Nm).
- Front lower arm bracket stay. Bolts: 33 ft. lbs. (44 Nm). Nut: 43 ft. lbs. (59 Nm).
- Theft deterrent horn
- Air cleaner and MAF meter assembly as follows:
- Air cleaner duct
- No. 1 air hose
- Cruise control actuator cable
- Undercover
- Negative battery cable

5. Fill the cooling system to the proper level.

6. Start the vehicle, check for leaks and repair if necessary.

Intake Manifold

REMOVAL & INSTALLATION

5E-FE Engine

1. Before servicing the vehicle, refer to the precautions in the beginning of this section.

2. Drain the cooling system.

3. Remove or disconnect the following:
- Negative battery cable. On vehicles equipped with an air bag, wait at least 90 seconds before proceeding.
- Accelerator cable
- With automatic transmission, throttle cable
- PCV hose
- Air cleaner
- All electrical wires and vacuum hoses that interfere with removal of the intake manifold
- EGR pipe and EGR valve
- Throttle body assembly

- Air intake chamber stay
- With distributor ignition, air pipe
- Engine wire clamps
- Intake manifold stay
- Intake manifold
- Intake manifold gasket

To install:

4. Install or connect the following:
- Using a new gasket, intake manifold. Nuts and bolts: 14 ft. lbs. (19 Nm).
- Intake manifold stay. Bolt(s) and nut(s): 15 ft. lbs. (20 Nm).
- With distributor ignition, vacuum hoses, then the air pipe. Bolts: 48 inch lbs. (6 Nm).
- Air intake chamber stay
- Throttle body, using a new gasket. Nuts and bolts: 108 inch lbs. (13 Nm).
- All components to the throttle body
- EGR pipe and EGR valve
- All electrical wires and vacuum hoses removed from the intake manifold
- Air cleaner hose
- Negative battery cable

5. Fill the cooling system.

6. Start the vehicle, check for leaks and repair if necessary.

1NZ-FE Engine

1. Before servicing the vehicle, refer to the precautions in the beginning of this section.

2. Drain the cooling system.

3. Remove or disconnect the following:
- Negative battery cable

- Water filler
- Outer front cowl top panel
- Alternator
- Air cleaner
- Accelerator cable
- Ignition coil
- PCV hoses
- Throttle body
- Intake manifold and discard the gasket

To install:

4. Install or connect the following:
- Intake manifold with a new gasket. Uniformly tighten the bolts and nuts, in several passes, from the ends, working towards the center, to 22 ft. lbs. (30 Nm).
- Engine wiring harness
- Throttle body
- PCV hoses
- Ignition coils
- Accelerator cable
- Air cleaner
- Alternator
- Water filler
- Negative battery cable

5. Fill the cooling system.

6. Start the vehicle, check for leaks and repair if necessary.

1ZZ-FE Engine

1. Before servicing the vehicle, refer to the precautions in the beginning of this section.

2. Drain the cooling system.

3. Remove or disconnect the following:
- Negative battery cable
- Drive belt and alternator

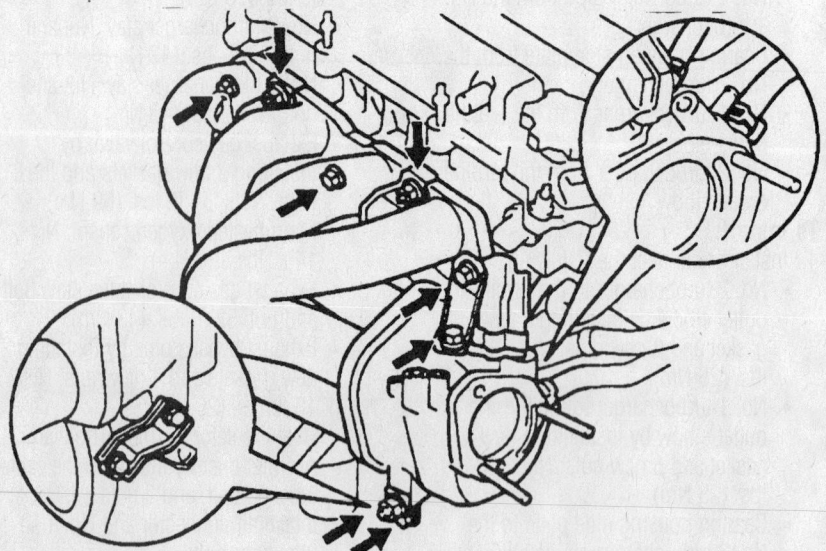

Intake manifold mounting fastener locations—1.8L (1ZZ-FE) engine

7923VG19

- Air intake duct
- Accelerator cable
- Exhaust pipe from the manifold.
- Exhaust manifold support bracket
- Spark plug wires, then ignition coils
- Spark plugs
- PCV hoses
- Throttle body assembly
- 2 bolts securing the wiring harness protector
- Wiring connectors and ground wires
- Intake manifold support bracket
- Intake manifold and gasket

To install:

4. Install or connect the following:
- Intake manifold with a new gasket. Torque the bolts to 14 ft. lbs. (18.5 Nm).
- Harness wiring to the cylinder head and harness protector
- Fuel injectors, throttle body and the PCV hoses
- Spark plugs and ignition coils. Bolts and nuts: 80 inch lbs. (9 Nm).
- Exhaust manifold and support bracket. Bolts: 37 ft. lbs. (49 Nm).
- Front exhaust pipe to the manifold. Bolts: 46 ft. lbs. (62 Nm).
- Oxygen sensor. Nuts: 14 ft. lbs. (20 Nm).
- Accelerator cable and air intake duct
- Alternator and drive belt
- Negative battery cable
5. Fill the cooling system.
6. Start the vehicle, check for leaks and repair if necessary.

2ZZ-GE Engine

1. Before servicing the vehicle, refer to the precautions in the beginning of this section.
2. Drain the cooling system.
3. Remove or disconnect the following:
- Negative battery cable
- Drive belt and alternator
- Air intake duct
- Accelerator cable
- Spark plug wires, then ignition coils
- Spark plugs
- PCV hoses
- Throttle body assembly
- Wiring harness
- Hoses and tubes connected to the head
- Intake manifold support bracket
- Intake manifold and gasket

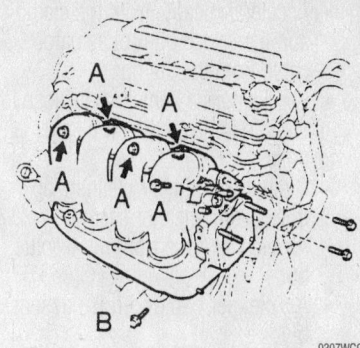

9307WG93

Intake manifold bolt installation—2ZZ-GE engine

To install:

4. Install or connect the following:
- Intake manifold with a new gasket. Bolts A: 20 ft. lbs. (27 Nm); bolt B: 34 ft. lbs. (46 Nm)
- Harness wiring to the cylinder head and harness protector
- Fuel injectors, throttle body and the PCV hoses
- Spark plugs and ignition coils. Bolts and nuts: 80 inch lbs. (9 Nm).
- Oxygen sensor. Nuts: 14 ft. lbs. (20 Nm).
- Accelerator cable and air intake duct
- Alternator and drive belt
- Negative battery cable
5. Fill the cooling system.
6. Start the vehicle, check for leaks and repair if necessary.

5S-FE Engine

CAMRY AND CAMRY SOLARA

1. Before servicing the vehicle, refer to the precautions in the beginning of this section.
2. Drain the cooling system.
3. Remove or disconnect the following:

- Negative battery cable. On vehicles equipped with an air bag, wait at least 90 seconds before proceeding
- Accelerator cable. With automatic transmission, throttle cable from the throttle body
- Intake air temperature sensor connector
- On California models, air cleaner hose
- Air cleaner hose clamp bolt, air cleaner cap clips, air hose from the throttle body, and the air cleaner cap together with the resonator and air cleaner hose
- Electrical connections and hoses from the throttle body
- Throttle body. Type A throttle bodies are secured with 4 bolts and Type B throttle bodies are secured with 2 bolts and 2 nuts
- Vacuum hose bracket and the engine wiring harness
- EGR valve
- No. 1 air intake chamber, and manifold stays
- Intake manifold and gasket

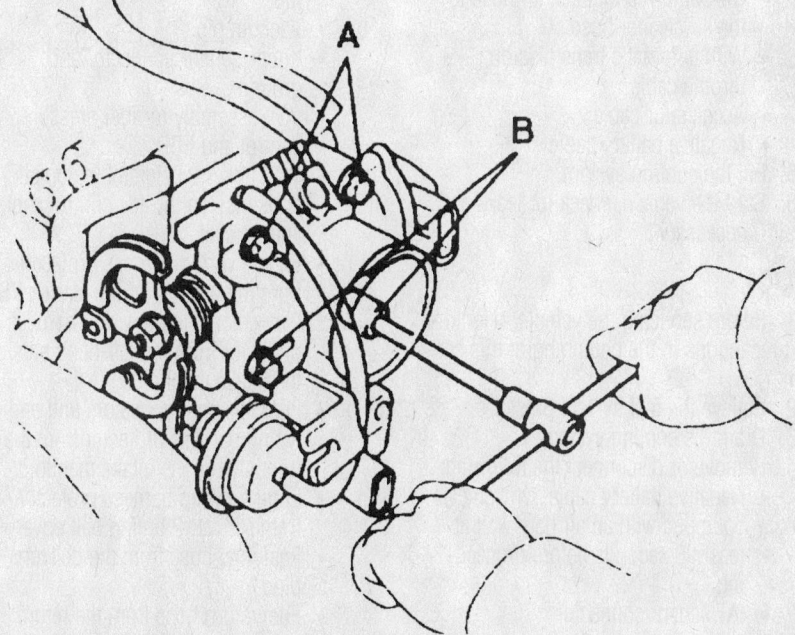

7923VG21

Mounting bolt length identification for the throttle body—2.2L (5S-FE) engine

To install:

4. Install or connect the following:
- Intake manifold with a new gasket. Torque the bolts to 14 ft. lbs. (19 Nm).
- Wire clamps to the wire brackets on the intake manifold
- Vacuum hose bracket and engine wiring harness
- No. 1 air intake chamber and manifold stays. 14mm bolts: 31 ft. lbs. (42 Nm). 12mm bolts: 16 ft. lbs. (22 Nm).
- EGR valve
- Throttle body with a new gasket
- Hoses and electrical connections to the throttle body

➡ **The protrusion on the gasket should be facing down and the water hose connections on the throttle body should also face down.**

- On type A throttle body, Bolts: 14 ft. lbs. (19 Nm). Bolt A is 45mm in length and bolt B is 55mm.
- On type B throttle body, bolts and nuts: 14 ft. lbs. (19 Nm).
- PCV hose
- 2 vacuum hoses to the EGR modulator
- Vacuum hose to the TVV for EVAP
- IAC valve connector
- Throttle position sensor connector
- Air cleaner hose to the throttle body
- Air cleaner cap with the resonator and air cleaner hose
- Intake air temperature sensor connector
- On California models, air hose to the air cleaner hose
- With automatic transmission, throttle cable
- Accelerator cable
- Negative battery cable
5. Fill the cooling system.
6. Start the vehicle, check for leaks and repair if necessary

CELICA

1. Before servicing the vehicle, refer to the precautions in the beginning of this section.
2. Relieve the fuel system pressure.
3. Drain the cooling system.
4. Remove or disconnect the following:
- Negative battery cable. On vehicles equipped with an air bag, wait at least 90 seconds before proceeding.
- IAT sensor connector
- High tension spark plug wire from air cleaner hose
- Accelerator cable from the clamp and the cruise control actuator cable from the clamps
- For California vehicles, air hose for the idle up from the air cleaner hose
- Clamps and the air cleaner cap from the air cleaner case
- Air cleaner hose from the throttle body
- Air cleaner cap and hose assembly
- Throttle body
- Vacuum sensor hose from the gas filter
- Brake booster vacuum line
- With air conditioning, air conditioning idle-up valve
- EGR temperature sensor connector and sensor connector from the wiring connector
- EVAP hose from the charcoal canister and hose clamp from the bracket on the air tube
- Vacuum hoses from the VSV for the EGR
- Vacuum modulator from the clamp on the intake manifold
- EGR valve, vacuum modulator, vacuum hoses assembly and gasket
- Intake manifold stay
- Automatic transmission throttle control cable
- 2 power steering air hose(s) from the air tube and the intake manifold
- Air hose from the fuel pressure regulator
- Hose bracket (for EGR) and the air tube
- Vacuum pipe
- Knock sensor connector and ground cables
- VSV assembly for fuel pressure control and EGR
- PCV hose, accelerator, automatic transmission throttle control cables and bracket
- Engine wiring harness protector from the bracket on the starter, VSS connector, engine wiring harness protector from the left-hand side of the intake manifold
- Fuel injector connectors and engine wiring harness protector from the front side of the intake manifold.
- Engine wiring harness protector from the No. 2 timing belt cover
- Fuel inlet hose from the delivery pipe
- Fuel return hose from the return pipe
- Engine wiring harness between the intake manifold and the cylinder head
- Intake manifold and gasket

To install:

5. Insert the intake manifold between the cylinder head and the firewall. Insert the engine wiring harness between the intake manifold and the cylinder head.
6. Install or connect the following:
- Intake manifold with a new gasket. Torque the bolts to 14 ft. lbs. (19 Nm).
- Fuel inlet pipe to the fuel delivery hose. Bolt: 25 ft. lbs. (34 Nm).
- Fuel return hose to the return hose
- Engine wiring harness protector to the No. 2 timing belt cover in reverse of removal sequence. Install the harness protector to the 2 brackets on the front side of the intake manifold
- Fuel injector connectors. The Nos. 1 and 3 injector connectors are brown; the No. 2 and 4 connectors are gray
- Engine wiring harness protector to the left-hand side of the intake manifold with the bolt and connect the VSS sensor connector
- Engine wiring harness protector to the starter bracket
- Cable bracket to the intake manifold, and the accelerator and automatic transmission throttle control cables
- PCV hose to the intake manifold
- VSV for fuel pressure control and EGR
- Knock sensor connector and ground cable to the intake manifold
- Air tube and hose bracket (for EGR)
- Power steering air hose(s) to the air tube and the intake manifold
- Vacuum sensing hose to the fuel pressure regulator
- Automatic transmission throttle control cable to the clamp on the rear side of the intake manifold.
- Intake manifold stay bolt. Bolt: 15 ft. lbs. (21 Nm). Nut: 32 ft. lbs. (44 Nm).
- EGR valve. Bolt: 43 ft. lbs. (59 Nm). Nuts: 108 inch lbs. (13 Nm).
- Vacuum modulator
- Vacuum hoses to the VSV for the EGR and hose clamp to the bracket on the air tube
- With A/C, the idle-up valve
- EGR gas temperature sensor connector and the sensor connector to the bracket on the intake manifold.

7923VG20

Intake manifold mounting fastener locations—2.2L (5S-FE) engine

- EVAP hose to the charcoal canister
- Vacuum sensor hose to the gas filter and the brake booster vacuum hose
- Throttle body assembly
- Air filter, air cleaner cap, hose assembly, air cleaner hose to the throttle body, and secure the air cleaner cap with the 4 clamps
- Accelerator and cruise control cables to the clamps
- High tension spark plug wire to the air cleaner hose
- IAT sensor connector
- Negative battery cable

7. Fill the cooling system.

8. Start the vehicle, check for leaks and repair if necessary.

1MZ-FE Engine

1. Before servicing the vehicle, refer to the precautions in the beginning of this section.

2. Relieve the fuel pressure from the fuel lines.

3. Remove or disconnect the following:
- Negative battery terminal. If equipped with an air bag system, wait at least 90 seconds or longer before performing any other work
- Air cleaner hose
- V-bank cover
- Air cleaner chamber assembly
- Accelerator cable
- automatic transmission throttle cable

- TPS connector
- IAC valve connector
- EGR gas temperature sensor connector
- A/C idle up valve connector
- VSV connector for the Acoustic Control Induction System (ACIS)
- VSV connector for the fuel pressure control
- Disconnect the VSV for the EVAP
- VSV connector for the EGR
- DLC1 from the bracket on the intake air control valve
- Power steering pressure tube from the No. 1 engine hanger by removing the 2 bolts
- Brake booster vacuum hose from the intake air control valve for the ACIS
- PCV hose from the PCV valve on the right-hand cylinder head
- Ground strap and cable from the air intake air control valve from the ACIS
- Ground cable from the air intake chamber
- Vacuum hose clamp from fuel pipe
- 2 bypass hoses from the throttle body
- 2 power steering air hoses to the air intake chamber
- Air assist hose from the throttle body
- Remove the EVAP hose from the pipe on emission control valve set
- 2 vacuum hoses from the pipes on the cylinder head rear plate

- Vacuum sensing hose from the fuel pressure regulator
- Engine wire clamp from emission control valve set
- 2 bolts and the No. 1 engine hanger
- 2 bolts and the air intake chamber stay
- No. 2 EGR pipe and 2 gaskets by removing the 4 nuts.
- Hose from the VSV from the EVAP
- Air intake chamber assembly and gasket
- Fuel injector connectors
- Air assist hoses and pipe
- Fuel return hose from the No. 1 fuel pipe
- Fuel inlet hose from the fuel filter
- Delivery pipes and injectors from the engine
- Heater hoses
- Intake manifold and gasket

To install:

4. Thoroughly clean the intake manifold and cylinder head surfaces.

5. Install or connect the following:
- Intake manifold with a new gasket. Torque the bolts to 11 ft. lbs. (15 Nm).
- Heater hoses to the intake manifold
- 2 new grommets to each injector
- 2 new O-rings, lightly oiled, on each injector
- Fuel injectors

➡ **Be sure to position the injector electrical connector outward**

- 4 spacers in position on the intake manifold
- Right-hand delivery pipe and the No. 1 fuel pipe together with the 3 injectors in position on the intake manifold
- 2 bolts, temporarily, holding the right-hand delivery pipe to the intake manifold
- Bolt, temporarily, holding the No. 1 fuel pipe to the intake manifold
- Left-hand delivery pipe and the No. 2 fuel pipe together with the 3 injectors in position
- Fuel return hose to the fuel pressure regulator
- 2 bolts, temporarily, holding the left-hand delivery pipe to the intake manifold
- No. 2 fuel pipe, temporarily, to the left and right-hand delivery pipes with the union bolts and 2 new gaskets

➡ **Check that the injectors rotate smoothly. If the injectors do not rotate smoothly, the probable cause is incorrect installation of the O-rings. Replace the O-rings.**

6. Tighten the 4 bolts holding the delivery pipes to the intake manifold. Bolts: 84 inch lbs. (10 Nm).

7. Tighten the bolt holding the No. 1 fuel pipe to the intake manifold. Bolts: 14 ft. lbs. (20 Nm).

8. Tighten the 2 union bolts holding the No. 2 fuel pipe to the delivery pipes. Bolts: 24 ft. lbs. (33 Nm).

9. Install or connect the following:
- Fuel inlet hose to the fuel filter. Use 2 new gaskets when installing the union bolt
- Fuel return hose to the No. 1 fuel pipe. When routing the fuel return hose, pass the hose under the heater hoses
- Air assist hoses to the intake manifold, then the air assist pipe to the bracket on the No. 1 fuel pipe
- Injector connectors
- Air intake chamber assembly. Bolts and nuts: 32 ft. lbs. (43 Nm).
- Hose to the VSV for the EVAP system
- 2 new gaskets and No. 2 EGR pipe with the 4 nuts. Nuts: 108 inch lbs. (12 Nm).
- No. 1 engine hanger with the 2 bolts. Bolts: 19 ft. lbs. (39 Nm).
- Air intake chamber stay with the 2 bolts. Bolts: 14 ft. lbs. (20 Nm).
- Brake booster vacuum hose to the intake air control valve for the ACIS
- PCV hose to the PCV valve on the right-hand cylinder head
- Ground strap and cable to the intake air control valve for the ACIS
- Connect the ground cable and strap with the nut. Nut: 10 ft. lbs. (15 Nm).
- Ground cable to air intake chamber
- Vacuum hose clamp to fuel pipe
- 2 water bypass hoses to the throttle body
- Air assist hose to the throttle body
- 2 power steering air hoses to the air intake chamber
- Connect the EVAP hose to the pipe on the emission control valve set
- 2 vacuum hoses to the pipes on the cylinder head rear plate
- Vacuum sensing hose to the fuel pressure regulator
- Engine wire clamp to the emission control valve set
- Power steering pressure tube with the 2 nuts
- TPS sensor connector
- IAC valve connector
- EGR gas temperature sensor connector
- A/C idle up valve connector
- VSV connector for the ACIS
- VSV connector for the fuel pressure control
- For California vehicles, install the VSV connector for the EVAP
- VSV connector for the EGR
- DLC1 to the bracket on the intake air control valve
- Accelerator cable
- automatic transmission throttle cable
- V-bank cover
- Air cleaner hose
- Negative battery cable

10. Fill the cooling system.

11. Start the vehicle, check for leaks and repair if necessary.

2JZ-GE Engine

1. Before servicing the vehicle, refer to the precautions in the beginning of this section.

2. Drain the cooling system.

3. Remove or disconnect the following:
- Negative battery cable. Wait at least 90 seconds before proceeding with any other work.
- VSV connector
- Vacuum sensing hose from the fuel pressure control and the air intake chamber
- With automatic transmissions, transmission dipstick and guide
- EGR pipe
- EGR gas temperature sensor connector from the No. 2 vacuum pipe and wiring connector
- No. 2 vacuum pipe from the air intake chamber and from the intake manifold
- 2 nuts holding the throttle body bracket to the cylinder head
- Intake air connector from the air intake chamber by removing the 4 bolts and 2 nuts
- Engine wire protector from the air intake chamber
- 3 vacuum hoses from the No. 1 vacuum pipe
- Vacuum hose from the air intake chamber
- Power steering air hose from the air intake chamber
- Brake booster vacuum hose from the air intake chamber by removing the union bolt and 2 gaskets
- Vacuum hose (from the actuator for ACIS) from the No. 1 vacuum pipe
- Fuel pressure regulator vacuum sensing hose from the air intake chamber, (except California)
- 2 nuts holding the air intake chamber stays to the cylinder head
- 2 bolts and disconnect the 2 air intake chamber stays from the air intake chamber
- Air intake chamber by removing the nut and 5 bolts

❉❉ CAUTION

Relieve the fuel pressure in the fuel line before disconnecting any fuel lines. Failure to do so may result in fire and/or explosion.

- Fuel return hose
- Oil dipstick and guide
- Engine wire bracket from the water pump
- Ground straps from the intake manifold
- Wire clamps from the intake manifold
- 6 injector connectors
- ECT sensor and the ECT sender gauge connectors
- Engine wire protector from the intake manifold
- Intake manifold stay
- Fuel inlet pipe clamp bolt from the intake manifold
- Union bolt and disconnect the fuel inlet pipe
- Intake manifold and gaskets

To install:

4. Thoroughly clean the contact surfaces of the head and manifold.

5. Install or connect the following:
- New gasket and intake manifold. Torque the bolts to 20 ft. lbs. (27 Nm).
- Inlet pipe with new gaskets. Bolt: 30 ft. lbs. (42 Nm).
- clamp bolt to the intake manifold
- Intake manifold stay. Bolts: 29 ft. lbs. (39 Nm).
- Engine wire protector to the intake manifold with the 3 nuts
- 6 injector connectors
- ECT sender gauge and connectors
- 2 wire clamps to the intake manifold with the bolts
- 2 ground straps to the intake manifold with the bolts

- Engine wire bracket to the water pump with the bolt
- New gasket on the air intake chamber

➡ **The protrusion on the gasket faces rearward**

- Intake air chamber, nut and 5 bolts
- 2 air intake chamber stays to the air intake chamber. Bolts: 13 ft. lbs. (18 Nm).
- 2 nuts holding the air intake chamber to the cylinder head. Nuts: 13 ft. lbs. (18 Nm).
- Except California vehicles, connect the vacuum sensing hose (from the fuel pressure regulator) to the air intake chamber
- Vacuum hose (from the actuator for ACIS) to the No. 1 vacuum pipe
- Brake booster vacuum hose to the air intake chamber. Tighten the union bolt to 22 ft. lbs. (29 Nm).
- Power steering air hose to the air intake chamber
- Vacuum hose (from the No. 2 vacuum pipe) to the air intake chamber
- 3 vacuum hoses (from the No. 2 vacuum pipe) to the No. 1 vacuum pipe
- Engine wire protector to the air intake chamber and install the bolt
- 4 bolts and 2 nuts holding the intake air connector to the air intake chamber
- Throttle body bracket to the cylinder head. Nuts: 15 ft. lbs. (21 Nm).
- No. 2 vacuum pipe to the air intake chamber and intake manifold. Nuts: 20 ft. lbs. (27 Nm).
- EGR gas temperature sensor connector
- EGR pipe. Union nut: 47 ft. lbs. (64 Nm). Bolts: 20 ft. lbs. (27 Nm).
- Oil dipstick guide and dipstick
- Fuel return hose
- With automatic transmission, transmission oil dipstick guide and dipstick
- VSV for the fuel pressure control
- Negative battery cable

6. Fill the cooling system.
7. Start the vehicle, check for leaks and repair if necessary.

2JZ-GTE Engine

1. Before servicing the vehicle, refer to the precautions in the beginning of this section.
2. Relieve fuel system pressure.
3. Drain the engine coolant.

4. Remove or disconnect the following:
- Negative battery cable. On vehicles equipped with an air bag, wait at least 90 seconds before proceeding
- Undercover
- Accelerator and cruise control actuator cables
- Throttle position sensor connector
- Sub throttle position sensor connector
- Sub throttle actuator connector
- Throttle body from the air intake chamber by removing the 2 bolts and 2 nuts
- Gasket from the throttle body
- EVAP hose
- Water bypass hose (from the No. 4 water bypass pipe)
- Power steering air hose
- Throttle body
- With automatic transmissions, transmission dipstick and guide
- Oil dipstick and guide
- Air intake chamber stay
- Control cable bracket from the air intake chamber
- IAC valve connector
- Turbo pressure sensor connector
- VSV connector for the fuel pressure control
- VSV connector for the EGR valve
- Engine wire protector
- IAC valve pipe from the clamp on the cylinder head cover and disconnect the air hose form the IAC valve
- Air hose (from the air intake chamber) from the vacuum pipe on the IAC valve pipe
- Air hose for the EGR from the valve pipe
- PCV hose from the PCV valve
- Vacuum sensing hose from the fuel pressure regulator
- Water bypass hose (from the IAC valve) from the No. 4 water bypass pipe
- EVAP hose (from the air intake chamber) from the vacuum pipe on the manifold stay
- EVAP hose (from the vacuum pipe on the No. 4 water bypass pipe) from the No. 2 vacuum pipe
- EVAP hose (from the charcoal canister) from the No. 2 vacuum pipe
- Power steering air hose from the air intake chamber
- Brake booster vacuum hose from the union on the air intake chamber
- EGR gas temperature sensor connector from the No. 2 vacuum pipe and wiring connector.

- EGR pipe
- No. 4 water bypass pipe
- Intake manifold stay
- Ground cable from the intake manifold
- Engine wire protector from the intake manifold
- Engine wire protector from the brackets
- Engine wire bracket.
- Air intake air chamber assembly from the intake manifold
- Water bypass hose from the IAC valve
- 6 injector connectors
- Camshaft position sensor connectors
- Engine wire clamps from the injector holders
- Fuel inlet pipe from the delivery pipe
- Fuel return pipe from the fuel pressure regulator
- Intake manifold and delivery pipe assembly

To install:

5. Install or connect the following:
- New gasket, the intake manifold and delivery pipe assembly. Bolts and nuts: 20 ft. lbs. (27 Nm).
- Fuel return pipe to the fuel pressure regulator. Union bolts: 20 ft. lbs. (27 Nm).
- Fuel inlet pipe to the delivery pipe. Union bolt: 30 ft. lbs. (41 Nm).
- Engine wire clamps to the injector holders
- Camshaft position sensor connectors
- Injector connectors. The No. 1, 3, and 5 injector connectors are dark gray; the No. 2, 4, and 6 injector connectors are gray
- Air intake chamber gasket
- Water bypass hose to the IAC valve
- Air intake chamber assembly to the intake manifold and install the 5 bolts and 2 nuts. Tighten the nuts and bolts in several passes to 20 ft. lbs. (27 Nm).
- Engine wire protector
- Ground cable to the intake manifold
- Intake manifold stay. Bolts: 29 ft. lbs. (39 Nm).
- No. 4 water bypass pipe
- EGR pipe and gasket with the 2 bolts. Bolts: 20 ft. lbs. (27 Nm).
- Union bolt holding the EGR pipe to the EGR valve. Bolt: 47 ft. lbs. (64 Nm).
- EGR gas temperature sensor connector

- Brake booster vacuum hose to the union on the air intake chamber
- Power steering air hose to the air intake chamber
- EVAP hose (from the charcoal canister) to the No. 2 vacuum pipe
- EVAP hose (from the vacuum pipe on the No. 4 water bypass pipe) to the No. 2 vacuum pipe
- EVAP hose (from the air intake chamber) to the vacuum pipe on the manifold stay
- Water bypass hose (from the IAC valve) to the No. 4 water bypass pipe
- Vacuum sensing hose to the fuel pressure regulator
- PCV hose to the PCV valve
- Air hose for the EGR to the valve pipe
- Air hose (from the air intake chamber) to the vacuum pipe on the IAC valve pipe
- Connect the IAC valve pipe to the clamp on the cylinder head cover
- Connect the air hose to the IAC valve
- Engine wire protector to the vehicle body with the bolt
- VSV connector for the EGR valve and the fuel pressure control
- Turbo pressure sensor and the IAC valve connectors
- Control cable bracket to the intake chamber. Bolts: 14 ft. lbs. (19 Nm).
- Intake chamber stay. Nut and bolt: 14 ft. lbs. (19 Nm).
- Oil dipstick guide and dipstick. Use a new O-ring for the dipstick guide
- Transmission oil dipstick guide and dipstick. Use a new O-ring for the dipstick guide
- Throttle body
- Power steering air hose
- Water bypass hose (from the cylinder head)
- Water bypass hose (from the No. 4 water bypass pipe)
- EVAP hose
- Gasket and throttle body to the air intake chamber. Nuts and bolts: 15 ft. lbs. (21 Nm).
- Sub throttle actuator connector
- Sub throttle position sensor connector
- Throttle position sensor connector
- Cruise control actuator and the accelerator cables
- Air hose
- Undercover
- Negative battery cable

6. Fill the cooling system.

7. Start the vehicle, check for leaks and repair if necessary.

Exhaust Manifold

REMOVAL & INSTALLATION

5E-FE Engine

1. Before servicing the vehicle, refer to the precautions in the beginning of this section.
2. Remove or disconnect the following:
   - Negative battery cable from the battery. On vehicles equipped with an air bag, wait at least 90 seconds before proceeding.
   - All electrical wires and vacuum hoses that interfere with removal of the exhaust manifold.
   - Exhaust heat insulator
   - Exhaust pipe stay
   - Exhaust pipe from the manifold by removing the 2 bolts and 2 compression springs.
   - Exhaust manifold and gasket

To install:
3. Clean the gasket surfaces
4. Install or connect the following:
   - Exhaust manifold with a new gasket. Bolts: 35 ft. lbs. (47 Nm).
   - Exhaust stay to the engine and exhaust manifold. Bolt and nuts: 29 ft. lbs. (40 Nm).
   - Exhaust manifold heat insulator. Bolts: 69 inch lbs. (8 Nm).
   - Exhaust pipe to the exhaust manifold with the 2 compression springs and 2 bolts. Bolts: 46 ft. lbs. (62 Nm).
   - All electrical wires and vacuum hoses that were disconnected for removal of the exhaust manifold.
   - Negative battery cable

1NZ-FE

1. Before servicing the vehicle, refer to the precautions in the beginning of this section.
2. Remove or disconnect the following:
   - Negative battery cable from the battery. On vehicles equipped with an air bag, wait at least 90 seconds before proceeding.
   - All electrical wires and vacuum hoses that interfere with removal of the exhaust manifold.
   - Exhaust heat insulator
   - Exhaust pipe stay
   - Exhaust pipe from the manifold by

removing the 2 bolts and 2 compression springs.
   - Exhaust manifold and gasket

To install:
3. Clean the gasket surfaces
4. Install or connect the following:
   - Exhaust manifold with a new gasket. Bolts: 20 ft. lbs. (27 Nm).
   - Exhaust stay to the engine and exhaust manifold. Bolt and nuts: 29 ft. lbs. (40 Nm).
   - Exhaust manifold heat insulator. Bolts: 71 inch lbs. (8 Nm).
   - Exhaust pipe to the exhaust manifold with the 2 compression springs and 2 bolts. Bolts: 46 ft. lbs. (62 Nm).
   - All electrical wires and vacuum hoses that were disconnected for removal of the exhaust manifold.
   - Negative battery cable

1ZZ-FE Engine

1. Before servicing the vehicle, refer to the precautions in the beginning of this section.
2. Drain the cooling system.
3. Remove or disconnect the following:
   - Negative battery cable
   - Drive belt and alternator
   - Air intake duct
   - Accelerator cable
   - Exhaust pipe from the manifold
   - Exhaust manifold support bracket
   - Heat insulator from the dash panel
   - Upper heat insulator
   - Exhaust manifold and gasket
   - If necessary, the lower heat insulator from the exhaust manifold.

To install:
4. Install or connect the following:
   - Lower heat insulator on the exhaust manifold. Bolts: 108 inch lbs. (12 Nm).
   - Exhaust manifold using a new gasket. Nuts, tightened several passes: 27 ft. lbs. (37 Nm).
   - Upper heat insulator. Bolts: 108 inch lbs. (12 Nm).
   - Heat insulator on the dash panel
   - Exhaust manifold support bracket. Bolts in an alternating pattern: 37 ft. lbs. (49 Nm).
   - Front exhaust pipe to the manifold. Bolts: 46 ft. lbs. (62 Nm).
   - Oxygen sensor, using new gasket and nuts. Nuts: 14 ft. lbs. (20 Nm).
   - Accelerator cable and air intake duct
   - Alternator and drive belt
   - Negative battery cable

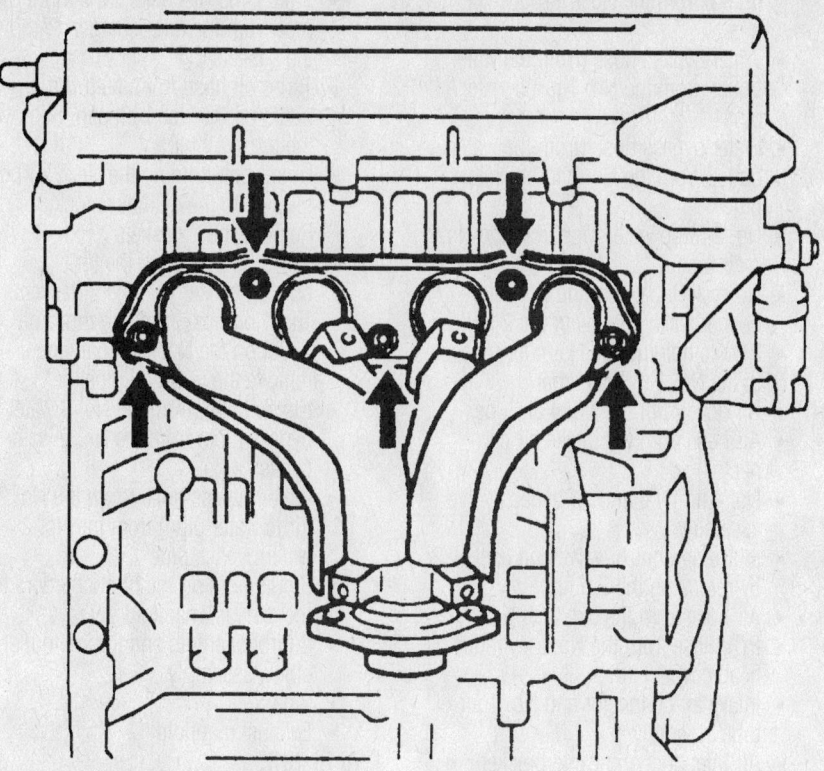

Exhaust manifold mounting nut locations—1.8L (1ZZ-FE) engine

7923VG22

5. Fill the cooling system.
6. Start the vehicle, check for leaks and repair if necessary.

2ZZ-GE Engine

1. Before servicing the vehicle, refer to the precautions in the beginning of this section.
2. Drain the cooling system.
3. Remove or disconnect the following:
   - Negative battery cable
   - Drive belt and alternator
   - Air intake duct
   - Accelerator cable
   - Exhaust pipe from the manifold
   - Exhaust manifold support bracket
   - Heat insulator from the dash panel.
   - Upper heat insulator
   - Exhaust manifold and gasket
   - If necessary, the lower heat insulator from the exhaust manifold.

To install:
4. Install or connect the following:
   - Lower heat insulator on the exhaust manifold. Bolts: 15 ft. lbs. (20 Nm).
   - Exhaust manifold using a new gasket. Nuts, tightened several passes: 37 ft. lbs. (50 Nm).
   - Upper heat insulator. Bolts: 15 ft. lbs. (20 Nm).
   - Heat insulator on the dash panel.

   - Exhaust manifold support bracket. Bolts: 37 ft. lbs. (49 Nm).
   - Front exhaust pipe to the manifold. Bolts: 46 ft. lbs. (62 Nm).
   - Oxygen sensor, using new gasket and nuts. Nuts: 14 ft. lbs. (20 Nm).
   - Accelerator cable and air intake duct.
   - Alternator and drive belt.
   - Negative battery cable
5. Fill the cooling system.
6. Start the vehicle, check for leaks and repair if necessary.

5S-FE Engine

1. Before servicing the vehicle, refer to the precautions in the beginning of this section.
2. Remove or disconnect the following:
   - Negative battery cable. On vehicles equipped with an air bag, wait at least 90 seconds before proceeding
   - Bolts holding the front exhaust pipe to the mounting bracket
   - Front exhaust pipe from the manifold
   - Main oxygen sensor connector and the sub oxygen sensor connector
   - Left-hand exhaust manifold stay
   - Upper manifold heat insulator
   - Right-hand exhaust manifold stay

   - Exhaust manifold, and the 3-way catalytic converter assembly
   - Oxygen sensor and gasket from the exhaust manifold
   - Sub oxygen sensor from the 3-way catalytic converter
   - Lower heat insulator from the exhaust manifold
   - 3-way catalytic converter heat insulators
   - 3-way Catalytic Converter (TWC), gasket, retainer and cushion from the exhaust manifold

To install:
3. Place the cushion, retainer, and a new gasket on the TWC and reinstall it to the exhaust manifold with the bolts and the nuts. Bolts and nuts: 22 ft. lbs. (30 Nm).
4. Install or connect the following:
   - Lower manifold heat insulator and TWC heat insulators with the bolts.
   - Main oxygen sensor to the exhaust manifold with a new gasket and new nuts. Nuts: 14 ft. lbs. (19 Nm).
   - Sub oxygen sensor to the front TWC and tighten to 33 ft. lbs. (45 Nm).
   - Exhaust manifold, using new gasket and front TWC assembly to the engine with the nuts. Nuts, tighten uniformly in several passes: 36 ft. lbs. (49 Nm).
   - Right-hand exhaust manifold stay. Bolts and nuts: 31 ft. lbs. (42 Nm).
   - Upper heat insulator
   - Left-hand exhaust manifold stay. Bolt: 29 ft. lbs. (39 Nm). Nut: 31 ft. lbs. (42 Nm).
   - Main and the sub oxygen sensors connectors.
   - Front exhaust pipe with a new gasket to the TWC. Nuts: 46 ft. lbs. (62 Nm).
   - Front exhaust pipe to the exhaust pipe bracket. Bolts: 14 ft. lbs. (19 Nm).
   - Negative battery cable

1MZ-FE Engine

1. Before servicing the vehicle, refer to the precautions in the beginning of this section.
2. Remove or disconnect the following:
   - Negative battery cable
   - Undercovers
   - Front exhaust pipe from the exhaust manifold
   - EGR pipe from the exhaust manifold
   - Heated oxygen sensor connector from the right exhaust manifold

- Exhaust manifold stay
- Exhaust manifold and gasket

To install:

3. Install or connect the following:
- Exhaust manifold using new gasket. Bolts, uniformly tightened: 36 ft. lbs. (49 Nm).
- Exhaust manifold stay. Bolt and nut: 15 ft. lbs. (20 Nm).
- Heated oxygen sensor connector to the right exhaust manifold
- EGR pipe, using new gaskets, to the exhaust manifold and the engine. Nuts: 108 inch lbs. (12 Nm).
- Front exhaust pipe, using new gasket, to the exhaust manifold. Nuts: 46 ft. lbs. (62 Nm).
- Undercovers
- Negative battery cable

2JZ-GTE Engine

1. Before servicing the vehicle, refer to the precautions in the beginning of this section.
2. Drain the cooling system.
3. Remove or disconnect the following:
- Negative battery cable. On vehicles equipped with an air bag, wait at least 90 seconds before proceeding.
- Cruise control actuator cable from the throttle body
- No. 1 air hose
- Air cleaner and MAF meter assembly
- Theft deterrent horn from the body.
- Undercover
- Front lower arm bracket stay
- Front upper crossmember extension
- No. 2 front exhaust pipe
- Heat insulator for the No. 2 front exhaust pipe
- With automatic transmission, automatic transmission oil cooler tubes from the engine
- Engine wire protector from the body
- Heater hose from the No. 3 water bypass pipe
- EVAP hose from the No. 1 vacuum pipe
- IAC valve pipe from the No. 2 air tube
- No. 1 vacuum pipe from the air tubes
- VSV assembly
- Crankshaft position sensor connector from the clamp
- Water bypass hose (from the water

pump) from the No.1 turbo water pipe
- Water bypass hose (from the water outlet) from the No. 1 turbo water pipe
- Water bypass hose (from the water outlet) from the No. 2 turbo water pipe
- No. 2 turbo water pipe from the No. 4 air tube
- No. 1 air tube from the No. 1 turbocharger by removing the 2 bolts
- 2 bolts holding the No. 4 air tube to the No. 1 turbocharger
- Air hose from the No. 4 air tube
- Air hose from the intake air connector
- No. 4 air tube and air bypass valve assembly
- Intake air control valve and gasket by removing the 2 nuts
- Air hose from the No. 2 air tube
- PCV hose from the No. 2 cylinder head cover
- Intake air connector and No. 1 air tube assembly
- Air inlet duct and cable bracket
- Heat insulator for the turbocharger
- Exhaust bypass pipe and gasket
- Exhaust gas control valve stay
- Main heated oxygen sensor by disconnecting the electrical connector and 2 nuts
- Exhaust gas control valve
- No. 1 turbocharger stay
- No. 2 turbocharger stay
- Union bolt holding the No. 1 turbo oil pipe to the cylinder block. Remove the 2 gaskets

- 2 nuts and disconnect the turbo oil pipe from the turbocharger. Remove the gaskets
- Turbo oil hose from the turbo oil outlet on the No. 1 oil pan. Remove the turbo oil pipe
- Union bolt holding the No. 2 turbo oil pipe to the cylinder block. Remove the 2 gaskets
- Turbo oil pipe from the turbocharger. Remove the 2 gaskets
- Turbo oil hose from the turbo oil outlet on the No. 1 oil pan and remove the turbo oil pipe
- Heater hose (from the No. 3 water bypass pipe) from the No. 2 water bypass pipe.
- Water bypass hose (from the No. 2 turbo water pipe) from the No. 2 water bypass pipe
- 8 nuts holding the turbochargers to the exhaust manifold
- 2 turbochargers and turbine outlet elbow assembly
- 2 gaskets
- Exhaust manifold

To install:

4. Install or connect the following:
- Exhaust manifold with 2 new gaskets. Torque the nuts, in several passes and in sequence: 29 ft. lbs. (39 Nm).
- 2 new gaskets and the turbochargers and turbine outlet elbow assembly to the exhaust manifold
- 8 **new** nuts and uniformly tighten the nuts in several passes. Nuts: 40 ft. lbs. (54 Nm).
- Water bypass hose (from the No. 2

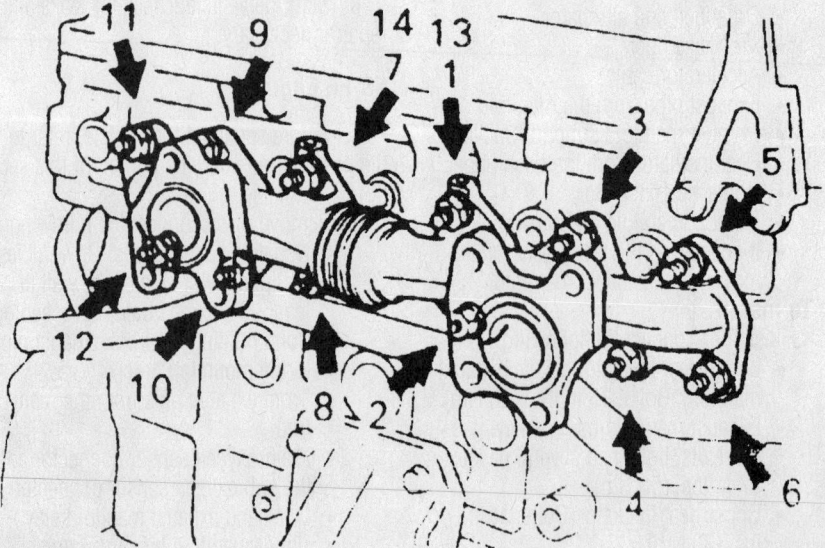

Exhaust manifold bolt installation sequence—3.0L (2JZ-GTE) engine

7923VG23

turbo water pipe) to the No. 2 water bypass pipe
- Heater hose (from the No. 3 water bypass pipe) to the No. 2 water bypass pipe
- No. 2 turbo oil pipe. Nuts: 15 ft. lbs. (21 Nm). Union bolt: 29 ft. lbs. (39 Nm).
- No. 1 turbo oil pipe. Nuts: 15 ft. lbs. (21 Nm). Union bolt: 29 ft. lbs. (39 Nm).
- No. 2 turbocharger stay. Nut and bolt: 32 ft. lbs. (43 Nm).
- No. 1 turbocharger stay. Nut and bolt: 32 ft. lbs. (43 Nm).
- Exhaust gas control valve by installing 2 new gaskets and the 3 nuts. Nuts: 51 ft. lbs. (69 Nm).
- Main heated oxygen sensor. Nuts: 14 ft. lbs. (20 Nm).
- Exhaust gas control valve stay. Bolt and nut: 32 ft. lbs. (43 Nm).
- Exhaust bypass pipe, with 2 new gaskets. Nuts: 18 ft. lbs. (25 Nm).
- Heat insulator for the turbocharger
- Air inlet duct
- Intake air connector and No. 1 air tube assembly
- PCV hose to the No. 2 cylinder head cover
- Air hose to the No. 2 air tube
- Intake air control valve and gasket. Nut: 15 ft. lbs. (21 Nm).
- No. 4 air tube and air bypass valve assembly
- Air hose to the intake air connector
- Air hose to the No. 4 air tube
- 2 bolts holding the No. 4 air tube to the No. 1 turbocharger. Bolt: 15 ft. lbs. (21 Nm).
- No. 1 air tube to the No. 1 turbocharger and install the 2 bolts. Bolts: 15 ft. lbs. (21 Nm).
- No. 2 turbo water pipe to the No. 4 air tube
- Water bypass hose (from the water outlet) to the No. 2 turbo water pipe
- Water hose (from the water outlet) to the No. 1 turbo water pipe
- Water bypass hose (from the water pump) to the No. 1 turbo water pipe
- Crankshaft position sensor connector to the clamp
- VSV assembly
- No. 1 vacuum pipe to the air tubes and install the 3 bolts
- 2 air hoses (from the vacuum pressure tank) to the vacuum pipe
- Air hose to the VSV for the intake air control valve

- Air hose (from the No. 2 air tube) from the vacuum pipe
- 2 air hoses (from the VSV for exhaust bypass valve) to the vacuum pipe
- Vacuum hose (from the air bypass valve) to the No. 1 air tube
- Air hose (from the VSV for exhaust gas control valve) to the vacuum pipe
- Air hose (from the VSV for waste gate valve) to the vacuum pipe
- Air hose to the No. 1 air tube
- Air hose to the No. 4 air tube
- Engine wire to the 3 clamps
- VSV connector for the exhaust bypass valve
- VSV connector for the intake air control valve
- IAC valve pipe to the No. 2 air tube
- EVAP hose to the No. 1 vacuum pipe
- Heater hose to the No. 3 water bypass pipe
- Engine wire protector to the body
- Automatic transmission oil cooler tubes to the engine
- Heat insulator for the No. 2 front exhaust pipe
- No. 2 front exhaust pipe, using new gasket. Nuts: 46 ft. lbs. (62 Nm).
- Front exhaust pipe to the No. 2 exhaust pipe and install the pipe support bracket. Bolts: 32 ft. lbs. (43 Nm).
- Bolts and nuts to hold the front exhaust pipe to the No. 2 front exhaust pipe. Tighten to 43 ft. lbs. (58 Nm).
- Upper front crossmember extension. Bolts to 22 ft. lbs. (29 Nm). Nuts: 25 ft. lbs. (33 Nm).
- Front lower arm bracket stay. Bolts: 33 ft. lbs. (44 Nm). Nut: 43 ft. lbs. (59 Nm).
- Theft deterrent horn
- Air cleaner and MAF meter
- Air cleaner duct
- No. 1 air hose
- Cruise control actuator cable to the throttle body
- Undercover
- Negative battery cable

5. Fill the cooling system to the proper level.

6. Start the vehicle, check for leaks and repair if necessary.

2JZ-GE Engine

1. Before servicing the vehicle, refer to the precautions in the beginning of this section.

2. Remove or disconnect the following:
- Negative battery cable. Wait at least 90 seconds before performing any other work.
- O$_2$ sensors at the exhaust manifolds.
- If equipped, outside heat insulator

3. Remove the 4 nuts to disconnect the manifold from the front exhaust pipe. Loosen the mounting nuts to the exhaust manifolds and remove the 2 manifolds and the gasket.

To install:

4. Scrape the mating surfaces of all old gasket material.

5. Install or connect the following:
- Manifolds with a new gasket. Nuts: 29 ft. lbs. (39 Nm).
- Front exhaust pipe, using new gasket, to the exhaust manifolds. Nuts: 46 ft. lbs. (62 Nm).
- Outer heat insulator. Nuts: 13 ft. lbs. (18 Nm).
- O$_2$ sensors
- Undercovers
- Negative battery cable

Front Crankshaft Seal

REMOVAL & INSTALLATION

➡ **The following procedures apply to engines using a timing belt. The procedures for front cover seals can be found later in this section.**

5E-FE Engine

➡ **The front oil seal can be removed from the engine without removing the oil pump.**

1. Before servicing the vehicle, refer to the precautions in the beginning of this section.

2. Remove or disconnect the following:
- Negative battery cable from the battery. On vehicles equipped with an air bag, wait at least 90 seconds before proceeding.
- Front covers and the timing belt
- Crankshaft timing belt sprocket

✳✳ WARNING

When removing the front seal, be extremely careful not to damage the crankshaft.

3. Using a knife, cut off the oil seal lip
4. Tape the end of a flat bladed tool to avoid damaging crankshaft. Pry out the oil seal using the taped end of the tool

5. Inspect the oil seal riding surface on the crankshaft for signs of wear or damage

To install:

6. Wipe the seal bore with a clean rag

7. Apply multipurpose grease the lip of a new oil seal

8. Drive the oil seal into place. Be sure the seal surface is flush with the oil pump case edge. Work from the front of the cover. Be extremely careful not to damage the seal

9. Install or connect the following:
- Sprocket without disturbing the Woodruff key
- Timing belt and front covers
- Negative battery cable

5S-FE Engine

1. Before servicing the vehicle, refer to the precautions in the beginning of this section.

2. Remove or disconnect the following:
- Negative battery cable. On vehicles equipped with an air bag, wait at least 90 seconds before proceeding

➡ **The front oil seal can be removed from the engine without removing the oil pump.**

- Timing belt covers and the timing belt
- Front crankshaft gear from the crankshaft. Be sure not to damage any part of the crankshaft

3. Using a knife, cut off the oil seal lip.

4. Pry out the oil seal. Wrap the edge of the tool with a rag or tape to prevent damaging the crankshaft. Be careful not to damage the crankshaft.

To install:

5. Using a new seal, apply a thin layer of liquid sealer to the outside of the seal.

6. Apply multi purpose grease to the new oil seal lip.

7. Oil seal until its surface is flush with the oil pump body edge.

8. Install or connect the following:
- Timing belt and the timing belt covers
- All other components
- Negative battery cable

1MZ-FE Engine

AVALON, CAMRY AND CAMRY SOLARA

1. Before servicing the vehicle, refer to the precautions in the beginning of this section.

2. Remove or disconnect the following:
- Negative battery cable. On vehicles equipped with an air bag, wait at

least 90 seconds before proceeding.
- Timing belt
- Crankshaft timing gear

3. Cut out the lip portion of the oil seal.

4. Tape the end of a suitable prybar to protect the crankshaft and carefully remove the oil seal.

✳✳ WARNING

Be careful not to damage the crankshaft sealing surface.

To install:

5. Apply multi-purpose grease to the lip of a new oil seal. Also apply a light coating of liquid sealant to the outside of the oil seal.

6. Oil seal, until its surface is flush with the oil pump case edge.

7. Install or connect the following:
- Crankshaft timing gear
- Timing belt
- Negative battery cable

2JZ-GE and 2JZ-GTE Engines

1. Before servicing the vehicle, refer to the precautions in the beginning of this section.

2. Remove or disconnect the following:
- Timing belt
- Crankshaft timing pulley

3. Cut the oil seal lip.

4. Tape the end of a small prying tool and remove the oil seal.

✳✳ WARNING

Be careful not to damage the crankshaft.

To install:

5. Apply multi-purpose grease to a new oil seal lip.

6. Oil seal, until its surface is flush with the oil pump case edge.

7. Install or connect the following:
- Crankshaft timing pulley
- Timing belt

Camshaft(s)

REMOVAL & INSTALLATION

5E-FE Engine

1. Before servicing the vehicle, refer to the precautions in the beginning of this section.

2. Remove or disconnect the following:
- Negative battery cable. On vehicles equipped with an air bag, wait at least 90 seconds before proceeding.
- Valve cover
- Timing belt assembly
- Camshaft timing sprocket

✳✳ WARNING

Due to the relatively small amount of camshaft thrust clearance, the camshaft must be kept level during removal. If the camshaft is not level on removal, the portion of the head receiving the thrust may crack or be damaged.

3. Set the intake camshaft so the service bolt holes of the intake camshaft gears are directly up.

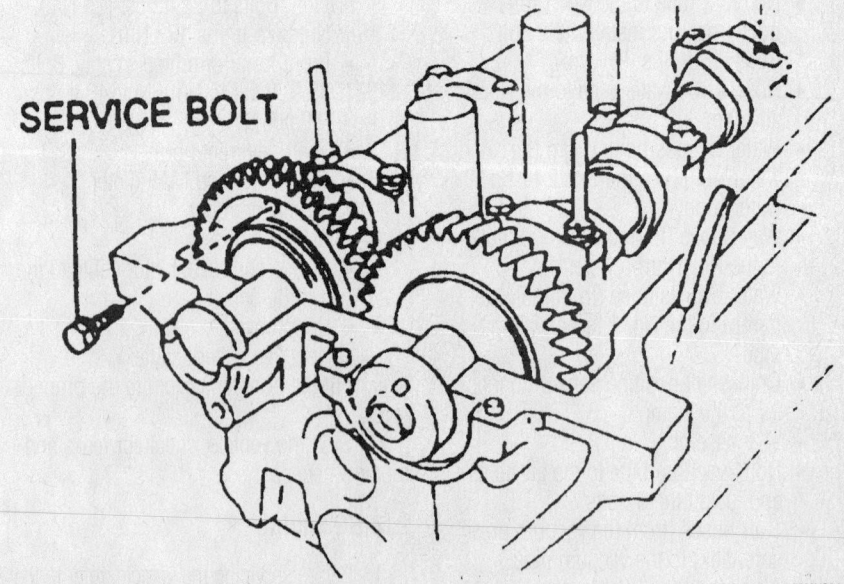

SERVICE BOLT

Securing the intake camshaft subgear to the main gear—1.5L (5E-FE) engine

7923VG24

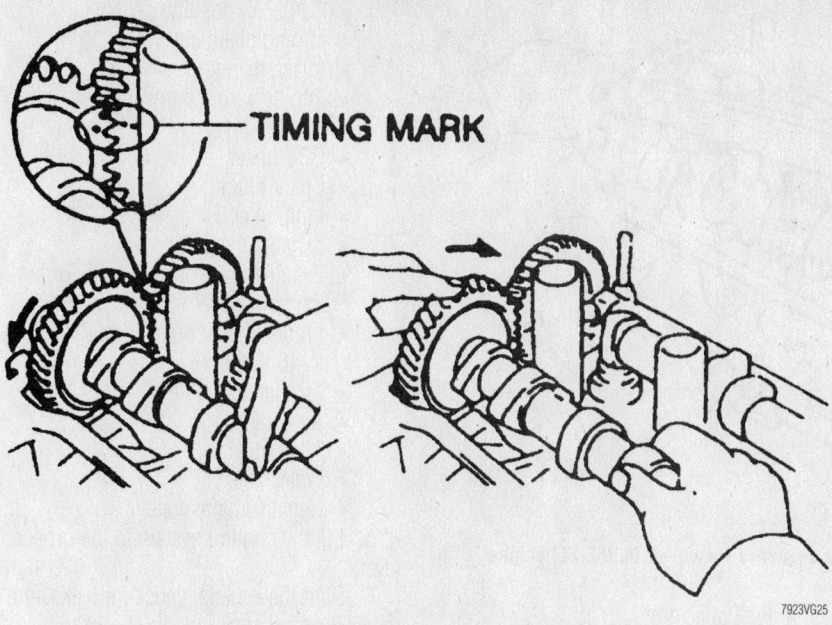

Exhaust and intake camshaft gear timing marks—1.5L (5E-FE) engine

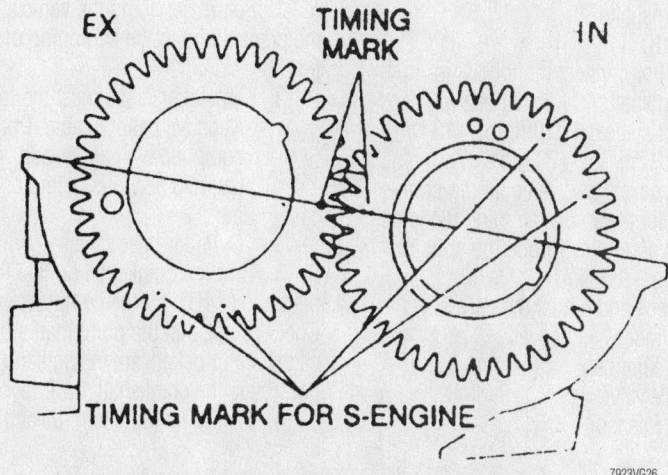

Correct exhaust and intake camshaft gear timing mark alignment—1.5L (5E-FE) engine

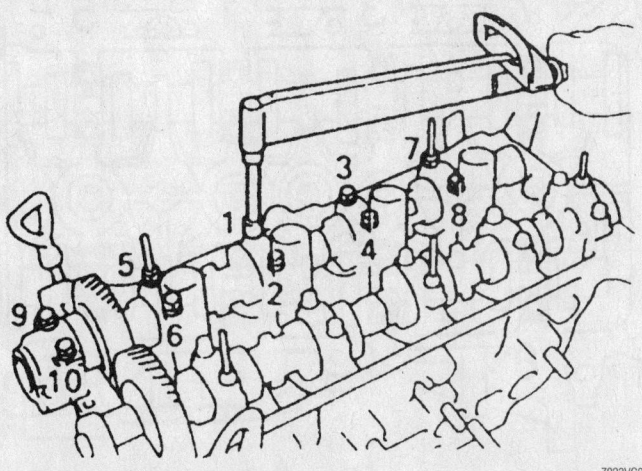

Tighten the intake camshaft bearing cap bolts in sequence—1.5L (5E-FE) engine

4. Remove each front bearing cap of the intake and exhaust camshafts.

5. Secure the intake camshaft sub-gear to the main gear with a 6mm diameter bolt, 16–20mm long and with a pitch of 1.0mm. Be sure that the torsional spring force of the subgear has been eliminated by the above operation.

6. In correct sequence, loosen and remove the 8 bolts of the 4 bearing caps of the exhaust camshaft and remove the exhaust camshaft. If the camshaft is not being lifted out straight, install the middle bearing cap and loosen it evenly to keep the camshaft straight.

7. In the reverse order of the installation sequence, loosen and remove the 8 bolts of the 4 bearing caps of the intake camshaft and remove the intake camshaft. If the camshaft is not being lifted out straight, install the middle bearing cap and loosen it evenly to keep the camshaft straight.

8. Remove the adjusting shim.

To install:

9. Install the valve shim to the engine.

10. Apply engine oil to the surface of the intake camshaft.

11. Place the intake camshaft on the cylinder head so the service bolt points directly up.

12. Install the 4 rearward bearing caps in their original order and temporarily tighten them evenly. Do not tighten the bolts at this time.

13. Apply engine oil to the portion of the exhaust camshaft.

14. Engage the exhaust camshaft gear to the intake camshaft gear by matching the proper timing marks on each gear.

15. Roll down the exhaust camshaft onto the bearing journals while engaging the gears with each other.

➡There are other marks present for the "S" engine. Do not use these marks.

16. Install the 4 rearward intake bearing caps in their original order and tighten evenly. Do not tighten the bolts at this time.

17. Remove the service bolt.

18. Clean the mating surfaces of the No. 2 bearing cap and apply sealer. Install and temporarily tighten the bolts. Install the camshaft housing plug.

19. Tighten the intake camshaft cap bolts to 108 inch lbs. (13 Nm), in the proper sequence.

20. Apply grease to a new camshaft oil seal lip and install it as far as the deepest part of the cylinder head.

21. Install the No. 1 bearing cap and temporarily tighten the bolts.

22. Tighten the exhaust camshaft bearing

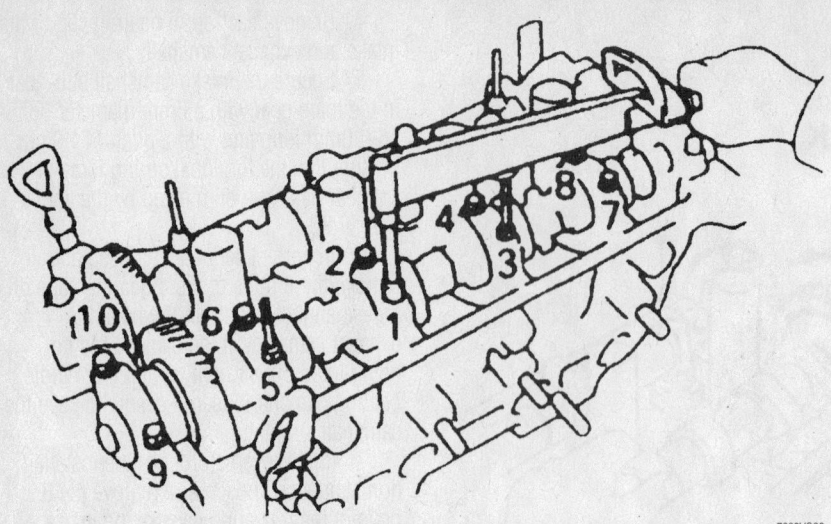

Tighten the exhaust bearing cap bolts according to the sequence shown—1.5L (5E-FE) engine

7923VG28

cap bolts to 108 inch lbs. (13 Nm) evenly and in the proper sequence.

23. Turn the camshaft 1 revolution and check that the timing marks of the camshaft gears are aligned.

24. Check and adjust the valve clearance.

25. Install or connect the following:
- Valve cover
- Camshaft timing sprocket. Bolt: 37 ft. lbs. (50 Nm).
- Timing belt
- Negative battery cable

1NZ-FE Engine

1. Before servicing the vehicle, refer to the precautions in the beginning of this section.

2. Drain the cooling system.

3. Remove or disconnect the following:
- Negative battery cable
- Water filler
- Outer front cowl top panel
- Alternator
- Air cleaner
- Accelerator cable
- Center exhaust pipe
- Exhaust manifold support
- Exhaust manifold
- Ignition coil
- Spark plugs
- PCV hoses
- Throttle body
- Engine wiring harness at the head
- Intake manifold
- Camshaft position sensor
- ECT sensor
- Oil control valve
- PCV valve
- Oil filer cap
- Cylinder head cover

- Fuel injectors
- Timing chain cover
- Camshaft sprockets and valve timing control assembly
- Camshafts

To install:

4. Install or connect the following:
- Camshafts. Camshaft bearing caps in 2 stages: 1st 10 ft. lbs. (13 Nm); 2nd 17 ft. lbs. (23 Nm).
- Sprockets and valve timing controller assembly, aligning the knock pin and hole. Torque the bolts to 47 ft. lbs. (64 Nm).
- Check and adjust the valves.
- Cylinder head cover
- Oil filler cap
- PCV valve
- ECT sensor

- Camshaft position sensor
- Timing chain cover
- Intake manifold
- Engine wiring harness
- Throttle body
- PCV hoses
- Spark plugs
- Ignition coils
- Exhaust manifold
- Exhaust manifold support. Torque the bolts to 27 ft. lbs. (37 Nm).
- Front exhaust pipe. Torque the nuts to 46 ft. lbs. (62 Nm).
- Accelerator cable
- Air cleaner
- Alternator
- Water filler
- Negative battery cable

5. Fill the cooling system to the proper level.

6. Start the vehicle, check for leaks and repair if necessary.

1ZZ-FE Engine

1. Before servicing the vehicle, refer to the precautions in the beginning of this section.

2. Remove or disconnect the following:
- Negative battery cable. On vehicles equipped with an air bag, wait at least 90 seconds before proceeding.
- Cylinder head cover

3. Turn the crankshaft so that the No. 1 piston is at TDC on the compression stroke. Check to see that the point marks on the camshaft sprockets are facing each other, if not, rotate the crankshaft 1 full revolution.

4. Tie the timing chain to each sprocket

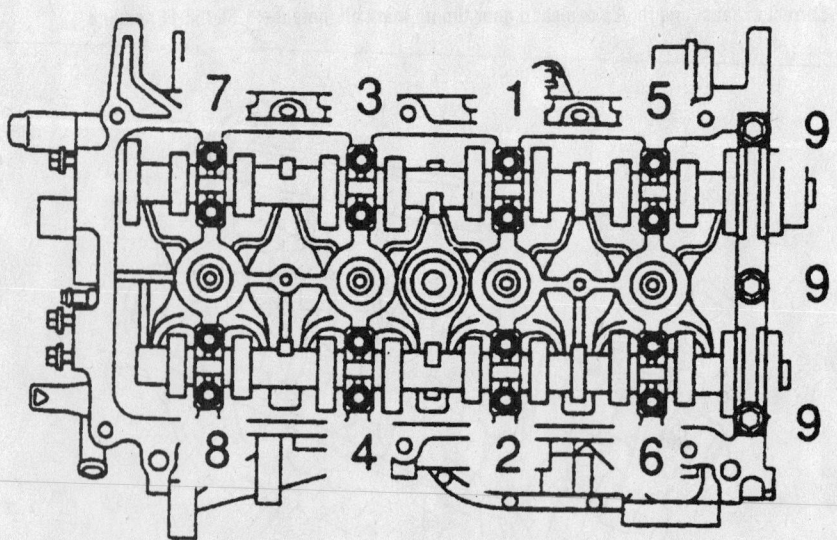

Camshaft bolt torque sequence—1NZ-FE engine

9307WG80

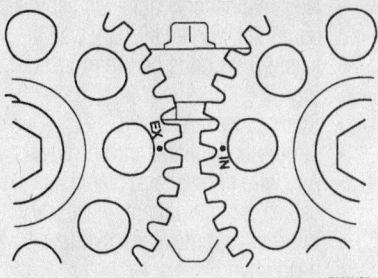

7923VG51

The sprocket marks will align when the No. 1 piston is at TDC on the compression stroke—1.8L (1ZZ-FE) engine

with string or wire to maintain correct valve timing.

5. Hold the camshafts with a wrench and remove the bolts securing the sprockets to the camshafts.

6. Using several passes, gradually remove the bearing cap bolts in the proper sequence. Then, remove the camshafts

To install:

7. Lubricate the camshafts with clean engine oil and place them on the cylinder head. Be sure to position the lobes for the No. 1 cylinder as shown in the illustration.

8. Install the bearing caps in their original positions. Apply clean engine oil to the threads and under the heads of the bearing cap bolts. After tightening the bolts on the No. 1 bearing cap to 17 ft. lbs. (23 Nm), tighten the remaining bolts in sequence using several passes to 10 ft. lbs. (13 Nm).

9. Check the valve clearance and make adjustments as needed.

10. Install or connect the following:
   • Camshaft sprockets and the chain
   • Cylinder head cover
   • Negative battery cable

2ZZ-GE Engine

1. Before servicing the vehicle, refer to the precautions in the beginning of this section.

2. Remove or disconnect the following:
   • Negative battery cable. On vehicles equipped with an air bag, wait at least 90 seconds before proceeding.
   • Cylinder head cover

3. Turn the crankshaft so that the No. 1 piston is at TDC on the compression stroke. Check to see that the point marks on the camshaft sprockets are facing each other, if not, rotate the crankshaft 1 full revolution.

4. Tie the timing chain to each sprocket with string or wire to maintain correct valve timing.

5. Hold the camshafts with a wrench and remove the bolts securing the sprockets to the camshafts.

6. Using several passes, gradually remove the bearing cap bolts in the proper sequence. Then, remove the camshafts

To install:

7. Lubricate the camshafts with clean engine oil and place them on the cylinder head. Be sure to position the lobes for the No. 1 cylinder as shown in the illustration.

8. Install the bearing caps in their original positions. Apply clean engine oil to the threads and under the heads of the bearing cap bolts. After tightening the bolts on the No. 1 bearing cap to 14 ft. lbs. (18 Nm), tighten the remaining bolts in sequence using several passes to 14 ft. lbs. (18 Nm).

9. Check the valve clearance and make adjustments as needed.

10. Install or connect the following:
   • Camshaft sprockets and the chain
   • Cylinder head cover
   • Negative battery cable

5S-FE Engine

1. Before servicing the vehicle, refer to the precautions in the beginning of this section.

2. Remove or disconnect the following:

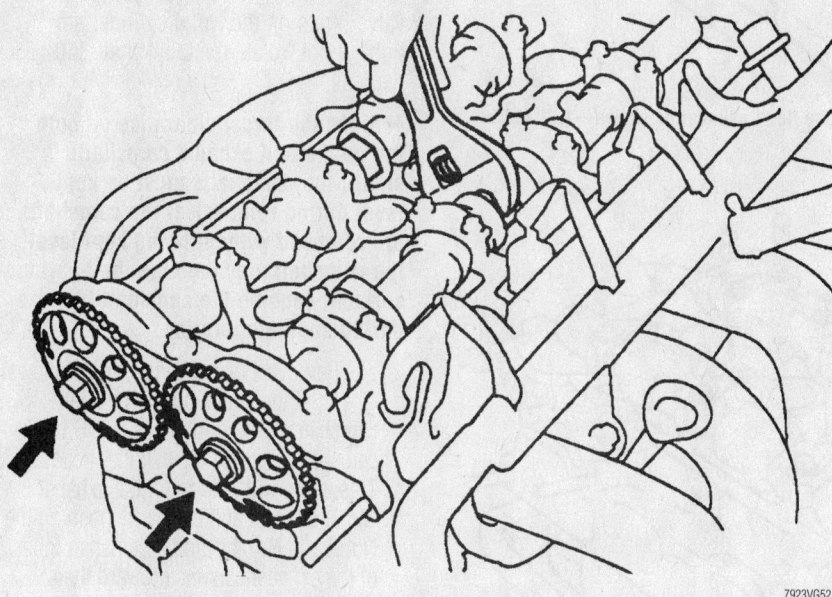

7923VG52

Hold the camshaft with a wrench while removing the sprocket bolt—1.8L (1ZZ-FE) engine

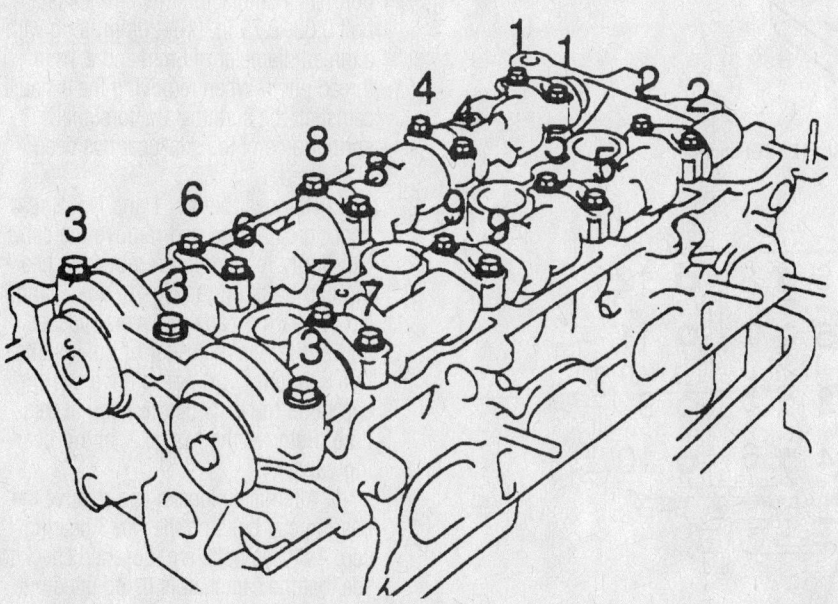

7923VG53

Camshaft bearing cap bolt removal sequence—1.8L (1ZZ-FE) engine

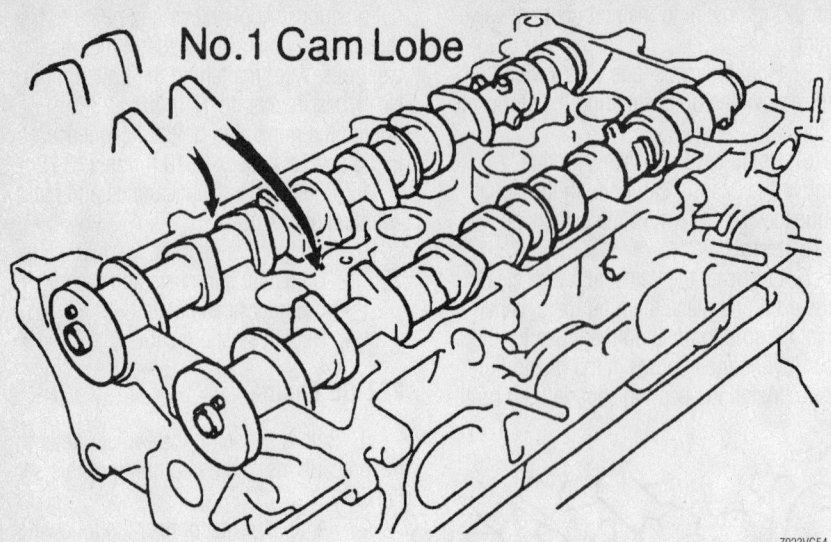

When installing the camshafts, position the lobes for the No. 1 cylinder as shown—1.8L (1ZZ-FE) engine

7923VG54

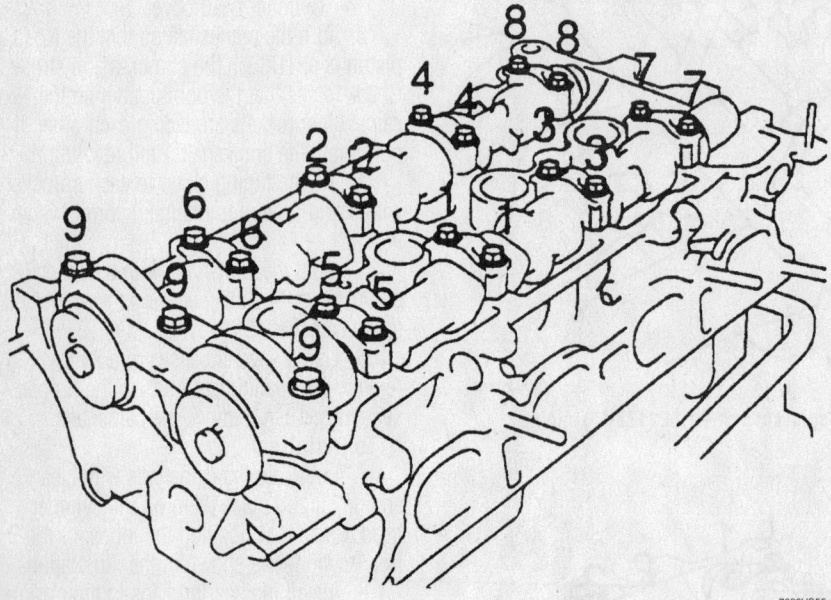

Camshaft bearing cap bolt tightening sequence—1.8L (1ZZ-FE) engine

7923VG55

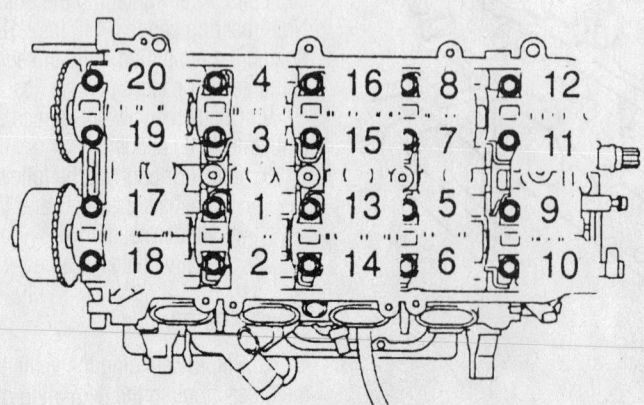

9307WG94

Camshaft bearing cap torque sequence—2ZZ-GE engines

- Negative battery cable. On vehicles equipped with an air bag, wait at least 90 seconds before proceeding.
- Spark plug wires
- Timing belt, gears, and the covers
- Any wire connectors, clamps, cables, or components necessary in order to remove the cylinder head cover.
- Cylinder head cover and gasket

3. Set the No. 1 cylinder to TDC. Turn the crankshaft pulley and align its groove with the timing mark **0** of the No. 1 timing belt cover. Check that the valve lifters on the No. 1 cylinder are loose and valve lifters on the No. 4 cylinder are tight. If not, rotate the crankshaft 360 degrees.

➡Since the thrust clearance on both the intake and exhaust camshafts is small, the camshafts must be kept level during removal. If the camshafts are removed without being kept level, the camshaft may damage the bearing surface, causing the camshaft to seize during engine operation.

4. Remove exhaust camshaft as follows:
   a. Set the knock pin of the intake camshaft at 10–45 degrees BTDC of camshaft angle on the cylinder head. This angle will help to lift the exhaust camshaft level and evenly by pushing the No. 2 and No. 4 cylinder camshaft lobes of the exhaust camshaft toward their valve lifters.
   b. Secure the exhaust camshaft sub-gear to the main gear using a service bolt. The manufacturer recommends a bolt 0.63–0.79 in. (16–20mm) long with a thread diameter of 6mm and a 1mm thread pitch. When removing the exhaust camshaft, be sure that the torsional spring force of the sub-gear has been eliminated.
   c. Remove the No. 1 and No. 2 rear bearing cap bolts and remove the cap. Uniformly loosen and remove the bearing cap bolts on the No. 1, No. 2, and No. 4 bearing caps in several passes and in the reverse order of the installation sequence. Do not remove bearing cap bolts to No. 3 bearing cap at this time. Remove the No. 1, 2, and 4 bearing caps.
   d. Alternately loosen and remove the bearing cap bolts on the No. 3 bearing cap. As these bolts are loosened check to see that the camshaft is being lifted out straight and level.

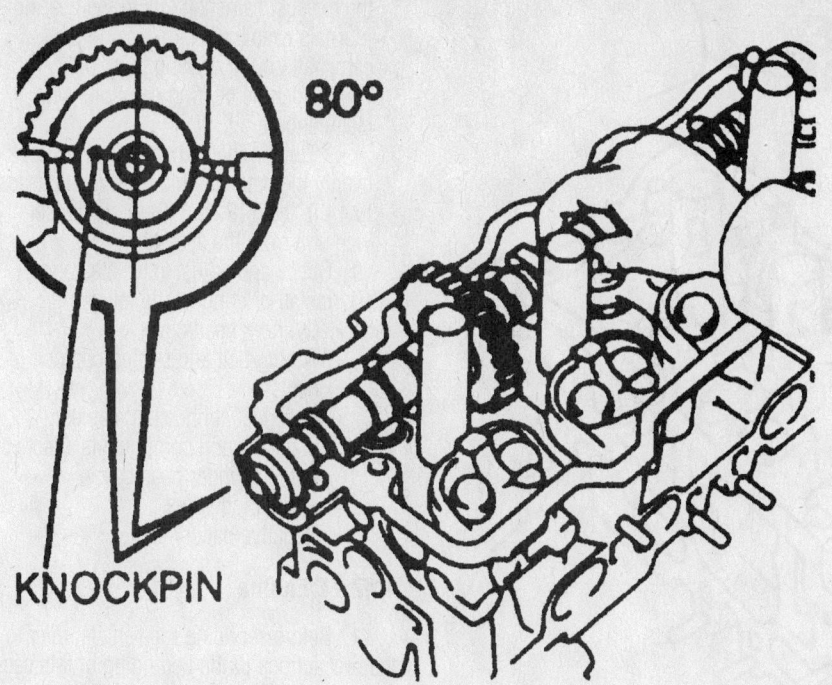

Intake camshaft removal and installation positioning—2.2L (5S-FE) engine

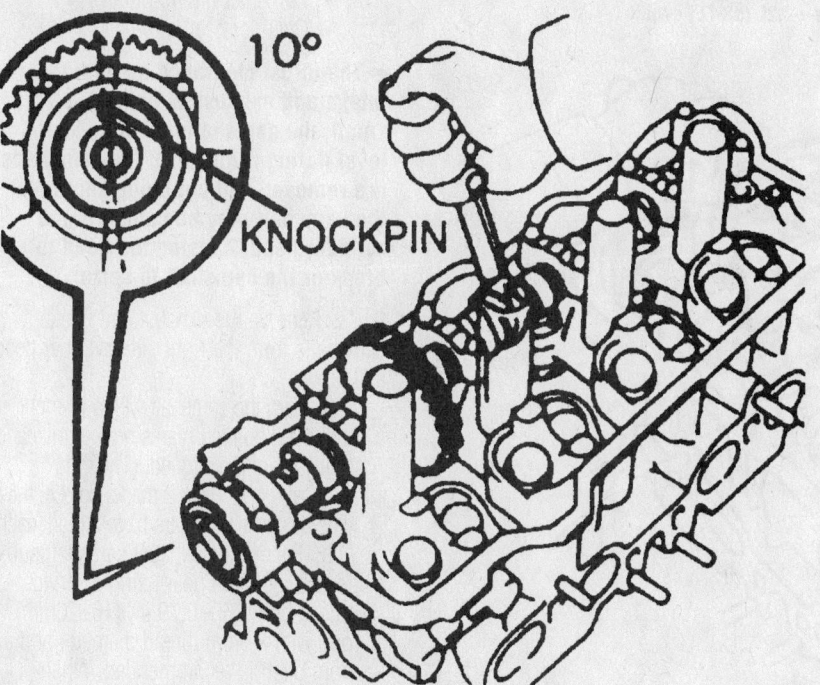

Exhaust camshaft removal and installation positioning—2.2L (5S-FE) engine

➡**If the camshaft is not lifted out straight and level, tighten the No. 3 bearing cap bolts. Reverse the order of Steps 7c through 7a and reset the intake camshaft knock pin to 10–45 degrees BTDC, then repeat Steps 7a through 7c. Do not attempt to pry the camshaft from its mounting.**

e. Remove the No. 3 bearing cap and exhaust camshaft from the engine.

5. Remove intake camshaft as follows:

a. Set the knock pin of the intake camshaft at 80–115 degrees BTDC of the camshaft angle on the cylinder head. This angle will help to lift the intake camshaft level and evenly by pushing

No. 1 and No. 3 cylinder camshaft lobes of the intake camshaft toward their valve lifters.

b. Remove the 2 front bearing cap bolts and remove the front bearing cap and oil seal. If the cap will not come apart easily, leave it in place without the bolts.

c. Uniformly loosen and remove the bearing cap bolts to No. 1, No. 3, and the No. 4 bearing caps in several phases and in the reverse order of the installation sequence. Do not remove bearing cap bolts to the No. 2 bearing cap at this time. Remove No. 1, 3, and 4 bearing caps.

d. Alternately loosen and remove bearing cap bolts to the No. 2 bearing cap. As these bolts are loosened and after breaking the adhesion on the front bearing cap, check to see that the camshaft is being lifted out straight and level.

➡**If the camshaft is not lifting out straight and level tighten the No. 2 bearing cap bolts. Reverse Steps 8b through 8d, then start over from Step 8b. Do not attempt to pry the camshaft from its mounting.**

e. Remove the No. 2 bearing cap with the intake camshaft from the engine.

6. Remove the valve lifter shims and hydraulic lifters. Identify each lifter and shim as it is removed so it can be reinstalled in the same position. If the lifters are to be reused, store them upside down in a sealed container.

To install:

7. Install the valve lifters into their original positions and shims.

➡**Before installing the intake camshaft, apply multi-purpose grease to the camshaft.**

8. Install the intake camshaft, as follows:

a. Position the camshaft at 80–115 degrees BTDC of camshaft angle on the cylinder head.

b. Apply sealant to the front bearing cap.

c. Coat the bearing cap bolts with clean engine oil.

d. Tighten the camshaft bearing caps evenly in sequence and in several passes to 14 ft. lbs. (19 Nm).

e. Apply MP grease to a new oil seal lip, and by using a suitable tool, tap a new oil seal into place.

• Exhaust camshaft as follows:

f. Set the knock pin of the camshaft

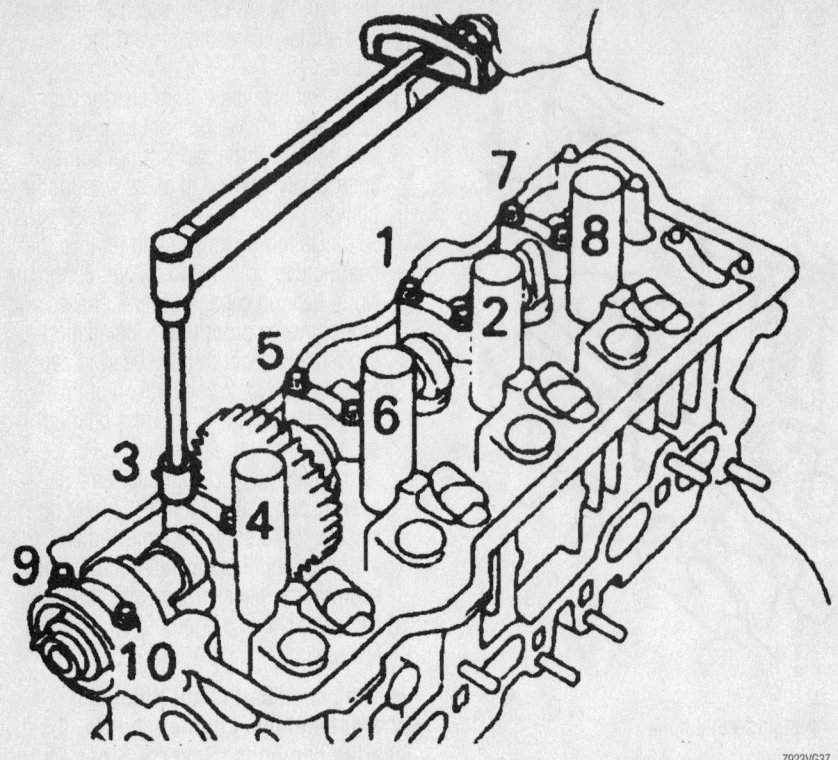

7923VG37

Intake camshaft bearing cap bolt tightening sequence—2.2L (5S-FE) engine

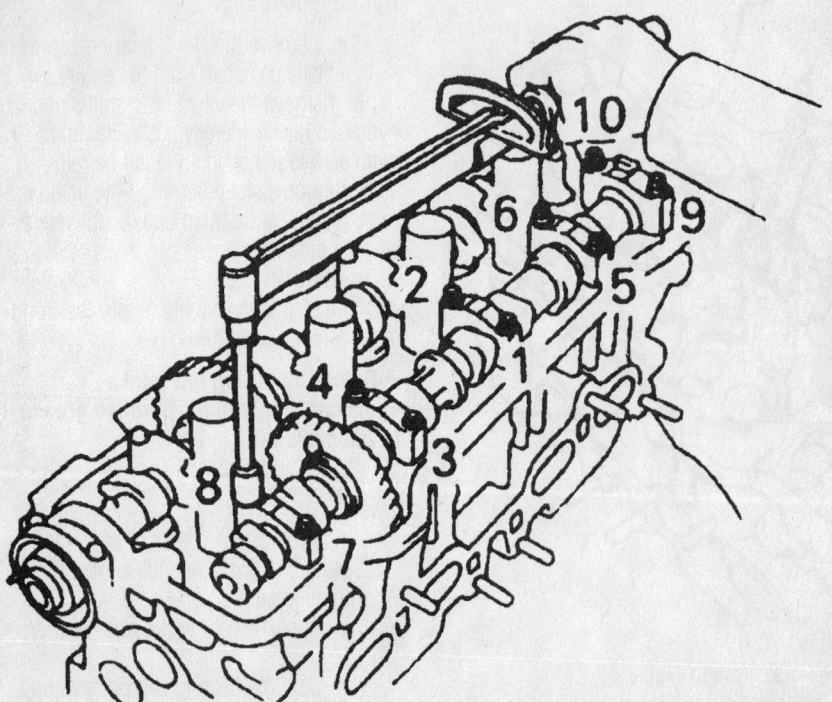

7923VG38

Exhaust camshaft bearing cap bolt tightening sequence—2.2L (5S-FE) engine

the exhaust camshaft sits in the bearing journals evenly without rocking the camshaft on the bearing journals.

j. Coat the bearing cap bolts with clean engine oil.

k. Tighten the camshaft bearing caps evenly in sequence and in several passes to 14 ft. lbs. (19 Nm). Remove the service bolt from the assembly.

9. Check and adjust valve clearance.

10. Install or connect the following:
- Cylinder head cover
- Timing belt and related components.
- Electrical connectors, cables, brackets, and components attached to the cylinder head cover.
- Spark plug wires
- Negative battery cable

1MZ-FE Engine

1. Before servicing the vehicle, refer to the precautions in the beginning of this section.

2. Remove or disconnect the following:
- Timing belt and idler pulley
- Camshaft timing pulleys
- Cylinder head covers

➡**The thrust clearance on both the intake and exhaust camshafts is very small; the camshafts must be kept level during removal. If the camshafts are removed without being kept level, the camshaft may be caught in the cylinder head, causing the head to break or the camshaft to seize.**

3. Remove the exhaust and intake camshafts from the right side cylinder head, as follows:

a. Turn the camshaft with a wrench until the 2 pointed marks on the drive and driven gears are aligned. (The right camshaft gears have 2 marks apiece; the left side camshaft gears have 1 mark each.)

b. Secure the exhaust camshaft sub-gear to the main gear using a service bolt. A bolt 0.63–0.79 in. (16–20mm) long with a 6mm thread diameter and a 1mm pitch is recommended. When removing the exhaust camshaft be sure the sub-gear is not loaded; all the force must be eliminated.

c. Uniformly loosen and remove the exhaust camshaft bearing cap bolts in several passes and in the proper sequence. Remove the 8 bearing cap bolts and remove the caps, keeping them in the correct order.

d. Remove the exhaust camshaft from the engine.

at 10–45 degrees BTDC of camshaft angle on the cylinder head.

g. Apply multipurpose grease to the camshaft.

h. Position the exhaust camshaft gear with the intake camshaft gear so that the timing marks are in alignment with one another. Be sure to use the proper alignment marks on the gears. Do not use the assembly reference marks.

i. Turn the intake camshaft clockwise or counterclockwise little by little until

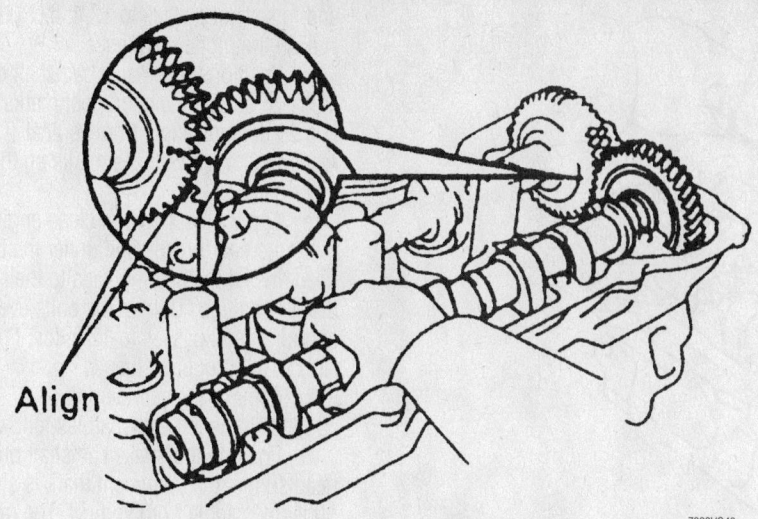

Aligning the camshaft gear timing marks for the right camshafts—3.0L (1MZ-FE) engine

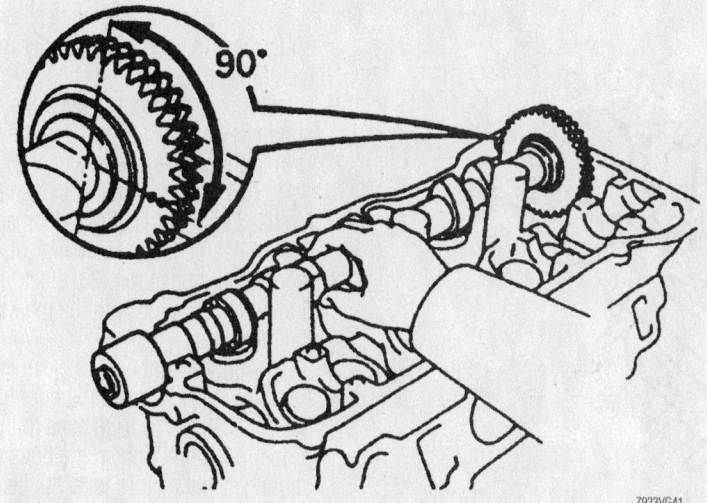

Camshaft installation for the right exhaust camshaft—3.0L (1MZ-FE) engine

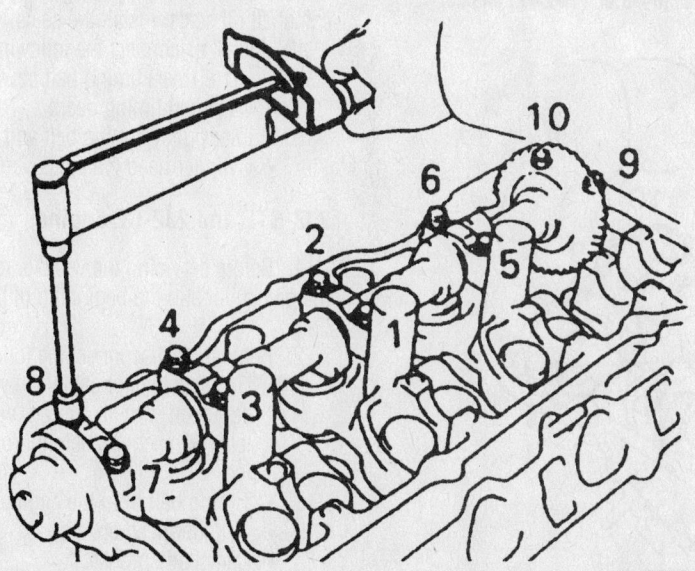

Bearing cap bolt tightening sequence for the right exhaust camshaft—3.0L (1MZ-FE) engine

e. Uniformly loosen and remove the 10 bearing cap bolts in several passes, in the proper sequence. Remove the bearing caps, keeping them in order, remove the oil seal, then, lift out the intake camshaft.

4. Remove the exhaust and intake camshafts from the left side cylinder head, as follows:

a. Turn the camshaft with a wrench until the pointed marks on the drive and driven gears are aligned. (The right camshaft gears have 2 marks apiece; the left side camshaft gears have 1 mark each.)

b. Secure the exhaust camshaft sub-gear to the main gear using a service bolt. A bolt 0.63–0.79 in. (16–20mm) long with a 6mm thread diameter and a 1mm pitch is recommended. When removing the exhaust camshaft be sure the sub-gear is not loaded; all the force must be eliminated.

c. Uniformly loosen and remove the exhaust camshaft bearing cap bolts in several passes and in the proper sequence. Remove the 8 bearing cap bolts and remove the caps. Keep the caps in the correct order.

d. Remove the exhaust camshaft from the engine.

e. Uniformly loosen and remove the 10 bearing cap bolts in several passes, in the reverse order of the installation sequence. Remove the bearing caps, keeping them in order, remove the oil seal, then lift out the intake camshaft.

5. Remove the valve lifter shims and hydraulic lifters. If the lifters are to be reused, store them upside down in a sealed container.

To install:

6. Install the valve lifters into their original positions and shims. Check the valve clearance and replace the shims as necessary.

➡Before installing the camshafts in either cylinder head, apply multi-purpose grease to each camshaft.

7. Install the right camshafts, as follows:

a. Position the intake camshaft on the head so that the alignment marks are at a 90 degrees angle from vertical. The mark should be at the "3 o'clock" position.

b. Apply sealant to the No. 1 bearing cap.

c. Apply a light coat of clean engine oil to the bolt threads and under the bolt head. Install the bearing caps to their proper position. Tighten the bolts evenly

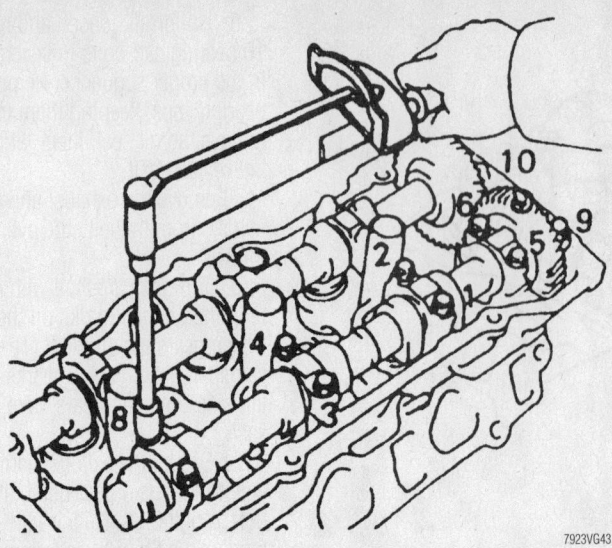

7923VG43

Bearing cap bolt tightening sequence for the right intake camshaft—3.0L (1MZ-FE) engine

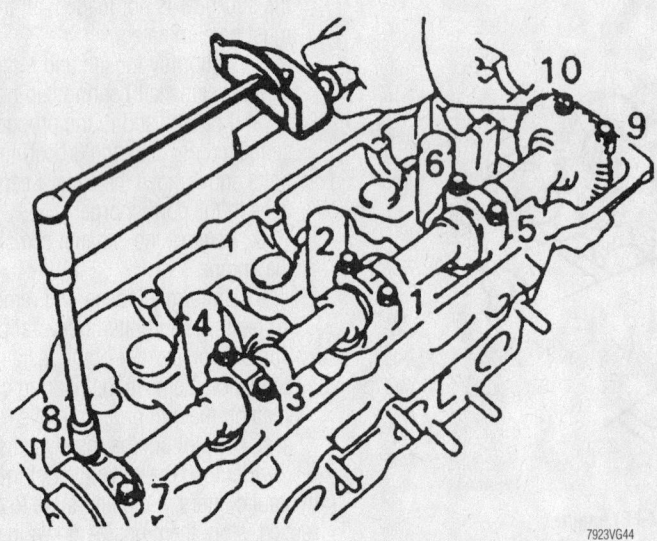

7923VG44

Bearing cap bolt tightening sequence for the left exhaust camshaft—3.0L (1MZ-FE) engine

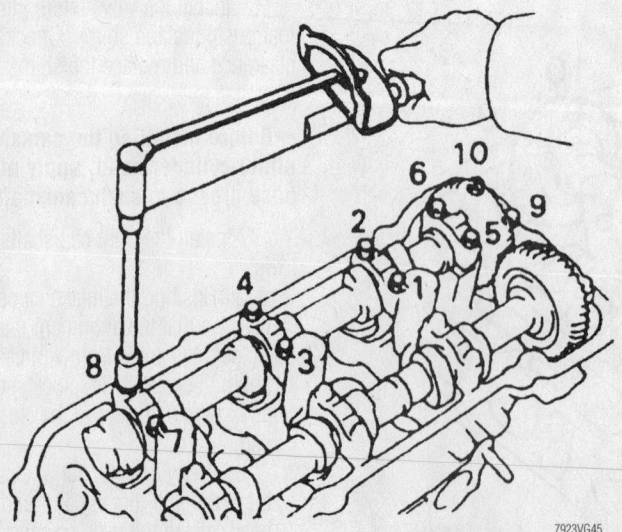

7923VG45

Bearing cap bolt tightening sequence for the left intake camshaft—3.0L (1MZ-FE) engine

and in several passes to 12 ft. lbs. (16 Nm) in the proper sequence.

d. Position the exhaust camshaft on the head so that the alignment marks are at a 90 degrees angle from vertical. The mark must align with the marks on the other gear.

e. Apply a light coat of clean engine oil to the bolt threads and under the bolt head. Install the bearing caps to their proper position. Tighten the bolts evenly and in several passes to 12 ft. lbs. (16 Nm) in the proper sequence.

f. Remove the service bolt.

8. Install the left camshaft, as follows:

a. Position the intake camshaft on the head so that the alignment mark is at a 90degrees angle from vertical. The mark should be at the "9 o'clock" position.

b. Apply sealant to the No. 1 bearing cap.

c. Apply a light coat of clean engine oil to the bolt threads and under the bolt head. Install the bearing caps to their proper position. Tighten the bolts evenly and in several passes to 12 ft. lbs. (16 Nm) in the proper sequence.

d. Position the exhaust camshaft on the head so that the alignment marks are at a 90 degree angle from vertical. The mark should be at the "3 o'clock" position and must align with the marks on the other gear.

e. Apply a light coat of clean engine oil to the bolt threads and under the bolt head. Install the bearing caps to their proper position. Tighten the bolts evenly and in several passes to 12 ft. lbs. (16 Nm) in the proper sequence.

f. Remove the service bolt.

9. Apply multi-purpose grease to new camshaft oil seals. Install the seals.

10. Install or connect the following:
- No. 3 (rear) timing belt cover
- Camshaft timing gears
- Idler pulley, timing belt and covers
- Cylinder head (valve) covers

2JZ-GTE and 2JZ-GE Engines

1. Before servicing the vehicle, refer to the precautions in the beginning of this section.

2. Remove or disconnect the following:
- Negative battery cable. On vehicles equipped with an air bag, wait at least 90 seconds before proceeding.
- Timing belt from the engine
- Cylinder head covers
- Camshaft sprocket
- No. 4 (inner) timing belt cover

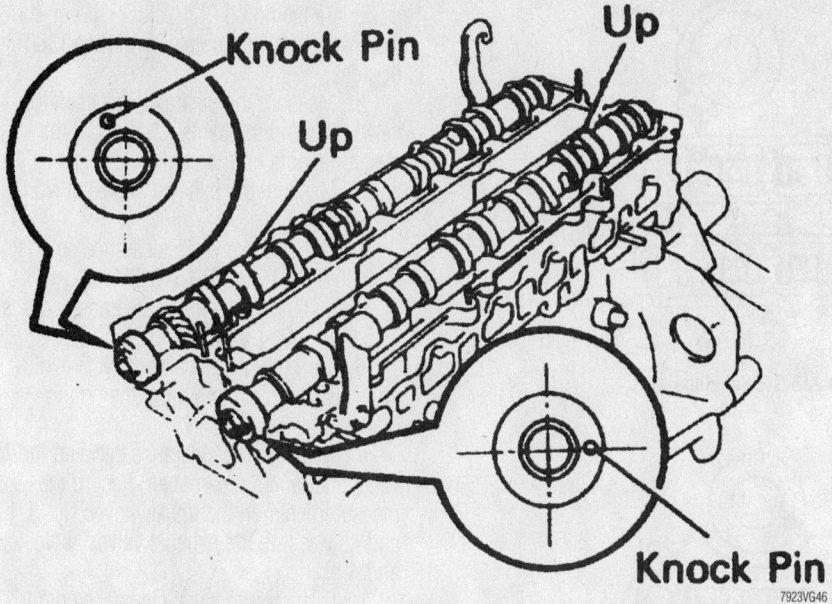

During installation, position the knock pins as shown—3.0L (2JZ-GE and 2JZ-GTE) engines

7923VG46

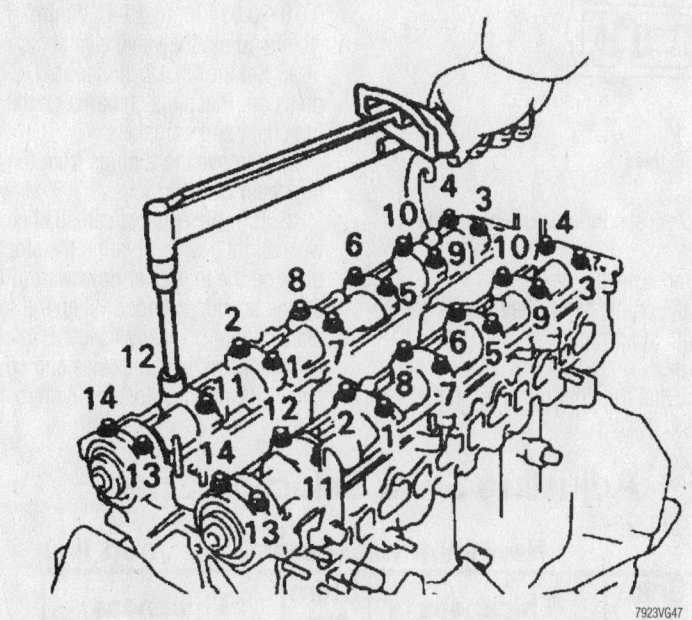

Camshaft bearing cap bolt tightening sequence—3.0L (2JZ-GE and 2JZ-GTE) engines

7923VG47

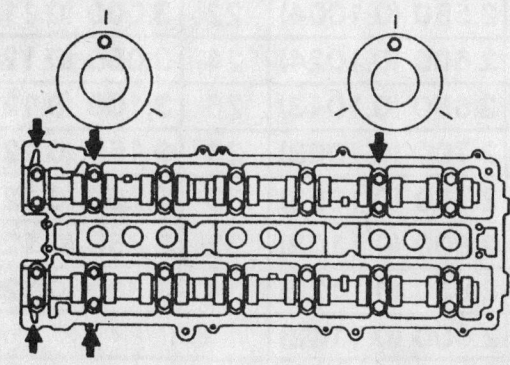

Tightening the camshafts (Step 1)—3.0L (2JZ-GE and 2JZ-GTE) engines

7923VG48

- No. 1 camshaft bearing cap bolts and bearing caps.
- All remaining bearing cap bolts. Note that there are separate sequences for the exhaust and intake camshafts. Lift off all 12 bearing caps.
- Exhaust and intake camshafts
- Valve lifter shims and hydraulic lifters. Identify each lifter and shim as it is removed so it can be reinstalled in the same position. If the lifters are to be reused, store them upside down in a sealed container.

To install:

3. Install or connect the following:
- Valve lifters into their original positions
- Shims. Check the valve clearance and replace the shims as necessary

➡ When reinstalling, remember that the camshafts must be handled carefully and kept straight and level to avoid damage.

- Camshafts, coated with engine oil
- No. 3 and No. 7 bearing caps in place
- Cap bolts, coated with oil. Then tighten them temporarily
- New oil seals, coated with multi-purpose grease, over the camshafts
- 2 No. 1 bearing caps. The mating surfaces should be coated with sealant. Install the bolts
- All remaining bearing caps
- All remaining cap bolts, coated with clean oil. Tighten them, in several passes, in the correct sequence, to 14 ft. lbs. (20 Nm). Note that there are separate sequences for the intake and exhaust sides

4. Install the oil seal, as far as it will go.

5. Rotate each camshaft until the forward straight (knock) pin is straight up. Loosen exhaust No. 1, 2 and 6 bearing cap bolts until they can be turned by hand; retighten them to 14 ft. lbs. (20 Nm). Loosen intake No. 1 and 2 bolts and retighten to 14 ft. lbs. (20 Nm).

6. Turn each camshaft 1/3 of a revolution (120 degrees). Loosen exhaust Nos. 4 and 7 bearing cap bolts; tighten them to 14 ft. lbs. (20 Nm). Loosen intake Nos. 4 and 6 bearing cap bolts; tighten them to 14 ft. lbs. (20 Nm).

7. Turn each camshaft an additional 1/3 of a revolution, loosen exhaust bearing cap bolts No. 3 and 5, then tighten them to 14 ft. lbs. (20 Nm). Loosen intake bearing cap

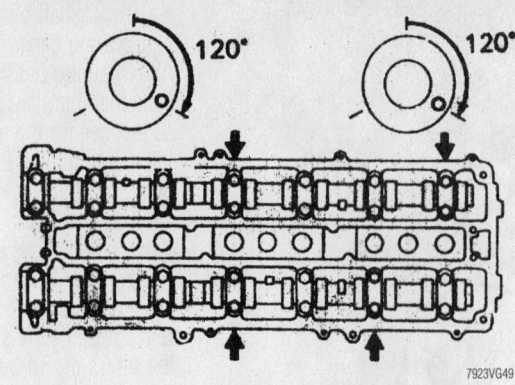

Tightening the camshafts (Step 2)—3.0L (2JZ-GE and 2JZ-GTE) engines

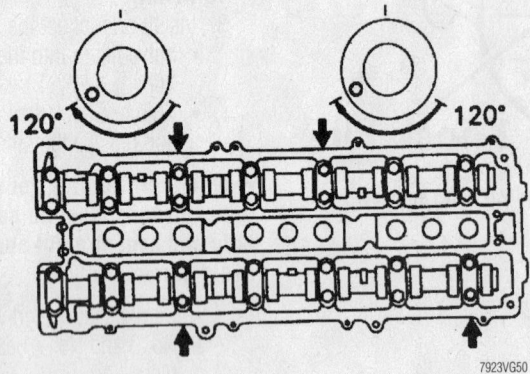

Tightening the camshafts (Step 3)—3.0L (2JZ-GE and 2JZ-GTE) engines

bolts No. 3 and 7, then tighten them to 14 ft. lbs. (20 Nm).

8. Check and adjust the valve clearance.

9. Install or connect the following:
- No 4. inside timing belt cover and the camshaft pulleys. Align the shaft pin with the pulley groove and slide the pulley on. Install the bolt temporarily. Hold the hex portion of the camshaft with a wrench; tighten the pulley bolt to 59 ft. lbs. (79 Nm).
- Cylinder head covers
- Timing belt
- Negative battery cable

Valve Lash

ADJUSTMENT

1NZ-FE Engine

➡ Adjust the valve clearance when the engine is cold.

1. Before servicing the vehicle, refer to the precautions in the beginning of this section.

2. Remove or disconnect the following:
- Negative battery cable. On vehicles equipped with an air bag, wait at least 90 seconds before proceeding.
- Cylinder head cover

3. Turn the crankshaft pulley and align its groove with the timing mark **0** of the No. 1 timing cover.

4. Check that the timing marks on the camshaft sprockets and valve timing controller are facing up (12 o'clock). If not, turn the crankshaft 1 complete revolution (360 degrees).

5. Measure the clearance between the valve lifter and the camshaft. Record the measurements on the intake valves No. 1 and 2. Measure the exhaust valves at No. 1 and 3.

 a. The intake valve clearance cold is 0.006–0.010 in. (0.15–0.25mm).

 b. The exhaust valve clearance cold is 0.010–0.014 in. (0.25–0.35mm).

6. Turn the crankshaft pulley 1 revolution (360 degrees) and align the timing mark as before.

7. Measure the clearance between the valve lifter and the camshaft. Record the measurements on the intake valves No. 3 and 4. Measure the exhaust valves at No. 2 and 4.

 a. The intake valve clearance cold is 0.006–0.010 in. (0.15–0.25mm).

 b. The exhaust valve clearance cold is 0.010–0.014 in. (0.25–0.35mm).

8. To adjust the valve clearance:

 a. Set the No.1 cylinder at TDC compression. Place matchmarks on the timing chain and sprockets.

 b. Remove the 2 plugs from the timing chain cover.

 c. Turn the exhaust camshaft clockwise slightly while rotating the stopper plate on the tensioner downward. Push in on the tension plunger. When the stopper plate cannot be easily lowered, rotate the exhaust camshaft clockwise and counterclockwise slightly. Insert a 3mm bar into

Adjusting Shim Selection Chart

Shim No.	Thickness	Shim No.	Thickness
02	2.500 (0.0984)	20	2.950 (0.1161)
04	2.550 (0.1004)	22	3.000 (0.1181)
06	2.600 (0.1024)	24	3.050 (0.1201)
08	2.650 (0.1043)	26	3.100 (0.1220)
10	2.700 (0.1063)	28	3.150 (0.1240)
12	2.750 (0.1083)	30	3.200 (0.1260)
14	2.800 (0.1102)	32	3.250 (0.1280)
16	2.850 (0.1122)	34	3.300 (0.1299)
18	2.900 (0.1142)		

New shim thickness — mm (in.)

Adjusting shim chart (intake and exhaust)—5E-FE engine

the holes in the stopper plate and tensioner to lock the tensioner. Remove the timing chain.

d. Remove the valve timing controller assembly.

e. Remove the lifters.

9. Determine the replacement adjusting shim size by either using the chart or the following formula:

- Intake: N = T + A—0.008 in. (0.20mm)
- Exhaust: N = T + A—0.012 in. (0.30mm)
- T = Thickness of removed shim
- A = Measured valve clearance
- N = Thickness of new shim

10. Install a new shim.
11. Recheck the valve clearance.
12. Install the cylinder head covers.
13. Connect the negative battery cable.

5E-FE

➡**Adjust the valve clearance when the engine is cold.**

1. Before servicing the vehicle, refer to the precautions in the beginning of this section.

2. Remove or disconnect the following:
- Negative battery cable. On vehicles equipped with an air bag, wait at

least 90 seconds before proceeding.
- Cylinder head covers

3. Turn the crankshaft pulley and align its groove with the timing mark **0** of the No. 1 timing cover.

4. Check that the timing marks on the camshaft sprockets are in alignment with the marks on the No. 4 timing cover. If not, turn the crankshaft 1 complete revolution (360 degrees).

5. Measure the clearance between the valve lifter and the camshaft. Record the measurements on the intake valves No. 1 and 2. Measure the exhaust valves at No. 1 and 3.

a. The intake valve clearance cold is 0.006–0.010 in. (0.15–0.25mm).

b. The exhaust valve clearance cold is 0.012–0.016 in. (0.31–0.41mm).

6. Turn the crankshaft pulley 1 revolution (360 degrees) and align the groove with the timing mark **0** of the No.1 timing belt cover.

7. Measure the clearance between the valve lifter and the camshaft. Record the measurements on the intake valves No. 3 and 4. Measure the exhaust valves at No. 2 and 4.

a. The intake valve clearance cold is 0.006–0.010 in. (0.15–0.25mm).

b. The exhaust valve clearance cold is 0.012–0.016 in. (0.31–0.41mm).

8. To adjust the valve clearance:

a. Remove the adjusting shim and turn the crankshaft to position the cam lobe of the camshaft on the valve to be adjusted upward.

b. Turn the valve lifter so that the notch is perpendicular to the camshaft and facing the spark plug side.

c. Using SST 09248-55040 (valve lifter press), or equivalent, hold the camshaft in place.

d. Using SST 09248-55040 (valve lifter press), or equivalent, press down the valve lifter and place SST 09248-05420 (valve lifter stopper), or equivalent between the camshaft and valve lifter.

e. Remove the SST 09248-44040 tool.

f. Using a small screwdriver and a magnetic finger, remove the adjusting shim.

9. Determine the replacement adjusting shim size by either using the chart or the following formula:

- Intake: N = T + A—0.008 in. (0.20mm)
- Exhaust: N = T + A—0.014 in. (0.36mm)
- T = Thickness of removed shim
- A = Measured valve clearance
- N = Thickness of new shim

10. Install a new shim.
11. Recheck the valve clearance.
12. Install the cylinder head covers.
13. Connect the negative battery cable.

➡**Adjust the valve clearance when the engine is cold.**

14. Before servicing the vehicle, refer to the precautions in the beginning of this section.

15. Remove or disconnect the following:
- Negative battery cable.
- Cylinder head covers

16. Turn the crankshaft pulley and align its groove with the timing mark **0** of the No. 1 timing cover.

17. Check that the timing marks on the camshaft sprockets are in alignment with the marks on the No. 4 timing cover. If not, turn the crankshaft 1 complete revolution (360 degrees).

18. Measure the clearance between the valve lifter and the camshaft. Record the measurements on the intake valves No. 1 and 2. Measure the exhaust valves at No. 1 and 3.

a. The intake valve clearance cold is 0.006–0.010 in. (0.15–0.25mm).

Adjusting Shim Selection Chart

New shim thickness   mm (in.)

Shim No.	Thickness	Shim No.	Thickness
1	2.500 (0.0984)	10	2.950 (0.1161)
2	2.550 (0.1004)	11	3.000 (0.1181)
3	2.600 (0.1024)	12	3.050 (0.1201)
4	2.650 (0.1043)	13	3.100 (0.1220)
5	2.700 (0.1063)	14	3.150 (0.1240)
6	2.750 (0.1083)	15	3.200 (0.1260)
7	2.800 (0.1102)	16	3.250 (0.1280)
8	2.850 (0.1122)	17	3.300 (0.1299)
9	2.900 (0.1142)		

HINT: New shims have the thickness in millimeters imprinted on the face.

7923VG57

Adjusting shim chart (intake and exhaust)— 5S-FE, 1MZ-FE, 2JZ-GE and 2JZ-GTE engines

1ZZ–FE: Valve Lifter Selection Chart (Intake)

New lifter thickness   mm (in.)

Lifter No.	Thickness	Lifter No.	Thickness	Lifter No.	Thickness
06	5.060 (0.1992)	30	5.300 (0.2087)	54	5.540 (0.2181)
08	5.080 (0.2000)	32	5.320 (0.2094)	56	5.560 (0.2189)
10	5.100 (0.2008)	34	5.340 (0.2102)	58	5.580 (0.2197)
12	5.120 (0.2016)	36	5.360 (0.2110)	60	5.600 (0.2205)
14	5.140 (0.2024)	38	5.380 (0.2118)	62	5.620 (0.2213)
16	5.160 (0.2031)	40	5.400 (0.2126)	64	5.640 (0.2220)
18	5.180 (0.2039)	42	5.420 (0.2134)	66	5.660 (0.2228)
20	5.200 (0.2047)	44	5.440 (0.2142)	68	5.680 (0.2236)
22	5.220 (0.2055)	46	5.460 (0.2150)	70	5.700 (0.2244)
24	5.240 (0.2063)	48	5.480 (0.2157)	72	5.720 (0.2252)
26	5.260 (0.2071)	50	5.500 (0.2165)	74	5.740 (0.2260)
28	5.280 (0.2079)	52	5.520 (0.2173)		

Intake valve clearance (Cold):
0.15 – 0.25 mm (0.006 – 0.010 in.)
EXAMPLE: The 5.250 mm (0.2067 in.) lifter is installed, and the measured clearance is 0.400 mm (0.0157 in.). Replace the 5.250 mm (0.2067 in.) lifter with a new No. 48 lifter.

(Valve Lifter Selection Chart — intake. The chart cross-references measured clearance mm (in.) against installed lifter thickness mm (in.) to yield the new lifter number. Installed lifter thickness values range 5.060 (0.1992) through 5.740 (0.2260); measured clearance rows range 0.000 – 0.030 (0.0000 – 0.0012) through 0.911 – 0.930 (0.0359 – 0.0366).)

Adjusting shim chart (intake)—1ZZ-FE engine

9307WG70

1ZZ–FE: Valve Lifter Selection Chart (Exhaust)

New lifter thickness   mm (in.)

Lifter No.	Thickness	Lifter No.	Thickness	Lifter No.	Thickness
06	5.060 (0.1992)	30	5.300 (0.2087)	54	5.540 (0.2181)
08	5.080 (0.2000)	32	5.320 (0.2094)	56	5.560 (0.2189)
10	5.100 (0.2008)	34	5.340 (0.2102)	58	5.580 (0.2197)
12	5.120 (0.2016)	36	5.360 (0.2110)	60	5.600 (0.2205)
14	5.140 (0.2024)	38	5.380 (0.2118)	62	5.620 (0.2213)
16	5.160 (0.2031)	40	5.400 (0.2126)	64	5.640 (0.2220)
18	5.180 (0.2039)	42	5.420 (0.2134)	66	5.660 (0.2228)
20	5.200 (0.2047)	44	5.440 (0.2142)	68	5.680 (0.2236)
22	5.220 (0.2055)	46	5.460 (0.2150)	70	5.700 (0.2244)
24	5.240 (0.2063)	48	5.480 (0.2157)	72	5.720 (0.2252)
26	5.260 (0.2071)	50	5.500 (0.2165)	74	5.740 (0.2260)
28	5.280 (0.2079)	52	5.520 (0.2173)		

Exhaust valve clearance (Cold):
0.25 – 0.35 mm (0.010 – 0.014 in.)
EXAMPLE: The 5.340 mm (0.2102 in.) lifter is installed, and the measured clearance is 0.440 mm (0.0173 in.).
Replace the 5.340 mm (0.2102 in.) lifter with a new No. 48 lifter.

Valve Lifter Selection Chart (Exhaust) — a triangular selection matrix. The horizontal axis is the installed lifter thickness mm (in.), ranging from 5.060 (0.1992) to 5.740 (0.2260). The vertical axis is the measured clearance mm (in.), ranging from 0.000 – 0.030 (0.0000 – 0.0012) to 1.011 – 1.030 (0.0398 – 0.0406). Intersection cells give the new lifter number (06–74).

Adjusting shim chart (exhaust)—1ZZ-FE engine

9307WG71

2ZZ–GE: Valve Shim Selection Chart (Intake)

Adjusting shim chart (intake)—2ZZ-GE engine

New Shim thickness mm (in.)

Shim No.	Thickness	Shim No.	Thickness	Shim No.	Thickness
00	2.000 (0.0787)	28	2.280 (0.0898)	56	2.560 (0.1008)
02	2.020 (0.0795)	30	2.300 (0.0906)	58	2.580 (0.1016)
04	2.040 (0.0803)	32	2.320 (0.0913)	60	2.600 (0.1024)
06	2.060 (0.0811)	34	2.340 (0.0921)	62	2.620 (0.1031)
08	2.080 (0.0819)	36	2.360 (0.0929)	64	2.640 (0.1039)
10	2.100 (0.0827)	38	2.380 (0.0937)	66	2.660 (0.1047)
12	2.120 (0.0835)	40	2.400 (0.0945)	68	2.680 (0.1055)
14	2.140 (0.0843)	42	2.420 (0.0953)	70	2.700 (0.1063)
16	2.160 (0.0850)	44	2.440 (0.0961)	72	2.720 (0.1071)
18	2.180 (0.0858)	46	2.460 (0.0969)	74	2.740 (0.1079)
20	2.200 (0.0866)	48	2.480 (0.0976)	76	2.760 (0.1087)
22	2.220 (0.0874)	50	2.500 (0.0984)	78	2.780 (0.1094)
24	2.240 (0.0882)	52	2.520 (0.0992)	80	2.800 (0.1102)
26	2.260 (0.0890)	54	2.540 (0.1000)		

Intake valve clearance (Cold):
0.15 – 0.25 mm (0.006 – 0.010 in.)
EXAMPLE: The 2.200 mm (0.0826 in.) shim is installed, and the measured clearance is 0.400 mm (0.0157 in.).
Replace the 2.400 mm (0.0945 in.) shim with a new No. 40 shim.

9307WG72

2ZZ–GE: Valve Shim Selection Chart (Exhaust)

New Shim thickness — mm (in.)

Shim No.	Thickness	Shim No.	Thickness	Shim No.	Thickness
00	2.000 (0.0787)	28	2.280 (0.0898)	56	2.560 (0.1008)
02	2.020 (0.0795)	30	2.300 (0.0906)	58	2.580 (0.1016)
04	2.040 (0.0803)	32	2.320 (0.0913)	60	2.600 (0.1024)
06	2.060 (0.0811)	34	2.340 (0.0921)	62	2.620 (0.1031)
08	2.080 (0.0819)	36	2.360 (0.0929)	64	2.640 (0.1039)
10	2.100 (0.0827)	38	2.380 (0.0937)	66	2.660 (0.1047)
12	2.120 (0.0835)	40	2.400 (0.0945)	68	2.680 (0.1055)
14	2.140 (0.0843)	42	2.420 (0.0953)	70	2.700 (0.1063)
16	2.160 (0.0850)	44	2.440 (0.0961)	72	2.720 (0.1071)
18	2.180 (0.0858)	46	2.460 (0.0969)	74	2.740 (0.1079)
20	2.200 (0.0866)	48	2.480 (0.0976)	76	2.760 (0.1087)
22	2.220 (0.0874)	50	2.500 (0.0984)	78	2.780 (0.1094)
24	2.240 (0.0882)	52	2.520 (0.0992)	80	2.800 (0.1102)
26	2.260 (0.0890)	54	2.540 (0.1000)		

Exhaust valve clearance (Cold):
0.35 – 0.45 mm (0.014 – 0.018 in.)

EXAMPLE: The 2.200 mm (0.0862 in.) shim is installed, and the measured clearance is 0.500 mm (0.0197 in.). Replace the 2.300 mm (0.0906 in.) shim with a new No. 30 shim.

The selection chart is a large matrix. The top axis lists the installed lifter thickness (new shim thickness) columns from 2.000 (0.0787) to 2.800 (0.1102) mm (in.). The left axis lists the measured clearance ranges as follows, and the body cells give the replacement shim numbers (00–80):

Measure clearance — mm (in.)

Measure clearance mm (in.)
0.000–0.030 (0.0000–0.0012)
0.031–0.050 (0.0012–0.0020)
0.051–0.070 (0.0020–0.0028)
0.071–0.090 (0.0028–0.0035)
0.091–0.110 (0.0036–0.0043)
0.111–0.130 (0.0044–0.0051)
0.131–0.150 (0.0052–0.0059)
0.151–0.170 (0.0059–0.0067)
0.171–0.190 (0.0067–0.0075)
0.191–0.210 (0.0075–0.0083)
0.211–0.230 (0.0083–0.0091)
0.231–0.250 (0.0091–0.0098)
0.251–0.270 (0.0099–0.0106)
0.271–0.290 (0.0107–0.0114)
0.291–0.310 (0.0115–0.0122)
0.311–0.330 (0.0122–0.0130)
0.331–0.349 (0.0130–0.0137)
0.350–0.450 (0.0138–0.0177)
0.451–0.470 (0.0178–0.0185)
0.471–0.490 (0.0185–0.0193)
0.491–0.510 (0.0193–0.0201)
0.511–0.530 (0.0201–0.0209)
0.531–0.550 (0.0209–0.0217)
0.551–0.570 (0.0217–0.0224)
0.571–0.590 (0.0225–0.0232)
0.591–0.610 (0.0232–0.0240)
0.611–0.630 (0.0241–0.0248)
0.631–0.650 (0.0248–0.0256)
0.651–0.670 (0.0256–0.0264)
0.671–0.690 (0.0264–0.0272)
0.691–0.710 (0.0272–0.0280)
0.711–0.730 (0.0280–0.0287)
0.731–0.750 (0.0288–0.0295)
0.751–0.770 (0.0296–0.0303)
0.771–0.790 (0.0304–0.0311)
0.791–0.810 (0.0311–0.0319)
0.811–0.830 (0.0319–0.0327)
0.831–0.850 (0.0327–0.0335)
0.851–0.870 (0.0335–0.0343)
0.871–0.890 (0.0343–0.0350)
0.891–0.910 (0.0351–0.0358)
0.911–0.930 (0.0359–0.0366)
0.931–0.950 (0.0367–0.0374)
0.951–0.970 (0.0374–0.0382)
0.971–0.990 (0.0382–0.0390)
0.991–1.010 (0.0390–0.0398)
1.011–1.030 (0.0398–0.0406)
1.031–1.050 (0.0406–0.0413)
1.051–1.070 (0.0414–0.0421)
1.071–1.090 (0.0422–0.0429)
1.091–1.110 (0.0430–0.0437)
1.111–1.130 (0.0437–0.0445)
1.131–1.150 (0.0445–0.0453)
1.151–1.170 (0.0453–0.0461)
1.171–1.190 (0.0461–0.0469)
1.191–1.210 (0.0469–0.0476)
1.211–1.230 (0.0477–0.0484)
1.231–1.250 (0.0485–0.0492)

Installed lifter thickness mm (in.) columns (top axis): 2.000 (0.0787), 2.020 (0.0795), 2.040 (0.0803), 2.060 (0.0811), 2.080 (0.0819), 2.100 (0.0827), 2.120 (0.0835), 2.140 (0.0843), 2.160 (0.0850), 2.180 (0.0858), 2.200 (0.0866), 2.210 (0.0870), 2.220 (0.0874), 2.230 (0.0878), 2.240 (0.0882), 2.250 (0.0886), 2.260 (0.0890), 2.270 (0.0894), 2.280 (0.0898), 2.290 (0.0902), 2.300 (0.0906), 2.310 (0.0909), 2.320 (0.0913), 2.330 (0.0917), 2.340 (0.0921), 2.350 (0.0925), 2.360 (0.0929), 2.370 (0.0933), 2.380 (0.0937), 2.390 (0.0941), 2.400 (0.0945), 2.410 (0.0949), 2.420 (0.0953), 2.430 (0.0957), 2.440 (0.0961), 2.450 (0.0965), 2.460 (0.0969), 2.470 (0.0972), 2.480 (0.0976), 2.490 (0.0980), 2.500 (0.0984), 2.510 (0.0988), 2.520 (0.0992), 2.530 (0.0996), 2.540 (0.1000), 2.550 (0.1004), 2.560 (0.1008), 2.580 (0.1016), 2.600 (0.1024), 2.620 (0.1031), 2.640 (0.1039), 2.660 (0.1047), 2.680 (0.1055), 2.700 (0.1063), 2.720 (0.1071), 2.740 (0.1079), 2.760 (0.1087), 2.780 (0.1094), 2.800 (0.1102).

Adjusting shim chart (exhaust)—2ZZ-GE engine

9307WG73

Adjusting Shim Selection Chart

New shim thickness mm (in.)

Shim No.	Thickness	Shim No.	Thickness
02	2.500 (0.0984)	20	2.950 (0.1161)
04	2.550 (0.1004)	22	3.000 (0.1181)
06	2.600 (0.1024)	24	3.050 (0.1201)
08	2.650 (0.1043)	26	3.100 (0.1220)
10	2.700 (0.1063)	28	3.150 (0.1240)
12	2.750 (0.1083)	30	3.200 (0.1260)
14	2.800 (0.1102)	32	3.250 (0.1280)
16	2.850 (0.1122)	34	3.300 (0.1299)
18	2.900 (0.1142)		

7923VG56

Adjusting shim chart (intake and exhaust)—1NZ-FE engine

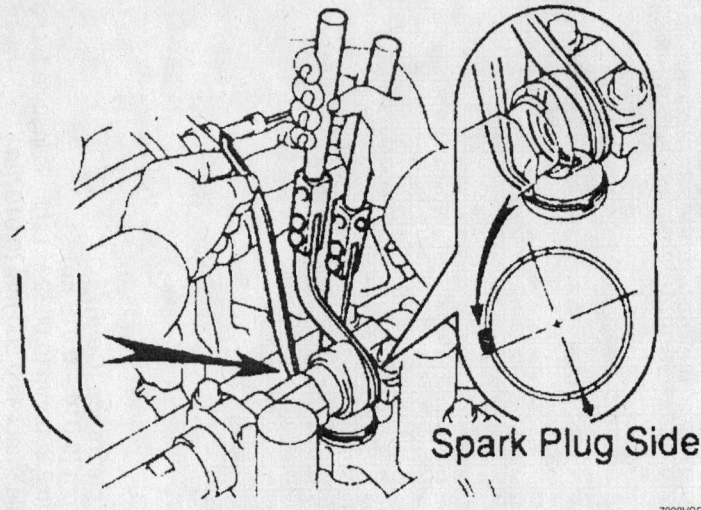

Spark Plug Side

7923VG58

Common method of removing valve shims

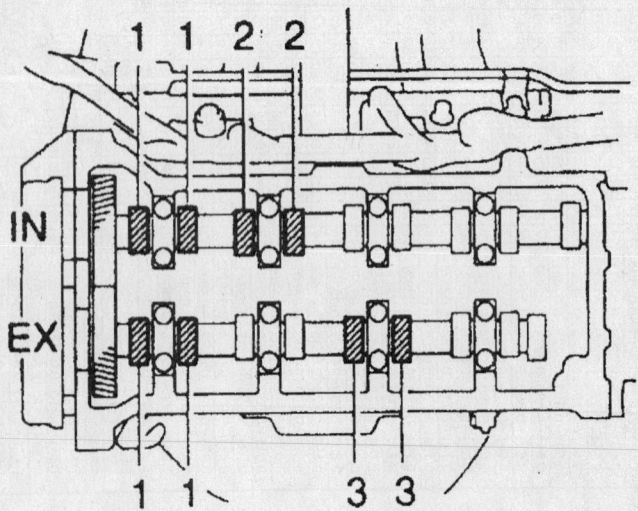

7923VG59

Intake valves (1 and 2) and exhaust valves (1 and 3)—1NZ-FE engine

b. The exhaust valve clearance cold is 0.010–0.014 in. (0.25–0.35mm).

19. Turn the crankshaft pulley 1 revolution (360 degrees) and align the groove with the timing mark **0** of the No.1 timing belt cover.

20. Measure the clearance between the valve lifter and the camshaft. Record the measurements on the intake valves No. 3 and 4. Measure the exhaust valves at No. 2 and 4.

    a. The intake valve clearance cold is 0.006–0.010 in. (0.15–0.25mm).

    b. The exhaust valve clearance cold is 0.010–0.014 in. (0.25–0.35mm).

21. To adjust the intake valve clearance:

    a. Remove the intake camshaft.

    b. Using a small screwdriver and a magnetic finger, remove the adjusting shim.

    c. Determine the replacement adjusting shim size by either using the chart or the following formula:

- Intake: $N = T + A - 0.008$ in. (0.20mm)
- T = Thickness of removed shim
- A = Measured valve clearance
- N = Thickness of new shim

    d. Install a new shim.

    e. Install intake camshaft.

    f. Recheck the valve clearance.

22. To adjust the exhaust valve clearance:

    a. Turn the crankshaft to position the cam lobe of the camshaft on the valve to be adjusted, upward.

    b. Turn the valve lifter so that the notch is perpendicular to the camshaft and facing the spark plug side.

    c. Using SST 09248–55040 (valve lifter press), or equivalent, hold the camshaft in place.

    d. Using SST 09248–55040 (valve lifter press), or equivalent, press down the valve lifter and place SST 09248–05420 (valve lifter stopper), or equivalent between the camshaft and valve lifter.

    e. Remove the SST 09248–44040 tool.

    f. Using a small screwdriver and a magnetic finger, remove the adjusting shim.

23. Determine the replacement adjusting shim size by either using the chart or the following formula:

- Exhaust: $N = T + A - 0.014$ in. (0.36mm)
- T = Thickness of removed shim
- A = Measured valve clearance
- N = Thickness of new shim

24. Install a new shim.

25. Recheck the valve clearance.

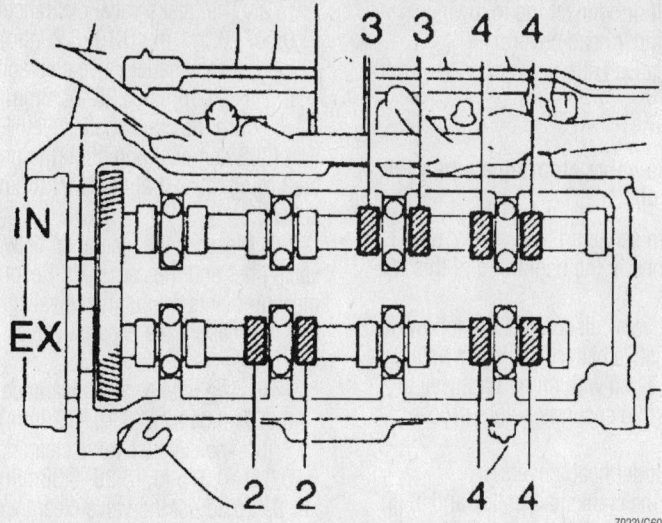

Intake valves (3 and 4) and exhaust valves (2 and 4)—5E-FE engine

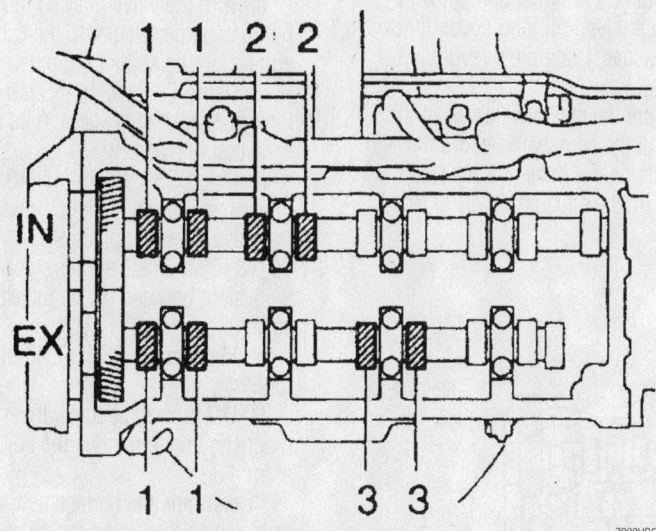

Intake valves (1 and 2) and exhaust valves (1 and 3)—5E-FE engine

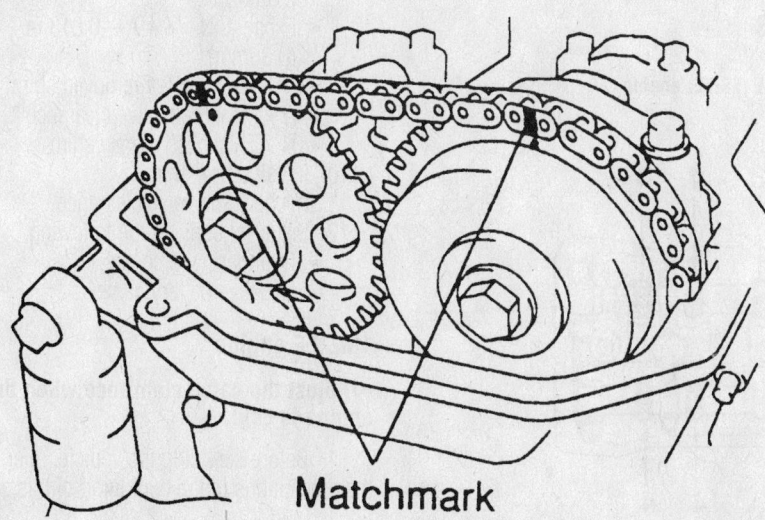

Timing marks at TDC compression—1ZZ-FE engine

26. Install or connect the following:
   • Cylinder head covers
   • Negative battery cable

2ZZ-GE Engine

➡**Adjust the valve clearance when the engine is cold.**

1. Before servicing the vehicle, refer to the precautions in the beginning of this section.

2. Remove or disconnect the following:
   • Negative battery cable.
   • Cylinder head covers

3. Turn the crankshaft pulley and align its groove with the timing mark **0** of the No. 1 timing cover.

4. Check that the timing marks on the camshaft sprockets are in alignment with the upper edge of the timing cover. If not, turn the crankshaft 1 complete revolution (360 degrees).

5. Measure the clearance between the valve lifter and the camshaft. Record the measurements on the intake valves No. 1 and 2. Measure the exhaust valves at No. 1 and 3.

   a. The intake valve clearance cold is 0.006–0.010 in. (0.15–0.25mm).

   b. The exhaust valve clearance cold is 0.014–0.018 in. (0.36–0.45mm).

6. Turn the crankshaft pulley 1 revolution (360 degrees) and align the groove with the timing mark **0** of the No.1 timing belt cover.

7. Measure the clearance between the valve lifter and the camshaft. Record the measurements on the intake valves No. 3 and 4. Measure the exhaust valves at No. 2 and 4.

   a. The intake valve clearance cold is 0.006–0.010 in. (0.15–0.25mm).

   b. The exhaust valve clearance cold is 0.014–0.018 in. (0.36–0.45mm).

8. To adjust the intake valve clearance:

   a. Remove the intake camshaft.

   b. Using a small screwdriver and a magnetic finger, remove the adjusting shim.

   c. Determine the replacement adjusting shim size by either using the chart or the following formula:

   • Intake: $N = T + A - 0.008$ in. (0.20mm)
   • T = Thickness of removed shim
   • A = Measured valve clearance
   • N = Thickness of new shim

   d. Install a new shim.

   e. Install intake camshaft.

   f. Recheck the valve clearance.

9. To adjust the exhaust valve clearance:

a. Turn the crankshaft to position the cam lobe of the camshaft on the valve to be adjusted, upward.

b. Turn the valve lifter so that the notch is perpendicular to the camshaft and facing the spark plug side.

c. Using SST 09248–55040 (valve lifter press), or equivalent, hold the camshaft in place.

d. Using SST 09248–55040 (valve lifter press), or equivalent, press down the valve lifter and place SST 09248–05420 (valve lifter stopper), or equivalent between the camshaft and valve lifter.

e. Remove the SST 09248–44040 tool.

f. Using a small screwdriver and a magnetic finger, remove the adjusting shim.

10. Determine the replacement adjusting shim size by either using the chart or the following formula:

- Exhaust: N = T + A—0.014 in. (0.36mm)
- T = Thickness of removed shim
- A = Measured valve clearance
- N = Thickness of new shim

11. Install a new shim.

12. Recheck the valve clearance.

Intake valves (1 and 2) and exhaust valves (1 and 3)—2.2L (5S-FE) engine

Intake valves (3 and 4) and exhaust valves (2 and 4)—2.2L (5S-FE) engine

13. Install or connect the following:
- Cylinder head covers
- Negative battery cable

5S-FE Engine

➡ **Adjust the valve clearance when the engine is cold.**

1. Before servicing the vehicle, refer to the precautions in the beginning of this section.

2. Remove or disconnect the following:
- Negative battery cable. On vehicles equipped with an air bag, wait at least 90 seconds before proceeding.
- Cylinder head covers

3. Turn the crankshaft pulley and align its groove with the timing mark **0** of the No. 1 timing cover.

4. Check that the timing marks on the camshaft sprockets are in alignment with the marks on the No. 4 timing cover. If not, turn the crankshaft 1 complete revolution (360 degrees).

5. Measure the clearance between the valve lifter and the camshaft. Record the measurements on the intake valves No. 1 and 2. Measure the exhaust valves at No. 1 and 3.

a. The intake valve clearance cold is 0.007–0.011 in. (0.19–0.29mm).

b. The exhaust valve clearance cold is 0.011–0.015 in. (0.28–0.38mm).

6. Turn the crankshaft pulley 1 revolution (360 degrees) and align the groove with the timing mark **0** of the No.1 timing belt cover.

7. Measure the clearance between the valve lifter and the camshaft. Record the measurements on the intake valves No. 3 and 4. Measure the exhaust valves at 2 and 4.

a. The intake valve clearance cold is 0.007–0.011 in. (0.19–0.29mm).

b. The exhaust valve clearance cold is 0.011–0.015 in. (0.29–0.38mm).

8. To adjust the valve clearance:

a. Turn the crankshaft to position the cam lobe of the camshaft on the valve to be adjusted, upward.

b. Turn the valve lifter so that the notch is perpendicular to the camshaft and facing the spark plug side.

c. Using SST 09248–55040 (valve lifter press), or equivalent, hold the camshaft in place.

d. Using SST 09248–55040 (valve lifter press), or equivalent, press down the valve lifter and place SST 09248–05420 (valve lifter stopper), or equivalent between the camshaft and valve lifter.

e. Remove the SST 09248–44040 tool.

f. Using a small screwdriver and a magnetic finger, remove the adjusting shim.

9. Determine the replacement adjusting shim size by either using the chart or the following formula:

- Intake: N = T + A—0.009 in. (24mm)
- Exhaust: N = T + A—0.013 in. (0.33mm)
- T = Thickness of removed shim
- A = Measured valve clearance
- N = Thickness of new shim

10. Install a new shim.

11. Recheck the valve clearance.

12. Install or connect the following:
- Cylinder head covers
- Negative battery cable

1MZ-FE engine

➡ **Adjust the valve clearance when the engine is cold.**

1. Before servicing the vehicle, refer to the precautions in the beginning of this section.

2. Remove or disconnect the following:

- Negative battery cable. On vehicles equipped with an air bag, wait at least 90 seconds before proceeding.
- Accelerator/throttle cable from the throttle linkage
- Air cleaner cover, air flow meter, and air duct assembly
- V-bank cover
- Emission control valve set
- Air intake chamber
- Engine harness from the injectors and the ignition coils
- Ignition coils and keep them in order for reassembly
- Spark plugs
- Cylinder head covers

3. Turn the crankshaft pulley and align its groove with the timing mark **0** of the No. 1 timing cover.

4. Check that the valve lifters on the No. 1 intake are loose and the No. 1 exhaust are tight. If not, turn the crankshaft 1 complete revolution (360 degrees).

➡ **All measurements should be written down. These recorded measurements will need to be used in conjunction with a mathematical formula to determine the thickness of the replacement shims.**

5. Measure the clearance between the valve lifters and the camshaft. Record the measurements on valves No. 1 and 6 intake; No. 2 and 3 exhaust.

    a. The intake valve clearance cold is 0.006–0.010 in. (0.15–0.25mm).

    b. The exhaust valve clearance cold is 0.010–0.014 in. (0.25–0.35mm).

6. Turn the crankshaft ⅔ of a revolution (240 degrees). Record the measurements on valves No. 2 and 3 intake; No. 4 and 5 exhaust.

7. Turn the crankshaft another ⅔ of a revolution. Record the measurements on valves No. 4 and 5 intake; No. 1 and 6 exhaust.

8. Remove the adjusting shim by turning the crankshaft to position the cam lobe of the camshaft in the up position on the valve to be adjusted. Using a small thin flat bladed tool, turn the valve lifter so that the notches are perpendicular to the camshaft. Press down the valve lifter with SST 09248–55010 part A, or equivalent. Place SST 09248–55010 part B between the camshaft and the valve lifter; remove part A.

9. Remove the adjusting shim with a magnet and a small screwdriver.

10. Determine the replacement adjusting shim size by either using the charts or the following formulas:

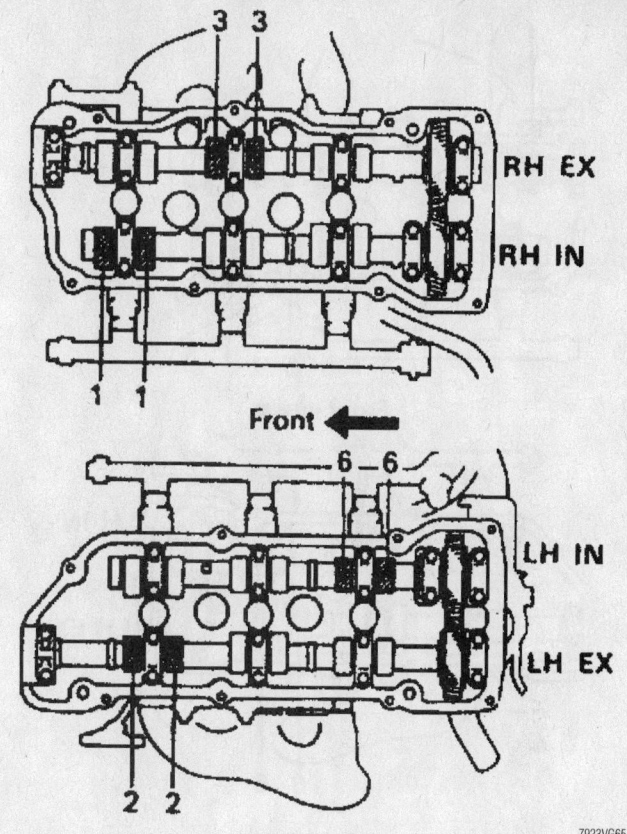

Adjust these valves during the 1st step—3.0L (1MZ-FE) engine

7923VG65

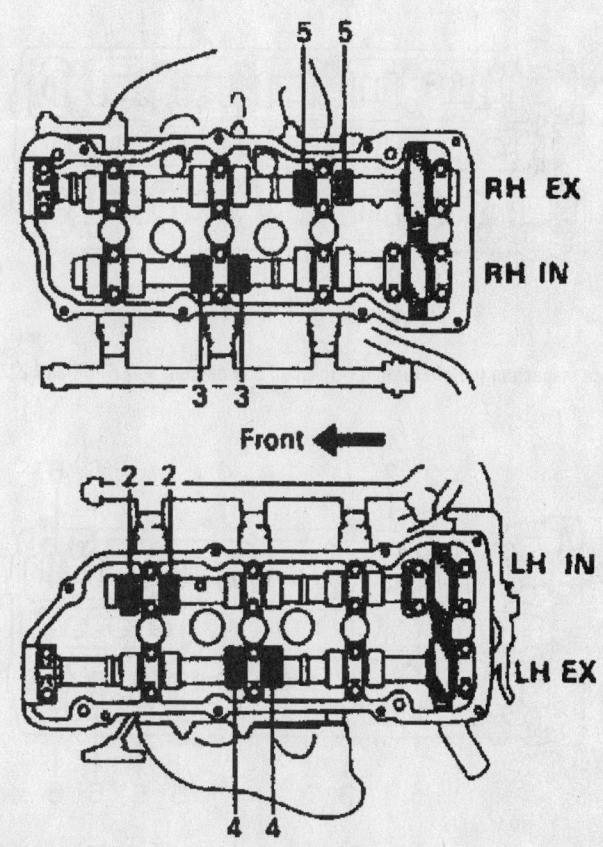

Adjust these valves during the 2nd step—3.0L (1MZ-FE) engine

7923VG66

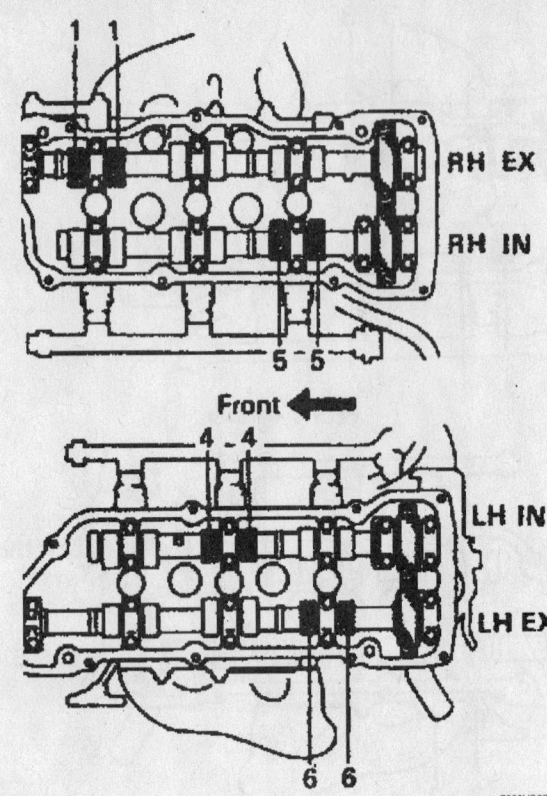

Adjust these valves during the 3rd step—3.0L (1MZ-FE) engine

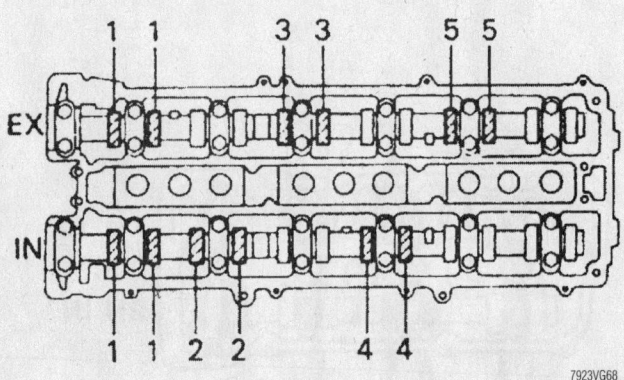

Valve clearance inspection (before turning crankshaft 360 degrees)—2JZ-GE and 2JZ-GTE engines

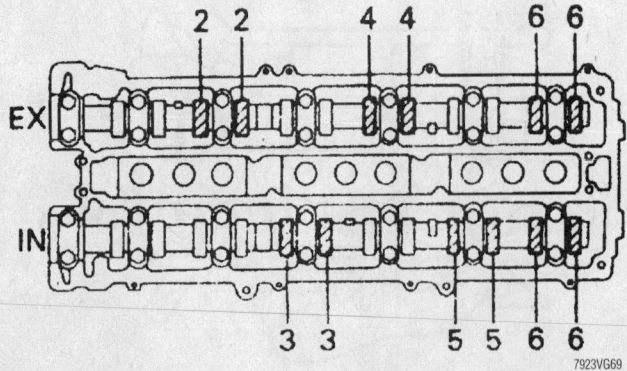

Valve clearance inspection (after turning crankshaft 360 degrees)—2JZ-GE and 2JZ-GTE engines

- Intake: $N = T + (A - 0.008$ in./0.020mm)
- Exhaust: $N = T + (A - 0.012$ in./0.30mm)
- T = Thickness of removed shim
- A = Measured valve clearance
- N = Thickness of new shim

11. Select a new shim with a thickness as close as possible to the calculated value. Install the new replacement shim.

➡ **Shims are available in 17 sizes in increments of 0.0020 in. (0.050mm), from 0.0984 in. (2.500mm) to 0.1299 in. (3.300mm).**

12. Recheck the valve clearance.
13. Install or connect the following:
- Cylinder head covers
- Spark plugs and the ignition coils
- Engine wiring harness to the injectors and the coils
- Intake chamber
- Emission control valve set
- V-bank cover
- Air flow meter, air duct, and air cleaner cover
- Negative battery cable

2JZ-GE and 2JZ-GTE Engines

➡ **Adjust the valve clearance when the engine is cold.**

1. Before servicing the vehicle, refer to the precautions in the beginning of this section.
2. Remove or disconnect the following:
- Negative battery cable. On vehicles equipped with an air bag, wait at least 90 seconds before proceeding.
- Cylinder head covers

3. Turn the crankshaft pulley and align its groove with the timing mark **0** of the No. 1 timing cover.
4. Check that the timing marks on the camshaft sprockets are in alignment with the marks on the No. 4 timing cover. If not, turn the crankshaft 1 complete revolution (360 degrees).
5. Measure the clearance between the valve lifter and the camshaft. Record the measurements on the intake valves No. 1, 2 and 4. Measure the exhaust valves at No. 1, 3 and 5.
   a. The intake valve clearance cold is 0.006–0.010 in. (0.15–0.25mm).
   b. The exhaust valve clearance cold is 0.010–0.014 in. (0.25–0.35mm).
6. Turn the crankshaft pulley 1 revolution (360 degrees) and align the groove with the timing mark **0** of the No.1 timing belt cover.

7. Measure the clearance between the valve lifter and the camshaft. Record the measurements on the intake valves No. 3, 5 and 6. Measure the exhaust valves at No. 2, 4, and 6.

    a. The intake valve clearance cold is 0.006–0.010 in. (0.15–0.25mm).

    b. The exhaust valve clearance cold is 0.010–0.014 in. (0.25–0.35mm).

8. To adjust the valve clearance:

    a. Remove the adjusting shim and turn the crankshaft to position the cam lobe of the camshaft on the adjusting valve upward.

    b. Turn the valve lifter so that the notches are perpendicular to the camshaft.

    c. Using SST 09248–55040 (valve lifter press), or equivalent, hold the camshaft in place.

    d. Using SST 09248–55040 (valve lifter press), or equivalent, press down the valve lifter and place SST 09248–05420 (valve lifter stopper), or equivalent between the camshaft and valve lifter.

    e. Remove the SST 09248–44040 tool.

    f. Using a small screwdriver and a magnetic finger, remove the adjusting shim.

9. Determine the replacement adjusting shim size by either using the chart or the following formula:

- Intake: $N = T + (A — 0.008$ in./0.20mm)
- Exhaust: $N = T + (A — 0.012$ in./0.30mm)
- T = Thickness of removed shim
- A = Measured valve clearance
- N = Thickness of new shim

10. Recheck the valve clearance.

11. Install or connect the following:
- Cylinder head covers
- Negative battery cable

Starter

REMOVAL & INSTALLATION

Echo

1. Before servicing the vehicle, refer to the precautions in the beginning of this section.

2. Remove or disconnect the following:
- Negative battery cable. On models equipped with an air bag, work must NOT be started until at least 90 seconds have passed from the time that both the ignition switch is turned to the LOCK position and the negative cable is disconnected from the battery
- Engine under covers
- Wiring from the starter terminals
- Starter mounting bolts
- Starter

To install:

3. Install or connect the following:
- Starter
- Starter wiring
- Engine under covers

Avalon and Camry

1. Before servicing the vehicle, refer to the precautions in the beginning of this section.

2. Remove or disconnect the following:
- Negative battery cable. On models equipped with an air bag, work must NOT be started until at least 90 seconds have passed from the time that both the ignition switch is turned to the LOCK position and the negative cable is disconnected from the battery.
- With cruise control, the battery
- With cruise control, the actuator and bracket
- Wiring from the starter
- Starter

To install:

3. Install or connect the following:
- Starter. Torque the bolts to 27 ft. lbs. (37 Nm).
- Starter cable
- Electrical connectors
- Actuator connector and clamp, if equipped with cruise control
- Battery and tray, if equipped with cruise control

Corolla

1. Before servicing the vehicle, refer to the precautions in the beginning of this section.

2. Remove or disconnect the following:
- Battery and tray. On models equipped with an air bag, work must NOT be started until at least 90 seconds have passed from the time that both the ignition switch is turned to the LOCK position and the negative cable is disconnected from the battery.
- Coolant reservoir
- Reservoir hose from the radiator
- Right side under cover
- Wiring clamp
- Wiring from the starter
- Starter

To install:

3. Install or connect the following:
- Starter
- Starter wiring and nut
- Starter connector and wire clamp
- Right side engine cover
- Radiator reservoir and hose
- Battery and tray

4. Check the cooling system and top off if necessary.

Supra

1. Before servicing the vehicle, refer to the precautions in the beginning of this section.

2. Remove or disconnect the following:
- Negative battery cable. On models equipped with an air bag, work must NOT be started until at least 90 seconds have passed from the time that both the ignition switch is turned to the LOCK position and the negative cable is disconnected from the battery.
- Air cleaner assembly
- On the 5S-FE, the battery if necessary
- With cruise control, the cruise control actuator from the body bracket
- The starter wiring
- The starter bolts
- The oxygen sensor wiring harness from the engine wire brackets and starter
- The starter

To install:

3. Install or connect the following:
- Starter and oxygen sensor wiring and brackets. Bolts: 29 ft. lbs. (39 Nm).
- Starter cable and tighten the nut
- All other starter wiring
- With cruise control, the actuator
- If removed, the battery
- Air cleaner
- Negative battery cable

GT-S and Celica

1. Before servicing the vehicle, refer to the precautions in the beginning of this section.

2. Remove or disconnect the following:
- Radiator upper support seal
- Air cleaner
- Engine under covers
- Starter wiring
- Starter

3. Installation is the reverse of removal. Torque the bolts to 28 ft. lbs. (37 Nm). Note that the bolts are 2 different lengths.

MR2

1. Before servicing the vehicle, refer to the precautions in the beginning of this section.

2. Remove or disconnect the following:
   - Negative battery cable
   - Front engine undercover
   - Starter wiring harness connectors
   - Starter mounting bolts
   - Starter motor

To install:

3. Install or connect the following:
   - Starter motor. Tighten the bolts to 28 ft. lbs. (37 Nm).
   - Starter wiring harness connectors
   - Front engine undercover
   - Negative battery cable

Oil Pan

REMOVAL & INSTALLATION

5E-FE

1. Before servicing the vehicle, refer to the precautions in the beginning of this section.

2. Drain the engine oil.

3. Remove or disconnect the following:
   - Negative battery cable. On vehicles equipped with an air bag, wait at least 90 seconds before proceeding.
   - Hood
   - Oil dipstick
   - Timing belt and suspend the engine with a hoist

➡ **Do not raise the engine more than necessary, since wiring and other components can be damaged.**

   - Crankshaft timing sprocket and oil pump sprocket
   - Air conditioning compressor and mounting bracket
   - With distributor ignition, exhaust pipe stay
   - Oxygen sensor
   - Front exhaust pipe
   - Oil pan

To install:

4. Clean all gasket mating surfaces.

5. Install or connect the following:
   - Oil pan, with new gasket and sealer. Torque the bolts to 10 ft. lbs. (13 Nm).
   - Front exhaust pipe using a new gasket. Bolts: 46 ft. lbs. (62 Nm).
   - Oxygen sensor
   - Exhaust pipe stay. Bolts: 14 ft. lbs. (19 Nm).

   - Crankshaft timing and oil pump sprockets
   - Timing belt
   - Oil dipstick
   - Oil
   - Negative battery cable
   - Hood

1NZ-FE

1. Before servicing the vehicle, refer to the precautions in the beginning of this section.

2. Drain the engine oil.

3. Remove or disconnect the following:
   - Negative battery cable
   - Oil filter
   - Front exhaust pipe
   - Engine under covers
   - Oil pan bolts

4. Using a thin blade, cut the sealer holding the oil pan and lower the pan.

5. Installation is the reverse of removal. Use RTV sealer. Torque the bolts, in a criss-cross pattern, to 80 inch lbs. (9 Nm).

1ZZ-FE Engine

1. Before servicing the vehicle, refer to the precautions in the beginning of this section.

2. Drain the engine oil.

3. Remove or disconnect the following:
   - Negative battery cable. On vehicles equipped with an air bag, wait at least 90 seconds before proceeding.
   - Undercovers
   - Front exhaust pipe
   - Oil pan mounting bolts and nuts
   - Oil pan, cutting off the applied sealer.

To install:

4. Remove any old sealant from the oil pan flange and thoroughly clean the sealing surface.

5. Install or connect the following:
   - Oil pan. Tighten the bolts and nuts in several passes. Bolts and nuts: 80 inch lbs. (9 Nm).
   - Front exhaust pipe
   - Negative battery cable
   - Undercovers

6. Fill the engine with clean oil.

7. Start the vehicle, check for leaks and repair if necessary.

2ZZ-GE Engine

1. Before servicing the vehicle, refer to the precautions in the beginning of this section.

2. Drain the engine oil.

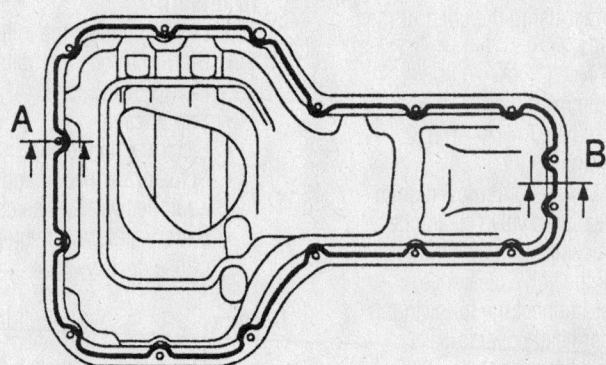

Seal Width
4 – 5 mm

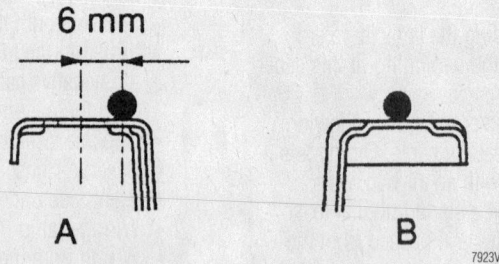

6 mm

A   B

Apply sealant to the oil pan as shown—1.8L (1ZZ-FE) engine

7923VG72

3. Remove or disconnect the following:
- Negative battery cable. On vehicles equipped with an air bag, wait at least 90 seconds before proceeding.
- Undercovers
- Front exhaust pipe
- Oil pan mounting bolts and nuts
- Oil pan, cutting off the applied sealer.

To install:

4. Remove any old sealant from the oil pan flange and thoroughly clean the sealing surface.

5. Install or connect the following:
- Oil pan. Tighten the bolts and nuts in several passes. Bolts and nuts: 80 inch lbs. (9 Nm).
- Front exhaust pipe
- Negative battery cable
- Undercovers

6. Fill the engine with clean oil.

7. Start the vehicle, check for leaks and repair if necessary.

5S-FE Engine

1. Before servicing the vehicle, refer to the precautions in the beginning of this section.

2. Drain the engine oil.

3. Remove or disconnect the following:
- Negative battery cable. On vehicles equipped with an air bag, wait at least 90 seconds before proceeding.

- Undercovers
- Front exhaust pipe
- Engine mounting center member
- On Celica, the 3-way Catalytic Converter (TWC)
- Rear end stiffener plate
- Oil dipstick
- Oil pan bolts and nuts. Cut off the applied sealer.

➡ **Do not use the cutting tool for the oil pump body side and rear oil seal retainer.**

To install:

4. Remove any old sealant from the oil pan flange and thoroughly clean both sealing surfaces.

5. Install or connect the following:
- Oil pan. Uniformly tighten the bolts and nuts in several passes. Bolts and nuts: 48 inch lbs. (5.4 Nm)
- Oil dipstick
- Rear end stiffener plate
- On Celica, the TWC.
- Engine mounting center member

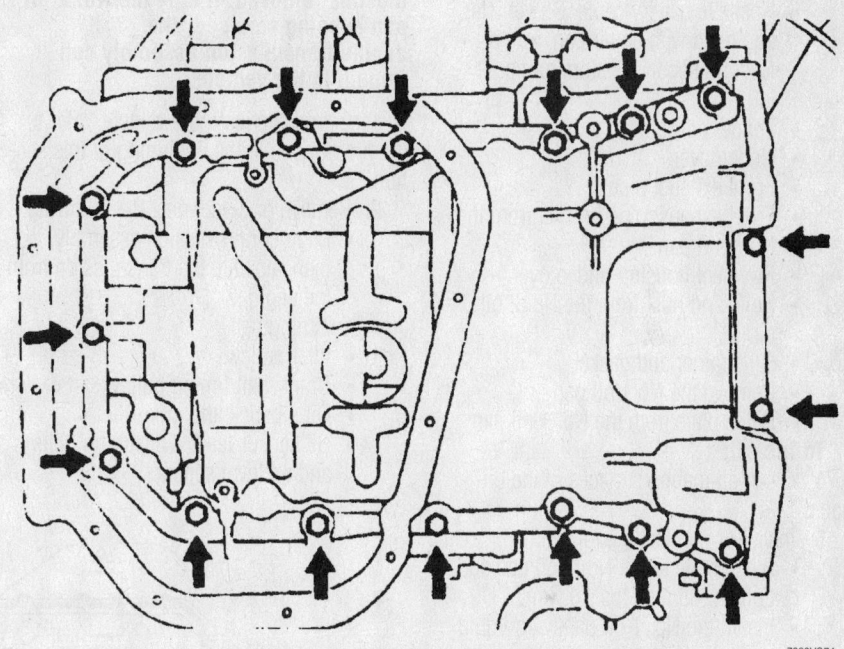

No. 1 oil pan mounting bolt locations—3.0L (1MZ-FE) engine

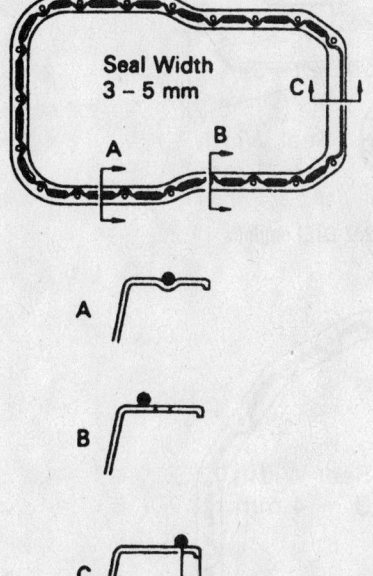

Seal Width 3 – 5 mm

A

B

C

5 mm (0.20 in.)

7923VG73

Oil pan sealing diagram—2.2L (5S-FE) engine

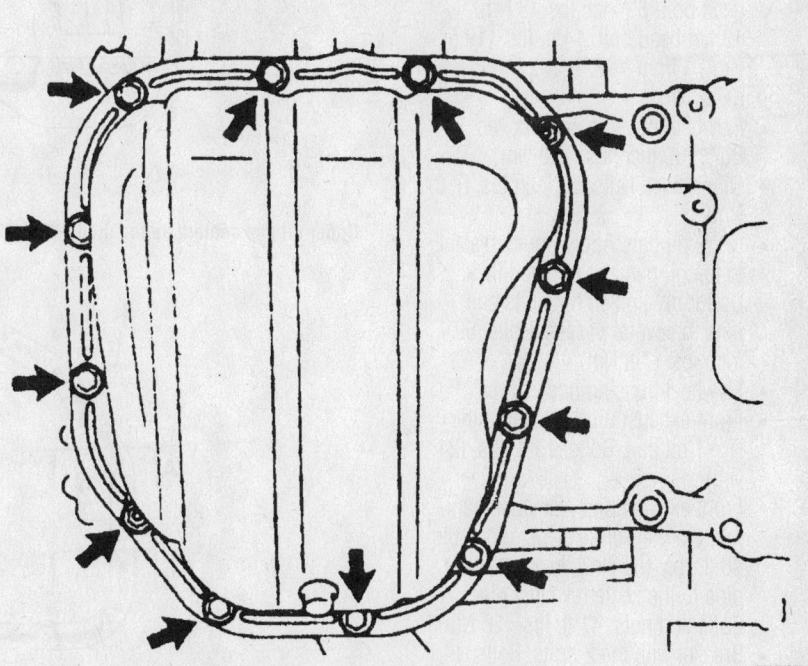

No. 2 oil pan mounting bolt locations—3.0L (1MZ-FE) engine

- Front exhaust pipe
- Negative battery cable
- Undercovers

6. Fill the engine with clean oil.

7. Start the vehicle, check for leaks and repair if necessary.

1MZ-FE Engine

1. Before servicing the vehicle, refer to the precautions in the beginning of this section.

2. Drain the engine oil.

3. Remove or disconnect the following:

- Negative battery cable. On vehicles equipped with an air bag, wait at least 90 seconds before proceeding.
- Fender apron seal
- Undercover
- Front exhaust pipe
- Front exhaust pipe bracket from the No. 1 oil pan.
- Flywheel housing undercover
- Bolts and nuts from the No. 2 oil pan.
- Oil strainer and gasket
- Remove the No.1 oil pan
- Baffle plate from the No. 1 oil pan

To install:

4. Clean all mating surfaces of the oil pans.

5. Install or connect the following:

- Baffle plate to the No. 1 oil pan and tighten: 69 inch lbs. (8 Nm).
- Install the No. 1 oil pan. with liquid sealant. Uniformly tighten the bolts and nuts in several passes. 10mm head bolt: 69 inch lbs. (8 Nm). 12mm head bolt: 14 ft. lbs. (19.5 Nm). 14mm head bolt-27 ft. lbs. (37.2 Nm)
- Flywheel housing undercover. Bolts: 69 inch lbs. (7.8 Nm).
- Oil strainer. Nuts: 69 inch lbs. (7.8 Nm).
- No. 2 oil pan. Apply liquid sealant to the oil pan and engine block. Uniformly tighten the bolts and nuts in several passes. Bolts: 69 inch lbs. (7.8 Nm).
- Flywheel housing undercover
- Front exhaust pipe bracket to the No. 1 oil pan. Bolts: 15 ft. lbs. (21 Nm).
- Front exhaust pipe. Exhaust manifolds to the front exhaust pipe nuts: 46 ft. lbs. (62 Nm). Front exhaust pipe to the center exhaust pipe. Bolts and nuts: 41 ft. lbs. (56 Nm).
- Bracket with the 2 bolts. Bolts: 14 ft. lbs. (19 Nm).

- Support stay with the 2 bolts. Bolts: 22 ft. lbs. (29 Nm).
- Undercover
- Right fender apron seal
- Negative battery cable

6. Fill the engine with clean oil.

7. Start the engine, check for leaks and repair if necessary.

2JZ-GE and 2JZ-GTE Engines

➡ **The No. 1 oil pan can not be removed with the engine in the vehicle. The engine/transmission assembly must be removed. If only the No. 2 oil pan is being serviced, the engine/transmission assembly can remain in the vehicle.**

1. Before servicing the vehicle, refer to the precautions in the beginning of this section.

2. Remove or disconnect the following:

- Engine/transmission assembly, then separate the transmission from the engine.
- Timing belt
- Idler pulley
- Crankshaft timing pulley
- Oil dipstick and guide
- Oil sensor lead, 4 attaching bolts and oil level sensor.

- Lower (No. 2) oil pan.
- Oil strainer and gasket
- Baffle plate
- On turbocharged engines, turbo oil outlet pipe
- Upper (No. 1) oil pan.
- O-ring

To install:

3. Install or connect the following:

- New O-ring in the block. Apply a 1/8 inch (3–4mm) sealant bead to the pan mating surface.
- Upper pan. 12mm bolts to 15 ft. lbs. (21 Nm). 14mm bolts to 29 ft. lbs. (39 Nm).
- On turbocharged engines, the turbo oil outlet pipe and gasket. Nuts: 20 ft. lbs. (27 Nm).
- 2 turbo oil outlet hoses.
- Baffle plate and oil strainer. Tighten both to 78 inch lbs. (9 Nm).
- Lower oil pan in the same manner as the upper pan. Bolts to 78 inch lbs. (9 Nm).
- Oil level sensor, using a new gasket. Tighten to 48 inch lbs. (5.4 Nm).
- Oil dipstick and guide, the timing pulleys and belt, and transmission to the engine.

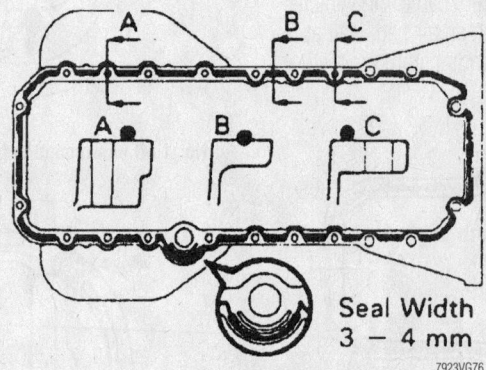

Upper oil pan sealant application—3.0L (2JZ-GE and 2JZ-GTE) engines

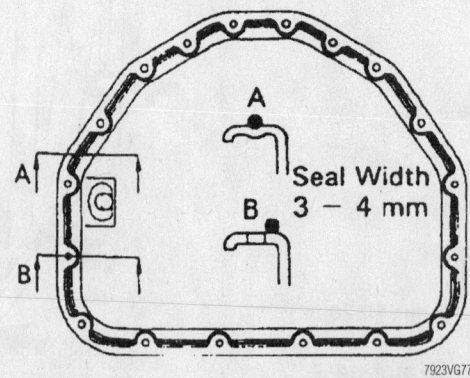

Lower oil pan sealant application—3.0L (2JZ-GE and 2JZ-GTE) engines

- Engine and transmission
- All fluids

Oil Pump

REMOVAL & INSTALLATION

5E-FE Engine

1. Before servicing the vehicle, refer to the precautions in the beginning of this section.
2. Drain the engine oil.
3. Remove or disconnect the following:
   - Negative battery cable. On vehicles equipped with an air bag, wait at least 90 seconds before proceeding.
   - Hood
   - Oil dipstick
   - Timing belt
   - Oil pan
   - Oil strainer with the O-ring
   - Pressure regulator valve
   - Oil pump and tension spring bracket
   - Nut and oil pump pulley

To install:
4. Clean all gasket mating surfaces.
5. Install or connect the following:
   - Oil pump pulley by first placing the driven rotors into the pump body with the marks facing the front
   - Oil pump pulley. Nut: 27 ft. lbs. (37 Nm).
   - Pressure regulator valve. Tighten to 22 ft. lbs. (30 Nm).
   - Oil strainer, using a new O-ring. Bolts: 84 inch lbs. (10 Nm).
   - Oil pan, using a new gasket and sealer
   - All remaining components in the reverse order they were removed.
   - Negative battery cable
   - Hood
6. Fill the engine with clean oil.
7. Start the vehicle, check for leaks and repair if necessary.

1NZ-FE

1. Before servicing the vehicle, refer to the precautions in the beginning of this section.
2. Drain the engine oil.
3. Remove or disconnect the following:
   - Negative battery cable
   - Timing chain cover
   - 2 bolts, 3 screws and the oil pump cover from the timing chain cover
   - Drive and driven rotors
   - Plug, spring and relief valve

4. Inspect the relief valve motion
5. Check rotor side clearance: 0.0012–0.0035 in. (0.03–0.09mm).
6. Check rotor tip clearance: 0.0024–0.0071 in. (0.06–0.18mm).
7. Check rotor-to-body clearance: 0.0098–0.0128 in. (0.250–0.325mm).

To install:
8. Install or connect the following:
   - Relief valve and spring. Torque the plug to 18 ft. lbs. (24 Nm).
   - Drive and driven rotors with the marks on the cover side
   - Cover. Torque the bolts to 80 inch lbs. (9 Nm); the screws to 96 inch lbs. (11 Nm).
   - Timing chain cover
   - Negative battery cable
9. Fill the engine with clean oil.
10. Start the vehicle, check for leaks and repair if necessary.

1ZZ-FE Engine

1. Before servicing the vehicle, refer to the precautions in the beginning of this section.
2. Drain the engine oil.
3. Remove or disconnect the following:
   - Negative battery cable
   - Timing chain and crankshaft sprocket
   - Oil pump and gasket

To install:
4. Clean the mounting surface.
5. Install or connect the following:
   - Oil pump, with new gasket. Bolts: 80 inch lbs. (9 Nm).
   - Crankshaft sprocket and timing chain
   - Negative battery cable
6. Fill the engine with clean oil.
7. Start the vehicle, check for leaks and repair if necessary.

2ZZ-GE Engine

1. Before servicing the vehicle, refer to the precautions in the beginning of this section.
2. Drain the engine oil.
3. Remove or disconnect the following:
   - Negative battery cable
   - Timing chain and crankshaft sprocket
   - Oil pump and gasket

To install:
4. Clean the mounting surface.
5. Install or connect the following:
   - Oil pump, with new gasket. Bolts: 80 inch lbs. (9 Nm).
   - Crankshaft sprocket and timing chain

- Negative battery cable
6. Fill the engine with clean oil.
7. Start the vehicle, check for leaks and repair if necessary.

5S-FE Engine

CAMRY AND CAMRY SOLARA

1. Before servicing the vehicle, refer to the precautions in the beginning of this section.
2. Drain the engine oil.
3. Remove or disconnect the following:
   - Negative battery cable. On vehicles equipped with an air bag, wait at least 90 seconds before proceeding.
   - Hood
   - Front exhaust pipe
   - Rear end stiffener plate
   - Oil dipstick and oil pan
   - Oil strainer and gasket
   - Timing belt and pulleys
   - Crankshaft position sensor
   - Oil pump and gasket

To install:
4. Install or connect the following:
   - Oil pump with new gasket. Bolts: 82 inch lbs. (9 Nm).

➡ **The long bolts are 1.38 in. (35mm) and all the others are 0.98 in. (25mm).**

   - Crankshaft position sensor
   - Timing belt and pulleys
   - Oil strainer with new gasket. Tighten to 48 inch lbs. (5 Nm).
   - Oil pan and dipstick

➡ **The pan must be installed within 5 minutes of sealant application or the procedure will have to be repeated.**

   - Rear end stiffener plate. Bolts: 27 ft. lbs. (37 Nm).
   - Front exhaust pipe
   - Negative battery cable
   - Hood
5. Fill the engine with clean oil.
6. Start the vehicle, check for leaks and repair if necessary.

CELICA

1. Before servicing the vehicle, refer to the precautions in the beginning of this section.
2. Drain the engine oil.
3. Remove or disconnect the following:
   - Negative battery cable. On vehicles equipped with an air bag, wait at least 90 seconds before proceeding.
   - Undercovers
   - Front exhaust pipe

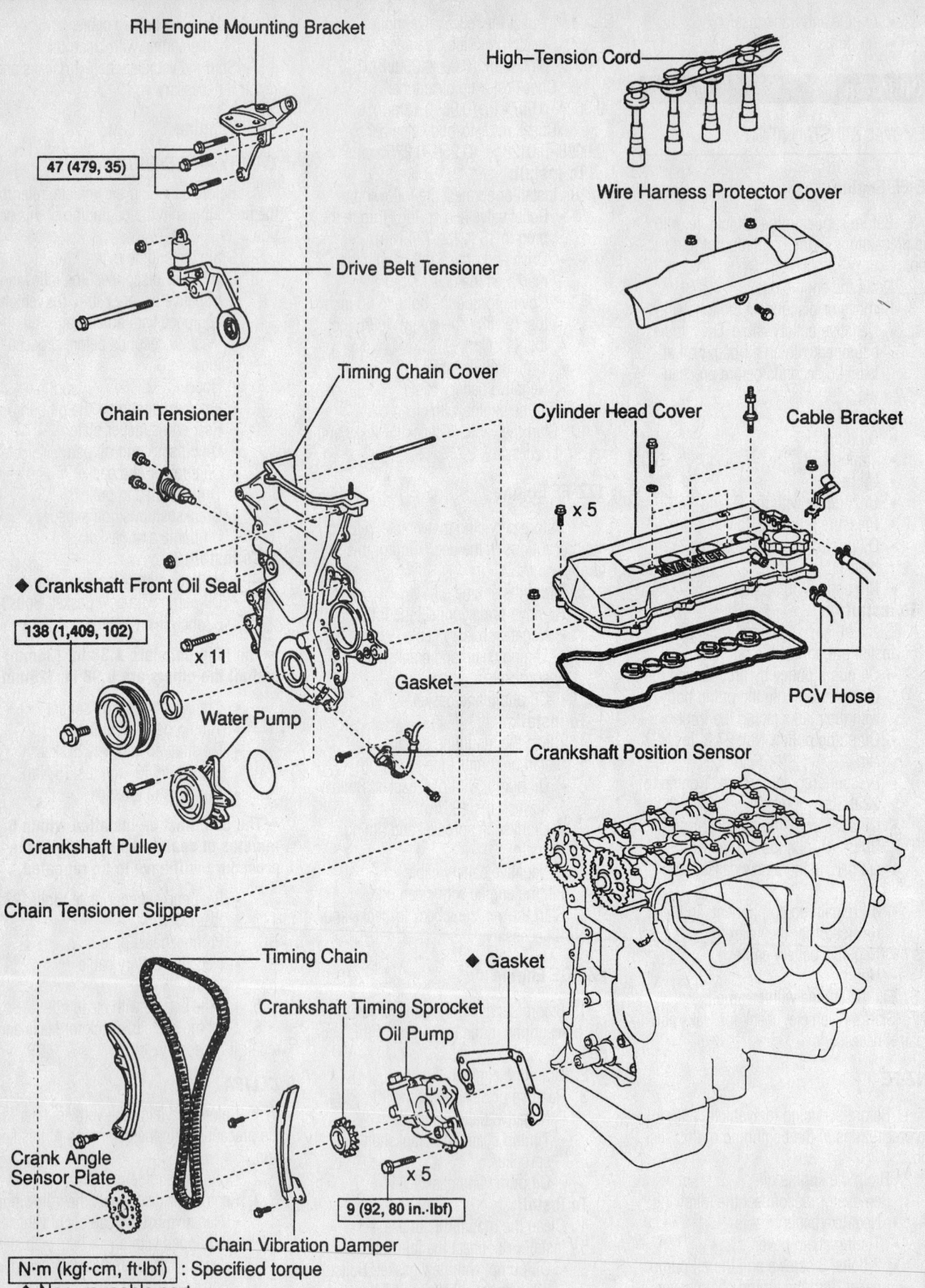

RH Engine Mounting Bracket

High—Tension Cord

47 (479, 35)

Wire Harness Protector Cover

Drive Belt Tensioner

Timing Chain Cover

Chain Tensioner

Cylinder Head Cover

Cable Bracket

x 5

◆ Crankshaft Front Oil Seal

138 (1,409, 102)

x 11

Gasket

PCV Hose

Water Pump

Crankshaft Position Sensor

Crankshaft Pulley

Chain Tensioner Slipper

Timing Chain

◆ Gasket

Crankshaft Timing Sprocket

Oil Pump

Crank Angle Sensor Plate

9 (92, 80 in.·lbf)

Chain Vibration Damper

N·m (kgf·cm, ft·lbf) : Specified torque

◆ Non—reusable part

7923VGB0

Exploded view of the oil pump mounting—1.8L (1ZZ-FE) engine

- Engine mounting center member
- Sub oxygen sensor wiring.
- Right-hand side exhaust manifold stay, TWC with gasket, retainer, and cushion.
- Rear end stiffener plate
- Oil dipstick
- Oil pan, cutting off the applied sealer.

➡ **Do not use the tool for the oil pump body side and rear oil seal retainer.**

- Oil strainer, baffle plate, and the gasket
- Timing belt
- No. 2 idler pulley and crankshaft timing pulley.
- Oil pump pulley
- Oil pump and gasket

To install:

4. Install or connect the following:
- Oil pump, with new gasket. Bolts, uniformly tightened in several passes: 78 inch lbs. (9 Nm). Bolt A is 0.98 in. (25mm) long and bolt B is 1.38 in. (35mm) long.
- Oil pump pulley. Nut: 18 ft. lbs. (24 Nm).
- Crankshaft timing pulley and the No. 2 idler pulley.
- Timing belt
- New gasket, oil strainer and baffle plate. Bolts and nuts: 48 inch lbs. (5 Nm).
- Oil pan and dipstick.
- Rear end stiffener plate
- Cushion, retainer, and a new gasket

to the front TWC. Bolts and nuts: 21 ft. lbs. (29 Nm).
- Right-hand side exhaust manifold stay. Bolts and nuts: 31 ft. lbs. (42 Nm)
- Sub oxygen sensor connector
- Engine mounting center member
- Front exhaust pipe
- Negative battery cable
- Undercovers

5. Fill the engine with clean oil.
6. Start the vehicle, check for leaks and repair if necessary.

1MZ-FE Engine

1. Before servicing the vehicle, refer to the precautions in the beginning of this section.
2. Drain the engine oil.
3. Remove or disconnect the following:
- Negative battery cable. On vehicles equipped with an air bag, wait at least 90 seconds before proceeding.
- Fender apron seal
- Undercover
- Front exhaust pipe
- Front exhaust pipe bracket from the No. 1 oil pan
- Alternator drive belt
- A/C compressor
- Power steering pump drive belt and adjusting strut
- Timing belt and belt pulleys
- Rear timing belt cover
- A/C compressor housing bracket

- No. 2 oil pan, oil strainer, No.1 oil pan and baffle plate
- Crankshaft position sensor
- 9 oil pump bolts. Make a note of the position of the each bolt.
- Oil pump body by prying between the oil pump and main bearing cap
- O-ring from the cylinder block
- Plug, gasket, spring, and relief valve from the oil pump body
- 9 screws, pump body cover, drive, and driven rotors

To install:

4. Install or connect the following:
- Driven rotors, drive, pump body cover
- 9 screws
- Oil pump relief valve, spring, gasket, and the plug to the oil pump body
- New O-ring on the cylinder block
- Liquid sealant to the oil pump and engine block
- Oil pump to the engine block
- Bolts, uniformly tightened in several passes: 10mm head: 69 inch lbs. (8 Nm). 12mm head: 14 ft. lbs. (20 Nm)
- Crankshaft position sensor. Bolt: 69 inch lbs. (8 Nm).
- Baffle plate to the No. oil pan. Tighten to 69 inch lbs. (8 Nm).
- No. 1 oil pan, oil strainer and No. 2 oil pan.
- A/C compressor housing bracket. Bolts: 18 ft. lbs. (25 Nm).
- Rear timing belt cover. Bolts: 74 inch lbs. (9 Nm).
- Timing belt pulleys
- Timing belt
- Adjusting strut and power steering drive belt. Bolt and nut: 32 ft. lbs. (43 Nm).
- A/C compressor
- Alternator drive belt
- Front exhaust pipe bracket to the No. 1 oil pan. Bolts: 15 ft. lbs. (21 Nm).
- Front exhaust pipe
- Undercover
- Right fender apron seal
- Negative battery cable

5. Fill the engine with clean oil.
6. Start the vehicle, check for leaks and repair if necessary.

2JZ-GE and 2JZ-GTE Engines

1. Before servicing the vehicle, refer to the precautions in the beginning of this section.
2. Remove or disconnect the following:

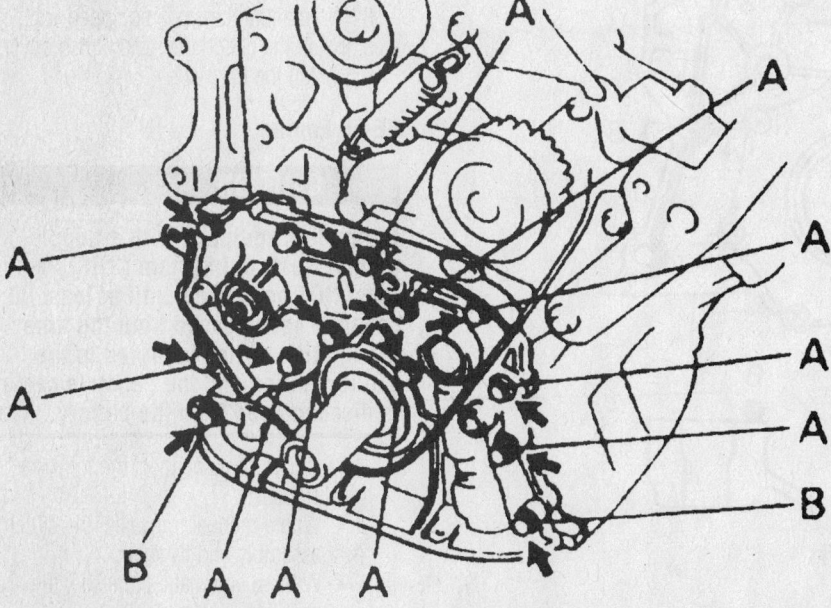

Oil pump mounting bolt identification—2.2L (5S-FE) engine

7923VG78

- Negative battery cable. On vehicles equipped with an air bag, wait at least 90 seconds before proceeding.
- Engine and transmission. Separate the transmission from the engine.
- Timing belt
- Idler pulley
- Crankshaft timing pulley
- Oil dipstick and tube, oil level sensor, No. 2 (lower) oil pan, oil strainer and oil baffle plate.
- With turbocharger, turbo oil outlet pipe.
- No.1 (upper) oil pan
- Oil pump, by driving the pump off the cylinder block using a brass drift.
- O-rings

To install:

3. Install or connect the following:
- 2 new O-rings in the cylinder block
- A ⅛ inch (3–4mm) bead of sealant around the pump mating surface
- Oil pump. Bolts: 15 ft. lbs. (21 Nm)
- O-ring on the block
- A bead of sealant around the No. 1 oil pan
- No.1 oil pan. Bolts: 12mm heads to 15 ft. lbs. (21 Nm). 14mm heads to 29 ft. lbs. (39 Nm).

- With turbocharger, turbo oil outlet pipe. Bolts: 20 ft. lbs. (27 Nm).
- Oil baffle plate, oil strainer, No. 2 oil pan, oil lever sensor and oil dipstick with a new O-ring.
- Crankshaft and idler pulley
- Timing belt
- Engine and transmission
- Negative battery cable

Rear Main Seal

REMOVAL & INSTALLATION

5S-FE, 1ZZ-FE, 2ZZ-GE and 1NZ-FE Engines

1. Remove or disconnect the following:
- Transaxle
- Clutch assembly
- Flywheel or flexplate

2. Use a small sharp knife to cut off the lip of the oil seal. Take great care not to score any metal with the knife.

3. Use a small prytool to pry the old seal from the retaining plate. Be careful not to damage the plate. Protect the tip of the tool with tape and pad the fulcrum point with cloth.

4. Inspect the crankshaft and seal lip contact surfaces for any sign of damage.

To install:

5. Apply a light coat of multi-purpose grease to the lip of a new oil seal. Loosely fit the seal into place by hand, making sure it is not crooked.

6. Use a seal driver of the correct size to install the seal. Tap it into place until the surface of the seal is flush with the edge of the housing.

1MZ-FE Engine

1. Remove or disconnect the following:
- Transaxle
- Clutch cover assembly and flywheel or driveplate
- Remove the rear end plate.
- Oil seal retainer and gasket. Discard the gasket or sealant.

2. Use a small prybar to pry the oil seal from the retaining plate. Be careful not to damage the plate.

To install:

3. Clean the retainer contact surfaces thoroughly and lubricate the new oil seal with multi-purpose grease.

4. Drive the oil seal into the retainer until its surface is flush with the edge of the retainer. Make sure that the seal is installed evenly in the retainer to ensure proper sealing.

5. Apply a ⅛ inch bead of sealant to the oil seal retainer. Install the retainer and install the dust seal. Tighten the bolts to 69 inch lbs. (8 Nm).

6. Install the rear end plate.

7. On automatic transaxle equipped vehicles, install the driveplate.

8. On manual transaxle equipped vehicles, install the clutch disc and clutch cover.

9. Install the transaxle.

5E-FE Engine

❄❄ CAUTION

On models equipped with a Supplemental Restraint System (SRS), work must NOT be started until at least 90 seconds have passed from the time the ignition switch is turned to the LOCK position and the negative cable is disconnected from the battery.

1. Remove or disconnect the following:
- Transaxle
- With a manual transaxle, the clutch assembly and flywheel
- With an automatic transaxle, the flexplate
- The rear endplate

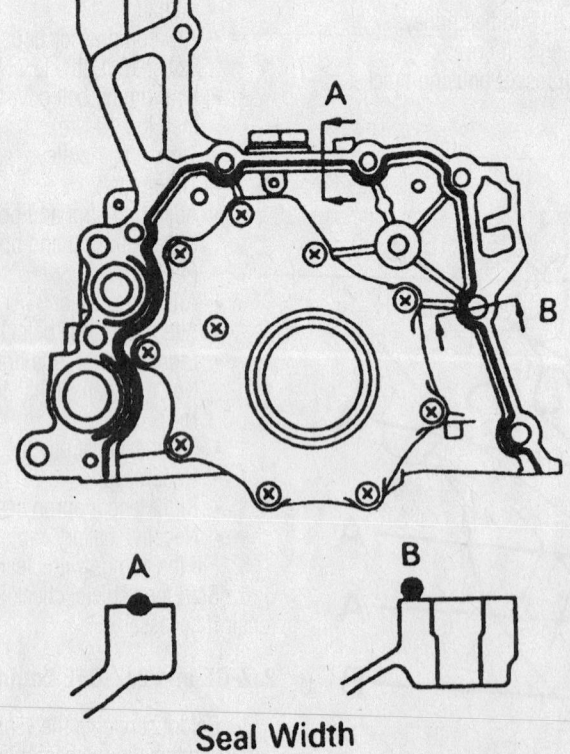

A

B

**Seal Width
2 – 3 mm**

7923VG79

Oil pump sealant application—3.0L (2JZ-GE and 2JZ-GTE) engines

Removing the rear endplate

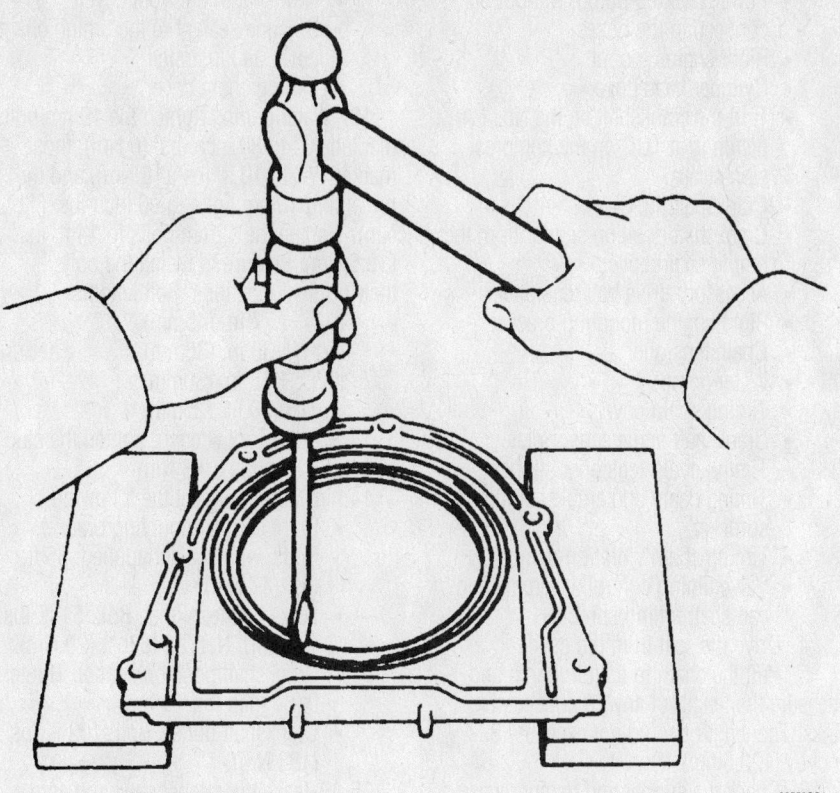

Always place the seal on blocks of wood, then tap the seal from the retainer—1MZ-FE engine

Loosening the rear main seal attaching bolts

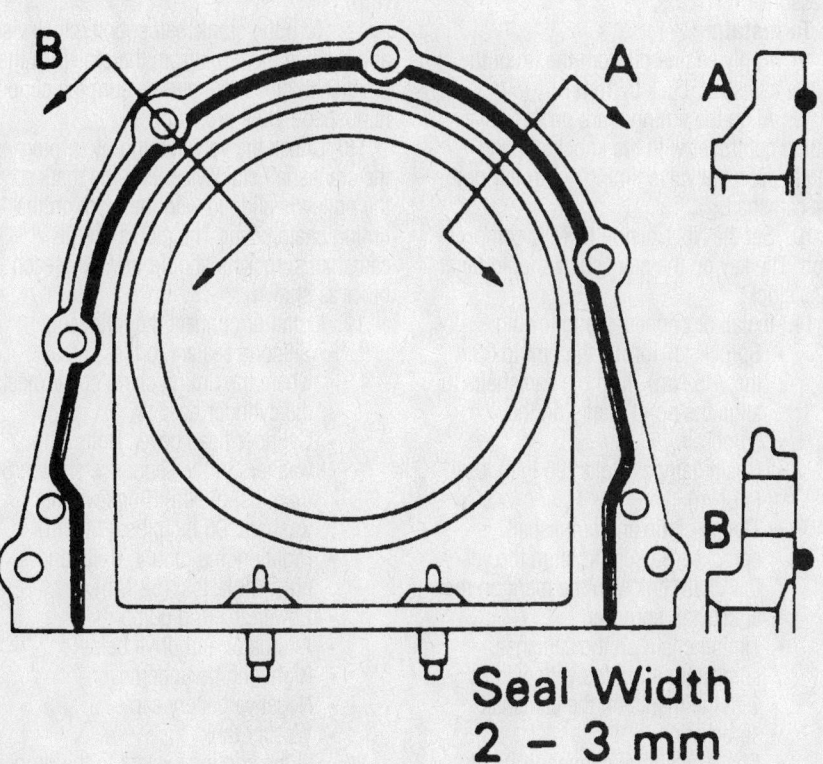

Seal Width 2 – 3 mm

Apply sealant to the rear oil seal retainer—1MZ-FE engine

Removing the rear main seal

• Rear oil seal retainer

2. Using a small prybar, pry the rear oil seal retainer from the mating surfaces.

3. Using a drive punch or a hammer and a small prytool, drive the oil seal from the rear bearing retainer.

➡ **When removing the rear oil seal, be careful not to damage the seal mounting surface.**

4. Using a putty knife, clean the gasket mounting surfaces. Make certain that the contact surfaces are completely free of oil and foreign matter.

To install:

5. Clean the oil seal mounting surface.

6. Using multi-purpose grease, lubricate the new seal lips.

7. Using a seal installation tool or a smooth, round driver, tap the seal straight into the bore of the retainer.

8. Position a new gasket on the retainer and coat it lightly with gasket sealer. Fit the seal retainer into place on the motor; be careful when installing the oil seal over the crankshaft.

9. Install the oil seal retainer bolts. Tighten them 82 inch lbs. (9.3 Nm).

10. Install the rear endplate. Tighten the bolts until snug.

11. Reinstall either the flexplate or the flywheel.

12. Install the torque converter (automatic) or the clutch disc and pressure plate (manual).

13. Reinstall the transaxle.

2JZ-GE and 2JZ-GTE Engines

1. Remove the transmission.

2. Remove the clutch cover assembly and flywheel (manual trans.) or the flexplate (auto. trans.).

3. Use a small, sharp knife to cut off the lip of the oil seal. Take great care not to score any metal with the knife.

4. Use a small prybar to pry the old seal from the retaining plate. Be careful not to damage the plate. Protect the tip of the tool with tape and pad the fulcrum point with cloth.

5. Inspect the crankshaft and seal lip contact surfaces for any sign of damage.

To install:

6. Apply a light coat of multi-purpose grease to the lip of a new oil seal. Loosely fit the seal into place by hand, making sure it is not crooked.

7. Use a seal driver. Tap it into place until the surface of the seal is flush with the edge of the housing.

Timing Chain, Sprockets, Front Cover and seal

REMOVAL & INSTALLATION

1ZZ-FE Engine

1. Before servicing the vehicle, refer to the precautions in the beginning of this section.

2. Drain the cooling system.

3. Remove or disconnect the following:

- Negative battery cable
- Right engine cover
- Accessory drive belt and generator

- Power steering pump, without disconnecting the hoses.
- Right engine mount
- Cylinder head cover
- Turn the crankshaft so the No. 1 piston is at TDC on the compression stroke.
- Crankshaft pulley
- Crankshaft position sensor from the timing chain cover.
- Accessory drive belt tensioner.
- Right engine mounting bracket
- Chain tensioner
- Water pump
- Timing chain cover
- Crankshaft angle sensor plate
- Timing chain tensioner slipper
- Timing chain and crankshaft timing sprocket.
- Timing chain vibration damper
- Valve timing control assembly and camshaft timing sprocket

4. Drive the seal from the cover.

5. Pull the chain to its full length and measure the length of any 16 consecutive links. The length should not exceed 4.827 inches (122.6mm).

6. Check the slipper and damper wear. Maximum wear should not exceed 0.039 in. (1mm).

7. The tensioner plunger should move smoothly and lock into place with finger pressure.

To install:

8. Apply engine oil from the tip of the intake camshaft, back to 16mm.

9. Align the timing mark on the valve timing controller with the knock pin and gently push the valve timing controller onto the camshaft.

10. Set the No.1 piston to TDC compression. The key on the crankshaft should be at 12 o'clock.

11. Install or connect the following:

- Sprockets. Torque the bolt to 33 ft. lbs. (45 Nm). Turn the camshafts to align the point marks on the sprockets.
- Chain damper. Bolts: 96 inch lbs. (11 Nm).
- Timing chain and crankshaft sprocket. Be sure to align the yellow chain link with the mark on the crankshaft sprocket.
- Timing chain on the camshaft sprockets. Align the yellow links with the marks on the camshaft sprockets.
- Chain tensioner slipper. Bolt: 14 ft. lbs. (18.5 Nm).
- Crankshaft angle sensor plate with the **F** mark forward

- New seal in the front cover
- Silicone sealant to the timing chain cover as illustrated
- Timing chain cover

12. Water pump. Tighten the 10mm bolts marked "C" to 80 inch lbs. (9 Nm), those marked "A" to 10 ft. lbs. (13 Nm), and the remaining 10mm bolts to 96 inch lbs. (11 Nm). Tighten the 12mm bolts to 14 ft. lbs. (18.5 Nm). Be sure to install the bolts in their original locations. Bolt lengths:

   a. A: 1.77 in. (45mm)
   b. B: 1.38 in. (35mm)
   c. C: 1.18 in. (30mm)
   d. D: 0.98 in. (25mm)

13. With a Torx wrench, tighten the stud bolt to 82 inch lbs. (9.3 Nm).

14. Install or connect the following:

- Right engine mounting bracket. Bolts, with sealant applied: 35 ft. lbs. (47 Nm).
- Drive belt tensioner. Bolt: 51 ft. lbs. (69 Nm). Nut: 21 ft. lbs. (29 Nm).
- Crankshaft position sensor. Tighten to 80 inch lbs. (9 Nm).
- Crankshaft pulley. Bolt: 102 ft. lbs. (138 Nm).

15. Release the ratchet pawl and compress the chain tensioner. Place the hook on the pin to keep the tensioner compressed.

16. Install the tensioner, using a new O-ring. Torque the bolts to 80 inch lbs. (9 Nm).

17. Turn the crankshaft counterclockwise and remove the hook from the pin. Turn the crankshaft clockwise and be sure the slipper is pushed by the plunger.

18. Check the valve timing by turning the crankshaft clockwise until the mark of the pulley is aligned with the mark on the timing chain cover. The marks on the camshaft sprockets should be facing each other as shown.

19. Install or connect the following:

- Silicone sealant to the 2 areas where the timing chain cover meets the cylinder head.
- Cylinder head cover. Bolts with washers in the sequence shown: 80 inch lbs. (9 Nm). Bolts without washers: 96 inch lbs. (11 Nm).
- Right engine mount. Bolts and nuts: 38 ft. lbs. (52 Nm).
- Power steering pump
- Alternator and drive belt
- Right engine undercover
- Negative battery cable
- Washer tank

20. Fill the cooling system to the proper level.

21. Start the vehicle, check for leaks and repair if necessary.

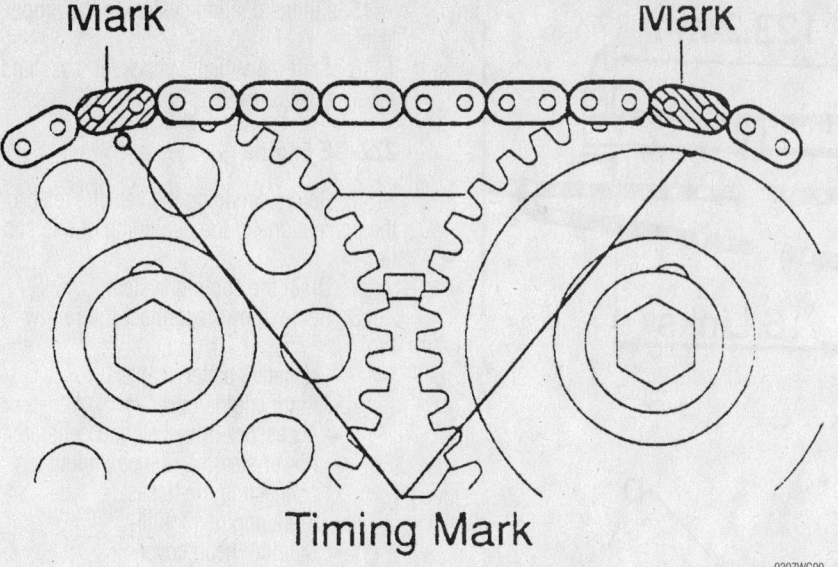

9307WG90

Timing chain link marks—1ZZ-FE and 2ZZ-GE engines

1NZ-FE

1. Before servicing the vehicle, refer to the precautions in the beginning of this section.

2. Drain the cooling system.

3. Remove or disconnect the following:
   - Negative battery cable
   - Right front wheel
   - Alternator
   - Power steering pump
   - Right engine mount insulator. Use a jack and wood block for support.
   - With A/C, the bolt holding the liquid tube to the insulator
   - Ignition coils
   - Cylinder head cover

4. Place No.1 cylinder on TDC compression. Make sure that the timing marks on the camshaft sprockets and valve timing controller assembly are facing UP (12 o'clock). If not, turn the crankshaft 360 degrees to align the marks.

5. Remove or disconnect the following:
   - Crankshaft pulley bolt
   - Pulley and pin
   - Crankshaft position sensor
   - Right engine mount bracket
   - Water pump
   - Oil control valve
   - 13 bolts, 1 nut and 1 stud bolt. Pry the cover off.
   - 2 O-rings from the block and pan
   - Chain tensioner
   - Tensioner slipper
   - Chain vibration damper
   - Chain

6. Drive the seal from the cover.

7. Pull the chain to its full length and measure the length of any 16 consecutive links. The length should not exceed 4.85 inches (123.2mm).

8. Check the slipper and damper wear. Maximum wear should not exceed 0.039 in. (1mm).

9. The tensioner plunger should move smoothly and lock into place with finger pressure.

To install:

10. Set the crankshaft at 140 degrees ATDC. Set the camshaft sprockets at 20 degrees ATDC; then, reset the crankshaft to 20 degrees ATDC.

11. Install or connect the following:
   - New seal, driven into place until flush with the cover edge
   - Vibration damper. Torque the bolts to 80 inch lbs. (9 Nm).
   - Timing chain

➡ **A new chain will have 3 marked links to align with the 3 sprockets, as shown in the accompanying illustration.**

   - Slipper
   - Tensioner. Torque the bolts to 80 inch lbs. (9 Nm).
   - Cover, using a 4–5mm bead of RTV sealer and new O-rings to the block and pan

12. Uniformly tighten the bolts in several passes, using the accompanying illustration, to:
   - A, C, E and G: 96 inch lbs. (11 Nm)
   - B, D and F: 18 ft. lbs. (24 Nm)

13. Bolt lengths are as follows:
   - A: 20mm
   - B: 30mm
   - C: 35mm
   - D: 20mm

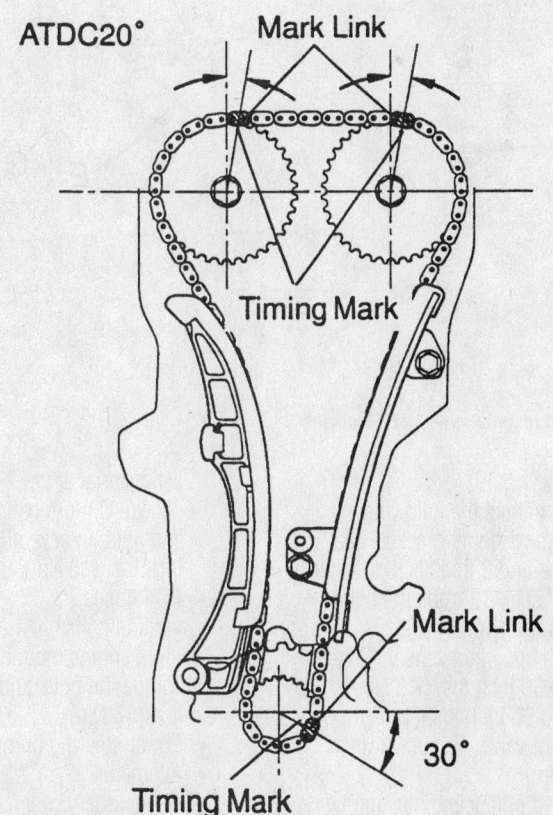

9307WG78

Timing chain installation—1NZ-FE engine

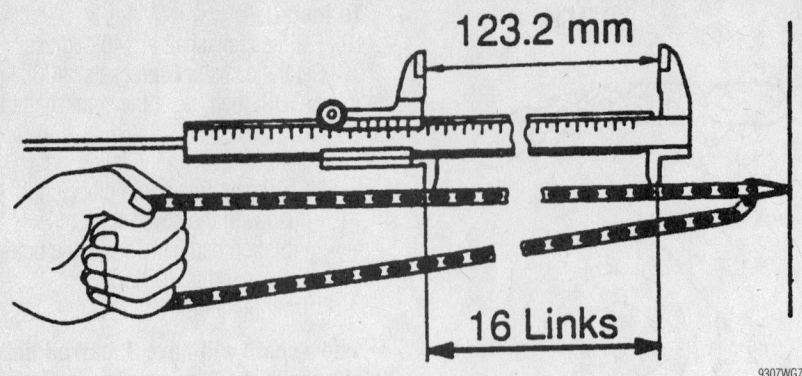

Measuring the timing chain—1NZ-FE engine

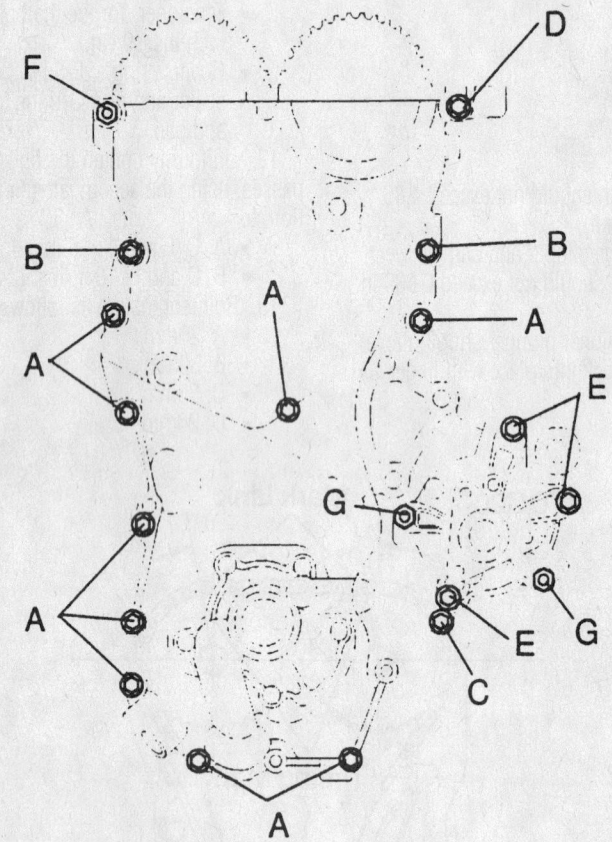

Timing cover bolt installation—1NZ-FE engine

- E: 35mm
14. Install or connect the following:
  - Right engine mount bracket. Coat all but the end 2 threads of the bolt with RTV sealer. Torque the bolt to 41 ft. lbs. (55 Nm).
  - Crankshaft position sensor. Torque bolt A to 66 inch lbs. (7.5 Nm); bolts B to 96 inch lbs. (11 Nm).
  - Oil control valve. Torque: 71 inch lbs. (8 Nm).
  - Crankshaft pulley and pin. Torque the bolt to 94 ft. lbs. (128 Nm).
  - Cylinder head cover with RTV gas-

ket material at the 2 locations shown. Uniformly tighten the bolts and nuts, in several passes, to 84 inch lbs. (10 Nm).
  - PCV hoses
  - Ignition coils
  - Right engine mount insulator. Torque the bolts and nuts to 35 ft. lbs. (47 Nm).
  - Power steering pump
  - Alternator
  - Right under cover
  - Wheel
  - Negative battery cable

15. Fill the cooling system to the proper level.

16. Start the vehicle, check for leaks and repair if necessary.

2ZZ-GE Engine

1. Before servicing the vehicle, refer to the precautions in the beginning of this section.

2. Drain the cooling system.

3. Remove or disconnect the following:

- Negative battery cable
- Right engine cover
- Accessory drive belt and generator
- Power steering pump, without disconnecting the hoses.
- Right engine mount
- Cylinder head cover
- Turn the crankshaft so the No. 1 piston is at TDC on the compression stroke.
- Crankshaft pulley
- Crankshaft position sensor from the timing chain cover.
- Accessory drive belt tensioner.
- Right engine mounting bracket
- Chain tensioner
- Water pump
- Timing chain cover
- Crankshaft angle sensor plate
- Timing chain tensioner slipper
- Timing chain and crankshaft timing sprocket.
- Timing chain vibration damper
- Valve timing control assembly and camshaft timing sprocket

4. Drive the seal from the cover.

5. Pull the chain to its full length and measure the length of any 16 consecutive links. The length should not exceed 4.827 inches (122.6mm).

6. Check the slipper and damper wear. Maximum wear should not exceed 0.039 in. (1mm).

7. The tensioner plunger should move smoothly and lock into place with finger pressure.

To install:

8. Apply engine oil from the tip of the intake camshaft, back to 16mm.

9. Align the timing mark on the valve timing controller with the knock pin and gently push the valve timing controller onto the camshaft.

10. Set the No.1 piston to TDC compression. The key on the crankshaft should be at 12 o'clock.

11. Install or connect the following:
- Sprockets. Torque the bolt to 33 ft. lbs. (45 Nm). Turn the camshafts to

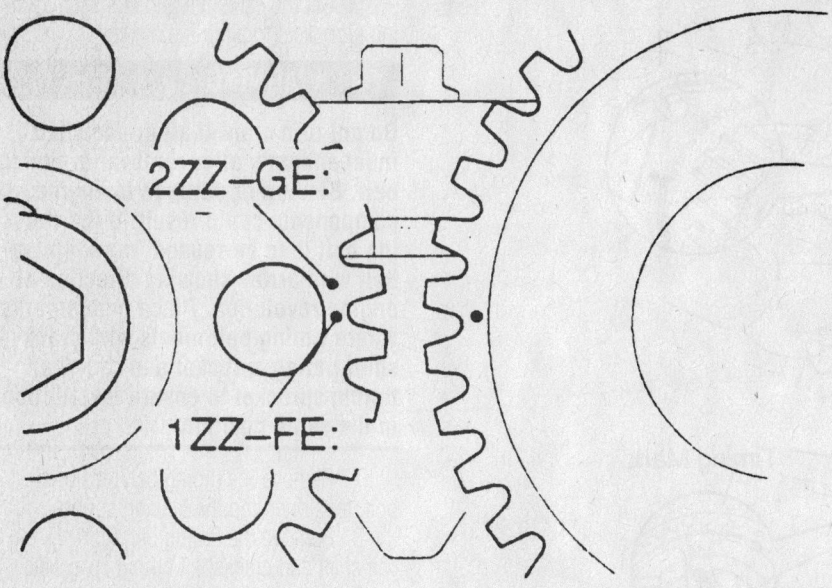

Timing mark identification at TDC compression—1ZZ-FE and 2ZZ-GE engines

9307WG89

align the point marks on the sprockets.

- Chain damper. Bolts: 96 inch lbs. (11 Nm).
- Timing chain and crankshaft sprocket. Be sure to align the yellow chain link with the mark on the crankshaft sprocket.
- Timing chain on the camshaft sprockets. Align the yellow links with the marks on the camshaft sprockets.
- Chain tensioner slipper. Bolt: 14 ft. lbs. (18.5 Nm).
- Crankshaft angle sensor plate with the **F** mark forward
- New seal in the front cover
- Silicone sealant to the timing chain cover as illustrated
- Timing chain cover

12. Install the water pump. Tighten the 10mm bolts marked "C" to 80 inch lbs. (9 Nm), those marked "A" to 10 ft. lbs. (13 Nm), and the remaining 10mm bolts to 96 inch lbs. (11 Nm). Tighten the 12mm bolts to 14 ft. lbs. (18.5 Nm). Be sure to install the bolts in their original locations. Bolt lengths:

   a. A: 1.77 in. (45mm)
   b. B: 1.38 in. (35mm)
   c. C: 1.18 in. (30mm)
   d. D: 0.98 in. (25mm)

13. With a Torx wrench, tighten the stud bolt to 82 inch lbs. (9.3 Nm).

14. Install or connect the following:
- Right engine mounting bracket. Bolts, with sealant applied: 35 ft. lbs. (47 Nm).

- Drive belt tensioner. Bolt: 51 ft. lbs. (69 Nm). Nut: 21 ft. lbs. (29 Nm).
- Crankshaft position sensor. Tighten to 80 inch lbs. (9 Nm).
- Crankshaft pulley. Bolt: 102 ft. lbs. (138 Nm).

15. Release the ratchet pawl and compress the chain tensioner. Place the hook on the pin to keep the tensioner compressed.

16. Install the tensioner, using a new O-ring. Torque the bolts to 80 inch lbs. (9 Nm).

17. Turn the crankshaft counterclockwise and remove the hook from the pin. Turn the crankshaft clockwise and be sure the slipper is pushed by the plunger.

18. Check the valve timing by turning the crankshaft clockwise until the mark of the pulley is aligned with the mark on the timing chain cover. The marks on the camshaft sprockets should be facing each other as shown.

19. Install or connect the following:
- Silicone sealant to the 2 areas where the timing chain cover meets the cylinder head.
- Cylinder head cover. Bolts with washers in the sequence shown: 80 inch lbs. (9 Nm). Bolts without washers: 96 inch lbs. (11 Nm).
- Right engine mount. Bolts and nuts: 38 ft. lbs. (52 Nm).
- Power steering pump
- Alternator and drive belt
- Right engine undercover
- Negative battery cable
- Washer tank

20. Fill the cooling system to the proper level.

21. Start the vehicle, check for leaks and repair if necessary.

Timing Belt

REMOVAL & INSTALLATION

1.5L (5E-FE) Engine

1. Before servicing the vehicle, refer to the precautions in the beginning of this section.

2. Remove all necessary components for access to the timing belt covers.

3. Remove the No. 2 timing belt cover.

4. Rotate the engine clockwise until the crankshaft pulley is aligned with the **0** mark on the No. 1 timing belt cover. Verify the hole in the camshaft timing pulley is aligned with the timing mark on the No. 1 bearing cap. If not as specified, rotate the crankshaft an additional 360 degrees.

5. For vehicles with A/C and/or power steering, remove the four bolts to the No. 2 crankshaft pulley. Then, remove the No. 2 crankshaft pulley.

6. Using Toyota tools SST 09213-14010 and SST 09330-00021, remove the No. 1 crankshaft pulley bolt.

7. Using a crankshaft pulley/damper puller (such as Toyota tool SST 09950-50010), remove the No. 1 crankshaft pulley from the crankshaft.

8. Remove the No. 3 timing belt cover (plug).

9. Remove the No. 1 timing belt cover and timing belt guide.

10. Place matchmarks on the timing belt on both sides of the cam and crankshaft gear timing marks. Also, place an arrow on the top surface of the belt to indicate the direction of travel.

11. Using pliers, remove the tension spring.

12. Loosen the No. 1 idler pulley bolt and push the pulley to the left as far as it will go, then temporarily tighten the bolt.

13. Remove the belt.

To install:

14. For vehicles with a distributor (distributor ignition), use the crankshaft bolt to turn the crankshaft until the timing marks on the sprocket and oil pump body align. This is method is used to set the piston at Top Dead Center (TDC) before the marks on the belt cover can be seen.

15. For vehicles with a crankshaft position sensor (distributorless ignition), use the crankshaft bolt to turn the crankshaft

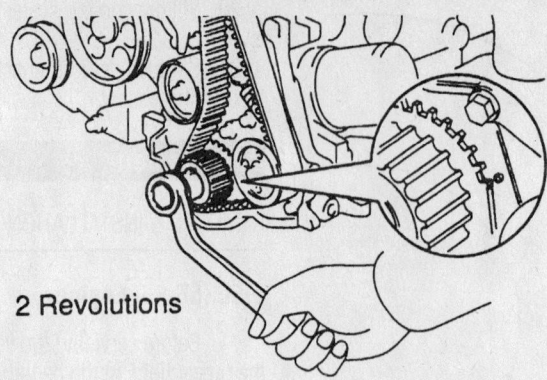

2 Revolutions

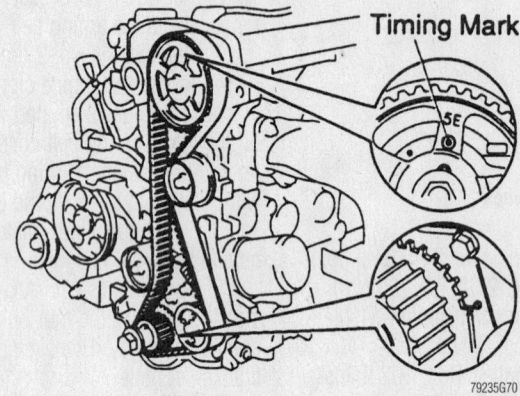

Timing Mark

5E

79235G70

Turn the engine two revolutions, then ensure the timing marks are still aligned—Toyota 1.5L (5E-FE) engine

until the rotor side of the crankshaft position sensor faces inward.

16. Turn the camshaft and align the hole of the camshaft timing pulley with the timing mark of the bearing cap. The matchmarks, if using the old belt should align. Place the belt over the crankshaft pulley and the idler pulleys.

17. Install the belt on the crankshaft gear (using the matchmarks if reinstalling the old belt) and install the timing belt guide with flange out.

18. Loosen the No. 1 idler pulley bolt until the pulley is moved slightly by the spring tension.

19. Turn the crankshaft pulley two revolutions from TDC-to-TDC.

➡ **Always rotate the crankshaft clockwise.**

20. Check that the pulleys align with the reference marks. If not, reinstall the belt.

21. When the timing is verified, tighten the adjuster pulley (No. 1 idler pulley) to 13 ft. lbs. (18 Nm).

22. Install No. 1 and No. 3 lower timing belt covers.

23. Install the No. 1 crankshaft pulley and tighten the pulley bolt to 112 ft. lbs. (152 Nm).

24. Tighten the four No. 2 crankshaft pulley bolts to 14 ft. lbs. (19 Nm).

25. Install the No. 2 timing belt cover with the four bolts.

26. Install the remaining components in the reverse order of the removal procedure. When installing the right-hand engine mounting insulator, tighten the bracket bolt to 47 ft. lbs. (64 Nm) and the through-bolt to 54 ft. lbs. (73 Nm).

1.6L (4A-FE) and 1.8L (7A-FE) Corolla engines

1. Before servicing the vehicle, refer to the precautions in the beginning of this section.

2. Remove all necessary components for access to the timing belt covers.

3. Turn the crankshaft to align the timing mark on crankshaft pulley at **0**, setting the piston in No. 1 cylinder at Top Dead Center (TDC) on the compression stroke. Check that the valve lifters on the No. 1 cylinder are loose. If not, turn crankshaft pulley 1 complete revolution (360 degrees).

4. Remove the nine bolts and timing belt covers from the engine.

5. Slide the timing belt guide from crankshaft.

6. Set the camshaft and crankshaft timing sprockets to align the marks.

✳✳ WARNING

Do not turn crankshaft or camshaft independently after removal of timing belt. Binding or damage to engine components could result. If the timing belt is to be reused, mark timing belt with arrow showing direction of engine revolution. Place matchmarks where timing belt meets with crankshaft timing sprocket and camshaft timing sprocket to ensure installation in the same position.

7. Remove the timing belt tensioner bolt, tensioner and the tension spring.

8. Remove the timing belt from camshaft and crankshaft timing sprockets.

9. Remove the timing belt.

✳✳ WARNING

Do not bend, twist or turn the timing belt inside out. Do not allow the belt to come in contact with oil, coolant or steam.

To install:

➡ **Inspect the camshaft timing sprocket to ensure mark is still aligned as indicated.**

10. Reinstall the timing belt tensioner and the tension spring. Pry the tensioner to the left as far as it will go and temporarily tighten the retaining bolt.

11. Install the timing belt. If reinstalling the old belt, observe the matchmarks made during removal. Be sure the belt is fully and squarely seated on the sprockets.

12. Loosen the retaining bolt for the timing belt tensioner and allow it to tension the belt.

13. Temporarily install the crankshaft pulley bolt and turn the crank clockwise 2 full revolutions from TDC to TDC. Insure that each timing mark realigns exactly.

14. Tighten the timing belt tensioner bolt to 27 ft. lbs. (37 Nm).

15. Measure the timing belt deflection at the **SIDE** point, looking for 1/4 in. (5–6mm) of deflection at 4.4 lbs. (2 kg) of pressure. If the deflection is not correct, adjust with the timing belt tensioner.

16. Install the timing belt guide, with the cup side facing outward, onto the crankshaft and install the timing belt covers from the lowest to the highest. Tighten the nine cover bolts to 62 inch lbs. (7 Nm).

17. Install all applicable remaining com-

ponents. During assembly, be sure to tighten the crankshaft pulley bolt to 87 ft. lbs. (118 Nm), the mounting bracket-to-engine mount bolt to 47 ft. lbs. (64 Nm), the mounting bracket-to-engine mount nuts to 38 ft. lbs. (52 Nm), the engine mount-to-body bolt **A** to 19 ft. lbs. (25 Nm), the engine mount-to-body bolts **B** to 19 ft. lbs. (25 Nm), the engine mount-to-body bolt **C** to 19 ft. lbs. (25 Nm), if equipped with cruise control.

2.2L (5S-FE) Engines

1. Before servicing the vehicle, refer to the precautions in the beginning of this section.

2. Remove all necessary components for access to the timing belt covers.

3. Remove the No. 2 timing cover.

4. Position the No. 1 cylinder to Top Dead Center (TDC) on the compression stroke by turning the crankshaft pulley and aligning its groove with the timing mark **O** of the No. 1 timing belt cover. Check that the hole of the camshaft timing pulley is aligned with the alignment mark of the bearing cap. If not, turn the crankshaft one revolution (360 degrees).

5. Remove the timing belt from the camshaft timing pulley, as follows:

a. If reusing the belt, place matchmarks on the timing belt and the camshaft pulley. Loosen the mount bolt of the No. 1 idler pulley and position the pulley toward the left as far as it will go. Tighten the bolt. Remove the belt from the camshaft pulley.

6. Remove the camshaft timing pulley as follows:

a. Using Toyota tools Nos. 09249-63010 and 09960-10010, remove the bolt and the camshaft pulley.

7. Remove the crankshaft pulley as follows:

a. Use Toyota tools Nos. 09213-54015 and 09330-00021 to hold the crankshaft pulley. Remove the pulley set bolt and remove the pulley using a puller.

8. Remove the No. 1 timing belt cover.

9. Remove the timing belt and the belt guide. If reusing the belt mark the belt and the crankshaft pulley in the direction of engine rotation and matchmark for correct installation.

To install:

10. Install the crankshaft timing pulley, as follows:

a. Align the timing pulley set key with the key groove of the pulley.

b. Slide on the timing pulley with the flange side facing inward.

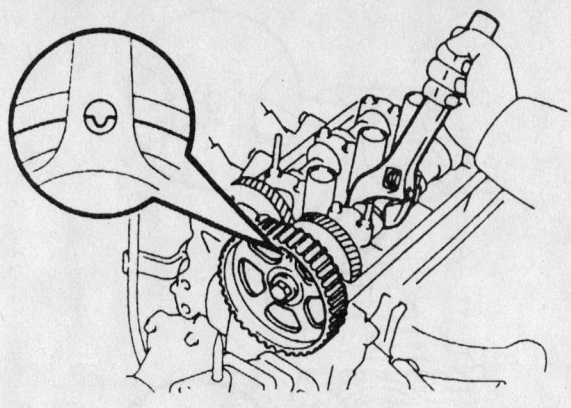

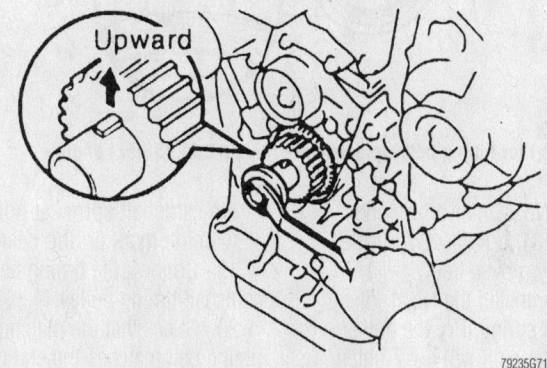

Sprocket alignment for timing belt replacement—Toyota 1.6L (4A-FE) and 1.8L (7A-FE) engines

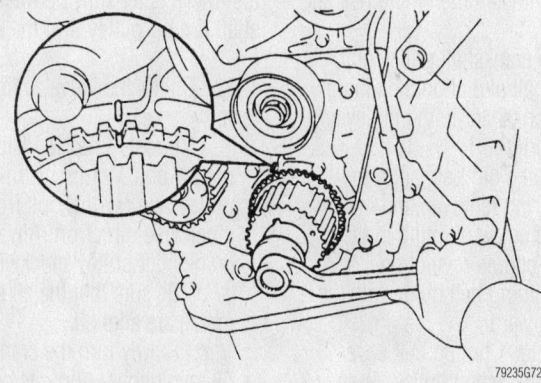

Crankshaft positioning for timing belt removal and installation—Toyota 2.2L (5S-FE) engine

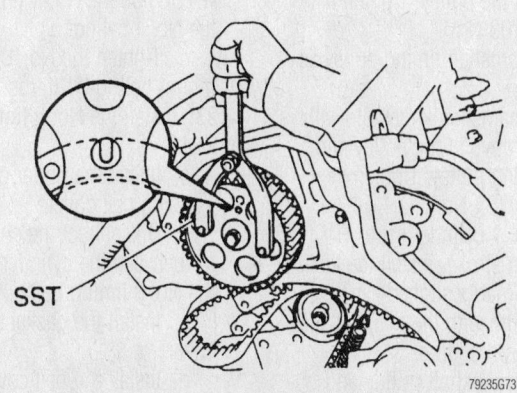

Using a spanner wrench, turn the camshaft into position so that the alignment mark is visible through the hole in the sprocket—Toyota 2.2L (5S-FE) engine

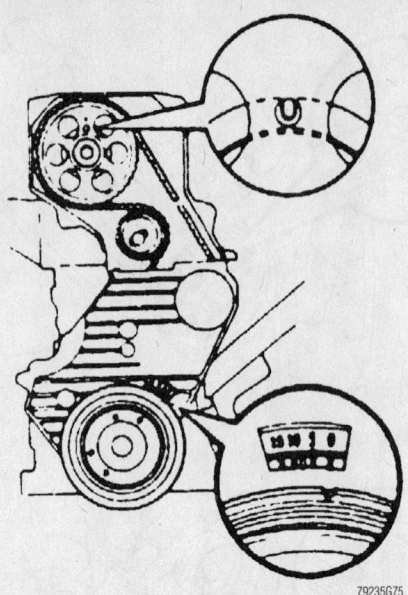

79235G75

Sprocket alignment for timing belt replacement—Toyota 2.2L (5S-FE) engine

11. Install the No. 2 idler pulley and tighten the bolt to 31 ft. lbs. (42 Nm). Be sure that the pulley moves freely.

12. Temporarily install the No. 1 idler pulley and tension spring. Pry the pulley toward the left as far as it will go. Tighten the bolt. Be sure that the pulley rotates freely.

13. Temporarily install the timing belt, as follows:

a. Using the crankshaft pulley bolt, turn the crankshaft and align the timing marks of the crankshaft timing pulley and the oil pump body.

b. If reusing the old belt, align the marks made during removal, and install the belt with the arrow pointing in the direction of the engine revolution.

14. Install the timing belt guide with the cup side facing outward.

15. Install the No. 1 timing belt cover.

16. Install the crankshaft pulley. Align the pulley set key with the key groove of the pulley and slide on the pulley. Tighten the bolt to 80 ft. lbs. (108 Nm).

17. Install the camshaft timing pulley as follows:

a. Align the camshaft knock pin with the knock pin groove of the pulley and slide on the timing pulley. Tighten the bolt to 40 ft. lbs. (54 Nm).

18. With the No. 1 cylinder set at TDC on the compression stroke, install the timing belt (all timing marks aligned). If reusing the belt, align with the marks made during the removal procedure:

a. Turn the crankshaft pulley and align its groove with the timing mark **0** of the No. 1 timing belt cover. Be sure

the camshaft sprocket hole is aligned with the mark on the bearing cap.

19. Connect the timing belt to the camshaft timing pulley.

20. Check that the matchmark on the timing belt matches the end of the No. 1 timing belt cover.

21. Once the belt is installed, be sure that there is tension between the crankshaft timing pulley and the camshaft pulley.

22. Check the valve timing as follows:

a. Loosen the No. 1 idler pulley mount bolt ½ turn. Turn the crankshaft pulley two revolutions from TDC in the clockwise direction. Always turn the crankshaft pulley clockwise.

b. Be sure that the all the timing marks are aligned.

c. Slowly turn the crankshaft pulley 1 ⅞ revolutions. Align its groove with the mark at 45° Before Top Dead Center (BTDC) on the No. 1 timing belt cover for the No. 1 cylinder.

d. Tighten the No. 1 idler pulley mount bolt to 31 ft. lbs. (42 Nm).

23. Install the No. 2 timing belt cover as follows:

a. Install the upper gasket to the No. 1 timing belt cover.

b. Disconnect the engine wire protector between the cylinder head cover and the No. 3 timing belt cover.

c. Install the gasket to the timing belt cover.

d. Install the belt covers and all remaining components. During assembly, tighten the right engine mount

bracket bolts to 38 ft. lbs. (52 Nm), the engine mount insulator bolt to 47 ft. lbs. (64 Nm), the through-bolt to 54 ft. lbs. (78 Nm), the power steering reservoir bracket bolt to 21 ft. lbs. (28 Nm), the power steering reservoir-to-bracket bolt to 27 ft. lbs. (37 Nm) and the nut to 38 ft. lbs. (52 Nm).

3.0L (1MZ-FE) Engine

1. Before servicing the vehicle, refer to the precautions in the beginning of this section.

2. Remove all necessary components for access to the timing belt covers.

✳✳ CAUTION

If equipped with an air bag, be sure to disconnect the negative battery cable and wait at least 90 seconds before proceeding.

3. Remove the lower timing belt cover by removing the four bolts.

4. Remove the No. 2 timing belt cover as follows:

a. Remove the bolt and disconnect the engine wire protector from the No. 3 (rear) timing belt cover.

b. Disconnect the engine wire protector clamp from the No. 3 timing belt cover.

c. Remove the five bolts from the No. 2 timing belt cover.

d. Remove the No. 2 cover from the engine.

5. Remove the right engine mounting bracket by removing the nut and two bolts.

6. Remove the crankshaft timing belt guide.

7. Temporarily install the crankshaft pulley bolt.

8. Turn the crankshaft and align the crankshaft timing pulley groove with the oil pump alignment mark. Always turn the engine clockwise.

9. Ensure the timing mark of the camshaft timing pulleys and rear timing belt covers are aligned. If not, turn the engine over an additional 360 degrees (one revolution).

10. Remove the crankshaft pulley bolt.

➡**If the belt is to be reused, align the installation marks on the belt to the marks on the pulleys. If the marks have worn off, make new ones.**

11. Alternately loosen the two timing belt tensioner bolts. Remove the tensioner and dust boot.

12. Remove the timing belt.

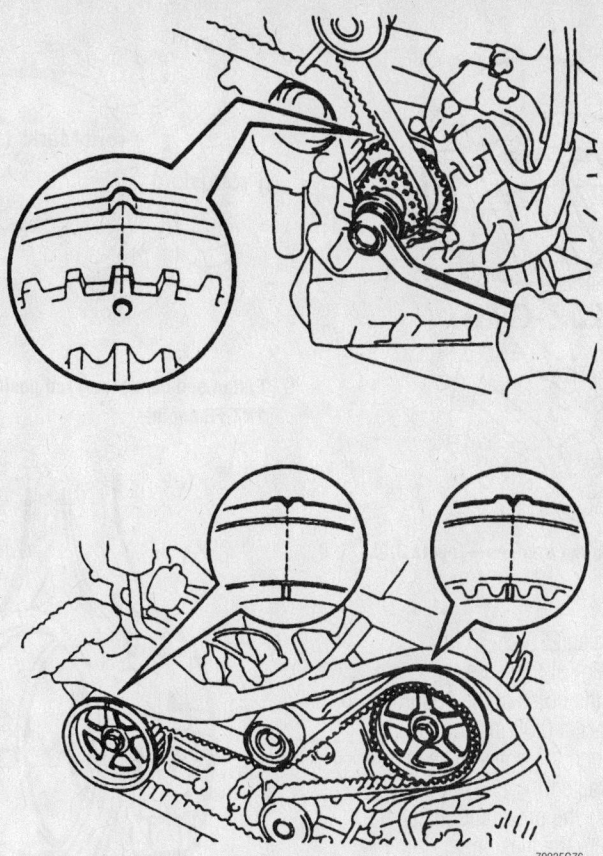

Camshaft and crankshaft timing belt sprocket alignment mark positioning for belt service—Toyota 3.0L (1MZ-FE) engine

To install:

13. Remove any oil or water from the pulleys.

14. Align the front mark of the timing belt with the dot mark of the crankshaft timing pulley.

15. Align the installation marks on the timing belt with the timing marks of the camshaft pulleys.

16. Install the timing belt in the following order:
 a. Crankshaft pulley.
 b. Water pump pulley.
 c. Left camshaft pulley.
 d. No. 2 idler pulley.
 e. Right camshaft pulley.
 f. No. 1 idler pulley.

17. Using a press, slowly press the timing belt tensioner until the holes of the pushrod and housing align. Insert a 0.05 in. (1.27mm) hexagonal Allen wrench through the holes to preserve the setting position.

18. Install the dust boot to the tensioner.

19. Install the tensioner with the two bolts. Alternately tighten the bolts to 20 ft. lbs. (27 Nm). Remove the Allen wrench.

20. Turn the crankshaft clockwise and align the crankshaft timing pulley groove with the oil pump alignment mark.

21. Ensure the camshaft timing marks align with the timing marks on the rear timing belt cover.

22. Install the timing belt guide.

23. Install the right engine mounting bracket and tighten the bolts to 21 ft. lbs. (28 Nm).

24. Install the upper timing belt cover with the five bolts. Tighten the bolts to 74 inch lbs. (8 Nm).

25. Install the engine wire protector clamp to the No. 3 timing belt cover.

26. Install the engine wire protector to the No. 3 timing belt cover with the bolt.

27. Install the lower timing belt cover by installing the four bolts. Tighten the bolts to 74 inch lbs. (8 Nm).

28. Install the remaining components. During installation be sure to tighten the crankshaft pulley bolt to 159 ft. lbs. (215 Nm) and the No. 2 alternator bracket nut to 21 ft. lbs. (28 Nm).

3.0L (2JZ-GTE and 2JZ-GE) Engines

1. Before servicing the vehicle, refer to the precautions in the beginning of this section.

2. Remove all necessary components for access to the timing belt covers.

If equipped with an air bag, be sure to disconnect the negative battery cable and wait at least 90 seconds before proceeding.

3. Remove the upper two timing belt covers (Nos. 2 and 3).

4. Remove the drive belt tensioner.

5. Set the No. 1 cylinder to Top Dead Center (TDC) on the compression stroke. Turn the crankshaft pulley clockwise to align the groove with the **0** mark on the lower (No. 1) timing belt cover. Check that the timing marks on the camshaft pulleys are aligned with the marks on the rear belt cover. If the camshaft marks do not align, turn the crankshaft another 360 degrees.

6. Alternately loosen the two bolts holding the timing belt tensioner. Remove the bolts and remove the tensioner.

7. Remove the timing belt from the camshaft pulleys. If the belt is to be reused, place matchmarks on the belt and gears before removing the belt. Mark the belt with an arrow to show direction of rotation.

8. Using Toyota tool SST 09960-10010, remove the bolts for the camshaft timing gears.

9. Remove the camshaft gears from the engine.

10. If necessary, disconnect the oil cooler tubes from the front of the engine by removing the two bolts and hose clamps.

11. Remove the crankshaft pulley by using Toyota tool 09330-0021 to hold the pulley and using tool 09213-70010 to remove the pulley bolt.

12. Remove the lower (No. 1) timing belt cover and the timing belt guide.

13. Remove the timing belt. If the belt is to be reused, protect it from contact with oil, grease or fluids.

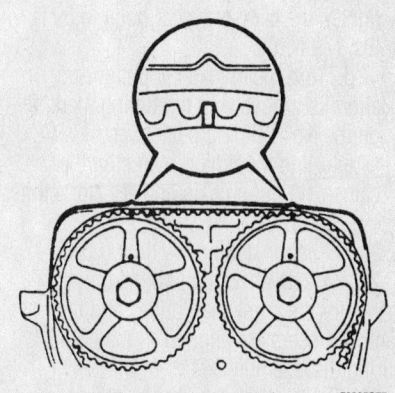

Camshaft timing mark alignment—Toyota 3.0L (2JZ-GTE and 2JZ-GE) engines

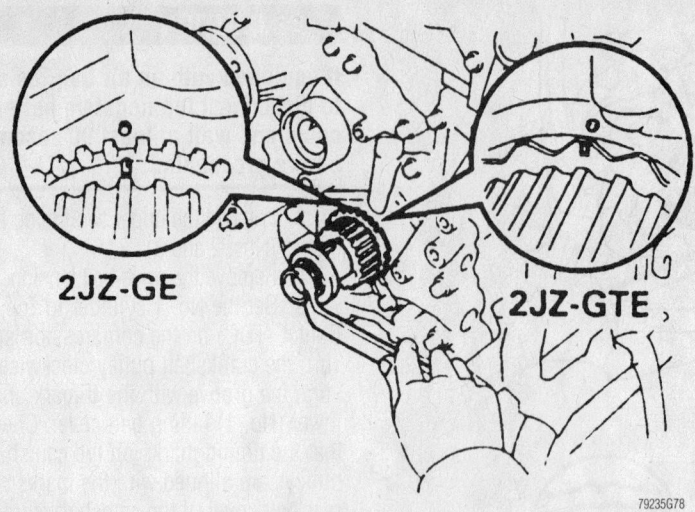

Crankshaft timing marks; notice the timing mark difference between the engines—Toyota 3.0L (2JZ-GTE and 2JZ-GE) engines

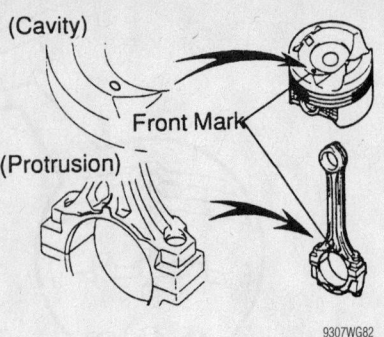

Piston and connecting rod positioning—1NZ-FE engine

To install:

14. Use the crankshaft pulley bolt to turn the crankshaft (clockwise) until the mark on the gear aligns with the oil pump body. Check all the pulleys and gears for cleanliness; remove any grease, oil or coolant. Install the timing belt onto the crankshaft gear and idler pulleys.

15. Install the timing belt guide with the cupped side facing outward.

16. Install the No. 1 timing belt cover.

17. Install the crankshaft pulley. Align the set key with the groove. Hold the pulley with the proper tool and tighten the pulley bolt to 239 ft. lbs. (324 Nm).

18. If equipped with automatic transmission, connect the oil cooler tubes with the clamps and two bolts.

19. Install the camshaft gears as follows:

a. Align the camshaft knock pin with the groove on the gear and slide on the timing gear.

b. Temporarily install the timing gear bolt.

c. Using the same tools as removal, tighten the camshaft gear bolts to 59 ft. lbs. (79 Nm).

d. Turn the crankshaft pulley and align its groove with the timing mark, **0** on the No. 1 timing belt cover.

e. Align the timing marks on the camshaft timing gears and the No. 4 timing belt cover.

20. Finish installing the timing belt.

21. Double check that all the timing marks for the crankshaft pulley and the camshaft gears are aligned as they were during disassembly.

22. Set the timing belt tensioner:

a. Use a press to slowly push in the pushrod on the tensioner. This will require between 220–2200 lbs. (100–1000 kg) of pressure.

b. Align the holes of the pushrod and housing. Place a 0.06 in. (1.5mm) hex wrench through the holes to keep the pushrod retracted.

c. Release the press and install the dust boot onto the tensioner.

23. Install the tensioner; alternately tighten the bolts to 20 ft. lbs. (26 Nm).

24. Remove the hex wrench from the tensioner with a pair of pliers.

25. Turn the crankshaft pulley two full turns clockwise. Check that each pulley's timing marks align correctly after the two turns. If any mark does not align, remove the timing belt and reinstall it.

26. Install the drive belt tensioner and tighten the bolts to 15 ft. lbs. (21 Nm).

27. Install the Nos. 2 and 3 timing belt covers.

28. Install all remaining components. During assembly be sure to tighten the drive belt tensioner damper nuts to 14 ft. lbs. (20 Nm).

Piston and Ring Positioning

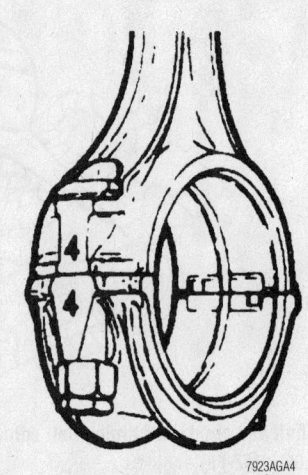

Before removing the caps from the connecting rods, be sure to matchmark them as shown

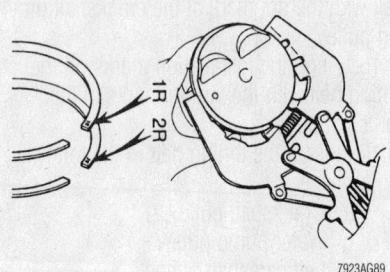

5E-FE, 1ZZ-FE, 2ZZ-GE engine—piston ring identification mark locations

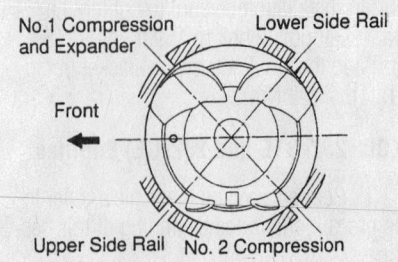

Piston ring positioning—1NZ-FE engine

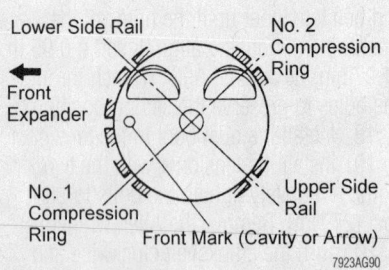

5E-FE, 1ZZ-FE, 2ZZ-GE engine—piston ring end-gap spacing

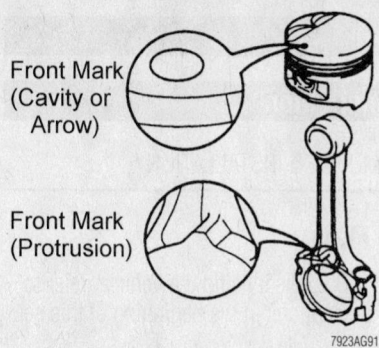

5E-FE, 1ZZ-FE, 2ZZ-GE engine—piston-to-connecting rod assembly

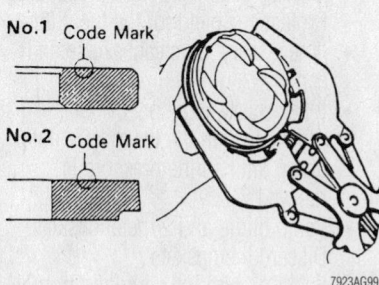

3S-GTE engine—compression ring positioning

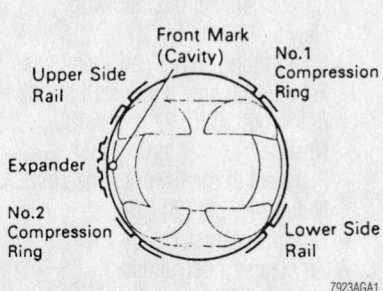

3S-GTE engine—piston ring end-gap spacing

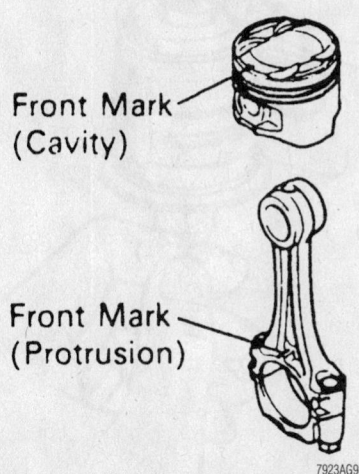

3S-GTE and 5S-FE engines—piston-to-connecting rod assembly

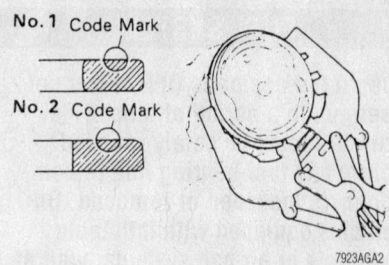

5S-FE engine—compression ring positioning

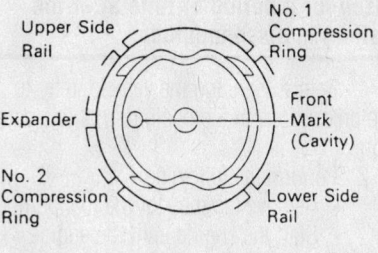

5S-FE engine—piston ring end-gap spacing

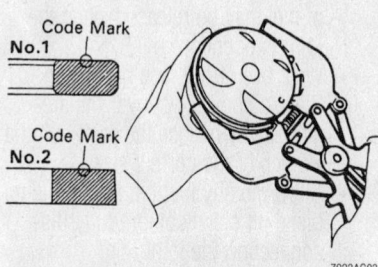

1MZ-FE, 2JZ-GE and 2JZ-GTE engines—compression ring positioning

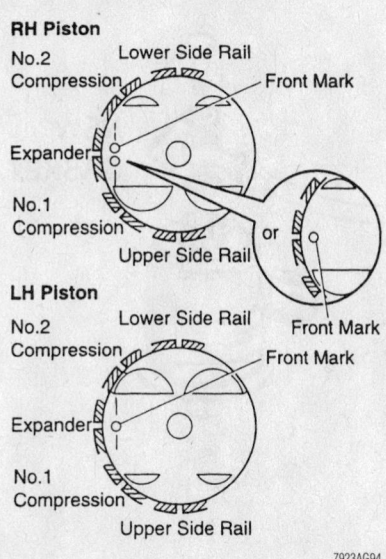

1MZ-FE engine—piston ring end-gap spacing

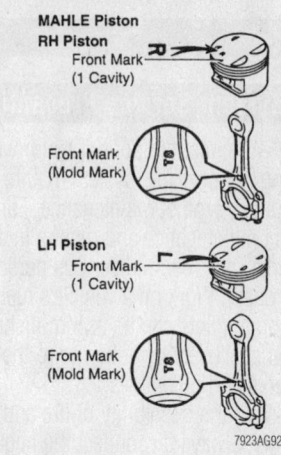

Avalon 1MZ-FE engine—piston-to-connecting rod assembly

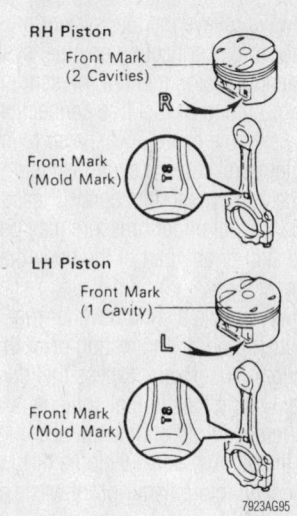

Camry 1MZ-FE engine—piston-to-connecting rod assembly

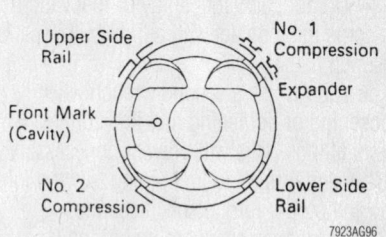

2JZ-GE and 2JZ-GTE engines—piston ring end-gap spacing

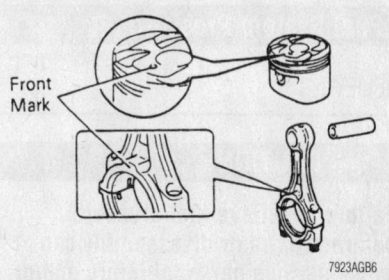

2JZ-GE and 2JZ-GTE engines—piston-to-connecting rod assembly

FUEL SYSTEM

Fuel System Service Precautions

Safety is the most important factor when performing not only fuel system maintenance, but any type of maintenance. Failure to conduct maintenance and repairs in a safe manner may result in serious personal injury or death. Work on a vehicle's fuel system components can be accomplished safely and effectively by adhering to the following rules and guidelines.

• To avoid the possibility of fire and personal injury, always disconnect the negative battery cable unless the repair or test procedure requires that battery voltage by applied.

• Always relieve the fuel system pressure prior to disconnecting any fuel system component (injector, fuel rail, pressure regulator, etc.) fitting or fuel line connection. Exercise extreme caution whenever relieving fuel system pressure, to avoid exposing skin, face and eyes to fuel spray. Please be advised that fuel under pressure may penetrate the skin or any part of the body that it contacts.

• Always place a shop towel or rag around the fitting or connection prior to loosening to absorb any excess fuel due to spillage. Ensure that all fuel spillage is quickly remove from engine surfaces. Ensure that all fuel-soaked cloths or towels are deposited into a flame-proof waste container with a lid.

• Always keep a dry chemical (Class B) fire extinguisher near the work area.

• Do not allow fuel spray or fuel vapors to come into contact with a light bulb, spark or open flame.

• Always use a second wrench when loosening or tightening fuel line connections fittings. This will prevent unnecessary stress and torsion to fuel piping. Always follow the proper torque specifications.

• Always replace worn fuel fitting O-rings with new ones. Do not substitute fuel hose where rigid pipe is installed.

Fuel System Pressure

RELIEVING

✳✳ CAUTION

Failure to relieve fuel pressure before repairs or disassembly can cause serious personal injury and/or property damage. Fuel pressure is maintained within the fuel lines,

even if the engine is OFF or has not been run in a period of time. This pressure must be safely relieved before any fuel-bearing line or component is loosened or removed. On vehicles equipped with inflatable restraints or air bag systems, wait at least 90 seconds after disconnecting the battery cable before performing any other work. The back-up power will keep the restraint system energized for a period of time after the battery is disconnected.

1. Before servicing the vehicle, refer to the precautions in the beginning of this section.
2. Perform the following:
   • Remove the fuse for the fuel pump
   • Start the engine until the engine stalls
   • Disconnect the negative battery cable
   • Place a catch-pan under the joint to be disconnected. A large quantity of fuel may be released when the joint is opened
   • Wear eye or full face protection
   • Place a shop towel over the area and slowly release the joint using a wrench of the correct size.
   • Allow the any fuel left in the line to bleed off slowly before fully disconnecting the joint.
   • Plug the opened lines
3. After connecting fuel lines, install the fuse for the fuel pump and start the engine.

Fuel Filter

REMOVAL & INSTALLATION

All Models

1. Before servicing the vehicle, refer to the precautions in the beginning of this section.
2. Remove or disconnect the following:
   • Negative battery cable. On vehicles equipped with an air bag, wait at least 90 seconds before proceeding
   • Protective shield for the fuel filter
   • If necessary, air cleaner hose and cap
   • If necessary, charcoal canister.
   • Slowly loosen the lower flare nut fitting until all the pressure is relieved
   • Banjo fitting and 2 metal gaskets. Discard the gaskets
   • Fuel line with the flared nut from the filter
   • Filter from the mounting bracket

To install:

3. Install or connect the following:
   • New fuel filter
   • Banjo fitting with a new metal gasket on each side and install the union bolt. Bolt: 22 ft. lbs. (30 Nm).
   • Flare nut to the lower connection. Nut: 22 ft. lbs. (30 Nm).
   • Charcoal canister
   • Air cleaner hose and cap

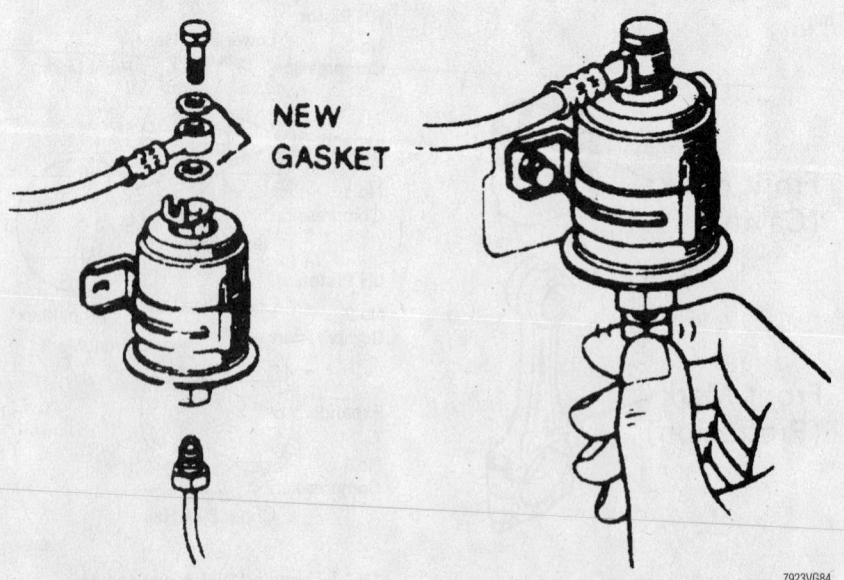

NEW GASKET

Always use new gaskets when replacing a fuel filter

7923VG84

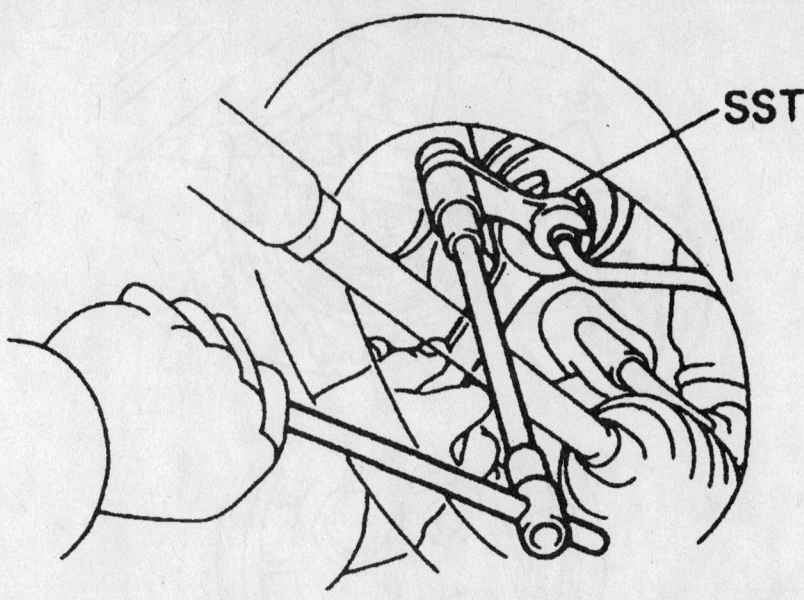

7923VG85

A line wrench with an extension may be needed to loosen the inlet line at the filter—Corolla

- Protective shield
- Negative battery cable

Fuel Pump

REMOVAL & INSTALLATION

Avalon, Camry and Camry Solara

1. Before servicing the vehicle, refer to the precautions in the beginning of this section.
2. Relieve the fuel system pressure.
3. Remove or disconnect the following:
   - Negative battery cable. On vehicles equipped with an air bag, wait at least 90 seconds before proceeding.
   - Fuel tank

➡**On some models it will be necessary to remove the rear seat cushion.**

- Floor service hole cover
- Electrical connector at the fuel pump assembly
- Fuel outlet pipe from the fuel pump bracket
- Return hose from the fuel pump bracket
- Fuel pump bracket assembly from the fuel tank by removing the bolts
- Pump bracket assembly
- Fuel pump from the fuel bracket

To install:
4. Install or connect the following:
   - Fuel pump to the fuel bracket
   - Pump bracket assembly

- Fuel pump to the fuel tank. Bolts: 35 inch lbs. (4 Nm).
- Return hose to the fuel pump bracket
- Outlet pipe to the fuel pump bracket. Tighten to 21 ft. lbs. (28 Nm).
- Service hole cover to the fuel tank

- Fuel pump connector
- Rear seat cushion
- Fuel tank
- Negative battery cable

Echo

1. Before servicing the vehicle, refer to the precautions in the beginning of this section.
2. Relieve the fuel system pressure.
3. Remove or disconnect the following:
   - Negative battery cable. On vehicles equipped with an air bag, wait at least 90 seconds before proceeding.
   - Rear seat cushion
   - Floor service hole cover
   - Electrical connector at the fuel pump assembly
   - Fuel outlet pipe from the fuel pump bracket
   - Return hose from the fuel pump bracket.
   - Fuel pump set plate from the fuel tank by removing the 8 bolts
   - Fuel pump from the fuel bracket

To install:
4. Install or connect the following:
   - Fuel pump to the fuel tank. Bolts: 35 inch lbs. (4 Nm).

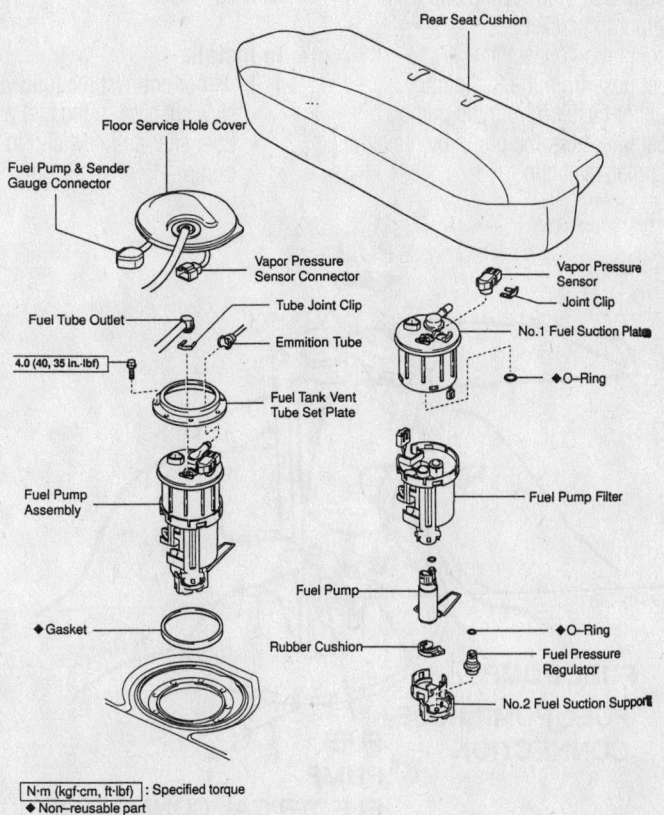

N·m (kgf·cm, ft·lbf) : Specified torque
◆ Non–reusable part

9307WG83

Fuel pump removal—Echo

- Return hose to the fuel pump bracket
- Outlet pipe to the fuel pump bracket. Tighten to 21 ft. lbs. (28 Nm).
- Service hole cover to the fuel tank
- Fuel pump connector
- Rear seat cushion
- Negative battery cable

Celica, GT-S, and Corolla

1. Before servicing the vehicle, refer to the precautions in the beginning of this section.
2. Relieve the fuel system pressure.
3. Remove or disconnect the following:
   - Negative battery cable. On vehicles equipped with an air bag, wait at least 90 seconds before proceeding
   - Rear seat cushion and floor service hole cover
   - Access plate-to-fuel tank bolts, then pull out the plate/fuel pump assembly
   - Fuel pump sender and fuel pump connector
   - Outlet pipe from the fuel pump bracket
   - Return hose from the pump bracket
   - Fuel pump bracket assembly from the fuel tank
   - Lower side of the fuel pump from the pump bracket
   - Fuel pump connector
   - Fuel hose from the fuel pump
   - Rubber cushion from the pump
   - Fuel filter from the pump by removing the small clip

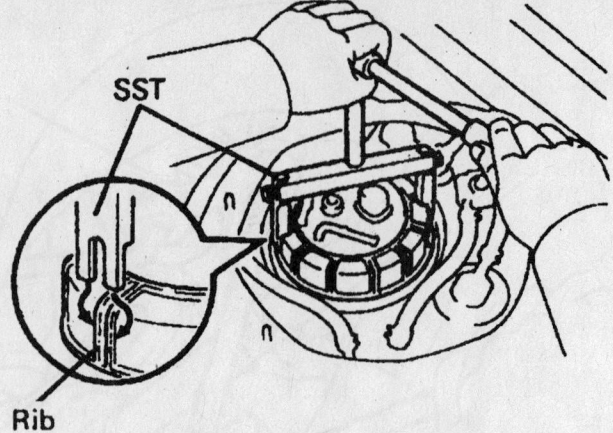

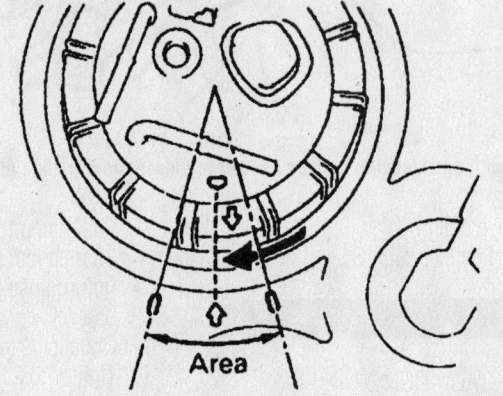

Tighten the fuel pump retainer until the arrow mark on the retainer is within the lines on the fuel tank—Supra

To install:

4. Install or connect the following:
   - New cushion to the fuel pump
   - Fuel filter and new clip to the fuel pump

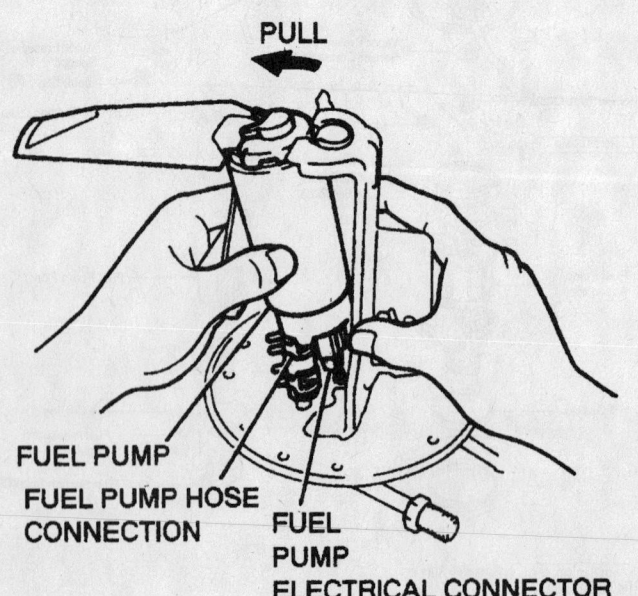

PULL

FUEL PUMP

FUEL PUMP HOSE CONNECTION

FUEL PUMP ELECTRICAL CONNECTOR

Pull the pump off the sender unit; the filter is still attached to the pump

- Fuel hose to the fuel pump, fuel pump connector and fuel pump to the bracket
- Fuel pump bracket assembly to the fuel tank using a new gasket. Bolts: 30 inch lbs. (3 Nm).
- Fuel return hose and the fuel outlet pipe to the fuel pump bracket
- Fuel pump and fuel pump sender connector
- Fuel tank
- Negative battery cable
- Floor service hole cover and rear seat cushion

Supra

1. Before servicing the vehicle, refer to the precautions in the beginning of this section.
2. Relieve the fuel system pressure.
3. Remove or disconnect the following:
   - Negative battery cable. On vehicles equipped with an air bag, wait at least 90 seconds before proceeding
   - Floor carpet, spare wheel cover, spare wheel, service hole cover
   - Fuel pump electrical connector from the fuel pump

- Outlet hose from the fuel pump
- Fuel return hose from the fuel pump
- Fuel breather clamp
- Loosen retainer to the fuel pump
- Fuel return hose from the return port of the fuel pump bracket
- Retainer, fuel pump, sender gauge assembly, and the gasket as a unit

To install:

4. Install or connect the following:
- New gasket to the fuel pump
- Fuel pump and sender gauge assembly into the fuel tank
- Align the arrow marks of the fuel pump bracket and the fuel tank
- Fuel pump retainer until the arrow mark on the retainer is within the lines on the fuel tank
- Retainer clamp to the fuel pump
- Fuel pump electrical connector to the fuel pump
- Outlet hose with 2 new gaskets. Bolt: 22 ft. lbs. (29 Nm).
- Fuel return hose to the fuel pump
- Fuel breather hose to the fuel pump
- Negative battery cable
- Service hole cover
- Spare wheel, spare wheel cover, floor carpet

MR2

1. Before servicing the vehicle, refer to the precautions in the beginning of this section.
2. Remove or disconnect the following:
- Negative battery cable
- Luggage compartment box
- Service access panel
- Fuel pump and gauge harness connector
- Vapor pressure sensor harness connector
- Fuel main tube
- Emissions tube
- Fuel pump module

To install:

3. Install or connect the following:
- Fuel pump module. Tighten the bolts to 30 inch lbs. (3.4 Nm).
- Emissions tube
- Fuel main tube
- Vapor pressure sensor harness connector
- Fuel pump and gauge harness connector
- Service access panel
- Luggage compartment box
- Negative battery cable
4. Start the engine and check for leaks.

Fuel Injectors

REMOVAL & INSTALLATION

Avalon and Camry

5S-FE ENGINE

1. Before servicing the vehicle, refer to the precautions in the beginning of this section.
2. Drain the cooling system.
3. Remove or disconnect the following:
- Negative battery cable
- Accelerator cable from the throttle body
- If equipped with automatic transmission, the throttle cable
- Air intake temperature sensor connector
- Cruise control actuator cable from the clamp on the resonator
- Air cleaner
- Wiring from the throttle position sensor and the ISC valve
- Hoses from the PCV, EGR vacuum modulator and EVAP VSV
- Throttle body
- PS vacuum hoses
- Hoses from the EVAP Bi-metal Vacuum Switching Valve (BVSV)
- EGR valve and the vacuum modulator
- Vacuum sensor hose at the air intake chamber
- Brake booster vacuum hose and the vacuum sensing hose
- With air conditioning, the magnet switch VSV wiring
- Ground straps from the intake manifold
- Knock sensor and EGR Vacuum Switching Valve (VSV) wiring
- Engine wire harness
- Stays or supports holding the air intake chamber and the intake manifold
- Intake manifold
- Wiring from each injector
- Fuel inlet pipe
- Fuel return hose
- Delivery pipe or fuel rail along with the injectors
- Insulators and rail spacers from the head
- Injectors from the fuel rail
- O-ring and grommet from each injector

To install:

4. Install or connect the following:
- New grommet on each injector
- New O-rings, coated with gasoline, on each injector
- Injectors into the fuel rail while turning each left and right. After installation, check that the injectors turn freely in place; if not, remove the injector and inspect the O-ring for damage or deformation.
- New insulators and spacers on the head
- Fuel rail and injectors; check that the injectors still turn freely in

Loosening and removing the pulsation damper—5S-FE engine

89555G10

Unbolt and remove the fuel delivery pipe and injectors as an assembly

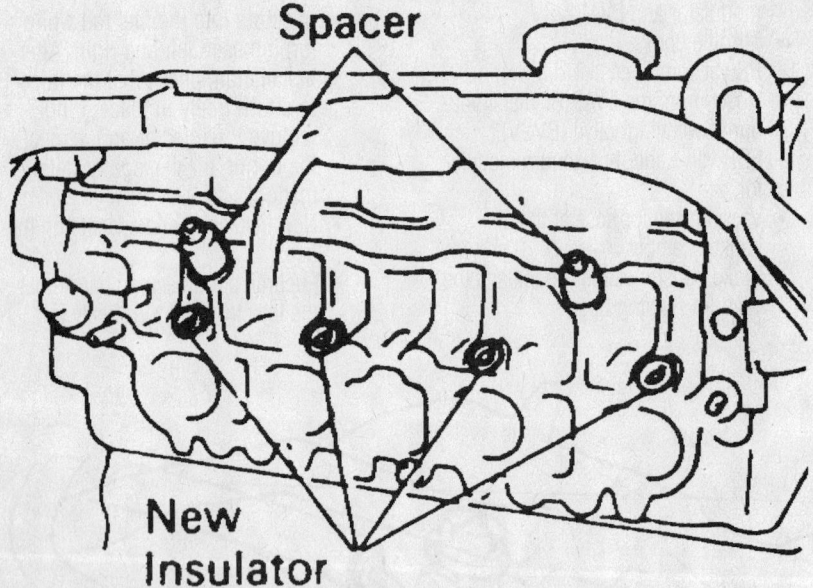

Place new injector seals and spacers in position on the cylinder head—5S-FE engine shown

position. Position the injector connectors upward
- Retaining bolts, tightening them to 9 ft. lbs. (13 Nm)
- Fuel return hose
- Fuel inlet pipe and pulsation damper to the delivery pipe. Use new gaskets; tighten the union bolt to 25 ft. lbs. (34 Nm).
- Wiring to each injector
- Intake manifold. Nuts and bolts evenly in several passes to 14 ft. lbs. (19 Nm).

- Air chamber and manifold stays. 14mm bolt: 31 ft. lbs. (42 Nm) and the 12mm bolt to 16 ft. lbs. (22 Nm).
- Engine wire harness
- Wiring to the knock sensor and the EGR VSV
- Both engine ground straps to the intake manifold
- A/C magnet switch wiring
- Hoses for the vacuum sensor, brake booster and vacuum sensing hose
- EGR valve and vacuum modulator.

Use new gaskets. Tighten the union nut to 43 ft. lbs. (59 Nm) and the bolt to 9 ft. lbs. (13 Nm).
- Hoses to the charcoal canister and EGR VSV
- Wiring to the EGR temperature sensor if it was removed
- Vacuum hoses to the EVAP BVSV
- PS vacuum hoses
- Throttle body. Bolts, evenly and alternately, to 14 ft. lbs. (19 Nm).

➡The upper mounting bolts are shorter than the lower mounting bolts. Make certain the bolts are correctly placed before tightening.

- PCV, EGR vacuum modulator and EGR VSV hoses to the throttle body
- Wiring for the throttle position sensor
- Air cleaner cap, resonator and intake hose
- Wiring to the air intake temperature sensor
- Cruise control actuator cable
- Throttle control cable
- Accelerator cable
- Negative battery cable

5. Fill the cooling system to the proper level.

6. Start the vehicle, check for leaks and repair if necessary.

1MZ-FE ENGINE

1. Before servicing the vehicle, refer to the precautions in the beginning of this section.

2. Properly relieve the fuel system pressure.

3. Drain the cooling system.

4. Remove or disconnect the following:
- Negative battery cable. Work must be started approximately 90 seconds or longer after the negative battery cable has been disconnected, if equipped with an air bag
- Accelerator and throttle cables
- Air cleaner assembly
- V-bank cover
- Emission valve control set
- No. 2 EGR pipe
- Hydraulic motor pressure pipe from the water inlet and air inlet chamber
- Air intake chamber assembly
- Injector wiring
- Air assist pipe from the bracket on the No. 1 fuel pipe
- Air assist hoses from the intake manifold
- Fuel return hose from the No. 1 fuel pipe
- Fuel inlet hose for the fuel filter

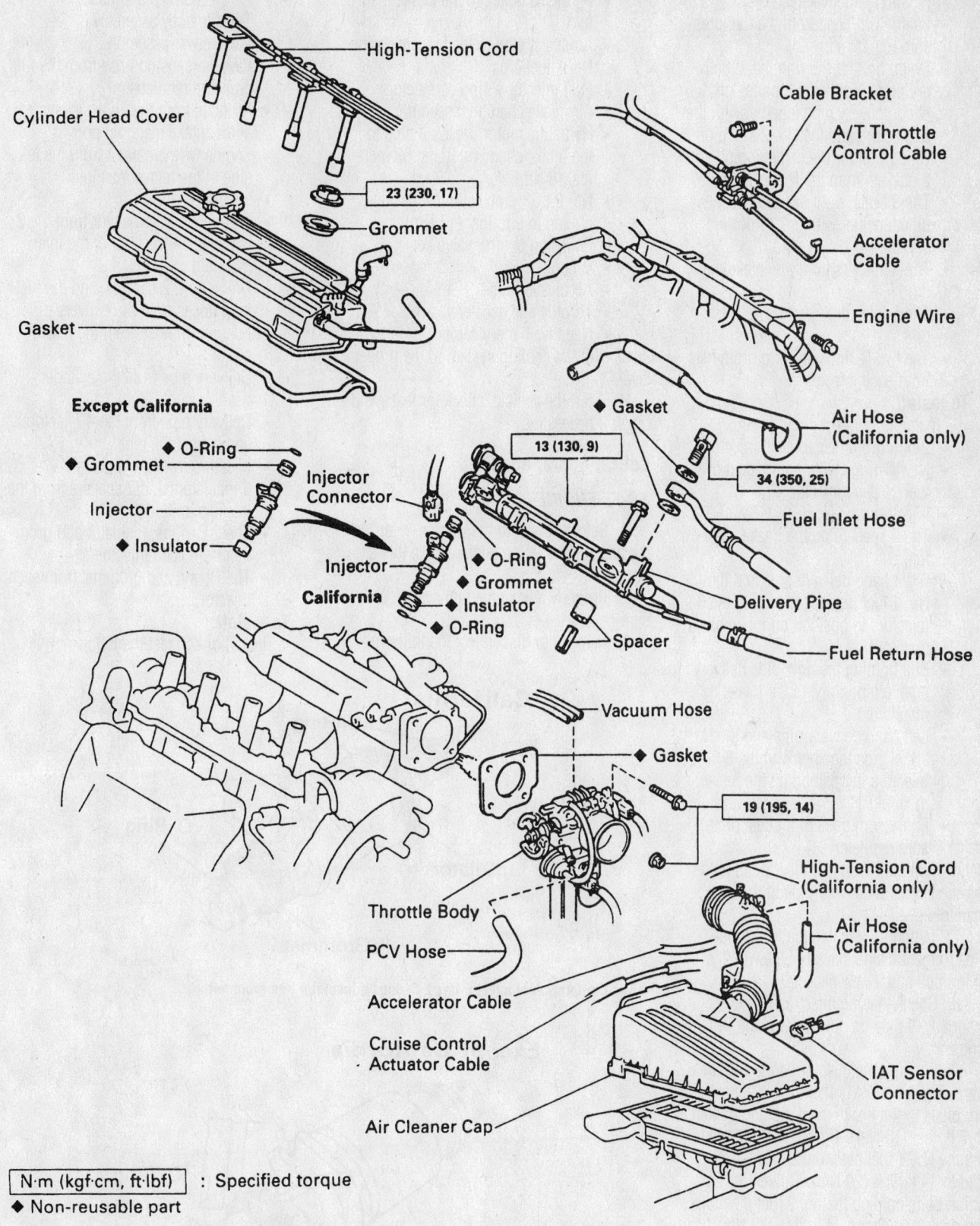

High-Tension Cord

Cable Bracket

A/T Throttle Control Cable

Cylinder Head Cover

23 (230, 17)

Grommet

Accelerator Cable

Gasket

Engine Wire

Except California

◆ Gasket

Air Hose (California only)

◆ O-Ring

◆ Grommet

Injector Connector

13 (130, 9)

34 (350, 25)

Injector

Fuel Inlet Hose

◆ Insulator

Injector

◆ O-Ring

◆ Grommet

California

◆ Insulator

◆ O-Ring

Delivery Pipe

Spacer

Fuel Return Hose

Vacuum Hose

◆ Gasket

19 (195, 14)

Throttle Body

PCV Hose

High-Tension Cord (California only)

Air Hose (California only)

Accelerator Cable

Cruise Control Actuator Cable

IAT Sensor Connector

Air Cleaner Cap

N·m (kgf·cm, ft·lbf) : Specified torque

◆ Non-reusable part

89595G28

View of the fuel injector, delivery pipe and related components—5S-FE engine

- 2 union bolts holding the No. 2 fuel pipe to the delivery pipes
- Fuel return hose from the fuel pressure regulator
- Union bolt for the right hand delivery pipe, 2 gaskets, 2 bolts, left hand delivery pipe together with the 3 injectors and the No. 2 fuel pipe
- Union bolt for the delivery pipe and 2 gaskets from the No. 2 fuel pipe
- The 3 bolts, right hand delivery pipe together with the 3 injectors and the No. 1 fuel pipe
- The 4 spacers from the intake manifold
- The 6 injectors from the delivery pipes
- The two O-rings and two grommets from each injector

To install:

5. Install or connect the following:
- 2 new grommets to each injector
- New O-rings, with a light coat of fuel, to each injector
- Injectors
- The 4 spacers on the intake manifold
- Right hand delivery pipe and the No. 1 fuel pipe together with the 3 injectors in position on the intake manifold
- Bolt holding the right side delivery pipe, temporarily, to the intake manifold
- Left hand delivery pipe and the No. 2 fuel pipe together with the 3 injectors in position on the intake manifold
- Fuel return hose to the fuel pressure regulator

6. Temporarily install the 2 bolts holding the left hand delivery pipe to the intake manifold.

7. Temporarily install the No. 2 fuel pipe to the left side delivery pipe with the union bolt and 2 new gaskets.

8. Check that the injectors rotate smoothly. If they do not, Replace the O-rings.

9. Position the injector connector outward. Tighten the 4 bolts holding the delivery pipes to the intake manifold and tighten to 7 ft. lbs. (10 Nm). Tighten the bolt holding the No. 1 fuel pipe to the intake manifold to 14 ft. lbs. (20 Nm). Tighten the 2 union bolts holding the no. 2 fuel pipe to the delivery pipes to 24 ft. lbs. (32 Nm).

10. Install or connect the following:
- Fuel inlet and return hoses. Union bolt: 22 ft. lbs. (30 Nm).
- Fuel return hose to the No. 1 fuel pipe. Pass the fuel return hose

under the heater hoses
- Air assist hoses to the intake manifold
- Air assist pipe to the bracket on the No. 1 fuel pipe
- Fuel injector wiring connectors
- Air intake chamber assembly
- Hydraulic motor pressure pipe to the intake chamber. Bolts: 69 inch lbs. (8 Nm)
- No. 2 EGR pipe with new gaskets, tighten to 9 ft. lbs. (12 Nm)
- Emission control valve set
- V-bank cover
- Air cleaner hose
- Throttle and accelerator cables
- Negative battery cable

11. Fill the cooling system to the proper level.

12. Start the vehicle, check for leaks and repair if necessary.

Celica, Supra, and GT-S

5S-FE ENGINE

1. Before servicing the vehicle, refer to the precautions in the beginning of this section.

2. Properly relieve the fuel system pressure.

3. Remove or disconnect the following:

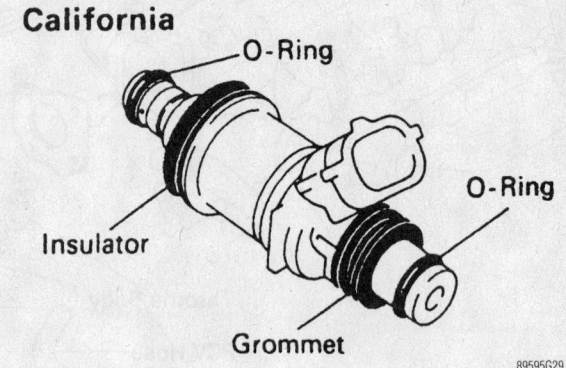

California

O-Ring

O-Ring

Insulator

Grommet

89595G29

The California fuel injector has 2 O-rings a insulator and grommet

Except California

O-Ring

Grommet

89595G30

View of the non-California fuel injector

- Negative battery cable
- Air cleaner cap and hose
- Throttle body assembly
- Valve cover assembly
- Vacuum sensing hose from the fuel pressure regulator
- Air hose from the air assist system at the intake manifold port
- Engine wire protector from the left side of the intake manifold
- Injector wiring
- Engine wiring protector from the 2 brackets on the front of the intake manifold
- Union bolt, 2 gaskets, and the fuel inlet hoses from the delivery pipe
- Fuel return hose from the return pipe
- Delivery pipe from the cylinder head
- Delivery pipe from the 4 injectors and extract the pipe
- Injectors from the pipe
- 4 insulators and 2 spacers from the intake manifold
- The 2 O-rings, insulator and grommet from each injector
- The O-ring and grommet from each injector

To install:

4. Install or connect the following:

- New insulator and grommet on each injector
- New grommet on each insulator
- O-ring(s) on each injector
- Spacers and insulators to the intake manifold
- Delivery pipe between the intake manifold and cylinder head
- Injectors into the delivery pipe

➡ **Position each injector in with the connector facing upward.**

- Delivery pipe assembly to the cylinder head
5. Temporarily install the bolts holding

the pipe to the cylinder head. Check that the injectors rotate smoothly. If they do not rotate smoothly, the O-ring(s) may have been install incorrectly. Replace the O-ring if this occurs.

6. Install or connect the following:
- Delivery pipe retaining bolts: 9 ft. lbs. (13 Nm).
- Fuel inlet hose to the delivery pipe with new gaskets and union bolt. Tighten the assembly to 25 ft. lbs. (34 Nm).

➡ **Be careful of the fuel inlet hose installation direction.**

- Fuel return hose to the return pipe
- Engine wire protector to the brackets on the front side of the intake manifold
- Injector wiring harnesses

➡ **The No. 1 and No. 3 injector wiring are brown and the No. 2 and No. 4 are gray.**

- Engine wire protector to the left side of the intake manifold
- Vacuum sensing hose to the fuel pressure regulator
- Air hose for the air assist system to the intake manifold port
- Valve cover assembly
- Throttle body
- Control cables to the throttle body
- Air cleaner cap and hose
- Negative battery cable

1ZZ-FE ENGINE

1. Before servicing the vehicle, refer to the precautions in the beginning of this section.
2. Properly relieve the fuel system pressure.
3. Remove or disconnect the following:
- No. 2 cylinder head cover
- PCV hose
- Fuel tube from the fuel pipe
- Injector connectors
- Delivery pipe and injectors
- Spacers from the head
- Injectors from the delivery pipe

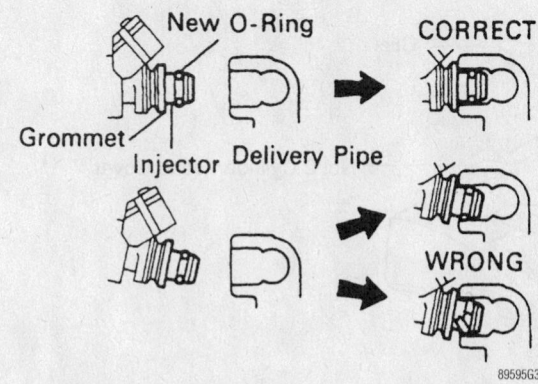

The correct way to install a new fuel injector O-ring

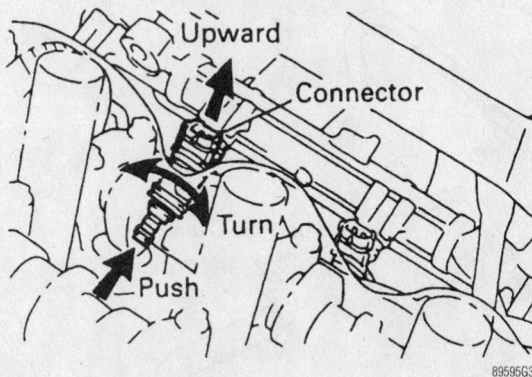

When inserting the injector, turn left and right making sure the connector is facing upward

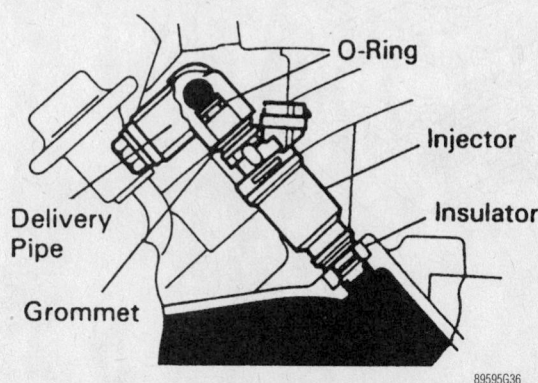

The injector will seat in the fuel rail and intake manifold as shown

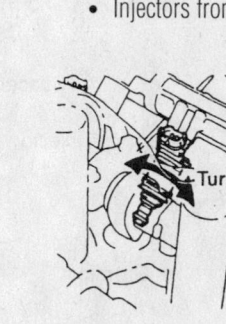

Once the injector is in check for it to rotate smoothly

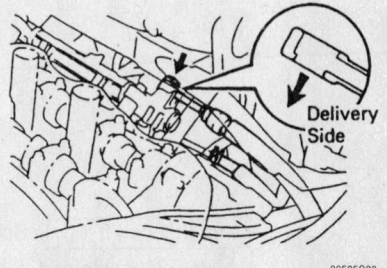

Place the fuel inlet hose on in the correct direction

- O-ring and grommet from each injector

To install:

4. Install or connect the following:
- New grommets
- New O-rings coated with light machine oil
- Injectors on the delivery pipe

➡ **Coat the contact point on the pipe with light machine oil and twist the injectors into place. The connector should face outward.**

- Spacers

➡ **Coat the seats in the head where the injectors contact, with light machine oil.**

- Delivery pipe and injectors

5. Loosely install the hold-down bolts and check that the injectors rotate smoothly. If they don't, the probable cause is incorrect O-ring installation.

6. Torque the hold-down bolts to 14 ft. lbs. (19 Nm).

7. Torque the fuel pipe bolt to 84 inch lbs. (9 Nm).

8. Connect the fuel line.
9. Install the PCV hose.
10. Install the No. 2 cover.

2ZZ-GE ENGINE

1. Before servicing the vehicle, refer to the precautions in the beginning of this section.

2. Properly relieve the fuel system pressure.

3. Remove or disconnect the following:
- No. 2 cylinder head cover
- Fuel tube from the fuel pipe

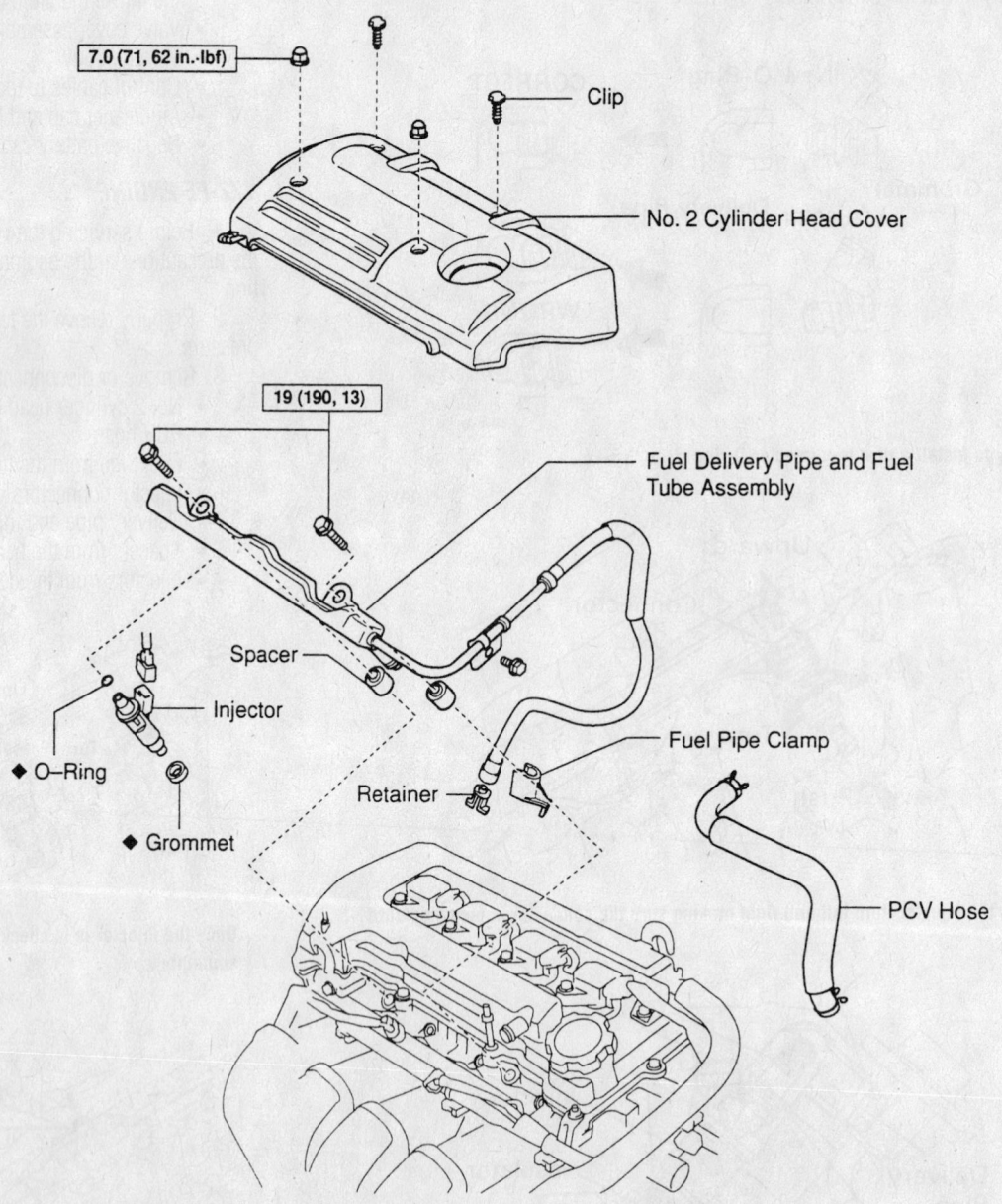

7.0 (71, 62 in.·lbf)

Clip

No. 2 Cylinder Head Cover

19 (190, 13)

Fuel Delivery Pipe and Fuel Tube Assembly

Spacer

Injector

◆ O-Ring

◆ Grommet

Retainer

Fuel Pipe Clamp

PCV Hose

N·m (kgf·cm, ft·lbf) : Specified torque

◆ Non–reusable part

9307WG95

Fuel injector removal and installation—1ZZ-FE engine

- Injector connectors
- Delivery pipe and injectors
- Spacers from the head
- Injectors from the delivery pipe
- O-ring and grommet from each injector

To install:

4. Install or connect the following:
   - New grommets
   - New O-rings coated with light machine oil

- Injectors on the delivery pipe

➡ **Coat the contact point on the pipe with light machine oil and twist the injectors into place. The connector should face outward.**

   - Spacers

➡ **Coat the seats in the head where the injectors contact, with light machine oil.**

   - Delivery pipe and injectors

5. Loosely install the hold-down bolts and check that the injectors rotate smoothly. If they don't, the probable cause is incorrect O-ring installation.

6. Torque the hold-down bolts to 21 ft. lbs. (29 Nm).

7. Torque the fuel pipe bolt to 84 inch lbs. (9 Nm).

8. Connect the fuel line.

9. Install the PCV hose.

10. Install the No. 2 cover.

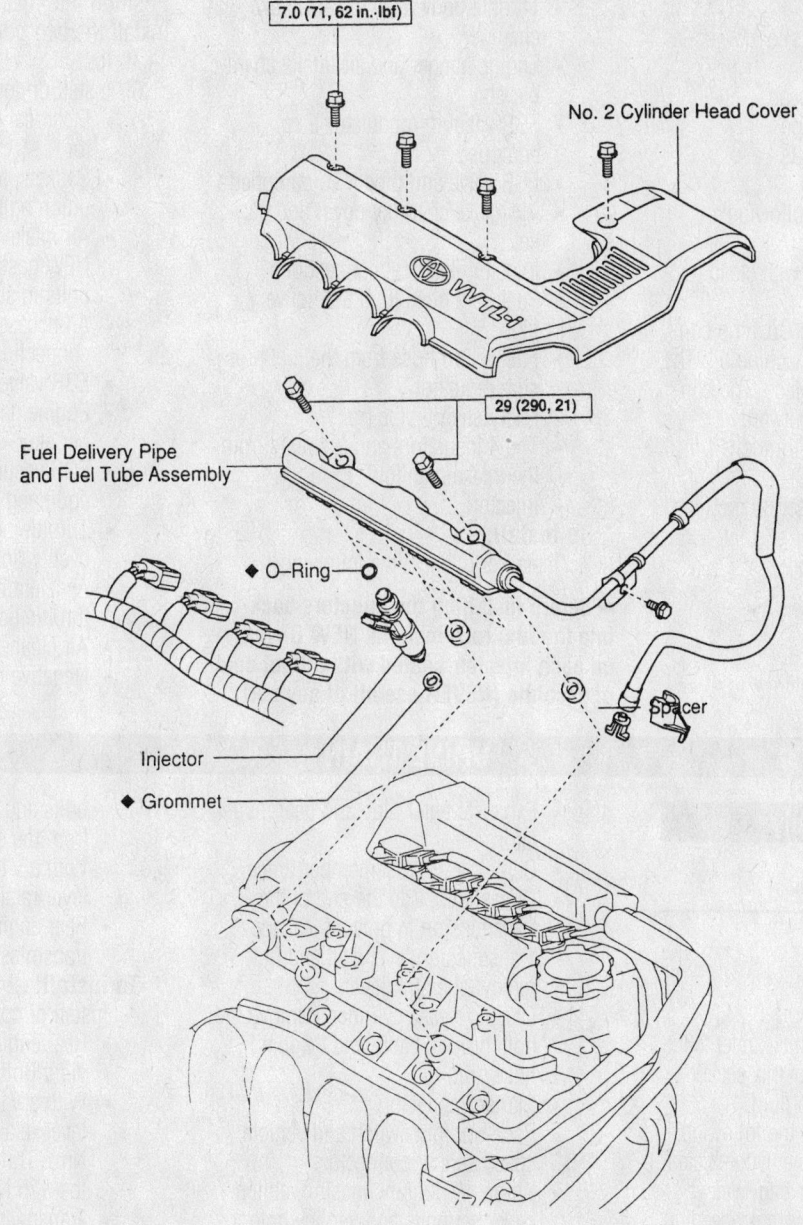

7.0 (71, 62 in.·lbf)

No. 2 Cylinder Head Cover

29 (290, 21)

Fuel Delivery Pipe and Fuel Tube Assembly

◆ O–Ring

Spacer

Injector

◆ Grommet

N·m (kgf·cm, ft·lbf) : Specified torque
◆ Non–reusable part

9307WG96

Fuel injector removal and installation—2ZZ-GE engine

Echo

1NZ-FE ENGINE

1. Before servicing the vehicle, refer to the precautions in the beginning of this section.

2. Remove or disconnect the following:

- Negative battery cable
- Cylinder head cover
- Fuel pipe clamp
- Fuel inlet line from the fuel pipe
- Injector connectors from the injectors
- Delivery pipe (3 bolts) and injectors
- 2 spacers from the head
- Injectors from the pipe
- O-rings and grommets

To install:

3. Install or connect the following:
- New grommets
- New O-rings coated with clean engine oil
- Injectors to the pipe. Coat the contact area with light machine oil. The injectors twist into place. The connector should face outward
- Delivery pipe bolts. Torque: 14 ft. lbs. (19 Nm).
- Fuel pipe bolt. Torque: 80 inch lbs. (9 Nm).
- Fuel hose to fuel pipe
- Pipe clamp
- Wire harness cover
- PCV hose
- Negative battery cable

Corolla and MR2

1ZZ-FE ENGINES

1. Before servicing the vehicle, refer to the precautions in the beginning of this section.

2. Properly relieve the fuel system pressure.

3. Remove or disconnect the following:
- Negative battery cable
- Air cleaner
- Accelerator cable bracket from the throttle body
- Throttle body from the air intake chamber
- Engine hanger and air intake chamber stay
- EGR vacuum modulator if so equipped
- EGR valve and pipe if so equipped
- Air intake chamber cover and gasket
- Injector electrical connections
- Fuel inlet hose from the delivery pipe
- Fuel return hose from the fuel pressure regulator
- Fuel delivery pipe (rail)
- The 4 insulators and 2 collars from the intake manifold
- Injectors

To install:

4. Install or connect the following:

➡ **Before installing the injectors back into the fuel rail, install a NEW O-ring on each injector, coated with a light coat of gasoline (NEVER use oil of any sort).**

- Injectors

➡ **Make certain each injector can be smoothly rotated. If they do not rotate smoothly, the O-ring is not in its correct position.**

- Insulators into each injector hole
- The two spacers on the delivery pipe mounting holes in the intake manifold

5. Place the delivery pipe and injectors on the intake manifold and again check that the injectors rotate smoothly. Position the injector connector upward. Install the two bolts and tighten them to 11 ft. lbs.

6. Install or connect the following:
- Electrical connectors to each injector
- Gaskets, the inlet pipe and fuel union bolt. Bolt to 22 ft. lbs.
- Air intake chamber cover with a NEW gasket. Torque the retaining bolts in steps to 14 ft. lbs.
- All necessary hoses and electrical connections
- EGR valve and pipe if so equipped
- Engine hanger and air intake chamber stay
- EGR vacuum modulator if so equipped
- Throttle body. Torque the bolts evenly (in a X-pattern) to 16 ft. lbs.
- Accelerator cable bracket to the throttle body
- Air cleaner hose and cap
- Negative battery cable

DRIVE TRAIN

Transmission Assembly

REMOVAL & INSTALLATION

Supra

MANUAL

1. Before servicing the vehicle, refer to the precautions in the beginning of this section.

2. Drain the transmission fluid.

3. Remove or disconnect the following:
- Negative battery cable. On vehicles equipped with an air bag, wait at least 90 seconds before proceeding
- Fan shroud set bolts
- Shift lever knob, upper console panel, shift and select lever boot No. 1 and No. 2
- Oxygen sensor, exhaust front pipe and pipe support bracket
- Exhaust center pipe and heat insulator
- Center floor crossmember brace
- Driveshafts. Cap the end of the transmission to prevent leakage
- Transmission lever bolt and nut, remove the shift lever
- Clutch release cylinder. Remove the bolt, ground cable and flexible hose bracket
- Starter connector
- Back-up light switch and vehicle speed sensor connectors
- With a V160 transmission, clutch cover set bolts and service hole cover. Place matchmarks on the flywheel and clutch cover. Remove the 6 bolts
- Rear engine mounting member
- Starter
- Remaining transmission mounting bolts and remove the transmission from the engine
- With a V160 transmission, shift lever retainer from the transmission
- Rear engine mounting from the transmission

To install:

4. Install or connect the following:
- Rear engine mount to the transmission. Bolts: 18 ft. lbs. (25 Nm).
- With a V160 transmission, shift lever retainer. Bolts: 14 ft. lbs. (19 Nm). Through-bolt and nut: 18 ft. lbs. (25 Nm).
- Transmission to the engine. Bolts: 53 ft. lbs. (72 Nm).
- Starter. Bolts: 29 ft. lbs. (39 Nm).
- Rear engine mounting member. Nuts: 10 ft. lbs. (13 Nm). Bolts: 19 ft. lbs. (25 Nm).
- With a V160 transmission, clutch

cover set bolts. Bolts: 14 ft. lbs. (19 Nm). Install the service hole cover. Bolts: 108 inch lbs. (12 Nm).
- Vehicle speed sensor and back-up light switch connectors
- Starter wire connector
- Clutch release cylinder. Bolts: 108 inch lbs. (12 Nm). Install bolt with the clamp and ground cable. Bolt: 53 ft. lbs. (72 Nm).
- Transmission shift lever. Bolts: 14 ft. lbs. (19 Nm).
- Driveshafts, with grease applied to the flexible coupling centering bushings
- Center floor crossmember brace. Bolts: 108 inch lbs. (13 Nm).
- Heat insulator. Bolts: 48 inch lbs. (5 Nm).
- Center pipe, using new gaskets. Bolts: 14 ft. lbs. (19 Nm).
- Front pipe, using new gaskets. Bolts: 43 ft. lbs. (58 Nm).
- Pipe support bracket. Bolts: 27 ft. lbs. (37 Nm).
- Front pipe set bolts and nuts. Bolts and nuts: 32 ft. lbs. (43 Nm).
- Oxygen sensor and cover using a new gasket. Tighten to 13 ft. lbs. (18 Nm).
- Shift lever mount bolts. Bolts: 69 inch lbs. (8 Nm).
- Shift and select lever boot No. 1 and No. 2.
- Upper console panel to the console box with the 4 clips.

- Shift lever knob
- Fan shroud set bolts
- Negative battery cable

5. Fill the transmission fluid to the proper level.

6. Start the vehicle, check for leaks and repair if necessary.

AUTOMATIC

1. Before servicing the vehicle, refer to the precautions in the beginning of this section.

2. Remove or disconnect the following:
- Negative battery cable. On vehicles equipped with an air bag, wait at least 90 seconds before proceeding.
- Transmission oil level gauge and filler pipe.
- Undercover
- Oxygen sensor
- Exhaust pipe
- Heat insulator by removing the 4 bolts (normal roof) or 6 bolts (sport roof).
- Rear center floor crossmember brace
- Driveshaft
- Control rod from the shift lever
- Shift control rod from the park/neutral position switch by removing the nut.
- No. 1 vehicle speed sensor
- No. 2 vehicle speed sensor
- Solenoid wire
- Park/neutral position switch

- Automatic transmission fluid temperature sensor
- With overdrive, overdrive direct clutch speed sensor
- Starter electrical connector to the starter, then the nut and cable from the starter
- Transmission oil cooler pipes
- With turbocharged models, intercooler pipe
- Torque converter clutch mounting bolts
- Rear mounting by removing the 4 outer bolts
- 4 wiring harness clamps from the retainer
- Starter and transmission set bolts
- Transmission

To install:

3. Install or connect the following:
- Transmission and starter. Bolts: 14mm bolts: 27 ft. lbs. (37 Nm.). 17mm bolts: 53 ft. lbs. (72 Nm)
- Starter cable and nut, then electrical connector to the starter
- 4 wiring harness clamps to the retainer
- Rear mounting bolts. Bolts: 19 ft. lbs. (25 Nm).
- Torque converter clutch mounting bolts
- With turbocharged models, intercooler pipe
- Transmission oil cooler pipes. Cooler union nuts: 25 ft. lbs. (34 Nm). Cooler pipe bracket bolt: 84 inch lbs. (10 Nm).
- Center and rear oil cooler pipe brackets. Bolts: 84 inch lbs. (34 Nm).
- Automatic transmission fluid temperature sensor
- Park/neutral position switch
- Solenoid wire
- No. 2 vehicle speed sensor
- No. 1 vehicle speed sensor
- Overdrive direct clutch speed sensor
- Shift control rod to the park/neutral position switch. Nut: 12 ft. lbs. (16 Nm).
- Shift control rod to the shift lever. Nut: 108 inch lbs. (13 Nm). Inspect and adjust the park/neutral position switch as needed
- Driveshaft
- Rear center floor crossmember brace. Bolts: 108 inch lbs. 913 Nm).
- Heat insulator with the bolts and tighten to 48 inch lbs. (5.4 Nm).
- Exhaust pipe. Bracket to transmis-

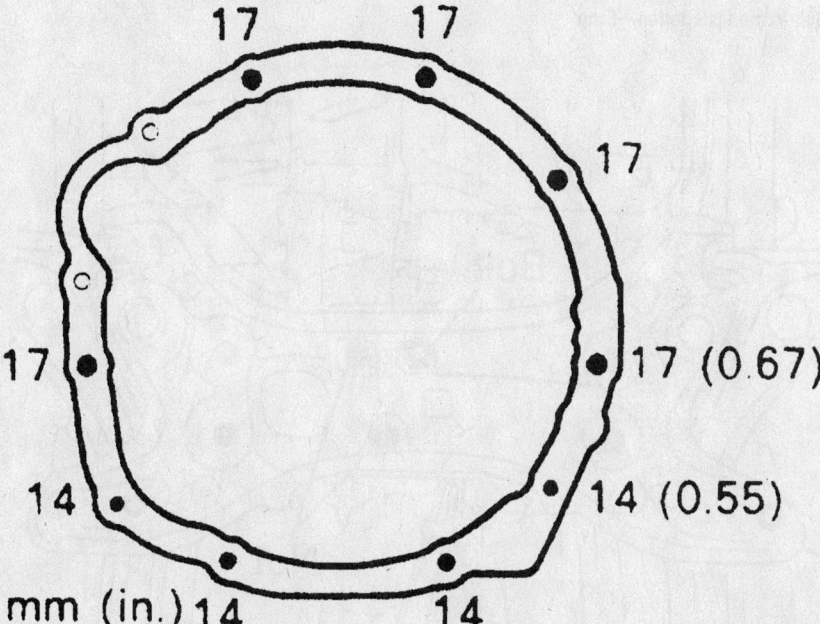

17 17
17
17
17 (0.67)
14 (0.55)
mm (in.) 14   14

Transmission bolt identification—Supra

7923VG99

sion housing: 27 ft. lbs. (37 Nm);
No. 2 exhaust pipe to center
exhaust pipe: 43 ft. lbs. (58 Nm).
• Oxygen sensor
• Undercover
• Transmission filler pipe and level
  gauge
• Negative battery cable

Manual Transaxle Assembly

REMOVAL & INSTALLATION

Echo

1. Before servicing the vehicle, refer to
the precautions in the beginning of this sec-
tion.
2. Drain the transaxle fluid.
3. Remove or disconnect the following:
• Hood
• Wiper arms
• Right and left cowl top ventilator
  covers
• No. 2 cylinder head cover
• Battery
• Air cleaner assembly
• Wiring harness from the transaxle
• Transaxle control cable
• Clutch release cylinder
• Ground cable from the left engine
  mount
• Back-up light switch wires
• Vehicle speed sensor wiring
• 2 transaxle upper side mounting
  bolts
• Starter
4. At this point, attach an engine crane
to support the engine.
5. Remove or disconnect the following:
• Left side engine under cover
• Both halfshafts
• 2 bolts and 1 nut securing the
  engine rear mount to the cross-
  member
• Sliding yoke
• Power steering hoses
• Support the transaxle
• Engine left mounting bracket
• Engine rear mount and bracket
• 5 transaxle lower side mount bolts
• Transaxle

To install:
6. Install or connect the following:
• Transaxle. Torque the 5 lower bolts
  to 25 ft. lbs. (33 Nm).
• Engine rear mount and bracket.
  Torque the mount bolt and nut to
  47 ft. lbs. (64 Nm); the bracket
  bolts to 36 ft. lbs. (49 Nm).
• Engine left mounting bracket.

Torque the bolts to 36 ft. lbs. (49
Nm).
• Power steering hoses
• Sliding yoke
• 2 bolts and 1 nut securing the
  engine rear mount to the crossmember. Torque the bolts to 36 ft.
  lbs. (49 Nm).
• Both halfshafts
• Left side engine under cover
• Starter. Torque the bolts to 29 ft.
  lbs. (39 Nm).
• 2 transaxle upper side mounting
  bolts. Torque the bolts to 25 ft. lbs.
  (33 Nm).
• Vehicle speed sensor wiring

• Back-up light switch wires
• Ground cable from the left engine
  mount
• Clutch release cylinder
• Transaxle control cable
• Wiring harness from the transaxle
• Air cleaner assembly
• Battery
• No. 2 cylinder head cover
• Right and left cowl top ventilator
  covers
• Wiper arms
• Hood
7. Fill the transaxle to the proper level.
8. Start the vehicle, check for leaks and
repair if necessary.

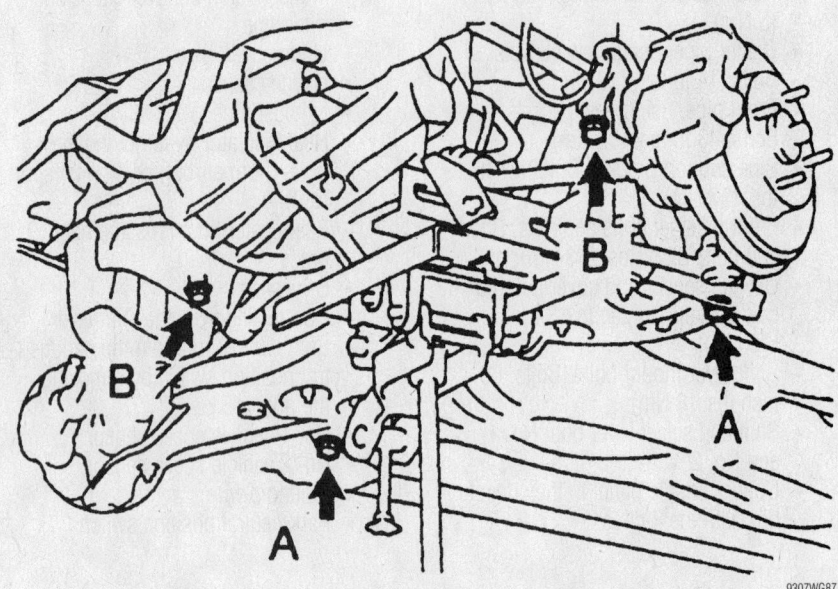

Sub-frame installation—Echo

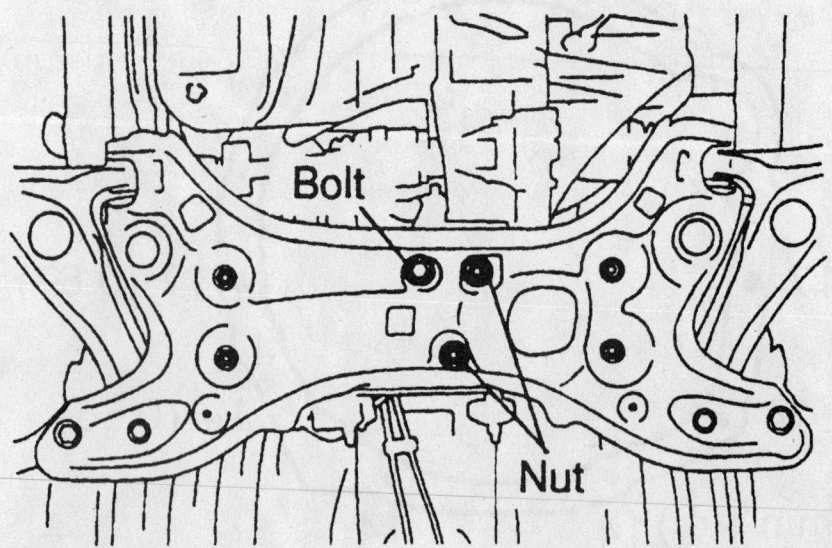

Crossmember installation—Echo

Camry and Camry Solara

1. Before servicing the vehicle, refer to the precautions in the beginning of this section.

2. Drain the transaxle fluid.

3. Remove or disconnect the following:

- Negative battery cable. On vehicles equipped with an air bag, wait at least 90 seconds before proceeding.
- Air cleaner
- With cruise control, cruise control actuator
- Clutch release cylinder and tube clamp
- Starter
- Back-up light switch connector and ground strap
- Wires clamp
- Clips and washers that attach the transaxle control cables to the control levers
- Transaxle control cables.
- Speed sensor connector
- Undercovers
- Left and right halfshafts
- 4 steering gear housing bolts.
- Stabilizer bar bushing bracket
- 2 set bolts and nuts
- Steering gear box from the suspension member and suspend it securely
- Exhaust pipe
- Stiffener plate
- Engine front mounting from the suspension member
- Engine rear mounting from the front suspension member
- Left engine mounting
- Steering cooler pipe from the suspension member
- 2 fender liner set screws
- The 2 bolts and 4 nuts located on the outside of the suspension member brackets
- The 4 larger bolts holding the suspension member to the vehicle body
- The 2 front lower braces, rear braces, and the front suspension member.
- Transaxle

To install:

4. Move the transaxle into position so that the input shaft spline is aligned with the clutch disc.

5. Install or connect the following:

- Transaxle into the engine and secure with the lower mounting bolts. Bolts: 10mm mounting bolts: 47 ft. lbs. (63 Nm). 12mm bolts to 34 ft. lbs. (46 Nm).
- Front suspension member and the 2 front lower braces and rear lower braces. 4 large bolts that hold the suspension member to the vehicle: 134 ft. lbs. (181 Nm); 2 outside bolts and 4 outside nuts: 24 ft. lbs. (32 Nm).
- 2 fender liner set screws
- Steering cooler pipe to the suspension member
- Engine left mount. Bolts: 38 ft. lbs. (52 Nm); 2 nuts and 2 grommets. Nuts: 59 ft. lbs. (80 Nm).
- Engine rear mounting to the front suspension member. Nuts: 59 ft. lbs. (80 Nm).
- Engine front mounting to the suspension member. Bolt: 59 ft. lbs. (80 Nm).
- Stiffener plate. Bolts: 27 ft. lbs. (37 Nm).
- Exhaust pipe
- Steering gear housing to the front suspension member. Bolts and nuts: 134 ft. lbs. (181 Nm).

- Stabilizer bar bushing bracket. 4 bolts: 14 ft. lbs. (19 Nm).
- Right and left halfshafts
- Undercovers
- Vehicle speed sensor
- Control cables by installing the washers and clips
- Clamp that retains the wires to the transaxle
- Back-up light switch connector and ground cables
- Starter. Bolts: 29 ft. lbs. (39 Nm).
- Pipe clamp and clutch release cylinder to the transaxle. Bolts: 108 inch lbs. (13 Nm).
- Cruise control actuator
- Air cleaner
- Negative battery cable

6. Fill the transaxle fluid to the proper level.

7. Start the vehicle, check for leaks and repair if necessary.

Celica

1. Before servicing the vehicle, refer to the precautions in the beginning of this section.

2. Drain the transaxle fluid.

3. Remove or disconnect the following:

- Negative battery cable. On vehicles equipped with an air bag, wait at least 90 seconds before proceeding
- Air cleaner case assembly with hose
- Release cylinder tube bracket
- Clutch release cylinder
- Back-up light switch connector
- Ground cable on the transaxle
- Shift cables from the transaxle
- Vehicle speed sensor connector or the speedometer cable
- Engine wire clamps

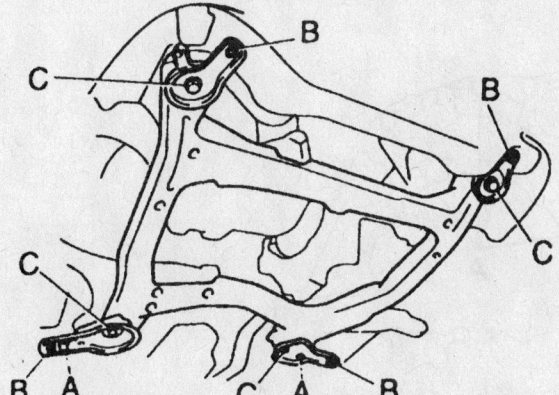

Bolt A: 32 N·m (330 kgf·cm, 24 ft·lbf)
Nut B: 36 N·m (370 kgf·cm, 27 ft·lbf)
Bolt C: 181 N·m (1,850 kgf·cm, 134 ft·lbf)

Front suspension member and fastener locations—Camry

7923VG93

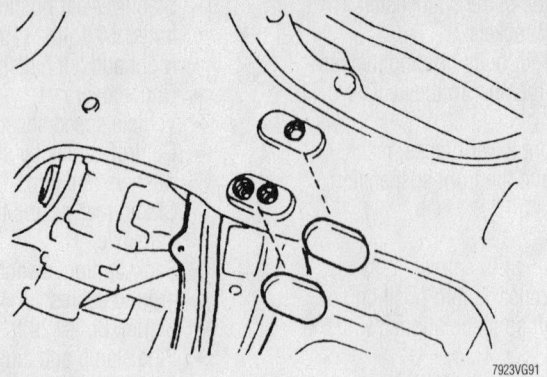

Rear mounting insulator set bolt locations—Celica

- Starter set bolt from the transaxle upper side
- Undercovers
- Halfshafts
- Front exhaust pipe and support bracket
- Starter
- Engine center support member
- Engine rear mounting
- Engine front mounting bracket and insulator
- Engine left mounting bracket
- Transaxle mounting bolts from the engine rear end plate side.
- Transaxle case protector
- Engine left side and remove the 3 upper transaxle bolts.
- Transaxle

To install:

4. Position the transaxle to the engine and raise the engine right side. Align the input shaft with the clutch disc

5. Install or connect the following:
- Transaxle to the engine. 3 upper transaxle bolts: 47 ft. lbs. (64 Nm).
- Transaxle case protector. Bolts: 108 inch lbs. (13 Nm).
- 4 transaxle lower bolts. Bolt A: 17 ft. lbs. (23 Nm); Bolt B: 34 ft. lbs. (46 Nm).
- Left engine mounting bracket to the engine left mounting insulator. Bolts: 47 ft. lbs. (64 Nm).
- Engine front mounting bracket and insulator. 2 bracket bolts: 57 ft. lbs. (77 Nm). Through-bolt: 64 ft. lbs. (87 Nm).
- Engine rear mounting bracket and insulator. Bracket bolts: 57 ft. lbs. (77 Nm). Through-bolt: 64 ft. lbs. (87 Nm).
- Engine center support member
- Starter
- Front exhaust pipe and support bracket
- Halfshafts

- Transaxle oil
- Undercovers
- Engine support fixture
- Starter set bolt to the transaxle upper side. Bolt: 29 ft. lbs. (39 Nm).
- Engine wire clamps
- Vehicle speed sensor connector or the speedometer cable
- Transaxle shift cables and ground cable
- Back-up light switch connector
- Release cylinder
- Air cleaner case assembly
- Negative battery cable

6. Fill the transaxle fluid to the proper level.

7. Start the vehicle, check for leaks and repair if necessary.

Celica GT-S

1. Before servicing the vehicle, refer to the precautions in the beginning of this section.

2. Drain the transaxle fluid.

3. Remove or disconnect the following:
- Hood
- No. 2 cylinder head cover
- Radiator overflow bottle
- Battery

- Air cleaner assembly
- ECM box
- Wiring harness
- Battery tray
- Transaxle control cable
- Engine ground cable
- Speed sensor connector
- Back-up light switch connector
- Clutch release cylinder
- Starter
- Transaxle control cable bracket
- 2 upper transaxle mounting bolts

4. Attach an engine crane to the engine.

5. Remove or disconnect the following:
- Left engine mount bracket
- Engine under covers
- Both halfshafts
- Front exhaust pipe

6. Unbolt the steering rack from the sub-frame and suspend it.

7. Remove or disconnect the following:
- Stabilizer bar
- Wheels
- Ball joints from the lower arms
- Lower arms and sub-frame
- Engine rear mount and bracket

8. Raise the engine slightly.

9. Remove or disconnect the following:
- 4 lower transaxle side bolts
- Transaxle from the engine

To install:

10. Install or connect the following:
- Transaxle to the engine. Install the 4 lower side bolts. Torque the 2 bottom bolts to 17 ft. lbs. (23 Nm); the 2 side bolts to 35 ft. lbs. (48 Nm).
- Rear mount and bracket. Torque the bracket bolts to 47 ft. lbs. (64 Nm); the mount bolt to 64 ft. lbs. (87 Nm).
- Lower control arms-to-frame. Bolts: 101 ft. lbs. (137 Nm); tie rod ball stud nuts: 36 ft. lbs. (49 Nm).

11. Install the sub-frame. Torque the bolts as illustrated:

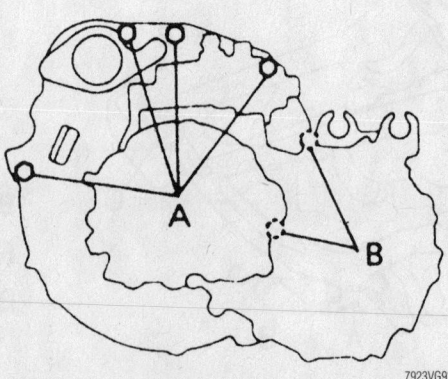

Upper transaxle mounting bolt locations—Celica

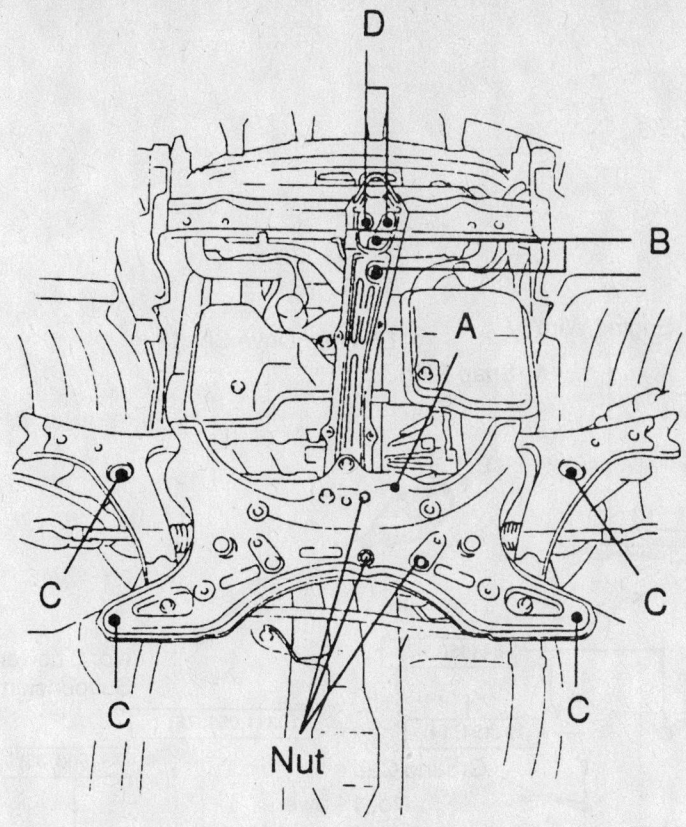

9307WG99

Celica sub-frame torques

a. A and B: 38 ft. lbs. (52 Nm)
b. C: 116 ft. lbs. (157 Nm)
c. D: 29 ft. lbs. (39 Nm)

12. Install or connect the following:
- Steering rack. Bolts: 43 ft. lbs. (58 Nm).
- Front exhaust pipe. Bolts: 32 ft. lbs. (43 Nm).
- Halfshafts
- Under covers
- Left mount bracket. Bolt: 44 ft. lbs. (60 Nm).
- Left mount. Bolts: 44 ft. lbs. (60 Nm); nut: 59 ft. lbs. (80 Nm).
- Transaxle upper side mount bolts. Torque: 47 ft. lbs. (64 Nm).
- Control cable bracket. Bolts: 18 ft. lbs. (25 Nm).
- Starter. Bolts: 28 ft. lbs. (37 Nm).
- Clutch release cylinder
- Wiring
- Control cable
- Battery tray
- ECM
- Air cleaner
- Battery
- Reservoir
- No. 2 cylinder head cover
- Hood

13. Fill the transaxle fluid to the proper level.

14. Start the vehicle, check for leaks and repair if necessary.

Corolla

1. Before servicing the vehicle, refer to the precautions in the beginning of this section.
2. Drain the transaxle fluid.
3. Remove or disconnect the following:
- Negative battery cable. On vehicles equipped with an air bag, wait at least 90 seconds before proceeding.
- Air cleaner case assembly with hose
- Coolant reservoir tank
- Release cylinder tube bracket
- Clutch release cylinder
- Back-up light switch connector
- Ground cable
- Shift cables from the transaxle
- Vehicle speed sensor connector or the speedometer cable
- Engine wire clamps
- Starter set bolt from the transaxle upper side
- 2 transaxle upper mounting bolts
- Engine left mounting stay
- Engine left mounting set bolt from the rear side

- Undercovers
- Lower ball joint from the lower arm.
- Halfshafts
- Front exhaust pipe
- Hole cover
- Engine front mounting set bolts
- Engine rear mounting
- Engine center support member
- Starter
- Transaxle mounting bolts from the engine rear end plate side
- Engine left mounting set bolts from the front side
- Transaxle mounting bolts from the engine front side, then engine rear side
- Transaxle

To install:
4. Align the input shaft with the clutch disc and install the transaxle to the engine. 12mm bolts: 47 ft. lbs. (64 Nm). 10mm bolts: 34 ft. lbs. (46 Nm).
5. Install or connect the following:
- Left engine mount. Bolts: 41 ft. lbs. (56 Nm).
- Transaxle mounting bolts to the engine rear end plate side. Bolts: 17 ft. lbs. (23 Nm).
- Starter, lower bolt and electrical connector to the starter. Bolt: 29 ft. lbs. (39 Nm).
- Engine center support member. Radiator support bolts: 45 ft. lbs. (61 Nm). Frame bolts: 152 ft. lbs. (206 Nm).
- Engine rear mounting. Bolts: 35 ft. lbs. (48 Nm).
- Engine front mounting. Bolts: 47 ft. lbs. (64 Nm).
- Hole covers
- Front exhaust pipe
- Halfshafts
- Lower ball joint to lower arm. Bolt and nuts: 105 ft. lbs. (142 Nm).
- Undercovers
- Engine left mounting set bolt to the rear side. Bolt: 41 ft. lbs. (56 Nm).
- Engine left mounting stay. Bolt: 15 ft. lbs. (21 Nm).
- 2 transaxle upper side mounting bolts. Bolts: 29 ft. lbs. (39 Nm).
- Starter set bolt to the transaxle upper side. Bolt: 29 ft. lbs. (39 Nm).
- Engine wire clamps
- Vehicle speed sensor connector or the speedometer cable.
- Transaxle shift cables and ground cable.
- Back-up light switch connector.
- Release cylinder and release cylinder tube bracket. Bolts: 108 inch

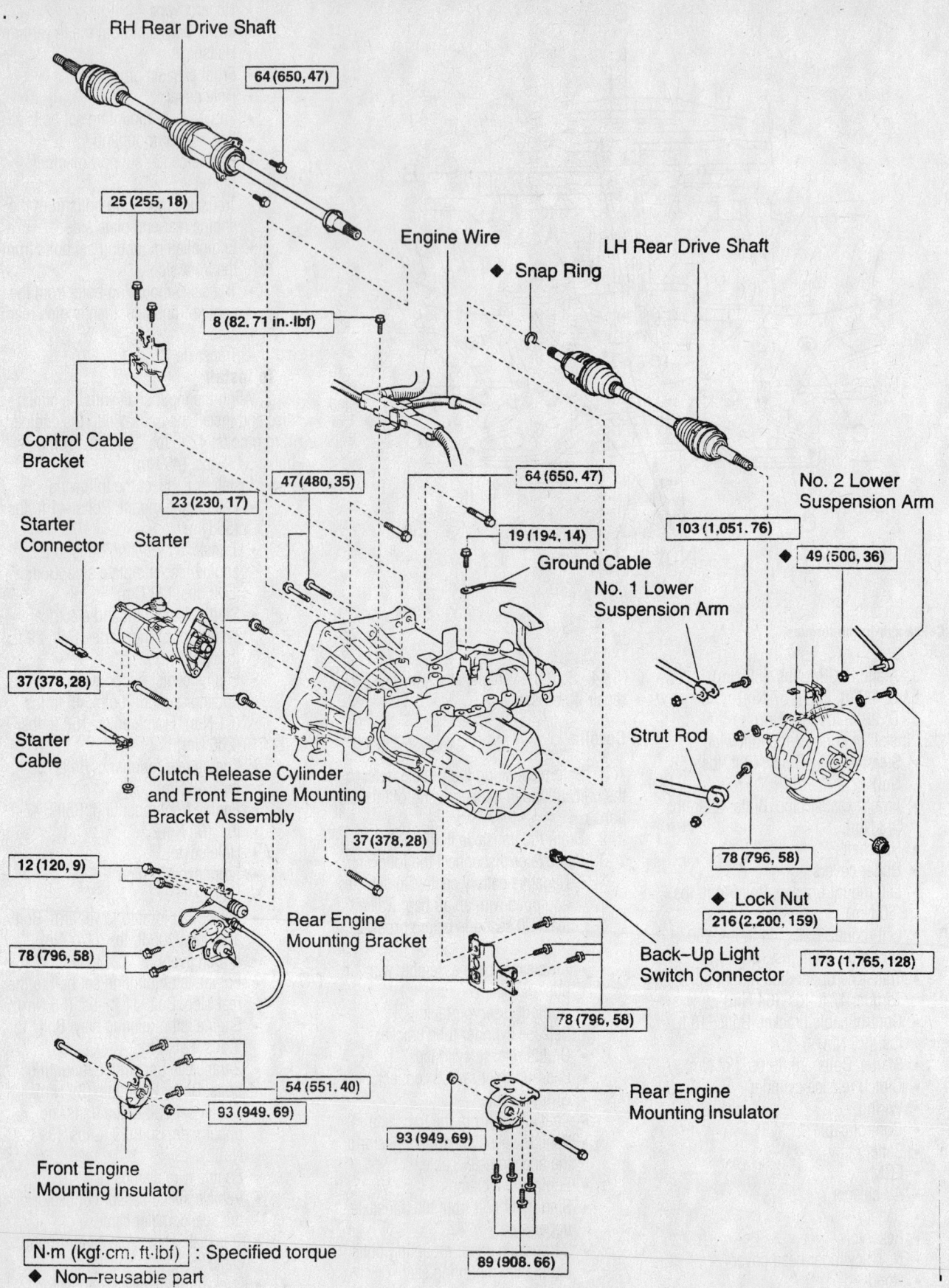

RH Rear Drive Shaft

64 (650, 47)

25 (255, 18)

8 (82. 71 in.·lbf)

Engine Wire

◆ Snap Ring

LH Rear Drive Shaft

Control Cable Bracket

47 (480, 35)

23 (230, 17)

64 (650, 47)

No. 2 Lower Suspension Arm

103 (1,051.76)

◆ 49 (500, 36)

Starter Connector

Starter

19 (194, 14)

Ground Cable

No. 1 Lower Suspension Arm

37 (378, 28)

Strut Rod

Starter Cable

Clutch Release Cylinder and Front Engine Mounting Bracket Assembly

37 (378, 28)

78 (796, 58)

◆ Lock Nut

216 (2,200, 159)

12 (120, 9)

Rear Engine Mounting Bracket

Back-Up Light Switch Connector

173 (1,765, 128)

78 (796, 58)

78 (796, 58)

54 (551. 40)

93 (949. 69)

93 (949. 69)

Rear Engine Mounting Insulator

Front Engine Mounting Insulator

N·m (kgf·cm. ft·lbf) : Specified torque

◆ Non-reusable part

89 (908. 66)

Manual transaxle mounting exploded view—MR2

9357WG04

lbs. (12 Nm).
- Coolant reservoir tank
- Air cleaner case assembly
- Negative battery cable

6. Fill the transaxle fluid to the proper level.

7. Start the vehicle, check for leaks and repair if necessary.

MR2

1. Before servicing the vehicle, refer to the precautions in the beginning of this section.

2. Drain the transaxle oil.

3. Remove or disconnect the following:
- Engine hood
- Rear suspension upper brace
- Air filter assembly
- Battery and tray
- Engine ground cable
- Back-up light switch connector
- Shift cables and bracket
- Transaxle upper flange bolts

4. Attach an engine support fixture.
- Left motor mount
- Engine undercovers
- Axle halfshafts
- Clutch slave cylinder
- Front motor mount and bracket
- Starter motor
- Rear motor mount and bracket
- Transaxle lower flange bolts
- Transaxle

To install:

5. Installation is the reverse of the removal procedure, while using the following torque values:
- Upper transaxle flange bolts: 47 ft. lbs. (64 Nm)
- Lower transaxle flange bolts: **A** bolts to 35 ft. lbs. (47 Nm) and **B** bolts to 17 ft. lbs. (23 Nm). Refer to the illustration.
- Rear engine mount bracket: 58 ft. lbs. (78 Nm)
- Rear engine mount bolts: 66 ft. lbs. (89 Nm)
- Rear engine mount through bolt: 69 ft. lbs. (93 Nm)
- Starter motor: 28 ft. lbs. (37 Nm)
- Front engine mount bracket bolts: 56 ft. lbs. (78 Nm)
- Front engine mount through bolt: 69 ft. lbs. (93 Nm)
- Clutch slave cylinder: 108 Inch lbs. (12 Nm)
- Left engine mount bracket: 38 ft. lbs. (52 Nm)
- Left engine mount through bolt: 64 ft. lbs. (87 Nm)

Upper transaxle flange bolts—MR2

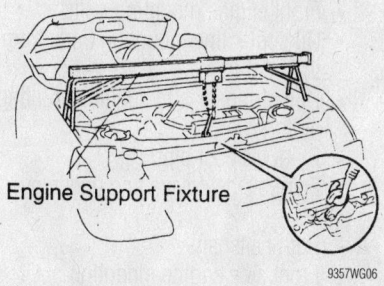

Engine support fixture—MR2

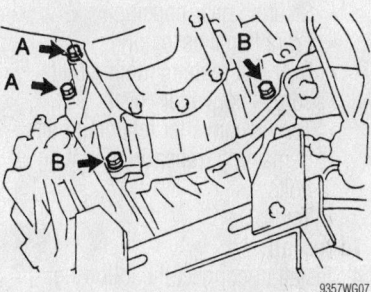

Lower transaxle flange bolts—MR2

- Suspension upper brace bolts: 59 ft. lbs. (80 Nm)
- Engine hood bolts: 15 ft. lbs. (20 Nm)

6. Fill the transaxle with the correct oil.

Automatic Transaxle

REMOVAL & INSTALLATION

Echo

1. Before servicing the vehicle, refer to the precautions in the beginning of this section.

2. Drain the transaxle fluid.

3. Remove or disconnect the following:
- Hood
- Right and left cowl top ventilator covers
- No. 2 cylinder head cover
- Battery
- Air cleaner bracket
- Wiring harness from the transaxle
- Transaxle shift cable
- Clutch release cylinder
- Ground cable from the left engine mount
- Park/Neutral switch wiring
- Solenoid wiring
- Direct Clutch Speed Sensor wiring
- Filler pipe hose
- 2 transaxle upper side mounting bolts
- Engine under covers
- Starter
- Oil cooler hose
- Both halfshafts
- Torque converter access plug
- Torque converter bolts and attach an engine crane to support the engine
- Front suspension subframe (lower arms, steering gear and stabilizer)
- 5 transaxle bolts
- 2 transaxle mount bolts
- Transaxle

To install:

4. Install or connect the following:

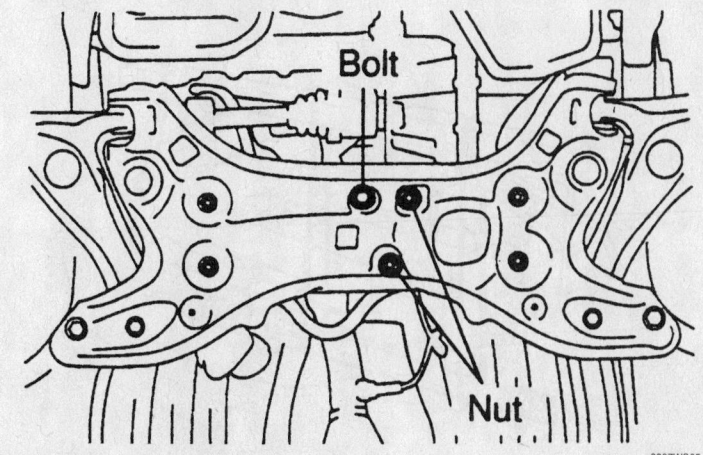

Rear mount installation—Echo

- Transaxle. Torque the 5 bolts to 22 ft. lbs. (30 Nm).
- 2 transaxle mount bolts. Torque the mount bolts to 36 ft. lbs. (49 Nm).
- Torque converter bolts: 20 ft. lbs. (27 Nm).
- Access plug
- Subframe assembly
- Oil cooler hoses
- Starter. Torque the bolts to 29 ft. lbs. (39 Nm).
- 2 transaxle upper side mounting bolts. Torque the bolts to 22 ft. lbs. (30 Nm).
- Both halfshafts
- Engine under covers
- Filler hose
- Transaxle control cable
- Ground cable from the left engine mount
- Wiring
- Air cleaner bracket
- No. 2 cylinder head cover
- Right and left cowl top ventilator covers
- Battery
- Hood

5. Fill the transaxle fluid to the proper level.

6. Start the vehicle, check for leaks and repair if necessary.

Avalon, Camry and Camry Solara

1. Before servicing the vehicle, refer to the precautions in the beginning of this section.

2. Drain the transaxle fluid.

3. Remove or disconnect the following:
- Negative battery cable. On vehicles equipped with an air bag, wait at least 90 seconds before proceeding
- Battery
- Air cleaner assembly
- Throttle cable from the throttle body

- Cruise control actuator cover and connector, if equipped
- Ground wire
- Starter
- Speed sensor connectors, direct clutch speed sensor, and the park/neutral position switch connector on the transaxle
- Solenoid connector on the transaxle
- Shift control cable
- Oil cooler hoses
- Front side transaxle mounting bolts
- Front engine mounting bolts
- Oil cooler line mounting bolts from the front frame
- Upper transaxle to engine mounting bolts
- Front exhaust pipe
- Engine side covers and undercovers
- Both halfshafts
- Front side engine mounting nut
- Rear side engine mounting bolts (remove hole plugs)
- Left side transaxle mounting bolts
- Steering gear housing
- Front frame assembly
- Rear end plate mounting bolts
- Torque converter cover
- Torque converter retaining bolts
- Remaining transaxle mounting bolts
- Transaxle

To install:

4. Install or connect the following:
- Transaxle aligning the 2 dowel pins on the block with the converter housing. 10mm bolts: 34 ft. lbs. (46 Nm); 12mm bolts: 47 ft. lbs. (64 Nm)
- Torque converter bolts coated with sealer. Install the bolts starting with the green bolt followed by the rest. Bolts: 20 ft. lbs. (27 Nm).

- Rear end plate. Bolts: 27 ft. lbs. (37 Nm).
- Front frame assembly. 12mm bolts: 24 ft. lbs. (32 Nm); 19mm bolts: 134 ft. lbs. (181 Nm); Nut: 27 ft. lbs. (36 Nm).
- Fender liner set screws
- Steering gear to the frame. Bolts and nuts: 134 ft. lbs. (181 Nm).
- Sway bar brackets. Bolts: 14 ft. lbs. (19 Nm).
- Left transaxle mounting bolts. Bolts: 38 ft. lbs. (52 Nm).
- Rear side mounting bolts and nuts. Bolts and nuts: 48 ft. lbs. (66 Nm). Install the plugs.
- Front engine mounting nut. Nut: 59 ft. lbs. (80 Nm).
- Halfshafts
- Right and left engine side covers
- Lower engine cover
- Exhaust pipe. Nuts: 46 ft. lbs. (62 Nm).
- Exhaust pipe to the converter. Nuts and bolts: 32 ft. lbs. (43 Nm).
- Upper transaxle mounting bolts. Bolts: 47 ft. lbs. (64 Nm).
- Oil cooler clamping bolts to the front frame.
- Front side engine mounting bolts. Bolts: 59 ft. lbs. (80 Nm).
- Front side transaxle mounting bolts. Bolts: 59 ft. lbs. (80 Nm).
- Oil cooler hoses
- Shift control cable
- Solenoid electrical connector
- Park/neutral switch electrical connector
- Speed sensor and the direct clutch speed sensor connectors.
- Starter
- Ground strap
- Cruise control actuator and cover
- Throttle cable to the engine. Nuts: 11 ft. lbs. (15 Nm).
- Air cleaner
- Battery and battery cables.

5. Fill the transaxle to the proper level.

6. Start the vehicle, check for leaks and repair if necessary.

Corolla

1. Before servicing the vehicle, refer to the precautions in the beginning of this section.

2. Drain the transaxle fluid.

3. Remove or disconnect the following:

- Negative battery cable. On vehicles equipped with an air bag, wait at least 90 seconds before proceeding

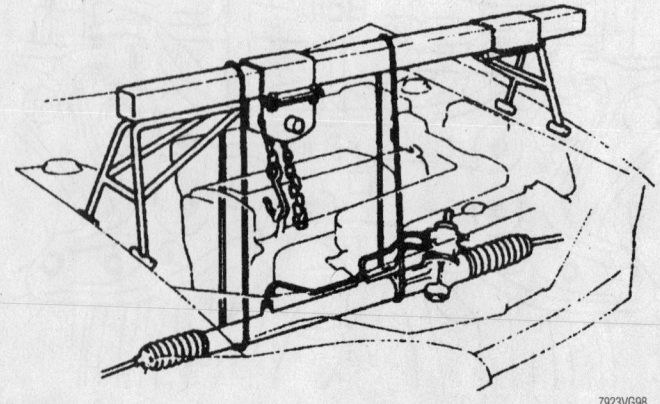

7923VG98

Tie the steering rack to the engine support fixture components, as shown—Avalon and Camry

- Negative battery cable from the transaxle
- Transaxle level gauge
- With A245E transaxle, reservoir tank and air cleaner assembly
- Throttle cable from the bracket
- Engine left mounting upper side bolts
- Engine left mounting stay
- Ground cable from the transaxle
- Wiring harness clamp and throttle cable clamp
- With A245E transaxle, starter upper side mounting bolt and the 2 transaxle mounting bolts from transaxle side
- Undercovers
- Left and right halfshafts
- Front exhaust pipe
- Engine support fixture
- Suspension member
- Starter
- Vehicle speed sensor connector
- Solenoid connector and park/neutral position switch connector. Remove the wiring harness clamps
- Nut from the manual shift lever, then the control cable from the bracket by removing the clip
- Oil cooler hoses
- Transaxle filler tube
- With A131L transaxle, stiffener plate
- With A245E transaxle, converter cover
- Torque converter bolts
- 5 (A245E) or 4 (A131L) transaxle mounting bolts
- Transaxle

To install:

4. Install or connect the following:
- A245E Transaxle. Install the 5 lower transaxle bolts. Tighten the bolts as follows: Bolt A: 17 ft. lbs. (23 Nm); Bolt B: 18 ft. lbs. (25 Nm); Bolt C: 34 ft. lbs. (46 Nm); Bolt D: 47 ft. lbs. (64 Nm)
- A131L transaxle. Install the 4 lower transaxle bolts. Bolts: 47 ft. lbs. (64 Nm).
- Torque converter bolts to the transaxle. Bolts: 18 ft. lbs. (25 Nm).
- With the A131L transaxle, stiffener plate. Bolts: 13 ft. lbs. (18 Nm).
- With the A245E transaxle, torque converter cover.
- Transaxle filler pipe
- Oil cooler hoses and replace the clips to their original positions
- Control cable for the transaxle to the bracket and install the clip
- Control cable to the manual shaft lever by installing the nut

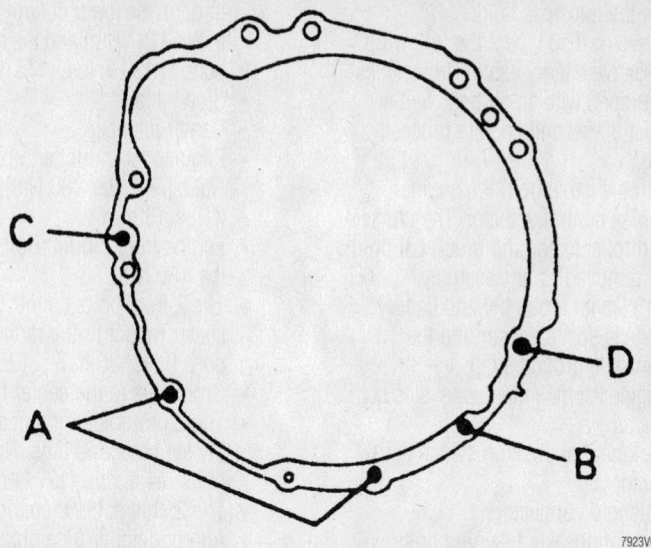

Transaxle mounting bolt torque specifications—Corolla A245E transaxle

- Solenoid connector and park/neutral position switch connector. Connect the wiring to the clamps
- Vehicle speed sensor wiring
- Starter. Bolt: 29 ft. lbs. (39 Nm).
- Suspension member. Bolt A: 45 ft. lbs. (61 Nm); Bolt B: 47 ft. lbs. (64 Nm); Bolt C: 152 ft. lbs. (206 Nm); Bolt D: 152 ft. lbs. (206 Nm); Nuts: 42 ft. lbs. (57 Nm).
- Front exhaust pipe
- Left and right halfshafts
- Undercovers
- Transaxle mounting bolts to the transaxle side
- Starter upper side mounting bolt. Bolt: 29 ft. lbs. (39 Nm).
- Wiring harness clamp and throttle cable clamp

- Ground cable. Bolt: 13 ft. lbs. (18 Nm).
- Engine left mounting stay. Bolts: 15 ft. lbs. (21 Nm).
- Engine left mounting upper side bolts. Bolts: 38 ft. lbs. (52 Nm).
- Throttle cable
- Air cleaner and the reservoir tank
- Transaxle level gauge
- Negative battery cable

5. Fill the transaxle fluid to the proper level.

6. Start the vehicle, check for leaks and repair if necessary.

Celica

1. Before servicing the vehicle, refer to the precautions in the beginning of this section.

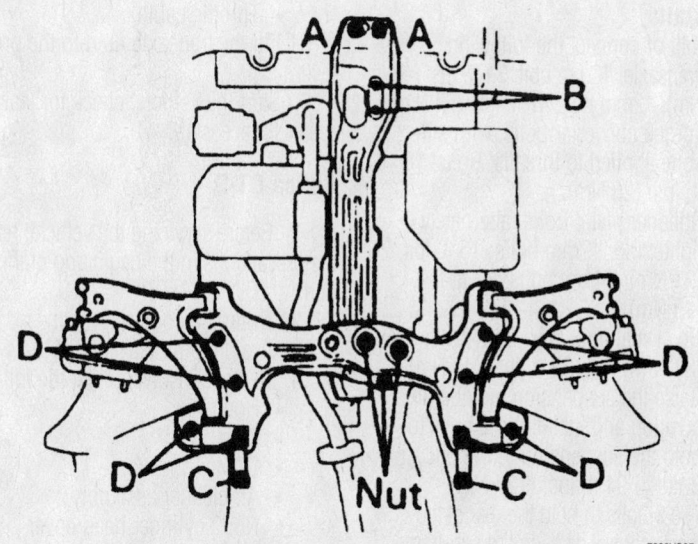

Suspension member fastener identification—Corolla

2. Drain the transaxle fluid.
3. Remove or disconnect the following:
- Negative battery cable. On vehicles equipped with an air bag, wait at least 90 seconds before proceeding.
- Throttle cable from the engine
- Cruise control actuator. The cruise control actuator and bracket should be removed as an assembly.
- Air cleaner assembly and battery.
- Vehicle speed sensor and the transaxle ground strap.
- Engine left mounting upper side bolt
- Starter
- Park/neutral position switch connector
- Solenoid connectors
- Upper transaxle retaining bolts
- Transaxle oil cooler hoses
- Undercover
- Both halfshafts
- Shift control cable from the control shaft lever and body bracket
- Engine rear mounting through-bolt
- Front exhaust pipe
- Air conditioner pipe bracket by removing the bolt
- Shift cable from the suspension member
- The 2 power steering gear assembly set bolts and nuts
- The 3 grommets from the center crossmember
- The 13 bolts and 2 nuts holding the suspension and center crossmembers
- Crossmembers from the vehicle
- No. 1 manifold stay
- Stiffener plate
- Torque converter bolts
- Transaxle

To install:

4. Install or connect the following:
- Transaxle. 10mm bolt: 34 ft. lbs. (46 Nm); 12mm bolt: 47 ft. lbs. (64 Nm).
- Torque converter bolts, with silicone applied to threads. Bolts: 18 ft. lbs. (25 Nm).
- Stiffener plate. Bolts, alternately tightening: 12mm bolts: 15 ft. lbs. (21 Nm); 14mm bolts: 32 ft. lbs. (43 Nm).
- No. 1 manifold stay. Bolt: 15 ft. lbs. (21 Nm). Nut: 32 ft. lbs. (43 Nm).
- Raise the suspension member into position and install the 2 bolts to hold the suspension to the body. Bolts to 94 ft. lbs. (127 Nm).
- The 3 bolts to hold the rear of the lower control arms to the sub-frame and body. Torque the bolt that goes through the lower control arm to 123 ft. lbs. (167 Nm) and the other 2 bolts to 130 ft. lbs. (175 Nm).
- Side bolts
- Center member
- Engine rear mount and bracket. Nuts: 59 ft. lbs. (80 Nm); Bolt: 65 ft. lbs. (88 Nm).
- Engine front mount. Bolts: 59 ft. lbs. (80 Nm).
- The 2 front bolts connecting the center mount to the radiator support. Bolts: 26 ft. lbs. (35 Nm).
- Grommets to the center member
- The 2 power steering gear assembly set bolts and nuts. Nuts and bolts: 94 ft. lbs. (127 Nm).
- The 2 shift cable mounting bolts
- Air conditioner pipe bracket
- Front exhaust pipe
- Engine rear mounting bolt. Bolt: 64 ft. lbs. (88 Nm).
- Shift control cable to the control shaft lever and body bracket. Install the clips
- Left and right halfshafts
- Undercovers
- Oil cooler hoses with the 2 clips
- Upper transaxle mounting bolts. Bolts: 47 ft. lbs. (64 Nm).
- Solenoid connectors
- Park/neutral position switch connector
- Vehicle speed sensor connector and the ground strap to the transaxle
- Starter. Bolts: 29 ft. lbs. (39 Nm).
- Left mounting upper side bolt. Bolt: 47 ft. lbs. (64 Nm).
- Air cleaner assembly
- Cruise control actuator
- Battery and cables
- Throttle cable

5. Fill the transaxle fluid to the proper level.
6. Start the vehicle, check for leaks and repair if necessary.

Celica GT-S

1. Before servicing the vehicle, refer to the precautions in the beginning of this section.
2. Drain the cooling system.
3. Drain the transaxle fluid.
4. Remove or disconnect the following:
- Hood
- Battery
- ECM
- Air cleaner assembly
- No. 2 cylinder head cover
- Ground cables
- Control cable bracket
- Input speed turbine sensor
- Vehicle speed sensor
- Solenoid connector
- Park/neutral switch connector
- Control cable
- Coolant reservoir
- Starter
- Engine under covers
- Halfshafts
- Stabilizer bar end links

5. Unbolt the steering rack and suspend it
- 9 bolts and 3 nuts and lower the suspension member
- Engine rear mount and attach and engine crane to support the engine.
- Upper left side engine
- Fluid cooler hoses and support the transaxle with a jack.
- Torque converter access plug
- Torque converter bolts
- Transaxle bolts
- Transaxle

To install:

6. Install or connect the following:
- Transaxle. 2 lower bolts: 17 ft. lbs. (23 Nm); 2 lower side bolts: 34 ft. lbs. (46 Nm).
- Torque converter bolts. Torque: 25 ft. lbs. (41 Nm).
- Oil cooler hoses
- Upper left side mount bolt and nut. Torque: 59 ft. lbs. (80 Nm).
- Engine rear mount insulator. Bolt: 64 ft. lbs. (87 Nm).

7. Install the suspension member. Torque the bolts as illustrated:
   a. A: 116 ft. lbs. (157 Nm).
   b. B: 38 ft. lbs. (52 Nm).
   c. C: 29 ft. lbs. (39 Nm).
   d. Nut: 38 ft. lbs. (52 Nm).

8. Install or connect the following:
- Steering rack. Bolts: 33 ft. lbs. (45 Nm).
- Stabilizer bar end links. Torque: 32 ft. lbs. (44 Nm).
- Halfshafts
- Under covers
- Starter. Torque: 28 ft. lbs. (37 Nm).
- Coolant reservoir
- Control cable
- Wiring
- Upper transaxle bolts. Torque: 47 ft. lbs. (64 Nm).
- Control cable clamp
- No. 2 cylinder head cover. Torque: 62 inch lbs. (7 Nm).
- Air cleaner
- ECM
- Battery
- Hood

9. Fill the cooling system.
10. Fill the transaxle to the proper level.
11. Start the vehicle, check for leaks and repair if necessary.

Clutch

ADJUSTMENTS

Hydraulic clutch actuating systems used in Toyota vehicles do not require adjustment.

REMOVAL & INSTALLATION

Camry, Camry, Echo, Solara, and Supra

1. Before servicing the vehicle, refer to the precautions in the beginning of this section.
2. Remove or disconnect the following:
   - Negative battery cable. On vehicles equipped with an air bag, wait at least 90 seconds before proceeding.
   - Remove the transaxle assembly from the vehicle
   - Clutch pressure plate retaining bolts
   - Clutch cover
   - Clutch disc

- Retaining clip and bearing from the transaxle
- Release fork and boot assembly

To install:
3. Install or connect the following:
   - Clutch disc onto the flywheel
   - Clutch cover, aligning the matchmarks
   - Clutch cover retaining bolts. Bolts, tightened in a crisscross pattern: 14 ft. lbs. (19 Nm).
   - Boot, release fork, hub and bearing assemblies
   - Transaxle
   - Negative battery cable

Corolla, Celica, and MR2

➡ **Do not allow grease or oil to get on any part of the disc, pressure plate, or flywheel surfaces.**

1. Before servicing the vehicle, refer to the precautions in the beginning of this section.
2. Remove or disconnect the following:

   - Negative battery cable. On vehicles equipped with an air bag, wait at least 90 seconds before proceeding
   - Transaxle assembly
3. Make matchmarks on the clutch cover (pressure plate) and flywheel so that the

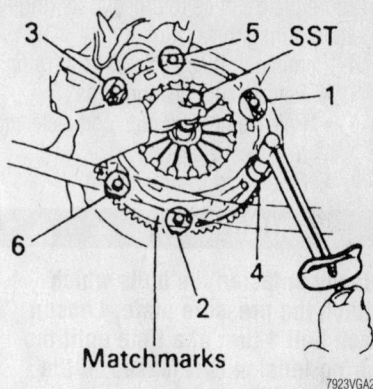

Tighten the pressure plate bolts according to the sequence shown—Camry

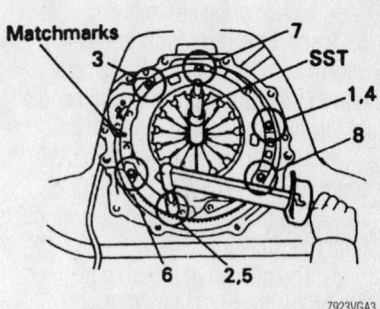

Be sure to tighten the pressure plate bolts in the correct order—Supra

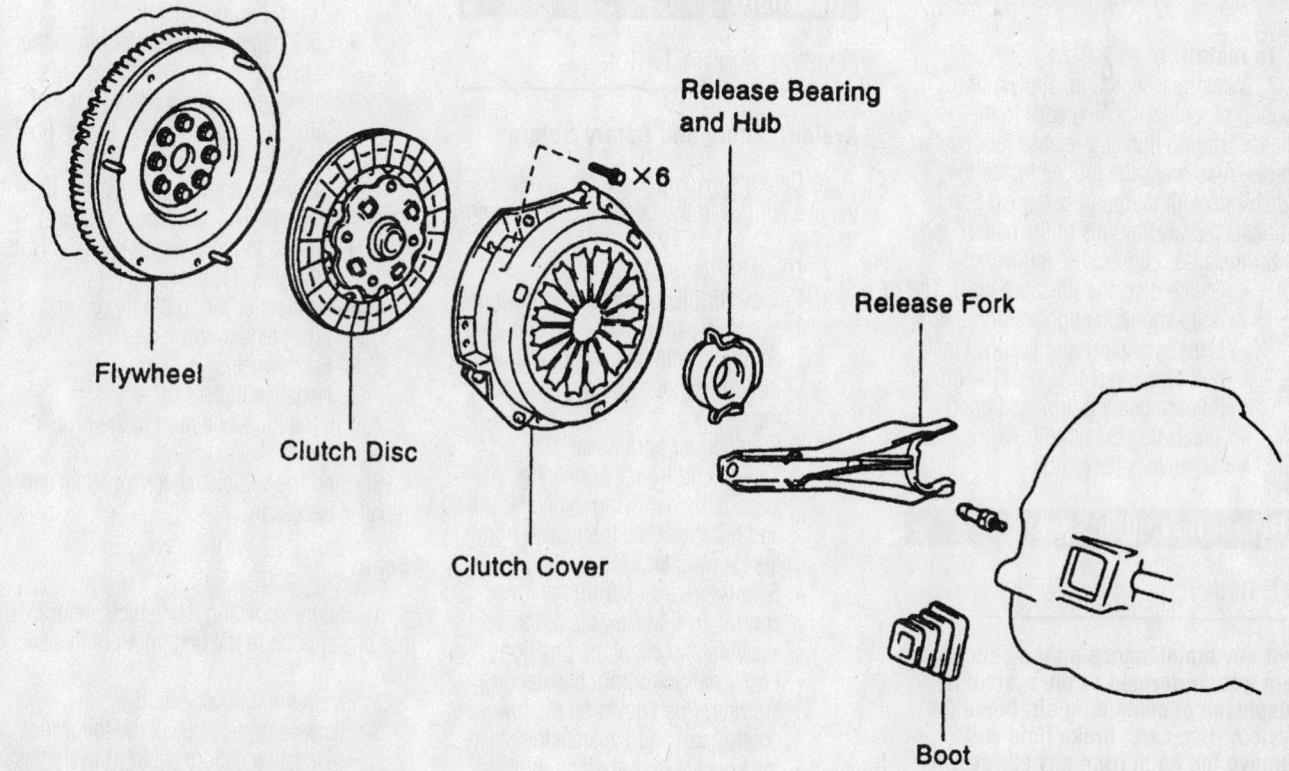

Clutch component assembly—Camry shown, others similar

pressure plate can be returned to its original position during installation.

4. Remove or disconnect the following:
- Release fork bearing clips
- Release bearing hub, complete with the release bearing
- Release fork and support

✳✳ CAUTION

Slowly unfasten the bolts which attach the pressure plate. Loosen each bolt 1 turn at a time until the spring tension is released. If the bolts are released improperly the clutch assembly could fly apart, causing possible injury.

- Pressure plate from the clutch cover/spring assembly

5. Inspect the disc, pressure plate and flywheel for damage and wear using a caliper to measure depth and width and a dial indicator to measure runout.

a. The minimum clutch disc rivet head depth is 0.012 in. (0.3mm).

b. The maximum clutch disc runout is 0.031 in. (0.8mm).

c. The maximum pressure plate spring depth is 0.024 in. (0.6mm).

d. The maximum pressure plate spring width is 0.197 in. (5.0mm).

e. The maximum flywheel runout is 0.004 in. (0.1mm).

6. Replace or machine parts as necessary.

To install:

7. When reassembling, apply a thin coating of multipurpose grease to the release bearing hub and release fork contact points. Also, pack the groove inside the clutch hub with multipurpose grease and lubricate the pivot points of the release fork.

8. Install or connect the following:
- Clutch disc and pressure plate. The bolts should be tightened in 2 or 3 steps, gradually and evenly. Final bolt torque is 14 ft. lbs. (19 Nm).
- Release bearing, fork and boot
- Transaxle assembly
- Negative battery cable

Hydraulic Clutch System

BLEEDING

➡**If any maintenance on the clutch system was performed or the system is suspected of containing air, bleed the system. Use care; brake fluid will remove the paint from any surface. If the brake fluid spills onto any painted surface, wash it off immediately with soap and water.**

1. Before servicing the vehicle, refer to the precautions in the beginning of this section.

2. Fill the clutch reservoir with brake fluid. Check the reservoir level frequently and add fluid as needed.

3. Connect one end of a vinyl tube to the bleeder plug on the slave cylinder and submerge the other end into a clear container half-filled with brake fluid.

4. Slowly pump the clutch pedal several times.

5. Have an assistant hold the clutch pedal down and loosen the bleeder plug until fluid and/or air starts to run out of the bleeder plug. Close the bleeder plug while the pedal is held to the floor.

➡**Do not allow the pedal to rise back-up while the bleeder is still open. If this happens, it will allow air to re-enter the slave cylinder and cause the clutch system not to work properly.**

6. Repeat Steps 2 and 3 until all the air bubbles are removed from the system.

7. Tighten the bleeder plug when all the air is gone.

8. Refill the master cylinder to the proper level as required.

9. Check the system for leaks.

Halfshaft

REMOVAL & INSTALLATION

Avalon, Camry and Camry Solara

1. Before servicing the vehicle, refer to the precautions in the beginning of this section.

2. Drain the transaxle fluid.

3. Remove or disconnect the following:
- Negative battery cable. On vehicles equipped with an air bag, wait at least 90 seconds before proceeding.
- Front fender apron seal
- Tie rod end from the steering knuckle by removing the cotter pin and nut. Separate the tie rod from the steering knuckle
- Stabilizer bar link from the lower control arm. Make note of the washers and cushions positions.
- Lower ball joint from the steering knuckle. Push down on the lower control arm and separate the steering knuckle from the ball joint
- Cotter pin, lock cap and locknut holding the halfshaft to the steering knuckle
- Halfshaft from the steering knuckle
- Left halfshaft from the transaxle
- Snapring from the halfshaft.
- Right halfshaft from the transaxle

➡**The lockbolt is located in the center of the halfshaft, near the dampener.**

To install:

4. Install the right halfshaft to the transaxle, as follows:

a. Coat the side gear shaft and differential case sliding surface with gear oil.

b. Using snapring pliers, install the snapring to the halfshaft.

c. Install the halfshaft and the bearing lockbolt. Lockbolt: 24 ft. lbs. (32 Nm).

5. Install the left halfshaft to the transaxle, as follows:

a. Install a new snapring to the inner spline of the halfshaft.

b. Coat the side gear shaft and differential case sliding surface with gear oil.

c. Install the halfshaft to the transaxle with the snapring opening facing down. The halfshaft should click into place when installing.

d. After installation of the halfshaft, check that the halfshaft cannot be removed by hand.

6. Install or connect the following:
- Halfshaft to the steering knuckle, then install the locknut. Locknut: 217 ft. lbs. (294 Nm).
- Lock cap and new cotter pin to the halfshaft
- Steering knuckle to the lower ball joint. Nuts and bolt: 94 ft. lbs. (127 Nm).
- Stabilizer bar link to the lower control arm. Nut: 29 ft. lbs. (39 Nm).
- Tie rod to the steering knuckle. Nut: 36 ft. lbs. (49 Nm).
- New cotter pin to the tie rod end
- Front fender apron seal
- Front wheels
- Negative battery cable

7. Fill the transaxle fluid to the proper level.

8. Start the vehicle, check for leaks and repair if necessary.

Echo

1. Before servicing the vehicle, refer to the precautions in the beginning of this section.

2. Drain the transaxle fluid.

3. Remove or disconnect the following:
- Negative battery cable. On vehicles equipped with an air bag, wait at

least 90 seconds before proceeding.
- Both front wheels
- Locknut holding the halfshaft to the steering knuckle
- Tie rod end from the steering knuckle by removing the cotter pin and nut. Separate the tie rod from the steering knuckle
- Stabilizer bar link from the lower control arm. Make note of the washers and cushions positions
- Lower ball joint from the steering knuckle. Push down on the lower control arm and separate the steering knuckle from the ball joint
- Halfshaft from the steering knuckle
- Left halfshaft from the transaxle
- Snapring from the halfshaft
- Right halfshaft from the transaxle

➡**The lockbolt is located in the center of the halfshaft, near the dampener.**

To install:

4. Install the right halfshaft to the transaxle, as follows:

   a. Coat the side gear shaft and differential case sliding surface with gear oil.

   b. Using snapring pliers, install the snapring to the halfshaft.

   c. Install the halfshaft

5. Install the left halfshaft to the transaxle, as follows:

   a. Install a new snapring to the inner spline of the halfshaft.

   b. Coat the side gear shaft and differential case sliding surface with gear oil.

   c. Install the halfshaft to the transaxle with the snapring opening facing down. The halfshaft should click into place when installing.

   d. After installation of the halfshaft,

check that the halfshaft cannot be removed by hand.

6. Install or connect the following:
- Halfshaft to the steering knuckle, then install the locknut. Locknut: 159 ft. lbs. (216 Nm).
- Lock cap and new cotter pin to the halfshaft
- Steering knuckle to the lower ball joint. Nut: 72 ft. lbs. (98 Nm).
- Stabilizer bar link to the lower control arm. Nut: 29 ft. lbs. (39 Nm).
- Tie rod to the steering knuckle. Nut: 36 ft. lbs. (49 Nm).
- New cotter pin to the tie rod end
- Both front wheels
- Negative battery cable

7. Fill the transaxle fluid to the proper level.

8. Start the vehicle, check for leaks and repair if necessary.

Celica and Celica GT-S

➡**The hub bearing could be damaged if subjected to the full weight of the vehicle, such as if the vehicle is moved without the halfshafts. If it is absolutely necessary to place the full vehicle weight on the hub bearing, first support the bearing with SST No. 09608–16041.**

1. Before servicing the vehicle, refer to the precautions in the beginning of this section.

2. Drain the transaxle fluid.

3. Remove or disconnect the following:
- Negative battery cable. On vehicles equipped with an air bag, wait at least 90 seconds before proceeding
- Both front wheels
- Cotter pin, locknut cap, and the bearing locknut

- Undercovers
- Tie rod ball joint from the steering knuckle
- Stabilizer bar link from the lower suspension arm
- Lower ball joint from the lower suspension arm
- Halfshaft from the knuckle

➡**Be careful not to damage the inner oil seal or the ABS sensor rotor on the halfshaft.**

4. To remove the left side halfshaft, separate the halfshaft from the transaxle.

5. To remove the right side halfshaft perform the following steps:
- Remove the 2 bolts of the center bearing bracket
- Pull the halfshaft out together with the center bearing case and the center halfshaft.
- Remove the center shaft with the right-hand halfshaft from the transaxle through the bearing bracket.

➡**Do not damage the oil seal lip.**

To install:

6. Install or connect the following:
- Snapring opening side facing downward, on the oiled inboard joint tulip
- Left side halfshaft into the transaxle
- Right side halfshaft, with the bearing case and center shaft, into the transaxle
- Center bearing case (right side). Bolts: 47 ft. lbs. (64 Nm)

7. After installing either halfshaft, check that there is 0.08–0.12 in. (2–3mm) of axial play. Check that the halfshaft is making contact with the pinion shaft and that the halfshaft cannot be pulled out.

8. Install or connect the following:
- Halfshaft into the knuckle
- Lower suspension arm to the lower ball joint. Bolt and nuts: 94 ft. lbs. (127 Nm).
- Tie rod end to the steering knuckle. Nut: 36 ft. lbs. (49 Nm).
- Stabilizer bar link to the lower suspension arm. Nuts: 33 ft. lbs. (44 Nm).
- Front wheels
- Locknut and washer. Locknut: 159 ft. lbs. (216 Nm).
- Negative battery cable
- Locknut cap and a new cotter pin.
- Undercover

9. Fill the transaxle fluid to the proper level

10. Start the vehicle, check for leaks and repair if necessary.

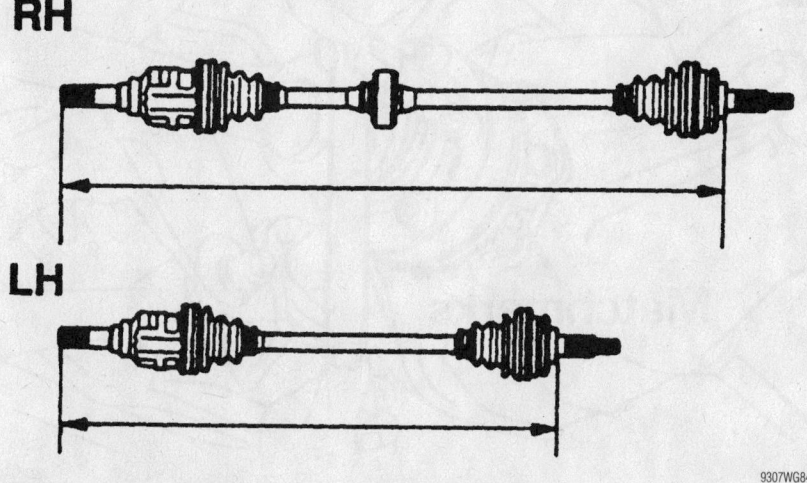

RH

LH

9307WG84

Measuring halfshaft length—Echo

Corolla

➡The hub bearing could be damaged if subjected to the full weight of the vehicle, such as if the vehicle is moved without the halfshafts. If it is absolutely necessary to place the full vehicle weight on the hub bearing, first support the bearing with SST No. 09608–16041.

1. Before servicing the vehicle, refer to the precautions in the beginning of this section.
2. Drain the transaxle fluid.
3. Remove or disconnect the following:
   - Negative battery cable. On vehicles equipped with an air bag, wait at least 90 seconds before proceeding
   - Cotter pin, locknut cap, and bearing locknut
   - Front wheels
   - Undercovers
   - With ABS, speed sensor
   - Tie rod ball joint from the steering knuckle
   - Lower ball joint from the lower suspension arm
4. Drive the halfshaft from the knuckle.

➡Most halfshafts can be separated from the knuckle using a brass or plastic hammer; some others may require the use of a puller.

5. Remove the halfshaft from the transaxle

To install:
6. Install or connect the following:
   - Snapring, opening side facing downward, to the inboard, oiled, joint tulip
   - Halfshaft into the transaxle. After installing the halfshaft to the transaxle, check that there is 0.08–0.12 in. (2–3mm) of axial play. Check that the halfshaft is making contact with the pinion shaft and that the halfshaft cannot be pulled out
   - Halfshaft into the knuckle
   - Lower suspension arm to the steering knuckle. Nuts and bolts: 105 ft. lbs. (142 Nm).
   - Tie rod end to the steering knuckle. Nut: 36 ft. lbs. (49 Nm).
   - ABS speed sensor
   - Hub locknut and washer
   - Negative battery cable
   - Wheels
   - Locknut cap and NEW cotter pin.
   - Undercovers
7. Fill the transaxle fluid to the proper level.

8. Start the vehicle, check for leaks and repair if necessary.

Supra

1. Before servicing the vehicle, refer to the precautions in the beginning of this section.
2. Remove or disconnect the following:
   - Negative battery cable. On vehicles equipped with an air bag, wait at least 90 seconds before proceeding.
   - Rear exhaust assembly
   - Cotter pin, locknut cap, and the locknut holding the halfshaft to the rear axle carrier.
   - Lower suspension arm brace
3. Place matchmarks on the halfshaft and the differential side gear shaft.
4. Remove or disconnect the following:
   - Hexagon bolts and washers
   - Halfshaft

To install:
5. Install or connect the following:
   - Halfshaft
   - Hexagon bolts, oiled. Bolts: 61 ft. lbs. (83 Nm).
   - Lower suspension arm brace. Bolts: 13 ft. lbs. (18 Nm).
   - Bearing locknut: 213 ft. lbs. (289 Nm).
   - Locknut cap and new cotter pin
   - Rear exhaust assembly
   - Negative battery cable

MR2

1. Before servicing the vehicle, refer to the precautions in the beginning of this section.

2. Remove or disconnect the following:
   - Rear wheel
   - Engine undercovers
   - Hub locknut
   - Brake fluid hose bracket
   - Strut rod
   - Lower suspension arms
   - Strut-to-spindle bolts
3. Drive the stub shaft from the axle hub with a plastic-faced hammer.
4. If removing the right halfshaft, remove the center bearing bracket bolts, then remove the axle halfshaft.
5. If removing the left halfshaft, drive the axle out of the transaxle with a hammer and brass punch.

To install:

➡Use new snap rings, circlips, and lock nuts for assembly.

➡Final tightening of suspension fasteners must take place with the suspension at curb height.

6. Install or connect the following:
   - Axle halfshaft. Tighten the center bearing bracket bolts to 47 ft. lbs. (64 Nm).
   - Strut-to-spindle-bolts and tighten them to 128 ft. lbs. (173 Nm).
   - Lower suspension arms. Tighten the No. 1 arm bolt to 76 ft. lbs. (103 Nm) and the No. 2 bolt to 36 ft. lbs. (49 Nm).
   - Strut rod. Tighten the bolt to 58 ft. lbs. (78 Nm).
   - Brake fluid hose bracket
   - Hub locknut: 159 ft. lbs. (216 Nm)

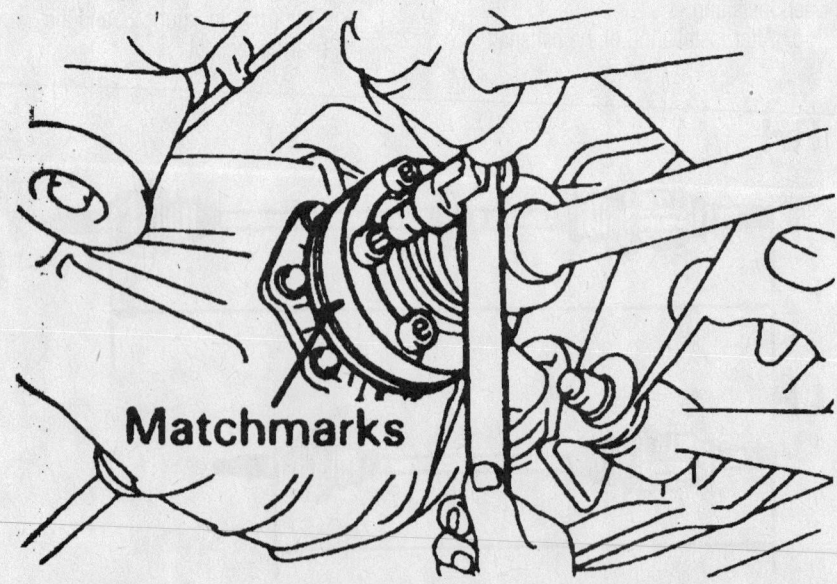

7923VGA4

Whenever removing the halfshaft, be sure to matchmark it to the differential side gear—Supra

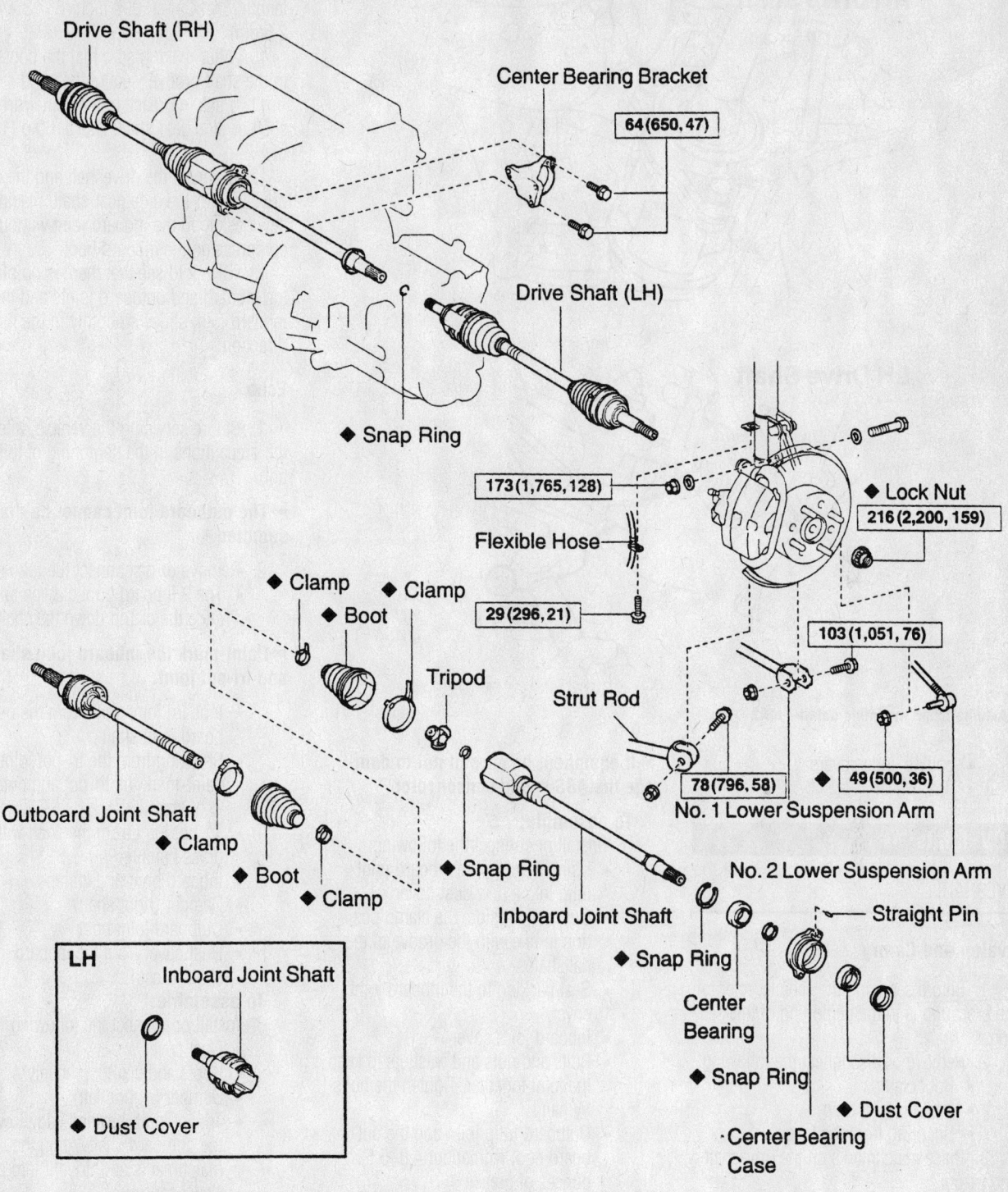

Drive Shaft (RH)

Center Bearing Bracket

64 (650, 47)

Drive Shaft (LH)

◆ Snap Ring

173 (1,765, 128)

Flexible Hose

◆ Lock Nut

216 (2,200, 159)

29 (296, 21)

103 (1,051, 76)

◆ Clamp ◆ Clamp

◆ Boot

Tripod

Strut Rod

Outboard Joint Shaft

◆ Clamp

◆ Boot

◆ Clamp

◆ Snap Ring

Inboard Joint Shaft

78 (796, 58)

49 (500, 36)

No. 1 Lower Suspension Arm

No. 2 Lower Suspension Arm

Straight Pin

◆ Snap Ring

Center Bearing

◆ Snap Ring

◆ Dust Cover

Center Bearing Case

LH

Inboard Joint Shaft

◆ Dust Cover

N·m (kgf·cm, ft·lbf) : Specified torque

◆ Non–reusable part

9357WG08

Axle halfshaft mounting exploded view—MR2

RH Drive Shaft

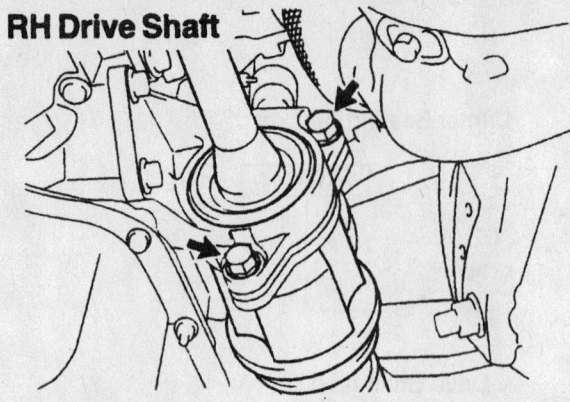

LH Drive Shaft

9357WG09

Axle halfshaft mounting detail—MR2

- Engine undercovers
- Rear wheel

CV-Joints

OVERHAUL

Avalon and Camry

1. Before servicing the vehicle, refer to the precautions in the beginning of this section.

2. Remove or disconnect the following:
- Boot clamps
- Inboard joint tulip
- Snapring from the tri-pot

3. Place matchmarks on the driveshaft and tri-pot

4. Remove or disconnect the following:
- Tri-pot joint off the driveshaft without hitting the joint roller
- Inboard joint boot
- On the right side, the clamp and dynamic damper
- Clamps and the outboard drive boot. DO NOT disassemble the outboard joint
- Dust cover from the inboard joint using a press

➡If equipped, be careful not to damage the ABS speed sensor rotor.

To assemble:

5. Install or connect the following:
- Using a press, the inboard joint tulip into a new dust cover
- On the right side, the clamp position in line with the groove of the halfshaft
- Seal packing to the inboard joint cover
- Inboard joint cover
- Bolts and nuts and washers to keep the joint together. Tighten the bolts by hand
- Outboard tulip joint and the outboard boot with about 4.8–5.5 ounces of grease
- Boot onto the outboard joint

6. Pack the inboard tulip joint and boot with grease that was supplied with the boot kit.

7. Install or connect the following:
- Inboard tulip joint onto the halfshaft
- Boot onto the halfshaft

8. Before checking the standard length, bend the band and lock it.

9. Make sure that the boot is not stretched or squashed when the driveshaft

is at standard length. Standard driveshaft length: 34.14 +/- 0.099 inch (867.3 +/- 2.5mm)

10. After making sure that the boots are in the shaft groove, secure the band

11. Pack in grease to the center driveshaft or side gear shaft. Use 5160 g (1.82.1 oz.)

12. Connect the driveshaft and the center driveshaft or the side gear shaft, placing a new gasket on the inboard joint without compressing the inboard boot.

13. Check to see that there is no play in the inboard and outboard joints and that the inboard joint slides smoothly in the thrust direction.

Echo

1. Before servicing the vehicle, refer to the precautions in the beginning of this section.

➡The outboard joint cannot be disassembled.

2. Remove or disconnect the following:
- The 2 inboard boot clamps and slide the clamp down the shaft

➡Paint-mark the inboard joint shaft and tri-pot joint.

- Inboard joint shaft from the outboard joint shaft
- Snapring from the tri-pot joint, and paint-mark the tri-pot and outboard joint shaft
- Tri-pot joint from the shaft with a brass hammer
- Inboard boot and clamps
- Damper (right shaft)
- Outboard joint boot
- Dust cover from the inboard joint shaft (press)

To assemble:

3. Install or connect the following:
- Dust cover
- Boots and clamps, loosely
- Damper (right shaft)
- Tri-pot joint, beveled edge toward the outboard shaft. Align the matchmarks and drive it into place
- New snapring
- Outboard boot. The grease capacity is 5.5–6 ounces (155–170 g).
- Inboard shaft to outboard shaft
- Inboard boot. The grease capacity is 4.4–4.8 ounces (125–135 g).

➡Toyota recommends different types of grease for the joints. OEM replacement boot kits have the grease color coded. The outboard grease is black; the inboard, yellow.

4. Check the boots at standard halfshaft length. The right shaft should be 32.02 inches +/- 0.197 in. (813.3mm +/- 5mm); the left shaft should be 22.61 inches +/- 0.197 in. (574.3mm +/- 5mm).

5. Check the position of the damper before installing the clamp. Distance from the outer face of the damper to the outer face of the outer joint should be 16.835 inches +/- 0.079 in. (427.6mm +/- 2mm).

Celica, Supra, GT-S, and MR2

5S-FE, 1ZZ-FE AND 2ZZ-GE ENGINES

1. Before servicing the vehicle, refer to the precautions in the beginning of this section.

2. Remove or disconnect the following:
• Halfshaft
• On the inboard joint tulip, the boot clamps. Slide the inboard joint boot toward the outboard joint.

➡**Place matchmarks on the tri-pot and inboard joint tulip or center driveshaft. Do not punch the marks.**

• Inboard joint tulip or center driveshaft from the driveshaft

3. On the tri-pot, use snapring expander, temporarily, slide the snapring toward the outboard joint side.

➡**Place matchmarks on the driveshaft and tri-pot. Do not punch the marks.**

4. Remove or disconnect the following:
• Tri-pot from the driveshaft
• Snapring
• Inboard joint boot
• On the right side with dynamic damper, the clamps from the damper and extract the unit
• Clamps from the outboard joint
• Boot from the outboard joint

➡**Do not disassemble the outboard joint.**

• On the left side, the dust cover from the inboard joint tulip
• On the right side, the dust cover from the center driveshaft
• On the right side, the snapring and bearing from the case
• The dust cover and snapring
• The bearing from the shaft and the other snapring
• The straight pin
• The No. 2 dust deflector

To assemble:

5. Install or connect the following:
• On the right side, assemble the center driveshaft by inserting the straight pin into the bearing case

Place matchmarks on the tri-pot and outboard joints

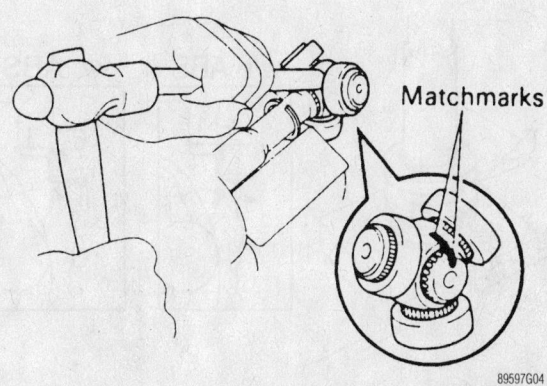

With matchmarks on the tri-pot, tap the joint for the driveshaft

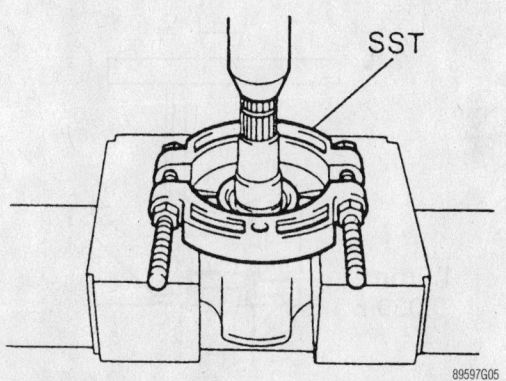

Using a press, remove the dust cover from the center driveshaft on the left side

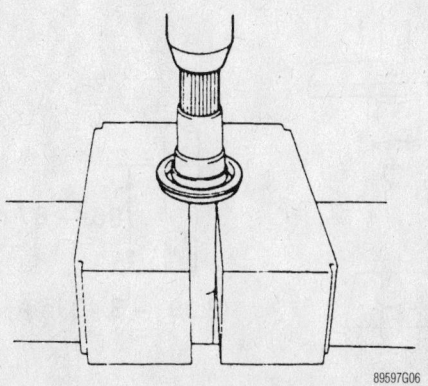

Use a press to drive the old dust cover out of the right side center driveshaft

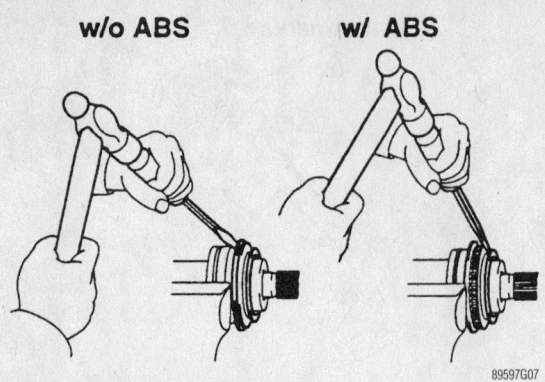

w/o ABS   w/ ABS

89597G07

Removing the no. 2 dust deflector on models with and without ABS

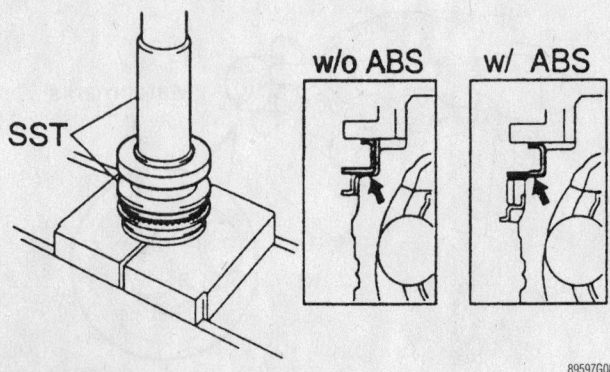

SST

w/o ABS   w/ ABS

89597G08

Press the new No. 2 deflector and seat it properly

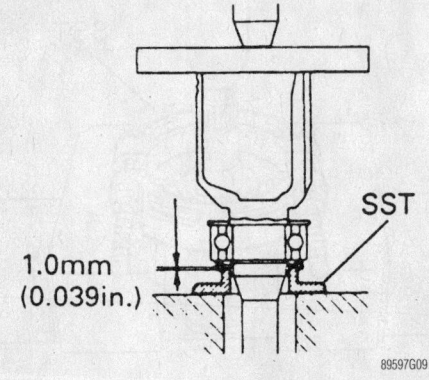

SST

1.0mm
(0.039in.)

89597G09

When installing the dust cover, the clearance should be within the ranges as specified

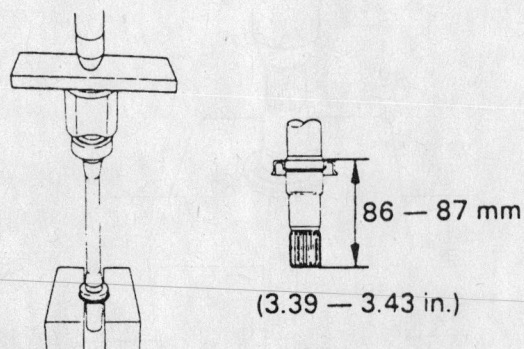

86 — 87 mm

(3.39 — 3.43 in.)

89597G10

On the right side driveshaft, press the dust cover till the tip of the shaft reaches specifications

- A new bearing into the case
- A new snapring
- A new bearing into the case assembly on the center driveshaft
- A new snapring
- New dust cover. The clearance between the dust cover and bearing should be kept in the ranges in the illustration.
- On the left side, use a press and insert the new dust cover
- On the right side, using a steel plate and press, press in a new dust cover until the distance from the tip of the center driveshaft to the cover reaches the specification as shown

6. Temporarily install boots and the damper. Before installing the outboard joint, wrap vinyl tape around the spline of the driveshaft to prevent damaging the boots and damper.

7. Install or connect the following:
- Temporarily, the new outboard boot onto the driveshaft
- On the dynamic damper, temporarily install the damper to the driveshaft.
- Temporarily, a new inboard joint to the driveshaft

8. Install as follows:
- A new snapring on the tri-pot joint
- Place the beveled side of the tri-pot joint axial spline toward the outboard joint. Align the matchmarks placed before removal
- The tri-pot joint to the driveshaft

➡ Do not tap the roller.

- A new snapring
- Boot to the outboard joint. Before assembling the boot, fill grease into the outboard joint and boot. Special grease (usually black) is supplied with the boot kits. Grease capacity is 4.2–4.6 oz (120–130 g)
- Inboard joint tulip to the front driveshaft. Pack grease (usually yellow) into the boot and inboard joint tulip.
- Inboard joint tulip onto the driveshaft
- Boot to the inboard joint tulip
- Boot clamps to both boots

The standard driveshaft length is as follows:
- Automatic left: 22.516–22.910 inch (571.7–581.7mm)
- Automatic right: 33.74–34.13 inch (857.0–867.0mm
- Manual left: 22.34–22.73 inch (525.5–577.4mm)

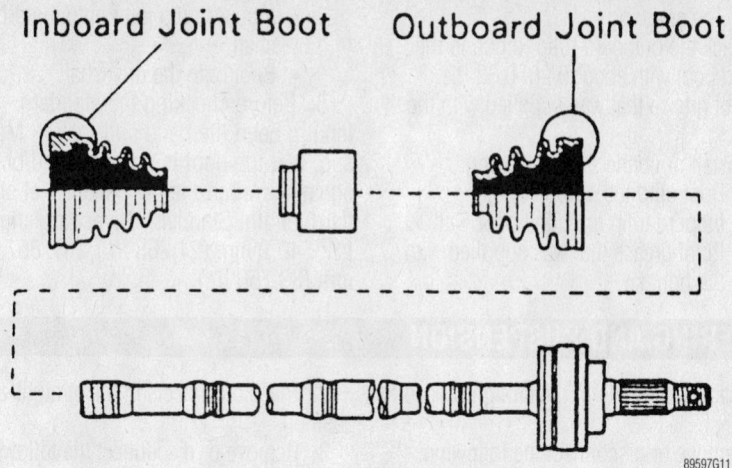

Inboard Joint Boot   Outboard Joint Boot

89597G11

Temporarily install the driveshaft boots on the inboards and outboard and dynamic damper

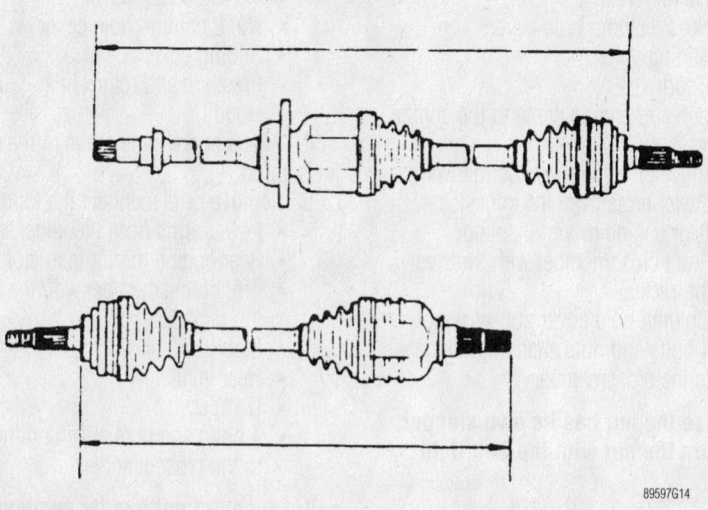

89597G14

Measure the driveshaft length at these points and make sure it reaches specifications prior to installation

- Manual right: 33.56–33.95 inch (852.5–862.5mm)

9. Bend the band of the clamp and lock onto the boot.

10. On the dynamic damper attach the clamp on the shaft groove. Check the distance as shown in the illustration. Correct distance should be 7.677–8.071 inch (195.0–205.0mm).

11. Using a screwdriver, tighten the band of the clamp and lock it.

2JZ-GE ENGINES

1. Before servicing the vehicle, refer to the precautions in the beginning of this section.

2. Remove or disconnect the following:
- Halfshaft
- Inboard joint boot clamps
- Inboard joint tulip
- Tri-pot joint
- Inboard joint boot

- On the dynamic damper (right side), the clamp from the damper
- Dynamic damper
- Outboard joint boot, but do not disassemble the joint.
- Dust cover from the inboard joint tulip using a press

To assemble:

3. Install or connect the following:
- New dust cover onto the inboard joint tulip
- Temporarily, boots and the right damper. Before installing the outboard joint boot, wrap vinyl tape around the spline of the driveshaft to prevent damaging the boots and damper
- Temporarily, a new outboard boot to the driveshaft
- On the right side, the dynamic damper to the driveshaft
- Temporarily, the new inboard joint boot to the driveshaft

4. Place the beveled side of the tri-pot joint axial spline toward the outboard joint. Align the matchmarks placed before removal.

5. Install or connect the following:
- Using a brass bar and hammer, the tri-pot joint to the driveshaft. Do not tap the roller.
- A new snapring
- Boot to the outboard joint

➡ **Before assembling the boot, fill it with the grease (black), supplied with the kit, into the outboard joint and boot.**

- The inboard joint tulip to the front driveshaft. Pack grease into the inboard joint tulip and boot. The grease should be yellow in color and supplied with the boot kit
- The inboard joint tulip to the driveshaft
- Temporarily, the boot to the inboard joint tulip
- The boot clamps on both boots

6. Be sure the boots are on the shaft groove. Set the driveshaft length to: left— 21.81–22.21 inch (554.2–564.2mm); right— 34.33–34.73 inch (872.25–882.2mm)

7. Using a screwdriver, bend the band and lock it.

8. On the right side driveshaft, place the clamp on in the shaft groove. Check that the distance is within specifications. Distance should be 15.25–15.65 inch (387.5–397.5mm).

9. Using a screwdriver, bend the band and lock it.

Corolla

1. Before servicing the vehicle, refer to the precautions in the beginning of this section.

2. Remove or disconnect the following:
- Inboard joint boot clips
- Inboard joint tulip from the driveshaft
- Snapring
- Using a brass rod and hammer, the tri-pot joint off the driveshaft without hitting the joint roller
- Inboard joint boot
- Clamp and driveshaft damper
- Clamps and the outboard drive boot. DO NOT disassemble the outboard joint.

To assemble:

3. Install or connect the following:

➡ **Before installing the boot, wrap the spline end of the shaft with masking tape to prevent damage to the boot.**

- Driveshaft damper with a new clamp
- Temporarily, the inboard boot with new clamp to the drive joint

➡**The inboard boot and clamp are larger than those of the outboard boot.**

- The tri-pot onto the driveshaft with a brass rod and hammer without hitting the joint roller

- The snapring
4. Pack the outboard tulip joint and the outboard boot with about 0.26–0.33 lbs. ounces of grease that was supplied with the boot kit.
5. Install or connect the following:
- Boot onto the outboard joint
- Inboard tulip joint and boot with ½ lb. of grease that was supplied with the boot kit

- Inboard tulip joint onto the drive-shaft
- Boot onto the driveshaft
6. Before checking the standard length, bend the band and lock it. Make sure that the boot is not stretched or squashed when the driveshaft is at standard length. Standard driveshaft length: LH: 540.2 mm (21.268 in.); RH: 857.4 mm (33.756 in.)

STEERING AND SUSPENSION

Air Bag

PRECAUTIONS

Several precautions must be observed when handling the inflator module to avoid accidental deployment and possible personal injury.
- Never carry the inflator module by the wires or connector on the underside of the module.
- When carrying a live inflator module, hold securely with both hands, and ensure that the bag and trim cover are pointed away.
- Place the inflator module on a bench or other surface with the bag and trim cover facing up.
- With the inflator module on the bench, never place anything on or close to the module that may be thrown in the event of an accidental deployment.

DISARMING

To avoid personal injury when working on vehicles equipped with an air bag, the negative battery cable must be disconnected and at least 90 seconds must elapse before working on the system. Failure to do so may result in deployment of the air bag.

REARMING

After vehicle service is completed, reattach the battery cables (positive cable first!) to rearm the air bag system.

Rack and Pinion Steering Gear

REMOVAL & INSTALLATION

Manual

ECHO

1. Before servicing the vehicle, refer to the precautions in the beginning of this section.

2. Place the wheels in the straight-ahead position.
3. Remove or disconnect the following:
- Steering wheel
- Engine under covers
- Tie rod ends
- No. 2 column hole cover
- Sliding yoke
- Hood
4. Attach and engine crane to the engine for support.
5. Remove or disconnect the following:
- Lower arms from the knuckles
- Rear engine mount insulator
- Front crossmember with the steering rack
- Column hole cover sub-assembly
- 4 bolts and nuts attaching the gear to the crossmember

➡**Because the nut has its own stopper, do not turn the nut with the bolt tight.**

To install:
6. Install or connect the following:
- Steering rack to crossmember. Torque the nuts to 54 ft. lbs. (74 Nm).
- Column hole cover sub-assembly
- Front suspension assembly. Front bolts: 52 ft. lbs. (70 Nm); rear bolts: 86 ft. lbs. (116 Nm).
- Rear mount insulator. Torque the bolt and 2 nuts to 59 ft. lbs. (80 Nm).
- Lower arm to knuckle. Horizontal bolt: 65 ft. lbs. (88 Nm); vertical bolt: 97 ft. lbs. (132 Nm).
- Hod
- Sliding yoke
- No. 2 column hole cover
- Tie rods
- Under covers
- Steering wheel. Nut: 25 ft. lbs. (34 Nm).

Power

ECHO

1. Before servicing the vehicle, refer to the precautions in the beginning of this section.

2. Place the wheels in the straight-ahead position.
3. Remove or disconnect the following:
- Steering wheel
- Engine under covers
- Tie rod ends
- No. 2 column hole cover
- Sliding yoke
- Pressure and return lines
- Hood
4. Attach and engine crane to the engine for support.
5. Remove or disconnect the following:
- Lower arms from the knuckles
- Rear engine mount insulator
- Front crossmember with the steering rack
- Stabilizer bar
- Heat insulator
- Damper
- 4 bolts and nuts attaching the gear to the crossmember

➡**Because the nut has its own stopper, do not turn the nut with the bolt tight.**

To install:
6. Install or connect the following:
- Steering rack to crossmember. Torque the nuts to 54 ft. lbs. (74 Nm).
- Damper. Bolts: 13 ft. lbs. (18 Nm).
- Heat insulator. Bolt: 26 ft. lbs. (35 Nm).
- Stabilizer bar
- Front suspension assembly. Front bolts: 52 ft. lbs. (70 Nm); rear bolts: 86 ft. lbs. (116 Nm).
- Rear mount insulator. Torque the bolt and 2 nuts to 59 ft. lbs. (80 Nm).
- Lower arm to knuckle. Horizontal bolt: 65 ft. lbs. (88 Nm); vertical bolt: 97 ft. lbs. (132 Nm).
- Hood
- Sliding yoke
- No. 2 column hole cover
- Tie rods
- Pressure and return lines
- Under covers

- Steering wheel. Nut: 25 ft. lbs. (34 Nm).

AVALON, CAMRY AND CAMRY SOLARA

1. Before servicing the vehicle, refer to the precautions in the beginning of this section.

2. Position the front wheels straight ahead.

3. Remove or disconnect the following:
   - Negative battery cable. On vehicles equipped with an air bag, wait at least 90 seconds before proceeding.

➡ **If equipped with an air bag, disable the system and secure the steering wheel.**

- Front wheels
- Left and right front fender apron seals.
- Cotter pin and nut holding the steering knuckle to the tie rod end.
- Tie rod end from the steering knuckle.

➡ **Place matchmarks on the intermediate shaft and the control valve shaft.**

- Lower bolt holding the control valve shaft to the intermediate shaft.
- Intermediate shaft from steering rack housing.
- Tube clamp
- Return line and the pressure line from the control valve housing.
- Stabilizer bar bolts and nuts. Do not remove the bar from the vehicle.
- If necessary, rear engine mounting and bracket for additional clearance.
- On the V6 engine, oxygen sensor.
- Steering gear mounting bolts and nuts
- Steering gear

To install:

4. Install or connect the following:
   - Steering gear. Nuts and bolts: 134 ft. lbs. (181 Nm).
   - On the V6, oxygen sensor.
   - Rear engine mounting bracket. Bolts: 38 ft. lbs. (52 Nm).
   - Stabilizer bar bolts and nuts.
   - Tube clamp. Nut: 84 inch lbs. (10 Nm).
   - Intermediate shaft to the steering rack. Bolts: 26 ft. lbs. (35 Nm).
   - Tie rods to the steering knuckles with the castellated nuts.
   - Front fender apron seals

- Front wheels
- Negative battery cable
- Power steering fluid

CELICA AND GT-S

1. Before servicing the vehicle, refer to the precautions in the beginning of this section.

2. Remove or disconnect the following:
   - Negative battery cable. On vehicles equipped with an air bag, wait at least 90 seconds before proceeding.
   - Right and left-hand engine undercovers.
   - Left and right-hand tie rod ends. Separate the tie rod using a puller.
   - Oxygen sensor
   - Front exhaust pipe and brackets.

➡ **Place matchmarks on the steering column intermediate shaft and the steering gear control valve shaft.**

- Lower bolt to the intermediate shaft
- Intermediate shaft from the control valve shaft.
- Pressure feed and return tubes from the steering rack.
- Tube clamp bracket

3. Support the engine and transaxle with a support fixture.

4. Remove or disconnect the following:
   - Lower control arms from the lower ball joints.
   - Through-bolts to the front and rear mounting insulators.
   - Bolts holding the rear of the lower control arm to the sub-frame and body. Remove the bolts on both sides of the sub-frame.

5. Support the front sub-frame with a jack.

6. Remove or disconnect the following:
   - Bolts holding the sub-frame to the body. Lower the front sub-frame with the lower suspension arms and steering gear.
   - Power steering gear assembly by removing the set bolts and nuts from the sub-frame.

To install:

7. Install or connect the following:
   - Power steering gear assembly to the sub-frame. Bolts and nuts: 94 ft. lbs. (127 Nm).
   - Sub-frame to the body. Bolts: 94 ft. lbs. (127 Nm).
   - Bolts to hold the rear of the lower control arms to the sub-frame and body. Bolt that goes through the lower control arm to 123 ft. lbs. (167 Nm) and the other 2 bolts to

130 ft. lbs. (175 Nm). Install the bolts to both sides
   - Through-bolts to the front and rear mounting insulators. Bolts: 64 ft. lbs. (88 Nm).
   - Lower control arms to the lower ball joints. Nuts and bolts: 94 ft. lbs. (127 Nm).
   - Pressure feed and return tubes to the steering rack. Tubes: 26 ft. lbs. (36 Nm).
   - Tube clamp bracket. Bolts: 108 inch lbs. (13 Nm).

8. Align the matchmarks on the intermediate shaft and control valve shaft.

9. Install or connect the following:
   - Lower bolt. Upper and lower bolts: 26 ft. lbs. (35 Nm).
   - Front exhaust pipe
   - Oxygen sensor
   - Tie rod ends to the steering knuckles
   - Right and left-hand engine undercovers
   - Negative battery cable
   - Power steering fluid

COROLLA

1. Before servicing the vehicle, refer to the precautions in the beginning of this section.

2. Position the front wheels straight ahead.

3. Remove or disconnect the following:
   - Negative battery cable. On vehicles equipped with an air bag, wait at least 90 seconds before proceeding

➡ **If equipped with an air bag, disable the system and secure the steering wheel.**

- Steering column hole cover and loosen the upper pinch bolt on the sliding yoke
- Lower pinch bolt at the pinion shaft
- Front wheels
- Left and right engine undercovers.
- Left and right tie rod ends.

4. Install an engine support and tension it to support the engine without raising it.

✷✷ CAUTION

The engine hoist is now in place and under tension. Use care when repositioning the vehicle and make necessary adjustments to the engine support.

5. Remove or disconnect the following:
   - Lower control arms from the ball joints

- If equipped with a stabilizer bar, stabilizer bar links from both lower control arms
- Right rear control arm bushing retaining bracket. Do this for both lower control arms
- Stabilizer bar
- Grommet in the crossmember
- Bolt and nuts holding in the middle of the crossmember and support the crossmember with a jack
- Bolts from the outer side of the suspension crossmember
- Suspension crossmember with the lower suspension arms
- Exhaust front pipe support
- Engine rear mount insulator
- Engine rear mount bracket
- Pressure feed and return tubes
- Brackets and grommets to the power steering rack

6. Slide the power steering gear assembly to the right side of the vehicle.

To install:

7. Install or connect the following:
- Power steering assembly
- Grommets and brackets. Nuts and bolts: 43 ft. lbs. (59 Nm).
- Pressure feed and return tubes. Nuts: 26 ft. lbs. (36 Nm).
- Engine rear mount bracket. Bolts: 57 ft. lbs. (77 Nm).
- Engine rear mount insulator. Bolt: 64 ft. lbs. (87 Nm).
- Exhaust front pipe support. Bolts: 14 ft. lbs. (19 Nm).
- Outer 6 bolts to hold the crossmember to the vehicle. Bolts: 152 ft. lbs. (206 Nm).
- Center crossmember-to-radiator support bolts: 45 ft. lbs. (61 Nm).
- Lower A frame-to-center bolts: 161 ft. lbs. (218 Nm).
- Lower A frame-to-outer bolts: 109 ft. lbs. (147 Nm).

- Front, center and rear mount bolts: 45 ft. lbs. (61 Nm).
- Grommet to the crossmember
- Stabilizer bar
- Lower control arm bushing retaining bracket. Do not tighten the bolts or nut at this time
- Lower control arm to the lower ball joint. Bolt and nuts: 105 ft. lbs. (142 Nm). Connect both lower control arms to the ball joints
- Stabilizer bar links to the lower control arms. Nuts: 33 ft. lbs. (44 Nm).
- Sliding yoke to the pinion shaft. Lower bolt: 26 ft. lbs. (35 Nm). Tighten upper bolt: 20 ft. lbs. (27 Nm).
- Steering column hole cover. Bolts: 43 inch lbs. (5 Nm).
- Left and right-hand tie rod ends. Nuts: 36 ft. lbs. (49 Nm).
- Front wheels

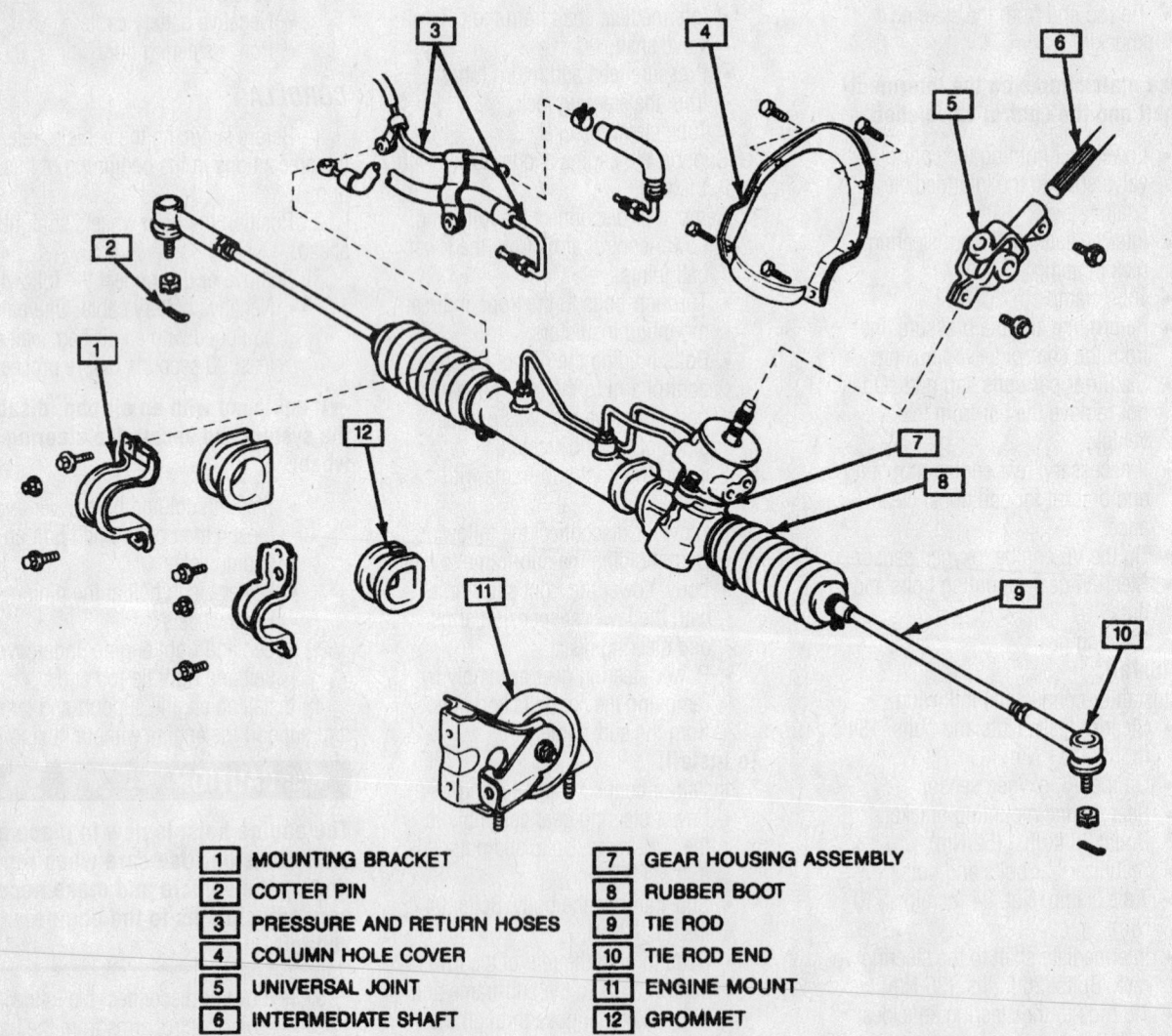

1 MOUNTING BRACKET		**7** GEAR HOUSING ASSEMBLY	
2 COTTER PIN		**8** RUBBER BOOT	
3 PRESSURE AND RETURN HOSES		**9** TIE ROD	
4 COLUMN HOLE COVER		**10** TIE ROD END	
5 UNIVERSAL JOINT		**11** ENGINE MOUNT	
6 INTERMEDIATE SHAFT		**12** GROMMET	

7923VGA5

Exploded view of a typical power rack and pinion steering gear unit

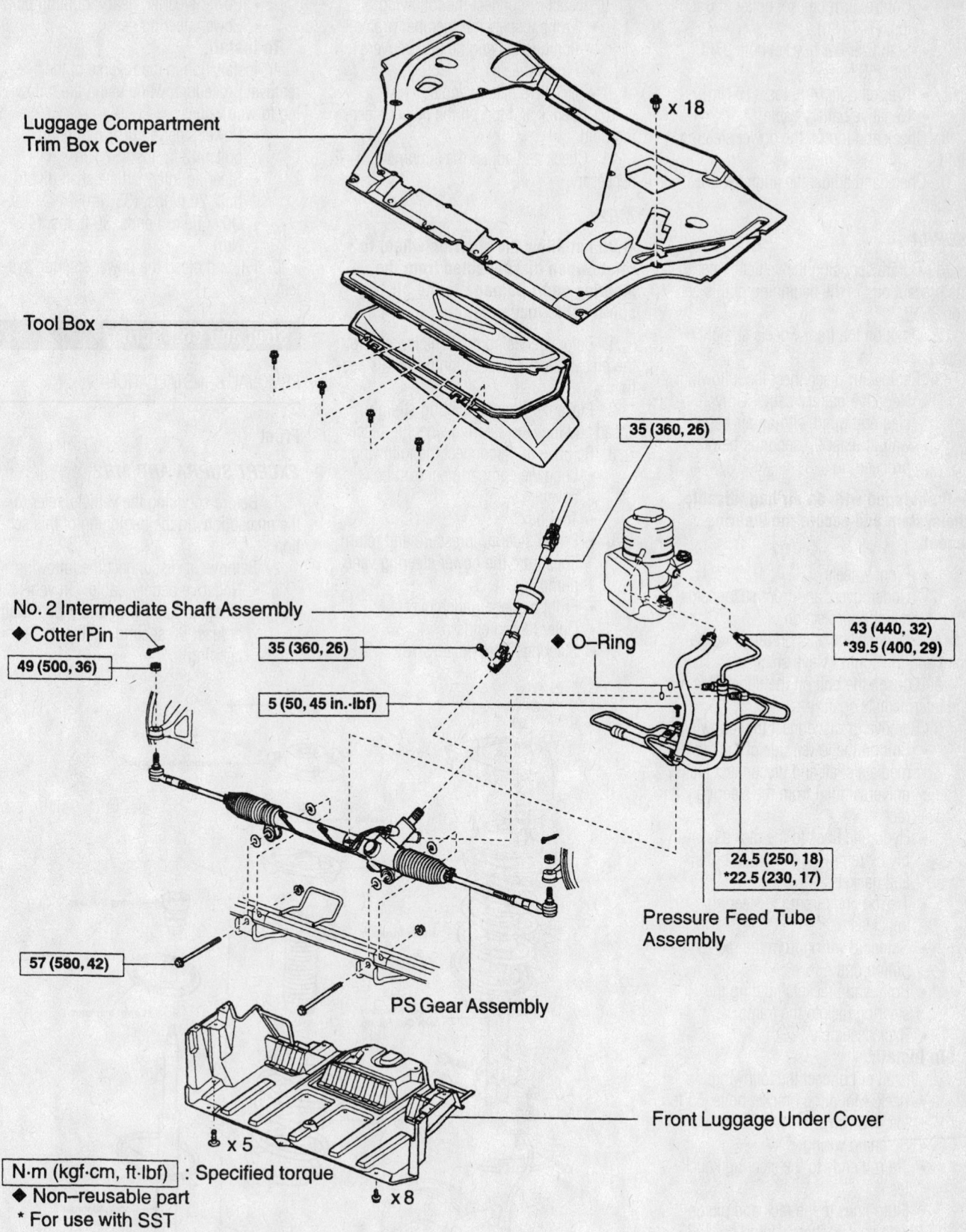

Luggage Compartment Trim Box Cover

x 18

Tool Box

35 (360, 26)

No. 2 Intermediate Shaft Assembly

◆ Cotter Pin

49 (500, 36)

35 (360, 26)

5 (50, 45 in.·lbf)

◆ O–Ring

43 (440, 32)
*39.5 (400, 29)

24.5 (250, 18)
*22.5 (230, 17)

Pressure Feed Tube Assembly

57 (580, 42)

PS Gear Assembly

Front Luggage Under Cover

x 5

N·m (kgf·cm, ft·lbf) : Specified torque
◆ Non–reusable part
* For use with SST

x 8

9357WG10

Power rack and pinion steering gear mounting exploded view—MR2

- Control arm bracket bolts: 108 ft. lbs. (147 Nm).
- Stabilizer bar bracket bolt: 37 ft. lbs. (50 Nm).
- Bracket nut: 14 ft. lbs. (19 Nm).
- Negative battery cable

8. Check and top off the power steering fluid.

9. Check and adjust the alignment, if needed.

SUPRA

1. Before servicing the vehicle, refer to the precautions in the beginning of this section.

2. Position the front wheels straight ahead.

3. Remove or disconnect the following:
- Negative battery cable. On vehicles equipped with an air bag, wait at least 90 seconds before proceeding

➡ **If equipped with an air bag, disable the system and secure the steering wheel.**

- Front wheels
- Undercovers and front suspension member protection

4. Place matchmarks on the universal joint and the control valve shaft.

5. Loosen the bolt on the upper side of the intermediate shaft.

6. Remove or disconnect the following:
- Bolt on the lower side of the intermediate shaft and disconnect the universal joint from the steering rack
- Hydraulic lines to the rack assembly by removing the union bolts and gaskets
- Tie rod ends from the steering knuckles
- Solenoid wiring from the rack and pinion unit
- Bolts and brackets holding the steering rack to the frame
- Rack assembly

To install:

7. Install or connect the following:
- Rack. Mounting bracket bolts: 55 ft. lbs. (75 Nm).
- Solenoid wiring
- Tie rod ends to the steering knuckles
- Fluid lines to the rack and pinion with new washers. Union bolts: 36 ft. lbs. (49 Nm).

8. Align the matchmarks on the universal joint and the control valve shaft. Tighten the upper and lower bolts to 26 ft. lbs. (35 Nm).

9. Install or connect the following:
- Front suspension member protection and the engine undercovers
- Front wheels
- Negative battery cable

10. Chedck and top off the power steering fluid.

11. Check and adjust the alignment, if necessary.

MR2

➡ **Do not allow the steering wheel to rotate when disconnected from the steering gear. Damage to the air bag spiral cable may result.**

1. Before servicing the vehicle, refer to the precautions in the beginning of this section.

2. Place the wheels pointing straight ahead and lock the steering wheel in place.

3. Remove or disconnect the following:
- Luggage compartment trim box cover
- Tool box
- Power steering pressure and return lines from the power steering vane pump
- Front luggage undercover
- Outer tie rod ends
- Steering intermediate shaft

- Power steering gear mounting bolts
- Power steering gear

To install:

4. Installation is the reverse of the removal procedure, while using the following torwue values:
- Power steering gear mounting bolts: 42 ft. .lbs. (57 Nm)
- Steering intermediate shaft pinch bolt: 26 ft. lbs. (35 Nm)
- Outer tie rod ends: 36 ft. lbs. (49 Nm)

5. Fill and bleed the power steering system.

Strut and Coil Spring

REMOVAL & INSTALLATION

Front

EXCEPT SUPRA AND MR2

1. Before servicing the vehicle, refer to the precautions in the beginning of this section.

2. Remove or disconnect the following:
- Negative battery cable. On vehicles equipped with an air bag, wait at least 90 seconds before proceeding

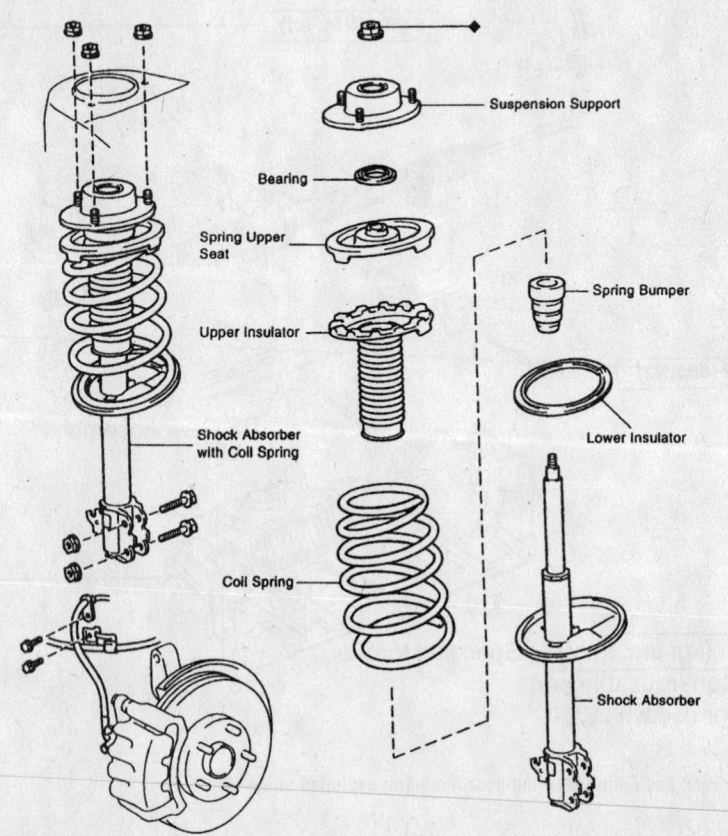

Suspension Support

Bearing

Spring Upper Seat

Upper Insulator

Shock Absorber with Coil Spring

Spring Bumper

Lower Insulator

Coil Spring

Shock Absorber

7923VGA6

Common coil spring and strut component assembly—except Supra

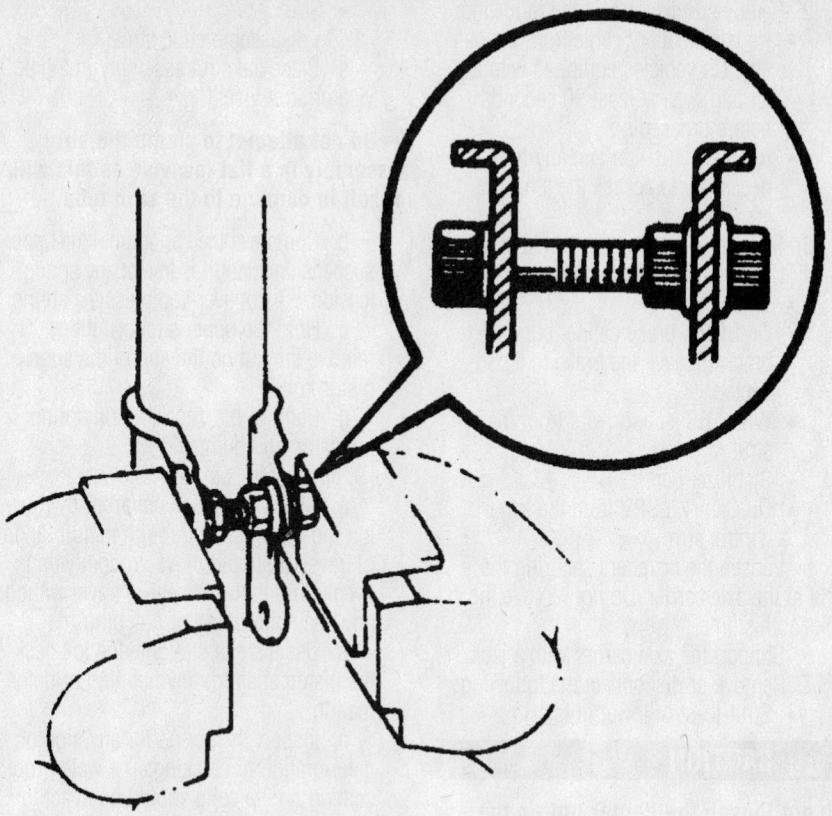

Proper method of supporting the strut in a vise

7923VGA7

✳✳ WARNING

Do not support the weight of the vehicle on the suspension arm; the arm will deform under its weight.

- Wheel
- Bolt, and disconnect the brake hose from the strut
- With ABS brakes, wiring harness from the strut
- Bolts and strut from the steering knuckle
- Strut

3. To disassemble the strut:
- Install a bolt and 2 nuts to the bracket at the lower portion of the strut shell and secure it in a vise
- Compress the coil spring
- Remove the dust cover and hold the spring seat so that it will not turn. Remove the nut on the top of the strut
- Remove the suspension support, bearing, dust seal, spring seat, spring, insulators and bumper

To install:
4. To assemble the strut:
- Install the spring bumper to piston
- Using a spring compressor, compress the spring

- Install the coil spring to the strut. Fit the lower end of the coil spring into the gap of the lower seat
- Install the spring seat with the insulator
- Install the dust seal on the spring seat
- Install the suspension support and tighten 35 ft. lbs. (47 Nm). After the nut has been tighten, release the compressor tool tension
- Pack multipurpose grease into the suspension support. Install the dust cover

➡ **Do not use an impact wrench to tighten the nut. Also, check that the bearing fits into the recess in the suspension support.**

5. Install or connect the following:
- Nuts holding the strut to the strut tower. Nuts to 29 ft. lbs. (39 Nm), except on Avalon, Camry and Celica: 59 ft. lbs. (80 Nm)
- Steering knuckle to the strut lower bracket

6. Insert the 2 bolts from the rear side and tighten the strut-to-steering knuckle arm bolts. Tighten as follows:
- Echo: 97 ft. lbs. (132 Nm).
- Corolla: 203 ft. lbs. (275 Nm).

- Celica: 113 ft. lbs. (153 Nm).
- Avalon and Camry: 156 ft. lbs. (211 Nm).

7. Install or connect the following:
- Brake line to the steering knuckle
- If equipped with ABS, secure the wiring harness
- Wheel
- Negative battery cable

8. Check and adjust the alignment, if needed.

SUPRA

1. Before servicing the vehicle, refer to the precautions in the beginning of this section.

2. Remove or disconnect the following:
- Negative battery cable. On vehicles equipped with an air bag, wait at least 90 seconds before proceeding
- Wheel
- Brake caliper support bracket. Suspend it with a piece of wire
- Fender apron, engine undercover, and the front fender wheel opening molding
- Windshield washer tank, if removing the left side strut
- ABS speed sensor at the steering knuckle and wiring harness clamp
- Plug from the upper strut mount. Do not remove the center bolt

✳✳ CAUTION

Do not remove the center bolt to the strut at this time. The spring on the strut is under high pressure and can cause serious injury or vehicle damage.

- Upper control arm through-bolt from the sub-frame
- Upper control arm and turn the control arm completely around. It is not necessary to remove the upper ball joint
- Strut at the lower control arm
- Upper mounting nuts and strut assembly with the coil spring

3. Compress the coil spring.
4. Remove or disconnect the following:
- Piston rod locknut
- Suspension support, coil spring and bumper

To install:
5. Match the bolt of the suspension support with the cut out portion of the insulator.

6. Install or connect the following:
- Spring bumper
- Compressed coil spring, matching

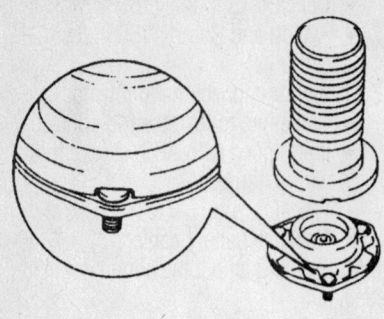

Aligning the insulator to the support—Supra

the end of the coil into the recess of the strut spring seat
- Suspension support to the rod and temporarily install a new nut
7. Turn the suspension support so one of the bolts on the support faces the same direction as shown in the illustration.

➡**Align the bolt so a line drawn between the rod and bolt would be at 90 degrees to the direction of the lower bushing.**

8. Remove the spring compressor.
9. Install or connect the following:
- Strut and tighten the 3 upper strut mount nuts to 26 ft. lbs. (35 Nm). Tighten the middle nut to 22 ft. lbs. (29 Nm) and install the plug
- Lower end of the strut to the lower control arm. Do not tighten the bolt at this time
- Upper control arm and through-bolt and nut. Do not tighten the bolt at this time
- Speed sensor, wiring harness, and the washer tank
- Fender apron and engine undercover
- Caliper support bracket. Bolts: 87 ft. lbs. (118 Nm).
- Wheel
10. Bounce the vehicle a few times to stabilize the suspension, then tighten the strut-to-lower arm bolt to 106 ft. lbs. (143 Nm). Tighten the upper arm to 121 ft. lbs. (164 Nm).
11. Reconnect negative battery cable.
12. Check and adjust the alignment, if needed.

MR2

Rear

EXCEPT MR2, CELICA AND GT-S

1. Before servicing the vehicle, refer to the precautions in the beginning of this section.

2. Remove or disconnect the following:
- Negative battery cable from the battery. On vehicles equipped with an air bag, wait at least 90 seconds before proceeding
- Rear seat cushion and any trim necessary to access the strut towers
3. Support the axle beam with a jack.
4. Remove or disconnect the following:
- Wheel
- On Supra, brake caliper support bracket. Leave the brake line connected
- With ABS, sensor wire from the strut
- Stabilizer bar
- On Camry, LSPV from the lower control arm
5. Loosen the fasteners securing the strut to the axle carrier. Do not remove the bolts at this time.
6. Support the axle carrier with a jack.
7. Remove or disconnect the following:
- Strut-to-strut tower nuts

❋❋ CAUTION

Do not loosen the center nut on the top of the strut piston.

- Strut
8. To disassemble the strut:
a. Place the strut assembly in a pipe vise or strut vise.

➡**Do not attempt to clamp the strut assembly in a flat jaw vise as this will result in damage to the strut tube.**

b. Compress the spring until the upper suspension support is free of any spring tension. Do not over-compress the spring.
c. Hold the upper support, then remove the nut on the end of the shock piston rod.
d. Remove the support, coil spring, insulator, and bumper.
9. Inspect the strut as follows:
a. Check the shock absorber by moving the piston shaft through its full range of travel. It should move smoothly and evenly throughout its entire travel without any trace of binding or notching.
b. Use a small straightedge to check the piston shaft for any bending or deformation.
c. Inspect the spring for any sign of deterioration or cracking. The waterproof coating on the coils should be intact to prevent rusting.

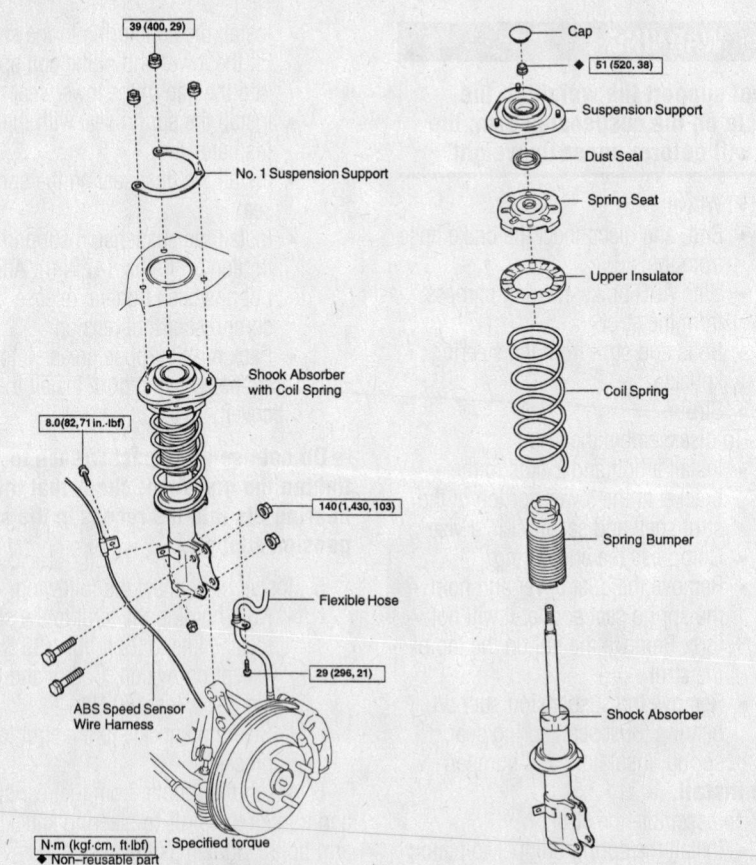

| 39 (400, 29) |
| No. 1 Suspension Support |
| Shock Absorber with Coil Spring |
| 8.0 (82, 71 in.-lbf) |
| 140 (1,430, 103) |
| Flexible Hose |
| 29 (296, 21) |
| ABS Speed Sensor Wire Harness |

| Cap |
| 51 (520, 38) ◆ |
| Suspension Support |
| Dust Seal |
| Spring Seat |
| Upper Insulator |
| Coil Spring |
| Spring Bumper |
| Shock Absorber |

N·m (kgf·cm, ft·lbf) : Specified torque
◆ Non–reusable part

Front strut and coil spring mounting exploded view—MR2

To install:

➡**Never reuse a self-locking nut. Always replace self-locking nuts and cotter pins as applicable.**

10. Assemble the strut as follows:

a. Loosely assemble all components onto the strut assembly. Be sure the spring end aligns with the hollow in the lower seat.

b. Align the upper suspension support with the piston rod and install the support.

c. Align the suspension support with the strut lower bracket. This assures the spring will be properly seated top and bottom.

d. Compress the spring to expose the strut piston rod threads.

e. Install a new strut piston nut and tighten to the following:

- Corolla, Celica, Camry and Avalon: 36 ft. lbs. (49 Nm).
- Supra: 20 ft. lbs. (27 Nm).

f. Remove the spring compressor. Be sure the paint mark on the upper support faces the outside of the strut.

11. Place the strut on the vehicle and install the nuts to hold the strut to the strut tower. Nuts: 29 ft. lbs. (39 Nm), except on Supra. On Supra models: 19 ft. lbs. (26 Nm).

12. Install or connect the following:

- Strut to the axle carrier and install the bolt and nut. Do not tighten at this time
- Stabilizer link to the strut
- On Camry, load sensing proportioning valve to the control arm. Tighten to 94 ft. lbs. (130 Nm)
- On Supra, brake caliper
- Wheel. Bounce the vehicle up and down to stabilize the suspension

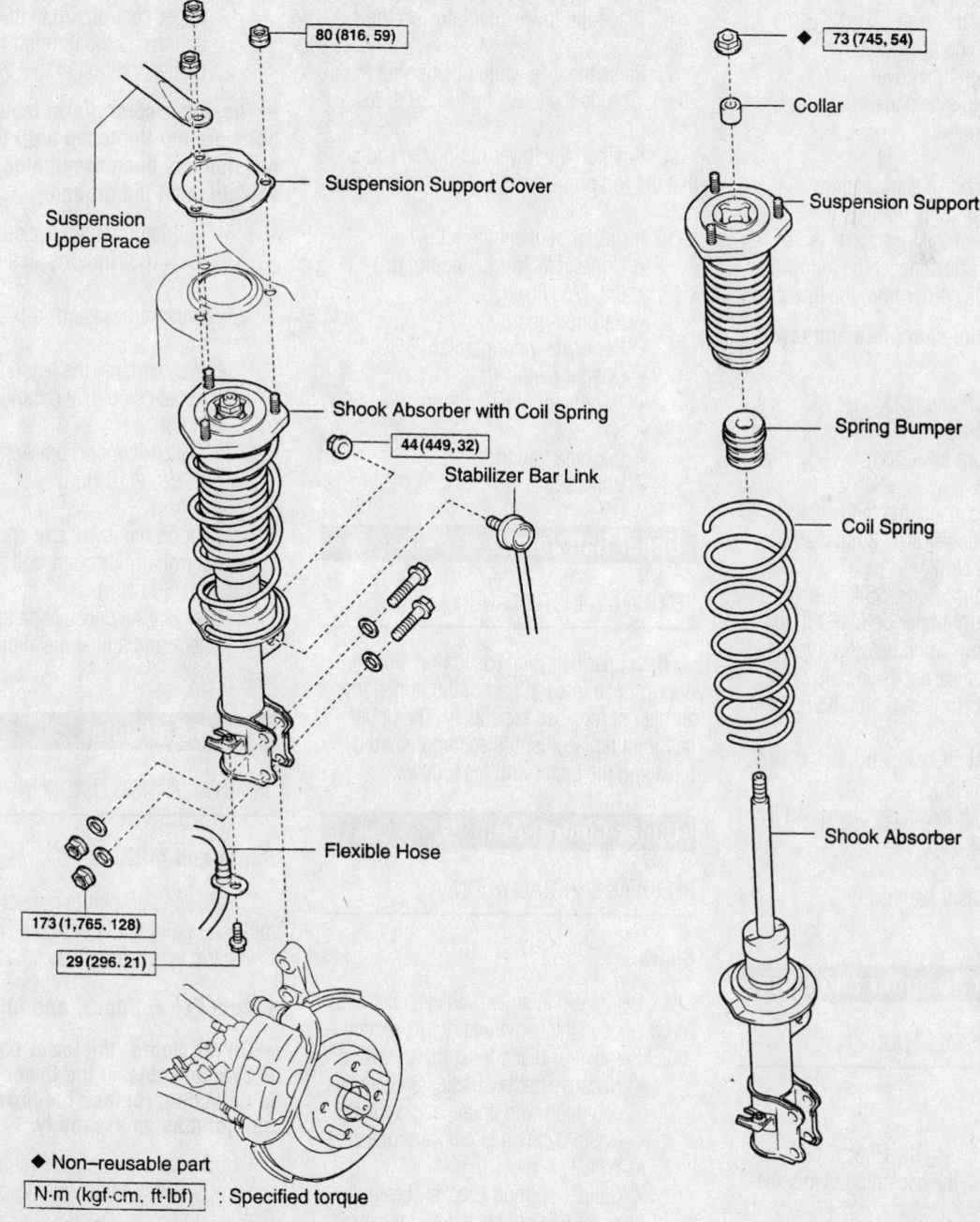

Suspension Upper Brace

80 (816, 59)

Suspension Support Cover

Shook Absorber with Coil Spring

44 (449, 32)

Stabilizer Bar Link

Flexible Hose

173 (1,765, 128)

29 (296, 21)

73 (745, 54)

Collar

Suspension Support

Spring Bumper

Coil Spring

Shook Absorber

◆ Non–reusable part

N·m (kgf·cm, ft·lbf) : Specified torque

Rear strut and coil spring assembly exploded view—MR2

9357WG12

13. With the vehicle weight on the suspension, tighten the bolt holding the strut to the axle carrier as follows:
- Corolla: 105 ft. lbs. (142 Nm).
- Celica, Camry and Avalon: 188 ft. lbs. (255 Nm)
- Supra: 106 ft. lbs. (143 Nm).

14. Install or connect the following:
- Rear seat cushion and any trim
- Negative battery cable

CELICA AND GT-S

1. Before servicing the vehicle, refer to the precautions in the beginning of this section.
2. Remove or disconnect the following:
- Wheel
- Package tray trim
- Luggage compartment mat
- Rear deck trim cover
- Deck trim side panel
- ABS harness
- Tailpipe
- Parking brake cable clamp

3. Loosen, but don't remove, the 3 bolts on the front side of the lower arm.
4. Remove or disconnect the following:
- Rear axle carrier from the lower arm

➡ Matchmark the cam plate and lower arm.

- Strut bolt and nut
- Lower arm
- Strut from lower arm

To install:

5. Install or connect the following:
- Strut to lower arm. 3 nuts: 59 ft. lbs. (80 Nm).
- Lower arm. Bolts: 55 ft. lbs. (74 Nm). Strut upper bolt: 103 ft. lbs. (140 Nm), while holding the nut.
- Rear axle carrier. Side bolt: 55 ft. lbs. (74 Nm); cam bolt: 55 ft. lbs. (Nm).
- Lower suspension arm bolts: 48 ft. lbs. (65 Nm).
- Parking brake cable clamp: 48 inch lbs. (5.4 Nm).
- Tailpipe
- ABS sensor harness
- Wheel

Rear Shock Absorber and Spring

REMOVAL & INSTALLATION

Echo

1. Before servicing the vehicle, refer to the precautions in the beginning of this section.
2. Support the rear axle beam.

3. Remove or disconnect the following:
- Wheels
- Package tray trim
- Rear seat
- Door scuff plate and door opening trim
- Quarter panel trim
- Roof side inner garnish
- Partition board
- Shock absorber upper nuts and lower bolt

4. Lower the axle slowly and remove the spring

To install:

5. Position the upper insulator so that its gap fits into the end of the spring.
6. Place the lower insulator on the axle.
7. Raise the axle while positioning the shock. Torque the lower bolt to 36 ft. lbs. (49 Nm).
8. Position the lower nut on the rod so that the rod protrudes 15–18mm above the nut.
9. Install or connect the following:
- Upper nut. Torque the nut to 18 ft. lbs. (25 Nm).
- Partition board
- Roof side inner garnish
- Quarter panel
- Door scuff plate and trim
- Rear seat
- Package tray trim
- Wheels

Upper Ball Joint

REMOVAL & INSTALLATION

The upper ball joint (used only on the Supra) is an integral part of the upper arm and is not replaced separately. The upper ball joint replacement is accomplished by replacing the upper arm, as follows:

Upper Control Arm

REMOVAL & INSTALLATION

Supra

1. Before servicing the vehicle, refer to the precautions in the beginning of this section.
2. Remove or disconnect the following:
- Negative battery cable. On vehicles equipped with an air bag, wait at least 90 seconds before proceeding
- Wheel
- Caliper support bracket. Leave the brake line connected and suspend it aside

- Rotor
- Front fender splash shield, fender liner, and wheel opening molding
- If removing the left side arm, the washer tank
- ABS speed sensor from the steering knuckle. Remove the 3 bolts and disconnect the wire harness clamp
- Cotter pin and the nut from the upper ball joint; press the upper ball joint from the knuckle
- Upper control arm

To install:

3. Install or connect the following:
- Upper control arm. Connect the upper control arm to the sub-frame and install the through-bolt. Do not tighten the bolt at this time

➡ The upper control arm mounting bolts are not tightened until the suspension has been assembled and vehicle is on the ground.

- Ball joint to the knuckle. Nut: 76 ft. lbs. (103 Nm). Install a new cotter pin
- Wire harness and ABS speed sensor
- Washer tank, the fender liner, splash shield, and molding
- Rotor
- Caliper support bracket. Bolts: 87 ft. lbs. (118 Nm).
- Wheel

4. Support the lower arm and tighten the upper control arm through-bolt and nut to 121 ft. lbs. (164 Nm).
5. Connect the negative battery cable
6. Check and adjust the alignment, if needed.

Lower Ball Joint

REMOVAL & INSTALLATION

Supra and MR2

The lower ball joint is not serviceable. If the lower ball joint is defective, replace the lower arm and ball joint as an assembly.

Except Echo, Supra, and MR2

➡ On the Supra, the lower ball joint is not replaceable. If the lower ball joint is defective, replace the lower arm and ball joint as an assembly.

1. Before servicing the vehicle, refer to the precautions in the beginning of this section.
2. Remove or disconnect the following:

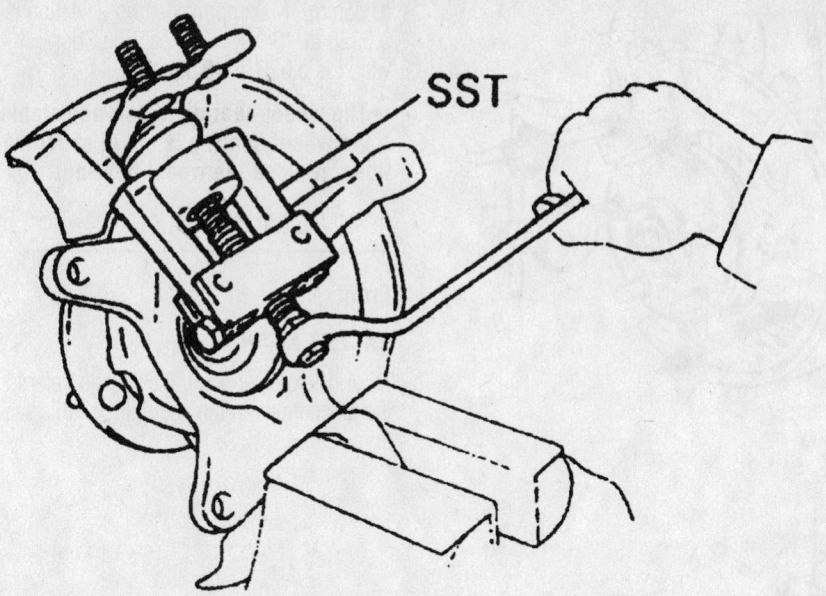

7923VGA9

Removing the ball joint from the knuckle

- Negative battery cable. On vehicles equipped with an air bag, wait at least 90 seconds before proceeding
- Front wheels
- Cotter pin from the bearing locknut cap, then remove the cap

3. Depress the brake pedal and loosen the axle nut

4. Remove or disconnect the following:
- Brake caliper attaching hardware, position the caliper aside with the hydraulic line still attached and suspend it with a wire
- ABS speed sensor, if equipped
- Rotor

5. Loosen the 2 nuts holding the strut to the steering knuckle assembly. Do not remove at this time.

6. Remove or disconnect the following:
- Cotter pin and nut from the tie rod end. Using a tie rod end removal tool, separate the tie rod end from the steering knuckle
- Steering knuckle from the strut assembly
- Axle nut and grasp the hub and knuckle assembly. With a plastic hammer tap the axle shaft to remove knuckle and hub

➡ **Cover the halfshaft boot with a shop rag to protect it from any damage.**

7. Clamp the steering knuckle in a vise and remove the dust deflector. Remove the nut holding the steering knuckle to the ball joint. Press the ball joint out of the steering knuckle.

8. Remove the ball joint from the arm.

To install:

9. Install the Lower ball joint to the lower arm. Tighten the fasteners to:
    a. Corolla, Avalon, Camry: 94 ft. lbs. (127 Nm).
    b. GT-S: 105 ft. lbs. (142 Nm).

10. Install the ball joint to the steering knuckle. Tighten the ball joint-to-steering knuckle nut to:
    a. Corolla: 87 ft. lbs. (118 Nm).
    b. GT-S: 76 ft. lbs. (103 Nm).
    c. Avalon and Camry: 90 ft. lbs. (123 Nm).

11. Install or connect the following:
- New cotter pin. Drive the deflector shield onto the knuckle
- Knuckle and hub assembly to the axle and temporarily tighten the axle nut
- Knuckle assembly to the lower strut bracket. Temporarily insert the mounting bolts from the rear and install the nuts
- Tie rod end to the knuckle

12. Tighten the bolts on the lower side of the strut assembly.

13. Install or connect the following:
- ABS speed sensor
- Brake disc and the caliper

14. Tighten the axle nut.

15. Connect the negative battery cable.

16. Check and adjust the alignment, if needed.

Echo

1. Before servicing the vehicle, refer to the precautions in the beginning of this section.

➡ **The ball joint is not replaceable.**

2. Remove or disconnect the following:
- Wheels
- Stabilizer bar end link
- Lower arm from the knuckle

3. For the right arm, jack up the engine slightly and disconnect the rear engine mount.

4. On either side, remove the 2 bolts and 1 nut and remove the arm.

5. Flip the ball stud back and forth a few times. Install the nut. Using a torque wrench, rotate the ball stud at the rate of 1 turn in 2–4 seconds. Take a torque reading on the 5th revolution. Turning torque should be at least 5 inch lbs. (0.59 Nm), and not more than 30 inch lbs. (3.4 Nm).

To install:

6. Install or connect the following:
- Lower control arm. Torque the bolts to A: 65 ft. lbs. (88 Nm); B: 97 ft. lbs. (132 Nm).

➡ **DO NOT turn the nut.**

- On the right side, the engine rear mount. Bolt and nuts: 59 ft. lbs. (80 Nm).
- Lower arm to knuckle. Torque the nut to 72 ft. lbs. (98 Nm).
- Stabilizer bar
- Wheel

Wheel Bearings

ADJUSTMENT

Front

1. Before servicing the vehicle, refer to the precautions in the beginning of this section.

2. All models use a non-adjustable wheel bearing. To determine the condition of the wheel bearing, check the backlash in bearing shaft direction and the axle hub deviation. Maximum for backlash should be as follows:
- Corolla, Echo, Supra, Avalon, Camry and Camry Solara: 0.0020 in. (0.05mm)
- Celica: 0.0031 in. (0.08mm)

3. Maximum axle hub deviation is:
- Corolla: 0.0028 in. (0.07mm)
- Celica: 0.0028 in. (0.07mm)
- Supra, Echo, Avalon and Camry: 0.0020 in. (0.05mm)

4. If the wheel bearing is out of specifications, replace the wheel bearing.

Rear

Check the backlash in bearing shaft direction and the axle hub deviation. Maximum for backlash should be 0.0020 in.

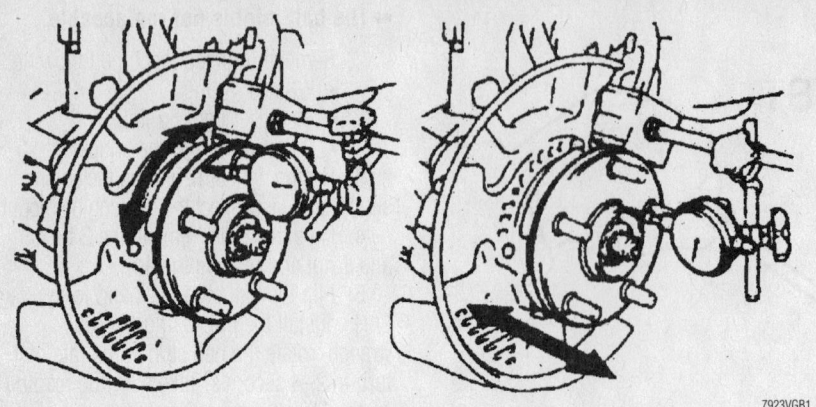

Checking the wheel bearings for deviation and free-play

7923VGB1

(0.05mm). Maximum axle hub deviation is 0.0028 in. (0.07mm), except on Supra, which is 0.020 in. (0.05mm).

➡ **The wheel bearing is non-adjustable. If the wheel bearing is out of specifications, replace the wheel bearing.**

REMOVAL & INSTALLATION

Front

EXCEPT SUPRA

1. Before servicing the vehicle, refer to the precautions in the beginning of this section.

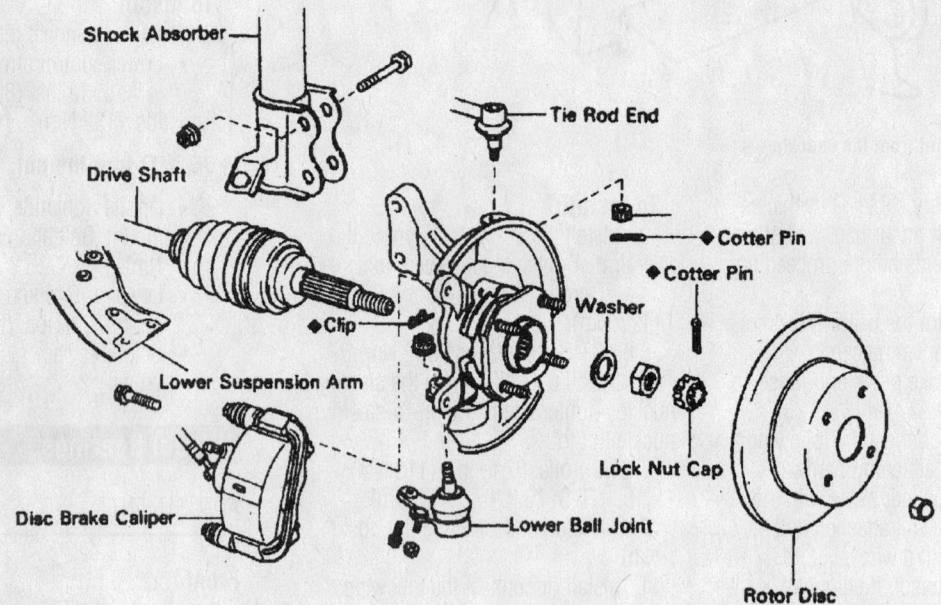

- Shock Absorber
- Tie Rod End
- Drive Shaft
- Cotter Pin
- Cotter Pin
- Clip
- Washer
- Lower Suspension Arm
- Lock Nut Cap
- Disc Brake Caliper
- Lower Ball Joint
- Rotor Disc

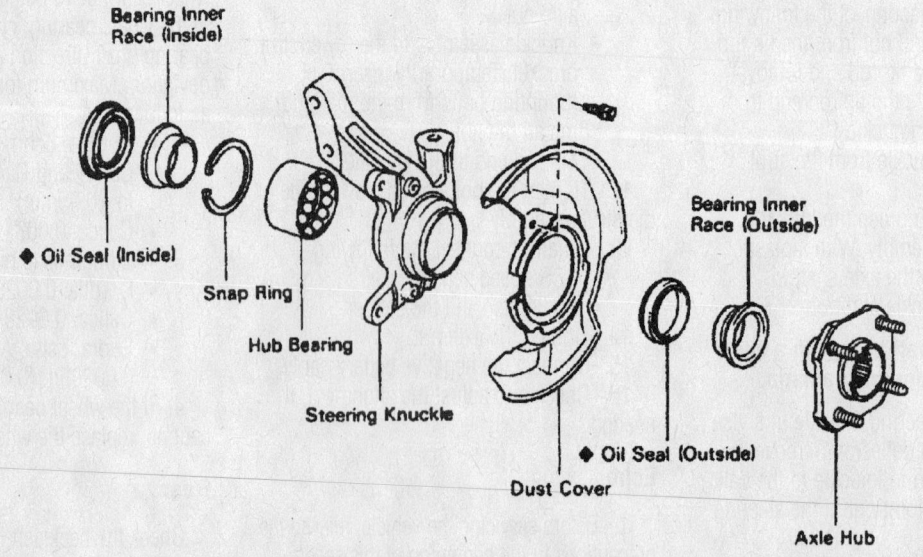

- Bearing Inner Race (Inside)
- Oil Seal (Inside)
- Snap Ring
- Hub Bearing
- Steering Knuckle
- Bearing Inner Race (Outside)
- Oil Seal (Outside)
- Dust Cover
- Axle Hub

7923VGB2

Steering knuckle and hub assembly—except Supra

2. Remove or disconnect the following:
- Negative battery cable. On vehicles equipped with an air bag, wait at least 90 seconds before proceeding
- Wheels
- Axle nut cap
- Axle nut
- Caliper. Position the caliper aside with the hydraulic line still attached and suspend it with a wire.
- ABS speed sensor
- Rotor

3. Loosen the nuts on the lower side of the strut assembly. Do not remove at this time.

4. Remove or disconnect the following:
- Tie rod end from the steering knuckle
- Steering knuckle from the lower control arm
- Knuckle from the strut assembly
- Hub

➡ **Cover the halfshaft boot with a shop rag to protect it from any damage.**

5. Clamp the steering knuckle in a vise and remove the dust deflector. Remove the nut holding the steering knuckle to the ball joint. Press the ball joint out of the steering knuckle.

6. Remove the inner axle seal.

7. Using a Torx® wrench, remove the bolts securing the dust cover.

8. Using hub puller, remove the hub and backing plate from the steering knuckle.

9. Using a proper sized driver and a press, remove the inner hub race from the axle hub.

10. Using seal removal tool, remove the outer axle seal.

11. Using snapring pliers, remove the snapring from the inner side of the steering knuckle.

12. Using a proper sized driver and a press, remove the bearing from the steering knuckle. The bearing is pressed from the front of the steering knuckle and is removed through the back of the steering knuckle.

To install:

13. Perform the following:

14. Using a proper sized driver and a press, install a new bearing to the steering knuckle.

15. Install the snapring to the steering knuckle using snapring pliers.

16. Using a seal driver and a hammer, install a new outer oil seal. Apply multipurpose grease to the oil seal lip.

17. Place the dust cover on the steering knuckle. Bolts: 78 inch lbs. (9 Nm).

18. Using a press and a proper sized

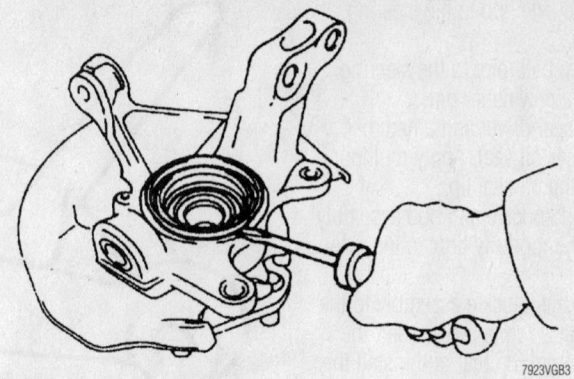

Removing the inner axle seal from the hub assembly

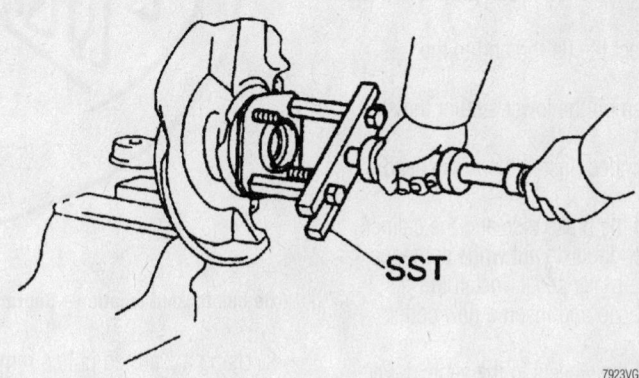

Removing the axle hub from the knuckle

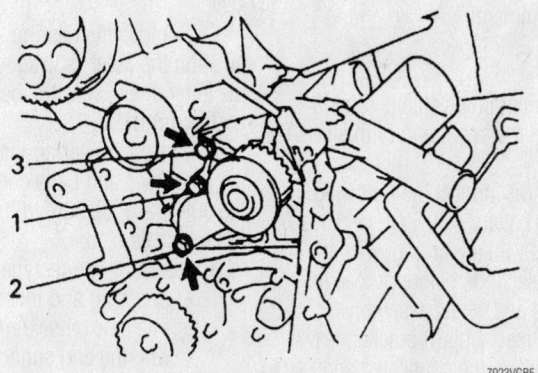

Removing the snapring from the knuckle before pressing out the bearing

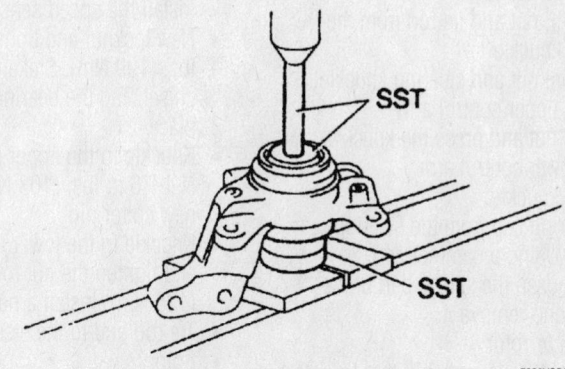

Removing the bearing from the steering knuckle using a press

driver, install the axle hub to the steering knuckle.

19. Attach the ball joint to the steering knuckle. Install a new cotter pin.

20. Using a seal driver and a hammer, install a new inner oil seal. Apply multipurpose grease to the oil seal lip.

21. Install the knuckle and hub assembly to the axle and temporarily tighten the axle nut.

22. Connect the knuckle assembly to the lower strut bracket. Temporarily insert the mounting bolts from the rear and install the nuts making sure the matchmarks made earlier are in alignment.

23. Connect the lower ball joint to lower arm.

24. Connect the tie rod end to the knuckle.

25. Tighten on the lower side of the strut assembly.

26. If equipped, install the ABS speed sensor.

27. Install the brake disc and the caliper.

28. Tighten the axle nut while someone depresses the brake pedal. Install the adjusting nut cap and insert a new cotter pin.

29. Install the wheels to the vehicle. Verify that the wheel turns freely.

30. Connect the negative battery cable to the battery.

31. Check alignment.

SUPRA

1. Before servicing the vehicle, refer to the precautions in the beginning of this section.

2. Remove or disconnect the following:
- Negative battery cable. On vehicles equipped with an air bag, wait at least 90 seconds before proceeding
- Front tire and wheel assembly
- Brake caliper support bracket, leaving the brake line connected and support it using a piece of wire
- Rotor
- ABS speed sensor
- Cotter pin, nut and tie rod from the steering knuckle
- Cotter pin, nut and steering knuckle from the upper control arm
- Clip and nut and press the knuckle off the lower control arm
- Steering knuckle
- Hub bearing cap from the steering knuckle. Using a hammer and chisel, loosen the staked part of the hub nut and remove it
- ABS sensor rotor

3. Remove the 4 bolts and shift the brake dust shield toward the hub (outside).

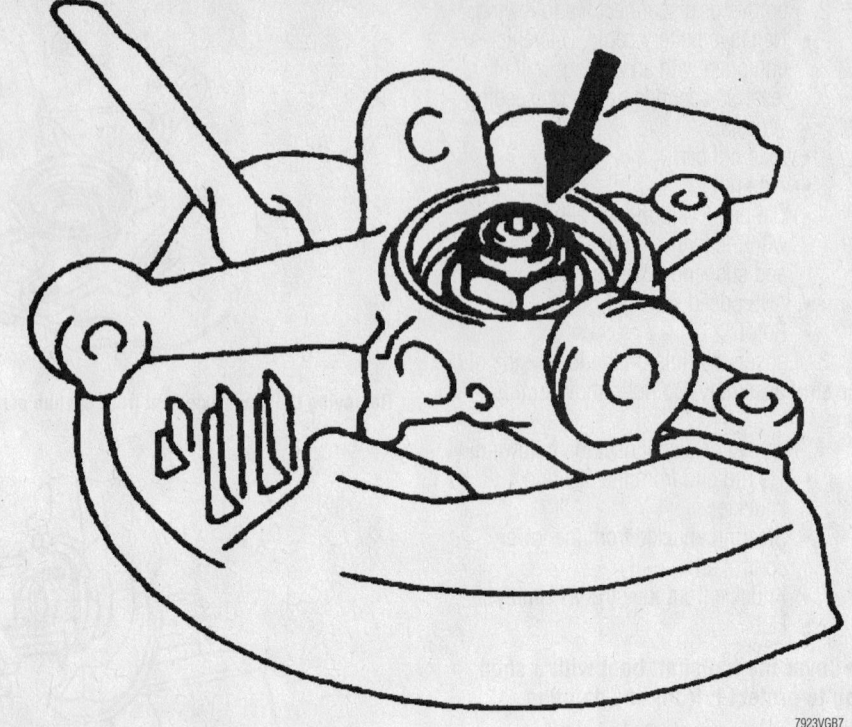

Axle hub locknut location—Supra

7923VGB7

4. Using a 2 armed puller, remove the axle hub from the knuckle.

5. With a puller, remove the inner bearing race from the axle hub. Pry out the oil seal.

6. Remove the bearing snapring, then position the inner race above the bearing on the inner side. Press the bearing out.

To install:

7. Press the bearing into the knuckle. If the inner race and balls come loose from the outer race, be sure to install them on the same side as before.

8. Install or connect the following:
- Snapring and inner race, then tap in a new oil seal until it is flush with the end surface of the knuckle
- Brake dust cover. Bolts: 74 inch lbs. (9 Nm).
- Press the hub into the knuckle and install the speed sensor
- New locknut and tighten it to 147 ft. lbs. (199 Nm). Stake the nut with a chisel. Tap the bearing cap into place
- Knuckle to the upper control arm. Nut: 76 ft. lbs. (103 Nm). Install a new cotter pin
- Knuckle to the lower control arm and tighten the nut to 92 ft. lbs. (125 Nm). Install a new clip
- Tie rod end to the steering knuckle
- Rotor
- Caliper support bracket

- Speed sensor to the knuckle
- Front wheel
- Negative battery cable

9. Check and adjust the alignment, if needed.

Rear

AVALON, CAMRY, CAMRY SOLARA, CELICA, COROLLA AND ECHO

1. Before servicing the vehicle, refer to the precautions in the beginning of this section.

2. Remove or disconnect the following:
- Negative battery cable. On vehicles equipped with an air bag, wait at least 90 seconds before proceeding
- Wheel
- Brake drum or rotor
- With ABS brakes, ABS wheel speed sensor
- Hub
- O-ring from the backing plate

To install:

3. Install or connect the following:
- New O-ring onto the backing plate. Coat the O-ring with multipurpose grease
- Hub to the knuckle. Bolts: Except Echo, 59 ft. lbs. (80 Nm); Echo, 38 ft. lbs. (52 Nm).
- With ABS brakes, ABS wheel speed sensor
- Brake drum or rotor
- Wheel

- Negative battery cable
4. Check and adjust the alignment, if needed.

SUPRA

1. Before servicing the vehicle, refer to the precautions in the beginning of this section.
2. Remove or disconnect the following:
- Negative battery cable. On vehicles equipped with an air bag, wait at least 90 seconds before proceeding
- Rear tire and wheel assembly
- Brake caliper support bracket from the rear axle carrier and support it with a piece of wire

➡**Matchmark the disc brake rotor and the axle hub.**

- Rotor
- Speed sensor
- Rear halfshaft
- Parking brake shoes
- Parking brake cable
- Strut rod at the axle carrier
3. Perform the following:
   a. Remove the nut, then press out the No. 1 lower suspension arm.
   b. Remove the nut, then press out the No. 2 lower suspension arm.
   c. Remove the nut, then press out the upper suspension arm. Remove the axle carrier.
   d. Remove the dust deflector and pull out the oil seal.
   e. Using a 2 arm puller, remove the axle hub from the carrier.
   f. Remove the backing plate.
   g. Press the inner race (outside) from the hub. Then, remove the oil seal and the snapring.
   h. Place the inner race (outside) over the bearing and tap out the bearing and inner race (inside).

To install:
4. Install or connect the following:
- Bearing to the axle carrier.

➡**If the inner races come loose from the bearing outer race, be sure to install them on the same side as before.**

- Snapring, the inner race (outside), and a new oil seal
- Backing plate. Install the inner race (inside) and press in the axle hub with the proper tools
- Inner oil seal. Align the holes for the speed sensor in the dust deflector and axle carrier. Install the dust deflector

- Upper arm to the axle carrier. Nut and bolt: 80 ft. lbs. (109 Nm).
- No. 2 lower arm to the carrier. Nut: 110 ft. lbs. (150 Nm).
- No. 1 lower arm to the carrier. Nut: 43 ft. lbs. (59 Nm).
- Strut rod to the carrier. Do not tighten the bolt at this time.
- Parking brake cable and slide the backing plate to the inside. Hex bolt: 132 ft. lbs. (180 Nm). Hub bolts: 19 ft. lbs. (26 Nm).
- Bolts at the parking brake cable: 69 inch lbs. (8 Nm).
- Parking brake shoes and the ABS sensor
- Halfshafts. Locknut: 213 ft. lbs. (289 Nm).
- Rotor
- Brake caliper to the rear axle carrier. Bolts: 77 ft. lbs. (104 Nm).
- Rear tire and wheel assembly. Lower the vehicle and bounce it a few times to stabilize the suspension. Raise the vehicle again, support the axle carrier and tighten the strut rod to 136 ft. lbs. (184 Nm).
- Negative battery cable

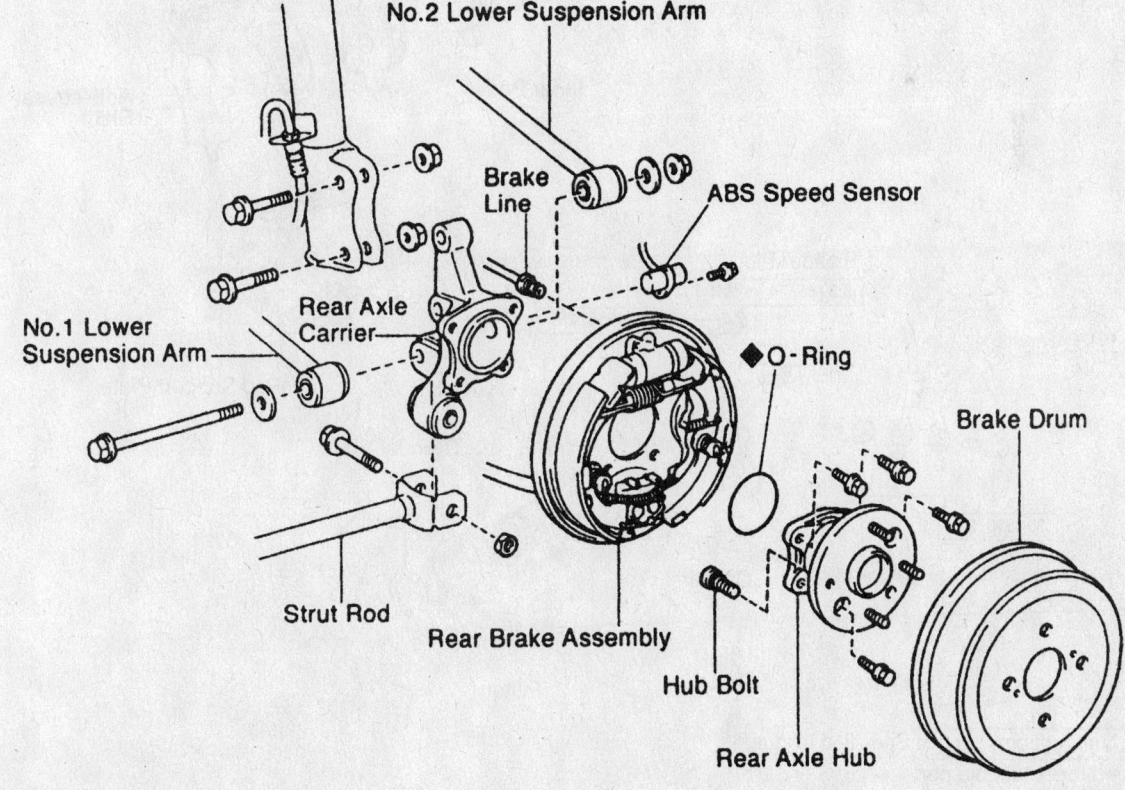

◆ **Non-reusable part**

Exploded view of the hub and wheel bearing assembly—Corolla shown, Celica similar

7923VGB9

BRKES

Brake Caliper

REMOVAL & INSTALLATION

All Models

FRONT AND REAR

1. Before servicing the vehicle, refer to the precautions in the beginning of this section.

2. Remove the wheels.
3. Disconnect the brake hose from the caliper.
4. Remove the bolts that attach the caliper to the torque plate. If applicable, hold the flats of the sliding pin with a wrench while loosening the caliper attaching bolts.
5. Lift up and remove the caliper assembly.

To install:

6. Install the caliper and loosely install the bolts.
7. Hold the flats of the sliding pin with a wrench, then tighten the bolts. On front calipers, tighten single piston types to 18 ft. lbs. (25 Nm) and dual piston types to 25 ft. lbs. (34 Nm). Tighten the bolts on rear calipers to 14 ft. lbs. (20 Nm).

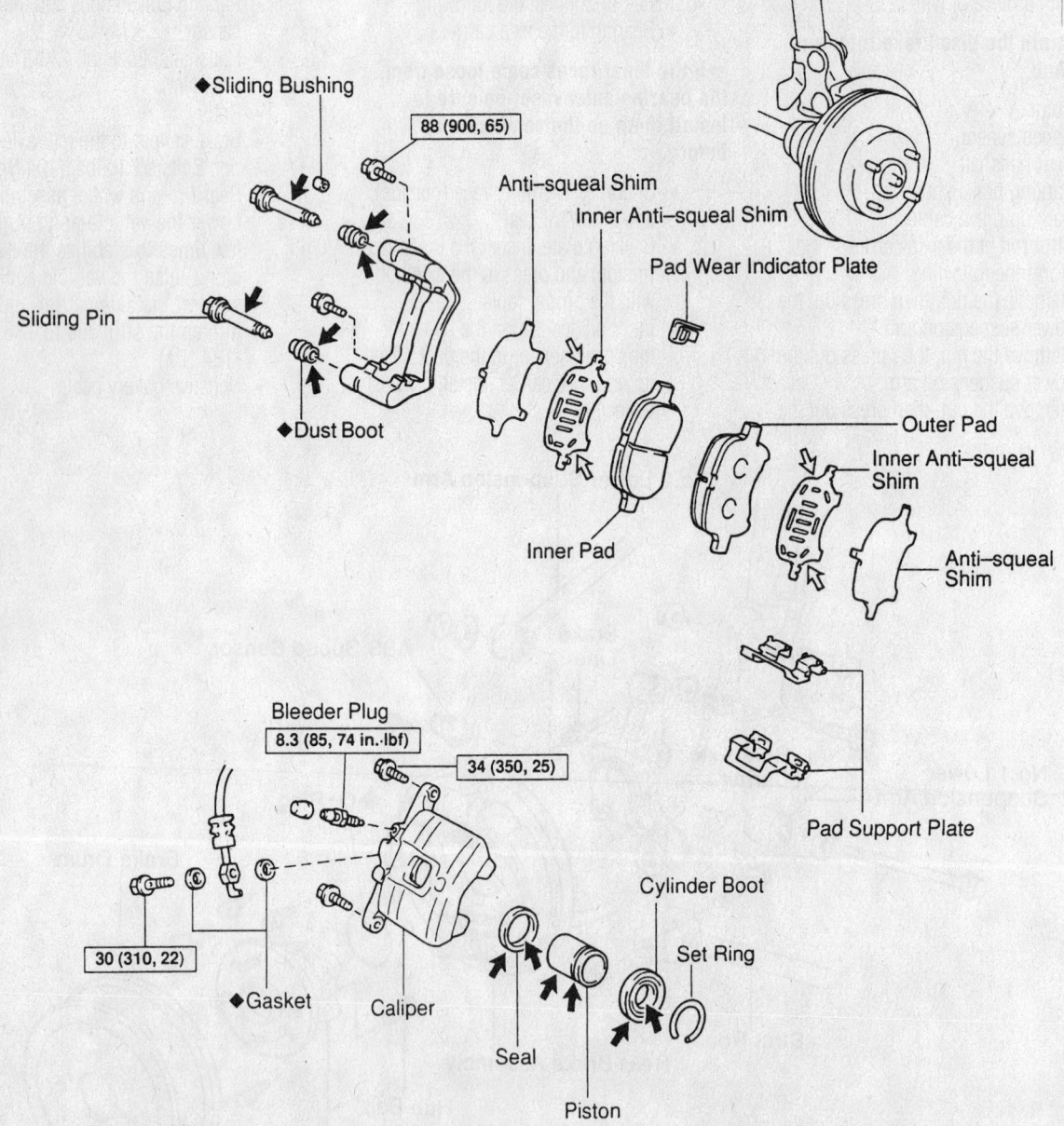

♦ Sliding Bushing

88 (900, 65)

Anti–squeal Shim

Inner Anti–squeal Shim

Pad Wear Indicator Plate

Sliding Pin

♦ Dust Boot

Outer Pad

Inner Anti–squeal Shim

Anti–squeal Shim

Inner Pad

Pad Support Plate

Bleeder Plug
8.3 (85, 74 in.·lbf)

34 (350, 25)

Cylinder Boot

Set Ring

30 (310, 22)

♦ Gasket

Caliper

Seal

Piston

N·m (kgf·cm, ft·lbf) : Specified torque
♦ Non–reusable part
➡ Lithium soap base glycol grease
⇨ Disc brake grease

Front caliper—Corolla

93016G66

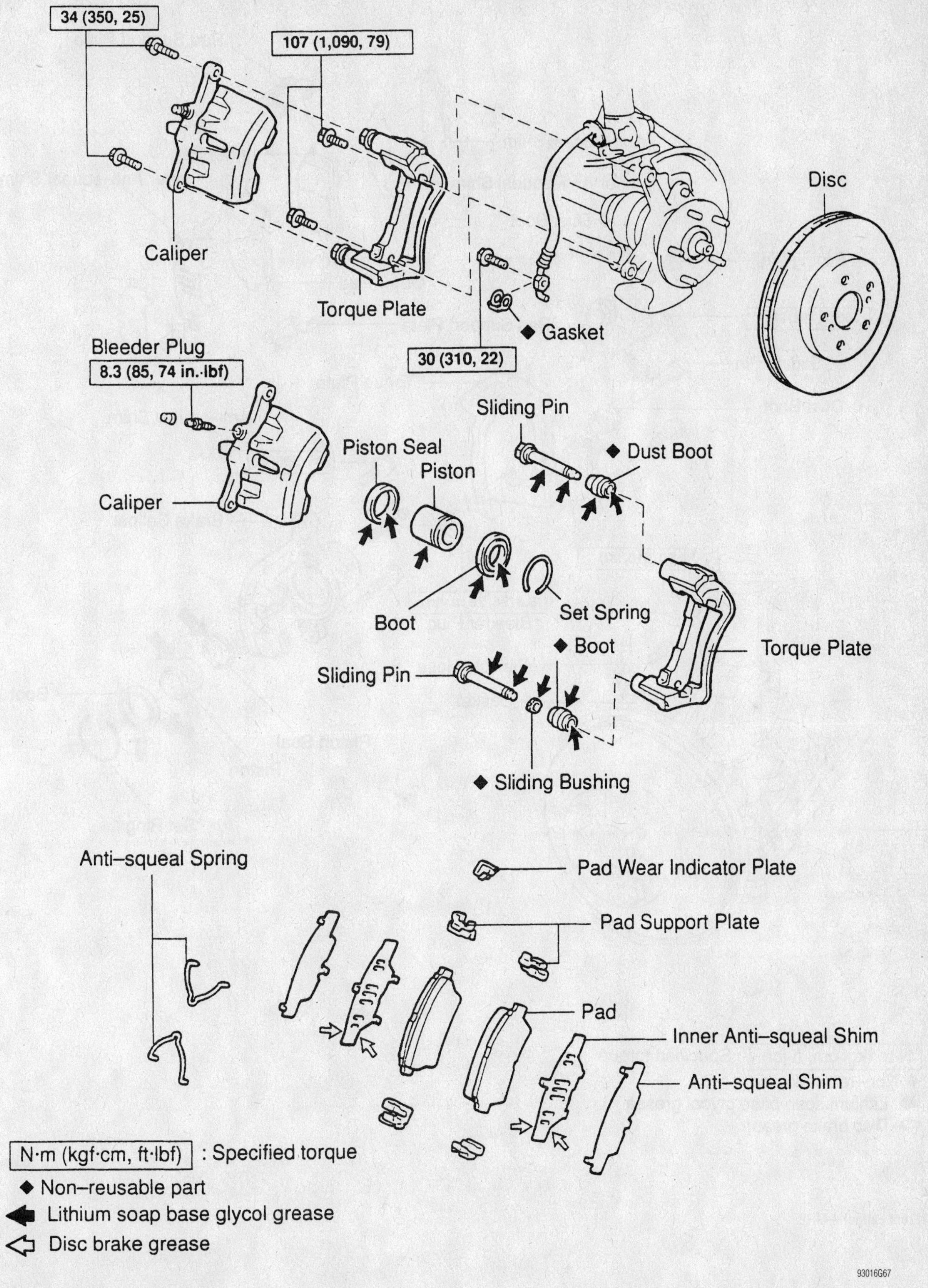

34 (350, 25)

107 (1,090, 79)

Disc

Caliper

Torque Plate

Gasket

30 (310, 22)

Bleeder Plug
8.3 (85, 74 in.·lbf)

Sliding Pin

Piston Seal
Piston

Dust Boot

Caliper

Boot

Set Spring

Boot

Torque Plate

Sliding Pin

Sliding Bushing

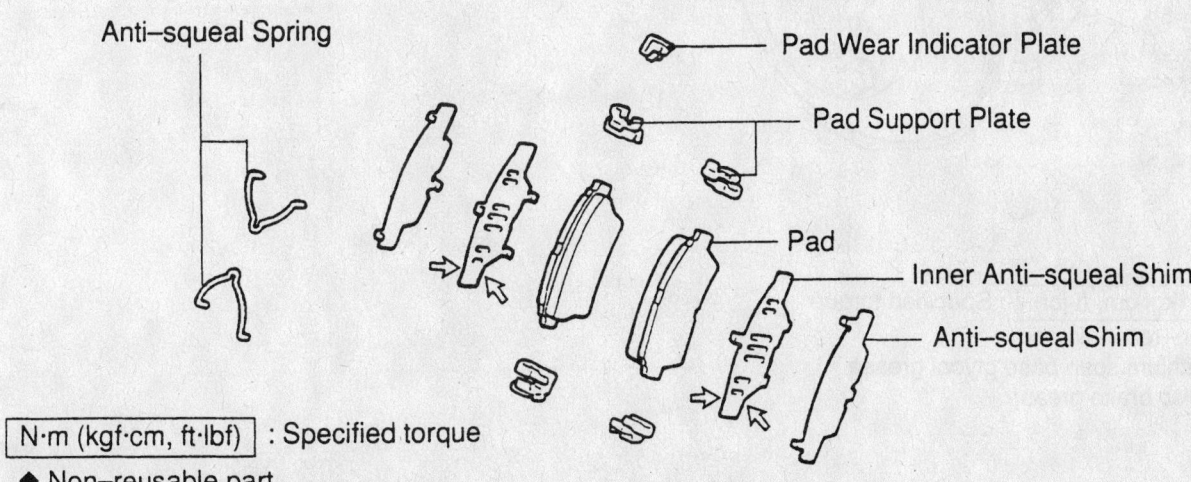

Anti–squeal Spring

Pad Wear Indicator Plate

Pad Support Plate

Pad

Inner Anti–squeal Shim

Anti–squeal Shim

N·m (kgf·cm, ft·lbf) : Specified torque

◆ Non–reusable part

◀ Lithium soap base glycol grease

◁ Disc brake grease

Front caliper—Celica, Camry, Avalon

93016G67

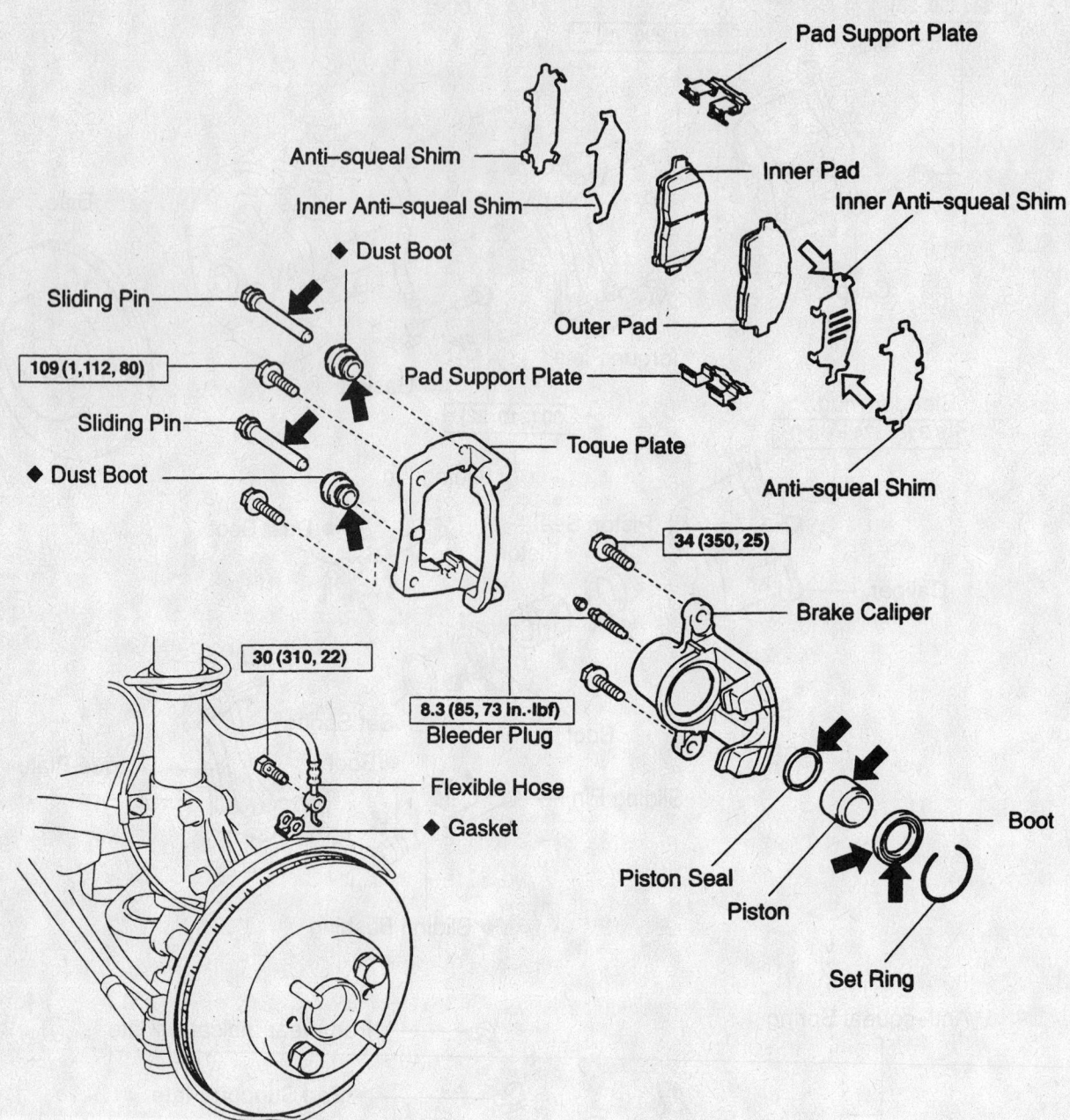

Pad Support Plate

Anti–squeal Shim

Inner Anti–squeal Shim

Inner Pad

Inner Anti–squeal Shim

Dust Boot

Sliding Pin

Outer Pad

109 (1,112, 80)

Pad Support Plate

Sliding Pin

Toque Plate

Anti–squeal Shim

◆ Dust Boot

34 (350, 25)

Brake Caliper

30 (310, 22)

8.3 (85, 73 In.·lbf)
Bleeder Plug

Flexible Hose

◆ Gasket

Piston Seal

Piston

Boot

Set Ring

N·m (kgf·cm, ft·lbf) : Specified torque
◆ Non–reusable part
➡ Lithium soap base glycol grease
⇨ Disc brake grease

9357WG13

Front caliper—MR2

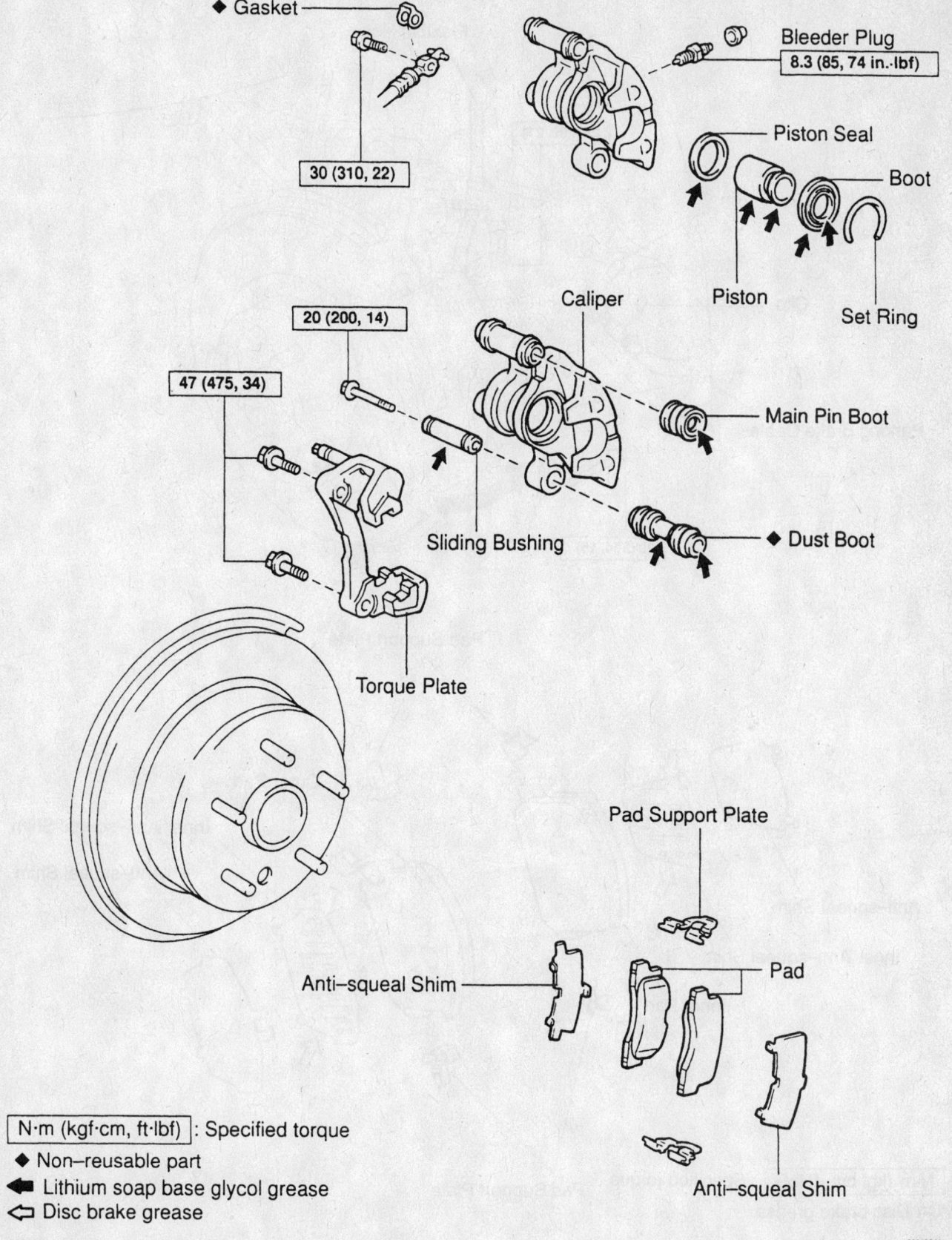

◆ Gasket

Bleeder Plug
8.3 (85, 74 in.·lbf)

Piston Seal

Boot

30 (310, 22)

Piston

Set Ring

Caliper

20 (200, 14)

47 (475, 34)

Main Pin Boot

Sliding Bushing

◆ Dust Boot

Torque Plate

Pad Support Plate

Anti-squeal Shim

Pad

N·m (kgf·cm, ft·lbf) : Specified torque

◆ Non-reusable part

◀ Lithium soap base glycol grease

◁ Disc brake grease

Anti-squeal Shim

93016G69

Rear caliper— Celica, Camry, Avalon

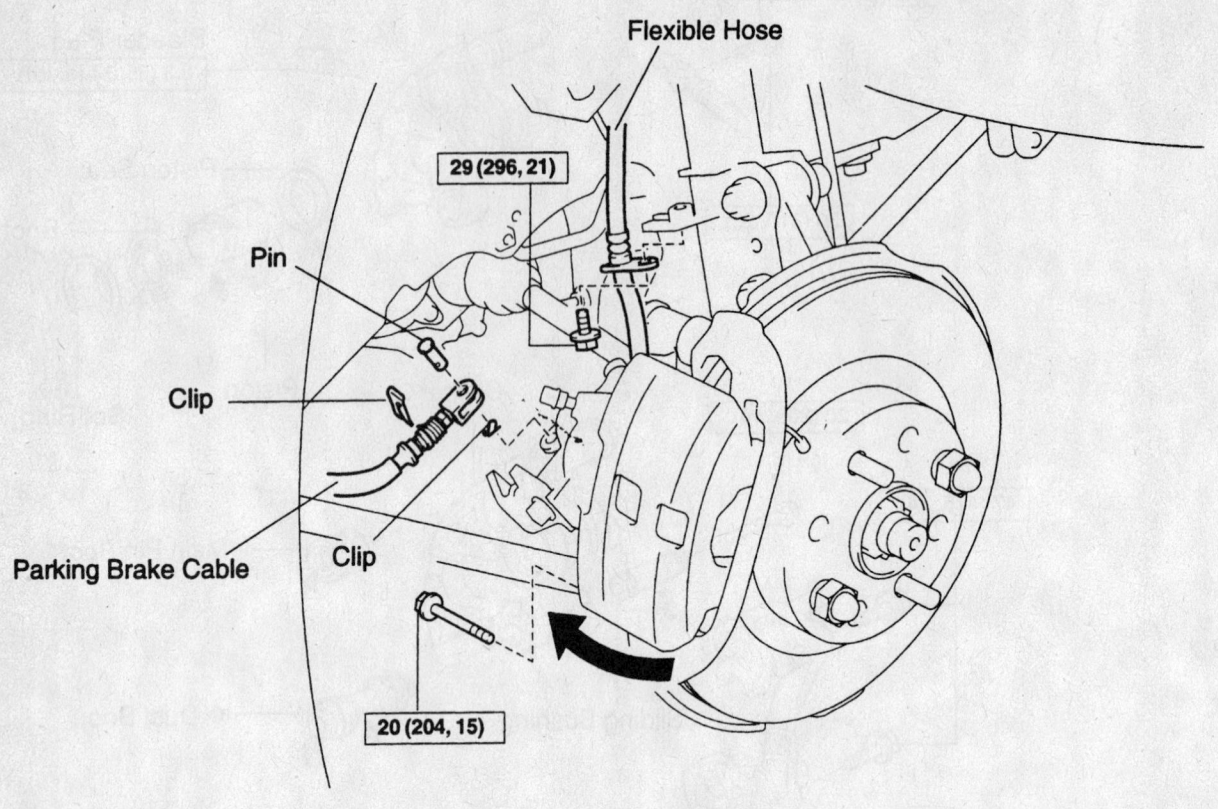

Flexible Hose

29 (296, 21)

Pin

Clip

Parking Brake Cable

Clip

20 (204, 15)

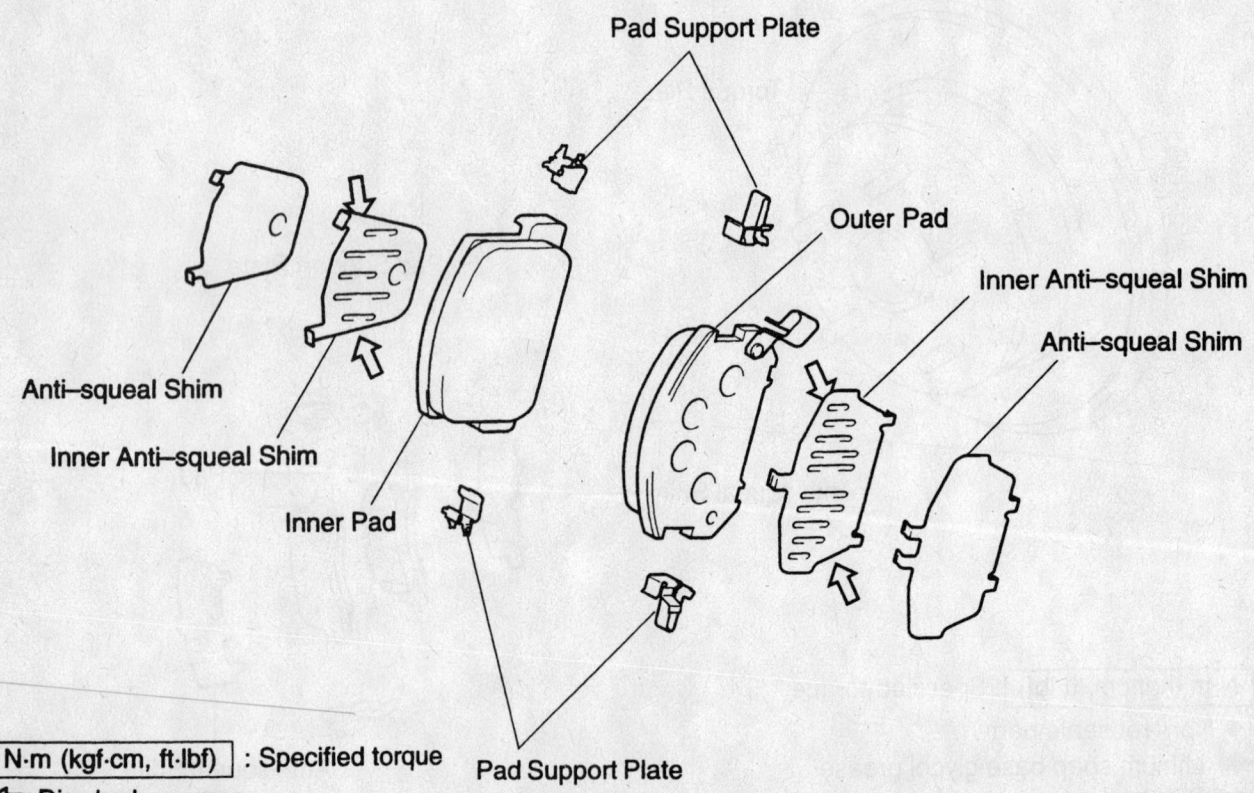

Pad Support Plate

Outer Pad

Inner Anti–squeal Shim

Anti–squeal Shim

Anti–squeal Shim

Inner Anti–squeal Shim

Inner Pad

Pad Support Plate

N·m (kgf·cm, ft·lbf) : Specified torque

⇐ Disc brake grease

9357WG14

Rear caliper—MR2

8. Connect the brake hose to the caliper, using 2 new washers.

9. Fill the brake system to the proper level and bleed the brake system.

10. Add brake fluid to the reservoir to fill to the correct level.

11. Lower the vehicle to the ground.

Disc Brake Pads

REMOVAL & INSTALLATION

All Models

FRONT AND REAR

1. Before servicing the vehicle, refer to the precautions in the beginning of this section.

2. Remove the wheels.

3. Loosen and remove the caliper mounting bolts, then remove the caliper assembly, without disconnecting the brake line. Position it aside.

4. Slide out the old brake pads along with any anti-squeal shims, springs, pad wear indicators and pad support plates.

To install:

5. Install the pad support plates into the torque plate.

6. Install the pad wear indicators onto the pads. Be sure the arrow on the indicator plate is pointing in the direction of rotation.

7. Install the anti-squeal shims on the outside of each pad and then install the pad assemblies into the torque plate.

8. Compress the caliper piston into the bore. For Corolla rear calipers, use tool SST 09719-14020, to rotate the piston clockwise while pressing it into the bore until it locks.

9. Position the caliper back down over the pads.

10. Install and tighten the caliper mounting bolts.

11. Install the wheels. Check the brake fluid level.

Brake Drums

REMOVAL & INSTALLATION

1. Before servicing the vehicle, refer to the precautions in the beginning of this section.

2. Remove the wheels.

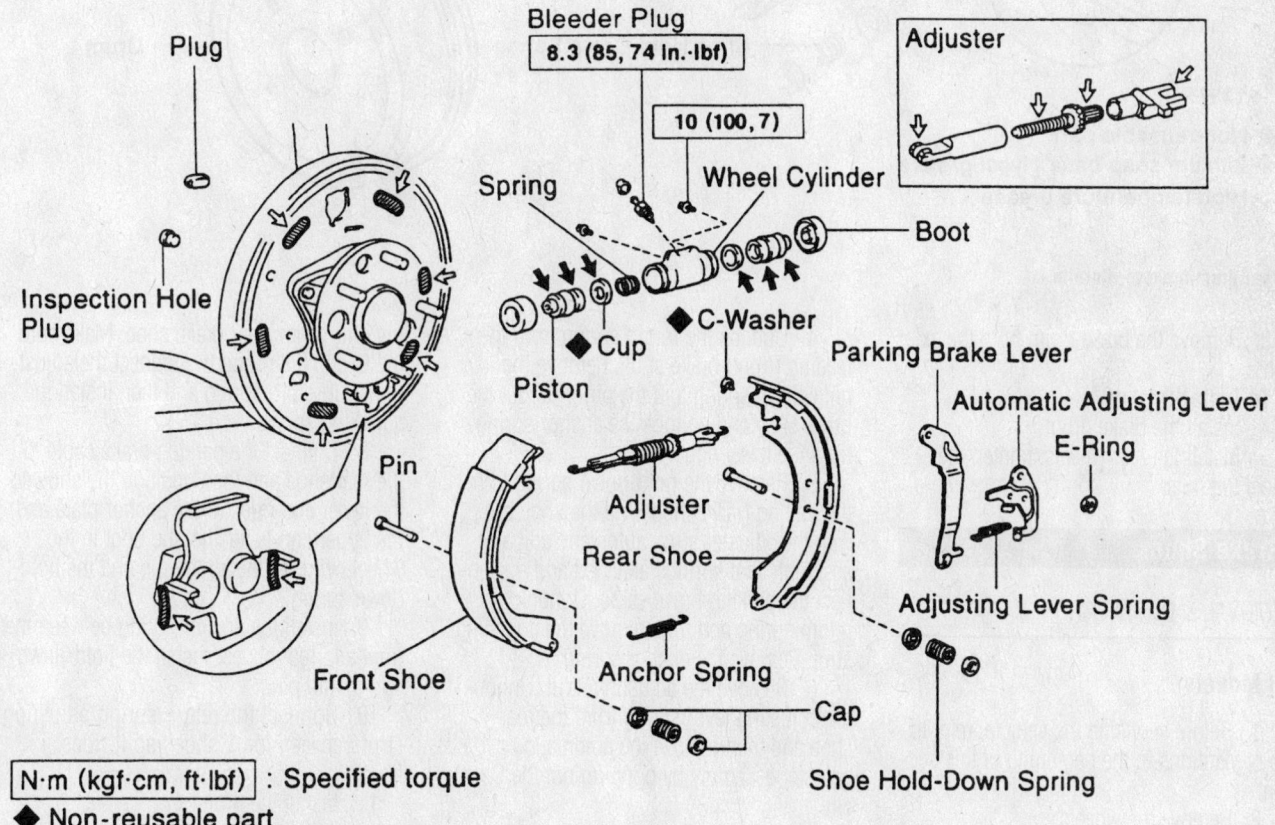

Bleeder Plug
8.3 (85, 74 In.·lbf)

10 (100, 7)

Plug

Inspection Hole Plug

Spring

Wheel Cylinder

Boot

◆ Cup

Piston

◆ C-Washer

Pin

Adjuster

Rear Shoe

Front Shoe

Anchor Spring

Cap

Shoe Hold-Down Spring

Adjuster

Parking Brake Lever

Automatic Adjusting Lever

E-Ring

Adjusting Lever Spring

N·m (kgf·cm, ft·lbf) : Specified torque

◆ Non-reusable part

◀ Lithium soap base glycol grease

◁ High temperature grease

93016G70

Rear drum brakes—Camry

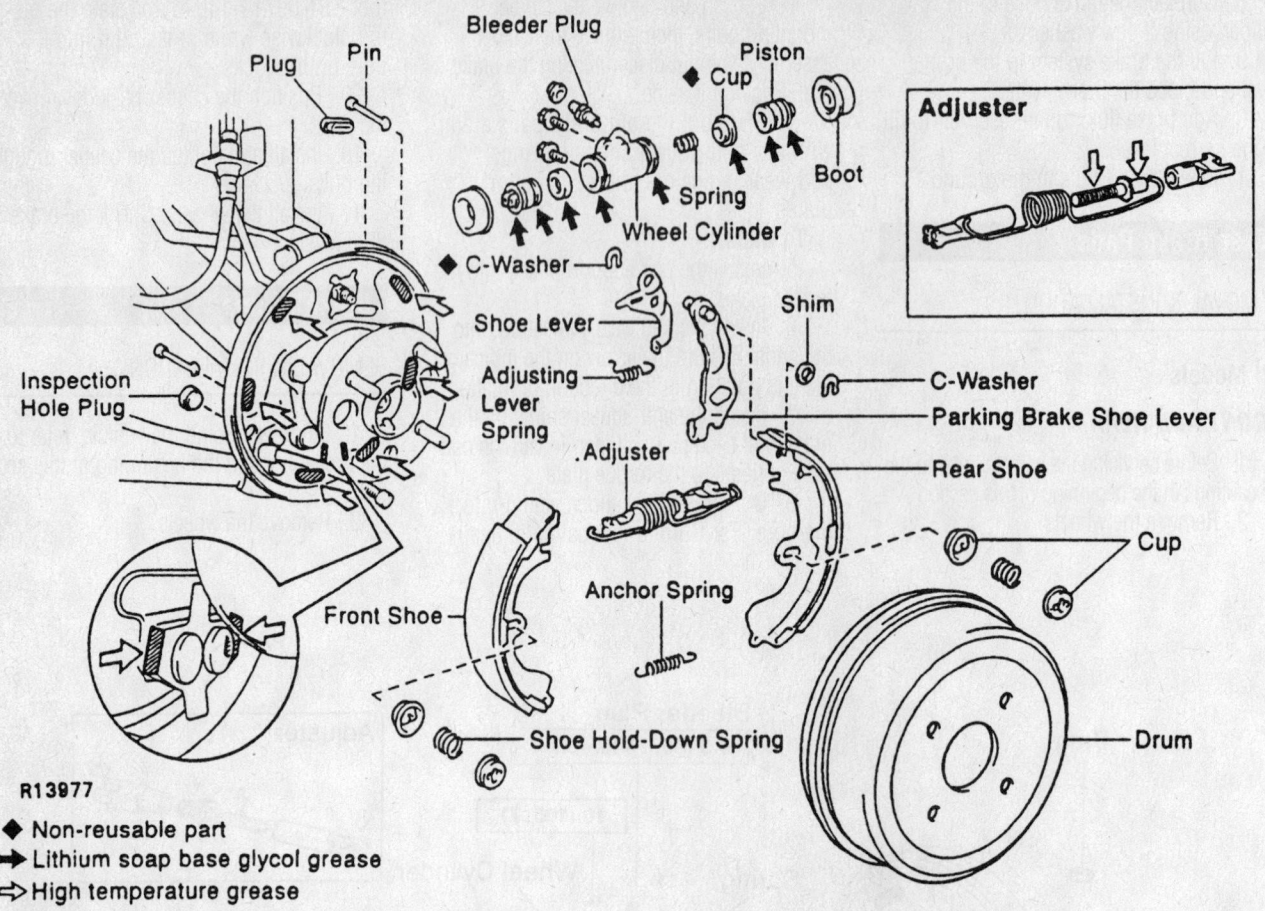

R13977

◆ Non-reusable part
➡ Lithium soap base glycol grease
⇨ High temperature grease

Rear drum brakes—Corolla

3. Remove the brake drum from the axle hub.
   To install:
4. Install the brake drum.
5. Install the rear wheels, tighten the wheel lug nuts.

Brake Shoes

REMOVAL & INSTALLATION

All Models

1. Before servicing the vehicle, refer to the precautions in the beginning of this section.
2. Remove the wheels.
3. Remove the brake drum.

4. Unhook the return spring from the leading (front) brake shoe. Remove the hold-down spring and the pin. Pull out the brake shoe and unhook the anchor spring from the lower edge.
5. Remove the hold-down spring from the trailing (rear) shoe. Pull the shoe out with the adjuster strut, automatic adjuster assembly and springs attached and disconnect the parking brake cable. Unhook the return spring and then remove the adjusting strut. Remove the anchor spring.
6. Remove the adjusting strut. Unhook the adjusting lever spring from the rear shoe and then remove the automatic adjuster assembly by popping out the C-clip.
   To install:
7. Mount the automatic adjuster assem-

bly onto a new rear brake shoe. Make sure the C-clip fits properly. Connect the adjusting strut/return spring and then install the adjusting spring.
8. Connect the parking brake cable to the rear shoe and then position the shoe so the lower end rides in the anchor plate and the upper end is against the boot in the wheel cylinder. Install the pin and the hold-down spring.
9. Install the anchor spring between the front and rear shoes. Install the hold-down spring and pin.
10. Connect the return spring/adjusting strut between the 2 shoes so it rides freely.
11. Install the drum.
12. Install the wheel.

SPECIFICATION CHARTS

ENGINE AND VEHICLE IDENTIFICATION

		Engine						Model Year	
Code ①	Liters (cc)	Cu. In.	Cyl.	Fuel Sys.	Engine Type	Eng. Mfg.		Code ②	Year
2UZ-FE	4.7 (4664)	285	8	SFI	DOHC	Toyota		Y	2000
5VZ-FE	3.4 (3378)	206	6	MFI	DOHC	Toyota		1	2001
								2	2002
								3	2003
								4	2004

SFI: Sequential Fuel Injection

MFI: Multi-port Fuel Injection

DOHC: Double Overhead Camshaft

① Stamped on the left side of the engine block

② 10th digit of the Vehicle Identification Number (VIN)

42356-TUND-C01

GENERAL ENGINE SPECIFICATIONS

Year	Model	Engine Displacement Liters (cc)	Engine Series (ID/VIN)	Fuel System	Net Horsepower @ rpm	Net Torque @ rpm (ft. lbs.)	Bore x Stroke (in.)	Com- pression Ratio	Oil Pressure @ rpm
2000	Tundra	3.4 (3378)	5VZ-FE	MFI	190@4800	220@3600	3.68x3.23	9.6:1	36-75@3000
		4.7 (4664)	2UZ-FE	MFI	245@4800	315@3400	3.70x3.30	9.6:1	45-65@3000
2001	Tundra	3.4 (3378)	5VZ-FE	MFI	190@4800	220@3600	3.68x3.23	9.6:1	36-75@3000
		4.7 (4664)	2UZ-FE	MFI	245@4800	315@3400	3.70x3.30	9.6:1	45-65@3000
2002	Tundra	3.4 (3378)	5VZ-FE	MFI	190@4800	220@3600	3.68x3.23	9.6:1	36-75@3000
		4.7 (4664)	2UZ-FE	MFI	245@4800	315@3400	3.70x3.30	9.6:1	45-65@3000
2003	Tundra	3.4 (3378)	5VZ-FE	MFI	190@4800	220@3600	3.68x3.23	9.6:1	36-75@3000
		4.7 (4664)	2UZ-FE	MFI	245@4800	315@3400	3.70x3.30	9.6:1	45-65@3000

NA: Not Available

SFI: Sequential Fuel Injection

MFI: Multi-port Fuel Injection

42356-TUND-C02

ENGINE TUNE-UP SPECIFICATIONS

Year	Engine Displacement Liters (cc)	Engine ID/VIN	Spark Plug Gap (in.)	Ignition Timing (deg.)*	Fuel Pump (psi)	Idle Speed (rpm)		Valve Clearance	
						MT	AT	Intake	Exhaust
2000	3.4 (3378)	5VZ-FE	0.031	5B	38-44	650-750	650-750	0.006-0.010	0.010-0.014
	4.7 (4664)	2UZ-FE	0.043	5B	38-44	650-750	650-750	0.006-0.009	0.011-0.014
2001	3.4 (3378)	5VZ-FE	0.031	5B	38-44	650-750	650-750	0.006-0.010	0.010-0.014
	4.7 (4664)	2UZ-FE	0.043	5B	38-44	650-750	650-750	0.006-0.009	0.011-0.014
2002	3.4 (3378)	5VZ-FE	0.043	8-12	38-44	650-750	650-750	0.006-0.010	0.010-0.014
	4.7 (4664)	2UZ-FE	0.043	8-12	38-44	650-750	650-750	0.006-0.009	0.011-0.014
2003	3.4 (3378)	5VZ-FE	0.043	8-12	38-44	650-750	650-750	0.006-0.010	0.010-0.014
	4.7 (4664)	2UZ-FE	0.043	8-12	38-44	650-750	650-750	0.006-0.009	0.011-0.014

NOTE: The Vehicle Emission Control Information label often reflects specification changes made during production.

The label figures must be used if they differ from those in this chart.

B: Before top dead center

*w/terminals TC and CG connected to DLC3

42356-TUND-C03

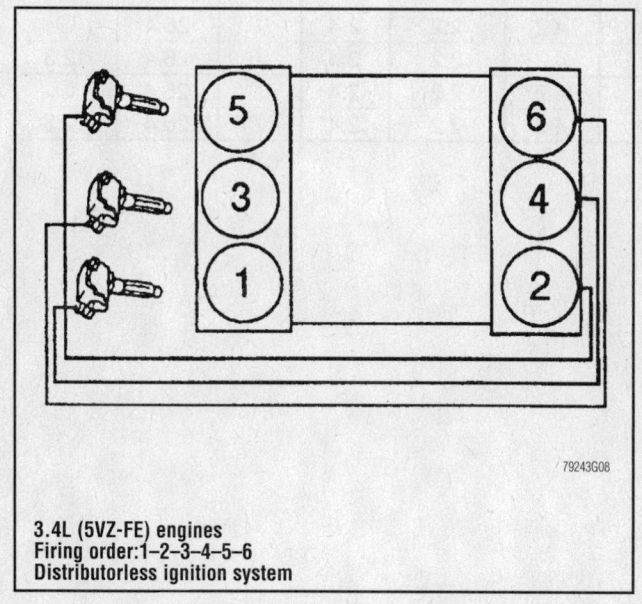

3.4L (5VZ-FE) engines
Firing order:1–2–3–4–5–6
Distributorless ignition system

79243G08

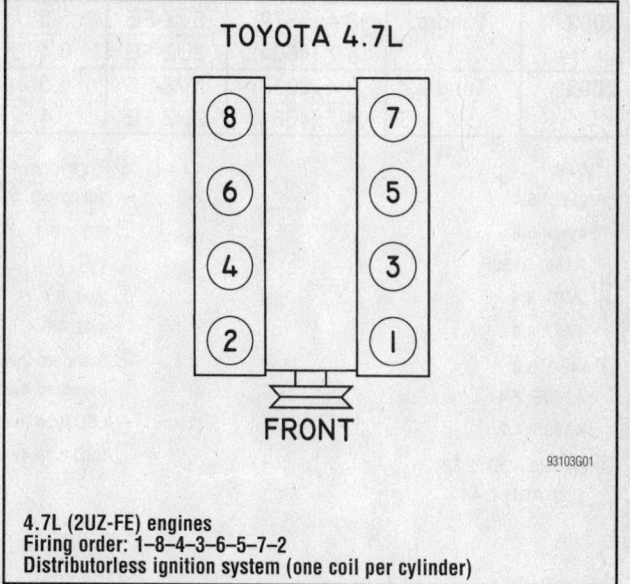

4.7L (2UZ-FE) engines
Firing order: 1–8–4–3–6–5–7–2
Distributorless ignition system (one coil per cylinder)

93103G01

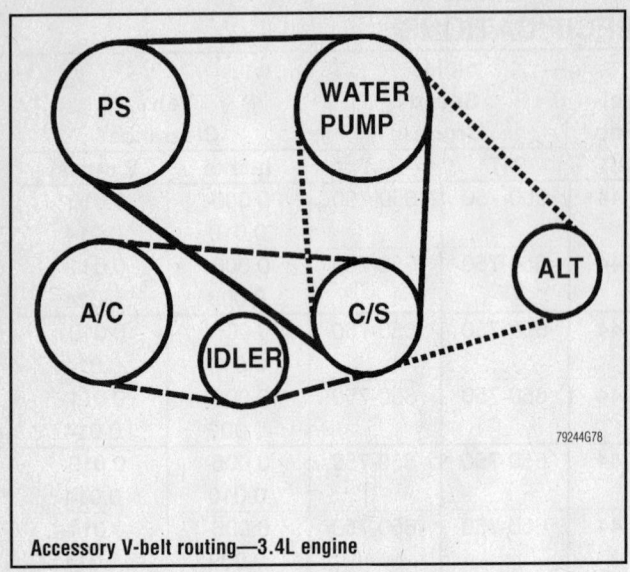

Accessory V-belt routing—3.4L engine

79244G78

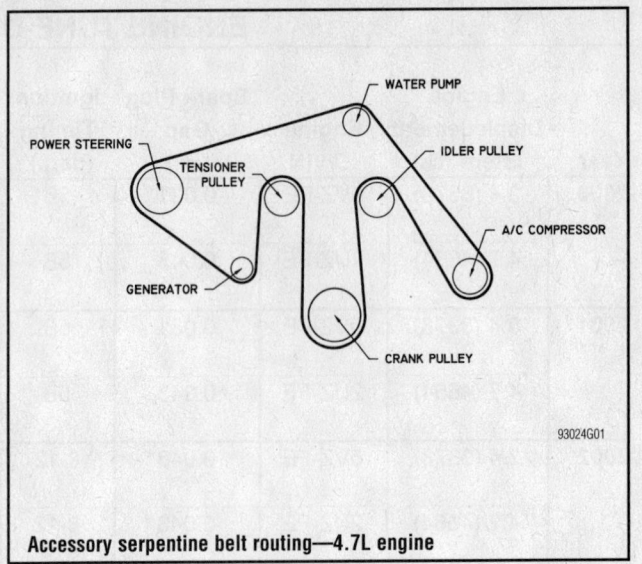

Accessory serpentine belt routing—4.7L engine

93024G01

CAPACITIES

Year	Model	Engine Displacement Liters (cc)	Engine ID/VIN	Engine Oil with Filter (qts.)	Transmission (pts.) 5-Spd	Transmission (pts.) Auto.	Transfer Case (pts.)	Drive Axle Front (pts.)	Drive Axle Rear (pts.)	Fuel Tank (gal.)	Cooling System (qts.)
2000	Tundra	3.4 (3378)	5VZ-FE	5.5	①	②	2.3	③	④	18.0	⑤
		4.7 (4664)	2UZ-FE	6.4	①	②	2.4	③	④	18.1	12.3
2001	Tundra	3.4 (3378)	5VZ-FE	5.5	①	②	2.3	③	④	18.0	⑤
		4.7 (4664)	2UZ-FE	6.4	①	②	2.4	③	④	18.1	12.3
2002	Tundra	3.4 (3378)	5VZ-FE	5.5	⑥	4.2	2.2	2.4	⑦	26.4	⑤
		4.7 (4664)	2UZ-FE	6.4	⑥	4.2	2.2	2.4	⑦	26.4	12.3
2003	Tundra	3.4 (3378)	5VZ-FE	5.5	⑥	4.2	2.2	2.4	⑦	26.4	⑤
		4.7 (4664)	2UZ-FE	6.4	⑥	4.2	2.2	2.4	⑦	26.4	12.3

① W59:
　2WD: 5.4
　4WD: 5.2
　R150, R150F:
　2WD: 5.4
　4WD: 4.6

② A43D: 5.0
　A340E: 3.4
　A340F: 4.2

③ Without ADD: 2.32
　With ADD: 2.44

④ Extra long: 4.4
　All others: 5.4

⑤ M/T: 10.3
　A/T: 10.0

⑥ 2wd: 5.4
　4wd: 4.6

⑦ Standard 2wd: 8.04
　Standard 4wd: 7.40
　LSD type 2wd: 6.66
　LSD type 4wd: 6.24

42356-TUND-C04

CRANKSHAFT AND CONNECTING ROD SPECIFICATIONS

All measurements are given in inches.

Year	Engine Displacement Liters (cc)	Engine ID/VIN	Crankshaft				Connecting Rod		
			Main Brg. Journal Dia.	Main Brg. Oil Clearance	Shaft End-play	Thrust on No.	Journal Diameter	Oil Clearance	Side Clearance
2000	3.4 (3378)	5VZ-FE	2.5191-2.5197	0.0008-0.0015	0.0008-0.0087	2	2.1648-2.1654	0.0009-0.0021	0.0059-0.0130
	4.7 (4665)	2UZ-FE	2.6373-2.6378	0.0016-0.0023	0.0008-0.0087	3	2.0465-2.0472	0.0011-0.0021	0.0063-0.0138
2001	3.4 (3378)	5VZ-FE	2.5191-2.5197	0.0008-0.0015	0.0008-0.0087	2	2.1648-2.1654	0.0009-0.0021	0.0059-0.0130
	4.7 (4665)	2UZ-FE	2.6373-2.6378	0.0016-0.0023	0.0008-0.0087	3	2.0465-2.0472	0.0011-0.0021	0.0063-0.0138
2002	3.4 (3378)	5VZ-FE	2.5191-2.5197	0.0008-0.0015	0.0008-0.0087	2	2.1648-2.1654	0.0009-0.0021	0.0059-0.0130
	4.7 (4665)	2UZ-FE	2.6373-2.6378	0.0016-0.0023	0.0008-0.0087	3	2.0465-2.0472	0.0011-0.0021	0.0063-0.0138
2003	3.4 (3378)	5VZ-FE	2.5191-2.5197	0.0008-0.0015	0.0008-0.0087	2	2.1648-2.1654	0.0009-0.0021	0.0059-0.0130
	4.7 (4665)	2UZ-FE	2.6373-2.6378	0.0016-0.0023	0.0008-0.0087	3	2.0465-2.0472	0.0011-0.0021	0.0063-0.0138

42356-TUND-C05

VALVE SPECIFICATIONS

Year	Engine Displacement Liters (cc)	Engine ID/VIN	Seat Angle (deg.)	Face Angle (deg.)	Spring Test Pressure (lbs. @ in.)	Spring Installed Height (in.)	Stem-to-Guide Clearance (in.)		Stem Diameter (in.)	
							Intake	Exhaust	Intake	Exhaust
2000	3.4 (3378)	5VZ-FE	45	44.5	41.9-46.3@ 1.311	1.311	0.0010-0.0024	0.0012-0.0026	0.2350-0.2356	0.2348-0.2354
	4.7 (4664)	2UZ-FE	45	44.5	45.9-50.7@ 1.378	1.380	0.0010-0.0024	0.0012-0.0026	0.2154-0.2159	0.2152-0.2157
2001	3.4 (3378)	5VZ-FE	45	44.5	41.9-46.3@ 1.311	1.311	0.0010-0.0024	0.0012-0.0026	0.2350-0.2356	0.2348-0.2354
	4.7 (4664)	2UZ-FE	45	44.5	45.9-50.7@ 1.378	1.380	0.0010-0.0024	0.0012-0.0026	0.2154-0.2159	0.2152-0.2157
2002	3.4 (3378)	5VZ-FE	45	44.5	41.9-46.3@ 1.311	1.311	0.0010-0.0024	0.0012-0.0026	0.2350-0.2356	0.2348-0.2354
	4.7 (4664)	2UZ-FE	45	44.5	45.9-50.7@ 1.378	1.380	0.0010-0.0024	0.0012-0.0026	0.2154-0.2159	0.2152-0.2157
2003	3.4 (3378)	5VZ-FE	45	44.5	41.9-46.3@ 1.311	1.311	0.0010-0.0024	0.0012-0.0026	0.2350-0.2356	0.2348-0.2354
	4.7 (4664)	2UZ-FE	45	44.5	45.9-50.7@ 1.378	1.380	0.0010-0.0024	0.0012-0.0026	0.2154-0.2159	0.2152-0.2157

42356-TUND-C06

PISTON AND RING SPECIFICATIONS
All measurements are given in inches.

Year	Engine Displ. Liters (cc)	Engine ID/VIN	Piston Clearance	Ring Gap			Ring Side Clearance		
				Top Compression	Bottom Compression	Oil Control	Top Compression	Bottom Compression	Oil Control
2000	3.4 (3378)	5VZ-FE	0.0053-0.0060	0.0118-0.0197	0.0157-0.0236	0.0059-0.0217	0.0016-0.0031	0.0012-0.0028	SNUG
	4.7 (4665)	2UZ-FE	0.0035-0.0044	0.0118-0.0197	0.0157-0.0256	0.0051-0.0189	0.0012-0.0031	0.0012-0.0028	SNUG
2001	3.4 (3378)	5VZ-FE	0.0053-0.0060	0.0118-0.0197	0.0157-0.0236	0.0059-0.0217	0.0016-0.0031	0.0012-0.0028	SNUG
	4.7 (4665)	2UZ-FE	0.0035-0.0044	0.0118-0.0197	0.0157-0.0256	0.0051-0.0189	0.0012-0.0031	0.0012-0.0028	SNUG
2002	3.4 (3378)	5VZ-FE	0.0053-0.0060	0.0118-0.0197	0.0157-0.0236	0.0059-0.0217	0.0016-0.0031	0.0012-0.0028	SNUG
	4.7 (4665)	2UZ-FE	0.0035-0.0044	0.0118-0.0197	0.0157-0.0256	0.0051-0.0189	0.0012-0.0031	0.0012-0.0028	SNUG
2003	3.4 (3378)	5VZ-FE	0.0053-0.0060	0.0118-0.0197	0.0157-0.0236	0.0059-0.0217	0.0016-0.0031	0.0012-0.0028	SNUG
	4.7 (4665)	2UZ-FE	0.0035-0.0044	0.0118-0.0197	0.0157-0.0256	0.0051-0.0189	0.0012-0.0031	0.0012-0.0028	SNUG

42356-TUND-C07

TORQUE SPECIFICATIONS
All readings in ft. lbs.

Year	Engine Displacement Liters (cc)	Engine ID/VIN	Cylinder Head Bolts	Main Bearing Bolts	Rod Bearing Bolts	Crankshaft Damper Bolts	Flywheel Bolts	Manifold		Spark Plugs	Lug Nuts
								Intake	Exhaust		
2000	3.4 (3378)	5VZ-FE	①	②	③	184	63-67	13	30	13	76
	4.7 (4664)	2UZ-FE	④	⑤	③	181	⑥	13	33	13	97
2001	3.4 (3378)	5VZ-FE	①	②	③	184	63-67	13	30	13	76
	4.7 (4664)	2UZ-FE	④	⑤	③	181	⑥	13	33	13	97
2002	3.4 (3378)	5VZ-FE	①	②	③	217	63	13	30	13	76
	4.7 (4664)	2UZ-FE	④	⑤	③	181	⑥	13	33	13	97
2003	3.4 (3378)	5VZ-FE	①	②	③	217	63	13	30	13	76
	4.7 (4664)	2UZ-FE	④	⑤	③	181	⑥	13	33	13	97

① Step 1: 25 ft. lbs.
   Step 2: Plus 90 degrees
   Step 3: Plus 90 degrees
   Recessed head: 13 ft. lbs.

② Step 1: 45 ft. lbs.
   Step 2: Plus 90 degrees

③ Step 1: 18 ft. lbs.
   Step 2: Plus 90 degrees

④ Step 1: 24 ft. lbs.
   Step 2: Plus 180 degrees

⑤ Step 1: 20 ft. lbs.
   Step 2: Plus 90 degrees

⑥ Step 1: 35 ft. lbs.
   Step 2: Plus 90 degrees

42356-TUND-C08

BRAKE SPECIFICATIONS

All measurements in inches unless noted

| Year | Model | Brake Disc | | | Brake Drum Diameter | | | Minimum Lining Thickness | | Brake Caliper | |
		Original Thickness	Minimum Thickness	Maximum Runout	Original Inside Diameter	Max. Wear Limit	Maximum Machine Diameter	Front	Rear	Bracket Bolts (ft. lbs.)	Mounting Bolts (ft. lbs.)
2000	2WD	0.866	0.787	0.0028	10.00	—	10.08	0.039	—	65	90
	4WD	0.866	0.787	0.0028	11.61	—	11.69	—	0.039	—	90
2001	2WD	1.102	1.024	0.0028	11.61	—	11.69	0.039	—	65	90
	4WD	1.102	1.024	0.0028	11.61	—	11.69	—	0.039	—	90
2002	2WD	1.102	1.024	0.0028	11.61	—	11.69	0.039	—	65	90
	4WD	1.102	1.024	0.0028	11.61	—	11.69	—	0.039	—	90
2003	2WD	1.102	1.024	0.0028	11.61	—	11.69	0.039	—	65	90
	4WD	1.102	1.024	0.0028	11.61	—	11.69	—	0.039	—	90

42356-TUND-C09

WHEEL ALIGNMENT

| Year | Model | Caster | | Camber | | Toe-in (in.) | Steering Axis Inclination (Deg.) |
		Range (+/-Deg.)	Preferred Setting (Deg.)	Range (+/-Deg.)	Preferred Setting (Deg.)		
2000	2WD	0.75	2.05	0.75	-0.12	0.06+/-0.08	10.87
	4WD	0.75	1.07	0.75	0.33	0.07+/-0.08	10.41
2001	2WD	0.75	2.05	0.75	-0.12	0.06+/-0.08	10.87
	4WD	0.75	1.07	0.75	0.33	0.07+/-0.08	10.41
2002	2WD	0.75	2.05	0.75	-0.12	0.06+/-0.08	10.87
	4WD	0.75	1.07	0.75	0.33	0.07+/-0.08	10.41
2003	2WD	0.75	2.05	0.75	-0.12	0.06+/-0.08	10.87
	4WD	0.75	1.07	0.75	0.33	0.07+/-0.08	10.41

All alignment figures based on nominal ride height and standard tires

42356-TUND-C10

TIRE, WHEEL AND BALL JOINT SPECIFICATIONS

| Year | Model | OEM Tires | | Tire Pressures (psi) | | Wheel Size | Ball Joint Inspection |
		Standard	Optional	Front	Rear		
2000	Tundra	P245/70R16	P265/70R16	26	Std:35/Opt:29	NA	NS
2001	Tundra	P245/70R16	P265/70R16	26	Std:35/Opt:29	NA	NS
2002	Tundra	P245/70R16	P265/70R16	26	Std:35/Opt:29	NA	NS
2003	Tundra	P245/70R16	P265/70R16	26	Std:35/Opt:29	NA	NS

OEM: Original Equipment Manufacturer

PSI: Pounds Per Square Inch

STD: Standard

OPT: Optional

NS: Not specified by manufacturer

NA: Not available

42356-TUND-C11

SCHEDULED MAINTENANCE INTERVALS
TOYOTA—TUNDRA

TO BE SERVICED	TYPE OF SERVICE	5	10	15	20	25	30	35	40	45	50	55	60	65	70	75	80	85	90	95
Automatic transmission and differential fluid	S/I			✓			✓			✓			✓			✓			✓	
Ball joints and boots	S/I			✓			✓			✓			✓			✓			✓	
Brake system	S/I			✓			✓			✓			✓			✓			✓	
Charcoal canister	S/I												✓							
Drive belts	S/I						✓						✓						✓	
Driveshaft bushing	L						✓						✓						✓	
Engine coolant	R						✓						✓						✓	
Engine oil & filter	R	✓	✓	✓	✓	✓	✓	✓	✓	✓	✓	✓	✓	✓	✓	✓	✓	✓	✓	✓
Exhaust system	S/I			✓			✓						✓			✓			✓	
Fuel lines	S/I						✓						✓						✓	
Fuel tank cap gasket	S/I						✓						✓						✓	
Halfshaft boots & flange bolts	S/I			✓			✓			✓			✓			✓			✓	
Limited slip differential fluid	R						✓						✓						✓	
Manual transmission and differential fluid	S/I						✓						✓						✓	
Non-platinum spark plugs	R						✓						✓						✓	
Platinum spark plugs	R												✓							
Propeller shaft (4WD)	L			✓			✓			✓			✓			✓			✓	
Propeller shaft bolts	S/I			✓			✓			✓			✓			✓			✓	
Steering gear	S/I			✓			✓			✓			✓			✓			✓	
Steering linkage	S/I			✓			✓			✓			✓			✓			✓	
Valves	S/I												✓							

R: Replace   S/I: Service or Inspect   L: Lubricate

FREQUENT OPERATION MAINTENANCE (SEVERE SERVICE)

If a vehicle is operated under any of the following conditions it is considered severe service:

- Towing a trailer or using a camper or car-top carrier.
- Repeated short trips of less than 5 miles in temperatures below freezing.
- Excessive idling or low-speed driving for long distances as in heavy commercial use, such as delivery, taxi or police cars.
- Operating on rough, muddy or salt-covered roads.
- Operating on unpaved or dusty roads.

Oil filter: service or inspect every 5000 miles or 4 months, whichever occurs first.

Brake linings and discs or drums: service or inspect every 5000 miles or 4 months, whichever occurs first.

Steering linkage: service or inspect every 5000 miles or 4 months, whichever occurs first.

Ball joints and boots: service or inspect every 5000 miles or 4 months, whichever occurs first.

Brake discs & pads (front): service or inspect every 6000 miles.

Halfshaft boots: service or inspect every 5000 miles or 4 months. Retighten the flange bolts, whichever occurs first.

Body chassis bolts and nuts: service or inspect every 5000 miles or 4 months, whichever occurs first.

Transmission and differential fluid: replace every 15,000 miles or 12 months, whichever occurs first.

Transfer case and differential fluid: replace every 15,000 miles or 12 months, whichever occurs first.

Timing belt: replace every 60,000 miles or 48 months, whichever occurs first.

PRECAUTIONS

Before servicing any vehicle, please be sure to read all of the following precautions, which deal with personal safety, prevention of component damage, and important points to take into consideration when servicing a motor vehicle:

• Never open, service or drain the radiator or cooling system when the engine is hot; serious burns can occur from the steam and hot coolant.

• Observe all applicable safety precautions when working around fuel. Whenever servicing the fuel system, always work in a well-ventilated area. Do not allow fuel spray or vapors to come in contact with a spark, open flame or excessive heat (a hot drop light, for example). Keep a dry chemical fire extinguisher near the work area. Always keep fuel in a container specifically designed for fuel storage; also, always properly seal fuel containers to avoid the possibility of fire or explosion. Refer to the additional fuel system precautions later in this section.

• Fuel injection systems often remain pressurized, even after the engine has been turned **OFF**. The fuel system pressure must be relieved before disconnecting any fuel lines. Failure to do so may result in fire and/or personal injury.

• Brake fluid often contains polyglycol ethers and polyglycols. Avoid contact with the eyes and wash your hands thoroughly after handling brake fluid. If you do get brake fluid in your eyes, flush your eyes with clean, running water for 15 minutes. If eye irritation persists, or if you have taken brake fluid internally, IMMEDIATELY seek medical assistance.

• The EPA warns that prolonged contact with used engine oil may cause a number of skin disorders, including cancer! You should make every effort to minimize your exposure to used engine oil. Protective gloves should be worn when changing oil. Wash your hands and any other exposed skin areas as soon as possible after exposure to used engine oil. Soap and water, or waterless hand cleaner should be used.

• All new vehicles are now equipped with an air bag system, often referred to as a Supplemental Restraint System (SRS) or Supplemental Inflatable Restraint (SIR) system. The system must be disabled before performing service on or around system components, steering column, instrument panel components, wiring and sensors. Failure to follow safety and disabling procedures could result in accidental air bag deployment, possible personal injury and unnecessary system repairs.

• Always wear safety goggles when working with, or around, the air bag system. When carrying a non-deployed air bag, be sure the bag and trim cover are pointed away from your body. When placing a non-deployed air bag on a work surface, always face the bag and trim cover upward, away from the surface. This will reduce the motion of the module if it is accidentally deployed. Refer to the additional air bag system precautions later in this section.

• Clean, high quality brake fluid from a sealed container is essential to the safe and proper operation of the brake system. You should always buy the correct type of brake fluid for your vehicle. If the brake fluid becomes contaminated, completely flush the system with new fluid. Never reuse any brake fluid. Any brake fluid that is removed from the system should be discarded. Also, do not allow any brake fluid to come in contact with a painted surface; it will damage the paint.

• Never operate the engine without the proper amount and type of engine oil; doing so WILL result in severe engine damage.

• Timing belt maintenance is extremely important! Many models utilize an interference type, non-freewheeling engine. If the timing belt breaks, the valves in the cylinder head may strike the pistons, causing potentially serious (also time consuming and expensive) engine damage. Refer to the maintenance interval charts in the front of this section for the recommended replacement interval for the timing belt, and to the timing belt procedure in this section for belt replacement and inspection.

• Disconnecting the negative battery cable on some vehicles may interfere with the functions of the on-board computer system(s) and may require the computer to undergo a relearning process once the negative battery cable is reconnected.

• When servicing drum brakes, only disassemble and assemble one side at a time, leaving the remaining side intact for reference.

ENGINE REPAIR

→Disconnecting the negative battery cable on some vehicles may interfere with the functions of the on board computer system. The computer may undergo a relearning process once the negative battery cable is reconnected.

Alternator

REMOVAL

3.4L Engine

1. Remove or disconnect the following:
   • Negative battery cable
   • Alternator wiring
   • Alternator locknut, pivot bolt, nut and adjusting bolt
   • Drive belt
   • Alternator

4.7L Engine

1. Before servicing the vehicle, refer to the precautions in the beginning of this section.
2. Drain the cooling system.
3. Remove or disconnect the following:
   • Negative battery cable
   • Accessory drive belt
   • Engine under cover
   • Radiator
   • Power steering pump pulley
   • Alternator harness connectors
   • Alternator

INSTALLATION

3.4L Engine

Install or connect the following:
   • Alternator

   • Drive belt. Tighten the locknut 25 ft. lbs. (33 Nm) and the pivot bolt 38 ft. lbs. (51 Nm).
   • Alternator wiring
   • Negative battery cable

4.7L Engine

1. Install or connect the following:
   • Alternator. Tighten the fasteners to 29 ft. lbs. (39 Nm).
   • Alternator harness connectors
   • Power steering pump pulley
   • Radiator
   • Engine under cover
   • Accessory drive belt
   • Negative battery cable
2. Fill the cooling system.
3. Start the engine and check for leaks.

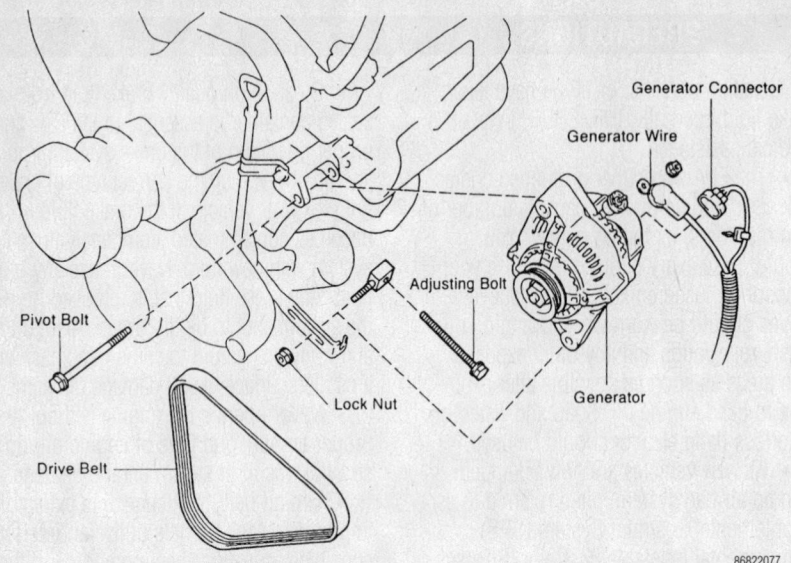

Exploded view of the alternator and drive belt—3.4L Engine

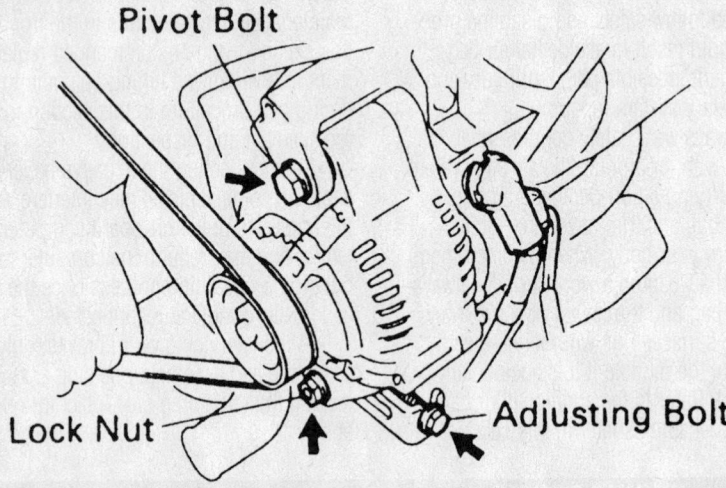

Locations of the adjusting and pivot bolts and the locknut—3.4L Engine

Ignition Timing

ADJUSTMENT

The engines are equipped with a Distributorless Ignition System (DIS). No timing adjustment is possible.

Engine Assembly

REMOVAL & INSTALLATION

3.4L Engine

2-WHEEL DRIVE

1. Before servicing the vehicle, refer to the precautions in the beginning of this section.

2. Properly relieve the fuel system pressure.
3. Remove or disconnect the following:
   - Hood
   - Battery
   - Engine under covers
4. Drain the engine coolant.
5. Drain the engine oil.
   - Radiator
   - Fan with the fluid coupling and fan pulleys
   - Air cleaner cap
   - Air cleaner case and filter
6. Disconnect the following hoses:
   - Heater hoses
   - Brake booster vacuum hose
   - Evaporative Emissions (EVAP) hose
   - Vacuum hose
   - Fuel return hose
   - Fuel inlet hose
7. Detach the starter wire and connectors, as follows:
   - Ground strap, by removing the bolt
   - Starter wires
8. Remove or disconnect the following:
   - Alternator connector and wire
   - Throttle cable, if equipped with an automatic transmission
   - Cruise control cable, if equipped with cruise control
9. Disconnect the engine wiring harness, as follows:
   - Glove box door
   - Lower the finish No. 2 panel
   - Heater to register duct
   - 3 Engine Control Module (ECM) connectors
   - 2 cassette connectors and the 2 wire clamps from the lower finish panel
   - Engine wiring harness clamp
10. Remove or disconnect the following:
    - Igniter connector
    - Ground strap
    - 2 engine wiring harness retainer-to-cowl panel nuts and pull out the engine wiring harness
11. If equipped with a manual transmission, remove or disconnect the following:
    - Shift lever knob
    - 4 shift lever boot screws
    - 6 shift lever assembly bolts, the assembly and gasket
12. Remove or disconnect the following:
    - Driveshaft from the transmission
    - Speedometer cable

➡ **Do not lose the felt protector and washers.**

   - Front exhaust pipe
   - Clutch release cylinder, if equipped with a manual transmission
   - Nut and the control cable
13. Place a jack under the transmission.
14. Remove or disconnect the following:
    - Transmission rear mounting bracket by removing the 8 bolts
    - Bolt and the air conditioning compressor wire clamp, if equipped with air conditioning
15. If necessary, install a No. 2 engine hanger with 2 bolts. Tighten the 2 bolts to 30 ft. lbs. (40 Nm).
16. Attach the engine hoist chain to the 2 engine hangers.
17. Remove or disconnect the following:
    - 4 engine front mounting insulators-to-frame bolts and nuts
    - Engine from the transmission

To install:

18. Install or connect the following:
    - Engine to the transmission
    - Engine mounts to the body mountings. Install the bolts and nuts but do not tighten at this time.
19. Remove the engine chain hoist the No. 2 engine hanger.
20. Install or connect the following:
    - Air conditioning wire with the bolt, if equipped with air conditioning
    - Transmission mounting bracket. Tighten the frame bolts to 43 ft. lbs. (58 Nm) and the mounting insulator bolts to 13 ft. lbs. (18 Nm).
    - Tighten the engine mounting nuts and bolts to 28 ft. lbs. (38 Nm).
    - Control cable
    - Clutch release cylinder, if equipped with a manual transmission. Torque the bolts to 9 ft. lbs. (12 Nm).
21. Install or connect the following:
    - Front exhaust pipe
    - Speedometer cable
    - Driveshaft
22. If equipped with a manual transmission, install or connect the following:
    - 6 shift lever assembly bolts, the assembly and gasket
    - 4 shift lever boot screws
    - Shift lever knob
23. Install or connect the following:
    - All engine wiring harness, hoses and cables
    - Fan with the fluid coupling and fan pulleys. Tighten the nuts to 48 inch lbs. (5.4 Nm).
    - Air cleaner case and air filter
    - Radiator
24. Install or connect the following hoses:
    - Fuel inlet hose
    - Fuel return hose
    - Vacuum hose
    - Evaporative Emissions (EVAP) hose
    - Brake booster vacuum hose
    - Heater hoses
25. Fill the engine with oil.
26. Fill the engine and radiator with coolant.
27. Install or connect the following:
    - Engine undercover
    - Battery
    - Hood
28. Start the engine and check for leaks.
29. Make any necessary adjustments and road test the vehicle.

4-WHEEL DRIVE

1. Before servicing the vehicle, refer to the precautions in the beginning of this section.
2. Remove or disconnect the following:
    - Transmission
    - Hood
3. Release the fuel system pressure.
4. Remove or disconnect the following:
    - Battery
    - Engine undercovers
5. Drain the engine coolant.
6. Drain the engine oil.
7. Remove or disconnect the following:
    - Radiator
    - Fan with the fluid coupling and fan pulleys
    - Air cleaner cap
    - Mass Air Flow (MAF) meter and the resonator
    - Cruise control cable, if equipped with cruise control
    - Throttle cable, if equipped with an automatic transmission
8. Disconnect the following hoses:
    - Heater hoses
    - Brake booster vacuum hose
    - Evaporative Emissions (EVAP) hose
    - Automatic Disconnecting Differential (ADD) vacuum hose
    - Vacuum hose
    - Fuel return hose
    - Fuel inlet hose
9. Detach the starter wire and connectors, as follows:
    - Ground strap, by removing the bolt
    - 3 starter wire clamps and connector
10. Detach the alternator connector and wire.
11. Disconnect the engine wiring harness, as follows:
    - Glove box door
    - Lower the finish No. 2 panel
    - 3 Engine Control Module (ECM) connectors
    - 2 cassette connectors and the 2 wire clamps from the lower finish panel
    - Igniter connector
    - Ground strap
    - Engine wiring harness clamp
12. Remove or disconnect the following:
    - 2 engine wiring harness retainer-to-cowl panel nuts and wiring harness
    - Air conditioning compressor wire clamp and compressor bracket, if equipped with air conditioning
13. If necessary, install a No. 2 engine hanger with 2 bolts. Tighten the 2 bolts to 30 ft. lbs. (40 Nm).
14. Attach the engine hoist chain to the 2 engine hangers.
15. Remove or disconnect the following:
    - 4 engine front mounting insulators-to-frame bolts and nuts
    - Engine

To install:

16. Install or connect the following:
    - Engine
    - Engine mounts-to-body mountings. Install the bolts and nuts but do not tighten at this time.
17. Remove the engine chain hoist the No. 2 engine hanger.
18. Install or connect the following:
    - Air conditioning wire with the bolt and the compressor bracket, if equipped with air conditioning
    - Tighten the engine mounting nuts and bolts to 28 ft. lbs. (38 Nm).
19. Install the engine wiring harness, as follows:
    - Engine wiring harness clamp
    - Ground strap
    - Igniter connector
    - 2 cassette connectors and the 2 wire clamps from the lower finish panel
    - 3 Engine Control Module (ECM) connectors
    - Lower the finish No. 2 panel
    - Glove box door
20. Install or connect the following:
    - Cruise control cable, if equipped with cruise control
    - Throttle cable, if equipped with an automatic transmission
21. Connect the following hoses:
    - Fuel inlet hose
    - Fuel return hose
    - Vacuum hose
    - Automatic Disconnecting Differential (ADD) vacuum hose
    - Evaporative Emissions (EVAP) hose
    - Brake booster vacuum hose
    - Heater hoses
22. Install or connect the following:
    - All wires, hoses and cables
    - Fan with the fluid coupling and fan pulleys. Tighten the nuts to 48 inch lbs. (5.4 Nm).
    - Air cleaner case and air filter
    - MAF meter, resonator and the air cleaner cap
    - Radiator
23. Fill the engine with oil.
24. Fill the engine and radiator with coolant.
25. Install or connect the following:
    - Transmission and refill it with transmission oil
    - Engine undercover
    - Battery
    - Hood
26. Start the engine, make any necessary adjustments and check for leaks.

4.7L Engine

1. Before servicing the vehicle, refer to the precautions in the beginning of this section.
2. Relieve the fuel system pressure.
3. Drain the cooling system.
4. Drain the engine oil.
5. Remove or disconnect the following:
   - Battery and tray
   - Hood
   - Engine appearance cover
   - Air intake pipe
   - Engine under covers
   - Coolant recovery tank
   - Radiator hoses
   - Radiator and fan shroud
   - Accessory drive belt
   - Cooling fan and pulley
   - Powertrain Control Module (PCM) harness connectors and pass the wiring harness through the firewall
   - Accelerator cable
   - Power steering vacuum hoses
   - Alternator harness connectors
   - Heater hoses
   - Engine control wiring harness and grommet at the firewall
   - Ground cable connector
   - Fuel lines
   - Evaporative Emissions (EVAP) canister hoses
   - Wire clamp at right inner fender
   - Negative battery cable at the relay box and right inner fender
   - Positive battery cable
   - Center console
   - Transmission shift lever assembly
   - Transfer case shift lever and rod
   - Exhaust front pipes
   - Stabilizer bar
   - Front and rear driveshafts
   - A/C compressor
   - Power steering pump
6. Attach a hoist to the engine lifting eyes.
7. Remove or disconnect the following:
   - Transfer case skid plate
   - Left and right motor mounts
   - Transmission mount crossmember
8. Attach a hoist to the engine lifting eyes and raise the powertrain out of the vehicle.

To install:

9. Lower the powertrain into the vehicle.
10. Install or connect the following:
    - Transmission mount crossmember. Tighten the bolts to 37 ft. lbs. (50 Nm) and the nuts to 55 ft. lbs. (74 Nm).
    - Transfer case skid plate
    - Left and right motor mounts. Tighten the fasteners to 22 ft. lbs. (30 Nm).
    - Power steering pump. Tighten the bolts to 13 ft. lbs. (17 Nm).
    - A/C compressor. Tighten the bolts to 36 ft. lbs. (49 Nm).
    - Front driveshaft. Tighten the fasteners to 59 ft. lbs. (80 Nm).
    - Rear driveshaft. Tighten the fasteners to 78 ft. lbs. (106 Nm).
    - Stabilizer bar. Tighten the bracket bolts to 13 ft. lbs. (18 Nm) and the link nuts to 18 ft. lbs. (25 Nm).
    - Exhaust front pipes
    - Transfer case shift lever and rod
    - Transmission shift lever assembly
    - Center console
    - Positive battery cable
    - Negative battery cable at the relay box and right inner fender
    - Wire clamp at right inner fender
    - EVAP canister hoses
    - Fuel lines
    - Ground cable connector
    - Engine control wiring harness and grommet at the firewall
    - Heater hoses
    - Alternator harness connectors
    - Power steering vacuum hoses
    - Accelerator cable
    - PCM harness connectors
    - Cooling fan and pulley
    - Accessory drive belt
    - Radiator and fan shroud
    - Radiator hoses
    - Coolant recovery tank
    - Engine under covers
    - Air intake pipe
    - Engine appearance cover
    - Hood
    - Battery and tray
11. Fill the crankcase to the correct level.
12. Fill the cooling system.
13. Start the engine and check for leaks.

Water Pump

REMOVAL & INSTALLATION

3.4L Engine

1. Before servicing the vehicle, refer to the precautions in the beginning of this section.
2. Remove or disconnect the following:
   - Negative battery cable
   - Engine undercover
3. Drain the engine coolant.
4. Remove the upper radiator hose.
5. Remove the power steering drive belt, as follows:
   - Stretch the belt and loosen the fan pulley mounting nuts
   - Loosen the lockbolt, pivot bolt and the adjusting bolt
   - Drive belt
6. Remove or disconnect the following:
   - Air conditioning drive belt, by loosening the idler pulley nut and adjusting bolt
   - Lockbolt, pivot bolt and the adjusting bolt
   - Alternator drive belt
   - No. 2 fan shroud, by removing the 2 clips
   - Fan with the fluid coupling and fan pulleys
   - Power steering pump and move it aside without disconnecting the lines from the pump
   - Compressor from the engine and move it aside without disconnecting the compressor lines, if equipped with air conditioning
   - Air conditioning bracket, if equipped with air conditioning
7. Remove the No. 2 timing belt cover, as follows:
   - Camshaft Position (CMP) sensor connector from the No. 2 timing belt cover
   - 3 spark plug wire clamps from the No. 2 timing belt cover
   - 6 bolts and the timing belt cover
8. Remove the fan bracket, as follows:
   - Power steering adjusting strut, by removing the nut
   - Fan bracket, by removing the bolt and nut
9. Set the No. 1 cylinder to Top Dead Center (TDC) of the compression stroke, as follows:
   a. Turn the crankshaft pulley and align its groove with the timing mark **0** of the No. 1 timing belt cover.
   b. Check that the timing marks of the camshaft timing pulleys and the No. 3 timing belt cover are aligned. If not, turn the crankshaft pulley 1 revolution (360 degrees).
10. Remove the camshaft timing pulleys, as follows:
    a. Remove the timing belt tensioner by alternately loosening the 2 bolts.
    b. Using Variable Wrench Set No. 09960-10010, remove the pulley bolt, the timing pulley and the knock pin.
    c. Remove the 2 timing pulleys with the timing belt.
11. Remove or disconnect the following:
    - Thermostat
    - No. 2 oil cooler hose, from the water pump

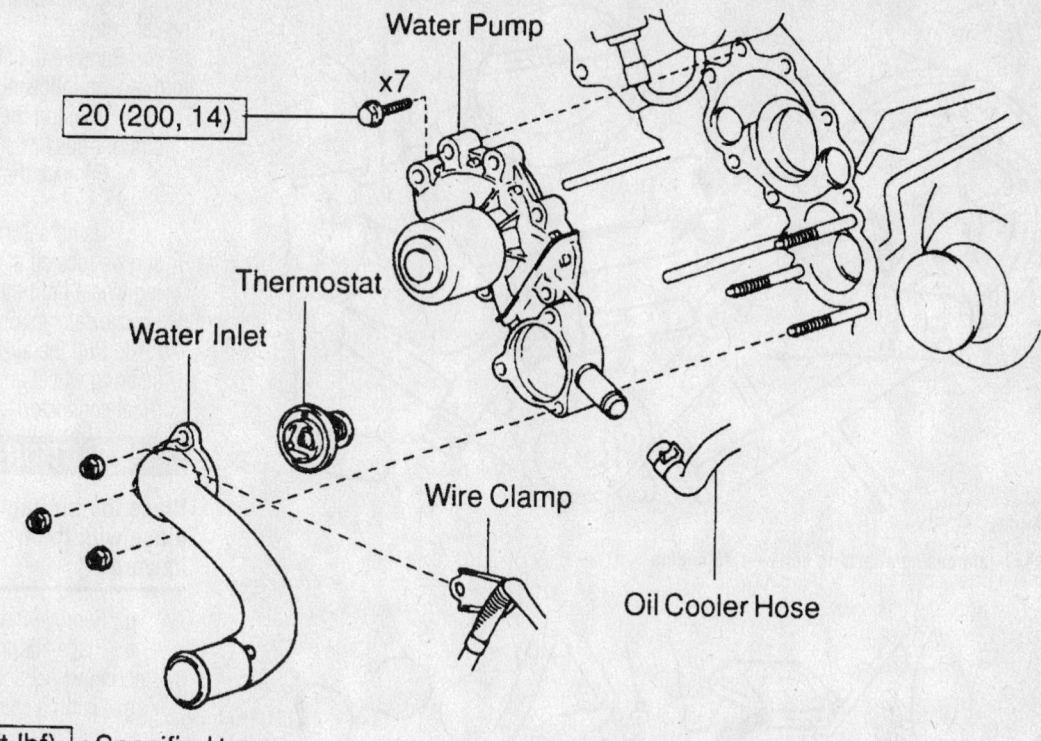

Exploded view of the water pump mounting—3.4L engine

20 (200, 14)

Water Pump

Thermostat

Water Inlet

Wire Clamp

Oil Cooler Hose

N·m(kgf·cm, ft·lbf) : Specified torque
◆ Non–Reusable part

7924YG08

- Water pump, by removing the 7 bolts
12. Thoroughly clean the mating surfaces.
To install:
13. Install or connect the following:
- Apply sealant (PN 08826-00100) to the water pump

❋❋ WARNING

Parts must be assembled within 5 minutes of application. Otherwise the material must be removed and reapplied.

- Water pump. Tighten the bolts to 14 ft. lbs. (20 Nm).
- No. 2 oil cooler hose
- Thermostat
- Left camshaft timing pulley. Tighten the pulley bolt to 81 ft. lbs. (110 Nm).
14. Set the No. 1 cylinder to TDC of the compression stroke.
15. Connect the timing belt to the left camshaft timing pulley. Check that the installation mark on the timing belt is aligned with the end of the No. 1 timing belt cover, as follows:
  a. Using Variable Pin Wrench Set

09960-01000, slightly turn the left camshaft timing pulley clockwise. Align the installation mark on the timing belt with the timing mark of the camshaft timing pulley and hang the timing belt on the left camshaft timing pulley.
  b. Align the timing marks of the left camshaft pulley and the No. 3 timing belt cover.
  c. Check that the timing belt has tension between the crankshaft timing pulley and the left camshaft timing pulley.
16. Install the right camshaft timing pulley and the timing belt.
17. Set the timing belt tensioner, as follows:
  a. Using a press, slowly press in the pushrod using 220–2,205 lbs. (981–9,807 N) of force.
  b. Align the holes of the pushrod and housing, pass a 1.5mm hexagon wrench through the holes to keep the setting position of the pushrod.
  c. Release the press and install the dust boot to the tensioner.
  d. Install the timing belt tensioner and alternately tighten the bolts to 20 ft. lbs. (28 Nm).
  e. Using pliers, remove the 1.5mm hexagon wrench from the belt tensioner.

18. Check the valve timing, as follows:
  a. Slowly turn the crankshaft pulley 2 revolutions from the TDC-to-TDC; always turn the crankshaft pulley clockwise.
  b. Check that each pulley aligns with the timing marks. If the timing marks do not align, remove the timing belt and reinstall it.
19. Install or connect the following:
- Fan bracket, with the bolt and nut
- Remaining components
- Negative battery cable
20. Fill with engine coolant.
21. Start the engine and check for leaks.

4.7L Engine

1. Before servicing the vehicle, refer to the precautions in the beginning of this section.
2. Drain the cooling system.
3. Remove or disconnect the following:
- Negative battery cable
- Timing belt
- No. 2 idler pulley
- Radiator hose
- Bypass hose
- Water inlet housing assembly
- Water pump
To install:
4. Install or connect the following:
- Water pump. Use a new gasket and

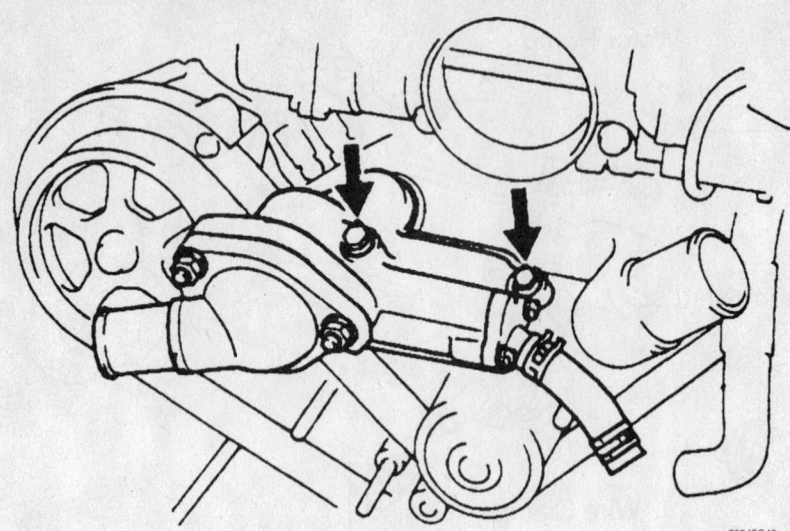

Water inlet housing attaching bolts—4.7L engine

7924SG40

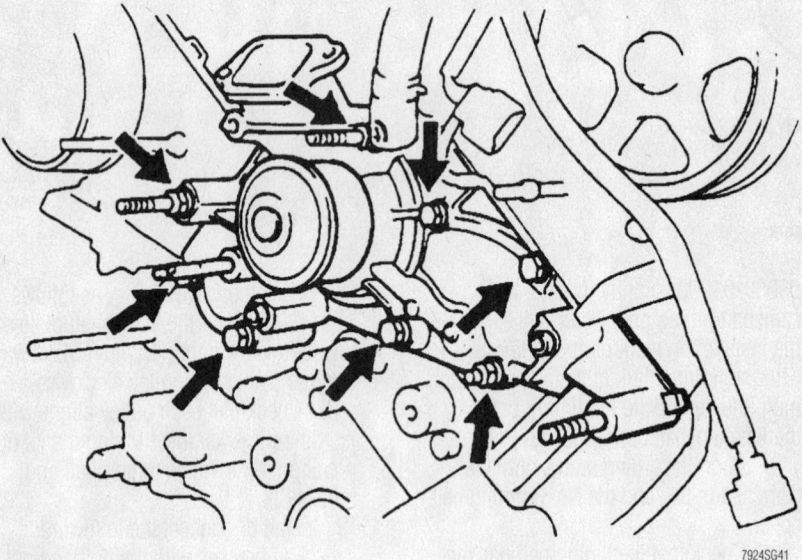

Water pump mounting bolts, stud bolts and nut locations—4.7L engine

7924SG41

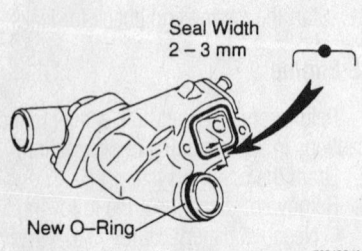

Seal Width
2 – 3 mm

New O–Ring

7924SG42

Water inlet housing sealant application—
4.7L engine

tighten the bolts to 15 ft. lbs. (21 Nm). Tighten the stud bolt and nut to 13 ft. lbs. (18 Nm).
- Water inlet housing assembly. Use a new O-ring and apply sealant as

shown. Tighten the bolts to 13 ft. lbs. (18 Nm).
- Bypass hose
- Radiator hose
- No. 2 idler pulley
- Timing belt
- Negative battery cable
5. Fill the cooling system.
6. Start the engine and check for leaks.

Heater Core

REMOVAL & INSTALLATION

1. Disconnect the negative battery cable.
2. Drain the cooling system into a clean container for reuse.

3. Disconnect the heater hoses from the heater core.
4. Remove the steering wheel by performing the following procedure:
   a. Position the front wheels facing straight-ahead.
   b. Remove the steering wheel side covers.
   c. Using a Torx® wrench, loosen the 2 screws located at each side of the steering wheel until the screw's circumference groove catches on the screw case.
   d. Pull the air bag module from the steering wheel and disconnect the electrical connector.

✳✳ CAUTION

Place the air bag module in a safe place with the front side facing upward.

   e. Remove the steering wheel nut.
   f. Place alignment marks on the steering wheel and the main shaft.
   g. Using a steering wheel puller, press the steering wheel from the steering column.
5. Remove the instrument panel and reinforcement by performing the following procedure:
   a. Remove the front door scuff plates, the cowl side trim and the front door opening trim.
   b. At the driver's side, remove the 2 assist grip plugs, the 2 screws and assist grip and the front pillar garnish.
   c. At the passenger's side, remove the 4 assist grip plugs, the 4 screws, the 2 assist grips and the front pillar garnish.
   d. Remove the instrument cluster finish panel.
   e. Remove the 2 screws and the hood lock control cable.
   f. Remove the 2 screws and the fuel lid control cable lever.
   g. Remove the lower No. 1 panel screw and the panel.
   h. Remove the lower left side panel.
   i. Remove the 3 steering column cover screws and the covers.
   j. At the steering column, disconnect the electrical connectors; then, remove the clamp, the 3 screws and the combination switch.
   k. Remove the No. 2 heater-to-register duct screw and the duct.
   l. Remove the steering column-to-instrument panel bolts and the steering column.
   m. At the combination meter, disconnect the electrical connectors; then,

remove the 4 screws and the combination meter.

n. Remove the glove compartment door stoppers, the 2 screws and the glove box door.

o. At the passenger's side air bag module, remove the No. 1 undercover, pull the air bag connector up from the undercover and disconnect it; then, remove the air bag.

✳✳ CAUTION

Place the air bag module in a safe place with the front side facing upward.

p. Remove the 3 lower No. 2 panel screws and the panel.

q. Remove the center cluster; then, pry the center cluster from the dash by prying the 8 clips in the following order:
- Left side
- Right side
- Top left side
- Top right side

r. Remove the 4 radio screws, pull the radio outward, disconnect the electrical connectors and remove the radio.

s. At the rear console panel, remove the transfer shift lever knob; then, pry the panel upward disengaging the 4 clips (2 on each side) and remove the panel.

t. At the rear of the console, remove the 2 rear end panel-to-console screws; then, pry the end panel rearward disengaging the 2 clips and remove the panel.

u. If not equipped with a rear air conditioning system, disconnect the connector and control cable; then, remove the 3 rear heater control panel screws and the panel.

v. Remove the 4 rear console box-to-chassis screws/bolts and the console box.

w. Remove the center lower cluster finish panel by prying panel rearward disengaging the 5 clips; then, disconnect the electrical connector.

x. Remove the 2 front console-to-chassis bolts/screws, disengage the 2 clips and remove the console.

y. At the instrument panel, disconnect the junction connectors (the connectors can be disconnected by loosening the bolts), the instrument panel-to-chassis 8 bolts and 2 nuts. Using an assistant, remove the instrument panel.

z. Disconnect the electrical connector and remove the ECM.

aa. Remove the No. 3 and No. 4 heater-to-register ducts.

bb. Remove the floor brace, the No. 1 brace and the reinforcement.

6. Remove the evaporator housing by performing the following procedure:

a. Discharge and recover the air conditioning system refrigerant.

b. Remove the air conditioning liquid line clamp.

c. Remove the air conditioning suction line clamp.

d. Disconnect both air conditioning lines and plug the openings to prevent contamination. Discard the 4 O-rings.

e. Remove the antenna relay electrical connector, the 2 screws and the relay.

f. Remove the evaporator housing-to-chassis 4 screws/2 nuts and the housing.

7. Remove the heater housing by performing the following procedure:

a. Remove the defroster nozzle.

b. Disconnect the electrical connector.

c. Remove the 4 nuts and the heater housing.

8. Remove the heater core-to-heater housing packing, the screw, the bracket, the clamp and the heater core.

To install:

9. Install the heater core, the clamp, the bracket, the screw and the heater core-to-heater housing packing.

10. Install the heater housing by performing the following procedure:

a. Install the heater housing and the 4 nuts.

b. Connect the electrical connector.

c. Install the defroster nozzle.

11. Install the evaporator housing by performing the following procedure:

a. Install the evaporator housing and the housing-to-chassis 4 screws and 2 nuts.

b. Install the antenna relay, the 2 screws and the electrical connector.

c. Using new O-rings, connect both air conditioning lines.

d. Install the air conditioning liquid line and suction line clamp.

12. Install the instrument panel and reinforcement by performing the following procedure:

a. Install the reinforcement, the No. 1 brace and the floor brace.

b. Install the No. 3 and No. 4 heater-to-register ducts.

c. Install the ECM and connect the electrical connector.

d. Using an assistant, install the instrument panel, connect the junction connectors, the instrument panel-to-chassis 8 bolts and 2 nuts.

e. Install the front the console, engage the 2 clips and install the 2 console-to-chassis bolts/screws.

f. Connect the electrical connector; then, install the center lower cluster finish panel by engaging the 5 clips.

g. Install the console box and the 4 rear console box-to-chassis screws/bolts.

h. If not equipped with a rear air conditioning system, install rear heater control panel, the 3 panel screws; then, connect the connector and control cable.

i. Install the rear of the console and engage the 2 clips; then, install the 2 rear end panel-to-console screws.

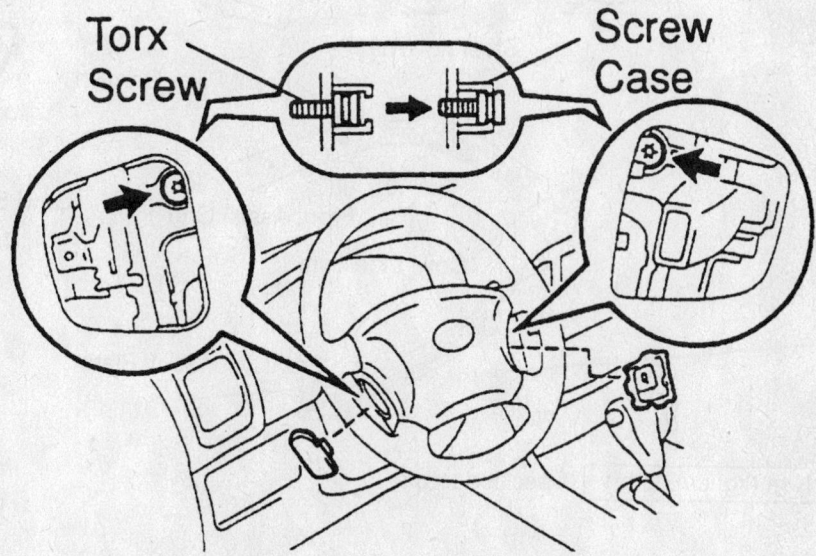

View the steering wheel's Torx® bolts

93113GG4

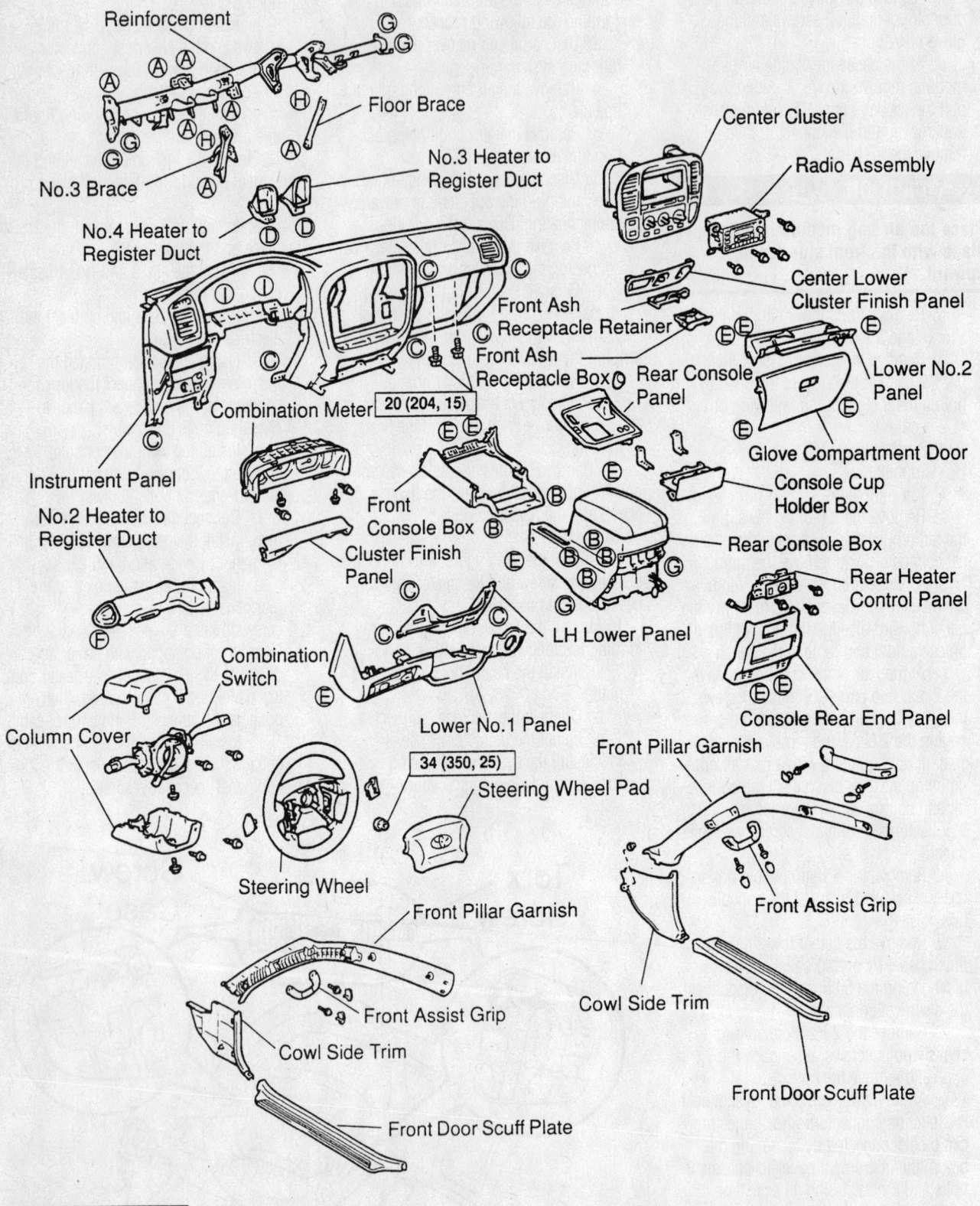

Reinforcement

Floor Brace

No.3 Heater to Register Duct

No.3 Brace

No.4 Heater to Register Duct

Center Cluster

Radio Assembly

Center Lower Cluster Finish Panel

Front Ash Receptacle Retainer

Front Ash Receptacle Box

Rear Console Panel

Lower No.2 Panel

Glove Compartment Door

Combination Meter 20 (204, 15)

Instrument Panel

No.2 Heater to Register Duct

Front Console Box

Cluster Finish Panel

Console Cup Holder Box

Rear Console Box

Rear Heater Control Panel

LH Lower Panel

Combination Switch

Column Cover

Lower No.1 Panel

Console Rear End Panel

Front Pillar Garnish

Front Pillar Garnish

34 (350, 25)

Steering Wheel Pad

Front Assist Grip

Steering Wheel

Front Pillar Garnish

Cowl Side Trim

Front Assist Grip

Cowl Side Trim

Front Door Scuff Plate

Front Door Scuff Plate

N·m (kgf·cm, ft·lbf) : Specified torque

93113GG7

Exploded view the instrument panel and related components

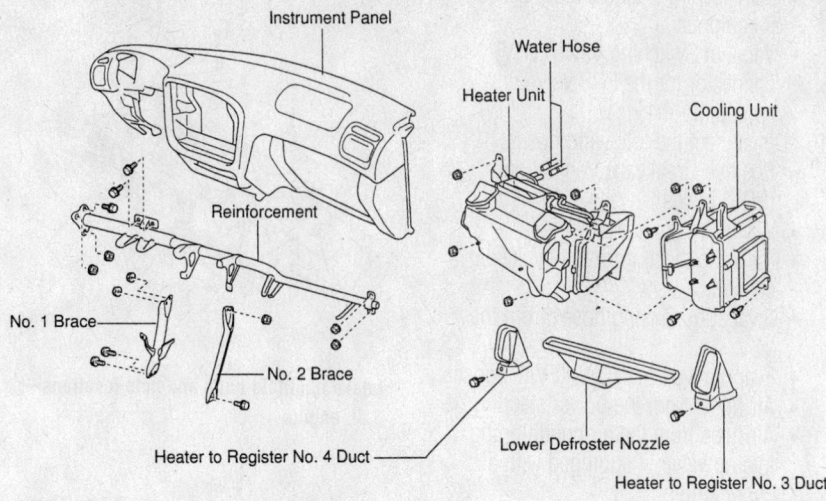

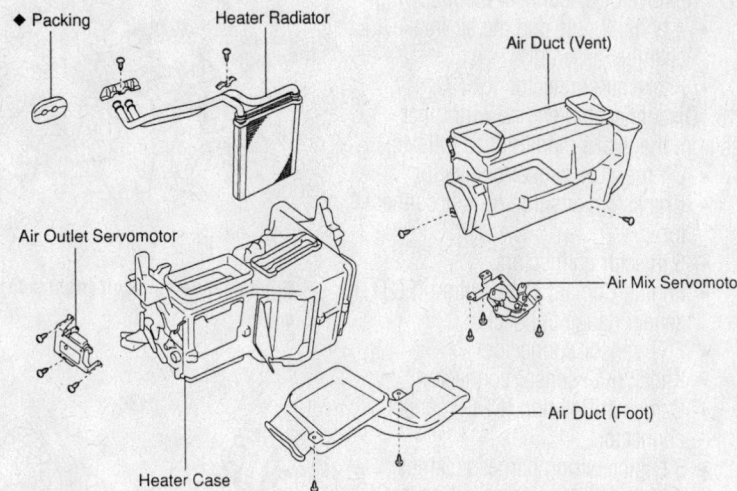

◆ Non-reusable part

93113GG0

Exploded view the heater core, heater housing, evaporator housing and related components

j. Install the rear console panel and engage the 4 clips (2 on each side); then, install the transfer shift lever knob.

k. Install the radio, connect the electrical connectors and the 4 radio screws.

l. Install the center cluster and engage the 8 center cluster clips.

m. Install the lower No. 2 panel and the 3 panel screws.

n. Install the passenger's side air bag module, connect it and install the No. 1 undercover.

o. Install the glove box door, the 2 screws and the glove compartment door stoppers.

p. Install the combination meter and the 4 screws; then, connect the electrical connectors.

q. Install the steering column and the steering column-to-instrument panel bolts.

r. Install the No. 2 heater-to-register duct and the duct screw.

s. At the steering column, install the combination switch, the 3 screws and the clamp; then, connect the electrical connectors.

t. Install the steering column covers and the 3 covers screws.

u. Install the lower left side panel.

v. Install the lower No. 1 panel and the panel screw.

w. Install the fuel lid control cable lever and the 2 screws.

x. Install the hood lock control cable and the 2 screws.

y. Install the instrument cluster finish panel.

z. At the passenger's side, install the front pillar garnish, the 2 assist grips, the 4 screws and the 4 assist grip plugs.

aa. At the driver's side, install the front pillar garnish, assist grip, the 2 screws and the 2 assist grip plugs.

bb. Install the front door scuff plates, the cowl side trim and the front door opening trim.

13. Install the steering wheel by performing the following procedure:

a. Install the steering wheel to the steering column.

b. Align the steering wheel-to-main shaft marks.

c. Install the steering wheel nut and torque to 25 ft. lbs. (34 Nm).

d. Install the air bag module to the steering wheel and connect the electrical connector.

e. Using a Torx® wrench, tighten the 2 screws located at each side of the steering wheel to 78 inch lbs. (8.8 Nm).

f. Install the steering wheel side covers.

14. Connect the heater hoses to the heater core.

15. Refill the cooling system.

16. Connect the negative battery cable.

a. Evacuate and charge the air conditioning system refrigerant.

17. Run the engine to normal operating temperatures; then, check the climate control operation and check for leaks.

Cylinder Head

REMOVAL & INSTALLATION

3.4L Engine

1. Before servicing the vehicle, refer to the precautions in the beginning of this section.

2. Disconnect the negative battery cable.

3. Relieve the fuel system pressure.

4. Remove the engine undercover.

5. Drain the cooling system.

6. Remove or disconnect the following:
   • Front exhaust pipe
   • Air cleaner cap
   • Mass Air Flow (MAF) meter and resonator

7. Disconnect the following cables:
   • Actuator cable from the bracket, if equipped with cruise control
   • Accelerator cable
   • Throttle cable, if equipped with an automatic transmission
   • Heater hose
   • Upper radiator hose
   • Power steering drive belt
   • Air conditioning drive belt, by loosening the idle pulley nut and adjusting bolt

- Loosen the lockbolt, pivot bolt and adjusting bolt and the alternator drive belt
- No. 2 fan shroud by removing the 2 clips
- Fan with the fluid coupling and fan pulleys
- Power steering pump and move it aside without disconnecting the pump lines
- Compressor and move it aside without disconnecting the compressor lines, if equipped with air conditioning
- Air conditioning bracket, if equipped with air conditioning
- Spark plug wires with the ignition coils
- Spark plugs
- No. 2 timing belt cover

8. Remove the fan bracket, as follows:

   a. Remove the power steering adjusting strut by removing the nut.

   b. Remove the fan bracket by removing the bolt and nut.

9. Set the No. 1 cylinder at Top Dead Center (TDC) of the compression stroke.

   a. Turn the crankshaft pulley and align its groove with the timing mark **0** on the No. 1 timing belt cover.

   b. Check that the timing marks of the camshaft timing pulleys and the No. 3 timing belt cover are aligned. If not, turn the crankshaft pulley 1 revolution (360 degrees).

10. Remove the timing belt tensioner by alternately loosening the 2 bolts.

11. Remove the camshaft timing pulleys, as follows:

    a. Using Variable Pin Wrench Set tool 09960-10010, remove the pulley bolt, the timing pulley and the knock pin.

    b. Remove the 2 timing pulleys with the timing belt.

12. Remove or disconnect the following:
- Bolt and the No. 2 idler pulley
- Alternator

13. Remove the Exhaust Gas Recirculation (EGR) pipe and 2 gaskets, if equipped with an EGR valve

14. Remove the intake chamber stay as follows:

    a. Remove the oil filler tube and No. 1 throttle cable clamp by removing the bolt and 2 nuts.

    b. Remove the intake chamber stay by removing the 2 bolts.

15. Remove the following connectors:
- VSV connector for the fuel pressure control.
- Throttle position sensor
- IAC valve connector

- EGR valve gas temperature sensor, if equipped
- Vacuum Switching Valve (VSV) connector for the EGR valve, if equipped

16. Disconnect the following hoses:
- Positive Crankcase Ventilation (PCV) hoses
- Water bypass hoses.
- Air assist hose from the intake air connector
- 2 vacuum sensing hoses from the VSV
- Evaporative Emissions (EVAP) hose
- Air hose, from the power steering
- Air hose from the air conditioning idle up valve, if equipped with air conditioning

17. Remove or disconnect the following:
- 4 bolts, 2 nuts and the air intake chamber assembly
- Intake air connector

18. Disconnect the engine wiring harness from the intake manifold, as follows:
- Oil pressure sensor connector
- Crankshaft position sensor connector
- 6 injector connectors
- Engine Coolant Temperature (ECT) sender gauge connector
- ECT sensor connector
- Knock (KS) sensor connector
- Camshaft Position (CMP) sensor connector
- 3 engine wiring harness clamps
- 3 bolts and the engine wiring harness from the cylinder head

19. Remove or disconnect the following:
- CMP sensor
- No. 3 (rear) timing belt cover, by removing the 6 bolts
- Fuel pressure regulator
- Intake manifold assembly
- Power steering pump bracket
- Oil dipstick and guide
- Exhaust crossover pipe and gaskets, by removing the 6 nuts

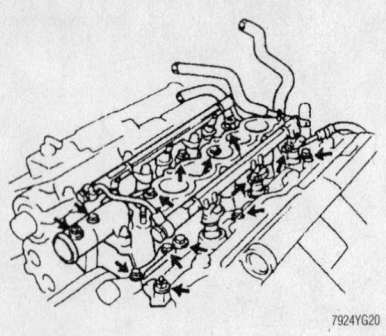

Intake manifold bolts and nuts locations—3.4L engine

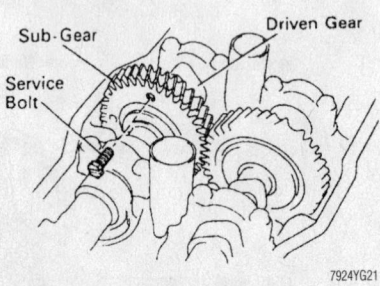

Drive gear service bolt (right side)—3.4L engine

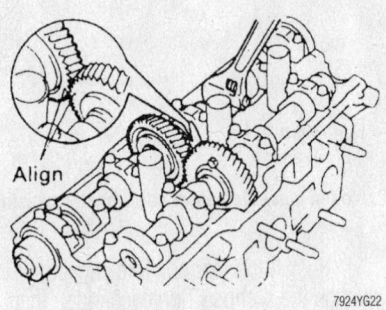

Aligning the timing mark (1 dot mark) of the left camshafts—3.4L engine

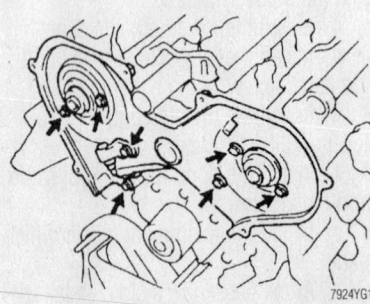

Rear timing belt cover bolt locations—3.4L engine

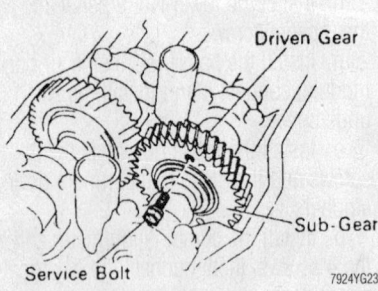

Drive gear service bolt (left side)—3.4L engine

- Left-hand exhaust manifold, by removing the heat insulator and 6 nuts
- Right-hand exhaust manifold, by removing the heat insulator and 6 nuts
- 8 bolts, seal washers, cylinder head cover and gasket
- Both cylinder head covers
- Semi-circular plugs
- Right exhaust camshafts
- Right-hand intake camshaft
- Left exhaust camshafts
- Left-hand intake camshaft
- Valve lifters and shims from the cylinder head; arrange the valve lifters and shims in correct order

20. Remove the cylinder heads, as follows:
- Bolt and disconnect the ground strap
- Cylinder head (recessed head) bolt on each cylinder head, using an 8mm hexagon wrench; then repeat for the other side
- 8 cylinder head (12-pointed head) bolts on each cylinder head, by loosening the bolts in several passes and in the reverse order of the tightening sequence
- 16 cylinder head bolts and plate washers.
- Cylinder head

To install:

21. Clean all surfaces.
22. Install or connect the following:
- New cylinder head gaskets
- Cylinder heads

23. Apply a light coat of engine oil on the threads and under the heads of the cylinder head bolts.

24. Tighten the cylinder head bolts, in sequence, using several passes, as follows:
- Step 1: 25 ft. lbs. (34 Nm)
- Step 2: Mark the front of the cylinder head bolt with paint
- Step 3: An additional 90 degrees
- Step 4: Check that the painted mark is now at a 90 degrees angle to the front

25. Install the recessed head cylinder head bolts, as follows:
- Apply a light coat of engine oil on the threads and under the heads of the cylinder head bolts
- Cylinder head bolt on each cylinder head, using a 8mm hexagon wrench; then, repeat for the other side, as shown. Tighten the bolts to 13 ft. lbs. (18 Nm).
- Bolt and the ground strap

26. Install or connect the following:
- Valve lifters and shims

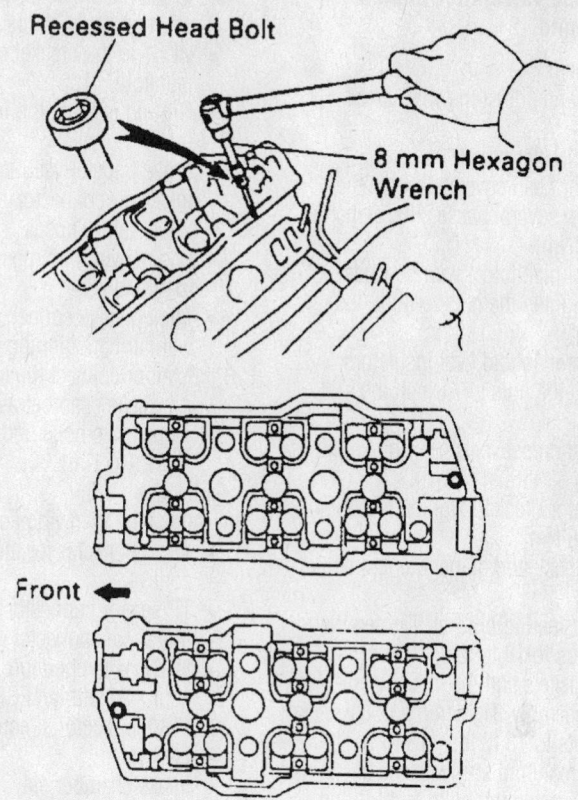

Front ◄—

7924YG24

Cylinder head recessed bolts—3.4L engine

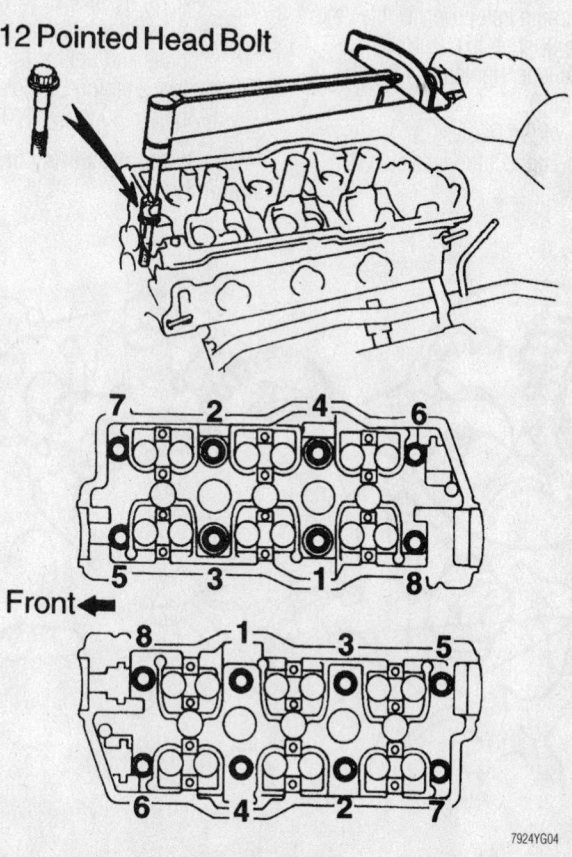

Front ◄—

7924YG04

Cylinder head torque sequence—3.4L (5VZ-FE) engine

➡️**Check that the valve lifter rotates smoothly by hand.**

- Camshafts
- Check and adjust the valve clearance
- Semi-circular plugs
- Cylinder head covers. Tighten the bolts, in several passes, to 53 inch lbs. (6 Nm).
- Exhaust manifolds, with new gaskets. Tighten the nuts to 30 ft. lbs. (40 Nm).
- Exhaust manifold heat insulators. Tighten the nuts to 71 inch lbs. (8 Nm).
- Exhaust crossover pipe. Tighten the nuts to 33 ft. lbs. (45 Nm).
- Alternator bracket. Tighten to 14 ft. lbs. (18 Nm).
- Oil dipstick and guide, using a new O-ring
- Power steering bracket. Tighten the fasteners to 14 ft. lbs. (18 Nm).
- New gaskets and the intake manifold assembly. Tighten the bolts and nuts to 13 ft. lbs. (18 Nm).
- Intake manifold stay with the 2 bolts. Tighten the bolts to 14 ft. lbs. (18 Nm).
- Fuel inlet hose
- Fuel pressure regulator
- No. 3 timing belt cover. Tighten the bolts to 80 inch lbs. (9 Nm).
- CMP sensor. Tighten to 71 inch lbs. (8 Nm).
- Engine wiring harness

27. Install or connect the intake air connector, as follows:

- Intake manifold. Tighten the bolts and nuts to 14 ft. lbs. (19 Nm).
- DLC1 to the bracket on the intake manifold
- Ground strap to the intake manifold
- Brake booster vacuum hose to the intake air connector
- 2 fuel return hoses
- Engine wiring harness to the intake manifold
- Idle up valve connector, if equipped with air conditioning

28. Install or connect the following:

- Air intake chamber assembly. Tighten the bolts and nuts to 14 ft. lbs. (18.5 Nm).
- Hoses

29. Attach the following connectors:

- VSV connector for the fuel pressure control
- TP sensor connector
- IAC valve connector
- EGR gas temperature connector, if equipped with an EGR valve
- VSV connector, if equipped with an EGR valve
- Intake chamber stay
- New gaskets and the EGR pipe. Tighten the clamp nuts to 71 inch lbs. (8 Nm) and the EGR pipe nuts to 14 ft. lbs. (18 Nm).
- Alternator but do not tighten the bolts and nuts at this time.
- No. 2 timing belt idler. Tighten the bolt to 30 ft. lbs. (40 Nm).

➡️**Check that the pulley bracket moves smoothly.**

- Left camshaft timing pulley

30. Set the No. 1 cylinder to TDC of the compression stroke, as follows:

a. Turn the crankshaft pulley and align its groove with the timing mark **0** on the No. 1 timing belt cover.

b. Turn the camshaft, align the knock pin hole of the camshaft with the timing mark of the No. 3 timing belt cover.

c. Turn the camshaft timing pulley, align the timing marks of the camshaft timing pulley and the No. 3 timing belt cover.

31. Connect the timing belt to the left camshaft timing pulley, as follows:

a. Check that the installation mark on the timing belt is aligned with the end of the No. 1 timing belt cover.

b. Using Variable Pin Wrench Set 09960-01000, slightly turn the left camshaft timing pulley clockwise. Align the installation mark on the timing belt with the timing mark of the camshaft timing pulley and hang the timing belt on the left camshaft timing pulley.

c. Align the timing marks of the left camshaft pulley and the No. 3 timing belt cover.

d. Check that the timing belt has tension between the crankshaft timing pulley and the left camshaft timing pulley.

32. Install the right camshaft timing pulley and the timing belt, as follows:

a. Align the installation mark on the timing belt with the timing mark of the right camshaft timing pulley, and hang the timing belt on the right camshaft timing pulley with the flange side facing inward.

b. Slide the right camshaft timing pulley on the camshaft. Align the timing marks on the right camshaft timing pulley and the No. 3 timing belt cover.

c. Align the knock pin hole of the camshaft with the knock pin groove of the pulley and install the knock pin. Install the bolt and tighten to 81 ft. lbs. (110 Nm).

33. Set the timing belt tensioner as follows:

a. Using a press, slowly press in the pushrod using 220–2,205 lbs. (981–9,807 N) of force.

b. Align the holes of the pushrod and housing, pass a 1.5mm hexagon wrench through the holes to keep the setting position of the pushrod.

c. Release the press and install the dust boot to the tensioner.

34. Install the timing belt tensioner. Tighten the bolts alternately to 20 ft. lbs. (28 Nm).

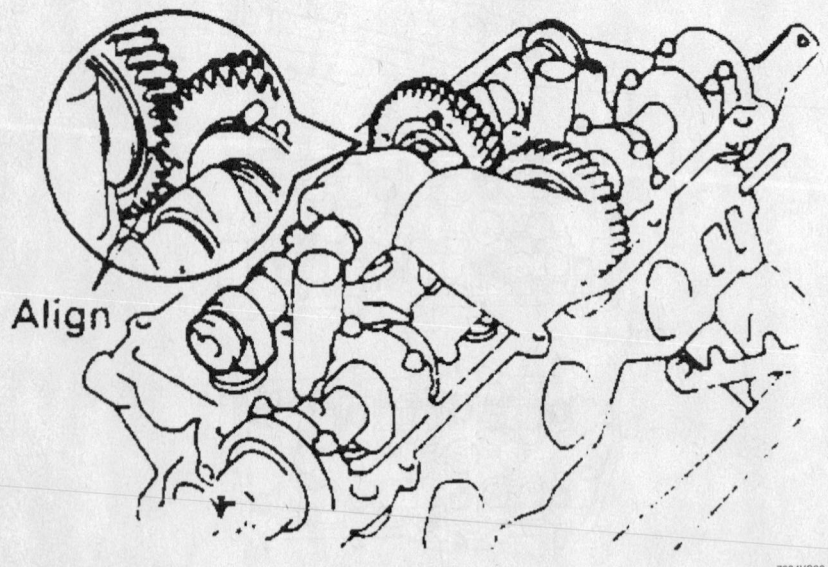

Align

Aligning the right camshafts for installation—3.4L engine

- Timing belt
- Camshaft sprockets
- Camshaft Position (CMP) sensor
- Power steering pump
- Exhaust front pipes
- Transmission dipstick tube
- Ignition coils

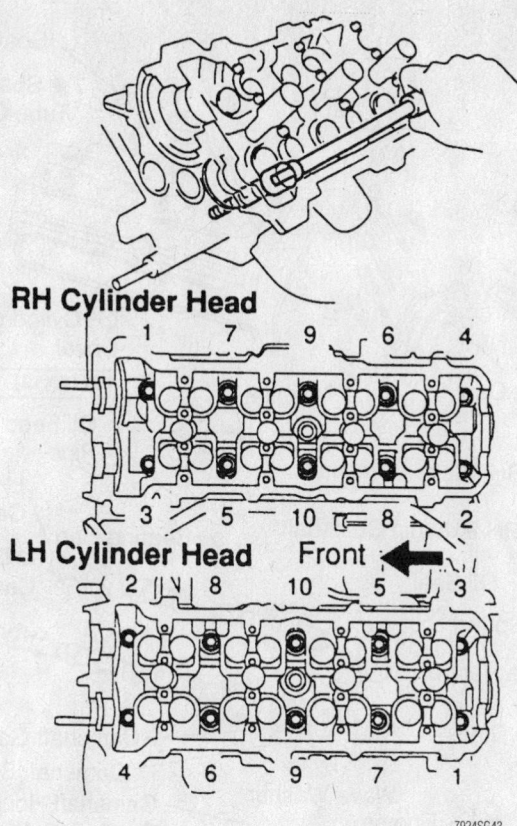

Cylinder head loosening sequence—4.7L engine

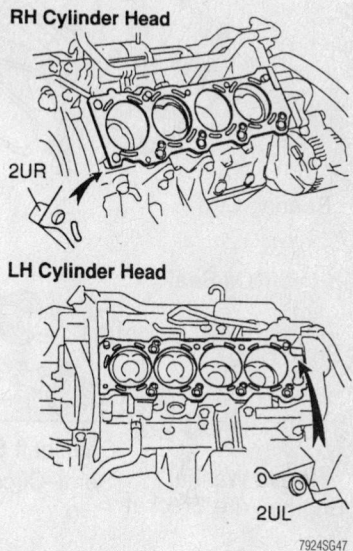

Cylinder head gasket identification—4.7L engine

35. Using pliers, remove the 1.5mm hexagon wrench from the belt tensioner.
36. Check the valve timing.
37. Install or connect the following:
   - Remaining components
   - Negative battery cable
38. Fill the radiator with engine coolant.
39. Start the engine and check for leaks.
40. Check the ignition timing.
41. Install the engine undercover.
42. Road test the vehicle.
43. Recheck all fluid levels.

4.7L Engine

1. Before servicing the vehicle, refer to the precautions in the beginning of this section.
2. Drain the cooling system.
3. Relieve the fuel system pressure.
4. Remove or disconnect the following:
   - Battery and tray
   - Engine appearance cover
   - Engine under covers
   - Air intake assembly
   - Accessory drive belt
   - A/C compressor and bracket
   - Cooling fan and bracket
   - Radiator
   - Idler pulley
   - Front covers

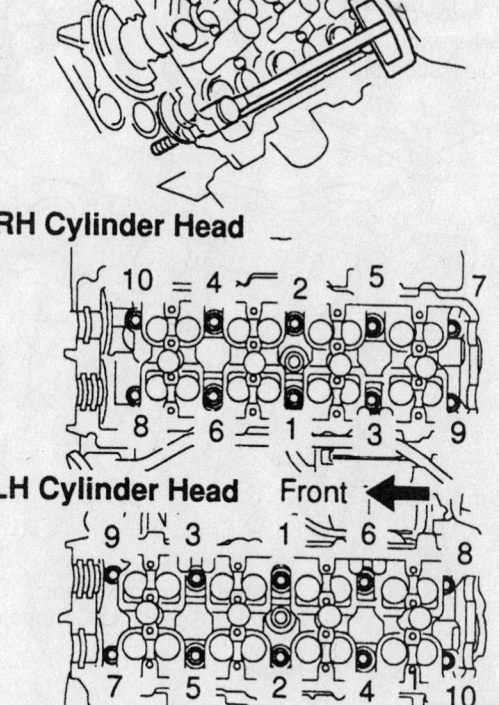

Cylinder head torque sequence—4.7L (2UZ-FE) engine

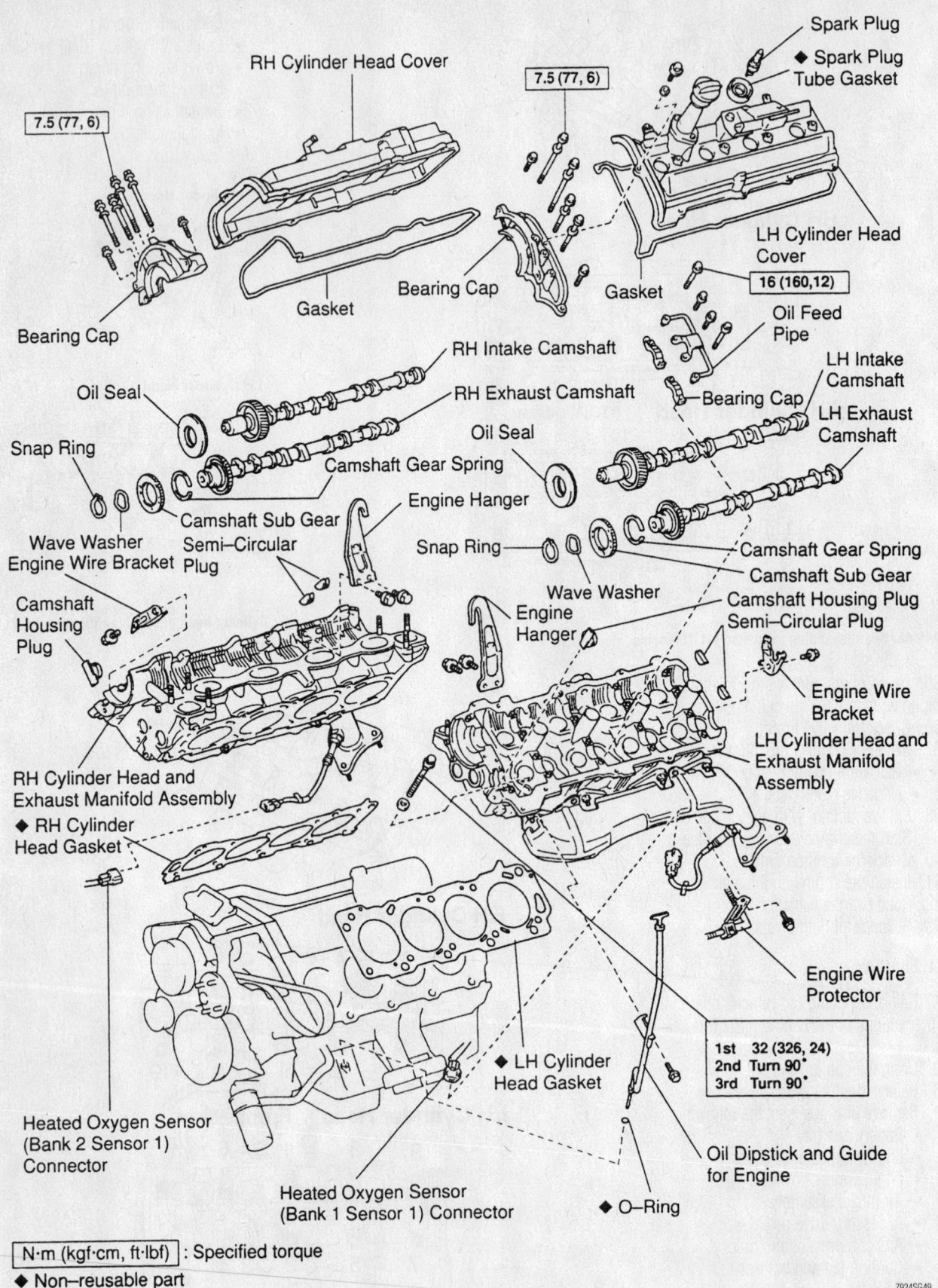

Spark Plug

◆ Spark Plug
Tube Gasket

RH Cylinder Head Cover

7.5 (77, 6)

7.5 (77, 6)

LH Cylinder Head
Cover

16 (160,12)

Bearing Cap

Gasket

Gasket

Bearing Cap

Oil Feed
Pipe

Bearing Cap

RH Intake Camshaft

RH Exhaust Camshaft

LH Intake
Camshaft

LH Exhaust
Camshaft

Oil Seal

Oil Seal

Snap Ring

Camshaft Gear Spring

Engine Hanger

Camshaft Sub Gear

Semi–Circular
Plug

Snap Ring

Camshaft Gear Spring

Camshaft Sub Gear

Wave Washer
Engine Wire Bracket

Wave Washer

Camshaft Housing Plug

Engine
Hanger

Semi–Circular Plug

Camshaft
Housing
Plug

Engine Wire
Bracket

LH Cylinder Head and
Exhaust Manifold
Assembly

RH Cylinder Head and
Exhaust Manifold Assembly

◆ RH Cylinder
Head Gasket

Engine Wire
Protector

1st 32 (326, 24)
2nd Turn 90°
3rd Turn 90°

◆ LH Cylinder
Head Gasket

Heated Oxygen Sensor
(Bank 2 Sensor 1)
Connector

Oil Dipstick and Guide
for Engine

Heated Oxygen Sensor
(Bank 1 Sensor 1) Connector

◆ O–Ring

N·m (kgf·cm, ft·lbf) : Specified torque

◆ Non–reusable part

7924SG49

Exploded view of the cylinder head mounting—4.7L engine

- Rear timing belt covers
- Fuel lines
- Intake manifold
- Water inlet housing assembly
- Front and rear water bypass joints
- Engine lifting eyes
- Oil dipstick tube
- Valve covers
- Camshafts
- Cylinder heads with the exhaust manifolds attached. Loosen the bolts in the sequence shown.

To install:

5. Install the cylinder heads with new gaskets. Tighten the bolts in sequence as follows:
   a. Step 1: 24 ft. lbs. (32 Nm)
   b. Step 2: Plus 180 degrees
6. Install or connect the following:
   - Camshafts
   - Valve covers
   - Oil dipstick tube
   - Engine lifting eyes
   - Front and rear water bypass joints
   - Water inlet housing assembly
   - Intake manifold
   - Fuel lines
   - Rear timing belt covers
   - Ignition coils
   - Transmission dipstick tube
   - Exhaust front pipes
   - Power steering pump
   - CMP sensor
   - Camshaft sprockets
   - Timing belt
   - Front covers
   - Idler pulley
   - Radiator
   - Cooling fan and bracket
   - A/C compressor and bracket
   - Accessory drive belt
   - Air intake assembly
   - Engine under covers
   - Engine appearance cover
   - Battery and tray
7. Fill the cooling system.
8. Start the engine and check for leaks.

Intake Manifold

REMOVAL & INSTALLATION

3.4L Engine

1. Before servicing the vehicle, refer to the precautions in the beginning of this section.
2. Disconnect the negative battery cable.
3. Relieve the fuel system pressure.
4. Remove the engine undercover.

5. Drain the cooling system.
6. Remove or disconnect the following:
   - Air cleaner cap
   - Mass Air Flow (MAF) meter and the resonator
   - Actuator cable with the bracket, if equipped with cruise control
   - Accelerator cable
   - Throttle cable, if equipped with automatic transmission
7. Disconnect the following hoses:
   - Heater hose
   - Brake booster vacuum hose
   - Evaporative Emissions (EVAP) hose
   - Automatic Disconnecting Differential (ADD) vacuum hose, for 4-Wheel drive
   - Fuel inlet and fuel return hose
8. Remove or disconnect the following:
   - Spark plug wires, with the ignition coils
   - Intake chamber stay
   - No. 2 timing belt cover
   - Air intake chamber assembly
   - Throttle Position (TP) sensor connector
   - Idle Air Control (IAC) valve connector
   - Positive Crankcase Ventilation (PCV) hoses
   - Water bypass hoses
   - Air assist hose from the throttle body
   - Intake air connector
   - Engine wiring harness
   - Fuel return hose
   - Vacuum hose, from the fuel pressure regulator
   - Ground strap, from the intake air connector

- Data Link Connector 1 (DLC1), from the bracket
- 6 injector connectors
- Engine Coolant Temperature (ECT) sensor and sender gauge connectors
- Engine wiring harness protector from the cylinder head
- Fuel pressure regulator
- Intake manifold assembly
- Intake manifold stay
- Intake manifold, delivery pipes and the injectors assembly with the gaskets

To install:

9. Install or connect the following:
   - New gaskets
   - Intake manifold assembly. Tighten the bolts and nuts to 13 ft. lbs. (18 Nm).
   - Intake manifold stay. Tighten the bolts to 14 ft. lbs. (18 Nm).
   - Fuel pressure regulator
   - Engine wiring harness to the cylinder head, by installing the 3 bolts
   - 3 engine wiring harness clamps
   - 6 injector connectors
   - ECT sender gauge connector
   - ECT sensor connector
   - Intake manifold. Tighten the bolts and nuts to 14 ft. lbs. (18.5 Nm).
   - DLC1 to the bracket on the intake manifold
   - Ground strap to the intake manifold, by installing the bolt
   - Brake booster vacuum hose, to the intake air connector
   - 2 fuel return hoses
   - Engine wiring harness to the intake manifold

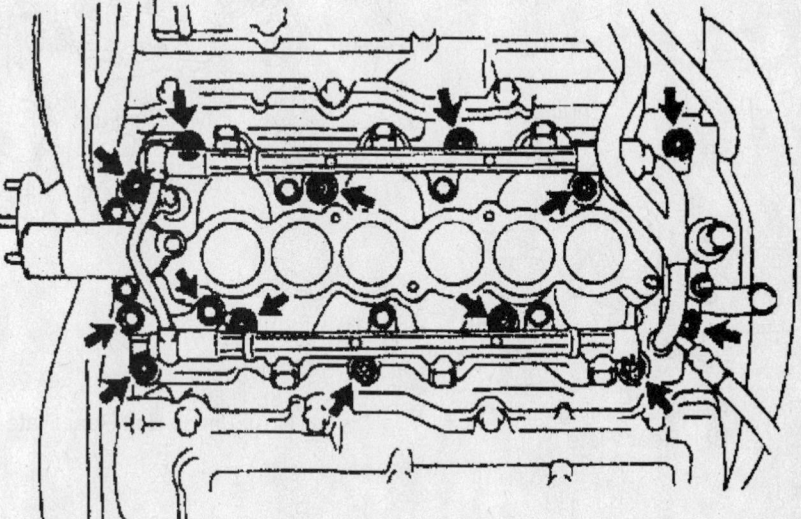

Intake manifold bolts and nuts—3.4L engine

7924YG38

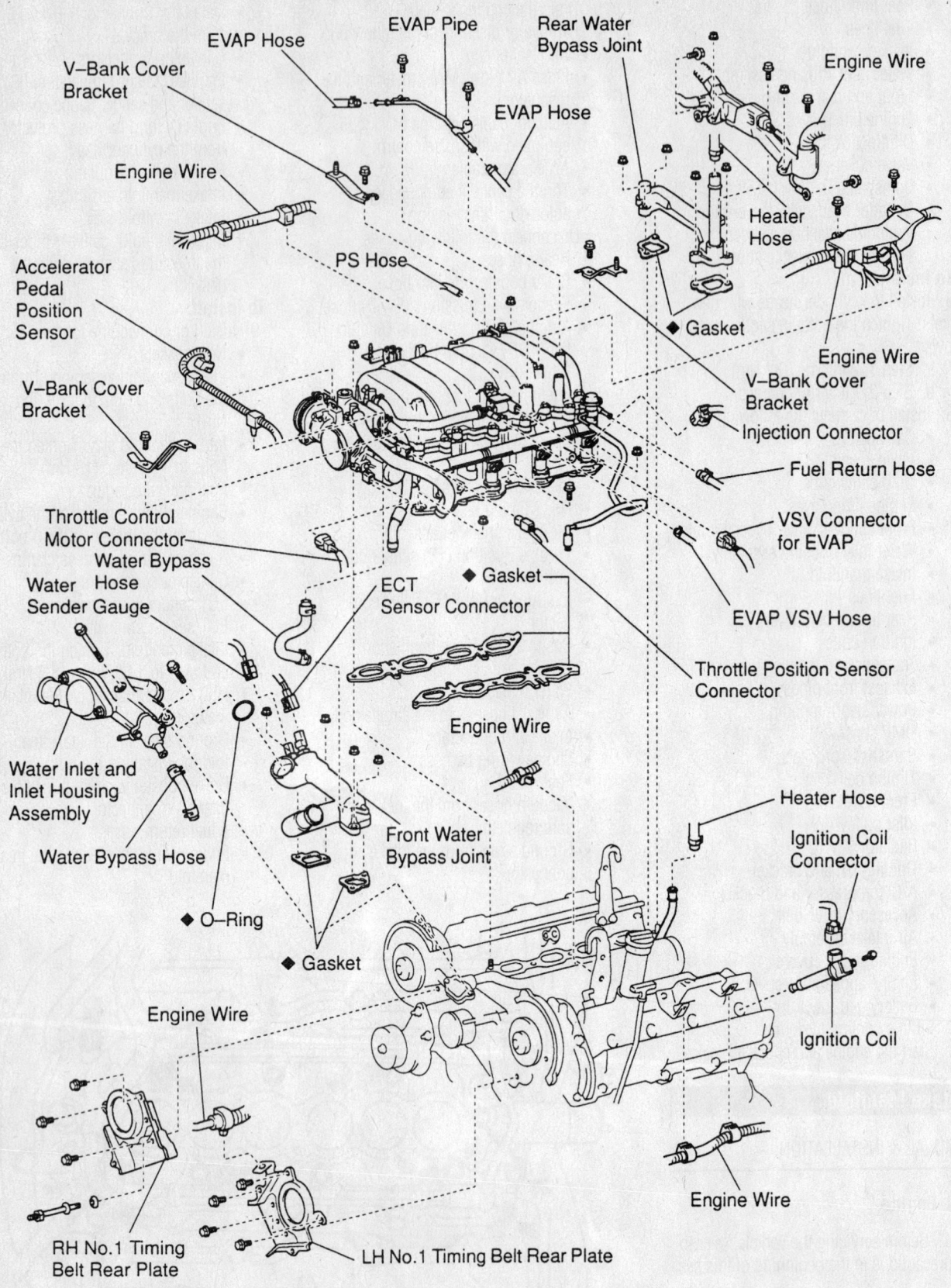

EVAP Hose

EVAP Pipe

Rear Water
Bypass Joint

Engine Wire

V-Bank Cover
Bracket

EVAP Hose

Engine Wire

Heater
Hose

Accelerator
Pedal
Position
Sensor

PS Hose

Gasket

Engine Wire
V-Bank Cover
Bracket

V-Bank Cover
Bracket

Injection Connector

Throttle Control
Motor Connector
Water Bypass
Hose

Fuel Return Hose

VSV Connector
for EVAP

Water
Sender Gauge

ECT
Sensor Connector

Gasket

EVAP VSV Hose

Throttle Position Sensor
Connector

Water Inlet and
Inlet Housing
Assembly

Engine Wire

Heater Hose

Water Bypass Hose

Front Water
Bypass Joint

Ignition Coil
Connector

O-Ring

Gasket

Ignition Coil

Engine Wire

Engine Wire

RH No.1 Timing
Belt Rear Plate

LH No.1 Timing Belt Rear Plate

◆ Non-reusable part

7924SG50

Exploded of the intake manifold mounting—4.7L engine

- Air intake chamber assembly to the engine. Tighten the bolts and nuts to 14 ft. lbs. (18.5 Nm).
- Intake chamber stay. Tighten the bolts to 30 ft. lbs. (40 Nm).
- New O-ring to the oil filler tube
- Oil filler tube end into the tube hole in the oil pan
- Oil filler tube and No. 1 throttle cable clamp
- No. 2 timing belt cover. Tighten the bolts to 80 inch lbs. (9 Nm).
- PCV hoses
- Water bypass hoses
- Air assist hose to the throttle body
- IAC valve connector
- TP sensor connector
- Brake booster vacuum hose
- EVAP hose
- Automatic Disconnecting Differential (ADD) vacuum hose, for 4-Wheel drive
- Fuel inlet and fuel return hose
- 3 spark plug wire clamps to the No. 2 timing belt cover
- CMP connector to the No. 2 timing belt cover
- Spark plug wires with the ignition coils
- Heater hose
- Actuator cable with the bracket, if equipped with cruise control
- Accelerator cable
- Throttle cable, if equipped with automatic transmission
- MAF meter, resonator and the air cleaner cap
- Negative battery cable

10. Fill the radiator with engine coolant.
11. Start the engine and check for leaks.
12. Install the engine undercover.
13. Road test the vehicle.
14. Recheck all fluid levels.

4.7L Engine

1. Before servicing the vehicle, refer to the precautions in the beginning of this section.
2. Drain the cooling system.
3. Relieve the fuel system pressure.
4. Remove or disconnect the following:
- Negative battery cable
- Engine appearance cover
- Accelerator cable
- Throttle Position (TP) sensor connector
- Accelerator pedal position sensor
- Throttle motor connector
- Evaporative Emissions (EVAP) vacuum switching valve connector
- Fuel injector connectors

- Engine Coolant Temperature (ECT) sensor connector
- ETC gauge sender connector
- Heated Oxygen (HO$_2$S) sensor connectors
- Fuel pressure regulator vacuum hose
- Positive Crankcase Ventilation (PCV) valve and hose
- EVAP hoses
- Power steering vacuum hoses
- Water bypass hose
- Engine control wiring harness clamps
- Cylinder head ground cables
- Intake manifold wire harness protector
- EVAP pipe
- Engine appearance cover brackets
- Intake manifold

To install:

5. Install or connect the following:
- Intake manifold. Tighten the fasteners to 13 ft. lbs. (18 Nm).
- Engine appearance cover brackets
- EVAP pipe
- Intake manifold wire harness protector
- Cylinder head ground cables
- Engine control wiring harness clamps
- Water bypass hose
- Power steering vacuum hoses
- EVAP hoses
- PCV valve and hose
- Fuel pressure regulator vacuum hose
- HO$_2$S sensor connectors
- ETC gauge sender connector
- ECT sensor connector
- Fuel injector connectors
- EVAP vacuum switching valve connector
- Throttle motor connector
- Accelerator pedal position sensor
- TP sensor connector
- Accelerator cable
- Engine appearance cover
- Negative battery cable

6. Fill the cooling system.
7. Start the engine and check for leaks.

Exhaust Manifold

REMOVAL & INSTALLATION

3.4L Engine

1. Before servicing the vehicle, refer to the precautions in the beginning of this section.

Exhaust crossover pipe mounting nut locations—3.4L engine

2. Remove or disconnect the following:
- Exhaust crossover pipe, from the exhaust manifold by removing the 3 nuts
- Exhaust Gas Recirculation (EGR) pipe, from the exhaust manifold, on the left manifold equipped with an EGR valve
- Exhaust manifold heat insulator, by removing the 3 nuts
- Exhaust manifold

To install:

3. Install or connect the following:
- Exhaust manifold, using a new gasket. Tighten the nuts to 30 ft. lbs. (40 Nm).
- Exhaust heat insulator. Tighten the nuts to 71 inch lbs. (8 Nm).
- EGR pipe to the exhaust manifold, if equipped with an EGR valve. Tighten the manifold nuts to 14 ft. lbs. (18 Nm) and the clamp nuts to 71 inch lbs. (8 Nm).
- Crossover pipe to the exhaust manifold, using a new gasket. Tighten the nuts to 33 ft. lbs. (45 Nm).

4.7L Engine

1. Before servicing the vehicle, refer to the precautions in the beginning of this section.
2. Attach a hoist to the engine lifting eyes.
3. Remove or disconnect the following:
- Negative battery cable
- Heated Oxygen (HO$_2$S) sensor connectors
- Exhaust manifold heat shield
- Exhaust front pipe
- Motor mount
- Motor mount bracket
- Exhaust manifold

To install:

➡ Use new exhaust manifold nuts for assembly.

4. Install or connect the following:
- Exhaust manifold. Tighten the nuts to 32 ft. lbs. (44 Nm).

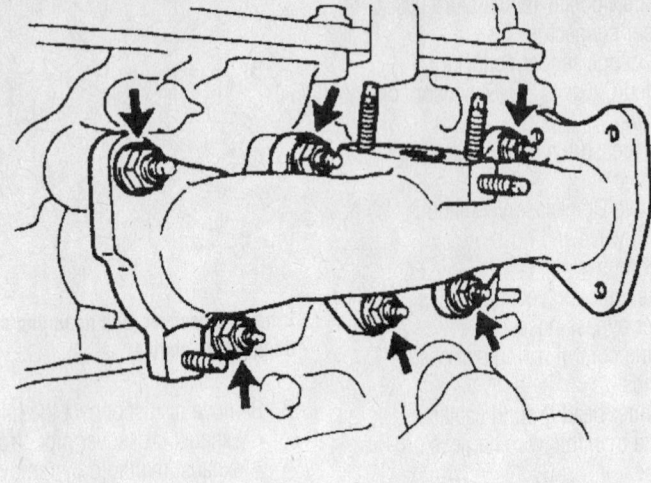

7924YG44

Exhaust manifold nuts—3.4L engine

- Motor mount bracket. Tighten the bolts to 27 ft. lbs. (36 Nm).
- Motor mount. Tighten the fasteners to 22 ft. lbs. (30 Nm).
- Exhaust front pipe. Tighten the nuts to 46 ft. lbs. (62 Nm).
- Exhaust manifold heat shield
- HO2S sensor connectors
- Negative battery cable

5. Start the engine and check for leaks.

Front Crankshaft Seal

REMOVAL & INSTALLATION

3.4L Engine

➡There are 2 methods to replace the oil seal, which are as follows:

OIL PUMP BODY INSTALLED

1. Before servicing the vehicle, refer to the precautions in the beginning of this section.

2. Remove or disconnect the following:

- Negative battery cable
- Timing belt and crankshaft pulley
- Cut off the oil seal lip, using a knife
- Pry out the oil seal, using a suitable tool

✳✳ WARNING

Be careful not to damage the crankshaft.

To install:

3. Install or connect the following:

- Apply multi-purpose grease to the new oil seal lip
- Tap in the new oil seal until its surface is flush with the oil pump case edge, using Seal Driver tool 09309-37010 and a mallet
- Crankshaft pulley and the timing belt
- Engine undercover, if removed
- Negative battery cable

OIL PUMP BODY REMOVED

1. Before servicing the vehicle, refer to the precautions in the beginning of this section.

2. Carefully pry out the seal using a suitable tool.

3. Apply multi-purpose grease to the new oil seal lip.

4. Using Seal Driver tool 09309-37010, drive the new seal into place.

4.7L Engine

1. Before servicing the vehicle, refer to the precautions in the beginning of this section.

2. Drain the cooling system.

3. Remove or disconnect the following:

- Negative battery cable
- Engine under cover
- Engine appearance cover
- Air intake assembly
- Accessory drive belt
- Cooling fan and pulley
- Radiator
- Drive belt idler pulley
- Camshaft Position (CMP) sensor connector
- Upper timing covers
- Oil cooler pipe
- Center timing cover
- A/C compressor
- Cooling fan bracket
- Crankshaft pulley
- Lower timing cover
- Timing belt
- Crankshaft timing sprocket
- Front crankshaft seal

To install:

4. Install the oil seal so that it is flush with the oil pump housing.

5. Install or connect the following:

- Crankshaft timing sprocket
- Timing belt
- Lower timing cover
- Crankshaft pulley. Tighten the bolt to 181 ft. lbs. (245 Nm).
- Cooling fan bracket. Tighten the 12mm bolts to 12 ft. lbs. (16 Nm) and the 14mm bolts to 24 ft. lbs. (32 Nm).
- A/C compressor
- Center timing cover
- Oil cooler pipe
- Upper timing covers
- CMP sensor connector
- Drive belt idler pulley. Tighten the bolt to 27 ft. lbs. (37 Nm).
- Radiator
- Cooling fan and pulley. Tighten the nuts to 16 ft. lbs. (21 Nm).
- Accessory drive belt
- Air intake assembly
- Engine appearance cover

- Engine under cover
- Negative battery cable

6. Fill the cooling system.
7. Start the engine and check for leaks.

Camshaft and Valve Lifters

REMOVAL & INSTALLATION

3.4L Engine

1. Before servicing the vehicle, refer to the precautions in the beginning of this section.
2. Remove or disconnect the following:
   - Negative battery cable
   - Engine undercover
3. Drain the cooling system.
4. Remove or disconnect the following:
   - Air cleaner cap
   - Mass Air Flow (MAF) meter and the resonator
   - Actuator cable with the bracket, if equipped with cruise control
   - Accelerator cable
   - Throttle cable, if equipped with an automatic transmission
   - Heater hose
   - Upper radiator hose from the engine
5. Remove the power steering drive belt, as follows:
   a. Stretch the belt and loosen the fan pulley mounting nuts.
   b. Loosen the lockbolt, pivot bolt and adjusting bolt and remove the drive belt from the engine.
6. Remove or disconnect the following:
   - Air conditioning drive belt, by loosening the idle pulley nut and adjusting bolt
   - Loosen the alternator lockbolt, pivot bolt and adjusting bolt
   - Alternator drive belt
   - No. 2 fan shroud, by removing the 2 clips
   - Fan with the fluid coupling and fan pulleys
   - Power steering pump and move it aside without disconnecting the lines
   - Compressor and move it aside without disconnecting the lines, if equipped with air conditioning
   - Air conditioning bracket, if equipped with air conditioning
   - Spark plug wires with the ignition coils
   - Spark plugs
   - Camshaft Position (CMP) sensor connector, from the No. 2 timing belt cover
   - 3 spark plug wire clamps from the No. 2 timing belt cover
   - 6 bolts and the timing belt cover
   - Power steering adjusting strut by removing the nut
   - Fan bracket by removing the bolt and nut

7. Set the No. 1 cylinder at Top Dead Center (TDC) of the compression stroke, as follows:
   a. Turn the crankshaft pulley and align its groove with the timing mark **0** on the No. 1 timing belt cover.
   b. Check that the timing marks of the camshaft timing pulleys and the No. 3 timing belt cover are aligned. If not, turn the crankshaft pulley 1 revolution (360 degrees).

8. Remove or disconnect the following:
   - Timing belt tensioner, by alternately loosening the 2 bolts
   - Pulley bolt, the timing pulley and the knock pin, using Variable Pin Wrench Set 09960-10010
   - Both timing pulleys with the timing belt
   - No. 2 idler pulley
   - Alternator
   - Positive Crankcase Ventilation PCV hoses
   - Water bypass hoses
   - Air assist hose from the intake air connector
   - 2 vacuum sensing hoses from the Vacuum Switching Valve (VSV)
   - Evaporative Emissions (EVAP) hose
   - Air hose from the power steering
   - Air hose from the air conditioning idle up valve, if equipped with air conditioning
   - 4 bolts, 2 nuts and the air intake chamber assembly
   - Intake air connector
   - Camshaft Position (CMP) sensor
   - No. 3 (rear) timing belt cover by removing the 6 bolts
   - 8 bolts, seal washers, both cylinder head cover and gaskets
   - Semi-circular plugs

9. Remove the right exhaust camshafts, as follows:
   a. Bring the service bolt hole of the driven sub-gear upward by turning the hexagon head portion of the exhaust camshaft with a wrench.
   b. Align the timing mark (2 dot marks) of the camshaft drive and driven gears by turning the camshaft with a wrench.
   c. Secure the exhaust camshaft sub- gear to the driven gear with a service bolt (6mm diameter, 16–20mm bolt length and 1.0mm in thread pitch).

➡ **When removing the camshaft, be sure the torsional spring force of the sub-gear has been eliminated by the above operation.**

   d. Uniformly loosen and remove the bearing cap bolts in several passes, in the sequence shown.
   e. Remove the bearing caps and camshaft. Make a note of the bearing cap positions for proper installation.

➡ **Do not pry on or attempt to force the camshaft with a tool or other object.**

10. Remove the right-hand intake camshaft, as follows:
    a. Uniformly loosen and remove the bearing cap bolts in several passes, in the sequence shown.
    b. Remove the bearing caps, oil seal and camshaft. Make a note of the bearing cap positions for proper installation.
11. Remove the left exhaust camshafts, as follows:
    a. Align the timing mark (1 dot mark) of the camshaft drive and driven gears by turning the camshaft with a wrench.
    b. Secure the exhaust camshaft sub- gear to the driven gear with a service bolt (6mm diameter, 16–20mm bolt length and 1.0mm in thread pitch).

➡ **When removing the camshaft, be sure that the torsional spring force of the sub-gear has been eliminated by the above operation.**

    c. Uniformly loosen and remove the bearing cap bolts in several passes, in the sequence shown.
    d. Remove the bearing caps and camshaft. Make a note of the bearing cap positions for proper installation.
12. Remove the left-hand intake camshaft, as follows:
    a. Uniformly loosen and remove the bearing cap bolts in several passes, in the sequence shown.
    b. Remove the bearing caps, oil seal and camshaft. Make a note of the bearing cap positions for proper installation.
13. Remove the valve lifters and shims from the cylinder head. Arrange the valve lifters and shims in correct order.

To install:
14. Clean all surfaces.
15. Install the valve lifters and shims. Check that the valve lifter rotates smoothly by hand.

16. Install the right intake camshaft, as follows:

    a. Apply engine oil to the thrust portion of the intake camshaft.

    b. Position the intake camshaft at 90 degrees angle of the timing mark (2 dot marks) on the cylinder head.

    c. Install the bearing caps in their proper locations. Apply a light coat of engine oil to the threads and install the cap bolts.

    d. Apply a light coat of engine oil on the threads and under the heads of the bearing cap bolts.

    e. Uniformly tighten the cap bolts in the sequence shown to 12 ft. lbs. (16 Nm).

17. Install the right exhaust camshaft, as follows:

    a. Apply engine oil to the thrust portion of the intake camshaft.

    b. Align the timing marks (2 dot marks) of the camshaft drive and driven gears.

    c. Roll down the exhaust camshaft onto the bearing journals while engaging the gears with each other. Install the bearing caps in their proper locations.

    d. Apply a light coat of engine oil to the threads and install the cap bolts.

    e. Apply a light coat of engine oil on the threads and under the heads of the bearing cap bolts.

    f. Uniformly tighten the cap bolts in the sequence shown to 12 ft. lbs. (16 Nm).

    g. Remove the service bolt from the driven sub-gear. Check that the intake and exhaust camshafts turns smoothly.

    h. Align the timing marks (2 dot marks) of the camshaft drive and driven gears by turning the camshaft with a wrench.

18. Install the left intake camshaft, as follows:

    a. Apply engine oil to the thrust portion of the intake camshaft.

    b. Position the intake camshaft at 90 degrees angle of the timing mark (1 dot mark) on the cylinder head.

    c. Install the bearing caps in their proper locations. Apply a light coat of engine oil to the threads and install the cap bolts.

    d. Apply a light coat of engine oil on the threads and under the heads of the bearing cap bolts.

    e. Uniformly tighten the cap bolts in the sequence shown to 12 ft. lbs. (16 Nm).

19. Install the left exhaust camshaft, as follows:

    a. Apply engine oil to the thrust portion of the intake camshaft.

    b. Align the timing marks (1 dot mark) of the camshaft drive and driven gears.

    c. Roll down the exhaust camshaft onto the bearing journals while engaging the gears with each other. Install the bearing caps in their proper locations.

    d. Apply a light coat of engine oil to the threads and install the cap bolts.

    e. Apply a light coat of engine oil on the threads and under the heads of the bearing cap bolts.

    f. Uniformly tighten the cap bolts in the sequence shown to 12 ft. lbs. (16 Nm).

    g. Remove the service bolt.

20. Check and adjust the valve clearance.

21. Install or connect the following:
- Semi-circular plugs
- Cylinder head covers. Tighten the bolts, in several passes, to 53 inch lbs. (6 Nm).
- Alternator bracket. Tighten the bolts to 14 ft. lbs. (18 Nm).
- No. 3 timing belt cover. Tighten the 6 bolts to 80 inch lbs. (9 Nm).
- CMP sensor. Tighten it to 71 inch lbs. (8 Nm).
- Intake air connector
- Hoses
- Alternator but do not tighten the bolts and nuts at this time
- No. 2 timing belt idler. Tighten the bolt to 30 ft. lbs. (40 Nm).

➡**Check that the pulley bracket moves smoothly.**

22. Install the left camshaft timing pulley, as follows:

    a. Install the knock pin to the camshaft.

    b. Align the knock pin hose of the camshaft with the knock pin groove of the timing pulley.

    c. Slide the timing pulley on the camshaft with the flange side facing outward. Tighten the pulley bolt to 81 ft. lbs. (110 Nm).

23. Set the No. 1 cylinder to TDC of the compression stroke.

    a. Turn the crankshaft pulley and align its groove with the timing mark **0** on the No. 1 timing belt cover.

    b. Turn the camshaft, align the knock pin hole of the camshaft with the timing mark of the No. 3 timing belt cover.

    c. Turn the camshaft timing pulley, align the timing marks of the camshaft timing pulley and the No. 3 timing belt cover.

24. Connect the timing belt to the left camshaft timing pulley, as follows:

    a. Check that the installation mark on the timing belt is aligned with the end of the No. 1 timing belt cover.

    b. Using Variable Pin Wrench Set 09960-01000 or equivalent, slightly turn the left camshaft timing pulley clockwise. Align the installation mark on the timing belt with the timing mark of the camshaft timing pulley, and hang the timing belt on the left camshaft timing pulley.

    c. Align the timing marks of the left camshaft pulley and the No. 3 timing belt cover.

    d. Check that the timing belt has tension between the crankshaft timing pulley and the left camshaft timing pulley.

25. Install the right camshaft timing pulley and the timing belt, as follows:

    a. Align the installation mark on the timing belt with the timing mark of the right camshaft timing pulley, and hang the timing belt on the right camshaft timing pulley with the flange side facing inward.

    b. Slide the right camshaft timing pulley on the camshaft. Align the timing marks on the right camshaft timing pulley and the No. 3 timing belt cover.

    c. Align the knock pin hole of the camshaft with the knock pin groove of the pulley.

    d. Install the knock pin. Tighten the bolt to 81 ft. lbs. (110 Nm).

26. Set the timing belt tensioner, as follows:

    a. Using a press, slowly press in the pushrod using 220–2,205 lbs. (981–9,807 N) of force.

    b. Align the holes of the pushrod and housing, pass a 1.5mm hexagon wrench through the holes to keep the setting position of the pushrod.

    c. Release the press and install the dust boot to the tensioner.

27. Install the timing belt tensioner and alternately tighten the bolts to 20 ft. lbs. (28 Nm). Using pliers, remove the 1.5mm hexagon wrench from the belt tensioner.

28. Check the valve timing, as follows:

    a. Slowly turn the crankshaft pulley 2 revolutions from the TDC-to-TDC. Always turn the crankshaft pulley clockwise.

    b. Check that each pulley aligns with the timing marks. If the timing marks do not align, remove the timing belt and reinstall it.

29. Install or connect the following:
- Fan bracket with the bolt and nut

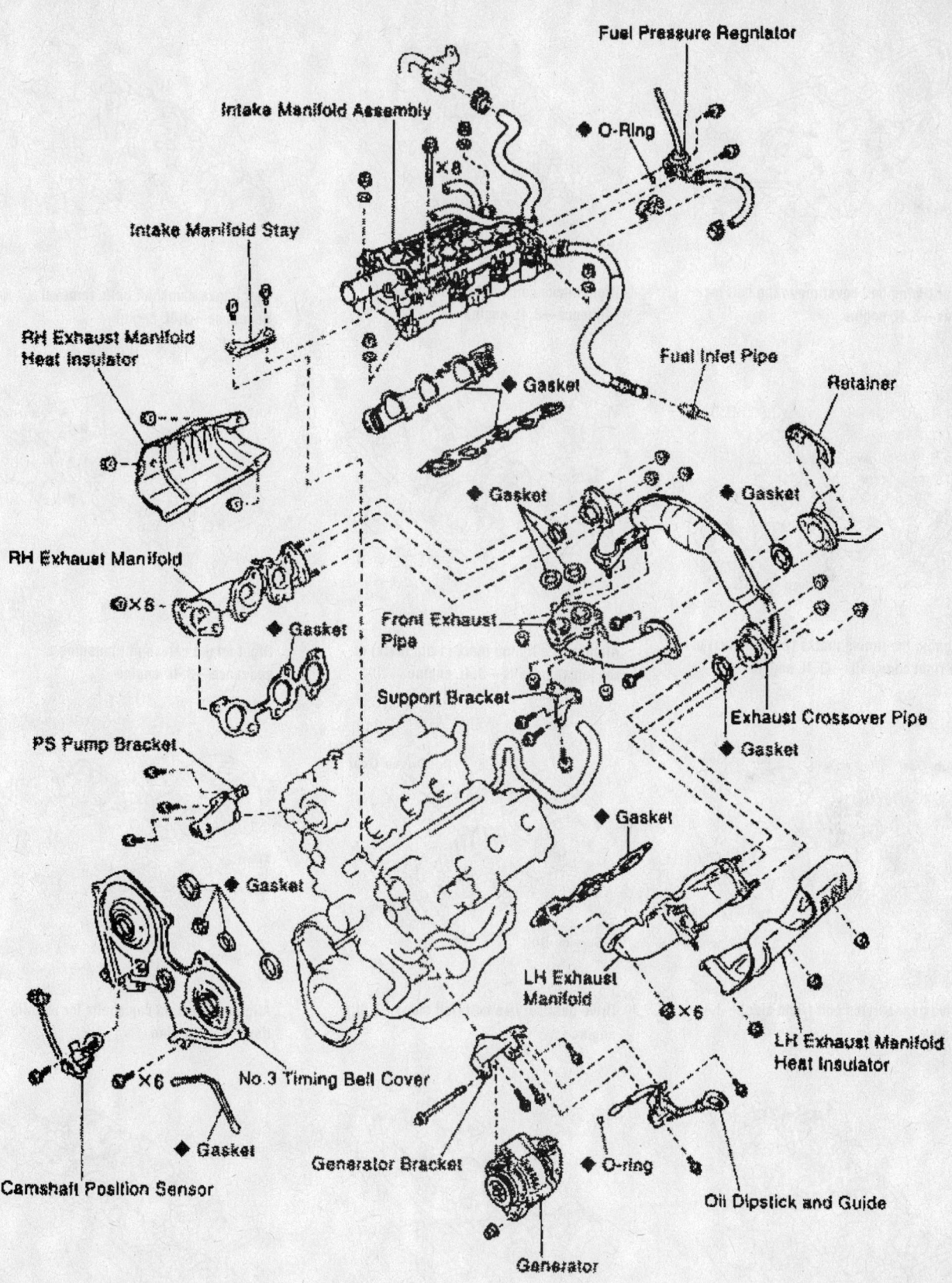

Fuel Pressure Regniator

Intake Manifold Assembly

◆ O-Ring

Intake Manifold Stay

×8

RH Exhaust Manifold
Heat Insulator

◆ Gasket

Fuel Inlet Pipe

Retainer

◆ Gasket

◆ Gasket

RH Exhaust Manifold

◆ Gasket

×6

Front Exhaust
Pipe

◆ Gasket

Exhaust Crossover Pipe

PS Pump Bracket

Support Bracket

◆ Gasket

◆ Gasket

LH Exhaust
Manifold

×6

LH Exhaust Manifold
Heat Insulator

No.3 Timing Bell Cover

×6

◆ Gasket

Camshaft Position Sensor

Generator Bracket

◆ O-ring

Oil Dipstick and Guide

Generator

◆ Non-reusable part

Exploded view of the cylinder head component assembly—3.4L engine

7924YG60

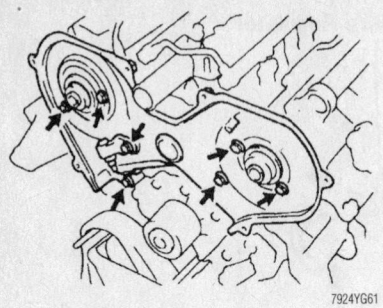

7924YG61

Rear timing belt cover mounting bolt locations—3.4L engine

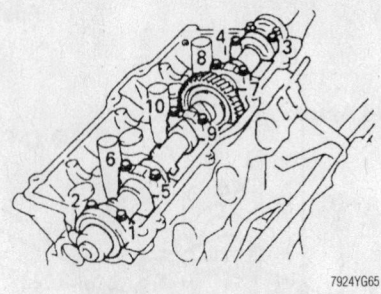

7924YG65

Right intake camshaft bolts removal sequence—3.4L engine

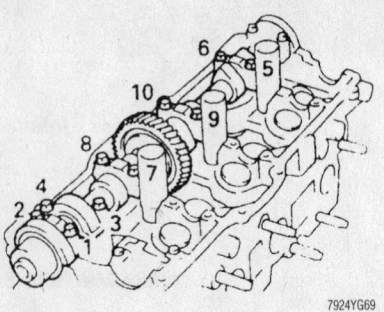

7924YG69

Left intake camshaft bolts removal sequence—3.4L engine

RH

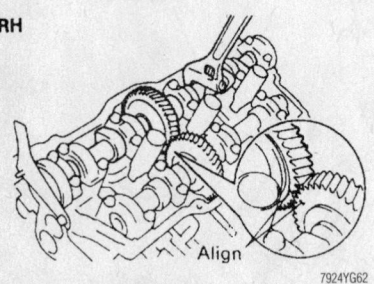

7924YG62

Aligning the timing marks (2 dot marks) of the right camshafts—3.4L engine

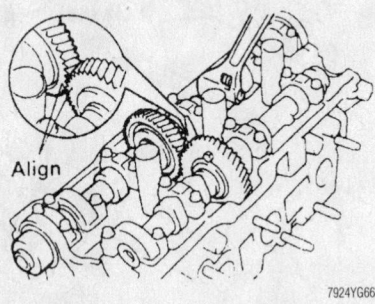

7924YG66

Aligning the timing mark (1 dot mark) of the left camshafts—3.4L engine

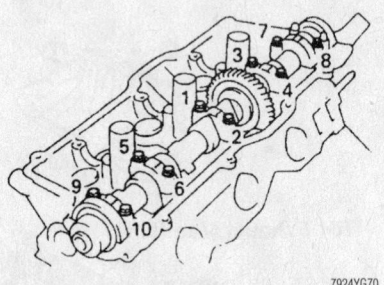

7924YG70

Right intake camshaft tightening sequence—3.4L engine

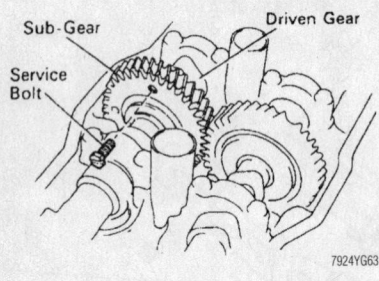

7924YG63

Drive gear service bolt (right side)—3.4L engine

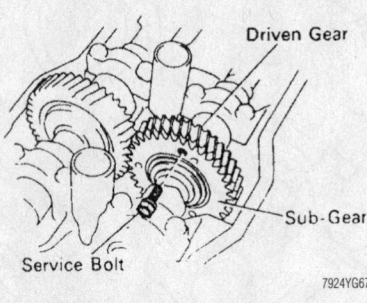

7924YG67

Drive gear service bolt (left side)—3.4L engine

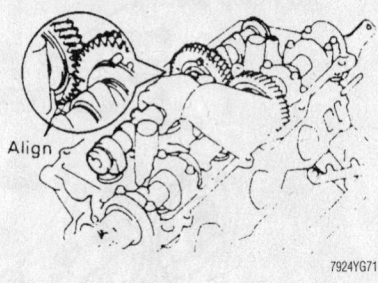

7924YG71

Aligning the right camshafts for installation—3.4L engine

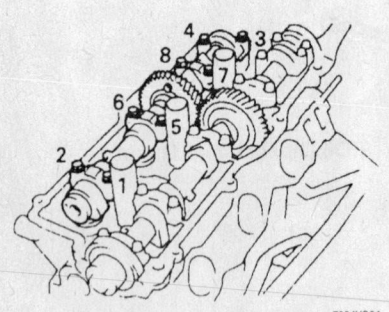

7924YG64

Right exhaust camshaft bolts removal sequence—3.4L engine

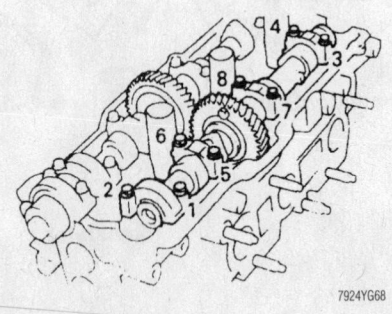

7924YG68

Left exhaust camshaft bolts removal sequence—3.4L engine

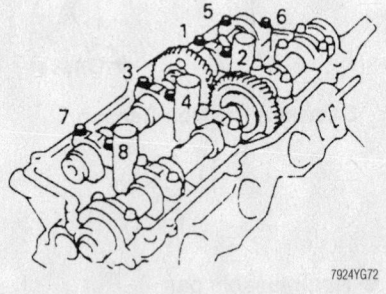

7924YG72

Right exhaust camshaft bolts tightening sequence—3.4L engine

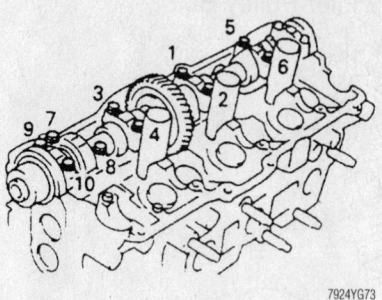

Left intake camshaft bolts tightening sequence—3.4L engine

Left exhaust camshaft bolts tightening sequence—3.4L engine

- Power steering adjusting strut with the nut
- No. 2 timing belt cover. Tighten the bolts to 80 inch lbs. (9 Nm).
- Negative battery cable
30. Fill the radiator with engine coolant.
31. Start the engine and check for leaks.
32. Check the ignition timing.
33. Install the engine undercover.
34. Road test the vehicle.
35. Recheck all fluid levels.

4.7L Engine

1. Before servicing the vehicle, refer to the precautions in the beginning of this section.

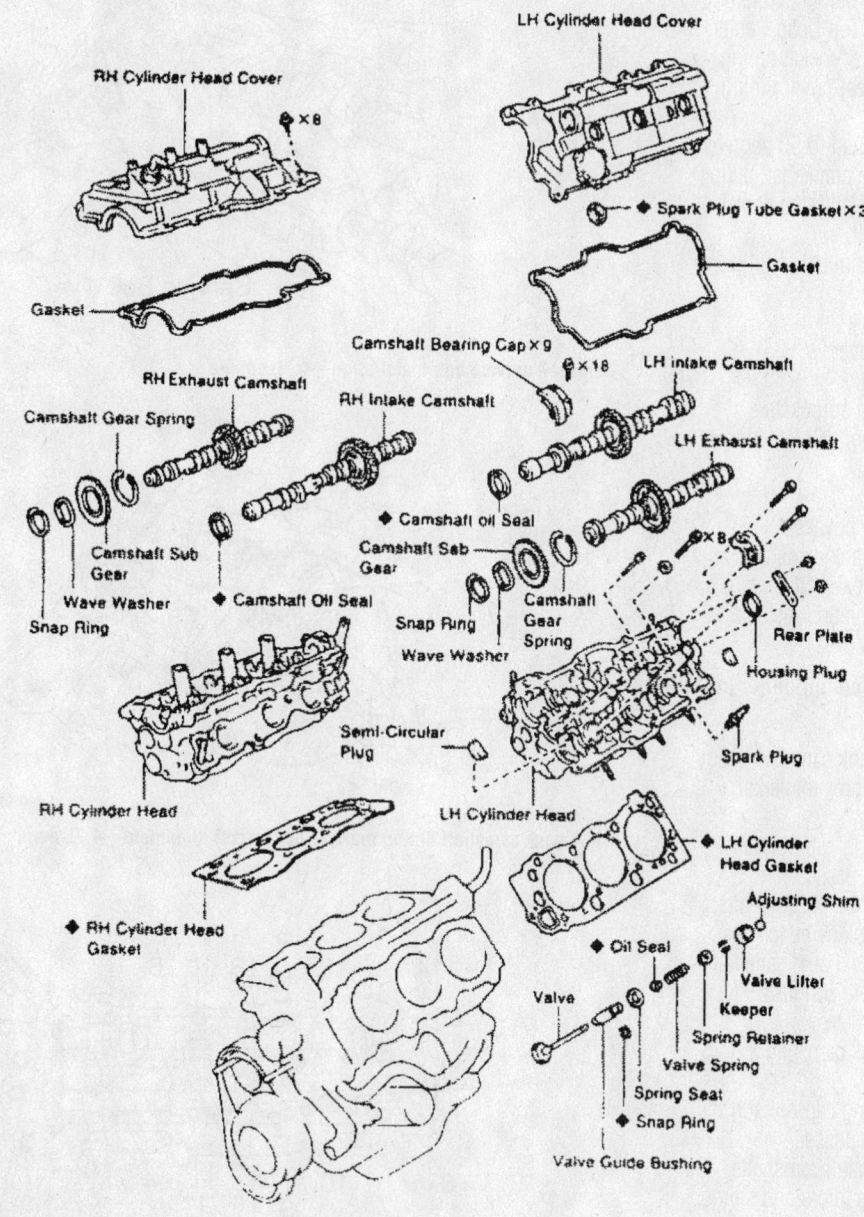

♦ Non-reusable part

Exploded view of the cylinder head component assembly—3.4L engine

2. Drain the cooling system.
3. Relieve the fuel system pressure.
4. Remove or disconnect the following:
- Negative battery cable
- Engine under covers
- Engine appearance cover
- Air intake hose
- Accessory drive belt
- Cooling fan
- Radiator
- Idler pulley
- Upper and middle timing belt covers
- A/C compressor
- Cooling fan bracket
- Alternator
- Accessory drive belt tensioner

5. Set the engine to Top Dead Center (TDC) with the camshaft sprocket timing marks aligned with the rear cover timing marks.

6. Rotate the crankshaft to 50 degrees After TDC as shown. The crankshaft pulley timing mark should align with the center of the No. 2 idler pulley bolt.

7. Remove or disconnect the following:
- Crankshaft pulley
- Lower timing cover
- Timing belt
- Camshaft timing sprockets
- Camshaft Position (CMP) sensor
- Ignition coils
- Valve cover
- Timing belt rear covers

8. Rotate the right bank camshafts as necessary to access the exhaust camshaft sub-gear service bolt hole and install a 6mm x 1.0mm bolt.

➡ **Keep all valvetrain components in order for assembly.**

9. Align the right bank camshaft 1 dot timing marks to a **10** degree angle as shown.

10. Loosen the bearing cap bolts in sequence and in several passes.

11. Remove the right bank camshafts.

12. Rotate the left bank camshafts as necessary to access the exhaust camshaft sub-gear service bolt hole and install a 6mm x 1.0mm bolt.

13. Align the left bank camshaft 2 dot timing marks as shown.

14. Loosen the bearing cap bolts in sequence and in several passes.

15. Remove the left bank camshafts.

16. Remove the valve lifters and shims.

To install:

17. Ensure that the crankshaft is at 50 degrees After TDC.

18. Install or connect the following:

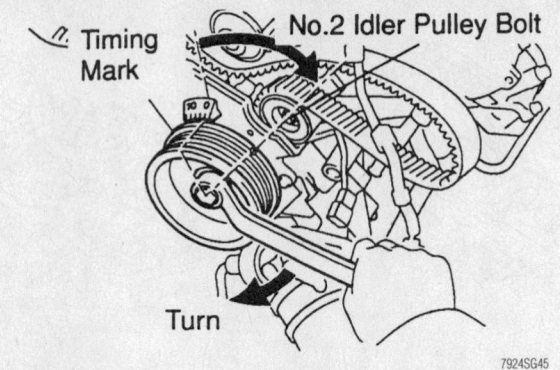

Setting the crankshaft to 50 degrees ATDC—4.7L engine

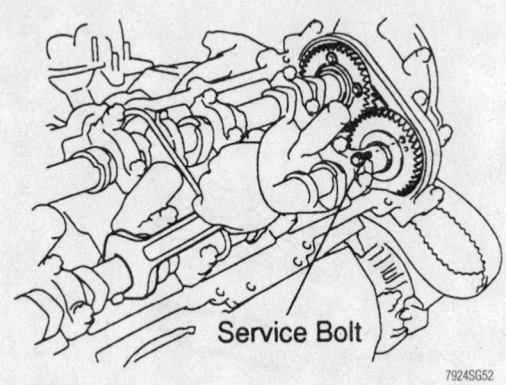

Camshaft service bolt installation—4.7L engine

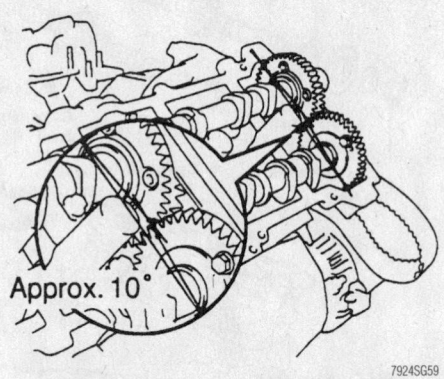

Right bank camshaft timing mark (1 dot marks) alignment—4.7L engine

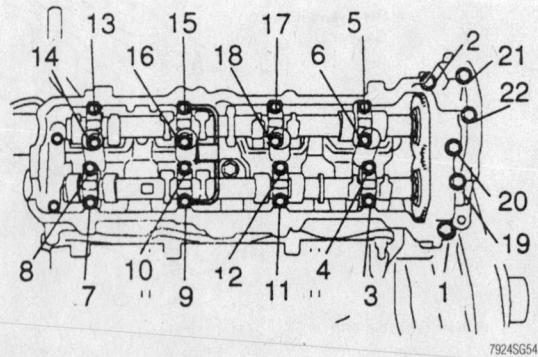

Right bank camshaft bearing cap loosening sequence—4.7L engine

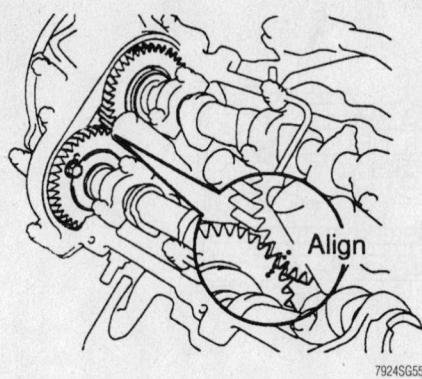

Left bank camshaft timing mark (2 dot marks) alignment—4.7L engine

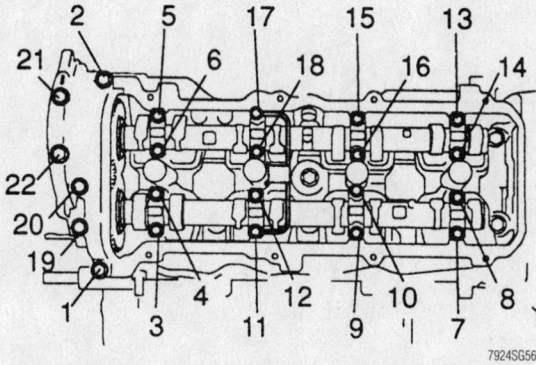

Left bank camshaft bearing cap loosening sequence—4.7L engine

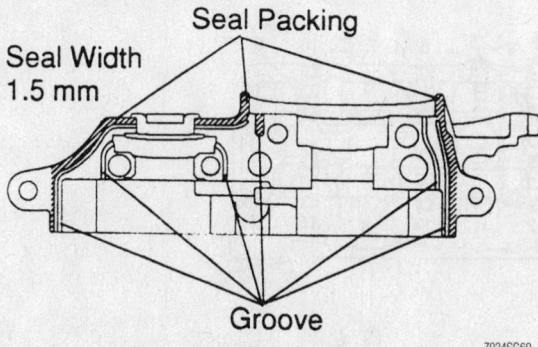

Apply a 1.5mm bead of sealant to the front bearing caps—4.7L engine

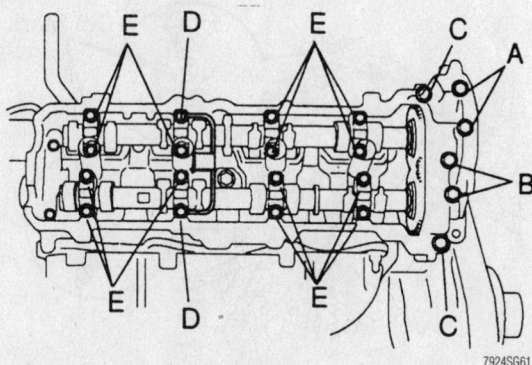

Right bank bearing cap bolt location—4.7L engine

- Valve lifters and shims in their original positions
- Right bank camshafts with the 1 dot timing marks at 10 degrees
- Left bank camshafts with the 2 dot timing marks aligned
- Left and right bank camshaft bearing caps in their original positions. Apply sealant to the front bearing caps as shown.
- Camshaft oil seals

19. The bearing cap bolts vary in length and are identified as follows:
- A: 3.70 inches (94mm)
- B: 2.83 inches (72mm)
- C: 0.98 inches (25mm)
- D: 2.05 inches (52mm)
- E: 1.50 inches (38mm)

20. Bolts in positions **A**, **B** and **C** are installed dry.

21. Lubricate the threads and under the contact flange for bolts in positions **D** and **E**.

22. Install oil feed pipes and the bearing cap bolts according to position in the illustrations.

23. Tighten the camshaft bearing bolts in sequence and in several passes to the following specifications:
- Bolt C: 66 inch lbs. (7.5 Nm)
- All others: 12 ft. lbs. (16 Nm)

24. Remove the service bolts from the exhaust camshaft gears.

25. Install or connect the following:
- Timing belt rear covers
- Valve cover
- Ignition coils
- CMP sensor
- Camshaft timing sprockets. Tighten the bolts to 80 ft. lbs. (108 Nm).
- Timing belt
- Lower timing cover
- Crankshaft pulley. Tighten the bolt to 181 ft. lbs. (245 Nm).
- Accessory drive belt tensioner
- Alternator
- Cooling fan bracket
- A/C compressor
- Upper and middle timing belt covers
- Idler pulley. Tighten the bolt to 27 ft. lbs. (37 Nm).
- Radiator
- Cooling fan
- Accessory drive belt
- Air intake hose
- Engine appearance cover
- Engine under covers
- Negative battery cable

26. Fill the cooling system.

27. Start the engine and check for leaks.

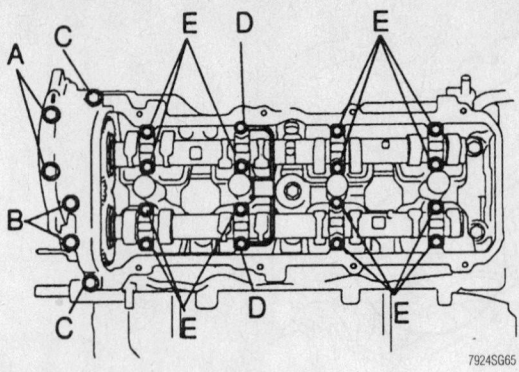

Left camshaft bearing cap bolt locations—4.7L engine

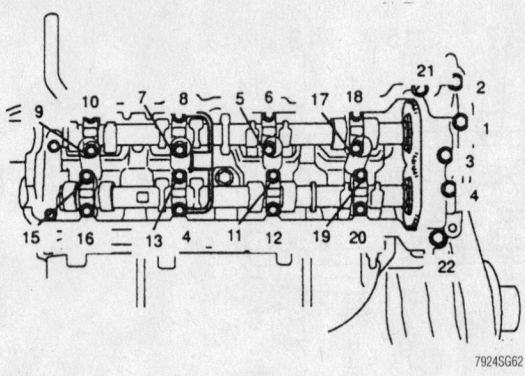

Right bank camshaft bearing cap bolt torque sequence—4.7L engine

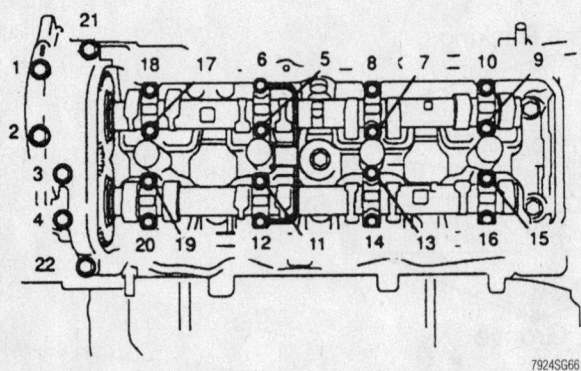

Left bank camshaft bearing cap bolt torque sequence—4.7L engine

Valve Lash

ADJUSTMENT

3.4L Engine

1. Before servicing the vehicle, refer to the precautions in the beginning of this section.
2. Disconnect the negative battery cable.
3. Drain the engine coolant.
4. Remove or disconnect the following:
   • Air intake connector

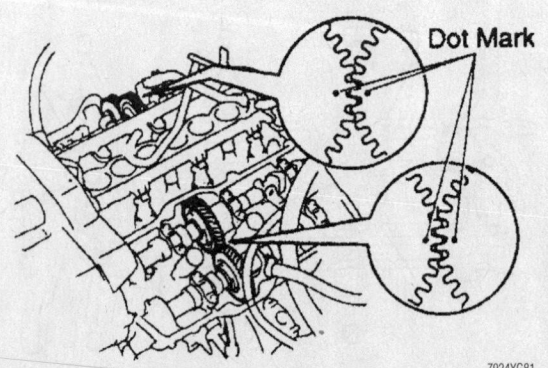

Aligning the timing marks—3.4L engine

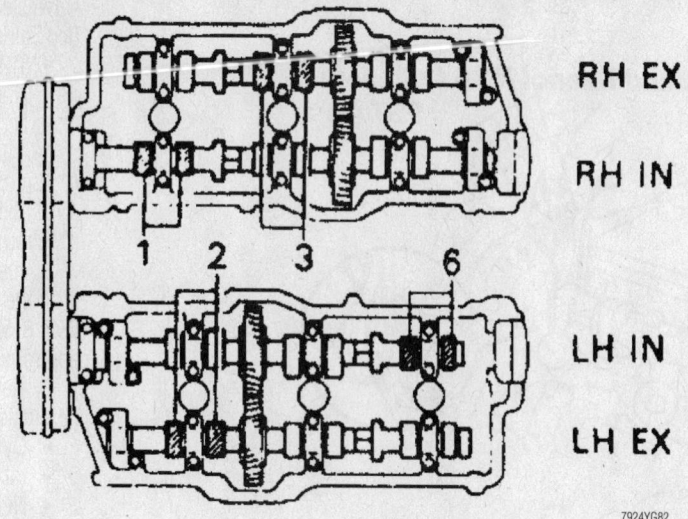

RH EX

RH IN

1   2   3   6

LH IN

LH EX

7924YG82

First valve adjustment—3.4L engine

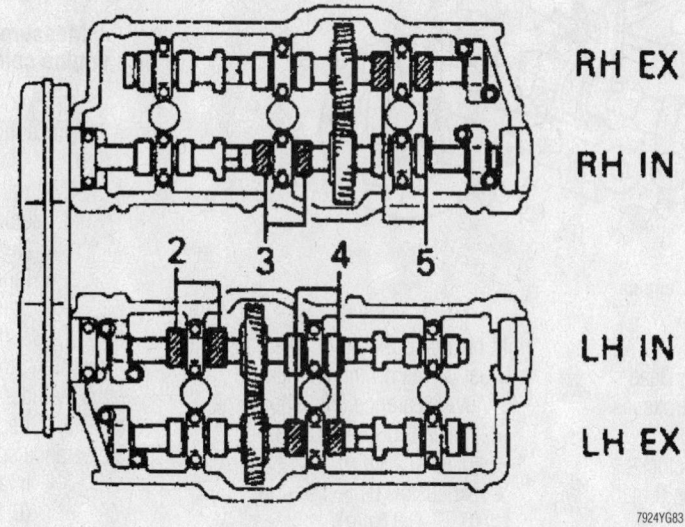

RH EX

RH IN

2   3   4   5

LH IN

LH EX

7924YG83

Second valve adjustment—3.4L engine

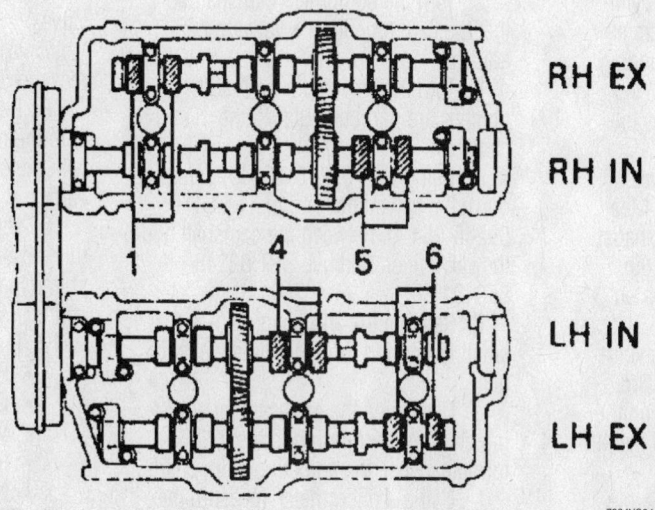

RH EX

RH IN

1   4   5   6

LH IN

LH EX

7924YG84

Third valve adjustment—3.4L engine

Front of No.1 and Rear of No.6 Cylinders

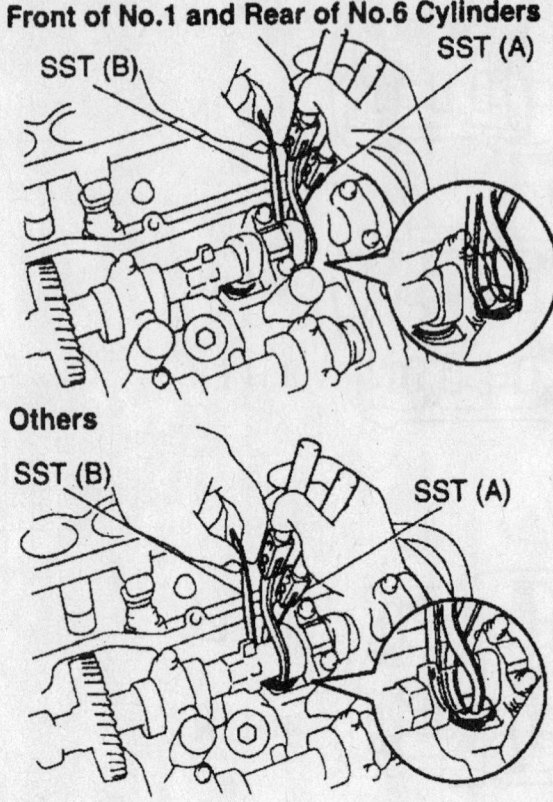

SST (B)   SST (A)

Others

SST (B)   SST (A)

7924YG85

Removing the adjusting shim—3.4L engine

• Cylinder head cover

5. Set the No. 1 cylinder to Top Dead Center (TDC) of the compression stroke, as follows:

   a. Turn the crankshaft pulley clockwise and align its groove with the **0** mark on the timing chain cover.

   b. Check that the timing marks (1 and 2 dots) of the camshaft drive and driven gears are in a straight line on the cylinder head surface. If not, turn the crankshaft 1 revolution (360 degrees) and align the marks.

6. Inspect the valve clearance, as follows:

   a. Measure the clearance between the valve lifter and the camshaft. Measure the 1st intake and the 3rd exhaust valves on the right head and the 6th intake and the 2nd exhaust valves on the left head.

   b. Turn the crankshaft ⅔ of a revolution (240 degrees) and adjust the 3rd intake and the 5th exhaust valves on the right head and the 2nd intake and the 4th exhaust valves on the left head.

   c. Turn the crankshaft ⅔ of a revolution (240 degrees) and adjust the 5th intake and the 1st exhaust valves on the right head and the 4th intake and the 6th exhaust valves on the left head.

7. Valve clearance cold should be:

   • Intake: 0.006–0.009 in. (0.13–0.23mm)
   • Exhaust: 0.011–0.014 in. (0.27–0.37mm)

8. Adjust the valve clearance by using adjusting shims, as follows:

   a. Turn the equipment camshaft so that the cam lobe for the valve to be adjusted faces up.

   b. Turn the valve lifter so that the notches are perpendicular to the camshaft.

   c. Using SST 09248-55040, press down the valve lifter and place SST 09248-05420, between the camshaft and the valve lifter. Remove SST 09248-55040.

   d. Remove the adjusting shim with a small flat prying tool and a magnetic finger.

   e. Determine the replacement adjusting shim size according to the following formula or use the adjusting shim charts.

   f. Using a micrometer, measure the thickness of the removed shim. Calculate the thickness of a new shim so that the valve clearance comes within the specified value.

   • T: Thickness of the removed shim
   • A: Measured valve clearance
   • N: Thickness of the new shim

   g. Intake: N = T + (A—0.007 in. (0.18mm))

   h. Exhaust: N = T + (A—0.013 in. (0.32mm))

   i. Install a new adjusting shim. Place it on the valve lifter. Using the SST 09248-55040, press down the valve lifter and remove SST 09248-05420.

   j. Recheck the valve clearance.

9. Install or connect the following:
   • Cylinder head cover
   • Intake air connector
   • Negative battery cable

10. Refill with engine coolant.

11. Start the engine and check for leaks.

4.7L Engine

➡ **Measure the valve clearance with the engine cold.**

1. Before servicing the vehicle, refer to the precautions in the beginning of this section.

2. Drain the cooling system.

3. Remove or disconnect the following:
   • Negative battery cable
   • Ignition coils
   • Valve covers

4. Set the engine to the top of the compression stroke with the valves closed for the cylinder to be measured.

5. Check the valve clearance. The valve clearance specifications are as follows:

   • Intake: 0.006–0.010 in. (0.15–0.25mm)
   • Exhaust: 0.010–0.014 in. (0.25–0.35mm)

6. Record the measurements for each valve.

7. When all valve clearances have been measured, remove the camshafts.

8. Remove the valve shims and measure them. Note this measurement along with the clearance measurement recorded earlier.

9. Using the valve clearance and shim thickness measurements, find replacement shims in the Adjusting Shim Selection charts.

10. Install or connect the following:
    • Replacement valve shims
    • Camshafts
    • Valve covers
    • Ignition coils
    • Negative battery cable

11. Fill the cooling system.

12. Start the engine and check for leaks.

New shim thickness

Shim No.	Thickness mm (in.)	Shim No.	Thickness mm (in.)	Shim No.	Thickness mm (in.)
00	2.000 (0.0787)	28	2.280 (0.0898)	56	2.560 (0.1008)
02	2.020 (0.0795)	30	2.300 (0.0906)	58	2.580 (0.1016)
04	2.040 (0.0803)	32	2.320 (0.0913)	60	2.600 (0.1024)
06	2.060 (0.0811)	34	2.340 (0.0921)	62	2.620 (0.1031)
08	2.080 (0.0819)	36	2.360 (0.0929)	64	2.640 (0.1039)
10	2.100 (0.0827)	38	2.380 (0.0937)	66	2.660 (0.1047)
12	2.120 (0.0835)	40	2.400 (0.0945)	68	2.680 (0.1055)
14	2.140 (0.0843)	42	2.420 (0.0953)	70	2.700 (0.1063)
16	2.160 (0.0850)	44	2.440 (0.0961)	72	2.720 (0.1071)
18	2.180 (0.0858)	46	2.460 (0.0969)	74	2.740 (0.1079)
20	2.200 (0.0866)	48	2.480 (0.0976)	76	2.760 (0.1087)
22	2.220 (0.0874)	50	2.500 (0.0984)	78	2.780 (0.1094)
24	2.240 (0.0882)	52	2.520 (0.0992)	80	2.800 (0.1102)
26	2.260 (0.0890)	54	2.540 (0.1000)		

Intake valve clearance (Cold):
0.15 – 0.25 mm (0.006 – 0.010 in.)

EXAMPLE:
The 2.300 mm (0.0906 in.) shim is installed, and the measured clearance is 0.440 mm (0.0173 in.). Replace the 2.300 mm (0.0906 in.) shim with a No. 54 shim.

Intake valve clearance shim selection chart—4.7L engine

7924SG71

New shim thickness

mm (in.)

Shim No.	Thickness	Shim No.	Thickness	Shim No.	Thickness
00	2.000 (0.0787)	28	2.280 (0.0898)	56	2.560 (0.1008)
02	2.020 (0.0795)	30	2.300 (0.0906)	58	2.580 (0.1016)
04	2.040 (0.0803)	32	2.320 (0.0913)	60	2.600 (0.1024)
06	2.060 (0.0811)	34	2.340 (0.0921)	62	2.620 (0.1031)
08	2.080 (0.0819)	36	2.360 (0.0929)	64	2.640 (0.1039)
10	2.100 (0.0827)	38	2.380 (0.0937)	66	2.660 (0.1047)
12	2.120 (0.0835)	40	2.400 (0.0945)	68	2.680 (0.1055)
14	2.140 (0.0843)	42	2.420 (0.0953)	70	2.700 (0.1063)
16	2.160 (0.0850)	44	2.440 (0.0961)	72	2.720 (0.1071)
18	2.180 (0.0858)	46	2.460 (0.0969)	74	2.740 (0.1079)
20	2.200 (0.0866)	48	2.480 (0.0976)	76	2.760 (0.1087)
22	2.220 (0.0874)	50	2.500 (0.0984)	78	2.780 (0.1094)
24	2.240 (0.0882)	52	2.520 (0.0992)	80	2.800 (0.1102)
26	2.260 (0.0890)	54	2.540 (0.1000)		

Exhaust valve clearance (Cold):
0.25 – 0.35 mm (0.010 – 0.014 in.)

EXAMPLE:
The 2.300 mm (0.0906 in.) shim is installed, and the measured clearance is 0.440 mm (0.0173 in.). Replace the 2.300 mm (0.0906 in.) shim with a No. 44 shim.

Exhaust valve clearance shim selection chart—4.7L engine

7924SG72

Starter Motor

REMOVAL & INSTALLATION

3.4L Engine

1. Before servicing the vehicle, refer to the precautions in the beginning of this section.
2. Remove or disconnect the following:
   - Negative battery cable
   - Starter electrical connectors
   - Starter

To install:

3. Install or connect the following:
   - Starter. Tighten the fasteners to 29 ft. lbs. (39 Nm).
   - Electrical connections
   - Negative battery cable

4.7L Engine

1. Before servicing the vehicle, refer to the precautions in the beginning of this section.
2. Drain the cooling system.
3. Relieve the fuel system pressure.
4. Remove or disconnect the following:
   - Negative battery cable
   - Engine appearance cover
   - Air intake tube
   - Intake manifold
   - Starter motor mounting bolts
   - Starter wiring connectors
   - Starter motor

To install:

5. Install or connect the following:
   - Starter motor

- Starter wiring connectors. Tighten the cable nut to 86 inch lbs. (10 Nm).
- Starter motor mounting bolts. Tighten the bolts to 29 ft. lbs. (39 Nm).
- Intake manifold
- Air intake tube
- Engine appearance cover
- Negative battery cable

6. Fill the cooling system.
7. Start the engine and check for leaks.

Timing Belt

REMOVAL & INSTALLATION

3.4L Engine

1. Disconnect the negative battery cable.

✲✲ CAUTION

Work must be started after 90 seconds from the time the ignition switch is turned to the LOCK position and the negative battery cable is disconnected.

2. Raise and safely support the vehicle.
3. Remove the engine undercover.
4. Drain the engine coolant.

✲✲ CAUTION

Never open, service or drain the radiator or cooling system when hot; serious burns can occur from the steam and hot coolant. Also, when draining engine coolant, keep in mind that cats and dogs are attracted to ethylene glycol antifreeze and could drink any that is left in an uncovered container or in puddles on the ground. This will prove fatal in sufficient quantities. Always drain coolant into a sealable container. Coolant should be reused unless it is contaminated or is several years old.

5. Disconnect the upper radiator hose from the engine.
6. Remove the power steering drive belt.
7. Remove the air conditioning drive belt by loosening the idler pulley nut and the adjusting bolt.
8. Loosen the lockbolt, pivot bolt, and the adjusting bolt and the alternator drive belt.
9. Remove the No. 2 fan shroud by removing the 2 clips.
10. Remove the fan with the fluid coupling and fan pulleys.
11. Disconnect the power steering pump from the engine and set aside. Do not disconnect the lines from the pump.
12. If equipped with air conditioning, disconnect the compressor from the engine and set aside. Do not disconnect the lines from the compressor.
13. If equipped with air conditioning, disconnect the air conditioning bracket.
14. Remove the No. 2 timing belt cover, as follows:
    a. Detach the camshaft position sensor connector from the No. 2 timing belt cover.
    b. Disconnect the 3 spark plug wire clamps from the No. 2 timing belt cover.
    c. Remove the 6 bolts and remove the timing belt cover.
15. Remove the fan bracket, as follows:
    a. Remove the power steering adjusting strut by removing the nut.
    b. Remove the fan bracket by removing the bolt and nut.
16. Set the No. 1 cylinder at Top Dead Center (TDC) of the compression stroke, as follows:
    a. Turn the crankshaft pulley and align its groove with the timing mark **0** of the No. 1 timing belt cover.
    b. Check that the timing marks of the camshaft timing pulleys and the No. 3 timing belt cover are aligned. If not, turn the crankshaft pulley one revolution (360 degrees).

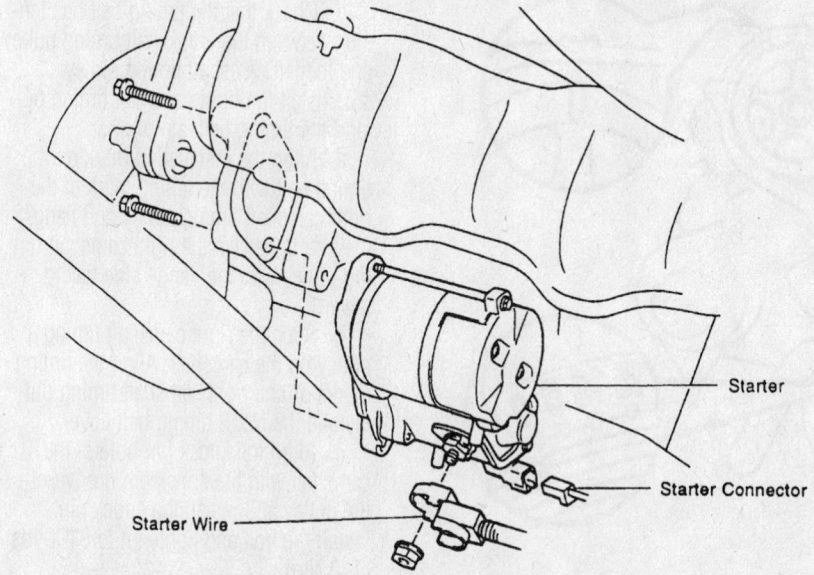

Exploded view of a common starter—3.4L engine

Starter

Starter Connector

Starter Wire

86822088

→ **If reusing the timing belt, be sure that you can still read the installation marks. If not, place new installation marks on the timing belt to match the timing marks of the camshaft timing pulleys.**

17. Remove the timing belt tensioner by alternately loosening the 2 bolts.

18. Remove the camshaft timing pulleys, as follows:

   a. Using SST 09960-10010, or equivalent, remove the pulley bolt, the timing pulley and the knock pin. Remove the 2 timing pulleys with the timing belt.

19. Remove the crankshaft pulley, as follows:

   a. Using SST 09213-54015 and 09330-00021 or their equivalents, loosen the pulley bolt.

   b. Remove the SST tool, the pulley bolt and the pulley.

20. Remove the starter wire bracket and the No. 1 timing belt cover.

21. Remove the timing belt guide and remove the timing belt.

22. Remove the bolt and the No. 2 idler pulley.

23. Remove the pivot bolt, the No. 1 idler pulley and the plate washer.

24. Remove the crankshaft gear.

To install:

25. Install the crankshaft timing gear.

   a. Align the timing pulley set key with the key groove of the gear.

   b. Using SST 09214-60010, or equivalent, and a hammer, tap in the timing gear with the flange side facing inward.

26. Install the plate washer and the No. 1 idler pulley with the pivot bolt and tighten it to 26 ft. lbs. (35 Nm). Check that the pulley bracket moves smoothly.

27. Install the No. 2 timing belt idler with the bolt. Tighten the bolt to 30 ft. lbs. (40 Nm). Check that the pulley bracket moves smoothly.

28. Temporarily install the timing belt, as follows:

   a. Using the crankshaft pulley bolt, turn the crankshaft and align the timing marks of the crankshaft timing pulley and the oil pump body.

   b. Align the installation mark on the timing belt with the dot mark of the crankshaft timing pulley.

   c. Install the timing belt on the crankshaft timing pulley, No. 1 idler pulley and the water pump pulleys.

29. Install the timing belt guide with the cup side facing outward.

30. Install the No. 1 timing belt cover and starter wire bracket. Tighten the timing belt cover bolts to 80 inch lbs. (9 Nm).

✳✳ WARNING

If any binding is felt when adjusting the timing belt tension by turning the crankshaft, STOP turning the engine, because the pistons may be hitting the valves.

31. Install the crankshaft pulley, as follows:

   a. Align the pulley set key with the key groove of the crankshaft pulley.

   b. Install the pulley bolt and tighten it to 184 ft. lbs. (250 Nm).

32. Install the left camshaft timing pulley.

   a. Install the knock pin to the camshaft.

   b. Align the knock pin hose of the camshaft with the knock pin groove of the timing pulley.

   c. Slide the timing belt pulley on the camshaft with the flange side facing outward. Tighten the pulley bolt to 81 ft. lbs. (110 Nm).

33. Set the No. 1 cylinder to TDC of the compression stroke, as follows:

   a. Turn the crankshaft pulley, and align its groove with the timing mark **0** of the No. 1 timing belt cover.

   b. Turn the camshaft to align the knock pin hole of the camshaft with the timing mark of the No. 3 timing belt cover.

   c. Turn the camshaft timing pulley, and align the timing marks of the camshaft timing pulley and the No. 3 timing belt cover.

34. Connect the timing belt to the left camshaft timing pulley, as follows:

→ **Check that the installation mark on the timing belt is aligned with the end of the No. 1 timing belt cover.**

   a. Using SST 09960-01000, or equivalent, slightly turn the left camshaft timing pulley clockwise. Align the installation mark on the timing belt with the timing mark of the camshaft timing pulley and hang the timing belt on the left camshaft timing pulley.

   b. Align the timing marks of the left camshaft pulley and the No. 3 timing belt cover.

   c. Check that the timing belt has tension between the crankshaft timing pulley and the left camshaft timing pulley.

35. Install the right camshaft timing pulley and the timing belt, as follows:

   a. Align the installation mark on the timing belt with the timing mark of the right camshaft timing pulley and hang the timing belt on the right camshaft timing pulley with the flange side facing inward.

   b. Slide the right camshaft timing pulley on the camshaft. Align the timing marks on the right camshaft timing pulley and the No. 3 timing belt cover.

   c. Align the knock pin hole of the camshaft with the knock pin groove of the pulley and install the knock pin. Install the bolt and tighten it to 81 ft. lbs. (110 Nm).

36. Set the timing belt tensioner, as follows:

Turn the crankshaft clockwise to align the timing marks before removing the timing belt—3.4L engine

a. Using a press, slowly press in the pushrod using 220–2205 lbs. (981–9807 N) of force.

b. align the holes of the pushrod and housing, pass a 1.5mm hex wrench through the holes to keep the setting position of the pushrod.

c. Release the press and install the dust boot on the tensioner.

37. Install the timing belt tensioner and alternately tighten the bolts to 20 ft. lbs. (28 Nm). Using pliers, remove the 1.5mm hex wrench from the belt tensioner.

38. Check the valve timing, as follows:

a. Slowly turn the crankshaft pulley 2 revolutions from TDC to TDC. Always turn the crankshaft pulley clockwise.

b. Check that each pulley aligns with the timing marks. If the timing marks do not align, remove the timing belt and reinstall it.

39. Install the fan bracket with the bolt and nut.

40. Install the power steering adjusting strut with the nut.

41. Install the No. 2 timing belt cover. Tighten the bolts to 80 inch lbs. (9 Nm). Install the remaining components.

42. Fill the cooling system with coolant.

43. Connect the negative battery cable.

44. Start the engine and check for leaks.

4.7L Engine

1. Disconnect the negative battery cable.

2. Raise and safely support the vehicle.

3. Remove the oil pan protector and the engine under cover.

4. Drain the cooling system and store the coolant for refilling purposes.

5. Lower the vehicle and remove the battery clamp cover.

6. From the top of the engine, remove the fuel return hose, the engine cover nuts/bolts and the cover.

7. Remove the air cleaner and the intake air connector assembly.

8. Remove the cooling fan pulley by performing the following procedures:

a. Loosen the 4 fan clutch-to-fan pulley nuts.

b. Using a box-end wrench on the serpentine drive belt tensioner bolt, rotate the tensioner counterclockwise and remove the drive belt.

➡ **The serpentine drive belt tensioner bolt is a left-hand thread.**

c. Remove the fan clutch-to-fan pulley nuts, the fan, the clutch assembly and the fan pulley.

9. Remove the radiator by performing the following procedures:

a. Disconnect the upper, lower and reservoir hoses from the radiator.

b. Disconnect and plug the automatic transmission oil cooler at the radiator. Disconnect the automatic transmission oil cooler hoses from the fan shroud clamp.

c. Remove the radiator reservoir tank.

d. Remove the fan shroud-to-radiator bolts and the shroud.

e. Remove the 2 upper radiator-to-chassis nuts.

f. Remove the middle radiator-to-chassis nut/bolts and brackets.

g. Carefully, lift the radiator from the vehicle.

10. Remove the serpentine drive belt idler pulley bolt, cover plate and pulley.

11. Remove the right side (No. 3) timing belt cover.

12. Remove the left side (No. 3) timing belt cover by performing the following procedures:

a. Disconnect the engine wire from both wire clamps.

b. Disconnect the camshaft position sensor wire from the wire clamp on the left-side (No.3) timing belt cover.

c. Disconnect the sensor connector from the connector bracket.

d. Disconnect the sensor connector.

e. Remove the wire grommet from the left-side (No. 3) timing belt cover.

f. Remove the oil cooler tube bolts and tube.

13. Remove the middle (No. 2) timing belt cover bolts and cover.

14. Remove the cooling fan bracket nuts/bolts and bracket.

➡ **If reusing the timing belt, make sure that there are 3 installation marks on the belt; if there are none, install them.**

15. Using the Crankshaft Pulley Holding tool 09213-70010, Bolt tool 90105-08076 and Companion Flange Holding tool 09330-00021, or equivalent, loosen the crankshaft pulley bolt.

16. Position the No. 1 cylinder to approximately 50 degrees After Top Dead Center (ATDC) of the compression stroke by performing the following procedures:

a. Rotate the crankshaft pulley (CLOCKWISE) to align its groove with the timing mark "0" on the lower (No. 1) timing belt cover.

b. Check that the camshaft sprocket timing marks are aligned with the rear timing belt plate marks; if not, rotate the crankshaft 1 revolution (360 degrees).

c. Rotate the crankshaft pulley approximately 50 degrees (CLOCKWISE) and align the crankshaft pulley timing mark between the centers of the crankshaft pulley bolt and the idler pulley bolt.

If the timing belt is disengaged, having the crankshaft pulley in the wrong angle can cause the valve to come into contact with the piston when removing the camshaft pulley.

17. Remove the crankshaft pulley bolt.

➡ **If reusing the timing belt and the installation marks have disappeared, place new installation marks on the timing belt to match the camshaft timing sprocket marks.**

➡ **To avoid meshing the timing sprocket and the timing belt, secure one with a string; then, place matchmarks on the timing belt and the right-side camshaft timing sprocket.**

18. Remove the timing belt tensioner bolts and the tensioner.

19. Using the Camshaft Holding tool 09960-10010, or equivalent, slightly turn the left-side camshaft sprocket clockwise to loosen the tension spring. Then, disconnect the timing belt from the camshaft sprockets.

20. Remove the alternator by performing the following procedures:

a. Disconnect the electrical connector from the alternator.

b. Remove the rubber cap/nut and disconnect the battery wire from the alternator.

c. Disconnect the wire clamp from the alternator cord clip.

d. Remove the alternator-to-engine nuts/bolts and the alternator.

21. Remove the serpentine drive belt tensioner nuts/bolts and the tensioner.

22. Using the Crankshaft Puller Assembly tool 09950-50012, or equivalent, press the crankshaft pulley from the crankshaft.

DO NOT rotate the crankshaft pulley.

23. Remove the lower (No. 1) timing belt cover bolts and the cover.

24. Remove the timing belt guide, spacer and the timing belt.

To install:

➡ **With the timing belt removed, this is a perfect opportunity to inspect and/or replace the water pump.**

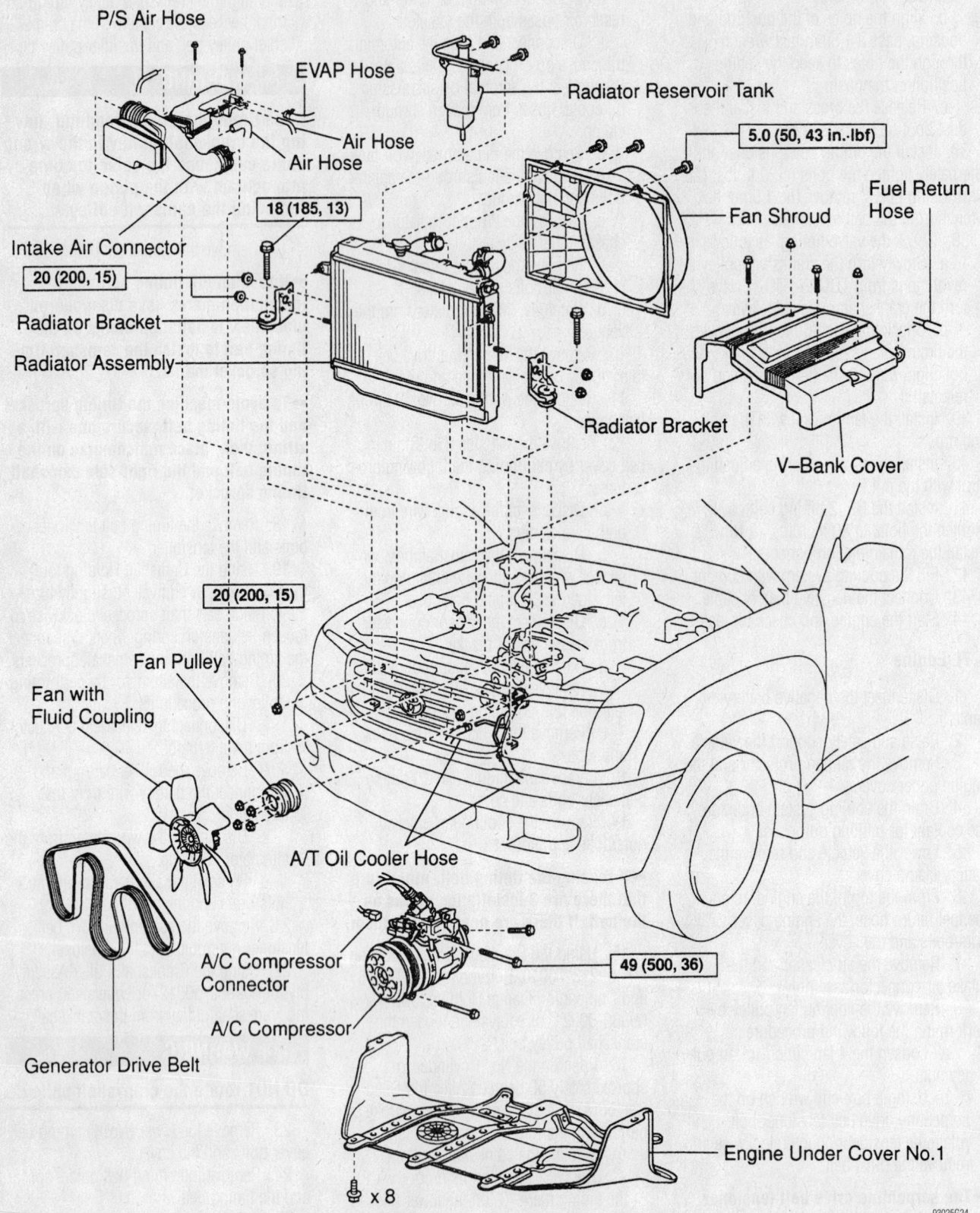

P/S Air Hose

EVAP Hose

Air Hose

Air Hose

Radiator Reservoir Tank

5.0 (50, 43 in.·lbf)

Fan Shroud

Fuel Return Hose

18 (185, 13)

Intake Air Connector

20 (200, 15)

Radiator Bracket

Radiator Assembly

Radiator Bracket

V–Bank Cover

20 (200, 15)

Fan Pulley

Fan with Fluid Coupling

A/T Oil Cooler Hose

A/C Compressor Connector

A/C Compressor

49 (500, 36)

Generator Drive Belt

Engine Under Cover No.1

x 8

93025G24

Exploded view of vehicle components for timing belt replacement—4.7L engine

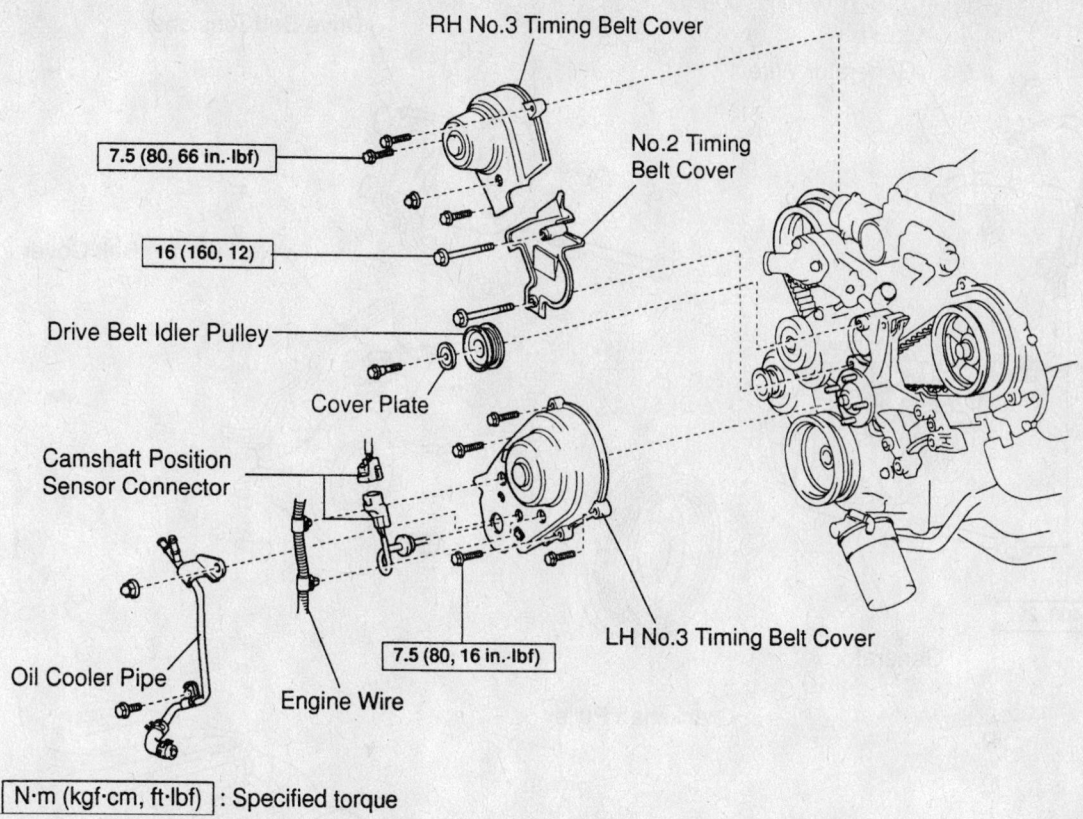

RH No.3 Timing Belt Cover

No.2 Timing Belt Cover

7.5 (80, 66 in.·lbf)

16 (160, 12)

Drive Belt Idler Pulley

Cover Plate

Camshaft Position Sensor Connector

LH No.3 Timing Belt Cover

7.5 (80, 16 in.·lbf)

Oil Cooler Pipe

Engine Wire

N·m (kgf·cm, ft·lbf) : Specified torque

93025G25

Exploded view of upper timing belt covers—4.7L engine

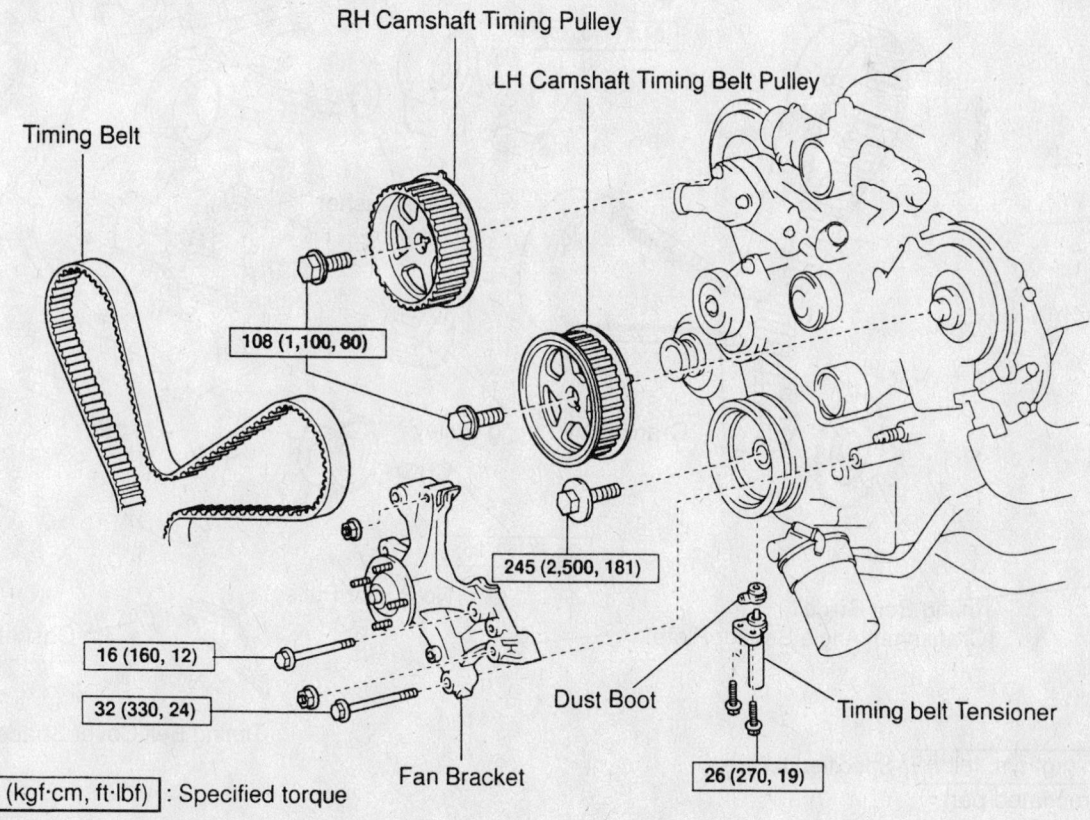

RH Camshaft Timing Pulley

LH Camshaft Timing Belt Pulley

Timing Belt

108 (1,100, 80)

245 (2,500, 181)

16 (160, 12)

32 (330, 24)

Dust Boot

Timing belt Tensioner

Fan Bracket

26 (270, 19)

N·m (kgf·cm, ft·lbf) : Specified torque

93025G26

Exploded view of upper timing sprockets and components—4.7L engine

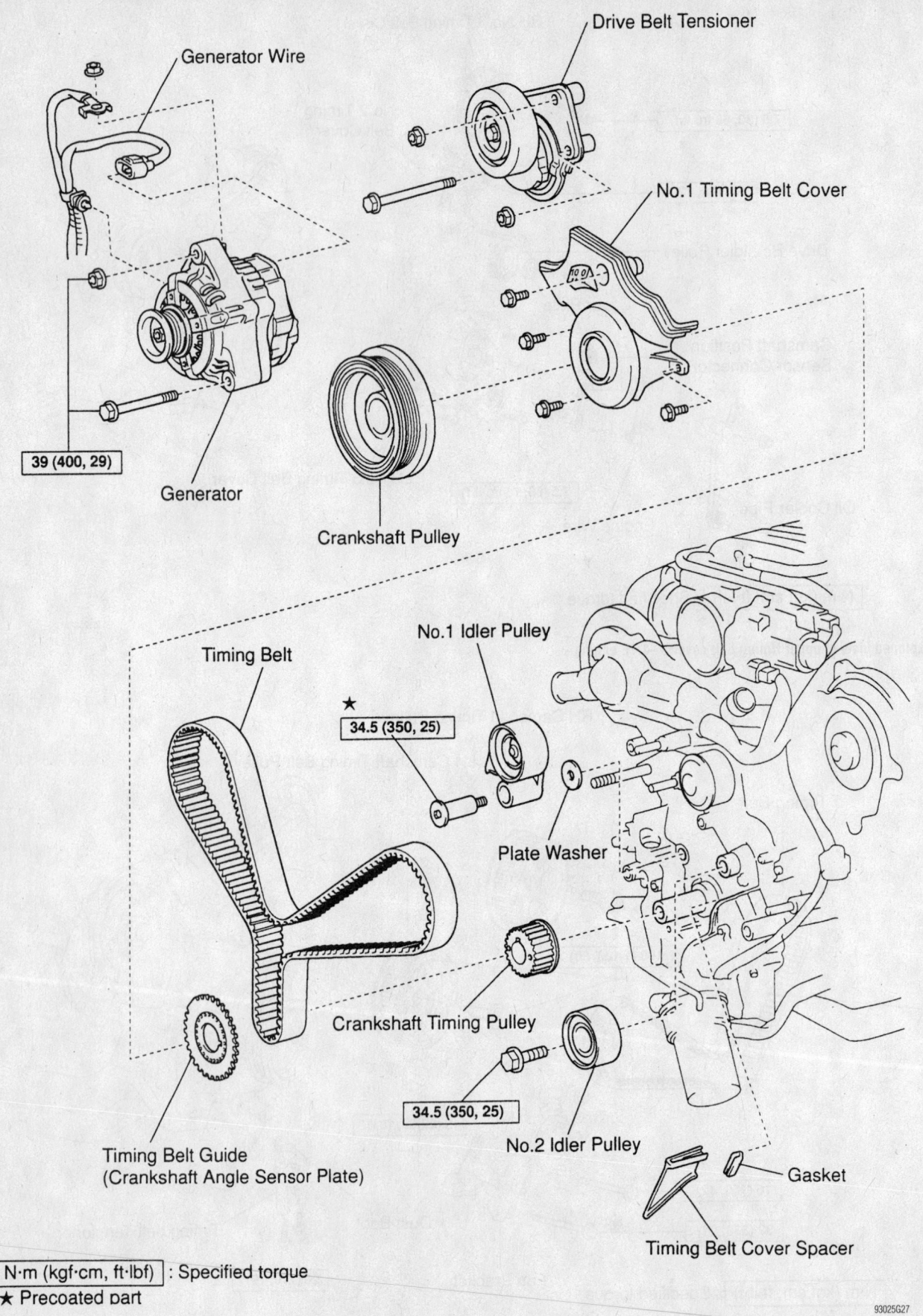

Generator Wire

Drive Belt Tensioner

No.1 Timing Belt Cover

39 (400, 29)

Generator

Crankshaft Pulley

Timing Belt

No.1 Idler Pulley

★
34.5 (350, 25)

Plate Washer

Crankshaft Timing Pulley

34.5 (350, 25)

No.2 Idler Pulley

Gasket

Timing Belt Cover Spacer

Timing Belt Guide
(Crankshaft Angle Sensor Plate)

N·m (kgf·cm, ft·lbf) : Specified torque

★ Precoated part

93025G27

Exploded view of lower timing belt cover, sprockets and components—4.7L engine

25. Inspect the timing belt tensioner by performing the following procedures:

   a. Inspect the seal for leakage; if leakage is suspected, replace the tensioner.

   b. Using both hands to hold the tensioner facing upward, strongly press the pushrod against a solid surface. If the pushrod moves, replace the tensioner.

✳✳ WARNING

Never hold the tensioner with the pushrod facing downward.

   c. Measure the pushrod protrusion from the housing end, it should be 0.413–0.453 in. (10.5–11.5mm). If the protrusion is not as specified, replace the tensioner.

26. Temporarily install the timing belt by performing the following procedures:

   a. Align the timing belt's installation mark with the crankshaft timing sprocket.

   b. Install the timing belt on the crankshaft timing sprocket, the No. 1 idler pulley and the No. 2 idler pulley.

27. Install the gasket to the timing belt cover spacer and install the cover spacer.

28. Install the timing belt guide with the cup side facing outward.

29. Install the lower (No. 1) timing belt cover.

30. Install the crankshaft pulley by performing the following procedures:

   a. Align the crankshaft pulley with the crankshaft key.

   b. Using the Crankshaft Installer tool 09223-46011, or equivalent, and a hammer, tap the crankshaft pulley into position.

31. Install the serpentine drive belt tensioner and torque the tensioner-to-engine bolts to 12 ft. lbs. (16 Nm).

➡ **To install the serpentine drive belt tensioner, use a bolt 4.18 in. (106mm) in length.**

93025G28

Alignment of timing belt with the timing sprockets—4.7L engine

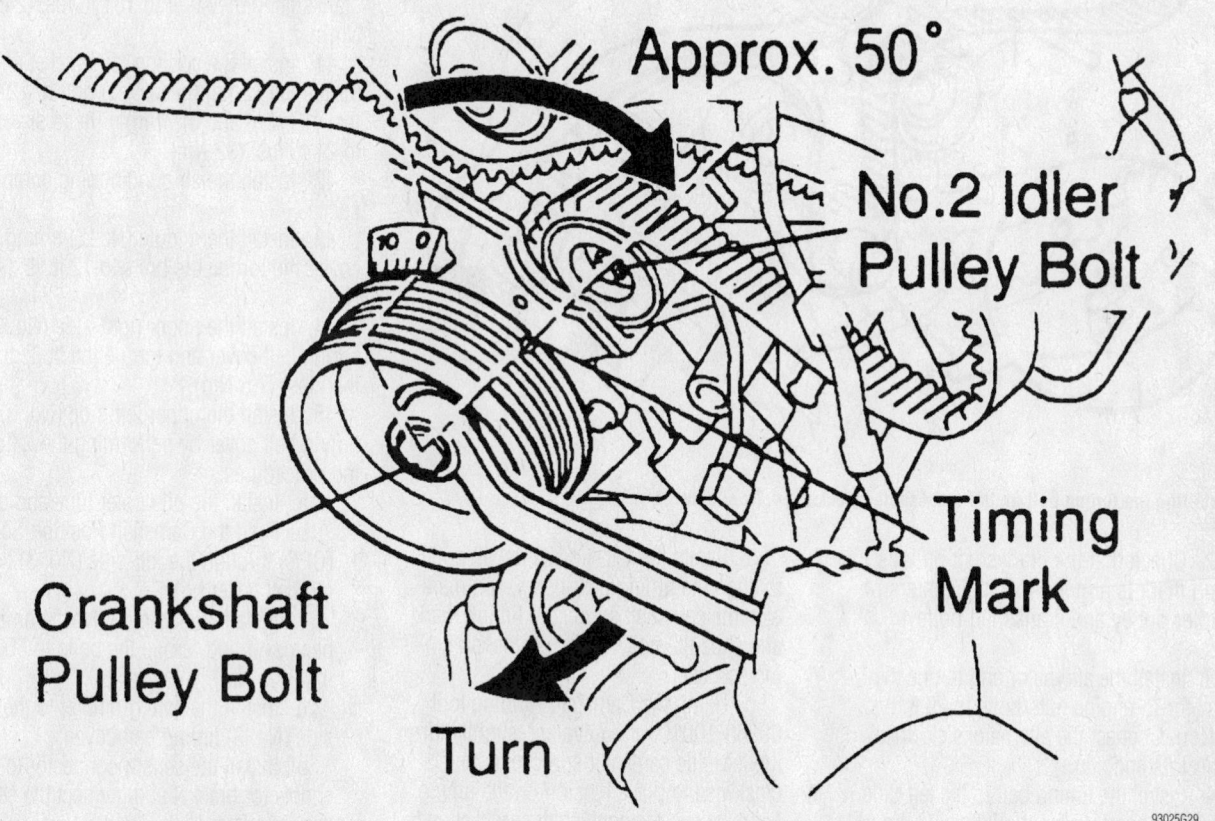

93025G29

Aligning of crankshaft pulley timing mark with the center line of the crankshaft pulley bolt and the idler pulley bolt—4.7L engine

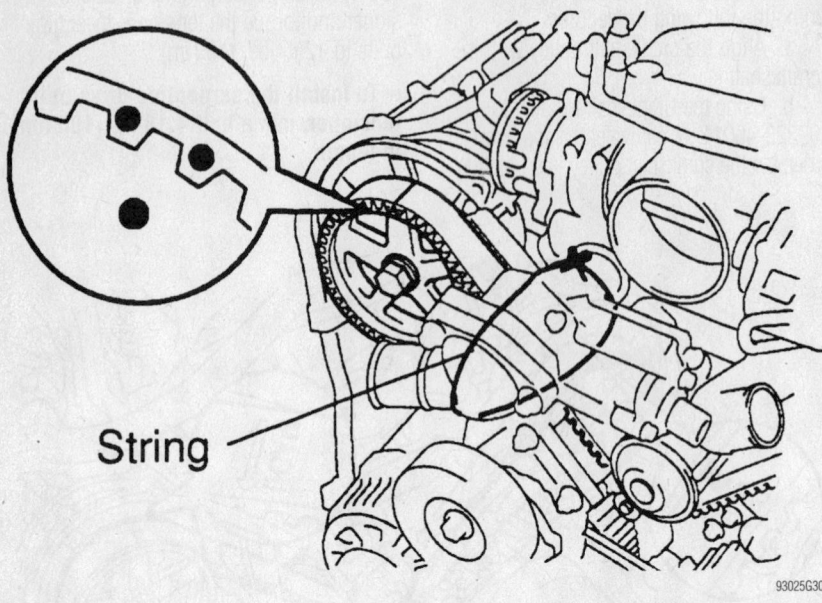

Securing the timing belt with string and matchmarking the camshaft with the timing belt—4.7L engine

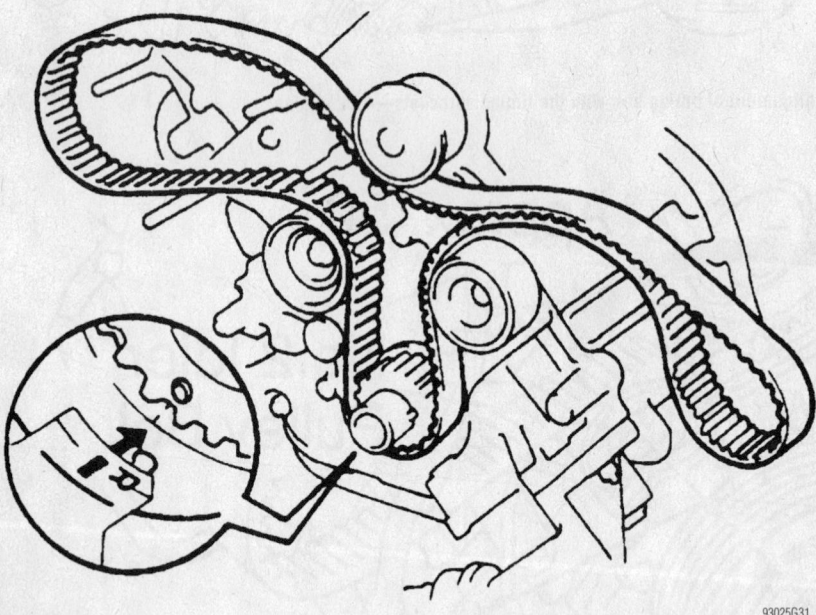

Installing the timing belt on the crankshaft sprocket—4.7L engine

32. Check that the crankshaft pulley's timing mark is aligned with the centers of the idler pulley and crankshaft pulley bolts.

33. Install the alternator and torque the alternator-to-engine nuts/bolts to 29 ft. lbs. (39 Nm). Connect the alternator's electrical connectors and clip.

34. Install the timing belt to the left-side camshaft by performing the following procedures:

a. Rotate the left-side camshaft pulley to align the timing belt installation mark with the camshaft sprocket's timing mark and slide the belt onto the camshaft timing sprocket.

b. Using the Camshaft Holding tool 09960-10010, or equivalent, slightly turn the left-side camshaft sprocket counterclockwise to place tension on the timing belt between the crankshaft sprocket and the camshaft sprocket.

35. Rotate the right-side camshaft pulley to align the timing belt installation mark with the camshaft sprocket's timing mark and slide the belt onto the camshaft timing sprocket.

36. Using a vertical press, slowly press the pushrod into the housing using 200–2205 lbs. (981–9807 N) until the holes align, then, install a 1.27mm Allen® wrench to secure the pushrod and release the press. Install the dust boot on the tensioner housing.

37. Install the timing belt tensioner and torque the bolts to 19 ft. lbs. (26 Nm).

38. Using a pair of pliers, remove the Allen® wrench from the tensioner housing.

39. Check the valve timing by performing the following procedure:

a. Temporarily install the crankshaft pulley bolt.

b. Slowly, rotate the crankshaft pulley 2 revolutions (CLOCKWISE) and realign the TDC marks.

➡ **If the pulley/sprocket timing marks do not realign, remove the timing belt and reinstall it.**

40. Using the Crankshaft Pulley Holding tool 09213-70010, Bolt tool 90105-08076 and Companion Flange Holding tool 09330-00021, or equivalent, torque the crankshaft pulley bolt to 181 ft. lbs. (245 Nm).

41. Install the cooling fan bracket and torque the 12mm (head size) bolt to 12 ft. lbs. (16 Nm) and the 14mm (head size) bolt to 24 ft. lbs. (32 Nm).

42. Install the air conditioning compressor.

43. Install the middle (No. 2) timing belt cover and torque the bolts to 12 ft. lbs. (16 Nm).

44. Install the upper right-side (No. 3) timing belt cover and torque the bolts to 66 inch lbs. (7.5 Nm).

45. Install the upper left-side (No. 3) timing belt cover by performing the following procedures:

a. Install the oil cooler tube and bolt.

b. Feed the Camshaft Position Sensor (CPS) through the left-side (No. 3) timing belt cover hole.

c. Install the left-side (No. 3) timing belt cover and torque the bolts to 66 inch lbs. (7.5 Nm).

d. Install the wire grommet to the left-side (No. 3) timing belt cover.

e. Install the sensor connector to the connector bracket and connect the sensor connector.

f. Install the sensor wire and the

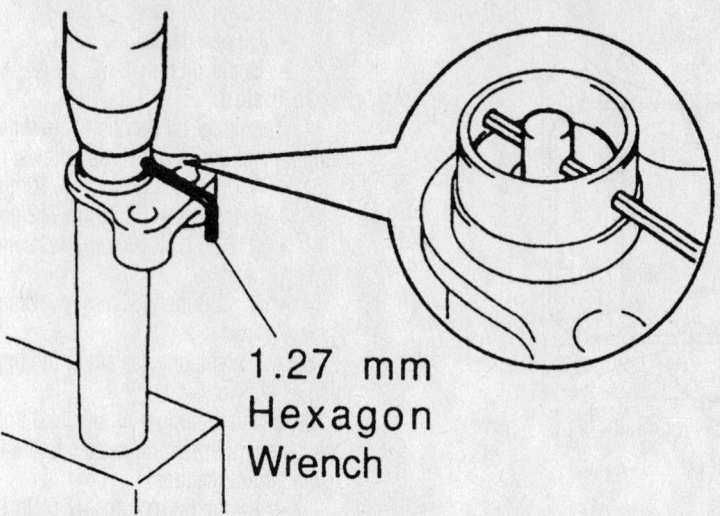

1.27 mm
Hexagon
Wrench

Securing the timing belt tensioner pushrod—4.7L engine

93025G32

93025G33

Checking the TDC alignment marks after rotating the crankshaft 2 revolutions—4.7L engine

engine wire to the clamps on the left-side (No. 3) timing belt cover.

46. Install the drive belt idler pulley and cover plate; then, torque the pulley bolt to 27 ft. lbs. (37 Nm).

47. To complete the installation, reverse the removal procedures.

48. Refill the cooling system and connect the negative battery cable.

Oil Pan

REMOVAL & INSTALLATION

3.4L Engine

1. Before servicing the vehicle, refer to the precautions in the beginning of this section.

2. Disconnect the negative battery cable.
3. Drain the engine oil.
4. Remove or disconnect the following:

- Engine undercover
- Front differential, if equipped with 4WD
- Oil pan, separate it from the engine using SST 09032-00100 and a brass bar

To install:

5. Apply seal packing to the oil pan.
6. Install the oil pan to the cylinder block. Tighten the nuts and bolts to: 67 inch lbs. (8 Nm)

❋❋ WARNING

If parts are not assembled within 5 minutes of applying time, the effectiveness of the seal packing is lost and must be removed and reapplied.

7. Install or connect the following:
- Front differential, if removed
- Engine undercover
- Negative battery cable
8. Fill with engine oil.
9. Start the engine and check for leaks.

4.7L Engine

1. Before servicing the vehicle, refer to the precautions in the beginning of this section.

2. Remove the engine from the vehicle and mount it on a stand.

3. Remove or disconnect the following:

- Oil dipstick tube

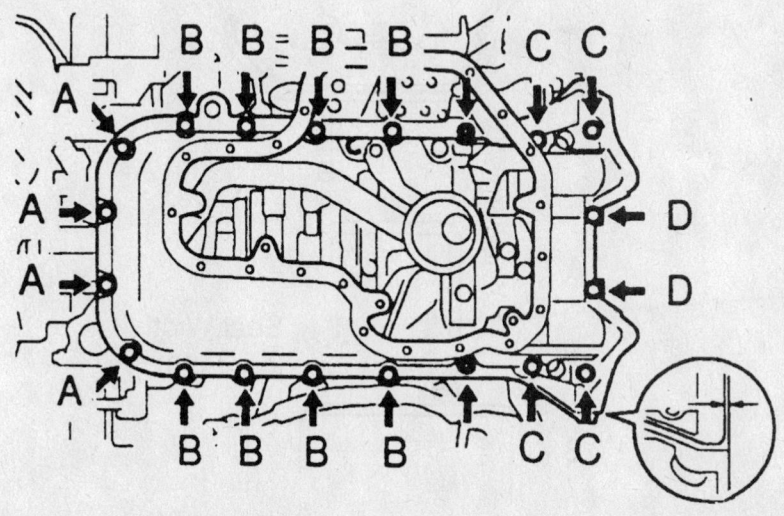

Upper oil pan bolt location—4.7L engine

7924SG75

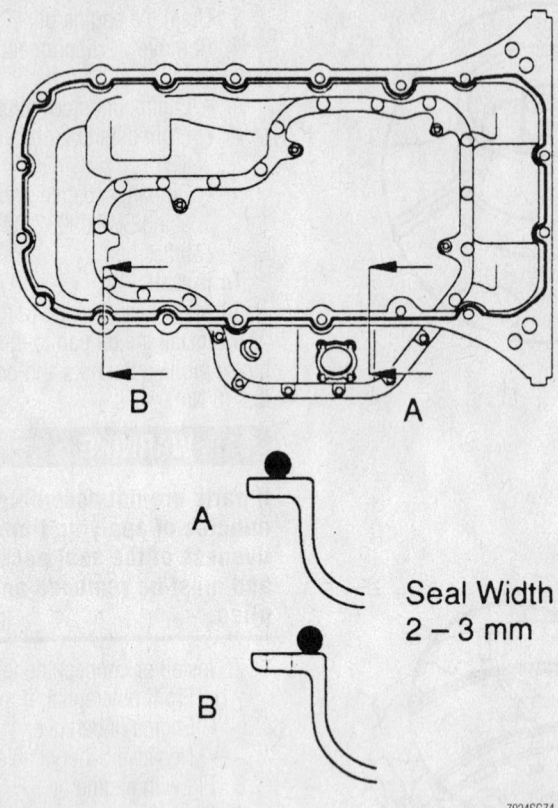

Seal Width
2 – 3 mm

7924SG74

Upper oil pan sealant application—4.7L engine

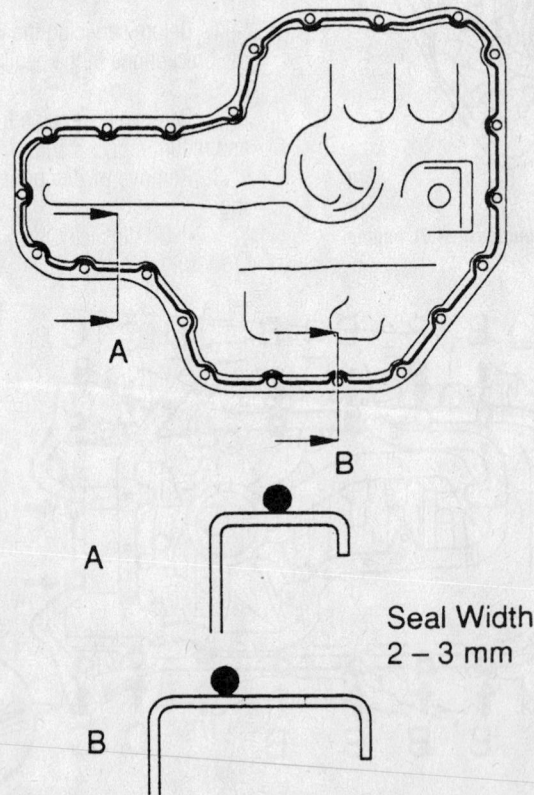

Seal Width
2 – 3 mm

7924SG76

Lower oil pan sealant application—4.7L engine

- Lower oil pan
- Oil pan baffle
- Upper oil pan

To install:

4. The upper oil pan bolts are different lengths and are identified as follows:
- A: 0.79 inch (20mm) w/10mm head
- B: 0.98 inch (25mm) w/12mm head
- C: 2.36 inch (60mm) w/12mm head
- D: 1.38 inch (35mm) w/10mm head

5. Apply silicone sealant to the upper oil pan as shown.

6. Install the upper oil pan and tighten the fasteners in several passes to the following specifications:
- 10mm: 66 inch lbs. (7.5 Nm)
- 12mm: 21 ft. lbs. (28 Nm)

7. Install or connect the following:
- Oil pan baffle. Tighten the fasteners to 66 inch lbs. (7.5 Nm).
- Lower oil pan. Tighten the fasteners in several passes to 66 inch lbs. (7.5 Nm).
- Oil dipstick tube

8. Install the engine.

Oil Pump

REMOVAL & INSTALLATION

3.4L Engine

1. Before servicing the vehicle, refer to the precautions in the beginning of this section.

2. Remove or disconnect the following:
- Negative battery cable
- Engine undercover
- Crankshaft timing pulley
- Front differential, if equipped with 4WD

3. Drain the engine oil from the engine.

4. Remove or disconnect the following:
- Timing belt and crankshaft gear
- Oil cooler tube and clamp, if equipped with automatic transmission
- Stiffener plate
- Flywheel housing undercover and dust cover
- Rear end cover and dust cover
- Starter wire clamp
- Crankshaft Position (CKP) sensor
- Oil pan

➡ Be careful not to damage the baffle plate flange.

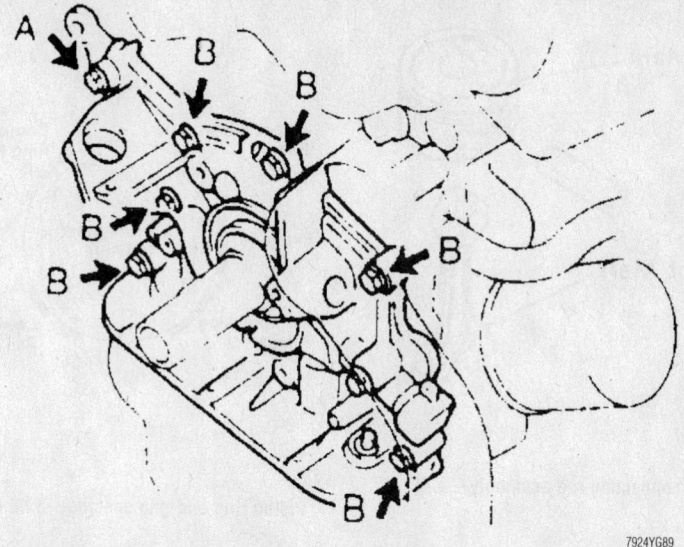

Oil pump bolt identification—3.4L engine

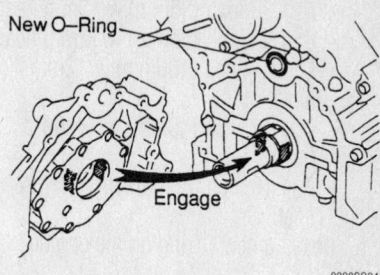

Location of the O-ring seal—4.7L engine

- Oil strainer
- Oil baffle plate
- Oil pump body by removing the 8 bolts.
- O-ring from the cylinder block

To install:

5. Install or connect the following:
- Apply Seal Packing PN 08826-00080 to the oil pump
- New O-ring into the groove of the cylinder block
- Oil pump to the crankshaft with the spline teeth of the drive rotor engaged with the large teeth of the crankshaft. Tighten the oil pump bolts "A" 15 ft. lbs. (20 Nm) and bolts "B" 31 ft. lbs. (42 Nm)
- CKP
- Oil pan baffle plate
- Oil strainer with a new gasket. Tighten the bolts to 13 ft. lbs. (18 Nm).
- Remaining components
- Negative battery cable

6. Fill with engine oil.
7. Start the engine and check for leaks.

4.7L Engine

1. Before servicing the vehicle, refer to the precautions in the beginning of this section.

2. Remove the engine from the vehicle and mount it on a stand.

3. Remove or disconnect the following:

- Front cover
- Timing belt
- Timing belt idler pulleys
- Crankshaft timing sprocket

- Oil dipstick tube
- Oil filter and bracket
- Crankshaft Position (CKP) sensor
- Oil pan and baffle
- Oil pump pickup tube
- Oil pump

To install:

4. The upper oil pan bolts are different lengths and are identified as follows:

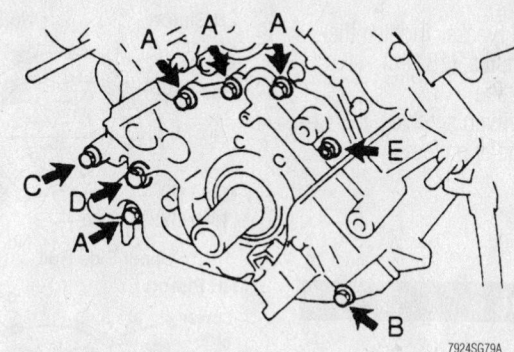

Oil pump bolt location—4.7L engine

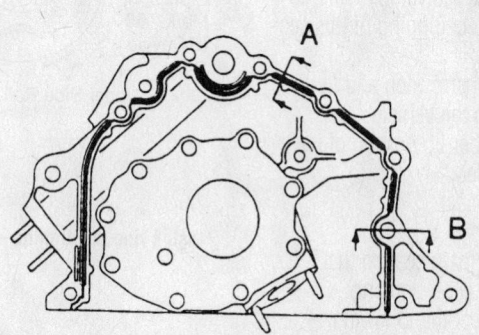

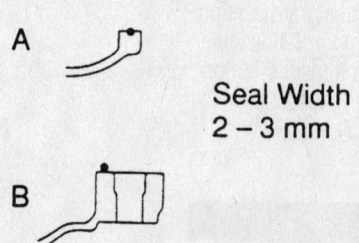

Seal Width
2 – 3 mm

Oil pump housing sealant application—4.7L engine

- A: 1.38 inch (35mm) w/12mm head
- B: 1.97 inch (50mm) w/12mm head
- C: 4.17 inch (106mm) w/12mm head
- D: 1.57 inch (40mm) w/14mm head
- E: 1.18 inch (30mm) w/6mm hex head

5. Install a new O-ring on the engine block.

6. Apply silicone sealant to the oil pump housing as shown.

7. Install the oil pump. Tighten the bolts in several passes to the following specifications:

- 12mm: 11 ft. lbs. (15.5 Nm)
- 14mm: 22 ft. lbs. (30.5 Nm)
- 6mm Hex: 11 ft. lbs. (15.5 Nm)

8. Install or connect the following:
- Oil pump pickup tube. Tighten the bolts to 66 inch lbs. (7.5 Nm).
- Oil pan and baffle
- CKP sensor
- Oil filter and bracket. Tighten the bolts to 13 ft. lbs. (18 Nm).
- Oil dipstick tube
- Crankshaft timing sprocket
- Timing belt idler pulleys
- Timing belt
- Front cover

9. Install the engine.

Rear Main Seal

REMOVAL & INSTALLATION

1. Before servicing the vehicle, refer to the precautions in the beginning of this section.

2. Remove the transmission and flywheel/driveplate from the vehicle.

3. Cut off the rubber lip portion of the seal with a sharp knife.

4. Pry out the oil seal.

To install:

5. Install the rear main seal so that it is flush with the seal retainer housing.

6. Install or connect the following:
- Flywheel/driveplate. Tighten the bolts to 28 ft. lbs. (38 Nm) for 3.4L engine with a manual transmission, 61 ft. lbs. (83 Nm) for 3.4L engine with an automatic transmission or 35 ft. lbs. (48 Nm) plus a 90 degree turn for 4.7L engine.
- Transmission

Piston and Ring

POSITIONING

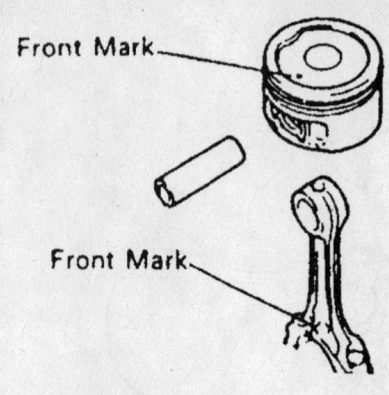

Piston to connecting rod assembly—3.4L engines

7924AG81

Piston ring end-gap spacing—3.4L engine

7924AG80

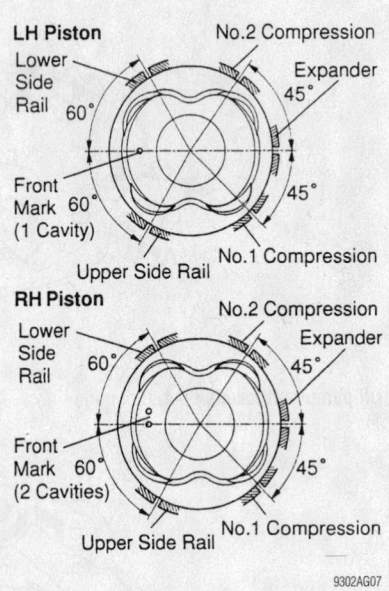
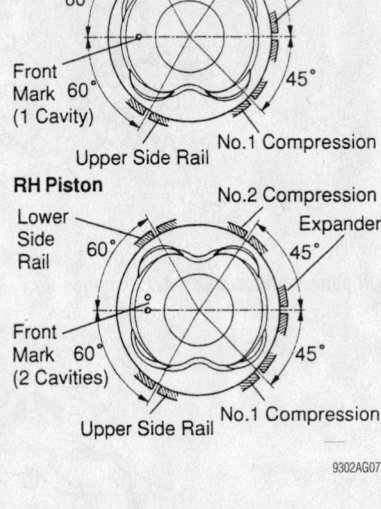

Piston ring positioning—4.7L engine

9302AG07

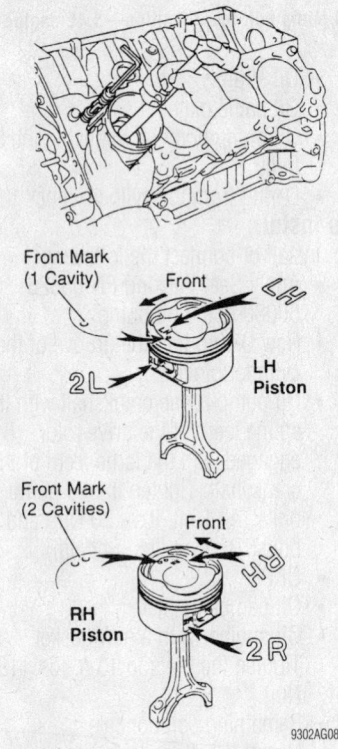

Piston positioning—4.7L engine

9302AG08

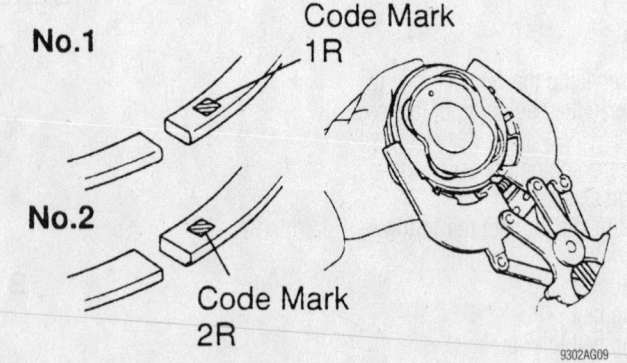

Piston ring identification—4.7L engine

9302AG09

FUEL SYSTEM

Fuel System Service Precautions

Safety is the most important factor when performing not only fuel system maintenance but any type of maintenance. Failure to conduct maintenance and repairs in a safe manner may result in serious personal injury or death. Maintenance and testing of the vehicle's fuel system components can be accomplished safely and effectively by adhering to the following rules and guidelines.

• To avoid the possibility of fire and personal injury, always disconnect the negative battery cable unless the repair or test procedure requires that battery voltage be applied.

• Always relieve the fuel system pressure prior to disconnecting any fuel system component (injector, fuel rail, pressure regulator, etc.), fitting or fuel line connection. Exercise extreme caution whenever relieving fuel system pressure, to avoid exposing skin, face and eyes to fuel spray. Please be advised that fuel under pressure may penetrate the skin or any part of the body that it contacts.

• Always place a shop towel or cloth around the fitting or connection prior to loosening to absorb any excess fuel due to spillage. Ensure that all fuel spillage (should it occur) is quickly removed from engine surfaces. Ensure that all fuel soaked cloths or towels are deposited into a suitable waste container.

• Always keep a dry chemical (Class B) fire extinguisher near the work area.

• Do not allow fuel spray or fuel vapors to come into contact with a spark or open flame.

• Always use a back-up wrench when loosening and tightening fuel line connection fittings. This will prevent unnecessary stress and torsion to fuel line piping.

• Always replace worn fuel fitting O-rings with new. Do not substitute fuel hose or equivalent, where fuel pipe is installed.

Fuel System Pressure

RELIEVING

1. Before servicing the vehicle, refer to the precautions in the beginning of this section.
2. Disconnect the fuel pump connector near the fuel tank.
3. Start the engine and allow it to run until it stalls. Crank the engine for a few seconds to relieve additional fuel pressure.
4. Disconnect the negative battery cable.
5. When repairs are complete, connect the negative battery cable.

Fuel Filter

REMOVAL & INSTALLATION

1. Before servicing the vehicle, refer to the precautions in the beginning of this section.
2. Relieve the fuel system pressure.
3. Remove or disconnect the following:

• Negative battery cable
• Fuel lines
• Fuel filter

To install:
4. Install the fuel filter.
5. Use new washers and tighten the fuel line bolts to the following specifications:
• Banjo bolt fittings: 21 ft. lbs. (29 Nm)
• Flare nut fitting: 28 ft. lbs. (38 Nm)
6. Connect the negative battery cable.

7. Start the engine and check for leaks.

Fuel Pump

REMOVAL & INSTALLATION

1. Before servicing the vehicle, refer to the precautions in the beginning of this section.
2. Relieve the fuel system pressure.
3. Remove or disconnect the following:

• Negative battery cable
• Fuel tank
• Fuel pump harness connector
• Fuel lines
• Fuel pump module

To install:
4. Install or connect the following:
• Fuel pump module. Tighten the bolts to 35 inch lbs. (4 Nm).
• Fuel lines
• Fuel pump harness connector
• Fuel tank
• Negative battery cable
5. Start the engine and check for leaks.

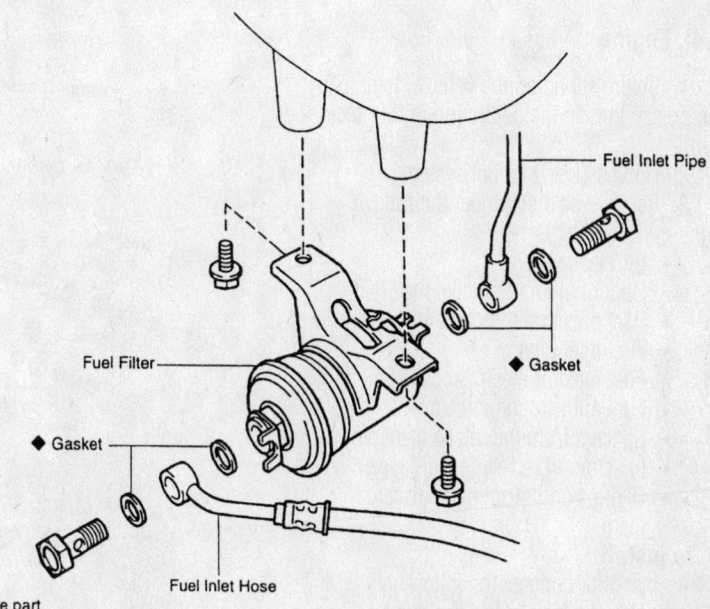

Fuel Inlet Pipe

Gasket

Fuel Filter

Gasket

Fuel Inlet Hose

◆ Non-reusable part

7924SG28

Always use new gaskets when replacing the fuel filter

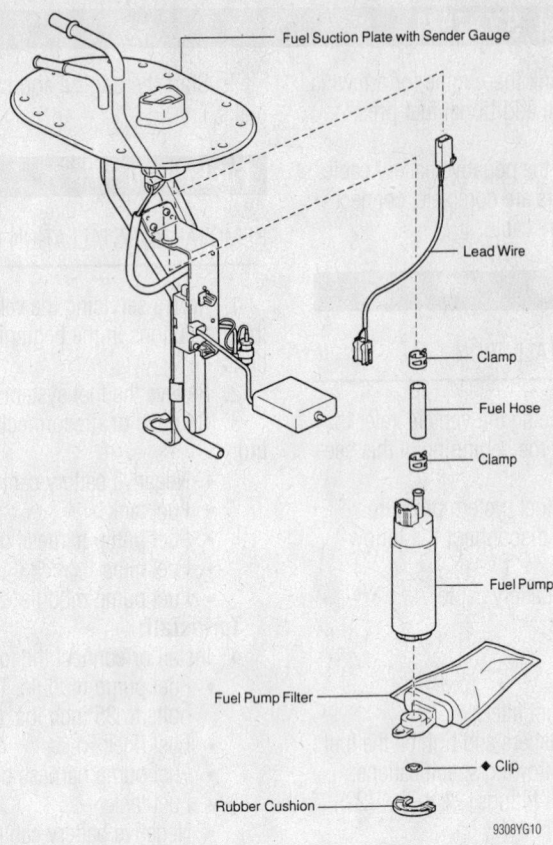

Exploded view of the fuel pump and related components

Fuel Injector

REMOVAL & INSTALLATION

3.4L Engine

1. Before servicing the vehicle, refer to the precautions in the beginning of this section.

2. Depressurize the fuel system.

3. Remove or disconnect the following:

- Air cleaner hose
- Upper half of the intake manifold
- Fuel pressure regulator
- Fuel inlet pipe
- Fuel injector electrical connections
- Fuel rail with the injectors
- Spacers from the intake manifold
- Injectors from the delivery pipes
- O-rings and grommets, discard them

To install:

4. Install or connect the following:

- New grommets and O-rings on each injector, lubricated with a light coat of gasoline
- Fuel injector with the electrical connector facing outward
- Spacers on the intake manifold

5. Temporarily install the bolts to hold the delivery pipes to the intake manifold.

6. Check that the injectors rotate smoothly. If they do not, the O-rings have probably been installed incorrectly.

7. Install or connect the following:

- Fuel injector electrical connectors
- Fuel pipe with new gaskets. Tighten the bolts to 25 ft. lbs. (34 Nm) and the delivery pipes-to-intake manifold bolts to 10 ft. lbs. (13 Nm).
- Fuel pipe union with new gaskets. Tighten the clamp bolt to 71 inch lbs. (8 Nm).
- Fuel pressure regulator

8. Inspect the vacuum lines and connections. Look for any loose connections, sharp bends or damage.

9. Install or connect the following:

- Air cleaner
- Air cleaner hose

10. Start the engine and check for vacuum and fuel leaks.

4.7L Engine

1. Before servicing the vehicle, refer to the precautions in the beginning of this section.

2. Relieve the fuel system pressure.

3. Remove or disconnect the following:

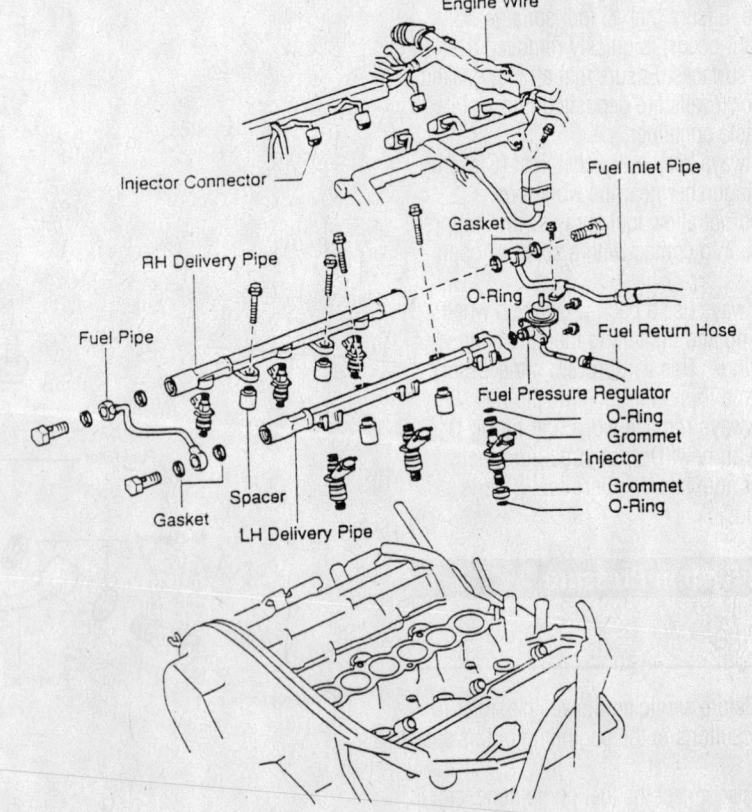

Fuel injector arrangement and related components—3.4L engine

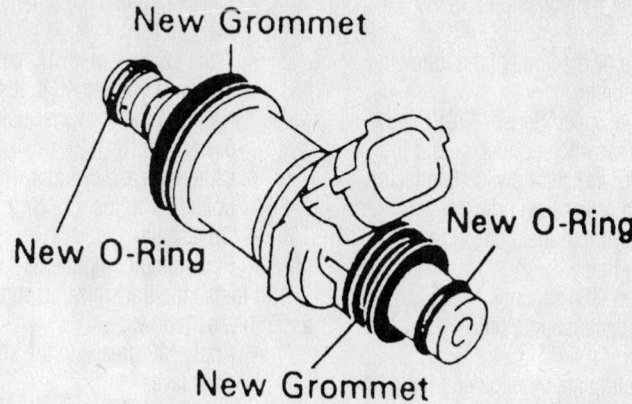

Install new O-rings and grommets on each injector—3.4L engine

- Negative battery cable
- Engine appearance cover
- Air intake tube
- Fuel lines
- Fuel pulsation damper
- Fuel pressure regulator vacuum line
- Accelerator cable and bracket
- Positive Crankcase Ventilation (PCV) valve and hose

- Evaporative Emissions (EVAP) vacuum switching valve
- Engine appearance cover brackets
- Fuel injector harness connectors
- Engine harness protector
- Fuel supply manifold crossover pipe
- Fuel supply manifolds with injectors attached
- Fuel injectors

To install:

4. Install the fuel injectors to the supply manifold with new O-ring seals and new grommets.

5. Install new injector insulators to the intake manifold.

6. Install or connect the following:
- Fuel supply manifolds with injectors attached. Tighten the bolts to 66 inch lbs. (7.5 Nm).
- Fuel supply manifold crossover pipe. Tighten the bolts to 29 ft. lbs. (39 Nm).
- Engine harness protector
- Fuel injector harness connectors
- Engine appearance cover brackets
- EVAP vacuum switching valve
- PCV valve and hose
- Accelerator cable and bracket
- Fuel pressure regulator vacuum line
- Fuel pulsation damper
- Fuel lines
- Air intake tube
- Engine appearance cover
- Negative battery cable

7. Start the engine and check for leaks.

DRIVE TRAIN

Manual Transmission Assembly

REMOVAL & INSTALLATION

2WD

1. Before servicing the vehicle, refer to the precautions in the beginning of this section.

2. Remove or disconnect the following:
- Negative battery cable

3. Drain the transmission oil.

4. Remove the shift lever assembly, as follows:
- Shift lever knob
- 4 screws, shift lever boot retainer and shift lever boot
- 6 bolts, shift lever assembly and baffle
- Turn over the dust boot
- Shift lever cap, cover it with a cloth
- Shift lever cap by pressing downward and rotating it counterclockwise
- Shift lever

5. Remove or disconnect the following:
- Driveshaft
- Vehicle Speed Sensor (VSS) and the back-up light switch connectors
- Oxygen (O₂) sensor connector

- Front exhaust pipe from the exhaust manifold and catalytic converter
- Clutch release cylinder
- Starter wires
- Starter

6. Position a jack and wooden block under the transmission.

7. Remove or disconnect the following:
- Rear endplate
- Rear engine mount bracket
- Crossmember

8. Attach a engine hoist to the engine hangers.

9. Remove or disconnect the following:
- Engine mounts
- Engine/transmission assembly out of the vehicle

10. Safely support the engine/transmission assembly.

11. Remove or disconnect the following:
- Transmission-to-engine bolts
- Transmission mount

To install:

12. Install or connect the following:
- Transmission
- Transmission mount. Tighten the 4 bolts to 48 ft. lbs. (65 Nm).

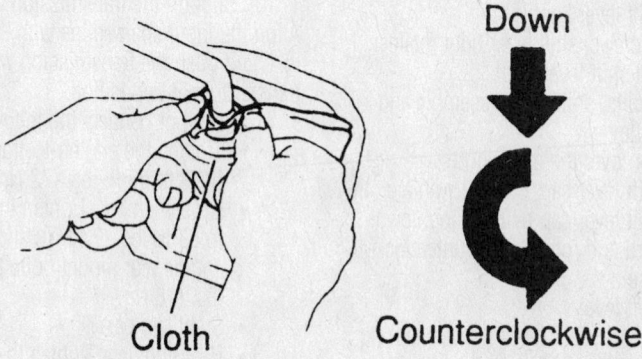

Removing the transfer shift lever

- Tighten the 6 transmission-to-engine bolts to 53 ft. lbs. (72 Nm).
- Crossmember. Torque the bolts to 53 ft. lbs. (72 Nm).
- Rear engine mount-to-crossmember bolts to 13 ft. lbs. (18 Nm).
- Starter. Tighten the 2 bolts to 29 ft. lbs. (39 Nm).
- Rear endplate. Tighten the 4 bolts to 27 ft. lbs. (37 Nm).

13. Install or connect the following:
- Tighten the engine mounts to 28 ft. lbs. (38 Nm).
- Starter wires
- Clutch release cylinder
- Driveshaft
- Front exhaust pipe. Tighten the exhaust pipe-to-manifold bolts to 46 ft. lbs. (62 Nm) and the exhaust pipe-to-catalytic converter bolts to 35 ft. lbs. (48 Nm).
- Remaining components
- Negative battery cable

14. Install the shift lever assembly, as follows:
- Shift lever
- Shift lever cap, cover it with a cloth
- Shift lever cap by pressing downward and rotating it clockwise
- Turn over the dust boot
- 6 bolts, shift lever assembly and baffle
- 4 screws, shift lever boot retainer and shift lever boot
- Shift lever knob

15. Fill the transmission with oil.
16. Start the engine and check for leaks.
17. Install the engine undercover.
18. Road test the vehicle and check all fluids.

4WD

1. Before servicing the vehicle, refer to the precautions in the beginning of this section.
2. Disconnect negative battery cable.
3. Remove the transmission shift lever assembly, as follows:
- Shift lever knob
- 4 screws, shift lever boot retainer and shift lever boot
- 6 bolts, shift lever assembly and baffle
- Turn over the dust boot
- Shift lever cap, cover it with a cloth
- Shift lever cap by pressing downward and rotating it counterclockwise
- Shift lever
- Transfer shift lever, using snapring pliers to pull it from the transfer case

4. Drain the transmission and the transfer oil.
5. Remove or disconnect the following:
- Driveshafts
- Vehicle Speed Sensor (VSS), back-up light switch connector and the transfer indicator switch connector
- Clutch release cylinder, move it aside without disconnecting the clutch line
- Oxygen (O_2) sensor
- Front exhaust pipe bracket
- Starter
- Rear endplate by removing the nuts and 2 bolts

6. Using a transmission jack, support the transmission.
7. Remove or disconnect the following:
- 4 engine rear mount bolts
- 8 bolts and the frame crossmember from the side frame
- 6 transmission-to-engine bolts
- 3 wire clamps from the transmission
- Transmission with the transfer case
- 4 engine rear mount bolts from the transfer case
- Transfer adapter rear mount bolts
- Transfer case from the transmission

To install:

8. Apply MP grease to the adapter oil seal and shift the 2 shift fork shafts to the high 4 position.
9. Install or connect the following:
- Transfer case to the transmission. Tighten the bolts to 17 ft. lbs. (24 Nm).

❋❋ WARNING

Be careful not to damage the oil seal by the input gear spline when installing the transfer.

- Engine rear mounting. Tighten the 4 bolts to 48 ft. lbs. (65 Nm).
- Transmission/transfer case assembly

10. Support the transmission with a jack. Align the input shaft spline with the clutch disc and push the transmission with the transfer fully into position.
11. Install or connect the following:
- Tighten the engine-to-transmission bolts to 53 ft. lbs. (72 Nm).
- Crossmember. Tighten the 4 bolts to 53 ft. lbs. (72 Nm) and the 4 engine rear mount bolts to 13 ft. lbs. (18 Nm).
- Stabilizer bar
- Rear endplate Tighten the 2 bolts and nuts to 27 ft. lbs. (37 Nm).

- Starter. Tighten the bolts to 29 ft. lbs. (39 Nm).
- Front exhaust pipe. Tighten the manifold bolts to 46 ft. lbs. (62 Nm) and the converter bolts to 35 ft. lbs. (48 Nm).
- Clutch release cylinder. Tighten the bolts to 9 ft. lbs. (12 Nm).
- Driveshafts
- Remaining components

12. Install the transmission shift lever assembly, as follows:
- Apply MP grease to the shift lever
- Shift lever
- Shift lever cap, cover it with a cloth
- Shift lever cap by pressing downward and rotating it clockwise
- Turn over the dust boot
- 6 bolts, shift lever assembly and baffle
- 4 screws, shift lever boot retainer and shift lever boot
- Shift lever knob

13. Connect the negative battery cable.
14. Start the engine and check for leaks.
15. Road test the vehicle for proper operation. Recheck all fluid levels.

Clutch Assembly

REMOVAL & INSTALLATION

1. Before servicing the vehicle, refer to the precautions in the beginning of this section.
2. Remove or disconnect the following:
- Negative battery cable
- Transmission assembly

3. Matchmark the clutch cover to the flywheel.
4. At the clutch cover, loosen each bolt 1 turn until spring tension is released.
5. Remove or disconnect the following:
- Clutch cover set bolts and the clutch cover with the clutch disc.
- Release bearing retaining clip and withdraw the it
- Release fork and boot assembly

To install:
6. Install or connect the following:
- Clutch disc onto the flywheel, using a clutch disc alignment tool
- Clutch cover, position it onto the flywheel and if reusing the old pressure plate, align the matchmarks.
- Clutch cover. Tighten the bolts in a crisscross pattern to 14 ft. lbs. (19 Nm).

7. Lubricate the release fork pivot and

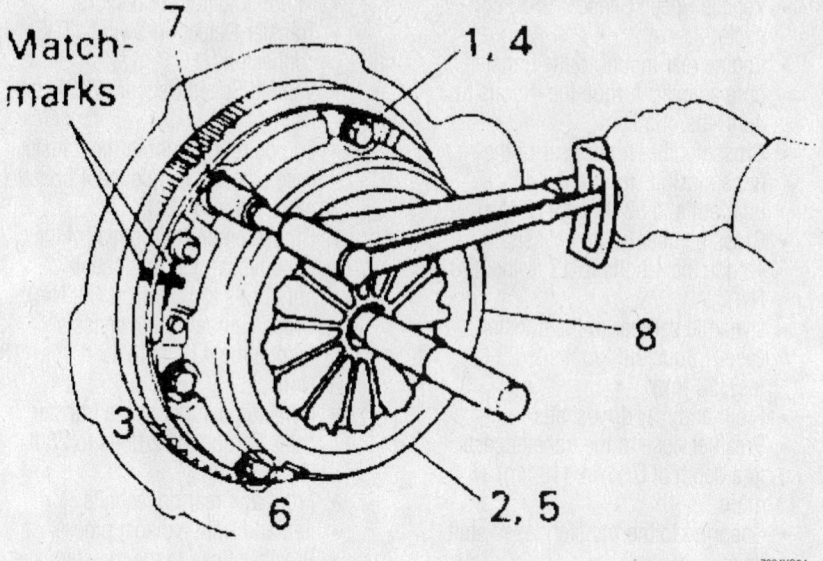

Match-marks

7   1, 4

8

3   6   2, 5

7924YG94

Bolt tightening sequence for the clutch cover—all engines

contact points, the release bearing, bearing hub and input shaft spline surfaces with a suitable molybdenum disulfide lithium based or multi-purpose grease.

8. Install or connect the following:
    • Boot, release fork, hub and the bearing assemblies
    • Transmission
    • Negative battery cable

Hydraulic Clutch System

BLEEDING

1. Before servicing the vehicle, refer to the precautions in the beginning of this section.

2. Fill the clutch reservoir with brake fluid. Check the reservoir level frequently and add fluid as needed.

3. Connect one end of a vinyl tube to the bleeder plug on the slave cylinder and submerge the other end into a clear container half-filled with brake fluid.

4. Slowly pump the clutch pedal several times.

5. Have an assistant hold the clutch pedal down and loosen the bleeder plug until fluid and/or air starts to run out of the bleeder plug. Close the bleeder plug while the pedal is held to the floor.

6. Repeat Steps 2 and 3 until all the air bubbles are removed from the system.

7. Tighten the bleeder plug when all the air is gone.

8. Refill the master cylinder to the proper level as required.

9. Check the system for leaks.

Automatic Transmission Assembly

REMOVAL & INSTALLATION

3.4L Engine

1. Before servicing the vehicle, refer to the precautions in the beginning of this section.

2. Remove or disconnect the following:

    • Negative battery cable
    • Throttle cable
    • Transmission dipstick
    • Oil filler tube and discard the O-ring

3. Shift the transfer case into **H4** position.

4. Remove or disconnect the following:

    • Transfer case shift lever knob
    • No. 1 engine under cover
    • Front and center exhaust pipes
    • Driveshaft(s)
    • No. 1 and 2 Vehicle Speed Sensor (VSS) connectors
    • Shift control cable
    • Oil cooler pipe
    • Automatic Transmission Fluid (ATF) sensor connector
    • Park/Neutral Position (PNP) switch
    • Rear endplate
    • Torque converter bolts
    • Crossmember
    • Engine rear mounting insulator
    • Starter
    • Transmission

To install:

5. Install or connect the following:

    • Transmission. Torque the bolts to 53 ft. lbs. (71 Nm).
    • Starter. Torque the bolts to 29 ft. lbs. (39 Nm).
    • Engine rear mounting insulator. Torque the bolts to 48 ft. lbs. (65 Nm).
    • Crossmember. Torque the nuts to 53 ft. lbs. (72 Nm) and the bolts to 13 ft. lbs. (18 Nm)..
    • Torque converter. Torque the bolts to 30 ft. lbs. (41 Nm).
    • Rear endplate. Torque the bolts to 13 ft. lbs. (18 Nm).
    • Park/Neutral Position (PNP) switch
    • Automatic Transmission Fluid (ATF) sensor connector
    • Oil cooler pipe. Torque the bolts to 25 ft. lbs. (34 Nm).
    • Shift control cable. Torque the bolts to 9 ft. lbs. (12 Nm).
    • No. 1 and 2 Vehicle Speed Sensor (VSS) connectors
    • Driveshaft(s)
    • Front and center exhaust pipes
    • No. 1 engine under cover
    • Transfer case shift lever knob
    • Oil filler tube using a new O-ring
    • Transmission dipstick
    • Throttle cable
    • Negative battery cable

4.7L Engine

1. Install or connect the following:

    • Transmission. Tighten the flange bolts to 53 ft. lbs. (72 Nm).
    • Transmission mount crossmember. Tighten the bolts to 37 ft. lbs. (50 Nm) and the nuts to 54 ft. lbs. (74 Nm).
    • Transmission oil cooler lines
    • Torque converter. Tighten the bolts to 35 ft. lbs. (48 Nm).
    • Motor actuator connector
    • L4 solenoid valve position switch connector
    • Center differential lock indicator switch connector
    • PNP switch connector
    • Transmission fluid temperature sensor connector
    • Solenoid harness connector
    • Overdrive clutch speed sensor connector
    • VSS sensor connectors
    • Front driveshaft. Tighten the fasteners to 59 ft. lbs. (80 Nm).
    • Rear driveshaft. Tighten the fasteners to 78 ft. lbs. (106 Nm).

- Exhaust front pipes
- Engine under covers
- Transfer case shift lever and rod
- Transmission gear select lever and rod
- Center console
- Transmission dipstick tube
- Coolant recovery reservoir
- Cooling fan and shroud
- Air intake assembly
- Battery and tray

2. Check the transmission and transfer case fluid levels and adjust as necessary.

Transfer Case Assembly

REMOVAL & INSTALLATION

VF2A Model

This transfer case is to be used with the 3.4L engine.

1. Before servicing the vehicle, refer to the precautions in the beginning of this section.
2. Drain the transfer case oil.
3. Shift the transfer shift lever into the **H4** position.
4. Remove or disconnect the following:
   - Transfer case shift lever knob
   - 4 screws, transfer case shift lever boot retainer and the shift lever boot
   - Snapring from the transfer case shift lever and the lever
   - Breather hose from the transfer case
   - Front and rear driveshafts
   - Dynamic damper from the transfer case
   - Crossmember from the rear of the transmission
   - 4 bolts and the engine rear mount from the transfer case adapter
   - Vehicle Speed Sensor (VSS) connector
   - Transfer Detection Switch (TDS) connector
5. Support the transfer case.
6. Remove or disconnect the following:
   - 8 transfer case-to-transfer adapter bolts
   - Transfer case

To install:

7. Install or connect the following:
   - Transfer case
   - Transfer case-to-transfer adapter bolts. Torque the 8 bolts to 17 ft. lbs. (24 Nm).
   - Transfer Detection Switch (TDS) connector

- Vehicle Speed Sensor (VSS) connector
- Engine rear mount to the transfer case adapter. Torque the 4 bolts to 48 ft. lbs. (65 Nm).
- Crossmember to the rear of the transmission. Torque the 4 nuts/bolts to 53 ft. lbs. (72 Nm).
- Crossmember to the chassis. Torque the 4 bolts to 13 ft. lbs. (18 Nm).
- Dynamic damper to the transfer case. Torque the 2 bolts to 28 ft. lbs. (38 Nm).
- Front and rear driveshafts
- Breather hose to the transfer case to a depth of 0.51 in. (13mm) or more
- Snapring to the transfer case's shift lever
- 4 screws, transfer case shift lever boot retainer and the shift lever boot
- Transfer case shift lever knob

8. Refill the transfer case to the correct level.
9. Test drive the vehicle.

VF2BM Model

This transfer case is to be used with the 4.7L engine.

1. Before servicing the vehicle, refer to the precautions in the beginning of this section.
2. Drain the transfer case oil.
3. Turn the touch select 2–4 switch **ON**.
4. Remove or disconnect the following:
   - Breather hose from the transfer case
   - Left and right exhaust pipes
   - Front and rear driveshafts
   - Crossmember from the rear of the transmission
   - 4 bolts and the engine rear mount from the transfer case adapter
   - Vehicle Speed Sensor (VSS) connector
   - Transfer Detection Switch (TDS) connectors
   - Motor actuator connectors
5. Support the transfer case.
6. Remove or disconnect the following:
   - 8 transfer case-to-transfer adapter bolts
   - Transfer case

To install:

7. Install or connect the following:
   - Transfer case
   - Transfer case-to-transfer adapter bolts. Torque the 8 bolts to 17 ft. lbs. (24 Nm).

- Motor actuator connectors
- Transfer Detection Switch (TDS) connector
- Vehicle Speed Sensor (VSS) connector
- Engine rear mount to the transfer case adapter. Torque the 4 bolts to 48 ft. lbs. (65 Nm).
- Crossmember to the rear of the transmission. Torque the 4 nuts/bolts to 53 ft. lbs. (72 Nm).
- Crossmember to the chassis. Torque the 4 bolts to 13 ft. lbs. (18 Nm).
- Dynamic damper to the transfer case. Torque the 2 bolts to 28 ft. lbs. (38 Nm).
- Front and rear driveshafts
- Left and right exhaust pipes
- Breather hose to the transfer case to a depth of 0.51 in. (13mm) or more

8. Refill the transfer case to the correct level.
9. Test drive the vehicle.

Halfshaft

REMOVAL & INSTALLATION

1. Before servicing the vehicle, refer to the precautions in the beginning of this section.
2. Remove or disconnect the following:
   - Front wheel
   - Under cover
3. Drain the differential oil.
4. Remove or disconnect the following:
   - Grease cap
   - Cotter pin and lock cap
   - Halfshaft locknut by applying the brakes
   - Lower control arm from the lower ball joint
   - Halfshaft from the steering knuckle, using a plastic hammer
   - Left strut, for the left halfshaft
   - Right halfshaft, using a brass bar and a hammer
   - Left halfshaft, using tools 09520-01010 and 09520-24010
   - Snapring from the inboard joint shaft

To install:

5. Install or connect the following:
   - New snapring, onto the inboard joint shaft with the opening facing downward
   - Halfshafts to the differential using a brass bar and a hammer
   - Halfshafts to the steering knuckles

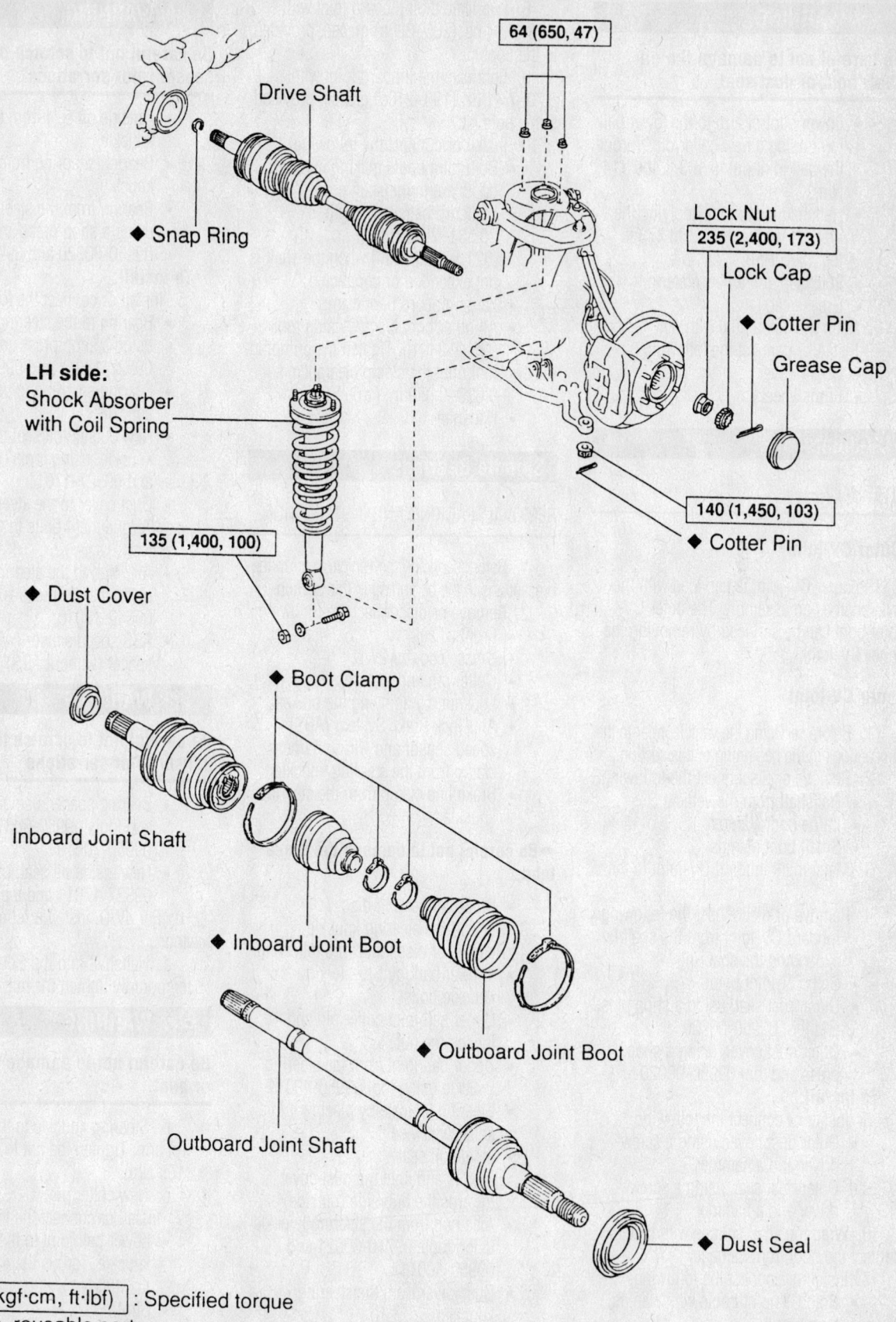

64 (650, 47)

Drive Shaft

◆ Snap Ring

Lock Nut
235 (2,400, 173)

Lock Cap

◆ Cotter Pin

Grease Cap

LH side:
Shock Absorber
with Coil Spring

135 (1,400, 100)

140 (1,450, 103)

◆ Cotter Pin

◆ Dust Cover

◆ Boot Clamp

Inboard Joint Shaft

◆ Inboard Joint Boot

◆ Outboard Joint Boot

Outboard Joint Shaft

◆ Dust Seal

N·m (kgf·cm, ft·lbf) : Specified torque

N ◆ Non–reusable part

9308YG12

View of the halfshaft and related components

Be careful not to damage the oil seal, boot or dust seal.

- Lower control arm to the lower ball joint using a new cotter pin. Torque the ball joint nut to 103 ft. lbs. (140 Nm).
- Halfshaft locknut by applying the brakes. Torque the nut to 173 ft. lbs. (235 Nm).
- Lock cap and a new cotter pin
- Grease cap
6. Refill the differential with oil.
7. Install or connect the following:
   - Under cover
   - Front wheel

CV-Joints

OVERHAUL

Outer CV-joint

The outer CV-joint is serviced with the axle shaft as an assembly. The outer CV-joint boot can be serviced by removing the inner CV-joint.

Inner CV-joint

1. Before servicing the vehicle, refer to the precautions in the beginning of this section.
2. Remove or disconnect the following:
   - Halfshaft from the vehicle
   - Large boot clamps
   - Small boot clamps
3. Matchmark inboard CV-joint to the shaft
4. Remove or disconnect the following:
   - Inboard CV-joint from the shaft by expanding the snapring
   - Both CV-joint boots
   - Outer dust seal, using a shop press and tool 09950-00020
   - Outer dust cover, using a shop press and tool 09950-00020

To install:
5. Install or connect the following:
   - Outer dust cover, using a screwdriver and a hammer
   - Outer dust seal, using a screwdriver and a hammer
6. Wrap the shaft splines with tape to protect the boot from damage.
7. Install or connect the following:
   - Both CV-joint boots with clamps, temporarily
   - Inboard CV-joint to the shaft by aligning the matchmarks and expanding the snapring

8. Lubricate the outboard joint with 7.23–7.94 oz. (205–225g) grease, provided in the boot kit.
9. Lubricate the inboard joint with 6.70–7.41 oz. (190–210g) grease, provided in the boot kit.
10. Install or connect the following:
   - Both joint boots making sure the boots are in the shaft groove
   - Standard halfshaft length is 20.531–20.689 in. (521.5–525.5mm) when the shaft is not expanded or contracted
   - Large inboard boot clamp
   - All other boot clamps using tool 09521-24010. Tighten the crimping tool until the clamp clearance is 0.039–0.059 in. (1.0–1.5mm)
   - Halfshaft

Spindle Bearings

REMOVAL, PACKING AND INSTALLATION

1. Before servicing the vehicle, refer to the precautions in the beginning of this section.
2. Remove or disconnect the following:
   - Front wheel
   - Grease cap, for 2WD
   - Cotter pin and lock cap, for 4WD
   - Locknut, by applying the brakes
   - Anti-lock Brake System (ABS) speed sensor and wiring harness clamp from the steering knuckle
   - Brake line clamp from the steering knuckle

➡ **Be careful not to damage the brake tube.**

   - Brake caliper and disc
3. Support the steering knuckle
4. Remove or disconnect the following:
   - 4 lower ball joint-to-steering knuckle bolts
   - Upper ball joint cotter pin and loosen the nut
   - Upper ball joint from the steering knuckle using tool 09950-40011
   - Steering knuckle by placing it in a soft-jawed vise
   - Inside oil seal
   - 4 bolts and shift the dust cover towards the hub side (outside)
   - Axle hub from the steering knuckle using tools 09710-30021 and 09950-40011
   - Dust cover from the steering knuckle
   - Bearing spacer and ABS speed sensor (with ABS) or spacer (without ABS)

Be careful not to scratch the speed sensor rotor serrations

   - Outside oil seal from the steering knuckle
   - Bearing snapring from the steering knuckle
   - Bearing from the steering knuckle, using a shop press and tools 09950-60020 and 09950-70010

To install:
5. Install or connect the following:
   - Bearing to the steering knuckle, using a shop press and tools 09527-17011 and 09950-60020
   - Bearing snapring to the steering knuckle
   - New outside oil seal to the steering knuckle, using tools 09223-15030 and 09527-17011
   - Dust cover to the steering knuckle. Torque the 4 bolts to 13 ft. lbs. (18 Nm).
   - Axle hub to the steering knuckle using a shop press and tool 09649-17010
   - ABS speed sensor (with ABS) or spacer (without ABS)

Be careful not to scratch the speed sensor rotor serrations

   - Bearing spacer, using a shop press and tools 09950-60010 and 09950-70010
   - New inside oil seal, using tool 09527-17011 and a plastic hammer
6. For 4WD, install or connect the following:
   a. Halfshaft into the axle hub and temporarily tighten the nut

Be careful not to damage the oil seal or boot.

   b. Steering knuckle to the upper control arm. Tighten the nut to 77 ft. lbs. (105 Nm).
   c. New cotter pin.
7. Install or connect the following:
   - Lower ball joint to the steering knuckle. Torque the 4 bolts to 59 ft. lbs. (80 Nm).
   - Strut
   - Brake disc and caliper. Torque both caliper bolts to 90 ft. lbs. (123 Nm).
   - Brake line clamp to the steering

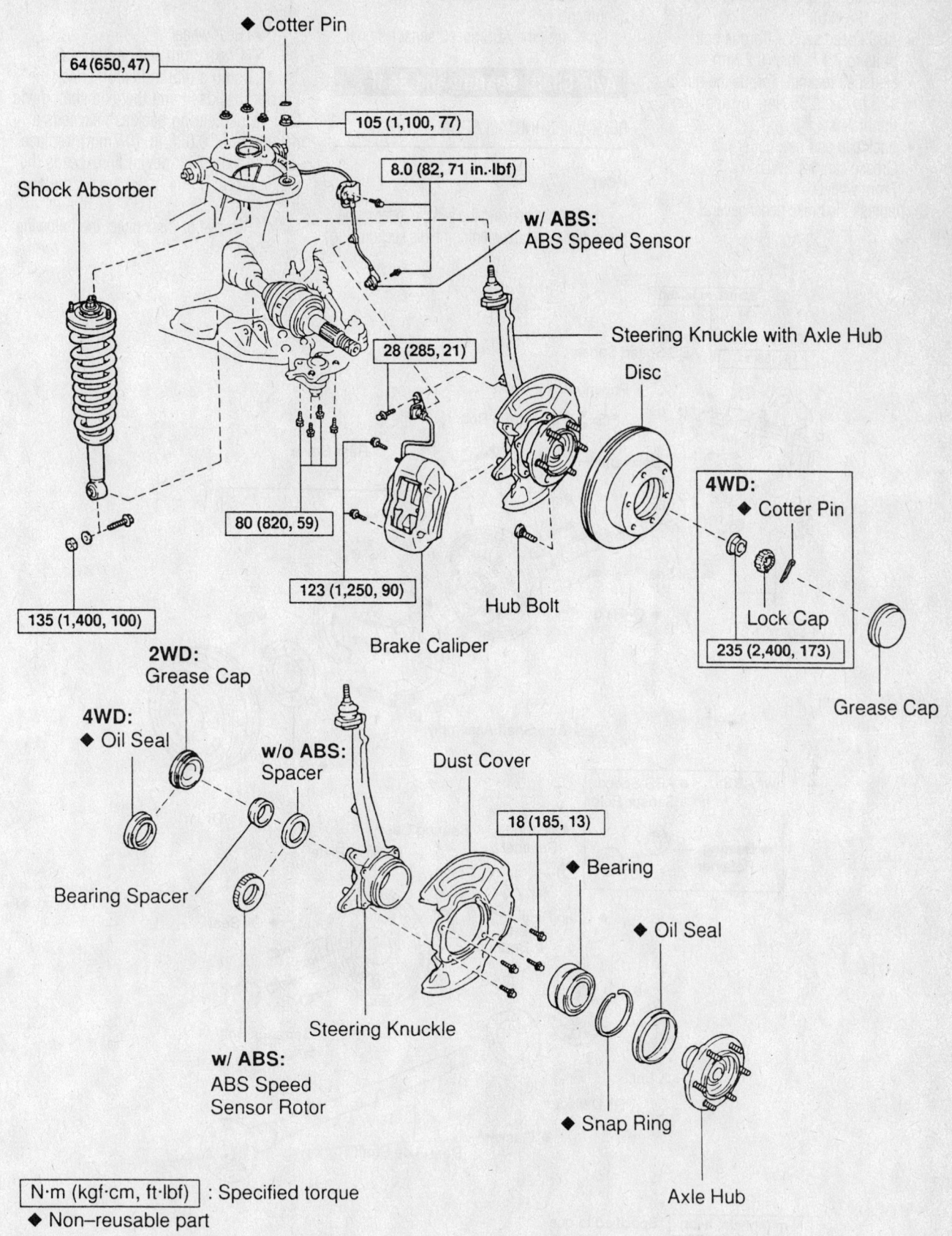

Cotter Pin

64 (650, 47)

105 (1,100, 77)

Shock Absorber

8.0 (82, 71 in.·lbf)

w/ ABS:
ABS Speed Sensor

Steering Knuckle with Axle Hub

Disc

28 (285, 21)

4WD:
Cotter Pin

Lock Cap

235 (2,400, 173)

80 (820, 59)

Grease Cap

135 (1,400, 100)

123 (1,250, 90)

Hub Bolt

Brake Caliper

2WD:
Grease Cap

4WD:
Oil Seal

w/o ABS:
Spacer

Dust Cover

18 (185, 13)

Bearing

Bearing Spacer

Oil Seal

Steering Knuckle

Snap Ring

w/ ABS:
ABS Speed
Sensor Rotor

Axle Hub

N·m (kgf·cm, ft·lbf) : Specified torque
◆ Non–reusable part

9308YG13

Exploded view of the front axle hub and related components

knuckle. Torque the bolt to 21 ft. lbs. (28 Nm).
- ABS speed sensor. Torque both bolts to 7.1 ft. lbs. (8.2 Nm).
- Halfshaft locknut. Torque the nut to 173 ft. lbs. (235 Nm), by applying the brakes.
- Lock cap and new cotter pin
- Grease cap, for 2WD
- Front wheel

8. Depress the brake pedal several times.

9. Check and/or adjust the front wheel alignment.
10. Check the ABS speed sensor signal.

Axle Shaft, Bearing and Seal

REMOVAL & INSTALLATION

Rear

1. Before servicing the vehicle, refer to the precautions in the beginning of this section.

2. Remove or disconnect the following:
- Rear wheel
- Brake drum and gasket

3. Using a dial indicator, check the bearing backlash and the axle shaft deviation. If the bearing backlash exceeds a maximum or 0.028 in. (0.7mm), replace it. If the axle shaft deviation exceeds the maximum of 0.004 in. (0.1mm), replace it.

4. Remove or disconnect the following:

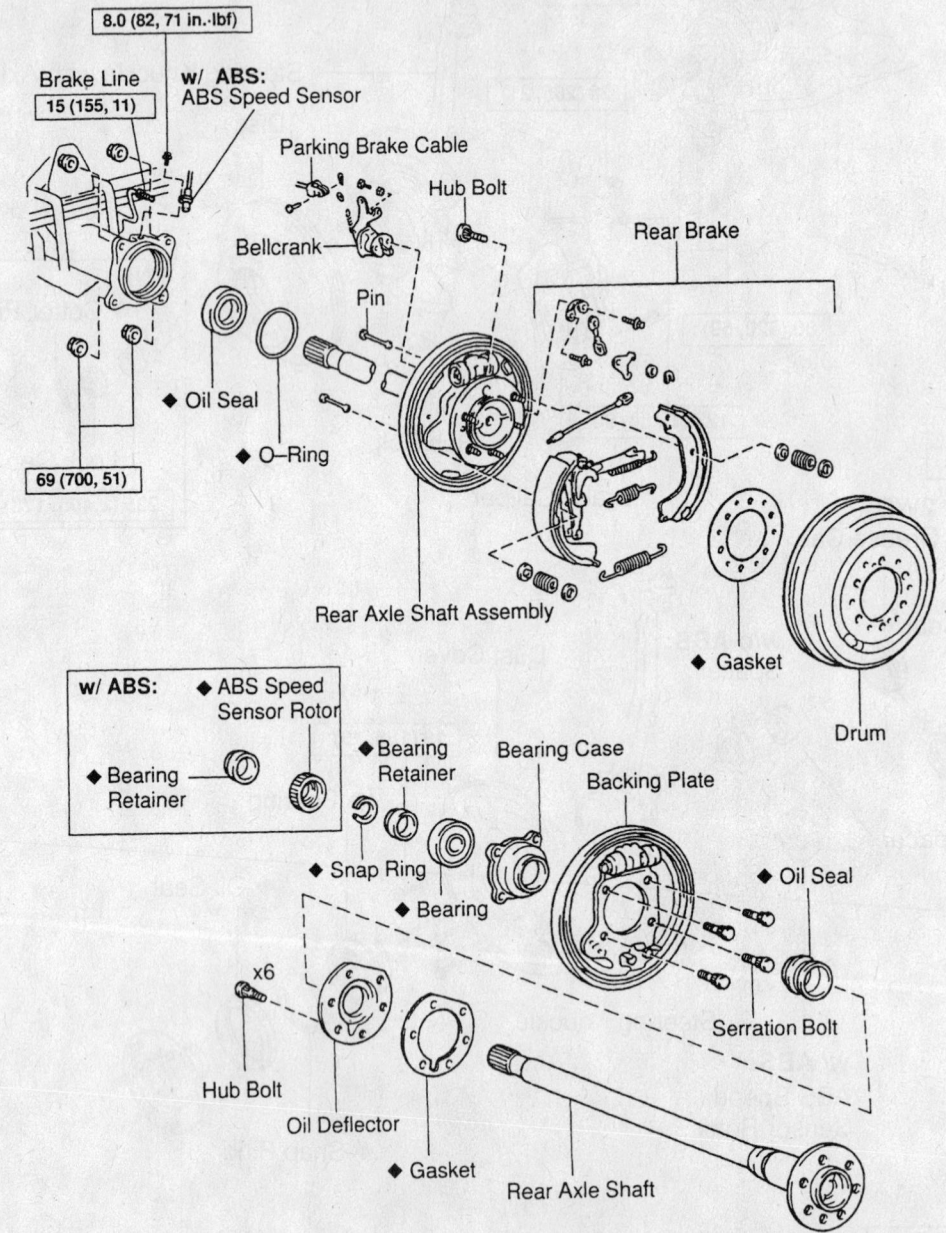

N·m (kgf·cm, ft·lbf) : Specified torque

◆ Non–reusable part

9308YG14

Exploded view of the rear axle

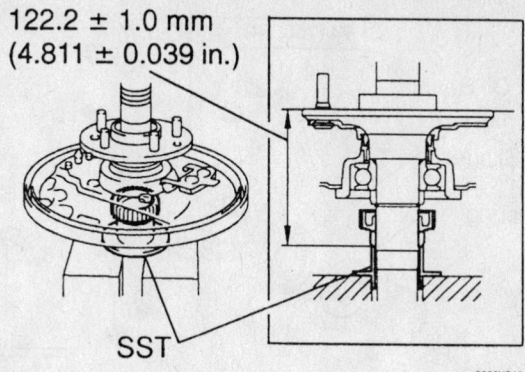

122.2 ± 1.0 mm
(4.811 ± 0.039 in.)

SST

9308YG15

Standard length of ABS speed sensor rotor and bearing retainer—Rear axle

- Anti-lock Brake System (ABS) speed sensor from the rear axle housing, if equipped
- Brake line from the wheel cylinder, using tool 09023-00100
- Parking brake cable
- 4 backing plate nuts
- Axle shaft assembly, by pulling it from the axle housing

✸✸ WARNING

Be careful not to damage the oil seal.

- O-ring from the rear axle housing
- Inner side oil seal using tool 09308-00010

5. If equipped with ABS, perform the following:

    a. Remove and discard the 4 serration bolt nuts; then, using a hammer, drive the bolts from the backing plate.

    b. Using a grinder, grind the retainer and sensor rotor surfaces; then, chisel them out.

6. Remove the snapring from the axle shaft.

7. Remove the axle shaft from the backing plate, as follows:

    a. Position tool 09521-25011 onto the backing plate with the 4 nuts.

    b. Using a shop press, remove the axle shaft and bearing retainer from the backing plate.

8. Using tool 09308-00010, pull the oil seal from the backing plate.

9. Using a shop press and tools 09223-56010 and 09950-60010, press the bearing from the backing plate.

To install:

10. Install or connect the following:
- Bearing into the backing plate, using a shop press and tools 09223-56010 and 09950-60010
- New O-ring to the rear axle housing

- New oil seal into the backing plate, using a hammer and tools 09950-70010 and 09950-60010

11. Install the axle shaft to the backing plate, as follows:
- New outer side seal, lubricate the oil seal lip with multi-purpose grease
- Backing plate and bearing retainer onto the rear axle shaft
- Axle shaft onto the backing plate, by pressing it using a shop press and tool 09316-60011
- New snapring

✸✸ WARNING

Be careful not to damage the oil seal.

12. Install or connect the following:
- New sensor rotor and new bearing retainer onto the axle shaft, using a shop press and tool 09316-60011 to a standard length of 4.77–4.85 in. (121.2–123.2mm), if equipped with ABS
- New inner side oil seal, using a hammer and tools 09950-60020 and 09950-70010
- Axle shaft assembly. Torque the bolts to 51 ft. lbs. (69 Nm).

✸✸ WARNING

Be careful not to damage the oil seal.

- Parking brake cable
- Brake line to the wheel cylinder, using tool 09023-00100. Torque the brake line to 11 ft. lbs. (15 Nm).
- Rear brake assembly
- ABS speed sensor to the rear axle housing. Torque it to 7.1 ft. lbs. (8.0 Nm).

13. Using a dial indicator, check the bearing backlash and the axle shaft deviation. If the bearing backlash exceeds a maximum or 0.028 in. (0.7mm), replace it. If the axle shaft deviation exceeds the maximum of 0.004 in. (0.1mm), replace it.

14. Install or connect the following:
- New gasket and brake drum
- Rear wheel. Torque the lug nuts to 81 ft. lbs. (110 Nm).

15. Bleed the brake system.

16. Check the ABS speed sensor signal.

Pinion Seal

REMOVAL & INSTALLATION

Front

1. Before servicing the vehicle, refer to the precautions in the beginning of this section.

2. Remove the under cover.

3. Drain the differential housing oil.

4. Remove the front driveshaft.

5. Remove the companion flange, as follows:
- Loosen the staked part of the nut, using a chisel and a hammer
- Companion flange nut, using tool 09330-00021
- Companion flange, using tools 09950-30011 and 09954-03010

6. Remove the oil seal and slinger, as follows:
- Oil seal, using tool 09308-10010
- Oil slinger

To install:

7. Install or connect the following:
- Oil slinger
- New oil seal, using a hammer and tool 09554-22010 to a depth of 0.153–0.189 in. (4.2–4.8mm).

8. Install the companion flange, as follows:
- Companion flange
- New nut, lubricated with hypoid gear oil
- Torque the nut to 80 ft. lbs. (108 Nm), using tool 09330-00021.

9. Adjust the drive pinion preload

10. Rotate the drive pinion, using a torque wrench while tightening the flange nut to make sure the bearing preload is 10.4–16.5 inch lbs. (1.2–1.9 Nm) for a new bearing or 5.2–8.7 inch lbs. (0.6–1.0 Nm) for a used bearing. Tighten the flange nut to achieve the preload torque readings originally recorded.

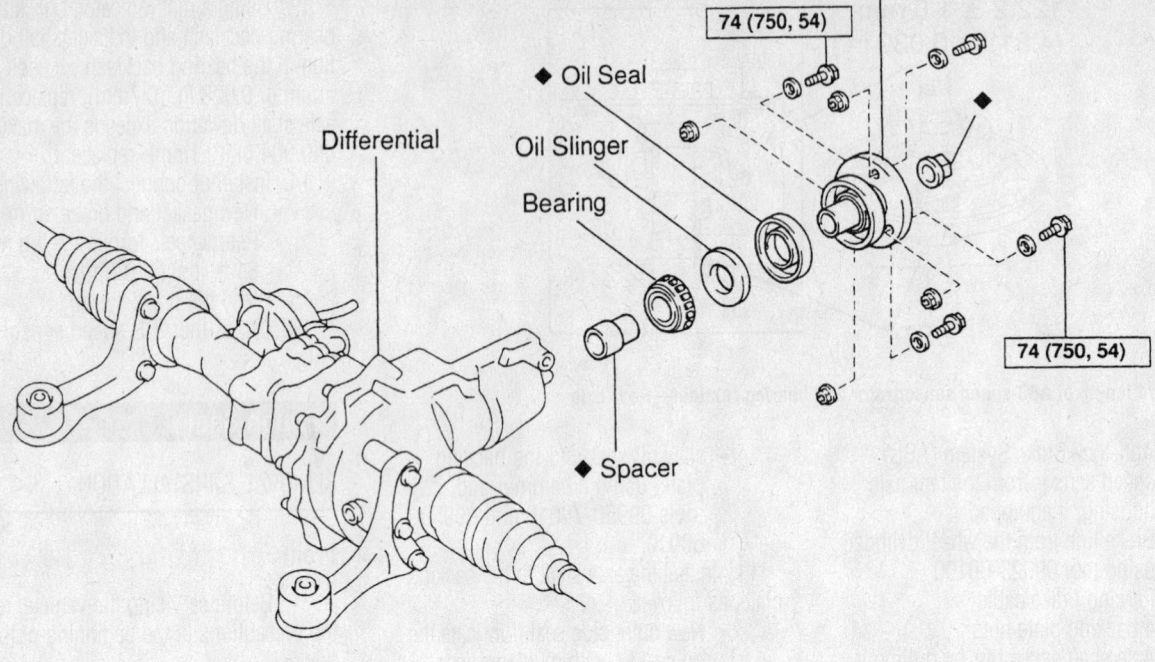

74 (750, 54)

◆ Oil Seal

Differential

Oil Slinger

Bearing

74 (750, 54)

◆ Spacer

N·m (kgf·cm, ft·lbf) : Specified torque

◆ Non–reusable part

9308YG16

Exploded view of the front differential assembly—Rear differential assembly is similar

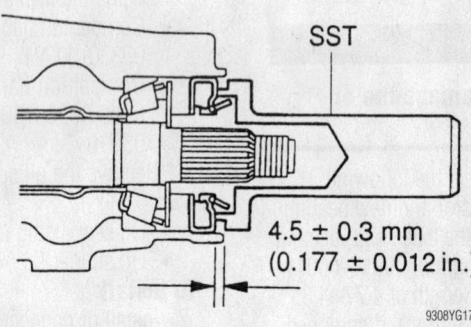

SST

4.5 ± 0.3 mm
(0.177 ± 0.012 in.)

9308YG17

Positioning the front pinion seal in the differential housing—Rear differential assembly is similar

※ CAUTION

Never loosen the pinion nut to reduce bearing preload.

11. Install or connect the following:
- Drive pinion nut, stake it
- Front driveshaft. Tighten the fasteners to 54 ft. lbs. (74 Nm).
- Under cover

12. Fill the differential with gear lubricant and check for leaks.

Rear

1. Before servicing the vehicle, refer to the precautions in the beginning of this section.

2. Drain the differential housing oil.

3. Remove the rear driveshaft.

4. Remove the companion flange, as follows:
- Loosen the staked part of the nut, using a chisel and a hammer
- Companion flange nut, using tool 09330-00021
- Companion flange, using tools 09950-30011 and 09954-03010
- Oil seal, using tool 09308-10010

To install:

5. Install the new oil seal until it is flush with the housing, using a plastic hammer and tools 09316-12010 and 09649-17010

➡ **Use vinyl tape to connect both oil seal installation tools.**

6. Install the companion flange, as follows:
- Companion flange
- New nut, lubricated with hypoid gear oil
- Torque the nut to 109 ft. lbs. (147 Nm), using tool 09330-00021.

7. Adjust the drive pinion preload

8. Rotate the drive pinion, using a torque wrench while tightening the flange nut to make sure the bearing preload is 11.4–16.7 inch lbs. (1.3–1.9 Nm) for a new bearing or 4.3–6.9 inch lbs. (0.5–0.8 Nm) for a used bearing. Tighten the flange nut to achieve the preload torque readings originally recorded.

※ CAUTION

Never loosen the pinion nut to reduce bearing preload.

9. Install or connect the following:
- Drive pinion nut, stake it
- Rear driveshaft. Tighten the fasteners to 54 ft. lbs. (74 Nm).

10. Refill the differential with gear lubricant and check for leaks; 3.33 qts. for 2WD or 3.12 qts. for 4WD.

STEERING AND SUSPENSION

Air Bag

✳✳ CAUTION

Some vehicles are equipped with an air bag system. The system must be disarmed before performing service on, or around, system components, the steering column, instrument panel components, wiring and sensors. Failure to follow the safety precautions and the disarming procedure could result in accidental air bag deployment, possible injury and unnecessary system repairs.

PRECAUTIONS

Several precautions must be observed when handling the inflator module to avoid accidental deployment and possible personal injury.

• Never carry the inflator module by the wires or connector on the underside of the module.

• When carrying a live inflator module, hold securely with both hands and ensure that the bag and trim cover are pointed away.

• Place the inflator module on a bench or other surface with the bag and trim cover facing up.

• With the inflator module on the bench, never place anything on or close to the module which may be thrown in the event of an accidental deployment.

DISARMING

To avoid personal injury when working on vehicles equipped with an air bag, the negative battery cable must be disconnected and at least 90 seconds must elapse before working on the system. Failure to do so may result in deployment of the air bag.

Power Rack And Pinion Steering Gear

REMOVAL & INSTALLATION

1. Before servicing the vehicle, refer to the precautions in the beginning of this section.

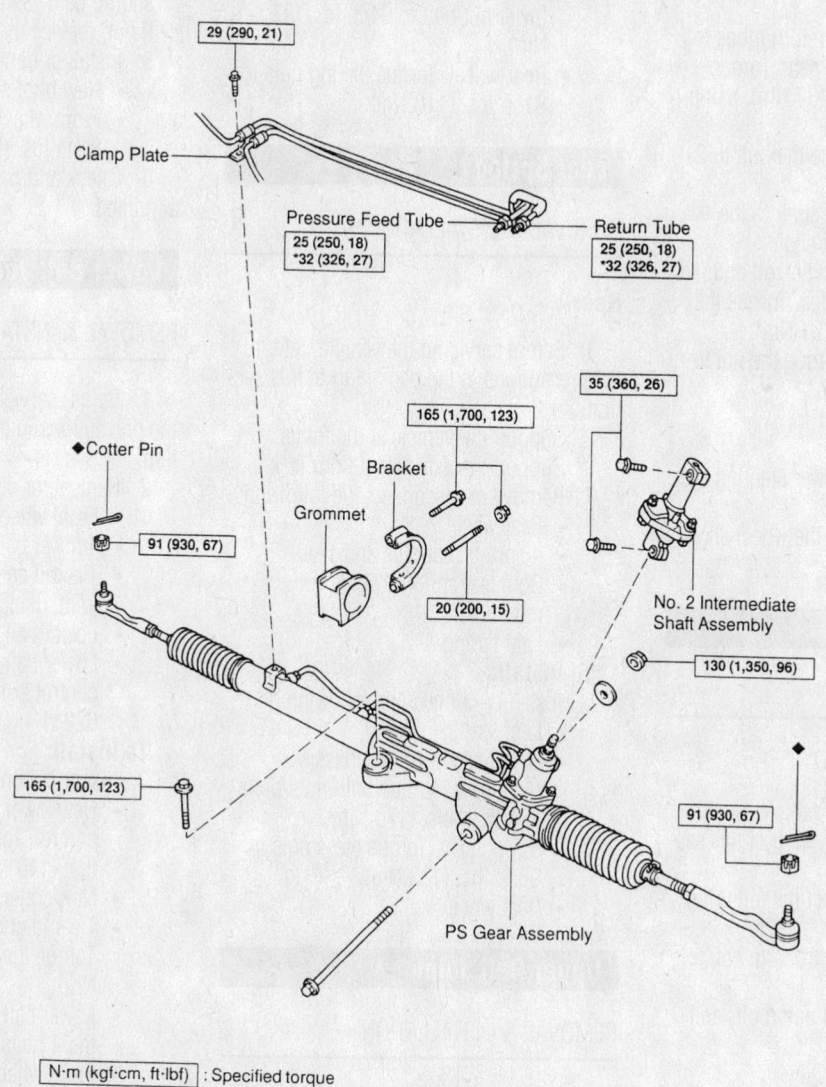

29 (290, 21)

Clamp Plate

Pressure Feed Tube
25 (250, 18)
*32 (326, 27)

Return Tube
25 (250, 18)
*32 (326, 27)

35 (360, 26)

165 (1,700, 123)

◆Cotter Pin

Bracket

Grommet

91 (930, 67)

20 (200, 15)

No. 2 Intermediate
Shaft Assembly

130 (1,350, 96)

165 (1,700, 123)

91 (930, 67)

PS Gear Assembly

N·m (kgf·cm, ft·lbf) : Specified torque
◆Non–reusable part
* For use with SST

9308YG18

Exploded view of the power rack and pinion steering gear mounting

2. Position the front wheels in the straight-ahead position.

3. Remove or disconnect the following:
- Engine under cover
- Steering wheel pad
- Steering wheel
- Left and right outer tie-rod ends from the steering knuckles

4. Matchmark the No. 2 intermediate shaft to the steering gear input shaft.

5. Remove or disconnect the following:
- Clamp plate
- Pressure feed and return tubes from the power steering gear, using tool 09631-22020
- Power steering gear assembly

To install:

6. Install or connect the following:
- Power steering gear assembly. Torque the set bolt to 123 ft. lbs. (165 Nm) and the set nut/bolt to 96 ft. lbs. (91 Nm).
- Pressure feed and return tubes to the power steering gear. Torque them to 27 ft. lbs. (32 Nm), using tool 09631-22020.
- Clamp plate. Torque the bolt to 21 ft. lbs. (29 Nm).
- No. 2 intermediate shaft to the steering gear input shaft
- Left and right outer tie-rod ends to the steering knuckles. Torque the nuts to 67 ft. lbs. (91 Nm).
- Steering wheel. Torque the nut to 26 ft. lbs. (35 Nm).
- Steering wheel pad
- Engine under cover

7. Fill and bleed the power steering system.

8. Check and/or adjust the wheel alignment, as necessary.

Strut

REMOVAL & INSTALLATION

Front

1. Before servicing the vehicle, refer to the precautions in the beginning of this section.

2. Remove or disconnect the following:
- Front wheel
- Strut-to-lower control arm nut/bolt and the strut
- Strut-to-chassis 3 nuts/bolts and the strut

To install:

3. Install or connect the following:
- Strut to the chassis. Torque the 3 nuts/bolts to 47 ft. lbs. (64 Nm).

- Strut to the lower control arm. Torque the nut/bolt to 100 ft. lbs. (135 Nm).
- Front wheel

Shock Absorber

REMOVAL & INSTALLATION

Rear

1. Before servicing the vehicle, refer to the precautions in the beginning of this section.

2. Remove or disconnect the following:
- Rear wheel
- Shock absorber

To install:

3. Install or connect the following:
- Shock absorber. Tighten the upper nut to 15 ft. lbs. (20 Nm) and the lower nut/bolt to 64 ft. lbs. (87 Nm).
- Rear wheel. Torque the lug nuts to 81 ft. lbs. (110 Nm).

Leaf Spring

REMOVAL & INSTALLATION

Rear

1. Before servicing the vehicle, refer to the precautions in the beginning of this section.

2. Support the vehicle at the frame.

3. Support the axle with a floor jack.

4. Remove or disconnect the following:
- Rear wheel
- 4 spring seat nuts and seat
- Both leaf spring-to-chassis nuts/bolts
- Leaf spring

To install:

5. Install or connect the following:
- Leaf spring
- Both leaf spring-to-chassis nuts/bolts. Torque both nuts/bolts to 125 ft. lbs. (170 Nm).
- Spring seat. Torque the 4 nuts to 98 ft. lbs. (133 Nm).
- Rear wheel

Upper Ball Joint

REMOVAL & INSTALLATION

1. Before servicing the vehicle, refer to the precautions in the beginning of this section.

2. Remove or disconnect the following:
- Front wheel
- Steering knuckle with the axle hub
- Wire and boot
- Snapring
- Upper ball joint from the steering knuckle, using a deep socket wrench and tool 09050-40011

To install:

3. Install or connect the following:
- New upper ball joint to the steering knuckle, using a deep socket and tool 09309-37010
- New snapring

4. Using a torque wrench, inspect the upper ball joint rotation, as follows:

a. Flip the ball joint back-and-forth 5 times.

b. Using a torque wrench, continuously turn the nut 1 turn in 2–4 seconds.

c. Take the reading on the 5th turn; it should be 6–39 inch lbs. (0.7–4.4 Nm). If not, replace the upper ball joint.

5. Install or connect the following:
- New boot secured with a wire
- Front wheel. Torque the lug nuts to 81 ft. lbs. (110 Nm).

6. Check and/or adjust the front wheel alignment.

Lower Ball Joint

REMOVAL & INSTALLATION

1. Before servicing the vehicle, refer to the precautions in the beginning of this section.

2. Remove or disconnect the following:
- Front wheel
- 4 lower ball joint set bolts
- Tie-rod end from the lower ball joint, using tool 09610-20012
- Lower ball joint nut.
- Lower ball joint from the lower control arm, using tool 09628-62011

To install:

3. Install or connect the following:
- New lower ball joint to the lower control. Torque the bolts to 103 ft. lbs. (140 Nm).
- New cotter pin
- Tie-rod end to the lower ball joint. Torque the nut to 67 ft. lbs. (91 Nm).
- Lower ball joint set bolts. Torque the 4 bolts to 59 ft. lbs. (80 Nm).
- Front wheel. Torque the lug nuts to 81 ft. lbs. (110 Nm).

4. Check and/or adjust the front wheel alignment.

Upper Control Arm

REMOVAL & INSTALLATION

1. Before servicing the vehicle, refer to the precautions in the beginning of this section.

2. Remove or disconnect the following:
   - Front wheel
   - Strut
   - Wheel speed sensor harness, if equipped with Anti-lock Brake System (ABS)
3. Upper ball joint, as follows:

   - Cotter pin and loosen the nut
   - Upper ball joint from the upper control arm, using tool 09950-40011
   - Steering knuckle, support it securely
   - Upper ball joint nut
4. Remove or disconnect the following:

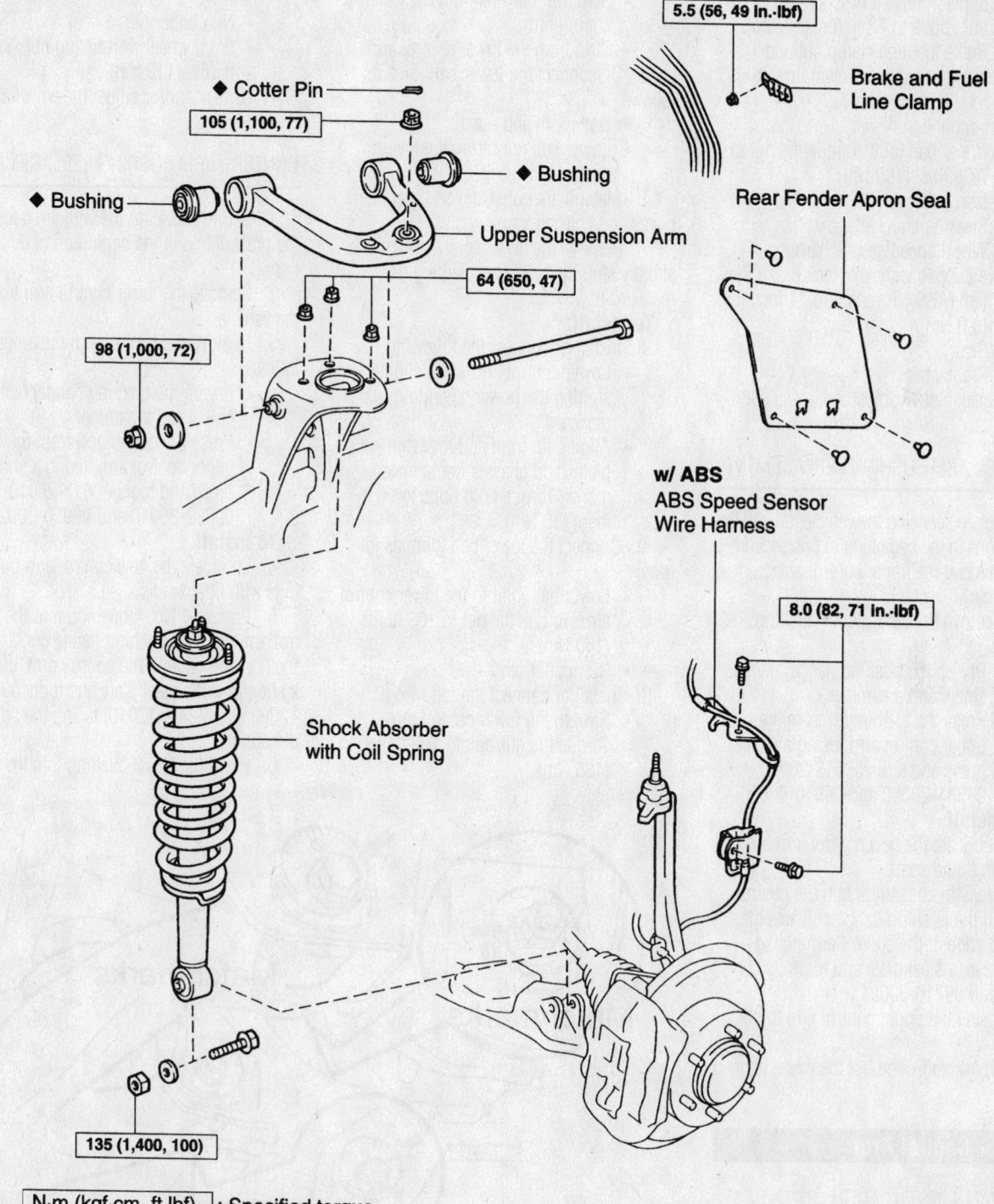

◆ Cotter Pin

105 (1,100, 77)

◆ Bushing

◆ Bushing

Upper Suspension Arm

64 (650, 47)

98 (1,000, 72)

5.5 (56, 49 in.·lbf)

Brake and Fuel Line Clamp

Rear Fender Apron Seal

w/ ABS
ABS Speed Sensor Wire Harness

8.0 (82, 71 in.·lbf)

Shock Absorber with Coil Spring

135 (1,400, 100)

N·m (kgf·cm, ft·lbf) : Specified torque
◆ Non–reusable part

9308YG19

Exploded view of the front suspension and related components

- 4 clips and the fender apron seal
- Brake/fuel line clamp nut and clamp
- Both upper control arm-to-chassis nuts/bolts
- Upper control arm

To install:

5. Install or connect the following:
   - Upper control arm. Torque both upper control arm-to-chassis nuts/bolts to 72 ft. lbs. (98 Nm).
   - Brake/fuel line clamp nut and clamp. Torque the clamp nut to 49 inch lbs. (5.5 Nm).
   - Fender apron seal
   - Upper ball joint. Torque the nut to 77 ft. lbs. (105 Nm).
   - New cotter pin
   - Steering knuckle
   - Wheel speed sensor harness, if equipped with Anti-lock Brake System (ABS). Torque it to 71 inch lbs. (8.0 Nm).
   - Strut
   - Front wheel

6. Check and/or adjust the wheel alignment.

CONTROL ARM BUSHING REPLACEMENT

1. Before servicing the vehicle, refer to the precautions in the beginning of this section.
2. Remove the upper control arm from the vehicle.
3. Remove the control arm bushings, as follows:
   - Pry up the bushing flange, using a chisel and a hammer
   - Press the bushing(s) from the upper control arm, using a shop press and tools 09613-26010, 09631-20060 and 09950-00020

To install:

4. Lubricate the new control arm bushings with liquid soap.
5. Press the bushings into the control arm until the bushing flange contacts the housing edge of the control arm, using a shop press, a steel plate and tools 09631-12090 and 09710-30021
6. Install the upper control arm to the vehicle.
7. Check and/or adjust the wheel alignment.

Lower Control Arm

REMOVAL & INSTALLATION

1. Before servicing the vehicle, refer to the precautions in the beginning of this section.

2. Remove front wheel.
3. Disconnect the tie-rod end, as follows:
   - Cotter pin and nut
   - Tie-rod end from the lower ball joint, using tool 09610-20012
4. Remove or disconnect the following:
   - Power steering gear set bolts and nuts
   - Stabilizer bar link from the lower control arm
   - Strut from the lower control arm
5. Disconnect the lower ball joint, as follows:
   - Cotter pin and nut
   - Lower ball joint from the lower control arm
6. Matchmark both front and rear cam plates and chassis frame.
7. Remove the lower control arm while slightly shifting the power steering gear rearward.

To install:

8. Install or connect the following:
   - Lower control arm while slightly shifting the power steering gear rearward
   - Align both front and rear cam plates and chassis frame matchmarks. Torque both bolts to 96 ft. lbs. (130 Nm).
9. Connect the lower ball joint, as follows:
   - Lower ball joint to the lower control arm. Torque the nut to 103 ft. lbs. (140 Nm).
   - New cotter pin
10. Install or connect the following:
    - Strut to the lower control arm. Torque the nut/bolt to 100 ft. lbs. (135 Nm).

- Stabilizer bar link to the lower control arm. Torque the nut to 51 ft. lbs. (69 Nm).
- Power steering gear set bolts and nuts. Torque the set bolt and clamp nut/bolt to 122 ft. lbs. and the set nut/bolt to 96 ft. lbs. (130 Nm).
- Tie-rod end to the lower ball joint. Torque the nut to 67 ft. lbs. (91 Nm).
- New cotter pin
- Front wheel. Torque lug nuts to 81 ft. lbs. (110 Nm).

11. Check and/or adjust the wheel alignment.

CONTROL ARM BUSHING REPLACEMENT

1. Before servicing the vehicle, refer to the precautions in the beginning of this section.
2. Remove the lower control arm from the vehicle.
3. Remove the control arm bushings, as follows:
   - Pry up the bushing flange, using a chisel and a hammer
   - Press the bushing(s) from the upper control arm, using a shop press and tools 09613-26010, 09632-36010 and 09950-00020

To install:

4. Lubricate the new control arm bushings with liquid soap.
5. Press the No. 1 bushing into the control arm until the bushing flange contacts the housing edge of the control arm, using a shop press, a steel plate and tools 09631-12090 and 09502-12010, facing the correct direction.
6. Press the No. 2 bushing into the con-

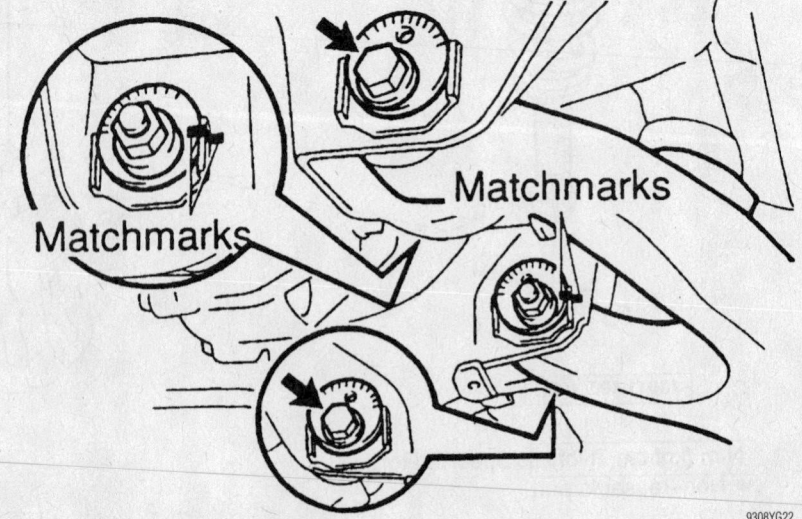

View of the lower control arm's cam plate alignment

9308YG22

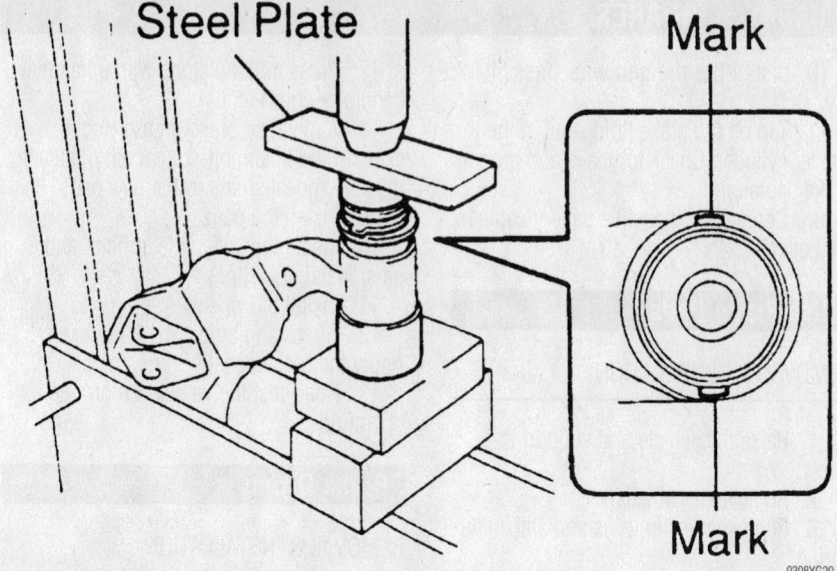

View of the No. 1 bushing's installed direction

9308YG20

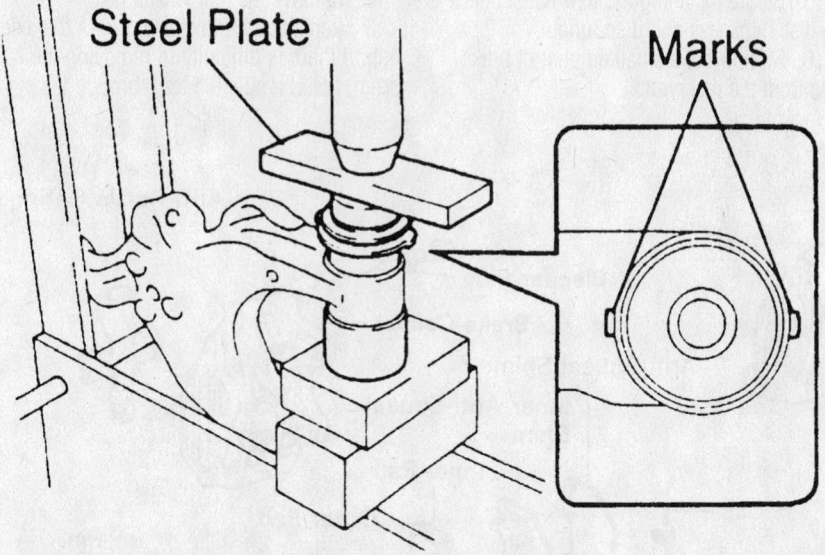

View of the No. 2 bushing's installed direction

9308YG21

trol arm until the bushing flange contacts the housing edge of the control arm, using a shop press, a steel plate and tools 09631-12090 and 09950-60020, facing the correct direction.

7. Install the lower control arm to the vehicle.

8. Check and/or adjust the wheel alignment.

Wheel Bearing

ADJUSTMENT

The wheel bearings are sealed unit; no adjustment is possible.

REMOVAL & INSTALLATION

1. Before servicing the vehicle, refer to the precautions in the beginning of this section.

2. Remove or disconnect the following:
- Front wheel
- Axle hub/steering knuckle assembly and place it in a vise
- Grease cap, for 2WD
- Inner grease seal, for 4WD

3. Remove the axle hub from the steering knuckle
- 4 bolts and shift the dust cover towards the outside (hub side)

- Axle hub from the steering knuckle, using tools 09710-30021 and 09950-40011
- Dust cover from the steering knuckle
- Bearing spacer and Anti-lock Brake System (ABS) speed sensor, if equipped with ABS
- Spacer, if not equipped with ABS

✳✳ WARNING

Be careful not to scratch the speed sensor rotor serrations.

4. Remove the outside oil seal from steering knuckle, using a small prybar

5. Remove the bearing from the steering knuckle, as follows:
- Snapring
- Bearing from the steering knuckle, using tools 09950-60020 and 09950-70010

To install:

6. Install the bearing to the steering knuckle, as follows:
- Bearing to the steering knuckle, using tools 09950-60020 and 09527-17011
- Snapring

7. Install the new outside oil seal to steering knuckle, using a plastic hammer and tools 09223-15030 and 09527-17011

8. Install the axle hub to the steering knuckle, as follows:
- Dust cover to the steering knuckle. Torque the 4 bolts to 13 ft. lbs. (18 Nm).
- Axle hub to the steering knuckle, using a shop press and tool 09649-17010

9. Install or connect the following:
- Bearing spacer and Anti-lock Brake System (ABS) speed sensor, if equipped with ABS

✳✳ WARNING

Be careful not to scratch the speed sensor rotor serrations.

- Bearing spacer, if not equipped with ABS, using a shop press and tools 09950-60010 and 09950-70010
- Grease cap, for 2WD
- Inner grease seal, for 4WD, using a plastic hammer and tool 09527-17011
- Axle hub/steering knuckle assembly
- Front wheel

BRAKES

Brake Caliper

REMOVAL & INSTALLATION

1. Disconnect the negative battery cable from the battery.
2. Raise and support the vehicle safely.
3. Remove the wheels.
4. Disconnect the brake hose from the caliper. Plug the end of the hose to prevent loss of fluid.
5. Remove the bolts that attach the caliper to the torque plate.
6. Lift the bottom of the caliper up and remove the caliper assembly.

To install:
7. Grease the caliper slides and bolts with lithium grease or equivalent. Install the caliper and secure with the bolts. Torque the bolts to 90 ft. lbs. (123 Nm).
8. Connect the brake hose to the caliper. Torque 11 ft. lbs. (15 Nm).
9. Fill the brake system to the proper level and bleed the brake system.

10. Install the tire and wheel assembly.
11. Top off the brake fluid level in the master cylinder. Check for leaks and proper brake operation.
12. Connect the negative battery cable to the battery.

Disc Brake Pads

REMOVAL & INSTALLATION

1. Raise the vehicle and support it safely.
2. Remove the wheels.
3. Remove the clip, pins and anti-rattle spring.
4. Withdraw the pads and remove the anti-squeal shims.

To install:
5. Before installing the new pads, check the disc thickness and disc runout.
6. Siphon out a small amount of brake fluid from the reservoir.

7. Press in the pistons with a hammer handle or equivalent.
8. Apply disc brake grease to both sides of the inner anti-squeal shim. Install the anti-squeal shims to the new pads.
9. Install the pads.
10. Install the anti-rattle springs and pins. Install the clip.
11. Install the wheels.
12. Check and adjust the fluid level. Apply the brake pedal several times.
13. Road-test the vehicle for proper operation.

Brake Drums

REMOVAL & INSTALLATION

1. Raise and safely support the vehicle.
2. Remove the rear wheel(s).
3. Remove the brake drum from the axle hub. If there is difficulty in removing the drum, insert a suitable tool through the

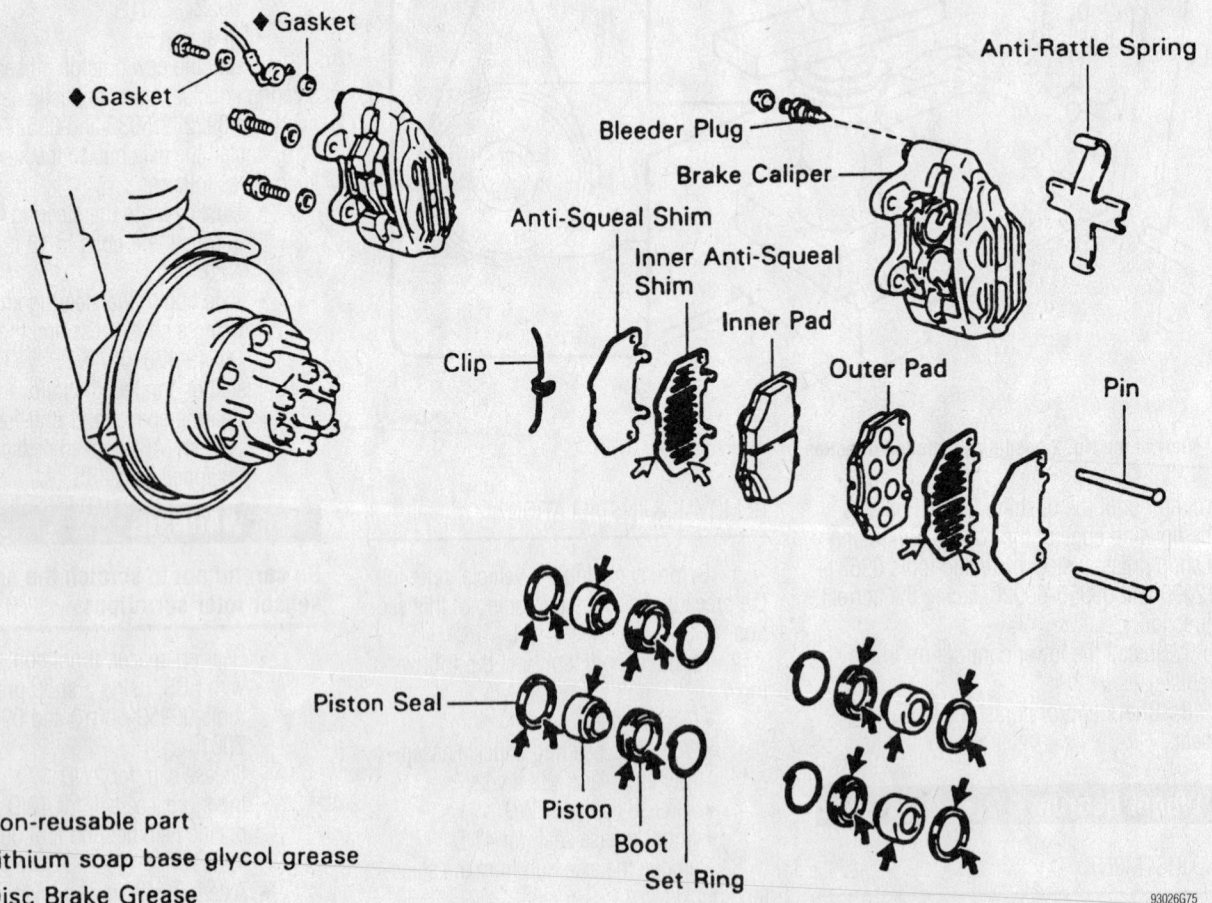

♦ Non-reusable part
← Lithium soap base glycol grease
⇐ Disc Brake Grease

Exploded view of the front disc brake components

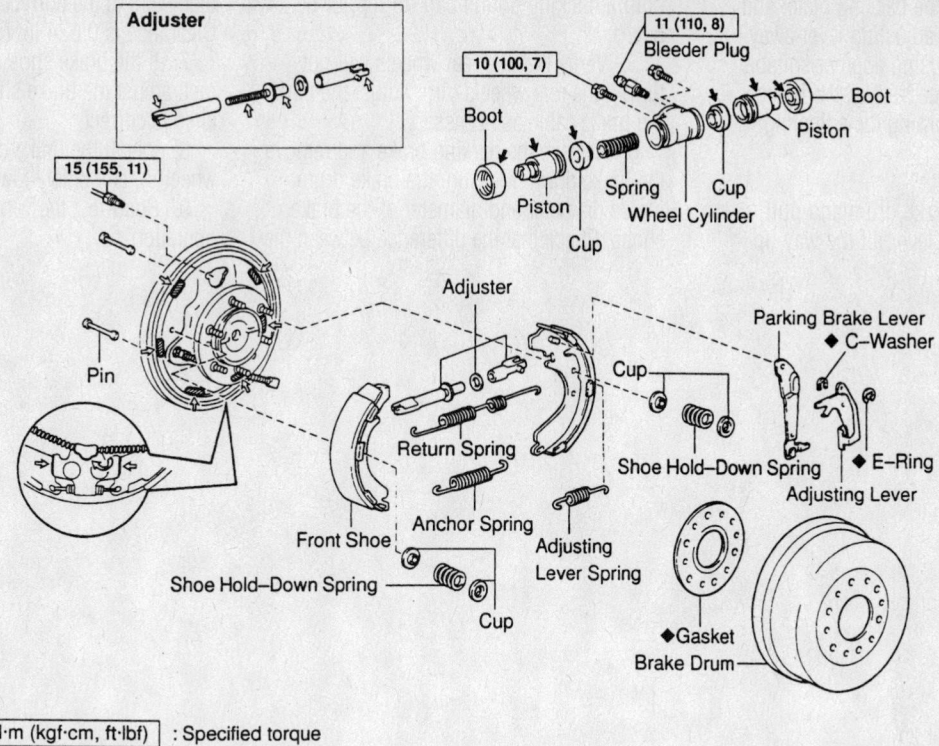

Adjuster

11 (110, 8)
Bleeder Plug

10 (100, 7)

Boot

15 (155, 11)

Boot

Piston

Piston

Spring

Cup

Cup

Wheel Cylinder

Pin

Adjuster

Cup

Parking Brake Lever
◆ C-Washer

Return Spring

Shoe Hold-Down Spring

◆ E-Ring

Adjusting Lever

Front Shoe

Anchor Spring

Adjusting
Lever Spring

Shoe Hold-Down Spring

Cup

◆Gasket

Brake Drum

N·m (kgf·cm, ft·lbf) : Specified torque
◆ Non-reusable part
➡ Lithium soap base glycol grease
⇨ High temperature grease

93026G77

Exploded view of the rear brake drums components

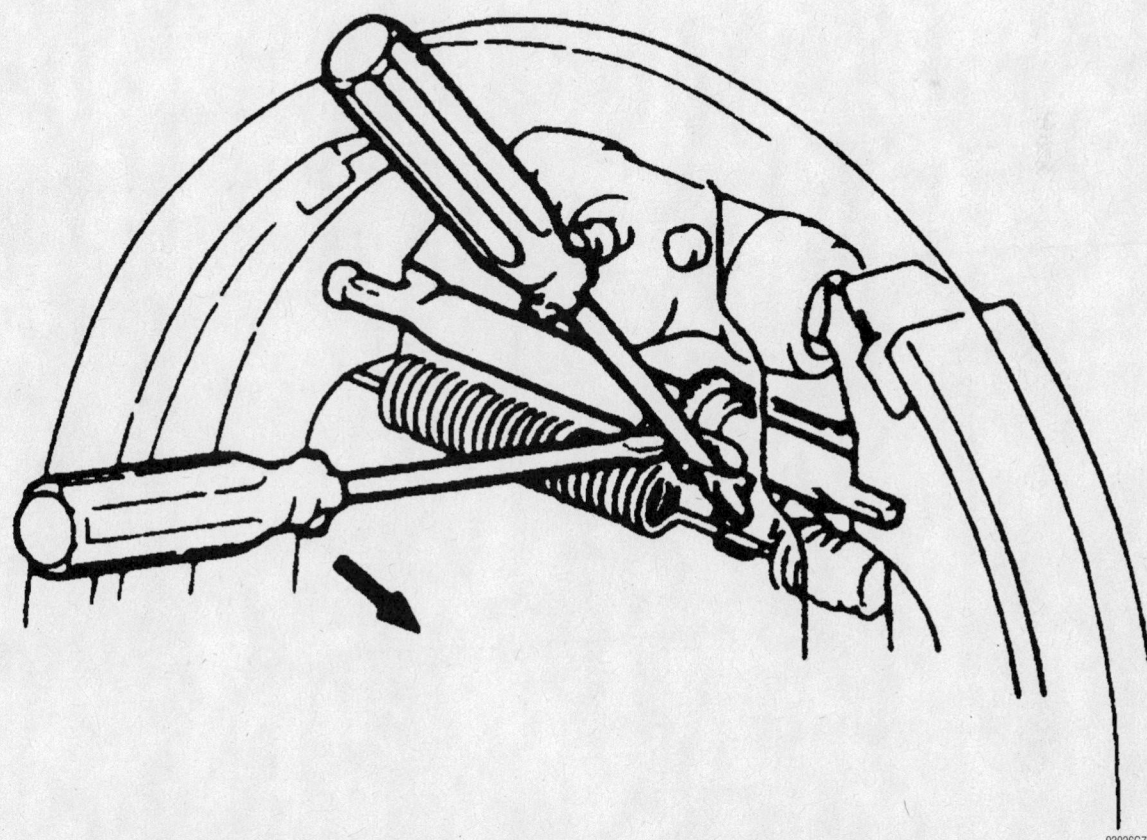

93026G78

Use a brake adjusting tool (brake spoon) and a prytool to adjust the brake shoes through the adjusting hole

hole in the rear of the backing plate, and hold the automatic adjusting lever away from the adjuster. Using another suitable tool at the same time, reduce the brake shoe adjuster by turning the adjusting wheel.

To install:

4. Install the brake drum and pull the parking brake lever all the way up until a clicking sound can no longer be heard.

5. Verify that the rear wheels will not turn. If the rear wheels turn, adjust the parking brake cable as necessary.

6. Release the parking brake and remove the brake drum. Measure the brake drum inside diameter and diameter of the brake shoes. Check that the difference between the diameters is the correct shoe clearance. Clearance is 0.024 in. (6mm).

7. If the brake shoe clearance is not correct, adjust the brake shoes until the clearance is correct.

8. Install the brake drum, replace the wheel(s), and safely lower the vehicle.

9. Road-test the vehicle for proper brake operation.

TOYOTA

Sienna

34

SPECIFICATION CHARTS

ENGINE AND VEHICLE IDENTIFICATION

		Engine					Model Year	
Code ①	Liters (cc)	Cu. In.	Cyl.	Fuel Sys.	Engine Type	Eng. Mfg.	Code ②	Year
1MZ-FE	3.0 (2995)	183	6	MFI	DOHC	Toyota	Y	2000
							1	2001
MFI: Multi-port Fuel Injection							2	2002
DOHC: Double Overhead Camshaft							3	2003
① Stamped on the left side of the engine block							4	2004
② 10th digit of the Vehicle Identification Number (VIN)								

42356-SIEN-C01

GENERAL ENGINE SPECIFICATIONS

Year	Model	Engine Displacemen Liters (cc)	Engine Series (ID/VIN)	Fuel System	Net Horsepower @ rpm	Net Torque @ rpm (ft. lbs.)	Bore x Stroke (in.)	Compression Ratio	Oil Pressure @ rpm
2000	Sienna	3.0 (2995)	1MZ-FE	MFI	194@5200	209@4400	3.44x3.27	10.5:1	43-78@3000
2001	Sienna	3.0 (2995)	1MZ-FE	MFI	194@5200	209@4400	3.44x3.27	10.5:1	43-78@3000
2002	Sienna	3.0 (2995)	1MZ-FE	MFI	210@5800	220@4400	3.44x3.27	10.5:1	43-78@3000
2003	Sienna	3.0 (2995)	1MZ-FE	MFI	210@5800	220@4400	3.44x3.27	10.5:1	43-78@3000

MFI: Multi-port Fuel Injection

42356-SIEN-C02

ENGINE TUNE-UP SPECIFICATIONS

Year	Engine Displacement Liters (cc)	Engine ID/VIN	Spark Plug Gap (in.)	Ignition Timing (deg.)*	Fuel Pump (psi)	Idle Speed (rpm) MT	Idle Speed (rpm) AT	Valve Clearance Intake	Valve Clearance Exhaust
2000	3.0 (2995)	1MZ-FE	0.043	10B	38-44	—	650-750	0.006-0.010	0.010-0.014
2001	3.0 (2995)	1MZ-FE	0.043	10B	38-44	—	650-750	0.006-0.010	0.010-0.014
2002	3.0 (2995)	1MZ-FE	0.043	10B	38-44	—	650-750	0.006-0.010	0.010-0.014
2003	3.0 (2995)	1MZ-FE	0.043	10B	38-44	—	650-750	0.006-0.010	0.010-0.014

NOTE: The Vehicle Emission Control Information label often reflects specification changes made during production.

The label figures must be used if they differ from those in this chart.

B: Before top dead center

* With terminals TE1 and E1 connected to DLC1

42356-SIEN-C03

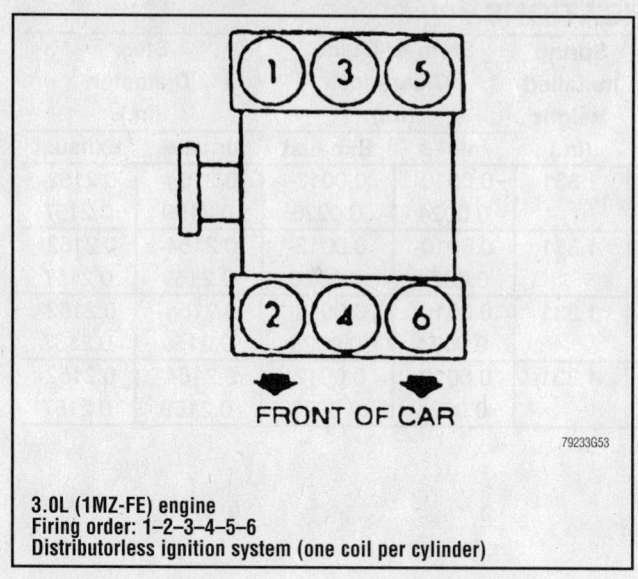

3.0L (1MZ-FE) engine
Firing order: 1–2–3–4–5–6
Distributorless ignition system (one coil per cylinder)

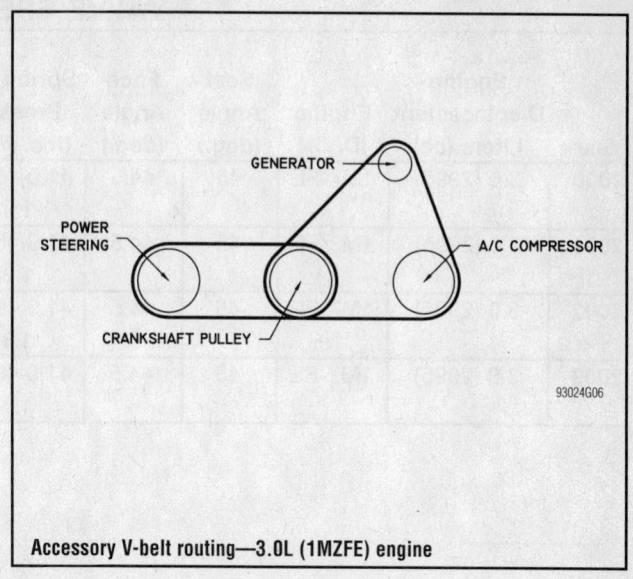

Accessory V-belt routing—3.0L (1MZFE) engine

CAPACITIES

Year	Model	Engine Displacement Liters (cc)	Engine ID/VIN	Engine Oil with Filter (qts.)	Transmission (pts.) 5-Spd	Transmission (pts.) Auto.	Transfer Case (pts.)	Drive Axle Front (pts.)	Drive Axle Rear (pts.)	Fuel Tank (gal.)	Cooling System (qts.)
2000	Sienna	3.0 (2995)	1MZ-FE	5.0	—	7.4	—	—	—	20.9	10.5
2001	Sienna	3.0 (2995)	1MZ-FE	5.0	—	7.4	—	—	—	20.9	10.5
2002	Sienna	3.0 (2995)	1MZ-FE	5.0	—	10.0	—	—	—	20.9	①
2003	Sienna	3.0 (2995)	1MZ-FE	5.0	—	10.0	—	—	—	20.9	①

① w/o rear heater: 10.0
With rear heater: 11.0

42356-SIEN-C04

CRANKSHAFT AND CONNECTING ROD SPECIFICATIONS

All measurements are given in inches.

Year	Engine Displacement Liters (cc)	Engine ID/VIN	Crankshaft Main Brg. Journal Dia.	Crankshaft Main Brg. Oil Clearance	Crankshaft Shaft End-play	Crankshaft Thrust on No.	Connecting Rod Journal Diameter	Connecting Rod Oil Clearance	Connecting Rod Side Clearance
2000	3.0 (2995)	1MZ-FE	2.4011-2.4016	①	0.0016-0.0095	2	2.0863-2.0866	0.0015-0.0025	0.0059-0.0188
2001	3.0 (2995)	1MZ-FE	2.4011-2.4016	①	0.0016-0.0095	2	2.0863-2.0866	0.0015-0.0025	0.0059-0.0188
2002	3.0 (2995)	1MZ-FE	2.4011-2.4016	①	0.0016-0.0095	2	2.0863-2.0866	0.0015-0.0025	0.0059-0.0118
2003	3.0 (2995)	1MZ-FE	2.4011-2.4016	①	0.0016-0.0095	2	2.0863-2.0866	0.0015-0.0025	0.0059-0.0118

① Journals 1 and 4: 0.0006 - 0.0013 in.
Journals 2 and 3: 0.0010 - 0.0018 in.

42356-SIEN-C05

VALVE SPECIFICATIONS

Year	Engine Displacement Liters (cc)	Engine ID/VIN	Seat Angle (deg.)	Face Angle (deg.)	Spring Test Pressure (lbs. @ in.)	Spring Installed Height (in.)	Stem-to-Guide Clearance (in.)		Stem Diameter (in.)	
							Intake	Exhaust	Intake	Exhaust
2000	3.0 (2995)	1MZ-FE	45	44.5	41.9-46.3@ 1.33	1.331	0.0010- 0.0024	0.0012- 0.0026	0.2154- 0.2159	0.2152- 0.2157
2001	3.0 (2995)	1MZ-FE	45	44.5	41.9-46.3@ 1.33	1.331	0.0010- 0.0024	0.0012- 0.0026	0.2154- 0.2159	0.2152- 0.2157
2002	3.0 (2995)	1MZ-FE	45	44.5	41.9-46.3@ 1.33	1.331	0.0010- 0.0024	0.0012- 0.0026	0.2154- 0.2159	0.2152- 0.2157
2003	3.0 (2995)	1MZ-FE	45	44.5	41.9-46.3@ 1.33	1.331	0.0010- 0.0024	0.0012- 0.0026	0.2154- 0.2159	0.2152- 0.2157

42356-SIEN-C06

PISTON AND RING SPECIFICATIONS

All measurements are given in inches.

Year	Engine Displ. Liters (cc)	Engine ID/VIN	Piston Clearance	Ring Gap			Ring Side Clearance		
				Top Comp.	Bottom Comp.	Oil Control	Top Comp.	Bottom Comp.	Oil Control
2000	3.0 (2995)	1MZ-FE	0.0033- 0.0042	0.0098- 0.0138	0.0138- 0.0177	0.0059- 0.0157	0.0008- 0.0028	0.0008- 0.0024	SNUG
2001	3.0 (2995)	1MZ-FE	0.0033- 0.0042	0.0098- 0.0138	0.0138- 0.0177	0.0059- 0.0157	0.0008- 0.0028	0.0008- 0.0024	SNUG
2002	3.0 (2995)	1MZ-FE	①	0.0098- 0.0138	0.0138- 0.0177	0.0059- 0.0157	0.0008- 0.0028	0.0008- 0.0024	SNUG
2003	3.0 (2995)	1MZ-FE	①	0.0098- 0.0138	0.0138- 0.0177	0.0059- 0.0157	0.0008- 0.0028	0.0008- 0.0024	SNUG

① AISIN piston: 0.0033-0.0042
   MAHLE piston: 0.0013-0.0023

42356-SIEN-C07

TORQUE SPECIFICATIONS
All readings in ft. lbs.

Year	Engine Displacement Liters (cc)	Engine ID/VIN	Cylinder Head Bolts	Main Bearing Bolts	Rod Bearing Bolts	Crankshaft Damper Bolts	Flywheel Bolts	Manifold Intake	Manifold Exhaust	Spark Plugs	Lug Nuts
2000	3.0 (2995)	1MZ-FE	①	②	③	159	61	11	36	13	76
2001	3.0 (2995)	1MZ-FE	①	②	③	159	61	11	36	13	76
2002	3.0 (2995)	1MZ-FE	①	②	④	159	61	11	36	13	76
2003	3.0 (2995)	1MZ-FE	①	②	④	159	61	11	36	13	76

① Step 1: 40 ft. lbs.
   Step 2: Plus 90 degrees
   Recessed bolt: 13 ft. lbs.

② 6-point bolts: 20 ft. lbs.
   12-point bolts:
       Step 1: 16 ft. lbs.
       Step 2: Plus 90 degrees

③ Step 1: 29 ft. lbs.
   Step 2: Plus 90 degrees

④ Step 1: 18 ft. lbs.
   Step 2: +90 degrees

42356-SIEN-C08

BRAKE SPECIFICATIONS
All measurements in inches unless noted

Year	Model		Brake Disc Original Thickness	Brake Disc Minimum Thickness	Brake Disc Maximum Runout	Brake Drum Diameter Original Inside Diameter	Brake Drum Diameter Maximum Machine Diameter	Minimum Lining Thickness	Brake Caliper Bracket Bolts (ft. lbs.)	Brake Caliper Mounting Bolts (ft. lbs.)
2000	Sienna	F	1.102	1.024	0.002	—	—	0.039	79	25
		R	—	—	—	9.84	9.921	0.039	—	—
2001	Sienna	F	1.102	1.024	0.002	—	—	0.039	79	25
		R	—	—	—	9.84	9.921	0.039	—	—
2002	Sienna	F	1.102	1.024	0.002	—	—	0.039	79	25
		R	—	—	—	9.84	9.921	0.039	—	—
2003	Sienna	F	1.102	1.024	0.002	—	—	0.039	79	25
		R	—	—	—	9.84	9.921	0.039	—	—

F: Front
R: Rear

42356-SIEN-C09

WHEEL ALIGNMENT

Year	Model		Caster Range (+/-Deg.)	Caster Preferred Setting (Deg.)	Camber Range (+/-Deg.)	Camber Preferred Setting (Deg.)	Toe-in (in.)	Inside Wheel Angle (Deg.)
2000	Sienna	F	0.75	1.53	0.75	-0.50	0.10+/-0.08	34.32
		R	—	—	0.75	-0.92	0.09+/-0.12	—
2001	Sienna	F	0.75	1.53	0.75	-0.50	0.10+/-0.08	34.32
		R	—	—	0.75	-0.92	0.09+/-0.12	—
2002	Sienna	F	0.75	1.53	0.75	-0.50	0.10+/-0.08	34.32
		R	—	—	0.75	-0.92	0.09+/-0.12	—
2003	Sienna	F	0.75	1.53	0.75	-0.50	0.10+/-0.08	34.32
		R	—	—	0.75	-0.92	0.09+/-0.12	—

42356-SIEN-C10

TIRE, WHEEL AND BALL JOINT SPECIFICATIONS

Year	Model	OEM Tires Standard	OEM Tires Optional	Tire Pressures (psi) Front	Tire Pressures (psi) Rear	Wheel Size	Ball Joint Inspection
2000	Sienna	P205/70R15 95S	P215/65R15 95S	35	35	6.5	①
2001	Sienna	P205/70R15 95S	P215/65R15 95S	35	35	6.5	①
2002	Sienna CE	P205/70R15 95S	none	35	35	6.5	①
	Sienna LE	P205/70R15 95S	P215/65R15 95S	35②	35②	6.5	①
	Sienna XLE	P215/65R15 95S	none	35②	35②	6.5	①
2003	Sienna CE	P205/70R15 95S	none	35	35	6.5	①
	Sienna LE	P205/70R15 95S	P215/65R15 95S	35②	35②	6.5	①
	Sienna XLE	P215/65R15 95S	none	35②	35②	6.5	①

OEM: Original Equipment Manufacturer

PSI: Pounds Per Square Inch

① Ball joint turning torque should be 30 inch lbs.

② P215/65R15
  Up to 6 passengers: 32
  Trailer towing or up to vehicle capacity weight: 35

42356-SIEN-C11

SCHEDULED MAINTENANCE INTERVALS
TOYOTA—SIENNA

TO BE SERVICED	TYPE OF SERVICE	VEHICLE MILEAGE INTERVAL (x1000)																		
		5	10	15	20	25	30	35	40	45	50	55	60	65	70	75	80	85	90	95
Automatic transmission and differential fluid	S/I			✓			✓			✓			✓			✓			✓	
Ball joints and boots	S/I			✓			✓			✓			✓			✓			✓	
Brake linings, discs/drums, lines & hoses	S/I			✓			✓			✓			✓			✓			✓	
Charcoal canister	S/I												✓							
Drive belts	S/I						✓						✓						✓	
Engine coolant	R						✓						✓						✓	
Engine oil & filter	R	✓	✓	✓	✓	✓	✓	✓	✓	✓	✓	✓	✓	✓	✓	✓	✓	✓	✓	✓
Exhaust pipes & mounts	S/I			✓			✓			✓			✓			✓			✓	
Fuel lines & connections, fuel tank vapor vent system hoses, fuel tank band	S/I						✓						✓						✓	
Fuel tank cap gasket	S/I						✓						✓						✓	
Halfshaft boots & flange bolts	S/I			✓			✓			✓			✓			✓			✓	
Non-platinum spark plugs	R						✓						✓						✓	
Platinum spark plugs	R												✓							
Rack and pinion assembly	S/I			✓			✓			✓			✓			✓			✓	
Steering linkage	S/I			✓			✓			✓			✓			✓			✓	
Valves	S/I												✓							

R: Replace   S/I: Service or Inspect   L: Lubricate

FREQUENT OPERATION MAINTENANCE (SEVERE SERVICE)

If a vehicle is operated under any of the following conditions it is considered severe service:

- Towing a trailer or using a camper or car-top carrier.
- Repeated short trips of less than 5 miles in temperatures below freezing.
- Excessive idling or low-speed driving for long distances as in heavy commercial use, such as delivery, taxi or police cars.
- Operating on rough, muddy or salt-covered roads.
- Operating on unpaved or dusty roads.

Oil filter: service or inspect every 5000 miles or 4 months, whichever occurs first.

Brake linings and discs or drums: service or inspect every 5000 miles or 4 months, whichever occurs first.

Steering linkage: service or inspect every 5000 miles or 4 months, whichever occurs first.

Ball joints and boots: service or inspect every 5000 miles or 4 months, whichever occurs first.

Brake discs & pads (front): service or inspect every 6000 miles.

Halfshaft boots: service or inspect every 5000 miles or 4 months. Retighten the flange bolts, whichever occurs first.

Body chassis bolts and nuts: service or inspect every 5000 miles or 4 months, whichever occurs first.

Transmission and differential fluid: replace every 15,000 miles or 12 months, whichever occurs first.

Timing belt: replace every 60,000 miles or 48 months, whichever occurs first.

42356-SIEN-C12

PRECAUTIONS

Before servicing any vehicle, please be sure to read all of the following precautions, which deal with personal safety, prevention of component damage, and important points to take into consideration when servicing a motor vehicle:

• Never open, service or drain the radiator or cooling system when the engine is hot; serious burns can occur from the steam and hot coolant.

• Observe all applicable safety precautions when working around fuel. Whenever servicing the fuel system, always work in a well-ventilated area. Do not allow fuel spray or vapors to come in contact with a spark, open flame or excessive heat (a hot drop light, for example). Keep a dry chemical fire extinguisher near the work area. Always keep fuel in a container specifically designed for fuel storage; also, always properly seal fuel containers to avoid the possibility of fire or explosion. Refer to the additional fuel system precautions later in this section.

• Fuel injection systems often remain pressurized, even after the engine has been turned **OFF**. The fuel system pressure must be relieved before disconnecting any fuel lines. Failure to do so may result in fire and/or personal injury.

• Brake fluid often contains polyglycol ethers and polyglycols. Avoid contact with the eyes and wash your hands thoroughly after handling brake fluid. If you do get brake fluid in your eyes, flush your eyes with clean, running water for 15 minutes. If

eye irritation persists, or if you have taken brake fluid internally, IMMEDIATELY seek medical assistance.

• The EPA warns that prolonged contact with used engine oil may cause a number of skin disorders, including cancer! You should make every effort to minimize your exposure to used engine oil. Protective gloves should be worn when changing oil. Wash your hands and any other exposed skin areas as soon as possible after exposure to used engine oil. Soap and water, or waterless hand cleaner should be used.

• All new vehicles are now equipped with an air bag system. The system must be disabled before performing service on or around system components, steering column, instrument panel components, wiring and sensors. Failure to follow safety and disabling procedures could result in accidental air bag deployment, possible personal injury and unnecessary system repairs.

• Always wear safety goggles when working with, or around, the air bag system. When carrying a non-deployed air bag, be sure the bag and trim cover are pointed away from your body. When placing a non-deployed air bag on a work surface, always face the bag and trim cover upward, away from the surface. This will reduce the motion of the module if it is accidentally deployed. Refer to the additional air bag system precautions later in this section.

• Clean, high quality brake fluid from a sealed container is essential to the safe and

proper operation of the brake system. You should always buy the correct type of brake fluid for your vehicle. If the brake fluid becomes contaminated, completely flush the system with new fluid. Never reuse any brake fluid. Any brake fluid that is removed from the system should be discarded. Also, do not allow any brake fluid to come in contact with a painted surface; it will damage the paint.

• Never operate the engine without the proper amount and type of engine oil; doing so WILL result in severe engine damage.

• Timing belt maintenance is extremely important! Many models utilize an interference type, non-freewheeling engine. If the timing belt breaks, the valves in the cylinder head may strike the pistons, causing potentially serious (also time consuming and expensive) engine damage. Refer to the maintenance interval charts in the front of this section for the recommended replacement interval for the timing belt, and to the timing belt section for belt replacement and inspection.

• Disconnecting the negative battery cable on some vehicles may interfere with the functions of the on-board computer system(s) and may require the computer to undergo a relearning process once the negative battery cable is reconnected.

• When servicing drum brakes, only disassemble and assemble one side at a time, leaving the remaining side intact for reference.

ENGINE REPAIR

Distributor

Sienna models are equipped with a distiburtorless ignition system.

Alternator

REMOVAL

1. Before servicing the vehicle, refer to the precautions in the beginning of this section.
2. Remove or disconnect the following:
• Alternator electrical connectors
• Wiring harness from the clip
• Pivot bolt
• Plate washer
• Adjusting lockbolt
• Drive belt
• Alternator

INSTALLATION

1. Install or connect the following:
• Alternator
• Drive belt. Tension the belt to 170–180 lbs. for a new belt or 95–135 lbs. for a used belt.
• Adjusting lockbolt. Tighten the bolt to 13 ft. lbs. (18 Nm).
• Plate washer
• Pivot bolt. Tighten the bolt to 41 ft. lbs. (56 Nm).
• Wiring harness from the clip
• Alternator electrical connectors

Ignition Timing

ADJUSTMENT

Ignition timing is controlled by the ECM and is not adjustable.

Engine Assembly

REMOVAL & INSTALLATION

1. Before servicing the vehicle, refer to the precautions in the beginning of this section.
2. Matchmark the hood position.
3. Remove or disconnect the following:
• Hood
• Wiper and blade assembly
• Top cowl seal and panel
• Window washer hoses from the ventilator louvers
• Left and right ventilator louvers
• Heater air duct
4. Properly relieve the fuel system pressure.
5. Remove or disconnect the following:
• Both battery cables
• Battery and tray

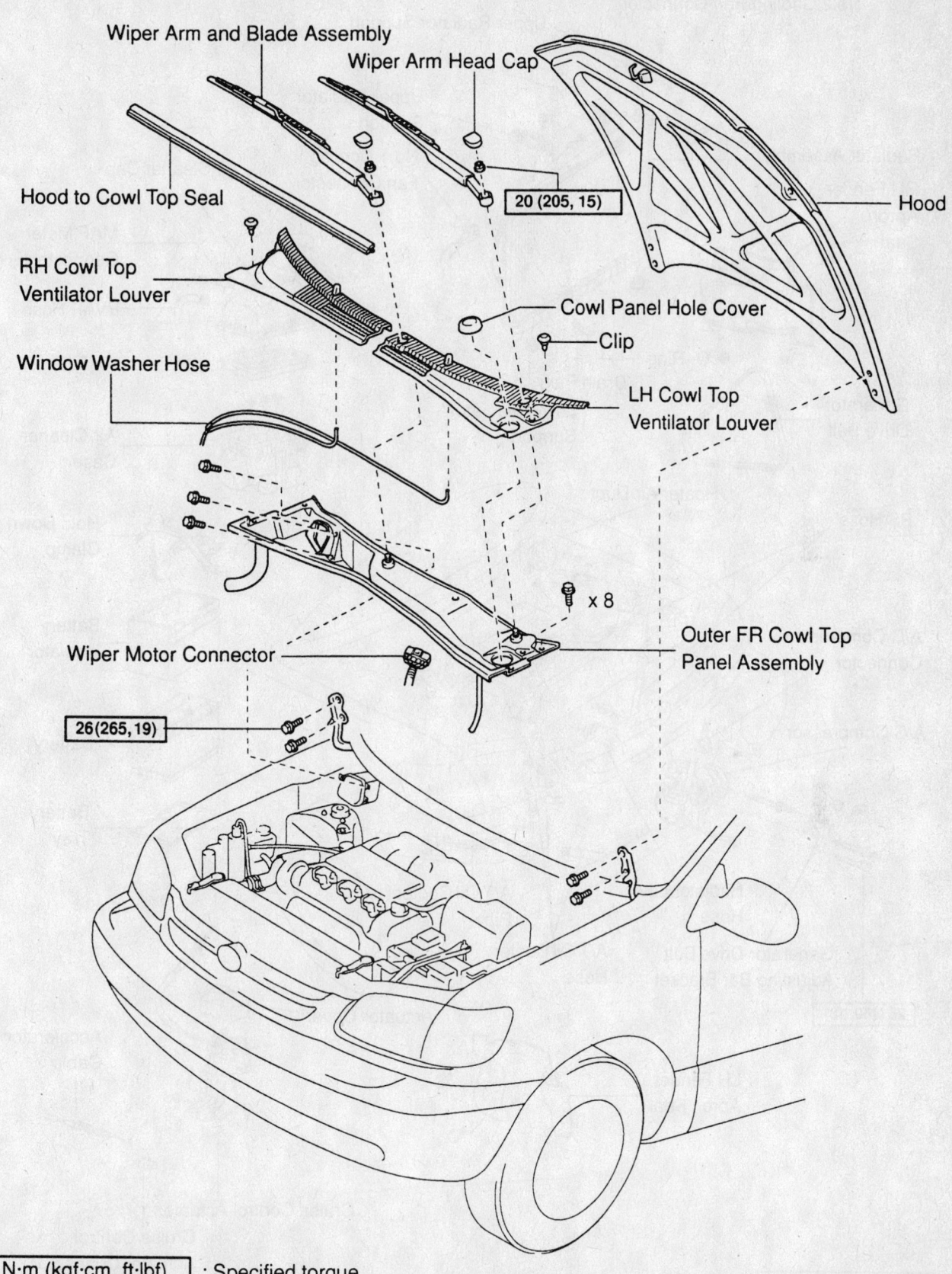

Wiper Arm and Blade Assembly

Wiper Arm Head Cap

Hood to Cowl Top Seal

20 (205, 15)

Hood

RH Cowl Top Ventilator Louver

Cowl Panel Hole Cover

Clip

Window Washer Hose

LH Cowl Top Ventilator Louver

x 8

Wiper Motor Connector

Outer FR Cowl Top Panel Assembly

26 (265, 19)

N·m (kgf·cm, ft·lbf) : Specified torque

7924ZG06

Exploded view of the top cowl and related components

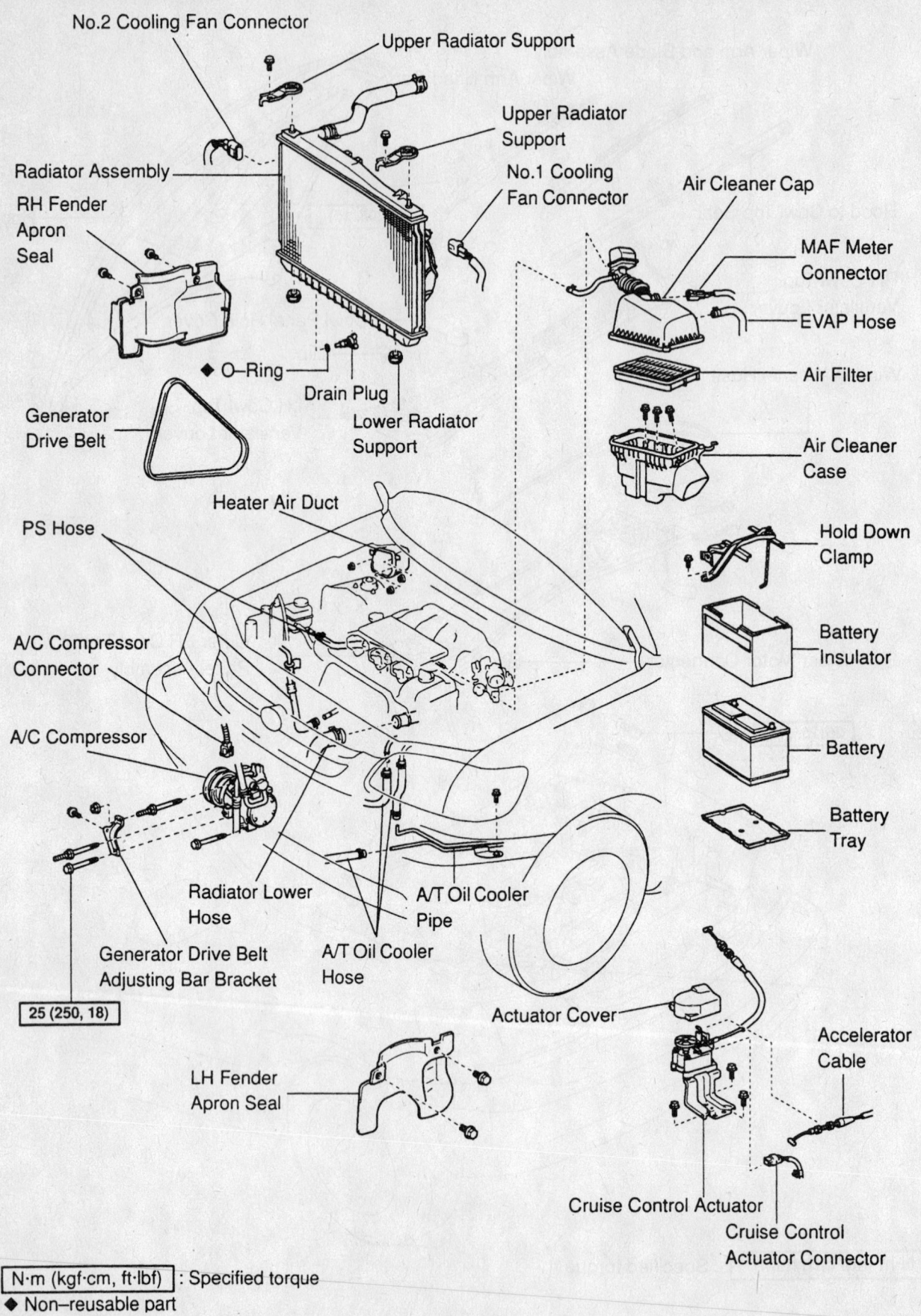

No.2 Cooling Fan Connector

Upper Radiator Support

Upper Radiator Support

No.1 Cooling Fan Connector

Air Cleaner Cap

MAF Meter Connector

EVAP Hose

Air Filter

Radiator Assembly

RH Fender Apron Seal

◆ O–Ring

Drain Plug

Lower Radiator Support

Generator Drive Belt

Air Cleaner Case

Heater Air Duct

PS Hose

Hold Down Clamp

Battery Insulator

A/C Compressor Connector

A/C Compressor

Battery

Battery Tray

Radiator Lower Hose

A/T Oil Cooler Pipe

Generator Drive Belt Adjusting Bar Bracket

A/T Oil Cooler Hose

25 (250, 18)

Actuator Cover

Accelerator Cable

LH Fender Apron Seal

Cruise Control Actuator

Cruise Control Actuator Connector

N·m (kgf·cm, ft·lbf) : Specified torque

◆ Non–reusable part

Exploded view of engine pre-removal components

7924ZG07

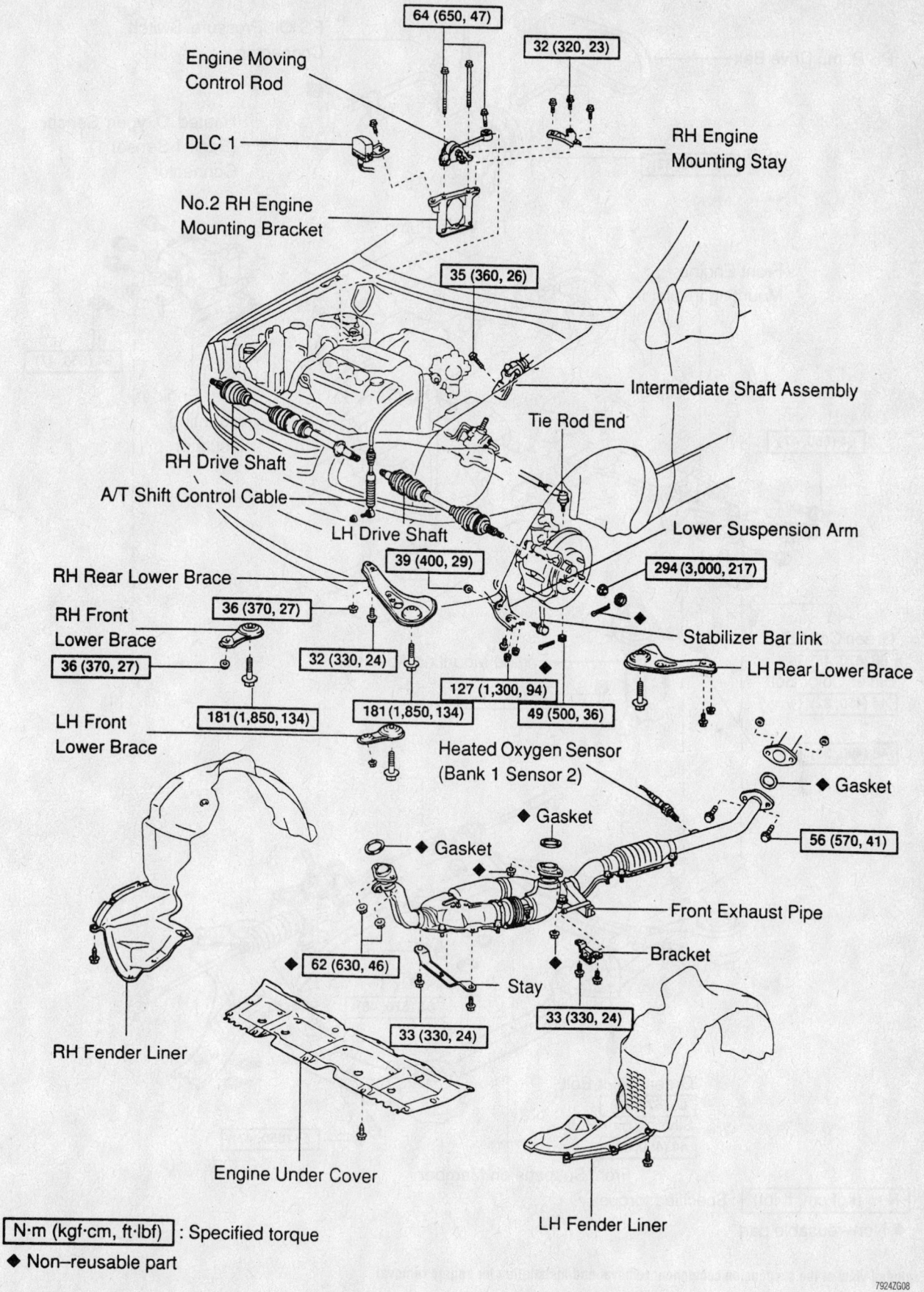

Engine Moving Control Rod

DLC 1

No.2 RH Engine Mounting Bracket

64 (650, 47)

32 (320, 23)

RH Engine Mounting Stay

35 (360, 26)

Intermediate Shaft Assembly

Tie Rod End

RH Drive Shaft

A/T Shift Control Cable

LH Drive Shaft

Lower Suspension Arm

RH Rear Lower Brace

RH Front Lower Brace

LH Front Lower Brace

39 (400, 29)

36 (370, 27)

36 (370, 27)

32 (330, 24)

181 (1,850, 134)

181 (1,850, 134)

294 (3,000, 217)

Stabilizer Bar link

LH Rear Lower Brace

127 (1,300, 94)

49 (500, 36)

Heated Oxygen Sensor (Bank 1 Sensor 2)

◆ Gasket

◆ Gasket

◆ Gasket

56 (570, 41)

Front Exhaust Pipe

Bracket

◆ 62 (630, 46)

Stay

33 (330, 24)

33 (330, 24)

RH Fender Liner

Engine Under Cover

LH Fender Liner

N·m (kgf·cm, ft·lbf) : Specified torque

◆ Non–reusable part

7924ZG08

Exploded view of engine removal and installation tightening specifications of the related components

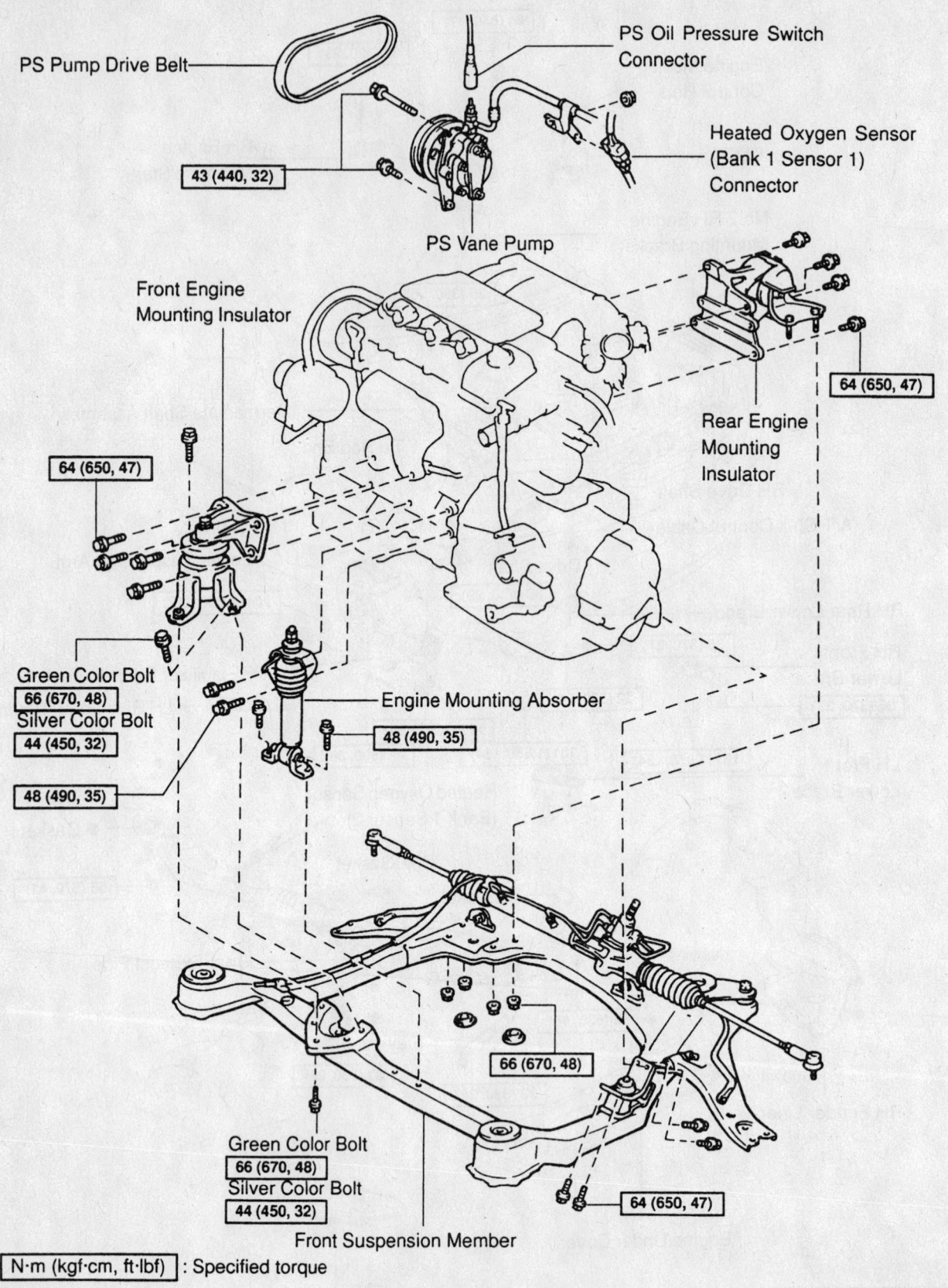

PS Pump Drive Belt

PS Oil Pressure Switch Connector

Heated Oxygen Sensor (Bank 1 Sensor 1) Connector

43 (440, 32)

PS Vane Pump

Front Engine Mounting Insulator

Rear Engine Mounting Insulator

64 (650, 47)

64 (650, 47)

Green Color Bolt
66 (670, 48)
Silver Color Bolt
44 (450, 32)

48 (490, 35)

Engine Mounting Absorber

48 (490, 35)

66 (670, 48)

Green Color Bolt
66 (670, 48)
Silver Color Bolt
44 (450, 32)

64 (650, 47)

Front Suspension Member

N·m (kgf·cm, ft·lbf) : Specified torque

◆ Non–reusable part

7924ZG09

Exploded view of the suspension component removal and installation for engine removal

6. Drain the engine coolant.
7. Drain the engine oil.
8. Remove or disconnect the following:
- Intake air cleaner and case assembly
- Cruise control actuator, if equipped
- Upper and lower radiator hoses
- Radiator
- Automatic transmission oil cooler lines
- Any connectors, hoses and sensors that would interfere with engine removal
- Engine Control Module (ECM) engine wiring harness from inside the glove box; then, pull the harness into the engine compartment
- Compressor

➡It may be necessary to remove the air conditioning compressor lines in order to remove the engine.

- Automatic transmission shifter cable from the transaxle
- Header pipes from the exhaust manifolds
- Left and right fender apron seals
- Halfshafts
- Stabilizer links and the steering intermediate shaft
- Power steering pump
- Engine undercover
- Engine hanger to the engine
- Engine sling device to the engine hangers
- Right-hand motor mount and moving control rod
- Front suspension lower braces

9. Lower the engine, transaxle and front suspension member as an assembly from the vehicle.

To install:

10. Raise the engine, transaxle and front suspension member as an assembly into the vehicle.
11. Install the front suspension lower braces, and tighten the fasteners, as follows:
- Bolt A: 134 ft. lbs. (181 Nm)
- Bolt B: 24 ft. lbs. (32 Nm)
- Nut C: 27 ft. lbs. (36 Nm)
12. Install or connect the following:
- Moving control rod. Tighten the bolts to 47 ft. lbs. (64 Nm).
- Right-hand motor mount. Tighten the bolts to 23 ft. lbs. (32 Nm).
- Engine sling device from the engine hangers
- Engine undercover
- Power steering pump hoses
- Stabilizer links and the steering intermediate shaft
- Halfshafts

- Left and right fender apron seals
- Header pipes to the exhaust manifolds
- Automatic transmission shifter cable to the transaxle
- Air conditioning compressor to the engine
13. Push the wiring harness into the glove box.
14. Install or connect the following:
- ECM
- Any connectors, hoses and sensors that were removed
- Automatic transmission oil cooler lines
- Upper and lower radiator hoses and fit the radiator
- Cruise control actuator, if removed
- Intake air cleaner and case assembly
15. Fill the engine oil to proper level.
16. Fill the engine with coolant.
17. Install or connect the following:
- Battery tray and battery
- Battery cables
- Heater air duct
- Left and right ventilator louvers
- Window washer hoses from the ventilator louvers
- Top cowl seal and panel
- Wiper and blade assembly
- Hood
- New oil filter
18. Refill the engine with oil.
19. Refill the engine with engine coolant.
20. Install the engine undercovers.
21. Start the engine and check for leaks.

Heater Core

REMOVAL & INSTALLATION

Front Heater

1. Disconnect the negative battery cable.
2. Drain the cooling system into a clean container for reuse.
3. Disconnect the heater hoses from the heater core.
4. Remove the steering wheel by performing the following procedure:
   a. Position the front wheels facing straight-ahead.
   b. Remove the steering wheel side covers.
   c. Using a Torx® wrench, loosen the 2 screws located at each side of the steering wheel until the screw's circumference groove catches on the screw case.
   d. Pull the air bag module from the

steering wheel and disconnect the electrical connector.

⁕⁕ CAUTION

Place the air bag module in a safe place with the front side facing upward.

   e. Remove the steering wheel nut.
   f. Place alignment marks on the steering wheel and the main shaft.
   g. Using a steering wheel puller, press the steering wheel from the steering column.
5. Remove the instrument panel and reinforcement by performing the following procedure:
   a. Remove the front door scuff plates.
   b. Remove the cowl side boards.
   c. Remove the front door trim covers.
   d. Remove the front pillar garnish by disengaging the 5 clips. If equipped with a tweeter speaker, disconnect the electrical connector.
   e. Remove the steering column covers-to-steering column screws and the covers.
   f. Remove the combination switch-to-steering column screws, disconnect the electrical connector(s) and remove the combination switch.
   g. Remove the 2 hood open lever screws and the hood open lever.
   h. Remove the 2 lower finish panel bolts and disengage the panel from the 3 clips.
   i. Remove the 2 No. 1 safety pad insert bolts and the insert.
   j. Remove the 2 No. 2 finish panel bolts and disengage the panel from the 4 clips.
   k. In the left side of the glove compartment, pry out the glove box door finish plate and disconnect the air bag module connector.
   l. Remove the glove box 3 nuts and 2 screws and the glove box.
   m. Remove the center cluster finish panel by disengaging the claw (bottom center) and 4 clips (one at each corner).
   n. Remove the ashtray, the 2 ashtray receptacle box screws.
   o. Remove the 4 lower center cluster finish panel screws and disconnect the connector.
   p. Remove the clock, the No. 1 and No. 2 registers from the panel.
   q. Remove the 3 cluster finish panel screws, disengage the 8 clips and remove the panel.
   r. Remove the combination meter.

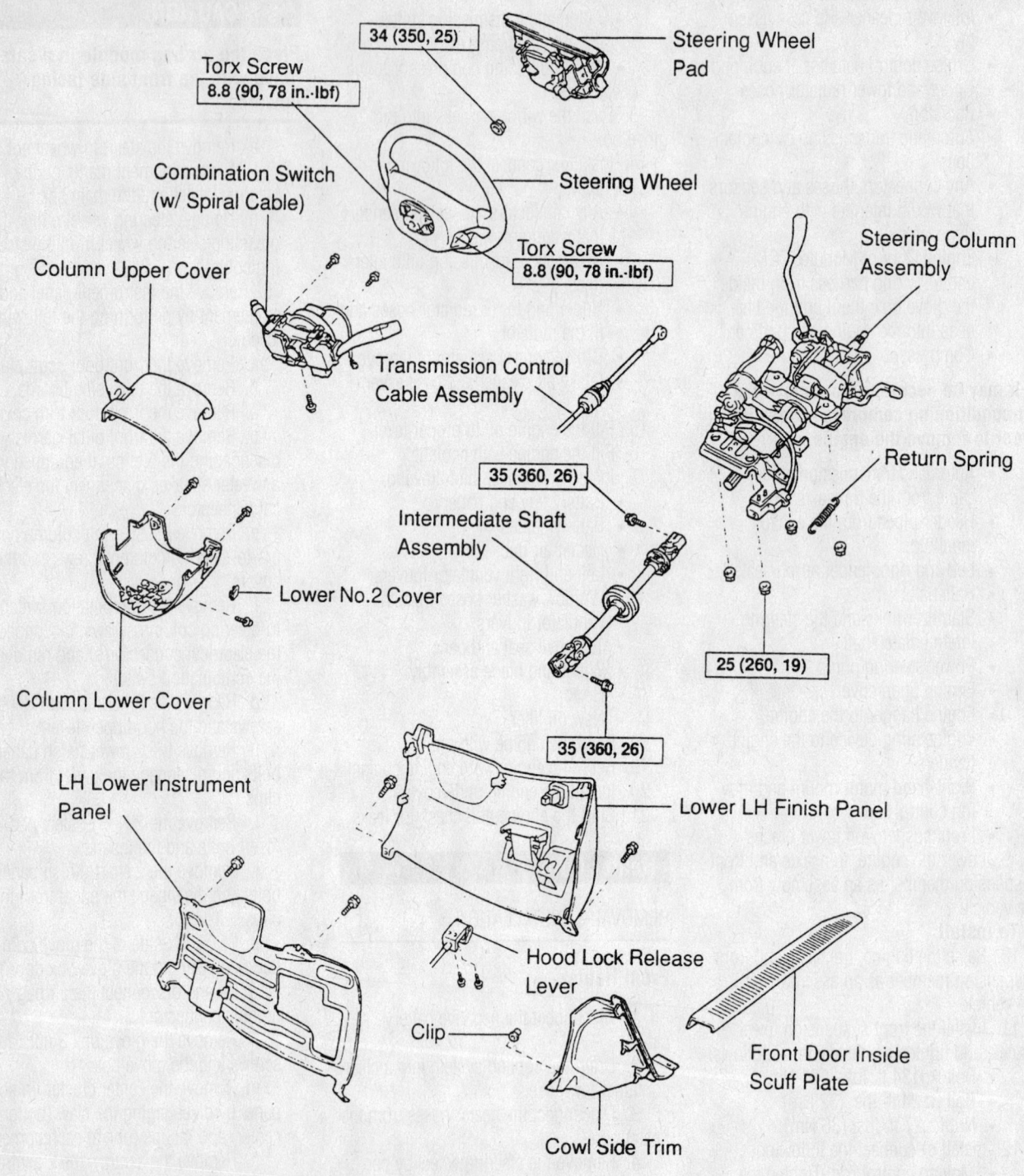

34 (350, 25)

Steering Wheel
Pad

Torx Screw
8.8 (90, 78 in.-lbf)

Combination Switch
(w/ Spiral Cable)

Steering Wheel

Torx Screw
8.8 (90, 78 in.-lbf)

Steering Column
Assembly

Column Upper Cover

Transmission Control
Cable Assembly

Return Spring

35 (360, 26)

Intermediate Shaft
Assembly

Lower No.2 Cover

25 (260, 19)

Column Lower Cover

35 (360, 26)

LH Lower Instrument
Panel

Lower LH Finish Panel

Hood Lock Release
Lever

Clip

Front Door Inside
Scuff Plate

Cowl Side Trim

N·m (kgf·cm, ft·lbf) : Specified torque

93113GH3

Exploded view of the steering wheel, steering column and related components

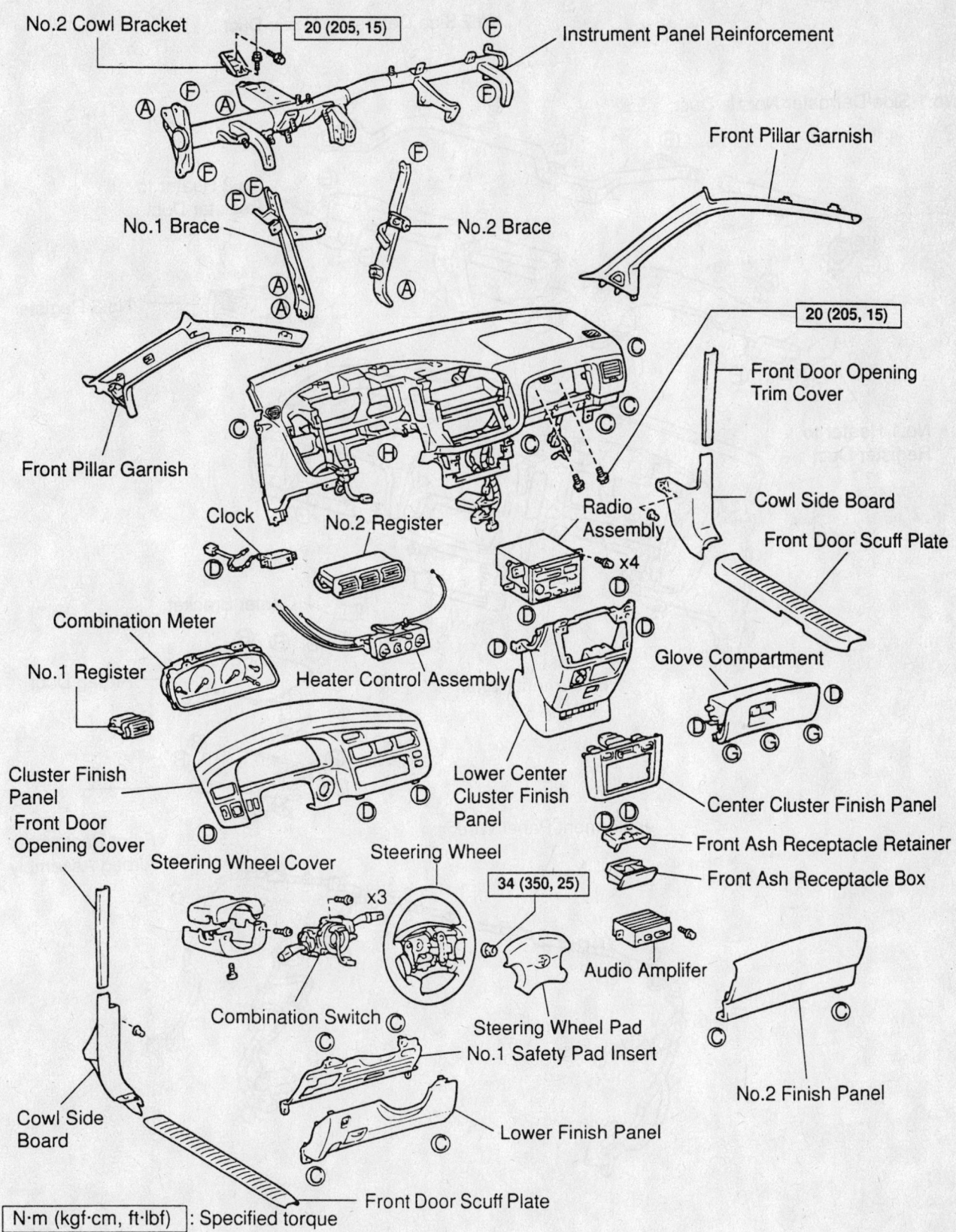

No.2 Cowl Bracket

20 (205, 15)

Instrument Panel Reinforcement

Front Pillar Garnish

No.1 Brace

No.2 Brace

20 (205, 15)

Front Door Opening Trim Cover

Front Pillar Garnish

Cowl Side Board

Front Door Scuff Plate

Clock

No.2 Register

Radio Assembly

Combination Meter

No.1 Register

Heater Control Assembly

Glove Compartment

Cluster Finish Panel

Front Door Opening Cover

Steering Wheel Cover

Lower Center Cluster Finish Panel

Center Cluster Finish Panel

Front Ash Receptacle Retainer

Front Ash Receptacle Box

Steering Wheel

34 (350, 25)

x3

Combination Switch

Audio Amplifer

Steering Wheel Pad

No.1 Safety Pad Insert

No.2 Finish Panel

Cowl Side Board

Lower Finish Panel

Front Door Scuff Plate

N·m (kgf·cm, ft·lbf) : Specified torque

93113GH4

Exploded view of the instrument panel and related components

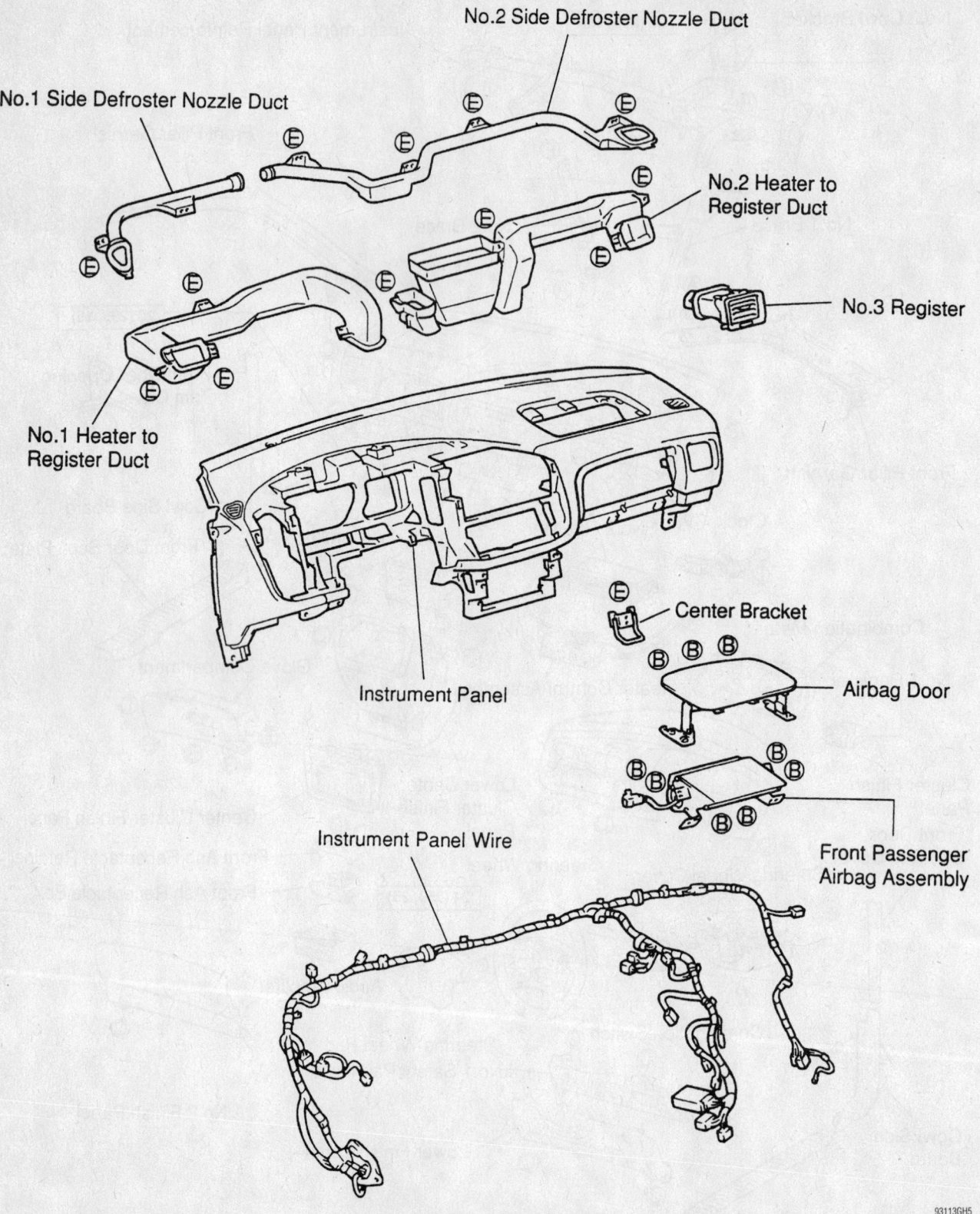

No.2 Side Defroster Nozzle Duct

No.1 Side Defroster Nozzle Duct

No.2 Heater to Register Duct

No.3 Register

No.1 Heater to Register Duct

Center Bracket

Airbag Door

Instrument Panel

Instrument Panel Wire

Front Passenger Airbag Assembly

93113GH5

Exploded view of the ventilation system and related components

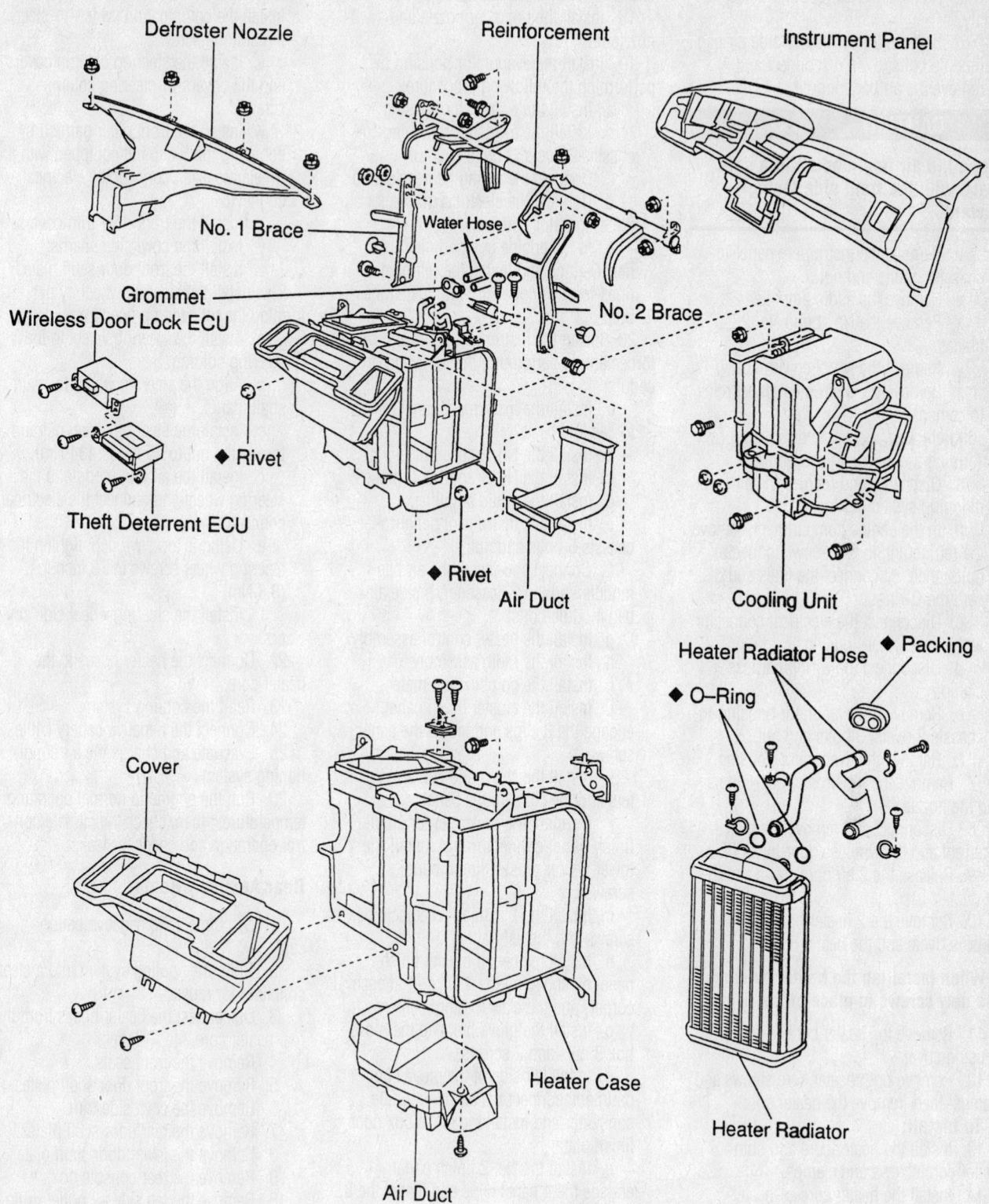

Defroster Nozzle

Reinforcement

Instrument Panel

No. 1 Brace

Water Hose

Grommet

Wireless Door Lock ECU

No. 2 Brace

◆ Rivet

Theft Deterrent ECU

◆ Rivet

Air Duct

Cooling Unit

Heater Radiator Hose   ◆ Packing

◆ O–Ring

Cover

Heater Case

Heater Radiator

Air Duct

◆ Non–reusable part

93113GH6

Exploded view of the heater core, heater housing, evaporator housing and related components

s. Remove the radio assembly.

t. Remove the heater control assembly.

u. Remove 2 passenger's side air bag module bolts; then, disconnect and remove the air bag module.

✳✳ CAUTION

Place the air bag module in a safe place with the front side facing upward.

v. Remove the instrument panel-to-chassis 5 bolts and nut.

w. Remove the audio amplifier.

x. Remove the No. 1 and No. 2 braces.

y. Remove the No. 2 cowl brace.

z. Remove the instrument panel reinforcement.

6. Remove the evaporator housing by performing the following procedure:

a. Discharge and recover the air conditioning system refrigerant.

b. In the engine compartment, remove the refrigerant lines-to-cowl connector bolts; then, disconnect the lines and discard the O-rings.

c. Disconnect the electrical connector at the evaporator housing.

d. Disconnect the wiring harness clamp.

e. Remove the evaporator housing-to-chassis 2 rivets, 3 bolts and nut.

f. Remove the evaporator housing.

7. Remove the 4 defroster nozzle nuts and the nozzle.

8. Disconnect and remove the theft deterrent and the wireless door lock ECUs.

9. Release the 2 air duct claws and the air duct.

10. Remove the 2 heater housing-to-chassis rivets and the heater housing.

➡ **When installing the heater housing, use new screws in place of the rivets.**

11. Remove the heater core-to-heater housing cover.

12. Remove both heater core screws and clamps; then, remove the heater core.

To install:

13. Install the heater core and both heater core screws and clamps.

14. Install the heater core-to-heater housing cover.

➡ **When installing the heater housing, use new screws in place of the rivets.**

15. Install the heater housing-to-chassis and the 2 heater housing screws.

16. Release the air duct and the air duct claws.

17. Connect and install the theft deterrent and the wireless door lock ECUs.

18. Install the defroster nozzle and the 4 nozzle nuts.

19. Install the evaporator housing by performing the following procedure:

a. Install the evaporator housing.

b. Install the evaporator housing-to-chassis 2 rivets, 3 bolts and nut.

c. Connect the wiring harness clamp.

d. Connect the electrical connector at the evaporator housing.

e. In the engine compartment, use new O-rings and install the refrigerant lines-to-cowl connector and install the bolts.

20. Install the instrument panel and reinforcement by performing the following procedure:

a. Install the instrument panel reinforcement.

b. Install the No. 2 cowl brace.

c. Install the No. 1 and No. 2 braces.

d. Install the audio amplifier.

e. Install the instrument panel-to-chassis 5 bolts and nut.

f. Connect and install the air bag module and the 2 passenger's side air bag module bolts.

g. Install the heater control assembly.

h. Install the radio assembly.

i. Install the combination meter.

j. Install the cluster finish panel, engage the 8 clips and install the panel screws.

k. Install the No. 1 and No. 2 registers and the clock to the panel.

l. Connect the lower center cluster finish panel connector and install the 4 lower center cluster finish panel screws.

m. Install the 2 ashtray receptacle box screws and the ashtray.

n. Install the center cluster finish panel by engaging the 4 clips (1 at each corner) and the claw (bottom center).

o. Install the glove box and the glove box 3 nuts and 2 screws.

p. In the left side of the glove compartment, connect the air bag module connector and install the glove box door finish plate.

q. Install the No. 2 finish panel, engage the 4 panel clips and install the 3 panel bolts.

r. Install the No. 1 safety pad insert and the 2 insert bolts.

s. Install the finish panel, engage the 3 finish panel clips and install 2 lower finish panel bolts.

t. Install the hood open lever and the 2 hood open lever screws.

u. Install the combination switch, connect the electrical connector(s) and install the combination switch-to-steering column screws.

v. Install the steering column covers and the covers-to-steering column screws.

w. Install the front pillar garnish by engaging the 5 clips. If equipped with a tweeter speaker, connect the electrical connector.

x. Install the front door trim covers.

y. Install the cowl side boards.

z. Install the front door scuff plates.

21. Install the steering wheel by performing the following procedure:

a. Install the steering wheel to the steering column.

b. Align the steering wheel-to-main shaft marks.

c. Install the steering wheel nut and torque the nut to 25 ft. lbs. (34 Nm).

d. Install the air bag module to the steering wheel and connect the electrical connector.

e. Using a Torx® wrench, tighten the steering wheel screws to 78 inch lbs. (8.8 Nm).

f. Install the steering wheel side covers.

22. Connect the heater hoses to the heater core.

23. Refill the cooling system.

24. Connect the negative battery cable.

25. Evacuate and charge the air conditioning system.

26. Run the engine to normal operating temperatures; then, check the climate control operation and check for leaks.

Rear Auxiliary Heater

1. Disconnect the negative battery cable.

2. Drain the cooling system into a clean container for reuse.

3. Disconnect the heater hoses from the rear heater core.

4. Remove the front seats.

5. Remove the front door scuff plates.

6. Remove the cowl side trim.

7. Remove the rear door scuff plates.

8. Remove the lower door scuff plates.

9. Remove the rear console box.

10. Remove the left side air outlet grille.

11. Pull the carpet rearward.

12. Remove the 3 clips and the air outlet grille.

13. Remove the rear air duct 2 bolts, 2 clips and the duct.

14. Disconnect the electrical connectors.

15. Remove the 3 rear heater housing bolts and the housing.

16. Remove both heater core-to-heater housing screws and clamps.

17. Remove the heater core-to-heater housing screw and plate.

18. Remove the heater core.

To install:

19. Install the heater core.

20. Install the heater core-to-heater housing screw and plate.

21. Install both heater core-to-heater housing screws and clamps.

22. Install the rear heater housing and the 3 housing bolts.

23. Connect the electrical connectors.

24. Install the rear air duct and the 2 bolts and 2 clips.

25. Install the 3 clips and the air outlet grille.

26. Move the carpet forward.

27. Install the left side air outlet grille.

28. Install the rear console box.

29. Install the lower door scuff plates.

30. Install the rear door scuff plates.

31. Install the cowl side trim.

32. Install the front door scuff plates.

33. Install the front seats.

34. Connect the heater hoses to the rear heater core.

35. Refill the cooling system.

36. Connect the negative battery cable.

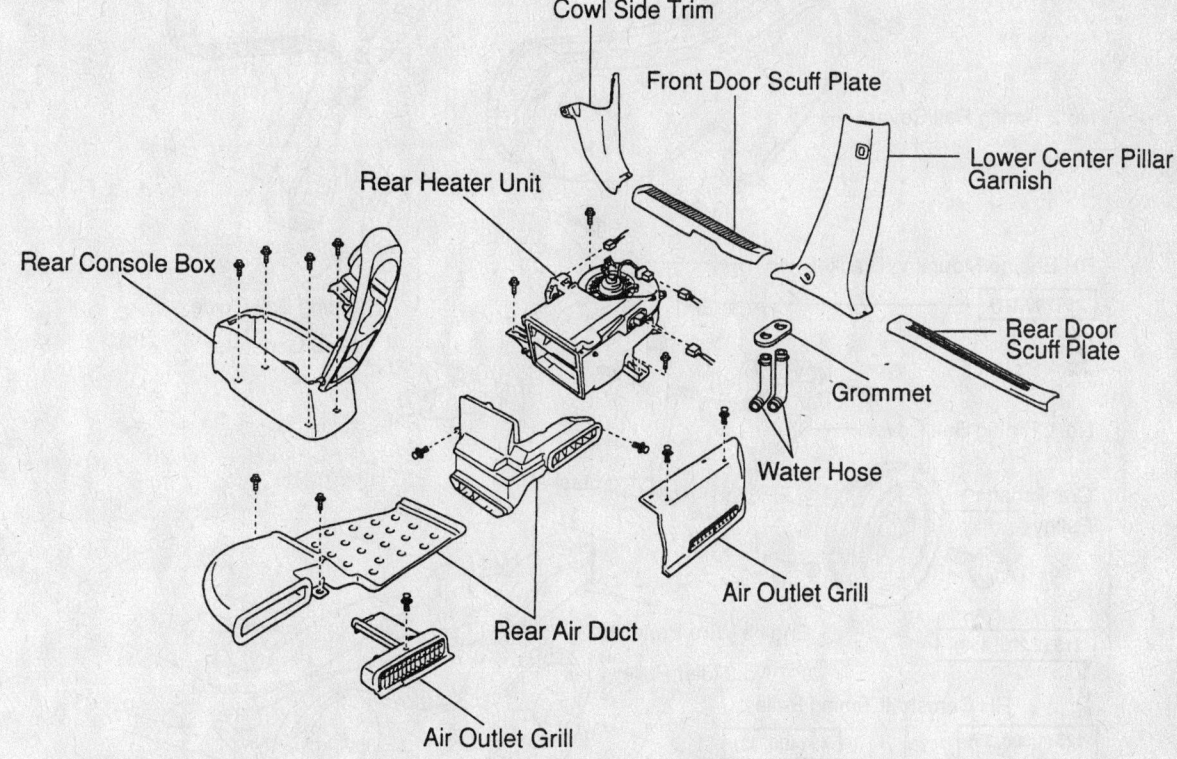

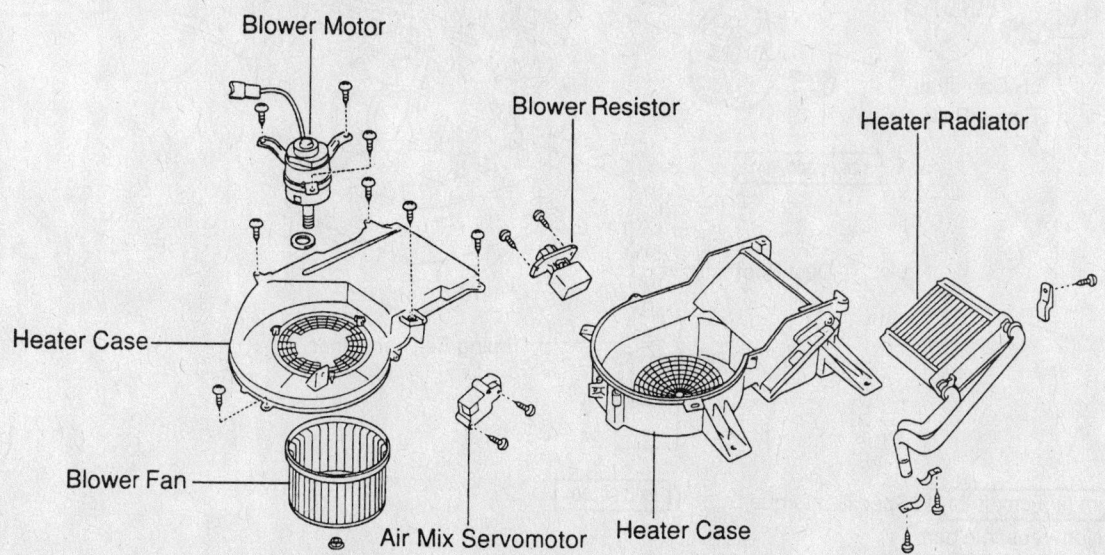

93113GH7

Exploded view of the rear heater core, the rear heater housing and related components

Water Pump

REMOVAL & INSTALLATION

1. Before servicing the vehicle, refer to the precautions in the beginning of this section.

2. Disconnect the negative battery cable.

3. Drain the engine coolant.

4. Remove or disconnect the following:

- Wiper and blade assembly
- Top cowl seal and panel
- Window washer hoses, from the ventilator louvers
- Left and right ventilator louvers
- Heater air duct
- Timing belt

5. Mark the left and right camshaft pulleys with a touch of paint.

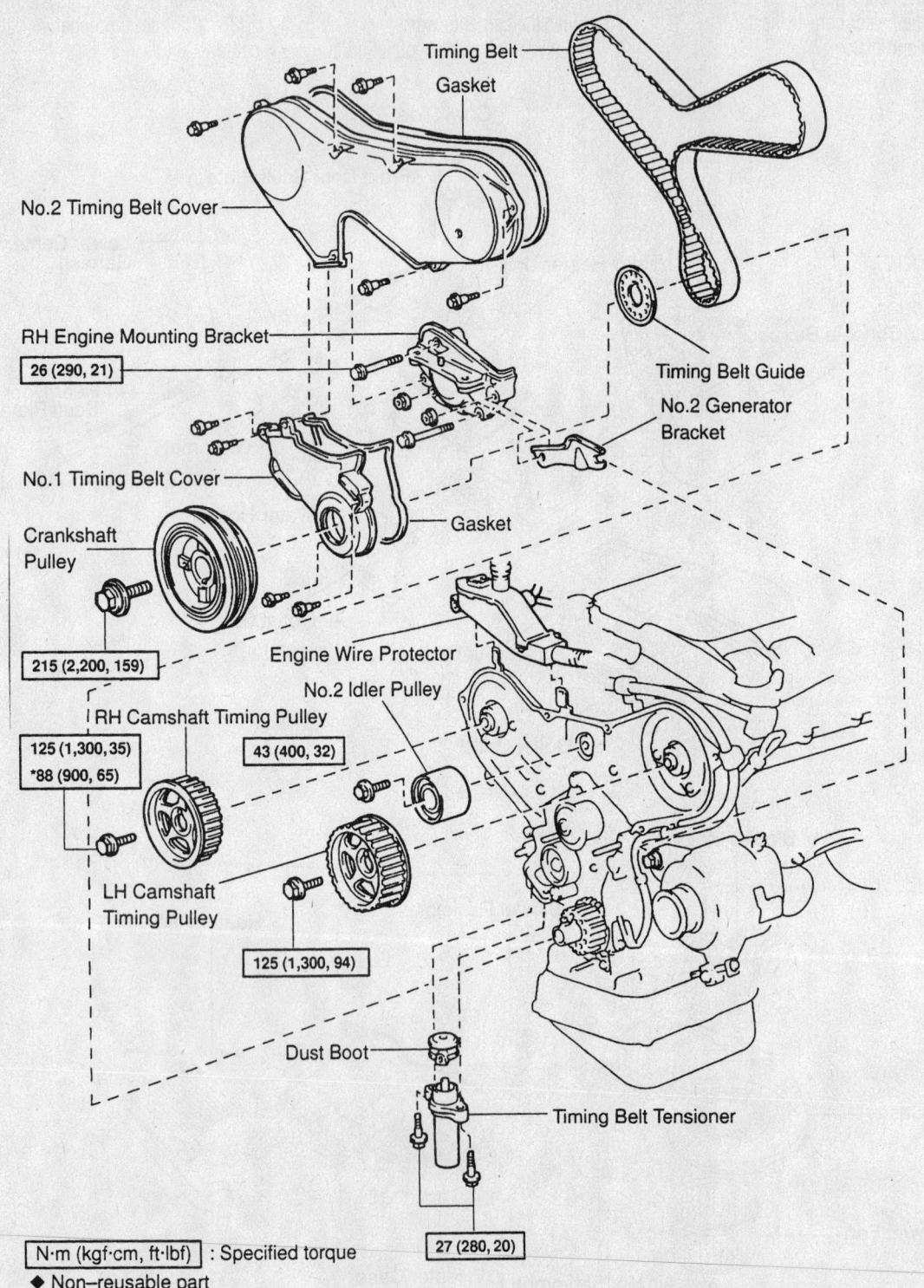

Timing Belt

Gasket

No.2 Timing Belt Cover

RH Engine Mounting Bracket
26 (290, 21)

Timing Belt Guide

No.2 Generator Bracket

No.1 Timing Belt Cover

Crankshaft Pulley

Gasket

215 (2,200, 159)

Engine Wire Protector

No.2 Idler Pulley

RH Camshaft Timing Pulley

125 (1,300, 35)
*88 (900, 65)

43 (400, 32)

LH Camshaft Timing Pulley

125 (1,300, 94)

Dust Boot

Timing Belt Tensioner

N·m (kgf·cm, ft·lbf) : Specified torque
27 (280, 20)

◆ Non-reusable part
*For use with SST

7924ZG15

Exploded view of the components to gain access to the water pump

6. Remove or disconnect the following:
- Right and left camshaft pulleys bolts
- Pulleys from the engine

➡ **Be sure not to mix up the pulleys.**

- No. 2 idler pulley by removing the bolt
- 3 clamps and engine wire from the rear timing belt cover
- 6 No. 3 timing belt cover-to-engine bolts
- Water pump nuts/bolts
- Water pump and gasket from the engine

To install:

7. Check that the water pump turns smoothly. Also check the air hole for coolant leakage.

8. Apply liquid sealer to the gasket, water pump and engine block.

9. Install or connect the following:
- Water pump, using a new gasket. Tighten the nuts/bolts to 53 inch lbs. (6 Nm).
- Rear timing belt cover. Tighten the 6 bolts to 74 inch lbs. (9 Nm).
- Engine wire with the 3 clamps to the rear timing belt cover

- No. 2 idler pulley. Tighten the bolt to 32 ft. lbs. (43 Nm).

➡ **After tightening the bolt, be sure the idler pulley moves smoothly.**

- Right-hand camshaft pulley, with the flange side **outward**.

➡ **Be sure to align the knock pin hole on the camshaft pulley with the knock pin on the camshaft.**

- Tighten the camshaft bolt to 65 ft. lbs. (88 Nm), using the removal tools
- Left-hand camshaft pulley, with the flange side **inward**.

➡ **Be sure to align the knock pin hole on the camshaft pulley with the knock pin on the camshaft.**

- Tighten the camshaft bolt to 94 ft. lbs. (125 Nm), using the removal tools
- Timing belt

10. Fill the engine coolant.

11. Install or connect the following:
- Heater air duct
- Left and right ventilator louvers
- Window washer hoses to the ventilator louvers
- Top cowl seal and panel
- Wiper and blade assembly
- Negative battery cable

12. Start the engine.

13. Top off the engine coolant and check for leaks.

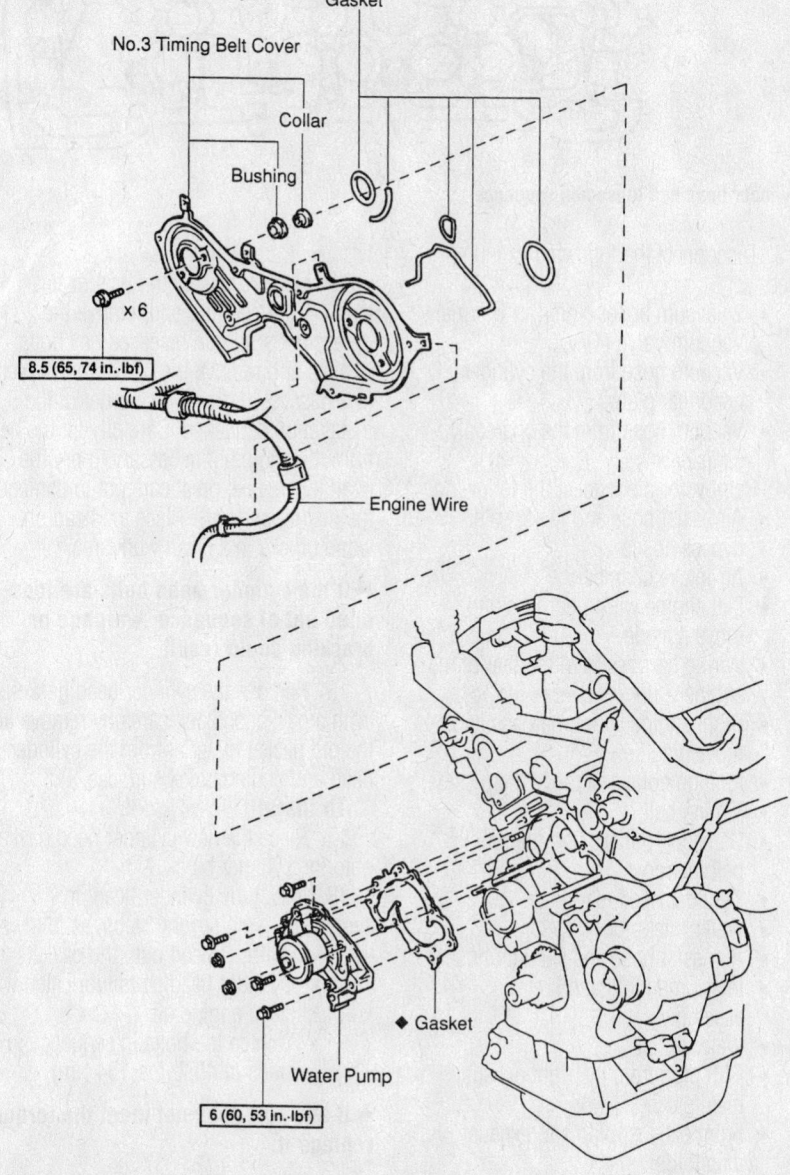

No.3 Timing Belt Cover

Gasket

Collar

Bushing

x 6

8.5 (65, 74 in.-lbf)

Engine Wire

◆ Gasket

Water Pump

6 (60, 53 in.-lbf)

N·m (kgf·cm, ft·lbf) : Specified torque
◆ Non–reusable part

7924ZG16

Exploded view of the water pump and related components

Cylinder Head

REMOVAL & INSTALLATION

1. Before servicing the vehicle, refer to the precautions in the beginning of this section.

2. Remove or disconnect the following:
- Wiper and blade assembly
- Top cowl seal and panel
- Window washer hoses from the ventilator louvers
- Left and right ventilator louvers
- Heater air duct

3. Relieve the fuel pressure.

4. Remove or disconnect the following:
- Turn the ignition key to the **OFF** position
- Negative battery cable

➡ **Wait at least 90 seconds from the time the negative battery was disconnected to start work.**

5. Drain the cooling system.

6. Remove or disconnect the following:
- Accelerator and throttle cables, if equipped with an automatic transaxle
- Air cleaner cover, air flow meter and the air duct
- Cruise control actuator and bracket, if equipped
- 2 engine ground straps
- Right engine mounting support
- Radiator hoses
- 2 heater hoses
- Fuel feed and return lines from the fuel rail assembly
- Pressure hose from the hydraulic motor
- V-bank cover

7. Disconnect the following vacuum hoses:
- Fuel pressure control Vacuum Switching Valve (VSV)
- Fuel pressure regulator
- Cylinder head rear plate
- Intake air control valve VSV
- Exhaust Gas Recirculation (EGR) vacuum modulator
- EGR valve

8. Disconnect the following wiring and hoses:
- Intake air control valve
- Fuel pressure regulator
- EGR VSV

9. Remove the 2 nuts and the emission control valve set.

10. Disconnect the following hoses;
- Brake booster vacuum hose
- PCV hose
- Intake air control valve vacuum hose

11. Remove or disconnect the following:
- Data Link Connector (DLC) from the mounting bracket
- 2 ground straps from the intake chamber
- Hydraulic motor pressure hose from the intake chamber
- Right Oxygen (O$_2$) sensor connector from the power steering pressure tube
- 2 nuts and the power steering pressure tube from the intake chamber
- Both power steering air hoses
- Engine hanger and the intake chamber support
- EGR pipe and gaskets

12. Disconnect the following wiring:
- Throttle Position (TP) sensor connector
- Idle Air Control (IAC) valve connector
- EGR gas temperature connector
- Air conditioning idle up connector

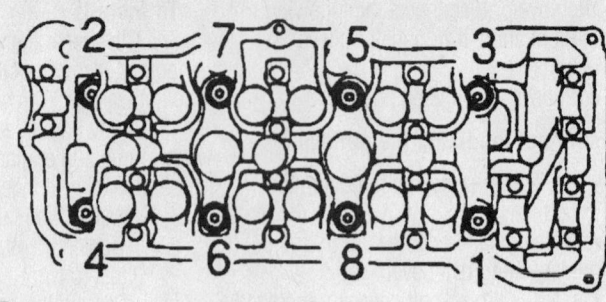

Front ←

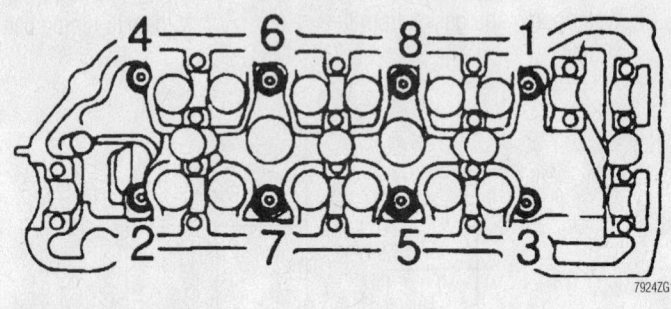

7924ZG19

Cylinder head bolt loosening sequence

13. Disconnect the following vacuum hoses:
- 2 vacuum hoses from the Thermal Vacuum Valve (TVV)
- Vacuum hose from the cylinder head rear plate
- Vacuum hose from the charcoal canister

14. Remove or disconnect the following:
- Air assist hose and the 2 water bypass hoses
- Air intake chamber
- Left engine wiring harness and move it aside
- Wiring harness from the rear of the engine
- Right engine wiring harness and move it aside
- Ignition coils and move them aside
- Timing belt
- Camshaft pulleys and the timing belt rear cover
- Cylinder head rear plate
- Water inlet pipe
- Air assist hose and vacuum hose
- Intake manifold and fuel rail assembly
- Water outlet
- EGR pipe from the right exhaust manifold
- Front exhaust pipe and exhaust manifolds
- Dipstick assembly and the power steering pump bracket
- Valve covers and the Camshaft Position (CMP) sensor

- Camshafts

15. Be sure the engine is at/or near ambient temperature and remove the 2 (1 on each head) 8mm recessed hex bolts. Loosen and remove the 8 head bolts evenly, in 3 passes, in the reverse order of the installation sequence. Carefully lift the head from the engine; if necessary to pry the head loose, take great care not to damage the mating surfaces. Place the head on wood blocks in a clean work area.

➡️**If the cylinder head bolts are loosened out of sequence, warpage or cracking could result.**

16. Remove the cylinder head gasket. With a gasket scraper, carefully remove all the old gasket material from the cylinder head and engine block surfaces.

To install:

17. Place the new cylinder head gasket onto the cylinder block.

18. Install the cylinder head, in sequence, using several steps, as follows:
- Cylinder head onto the gasket
- Cylinder head bolts lubricated with clean engine oil
- Tighten the bolts in sequence in 3 steps to 40 ft. lbs. (54 Nm).

➡️**If any bolt does not meet the torque, replace it.**

- Mark the forward edge of each bolt with paint, then tighten each bolt, in proper sequence, an additional 90 degrees.
- Check that each painted mark is

now at a 90 degrees angle to the front

➡ **The paint mark applied to the bolt in the 9 o'clock position and should now be in the 12 o'clock position.**

- Remaining 8mm bolts, lubricated with engine oil. Tighten both bolts to 13 ft. lbs. (18 Nm).

19. Install the camshafts.
20. Check and adjust the valves.
21. Apply sealant to the cylinder heads where the camshaft supports meet the cylinder heads.
22. Install or connect the following:
- Cylinder head covers, using new gaskets
- Dipstick and power steering pump bracket
- Exhaust manifolds. Tighten the nuts to 36 ft. lbs. (49 Nm).
- EGR pipe to the right exhaust manifold
- Water outlet
- Intake manifold and the fuel rail assembly. Tighten the intake manifold nuts/bolts to 11 ft. lbs. (15 Nm).
- Air assist hose and the 2 water bypass hoses
- Water inlet pipe and cylinder head rear plate
- Timing belt rear cover and camshaft pulleys
- Timing belt
- Spark plugs and ignition coils
- Right engine wiring harness
- Wiring harness to the rear of the engine
- Left engine wiring harness
- Air intake chamber
- EGR pipe, using new gaskets

23. Connect the following vacuum hoses:
- The 2 TVV vacuum hoses
- The vacuum hose to the rear cylinder head plate
- Charcoal canister vacuum hose

24. Connect the following electrical wiring:
- TP sensor connector
- IAC valve connector
- EGR gas temperature connector
- Air conditioning idle up connector

25. Install or connect the following:
- Engine hanger and the intake chamber support
- Both power steering air hoses
- Power steering pressure tube to the intake chamber
- O$_2$ sensor connector to the pressure tube.

- Both ground straps, to the intake chamber
- DLC to the bracket

26. Connect the following hoses:
- Power brake booster vacuum hose
- PCV hose
- IAC valve vacuum hose

27. Install or connect the following:
- Emission control valve set and related vacuum hoses and connectors
- V-bank cover
- Pressure hose to the hydraulic motor
- Fuel lines to the fuel rail assembly
- Heater and radiator hoses
- Right engine mounting support
- Both engine ground straps
- Upper front suspension brace, if removed. Tighten the nuts to 59 ft. lbs. (80 Nm).
- Cruise control actuator and bracket
- Air cleaner, air flow meter and air duct assembly
- Accelerator and throttle cables, if equipped with an automatic transaxle

28. Fill the cooling system.
29. Install or connect the following:
- Negative battery cable
- Heater air duct
- Left and right ventilator louvers
- Window washer hoses from the ventilator louvers
- Top cowl seal and panel
- Wiper and blade assembly

30. Start the engine and check for leaks.
31. Bleed the air from the cooling system.
32. Road test the vehicle and check for unusual noise, shock, slippage, correct shift points and smooth operation.
33. Recheck the coolant and engine oil levels.

Intake Manifold

REMOVAL & INSTALLATION

1. Before servicing the vehicle, refer to the precautions in the beginning of this section.
2. Remove or disconnect the following:
- Wiper and blade assembly
- Top cowl seal and panel
- Window washer hoses from the ventilator louvers
- Left and right ventilator louvers
- Heater air duct

3. Properly relieve the fuel system pressure.

4. Remove the battery and battery tray.
5. Drain and recycle the engine coolant.
6. Remove or disconnect the following:
- Accelerator cable, on automatic transaxles
- Throttle cable
- Air cleaner cap assembly
- Any wiring or hoses interfering with removal
- Right side engine mount stay
- Radiator and heater hoses in the way of the intake manifold removal
- V-bank cover
- All the vacuum hose and wiring for the emission control valve set
- Air intake chamber and discard the gasket
- Exhaust Gas Recirculation (EGR) pipe and discard the gaskets
- Hydraulic motor pressure hose from the air intake chamber
- Engine wiring harnesses from the left side, right side, rear and No. 3 timing belt cover
- Front exhaust pipe, if necessary
- Timing belt, camshaft timing pulleys, No. 2 idler pulley and No. 3 timing belt cover
- Cylinder head rear plate
- 2 bolts, nuts and plate washers with the intake manifold.

➡ **The delivery pipes with injectors will be attached to the manifold.**

- Other fuel related components such as the No. 2 fuel pipe and pulsation damper, if needed
- Delivery pipes from the intake manifold

7. Clean and inspect the intake manifold mating surfaces. Scrape all old gasket martial off.

To install:

8. Install or connect the following:
- Delivery pipes with injectors to the intake manifold.

➡ **Be sure to place 4 spacers in position on the manifold. Temporarily install 4 bolts to retain the delivery pipes to the manifold. Inspect the injectors for smooth rotation.**

- Tighten the delivery pipes bolts to 84 inch lbs. (10 Nm), once the injectors are properly seated
- No. 2 fuel pipe with union bolts and gaskets. Tighten the bolts to 24 ft. lbs. (32 Nm).
- No. 1 fuel pipe with pulsation damper, using 4 new gaskets. Tighten the damper to 35 ft. lbs.

(32 Nm) and the bolt to 11 ft. lbs. (15 Nm).

- Fuel pressure regulator, if removed
- Intake manifold. Tighten the 9 bolts and 2 nuts in a crisscross pattern to 11 ft. lbs. (15 Nm).

➡ **Be sure the gasket is in place properly prior to tightening.**

9. Retighten the water outlet mounting nuts/bolts to 11 ft. lbs. (15 Nm), if loosened.

10. Install or connect the following:
- Air assist hose and water inlet pipe, using a new O-ring, by applying a small amount of soapy water. Tighten the fastener(s) to 14 ft. lbs. (20 Nm).
- Ground strap
- Vacuum hoses removed to the air intake chamber and vacuum tank
- Any remaining components, using new gaskets. Tighten the air intake chamber nuts/bolts to 32 ft. lbs. (43 Nm), the EGR pipe nuts to 108 inch lbs. (12 Nm) and the emission control valve set to 69 inch lbs. (8 Nm).
- Air cleaner assembly
- Heater hoses
- Battery and tray
- Throttle cable with bracket onto the throttle body
- Accelerator cable, by adjusting it, if equipped with an automatic transaxle

11. Refill the cooling system
12. Install or connect the following:
- Negative battery cable
- Heater air duct
- Left and right ventilator louvers
- Window washer hoses from the ventilator louvers
- Top cowl seal and panel
- Wiper and blade assembly

13. Start the engine and inspect for leaks.

Exhaust Manifold

REMOVAL & INSTALLATION

Front Manifold

➡ **Removing the oil filter helps gain access to a lower bolt in the front exhaust manifold.**

1. Before servicing the vehicle, refer to the precautions in the beginning of this section.
2. Remove or disconnect the following:

- Negative battery cable
- Engine undercovers
- Front exhaust pipe from the exhaust manifolds, by removing the nuts

➡ **Check for access to some of the manifold lower bolts, if so remove any possible.**

- Heated Oxygen (HO2) sensor
- Exhaust manifold stay, by removing the bolt and nut
- Remaining exhaust manifold nuts; then, separate the exhaust manifold from the engine

To install:
3. Install or connect the following:
- Exhaust manifold, using a new gasket. Uniformly, tighten the bolts to 36 ft. lbs. (49 Nm).
- Exhaust manifold stay. Tighten the nut/bolt to 15 ft. lbs. (20 Nm).
- Heated Oxygen (HO2) sensor to the exhaust manifold
- Front exhaust pipe to the exhaust manifold, using a new gasket. Tighten both nuts to 46 ft. lbs. (62 Nm).
- Engine undercovers
- Negative battery cable

Rear Manifold

1. Before servicing the vehicle, refer to the precautions in the beginning of this section.
2. Remove or disconnect the following:
- Negative battery cable
- Engine undercovers
- Front exhaust pipe from both exhaust manifolds, from below the engine
- Exhaust Gas Recirculation (EGR) pipe from the rear exhaust manifold, by removing the 4 nuts
- Heated Oxygen (HO2) sensor wiring, from the right exhaust manifold
- Exhaust manifold stay
- 6 exhaust manifold nuts and the exhaust manifold

To install:
3. Install or connect the following:
- Exhaust manifold to the engine, using a new gasket. Tighten the 6 nuts to 36 ft. lbs. (49 Nm).
- Exhaust manifold stay. Tighten the nut/bolt to 15 ft. lbs. (20 Nm).
- HO2 sensor wiring to the exhaust manifold
- EGR pipe to the exhaust manifold and the engine, using new gaskets.

Tighten the 4 nuts to 108 inch lbs. (12 Nm).
- Front exhaust pipe to the exhaust manifold, use a new gasket. Tighten both nuts to 46 ft. lbs. (62 Nm).
- Engine undercovers
- Negative battery cable

Front Crankshaft Seal

REMOVAL & INSTALLATION

1. Before servicing the vehicle, refer to the precautions in the beginning of this section.
2. Remove or disconnect the following:
- Engine coolant reservoir tank and the alternator belt
- Right front wheel and the splash shield
- Power steering pump drive belt, by loosening both bolts
- Both ground wire connectors
- Right engine mounting stay
- Engine moving control rod and the No. 2 right engine mount bracket

➡ **To extract the engine bracket and control rod, raise the engine slightly.**

- No. 2 alternator bracket
- Crankshaft pulley bolt, using a prybar and wrench or Crankshaft Pulley Holding tool 09213-54015 and Flange Holding tool 09330-00021
- Crankshaft pulley, using a puller
- No. 1 timing belt cover

3. Remove the No. 2 timing belt cover, as follows:
- Engine wire protector from the No. 3 (rear) timing belt cover
- Engine wire protector clamp from the No. 3 timing belt cover
- 5 bolts from the No. 2 timing belt cover
- No. 2 cover

To install:
4. Install or connect the following:
- No. 2 timing belt cover, using a new gasket

➡ **Install it evenly to the part of the belt cover shaded black. After installation, press down on it so that the adhesive sticks to the belt cover firmly.**

- No. 2 timing belt cover. Tighten the 5 bolts to 74 inch lbs. (8 Nm).
- Engine wire protector clamp to the No. 3 timing belt cover
- Engine wire protector to the No. 3 timing belt cover with the bolt

- No. 3 timing belt cover, using a new gasket
- Tighten the 4 No. 1 timing belt cover bolts to 74 inch lbs. (8 Nm).
- Crankshaft pulley. Tighten the bolt to 159 ft. lbs. (215 Nm).
- No. 2 alternator bracket. Tighten the nut to 21 ft. lbs. (28 Nm). Do not tighten the pivot bolt at this time.
- No. 2 right engine mounting bracket and the moving control rod
- Right engine mount stay
- Both ground wire connectors
- Drive belts by adjusting them
- Coolant reservoir
- Right front splash shield and wheel
- Negative battery cable

5. Start the vehicle and check for any leaks.

6. Recheck the ignition timing.

Camshaft and Valve Lifters

REMOVAL & INSTALLATION

1. Before servicing the vehicle, refer to the precautions in the beginning of this section.

2. Remove or disconnect the following:
- Timing belt and idler pulley
- Camshaft timing pulleys
- Cylinder head covers

➡ The thrust clearance on both the intake and exhaust camshafts is very small; the camshafts must be kept level during removal. If the camshafts are removed without being kept level, the camshaft may be caught in the cylinder head, causing the head to break or the camshaft to seize.

3. Remove the exhaust and intake camshafts from the right side cylinder head, as follows:

a. Turn the camshaft with a wrench until the 2 pointed marks drive and driven gears are aligned. (The right camshaft gears have 2 marks apiece; the left side camshaft gears have 1 mark each.)

b. Secure the exhaust camshaft sub-gear to the main gear using a service bolt. A bolt 0.63–0.79 in. (16–20mm) long with a 6mm thread diameter and a 1mm pitch is recommended. When removing the exhaust camshaft be sure the sub-gear is not loaded; all the force must be eliminated.

c. Uniformly loosen and remove the exhaust camshaft bearing cap bolts in several passes and in the proper

sequence. Remove the 8 bearing cap bolts and remove the caps, keeping them in the correct order.

d. Remove the exhaust camshaft from the engine.

e. Uniformly loosen and remove the 10 bearing cap bolts in several passes, in the proper sequence. Remove the bearing caps, keeping them in order, remove the oil seal, then lift out the intake camshaft.

4. Remove the exhaust and intake camshafts from the left side cylinder head, as follows:

a. Turn the camshaft with a wrench until the pointed marks on the drive and driven gears are aligned. (The right camshaft gears have 2 marks apiece; the

left side camshaft gears have 1 mark each.)

b. Secure the exhaust camshaft sub-gear to the main gear using a service bolt. A bolt 16–20mm long with a 6mm thread diameter and a 1mm pitch is recommended. When removing the exhaust camshaft be sure the sub-gear is not loaded; all the force must be eliminated.

c. Uniformly loosen and remove the exhaust camshaft bearing cap bolts in several passes and in the proper sequence. Remove the 8 bearing cap bolts and remove the caps. Keep the caps in the correct order.

d. Remove the exhaust camshaft from the engine.

e. Uniformly loosen and remove the

Intake

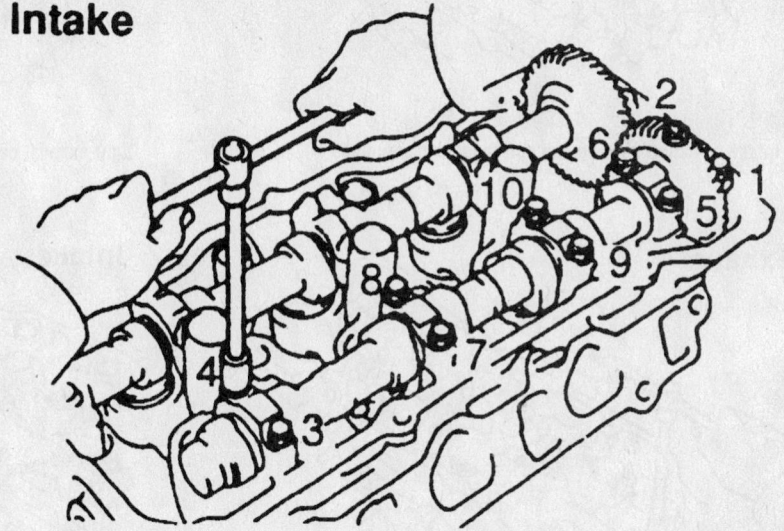

Right intake camshaft bearing cap bolt loosening sequence

Exhaust

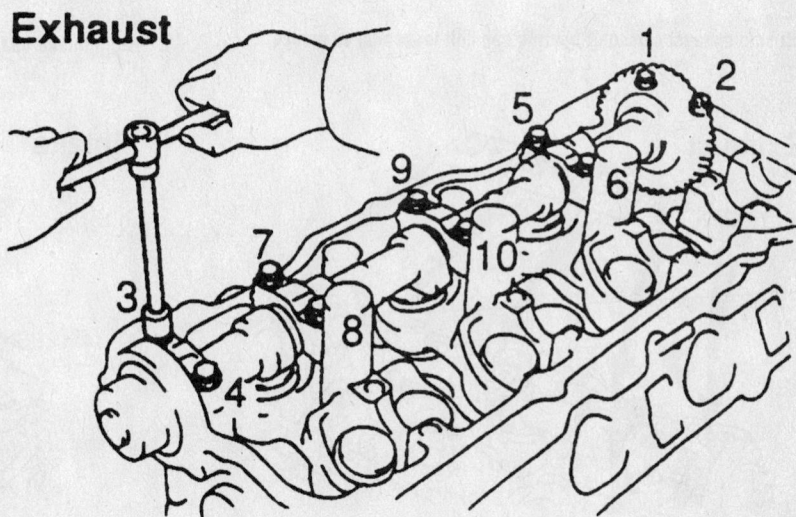

Right side exhaust camshaft bearing cap bolt loosening sequence

10 bearing cap bolts in several passes, in the reverse order of the installation sequence. Remove the bearing caps, keeping them in order, remove the oil seal, then lift out the intake camshaft.

5. Remove the valve lifter shims and hydraulic lifters. Identify each lifter and shim as it is removed so it can be rein-

stalled in the same position. If the lifters are to be reused, store them upside down in a sealed container.

To install:

6. Install the valve lifters into their original positions and install the shims. Check valve clearance and replace the shims as necessary.

7. When reinstalling, remember that the camshafts must be handled carefully and kept straight and level to avoid damage.

8. Before installing the camshafts in either cylinder head, apply multi-purpose grease to each camshaft.

9. Install the right camshafts, as follows:

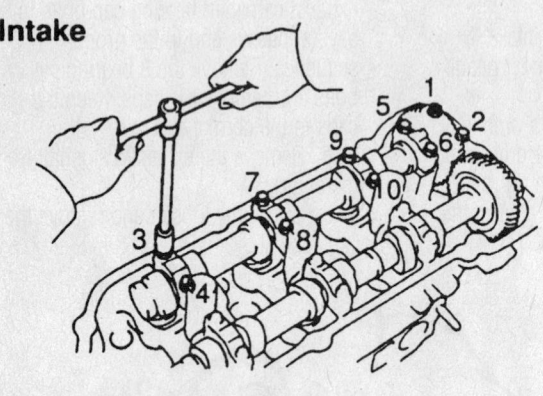

Intake

Left intake camshaft bearing cap bolt loosening sequence

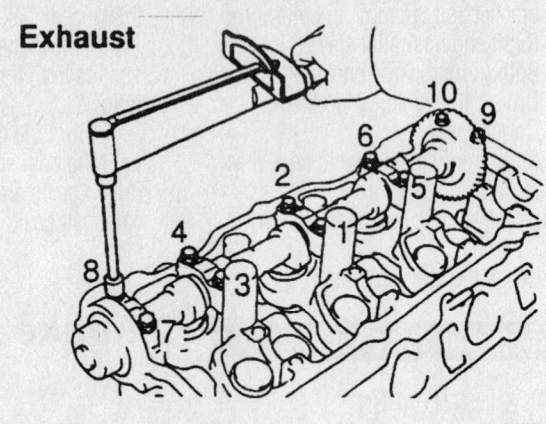

Exhaust

Right exhaust camshaft bearing cap bolt tightening sequence

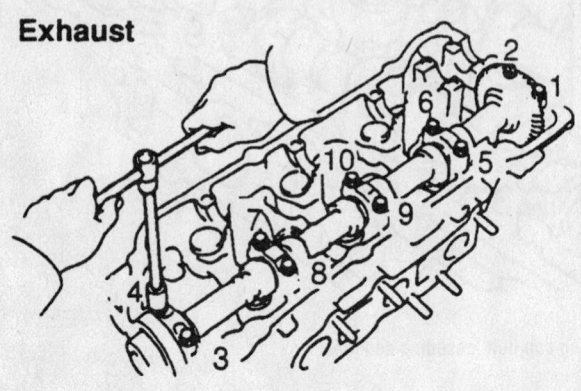

Exhaust

Left side exhaust camshaft bearing cap bolt loosening sequence

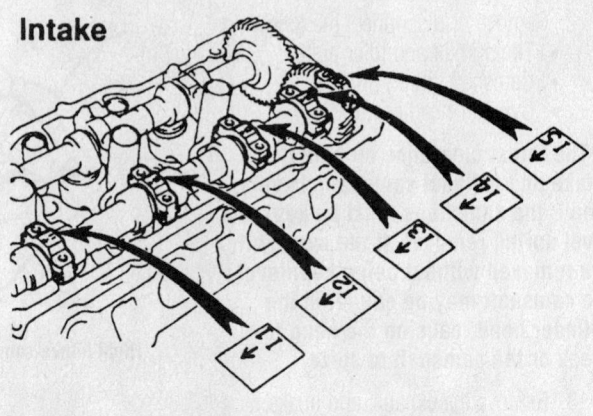

Intake

Right intake bearing caps must be placed in their proper locations

Exhaust

Right exhaust bearing caps must be placed in their proper locations

Intake

Right intake camshaft bearing cap bolt tightening sequence

Exhaust

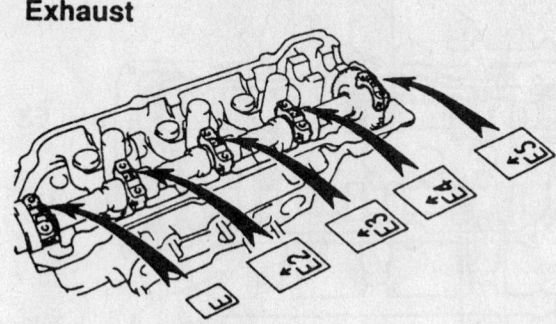

Exhaust

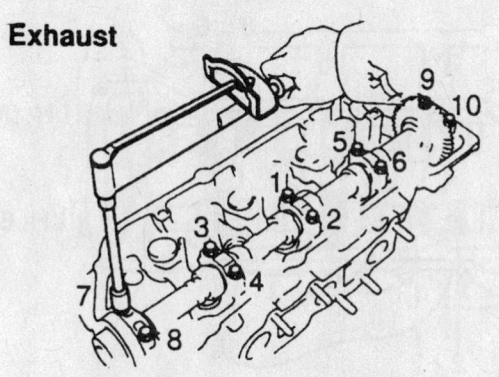

Left exhaust bearing caps locations and bolt tightening sequence

Intake

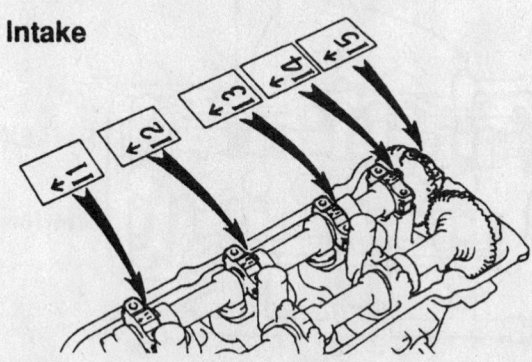

Intake

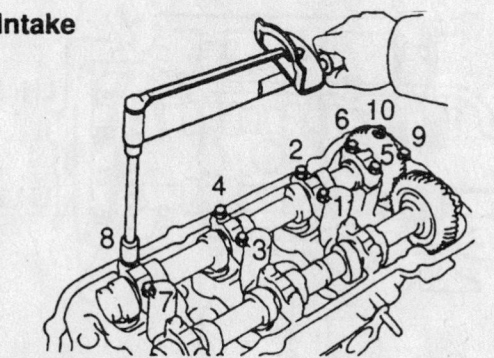

Left intake camshaft bearing cap locations and bolt tightening sequence

a. Position the intake camshaft on the head so that the alignment marks are at a 90 degrees angle from vertical. The mark should be at the "3 o'clock" position.

b. Apply sealant to the No. 1 bearing cap.

c. Apply a light coat of clean engine oil to the bolt threads and under the bolt head. Install the bearing caps to their proper position. Tighten the bolts evenly and in several passes to 12 ft. lbs. (16 Nm) in the proper sequence.

d. Position the exhaust camshaft on the head so that the alignment marks are at a 90 degrees angle from vertical. The mark should be at the "9 o'clock" position and must align with the marks on the other gear.

e. Apply a light coat of clean engine oil to the bolt threads and under the bolt head. Install the bearing caps to their proper position. Tighten the bolts evenly and in several passes to 12 ft. lbs. (16 Nm) in the proper sequence.

f. Remove the service bolt.

10. Install the left camshafts, as follows:

a. Position the intake camshaft on the head so that the alignment mark is at a 90 degrees angle from vertical. The mark should be at the "9 o'clock" position.

b. Apply sealant to the No. 1 bearing cap.

c. Apply a light coat of clean engine oil to the bolt threads and under the bolt head. Install the bearing caps to their proper position. Tighten the bolts evenly and in several passes to 12 ft. lbs. (16 Nm) in the proper sequence.

d. Position the exhaust camshaft on the head so that the alignment marks are at a 90 degrees angle from vertical. The mark should be at the "3 o'clock" position and must align with the marks on the other gear.

e. Apply a light coat of clean engine oil to the bolt threads and under the bolt head. Install the bearing caps to their proper position. Tighten the bolts evenly and in several passes to 12 ft. lbs. (16 Nm) in the proper sequence.

f. Remove the service bolt.

11. Install or connect the following:
- New camshaft oil seals, lubricated with multi-purpose grease
- No. 3 (rear) timing belt cover
- Camshaft timing gears
- Idler pulley, timing belt and covers

12. Check and adjust the valve clearance.

13. Install the cylinder head (valve) covers.

14. Start the engine. Check the ignition timing.

15. Test drive the vehicle.

16. Check all fluid levels.

Valve Lash

ADJUSTMENT

➡ **Adjust the valve clearance when the engine is cold.**

1. Before servicing the vehicle, refer to the precautions in the beginning of this section.

2. Remove or disconnect the following:
- Negative battery cable. If equipped with an air bag, wait at least 90 seconds before proceeding.
- Accelerator/throttle cable from the throttle linkage
- Air cleaner cover, air flow meter and air duct assembly
- V-bank cover
- Emission control valve set
- Air intake chamber
- Engine harness from the injectors and the ignition coils
- Ignition coils and keep them in order for reassembly
- Spark plugs
- Cylinder head covers

3. Turn the crankshaft pulley and align its groove with the timing mark **0** of the No. 1 timing cover.

4. Check that the valve lifters on the No. 1 intake are loose and the No. 1 exhaust are tight. If not, turn the crankshaft 1 complete revolution (360 degrees).

➡ **All measurements should be written down. These recorded measurements will need to be used in conjunction with a mathematical formula to determine the thickness of the replacement shims.**

5. Measure the clearance between the valve lifters and the camshaft. Record the measurements on valves No. 1 and 6 intake; No. 2 and 3 exhaust.

 a. The intake valve clearance cold is 0.006–0.010 in. (0.15–0.25mm).

 b. The exhaust valve clearance cold is 0.010–0.014 in. (0.25–0.35mm).

6. Turn the crankshaft ⅔ of a revolution (240 degrees). Record the measurements on valves No. 2 and 3 intake; No. 4 and 5 exhaust.

7. Turn the crankshaft another ⅔ of a revolution. Record the measurements on valves No. 4 and 5 intake; No. 1 and 6 exhaust.

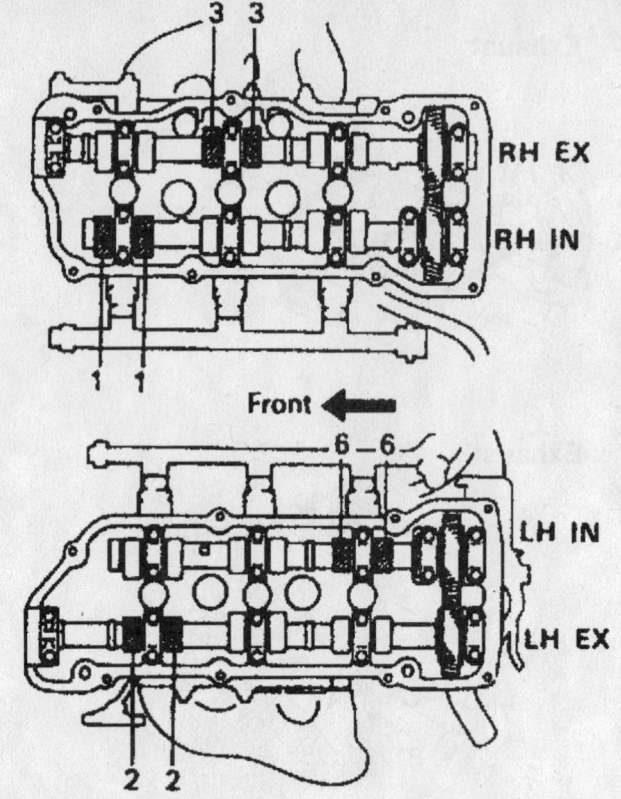

Adjust these valves during the 1st step

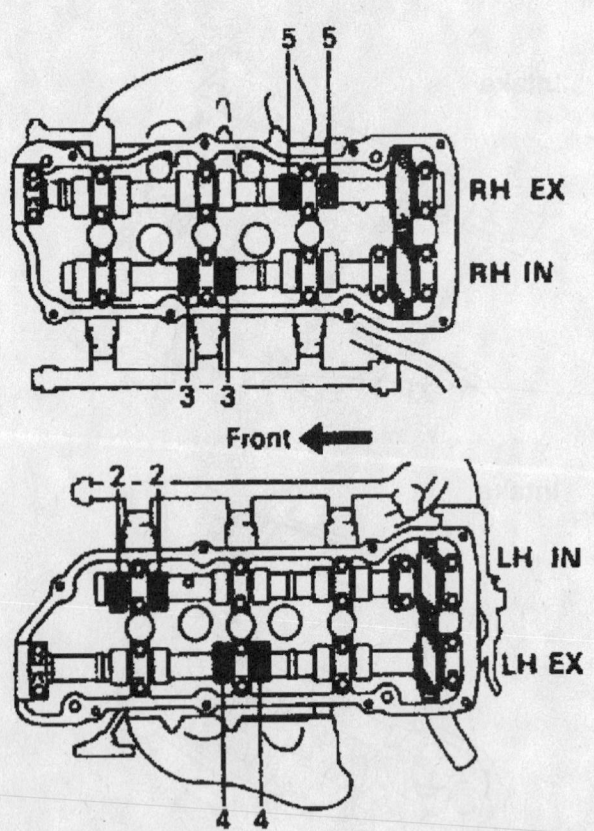

Adjust these valves during the 2nd step

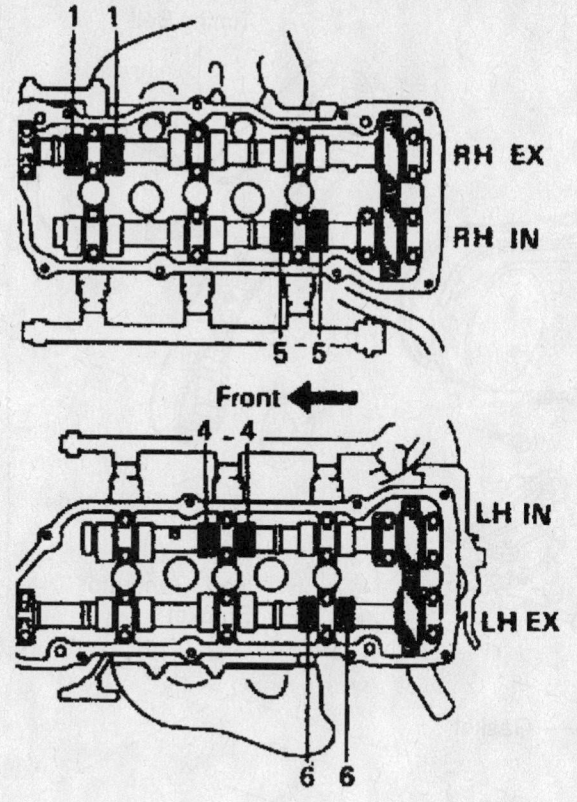

RH EX

RH IN

5 5

Front ←

4 4

LH IN

LH EX

6 6

7923VG67

Adjust these valves during the 3rd step

8. Remove the adjusting shim by turning the crankshaft to position the cam lobe of the camshaft in the up position on the valve to be adjusted. Using a small thin flat bladed tool, turn the valve lifter so that the notches are perpendicular to the camshaft. Press down the valve lifter with tool 09248-55010 part A. Place too 09248-55010 part B between the camshaft and the valve lifter; remove part A.

9. Remove the adjusting shim with a magnet and a small screwdriver.

10. Determine the replacement adjusting shim size by either using the charts or the following formulas:
- Intake: $N = T + (A-0.008$ in./0.020mm)
- Exhaust: $N = T + (A-0.012$ in./0.30mm)
- T = Thickness of removed shim
- A = Measured valve clearance
- N = Thickness of new shim

11. Select a new shim with a thickness as close as possible to the calculated value. Install the new replacement shim.

➡**Shims are available in 17 sizes in increments of 0.0020 in. (0.050mm), from 0.0984 in. (2.500mm) to 0.1299 in. (3.300mm).**

12. Recheck the valve clearance.

13. Install or connect the following:
- Cylinder head covers
- Spark plugs and the ignition coils
- Engine wiring harness to the injectors and the coils
- Intake chamber
- Emission control valve set
- V-bank cover
- Air flow meter, air duct and air cleaner cover
- Negative battery cable

Starter Motor

REMOVAL & INSTALLATION

1. Before servicing the vehicle, refer to the precautions in the beginning of this section.

2. Remove or disconnect the following:
- Battery
- Battery tray

3. Remove or disconnect the cruise control actuator, if equipped, as follows:
- Actuator connector and clamp
- 3 bolts and the actuator with the bracket

4. Remove or disconnect the following:
- Automatic transaxle shift control cable

- Engine wiring
- Starter electrical connectors
- Both bolts, shift control cable clamp and the starter

To install:

5. Install or connect the following:
- Starter and the shift control cable clamp. Tighten the bolts to 27 ft. lbs. (37 Nm).
- Starter electrical connectors
- Engine wiring
- Automatic transaxle shift control cable

6. Install or connect the following, if equipped with cruise control:
- 3 bolts and the actuator with the bracket
- Actuator connector and clamp

7. Install or connect the following:
- Battery tray
- Battery

Timing Belt

REMOVAL & INSTALLATION

1. Disconnect the negative battery cable.

2. Remove the outer front cowl top panel assembly by performing the following procedure:

a. Remove the wiper arm/blade assemblies head caps, nuts and assemblies.

b. Remove the head-to-cowl seal and the cowl panel hole cover.

c. Disconnect the windshield washer clip and hose.

d. Remove both (right and left) cowl top ventilator louvers.

e. Disconnect the electrical connector from the windshield wiper motor.

f. Remove the outer front cowl top panel assembly-to-cowl bolts and the panel.

3. Raise and safely support the vehicle.

4. Remove the right front wheel assembly and apron seal.

5. Remove the alternator by performing the following procedure:

a. Loosen the pivot bolt, adjusting lockbolt and adjusting bolt, then, remove the drive belt.

b. Disconnect the alternator's electrical connector.

c. Remove the nut and the alternator wire.

d. Disconnect the wiring harness from the clip.

e. Remove the alternator-to-bracket

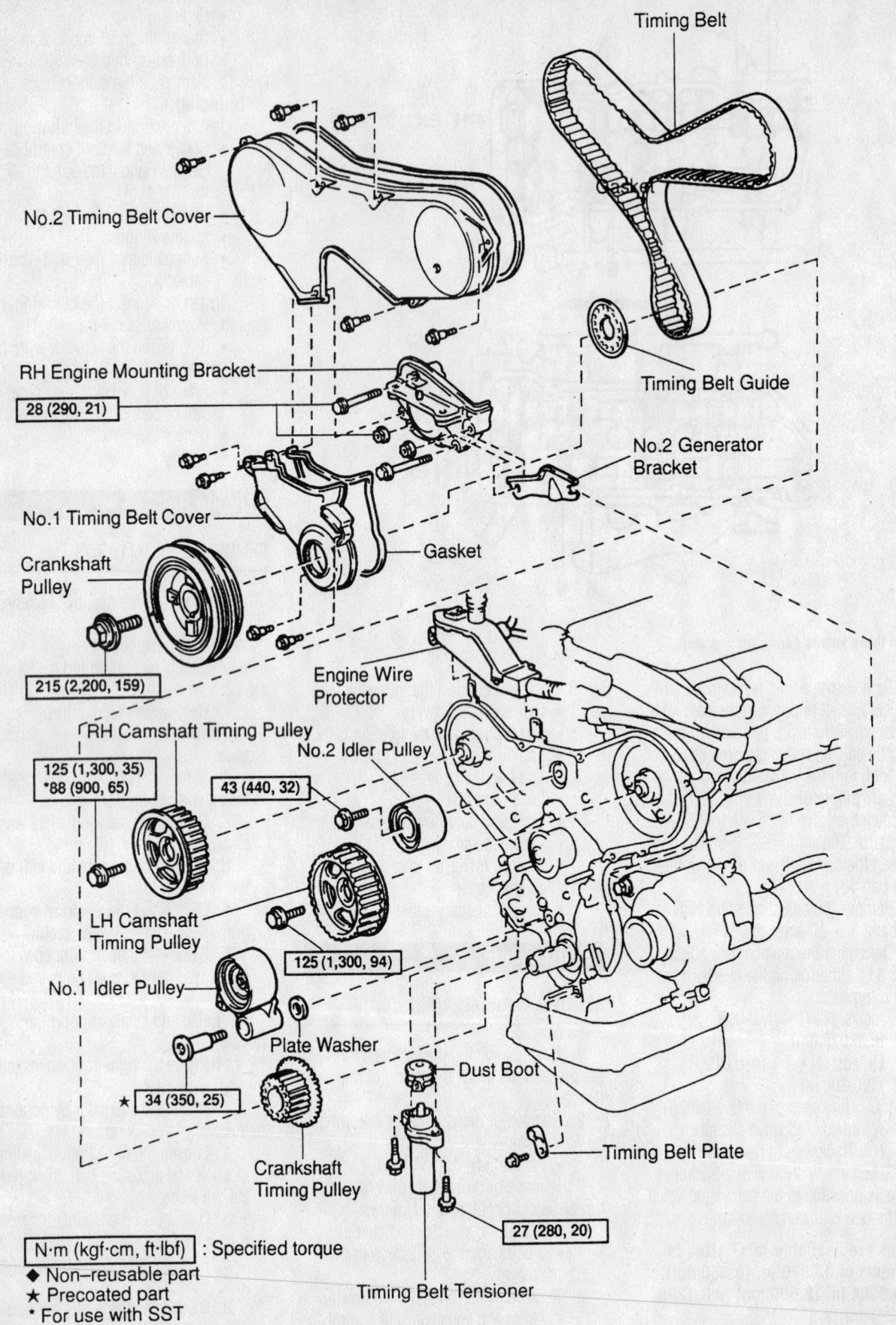

Timing Belt

No.2 Timing Belt Cover

Timing Belt Guide

RH Engine Mounting Bracket

28 (290, 21)

No.2 Generator Bracket

No.1 Timing Belt Cover

Crankshaft Pulley

Gasket

215 (2,200, 159)

Engine Wire Protector

RH Camshaft Timing Pulley

No.2 Idler Pulley

125 (1,300, 35)
*88 (900, 65)

43 (440, 32)

LH Camshaft Timing Pulley

125 (1,300, 94)

No.1 Idler Pulley

Plate Washer

Dust Boot

★ 34 (350, 25)

Crankshaft Timing Pulley

Timing Belt Plate

27 (280, 20)

Timing Belt Tensioner

N·m (kgf·cm, ft·lbf) : Specified torque
◆ Non–reusable part
★ Precoated part
* For use with SST

93025G19

Exploded view of the timing belt assembly

pivot bolt, the washer, adjusting lockbolt and alternator.

6. Loosen the power steering pump's mount and adjusting bolt, then, remove the drive belt.

7. Disconnect the hose from the engine coolant reservoir.

8. Disconnect the Diagnostic Link Connector 1 (DLC1) from the No. 2 right side engine mounting bracket.

9. Remove the right side engine mounting stay, the engine moving control rod and the No. 2 right side engine mounting bracket.

10. Loosen the alternator's pivot bolt, the nut and the No. 2 alternator bracket.

11. Using the Crankshaft Pulley Holding tool 09213-54015, Bolt tool 91651-60855 and Companion Flange Holding tool 09330-00021, or equivalent, remove the crankshaft pulley bolt.

12. Using a Puller "C" Set 09950-50011 (Hanger 150 tool 09951-05010, Slide Arm tool 09952-5010, Center Bolt 100 tool 09953-05010, Center Bolt 150 tool 09953-05020 and 2 No. 2 Claw tools 09954-05020), pull the crankshaft pulley from the crankshaft.

13. Remove the lower (No. 1) timing belt cover. Remove the timing belt guide from the crankshaft.

14. Remove the engine wire protector clamps from the upper (No. 2) timing belt cover and remove the upper (No. 2) timing belt cover.

15. Remove the right side engine mounting brace

➡**If reusing the timing belt, be sure that you can still read the installation marks. If not, place new installation marks on the timing belt to match the timing marks of the camshaft timing pulleys.**

16. Temporarily install the crankshaft pulley bolt.

17. Set the No. 1 cylinder to Top Dead Center (TDC) of the compression stroke, as follows:

a. Rotate the crankshaft (CLOCK-WISE) to align the timing marks: dimple on the crankshaft timing sprocket with the notch on the oil pump body.

b. Check that the timing marks on the camshaft sprockets and the rear timing belt cover are aligned; if not, rotate the crankshaft 360 degrees (1 revolution) and align the marks.

18. Remove the timing belt tensioner and the timing belt.

To install:

19. Inspect the timing belt tensioner by performing the following procedures:

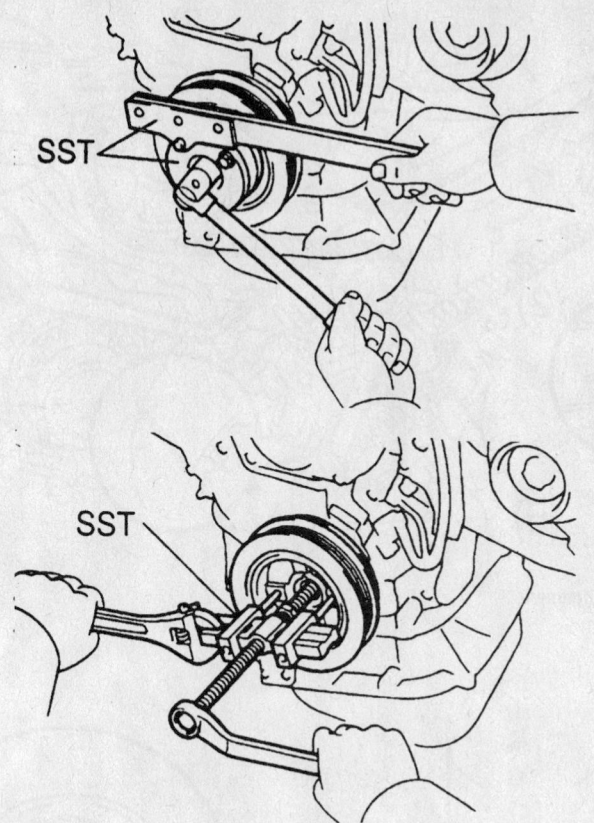

Removing/installing the crankshaft pulley

93025G20

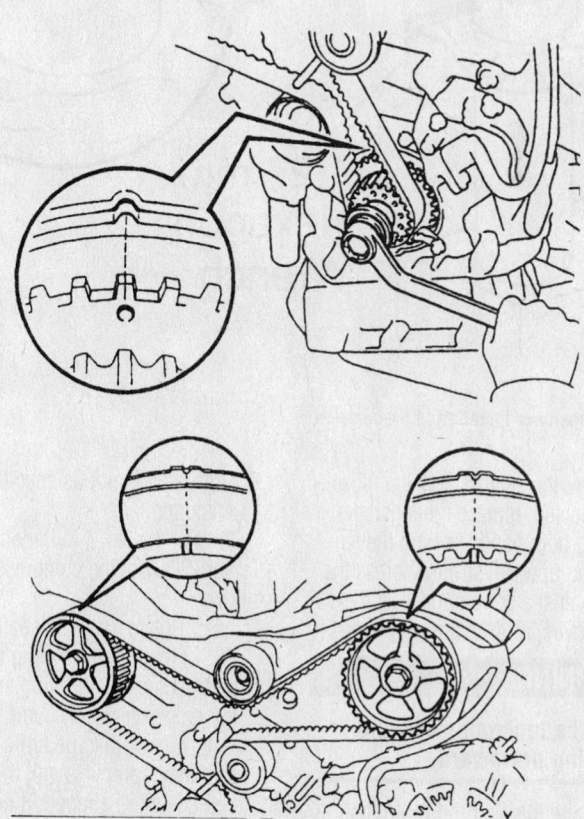

93025G21

View of the timing mark locations

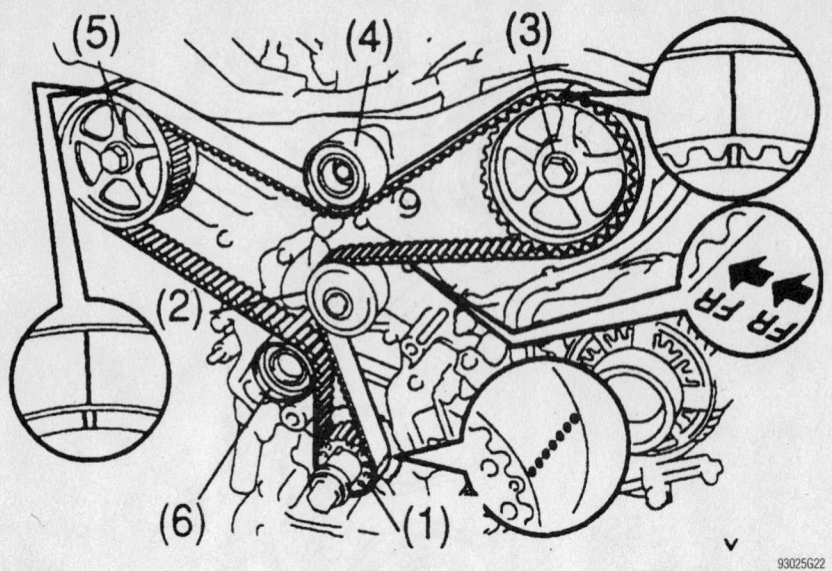

Timing belt alignment

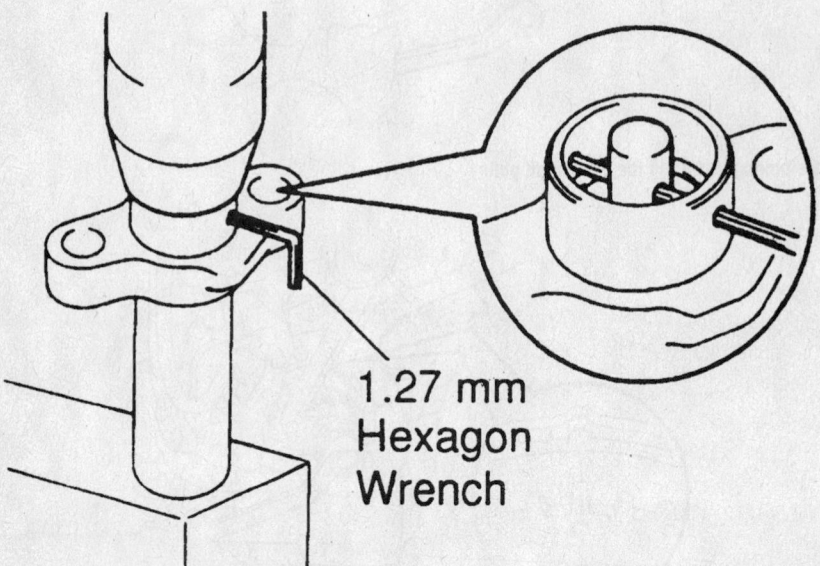

1.27 mm Hexagon Wrench

Timing belt tensioner installation preparation

a. Inspect the seal for leakage; if leakage is suspected, replace the tensioner.

b. Using both hands to hold the tensioner facing upward, strongly press the pushrod against a solid surface. If the pushrod moves, replace the tensioner.

✳✳ WARNING

Never hold the tensioner with the pushrod facing downward.

c. Measure the pushrod protrusion from the housing end, it should be 0.394–0.425 in. (10.0–10.8mm); if the protrusion is not as specified, replace the tensioner.

20. Set the No. 1 cylinder to Top Dead Center (TDC) of the compression stroke, as follows:

a. Rotate the crankshaft (CLOCKWISE) to align the timing marks: dimple on the crankshaft timing sprocket with the notch on the oil pump body.

b. Check that the timing marks on the camshaft sprockets and the rear timing belt cover are aligned; if not, rotate the crankshaft 1 revolution (360 degrees) and align the marks.

21. Install the timing belt in the following order:

a. Crankshaft timing sprocket
b. Water pump pulley
c. Left camshaft timing sprocket
d. No. 2 idler pulley
e. Right camshaft sprocket
f. No. 1 idler pulley

22. Using a vertical press, slowly press the pushrod into the housing using 200–2205 lbs. (981–9807 N) until the holes align, then, install a 1.27mm Allen® wrench to secure the pushrod and release the press. Install the dust boot on the tensioner housing.

23. Install the timing belt tensioner and torque the bolts to 20 ft. lbs. (27 Nm).

24. Remove the Allen® wrench from the tensioner housing.

25. Slowly, rotate the crankshaft (CLOCKWISE) 2 complete revolutions and realign the timing marks. If the timing marks do not align, remove the timing belt and reinstall it.

26. Remove the crankshaft pulley bolt.

27. Install the right side engine mounting bracket and torque the bolts to 21 ft. lbs. (28 Nm).

28. Clean and install the upper (No. 2) timing belt cover.

➡ **If the gasket material on the timing belt covers is cracked, peeling or etc., replace it.**

29. Install the timing belt guide on the crankshaft with the cup side facing outward.

30. Clean and install the lower (No. 1) timing belt cover.

➡ **If the gasket material on the timing belt covers is cracked, peeling or etc., replace it.**

31. Install the crankshaft pulley.

32. Using the Crankshaft Pulley Holding tool 09213-54015, Bolt tool 91651-60855 and Companion Flange Holding tool 09330-00021, or equivalent, install the crankshaft pulley bolt and torque the bolt to 159 ft. lbs. (215 Nm).

33. To complete the installation, reverse the removal procedures.

34. Connect the negative battery cable.

35. Start the engine and check for leaks.

Oil Pan

REMOVAL & INSTALLATION

1. Before servicing the vehicle, refer to the precautions in the beginning of this section.

2. Remove or disconnect the following:
- Right front wheel
- Fender apron seal
- Engine undercover

3. Drain the engine oil from the engine.

4. Remove or disconnect the following:
- Front exhaust pipe
- Front exhaust pipe bracket from the No. 1 oil pan
- Flywheel housing undercover
- 10 bolts and 2 nuts to the No. 2 oil pan

5. Insert the blade of the Oil Pan Seal Cutting tool 09032-00100 between the No. 1 and No. 2 oil pans. Clean the surfaces of the oil pans.

6. Remove or disconnect the following:
- 3 oil strainer nuts and gasket

7. Remove the No. 1 oil pan, as follows:
- 2 bolts and the flywheel housing undercover
- 17 bolts and 2 nuts to the No. 1 oil pan

➡ **Make a note of the position of the each bolt. When replacing the bolts into the oil pan, place each bolt in the position from which it was removed.**

- Oil pan, by prying the portions between the cylinder block and the oil pan

➡ **Be careful not to damage the contact surfaces.**

- Baffle plate from the No. 1 oil pan

To install:

8. Clean all mating surfaces of the oil pans.

9. Install the baffle plate to the No. 1 oil pan and tighten to 69 inch lbs. (8 Nm).

10. Install the No. 1 oil pan, as follows:
a. Using a non residue solvent, clean both sealing surfaces to the oil pan.
b. Apply liquid sealant to the oil pan and engine block.
c. Install the oil pan with the 17 bolts and 2 nuts. Uniformly tighten the bolts and nuts in several passes.
d. Tighten the No. 1 oil pan bolts, as follows:
- 10mm head bolt: 69 inch lbs. (8 Nm)
- 12mm head bolt: 14 ft. lbs. (20 Nm)
- 14mm head bolt: 27 ft. lbs. (37 Nm)
e. Install the flywheel housing undercover with the 2 bolts. Tighten the bolts to 69 inch lbs. (8 Nm).

11. Install the oil strainer with the 3 nuts. Tighten the nuts to 69 inch lbs. (8 Nm).

12. Install the No. 2 oil pan, as follows:
a. Using a non residue solvent, clean both sealing surfaces to the oil pan.
b. Apply liquid sealant to the oil pan and engine block.
c. Install the No. 2 oil pan with the 10 bolts and 2 nuts. Uniformly tighten the bolts and nuts in several passes. Tighten the bolts to 69 inch lbs. (8 Nm).

13. Install or connect the following:
- Flywheel housing undercover
- Front exhaust pipe bracket to the No. 1 oil pan. Tighten the bolts to 15 ft. lbs. (21 Nm).

14. Install the front exhaust pipe, as follows:
- Temporarily install the 3 new gaskets and the front exhaust pipe with the 2 bolts and 6 nuts
- Tighten the 4 exhaust manifolds-to-front exhaust pipe nuts to 46 ft. lbs. (62 Nm).
- Tighten the both front exhaust pipe-to-center exhaust pipe nuts/bolts to 41 ft. lbs. (56 Nm).
- Bracket. Tighten both bolts to 14 ft. lbs. (19 Nm).
- Support stay. Tighten both bolts to 22 ft. lbs. (29 Nm).

15. Install or connect the following:
- Engine undercover
- Right fender apron seal
- Right front wheel

16. Fill the engine with oil.

17. Start the engine and check for leaks.

Oil Pump

REMOVAL & INSTALLATION

1. Before servicing the vehicle, refer to the precautions in the beginning of this section.

2. Remove or disconnect the following:
- Oil pan
- Crankshaft Position (CKP) sensor
- 9 oil pump bolts

➡ **Make a note of the position of the each bolt. When replacing the bolts into the oil pump body, place each bolt in the position from which it was removed.**

- Oil pump body, by prying between the oil pump and main bearing cap
- O-ring from the cylinder block
- Plug, gasket, spring and relief valve from the oil pump body

- 9 screws, pump body cover, drive and driven rotors

To install:

3. Install or connect the following:
- Driven rotors, drive, pump body cover, using the 9 screws
- Oil pump relief valve, spring, gasket and the plug to the oil pump body
- New O-ring on the cylinder block

4. Using a non residue solvent, clean both sealing surfaces to the oil pump.

5. Apply liquid sealant to the oil pump and engine block.

6. Install or connect the following:
- Oil pump

➡ **Be sure to engage the spline teeth of the oil pump drive gear with the large teeth of the crankshaft.**

- 9 oil pump bolts. Tighten the bolts in several passes to 69 inch lbs. (8 Nm), for 10mm or to 14 ft. lbs. (20 Nm), for 12mm.
- CKP sensor. Tighten the bolt to 69 inch lbs. (8 Nm).
- Baffle plate to the No. oil pan. Tighten to 69 inch lbs. (8 Nm).
- No. 1 oil pan, oil strainer and No. 2 oil pan

7. Refill the engine with oil.

8. Start the engine and inspect for leaks.

9. Recheck the engine oil level.

Rear Main Seal

REMOVAL & INSTALLATION

If the rear oil seal retainer is not installed to the block, use a tapered ended screwdriver and hammer to remove the oil seal. Apply multi-purpose grease to the new oil seal lip. Using a seal driver, tap the seal into place. Be careful not to install it slantwise.

1. Before servicing the vehicle, refer to the precautions in the beginning of this section.

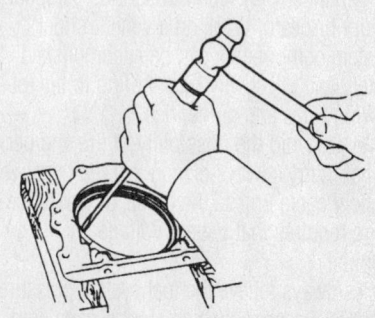

Carefully tap the old seal from the retainer

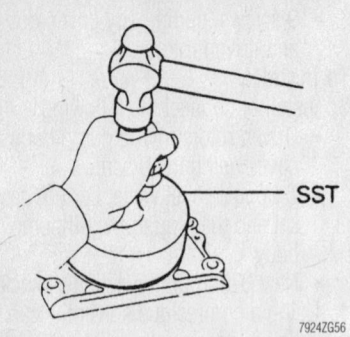

Use the proper sized driver to seat the

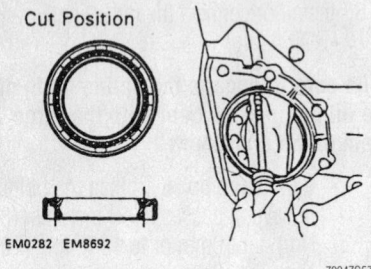

Cut off the oil seal lip, then pry the seal out of the retaining plate

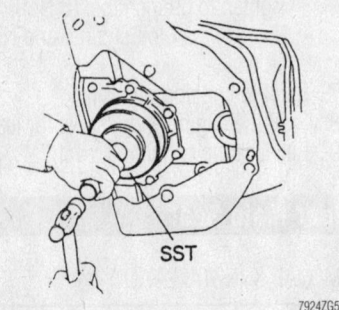

Tap a new seal into place

If the rear oil seal retainer is installed on the cylinder block, using a knife, cut off the lip of the seal. Using a taped ended prytool, pry the old seal out of the retainer. Inspect the oil seal lip contacting surface of the crankshaft for cracks or damage. Apply multipurpose grease to the new oil seal, then tap the seal in place with a seal installer. Be careful not to install the seal slantwise.

Piston and Rings

POSITIONING

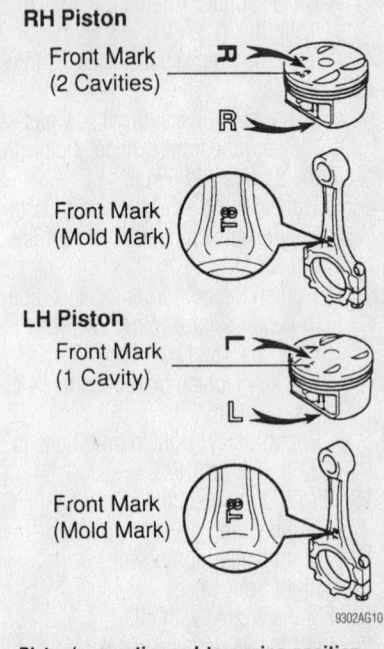

Piston/connecting rod-to-engine positioning

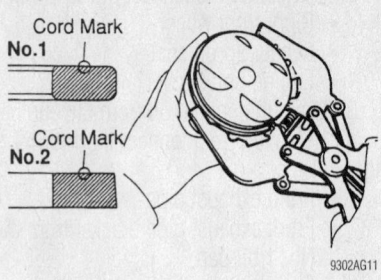

Piston ring positioning

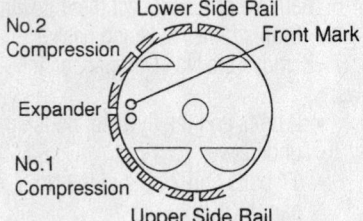

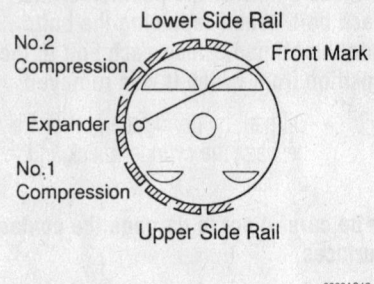

Piston ring identification

FUEL SYSTEM

Fuel System Service Precautions

Safety is the most important factor when performing not only fuel system maintenance but any type of maintenance. Failure to conduct maintenance and repairs in a safe manner may result in serious personal injury or death. Work on a vehicle's fuel system components can be accomplished safely and effectively by adhering to the following rules and guidelines.

• To avoid the possibility of fire and personal injury, always disconnect the negative battery cable unless the repair or test procedure requires that battery voltage be applied.

• Always relieve the fuel system pressure prior to disconnecting any fuel system component (injector, fuel rail, pressure regulator,

etc.) fitting or fuel line connection. Exercise extreme caution whenever relieving fuel system pressure, to avoid exposing skin, face and eyes to fuel spray. Please be advised that fuel under pressure may penetrate the skin or any part of the body that it contacts.

• Always place a shop towel or cloth around the fitting or connection prior to loosening to absorb any excess fuel due to spillage. Ensure that all fuel spillage is quickly remove from engine surfaces. Ensure that all fuel-soaked cloths or towels are deposited into a flame-proof waste container with a lid.

• Always keep a dry chemical (Class B) fire extinguisher near the work area.

• Do not allow fuel spray or fuel vapors to come into contact with a light bulb, spark or open flame.

• Always use a second wrench when loosening or tightening fuel line connection fittings. This will prevent unnecessary stress and torsion to fuel piping. Always follow the proper torque specifications.

• Always replace worn fuel fitting O-rings with new ones. Do not substitute fuel hose where rigid pipe is installed.

Fuel System Pressure

RELIEVING

1. Before servicing the vehicle, refer to the precautions in the beginning of this section.

2. Disconnect the negative battery terminal. Wait at least 90 seconds prior to working on models equipped with an airbag.

3. Place a catch-pan under the joint to be disconnected. A large quantity of fuel may be released when the joint is opened.

➡ **Wear eye or full-face protection.**

4. Place a shop towel over the area and slowly loosen the joint using a wrench of the correct size. Use a back-up wrench if needed.

5. Allow the fuel left in the line to bleed off slowly before fully disconnecting the joint.

6. Plug the opened lines immediately to prevent fuel spillage or the entry of dirt.

7. Dispose of the released fuel properly.

8. After adjoining fuel lines, connect the negative battery cable and start the engine.

9. Check for leaks and repair as needed.

Fuel Filter

REMOVAL & INSTALLATION

1. Before servicing the vehicle, refer to the precautions in the beginning of this section.

2. Disconnect the negative battery cable.

3. Relieve the fuel system pressure.

➡ **The fuel filter is located in the engine compartment, at the inlet line to the fuel rail.**

4. Remove or disconnect the following:
    - Inlet and outlet lines from the filter
    - Fuel filter

To install:

5. Install or connect the following:
    - Fuel filter, using new O-rings. Tighten the lines to 22 ft. lbs. (29 Nm).
    - Negative battery cable

6. Start the engine and check for leaks.

Fuel Pump

REMOVAL & INSTALLATION

1. Before servicing the vehicle, refer to the precautions in the beginning of this section.

2. Relieve the fuel pressure.

3. Disconnect the negative battery cable.

Wait at least 90 seconds before proceeding on models with an airbag.

4. Drain the fuel tank.

5. Remove or disconnect the following:
    - Fuel tank
    - Access plate bolts
    - Fuel pump assembly
    - Fuel pump electrical connectors.

6. Pull the bracket from the lower side of the fuel pump.

7. Remove or disconnect the following:
    - Fuel pump from the fuel hose
    - Rubber cushion, the clip and the fuel filter from the bottom of the fuel pump.

To install:

8. Install or connect the following:
    - Fuel pump filter to the fuel pump using a new clip
    - Fuel pump to the bracket using new gaskets
    - Fuel hose to the outlet port of the fuel pump
    - Fuel pump bracket. Tighten the bolts to 26 inch lbs. (3 Nm).
    - Fuel tank
    - All electrical and fuel harness

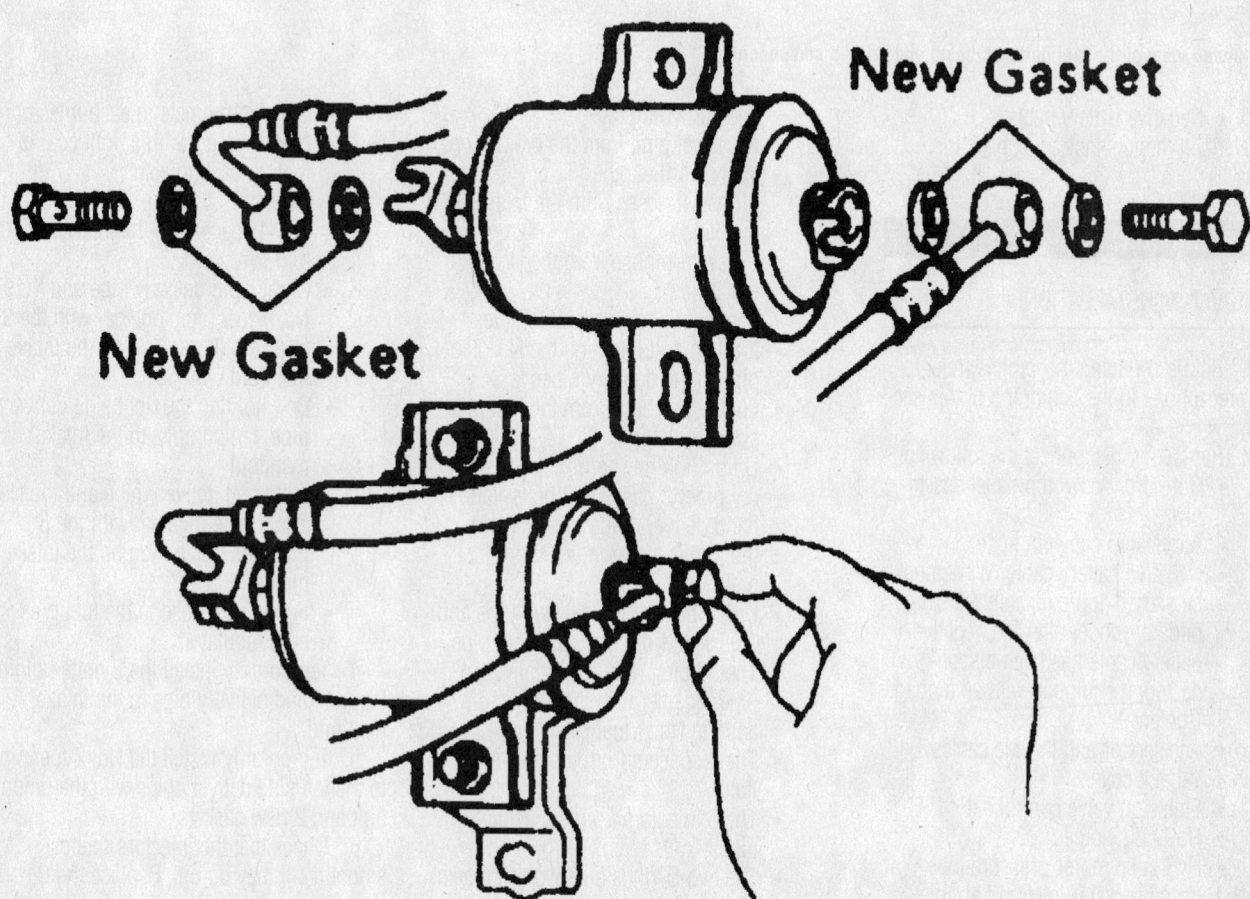

Exploded view of the fuel filter

79242G59

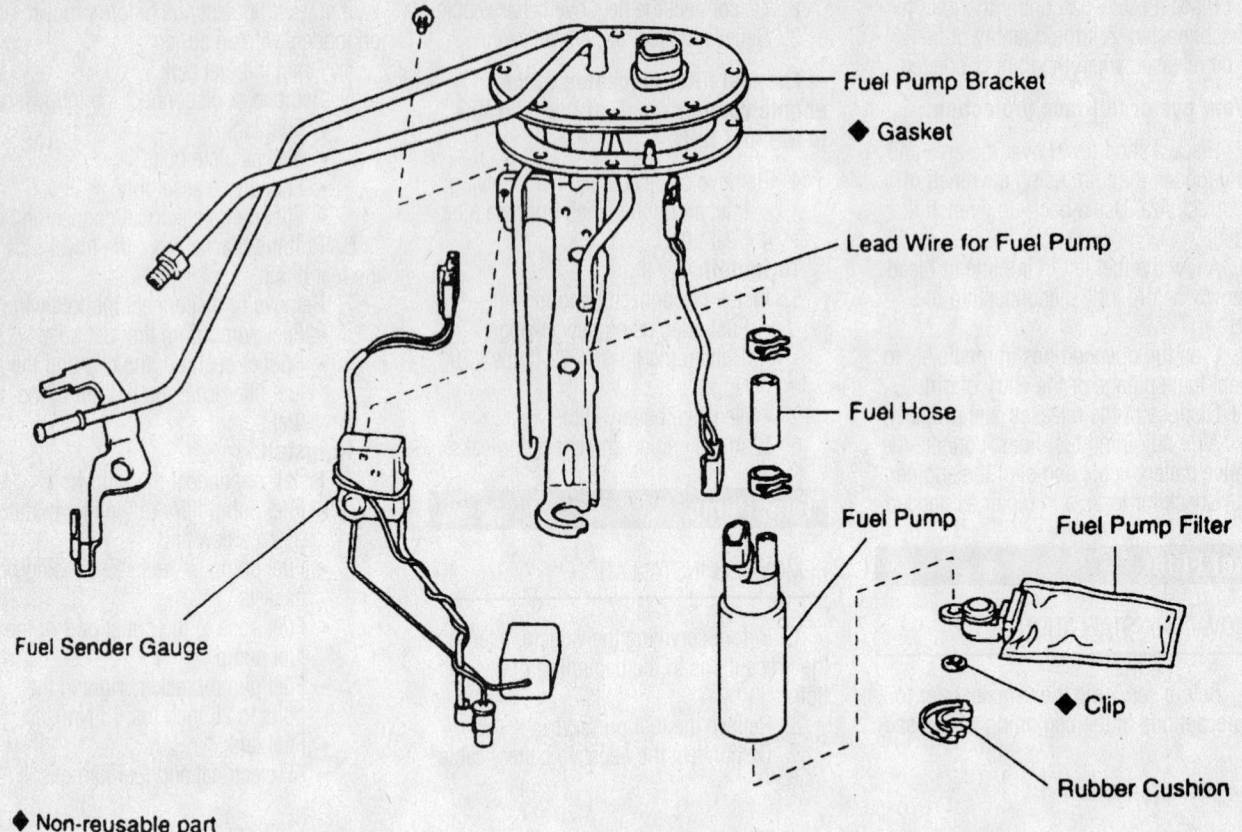

Fuel Pump Bracket

◆ Gasket

Lead Wire for Fuel Pump

Fuel Hose

Fuel Pump

Fuel Pump Filter

◆ Clip

Rubber Cushion

Fuel Sender Gauge

7924ZG60

◆ Non-reusable part

Exploded view of the fuel pump, bracket and related components

- Negative battery cable
9. Refill the fuel tank and check for leaks.

Fuel Injector

REMOVAL & INSTALLATION

1. Before servicing the vehicle, refer to the precautions in the beginning of this section.
2. Remove or disconnect the following:
   - Outer front cowl top panel assembly
   - Air cleaner cap with hose
   - Negative battery cable. Work must be started approximately 90 seconds or longer after the negative battery cable has been disconnected, if equipped with an air bag.
   - Coolant
   - Accelerator and throttle cables
   - V-bank cover
   - Emission valve control set
   - No. 2 EGR pipe
   - Hydraulic motor pressure pipe from the water inlet and air inlet chamber
   - Air intake chamber assembly

- Injector wiring
- Air assist pipe from the bracket on the No. 1 fuel pipe
- Air assist hoses from the intake manifold
- Fuel return hose from the No. 1 fuel pipe
- Fuel inlet hose for the fuel filter
- 2 union bolts holding the No. 2 fuel pipe to the delivery pipes
- Fuel return hose from the fuel pressure regulator
- Union bolt for the right hand delivery pipe, 2 gaskets, 2 bolts, left hand delivery pipe together with the 3 injectors and the No. 2 fuel pipe
- Union bolt for the delivery pipe and 2 gaskets from the No. 2 fuel pipe
- The 3 bolts, right hand delivery pipe together with the 3 injectors and the No. 1 fuel pipe
- The 4 spacers from the intake manifold
- The 6 injectors from the delivery pipes
- The two O-rings and two grommets from each injector

To install:
3. Install or connect the following:

- 2 new grommets to each injector
- New O-rings, with a light coat of fuel, to each injector
- Injectors
- The 4 spacers on the intake manifold
- Right hand delivery pipe and the No. 1 fuel pipe together with the 3 injectors in position on the intake manifold
- Bolt holding the right side delivery pipe, temporarily, to the intake manifold
- Left hand delivery pipe and the No. 2 fuel pipe together with the 3 injectors in position on the intake manifold
- Fuel return hose to the fuel pressure regulator

4. Temporarily install the 2 bolts holding the left hand delivery pipe to the intake manifold.

5. Temporarily install the No. 2 fuel pipe to the left side delivery pipe with the union bolt and 2 new gaskets.

6. Check that the injectors rotate smoothly. If they do not, Replace the O-rings.

7. Position the injector connector outward. Tighten the 4 bolts holding the

delivery pipes to the intake manifold and tighten to 7 ft. lbs. (10 Nm). Tighten the bolt holding the No. 1 fuel pipe to the intake manifold to 14 ft. lbs. (20 Nm). Tighten the 2 union bolts holding the no. 2 fuel pipe to the delivery pipes to 24 ft. lbs. (32 Nm).

8. Install or connect the following:
  • Fuel inlet and return hoses. Union bolt: 22 ft. lbs. (30 Nm)
  • Fuel return hose to the No. 1 fuel

pipe. Pass the fuel return hose under the heater hoses.
  • Air assist hoses to the intake manifold
  • Air assist pipe to the bracket on the No. 1 fuel pipe
  • Fuel injector wiring connectors
  • Air intake chamber assembly
  • Hydraulic motor pressure pipe to the intake chamber. Bolts: 69 inch lbs. (8 Nm)

  • No. 2 EGR pipe with new gaskets, tighten to 9 ft. lbs. (12 Nm)
  • Emission control valve set
  • V-bank cover
  • Air cleaner hose
  • Throttle and accelerator cables
  • Coolant
  • Air cleaner cap with hose
  • Outer front cowl top panel assembly
  • Negative battery cable

DRIVE TRAIN

Transmission Assembly

REMOVAL & INSTALLATION

1. Before servicing the vehicle, refer to the precautions in the beginning of this section.

2. Remove or disconnect the following:
  • Hood
  • Wiper and blade assembly
  • Top cowl seal and panel
  • Window washer hoses, from the ventilator louvers
  • Left and right ventilator louvers
  • Heater air duct
  • Battery and tray
  • Throttle cable
  • Cruise control actuator with its bracket, if equipped
  • Starter
  • Shift control cable
  • Body-to-engine ground strap
  • Park/Neutral Position (PNP) switch, solenoid and ATF temperature connectors
  • 5 upper transaxle-to-engine mounting bolts
  • Front wheel
  • Engine undercover
  • Halfshafts
  • Front exhaust pipe
  • Stabilizer bar
  • Both steering gear mounting bolts and support it in the vehicle
  • Shift control cable from its bracket
  • Power steering pipe and the oil cooler clamps from the frame
  • Both left-side transaxle mounting nuts
  • Rear-side engine mounting nuts
  • Engine shock absorber mounting bolts
  • 3 front-side engine mounting bolts

3. Attach an engine sling to the engine hangers in order to support the engine weight.

4. Remove or disconnect the following:
  • Front frame mounting bolts and the frame
  • Transaxle oil cooler lines

5. Support the transaxle with a transmission jack.

  • Torque converter access cover
  • 6 torque converter mounting bolts
  • 3 lower transaxle-to-engine mounting bolts
  • Engine from the transaxle

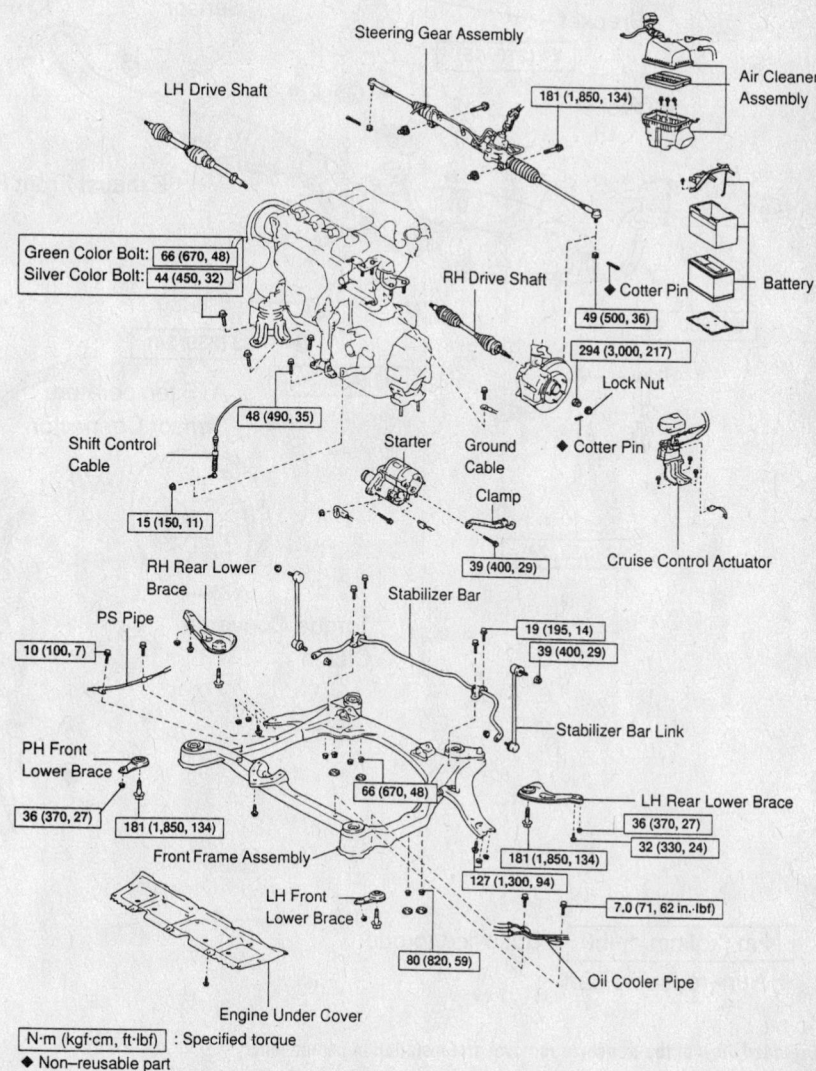

Exploded view of the transaxle removal and installation components

7924ZG65

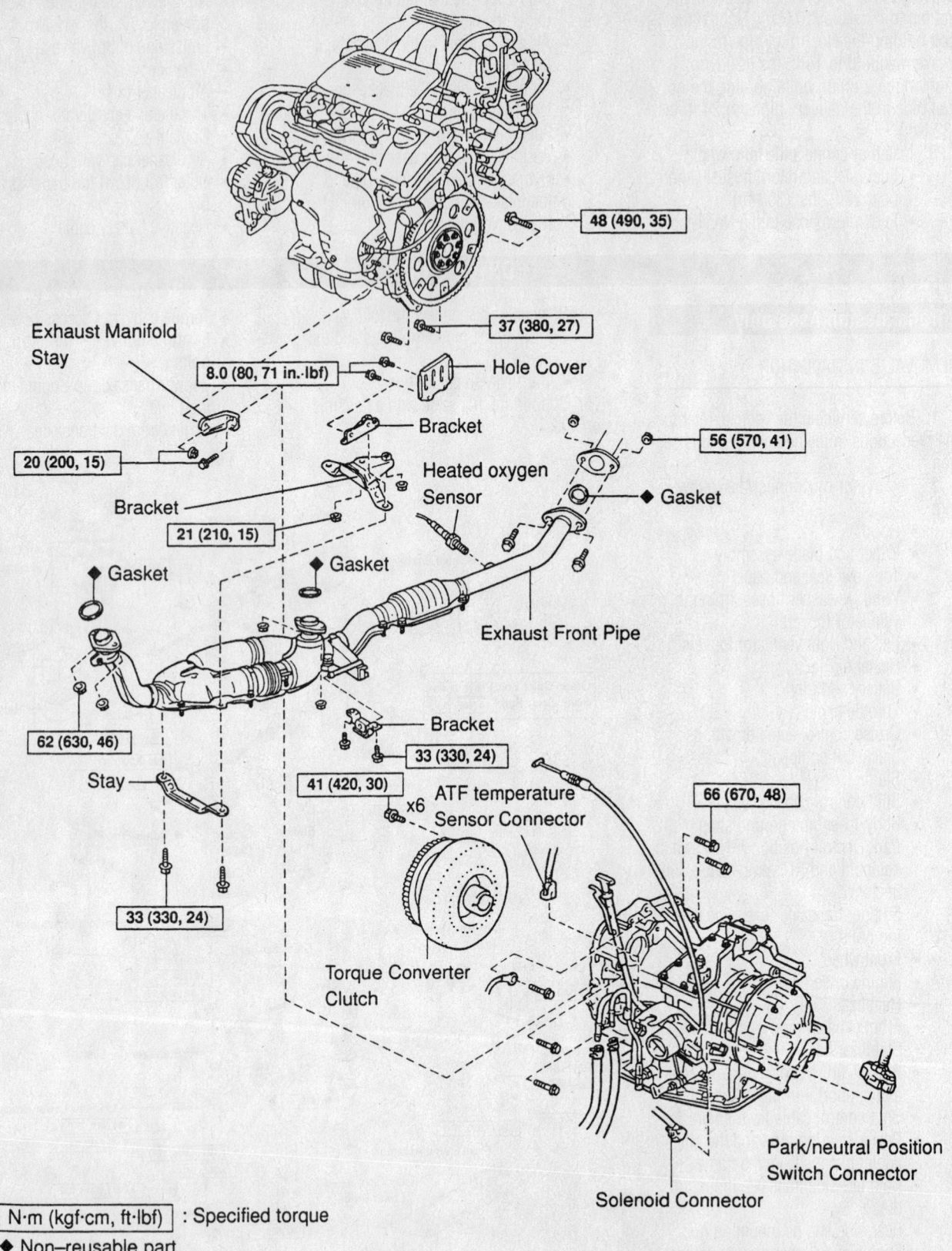

48 (490, 35)

37 (380, 27)

Exhaust Manifold
Stay

8.0 (80, 71 in.·lbf)

Hole Cover

20 (200, 15)

Bracket

56 (570, 41)

Bracket

Heated oxygen
Sensor

◆ Gasket

21 (210, 15)

◆ Gasket

◆ Gasket

Exhaust Front Pipe

62 (630, 46)

Bracket

Stay

33 (330, 24)

41 (420, 30)

ATF temperature
Sensor Connector

66 (670, 48)

x6

33 (330, 24)

Torque Converter
Clutch

Park/neutral Position
Switch Connector

Solenoid Connector

N·m (kgf·cm, ft·lbf) : Specified torque

◆ Non–reusable part

7924ZG66

Exploded view of the transaxle removal and installation components

To install:

6. Install or connect the following:
- Transaxle
- 3 lower transaxle-to-engine mounting bolts and tighten to the illustrated value.
- Torque converter-to-flexplate bolts, starting with the black bolt, then the other 5.

7. The rest of installation is the reverse of the removal referring to the illustrations for the tightening specifications.

Halfshaft

REMOVAL & INSTALLATION

1. Before servicing the vehicle, refer to the precautions in the beginning of this section.

2. Remove or disconnect the following:
- Front wheels
- Cotter pin and locknut cap

➡Have an assistant depress the brake pedal and loosen the bearing locknut.

- Engine undercover
- Fender apron seal
- Tie rod end, from the steering knuckle
- Steering knuckle, from the lower control arm
- Halfshaft from the axle hub, using a plastic hammer
- Cover the outer boot with a rag

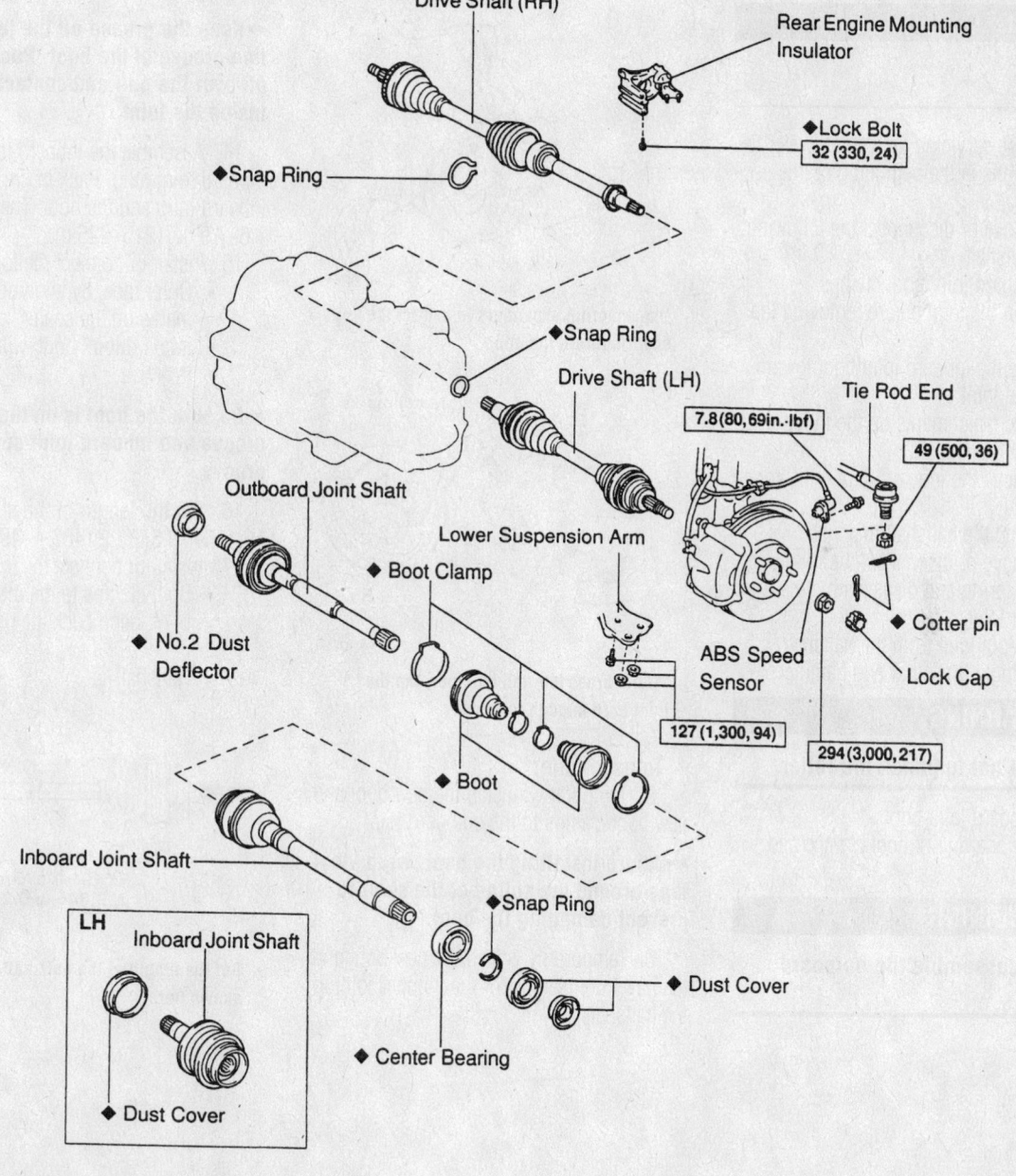

N·m (kgf·cm, ft·lbf) : Specified torque
◆ Non–reusable part

Exploded view of halfshaft

7924ZG73

- Halfshaft from the transaxle, using the proper tools

To install:

3. Reverse the removal procedures to complete installation, tightening fasteners to specifications.

4. Fill the transaxle with gear oil, install the fender apron, check front end alignment and test drive.

➡**If the cotter pin holes do not align, always correct by tightening the nut until the next hole aligns.**

5. Install a new cotter pin.

CV-Joints

OVERHAUL

1. Before servicing the vehicle, refer to the precautions in the beginning of this section.

2. Remove or disconnect the following:
- Halfshaft
- Inboard joint boot clamps

3. Clean the joint before removing the boot.

4. Slide the inboard joint boot toward the outboard joint.

5. Place matchmarks on the inboard joint tulip and the shaft.

6. Remove the inboard joint tulip from the driveshaft.

7. Clamp the halfshaft in a vise.

8. Remove or disconnect the following:
- Snapring and disassemble the tri-pot joint
- Tri-pot joint from the halfshaft, using a brass bar and hammer

※ WARNING

Be careful not to punch the roller.

- Inboard joint boot
- Outboard joint boot clamps and boot

※ WARNING

Do not disassemble the outboard joint.

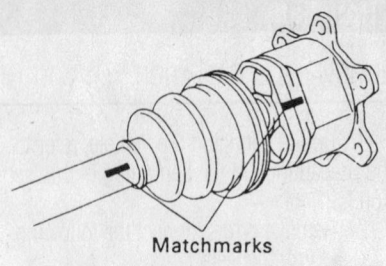

Matchmarks

90917G47

Place matchmarks on the inboard joint outer race and shaft

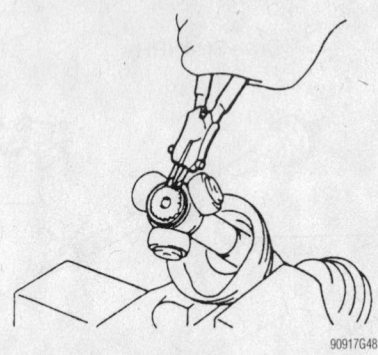

90917G48

Use snapring expanders to extract the snapring from the end

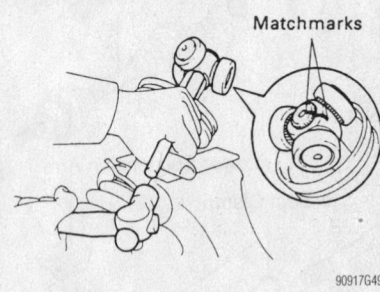

Matchmarks

90917G49

With a brass bar and hammer, tap the joint hard enough to remove

To assemble:

9. Temporarily, install the new boot and new boot clamps to the outboard joint.

➡**Before installing the boot, wrap vinyl tape around the spline of the shaft to prevent damaging the boot.**

10. Temporarily, install the new boot and the new boot clamps for the inboard joint to the halfshaft.

11. Assemble the tri-pot joint, as follows:

a. Place the beveled side of the tri-pot axial spline toward the outboard joint.

b. Align the matchmarks placed before disassembly.

c. Using a brass bar and hammer, tap in the tri-pot joint onto the driveshaft. Do not punch the roller.

12. Install a new snapring.

13. Before assembling the boot to the outboard joint, pack the boot with grease. The capacity is 4.2–4.6 oz. (120–130 g).

➡**Keep the grease off the joint connection groove of the boot. Pack in grease all over the ball and contact surface inside the joint.**

14. Assemble the inboard joint to the inboard joint tulip. Pack in grease to the inboard tulip and the boot. The capacity is 7.6–7.9 oz. (215–225 g).

15. Install or connect the following:
- Outer race, by aligning the matchmarks on the shaft
- Inboard joint boot, without twisting it

➡**Be sure the boot is on the shaft groove and inboard joint outer race groove.**

16. Set the length of the shaft to 19.146–19.546 in. (486.4–496.41mm).

17. Install or connect the following:
- Both clamps to the inboard joint boot; bend back the band and lock it
- Halfshaft

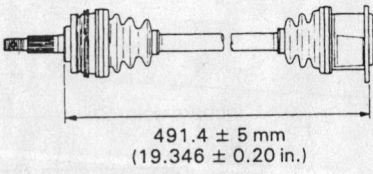

491.4 ± 5 mm (19.346 ± 0.20 in.)

90917G50

Set the length of the halfshaft at the points shown here

STEERING & SUSPENSION

Air Bag

�֎ CAUTION

Some vehicles are equipped with an air bag system. The system must be disabled before performing service on or around system components, steering column, instrument panel components, wiring and sensors. Failure to follow safety and disabling procedures could result in accidental air bag deployment, possible personal injury and unnecessary system repairs.

PRECAUTIONS

Several precautions must be observed when handling the inflator module to avoid accidental deployment and possible personal injury.

• Never carry the inflator module by the wires or connector on the underside of the module.

• When carrying a live inflator module, hold securely with both hands, and ensure that the bag and trim cover are pointed away.

• Place the inflator module on a bench or other surface with the bag and trim cover facing up.

• With the inflator module on the bench, never place anything on or close to the module, which may be thrown in the event of an accidental deployment.

DISARMING

To avoid personal injury when working on vehicles equipped with an air bag, the negative battery cable must be disconnected and at least 90 seconds must elapse before working on the system. Failure to do so may result in deployment of the air bag.

Power Rack and Pinion Steering Gear

REMOVAL & INSTALLATION

1. Before servicing the vehicle, refer to the precautions in the beginning of this section.
2. Remove or disconnect the following:
   • Negative battery cable

➡**Wait at least 90 seconds before working on the vehicle to allow the Supplemental Restraint System (SRS) system to disarm.**

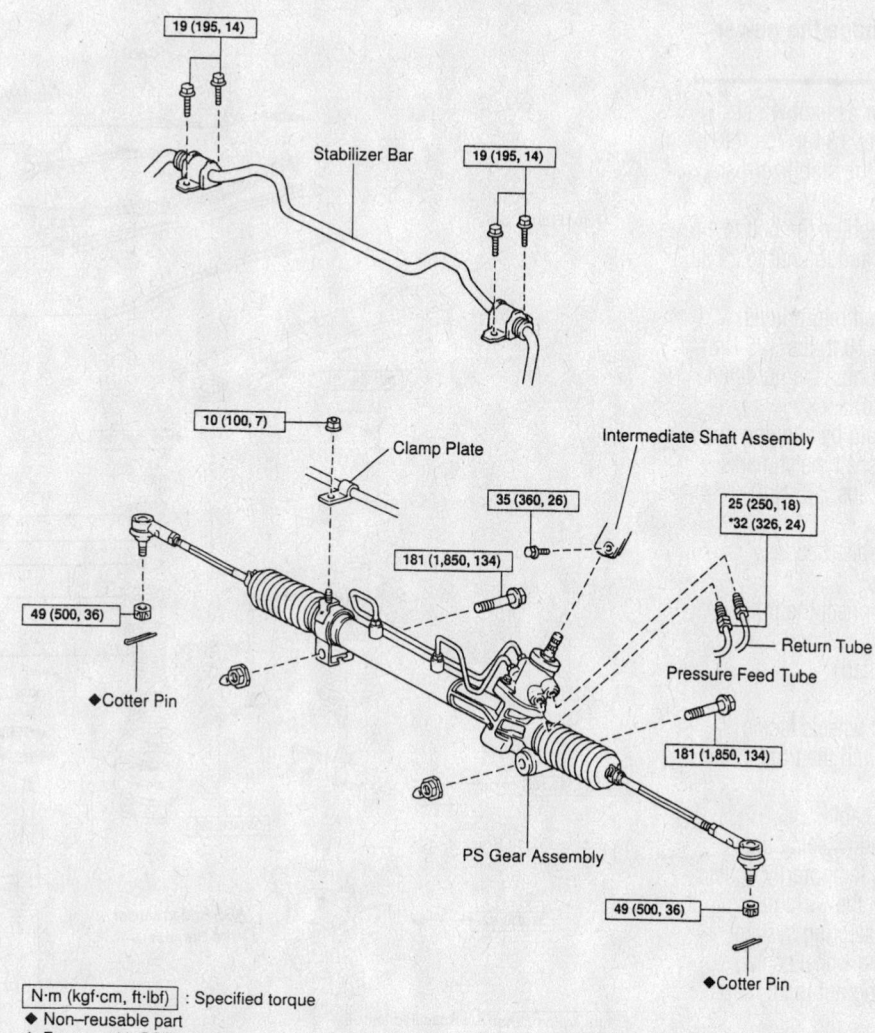

N·m (kgf·cm, ft·lbf) : Specified torque
◆ Non–reusable part
* For use with SST

Exploded view of the power steering gear and related components

7924ZG76

- Right and left side fender apron seals
- Right and left tie rod ends

3. Place matchmarks on the intermediate shaft.

4. Remove or disconnect the following:

- Pinch bolt and the intermediate shaft out from under the vehicle
- Power steering line clamp
- Pressure and feed lines
- Stabilizer bar, unbolt it but do not remove it
- Heated Oxygen (HO2) sensor
- Both gear assembly set bolts and nuts, by lifting the stabilizer bar
- Gear assembly from the left side of the vehicle

To install:

5. Install or connect the following:

- Gear assembly to the left side of the vehicle

✴✴ WARNING

Be careful not to damage the power steering lines.

- Tighten the gear assembly set bolts and nuts to 134 ft. lbs. (181 Nm), by lifting the stabilizer bar
- HO2 sensor
- Stabilizer bar. Tighten the bolt to 14 ft. lbs. (19 Nm) and the nut to 29 ft. lbs. (39 Nm).
- Pressure and feed return lines. Tighten them to 18 ft. lbs. (25 Nm).
- Line clamps. Tighten the nut to 84 inch lbs. (10 Nm).
- Intermediate shaft, by aligning the joint and main shaft matchmarks. Tighten to 26 ft. lbs. (35 Nm).
- Tie rod ends
- Fender apron seals. Securely tighten the bolts.

6. Remove or disconnect the following:

- Steering wheel pad
- Steering wheel

7. Position the front wheels facing straight-ahead. Do this with the front of the vehicle on jackstands.

8. Center the spiral cable.

9. Install the steering wheel at the straight-ahead position. Temporarily tighten the wheel set nut. Attach the wiring.

10. Bleed the power steering system.

11. Check the steering wheel center point. Tighten the steering nut to 26 ft. lbs. (35 Nm).

12. Check and/or adjust the front wheel alignment.

Strut

REMOVAL & INSTALLATION

The Sienna is equipped with front struts and rear shock absorbers.

Front

1. Before servicing the vehicle, refer to the precautions in the beginning of this section.

➡**Do not support the weight of the vehicle on the suspension arm; the arm will deform under its weight.**

2. Remove or disconnect the following:

- Wheel
- Brake hose and the Anti-lock Brake System (ABS) speed sensor wire from the strut
- Sway bar link from the strut
- Outer front cowl top panel

3. Matchmark the strut lower bracket and camber adjust cam, if equipped.

4. Remove or disconnect the following:

- Lower strut end from the steering knuckle's lower arm
- 3 upper strut mounting plate to the upper wheel arch nuts
- Strut

To install:

5. Align the upper suspension support hole with the strut piston or end, so they fit properly.

6. Install or connect the following:

- Strut piston rod end to the upper suspension support. Tighten the new nut to 29–40 ft. lbs. (39–54 Nm).

✴✴ WARNING

Do not use an impact wrench to tighten the nut.

- Lubricate the suspension support bearing with multi-purpose grease.

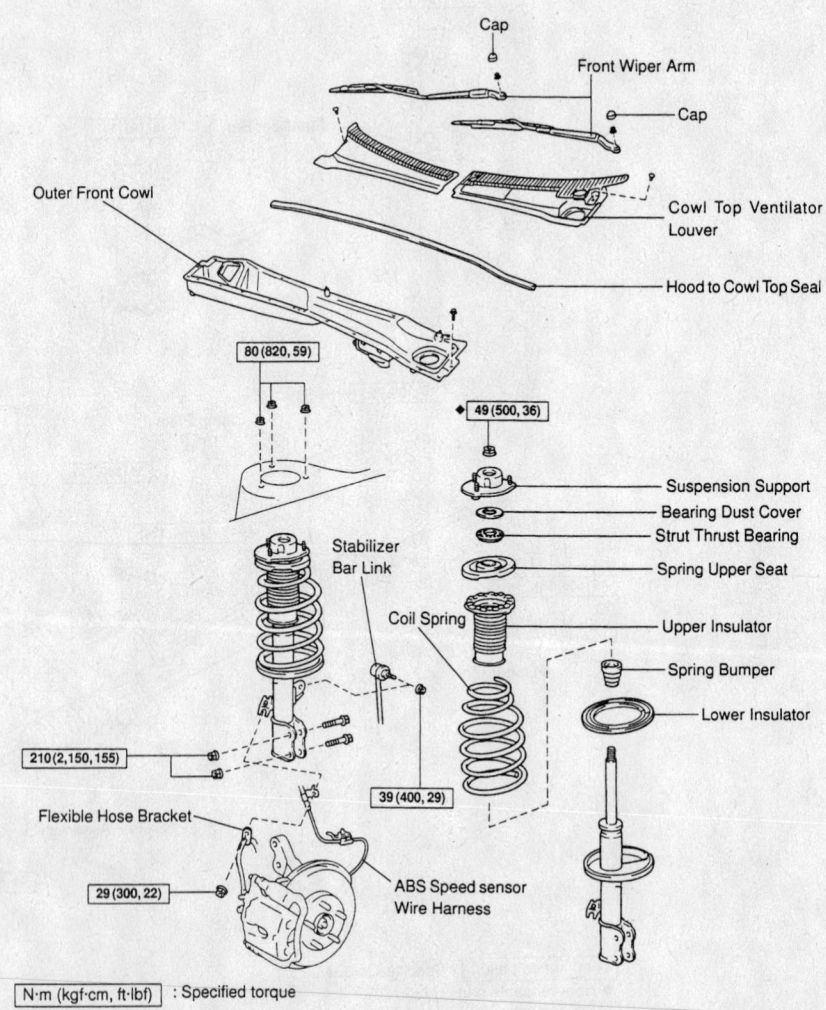

View of the front strut assembly and related components

79242ZG79

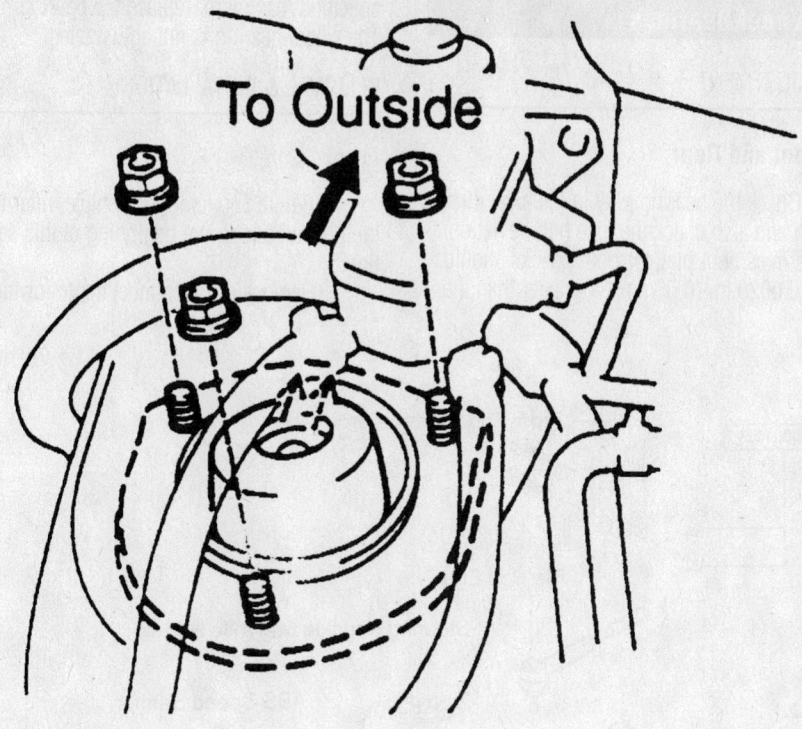

To Outside

7924ZG80

Cutaway view of the upper strut bearing position for installation

Pack the upper support space with multi-purpose grease, after installation.

- Tighten the 3 suspension support-to-wheel arch nuts to 47 ft. lbs. (64 Nm).
- Tighten the strut-to-steering knuckle arm bolts to 156 ft. lbs. (211 Nm).
- Sway bar link to the strut. Tighten the nut to 29 ft. lbs. (39 Nm).
- Outer front cowl top panel
- ABS speed sensor and the brake hose to the strut, if equipped.
- Wheel

Shock Absorber

REMOVAL & INSTALLATION

1. Before servicing the vehicle, refer to the precautions in the beginning of this section.
2. Support the axle beam with jacks.
3. Remove or disconnect the following:
   - Rear wheels
   - Interior service covers to access the upper shock mounts
   - Both nuts and retainers from the upper mount
   - Shock absorber

To install:

4. Install the shock absorber. Tighten the lower mounting bolt to 27 ft. lbs. (37 Nm).

5. If the upper cushion is showing signs of wear, replace it.
6. Install or connect the following:
   - Upper shock absorber. Tighten the nuts to 18 ft. lbs. (25 Nm).
   - Wheels

Coil Spring

REMOVAL & INSTALLATION

Front

1. Before servicing the vehicle, refer to the precautions in the beginning of this section.
2. Remove or disconnect the following:
   - Wheel

➡ If equipped, be careful not to damage the oil seal, driveshaft boot and/or speed sensor rotor when removing the steering knuckle.

   - Shock absorber (strut assembly)
3. Install a nut/bolt to the bracket at the lower portion of the strut assembly and secure it in a vise.
4. Compress the coil spring with a spring compressor.

❉❉ CAUTION

The proper tools must be used for this procedure. The spring on the

strut is under high pressure and can cause serious injury if not properly removed and installed.

5. Remove or disconnect the following:
   - Center retaining nut, by holding the spring seat
   - Support, dust seal, spring seat, insulator and spring from the strut assembly

To install:

6. Install the spring bumper and lower insulator to the strut assembly.
7. Compress the coil spring and fit the lower end of the spring into the spring seat gap.
8. Install or connect the following:
   - Upper insulator, spring seat, dust seal, support and spring seat. Tighten the new retaining nut to 34 ft. lbs. (47 Nm).
9. Rotate the spring seat so that the OUT mark of the spring seat faces the outside of the vehicle.
   - Strut
   - Wheel
10. If required, bleed the brake system and check for leaks.
11. Check and/or adjust the front wheel alignment.

Rear

1. Before servicing the vehicle, refer to the precautions in the beginning of this section.
2. Remove or disconnect the following:
   - Shock absorbers
   - Coil springs

To install:

3. Install or connect the following:
   - Coil springs
   - Raise the axle beam enough to apply tension on the springs
   - Shock absorbers

Lower Ball Joint

REMOVAL & INSTALLATION

1. Before servicing the vehicle, refer to the precautions in the beginning of this section.
2. Remove or disconnect the following:
   - Wheel
   - Steering knuckle with the axle hub
   - Dust deflector, by prying it from the knuckle
   - Cotter pin and nut from the ball joint

- Ball joint from the steering knuckle, by removing the 2 bolts
- Lower ball joint, using a Ball Joint Separator tool 09628-62011

To install:

3. Install or connect the following:
- Lower ball joint. Tighten the nut to 76 ft. lbs. (103 Nm) and both bolts to 94 ft. lbs. (127 Nm).
- New cotter pin
- Wheel

Wheel Bearings

ADJUSTMENT

Front and Rear

Check the bearing play in the axial direction and also check the axle hub deviation. The maximum play for both checks should be 0.0020 in. (0.05mm). If greater than the specified maximum, replace the bearing. The wheel bearing is not adjustable.

REMOVAL & INSTALLATION

Front

1. Before servicing the vehicle, refer to the precautions in the beginning of this section.

2. Remove or disconnect the following:

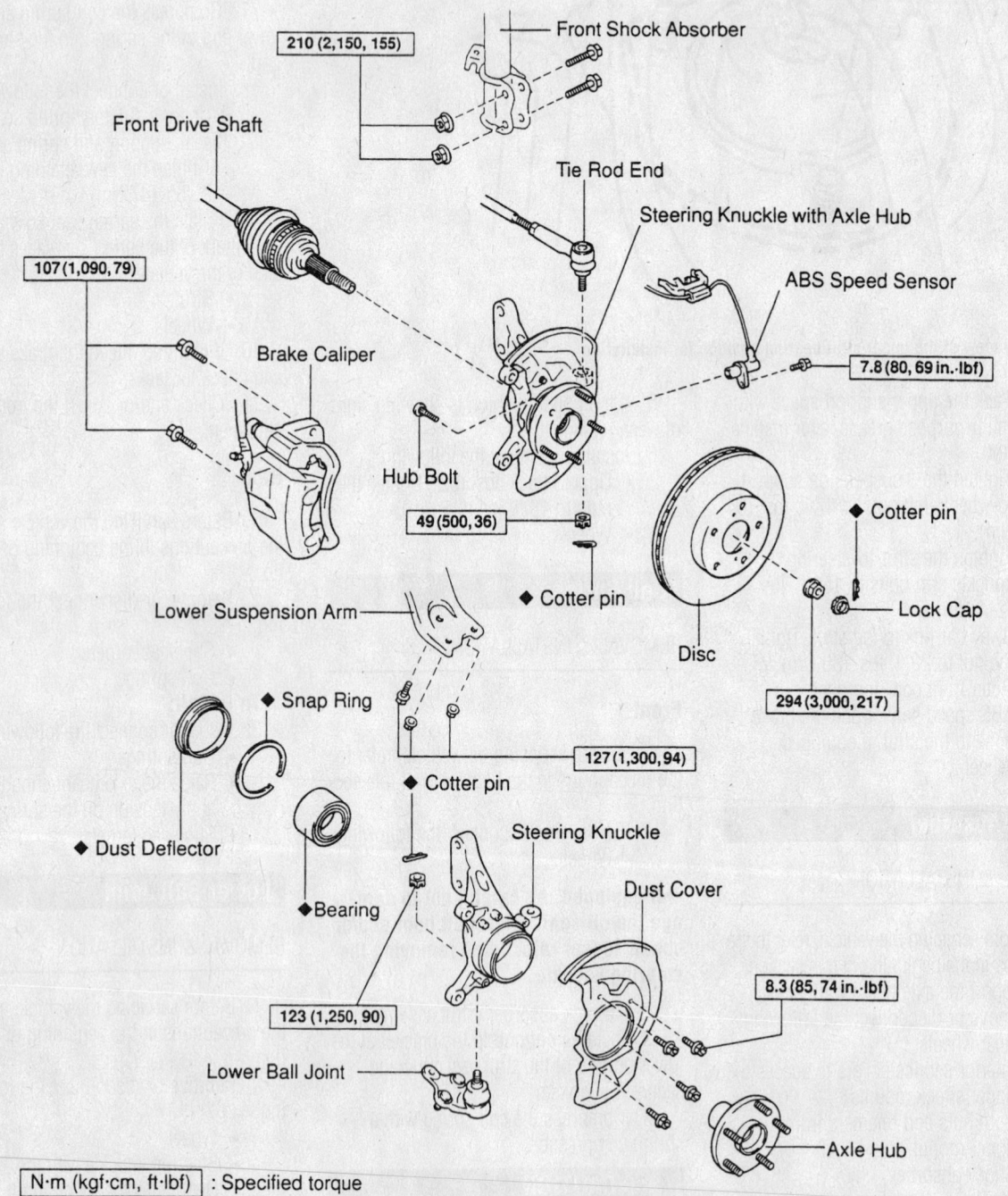

210 (2,150, 155) — Front Shock Absorber
Front Drive Shaft
Tie Rod End
Steering Knuckle with Axle Hub
107 (1,090, 79)
ABS Speed Sensor
Brake Caliper
7.8 (80, 69 in.·lbf)
Hub Bolt
◆ Cotter pin
49 (500, 36)
◆ Cotter pin
Lower Suspension Arm
Disc
Lock Cap
◆ Snap Ring
294 (3,000, 217)
127 (1,300, 94)
◆ Cotter pin
◆ Dust Deflector
◆ Bearing
Steering Knuckle
Dust Cover
123 (1,250, 90)
8.3 (85, 74 in.·lbf)
Lower Ball Joint
Axle Hub

N·m (kgf·cm, ft·lbf) : Specified torque
◆ Non–reusable part

7924ZG82

Exploded view of the front hub, bearing and steering knuckle assembly

- Front wheels
- Fender apron seal

3. Check the bearing backlash and axle hub deviation, as follows:

　a. Remove the 2 brake caliper set bolts.

　b. Hang the caliper using stiff wire on the shock absorber assembly.

　c. Remove the rotor.

　d. Place a dial indicator near the center of the axle hub and check the backlash in the bearing shaft direction.

　e. Backlash maximum should read 0.0020 inch (0.05mm). If greater than specified, replace the bearing.

　f. Using the dial indicator, check the deviation at the surface of the axle hub outside and hub bolt. Maximum is 0.0020 inch (0.05mm). If greater than specified, replace the axle hub.

4. Install the rotor and caliper assembly.

5. Remove or disconnect the following:
- Cotter pin (discard it) and lockcap off the center hub nut
- Driveshaft locknut, by applying the front brakes
- Tie rod end, from the steering knuckle
- Left and right stabilizer end brackets, from the lower arms
- Both nuts and the lower arm from the ball joint
- Driveshaft from the axle hub. Secure the shaft aside using wire.

✳✳ WARNING

Be careful not to damage the shaft boot or Anti-lock Brake System (ABS) sensor rotor.

- Both brake caliper mounting bolts and the caliper.

➡**Support caliper from the vehicle using wire.**

- Brake rotor
- Sensor from the steering knuckle, if equipped with ABS
- Both nuts from the lower end of the shock
- Steering knuckle and hub assembly

6. Clamp the steering knuckle in a vise with soft jaws to protect the knuckle.

7. Remove or disconnect the following:
- Dust deflector from the hub, using a screwdriver
- Bearing inner oil seal, by prying it from the knuckle
- Snapring from the knuckle bore
- Dust deflector from the steering knuckle
- Axle hub, by pulling it from the dust deflector, using a 2-armed mechanical puller
- Inner (inside) bearing race from the bearing, using the puller
- Sensor control rotor from the axle hub, using Torx® wrench
- Outer bearing race, using the puller
- Outer bearing seal, using the puller

8. Position the inner (outside) race inside the bearing.

9. Using a brass rod, tap the bearing from the steering knuckle.

To install:

10. Clean all the oil seal and bearing seating surfaces with a clean, dry rag.

11. Install or connect the following:
- Bearing into the bore, using a Bearing Driver tool 09608-32010 and a press
- New outer oil seal, driving it into the steering knuckle, by inserting the seal side lip into the factory tool
- Brake disc cover to the steering knuckle with the bolts

12. Apply multi-purpose grease between the oil seal lip, oil seal and bearing.

13. Install or connect the following:
- Hub, by pressing it into the knuckle
- New snapring into the knuckle
- New oil seal, by pressing it into the knuckle once lubricated with multi-purpose grease
- Dust deflector, by pressing it into the knuckle.

✳✳ WARNING

Align the speed sensor holes in the dust deflector and steering knuckle, if equipped with ABS.

- Ball joint to the steering knuckle. Tighten the bolts to 94 ft. lbs. (127 Nm).
- Steering knuckle/hub assembly and temporarily install the lower shock bolts
- Lower ball joint to the lower arm. Tighten the bolt and nuts to 94 ft. lbs. (127 Nm).
- Tie rod to the knuckle. Tighten the nut to 36 ft. lbs. (49 Nm).
- New cotter pin
- Tighten the lower shock nuts to 156 ft. lbs. (211 Nm).
- Both side stabilizer end brackets to the lower arm. Tighten the fasteners to 43 ft. lbs. (58 Nm).
- Front ABS sensor. Tighten it to 69 inch lbs. (8 Nm).
- Front brake rotor and caliper. Tighten the caliper bolts to 79 ft. lbs. (107 Nm).
- Driveshaft locknut, by applying the brakes. Tighten it to 217 ft. lbs. (294 Nm).
- Lockcap and new cotter pin
- Front fender apron seal
- Front wheel. Tighten the lug nuts to 76 ft. lbs. (103 Nm).

Rear

1. Before servicing the vehicle, refer to the precautions in the beginning of this section.

2. Remove or disconnect the following:
- Rear wheel
- Brake drum, if equipped

3. If equipped with disk brakes, remove the following:
- Flexible brake hose from the rear strut assembly
- Brake caliper and support it on using wire
- Brake rotor

4. Remove or disconnect the following:
- Anti-lock Brake System (ABS) speed sensor connector
- 4 rear axle hub assembly nuts
- Hub assembly

To install:

5. Install or connect the following:
- New hub assembly. Tighten the nuts to 59 ft. lbs. (80 Nm).
- ABS speed sensor

6. If equipped with disc brakes, install the following:
- Brake rotor
- Brake caliper. Tighten the mounting bolts to 34 ft. lbs. (47 Nm).
- Flexible brake hose to the rear strut assembly. Tighten the mounting bolt to 21 ft. lbs. (29 Nm).

7. Install or connect the following:
- Brake drum, if equipped
- Wheels

8. Test drive the vehicle.

BRAKES

Brake Caliper

REMOVAL & INSTALLATION

1. Disconnect the negative battery cable from the battery.
2. Raise and support the vehicle safely.
3. Remove the wheels.
4. Disconnect the brake hose from the caliper by removing the union bolt and 2 gaskets. Plug the end of the hose to prevent loss of fluid.

5. Remove the bolts that attach the caliper to the torque plate.
6. Lift the bottom of the caliper up and remove the caliper assembly.

To install:

7. Grease the caliper slides and bolts with lithium grease or equivalent. Install the caliper and secure with the bolts. Torque the bolts to 27 ft. lbs. (36 Nm).
8. Reconnect the brake hose to the caliper, using 2 new washers. Make sure the flexible hose lock is securely in the lock

hole of the caliper. Torque the union bolt to 22 ft. lbs. (30 Nm). Also, verify that the brake hose is not twisted.
9. Fill the brake system to the proper level and bleed the brake system.
10. Install the tire and wheel assembly.
11. Top off the brake fluid level in the master cylinder. Check for leaks and proper brake operation.
12. Connect the negative battery cable to the battery.

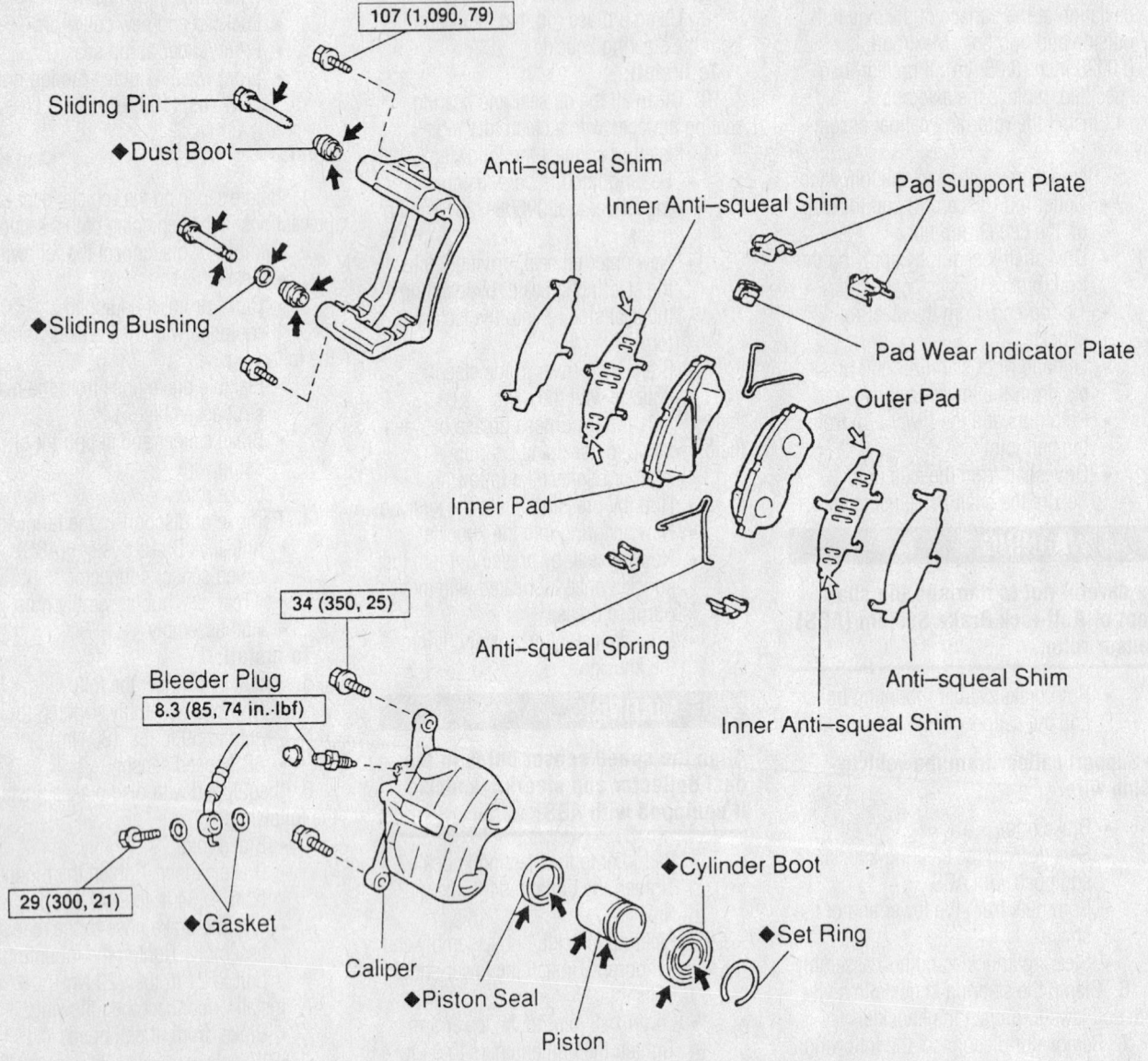

107 (1,090, 79)
Sliding Pin
◆Dust Boot
◆Sliding Bushing
Anti–squeal Shim
Inner Anti–squeal Shim
Pad Support Plate
Pad Wear Indicator Plate
Outer Pad
Inner Pad
Anti–squeal Spring
Anti–squeal Shim
Inner Anti–squeal Shim
34 (350, 25)
Bleeder Plug
8.3 (85, 74 in.·lbf)
29 (300, 21)
◆Gasket
Caliper
◆Piston Seal
Piston
◆Cylinder Boot
◆Set Ring

N·m (kgf·cm, ft·lbf) : Specified torque
◆ Non–reusable part
➡ Lithium soap base glycol grease
⇨ Disc brake grease

93026G72

Exploded view of the front disc brake caliper assembly—Sienna

Disc Brake Pads

REMOVAL & INSTALLATION

FRONT

1. Raise and safely support the front of the vehicle.

2. Remove the front wheels and temporarily fasten the rotor disc with the hub nuts.

3. Hold the sliding pin on the bottom of the caliper and loosen the installation bolt.

4. Remove the lower installation bolt.

5. Lift up the caliper and suspend it securely. Do not remove the upper installation bolt.

6. Remove the following parts:
   - The 2 anti-squeal springs.
   - The 2 brake pads.
   - The 4 anti-squeal shims.
   - The 4 pad support plates.

To install:

7. Install the pad support plates.

8. Install a pad wear indicator plate to the pad. Install the anti-squeal shims and support plates to each pad.

➡️**It recommended that a suitable anti-squeal compound be applied to both sides of the inner anti-squeal shim.**

9. Draw out a small amount of brake fluid from the brake reservoir. Press in the caliper piston with a suitable tool.

10. Press the brake piston in carefully so the boot will not become wedged.

11. Install the 2 pads so that the wear indicator plate is facing upward. Do not allow oil or grease to get in the rubbing face of the pads.

12. Lower and install the caliper. Torque the sliding main pin to 27 ft. lbs. (36 Nm).

➡️**When installing the sliding main pin, be careful that the plug installed in the torque plate does not come loose.**

13. Install the front wheels and lower the vehicle.

14. Check the fluid level in the master cylinder and add as necessary. Be sure to pump the brake pedal a few times before road-testing the vehicle.

REAR

1. Raise and safely support the rear of the vehicle.

2. Remove the rear wheels and temporarily fasten the rotor disc with the hub nuts.

3. Hold the sliding pin on the bottom of the caliper and loosen the installation bolt.

4. Remove the lower installation bolt.

5. Lift up the caliper and suspend it securely. Do not remove the upper installation bolt.

6. Remove the following parts:
   - The 2 anti-squeal springs.
   - The 2 brake pads.
   - The 4 anti-squeal shims.
   - The 4 pad support plates.

To install:

7. Install the pad support plates.

8. Install a pad wear indicator plate to the pad. Install the anti-squeal shims and support plates to each pad.

➡️**It recommended that a suitable anti-squeal compound (available at your local parts house) be applied to both sides of the inner anti-squeal shim.**

9. Draw out a small amount of brake fluid from the brake reservoir. Press in the caliper piston with a suitable tool.

10. Press the brake piston in carefully so the boot will not become wedged.

11. Install the 2 pads so that the wear indicator plate is facing upward. Do not allow oil or grease to get in the rubbing face of the pads.

12. Lower and install the caliper. Torque the sliding main pin to 25 ft. lbs. (34 Nm).

➡️**When installing the sliding main pin, be careful that the plug installed in the torque plate does not come loose.**

13. Install the rear wheels and lower the vehicle.

14. Check the fluid level in the master cylinder and add as necessary. Be sure to pump the brake pedal a few times before road-testing the vehicle.

Brake Shoes

REMOVAL & INSTALLATION

1. Disconnect the negative battery cable from the battery.

2. Loosen the rear wheel lug nuts slightly. Release the parking brake.

3. Block the front wheels, raise the rear of the vehicle, and safely support it with jackstands.

4. Remove the wheel lug nuts and the wheel.

5. Remove the brake drum retaining screws, if equipped. Remove the brake drum.

6. If the drum is difficult to remove, perform the following:

   a. Insert the end of a bent wire (a coat hanger will do nicely) through the hole in the brake drum and hold the automatic adjusting lever away from the adjuster.

   b. Reduce the brake shoe adjustment by turning the adjuster bolt with a brake tool.

   c. The drum should now be loose enough to remove without much effort.

7. Carefully unhook the return spring from the leading (front) brake shoe.

8. Press the hold down spring retainer in and turn the pin on the front brake shoe.

9. Remove the hold down spring, retainers and the pin for the front brake shoe.

10. Pull out the brake shoe and unhook the anchor spring from the lower edge.

11. Remove the hold down spring from the trailing (rear) shoe. Pull the shoe out with the adjuster, automatic adjuster assembly and springs attached. Disconnect the parking brake cable. Remove the tension/return and anchor springs from the rear shoe.

12. Unhook the adjusting lever spring from the rear shoe and then remove the automatic adjuster assembly.

To install:

13. Inspect the shoes for signs of unusual wear or scoring.

14. Check the wheel cylinder for any sign of fluid seepage or frozen pistons.

15. Clean and inspect the brake backing plate and all other components. Check that the brake drum inner diameter is within specified limits. Lubricate the backing plate at the positions the brakes come in contact with the backing plate. Also lubricate the anchor plate.

16. Mount the automatic adjuster assembly onto a new rear brake shoe.

17. Connect the parking brake cable to the rear shoe and then install the automatic adjusting lever, spring and E-ring. Position the rear shoe so the lower end rides in the anchor plate and the upper end is against the boot of the wheel cylinder.

18. Install the pin and the hold down spring. Press the retainer down over the pin and rotate the pin so the crimped edge is held by the retainer.

19. Place the front brake into position and install the anchor spring between the front and rear shoes. Stretch the spring enough so the front shoe will fit as the rear did. Install the hold down spring, pin and retainer to the front brake shoe.

20. Connect the return spring to the front brake shoe.

21. Check the operation of the automatic adjuster mechanism:

   a. Apply the parking brake lever and verifying the adjusting bolt turns.

b. Adjust the strut to where it is the shortest possible length.

c. Install the brake drum.

d. Apply the parking brake lever until the clicking sound can no longer be heard.

22. Check the clearance between the brake shoes and drum:

a. Remove the brake drum.

b. Measure the brake drum inside diameter and diameter of the brake shoes. The difference is "Shoe-to-drum clearance" and should be approximately 0.024 inch (0.6mm). If incorrect, check the parking brake system.

➡ **A special brake caliper tool is required to gauge the brake drum inside diameter and shoe-to-drum clearance. However it is not required to perform brake shoe adjustment.**

23. Install the brake drum.

24. Adjust the brake pedal until a slight drag is felt when the drum is spun by hand.

25. Pull the parking lever all the way up until a clicking sound can no longer be heard. Check the clearance between brake shoes and brake drum.

26. Install the rear wheels, tighten the wheel lug nuts and lower the vehicle.

27. Retighten the wheel lug nuts and pump the brake pedal a few times before moving the vehicle. Adjust the rear brakes again if necessary.

28. Check the level of brake fluid in the master cylinder, then perform a test drive.

29. Connect the negative battery cable to the battery.

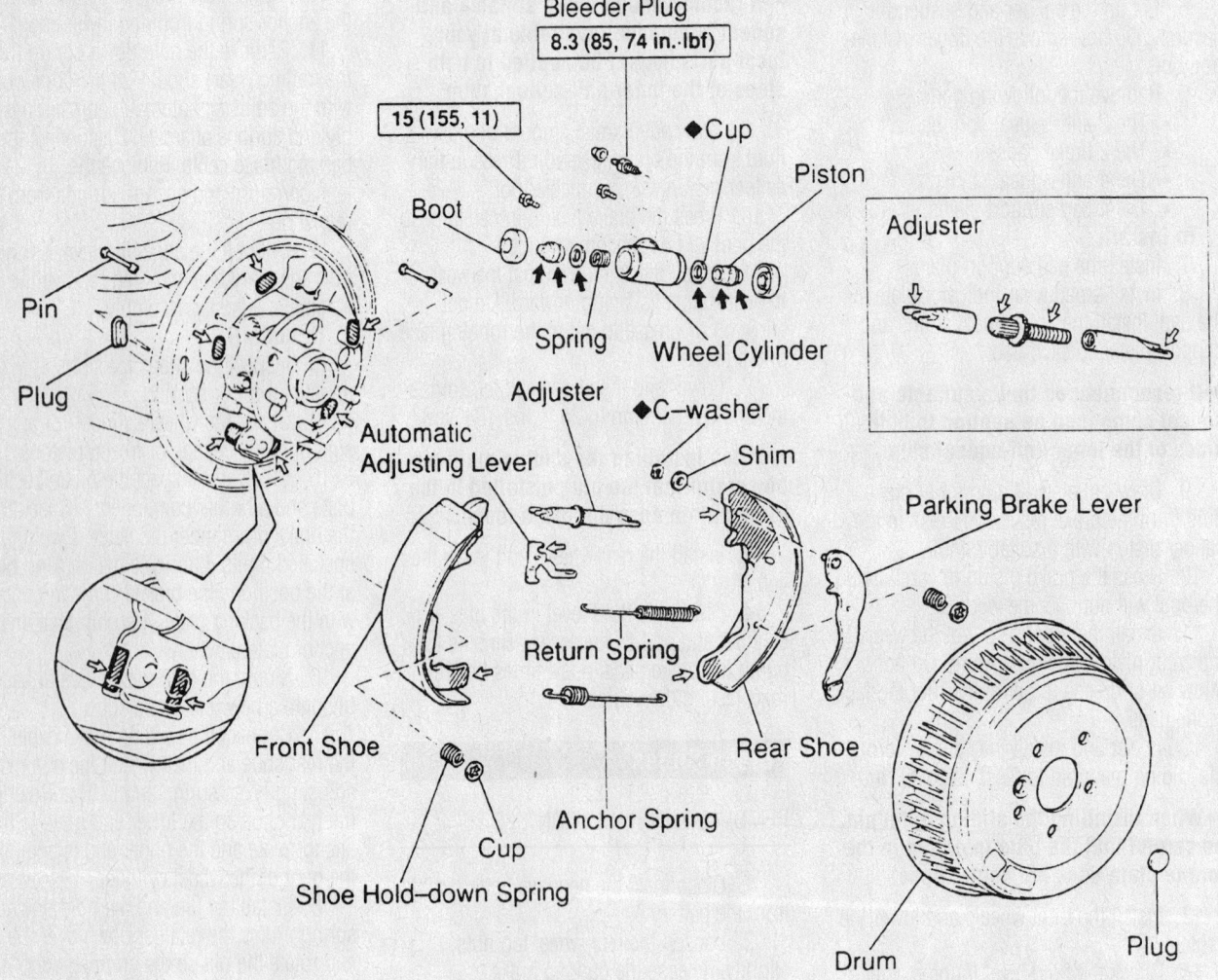

N·m (kgf·cm, ft·lbf) : Specified torque

◆ Non–reusable part

➡ Lithium soap base glycol grease

⇨ High temperature grease

Exploded view of the rear drum brake components—Sienna

93026G80

SPECIFICATION CHARTS

ENGINE AND VEHICLE IDENTIFICATION

		Engine						Model Year	
Code ①	Liters (cc)	Cu. In.	Cyl.	Fuel Sys.	Engine Type	Eng. Mfg.		Code ②	Year
2RZ-FE	2.4 (2438)	149	4	MFI	DOHC	Toyota		Y	2000
3RZ-FE	2.7 (2693)	164	4	MFI	DOHC	Toyota		1	2001
5VZ-FE	3.4 (3378)	206	6	MFI	DOHC	Toyota		2	2002
								3	2003
								4	2004

SFI: Sequential Fuel Injection

MFI: Multi-port Fuel Injection

DOHC: Double Overhead Camshaft

① Stamped on the left side of the engine block

② 10th digit of the Vehicle Identification Number (VIN)

42356-TACO-C01

GENERAL ENGINE SPECIFICATIONS

Year	Model	Engine Displacement Liters (cc)	Engine Series (ID/VIN)	Fuel System	Net Horsepower @ rpm	Net Torque @ rpm (ft. lbs.)	Bore x Stroke (in.)	Com- pression Ratio	Oil Pressure @ rpm
2000	Tacoma	2.4 (2438)	2RZ-FE	MFI	142@5000	160@4000	3.74x3.38	9.5:1	36-71@3000
		2.7 (2693)	3RZ-FE	MFI	150@4800	177@4000	3.74x3.74	9.5:1	36-71@3000
		3.4 (3378)	5VZ-FE	MFI	190@4800	220@3600	3.68x3.23	9.6:1	NA
2001	Tacoma	2.4 (2438)	2RZ-FE	MFI	142@5000	160@4000	3.74x3.38	9.5:1	36-71@3000
		2.7 (2693)	3RZ-FE	MFI	150@4800	177@4000	3.74x3.74	9.5:1	36-71@3000
		3.4 (3378)	5VZ-FE	MFI	190@4800	220@3600	3.68x3.23	9.6:1	NA
2002	Tacoma	2.4 (2438)	2RZ-FE	MFI	142@5000	160@4000	3.74x3.38	9.5:1	36-71@3000
		2.7 (2693)	3RZ-FE	MFI	150@4800	177@4000	3.74x3.74	9.5:1	36-71@3000
		3.4 (3378)	5VZ-FE	MFI	190@4800	220@3600	3.68x3.23	9.6:1	NA
2003	Tacoma	2.4 (2438)	2RZ-FE	MFI	142@5000	160@4000	3.74x3.38	9.5:1	36-71@3000
		2.7 (2693)	3RZ-FE	MFI	150@4800	177@4000	3.74x3.74	9.5:1	36-71@3000
		3.4 (3378)	5VZ-FE	MFI	190@4800	220@3600	3.68x3.23	9.6:1	NA

NA: Not Available

MFI: Multi-port Fuel Injection

42356-TACO-C02

ENGINE TUNE-UP SPECIFICATIONS

Year	Engine Displacement Liters (cc)	Engine ID/VIN	Spark Plug Gap (in.)	Ignition Timing (deg.)*	Fuel Pump (psi)	Idle Speed (rpm)		Valve Clearance	
						MT	AT	Intake	Exhaust
2000	2.4 (2438)	2RZ-FE	0.031	5B	38-44	650-750	—	0.006-0.010	0.010-0.014
	2.7 (2693)	3RZ-FE	0.031	5B	38-44	650-750	650-750	0.006-0.010	0.010-0.014
	3.4 (3378)	5VZ-FE	0.031	10B	38-44	650-750	650-750	0.006-0.010	0.010-0.014
2001	2.4 (2438)	2RZ-FE	0.031	5B	38-44	650-750	—	0.006-0.010	0.010-0.014
	2.7 (2693)	3RZ-FE	0.031	5B	38-44	650-750	650-750	0.006-0.010	0.010-0.014
	3.4 (3378)	5VZ-FE	0.031	10B	38-44	650-750	650-750	0.006-0.010	0.010-0.014
2002	2.4 (2438)	2RZ-FE	0.043	5B	38-44	650-750	—	0.006-0.010	0.010-0.014
	2.7 (2693)	3RZ-FE	0.043	5B	38-44	650-750	650-750	0.006-0.010	0.010-0.014
	3.4 (3378)	5VZ-FE	0.043	10B	38-44	650-750	650-750	0.006-0.010	0.010-0.014
2003	2.4 (2438)	2RZ-FE	0.043	5B	38-44	650-750	—	0.006-0.010	0.010-0.014
	2.7 (2693)	3RZ-FE	0.043	5B	38-44	650-750	650-750	0.006-0.010	0.010-0.014
	3.4 (3378)	5VZ-FE	0.043	10B	38-44	650-750	650-750	0.006-0.010	0.010-0.014

NOTE: The Vehicle Emission Control Information label often reflects specification changes made during production.

The label figures must be used if they differ from those in this chart.

B: Before top dead center

*with Terminals TE1 and E1 connected to DLC1

42356-TACO-C03

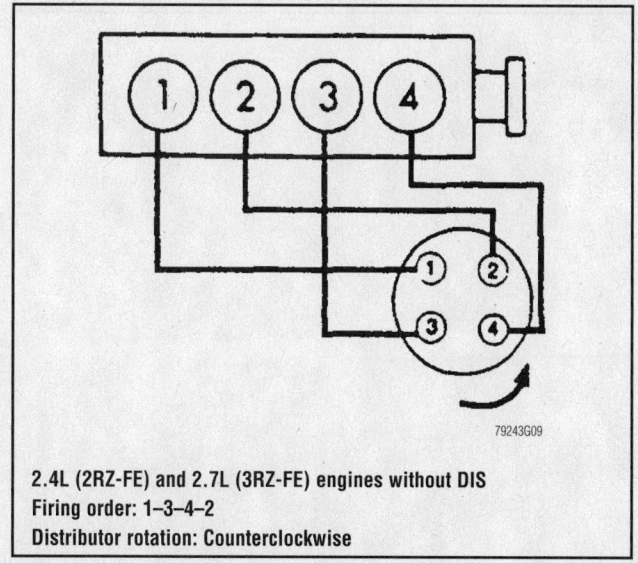

2.4L (2RZ-FE) and 2.7L (3RZ-FE) engines without DIS
Firing order: 1-3-4-2
Distributor rotation: Counterclockwise

79243G09

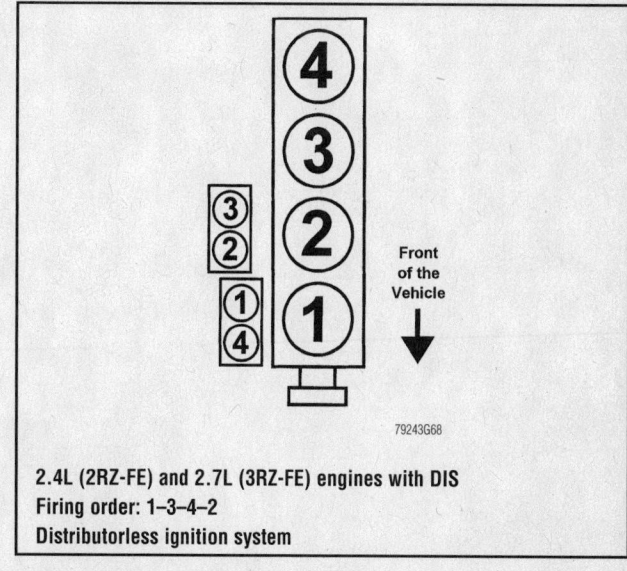

2.4L (2RZ-FE) and 2.7L (3RZ-FE) engines with DIS
Firing order: 1-3-4-2
Distributorless ignition system

79243G68

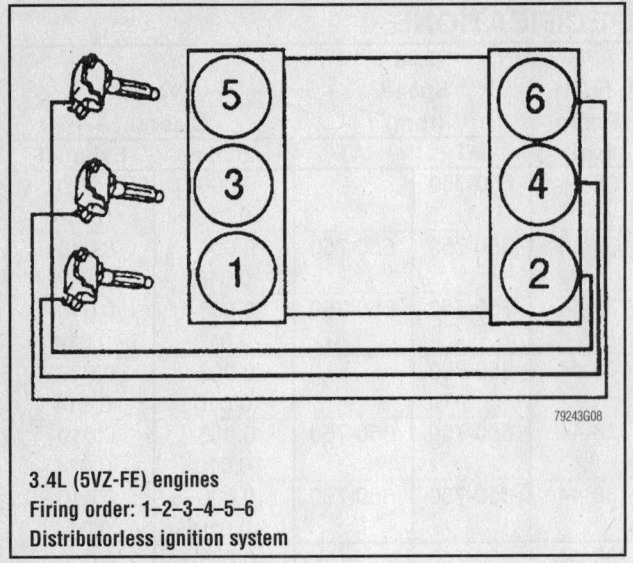

3.4L (5VZ-FE) engines
Firing order: 1–2–3–4–5–6
Distributorless ignition system

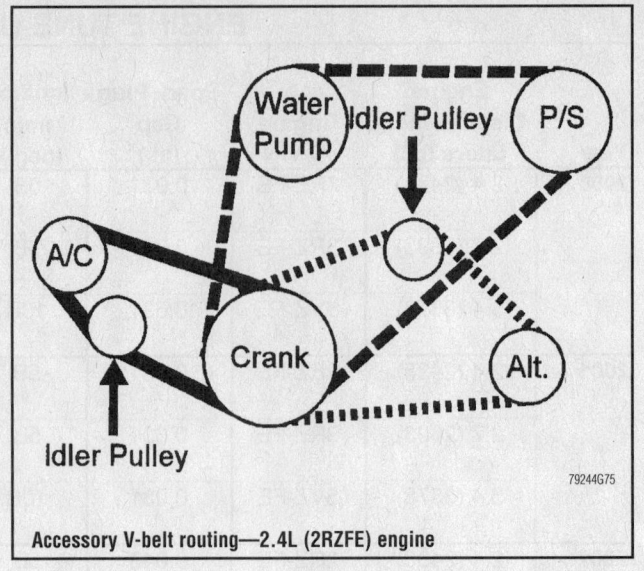

Accessory V-belt routing—2.4L (2RZFE) engine

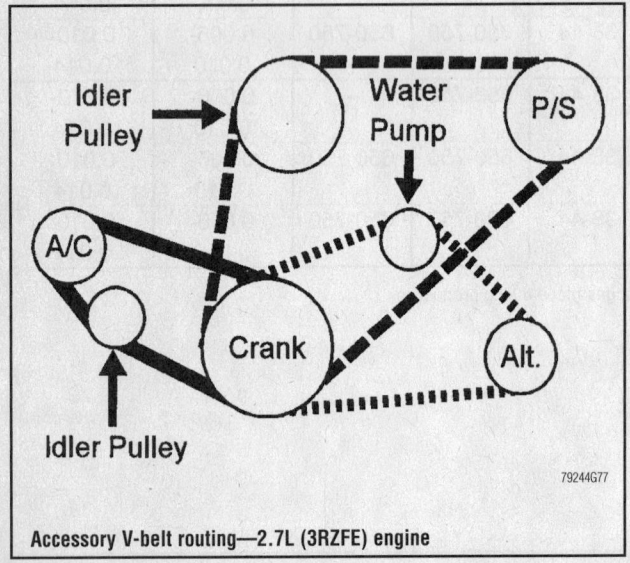

Accessory V-belt routing—2.7L (3RZFE) engine

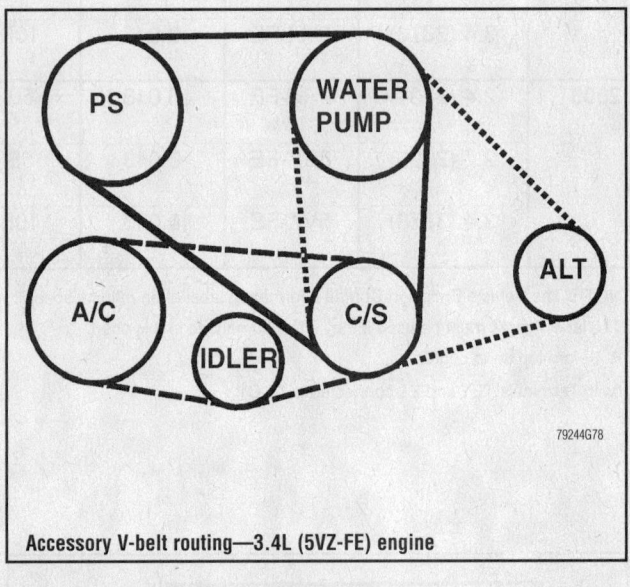

Accessory V-belt routing—3.4L (5VZ-FE) engine

CAPACITIES

| Year | Model | Engine Displacement Liters (cc) | Engine ID/VIN | Engine Oil with Filter (qts.) | Transmission (pts.) | | Transfer Case (pts.) | Drive Axle | | Fuel Tank (gal.) | Cooling System (qts.) |
					5-Spd	Auto.		Front (pts.)	Rear (pts.)		
2000	Tacoma	2.4 (2438)	2RZ-FE	5.8	①	②	2.2	③	2.9	15.1	④
		2.7 (2693)	3RZ-FE	5.8	①	②	2.2	③	⑤	18.0	④
		3.4 (3378)	5VZ-FE	⑥	①	②	2.2	③	⑤	18.1	⑦
2001	Tacoma	2.4 (2438)	2RZ-FE	5.8	①	②	2.2	③	2.9	15.1	④
		2.7 (2693)	3RZ-FE	5.8	①	②	2.2	③	⑤	18.0	④
		3.4 (3378)	5VZ-FE	⑥	①	②	2.2	③	⑤	18.1	⑦
2002	Tacoma	2.4 (2438)	2RZ-FE	5.8	⑧	2.4	2.2	2.7	2.9	18.5	④
		2.7 (2693)	3RZ-FE	5.8	⑧	2.4	2.2	⑨	⑤	18.5	④
		3.4 (3378)	5VZ-FE	5.5	①	2.4	2.2	⑨	⑤	18.5	⑩
2003	Tacoma	2.4 (2438)	2RZ-FE	5.8	⑧	2.4	2.2	2.7	2.9	18.5	④
		2.7 (2693)	3RZ-FE	5.8	⑧	2.4	2.2	⑨	⑤	18.5	④
		3.4 (3378)	5VZ-FE	5.5	①	2.4	2.2	⑨	⑤	18.5	⑩

① W59:
  2WD: 5.4
  4WD: 5.2
  R150, R150F:
  2WD: 5.4
  4WD: 4.6

② A43D: 5.0
  A44D: 5.0
  A340E: 3.4
  A340F: 4.2

③ Without ADD: 2.32
  With ADD: 2.44

④ 2WD M/T: 8.5
  2WD A/T: 8.2
  4WD M/T: 8.8
  4WD A/T: 8.7

⑤ Extra long: 4.4
  All others: 5.4

⑥ 2WD: 5.7
  4WD: 5.5

⑦ M/T: 10.3
  A/T: 10.0

⑧ 2wd: 5.4
  4wd: 4.6

⑨ 4wd extra long: 5.16
  Short models w/diff. Lock: 5.60
  Extra long models w/diff. Lock: 6.36
  w/o diff. Lock: 5.38

⑩ w/rear heater: 9.5
  w/o rear heater: 8.5

42356-TACO-C04

CRANKSHAFT AND CONNECTING ROD SPECIFICATIONS

All measurements are given in inches.

Year	Engine Displacement Liters (cc)	Engine ID/VIN	Crankshaft				Connecting Rod		
			Main Brg. Journal Dia.	Main Brg. Oil Clearance	Shaft End-play	Thrust on No.	Journal Diameter	Oil Clearance	Side Clearance
2000	2.4 (2438)	2RZ-FE	2.3617-2.3622	0.0009-0.0022	0.0008-0.0087	2	2.0861-2.0866	0.0012-0.0022	0.0063-0.0123
	2.7 (2693)	3RZ-FE	2.2615-2.3620	0.0012-0.0022	0.0008-0.0087	3	2.0861-2.0866	0.0009-0.0022	0.0063-0.0123
	3.4 (3378)	5VZ-FE	2.5191-2.5197	0.0008-0.0015	0.0008-0.0087	2	2.1648-2.1654	0.0009-0.0021	0.0059-0.0130
2001	2.4 (2438)	2RZ-FE	2.3617-2.3622	0.0009-0.0022	0.0008-0.0087	2	2.0861-2.0866	0.0012-0.0022	0.0063-0.0123
	2.7 (2693)	3RZ-FE	2.2615-2.3620	0.0012-0.0022	0.0008-0.0087	3	2.0861-2.0866	0.0009-0.0022	0.0063-0.0123
	3.4 (3378)	5VZ-FE	2.5191-2.5197	0.0008-0.0015	0.0008-0.0087	2	2.1648-2.1654	0.0009-0.0021	0.0059-0.0130
2002	2.4 (2438)	2RZ-FE	2.3617-2.3622	0.0009-0.0022	0.0008-0.0087	2	2.0861-2.0866	0.0012-0.0022	0.0063-0.0123
	2.7 (2693)	3RZ-FE	2.2615-2.3620	0.0012-0.0022	0.0008-0.0087	3	2.0861-2.0866	0.0009-0.0022	0.0063-0.0123
	3.4 (3378)	5VZ-FE	2.5191-2.5197	0.0008-0.0015	0.0008-0.0087	2	2.1648-2.1654	0.0009-0.0021	0.0059-0.0130
2003	2.4 (2438)	2RZ-FE	2.3617-2.3622	0.0009-0.0022	0.0008-0.0087	2	2.0861-2.0866	0.0012-0.0022	0.0063-0.0123
	2.7 (2693)	3RZ-FE	2.2615-2.3620	0.0012-0.0022	0.0008-0.0087	3	2.0861-2.0866	0.0009-0.0022	0.0063-0.0123
	3.4 (3378)	5VZ-FE	2.5191-2.5197	0.0008-0.0015	0.0008-0.0087	2	2.1648-2.1654	0.0009-0.0021	0.0059-0.0130

42356-TACO-C05

VALVE SPECIFICATIONS

Year	Engine Displacement Liters (cc)	Engine ID/VIN	Seat Angle (deg.)	Face Angle (deg.)	Spring Test Pressure (lbs. @ in.)	Spring Installed Height (in.)	Stem-to-Guide Clearance (in.)		Stem Diameter (in.)	
							Intake	Exhaust	Intake	Exhaust
2000	2.4 (2438)	2RZ-FE	45	44.5	40.0-46.0@ 1.406	1.406	0.0010-0.0024	0.0012-0.0026	0.2350-0.2356	0.2348-0.2354
	2.7 (2693)	3RZ-FE	45	44.5	40.0-46.0@ 1.406	1.406	0.0010-0.0024	0.0012-0.0026	0.2350-0.2356	0.2348-0.2354
	3.4 (3378)	5VZ-FE	45	44.5	41.9-46.3@ 1.311	1.311	0.0010-0.0024	0.0012-0.0026	0.2350-0.2356	0.2348-0.2354
2001	2.4 (2438)	2RZ-FE	45	44.5	40.0-46.0@ 1.406	1.406	0.0010-0.0024	0.0012-0.0026	0.2350-0.2356	0.2348-0.2354
	2.7 (2693)	3RZ-FE	45	44.5	40.0-46.0@ 1.406	1.406	0.0010-0.0024	0.0012-0.0026	0.2350-0.2356	0.2348-0.2354
	3.4 (3378)	5VZ-FE	45	44.5	41.9-46.3@ 1.311	1.311	0.0010-0.0024	0.0012-0.0026	0.2350-0.2356	0.2348-0.2354
2002	2.4 (2438)	2RZ-FE	45	44.5	40.0-46.0@ 1.406	1.406	0.0010-0.0024	0.0012-0.0026	0.2350-0.2356	0.2348-0.2354
	2.7 (2693)	3RZ-FE	45	44.5	40.0-46.0@ 1.406	1.406	0.0010-0.0024	0.0012-0.0026	0.2350-0.2356	0.2348-0.2354
	3.4 (3378)	5VZ-FE	45	44.5	41.9-46.3@ 1.311	1.311	0.0010-0.0024	0.0012-0.0026	0.2350-0.2356	0.2348-0.2354
2003	2.4 (2438)	2RZ-FE	45	44.5	40.0-46.0@ 1.406	1.406	0.0010-0.0024	0.0012-0.0026	0.2350-0.2356	0.2348-0.2354
	2.7 (2693)	3RZ-FE	45	44.5	40.0-46.0@ 1.406	1.406	0.0010-0.0024	0.0012-0.0026	0.2350-0.2356	0.2348-0.2354
	3.4 (3378)	5VZ-FE	45	44.5	41.9-46.3@ 1.311	1.311	0.0010-0.0024	0.0012-0.0026	0.2350-0.2356	0.2348-0.2354

42356-TACO-C06

PISTON AND RING SPECIFICATIONS

All measurements are given in inches.

Year	Engine Displacement Liters (cc)	Engine ID/VIN	Piston Clearance	Ring Gap			Ring Side Clearance		
				Top Compression	Bottom Compression	Oil Control	Top Compression	Bottom Compression	Oil Control
2000	2.4 (2438)	2RZ-FE	0.0012-0.0020	0.0118-0.0169	0.0177-0.0236	0.0051-0.0150	0.0008-0.0028	0.0012-0.0028	SNUG
	2.7 (2693)	3RZ-FE	0.0019-0.0028	0.0118-0.0157	0.0157-0.0194	0.0051-0.0150	0.0008-0.0028	0.0012-0.0028	SNUG
	3.4 (3378)	5VZ-FE	0.0053-0.0060	0.0118-0.0197	0.0157-0.0236	0.0059-0.0217	0.0016-0.0031	0.0012-0.0028	SNUG
2001	2.4 (2438)	2RZ-FE	0.0012-0.0020	0.0118-0.0169	0.0177-0.0236	0.0051-0.0150	0.0008-0.0028	0.0012-0.0028	SNUG
	2.7 (2693)	3RZ-FE	0.0019-0.0028	0.0118-0.0157	0.0157-0.0194	0.0051-0.0150	0.0008-0.0028	0.0012-0.0028	SNUG
	3.4 (3378)	5VZ-FE	0.0053-0.0060	0.0118-0.0197	0.0157-0.0236	0.0059-0.0217	0.0016-0.0031	0.0012-0.0028	SNUG
2002	2.4 (2438)	2RZ-FE	0.0012-0.0020	0.0118-0.0169	0.0177-0.0236	0.0051-0.0150	0.0008-0.0028	0.0012-0.0028	SNUG
	2.7 (2693)	3RZ-FE	0.0019-0.0028	0.0118-0.0157	0.0157-0.0194	0.0051-0.0150	0.0008-0.0028	0.0012-0.0028	SNUG
	3.4 (3378)	5VZ-FE	0.0053-0.0060	0.0118-0.0197	0.0157-0.0236	0.0059-0.0217	0.0016-0.0031	0.0012-0.0028	SNUG
2003	2.4 (2438)	2RZ-FE	0.0012-0.0020	0.0118-0.0169	0.0177-0.0236	0.0051-0.0150	0.0008-0.0028	0.0012-0.0028	SNUG
	2.7 (2693)	3RZ-FE	0.0019-0.0028	0.0118-0.0157	0.0157-0.0194	0.0051-0.0150	0.0008-0.0028	0.0012-0.0028	SNUG
	3.4 (3378)	5VZ-FE	0.0053-0.0060	0.0118-0.0197	0.0157-0.0236	0.0059-0.0217	0.0016-0.0031	0.0012-0.0028	SNUG

42356-TACO-C07

TORQUE SPECIFICATIONS
All readings in ft. lbs.

Year	Engine Displacement Liters (cc)	Engine ID/VIN	Cylinder Head Bolts	Main Bearing Bolts	Rod Bearing Bolts	Crankshaft Damper Bolts	Flywheel Bolts	Manifold		Spark Plugs	Lug Nuts
								Intake	Exhaust		
2000	2.4 (2438)	2RZ-FE	①	②	③	193	④	22	36	14	83
	2.7 (2693)	3RZ-FE	①	②	③	193	⑤	22	36	14	83
	3.4 (3378)	5VZ-FE	⑥	⑦	⑧	184	63-67	13	30	13	83
2001	2.4 (2438)	2RZ-FE	①	②	③	193	④	22	36	14	83
	2.7 (2693)	3RZ-FE	①	②	③	193	⑤	22	36	14	83
	3.4 (3378)	5VZ-FE	⑥	⑦	⑧	184	63-67	13	30	13	83
2002	2.4 (2438)	2RZ-FE	①	②	③	193	④	22	36	14	83
	2.7 (2693)	3RZ-FE	①	②	③	193	⑨	22	36	14	83
	3.4 (3378)	5VZ-FE	⑩	⑦	⑧	184	63	13	30	13	83
2003	2.4 (2438)	2RZ-FE	①	②	③	193	④	22	36	14	83
	2.7 (2693)	3RZ-FE	①	②	③	193	⑨	22	36	14	83
	3.4 (3378)	5VZ-FE	⑩	⑦	⑧	184	63	13	30	13	83

① Step 1: 29 ft. lbs.
  Step 2: Plus 90 degrees
  Step 3: Plus 90 degrees

② Step 1: 29 ft. lbs.
  Step 2: Plus 90 degrees

③ Step 1: 33 ft. lbs.
  Step 2: Plus 90 degrees

④ Manual transmission: 65 ft. lbs.
  Automatic transmission: 54 ft. lbs.

⑤ Manual transmission: 19 ft. lbs. + 90°
  Automatic transmission: 54 ft. lbs.

⑥ Step 1: 25 ft. lbs.
  Step 2: Plus 90 degrees
  Recessed head: 13 ft. lbs.

⑦ Step 1: 45 ft. lbs.
  Step 2: Plus 90 degrees

⑧ Step 1: 18 ft. lbs.
  Step 2: Plus 90 degrees

⑨ MT: 19 ft. lbs. +90 degrees
  AT: 54 ft. lbs.

⑩ Step 1: 25 ft. lbs.
  Step 2: Plus 90 degrees
  Step 3: Plus 90 degrees
  Recessed head: 13 ft. lbs.

42356-TACO-C08

BRAKE SPECIFICATIONS
All measurements in inches unless noted

| Year | Model | Brake Disc | | | Brake Drum Diameter | | Minimum Lining Thickness | | Brake Caliper | |
		Original Thickness	Minimum Thickness	Maximum Runout	Original Inside Diameter	Maximum Machine Diameter	Front	Rear	Bracket Bolts (ft. lbs.)	Mounting Bolts (ft. lbs.)
2000	2WD	0.866	0.787	0.0028	10.00	10.08	0.039	—	80	65
	4WD	0.866	0.787	0.0028	11.61	11.69	—	0.039	—	90
2001	2WD	0.866	0.787	0.0028	10.00	10.08	0.039	—	80	65
	4WD	0.866	0.787	0.0028	11.61	11.69	—	0.039	—	90
2002	2WD	0.866	0.787	0.0028	10.00	10.08	0.039	—	80	65
	4WD	0.866	0.787	0.0028	11.61	11.69	—	0.039	—	90
2003	2WD	0.866	0.787	0.0028	10.00	10.08	0.039	—	80	65
	4WD	0.866	0.787	0.0028	11.61	11.69	—	0.039	—	90

F: Front

R: Rear

42356-TACO-C09

WHEEL ALIGNMENT

| Year | Model | Caster | | Camber | | Toe-in (in.) | Steering Axis Inclination (Deg.) |
		Range (+/-Deg.)	Preferred Setting (Deg.)	Range (+/-Deg.)	Preferred Setting (Deg.)		
2000	2WD	0.75	0	0.75	0.67	0.06+/-0.08	10.00
	4WD	0.75	1.62	0.75	0.30	0.06+/-0.08	10.40
2001	2WD	0.75	0	0.75	0.67	0.06+/-0.08	10.00
	4WD	0.75	1.62	0.75	0.30	0.06+/-0.08	10.40
2002	2WD	0.75	0	0.75	0.67	0.06+/-0.08	10.00
	4WD	0.75	1.62	0.75	0.30	0.06+/-0.08	10.40
2003	2WD	0.75	0	0.75	0.67	0.06+/-0.08	10.00
	4WD	0.75	1.62	0.75	0.30	0.06+/-0.08	10.40

All alignment figures based on nominal ride height and standard tires

42356-TACO-C10

TIRE, WHEEL AND BALL JOINT SPECIFICATIONS

Year	Model	OEM Tires		Tire Pressures (psi)		Wheel Size	Ball Joint Inspection ①
		Standard	Optional	Front	Rear		
2000	Tacoma 2wd	P195/75R14	None	29	35	5-J	4-30
	Tacoma 2wd Xtracab	P215/70R14	None	29	29	6-JJ	4-30
	Tacoma 4wd & Prerunner	P225/75R15	31x10.5R15 LT	26	29	7-JJ	4-30
2001	Tacoma 2wd	P205/75R15	P225/75R15	Std: 29 Opt: 26	Std: 29 Opt: 29	6-JJ	4-30
	Tacoma Double Cab	P265/70R16	None	26	26	7-JJ	4-30
	Tacoma Prerunner	P235/55R16	None	26	29	7-JJ	4-30
	Tacoma 4wd	P225/75R15	P265/70R16	Std: 26 Opt: 26	Std: 29 Opt: 26	Std: 6-JJ Opt: 7-JJ	4-30
2002	Tacoma 2wd Reg. Cab	P205/75R15	P235/55R16	Std: 29 Opt: 29	Std: 29 Opt: 32	6-JJ Opt: 6.5-J	Upper 4-30 Lower 0.8-30
	Tacoma 4wd Reg. Cab	P225/75SR15	P265/70R16	Std: 26 Opt: 26	Std: 29 Opt: 26	6-JJ Opt.: 7-JJ	Upper 6-39 Lower 1-22
	Tacoma 4wd xtracab	P225/75SR15	P265/70R16	26	29	7-JJ	Upper 6-39 Lower 1-22
	Tacoma 2wd xtracab S-Runner	P235/55R16	None	29	32	7-JJ	Upper 4-30 Lower 0.8-30
	Tacoma 2wd xtracab PreRunner	P225/75R15	P265/70R16	Std: 26 Opt: 26	Std: 29 Opt: 26	7-JJ	Upper 6-39 Lower 1-22
	Tacoma 4wd Crew Cab	P225/75SR15	P265/70R16	Std: 26 Opt: 26	Std: 29 Opt: 26	7-JJ	Upper 6-39 Lower 1-22
	Tacoma 2wd Crew Cab PreRunner	P225/75R15	P265/70R16	Std: 26 Opt: 26	Std: 29 Opt: 26	7-JJ	Upper 6-39 Lower 1-22
2003	Tacoma 2wd Reg. Cab	P205/75R15	P235/55R16	Std: 29 Opt: 29	Std: 29 Opt: 32	6-JJ Opt: 6.5J	Upper 4-30 Lower 0.8-30
	Tacoma 4wd Reg. Cab	P225/75SR15	P265/70R16	Std: 26 Opt: 26	Std: 29 Opt: 26	6-JJ Opt.: 7-JJ	Upper 6-39 Lower 1-22
	Tacoma 4wd xtracab	P225/75SR15	P265/70R16	26	29	7-JJ	Upper 6-39 Lower 1-22
	Tacoma 2wd xtracab S-Runner	P235/55R16	None	29	32	7-JJ	Upper 4-30 Lower 0.8-30
	Tacoma 2wd xtracab PreRunner	P225/75R15	P265/70R16	Std: 26 Opt: 26	Std: 29 Opt: 26	7-JJ	Upper 6-39 Lower 1-22
	Tacoma 4wd Crew Cab	P225/75SR15	P265/70R16	Std: 26 Opt: 26	Std: 29 Opt: 26	7-JJ	Upper 6-39 Lower 1-22
	Tacoma 2wd Crew Cab PreRunner	P225/75R15	P265/70R16	Std: 26 Opt: 26	Std: 29 Opt: 26	7-JJ	Upper 6-39 Lower 1-22

OEM: Original Equipment Manufacturer

PSI: Pounds Per Square Inch

STD: Standard

OPT: Optional

① Torque required in inch lbs. to rotate ball joint when removed from the knuckle

42356-TACO-C11

SCHEDULED MAINTENANCE INTERVALS
TOYOTA—TACOMA

TO BE SERVICED	TYPE OF	VEHICLE MILEAGE INTERVAL (x1000)																		
		5	10	15	20	25	30	35	40	45	50	55	60	65	70	75	80	85	90	95
Automatic transmission and differential fluid	S/I			✓			✓			✓			✓			✓			✓	
Ball joints and boots	S/I			✓			✓			✓			✓			✓			✓	
Brake linings, discs/drums, lines & hoses	S/I			✓			✓			✓			✓			✓			✓	
Charcoal canister	S/I												✓							
Drive belts	S/I						✓						✓						✓	
Driveshaft bushing (4WD)	L						✓						✓						✓	
Engine coolant	R						✓						✓						✓	
Engine oil & filter	R	✓	✓	✓	✓	✓	✓	✓	✓	✓	✓	✓	✓	✓	✓	✓	✓	✓	✓	✓
Exhaust pipes & mounts	S/I			✓			✓			✓			✓			✓			✓	
Fuel lines & connections, fuel tank vapor vent system hoses, fuel tank band	S/I						✓						✓						✓	
Fuel tank cap gasket	S/I						✓						✓						✓	
Halfshaft boots & flange bolts	S/I			✓			✓			✓			✓			✓			✓	
Limited slip differential fluid	R						✓						✓						✓	
Manual transmission and differential fluid	S/I						✓						✓						✓	
Non-platinum spark plugs	R						✓												✓	
Platinum spark plugs	R												✓							
Propeller shaft (4WD)	L			✓			✓			✓			✓			✓			✓	
Propeller shaft bolts	S/I			✓			✓			✓			✓			✓			✓	
Rack and pinion assembly	S/I			✓			✓			✓			✓			✓			✓	
Rear wheel bearing	L						✓						✓						✓	
Steering linkage	S/I			✓			✓			✓			✓			✓			✓	
Valves	S/I												✓							

R: Replace   S/I: Service or Inspect   L: Lubricate

FREQUENT OPERATION MAINTENANCE (SEVERE SERVICE)

If a vehicle is operated under any of the following conditions it is considered severe service:

- Towing a trailer or using a camper or car-top carrier.
- Repeated short trips of less than 5 miles in temperatures below freezing.
- Excessive idling or low-speed driving for long distances as in heavy commercial use, such as delivery, taxi or police cars.
- Operating on rough, muddy or salt-covered roads.
- Operating on unpaved or dusty roads.

Oil filter: service or inspect every 5000 miles or 4 months, whichever occurs first.

Brake linings and discs or drums: service or inspect every 5000 miles or 4 months, whichever occurs first.

Steering linkage: service or inspect every 5000 miles or 4 months, whichever occurs first.

Ball joints and boots: service or inspect every 5000 miles or 4 months, whichever occurs first.

Brake discs & pads (front): service or inspect every 6000 miles.

Halfshaft boots: service or inspect every 5000 miles or 4 months. Retighten the flange bolts, whichever occurs first.

Body chassis bolts and nuts: service or inspect every 5000 miles or 4 months, whichever occurs first.

Transmission and differential fluid: replace every 15,000 miles or 12 months, whichever occurs first.

Transfer case and differential fluid: replace every 15,000 miles or 12 months, whichever occurs first.

Timing belt: replace every 60,000 miles or 48 months, whichever occurs first.

42356-TACO-C12

PRECAUTIONS

Before servicing any vehicle, please be sure to read all of the following precautions, which deal with personal safety, prevention of component damage, and important points to take into consideration when servicing a motor vehicle:

• Never open, service or drain the radiator or cooling system when the engine is hot; serious burns can occur from the steam and hot coolant.

• Observe all applicable safety precautions when working around fuel. Whenever servicing the fuel system, always work in a well-ventilated area. Do not allow fuel spray or vapors to come in contact with a spark, open flame or excessive heat (a hot drop light, for example). Keep a dry chemical fire extinguisher near the work area. Always keep fuel in a container specifically designed for fuel storage; also, always properly seal fuel containers to avoid the possibility of fire or explosion. Refer to the additional fuel system precautions later in this section.

• Fuel injection systems often remain pressurized, even after the engine has been turned **OFF**. The fuel system pressure must be relieved before disconnecting any fuel lines. Failure to do so may result in fire and/or personal injury.

• Brake fluid often contains polyglycol ethers and polyglycols. Avoid contact with the eyes and wash your hands thoroughly after handling brake fluid. If you do get brake fluid in your eyes, flush your eyes with clean, running water for 15 minutes. If eye irritation persists, or if you have taken brake fluid internally, IMMEDIATELY seek medical assistance.

• The EPA warns that prolonged contact with used engine oil may cause a number of skin disorders, including cancer! You should make every effort to minimize your exposure to used engine oil. Protective gloves should be worn when changing oil. Wash your hands and any other exposed skin areas as soon as possible after exposure to used engine oil. Soap and water, or waterless hand cleaner should be used.

• All new vehicles are now equipped with an air bag system. The system must be disabled before performing service on or around system components, steering column, instrument panel components, wiring and sensors. Failure to follow safety and disabling procedures could result in accidental air bag deployment, possible personal injury and unnecessary system repairs.

• Always wear safety goggles when working with, or around, the air bag system. When carrying a non-deployed air bag, be sure the bag and trim cover are pointed away from your body. When placing a non-deployed air bag on a work surface, always face the bag and trim cover upward, away from the surface. This will reduce the motion of the module if it is accidentally deployed. Refer to the additional air bag system precautions later in this section.

• Clean, high quality brake fluid from a sealed container is essential to the safe and proper operation of the brake system. You should always buy the correct type of brake fluid for your vehicle. If the brake fluid becomes contaminated, completely flush the system with new fluid. Never reuse any brake fluid. Any brake fluid that is removed from the system should be discarded. Also, do not allow any brake fluid to come in contact with a painted surface; it will damage the paint.

• Never operate the engine without the proper amount and type of engine oil; doing so WILL result in severe engine damage.

• Timing belt maintenance is extremely important! Many models utilize an interference-type, non-freewheeling engine. If the timing belt breaks, the valves in the cylinder head may strike the pistons, causing potentially serious (also time-consuming and expensive) engine damage. Refer to the maintenance interval charts for the recommended replacement interval for the timing belt, and to the timing belt section for belt replacement and inspection.

• Disconnecting the negative battery cable on some vehicles may interfere with the functions of the on-board computer system(s) and may require the computer to undergo a relearning process once the negative battery cable is reconnected.

• When servicing drum brakes, only disassemble and assemble one side at a time, leaving the remaining side intact for reference.

ENGINE REPAIR

Distributor

All engines are equipped with distributorless ignition systems.

Alternator

REMOVAL

On some models, the alternator is mounted very low on the engine. On these models, it may be necessary to remove the gravel shield and work from beneath the vehicle in order to gain access to the alternator. Replacing the alternator while the engine is cold is recommended.

2.4L and 2.7L Engines

Remove or disconnect the following:
• Negative battery cable
• Alternator wiring

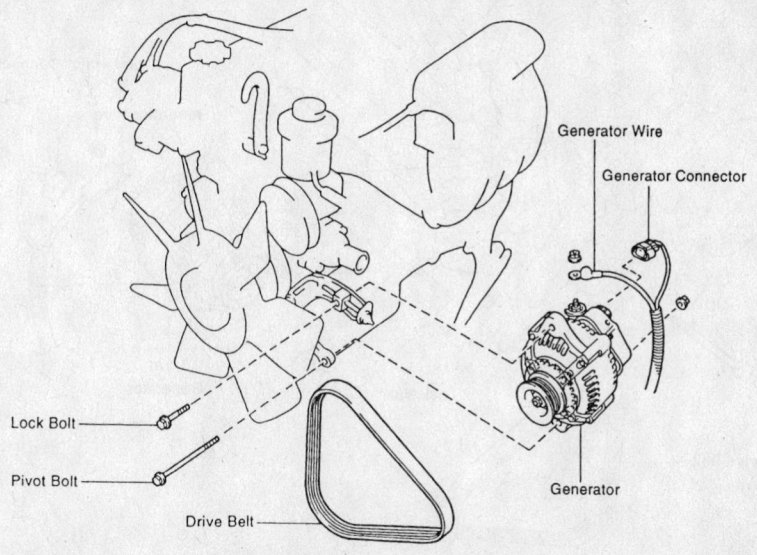

Exploded view of the alternator and drive belt—2.4L and 2.7L engines

86822073

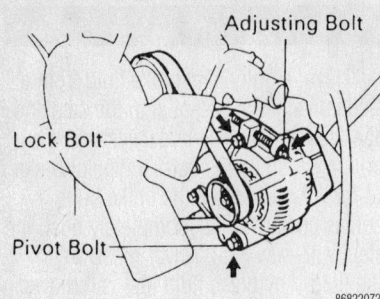

Locations of the adjusting, pivot and lock-bolts—2.4L and 2.7L engines

- Alternator lockbolt, pivot bolt, nut and adjusting bolt
- Drive belt
- Wiring harness with the clip
- Alternator

3.4L Engine

1. Remove or disconnect the follow-ing:
   - Negative battery cable
   - Alternator wiring
   - Alternator locknut, pivot bolt, nut and adjusting bolt
   - Drive belt
   - Alternator

INSTALLATION

2.4L and 2.7L Engines

Install or connect the following:
- Alternator
- Drive belt; adjust it to the proper tension. Tighten the lockbolt to 21 ft. lbs. (29

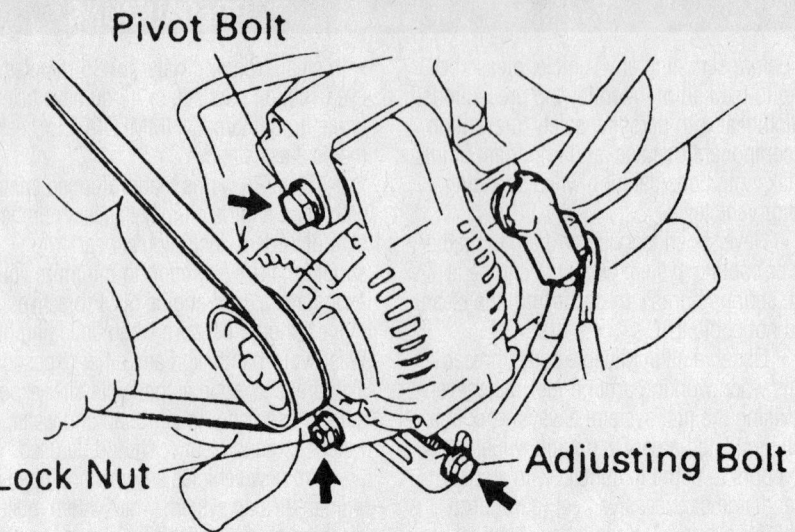

Locations of the adjusting and pivot bolts and the locknut—3.4L Engine

Nm) and the pivot bolt to 43 ft. lbs. (59 Nm).
- Wire harness with clip
- Alternator wiring. Tighten the nut to 7 ft. lbs. (10 Nm).
- Rubber boot over the terminal
- Wiring harness connector
- Negative battery cable

3.4L Engine

Install or connect the following:
- Alternator
- Drive belt. Tighten the locknut 25 ft. lbs. (33 Nm) and the pivot bolt 38 ft. lbs. (51 Nm).
- Alternator wiring
- Negative battery cable

Ignition Timing

ADJUSTMENT

All engines use a distributorless ignition system referred to as Direct Ignition System (DIS). All spark advance is permanently set by the PCM.

2.4L (2RZ-FE) Engine

➡**The ignition timing is not adjustable but can be checked.**

1. Before servicing the vehicle, refer to the precautions in the beginning of this section.
2. Warm the engine to normal operating temperature.
3. Attach a hand-held tester to the Data Link Connector 3 (DLC3) under the dashboard on the driver's side.
4. Jumper terminals T_{E1} and E_1 of the DLC1.
5. Check the idle speed.
6. Aim the timing light at the timing indicator and check the ignition timing. Timing should be between 3–7 degrees BTDC at idle.
7. For a further check on ignition timing, disconnect the hand-held tester from the DLC3 and disconnect the jumper wire from the DLC1.
8. Point the timing light at the crankshaft pulley and read the timing. Timing should be between 7–18 degrees BTDC at idle.
9. Remove timing light from the engine.

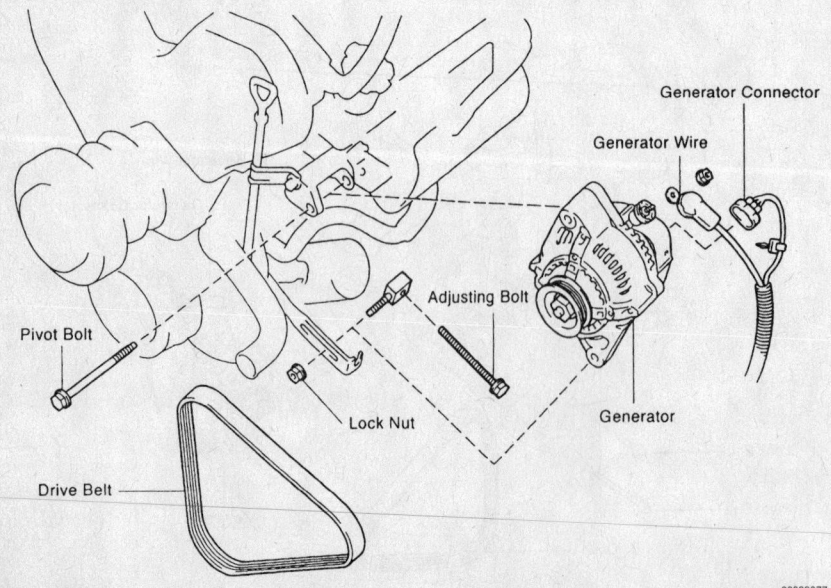

Exploded view of the alternator and drive belt—3.4L Engine

Engine Assembly

REMOVAL & INSTALLATION

2.4L (2RZ-FE) Engine

1. Before servicing the vehicle, refer to the precautions in the beginning of this section.

2. Properly relieve the fuel system pressure.

3. Turn the ignition switch **OFF**.

4. Remove or disconnect the following:
- Battery cables; negative cable first
- Hood by matchmarking the hood hinges
- Engine undercover

5. Drain the engine oil, transmission oil and cooling system.

6. Remove or disconnect the following:
- Radiator
- Drive belts
- Loosen the lockbolt and adjusting bolt to the idler pulley, if equipped with power steering
- Loosen the idler pulley nut and adjusting bolt, if equipped with air conditioning
- Fan (with fan clutch), water pump pulley and fan shroud
- Accelerator cable from the throttle body, if equipped with a manual transaxle
- Accelerator and throttle cables from the throttle body, if equipped with an automatic transaxle
- Actuator cover and cruise control cable from the actuator, if equipped with cruise control
- Air cleaner cap, Mass Air Flow (MAF) and resonator
- Air cleaner case
- Intake air connector
- Air conditioning compressor and bracket, if equipped with air conditioning
- Alternator wires from the alternator
- Heater hoses at the cowl panel
- Brake booster vacuum hose
- EVAP hose
- Vacuum hose, if equipped with 4WD with Automatic Disconnecting Differential (ADD)
- Both power steering hoses, if equipped with power steering
- Fuel return hose
- Fuel inlet hose

7. Remove the power steering pump as follows:
- Nut and power steering pulley

- Both bolts and the power steering pump

8. Disconnect the Engine Control Module (ECM) wiring from the ECM as follows:
- Right front door scuff plate
- Cowl panel side trim by removing the clip
- 4 ECM electrical connectors.

9. detach the engine wiring harness and connectors as follows:
- Igniter connector
- Ground strap from the cowl top panel
- Both engine wiring harness clamps
- Engine wiring harness retainer to the cowl panel nuts and pull out the engine wiring harness from the vehicle

10. Disconnect the front exhaust pipe from the exhaust manifold and catalytic converter.

11. If equipped with manual transmission, remove the shift lever assembly as follows:
- Shift lever knob
- 4 screws and shift lever boot
- 6 bolts, shift lever assembly and baffle

12. Remove or disconnect the following:
- Driveshaft
- Speedometer cable from the transmission
- Clutch release cylinder, if equipped with manual transmission
- Cross-shaft, if equipped with automatic transmission
- Wires at the starter

13. Position a jack and wooden block under the transmission and remove the rear engine mounting bracket.

14. Attach an engine hoist to the engine hangers.

15. Remove or disconnect the following:
- Nuts and bolts from the engine mounts
- Engine/transmission assembly

To install:

16. Install or connect the following:
- Engine/transmission assembly, by keeping the engine level, while aligning the engine mounts
- Engine mount fasteners but do not fully tighten them

17. Position a jack and wooden block under the transmission.

18. Install the rear engine mounting bracket. Tighten the frame bolts to 43 ft. lbs. (58 Nm) and the mount bolts to 13 ft. lbs. (18 Nm).

19. Remove the jack and engine hoist.

20. Install or connect the following:

- Tighten the engine mounts to 28 ft. lbs. (38 Nm).
- Starter wires to the starter
- Clutch release cylinder, if equipped with manual transmission
- Cross-shaft, if equipped with automatic transmission
- Speedometer to the transmission
- Driveshaft

21. If equipped with manual transmission, install the shift lever assembly as follows:
- Baffle and shift lever assembly with the 6 bolts
- Shift lever boot with the 4 screws
- Shift lever knob
- Front exhaust pipe to the exhaust manifold
- All wires and connectors
- Cowl side trim and clip
- Front door scuff plate
- Alternator wires to the alternator

22. Install the power steering pump as follows:
- Power steering pump to the bracket with the 2 bolts. Tighten the bolts to 43 ft. lbs. (58 Nm).
- Power steering pulley with the nut. Tighten the nut to 32 ft. lbs. (43 Nm).

23. Install or connect the following:
- All hoses
- Compressor, if equipped with air conditioning
- Intake air connector. Tighten the 2 bolts to 13 ft. lbs. (18 Nm).
- Water pump pulley, fan shroud, fan (with fan clutch) and alternator drive belt.
- Drive belt, if equipped with air conditioning
- Power steering drive belt
- Accelerator cable to the throttle body, if equipped with manual transmission
- Accelerator and throttle cables to the throttle body, if equipped with automatic transmission
- Air cleaner case
- MAF meter, resonator and air cleaner cap
- Radiator with the support tabs through the radiator service holes. Tighten the bolts to 108 inch lbs. (13 Nm).
- Lower radiator hose to the radiator
- Oil cooler hoses to the radiator, if equipped with an automatic transmission
- No. 2 fan shroud.
- Radiator reservoir hose to the radiator

- Upper radiator hose to the radiator
- Air pipe
- Radiator grille to the vehicle with the 11 clips
- Both fillers
- Clearance lights to the grille with the 4 bolts and 2 clips
- Both battery cables

24. Fill the engine oil, engine coolant and transmission oil.

25. Start the engine and check for leaks.

26. Check ignition timing.

27. Install or connect the following:
- Engine undercover
- Hood

28. Road test the vehicle and check all fluids.

2.7L (3RZ-FE) Engine

1. Before servicing the vehicle, refer to the precautions in the beginning of this section.

2. Properly relieve the fuel system pressure.

3. Turn the ignition switch **OFF**.

4. Remove or disconnect the following:
- Battery cables; negative cable first
- Hood
- Engine undercover

5. Drain the engine oil, transmission oil and cooling system.

6. Remove or disconnect the following:
- Radiator
- Idler pulley and drive belt, if equipped with power steering
- Idler pulley nut/bolt and drive belt, if equipped with air conditioning
- Alternator drive belt, fan (with fan clutch), water pump pulley and fan shroud
- Accelerator cable from the throttle body, if equipped with a manual transaxle
- Accelerator and throttle cables from the throttle body, if equipped with a automatic transaxle
- Actuator cover and the cruise control cable from the actuator, if equipped with cruise control
- Air cleaner cap
- Mass Air Flow (MAF) meter and resonator
- Air cleaner case
- Intake air connector
- Air conditioning compressor and bracket, if equipped with air conditioning
- Alternator wires
- Heater hoses at the cowl panel

7. Disconnect the following hoses:
- Brake booster vacuum hose
- Evaporative Emissions (EVAP) hose
- Vacuum hose, if equipped with 4WD with Automatic Disconnecting Differential (ADD)
- Both power steering hoses, if equipped with power steering
- Fuel return hose
- Fuel inlet hose

8. Remove the power steering pump as follows:
- Nut and power steering pulley
- Power steering pump

9. Disconnect the Engine Control Module (ECM) wiring from the ECM as follows:
- 4 screws to the right front door scuff plate
- Scuff plate
- Cowl panel side trim by removing the clip
- 4 ECM electrical connectors

10. Detach the engine wiring harness and connectors from the vehicle as follows:
- Igniter
- Ground strap from the cowl top panel
- 2 engine wiring harness clamps
- Engine wiring harness retainer to cowl panel nut and wiring harness

11. Disconnect the front exhaust pipe from the exhaust manifold and catalytic converter.

12. If equipped with manual transmission, remove the shift lever assembly as follows:

13. Remove or disconnect the following:
- Shift lever knob
- 4 screws and shift lever boot
- 6 bolts, shift lever assembly and baffle

14. Remove or disconnect the following:
- Driveshaft
- Speedometer cable from the transmission
- Clutch release cylinder, if equipped with manual transmission
- Cross-shaft, If equipped with automatic transmission
- Wires at the starter

15. Position a jack and wooden block under the transmission.

16. Remove the rear engine mounting bracket.

17. Attach an engine hoist to the engine hangers.

18. Remove or disconnect the following:
- Nuts and bolts from the engine mounts
- Engine/transmission

To install:

19. Attach the engine hoist to the engine hangers.

20. Install or connect the following:
- Engine/transmission assembly

➡️**Keep the engine level, while aligning the engine mounts.**

- Engine mount fasteners but do not fully tighten them
- Position a jack and wooden block under the transmission
- Rear engine mounting bracket. Tighten the frame bolts to 19 ft. lbs. (26 Nm) and the mount bolts to 13 ft. lbs. (18 Nm).

21. Remove the jack and engine hoist.

22. Install or connect the following:
- Tighten the engine mounts to 28 ft. lbs. (38 Nm).
- Starter wires
- Clutch release cylinder, if equipped with manual transmission
- Cross-shaft, if equipped with automatic transmission
- Speedometer to the transmission
- Driveshaft

23. If equipped with manual transmission, install the shift lever assembly as follows:
- Baffle and shift lever assembly with the 6 bolts
- Shift lever boot with the 4 screws
- Shift lever knob

24. Install or connect the following:
- Front exhaust pipe to the exhaust manifold
- Engine wiring harness
- Cowl side trim and clip
- Front door scuff plate
- Alternator wires

25. Install the power steering pump as follows:
- Power steering pump to the bracket. Tighten the 2 bolts to 43 ft. lbs. (58 Nm).
- Power steering pulley. Tighten the nut to 32 ft. lbs. (43 Nm).

26. Install or connect the following:
- All hoses
- Heater hoses at the cowl panel
- Compressor, if equipped with air conditioning
- Intake air connector. Tighten the 2 bolts to 13 ft. lbs. (18 Nm).

27. Install the water pump pulley, fan shroud, fan (with fan clutch) and alternator drive belt as follows:
- Fan (with the fan clutch), water pump pulley and fan shroud in position
- Water pump pulley but do not tighten the nuts
- Alternator drive belt
- Stretch the alternator belt tight. Tighten the fan nuts to 16 ft. lbs. (21 Nm).

- Adjust the alternator drive belt
28. Install or connect the following:
  - Adjust the drive belt, if equipped with air conditioning
  - Adjust the power steering drive belt
  - Accelerator cable to the throttle body, if equipped with manual transmission
  - Accelerator and throttle cables to the throttle body, if equipped with automatic transmission
  - Air cleaner case
  - MAF meter, resonator and air cleaner cap
  - Radiator with the tabs on the supports through the radiator service holes. Tighten the bolts to 108 inch lbs. (13 Nm).
  - Lower radiator hose to the radiator
  - Oil cooler hoses to the radiator, if equipped with automatic transmission
  - No. 2 fan shroud
  - Radiator reservoir hose to the radiator
  - Upper radiator hose to the radiator
  - Air pipe with the 2 bolts, if removed
  - Radiator grille with the 11 clips
  - 2 fillers
  - Clearance lights to the grille with the 4 bolts and 2 clips
  - Negative and positive cables to the battery
29. Fill the engine oil, engine coolant and transmission oil.
30. Start the engine and check for leaks.
31. Check ignition timing.
32. Install or connect the following:
  - Engine undercover
  - Hood
33. Road test the vehicle and check all fluids.

3.4L (5VZ-FE) Engine

2WD

1. Before servicing the vehicle, refer to the precautions in the beginning of this section.
2. Properly relieve the fuel system pressure.
3. Remove or disconnect the following:
  - Hood
  - Battery
  - Engine under covers
4. Drain the engine coolant.
5. Drain the engine oil.
6. Remove or disconnect the following:
  - Radiator
7. Remove the power steering drive belt as follows:

- Stretch the belt and loosen the fan pulley mounting nuts
- Loosen the lockbolt, pivot bolt and adjusting bolt and remove the drive belt
8. Remove or disconnect the following:
  - Air conditioning drive belt by loosening the idle pulley nut and adjusting bolt, if equipped with air conditioning
  - Loosen the lockbolt, pivot bolt and adjusting bolt and the alternator drive belt
  - Fan with the fluid coupling and fan pulleys
  - Power steering pump, do not disconnect the lines from the pump
  - Compressor, if equipped with air conditioning; do not disconnect the lines from the compressor
  - Air cleaner cap
  - Mass Air Flow (MAF) meter and resonator
  - Air cleaner case and filter
9. Disconnect the following cables:
  - Actuator cable with the bracket, if equipped with cruise control
  - Accelerator cable
  - Throttle cable, if equipped with an automatic transmission
10. Disconnect the following hoses:
  - Heater hoses
  - Brake booster vacuum hose
  - Evaporative Emission (EVAP) hose
  - Fuel return hose
  - Fuel inlet hose
11. Detach the starter wire and connectors as follows:
  - Ground strap by removing the bolt
  - Positive cable from the battery
  - 3 starter wire clamps and connector
12. Detach the alternator connector and wire.
13. Detach the engine wiring harness and connectors as follows:
  - Right front door scuff plate.
  - Cowl panel side trim, by removing the clip
  - Engine Control Module (ECM)
  - 2 connectors from the cowl wire
  - Igniter
  - Ground strap
  - 6 engine wiring harness clamps
  - Engine wiring harness
14. If equipped with manual transmission, remove the shift lever assembly as follows:
  - Shift lever knob
  - 4 screws and the shift lever boot
  - Shift lever assembly and gasket, by removing the 6 bolts
15. Remove or disconnect the following:

- Stabilizer bar
- Driveshaft from the transmission
- Speedometer cable
- Front exhaust pipe
- Clutch release cylinder, if equipped with a manual transmission
- Cross-shaft, if equipped with an automatic transmission
16. Place a jack under the transmission.
17. Remove or disconnect the following:
  - Transmission rear mounting bracket, by removing the 8 bolts
  - Air conditioning compressor wire clamp, if equipped with air conditioning
18. If necessary, install a No. 2 engine hanger with 2 bolts. Tighten the 2 bolts to 30 ft. lbs. (40 Nm).
19. Attach the engine hoist chain to the 2 engine hangers.
20. Remove or disconnect the following:
  - 4 bolts/nuts holding the engine front mounting insulators to the frame
  - Engine/transmission assembly

To install:

21. Install or connect the following:
  - Engine
  - Engine mounts to the body mountings. Install the bolts and nuts but do not tighten at this time.
22. Remove the engine chain hoist the No. 2 engine hanger.
23. Install or connect the following:
  - Air conditioning wire, if equipped with air conditioning
  - Transmission mounting bracket. Tighten the frame bolts to 43 ft. lbs. (58 Nm) and the mounting insulator bolts to 13 ft. lbs. (18 Nm).
  - Tighten the engine mounting nuts and bolts to 28 ft. lbs. (38 Nm).
  - Cross-shaft, if equipped with an automatic transmission
  - Clutch release cylinder, if equipped with a manual transmission. Tighten the bolts to 108 inch lbs. (13 Nm).
  - Front exhaust pipe
  - Speedometer cable
  - Driveshaft
  - Stabilizer bar
24. Install the shift lever assembly, as follows:
25. Install or connect the following:
  - New gasket and shift lever assembly with the 6 bolts
  - Shift lever boot with the 4 screws
  - Shift lever knob
26. Install or connect the following:
  - All engine wiring harness, hoses and cables

- Air cleaner case and air filter
- MAF meter, resonator and air cleaner cap
- Air conditioning compressor, if equipped
- Remaining components

27. Fill the engine with oil.
28. Fill the engine and radiator with coolant.
29. Install the engine undercover.
30. Start the engine and check for leaks.

4WD

1. Before servicing the vehicle, refer to the precautions in the beginning of this section.
2. Properly relieve the fuel system pressure.
3. Remove or disconnect the following:
   - Transmission
   - Hood
   - Battery from the vehicle
   - Engine under covers
4. Drain the engine coolant.
5. Drain the engine oil.
6. Remove or disconnect the following:
   - Radiator
7. Remove the power steering drive belt, as follows:
   - Stretch the belt and loosen the fan pulley mounting nuts
   - Loosen the lockbolt, pivot bolt and adjusting bolt
   - Drive belt from the engine
8. Remove or disconnect the following:
   - Air conditioning drive belt by loosening the idle pulley nut and adjusting bolt, if equipped with air conditioning
   - Alternator drive belt
   - Fan with the fluid coupling and fan pulleys
   - Power steering pump; do not disconnect the lines from the pump
   - Compressor, if equipped with air conditioning. Do not disconnect the lines from the compressor.
   - Air cleaner cap, Mass Air Flow Meter (MAF) meter and resonator
   - Air cleaner case and filter
9. Disconnect the following cables:
   - Actuator cable with the bracket, if equipped with cruise control
   - Accelerator cable
   - Throttle cable, if equipped with an automatic transmission
10. Disconnect the following hoses:
    - Heater hoses
    - Brake booster vacuum hose
    - Evaporative Emissions (EVAP) hose
    - Automatic Disconnecting Differential (ADD) vacuum hose

- Fuel return hose
- Fuel inlet hose

11. Detach the starter wire and connectors as follows:
12. Remove or disconnect the following:
    - Ground strap by removing the bolt
13. Disconnect the positive cable from the battery, as follows:
    - 3 starter wire clamps and connector
    - Automatic Disconnecting Differential (ADD) indicator switch connector
14. Detach the alternator connector and wire.
15. Detach the engine wiring harness and connectors, as follows:
    - Right front door scuff plate
    - Cowl panel side trim, by removing the clip
    - Engine Control Module (ECM)
    - 2 connectors from the cowl wire
    - Igniter connector
    - Ground strap
    - 6 engine wiring harness clamps
    - Engine wiring harness
16. Remove or disconnect the following:
    - Air conditioning compressor wire clamp, if equipped with air conditioning
17. If necessary, install a No. 2 engine hanger with 2 bolts. Tighten the 2 bolts to 30 ft. lbs. (40 Nm).
18. Attach the engine hoist chain to the 2 engine hangers.
19. Remove or disconnect the following:
    - 4 Engine front mounting insulators-to-frame bolts/nuts
    - Engine

To install:
20. Install or connect the following:
    - Engine
    - Engine mounts to the body mountings. Install the bolts and nuts but do not tighten at this time.
21. Remove the engine chain hoist the No. 2 engine hanger.
22. Install or connect the following:
    - Air conditioning wire with the bolt, if equipped with air conditioning
    - Tighten the engine mounting nuts and bolts to 28 ft. lbs. (38 Nm).
    - All engine wiring harness, hoses and cables
    - Air cleaner case and air filter
    - MAF meter, resonator and air cleaner cap
    - Air conditioning compressor, if equipped
    - Fan with the fluid coupling and fan pulleys. Tighten the nuts to 48 inch lbs. (5.4 Nm).
    - Alternator drive belt

- Adjust the air conditioning drive belt, if equipped
- Power steering pump, pump pulley and the drive belt
- Radiator

23. Fill the engine with oil.
24. Fill the engine and radiator with coolant.
25. Install or connect the following:
    - Hood
    - Engine undercover
    - Transmission
26. Start the engine and check for leaks.

Heater Core

REMOVAL & INSTALLATION

1. Disconnect the negative battery cable.

❊❊ CAUTION

After the negative battery cable has been disconnected, wait at least 1½ minutes for the air bag module to deplete its energy.

2. Drain the cooling system into a clean container for reuse.
3. Disconnect the heater hoses from the heater core.
4. Remove the steering wheel by performing the following procedure:
   a. Position the front wheels in the straight-ahead position.
   b. At both sides of the steering wheel, remove the side covers.
   c. Using a Torx® wrench, loosen the steering wheel Torx® screws until the screw's circumference ring catches on the screw case.
   d. Carefully, lift the air bag module, disconnect the electrical connector and remove the air bag.

❊❊ CAUTION

Place the air bag module in a safe location with the front facing upward.

   e. Remove the steering wheel nut.
   f. Using a steering wheel puller, press the steering wheel from the steering column.
5. Remove the instrument panel and reinforcement by performing the following procedure:
   a. Remove the steering column cover screws and the covers.
   b. Remove the combination switch-to-steering column screws and the combination switch.

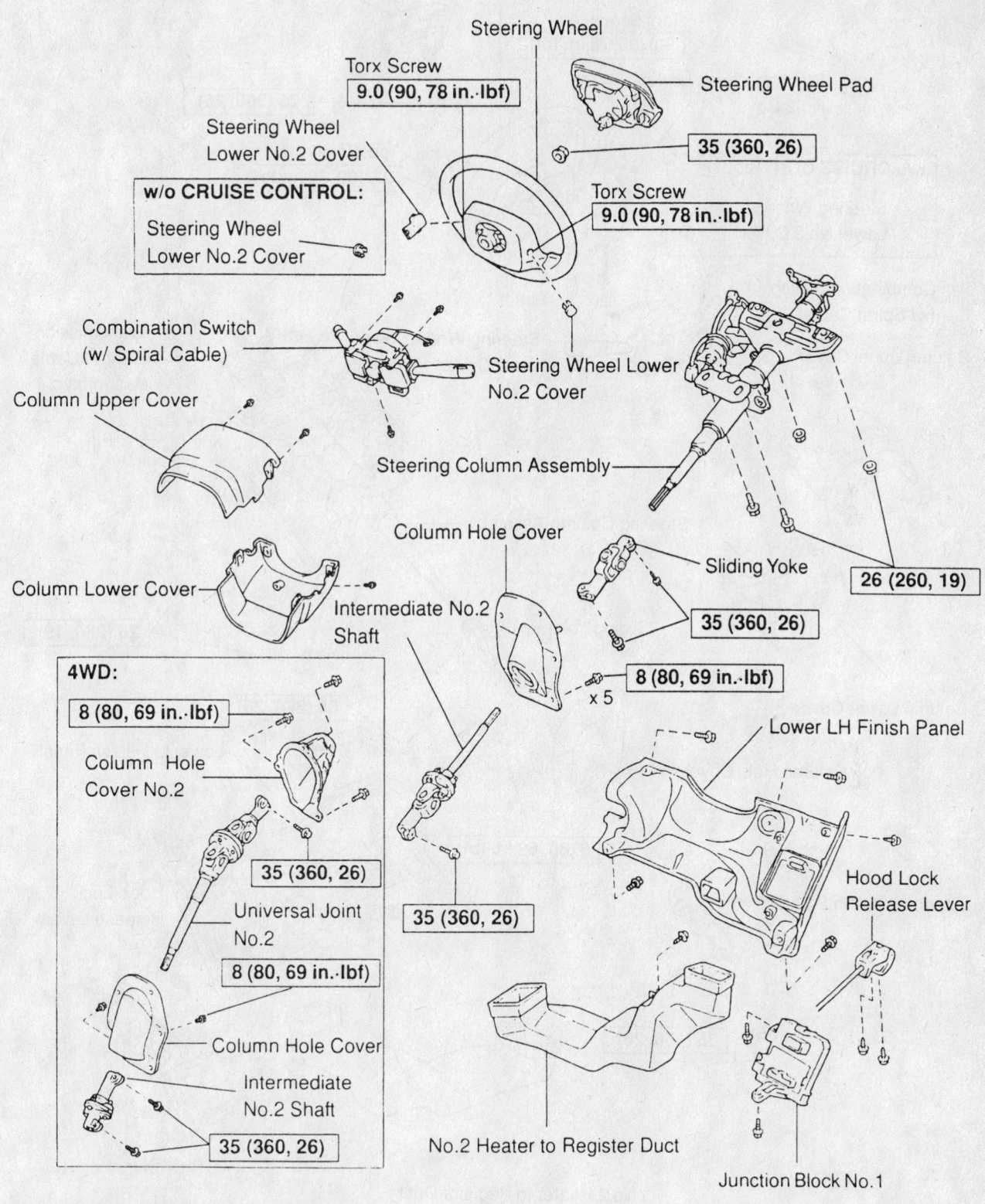

Steering Wheel

Torx Screw
9.0 (90, 78 in.·lbf)

Steering Wheel Pad

35 (360, 26)

Steering Wheel
Lower No.2 Cover

w/o CRUISE CONTROL:

Steering Wheel
Lower No.2 Cover

Torx Screw
9.0 (90, 78 in.·lbf)

Combination Switch
(w/ Spiral Cable)

Steering Wheel Lower
No.2 Cover

Column Upper Cover

Steering Column Assembly

Column Hole Cover

Sliding Yoke

26 (260, 19)

Column Lower Cover

35 (360, 26)

Intermediate No.2
Shaft

8 (80, 69 in.·lbf)

x 5

4WD:

8 (80, 69 in.·lbf)

Lower LH Finish Panel

Column Hole
Cover No.2

35 (360, 26)

Universal Joint
No.2

8 (80, 69 in.·lbf)

35 (360, 26)

Hood Lock
Release Lever

Column Hole Cover

Intermediate
No.2 Shaft

35 (360, 26)

No.2 Heater to Register Duct

Junction Block No.1

N·m (kgf·cm, ft·lbf) : Specified torque

93113GI9

Exploded view of the steering wheel, air bag module, floor shift steering column and related components—Toyota Tacoma

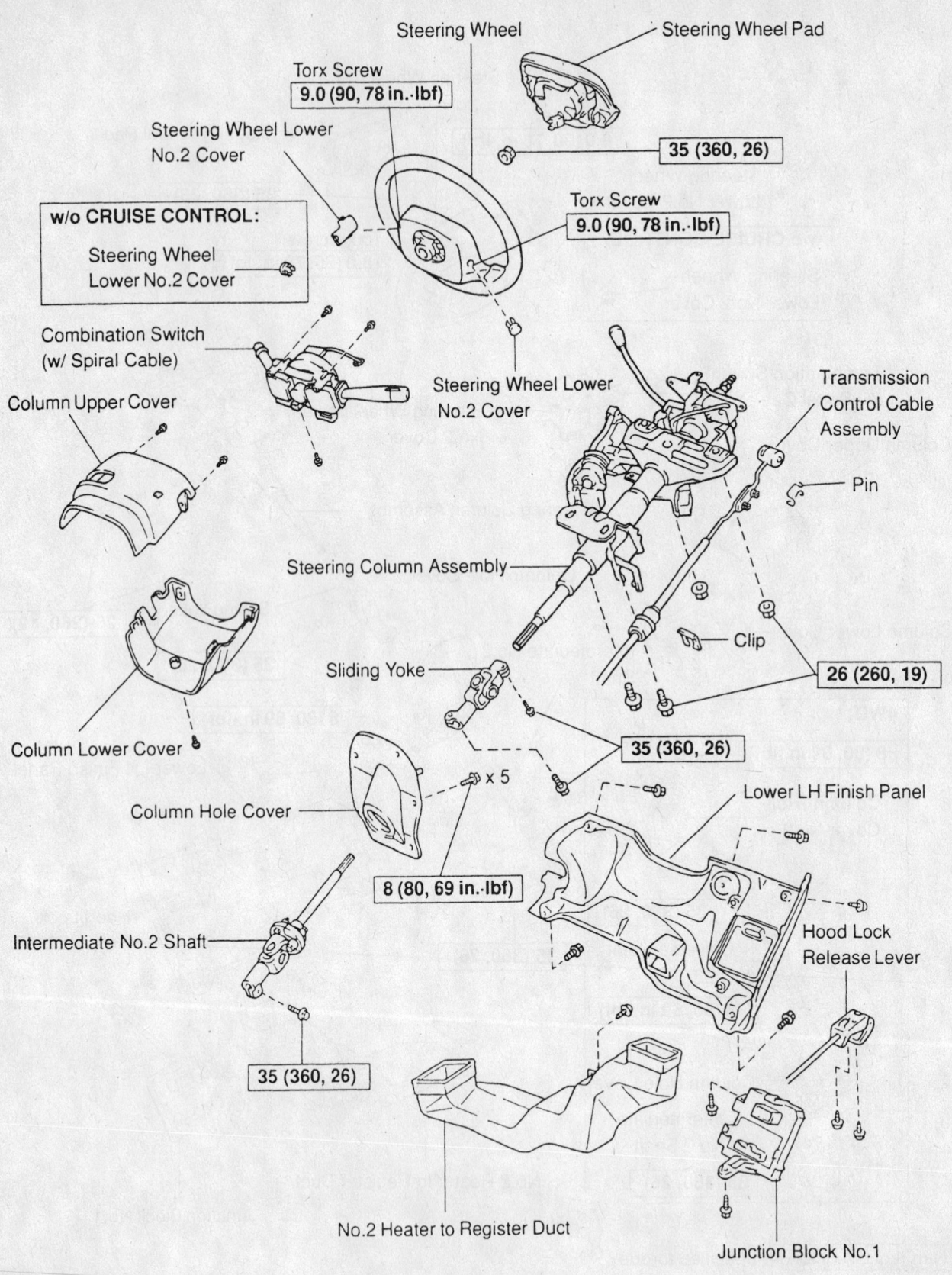

Steering Wheel

Steering Wheel Pad

Torx Screw
9.0 (90, 78 in.·lbf)

Steering Wheel Lower
No.2 Cover

35 (360, 26)

w/o CRUISE CONTROL:

Steering Wheel
Lower No.2 Cover

Torx Screw
9.0 (90, 78 in.·lbf)

Combination Switch
(w/ Spiral Cable)

Steering Wheel Lower
No.2 Cover

Transmission
Control Cable
Assembly

Column Upper Cover

Pin

Steering Column Assembly

Clip

Column Lower Cover

Sliding Yoke

35 (360, 26)

26 (260, 19)

Column Hole Cover

x 5

Lower LH Finish Panel

8 (80, 69 in.·lbf)

Hood Lock
Release Lever

Intermediate No.2 Shaft

35 (360, 26)

No.2 Heater to Register Duct

Junction Block No.1

N·m (kgf·cm, ft·lbf) : Specified torque

93113GI0

Exploded view of the steering wheel, air bag module, column shift steering column and related components

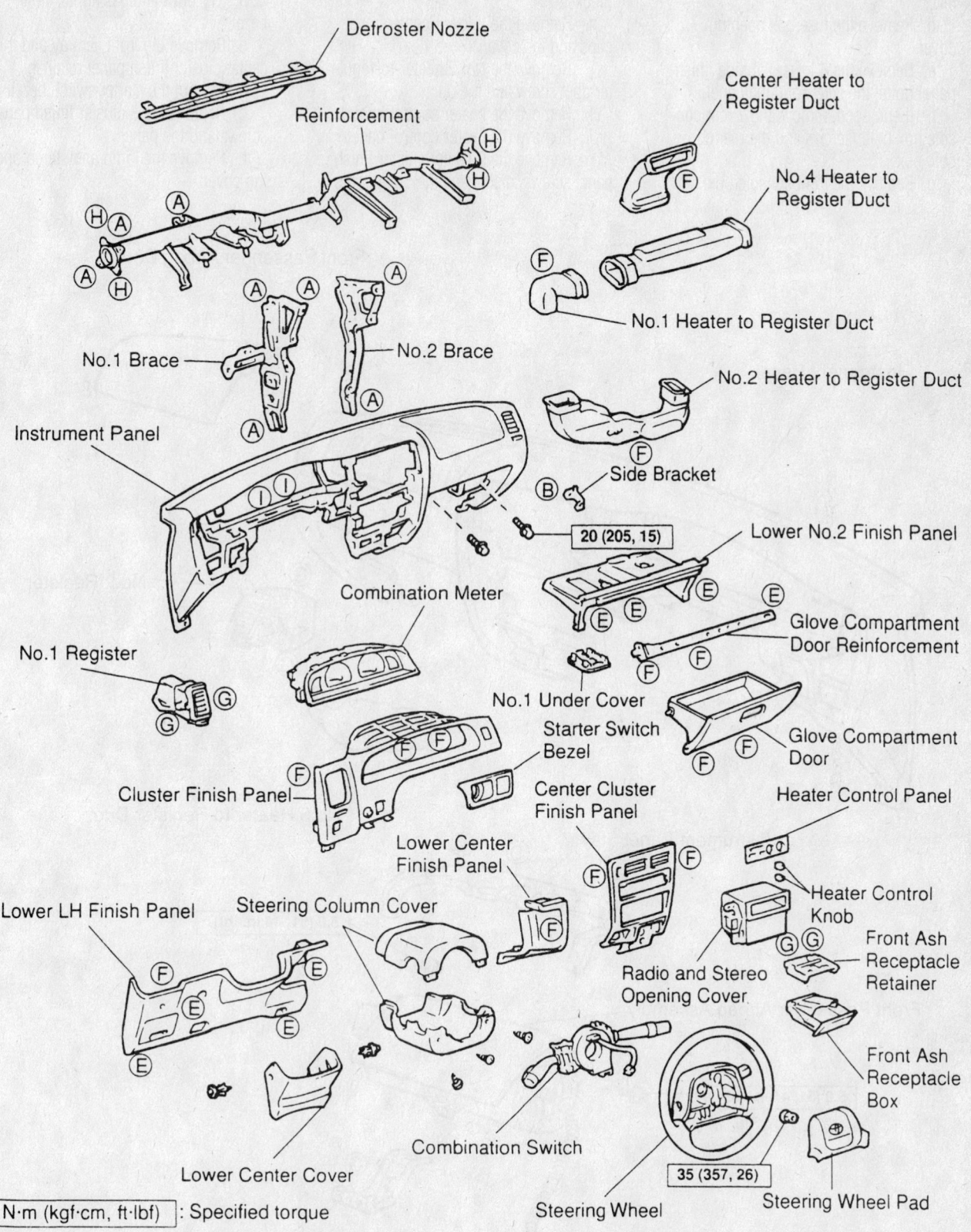

Defroster Nozzle

Reinforcement

Center Heater to Register Duct

No.4 Heater to Register Duct

No.1 Heater to Register Duct

No.1 Brace

No.2 Brace

No.2 Heater to Register Duct

Instrument Panel

Side Bracket

20 (205, 15)

Lower No.2 Finish Panel

Combination Meter

Glove Compartment Door Reinforcement

No.1 Register

No.1 Under Cover

Starter Switch Bezel

Glove Compartment Door

Cluster Finish Panel

Center Cluster Finish Panel

Heater Control Panel

Lower Center Finish Panel

Heater Control Knob

Lower LH Finish Panel

Steering Column Cover

Radio and Stereo Opening Cover

Front Ash Receptacle Retainer

Front Ash Receptacle Box

Combination Switch

35 (357, 26)

Lower Center Cover

Steering Wheel

Steering Wheel Pad

N·m (kgf·cm, ft·lbf) : Specified torque

93113GJ1

Exploded view of the instrument panel and related components

c. Remove the 2 hood lock release lever screws and the hood lock release lever.

d. Remove the fuse box opening cover.

e. Remove the 4 lower left side finish panel bolts, the screw and the panel.

f. If equipped, remove the 2 rear console box bolts/screws and the rear console box.

g. Remove the front console box.

h. Remove the 2 upper console box mounting bracket screws and the bracket.

i. Remove the 2 lower center cover clips and the cover.

j. Remove the No. 2 heater-to-register duct screw and the duct.

k. Remove the heater control knobs.

l. Remove the heater control panel.

m. Remove the 2 center cluster finish panel screws, disengage the 5 clips, dis-

connect the electrical connectors and remove the panel.

n. Pry out the cigar lighter hole bezel.

o. Remove the front ashtray and the center cluster finish panel retainer.

p. Pry out the starter switch bezel.

q. Remove the 3 cluster finish panel screws and the panel.

r. Remove the radio and stereo opening cover.

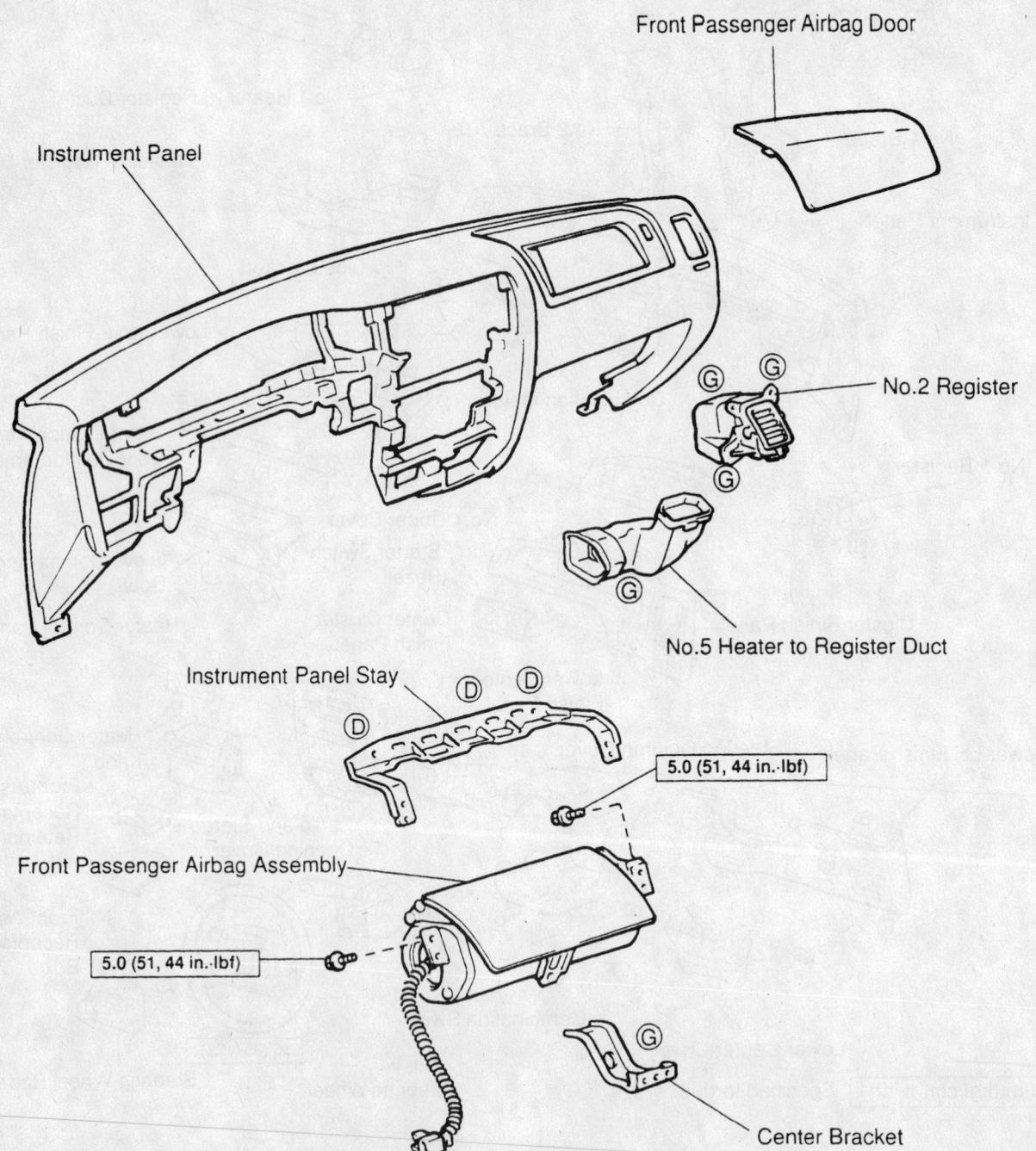

Exploded view of the instrument panel air bag module, ventilation components and brackets

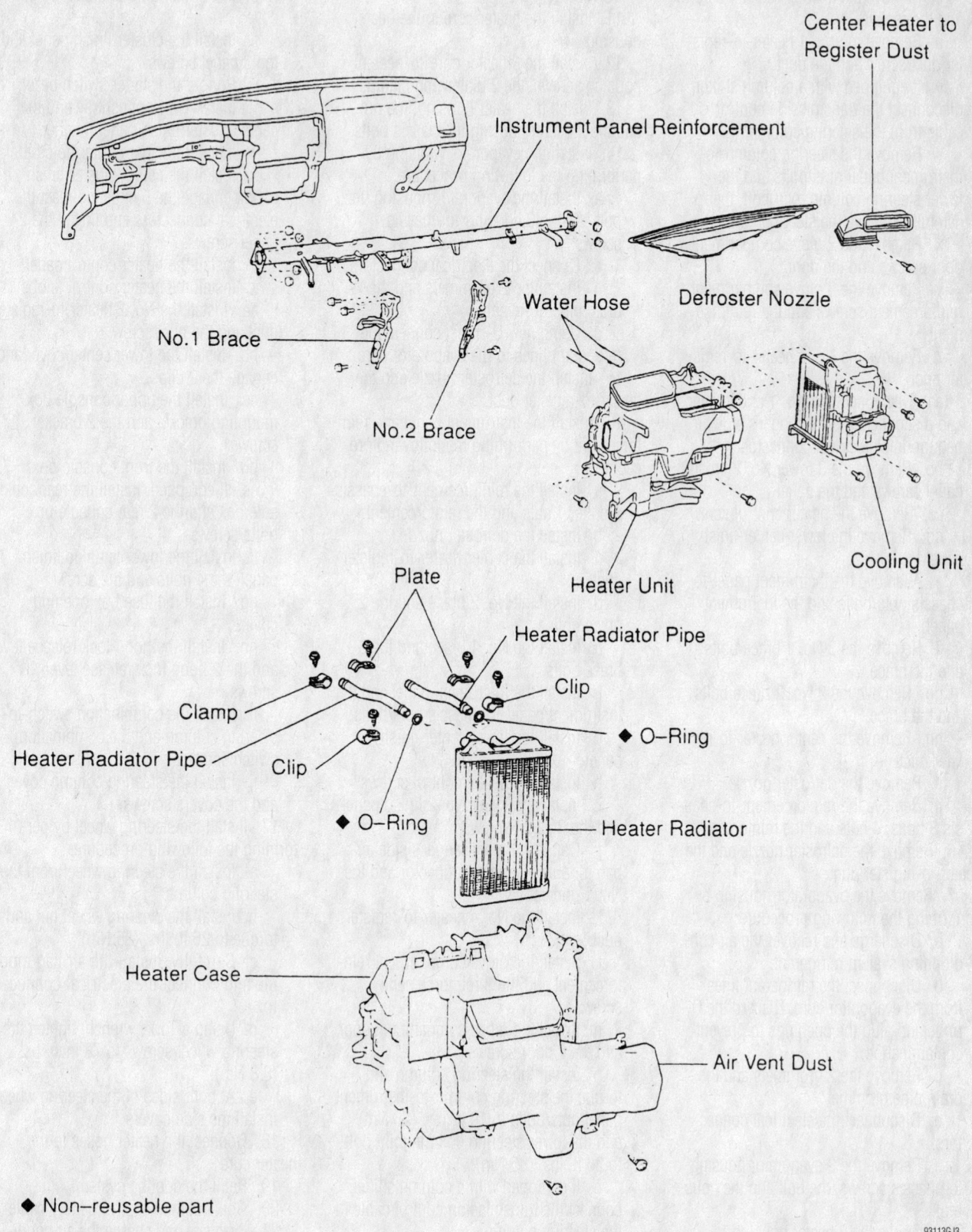

Center Heater to
Register Dust

Instrument Panel Reinforcement

Defroster Nozzle

Water Hose

No.1 Brace

No.2 Brace

Cooling Unit

Heater Unit

Plate

Heater Radiator Pipe

Clamp

Clip

Heater Radiator Pipe

Clip

◆ O–Ring

◆ O–Ring

Heater Radiator

Heater Case

Air Vent Dust

◆ Non–reusable part

93113GJ3

Exploded view of the front heater core, heater housing, evaporator housing and related components

s. Remove the combination meter and disconnect the electrical connectors.

t. Remove the 2 No. 1 register screws and the register.

u. Remove the No. 1 heater-to-register duct screw and the duct.

v. If equipped with a column shifter, disconnect the transmission control cable from the steering column.

w. Remove the steering column-to-instrument panel nuts/bolts and the lower steering column joint bolt; then, carefully, remove the steering column.

x. Remove the 2 glove compartment door screws and the door.

y. Remove the 3 glove compartment reinforcement screws and the reinforcement.

z. Remove the No. 4 heater-to-register duct.

aa. Disconnect the No. 1 undercover and disconnect the passenger's side air bag module electrical connector.

bb. Remove the 3 lower No. 2 finish panel screws and the panel.

cc. Remove the heater control screws.

dd. Remove the lower center finish panel.

ee. Remove the instrument panel-to-chassis nuts/bolts and the instrument panel.

ff. Remove the 3 No. 1 brace bolts and the brace.

gg. Remove the 2 No. 2 brace bolts and the brace.

hh. Remove the center heater-to-register duct.

ii. Remove the defroster nozzle.

jj. Remove the reinforcement-to-chassis 3 bolts, 4 nuts and the reinforcement.

6. Remove the defroster nozzle and the heater-to-register duct.

7. Remove the evaporator housing by performing the following procedure:

a. Discharge and recover the air conditioning system refrigerant.

b. Disconnect the refrigerant lines from the evaporator core. Discard the O-rings and plug the openings to prevent contamination.

c. Remove the 2 grommets and the drain pipe grommet.

d. Disconnect the electrical connectors.

e. Remove the 3 evaporator housing-to-chassis screws, the bolt and the housing.

8. Remove the 2 heater housing-to-chassis bolts, the nut and the heater housing.

9. Remove the 3 heater core-to-heater housing screws, the 2 plates and clamp.

10. Remove the heater core from the heater housing.

To install:

11. Install the heater core to the heater housing.

12. Install the 3 heater core-to-heater housing screws, the 2 plates and clamp.

13. Install the heater housing, the nut and the 2 heater housing to-chassis bolts.

14. Install the evaporator housing by performing the following procedure:

a. Install the evaporator housing, the bolt and the 3 housing-to-chassis screws.

b. Connect the electrical connectors.

c. Install the 2 grommets and the drain pipe grommet.

d. Using new O-rings, connect the refrigerant lines to the evaporator core.

15. Install the defroster nozzle and the heater-to-register duct.

16. Install the instrument panel and reinforcement by performing the following procedure:

a. Install the reinforcement-to-chassis 3 bolts, 4 nuts and the reinforcement.

b. Install the defroster nozzle.

c. Install the center heater-to-register duct.

d. Install the No. 2 brace and the 2 brace bolts.

e. Install the No. 1 brace and the 3 brace bolts.

f. Install the instrument panel and the instrument panel-to-chassis nuts/bolts.

g. Install the lower center finish panel.

h. Install the heater control screws.

i. Install the lower No. 2 finish panel and the 3 panel screws.

j. Connect the passenger's side air bag module electrical connector and the No. 1 undercover.

k. Install the No. 4 heater-to-register duct.

l. Install the glove compartment reinforcement and the 3 reinforcement screws.

m. Install the glove compartment door and the 2 door screws.

n. Install the steering column and torque the steering column-to-instrument panel nuts/bolts to 19 ft. lbs. (26 Nm) and the lower steering column joint bolt to 26 ft. lbs. (35 Nm).

o. If equipped with a column shifter, connect the transmission control cable to the steering column.

p. Install the No. 1 heater-to-register duct and the duct screw.

q. Install the No. 1 register and the 2 register screws.

r. Install the combination meter and connect the electrical connectors.

s. Install the radio and stereo opening cover.

t. Install the cluster finish panel and the 3 panel screws.

u. Pry out the starter switch bezel.

v. Install the front ashtray and the center cluster finish panel retainer.

w. Install the cigar lighter hole bezel.

x. Install the center cluster finish panel, engage the 5 clips, connect the electrical connectors and install the 2 panel screws.

y. Install the heater control panel.

z. Install the heater control knobs.

aa. Install the No. 2 heater-to-register duct and the duct screw.

bb. Install the lower center cover and engage the 2 clips.

cc. Install the upper console box mounting bracket and the 2 bracket screws.

dd. Install the front console box.

ee. If equipped, install the rear console box and the 2 rear console box bolts/screws.

ff. Install the lower left side finish panel, the 4 bolts and the screw.

gg. Install the fuse box opening cover.

hh. Install the hood lock release lever and the 2 hood lock release lever screws.

ii. Install the combination switch-to-steering column and the combination switch screws.

jj. Install the steering column cover and the covers screws.

17. Install the steering wheel by performing the following procedure:

a. Install the steering wheel from the steering column.

b. Install the steering wheel nut and torque to 26 ft. lbs. (35 Nm).

c. Carefully, install the air bag module and connect the electrical connector.

d. Using a Torx® wrench, tighten the steering wheel screws to 78 inch lbs. (8.8 Nm).

e. At both sides of the steering wheel, install the side covers.

18. Connect the heater hoses to the heater core.

19. Refill the cooling system.

20. Connect the negative battery cable.

21. Evacuate and charge the air conditioning system refrigerant.

22. Run the engine to normal operating temperatures; then, check the climate control operation and check for leaks.

Water Pump

REMOVAL & INSTALLATION

2.4L (2RZ-FE) and 2.7L (3RZ-FE) Engines

1. Before servicing the vehicle, refer to the precautions in the beginning of this section.
2. Remove or disconnect the following:
   - Negative battery cable
   - Engine undercover
3. Drain the cooling system.
4. Remove or disconnect the following:
   - 2 bolts and the air pipe, for the California vehicles with 3RZ-FE engine
   - Upper radiator hose from the radiator
   - Oil dipstick guide, by removing the bolt
   - Power steering drive belt, by loosening the lockbolt and adjusting bolt to the idler pulley, if equipped with power steering
   - No. 2 fan shroud, by removing the 2 clips
   - No. 1 fan shroud, by removing the 4 bolts
   - Loosen the idler pulley nut and adjusting bolt and remove the air conditioning drive belt, if equipped with air conditioning
5. Remove the alternator drive belt, fan (with fan clutch), water pump pulley and the fan shroud, as follows:
   - Stretch the belt and loosen the water pump pulley mounting nuts
   - Loosen the lock, pivot and the adjusting bolts for the alternator
   - Alternator drive belt
   - 4 water pump pulley mounting nuts
   - Fan (with fan clutch) and the water pump pulley
6. Remove the water pump and discard the gasket.

To install:

7. Clean all gasket mounting surfaces.
8. Install or connect the following:
   - Apply a thin layer of liquid sealant to a new gasket
   - Place the gasket and water pump into position. Tighten the 14mm head bolts **A** to 18 ft. lbs. (25 Nm) and the 12mm head bolts to 78 inch lbs. (9 Nm).
9. Install the water pump pulley, fan shroud, fan (with fan clutch) and the alternator drive belt, as follows:
   - Fan (with the fan clutch), water pump pulley and the fan shroud in position
   - Water pump pulley mounting nuts but do not tighten the nuts at this time
   - Alternator drive belt
   - Stretch the alternator belt tight. Tighten the fan nuts to 16 ft. lbs. (21 Nm).
   - Adjust the alternator drive belt
10. Install or connect the following:
    - Adjust the drive belt, if equipped with air conditioning
    - No. 1 fan shroud, by installing the 4 bolts
    - No. 2 fan shroud, with the 2 clips
    - Adjust the power steering drive belt
    - Oil dipstick guide, with the bolt
    - Upper radiator hose to the radiator
    - Air pipe, If removed
    - Negative battery cable
11. Fill and bleed the cooling system.
12. Start the engine and check for leaks.
13. Install the engine undercover.

3.4L (5VZ-FE) Engine

1. Before servicing the vehicle, refer to the precautions in the beginning of this section.

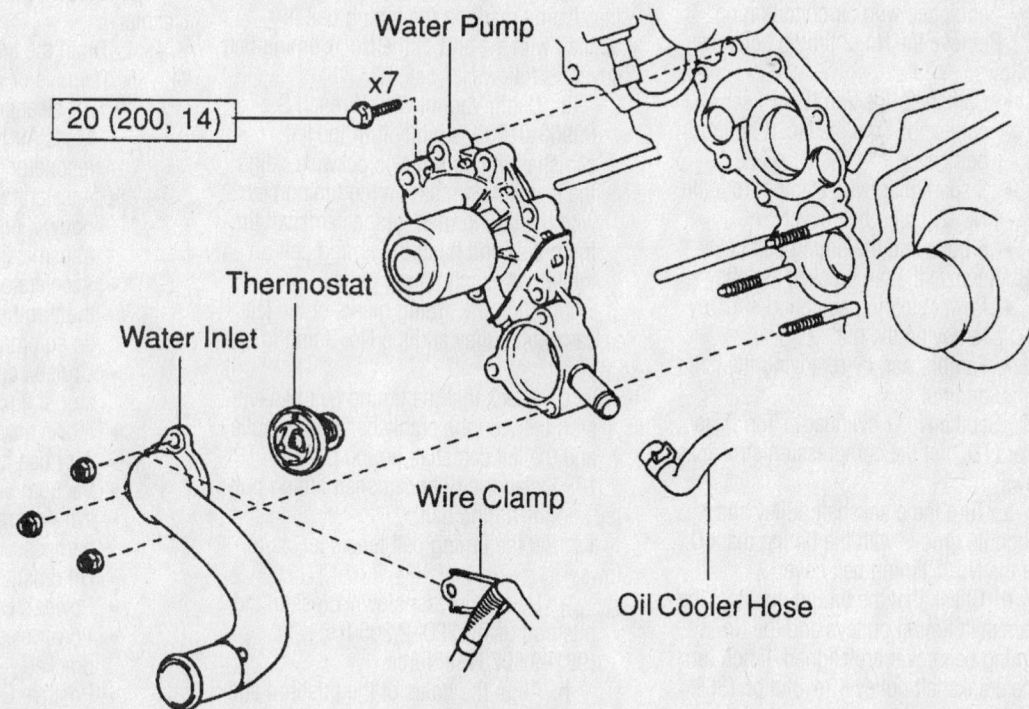

Exploded view of the water pump mounting—3.4L (5VZ-FE) engine

N·m(kgf·cm, ft·lbf) : Specified torque
◆ Non–Reusable part

7924YG08

2. Remove or disconnect the following:
- Negative battery cable
- Engine undercover

3. Drain the engine coolant.

4. Remove the upper radiator hose.

5. Remove the power steering drive belt, as follows:
- Stretch the belt and loosen the fan pulley mounting nuts
- Loosen the lockbolt, pivot bolt and the adjusting bolt
- Drive belt

6. Remove or disconnect the following:
- Air conditioning drive belt, by loosening the idler pulley nut and adjusting bolt
- Lockbolt, pivot bolt and the adjusting bolt
- Alternator drive belt
- No. 2 fan shroud, by removing the 2 clips
- Fan with the fluid coupling and fan pulleys
- Power steering pump and move it aside without disconnecting the lines from the pump
- Compressor from the engine and move it aside without disconnecting the compressor lines, if equipped with air conditioning
- Air conditioning bracket, if equipped with air conditioning

7. Remove the No. 2 timing belt cover, as follows:
- Camshaft Position (CMP) sensor connector from the No. 2 timing belt cover
- 3 spark plug wire clamps from the No. 2 timing belt cover
- 6 bolts and the timing belt cover

8. Remove the fan bracket, as follows:
- Power steering adjusting strut, by removing the nut
- Fan bracket, by removing the bolt and nut

9. Set the No. 1 cylinder to Top Dead Center (TDC) of the compression stroke, as follows:

  a. Turn the crankshaft pulley and align its groove with the timing mark **0** of the No. 1 timing belt cover.

  b. Check that the timing marks of the camshaft timing pulleys and the No. 3 timing belt cover are aligned. If not, turn the crankshaft pulley 1 revolution (360 degrees).

10. Remove the camshaft timing pulleys, as follows:

  a. Remove the timing belt tensioner by alternately loosening the 2 bolts.

  b. Using Variable Wrench Set No.

09960-10010, remove the pulley bolt, the timing pulley and the knock pin.

  c. Remove the 2 timing pulleys with the timing belt.

11. Remove or disconnect the following:
- Thermostat
- No. 2 oil cooler hose, from the water pump
- Water pump, by removing the 7 bolts

12. Thoroughly clean the mating surfaces.

To install:

13. Install or connect the following:
- Apply sealant (PN 08826-00100) to the water pump

❄❄ WARNING

Parts must be assembled within 5 minutes of application. Otherwise the material must be removed and reapplied.

- Water pump. Tighten the bolts to 14 ft. lbs. (20 Nm).
- No. 2 oil cooler hose
- Thermostat
- Left camshaft timing pulley. Tighten the pulley bolt to 81 ft. lbs. (110 Nm).

14. Set the No. 1 cylinder to TDC of the compression stroke.

15. Connect the timing belt to the left camshaft timing pulley. Check that the installation mark on the timing belt is aligned with the end of the No. 1 timing belt cover, as follows:

  a. Using Variable Pin Wrench Set 09960-01000, slightly turn the left camshaft timing pulley clockwise. Align the installation mark on the timing belt with the timing mark of the camshaft timing pulley and hang the timing belt on the left camshaft timing pulley.

  b. Align the timing marks of the left camshaft pulley and the No. 3 timing belt cover.

  c. Check that the timing belt has tension between the crankshaft timing pulley and the left camshaft timing pulley.

16. Install the right camshaft timing pulley and the timing belt.

17. Set the timing belt tensioner, as follows:

  a. Using a press, slowly press in the pushrod using 220–2,205 lbs. (981–9,807 N) of force.

  b. Align the holes of the pushrod and housing, pass a 1.5mm hexagon wrench through the holes to keep the setting position of the pushrod.

  c. Release the press and install the dust boot to the tensioner.

  d. Install the timing belt tensioner and alternately tighten the bolts to 20 ft. lbs. (28 Nm).

  e. Using pliers, remove the 1.5mm hexagon wrench from the belt tensioner.

18. Check the valve timing, as follows:

  a. Slowly turn the crankshaft pulley 2 revolutions from the TDC-to-TDC; always turn the crankshaft pulley clockwise.

  b. Check that each pulley aligns with the timing marks. If the timing marks do not align, remove the timing belt and reinstall it.

19. Install or connect the following:
- Fan bracket, with the bolt and nut
- Remaining components
- Negative battery cable

20. Fill with engine coolant.

21. Start the engine and check for leaks.

Cylinder Head

REMOVAL & INSTALLATION

2.4L (2RZ-FE) and 2.7L (3RZ-FE) Engines

1. Before servicing the vehicle, refer to the precautions in the beginning of this section.

2. Release the fuel system pressure.

3. Disconnect the negative battery cable.

4. Drain the engine coolant.

5. Remove or disconnect the following:
- Air cleaner cap
- Mass Air Flow (MAF) meter and resonator
- Accelerator cable from the throttle body, if equipped with a manual transmission
- Accelerator and throttle cables from the throttle body, if equipped with an automatic transmission
- Cruise control cable from the actuator, if equipped with cruise control
- Intake air connector
- Air hose for Idle Air Control (IAC)
- Vacuum sensing hose
- Wire clamp for the engine wiring harness
- Oil dipstick guide
- Power steering belt
- Power steering pulley, pump and bracket
- Positive Crankcase Ventilation (PCV) hoses
- Distributor
- Spark plug wires from the spark plugs
- Engine wiring harness

- Air conditioning compressor, if equipped with air conditioning
- Oil pressure sensor
- Engine Coolant Temperature (ECT) sensor connector
- ECT sender gauge connector
- Exhaust Gas Recirculation (EGR) gas temperature sensor connector
- Vacuum Switching Valve (VSV) connector
- 2 vacuum hose from the VSV
- Ground strap from the cowl top panel
- Engine wiring harness from the air intake chamber
- Throttle Position (TP) sensor connector
- IAC valve connector
- Crankshaft Position (CKP) sensor connector
- Knock Sensor (KS) connector
- Data Link Connector 1 (DLC1) from the bracket
- Engine wiring harness clamp
- EGR pipe
- Intake chamber stay
- Air intake chamber assembly

6. Disconnect the following hoses:
- Evaporative Emissions (EVAP) hose from the throttle body
- Brake booster vacuum hose from the union
- Water bypass hose from the water bypass pipe
- Water bypass hose from the cylinder head rear cover

7. Remove or disconnect the following:
- Injector connectors
- Fuel inlet pipe

- Hoses and the fuel return pipe
- Delivery pipe and injectors
- Intake manifold
- Front exhaust pipe
- Exhaust manifold and gasket
- Water outlet
- Cylinder head rear cover
- Spark plugs
- Front engine hanger
- Engine wiring harness brackets
- Cylinder head cover

8. Set No. 1 cylinder to Top Dead Center (TDC) of the compression stroke. The groove on the crankshaft pulley should align with the **0** mark on the timing chain cover and the timing marks (1 and 2 dots) of the camshaft gears should form a straight line in respect to the cylinder head surface. If not, turn the crankshaft 1 revolution (360 degrees).

9. Remove or disconnect the following:
- Chain tensioner and gasket
- Camshaft timing gear
- Exhaust camshafts

10. Remove the intake camshaft, as follows:

a. Uniformly, loosen and remove the bearing cap bolts in the reverse order of the tightening in several passes, in sequence.

b. Remove the bearing caps and camshaft. Make a note of the bearing cap positions for proper installation.

➡**If the camshaft is not being lifted out straight and level, reinstall the No. 3 bearing cap with the 2 bolts. Then, alternately loosen and remove the 2 bearing cap bolts with the camshaft gear pulled up.**

11. Remove or disconnect the following:
- Valve lifters and shims

➡**Arrange the valve lifters and shims in correct order.**

- Cylinder head, by uniformly loosen and remove the cylinder head bolts in the reverse order of the tightening, in sequence, using several passes

To install:

12. Before installing, thoroughly clean the gasket mating surfaces and check for warpage.

13. Apply sealant (PN 08826-00080) to the 2 locations. Place a new head gasket on the block and install the cylinder head.

14. Install the cylinder head as follows:

a. Lightly coat the cylinder head bolts with engine oil.

b. Install the bolts and tighten, in several passes, in the sequence. Tighten all bolts to 29 ft. lbs. (39 Nm).

c. Mark the front of the bolt with paint and retighten bolts 90 degrees in the proper sequence.

d. Retighten an additional 90 degrees. Check that the painted mark is now facing rearward.

15. Install or connect the following:
- Tighten the 2 front mounting bolts to 15 ft. lbs. (21 Nm).
- Valve lifters and shims in their proper locations. Check that the valve lifter rotates smoothly by hand.
- Intake and exhaust camshafts

16. Set No. 1 cylinder to TDC compression stroke. The groove on the crankshaft pulley should align with the **0** mark on the timing chain cover and the timing marks (1 and 2 dots) of the camshaft gears should form a straight line in respect to the cylinder head surface. If not, turn the crankshaft 1 revolution (360 degrees).

17. Install the timing gear, as follows:

a. Place the gear over the straight pin of the intake camshaft.

b. Hold the intake camshaft with a wrench. Install and tighten the bolt to 54 ft. lbs. (74 Nm).

c. Hold the exhaust camshaft and install the bolt and distributor gear. Tighten the bolt to 34 ft. lbs. (46 Nm).

18. Install or connect the following:
- Chain tensioner, using a new gasket (mark toward the front)
- Recheck the valve timing
- Check and adjust the valve clearance
- Spark plugs
- Semi-circular plug

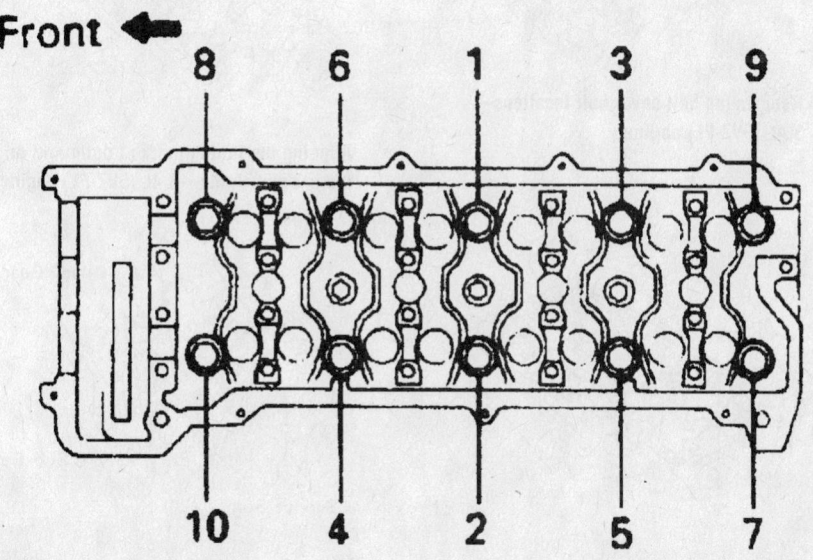

Cylinder head bolt tightening sequence—2.4L (2RZ-FE) and 2.7L (3RZ-FE) engines

19. Recheck the engine for proper valve timing.

20. Install or connect the following:
- Cylinder head cover, using a new gasket
- Engine wiring harness brackets
- Front engine hanger. Tighten the bolts to 30 ft. lbs. (42 Nm).
- Cylinder head rear cover. Tighten the bolts to 10 ft. lbs. (13 Nm).
- Water outlet, using a new gasket. Tighten the bolts to 14 ft. lbs. (20 Nm).
- Upper radiator hose
- Exhaust manifold. Tighten the bolts to 36 ft. lbs. (49 Nm).
- Remaining components
- Negative battery cable

21. Fill the engine and radiator with engine coolant.

22. Start the engine and check for leaks.

23. Check the ignition timing. Road test the vehicle for proper operation.

24. Recheck all fluid levels.

3.4L (5VZ-FE) Engine

1. Before servicing the vehicle, refer to the precautions in the beginning of this section.

2. Disconnect the negative battery cable.

3. Relieve the fuel system pressure.

4. Remove the engine undercover.

5. Drain the cooling system.

6. Remove or disconnect the following:
- Front exhaust pipe
- Air cleaner cap
- Mass Air Flow (MAF) meter and resonator

7. Disconnect the following cables:
- Actuator cable from the bracket, if equipped with cruise control
- Accelerator cable
- Throttle cable, if equipped with an automatic transmission
- Heater hose
- Upper radiator hose
- Power steering drive belt
- Air conditioning drive belt, by loosening the idle pulley nut and adjusting bolt
- Loosen the lockbolt, pivot bolt and adjusting bolt and the alternator drive belt
- No. 2 fan shroud by removing the 2 clips
- Fan with the fluid coupling and fan pulleys
- Power steering pump and move it aside without disconnecting the pump lines
- Compressor and move it aside without disconnecting the compressor lines, if equipped with air conditioning
- Air conditioning bracket, if equipped with air conditioning
- Spark plug wires with the ignition coils
- Spark plugs
- No. 2 timing belt cover

8. Remove the fan bracket, as follows:

a. Remove the power steering adjusting strut by removing the nut.

b. Remove the fan bracket by removing the bolt and nut.

9. Set the No. 1 cylinder at Top Dead Center (TDC) of the compression stroke.

a. Turn the crankshaft pulley and align its groove with the timing mark **0** on the No. 1 timing belt cover.

b. Check that the timing marks of the camshaft timing pulleys and the No. 3 timing belt cover are aligned. If not, turn the crankshaft pulley 1 revolution (360 degrees).

10. Remove the timing belt tensioner by alternately loosening the 2 bolts.

11. Remove the camshaft timing pulleys, as follows:

a. Using Variable Pin Wrench Set tool 09960-10010, remove the pulley bolt, the timing pulley and the knock pin.

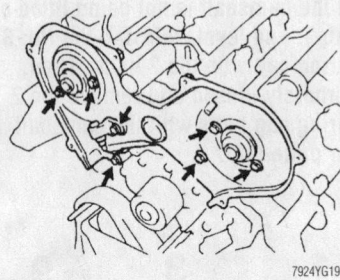

Rear timing belt cover bolt locations—3.4L (5VZ-FE) engine

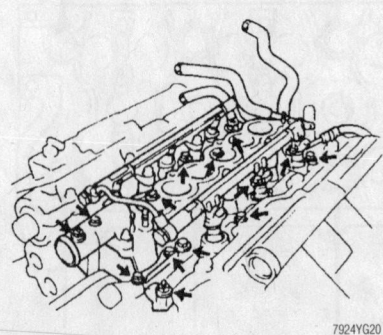

Intake manifold bolts and nuts locations—3.4L (5VZ-FE) engine

b. Remove the 2 timing pulleys with the timing belt.

12. Remove or disconnect the following:
- Bolt and the No. 2 idler pulley
- Alternator

13. Remove the Exhaust Gas Recirculation (EGR) pipe and 2 gaskets, if equipped with an EGR valve

14. Remove the intake chamber stay as follows:

a. Remove the oil filler tube and No. 1 throttle cable clamp by removing the bolt and 2 nuts.

b. Remove the intake chamber stay by removing the 2 bolts.

15. Remove the following connectors:
- VSV connector for the fuel pressure control.
- Throttle position sensor
- IAC valve connector

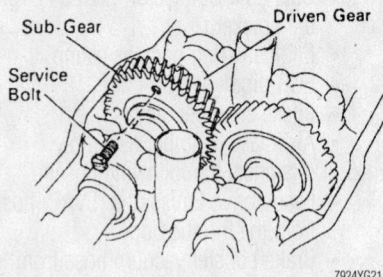

Drive gear service bolt (right side)—3.4L (5VZ-FE) engine

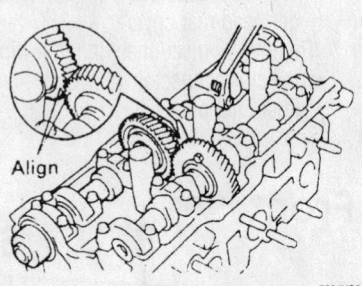

Aligning the timing mark (1 dot mark) of the left camshafts—3.4L (5VZ-FE) engine

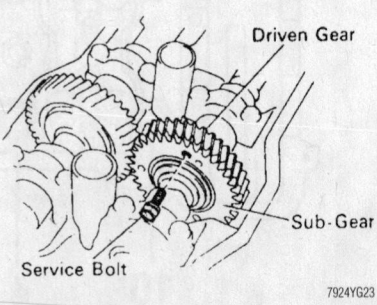

Drive gear service bolt (left side)—3.4L (5VZ-FE) engine

- EGR valve gas temperature sensor, if equipped
- Vacuum Switching Valve (VSV) connector for the EGR valve, if equipped

16. Disconnect the following hoses:
- Positive Crankcase Ventilation (PCV) hoses
- Water bypass hoses.
- Air assist hose from the intake air connector
- 2 vacuum sensing hoses from the VSV
- Evaporative Emissions (EVAP) hose
- Air hose, from the power steering
- Air hose from the air conditioning idle up valve, if equipped with air conditioning

17. Remove or disconnect the following:
- 4 bolts, 2 nuts and the air intake chamber assembly
- Intake air connector

18. Disconnect the engine wiring harness from the intake manifold, as follows:
- Oil pressure sensor connector
- Crankshaft position sensor connector
- 6 injector connectors
- Engine Coolant Temperature (ECT) sender gauge connector
- ECT sensor connector
- Knock (KS) sensor connector
- Camshaft Position (CMP) sensor connector
- 3 engine wiring harness clamps
- 3 bolts and the engine wiring harness from the cylinder head

19. Remove or disconnect the following:
- CMP sensor
- No. 3 (rear) timing belt cover, by removing the 6 bolts
- Fuel pressure regulator
- Intake manifold assembly
- Power steering pump bracket
- Oil dipstick and guide
- Exhaust crossover pipe and gaskets, by removing the 6 nuts
- Left-hand exhaust manifold, by removing the heat insulator and 6 nuts
- Right-hand exhaust manifold, by removing the heat insulator and 6 nuts
- 8 bolts, seal washers, cylinder head cover and gasket
- Both cylinder head covers
- Semi-circular plugs
- Right exhaust camshafts
- Right-hand intake camshaft
- Left exhaust camshafts
- Left-hand intake camshaft
- Valve lifters and shims from the cylinder head; arrange the valve lifters and shims in correct order

20. Remove the cylinder heads, as follows:
- Bolt and disconnect the ground strap
- Cylinder head (recessed head) bolt on each cylinder head, using an 8mm hexagon wrench; then repeat for the other side
- 8 cylinder head (12-pointed head) bolts on each cylinder head, by loosening the bolts in several passes and in the reverse order of the tightening sequence
- 16 cylinder head bolts and plate washers.
- Cylinder head

To install:

21. Clean all surfaces.
22. Install or connect the following:
- New cylinder head gaskets
- Cylinder heads

23. Apply a light coat of engine oil on the threads and under the heads of the cylinder head bolts.

24. Tighten the cylinder head bolts, in sequence, using several passes, as follows:
- Step 1: 25 ft. lbs. (34 Nm)
- Step 2: Mark the front of the cylinder head bolt with paint
- Step 3: An additional 90 degrees
- Step 4: Check that the painted mark

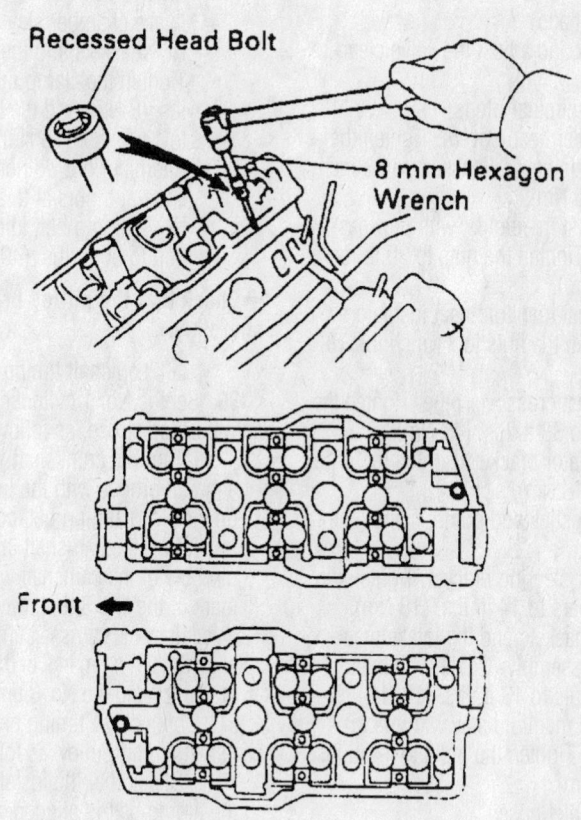

Cylinder head recessed bolts—3.4L (5VZ-FE) engine

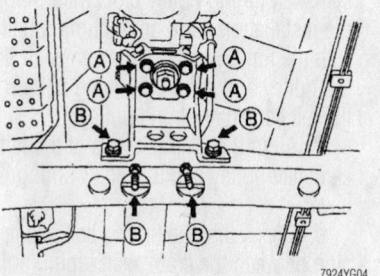

Cylinder head torque sequence—3.4L (5VZ-FE) engine

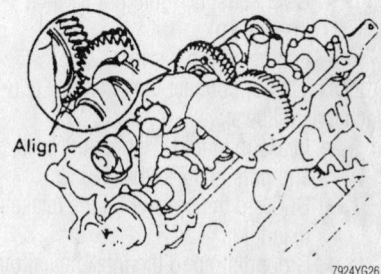

Aligning the right camshafts for installation—3.4L (5VZ-FE) engine

is now at a 90 degrees angle to the front

25. Install the recessed head cylinder head bolts, as follows:
- Apply a light coat of engine oil on the threads and under the heads of the cylinder head bolts
- Cylinder head bolt on each cylinder head, using a 8mm hexagon wrench; then, repeat for the other side, as shown. Tighten the bolts to 13 ft. lbs. (18 Nm).
- Bolt and the ground strap

26. Install or connect the following:
- Valve lifters and shims

➡ **Check that the valve lifter rotates smoothly by hand.**

- Camshafts
- Check and adjust the valve clearance
- Semi-circular plugs
- Cylinder head covers. Tighten the bolts, in several passes, to 53 inch lbs. (6 Nm).
- Exhaust manifolds, with new gaskets. Tighten the nuts to 30 ft. lbs. (40 Nm).
- Exhaust manifold heat insulators. Tighten the nuts to 71 inch lbs. (8 Nm).
- Exhaust crossover pipe. Tighten the nuts to 33 ft. lbs. (45 Nm).
- Alternator bracket. Tighten to 14 ft. lbs. (18 Nm).
- Oil dipstick and guide, using a new O-ring
- Power steering bracket. Tighten the fasteners to 14 ft. lbs. (18 Nm).
- New gaskets and the intake manifold assembly. Tighten the bolts and nuts to 13 ft. lbs. (18 Nm).
- Intake manifold stay with the 2 bolts. Tighten the bolts to 14 ft. lbs. (18 Nm).
- Fuel inlet hose
- Fuel pressure regulator
- No. 3 timing belt cover. Tighten the bolts to 80 inch lbs. (9 Nm).
- CMP sensor. Tighten to 71 inch lbs. (8 Nm).
- Engine wiring harness

27. Install or connect the intake air connector, as follows:
- Intake manifold. Tighten the bolts and nuts to 14 ft. lbs. (19 Nm).
- DLC1 to the bracket on the intake manifold
- Ground strap to the intake manifold
- Brake booster vacuum hose to the intake air connector
- 2 fuel return hoses

- Engine wiring harness to the intake manifold
- Idle up valve connector, if equipped with air conditioning

28. Install or connect the following:
- Air intake chamber assembly. Tighten the bolts and nuts to 14 ft. lbs. (18.5 Nm).
- Hoses

29. Attach the following connectors:
- VSV connector for the fuel pressure control
- TP sensor connector
- IAC valve connector
- EGR gas temperature connector, if equipped with an EGR valve
- VSV connector, if equipped with an EGR valve
- Intake chamber stay
- New gaskets and the EGR pipe. Tighten the clamp nuts to 71 inch lbs. (8 Nm) and the EGR pipe nuts to 14 ft. lbs. (18 Nm).
- Alternator but do not tighten the bolts and nuts at this time.
- No. 2 timing belt idler. Tighten the bolt to 30 ft. lbs. (40 Nm).

➡ **Check that the pulley bracket moves smoothly.**

- Left camshaft timing pulley

30. Set the No. 1 cylinder to TDC of the compression stroke, as follows:
a. Turn the crankshaft pulley and align its groove with the timing mark **0** on the No. 1 timing belt cover.
b. Turn the camshaft, align the knock pin hole of the camshaft with the timing mark of the No. 3 timing belt cover.
c. Turn the camshaft timing pulley, align the timing marks of the camshaft timing pulley and the No. 3 timing belt cover.

31. Connect the timing belt to the left camshaft timing pulley, as follows:
a. Check that the installation mark on the timing belt is aligned with the end of the No. 1 timing belt cover.
b. Using Variable Pin Wrench Set 09960-01000, slightly turn the left camshaft timing pulley clockwise. Align the installation mark on the timing belt with the timing mark of the camshaft timing pulley and hang the timing belt on the left camshaft timing pulley.
c. Align the timing marks of the left camshaft pulley and the No. 3 timing belt cover.
d. Check that the timing belt has tension between the crankshaft timing pulley and the left camshaft timing pulley.

32. Install the right camshaft timing pulley and the timing belt, as follows:

a. Align the installation mark on the timing belt with the timing mark of the right camshaft timing pulley, and hang the timing belt on the right camshaft timing pulley with the flange side facing inward.
b. Slide the right camshaft timing pulley on the camshaft. Align the timing marks on the right camshaft timing pulley and the No. 3 timing belt cover.
c. Align the knock pin hole of the camshaft with the knock pin groove of the pulley and install the knock pin. Install the bolt and tighten to 81 ft. lbs. (110 Nm).

33. Set the timing belt tensioner as follows:
a. Using a press, slowly press in the pushrod using 220–2,205 lbs. (981–9,807 N) of force.
b. Align the holes of the pushrod and housing, pass a 1.5mm hexagon wrench through the holes to keep the setting position of the pushrod.
c. Release the press and install the dust boot to the tensioner.

34. Install the timing belt tensioner. Tighten the bolts alternately to 20 ft. lbs. (28 Nm).

35. Using pliers, remove the 1.5mm hexagon wrench from the belt tensioner.

36. Check the valve timing.

37. Install or connect the following:
- Remaining components
- Negative battery cable

38. Fill the radiator with engine coolant.

39. Start the engine and check for leaks.

40. Check the ignition timing.

41. Install the engine undercover.

42. Road test the vehicle.

43. Recheck all fluid levels.

Intake Manifold

REMOVAL & INSTALLATION

2.4L (2RZ-FE) Engine

1. Relieve the fuel system pressure.
2. Disconnect the negative battery cable.
3. Drain the engine coolant.
4. Remove or disconnect the following:
- Air cleaner cap
- Mass Air Flow (MAF) meter and the resonator
- Accelerator cable from the throttle body, if equipped with a manual transaxle
- Accelerator and throttle cables from the throttle body, if equipped with an automatic transaxle

- Intake air connector
- Air conditioning idle-up valve, if equipped with air conditioning
- No. 1 and No. 2 Positive Crankcase Ventilation (PCV) hoses
- Spark plug wires from the spark plugs
- Throttle body
- Air conditioning compressor connector, if equipped with air conditioning
- Oil pressure sensor connector
- Engine Coolant Temperature (ECT) sensor connector

- Exhaust Gas Recirculation (EGR) gas temperature sensor connector
- EGR Vacuum Switching Valve (VSV) connector

5. Disconnect the engine wiring harness, as follows:
- 2 bolts and the harness from the intake chamber
- 5 engine harness clamps and harness

6. Remove or disconnect the following:
- Knock (KS) sensor connector
- Crankshaft Position (CKP) sensor connector

- Fuel pressure control VSV connector
- Data Link Connector 1 (DLC1) from the bracket
- 2 engine wiring harness clamps
- Engine wiring harness
- Fuel injectors
- EGR valve and vacuum modulator
- Intake chamber stay, by removing the 2 bolts
- Fuel return pipe, by removing the hoses and 2 bolts

7. Remove the intake chamber, as follows:

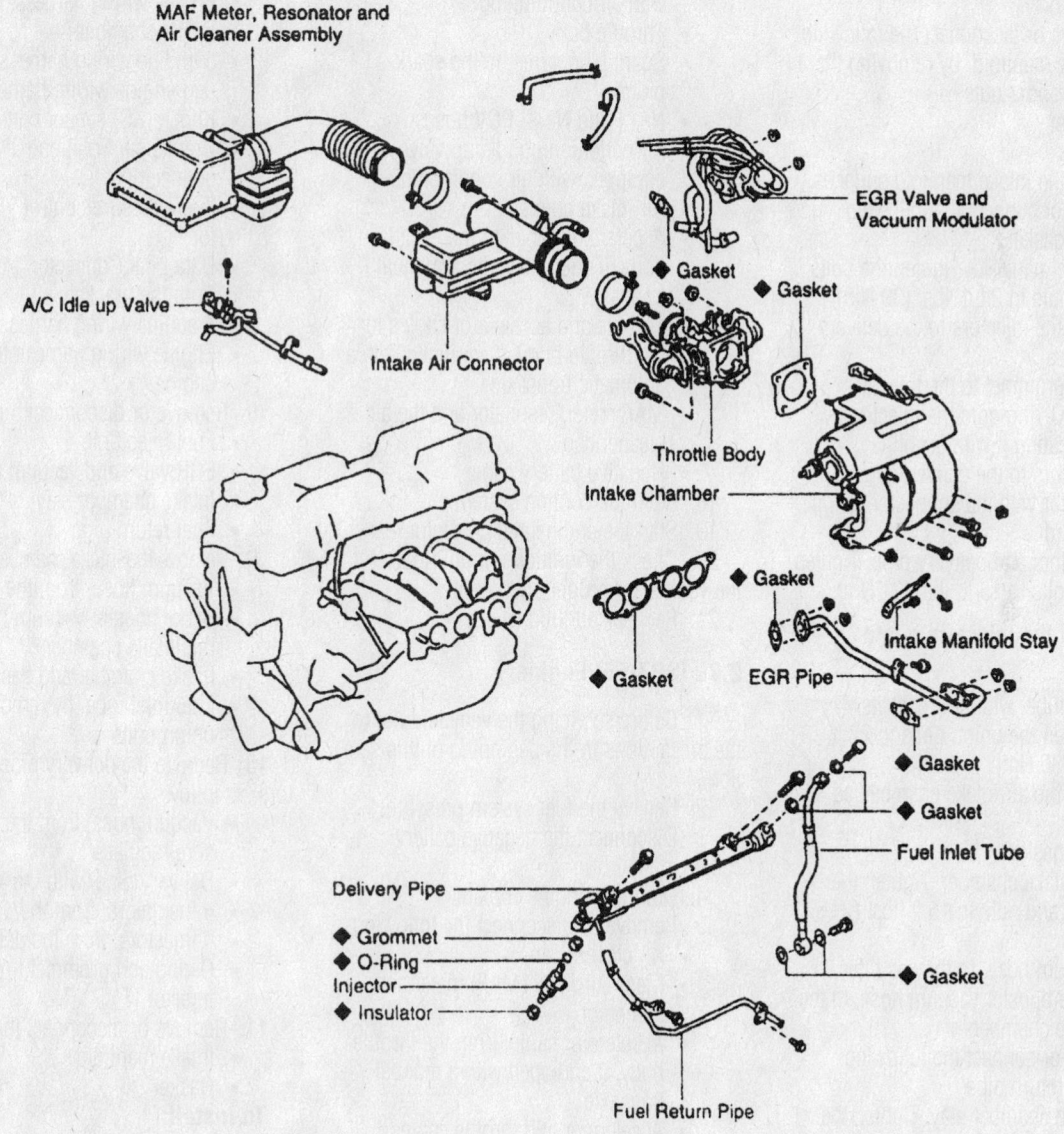

◆ Non-reusable part

7924YG32

Exploded view of the intake manifold assembly—2.4L (2RZ-FE) and 2.7L (3RZ-FE) engines

- Vacuum hose, from the gas filter
- Brake booster vacuum hose, from the intake chamber
- 3 bolts, 2 nuts, air intake chamber and gasket

8. Remove the fuel inlet tube, by removing the union bolts

9. Remove the delivery pipe and injectors, as follows:
  - Vacuum hose, from the fuel pressure regulator
  - Bolts and delivery pipe together with the 4 injectors
  - 4 insulators, from the 4 spacers
  - Injectors from the delivery pipe
  - O-ring and grommet, from each injector

10. Remove or disconnect the following:
  - Intake manifold, by removing the 3 bolts and 2 nuts
  - Gasket

To install:

11. Clean the intake manifold surfaces.

12. Install or connect the following:
  - New gasket
  - Intake manifold. Tighten the bolts and nuts to 22 ft. lbs. (30 Nm).

13. Install the injectors to the delivery pipe, as follows:
  - New grommet to the injector
  - New O-ring onto the injector, by lubricating it with gasoline
  - Injectors to the delivery pipe
  - Injector with the connector facing upward
  - Injectors and delivery pipe. Tighten the bolts to 15 ft. lbs. (21 Nm).

➡**Check that the injectors rotate smoothly.**

  - Fuel tube, with new gaskets. Tighten the union bolts to 22 ft. lbs. (30 Nm).

14. Install the air intake chamber, as follows:
  - New gasket
  - Air intake chamber. Tighten the bolts and nuts to 15 ft. lbs. (21 Nm).
  - Vacuum hose, to the gas filter
  - Brake booster vacuum hose, to the intake chamber

15. Install or connect the following:
  - Fuel return pipe
  - Intake chamber stay. Tighten the bolts to 14 ft. lbs. (20 Nm).
  - EGR valve and vacuum modulator
  - Injector connectors

16. Connect the engine wiring harness to the engine, as follows:
  - Engine wiring harness to the intake manifold

- Engine wiring harness clamps
- DLC1 to the bracket
- Fuel pressure control VSV connector
- KS sensor connector
- CKP sensor connector
- 5 engine wiring harness clamps
- Engine wiring harness, to the intake chamber

17. Install or connect the following:
- EGR VSV connector
- EGR gas temperature sensor connector
- ECT sensor connector
- Oil pressure sensor connector
- Compressor connector, if equipped with air conditioning
- Throttle body
- Spark plug wires, to the spark plugs
- No. 1 and No. 2 PCV hoses
- Air conditioning idle up valve, if equipped with air conditioning
- Air intake connector
- Accelerator cable, to the throttle body, if equipped with a manual transaxle
- Throttle and accelerator cables to the throttle body, if equipped with a automatic transaxle
- MAF meter, resonator and the air cleaner cap
- Negative battery cable

18. Refill the cooling system.

19. Start the engine and check for leaks.

20. Check the ignition timing. Road test the vehicle for proper operation.

21. Recheck all fluid levels.

2.7L (3RZ-FE) Engine

1. Before servicing the vehicle, refer to the precautions in the beginning of this section.

2. Relieve the fuel system pressure.

3. Disconnect the negative battery cable.

4. Drain the engine coolant.

5. Remove or disconnect the following:
- Air cleaner cap
- Mass Air Flow (MAF) meter and resonator
- Accelerator cable, from the throttle body, if equipped with a manual transaxle
- Accelerator and throttle cables, from the throttle body, if equipped with an automatic transaxle
- Intake air connector
- Air conditioning idle-up valve, if equipped with air conditioning
- No. 1 and No. 2 PCV hoses

- Spark plug wires, from the spark plugs
- Throttle body

6. Detach the following connectors:
- Air conditioning compressor connector, if equipped with air conditioning
- Oil pressure sensor connector
- Engine Coolant Temperature (ECT) sensor connector
- Exhaust Gas Recirculation (EGR) gas temperature sensor connector
- EGR Vacuum Switching Valve (VSV) connector

7. Disconnect the engine wiring harness, as follows:
- Engine wiring harness, from the intake chamber
- 5 engine wiring harness clamps and engine wiring harness
- Knock (KS) sensor connector
- Crankshaft Position (CKP) sensor connector
- Fuel pressure control VSV connector
- Data Link Connector 1 (DLC1), from the bracket
- 2 engine wiring harness clamps.
- Engine wiring harness from the engine

8. Remove or disconnect the following:
- Fuel injectors
- EGR valve and vacuum modulator
- Intake chamber stay
- Fuel return pipe

9. Remove the intake chamber, as follows:
- Vacuum hose, from the gas filter
- Brake booster vacuum hose, from the intake chamber
- Intake chamber and gasket
- Fuel inlet tube, by removing the union bolts

10. Remove the delivery pipe and injectors, as follows:
- Vacuum hose, from the fuel pressure regulator
- Delivery pipe, with the 4 injectors
- 4 insulators, from the 4 spacers
- 4 injectors, from the delivery pipe
- O-ring and grommet, from each injector

11. Remove or disconnect the following:
- Intake manifold
- Gasket

To install:

12. Clean the intake manifold surfaces.

13. Install or connect the following:
- New gasket
- Intake manifold. Tighten the bolts and nuts to 22 ft. lbs. (30 Nm).

14. Install the injectors to the delivery pipe, as follows:

- New grommet onto the injector
- New O-ring onto the injector, lubricated with gasoline
- Injectors onto the delivery pipe, with the electrical connector upward

15. Install or connect the following:
- Delivery pipe. Tighten the bolts to 15 ft. lbs. (21 Nm).

➡ **Check that the injectors rotate smoothly.**

- Fuel tube, with new gaskets. Tighten the union bolts to 22 ft. lbs. (30 Nm).

16. Install the air intake chamber, as follows:
- Air intake chamber with a new gasket. Tighten the bolts and nuts to 15 ft. lbs. (21 Nm).
- Vacuum hose, to the gas filter
- Brake booster vacuum hose, to the intake chamber

17. Install or connect the following:
- Fuel return pipe
- Intake chamber stay. Tighten the bolts to 14 ft. lbs. (20 Nm).

18. Install the EGR valve and vacuum modulator, as follows:
- New gasket, EGR valve and vacuum modulator. Tighten the bolt to 74 inch lbs. (9 Nm) and the nuts to 14 ft. lbs. (19 Nm).
- Vacuum hoses, to the EGR VSV
- Water bypass hose
- EGR pipe, using new gaskets. Tighten the bolts to 14 ft. lbs. (18 Nm), intake manifold nuts to 14 ft. lbs. (19 Nm) and the cylinder head nuts to 15 ft. lbs. (20 Nm).
- Injector connectors

19. Connect the engine wiring harness to the engine, as follows:
- Engine wiring harness, to the intake manifold
- 2 engine wiring harness clamps
- DLC1, to the bracket
- Fuel pressure control VSV connector
- KS sensor connector
- CKP sensor connector
- 5 engine wiring harness clamps.
- Engine wiring harness, to the intake chamber

20. Attach the following connectors to the engine:
- EGR VSV connector
- EGR gas temperature sensor connector
- ECT sensor connector
- Oil pressure sensor connector
- Compressor connector, if equipped with air conditioning

21. Install or connect the following:
- Throttle body
- Spark plug wires, to the spark plugs
- No. 1 and No. 2 PCV hoses
- Air conditioning idle up valve, if equipped with air conditioning
- Air intake connector, by installing the 2 bolts, hose clamp and 2 air hoses
- Accelerator cable, if equipped with a manual transaxle
- Throttle and accelerator cables, if equipped with an automatic transaxle
- MAF meter, resonator and the air cleaner cap
- Negative battery cable

22. Refill the cooling system.
23. Start the engine and check for leaks.
24. Check the ignition timing. Road test the vehicle for proper operation.
25. Recheck all fluid levels.

3.4L (5VZ-FE) Engine

1. Before servicing the vehicle, refer to the precautions in the beginning of this section.
2. Disconnect the negative battery cable.
3. Relieve the fuel system pressure.
4. Drain the engine coolant.
5. Remove or disconnect the following:
- Spark plug wires from the spark plugs
- Air cleaner cap
- Mass Air Flow (MAF) meter and resonator
- Actuator cable with the bracket, if equipped with cruise control
- Accelerator cable
- Throttle cable, if equipped with an automatic transmission
- Exhaust Gas Recirculation (EGR) pipe and gaskets, if equipped with an EGR valve
- Oil filler tube and No. 1 throttle cable clamp, by removing the bolt and 2 nuts
- Intake chamber stay, by removing the 2 bolts
- Vacuum Switching Valve (VSV) connector, for the fuel pressure control
- Throttle Position (TP) sensor connector
- Idle Air Control (IAC) valve connector
- EGR gas temperature connector, if equipped with an EGR valve
- VSV connector for the EGR valve, if equipped with an EGR valve

- Disconnect the Positive Crankcase Ventilation (PCV) hoses
- Water bypass hoses
- Air assist hose from the intake air connector
- 2 vacuum sensing hoses from the VSV
- Evaporative Emission (EVAP) hose
- Air hose from the power steering
- Air hose from the air conditioning idle up valve, if equipped with air conditioning
- Air intake chamber assembly
- Engine wiring harness from the intake air connector
- 2 fuel return hoses
- Brake booster vacuum hose, from the intake air connector
- Ground strap, from the intake air connector
- Data Link Connector 1 (DLC1) from the intake air connector bracket
- Idle up valve connector, if equipped with air conditioning
- Intake air connector
- Upper radiator hose, from the engine
- Oil pressure sensor connector
- Crankshaft Position (CKP) sensor connector
- 6 injector connectors
- Engine Coolant Temperature (ECT) sender gauge connector
- ECT sensor connector
- Knock (KS) sensor connector
- Camshaft Position (CMP) sensor connector
- 3 engine wiring harness clamps
- Engine wiring harness, from the cylinder head
- Fuel pressure regulator
- Heater hose
- Camshaft Position sensor.
- Fuel inlet hose
- Intake manifold stay
- Intake manifold assembly

To install:
6. Clean all surfaces.
7. Install or connect the following:
- New gaskets
- Intake manifold assembly. Tighten the bolts and nuts to 13 ft. lbs. (18 Nm).
- Intake manifold stay. Tighten the bolts to 13 ft. lbs. (18 Nm):
- Fuel inlet hose
- CMP sensor. Tighten it to 71 inch lbs. (8 Nm).
- Engine wiring harness to the cylinder head
- 3 engine wiring harness clamps
- Oil pressure sensor connector

- CKP sensor connector
- 6 injector connectors
- ECT sender gauge connector
- ECT sensor connector
- KS sensor connector
- CMP sensor connector
- Heater hose
- Intake manifold. Tighten the bolts and nuts to 14 ft. lbs. (18.5 Nm).
- DLC1 to the bracket on the intake manifold
- Ground strap to the intake manifold
- Brake booster vacuum hose, to the intake air connector
- 2 fuel return hoses
- Engine wiring harness to the intake manifold
- Idle up valve connector, if equipped with air conditioning
- Air intake chamber assembly to the engine. Tighten the bolts and nuts to 14 ft. lbs. (18.5 Nm).
- PCV hoses
- Water bypass hoses
- Air assist hose to the intake manifold
- 2 vacuum sensing hoses to the VSV
- EVAP hose
- Air hose to the power steering
- Air hose to the air conditioning idle up valve, if equipped with air conditioning
- VSV connector for the fuel pressure control
- TP sensor connector
- IAC valve connector
- EGR gas temperature connector, if equipped with an EGR valve
- VSV connector for the EGR valve, if equipped with an EGR valve
- Intake chamber stay. Tighten the bolts to 30 ft. lbs. (40 Nm).
- New O-ring to the oil filler tube
- Oil filler tube end into the tube hole in the oil pan
- Oil filler tube and No. 1 throttle cable clamp
- New gaskets and the EGR pipe. Tighten the clamp nuts to 71 inch lbs. (8 Nm) and the EGR pipe nuts to 14 ft. lbs. (18 Nm).
- Fuel pressure regulator
- 3 clamps for the spark plug wires, to the No. 2 timing belt cover
- CMP connector to the No. 2 timing belt cover
- Upper radiator hose

8. Fill with engine coolant.
9. Install or connect the following:
   - Spark plug wires to the spark plugs
   - Actuator cable with the bracket, if equipped with cruise control

- Accelerator cable
- Throttle cable, if equipped with an automatic transmission
- Air cleaner hose
- Negative battery cable

10. Fill the radiator with engine coolant.
11. Start the engine and check for leaks.

Exhaust Manifold

REMOVAL & INSTALLATION

2.4L (2RZ-FE) and 2.7L (3RZ-FE) Engines

1. Before servicing the vehicle, refer to the precautions in the beginning of this section.
2. Remove or disconnect the following:
   - Clamp from the support bracket
   - Support bracket
   - Front exhaust pipe and gaskets from the exhaust manifold
   - Heat insulator
   - Exhaust manifold and gasket

To install:
3. Install or connect the following:
   - Exhaust manifold and gasket. Tighten the nuts to 36 ft. lbs. (49 Nm).

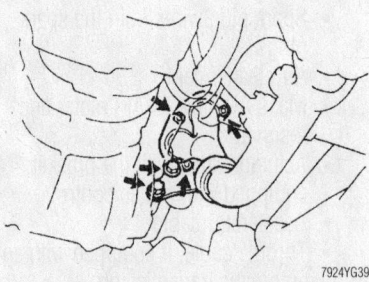

7924YG39

Front exhaust pipe to exhaust manifold nut and bolt locations—2.4L (2RZ-FE) and 2.7L (3RZ-FE) engines

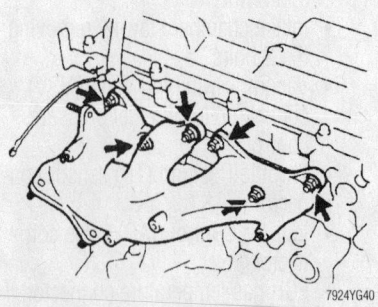

7924YG40

Exhaust manifold nuts—2.4L (2RZ-FE) and 2.7L (3RZ-FE) engines

- Heat insulator. Tighten the bolts and nuts to 48 inch lbs. (5.5 Nm).
- Front exhaust pipe assembly to the exhaust manifold. Tighten the nuts to 46 ft. lbs. (62 Nm).
- Support bracket. Tighten the bolts to 29 ft. lbs. (39 Nm).
- Clamp. Tighten the bolt to 14 ft. lbs. (19 Nm).
4. Start the engine.
5. Check for exhaust leaks.

3.4L (5VZ-FE) Engine

1. Before servicing the vehicle, refer to the precautions in the beginning of this section.
2. Remove or disconnect the following:
   - Exhaust crossover pipe, from the exhaust manifold by removing the 3 nuts
   - Exhaust Gas Recirculation (EGR)

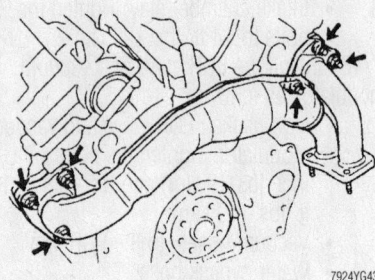

7924YG43

Exhaust crossover pipe mounting nut locations—3.4L (5VZ-FE) engine

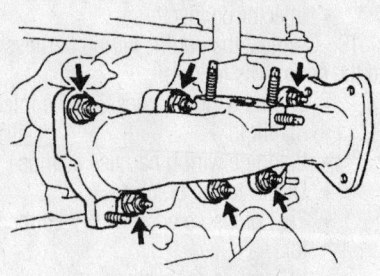

7924YG44

Exhaust manifold nuts—3.4L (5VZ-FE) engine

pipe, from the exhaust manifold, on the left manifold equipped with an EGR valve
- Exhaust manifold heat insulator, by removing the 3 nuts
- Exhaust manifold

To install:

3. Install or connect the following:
- Exhaust manifold, using a new gasket. Tighten the nuts to 30 ft. lbs. (40 Nm).
- Exhaust heat insulator. Tighten the nuts to 71 inch lbs. (8 Nm).
- EGR pipe to the exhaust manifold, if equipped with an EGR valve. Tighten the manifold nuts to 14 ft. lbs. (18 Nm) and the clamp nuts to 71 inch lbs. (8 Nm).
- Crossover pipe to the exhaust manifold, using a new gasket. Tighten the nuts to 33 ft. lbs. (45 Nm).

Front Crankshaft Seal

REMOVAL & INSTALLATION

3.4L (5VZ-FE) Engine

➡There are 2 methods to replace the oil seal, which are as follows:

OIL PUMP BODY INSTALLED

1. Before servicing the vehicle, refer to the precautions in the beginning of this section.
2. Remove or disconnect the following:
- Negative battery cable
- Timing belt and crankshaft pulley
- Cut off the oil seal lip, using a knife
- Pry out the oil seal, using a suitable tool

✳✳ WARNING

Be careful not to damage the crankshaft.

To install:

3. Install or connect the following:
- Apply multi-purpose grease to the new oil seal lip
- Tap in the new oil seal until its surface is flush with the oil pump case edge, using Seal Driver tool 09309-37010 and a mallet
- Crankshaft pulley and the timing belt
- Engine undercover, if removed
- Negative battery cable

OIL PUMP BODY REMOVED

1. Before servicing the vehicle, refer to the precautions in the beginning of this section.

2. Carefully pry out the seal using a suitable tool.
3. Apply multi-purpose grease to the new oil seal lip.
4. Using Seal Driver tool 09309-37010, drive the new seal into place.

Camshaft and Valve Lifters

REMOVAL & INSTALLATION

2.4L (2RZ-FE) Engine

1. Before servicing the vehicle, refer to the precautions in the beginning of this section.
2. Remove or disconnect the following:
- Timing chain
- Exhaust camshaft by bringing the service bolt hole of the driven sub-gear upwards. Turn the hexagon wrench head portion of the exhaust camshaft with a wrench.

3. Secure the exhaust camshaft sub-gear to the main gear with a service bolt. The thread diameter should be 0.23 in. (6mm) with a thread pitch of 0.04 in. (1.0mm) and a bolt length of 0.63–0.79 in. (16–20mm).

➡When removing the camshaft, be sure that the torsional spring force of the sub-gear has been eliminated by the above operation.

4. Uniformly loosen and remove the exhaust bearing cap bolts (10 of them), in several passes. Use the reverse order of the tightening sequence. Remove the 5 bearing caps and the camshaft. Do the same for the intake camshafts.

➡If the camshaft is not being lifted out straight and level, reinstall the No. 3 cap with the 2 bolts. Alternately loosen, then remove the bearing cap bolts with the camshaft pulled up. Do not pry on or force the camshaft.

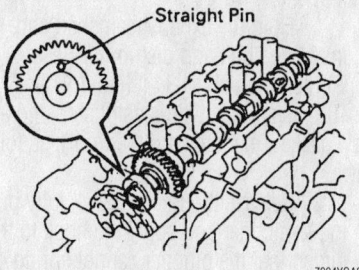

Install the camshaft with the pin facing upwards—2.4L (2RZ-FE) engine

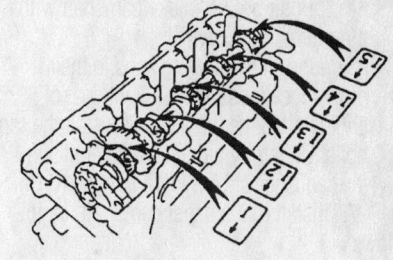

Intake camshaft bearing cap locations— 2.4L (2RZ-FE) engine

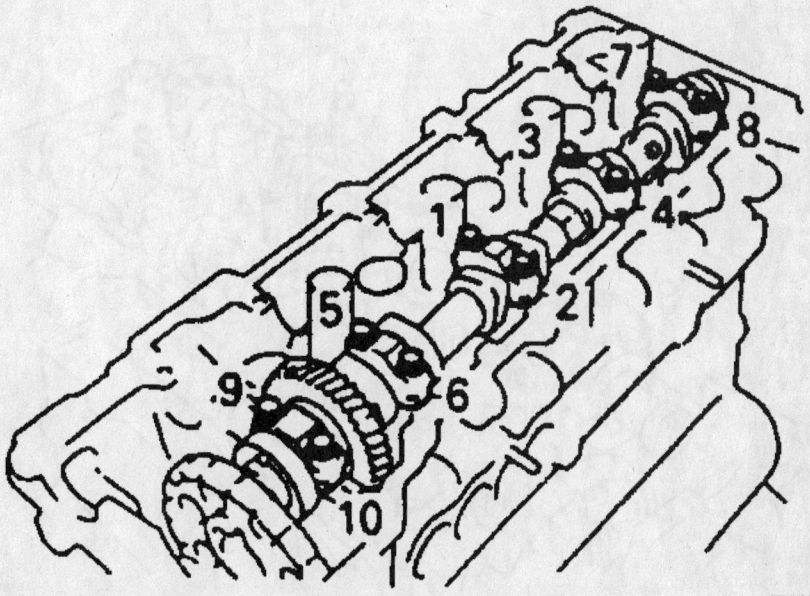

Tighten the intake bearing caps following this order—2.4L (2RZ-FE) engine

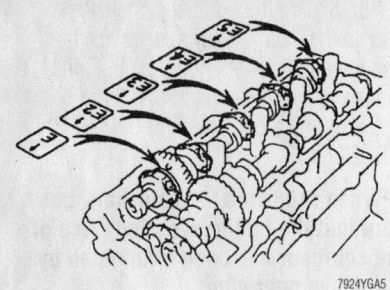

Position the exhaust camshaft bearing caps as shown—2.4L (2RZ-FE) engine

5. Inspect the camshafts for excessive runout. Inspect the cam lobes and journals. The bearings are part of the cam and should be inspected for flaking or scoring. If the bearings are damaged, replace the caps and the cylinder head as a set. The camshaft journal oil and thrust clearances should be checked.

To install:

6. Install the intake camshaft, as follows:

a. Apply multi-purpose grease to the thrust portion of the intake camshaft.

b. Position the intake camshaft with the pin facing upward.

c. Install the bearing caps in their proper locations. Apply a light coat of engine oil to the threads and install the cap bolts. Uniformly tighten the cap bolts in the sequence shown to 12 ft. lbs. (16 Nm).

7. Install the exhaust camshaft, as follows:

a. Apply engine oil to the thrust portion of the intake camshaft.

b. Engage the exhaust camshaft gear to the intake camshaft gear by matching the timing marks (1 and 2 dots) on each other.

c. Roll down the exhaust camshaft onto the bearing journals while engaging the gears with each other. Install the bearing caps in their proper locations.

d. Apply a light coat of engine oil to the threads and install the cap bolts. Uniformly tighten the cap bolts in the sequence shown to 12 ft. lbs. (16 Nm).

e. Remove the service bolt from the driven sub-gear. Check that the intake and exhaust camshafts turn smoothly.

8. Set No. 1 cylinder to Top Dead Center (TDC) of the compression stroke. The crankshaft pulley groove aligns with the **0** mark on timing cover and camshaft timing marks with 1 dot and 2 dots will be in a straight line on the cylinder head surface.

9. Install the timing gear, as follows:

a. Place the gear over the straight pin of the intake camshaft.

b. Hold the intake camshaft with a wrench. Install and tighten the bolt to 54 ft. lbs. (74 Nm).

c. Hold the exhaust camshaft and install the bolt and distributor gear. Tighten the bolt to 34 ft. lbs. (46 Nm).

10. Install the chain tensioner, using a new gasket (mark toward the front), as follows:

a. Release the ratchet pawl, fully push in the plunger and apply the hook to the pin so that the plunger cannot spring out.

b. Turn the crankshaft pulley clockwise to provide some slack for the chain on the tensioner side.

c. Push the tensioner by hand until it touches the head installation surface, then install the 2 nuts. Tighten the nuts to 13 ft. lbs. (18 Nm). Check that the hook of the tensioner is not released.

d. Turn the crankshaft to the left so that the hook of the chain tensioner is released from the pin of the plunger, allowing the plunger to spring out and the slipper to be pushed into the chain.

11. Check and adjust the valve clearance. Intake valve clearance is 0.006–0.010 inch (0.15–0.25mm) and exhaust valve clearance is 0.010–0.014 inch (0.25–0.35mm).

12. Recheck the engine for proper valve timing. Check and adjust the valve clearance.

13. Install the spark plugs and the semi-circular plug.

14. Recheck the engine for proper valve timing. Install the valve cover and engine hangers. Tighten the engine hanger bolts to 30 ft. lbs. (42 Nm).

15. Reinstall all other parts from the timing chain removal. Fill any fluids, start the engine, top off the fluids.

2.7L (3RZ-FE) Engine

1. Before servicing the vehicle, refer to the precautions in the beginning of this section.

2. Disconnect the negative battery cable.

3. Drain the engine coolant.

4. Remove or disconnect the following:

- Air cleaner cap
- Mass Air Flow (MAF) meter and the resonator
- Accelerator cable from the throttle body, if equipped with a manual transmission
- Accelerator and throttle cables from the throttle body, if equipped with an automatic transmission
- Cruise control cable from the actuator, if equipped with cruise control
- Intake air connector
- Air hose for Idle Air Control (IAC)
- Vacuum sensing hose
- Wire clamp for the engine wiring harness
- Positive Crankcase Ventilation (PCV) hoses
- Spark plug wires from the spark plugs
- Engine wiring harness clamps and harness
- Air conditioning compressor connector, if equipped with air conditioning

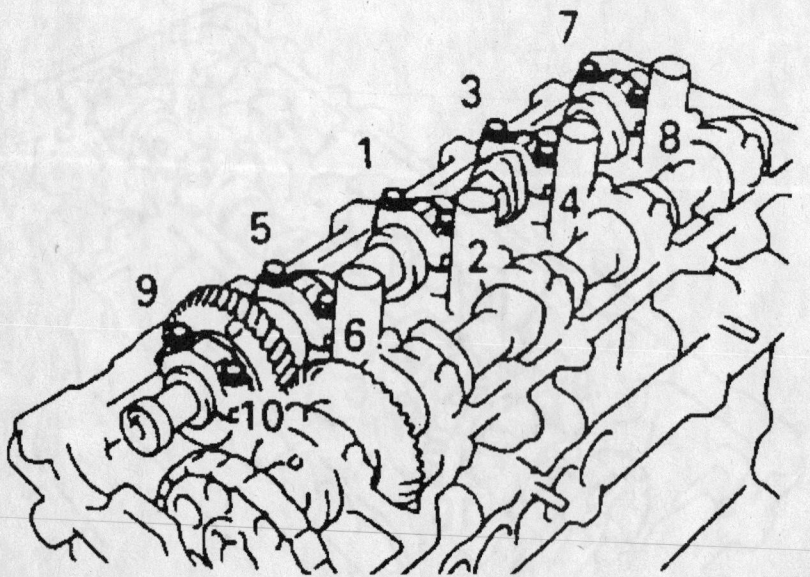

Tighten the exhaust bearing cap bolts in this order—2.4L (2RZ-FE) engine

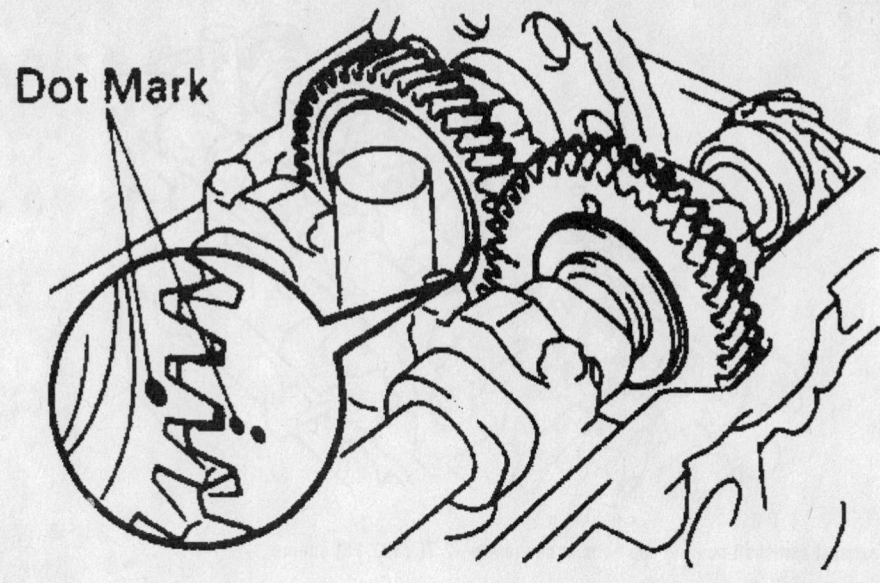

Dot Mark

Camshafts TDC/compression timing marks. Marks with 1 and 2 dots will be in straight line on cylinder head surface—2.7L (3RZ-FE) engine

- Oil pressure sensor connector
- Engine Coolant Temperature (ECT) sensor connector
- Engine coolant temperature sender gauge connector
- Exhaust Gas Recirculation (EGR) gas temperature sensor connector
- Vacuum Switching Valve (VSV) connector for the EGR
- 2 vacuum hoses from the VSV for the EGR
- Ground strap from the cowl top panel

- Engine wiring harness from the air intake chamber
- Throttle Position (TP) sensor connector
- IAC valve connector
- Crankshaft Position (CKP) sensor connector
- Knock (KS) sensor connector
- Data Link Connector 1 (DLC1) from the bracket
- Engine wiring harness clamp
- EGR pipe
- Intake chamber stay

- Air intake chamber assembly.
- Evaporative Emission (EVAP) hose from the throttle body
- Brake booster vacuum hose from the union
- Water bypass hose from the water bypass pipe
- Water bypass hose from the cylinder head rear cover
- Front engine hanger
- Engine wiring harness brackets
- Cylinder head cover

5. Set No. 1 cylinder to Top Dead Center (TDC) compression stroke. The groove on the crankshaft pulley should align with the **0** mark on the timing chain cover and the timing marks (1 and 2 dots) of the camshaft gears should form a straight line in respect to the cylinder head surface. If not, turn the crankshaft 1 revolution (360 degrees).

6. Remove the chain tensioner and gasket.

7. Remove the camshaft timing gear as follows:

a. Remove the 2 semi-circular plugs.

b. Place matchmarks on the camshaft timing gear and No. 1 timing chain.

c. Hold the hexagon head portion of the exhaust camshaft with a wrench and remove the fastener and distributor gear.

d. Hold the hexagon head portion of the intake camshaft with a wrench and remove the bolt.

e. Remove the camshaft timing gear and chain from the intake camshaft and leave on the slipper and damper.

8. Remove exhaust camshafts:

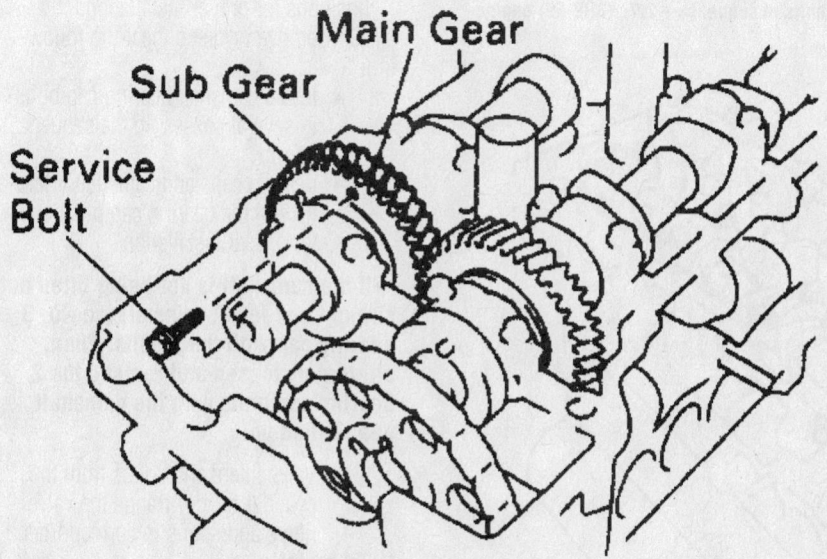

Main Gear
Sub Gear
Service Bolt

Secure the exhaust camshaft sub-gear to the main gear with a service bolt—2.7L (3RZ-FE) engine

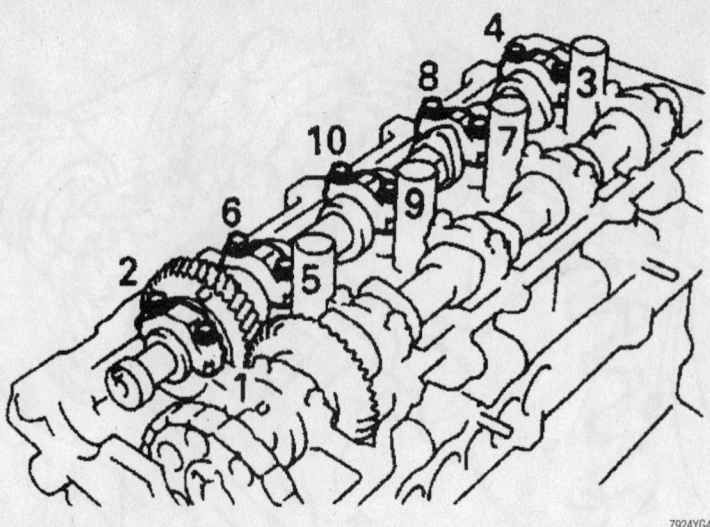

Loosen and remove the exhaust camshaft bearing cap bolts in sequence—2.7L (3RZ-FE) engine

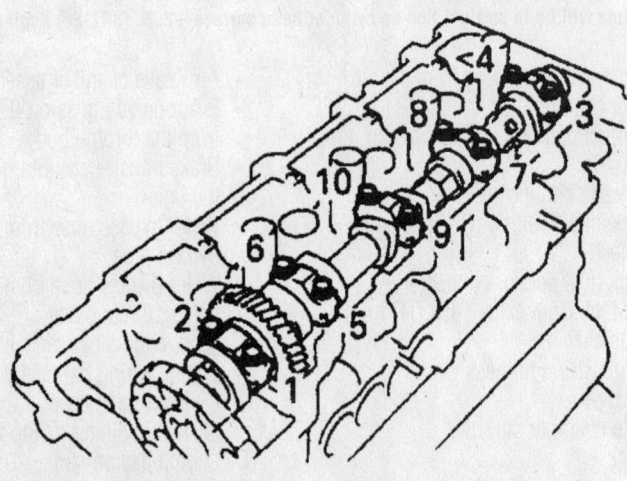

Loosen and remove the intake camshaft bearing cap bolts in sequence—2.7L (3RZ-FE) engine

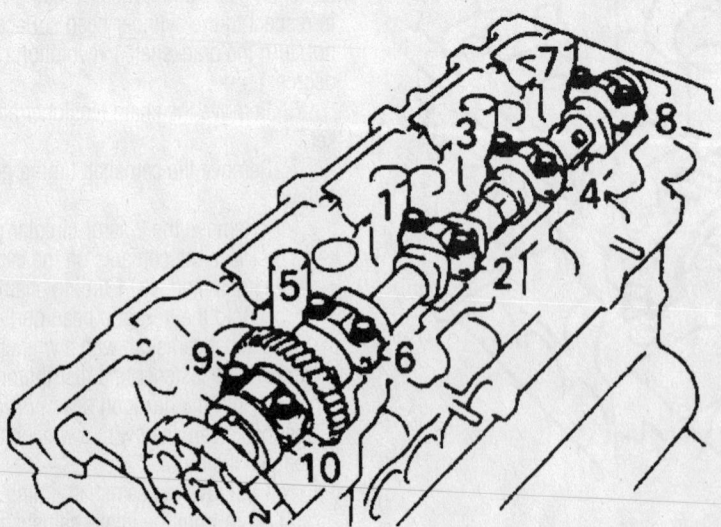

Tighten the intake camshaft bearing cap bolts in sequence—2.7L (3RZ-FE) engine

a. Bring the service bolt hole of the driven sub-gear upward by turning the hexagon head portion of the exhaust camshaft with a wrench.

b. Secure the exhaust camshaft sub-gear to the driven gear with a service bolt (6mm diameter, 0.63–0.79 inches in length and 1.0mm in thread pitch).

➡ When removing the camshaft, be sure that the torsional spring force of the sub-gear has been eliminated by the above operation.

c. Uniformly loosen and remove the bearing cap bolts in several passes, in the sequence shown.

d. Remove the bearing caps and camshaft. Make a note of the bearing cap positions for proper installation.

9. Remove or disconnect the following:

- Intake camshaft bearing cap bolts in several passes, in the sequence shown
- Bearing caps and camshaft. Make a note of the bearing cap positions for proper installation.

➡ If the camshaft is not being lifted out straight and level, reinstall the No. 3 bearing cap with the 2 bolts. Then, alternately loosen and remove the 2 bearing cap bolts with the camshaft gear pulled up.

- Valve lifters and shims from the cylinder head. Arrange the valve lifters and shims in correct order.

To install:

10. Install the valve lifters and shims in their proper locations. Check that the valve lifter rotates smoothly by hand.

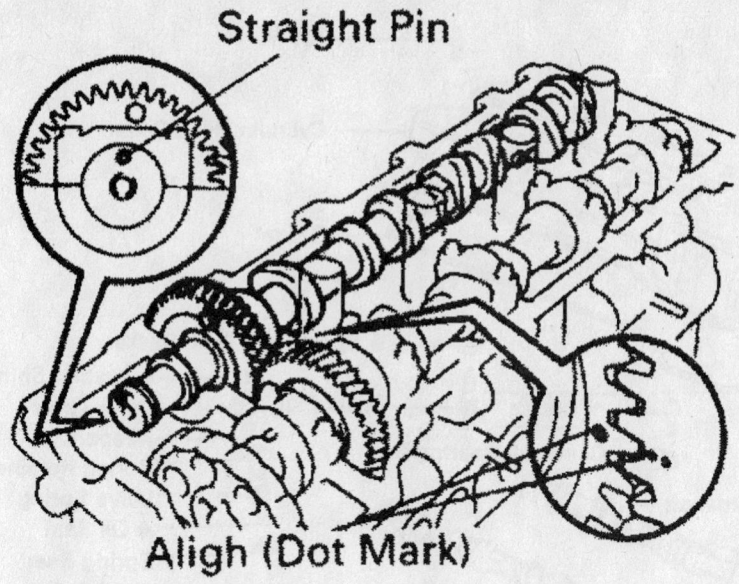

Engage both camshaft gears while matching the timing marks—2.7L (3RZ-FE) engine

11. Install the intake camshaft, as follows:

a. Apply engine oil to the thrust portion of the intake camshaft.

b. Position the intake camshaft with the knock pin facing upward.

c. Install the bearing caps in their proper locations. Apply a light coat of engine oil to the threads and install the cap bolts. Uniformly tighten the cap bolts, in sequence, to 12 ft. lbs. (16 Nm).

12. Install the exhaust camshaft, as shown:

a. Apply engine oil to the thrust portion of the intake camshaft.

b. Engage the exhaust camshaft gear to the intake camshaft gear by matching the timing marks (1 and 2 dots) on each other.

c. Roll down the exhaust camshaft onto the bearing journals while engaging the gears with each other. Install the bearing caps in their proper locations.

d. Apply a light coat of engine oil to the threads and install the cap bolts. Uniformly tighten the cap bolts in the sequence shown to 12 ft. lbs. (16 Nm).

e. Remove the service bolt from the driven sub-gear. Check that the intake and exhaust camshafts turns smoothly.

13. Set No. 1 cylinder to Top Dead Center (TDC) of the compression stroke: Crankshaft pulley groove align with **0** mark on timing cover and camshafts timing marks with 1 dot and 2 dots will be straight line on the cylinder head surface.

14. Install the timing gear. Place the gear over the straight pin of the intake camshaft.

a. Hold the intake camshaft with a wrench. Install and tighten the bolt to 54 ft. lbs. (74 Nm).

b. Hold the exhaust camshaft and install the bolt and distributor gear. Tighten the bolt to 34 ft. lbs. (46 Nm).

15. Install or connect the following:

- Chain tensioner, using a new gasket (mark toward the front)
- Recheck the engine for proper valve timing. Check and adjust the valve clearance.
- Semi-circular plug
- Recheck the engine for proper valve timing.
- Cylinder head cover with a new gasket
- Engine wiring harness brackets
- Front engine hanger. Tighten the bolts to 30 ft. lbs. (42 Nm).
- Air intake chamber assembly. Tighten the bolts to 15 ft. lbs. (20 Nm).
- Hoses
- Intake chamber stay
- Air intake chamber stay. Tighten the bolts to 15 ft. lbs. (20 Nm).
- EGR pipe. Tighten the bolts to 13 ft. lbs. (18 Nm), nut "A" to 14 ft. lbs. (19 Nm) and nut "B" to 15 ft. lbs. (20 Nm).
- Engine wiring harness
- Spark plug wires to the spark plugs
- PCV hoses
- Intake air connector. Tighten the bolts to 13 ft. lbs. (18 Nm).
- Air hose for the IAC
- Vacuum sensing hose
- Wire clamp for the engine wiring harness
- Cruise control cable to the actuator, if equipped with cruise control
- Accelerator cable to the throttle body, if equipped with a manual transmission
- Accelerator and throttle cables to the throttle body, if equipped with an automatic transmission
- Air cleaner cap, MAF meter and the resonator assembly
- Negative battery cable

16. Refill the cooling system.

17. Start the engine and check for leaks.

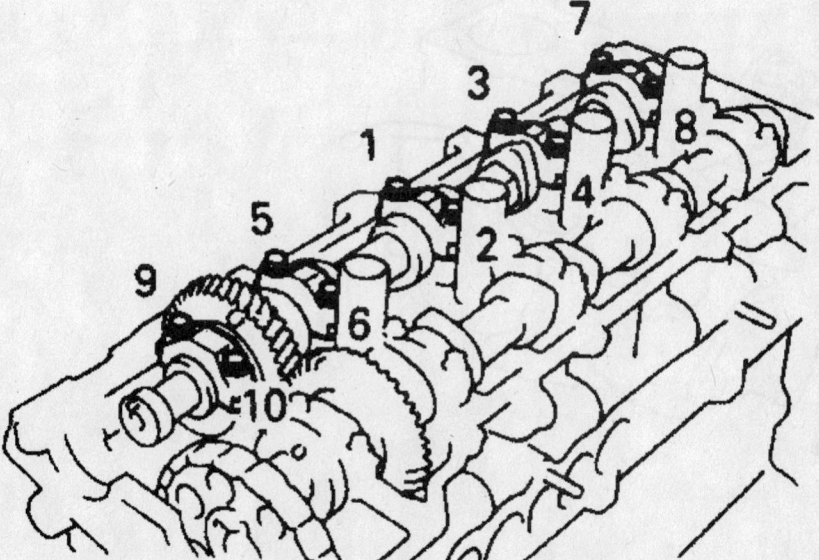

Tighten the exhaust camshaft bearing cap bolts in sequence—2.7L (3RZ-FE) engine

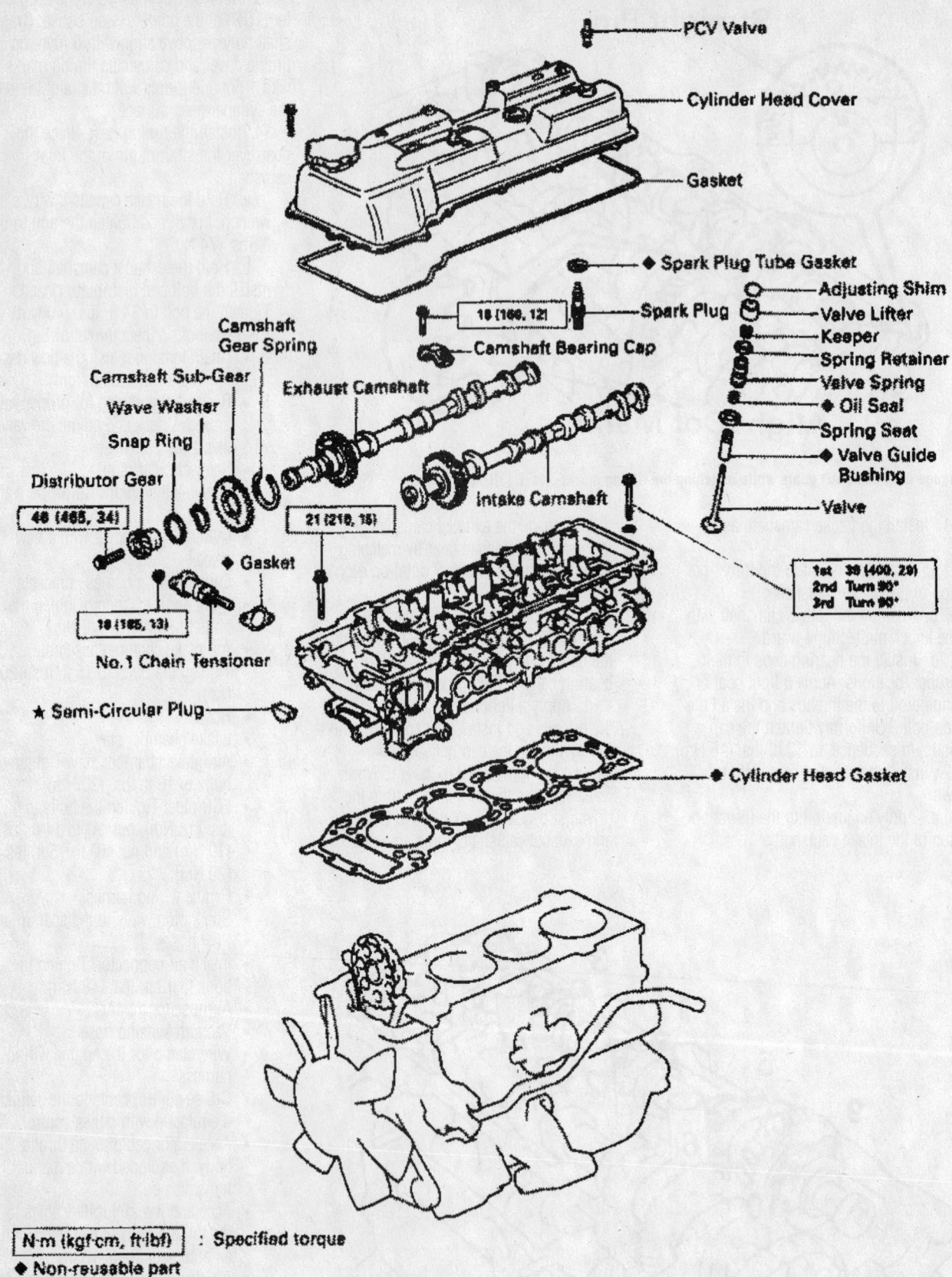

PCV Valve

Cylinder Head Cover

Gasket

◆ Spark Plug Tube Gasket

Adjusting Shim

Valve Lifter

18 (180, 12) Spark Plug

Keeper

Camshaft Bearing Cap

Spring Retainer

Camshaft
Gear Spring

Valve Spring

Exhaust Camshaft

◆ Oil Seal

Camshaft Sub-Gear

Spring Seat

Wave Washer

◆ Valve Guide
Bushing

Snap Ring

Distributor Gear

Intake Camshaft

Valve

48 (485, 34)

21 (210, 15)

1st 38 (400, 29)
2nd Turn 90°
3rd Turn 90°

◆ Gasket

18 (185, 13)

No. 1 Chain Tensioner

★ Semi-Circular Plug

◆ Cylinder Head Gasket

N·m (kgf·cm, ft·lbf) : Specified torque

◆ Non-reusable part

★ Precoated part

7924YG54

Exploded view of the cylinder head components—2.7L (3RZ-FE) engine

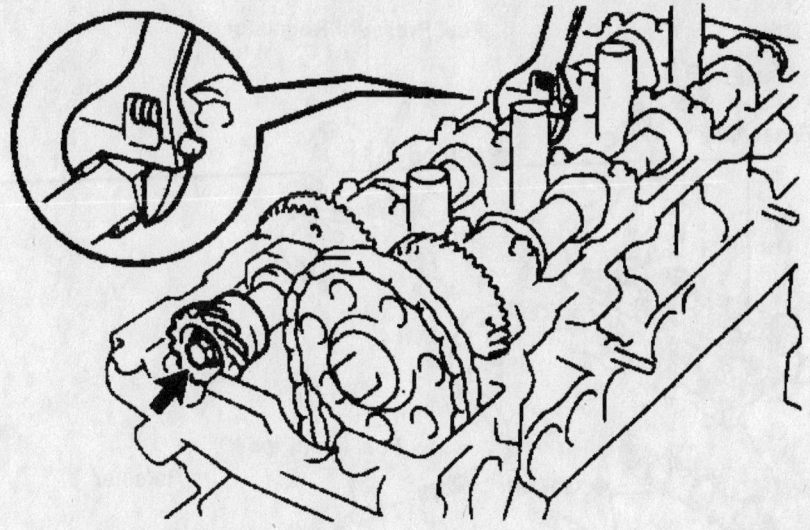

7924YG55

Using a wrench to hold the camshaft—2.7L (3RZ-FE) engine

18. Check the ignition timing. Road test the vehicle for proper operation.

19. Recheck all fluid levels.

3.4L (5VZ-FE) Engine

1. Before servicing the vehicle, refer to the precautions in the beginning of this section.

2. Remove or disconnect the following:
- Negative battery cable
- Engine undercover

3. Drain the cooling system.

4. Remove or disconnect the following:
- Air cleaner cap
- Mass Air Flow (MAF) meter and the resonator
- Actuator cable with the bracket, if equipped with cruise control
- Accelerator cable
- Throttle cable, if equipped with an automatic transmission
- Heater hose
- Upper radiator hose from the engine

5. Remove the power steering drive belt, as follows:

a. Stretch the belt and loosen the fan pulley mounting nuts.

b. Loosen the lockbolt, pivot bolt and adjusting bolt and remove the drive belt from the engine.

6. Remove or disconnect the following:
- Air conditioning drive belt, by loosening the idle pulley nut and adjusting bolt
- Loosen the alternator lockbolt, pivot bolt and adjusting bolt
- Alternator drive belt

- No. 2 fan shroud, by removing the 2 clips
- Fan with the fluid coupling and fan pulleys
- Power steering pump and move it aside without disconnecting the lines
- Compressor and move it aside without disconnecting the lines, if equipped with air conditioning
- Air conditioning bracket, if equipped with air conditioning
- Spark plug wires with the ignition coils
- Spark plugs
- Camshaft Position (CMP) sensor connector, from the No. 2 timing belt cover
- 3 spark plug wire clamps from the No. 2 timing belt cover
- 6 bolts and the timing belt cover
- Power steering adjusting strut by removing the nut
- Fan bracket by removing the bolt and nut

7. Set the No. 1 cylinder at Top Dead Center (TDC) of the compression stroke, as follows:

a. Turn the crankshaft pulley and align its groove with the timing mark **0** on the No. 1 timing belt cover.

b. Check that the timing marks of the camshaft timing pulleys and the No. 3 timing belt cover are aligned. If not, turn the crankshaft pulley 1 revolution (360 degrees).

8. Remove or disconnect the following:
- Timing belt tensioner, by alternately loosening the 2 bolts
- Pulley bolt, the timing pulley and

the knock pin, using Variable Pin Wrench Set 09960-10010
- Both timing pulleys with the timing belt
- No. 2 idler pulley
- Alternator
- Positive Crankcase Ventilation PCV hoses
- Water bypass hoses
- Air assist hose from the intake air connector
- 2 vacuum sensing hoses from the Vacuum Switching Valve (VSV)
- Evaporative Emissions (EVAP) hose
- Air hose from the power steering
- Air hose from the air conditioning idle up valve, if equipped with air conditioning
- 4 bolts, 2 nuts and the air intake chamber assembly
- Intake air connector
- Camshaft Position (CMP) sensor
- No. 3 (rear) timing belt cover by removing the 6 bolts
- 8 bolts, seal washers, both cylinder head cover and gaskets
- Semi-circular plugs

9. Remove the right exhaust camshafts, as follows:

a. Bring the service bolt hole of the driven sub-gear upward by turning the hexagon head portion of the exhaust camshaft with a wrench.

b. Align the timing mark (2 dot marks) of the camshaft drive and driven gears by turning the camshaft with a wrench.

c. Secure the exhaust camshaft sub-gear to the driven gear with a service bolt (6mm diameter, 16–20mm bolt length and 1.0mm in thread pitch).

➡ **When removing the camshaft, be sure the torsional spring force of the sub-gear has been eliminated by the above operation.**

d. Uniformly loosen and remove the bearing cap bolts in several passes, in the sequence shown.

e. Remove the bearing caps and camshaft. Make a note of the bearing cap positions for proper installation.

➡ **Do not pry on or attempt to force the camshaft with a tool or other object.**

10. Remove the right-hand intake camshaft, as follows:

a. Uniformly loosen and remove the bearing cap bolts in several passes, in the sequence shown.

b. Remove the bearing caps, oil seal and camshaft. Make a note of the bearing cap positions for proper installation.

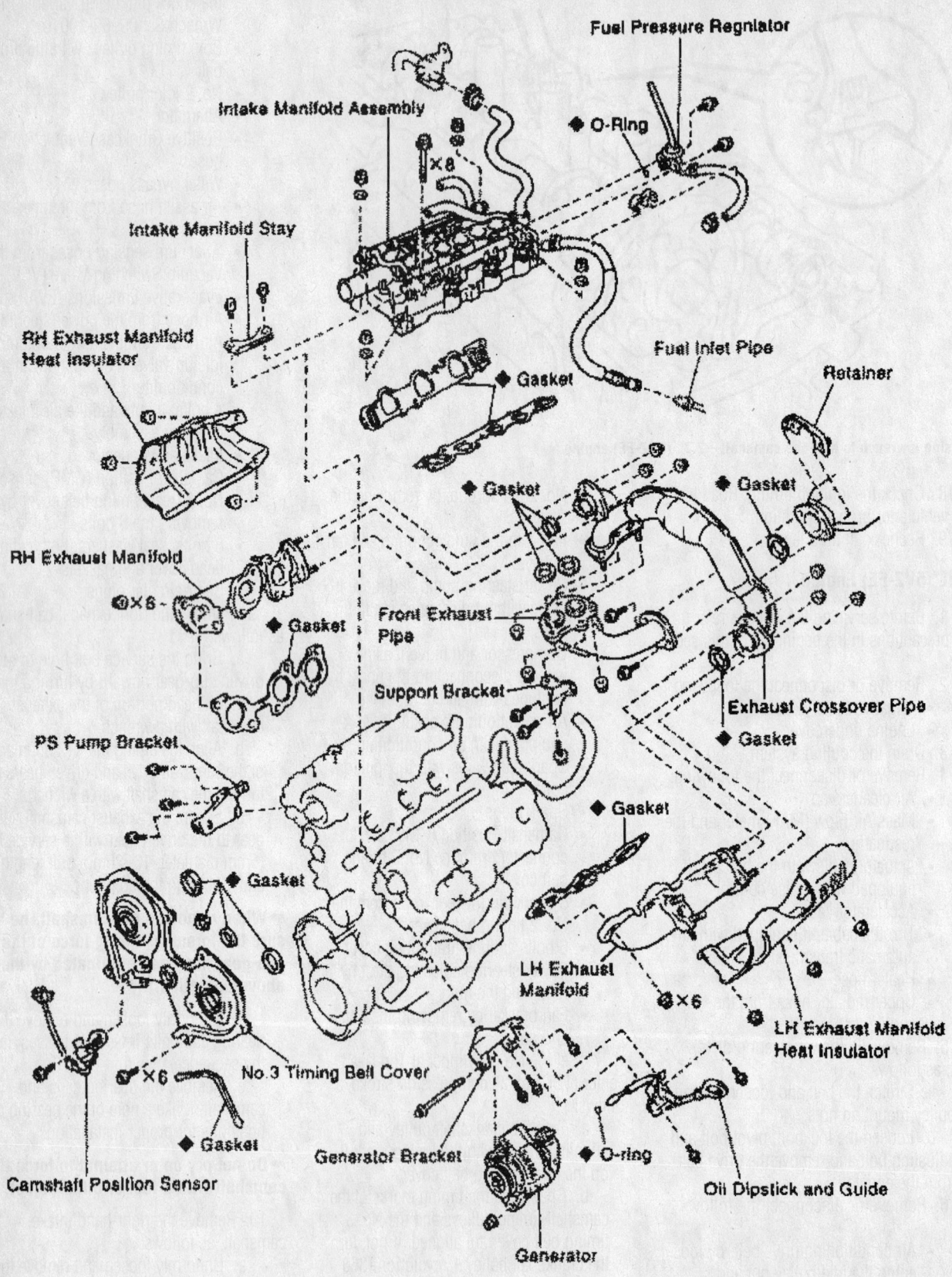

Fuel Pressure Regulator

Intake Manifold Assembly

◆ O-Ring

Intake Manifold Stay

RH Exhaust Manifold Heat Insulator

Fuel Inlet Pipe

Retainer

◆ Gasket

◆ Gasket

◆ Gasket

RH Exhaust Manifold

×6

◆ Gasket

Front Exhaust Pipe

Exhaust Crossover Pipe

◆ Gasket

PS Pump Bracket

Support Bracket

◆ Gasket

◆ Gasket

◆ Gasket

LH Exhaust Manifold

×6

LH Exhaust Manifold Heat Insulator

×6

No.3 Timing Bell Cover

◆ Gasket

Generator Bracket

◆ O-ring

Oil Dipstick and Guide

Camshaft Position Sensor

Generator

◆ Non-reusable part

7924YG60

Exploded view of the cylinder head component assembly—3.4L (5VZ-FE) engine

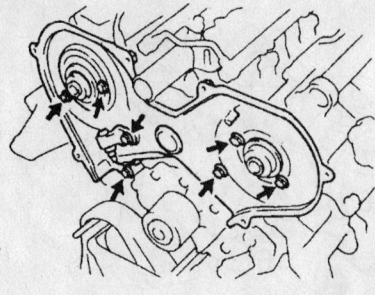

7924YG61

Rear timing belt cover mounting bolt locations—3.4L (5VZ-FE) engine

11. Remove the left exhaust camshafts, as follows:

a. Align the timing mark (1 dot mark) of the camshaft drive and driven gears by turning the camshaft with a wrench.

b. Secure the exhaust camshaft sub-gear to the driven gear with a service bolt (6mm diameter, 16–20mm bolt length and 1.0mm in thread pitch).

➡**When removing the camshaft, be sure that the torsional spring force of the sub-gear has been eliminated by the above operation.**

c. Uniformly loosen and remove the bearing cap bolts in several passes, in the sequence shown.

d. Remove the bearing caps and camshaft. Make a note of the bearing cap positions for proper installation.

12. Remove the left-hand intake camshaft, as follows:

a. Uniformly loosen and remove the bearing cap bolts in several passes, in the sequence shown.

b. Remove the bearing caps, oil seal and camshaft. Make a note of the bearing cap positions for proper installation.

13. Remove the valve lifters and shims from the cylinder head. Arrange the valve lifters and shims in correct order.

To install:

14. Clean all surfaces.

15. Install the valve lifters and shims. Check that the valve lifter rotates smoothly by hand.

16. Install the right intake camshaft, as follows:

a. Apply engine oil to the thrust portion of the intake camshaft.

b. Position the intake camshaft at 90 degrees angle of the timing mark (2 dot marks) on the cylinder head.

c. Install the bearing caps in their proper locations. Apply a light coat of engine oil to the threads and install the cap bolts.

RH

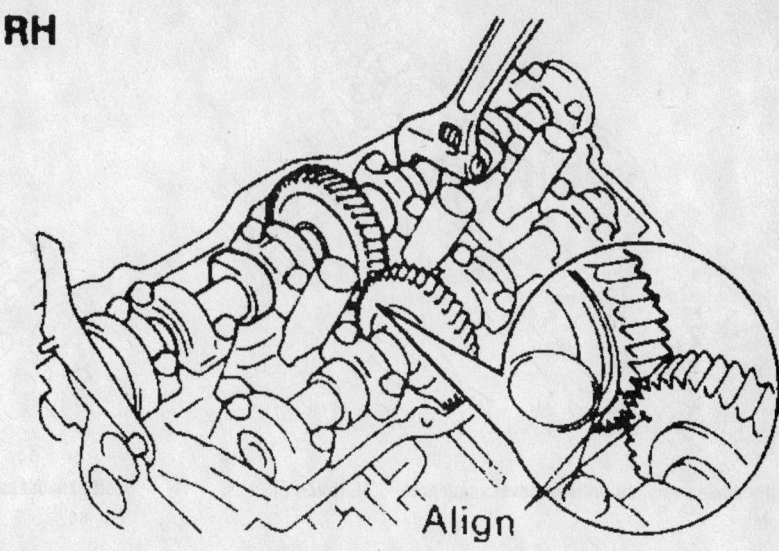

7924YG62

Aligning the timing marks (2 dot marks) of the right camshafts—3.4L (5VZ-FE) engine

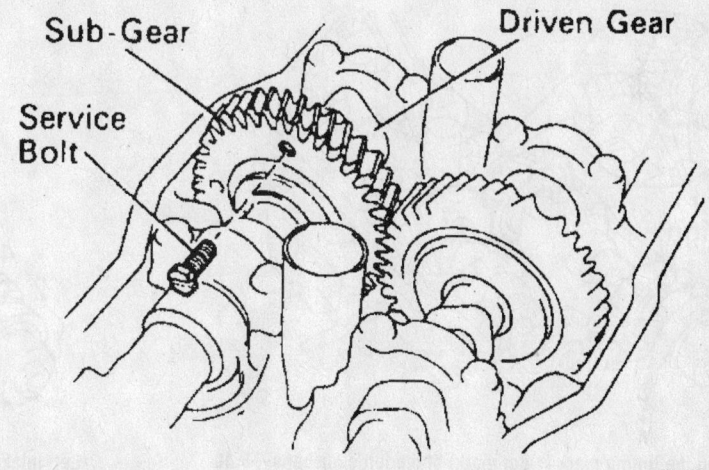

7924YG63

Drive gear service bolt (right side)—3.4L (5VZ-FE) engine

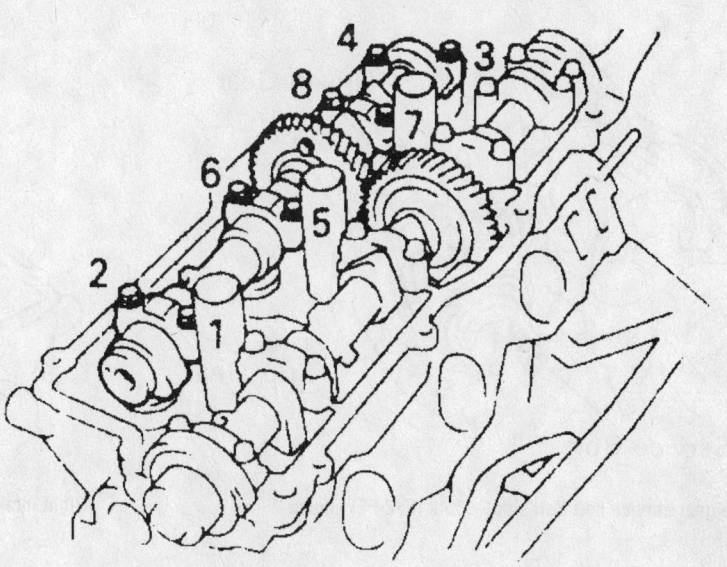

7924YG64

Right exhaust camshaft bolts removal sequence—3.4L (5VZ-FE) engine

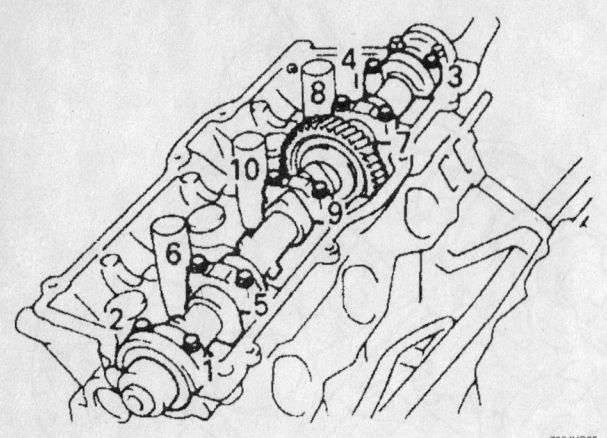

Right intake camshaft bolts removal sequence—3.4L (5VZ-FE) engine

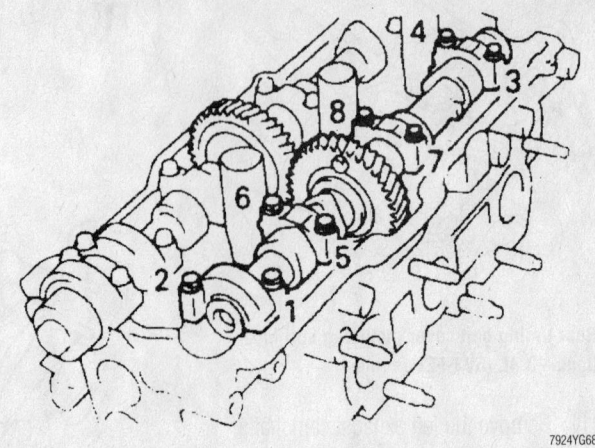

Left exhaust camshaft bolts removal sequence—3.4L (5VZ-FE) engine

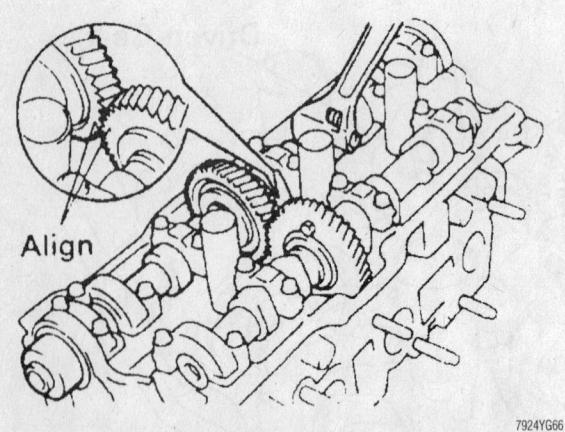

Aligning the timing mark (1 dot mark) of the left camshafts—3.4L (5VZ-FE) engine

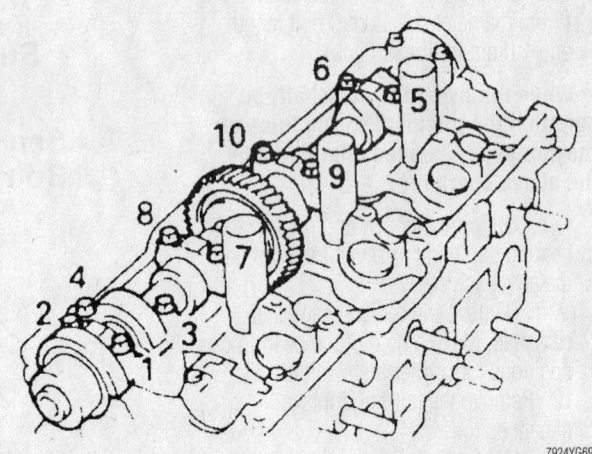

Left intake camshaft bolts removal sequence—3.4L (5VZ-FE) engine

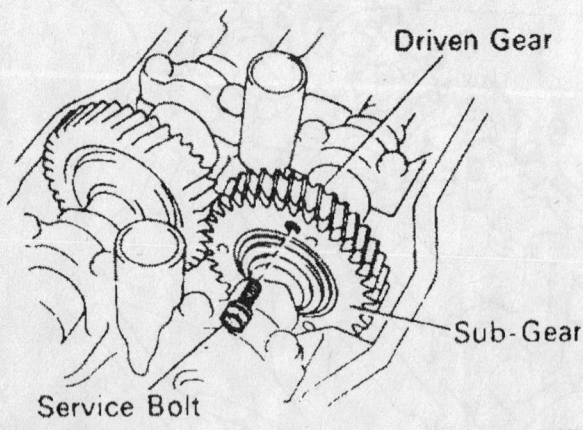

Drive gear service bolt (left side)—3.4L (5VZ-FE) engine

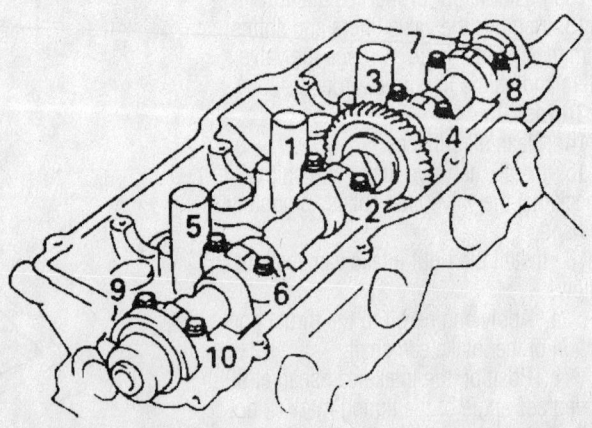

Right intake camshaft tightening sequence—3.4L (5VZ-FE) engine

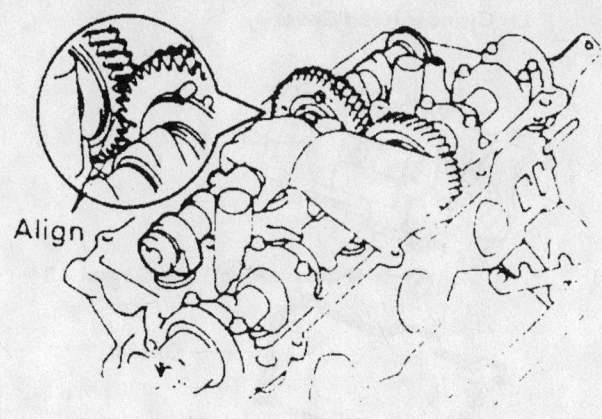

Aligning the right camshafts for installation—3.4L (5VZ-FE) engine

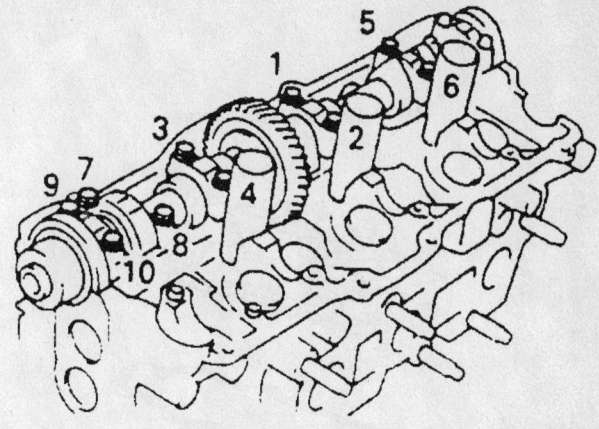

Left intake camshaft bolts tightening sequence—3.4L (5VZ-FE) engine

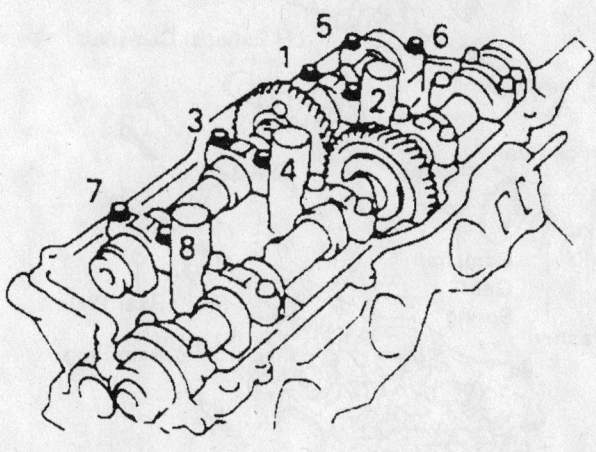

Right exhaust camshaft bolts tightening sequence—3.4L (5VZ-FE) engine

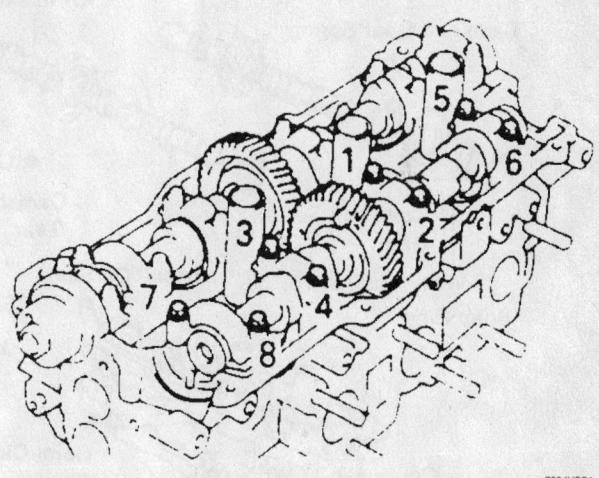

Left exhaust camshaft bolts tightening sequence—3.4L (5VZ-FE) engine

   d. Apply a light coat of engine oil on the threads and under the heads of the bearing cap bolts.

   e. Uniformly tighten the cap bolts in the sequence shown to 12 ft. lbs. (16 Nm).

17. Install the right exhaust camshaft, as follows:

   a. Apply engine oil to the thrust portion of the intake camshaft.

   b. Align the timing marks (2 dot marks) of the camshaft drive and driven gears.

   c. Roll down the exhaust camshaft onto the bearing journals while engaging the gears with each other. Install the bearing caps in their proper locations.

   d. Apply a light coat of engine oil to the threads and install the cap bolts.

   e. Apply a light coat of engine oil on the threads and under the heads of the bearing cap bolts.

   f. Uniformly tighten the cap bolts in the sequence shown to 12 ft. lbs. (16 Nm).

   g. Remove the service bolt from the driven sub-gear. Check that the intake and exhaust camshafts turns smoothly.

   h. Align the timing marks (2 dot marks) of the camshaft drive and driven gears by turning the camshaft with a wrench.

18. Install the left intake camshaft, as follows:

   a. Apply engine oil to the thrust portion of the intake camshaft.

   b. Position the intake camshaft at 90 degrees angle of the timing mark (1 dot mark) on the cylinder head.

   c. Install the bearing caps in their proper locations. Apply a light coat of engine oil to the threads and install the cap bolts.

   d. Apply a light coat of engine oil on

the threads and under the heads of the bearing cap bolts.

   e. Uniformly tighten the cap bolts in the sequence shown to 12 ft. lbs. (16 Nm).

19. Install the left exhaust camshaft, as follows:

   a. Apply engine oil to the thrust portion of the intake camshaft.

   b. Align the timing marks (1 dot mark) of the camshaft drive and driven gears.

   c. Roll down the exhaust camshaft onto the bearing journals while engaging the gears with each other. Install the bearing caps in their proper locations.

   d. Apply a light coat of engine oil to the threads and install the cap bolts.

   e. Apply a light coat of engine oil on the threads and under the heads of the bearing cap bolts.

   f. Uniformly tighten the cap bolts in

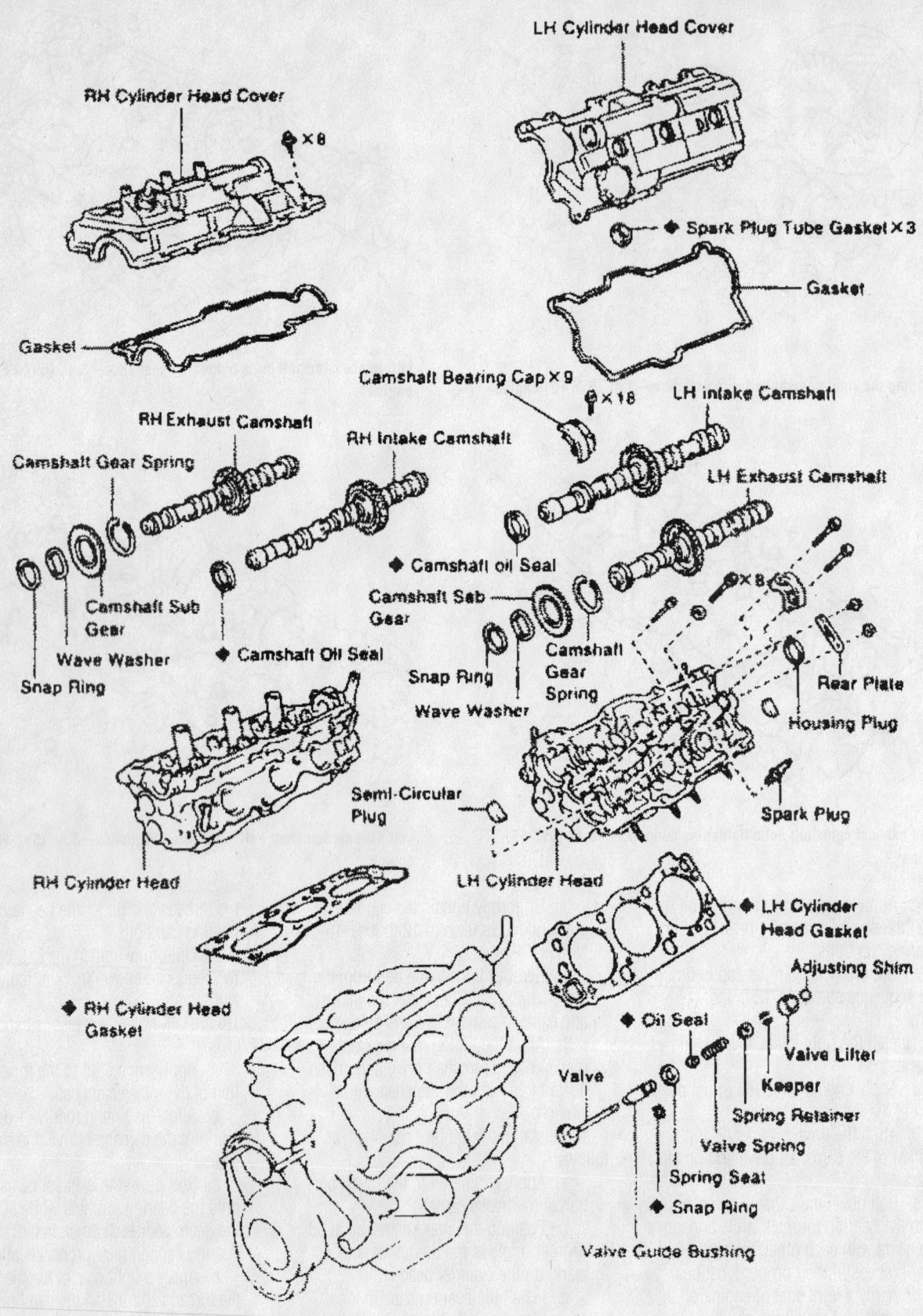

RH Cylinder Head Cover

LH Cylinder Head Cover

×8

Spark Plug Tube Gasket×3

Gasket

Gasket

Camshaft Bearing Cap×9

×18

RH Exhaust Camshaft

LH Intake Camshaft

RH Intake Camshaft

LH Exhaust Camshaft

Camshaft Gear Spring

Camshaft Sub Gear

Camshaft Oil Seal

Camshaft Sub Gear

Wave Washer

Camshaft Gear Spring

×8

Snap Ring

Camshaft Oil Seal

Snap Ring

Camshaft Oil Seal

Wave Washer

Rear Plate

Housing Plug

Semi-Circular Plug

Spark Plug

RH Cylinder Head

LH Cylinder Head

LH Cylinder Head Gasket

Adjusting Shim

RH Cylinder Head Gasket

Oil Seal

Valve Lifter

Valve

Keeper

Spring Retainer

Valve Spring

Spring Seat

Snap Ring

Valve Guide Bushing

◆ Non-reusable part

Exploded view of the cylinder head component assembly—3.4L (5VZ-FE) engine

7924YG75

the sequence shown to 12 ft. lbs. (16 Nm).

    g. Remove the service bolt.

20. Check and adjust the valve clearance.

21. Install or connect the following:
- Semi-circular plugs
- Cylinder head covers. Tighten the bolts, in several passes, to 53 inch lbs. (6 Nm).
- Alternator bracket. Tighten the bolts to 14 ft. lbs. (18 Nm).
- No. 3 timing belt cover. Tighten the 6 bolts to 80 inch lbs. (9 Nm).
- CMP sensor. Tighten it to 71 inch lbs. (8 Nm).
- Intake air connector
- Hoses
- Alternator but do not tighten the bolts and nuts at this time
- No. 2 timing belt idler. Tighten the bolt to 30 ft. lbs. (40 Nm).

➡**Check that the pulley bracket moves smoothly.**

22. Install the left camshaft timing pulley, as follows:

    a. Install the knock pin to the camshaft.

    b. Align the knock pin hose of the camshaft with the knock pin groove of the timing pulley.

    c. Slide the timing pulley on the camshaft with the flange side facing outward. Tighten the pulley bolt to 81 ft. lbs. (110 Nm).

23. Set the No. 1 cylinder to TDC of the compression stroke.

    a. Turn the crankshaft pulley and align its groove with the timing mark **0** on the No. 1 timing belt cover.

    b. Turn the camshaft, align the knock pin hole of the camshaft with the timing mark of the No. 3 timing belt cover.

    c. Turn the camshaft timing pulley, align the timing marks of the camshaft timing pulley and the No. 3 timing belt cover.

24. Connect the timing belt to the left camshaft timing pulley, as follows:

    a. Check that the installation mark on the timing belt is aligned with the end of the No. 1 timing belt cover.

    b. Using Variable Pin Wrench Set 09960-01000 or equivalent, slightly turn the left camshaft timing pulley clockwise. Align the installation mark on the timing belt with the timing mark of the camshaft timing pulley, and hang the timing belt on the left camshaft timing pulley.

    c. Align the timing marks of the left

camshaft pulley and the No. 3 timing belt cover.

    d. Check that the timing belt has tension between the crankshaft timing pulley and the left camshaft timing pulley.

25. Install the right camshaft timing pulley and the timing belt, as follows:

    a. Align the installation mark on the timing belt with the timing mark of the right camshaft timing pulley, and hang the timing belt on the right camshaft timing pulley with the flange side facing inward.

    b. Slide the right camshaft timing pulley on the camshaft. Align the timing marks on the right camshaft timing pulley and the No. 3 timing belt cover.

    c. Align the knock pin hole of the camshaft with the knock pin groove of the pulley.

    d. Install the knock pin. Tighten the bolt to 81 ft. lbs. (110 Nm).

26. Set the timing belt tensioner, as follows:

    a. Using a press, slowly press in the pushrod using 220–2,205 lbs. (981–9,807 N) of force.

    b. Align the holes of the pushrod and housing, pass a 1.5mm hexagon wrench through the holes to keep the setting position of the pushrod.

    c. Release the press and install the dust boot to the tensioner.

27. Install the timing belt tensioner and alternately tighten the bolts to 20 ft. lbs. (28 Nm). Using pliers, remove the 1.5mm hexagon wrench from the belt tensioner.

28. Check the valve timing, as follows:

    a. Slowly turn the crankshaft pulley 2 revolutions from the TDC-to-TDC.

Always turn the crankshaft pulley clockwise.

    b. Check that each pulley aligns with the timing marks. If the timing marks do not align, remove the timing belt and reinstall it.

29. Install or connect the following:
- Fan bracket with the bolt and nut
- Power steering adjusting strut with the nut
- No. 2 timing belt cover. Tighten the bolts to 80 inch lbs. (9 Nm).
- Negative battery cable

30. Fill the radiator with engine coolant.
31. Start the engine and check for leaks.
32. Check the ignition timing.
33. Install the engine undercover.
34. Road test the vehicle.
35. Recheck all fluid levels.

Valve Lash

ADJUSTMENT

2.4L (2RZ-FE) Engine

1. Before servicing the vehicle, refer to the precautions in the beginning of this section.

2. Remove or disconnect the following:
- Negative battery cable
- Air intake connector
- Positive Crankcase Ventilation (PCV) hoses
- Spark plug wires
- 4 clamps and the engine wiring harness
- Air conditioning compressor con-

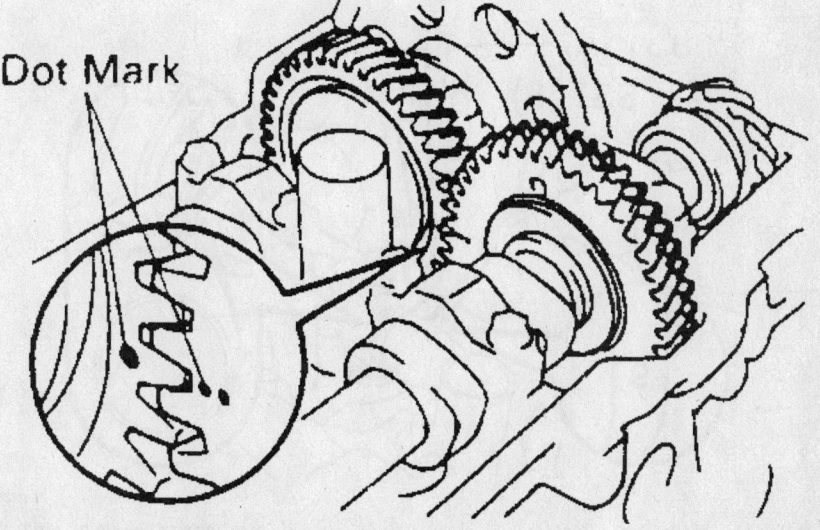

Aligning the timing marks—2.4L (2RZ-FE) and 2.7L (3RZ-FE) engines

7924YG76

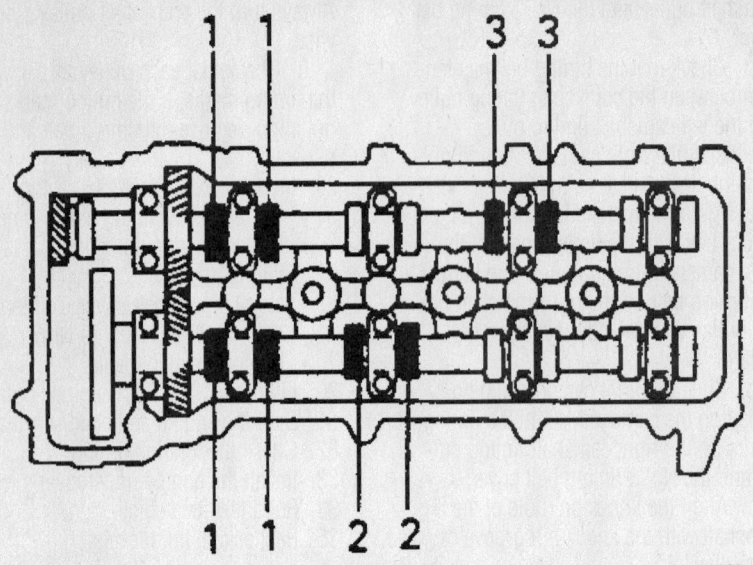

First valve adjustment—2.4L (2RZ-FE) and 2.7L (3RZ-FE) engines

7924YG77

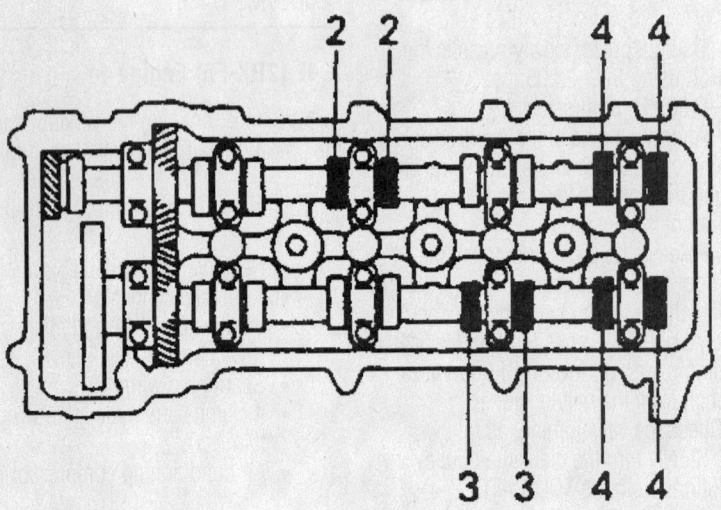

Second valve adjustment—2.4L (2RZ-FE) and 2.7L (3RZ-FE) engines

7924YG78

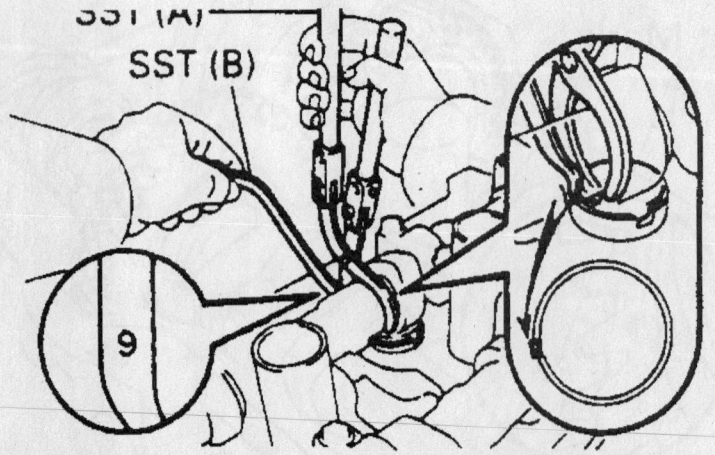

7924YG79

Removing adjusting shim using the special tools shown above—2.4L (2RZ-FE) and 2.7L (3RZ-FE) engines

nector, if equipped with air conditioning
- Oil pressure sensor connector
- Engine Coolant Temperature (ECT) sensor connector
- Distributor connector
- Cylinder head cover

3. Set the No. 1 cylinder to Top Dead Center (TDC) of the compression stroke, as follows:

   a. Turn the crankshaft pulley clockwise and align its groove with the **0** mark on the timing chain cover.

   b. Check that the timing marks (1 and 2 dots) of the camshaft drive and driven gears are in a straight line on the cylinder head surface. If not, turn the crankshaft 1 revolution (360 degrees) and align the marks.

4. Inspect the valve clearance, as follows:

   a. Measure the clearance between the valve lifter and the camshaft. Measure the 1st and 2nd intake and the 1st and 3rd exhaust valves.

   b. Turn the crankshaft pulley 1 revolution (360 degrees) and align the marks as above. Measure the 3rd and 4th intake and the 2nd and 4th exhaust valves.

5. Valve clearance "cold" should be:
- Intake: 0.006–0.010 in. (0.15–0.25mm)
- Exhaust: 0.010–0.014 in. (0.25–0.35mm)

6. Adjust the valve clearance by using adjusting shims, as follows:

   a. Turn the equipment driveshaft so that the cam lobe for the valve to be adjusted faces up.

   b. Using SST 09248-55040, press down the valve lifter and place SST 09248-05420, between the camshaft and the valve lifter.

   c. Remove SST 09248-55040.

   d. Remove the adjusting shim with a small flat prying tool and a magnetic finger.

   e. Determine the replacement adjusting shim size according to the following formula or use the adjusting shim charts.

   f. Using a micrometer, measure the thickness of the removed shim. Calculate the thickness of a new shim so that the valve clearance comes within the specified value:
- T: Thickness of the removed shim
- A: Measured valve clearance
- N: Thickness of the new shim

   g. Intake: N = T + (A—0.008 in. (0.20mm))

   h. Exhaust: N = T + (A—0.012 in. (0.30mm))

i. Install a new adjusting shim. Place it on the valve lifter. Using the SST 09248-55040, press down the valve lifter.

j. Remove the SST 09248-05420.

k. Recheck the valve clearance.

7. Install or connect the following:
- Cylinder head cover
- Engine wiring harness and clamps
- Distributor connector
- ECT sensor connector
- Oil pressure sensor connector
- Air conditioning compressor connector, if disconnected
- Spark plug wires
- PCV hoses
- Air intake connector
- Negative battery cable

8. Check the ignition timing.

2.7L (3RZ-FE) Engine

1. Before servicing the vehicle, refer to the precautions in the beginning of this section.

2. Disconnect the negative battery cable.

3. Drain the engine coolant.

4. Remove or disconnect the following:
- Intake air connector
- Positive Crankcase Ventilation (PCV) hoses
- Spark plug wires
- Engine wiring harness clamps and harness
- Air conditioning compressor connector, if equipped with air conditioning
- Oil pressure sensor connector
- Engine Coolant Temperature (ECT) sensor connector
- Distributor connector
- Cylinder head cover

5. Set the No. 1 cylinder to Top Dead Center (TDC) of the compression stroke, as follows:

a. Turn the crankshaft pulley clockwise and align its groove with the **0** mark on the timing chain cover.

b. Check that the timing marks (1 and 2 dots) of the camshaft drive and driven gears are in a straight line on the cylinder head surface. If not, turn the crankshaft 1 revolution (360 degrees) and align the marks.

6. Inspect the valve clearance, as follows:

a. Measure the clearance between the valve lifter and the camshaft. Measure the 1st and 2nd intake and the 1st and 3rd exhaust valves.

b. Turn the crankshaft pulley 1 revolution (360 degrees) and align the marks

as above. Measure the 3rd and 4th intake and the 2nd and 4th exhaust valves.

7. Valve clearance cold should be:
- Intake: 0.006–0.010 in. (0.15–0.25mm)
- Exhaust: 0.010–0.014 in. (0.25–0.35mm)

8. Adjust the valve clearance by using adjusting shims, as follows:

a. Turn the camshaft so the cam lobe for the valve to be adjusted faces up.

b. Using SST 09248-55040, press down the valve lifter and place SST 09248-05420, between the camshaft and the valve lifter. Remove SST 09248-55040.

c. Remove the adjusting shim with a small flat prying tool and a magnetic finger.

d. Determine the replacement adjusting shim size according to the following formula or use the adjusting shim charts.

e. Using a micrometer, measure the thickness of the removed shim. Calculate the thickness of a new shim so the valve clearance comes within the specified value.
- T: Thickness of the removed shim
- A: Measured valve clearance
- N: Thickness of the new shim

f. Intake: $N = T + (A — 0.008$ in. $(0.20mm))$

g. Exhaust: $N = T + (A — 0.012$ in. $(0.30mm))$

h. Install a new adjusting shim. Place it on the valve lifter. Using the SST 09248-55040, press down the valve lifter and remove SST 09248-05420.

i. Recheck the valve clearance.

9. Install or connect the following:
- Cylinder head cover
- Engine wiring harness and clamps
- Distributor connector
- ECT sensor connector
- Oil pressure sensor connector
- Air conditioning compressor connector, if disconnected
- Spark plug wires
- PCV hoses
- Intake air connector
- Negative battery cable

10. Refill with engine coolant.

11. Check the ignition timing.

3.4L (5VZ-FE) Engine

1. Before servicing the vehicle, refer to the precautions in the beginning of this section.

2. Disconnect the negative battery cable.

3. Drain the engine coolant.

4. Remove or disconnect the following:
- Air intake connector
- Cylinder head cover

5. Set the No. 1 cylinder to Top Dead Center (TDC) of the compression stroke, as follows:

a. Turn the crankshaft pulley clockwise and align its groove with the **0** mark on the timing chain cover.

b. Check that the timing marks (1 and 2 dots) of the camshaft drive and driven gears are in a straight line on the cylinder head surface. If not, turn the crankshaft 1 revolution (360 degrees) and align the marks.

6. Inspect the valve clearance, as follows:

a. Measure the clearance between the

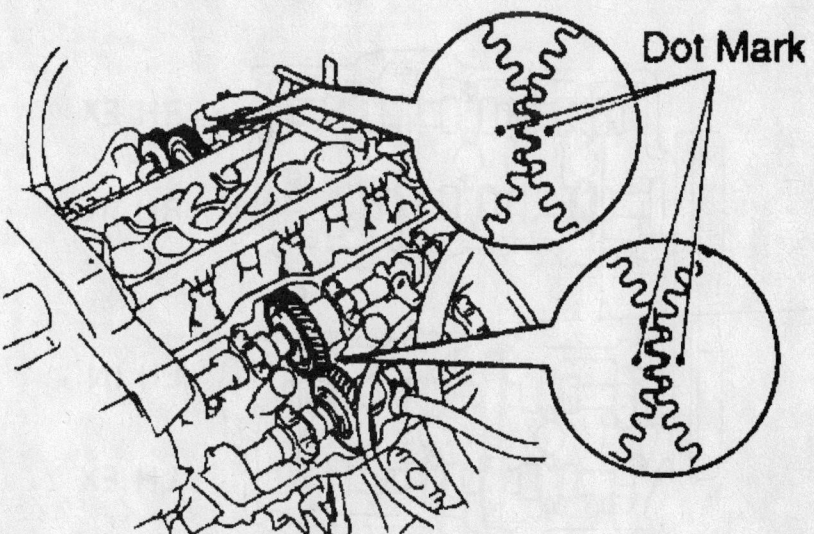

Aligning the timing marks—3.4L (5VZ-FE) engine

7924YG81

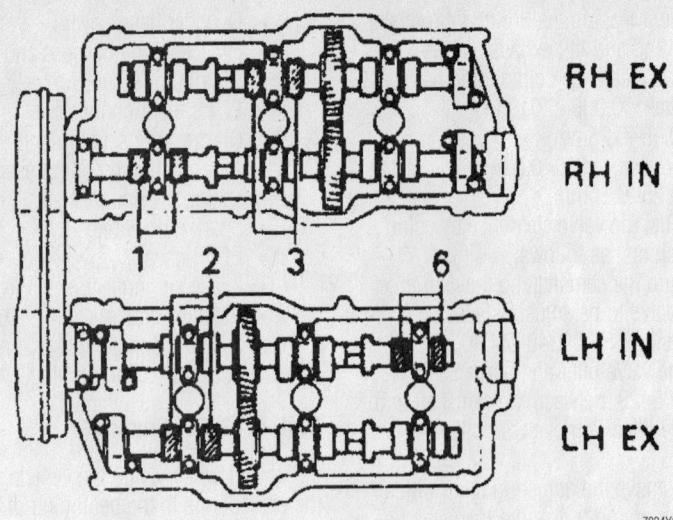

RH EX

RH IN

1 2 3 6

LH IN

LH EX

7924YG82

First valve adjustment—3.4L (5VZ-FE) engine

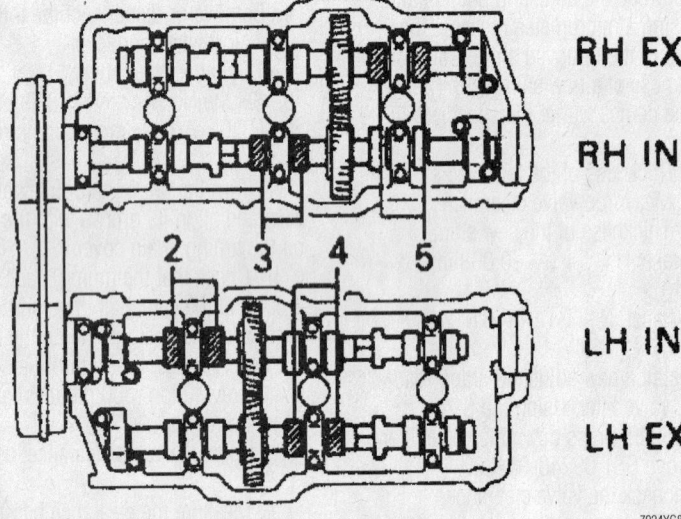

RH EX

RH IN

2 3 4 5

LH IN

LH EX

7924YG83

Second valve adjustment—3.4L (5VZ-FE) engine

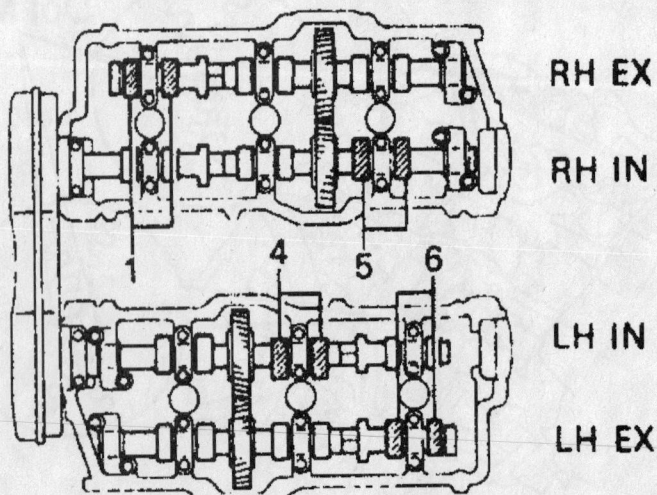

RH EX

RH IN

1 4 5 6

LH IN

LH EX

7924YG85

Removing the adjusting shim—3.4L (5VZ-FE) engine

valve lifter and the camshaft. Measure the 1st intake and the 3rd exhaust valves on the right head and the 6th intake and the 2nd exhaust valves on the left head.

b. Turn the crankshaft ⅔ of a revolution (240 degrees) and adjust the 3rd intake and the 5th exhaust valves on the right head and the 2nd intake and the 4th exhaust valves on the left head.

c. Turn the crankshaft ⅔ of a revolution (240 degrees) and adjust the 5th intake and the 1st exhaust valves on the right head and the 4th intake and the 6th exhaust valves on the left head.

7. Valve clearance cold should be:
- Intake: 0.006–0.009 in. (0.13–0.23mm)
- Exhaust: 0.011–0.014 in. (0.27–0.37mm)

8. Adjust the valve clearance by using adjusting shims, as follows:

a. Turn the equipment camshaft so that the cam lobe for the valve to be adjusted faces up.

b. Turn the valve lifter so that the notches are perpendicular to the camshaft.

c. Using SST 09248-55040, press down the valve lifter and place SST 09248-05420, between the camshaft and the valve lifter. Remove SST 09248-55040.

d. Remove the adjusting shim with a small flat prying tool and a magnetic finger.

e. Determine the replacement adjusting shim size according to the following formula or use the adjusting shim charts.

f. Using a micrometer, measure the thickness of the removed shim. Calculate the thickness of a new shim so that the valve clearance comes within the specified value.
- T: Thickness of the removed shim
- A: Measured valve clearance
- N: Thickness of the new shim

g. Intake: $N = T + (A—0.007$ in. $(0.18mm))$

h. Exhaust: $N = T + (A—0.013$ in. $(0.32mm))$

i. Install a new adjusting shim. Place it on the valve lifter. Using the SST 09248-55040, press down the valve lifter and remove SST 09248-05420.

j. Recheck the valve clearance.

9. Install or connect the following:
- Cylinder head cover
- Intake air connector
- Negative battery cable

10. Refill with engine coolant.

11. Start the engine and check for leaks.

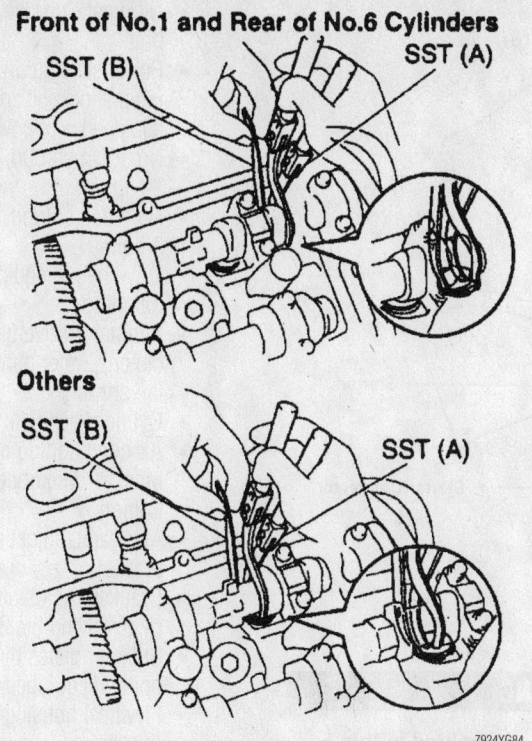

Front of No.1 and Rear of No.6 Cylinders

SST (B)   SST (A)

Others

SST (B)   SST (A)

7924YG84

Third valve adjustment—3.4L (5VZ-FE) engine

REMOVAL & INSTALLATION

2.4L Engine

1. Before servicing the vehicle, refer to the precautions in the beginning of this section.

2. Remove or disconnect the following:
   - Negative battery cable
   - Engine cover
   - Accelerator cable and intake air connector
   - Intake manifold assembly
   - Starter motor
   - Starter electrical connectors

To install:

3. Install or connect the following:
   - Starter. Tighten both bolts to 29 ft. lbs. (39 Nm).
   - Intake manifold
   - Accelerator cable and intake air connector
   - Engine cover
   - Negative battery cable

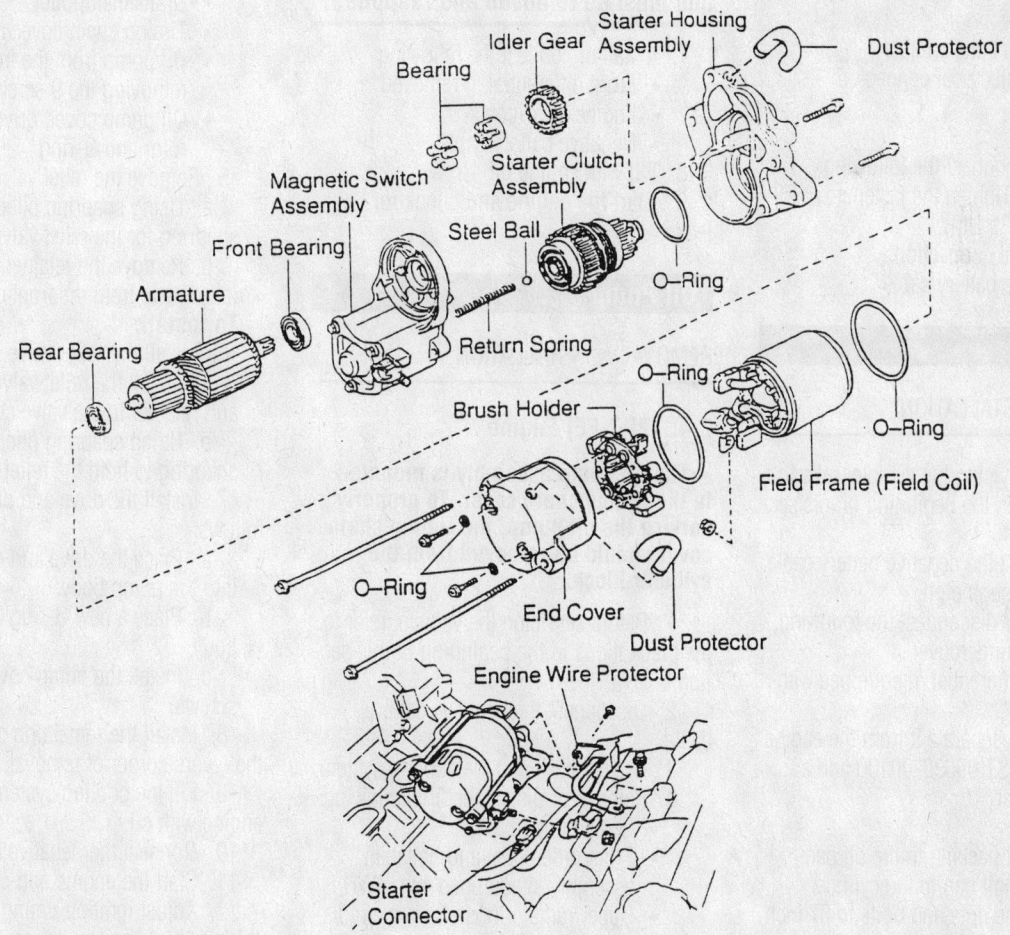

Idler Gear   Starter Housing Assembly   Dust Protector

Bearing

Magnetic Switch Assembly   Starter Clutch Assembly

Front Bearing   Steel Ball   O-Ring

Armature

Rear Bearing   Return Spring   O-Ring

Brush Holder   O-Ring

Field Frame (Field Coil)

O-Ring

O-Ring   End Cover   Dust Protector

Engine Wire Protector

Starter Connector

93162G18

This starter motor location—2.4L engine

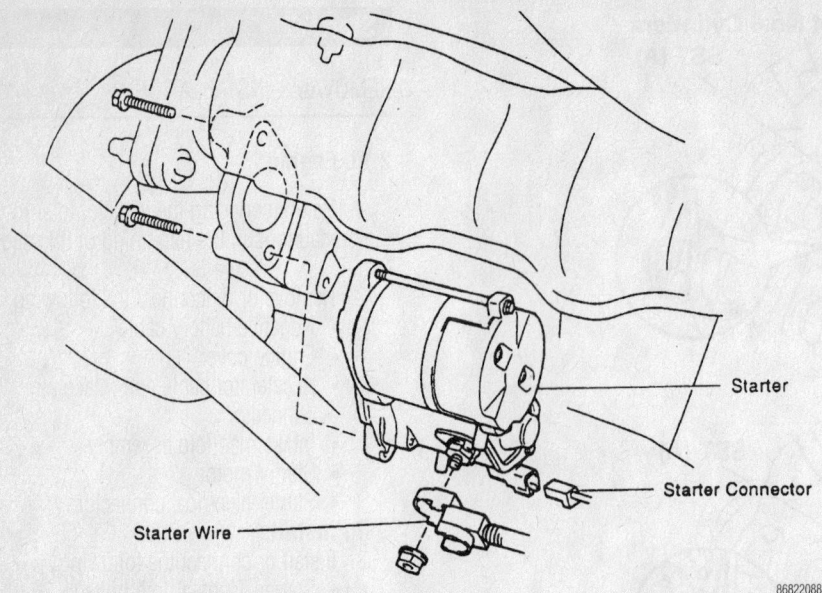

Exploded view of the starter—except 2.4L engine

Except 2.4L Engine

1. Before servicing the vehicle, refer to the precautions in the beginning of this section.

2. Remove or disconnect the following:

- Negative battery cable
- Starter electrical connectors
- Starter

To install:

3. Install or connect the following:

- Starter. Tighten the fasteners to 29 ft. lbs. (39 Nm).
- Electrical connections
- Negative battery cable

Oil Pan

REMOVAL & INSTALLATION

1. Before servicing the vehicle, refer to the precautions in the beginning of this section.

2. Disconnect the negative battery cable.

3. Drain the engine oil.

4. Remove or disconnect the following:

- Engine undercover
- Front differential, if equipped with 4WD
- Oil pan, separate it from the engine using SST 09032-00100 and a brass bar

To install:

5. Apply seal packing to the oil pan.

6. Install the oil pan to the cylinder block. Tighten the nuts and bolts to 67 inch lbs. (8 Nm)

※※ WARNING

If parts are not assembled within 5 minutes of applying time, the effectiveness of the seal packing is lost and must be removed and reapplied.

7. Install or connect the following:

- Front differential, if removed
- Engine undercover
- Negative battery cable

8. Fill with engine oil.

9. Start the engine and check for leaks.

Oil Pump

REMOVAL & INSTALLATION

2.4L (2RZ-FE) Engine

➡The oil pump assembly is mounted in the timing chain cover. To properly service the oil pump, the timing chain cover should be removed from the cylinder block.

1. Before servicing the vehicle, refer to the precautions in the beginning of this section.

2. Disconnect the negative battery cable.

3. Drain the oil and the cooling system.

4. Remove or disconnect the following:

- Engine undercover
- Front differential and halfshaft assembly, if equipped with 4WD
- Upper radiator hose from the radiator

- Oil dipstick guide, by removing the bolt
- Power steering drive belt, by loosening the lockbolt and adjusting bolt, if equipped with power steering
- No. 2 fan shroud, by removing the 2 clips
- No. 1 fan shroud, by removing the 4 bolts
- Drive belt, if equipped with air conditioning
- Alternator drive belt, fan (with fan clutch), water pump pulley and the fan shroud
- Cylinder head
- Air conditioning compressor and bracket, if equipped with air conditioning
- Alternator, adjusting bar and bracket
- Crankshaft Position (CKP) sensor, by removing the 2 bolts
- Stiffener plates by removing the 8 bolts, if equipped with 2WD
- Flywheel housing undercover and dust seal
- Oil pan.
- Oil strainer and gasket
- Crankshaft pulley
- Timing chain cover
- Oil pump from the front cover, by removing the 9 screws
- Oil pump cover, drive rotor, driven rotor and O-ring

5. Remove the relief valve, as follows:

a. Using snapring pliers, remove the snapring for the relief valve.

b. Remove the retainer, spring(s) and relief valve from the front cover.

To install:

6. Install the relief valve, as follows:

a. Install the relief valve, spring(s) and retainer to the valve cover.

b. Using snapring pliers, install the snapring to hold the relief valve.

7. Install the drive and driven rotors as follows:

a. Place the drive and driven rotors into the pump body.

b. Place a new O-ring to the pump body.

c. Install the pump cover with the 9 screws.

8. Install the remaining components in the reverse order of removal.

9. Fill the cooling system and fill the engine with oil.

10. Connect the negative battery cable.

11. Start the engine and check for leaks.

12. Adjust ignition timing. Road test the vehicle for proper operation.

13. Recheck all fluid levels.

2.7L (3RZ-FE) Engine

➡ **The oil pump assembly is mounted in the timing chain cover. To properly service the oil pump, the timing chain cover should be removed from the cylinder block.**

1. Before servicing the vehicle, refer to the precautions in the beginning of this section.

2. Disconnect the negative battery cable.

3. Drain the oil and the cooling system.

4. Remove or disconnect the following:

- Engine undercover
- Front differential and halfshaft assembly, if equipped with 4WD
- 2 bolts and the air pipe, for California vehicles with 3RZ-FE engine
- Upper radiator hose from the radiator
- Oil dipstick guide
- Power steering drive belt, by loosening the lockbolt and adjusting bolt, if equipped with power steering
- No. 2 fan shroud by removing the 2 clips
- No. 1 fan shroud by removing the 4 bolts
- Drive belt, by loosening the idler pulley nut and adjusting bolt, if equipped with air conditioning
- Alternator drive belt, fan (with fan clutch), water pump pulley and the fan shroud
- Cylinder head
- Air conditioning compressor and bracket with the lines attached, if equipped with air conditioning
- Alternator, adjusting bar and bracket
- Crankshaft Position (CKP) sensor by removing the 2 bolts
- Stiffener plates by removing the 8 bolts, if equipped with 2WD
- Flywheel housing undercover and dust seal
- Oil pan
- Oil strainer and gasket
- Crankshaft pulley
- Timing chain cover
- Oil pump from the front cover by removing the 9 screws
- Oil pump cover, drive rotor, driven rotor and O-ring

5. Remove the relief valve, as follows:

a. Using snapring pliers, remove the snapring for the relief valve.

b. Remove the retainer, spring(s) and relief valve from the front cover.

To install:

6. Install the relief valve, as follows:

a. Install the relief valve, spring(s) and retainer to the valve cover.

b. Using snapring pliers, install the snapring to hold the relief valve.

7. Install the drive and driven rotors, as follows:

a. Place the drive and driven rotors into the pump body.

b. Place a new O-ring to the pump body.

c. Install the pump cover with the 9 screws.

8. Install the remaining components in the reverse order of removal.

9. Fill the cooling system and fill the engine with oil.

10. Connect the negative battery cable.

11. Start the engine and check for leaks.

12. Adjust ignition timing. Road test the vehicle for proper operation.

13. Recheck all fluid levels.

3.4L (5VZ-FE) Engine

1. Before servicing the vehicle, refer to the precautions in the beginning of this section.

2. Remove or disconnect the following:

- Negative battery cable
- Engine undercover
- Crankshaft timing pulley
- Front differential, if equipped with 4WD

3. Drain the engine oil from the engine.

4. Remove or disconnect the following:

- Timing belt and crankshaft gear
- Oil cooler tube and clamp, if equipped with automatic transmission
- Stiffener plate
- Flywheel housing undercover and dust cover
- Rear end cover and dust cover
- Starter wire clamp
- Crankshaft Position (CKP) sensor
- Oil pan

➡ **Be careful not to damage the baffle plate flange.**

- Oil strainer
- Oil baffle plate
- Oil pump body by removing the 8 bolts.
- O-ring from the cylinder block

To install:

5. Install or connect the following:

- Apply Seal Packing PN 08826-00080 to the oil pump
- New O-ring into the groove of the cylinder block
- Oil pump to the crankshaft with the spline teeth of the drive rotor engaged with the large teeth of the crankshaft. Tighten the oil pump bolts "A" 15 ft. lbs. (20 Nm) and bolts "B" 31 ft. lbs. (42 Nm)
- CKP
- Oil pan baffle plate
- Oil strainer with a new gasket.

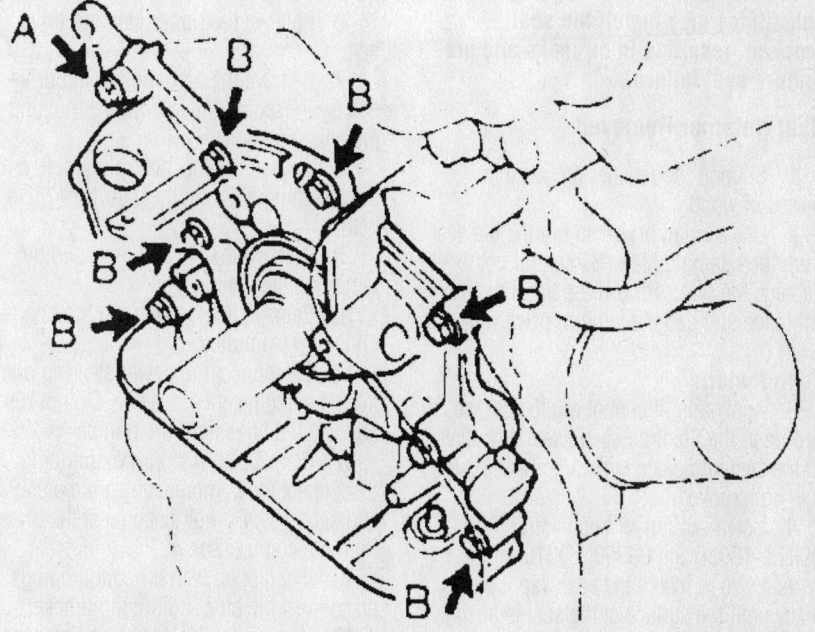

Oil pump bolt identification—3.4L (5VZ-FE) engine

7924YG89

Tighten the bolts to 13 ft. lbs. (18 Nm).
- Remaining components
- Negative battery cable

6. Fill with engine oil.
7. Start the engine and check for leaks.

Rear Main Seal

REMOVAL & INSTALLATION

Seal Retainer On Engine

1. Remove the transmission.
2. Remove the clutch cover assembly and flywheel (manual trans.) or the flexplate (automatic trans.).
3. Use a small, sharp knife to cut off the lip of the oil seal. Take great care not to score any metal with the knife.
4. Use a small prybar to pry the old seal from the retaining plate. Be careful not to damage the plate. Protect the tip of the tool with tape and pad the fulcrum point with cloth.
5. Inspect the crankshaft and seal lip contact surfaces for any sign of damage.

To install:

6. Apply a light coat of multi-purpose grease to the lip of a new oil seal. Loosely fit the seal into place by hand, making sure it is not crooked.
7. Use a seal driver such as (SST 09223–15030 and 09950–70010) of the correct size to install the seal. Tap it into place until the surface of the seal is flush with the edge of the housing.

➡**Use the correct tools. Homemade substitutes may install the seal crooked, resulting in oil leaks and premature seal failure.**

Seal Retainer Removed

1. Support the retainer on two thin pieces of wood.
2. Use a small prybar to pry the old seal from the retaining plate. Be careful not to damage the plate. Protect the tip of the tool with tape and pad the fulcrum point with cloth.

To install:

3. Apply a light coat of multi-purpose grease to the lip of a new oil seal. Loosely fit the seal into place by hand, making sure it is not crooked.
4. Use a seal driver such as (SST 09223–15030 and 09950–70010) of the correct size to install the seal. Tap it into place until the surface of the seal is flush with the edge of the housing.

Timing Belt

REMOVAL & INSTALLATION

3.4L Engine

1. Disconnect the negative battery cable.

✳✳ CAUTION

Work must be started after 90 seconds from the time the ignition switch is turned to the LOCK position and the negative battery cable is disconnected.

2. Raise and safely support the vehicle.
3. Remove the engine undercover.
4. Drain the engine coolant.

✳✳ CAUTION

Never open, service or drain the radiator or cooling system when hot; serious burns can occur from the steam and hot coolant. Also, when draining engine coolant, keep in mind that cats and dogs are attracted to ethylene glycol antifreeze and could drink any that is left in an uncovered container or in puddles on the ground. This will prove fatal in sufficient quantities. Always drain coolant into a sealable container. Coolant should be reused unless it is contaminated or is several years old.

5. Disconnect the upper radiator hose from the engine.
6. Remove the power steering drive belt.
7. Remove the air conditioning drive belt by loosening the idler pulley nut and the adjusting bolt.
8. Loosen the lockbolt, pivot bolt, and the adjusting bolt and the alternator drive belt.
9. Remove the No. 2 fan shroud by removing the 2 clips.
10. Remove the fan with the fluid coupling and fan pulleys.
11. Disconnect the power steering pump from the engine and set aside. Do not disconnect the lines from the pump.
12. If equipped with air conditioning, disconnect the compressor from the engine and set aside. Do not disconnect the lines from the compressor.
13. If equipped with air conditioning, disconnect the air conditioning bracket.
14. Remove the No. 2 timing belt cover, as follows:

a. Detach the camshaft position sensor connector from the No. 2 timing belt cover.
b. Disconnect the 3 spark plug wire clamps from the No. 2 timing belt cover.
c. Remove the 6 bolts and remove the timing belt cover.

15. Remove the fan bracket, as follows:

a. Remove the power steering adjusting strut by removing the nut.
b. Remove the fan bracket by removing the bolt and nut.

16. Set the No. 1 cylinder at Top Dead Center (TDC) of the compression stroke, as follows:

a. Turn the crankshaft pulley and align its groove with the timing mark **0** of the No. 1 timing belt cover.
b. Check that the timing marks of the camshaft timing pulleys and the No. 3 timing belt cover are aligned. If not, turn the crankshaft pulley one revolution (360 degrees).

➡**If reusing the timing belt, be sure that you can still read the installation marks. If not, place new installation marks on the timing belt to match the timing marks of the camshaft timing pulleys.**

17. Remove the timing belt tensioner by alternately loosening the 2 bolts.
18. Remove the camshaft timing pulleys, as follows:

a. Using SST 09960-10010, or equivalent, remove the pulley bolt, the timing pulley and the knock pin. Remove the 2 timing pulleys with the timing belt.

19. Remove the crankshaft pulley, as follows:

a. Using SST 09213-54015 and 09330-00021 or their equivalents, loosen the pulley bolt.
b. Remove the SST tool, the pulley bolt and the pulley.

20. Remove the starter wire bracket and the No. 1 timing belt cover.
21. Remove the timing belt guide and remove the timing belt.
22. Remove the bolt and the No. 2 idler pulley.
23. Remove the pivot bolt, the No. 1 idler pulley and the plate washer.
24. Remove the crankshaft gear.

To install:

25. Install the crankshaft timing gear.

a. Align the timing pulley set key with the key groove of the gear.
b. Using SST 09214-60010, or equivalent, and a hammer, tap in the timing gear with the flange side facing inward.

26. Install the plate washer and the No. 1

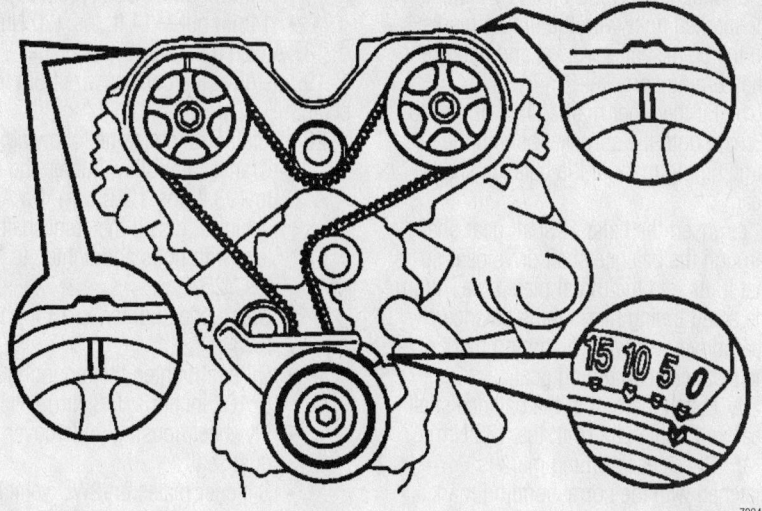

79245G37

Turn the crankshaft clockwise to align the timing marks before removing the timing belt—3.4L engine

idler pulley with the pivot bolt and tighten it to 26 ft. lbs. (35 Nm). Check that the pulley bracket moves smoothly.

27. Install the No. 2 timing belt idler with the bolt. Tighten the bolt to 30 ft. lbs. (40 Nm). Check that the pulley bracket moves smoothly.

28. Temporarily install the timing belt, as follows:

a. Using the crankshaft pulley bolt, turn the crankshaft and align the timing marks of the crankshaft timing pulley and the oil pump body.

b. Align the installation mark on the timing belt with the dot mark of the crankshaft timing pulley.

c. Install the timing belt on the crankshaft timing pulley, No. 1 idler pulley and the water pump pulleys.

29. Install the timing belt guide with the cup side facing outward.

30. Install the No. 1 timing belt cover and starter wire bracket. Tighten the timing belt cover bolts to 80 inch lbs. (9 Nm).

✳✳ WARNING

If any binding is felt when adjusting the timing belt tension by turning the crankshaft, STOP turning the engine, because the pistons may be hitting the valves.

31. Install the crankshaft pulley, as follows:

a. Align the pulley set key with the key groove of the crankshaft pulley.

b. Install the pulley bolt and tighten it to 184 ft. lbs. (250 Nm).

32. Install the left camshaft timing pulley.

a. Install the knock pin to the camshaft.

b. Align the knock pin hose of the camshaft with the knock pin groove of the timing pulley.

c. Slide the timing belt pulley on the camshaft with the flange side facing outward. Tighten the pulley bolt to 81 ft. lbs. (110 Nm).

33. Set the No. 1 cylinder to TDC of the compression stroke, as follows:

a. Turn the crankshaft pulley, and align its groove with the timing mark **0** of the No. 1 timing belt cover.

b. Turn the camshaft to align the knock pin hole of the camshaft with the timing mark of the No. 3 timing belt cover.

c. Turn the camshaft timing pulley, and align the timing marks of the camshaft timing pulley and the No. 3 timing belt cover.

34. Connect the timing belt to the left camshaft timing pulley, as follows:

➡ **Check that the installation mark on the timing belt is aligned with the end of the No. 1 timing belt cover.**

a. Using SST 09960-01000, or equivalent, slightly turn the left camshaft timing pulley clockwise. Align the installation mark on the timing belt with the timing mark of the camshaft timing pulley and hang the timing belt on the left camshaft timing pulley.

b. Align the timing marks of the left camshaft pulley and the No. 3 timing belt cover.

c. Check that the timing belt has tension between the crankshaft timing pulley and the left camshaft timing pulley.

35. Install the right camshaft timing pulley and the timing belt, as follows:

a. Align the installation mark on the timing belt with the timing mark of the right camshaft timing pulley and hang the timing belt on the right camshaft timing pulley with the flange side facing inward.

b. Slide the right camshaft timing pulley on the camshaft. Align the timing marks on the right camshaft timing pulley and the No. 3 timing belt cover.

c. Align the knock pin hole of the camshaft with the knock pin groove of the pulley and install the knock pin. Install the bolt and tighten it to 81 ft. lbs. (110 Nm).

36. Set the timing belt tensioner, as follows:

a. Using a press, slowly press in the pushrod using 220–2205 lbs. (981–9807 N) of force.

b. Align the holes of the pushrod and housing, pass a 1.5mm hex wrench through the holes to keep the setting position of the pushrod.

c. Release the press and install the dust boot on the tensioner.

37. Install the timing belt tensioner and alternately tighten the bolts to 20 ft. lbs. (28 Nm). Using pliers, remove the 1.5mm hex wrench from the belt tensioner.

38. Check the valve timing, as follows:

a. Slowly turn the crankshaft pulley 2 revolutions from TDC to TDC. Always turn the crankshaft pulley clockwise.

b. Check that each pulley aligns with the timing marks. If the timing marks do not align, remove the timing belt and reinstall it.

39. Install the fan bracket with the bolt and nut.

40. Install the power steering adjusting strut with the nut.

41. Install the No. 2 timing belt cover. Tighten the bolts to 80 inch lbs. (9 Nm). Install the remaining components.

42. Fill the cooling system with coolant.

43. Connect the negative battery cable.

44. Start the engine and check for leaks.

Timing Chain Cover, Seal and Timing Chain

REMOVAL & INSTALLATION

2RZ-FE and 3RZ-FE

1. Before servicing the vehicle, refer to the precautions in the beginning of this section.

2. Disconnect the negative battery cable.

3. Drain the engine coolant.

4. Remove or disconnect the following:
- Engine undercover
- Engine oil
- On 4WD vehicles, the front differential
- Alternator belt, fan with coupling and the water pump pulley
- Cylinder head
- A/C belt, compressor, and the bracket
- Alternator adjusting bar and bracket
- Crankshaft position sensor and O-ring
- On 2WD vehicles, the stiffener plates
- Flywheel housing undercover and dust seal
- Oil pan
- Oil strainer and gasket
- Crankshaft pulley
- Water bypass pipe
- Chain cover assembly. Remove the bolts shown by the arrows.
- No. 1 timing chain and camshaft gear
- Crankshaft timing gear
- No. 1 timing chain tensioner slipper and No. 1 vibration damper
- No. 1 damper

5. On the 2RZ-FE remove the crankshaft position sensor rotor and the timing chain oil jet.

6. On the 3RZ-FE, remove the No. 2 and No. 3 vibration dampers and the No. 2 chain tensioner as follows:

a. Install a pin in the No. 2 tensioner and lock the plunger.

b. Remove the bolt and the No. 2 damper.

c. Remove the 2 bolts and the No. 3 damper.

d. Remove the nut and the No. 2 tensioner.

7. Remove the balance shaft driven gear, shaft, No. 2 timing chain and the No. 2 crankshaft sprocket, as follows:

a. Unbolt the balance shaft driven gear.

b. Remove the balance shaft gear with the shaft.

c. Remove the No. 2 timing chain with the No. 2 crankshaft timing sprocket.
- Remove the No. 4 vibration damper.

To install:

8. Install the No. 4 vibration dampener.

9. Install the No. 2 timing chain, No. 2 crankshaft timing sprocket, balance shaft drive gear and shaft as follows:

a. Install the No. 2 chain by matching the marked links with the timing marks on the crankshaft sprocket and balance shaft timing sprocket.

b. Fit the other marked link of the No. 2 chain onto the sprocket behind the large timing mark of the balance shaft gear.

c. Insert the balance shaft gear shaft through the balance shaft drive gear so that it fits into the thrust plate hole. Align the small timing mark of the balance shaft drive gear with the timing mark of the balance shaft timing gear.

d. Install the bolt to the balance shaft gear and tighten to 18 ft. lbs. (25 Nm).

e. Check each timing mark is matched with the corresponding mark link.

10. Install the No. 2, No. 3 vibration dampers and the No. 2 chain tensioner, as follows:

➡**Assemble the chain tensioner with the pin installed, then remove the pin after assembly.**

a. Install the No. 2 chain tensioner with the nut, tighten to 13 ft. lbs. (18 Nm).

b. Install the No. 3 damper with the bolts, tighten to 13 ft. lbs. (18 Nm).

c. Install the No. 2 damper, tighten to 20 ft. lbs. (27 Nm).

d. Remove the pin from the No. 2 chain tensioner and free the plunger.

11. On the 2RZ-FE remove the crankshaft position sensor rotor and the timing chain oil jet.

12. Install or connect the following:
- No. 1 timing chain tensioner slipper and the No. 1 vibration damper. Torque the No. 1 damper to 22 ft. lbs. (29 Nm); torque the slipper to 20 ft. lbs. (27 Nm). Check that the slipper moves smoothly.
- Crankshaft timing gear.
- No. 1 timing chain and camshaft timing gear.

13. Align the timing mark between the marked link of the No. 1 timing chain, and install the No. 1 timing chain to the gear.

14. Align the timing mark of the crankshaft timing gear with the mark of the No. 1 timing chain, then install the No. 1 timing chain.

15. Tie the No. 1 chain with a wire or cord, make sure it does not come loose.

16. Install or connect the following:
- Timing chain cover assembly

17. Tighten the following:
- 12mm **A** bolts—14 ft. lbs. (20 Nm)
- 12mm **B** bolts—18 ft. lbs. (25 Nm)
- 14mm bolts—32 ft. lbs. (44 Nm)
- 14mm nut—14 ft. lbs. (20 Nm)

18. Attach the water bypass pipe.

19. Remove the cord or wire from the chain.

20. Install or connect the following:
- Crankshaft pulley, tighten the bolt to 193 ft. lbs. (260 Nm). On A/C vehicles, install the crankshaft pulleys with bolts and tighten to 18 ft. lbs. (25 Nm).
- Oil strainer, tighten to 13 ft. lbs. (18 Nm)
- Oil pan, tighten the mounting bolts to 108 inch lbs. (13 Nm)
- Flywheel housing undercover and dust seal
- Stiffener plates on 2WD vehicles, tighten to 27 ft. lbs. (37 Nm)
- Crankshaft position sensor with a new O-ring
- Alternator, adjusting bar and bracket
- A/C compressor and bracket
- Cylinder head
- Water pump pulley and the fluid coupling with the fan
- On 4WD vehicles, the front differential and driveshaft assemblies

21. Adjust the drive belt tension.

22. Fill with engine coolant and engine oil.

23. Install the engine undercover.

24. Connect the negative battery cable.

SEAL REPLACEMENT

Cover Removed

1. Unbolt the timing chain cover assembly. Be careful to loosen only the correct bolts.

2. Pry out the seal from the cover with a flat-bladed tool.

3. It is a good idea to remove the oil pump from the timing cover and replace the O-ring.

To install:

4. Clean and inspect the timing cover area. Install new gaskets around the dowel areas and pump spline.

5. Apply multi-purpose grease to the new oil seal lip.

6. Tap the seal into place with SST 09223–50010/60010 or equivalent, and a hammer. Do this until the seal surface is flush with the cover edge.

7. Install the cover, tighten the bolts as specified for your engine.

8. If the oil pump was removed, install a new O-ring behind the pump prior to installation.

Cover Installed

1. Unbolt and remove the oil pump.

2. Using a knife, carefully cut off the oil seal lip. With a flat-bladed tool, (preferably with tape around it) pry the seal from the cover.

To install:

3. Apply multi-purpose grease to the new oil seal lip.

4. Tap the seal into place with SST 09223–50010/60011 or equivalent seal driver, and a hammer. Do this until the seal surface is flush with the cover edge.

5. Install the oil pump with a new O-ring.

Piston and Ring

POSITIONING

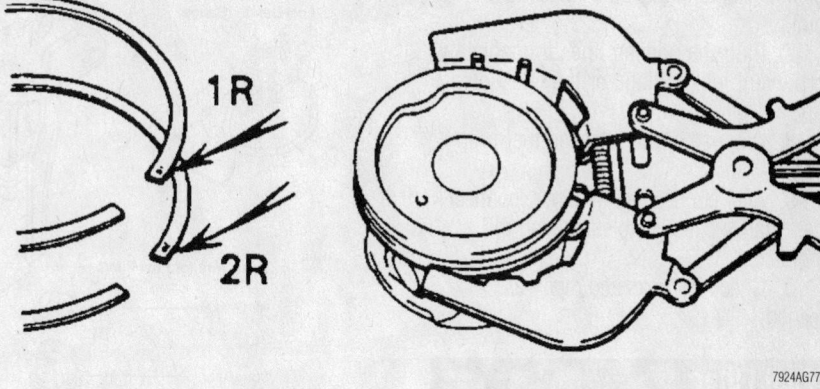

Compression ring identification mark locations—2.7L engine

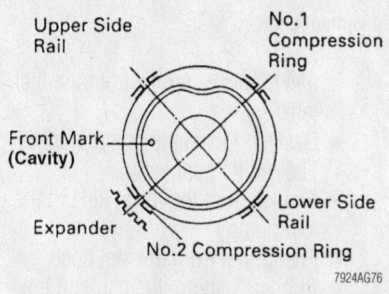

Piston ring end-gap spacing—2.4L (2RZ-FE) and 2.7L engines

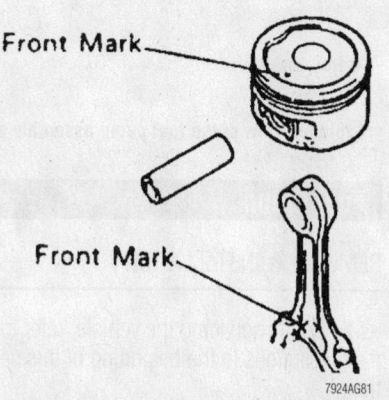

Piston to connecting rod assembly—2.4L (2RZ-FE), 2.7L and 3.4L engines

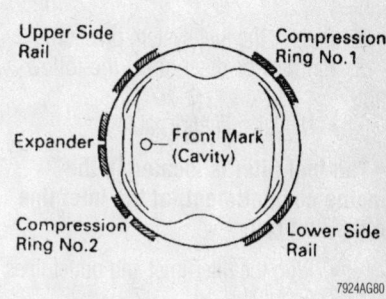

Piston ring end-gap spacing—3.4L engine

FUEL SYSTEM

Fuel System Service Precautions

Safety is the most important factor when performing not only fuel system maintenance, but any type of maintenance. Failure to conduct maintenance and repairs in a safe manner may result in serious personal injury or death. Work on a vehicle's fuel system components can be accomplished safely and effectively by adhering to the following rules and guidelines.

• To avoid the possibility of fire and personal injury, always disconnect the negative battery cable unless the repair or test procedure requires that battery voltage be applied.

• Always relieve the fuel system pressure prior to disconnecting any fuel system component (injector, fuel rail, pressure regulator, etc.) fitting or fuel line connection. Exercise extreme caution whenever relieving fuel system pressure, to avoid exposing

skin, face and eyes to fuel spray. Please be advised that fuel under pressure may penetrate the skin or any part of the body that it contacts.

• Always place a shop towel or cloth around the fitting or connection prior to loosening to absorb any excess fuel due to spillage. Ensure that all fuel spillage is quickly remove from engine surfaces. Ensure that all fuel-soaked cloths or towels are deposited into a flame-proof waste container with a lid.

• Always keep a dry chemical (Class B) fire extinguisher near the work area.

• Do not allow fuel spray or fuel vapors to come into contact with a light bulb, spark or open flame.

• Always use a second wrench when loosening or tightening fuel line connection fittings. This will prevent unnecessary stress and torsion to fuel piping. Always follow the proper torque specifications.

• Always replace worn fuel fitting O-rings with new ones. Do not substitute fuel hose where rigid pipe is installed.

Fuel System Pressure

RELIEVING

1. Before servicing the vehicle, refer to the precautions in the beginning of this section.

2. Disconnect the negative battery terminal.

3. Place a catch-pan under the joint to be disconnected. A large quantity of fuel may be released when the joint is opened.

4. Wear eye or full face protection.

5. Place a shop towel over the area and slowly loosen the joint using a wrench of the correct size. Use a back-up wrench if needed.

6. Allow the fuel left in the line to bleed off slowly before fully disconnecting the joint.

7. Plug the opened lines immediately to prevent fuel spillage or the entry of dirt.

8. Dispose of the released fuel properly.

9. After connecting fuel lines, connect the negative battery cable and start the engine.

10. Check for leaks and repair as needed.

Fuel Filter

REMOVAL & INSTALLATION

1. Before servicing the vehicle, refer to the precautions in the beginning of this section.

2. Relieve the fuel system pressure.

3. Remove or disconnect the following:

- Negative battery cable

➡ The fuel filter is located in the engine compartment, at the inlet line to the fuel rail.

- Plug the filter inlet and outlet lines
- Fuel filter
- Bracket from the fuel filter

To install:

4. Install or connect the following:

- Fuel filter bracket to the fuel filter
- Fuel filter. Tighten the 2 bolts to 14 ft. lbs. (20 Nm).
- New gaskets. Tighten the union bolts to 22 ft. lbs. (30 Nm).
- Negative battery cable

5. Start the engine and check for leaks.

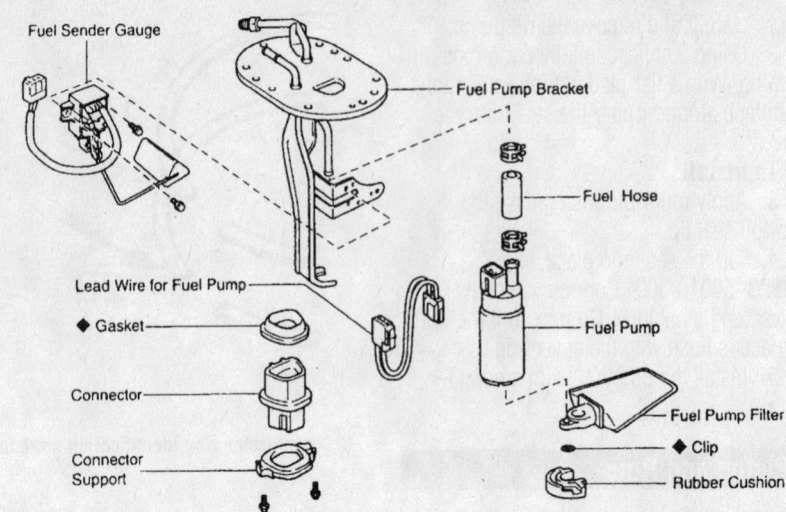

Reference (2WD)

Fuel Sender Gauge — Fuel Pump Bracket — Fuel Hose — Lead Wire for Fuel Pump — Fuel Pump — ◆ Gasket — Connector — Fuel Pump Filter — Connector Support — ◆ Clip — Rubber Cushion

◆ Non-reusable part

7924YG91

Exploded view of the fuel pump assembly and related components

Fuel Pump

REMOVAL & INSTALLATION

1. Before servicing the vehicle, refer to the precautions in the beginning of this section.

2. Relieve the fuel pressure.

3. Disconnect the negative battery cable from the battery.

4. Drain the fuel from the fuel tank.

5. Remove or disconnect the following:

- Fuel tank
- Fuel pump connector from the clamp
- Access plate bolts, then pull out the fuel pump assembly from the fuel tank
- Gasket(s) from the pump bracket
- Fuel pump connector
- Bracket from the lower side of the fuel pump
- Fuel pump from the fuel hose
- Rubber cushion, the clip and the fuel filter at the bottom of the fuel pump

To install:

6. Install or connect the following:

- Fuel pump filter to the fuel pump with a new clip
- Fuel pump to the fuel pump bracket
- Fuel hose to the outlet port of the fuel pump
- Fuel pump connector
- Fuel pump assembly with a new gasket(s). Tighten the bolts to 31 inch lbs. (4 Nm).
- Fuel pump connector to the clamp
- Fuel tank
- All electrical and fuel connections
- Negative battery cable

7. Refill the fuel tank and check for leaks.

Fuel Injector

REMOVAL & INSTALLATION

2.4L and 2.7L Engines

1. Before servicing the vehicle, refer to the precautions in the beginning of this section.

Pressure Regulator — SST — SST — SST — Injector — SST — SST — SST — Fuel Filter

7924YG90

Exploded view of the fuel delivery components—2.4L (2RZ-FE) and 2.7L (3RZ-FE) engines

2. Relieve the fuel system pressure.
3. Remove or disconnect the following:
   - Throttle body
   - Fuel injector electrical connectors
   - Crankshaft Position (CKP) sensor connector
   - Knock Sensor (KS) connector
   - Data Link Connector 1 (DLC1) and wire clamp from the brackets
   - Vacuum line from the fuel pressure regulator
   - Fuel return hose from the pressure regulator
   - Union bolt and gaskets
   - Fuel inlet pipe from the fuel rail
   - Fuel rail with the injectors attached

✳✳ WARNING

The injectors are only retained by their O-rings and will tend to drop out of the fuel rail.

   - 4 insulators from the four spacers
   - Fuel injectors from the fuel rail
   - O-ring and grommet, discard them

To install:

4. Install or connect the following:
   - New grommets and O-rings on each injector, lubricated with a light coat of gasoline
   - Fuel injectors, with the electrical connector facing upwards
   - New insulators and spacers on the intake manifold
5. Temporarily install the bolts holding the delivery pipe to the intake manifold.
6. Check that the injectors rotate smoothly.
7. Install or connect the following:
   - Tighten the delivery pipe-to-intake manifold bolts to 15 ft. lbs. (21 Nm).
   - Injector electrical connectors
   - Fuel inlet pipe with new gaskets. Tighten the union bolt to 22 ft. lbs. (29 Nm) and the bolt to 14 ft. lbs. (20 Nm).
   - Fuel return pipe to the fuel pressure regulator
   - Vacuum line to the pressure regulator
   - Throttle body
   - Negative battery cable

3.4L Engine

1. Before servicing the vehicle, refer to the precautions in the beginning of this section.
2. Depressurize the fuel system.
3. Remove or disconnect the following:
   - Air cleaner hose

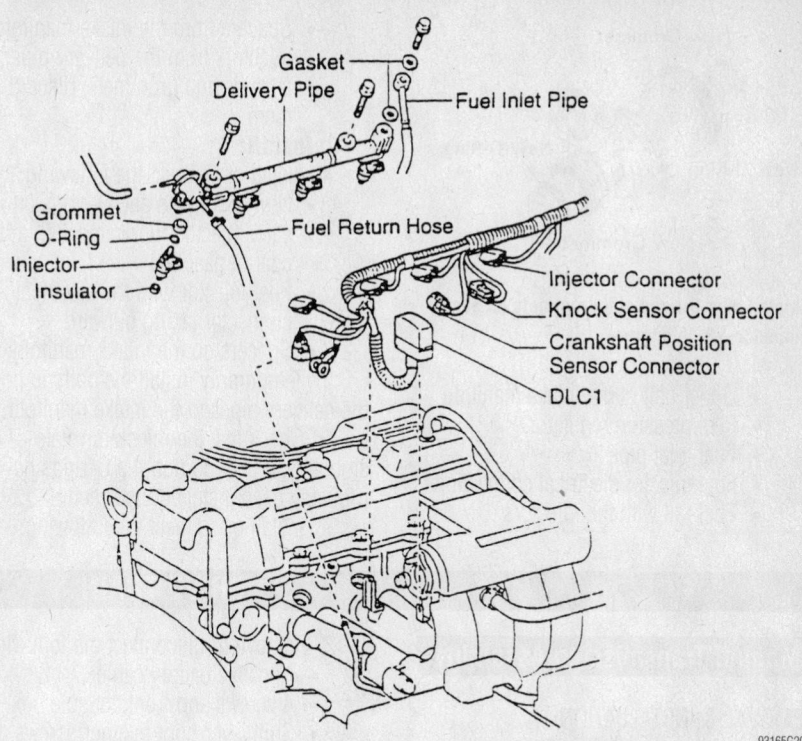

Fuel injector arrangement and related components—2.4L and 2.7L engines

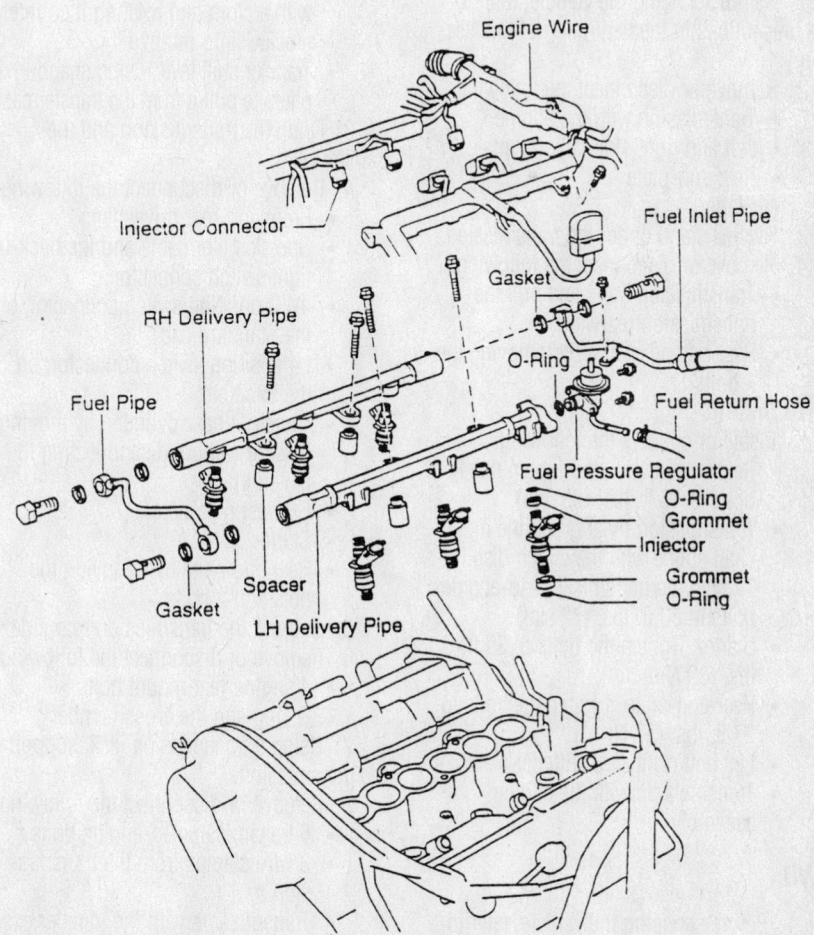

Fuel injector arrangement and related components—3.4L engine

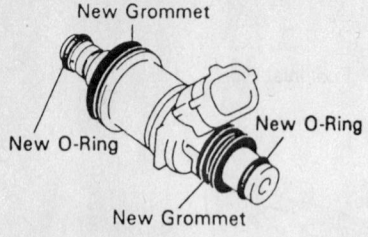

Install new O-rings and grommets on each injector—3.4L engine

- Upper half of the intake manifold
- Fuel pressure regulator
- Fuel inlet pipe
- Fuel injector electrical connections
- Fuel rail with the injectors

- Spacers from the intake manifold
- Injectors from the delivery pipes
- O-rings and grommets, discard them

To install:

4. Install or connect the following:
- New grommets and O-rings on each injector, lubricated with a light coat of gasoline
- Fuel injector with the electrical connector facing outward
- Spacers on the intake manifold

5. Temporarily install the bolts to hold the delivery pipes to the intake manifold.

6. Check that the injectors rotate smoothly. If they do not, the O-rings have probably been installed incorrectly.

7. Install or connect the following:

- Fuel injector electrical connectors
- Fuel pipe with new gaskets. Tighten the bolts to 25 ft. lbs. (34 Nm) and the delivery pipes-to-intake manifold bolts to 10 ft. lbs. (13 Nm).
- Fuel pipe union with new gaskets. Tighten the clamp bolt to 71 inch lbs. (8 Nm).
- Fuel pressure regulator

8. Inspect the vacuum lines and connections. Look for any loose connections, sharp bends or damage.

9. Install or connect the following:
- Air cleaner
- Air cleaner hose

10. Start the engine and check for vacuum and fuel leaks.

DRIVE TRAIN

Manual Transmission Assembly

REMOVAL & INSTALLATION

2WD

1. Before servicing the vehicle, refer to the precautions in the beginning of this section.

2. Remove or disconnect the following:
- Transmission with the engine
- Left and right side stiffener plates
- Rear end-plate
- Starter

3. Place a stand under the transmission.

4. Remove or disconnect the following:
- Transmission bolts and pull the transmission rearward
- Rear engine mount by removing the 4 bolts

To install:

5. Install or connect the following:
- Rear engine mount. Tighten the 4 bolts to 48 ft. lbs. (65 Nm).
- Transmission by aligning the input shaft spline with the clutch disc. Tighten the transmission-to-engine bolts to 53 ft. lbs. (72 Nm).
- Starter. Tighten the bolts to 29 ft. lbs. (39 Nm).
- Rear end-plate. Tighten the bolts to 27 ft. lbs. (37 Nm).
- Left and right side stiffener plates
- Transmission with the engine assembly.

4WD

1. Before servicing the vehicle, refer to the precautions in the beginning of this section.

2. Remove or disconnect the following:
- Negative battery cable
- 4 screws and front console box
- Shift lever boot retainer screws and the shift lever boot
- Shift lever cap, by pressing downward on the shift lever cap covered with a cloth and rotating it counterclockwise to remove it
- Transfer shift lever, using snapring pliers to pull it from the transfer case

3. Drain the transmission and the transfer oil.

4. Remove or disconnect the following:
- Front and rear driveshafts
- Speedometer cable and the back-up light switch connector
- 4WD position switch connector, on the Standard cab
- L4 position switch connector, on the Extra cab
- Clutch release cylinder, by moving it aside without disconnecting the clutch line
- Exhaust pipe bracket
- Starter
- Rear end plate by removing the nuts and 2 bolts

5. Support the transmission rear side.

6. Remove or disconnect the following:
- 4 engine rear mount bolts
- O-ring and the crossmember

7. Using a transmission jack, support the transmission.

8. Remove or disconnect the following:
- 6 transmission-to-engine bolts
- 3 wire clamps from the transmission
- Transmission with the transfer case
- Engine rear mounting from the transfer case

- Transfer adapter rear mount bolts
- Transfer case from the transmission

To install:

9. Apply MP grease to the adapter oil seal and shift the 2 shift fork shafts to the high 4 position.

10. Install or connect the following:
- Transfer to the transmission. Tighten the bolts to 17 ft. lbs. (24 Nm).

➡ **Be careful not to damage the oil seal by the input gear spline when installing the transfer.**

- Transmission/transfer case assembly, by aligning the input shaft spline with the clutch disc.

11. Support the transmission with a jack.

12. Install or connect the following:
- Tighten the engine to transmission bolts to 53 ft. lbs. (72 Nm)
- Engine rear mount. Tighten the 4 bolts to 48 ft. lbs. (65 Nm)

13. Raise the transmission slightly with a jack.

14. Install or connect the following:
- Crossmember. Tighten the 4 bolts to 48 ft. lbs. (65 Nm).
- Engine rear mount. Tighten the bolts to 14 ft. lbs. (19 Nm).
- Rear end-plate. Tighten the 4 bolts and nuts to 13 ft. lbs. (18 Nm) on R150 and R150F transmissions or to 27 ft. lbs. (37 Nm) on W59 transmissions.
- Starter. Tighten both bolts to 29 ft. lbs. (39 Nm).
- Front exhaust pipe. Tighten the exhaust pipe-to-manifold bolts to 46 ft. lbs. (62 Nm), the exhaust bracket bolts to 33 ft. lbs. (44 Nm)

and the exhaust pipe-to-catalytic converter bolts to 35 ft. lbs. (48 Nm).
- Clutch release cylinder. Tighten the 2 bolts to 108 inch lbs. (13 Nm).
- L4 position switch connector on the extra cab or the 4WD position switch connector on the standard cab.
- VSS and the back-up light switch connector.
- Front and rear driveshafts.

15. Refill the transmission to the correct level.

16. Apply MP grease to the transfer shift lever.

17. Install or connect the following:
- Transfer shift lever.
- Snapring, using pliers

18. Install the transmission shift lever, as follows:

a. Apply MP grease to the transmission shift lever.

b. Align the groove of the shift lever cap and the pin par of the case cover. Cover the shift lever cap with a cloth. Pressing down on the shift lever cap, rotate it clockwise to install.

c. Install shift lever boot retainer with the 4 screws.

d. Install the front console box with the 4 screws.

19. Connect the negative battery cable. Start the engine and check for leaks.

20. Road test the vehicle for proper operation. Recheck all fluid levels.

Clutch Assembly

REMOVAL & INSTALLATION

1. Before servicing the vehicle, refer to the precautions in the beginning of this section.

2. Remove or disconnect the following:
- Negative battery cable
- Transmission assembly

3. Matchmark the clutch cover to the flywheel.

4. At the clutch cover, loosen each bolt 1 turn until spring tension is released.

5. Remove or disconnect the following:
- Clutch cover set bolts and the clutch cover with the clutch disc.
- Release bearing retaining clip and withdraw the it
- Release fork and boot assembly

To install:

6. Install or connect the following:
- Clutch disc onto the flywheel, using a clutch disc alignment tool

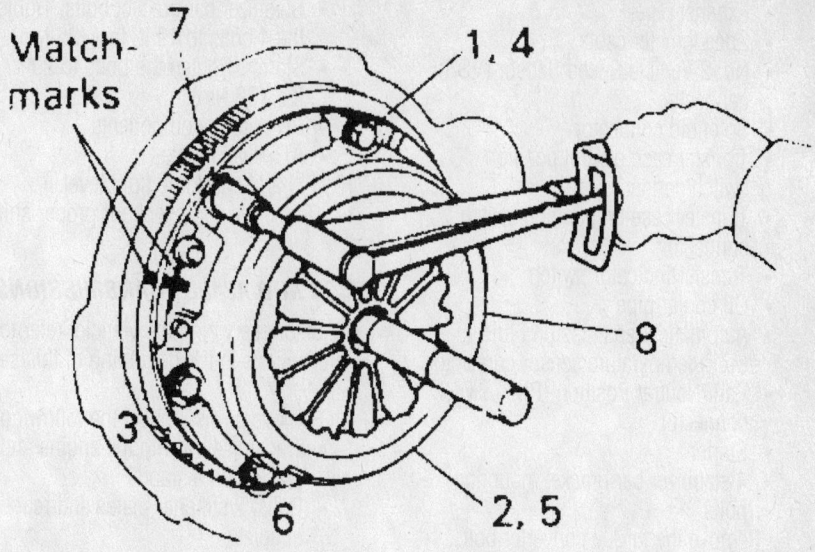

Bolt tightening sequence for the clutch cover—all engines

- Clutch cover, position it onto the flywheel and if reusing the old pressure plate, align the matchmarks.
- Clutch cover. Tighten the bolts in a crisscross pattern to 14 ft. lbs. (19 Nm).

7. Lubricate the release fork pivot and contact points, the release bearing, bearing hub and input shaft spline surfaces with a suitable molybdenum disulfide lithium based or multi-purpose grease.

8. Install or connect the following:
- Boot, release fork, hub and the bearing assemblies
- Transmission
- Negative battery cable

Hydraulic Clutch System

BLEEDING

1. Before servicing the vehicle, refer to the precautions in the beginning of this section.

2. Fill the clutch reservoir with brake fluid. Check the reservoir level frequently and add fluid as needed.

3. Connect one end of a vinyl tube to the bleeder plug on the slave cylinder and submerge the other end into a clear container half-filled with brake fluid.

4. Slowly pump the clutch pedal several times.

5. Have an assistant hold the clutch pedal down and loosen the bleeder plug until fluid and/or air starts to run out of the bleeder plug. Close the bleeder plug while the pedal is held to the floor.

6. Repeat Steps 2 and 3 until all the air bubbles are removed from the system.

7. Tighten the bleeder plug when all the air is gone.

8. Refill the master cylinder to the proper level as required.

9. Check the system for leaks.

Automatic Transmission Assembly

REMOVAL & INSTALLATION

Model A340F Transmission

1. Before servicing the vehicle, refer to the precautions in the beginning of this section.

2. Remove or disconnect the following:
- Automatic Transmission Fluid (ATF) level gauge
- Engine undercover

3. Drain the transmission fluid.

4. Remove or disconnect the following:
- Throttle cable
- No. 1 fan shroud

5. Remove the transmission shift lever assembly and the transfer shift lever, as follows:
- Rear console box
- Front console box with the transfer shift lever knob
- Connectors
- Shift control rod
- Transmission shift lever assembly
- Snapring and pull it from the transfer case
- Oil filler pipe, with the O-ring
- Front and rear driveshaft

- Exhaust pipe
- Speedometer cable
- No. 2 Vehicle Speed Sensor (VSS) connector
- Solenoid connector
- Transfer case neutral position switch connector
- Transfer case L4 position switch connector
- Transfer indicator switch
- Oil cooler pipe
- Automatic Transmission Fluid (ATF) temperature sensor connector
- Park/Neutral Position (PNP) switch connector
- Starter
- 4 stabilizer bar bracket mounting bolts

6. Remove the torque converter bolts, as follows:
- Flywheel housing undercover
- Torque converter clutch mounting bolts, while turning the crankshaft to gain access

7. Remove the front differential rear mounting cushion, as follows:
- Nut, using a hexagon wrench
- Front differential by lifting it

➡Be careful not to touch the torque converter clutch housing and the front differential companion flange

- 2 rear mount cushion bolts

8. Remove or disconnect the following:
- Support the transmission's rear side
- 4 engine rear mount bolts
- 4 nuts, bolts and the crossmember, by supporting the transmission
- Transmission

To install:
9. Install or connect the following:
- Transmission. Tighten the engine-to-transmission bolts to 53 ft. lbs. (71 Nm).
- Crossmember. Tighten the bolts to 48 ft. lbs. (65 Nm).
- Engine rear mount. Tighten the bolts to 14 ft. lbs. (19 Nm).
- Front differential rear mount cushion. Tighten the nut to 64 ft. lbs. (41 Nm).
- Torque converter clutch mount bolt

➡Install the green colored bolt, then the 5 others. Tighten the bolts to 30 ft. lbs. (41 Nm).

- Flywheel housing undercover. Tighten the bolts to 13 ft. lbs. (18 Nm) for 3.4L (5VZ-FE) engine or to 27 ft. lbs. (37 Nm) for 2.7L (3RZ-FE) engine.

- Stabilizer bar bracket bolts. Tighten the 4 bolts to 19 ft. lbs. (25 Nm).
- Starter. Tighten the bolts to 29 ft. lbs. (39 Nm).
- Remaining components
- ATF level gauge

10. Fill and check the fluid level.
11. Test drive and check for proper shifting.

A340D AND A340E TRANSMISSIONS

1. Before servicing the vehicle, refer to the precautions in the beginning of this section.
2. Remove or disconnect the following:
- Transmission with the engine and place it on a stand
- Bolts, 2 stiffener plates and rear endplate

3. Turn the crankshaft to gain access to the torque converter bolts.
4. Remove or disconnect the following:
- Torque converter bolts
- Starter
- Transmission-to-engine bolts
- Transmission

To install:
5. Install or connect the following:
- Transmission to the engine. Tighten the bolts to 53 ft. lbs. (71 Nm).
- Starter. Tighten both bolts to 29 ft. lbs. (39 Nm).
- Torque converter. Tighten the bolts to 30 ft. lbs. (41 Nm).
- Stiffener plate and rear endplate. Tighten the bolts to 27 ft. lbs. (37 Nm).
- Starter wires
- Transmission with the engine

Transfer Case Assembly

REMOVAL & INSTALLATION

1. Before servicing the vehicle, refer to the precautions in the beginning of this section.
2. Disconnect the negative battery cable.
3. Drain the transmission and the transfer case.
4. Remove or disconnect the following:
- Transfer case with the transmission
- Breather hose from the transfer upper cover and the transmission control retainer, if equipped with an automatic transmission
- Rear engine mounting
- Dynamic damper

5. Remove the driveshaft upper dust cover and the transfer from the transmissions, as follows:

- Dust cover bolt from the bracket
- Transfer case adapter rear mounting bolts
- Transfer case, by pulling it straight up and away from the transmission.

✳✳ WARNING

Be careful not to damage the adapter rear oil seal with the transfer input gear spline.

To install:
6. Install the transfer case and the driveshaft upper dust cover to the transmission with a new gasket, as follows:
- Shift the 2 shift fork shafts to the high 4 position
- Apply MP grease to the adapter oil seal
- New gasket to the transfer adapter
- Transfer case to the transmission.

✳✳ WARNING

Take care not to damage the oil seal by the input gear spline.

- Transfer case adapter. Tighten the rear bolts to 27 ft. lbs. (37 Nm).
- Dust cover to the bracket. Tighten the bolt to 17 ft. lbs. (23 Nm).

7. Install or connect the following:
- Engine rear mount. Tighten the bolts to 19 ft. lbs. (25 Nm).
- Dynamic damper. Tighten the bolts to 27 ft. lbs. (37 Nm).
- Breather hose, if equipped with an automatic transmission
- Transfer case with the transmission to the engine

8. Fill the transmission and the transfer case with oil.
9. Test drive the vehicle and check the abnormal noise and smooth operation.
10. Recheck the fluid levels.

Halfshaft

REMOVAL & INSTALLATION

1. Before servicing the vehicle, refer to the precautions in the beginning of this section.
2. Drain the differential oil from the differential.
3. If not equipped with a free-wheeling hub, disconnect the halfshaft from the steering knuckle, as follows:
- Grease cap
- Cotter pin and lockcap, from the halfshaft

- Locknut from the halfshaft, while having an assistant apply the brakes
4. If equipped with free-wheeling hub, remove the free wheel hub, as follows:
  - Set the control handle to FREE
  - Cover bolts and pull off the cover

- Center bolt with washer
- Mounting nuts and washer to the hub body
- Cone washer, using a brass bar and hammer to tap on the bolt heads
- Free wheel hub body and gasket
- Snapring from the end of the half-shaft, using a snapring expander

5. Remove or disconnect the following:
  - Halfshaft from the differential, using a brass bar and hammer
  - Cotter pin and nut, from the lower ball joint
  - Lower control arm from the lower ball joint
  - Halfshaft

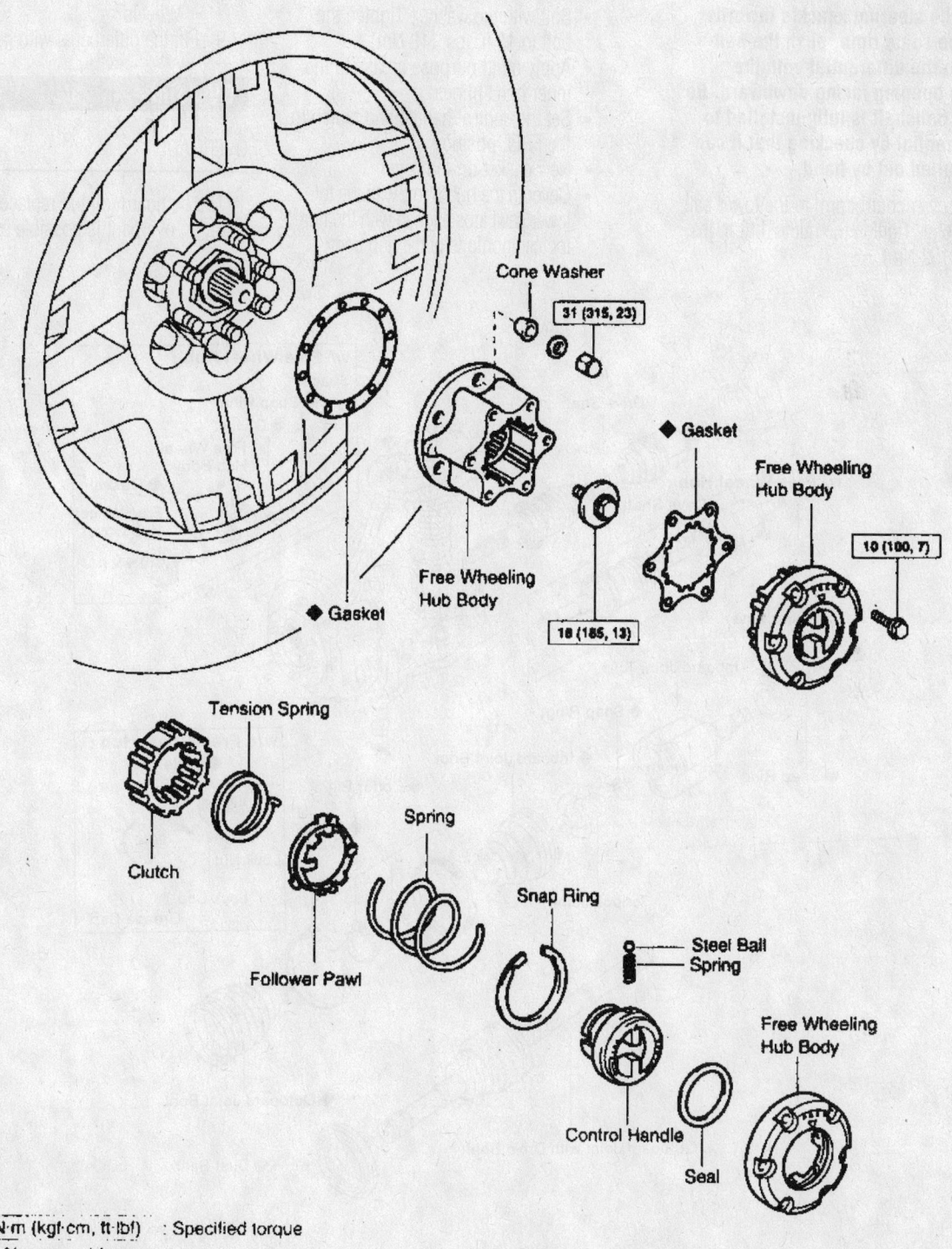

Cone Washer

31 (315, 23)

◆ Gasket

Free Wheeling Hub Body

◆ Gasket

Free Wheeling Hub Body

19 (190, 7)

18 (185, 13)

Free Wheeling Hub Body

Tension Spring

Clutch

Follower Pawl

Spring

Snap Ring

Steel Ball

Spring

Control Handle

Seal

Free Wheeling Hub Body

N·m (kgf·cm, ft·lbf) : Specified torque

◆ Non-reusable part

Exploded view of the free wheeling hub assembly

7924YG96

➡ **If it is difficult to remove the half-shaft from the steering knuckle, use a rubber hammer and tap the halfshaft from the steering knuckle.**

- Snapring from the inboard shaft

To install:

6. Install or connect the following:
- New snapring to the inboard shaft
- Halfshaft to the steering knuckle

➡ **Push the steering knuckle inwards and at the same time, push the half-shaft into the differential with the snapring opening facing downward. Be sure the halfshaft is fully installed to the differential by checking that it can-not be pulled out by hand.**

- Lower control arm to the lower ball joint. Tighten the nut to 112 ft. lbs. (152 Nm).

- New cotter pin

7. If equipped with a free wheeling hub, install the hub, as follows:
- Spacer
- Snapring to the halfshaft, using a snapring expander
- New gasket on the front axle hub
- Fee wheeling hub body, with the 6 cone washers and nuts. Tighten the 6 nuts to 23 ft. lbs. (31 Nm).
- Bolt with the washer. Tighten the bolt to 13 ft. lbs. (18 Nm).
- Apply multi purpose grease to the inner hub splines
- Set the control handle and clutch to the FREE position
- New gasket on the cover
- Cover to the hub body, with the fol-lower pawl tabs aligned with the non-toothed portions of the hub body.

- Tighten the cover bolts to 84 inch lbs. (10 Nm).

8. If equipped without a free wheeling hub, install the halfshaft to the steering knuckle, as follows:
- Locknut to the halfshaft. Tighten the locknut to 174 ft. lbs. (235 Nm).
- Lockcap and cotter pin to the half-shaft
- Grease cap
- Wheels

9. Fill the differential with gear oil.

CV-Joints

OVERHAUL

The outboard joint is replaced with half-shaft; no overhaul is possible or necessary.

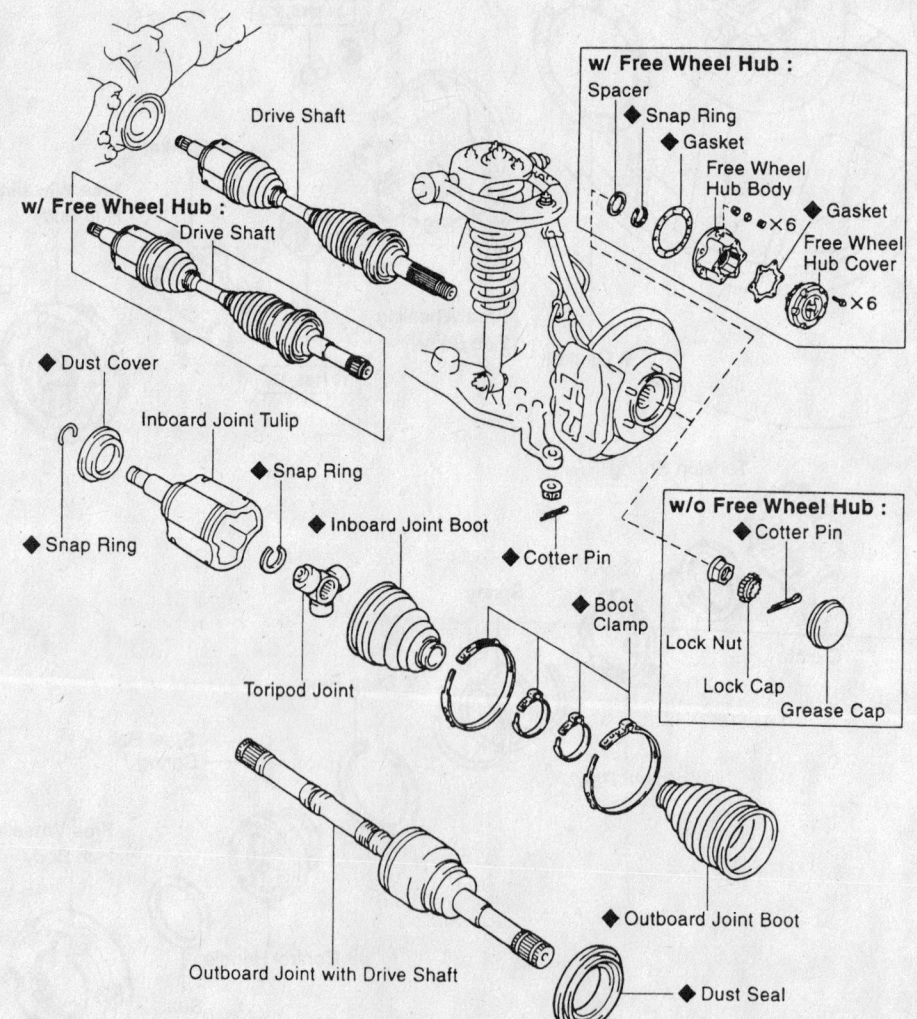

◆ Non-reusable part

Exploded view of the halfshaft assembly—4WD

9308YG07

Inboard Joint

1. Before servicing the vehicle, refer to the precautions in the beginning of this section.

2. Remove the halfshaft from the vehicle.

3. Remove the large clamp from the inboard joint.

4. Remove the small clamp, using side cutters, from the inboard joint.

5. Slide the inboard joint boot toward the outboard joint.

6. Matchmark the inboard joint to the halfshaft.

7. Remove the inboard joint housing from the halfshaft.

8. Remove the snapring from the end of the halfshaft.

9. Matchmark the halfshaft to the tri-pot joint.

10. Remove the tri-pot from the halfshaft, using a brass bar and a hammer.

✳✳ WARNING

Do not tap on the tri-pot joint.

11. Remove the inboard and outboard boots from the halfshaft.

✳✳ WARNING

Do not disassemble the outboard joint.

To assemble:

12. Wrap vinyl tape around the halfshaft splines to prevent damaging the boots.

13. Install the outboard and inboard boots to the halfshaft with the small end clamps.

14. Assemble the tri-pot joint to the halfshaft with the beveled side facing the outboard joint and align the matchmarks.

15. Install the tri-pot joint, using a brass bar and a hammer.

✳✳ WARNING

Do not tap on the roller.

16. Install the snapring.

17. Lubricate the outboard joint with ½ of the grease supplied with the kit.

18. Assemble the boot to the outboard joint

19. Assemble the inboard joint housing to the halfshaft by aligning the matchmarks.

20. Temporarily install the boot onto the tri-pot housing.

21. Make sure the boots are positioned in the shaft grooves.

22. With the halfshaft positioned at the standard length of 17.094–17.252 in. (434.2–438.2mm), make sure that the boots are not stretched or contracted.

23. Install a new inboard joint clamp.

24. Crimp the large clamp with tool 09521-24010 so that the crimp clearance is 0.039–0.059 in. (1.0–1.5mm).

25. Install the halfshaft.

Spindle Bearings

REMOVAL, PACKING AND INSTALLATION

1. Before servicing the vehicle, refer to the precautions in the beginning of this section.

2. Remove or disconnect the following:
   - Front wheel
   - Shock absorber
   - Grease cap
   - Driveshaft
   - Cotter pin and lockcap
   - Locknut, with an assistant applying the brakes
   - Speed sensor and harness from the steering knuckle, if equipped with Anti-lock Brake System (ABS)
   - Brake line from the steering knuckle
   - Caliper and rotor
   - Lower ball joint bolts and the joint from the steering knuckle
   - Cotter pin and axle hub nut
   - Steering knuckle
   - Bearings from the steering knuckle

To install:

3. Install or connect the following:
   - Bearings to the steering knuckle
   - Steering knuckle
   - Cotter pin and axle hub nut. Tighten the nut to 80 ft. lbs. (108 Nm).
   - Lower ball joint to the steering knuckle
   - Caliper and rotor
   - Brake line to the steering knuckle
   - Speed sensor and harness to the steering knuckle, if equipped with Anti-lock Brake System (ABS)
   - Locknut, with an assistant applying the brakes. Torque the locknut to 174 ft. lbs. (235 Nm).
   - Cotter pin and lockcap
   - Driveshaft
   - Grease cap
   - Shock absorber
   - Front wheel

Axle Shaft, Bearing and Seal

REMOVAL & INSTALLATION

Front

1. Before servicing the vehicle, refer to the precautions in the beginning of this section.

2. Remove or disconnect the following:
   - Front wheel
   - Shock absorber
   - Grease cap
   - Axle shaft's cotter pin and lock cap
   - Locknut, using an assistant to apply the brakes
   - Speed sensor and harness from the steering knuckle, if equipped with Anti-lock Brake System (ABS)
   - Brake line from the steering knuckle
   - Caliper and rotor
   - Lower ball joint bolts
   - Cotter pin and loosen the axle hub nut
   - Steering knuckle
   - Axle shaft

To install:

3. Install or connect the following:
   - Axle shaft
   - Steering knuckle
   - Tighten the axle hub nut to 80 ft. lbs. (108 Nm) and the locknut to 174 ft. lbs. (235 Nm).
   - Cotter pin
   - Lower ball joint bolts
   - Caliper and rotor
   - Brake line to the steering knuckle
   - Speed sensor and harness to the steering knuckle, if equipped with Anti-lock Brake System (ABS)
   - Locknut, using an assistant to apply the brakes
   - Axle shaft's cotter pin and lock cap
   - Grease cap
   - Shock absorber
   - Front wheel

Rear

1. Before servicing the vehicle, refer to the precautions in the beginning of this section.

2. Remove or disconnect the following:
   - Rear wheel
   - Brake drum

3. Check the bearing backlash and axle shaft deviation, as follows:

   a. Using a dial indicator, check that the backlash in the bearing shaft direction. The maximum is 0.027 in. (0.7mm).

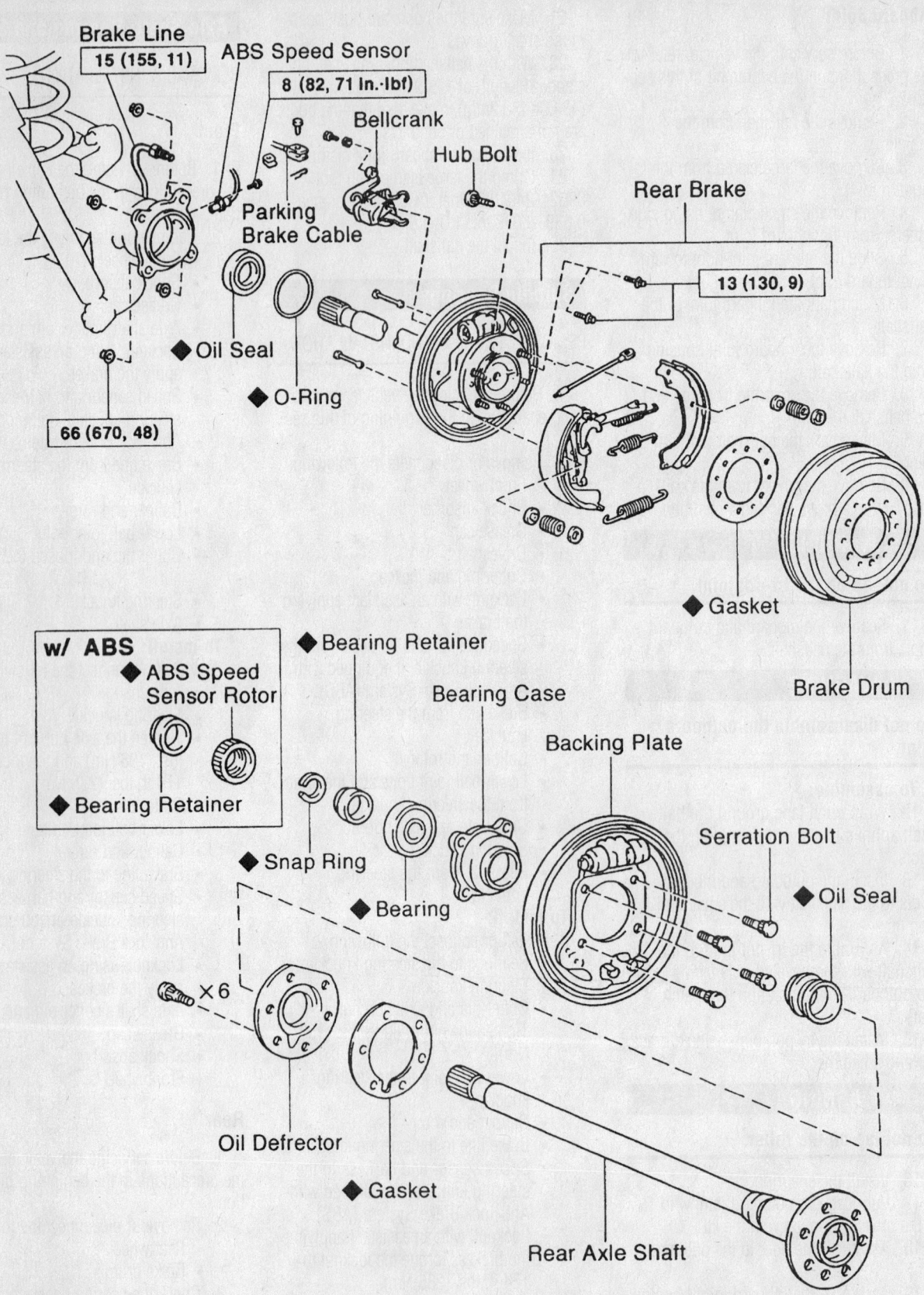

Brake Line 15 (155, 11)

ABS Speed Sensor 8 (82, 71 In.·lbf)

Bellcrank

Hub Bolt

Rear Brake

Parking Brake Cable

◆ Oil Seal

◆ O-Ring

13 (130, 9)

66 (670, 48)

◆ Gasket

Brake Drum

w/ ABS

◆ ABS Speed Sensor Rotor

◆ Bearing Retainer

◆ Bearing Retainer

Bearing Case

Backing Plate

Serration Bolt

◆ Snap Ring

◆ Bearing

◆ Oil Seal

×6

Oil Deflector

◆ Gasket

Rear Axle Shaft

N·m (kgf·cm, ft·lbf) : Specified torque

◆ Non-reusable part

Exploded view of the rear axle shaft and components—typical

86827G97

b. If the backlash exceeds the maximum, replace the bearing.

c. Using a dial indicator, check the deviation at the surface of the axle shaft outside the hub bolt. Maximum is 0.0039 in. (0.1mm).

d. If the deviation exceeds the maximum, replace the axle shaft.

4. Remove or disconnect the following:
- Anti-lock Brake System (ABS) speed sensor from the axle housing, if equipped
- Axle shaft assembly by removing the 4 nuts from the backing plate
- O-ring from the axle housing
- Bearing and retainer (differential side) and ABS speed sensor rotor, if equipped
- Snapring from the axle shaft
- Axle shaft

➥**Inspect the axle shaft and flange runouts. The axle shaft run-out should be 0.079 in. (2.0mm) and the flange runout should be 0.004 in. (0.1mm).**

- Outer seal
- Bearing from the axle shaft

To install:

5. Install or connect the following:
- Bearing from the axle shaft
- Outer seal
- Axle shaft
- Snapring to the axle shaft
- Bearing and retainer (differential side) and ABS speed sensor rotor, if equipped
- O-ring to the axle housing
- Axle shaft assembly. Tighten the 4 backing plate nuts to 48 ft. lbs. (66 Nm).
- Anti-lock Brake System (ABS) speed sensor to the axle housing, if equipped
- Brake drum
- Rear wheel

Pinion Seal

REMOVAL & INSTALLATION

Front

1. Before servicing the vehicle, refer to the precautions in the beginning of this section.
2. Remove the engine undercover.
3. Drain the differential oil.
4. Remove or disconnect the following:
- Front driveshaft
- Companion flange nut, by unstaking it

- Companion flange
- Pinion seal, using an extractor

To install:

5. Install a new oil seal, to a depth of 0.059 in. (1.5mm) below the lip, using a seal driver.
6. Lubricate the seal lip with multi-purpose grease.
7. Install or connect the following:
- Companion flange, coat the threads with multi-purpose grease
- New companion flange nut. Tighten it to 89 ft. lbs. (120 Nm).
8. Measure the bearing preload, using a torque wrench. The correct preload should be 5–9 inch lbs. (0.6–1.0 Nm) for a used bearing or 10–17 inch lbs. (1–2 Nm) for a new bearing.

➥**If the preload is greater that specified, replace the bearing spacer. If the preload is less than specified, tighten the companion flange nut in 108 inch lbs. (13 Nm) increments until the correct preload is achieved. Maximum torque for the nut is 173 ft. lbs. (235 Nm). If the value is exceeded, the bearing spacer must be replaced; do not back off the flange nut to lower the torque or preload.**

9. Install the front driveshaft by aligning the matchmarks.
10. Check the companion flange run-out; maximum allowable run-out is 0.003 in. (0.10mm).
11. Stake the pinion flange nut.
12. Refill the differential with oil.

Rear

1. Before servicing the vehicle, refer to the precautions in the beginning of this section.
2. Remove or disconnect the following:
- Rear driveshaft by matchmarking it
- Companion flange nut, by loosen the staked portion
- Companion flange, using a screw-type extractor
- Oil seal, using an extractor

To install:

3. Install a new oil seal, to a depth of 0.039 in. (1.0mm) below the lip, using a seal driver.
4. Lubricate the seal lip with multi-purpose grease.
5. Install or connect the following:
- Companion flange, coat the threads with multi-purpose grease
- New companion flange nut. Tighten it to 109 ft. lbs. (147 Nm).

Using a chisel and hammer, loosen the staked part of the nut. Hold the flange with SST 09950-30010 or equivalent and remove the nut

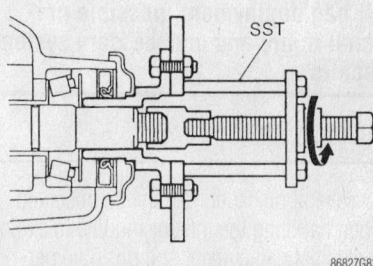

Screw-type extractor from Tool 09950-30010

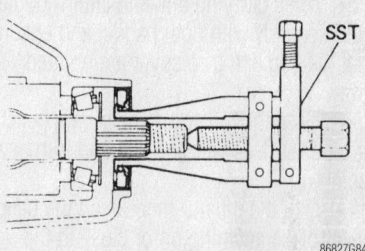

Extractor fits into the Seal Removal Tool 09308-10010

6. Measure the bearing preload, using a torque wrench. The correct preload should be 8–11 inch lbs. (0.9–1.2 Nm) for a 2 spider gear differential or to 4–7 inch lbs. (0.4–0.8 Nm) for a 4 spider gear differential.

➥**If the preload is greater that specified, replace the bearing spacer. If the preload is less than specified, tighten the companion flange nut in 9 ft. lbs. (13 Nm) increments until the correct preload is achieved. Maximum torque for the nut is 325 ft. lbs. (441 Nm). If the value is exceeded, the bearing spacer must be replaced; do not back off the flange nut to lower the torque or preload.**

7. Stake the pinion flange nut.
8. Install the rear driveshaft by aligning the matchmarks.

STEERING AND SUSPENSION

Air Bag

❋❋ CAUTION

Some vehicles are equipped with an air bag system. The system must be disabled before performing service on or around system components, steering column, instrument panel components, wiring and sensors. Failure to follow safety and disabling procedures could result in accidental air bag deployment, possible personal injury and unnecessary system repairs.

PRECAUTIONS

Several precautions must be observed when handling the inflator module to avoid accidental deployment and possible personal injury.

• Never carry the inflator module by the wires or connector on the underside of the module.

• When carrying a live inflator module, hold securely with both hands, and ensure that the bag and trim cover are pointed away.

• Place the inflator module on a bench or other surface with the bag and trim cover facing up.

• With the inflator module on the bench, never place anything on or close to the module which may be thrown in the event of an accidental deployment.

DISARMING

To avoid personal injury when working on vehicles equipped with an air bag, the negative battery cable must be disconnected and at least 90 seconds must elapse before working on the system. Failure to do so may result in deployment of the air bag.

Power Rack and Pinion Steering Gear

REMOVAL & INSTALLATION

1. Before servicing the vehicle, refer to the precautions in the beginning of this section.
2. Remove or disconnect the following:
   • Negative battery cable
   • Right and left tie rod ends from the knuckle

• Intermediate No. 2 shaft from the steering rack
• Pressure feed and the return tubes, using SST 09631-22020
• Power steering rack

To install:

3. Install the power steering rack. Tighten the bolts to 148 ft. lbs. (201 Nm) for 2WD or to the following values for 4WD:
   • Rack assembly bolt: 123 ft. lbs. (167 Nm)
   • Rack assembly nut: 141 ft. lbs. (191 Nm)
   • Bracket nut and bolt: 123 ft. lbs. (167 Nm)
4. Install or connect the following:
   • New O-ring
   • Pressure feed tube. Tighten it to 33 ft. lbs. (45 Nm).
   • Return tube. Tighten it to 36 ft. lbs. (49 Nm) for 2WD or to 29 ft. lbs. (40 Nm) for 4WD
   • Intermediate No. 2 shaft to the steering rack
   • Right and left tie rod ends to the steering knuckle
   • Tighten the castle nuts to specification
   • New cotter pins
   • Negative battery cable
5. Check the steering wheel center point.
6. Bleed the power steering system.
7. Check the front wheel alignment. Tighten the tie rod end locknuts to 67 ft. lbs. (90 Nm)

Shock Absorbers

REMOVAL & INSTALLATION

Front

2WD

1. Before servicing the vehicle, refer to the precautions in the beginning of this section.
2. Remove or disconnect the following:
   • Front wheel
   • Shock absorber from the lower control arm
   • Nut, retainers and the cushion from the top of the shock absorber
   • Shock absorber
   • Retainers and cushion from the shock absorber

To install:

3. Install or connect the following:
   • Retainers and cushion to the shock absorber

• Shock absorber
• Retainers, cushion and nut to the top of the shock absorber. Tighten the nut to 18 ft. lbs. (25 Nm).
• Lower shock absorber-to-lower control arm. Tighten the bolts to 13 ft. lbs. (19 Nm)
• Wheels

Rear

1. Before servicing the vehicle, refer to the precautions in the beginning of this section.
2. Remove the wheel.
3. Lower the floor jack to take tension off of the spring.
4. Remove or disconnect the following:
   • Shock absorber from the rear axle housing
   • Nut, retainers and the cushions holding the shock absorber to the frame
   • Shock absorber with the washers and bushings

To install:

5. Install the shock absorber to the frame with the washers and bushings.
6. Tighten the shock absorber-to-frame nut to the following values:
   • 2WD: 19 ft. lbs. (25 Nm)
   • 4WD: 53 ft. lbs. (72 Nm)
7. Connect the shock absorber to the rear axle housing. Tighten the bolt to the following specifications:
   • 2WD: 19 ft. lbs. (25 Nm)
   • 4WD: 53 ft. lbs. (72 Nm)
8. Install the wheels.

Struts

REMOVAL & INSTALLATION

4WD

1. Before servicing the vehicle, refer to the precautions in the beginning of this section.
2. Remove the wheel.
3. Remove the front wheel.
4. Remove the strut absorber nut and washer from the lower control arm.

❋❋ WARNING

Do not remove the bolt at this time.

5. While slowly lowering the front suspension, remove the lower bolt from the strut absorber assembly.

6. Support the strut.

7. Remove the three nuts from the top of the strut tower.

8. Lower the strut out of the wheel well.

To install:

9. The remainder of installation is the reverse of removal. Please note the following torque specifications:

- 3 upper strut to body nuts: 47 ft. lbs. (64 Nm)
- Lower strut absorber mounting bolt: 101 ft. lbs. (135 Nm)

Coil Spring

REMOVAL & INSTALLATION

2WD

1. Before servicing the vehicle, refer to the precautions in the beginning of this section.

2. Remove or disconnect the following:

- Shock absorber from the suspension, by removing the 2 bottom bolts and top nut
- Compress the coil spring, using a Spring Compressor
- Nut and sway bar link from the lower control arm
- 2 sway bar bracket bolts on the side of the suspension that the lower control arm is being removed.

➡**This will allow access to the lower control arm through-bolt.**

3. Support the steering knuckle and upper control arm.

4. Remove or disconnect the following:

- Cotter pin and nut from the lower ball joint
- Lower ball joint from the lower control arm, using SST 09628-62011
- Nut from the lower control arm set bolt
- Nut from the strut bar front set bolt
- Lower control arm and strut bar as an assembly, pulling out the 2 bolts

➡**When the lower control arm is removed, set the coil spring aside.**

To install:

5. Install or connect the following:

- Place the end of the coil spring in contact with the lower control arm seat
- Lower control arm, spring, and strut arm to the suspension.

- Strut arm bolt and lower control arm bolt
- Nuts for the strut arm bolt and lower control arm bolt; do not tighten the bolts at this time
- Lower control arm to the lower ball joint. Tighten the nut to 80 ft. lbs. (110 Nm).
- New cotter pin

6. Remove the support from the upper control arm and steering knuckle.

7. Install or connect the following:

- Sway bar bracket to the suspension. Tighten the bolts to 22 ft. lbs. (29 Nm).
- Sway bar link to the lower control arm. Tighten the nut to 29 ft. lbs. (39 Nm).

8. Making sure the coil spring is in its correct position, slowly remove the spring compressor from the coil.

9. Install or connect the following:

- Shock absorber. Tighten the top nut to 18 ft. lbs. (25 Nm) and the bottom 2 bolts to 29 ft. lbs. (39 Nm).
- Wheel
- Stabilize the suspension by pushing up and down on the vehicle
- Tighten the strut bar nut/bolt to 221 ft. lbs. (300 Nm) and the lower control arm bolt/nut to 148 ft. lbs. (200 Nm).

10. Check the front wheel alignment.

4WD

1. Before servicing the vehicle, refer to the precautions in the beginning of this section.

2. Remove or disconnect the following:

- Strut to the lower control arm nut/bolt.
- 3 strut-to-strut tower nuts/bolts
- Strut

3. Compress the coil spring until there is a clearance on both ends, using SST 09727-30030

4. Remove or disconnect the following:

- Strut center nut
- Suspension support and coil spring
- Insulator from the suspension support

To install:

5. Install or connect the following:

- Insulator to the suspension support

➡**Match the bolt of the suspension support with the cut out part of the insulator.**

- Coil spring to the strut, by compressing it with a coil spring compressor

➡**Fit the lower end of the coil spring into the gap of the spring seat of the strut.**

- Suspension support to the strut rod
- Temporarily tighten a new suspension support center nut

6. Position the suspension support so that a line drawn between the 2 bolts would be parallel to the direction of the lower bushing.

7. Remove the compressor from the spring.

8. Install or connect the following:

- Tighten the strut center nut to 22 ft. lbs. (29 Nm).
- Strut
- Strut-to-strut tower. Tighten the 3 nuts to 47 ft. lbs. (64 Nm).
- Strut-to-lower control arm. Tighten the nut/bolt to 101 ft. lbs. (135 Nm).
- Front wheels

Leaf Springs

REMOVAL & INSTALLATION

2WD

1. Loosen the rear wheel lug nuts.

2. Raise the rear of the vehicle. Support the frame and rear axle housing with stands.

3. Remove the lug nuts and the wheel.

4. Remove the cotter pin, nut, and washer from the lower end of the shock absorber.

5. Detach the shock absorber from the spring seat.

6. Remove the parking brake cable clamp.

➡**Remove the parking brake equalizer, if necessary.**

7. Unfasten the U-bolt nuts and remove the spring seat assemblies.

8. Adjust the height of the rear axle housing so that the weight of the rear axle is removed from the rear springs.

9. Unfasten the spring shackle retaining nuts. Withdraw the spring shackle inner plate. Carefully pry out the spring shackle with a bar.

10. Remove the spring bracket pin from the front end of the spring hanger and remove the rubber bushing.

11. Remove the spring. Use care not to damage the hydraulic brake line or the parking brake cable.

To install:

12. Install the rubber bushing in the eye of the spring.

2WD

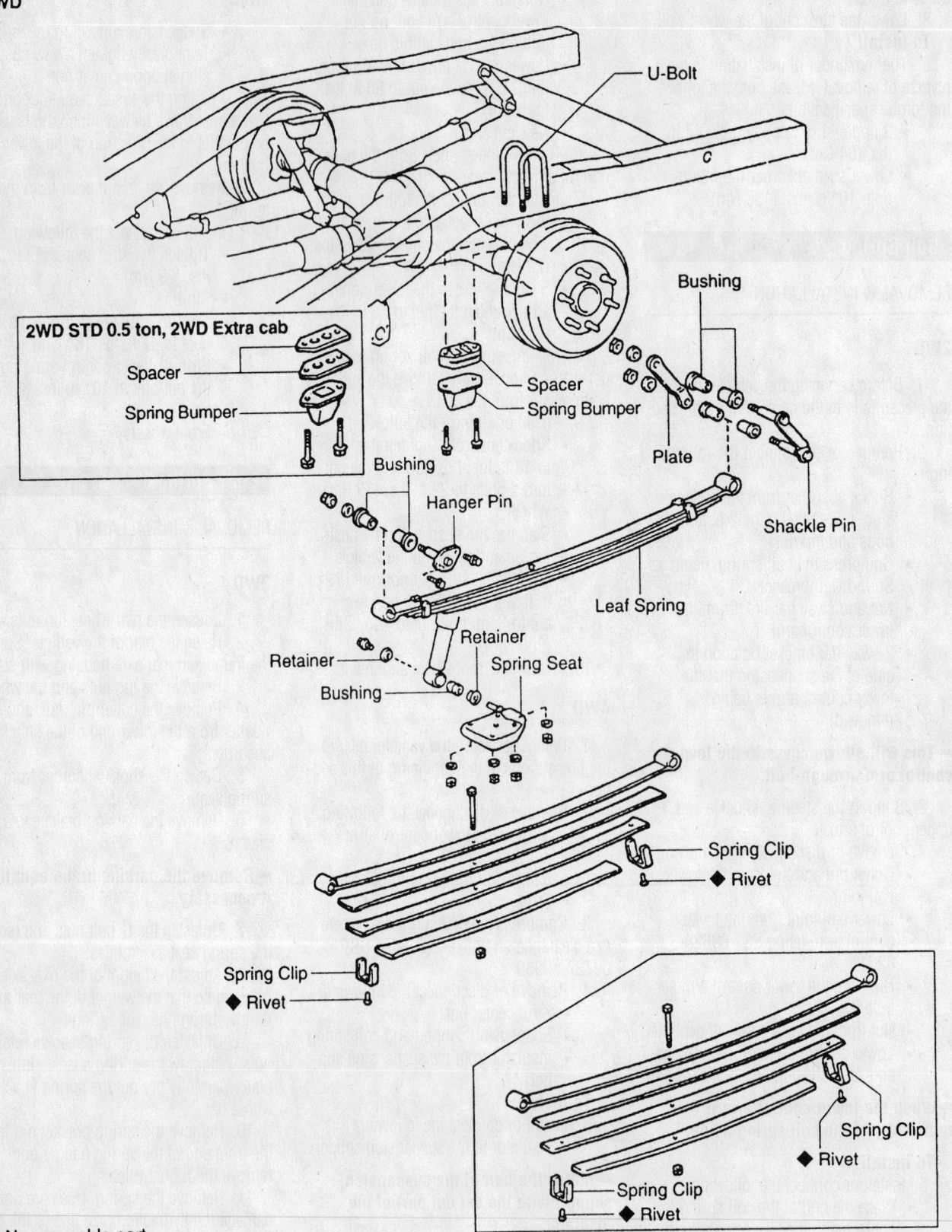

U-Bolt

2WD STD 0.5 ton, 2WD Extra cab

Spacer

Spring Bumper

Bushing

Spacer

Spring Bumper

Bushing

Hanger Pin

Bushing

Plate

Shackle Pin

Leaf Spring

Retainer

Retainer

Bushing

Spring Seat

Spring Clip

◆ Rivet

Spring Clip

◆ Rivet

Spring Clip

◆ Rivet

Spring Clip

◆ Rivet

◆ Non-reusable part

86828G55

Exploded view of the rear leaf spring and related components—2WD

4WD

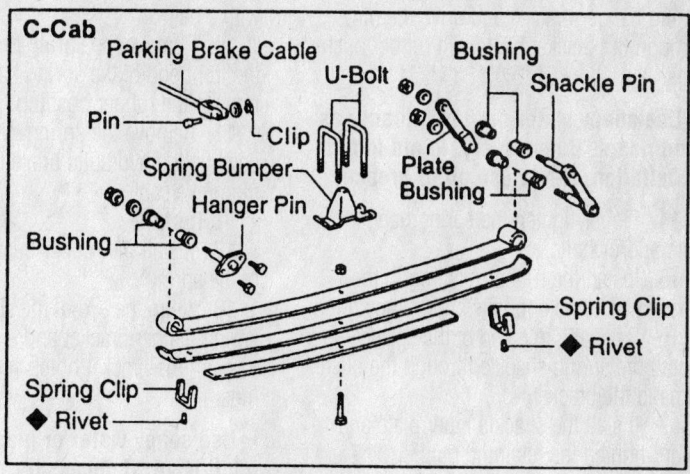

C-Cab

Parking Brake Cable

Pin

Clip

U-Bolt

Spring Bumper

Hanger Pin

Bushing

Bushing

Plate
Bushing

Shackle Pin

Spring Clip

◆ Rivet

Spring Clip

◆ Rivet

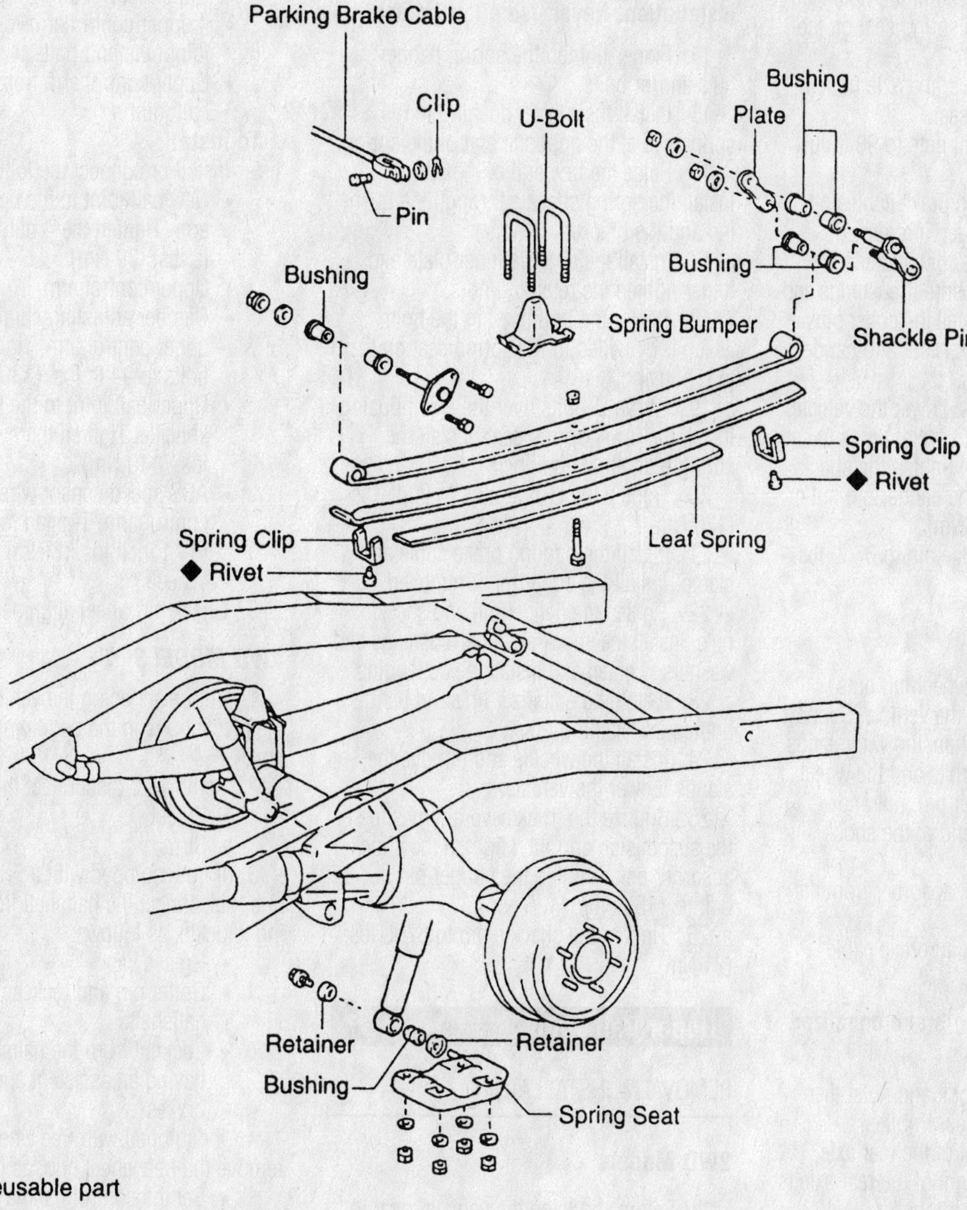

Parking Brake Cable

Clip

U-Bolt

Plate

Bushing

Pin

Bushing

Bushing

Spring Bumper

Bushing

Shackle Pin

Spring Clip

◆ Rivet

Leaf Spring

Spring Clip

◆ Rivet

Retainer

Retainer

Bushing

Spring Seat

◆ Non-reusable part

Exploded view of the rear leaf spring and components—4WD

86828G59

13. Align the eye of the spring with the spring hanger bracket and drive the pin through the bracket holes and rubber bushings.

➡**Use soapy water or glass cleaner as a lubricant, if necessary, to aid in pin installation. Never use oil or grease.**

14. Finger-tighten the spring hanger nuts and/or bolts.

15. Install the rubber bushing in the spring eye at the opposite end of the spring.

16. Raise the free end of the spring. Install the spring shackle through the bushing and the bracket.

17. Install the shackle inner plate and finger-tighten the retaining nuts.

18. Center the bolt head in the hole which is provided in the spring seat on the axle housing.

19. Fit the U-bolts over the axle housing. Install the lower spring seat.

20. Tighten the U-bolt nuts to 90 ft. lbs. (120 Nm)

21. Install the parking brake cable and clamp. Install the equalizer, if removed.

22. Tighten the hanger pin and shackle nuts. Install the shock absorber bushings and washers. Tighten and install the cotter pins.

23. Install the stabilizer link and hand-tighten its retaining nuts.

24. Install the wheels. Lower the vehicle.

25. Bounce the truck several times to set the suspension and then tighten the shock absorber bolt. Tighten the hanger pin nut or bolt to 115 ft. lbs. (120 Nm).

26. Tighten the shackle pin to 67 ft. lbs. (91 Nm).

4WD

1. Loosen the rear wheel lug nuts.

2. Raise the rear of the vehicle. Support the frame and rear axle housing with stands.

3. Remove the lug nuts and the wheel.

4. Remove the cotter pin, nut and washer from the lower end of the shock absorber.

5. Detach the shock absorber from the spring seat.

6. Remove the parking brake cable clamp.

➡**Remove the parking brake equalizer, if necessary.**

7. Unfasten the U-bolt and nuts, then remove the spring seat assemblies.

8. Adjust the height of the rear axle housing so that the weight of the rear axle is removed from the rear springs.

9. Unfasten the spring shackle retaining nuts. Withdraw the spring shackle inner

plate. Carefully pry out the spring shackle with a bar.

10. Remove the spring bracket pin from the front end of the spring hanger and remove the rubber bushing.

11. Remove the spring. Use care not to damage the hydraulic brake line or the parking brake cable.

To install:

12. Install the rubber bushing in the eye of the spring.

13. Align the eye of the spring with the spring hanger bracket and drive the pin through the bracket holes and rubber bushings.

➡**Use soapy water or glass cleaner as a lubricant, if necessary, to aid in pin installation. Never use oil or grease.**

14. Finger-tighten the spring hanger nuts and/or bolts.

15. Install the rubber bushing in the spring eye at the opposite end of the spring.

16. Raise the free end of the spring. Install the spring shackle through the bushing and the bracket.

17. Install the shackle inner plate and finger-tighten the retaining nuts.

18. Center the bolt head in the hole which is provided in the spring seat on the axle housing.

19. Fit the U-bolts over the axle housing. Install the lower spring seat. Install the spring bumper, if equipped.

20. Tighten the U-bolt nuts to 90 ft. lbs. (120 Nm)

21. Install the parking brake cable and clamp. Install the equalizer, if removed.

22. Tighten the hanger pin and shackle nuts. Install the shock absorber bushings and washers. Tighten and install the cotter pins.

23. Install the stabilizer link and hand-tighten its retaining nuts.

24. Install the wheels and remove the stands. Lower the vehicle.

25. Bounce the truck several times to set the suspension and then tighten the shock absorber bolt. Tighten the hanger pin nut or bolt to 115 ft. lbs. (120 Nm).

26. Tighten the shackle pin to 67 ft. lbs. (91 Nm).

Upper Ball Joint

REMOVAL & INSTALLATION

2WD Models

1. Before servicing the vehicle, refer to the precautions in the beginning of this section.

2. Remove the wheels.

3. Support the lower control arm with a floor jack.

4. Remove or disconnect the following:
- Anti-lock Brake System (ABS) speed sensor wire from the upper control arm
- 2 bolts and camber adjusting shims from the upper control arm

➡**Before removing the shims from the upper control arm, make a note of each shim size and position.**

- Upper control arm cotter pin and nut.
- Upper ball joint from the steering knuckle, using SST 09628-62011
- Upper control arm
- 4 upper control arm-to-upper ball joint nuts and bolts.
- Upper control arm from the upper ball joint

To install:

5. Install or connect the following:
- New ball joint to the upper control arm. Tighten the 4 nuts/bolts to 29 ft. lbs. (39 Nm).
- Upper control arm
- Camber adjusting shims to the upper control arm. Tighten the 2 bolts to 94 ft. lbs. (130 Nm).
- Upper ball joint to the steering knuckle. Tighten the nut to 80 ft. lbs. (110 Nm).
- ABS speed sensor wire to the upper control arm. Tighten the ABS bolt to 71 inch lbs. (8 Nm).
- Wheels

6. Check the wheel alignment.

4WD MODELS

1. Before servicing the vehicle, refer to the precautions in the beginning of this section.

2. Remove or disconnect the following:
- Wheel
- Strut

3. If not equipped with a FREE wheeling hub, disconnect the halfshaft from the steering knuckle, as follows:
- Grease cap
- Cotter pin and lockcap from the halfshaft
- Locknut from the halfshaft, while having an assistant apply the brakes

4. If equipped with free wheeling hub, remove the free wheel hub, as follows:
- Set the control handle to FREE
- Cover bolts and pull off the cover
- Center bolt with washer
- Hub body nuts and washer

- Cone washer, using a brass bar and hammer to tap on the bolt heads
- Free wheel hub body and gasket
- Snapring and spacer from the half-shaft end, using a snapring expander
- Anti-lock Brake System (ABS) speed sensor from the steering knuckle, if equipped with ABS
- Brake hose from the steering knuckle
- Brake caliper support bracket and support it on a wire

✳✳ WARNING

Do not allow the caliper to hang from the brake hose.

5. Remove or disconnect the following:
- Rotor
- Lower ball joint from the steering knuckle
- Upper control arm cotter pin and nut
- Steering knuckle from the upper control arm, using SST 09950-40010
- Steering knuckle from the vehicle

➡ **If it is difficult to remove the half-shaft from the steering knuckle, use a rubber hammer to tap the halfshaft from the steering knuckle.**

- Wire and boot from the upper ball joint
- Snapring from the ball joint, using a snapring expander
- Upper ball joint from the steering knuckle, using SST 09950-40010 (puller set) and a deep socket wrench

To install:
6. Install or connect the following:
- Press in a new upper ball joint, using SST 09309-37010 and a socket wrench
- New snapring, using a snapring expander
- New boot, secured with a new piece of wire
- Steering knuckle to the halfshaft
- Steering knuckle to the lower ball joint by installing the 4 bolts; do not tighten the bolts at this time.
- Upper control arm
- Upper ball joint to the steering knuckle. Tighten the nut to 80 ft. lbs. (105 Nm).
- New cotter pin
- Tighten the lower ball joint-to-steering knuckle bolts to 59 ft. lbs. (80 Nm).

- Brake rotor
- Caliper support bracket to the steering knuckle. Tighten both bolts to 90 ft. lbs. (123 Nm).
- Brake hose clamp to the steering knuckle. Tighten the bolt to 13 ft. lbs. (18 Nm).
- ABS speed sensor and wiring harness to the steering knuckle, if equipped with ABS
- Spacer and snapring to the half-shaft, using a snapring expander

7. If equipped with a free wheeling hub, install the hub, as follows:
- New front axle hub gasket
- Free wheeling hub body with the 6 cone washers and nuts. Tighten the 6 nuts to 23 ft. lbs. (31 Nm).
- Bolt with the washer. Tighten the bolt to 13 ft. lbs. (18 Nm).
- Apply multi-purpose grease to the inner hub splines
- Set the control handle and clutch to the FREE position
- New gasket on the cover
- Cover to the hub body with the follower pawl tabs aligned with the non-toothed portions of the hub body
- Tighten the cover bolts to 84 inch lbs. (10 Nm).

8. If equipped without a free wheeling hub, install the halfshaft to the steering knuckle, as follows:
- Locknut to the halfshaft. Tighten the nut to 174 ft. lbs. (235 Nm).
- Halfshaft lockcap and cotter pin
- Grease cab
- Strut. Tighten the strut-to-lower control arm nut to 101 ft. lbs. (135 Nm) and the upper 3 nuts to 47 ft. lbs. (64 Nm).
- Front wheels

9. Check the wheel alignment.

Lower Ball Joint

REMOVAL & INSTALLATION

2WD Models

1. Before servicing the vehicle, refer to the precautions in the beginning of this section.
2. Remove the wheel.
3. Support the lower control with a floor jack.
4. Remove or disconnect the following:
- Loosen the 2 lower ball joint set bolts
- Cotter pin and nut from the tie rod

- Tie rod from the ball joint bracket
- Cotter pin and nut from the lower ball joint
- Lower ball joint from the lower control arm
- Both lower ball joint set bolts
- Ball joint from the suspension

To install:
5. Install or connect the following:
- Lower ball joint to the steering knuckle and lower control arm
- Both lower ball joint set bolts; do not tighten the bolts at this time
- Lower ball joint nut to hold the lower ball joint to the lower control arm. Tighten the nut to 80 ft. lbs. (110 Nm).
- New cotter pin to the lower ball joint
- Tie rod end to the ball joint bracket. Tighten the nut to 53 ft. lbs. (72 Nm).
- New cotter pin to the tie rod end
- Tighten the both lower ball joint set bolts to 116 ft. lbs. (160 Nm).
- Wheel

6. Check the wheel alignment.

4WD MODELS

1. Before servicing the vehicle, refer to the precautions in the beginning of this section.
2. Remove or disconnect the following:
- Wheel
- Loosen the 4 lower ball joint set bolts
- Cotter pin and nut from the tie rod
- Tie rod from the ball joint bracket
- Cotter pin and nut from the lower ball joint
- Lower ball joint from the lower control arm
- 4 lower ball joint set bolts
- Ball joint from the suspension

To install:
3. Install or connect the following:
- Lower ball joint to the steering knuckle and lower control arm
- 4 lower ball joint set bolts; do not tighten the bolts at this time
- Lower ball joint-to-lower control arm nut. Tighten the nut to 112 ft. lbs. (152 Nm).
- New cotter pin to the lower ball joint
- Tie rod end to the ball joint bracket. Tighten the nut to 67 ft. lbs. (90 Nm).
- New cotter pin to the tie rod end
- Tighten the 2 lower ball joint set bolts to 83 ft. lbs. (113 Nm).

- Wheel
4. Check the wheel alignment.

Upper Control Arm

REMOVAL & INSTALLATION

2WD

1. Before servicing the vehicle, refer to the precautions in the beginning of this section.
2. Remove or disconnect the following:
   - Front wheel
   - Anti-lock Brake System (ABS) speed sensor and wire harness
   - Stabilizer bar link
   - Steering knuckle from the upper bar joint
3. Loosen the 2 bolts; then, remove the front and rear alignment adjusting shims.
4. Make note of the number and thickness of the front and rear shims.
5. Remove or disconnect the following:
   - Upper control arm
   - Upper ball joint from the arm

To install:
6. Install or connect the following:
   - Upper ball joint to the arm. Tighten the fasteners to 29 ft. lbs. (39 Nm).

➡**Do not lose the camber adjusting shims. Record the position and thickness of the camber shims so that these can be reinstalled to there original locations. Install the equal number and thickness of shims into there locations.**

   - Upper control arm with the shims. Tighten the mounting bolts to 94 ft. lbs. (130 Nm).
   - Steering knuckle to the upper ball joint
   - Stabilizer bar link
   - ABS speed sensor and wire harness
   - Front wheel. Tighten the lug nuts to 83 ft. lbs. (110 Nm).

4WD

1. Before servicing the vehicle, refer to the precautions in the beginning of this section.
2. Remove or disconnect the following:
   - Front wheel
   - Shock and coil spring assembly
   - Anti-lock Brake System (ABS) speed sensor wire harness clamp
3. Upper ball joint, as follows:
   - Cotter pins and loosen the nut

- Upper ball joint from the control arm, using a ball joint separator
- Support the steering knuckle
- Nut
4. Detach the control arm, by removing the nut, bolt, washers and lowering the arm.

To install:
5. Install or connect the following:
   - Upper control arm with the washer, bolt and nut. Tighten the nut to 87 ft. lbs. (115 Nm).
   - Upper ball joint to the control arm. Tighten the mounting nut to 80 ft. lbs. (105 Nm).
   - New cotter pin
   - ABS speed sensor wire harness

clamp. Tighten it to 71 inch lbs. (8 Nm).
   - Shock and coil spring assembly
6. Check and/or adjust the alignment.

CONTROL ARM BUSHING REPLACEMENT

4WD

1. Before servicing the vehicle, refer to the precautions in the beginning of this section.
2. Remove the upper control arm from the vehicle.
3. Pry up the bushing flange, using a chisel and a hammer.

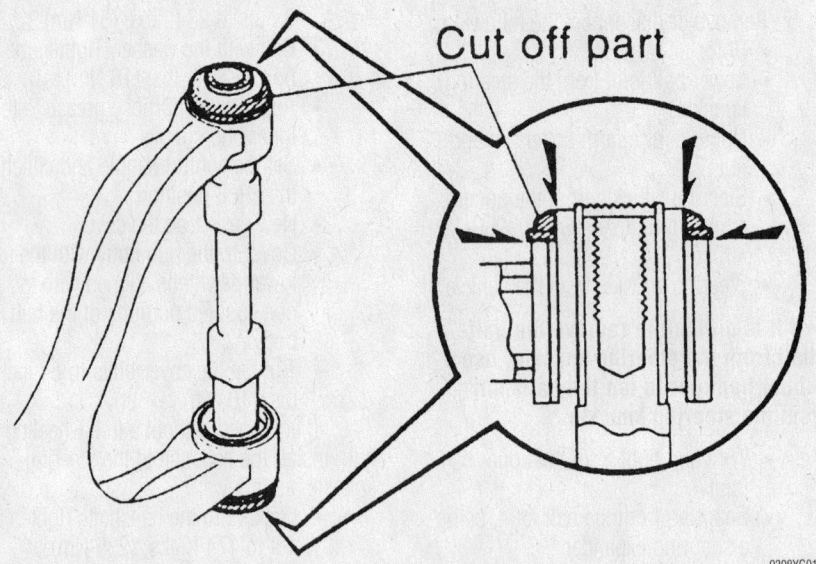

9308YG01

Cut the bushing flush with the upper control arm tube and shaft

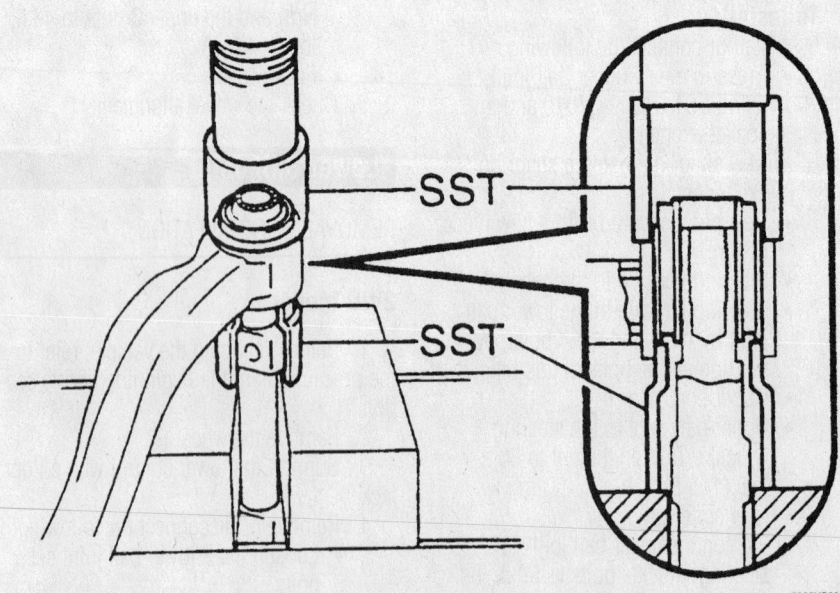

9308YG02

Position the upper control arm bushings with the tools

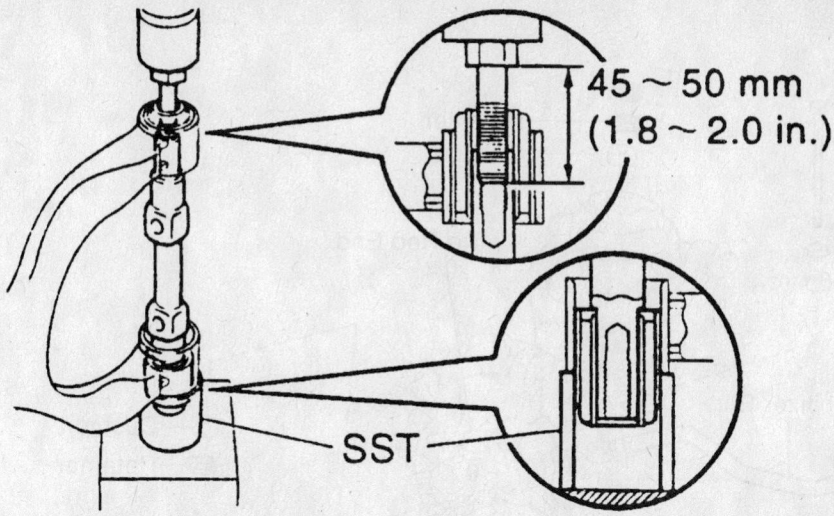

Positioning the upper control arm tube and bolt

45 ~ 50 mm
(1.8 ~ 2.0 in.)

SST

9308YG03

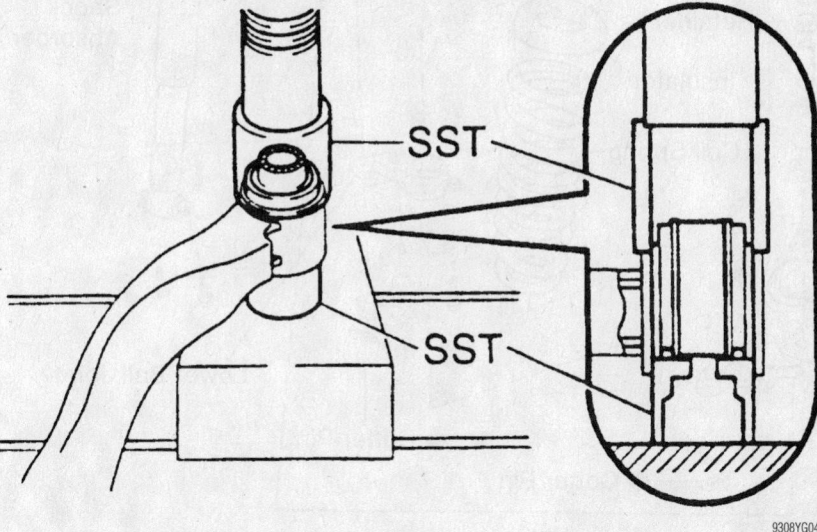

Removing the bushings from the upper control arm

SST

SST

9308YG04

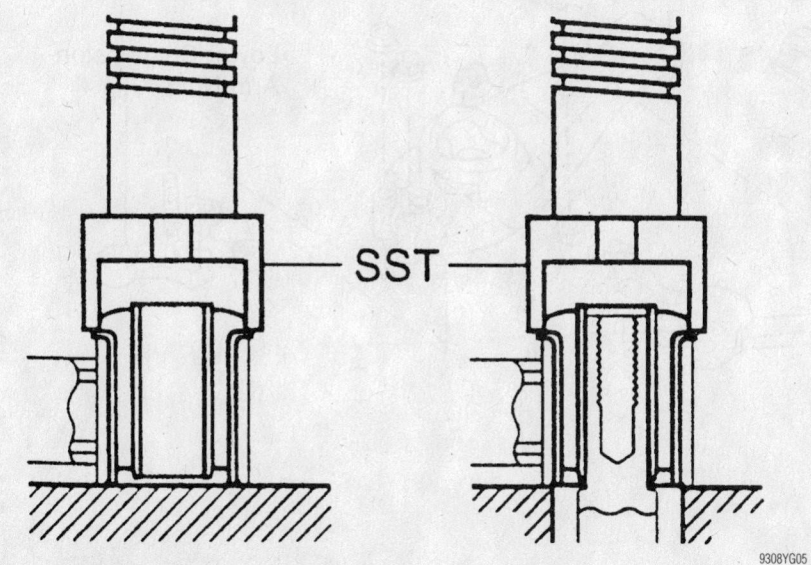

Installing the bushings to the upper control arm

SST

9308YG05

4. Using tools 09613-26010, 09613-20060 and 09950-00020 and a shop press, remove the bushing.

To install:

5. Using tools 09223-00010, 09506-35010 and a shop press, press the new bushing into the upper control arm.

6. Install the upper control arm to the chassis. Torque the nuts to 87 ft. lbs. (115 Nm).

7. Check and/or adjust the alignment.

2WD

1. Before servicing the vehicle, refer to the precautions in the beginning of this section.

2. Remove the upper control arm.

3. Cut off the outer edges of the bushing so it is flush with arm tube and the shaft.

➡**Be careful not to damage the edge of the arm tubes.**

4. Using a shop press with tool 09710-03031, press down the suspension arm tube until it touches tool 09710-03141.

➡**Do not press the tube excessively.**

5. Temporarily install a 1.8–2.0 in. (45–50mm) bolt to the arm shaft on the other side.

6. Using a shop press with tool 09710-03141, remove the bushing from the arm shaft.

7. Repeat this procedure for the other bushing.

To install:

8. Using a shop press and tool 09710-03101, install the new bushing.

9. Place the arm shaft to the bushing.

10. Using a shop press and tool 09710-03101, install the other new bushing.

➡**Pass the arm shaft through the bushing to make sure that the shaft turns freely and there is no axial play**

11. Install the upper control arm. Torque the lockbolts to 92 ft. lbs. (125 Nm).

12. Check and/or adjust the alignment.

Lower Control Arm

REMOVAL & INSTALLATION

2WD

1. Before servicing the vehicle, refer to the precautions in the beginning of this section.

2. Remove or disconnect the following:
   • Front wheel

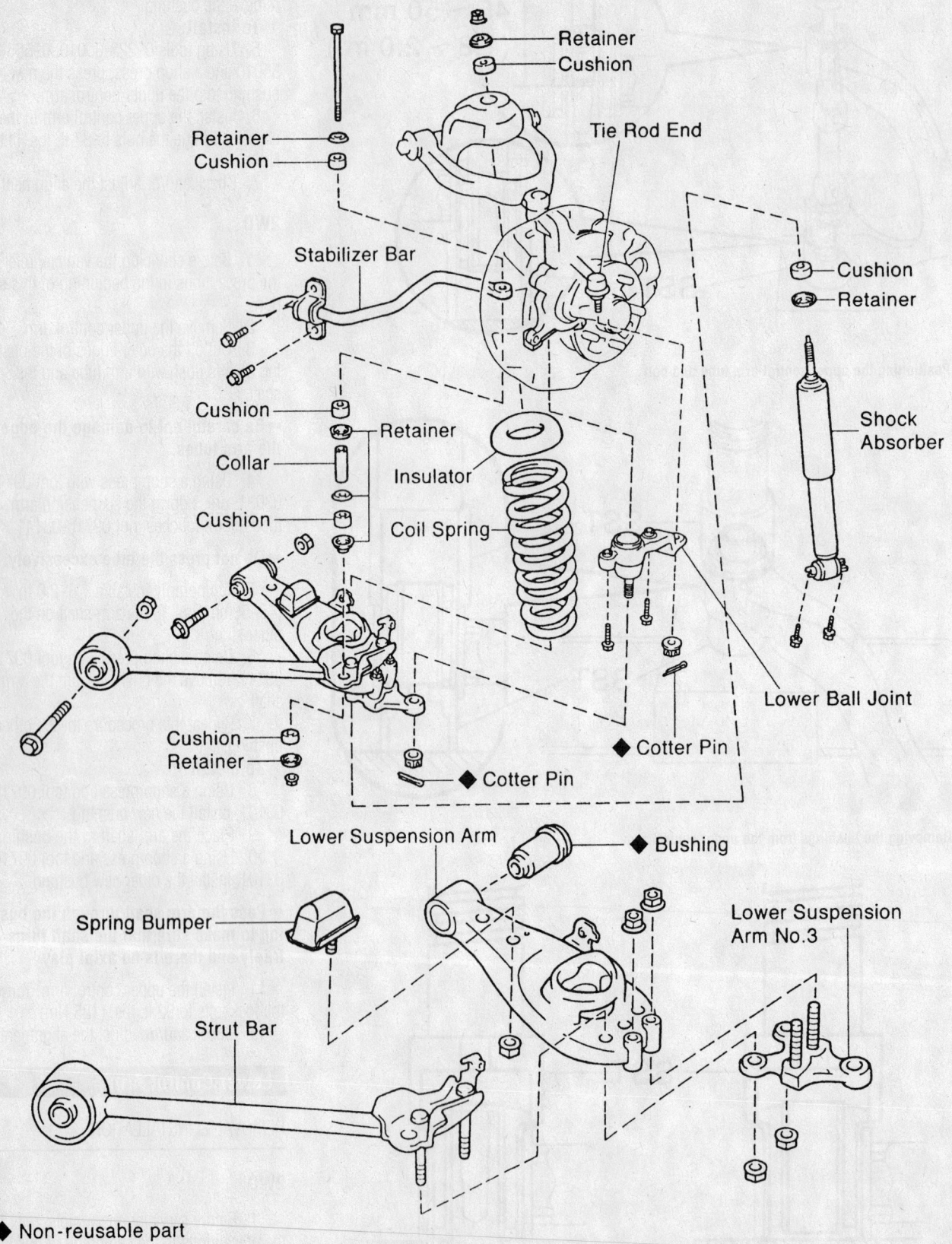

Retainer
Cushion
Retainer
Cushion
Tie Rod End
Stabilizer Bar
Cushion
Retainer
Cushion
Retainer
Collar
Insulator
Cushion
Coil Spring
Shock Absorber
Lower Ball Joint
◆ **Cotter Pin**
Cushion
Retainer
◆ **Cotter Pin**
Lower Suspension Arm
◆ **Bushing**
Lower Suspension Arm No.3
Spring Bumper
Strut Bar

◆ Non-reusable part

86828G01

Exploded view of the lower control arm and related front suspension components

- Shock absorber

3. Compress the spring using a spring compressor, following the manufacturer's instructions.

4. Remove the stabilizer bar, as follows:
- Stabilizer bar link from the lower control arm
- 2 stabilizer bar bracket set bolts

5. Remove the lower control arm and strut bar, as follows:
- Support the upper control arm and steering knuckle assembly
- Cotter pin and nut
- Lower ball joint from the lower control arm
- Loosen the lower control arm set bolt and remove the nut
- Loosen the strut bar front set bolt and remove the nut
- Pull out the bolts and remove the lower control arm along with the strut bar

6. Remove or disconnect the following:
- Coil spring compressor tool and coil
- Strut bar from the lower control arm. Separate the nut and spring bumper.
- Lower suspension arm No. 3

To install:

7. Install or connect the following:
- Lower suspension arm No. 3. Tighten the bolts to 111 ft. lbs. (150 Nm).
- Spring bumper. Tighten it to 32 ft. lbs. (43 Nm) and the strut bar-to-lower control arm to 111 ft. lbs. (150 Nm).

8. Place each end of the coil spring and lower control arm seat in contact when applying the coil spring expander.

9. Install the lower control arm and strut bar, as follows:
- Attach the strut bar front set bolt. Tighten the set bolt to 221 ft. lbs. (300 Nm).

➡**Make sure the suspension is stabilized prior to tightening the bolt.**

- Lower control arm set bolt. Tighten the nut to 148 ft. lbs. (200 Nm).

➡**Make sure the suspension is stabilized prior to tightening the bolt.**

- Lower ball joint with a ball joint installer tool. Tighten the nut to 80 ft. lbs. (110 Nm).
- Cotter pin.

10. Install or connect the following:
- Stabilizer bar bracket. Tighten the set bolts to 22 ft. lbs. (29 Nm).
- Stabilizer link to the lower control

arm. Tighten the fasteners to 29 ft. lbs. (39 Nm).

11. Remove the spring compressing tool.

12. Install or connect the following:
- Shock absorber
- Wheel. Tighten the lug nuts.

13. Check and/or adjust the alignment.

4WD

1. Before servicing the vehicle, refer to the precautions in the beginning of this section.

2. Remove or disconnect the following:
- Front wheel
- Steering gear assembly
- Stabilizer bar link
- Shock absorber from lower control arm

3. Support the upper control and steering knuckle securely.

4. Remove or disconnect the following:
- Cotter pin and nut from the lower ball joint
- Lower ball joint from the lower control arm

5. Place matchmarks on the front and rear adjusting cams.

6. Remove or disconnect the following:
- 2 bolts, nuts, adjusting cams and lower control arm
- Spring bumpers, with a special tool 09922-10010

To install:

7. Install or connect the following:
- Spring bumpers. Tighten to 17 ft. lbs. (23 Nm).
- Lower control arm, placing it in the appropriate position with the matchmarks. Tighten the arm to 96 ft. lbs. (130 Nm).

8. Install or connect the following:
- Lower ball joint. Tighten the nut to 112 ft. lbs. (152 Nm).
- Shock absorber to the lower control arm
- Stabilizer bar link
- Steering gear assembly
- Front wheel. Tighten the lug nuts.

9. Check and/or adjust the alignment.

CONTROL ARM BUSHING REPLACEMENT

4WD

1. Before servicing the vehicle, refer to the precautions in the beginning of this section.

2. Remove the lower control arm from the vehicle.

3. Pry up the bushing flange, using a chisel and a hammer.

4. Using tools 09613-26010, 09632-36010 and 09950-00020 and a shop press, remove the bushing.

To install:

5. Using tools 09316-20011, 09710-30021 and a shop press, press the new bushing into the lower control arm.

6. Install the lower control arm to the chassis. Torque the bolts to 96 ft. lbs. (130 Nm).

7. Check and/or adjust the alignment.

2WD

The lower control arm is equipped with a single bushing.

1. Before servicing the vehicle, refer to the precautions in the beginning of this section.

2. Remove the lower control arm from the vehicle.

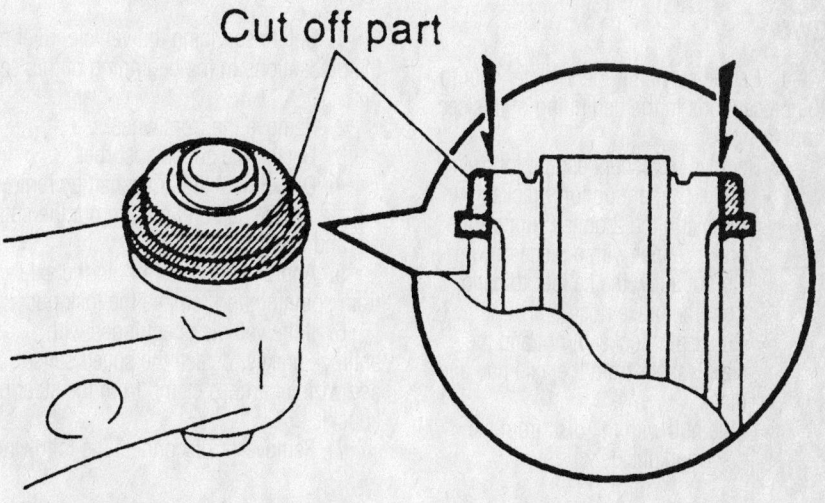

Cut off part

Cutting off part of the bushing—2WD

9308YG06

3. Cut off a portion of the bushing to expose the edge of the arm tube.

4. Position the lower control arm on a shop press with the cut side facing downward, resting on tool 09710-30021; then, press the bushing from the arm.

To install:

5. Position a new bushing onto the lower control arm.

6. Position the lower control arm on a shop press, resting on tool 09710-30021.

7. Press the bushing into the lower control arm.

8. Install the lower control arm.

9. Check and/or adjust the alignment.

Front Wheel Bearings

ADJUSTMENT

The bearings on the 4WD are not adjustable.

2WD

1. Tighten the adjusting nut to 26 ft. lbs. (35 Nm).

2. Turn the disc/hub assembly 2–3 times, from the left to the right.

3. Loosen the adjusting nut until it can be turned by hand.

4. Attach a spring tension gauge to 1 lug on the hub assembly. Pull on the gauge and measure the frictional force. Frictional force should be 1–3 lbs. (5.0–14.0 N).

5. Adjust the preload by tightening the nut.

6. Measure the hub axial play. The limit is 0.0020 in. (0.05mm).

REMOVAL & INSTALLATION

2WD

1. Before servicing the vehicle, refer to the precautions in the beginning of this section.

2. Remove or disconnect the following:
- Brake caliper support bracket, by removing the 2 bolts. Support the brake caliper with a piece of wire. Do not allow the caliper to hang from the brake hose.
- Cotter pin, lockcap, nut and the claw washer from the axle hub and disc
- Axle hub with the disc from the steering knuckle.

✳✳ WARNING

Do not drop the outer bearing when removing the hub.

- Inner oil seal
- Inner bearing
- Bearing outer races, using SST 09527-17011, a brass bar and a hammer

3. If it is necessary to separate the hub and rotor, place matchmarks on the hub and rotor.

4. Remove the 5 bolts and remove the hub from the rotor.

To install:

5. Install or connect the following:
- New bearing races, using SST 09527-17011 and a press.
- Hub to the rotor. Tighten the 5 bolts to 47 ft. lbs. (64 Nm).

6. Clean all parts.

7. Repack the bearings with multi purpose grease and apply the same grease to the outer bearings.

8. Install or connect the following:
- Inner bearing and seal to the hub. Coat the inner seal with multi purpose grease.
- Outer bearing to the hub
- Hub to the steering knuckle
- Axle hub to the steering knuckle claw washer and nut

9. Adjust the bearing preload as described above.

10. Install or connect the following:
- Locknut, cotter pin and the grease cap
- Disc brake caliper. Tighten the 2 bolts to 80 ft. lbs. (108 Nm).
- Wheel

4WD

1. Before servicing the vehicle, refer to the precautions in the beginning of this section.

2. Remove the front wheel.

3. Detach the shock absorber.

4. Disconnect the driveshaft by removing the grease cap and pulling out the cotter pin and lock cap.

5. Apply the brakes to hold the axle from spinning and remove the lock nut.

6. If the vehicle is equipped with antilock brakes, detach the speed sensor and wiring harness clamp from the steering knuckle.

7. Remove or disconnect the following:

- Banjo bolt and 2 gaskets from the caliper
- Flexible brake hose from the caliper
- Brake caliper and then the rotor
- Lower ball joint
- Steering knuckle

8. Clamp the axle hub in a soft jaw vise.

➡**Close the vise until it holds the hub bolts.**

9. Using the proper seal puller or pry-tool, remove the oil seal.

10. On vehicles equipped with a free wheel hub and on the Pre Runner, use a chisel and hammer to loosen the staked part of the lock nut.

11. Remove the lock nut. A special service tool may be required.

12. Remove the Antilock Brake System (ABS) speed sensor rotor/spacer.

➡**Do not scratch the speed sensor rotor.**

13. Detach the bolts to the dust shield and shift the shield towards the outside of the hub.

14. Remove the axle from the steering knuckle, a special service tool may be required.

15. Remove the dust cover from the steering knuckle.

16. On vehicles without a free wheeling hub, that are 4WD only, remove the bearing spacer and ABS speed sensor rotor spacer.

17. Remove the outside seal by prying it out with a seal puller.

18. Remove the bearing from the steering knuckle by removing the snapring with a pair of snapring pliers.

19. Press the bearing from the steering knuckle.

To install:

20. Install or connect the following:
- New bearing
- New oil seal
- Axle hub to the steering knuckle. Torque the bolts to 13 ft. lbs. (18 Nm).
- ABS speed sensor rotor

21. On vehicles that are equipped with a free wheel hub, install the bearing spacer.

22. Except Pre-Runner install a new inside oil seal.

23. On the Pre-Runner, install the grease cap.

24. The remainder of the installation procedure is the reverse of removal. New lock nut torque is 203 ft. lbs. (274 Nm).

BRAKES

Brake Caliper

REMOVAL & INSTALLATION

1. Disconnect the negative battery cable from the battery.
2. Raise and support the vehicle safely.
3. Remove the wheels.
4. Disconnect the brake hose from the caliper by removing the union bolt and 2 gaskets. Plug the end of the hose to prevent loss of fluid.
5. Remove the bolts that attach the caliper to the torque plate.
6. Lift the bottom of the caliper up and remove the caliper assembly.

To install:

7. Grease the caliper slides and bolts with lithium grease or equivalent. Install the caliper and secure with the bolts. Torque the bolts to 65 ft. lbs. (88 Nm).
8. Connect the brake hose to the caliper, using 2 new washers. Make sure the flexible hose lock is securely in the lock hole of the caliper. Torque the union bolt to 22 ft. lbs. (30 Nm).
9. Fill the brake system to the proper level and bleed the brake system.
10. Install the tire and wheel assembly.

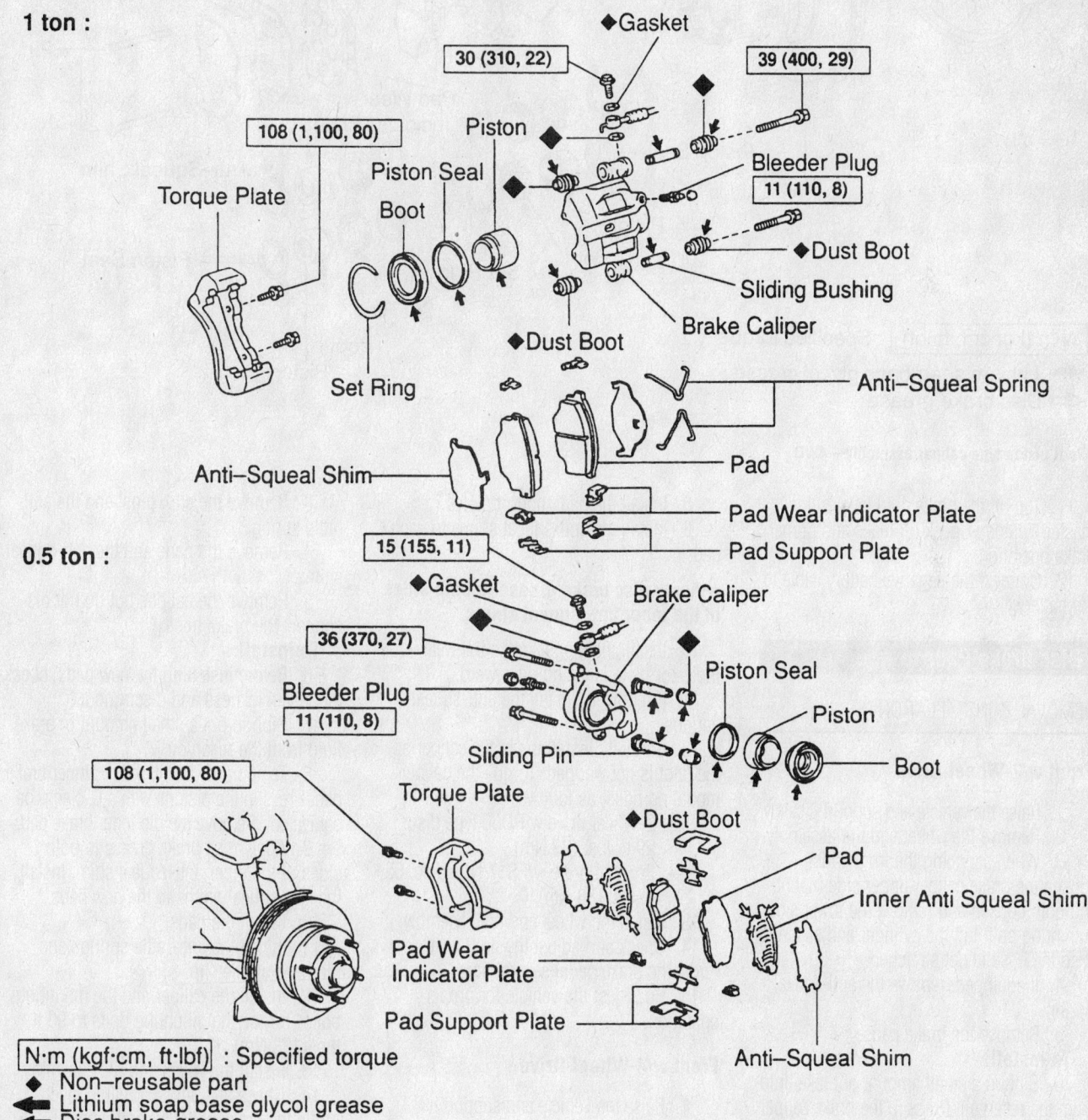

1 ton:

Gasket • 30 (310, 22) • 39 (400, 29) • Piston • 108 (1,100, 80) • Torque Plate • Piston Seal • Boot • Bleeder Plug • 11 (110, 8) • Dust Boot • Sliding Bushing • Brake Caliper • Dust Boot • Set Ring • Anti-Squeal Spring • Anti-Squeal Shim • Pad • Pad Wear Indicator Plate • Pad Support Plate • 15 (155, 11)

0.5 ton:

Gasket • 36 (370, 27) • Brake Caliper • Piston Seal • Piston • Bleeder Plug • 11 (110, 8) • Boot • Sliding Pin • Dust Boot • Pad • Inner Anti Squeal Shim • 108 (1,100, 80) • Torque Plate • Pad Wear Indicator Plate • Pad Support Plate • Anti-Squeal Shim

N·m (kgf·cm, ft·lbf) : Specified torque
◆ Non-reusable part
◀ Lithium soap base glycol grease
◁ Disc brake grease

Single piston type caliper assembly—2WD

93026G70

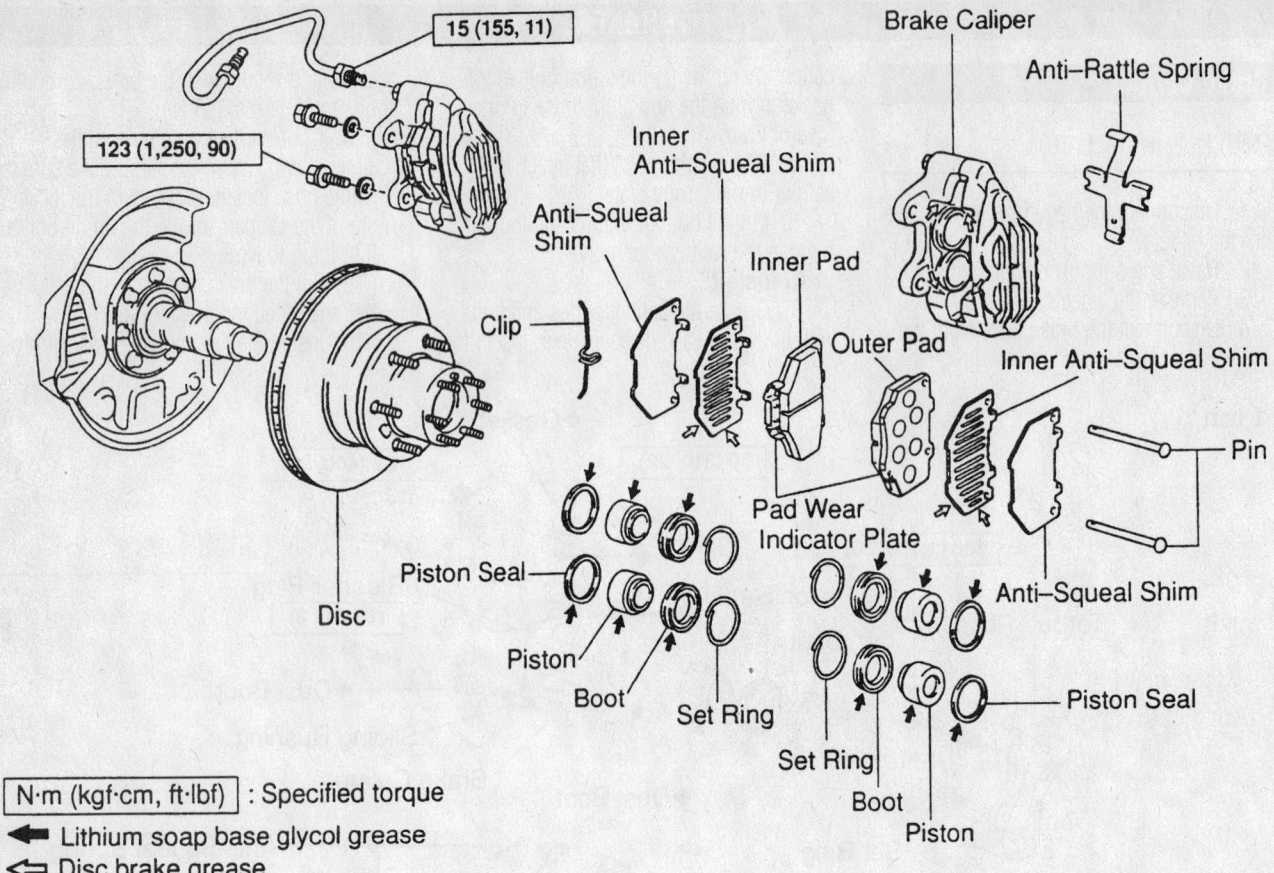

15 (155, 11)

123 (1,250, 90)

Brake Caliper

Anti–Rattle Spring

Inner Anti–Squeal Shim

Anti–Squeal Shim

Inner Pad

Clip

Outer Pad

Inner Anti–Squeal Shim

Pin

Pad Wear Indicator Plate

Anti–Squeal Shim

Piston Seal

Piston

Boot

Set Ring

Set Ring

Boot

Piston

Piston Seal

Disc

N·m (kgf·cm, ft·lbf) : Specified torque
◄ Lithium soap base glycol grease
⇐ Disc brake grease

93026G71

Dual piston type caliper assembly—4WD

11. Top off the brake fluid level in the master cylinder. Check for leaks and proper brake operation.

12. Connect the negative battery cable to the battery.

Disc Brake Pads

REMOVAL & INSTALLATION

Front w/2-Wheel Drive

1. Raise the vehicle and support it safely.
2. Remove the wheel and tire assembly.
3. When servicing the front pads, loosen the brake caliper upper side mounting bolt. Loosen and remove the lower side mounting bolt. Lift the cylinder and suspend it so the hose is not stretched.
4. If equipped, remove the anti-squeal spring.
5. Remove the brake pads.
To install:
6. Siphon a small amount of brake fluid from the reservoir. Press in the brake caliper piston with a hammer handle or equivalent.
7. Before installing the new pads, check the disc thickness and disc runout.

8. Install the pad support plates.
9. Install the anti-squeal shims to each pad.

➡**Apply disc brake grease to both sides of the inner anti-squeal shims.**

10. Install the disc pads so the wear indicator plate is facing downward.
11. If removed, install the anti-squeal springs.
12. Carefully install the brake caliper so the boot is not wedged. Torque the caliper mounting bolts, as follows:
   - 2-Wheel drive w/PD60 type disc: 29 ft. lbs. (39 Nm)
   - 2-wheel drive w/FS17 type disc: 65 ft. lbs. (88 Nm)
13. Install the wheel and tire assembly.
14. Check and adjust the fluid level. Apply the brake pedal several times.
15. Road-test the vehicle for proper operation.

Front w/4-Wheel Drive

1. Raise the vehicle and support it safely.
2. Remove the wheel and tire assembly.

3. Remove the clip, pins, and the anti-rattle spring.
4. Remove the pads and the anti-squeal shims.
5. Remove the caliper, but do not disconnect the brake hose.
To install:
6. Before installing the new pads, check the disc thickness and disc runout.
7. Siphon out a small amount of brake fluid from the reservoir.
8. Temporarily install the old inner brake pad. Press in the pistons with a C-clamp or equivalent. Remove the old inner brake pad.
9. Apply disc brake grease to both sides of the inner anti-squeal shim. Install the anti-squeal shims to the new pads.
10. Install the pads.
11. Install the anti-rattle springs and pins. Install the clip.
12. Install the caliper and the mounting bolts. Torque the mounting bolts to 90 ft. lbs. (123 Nm).
13. Install the wheel and tire assembly.
14. Check and adjust the fluid level. Apply the brake pedal several times.
15. Road-test the vehicle for proper operation.

Rear

1. Raise the vehicle and support it safely.
2. Remove the wheel and tire assembly.
3. Remove the brake caliper and suspend it with a wire so the hose is not stretched or stressed.
4. Remove the brake pads, anti-squeal shim, pad support plates and wear indicators.

To install:

5. Before installing the new pads, check the disc thickness and disc runout.
6. Temporarily install the old inner brake pad. Press in the piston with a C-clamp or equivalent. Remove the old inner brake pad.
7. Install the pad support plates.
8. Install the pad wear indicator plate to each pads.
9. Install the anti-squeal shim to the outer pad. Install the pads so the wear indicator plate is facing upward.
10. Install the brake caliper. Torque the main sliding pin and the sub pin to 65 ft. lbs. (88 Nm).

11. Install the wheel and tire assembly.
12. Apply the brake pedal several times.
13. Road-test the vehicle for proper operation.

Brake Drums

REMOVAL & INSTALLATION

1. Raise and safely support the vehicle.
2. Remove the rear wheel(s).
3. Remove the brake drum from the axle hub. If there is difficulty in removing the drum, insert a suitable tool through the hole in the rear of the backing plate, and hold the automatic adjusting lever away from the adjuster. Using another suitable tool at the same time, reduce the brake shoe adjuster by turning the adjusting wheel.

To install:

4. Install the brake drum and pull the parking brake lever all the way up until a clicking sound can no longer be heard.
5. Verify that the rear wheels will not

turn. If the rear wheels turn, adjust the parking brake cable as necessary.
6. Release the parking brake and remove the brake drum. Measure the brake drum inside diameter and diameter of the brake shoes. Check that the difference between the diameters is the correct shoe clearance. Clearance is 0.024 in. (6mm).
7. If the brake shoe clearance is not correct, adjust the brake shoes until the clearance is correct.
8. Install the brake drum, replace the wheel(s), and safely lower the vehicle.
9. Road-test the vehicle for proper brake operation.

Brake Shoes

REMOVAL & INSTALLATION

1. Loosen the rear wheel lug nuts slightly.
2. Block the front wheels, raise the rear of the vehicle, and safely support it with jackstands.

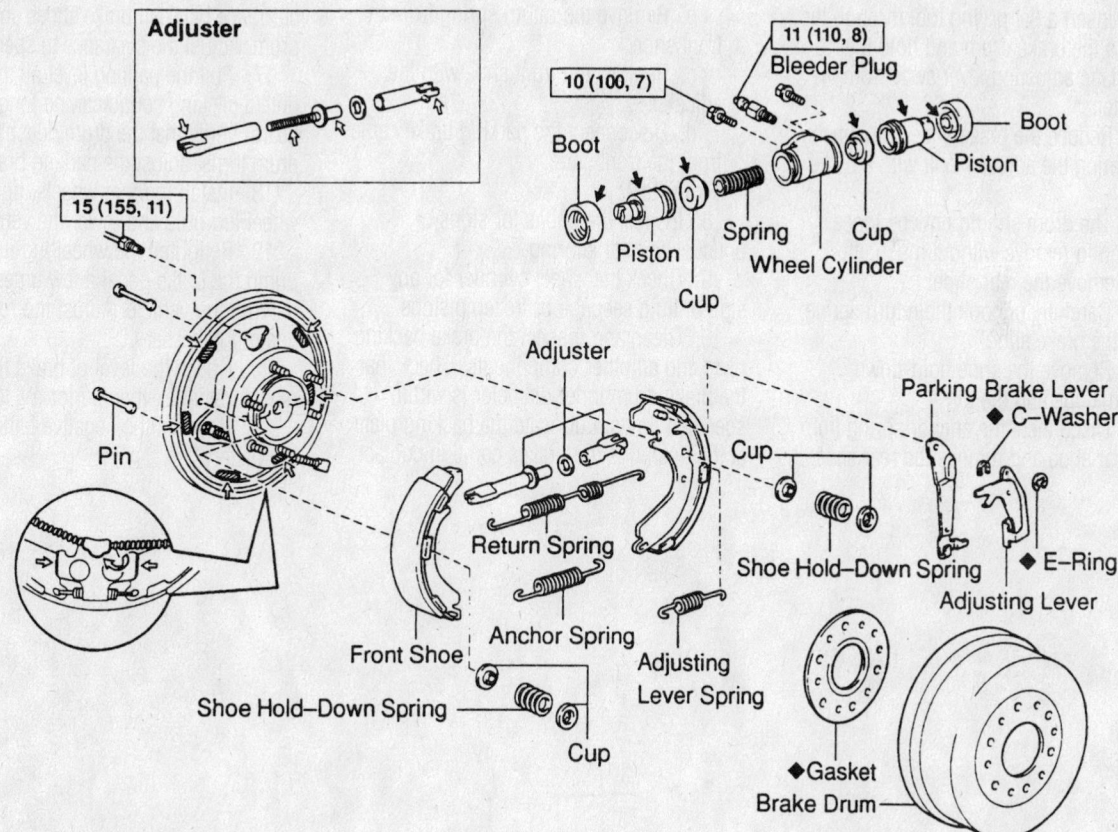

N·m (kgf·cm, ft·lbf) : Specified torque
◆ Non–reusable part
➡ Lithium soap base glycol grease
⇨ High temperature grease

93026G77

Exploded view of the rear brake drums components—2WD

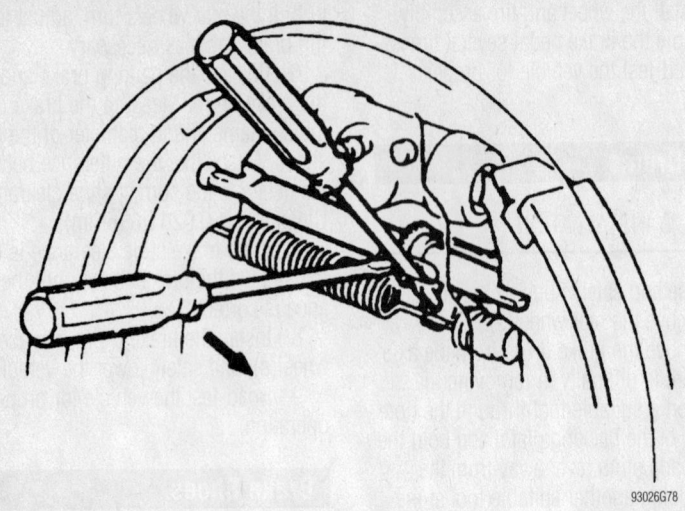

93026G78

Use a brake adjusting tool (brake spoon) and a prytool to adjust the brake shoes through the adjusting hole

3. Remove the wheel lug nuts and the wheel.

4. Remove the brake drum.

5. If the drum is difficult to remove, perform the following:

a. Insert a flat prying tool through the hole in the brake drum and hold the automatic adjusting lever away from the adjuster.

b. Reduce the brake shoe adjustment by turning the adjuster bolt with a brake tool.

c. The drum should now be loose enough to remove without much effort.

6. Remove the rear shoe.

a. Carefully unhook the return spring from the brake shoe.

b. Remove the shoe hold-down spring, cups and the pin.

c. Disconnect the anchor spring from the rear shoe and remove the rear shoe.

d. Disconnect the anchor spring from the front shoe.

7. Remove the front shoe.

a. Remove the shoe hold-down spring, cups and pin.

b. Remove the return spring from the front shoe.

c. Remove the front shoe with the adjuster.

d. Disconnect the parking brake cable from the front shoe.

To install:

8. Inspect the shoes for signs of unusual wear or scoring.

9. Check the wheel cylinder for any sign of fluid seepage or frozen pistons.

10. Clean and inspect the brake backing plate and all other components. Check that the brake drum inner diameter is within specified limits. Lubricate the backing plate at the positions the brakes come in contact with the backing plate. Also lubricate the anchor plate.

11. Mount the automatic adjuster assembly onto a new rear brake shoe.

12. Install the front shoe.

a. Install the parking brake cable to the front shoe.

b. Install the front shoe with the adjuster.

c. Install the return spring to the front shoe.

d. Install the shoe hold-down spring, cups and pin.

13. Install the rear shoe.

a. Install the anchor spring to the front shoe.

b. Install the anchor spring to the rear shoe and install the rear shoe.

c. Install the shoe hold-down spring, cups and the pin.

d. Hook the return spring to the brake shoe.

14. Install the brake drum.

15. Adjust the brake shoes until a slight drag is felt when the drum is spun by hand.

16. Remove the brake drum and check the clearance between brake shoes and brake drum. Adjust the clearance to specification.

17. Pull the parking lever all the way up until a clicking sound can no longer be heard. Verify that the drum doesn't turn. If the drum turns, adjust the parking brake cable.

18. Install the rear wheels, tighten the wheel lug nuts and lower the vehicle.

19. Retighten the wheel lug nuts and pump the brake pedal a few times before moving the vehicle. Adjust the rear brakes again if necessary.

20. Check the level of brake fluid in the master cylinder, then perform a test drive.

21. Connect the negative battery cable to the battery.

SPECIFICATION CHARTS

ENGINE AND VEHICLE IDENTIFICATION

Code ①	Liters (cc)	Cu. In.	Cyl.	Fuel Sys.	Engine Type	Eng. Mfg.	Code ②	Year
1AZ-FE	2.0 (1998)	122	4	SFI	DOHC	Toyota	Y	2000
1MZ-FE	3.0 (2995)	183	6	SFI	DOHC	Toyota	1	2001
3S-FE	2.0 (1998)	122	4	MFI	DOHC	Toyota	2	2002
3RZ-FE	2.7 (2693)	164	4	MFI	DOHC	Toyota	3	2003
5VZ-FE	3.4 (3378)	206	6	MFI	DOHC	Toyota	4	2004
2AZ-FE	2.4 (2362)	144	4	SFI	DOHC	Toyota		

SFI: Sequential Fuel Injection

MFI: Multi-port Fuel Injection

DOHC: Double Overhead Camshaft

① Stamped on the left side of the engine block

② 10th digit of the Vehicle Identification Number (VIN)

42356-RAV-C01

GENERAL ENGINE SPECIFICATIONS

Year	Model	Engine Displacement Liters (cc)	Engine Series (ID/VIN)	Fuel System	Net Horsepower @ rpm	Net Torque @ rpm (ft. lbs.)	Bore x Stroke (in.)	Compression Ratio	Oil Pressure @ rpm
2000	RAV4	2.0 (1998)	3S-FE	MFI	127@5400	132@4600	3.40x3.40	9.5:1	NA
	RX300	3.0 (2995)	1MZ-FE	SFI	220@5800	222@4400	3.44x3.27	10.5:1	43-78@3000
	4Runner	2.7 (2693)	3RZ-FE	MFI	150@4800	177@4000	3.74x3.74	9.5:1	36-71@3000
		3.4 (3378)	5VZ-FE	MFI	183@4800	217@3600	3.68x3.23	9.6:1	NA
2001	RAV4	2.0 (1998)	1AZ-FE	SFI	127@5400	132@4600	3.40x3.40	9.5:1	NA
	4Runner	3.4 (3378)	5VZ-FE	MFI	183@4800	217@3600	3.68x3.23	9.6:1	NA
	RX300	3.0 (2995)	1MZ-FE	SFI	220@5800	222@4400	3.44x3.27	10.5:1	43-78@3000
	Highlander	2.4 (2362)	2AZ-FE	SFI	155@5600	163@4000	3.48x3.78	NA	36@3000
		3.0 (2995)	1MZ-FE	SFI	220@5800	222@4400	3.44x3.27	10.5:1	43-78@3000
2002	RAV4	2.0 (1998)	1AZ-FE	SFI	127@5400	132@4600	3.40x3.40	9.5:1	NA
	4Runner	3.4 (3378)	5VZ-FE	MFI	183@4800	217@3600	3.68x3.23	9.6:1	NA
	RX300	3.0 (2995)	1MZ-FE	SFI	220@5800	222@4400	3.44x3.27	10.5:1	43-78@3000
	Highlander	2.4 (2362)	2AZ-FE	SFI	155@5600	163@4000	3.48x3.78	NA	36@3000
		3.0 (2995)	1MZ-FE	SFI	220@5800	222@4400	3.44x3.27	10.5:1	43-78@3000
2003	RAV4	2.0 (1998)	1AZ-FE	SFI	127@5400	132@4600	3.40x3.40	9.5:1	NA
	4Runner	3.4 (3378)	5VZ-FE	MFI	183@4800	217@3600	3.68x3.23	9.6:1	NA
	RX300	3.0 (2995)	1MZ-FE	SFI	220@5800	222@4400	3.44x3.27	10.5:1	43-78@3000
	Highlander	2.4 (2362)	2AZ-FE	SFI	155@5600	163@4000	3.48x3.78	NA	36@3000
		3.0 (2995)	1MZ-FE	SFI	220@5800	222@4400	3.44x3.27	10.5:1	43-78@3000

NA: Not Available

SFI: Sequential Fuel Injection

MFI: Multi-port Fuel Injection

42356-RAV-C02

ENGINE TUNE-UP SPECIFICATIONS

Year	Engine Displacement Liters (cc)	Engine ID/VIN	Spark Plug Gap (in.)	Ignition Timing (deg.)*	Fuel Pump (psi)	Idle Speed (rpm)		Valve Clearance	
						MT	AT	Intake	Exhaust
2000	2.0 (1998)	3S-FE	0.043	8-12B	44-50	700-800	700-800	0.007-0.011	0.011-0.015
	3.0 (2995)	1MZ-FE	0.043	8-12B	44-50	—	650-750	0.006-0.010	0.010-0.014
	2.7 (2693)	3RZ-FE	0.031	8-12B	38-44	650-750	650-750	0.006-0.010	0.010-0.014
	3.4 (3378)	5VZ-FE	0.031	8-12B	38-44	650-750	650-750	0.006-0.010	0.010-0.014
2001	2.0 (1998)	1AZ-FE	0.043	8-12B	44-50	700-800	700-800	0.007-0.011	0.011-0.015
	2.4 (2362)	2AZ-FE	0.041	8-12B	44-50	—	600-700	0.007-0.011	0.012-0.016
	3.0 (2995)	1MZ-FE	0.043	8-12B	44-50	—	650-750	0.006-0.010	0.010-0.014
	3.4 (3378)	5VZ-FE	0.031	8-12B	38-44	—	650-750	0.006-0.010	0.010-0.014
2002	2.0 (1998)	1AZ-FE	0.043	8-12B	44-50	700-800	700-800	0.008-0.011	0.012-0.016
	2.4 (2362)	2AZ-FE	0.041	8-12B	44-50	—	600-700	0.007-0.011	0.012-0.016
	3.0 (2995)	1MZ-FE	0.039-0.043	8-12B	44-50	—	650-750	0.006-0.010	0.010-0.014
	3.4 (3378)	5VZ-FE	0.039-0.043	8-12B	38-44	—	650-750	0.006-0.009	0.011-0.014
2003	2.0 (1998)	1AZ-FE	0.043	8-12B	44-50	700-800	700-800	0.008-0.011	0.012-0.016
	2.4 (2362)	2AZ-FE	0.041	8-12B	44-50	—	600-700	0.007-0.011	0.012-0.016
	3.0 (2995)	1MZ-FE	0.039-0.043	8-12B	44-50	—	650-750	0.006-0.010	0.010-0.014
	3.4 (3378)	5VZ-FE	0.039-0.043	8-12B	38-44	—	650-750	0.006-0.009	0.011-0.014

NOTE: The Vehicle Emission Control Information label often reflects specification changes made during production. The label figures must be used if they differ

B: Before top dead center

* With terminals TC and CG connected to DLC3

42356-RAV-C03

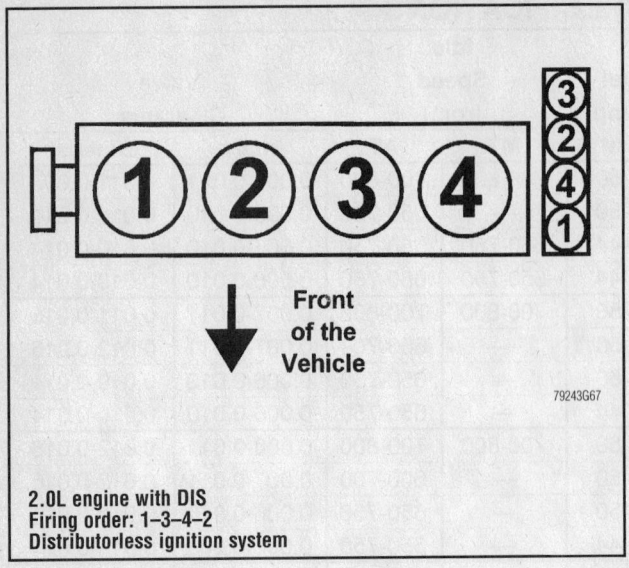

2.0L engine with DIS
Firing order: 1–3–4–2
Distributorless ignition system

79243G67

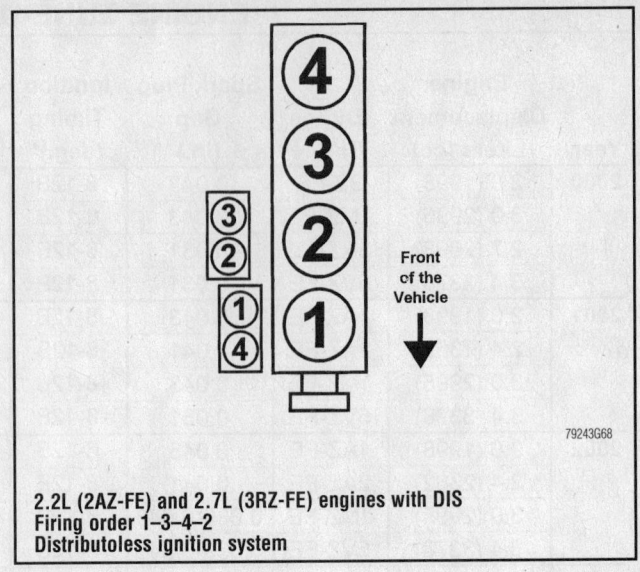

Front of the Vehicle

2.2L (2AZ-FE) and 2.7L (3RZ-FE) engines with DIS
Firing order 1–3–4–2
Distributoless ignition system

79243G68

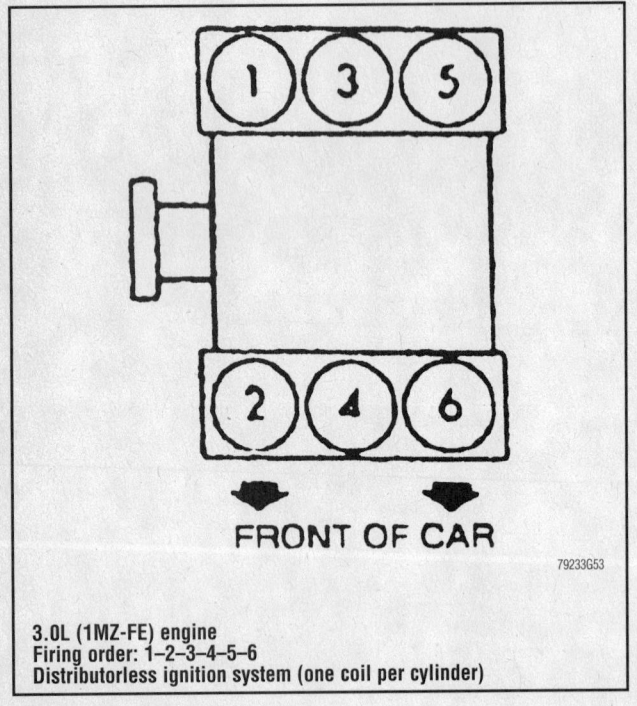

FRONT OF CAR

3.0L (1MZ-FE) engine
Firing order: 1–2–3–4–5–6
Distributorless ignition system (one coil per cylinder)

79233G53

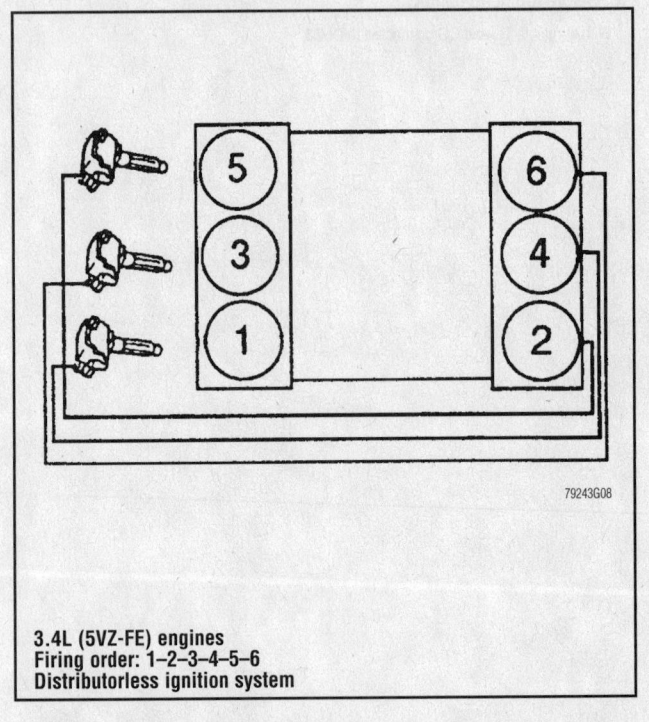

3.4L (5VZ-FE) engines
Firing order: 1–2–3–4–5–6
Distributorless ignition system

79243G08

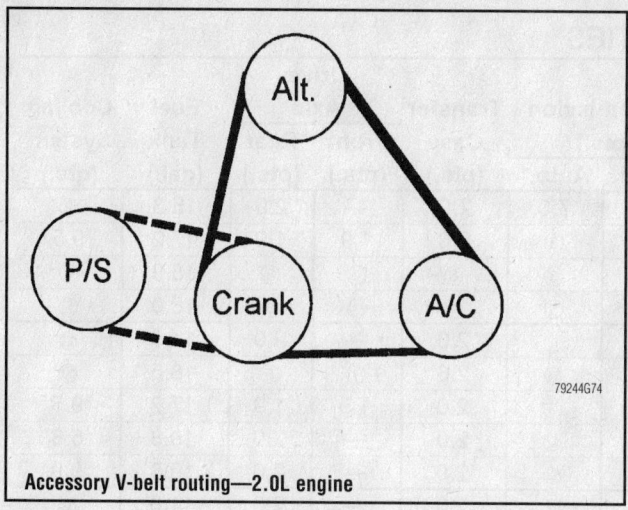

Accessory V-belt routing—2.0L engine

79244G74

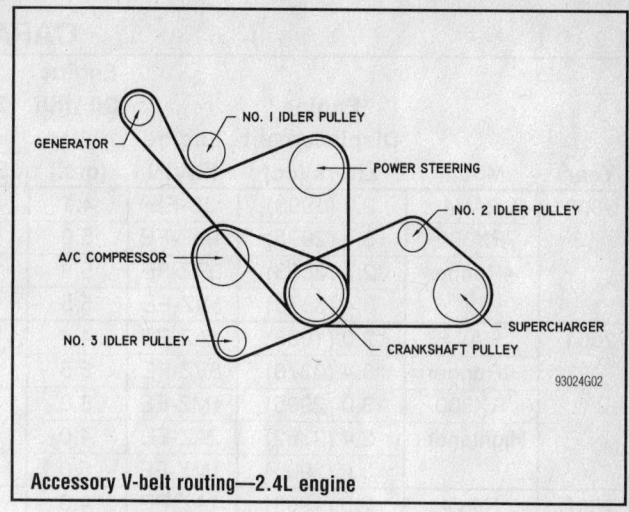

Accessory V-belt routing—2.4L engine

93024G02

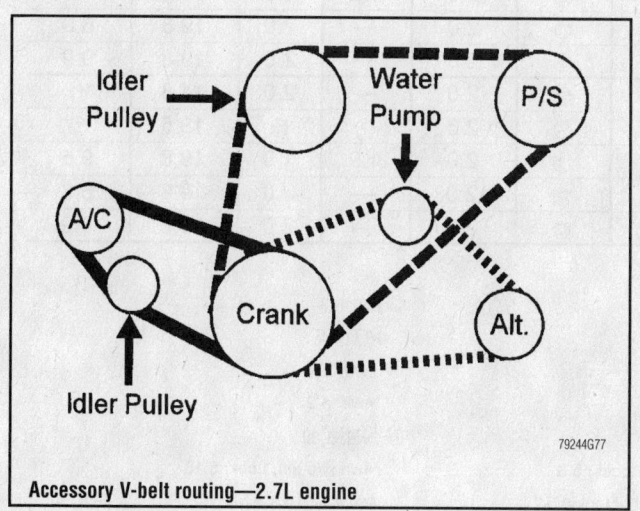

Accessory V-belt routing—2.7L engine

79244G77

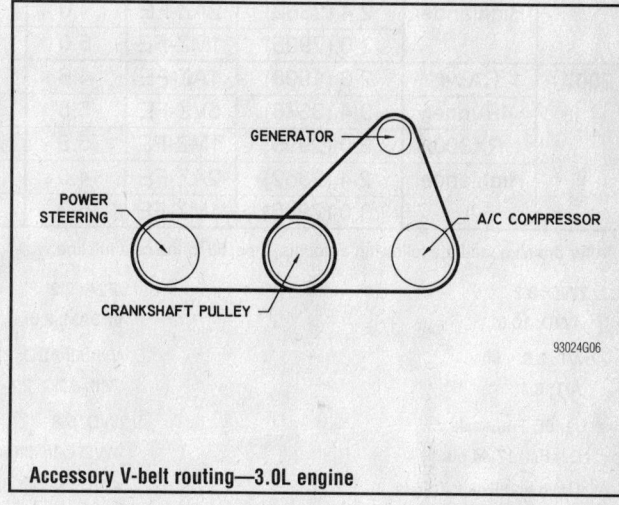

Accessory V-belt routing—3.0L engine

93024G06

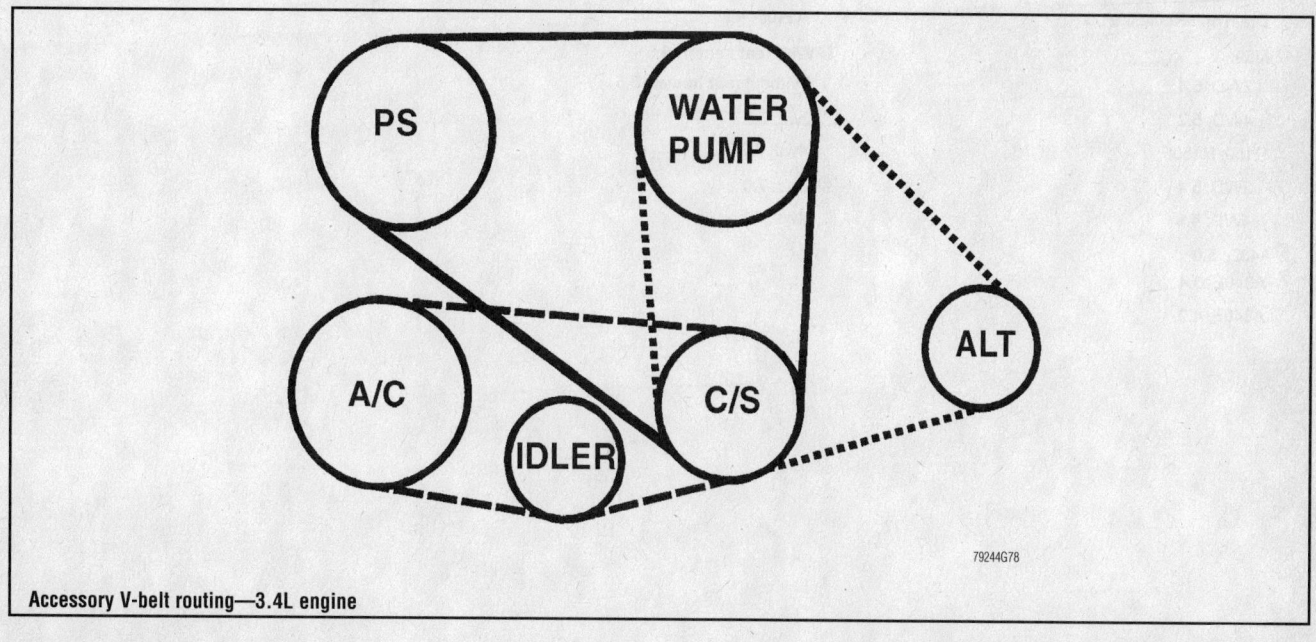

Accessory V-belt routing—3.4L engine

79244G78

CAPACITIES

Year	Model	Engine Displacement Liters (cc)	Engine ID/VIN	Engine Oil with Filter (qts.)	Transmission (pts.) 5-Spd	Transmission (pts.) Auto.*	Transfer Case (pts.)	Drive Axle Front (pts.)	Drive Axle Rear (pts.)	Fuel Tank (gal.)	Cooling System (qts.)
2000	RAV4	2.0 (1998)	3S-FE	4.1	①	7.0	2.0	—	2.0	15.3	②
	RX300	3.0 (2995)	1MZ-FE	5.0	—	③	2.0	1.9	1.9	17.2	9.5
	4Runner	2.7 (2693)	3RZ-FE	5.8	④	⑤	⑥	⑦	⑧	18.0	⑨
		3.4 (3378)	5VZ-FE	5.5	④	⑤	⑥	⑦	⑧	18.0	⑩
2001	RAV4	2.0 (1998)	1AZ-FE	4.4	⑪	⑫	2.0	—	2.0	14.8	⑬
	4Runner	3.4 (3378)	5VZ-FE	5.5	—	⑭	2.6	⑦	⑧	18.5	⑩
	RX300	3.0 (2995)	1MZ-FE	5.0	—	③	2.0	1.9	1.9	17.2	9.5
	Highlander	2.4 (2362)	2AZ-FE	4.0	—	⑫	2.0	—	2.0	19.8	6.8
		3.0 (2995)	1MZ-FE	5.0	—	⑫	2.0	—	2.0	19.8	9.9
2002	RAV4	2.0 (1998)	1AZ-FE	4.6	⑪	⑫	2.0	—	2.0	14.8	⑬
	4Runner	3.4 (3378)	5VZ-FE	5.5	—	⑭	2.6	⑦	⑮	18.5	⑩
	RX300	3.0 (2995)	1MZ-FE	5.0	—	③	2.0	1.9	1.9	17.2	9.5
	Highlander	2.4 (2362)	2AZ-FE	4.0	—	⑫	2.0	—	2.0	19.8	6.8
		3.0 (2995)	1MZ-FE	5.0	—	⑫	2.0	—	2.0	19.8	9.9
2003	RAV4	2.0 (1998)	1AZ-FE	4.6	⑪	⑫	2.0	—	2.0	14.8	⑬
	4Runner	3.4 (3378)	5VZ-FE	5.5	—	⑭	2.6	⑦	⑮	18.5	⑩
	RX300	3.0 (2995)	1MZ-FE	5.0	—	③	2.0	1.9	1.9	19.8	9.5
	Highlander	2.4 (2362)	2AZ-FE	4.0	—	⑫	2.0	—	2.0	19.8	6.8
		3.0 (2995)	1MZ-FE	5.0	—	⑫	2.0	—	2.0	19.8	9.9

*After draining, add the following amounts, then, fill to the cold full line.

① 2WD: 8.2
 4WD: 10.6

② M/T: 8.5
 A/T: 8.1

③ U140E Transaxle:
 Dry Fill: 17.44 pts.
 Drain and Refill: 7.4 pts.
 U140F Transaxle:
 Dry Fill: 19.34 pts.
 Drain and Refill: 8.6 pts.

④ W59:
 2WD: 5.4
 4WD: 5.2
 R150, R150F:
 2WD: 5.4
 4WD: 4.6

⑤ A43D: 5.0
 A340E: 3.4
 A340F: 4.2

⑥ VF2A: 2.2
 VF3AM: 2.6

⑦ Without ADD: 2.32
 With ADD: 2.44

⑧ 2WD: 5.8
 4WD with differential locks: 5.8
 4WD without differential locks: 5.2

⑨ With rear heater: 11.6
 Without rear heater: 10.6
 A340F: 4.2

⑩ With rear heater: 9.5
 Without rear heater: 8.5

⑪ 2wd: 5.2
 4wd: 7.2

⑫ 2wd: 7.0
 4wd: 8.2

⑬ MT: 6.7
 AT: 6.6

⑭ 2wd: 3.4
 4wd: 4.2

⑮ 2wd: 5.82
 4wd w/o diff. Lock: 5.18
 4wd w/diff. Lock: 5.82

42356-RAV4-C04

CRANKSHAFT AND CONNECTING ROD SPECIFICATIONS

All measurements are given in inches.

Year	Engine Displacement Liters (cc)	Engine ID/VIN	Crankshaft				Connecting Rod		
			Main Brg. Journal Dia.	Main Brg. Oil Clearance	Shaft End-play	Thrust on No.	Journal Diameter	Oil Clearance	Side Clearance
2000	2.0 (1998)	3S-FE	2.1653-2.1655	0.0010-0.0017	0.0008-0.0087	3	2.0466-2.0472	0.0009-0.0022	0.0063-0.0123
	3.0 (2995)	1MZ-FE	2.4011-2.4016	①	0.0016-0.0095	2	2.0863-2.0866	0.0015-0.0025	0.0059-0.0188
	2.7 (2693)	3RZ-FE	2.2615-2.3620	0.0012-0.0022	0.0008-0.0087	3	2.0861-2.0866	0.0009-0.0022	0.0063-0.0123
	3.4 (3378)	5VZ-FE	2.5191-2.5197	0.0008-0.0015	0.0008-0.0087	2	2.1648-2.1654	0.0009-0.0021	0.0059-0.0130
2001	2.0 (1998)	1AZ-FE	2.1653-2.1655	0.0010-0.0017	0.0008-0.0087	3	2.0466-2.0472	0.0009-0.0022	0.0063-0.0123
	2.4 (2362)	2AZ-FE	2.0654-2.1648	0.0009-0.0019	0.0016-0.0094	2	1.8894-1.8898	0.0009-0.0019	0.0063-0.0143
	3.0 (2995)	1MZ-FE	2.4011-2.4016	①	0.0016-0.0095	2	2.0863-2.0866	0.0015-0.0025	0.0059-0.0188
	3.4 (3378)	5VZ-FE	2.5191-2.5197	0.0008-0.0015	0.0008-0.0087	2	2.1648-2.1654	0.0009-0.0021	0.0059-0.0130
2002	2.0 (1998)	1AZ-FE	2.1649-2.1655	0.0010-0.0016	0.0008-0.0087	3	1.8894-1.8898	0.0009-0.0019	0.0063-0.0143
	2.4 (2362)	2AZ-FE	2.0654-2.1648	0.0009-0.0019	0.0016-0.0094	2	1.8894-1.8898	0.0009-0.0019	0.0063-0.0143
	3.0 (2995)	1MZ-FE	2.4011-2.4016	①	0.0016-0.0095	2	2.0863-2.0866	0.0015-0.0025	0.0059-0.0188
	3.4 (3378)	5VZ-FE	2.5191-2.5197	②	0.0008-0.0087	2	2.1648-2.1654	0.0009-0.0021	0.0059-0.0130
2003	2.0 (1998)	1AZ-FE	2.1649-2.1655	0.0010-0.0016	0.0008-0.0087	3	1.8894-1.8898	0.0009-0.0019	0.0063-0.0143
	2.4 (2362)	2AZ-FE	2.0654-2.1648	0.0009-0.0019	0.0016-0.0094	2	1.8894-1.8898	0.0009-0.0019	0.0063-0.0143
	3.0 (2995)	1MZ-FE	2.4011-2.4016	①	0.0016-0.0095	2	2.0863-2.0866	0.0015-0.0025	0.0059-0.0188
	3.4 (3378)	5VZ-FE	2.5191-2.5197	②	0.0008-0.0087	2	2.1648-2.1654	0.0009-0.0021	0.0059-0.0130

① Journals 1 and 4: 0.0006 - 0.0013 in.
   Journals 2 and 3: 0.0010 - 0.0018 in.

② No. 1: 0.0008-0.0015 in.
   All others: 0.0009-0.0017 in.

42356-RAV4-C05

VALVE SPECIFICATIONS

Year	Engine Displacement Liters (cc)	Engine ID/VIN	Seat Angle (deg.)	Face Angle (deg.)	Spring Test Pressure (lbs. @ in.)	Spring Installed Height (in.)	Stem-to-Guide Clearance (in.)		Stem Diameter (in.)	
							Intake	Exhaust	Intake	Exhaust
2000	2.0 (1998)	3S-FE	45	44.5	36.8-42.5@ 1.366	1.366	0.0010-0.0024	0.0012-0.0026	0.2350-0.2356	0.2348-0.2354
	3.0 (2995)	1MZ-FE	45	40.5	41.9-46.3@ 1.437	1.331	0.0010-0.0024	0.0012-0.0026	0.2154-0.2159	0.2152-0.2156
	2.7 (2693)	3RZ-FE	45	44.5	40.0-46.0@ 1.406	1.406	0.0010-0.0024	0.0012-0.0026	0.2350-0.2356	0.2348-0.2354
	3.4 (3378)	5VZ-FE	45	44.5	41.9-46.3@ 1.311	1.311	0.0010-0.0024	0.0012-0.0026	0.2350-0.2356	0.2348-0.2354
2001	2.0 (1998)	1AZ-FE	45	44.5	36.8-42.5@ 1.366	1.366	0.0010-0.0024	0.0012-0.0026	0.2350-0.2356	0.2348-0.2354
	2.4 (2362)	2AZ-FE	45	44.5	NA	NA	0.0010-0.0024	0.0012-0.0026	0.2154-0.2159	0.2152-0.2157
	3.0 (2995)	1MZ-FE	45	40.5	41.9-46.3@ 1.437	1.331	0.0010-0.0024	0.0012-0.0026	0.2154-0.2159	0.2152-0.2156
	3.4 (3378)	5VZ-FE	45	44.5	41.9-46.3@ 1.311	1.311	0.0010-0.0024	0.0012-0.0026	0.2350-0.2356	0.2348-0.2354
2002	2.0 (1998)	1AZ-FE	45	44.5	41.4-45.9@ 1.339	1.339	0.0010-0.0024	0.0012-0.0026	0.2154-0.2159	0.2152-0.2157
	2.4 (2362)	2AZ-FE	45	44.5	NA	NA	0.0010-0.0024	0.0012-0.0026	0.2154-0.2159	0.2152-0.2157
	3.0 (2995)	1MZ-FE	45	40.5	41.9-46.3@ 1.437	1.331	0.0010-0.0024	0.0012-0.0026	0.2154-0.2159	0.2152-0.2156
	3.4 (3378)	5VZ-FE	45	44.5	41.9-46.3@ 1.311	1.311	0.0010-0.0024	0.0012-0.0026	0.2350-0.2356	0.2348-0.2354
2003	2.0 (1998)	1AZ-FE	45	44.5	41.4-45.9@ 1.339	1.339	0.0010-0.0024	0.0012-0.0026	0.2154-0.2159	0.2152-0.2157
	2.4 (2362)	2AZ-FE	45	44.5	NA	NA	0.0010-0.0024	0.0012-0.0026	0.2154-0.2159	0.2152-0.2157
	3.0 (2995)	1MZ-FE	45	40.5	41.9-46.3@ 1.437	1.331	0.0010-0.0024	0.0012-0.0026	0.2154-0.2159	0.2152-0.2156
	3.4 (3378)	5VZ-FE	45	44.5	41.9-46.3@ 1.311	1.311	0.0010-0.0024	0.0012-0.0026	0.2350-0.2356	0.2348-0.2354

42356-RAV4-C06

PISTON AND RING SPECIFICATIONS
All measurements are given in inches.

Year	Engine Displ. Liters (cc)	Engine ID/VIN	Piston Clearance	Ring Gap			Ring Side Clearance		
				Top Comp.	Bottom Comp.	Oil Control	Top Comp.	Bottom Comp.	Oil Control
2000	2.0 (1998)	3S-FE	0.0056-0.0064	0.0106-0.0185	0.0177-0.0256	0.0039-0.0177	0.0012-0.0028	0.0012-0.0028	SNUG
	3.0 (2995)	1MZ-FE	0.0033-0.0042	0.0098-0.0138	0.0138-0.0177	0.0059-0.0157	0.0008-0.0028	0.0008-0.0024	SNUG
	2.7 (2693)	3RZ-FE	0.0019-0.0028	0.0118-0.0157	0.0157-0.0194	0.0051-0.0150	0.0008-0.0028	0.0012-0.0028	SNUG
	3.4 (3378)	5VZ-FE	0.0053-0.0060	0.0118-0.0197	0.0157-0.0236	0.0059-0.0217	0.0016-0.0031	0.0012-0.0028	SNUG
2001	2.0 (1998)	1AZ-FE	0.0056-0.0064	0.0106-0.0185	0.0177-0.0256	0.0039-0.0177	0.0012-0.0028	0.0012-0.0028	SNUG
	2.4 (2362)	2AZ-FE	0.0020-0.0029	0.0087-0.0126	0.0197-0.0236	0.0039-0.0138	0.0012-0.0028	0.0012-0.0028	SNUG
	3.0 (2995)	1MZ-FE	0.0033-0.0042	0.0098-0.0138	0.0138-0.0177	0.0059-0.0157	0.0008-0.0028	0.0008-0.0024	SNUG
	3.4 (3378)	5VZ-FE	0.0053-0.0060	0.0118-0.0197	0.0157-0.0236	0.0059-0.0217	0.0016-0.0031	0.0012-0.0028	SNUG
2002	2.0 (1998)	1AZ-FE	0.0025-0.0034	0.0118-0.0157	0.0185-0.0244	0.0039-0.0138	0.0008-0.0028	0.0008-0.0024	0.0028-0.0059
	2.4 (2362)	2AZ-FE	0.0020-0.0029	0.0087-0.0126	0.0197-0.0236	0.0039-0.0138	0.0012-0.0028	0.0012-0.0028	SNUG
	3.0 (2995)	1MZ-FE	0.0033-0.0042	0.0098-0.0138	0.0138-0.0177	0.0059-0.0157	0.0008-0.0028	0.0008-0.0024	SNUG
	3.4 (3378)	5VZ-FE	0.0053-0.0060	0.0118-0.0197	0.0157-0.0236	0.0059-0.0217	0.0016-0.0031	0.0012-0.0028	SNUG
2003	2.0 (1998)	1AZ-FE	0.0025-0.0034	0.0118-0.0157	0.0185-0.0244	0.0039-0.0138	0.0008-0.0028	0.0008-0.0024	0.0028-0.0059
	2.4 (2362)	2AZ-FE	0.0020-0.0029	0.0087-0.0126	0.0197-0.0236	0.0039-0.0138	0.0012-0.0028	0.0012-0.0028	SNUG
	3.0 (2995)	1MZ-FE	0.0033-0.0042	0.0098-0.0138	0.0138-0.0177	0.0059-0.0157	0.0008-0.0028	0.0008-0.0024	SNUG
	3.4 (3378)	5VZ-FE	0.0053-0.0060	0.0118-0.0197	0.0157-0.0236	0.0059-0.0217	0.0016-0.0031	0.0012-0.0028	SNUG

42356-RAV4-C07

TORQUE SPECIFICATIONS
All readings in ft. lbs.

Year	Engine Displacement Liters (cc)	Engine ID/VIN	Cylinder Head Bolts	Main Bearing Bolts	Rod Bearing Bolts	Crankshaft Damper Bolts	Flywheel Bolts	Manifold Intake	Manifold Exhaust	Spark Plugs	Lug Nuts
2000	2.0 (1998)	3S-FE	①	43	②	80	③	14	36	13	76
	3.0 (2995)	1MZ-FE	④	⑤	②	159	61	32	36	13	70
	2.7 (2693)	3RZ-FE	⑥	⑦	⑧	116	⑨	22	36	14	83
	3.4 (3378)	5VZ-FE	⑩	⑪	②	184	63-67	13	30	13	76
2001	2.0 (1998)	1AZ-FE	①	43	②	80	③	14	36	13	76
	2.4 (2362)	2AZ-FE	⑫	29	②	125	72	22	27	14	76
	3.0 (2995)	1MZ-FE	④	⑤	②	159	61	32	36	13	70
	3.4 (3378)	5VZ-FE	⑩	⑪	②	184	63-67	13	30	13	76
2002	2.0 (1998)	1AZ-FE	⑫	⑬	②	125	⑭	22	25	15	76
	2.4 (2362)	2AZ-FE	⑫	29	②	125	72	22	27	14	76
	3.0 (2995)	1MZ-FE	④	⑤	②	159	61	32	36	13	70
	3.4 (3378)	5VZ-FE	⑮	⑪	②	213	61	13	30	13	76
2003	2.0 (1998)	1AZ-FE	⑫	⑬	②	125	⑭	22	25	15	76
	2.4 (2362)	2AZ-FE	⑫	29	②	125	72	22	27	14	76
	3.0 (2995)	1MZ-FE	④	⑤	②	159	61	32	36	13	70
	3.4 (3378)	5VZ-FE	⑮	⑪	②	213	61	13	30	13	76

① Step 1: 35 ft. lbs.
   Step 2: Plus 90 degrees

② Step 1: 18 ft. lbs.
   Step 2: Plus 90 degrees

③ Manual transmission: 65 ft. lbs.
   Automatic transmission: 61 ft. lbs.

④ Step 1: 12 point bolts to 40 ft. lbs.
   Step 2: 12 point bolts plus 90 degrees
   Step 3: Hex head recessed bolt to 13 ft. lbs.

⑤ Step 1: 12 point cap bolts to 16 ft. lbs.
   Step 2: 12 point cap bolts plus 90 degrees
   Step 3: Hex head side bolts to 20 ft. lbs.

⑥ Step 1: 29 ft. lbs.
   Step 2: Plus 90 degrees
   Step 3: Plus 90 degrees

⑦ Step 1: 29 ft. lbs.
   Step 2: Plus 90 degrees

⑧ Step 1: 33 ft. lbs.
   Step 2: Plus 90 degrees

⑨ Manual transmission: 19 ft. lbs. + 90°
   Automatic transmission: 54 ft. lbs.

⑩ Step 1: 25 ft. lbs.
   Step 2: Plus 90 degrees
   Recessed head: 13 ft. lbs.

⑪ Step 1: 45 ft. lbs.
   Step 2: Plus 90 degrees

⑫ Step 1: 58
   Step 2: plus 90 degrees

⑬ Step 1: 15 ft. lbs.
   Step 2: 30 ft. lbs.
   Step 3: +90 degrees

⑭ MT: 96 ft. lbs.
   AT: 72 ft. lbs.

⑮ 12-pointed bolts
   Step 1: 25 ft. lbs.
   Step 2: +90 degrees
   Step 3: +90 degrees
   Recessed heads: 13 ft. lbs.

42356-RAV4-C08

WHEEL ALIGNMENT

Year	Model		Caster Range (+/-Deg.)	Caster Preferred Setting (Deg.)	Camber Range (+/-Deg.)	Camber Preferred Setting (Deg.)	Toe-in (in.)	Steering Axis Inclination (Deg.)
2000	RAV4	2WD	0.75	+1.42	0.75	-0.33	0+/-0.08	11+/-0.75
		4WD	0.75	+1.33	0.75	-0.25	0+/-0.08	11+/-0.75
	RX300	F	0.75	+2.08	0.75	-0.33	0.04+/-0.08	12.16+/-0.75
		2WD R	—	—	0.75	-0.33	0.08+/-0.08	—
		4WD R	—	—	0.75	-0.35	0.12+/-0.08	—
	4Runner	2WD	0.75	+3.25	0.75	-0.25	0.08+/-0.08	11+/-0.75
		4WD	0.75	+3.06	0.75	-0.25	0.08+/-0.08	11+/-0.75
2001	RAV4	2WD	0.75	+1.42	0.75	-0.33	0+/-0.08	11+/-0.75
		4WD	0.75	+1.33	0.75	-0.25	0+/-0.08	11+/-0.75
	Highlander	2WD F	0.75	+2.75	0.75	-0.67	0+/-0.08	10.75+/-0.75
		4WD F	0.75	+2.75	0.75	-0.58	0+/-0.08	10.58+/-0.75
		2WD R	—	—	0.75	-1.33	0.12+/-0.08	—
		4WD R	—	—	0.75	-0.75	0.12+/-0.08	—
	RX300	F	0.75	+2.08	0.75	-0.33	0.04+/-0.08	12.16+/-0.75
		2WD R	—	—	0.75	-0.33	0.08+/-0.08	—
		4WD R	—	—	0.75	-0.35	0.12+/-0.08	—
	4Runner	2WD	0.75	+3.25	0.75	-0.25	0.08+/-0.08	11+/-0.75
		4WD	0.75	+3.06	0.75	-0.25	0.08+/-0.08	11+/-0.75
2002	RAV4	2WD	0.75	+2.00	0.75	-0.42	0+/-0.08	11+/-0.75
		4WD	0.75	+1.92	0.75	-0.33	0.04+/-0.08	10.75+/-0.75
	Highlander	2WD F	0.75	+2.75	0.75	-0.67	0+/-0.08	10.75+/-0.75
		4WD F	0.75	+2.75	0.75	-0.58	0+/-0.08	10.58+/-0.75
		2WD R	—	—	0.75	-1.33	0.12+/-0.08	—
		4WD R	—	—	0.75	-0.75	0.12+/-0.08	—
	RX300	F	0.75	+2.08	0.75	-0.33	0.04+/-0.08	12.16+/-0.75
		2WD R	—	—	0.75	-0.33	0.08+/-0.08	—
		4WD R	—	—	0.75	-0.35	0.12+/-0.08	—
	4Runner	2WD	0.75	+3.25	0.75	-0.25	0.08+/-0.08	11+/-0.75
		4WD	0.75	+3.06	0.75	-0.25	0.08+/-0.08	11+/-0.75
2003	RAV4	2WD	0.75	+2.00	0.75	-0.42	0+/-0.08	11+/-0.75
		4WD	0.75	+1.92	0.75	-0.33	0.04+/-0.08	10.75+/-0.75
	Highlander	2WD F	0.75	+2.75	0.75	-0.67	0+/-0.08	10.75+/-0.75
		4WD F	0.75	+2.75	0.75	-0.58	0+/-0.08	10.58+/-0.75
		2WD R	—	—	0.75	-1.33	0.12+/-0.08	—
		4WD R	—	—	0.75	-0.75	0.12+/-0.08	—
	RX300	F	0.75	+2.08	0.75	-0.33	0.04+/-0.08	12.16+/-0.75
		2WD R	—	—	0.75	-0.33	0.08+/-0.08	—
		4WD R	—	—	0.75	-0.35	0.12+/-0.08	—
	4Runner	2WD	0.75	+3.25	0.75	-0.25	0.08+/-0.08	11+/-0.75
		4WD	0.75	+3.06	0.75	-0.25	0.08+/-0.08	11+/-0.75

TIRE, WHEEL AND BALL JOINT SPECIFICATIONS

Year	Model	OEM Tires		Tire Pressures (psi)		Wheel Size	Ball Joint Inspection
		Standard	Optional	Front	Rear		
2000	RAV4	P215/70R16	P235/60HR15	28	28	6.5-JJ	4-30 ①
	RX300	P255/70HR16	None	30	30	6.5-JJ	③
	4Runner	P225/75R15	P265/70R16	Std: 29 Opt: 32	Std: 29 Opt: 32	7-JJ	6-39 ①
2001	RAV4	P215/70R16	P235/60HR16	29	29	②	4-30 ①
	RX300	P255/70HR16	None	30	30	6.5-JJ	③
	4Runner	P225/75R15	P265/70R16	Std: 29 Opt: 32	Std: 29 Opt: 32	7-JJ	6-39 ①
2002	RAV4	P215/70R16	P235/60HR16	29	29	③	9-43 ②
	Highlander	P225/70R16	None	30	30	6.5-JJ	③
	4Runner SR5	P225/75R15	P265/70R16	Std: 29 Opt: 32	Std: 29 Opt: 32	7-JJ	④
	4Runner Limited	P265/70R16	None	32	32	7-JJ	④
2003	RAV4	P215/70R16	P235/60HR16	29	29	②	9-43 ①
	Highlander	P225/70R16	None	30	30	6.5-JJ	③
	RX300	P255/70HR16	None	30	30	6.5-JJ	③
	4Runner SR5	P225/75R15	P265/70R16	Std: 29 Opt: 32	Std: 29 Opt: 32	7-JJ	④
	4Runner Limited	P265/70R16	None	32 Opt: 32	32 Opt: 32	7-JJ	④

OEM: Original Equipment Manufacturer

PSI: Pounds Per Square Inch

STD: Standard

OPT: Optional

① Torque required in inch lbs. to rotate ball joint when removed from the knuckle

② Steel wheel: 6.5J; aluminum wheel: 7JJ

③ Replace if any measurable movement is found.

④ Turning torque: upper 6-39 in. lbs.; lower 0.8-21.7 in. lbs.

42356-RAV4-C10

BRAKE SPECIFICATIONS
All measurements in inches unless noted

Year	Model		Brake Disc Original Thickness	Brake Disc Minimum Thickness	Brake Disc Maximum Runout	Brake Drum Diameter Original Inside Diameter	Brake Drum Diameter Max. Wear Limit	Brake Drum Diameter Maximum Machine Diameter	Minimum Lining Thickness	Brake Caliper Bracket Bolts (ft. lbs.)	Brake Caliper Mounting Bolts (ft. lbs.)
2000	RAV4		0.709	0.630	0.0020	9.00	—	9.08	0.039	78	20
	RX300	F	1.020	1.024	0.0020	—	—	—	0.039	79	25
		R	0.394	0.354	0.0059	—	—	—	0.039	34	14
	Highlander	F	1.102	1.024	0.0020	—	—	—	0.039	79	25
		R	0.394	0.354	0.0059	—	—	—	0.039	43	25
	4Runner		0.866	0.787	0.0028	11.61	—	11.69	0.039	—	90
2001	RAV4		0.709	0.630	0.0020	9.00	—	9.08	0.039	78	20
	RX300	F	1.020	1.024	0.0020	—	—	—	0.039	79	25
		R	0.394	0.354	0.0059	—	—	—	0.039	34	14
	Highlander	F	1.102	1.024	0.0020	—	—	—	0.039	79	25
		R	0.394	0.354	0.0059	—	—	—	0.039	43	25
	4Runner		0.866	0.787	0.0028	11.61	—	11.69	0.039	—	90
2002	RAV4		0.984	0.906	0.0020	9.00	—	9.08	0.039	78	20
	RX300	F	1.020	1.024	0.0020	—	—	—	0.039	79	25
		R	0.394	0.354	0.0059	—	—	—	0.039	34	14
	Highlander	F	1.102	1.024	0.0020	—	—	—	0.039	79	25
		R	0.394	0.354	0.0059	—	—	—	0.039	43	25
	4Runner		0.866	0.787	0.0028	11.61	—	11.69	0.039	—	90
2003	RAV4		0.984	0.906	0.0020	9.00	—	9.08	0.039	78	20
	RX300	F	1.020	1.024	0.0020	—	—	—	0.039	79	25
		R	0.394	0.354	0.0059	—	—	—	0.039	34	14
	Highlander	F	1.102	1.024	0.0020	—	—	—	0.039	79	25
		R	0.394	0.354	0.0059	—	—	—	0.039	43	25
	4Runner		0.866	0.787	0.0028	11.61	—	11.69	0.039	—	90

F: Front

R: Rear

42356-RAV4-C11

SCHEDULED MAINTENANCE INTERVALS

LEXUS—RX300

TO BE SERVICED	TYPE OF SERVICE	VEHICLE MILEAGE INTERVAL (x1000)												
		7.5	15	22.5	30	37.5	45	52.5	60	67.5	75	82.5	90	97.5
Engine oil & filter	R	✓	✓	✓	✓	✓	✓	✓	✓	✓	✓	✓	✓	✓
Automatic transmission fluid	S/I		✓		✓		✓		✓		✓		✓	
Ball joints & dust covers	S/I		✓		✓		✓		✓		✓		✓	
Bolts & nuts on chassis & body	S/I		✓		✓		✓		✓		✓		✓	
Brake linings & drums	S/I		✓		✓		✓		✓		✓		✓	
Brake line pipes & hoses	S/I		✓		✓		✓		✓		✓		✓	
Brake pads & discs (front & rear)	S/I		✓		✓		✓		✓		✓		✓	
Propeller shaft grease	S/I		✓		✓		✓		✓		✓		✓	
Steering knuckle & chassis grease	S/I		✓		✓		✓		✓		✓		✓	
Steering linkage	S/I		✓		✓		✓		✓		✓		✓	
Air cleaner filter	R				✓				✓				✓	
Spark plugs	R				✓				✓				✓	
Drive belts	S/I				✓				✓				✓	
Exhaust pipes & mountings	S/I				✓				✓				✓	
Fuel lines & connections	S/I				✓				✓				✓	
Engine coolant	R						✓				✓			
Charcoal canister	R								✓					
Fuel tank cap gasket	R								✓					
Oxygen sensors (exc. Calif.)①	R													

R: Replace   S/I: Service or Inspect

① Heated oxygen sensors (except Calif.): replace every 80,000 miles.

FREQUENT OPERATION MAINTENANCE (SEVERE SERVICE)

If a vehicle is operated under any of the following conditions it is considered severe service:

- Extremely dusty areas.

- 50% or more of the vehicle operation is in 32°C (90°F) or higher temperatures, or constant operation in temperatures below 0°C (32°F).

- Prolonged idling (vehicle operation in stop and go traffic).

- Frequent short running periods (engine does not warm to normal operating temperatures).

- Police, taxi, delivery usage or trailer towing usage.

Air cleaner filter: service or inspect every 3750 miles

Engine oil & filter: replace every 3750 miles.

Ball joints & dust covers: service or inspect every 7500 miles.

Bolts & nuts on chassis & body: service or inspect every 7500 miles.

Brake pads & discs (front & rear): service or inspect every 7500 miles.

Steering knuckle & chassis grease: service or inspect every 7500 miles.

Steering linkage: service or inspect every 7500 miles.

Exhaust pipes & mountings: service or inspect every 15,000 miles.

42356-RAV4-C12

SCHEDULED MAINTENANCE INTERVALS
TOYOTA—4RUNNER

TO BE SERVICED	TYPE OF SERVICE	VEHICLE MILEAGE INTERVAL (x1000)																		
		5	10	15	20	25	30	35	40	45	50	55	60	65	70	75	80	85	90	95
Automatic transmission and differential fluid	S/I			✓			✓			✓			✓			✓			✓	
Ball joints and boots	S/I			✓			✓			✓			✓			✓			✓	
Brake system	S/I			✓			✓			✓			✓			✓			✓	
Charcoal canister	S/I												✓							
Drive belts	S/I						✓						✓						✓	
Driveshaft bushing	L						✓						✓						✓	
Engine coolant	R						✓						✓						✓	
Engine oil & filter	R	✓	✓	✓	✓	✓	✓	✓	✓	✓	✓	✓	✓	✓	✓	✓	✓	✓	✓	✓
Exhaust system	S/I			✓			✓			✓			✓			✓			✓	
Fuel tank cap gasket	S/I	·					✓						✓						✓	
Halfshaft boots & flange bolts	S/I			✓			✓			✓			✓			✓			✓	
Limited slip differential fluid	R						✓												✓	
Manual transmission and differential fluid	S/I						✓						✓						✓	
Platinum spark plugs	R												✓							
Propeller shaft (4WD)	L			✓			✓			✓			✓			✓			✓	
Propeller shaft bolts	S/I			✓			✓			✓			✓			✓			✓	
Rack and pinion assembly	S/I			✓			✓			✓			✓			✓			✓	
Rear wheel bearing	L						✓						✓						✓	
Steering linkage	S/I			✓			✓			✓			✓			✓			✓	
Valves	S/I												✓							

R: Replace   S/I: Service or Inspect   L: Lubricate

FREQUENT OPERATION MAINTENANCE (SEVERE SERVICE)

If a vehicle is operated under any of the following conditions it is considered severe service:

- Towing a trailer or using a camper or car-top carrier.

- Repeated short trips of less than 5 miles in temperatures below freezing.

- Excessive idling or low-speed driving for long distances as in heavy commercial use, such as delivery, taxi or police cars.

- Operating on rough, muddy or salt-covered roads.

- Operating on unpaved or dusty roads.

Oil filter: service or inspect every 5000 miles or 4 months, whichever occurs first.

Brake linings and discs or drums: service or inspect every 5000 miles or 4 months, whichever occurs first.

Steering linkage: service or inspect every 5000 miles or 4 months, whichever occurs first.

Ball joints and boots: service or inspect every 5000 miles or 4 months, whichever occurs first.

Brake discs & pads (front): service or inspect every 6000 miles.

Halfshaft boots: service or inspect every 5000 miles or 4 months. Retighten the flange bolts, whichever occurs first.

Body chassis bolts and nuts: service or inspect every 5000 miles or 4 months, whichever occurs first.

Transmission and differential fluid: replace every 15,000 miles or 12 months, whichever occurs first.

Transfer case and differential fluid: replace every 15,000 miles or 12 months, whichever occurs first.

Timing belt: replace every 60,000 miles or 48 months, whichever occurs first.

42356-RAV4-C13

SCHEDULED MAINTENANCE INTERVALS

TOYOTA—HIGHLANDER

TO BE SERVICED	TYPE OF SERVICE	VEHICLE MILEAGE INTERVAL (x1000)												
		7.5	15	22.5	30	37.5	45	52.5	60	67.5	75	82.5	90	97.5
Engine oil & filter	R	✓	✓	✓	✓	✓	✓	✓	✓	✓	✓	✓	✓	✓
Automatic transmission fluid	S/I		✓		✓		✓		✓		✓		✓	
Ball joints & dust covers	S/I		✓		✓		✓		✓		✓		✓	
Bolts & nuts on chassis & body	S/I		✓		✓		✓		✓		✓		✓	
Brake linings & drums	S/I		✓		✓		✓		✓		✓		✓	
Brake line pipes & hoses	S/I		✓		✓		✓		✓		✓		✓	
Brake pads & discs (front & rear)	S/I		✓		✓		✓		✓		✓		✓	
Propeller shaft grease	S/I		✓		✓		✓		✓		✓		✓	
Steering knuckle & chassis grease	S/I		✓		✓		✓		✓		✓		✓	
Steering linkage	S/I		✓		✓		✓		✓		✓		✓	
Air cleaner filter	R				✓				✓				✓	
Spark plugs ①	R				✓				✓				✓	
Drive belts	S/I				✓				✓				✓	
Exhaust pipes & mountings	S/I				✓				✓				✓	
Fuel lines & connections	S/I				✓				✓				✓	
Engine coolant	R						✓				✓			
Charcoal canister	R								✓					
Fuel tank cap gasket	R								✓					
Heated oxygen sensors (except Calif.) ②	R													

R: Replace   S/I: Service or Inspect

① Platinum plugs are replaced at 100,000 mile intervals.

② Heated oxygen sensors (except Calif.): replace every 80,000 miles.

FREQUENT OPERATION MAINTENANCE (SEVERE SERVICE)

If a vehicle is operated under any of the following conditions it is considered severe service:

- Extremely dusty areas.

- 50% or more of the vehicle operation is in 32°C (90°F) or higher temperatures, or constant operation in temperatures below 0°C (32°F).

- Prolonged idling (vehicle operation in stop and go traffic).

- Frequent short running periods (engine does not warm to normal operating temperatures).

- Police, taxi, delivery usage or trailer towing usage.

Air cleaner filter: service or inspect every 3750 miles

Engine oil & filter: replace every 3750 miles.

Ball joints & dust covers: service or inspect every 7500 miles.

Bolts & nuts on chassis & body: service or inspect every 7500 miles.

Brake pads & discs (front & rear): service or inspect every 7500 miles.

Steering knuckle & chassis grease: service or inspect every 7500 miles.

Steering linkage: service or inspect every 7500 miles.

Exhaust pipes & mountings: service or inspect every 15,000 miles.

42356-RAV4-C14

SCHEDULED MAINTENANCE INTERVALS
TOYOTA—RAV4

TO BE SERVICED	TYPE OF SERVICE	5	10	15	20	25	30	35	40	45	50	55	60	65	70	75	80	85	90	95
		VEHICLE MILEAGE INTERVAL (x1000)																		
Automatic transmission and differential fluid	S/I			✓			✓			✓			✓			✓			✓	
Ball joints and boots	S/I			✓			✓			✓			✓			✓			✓	
Brake system	S/I			✓			✓			✓			✓			✓			✓	
Charcoal canister	S/I												✓							
Drive belts	S/I						✓						✓						✓	
Driveshaft bushing	L						✓						✓						✓	
Engine coolant	R						✓						✓						✓	
Engine oil & filter	R	✓	✓	✓	✓	✓	✓	✓	✓	✓	✓	✓	✓	✓	✓	✓	✓	✓	✓	✓
Exhaust pipes & mounts	S/I			✓			✓			✓			✓			✓				
Fuel tank cap gasket	S/I						✓						✓						✓	
Halfshaft boots & flange bolts	S/I			✓			✓			✓			✓			✓			✓	
Limited slip differential fluid	R						✓						✓						✓	
Manual transmission and differential fluid	S/I						✓						✓						✓	
Platinum spark plugs	R												✓							
Propeller shaft bolts	S/I			✓			✓			✓			✓			✓			✓	
Steering linkage	S/I			✓			✓			✓			✓			✓			✓	
Transfer case and differential fluid	S/I			✓			✓			✓			✓			✓			✓	
Valves	S/I												✓							

R: Replace   S/I: Service or Inspect   L: Lubricate

FREQUENT OPERATION MAINTENANCE (SEVERE SERVICE)

If a vehicle is operated under any of the following conditions it is considered severe service:

- Towing a trailer or using a camper or car-top carrier.

- Repeated short trips of less than 5 miles in temperatures below freezing.

- Excessive idling or low-speed driving for long distances as in heavy commercial use, such as delivery, taxi or police cars.

- Operating on rough, muddy or salt-covered roads.

- Operating on unpaved or dusty roads.

Oil filter: service or inspect every 5000 miles or 4 months, whichever occurs first.

Brake linings and discs or drums: service or inspect every 5000 miles or 4 months, whichever occurs first.

Steering linkage: service or inspect every 5000 miles or 4 months, whichever occurs first

Ball joints and boots: service or inspect every 5000 miles or 4 months, whichever occurs first.

Brake discs & pads (front): service or inspect every 6000 miles.

Halfshaft boots: service or inspect every 5000 miles or 4 months. Retighten the flange bolts, whichever occurs first.

Body chassis bolts and nuts: service or inspect every 5000 miles or 4 months, whichever occurs first.

Transmission and differential fluid: replace every 15,000 miles or 12 months, whichever occurs first.

Transfer case and differential fluid: replace every 15,000 miles or 12 months, whichever occurs first.

Timing belt: replace every 60,000 miles or 48 months, whichever occurs first.

42356-RAV4-C15

PRECAUTIONS

Before servicing any vehicle, please be sure to read all of the following precautions, which deal with personal safety, prevention of component damage, and important points to take into consideration when servicing a motor vehicle:

• Never open, service or drain the radiator or cooling system when the engine is hot; serious burns can occur from the steam and hot coolant.

• Observe all applicable safety precautions when working around fuel. Whenever servicing the fuel system, always work in a well-ventilated area. Do not allow fuel spray or vapors to come in contact with a spark, open flame, or excessive heat (a hot drop light, for example). Keep a dry chemical fire extinguisher near the work area. Always keep fuel in a container specifically designed for fuel storage; also, always properly seal fuel containers to avoid the possibility of fire or explosion. Refer to the additional fuel system precautions later in this section.

• Fuel injection systems often remain pressurized, even after the engine has been turned **OFF**. The fuel system pressure must be relieved before disconnecting any fuel lines. Failure to do so may result in fire and/or personal injury.

• Brake fluid often contains polyglycol ethers and polyglycols. Avoid contact with the eyes and wash your hands thoroughly after handling brake fluid. If you do get brake fluid in your eyes, flush your eyes with clean, running water for 15 minutes. If eye irritation persists, or if you have taken brake fluid internally, IMMEDIATELY seek medical assistance.

• The EPA warns that prolonged contact with used engine oil may cause a number of skin disorders, including cancer. You should make every effort to minimize your exposure to used engine oil. Protective gloves should be worn when changing oil. Wash your hands and any other exposed skin areas as soon as possible after exposure to used engine oil. Soap and water, or waterless hand cleaner should be used.

• All new vehicles are now equipped with an air bag system. The system must be disabled before performing service on or around system components, steering column, instrument panel components, wiring and sensors. Failure to follow safety and disabling procedures could result in accidental air bag deployment, possible personal injury and unnecessary system repairs.

• Always wear safety goggles when working with, or around, the air bag system. When carrying a non-deployed air bag, be sure the bag and trim cover are pointed away from your body. When placing a non-deployed air bag on a work surface, always face the bag and trim cover upward, away from the surface. This will reduce the motion of the module if it is accidentally deployed. Refer to the additional air bag system precautions later in this section.

• NEVER disconnect the negative battery cable with the ignition **ON** or the engine running. Removing power from the computer control module with the ignition **ON** may destroy the module.

• Clean, high quality brake fluid from a sealed container is essential to the safe and proper operation of the brake system. You should always buy the correct type of brake fluid for your vehicle. If the brake fluid becomes contaminated, completely flush the system with new fluid. Never reuse any brake fluid. Any brake fluid that is removed from the system should be discarded. Also, do not allow any brake fluid to come in contact with a painted surface; it will damage the paint.

• Never operate the engine without the proper amount and type of engine oil; doing so WILL result in severe engine damage.

• Timing belt maintenance is extremely important. Many models utilize an interference-type, non-freewheeling engine. If the timing belt breaks, the valves in the cylinder head may strike the pistons, causing potentially serious (also time-consuming and expensive) engine damage. Refer to the maintenance interval charts in the front of this section for the recommended replacement interval for the timing belt, and to the timing belt section for belt replacement and inspection.

• Disconnecting the negative battery cable on some vehicles may interfere with the functions of the on-board computer system(s) and may require the computer to undergo a relearning process once the negative battery cable is reconnected.

• When servicing drum brakes, only disassemble and assemble one side at a time, leaving the remaining side intact for reference.

• Only an MVAC-trained, EPA-certified automotive technician should service the air conditioning system or its components.

ENGINE REPAIR

➡**Disconnecting the negative battery cable on some vehicles may interfere with the functions of the on board computer system. The computer may undergo a relearning process once the negative battery cable is reconnected.**

Distributor

All models are equipped with a distiburtorless ignition system.

Alternator

REMOVAL

On some models, the alternator is mounted very low on the engine. On these models, it may be necessary to remove the gravel shield and work from beneath the vehicle in order to gain access to the alternator. Replacing the alternator while the engine is cold is recommended. A hot engine can result in personal injury.

2.0L RAV4

1. Before servicing the vehicle, refer to the precautions in the beginning of this section.

2. Remove or disconnect the following:

• Electrical wiring from the alternator
• Loosen the adjusting lockbolt and the pivot bolt.
• Loosen the adjusting bolt to relieve tension on the drive belt, if equipped with air conditioning
• Drive belt

➡**It may be necessary to remove other belts for access.**

• Pivot bolt and the adjusting lockbolt
• Alternator

2.4L Highlander

1. Before servicing the vehicle, refer to the precautions in the beginning of this section.

2. Remove or disconnect the following:
• Electrical wiring from the alternator
• Drive belt
• 1 adjusting and 2 mounting bolts
• Alternator

3.0L Highlander

1. Before servicing the vehicle, refer to the precautions in the beginning of this section.
2. Remove or disconnect the following:
   - Alternator electrical connectors
   - Wiring harness from the clip
   - Pivot bolt
   - Adjuster lockbolt
   - Drive belt
   - Alternator

3.0L RX300

1. Before servicing the vehicle, refer to the precautions in the beginning of this section.
2. Remove or disconnect the following:
   - Alternator electrical connectors
   - Wiring harness from the clip
   - Pivot bolt
   - Plate washer
   - Adjusting lockbolt
   - Drive belt
   - Alternator

2.7L 4Runner

Remove or disconnect the following:
   - Negative battery cable
   - Alternator wiring
   - Alternator lockbolt, pivot bolt, nut and adjusting bolt
   - Drive belt
   - Wiring harness with the clip
   - Alternator

3.4L 4Runner

1. Remove or disconnect the following:
   - Negative battery cable
   - Alternator wiring
   - Alternator locknut, pivot bolt, nut and adjusting bolt
   - Drive belt
   - Alternator

INSTALLATION

2.0L RAV4

1. Install or connect the following:
   - Alternator
   - Adjusting lockbolt and the pivot bolt
   - Drive belt
2. Adjust the drive belt tension to:
   - New belt with A/C: 140–190 lbs.
   - Used belt with A/C: 100–120 lbs.
   - New belt without A/C: 100–150 lbs.

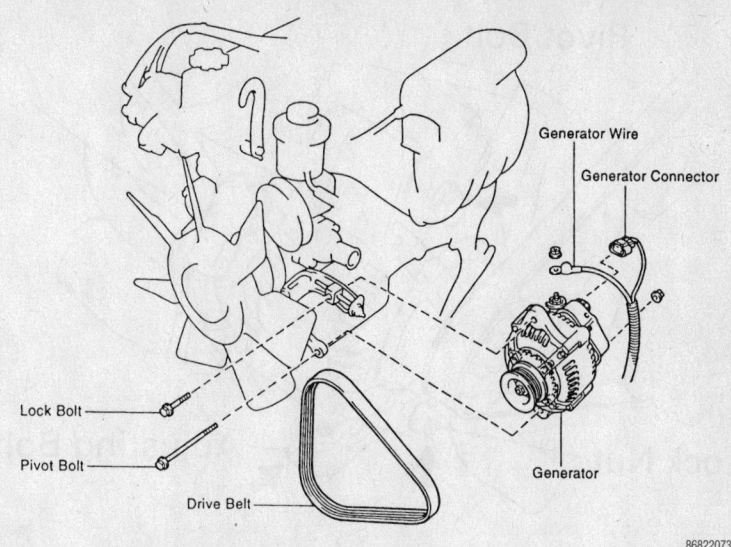

Exploded view of the alternator and drive belt—2.7L 4Runner

86822073

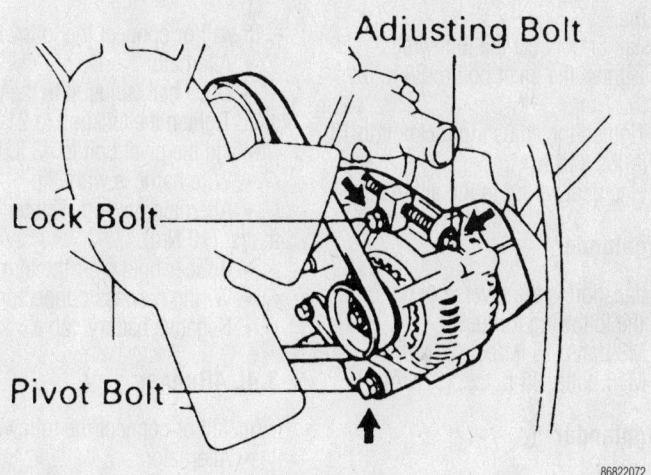

Locations of the adjusting, pivot and lockbolts—2.7L 4Runner

86822072

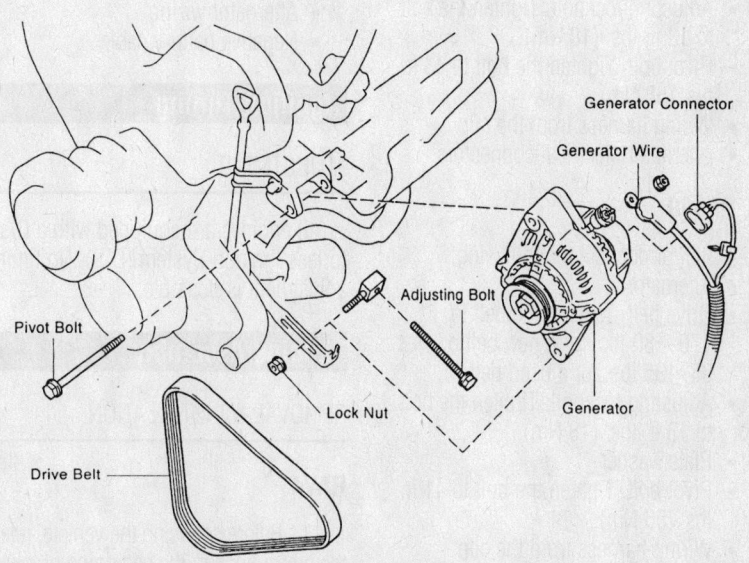

Exploded view of the alternator and drive belt—3.4L 4Runner

86822077

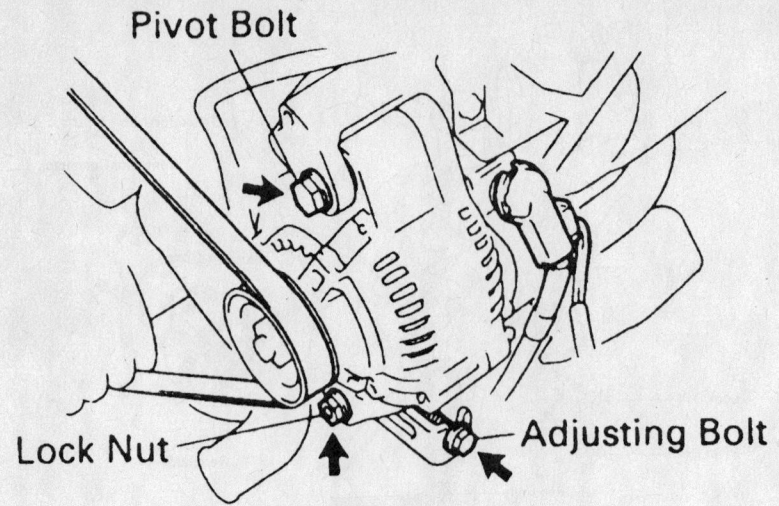

Pivot Bolt

Lock Nut

Adjusting Bolt

86822078

Locations of the adjusting and pivot bolts and the locknut—3.4L 4Runner

- Used belt without A/C: 75–115 lbs.
3. Install or connect the following:
  - Tighten the pivot bolt to 38 ft. lbs. (52 Nm).
  - Tighten the adjusting lockbolt to 13 ft. lbs. (18 Nm).
  - Electrical wiring to the alternator

2.4L Highlander

1. Installation is the reverse of removal. Observe the following torques:
   - M8 bolts: 15 ft. lbs. (21Nm)
   - M10 bolts: 38 ft. lbs. (52Nm)

3.0L Highlander

1. Install or connect the following:
   - Alternator
   - Drive belt
   - Adjusting lockbolt. Tighten the bolt to 13 ft. lbs. (18 Nm).
   - Pivot bolt. Tighten the bolt to 43 ft. lbs. (58 Nm).
   - Wiring harness from the clip
   - Alternator electrical connectors

3.0L RX300

1. Install or connect the following:
   - Alternator
   - Drive belt. Tension the belt to 170–180 lbs. for a new belt or 95–135 lbs. for a used belt.
   - Adjusting lockbolt. Tighten the bolt to 13 ft. lbs. (18 Nm).
   - Plate washer
   - Pivot bolt. Tighten the bolt to 41 ft. lbs. (56 Nm).
   - Wiring harness from the clip
   - Alternator electrical connectors

2.7L 4Runner

Install or connect the following:
- Alternator
- Drive belt; adjust it to the proper tension. Tighten the lockbolt to 21 ft. lbs. (29 Nm) and the pivot bolt to 43 ft. lbs. (59 Nm).
- Wire harness with clip
- Alternator wiring. Tighten the nut to 7 ft. lbs. (10 Nm).
- Rubber boot over the terminal
- Wiring harness connector
- Negative battery cable

3.4L 4Runner

Install or connect the following:
- Alternator
- Drive belt. Tighten the locknut 25 ft. lbs. (33 Nm) and the pivot bolt 38 ft. lbs. (51 Nm).
- Alternator wiring
- Negative battery cable

Ignition Timing

ADJUSTMENT

All engines are equipped with a Distributorless Ignition System (DIS). No timing adjustment is possible.

Engine Assembly

REMOVAL & INSTALLATION

RAV4

1. Before servicing the vehicle, refer to the precautions in the beginning of this section.

2. Relieve the fuel system pressure.
3. Remove or disconnect the following:
   - Negative battery cable
   - Battery
   - Hood
   - Engine undercover
4. Drain the engine coolant and oil.
5. Drain the transaxle assembly.
6. Remove or disconnect the following:
   - Air cleaner and case
   - Accelerator cable from the throttle body, bracket and clamps
7. Disconnect and remove the engine wire from the No. 2 relay block, as follows:
   - No. 2 relay block from the body by removing the 2 bolts
   - Upper cover to the relay block
   - Electrical connectors
   - Engine wire, by removing the 2 nuts
   - Charcoal canister
   - Alternator
   - Upper and lower radiator hoses
   - Water inlet from the engine by removing the 2 nuts
   - Heater hoses
   - Fuel hose, by placing a rag under the fuel inlet hose
   - Starter by disconnecting the electrical connectors and 2 bolts, if equipped with manual transmission
   - Ground cable from the transaxle by removing the bolt
   - Clutch release cylinder from the transaxle, if equipped with manual transmission
   - Transaxle control cables (2 cable for manual transmission or 1 for automatic transmission) from the transaxle.
   - Transaxle cable from the front suspension crossmember and engine mounting centermember by removing the 2 bolts, if equipped with an automatic transmission
   - Transaxle oil cooler hoses, if equipped with an automatic transmission or 4WD with manual transmission
8. Detach the following:
   - Vapor pressure sensor connector
   - Igniter connector
   - Ignition coil connector
   - Noise filter connector
   - Ignition coil wire
   - Manifold Absolute Pressure (MAP) sensor connector
   - MAP sensor vacuum hose from the gas filter on the intake manifold
   - Brake booster hose from the intake manifold

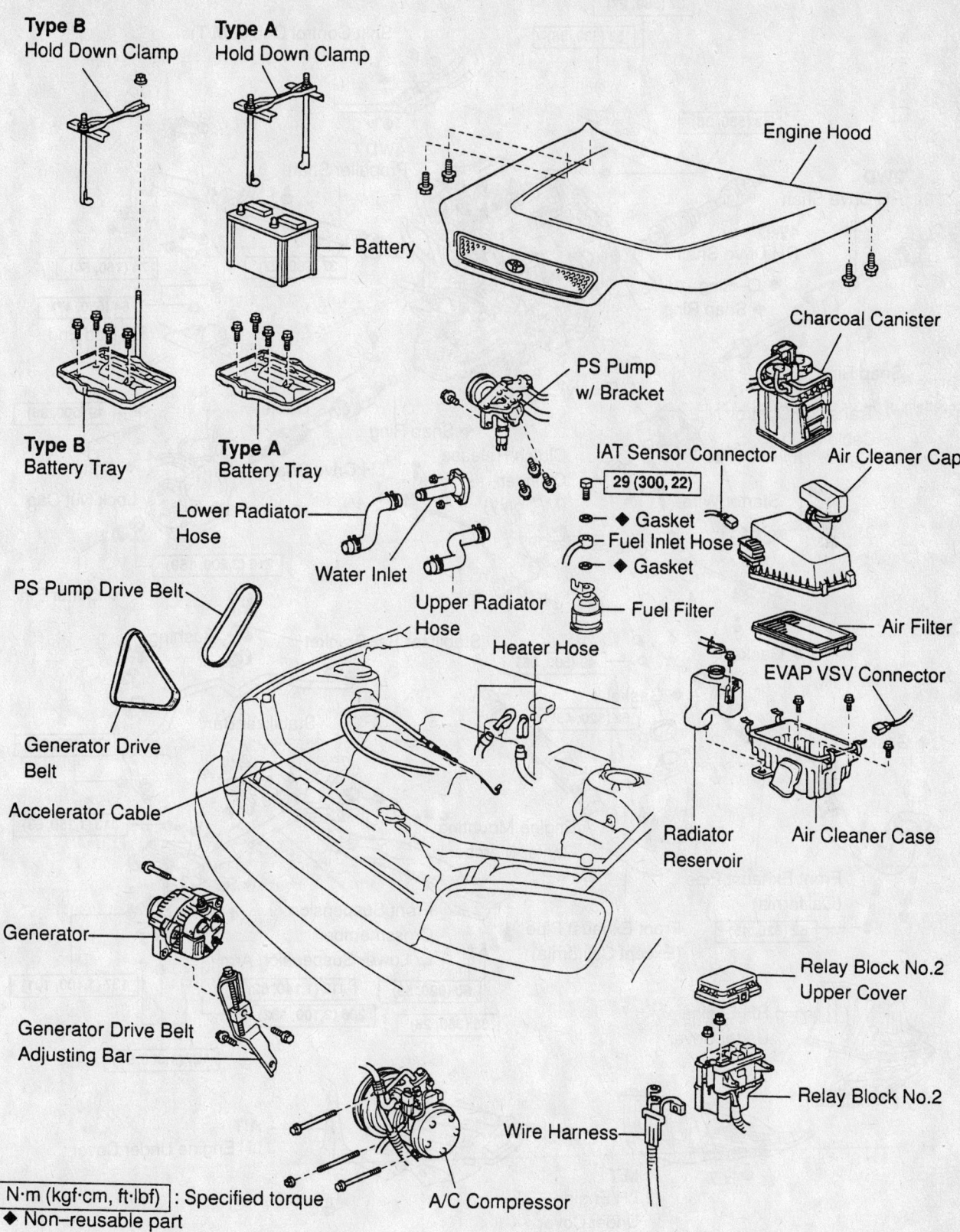

Type B
Hold Down Clamp

Type A
Hold Down Clamp

Engine Hood

Battery

Charcoal Canister

Type B
Battery Tray

Type A
Battery Tray

PS Pump
w/ Bracket

IAT Sensor Connector

Air Cleaner Cap

Lower Radiator
Hose

29 (300, 22)

◆ Gasket

Fuel Inlet Hose

◆ Gasket

Air Filter

Water Inlet

Fuel Filter

PS Pump Drive Belt

Upper Radiator
Hose

EVAP VSV Connector

Heater Hose

Generator Drive
Belt

Accelerator Cable

Radiator
Reservoir

Air Cleaner Case

Generator

Relay Block No.2
Upper Cover

Generator Drive Belt
Adjusting Bar

Relay Block No.2

Wire Harness

N·m (kgf·cm, ft·lbf) : Specified torque
◆ Non−reusable part

A/C Compressor

7924ZG04

Exploded view of the engine accessory removal components—RAV4 model

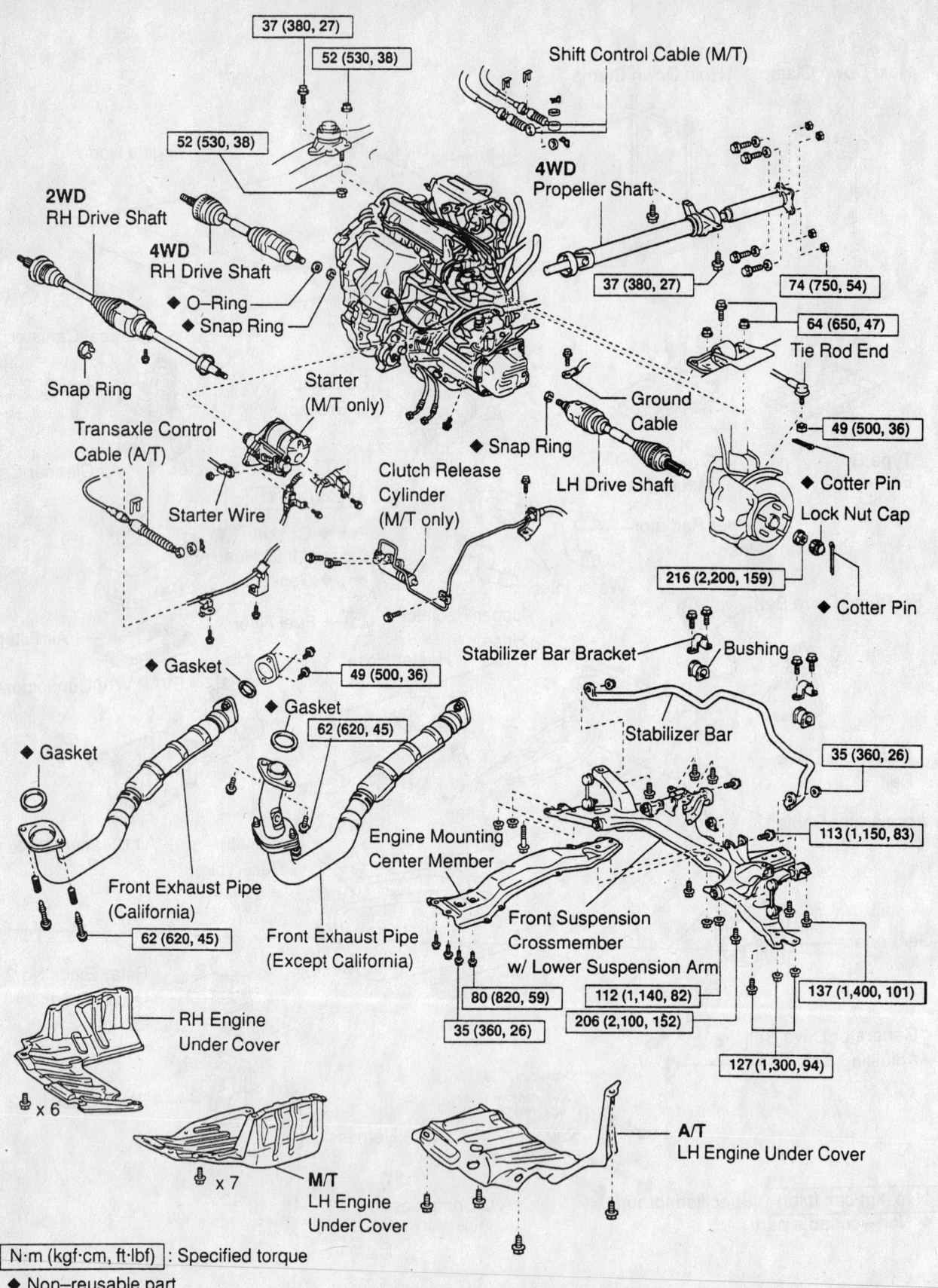

37 (380, 27)

52 (530, 38)

Shift Control Cable (M/T)

52 (530, 38)

2WD
RH Drive Shaft

4WD
RH Drive Shaft

◆ O–Ring

◆ Snap Ring

Snap Ring

Transaxle Control
Cable (A/T)

Starter Wire

Starter
(M/T only)

Clutch Release
Cylinder
(M/T only)

4WD
Propeller Shaft

37 (380, 27)

74 (750, 54)

64 (650, 47)

Tie Rod End

Ground
Cable

◆ Snap Ring

LH Drive Shaft

49 (500, 36)

◆ Cotter Pin

Lock Nut Cap

216 (2,200, 159)

◆ Cotter Pin

◆ Gasket

49 (500, 36)

◆ Gasket

62 (620, 45)

◆ Gasket

Front Exhaust Pipe
(California)

62 (620, 45)

Front Exhaust Pipe
(Except California)

Stabilizer Bar Bracket

Bushing

Stabilizer Bar

35 (360, 26)

113 (1,150, 83)

Engine Mounting
Center Member

Front Suspension
Crossmember
w/ Lower Suspension Arm

80 (820, 59)

35 (360, 26)

112 (1,140, 82)

206 (2,100, 152)

137 (1,400, 101)

127 (1,300, 94)

RH Engine
Under Cover

x 6

x 7 **M/T**
LH Engine
Under Cover

A/T
LH Engine Under Cover

N·m (kgf·cm, ft·lbf) : Specified torque

◆ Non–reusable part

Exploded view of the engine removal—RAV4 model

7924ZG05

- Differential lock control solenoid connector, if equipped with a 4WD manual transmission
- Ground strap from cowl

9. Detach the engine wire from the passenger compartment, as follows:
  - Right-hand scuff plate
  - Right-hand side trim
  - Right-hand carpet center cover
  - 2 ECM connectors
  - 2 connectors from the bracket connectors
  - No. 4 junction block connector
  - Wire clamp from the bracket
  - Engine wire from the passenger compartment

10. Remove the front exhaust pipe, as follows:
  - 3 nuts and the front exhaust pipe from the exhaust manifold, using a 14mm deep socket wrench; discard the gasket
  - 2 bolts and 2 nuts holding the front exhaust pipe to the catalytic converter
  - Front exhaust pipe and 2 gaskets

11. Remove the compressor from the engine and suspend the compressor securely.

➡**It is not necessary to remove the air conditioning compressor lines in order to remove the engine.**

12. Remove or disconnect the following:
  - Driveshaft, if equipped with 4WD
  - Halfshaft
  - Sway bar

13. Remove the front suspension crossmember assembly, as follows:
  - 2 centermember set nuts holding the centermember to the middle of the crossmember.
  - 2 rack and pinion assembly set bolts/nuts from the crossmember. Securely suspend the steering gear assembly.
  - Catalytic converter with pipe from the ring
  - Support the suspension crossmember with a jack
  - 6 bolts from the suspension crossmember
  - Suspension crossmember with the lower suspension arms

14. Remove the engine mounting centermember, as follows:
  - 2 bolts holding the centermember to the front engine mounting insulator
  - 2 bolts holding the centermember to the body

- Centermember

15. Disconnect the power steering pump from the engine, as follows:
  - 2 vacuum hoses from the steering pump
  - Adjusting bolt for the power steering unit. Loosen the pivot bolt to the power steering pump and remove the drive belt. Use Torque Wrench Adapter tool 09249-63010 and a deep socket to loosen the pivot bolt.
  - Power steering pump from the engine by removing the 3 bracket bolts

16. Install a engine hanger to the engine.

17. Attach the engine sling device to the engine hangers.

18. Remove or disconnect the following:
  - Left-hand engine mounting bracket from the mounting insulator by removing the 2 nuts and 2 bolts
  - Ground connector next to the right-hand engine mount
  - Right-hand engine mounting bracket from the mounting insulator by removing the bolt and 2 nuts

19. Lower the engine and transaxle and at the same time, raise the vehicle to gain clearance to the remove the engine.

20. Place the assembly on a stand and separate the engine from the transaxle.

To install:

21. Install or connect the following:
  - Engine and transaxle assembly
  - Left-hand engine mounting bracket to the mounting insulator. Tighten both nuts/bolts to 47 ft. lbs. (64 Nm).
  - Bolt and 2 nuts to hold the right-hand engine mounting bracket to the mounting insulator. Tighten the bolt to 27 ft. lbs. (37 Nm) and both nuts to 38 ft. lbs. (52 Nm).
  - Ground connector next to the right-hand engine mount
  - Engine sling and hanger

22. Install the power steering pump, as follows:
  - Pump with the bracket. Tighten the 3 bolts to 32 ft. lbs. (43 Nm).
  - Pivot and adjusting bolts. Tighten the pivot bolt to 32 ft. lbs. (43 Nm) and the adjusting bolt to 29 ft. lbs. (39 Nm).
  - Drive belt. Adjust the tension.
  - Both air hoses to the power steering pump

23. Install or connect the following:
  - Engine mounting centermember to the body; install the 4 bolts but do not tighten the bolts at this time

24. Install or connect the front crossmember, as follows:
  - Suspension crossmember with the lower control arms. Torque both bolts crossmember-to-chassis bolts to 152 ft. lbs. (206 Nm).
  - Rack and pinion. Tighten both nuts/bolts to 83 ft. lbs. (113 Nm).
  - Centermember to the crossmember. Tighten both nuts to 82 ft. lbs. (112 Nm).
  - Tighten the lower control arm rear brackets to 101 ft. lbs. (137 Nm).
  - Tighten both engine mounting centermember-to-front engine mounting insulator bolts to 59 ft. lbs. (80 Nm).
  - Tighten both engine mounting centermember-to-body bolts to 26 ft. lbs. (35 Nm).

25. Install or connect the following:
  - Sway bar
  - Halfshafts
  - Driveshaft, if equipped with 4WD

26. Install the air conditioning compressor. Tighten nut/bolts, as follows:
  - Stud bolt to 34 ft. lbs. (47 Nm)
  - Bolt to 27 ft. lbs. (37 Nm)
  - Nut to 20 ft. lbs. (27 Nm)

27. Install or connect the following:
  - Air conditioning compressor connector
  - Front exhaust pipe with new gaskets. Tighten the 3 nuts to 46 ft. lbs. (62 Nm) and 2 bolts to 35 ft. lbs. (48 Nm).

28. Attach the engine wire to the passenger compartment, as follows:
  - Engine wire through the cowl panel
  - Wire clamp to the bracket
  - Both ECM connectors
  - Both connectors on the bracket
  - No. 4 junction block
  - Right-hand floor carpet center cover
  - Right-hand cowl side trim
  - Right-hand scuff plate

29. Connect the following:
  - Vapor pressure sensor connector
  - Igniter connector
  - Ignition coil connector
  - Noise filter connector
  - Ignition coil wire
  - MAP sensor connector
  - MAP sensor vacuum hose to the gas filter on the intake manifold
  - Brake booster hose to the intake manifold
  - Differential lock control solenoid connector, if equipped with 4WD manual transmission
  - Ground strap to cowl

30. Install or connect the following:
- Transaxle oil cooler hoses, if equipped with an automatic transmission or 4WD with manual transmission
- Transaxle control cable(s) to the transaxle
- Transaxle control cable to the front crossmember and engine mounting centermember, for an automatic transmission
- Clutch release cylinder, for manual transmission Tighten both bolts to 108 inch lbs. (12 Nm).
- Ground cable to the transaxle
- Starter, for a manual transmission
- Fuel inlet hose to the fuel filter, using new gaskets. Tighten the union bolt to 22 ft. lbs. (29 Nm).
- Heater hoses
- Water inlet to the engine. Tighten both nuts to 78 inch lbs. (8.8 Nm).
- Upper and lower radiator hoses
- Alternator
- Charcoal canister

31. Connect the engine wire to the No. 2 relay box, as follows:
- Engine wire to the No. 2 relay block with the 2 nuts
- Connector
- Upper cover
- No. 2 relay block to the body with the 2 bolts

32. Install or connect the following:
- Accelerator cable to the throttle body, cable bracket and clamps
- Air cleaner case and cap
- Battery
- Negative battery cables

33. Fill the transaxle with oil.
34. Fill the engine with oil.
35. Fill the engine coolant.
36. Start the engine and check for leaks.
37. Install or connect the following:
- Engine undercovers
- Hood

38. Recheck all fluid levels.
39. Check and/or adjust the front wheel alignment.

2.4L Highlander

1. Before servicing the vehicle, refer to the precautions in the beginning of this section.
2. Matchmark the hood position.
3. Remove or disconnect the following:
- Front wheels
- No.1 engine undercover
- Right and left fender splash shields
- Right fender apron seal
- Engine oil
- Coolant

- Transaxle fluid
- Transfer case oil
- Battery
- Air cleaner
- Radiator hoses
- Oil cooler hoses
- Upper engine stay
- Upper engine mount bracket
- Accessory drive belts
- Steering pump reservoir
- Steering pump hoses
- All cables and wires connected to the engine
- Exhaust pipe
- Front drive shaft
- Stabilizer links
- Left and right axle hub nuts
- Left and right speed sensors
- Left and right tie rods
- Left and right lower control arms
- Torque converter-to-drive plate bolts
- Intermediate steering shaft
- AC compressor

4. Attach a crane, remove the 6 side rail plate subassembly bolts (3 each side) and the front suspension member rear brace.
5. Lift the engine out of the vehicle.
6. Installation is the reverse of removal. Observe the following torques:
- Frame side plate bolts: Large 63 ft. lbs. (85Nm); small 24 ft. lbs. (32Nm)
- Suspension member rear brace: Large 63 ft. lbs. (85Nm); small 24 ft. lbs. (32Nm)
- Intermediate shaft bolt: 26 ft. lbs. (35Nm)
- Torque converter bolts: 30 ft. lbs. (41Nm)
- Lower control arms, bolts and nuts: 94 ft. lbs. (127Nm)
- Tie rod nuts: 36 ft. lbs. (49Nm)
- Speed sensors: 71 inch lbs. (8Nm)
- Hub nuts: 217 ft. lbs. (294Nm)
- Stabilizer link nuts: 55 ft. lbs. (74Nm)
- Driveshaft nuts: 55 ft. lbs. (74Nm)
- Engine mount bracket: 15 ft. lbs. (20Nm)
- Engine mount stay: 47 ft. lbs. (64Nm)

3.0L Highlander

1. Before servicing the vehicle, refer to the precautions in the beginning of this section.
2. Matchmark the hood position.
3. Remove or disconnect the following:
- Front wheels
- No.1 engine undercover
- Right and left fender splash shields

- Right fender apron seal
- Coolant
- Engine oil
- Transaxle fluid
- Transfer case oil
- Battery
- Air cleaner
- Radiator and heater hoses
- Oil cooler hoses
- Upper engine stay
- Upper engine mount bracket
- Accessory drive belts
- Steering pump reservoir
- Steering pump hoses
- All cables and wires connected to the engine
- Exhaust pipes
- Front drive shaft with center bearing
- Stabilizer links
- Starter
- Alternator and brackets
- Left and right axle hub nuts
- Left and right speed sensors
- Left and right tie rods
- Left and right lower control arms
- Torque converter-to-drive plate bolts
- Intermediate steering shaft
- AC compressor

4. Attach a crane, remove the 6 side rail plate subassembly bolts (3 each side) and the front suspension member rear brace.
5. Lift the engine out of the vehicle.
6. Installation is the reverse of removal. Observe the following torques:
- Frame side plate bolts: Large 63 ft. lbs. (85Nm); small 24 ft. lbs. (32Nm)
- Suspension member rear brace: Large 63 ft. lbs. (85Nm); small 24 ft. lbs. (32Nm)
- Intermediate shaft bolt: 26 ft. lbs. (35Nm)
- Torque converter bolts: 30 ft. lbs. (41Nm)
- Lower control arms, bolts and nuts: 94 ft. lbs. (127Nm)
- Tie rod nuts: 36 ft. lbs. (49Nm)
- Speed sensors: 71 inch lbs. (8Nm)
- Hub nuts: 217 ft. lbs. (294Nm)
- Stabilizer link nuts: 55 ft. lbs. (74Nm)
- Driveshaft nuts: 55 ft. lbs. (74Nm)
- Engine mount bracket: 15 ft. lbs. (20Nm)
- Engine mount stay: 47 ft. lbs. (64Nm)

RX 300

1. Before servicing the vehicle, refer to the precautions in the beginning of this section.

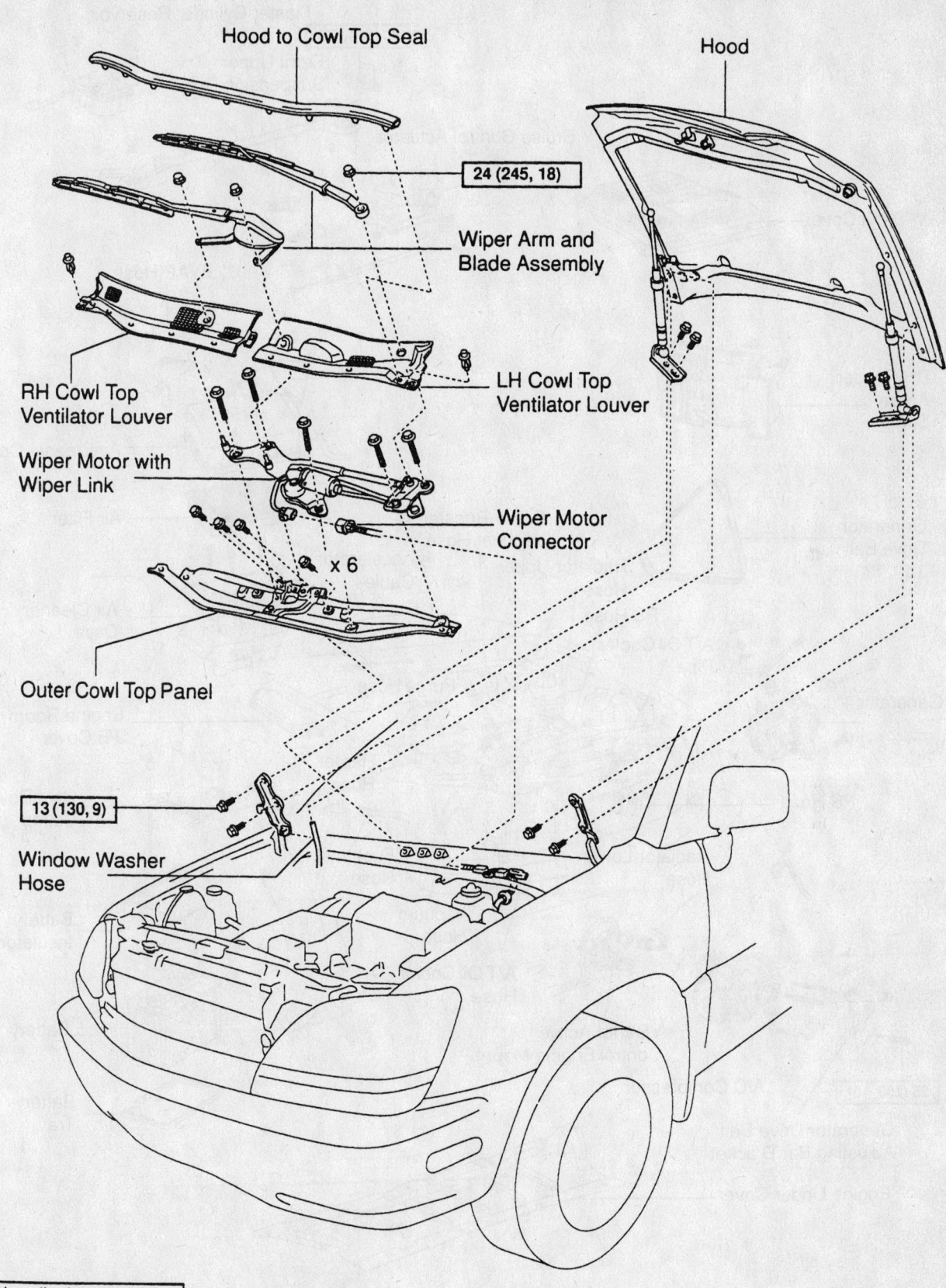

Hood to Cowl Top Seal

Hood

24 (245, 18)

Wiper Arm and
Blade Assembly

RH Cowl Top
Ventilator Louver

LH Cowl Top
Ventilator Louver

Wiper Motor with
Wiper Link

Wiper Motor
Connector

x 6

Outer Cowl Top Panel

13 (130, 9)

Window Washer
Hose

N·m (kgf·cm, ft·lbf) : Specified torque

7924ZG83

Exploded view of the top cowl and related components—RX 300

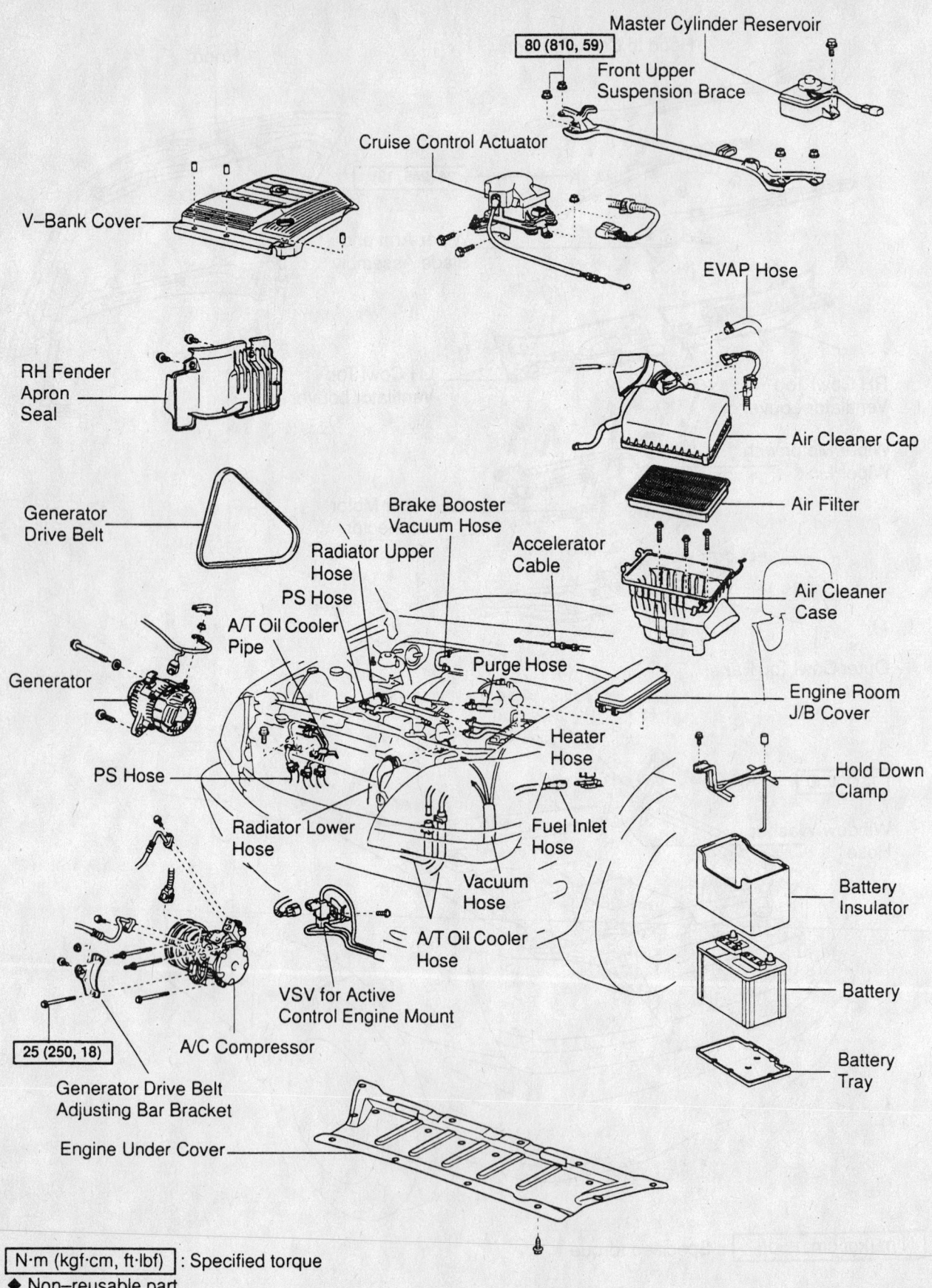

Master Cylinder Reservoir

80 (810, 59)

Front Upper Suspension Brace

Cruise Control Actuator

V–Bank Cover

EVAP Hose

RH Fender Apron Seal

Air Cleaner Cap

Air Filter

Generator Drive Belt

Brake Booster Vacuum Hose

Radiator Upper Hose

PS Hose

Accelerator Cable

Air Cleaner Case

A/T Oil Cooler Pipe

Purge Hose

Generator

Engine Room J/B Cover

PS Hose

Heater Hose

Hold Down Clamp

Radiator Lower Hose

Fuel Inlet Hose

Vacuum Hose

Battery Insulator

A/T Oil Cooler Hose

VSV for Active Control Engine Mount

Battery

25 (250, 18)

A/C Compressor

Battery Tray

Generator Drive Belt Adjusting Bar Bracket

Engine Under Cover

N·m (kgf·cm, ft·lbf) : Specified torque

◆ Non–reusable part

7924ZG84

Exploded view of engine pre-removal components—RX 300

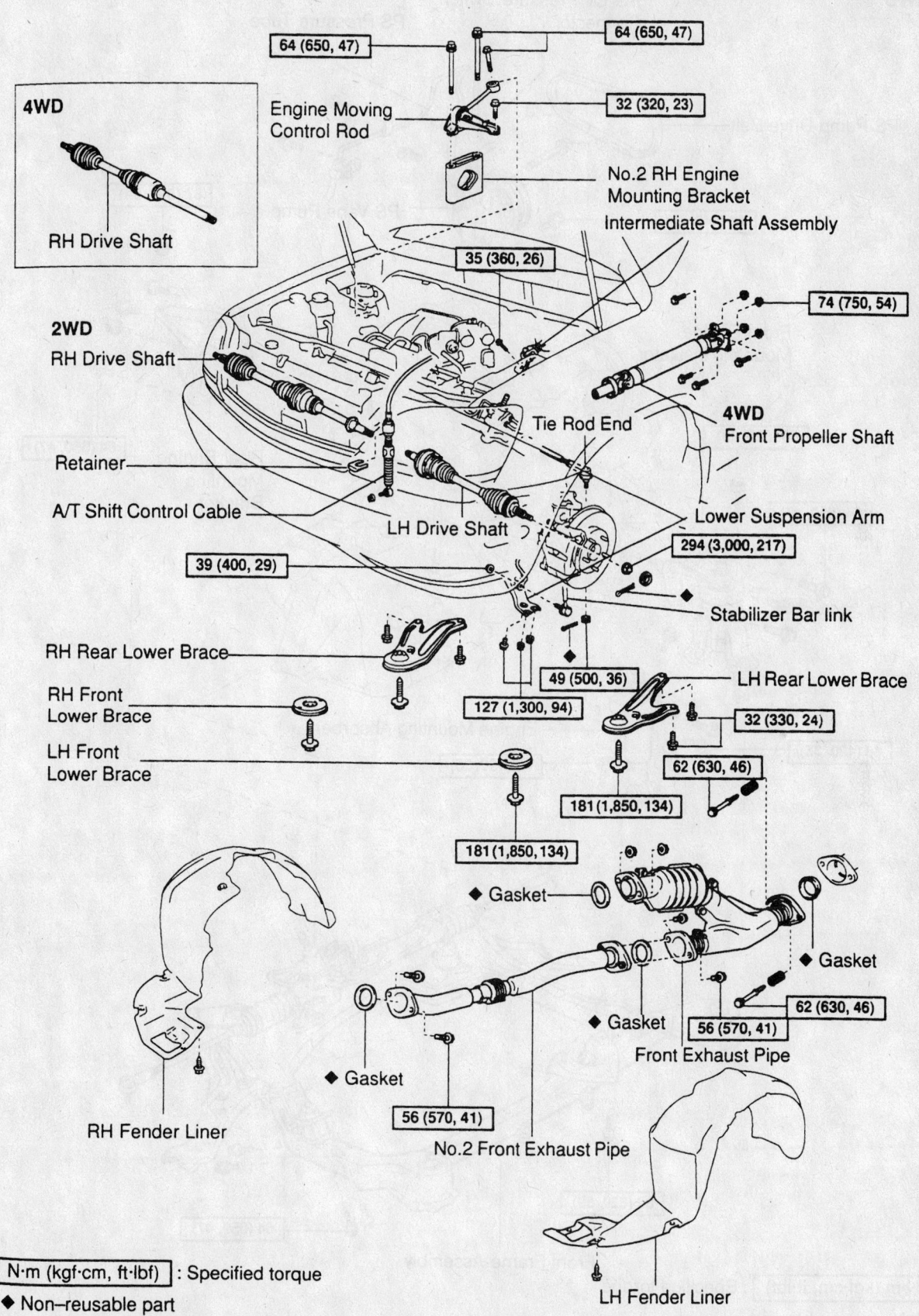

4WD

RH Drive Shaft

64 (650, 47)

64 (650, 47)

Engine Moving
Control Rod

32 (320, 23)

No.2 RH Engine
Mounting Bracket

Intermediate Shaft Assembly

35 (360, 26)

74 (750, 54)

2WD

RH Drive Shaft

Tie Rod End

4WD
Front Propeller Shaft

Retainer

A/T Shift Control Cable

LH Drive Shaft

Lower Suspension Arm

294 (3,000, 217)

Stabilizer Bar link

39 (400, 29)

RH Rear Lower Brace

49 (500, 36)

LH Rear Lower Brace

RH Front
Lower Brace

127 (1,300, 94)

32 (330, 24)

LH Front
Lower Brace

62 (630, 46)

181 (1,850, 134)

181 (1,850, 134)

◆ Gasket

62 (630, 46)

◆ Gasket

RH Fender Liner

◆ Gasket

◆ Gasket

56 (570, 41)

Front Exhaust Pipe

◆ Gasket

56 (570, 41)

No.2 Front Exhaust Pipe

LH Fender Liner

N·m (kgf·cm, ft·lbf) : Specified torque

◆ Non-reusable part

7924ZG85

Exploded view of engine removal and installation tightening specifications of the related components—RX 300

2WD

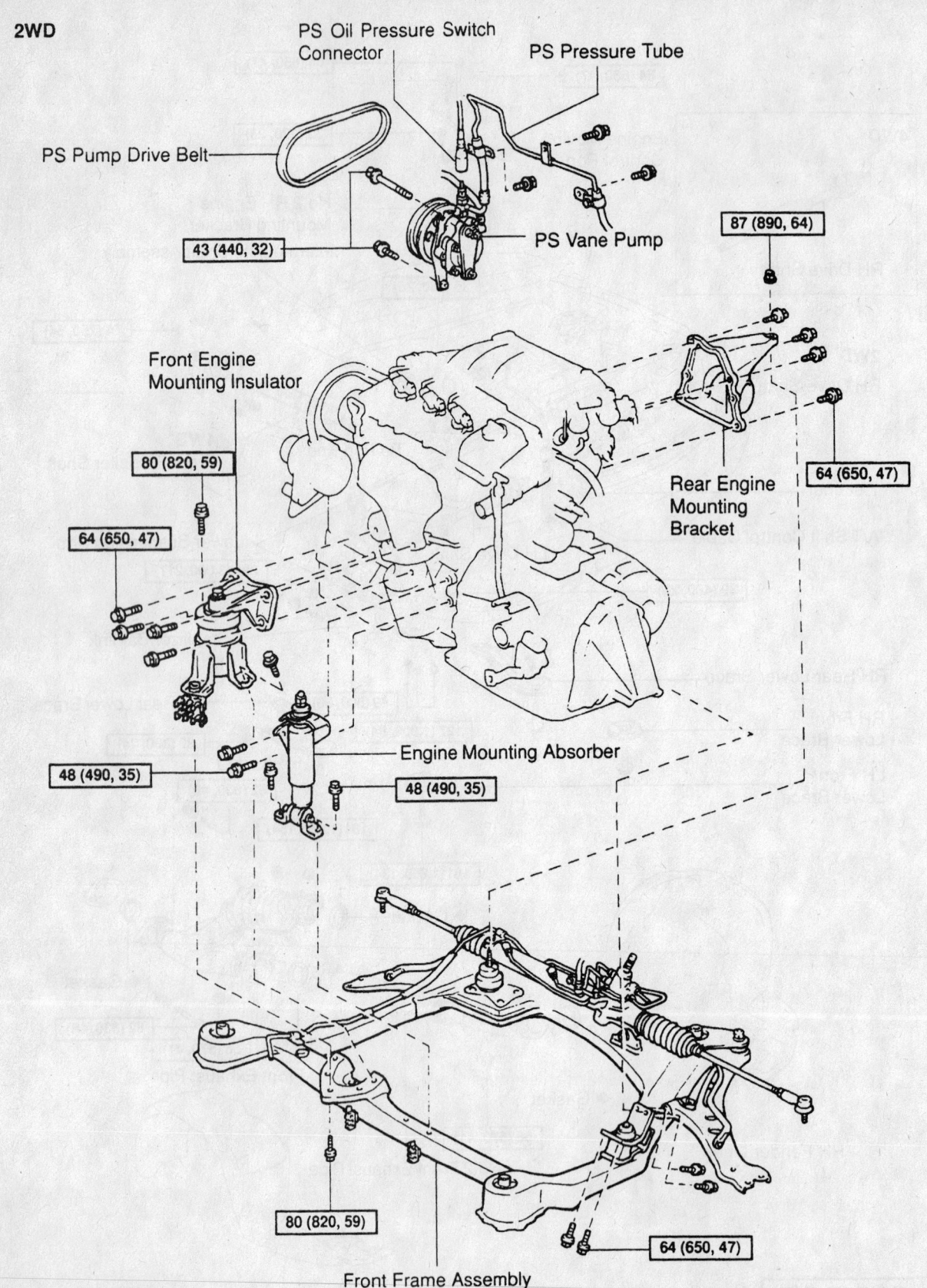

PS Oil Pressure Switch Connector

PS Pressure Tube

PS Pump Drive Belt

43 (440, 32)

PS Vane Pump

87 (890, 64)

Front Engine Mounting Insulator

80 (820, 59)

64 (650, 47)

Rear Engine Mounting Bracket

64 (650, 47)

Engine Mounting Absorber

48 (490, 35)

48 (490, 35)

80 (820, 59)

64 (650, 47)

Front Frame Assembly

N·m (kgf·cm, ft·lbf) : Specified torque

◆ Non–reusable part

7924ZG86

Exploded view of the suspension component removal and installation for engine removal—2WD RX 300

4WD

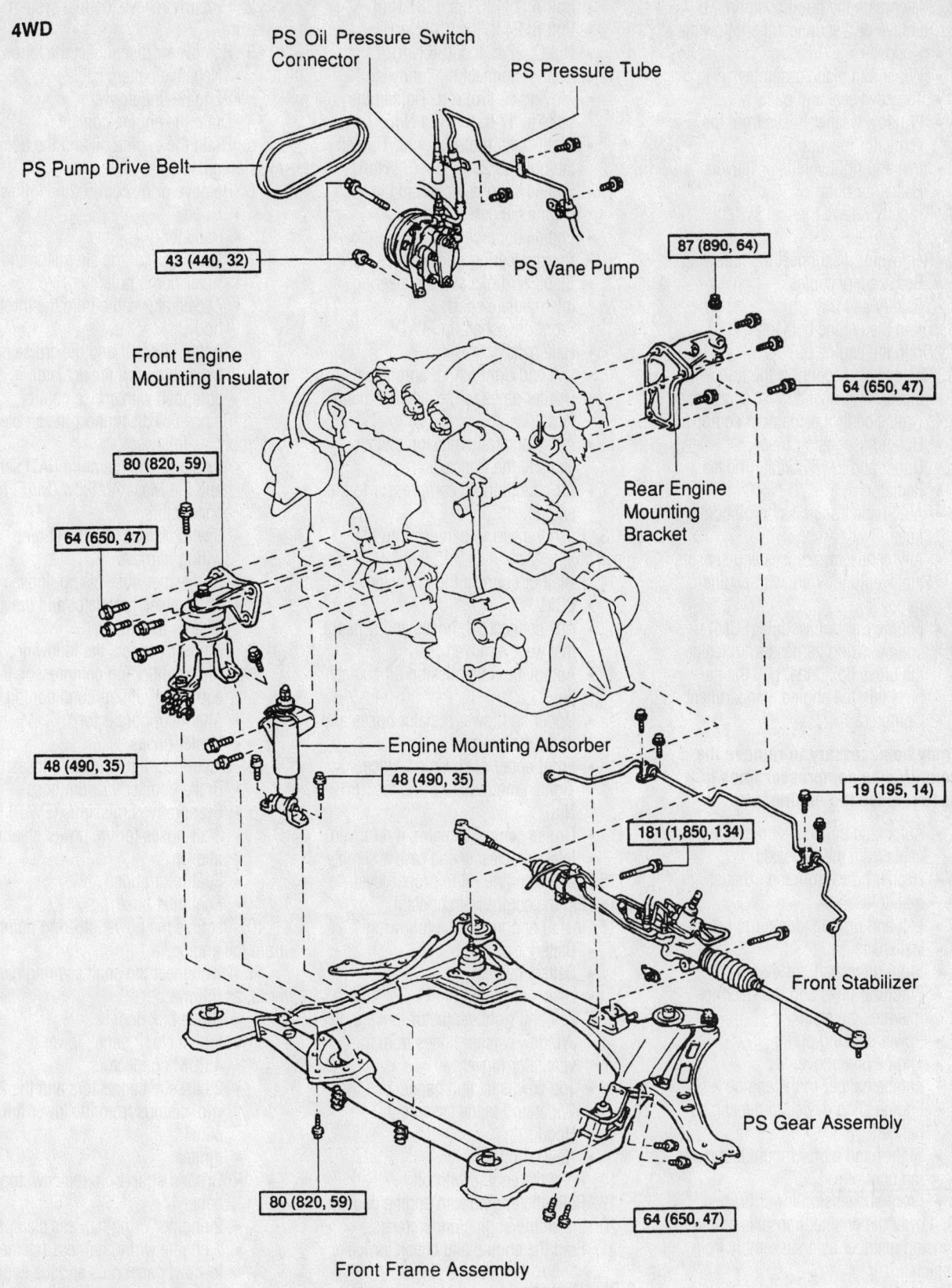

PS Oil Pressure Switch Connector

PS Pressure Tube

PS Pump Drive Belt

43 (440, 32)

PS Vane Pump

87 (890, 64)

64 (650, 47)

Front Engine Mounting Insulator

Rear Engine Mounting Bracket

80 (820, 59)

64 (650, 47)

Engine Mounting Absorber

48 (490, 35)

48 (490, 35)

19 (195, 14)

181 (1,850, 134)

Front Stabilizer

PS Gear Assembly

80 (820, 59)

64 (650, 47)

Front Frame Assembly

N·m (kgf·cm, ft·lbf) : Specified torque

◆ Non–reusable part

7924ZG87

Exploded view of the suspension component removal and installation for engine removal—4WD RX 300

2. Matchmark the hood position.

3. Remove or disconnect the following:
- Hood
- Wiper and blade assembly
- Top cowl seal and panel
- Window washer hoses from the ventilator louvers
- Left and right ventilator louvers
- Heater air duct

4. Properly relieve the fuel system pressure.

5. Remove or disconnect the following:
- Both battery cables
- Battery and tray

6. Drain the engine coolant.

7. Drain the engine oil.

8. Remove or disconnect the following:
- Intake air cleaner and case assembly
- Cruise control actuator, if equipped
- Upper suspension brace
- Upper and lower radiator hoses
- Radiator
- Automatic transmission oil cooler lines
- Any connectors, hoses and sensors that would interfere with engine removal
- Engine Control Module (ECM) engine wiring harness from inside the glove box; then, pull the harness into the engine compartment
- Compressor

➡It may be necessary to remove the air conditioning compressor lines in order to remove the engine.

- Automatic transmission shifter cable from the transaxle
- Header pipes from the exhaust manifolds
- Left and right fender apron seals
- Halfshafts
- Front driveshaft, for 4WD
- Stabilizer links and the steering intermediate shaft
- Power steering pump
- Engine undercover
- Engine hanger to the engine
- Engine sling device to the engine hangers
- Right-hand motor mount and moving control rod
- Front suspension lower braces

9. Lower the engine, transaxle and front suspension member as an assembly from the vehicle.

To install:

10. Raise the engine, transaxle and front suspension member as an assembly into the vehicle.

11. Install the front suspension lower braces, and tighten the fasteners, as follows:

- Bolt A: 134 ft. lbs. (181 Nm)
- Bolt B: 24 ft. lbs. (32 Nm)
- Nut C: 27 ft. lbs. (36 Nm)

12. Install or connect the following:
- Moving control rod. Tighten the bolts to 47 ft. lbs. (64 Nm).
- Right-hand motor mount. Tighten the bolts to 23 ft. lbs. (32 Nm).
- Engine sling device from the engine hangers
- Engine undercover
- Power steering pump hoses
- Stabilizer links and the steering intermediate shaft
- Front driveshaft, for 4WD
- Halfshafts
- Left and right fender apron seals
- Header pipes to the exhaust manifolds
- Automatic transmission shifter cable to the transaxle
- Air conditioning compressor to the engine

13. Push the wiring harness into the glove box.

14. Install or connect the following:
- ECM
- Any connectors, hoses and sensors that were removed
- Automatic transmission oil cooler lines
- Upper and lower radiator hoses and fit the radiator
- Front upper suspension brace. Tighten the nuts to 59 ft. lbs. (80 Nm).
- Cruise control actuator, if removed
- Intake air cleaner and case assembly

15. Fill the engine oil to proper level.

16. Fill the engine with coolant.

17. Install or connect the following:
- Battery tray and battery
- Battery cables
- Heater air duct
- Left and right ventilator louvers
- Window washer hoses from the ventilator louvers
- Top cowl seal and panel
- Wiper and blade assembly
- Hood
- New oil filter

18. Refill the engine with oil.

19. Refill the engine with engine coolant.

20. Install the engine undercovers.

21. Start the engine and check for leaks.

2.7L 4Runner

2WD

1. Before servicing the vehicle, refer to the precautions in the beginning of this section.

2. Properly relieve the fuel system pressure.

3. Remove or disconnect the following:
- Negative battery cable
- Engine undercover

4. Drain the engine coolant.

5. Drain the engine oil and the transmission oil.

6. Remove or disconnect the following:
- Hood
- Radiator
- Drive belt for the alternator and the water pump pulley
- Accelerator cable from the throttle body
- Actuator cover and the cruise control cable from the actuator, if equipped with cruise control

7. Remove or disconnect the air cleaner assembly, as follows:
- Intake Air Temperature (IAT) sensor and the Mass Air Flow (MAF) meter connectors
- 3 wire clamps and the engine wiring harness
- Air cleaner hose clamp, loosen it
- MAF meter, resonator and the air cleaner assembly

8. Install or connect the following:
- Air conditioning compressor, if equipped with air conditioning
- Alternator connector
- Heater hoses

9. Disconnect the following hoses:
- Brake booster vacuum hose
- Evaporative Emissions (EVAP) hose
- 2 air hoses for the power steering idle-up
- Fuel return hose
- Fuel inlet hose

10. Remove the power steering pump from the engine.

11. Disconnect the engine wiring harness, as follows:
- Glove box door
- Finish No. 2 panel, lower it
- 4 ECM connectors
- 2 cassette connectors and the 2 wire clamps from the lower finish panel
- Igniter
- Ground strap from the cowl top panel
- 2 engine wiring harness clamps.
- 2 engine wiring harness retainer-to-cowl panel nuts and the engine wiring harness

12. Install or connect the following:
- Heated Oxygen (HO$_2$S) sensor
- Front exhaust pipe

13. Remove the shift lever assembly for a manual transmission, as follows:

14. Remove or disconnect the following:
   • Shift lever knob
   • 4 screws and the shift lever boot
   • 6 bolts, the shift lever assembly and baffle

15. Remove or disconnect the following:
   • Driveshaft
   • Speedometer cable
   • Clutch release cylinder, if equipped with a manual transmission
   • Cross-shaft, if equipped with an automatic transmission
   • Starter wire

16. Place a jack under the transmission and remove the engine rear mounting bracket.

17. Install a rear engine hanger in the correct direction.

18. Attach the engine hoist chain to the 2 engine hangers.

19. Remove or disconnect the following:
   • Engine front mounting insulators-to-frame bolts/nuts
   • Engine/transmission assembly
   • Engine

To install:

20. Install or connect the following:
   • Transmission to the engine
   • Chain hoist to the engine hangers
   • Engine/transmission assembly into the engine compartment

➡**Keep the engine level and align the right and left mounting and body mountings.**

   • Right and left mounting insulators to the body mountings and temporarily install the bolts/nuts

21. Raise the transmission onto the frame

22. Remove the chain hoist.

23. Remove the bolt and the rear engine hanger.

24. Install the engine rear mounting bracket and tighten to:
   • Bolt A: 13 ft. lbs. (19 Nm)
   • Bolt B: 19 ft. lbs. (26 Nm)

25. Install or connect the following:
   • Tighten the left and right engine mounting insulator bolts and nuts to 28 ft. lbs. (38 Nm).
   • Starter wire
   • Clutch release cylinder, for a manual transmission. Tighten the clutch line bolt to 29 ft. lbs. (39 Nm) and the clutch release cylinder bolts to 108 inch lbs. (13 Nm).
   • Cross-shaft, for an automatic transmission. Tighten the bolt to 29 ft. lbs. (39 Nm) and the nut to 13 ft. lbs. (18 Nm).
   • Speedometer cable

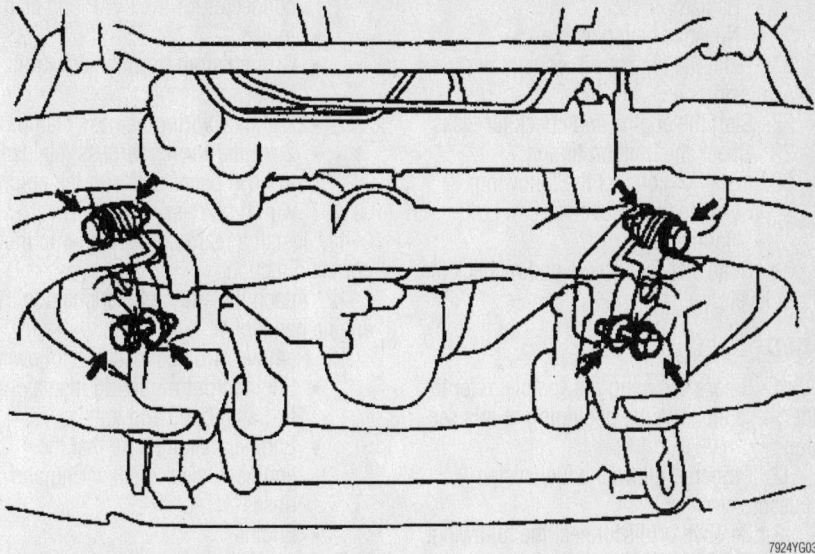

Be sure to support the engine before removing the right and left engine mounts—4Runner 2-wheel drive with 2.7L engine

   • Driveshaft
   • Shift lever assembly, for a manual transmission

26. Install the front exhaust pipe, as follows:
   • New gaskets and the front exhaust pipe assembly. Tighten the 3 new nuts to 46 ft. lbs. (62 Nm).
   • Support bracket. Tighten the bolts to 29 ft. lbs. (39 Nm).
   • 3-way catalytic converter with a new gasket to the tail pipe. Tighten to 29 ft. lbs. (39 Nm).
   • HO_2S sensor
   • Engine wiring harness

   • Power steering pump
   • All hoses previously removed
   • Alternator wire
   • Air conditioning compressor, if removed. Tighten the bolts to 18 ft. lbs. (25 Nm).
   • Intake air connector. Tighten the bolts to 13 ft. lbs. (18 Nm).
   • Air cleaner assembly
   • Throttle cable to the throttle body
   • Cruise control cable to the actuator and the actuator cover, if disconnected
   • Drive belt for the alternator and water pump pulley

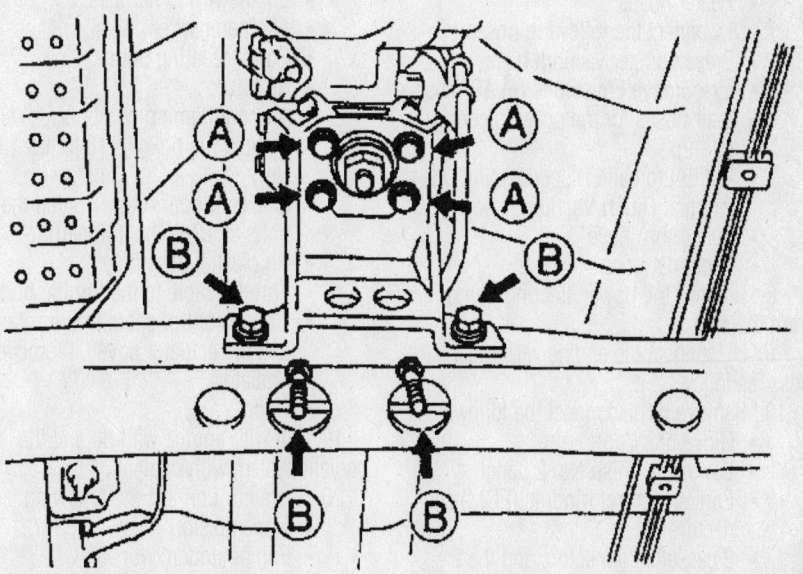

Bolt tightening pattern for the engine rear mounting bracket—4Runner 2-wheel drive with 2.7L engine

- Radiator
- Negative battery cable

27. Refill the engine oil, coolant and transmission oil.

28. Start the engine and check for leaks.

29. Check the ignition timing.

30. Install or connect the following:
- Engine undercover
- Hood

31. Road test the vehicle and recheck the fluid levels.

4WD

1. Before servicing the vehicle, refer to the precautions in the beginning of this section.

2. Properly relieve the fuel system pressure.

3. Remove or disconnect the following:
- Negative battery cable
- Transmission
- Engine undercover

4. Drain the engine coolant.

5. Drain the engine oil.

6. Remove or disconnect the following:
- Hood
- Radiator
- Drive belt for the alternator and the water pump pulley
- Accelerator cable from the throttle body
- Actuator cover and the cruise control cable from the actuator, if equipped with cruise control
- Air cleaner assembly
- Intake air connector
- Air conditioning compressor, if equipped with air conditioning
- Alternator connector
- Heater hoses

7. Disconnect the following hoses:
- Brake booster vacuum hose
- Evaporative Emissions (EVAP) hose
- 2 air hoses for the power steering idle-up
- With Automatic Disconnecting Differential (ADD) Vacuum hose
- Fuel return hose
- Fuel inlet hose

8. Remove the power steering pump from the engine.

9. Disconnect the engine wiring harness, as follows:

10. Remove or disconnect the following:
- Glove box door
- Lower the finish No. 2 panel
- Engine Control Module (ECM) connectors
- 2 cassette connectors and the 2 wire clamps from the lower finish panel
- Vacuum Switching Valve (VSV)

connector for the EVAP and clamp
- Igniter
- Ground strap from the cowl top panel
- 2 engine wiring harness clamps
- 2 engine wiring harness retainer-to-cowl panel nuts and the engine wiring harness

11. Install a rear engine hanger in the correct direction.

12. Attach the engine hoist chain to the 2 engine hangers.

13. Remove or disconnect the following:
- Engine front mounting insulators-to-frame bolts and nuts
- Engine, making sure that the engine is clear of all wiring and hoses
- Engine

To install:

14. Attach a chain hoist to the engine hangers.

15. Install or connect the following:
- Engine into the engine compartment

➡ Keep the engine level and align the right and left mounting and body mountings

- Right and left mounting insulators to the body mountings and temporarily install the bolts and nuts

16. Remove the chain hoist.

17. Remove the bolt and the rear engine hanger.

18. Install or connect the following:
- Tighten the right and left engine mounting insulator bolts/nuts to 28 ft. lbs. (38 Nm).
- Engine wiring harness
- Alternator wire
- Power steering pump
- All hoses
- Air conditioning compressor. Tighten the bolts to 18 ft. lbs. (25 Nm).
- Intake air connector. Tighten the bolts to 13 ft. lbs. (18 Nm).
- Air cleaner assembly
- Throttle cable to the throttle body
- Cruise control cable to the actuator and the actuator cover, if removed
- Radiator
- Hood

19. Fill with engine with oil and the coolant system with coolant.

20. Install or connect the following:
- Transmission
- Engine undercover
- Negative battery cable

21. Fill the transmission fluid.

22. Check the ignition timing.

23. Test drive the vehicle and check for leaks.

24. Recheck fluid levels.

3.4L 4Runner

2WD

1. Before servicing the vehicle, refer to the precautions in the beginning of this section.

2. Properly relieve the fuel system pressure.

3. Remove or disconnect the following:
- Hood
- Battery
- Engine under covers

4. Drain the engine coolant.

5. Drain the engine oil.
- Radiator
- Fan with the fluid coupling and fan pulleys
- Air cleaner cap
- Mass Air Flow (MAF) meter and the resonator
- Air cleaner case and filter

6. Disconnect the following hoses:
- Heater hoses
- Brake booster vacuum hose
- Evaporative Emissions (EVAP) hose
- Fuel return hose
- Fuel inlet hose

7. Detach the starter wire and connectors, as follows:
- Ground strap, by removing the bolt
- 3 starter wire clamps and connector

8. Detach the alternator connector and wire.

9. Disconnect the engine wiring harness, as follows:
- Glove box door
- Lower the finish No. 2 panel
- 4 ECM connectors
- 2 cassette connectors and the 2 wire clamps from the lower finish panel
- Engine wiring harness clamp

10. Remove or disconnect the following:
- Igniter connector
- Ground strap
- Vacuum Switching Valve (VSV) connector for the Evaporative Emissions (EVAP)
- Vapor pressure sensor connector and clamp
- Vapor connector for the vapor pressure sensor and clamp
- 2 engine wiring harness retainer-to-cowl panel nuts and pull out the engine wiring harness
- Driveshaft from the transmission
- Speedometer cable
- Front exhaust pipe

- Nut and the control cable
11. Place a jack under the transmission.
12. Remove or disconnect the following:
    - Transmission rear mounting bracket by removing the 8 bolts
    - Bolt and the air conditioning compressor wire clamp, if equipped with air conditioning
13. If necessary, install a No. 2 engine hanger with 2 bolts. Tighten the 2 bolts to 30 ft. lbs. (40 Nm).
14. Attach the engine hoist chain to the 2 engine hangers.
15. Remove or disconnect the following:
    - 4 engine front mounting insulators-to-frame bolts and nuts
    - Engine and transmission

To install:
16. Install or connect the following:
    - Engine
    - Engine mounts to the body mountings. Install the bolts and nuts but do not tighten at this time.
17. Remove the engine chain hoist the No. 2 engine hanger.
18. Install or connect the following:
    - Air conditioning wire with the bolt, if equipped with air conditioning
    - Transmission mounting bracket. Tighten the frame bolts to 43 ft. lbs. (58 Nm) and the mounting insulator bolts to 13 ft. lbs. (18 Nm).
    - Tighten the engine mounting nuts and bolts to 28 ft. lbs. (38 Nm).
    - Control cable
    - Front exhaust pipe
    - Speedometer cable
    - Driveshaft
    - All engine wiring harness, hoses and cables
    - Fan with the fluid coupling and fan pulleys. Tighten the nuts to 48 inch lbs. (5.4 Nm).
    - Air cleaner case and air filter
    - MAF meter, resonator and the air cleaner cap
    - Radiator
19. Fill the engine with oil.
20. Fill the engine and radiator with coolant.
21. Install or connect the following:
    - Engine undercover
    - Battery
    - Hood
22. Start the engine and check for leaks.
23. Make any necessary adjustments and road test the vehicle.

4WD

1. Before servicing the vehicle, refer to the precautions in the beginning of this section.

2. Remove or disconnect the following:
    - Transmission
    - Hood
3. Release the fuel system pressure.
4. Remove or disconnect the following:
    - Battery
    - Engine undercovers
5. Drain the engine coolant.
6. Drain the engine oil.
7. Remove or disconnect the following:
    - Radiator
    - Fan with the fluid coupling and fan pulleys
    - Air cleaner cap
    - Mass Air Flow (MAF) meter and the resonator
8. Disconnect the following hoses:
    - Heater hoses
    - Brake booster vacuum hose
    - Evaporative Emissions (EVAP) hose
    - Automatic Disconnecting Differential (ADD) vacuum hose
    - Fuel return hose
    - Fuel inlet hose
9. Detach the starter wire and connectors, as follows:
    - Ground strap, by removing the bolt
    - 3 starter wire clamps and connector
10. Detach the alternator connector and wire.
11. Disconnect the engine wiring harness, as follows:
    - Glove box door
    - Lower the finish No. 2 panel
    - 4 ECM connectors
    - 2 cassette connectors and the 2 wire clamps from the lower finish panel
    - Engine wiring harness clamp
12. Disconnect the following:
    - Igniter connector
    - Ground strap
    - Vacuum Switching Valve (VSV) connector for the EVAP
    - Vapor pressure sensor connector and clamp
    - Vapor connector for the vapor pressure sensor and clamp
13. Remove or disconnect the following:
    - 2 engine wiring harness retainer-to-cowl panel nuts and wiring harness
    - Air conditioning compressor wire clamp, if equipped with air conditioning
14. If necessary, install a No. 2 engine hanger with 2 bolts. Tighten the 2 bolts to 30 ft. lbs. (40 Nm).
15. Attach the engine hoist chain to the 2 engine hangers.
16. Remove or disconnect the following:

- 4 engine front mounting insulators-to-frame bolts and nuts
- Engine

To install:
17. Install or connect the following:
    - Engine
    - Engine mounts-to-body mountings. Install the bolts and nuts but do not tighten at this time.
18. Remove the engine chain hoist the No. 2 engine hanger.
19. Install or connect the following:
    - Air conditioning wire with the bolt, if equipped with air conditioning
    - Tighten the engine mounting nuts and bolts to 28 ft. lbs. (38 Nm).
    - Engine wiring harness
    - Engine wiring harness clamp
    - All wires, hoses and cables
    - Fan with the fluid coupling and fan pulleys. Tighten the nuts to 48 inch lbs. (5.4 Nm).
    - Air cleaner case and air filter
    - MAF meter, resonator and the air cleaner cap
    - Radiator
20. Fill the engine with oil.
21. Fill the engine and radiator with coolant.
22. Install or connect the following:
    - Transmission and refill it with transmission oil
    - Engine undercover
    - Battery
    - Hood
23. Start the engine, make any necessary adjustments and check for leaks.

Heater Core

REMOVAL & INSTALLATION

RAV4

1. Disconnect the negative battery cable.

CAUTION

After the negative battery cable has been disconnected, wait at least 1½ minutes for the air bag module to deplete its energy.

2. Drain the cooling system into a clean container for reuse.
3. Disconnect the heater hoses from the heater core.
4. Remove the steering wheel by performing the following procedure:
   a. Position the front wheels in the straight-ahead position.

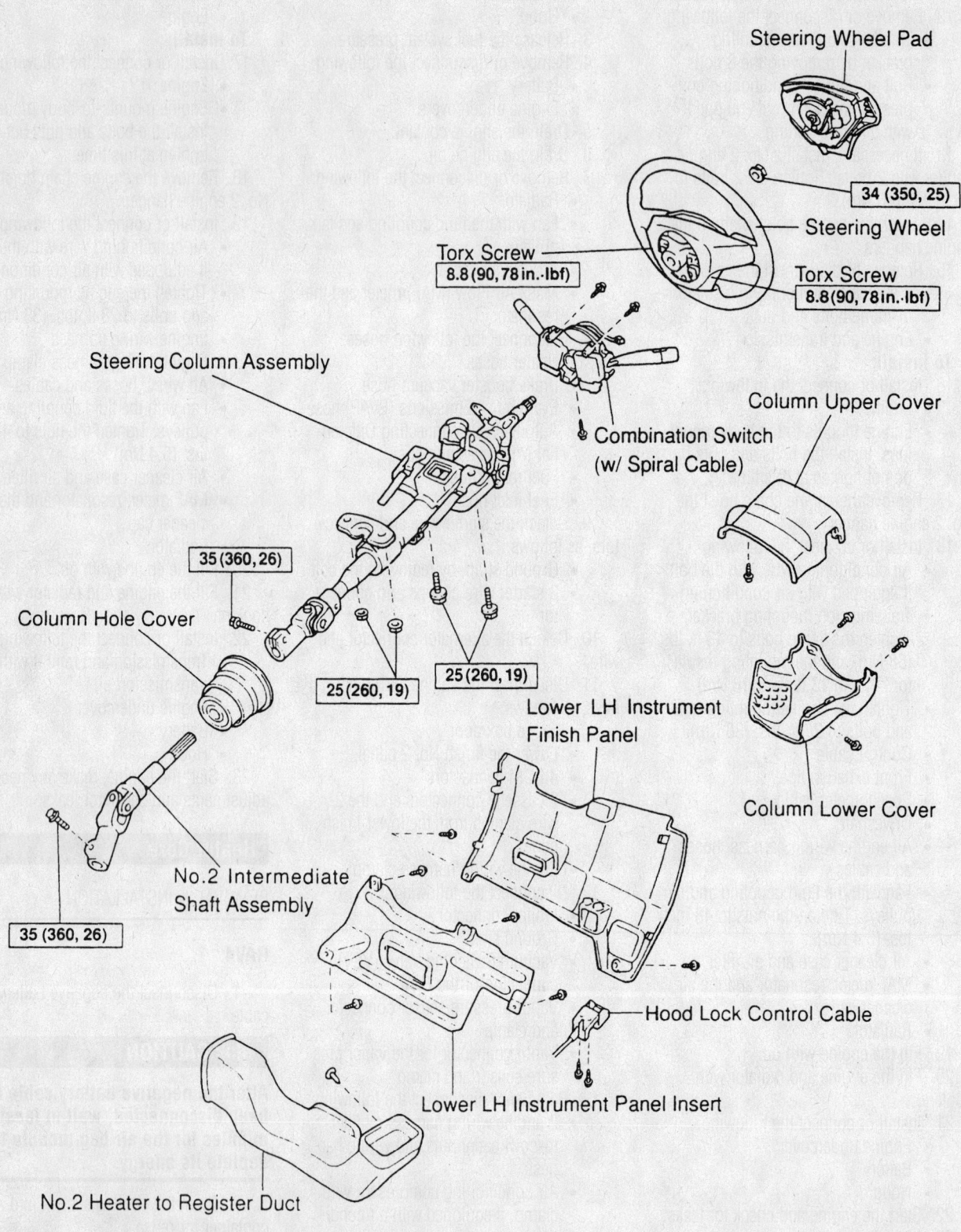

Steering Wheel Pad

34 (350, 25)

Steering Wheel

Torx Screw
8.8 (90, 78 in.·lbf)

Torx Screw
8.8 (90, 78 in.·lbf)

Steering Column Assembly

Combination Switch
(w/ Spiral Cable)

Column Upper Cover

35 (360, 26)

Column Hole Cover

25 (260, 19)

25 (260, 19)

Column Lower Cover

Lower LH Instrument
Finish Panel

No.2 Intermediate
Shaft Assembly

35 (360, 26)

Hood Lock Control Cable

Lower LH Instrument Panel Insert

No.2 Heater to Register Duct

N·m (kgf·cm, ft·lbf) : Specified torque

93113GK4

Exploded view of the steering wheel, air bag module, steering column and related components—Toyota RAV4

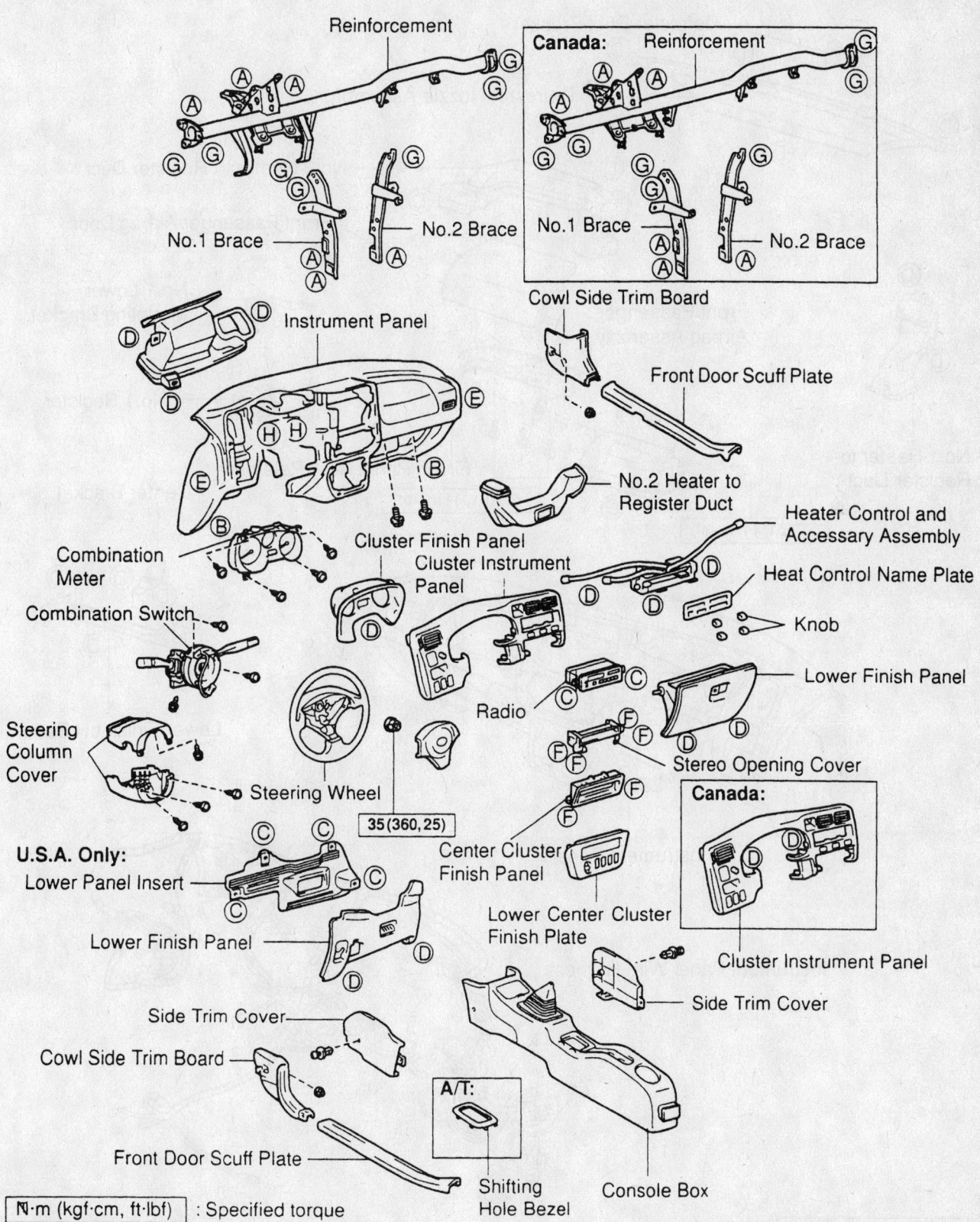

Reinforcement

Canada: Reinforcement

No.1 Brace

No.2 Brace

No.1 Brace

No.2 Brace

Cowl Side Trim Board

Instrument Panel

Front Door Scuff Plate

No.2 Heater to Register Duct

Heater Control and Accessary Assembly

Heat Control Name Plate

Knob

Combination Meter

Cluster Finish Panel
Cluster Instrument Panel

Lower Finish Panel

Combination Switch

Radio

Steering Column Cover

Steering Wheel

Stereo Opening Cover

35 (360, 25)

Canada:

U.S.A. Only:
Lower Panel Insert

Center Cluster Finish Panel

Cluster Instrument Panel

Lower Finish Panel

Lower Center Cluster Finish Plate

Side Trim Cover

Side Trim Cover

Cowl Side Trim Board

A/T:

Front Door Scuff Plate

Shifting Hole Bezel

Console Box

N·m (kgf·cm, ft·lbf) : Specified torque

93113GK5

Exploded view of the instrument panel and related components—Toyota RAV4

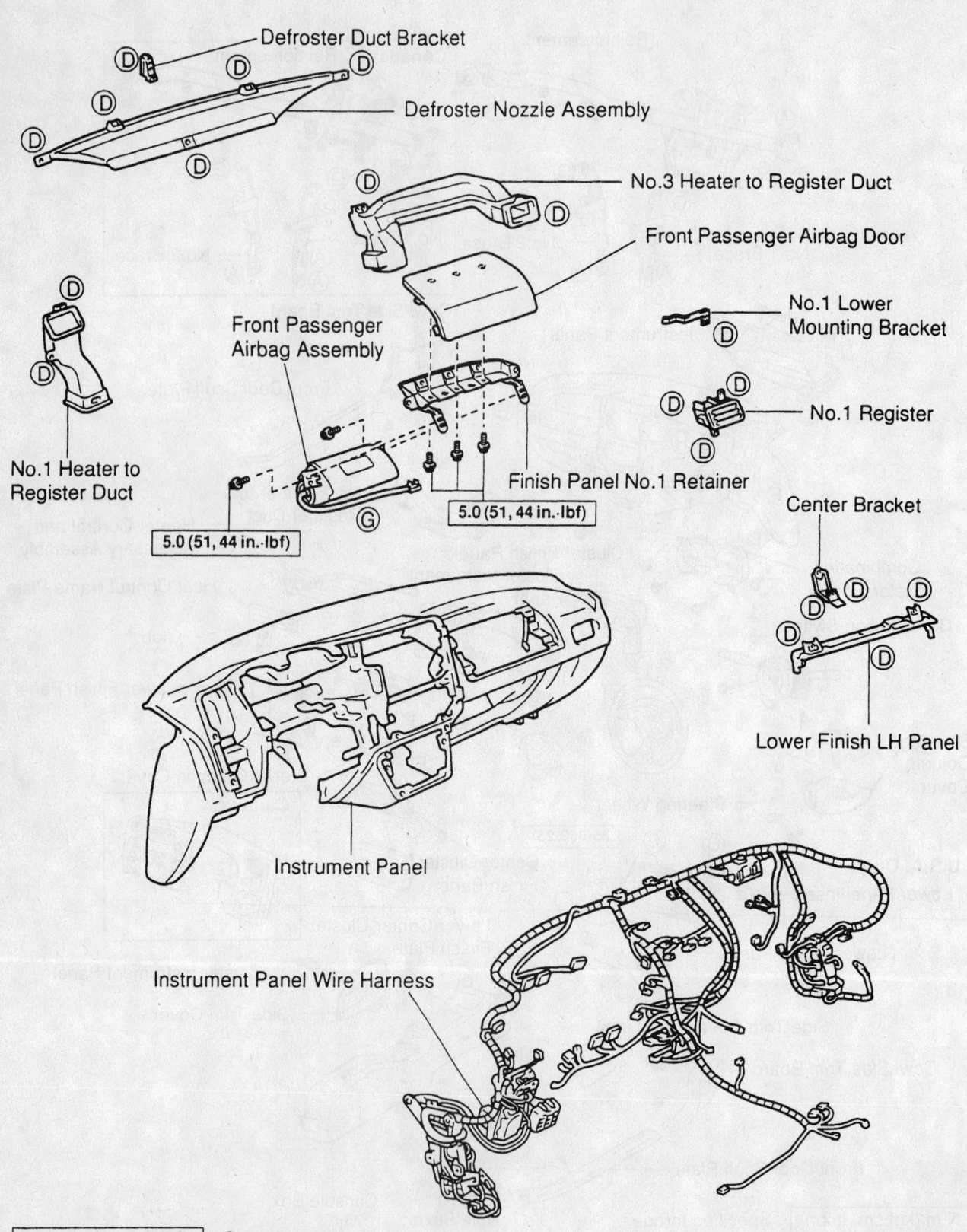

Defroster Duct Bracket

Defroster Nozzle Assembly

No.3 Heater to Register Duct

Front Passenger Airbag Door

No.1 Lower Mounting Bracket

Front Passenger Airbag Assembly

No.1 Register

No.1 Heater to Register Duct

Finish Panel No.1 Retainer

Center Bracket

5.0 (51, 44 in.·lbf)

5.0 (51, 44 in.·lbf)

Lower Finish LH Panel

Instrument Panel

Instrument Panel Wire Harness

N·m (kgf·cm, ft·lbf) : Specified torque

93113GK6

Exploded view of the instrument panel air bag module, ventilation components and wiring harness—Toyota RAV4

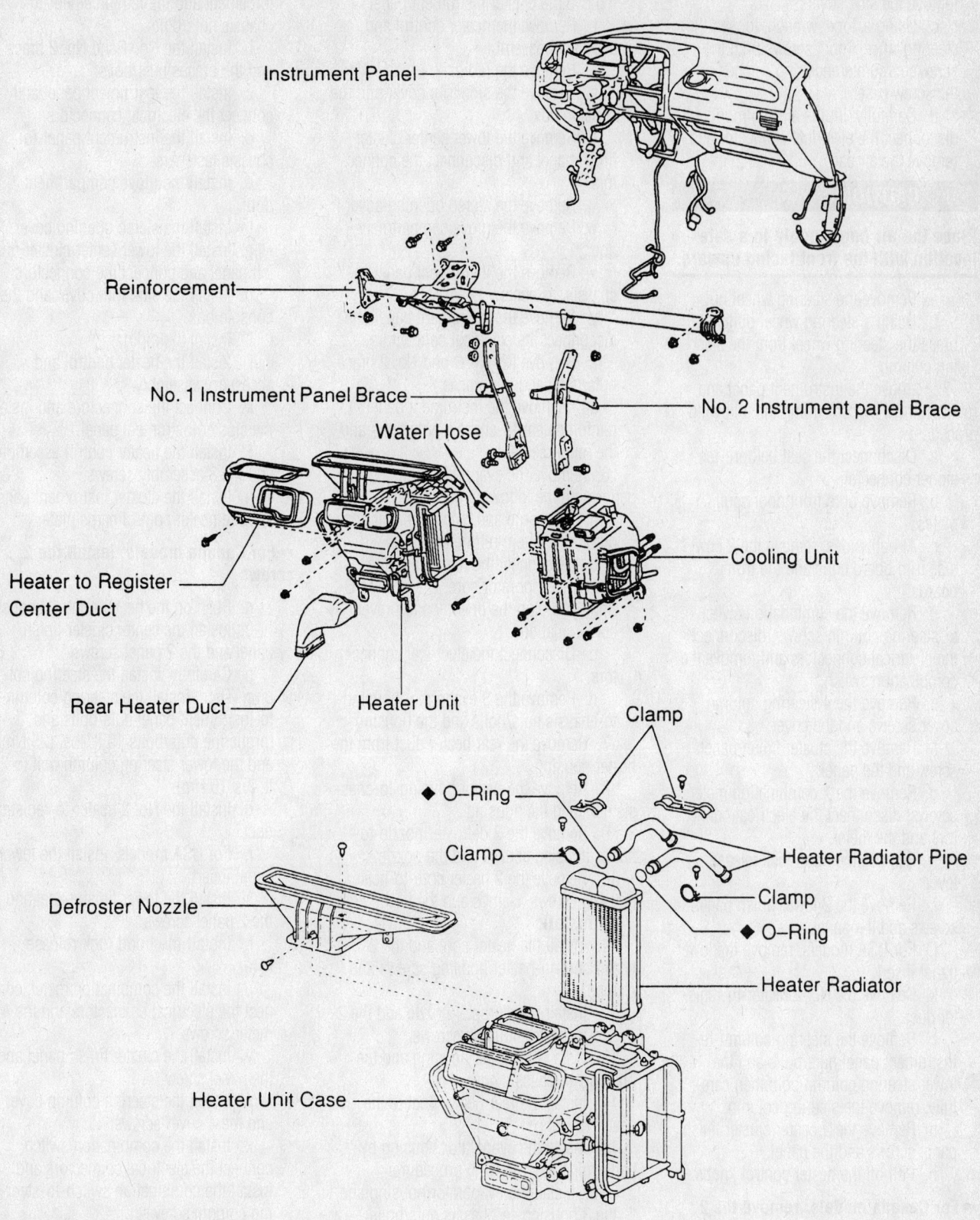

Instrument Panel

Reinforcement

No. 1 Instrument Panel Brace

No. 2 Instrument panel Brace

Water Hose

Heater to Register Center Duct

Cooling Unit

Rear Heater Duct

Heater Unit

Clamp

◆ O–Ring

Clamp

Heater Radiator Pipe

Clamp

◆ O–Ring

Defroster Nozzle

Heater Radiator

Heater Unit Case

◆ Non–reusable part

93113GK7

Exploded view of the heater core, heater housing, evaporator housing and related components—Toyota RAV4

b. At both sides of the steering wheel, remove the side covers.

c. Using a Torx® wrench, loosen the steering wheel Torx® screws until the screw's circumference ring catches on the screw case.

d. Carefully, lift the air bag module, disconnect the electrical connector and remove the air bag.

❄❄ CAUTION

Place the air bag module in a safe location with the front facing upward.

e. Remove the steering wheel nut.

f. Using a steering wheel puller, press the steering wheel from the steering column.

5. Remove the instrument panel and reinforcement by performing the following procedure:

a. Disconnect the seat belt pre-tensioner connector.

b. Remove both front door scuff plates.

c. At both sides, remove the 2 cowl side trim board clips and the trim boards.

d. Remove the combination switch-to-steering column screws, disconnect the electrical connectors and remove the combination switch.

e. Remove the 4 steering column cover screws and the cover.

f. Remove the cluster finish panel screw and the panel.

g. Remove the 4 combination meter screws, disconnect the electrical connectors and the meter.

h. Remove the hood lock release lever.

i. Remove the 2 lower finish panel screws and the panel.

j. For USA models, remove the lower panel insert.

k. Remove the No. 2 heater-to-register duct.

l. Remove the steering column-to-instrument panel nuts/bolts and the lower steering column bolt; then carefully, remove the steering column.

m. Remove the 2 center cluster finish panel screws and the panel.

n. Pull off the heater control knobs.

➡**For Canada models, remove the 2 screws.**

o. Pry off the heater control name plate and the cluster instrument panel.

p. Remove the 3 heater control assembly screws and the assembly.

q. Disconnect the connectors and remove the cluster instrument panel.

r. Remove the heater control and accessory assembly.

s. Remove the radio.

t. Remove the side trim cover and the console box.

u. Remove the lower center cluster finish panel and disconnect the connectors.

v. Remove the stereo opening cover.

w. Remove the glove compartment door.

x. Remove the instrument panel-to-chassis fasteners.

y. Remove the instrument panel and disconnect the electrical connectors.

z. Remove the No. 1 and No. 2 brace nuts/bolts and the braces.

aa. Remove the instrument panel reinforcement-to-chassis nuts/bolts and the reinforcement.

6. Remove the evaporator housing by performing the following procedure:

a. Discharge and recover the air conditioning system refrigerant.

b. Disconnect the refrigerant lines from the evaporator core. Discard the O-rings and plug the openings to prevent contamination.

c. Disconnect the electrical connectors.

d. Remove the 3 evaporator housing-to-chassis nuts/bolts and the housing.

7. Remove the rear heater duct from the heater housing.

8. Remove the heater housing-to-chassis nuts and the housing.

9. Remove the 2 defroster nozzle-to-heater housing screws and the nozzle.

10. Remove the 2 heater core-to-heater housing screws, clamps and the heater core.

To install:

11. Install the heater core and the 2 heater core-to-heater housing screws and clamps.

12. Install the defroster nozzle and the 2 nozzle-to-heater housing screws.

13. Install the heater housing and the housing-to-chassis nuts.

14. Install the rear heater duct to the heater housing.

15. Install the evaporator housing by performing the following procedure:

a. Install the evaporator housing and the 3 housing-to-chassis nuts/bolts.

b. Connect the electrical connectors.

c. Using new O-rings, connect the refrigerant lines to the evaporator core.

16. Install the instrument panel and reinforcement by performing the following procedure:

a. Install the instrument panel reinforcement and the reinforcement-to-chassis nuts/bolts.

b. Install the No. 1 and No. 2 brace and the braces nuts/bolts.

c. Install the instrument panel and connect the electrical connectors.

d. Install the instrument panel-to-chassis fasteners.

e. Install the glove compartment door.

f. Install the stereo opening cover.

g. Install the lower center cluster finish panel and connect the connectors.

h. Install the side trim cover and the console box.

i. Install the radio.

j. Install the heater control and accessory assembly.

k. Connect the connectors and install the cluster instrument panel.

l. Install the heater control assembly and the 3 assembly screws.

m. Install the cluster instrument panel and the heater control name plate.

➡**For Canada models, install the 2 screws.**

n. Push on the heater control knobs.

o. Install the center cluster finish panel and the 2 panel screws.

p. Carefully, install the steering column. Then, install the steering column-to-instrument panel nuts/bolts and torque the nuts/bolts 19 ft. lbs. (25 Nm) and the lower steering column bolt to 26 ft. lbs. (5 Nm).

q. Install the No. 2 heater-to-register duct.

r. For USA models, install the lower panel insert.

s. Install the lower finish panel and the 2 panel screws.

t. Install the hood lock release lever.

u. Install the combination meter, connect the electrical connectors and the 4 meter screws.

v. Install the cluster finish panel and the panel screw.

w. Install the steering column cover and the 4 cover screws.

x. Install the combination switch, connect the electrical connectors and install the combination switch-to-steering column screws.

y. At both sides, install the cowl side trim board and the 2 trim boards clips.

z. Install both front door scuff plates.

aa. Connect the seat belt pre-tensioner connector.

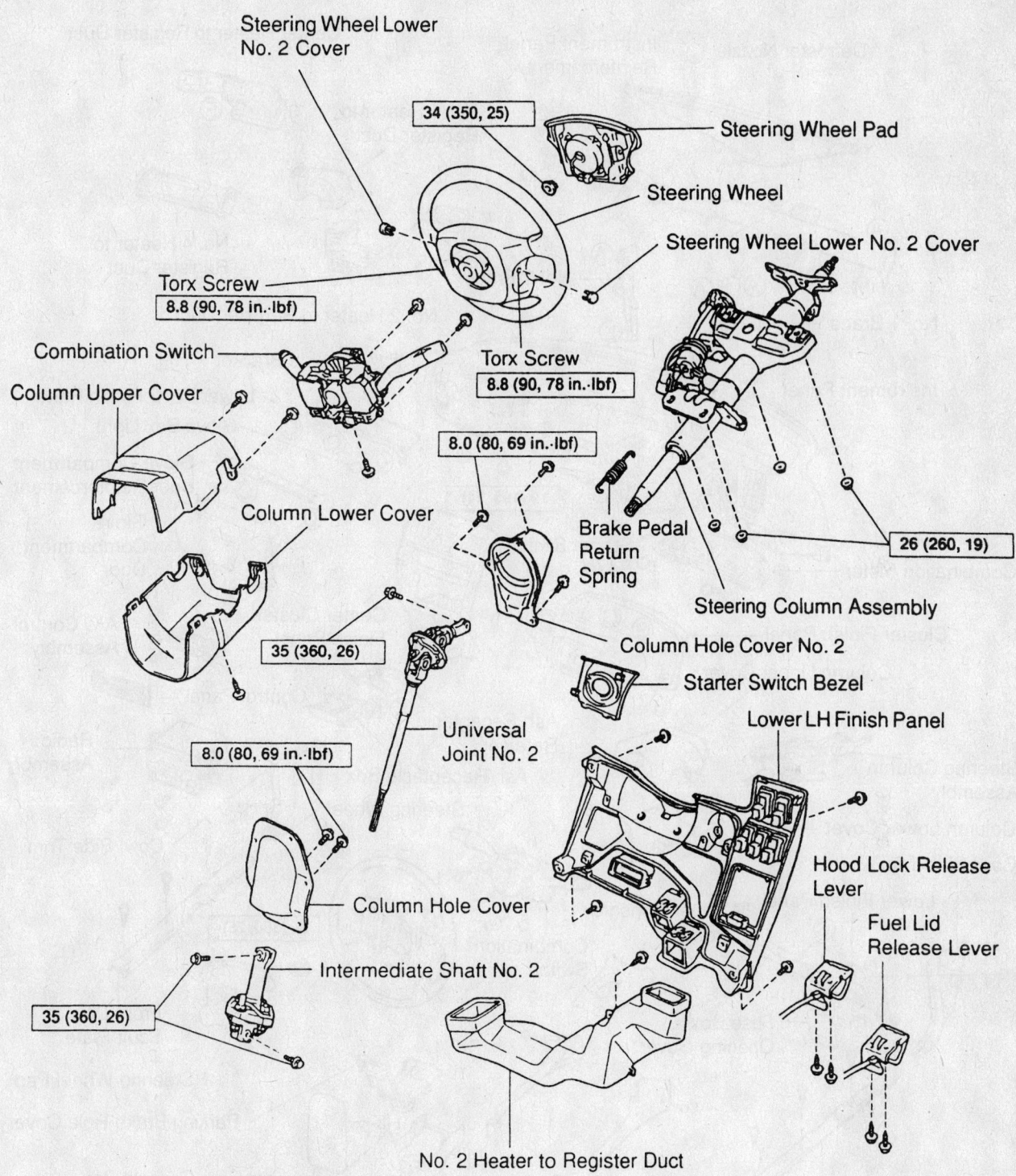

Steering Wheel Lower No. 2 Cover

34 (350, 25)

Steering Wheel Pad

Steering Wheel

Steering Wheel Lower No. 2 Cover

Torx Screw
8.8 (90, 78 in.·lbf)

Combination Switch

Column Upper Cover

Torx Screw
8.8 (90, 78 in.·lbf)

8.0 (80, 69 in.·lbf)

Column Lower Cover

Brake Pedal Return Spring

26 (260, 19)

Steering Column Assembly

Column Hole Cover No. 2

35 (360, 26)

Starter Switch Bezel

Lower LH Finish Panel

Universal Joint No. 2

8.0 (80, 69 in.·lbf)

Column Hole Cover

Hood Lock Release Lever

Fuel Lid Release Lever

Intermediate Shaft No. 2

35 (360, 26)

No. 2 Heater to Register Duct

N·m (kgf·cm, ft·lbf) : Specified torque

93113GJ9

Exploded view of the steering wheel, air bag module, steering column and related components—Toyota 4Runner

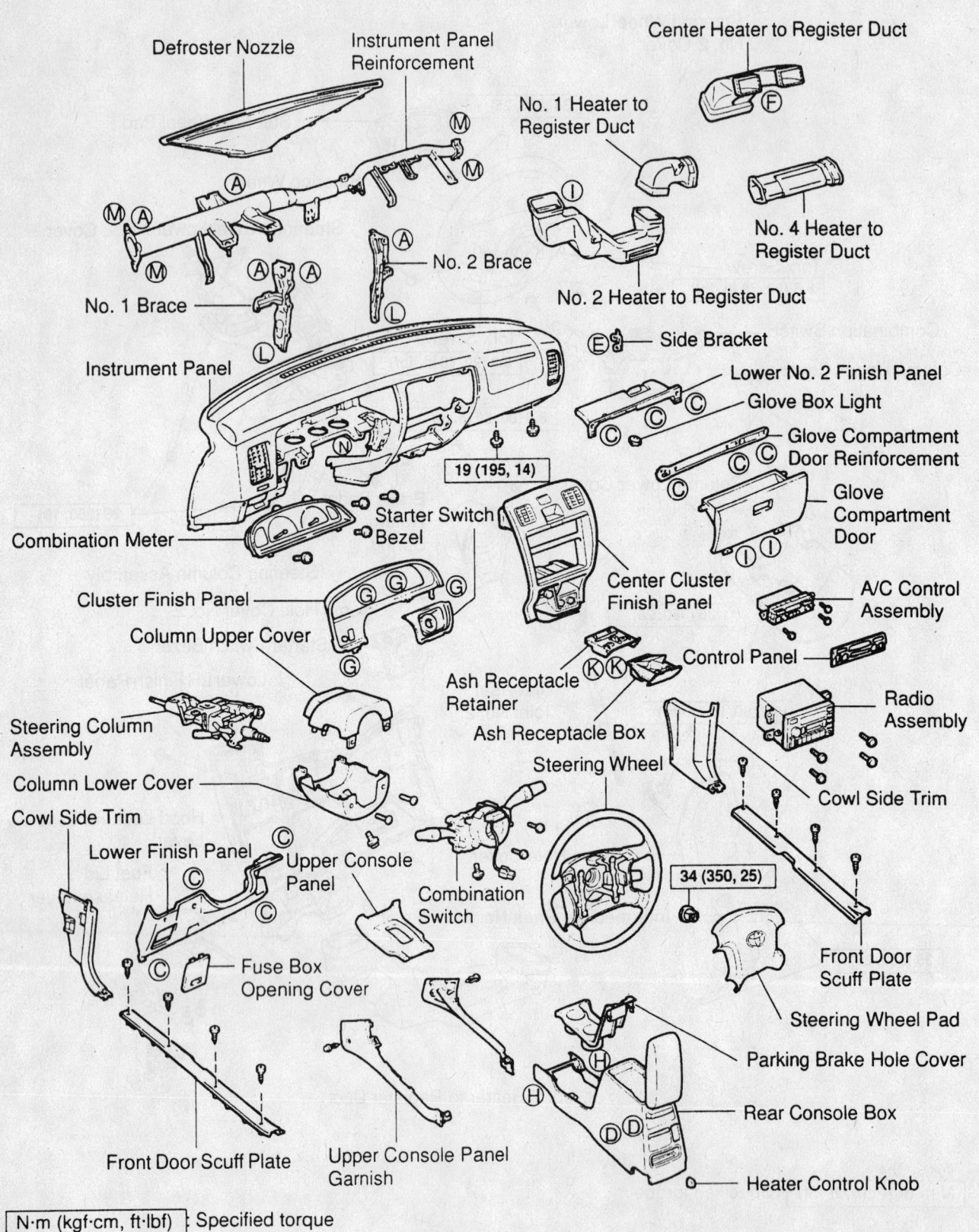

Defroster Nozzle

Instrument Panel Reinforcement

Center Heater to Register Duct

No. 1 Heater to Register Duct

No. 4 Heater to Register Duct

No. 2 Heater to Register Duct

No. 2 Brace

No. 1 Brace

Instrument Panel

Side Bracket

Lower No. 2 Finish Panel

Glove Box Light

Glove Compartment Door Reinforcement

Glove Compartment Door

Combination Meter

Starter Switch Bezel

19 (195, 14)

Center Cluster Finish Panel

A/C Control Assembly

Control Panel

Cluster Finish Panel

Column Upper Cover

Radio Assembly

Steering Column Assembly

Ash Receptacle Retainer

Ash Receptacle Box

Steering Wheel

Cowl Side Trim

Column Lower Cover

Cowl Side Trim

Lower Finish Panel

Upper Console Panel

Combination Switch

34 (350, 25)

Front Door Scuff Plate

Fuse Box Opening Cover

Steering Wheel Pad

Parking Brake Hole Cover

Rear Console Box

Front Door Scuff Plate

Upper Console Panel Garnish

Heater Control Knob

N·m (kgf·cm, ft·lbf) : Specified torque

93113GJ0

Exploded view of the instrument panel and related components—Toyota 4Runner

17. Install the steering wheel by performing the following procedure:

a. Install the steering wheel to the steering column.

b. Install the steering wheel nut and torque the nut to 25 ft. lbs. (34 Nm).

c. Connect the electrical connector and carefully, install the air bag module.

d. Using a Torx® wrench, tighten the steering wheel screws to 78 inch lbs. (8.8 Nm).

e. At both sides of the steering wheel, install the side covers.

18. Connect the heater hoses to the heater core.

19. Refill the cooling system.

20. Connect the negative battery cable.

21. Evacuate and charge the air conditioning system.

22. Run the engine to normal operating temperatures. Check the climate control operation and check for leaks.

4Runner

FRONT HEATER

1. Disconnect the negative battery cable.

✳✳ CAUTION

After the negative battery cable has been disconnected, wait at least 1½ minutes for the air bag module to deplete its energy.

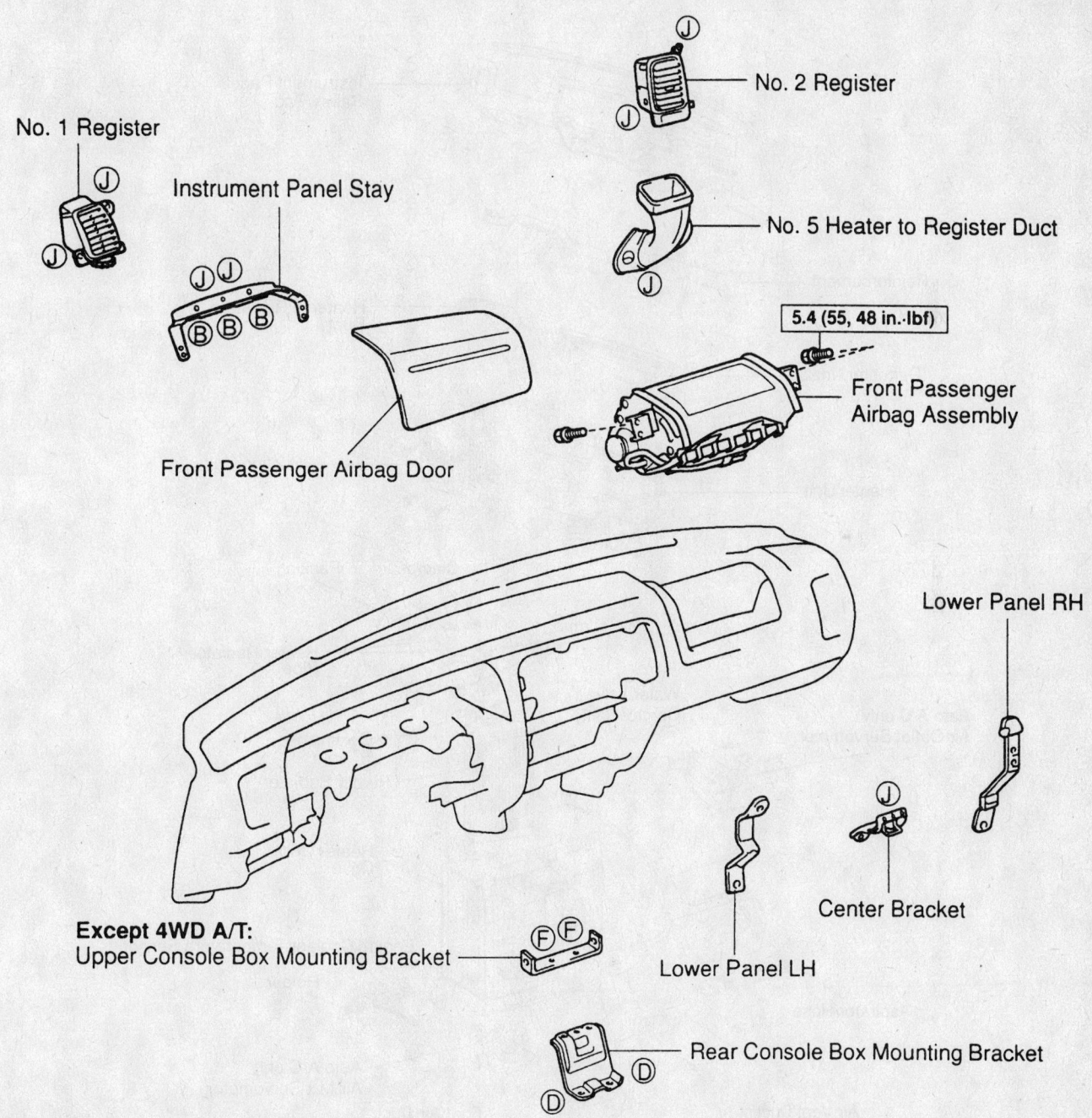

No. 1 Register

Instrument Panel Stay

No. 2 Register

No. 5 Heater to Register Duct

5.4 (55, 48 in.·lbf)

Front Passenger Airbag Assembly

Front Passenger Airbag Door

Lower Panel RH

Center Bracket

Lower Panel LH

Except 4WD A/T:
Upper Console Box Mounting Bracket

Rear Console Box Mounting Bracket

N·m (kgf·cm, ft·lbf) : Specified torque

93113GK1

Exploded view of the instrument panel air bag module, ventilation components and brackets—Toyota 4Runner

2. Drain the cooling system into a clean container for reuse.

3. Disconnect the heater hoses from the heater core.

4. Remove the steering wheel by performing the following procedure:

a. Position the front wheels in the straight-ahead position.

b. At both sides of the steering wheel, remove the side covers.

c. Using a Torx® wrench, loosen the steering wheel screws until the screw's circumference ring catches on the screw case.

d. Carefully, lift the air bag module, disconnect the electrical connector and remove the air bag.

☀ CAUTION

Place the air bag module in a safe location with the front facing upward.

e. Remove the steering wheel nut.

f. Using a steering wheel puller, press the steering wheel from the steering column.

5. Remove the instrument panel and reinforcement by performing the following procedure:

a. Remove both front door scuff plates.

b. Remove both cowl side trims.

c. Remove the 2 hood lock release lever screws and the hood lock release lever.

d. Remove the 2 fuel lid release lever screws and the fuel lid release lever.

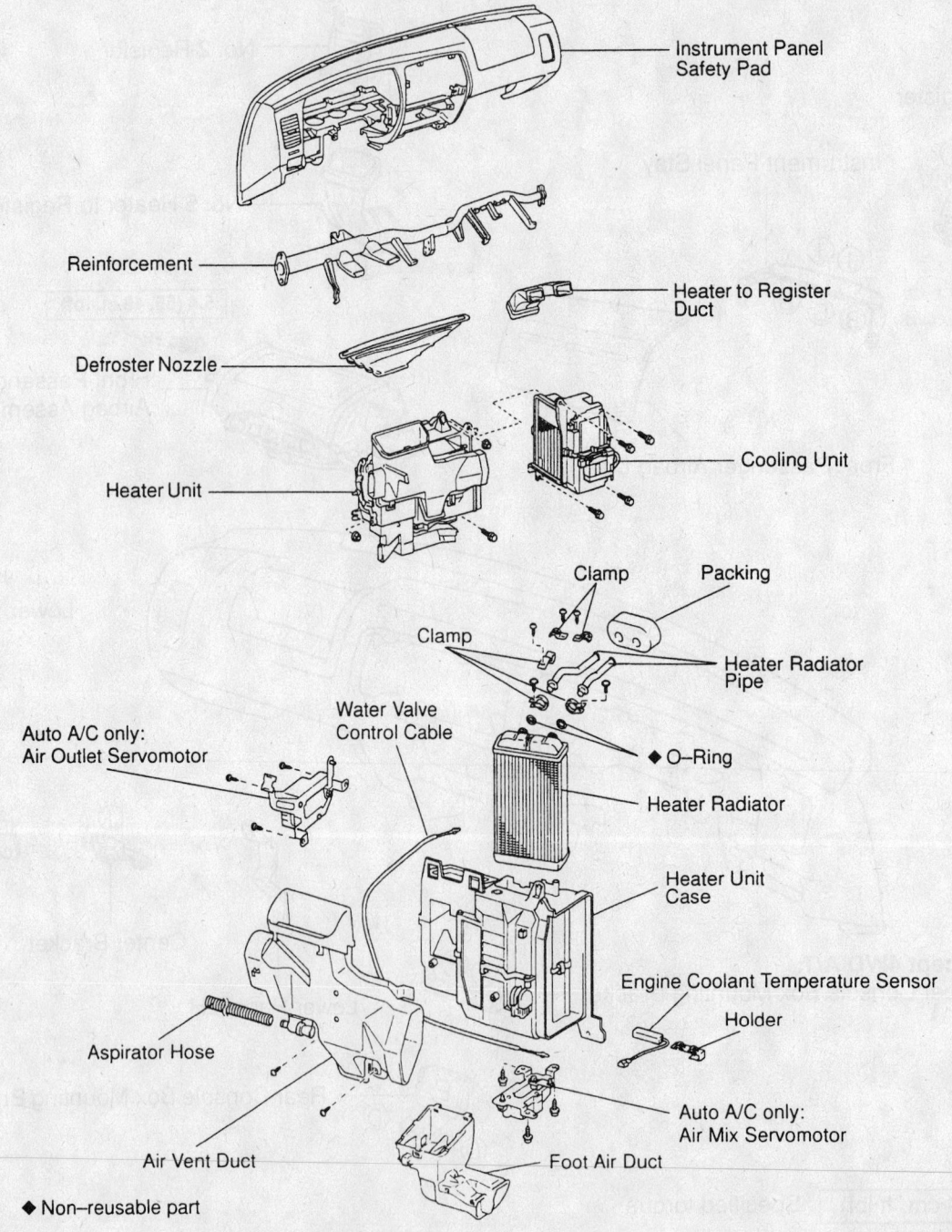

Instrument Panel Safety Pad

Reinforcement

Heater to Register Duct

Defroster Nozzle

Cooling Unit

Heater Unit

Clamp

Packing

Clamp

Heater Radiator Pipe

Auto A/C only:
Air Outlet Servomotor

Water Valve Control Cable

◆ O–Ring

Heater Radiator

Heater Unit Case

Engine Coolant Temperature Sensor

Holder

Aspirator Hose

Auto A/C only:
Air Mix Servomotor

Air Vent Duct

Foot Air Duct

◆ Non–reusable part

93113GK2

Exploded view of the front heater core, heater housing, evaporator housing and related components—Toyota 4Runner

e. Remove the 4 lower finish panel bolts and the panel.

f. Remove the No. 1 and No. 2 heater-to-register duct screw and the ducts.

g. Pry out the starter switch bezel.

h. Remove the steering column cover screws and the covers.

i. Remove the combination switch-to-steering column screws, disconnect the electrical connector and the combination switch.

j. Remove the steering column-to-instrument panel nuts/bolts and the lower steering column bolt; then carefully, remove the steering column.

k. Remove the 4 cluster finish panel screws and the panel.

l. Remove the 4 combination meter screws, disconnect the electrical connectors and remove the combination meter.

m. Pry out the parking brake hole cover.

n. Pry out the upper console panel.

o. Disengage the 7 center cluster finish panel clips and remove the panel.

➡**Remove the center cluster finish panel clips by starting at the bottom and working toward the top.**

p. Remove the heater control knobs.

q. Remove the 2 rear console box bolts/screws and the rear console box.

r. Remove the upper console panel garnish.

s. Remove the 2 glove compartment door screws and the door.

t. Disconnect the passenger's side air bag module electrical connector.

u. Remove the glove box light.

v. Remove the 3 lower No. 2 finish panel bolts and the panel.

w. Remove the 3 glove compartment door reinforcement bolts and the reinforcement.

x. Remove the No. 4 heater-to-register duct.

y. Remove the radio assembly.

z. Remove the side bracket bolt and the bracket.

aa. If equipped with manual air conditioning, remove the heater control assembly.

bb. If equipped with automatic air conditioning, remove the air conditioning control assembly.

cc. Remove the instrument panel-to-chassis nut and 2 bolts; then, remove the instrument panel.

dd. Remove the instrument panel reinforcement-to-chassis nuts/bolts and the reinforcement.

6. Remove the defroster nozzle and heater-to-register duct.

7. Remove the evaporator housing by performing the following procedure:

a. Discharge and recover the air conditioning system refrigerant.

b. Disconnect the refrigerant lines from the evaporator core. Discard the O-rings and plug the openings to prevent contamination.

c. Disconnect the electrical connectors.

d. Remove the 3 evaporator housing-to-chassis screws and the housing.

8. Disconnect the mode control servomotor connector.

9. Disconnect the aspirator hose from the room temperature sensor.

10. Disconnect the heater valve control cable.

11. Remove the heater housing-to-chassis nuts and the heater housing.

12. Remove the 3 heater core-to-heater housing screws, the 2 clips and clamp.

13. Remove the heater core from the heater housing.

To install:

14. Install the 3 heater core-to-heater housing screws, the 2 clips and clamp.

15. Install the heater housing and the heater housing-to-chassis nuts.

16. Connect the heater valve control cable.

17. Connect the aspirator hose to the room temperature sensor.

18. Connect the mode control servomotor connector.

19. Install the defroster nozzle and heater-to-register duct.

20. Install the evaporator housing by performing the following procedure:

a. Install the evaporator housing and the 3 housing-to-chassis screws.

b. Connect the electrical connectors.

c. Using new O-rings, connect the refrigerant lines to the evaporator core.

21. Install the instrument panel and reinforcement by performing the following procedure:

a. Install the instrument panel reinforcement and the reinforcement-to-chassis nuts/bolts.

b. Install the instrument panel and the instrument panel-to-chassis nut and 2 bolts.

c. If equipped with automatic air conditioning, install the air conditioning control assembly.

d. If equipped with manual air conditioning, install the heater control assembly.

e. Install the side bracket and the bracket bolt.

f. Install the radio assembly.

g. Install the No. 4 heater-to-register duct.

h. Install the glove compartment door reinforcement and the 3 reinforcement bolts.

i. Install the lower No. 2 finish panel and the 3 panel bolts.

j. Install the glove box light.

k. Connect the passenger's side air bag module electrical connector.

l. Install the glove compartment door and the 2 door screws.

m. Install the upper console panel garnish.

n. Install the rear console box and the 2 rear console box bolts/screws.

o. Install the heater control knobs.

p. Install the center cluster finish panel and engage the 7 panel clips.

q. Install the upper console panel.

r. Install the parking brake hole cover.

s. Install the combination meter, connect the electrical connectors and install the 4 combination meter screws.

t. Install the cluster finish panel and the 4 panel screws.

u. Install the steering column and torque the steering column-to-instrument panel nuts to 19 ft. lbs. (26 Nm) and the lower steering column bolt to 26 ft. lbs. (35 Nm).

v. Install the combination switch-to-steering column, connect the electrical connector and the combination switch screws.

w. Install the steering column cover and the cover screws.

x. Pry out the starter switch bezel.

y. Install the No. 1 and No. 2 heater-to-register duct and the duct screws.

z. Install the lower finish panel and the 4 panel bolts.

aa. Install the fuel lid release lever and the 2 fuel lid release lever screws.

bb. Install the hood lock release lever and the 2 hood lock release lever screws.

cc. Install both cowl side trims.

dd. Install both front door scuff plates.

22. Install the steering wheel by performing the following procedure:

a. Install the steering wheel to the steering column.

b. Install the steering wheel nut and torque the nut to 25 ft. lbs. (34 Nm).

c. Carefully, install the air bag module and connect the electrical connector.

d. Using a Torx® wrench, torque the steering wheel screws to 78 inch lbs. (8.8 Nm).

e. At both sides of the steering wheel, install the side covers.

23. Connect the heater hoses to the heater core.

24. Refill the cooling system.

25. Connect the negative battery cable.

26. Evacuate and charge the air conditioning system.

27. Run the engine to normal operating temperatures; then, check the climate control operation and check for leaks.

REAR AUXILIARY HEATER

1. Disconnect the negative battery cable.

2. Drain the cooling system into a clean container for reuse.

3. Remove the front seats.

4. Remove the center console box.

5. Move the floor carpet backward.

6. Disconnect the rear heater hoses from the rear heater core.

7. Remove the rear heater duct bolt, screw and duct.

8. Remove the rear heater control assembly.

9. Disconnect the electrical connectors.

10. Remove the 3 rear heater housing-to-chassis screws and the housing.

11. Remove the blower resistor, the rear heater relay and with wiring harness from the rear heater housing.

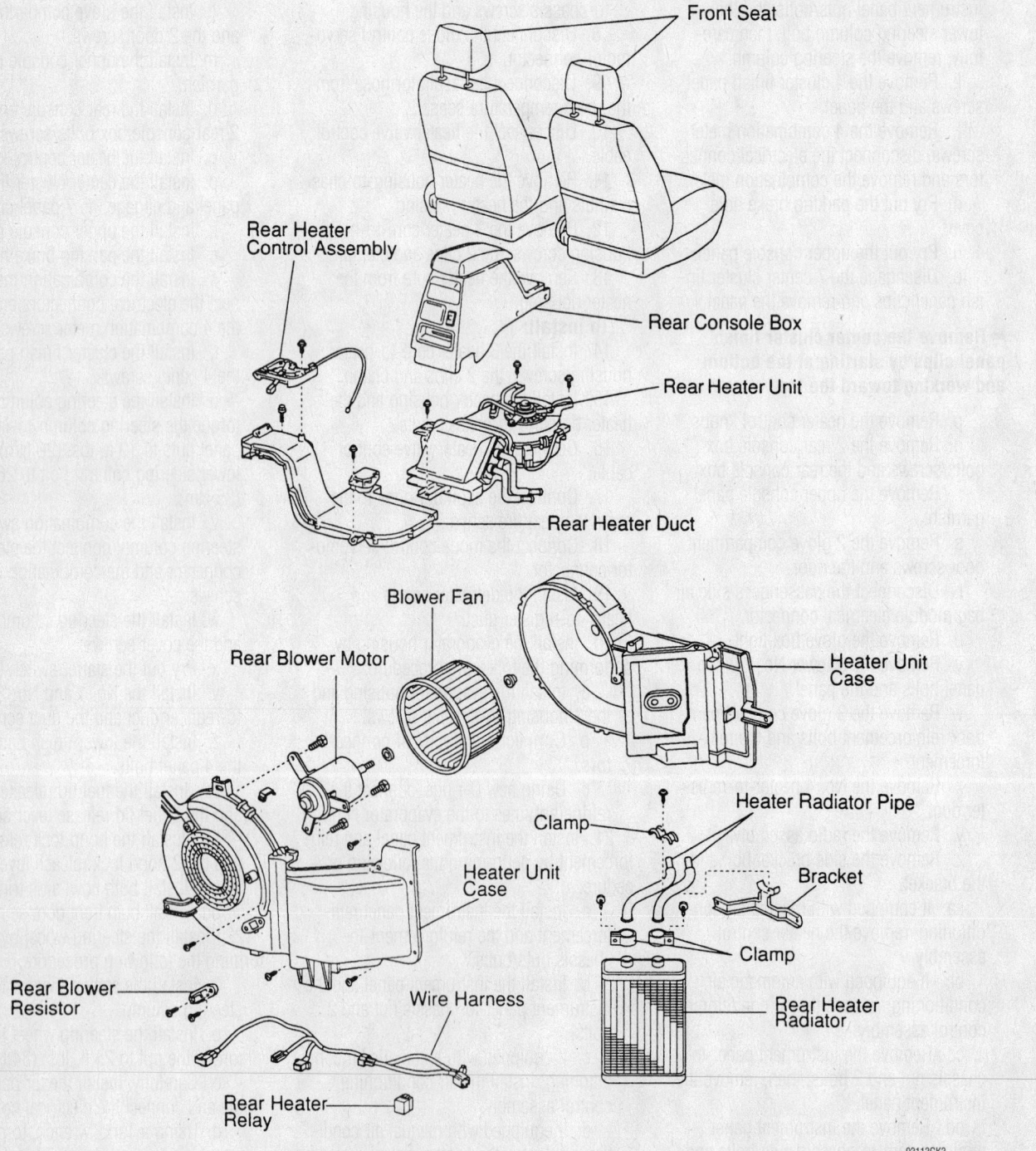

Exploded view of the rear heater housing and heater core—Toyota 4Runner

93113GK3

12. Remove the rear heater housing case screws and separate the cases.

13. Remove the rear heater core.

To install:

14. Install the rear heater core.

15. Assemble the rear heater housing case and install the case screws.

16. Install the blower resistor, the rear heater relay and wiring harness to the rear heater housing.

17. Install the rear heater housing and the 3 housing-to-chassis screws.

18. Connect the electrical connectors.

19. Install the rear heater control assembly.

20. Install the rear heater duct, bolt and screw.

21. Connect the rear heater hoses to the rear heater core.

22. Move the floor carpet foreword.

23. Install the center console box.

24. Install the front seats.

25. Refill the cooling system.

26. Connect the negative battery cable.

27. Run the engine to normal operating temperatures; then, check the climate control operation and check for leaks.

RX 300 and Highlander

FRONT HEATER

1. Disconnect the negative battery cable.

2. Drain the cooling system into a clean container for reuse.

3. Disconnect the heater hoses from the heater core.

4. Remove the steering wheel by performing the following procedure:

    a. Position the front wheels facing straight-ahead.

    b. Remove the steering wheel side covers.

    c. Using a Torx® wrench, loosen the 2 screws located at each side of the steering wheel until the screw's circumference groove catches on the screw case.

    d. Pull the air bag module from the steering wheel and disconnect the electrical connector.

❊❊ CAUTION

Place the air bag module in a safe place with the front side facing upward.

    e. Remove the steering wheel nut.

    f. Place alignment marks on the steering wheel and the main shaft.

    g. Using a steering wheel puller, press the steering wheel from the steering column.

5. Remove the instrument panel and reinforcement by performing the following procedure:

    a. Remove the front door scuff plates.

    b. Remove the cowl side boards.

    c. Remove the front door trim covers.

    d. Remove the front pillar garnish by disengaging the 5 clips. If equipped with a tweeter speaker, disconnect the electrical connector.

    e. Remove the steering column covers-to-steering column screws and the covers.

    f. Remove the combination switch-to-steering column screws, disconnect the electrical connector(s) and remove the combination switch.

    g. Remove the 2 hood open lever screws and the hood open lever.

    h. Remove the 2 lower finish panel bolts and disengage the panel from the 3 clips.

    i. Remove the 2 No. 1 safety pad insert bolts and the insert.

    j. Remove the 2 No. 2 finish panel bolts and disengage the panel from the 4 clips.

    k. In the left side of the glove compartment, pry out the glove box door finish plate and disconnect the air bag module connector.

    l. Remove the glove box 3 nuts and 2 screws and the glove box.

    m. Remove the center cluster finish panel by disengaging the claw (bottom center) and 4 clips (1 at each corner).

    n. Remove the ashtray, the 2 ashtray receptacle box screws.

    o. Remove the 4 lower center cluster finish panel screws and disconnect the connector.

    p. Remove the clock, the No. 1 and No. 2 registers from the panel.

    q. Remove the 3 cluster finish panel screws, disengage the 8 clips and remove the panel.

    r. Remove the combination meter.

    s. Remove the radio assembly.

    t. Remove the heater control assembly.

    u. Remove 2 passenger's side air bag module bolts; then, disconnect and remove the air bag module.

❊❊ CAUTION

Place the air bag module in a safe place with the front side facing upward.

    v. Remove the instrument panel-to-chassis 5 bolts and nut.

    w. Remove the audio amplifier.

    x. Remove the No. 1 and No. 2 braces.

    y. Remove the No. 2 cowl brace.

    z. Remove the instrument panel reinforcement.

6. Remove the evaporator housing by performing the following procedure:

    a. Discharge and recover the air conditioning system refrigerant.

    b. In the engine compartment, remove the refrigerant lines-to-cowl connector bolts; then, disconnect the lines and discard the O-rings.

    c. Disconnect the electrical connector at the evaporator housing.

    d. Disconnect the wiring harness clamp.

    e. Remove the evaporator housing-to-chassis 2 rivets, 3 bolts and nut.

    f. Remove the evaporator housing.

7. Remove the 4 defroster nozzle nuts and the nozzle.

8. Disconnect and remove the theft deterrent and the wireless door lock ECUs.

9. Release the 2 air duct claws and the air duct.

10. Remove the 2 heater housing-to-chassis rivets and the heater housing.

➡**When installing the heater housing, use new screws in place of the rivets.**

11. Remove the heater core-to-heater housing cover.

12. Remove both heater core screws and clamps; then, remove the heater core.

To install:

13. Install the heater core and both heater core screws and clamps.

14. Install the heater core-to-heater housing cover.

➡**When installing the heater housing, use new screws in place of the rivets.**

15. Install the heater housing-to-chassis and the 2 heater housing screws.

16. Release the air duct and the air duct claws.

17. Connect and install the theft deterrent and the wireless door lock ECUs.

18. Install the defroster nozzle and the 4 nozzle nuts.

19. Install the evaporator housing by performing the following procedure:

    a. Install the evaporator housing.

    b. Install the evaporator housing-to-chassis 2 rivets, 3 bolts and nut.

    c. Connect the wiring harness clamp.

    d. Connect the electrical connector at the evaporator housing.

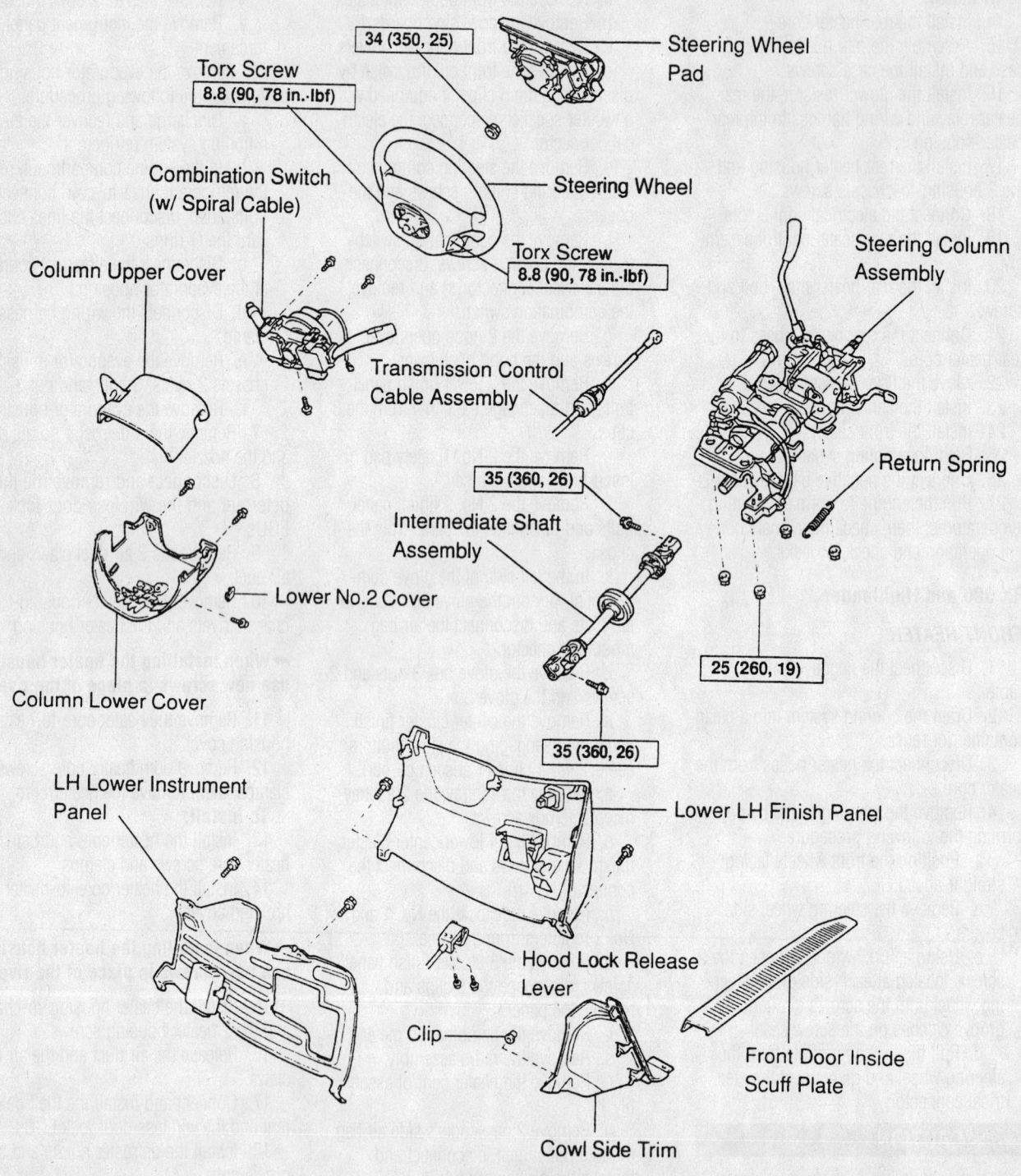

34 (350, 25)

Steering Wheel Pad

Torx Screw
8.8 (90, 78 in.·lbf)

Combination Switch
(w/ Spiral Cable)

Steering Wheel

Column Upper Cover

Torx Screw
8.8 (90, 78 in.·lbf)

Steering Column Assembly

Transmission Control Cable Assembly

Return Spring

35 (360, 26)

Intermediate Shaft Assembly

Lower No.2 Cover

Column Lower Cover

25 (260, 19)

35 (360, 26)

LH Lower Instrument Panel

Lower LH Finish Panel

Hood Lock Release Lever

Clip

Front Door Inside Scuff Plate

Cowl Side Trim

N·m (kgf·cm, ft·lbf) : Specified torque

93113GH3

Exploded view of the steering wheel, steering column and related components—Lexus RX 300

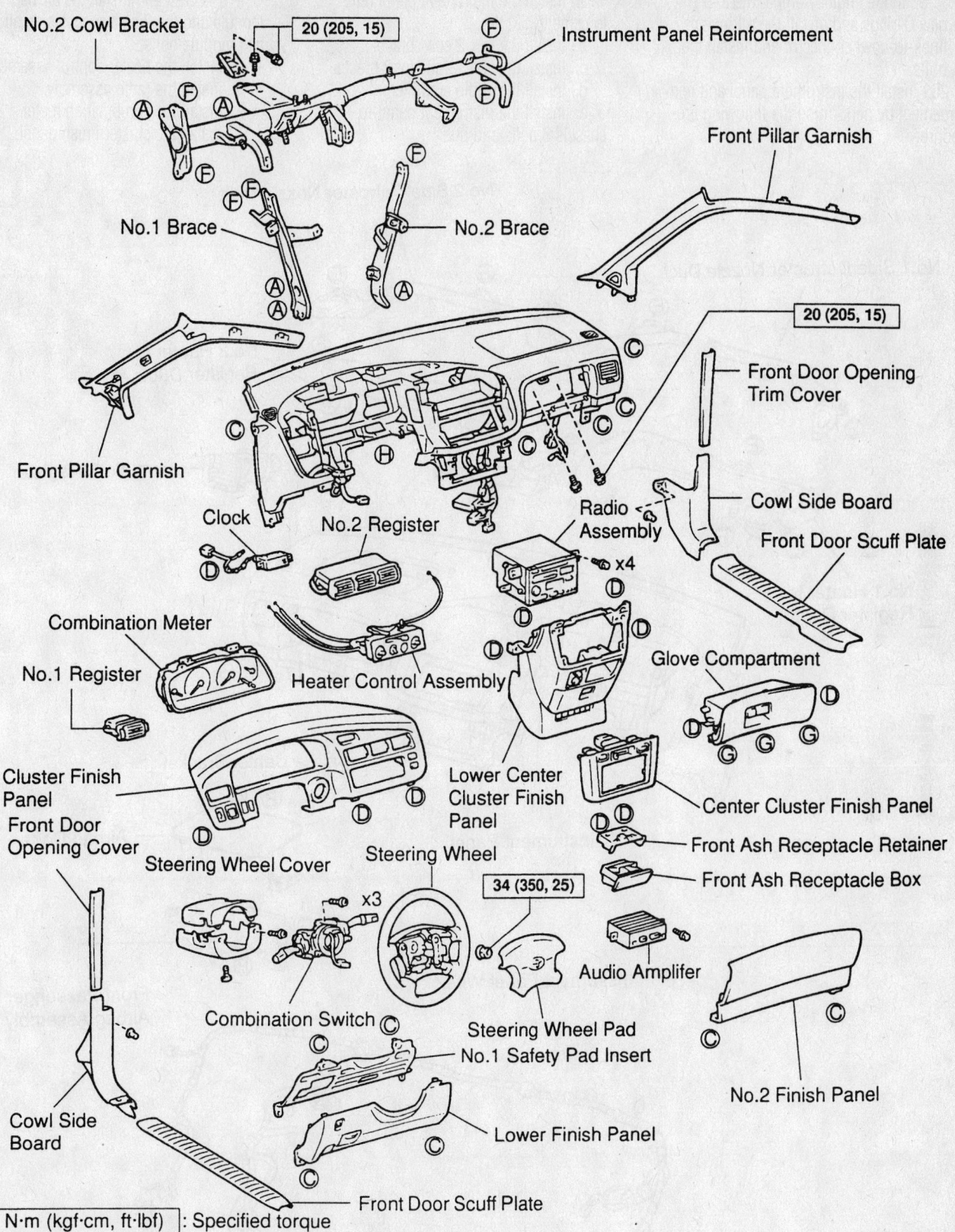

No.2 Cowl Bracket

20 (205, 15)

Instrument Panel Reinforcement

Front Pillar Garnish

No.1 Brace

No.2 Brace

20 (205, 15)

Front Door Opening Trim Cover

Front Pillar Garnish

Cowl Side Board

Front Door Scuff Plate

Clock

No.2 Register

Radio Assembly

x4

Combination Meter

No.1 Register

Heater Control Assembly

Glove Compartment

Cluster Finish Panel

Front Door Opening Cover

Lower Center Cluster Finish Panel

Center Cluster Finish Panel

Front Ash Receptacle Retainer

Front Ash Receptacle Box

Steering Wheel Cover

Steering Wheel

34 (350, 25)

Audio Amplifer

x3

Combination Switch

Steering Wheel Pad

No.1 Safety Pad Insert

No.2 Finish Panel

Cowl Side Board

Lower Finish Panel

Front Door Scuff Plate

N·m (kgf·cm, ft·lbf) : Specified torque

93113GH4

Exploded view of the instrument panel and related components—Lexus RX 300

e. In the engine compartment, use new O-rings and install the refrigerant lines-to-cowl connector and install the bolts.

20. Install the instrument panel and reinforcement by performing the following procedure:

a. Install the instrument panel reinforcement.

b. Install the No. 2 cowl brace.

c. Install the No. 1 and No. 2 braces.

d. Install the audio amplifier.

e. Install the instrument panel-to-chassis 5 bolts and nut.

f. Connect and install the air bag module and the 2 passenger's side air bag module bolts.

g. Install the heater control assembly.

h. Install the radio assembly.

i. Install the combination meter.

j. Install the cluster finish panel,

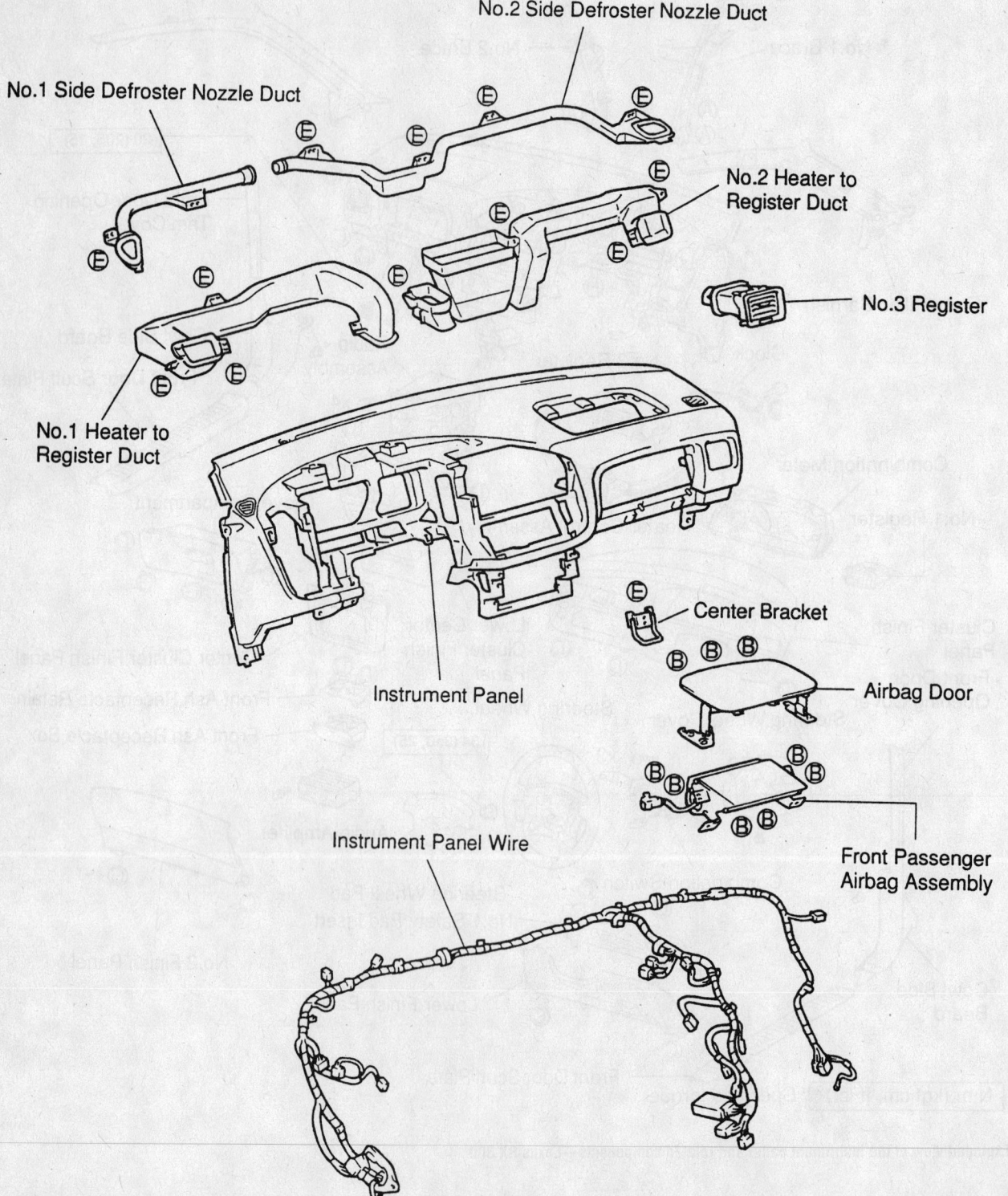

Exploded view of the ventilation system and related components—Lexus RX 300

93113GH5

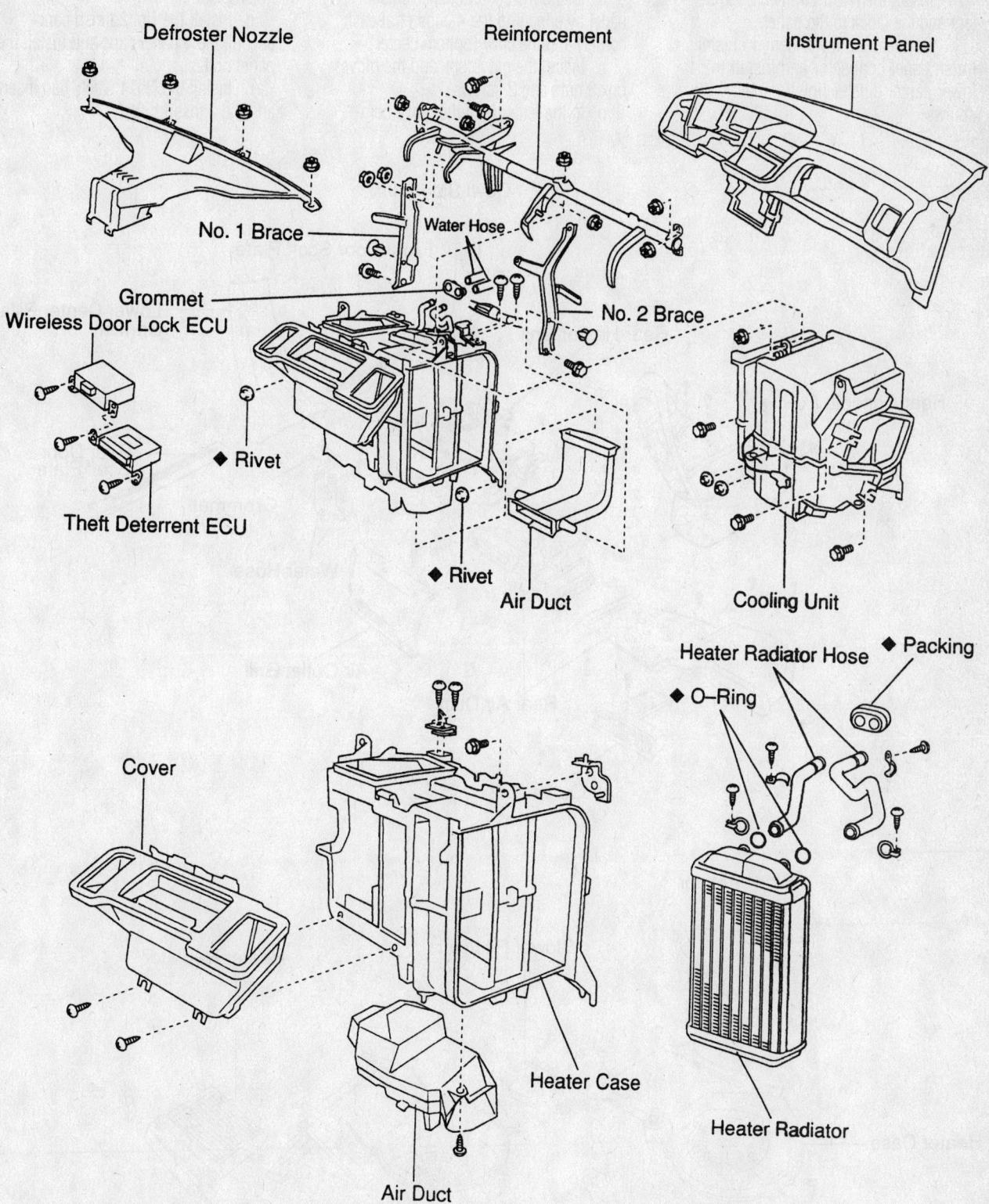

Defroster Nozzle

Reinforcement

Instrument Panel

No. 1 Brace

Water Hose

Grommet

Wireless Door Lock ECU

No. 2 Brace

◆ Rivet

Theft Deterrent ECU

◆ Rivet

Air Duct

Cooling Unit

Heater Radiator Hose ◆ Packing

◆ O–Ring

Cover

Heater Case

Heater Radiator

Air Duct

◆ Non–reusable part

93113GH6

Exploded view of the heater core, heater housing, evaporator housing and related components—Lexus RX 300

engage the 8 clips and install the panel screws.

k. Install the No. 1 and No. 2 registers and the clock to the panel.

l. Connect the lower center cluster finish panel connector and install the 4 lower center cluster finish panel screws.

m. Install the 2 ashtray receptacle box screws and the ashtray.

n. Install the center cluster finish panel by engaging the 4 clips (1 at each corner) and the claw (bottom center).

o. Install the glove box and the glove box 3 nuts and 2 screws.

p. In the left side of the glove com-

partment, connect the air bag module connector and install the glove box door finish plate.

q. Install the No. 2 finish panel, engage the 4 panel clips and install the 3 panel bolts.

r. Install the No. 1 safety pad insert and the 2 insert bolts.

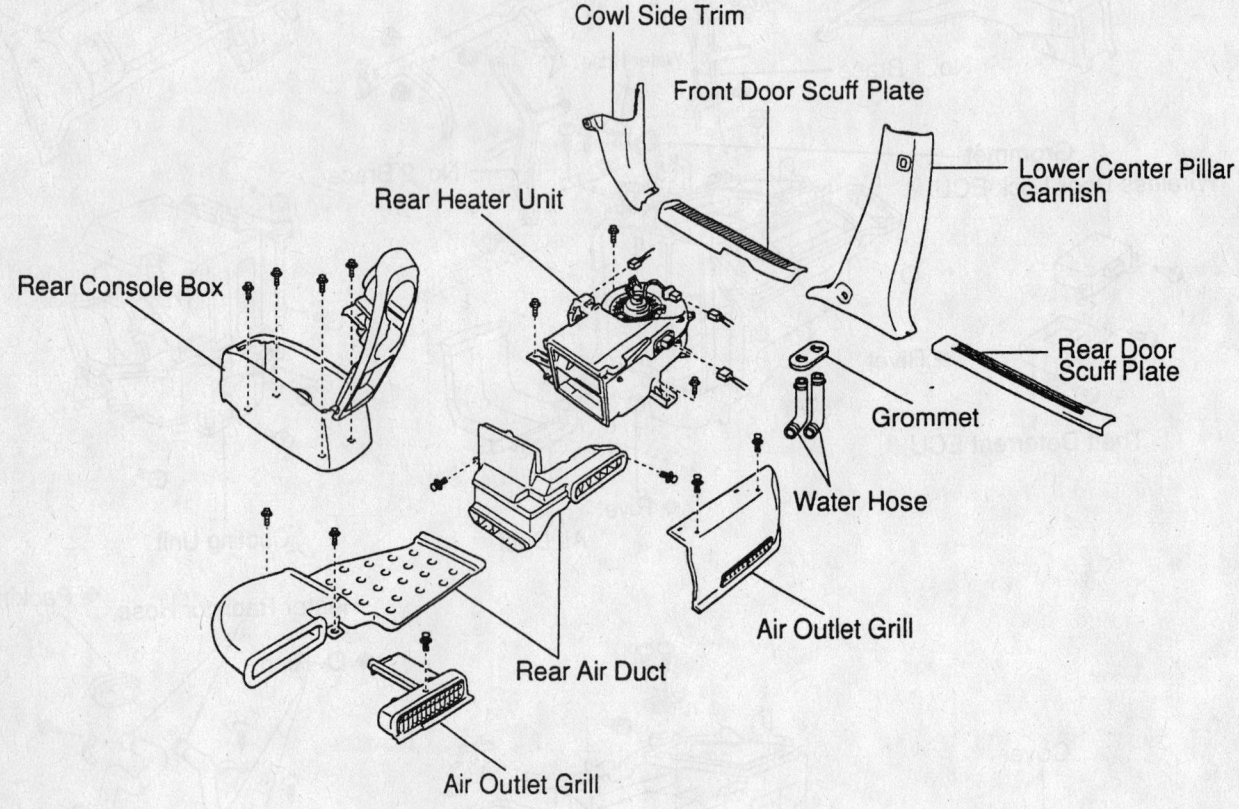

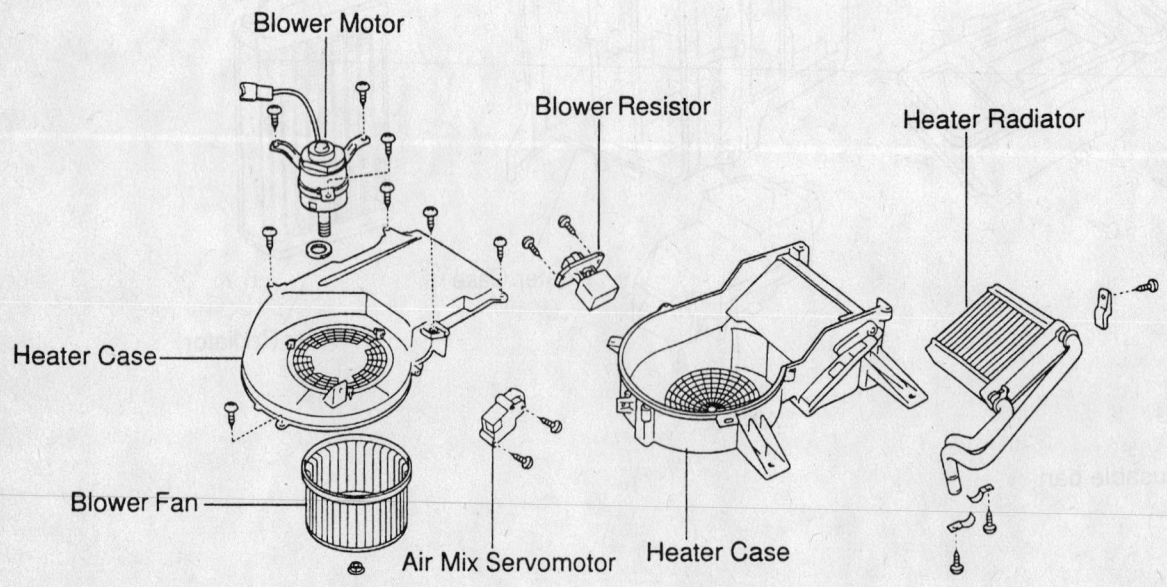

Exploded view of the rear heater core, the rear heater housing and related components—Lexus RX 300

93113GH7

s. Install the finish panel, engage the 3 finish panel clips and install 2 lower finish panel bolts.

t. Install the hood open lever and the 2 hood open lever screws.

u. Install the combination switch, connect the electrical connector(s) and install the combination switch-to-steering column screws.

v. Install the steering column covers and the covers-to-steering column screws.

w. Install the front pillar garnish by engaging the 5 clips. If equipped with a tweeter speaker, connect the electrical connector.

x. Install the front door trim covers.

y. Install the cowl side boards.

z. Install the front door scuff plates.

21. Install the steering wheel by performing the following procedure:

a. Install the steering wheel to the steering column.

b. Align the steering wheel-to-main shaft marks.

c. Install the steering wheel nut and torque the nut to 25 ft. lbs. (34 Nm).

d. Install the air bag module to the steering wheel and connect the electrical connector.

e. Using a Torx® wrench, tighten the steering wheel screws to 78 inch lbs. (8.8 Nm).

f. Install the steering wheel side covers.

22. Connect the heater hoses to the heater core.

23. Refill the cooling system.

24. Connect the negative battery cable.

25. Evacuate and charge the air conditioning system.

26. Run the engine to normal operating temperatures; then, check the climate control operation and check for leaks.

REAR AUXILIARY HEATER

1. Disconnect the negative battery cable.

2. Drain the cooling system into a clean container for reuse.

3. Disconnect the heater hoses from the rear heater core.

4. Remove the front seats.

5. Remove the front door scuff plates.

6. Remove the cowl side trim.

7. Remove the rear door scuff plates.

8. Remove the lower door scuff plates.

9. Remove the rear console box.

10. Remove the left side air outlet grille.

11. Pull the carpet rearward.

12. Remove the 3 clips and the air outlet grille.

13. Remove the rear air duct 2 bolts, 2 clips and the duct.

14. Disconnect the electrical connectors.

15. Remove the 3 rear heater housing bolts and the housing.

16. Remove both heater core-to-heater housing screws and clamps.

17. Remove the heater core-to-heater housing screw and plate.

18. Remove the heater core.

To install:

19. Install the heater core.

20. Install the heater core-to-heater housing screw and plate.

21. Install both heater core-to-heater housing screws and clamps.

22. Install the rear heater housing and the 3 housing bolts.

23. Connect the electrical connectors.

24. Install the rear air duct and the duct 2 bolts and 2 clips.

25. Install the 3 clips and the air outlet grille.

26. Move the carpet forward.

27. Install the left side air outlet grille.

28. Install the rear console box.

29. Install the lower door scuff plates.

30. Install the rear door scuff plates.

31. Install the cowl side trim.

32. Install the front door scuff plates.

33. Install the front seats.

34. Connect the heater hoses to the rear heater core.

35. Refill the cooling system.

36. Connect the negative battery cable.

Water Pump

REMOVAL & INSTALLATION

RAV4

1. Before servicing the vehicle, refer to the precautions in the beginning of this section.

2. Remove or disconnect the following:
- Negative battery cable
- Right-hand engine undercover

3. Drain the engine coolant from the radiator and engine.

4. Remove or disconnect the following:
- Timing belt
- Lower radiator hose from the water inlet
- Timing belt tension spring and the No. 2 idler pulley
- Crankshaft Position (CKP) sensor connector clamp
- Alternator drive belt adjusting bar
- 2 water pump-to-water bypass pipe nuts

- 3 water pump bolts in the sequence
- Water pump cover from the water bypass pipe
- Water pump and water pump cover assembly
- Gasket and 2 O-rings from the water pump and water bypass pipe
- 3 bolts, water pump and gasket, from the water pump cover

To install:

5. Install or connect the following:
- Water pump to the water pump cover, using a new gasket. Tighten the 3 bolts to 78 inch lbs. (9 Nm).
- New O-ring and gasket to the water pump cover
- New O-ring to the water bypass pipe, by applying soapy water to the O-ring
- Water pump cover to the water bypass pipe; do not install the nuts at this time
- Water pump. Tighten the 3 bolts, in sequence, to 78 inch lbs. (9 Nm).
- Water pump cover to the water pump pipe. Tighten the 2 bolts to 82 inch lbs. (9 Nm).
- Alternator drive belt adjusting bar. Tighten the bolt to 20 ft. lbs. (27 Nm).
- CKP connector clamp
- No. 2 idler pulley and timing belt tension spring
- Lower radiator hose
- Timing belt
- Negative battery cable

6. Fill the engine and radiator with engine coolant.

7. Start the engine and check for leaks.

8. Install the right-hand engine undercover.

2.4L Highlander

1. Before servicing the vehicle, refer to the precautions in the beginning of this section.

2. Disconnect the negative battery cable.

3. Drain the engine coolant.

4. Remove or disconnect the following:
- Alternator
- Water pump pulley
- Water pump

5. Installation is the reverse of removal. Torque the pump bolts and nuts to 80 inch lbs. (9Nm) and the pulley bolts to 19 ft. lbs. (26Nm).

3.0L Highlander

1. Before servicing the vehicle, refer to the precautions in the beginning of this section.

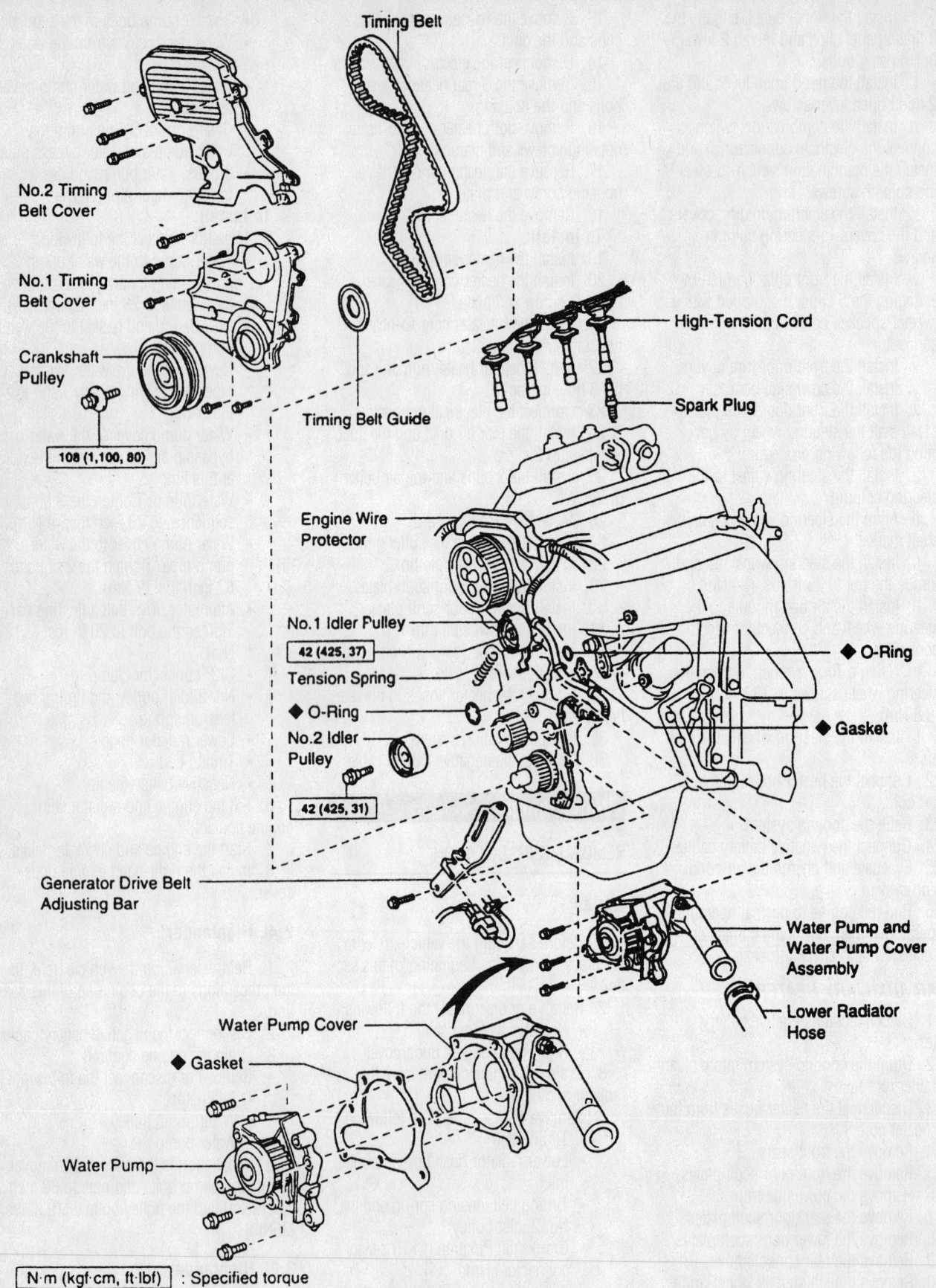

No.2 Timing Belt Cover

No.1 Timing Belt Cover

Crankshaft Pulley

108 (1,100, 80)

Timing Belt

Timing Belt Guide

High-Tension Cord

Spark Plug

Engine Wire Protector

No.1 Idler Pulley

42 (425, 37)

Tension Spring

◆ O-Ring

No.2 Idler Pulley

42 (425, 31)

◆ O-Ring

◆ Gasket

Generator Drive Belt Adjusting Bar

Water Pump and Water Pump Cover Assembly

Lower Radiator Hose

Water Pump Cover

◆ Gasket

Water Pump

N·m (kgf·cm, ft·lbf) : Specified torque

◆ Non-reusable part

7924ZG10

Exploded view of the water pump and related components—RAV4

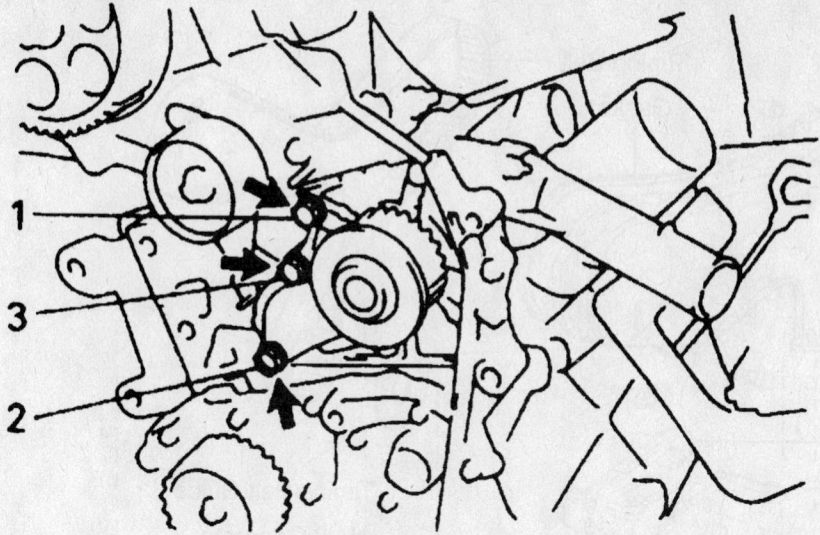

Loosening sequence for the water pump bolts—RAV4

7924ZG11

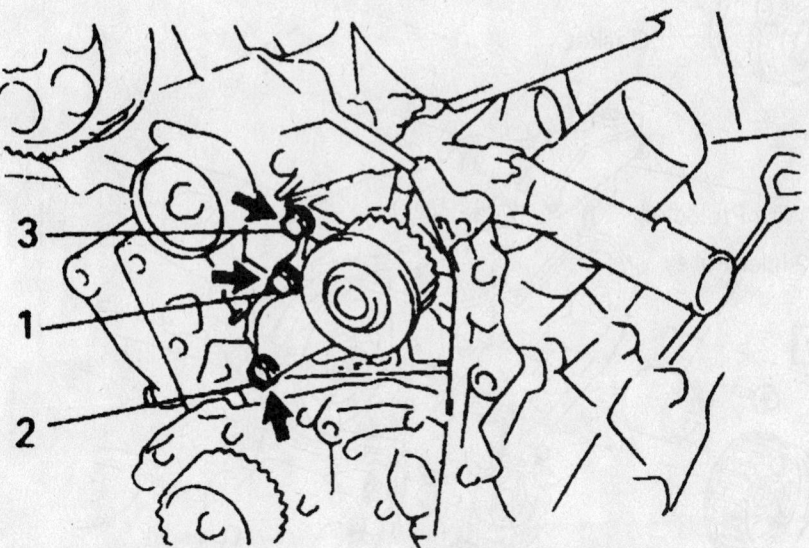

Tightening sequence for the water pump bolts—RAV4

7924ZG12

2. Disconnect the negative battery cable.
3. Drain the engine coolant.
4. Remove or disconnect the following:
- Right front wheel
- Right fender apron seal
- Accessory drive belts
- Upper engine mount and stay
- Alternator and bracket
- Crankshaft pulley
- Timing belt covers
- Transverse engine mounting bracket
- Timing belt
- Timing belt idler
- Camshaft pulley
- Water pump

5. Installation is the reverse of removal.

Torque the water pump bolts and nuts to 71 inch lbs. (8Nm).

RX 300

1. Before servicing the vehicle, refer to the precautions in the beginning of this section.
2. Disconnect the negative battery cable.
3. Drain the engine coolant.
4. Remove or disconnect the following:
- Wiper and blade assembly
- Top cowl seal and panel
- Window washer hoses, from the ventilator louvers
- Left and right ventilator louvers
- Heater air duct
- Front upper suspension brace
- Timing belt

5. Mark the left and right camshaft pulleys with a touch of paint.
6. Remove or disconnect the following:
- Right and left camshaft pulleys bolts
- Pulleys from the engine

➡**Be sure not to mix up the pulleys.**

- No. 2 idler pulley by removing the bolt
- 3 clamps and engine wire from the rear timing belt cover
- 6 No. 3 timing belt cover-to-engine bolts
- Water pump nuts/bolts
- Water pump and gasket from the engine

To install:

7. Check that the water pump turns smoothly. Also check the air hole for coolant leakage.
8. Apply liquid sealer to the gasket, water pump and engine block.
9. Install or connect the following:
- Water pump, using a new gasket. Tighten the nuts/bolts to 53 inch lbs. (6 Nm).
- Rear timing belt cover. Tighten the 6 bolts to 74 inch lbs. (9 Nm).
- Engine wire with the 3 clamps to the rear timing belt cover
- No. 2 idler pulley. Tighten the bolt to 32 ft. lbs. (43 Nm).

➡**After tightening the bolt, be sure the idler pulley moves smoothly.**

- Right-hand camshaft pulley, with the flange side **outward**.

➡**Be sure to align the knock pin hole on the camshaft pulley with the knock pin on the camshaft.**

- Tighten the camshaft bolt to 65 ft. lbs. (88 Nm), using the removal tools
- Left-hand camshaft pulley, with the flange side **inward**.

➡**Be sure to align the knock pin hole on the camshaft pulley with the knock pin on the camshaft.**

- Tighten the camshaft bolt to 94 ft. lbs. (125 Nm), using the removal tools
- Timing belt
- Front upper suspension brace, for RX 300. Tighten the nuts to 59 ft. lbs. (80 Nm).

10. Fill the engine coolant.
11. Install or connect the following:

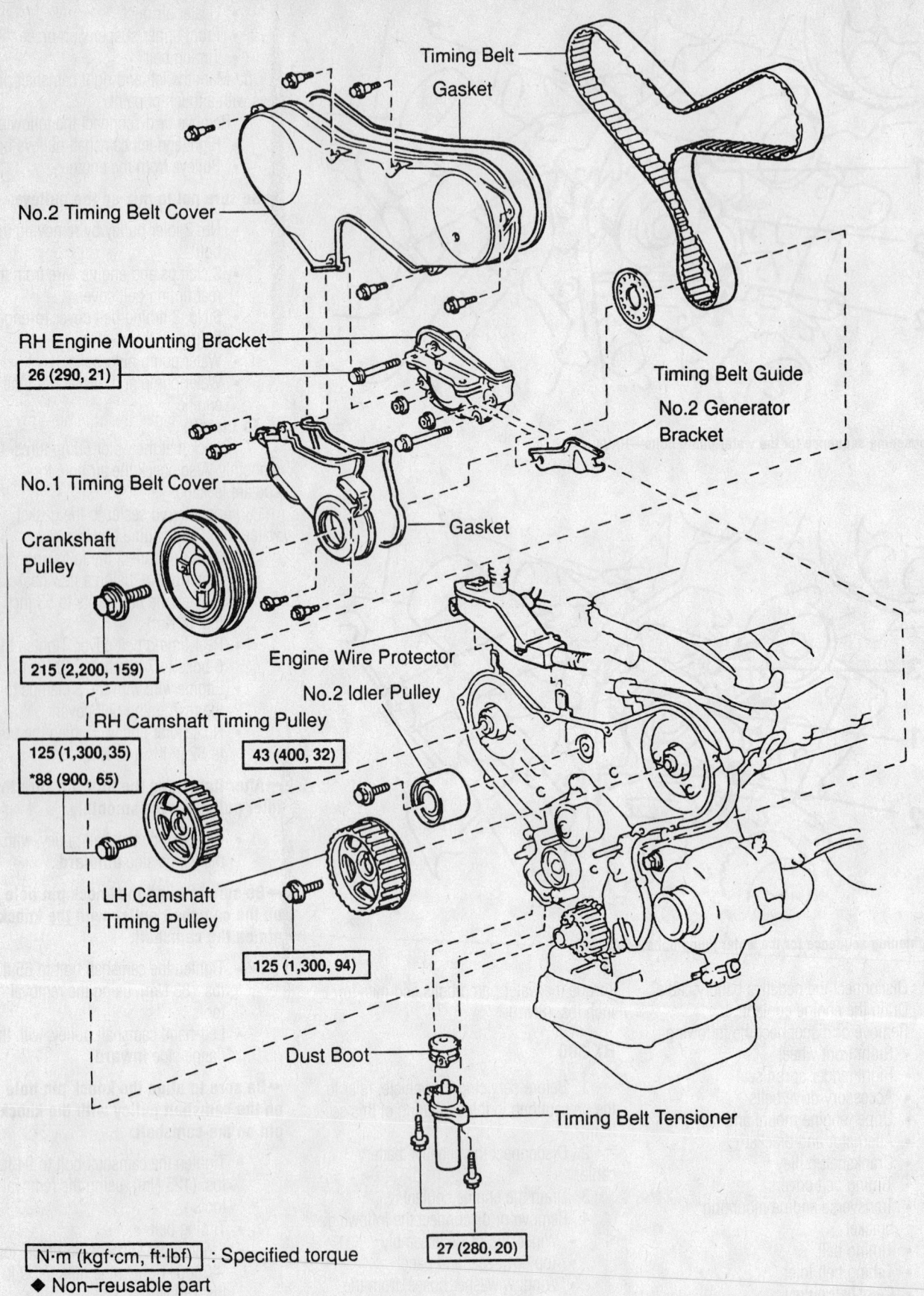

Timing Belt

Gasket

No.2 Timing Belt Cover

Timing Belt Guide

No.2 Generator Bracket

RH Engine Mounting Bracket

26 (290, 21)

No.1 Timing Belt Cover

Gasket

Crankshaft Pulley

Engine Wire Protector

215 (2,200, 159)

No.2 Idler Pulley

RH Camshaft Timing Pulley

125 (1,300, 35)
*88 (900, 65)

43 (400, 32)

LH Camshaft Timing Pulley

125 (1,300, 94)

Dust Boot

Timing Belt Tensioner

N·m (kgf·cm, ft·lbf) : Specified torque

27 (280, 20)

◆ Non–reusable part
*For use with SST

7924ZG15

Exploded view of the components to gain access to the water pump—RX 300

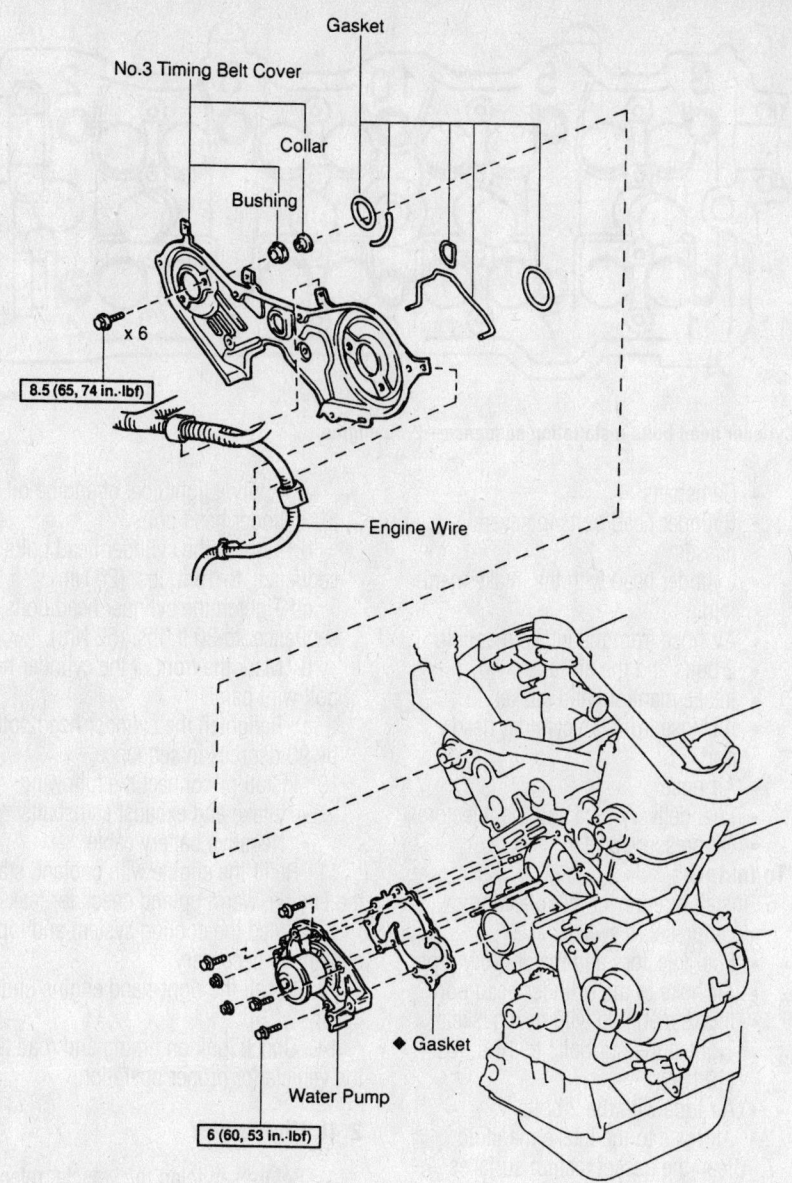

Exploded view of the water pump and related components—RX 300

No.3 Timing Belt Cover

Gasket

Collar

Bushing

Engine Wire

8.5 (65, 74 in.-lbf)

x 6

◆ Gasket

Water Pump

6 (60, 53 in.-lbf)

N·m (kgf·cm, ft·lbf) : Specified torque
◆ Non–reusable part

7924ZG16

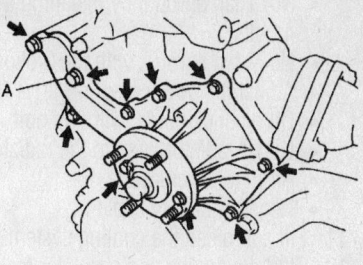

7924YG07

Water pump mounting bolt locations— 2.7L 4Runner

- Heater air duct
- Left and right ventilator louvers
- Window washer hoses to the ventilator louvers
- Top cowl seal and panel
- Wiper and blade assembly
- Negative battery cable

12. Start the engine.

13. Top off the engine coolant and check for leaks.

2.7L 4Runner

1. Before servicing the vehicle, refer to the precautions in the beginning of this section.

2. Remove or disconnect the following:
- Negative battery cable
- Engine undercover

3. Drain the cooling system.

4. Remove or disconnect the following:
- 2 bolts and the air pipe, for the California vehicles
- Upper radiator hose from the radiator
- Oil dipstick guide, by removing the bolt
- Power steering drive belt, by loosening the lockbolt and adjusting bolt to the idler pulley, if equipped with power steering

- No. 2 fan shroud, by removing the 2 clips
- No. 1 fan shroud, by removing the 4 bolts
- Loosen the idler pulley nut and adjusting bolt and remove the air conditioning drive belt, if equipped with air conditioning

5. Remove the alternator drive belt, fan (with fan clutch), water pump pulley and the fan shroud, as follows:
- Stretch the belt and loosen the water pump pulley mounting nuts
- Loosen the lock, pivot and the adjusting bolts for the alternator
- Alternator drive belt
- 4 water pump pulley mounting nuts
- Fan (with fan clutch) and the water pump pulley

6. Remove the water pump and discard the gasket.

To install:

7. Clean all gasket mounting surfaces.

8. Install or connect the following:
- Apply a thin layer of liquid sealant to a new gasket
- Place the gasket and water pump into position. Tighten the 14mm head bolts **A** to 18 ft. lbs. (25 Nm) and the 12mm head bolts to 78 inch lbs. (9 Nm).

9. Install the water pump pulley, fan shroud, fan (with fan clutch) and the alternator drive belt, as follows:
- Fan (with the fan clutch), water pump pulley and the fan shroud in position
- Water pump pulley mounting nuts but do not tighten the nuts at this time
- Alternator drive belt
- Stretch the alternator belt tight. Tighten the fan nuts to 16 ft. lbs. (21 Nm).
- Adjust the alternator drive belt

10. Install or connect the following:
- Adjust the drive belt, if equipped with air conditioning

- No. 1 fan shroud, by installing the 4 bolts
- No. 2 fan shroud, with the 2 clips
- Adjust the power steering drive belt
- Oil dipstick guide, with the bolt
- Upper radiator hose to the radiator
- Air pipe, If removed
- Negative battery cable

11. Fill and bleed the cooling system.
12. Start the engine and check for leaks.
13. Install the engine undercover.

3.4L 4Runner

1. Before servicing the vehicle, refer to the precautions in the beginning of this section.
2. Disconnect the negative battery cable.
3. Drain the cooling system.
4. Remove or disconnect the following:
- Timing belt
- Thermostat
- No. 2 oil cooler hose from the water pump
- Water pump
5. Thoroughly clean the mating surfaces.

To install:

6. Apply sealant (PN 08826-00100) to the water pump.

※※ WARNING

Parts must be assembled within 5 minutes of application. Otherwise the material must be removed and reapplied.

7. Install or connect the following:
- Water pump. Tighten the bolts to 14 ft. lbs. (20 Nm).
- No. 2 oil cooler hose
- Thermostat
- Timing belt
- Negative battery cable
8. Fill the cooling system.
9. Start the engine and check for leaks.

Cylinder Head

REMOVAL & INSTALLATION

Rav4

1. Before servicing the vehicle, refer to the precautions in the beginning of this section.
2. Release the fuel system pressure.
3. Remove or disconnect the following:
- Negative battery cable
- Right-hand engine undercover
4. Drain the engine coolant.
5. Remove or disconnect the following:

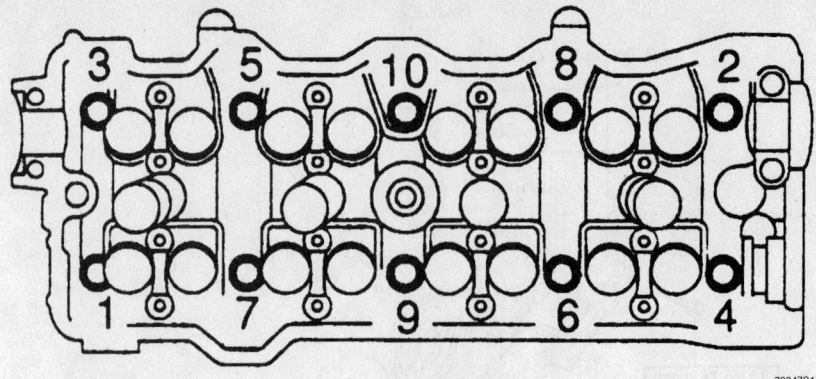

Cylinder head bolts installation sequence—2.0L engine

- Camshafts
- Cylinder head bolts in several passes
- Cylinder head with the intake manifold
- Air hose from the intake manifold
- 2 bolts and the air tube
- Intake manifold and gasket
- Air hose from the cylinder head port
- Air hose
- Fuel delivery pipe and the injectors
- Oil pressure switch

To install:

6. Install or connect the following:
- Oil pressure switch
- Fuel injectors and the delivery pipe
- Air hose to the cylinder head port
- Intake manifold with new gaskets. Tighten the nut/bolts to 14 ft. lbs. (19 Nm).
- Air tube with the 2 bolts
- Air hose to the intake manifold
7. Clean the gasket mating surfaces using care not to damage the aluminum components, replace the gasket; then, lower the cylinder head onto the engine. Be sure the dowel pins are aligned and no hoses or wires are between the head and cylinder block.
8. For 2000–01, tighten the cylinder head bolts as follows:
   a. Apply a light coat of engine oil to the cylinder head bolts.
   b. Tighten the cylinder head bolts, in several passes, in sequence, to 36 ft. lbs. (49 Nm).
   c. Mark the front of the cylinder head bolt with paint.
   d. Retighten the cylinder head bolts by 90 degrees in sequence.
   e. Retighten an additional 90 degrees and be sure that the paint mark is now positioned toward the rear.
9. For 2002, tighten the cylinder head bolts in 3 progressive steps, as follows:

   a. Apply a light coat of engine oil to the cylinder head bolts.
   b. Tighten the cylinder head bolts, in sequence, to 15 ft. lbs. (26 Nm).
   c. Tighten the cylinder head bolts, in sequence, to 30 ft. lbs. (52 Nm).
   d. Mark the front of the cylinder head bolt with paint.
   e. Retighten the cylinder head bolts by 90 degrees in sequence.
10. Install or connect the following:
- Intake and exhaust camshafts
- Negative battery cable
11. Refill the engine with coolant, start the engine, warm up and check for leaks.
12. Bleed the cooling system and top off coolant as necessary.
13. Install the right-hand engine undercover.
14. Check ignition timing and road test the vehicle for proper operation.

2.4L Highlander

1. Before servicing the vehicle, refer to the precautions in the beginning of this section.
2. Remove or disconnect the following:
- Front center suspension brace
- Timing chain
- Coolant
- Transfer case oil
- Radiator hoses
- Power steering hoses
- Heater hoses
- Fuel rail lines
- Camshaft timing oil control valve
- Front driveshaft
- Rear engine mount insulator (4wd)
- Transverse engine mount bracket (4wd)
- Intake manifold
- All wires and cables connected to the head
- Exhaust manifold
- Camshafts

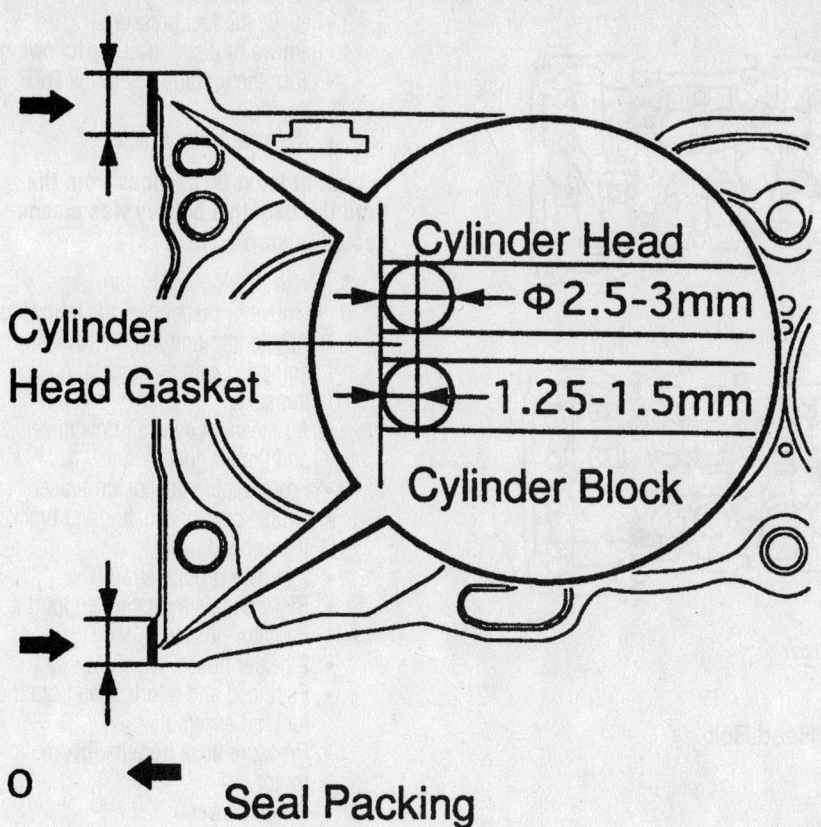

Apply a bead of RTV sealant as shown—2.4L engine cylinder head

9355ZG01

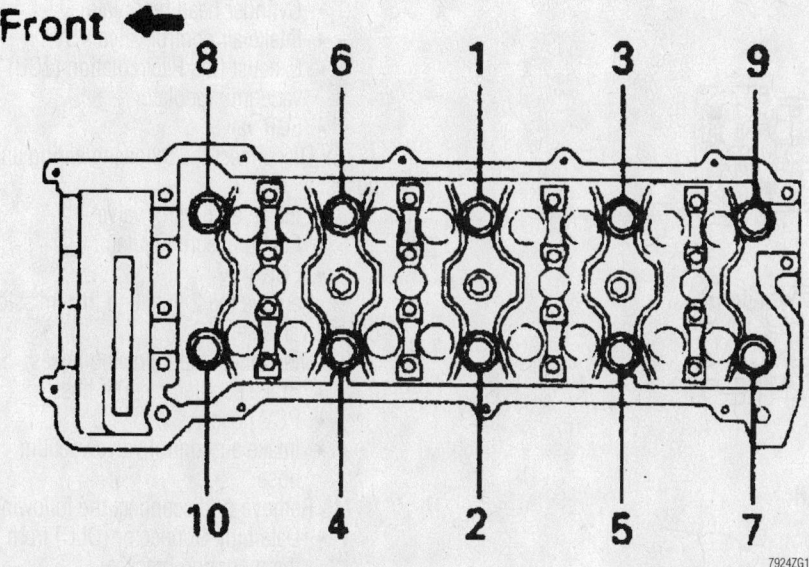

Cylinder head bolt tightening sequence—2.4L engines

7924ZG18

3. Loosen the 10 head bolts evenly, a little at a time in several passes and lift off the head. Check the head bolt length. Any bolt longer than 6.465 in. (164.2mm) should be replaced.

4. Installation is the reverse of removal. Install the head gasket with the lot number

stamp upward. Apply a bead of RTV sealer as shown. The head must be installed within 3 minutes of applying the sealer, and the head bolts must be tightened with 15 minutes. The head bolts must be tightened in sequence, in several passes, to 58 ft. lbs. (79Nm), then, an additional 90 degree turn each.

3.0L Highlander

1. Before servicing the vehicle, refer to the precautions in the beginning of this section.

2. Remove or disconnect the following:
- Coolant
- Engine oil
- Exhaust pipes
- Exhaust manifold
- Camshaft cover
- Upper center front suspension brace
- Air cleaner
- Intake air surge tank
- Fuel rail
- Heater hoses
- Intake manifold
- Radiator hose
- Water outlet
- Right front wheel
- Right fender apron seal
- Accessory drive belts
- Engine roll stopper rod
- Right engine mount
- Alternator and brackets
- Crankshaft pulley
- Timing belt covers
- Transverse engine mounting bracket
- Timing belt
- Timing belt tensioner
- Power steering pump
- Ignition coil pack
- Camshafts
- The hexagonal bolt, using an 8mm hex wrench, then the 8 head bolts
- Cylinder head

3. Installation is the reverse of removal. Measure the head bolts. The minimum diameter of the stretch portion of each bolt should be at least 8.75mm (0.3775 in.). Replace any bolt that does not measure up. The head gasket is installed with the **R** mark upwards. Install the 8 head bolts first, tightened in sequence, in 2 equal steps, to 40 ft. lbs. (54Nm). Then tighten each an additional 90 degrees. Finally, install the hex bolt, torqued to 14 ft. lbs. (19 Nm).

RX 300

1. Before servicing the vehicle, refer to the precautions in the beginning of this section.

2. Remove or disconnect the following:
- Wiper and blade assembly
- Top cowl seal and panel
- Window washer hoses from the ventilator louvers
- Left and right ventilator louvers
- Heater air duct

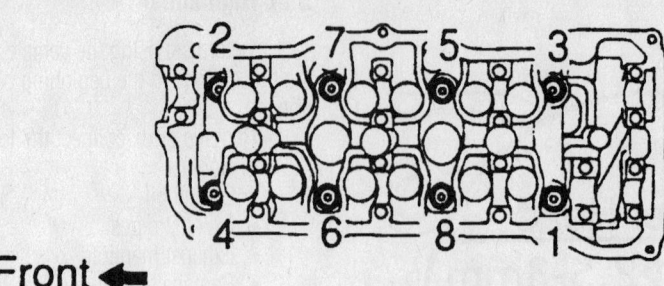

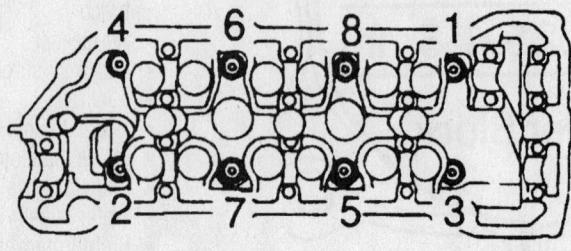

Cylinder head bolt loosening sequence—3.0L Highlander

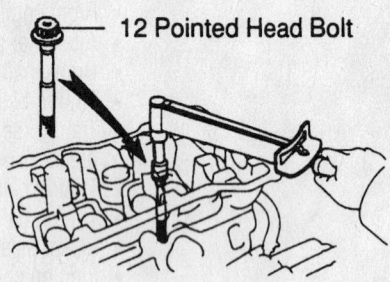

12 Pointed Head Bolt

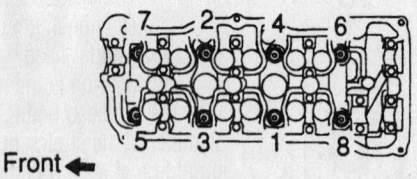

Front ←

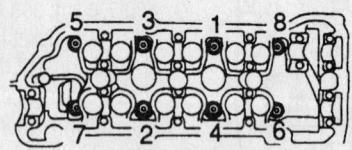

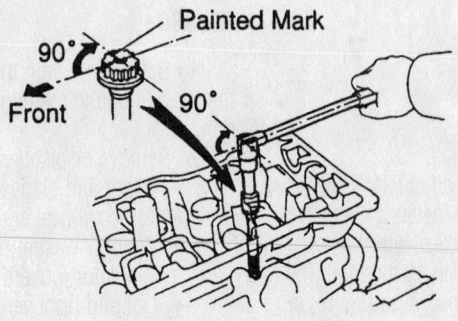

Painted Mark

90°

Front 90°

Cylinder head bolt tightening sequence—3.0L Highlander

3. Relieve the fuel pressure.
4. Remove or disconnect the following:
   • Turn the ignition key to the **OFF** position
   • Negative battery cable

➡ **Wait at least 90 seconds from the time the negative battery was disconnected to start work.**

5. Drain the cooling system.
6. Remove or disconnect the following:
   • Accelerator and throttle cables, if equipped with an automatic transaxle
   • Air cleaner cover, air flow meter and the air duct
   • Front upper suspension brace
   • Cruise control actuator and bracket, if equipped
   • 2 engine ground straps
   • Right engine mounting support
   • Radiator hoses
   • 2 heater hoses
   • Fuel feed and return lines from the fuel rail assembly
   • Pressure hose from the hydraulic motor
   • V-bank cover
7. Disconnect the following vacuum hoses:
   • Fuel pressure control Vacuum Switching Valve (VSV)
   • Fuel pressure regulator
   • Cylinder head rear plate
   • Intake air control valve VSV
   • Exhaust Gas Recirculation (EGR) vacuum modulator
   • EGR valve
8. Disconnect the following wiring and hoses:
   • Intake air control valve
   • Fuel pressure regulator
   • EGR VSV
9. Remove the 2 nuts and the emission control valve set.
10. Disconnect the following hoses;
    • Brake booster vacuum hose
    • PCV hose
    • Intake air control valve vacuum hose
11. Remove or disconnect the following:
    • Data Link Connector (DLC) from the mounting bracket
    • 2 ground straps from the intake chamber
    • Hydraulic motor pressure hose from the intake chamber
    • Right Oxygen (O₂) sensor connector from the power steering pressure tube
    • 2 nuts and the power steering pressure tube from the intake chamber

- Both power steering air hoses
- Engine hanger and the intake chamber support
- EGR pipe and gaskets

12. Disconnect the following wiring:
- Throttle Position (TP) sensor connector
- Idle Air Control (IAC) valve connector
- EGR gas temperature connector
- Air conditioning idle up connector

13. Disconnect the following vacuum hoses:
- 2 vacuum hoses from the Thermal Vacuum Valve (TVV)
- Vacuum hose from the cylinder head rear plate
- Vacuum hose from the charcoal canister

14. Remove or disconnect the following:
- Air assist hose and the 2 water bypass hoses
- Air intake chamber
- Left engine wiring harness and move it aside
- Wiring harness from the rear of the engine
- Right engine wiring harness and move it aside
- Ignition coils and move them aside
- Timing belt
- Camshaft pulleys and the timing belt rear cover
- Cylinder head rear plate
- Water inlet pipe
- Air assist hose and vacuum hose
- Intake manifold and fuel rail assembly
- Water outlet
- EGR pipe from the right exhaust manifold
- Front exhaust pipe and exhaust manifolds
- Dipstick assembly and the power steering pump bracket
- Valve covers and the Camshaft Position (CMP) sensor
- Camshafts

15. Be sure the engine is at/or near ambient temperature and remove the 2 (1 on each head) 8mm recessed hex bolts. Loosen and remove the 8 head bolts evenly, in 3 passes, in the reverse order of the installation sequence. Carefully lift the head from the engine; if necessary to pry the head loose, take great care not to damage the mating surfaces. Place the head on wood blocks in a clean work area.

➡**If the cylinder head bolts are loosened out of sequence, warpage or cracking could result.**

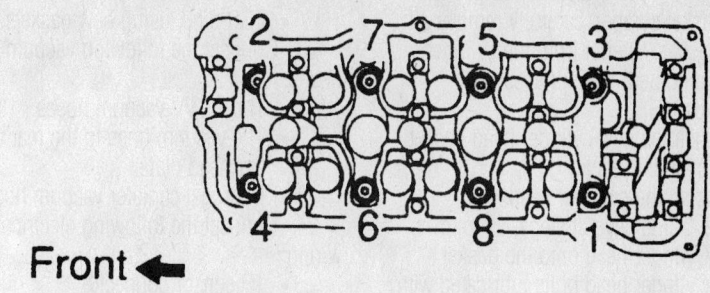

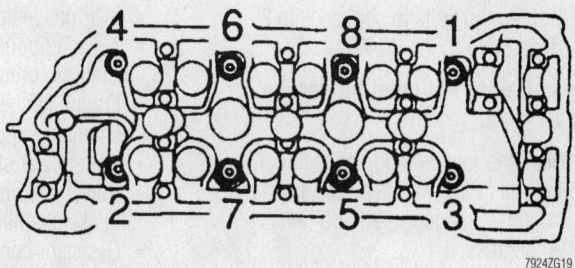

Cylinder head bolt loosening sequence—RX 300

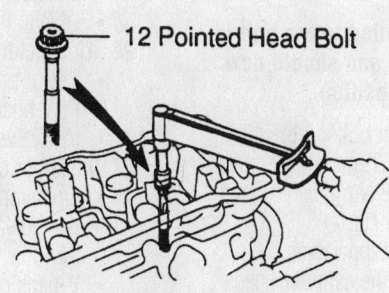

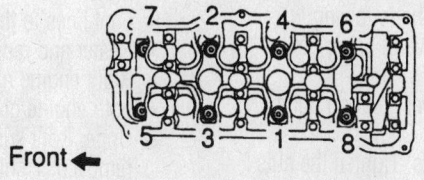

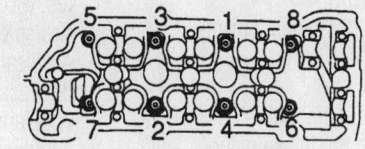

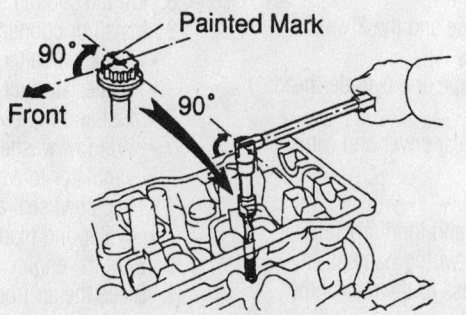

Cylinder head bolt tightening sequence—3.0L (1MZ-FE) engine

16. Remove the cylinder head gasket. With a gasket scraper, carefully remove all the old gasket material from the cylinder head and engine block surfaces.

To install:

17. Place the new cylinder head gasket onto the cylinder block.

18. Install the cylinder head, in sequence, using several steps, as follows:
- Cylinder head onto the gasket
- Cylinder head bolts lubricated with clean engine oil
- Tighten the bolts in sequence in 3 steps to 40 ft. lbs. (54 Nm).

➡️**If any bolt does not meet the torque, replace it.**

- Mark the forward edge of each bolt with paint, then tighten each bolt, in proper sequence, an additional 90 degrees.
- Check that each painted mark is now at a 90 degrees angle to the front

➡️**The paint mark applied to the bolt in the 9 o'clock position and should now be in the 12 o'clock position.**

- Remaining 8mm bolts, lubricated with engine oil. Tighten both bolts to 13 ft. lbs. (18 Nm).

19. Install the camshafts.

20. Check and adjust the valves.

21. Apply sealant to the cylinder heads where the camshaft supports meet the cylinder heads.

22. Install or connect the following:
- Cylinder head covers, using new gaskets
- Dipstick and power steering pump bracket
- Exhaust manifolds. Tighten the nuts to 36 ft. lbs. (49 Nm).
- EGR pipe to the right exhaust manifold
- Water outlet
- Intake manifold and the fuel rail assembly. Tighten the intake manifold nuts/bolts to 11 ft. lbs. (15 Nm).
- Air assist hose and the 2 water bypass hoses
- Water inlet pipe and cylinder head rear plate
- Timing belt rear cover and camshaft pulleys
- Timing belt
- Spark plugs and ignition coils
- Right engine wiring harness
- Wiring harness to the rear of the engine
- Left engine wiring harness

- Air intake chamber
- EGR pipe, using new gaskets

23. Connect the following vacuum hoses:
- The 2 TVV vacuum hoses
- The vacuum hose to the rear cylinder head plate
- Charcoal canister vacuum hose

24. Connect the following electrical wiring:
- TP sensor connector
- IAC valve connector
- EGR gas temperature connector
- Air conditioning idle up connector

25. Install or connect the following:
- Engine hanger and the intake chamber support
- Both power steering air hoses
- Power steering pressure tube to the intake chamber
- O_2 sensor connector to the pressure tube.
- Both ground straps, to the intake chamber
- DLC to the bracket

26. Connect the following hoses:
- Power brake booster vacuum hose
- PCV hose
- IAC valve vacuum hose

27. Install or connect the following:
- Emission control valve set and related vacuum hoses and connectors
- V-bank cover
- Pressure hose to the hydraulic motor
- Fuel lines to the fuel rail assembly
- Heater and radiator hoses
- Right engine mounting support
- both engine ground straps
- Upper front suspension brace, if removed. Tighten the nuts to 59 ft. lbs. (80 Nm).
- Cruise control actuator and bracket
- Air cleaner, air flow meter and air duct assembly
- Accelerator and throttle cables, if equipped with an automatic transaxle

28. Fill the cooling system.

29. Install or connect the following:
- Negative battery cable
- Heater air duct
- Left and right ventilator louvers
- Window washer hoses from the ventilator louvers
- Top cowl seal and panel
- Wiper and blade assembly

30. Start the engine and check for leaks.

31. Bleed the air from the cooling system.

32. Road test the vehicle and check for

unusual noise, shock, slippage, correct shift points and smooth operation.

33. Recheck the coolant and engine oil levels.

2.7L 4Runner

1. Before servicing the vehicle, refer to the precautions in the beginning of this section.

2. Release the fuel system pressure.

3. Disconnect the negative battery cable.

4. Drain the engine coolant.

5. Remove or disconnect the following:
- Air cleaner cap
- Mass Air Flow (MAF) meter and resonator
- Accelerator cable from the throttle body, if equipped with a manual transmission
- Accelerator and throttle cables from the throttle body, if equipped with an automatic transmission
- Cruise control cable from the actuator, if equipped with cruise control
- Intake air connector
- Air hose for Idle Air Control (IAC)
- Vacuum sensing hose
- Wire clamp for the engine wiring harness
- Oil dipstick guide
- Power steering belt
- Power steering pulley, pump and bracket
- Positive Crankcase Ventilation (PCV) hoses
- Distributor
- Spark plug wires from the spark plugs
- Engine wiring harness
- Air conditioning compressor, if equipped with air conditioning
- Oil pressure sensor
- Engine Coolant Temperature (ECT) sensor connector
- ECT sender gauge connector
- Exhaust Gas Recirculation (EGR) gas temperature sensor connector
- Vacuum Switching Valve (VSV) connector
- 2 vacuum hose from the VSV
- Ground strap from the cowl top panel
- Engine wiring harness from the air intake chamber
- Throttle Position (TP) sensor connector
- IAC valve connector
- Crankshaft Position (CKP) sensor connector
- Knock Sensor (KS) connector

- Data Link Connector 1 (DLC1) from the bracket
- Engine wiring harness clamp
- EGR pipe
- Intake chamber stay
- Air intake chamber assembly

6. Disconnect the following hoses:
- Evaporative Emissions (EVAP) hose from the throttle body
- Brake booster vacuum hose from the union
- Water bypass hose from the water bypass pipe
- Water bypass hose from the cylinder head rear cover

7. Remove or disconnect the following:
- Injector connectors
- Fuel inlet pipe
- Hoses and the fuel return pipe
- Delivery pipe and injectors
- Intake manifold
- Front exhaust pipe
- Exhaust manifold and gasket
- Water outlet
- Cylinder head rear cover
- Spark plugs
- Front engine hanger
- Engine wiring harness brackets
- Cylinder head cover

8. Set No. 1 cylinder to Top Dead Center (TDC) of the compression stroke. The groove on the crankshaft pulley should align with the **0** mark on the timing chain cover and the timing marks (1 and 2 dots) of the camshaft gears should form a straight line in respect to the cylinder head surface. If not, turn the crankshaft 1 revolution (360 degrees).

9. Remove or disconnect the following:
- Chain tensioner and gasket
- Camshaft timing gear
- Exhaust camshafts

10. Remove the intake camshaft, as follows:

a. Uniformly, loosen and remove the bearing cap bolts in the reverse order of the tightening in several passes, in sequence.

b. Remove the bearing caps and camshaft. Make a note of the bearing cap positions for proper installation.

➡**If the camshaft is not being lifted out straight and level, reinstall the No. 3 bearing cap with the 2 bolts. Then, alternately loosen and remove the 2 bearing cap bolts with the camshaft gear pulled up.**

11. Remove or disconnect the following:
- Valve lifters and shims

➡**Arrange the valve lifters and shims in correct order.**

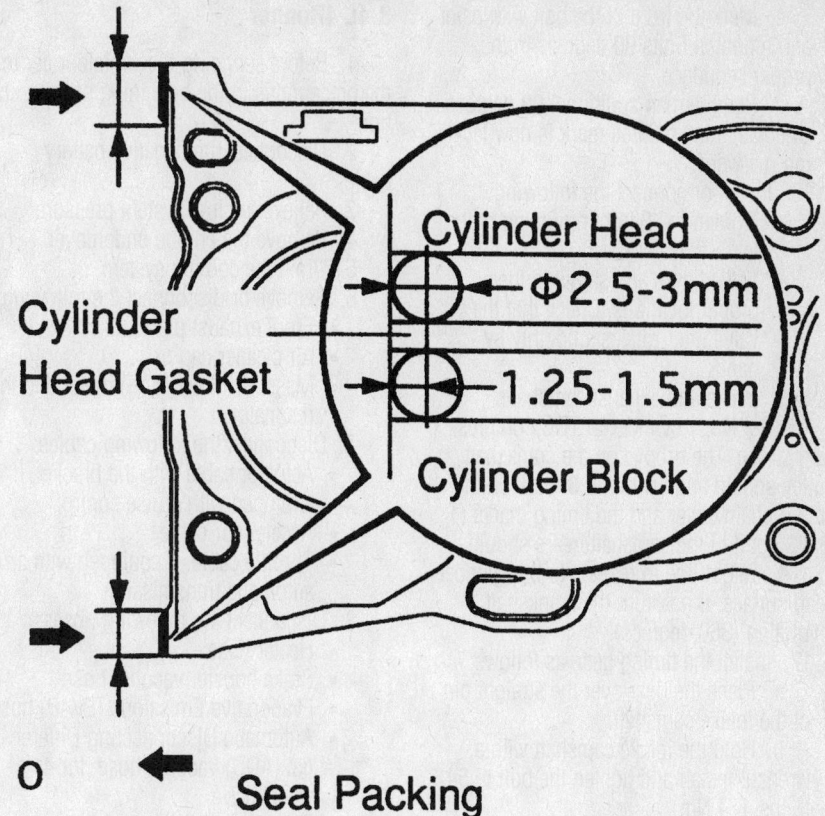

Apply a bead of RTV sealant as shown—2.7L engine cylinder head

- Cylinder head, by uniformly loosen and remove the cylinder head bolts in the reverse order of the tightening, in sequence, using several passes

To install:

12. Before installing, thoroughly clean the gasket mating surfaces and check for warpage.

13. Apply sealant (PN 08826-00080) as shown. Place a new head gasket on the block and install the cylinder head.

14. Install the cylinder head as follows:

a. Lightly coat the cylinder head bolts with engine oil.

b. Install the bolts and tighten, in several passes, in the sequence. Tighten all bolts to 29 ft. lbs. (39 Nm).

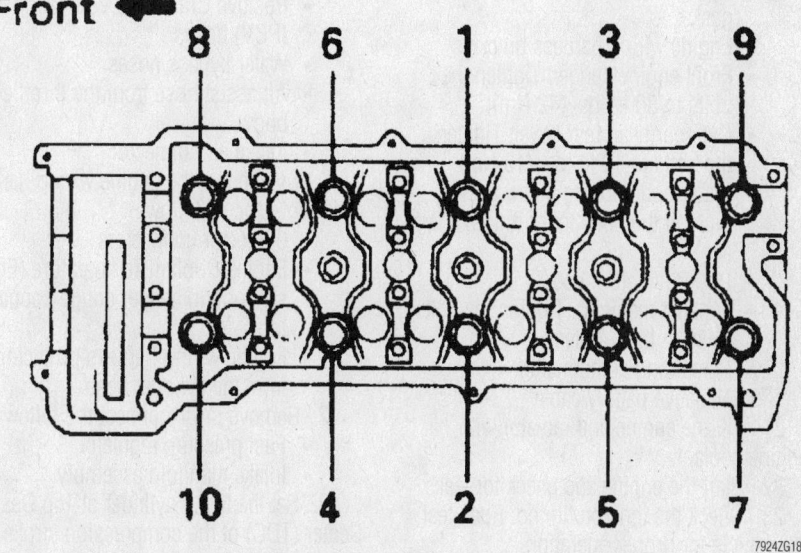

Cylinder head bolt tightening sequence—2.7L engines

c. Mark the front of the bolt with paint and retighten bolts 90 degrees in the proper sequence.

d. Retighten an additional 90 degrees. Check that the painted mark is now facing rearward.

15. Install or connect the following:
- Tighten the 2 front mounting bolts to 15 ft. lbs. (21 Nm).
- Valve lifters and shims in their proper locations. Check that the valve lifter rotates smoothly by hand.
- Intake and exhaust camshafts

16. Set No. 1 cylinder to TDC compression stroke. The groove on the crankshaft pulley should align with the **0** mark on the timing chain cover and the timing marks (1 and 2 dots) of the camshaft gears should form a straight line in respect to the cylinder head surface. If not, turn the crankshaft 1 revolution (360 degrees).

17. Install the timing gear, as follows:
a. Place the gear over the straight pin of the intake camshaft.

b. Hold the intake camshaft with a wrench. Install and tighten the bolt to 54 ft. lbs. (74 Nm).

c. Hold the exhaust camshaft and install the bolt and distributor gear. Tighten the bolt to 34 ft. lbs. (46 Nm).

18. Install or connect the following:
- Chain tensioner, using a new gasket (mark toward the front)
- Recheck the valve timing
- Check and adjust the valve clearance
- Spark plugs
- Semi-circular plug

19. Recheck the engine for proper valve timing.

20. Install or connect the following:
- Cylinder head cover, using a new gasket
- Engine wiring harness brackets
- Front engine hanger. Tighten the bolts to 30 ft. lbs. (42 Nm).
- Cylinder head rear cover. Tighten the bolts to 10 ft. lbs. (13 Nm).
- Water outlet, using a new gasket. Tighten the bolts to 14 ft. lbs. (20 Nm).
- Upper radiator hose
- Exhaust manifold. Tighten the bolts to 36 ft. lbs. (49 Nm).
- Remaining components
- Negative battery cable

21. Fill the engine and radiator with engine coolant.

22. Start the engine and check for leaks.

23. Check the ignition timing. Road test the vehicle for proper operation.

24. Recheck all fluid levels.

3.4L 4Runner

1. Before servicing the vehicle, refer to the precautions in the beginning of this section.

2. Disconnect the negative battery cable.

3. Relieve the fuel system pressure.

4. Remove the engine undercover.

5. Drain the cooling system.

6. Remove or disconnect the following:
- Front exhaust pipe
- Air cleaner cap
- Mass Air Flow (MAF) meter and the resonator

7. Disconnect the following cables:
- Actuator cable with the bracket, if equipped with cruise control
- Accelerator cable
- Throttle cable, if equipped with an automatic transmission

8. Disconnect the following hoses:
- Heater hose
- Brake booster vacuum hose
- Evaporative Emissions (EVAP) hose
- Automatic Disconnecting Differential (ADD) vacuum hose, for 4-wheel drive
- Fuel inlet and fuel return hose

9. Remove or disconnect the following:
- Spark plug wires with the ignition coils
- Spark plugs
- Intake chamber stay
- No. 2 timing belt cover
- Air intake chamber assembly

10. Remove the following connectors and hoses:
- Throttle Position (TP) sensor connector
- Idle Air Control (IAC) valve connector
- Positive Crankcase Ventilation (PCV) hoses
- Water bypass hoses
- Air assist hose from the throttle body
- Intake air connector

11. Disconnect the engine wiring harness protector, as follows:
- 6 injector connectors
- Engine Coolant Temperature (ECT) sensor and sender gauge connectors
- Engine wiring harness protector from the cylinder head

12. Remove or disconnect the following:
- Fuel pressure regulator
- Intake manifold assembly

13. Set the No. 1 cylinder at Top Dead Center (TDC) of the compression stroke, as follows:

a. Turn the crankshaft pulley and align its groove with the timing mark **0** of the No. 1 timing belt cover.

b. Check that the timing marks of the camshaft timing pulleys and the No. 3 timing belt cover are aligned. If not, turn the crankshaft pulley 1 revolution (360 degrees).

14. Remove or disconnect the following:
- Timing belt tensioner, by alternately loosening the 2 bolts
- Timing belt

15. Remove the camshaft timing pulleys, as follows:

a. Using Variable Pin Wrench Set 09960-10010, remove the pulley bolt, the timing pulley and the knock pin.

b. Remove the 2 timing pulleys with the timing belt.

16. Remove or disconnect the following:
- Bolt and the No. 2 idler pulley
- Camshaft Position (CMP) sensor
- No. 3 timing belt cover
- Alternator from the engine
- Alternator bracket
- Power steering pump and move it aside without disconnecting the pump lines
- Exhaust crossover pipe and gaskets, by removing the 6 nuts
- Left-hand exhaust manifold, by removing the heat insulator and 6 nuts
- Right-hand exhaust manifold, by removing the heat insulator and 6 nuts
- 8 bolts, seal washers, cylinder head cover and gasket.

➡ **Remove both cylinder head covers**

- Semi-circular plugs
- Right exhaust and intake camshafts
- Left exhaust and intake camshafts
- Valve lifters and shims from the cylinder head; arrange the valve lifters and shims in correct order

17. Remove the cylinder heads, as follows:
- Bolt and ground strap
- Cylinder head (recessed head) bolt on the cylinder head, using an 8mm hexagon wrench; then, repeat the procedure for the other side.
- 8 cylinder head (12-pointed head) bolts, on each cylinder head.

➡ **Loosen the bolts in several passes, in the reverse order of the tightening sequence.**

- 16 cylinder head bolts and plate washers
- Cylinder head

To install:
18. Clean all surfaces.
19. Install or connect the following:
- New cylinder head gaskets
- Cylinder heads

20. Apply a light coat of engine oil on the threads and under the heads of the cylinder head bolts.

21. Tighten the cylinder head bolts using several passes, in sequence, as follows:
- Step 1: 25 ft. lbs. (34 Nm)
- Step 2: Mark the front of the cylinder head bolt with paint
- Step 3: Turn 90 degrees
- Step 4: Check that the painted mark is now at a 90 degrees angle to the front

22. Install the recessed head cylinder head bolts, as follows:
- Step 1: Apply a light coat of engine oil on the threads and under the heads of the cylinder head bolts
- Step 2: Tighten the cylinder head bolts, using a 8mm hexagon wrench, to 13 ft. lbs. (18 Nm).
- Bolt and ground strap

23. Install or connect the following:
- Valve lifters and shims

➡**Check that the valve lifter rotates smoothly by hand.**

- Right intake and exhaust camshafts
- Left intake and exhaust camshafts

24. Check and adjust the valve clearance.
25. Install or connect the following:
- Semi-circular plugs
- Cylinder head covers. Uniformly, tighten the bolts, in several passes, to 53 inch lbs. (6 Nm).

- Exhaust manifolds with new gaskets. Tighten the nuts to 30 ft. lbs. (40 Nm).
- Exhaust manifold heat insulators. Tighten the nuts to 71 inch lbs. (8 Nm).
- Exhaust crossover pipe. Tighten the nuts to 33 ft. lbs. (45 Nm).
- Power steering pump
- Alternator bracket. Tighten the fasteners to 14 ft. lbs. (18 Nm).
- Alternator
- No. 3 timing belt cover. Tighten the bolts to 80 inch lbs. (9 Nm).
- CMP sensor. Tighten it to 71 inch lbs. (8 Nm).
- Timing belt
- No. 2 timing belt idler bolt. Tighten the bolt to 30 ft. lbs. (40 Nm).

➡**Check that the pulley bracket moves smoothly.**

- Left camshaft timing pulley

26. Set the No. 1 cylinder to TDC of the compression stroke, as follows:
   a. Connect the timing belt to the left camshaft timing pulley.
   b. Check that the installation mark on the timing belt is aligned with the end of the No. 1 timing belt cover.
   c. Install the right camshaft timing pulley and the timing belt.
   d. Set the timing belt tensioner. Alternately, tighten the bolts to 20 ft. lbs. (28 Nm).

27. Using pliers, remove the 1.5mm hexagon wrench from the belt tensioner.
28. Check the valve timing.
29. Install or connect the following:
- New gaskets and the intake mani-

fold assembly. Tighten the bolts and nuts to 13 ft. lbs. (18 Nm).
- Intake manifold stay. Tighten the bolts to 14 ft. lbs. (18 Nm).
- Fuel pressure regulator

30. Connect the engine wiring harness to the intake manifold, as follows:
- Engine wiring harness to the cylinder head
- 3 engine wiring harness clamps

31. Install or connect the following:
- 6 injector connectors
- ECT sender gauge connector
- ECT sensor connector
- Intake air connector
- Air intake chamber assembly. Tighten the bolts and nuts to 13 ft. lbs. (18 Nm).
- Intake chamber stay
- No. 2 timing belt cover. Tighten the bolts to 80 inch lbs. (9 Nm).
- PCV hoses
- Water bypass hoses
- Air assist hose to the throttle body
- IAC valve connector
- TP sensor connector
- CMP sensor connector to the No. 2 timing belt cover
- 3 spark plug wire clamps

32. Connect the following hoses:
- Brake booster vacuum hose
- EVAP hose
- Automatic Disconnecting Differential (ADD) vacuum hose, for 4-wheel drive
- Fuel inlet and fuel return hose
- Heater hose

33. Install or connect the following:
- Oil dipstick and guide, using a new O-ring
- Spark plugs
- Spark plug wires, with the ignition coils
- Alternator drive belt

34. Connect the following cables:
- Actuator cable with the bracket, if equipped with cruise control
- Accelerator cable
- Throttle cable, if equipped with an automatic transmission
- MAF meter, resonator and air cleaner cap
- Front exhaust pipe
- Negative battery cable

35. Fill the radiator with engine coolant.
36. Start the engine and check for leaks.
37. Check the ignition timing.
38. Install the engine undercover.
39. Road test the vehicle.
40. Recheck all fluid levels.

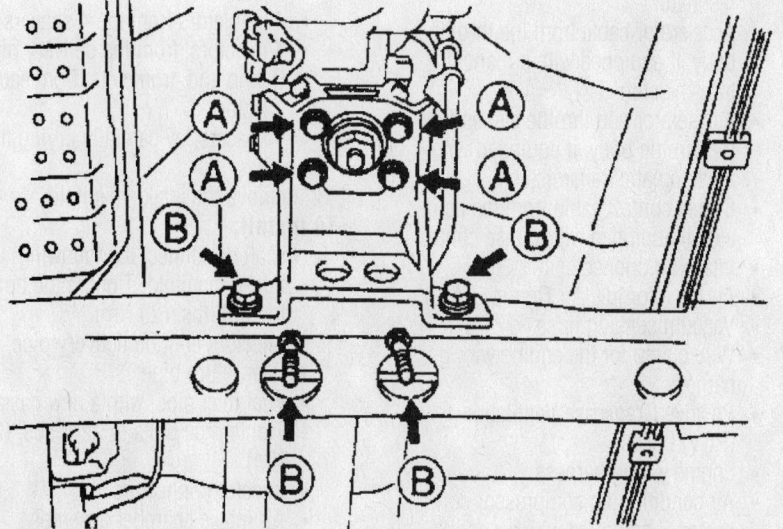

7924YG04

Cylinder head torque sequence—3.4L (5VZ-FE) engine

Intake Manifold

REMOVAL & INSTALLATION

RAV4

1. Before servicing the vehicle, refer to the precautions in the beginning of this section.
2. Properly relieve the fuel system pressure.
3. Remove or disconnect the following:
   - Negative battery cable
   - Air cleaner assembly
   - Throttle body from the intake manifold
4. Disconnect the engine wire from the intake manifold, as follows:
   - 4 injector connectors
   - 2 engine wire clamps from the intake manifold wire brackets
   - Engine wire protector from the right-hand side of the intake manifold by removing the bolt
   - Engine wire from the wire clamp
5. Remove the EGR valve, EGR pipe and modulator, as follows:
   - Both vacuum hoses from the Exhaust Gas Recirculation (EGR) Vacuum Switching Valve (VSV)
   - Vacuum modulator from the clamp on the intake manifold
   - Loosen the cylinder head side of the EGR pipe union nut
   - Both nuts, the EGR valve, pipe assembly and gasket
   - Vacuum modulator
6. Disconnect the following hoses:
   - Fuel filter vacuum sensor hose on the intake manifold
   - Brake booster vacuum hose from the intake manifold
   - Ground strap from the intake manifold
7. Remove or disconnect the following:
   - Intake manifold stay by removing the 2 bolts
   - Control cable from the clamp on the rear side of the intake manifold, if equipped with automatic transmission
   - Air hose from the intake manifold
   - Air tube from the intake manifold, by removing the 2 bolts
   - 6 bolts and 2 nuts from the intake manifold
   - Intake manifold

To install:

8. Install or connect the following:
   - Intake manifold. Tighten the 6 bolts and 2 nuts to 14 ft. lbs. (19 Nm).
   - Air tube with the 2 bolts
   - Air hose to the intake manifold
   - Control cable to the clamp on the rear side of the intake manifold, if equipped with an automatic transmission
   - Intake manifold stay. Tighten both bolts to 31 ft. lbs. (42 Nm).
9. Connect the following hoses:
   - Ground strap to the intake manifold
   - Brake booster vacuum hose to the intake manifold
   - Fuel filter vacuum sensor hose to the intake manifold
10. Install the EGR valve, EGR pipe and the vacuum modulator, as follows:
   - Vacuum modulator
   - EGR valve and pipe. Tighten both nuts to 108 inch lbs. (13 Nm) and the union nut to 43 ft. lbs. (59 Nm).
   - Vacuum hoses
11. Install or connect the following:
   - Engine wire and injectors

➡ **The No. 1 and No. 3 injector connectors are brown, and the No. 2 and No. 4 injector connectors are gray.**

   - Throttle body to the intake manifold
   - Air cleaner assembly
   - Negative battery cable

2.4L Highlander

1. Before servicing the vehicle, refer to the precautions in the beginning of this section.
2. Disconnect the negative battery cable.
3. Release the fuel system pressure.
4. Drain the engine coolant.
5. Remove or disconnect the following:
   - Air cleaner cap
   - Mass Air Flow (MAF) meter and the resonator
   - Accelerator cable from the throttle body, if equipped with a manual transmission
   - Accelerator and throttle cables from the throttle body, if equipped with an automatic transmission
   - Cruise control cable from the actuator, if equipped with cruise control
   - Intake air connector
   - Air hose for Idle Air Control (IAC)
   - Vacuum sensing hose
   - Wire clamp for the engine wiring harness
   - Positive Crankcase Ventilation (PCV) hoses.
   - Engine wiring harness
   - Air conditioning compressor connector, if equipped with air conditioning
   - Oil pressure sensor connector
   - Engine Coolant Temperature (ECT) sensor connector
   - ECT sender gauge connector
   - Exhaust Gas Recirculation (EGR) gas temperature sensor connector
   - Vacuum Switching Valve (VSV) connector, for the EGR
   - 2 vacuum hoses, from the VSV for the EGR
   - Ground strap, from the cowl top panel
   - Engine wiring harness, from the air intake chamber
   - Throttle Position (TP) sensor connector
   - IAC valve connector
   - Crankshaft Position (CKP) sensor connector
   - Knock (KS) sensor connector
   - Data Link Connector 1 (DLC1), from the bracket
   - Engine wiring harness clamp
   - EGR pipe
   - Intake chamber stay
   - Air intake chamber assembly
6. Disconnect the following hoses:
   - Evaporative Emission (EVAP) hose, from the throttle body
   - Brake booster vacuum hose, from the union
   - Water bypass hose, from the water bypass pipe
   - Water bypass hose, from the cylinder head rear cover
   - Injector connectors
   - Fuel inlet pipe
   - Hoses and the fuel return pipe.
7. Remove the delivery pipe and injectors, as follows:
   - Delivery pipe, together with the 4 injectors
   - 4 insulators from the 4 spacers
   - 4 injectors, from the delivery pipe
   - O-ring and grommets, from each injector
   - 4 spacers, by carefully prying them out
8. Remove the intake manifold.

To install:

9. Install or connect the following:
   - Intake manifold. Tighten the bolts to 22 ft. lbs. (29 Nm).
   - Injectors and the delivery pipe
   - Fuel return pipe
   - Fuel inlet pipe, with a new gasket. Tighten the bolts to 22 ft. lbs. (29 Nm).
   - Injector connectors
   - Air intake chamber assembly. Tighten the bolts to 15 ft. lbs. (21 Nm).

10. Connect the following hoses:
- Evaporative Emissions (EVAP) hose, to the throttle body
- Brake booster vacuum hose, to the union
- Water bypass hose, to water bypass pipe
- Water bypass hose, to cylinder head rear cover

11. Install or connect the following:
- Air intake chamber stay. Tighten the bolts to 15 ft. lbs. (20 Nm).
- EGR pipe. Tighten bolts to 13 ft. lbs. (18 Nm), nut "A" to 14 ft. lbs. (19 Nm) and nut B to 15 ft. lbs. (20 Nm).
- Air conditioning compressor connector
- Oil pressure sensor connector
- ECT sensor connector
- ECT sender gauge connector
- EGR gas temperature sensor connector
- VSV connector for the EGR
- 2 vacuum hose to the VSV for the EGR

- Ground strap to the cowl top panel
- Engine wiring harness to the air intake chamber
- TP sensor connector
- IAC valve connector
- CKP sensor connector
- KS sensor connector
- DLC1 to the bracket
- Engine wiring harness clamp
- PCV hoses
- Intake air connector. Tighten the bolts to 13 ft. lbs. (18 Nm).
- Cruise control cable to the actuator, if equipped with cruise control
- Accelerator cable to the throttle body, if equipped with a manual transmission
- Accelerator and throttle cables to the throttle body, if equipped with an automatic transmission

12. Fill the engine and radiator with engine coolant.

13. Install or connect the following:
- Air cleaner cap, MAF meter and resonator assembly
- Negative battery cable

14. Start the engine and check for leaks.

15. Road test the vehicle for proper operation.

16. Recheck all fluid levels.

3.0L Highlander

1. Before servicing the vehicle, refer to the precautions in the beginning of this section.

2. Remove or disconnect the following:

3. Properly relieve the fuel system pressure.

4. Drain and recycle the engine coolant.

5. Remove or disconnect the following:
- Accelerator cable
- Throttle cable
- Air cleaner
- Any wiring or hoses interfering with removal
- Right side engine mount stay
- Radiator and heater hoses in the way of the intake manifold removal
- V-bank cover
- All the vacuum hose and wiring for the emission control valve set

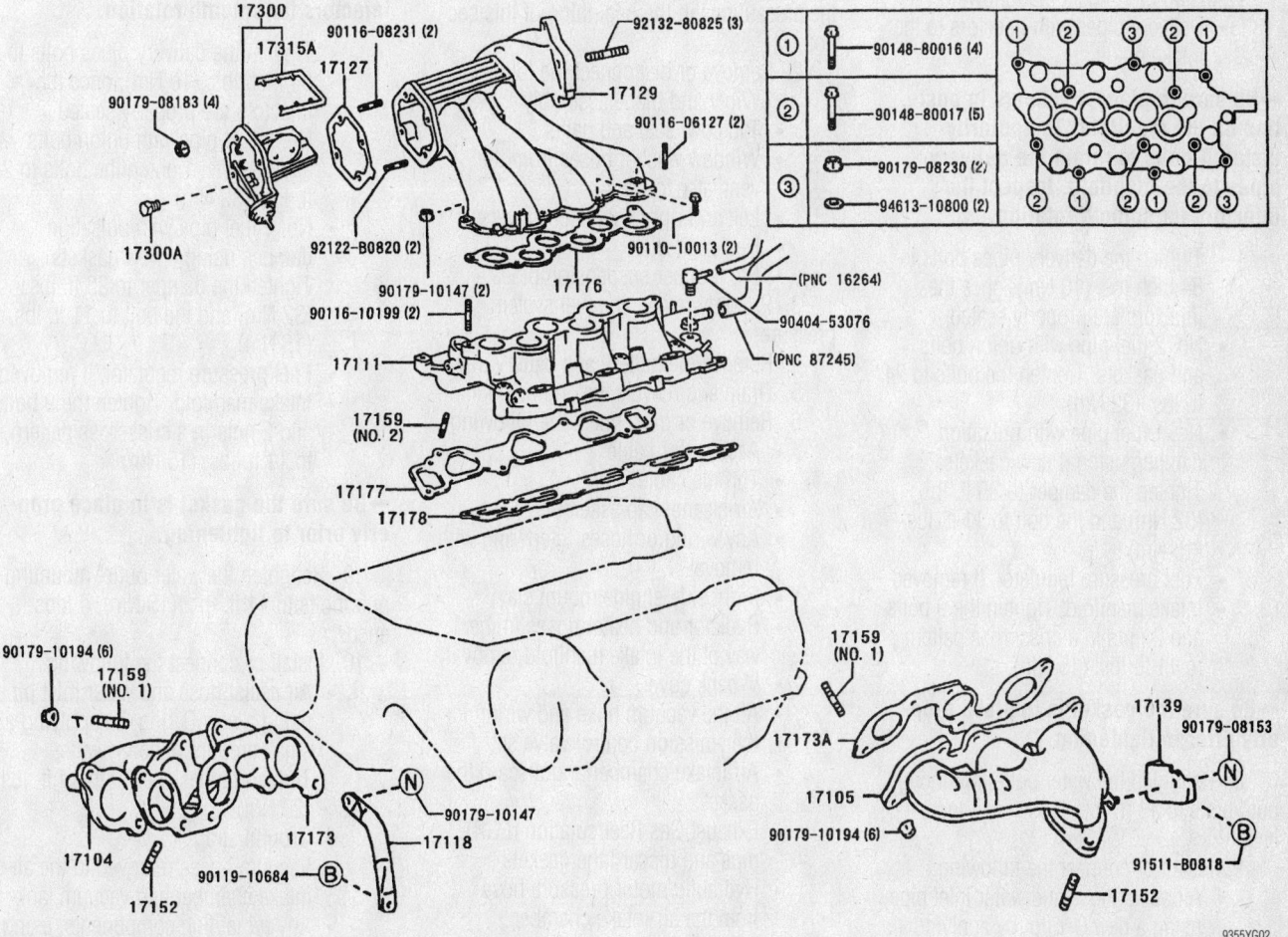

Intake manifold and related components—2.4L Highlander

9355YG02

- Air intake chamber and discard the gasket
- Exhaust Gas Recirculation (EGR) pipe and discard the gaskets
- Hydraulic motor pressure hose from the air intake chamber
- Engine wiring harnesses from the left side, right side, rear and No. 3 timing belt cover
- Front exhaust pipe, if necessary
- Timing belt, camshaft timing pulleys, No. 2 idler pulley and No. 3 timing belt cover
- Cylinder head rear plate
- 2 bolts, nuts and plate washers with the intake manifold.

➡**The delivery pipes with injectors will be attached to the manifold.**

- Other fuel related components such as the No. 2 fuel pipe and pulsation damper, if needed
- Delivery pipes from the intake manifold

6. Clean and inspect the intake manifold mating surfaces. Scrape all old gasket material off.

To install:

7. Install or connect the following:
- Delivery pipes with injectors to the intake manifold.

➡**Be sure to place 4 spacers in position on the manifold. Temporarily install 4 bolts to retain the delivery pipes to the manifold. Inspect the injectors for smooth rotation.**

- Tighten the delivery pipes bolts to 84 inch lbs. (10 Nm), once the injectors are properly seated
- No. 2 fuel pipe with union bolts and gaskets. Tighten the bolts to 24 ft. lbs. (32 Nm).
- No. 1 fuel pipe with pulsation damper, using 4 new gaskets. Tighten the damper to 35 ft. lbs. (32 Nm) and the bolt to 11 ft. lbs. (15 Nm).
- Fuel pressure regulator, if removed
- Intake manifold. Tighten the 9 bolts and 2 nuts in a crisscross pattern to 11 ft. lbs. (15 Nm).

➡**Be sure the gasket is in place properly prior to tightening.**

8. Retighten the water outlet mounting nuts/bolts to 11 ft. lbs. (15 Nm), if loosened.

9. Install or connect the following:
- Air assist hose and water inlet pipe, using a new O-ring, by applying a small amount of soapy water.

Tighten the fastener(s) to 14 ft. lbs. (20 Nm).
- Ground strap
- Vacuum hoses removed to the air intake chamber and vacuum tank
- Any remaining components, using new gaskets. Tighten the air intake chamber nuts/bolts to 32 ft. lbs. (43 Nm), the EGR pipe nuts to 108 inch lbs. (12 Nm) and the emission control valve set to 69 inch lbs. (8 Nm).
- Air cleaner assembly
- Heater hoses
- Throttle cable with bracket onto the throttle body
- Accelerator cable, by adjusting it, if equipped with an automatic transaxle

10. Refill the cooling system
11. Install or connect the following:
- Negative battery cable
- Heater air duct
12. Start the engine and inspect for leaks.

RX 300

1. Before servicing the vehicle, refer to the precautions in the beginning of this section.

2. Remove or disconnect the following:
- Wiper and blade assembly
- Top cowl seal and panel
- Window washer hoses from the ventilator louvers
- Left and right ventilator louvers
- Heater air duct
- Front upper suspension brace
3. Properly relieve the fuel system pressure.
4. Remove the battery and battery tray.
5. Drain and recycle the engine coolant.
6. Remove or disconnect the following:
- Accelerator cable
- Throttle cable
- Air cleaner cap assembly
- Any wiring or hoses interfering with removal
- Right side engine mount stay
- Radiator and heater hoses in the way of the intake manifold removal
- V-bank cover
- All the vacuum hose and wiring for the emission control valve set
- Air intake chamber and discard the gasket
- Exhaust Gas Recirculation (EGR) pipe and discard the gaskets
- Hydraulic motor pressure hose from the air intake chamber
- Engine wiring harnesses from the

left side, right side, rear and No. 3 timing belt cover
- Front exhaust pipe, if necessary
- Timing belt, camshaft timing pulleys, No. 2 idler pulley and No. 3 timing belt cover
- Cylinder head rear plate
- 2 bolts, nuts and plate washers with the intake manifold.

➡**The delivery pipes with injectors will be attached to the manifold.**

- Other fuel related components such as the No. 2 fuel pipe and pulsation damper, if needed
- Delivery pipes from the intake manifold

7. Clean and inspect the intake manifold mating surfaces. Scrape all old gasket material off.

To install:

8. Install or connect the following:
- Delivery pipes with injectors to the intake manifold.

➡**Be sure to place 4 spacers in position on the manifold. Temporarily install 4 bolts to retain the delivery pipes to the manifold. Inspect the injectors for smooth rotation.**

- Tighten the delivery pipes bolts to 84 inch lbs. (10 Nm), once the injectors are properly seated
- No. 2 fuel pipe with union bolts and gaskets. Tighten the bolts to 24 ft. lbs. (32 Nm).
- No. 1 fuel pipe with pulsation damper, using 4 new gaskets. Tighten the damper to 35 ft. lbs. (32 Nm) and the bolt to 11 ft. lbs. (15 Nm).
- Fuel pressure regulator, if removed
- Intake manifold. Tighten the 9 bolts and 2 nuts in a crisscross pattern to 11 ft. lbs. (15 Nm).

➡**Be sure the gasket is in place properly prior to tightening.**

9. Retighten the water outlet mounting nuts/bolts to 11 ft. lbs. (15 Nm), if loosened.

10. Install or connect the following:
- Air assist hose and water inlet pipe, using a new O-ring, by applying a small amount of soapy water. Tighten the fastener(s) to 14 ft. lbs. (20 Nm).
- Ground strap
- Vacuum hoses removed to the air intake chamber and vacuum tank
- Any remaining components, using new gaskets. Tighten the air intake

chamber nuts/bolts to 32 ft. lbs. (43 Nm), the EGR pipe nuts to 108 inch lbs. (12 Nm) and the emission control valve set to 69 inch lbs. (8 Nm).
- Air cleaner assembly
- Heater hoses
- Battery and tray
- Throttle cable with bracket onto the throttle body
- Accelerator cable, by adjusting it, if equipped with an automatic transaxle
- Front upper suspension brace. Tighten the nuts to 59 ft. lbs. (80 Nm).

11. Refill the cooling system
12. Install or connect the following:
- Negative battery cable
- Heater air duct
- Left and right ventilator louvers
- Window washer hoses from the ventilator louvers
- Top cowl seal and panel
- Wiper and blade assembly

13. Start the engine and inspect for leaks.

2.7L 4Runner

1. Before servicing the vehicle, refer to the precautions in the beginning of this section.
2. Disconnect the negative battery cable.
3. Release the fuel system pressure.
4. Drain the engine coolant.
5. Remove or disconnect the following:
- Air cleaner cap
- Mass Air Flow (MAF) meter and the resonator
- Accelerator cable from the throttle body, if equipped with a manual transmission
- Accelerator and throttle cables from the throttle body, if equipped with an automatic transmission
- Cruise control cable from the actuator, if equipped with cruise control
- Intake air connector
- Air hose for Idle Air Control (IAC)
- Vacuum sensing hose
- Wire clamp for the engine wiring harness
- Positive Crankcase Ventilation (PCV) hoses.
- Engine wiring harness
- Air conditioning compressor connector, if equipped with air conditioning
- Oil pressure sensor connector
- Engine Coolant Temperature (ECT) sensor connector

- ECT sender gauge connector
- Exhaust Gas Recirculation (EGR) gas temperature sensor connector
- Vacuum Switching Valve (VSV) connector, for the EGR
- 2 vacuum hoses, from the VSV for the EGR
- Ground strap, from the cowl top panel
- Engine wiring harness, from the air intake chamber
- Throttle Position (TP) sensor connector
- IAC valve connector
- Crankshaft Position (CKP) sensor connector
- Knock (KS) sensor connector
- Data Link Connector 1 (DLC1), from the bracket
- Engine wiring harness clamp
- EGR pipe
- Intake chamber stay
- Air intake chamber assembly

6. Disconnect the following hoses:
- Evaporative Emission (EVAP) hose, from the throttle body
- Brake booster vacuum hose, from the union
- Water bypass hose, from the water bypass pipe
- Water bypass hose, from the cylinder head rear cover
- Injector connectors
- Fuel inlet pipe
- Hoses and the fuel return pipe.

7. Remove the delivery pipe and injectors, as follows:
- Delivery pipe, together with the 4 injectors
- 4 insulators from the 4 spacers
- 4 injectors, from the delivery pipe
- O-ring and grommets, from each injector
- 4 spacers, by carefully prying them out

8. Remove the intake manifold.

To install:

9. Install or connect the following:
- Intake manifold. Tighten the bolts to 22 ft. lbs. (29 Nm).
- Injectors and the delivery pipe
- Fuel return pipe
- Fuel inlet pipe, with a new gasket. Tighten the bolts to 22 ft. lbs. (29 Nm).
- Injector connectors
- Air intake chamber assembly. Tighten the bolts to 15 ft. lbs. (21 Nm).

10. Connect the following hoses:
- Evaporative Emissions (EVAP) hose, to the throttle body

- Brake booster vacuum hose, to the union
- Water bypass hose, to water bypass pipe
- Water bypass hose, to cylinder head rear cover

11. Install or connect the following:
- Air intake chamber stay. Tighten the bolts to 15 ft. lbs. (20 Nm).
- EGR pipe. Tighten bolts to 13 ft. lbs. (18 Nm), nut "A" to 14 ft. lbs. (19 Nm) and nut B to 15 ft. lbs. (20 Nm).
- Air conditioning compressor connector
- Oil pressure sensor connector
- ECT sensor connector
- ECT sender gauge connector
- EGR gas temperature sensor connector
- VSV connector for the EGR
- 2 vacuum hose to the VSV for the EGR
- Ground strap to the cowl top panel
- Engine wiring harness to the air intake chamber
- TP sensor connector
- IAC valve connector
- CKP sensor connector
- KS sensor connector
- DLC1 to the bracket
- Engine wiring harness clamp
- PCV hoses
- Intake air connector. Tighten the bolts to 13 ft. lbs. (18 Nm).
- Cruise control cable to the actuator, if equipped with cruise control
- Accelerator cable to the throttle body, if equipped with a manual transmission
- Accelerator and throttle cables to the throttle body, if equipped with an automatic transmission

12. Fill the engine and radiator with engine coolant.
13. Install or connect the following:
- Air cleaner cap, MAF meter and resonator assembly
- Negative battery cable

14. Start the engine and check for leaks.
15. Road test the vehicle for proper operation.
16. Recheck all fluid levels.

3.4L 4Runner

1. Before servicing the vehicle, refer to the precautions in the beginning of this section.
2. Disconnect the negative battery cable.
3. Relieve the fuel system pressure.
4. Remove the engine undercover.

5. Drain the cooling system.
6. Remove or disconnect the following:
   - Air cleaner cap
   - Mass Air Flow (MAF) meter and the resonator
   - Actuator cable with the bracket, if equipped with cruise control
   - Accelerator cable
   - Throttle cable, if equipped with automatic transmission
7. Disconnect the following hoses:
   - Heater hose
   - Brake booster vacuum hose
   - Evaporative Emissions (EVAP) hose
   - Automatic Disconnecting Differential (ADD) vacuum hose, for 4-Wheel drive
   - Fuel inlet and fuel return hose
8. Remove or disconnect the following:
   - Spark plug wires, with the ignition coils
   - Intake chamber stay
   - No. 2 timing belt cover
   - Air intake chamber assembly
   - Throttle Position (TP) sensor connector
   - Idle Air Control (IAC) valve connector
   - Positive Crankcase Ventilation (PCV) hoses
   - Water bypass hoses
   - Air assist hose from the throttle body
   - Intake air connector
   - Engine wiring harness
   - Fuel return hose
   - Vacuum hose, from the fuel pressure regulator
   - Ground strap, from the intake air connector

   - Data Link Connector 1 (DLC1), from the bracket
   - 6 injector connectors
   - Engine Coolant Temperature (ECT) sensor and sender gauge connectors
   - Engine wiring harness protector from the cylinder head
   - Fuel pressure regulator
   - Intake manifold assembly
   - Intake manifold stay
   - Intake manifold, delivery pipes and the injectors assembly with the gaskets

To install:

9. Install or connect the following:
   - New gaskets
   - Intake manifold assembly. Tighten the bolts and nuts to 13 ft. lbs. (18 Nm).
   - Intake manifold stay. Tighten the bolts to 14 ft. lbs. (18 Nm).
   - Fuel pressure regulator
   - Engine wiring harness to the cylinder head, by installing the 3 bolts
   - 3 engine wiring harness clamps
   - 6 injector connectors
   - ECT sender gauge connector
   - ECT sensor connector
   - Intake manifold. Tighten the bolts and nuts to 14 ft. lbs. (18.5 Nm).
   - DLC1 to the bracket on the intake manifold
   - Ground strap to the intake manifold, by installing the bolt
   - Brake booster vacuum hose, to the intake air connector
   - 2 fuel return hoses
   - Engine wiring harness to the intake manifold

   - Air intake chamber assembly to the engine. Tighten the bolts and nuts to 14 ft. lbs. (18.5 Nm).
   - Intake chamber stay. Tighten the bolts to 30 ft. lbs. (40 Nm).
   - New O-ring to the oil filler tube
   - Oil filler tube end into the tube hole in the oil pan
   - Oil filler tube and No. 1 throttle cable clamp
   - No. 2 timing belt cover. Tighten the bolts to 80 inch lbs. (9 Nm).
   - PCV hoses
   - Water bypass hoses
   - Air assist hose to the throttle body
   - IAC valve connector
   - TP sensor connector
   - Brake booster vacuum hose
   - EVAP hose
   - Automatic Disconnecting Differential (ADD) vacuum hose, for 4-Wheel drive
   - Fuel inlet and fuel return hose
   - 3 spark plug wire clamps to the No. 2 timing belt cover
   - CMP connector to the No. 2 timing belt cover
   - Spark plug wires with the ignition coils
   - Heater hose
   - Actuator cable with the bracket, if equipped with cruise control
   - Accelerator cable
   - Throttle cable, if equipped with automatic transmission
   - MAF meter, resonator and the air cleaner cap
   - Negative battery cable
10. Fill the radiator with engine coolant.
11. Start the engine and check for leaks.
12. Install the engine undercover.
13. Road test the vehicle.
14. Recheck all fluid levels.

Exhaust Manifold

REMOVAL & INSTALLATION

RAV4

1. Before servicing the vehicle, refer to the precautions in the beginning of this section.
2. Remove or disconnect the following:
   - Negative battery cable
   - Front exhaust pipe from the exhaust manifold, using a 14mm deep socket wrench; discard the gasket
   - Main Oxygen (O_2) sensor and the sub Oxygen (O_2) sensor connectors

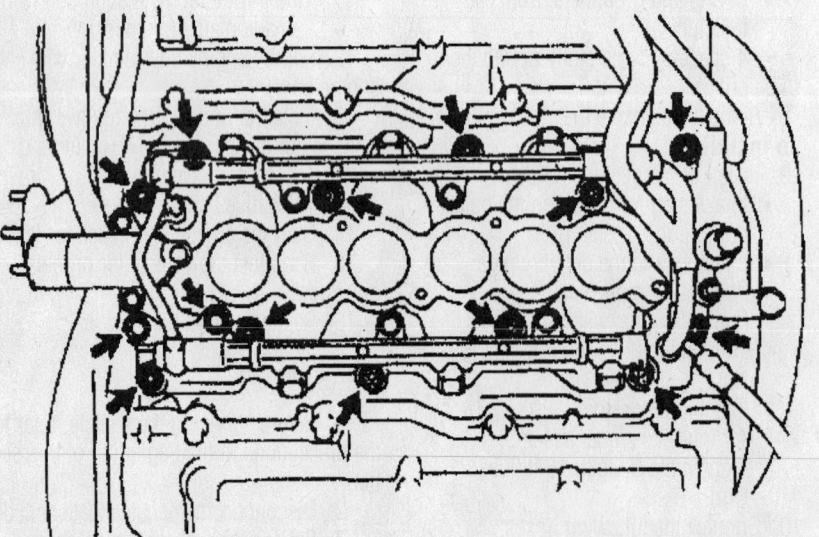

Intake manifold bolts and nuts—4Runner with 3.4L engine

7924YG38

- 6 bolts and the upper manifold heat insulator
- 2 right-hand exhaust manifold stay-to-cylinder block bolts
- 6 nuts, the exhaust manifold and the Three-Way Catalytic (TWC) converter assembly
- Exhaust manifold and front catalytic converter

To install:
3. Install or connect the following:
- Catalytic converter to the exhaust manifold. Tighten the nuts/bolts to 22 ft. lbs. (29 Nm).
- Exhaust manifold and the front TWC assembly. Tighten the 6 nuts, in several passes to 36 ft. lbs. (49 Nm).

- Right-hand manifold stay. Tighten both bolts to 31 ft. lbs. (42 Nm).
- Manifold upper heat insulator with the 6 bolts and attach the main Oxygen (O₂) and the sub Oxygen (O₂) sensor connectors
- Front exhaust pipe to the TWC, using a new gasket. Tighten the 3 nuts to 46 ft. lbs. (62 Nm).

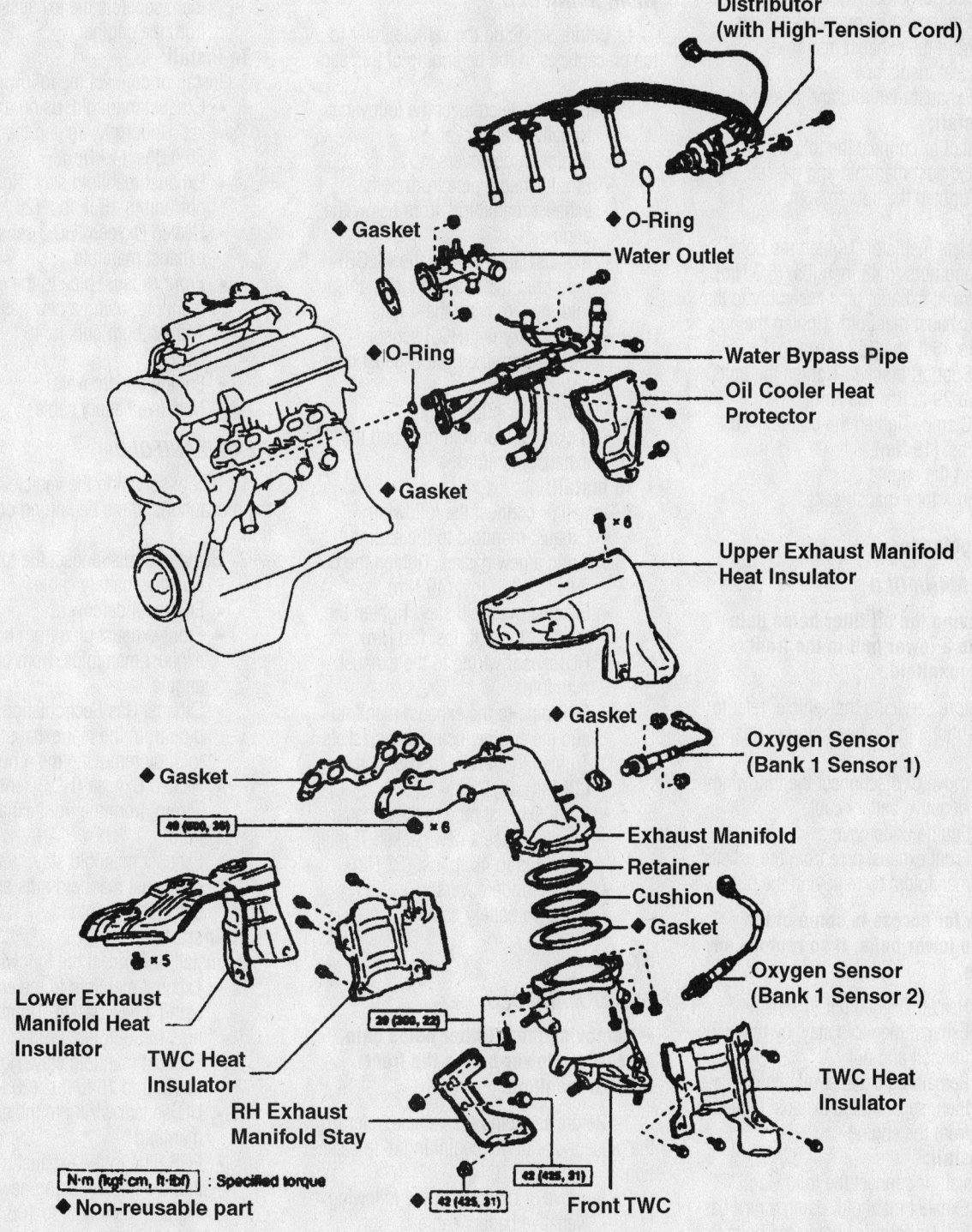

Distributor (with High-Tension Cord)

◆ O-Ring

◆ Gasket

Water Outlet

◆ O-Ring

Water Bypass Pipe

Oil Cooler Heat Protector

◆ Gasket

Upper Exhaust Manifold Heat Insulator

◆ Gasket

Oxygen Sensor (Bank 1 Sensor 1)

◆ Gasket

49 (500, 36) ×6

Exhaust Manifold

Retainer

Cushion

◆ Gasket

Oxygen Sensor (Bank 1 Sensor 2)

Lower Exhaust Manifold Heat Insulator

×5

TWC Heat Insulator

29 (300, 22)

RH Exhaust Manifold Stay

TWC Heat Insulator

N·m (kgf·cm, ft·lbf) : Specified torque

◆ Non-reusable part

42 (425, 31)

42 (425, 31)

Front TWC

7924ZG22

Exploded view of the exhaust manifold and components—rav4

- Negative battery cable
4. Start the engine and be sure that there are no exhaust leaks.

2.4L Highlander

1. Before servicing the vehicle, refer to the precautions in the beginning of this section.
2. Remove or disconnect the following:
   - Clamp from the support bracket
   - Support bracket
   - Front exhaust pipe and gaskets from the exhaust manifold
   - Heat insulator
   - Exhaust manifold and gasket

To install:
3. Install or connect the following:
   - Exhaust manifold and gasket. Tighten the nuts to 36 ft. lbs. (49 Nm).
   - Heat insulator. Tighten the bolts and nuts to 48 inch lbs. (5.5 Nm).
   - Front exhaust pipe assembly to the exhaust manifold. Tighten the nuts to 46 ft. lbs. (62 Nm).
   - Support bracket. Tighten the bolts to 29 ft. lbs. (39 Nm).
   - Clamp. Tighten the bolt to 14 ft. lbs. (19 Nm).
4. Start the engine.
5. Check for exhaust leaks.

3.0L Highlander

FRONT MANIFOLD

➡**Removing the oil filter helps gain access to a lower bolt in the front exhaust manifold.**

1. Before servicing the vehicle, refer to the precautions in the beginning of this section.
2. Remove or disconnect the following:
   - Negative battery cable
   - Engine undercovers
   - Front exhaust pipe from the exhaust manifolds, by removing the nuts

➡**Check for access to some of the manifold lower bolts, if so remove any possible.**

   - Heated Oxygen (HO$_2$) sensor
   - Exhaust manifold stay, by removing the bolt and nut
   - Remaining exhaust manifold nuts; then, separate the exhaust manifold from the engine

To install:
3. Install or connect the following:
   - Exhaust manifold, using a new gasket. Uniformly, tighten the bolts to 36 ft. lbs. (49 Nm).

   - Exhaust manifold stay. Tighten the nut/bolt to 15 ft. lbs. (20 Nm).
   - Heated Oxygen (HO$_2$) sensor to the exhaust manifold
   - Front exhaust pipe to the exhaust manifold, using a new gasket. Tighten both nuts to 46 ft. lbs. (62 Nm).
   - Engine undercovers
   - Negative battery cable

REAR MANIFOLD

1. Before servicing the vehicle, refer to the precautions in the beginning of this section.
2. Remove or disconnect the following:
   - Negative battery cable
   - Engine undercovers
   - Front exhaust pipe from both exhaust manifolds, from below the engine
   - Exhaust Gas Recirculation (EGR) pipe from the rear exhaust manifold, by removing the 4 nuts
   - Heated Oxygen (HO$_2$) sensor wiring, from the right exhaust manifold
   - Exhaust manifold stay
   - 6 exhaust manifold nuts and the exhaust manifold

To install:
3. Install or connect the following:
   - Exhaust manifold to the engine, using a new gasket. Tighten the 6 nuts to 36 ft. lbs. (49 Nm).
   - Exhaust manifold stay. Tighten the nut/bolt to 15 ft. lbs. (20 Nm).
   - HO$_2$ sensor wiring to the exhaust manifold
   - EGR pipe to the exhaust manifold and the engine, using new gaskets. Tighten the 4 nuts to 108 inch lbs. (12 Nm).
   - Front exhaust pipe to the exhaust manifold, use a new gasket. Tighten both nuts to 46 ft. lbs. (62 Nm).
   - Engine undercovers
   - Negative battery cable

RX 300

FRONT MANIFOLD

➡**Removing the oil filter helps gain access to a lower bolt in the front exhaust manifold.**

1. Before servicing the vehicle, refer to the precautions in the beginning of this section.
2. Remove or disconnect the following:
   - Negative battery cable
   - Engine undercovers

   - Front exhaust pipe from the exhaust manifolds, by removing the nuts

➡**Check for access to some of the manifold lower bolts, if so remove any possible.**

   - Heated Oxygen (HO$_2$) sensor
   - Exhaust manifold stay, by removing the bolt and nut
   - Remaining exhaust manifold nuts; then, separate the exhaust manifold from the engine

To install:
3. Install or connect the following:
   - Exhaust manifold, using a new gasket. Uniformly, tighten the bolts to 36 ft. lbs. (49 Nm).
   - Exhaust manifold stay. Tighten the nut/bolt to 15 ft. lbs. (20 Nm).
   - Heated Oxygen (HO$_2$) sensor to the exhaust manifold
   - Front exhaust pipe to the exhaust manifold, using a new gasket. Tighten both nuts to 46 ft. lbs. (62 Nm).
   - Engine undercovers
   - Negative battery cable

REAR MANIFOLD

1. Before servicing the vehicle, refer to the precautions in the beginning of this section.
2. Remove or disconnect the following:
   - Negative battery cable
   - Engine undercovers
   - Front exhaust pipe from both exhaust manifolds, from below the engine
   - Exhaust Gas Recirculation (EGR) pipe from the rear exhaust manifold, by removing the 4 nuts
   - Heated Oxygen (HO$_2$) sensor wiring, from the right exhaust manifold
   - Exhaust manifold stay
   - 6 exhaust manifold nuts and the exhaust manifold

To install:
3. Install or connect the following:
   - Exhaust manifold to the engine, using a new gasket. Tighten the 6 nuts to 36 ft. lbs. (49 Nm).
   - Exhaust manifold stay. Tighten the nut/bolt to 15 ft. lbs. (20 Nm).
   - HO$_2$ sensor wiring to the exhaust manifold
   - EGR pipe to the exhaust manifold and the engine, using new gaskets. Tighten the 4 nuts to 108 inch lbs. (12 Nm).
   - Front exhaust pipe to the exhaust

manifold, use a new gasket. Tighten both nuts to 46 ft. lbs. (62 Nm).

- Engine undercovers
- Negative battery cable

2.7L 4Runner

1. Before servicing the vehicle, refer to the precautions in the beginning of this section.

2. Remove or disconnect the following:

- Clamp from the support bracket
- Support bracket
- Front exhaust pipe and gaskets from the exhaust manifold
- Heat insulator
- Exhaust manifold and gasket

To install:

3. Install or connect the following:

- Exhaust manifold and gasket. Tighten the nuts to 36 ft. lbs. (49 Nm).
- Heat insulator. Tighten the bolts and nuts to 48 inch lbs. (5.5 Nm).
- Front exhaust pipe assembly to the exhaust manifold. Tighten the nuts to 46 ft. lbs. (62 Nm).
- Support bracket. Tighten the bolts to 29 ft. lbs. (39 Nm).
- Clamp. Tighten the bolt to 14 ft. lbs. (19 Nm).

4. Start the engine.
5. Check for exhaust leaks.

Front exhaust pipe to exhaust manifold nut and bolt locations—4Runner 2.7L

7924YG39

Exhaust manifold nuts—2.7L 4Runner

7924YG40

3.4L 4Runner

1. Before servicing the vehicle, refer to the precautions in the beginning of this section.

2. Remove or disconnect the following:

- Exhaust crossover pipe, from the exhaust manifold by removing the 3 nuts
- Exhaust Gas Recirculation (EGR) pipe, from the exhaust manifold, on the left manifold equipped with an EGR valve
- Exhaust manifold heat insulator, by removing the 3 nuts
- Exhaust manifold

To install:

3. Install or connect the following:

- Exhaust manifold, using a new gasket. Tighten the nuts to 30 ft. lbs. (40 Nm).
- Exhaust heat insulator. Tighten the nuts to 71 inch lbs. (8 Nm).
- EGR pipe to the exhaust manifold, if equipped with an EGR valve. Tighten the manifold nuts to 14 ft.

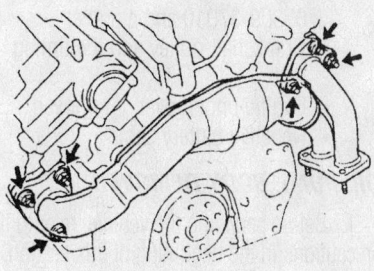

Exhaust crossover pipe mounting nut locations—3.4L 4Runner

7924YG43

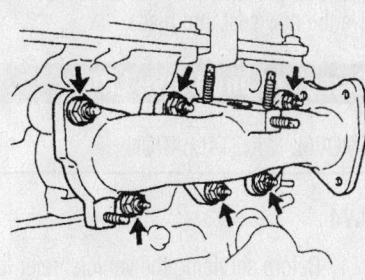

Exhaust manifold nuts—3.4L 4Runner

7924YG44

lbs. (18 Nm) and the clamp nuts to 71 inch lbs. (8 Nm).

- Crossover pipe to the exhaust manifold, using a new gasket. Tighten the nuts to 33 ft. lbs. (45 Nm).

Front Crankshaft Seal

REMOVAL & INSTALLATION

RAV4

1. Before servicing the vehicle, refer to the precautions in the beginning of this section.

➡ **The front oil seal can be removed from the engine without removing the oil pump.**

2. Remove or disconnect the following:

- Negative battery cable
- Timing belt covers and the timing belt
- Front crankshaft gear from the crankshaft, using Crankshaft Gear Puller tool 09950-50010

✳✳ WARNING

Be sure not to damage any part of the crankshaft.

- Cut off the oil seal lip
- Oil seal, using a suitable tool. Wrap the edge of the tool with a rag or tape to prevent damaging the crankshaft.

To install:

3. Install or connect the following:

- New seal, by applying a thin layer of liquid sealer to the outside of the seal
- Apply multi purpose grease to the new oil seal lip
- New oil seal, by tapping it in until its surface is flush with the oil pump body edge, using the Oil Seal Installer tool 09226-00010 and a hammer
- Timing belt and timing belt covers
- All other components
- Negative battery cable

4. Start the engine and check for leaks.

3.0L Highlander and RX 300

1. Before servicing the vehicle, refer to the precautions in the beginning of this section.

2. Remove or disconnect the following:

- Engine coolant reservoir tank and the alternator belt

- Right front wheel and the splash shield
- Power steering pump drive belt, by loosening both bolts
- Both ground wire connectors
- Right engine mounting stay
- Engine moving control rod and the No. 2 right engine mount bracket

➡ **To extract the engine bracket and control rod, raise the engine slightly.**

- No. 2 alternator bracket
- Crankshaft pulley bolt, using a pry-bar and wrench or Crankshaft Pulley Holding tool 09213-54015 and Flange Holding tool 09330-00021
- Crankshaft pulley, using a puller
- No. 1 timing belt cover

3. Remove the No. 2 timing belt cover, as follows:

- Engine wire protector from the No. 3 (rear) timing belt cover
- Engine wire protector clamp from the No. 3 timing belt cover
- 5 bolts from the No. 2 timing belt cover
- No. 2 cover

To install:

4. Install or connect the following:

- No. 2 timing belt cover, using a new gasket

➡ **Install it evenly to the part of the belt cover shaded black. After installation, press down on it so that the adhesive sticks to the belt cover firmly.**

- No. 2 timing belt cover. Tighten the 5 bolts to 74 inch lbs. (8 Nm).
- Engine wire protector clamp to the No. 3 timing belt cover
- Engine wire protector to the No. 3 timing belt cover with the bolt
- No. 3 timing belt cover, using a new gasket
- Tighten the 4 No. 1 timing belt cover bolts to 74 inch lbs. (8 Nm).
- Crankshaft pulley. Tighten the bolt to 159 ft. lbs. (215 Nm).
- No. 2 alternator bracket. Tighten the nut to 21 ft. lbs. (28 Nm). Do not tighten the pivot bolt at this time.
- No. 2 right engine mounting bracket and the moving control rod
- Right engine mount stay
- Both ground wire connectors
- Drive belts by adjusting them
- Coolant reservoir
- Right front splash shield and wheel
- Negative battery cable

5. Start the vehicle and check for any leaks.
6. Recheck the ignition timing.

3.4L 4Runner

➡ **There are 2 methods to replace the oil seal, which are as follows:**

OIL PUMP BODY INSTALLED

1. Before servicing the vehicle, refer to the precautions in the beginning of this section.
2. Remove or disconnect the following:

- Negative battery cable
- Timing belt and crankshaft pulley
- Cut off the oil seal lip, using a knife
- Pry out the oil seal, using a suitable tool

❊❊ WARNING

Be careful not to damage the crankshaft.

To install:

3. Install or connect the following:

- Apply multi-purpose grease to the new oil seal lip
- Tap in the new oil seal until its surface is flush with the oil pump case edge, using Seal Driver tool 09309-37010 and a mallet
- Crankshaft pulley and the timing belt
- Engine undercover, if removed
- Negative battery cable

OIL PUMP BODY REMOVED

1. Before servicing the vehicle, refer to the precautions in the beginning of this section.
2. Carefully pry out the seal using a suitable tool.
3. Apply multi-purpose grease to the new oil seal lip.
4. Using Seal Driver tool 09309-37010, drive the new seal into place.

Camshaft and Valve Lifters

REMOVAL & INSTALLATION

RAV4

1. Before servicing the vehicle, refer to the precautions in the beginning of this section.
2. Remove or disconnect the following:

- Negative battery cable
- Cylinder head cover and the upper timing belt cover

3. Rotate the crankshaft to set the engine at Top Dead Center (TDC)/compression for the No. 1 cylinder.

➡ **Due to the small thrust clearance on both the intake and exhaust camshafts,**

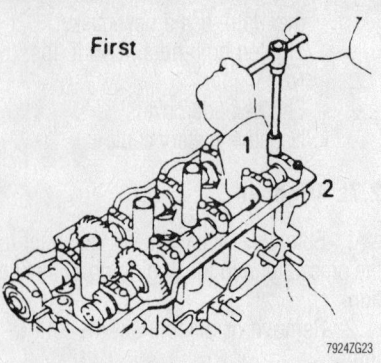

First

Exhaust camshaft bolt removal: Step 1—RAV4 2.0L

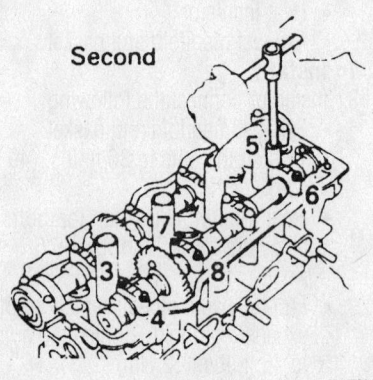

Second

Exhaust camshaft bolt removal: Step 2—RAV4 2.0L

the camshafts must be kept level during removal. If the camshafts are removed without being kept level, the camshaft may be caught in the cylinder head causing the head to break or the camshaft to seize.

4. Remove the camshaft timing sprocket and the timing belt.
5. Set the knock pin of the intake camshaft at 10–45 degrees Before Top Dead Center (BTDC) of camshaft angle. This angle will help to lift the exhaust camshaft level and evenly by pushing No. 2 and No. 4 cylinder camshaft lobes of the exhaust camshaft toward their valve lifters.
6. Secure the exhaust camshaft sub-gear to the main gear using a service bolt. The manufacturer recommends a bolt 0.63–0.79 in. (16–20mm) long with a thread diameter of 6mm and a 1mm thread pitch. When removing the exhaust camshaft, be sure that the torsional spring force of the sub-gear has been eliminated.
7. Remove the No. 1 and No. 2 rear bearing cap bolts and remove the cap. Uniformly loosen and remove bearing cap bolts No. 3 to No. 8 in several passes and in the proper sequence. Do not remove bearing

cap bolts No. 9 and 10 at this time. Remove the No. 1, 2 and 4 bearing caps.

8. Alternately loosen and remove bearing cap bolts No. 9 and 10. As these bolts are loosened check to see that the camshaft is being lifted out straight and level.

➡ **If the camshaft is not lifting out straight and level retighten No. 9 and 10 bearing cap bolts. Reverse the order of steps 5 through 7 and reset the intake camshaft knock pin to 10–45 degrees BTDC and repeat steps 5 through 7 again. Do not attempt to pry the camshaft from its mounting.**

9. Remove the No. 3 bearing cap and exhaust camshaft from the engine.

10. Set the knock pin of the intake camshaft at 80–115 degrees BTDC of camshaft angle. This angle will help to lift the intake camshaft level and evenly by pushing No. 1 and No. 3 cylinder camshaft lobes of the intake camshaft toward their valve lifters.

11. Remove the No. 1 and No. 2 front bearing cap bolts and remove the front bearing cap and oil seal. If the cap will not come apart easily, leave it in place without the bolts.

12. Uniformly loosen and remove bearing cap bolts No. 3 to No. 8 in several phases and in the proper sequence. Do not remove bearing cap bolts No. 9 and 10 at this time. Remove No. 1, 3 and 4 bearing caps.

13. Alternately loosen and remove bearing cap bolts No. 9 and 10. As these bolts are loosened and after breaking the adhesion on the front bearing cap, check to see that the camshaft is being lifted out straight and level.

➡ **If the camshaft is not lifting out straight and level retighten No. 9 and 10 bearing cap bolts. Reverse steps 10 through 12, than start over from step 10. Do not attempt to pry the camshaft from its mounting.**

14. Remove the No. 2 bearing cap with the intake camshaft from the engine.

15. Remove the valve adjusting shims from the engine. Be sure to replace the shims to their original location.

To install:

16. Install the valve adjusting shims to the engine.

17. Before installing the intake camshaft, apply multi-purpose grease to the thrust portion of the camshaft.

18. Position the camshaft at 80–115 degrees BTDC of camshaft angle on the cylinder head.

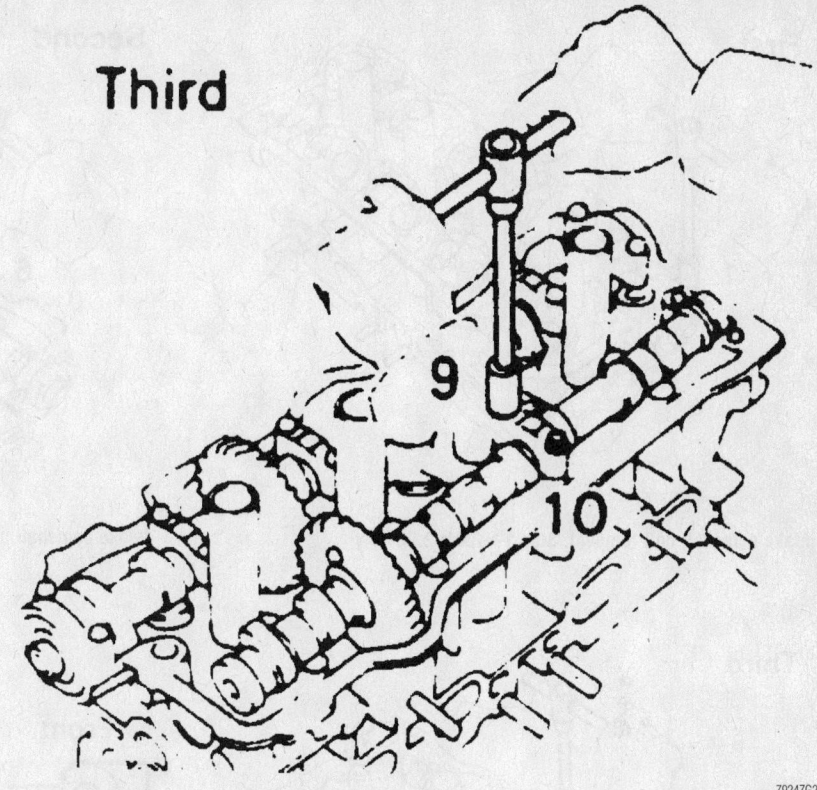

Exhaust camshaft bolt removal: Step 3—RAV4 2.0L

19. Apply sealant to the front bearing cap.

20. Coat the bearing cap bolts with clean engine oil.

21. Tighten the camshaft bearing caps evenly and in several passes to 14 ft. lbs. (19 Nm) in the proper sequence.

22. Set the knock pin of the camshaft at 10–45 degrees BTDC of camshaft angle.

23. Apply multipurpose grease to the thrust portion of the camshaft.

24. Position the exhaust camshaft gear with the intake camshaft gear so that the timing marks are in alignment with one another. Be sure to use the proper alignment marks on the gears. Do not use the assembly reference marks.

25. Turn the intake camshaft clockwise or counterclockwise little by little until the

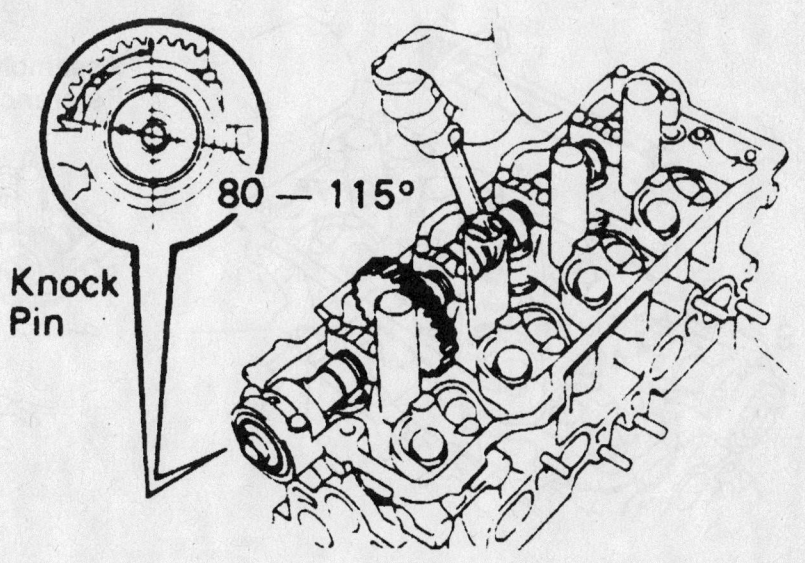

Intake camshaft knock pin alignment—RAV4 2.0L

First

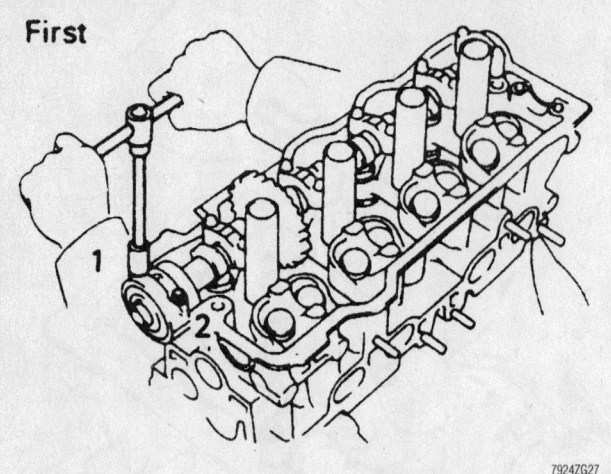

7924ZG27

Intake camshaft bolt removal: Step 1—RAV4 2.0L

Second

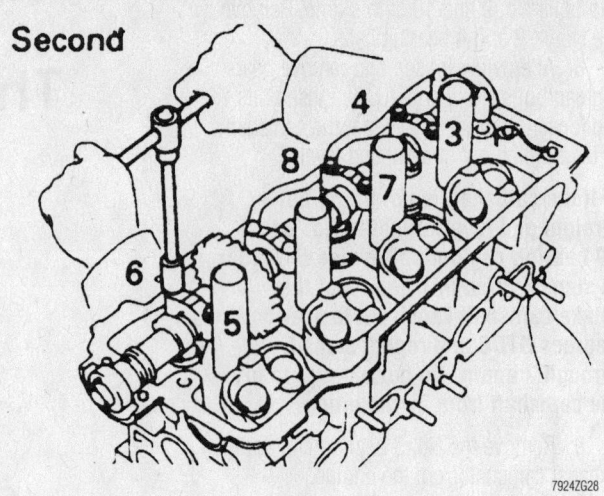

7924ZG28

Intake camshaft bolt removal: Step 2—RAV4 2.0L

Third

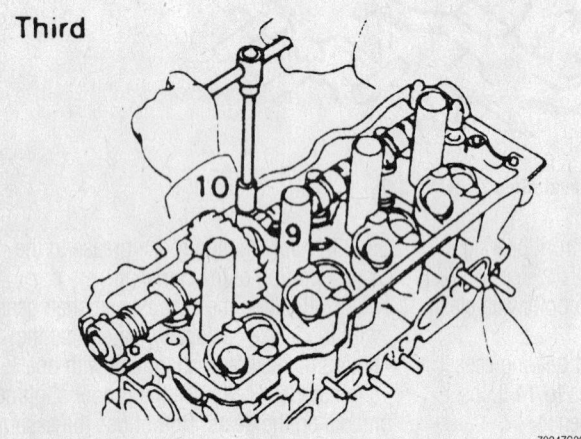

7924ZG29

Intake camshaft bolt removal: Step 3—RAV4 2.0L

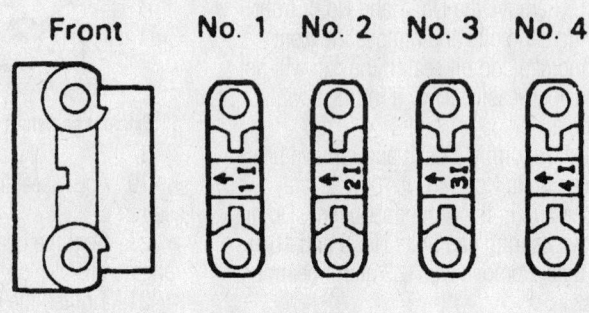

7924ZG30

Intake camshaft bearing cap positioning—RAV4 2.0L

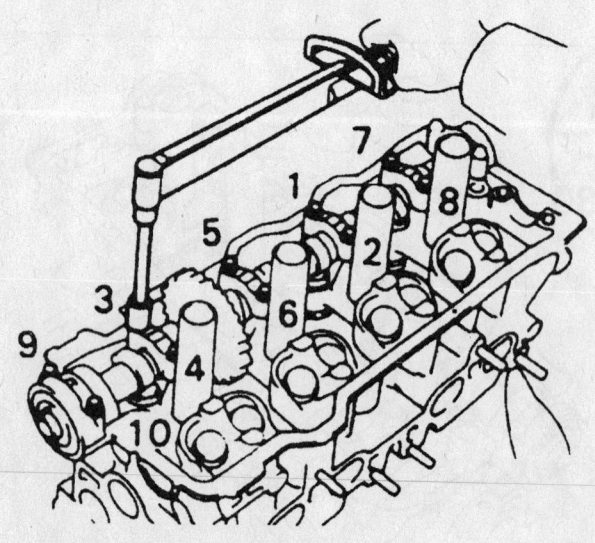

7924ZG31

Intake camshaft bolt tightening sequence—RAV4 2.0L

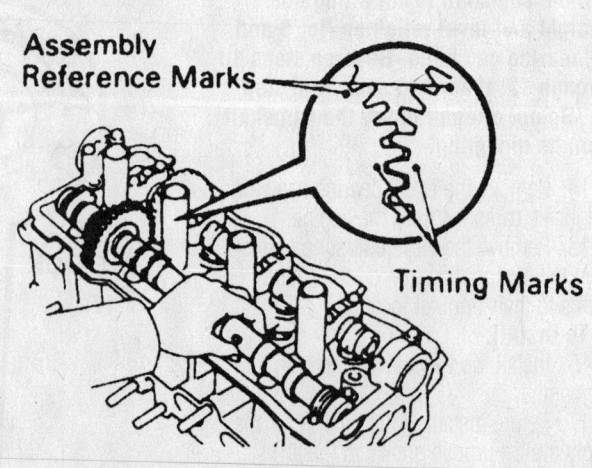

7924ZG32

Camshaft timing mark alignment—RAV4 2.0L

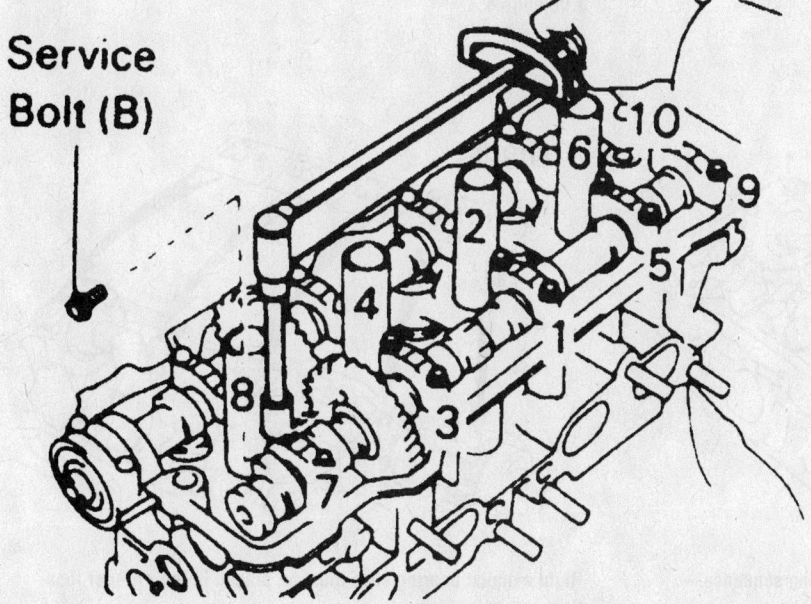

No. 1   No. 2   No. 3   No. 4   Rear

Exhaust camshaft bearing cap positioning—RAV4 2.0L

Service Bolt (B)

Exhaust camshaft bolt tightening sequence—RAV4 2.0L

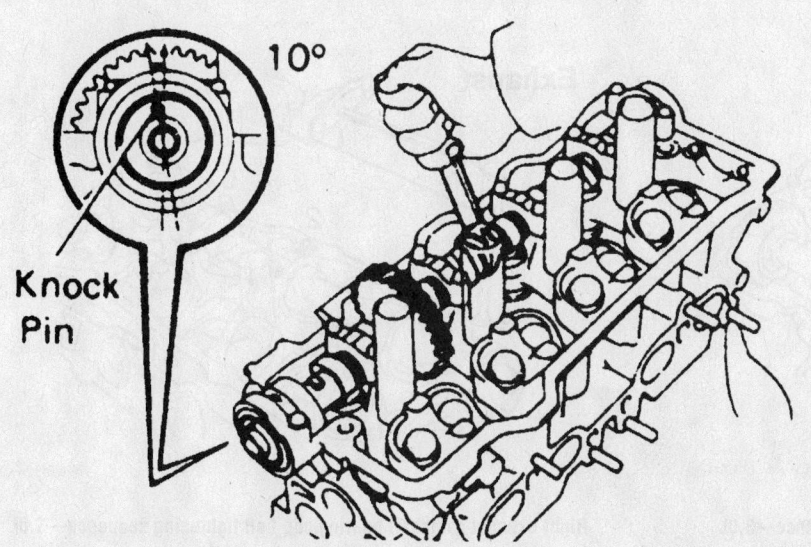

10°

Knock Pin

Exhaust camshaft knock pin alignment—RAV4 2.0L

exhaust camshaft sits in the bearing journals evenly without rocking the camshaft on the bearing journals.

26. Coat the bearing cap bolts with clean engine oil.

27. Tighten the camshaft bearing caps evenly and in several passes to 14 ft. lbs. (19 Nm). Remove the service bolt from the assembly.

28. Install the camshaft timing pulleys and the timing belt.

29. Adjust the valve clearance.

30. Install the head cover and the upper timing cover. Reconnect the negative battery cable.

31. Start the engine and check for leaks.

32. Check and adjust the ignition timing.

3.0L Highlander and RX 300

1. Before servicing the vehicle, refer to the precautions in the beginning of this section.

2. Remove or disconnect the following:
   • Timing belt and idler pulley
   • Camshaft timing pulleys
   • Cylinder head covers

➡ The thrust clearance on both the intake and exhaust camshafts is very small; the camshafts must be kept level during removal. If the camshafts are removed without being kept level, the camshaft may be caught in the cylinder head, causing the head to break or the camshaft to seize.

3. Remove the exhaust and intake camshafts from the right side cylinder head, as follows:

   a. Turn the camshaft with a wrench until the 2 pointed marks drive and driven gears are aligned. (The right camshaft gears have 2 marks apiece; the left side camshaft gears have 1 mark each.)

   b. Secure the exhaust camshaft sub-gear to the main gear using a service bolt. A bolt 0.63–0.79 in. (16–20mm) long with a 6mm thread diameter and a 1mm pitch is recommended. When removing the exhaust camshaft be sure the sub-gear is not loaded; all the force must be eliminated.

   c. Uniformly loosen and remove the exhaust camshaft bearing cap bolts in several passes and in the proper sequence. Remove the 8 bearing cap bolts and remove the caps, keeping them in the correct order.

   d. Remove the exhaust camshaft from the engine.

   e. Uniformly loosen and remove the

Intake

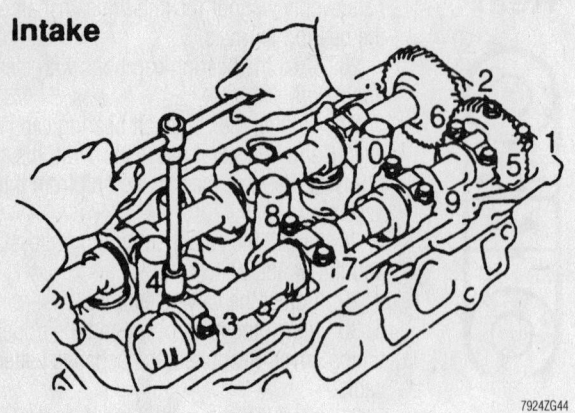

Right intake camshaft bearing cap bolt loosening sequence—3.0L engine

Exhaust

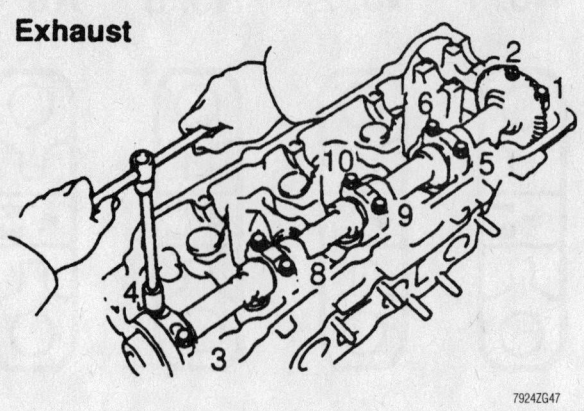

Left side exhaust camshaft bearing cap bolt loosening sequence—3.0L engine

Exhaust

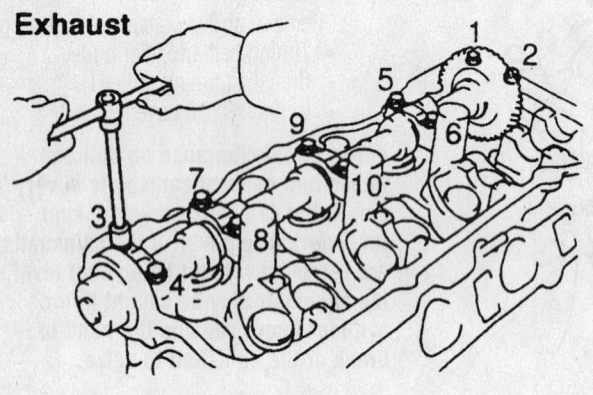

Right side exhaust camshaft bearing cap bolt loosening sequence—3.0L engine

Exhaust

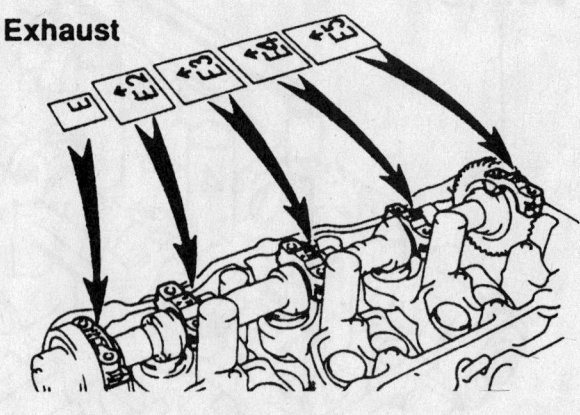

Right exhaust bearing caps must be placed in their proper locations—3.0L engine

Intake

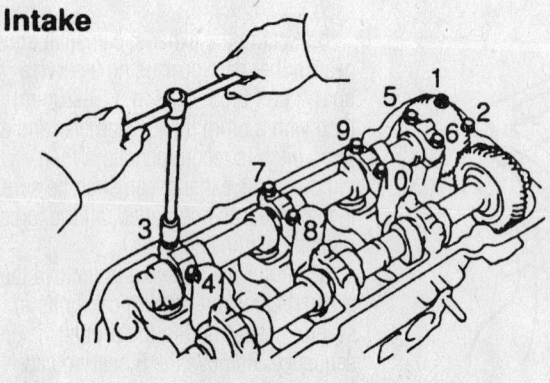

Left intake camshaft bearing cap bolt loosening sequence—3.0L engine

Exhaust

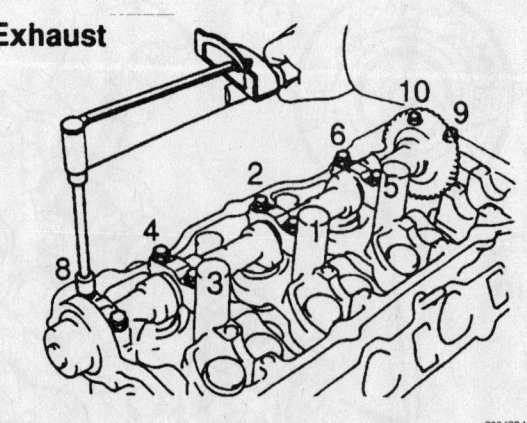

Right exhaust camshaft bearing cap bolt tightening sequence—3.0L engine

10 bearing cap bolts in several passes, in the proper sequence. Remove the bearing caps, keeping them in order, remove the oil seal, then lift out the intake camshaft.

4. Remove the exhaust and intake camshafts from the left side cylinder head, as follows:

a. Turn the camshaft with a wrench until the pointed marks on the drive and driven gears are aligned. (The right camshaft gears have 2 marks apiece; the left side camshaft gears have 1 mark each.)

b. Secure the exhaust camshaft sub-gear to the main gear using a service bolt. A bolt 16–20mm long with a 6mm thread diameter and a 1mm pitch is recommended. When removing the exhaust camshaft be sure the sub-gear is not loaded; all the force must be eliminated.

c. Uniformly loosen and remove the exhaust camshaft bearing cap bolts in several passes and in the proper sequence. Remove the 8 bearing cap bolts and remove the caps. Keep the caps in the correct order.

d. Remove the exhaust camshaft from the engine.

e. Uniformly loosen and remove the 10 bearing cap bolts in several passes, in the reverse order of the installation sequence. Remove the bearing caps, keeping them in order, remove the oil seal, then lift out the intake camshaft.

5. Remove the valve lifter shims and hydraulic lifters. Identify each lifter and shim as it is removed so it can be reinstalled in the same position. If the lifters are to be reused, store them upside down in a sealed container.

To install:

6. Install the valve lifters into their original positions and install the shims. Check valve clearance and replace the shims as necessary.

7. When reinstalling, remember that the camshafts must be handled carefully and kept straight and level to avoid damage.

8. Before installing the camshafts in either cylinder head, apply multi-purpose grease to each camshaft.

9. Install the right camshafts, as follows:

a. Position the intake camshaft on the head so that the alignment marks are at a 90 degrees angle from vertical. The mark should be at the "3 o'clock" position.

b. Apply sealant to the No. 1 bearing cap.

c. Apply a light coat of clean engine oil to the bolt threads and under the bolt

head. Install the bearing caps to their proper position. Tighten the bolts evenly and in several passes to 12 ft. lbs. (16 Nm) in the proper sequence.

d. Position the exhaust camshaft on the head so that the alignment marks are at a 90 degrees angle from vertical. The mark should be at the "9 o'clock" position and must align with the marks on the other gear.

e. Apply a light coat of clean engine oil to the bolt threads and under the bolt head. Install the bearing caps to their proper position. Tighten the bolts evenly and in several passes to 12 ft. lbs. (16 Nm) in the proper sequence.

f. Remove the service bolt.

10. Install the left camshafts, as follows:

a. Position the intake camshaft on the head so that the alignment mark is at a 90 degrees angle from vertical. The mark should be at the "9 o'clock" position.

b. Apply sealant to the No. 1 bearing cap.

c. Apply a light coat of clean engine oil to the bolt threads and under the bolt head. Install the bearing caps to their proper position. Tighten the bolts evenly and in several passes to 12 ft. lbs. (16 Nm) in the proper sequence.

d. Position the exhaust camshaft on the head so that the alignment marks are at a 90 degrees angle from vertical. The mark should be at the "3 o'clock" position and must align with the marks on the other gear.

Intake

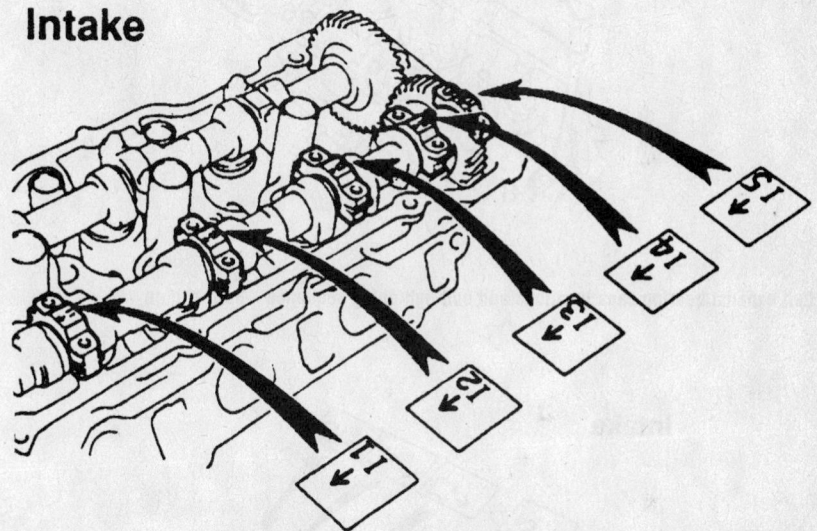

7924ZG50

Right intake bearing caps must be placed in their proper locations—3.0L engine

Intake

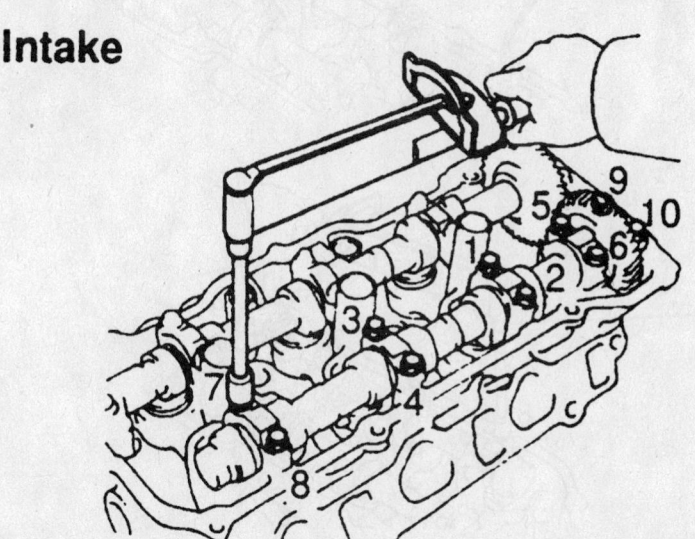

7924ZG51

Right intake camshaft bearing cap bolt tightening sequence—3.0L engine

Exhaust

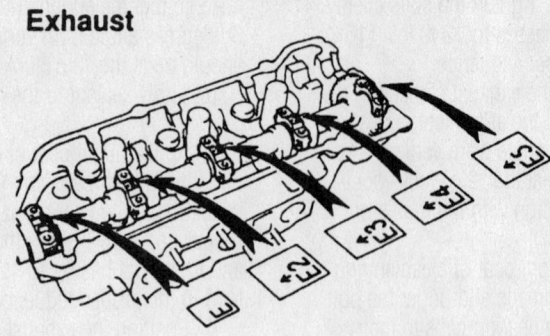

Exhaust

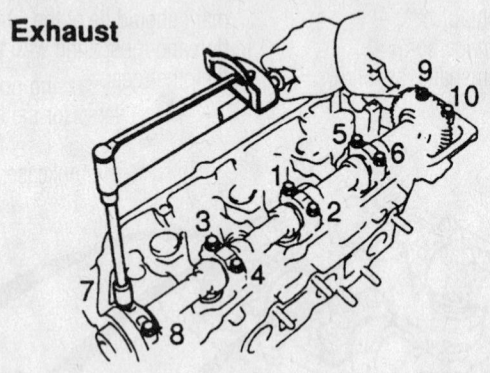

Left exhaust bearing caps locations and bolt tightening sequence—3.0L engine

Intake

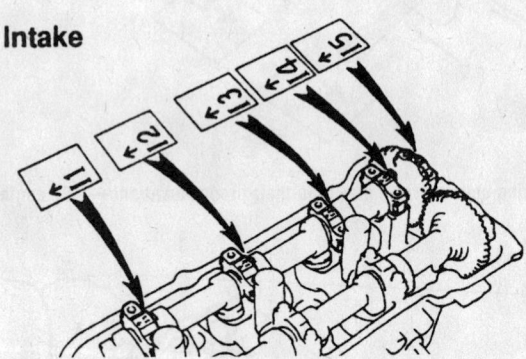

Intake

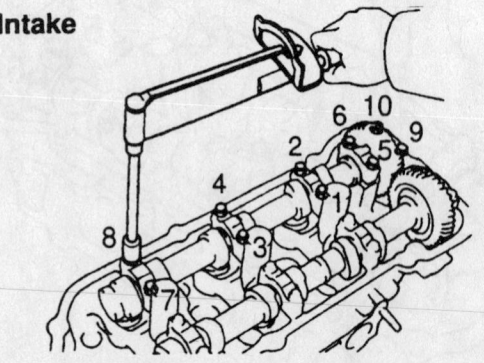

Left intake camshaft bearing cap locations and bolt tightening sequence—3.0L engine

e. Apply a light coat of clean engine oil to the bolt threads and under the bolt head. Install the bearing caps to their proper position. Tighten the bolts evenly and in several passes to 12 ft. lbs. (16 Nm) in the proper sequence.

f. Remove the service bolt.

11. Install or connect the following:
- New camshaft oil seals, lubricated with multi-purpose grease
- No. 3 (rear) timing belt cover
- Camshaft timing gears
- Idler pulley, timing belt and covers

12. Check and adjust the valve clearance.

13. Install the cylinder head (valve) covers.

14. Start the engine. Check the ignition timing.

15. Test drive the vehicle.

16. Check all fluid levels.

2.4L Highlander

1. Before servicing the vehicle, refer to the precautions in the beginning of this section.

2. Disconnect the negative battery cable.

3. Drain the engine coolant.

4. Remove or disconnect the following:
- Right front wheel
- Right fender splash shield
- Right fender apron seal
- No.1 engine undercover
- Coil pack
- Cylinder head cover

5. Set the No.1 piston at TDC compression.

6. Remove or disconnect the following:
- Timing chain tensioner No.1

7. Loosen the camshaft timing gear set bolt.

8. Raise the camshaft and remove the set bolt.

9. Remove or disconnect the following:
- Timing gear and chain
- Exhaust camshaft

10. Loosen the intake camshaft cap bolts in several passes, in the sequence shown. Remove the caps. Remove the camshaft.

11. Installation is the reverse of removal. Tighten the cap bolt, in several passes, in the sequences shown, to:
- Front caps: 22 ft. lbs. (30Nm)
- All other caps: 80 inch lbs. (9Nm)

12. See the Timing Chain Removal and Installation procedure.

2.7L 4Runner

1. Before servicing the vehicle, refer to the precautions in the beginning of this section.

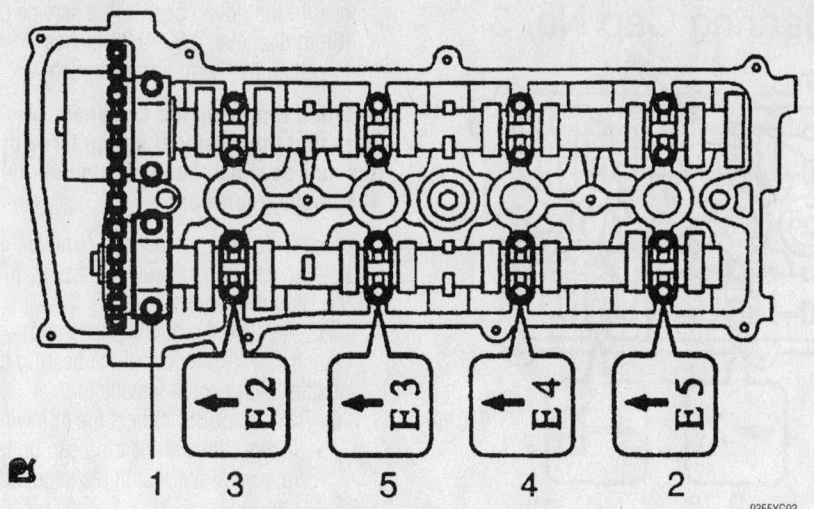

Exhaust camshaft cap bolt loosening sequence—2.4L Highlander

9355YG03

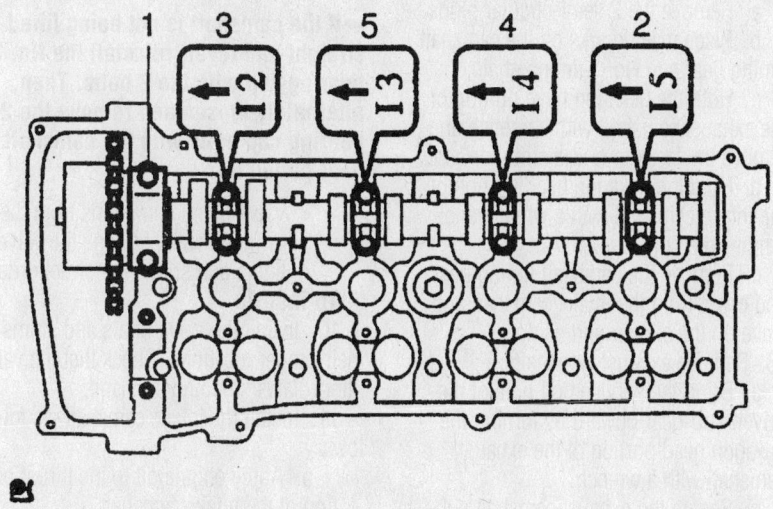

Intake camshaft cap bolt loosening sequence—2.4L Highlander

9355YG04

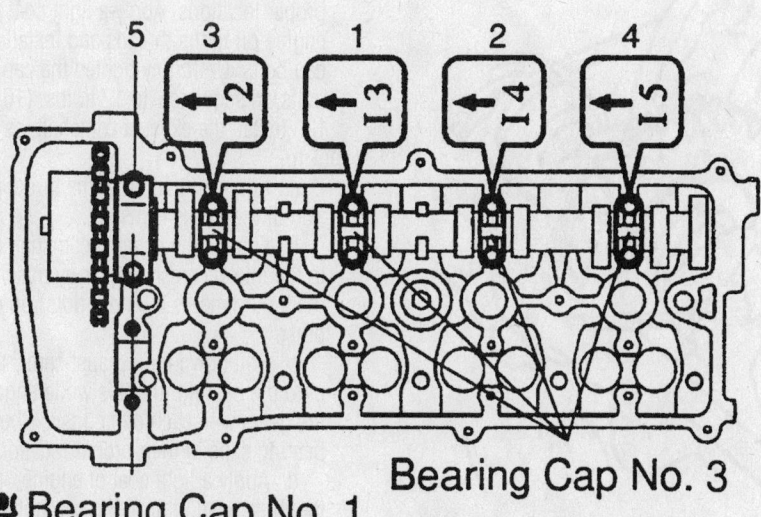

Bearing Cap No. 3

Bearing Cap No. 1

9355YG05

Intake camshaft cap bolt tightening sequence—2.4L Highlander

2. Disconnect the negative battery cable.
3. Drain the engine coolant.
4. Remove or disconnect the following:
- Air cleaner cap
- Mass Air Flow (MAF) meter and the resonator
- Accelerator cable from the throttle body, if equipped with a manual transmission
- Accelerator and throttle cables from the throttle body, if equipped with an automatic transmission
- Cruise control cable from the actuator, if equipped with cruise control
- Intake air connector
- Air hose for Idle Air Control (IAC)
- Vacuum sensing hose
- Wire clamp for the engine wiring harness
- Positive Crankcase Ventilation (PCV) hoses
- Spark plug wires from the spark plugs
- Engine wiring harness clamps and harness
- Air conditioning compressor connector, if equipped with air conditioning
- Oil pressure sensor connector
- Engine Coolant Temperature (ECT) sensor connector
- Engine coolant temperature sender gauge connector
- Exhaust Gas Recirculation (EGR) gas temperature sensor connector
- Vacuum Switching Valve (VSV) connector for the EGR
- 2 vacuum hoses from the VSV for the EGR
- Ground strap from the cowl top panel
- Engine wiring harness from the air intake chamber
- Throttle Position (TP) sensor connector
- IAC valve connector
- Crankshaft Position (CKP) sensor connector
- Knock (KS) sensor connector
- Data Link Connector 1 (DLC1) from the bracket
- Engine wiring harness clamp
- EGR pipe
- Intake chamber stay
- Air intake chamber assembly.
- Evaporative Emission (EVAP) hose from the throttle body
- Brake booster vacuum hose from the union
- Water bypass hose from the water bypass pipe

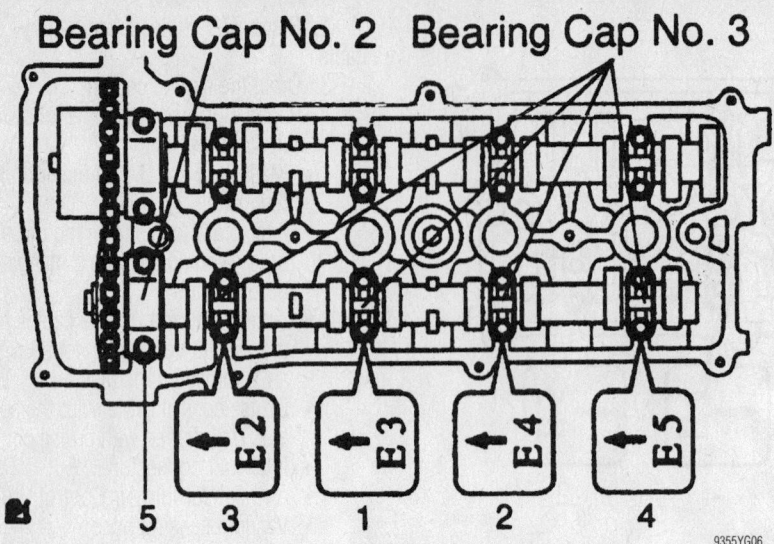

Exhaust camshaft cap bolt tightening sequence—2.4L Highlander

- Water bypass hose from the cylinder head rear cover
- Front engine hanger
- Engine wiring harness brackets
- Cylinder head cover

5. Set No. 1 cylinder to Top Dead Center (TDC) compression stroke. The groove on the crankshaft pulley should align with the **0** mark on the timing chain cover and the timing marks (1 and 2 dots) of the camshaft gears should form a straight line in respect to the cylinder head surface. If not, turn the crankshaft 1 revolution (360 degrees).

6. Remove the chain tensioner and gasket.

7. Remove the camshaft timing gear as follows:

a. Remove the 2 semi-circular plugs.

b. Place matchmarks on the camshaft timing gear and No. 1 timing chain.

c. Hold the hexagon head portion of the exhaust camshaft with a wrench and remove the fastener and distributor gear.

d. Hold the hexagon head portion of the intake camshaft with a wrench and remove the bolt.

e. Remove the camshaft timing gear and chain from the intake camshaft and leave on the slipper and damper.

8. Remove exhaust camshafts:

a. Bring the service bolt hole of the driven sub-gear upward by turning the hexagon head portion of the exhaust camshaft with a wrench.

b. Secure the exhaust camshaft sub-gear to the driven gear with a service bolt (6mm diameter, 0.63–0.79 inches in length and 1.0mm in thread pitch).

➡**When removing the camshaft, be sure that the torsional spring force of the sub-gear has been eliminated by the above operation.**

c. Uniformly loosen and remove the bearing cap bolts in several passes, in the sequence shown.

d. Remove the bearing caps and camshaft. Make a note of the bearing cap positions for proper installation.

9. Remove or disconnect the following:

- Intake camshaft bearing cap bolts in several passes, in the sequence shown
- Bearing caps and camshaft. Make a note of the bearing cap positions for proper installation.

➡**If the camshaft is not being lifted out straight and level, reinstall the No. 3 bearing cap with the 2 bolts. Then, alternately loosen and remove the 2 bearing cap bolts with the camshaft gear pulled up.**

- Valve lifters and shims from the cylinder head. Arrange the valve lifters and shims in correct order.

To install:

10. Install the valve lifters and shims in their proper locations. Check that the valve lifter rotates smoothly by hand.

11. Install the intake camshaft, as follows:

a. Apply engine oil to the thrust portion of the intake camshaft.

b. Position the intake camshaft with the knock pin facing upward.

c. Install the bearing caps in their proper locations. Apply a light coat of engine oil to the threads and install the cap bolts. Uniformly tighten the cap bolts, in sequence, to 12 ft. lbs. (16 Nm).

12. Install the exhaust camshaft, as shown:

a. Apply engine oil to the thrust portion of the intake camshaft.

b. Engage the exhaust camshaft gear to the intake camshaft gear by matching the timing marks (1 and 2 dots) on each other.

c. Roll down the exhaust camshaft onto the bearing journals while engaging the gears with each other. Install the bearing caps in their proper locations.

d. Apply a light coat of engine oil to the threads and install the cap bolts. Uniformly tighten the cap bolts in the sequence shown to 12 ft. lbs. (16 Nm).

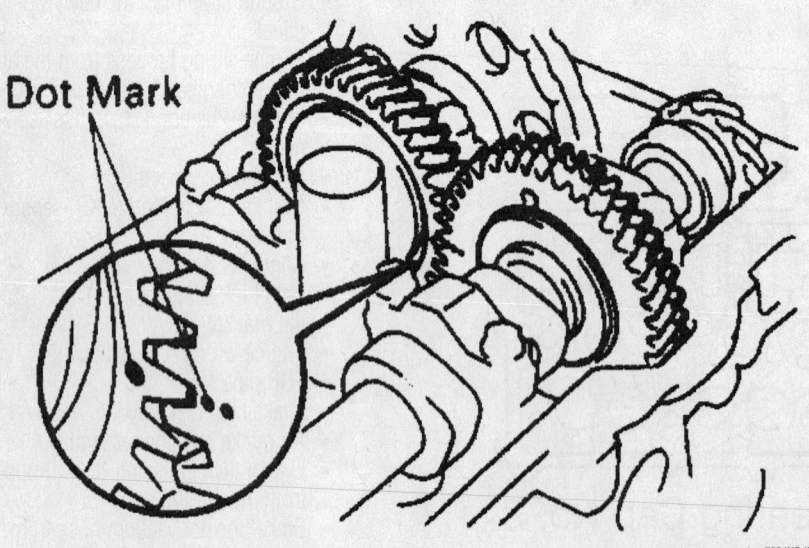

Camshafts TDC/compression timing marks. Marks with 1 and 2 dots will be in straight line on cylinder head surface—2.7L 4Runner

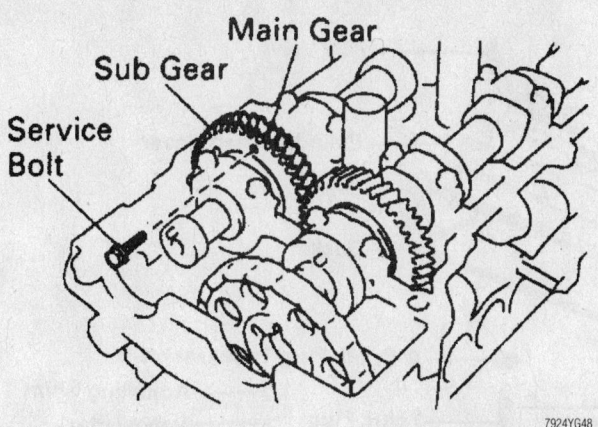

Secure the exhaust camshaft sub-gear to the main gear with a service bolt—2.7L 4Runner

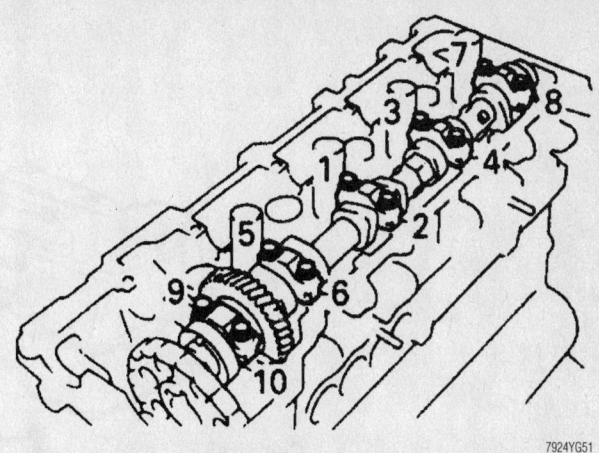

Tighten the intake camshaft bearing cap bolts in sequence—2.7L 4Runner

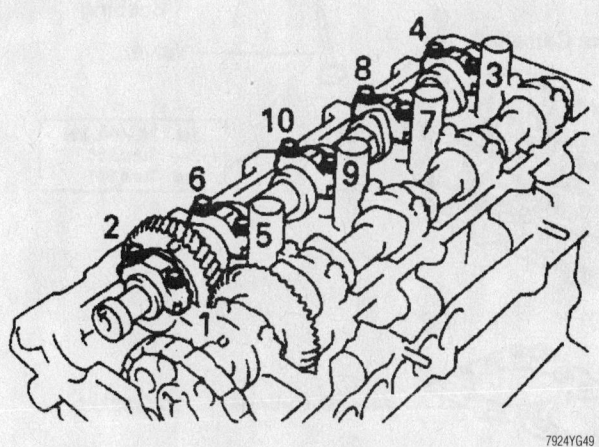

Loosen and remove the exhaust camshaft bearing cap bolts in sequence—2.7L 4Runner

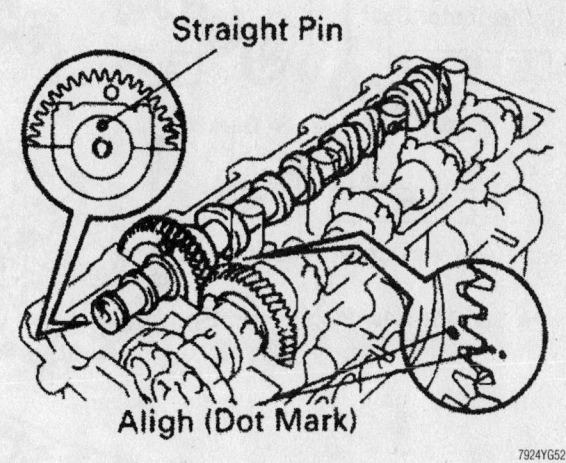

Engage both camshaft gears while matching the timing marks—2.7L 4Runner

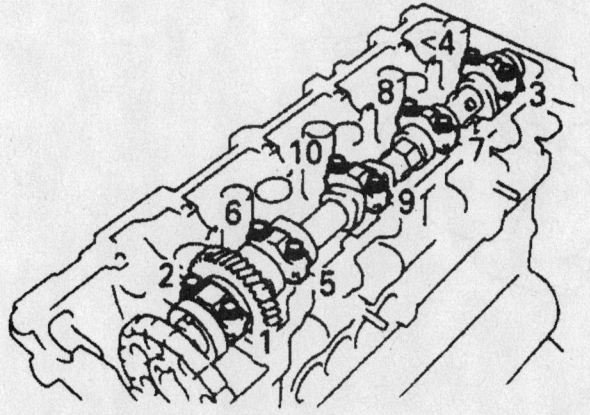

Loosen and remove the intake camshaft bearing cap bolts in sequence—2.7L 4Runner

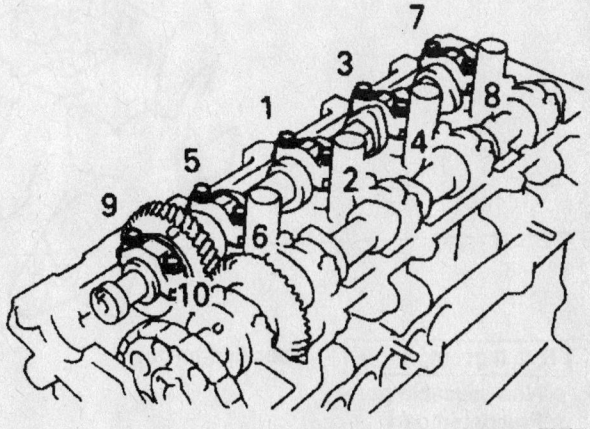

Tighten the exhaust camshaft bearing cap bolts in sequence—2.7L 4Runner

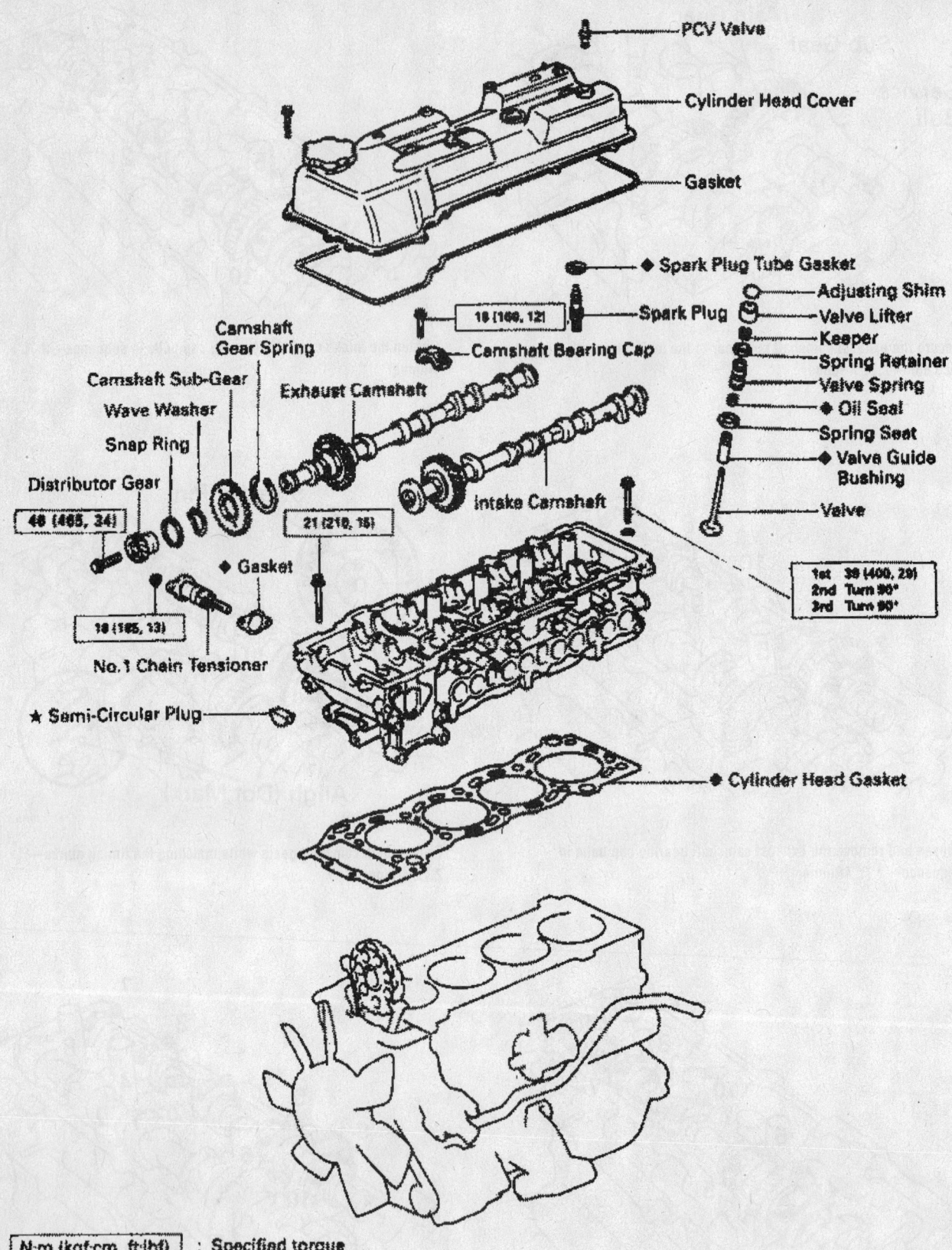

- PCV Valve
- Cylinder Head Cover
- Gasket
- ◆ Spark Plug Tube Gasket
- 18 (160, 12) Spark Plug
- Camshaft Bearing Cap
- Adjusting Shim
- Valve Lifter
- Keeper
- Spring Retainer
- Valve Spring
- ◆ Oil Seal
- Spring Seat
- ◆ Valve Guide Bushing
- Valve

Camshaft Gear Spring
Camshaft Sub-Gear
Wave Washer
Snap Ring
Distributor Gear
Exhaust Camshaft
Intake Camshaft

48 (465, 34)
21 (210, 15)
18 (185, 13)

1st   39 (400, 29)
2nd   Turn 90°
3rd   Turn 90°

- ◆ Gasket
- No.1 Chain Tensioner
- ★ Semi-Circular Plug
- ◆ Cylinder Head Gasket

N·m (kgf·cm, ft·lbf) : Specified torque
◆ Non-reusable part
★ Precoated part

7924YG54

Exploded view of the cylinder head components—2.7L 4Runner

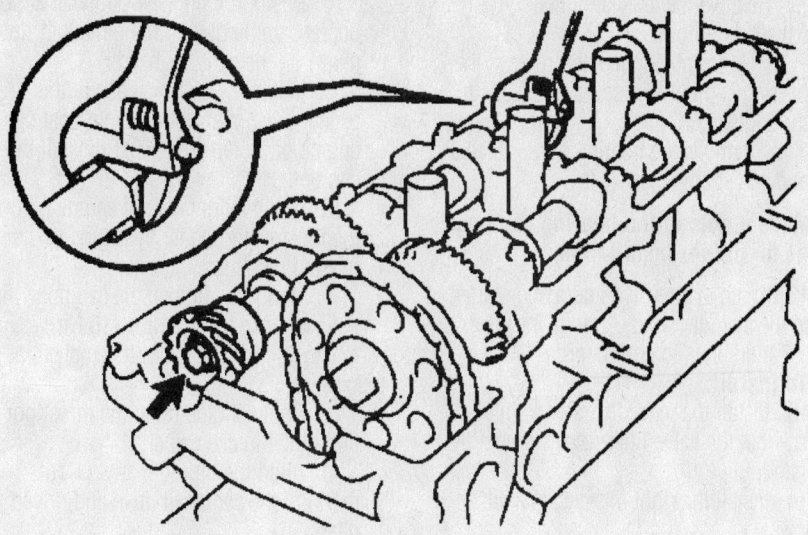

7924YG55

Using a wrench to hold the camshaft—2.7L 4Runner

e. Remove the service bolt from the driven sub-gear. Check that the intake and exhaust camshafts turns smoothly.

13. Set No. 1 cylinder to Top Dead Center (TDC) of the compression stroke: Crankshaft pulley groove align with **0** mark on timing cover and camshafts timing marks with 1 dot and 2 dots will be straight line on the cylinder head surface.

14. Install the timing gear. Place the gear over the straight pin of the intake camshaft.

   a. Hold the intake camshaft with a wrench. Install and tighten the bolt to 54 ft. lbs. (74 Nm).

   b. Hold the exhaust camshaft and install the bolt and distributor gear. Tighten the bolt to 34 ft. lbs. (46 Nm).

15. Install or connect the following:
   • Chain tensioner, using a new gasket (mark toward the front)
   • Recheck the engine for proper valve timing. Check and adjust the valve clearance.
   • Semi-circular plug
   • Recheck the engine for proper valve timing.
   • Cylinder head cover with a new gasket
   • Engine wiring harness brackets
   • Front engine hanger. Tighten the bolts to 30 ft. lbs. (42 Nm).
   • Air intake chamber assembly. Tighten the bolts to 15 ft. lbs. (20 Nm).
   • Hoses
   • Intake chamber stay
   • Air intake chamber stay. Tighten the bolts to 15 ft. lbs. (20 Nm).
   • EGR pipe. Tighten the bolts to 13 ft. lbs. (18 Nm), nut "A" to 14 ft. lbs.

(19 Nm) and nut "B" to 15 ft. lbs. (20 Nm).
   • Engine wiring harness
   • Spark plug wires to the spark plugs
   • PCV hoses
   • Intake air connector. Tighten the bolts to 13 ft. lbs. (18 Nm).
   • Air hose for the IAC
   • Vacuum sensing hose
   • Wire clamp for the engine wiring harness
   • Cruise control cable to the actuator, if equipped with cruise control
   • Accelerator cable to the throttle body, if equipped with a manual transmission
   • Accelerator and throttle cables to the throttle body, if equipped with an automatic transmission
   • Air cleaner cap, MAF meter and the resonator assembly
   • Negative battery cable

16. Refill the cooling system.

17. Start the engine and check for leaks.

18. Check the ignition timing. Road test the vehicle for proper operation.

19. Recheck all fluid levels.

3.4L 4Runner

1. Before servicing the vehicle, refer to the precautions in the beginning of this section.

2. Release the fuel pressure.

3. Remove or disconnect the following:
   • Negative battery cable
   • Engine undercover

4. Drain the cooling system.

5. Remove or disconnect the following:
   • Air cleaner cap
   • Mass Air Flow (MAF) meter and the resonator
   • Actuator cable with the bracket, if equipped with cruise control
   • Accelerator cable
   • Throttle cable, if equipped with an automatic transmission
   • Heater hose
   • Brake booster vacuum hose
   • Evaporative Emission (EVAP) hose
   • Automatic Disconnecting Differential (ADD) vacuum hose, for 4-wheel drive
   • Fuel inlet and fuel return hose
   • Spark plug wires, with the ignition coils
   • Intake chamber stay
   • Camshaft Position (CMP) sensor connector from the No. 2 timing belt cover
   • 3 spark plug wire clamps from the No. 2 timing belt cover
   • 6 bolts and the No. 2 timing belt cover
   • Air intake chamber assembly
   • Throttle Position (TP) sensor connector
   • Idle Air Control (IAC) valve connector
   • Positive Crankcase Ventilation (PCV) hoses
   • Water bypass hoses
   • Air assist hose from the throttle body
   • Intake air connector
   • 6 injector connectors
   • Engine Coolant Temperature (ECT) sensor and sender gauge connectors
   • Engine wiring harness protector, from the cylinder head

6. Set the No. 1 cylinder at Top Dead Center (TDC) of the compression stroke, as follows:

   a. Turn the crankshaft pulley and align its groove with the timing mark **0** on the No. 1 timing belt cover.

   b. Check that the timing marks of the camshaft timing pulleys and the No. 3 timing belt cover are aligned. If not, turn the crankshaft pulley 1 revolution (360 degrees).

7. Remove or disconnect the following:
   • Timing belt tensioner, by alternately loosening the 2 bolts
   • Timing belt
   • Camshaft timing pulley bolt, the timing pulley and the knock pin, using Variable Pin Wrench Set 09960-10010
   • Both timing pulleys

- Bolt and the No. 2 idler pulley
- Camshaft Position (CMP) sensor
- Timing belt cover
- 8 bolts, seal washers, cylinder head covers and gasket
- Semi-circular plugs

8. Remove the right exhaust camshafts, as follows:

a. Bring the service bolt hole of the driven sub-gear upward by turning the hexagon head portion of the exhaust camshaft with a wrench.

b. Align the timing mark (2 dot marks) of the camshaft drive and driven gears by turning the camshaft with a wrench.

c. Secure the exhaust camshaft sub-gear to the driven gear with a service bolt (6mm diameter, 16–20mm bolt length and 1.0mm in thread pitch).

➡ **When removing the camshaft, be sure that the torsional spring force of the sub-gear has been eliminated by the above operation.**

d. Uniformly loosen and remove the bearing cap bolts in several passes, in the sequence.

e. Remove the bearing caps and camshaft. Make a note of the bearing cap positions for proper installation.

9. Remove the right-hand intake camshaft, as follows:

a. Uniformly loosen and remove the bearing cap bolts in several passes, in the sequence.

b. Remove the bearing caps, oil seal and camshaft. Make a note of the bearing cap positions for proper installation.

10. Remove the left exhaust camshafts, as follows:

a. Align the timing mark (1 dot mark) of the camshaft drive and driven gears by turning the camshaft with a wrench.

b. Secure the exhaust camshaft sub-gear to the driven gear with a service bolt (6mm diameter, 16–20mm bolt length and 1.0mm in thread pitch).

➡ **When removing the camshaft, be sure the torsional spring force of the sub-gear has been eliminated by the above operation.**

c. Uniformly loosen and remove the bearing cap bolts in several passes, in the sequence.

d. Remove the bearing caps and camshaft. Make a note of the bearing cap positions for proper installation.

➡ **Do not pry on or attempt to force the camshaft with a tool or other object.**

11. Remove the left-hand intake camshaft, as follows:

a. Uniformly loosen and remove the bearing cap bolts in several passes, in the sequence.

b. Remove the bearing caps, oil seal and camshaft.

➡ **Make a note of the bearing cap positions for proper installation.**

12. Remove the valve lifters and shims from the cylinder head. Arrange the valve lifters and shims in correct order.

To install:

13. Install the valve lifters and shims. Check that the valve lifter rotates smoothly by hand.

14. Install the right intake camshaft, as follows:

a. Apply engine oil to the thrust portion of the intake camshaft.

b. Position the intake camshaft at 90 degrees angle of the timing mark (2 dot marks) on the cylinder head.

c. Install the bearing caps in their proper locations. Apply a light coat of engine oil to the threads and install the cap bolts.

d. Apply a light coat of engine oil on the threads and under the heads of the bearing cap bolts.

e. Uniformly tighten the cap bolts in the sequence to 12 ft. lbs. (16 Nm).

15. Install the right exhaust camshaft, as follows:

a. Apply engine oil to the thrust portion of the intake camshaft.

b. Align the timing marks (2 dot marks) of the camshaft drive and driven gears.

c. Roll down the exhaust camshaft onto the bearing journals while engaging the gears with each other. Install the bearing caps in their proper locations.

d. Apply a light coat of engine oil to the threads and install the cap bolts.

e. Apply a light coat of engine oil on the threads and under the heads of the bearing cap bolts.

f. Uniformly tighten the cap bolts in the sequence to 12 ft. lbs. (16 Nm).

g. Remove the service bolt from the driven sub-gear. Check that the intake and exhaust camshafts turn smoothly.

h. Align the timing marks (2 dot marks) of the camshaft drive and driven gears by turning the camshaft with a wrench.

16. Install the left intake camshaft, as follows:

a. Apply engine oil to the thrust portion of the intake camshaft.

b. Position the intake camshaft at 90 degrees angle of the timing mark (1 dot mark) on the cylinder head.

c. Install the bearing caps in their proper locations. Apply a light coat of engine oil to the threads and install the cap bolts.

d. Apply a light coat of engine oil on the threads and under the heads of the bearing cap bolts.

e. Uniformly tighten the cap bolts in the sequence to 12 ft. lbs. (16 Nm).

17. Install the left exhaust camshaft, as follows:

a. Apply engine oil to the thrust portion of the intake camshaft.

b. Align the timing marks (1 dot mark) of the camshaft drive and driven gears.

c. Roll down the exhaust camshaft onto the bearing journals while engaging the gears with each other. Install the bearing caps in their proper locations.

d. Apply a light coat of engine oil to the threads and install the cap bolts.

e. Apply a light coat of engine oil on the threads and under the heads of the bearing cap bolts.

f. Uniformly tighten the cap bolts in the sequence to 12 ft. lbs. (16 Nm).

g. Remove the service bolt.

18. Check and adjust the valve clearance.

19. Install or connect the following:

- Semi-circular plugs
- Cylinder head covers. Tighten the bolts, in several passes, to 53 inch lbs. (6 Nm).
- No. 3 timing belt cover. Tighten the 6 bolts to 80 inch lbs. (9 Nm).
- CMP sensor. Tighten it to 71 inch lbs. (8 Nm).
- No. 2 timing belt idler. Tighten the bolt to 30 ft. lbs. (40 Nm).

➡ **Check that the pulley bracket moves smoothly.**

20. Install the left camshaft timing pulley, as follows:

a. Install the knock pin to the camshaft.

b. Align the knock pin hose of the camshaft with the knock pin groove of the timing pulley.

c. Slide the timing pulley on the camshaft with the flange side facing outward. Tighten the pulley bolt to 81 ft. lbs. (110 Nm).

21. Set the No. 1 cylinder to Top Dead Center (TDC) of the compression stroke, as follows:

a. Turn the crankshaft pulley, and

align its groove with the timing mark **0** on the No. 1 timing belt cover.

b. Turn the camshaft, align the knock pin hole of the camshaft with the timing mark of the No. 3 timing belt cover.

c. Turn the camshaft timing pulley, align the timing marks of the camshaft timing pulley and the No. 3 timing belt cover.

22. Install or connect the following:
- Timing belt to the left camshaft timing pulley. Check that the installation mark on the timing belt is aligned with the end of the No. 1 timing belt cover.
- Right camshaft timing pulley
- Timing belt

23. Set the timing belt tensioner, as follows:

a. Using a press, slowly press in the pushrod using 220–2,205 lbs. (981–9,807 N) of force.

b. Align the holes of the pushrod and housing, pass a 1.5mm hexagon wrench through the holes to keep the setting position of the pushrod.

c. Release the press and install the dust boot to the tensioner.

24. Install the timing belt tensioner and alternately tighten the bolts to 20 ft. lbs. (28 Nm). Using pliers, remove the 1.5mm hexagon wrench from the belt tensioner.

25. Check the valve timing, as follows:

a. Slowly turn the crankshaft pulley 2 revolutions from the TDC-to-TDC. Always turn the crankshaft pulley clockwise.

b. Check that each pulley aligns with the timing marks. If the timing marks do not align, remove the timing belt and reinstall it.

26. Install or connect the following:
- Engine wiring harness to the cylinder head
- 3 engine wiring harness clamps.
- 6 injector connectors
- ECT sender gauge connector
- ECT sensor connector
- Intake air connector
- Air intake chamber assembly. Tighten the 4 bolts and 2 nuts to 13 ft. lbs. (18 Nm).
- Intake chamber stay. Tighten the 2 bolts to 30 ft. lbs. (40 Nm).
- New O-ring to the oil filler tube
- Oil filler tube end into the oil pan tube hole
- Oil filler tube and No. 1 throttle cable clamp
- No. 2 timing belt cover. Tighten the bolts to 80 inch lbs. (9 Nm).
- Remaining components
- Negative battery cable

27. Fill the cooling system.
28. Start the engine and check for leaks.
29. Check the ignition timing.
30. Install the engine undercover.
31. Road test the vehicle.
32. Recheck all fluid levels.

Valve Lash

ADJUSTMENT

2.0L

1. Before servicing the vehicle, refer to the precautions in the beginning of this section.

2. Remove the cylinder head covers.

3. Use a wrench to turn the crankshaft until the notch in the pulley aligns with timing mark **0** of the No. 1 timing belt cover. This will ensure that the No. 1 piston is at Top Dead Center (TDC) of the compression stroke.

➡Check that the valve lifters on the No. 1 cylinder are loose and those on the No. 4 cylinder are tight. If not, rotate the crankshaft 1 complete revolution (360 degrees) and then realign the marks.

4. Using a flat feeler gauge measure the clearance between the camshaft lobe and the valve lifter on the first set of valves

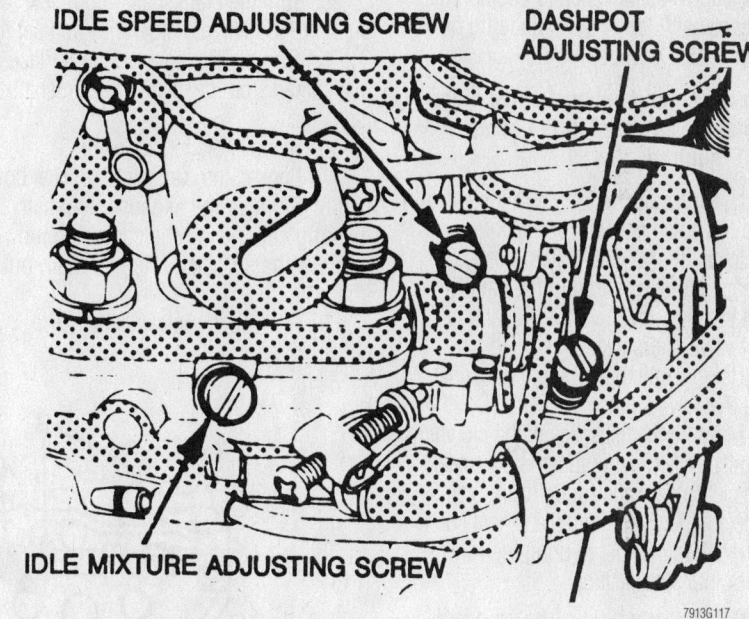

Adjust these valve first—2.0L

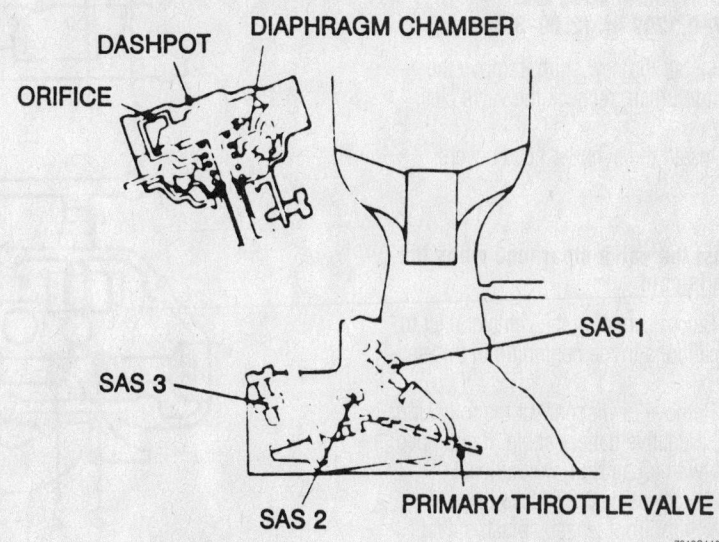

Adjust these valve second—2.0L

shown. This measurement should correspond to specifications.

➡ **If the measurement is within specifications, go on to the next step. If not, record the measurement taken for each individual valve.**

5. Rotate the crankshaft 1 complete revolution and realign the timing marks.

6. Measure the clearance of the second set of valves.

➡ **If the measurement for this set of valves (and also the previous one) is within specifications, go no further, the procedure is finished. If not, record the measurements and proceed to the next step.**

7. Rotate the crankshaft to position the intake camshaft lobe of the cylinder to be adjusted, facing upward.

➡ **Both intake and exhaust valve clearance may be adjusted at the same time, if required.**

8. Using a suitable tool, turn the valve lifter so the notch is easily accessible.

9. Install tool 09248-55010 between both camshaft lobes and turn the handle so the tool presses down both intake and exhaust valve lifters evenly.

10. Using a suitable tool and a magnet, remove the valve shims.

11. Measure the thickness of the old shim with a micrometer. Using this measurement and the clearance made earlier (from Step 3 or 5), determine what size replacement shim will be required in order to bring the valve clearance into specification.

➡ **Replacement shims are available in 27 sizes, in increments of 0.0020 in. (0.05mm). Shim sizes are 0.0787–0.1299 in. (2.00–3.30mm).**

12. Install the new shim, remove the special tool; then, recheck the valve clearances.

13. Install the cylinder head covers.

2.4L

➡ **Adjust the valve clearance when the engine is cold.**

1. Before servicing the vehicle, refer to the precautions in the beginning of this section.

2. Remove or disconnect the following:
- Negative battery cable. If equipped with an air bag, wait at least 90 seconds before proceeding.
- Right front wheel, splash shield and apron seal

- engine undercover
- Coil pack
- Air intake hoses
- Cylinder head cover

3. Place the No.1 piston at TDC compression. Check only those valves shown. Record the clearance. If out of clearance, the measurement will be used to calculate the adjusting shims.

4. Place the No.4 piston at TDC compression. Check only those valves shown. Record the clearance. If out of clearance, the measurement will be used to calculate the adjusting shims.

Clearance range is:
- Intake: 0.19–0.29mm
- Exhaust: 0.30–0.40mm

To adjust the valves:

5. Turn the crankshaft 1 complete revolution (360 degrees) clockwise and set the No.1 piston at TDC compression. Place matchmarks on the chain and camshaft sprocket.

6. Remove the tensioner.

7. Loosen the camshaft sprocket bolt.

8. Remove the exhaust camshaft bearing caps, raise the camshaft and remove the sprocket. Tie the chain out of the way.

9. Remove the intake camshaft.

10. Remove the lifters and keep them in order.

11. Measure the thickness of any lifter on which the clearance was out of range. Calculate the thickness of the necessary replacement lifter. Lifters are available in 0.020mm increments from 5.060mm to 5.740mm.

12. For Camshaft and Timing Chain installation, see the respective procedures in this section.

3.0L

➡ **Adjust the valve clearance when the engine is cold.**

1. Before servicing the vehicle, refer to the precautions in the beginning of this section.

2. Remove or disconnect the following:
- Negative battery cable. If equipped with an air bag, wait at least 90 seconds before proceeding.
- Accelerator/throttle cable from the throttle linkage
- Air cleaner cover, air flow meter and air duct assembly
- V-bank cover
- Emission control valve set
- Air intake chamber
- Engine harness from the injectors and the ignition coils

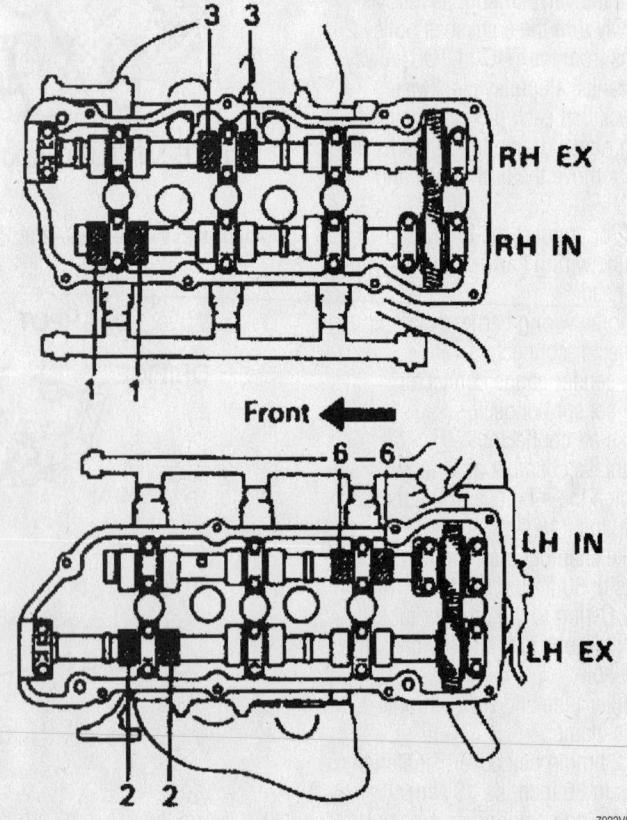

Adjust these valves during the 1st step—3.0L

7923VG65

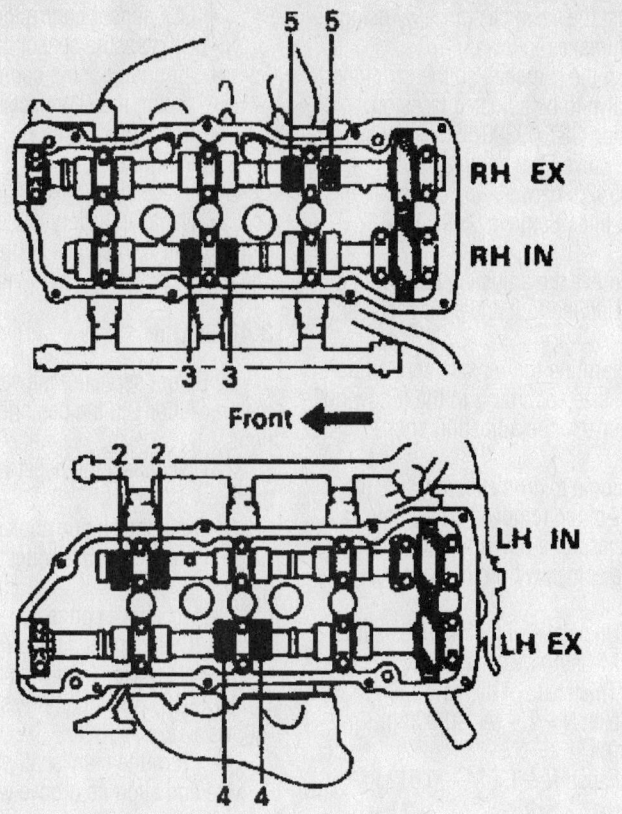

Adjust these valves during the 2nd step—3.0L

7923VG66

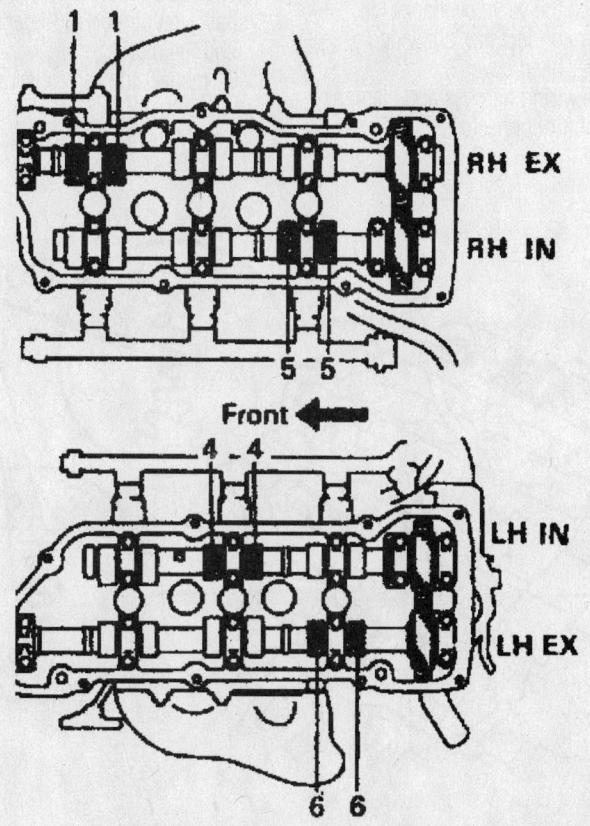

Adjust these valves during the 3rd step—3.0L

7923VG67

- Ignition coils and keep them in order for reassembly
- Spark plugs
- Cylinder head covers

3. Turn the crankshaft pulley and align its groove with the timing mark **0** of the No. 1 timing cover.

4. Check that the valve lifters on the No. 1 intake are loose and the No. 1 exhaust are tight. If not, turn the crankshaft 1 complete revolution (360 degrees).

➡**All measurements should be written down. These recorded measurements will need to be used in conjunction with a mathematical formula to determine the thickness of the replacement shims.**

5. Measure the clearance between the valve lifters and the camshaft. Record the measurements on valves No. 1 and 6 intake; No. 2 and 3 exhaust.

   a. The intake valve clearance cold is 0.006–0.010 in. (0.15–0.25mm).

   b. The exhaust valve clearance cold is 0.010–0.014 in. (0.25–0.35mm).

6. Turn the crankshaft ⅔ of a revolution (240 degrees). Record the measurements on valves No. 2 and 3 intake; No. 4 and 5 exhaust.

7. Turn the crankshaft another ⅔ of a revolution. Record the measurements on valves No. 4 and 5 intake; No. 1 and 6 exhaust.

8. Remove the adjusting shim by turning the crankshaft to position the cam lobe of the camshaft in the up position on the valve to be adjusted. Using a small thin flat bladed tool, turn the valve lifter so that the notches are perpendicular to the camshaft. Press down the valve lifter with tool 09248-55010 part A. Place too 09248-55010 part B between the camshaft and the valve lifter; remove part A.

9. Remove the adjusting shim with a magnet and a small screwdriver.

10. Determine the replacement adjusting shim size by either using the charts or the following formulas:

- Intake: N = T + (A—0.008 in./0.020mm)
- Exhaust: N = T + (A—0.012 in./0.30mm)
- T = Thickness of removed shim
- A = Measured valve clearance
- N = Thickness of new shim

11. Select a new shim with a thickness as close as possible to the calculated value. Install the new replacement shim.

➡**Shims are available in 17 sizes in increments of 0.0020 in. (0.050mm), from 0.0984 in. (2.500mm) to 0.1299 in. (3.300mm).**

12. Recheck the valve clearance.
13. Install or connect the following:
- Cylinder head covers
- Spark plugs and the ignition coils
- Engine wiring harness to the injectors and the coils
- Intake chamber
- Emission control valve set
- V-bank cover
- Air flow meter, air duct and air cleaner cover
- Negative battery cable

2.7L Engines

1. Before servicing the vehicle, refer to the precautions in the beginning of this section.
2. Disconnect the negative battery cable.
3. Drain the engine coolant.
4. Remove or disconnect the following:
- Intake air connector
- Positive Crankcase Ventilation (PCV) hoses
- Spark plug wires
- Engine wiring harness clamps and harness
- Air conditioning compressor connector, if equipped with air conditioning
- Oil pressure sensor connector
- Engine Coolant Temperature (ECT) sensor connector
- Distributor connector
- Cylinder head cover

5. Set the No. 1 cylinder to Top Dead Center (TDC) of the compression stroke, as follows:

a. Turn the crankshaft pulley clockwise and align its groove with the **0** mark on the timing chain cover.

b. Check that the timing marks (1 and 2 dots) of the camshaft drive and driven gears are in a straight line on the cylinder head surface. If not, turn the crankshaft 1 revolution (360 degrees) and align the marks.

6. Inspect the valve clearance, as follows:

a. Measure the clearance between the valve lifter and the camshaft. Measure the 1st and 2nd intake and the 1st and 3rd exhaust valves.

b. Turn the crankshaft pulley 1 revolution (360 degrees) and align the marks as above. Measure the 3rd and 4th intake and the 2nd and 4th exhaust valves.

7. Valve clearance cold should be:
- Intake: 0.006–0.010 in. (0.15–0.25mm)
- Exhaust: 0.010–0.014 in. (0.25–0.35mm)

8. Adjust the valve clearance by using adjusting shims, as follows:

a. Turn the camshaft so the cam lobe for the valve to be adjusted faces up.

b. Using SST 09248-55040, press down the valve lifter and place SST 09248-05420, between the camshaft and the valve lifter. Remove SST 09248-55040.

c. Remove the adjusting shim with a small flat prying tool and a magnetic finger.

d. Determine the replacement adjusting shim size according to the following formula or use the adjusting shim charts.

e. Using a micrometer, measure the thickness of the removed shim. Calculate the thickness of a new shim so the valve clearance comes within the specified value.
- T: Thickness of the removed shim
- A: Measured valve clearance
- N: Thickness of the new shim

f. Intake: $N = T + (A - 0.008 \text{ in.} (0.20\text{mm}))$

g. Exhaust: $N = T + (A - 0.012 \text{ in.} (0.30\text{mm}))$

h. Install a new adjusting shim. Place it on the valve lifter. Using the SST 09248-55040, press down the valve lifter and remove SST 09248-05420.

i. Recheck the valve clearance.

9. Install or connect the following:
- Cylinder head cover
- Engine wiring harness and clamps
- Distributor connector

- ECT sensor connector
- Oil pressure sensor connector
- Air conditioning compressor connector, if disconnected
- Spark plug wires
- PCV hoses
- Intake air connector
- Negative battery cable

10. Refill with engine coolant.
11. Check the ignition timing.

3.4L Engine

1. Before servicing the vehicle, refer to the precautions in the beginning of this section.
2. Disconnect the negative battery cable.
3. Drain the engine coolant.
4. Remove or disconnect the following:
- Air intake connector
- Cylinder head cover

5. Set the No. 1 cylinder to Top Dead Center (TDC) of the compression stroke, as follows:

a. Turn the crankshaft pulley clockwise and align its groove with the **0** mark on the timing chain cover.

b. Check that the timing marks (1 and 2 dots) of the camshaft drive and driven gears are in a straight line on the cylinder head surface. If not, turn the crankshaft 1 revolution (360 degrees) and align the marks.

6. Inspect the valve clearance, as follows:

a. Measure the clearance between the

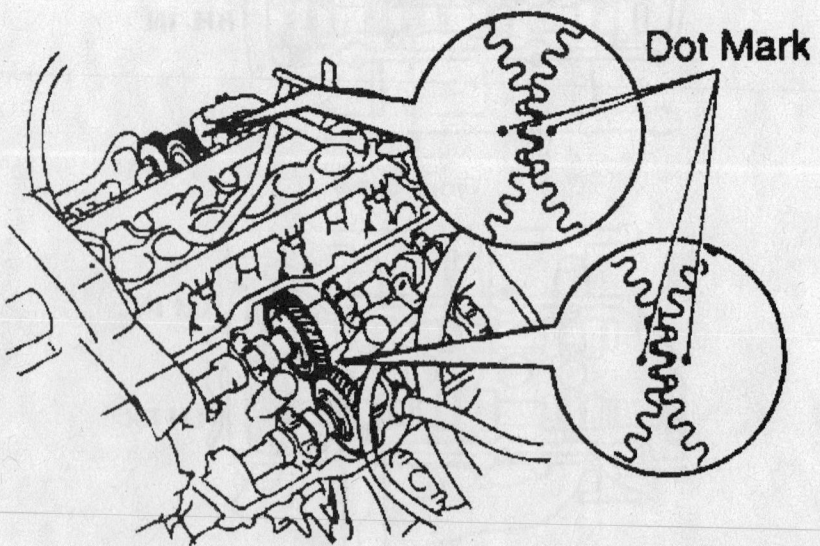

Aligning the timing marks—3.4L engine

7924YG81

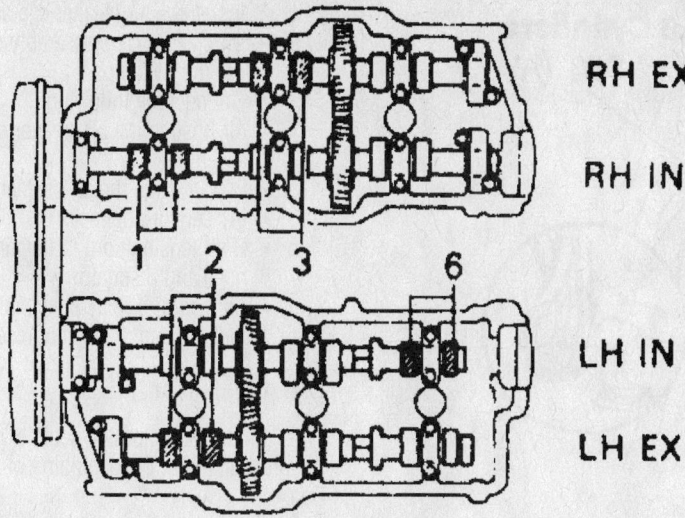

RH EX

RH IN

1 2 3 6

LH IN

LH EX

7924YG82

First valve adjustment—3.4L engine

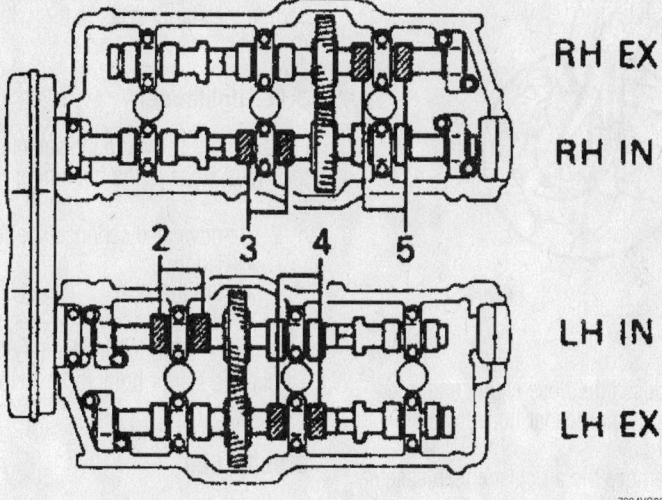

RH EX

RH IN

2 3 4 5

LH IN

LH EX

7924YG83

Second valve adjustment—3.4L engine

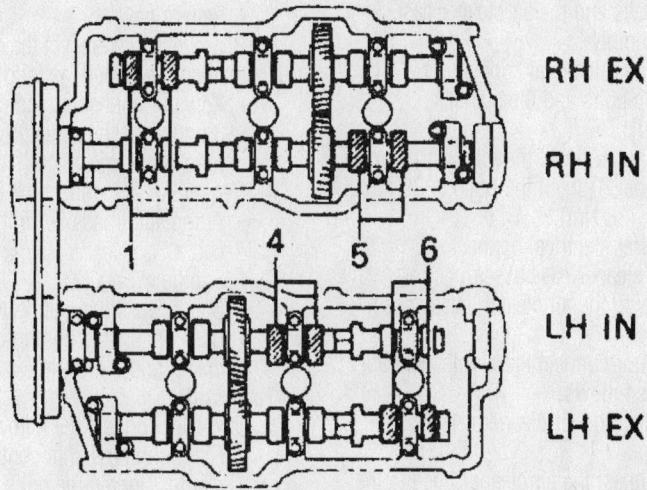

RH EX

RH IN

1 4 5 6

LH IN

LH EX

7924YG84

Third valve adjustment—3.4L engine

valve lifter and the camshaft. Measure the 1st intake and the 3rd exhaust valves on the right head and the 6th intake and the 2nd exhaust valves on the left head.

b. Turn the crankshaft ⅔ of a revolution (240 degrees) and adjust the 3rd intake and the 5th exhaust valves on the right head and the 2nd intake and the 4th exhaust valves on the left head.

c. Turn the crankshaft ⅔ of a revolution (240 degrees) and adjust the 5th intake and the 1st exhaust valves on the right head and the 4th intake and the 6th exhaust valves on the left head.

7. Valve clearance cold should be:
- Intake: 0.006–0.009 in. (0.13–0.23mm)
- Exhaust: 0.011–0.014 in. (0.27–0.37mm)

8. Adjust the valve clearance by using adjusting shims, as follows:

a. Turn the equipment camshaft so that the cam lobe for the valve to be adjusted faces up.

b. Turn the valve lifter so that the notches are perpendicular to the camshaft.

c. Using SST 09248-55040, press down the valve lifter and place SST 09248-05420, between the camshaft and the valve lifter. Remove SST 09248-55040.

d. Remove the adjusting shim with a small flat prying tool and a magnetic finger.

e. Determine the replacement adjusting shim size according to the following formula or use the adjusting shim charts.

f. Using a micrometer, measure the thickness of the removed shim. Calculate the thickness of a new shim so that the valve clearance comes within the specified value.
- T: Thickness of the removed shim
- A: Measured valve clearance
- N: Thickness of the new shim

g. Intake: $N = T + (A—0.007$ in. $(0.18mm))$

h. Exhaust: $N = T + (A—0.013$ in. $(0.32mm))$

i. Install a new adjusting shim. Place it on the valve lifter. Using the SST 09248-55040, press down the valve lifter and remove SST 09248-05420.

j. Recheck the valve clearance.

9. Install or connect the following:
- Cylinder head cover
- Intake air connector
- Negative battery cable

10. Refill with engine coolant.

11. Start the engine and check for leaks.

Front of No.1 and Rear of No.6 Cylinders

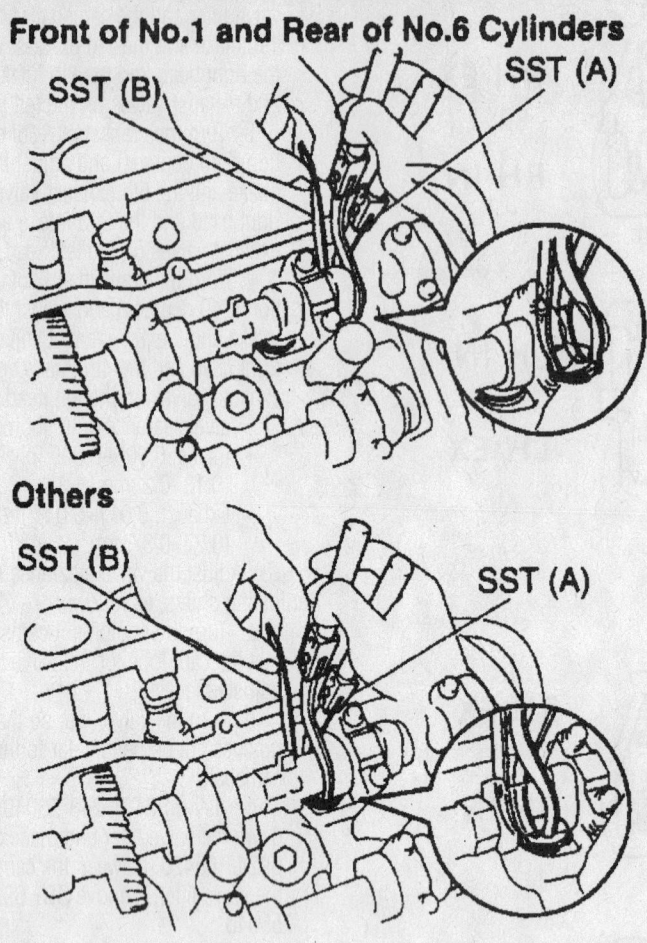

SST (B)   SST (A)

Others

SST (B)   SST (A)

7924YG85

Removing the adjusting shim—3.4L engine

Starter Motor

REMOVAL & INSTALLATION

2.0L RAV4

1. Before servicing the vehicle, refer to the precautions in the beginning of this section.

2. Remove the engine coolant reservoir.

3. Remove the air cleaner cap assembly, as follows:

   a. Disconnect the following:
   • Skid control relay connectors
   • High tension cord from the air cleaner hose and resonator
   • Intake Air Temperature (IAT) sensor connector
   • Positive Crankcase Ventilation (PCV) hose from the air cleaner hose
   • Air hose from the air cleaner cap assembly

   b. The air cleaner cap assembly from the air cleaner case assembly by removing the 4 clamps.

   c. Loosen the hose clamp and disconnect the air cleaner hose from the throttle body.

   d. Remove the air cleaner cap assembly.

4. Remove or disconnect the following:
   • Vacuum Switching Valve (VSV) from the air cleaner case assembly
   • 3 bolts and the air cleaner case assembly
   • Starter electrical connectors
   • Both bolts and the starter

To install:

5. Install or connect the following:
   • Starter. Tighten the bolts to 29 ft. lbs. (38 Nm).
   • Starter electrical connectors
   • Air cleaner case assembly
   • VSV to the air cleaner case assembly

6. Install or connect the air cleaner cap assembly, as follows:

   a. Install the air cleaner cap assembly.

   b. Connect the air cleaner hose to the throttle body and tighten the hose clamp.

   c. Install the air cleaner cap assembly to the air cleaner case assembly by installing the 4 clamps.

   d. Connect the following:
   • Air hose to the air cleaner cap assembly
   • PCV hose to the air cleaner hose
   • IAT sensor connector
   • High tension cord to the air cleaner hose and resonator
   • Skid control relay connectors

7. Install the engine coolant reservoir.

2.4L Highlander

1. Before servicing the vehicle, refer to the precautions in the beginning of this section.

2. Remove or disconnect the following:
   • Air cleaner
   • starter wiring
   • Starter

3. Installation is the reverse of removal. Torque the starter bolts to 29 ft. lbs. (39Nm).

3.0L Highlander

1. Before servicing the vehicle, refer to the precautions in the beginning of this section.

2. Remove or disconnect the following:
   • Battery
   • Air cleaner
   • Starter

3. Installation is the reverse of removal. Torque the starter bolts to 31 ft. lbs. (42Nm).

3.0L RX300

1. Before servicing the vehicle, refer to the precautions in the beginning of this section.

2. Remove or disconnect the following:
   • Battery
   • Battery tray

3. Remove or disconnect the cruise control actuator, if equipped, as follows:
   • Actuator connector and clamp
   • 3 bolts and the actuator with the bracket

4. Remove or disconnect the following:
   • Automatic transaxle shift control cable
   • Engine wiring
   • Starter electrical connectors
   • Both bolts, shift control cable clamp and the starter

To install:

5. Install or connect the following:
   • Starter and the shift control cable clamp. Tighten the bolts to 27 ft. lbs. (37 Nm).

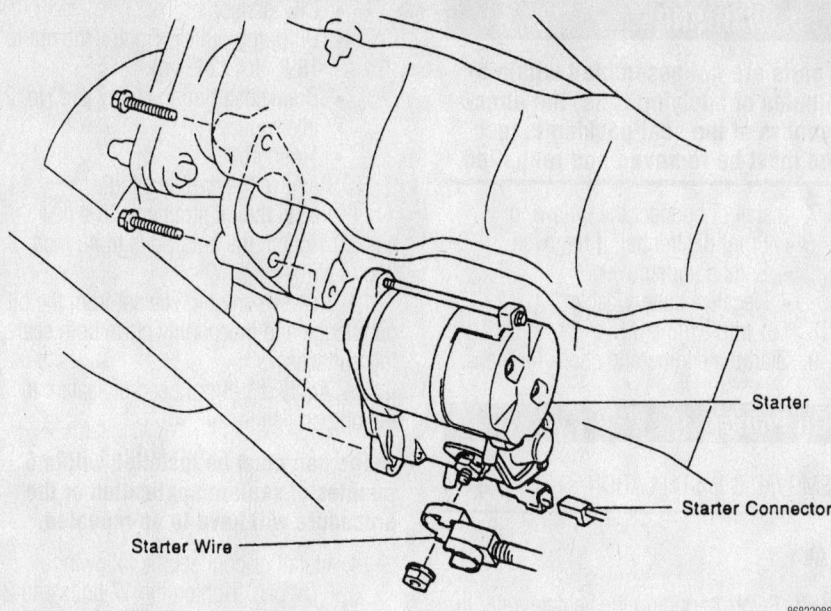

Exploded view of a 2.7L engine starter

- Starter electrical connectors
- Engine wiring
- Automatic transaxle shift control cable

6. Install or connect the following, if equipped with cruise control:
- 3 bolts and the actuator with the bracket
- Actuator connector and clamp

7. Install or connect the following:
- Battery tray
- Battery

2.7L & 3.4L 4Runner

1. Before servicing the vehicle, refer to the precautions in the beginning of this section.
2. Remove or disconnect the following:
- Negative battery cable
- Starter electrical connectors
- Starter

To install:

3. Install or connect the following:
- Starter. Tighten the fasteners to 29 ft. lbs. (39 Nm).
- Electrical connections
- Negative battery cable

Oil Pan

REMOVAL & INSTALLATION

RAV4

1. Before servicing the vehicle, refer to the precautions in the beginning of this section.

2. Remove the right-hand engine undercover.
3. Drain the crankcase oil.
4. Remove or disconnect the following:
- Dipstick
- Front exhaust pipe
- Stiffener plate from the engine by removing the 2 (manual transmission) or 3 (automatic transmission) bolts.
- 2 nuts and 17 bolts from the oil pan
- Oil pan and discard the gasket

To install:

5. Clean all gasket surfaces completely.
6. Apply a thin bead of sealer to the oil pan mounting surfaces.
7. Install or connect the following:
- Oil pan. Tighten the nuts/bolts to 48 inch lbs. (5 Nm).
- Stiffener plate. Tighten the bolts to 27 ft. lbs. (37 Nm).
- Front exhaust pipe
8. Fill the engine with oil to the proper level.
9. Start the engine and check for leaks. Recheck the engine oil level.
10. Install the right engine cover.

2.4L Highlander

1. Before servicing the vehicle, refer to the precautions in the beginning of this section.
2. Remove or disconnect the following:
- Engine undercover
- Engine oil
- Oil pan bolts, nuts and pan

3. Installation is the reverse of removal. Torque the 12 bolts and 2 nuts to 80 inch lbs. (9Nm).

3.0L Highlander and RX 300

1. Before servicing the vehicle, refer to the precautions in the beginning of this section.
2. Remove or disconnect the following:
- Right front wheel
- Fender apron seal
- Engine undercover
3. Drain the engine oil from the engine.
4. Remove or disconnect the following:
- Front exhaust pipe
- Front exhaust pipe bracket from the No. 1 oil pan
- Flywheel housing undercover
- 10 bolts and 2 nuts to the No. 2 oil pan
5. Insert the blade of the Oil Pan Seal Cutting tool 09032-00100 between the No. 1 and No. 2 oil pans. Clean the surfaces of the oil pans.
6. Remove or disconnect the following:
- 3 oil strainer nuts and gasket
7. Remove the No. 1 oil pan, as follows:
- 2 bolts and the flywheel housing undercover
- 17 bolts and 2 nuts to the No. 1 oil pan

➡ **Make a note of the position of the each bolt. When replacing the bolts into the oil pan, place each bolt in the position from which it was removed.**

- Oil pan, by prying the portions between the cylinder block and the oil pan

➡ **Be careful not to damage the contact surfaces.**

- Baffle plate from the No. 1 oil pan

To install:

8. Clean all mating surfaces of the oil pans.
9. Install the baffle plate to the No. 1 oil pan and tighten to 69 inch lbs. (8 Nm).
10. Install the No. 1 oil pan, as follows:
a. Using a non residue solvent, clean both sealing surfaces to the oil pan.
b. Apply liquid sealant to the oil pan and engine block.
c. Install the oil pan with the 17 bolts and 2 nuts. Uniformly tighten the bolts and nuts in several passes.
d. Tighten the No. 1 oil pan bolts, as follows:
- 10mm head bolt: 69 inch lbs. (8 Nm)

- 12mm head bolt: 14 ft. lbs. (20 Nm)
- 14mm head bolt: 27 ft. lbs. (37 Nm)

    e. Install the flywheel housing undercover with the 2 bolts. Tighten the bolts to 69 inch lbs. (8 Nm).

11. Install the oil strainer with the 3 nuts. Tighten the nuts to 69 inch lbs. (8 Nm).

12. Install the No. 2 oil pan, as follows:

    a. Using a non residue solvent, clean both sealing surfaces to the oil pan.

    b. Apply liquid sealant to the oil pan and engine block.

    c. Install the No. 2 oil pan with the 10 bolts and 2 nuts. Uniformly tighten the bolts and nuts in several passes. Tighten the bolts to 69 inch lbs. (8 Nm).

13. Install or connect the following:

- Flywheel housing undercover
- Front exhaust pipe bracket to the No. 1 oil pan. Tighten the bolts to 15 ft. lbs. (21 Nm).

14. Install the front exhaust pipe, as follows:

- Temporarily install the 3 new gaskets and the front exhaust pipe with the 2 bolts and 6 nuts
- Tighten the 4 exhaust manifolds-to-front exhaust pipe nuts to 46 ft. lbs. (62 Nm).
- Tighten the both front exhaust pipe-to-center exhaust pipe nuts/bolts to 41 ft. lbs. (56 Nm).
- Bracket. Tighten both bolts to 14 ft. lbs. (19 Nm).
- Support stay. Tighten both bolts to 22 ft. lbs. (29 Nm).

15. Install or connect the following:

- Engine undercover
- Right fender apron seal
- Right front wheel

16. Fill the engine with oil.

17. Start the engine and check for leaks.

2.7L & 3.4L 4Runner

1. Before servicing the vehicle, refer to the precautions in the beginning of this section.

2. Disconnect the negative battery cable.

3. Drain the engine oil.

4. Remove or disconnect the following:

- Engine undercover
- Front differential, if equipped with 4WD
- Oil pan, separate it from the engine using SST 09032-00100 and a brass bar

To install:

5. Apply seal packing to the oil pan.

6. Install the oil pan to the cylinder block. Tighten the nuts and bolts to:

- 4Runner: 108 inch lbs. (13 Nm)

If parts are not assembled within 5 minutes of applying time, the effectiveness of the seal packing is lost and must be removed and reapplied.

7. Install or connect the following:

- Front differential, if removed
- Engine undercover
- Negative battery cable

8. Fill with engine oil.

9. Start the engine and check for leaks.

Oil Pump

REMOVAL & INSTALLATION

RAV4

1. Before servicing the vehicle, refer to the precautions in the beginning of this section.

2. Remove or disconnect the following:

- Negative battery cable
- Hood
- Right-hand engine undercover

3. Drain the engine oil.

4. Remove or disconnect the following:

- Front exhaust pipe
- Rear end stiffener plate
- Oil dipstick
- 17 bolts and 2 nuts from the oil pan

5. Insert the blade of the Oil Pan Seal Cutting tool 09032-00100 between the oil pan and the cylinder block; then, cut off the applied sealer and remove the oil pan

➡**Do not use the tool for the oil pump body side and rear oil seal retainer.**

6. Remove the bolts, nuts, oil strainer and gasket.

7. Carefully suspend the engine with a sling device.

8. Remove or disconnect the following:

- Timing belt
- No. 2 idler pulley and crankshaft timing pulley
- Oil pump's pulley, using the Variable Pin Wrench Set 09960-10010
- Crankshaft Position (CKP) sensor
- Oil pump, by discarding the gasket

To install:

9. Install or connect the following:

- Oil pump, using a new gasket. Tighten the 12 bolts to 82 inch lbs. (9 Nm).

➡**The long bolts are 35mm and all the others are 25mm.**

- CKP sensor
- Oil pump pulley. Tighten the nut to 18 ft. lbs. (24 Nm).
- Crankshaft timing pulley and No. 2 idler pulley
- Timing belt

10. Remove the engine sling.

11. Install the oil strainer with a new gasket. Tighten the nuts/bolts to 48 inch lbs. (5 Nm).

12. Remove any old sealant from the oil pan flange and thoroughly clean both sealing surfaces.

13. Apply a 3–5mm bead of sealant to the oil pan flange.

➡**The pan must be installed within 5 minutes of sealant application or the procedure will have to be repeated.**

14. Install or connect the following:

- Oil pan. Tighten the 17 bolts and 2 nuts to 48 inch lbs. (5 Nm).
- Dipstick
- Rear end stiffener plate. Tighten the bolts to 27 ft. lbs. (37 Nm).
- Front exhaust pipe
- Negative battery cable
- Hood

15. Refill the engine with oil.

Be sure to prime the oil pump prior to initial engine start-up or engine damage may occur because of low oil pressure.

16. Start the engine and check for leaks.

17. Recheck the engine oil level.

18. Install the right-hand engine undercover.

2.4L Highlander

1. Before servicing the vehicle, refer to the precautions in the beginning of this section.

2. Remove or disconnect the following:

- Timing chain
- Oil pump

3. Installation is the reverse of removal. Torque the pump bolts to 14 ft. lbs. (19Nm).

3.0L Highlander and RX 300

1. Before servicing the vehicle, refer to the precautions in the beginning of this section.

2. Remove or disconnect the following:

- Oil pan
- Crankshaft Position (CKP) sensor
- 9 oil pump bolts

➡Make a note of the position of the each bolt. When replacing the bolts into the oil pump body, place each bolt in the position from which it was removed.

- Oil pump body, by prying between the oil pump and main bearing cap
- O-ring from the cylinder block
- Plug, gasket, spring and relief valve from the oil pump body
- 9 screws, pump body cover, drive and driven rotors

To install:

3. Install or connect the following:
- Driven rotors, drive, pump body cover, using the 9 screws
- Oil pump relief valve, spring, gasket and the plug to the oil pump body
- New O-ring on the cylinder block

4. Using a non residue solvent, clean both sealing surfaces to the oil pump.

5. Apply liquid sealant to the oil pump and engine block.

6. Install or connect the following:
- Oil pump

➡Be sure to engage the splined teeth of the oil pump drive gear with the large teeth of the crankshaft.

- 9 oil pump bolts. Tighten the bolts in several passes to 69 inch lbs. (8 Nm), for 10mm or to 14 ft. lbs. (20 Nm), for 12mm.
- CKP sensor. Tighten the bolt to 69 inch lbs. (8 Nm).
- Baffle plate to the No. oil pan. Tighten to 69 inch lbs. (8 Nm).
- No. 1 oil pan, oil strainer and No. 2 oil pan

7. Refill the engine with oil.
8. Start the engine and inspect for leaks.
9. Recheck the engine oil level.

2.7L 4Runner

1. Before servicing the vehicle, refer to the precautions in the beginning of this section.

2. Remove or disconnect the following:
- Negative battery cable
- Cylinder head assembly
- Water inlet and housing
- Timing chain cover
- 9 screws and separate the oil pump from the timing chain cover

To install:

3. Install or connect the following:
- Oil pump assembly to the timing chain cover
- Timing chain cover

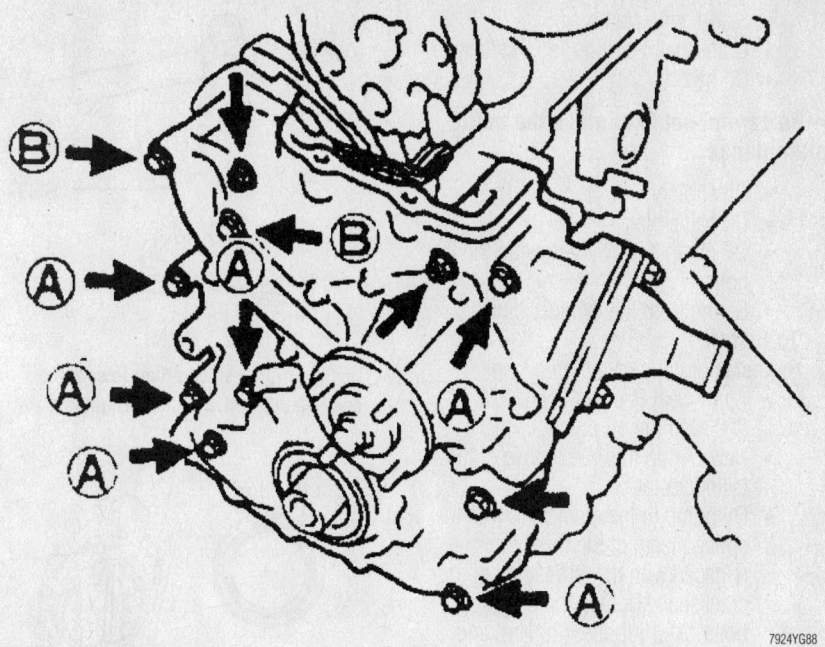

Timing cover bolt pattern—4Runner with 2.7L engine

- Water bypass pipe. Tighten both nuts to 14 ft. lbs. (20 Nm).
- Cylinder head assembly

3.4L 4Runner

1. Before servicing the vehicle, refer to the precautions in the beginning of this section.

2. Remove or disconnect the following:
- Negative battery cable
- Engine undercover
- Crankshaft timing pulley

- Front differential, if equipped with 4WD

3. Drain the engine oil from the engine.

4. Remove or disconnect the following:
- Timing belt and crankshaft gear
- Oil cooler tube and clamp, if equipped with automatic transmission
- Stiffener plate
- Flywheel housing undercover and dust cover
- Rear end cover and dust cover

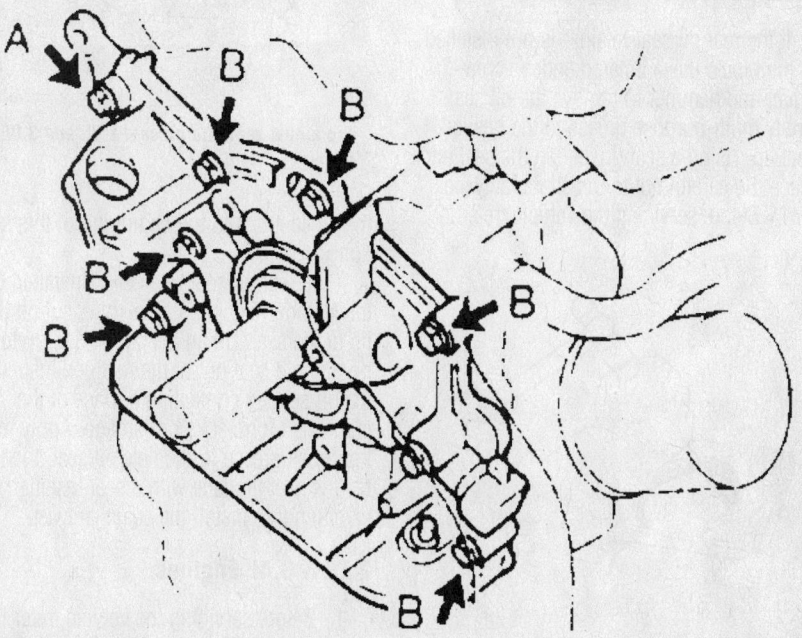

Oil pump bolt identification—4Runner, with 3.4L engine

- Starter wire clamp
- Crankshaft Position (CKP) sensor
- Oil pan

➡ Be careful not to damage the baffle plate flange.

- Oil strainer
- Oil baffle plate
- Oil pump body by removing the 8 bolts.
- O-ring from the cylinder block

To install:

5. Install or connect the following:
- Apply Seal Packing PN 08826-00080 to the oil pump
- New O-ring into the groove of the cylinder block
- Oil pump to the crankshaft with the splined teeth of the drive rotor engaged with the large teeth of the crankshaft. Tighten the oil pump bolts "A" 15 ft. lbs. (20 Nm) and bolts "B" 31 ft. lbs. (42 Nm)
- CKP
- Oil pan baffle plate
- Oil strainer with a new gasket. Tighten the bolts to 13 ft. lbs. (18 Nm).
- Remaining components
- Negative battery cable

6. Fill with engine oil.
7. Start the engine and check for leaks.

Rear Main Seal

REMOVAL & INSTALLATION

2.0L, 2.4L and 3.0L Engines

If the rear oil seal retainer is not installed to the block, use a tapered ended screwdriver and hammer to remove the oil seal. Apply multi-purpose grease to the new oil seal lip. Using a seal driver, tap the seal into place. Be careful not to install it slantwise.

1. Before servicing the vehicle, refer to

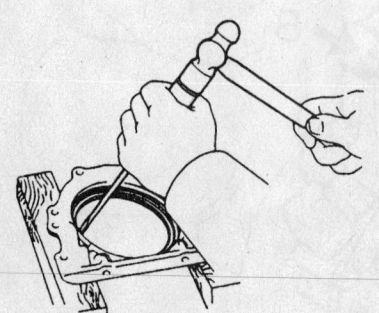

Carefully tap the old seal from the retainer—2.0L, 2.4L and 3.0L Engines

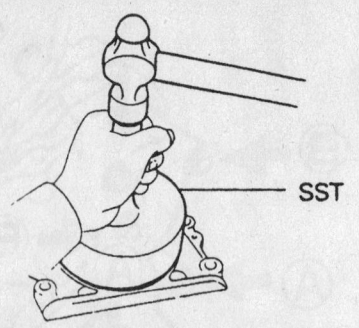

Use the proper sized driver to seat the seal—2.0L, 2.4L and 3.0L Engines

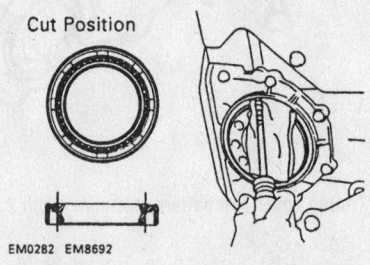

Cut off the oil seal lip, then pry the seal out of the retaining plate—2.0L and 3.0L Engines

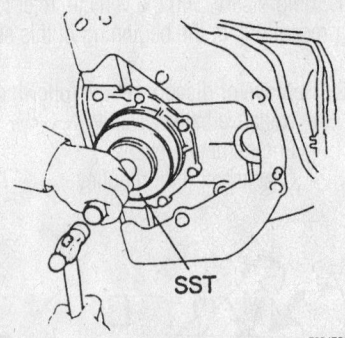

Tap a new seal into place—2.0L and 3.0L Engines

the precautions in the beginning of this section.

If the rear oil seal retainer is installed on the cylinder block, using a knife, cut off the lip of the seal. Using a taped ended prytool, pry the old seal out of the retainer. Inspect the oil seal lip contacting surface of the crankshaft for cracks or damage. Apply multipurpose grease to the new oil seal, then tap the seal in place with a seal installer. Be careful not to install the seal slantwise.

2.7L & 3.4L Engines

1. Before servicing the vehicle, refer to the precautions in the beginning of this section.

2. Remove the transmission and flywheel from the vehicle.
3. Cut off the rubber lip portion of the seal with a sharp knife.
4. Pry out the oil seal.

To install:

5. Install the rear main seal so that it is flush with the seal retainer housing.
6. Install or connect the following:
- Flywheel/driveplate. Tighten the bolts to 28 ft. lbs. (38 Nm) for engines with a manual transmission, 61 ft. lbs. (83 Nm) for engines with an automatic.
- Transmission

Timing Belt

REMOVAL & INSTALLATION

3.4L Engine

1. Disconnect the negative battery cable.

✷✷ CAUTION

Wait 90 seconds from the time the key is turned to LOCK and the negative battery cable is disconnected to begin work. This allows the SRS capacitor to discharge and prevent deployment of the air bag(s).

2. Raise and safely support the vehicle.
3. Remove the engine undercover.
4. Drain the engine coolant.

✷✷ CAUTION

Never open, service or drain the radiator or cooling system when hot; serious burns can occur from the steam and hot coolant. Also, when draining engine coolant, keep in mind that cats and dogs are attracted to ethylene glycol antifreeze and could drink any that is left in an uncovered container or in puddles on the ground. This will prove fatal in sufficient quantities. Always drain coolant into a sealable container. Coolant should be reused unless it is contaminated or is several years old.

5. Disconnect the upper radiator hose from the engine.
6. Remove the power steering drive belt.
7. Remove the air conditioning drive belt by loosening the idler pulley nut and the adjusting bolt.
8. If equipped with air conditioning,

disconnect the compressor from the engine and set aside. Do not disconnect the lines from the compressor.

9. If equipped with air conditioning, disconnect the air conditioning bracket.

10. Remove the fan with the fluid coupling and fan pulleys.

11. Loosen the lockbolt, pivot bolt, and the adjusting bolt and the alternator drive belt.

12. Remove the No. 2 fan shroud by removing the 2 clips.

13. Disconnect the power steering pump from the engine and set aside. Do not disconnect the lines from the pump.

14. Remove the oil dipstick and the guide.

15. Remove the No. 2 timing belt cover as follows:

  a. Detach the camshaft position sensor connector from the No. 2 timing belt cover.

  b. Disconnect the 4 spark plug wire clamps from the No. 2 timing belt cover.

  c. Remove the 6 bolts and remove the timing belt cover.

16. Remove the fan bracket as follows:

  a. Remove the power steering adjusting strut by removing the nut.

  b. Remove the fan bracket by removing the bolt and nut.

17. Using SST 09213-54015, or equivalent, remove the crankshaft pulley.

18. Remove the starter wire bracket and the No. 1 timing belt cover.

19. Remove the timing belt guide.

20. Set the No. 1 cylinder at Top Dead Center (TDC) of the compression stroke, as follows:

  a. Temporarily install the crankshaft pulley bolt to the crankshaft.

  b. Turn the crankshaft and align the timing marks of the crankshaft timing pulley and the oil pump body.

  c. Check that the timing marks of the camshaft timing pulleys and the No. 3 timing belt cover are aligned. If not, turn the crankshaft pulley one revolution (360 degrees).

➡**If reusing the timing belt, be sure that you can still read the installation marks. If not, place new installation marks on the timing belt to match the timing marks of the camshaft timing pulleys.**

21. Remove the timing belt tensioner by alternately loosening the 2 bolts.

22. Remove the right and left camshaft pulleys.

23. Remove the No. 2 idler pulley.

24. Using a 10mm hex wrench, remove

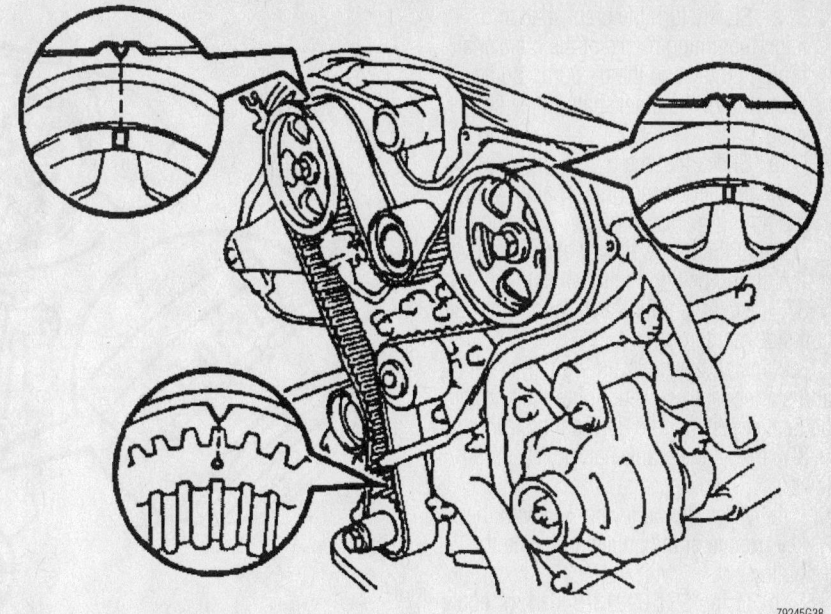

Crankshaft and camshaft timing mark locations—3.4L engine

79245G38

the pivot bolt, No. 1 idler pulley and the plate washer.

25. Remove the timing belt guide and remove the timing belt.

26. Remove the crankshaft timing pulley.

To install:

27. Install the crankshaft timing belt pulley, as follows:

  a. Align the timing belt pulley set key with the key groove of the timing pulley and slide on the timing pulley.

  b. Slide on the timing belt pulley with the flange side facing inward.

28. Install the plate washer and the No. 1 idler pulley with the pivot bolt and tighten it to 26 ft. lbs. (35 Nm). Check that the pulley bracket moves smoothly.

29. Install the No. 2 timing belt idler with the bolt. Tighten the bolt to 30 ft. lbs. (40 Nm). Check that the pulley bracket moves smoothly.

30. Install the left and right camshaft timing pulleys.

31. Set the No. 1 cylinder to TDC of the compression stroke, as follows:

  a. Using the crankshaft pulley bolt, turn the crankshaft and align the timing marks of the crankshaft timing pulley and the oil pump body.

  b. Using SST 09960-10010, or equivalent, to turn the camshaft pulley to align the marks of the camshaft timing belt pulley and the No. 3 timing belt cover.

32. Install the timing belt, as follows:

➡**The engine should be cold.**

  a. Face the front mark on the timing belt forward.

  b. Align the installation mark on the timing belt with the timing mark of the crankshaft timing pulley.

  c. Align the installation marks on the timing belt with the timing marks of the camshaft pulleys.

33. Install the timing belt in the following order:

  • Left camshaft pulley
  • No. 2 idler pulley
  • Right camshaft pulley
  • Water pump pulley
  • Crankshaft pulley
  • No. 1 idler pulley

✷✷ WARNING

If any binding is felt when adjusting the timing belt tension by turning the crankshaft, STOP turning the engine, because the pistons may be hitting the valves.

34. Set the timing belt tensioner as follows:

  a. Using a press, slowly press in the pushrod using 220 – 2205 lbs. (981 – 9807 N) of force.

  b. Align the holes of the pushrod and housing, pass a 1.27mm wrench through the holes to keep the setting position of the pushrod.

  c. Release the press and install the dust boot to the tensioner.

35. Install the timing belt tensioner and alternately tighten the bolts to 20 ft. lbs. (27 Nm). Using pliers, remove the 1.27mm wrench from the belt tensioner.

36. Check the valve timing, as follows:

a. Slowly turn the crankshaft and align the timing marks of the crankshaft timing pulley and the oil pump body. Always turn the crankshaft pulley clockwise.

b. Check that the timing marks of the right and left timing pulleys align with the timing marks of the No. 3 timing belt cover. If the marks do not align, remove the timing belt and reinstall it.

37. Install the timing belt guide with the cup side facing outward.

38. Install the No. 1 timing belt cover and starter wire bracket. Tighten the timing belt cover fasteners to 80 inch lbs. (9 Nm).

39. Install the crankshaft pulley, as follows:

a. Align the pulley set key with the key groove of the pulley and slide the pulley.

b. Using SST 09213-54014, or equivalent, tighten the bolt to 184 ft. lbs. (250 Nm).

40. Install the fan bracket with the bolt and nut.

41. Install the No. 2 timing belt cover, and tighten the bolts to 80 inch lbs. (9 Nm). Install the remaining components.

42. Fill the cooling system with coolant.

43. Connect the negative battery cable.

44. Start the engine and check for leaks.

45. Check the ignition timing.

2.0L Engine

The timing belt is not adjustable.

1. Disconnect the negative battery cable.

※※ CAUTION

To avoid air bag deployment, if equipped, work must be started after approximately 90 seconds or longer from the time the ignition switch is turned to the LOCK position and the negative battery cable is disconnected from the battery.

2. Disconnect the power steering reservoir tank and remove the reservoir bracket.

3. Detach the wiring harness bracket for the Data Link Connector 1 (DLC1).

4. Remove the alternator and alternator bracket.

5. If equipped with ABS brakes, remove the ABS actuator.

6. Remove the right front wheel and the fender apron seal.

7. Remove the power steering drive belt.

8. Slightly raise the engine using a

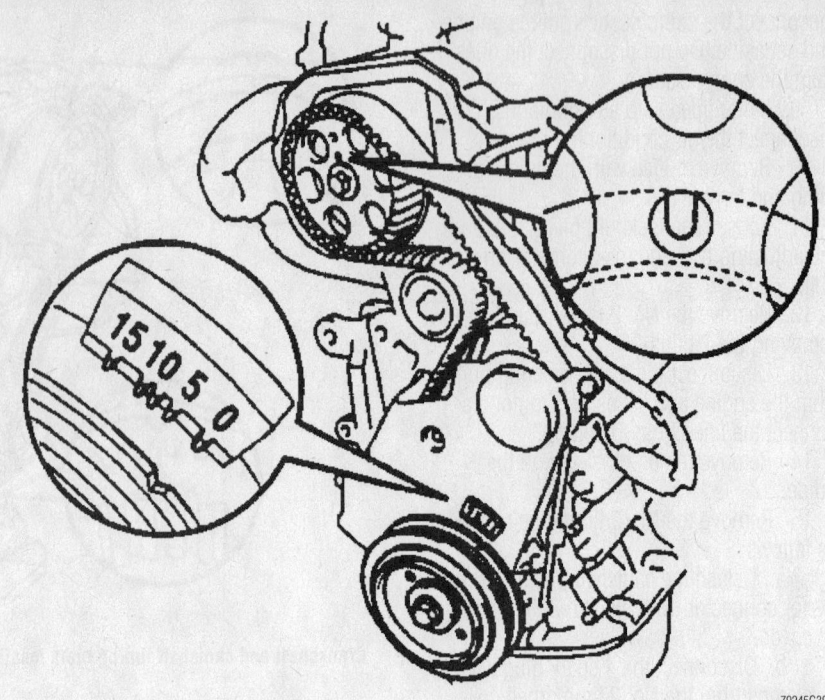

79245G39

It is necessary to align the timing reference indicators prior to removing the timing belt—2.0L engine

block of wood and floor jack under the oil pan to prevent damage.

9. Remove the 4 bolts, 2 nuts, and right-hand mounting bracket.

10. Remove the spark plugs.

11. Using SST 09213-54015, or equivalent, loosen the crankshaft pulley bolt and remove it by pulling it straight off the crankshaft.

12. Using SST 09249-63010, or equivalent, loosen the retaining bolts and remove the right engine mounting bracket.

13. Remove the upper (No. 2) timing belt cover.

14. Install the crankshaft pulley to the crankshaft and temporarily install the retaining bolt.

15. Turn the crankshaft pulley and align its groove with the timing mark **0** of the No. 1 timing belt cover. Check that the hole of the camshaft timing pulley is aligned with the timing mark of the bearing cap. If not, turn the crankshaft 360 degrees and align the marks.

➡**If the timing belt is to be reused, matchmark the timing belt to the timing pulleys and timing belt covers so the belt can be reinstalled in its original position. Also, be sure to mark an arrow on the belt to indicate which direction it was turning.**

16. Remove the timing belt from the camshaft timing pulley.

17. Hold the camshaft sprocket with a spanner wrench and remove the mounting bolt. Remove the camshaft pulley.

18. Remove the crankshaft pulley bolt and remove the crankshaft pulley.

19. Remove the No. 1 timing belt cover.

20. Remove the timing belt guide and the timing belt.

21. Remove the No. 1 idler pulley and tension spring.

22. Remove the No. 2 idler pulley.

23. Remove the crankshaft timing pulley.

24. Support the oil pump sprocket with a spanner wrench, then remove the mounting bolt and remove the sprocket.

To install:

25. Install the oil pump pulley. Tighten the nut to 18 ft. lbs. (24 Nm).

26. Install the crankshaft timing pulley. Align the pulley set key with the key groove of the pulley. Slide on the pulley facing the flange side inward.

27. Install the No. 2 idler pulley and tighten the mounting bolt to 31 ft. lbs. (42 Nm). Be sure that the pulley moves smoothly.

28. Install the No. 1 idler pulley with the bolt and the tension spring. Pry the pulley toward the left as far as it will go and tighten the bolt. Make sure that the pulley moves smoothly.

29. Temporarily install the timing belt. Using the crankshaft pulley bolt, turn the

crankshaft and position the key groove of the crankshaft timing pulley upward. If reusing the timing belt, align the points marked during removal.

30. Install the timing belt on the crankshaft timing pulley, oil pump pulley, No. 1 idler pulley, water pump pulley and the No. 2 idler pulley.

31. Install the timing belt guide.

➡ **If the old timing belt is being reinstalled, be sure the directional arrow is facing in the original direction and that the belt and sprocket/cover matchmarks are properly aligned.**

32. Install the lower (No. 1) timing belt cover and new gasket with the 4 bolts.

33. Align the crankshaft pulley set key with the pulley key groove. Temporarily install the crankshaft pulley and bolt.

34. Align the camshaft knock pin with the groove of the pulley, and slide the timing pulley onto the camshaft with the plate washer and set bolt.

35. Tighten the pulley set bolt to 40 ft. lbs. (54 Nm).

✳✳ WARNING

If any binding is felt when adjusting the timing belt tension by turning the crankshaft, STOP turning the engine, because the pistons may be hitting the valves.

36. Turn the crankshaft pulley and align the **0** mark on the lower (No. 1) timing belt cover.

37. Finish installing the timing belt and check the valve timing, as follows:

   a. If reusing the old timing belt, align the matchmarks made previously and install the timing belt onto the camshaft pulley.

   b. Align the marks on the timing belt with the marks on the camshaft pulley.

   c. Loosen the No. 1 idler pulley set bolt $1/2$ turn.

   d. Turn the crankshaft pulley 2 complete revolutions TDC to TDC. ALWAYS turn the crankshaft CLOCKWISE. Check that the pulleys are still in alignment with the timing marks.

   e. If the No. 1 idler pulley uses a green tension spring, slowly turn the crankshaft pulley 1 $7/8$ revolutions, and align its groove with the mark at 45 degrees BTDC (for the No. 1 cylinder) of the No. 1 timing belt cover.

   f. Tighten the No. 1 idler pulley set bolt to 31 ft. lbs. (42 Nm).

   g. Be sure there is belt tension

between the crankshaft and camshaft timing pulleys.

38. Place the right-hand engine mounting bracket in position but do not install the bolts.

39. Install the upper (No. 2) timing cover with a new gasket(s).

40. Remove the engine crankshaft pulley bolt and pulley.

41. Using SST 09249-63010, or equivalent, install the mounting bolts for the right-hand mounting bracket. Tighten the mounting bolts to 38 ft. lbs. (52 Nm).

42. Align the crankshaft pulley set key with the pulley key groove. Install the pulley. Tighten the pulley bolt to 80 ft. lbs. (108 Nm).

43. Install the spark plugs.

44. Install the right-hand mounting insulator, as follows:

   a. Attach the mounting insulator to the body and mounting bracket with the 4 bolts and 2 nuts.

   b. Tighten the 3 bolts to hold the mounting insulator to the body. Tighten the bolts to 47 ft. lbs. (64 Nm).

   c. Tighten the 2 nuts and bolt to hold the mounting insulator to the mounting bracket. Tighten the bolt to 27 ft. lbs. (37 Nm) and the nut to 38 ft. lbs. (52 Nm).

45. Install and adjust the power steering pump drive belt.

46. Install the right-hand engine undercover.

47. Install the right front wheel.

48. Lower the engine.

49. If equipped, install the ABS actuator.

50. Install the alternator and alternator bracket.

51. Install the wiring harness bracket for the DLC1.

52. Install the power steering reservoir bracket and reservoir.

53. Connect the negative battery cable.

54. Start the engine and check the timing.

Timing Chain, Sprockets, Front Cover and Seal

REMOVAL & INSTALLATION

2.4L Highlander

1. Before servicing the vehicle, refer to the precautions in the beginning of this section.

2. Disconnect the negative battery cable.

3. Drain the engine coolant.

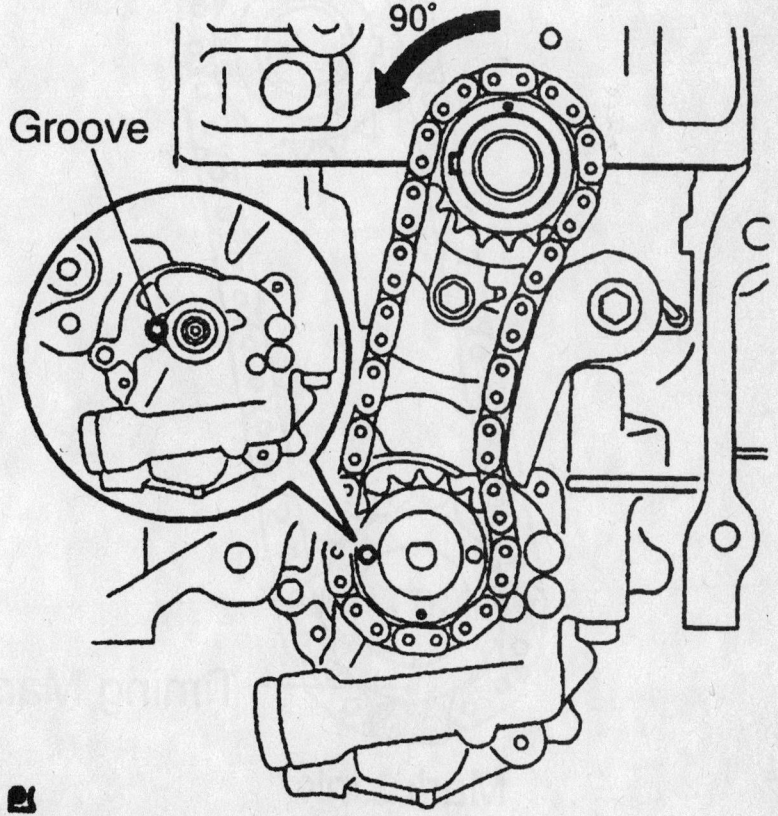

Aligning the adjusting hole and groove—2.4L engine

9355YG14

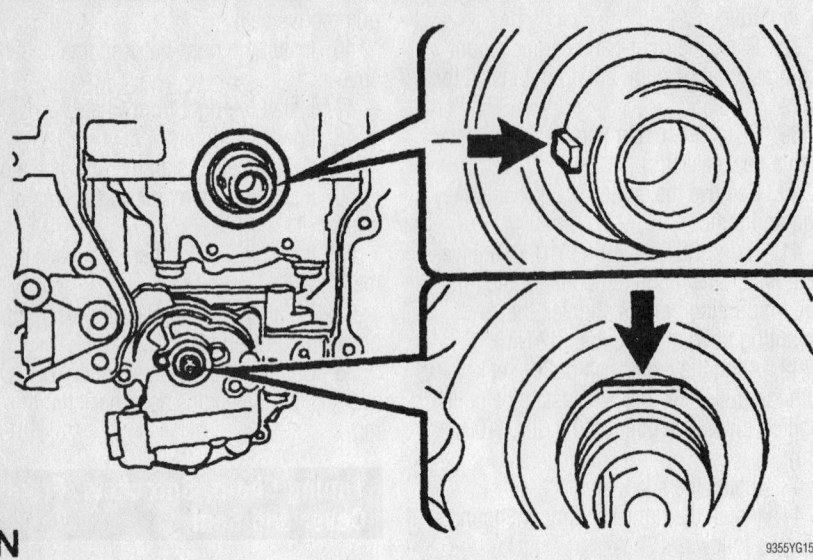

N

9355YG15

Aligning the crankshaft with the key in the left horizontal position—2.4L engine

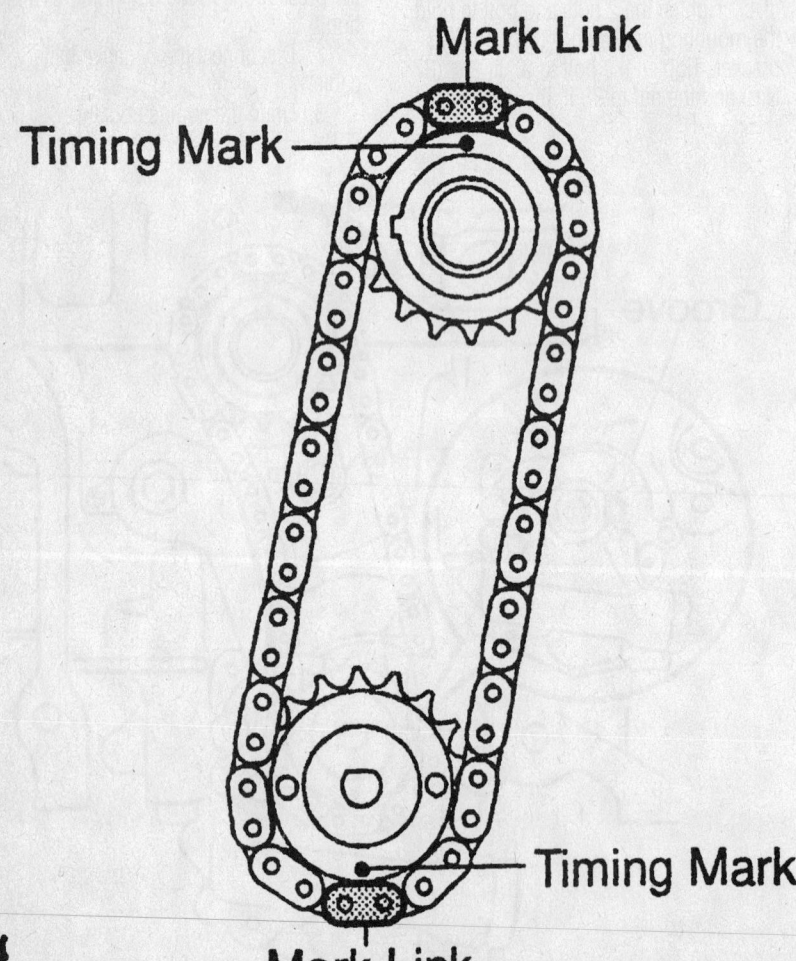

Mark Link

Timing Mark

Timing Mark

Mark Link

9355YG07

Install the secondary chain and gears with the timing marks aligned as shown—2.4L engine

4. Remove or disconnect the following:
- Hood
- Engine oil
- Right front wheel
- Right fender splash shield
- Right fender apron seal
- No.1 engine undercover
- Engine roll stopper and bracket
- Exhaust pipe
- Upper engine mount, right side
- Accessory drive belts
- Alternator
- Power steering pump

5. Set the No.1 piston at TDC compression.

- Crankshaft pulley
- Oil pan
- CKP sensor
- Chain tensioner assembly No.1
- V-belt tensioner

6. Take up the weight of the engine with a crane.

7. Remove or disconnect the following:

- Transverse engine mount insulator
- Transverse engine mount bracket
- Timing chain cover (14 bolts and 2 nuts
- CKP sensor plate
- Chain tensioner slipper
- Primary chain vibration damper
- Primary chain and crankshaft sprocket

8. Turn the crankshaft 90 degrees counterclockwise and align the adjusting hole of the oil pump drive shaft gear with the groove in the pump.

9. Insert a 4mm diameter bar into the hole to lock the gear in position and remove the nut.

10. Remove the bolt, tensioner plate, spring, tensioner oil pump driveshaft gear and the chain.

To install:

11. Turn the crankshaft so that the key is in the left horizontal position.

12. Install the secondary chain and gears with the timing marks aligned as shown.

13. Install the damper spring and tensioner plate. Torque the nut to 10 ft. lbs. (13Nm).

14. Align the oil pump adjusting hole and groove, lock it with the bar and torque the nut to 22 ft. lbs. (30Nm).

15. Rotate the crankshaft counterclockwise 90 degrees so the crankshaft key is at the 12 o'clock position and shown.

16. Install the primary chain damper. Torque the bolts to 80 inch lbs. (9Nm).

17. Set the No.1 piston at TDC compression with the timing marks aligned as shown.

Aligning the crankshaft with the key in the 12 o'clock position—2.4L engine

9355YG08

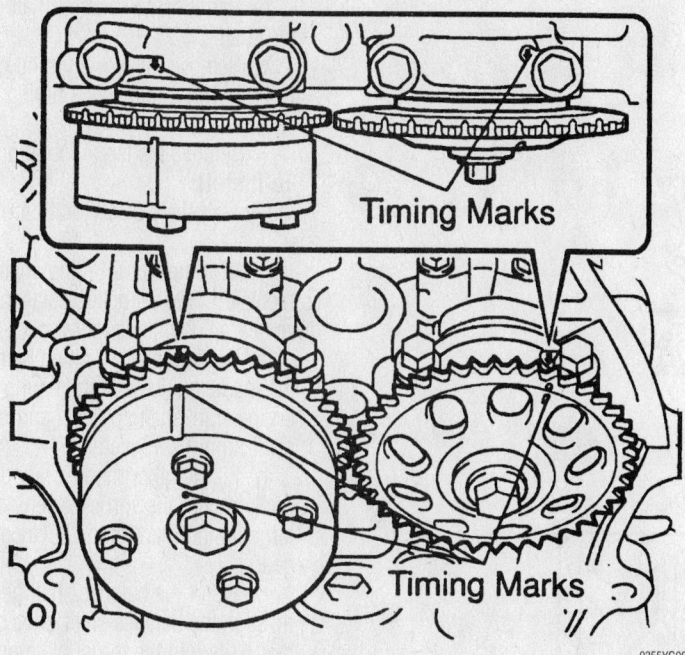

Aligning the timing marks with the No.1 piston at TDC compression—2.4L engine

9355YG09

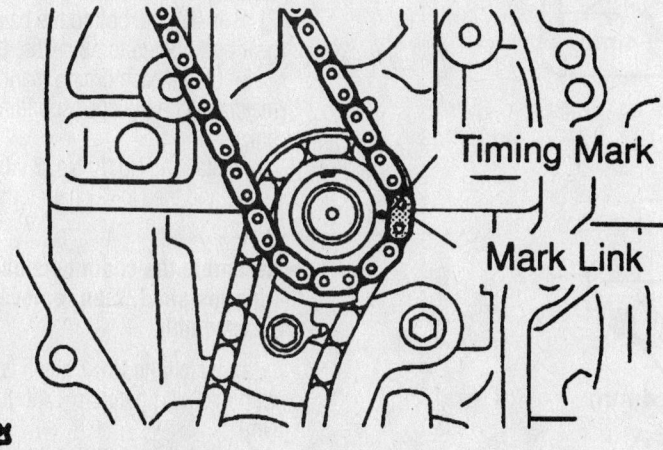

Aligning the timing chain bottom end marks—2.4L engine

9355YG10

18. Turn the crankshaft, using the pulley bolt, until the key is at the 12 o'clock position.

19. Install the bottom end of the chain, with sprocket, so that the colored links are aligned as shown.

20. Align the upper end timing marks as shown and install the chain.

21. Install the tensioner slipper. Torque the bolt to 14 ft. lbs. (19Nm).

22. Install the CKP sensor plate with the **F** mark outwards.

➡ **When installing the cover, use RTV sealant in the positions shown. The cover must be installed within 3 minutes of seal application. Do not start the engine within 2 hours of seal application.**

23. Apply the sealant and install the cover. Torque the cover bolts as follows:
- Bolt A: 80 inch lbs. (9Nm)
- Bolts B: 15 ft. lbs. (21 Nm)
- Bolts C: 32 ft. lbs. (43Nm)
- Nuts: 80 inch lbs. (9Nm)

24. The remainder of installation is the reverse of removal.

2.7L 4Runner

1. Before servicing the vehicle, refer to the precautions in the beginning of this section.

2. Disconnect the negative battery cable.

3. Drain the engine coolant.

4. Remove or disconnect the following:
- Engine undercover
- Engine oil
- On 4WD vehicles, the front differential
- Alternator belt, fan with coupling and the water pump pulley
- Cylinder head
- A/C belt, compressor, and the bracket
- Alternator adjusting bar and bracket
- Crankshaft position sensor and O-ring
- On 2WD vehicles, the stiffener plates
- Flywheel housing undercover and dust seal
- Oil pan
- Oil strainer and gasket
- Crankshaft pulley
- Water bypass pipe
- Chain cover assembly. Remove the bolts shown by the arrows.
- No. 1 timing chain and camshaft gear
- Crankshaft timing gear

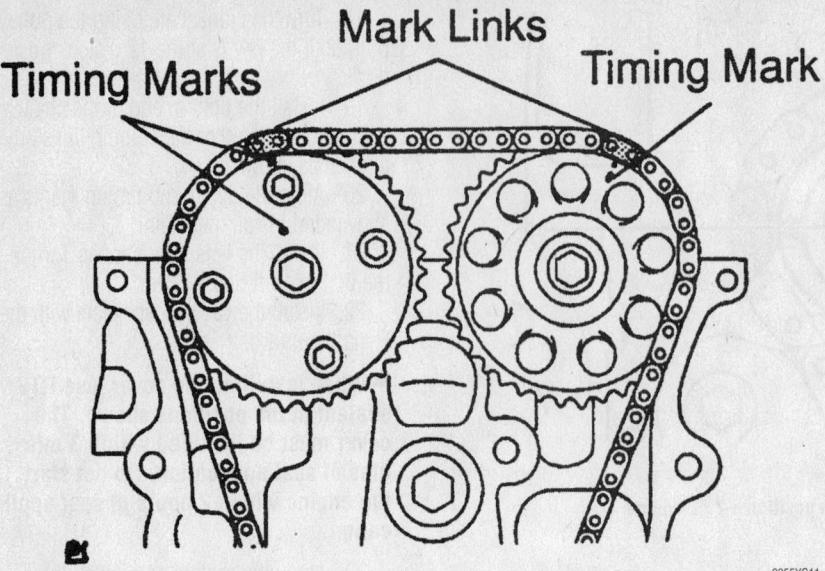

Aligning the timing chain upper end marks—2.4L engine

← Seal Packing

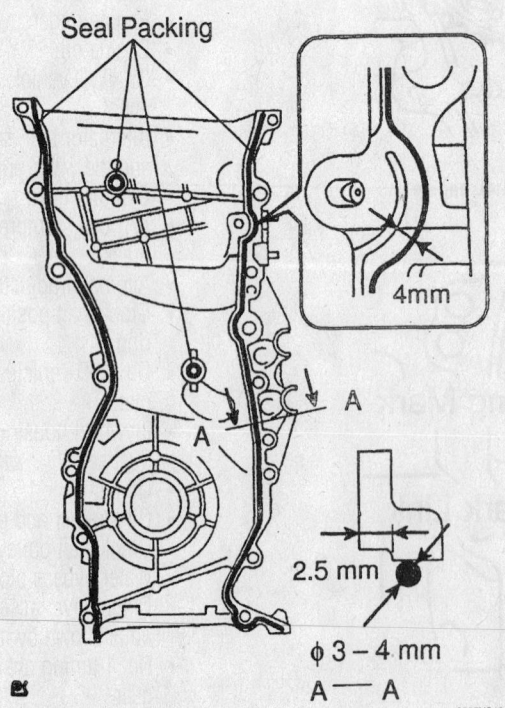

Timing cover sealant application—2.4L engine

- No. 1 timing chain tensioner slipper and No. 1 vibration damper
- No. 1 damper

5. Remove the No. 2 and No. 3 vibration dampers and the No. 2 chain tensioner as follows:

   a. Install a pin in the No. 2 tensioner and lock the plunger.

   b. Remove the bolt and the No. 2 damper.

   c. Remove the 2 bolts and the No. 3 damper.

   d. Remove the nut and the No. 2 tensioner.

6. Remove the balance shaft driven gear, shaft, No. 2 timing chain and the No. 2 crankshaft sprocket, as follows:

   a. Unbolt the balance shaft driven gear.

   b. Remove the balance shaft gear with the shaft.

   c. Remove the No. 2 timing chain with the No. 2 crankshaft timing sprocket.

- Remove the No. 4 vibration damper.

To install:

7. Install the No. 4 vibration dampener.

8. Install the No. 2 timing chain, No. 2 crankshaft timing sprocket, balance shaft drive gear and shaft as follows:

   a. Install the No. 2 chain by matching the marked links with the timing marks on the crankshaft sprocket and balance shaft timing sprocket.

   b. Fit the other marked link of the No. 2 chain onto the sprocket behind the large timing mark of the balance shaft gear.

   c. Insert the balance shaft gear shaft through the balance shaft drive gear so that it fits into the thrust plate hole. Align the small timing mark of the balance shaft drive gear with the timing mark of the balance shaft timing gear.

   d. Install the bolt to the balance shaft gear and tighten to 18 ft. lbs. (25 Nm).

   e. Check each timing mark is matched with the corresponding mark link.

9. Install the No. 2, No. 3 vibration dampers and the No. 2 chain tensioner, as follows:

➡**Assemble the chain tensioner with the pin installed, then remove the pin after assembly.**

   a. Install the No. 2 chain tensioner with the nut, tighten to 13 ft. lbs. (18 Nm).

   b. Install the No. 3 damper with the bolts, tighten to 13 ft. lbs. (18 Nm).

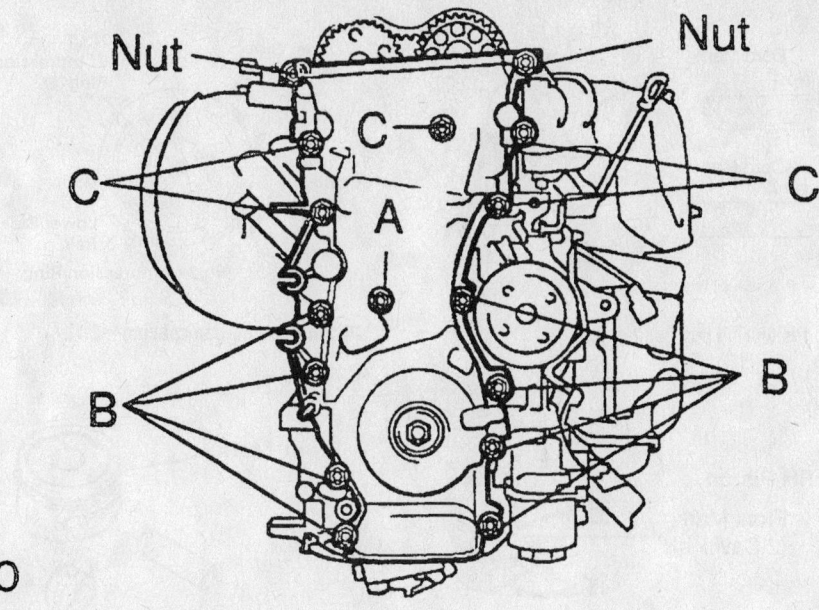

Timing cover bolt positions—2.4L engine

c. Install the No. 2 damper, tighten to 20 ft. lbs. (27 Nm).

d. Remove the pin from the No. 2 chain tensioner and free the plunger.

10. Install or connect the following:
- No. 1 timing chain tensioner slipper and the No. 1 vibration damper. Torque the No. 1 damper to 22 ft. lbs. (29 Nm); torque the slipper to 20 ft. lbs. (27 Nm). Check that the slipper moves smoothly.
- Crankshaft timing gear.
- No. 1 timing chain and camshaft timing gear.

11. Align the timing mark between the marked link of the No. 1 timing chain, and install the No. 1 timing chain to the gear.

12. Align the timing mark of the crankshaft timing gear with the mark of the No. 1 timing chain, then install the No. 1 timing chain.

13. Tie the No. 1 chain with a wire or cord, make sure it does not come loose.

14. Install or connect the following:
- Timing chain cover assembly

15. Tighten the following:
- 12mm **A** bolts—14 ft. lbs. (20 Nm)
- 12mm **B** bolts—18 ft. lbs. (25 Nm)
- 14mm bolts—32 ft. lbs. (44 Nm)
- 14mm nut—14 ft. lbs. (20 Nm)

16. Attach the water bypass pipe.

17. Remove the cord or wire from the chain.

18. Install or connect the following:
- Crankshaft pulley, tighten the bolt to 193 ft. lbs. (260 Nm). On A/C vehicles, install the crankshaft pul-

leys with bolts and tighten to 18 ft. lbs. (25 Nm).
- Oil strainer, tighten to 13 ft. lbs. (18 Nm)
- Oil pan, tighten the mounting bolts to 108 inch lbs. (13 Nm)
- Flywheel housing undercover and dust seal
- Stiffener plates on 2WD vehicles, tighten to 27 ft. lbs. (37 Nm)
- Crankshaft position sensor with a new O-ring
- Alternator, adjusting bar and bracket
- A/C compressor and bracket
- Cylinder head
- Water pump pulley and the fluid coupling with the fan
- On 4WD vehicles, the front differential and driveshaft assemblies

19. Adjust the drive belt tension.

20. Fill with engine coolant and engine oil.

21. Install the engine undercover.

22. Connect the negative battery cable.

SEAL REPLACEMENT

Cover Removed

1. Unbolt the timing chain cover assembly. Be careful to loosen only the correct bolts.

2. Pry out the seal from the cover with a flat-bladed tool.

3. It is a good idea to remove the oil pump from the timing cover and replace the O-ring.

To install:

4. Clean and inspect the timing cover area. Install new gaskets around the dowel areas and pump spline.

5. Apply multi-purpose grease to the new oil seal lip.

6. Tap the seal into place with SST 09223–50010/60010 or equivalent, and a hammer. Do this until the seal surface is flush with the cover edge.

7. Install the cover, tighten the bolts as specified for your engine.

8. If the oil pump was removed, install a new O-ring behind the pump prior to installation.

Cover Installed

1. Unbolt and remove the oil pump.

2. Using a knife, carefully cut off the oil seal lip. With a flat-bladed tool, (preferably with tape around it) pry the seal from the cover.

To install:

3. Apply multi-purpose grease to the new oil seal lip.

4. Tap the seal into place with SST 09223–50010/60011 or equivalent seal driver, and a hammer. Do this until the seal surface is flush with the cover edge.

5. Install the oil pump with a new O-ring.

Piston and Ring

POSITIONING

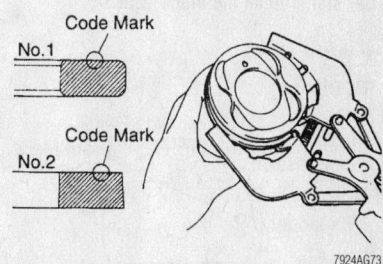

Compression ring identification mark locations—2.0L

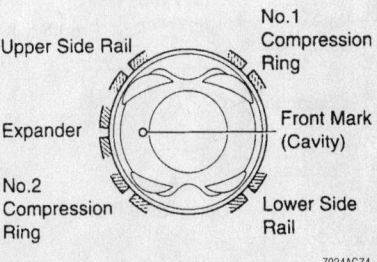

Piston ring end-gap spacing—2.0L

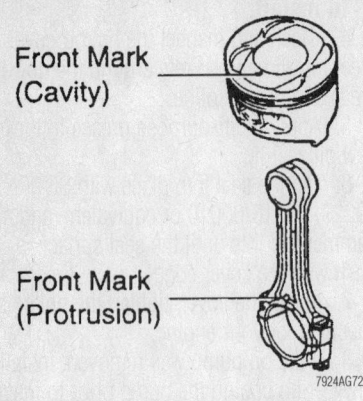

Front Mark (Cavity)

Front Mark (Protrusion)

7924AG72

Piston-to-connecting rod assembly–2.0L

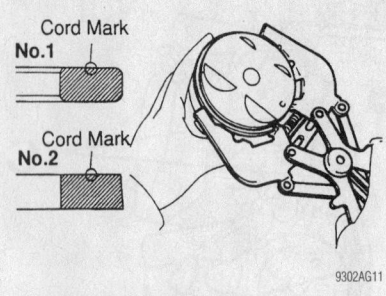

Cord Mark No.1

Cord Mark No.2

9302AG11

Piston ring positioning–3.0L

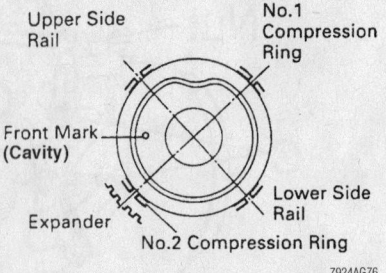

Upper Side Rail

No.1 Compression Ring

Front Mark (Cavity)

Lower Side Rail

Expander

No.2 Compression Ring

7924AG76

Piston ring end-gap spacing–2.7L

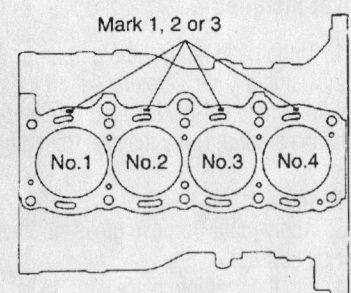

Mark 1, 2 or 3

No.1   No.2   No.3   No.4

Mark 1, 2 or 3

Front Mark (Cavity)

7924AG71

Piston-to-engine installation. Match the number on the piston crown with the number stamped on the block–2.0L

RH Piston

Front Mark (2 Cavities)

Front Mark (Mold Mark)

LH Piston

Front Mark (1 Cavity)

Front Mark (Mold Mark)

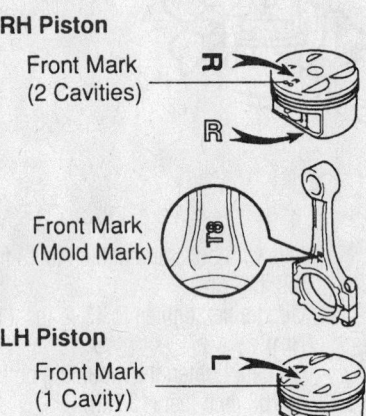

9302AG10

Piston/connecting rod-to-engine positioning–3.0L

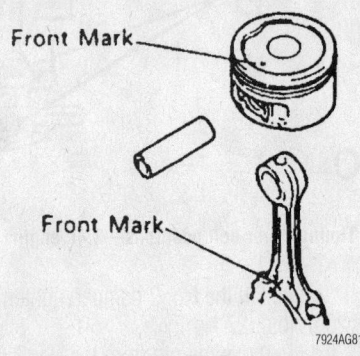

Front Mark

Front Mark

7924AG81

Piston to connecting rod assembly—2.7L & 3.4L engines

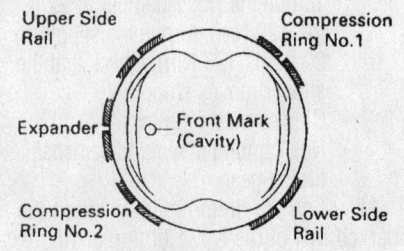

Upper Side Rail

Compression Ring No.1

Expander

Front Mark (Cavity)

Compression Ring No.2

Lower Side Rail

7924AG80

Piston ring end-gap spacing—3.4L engine

RH Piston

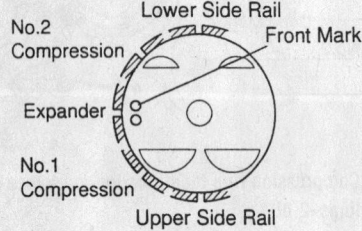

No.2 Compression

Lower Side Rail

Front Mark

Expander

No.1 Compression

Upper Side Rail

LH Piston

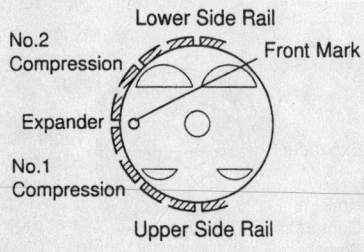

No.2 Compression

Lower Side Rail

Front Mark

Expander

No.1 Compression

Upper Side Rail

9302AG12

Piston ring identification–3.0L

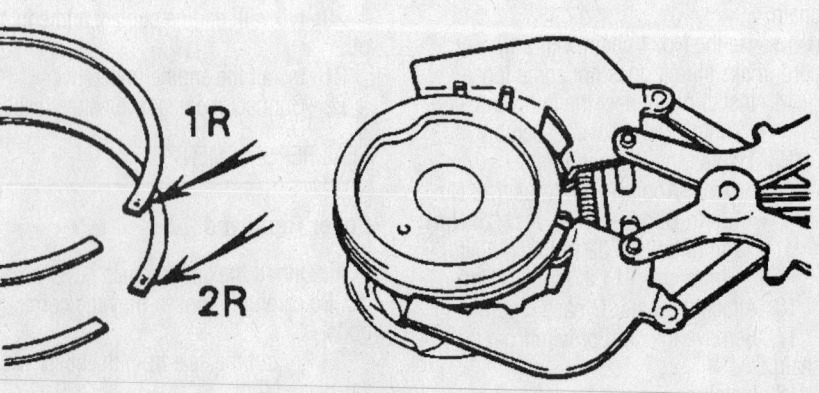

1R

2R

7924AG77

Compression ring identification mark locations—2.7L engine

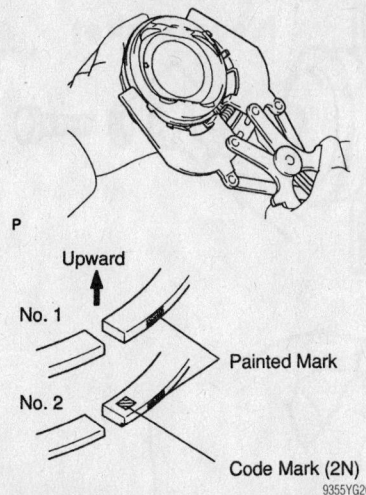

Piston ring identification—2.4L engine

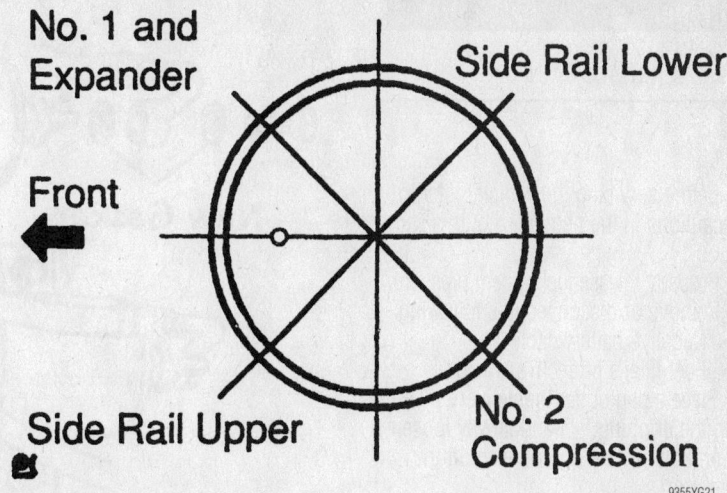

Piston ring installation—2.4L engine

FUEL SYSTEM

°Fuel System Service

Safety is the most important factor when performing not only fuel system maintenance but any type of maintenance. Failure to conduct maintenance and repairs in a safe manner may result in serious personal injury or death. Maintenance and testing of the vehicle's fuel system components can be accomplished safely and effectively by adhering to the following rules and guidelines.

• To avoid the possibility of fire and personal injury, always disconnect the negative battery cable unless the repair or test procedure requires that battery voltage be applied.

• Always relieve the fuel system pressure prior to disconnecting any fuel system component (injector, fuel rail, pressure regulator, etc.), fitting or fuel line connection. Exercise extreme caution whenever relieving fuel system pressure, to avoid exposing skin, face and eyes to fuel spray. Please be advised that fuel under pressure may penetrate the skin or any part of the body that it contacts.

• Always place a shop towel or cloth around the fitting or connection prior to loosening to absorb any excess fuel due to spillage. Ensure that all fuel spillage (should it occur) is quickly removed from engine surfaces. Ensure that all fuel soaked cloths or towels are deposited into a suitable waste container.

• Always keep a dry chemical (Class B) fire extinguisher near the work area.

• Do not allow fuel spray or fuel vapors to come into contact with a spark or open flame.

• Always use a back-up wrench when loosening and tightening fuel line connection fittings. This will prevent unnecessary stress and torsion to fuel line piping.

• Always replace worn fuel fitting O-rings with new. Do not substitute fuel hose or equivalent, where fuel pipe is installed.

Fuel System Pressure

RELIEVING

RAV4 and RX300

1. Before servicing the vehicle, refer to the precautions in the beginning of this section.
2. Disconnect the negative battery terminal.
3. Place a catch-pan under the joint to be disconnected. A large quantity of fuel may be released when the joint is opened.
4. Wear eye or full face protection.
5. Place a shop towel over the area and slowly loosen the joint using a wrench of the correct size. Use a back-up wrench if needed.
6. Allow the fuel left in the line to bleed off slowly before fully disconnecting the joint.
7. Plug the opened lines immediately to prevent fuel spillage or the entry of dirt.
8. Dispose of the released fuel properly.
9. After connecting fuel lines, connect the negative battery cable and start the engine.
10. Check for leaks and repair as needed.

4Runner

1. Before servicing the vehicle, refer to the precautions in the beginning of this section.
2. Disconnect the negative battery terminal.
3. Place a catch-pan under the joint to be disconnected. A large quantity of fuel may be released when the joint is opened.
4. Wear eye or full face protection.
5. Place a shop towel over the area and slowly loosen the joint using a wrench of the correct size. Use a back-up wrench if needed.
6. Allow the fuel left in the line to bleed off slowly before fully disconnecting the joint.
7. Plug the opened lines immediately to prevent fuel spillage or the entry of dirt.
8. Dispose of the released fuel properly.
9. After connecting fuel lines, connect the negative battery cable and start the engine.
10. Check for leaks and repair as needed.

Highlander

Disconnect the fuel pump connector at the tank. Start the engine and let it run out of fuel. Turn the switch off. Disconnect the battery ground. Reconnect the fuel pump connector.

Fuel Filter

REMOVAL & INSTALLATION

RAV4

1. Before servicing the vehicle, refer to the precautions in the beginning of this section.
2. Properly release fuel system pressure.
3. Remove or disconnect the following:
   - Negative battery cable
   - Fuel filter's protective shield
4. Place a pan under the delivery pipe to catch the dripping fuel and slowly loosen the union bolt or flare nut to bleed off the fuel pressure.
5. Drain the remaining fuel.
6. Remove or disconnect the following:
   - Inlet and outlet lines
   - Fuel filter

To install:

7. Coat the flare nut, union nut and bolt threads with engine oil.
8. Hand-tighten the inlet line to the fuel filter.

➡**When tightening the fuel line bolts to the fuel filter, use a torque wrench. The tightening torque is very important, as under or over tightening may cause fuel leakage. Insure that there is no fuel line interference and that there is sufficient clearance between it and any other parts.**

9. Install or connect the following:
   - Fuel filter. Tighten the inlet bolts to 22 ft. lbs. (30 Nm).
   - Delivery pipe using new gaskets. Tighten the union bolt to 22 ft. lbs. (30 Nm).
10. Run the engine for a few minutes and check for any fuel leaks.
11. Install the protective shield.

RX300

1. Before servicing the vehicle, refer to the precautions in the beginning of this section.
2. Disconnect the negative battery cable.
3. Relieve the fuel system pressure.

➡**The fuel filter is located in the engine compartment, at the inlet line to the fuel rail.**

4. Remove or disconnect the following:
   - Inlet and outlet lines from the filter
   - Fuel filter

To install:

5. Install or connect the following:
   - Fuel filter, using new O-rings.

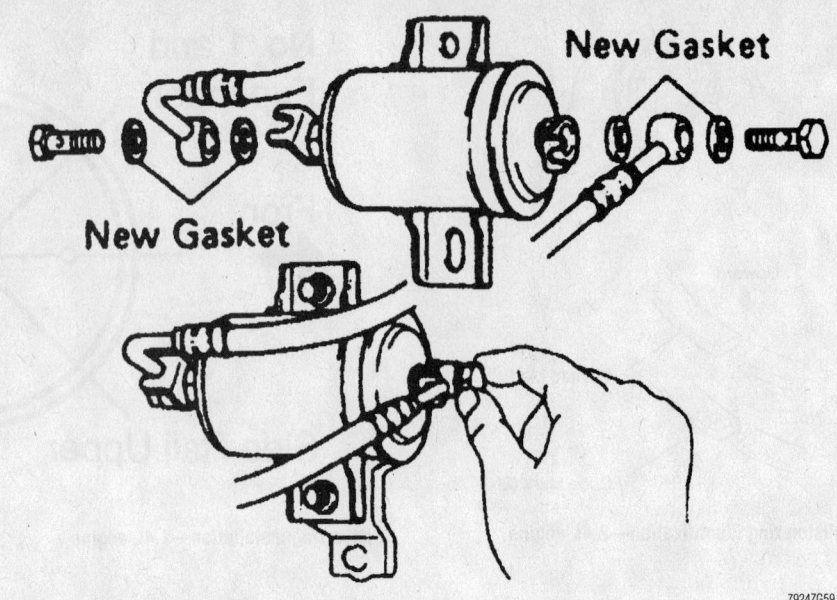

Exploded view of the fuel filter—RX300

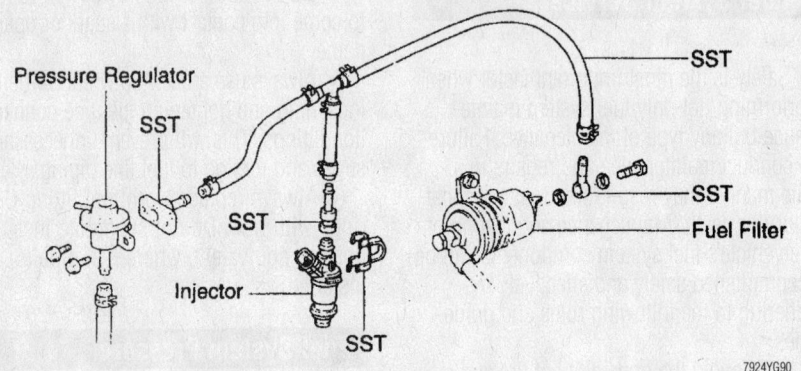

Exploded view of the fuel delivery components—2.7L engines

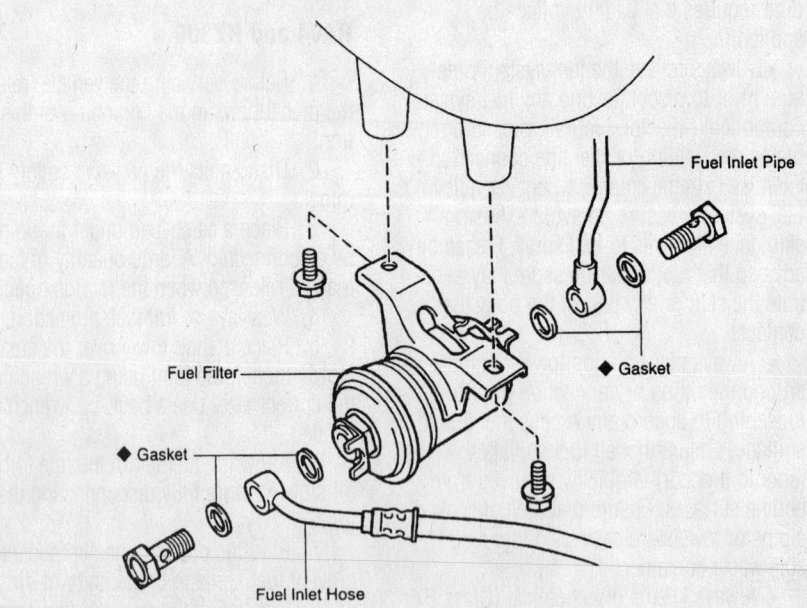

Always use new gaskets when replacing the fuel filter—3.4L

Tighten the lines to 22 ft. lbs. (29 Nm).
- Negative battery cable
6. Start the engine and check for leaks.

Highlander

The fuel filter is in the tank as part of the pump assembly and is not a normal maintenance item.

4Runner

1. Before servicing the vehicle, refer to the precautions in the beginning of this section.
2. Relieve the fuel system pressure.
3. Remove or disconnect the following:
- Negative battery cable

➡**The fuel filter is located in the engine compartment, at the inlet line to the fuel rail.**

- Plug the filter inlet and outlet lines
- Fuel filter
- Bracket from the fuel filter

To install:
4. Install or connect the following:
- Fuel filter bracket to the fuel filter
- Fuel filter. Tighten the 2 bolts to 14 ft. lbs. (20 Nm).
- New gaskets. Tighten the union bolts to 22 ft. lbs. (30 Nm).
- Negative battery cable
5. Start the engine and check for leaks.

Fuel Pump

REMOVAL & INSTALLATION

RAV4 and RX 300

1. Before servicing the vehicle, refer to the precautions in the beginning of this section.
2. Relieve the fuel system pressure.
3. Remove or disconnect the following:
- Negative battery cable
- Left-hand rear seat assembly
- Floor service hole by pulling back the carpet; then, remove the 4 screws
- Fuel pump and sender gauge connector

➡**Loosen the fuel cap to relieve any fuel pressure within the tank.**

- Fuel pipe union bolt and both gaskets
- Fuel pump outlet pipe
- Return vent hose from the fuel pump

- 8 fuel pump bolts and the pump assembly from the tank

To install:
4. Install or connect the following:
- Fuel pump to the fuel tank. Tighten the 8 bolts to 31 inch lbs. (3.5 Nm).
- Return vent hose to the fuel pump
- Outlet pipe to the fuel pump, using new gaskets. Tighten the union bolts to 22 ft. lbs. (29 Nm).
- Fuel pump and sender gauge connector
- Floor hole cover with the 4 screws
- Carpet
- Left rear seat assembly
- Negative battery cable
- Fuel cap
5. Start the vehicle and check for leaks.

4Runner

1. Before servicing the vehicle, refer to the precautions in the beginning of this section.
2. Relieve the fuel pressure.
3. Disconnect the negative battery cable from the battery.
4. Drain the fuel from the fuel tank.
5. Remove or disconnect the following:
- Fuel tank
- Fuel pump connector from the clamp
- Access plate bolts, then pull out the fuel pump assembly from the fuel tank
- Gasket(s) from the pump bracket
- Fuel pump connector
- Bracket from the lower side of the fuel pump
- Fuel pump from the fuel hose
- Rubber cushion, the clip and the fuel filter at the bottom of the fuel pump

To install:
6. Install or connect the following:
- Fuel pump filter to the fuel pump with a new clip
- Fuel pump to the fuel pump bracket
- Fuel hose to the outlet port of the fuel pump
- Fuel pump connector
- Fuel pump assembly with a new gasket(s). Tighten the bolts to 31 inch lbs. (4 Nm).
- Fuel pump connector to the clamp
- Fuel tank
- All electrical and fuel connections
- Negative battery cable
7. Refill the fuel tank and check for leaks.

Highlander

1. Before servicing the vehicle, refer to the precautions in the beginning of this section.
2. Relieve the fuel pressure.
3. Remove or disconnect the following:
- Both rear seats
- Carpet
- Service hole cover
- Connector
- Joint clip and pull out the fuel tube
- Vent tube set plate
- Pump and gauge assembly
4. Installation is the reverse of removal.

Fuel Injector

REMOVAL & INSTALLATION

2.0L

1. Before servicing the vehicle, refer to the precautions in the beginning of this section.
2. Remove or disconnect the following:
- Air cleaner assembly
- Cylinder head cover
- Throttle body from the intake manifold
- Distributor
3. Remove or disconnect the engine wire from the intake manifold, as follows:
- 4 injector connectors
- Both engine wire clamps from the intake manifold wire brackets
- Engine wire protector from the right side of the intake manifold
- Engine wire clamp
4. Remove or disconnect the Exhaust Gas Recirculation (EGR) valve, as follows:
- Vacuum hose from port **E** of the Vacuum Switching Valve (VSV)
- EGR hose from the vacuum modulator
- Loosen the EGR pipe nut from the cylinder head
- Both nuts, EGR valve, pipe assembly and gasket
5. Disconnect the engine compartment R/B No. 2
6. Remove or disconnect the fuel inlet hose and delivery pipe, as follows:
- Union bolt, both gaskets and the fuel inlet hose from the fuel filter outlet
- Air assist hose from the intake manifold port
- Air assist hose
- Loosen both delivery pipe-to-cylinder head bolts
- Delivery pipe from the 4 injectors

- Delivery pipe and fuel inlet hose assembly

7. Remove or disconnect the following:
- 4 injectors and spacers

※ WARNING

Be careful not to drop the injectors and spacers.

- O-rings, insulator and grommet from each injector

To install:

8. Install or connect the following:
- New O-rings, insulator and grommet, lubricated with gasoline, to each injector
- 4 injectors and spacers

9. Install or connect the fuel inlet hose and delivery pipe, as follows:
- Delivery pipe and fuel inlet hose assembly
- Delivery pipe to the 4 injectors. Tighten both delivery pipe-to-cylinder head bolts to 9 ft. lbs. (13 Nm).
- Air assist hose
- Air assist hose to the intake manifold port
- Union bolt, new gaskets and the fuel inlet hose to the fuel filter outlet. Tighten the union bolt to 22 ft. lbs. (29 Nm).

10. Connect the engine compartment R/B No. 2

11. Install or connect the EGR valve, as follows:
- New gasket, pipe assembly and EGR valve. Tighten the nut to 9 ft. lbs. (13 Nm) and the union nut to 43 ft. lbs. (59 Nm).
- EGR hose to the vacuum modulator
- Vacuum hose from port **E** of the VSV

12. Install or connect the engine wire to the intake manifold, as follows:
- Engine wire clamp
- Engine wire protector to the right side of the intake manifold
- Both engine wire clamps to the intake manifold wire brackets
- 4 injector connectors

13. Install or connect the following:
- Distributor
- Throttle body to the intake manifold
- Cylinder head cover
- Air cleaner assembly

2.4L

1. Before servicing the vehicle, refer to the precautions in the beginning of this section.
2. Relieve the fuel system pressure.

3. Remove or disconnect the following:
- Air cleaner and hoses
- Fuel line from the rail
- Injector connectors
- Injector wiring harness
- Fuel rail with injectors
- Injector spacers from the head

4. Installation is the reverse of removal. Coat the new o-rings with clean fuel. Before tightening the fuel rail bolts, make sure that each injector rotates smoothly. Tighten the bolts to 15 ft. lbs. (20Nm).

3.0L Highlander

1. Before servicing the vehicle, refer to the precautions in the beginning of this section.
2. Relieve the fuel system pressure.
3. Remove or disconnect the following:
- Coolant
- V-bank cover
- Battery
- Air cleaner assembly
- Upper front suspension brace
- Intake air surge tank
- Fuel supply line
- Injector connectors
- Fuel rail with injectors

4. Installation is the reverse of removal. Coat the new o-rings with clean fuel. Before tightening the fuel rail bolts, make sure that each injector rotates smoothly. Tighten the bolts to 84 inch lbs. (10Nm).

RX300

1. Before servicing the vehicle, refer to the precautions in the beginning of this section.

2. Remove or disconnect the following:
- Outer front cowl top panel assembly
- Air cleaner cap with hose
- Negative battery cable. Work must be started approximately 90 seconds or longer after the negative battery cable has been disconnected, if equipped with an air bag.
- Coolant
- Accelerator and throttle cables
- V-bank cover
- Emission valve control set
- No. 2 EGR pipe
- Hydraulic motor pressure pipe from the water inlet and air inlet chamber
- Air intake chamber assembly
- Injector wiring
- Air assist pipe from the bracket on the No. 1 fuel pipe
- Air assist hoses from the intake manifold

- Fuel return hose from the No. 1 fuel pipe
- Fuel inlet hose for the fuel filter
- 2 union bolts holding the No. 2 fuel pipe to the delivery pipes
- Fuel return hose from the fuel pressure regulator
- Union bolt for the right hand delivery pipe, 2 gaskets, 2 bolts, left hand delivery pipe together with the 3 injectors and the No. 2 fuel pipe
- Union bolt for the delivery pipe and 2 gaskets from the No. 2 fuel pipe
- The 3 bolts, right hand delivery pipe together with the 3 injectors and the No. 1 fuel pipe
- The 4 spacers from the intake manifold
- The 6 injectors from the delivery pipes
- The two O-rings and two grommets from each injector

To install:

3. Install or connect the following:
- 2 new grommets to each injector
- New O-rings, with a light coat of fuel, to each injector
- Injectors
- The 4 spacers on the intake manifold
- Right hand delivery pipe and the No. 1 fuel pipe together with the 3 injectors in position on the intake manifold
- Bolt holding the right side delivery pipe, temporarily, to the intake manifold
- Left hand delivery pipe and the No. 2 fuel pipe together with the 3 injectors in position on the intake manifold
- Fuel return hose to the fuel pressure regulator

4. Temporarily install the 2 bolts holding the left hand delivery pipe to the intake manifold.

5. Temporarily install the No. 2 fuel pipe to the left side delivery pipe with the union bolt and 2 new gaskets.

6. Check that the injectors rotate smoothly. If they do not, replace the O-rings.

7. Position the injector connector outward. Tighten the 4 bolts holding the delivery pipes to the intake manifold and tighten to 7 ft. lbs. (10 Nm). Tighten the bolt holding the No. 1 fuel pipe to the intake manifold to 14 ft. lbs. (20 Nm). Tighten the 2 union bolts holding the no. 2 fuel pipe to the delivery pipes to 24 ft. lbs. (32 Nm).

8. Install or connect the following:

- Fuel inlet and return hoses. Union bolt: 22 ft. lbs. (30 Nm)
- Fuel return hose to the No. 1 fuel pipe. Pass the fuel return hose under the heater hoses.
- Air assist hoses to the intake manifold
- Air assist pipe to the bracket on the No. 1 fuel pipe
- Fuel injector wiring connectors
- Air intake chamber assembly
- Hydraulic motor pressure pipe to the intake chamber. Bolts: 69 inch lbs. (8 Nm)
- No. 2 EGR pipe with new gaskets, tighten to 9 ft. lbs. (12 Nm)
- Emission control valve set
- V-bank cover
- Air cleaner hose
- Throttle and accelerator cables
- Coolant
- Air cleaner cap with hose
- Outer front cowl top panel assembly
- Negative battery cable

2.7L Engines

1. Before servicing the vehicle, refer to the precautions in the beginning of this section.

2. Relieve the fuel system pressure.
3. Remove or disconnect the following:
   - Throttle body
   - Fuel injector electrical connectors
   - Crankshaft Position (CKP) sensor connector
   - Knock Sensor (KS) connector
   - Data Link Connector 1 (DLC1) and wire clamp from the brackets
   - Vacuum line from the fuel pressure regulator
   - Fuel return hose from the pressure regulator
   - Union bolt and gaskets
   - Fuel inlet pipe from the fuel rail
   - Fuel rail with the injectors attached

❊❊ WARNING

The injectors are only retained by their O-rings and will tend to drop out of the fuel rail.

- 4 insulators from the four spacers
- Fuel injectors from the fuel rail
- O-ring and grommet, discard them

To install:
4. Install or connect the following:
   - New grommets and O-rings on

each injector, lubricated with a light coat of gasoline
   - Fuel injectors, with the electrical connector facing upwards
   - New insulators and spacers on the intake manifold

5. Temporarily install the bolts holding the delivery pipe to the intake manifold.
6. Check that the injectors rotate smoothly.
7. Install or connect the following:
   - Tighten the delivery pipe-to-intake manifold bolts to 15 ft. lbs. (21 Nm).
   - Injector electrical connectors
   - Fuel inlet pipe with new gaskets. Tighten the union bolt to 22 ft. lbs. (29 Nm) and the bolt to 14 ft. lbs. (20 Nm).
   - Fuel return pipe to the fuel pressure regulator
   - Vacuum line to the pressure regulator
   - Throttle body
   - Negative battery cable

3.4L Engine

1. Before servicing the vehicle, refer to the precautions in the beginning of this section.

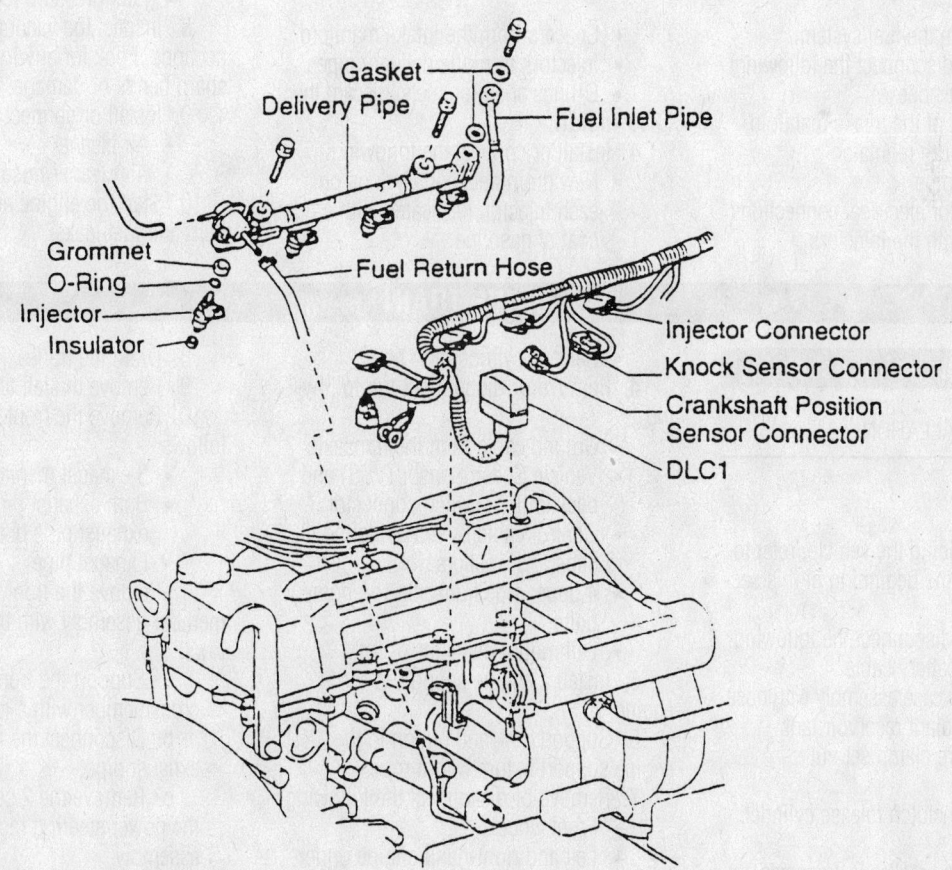

Fuel injector arrangement and related components—2.7L engines

93165G20

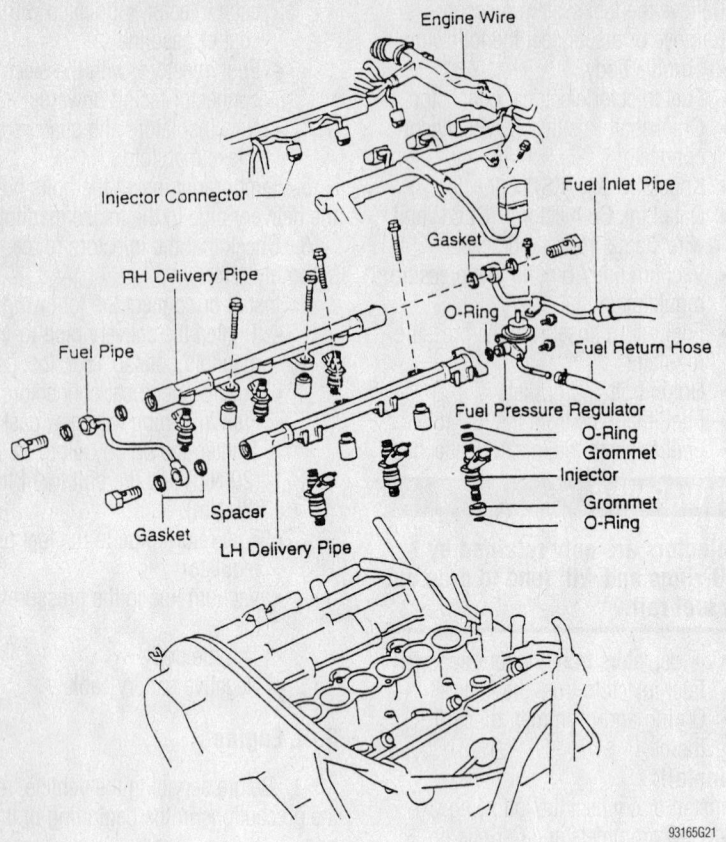

Fuel injector arrangement and related components—3.4L engine

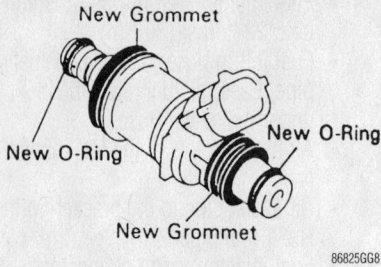

Install new O-rings and grommets on each injector—3.4L engine

2. Depressurize the fuel system.
3. Remove or disconnect the following:
   - Air cleaner hose
   - Upper half of the intake manifold
   - Fuel pressure regulator
   - Fuel inlet pipe
   - Fuel injector electrical connections
   - Fuel rail with the injectors
   - Spacers from the intake manifold
   - Injectors from the delivery pipes
   - O-rings and grommets, discard them

To install:
4. Install or connect the following:
   - New grommets and O-rings on each injector, lubricated with a light coat of gasoline
   - Fuel injector with the electrical connector facing outward
   - Spacers on the intake manifold
5. Temporarily install the bolts to hold the delivery pipes to the intake manifold.
6. Check that the injectors rotate smoothly. If they do not, the O-rings have probably been installed incorrectly.
7. Install or connect the following:
   - Fuel injector electrical connectors
   - Fuel pipe with new gaskets. Tighten the bolts to 25 ft. lbs. (34 Nm) and the delivery pipes-to-intake manifold bolts to 10 ft. lbs. (13 Nm).
   - Fuel pipe union with new gaskets. Tighten the clamp bolt to 71 inch lbs. (8 Nm).
   - Fuel pressure regulator
8. Inspect the vacuum lines and connections. Look for any loose connections, sharp bends or damage.
9. Install or connect the following:
   - Air cleaner
   - Air cleaner hose
10. Start the engine and check for vacuum and fuel leaks.

DRIVE TRAIN

Manual Transmission Assembly

REMOVAL & INSTALLATION

2WD RAV4

1. Before servicing the vehicle, refer to the precautions in the beginning of this section.
2. Remove or disconnect the following:
   - Negative battery cable
   - Air cleaner case assembly with hose
   - Engine coolant reservoir tank
   - Engine wire clamp set nut
   - Starter
3. Remove the clutch release cylinder, as follows:
   - Clutch line bracket-to-transaxle set bolts
   - Release cylinder and line
4. Remove or disconnect the following:
   - Ground cable from the transaxle
   - Vehicle Speed Sensor (VSS) and backup light switch connector
   - Control cable by removing the 4 clips and washers
   - 4 upper side transaxle-to-engine bolts
   - Left mount insulator
5. Install a engine support to the engine.
6. Support rack and pinion to the engine support fixture with a rope.
7. Remove or disconnect the following:
   - Front wheels
   - Left and right-hand engine under-covers
8. Drain the transaxle oil.
9. Remove the left and right halfshafts.
10. Remove the front exhaust pipe, as follows:
    - 3 exhaust manifold nuts and gasket
    - Both exhaust pipe-to-center exhaust pipe bolts
    - Exhaust pipe
11. Remove the front suspension crossmember assembly with the sway bar, as follows:
    a. Support the front suspension crossmember with a jack.
    b. Disconnect the ring from the center exhaust pipe.
    c. Remove the 2 set bolts and nuts of the power steering rack and pinion assembly.
    d. Remove the suspension cross-

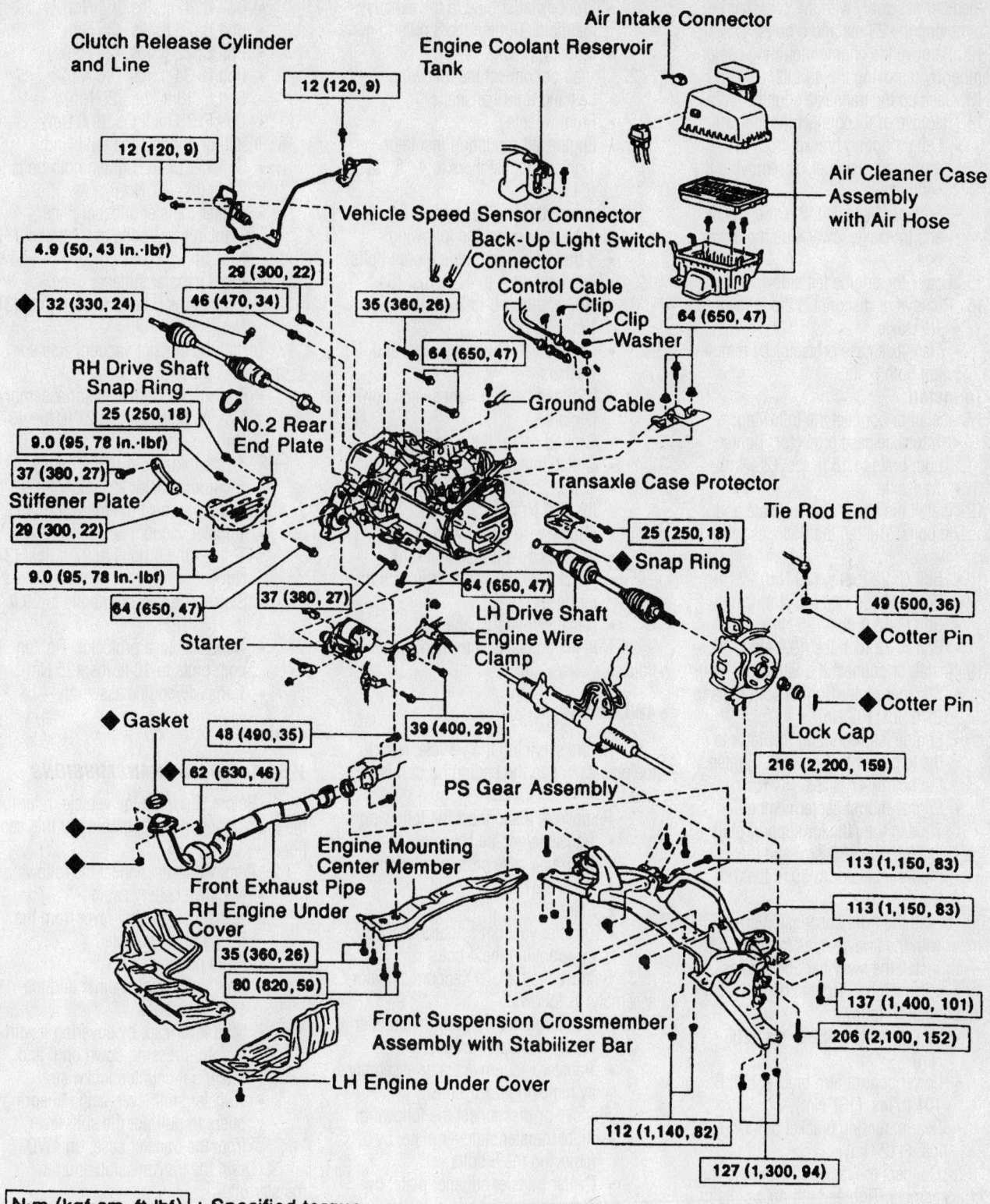

Clutch Release Cylinder and Line

Engine Coolant Reservoir Tank

Air Intake Connector

12 (120, 9)

12 (120, 9)

Air Cleaner Case Assembly with Air Hose

4.9 (50, 43 in.·lbf)

Vehicle Speed Sensor Connector

Back-Up Light Switch Connector

29 (300, 22)

◆ 32 (330, 24)

46 (470, 34)

35 (360, 26)

Control Cable Clip

Clip

Washer

64 (650, 47)

RH Drive Shaft Snap Ring

64 (650, 47)

25 (250, 18)

No.2 Rear End Plate

Ground Cable

Transaxle Case Protector

9.0 (95, 78 in.·lbf)

37 (380, 27)

Tie Rod End

Stiffener Plate

25 (250, 18)

29 (300, 22)

◆ Snap Ring

49 (500, 36)

9.0 (95, 78 in.·lbf)

64 (650, 47)

37 (380, 27)

64 (650, 47)

◆ Cotter Pin

Starter

LH Drive Shaft

Engine Wire Clamp

◆ Cotter Pin

Lock Cap

216 (2,200, 159)

◆ Gasket

48 (490, 35)

39 (400, 29)

◆ 62 (630, 46)

PS Gear Assembly

◆

◆

113 (1,150, 83)

113 (1,150, 83)

Engine Mounting Center Member

Front Exhaust Pipe

RH Engine Under Cover

35 (360, 26)

80 (820, 59)

137 (1,400, 101)

206 (2,100, 152)

Front Suspension Crossmember Assembly with Stabilizer Bar

LH Engine Under Cover

112 (1,140, 82)

127 (1,300, 94)

N·m (kgf·cm, ft·lbf) : Specified torque

◆ Non-reusable part

7924ZG61

Transaxle exploded view—2WD RAV4

member assembly with the sway bar by removing the 2 nuts and 6 bolts.

12. Remove the engine mounting center-member by removing the 4 bolts.

13. Jack up the transaxle slightly.

14. Remove or disconnect the following:
- Left mounting bracket from the mounting insulator by removing the set bolt
- Stiffener plate, No. 2 rear endplate and transaxle lower side mounting bolt

15. Lower the engine left side

16. Remove or disconnect the following:
- Transaxle
- Transaxle case protector by removing both bolts

To install:

17. Install or connect the following:
- Transaxle case protector. Tighten both bolts to 18 ft. lbs. (25 Nm).
- Transaxle

18. Install the No. 2 rear endplate and transaxle bolts. Tighten the bolts, as follows:
- Bolt C: 22 ft. lbs. (29 Nm)
- Bolt D: 34 ft. lbs. (46 Nm)
- Bolt E: 18 ft. lbs. (25 Nm)
- Bolt F: 78 inch lbs. (9.0 Nm)

19. Install or connect the following:
- Stiffener plate. Tighten both bolts to 27 ft. lbs. (37 Nm).
- Engine left mounting insulator to the left mounting bracket. Tighten the bolt to 47 ft. lbs. (64 Nm).
- Engine mount centermember. Tighten the radiator support bolts to 26 ft. lbs. (35 Nm) and the mount insulator to 59 ft. lbs. (80 Nm).

20. Install the front suspension cross-member with the sway bar, as follows:
- a. Install the sway bar and suspension crossmember. Tighten the nuts/bolts, as follows:
  - Vehicle bolt A: 152 ft. lbs. (206 Nm)
  - Lower control arm bracket bolt B: 101 ft. lbs. (137 Nm)
  - Rear mounting bracket bolt C: 82 ft. lbs. (112 Nm)
- b. Connect the rack and pinion to the crossmember. Tighten both nuts/bolts to 83 ft. lbs. (113 Nm).
- c. Connect the ring for the center exhaust pipe.

21. Install the front exhaust pipe, as follow:
- Pipe with new gaskets
- Front pipe to the center exhaust pipe. Tighten both bolts to 35 ft. lbs. (48 Nm).

- Front exhaust pipe to the exhaust manifold. Tighten the 3 nuts to 46 ft. lbs. (62 Nm).

22. Install or connect the following:
- Left and right halfshafts
- Front wheels
- Engine left mounting insulator. Tighten the fasteners to 47 ft. lbs. (64 Nm).

23. Remove the engine support fixture.

24. Install or connect the following:
- 4 transaxle upper side mount bolts. Tighten bolt A to 47 ft. lbs. (64 Nm) and bolt B to 26 ft. lbs. (35 Nm).
- Ground cable with the clips and washers
- VSS and backup light switch connectors
- Ground cable to the transaxle
- Clutch release cylinder and line
- Starter. Tighten both bolts to 29 ft. lbs. (39 Nm).
- Engine wire clamp with the nut
- Engine coolant reservoir tank
- Air cleaner case assembly with the air hose
- Negative battery cable

25. Fill the transaxle with fluid. Check all fluids.

4WD Rav4

1. Before servicing the vehicle, refer to the precautions in the beginning of this section.

2. Remove or disconnect the following:
- Transaxle/engine assembly
- Transaxle case protector, by removing the 2 bolts
- Starter
- Transfer vacuum actuator bracket, by removing the 4 bolts

3. Remove the transfer vacuum actuator assembly, as follows:
- 4 solenoid hoses from the transfer vacuum actuator assembly
- Transfer vacuum actuator assembly, by removing the 2 bolts

4. Remove or disconnect the following:
- Right transfer stiffener plate, by removing the 5 bolts
- Center transfer stiffener plate, by removing the 3 bolts
- Stiffener plate by removing the 2 bolts
- Transaxle from the engine, by removing the 9 transaxle mount bolts

To install:

5. Connect the transaxle to the engine. Tighten the 9 bolts, as follows:

- Bolt A: 47 ft. lbs. (64 Nm)
- Bolt B: 26 ft. lbs. (35 Nm)
- Bolt C: 22 ft. lbs. (29 Nm)
- Bolt D: 34 ft. lbs. (46 Nm)
- Bolt E: 18 ft. lbs. (25 Nm)
- Bolt F: 78 inch lbs. (9.0 Nm)

6. Install or connect the following:
- Stiffener plate. Tighten both bolts to 27 ft. lbs. (37 Nm).
- Center transfer stiffener plate. Tighten the 3 bolts to 27 ft. lbs. (37 Nm).
- Right transfer stiffener plate. Tighten the 5 bolts to 27 ft. lbs. (37 Nm).

7. Install the transfer vacuum actuator assembly, as follows:
- Transfer vacuum actuator assembly. Tighten both bolts to 27 ft. lbs. (37 Nm).
- 4 solenoid hoses to the transfer vacuum actuator assembly

8. Install or connect the following:
- Transfer vacuum actuator bracket. Tighten the 4 bolts to 27 ft. lbs. (37 Nm).
- Starter. Tighten both bolts to 29 ft. lbs. (39 Nm).
- Transaxle case protector. Tighten both bolts to 18 ft. lbs. (25 Nm).
- Transaxle/engine assembly

4Runner

W59 AND R150 TRANSMISSIONS

1. Before servicing the vehicle, refer to the precautions in the beginning of this section.

2. Remove or disconnect the following:
- Negative battery cable
- Transmission shift lever from the inside of the vehicle
- Front console box
- Shift lever boot retainer and the shift lever boot
- Shift lever cap, by covering it with a cloth, pressing down on it and rotating it counterclockwise
- Transfer shift lever, using snapring pliers to pull out the shift lever from the transfer case, on 4WD

3. Drain the transmission and the transfer oil.

4. Remove or disconnect the following:
- No. 1 and No. 2 engine undercover
- Front and rear driveshafts
- Vehicle Speed Sensor (VSS), back-up light switch and the 4WD position switch connectors
- L4 position switch connector, if equipped with Anti-lock Brake System (ABS) and/or differential lock

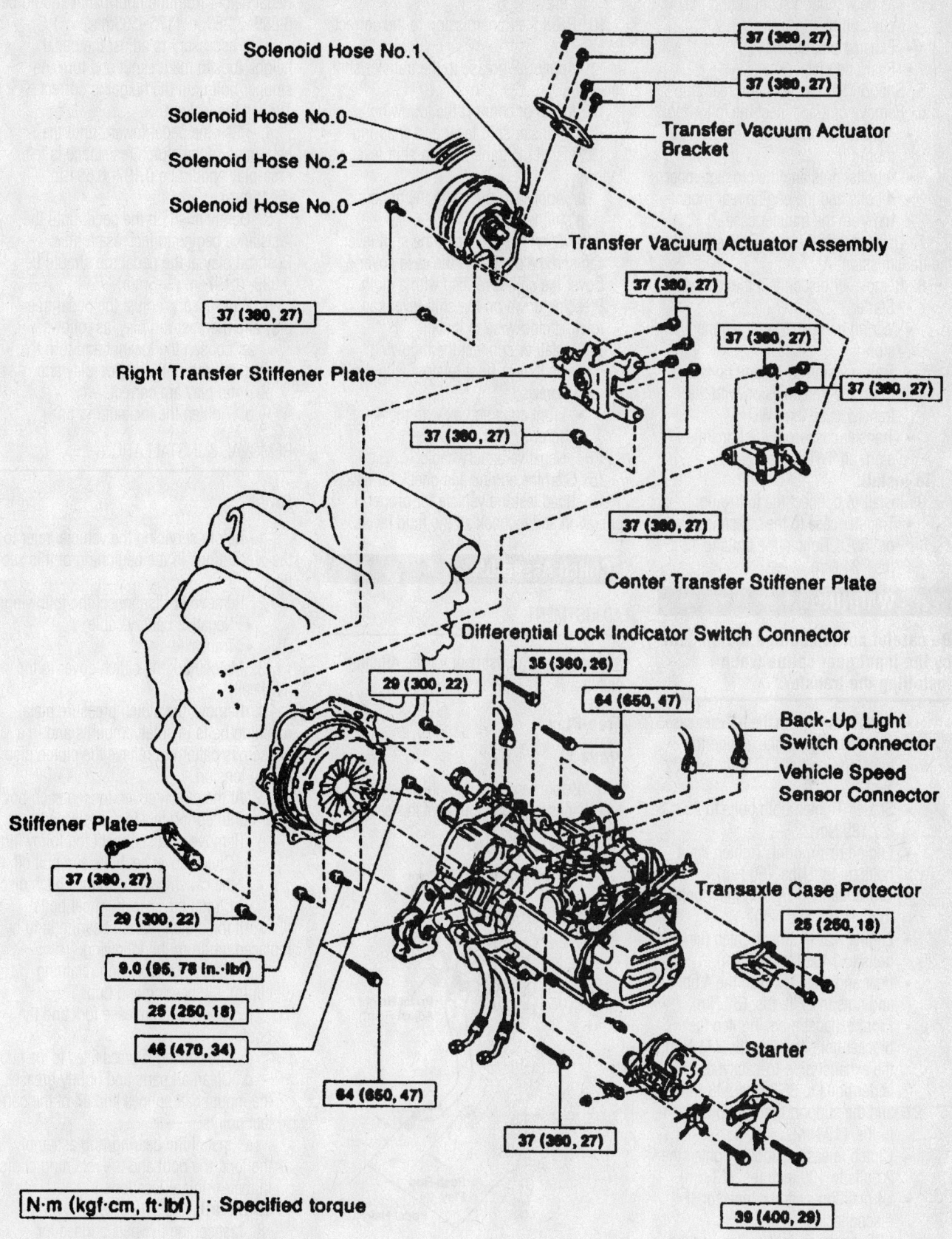

Solenoid Hose No.1

Solenoid Hose No.0

Solenoid Hose No.2

Solenoid Hose No.0

37 (380, 27)

37 (380, 27)

Transfer Vacuum Actuator Bracket

Transfer Vacuum Actuator Assembly

37 (380, 27)

37 (380, 27)

37 (380, 27)

37 (380, 27)

37 (380, 27)

Right Transfer Stiffener Plate

37 (380, 27)

37 (380, 27)

Center Transfer Stiffener Plate

Differential Lock Indicator Switch Connector

35 (360, 26)

29 (300, 22)

64 (650, 47)

Back-Up Light Switch Connector

Vehicle Speed Sensor Connector

Stiffener Plate

37 (380, 27)

29 (300, 22)

9.0 (95, 78 in.·lbf)

25 (250, 18)

46 (470, 34)

64 (650, 47)

Transaxle Case Protector

25 (250, 18)

Starter

37 (380, 27)

39 (400, 29)

N·m (kgf·cm, ft·lbf) : Specified torque

7924ZG62

Manual transaxle exploded view—4WD RAV4

- Clutch release cylinder and move it aside without disconnecting the clutch line
- Exhaust pipe bracket
- Rear end plate

5. Support the transmission rear side.

6. Remove or disconnect the following:
- 4 bolts from the engine rear mount
- 4 bolts, nuts and the crossmember
- 4 bolts and the engine rear mounting from the transfer case

7. Using a transmission jack, support the transmission.

8. Remove or disconnect the following:
- Starter
- Wiring harness from the transmission
- Transmission-to-engine bolts and lower the transmission with the transfer case, on 4WD
- Transfer case from the transmission, on 4WD

To install:

9. Install or connect the following:
- Transfer case to the transmission, on 4WD. Tighten the bolts to 17 ft. lbs. (24 Nm).

✳✳ WARNING

Be careful not to damage the oil seal by the input gear spline when installing the transfer.

- Transmission with the transfer case, on 4WD. Tighten the engine-to-transmission bolts to 53 ft. lbs. (72 Nm).
- Starter. Tighten both bolts to 29 ft. lbs. (39 Nm).
- Engine rear mount. Tighten the 4 bolts to 48 ft. lbs. (65 Nm).
- Crossmember. Tighten the 4 bolts to 48 ft. lbs. (65 Nm).
- Engine rear mount. Tighten the 4 bolts to 14 ft. lbs. (19 Nm).
- Rear end plate. Tighten the 4 bolts and nuts to 27 ft. lbs. (37 Nm).
- Front exhaust pipe. Tighten the bracket bolts to 52 ft. lbs. (71 Nm), the exhaust pipe-to-catalytic converter bolts to 35 ft. lbs. (48 Nm) and the support bracket bolts to 14 ft. lbs. (19 Nm).
- Clutch release cylinder. Tighten the 2 bolts to 108 inch lbs. (13 Nm).
- L4 position switch connector, if disconnected
- VSS, back-up light switch and the 4WD position switch connectors
- Front and rear driveshafts

- No. 1 and No. 2 engine under covers

10. Refill the transmission to the correct level.

11. Apply MP grease to the transfer shift lever.

12. Install or connect the following:
- Transfer shift lever and snapring

13. Install the transmission shift lever, as follows:

a. Apply MP grease to the transmission shift lever.

b. Align the groove of the shift lever cap and the pin par of the case cover. Cover the shift lever cap with a cloth. Pressing down on the shift lever cap, rotate it clockwise to install.

14. Install or connect the following:
- Shift lever boot retainer with the 4 screws
- Front console box with the 4 screws
- Negative battery cable

15. Start the engine and check for leaks.

16. Road test the vehicle for proper operation and recheck all the fluid levels.

Clutch Assembly

ADJUSTMENT

There is no adjustment for the 4Runner clutch.

Free-Play

RAV4

1. Before servicing the vehicle, refer to the precautions in the beginning of this section.

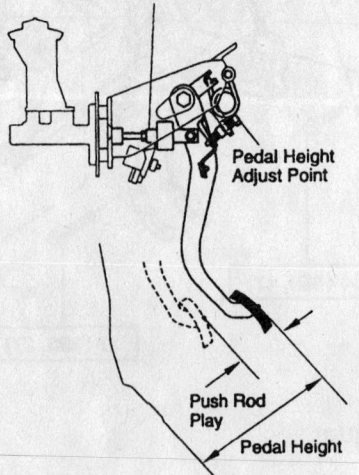

Push Rod Play Adjust Point

Pedal Height Adjust Point

Push Rod Play

Pedal Height

7924ZG67

Clutch pedal height measurement location—RAV4

2. Check that the pedal height is correct. Pedal height from the floor panel should be: 6.889–7.283 in. (175–185mm)

3. If necessary to adjust the pedal height, loosen the locknut and turn the stopper bolt until the height is correct. Tighten the locknut.

4. Push the pedal inward until the beginning of the clutch resistance is felt. Free-play should be 0.197–0.591 in. (5–15mm).

5. Gently push on the pedal until the resistance begins to increase a little. Pushrod play at the pedal top should be 0.039–0.197 in. (1–5mm).

6. If necessary, adjust the pedal free-play and the pushrod play, as follows:

a. Loosen the locknut and turn the push the rod until the free-play and pushrod play are correct.

b. Tighten the locknut.

REMOVAL & INSTALLATION

RAV4

1. Before servicing the vehicle, refer to the precautions in the beginning of this section.

2. Remove or disconnect the following:
- Negative battery cable
- Transaxle

3. Matchmark the clutch cover to the flywheel.

4. Remove the clutch pressure plate retaining bolts in small amounts and in a crisscross pattern to relieve the clutch disc spring tension.

5. At the clutch cover, loosen each bolt 1 turn until spring tension is released.

6. Remove or disconnect the following:
- Clutch cover set bolts and pull off the clutch cover with the clutch disc
- Clutch cover-to-flywheel bolts

7. If the clutch release bearing is to be replaced, perform the following:

a. Remove the bearing retaining clip(s), the bearing and hub.

b. Remove the release fork and the boot.

c. The bearing is press fitted to the hub.

d. Clean all parts and lightly grease the input shaft splines and all of the contact points.

e. Install the bearing/hub assembly, the fork, the boot and the retaining clip(s) in their original locations.

To install:

8. Inspect the flywheel surface for cracks, heat scoring (blue marks) and warpage. Replace or resurface the flywheel, if any damage is present.

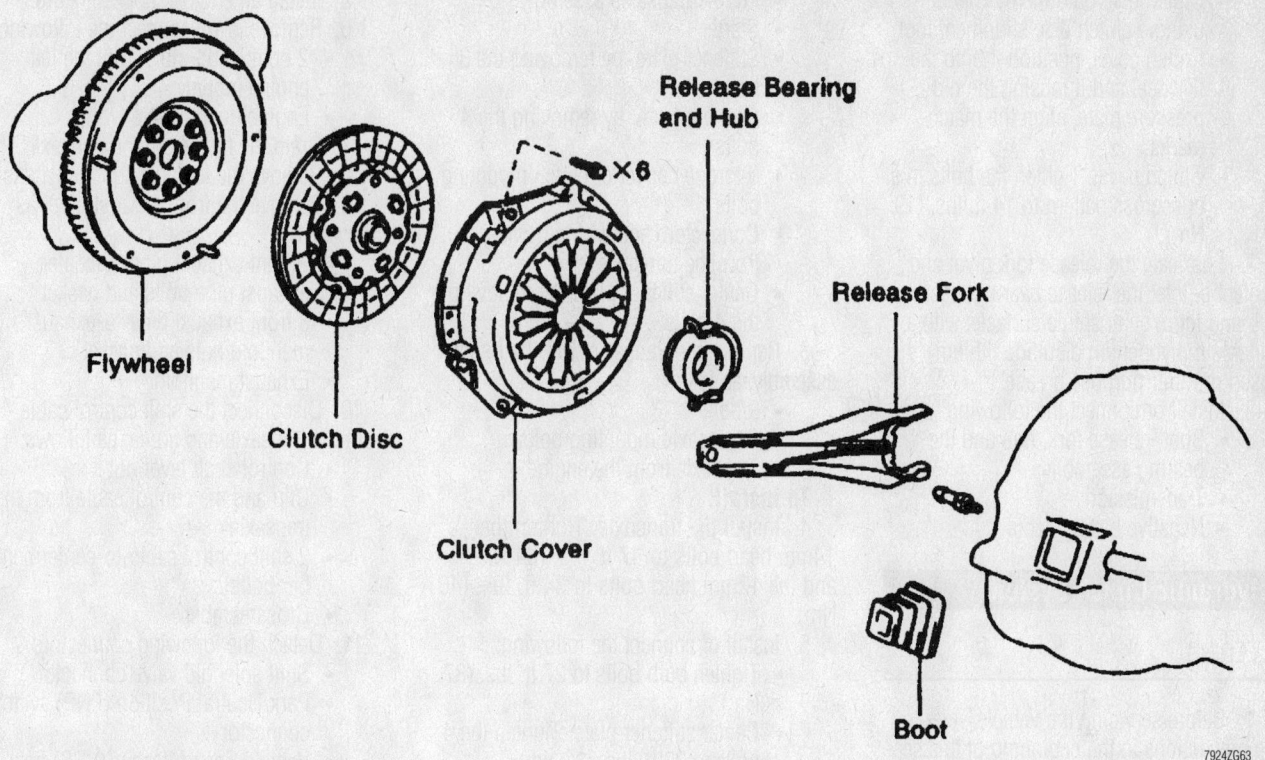

Clutch component assembly—Exploded view

➡ **Before installing any new parts, be sure they are clean. During installation, do not get grease or oil on any of the components, as this will shorten clutch life considerably.**

9. Using a clutch alignment tool, position the clutch disc against the flywheel. The raised center section of the disc faces the transaxle.

10. Install or connect the following:
- Clutch cover onto the flywheel by aligning the matchmarks

- Clutch cover. Tighten the bolts in a crisscross pattern to 14 ft. lbs. (19 Nm).

11. Lubricate the release fork pivot and contact points, release bearing, bearing hub and input shaft spline surfaces with a suitable molybdenum disulfide lithium based or multi-purpose grease.

12. Install or connect the following:
- Boot, release fork, hub and the bearing assemblies
- Transaxle
- Negative battery cable

4Runner

1. Before servicing the vehicle, refer to the precautions in the beginning of this section.

2. Remove or disconnect the following:
- Negative battery cable
- Transmission assembly

3. Matchmark the clutch cover to the flywheel.

4. At the clutch cover, loosen each bolt 1 turn until spring tension is released.

5. Remove or disconnect the following:
- Clutch cover set bolts and the clutch cover with the clutch disc.
- Release bearing retaining clip and withdraw the it
- Release fork and boot assembly

To install:

6. Install or connect the following:

Torque sequence for the clutch cover—RAV4

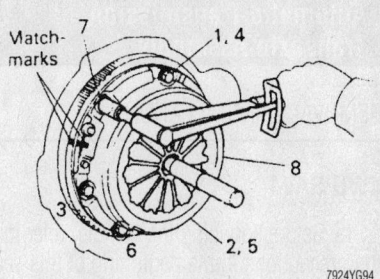

Bolt tightening sequence for the clutch cover—4Runner

- Clutch disc onto the flywheel, using a clutch disc alignment tool
- Clutch cover, position it onto the flywheel and if reusing the old pressure plate, align the match-marks.
- Clutch cover. Tighten the bolts in a crisscross pattern to 14 ft. lbs. (19 Nm).

7. Lubricate the release fork pivot and contact points, the release bearing, bearing hub and input shaft spline surfaces with a suitable molybdenum disulfide lithium based or multi-purpose grease.

8. Install or connect the following:
- Boot, release fork, hub and the bearing assemblies
- Transmission
- Negative battery cable

Hydraulic Clutch System

BLEEDING

1. Before servicing the vehicle, refer to the precautions in the beginning of this section.

2. Fill the clutch reservoir with brake fluid. Check the reservoir level frequently and add fluid as needed.

3. Connect one end of a vinyl tube to the bleeder plug on the slave cylinder and submerge the other end into a clear container half-filled with brake fluid.

4. Slowly pump the clutch pedal several times.

5. Have an assistant hold the clutch pedal down and loosen the bleeder plug until fluid and/or air starts to run out of the bleeder plug. Close the bleeder plug while the pedal is held to the floor.

6. Repeat Steps 2 and 3 until all the air bubbles are removed from the system.

7. Tighten the bleeder plug when all the air is gone.

8. Refill the master cylinder to the proper level as required.

9. Check the system for leaks.

Automatic Transmission/ Transaxle Assembly

REMOVAL & INSTALLATION

4WD RAV4

1. Before servicing the vehicle, refer to the precautions in the beginning of this section.

2. Remove or disconnect the following:
- Negative battery cable

- Engine/transaxle assembly
- Starter
- Stiffener plate, by removing the 3 bolts
- Rear endplate, by removing the 4 bolts
- 6 torque converter clutch mounting bolts
- Connectors and wiring harness, from the transaxle
- Center stiffener plate, by removing the 4 bolts

3. Remove the transaxle with the transfer assembly, as follows:
- 2 bolts
- 5 transaxle mounting bolts
- Transaxle from the engine

To install:

4. Install the transaxle. Tighten the 14mm head bolts to 47 ft. lbs. (64 Nm) and the 12mm head bolts to 34 ft. lbs. (46 Nm).

5. Install or connect the following:
- Tighten both bolts to 27 ft. lbs. (37 Nm).
- Center stiffener plate. Tighten the 4 bolts to 27 ft. lbs. (37 Nm).
- Connectors and the wiring harness to the transaxle
- Torque converter clutch mounting bolts. Tighten each bolt to 20 ft. lbs. (27 Nm).

➡ **Coat the threads of the bolts with an approved locking compound.**

- Rear endplate. Tighten the 4 bolts to 80 inch lbs. (9.0 Nm).
- Stiffener plate. Tighten the 3 bolts to 27 ft. lbs. (37 Nm).
- Starter. Tighten both bolts to 29 ft. lbs. (39 Nm).
- Engine/transaxle assembly
- Negative battery cable

2WD RAV4

1. Before servicing the vehicle, refer to the precautions in the beginning of this section.

2. Remove or disconnect the following:
- Negative battery cable
- Throttle cable
- Engine coolant reservoir tank
- Air cleaner assembly
- Ground cable from the transaxle
- Set nut of the engine wire clamp

3. Remove the starter, as follows:
- Connector and nut from the starter
- 2 bolts and the engine wire
- Starter

4. Remove the 3 upper side transaxle mounting bolts.

5. Install an engine support fixture.

6. Remove or disconnect the following:
- 2 bolts and 2 nuts from the left engine mount
- Engine undercovers

7. Drain the fluid from the transaxle.

8. Remove the left and right halfshafts.

9. Remove the front exhaust pipe, as follows:
- 2 front exhaust pipe-to-center exhaust pipe bolts and gasket
- 3 front exhaust pipe-to-exhaust manifold nuts and gasket
- Exhaust manifold

10. Disconnect the shift control cable from the transaxle and frame, as follows:
- Control shaft lever nut
- Clip and the control cable from the transaxle
- 2 shift control cable-to-centermember bolts
- Crossmember

11. Detach the following connectors:
- Shift solenoid valve connector
- Park/Neutral Position (PNP) switch connector
- Vehicle Speed Sensor (VSS) connector

12. Remove or disconnect the following:
- Oil cooler hoses from the transaxle
- Rack and pinion from the crossmember by removing both nuts/bolts

13. Support the rack and pinion.

14. Support the suspension crossmember with a floor jack.

15. Remove or disconnect the following:
- Crossmember-to-centermember fasteners
- Crossmember with the sway bar
- Stiffener plate by removing the 3 bolts
- Rear endplate by removing the 4 bolts
- 6 torque converter bolts
- Both rear side transaxle mounting bolts
- Transaxle

To install:

16. Install or connect the following:
- Transaxle
- Both rear side transaxle mounting bolts. Tighten the top bolt to 18 ft. lbs. (25 Nm) and the lower bolt to 34 ft. lbs. (46 Nm).
- Torque converter bolts. Tighten the bolts to 20 ft. lbs. (27 Nm).

➡ **First install the gray bolt; then, install the 5 black bolts.**

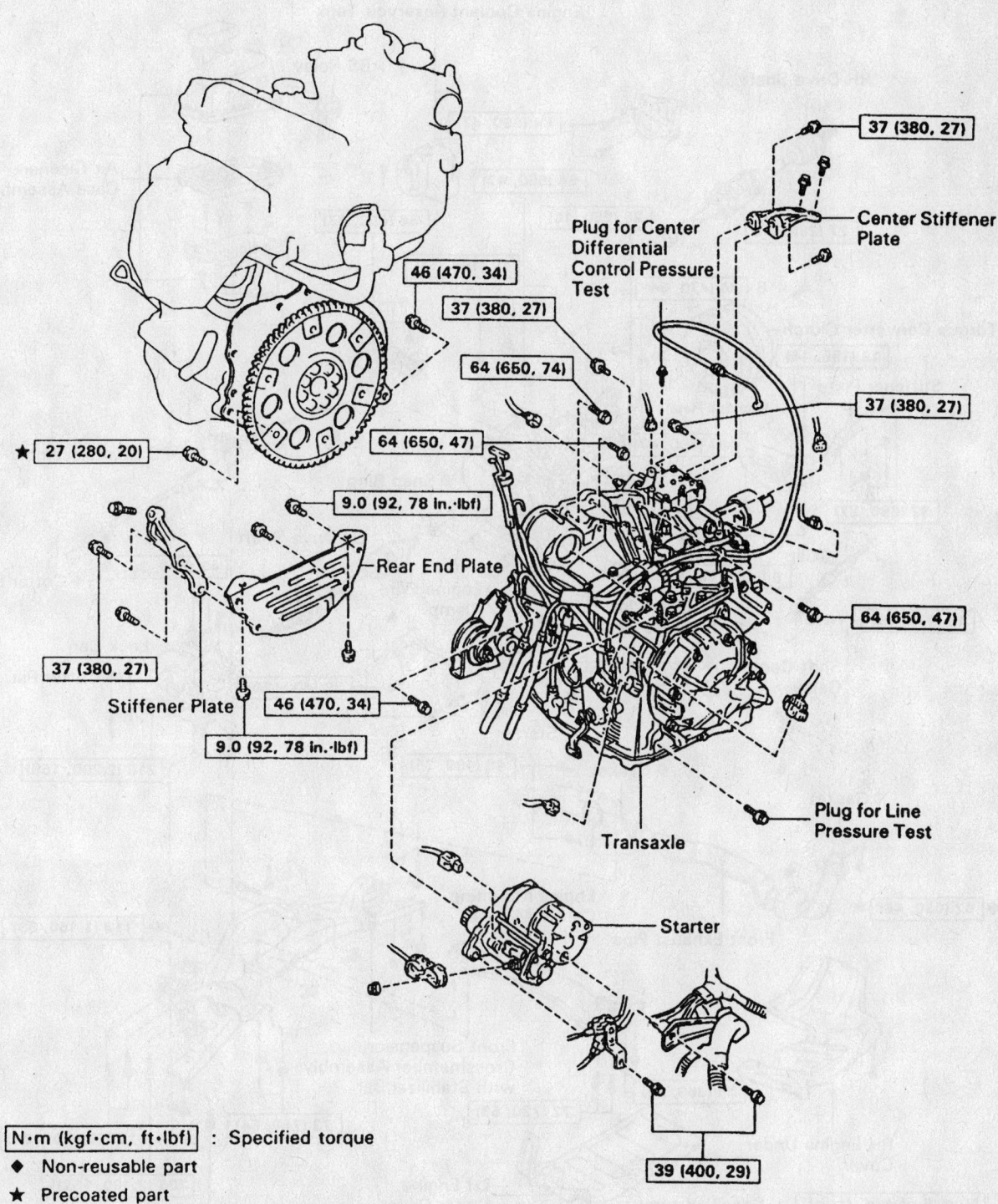

37 (380, 27)

Center Stiffener
Plate

Plug for Center
Differential
Control Pressure
Test

46 (470, 34)

37 (380, 27)

64 (650, 74)

37 (380, 27)

64 (650, 47)

★ 27 (280, 20)

9.0 (92, 78 in.·lbf)

Rear End Plate

64 (650, 47)

37 (380, 27)

Stiffener Plate

46 (470, 34)

9.0 (92, 78 in.·lbf)

Transaxle

Plug for Line
Pressure Test

Starter

39 (400, 29)

N·m (kgf·cm, ft·lbf) : Specified torque
◆ Non-reusable part
★ Precoated part

7924ZG62A

Automatic transaxle exploded view—4WD RAV4

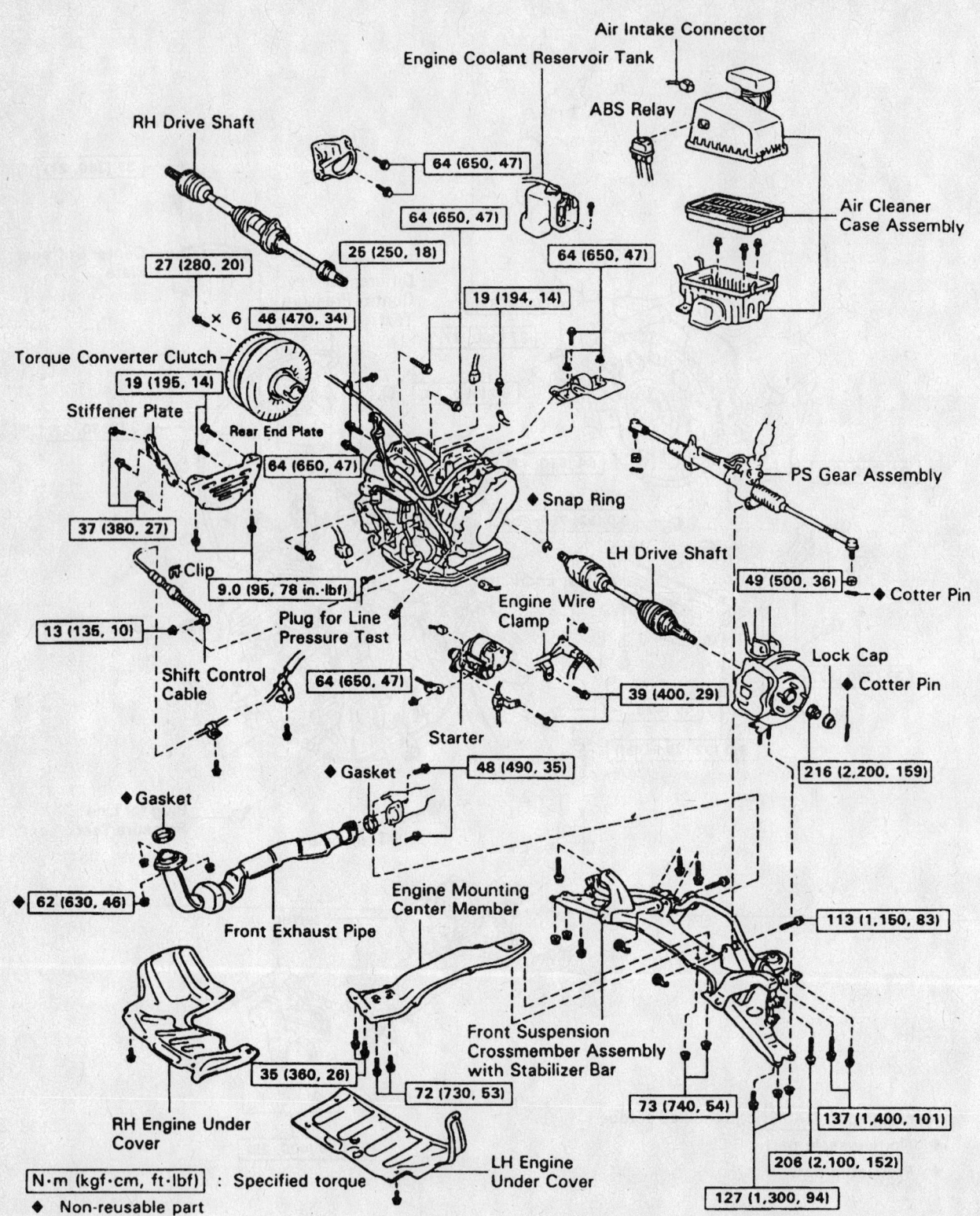

Air Intake Connector

Engine Coolant Reservoir Tank

ABS Relay

RH Drive Shaft

Air Cleaner Case Assembly

64 (650, 47)

64 (650, 47)

25 (250, 18)

64 (650, 47)

19 (194, 14)

27 (280, 20)

× 6

46 (470, 34)

Torque Converter Clutch

19 (195, 14)

Rear End Plate

Stiffener Plate

64 (650, 47)

PS Gear Assembly

37 (380, 27)

◆ Snap Ring

LH Drive Shaft

49 (500, 36)   ◆ Cotter Pin

Clip

9.0 (95, 78 in.·lbf)

Plug for Line Pressure Test

Engine Wire Clamp

Lock Cap

13 (135, 10)

◆ Cotter Pin

Shift Control Cable

64 (650, 47)

39 (400, 29)

216 (2,200, 159)

Starter

48 (490, 35)

◆ Gasket

◆ Gasket

62 (630, 46)

Engine Mounting Center Member

113 (1,150, 83)

Front Exhaust Pipe

Front Suspension Crossmember Assembly with Stabilizer Bar

35 (360, 26)

72 (730, 53)

73 (740, 54)

137 (1,400, 101)

RH Engine Under Cover

206 (2,100, 152)

LH Engine Under Cover

127 (1,300, 94)

N·m (kgf·cm, ft·lbf) : Specified torque

◆ Non-reusable part

7924ZG64

Transaxle exploded view—2WD RAV4

- Rear endplate. Tighten the engine bolts to 78 inch lbs. (9.0 Nm) and the transaxle bolts to 14 ft. lbs. (19 Nm).
- Stiffener plate. Tighten the 3 bolts to 27 ft. lbs. (37 Nm).

17. Install the front suspension crossmember and centermember with the sway bar. Tighten the nuts/bolts, as follows:
- Bolt A: 152 ft. lbs. (206 Nm)
- Bolt B: 101 ft. lbs. (137 Nm)
- Bolt C: 26 ft. lbs. (35 Nm)
- Bolt D: 53 ft. lbs. (72 Nm)
- Nut: 54 ft. lbs. (73 Nm)

18. Install or connect the following:
- Rack and pinion to the crossmember. Tighten the nuts to 83 ft. lbs. (113 Nm).
- Oil cooler hoses with both clips

19. Connect the following connectors:
- Shift solenoid valve connector
- PNP switch connector
- VSS connector

20. Install or connect the following:
- Shift control cable to the transaxle. Tighten the nut to 10 ft. lbs. (13 Nm).

21. Install the front exhaust pipe, as follows:
- Front exhaust pipe, using new gaskets
- Front exhaust pipe to the exhaust manifold. Tighten the 3 nuts to 46 ft. lbs. (62 Nm).
- Front exhaust pipe to the center exhaust pipe. Tighten both bolts to 35 ft. lbs. (48 Nm).

22. Install or connect the following:
- Left and right halfshafts
- Engine undercovers
- Left engine mount. Tighten the both nuts/bolts to 47 ft. lbs. (64 Nm).

23. Remove the engine fixture.

24. Install or connect the following:
- Upper side transaxle mount. Tighten the 3 bolts to 47 ft. lbs. (64 Nm).
- Starter. Tighten both bolts to 29 ft. lbs. (39 Nm).
- Engine wire
- Starter wire with the nut
- Engine wire clamp set nut
- Ground cable to the transaxle. Tighten the bolt to 14 ft. lbs. (19 Nm).
- Air cleaner assembly
- Engine coolant reservoir tank
- Throttle cable
- Negative battery cable

25. Check all fluids.

Highlander

The transaxle is removed along with the engine. See Engine Removal and Installation, in this section.

RX 300

1. Before servicing the vehicle, refer to the precautions in the beginning of this section.

2. Remove or disconnect the following:
- Hood
- Wiper and blade assembly
- Top cowl seal and panel
- Window washer hoses, from the ventilator louvers
- Left and right ventilator louvers
- Heater air duct
- Battery and tray
- Throttle cable
- Front upper suspension brace
- Cruise control actuator with its bracket, if equipped
- Starter
- Shift control cable
- Driveshaft, for 4WD
- Body-to-engine ground strap
- Park/Neutral Position (PNP) switch, solenoid and ATF temperature connectors
- 5 upper transaxle-to-engine mounting bolts
- Front wheel
- Engine undercover
- Halfshafts
- Front exhaust pipe
- Stabilizer bar
- Both steering gear mounting bolts and support it in the vehicle
- Shift control cable from its bracket
- Power steering pipe and the oil cooler clamps from the frame
- Both left-side transaxle mounting nuts
- Rear-side engine mounting nuts
- Engine shock absorber mounting bolts
- 3 front-side engine mounting bolts

3. Attach an engine sling to the engine hangers in order to support the engine weight.

4. Remove or disconnect the following:
- Front frame mounting bolts and the frame
- Transaxle oil cooler lines

5. Support the transaxle with a transmission jack.
- Torque converter access cover
- 6 torque converter mounting bolts
- 3 lower transaxle-to-engine mounting bolts

- Engine from the transaxle

To install:

6. Install or connect the following:
- Transaxle
- 3 lower transaxle-to-engine mounting bolts and tighten to the illustrated value.
- Torque converter-to-flexplate bolts, starting with the black bolt, then the other 5.

7. The rest of installation is the reverse of the removal referring to the illustrations for the tightening specifications.

4Runner

MODEL A340D, A340E AND A340H TRANSMISSIONS

➡**The transfer case and the transmission should be removed as an assembly.**

1. Before servicing the vehicle, refer to the precautions in the beginning of this section.

2. Remove or disconnect the following:
- Negative battery cable
- Air cleaner assembly, if necessary
- Transmission throttle cable from the throttle body
- Engine undercover

3. Drain the transmission and transfer case (if applicable) fluid.

4. Remove or disconnect the following:
- Wiring connectors from the transmission and transfer case, if applicable.
- Starter

5. Matchmarks on the front and rear driveshaft flanges and the differential pinion flanges. These marks must be aligned during installation.

6. Remove or disconnect the following:
- Front and rear driveshaft flanges.
- Center bearing bracket bolts, if equipped with a 2-piece driveshaft
- Driveshaft
- Speedometer cable
- Front exhaust pipe and bracket
- Transmission oil cooler lines, at the transmission
- Oil cooler lines bracket and the transmission oil filler tube, as required

7. Support the transmission, using a jack with a wooden block placed between the jack and the transmission pan. Raise the transmission, just enough to take the weight off of the rear mount.

8. Remove or disconnect the following:
- Rear engine mount with the bracket, the rear crossmember and

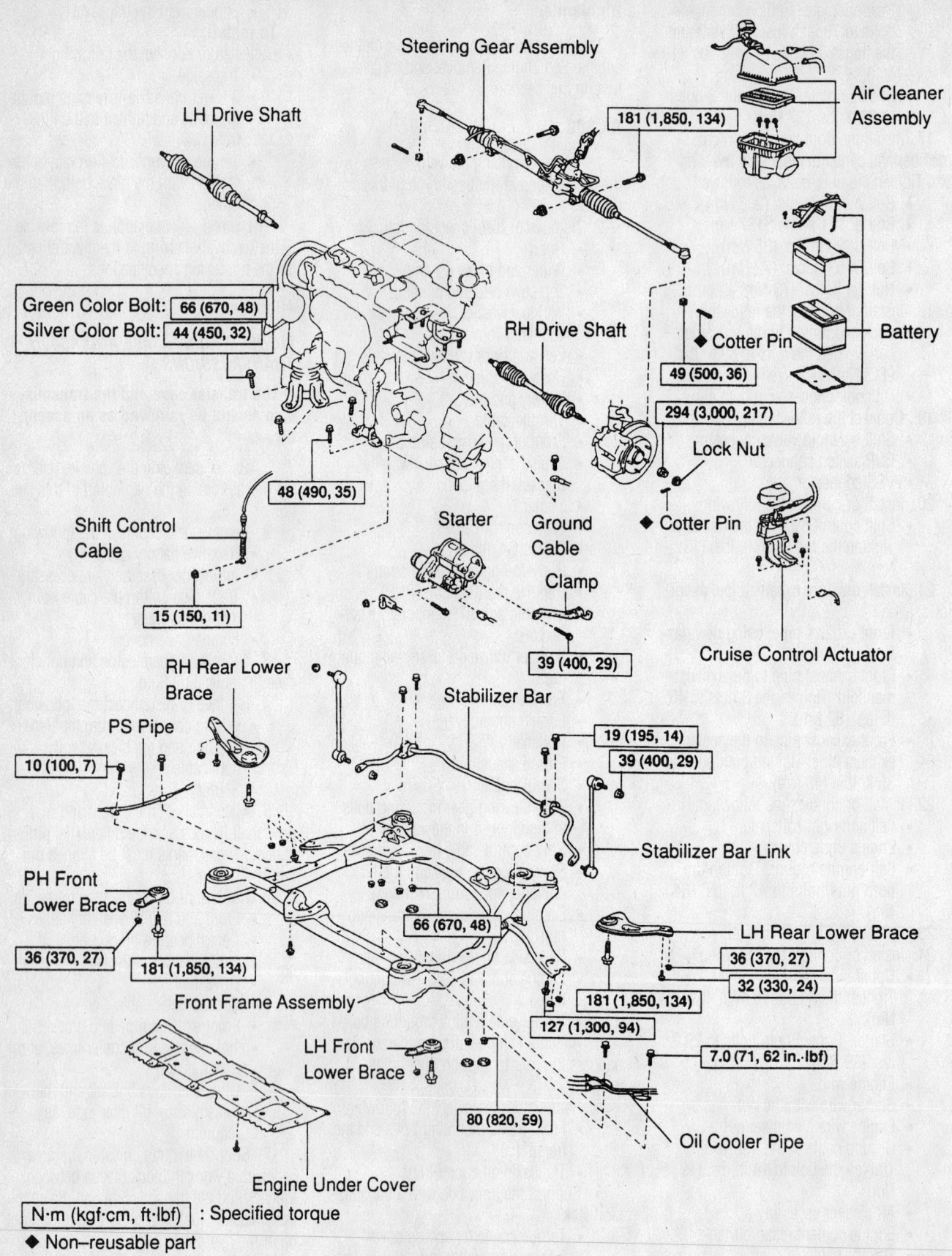

Steering Gear Assembly

LH Drive Shaft

Air Cleaner Assembly

181 (1,850, 134)

Green Color Bolt: 66 (670, 48)
Silver Color Bolt: 44 (450, 32)

RH Drive Shaft

Cotter Pin

49 (500, 36)

Battery

294 (3,000, 217)

Lock Nut

48 (490, 35)

Shift Control Cable

Starter

Ground Cable

Cotter Pin

Cruise Control Actuator

15 (150, 11)

Clamp

39 (400, 29)

RH Rear Lower Brace

Stabilizer Bar

PS Pipe

19 (195, 14)

39 (400, 29)

10 (100, 7)

Stabilizer Bar Link

PH Front Lower Brace

66 (670, 48)

LH Rear Lower Brace

36 (370, 27)

36 (370, 27)

181 (1,850, 134)

32 (330, 24)

Front Frame Assembly

181 (1,850, 134)

127 (1,300, 94)

LH Front Lower Brace

7.0 (71, 62 in.·lbf)

80 (820, 59)

Oil Cooler Pipe

Engine Under Cover

N·m (kgf·cm, ft·lbf) : Specified torque

◆ Non–reusable part

7924ZG65

Exploded view of the transaxle removal and installation components—RX 300 models

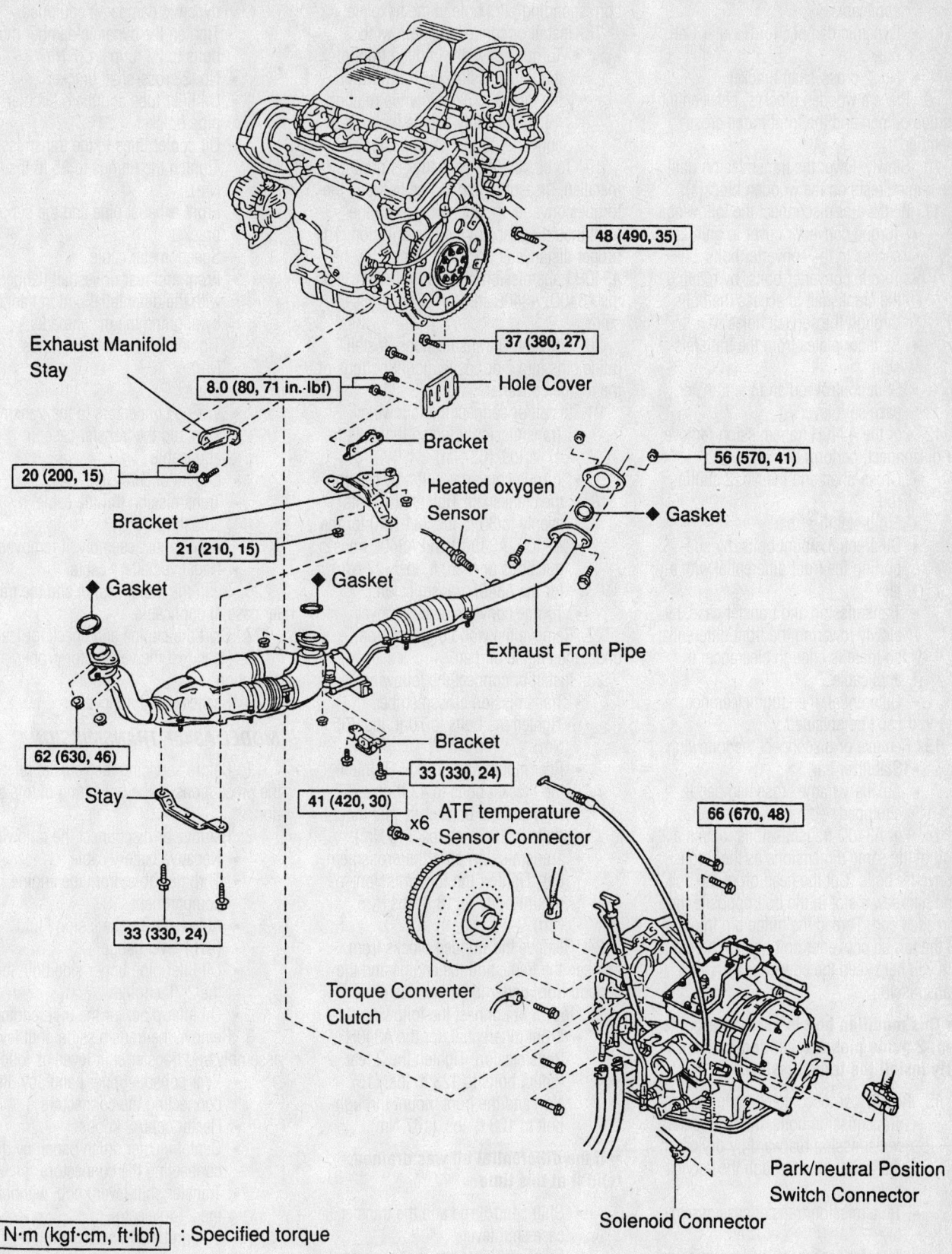

Exhaust Manifold Stay

48 (490, 35)

37 (380, 27)

8.0 (80, 71 in.·lbf)

Hole Cover

Bracket

20 (200, 15)

56 (570, 41)

Bracket

Heated oxygen Sensor

◆ Gasket

21 (210, 15)

◆ Gasket

◆ Gasket

Exhaust Front Pipe

Bracket

62 (630, 46)

33 (330, 24)

Stay

41 (420, 30)

ATF temperature Sensor Connector

66 (670, 48)

×6

33 (330, 24)

Torque Converter Clutch

Solenoid Connector

Park/neutral Position Switch Connector

N·m (kgf·cm, ft·lbf) : Specified torque

◆ Non–reusable part

7924ZG66

Exploded view of the transaxle removal and installation components—RX 300 models, Cont.

the transfer case undercover, if applicable
- Dynamic damper, for Regular Cab only
- No. 2 cross-shaft bracket

9. Place a wooden block(s) between the engine oil pan and the front frame cross-member.

10. Slowly, lower the transmission until the engine rests on the wooden block(s).

11. Remove or disconnect the following:
- Torque converter cover to gain access to the converter bolts
- Torque converter bolts, by rotating the crankshaft to access the bolts through the service holes
- Stiffener plates from the transmission
- Shift control rod and the transfer case shift lever

12. For the A340H transmission remove or disconnect, perform the following:
- Cross-shaft and the No. 2 shifting rod
- Front stabilizer bar
- Differential mount bolts, by supporting the front differential with a jack
- Transmission and transfer case, by slowly lowering the front differential so there is enough clearance, if applicable
- Differential, if enough clearance can't be obtained

13. Remove or disconnect the following:
- Stabilizer bar
- Auxiliary frame crossmember, if equipped

14. For A340D transmissions, obtain a bolt of the same dimensions as the torque converter bolts. Cut the head off of the bolt and hacksaw a slot in the bolt opposite the threaded end. Thread the guide pin into one of the torque converter bolt holes. The guide pin will help keep the converter with the transmission.

➡ **This modified bolt is used as a guide pin. 2 guide pins are needed to properly install the transmission.**

15. Remove or disconnect the following:
- Transmission bolts, then move the transmission rearward by prying on the dowel pins through the service hole
- Transmission/transfer case assembly
- Transfer case from the transmission

To install:

16. Connect the transfer case to the transmission.

17. Apply a coat of multi-purpose grease

to the torque converter stub shaft and the corresponding pilot hole in the flexplate.

18. Install or connect the following:
- Torque converter into the front of the transmission Push inward on the torque converter while rotating it to completely couple the torque converter to the transmission.

19. To be sure the converter is properly installed, measure the distance between the torque converter mounting lugs and the front mounting face of the transmission. The proper distance is 0.71 in. (18mm) for the A340H transmission or 0.79 in. (20mm) for the A340D, A340E and A340F transmissions.

20. For A340D transmissions, install guide pins into 2 opposite mounting lugs of the torque converter.

21. Install or connect the following:
- Transmission. Tighten the bolts to 47 ft. lbs. (63 Nm).
- Torque converter bolts, by rotating the crankshaft. Tighten the bolts evenly to 30 ft. lbs. (41 Nm) for the A340H, A3430D and A340E transmissions or to 20 ft. lbs. (27 Nm) for the A340F transmission.
- Torque converter access cover

22. Remove the wood block(s) from under the engine oil pan.

23. Install or connect the following:
- Transmission crossmember. Tighten the bolts to 70 ft. lbs. (95 Nm).
- Rear mount and bracket. Tighten the bracket bolts to 43 ft. lbs. (58 Nm) and the bracket-to-rear mount bolts to 108 inch lbs. (13 Nm).
- Transmission onto the crossmember. Tighten the transmission-to-mount bolts to 18 ft. lbs. (25 Nm).

24. Remove the wooden blocks from between the frame and the engine and the support from under the transmission.

25. Install or connect the following:
- Front differential, for the A340H transmission. Tighten the 2 rear mount bolts to 123 ft. lbs. (167 Nm) and the front mount through-bolt to 108 ft. lbs. (147 Nm).

➡ **If the differential oil was drained, refill it at this time.**

- Shift control rod and the transfer case shift lever
- Front stabilizer bar, if applicable
- Cross-shaft and the No. 2 shifting rod, if applicable
- Stiffener plates. Tighten the bolts to 27 ft. lbs. (37 Nm).

- Transfer case undercover and the dynamic damper, if equipped. Tighten the dynamic damper mount bolts to 27 ft. lbs. (37 Nm).
- No. 2 cross-shaft bracket
- Oil filler tube and the oil cooler pipe bracket
- Oil cooler lines to the transmission. Tighten the fittings to 25 ft. lbs. (34 Nm).
- Front exhaust pipe and the support bracket
- Speedometer cable
- Front and rear driveshaft flanges with the differential pinion flanges, by aligning the matchmarks. Tighten the bolts to 54 ft. lbs. (74 Nm).
- Starter
- Wiring connectors to the transmission and the transfer case, if applicable
- Engine undercover
- Transmission throttle cable, by adjusting it
- Air cleaner assembly, if removed
- Negative battery cable

26. Refill the transmission and the transfer case, if applicable.

27. Start the engine and check for leaks.

28. Road test the vehicle for proper operation.

29. Recheck all fluid levels.

MODEL A340F TRANSMISSION

1. Before servicing the vehicle, refer to the precautions in the beginning of this section.

2. Remove or disconnect the following:
- Negative battery cable
- Throttle cable, from the engine compartment
- Automatic Transmission Fluid (ATF) level gauge
- Oil filler pipe upper side bolt, for the 2.7L engine
- Oil filler pipe, for the 3.4L engine

3. Remove the transmission shift lever assembly and transfer shift lever, as follows:
- Rear console upper panel, by disconnecting the connectors
- Heater control knobs
- Center cluster finish panel, by disconnecting the connectors
- Transfer shift lever knob, without the 2–4 selector
- Bolt and the transfer shift lever knob, with the 2–4 selector
- Front console upper panel
- 2–4 selector connector, if equipped
- Transfer shift lever knob
- Shift control rod

- Transmission shift lever assembly connector and the 8 screws
- Shift lever snapring, using pliers and pull out it from the transfer case
- Engine undercover
- Front and rear driveshafts
- Exhaust pipe
- Oil filler pipe, for 2.7L engine

4. Disconnect the following connectors from the transmission:
   - No. 2 Vehicle Speed Sensor (VSS) connector
   - Solenoid connector
   - Automatic Transmission Fluid (ATF) temperature sensor connector
   - Park/neutral position switch connector
5. Detach the following connectors from the transfer case:
   - No. 1 Vehicle Speed Sensor (VSS) connector, for 2.7L engine
   - Transfer neutral position switch connector
   - Transfer L4 position switch connector
   - Transfer 4WD position switch connector
   - Actuator connector (2–4 selector only)
6. Remove or disconnect the following:
   - Wiring harness from the transmission and the transfer case
   - Both oil cooler pipes
   - Rear end-plate and torque converter clutch mounting bolt
7. Support the transmission with a jackstand.
8. Remove or disconnect the following:
   - Engine rear mount bolts
   - 4 bolts and the crossmember
   - Starter
   - Transmission

To install:

9. Install or connect the following:
   - Transmission. Tighten the bolts to 53 ft. lbs. (71 Nm).
   - Starter. Tighten the bolts to 29 ft. lbs. (39 Nm).
   - Crossmember. Tighten the 4 bolts to 48 ft. lbs. (65 Nm).
   - Engine rear mount. Tighten the 4 bolts to 14 ft. lbs. (19 Nm).
   - Clutch converter bolts, by installing the green colored bolt before the other 5. Tighten the bolts to 30 ft. lbs. (41 Nm).
   - Rear end-plate. Tighten the bolts to 13 ft. lbs. (18 Nm).
   - Both oil cooler pipes. Tighten to 25 ft. lbs. (34 Nm).
   - Oil cooler pipe clamps. Tighten the 10mm head bolt to 48 inch lbs. (5 Nm) and the 12mm head bolt to 108 inch lbs. (13 Nm).
   - Wiring harness to the transmission and the transfer case
   - Remaining components
10. Fill the transmission and transfer case with transmission fluid.
    - Throttle cable
    - Negative battery cable

Transfer Case Assembly

REMOVAL & INSTALLATION

The transfer case for the RAV4, Highlander and RX300 is part of the transmission/transaxle assembly and is serviced with those units.

4Runner

1. Before servicing the vehicle, refer to the precautions in the beginning of this section.
2. Disconnect the negative battery cable.
3. Drain the transmission and the transfer case.
4. Remove or disconnect the following:
   - Transfer case with the transmission
   - Breather hose from the transfer upper cover and the transmission control retainer, if equipped with an automatic transmission
   - Rear engine mounting
   - Dynamic damper
5. Remove the driveshaft upper dust cover and the transfer from the transmissions, as follows:
   - Dust cover bolt from the bracket
   - Transfer case adapter rear mounting bolts
   - Transfer case, by pulling it straight up and away from the transmission.

⁂ WARNING

Be careful not to damage the adapter rear oil seal with the transfer input gear spline.

To install:

6. Install the transfer case and the driveshaft upper dust cover to the transmission with a new gasket, as follows:
   - Shift the 2 shift fork shafts to the high 4 position
   - Apply MP grease to the adapter oil seal
   - New gasket to the transfer adapter
   - Transfer case to the transmission.

⁂ WARNING

Take care not to damage the oil seal by the input gear spline.

   - Transfer case adapter. Tighten the rear bolts to 27 ft. lbs. (37 Nm).
   - Dust cover to the bracket. Tighten the bolt to 17 ft. lbs. (23 Nm).
7. Install or connect the following:
   - Engine rear mount. Tighten the bolts to 19 ft. lbs. (25 Nm).
   - Dynamic damper. Tighten the bolts to 27 ft. lbs. (37 Nm).
   - Breather hose, if equipped with an automatic transmission
   - Transfer case with the transmission to the engine
8. Fill the transmission and the transfer case with oil.
9. Test drive the vehicle and check the abnormal noise and smooth operation.
10. Recheck the fluid levels.

Halfshaft

REMOVAL & INSTALLATION

RAV4 Front

1. Before servicing the vehicle, refer to the precautions in the beginning of this section.
2. Remove or disconnect the following:
   - Negative battery cable
   - Engine undercover
3. Drain the transaxle.
4. Remove or disconnect the following:
   - Anti-lock Brake System (ABS) sensor by removing the bolt, if equipped
   - Cotter pin, lock cap and the locknut holding the halfshaft to the steering knuckle
   - Tie rod ends, from the steering knuckle
   - Sway bar link, from the lower control arm
   - Lower ball joint, from the lower control arm
   - Halfshaft from the axle hub, using a plastic hammer
5. If working on a 2WD right-hand halfshaft and the vehicle is equipped with a manual transaxle, perform the following to remove the halfshaft:
   - Snapring from the center bearing bracket, using a brass bar and hammer
   - Bolt and the center bearing bracket
   - Halfshaft with the center halfshaft

2WD M/T

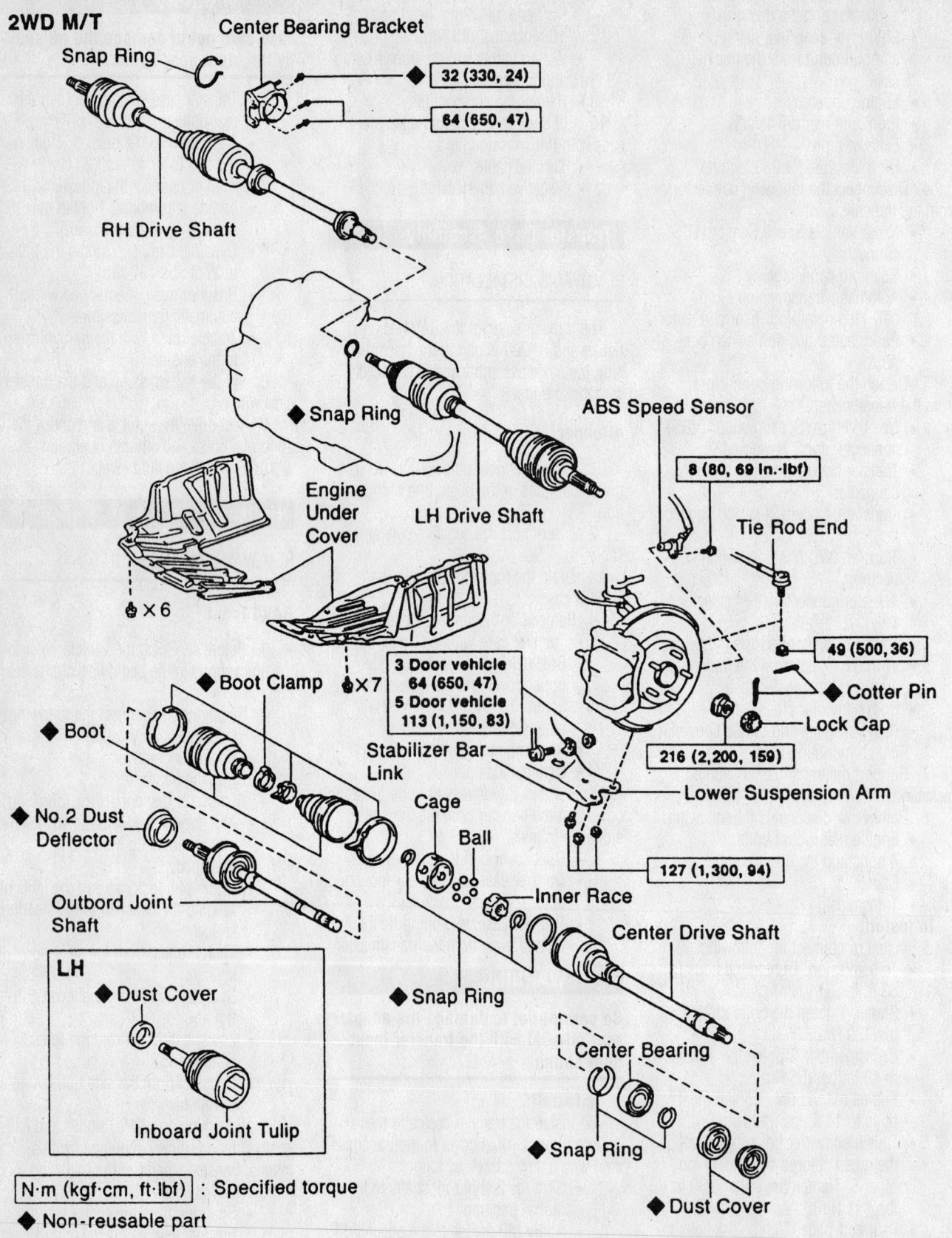

Snap Ring

Center Bearing Bracket

◆ 32 (330, 24)

64 (650, 47)

RH Drive Shaft

◆ Snap Ring

Engine Under Cover

LH Drive Shaft

ABS Speed Sensor

8 (80, 69 In.·lbf)

Tie Rod End

49 (500, 36)

◆ Cotter Pin

Lock Cap

216 (2,200, 159)

Lower Suspension Arm

127 (1,300, 94)

◆ Boot Clamp

×7

3 Door vehicle
64 (650, 47)
5 Door vehicle
113 (1,150, 83)

Stabilizer Bar Link

Cage

Ball

Inner Race

×6

◆ Boot

◆ No.2 Dust Deflector

Outboard Joint Shaft

Center Drive Shaft

◆ Snap Ring

LH

◆ Dust Cover

Center Bearing

Inboard Joint Tulip

◆ Snap Ring

◆ Dust Cover

N·m (kgf·cm, ft·lbf) : Specified torque

◆ Non-reusable part

7924ZG70

Front halfshaft exploded view (2WD with manual transmission only)—RAV4

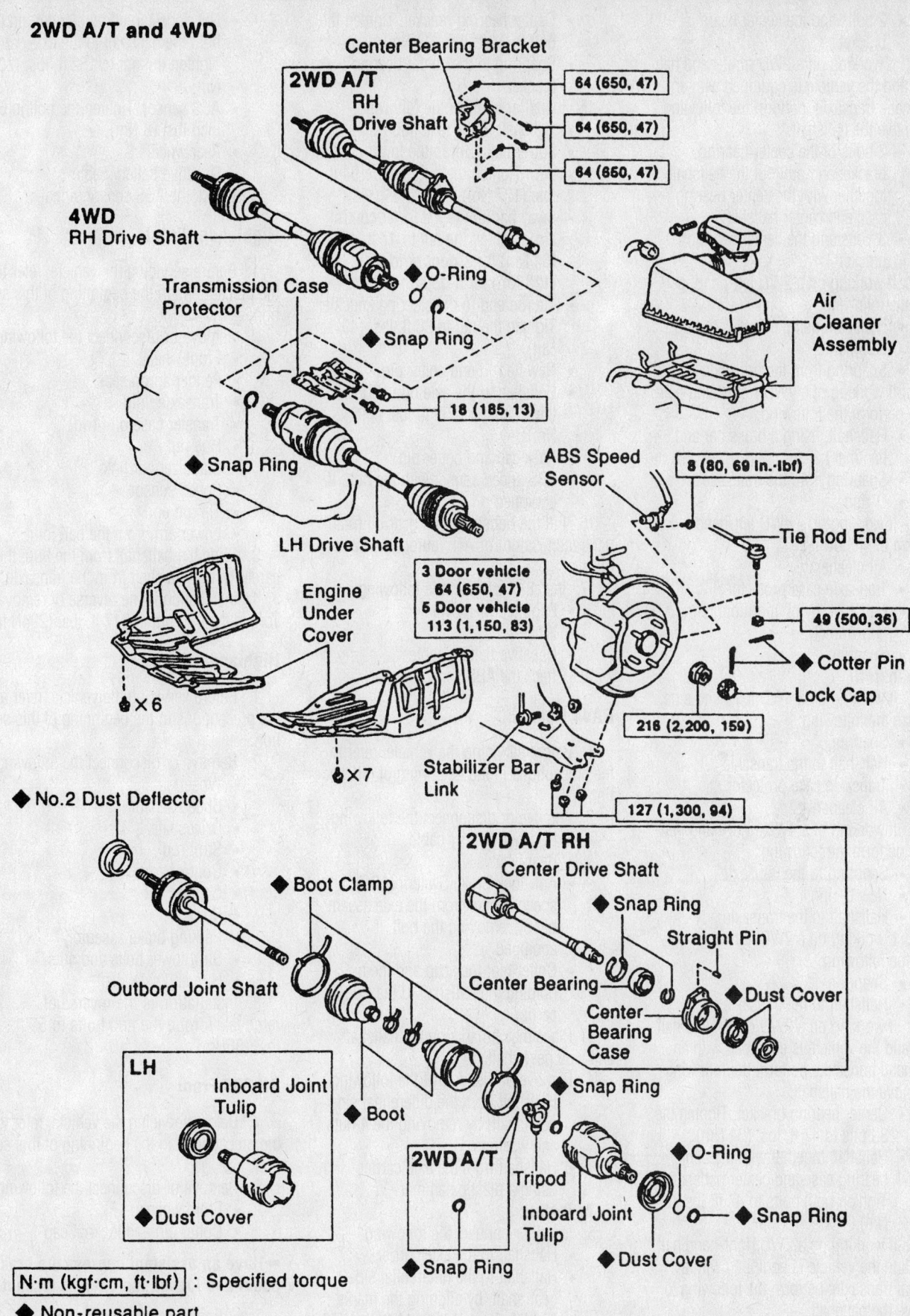

2WD A/T and 4WD

Center Bearing Bracket

2WD A/T RH Drive Shaft

64 (650, 47)
64 (650, 47)
64 (650, 47)

4WD RH Drive Shaft

◆ O-Ring

Transmission Case Protector

◆ Snap Ring

Air Cleaner Assembly

◆ Snap Ring

18 (185, 13)

LH Drive Shaft

ABS Speed Sensor

8 (80, 69 in.·lbf)

Tie Rod End

Engine Under Cover

3 Door vehicle 64 (650, 47)
5 Door vehicle 113 (1,150, 83)

49 (500, 36)

◆ Cotter Pin

Lock Cap

×6

216 (2,200, 159)

×7

Stabilizer Bar Link

127 (1,300, 94)

◆ No.2 Dust Deflector

◆ Boot Clamp

2WD A/T RH

Center Drive Shaft

◆ Snap Ring

Straight Pin

Outbord Joint Shaft

Center Bearing

Center Bearing Case

◆ Dust Cover

LH

Inboard Joint Tulip

◆ Boot

◆ Snap Ring

◆ Dust Cover

2WD A/T

Tripod

◆ O-Ring

Inboard Joint Tulip

◆ Snap Ring

◆ Snap Ring

◆ Dust Cover

N·m (kgf·cm, ft·lbf) : Specified torque

◆ Non-reusable part

7924ZG71

Front halfshaft exploded view (2WD with automatic transmission and 4WD)—RAV4

- 2 bolts and the center bearing bracket

6. If working on a 2WD right-hand halfshaft and the vehicle is equipped with an automatic transaxle, perform the following to remove the halfshaft:
  - 2 bolts of the center bearing bracket and pull out the halfshaft together with the center bearing case and center halfshaft
  - 3 bolts and the center bearing bracket

7. If working on a 2WD left-hand, perform the following:
  - Halfshaft, using a brass bar and hammer
  - Snapring from the transaxle

8. If working of a 4WD right-hand halfshaft, perform the following:
  - Halfshaft, using a brass bar and hammer
  - Snapring from the transaxle
  - O-ring

9. If working on a 4WD left-hand side, perform the following:
  - Air cleaner
  - Transaxle case protector
  - Halfshaft, by prying it out using a hub wrench
  - Snapring

To install:

10. If working on a 4WD left-hand side, perform the following:
  - Snapring
  - Halfshaft to the transaxle
  - Transaxle case protector
  - Air cleaner

11. If working of a 4WD right-hand halfshaft, perform the following:
  - Snapring to the transaxle
  - New O-ring
  - Halfshaft to the transaxle

12. If working on a 2WD left-hand, perform the following:
  - Snapring
  - Halfshaft to the transaxle

13. If working on a 2WD right-hand halfshaft and the vehicle is equipped with an automatic transaxle, perform the following to remove the halfshaft:
  - Center bearing bracket. Tighten the 3 bolts to 47 ft. lbs. (64 Nm).
  - Halfshaft together with the center bearing case and center halfshaft. Tighten both bolts to 47 ft. lbs. (64 Nm).

14. If working on a 2WD right-hand halfshaft and the vehicle is equipped with a manual transaxle, perform the following to remove the halfshaft:
  - Center bearing bracket
  - Halfshaft with the center halfshaft

- Center bearing bracket. Tighten the bolt to 24 ft. lbs. (32 Nm).
- Snapring to the center bearing bracket

15. Install or connect the following:
  - Halfshaft to the axle hub
  - Lower ball joint to the lower control arm. Tighten the nuts/bolt to 94 ft. lbs. (127 Nm).
  - Sway bar link to the lower control arm. Tighten the nut to 47 ft. lbs. (64 Nm) for 3-door or to 83 ft. lbs. (113 Nm) for 5-door.
  - Tie rod end to the steering knuckle. Tighten the nut to 36 ft. lbs. (49 Nm).
  - New tie rod end cotter pin
  - Halfshaft to the axle hub. Tighten the locknut to 159 ft. lbs. (216 Nm).
  - Lock cap and cotter pin
  - ABS speed sensor with the bolt, if equipped

16. Fill the transaxle with gear oil (manual transmission) or ATF (automatic transmission).

17. Install or connect the following:
  - Engine undercover
  - Wheels
  - Negative battery cable

18. Check the ABS sensor signal.

RAV4 Rear

1. Before servicing the vehicle, refer to the precautions in the beginning of this section.

2. Remove or disconnect the following:
  - Negative battery cable
  - Rear wheels
  - Anti-lock Brake System (ABS) speed sensor from the axle assembly by removing the bolt, if equipped
  - Cotter pin, lock cap and the nut holding the halfshaft to the axle carrier

3. Place matchmarks on the halfshaft and side gear shaft.

4. Remove or disconnect the following:
  - Halfshaft from the differential side gear shaft, by removing the 4 nuts and washers
  - Halfshaft from the axle carrier, using a plastic hammer

To install:

5. Install or connect the following:
  - Halfshaft to the axle carrier
  - Halfshaft to the differential side gear shaft, by aligning the marks. Tighten the 4 nuts to 41 ft. lbs. (56 Nm).

- Nut, lock cap and the cotter pin to hold the halfshaft to the axle carrier. Tighten the nut to 152 ft. lbs. (206 Nm).
- ABS sensor. Tighten the bolt to 69 inch lbs. (8 Nm).
- Rear wheels
- Negative battery cable

6. Check the ABS sensor signal.

Highlander Front

1. Before servicing the vehicle, refer to the precautions in the beginning of this section.

2. Remove or disconnect the following:
  - Front wheels
  - Fender apron seal
  - Transaxle fluid
  - Transfer case oil (4wd)
  - Hub nut
  - Stabilizer bar link
  - Speed sensor
  - Tie rod end
  - Lower arm from the ball joint

3. Slide the halfshaft from the hub, then, carefully, pry the shaft from the transaxle.

4. Installation is the reverse of removal. Torque the hub nut to 217 ft. lbs. (294Nm).

Highlander Rear

1. Before servicing the vehicle, refer to the precautions in the beginning of this section.

2. Remove or disconnect the following:
  - Wheel
  - Speed sensor
  - Driveshaft
  - Strut rod
  - Control arms
  - Caliper
  - Rotor
  - Parking brake assembly
  - Strut lower bolts and nuts
  - Hub and axle shaft

3. Installation is the reverse of removal. Torque the hub bolts to 59 ft. lbs. (80Nm).

RX 300 Front

1. Before servicing the vehicle, refer to the precautions in the beginning of this section.

2. Remove or disconnect the following:
  - Front wheels
  - Cotter pin and locknut cap

➡**Have an assistant depress the brake pedal and loosen the bearing locknut.**

  - Engine undercover
  - Fender apron seal

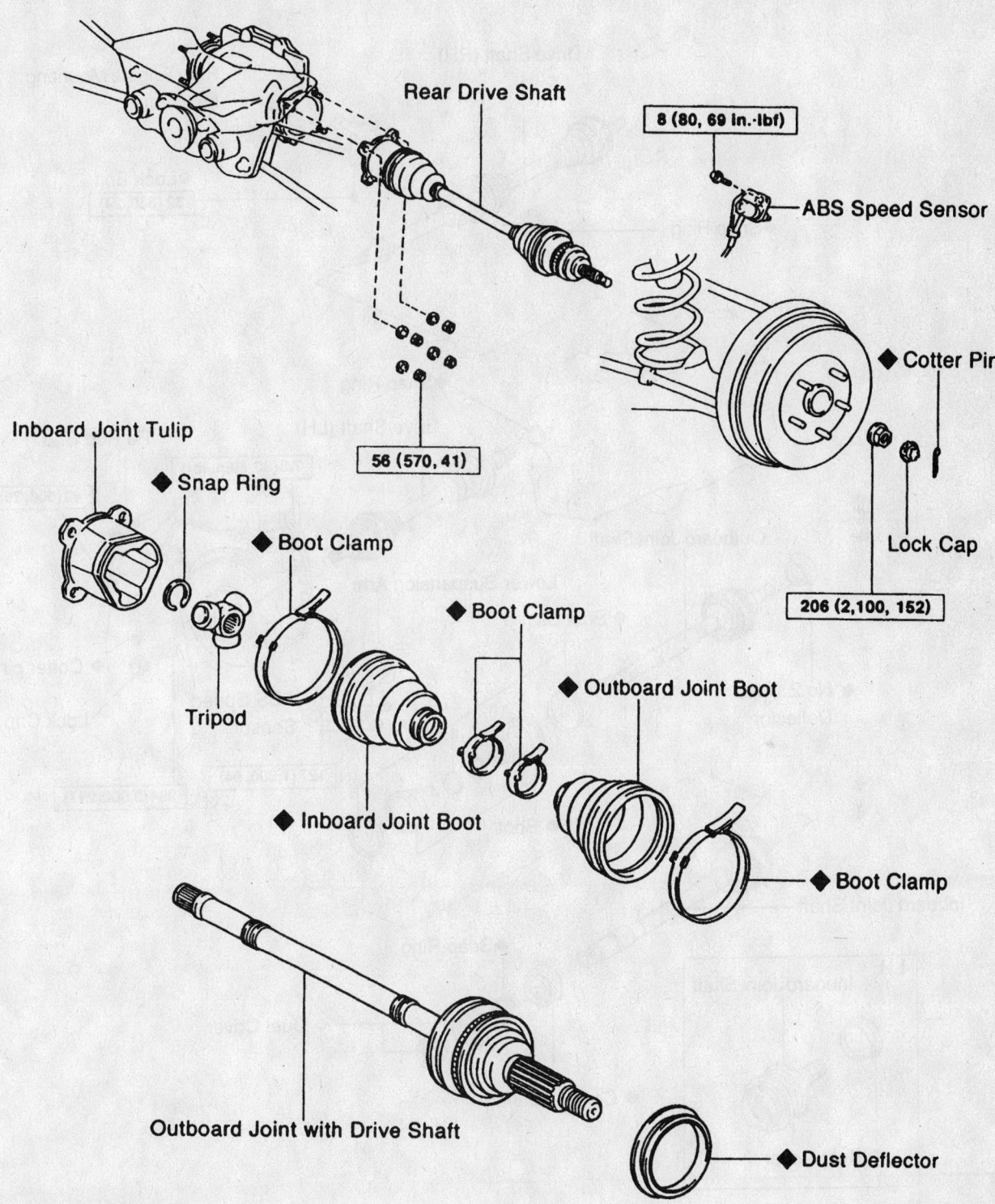

Rear Drive Shaft

8 (80, 69 in.·lbf)

ABS Speed Sensor

Cotter Pin

Inboard Joint Tulip

Snap Ring

Boot Clamp

Tripod

56 (570, 41)

Lock Cap

206 (2,100, 152)

Boot Clamp

Outboard Joint Boot

Inboard Joint Boot

Boot Clamp

Outboard Joint with Drive Shaft

Dust Deflector

N·m (kgf·cm, ft·lbf) : Specified torque

◆ Non-reusable part

7924ZG72

Rear halfshaft removal and installation (4WD only)—RAV4

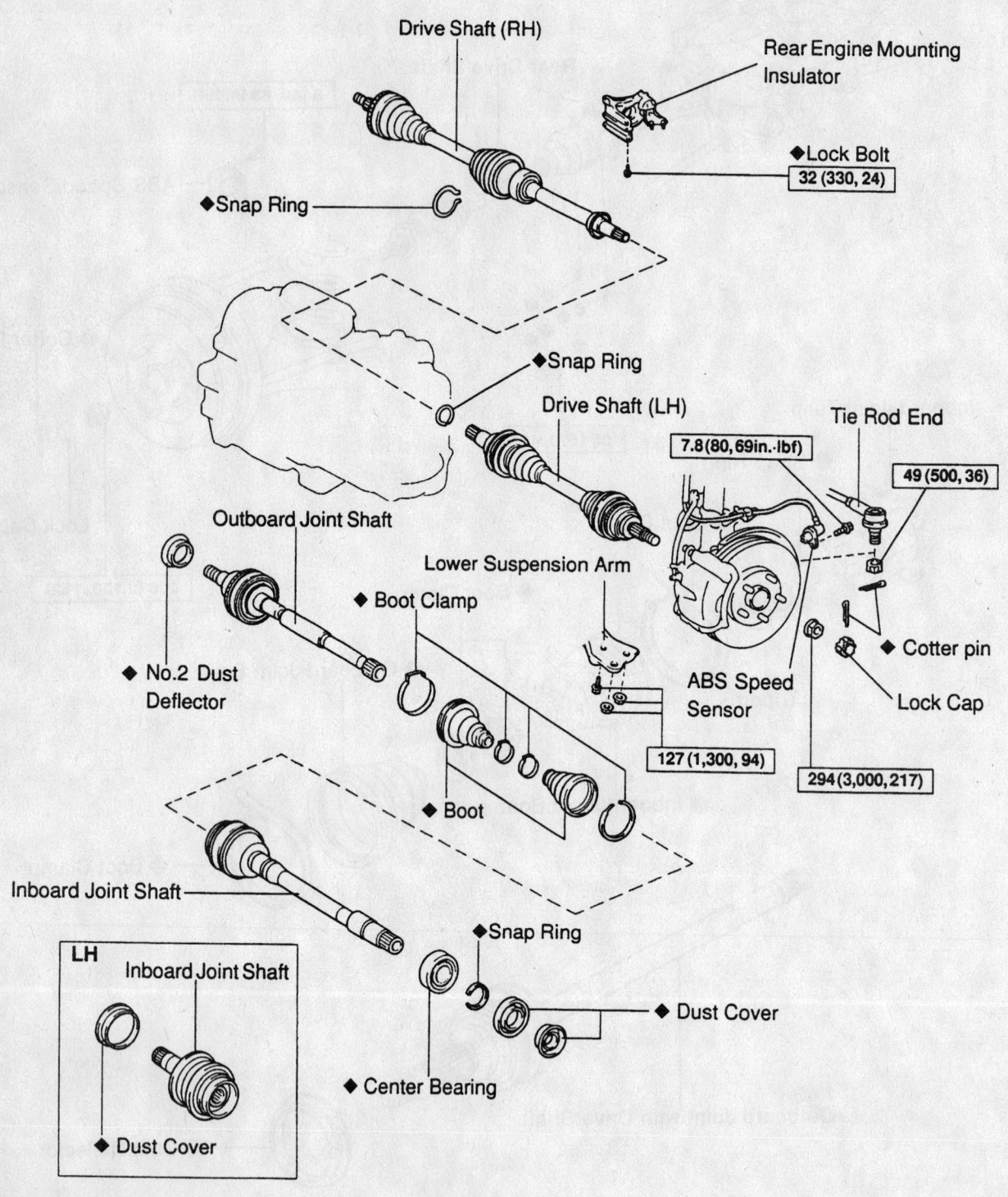

Drive Shaft (RH)

Rear Engine Mounting
Insulator

◆Lock Bolt
32 (330, 24)

◆Snap Ring

◆Snap Ring

Drive Shaft (LH)

Tie Rod End

7.8 (80, 69 in.·lbf)

49 (500, 36)

Outboard Joint Shaft

Lower Suspension Arm

◆ Boot Clamp

◆ Cotter pin

◆ No.2 Dust
Deflector

ABS Speed
Sensor

Lock Cap

127 (1,300, 94)

294 (3,000, 217)

◆ Boot

Inboard Joint Shaft

◆Snap Ring

◆ Dust Cover

LH
Inboard Joint Shaft

◆ Center Bearing

◆ Dust Cover

N·m (kgf·cm, ft·lbf) : Specified torque

◆ Non–reusable part

7924ZG73

Exploded view of front halfshaft—RX 300

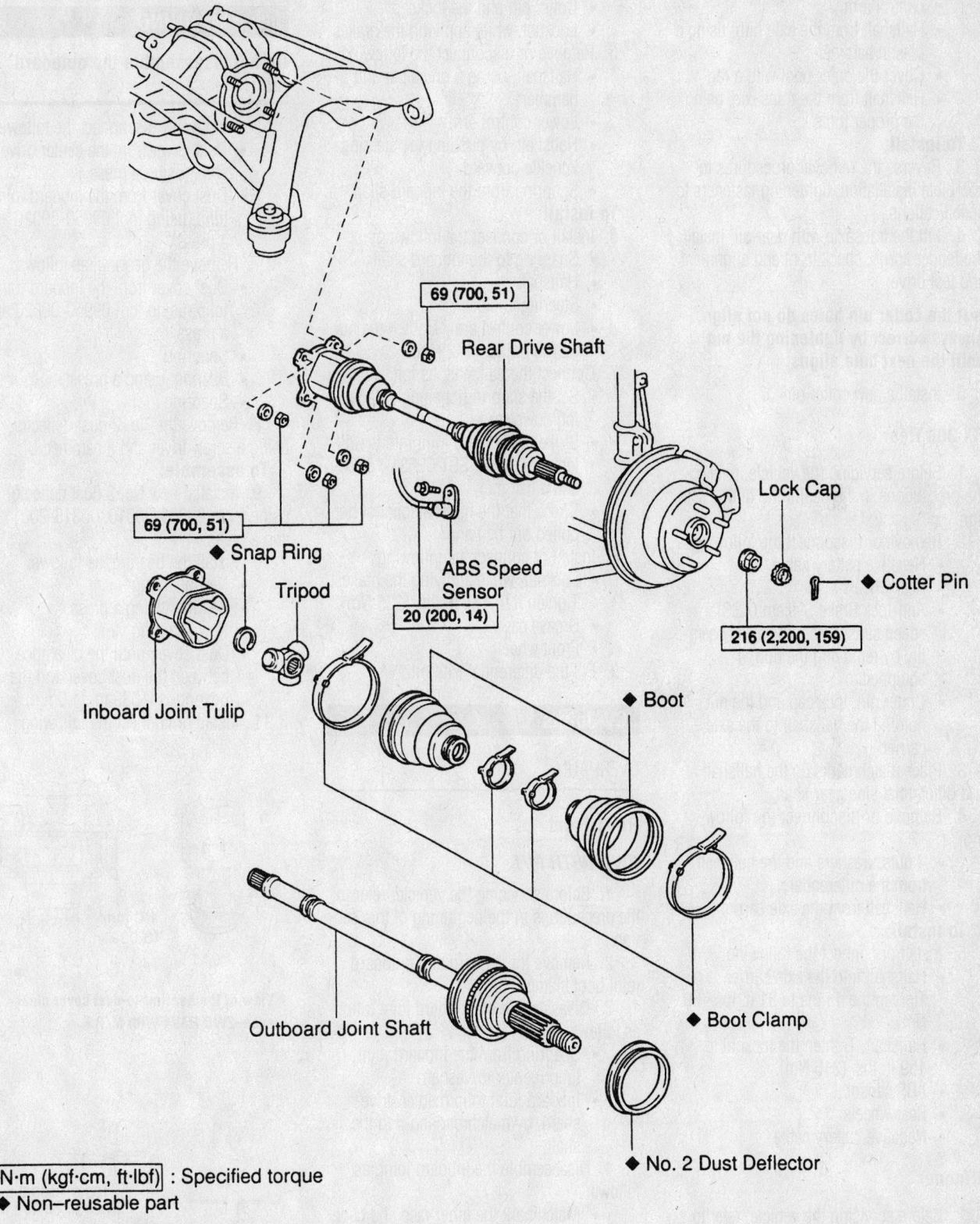

69 (700, 51)

Rear Drive Shaft

69 (700, 51)

◆ Snap Ring

Tripod

Inboard Joint Tulip

ABS Speed
Sensor

20 (200, 14)

Lock Cap

◆ Cotter Pin

216 (2,200, 159)

◆ Boot

Outboard Joint Shaft

◆ Boot Clamp

◆ No. 2 Dust Deflector

N·m (kgf·cm, ft·lbf) : Specified torque
◆ Non–reusable part

7924ZG88

Exploded view of the rear halfshaft—RX 300 model with 4WD

- Tie rod end, from the steering knuckle
- Steering knuckle, from the lower control arm
- Halfshaft from the axle hub, using a plastic hammer
- Cover the outer boot with a rag
- Halfshaft from the transaxle, using the proper tools

To install:

3. Reverse the removal procedures to complete installation, tightening fasteners to specifications.

4. Fill the transaxle with gear oil, install the fender apron, check front end alignment and test drive.

➡**If the cotter pin holes do not align, always correct by tightening the nut until the next hole aligns.**

5. Install a new cotter pin.

RX 300 Rear

1. Before servicing the vehicle, refer to the precautions in the beginning of this section.

2. Remove or disconnect the following:
- Negative battery cable
- Rear wheels
- Anti-lock Brake System (ABS) speed sensor from the axle assembly by removing the bolt, if equipped
- Cotter pin, lock cap and the nut holding the halfshaft to the axle carrier

3. Place matchmarks on the halfshaft and differential side gear shaft.

4. Remove or disconnect the following:
- 4 nuts, washers and the halfshaft from the differential
- Halfshaft from the axle carrier

To install:

5. Install or connect the following:
- Halfshaft into the axle carrier. Tighten the 4 nuts to 51 ft. lbs. (69 Nm).
- Halfshaft. Tighten the locknut to 159 ft. lbs. (216 Nm).
- ABS sensor
- Rear wheels
- Negative battery cable

4Runner

1. Before servicing the vehicle, refer to the precautions in the beginning of this section.

2. Remove the front wheel.

3. Drain the differential oil from the differential.

4. Remove the halfshaft locknut, as follows:
- Grease cap
- Cotter pin and the lockcap
- Locknut, while applying the brakes

5. Remove or disconnect the following:
- Halfshaft, using a brass bar and a hammer
- Lower control arm
- Halfshaft, by pushing the steering knuckle outward
- Snapring from the inboard shaft

To install:

6. Install or connect the following:
- Snapring to the inboard shaft
- Halfshaft
- Steering knuckle
- Lower control arm. Tighten the nut to 105 ft. lbs. (142 Nm).

7. Connect the halfshaft, as follows:
- Set the snapring opening side facing downward
- Strike the inboard joint into the differential, using SST 09631-10030 and a hammer
- Check that the halfshaft cannot be pulled out by hand

8. Install or connect the following:
- Locknut, while applying the brakes. Tighten it to 174 ft. lbs. (235 Nm).
- Grease cap
- Front wheel

9. Fill the differential with oil.

CV-Joints

OVERHAUL

RAV4 Front

2WD WITH M/T

1. Before servicing the vehicle, refer to the precautions in the beginning of this section.

2. Remove the inboard and outboard joint boot clamps.

3. Disassemble the inboard joint tulip, as follows:
- Snapring from the inboard joint tulip (center driveshaft)
- Inboard joint tulip (center driveshaft), by matchmarking it to the shaft

4. Disassemble the inboard joint, as follows:
- Matchmark the inner race and cage to the driveshaft
- 6 balls and cage
- Snapring
- Inner race, using a brass bar and a hammer

- Snapring

5. Remove the inboard and outboard joint boots and inboard joint clamps.

✳✳ WARNING

Do not disassemble the outboard joint.

6. Remove or disconnect the following:
- Dust cover from the center driveshaft, using a press
- Dust cover from the inboard joint tulip, using tool 09950-00020 and a press

7. Remove the bearing, as follows:
- Dust cover from the inboard joint tulip, using tool 09950-00020 and a press
- Snapring
- Bearing, using a press
- Snapring

8. Remove the No. 2 dust deflector, using a screwdriver and a hammer.

To assemble:

9. Install a new No. 2 dust deflector, using tools 09309-36010, 09316-20011 and a press.

10. Install the bearing, as follows:
- New snapring
- Bearing, using a press
- New snapring
- Dust cover, until the clearance between the dust cover and the bearing is 0.039 in. (1.0mm)

11. Install or connect the following:

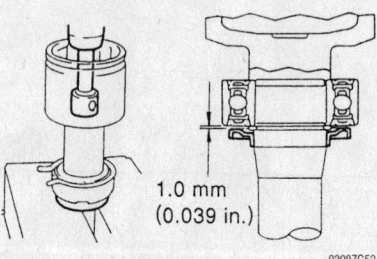

1.0 mm (0.039 in.)

9308ZG52

View of the bearing-to-dust cover clearance–2WD RAV4 With M/T

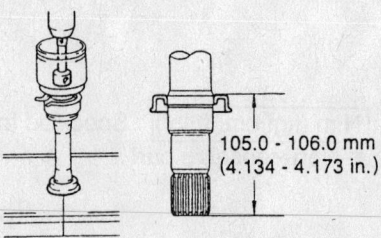

105.0 - 106.0 mm (4.134 - 4.173 in.)

9308ZG53

View of the dust cover-to-center drive distance–2WD RAV4 With M/T

RH

LH

9308ZG54

Measuring the front halfshaft lengths–2WD RAV4 with M/T

- Right dust cover, until the distance from the tip of the center drive is 4.134–4.173 in. (105.0–106.0mm) to the inner edge of the dust cover
- Left side dust cover, using a press

12. Temporarily install new outboard/inboard joint boots using new clamps, as follows:

a. Warp tape around the driveshaft splines.

b. Install the new outboard joint boot onto the driveshaft with both new clamps.

c. Install the new inboard joint boot onto the driveshaft.

13. Assemble the inboard joint onto the driveshaft, as follows:
- New snapring
- Cage

➡**The smaller diameter side must face outboard.**

- Inner race, using a brass bar and hammer by aligning the match-marks

※※ WARNING

Be careful not to damage the inner race.

- New snapring

14. Install the outboard joint boot packed with grease from the boot kit.

15. Install the inboard joint tulip, as follows:

- Cage to the inner race by aligning the matchmarks
- 6 cage balls

➡**Lubricate the balls with grease to keep them from falling.**

- Inboard joint tulip, by aligning the matchmarks
- New snapring
- Temporarily, install the inboard joint boot packed with grease from the kit

16. Install the boot clamps to both boots, as follows:
- Both boots to the shaft grooves
- Halfshaft length should be 32.988–33.382 in. (837.9–847.9mm) for the right side or 21.165–21.559 in. (537.6–547.6mm) for the left side
- Both new clamps on the inboard joint boot
- Crimp the new clamps using tool 09521-24010
- Adjust the crimp clearance to 0.047–0.157 in. (1.2–4.0mm)

2WD WITH A/T AND 4WD; HIGHLANDER AND RX300

1. Before servicing the vehicle, refer to the precautions in the beginning of this section.

2. Remove the inboard and outboard joint boot clamps.

3. Disassemble the inboard joint tulip, as follows:
- Matchmark the tri-pot, inboard joint tulip or center driveshaft to the driveshaft

※※ WARNING

Do not use punch marks.

- Inboard joint tulip from the driveshaft

4. Remove the inboard and outboard joint clamps.

5. Remove the tri-pot joint, as follows:
- Snapring
- Matchmark the tri-pot joint to the driveshaft
- Tri-pot joint, using a brass bar and hammer

※※ WARNING

Do not tap the roller.

6. Remove or disconnect the following:
- Inboard and outboard joint boots

➡**Do not disassemble the outboard joint.**

- Dust cover from the center driveshaft, using a press, for 2WD on the right side
- Dust cover from the inboard joint tulip, using tool 09950-00020 and a press, for 2WD on the left side and 4WD

7. Disassemble the center driveshaft, as follows:
- Snapring
- Bearing case, using a press

- Straight pin from the bearing case, using a pin punch and hammer
- Dust cover, using tool 09950-00020 and a press
- Snapring
- Bearing, using a press

8. Remove the No. 2 dust deflector, using a screwdriver and hammer.

To assemble:

9. Install a new No. 2 dust deflector, using a press.

10. Assemble the center driveshaft, as follows:
- Straight pin into the bearing case, using a pin punch and hammer
- New bearing, using tools 09959-60010, 09950-70010 and a press

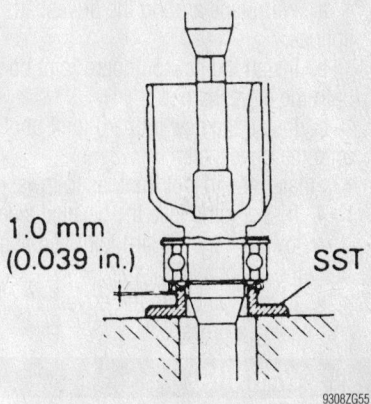

9308ZG55

View of the bearing-to-dust cover clearance–RAV4 with 2WD A/T and 4WD

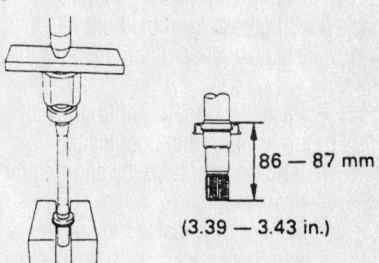

9308ZG56

View of the dust cover-to-center drive distance–RAV4 with 2WD A/T and 4WD

2WD A/T RH

Others

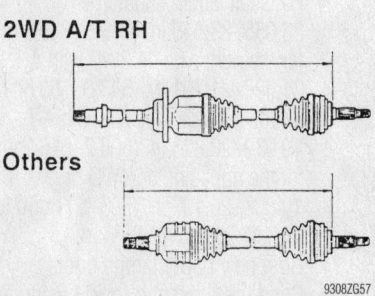

9308ZG57

Measuring the front halfshaft lengths–RAV4 with 2WD A/T and 4WD

- New snapring
- Bearing with the bearing case assembly to the center driveshaft, using tool 09710-30021 and a press
- New snapring
- New dust cover, until the clearance between the dust cover and the bearing is 0.039 in. (1.0mm)

11. Install or connect the following:
- Right dust cover (2WD), until the distance from the tip of the center drive is 3.39–3.34 in. (86–87mm) to the inner edge of the dust cover
- Left side dust cover (2WD and 4WD), using a press

12. Temporarily install new outboard/inboard joint boots using new clamps, as follows:
   a. Warp tape around the driveshaft splines.
   b. Install the new outboard joint boot onto the driveshaft.
   c. Install the new inboard joint boot onto the driveshaft.

13. Install the tri-pot joint, as follows:
- Tri-pot joint, face the beveled side toward the outboard joint and align the matchmarks
- Tri-pot joint onto the driveshaft, using a press

✷✷ WARNING

Be careful not to tap the roller.

- New snapring

14. Install the outboard joint boot packed with grease from the boot kit.

15. Install the inboard joint tulip, as follows:
- Pack the inboard joint boot with grease from the boot kit
- Inboard joint tulip, by aligning the matchmarks
- Temporarily, install the inboard joint boot packed with grease from the kit

16. Install the boot clamps to both boots, as follows:
- Both boots to the shaft grooves
- Halfshaft length should be 33.055–33.449 in. (839.6–849.6mm) for the right side on 2WD with A/T, 21.397–21.791 in. (543.5–553.5mm) for the left side on 2WD with A/T, 19.929–20.323 in. (506.2–516.2mm) for the right side on 4WD or 19.803–20.197 in. (503–511mm) for the left side on 4WD
- Both new boot clamps boot
- Bend the band and lock it using a screwdriver

RAV4 Rear

1. Before servicing the vehicle, refer to the precautions in the beginning of this section.

2. Remove the inboard and outboard joint boot clamps.

3. Disassemble the inboard joint tulip, as follows:
- Matchmark the tri-pot, inboard joint tulip or center driveshaft to the driveshaft

✷✷ WARNING

Do not use punch marks.

- Inboard joint tulip from the driveshaft

4. Remove the tri-pot joint, as follows:
- Snapring
- Matchmark the tri-pot joint to the driveshaft

✷✷ WARNING

Do not use punch marks.

- Tri-pot joint, using a brass bar and hammer

✷✷ WARNING

Do not tap the roller.

5. Remove or disconnect the following:
- Inboard and outboard joint boots

➡ **Do not disassemble the outboard joint.**

- No. 2 dust deflector from the center driveshaft, using a screwdriver and hammer

To assemble:

6. Install a new No. 2 dust deflector, using tools 09309-36010, 09316-20011 and a press.

7. Temporarily install new outboard/inboard joint boots using new clamps, as follows:
   a. Warp tape around the driveshaft splines.
   b. Install the new outboard joint boot onto the driveshaft.

c. Install the new inboard joint boot onto the driveshaft.

8. Install the tri-pot joint, as follows:
- Tri-pot joint, face the beveled side toward the outboard joint and align the matchmarks
- Tri-pot joint onto the driveshaft, using a brass bar and hammer

✷✷ WARNING

Be careful not to tap the roller.

- New snapring

9. Install the outboard joint boot packed with grease from the boot kit.

10. Install the inboard joint tulip, as follows:
- Pack the inboard joint boot with grease from the boot kit
- Inboard joint tulip, by aligning the matchmarks
- Inboard joint boot packed with grease from the kit

11. Install the boot clamps to both boots, as follows:
- Both boots to the shaft grooves
- Halfshaft length should be 23.392–23.795 in. (594.4–604.4mm) for the right side or 21.590–21.984 (548.4–558.4mm) for the left side
- New boot clamps boot
- Bend the band and lock it using a screwdriver

4Runner

OUTBOARD JOINT

The outboard joint is replaced with half-shaft; no overhaul is possible or necessary.

INBOARD (TRI-POT) JOINT

1. Before servicing the vehicle, refer to the precautions in the beginning of this section.

2. Remove the halfshaft from the vehicle.

3. Remove the large clamp from the inboard joint.

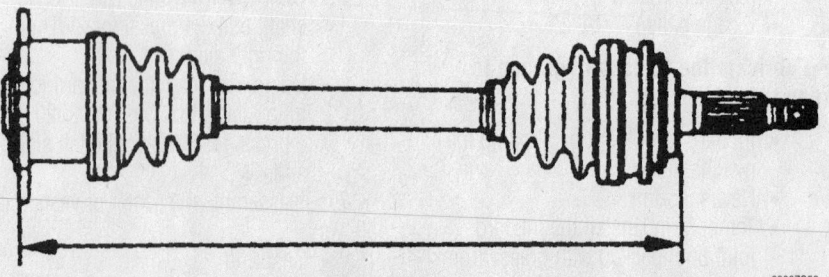

Measuring the rear halfshaft lengths–RAV4

9308ZG58

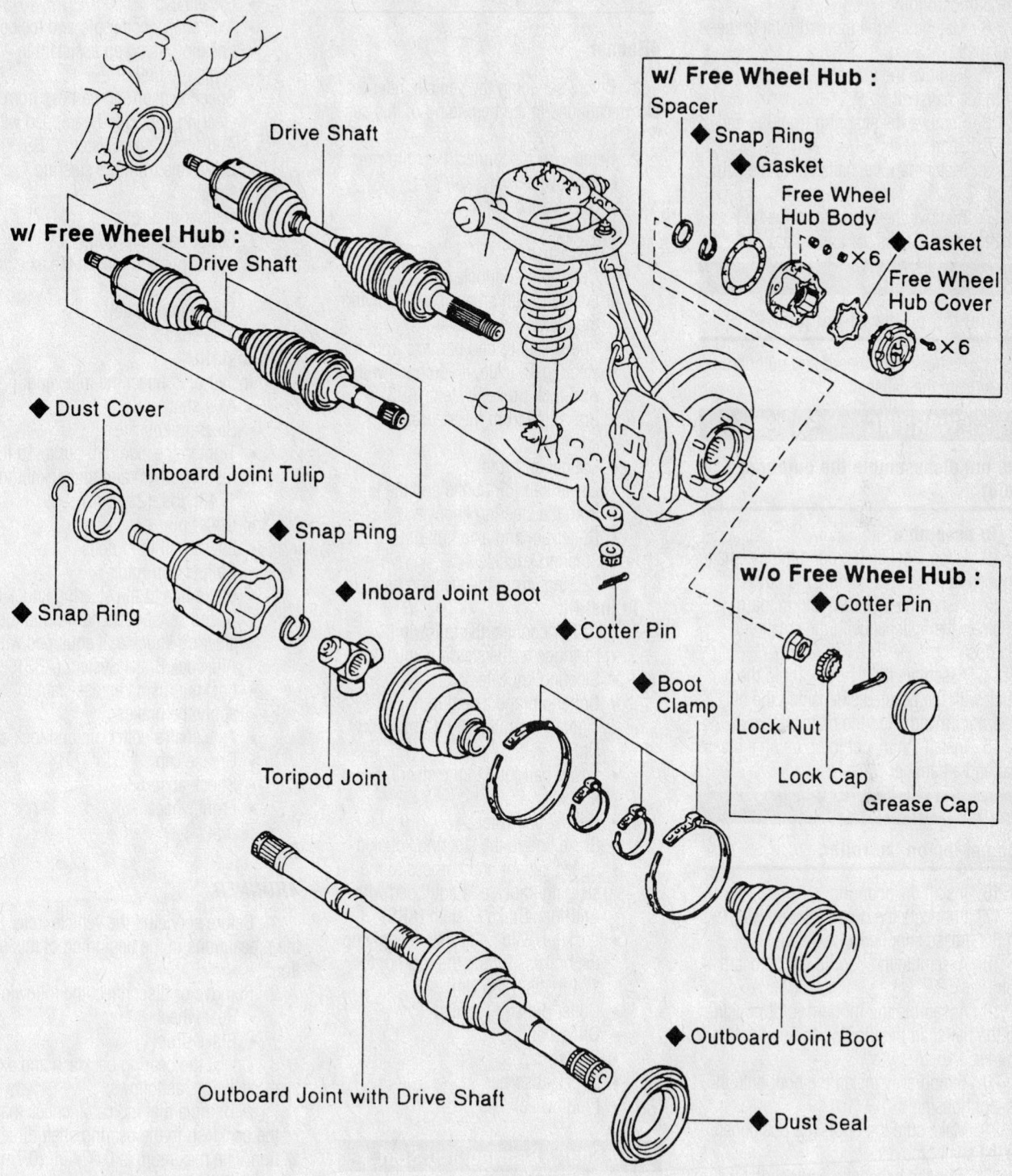

w/ Free Wheel Hub :

Drive Shaft

w/ Free Wheel Hub :
Drive Shaft

w/ Free Wheel Hub :
Spacer
◆ Snap Ring
◆ Gasket
Free Wheel
Hub Body
◆ Gasket
Free Wheel
Hub Cover
×6
×6

◆ Dust Cover

Inboard Joint Tulip

◆ Snap Ring

◆ Snap Ring

◆ Inboard Joint Boot

Toripod Joint

◆ Cotter Pin

◆ Boot Clamp

w/o Free Wheel Hub :
◆ Cotter Pin
Lock Nut
Lock Cap
Grease Cap

Outboard Joint with Drive Shaft

◆ Outboard Joint Boot

◆ Dust Seal

◆ Non-reusable part

9308YG07

Exploded view of the halfshaft assembly—4Runner

4. Remove the small clamp, using side cutters, from the inboard joint.

5. Slide the inboard joint boot toward the outboard joint.

6. Matchmark the inboard joint to the halfshaft.

7. Remove the inboard joint housing from the halfshaft.

8. Remove the snapring from the end of the halfshaft.

9. Matchmark the halfshaft to the tri-pot joint.

10. Remove the tri-pot from the half-shaft, using a brass bar and a hammer.

❋❋ WARNING

Do not tap on the tri-pot joint.

11. Remove the inboard and outboard boots from the halfshaft.

❋❋ WARNING

Do not disassemble the outboard joint.

To assemble:

12. Wrap vinyl tape around the halfshaft splines to prevent damaging the boots.

13. Install the outboard and inboard boots to the halfshaft with the small end clamps.

14. Assemble the tri-pot joint to the half-shaft with the beveled side facing the out-board joint and align the matchmarks.

15. Install the tri-pot joint, using a brass bar and a hammer.

❋❋ WARNING

Do not tap on the roller.

16. Install the snapring.

17. Lubricate the outboard joint with ½ of the grease supplied with the kit.

18. Assemble the boot to the outboard joint

19. Assemble the inboard joint housing to the halfshaft by aligning the match-marks.

20. Temporarily install the boot onto the tri-pot housing.

21. Make sure the boots are positioned in the shaft grooves.

22. With the halfshaft positioned at the standard length of 20.898–21.095 in. (525.8–535.8mm), make sure that the boots are not stretched or contracted.

23. Install a new inboard joint clamp.

24. Crimp the large clamp with tool 09521-24010 so that the crimp clearance is 0.039–0.059 in. (1.0–1.5mm).

25. Install the halfshaft.

Spindle Bearings

REMOVAL, PACKING AND INSTALLATION

4Runner

1. Before servicing the vehicle, refer to the precautions in the beginning of this section.

2. Remove or disconnect the following:
- Front wheel
- Shock absorber
- Grease cap
- Driveshaft
- Cotter pin and lockcap
- Locknut, with an assistant applying the brakes
- Speed sensor and harness from the steering knuckle, if equipped with Anti-lock Brake System (ABS)
- Brake line from the steering knuckle
- Caliper and rotor
- Lower ball joint bolts and the joint from the steering knuckle
- Cotter pin and axle hub nut
- Steering knuckle
- Bearings from the steering knuckle

To install:

3. Install or connect the following:
- Bearings to the steering knuckle
- Steering knuckle
- Cotter pin and axle hub nut. Tighten the nut to 80 ft. lbs. (108 Nm).
- Lower ball joint to the steering knuckle
- Caliper and rotor
- Brake line to the steering knuckle
- Speed sensor and harness to the steering knuckle, if equipped with Anti-lock Brake System (ABS)
- Locknut, with an assistant applying the brakes. Torque the locknut to 174 ft. lbs. (235 Nm).
- Cotter pin and lockcap
- Driveshaft
- Grease cap
- Shock absorber
- Front wheel

Axle Shaft, Bearing and Seal

REMOVAL & INSTALLATION

Front

4RUNNER

1. Before servicing the vehicle, refer to the precautions in the beginning of this section.

2. Remove or disconnect the following:
- Front wheel
- Shock absorber
- Grease cap
- Axle shaft's cotter pin and lock cap
- Locknut, using an assistant to apply the brakes
- Speed sensor and harness from the steering knuckle, if equipped with Anti-lock Brake System (ABS)
- Brake line from the steering knuckle
- Caliper and rotor
- Lower ball joint bolts
- Cotter pin and loosen the axle hub nut
- Steering knuckle
- Axle shaft

To install:

3. Install or connect the following:
- Axle shaft
- Steering knuckle
- Tighten the axle hub nut to 80 ft. lbs. (108 Nm) and the locknut to 174 ft. lbs. (235 Nm).
- Cotter pin
- Lower ball joint bolts
- Caliper and rotor
- Brake line to the steering knuckle
- Speed sensor and harness to the steering knuckle, if equipped with Anti-lock Brake System (ABS)
- Locknut, using an assistant to apply the brakes
- Axle shaft's cotter pin and lock cap
- Grease cap
- Shock absorber
- Front wheel

Rear

4RUNNER

1. Before servicing the vehicle, refer to the precautions in the beginning of this section.

2. Remove or disconnect the following:
- Rear wheel
- Brake drum

3. Check the bearing backlash and axle shaft deviation, as follows:

a. Using a dial indicator, check that the backlash in the bearing shaft direction. The maximum is 0.027 in. (0.7mm).

b. If the backlash exceeds the maximum, replace the bearing.

c. Using a dial indicator, check the deviation at the surface of the axle shaft outside the hub bolt. Maximum is 0.0039 in. (0.1mm).

d. If the deviation exceeds the maximum, replace the axle shaft.

4. Remove or disconnect the following:

- Anti-lock Brake System (ABS) speed sensor from the axle housing, if equipped
- Axle shaft assembly by removing the 4 nuts from the backing plate
- O-ring from the axle housing
- Bearing and retainer (differential side) and ABS speed sensor rotor, if equipped
- Snapring from the axle shaft
- Axle shaft

➡ Inspect the axle shaft and flange run-outs. The axle shaft run-out should be 0.079 in. (2.0mm) and the flange run-out should be 0.004 in. (0.1mm).

- Outer seal
- Bearing from the axle shaft

To install:

5. Install or connect the following:
- Bearing from the axle shaft
- Outer seal
- Axle shaft
- Snapring to the axle shaft

- Bearing and retainer (differential side) and ABS speed sensor rotor, if equipped
- O-ring to the axle housing
- Axle shaft assembly. Tighten the 4 backing plate nuts to 48 ft. lbs. (66 Nm).
- Anti-lock Brake System (ABS) speed sensor to the axle housing, if equipped
- Brake drum
- Rear wheel

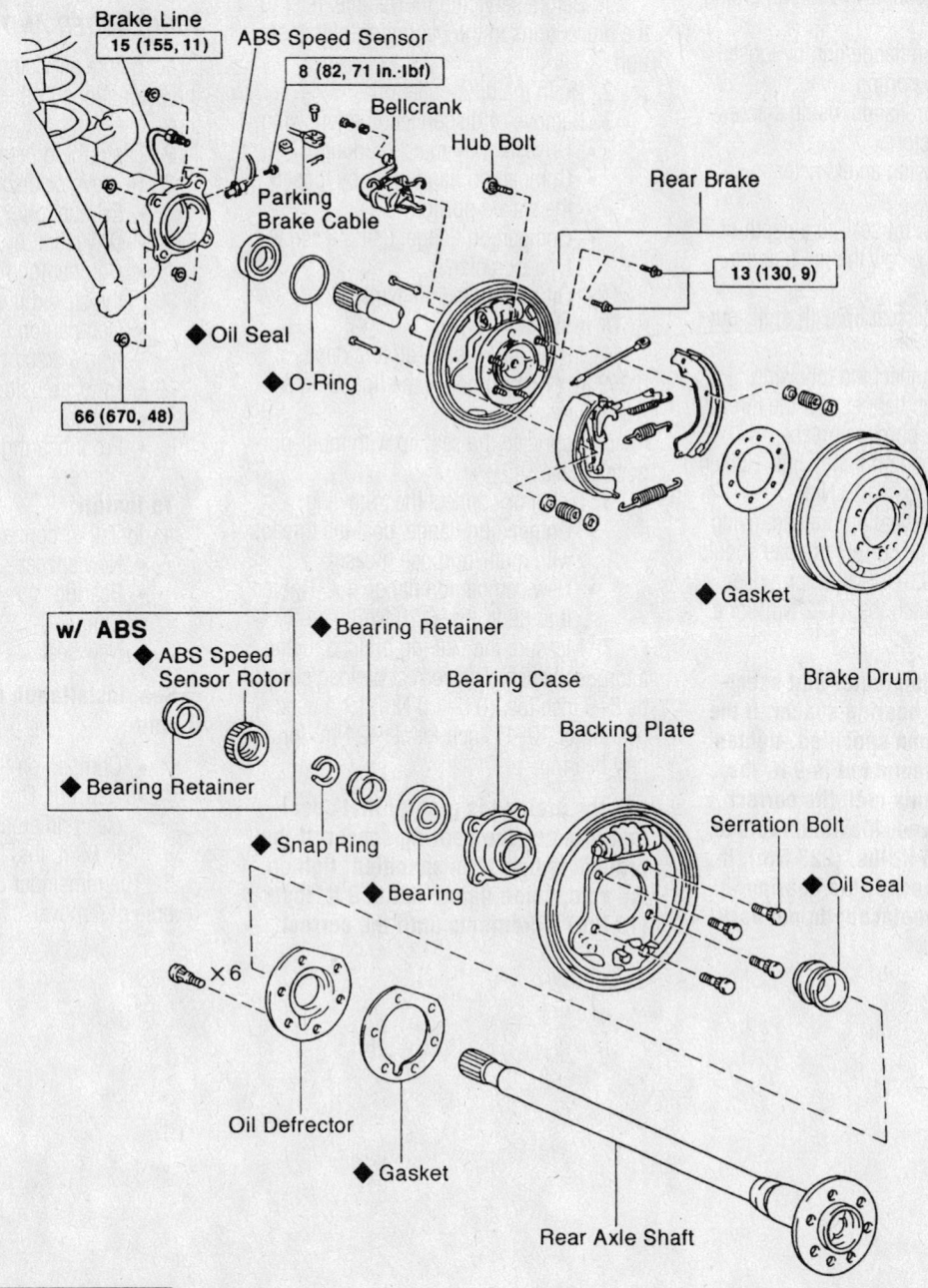

| N·m (kgf·cm, ft·lbf) | : Specified torque
◆ Non-reusable part

86827G97

Exploded view of the rear axle shaft and components—4Runner

Pinion Seal

REMOVAL & INSTALLATION

Front

4RUNNER

1. Before servicing the vehicle, refer to the precautions in the beginning of this section.
2. Drain the differential oil.
3. Remove or disconnect the following:
   - Front driveshaft by matchmarking it
   - Companion flange nut, by loosen the staked portion
   - Companion flange, using a screw-type extractor
   - Oil seal, using an extractor

To install:

4. Install a new oil seal, to a depth of 0.059 in. (1.5mm) below the lip, using a seal driver.
5. Lubricate the seal lip with multi-purpose grease.
6. Install or connect the following:
   - Companion flange, coat the threads with multi-purpose grease
   - New companion flange nut. Tighten it to 89 ft. lbs. (120 Nm).
7. Measure the bearing preload, using a torque wrench. The correct preload should be 5–9 inch lbs. (0.6–1.0 Nm) for a used bearing or 10–17 inch lbs. (1–2 Nm) for a new bearing.

➡**If the preload is greater that specified, replace the bearing spacer. If the preload is less than specified, tighten the companion flange nut in 9 ft. lbs. (13 Nm) increments until the correct preload is achieved. Maximum torque for the nut is 165 ft. lbs. (223 Nm). If the value is exceeded, the bearing spacer must be replaced; do not back off the flange nut to lower the torque or preload.**

8. Install the front driveshaft by aligning the matchmarks.
9. Check the companion flange run-out; maximum allowable run-out is 0.003 in. (0.10mm).
10. Stake the pinion flange nut.
11. Refill the differential with oil.

Rear

4RUNNER

1. Before servicing the vehicle, refer to the precautions in the beginning of this section.
2. Drain the differential oil.
3. Remove or disconnect the following:
   - Driveshaft by matchmarking it
   - Companion flange nut, by loosen the staked portion
   - Companion flange, using a screw-type extractor
   - Oil seal, using an extractor

To install:

4. Install a new oil seal, to a depth of 0.059 in. (1.5mm) below the lip, using a seal driver.
5. Lubricate the seal lip with multi-purpose grease.
6. Install or connect the following:
   - Companion flange, coat the threads with multi-purpose grease
   - New companion flange nut. Tighten it to 89 ft. lbs. (120 Nm).
7. Measure the bearing preload, using a torque wrench. The correct preload should be 5–9 inch lbs. (0.6–1.0 Nm) for a used bearing or 10–17 inch lbs. (1–2 Nm) for a new bearing.

➡**If the preload is greater that specified, replace the bearing spacer. If the preload is less than specified, tighten the companion flange nut in 9 ft. lbs. (13 Nm) increments until the correct preload is achieved. Maximum torque for the nut is 165 ft. lbs. (223 Nm). If the value is exceeded, the bearing spacer must be replaced; do not back off the flange nut to lower the torque or preload.**

8. Install the driveshaft by aligning the matchmarks.
9. Check the companion flange run-out; maximum allowable run-out is 0.003 in. (0.10mm).
10. Stake the pinion flange nut.
11. Refill the differential with oil.

HIGHLANDER AND RX300

1. Before servicing the vehicle, refer to the precautions in the beginning of this section.
2. Drain the differential oil.
3. Remove or disconnect the following:
   - Exhaust pipe
   - Driveshaft by matchmarking it
   - Companion flange nut, by loosen the staked portion
   - Companion flange, using a screw-type extractor
   - Oil seal, using an extractor
   - Slinger
   - Front bearing
   - Spacer

To install:

4. Install or connect the following
   - New spacer
   - Bearing
   - Slinger
   - New seal

➡**Seal installation depth: 2.0mm +/- 0.3mm**

   - Companion flange
   - New nut. Coat the threads with clean differential oil. Torque the nut to 80 ft. lbs. (108Nm).

5. The remainder of installation is the reverse of removal.

STEERING AND SUSPENSION

Air Bag

✳✳ CAUTION

Some vehicles are equipped with an air bag system. The system must be disarmed before performing service on, or around, system components, the steering column, instrument panel components, wiring and sensors. Failure to follow the safety precautions and the disarming procedure could result in accidental air bag deployment, possible injury and unnecessary system repairs.

PRECAUTIONS

Several precautions must be observed when handling the inflator module to avoid accidental deployment and possible personal injury.

• Never carry the inflator module by the wires or connector on the underside of the module.

• When carrying a live inflator module, hold securely with both hands and ensure that the bag and trim cover are pointed away.

• Place the inflator module on a bench or other surface with the bag and trim cover facing up.

• With the inflator module on the bench, never place anything on or close to the module which may be thrown in the event of an accidental deployment.

DISARMING

To avoid personal injury when working on vehicles equipped with an air bag, the negative battery cable must be disconnected and at least 90 seconds must elapse before working on the system. Failure to do so may result in deployment of the air bag.

Power Rack And Pinion Steering Gear

REMOVAL & INSTALLATION

RAV4

1. Before servicing the vehicle, refer to the precautions in the beginning of this section.

2. Disconnect the negative battery cable.

✳✳ CAUTION

To avoid personal injury when working on air bag equipped vehicles, work must be started after 90 seconds or longer from the time the ignition switch is turned to the LOCK position and the negative battery terminal is disconnected. If the air bag system is disconnected with the ignition switch at the ON or ACC, diagnostic codes will be set. When removing the air bag, take care not to pull the air bag wiring harness. When carrying the wheel pad, carry it with the upper surface facing away. When storing it, keep the upper surface of the pad facing upward.

3. Turn the key to the **LOCK** position and lock the steering wheel in place.

4. Place a drain pan under the steering rack.

5. Remove or disconnect the following:
   • Front wheels
   • Right and left-hand engine undercovers
   • Right and left-hand tie rod ends from the steering knuckle
   • Front exhaust pipe
   • Sway bar with the links

6. Disconnect the No. 2 intermediate shaft from the rack and pinion, as follows:
   a. Loosen the top bolt.
   b. Remove the lower bolt holding the No. 2 intermediate shaft to the rack and pinion.
   c. Shift the No. 2 intermediate shaft and place matchmarks on the control valve shaft and the No. 2 intermediate shaft.
   d. Disconnect the No. 2 shaft from the rack and pinion.

7. Install or connect the following:
   • Pressure feed and return tubes from the rack and pinion, using a line wrench
   • Pressure feed and return tube clamps, by removing the bolt
   • Right and left lower control arms, from the steering knuckle

8. Remove the front suspension crossmember assembly, as follows:
   • Both centermember set nuts, holding the centermember to the middle of the crossmember.
   • Both rack and pinion assembly set bolts and nuts from the crossmember.

• Securely suspend the steering gear assembly.
• Support the suspension crossmember with a jack.
• Both bolts from the suspension crossmember
• Suspension crossmember with the lower suspension arms

9. Remove the rack and pinion.

To install:

10. Install or connect the following:
   • Rack and pinion

11. Install the crossmember to the vehicle, as follows:
   • Suspension crossmember with the lower control arms. Tighten both bolts to 152 ft. lbs. (206 Nm).
   • Rack and pinion. Tighten the nuts/bolts to 83 ft. lbs. (113 Nm).
   • Centermember to the crossmember. Tighten both set nuts to 82 ft. lbs. (112 Nm).

12. Install or connect the following:
   • Right and left lower control arms
   • Pressure feed and return tubes clamps
   • Pressure feed and return tubes to the rack and pinion. Tighten the tubes to 26 ft. lbs. (36 Nm), using a torque wrench with a fulcrum length of 11.81 inches (300mm).
   • Steering column No. 2 intermediate shaft to the rack and pinion. Align the marks and tighten the upper and lower pinch bolts to 26 ft. lbs. (35 Nm).
   • Stabilizer bar links. Tighten the nuts to 22 ft. lbs. (29 Nm).
   • Front exhaust pipe with new gaskets. Tighten the bolts to 35 ft. lbs. (48 Nm) and the nuts to 46 ft. lbs. (62 Nm).
   • Right and left-hand tie rod ends to the steering knuckle. Tighten the nuts to 36 ft. lbs. (49 Nm) and install new cotter pins.
   • Right and left-hand engine undercovers
   • Front wheels

13. Fill the power steering unit and bleed the system. Check for leaks.

14. Check and/or adjust the front wheel alignment.

Highlander

1. Before servicing the vehicle, refer to the precautions in the beginning of this section.

2. Remove or disconnect the following:

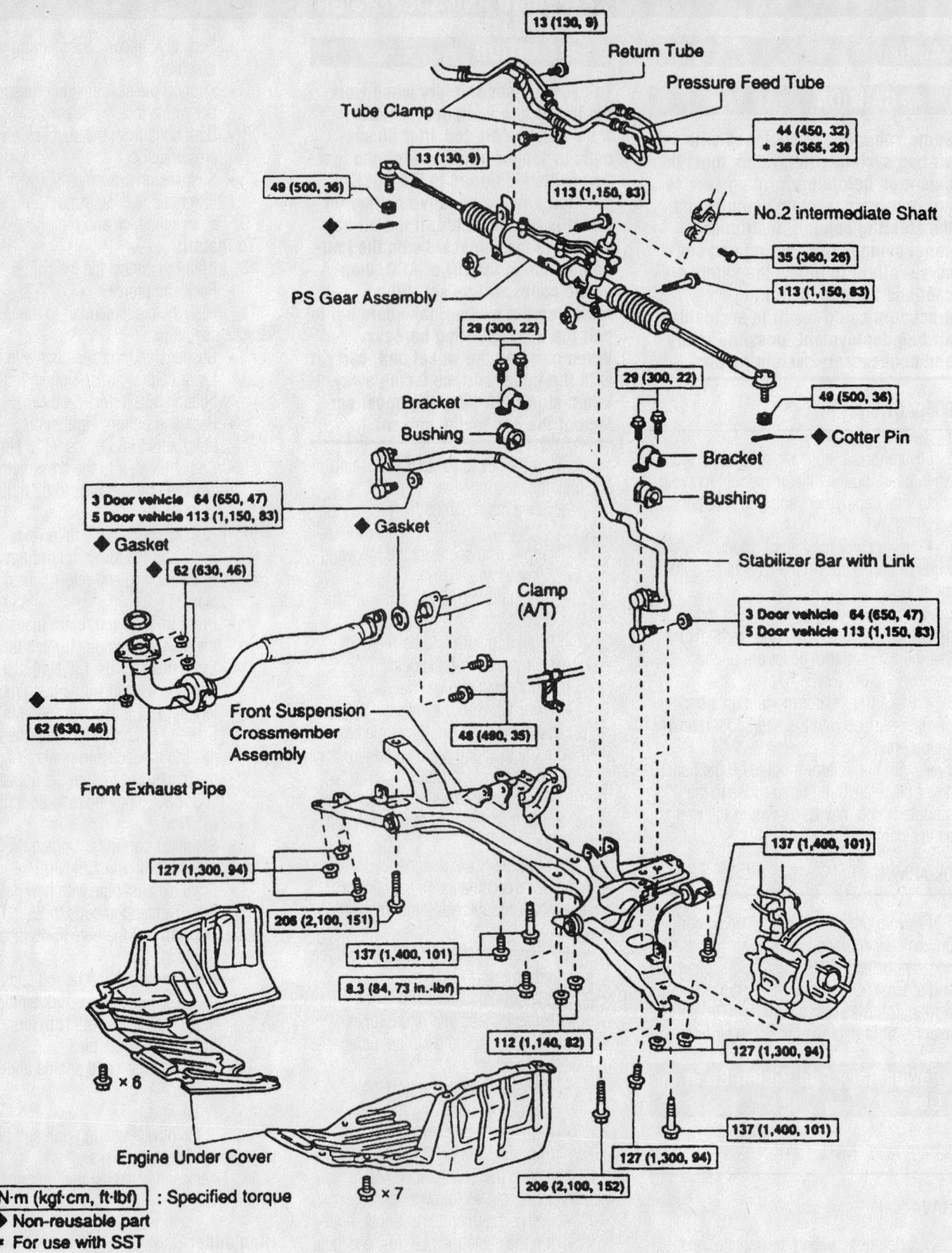

13 (130, 9)
Return Tube
Pressure Feed Tube
Tube Clamp
44 (450, 32)
* 36 (365, 26)
13 (130, 9)
49 (500, 36)
113 (1,150, 83)
No.2 intermediate Shaft
35 (360, 26)
113 (1,150, 83)
PS Gear Assembly
29 (300, 22)
29 (300, 22)
49 (500, 36)
Bracket
Bushing
◆ Cotter Pin
Bracket
Bushing
3 Door vehicle 64 (650, 47)
5 Door vehicle 113 (1,150, 83)
◆ Gasket
62 (630, 46)
◆ Gasket
Stabilizer Bar with Link
Clamp
(A/T)
3 Door vehicle 64 (650, 47)
5 Door vehicle 113 (1,150, 83)
62 (630, 46)
Front Suspension
Crossmember
Assembly
48 (490, 35)
Front Exhaust Pipe
137 (1,400, 101)
127 (1,300, 94)
206 (2,100, 151)
137 (1,400, 101)
137 (1,400, 101)
8.3 (84, 73 in.·lbf)
112 (1,140, 82)
127 (1,300, 94)
× 6
137 (1,400, 101)
Engine Under Cover
127 (1,300, 94)
N·m (kgf·cm, ft·lbf) : Specified torque
× 7
206 (2,100, 152)
◆ Non-reusable part
* For use with SST

7924ZG75

- Negative battery cable

➡**Wait at least 90 seconds before working on the vehicle to allow the Supplemental Restraint System (SRS) system to disarm.**

- Steering wheel
- Front wheels
- Tie rod ends
- Intermediate shaft

➡**Matchmark the shaft and gear.**

- Stabilizer bar end links
- Pressure and return lines
- Steering gear
- Installation is the reverse of

removal. Observe the following torques:
- Rack mounting bolts: 52 ft. lbs. (70Nm)
- Stabilizer bar end links: 55 ft. lbs. (74Nm)
- Intermediate shaft bolt: 26 ft. lbs. (35Nm)
- Tie rod end nuts: 36 ft. lbs. (49Nm)

RX 300

1. Before servicing the vehicle, refer to the precautions in the beginning of this section.
2. Remove or disconnect the following:

- Negative battery cable

➡**Wait at least 90 seconds before working on the vehicle to allow the Supplemental Restraint System (SRS) system to disarm.**

- Right and left side fender apron seals
- Right and left tie rod ends
3. Place matchmarks on the intermediate shaft.
4. Remove or disconnect the following:
- Pinch bolt and the intermediate shaft out from under the vehicle
- Power steering line clamp
- Pressure and feed lines

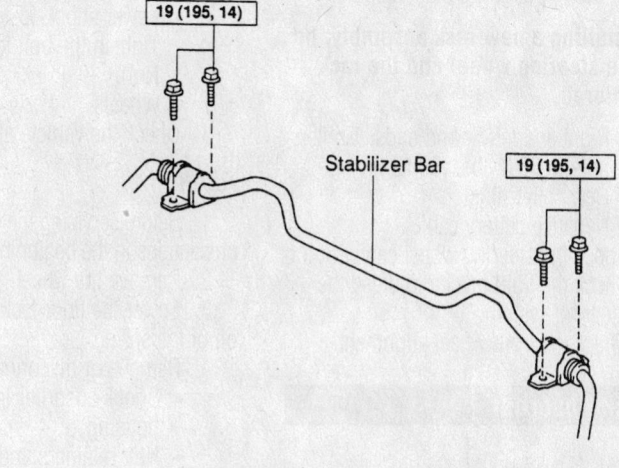

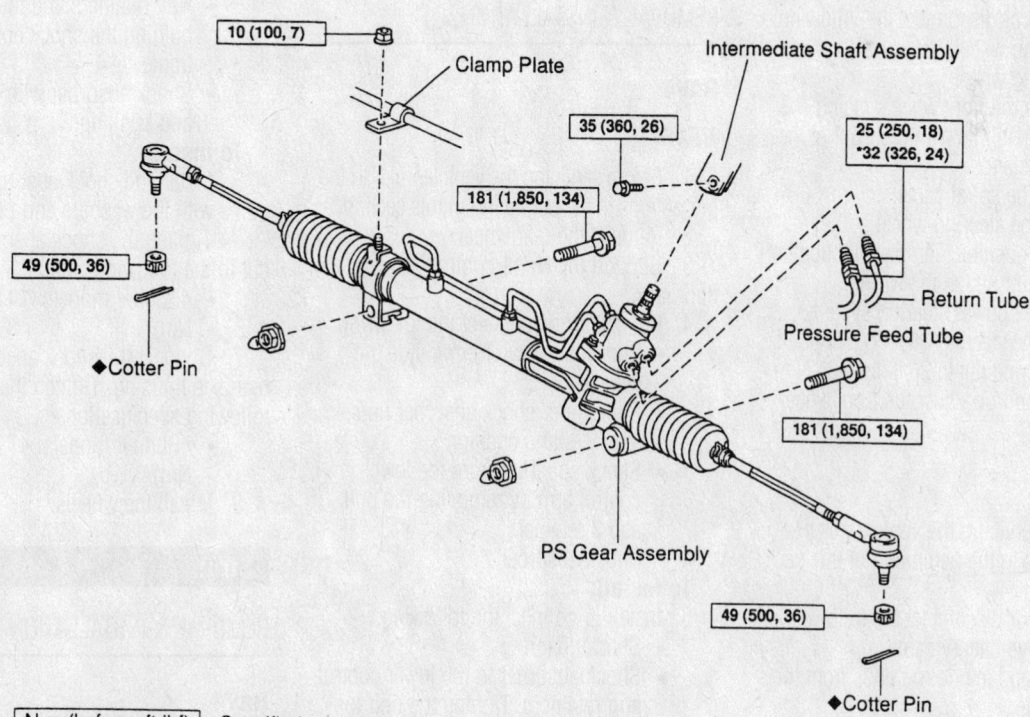

N·m (kgf·cm, ft·lbf) : Specified torque
◆ Non–reusable part
* For use with SST

Exploded view of the power steering gear and related components—RX 300 models

7924ZG76

- Stabilizer bar, unbolt it but do not remove it
- Heated Oxygen (HO2) sensor
- Both gear assembly set bolts and nuts, by lifting the stabilizer bar
- Gear assembly from the left side of the vehicle

To install:

5. Install or connect the following:
- Gear assembly to the left side of the vehicle

✳✳ WARNING

Be careful not to damage the power steering lines.

- Tighten the gear assembly set bolts and nuts to 134 ft. lbs. (181 Nm), by lifting the stabilizer bar
- HO2sensor
- Stabilizer bar. Tighten the bolt to 14 ft. lbs. (19 Nm) and the nut to 29 ft. lbs. (39 Nm).
- Pressure and feed return lines. Tighten them to 18 ft. lbs. (25 Nm).
- Line clamps. Tighten the nut to 84 inch lbs. (10 Nm).
- Intermediate shaft, by aligning the joint and main shaft matchmarks. Tighten to 26 ft. lbs. (35 Nm).
- Tie rod ends
- Fender apron seals. Securely tighten the bolts.

6. Remove or disconnect the following:
- Steering wheel pad
- Steering wheel

7. Position the front wheels facing straight-ahead. Do this with the front of the vehicle on jackstands.

8. Center the spiral cable.

9. Install the steering wheel at the straight-ahead position. Temporarily tighten the wheel set nut. Attach the wiring.

10. Bleed the power steering system.

11. Check the steering wheel center point. Tighten the steering nut to 26 ft. lbs. (35 Nm).

12. Check and/or adjust the front wheel alignment.

4Runner

1. Before servicing the vehicle, refer to the precautions in the beginning of this section.

2. Remove or disconnect the following:
- Negative battery cable
- Right and left tie rod ends from the knuckle
- Intermediate shaft from the steering rack, by matchmarking it
- Pressure feed and the return tubes, using SST 09631-22020

- Mount bracket and the grommet, from the power steering rack assembly
- Power steering rack and pinion

To install:

3. Install or connect the following:
- Power steering rack and pinion. Tighten the mounting bolts to 65 ft. lbs. (88 Nm).
- Grommet and mount bracket to the gear assembly. Tighten the bolts to 65 ft. lbs. (88 Nm).
- New O-ring
- Pressure feed and return tubes. Tighten the line fittings to 14 ft. lbs. (19 Nm).
- Intermediate shaft to the steering rack, by aligning the matchmarks

➡ **If installing a new rack assembly, be sure the steering wheel and the rack are centered.**

- Right and left tie rod ends. Tighten nuts to 67 ft. lbs. (90 Nm).
- New cotter pins
- Negative battery cable

4. Check the steering wheel center point.

5. Check the fluid level and bleed the power steering system.

6. Check the front wheel alignment.

Shock Absorber

REMOVAL & INSTALLATION

RAV4

REAR

1. Before servicing the vehicle, refer to the precautions in the beginning of this section.

2. Remove the rear wheel.

3. Support the No. 1 control arm with a floor jack.

4. Remove or disconnect the following:
- Suspension cap from inside the vehicle
- Both upper shock absorber nuts, retainers and cushion
- Shock absorber from the lower control arm by removing the bolt and 2 retainers
- Shock absorber

To install:

5. Install or connect the following:
- Shock absorber
- Shock absorber to the lower control arm retainers. Tighten the bolt to 27 ft. lbs. (37 Nm).
- Shock absorber to the chassis cushion and retainers. Tighten both nuts to 18 ft. lbs. (25 Nm).

- Suspension cap
- Wheel

4Runner

FRONT

1. Before servicing the vehicle, refer to the precautions in the beginning of this section.

2. Remove or disconnect the following:
- Front wheel
- Shock from the lower control arm
- 3 upper nuts
- Shock absorber

To install:

3. Install or connect the following:
- Shock absorber. Tighten the 3 nuts to 47 ft. lbs. (64 Nm).
- Lower shock-to-lower control arm. Tighten the bolt to 101 ft. lbs. (135 Nm).
- Wheels

4. Check the vehicle alignment.

REAR

1. Before servicing the vehicle, refer to the precautions in the beginning of this section.

2. Remove the wheel.

3. Lower the floor jack to take tension off of the spring.

4. Remove or disconnect the following:
- Shock absorber from the rear axle housing
- Nut, retainers and the cushions holding the shock absorber to the frame
- Shock absorber with the washers and bushings

To install:

5. Install the shock absorber to the frame with the washers and bushings.

6. Tighten the shock absorber-to-frame nut to the following values:
- 4Runner models: 14 ft. lbs. (20 Nm)

7. Connect the shock absorber to the rear axle housing. Tighten the bolt to the following specifications:
- 4Runner models: 47 ft. lbs. (64 Nm)

8. Install the wheels.

Strut

REMOVAL & INSTALLATION

RAV4

RAV4 is equipped with front strut and rear shock absorber type suspension arrangement.

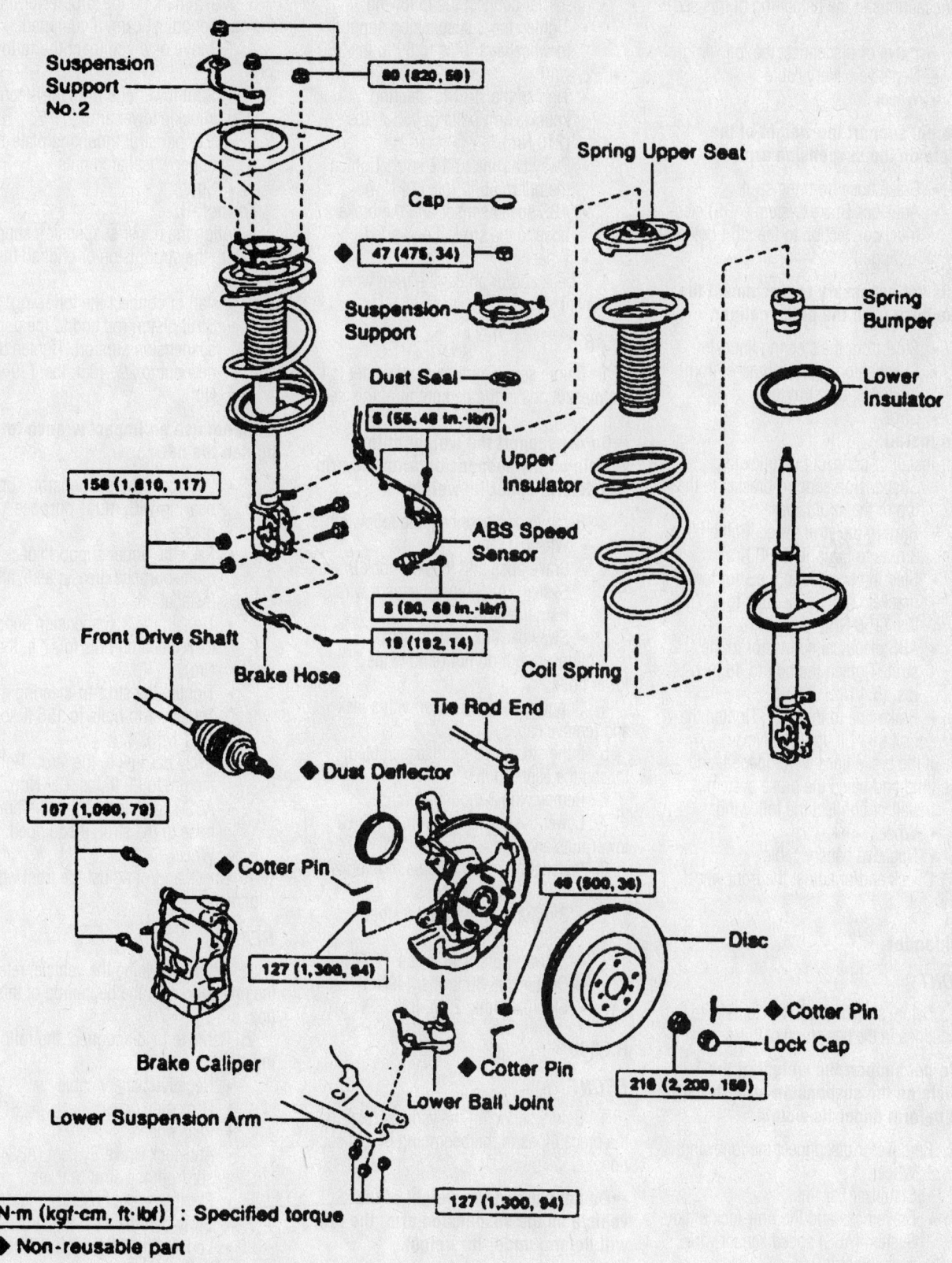

Suspension Support No.2

80 (820, 59)

Cap

47 (475, 34)

Suspension Support

Dust Seal

6 (55, 48 in.·lbf)

158 (1,610, 117)

ABS Speed Sensor

8 (80, 69 in.·lbf)

Front Drive Shaft

19 (192, 14)

Brake Hose

Spring Upper Seat

Spring Bumper

Lower Insulator

Upper Insulator

Coil Spring

Tie Rod End

Dust Deflector

167 (1,090, 79)

Cotter Pin

49 (500, 36)

Disc

Cotter Pin

Lock Cap

127 (1,300, 94)

Cotter Pin

216 (2,200, 159)

Brake Caliper

Lower Ball Joint

Lower Suspension Arm

127 (1,300, 94)

N·m (kgf·cm, ft·lbf) : Specified torque
◆ Non-reusable part

7924ZG78

Strut assembly exploded view—RAV4

1. Before servicing the vehicle, refer to the precautions in the beginning of this section.

2. Remove or disconnect the following:
- Negative battery cable
- Wheel

➡ **Do not support the weight of the vehicle on the suspension arm.**

- Brake hose from the strut
- Anti-lock Brake System (ABS) electrical connection to the strut bolt, if equipped

➡ **It is not necessary to disconnect the brake hose from the brake caliper.**

- Strut from the steering knuckle
- Suspension support bracket from the top of the strut tower
- Strut

To install:

3. Install or connect the following:
- Suspension support bracket to the top of the strut tower
- Strut to the strut tower. Tighten the 3 nuts to 59 ft. lbs. (80 Nm).
- Steering knuckle to the strut lower bracket. Tighten the nuts to 117 ft. lbs. (158 Nm).
- ABS electrical connector to the strut. Tighten the bolt to 48 inch lbs. (5.4 Nm).
- Brake line to the strut. Tighten the bolt to 14 ft. lbs. (19 Nm).

4. If the brake lines were opened, add brake fluid and bleed the brake system.

5. Install or connect the following:
- Wheel
- Negative battery cable

6. Check and/or adjust the front wheel alignment.

Highlander

FRONT

1. Before servicing the vehicle, refer to the precautions in the beginning of this section.

➡ **Do not support the weight of the vehicle on the suspension arm; the arm will deform under its weight.**

2. Remove or disconnect the following:
- Wheel
- Stabilizer bar link
- Brake hose and the Anti-lock Brake System (ABS) speed sensor wire from the strut
- Strut lower end from the steering knuckle lower arm
- 3 upper strut mounting plate-to-upper wheel arch nuts
- Strut

To install:

3. Install or connect the following:
- Tighten the 3 suspension support-to-wheel arch nuts to 59 ft. lbs. (80 Nm).
- Tighten the strut-to-steering knuckle arm bolts to 155 ft. lbs. (210 Nm).
- Sway bar link to the strut. Tighten the nut to 55 ft. lbs. (74 Nm).
- ABS speed sensor and the brake hose to the strut, if equipped
- Wheel

4. Check and/or adjust the front wheel alignment.

REAR

1. Before servicing the vehicle, refer to the precautions in the beginning of this section.

➡ **Do not support the weight of the vehicle on the suspension arm; the arm will deform under its weight.**

2. Remove or disconnect the following:
- Wheel
- Brake hose and the Anti-lock Brake System (ABS) speed sensor wire from the strut
- Sway bar link from the strut

3. Loosen, but do not remove the 2 lower bolts.

4. Support the axle carrier with a jack and remove cap.

5. If the strut is being disassembled, loosen the center nut.

6. Remove the 3 mounting nuts.

7. Lower the carrier and remove the 2 lower nuts and bolts.

8. Installation is the reverse of removal. Observe the following torques:
- 3 mounting nuts: 29 ft. lbs. (39Nm)
- 2 lower nuts: 188 ft. lbs. (255Nm)
- Center nut: 36 ft. lbs. (49Nm)
- Stabilizer link: 29 ft. lbs. (39Nm)

RX 300

FRONT

1. Before servicing the vehicle, refer to the precautions in the beginning of this section.

➡ **Do not support the weight of the vehicle on the suspension arm; the arm will deform under its weight.**

2. Remove or disconnect the following:
- Wheel
- Brake hose and the Anti-lock Brake System (ABS) speed sensor wire from the strut

- Sway bar link from the strut

3. Matchmark on the strut lower bracket and camber adjust cam, if equipped.

4. Remove or disconnect the following:
- Strut lower end from the steering knuckle lower arm
- 3 upper strut mounting plate-to-upper wheel arch nuts
- Strut

To install:

5. Align the upper suspension support hole with the strut piston or end, so they fit properly.

6. Install or connect the following:
- Strut piston rod end to the upper suspension support. Tighten the new nut to 29–40 ft. lbs. (39–54 Nm).

➡ **Do not use an impact wrench to tighten the nut.**

- Lubricate the suspension support bearing with multi-purpose grease.
- Pack the upper support space with multi-purpose grease, also, after installation.
- Tighten the 3 suspension support-to-wheel arch nuts to 47 ft. lbs. (64 Nm).
- Tighten the strut-to-steering knuckle arm bolts to 156 ft. lbs. (211 Nm).
- Sway bar link to the strut. Tighten the nut to 29 ft. lbs. (39 Nm).
- ABS speed sensor and the brake hose to the strut, if equipped
- Wheel

7. Check and/or adjust the front wheel alignment.

REAR

1. Before servicing the vehicle, refer to the precautions in the beginning of this section.

2. Remove or disconnect the following:
- Negative battery cable
- Deck side cover
- Rear wheels
- Anti-lock Brake System (ABS) sensor from the strut bracket
- Flexible brake hose from the strut
- Sway bar link from the strut
- Loosen the 2 lower strut mounting bolts

3. Support the rear axle carrier with a jack.

4. Remove or disconnect the following:
- 3 upper strut mounting nuts
- Strut, by lower the rear axle

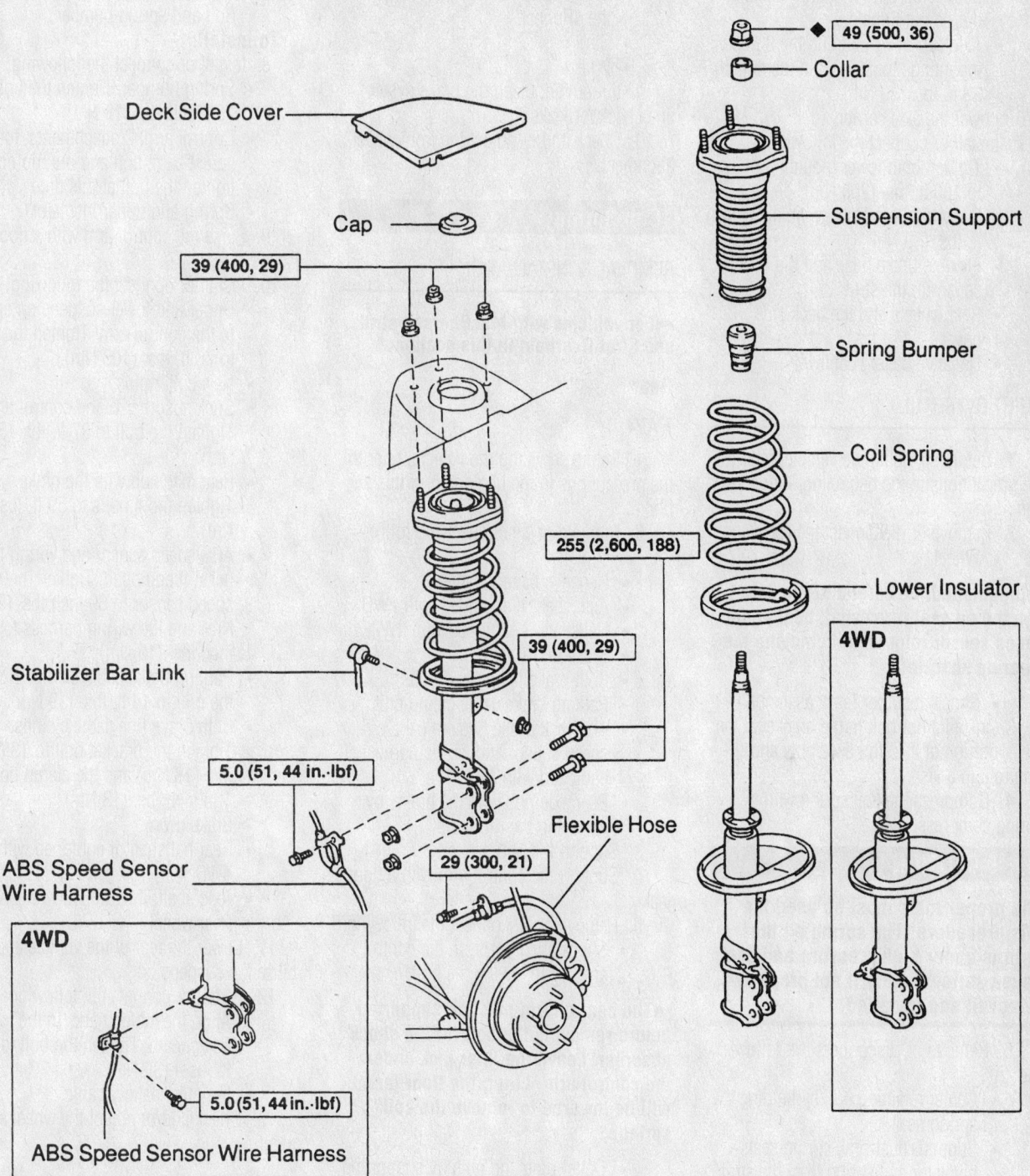

Deck Side Cover

Cap

39 (400, 29)

Stabilizer Bar Link

5.0 (51, 44 in.·lbf)

ABS Speed Sensor
Wire Harness

39 (400, 29)

255 (2,600, 188)

Flexible Hose

29 (300, 21)

49 (500, 36)

Collar

Suspension Support

Spring Bumper

Coil Spring

Lower Insulator

4WD

4WD

5.0 (51, 44 in.·lbf)

ABS Speed Sensor Wire Harness

N·m (kgf·cm, ft·lbf) : Specified torque

◆ Non–reusable part

7924ZG89

Exploded view of the rear strut assembly—RX 300

To install:

5. Install or connect the following:
   - Strut
   - Both lower strut mounting bolts, but do not tighten
   - Axle carrier by aligning the 3 upper mounting studs. Tighten the nuts to 29 ft. lbs. (39 Nm).
6. Lower the axle carrier.
7. Install or connect the following:
   - Tighten both lower mounting bolts to 188 ft. lbs. (255 Nm).
   - Sway bar link. Tighten the nut to 29 ft. lbs. (39 Nm).
   - Flexible brake hose and the ABS sensor to the strut
   - Rear wheels and the deck side cover
   - Negative battery cable

STRUT OVERHAUL

1. Before servicing the vehicle, refer to the precautions in the beginning of this section.
2. Remove or disconnect the following:
   - Wheel

→**If equipped, be careful not to damage the oil seal, driveshaft boot and/or speed sensor rotor when removing the steering knuckle.**

   - Shock absorber (strut assembly)
3. Install a nut/bolt to the bracket at the lower portion of the strut assembly and secure it in a vise.
4. Compress the coil spring with a spring compressor.

※※ CAUTION

The proper tools must be used for this procedure. The spring on the strut is under high pressure and can cause serious injury if not properly removed and installed.

5. Remove or disconnect the following:
   - Center retaining nut, by holding the spring seat
   - Support, dust seal, spring seat, insulator and spring from the strut assembly

To install:

6. Install the spring bumper and lower insulator to the strut assembly.
7. Compress the coil spring and fit the lower end of the spring into the spring seat gap.
8. Install or connect the following:
   - Upper insulator, spring seat, dust seal, support and spring seat.

Tighten the new retaining nut to 34 ft. lbs. (47 Nm) for the RAV4 and RX300; 36 ft. lbs. (49Nm) for the Highlander; 18 ft. lbs. (25Nm) for the 4Runner.
   - Strut
   - Wheel
9. If required, bleed the brake system and check for leaks.
10. Check and/or adjust the front wheel alignment.

Coil Spring

REMOVAL & INSTALLATION

→**For vehicles with MacPherson struts, see Strut Overhaul in this section.**

Rear

RAV4

1. Before servicing the vehicle, refer to the precautions in the beginning of this section.
2. Remove or disconnect the following:
   - Negative battery cable
   - Axle shaft, if equipped with 2WD
   - Halfshaft, if equipped with 4WD
   - Brake drum
   - Both brake line clamp bolts
   - Parking brake cable clamp bolt
   - Anti-lock Brake System (ABS) speed sensor and wiring harness, if equipped with ABS
   - Rear axle hub with the brake, by removing the 4 bolts
3. Support the hub securely.
4. Support the control arm with a floor jack.
5. Remove or disconnect the following:
   - Shock absorber from the control arm, by removing the bolt

→**The control arm must be supported before removing the bolt for the shock absorber. Leave the floor jack under the control arm. Later, the floor jack will be lowered to remove the coil spring.**

   - Cotter pins and nuts by supporting the lower and upper suspension arms
6. Disconnect the upper and lower control arms from the control arm, using tool 09628-62011
7. Remove the coil spring and control arm, as follows:
   - Matchmark the toe adjust cam and body.
   - Coil spring and upper insulator, by

loosening the bolt and lowering the control arm.
   - Bolt, toe-adjust cam, 2 attachments, nut and control arm
   - Bolt and spring bumper

To install:

8. Install or connect the following:
   - Spring bumper. Tighten the bolt to 108 inch lbs. (13 Nm).
   - Control arm, 2 attachments, toe-adjust cam, bolt and nut; do not tighten the bolt at this time
   - Spring and upper insulator
9. Raise the control arm with a floor jack.
10. Install or connect the following:
   - Upper and lower suspension arms to the control arm. Tighten the nuts to 76 ft. lbs. (103 Nm).
   - New cotter pins
   - Sock absorber to the control arm. Tighten the bolt to 27 ft. lbs. (37 Nm).
   - Rear axle hub with the brake. Tighten the 4 bolts to 59 ft. lbs. (80 Nm).
   - ABS speed sensor and wiring harness, if equipped. Tighten the ABS speed sensor to 69 inch lbs. (8 Nm) and the wiring harness to 108 inch lbs. (13 Nm).
   - Parking brake cable clamp. Tighten the bolt to 14 ft. lbs. (19 Nm).
   - Both brake line cable clamps. Tighten the bracket bolt to 13 ft. lbs. (18 Nm) and the clamp bolt to 108 inch lbs. (13 Nm).
   - Brake drum
   - Rear halfshaft, if equipped with 4WD
   - Axle shaft, if equipped with 2WD
   - Rear wheel
11. Lower the rear of the vehicle and stabilize the suspension.
12. Install or connect the following:
   - Align the matchmarks to the toe-adjust cam. Tighten the bolt to 98 ft. lbs. (132 Nm).
   - Negative battery cable
13. Check and/or adjust the wheel alignment

4RUNNER

1. Before servicing the vehicle, refer to the precautions in the beginning of this section.
2. Remove the wheel assemblies.
3. Support the axle housing with a floor jack.
4. Remove or disconnect the following:
   - Brake drum

- Parking brake cable from the brake shoe
- Parking brake cable from the axle housing.
5. Place matchmarks on the flanges for the driveshaft and differential.
6. Remove or disconnect the following:
- Driveshaft from the differential
- Brake hose line from the brake hose
- Brake hose-to-brake bracket clip
- Brake hose from the body
- Anti-lock Brake System (ABS) wiring harness bracket, if equipped with ABS
- Shock absorbers from the axle housing
- Lateral control rod nuts/bolts
- Control rod from the suspension
- Coil spring, by lower the rear axle housing

To install:
7. Install or connect the following:
- Coil springs into position and raise the axle housing.

※※ CAUTION

Be sure to fit the lower end of the coil spring into the gap of the spring seat on the lower control arm.

- Lateral control rod to the suspension. Tighten the bolts/nuts to 64 ft. lbs. (86 Nm).
- Shock absorbers to the axle housing. Tighten the bolt to 47 ft. lbs. (64 Nm).
- ABS wiring harness bracket, if equipped
- Brake hose to the bracket
- Clip
- Brake line to the brake hose. Tighten the tube.
- Parking brake cable bracket to the axle housing. Tighten to 108 inch lbs. (13 Nm).
- Parking brake cable to the brake shoes
- Driveshaft to the differential, by aligning the matchmarks. Tighten bolts/nuts to 54 ft. lbs. (73 Nm).
8. Fill the differential with the proper amount and type of oil.
9. Install the brake drum.
10. Bleed the brake system.
11. Install the wheel assemblies.
12. Lower the vehicle and bounce the vehicle several times to stabilize the suspension.
13. Tighten the lower control arm to 107 ft. lbs. (145 Nm).

Upper Ball Joint

REMOVAL & INSTALLATION

4Runner

1. Before servicing the vehicle, refer to the precautions in the beginning of this section.
2. Remove or disconnect the following:
- Front wheels
- Strut assembly
- Grease cap
3. If equipped with 4WD, disconnect the halfshaft, as follows:
- Cotter pin and lockcap
- Locknut, while applying the brakes
4. Remove or disconnect the following:
- Anti-lock Brake System (ABS) speed sensor and wiring harness clamp from the steering knuckle, if equipped with ABS
- Brake line bracket from the steering knuckle
- Front brake caliper and the rotor
- Lower ball joint
5. Remove the steering knuckle with the axle hub, as follows:
- Cotter pin and loosen the nut
- Steering knuckle from the upper control arm, using SST 09950-40010
- Steering knuckle
- Upper ball joint
- Wire and the boot
- Snapring
- Upper ball joint, using SST 09950-40010 and a deep socket wrench.

To install:
6. Install the upper ball joint, as follows:
- New ball joint with a new snapring
- New boot secured with a new wire
7. Install or connect the following:
- Steering knuckle with the axle hub to the upper control arm. Tighten the nut to 80 ft. lbs. (108 Nm).
- New cotter pin
- Lower ball joint. Tighten the 4 bolts to 59 ft. lbs. (80 Nm).
- Rotor and caliper. Tighten the caliper bolts to 90 ft. lbs. (123 Nm).
- Brake line bracket to the steering knuckle. Tighten it to 21 ft. lbs. (28 Nm).
- ABS speed sensor and wiring harness clamp to the steering knuckle, if equipped. Tighten the bolts to 72 inch lbs. (8 Nm).
- Halfshaft, if disconnected. Tighten the nut to 174 ft. lbs. (235 Nm).

- Grease cap
- Strut
- Front wheel
8. Check the alignment

Lower Ball Joint

REMOVAL & INSTALLATION

RAV4

1. Before servicing the vehicle, refer to the precautions in the beginning of this section.
2. Remove or disconnect the following:
- Negative battery cable
- Front wheel(s)
- Steering knuckle with the axle hub
- Dust deflector, by prying it from the knuckle
- Cotter pin and nut from the ball joint stud
- Lower ball joint from the steering knuckle, using a 2-jaw puller

To install:
3. Install or connect the following:
- Lower ball joint onto the steering knuckle. Tighten nut to 94 ft. lbs. (127 Nm).
- New cotter pin
- ABS speed sensor, by aligning it the dust deflector hole
- New dust deflector, using a driver
- Steering knuckle and hub
- Front wheel(s)
- Negative battery cable

Highlander

1. Before servicing the vehicle, refer to the precautions in the beginning of this section.
2. Remove or disconnect the following:
- Wheel
- Hub nut
- Caliper and rotor
- Lower control arm from the ball joint

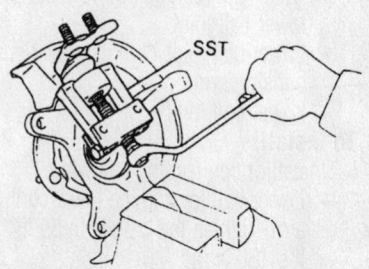

Use a 2-jaw puller to remove the lower ball joint—RAV4

- Tie rod end
- Halfshaft
- Ball joint from the knuckle

3. Installation is the reverse of removal. Observe the following torques:
- Ball stud nut: 90 ft. lbs. (123Nm)
- Lower arm-to-ball joint: 94 ft. lbs. (127Nm)
- Tie rod end: 36 ft. lbs. (49Nm)

RX300

1. Before servicing the vehicle, refer to the precautions in the beginning of this section.
2. Remove or disconnect the following:
- Wheel
- Steering knuckle with the axle hub
- Dust deflector, by prying it from the knuckle
- Cotter pin and nut from the ball joint
- Ball joint from the steering knuckle, by removing the 2 bolts
- Lower ball joint, using a Ball Joint Separator tool 09628-62011

To install:
3. Install or connect the following:
- Lower ball joint. Tighten the nut to 76 ft. lbs. (103 Nm) and both bolts to 94 ft. lbs. (127 Nm).
- New cotter pin
- Wheel

4Runner

1. Before servicing the vehicle, refer to the precautions in the beginning of this section.
2. Remove the wheel.
3. Disconnect the tie rod end, as follows:
- Loosen the 4 bolts
- Cotter pin and nut from the tie rod end
- Tie rod end from the steering knuckle, using SST 09610-20012
4. Remove the lower ball joint, as follows:
- Cotter pin and the nut from the lower ball joint
- Lower ball joint from the lower suspension arm
- Lower ball joint

To install:
5. Install or connect the following:
- Lower ball joint to the lower control arm. Tighten the 4 bolts to 59 ft. lbs. (80 Nm).
- Nut. Tighten it to 105 ft. lbs. (142 Nm).
- Tie rod end to the steering knuckle.

Tighten the nut to 66 ft. lbs. (90 Nm).
- Wheel

Upper Control Arm

REMOVAL & INSTALLATION

4Runner

4WD

1. Before servicing the vehicle, refer to the precautions in the beginning of this section.
2. Remove or disconnect the following:
- Shock and coil spring assembly
- Anti-lock Brake System (ABS) speed sensor wire harness clamp
3. Disconnect the upper ball joint, as follows:
- Cotter pins and loosen the nut
- Upper ball joint from the control arm, using a ball joint separator
- Support the steering knuckle
- Nut
4. Detach the control arm, by removing the nut, bolt, washers and lowering the arm.

To install:
5. Install or connect the following:
- Upper control arm with the washer, bolt and nut. Tighten the nut to 87 ft. lbs. (115 Nm).
- Upper ball joint to the control arm. Tighten the mounting nut to 80 ft. lbs. (105 Nm).
- New cotter pin
- ABS speed sensor wire harness clamp. Tighten it to 71 inch lbs. (8 Nm).
- Shock and coil spring assembly
6. Check and/or adjust the alignment.

CONTROL ARM BUSHING REPLACEMENT

4Runner

1. Before servicing the vehicle, refer to the precautions in the beginning of this section.
2. Remove the upper control arm from the vehicle.
3. Pry up the bushing flange, using a chisel and a hammer.
4. Using tools 09613-26010, 09613-20060 and 09950-00020 and a shop press, remove the bushing.

To install:
5. Using tools 09223-00010, 09506-35010 and a shop press, press the new bushing into the upper control arm.
6. Install the upper control arm to the

chassis. Torque the nuts to 87 ft. lbs. (115 Nm).
7. Check and/or adjust the alignment.

Lower Control Arm

REMOVAL & INSTALLATION

4Runner

2WD

1. Before servicing the vehicle, refer to the precautions in the beginning of this section.
2. Remove or disconnect the following:
- Front wheel
- Steering gear assembly
- Stabilizer bar link
- Shock absorber from the lower control arm
3. Support the upper control arm and the steering knuckle securely.
4. Remove or disconnect the following:
- Cotter pin and nut from the lower ball joint
- Lower ball joint from the control arm
- Bolts, nuts, adjusting cams and lower control arm, by placing matchmarks on the front and rear adjusting cams
- 2 spring bumpers, using tool SST 09922-10010

To install:
5. Install or connect the following:
- 2 spring bumpers. Tighten the spring bumpers to 17 ft. lbs. (23 Nm).
- Lower control arm and adjusting cams. Tighten the nuts and bolts to 96 ft. lbs. (130 Nm),
- Lower ball joint to the control arm
- Cotter pin and nut to the lower ball joint. Tighten the nut to 105 ft. lbs. (142 Nm).
- Shock absorber to the lower control arm
- Stabilizer bar link
- Steering gear assembly
- Front wheel
6. Check and/or adjust the alignment.

4WD

1. Before servicing the vehicle, refer to the precautions in the beginning of this section.
2. Remove or disconnect the following:
- Front wheel
- Steering gear assembly
- Stabilizer bar link
- Shock absorber from lower control arm

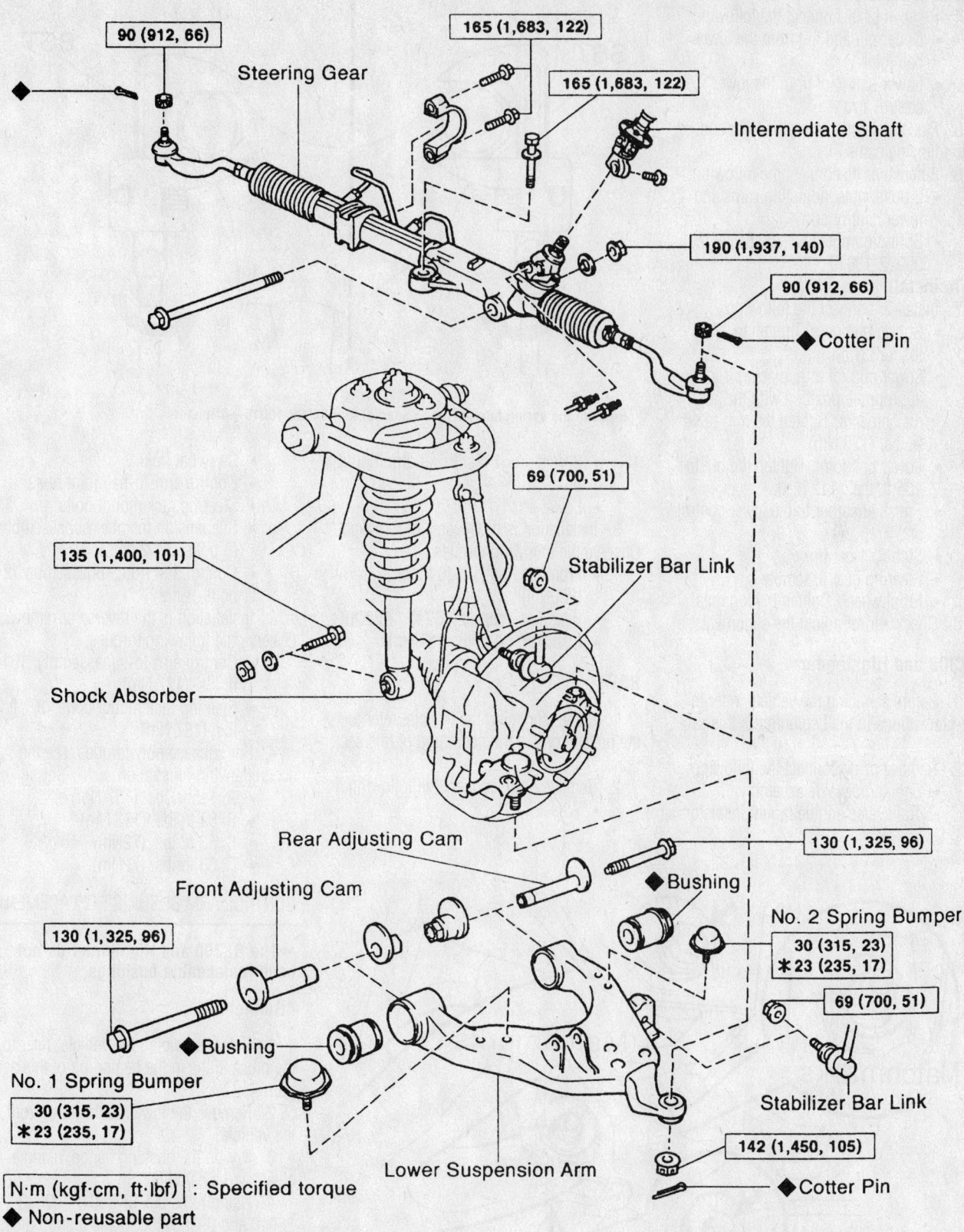

90 (912, 66)

Steering Gear

165 (1,683, 122)

165 (1,683, 122)

Intermediate Shaft

190 (1,937, 140)

90 (912, 66)

◆ Cotter Pin

69 (700, 51)

135 (1,400, 101)

Stabilizer Bar Link

Shock Absorber

130 (1, 325, 96)

Rear Adjusting Cam

◆ Bushing

No. 2 Spring Bumper

Front Adjusting Cam

30 (315, 23)
✳ 23 (235, 17)

130 (1, 325, 96)

69 (700, 51)

◆ Bushing

Stabilizer Bar Link

No. 1 Spring Bumper

30 (315, 23)
✳ 23 (235, 17)

142 (1,450, 105)

Lower Suspension Arm

◆ Cotter Pin

N·m (kgf·cm, ft·lbf) : Specified torque

◆ Non-reusable part

✳ For use with SST

86828G20

Exploded view of the lower control arm and associated components—4Runner

3. Support the upper control and steering knuckle securely.

4. Remove or disconnect the following:
- Cotter pin and nut from the lower ball joint
- Lower ball joint from the lower control arm

5. Place matchmarks on the front and rear adjusting cams.

6. Remove or disconnect the following:
- 2 bolts, nuts, adjusting cams and lower control arm
- Spring bumpers with a special tool 09922-10010.

To install:

7. Install or connect the following:
- Spring bumpers. Tighten to 17 ft. lbs. (23 Nm).
- Lower control arm, placing it in the appropriate position with the matchmarks. Tighten the arm to 96 ft. lbs. (130 Nm).
- Lower ball joint. Tighten the nut to 105 ft. lbs. (142 Nm).
- Shock absorber to the lower control arm
- Stabilizer bar link
- Steering gear assembly
- Front wheel. Tighten the lug nuts.

8. Check and/or adjust the alignment.

RX300 and Highlander

1. Before servicing the vehicle, refer to the precautions in the beginning of this section.

2. Remove or disconnect the following:
- Engine/transaxle assembly
- Transverse engine mount insulator

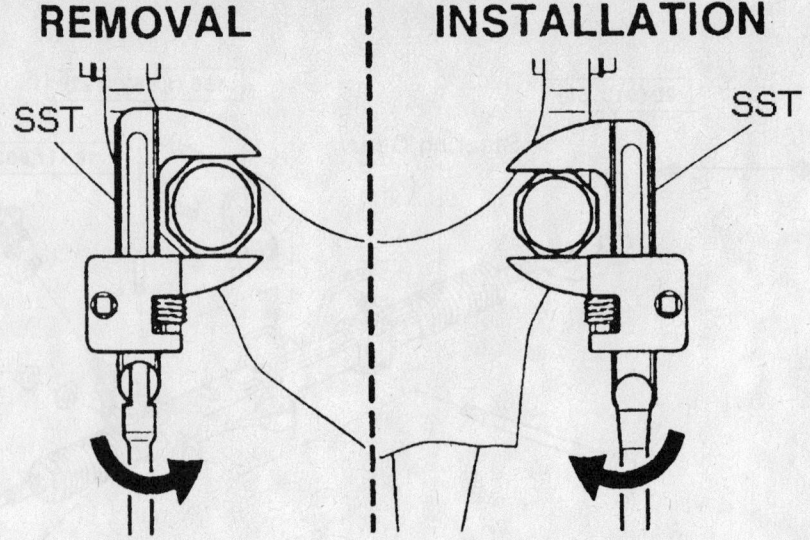

REMOVAL | **INSTALLATION**

SST

86828G22

Remove the spring bumpers with special tool 09922-10010—4Runner

- 2 front and 1 rear lower arm mount bolts
- Lower arm

3. Installation is the reverse of removal. Observe the following torques:
- Front side bolts: 148 ft. lbs. (200Nm)
- Rear arm bolt/nut: 152 ft. (206Nm)
- Insulator: 64 ft. lbs. (87Nm)

RAV4

1. Before servicing the vehicle, refer to the precautions in the beginning of this section.

2. Remove or disconnect the following:
- Wheel

- Sway bar link
- Control arm-to-ball joint bolts
- Steering rack mount bolts
- Suspension member subassembly (5 bolts and 2 nuts)
- Control arm from subassembly (2 bolts, 1 nut)

3. Installation is the reverse of removal. Observe the following torques:
- Control arm-to-subassembly: 101 ft. lbs. (137 Nm)
- Steering rack mount bolts: 101 ft. lbs. (137 Nm)

4. For subassembly torques, see the accompanying illustration.
- A: 115 ft. lbs. (157 Nm)
- B: 82 ft. lbs. (113 Nm)
- C: 53 ft. lbs. (72Nm)
- D: 53 ft. lbs. (72Nm)

CONTROL ARM BUSHING REPLACEMENT

➡The RX300 and Highlander do not have replaceable bushings.

4Runner

1. Before servicing the vehicle, refer to the precautions in the beginning of this section.

2. Remove the lower control arm from the vehicle.

3. Pry up the bushing flange, using a chisel and a hammer.

4. Using tools 09613-26010, 09632-36010 and 09950-00020 and a shop press, remove the bushing.

To install:

5. Using tools 09316-20011, 09710-30021 and a shop press, press the new bushing into the lower control arm.

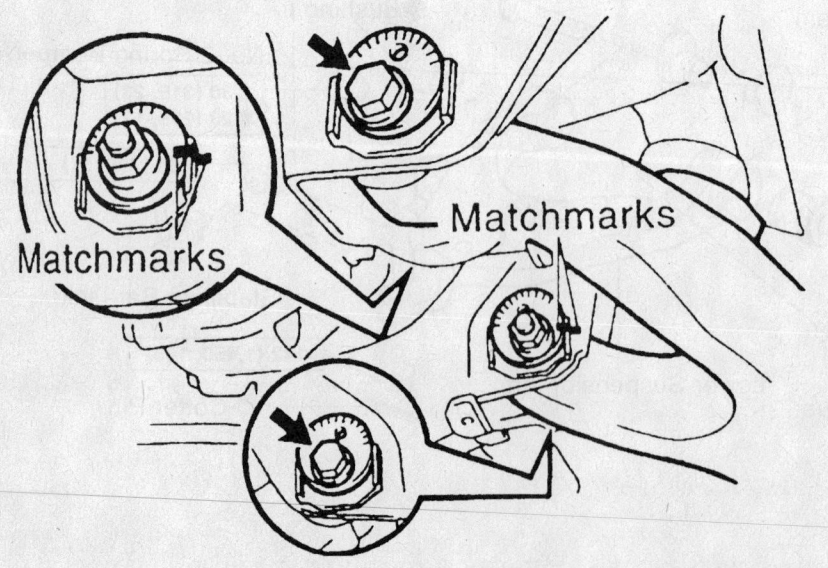

Matchmarks

Matchmarks

Matchmarks

86828G21

Place matchmarks on the front and rear adjusting cams—4Runner

RH Side

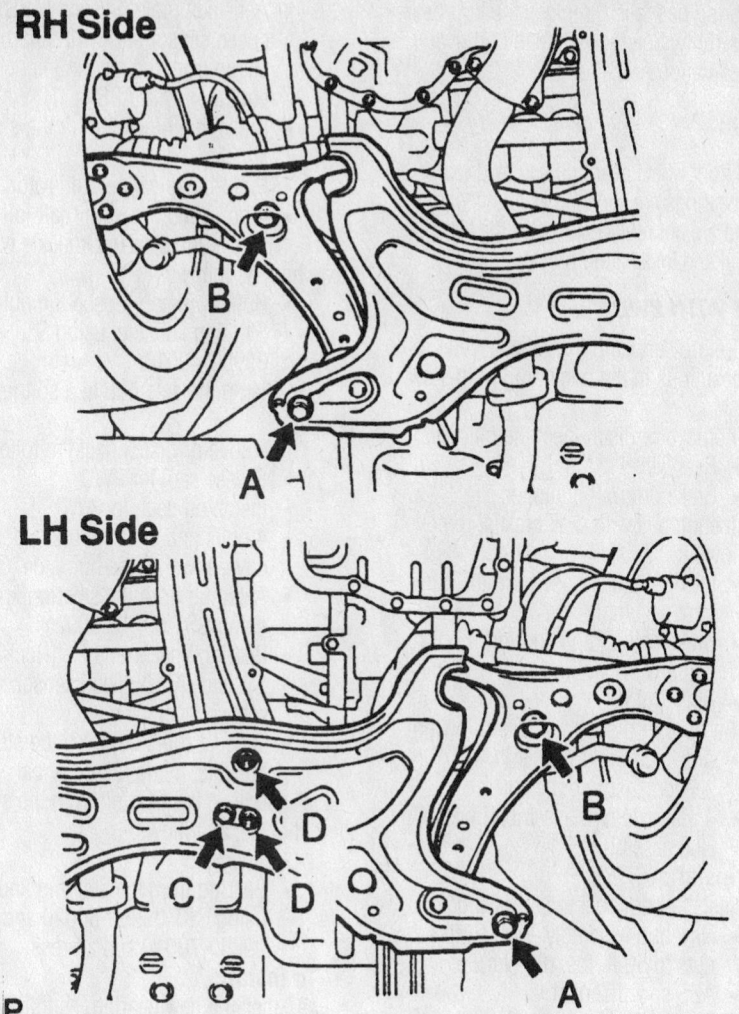

LH Side

9355YG25

RAV4 suspension subassembly bolt torques

6. Install the lower control arm to the chassis. Torque the bolts to 96 ft. lbs. (130 Nm).

7. Check and/or adjust the alignment.

RAV4

1. Matchmark the control arm with the triangle mark on the bushing.
2. Press out the old bushing.
3. Press in the new bushing, aligning the matchmarks.

Wheel Bearing

ADJUSTMENT

RAV4 and RX300

FRONT AND REAR

Check the bearing play in the axial direction and also check the axle hub deviation. The maximum play for both checks should be 0.0020 in. (0.05mm). If greater than the specified maximum, replace the bearing. The wheel bearing is not adjustable.

4Runner

The wheel bearings are sealed unit; no adjustment is possible.

REMOVAL & INSTALLATION

RAV4

FRONT

1. Before servicing the vehicle, refer to the precautions in the beginning of this section.
2. Remove or disconnect the following:
- Negative battery cable
- Front wheels
- Cotter pin and lockcap from the halfshaft end
- Halfshaft locknut, by applying the front brakes
- Brake caliper and support it on a wire
3. Matchmark the rotor to the hub.
4. Remove or disconnect the following:
- Rotor
- Anti-lock Brake System (ABS) speed sensor from the steering knuckle, if equipped
- Loosen the strut's lower end nuts
- Tie rod end from the steering knuckle
- Lower control arm from the ball joint, by removing the bolt and 2 nuts
- Halfshaft from the axle hub

➡ **Secure the halfshaft aside using a wire. Be careful not to damage the shaft boot or ABS sensor rotor.**

- Both strut's lower end nuts
- Steering knuckle
5. Clamp the steering knuckle in a vise with soft jaws to protect the knuckle.
6. Remove or disconnect the following:
- Dust deflector, by prying it from the hub
- Ball joint from the steering knuckle
- Hub from the knuckle, using slide hammer
- Inner race from the hub, using press and arbor tool
- 4 bolts and the dust cover
- Inner oil seal, using Seal Removal tool 09308-00010
- Outer oil seal, using Seal Removal tool 09308-00010
- Snapring
7. Install inner race (removed from the hub) on the outside of the bearing
8. Remove the steering knuckle bearing, using a bearing driver

To install:

9. Clean bearing seating surfaces with a clean, dry rag.
10. Install or connect the following:
- Bearing into the knuckle, using a press and Bearing Installer tool 09608-32010
- Snapring
- Dust cover. Tighten the 4 bolts to 74 inch lbs. (8 Nm).
- New outer oil seal, using a seal driver

➡ **Apply multi-purpose grease to the oil seal lip.**

- Hub into the steering knuckle
- New inner oil seal, using a seal driver

➡ **Apply multi-purpose grease to the oil seal lip.**

- Lower ball joint to the steering knuckle. Tighten the nut to 94 ft. lbs. (127 Nm).
- New cotter pin
- Dust deflector, by aligning it with the ABS speed sensor hole
- Knuckle to the lower strut and install the bolts
- Lower ball joint to the lower arm. Tighten the bolts to 94 ft. lbs. (127 Nm).
- Tie rod end to the steering knuckle. Tighten the nut to 36 ft. lbs. (49 Nm).
- Halfshaft to the hub and knuckle
- Tighten the lower strut nuts to 117 ft. lbs. (158 Nm).
- ABS speed sensor. Tighten the bolt to 69 inch lbs. (8 Nm).
- Rotor to the hub, by aligning the matchmark
- Brake caliper. Tighten the mounting bolts to 79 ft. lbs. (107 Nm).
- Axle locknut, using an assistant to apply the brakes. Tighten the nut to 159 ft. lbs. (216 Nm).
- Lockcap and a new cotter pin
- Wheel
- Negative battery cable

11. Turn the wheel by hand, verify that the wheel turns without noise and without binding.

12. Check the signal from the ABS sensor.

REAR

The wheel bearings are not replaceable. If the bearing has failed, replace the hub/bearing assembly. See the Halfshaft Removal and Installation procedure.

Highlander
FRONT

1. Before servicing the vehicle, refer to the precautions in the beginning of this section.

2. Remove the hub/knuckle assembly. See the Halfshaft Removal and Installation procedure. Mount the assembly in a vise.

3. Remove or disconnect the following:
- Dust deflector
- Snapring
- Hub from the spindle and mount the hub in a vise.
- Dust cover

4. Press out the bearings, inner bearing first.

5. Press in the new bearings.

6. Installation is the reverse of removal.

REAR

The wheel bearings are not replaceable. If the bearing has failed, replace the hub/bearing assembly. See the Halfshaft Removal and Installation procedure.

RX 300

The front wheel bearings and 4wd rear wheel bearings are serviced as a unit with the hub and are not replaceable. See the Halfshaft Removal and Installation procedure.

REAR WITH 2WD

1. Before servicing the vehicle, refer to the precautions in the beginning of this section.

2. Remove or disconnect the following:
- Rear wheel
- Brake drum, if equipped

3. If equipped with disk brakes, remove the following:
- Flexible brake hose from the rear strut assembly
- Brake caliper and support it on using wire
- Brake rotor

4. Remove or disconnect the following:
- Anti-lock Brake System (ABS) speed sensor connector
- 4 rear axle hub assembly nuts
- Hub assembly

To install:

5. Install or connect the following:
- New hub assembly. Tighten the nuts to 59 ft. lbs. (80 Nm).
- ABS speed sensor

6. If equipped with disc brakes, install the following:
- Brake rotor
- Brake caliper. Tighten the mounting bolts to 34 ft. lbs. (47 Nm).
- Flexible brake hose to the rear strut assembly. Tighten the mounting bolt to 21 ft. lbs. (29 Nm).

7. Install or connect the following:
- Brake drum, if equipped
- Wheels

8. Test drive the vehicle.

4Runner

1. Before servicing the vehicle, refer to the precautions in the beginning of this section.

2. Remove or disconnect the following:
- Front wheels
- Shock absorber
- Grease cap

3. On 4WD, remove the halfshaft, as follows:
- Cotter pin and lockcap
- Locknut, while applying the brakes

4. Remove or disconnect the following:

- Anti-lock Brake System (ABS) speed sensor and wiring harness clamp from the steering knuckle, if equipped with ABS
- Brake line bracket from the steering knuckle
- Front brake caliper and rotor
- 4 bolts and the lower ball joint

5. Remove the steering knuckle with the axle hub, as follows:
- Cotter pin and loosen the nut
- Steering knuckle, using SST 09950-40010

6. Clamp the axle hub in a soft jaw vise.

7. Remove or disconnect the following:
- Grease cap, for 2WD
- Inside oil seal, for 4WD
- 4 bolts and shift the brake dust cover towards the hub side
- Axle hub from the steering knuckle, using SST 09710-30021
- Bearing spacer and Anti-lock Brake System (ABS) speed sensor rotor/spacer
- Oil seal (outside) from the steering knuckle, using a flat pry bar

8. Remove the bearing from the steering knuckle, as follows:
- Snapring
- Bearing from the steering knuckle, using SST 09950-60020 and 09950-70010 and a press

To install:

9. Install a new bearing, as follows:
- New bearing to the steering knuckle, using SST 09527-17011 and 09950-60020 and a press
- New snapring

10. Install or connect the following:
- New outside oil seal, using SST 09223-15030 and a plastic hammer. Coat MP grease to the oil seal lip.
- Brake dust cover to the steering knuckle. Tighten the 4 bolts to 13 ft. lbs. (18 Nm).
- Axle hub to the steering knuckle, using a press
- ABS speed sensor rotor/spacer.

✳✳ WARNING

Be careful not to scratch the serration of the speed sensor rotor.

- Bearing spacer, using a press
- Grease cap, if removed
- New inside oil seal, if removed, using SST 09527-17011 and a plastic hammer
- Steering knuckle with the axle hub.

Tighten the nut to 80 ft. lbs. (108 Nm).
- New cotter pin
- Lower ball joint. Tighten the 4 bolts to 59 ft. lbs. (80 Nm).
- Rotor and the caliper. Tighten the caliper bolts to 90 ft. lbs. (123 Nm).

- Brake line bracket to the steering knuckle. Tighten the fasteners to 21 ft. lbs. (28 Nm).
- ABS speed sensor and wiring harness clamp to the steering knuckle, if removed. Tighten the bolts to 72 inch lbs. (8 Nm).

- Halfshaft, if disconnected. Tighten the nut to 174 ft. lbs. (235 Nm).
- Grease cap
- Shock absorber
- Front wheel
- Negative battery cable

BRAKES

Brake Caliper

REMOVAL & INSTALLATION

Highlander

FRONT

1. Before servicing the vehicle, refer to the precautions in the beginning of this section.

2. Remove or disconnect the following:
   - Front wheel
   - 2 mounting bolts and caliper
3. If the caliper is being replaced, disconnect the brake line and plug both openings.

➡ **Depending on the brake type, there may be either 1 or 2 sealing washers.**

4. Installation is the reverse of removal.

Torque the mounting bolts to 25 ft. lbs. (34Nm). If the brake hose was removed, torque the union bolt to 21 ft. lbs. (29Nm).

REAR

1. Before servicing the vehicle, refer to the precautions in the beginning of this section.

2. Remove or disconnect the following:
   - Rear wheel

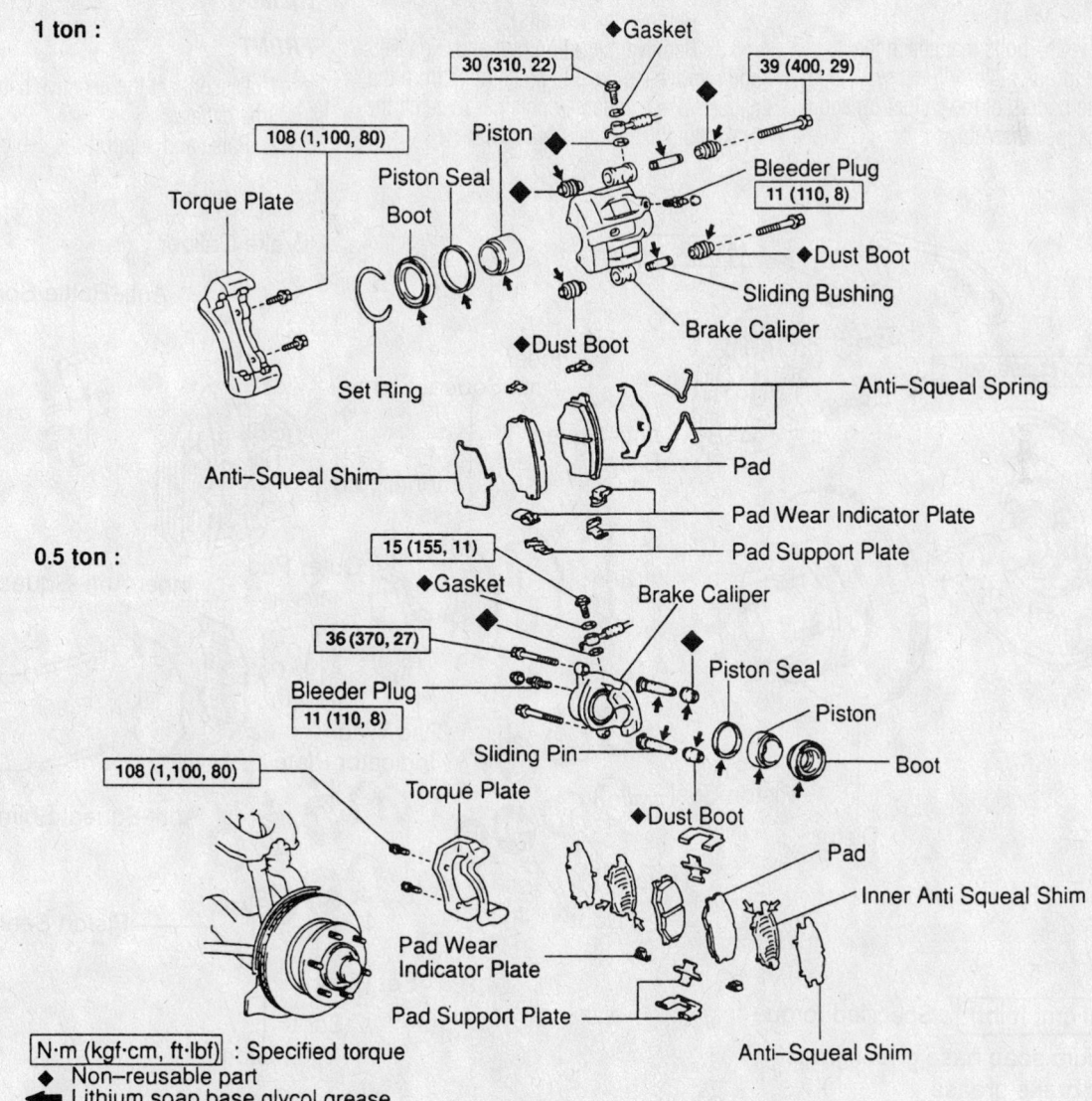

1 ton :

◆Gasket
30 (310, 22)   39 (400, 29)
108 (1,100, 80)
Piston
Torque Plate
Piston Seal
Boot   Bleeder Plug
              11 (110, 8)
              ◆Dust Boot
              Sliding Bushing
              Brake Caliper
Set Ring   ◆Dust Boot
              Anti–Squeal Spring
Anti–Squeal Shim   Pad
              Pad Wear Indicator Plate
              Pad Support Plate

0.5 ton :
15 (155, 11)
◆Gasket   Brake Caliper
36 (370, 27)   Piston Seal
Bleeder Plug   Piston
11 (110, 8)   Boot
              Sliding Pin
108 (1,100, 80)   ◆Dust Boot
              Torque Plate   Pad
              Inner Anti Squeal Shim
Pad Wear Indicator Plate
Pad Support Plate
              Anti–Squeal Shim

N·m (kgf·cm, ft·lbf) : Specified torque
◆ Non–reusable part
◀ Lithium soap base glycol grease
◁ Disc brake grease

93026G70

Single piston type caliper assembly—4Runner

• Slide pin and caliper

3. If the caliper is being replaced, disconnect the brake line and plug both openings.

→ **Depending on the brake type, there may be either 1 or 2 sealing washers.**

4. Installation is the reverse of removal. Torque the slide pin to 25 ft. lbs. (34Nm). If the brake hose was removed, torque the union bolt to 21 ft. lbs. (29Nm).

4Runner

1. Disconnect the negative battery cable from the battery.
2. Raise and support the vehicle safely.
3. Remove the wheels.
4. Disconnect the brake hose from the caliper by removing the union bolt and 2 gaskets. Plug the end of the hose to prevent loss of fluid.
5. Remove the bolts that attach the caliper to the torque plate.
6. Lift the bottom of the caliper up and remove the caliper assembly.

To install:

7. Grease the caliper slides and bolts with lithium grease or equivalent. Install the caliper and secure with the bolts. Torque the bolts to 90 ft. lbs. (123 Nm).
8. Connect the brake hose to the caliper, using 2 new washers. Make sure the flexible hose lock is securely in the lock hole of the caliper. Torque the union bolt to 11 ft. lbs. (15 Nm).
9. Fill the brake system to the proper level and bleed the brake system.
10. Install the tire and wheel assembly.
11. Top off the brake fluid level in the master cylinder. Check for leaks and proper brake operation.
12. Connect the negative battery cable to the battery.

RAV4

1. Raise and safely support the vehicle.
2. Remove the wheel(s).
3. Remove the union bolt and 2 gaskets and remove the flexible brake hose from the caliper. Use a suitable container to catch the brake fluid as it drains out.

4. Hold the sliding pin and loosen the 2 caliper mounting bolts. Remove the bolts and remove the caliper from the torque plate.
5. Remove the brake pads and brake hardware.

To install:

6. Install the brake pads and brake hardware.
7. Install the caliper to the torque plate with the 2 mounting bolts. Torque the bolts to 25 ft. lbs. (34 Nm).
8. Reconnect the flexible brake hose to the caliper with the 2 gaskets and the union bolt. Torque the union bolt to 22 ft. lbs. (30 Nm).
9. Refill the master cylinder with brake fluid and bleed the brake system.
10. Check for proper operation and make sure there are no leaks.

RX300

FRONT

1. Disconnect the negative battery cable from the battery.
2. Raise and support the vehicle safely.

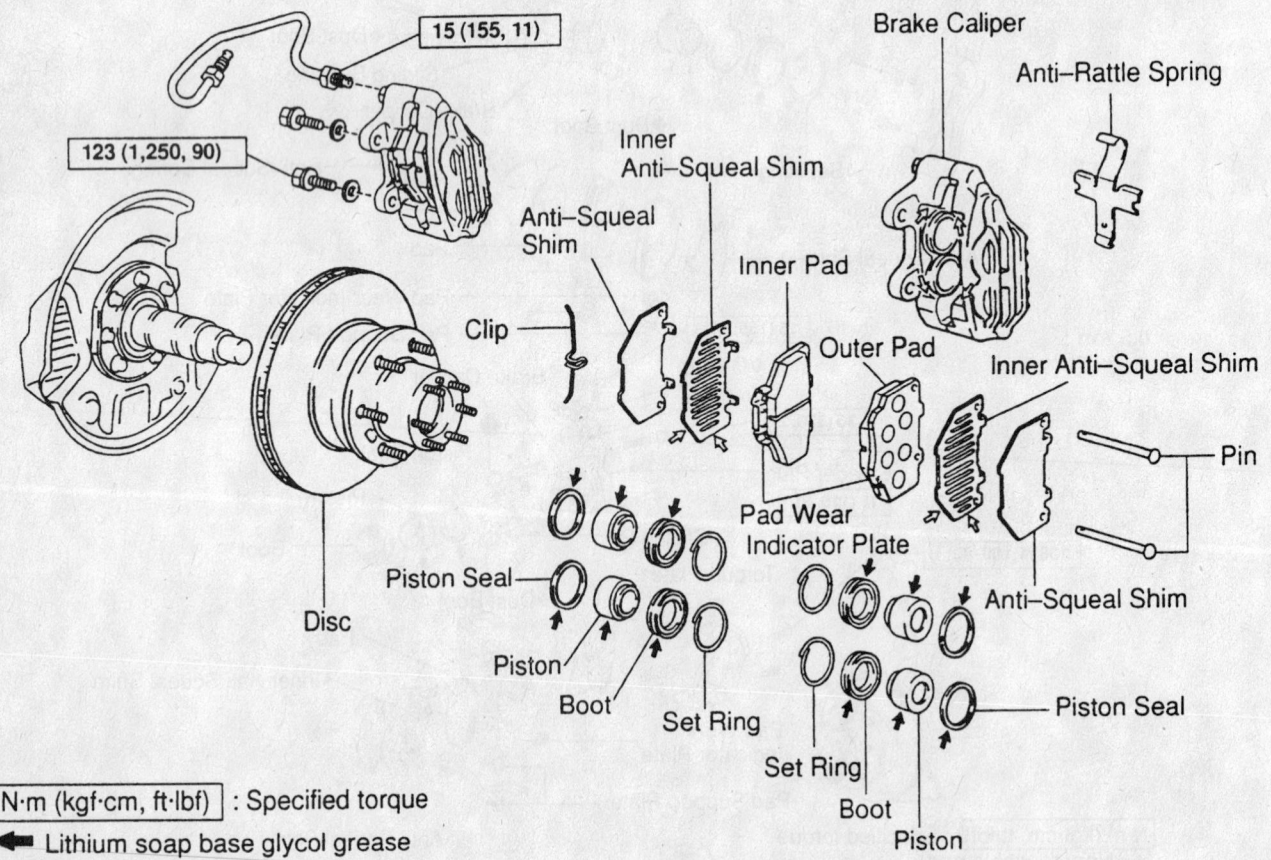

15 (155, 11)

123 (1,250, 90)

Disc

Clip

Anti–Squeal Shim

Inner Anti–Squeal Shim

Inner Pad

Piston Seal

Piston

Boot

Set Ring

Brake Caliper

Anti–Rattle Spring

Outer Pad

Pad Wear Indicator Plate

Inner Anti–Squeal Shim

Pin

Anti–Squeal Shim

Piston Seal

Set Ring

Boot

Piston

N·m (kgf·cm, ft·lbf) : Specified torque

◀ Lithium soap base glycol grease

⇐ Disc brake grease

93026G71

Dual piston type caliper assembly—4Runner

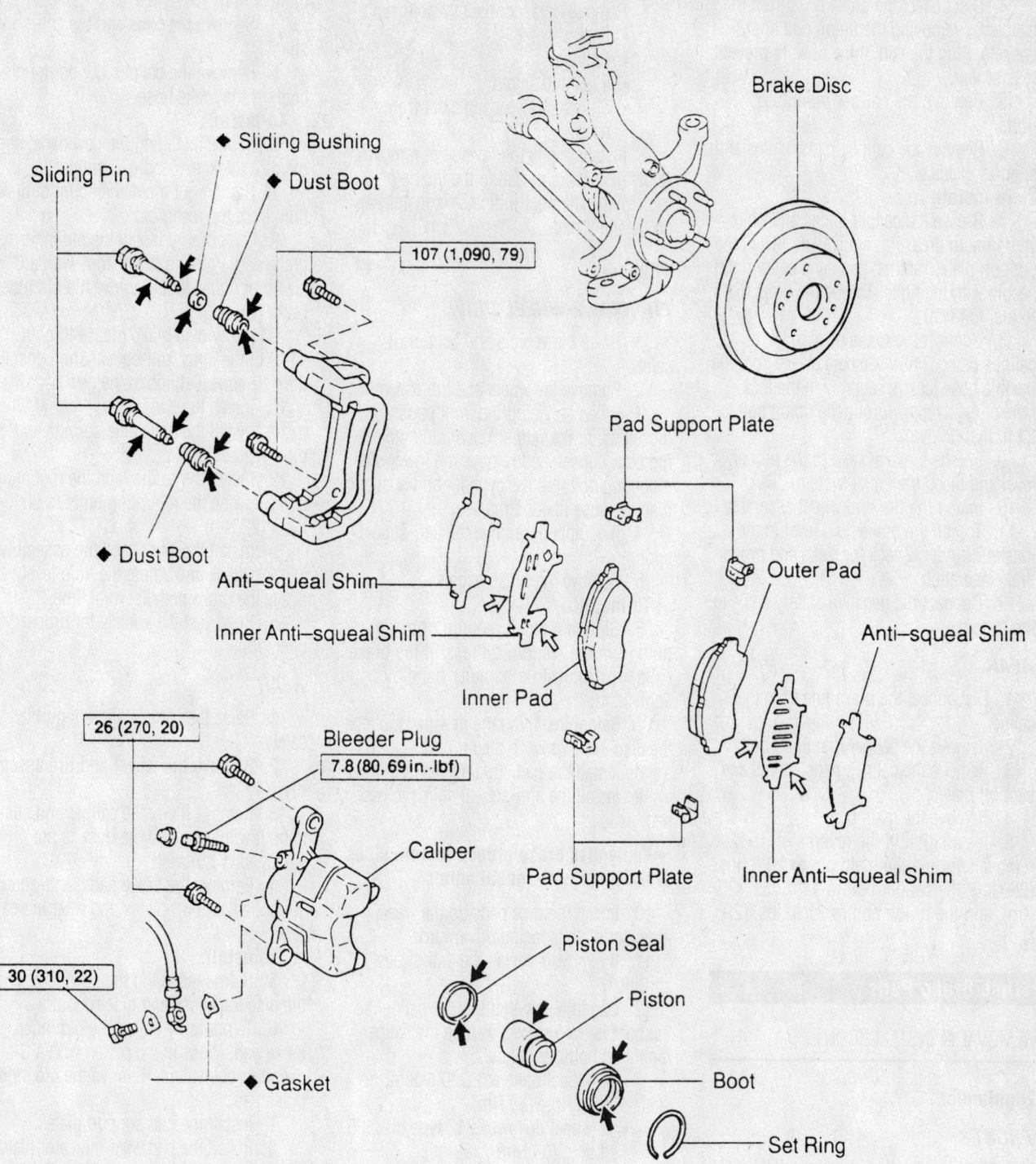

Sliding Pin

◆ Sliding Bushing

◆ Dust Boot

Brake Disc

107 (1,090, 79)

Pad Support Plate

Outer Pad

Anti–squeal Shim

◆ Dust Boot

Anti–squeal Shim

Inner Anti–squeal Shim

Inner Pad

Pad Support Plate

Inner Anti–squeal Shim

26 (270, 20)

Bleeder Plug

7.8 (80, 69 in.·lbf)

Caliper

Piston Seal

Piston

30 (310, 22)

Boot

◆ Gasket

Set Ring

N·m (kgf·cm, ft·lbf) : Specified torque

◆ Non–reusable part

➡ Lithium soap base glycol grease

⇨ Disc brake grease

93026G73

Exploded view of the front brake components—RAV4

3. Remove the wheels.

4. Disconnect the brake hose from the caliper by removing the union bolt and 2 gaskets. Plug the end of the hose to prevent loss of fluid.

5. Remove the caliper mounting bolts.

6. Remove the caliper, pads, shims and support plates.

To install:

7. Grease the caliper slides and bolts with lithium grease or equivalent. Install the support plates, shims, pads and caliper and secure with the bolts. Torque the bolts to 25 ft. lbs. (34 Nm).

8. Connect the brake hose to the caliper, using 2 new washers. Make sure the flexible hose lock is securely in the lock hole of the caliper. Torque the union bolt to 21 ft. lbs. (29 Nm).

9. Fill the brake system to the proper level and bleed the brake system.

10. Install the tire and wheel assembly.

11. Top off the brake fluid level in the master cylinder. Check for leaks and proper brake operation.

12. Connect the negative battery cable to the battery.

REAR

1. Disconnect the brake line from the caliper.

2. Remove the caliper mounting bolt.

3. Remove the caliper, pads, shims and support plates.

4. Remove the main pin.

5. Installation is the reverse of removal. Torque the main pin to 20 ft. lbs. (26 Nm), the caliper bolt to 14 ft. lbs. (20 Nm), and the union bolt to 21 ft. lbs. (29 Nm).

Disc Brake Pads

REMOVAL & INSTALLATION

Highlander

FRONT

1. Remove or disconnect the following:
- Front wheel
- 2 mounting bolts and caliper
- Pads and shims
- Shims and wear indicator from the pads

2. Installation is the reverse of removal. Apply disc brake grease to the inside of each shim. The wear indicator is installed on the inner pad.

REAR

1. Remove or disconnect the following:
- Rear wheel
- Caliper
- Pads and shims
- Shims and wear indicator from the pads

2. Installation is the reverse of removal. Apply disc brake grease to the inside of each shim. The wear indicator is installed on the inner pad.

4Runner

FRONT W/2-WHEEL DRIVE

1. Raise the vehicle and support it safely.

2. Remove the wheel and tire assembly.

3. When servicing the front pads, loosen the brake caliper upper side mounting bolt. Loosen and remove the lower side mounting bolt. Lift the cylinder and suspend it so the hose is not stretched.

4. If equipped, remove the anti-squeal spring.

5. Remove the brake pads.

To install:

6. Siphon a small amount of brake fluid from the reservoir. Press in the brake caliper piston with a hammer handle or equivalent.

7. Before installing the new pads, check the disc thickness and disc runout.

8. Install the pad support plates.

9. Install the anti-squeal shims to each pad.

➡ Apply disc brake grease to both sides of the inner anti-squeal shims.

10. Install the disc pads so the wear indicator plate is facing downward.

11. If removed, install the anti-squeal springs.

12. Carefully install the brake caliper so the boot is not wedged. Torque the caliper mounting bolts, as follows:
- 2-Wheel drive w/PD60 type disc: 29 ft. lbs. (39 Nm)
- 2-wheel drive w/FS17 type disc: 65 ft. lbs. (88 Nm)

13. Install the wheel and tire assembly.

14. Check and adjust the fluid level. Apply the brake pedal several times.

15. Road-test the vehicle for proper operation.

FRONT W/4-WHEEL DRIVE

1. Raise the vehicle and support it safely.

2. Remove the wheel and tire assembly.

3. Remove the clip, pins, and the anti-rattle spring.

4. Remove the pads and the anti-squeal shims.

5. Remove the caliper, but do not disconnect the brake hose.

To install:

6. Before installing the new pads, check the disc thickness and disc runout.

7. Siphon out a small amount of brake fluid from the reservoir.

8. Temporarily install the old inner brake pad. Press in the pistons with a C-clamp or equivalent. Remove the old inner brake pad.

9. Apply disc brake grease to both sides of the inner anti-squeal shim. Install the anti-squeal shims to the new pads.

10. Install the pads.

11. Install the anti-rattle springs and pins. Install the clip.

12. Install the caliper and the mounting bolts. Torque the mounting bolts to 90 ft. lbs. (123 Nm).

13. Install the wheel and tire assembly.

14. Check and adjust the fluid level. Apply the brake pedal several times.

15. Road-test the vehicle for proper operation.

REAR

1. Raise the vehicle and support it safely.

2. Remove the wheel and tire assembly.

3. Remove the brake caliper and suspend it with a wire so the hose is not stretched or stressed.

4. Remove the brake pads, anti-squeal shim, pad support plates and wear indicators.

To install:

5. Before installing the new pads, check the disc thickness and disc runout.

6. Temporarily install the old inner brake pad. Press in the piston with a C-clamp or equivalent. Remove the old inner brake pad.

7. Install the pad support plates.

8. Install the pad wear indicator plate to each pads.

9. Install the anti-squeal shim to the outer pad. Install the pads so the wear indicator plate is facing upward.

10. Install the brake caliper. Torque the main sliding pin and the sub pin to 65 ft. lbs. (88 Nm).

11. Install the wheel and tire assembly.

12. Apply the brake pedal several times.

13. Road-test the vehicle for proper operation.

RAV4

1. Raise and safely support the vehicle.
2. Remove the wheel(s).
3. Temporarily install 2 wheel stud nuts to hold the brake rotor in place.
4. If necessary, siphon a sufficient quantity of brake fluid from the master cylinder reservoir to prevent any brake fluid from overflowing the master cylinder when removing or installing new pads. This may be necessary, as the piston must be forced into the caliper bore to provide sufficient clearance when installing the pads.
5. Grasp the caliper from behind and carefully pull it towards you. This will start to seat the piston(s) in its bore. Using a C-clamp or other suitable tool, press the piston the remaining way into the caliper. Be careful not to cock the piston in the bore. Also, do not force the piston or the caliper and piston may be damaged.
6. Hold the sliding pin and loosen the 2 caliper mounting bolts. Remove the bolts and remove the caliper from the torque plate.
7. Secure the caliper assembly out of the way with a wire; so as not to stress the flexible hose.
8. Slide out the old brake pads along with any anti-squeal shims, springs, pad wear indicators and pad support plates. Make sure to note the position of all assorted pad hardware.
   To install:
9. Check the brake disc (rotor) for thickness and run-out. Inspect the caliper and piston assembly for breaks, cracks, fluid seepage or other damage. Overhaul or replace as necessary.
10. Install the pad support plates into the torque plate.
11. Install the pad wear indicators onto the pads. Be sure the arrow on the indicator plate is pointing in the direction of rotation.
12. Install the anti-squeal shims on the outside of each pad and then install the pad assemblies into the torque plate.
13. Install the caliper to the torque plate with the 2 mounting bolts. Torque the bolts to 20 ft. lbs. (26 Nm).
14. Remove the 2 temporary wheel stud nuts and check that the rotor turns freely.
15. Reinstall the wheel(s), safely lower the vehicle, and road-test for proper brake operation.
16. Be sure to pump the brakes several times prior to moving the vehicle.

RX300

FRONT

1. Hold the sliding pin and remove the lower bolt.
2. Lift the caliper up and secure it.
3. Remove the pads, 4 shims and wear indicator plate. Remove the 2 pad support plates.

➡**The support plates can be reused, provided they have sufficient rebound, are not deformed or cracked, show no signs of wear and are cleaned of all rust and debris.**

   To install:

➡**Always use new shims and wear indicators, even when re-installing the original pads.**

4. Install a wear indicator plate on the inner pad.
5. Apply disc brake grease to both sides of the inner anti-squeal shims and install the shims.
6. Install the inner pad with the wear indicator plate facing upwards.
7. Install the outer pad.
8. Install the caliper. Torque the bolt to 25 ft. lbs.

REAR

1. On 2WD, unbolt the brake hose from the shock absorber.
2. Remove the caliper installation bolt from the torque plate.
3. Lift the caliper up and secure it.
4. Remove the pads, 4 shims and 4 support plates.

➡**The support plates can be reused, provided they have sufficient rebound, are not deformed or cracked, show no signs of wear and are cleaned of all rust and debris.**

   To install:

➡**Always use new shims and wear indicators, even when re-installing the original pads.**

5. Install the support plates.
6. Apply disc brake grease to both sides of the anti-squeal shims.
7. Install the anti-squeal shims.
8. Install the inner pad with the wear indicator plate facing upwards.
9. Install the outer pad.
10. Install the caliper. Torque the bolt to 14 ft. lbs. (20 Nm).
11. On 2WD, connect the line to the shock absorber. Torque the bolt to 21 ft. lbs. (29 Nm).

Brake Drums

REMOVAL & INSTALLATION

4Runner

1. Raise and safely support the vehicle.
2. Remove the rear wheel(s).
3. Remove the brake drum from the axle hub. If there is difficulty in removing the drum, insert a suitable tool through the hole in the rear of the backing plate, and hold the automatic adjusting lever away from the adjuster. Using another suitable tool at the same time, reduce the brake shoe adjuster by turning the adjusting wheel.
   To install:
4. Install the brake drum and pull the parking brake lever all the way up until a clicking sound can no longer be heard.
5. Verify that the rear wheels will not turn. If the rear wheels turn, adjust the parking brake cable as necessary.
6. Release the parking brake and remove the brake drum. Measure the brake drum inside diameter and diameter of the brake shoes. Check that the difference between the diameters is the correct shoe clearance. Clearance is 0.024 in. (6mm).
7. If the brake shoe clearance is not correct, adjust the brake shoes until the clearance is correct.
8. Install the brake drum, replace the wheel(s), and safely lower the vehicle.
9. Road-test the vehicle for proper brake operation.

RAV4

1. Raise and safely support the vehicle.
2. Remove the wheel(s).
3. Temporarily install 2 wheel stud nuts to hold the brake rotor in place.
4. If necessary, siphon a sufficient quantity of brake fluid from the master cylinder reservoir to prevent any brake fluid from overflowing the master cylinder when removing or installing new pads. This may be necessary, as the piston must be forced into the caliper bore to provide sufficient clearance when installing the pads.
5. Grasp the caliper from behind and carefully pull it towards you. This will start to seat the piston(s) in its bore. Using a C-clamp or other suitable tool, press the piston the remaining way into the caliper. Be careful not to cock the piston in the bore. Also, do not force the piston or the caliper and piston may be damaged.
6. Hold the sliding pin and loosen the 2

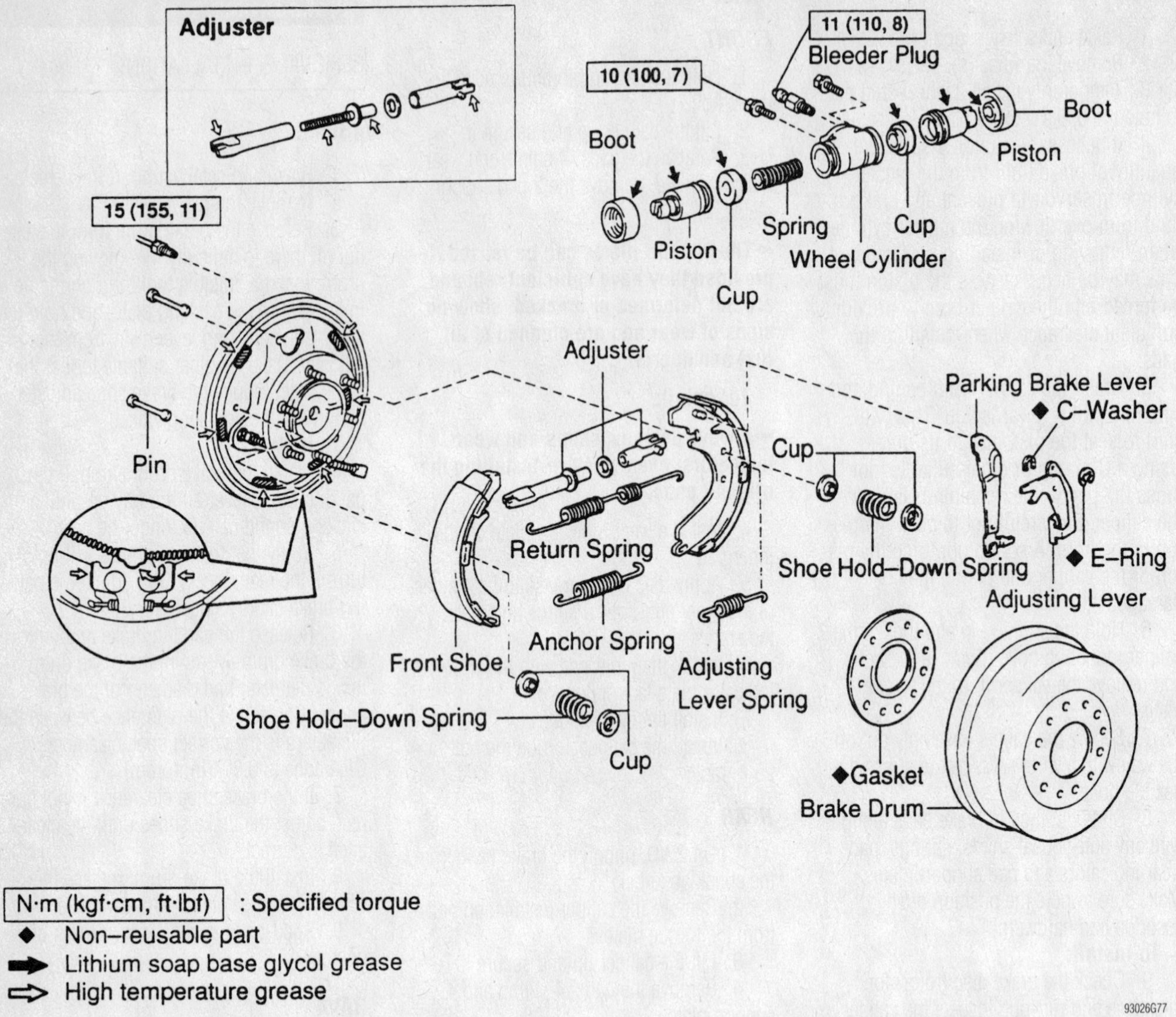

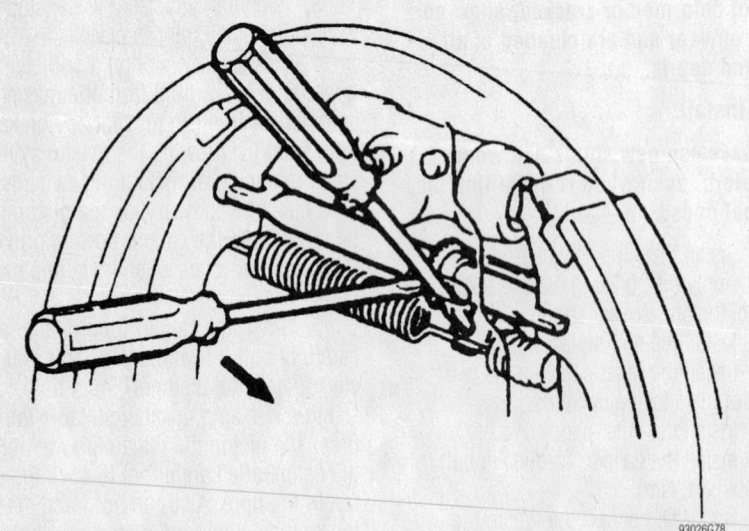

Adjuster

11 (110, 8)
Bleeder Plug

10 (100, 7)

Boot

Boot

Piston

Piston

Spring

Cup

Cup

Wheel Cylinder

15 (155, 11)

Adjuster

Parking Brake Lever

◆ C–Washer

Cup

Pin

Return Spring

Shoe Hold–Down Spring

◆ E–Ring

Adjusting Lever

Front Shoe

Anchor Spring

Adjusting Lever Spring

Shoe Hold–Down Spring

◆ Gasket

Brake Drum

Cup

N·m (kgf·cm, ft·lbf) : Specified torque
◆ Non–reusable part
➡ Lithium soap base glycol grease
⇨ High temperature grease

93026G77

Exploded view of the rear brake drums components

93026G78

Use a brake adjusting tool (brake spoon) and a prytool to adjust the brake shoes through the adjusting hole

caliper mounting bolts. Remove the bolts and remove the caliper from the torque plate.

7. Secure the caliper assembly out of the way with a wire, to avoid stressing the flexible hose.

8. Slide out the old brake pads along with any anti-squeal shims, springs, pad wear indicators and pad support plates. Make sure to note the position of all assorted pad hardware.

To install:

9. Check the brake disc (rotor) for thickness and run-out. Inspect the caliper and piston assembly for breaks, cracks, fluid seepage or other damage. Overhaul or replace as necessary.

10. Install the pad support plates into the torque plate.

11. Install the pad wear indicators onto the pads. Be sure the arrow on the indicator plate is pointing in the direction of rotation.

12. Install the anti-squeal shims on the outside of each pad and then install the pad assemblies into the torque plate.

13. Install the caliper to the torque plate with the 2 mounting bolts. Torque the bolts to 20 ft. lbs. (26 Nm).

14. Remove the 2 temporary wheel stud nuts and check that the rotor turns freely.

15. Reinstall the wheel(s), safely lower the vehicle, and road-test for proper brake operation.

16. Be sure to pump the brakes several times prior to moving the vehicle.

Brake Shoes

REMOVAL & INSTALLATION

4Runner

1. Loosen the rear wheel lug nuts slightly.
2. Block the front wheels, raise the rear of the vehicle, and safely support it with jackstands.
3. Remove the wheel lug nuts and the wheel.
4. Remove the brake drum.
5. If the drum is difficult to remove, perform the following:
   a. Insert a flat prying tool through the hole in the brake drum and hold the automatic adjusting lever away from the adjuster.
   b. Reduce the brake shoe adjustment by turning the adjuster bolt with a brake tool.
   c. The drum should now be loose enough to remove without much effort.
6. Remove the rear shoe.
   a. Carefully unhook the return spring from the brake shoe.
   b. Remove the shoe hold-down spring, cups and the pin.
   c. Disconnect the anchor spring from the rear shoe and remove the rear shoe.
   d. Disconnect the anchor spring from the front shoe.

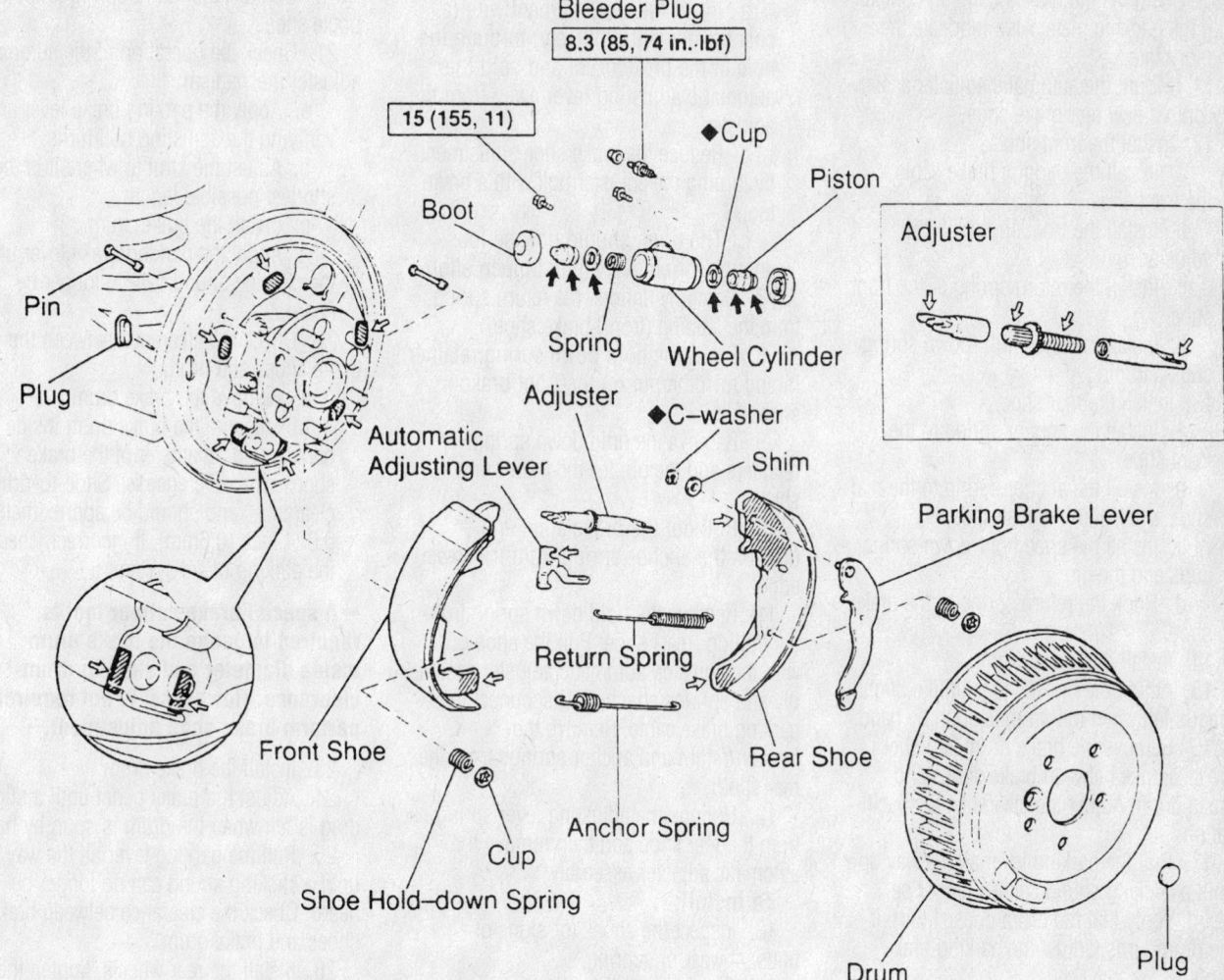

N·m (kgf·cm, ft·lbf) : Specified torque

◆ Non–reusable part
➡ Lithium soap base glycol grease
⇨ High temperature grease

93026G80

Exploded view of the rear drum brake components—RAV4

7. Remove the front shoe.

a. Remove the shoe hold-down spring, cups and pin.

b. Remove the return spring from the front shoe.

c. Remove the front shoe with the adjuster.

d. Disconnect the parking brake cable from the front shoe.

To install:

8. Inspect the shoes for signs of unusual wear or scoring.

9. Check the wheel cylinder for any sign of fluid seepage or frozen pistons.

10. Clean and inspect the brake backing plate and all other components. Check that the brake drum inner diameter is within specified limits. Lubricate the backing plate at the positions the brakes come in contact with the backing plate. Also lubricate the anchor plate.

11. Mount the automatic adjuster assembly onto a new rear brake shoe.

12. Install the front shoe.

a. Install the parking brake cable to the front shoe.

b. Install the front shoe with the adjuster.

c. Install the return spring to the front shoe.

d. Install the shoe hold-down spring, cups and pin.

13. Install the rear shoe.

a. Install the anchor spring to the front shoe.

b. Install the anchor spring to the rear shoe and install the rear shoe.

c. Install the shoe hold-down spring, cups and the pin.

d. Hook the return spring to the brake shoe.

14. Install the brake drum.

15. Adjust the brake shoes until a slight drag is felt when the drum is spun by hand.

16. Remove the brake drum and check the clearance between brake shoes and brake drum. Adjust the clearance to specification.

17. Pull the parking lever all the way up until a clicking sound can no longer be heard. Verify that the drum doesn't turn. If the drum turns, adjust the parking brake cable.

18. Install the rear wheels, tighten the wheel lug nuts and lower the vehicle.

19. Retighten the wheel lug nuts and pump the brake pedal a few times before moving the vehicle. Adjust the rear brakes again if necessary.

20. Check the level of brake fluid in the master cylinder, then perform a test drive.

21. Connect the negative battery cable to the battery.

RAV4

1. Disconnect the negative battery cable from the battery.

2. Loosen the rear wheel lug nuts slightly. Release the parking brake.

3. Block the front wheels, raise the rear of the vehicle, and safely support it with jackstands.

4. Remove the wheel lug nuts and the wheel.

5. Remove the brake drum retaining screws, if equipped. Remove the brake drum.

6. If the drum is difficult to remove, perform the following:

a. Insert the end of a bent wire (a coat hanger will do nicely) through the hole in the brake drum and hold the automatic adjusting lever away from the adjuster.

b. Reduce the brake shoe adjustment by turning the adjuster bolt with a brake tool.

c. The drum should now be loose enough to remove without much effort.

7. Carefully unhook the return spring from the leading (front) brake shoe.

8. Press the hold down spring retainer in and turn the pin on the front brake shoe.

9. Remove the hold down spring, retainers and the pin for the front brake shoe.

10. Pull out the brake shoe and unhook the anchor spring from the lower edge.

11. Remove the hold down spring from the trailing (rear) shoe. Pull the shoe out with the adjuster, automatic adjuster assembly and springs attached. Disconnect the parking brake cable. Remove the tension/return and anchor springs from the rear shoe.

12. Unhook the adjusting lever spring from the rear shoe and then remove the automatic adjuster assembly.

To install:

13. Inspect the shoes for signs of unusual wear or scoring.

14. Check the wheel cylinder for any sign of fluid seepage or frozen pistons.

15. Clean and inspect the brake backing plate and all other components. Check that the brake drum inner diameter is within specified limits. Lubricate the backing plate at the positions the brakes come in contact with the backing plate. Also lubricate the anchor plate.

16. Mount the automatic adjuster assembly onto a new rear brake shoe.

17. Connect the parking brake cable to the rear shoe and then install the automatic adjusting lever, spring and E-ring. Position the rear shoe so the lower end rides in the anchor plate and the upper end is against the boot of the wheel cylinder.

18. Install the pin and the hold down spring. Press the retainer down over the pin and rotate the pin so the crimped edge is held by the retainer.

19. Place the front brake into position and install the anchor spring between the front and rear shoes. Stretch the spring enough so the front shoe will fit as the rear did. Install the hold down spring, pin and retainer to the front brake shoe.

20. Connect the return spring to the front brake shoe.

21. Check the operation of the automatic adjuster mechanism:

a. Apply the parking brake lever and verifying the adjusting bolt turns.

b. Adjust the strut to where it is the shortest possible length.

c. Install the brake drum.

d. Apply the parking brake lever until the clicking sound can no longer be heard.

22. Check the clearance between the brake shoes and drum:

a. Remove the brake drum.

b. Measure the brake drum inside diameter and diameter of the brake shoes. The difference is "Shoe-to-drum clearance" and should be approximately 0.024 inch (0.6mm). If incorrect, check the parking brake system.

➡**A special brake caliper tool is required to gauge the brake drum inside diameter and shoe-to-drum clearance. However it is not required to perform brake shoe adjustment.**

23. Install the brake drum.

24. Adjust the brake pedal until a slight drag is felt when the drum is spun by hand.

25. Pull the parking lever all the way up until a clicking sound can no longer be heard. Check the clearance between brake shoes and brake drum.

26. Install the rear wheels, tighten the wheel lug nuts and lower the vehicle.

27. Retighten the wheel lug nuts and pump the brake pedal a few times before moving the vehicle. Adjust the rear brakes again if necessary.

28. Check the level of brake fluid in the master cylinder, then perform a test drive.

29. Connect the negative battery cable to the battery.

TOYOTA

Matrix

37

SPECIFICATION CHARTS

ENGINE AND VEHICLE IDENTIFICATION

	Engine							Model Year	
Code ①	Liters (cc)	Cu. In.	Cyl.	Fuel Sys.	Engine Type	Eng. Mfg.		Code ②	Year
1ZZ-FE	1.8 (1794)	109	4	EFI	DOHC	Toyota		3	2003
2ZZ-GE	1.8 (1796)	109.5	4	EFI	DOHC	Toyota		4	2004

EFI: Electronic Fuel Injection

DOHC: Double Overhead Camshaft

① 8th digit of VIN

② 10th digit of VIN

42356-TMAT-C01

GENERAL ENGINE SPECIFICATIONS

Year	Model	Engine Displacement Liters (cc)	Engine Series (ID/VIN)	Fuel System	Net Horsepower @ rpm	Net Torque @ rpm (ft. lbs.)	Bore x Stroke (in.)	Com-pression Ratio	Oil Pressure @ idle
2003	Matrix	1.8 (1794)	1ZZ-FE	EFI	①	②	3.11x3.60	10.0:1	4.2
	Matrix	1.8 (1796)	2ZZ-GE	EFI	180@7600	130@6800	3.23x3.35	11.5:1	2.8
2004	Matrix	1.8 (1794)	1ZZ-FE	EFI	①	②	3.11x3.60	10.0:1	4.2
	Matrix	1.8 (1796)	2ZZ-GE	EFI	180@7600	130@6800	3.23x3.35	11.5:1	2.8

EFI: Electronic Fuel Injection

① 2WD models: 130@6000

  4WD models: 123@6000

② 2WD models: 126@4200

  4WD models: 119@4200

42356-TMAT-C02

ENGINE TUNE-UP SPECIFICATIONS

Year	Engine Displacement Liters (cc)	Engine ID/VIN	Spark Plug Gap (in.)	Ignition Timing (deg.)	Fuel Pump (psi)	Idle Speed (rpm) MT	Idle Speed (rpm) AT	Valve Clearance In.	Valve Clearance Ex.
2003	1.8 (1794)	1ZZ-FE	0.043	①	44-50	650-750	650-750	0.0059-0.0098	0.0098-0.0138
	1.8 (1796)	2ZZ-GE	0.043	②	44-50	750-850	700-800	0.0031-0.0071	0.0087-0.0126
2004	1.8 (1794)	1ZZ-FE	0.043	①	44-50	650-750	650-750	0.0059-0.0098	0.0098-0.0138
	1.8 (1796)	2ZZ-GE	0.043	②	44-50	750-850	700-800	0.0031-0.0071	0.0087-0.0126

Note: The Vehicle Emission Control Information label often reflects specification changes made during production. The label figures must be used if they differ from those in this chart.

① With terminal TC and CG of DLC3 connected: 8-12 degrees BTDC

  With terminal TC and CG of DLC3 disconnected: 10-18 degrees BTDC

② With terminal TC and CG of DLC3 connected: 8-12 degrees BTDC

  With terminal TC and CG of DLC3 disconnected:

A/T: 10-18 degrees BTDC

M/T: 4-12 degrees BTDC

42356-TMAT-C03

CAPACITIES

Year	Model	Engine Displacement Liters (cc)	Engine ID/VIN	Engine Oil with Filter	Transmission (pts.)			Drive Axle		Fuel Tank (gal.)	Cooling System (qts.)
					5-Spd	6-Spd	Auto.	Front (pts.)	Rear (pts.)		
2003	Matrix	1.8 (1794)	1ZZ-FE	3.9	4.0	4.0	5.2	NA	1.04	①	6.9
	Matrix	1.8 (1796)	2ZZ-GE	4.7	4.8	4.8	6.1	NA	—	13.2	7.1
2004	Matrix	1.8 (1794)	1ZZ-FE	3.9	4.0	4.0	5.2	NA	1.04	①	6.9
	Matrix	1.8 (1796)	2ZZ-GE	4.7	4.8	4.8	6.1	NA	—	13.2	7.1

Note: All capacities are approximate. Add fluid gradually and check to be sure a proper fluid level is obtained.

NA - Not available

① 2WD: 13.2 gallons
   4WD: 11.9 gallons

42356-TMAT-C04

VALVE SPECIFICATIONS

Year	Engine Displacement Liters (cc)	Engine ID/VIN	Seat Angle (deg.)	Face Angle (deg.)	Spring Test Pressure (lbs. @ in.)	Spring Installed Height (in.)	Stem-to-Guide Clearance (in.)		Stem Diameter (in.)	
							Intake	Exhaust	Intake	Exhaust
2003	1.8 (1794)	1ZZ-FE	45	44.5	31.3-34.8 @ 1.252	1.323	0.0010-0.0024	0.0012-0.0026	0.2154-0.2159	0.2152-0.2158
	1.8 (1796)	2ZZ-GE	45	44.5	①	1.516	0.0010-0.0023	0.0012-0.0025	0.2145-0.2156	0.2144-0.2154
2004	1.8 (1794)	1ZZ-FE	45	44.5	31.3-34.8 @ 1.252	1.323	0.0010-0.0024	0.0012-0.0026	0.2154-0.2159	0.2152-0.2158
	1.8 (1796)	2ZZ-GE	45	44.5	①	1.516	0.0010-0.0023	0.0012-0.0025	0.2145-0.2156	0.2144-0.2154

① Intake: 49.6-55.5 @ 1.516
   Exhaust: 47.6-52.6 @ 1.516

42356-TMAT-C05

CRANKSHAFT AND CONNECTING ROD SPECIFICATIONS
All measurements are given in inches.

Year	Engine Displacement Liters (cc)	Engine ID/VIN	Crankshaft				Connecting Rod		
			Main Brg. Journal Dia.	Main Brg. Oil Clearance	Shaft End-play	Thrust on No.	Journal Diameter	Oil Clearance	Side Clearance
2003	1.8 (1762)	1ZZ-FE	1.8893-1.8898	0.0006-0.0013	0.0008-0.0087	3	1.7320-1.7323	0.0011-0.0024	0.0063-0.0135
	1.8 (1796)	2ZZ-GE	1.8893-1.8898	0.0006-0.0013	0.0016-0.0094	3	1.7713-1.7717	0.0011-0.0020	0.0063-0.0135
2004	1.8 (1762)	1ZZ-FE	1.8893-1.8898	0.0006-0.0013	0.0008-0.0087	3	1.7320-1.7323	0.0011-0.0024	0.0063-0.0135
	1.8 (1796)	2ZZ-GE	1.8893-1.8898	0.0006-0.0013	0.0016-0.0094	3	1.7713-1.7717	0.0011-0.0020	0.0063-0.0135

42356-TMAT-C06

PISTON AND RING SPECIFICATIONS
All measurements are given in inches.

Year	Engine Displacement Liters (cc)	Engine ID/VIN	Piston Clearance	Ring Gap			Ring Side Clearance		
				Top Compression	Bottom Compression	Oil Control	Top Compression	Bottom Compression	Oil Control
2003	1.8 (1762)	1ZZ-FE	0.0026-0.0035	0.0098-0.0138	0.0138-0.0197	0.0059-0.0197	0.0008-0.0028	0.0012-0.0028	0.0012-0.0043
	1.8 (1796)	2ZZ-GE	0.0003-0.0015	0.0098-0.0138	0.0138-0.0197	NA	0.0009-0.0028	0.0012-0.0028	NA
2004	1.8 (1762)	1ZZ-FE	0.0026-0.0035	0.0098-0.0138	0.0138-0.0197	0.0059-0.0197	0.0008-0.0028	0.0012-0.0028	0.0012-0.0043
	1.8 (1796)	2ZZ-GE	0.0003-0.0015	0.0098-0.0138	0.0138-0.0197	NA	0.0009-0.0028	0.0012-0.0028	NA

NA - Not available

42356-TMAT-C07

TORQUE SPECIFICATIONS
All readings in ft. lbs.

Year	Engine Displacement Liters (cc)	Engine ID/VIN	Cylinder Head Bolts	Main Bearing Bolts	Rod Bearing Bolts	Crankshaft Damper Bolts	Flywheel Bolts	Manifold		Spark Plugs	Lug Nuts
								Intake	Exhaust		
2003	1.8 (1794)	1ZZ-FE	①	②	③	102	①	22	27	18	76
	1.8 (1796)	2ZZ-GE	④	⑤	⑥	87	①	⑦	37	13	76
2004	1.8 (1794)	1ZZ-FE	①	②	③	102	①	22	27	18	76
	1.8 (1796)	2ZZ-GE	④	⑤	⑥	87	①	⑦	37	13	76

① Step 1: 36 ft. lbs.
  Step 2: 90 degree turn

② 12 pointed bolts:
  Step 1: 33 ft. lbs.
  Step 2: 90 degree turn
  Hex head bolts: 14 ft. lbs.

③ Step 1: 15 ft. lbs.
  Step 2: 90 degree turn

④ Step 1: 26 ft. lbs.
  Step 2: 180 degree turn

⑤ 12 pointed bolts:
  Step 1: 16 ft. lbs.
  Step 2: 32 ft. lbs.
  Step 3: 45 degree turn
  Step 4: 45 degree turn
  Hex head bolts: 13 ft. lbs.

⑥ Step 1: 22 ft. lbs.
  Step 2: 90 degree turn

⑦ Bolt A: 25 ft. lbs.
  Bolt B: 34 ft. lbs.

42356-TMAT-C08

WHEEL ALIGNMENT

Year	Model		Caster Range (+/-Deg.)	Caster Preferred Setting (Deg.)	Camber Range (+/-Deg.)	Camber Preferred Setting (Deg.)	Toe-in (in.)	Steering Axis Inclination (Deg.)
2003	Matrix - 2WD	F	0.75	+2.78	0.75	-0.77	0+/-0.08	12.47+/-0.75
		R	—	—	0.50	-1.45	0.11+/-0.11	—
	Matrix - 4WD	F	0.75	+2.77	0.75	-0.48	0+/-0.08	12.22+/-0.75
		R	—	—	0.75	-0.73	0.08+/-0.08	—
2004	Matrix - 2WD	F	0.75	+2.78	0.75	-0.77	0+/-0.08	12.47+/-0.75
		R	—	—	0.50	-1.45	0.11+/-0.11	—
	Matrix - 4WD	F	0.75	+2.77	0.75	-0.48	0+/-0.08	12.22+/-0.75
		R	—	—	0.75	-0.73	0.08+/-0.08	—

42356-TMAT-C09

TIRE, WHEEL AND BALL JOINT SPECIFICATIONS

Year	Model	OEM Tires Standard	OEM Tires Optional	Tire Pressures (psi) Front	Tire Pressures (psi) Rear	Wheel Size	Ball Joint Inspection
2003	Matrix	205/55R16	—	33	33	6.5-JJ	9-26 in. ①
2004	Matrix	205/55R16	—	33	33	6.5-JJ	9-26 in. ①

OEM: Original Equipment Manufacturer

PSI: Pounds Per Square Inch

STD: Standard

OPT: Optional

① Torque required in inch lbs. to rotate ball joint when removed from the knuckle

42356-TMAT-C10

BRAKE SPECIFICATIONS

All measurements in inches unless noted

Year	Model		Brake Disc Original Thickness	Brake Disc Minimum Thickness	Brake Disc Maximum Runout	Brake Drum Diameter Original Inside Diameter	Brake Drum Diameter Max. Wear Limit	Brake Drum Diameter Maximum Machine Diameter	Minimum Lining Thickness	Brake Caliper Bracket Bolts (ft. lbs.)	Brake Caliper Mounting Bolts (ft. lbs.)
2003	Matrix	F	0.984	0.906	0.0020	—	—	—	0.039	79	25
		R	0.354	0.295	0.0059	9.00	—	9.04	0.039	—	34
2004	Matrix	F	0.984	0.906	0.0020	—	—	—	0.039	79	25
		R	0.354	0.295	0.0059	9.00	—	9.04	0.039	—	34

F: Front

R: Rear

42356-TMAT-C11

SCHEDULED MAINTENANCE INTERVALS
TOYOTA—MATRIX

TO BE SERVICED	TYPE OF SERVICE	VEHICLE MILEAGE INTERVAL (x1000)												
		7.5	15	22.5	30	37.5	45	52.5	60	67.5	75	82.5	90	97.5
Engine oil & filter	R	✓	✓	✓	✓	✓	✓	✓	✓	✓	✓	✓	✓	✓
Drive belts	S/I								✓	✓	✓	✓	✓	✓
Automatic transaxle fluid & filter	S/I		✓		✓		✓		✓		✓		✓	
Ball joints & dust covers	S/I		✓		✓		✓		✓		✓		✓	
Bolts & nuts on body & chassis	S/I		✓		✓		✓		✓		✓		✓	
Brake line pipes & hoses	S/I		✓		✓		✓		✓		✓		✓	
Brake linings & drums	S/I		✓		✓		✓		✓		✓		✓	
Brake pads & discs (front & rear if equipped)	S/I		✓		✓		✓		✓		✓		✓	
Differential oil	S/I		✓		✓		✓		✓		✓		✓	
Drive shaft boots (except Supra)	S/I		✓		✓		✓		✓		✓		✓	
Manual transaxle oil	S/I		✓		✓		✓		✓		✓		✓	
Steering gear housing oil	S/I		✓		✓		✓		✓		✓		✓	
Steering linkage	S/I		✓		✓		✓		✓		✓		✓	
Air filter	R				✓				✓				✓	
Spark plugs	R				✓				✓				✓	
Spark plugs (platinum tip)	R								✓					
Exhaust system	S/I				✓				✓				✓	
Fuel lines & connections	S/I				✓				✓				✓	
Valve clearance	S/I				✓				✓				✓	
Engine coolant	R						✓				✓			
Fuel tank cap gasket	R								✓					
Charcoal canister	S/I								✓					

R: Replace   S/I: Service or Inspect

FREQUENT OPERATION MAINTENANCE (SEVERE SERVICE)

If a vehicle is operated under any of the following conditions it is considered severe service:

- Extremely dusty areas.

- 50% or more of the vehicle operation is in 32°C (90°F) or higher temperatures, or constant operation in temperatures below 0°C (32°F).

- Prolonged idling (vehicle operation in stop and go traffic).

- Frequent short running periods (engine does not warm to normal operating temperatures).

- Police, taxi, delivery usage or trailer towing usage.

Oil & oil filter: change every 6000 miles.

Bolts & nuts on chassis & body: tighten every 7500 miles.

Ball joints & dust covers: service or inspect every 12,000 miles.

Brake linings & drums: service or inspect ever 12,000 miles.

Brake pads & discs (front & rear if equipped): service or inspect every 12,000 miles.

Drive shaft boots & except Supra): service or inspect every 12,000 miles.

Steering linkage: service or inspect every 12,000 miles.

Air filter: service or inspect every 15,000 miles.

Exhaust system: service or inspect every 15,000 miles.

Timing belt: replace every 60,000 miles.

42356-TMAT-C12

PRECAUTIONS

Before servicing any vehicle, please be sure to read all of the following precautions, which deal with personal safety, prevention of component damage, and important points to take into consideration when servicing a motor vehicle:

• Never open, service or drain the radiator or cooling system when the engine is hot; serious burns can occur from the steam and hot coolant.

• Observe all applicable safety precautions when working around fuel. Whenever servicing the fuel system, always work in a well-ventilated area. Do not allow fuel spray or vapors to come in contact with a spark, open flame or excessive heat (a hot drop light, for example). Keep a dry chemical fire extinguisher near the work area. Always keep fuel in a container specifically designed for fuel storage; also, always properly seal fuel containers to avoid the possibility of fire or explosion. Refer to the additional fuel system precautions later in this section.

• Fuel injection systems often remain pressurized, even after the engine has been turned **OFF**. The fuel system pressure must be relieved before disconnecting any fuel lines. Failure to do so may result in fire and/or personal injury.

• Brake fluid often contains polyglycol ethers and polyglycols. Avoid contact with the eyes and wash your hands thoroughly after handling brake fluid. If you do get brake fluid in your eyes, flush your eyes with clean, running water for 15 minutes. If eye irritation persists, or if you have taken brake fluid internally, IMMEDIATELY seek medical assistance.

• The EPA warns that prolonged contact with used engine oil may cause a number of skin disorders, including cancer! You should make every effort to minimize your exposure to used engine oil. Protective gloves should be worn when changing oil. Wash your hands and any other exposed skin areas as soon as possible after exposure to used engine oil. Soap and water, or waterless hand cleaner should be used.

• All new vehicles are now equipped with an air bag system. The system must be disabled before performing service on or around system components, steering column, instrument panel components, wiring and sensors. Failure to follow safety and disabling procedures could result in accidental air bag deployment, possible personal injury and unnecessary system repairs.

• Always wear safety goggles when working with, or around, the air bag system. When carrying a non-deployed air bag, be sure the bag and trim cover are pointed away from your body. When placing a non-deployed air bag on a work surface, always face the bag and trim cover upward, away from the surface. This will reduce the motion of the module if it is accidentally deployed. Refer to the additional air bag system precautions later in this section.

• Clean, high quality brake fluid from a sealed container is essential to the safe and proper operation of the brake system. You should always buy the correct type of brake fluid for your vehicle. If the brake fluid becomes contaminated, completely flush the system with new fluid. Never reuse any brake fluid. Any brake fluid that is removed from the system should be discarded. Also, do not allow any brake fluid to come in contact with a painted surface; it will damage the paint.

• Never operate the engine without the proper amount and type of engine oil; doing so WILL result in severe engine damage.

• Timing belt maintenance is extremely important! Many models utilize an interference-type, non-freewheeling engine. If the timing belt breaks, the valves in the cylinder head may strike the pistons, causing potentially serious (also time-consuming and expensive) engine damage.

• Disconnecting the negative battery cable on some vehicles may interfere with the functions of the on-board computer system(s) and may require the computer to undergo a relearning process once the negative battery cable is reconnected.

• When servicing drum brakes, only disassemble and assemble one side at a time, leaving the remaining side intact for reference.

• Only an MVAC-trained, EPA-certified automotive technician should service the air conditioning system or its components.

ENGINE REPAIR

Alternator

REMOVAL & INSTALLATION

1. Before servicing the vehicle, refer to the precautions in the beginning of this section.
2. Remove or disconnect the following:
   • Negative battery cable
   • Drive belt
   • Wire clamp from the clip on the rectifier end frame
   • Rubber clamp and nut
   • Alternator wiring and connector
   • Alternator

To install:
3. Install or connect the following:
   • Alternator. Torque the 12mm bolt to 18 ft. lbs. (25 Nm) and the 14mm bolt to 39 ft. lbs. (54 Nm).
   • Alternator connector and wiring

   • Rubber clamp and nut
   • Wire clamp
   • Drive belt
   • Negative battery cable

Ignition Timing

ADJUSTMENT

➡The timing on engines equipped with Distributorless Ignition Systems (DIS) is not adjustable.

Engine Assembly

REMOVAL & INSTALLATION

1. Before servicing the vehicle, refer to the precautions in the beginning of this section.
2. Relieve the fuel system pressure.

3. Drain the cooling system.
4. Drain the engine oil.
5. Drain the transaxle fluid and transfer fluid, if equipped.
6. Remove or disconnect the following:
   • Negative battery cable. Wait at least 90 seconds before proceeding.
   • Battery
   • Hood
   • Undercovers
   • Radiator inlet and outlet hoses
   • Radiator hose outlet
   • Oil cooler inlet and outlet tubes
   • Upper radiator support and radiator, if equipped with A/C
   • Battery
   • Air cleaner assembly
   • Fuel pipe clamp
   • Fuel tube sub-assembly
   • Accelerator control cable
   • Cruise control actuator, if equipped
   • Union-to-connector tube hose

- Heater inlet and outlet hoses
- Transmission shift cable(s)
- Clutch release cylinder, on manual transaxle
- Glove compartment door
- Engine relay block cover
- 3 connectors from the relay block
- 2 ground cables
- Engine wire from the Engine Control Module (ECM) and junction block
- Engine wire from the cabin
- Drive belt
- Compressor and magnetic clutch, if equipped with A/C. Unbolt and position aside, DO NOT disconnect the lines.
- Vane pump oil reservoir from the bracket
- Return tube
- Right side front door scuff plate
- Right side cowl side trim plate
- Right side rear door scuff plate, AWD
- Lower right side center pillar garnish, AWD
- Right front seat, AWD
- Column hole cover silencer sheet
- Steering intermediate shaft
- Front floor panel brace, FWD
- Center exhaust pipe, AWD
- Propeller shaft with center bearing shaft, AWD
- Front exhaust pipe
- Front hub nuts
- Tie rod ends from the steering knuckles

7. Separate the front stabilizer links and lower control arm ball joints
- Front halfshafts

8. Remove the engine from the vehicle, as follows:
    a. Set the engine lifter.
    b. Remove the bolts and nuts, then remove the engine mounting insulator.
    c. Remove the through bolt and nut,

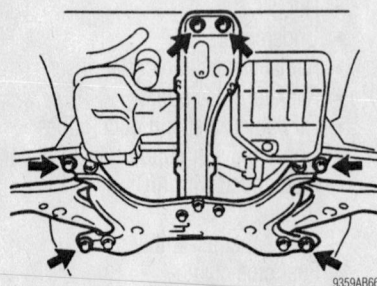

Remove the 6 bolts, as indicated by arrows

then detach the engine mounting insulator from the vehicle.
    d. Remove the 6 bolts as shown.
    e. Use a suitable tool to suspend the engine assembly, as shown in the figure.
    f. No. 1 engine hanger: P/N 12281-15040 (1ZZ-FE), 12281-88600 (2ZZ-GE).
    g. No. 2 engine hanger: P/N 12281-22021 (1ZZ-FE), 12281-88600 (2ZZ-GE).
    h. Bolt: P/N 91512-B1016.
    i. Torque the bolts to 28 ft. lbs. (38 Nm).

✳✳ CAUTION

Do not try to suspend the engine by hooking the chain to any other part.

    j. Attach an engine chain hoist to the hangers.

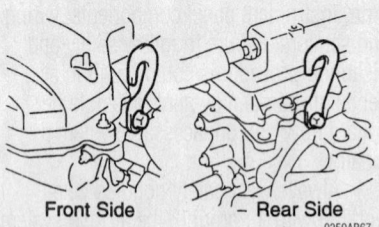

Front Side — **Rear Side**

9359AB67

Install the engine hangers—1ZZ-FE shown, 2ZZ-GE similar

    k. Using the chain block and sling device, suspend the engine.
    l. Remove the engine and transaxle assembly from the vehicle.

9. Remove or disconnect the following components, as necessary:
- Vane pump
- Steering gear
- Crossmember
- Manifold stay
- Oxygen (O$_2$) sensor
- Exhaust manifold
- Starter
- Transaxle
- Transfer case
- Clutch
- Flywheel
- Alternator
- Ignition coil
- Fuel delivery pipe
- Intake manifold
- Oil level gauge
- Water inlet and bypass pipes
- Thermostat
- Oil pressure switch
- Crankshaft Position (CKP) sensor
- Knock Sensor (KS)
- Drive belt tensioner
- Engine mounts and brackets
- Coolant Temperature Sensor (CTS)

To install:
10. Install any removed components to the engine and transaxle assembly.
11. To install the engine:
    a. Place the engine and transaxle on an engine lifter.
    b. Install the engine with the transaxle to the vehicle.
    c. Temporarily install the crossmember and 6 bolts.
    d. Install the left engine mounting insulator. Tighten the bolts to 59 ft. lbs. (80 Nm).
    e. Install the right engine mounting insulator. Tighten the bolts to 38 ft. lbs. (52 Nm).
    f. Insert SST 09670-00010 to the positioning holes of the right handle crossmember and on the right handle of the vehicle. Temporarily tighten bolt A, then bolt B.

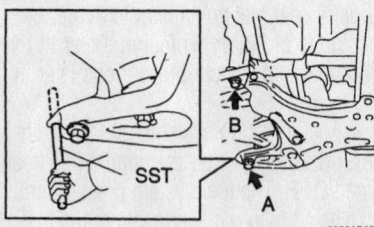

9359AB68

Insert the SST to the positioning holes of the right handle crossmember and on the right handle of the vehicle. Temporarily tighten bolt A, then bolt B

    g. Insert SST 09670-00010 to the positioning holes of the left handle crossmember and on the left handle of the vehicle. Temporarily tighten bolt A, then bolt B.

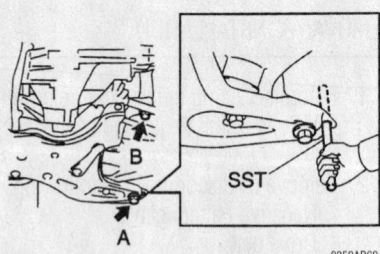

9359AB69

Insert the SST to the positioning holes of the left handle crossmember and on the left handle of the vehicle. Temporarily tighten bolt A, then bolt B

    h. Insert the SST to the positioning holes on the right-handle crossmember and right handle. Tighten bolt A to 116 ft. lbs. (157 Nm) and bolt B to 83 ft. lbs. (113 Nm).

i. Insert the SST to the positioning holes on the left-handle crossmember and left handle. Tighten bolt A to 116 ft. lbs. (157 Nm) and bolt B to 83 ft. lbs. (113 Nm).

j. Tighten the 2 crossmember bolts, shown in the figure, to 29 ft. lbs. (39 Nm).

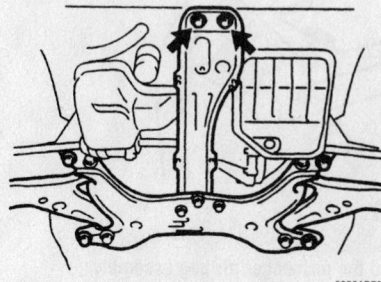

Tighten the 2 crossmember bolts, indicated by arrows

12. Installation of the remaining components is the reverse of the removal procedure.

13. Make sure all fluid levels are accurate, then start the engine check for leaks.

Water Pump

REMOVAL & INSTALLATION

1ZZ-FE Engine

1. Before servicing the vehicle, refer to the precautions in the beginning of this section.
2. Drain the cooling system.
3. Remove or disconnect the following:
   • Negative battery cable

• Right-hand engine under cover
• Drive belt
• Alternator
• Water pump

To install:
4. Install or connect the following:
   • Water pump. Torque bolts marked **A** (short) to 80 inch lbs. (9 Nm) and bolts marked **B** (long) to 96 inch lbs. (11 Nm).
   • Alternator
   • Drive belt
   • Right engine under cover
   • Negative battery cable
5. Fill the cooling system to the proper level.
6. Start the vehicle, check for leaks and repair if necessary.

2ZZ-GE Engine

1. Before servicing the vehicle, refer to the precautions in the beginning of this section.
2. Drain the cooling system.
3. Remove or disconnect the following:
   • Negative battery cable

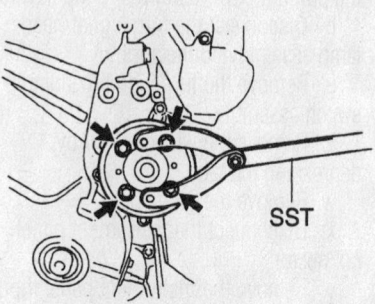

View of the special tool needed to remove and install the water pump pulley—2ZZ-GE engine

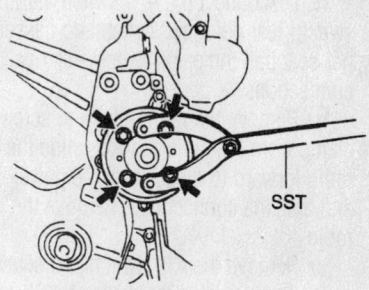

Water pump mounting and bolt locations—2ZZ-GE engine

• Right-hand engine under cover
• Drive belt
• Alternator
• Water pump pulley, using SST 09960-10010
• Water pump and O-ring

To install:
4. Install or connect the following:
   • Water pump with new O-ring. Torque the bolts to 80 inch lbs. (9 Nm).
   • Water pump pulley, using SST 09960-10010. Torque the bolts to 11 ft. lbs. (15 Nm).
   • Alternator
   • Drive belt
   • Right engine under cover
   • Negative battery cable
5. Fill the cooling system to the proper level.
6. Start the vehicle, check for leaks and repair if necessary.

Heater Core

REMOVAL & INSTALLATION

1. Before servicing the vehicle, refer to the precautions in the beginning of this section.
2. Drain the cooling system.
3. Discharge and recover the A/C system refrigerant using approved equipment.
4. Remove or disconnect the following:
   • Negative battery cable
   • Heater hoses from the core

Water pump bolt identification—1.8L (1ZZ-FE) engine

- Evaporator inlet and outlet tubes from the evaporator and cap the lines to avoid system contamination

5. Remove the instrument panel as follows:

a. Disable the air bag system.

b. Using a taped flat–bladed tool, carefully pry the retaining clips attaching the center trim plate to the instrument panel.

c. Disconnect the A/C switch, hazard switch; rear defogger switch and passenger seat belt indicator switch electrical connections.

d. Remove the radio retaining screws, clamp from the radio bracket, slide the radio forward to disconnect the power and antenna connections. Remove the radio.

e. Remove the A/C switch and screw.

f. Remove the hazard switch.

g. Remove the rear defogger switch.

h. Remove the manual transmission shift knob.

i. Using a taped flat–bladed tool, carefully pry the retaining clips attaching the front floor console trim plate to the floor console assembly.

j. Disconnect the 2 cigar lighter connectors.

k. Disconnect the accessory power receptacle connectors.

l. Remove the cigar lighters and power receptacle.

m. Place both wheels in the straight ahead position.

n. Remove the bolts from the steering wheel module.

o. Release the Connector Position Assurance (CPA) from the inflator module.

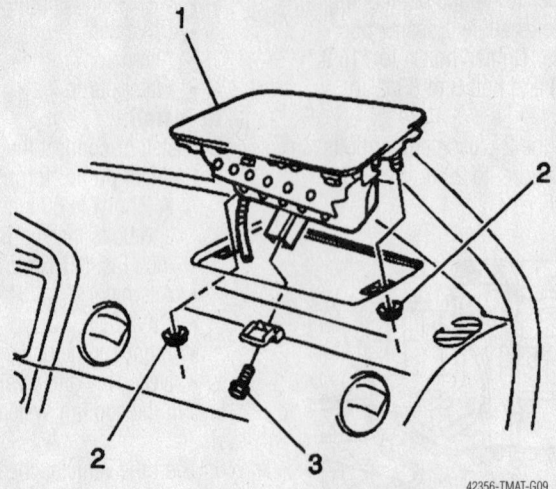

Remove the instrument panel module connectors and the passenger air bag assembly

p. Disconnect the steering wheel module connectors.

q. Remove the steering wheel module.

r. Matchmark the steering wheel nut–to–shaft position, then remove the steering wheel nut and the wheel.

s. Remove the upper and lower steering column cover screws and the covers.

t. Disconnect the turn signal/headlamp assembly connectors.

u. Remove the turn signal/headlamp switch assembly

v. Remove the wiper switch by depressing the tab.

w. Remove the glove box.

x. Disconnect the instrument panel connector.

y. Remove the instrument panel module connectors and the passenger air bag assembly.

z. Remove the cluster trim plate by disengaging the clips.

aa. Remove the cluster screw and disengage the 2 lower clips.

bb. Disconnect the cluster electrical connectors and remove the cluster.

cc. Remove the windshield garnish moldings.

dd. Using a taped flat–bladed tool, carefully pry the retaining clips attaching the instrument panel left trim plate to the instrument panel.

ee. Disconnect the power mirror and dimmer switch connectors.

ff. Remove the power mirror and dimmer switches.

gg. Disconnect any remaining electrical connections.

hh. Remove the upper instrument panel screws and the panel by pulling towards the rear to disengage the tabs.

ii. Disconnect the steering wheel coil connector.

jj. Release the 3 claws and remove the coil assembly.

kk. If the vehicle is equipped with an automatic transmission, insert the key into the cylinder, turn to the ACC position, push in the release button, disconnect the park lock cable, remove the key from the cylinder and lock the steering wheel.

ll. Move the silencer pad from the column.

mm. Matchmark the steering shaft coupling to the shaft.

nn. Loosen the upper bolt on the coupling.

oo. Remove the lower bolt from the coupling.

pp. Move the coupling onto the column shaft.

qq. Disconnect the wiring harness clamps from the column.

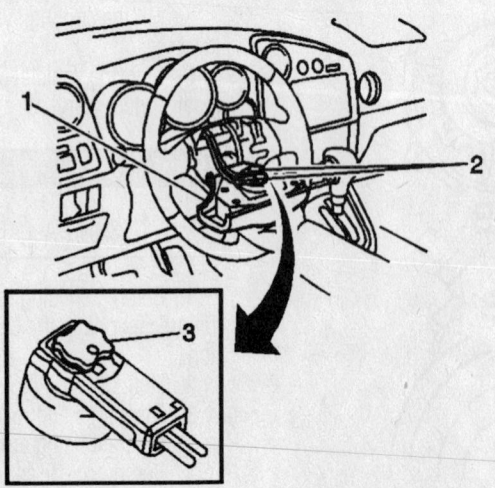

Exploded view of the CPA assembly

rr. Remove the 3 bolts and the column.

ss. Remove the body hinge trim panels.

tt. Remove the sill plates.

uu. Remove the front floor console storage door.

vv. Remove the screws attaching the console to the instrument panel, pull the console rewards and up and remove the front floor console.

ww. Remove the HVAC retaining screw; disconnect the electrical connectors and module control, temperature control and A/C cables. Remove the unit.

xx. Push in the clip and disconnect the cable from the manual selector shifter assembly.

yy. Using a suitable prytool, discon-

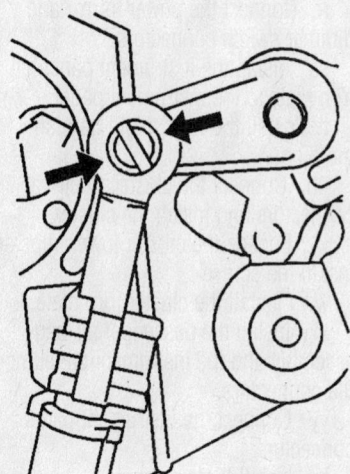

Push in the clip and disconnect the cable from the manual selector shifter assembly

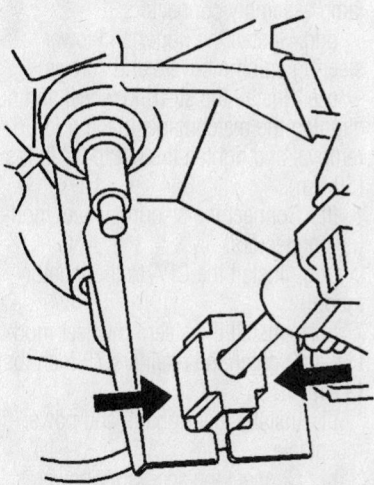

Using a suitable prytool, disconnect the park lock cable from the bracket

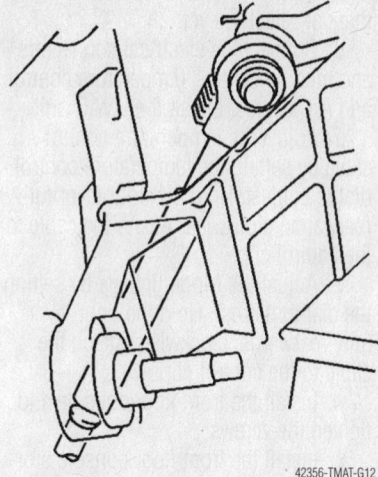

Disconnect the shift select cable from the manual selector lever

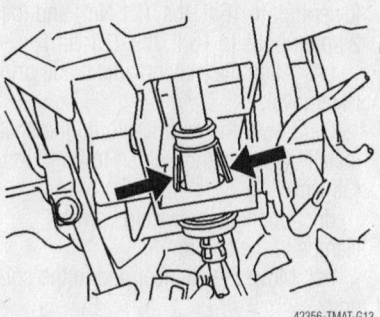

Using a suitable prytool, disconnect the shift select cable from the shift lever plate

nect the park lock cable from the bracket.

zz. Disconnect the shift select cable from the manual selector lever.

aaa. Using a suitable prytool, disconnect the shift select cable from the shift lever plate.

bbb. Disconnect the electrical connectors and the wire harness clip.

ccc. Remove the nuts from the selector and remove the selector.

ddd. Disconnect the hood release cable from the release handle.

eee. Remove the 8 bolts, 4 push retainers and the wire harness clamps from the lower instrument panel and remove the panel.

fff. Disengage the wiring harness clips, remove the bolts retaining the lower instrument panel pad and the pad.

ggg. Remove the ground cable.

hhh. Remove the connector housing bracket from the right instrument panel center support brace.

iii. Remove the left brace nut, right brace nut, left brace bolt, right brace bolt,

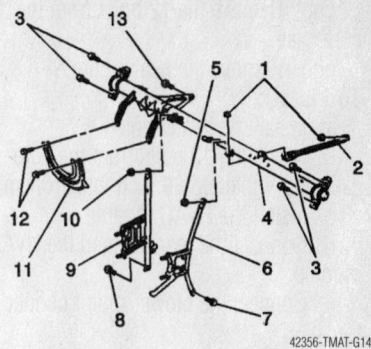

Exploded view of the instrument panel reinforcement

left center support brace and right center support brace.

jjj. Remove the windshield defroster nozzle duct from the heater case.

kkk. Remove the 5 bolts and the nuts from the instrument panel reinforcement at the hinge pillars.

lll. Remove the instrument panel reinforcement.

mmm. Disconnect the blower motor connector.

nnn. Disconnect the rear ducts from the HVAC module.

ooo. Remove the HVAC module.

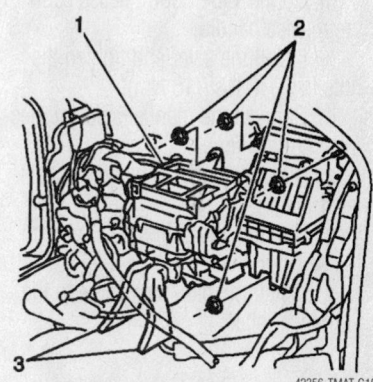

Remove the HVAC module

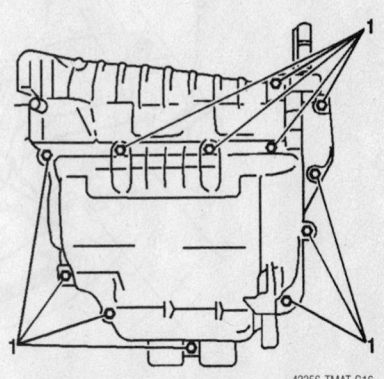

Remove the 12 bolts from the core case to access the heater core

ppp. Remove the 12 bolts from the core case.

qqq. Remove the heater core.

To install:

a. Install the heater core.

b. Install the 12 bolts from the core case and tighten to 89 inch lbs. (10 Nm).

c. Install the HVAC module.

d. Connect the rear ducts to the HVAC module.

e. Connect the blower motor connector.

f. Install the instrument panel reinforcement.

g. Install the 5 bolts and the nuts to the instrument panel reinforcement at the hinge pillars. Tighten to 21 ft. lbs. (28 Nm).

h. Install the windshield defroster nozzle duct to the heater case.

i. Install left center support brace and right center support brace. Tighten the nuts and bolt to 15 ft. lbs. (20 Nm).

j. Install the connector housing bracket.

k. Connect the ground cable.

l. Install the lower instrument panel pad and the bolts and attach the wiring harness clips.

m. Connect the hood release cable to the release handle.

n. Install the lever and tighten the nuts to 12 ft. lbs. (18 Nm).

o. Connect the manual selector electrical connections.

p. Attach the shift cable to the shift lever plate.

q. Connect the shift select cable to the selector lever.

r. Install the park lock cable to the shift lever plate.

s. Connect the park lock cable to the manual selector lever.

t. Connect the electrical connectors and module control, temperature control and A/C cables. Install the HVAC unit.

u. Adjust the temperature control cable by setting the temperature control dial to coldest. Hold the door lever fully rearwards, clockwise. Attach the cable to the control clip.

v. Adjust the Mode linkage by setting the dial to defrost. Hold the door lever fully rearwards, clockwise. Attach the cable to the control clip.

w. Install the front floor console and tighten the screws.

x. Install the front floor console storage door.

y. Install the sill plates.

z. Install the column.

aa. Install the 3 bolts and tighten the lower bolt to 16 ft. lbs. (21 Nm) and the 2 upper bolts to 16 ft. lbs. (21 Nm).

bb. Align the matchmarks made prior to removal.

cc. Lower the coupling onto the shaft. Install the bolts and tighten to 26 ft. lbs. (35 Nm).

dd. Connect the wiring harness clamps to the column.

ee. Move the silencer pad to the column.

ff. If the vehicle is equipped with an automatic transmission, insert the key into the cylinder, turn to the ACC position, insert the park lock cable making sure the release button engages. Make sure the key will not rotate to the lock position unless the shifter is in the park position, remove the key from the cylinder and lock the steering wheel.

gg. Install the body hinge trim panels.

hh. Make sure the turn signal switch is in the neutral position.

ii. If installing a new coil, remove the lock pin.

jj. Install the coil making sure the 3 claws engage.

kk. While holding the coil casing, turn the coil center casing counterclockwise until the coil reaches its stop.

ll. Turn the coil center casing clockwise 2 ½ turns.

mm. Align the center casing with the arrow on the outer casing.

nn. Connect the coil electrical connector.

oo. Install the upper instrument panel and screws.

pp. Connect the electrical connections.

qq. Install the power mirror and dimmer switches.

rr. Connect the power mirror and dimmer switch connectors.

ss. Install the instrument panel left trim plate to the instrument panel.

tt. Install the windshield garnish moldings.

uu. Connect the cluster electrical connectors and install the cluster.

vv. Engage the cluster lower clips and install the screw.

ww. Install the cluster trim plate.

xx. Install the passenger air bag assembly and the instrument panel module connectors.

yy. Connect the instrument panel connector.

zz. Install the glove box.

aaa. Install the wiper switch.

bbb. Install the turn signal/headlamp switch assembly

ccc. Connect the turn signal/headlamp assembly connectors.

ddd. Install the upper and lower steering column covers and screws.

eee. Install the steering wheel and nut aligning the matchmarks made prior to removal and tighten the nut to 37 ft. lbs. (50 Nm).

fff. Connect the steering wheel module connectors.

ggg. Install the CPA to the inflator module.

hhh. Install the steering wheel module and tighten the retainers 78 inch lbs. (9 Nm)

iii. Install cigar lighters and power receptacle.

jjj. Connect the accessory power receptacle connectors.

kkk. Connect the 2 cigar lighter connectors.

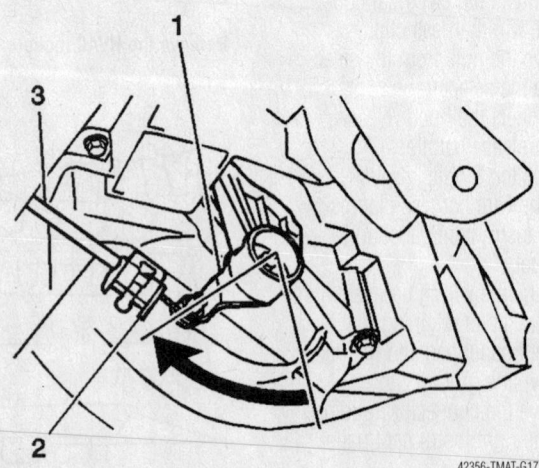

42356-TMAT-G17

Location of the components used to adjust the temperature control cable (3) clip (2), door lever (1).

lll. Install the front floor console trim plate to the floor console assembly.

mmm. Install the manual transmission shift knob.

nnn. Install the rear defogger switch.

ooo. Install the hazard switch.

ppp. Install the A/C switch and screw.

qqq. Install the radio.

rrr. Connect the A/C switch, hazard switch, rear defogger switch and passenger seat belt indicator switch electrical connections.

sss. Install the center trim plate to the instrument panel.

ttt. Connect the evaporator inlet and outlet tubes.

uuu. Connect the heater hoses to the core.

vvv. Connect the negative battery cable.

www. Recharge the A/C system and fill the cooling system.

Cylinder Head

REMOVAL & INSTALLATION

1. Before servicing the vehicle, refer to the precautions in the beginning of this section.

2. Drain the cooling system.

3. Remove or disconnect the following:
- Right side engine under cover
- Right front wheel and tire
- Cylinder head cover
- Air cleaner assembly with hose
- Accelerator control cable
- Wire harness clamp and suction hose assembly, 2ZZ-GE engine only
- Water bypass hoses
- Fuel pipe clamp
- Fuel tube sub-assembly
- Union-to-connector tube hose
- Radiator and heater inlet hoses
- Drive belt

4. Separate the vane pipe assembly, but do not disconnect the hose, 1ZZ-FE engine.
- Alternator bracket, 2ZZ-GE
- Alternator

5. Separate the compressor and magnetic clutch, on 2ZZ-GE engines with air conditioning.
- Front exhaust pipe assembly
- Power steering pump reservoir and position it aside, 1ZZ-FE engine

6. Place a jack with a wooden block under the vehicle for support, then remove the 4 bolts and 2 nuts and remove the right side engine mount.

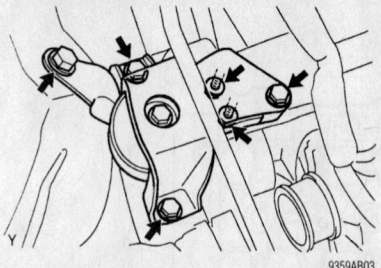

With the engine supported, remove the right side engine mount—1ZZ-FE engine shown, 2ZZ-GE similar

7. Remove the engine wire, on 1ZZ-FE engines as follows:
a. Remove the 5 clamps from the brackets.
b. Detach the connectors.
c. Remove the ignition coil connectors.
d. Bolt and nut holding the engine wire.

8. Remove or disconnect the following:
- Ignition coil assembly
- Positive Crankcase Ventilation (PCV) hoses
- Valve (cylinder head) cover sub-assembly

9. Set the No. 1 cylinder to Top Dead Center (TDC) of the compressor stroke as follows:
a. Turn the crankshaft pulley, and align its groove with the "0" timing mark of the timing chain cover.
b. Make sure the point marks of the camshaft timing sprockets and Variable Valve Timing (VVT) timing sprockets are in a straight line as shown. If not, turn the crankshaft 1 complete revolution (360°) and align the marks.

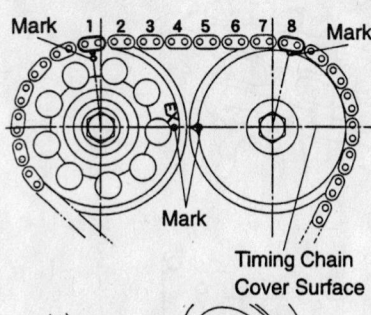

Timing Chain Cover Surface

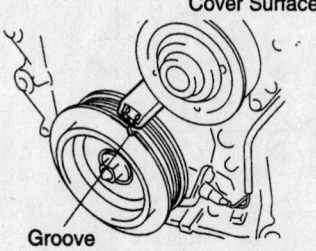

Proper timing mark alignment for TDC

10. Remove or disconnect the following:
- Crankshaft pulley, using SST 09960-10010
- Belt tensioner
- Exhaust manifold stay and head insulator, 2ZZ-GE engine
- Water pump pulley and pump
- Transverse engine mounting bracket
- Crankshaft Position (CKP) sensor
- No. 1 chain tensioner assembly, making sure not to revolve the crankshaft without the tensioner
- Timing chain or belt cover
- Timing gear cover oil seal
- CKP sensor plate No. 1
- Timing chain tensioner slipper
- Timing chain vibration damper No. 1

➡ **In case you turn the camshafts with the timing chain removed, turn the crankshaft ¼ turn for the valve to avoid contact with the pistons.**

- Timing chain sub-assembly. Remove the chain with the crankshaft gear, using screwdrivers as shown.

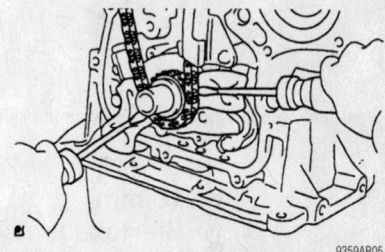

Remove the timing chain with the crankshaft gear

- Surge tank stay, 2ZZ-GE engine
- Intake manifold
- Oil level gauge
- Water bypass pipe bolts and pipe, 1ZZ-FE engine
- Camshafts
- Camshaft timing oil control valve, 1ZZ-FE engine
- Manifold stay, 1ZZ-FE engine
- Cylinder head bolts in sequence. To prevent damage to the cylinder head, loosen each bolt about ¼ of a turn during each pass until the bolts are loose.
- Cylinder head

To install:

11. Clean and degrease the surface of the cylinder head and engine block.

12. Check the length of the cylinder head bolts. They should be 5.780–5.835 in.

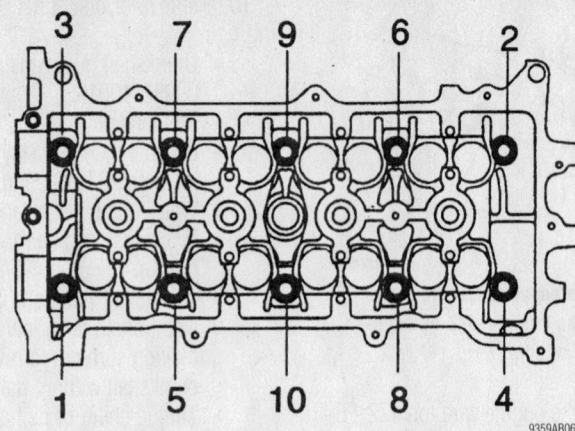

Cylinder head bolt loosening sequence

9359AB06

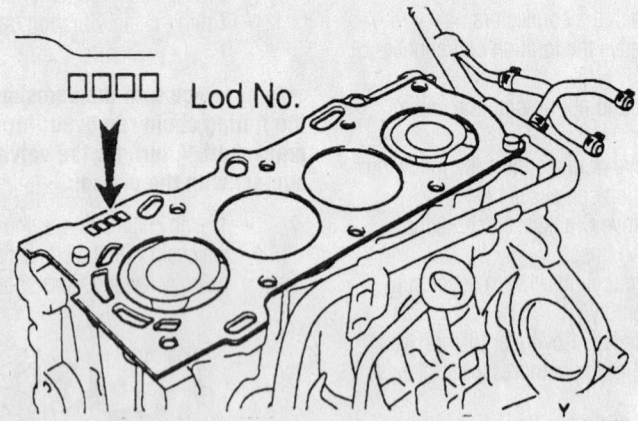

Position the head gasket correctly on the cylinder head—1.8L (1ZZ-FE) engine

7923VG12

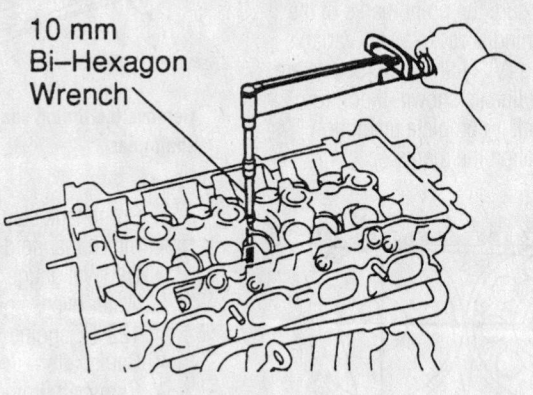

10 mm Bi–Hexagon Wrench

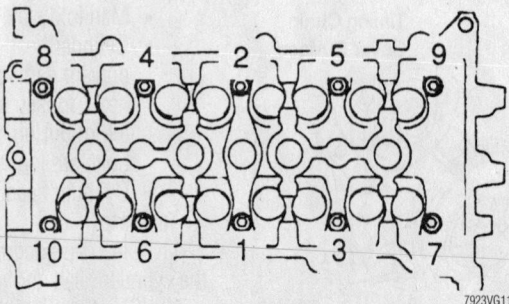

Cylinder head bolt tightening sequence—1ZZ-FE and 2ZZ-GE engines

7923VG11

(146.8–148.2mm) long. If they are longer than 5.846 in. (148.5mm), they must be replaced.

13. Install or connect the following:
- New gasket on the engine block with the Lot No. stamp facing up.
- Cylinder head
- Apply a light coat of oil to cylinder head bolt threads and tighten in sequence, in several passes, to 36 ft. lbs. (49 Nm) for 1ZZ-FE engines or to 26 ft. lbs. (35 Nm) for 2ZZ-GE engines.
- Tighten each head bolt, in sequence, an additional 90 degree turn for 1ZZ-FE engines, or 180 degree turn for 2ZZ-GE engines.
- Manifold stay, 1ZZ-FE engine. Tighten the bolts to 36 ft. lbs. (49 Nm).
- Camshaft timing oil control valve, on 1ZZ-FE engines, and tighten to 80 inch lbs. (9 Nm)
- Camshaft
- Water by-pass pipe, on 1ZZ-FE engines, and tighten to 80 inch lbs. (9 Nm)
- Oil level gauge
- Intake manifold
- Surge tank stay, 2ZZ-GE engine. Tighten to 18 ft. lbs. (24 Nm).
- Timing chain
- Timing chain vibration damper. Tighten the bolts to 80 inch lbs. (9 m).
- Timing chain tensioner slipper and tighten the bolt to 14 ft. lbs. (19 Nm).
- Crankshaft position sensor plate, with the "F" mark facing forward.
- Timing gear cover oil seal
- Timing cover. For 1ZZ-FE engine, tighten the "A" bolts to 10 ft. lbs. (13 Nm), the "B" bolts to 14 ft. lbs. (19 Nm) and the stud bolt to 84 inch lbs. (9.5 Nm), using a Torx® wrench. For 2ZZ-GE engines, tighten the M8 bolts to 15 ft. lbs. (21 m), the M6 bolts to 8 ft. lbs. (11 Nm) and the stud bolt to 84 inch lbs. (9.5 Nm).

➡**When installing the tensioner, make sure to set the hook again if the hook releases the plunger.**

- Timing chain tensioner. Torque the nuts to 80 inch lbs. (9 Nm).
- CKP sensor and tighten the bolts to 80 inch lbs. (9 Nm)
- Transverse engine mounting bracket. Tighten the bolts to 35 ft. lbs. (47 Nm).

- Water pump and pulley
- Exhaust manifold stay and heat insulator, 2ZZ-GE engine
- Belt tensioner. Tighten the nut to 21 ft. lbs. (29 Nm) and the bolt to 51 ft. lbs. (69 Nm) on 1ZZ-FE engines or to 74 ft. lbs. (100 Nm) on 2ZZ-GE engines.

14. Install the crankshaft pulley, as follows:

a. Align the pulley set key with the key groove of the pulley and slide on the pulley.

b. Use SST 09960-11010 to install the bolt and tighten to 102 ft. lbs. (138 Nm) for 1ZZ-FE engine or to 87 ft. lbs. (118 Nm) on 2ZZ-GE engines.

c. Turn the crankshaft counterclockwise and disconnect the plunger knock pin from the hook.

d. Turn the crankshaft clockwise and check that the slipper is pushed by the plunger. If the plunger does not spring out, press the slipper into the chain tensioner with a screwdriver so that the hook is released from the knock pin and the plunger springs out.

15. Install or connect the following:

- Cylinder head sub-assembly cover. Install seal packing into the locations shown and install within 3 minutes. Tighten the "A" bolts to 8 ft. lbs. (11 Nm) and the "B" bolts to 80 inch lbs. (9 Nm) for 1ZZ-FE engines. For 2ZZ-GE engines, tighten the bolts to 7 ft. lbs. (10 Nm).
- Ignition coil assembly. Torque the bolts to 80 inch lbs. (9 Nm).
- Engine wire and tighten to 80 inch lbs. (9 Nm), 1ZZ-FE
- Right side engine mount. Tighten to 38 ft. lbs. (52 Nm).
- Front exhaust pipe
- Vane pump, 1ZZ-FE
- Compressor and magnetic clutch, 2ZZ-GE
- Alternator bracket, 2ZZ-GE engine
- Alternator
- Suction hose and wire harness clamp, 2ZZ-GE engine
- Air cleaner and hose
- Main cylinder head cover and tighten to 62 inch lbs. (7 Nm)
- Right front wheel and tire. Tighten the lug nuts to 76 ft. lbs. (103 Nm).

16. Fill the cooling system to the proper level.

17. Start the vehicle, check for leaks and repair if necessary.

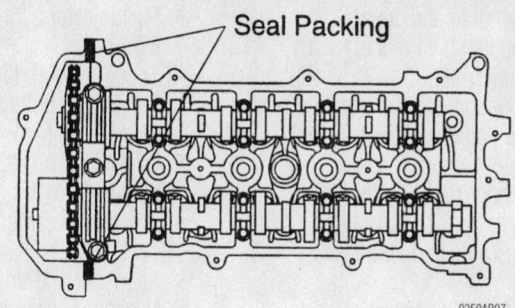

Seal packing installation locations

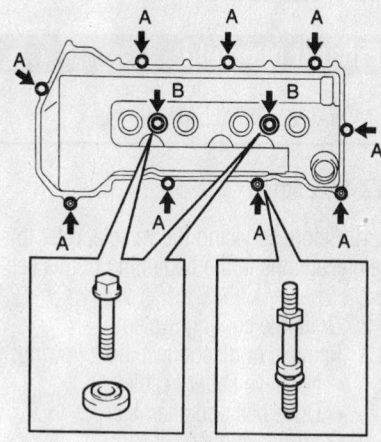

Cylinder head (valve) cover bolt locations—1ZZ-FE engine

Intake Manifold

REMOVAL & INSTALLATION

1ZZ-FE Engine

1. Before servicing the vehicle, refer to the precautions in the beginning of this section.

2. Drain the cooling system.

3. Remove or disconnect the following:
- Negative battery cable
- Drive belt and alternator
- Air intake duct
- Accelerator cable
- Exhaust pipe from the manifold.
- Exhaust manifold support bracket
- Spark plug wires, then ignition coils
- Spark plugs
- Positive Crankcase Ventilation (PCV) hoses
- Throttle body assembly
- 2 bolts securing the wiring harness protector
- Wiring connectors and ground wires
- Intake manifold support bracket
- Intake manifold and gasket

To install:

4. Install or connect the following:
- Intake manifold with a new gasket. Torque the bolts to 22 ft. lbs. (30 Nm).
- Harness wiring to the cylinder head and harness protector
- Fuel injectors, throttle body and the PCV hoses
- Spark plugs and ignition coils. Tighten the bolts and nuts to 80 inch lbs. (9 Nm).

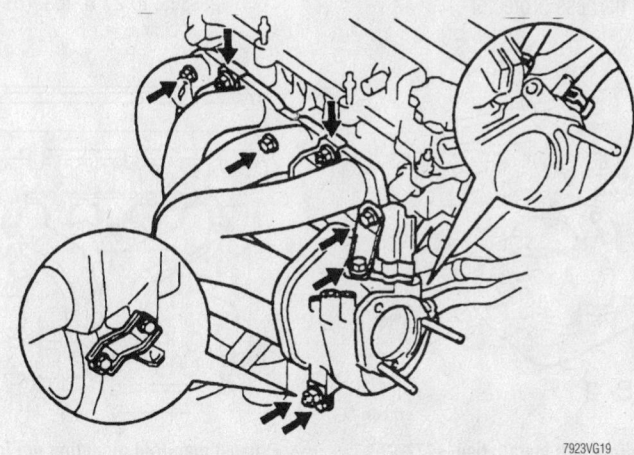

Intake manifold mounting fastener locations—1.8L (1ZZ-FE) engine

- Exhaust manifold and support bracket. Tighten the bolts to 37 ft. lbs. (49 Nm).
- Front exhaust pipe to the manifold. Tighten the bolts to 46 ft. lbs. (62 Nm).
- Oxygen Sensor (O₂S). Tighten the nuts to 14 ft. lbs. (20 Nm).
- Accelerator cable and air intake duct
- Alternator and drive belt
- Negative battery cable

5. Fill the cooling system.

6. Start the vehicle, check for leaks and repair if necessary.

2ZZ-GE Engine

1. Before servicing the vehicle, refer to the precautions in the beginning of this section.

2. Drain the cooling system.

3. Remove or disconnect the following:
- Negative battery cable
- Drive belt and alternator
- Air intake duct
- Accelerator cable
- Spark plug wires, then ignition coils
- Spark plugs
- Positive Crankcase Ventilation (PCV) hoses
- Throttle body assembly
- Wiring harness
- Hoses and tubes connected to the head
- Intake manifold support bracket
- Intake manifold and gasket

To install:

4. Install or connect the following:
- Intake manifold with a new gasket. Tighten bolts A to 25 ft. lbs. (34 Nm) and bolt B to 34 ft. lbs. (46 Nm)
- Harness wiring to the cylinder head and harness protector

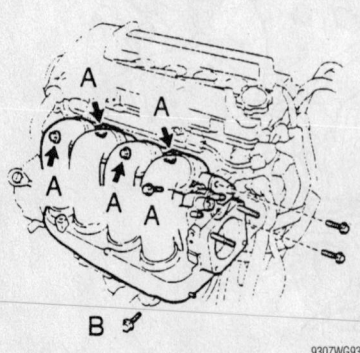

Intake manifold bolt installation—2ZZ-GE engine

9307WG93

- Fuel injectors, throttle body and the PCV hoses
- Spark plugs and ignition coils. Tighten the bolts and nuts to 80 inch lbs. (9 Nm).
- Oxygen Sensor (O₂S). Tighten the nuts to 14 ft. lbs. (20 Nm).
- Accelerator cable and air intake duct
- Alternator and drive belt
- Negative battery cable

5. Fill the cooling system.

6. Start the vehicle, check for leaks and repair if necessary.

Exhaust Manifold

REMOVAL & INSTALLATION

1ZZ-FE Engine

1. Before servicing the vehicle, refer to the precautions in the beginning of this section.

2. Drain the cooling system.

3. Remove or disconnect the following:
- Negative battery cable
- Drive belt and alternator
- Air intake duct
- Accelerator cable
- Exhaust pipe from the manifold
- Exhaust manifold support bracket
- Heat insulator from the dash panel
- Upper heat insulator
- Exhaust manifold and gasket
- If necessary, the lower heat insulator from the exhaust manifold.

To install:

4. Install or connect the following:
- Lower heat insulator on the exhaust manifold. Tighten the bolts to 108 inch lbs. (12 Nm).
- Exhaust manifold using a new gasket. Tighten the nuts, in several passes, to 27 ft. lbs. (37 Nm).

Exhaust manifold mounting nut locations—1.8L (1ZZ-FE) engine

7923VG22

- Upper heat insulator. Tighten the bolts to 108 inch lbs. (12 Nm).
- Heat insulator on the dash panel
- Exhaust manifold support bracket. Tighten the bolts, in an alternating pattern, to 37 ft. lbs. (49 Nm).
- Front exhaust pipe to the manifold. Tighten the bolts to 46 ft. lbs. (62 Nm).
- Oxygen Sensor (O₂S). Tighten the nuts to 14 ft. lbs. (20 Nm).
- Accelerator cable and air intake duct
- Alternator and drive belt
- Negative battery cable

5. Fill the cooling system.

6. Start the vehicle, check for leaks and repair if necessary.

2ZZ-GE Engine

1. Before servicing the vehicle, refer to the precautions in the beginning of this section.

2. Drain the cooling system.

3. Remove or disconnect the following:
- Negative battery cable
- Drive belt and alternator
- Air intake duct
- Accelerator cable
- Exhaust pipe from the manifold
- Exhaust manifold support bracket
- Heat insulator from the dash panel.
- Upper heat insulator
- Exhaust manifold and gasket
- If necessary, the lower heat insulator from the exhaust manifold.

To install:

4. Install or connect the following:
- Lower heat insulator on the exhaust manifold. Tighten the bolts to 15 ft. lbs. (20 Nm).
- Exhaust manifold using a new gasket. Tighten the nuts, in several passes to 37 ft. lbs. (50 Nm).
- Upper heat insulator. Tighten the bolts to 15 ft. lbs. (20 Nm).
- Heat insulator on the dash panel.
- Exhaust manifold support bracket. Tighten the bolts to 37 ft. lbs. (49 Nm).
- Front exhaust pipe to the manifold. Tighten the bolts to 46 ft. lbs. (62 Nm).
- Oxygen Sensor (O₂S). Tighten the nuts to 14 ft. lbs. (20 Nm).
- Accelerator cable and air intake duct.
- Alternator and drive belt.
- Negative battery cable

5. Fill the cooling system.

6. Start the vehicle, check for leaks and repair if necessary.

Camshaft(s)

REMOVAL & INSTALLATION

1ZZ-FE Engine

1. Before servicing the vehicle, refer to the precautions in the beginning of this section.

2. Remove or disconnect the following:
- Negative battery cable
- Right side engine under cover
- Cylinder head cover
- Suction hose sub-assembly, 2ZZ-GE engine
- Drive belt
- Power steering pump reservoir and position it aside, 1ZZ-FE engine

3. Place a jack with a wooden block under the vehicle for support, then remove the 4 bolts and 2 nuts and remove the right side engine mount.

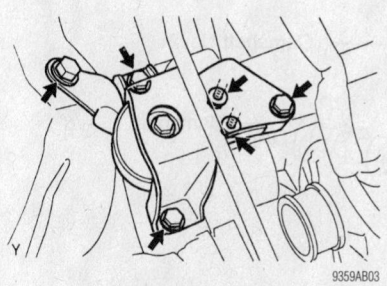

With the engine supported, remove the right side engine mount—1ZZ-FE engine shown, 2ZZ-GE similar

4. Remove the engine wire, on 1ZZ-FE engines:
a. Remove the 5 clamps from the brackets.
b. Detach the connectors.
c. Remove the ignition coil connectors.
d. Bolt and nut holding the engine wire.

5. Remove or disconnect the following:
- Ignition coil assembly
- Positive Crankcase Ventilation (PCV) hoses from the valve cover
- Valve (cylinder head) cover sub-assembly

6. Set the No. 1 cylinder to Top Dead Center (TDC) of the compressor stroke as follows:
a. Turn the crankshaft pulley, and align its groove with the "0" timing mark of the timing chain cover.
b. Make sure the point marks of the

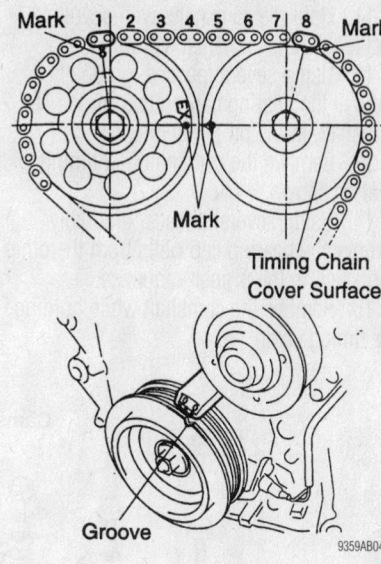

Proper timing mark alignment for TDC

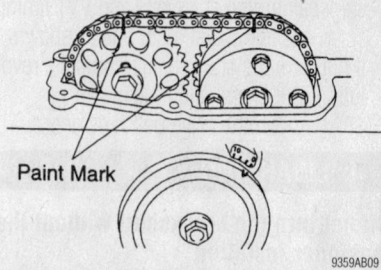

Matchmark the timing chain and cam sprockets

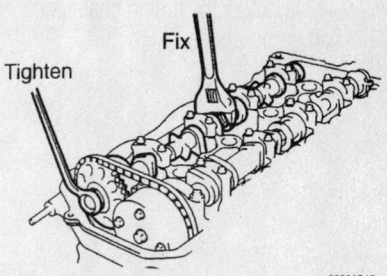

Hold the camshaft with a wrench while removing the set bolt

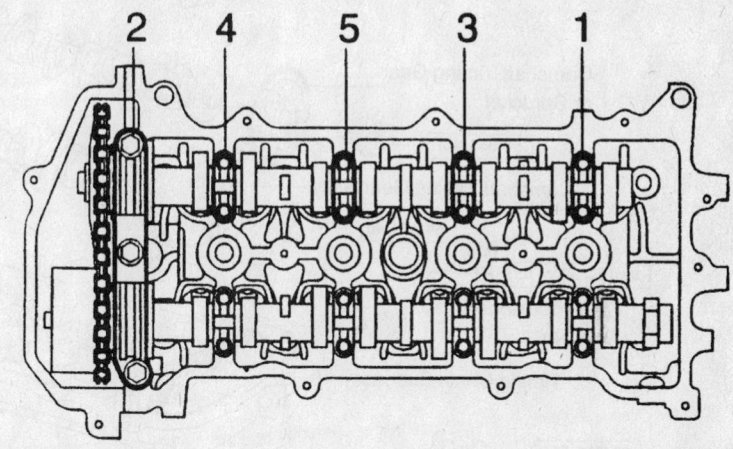

Camshaft bearing cap bolt removal sequence—1ZZ-FE engine

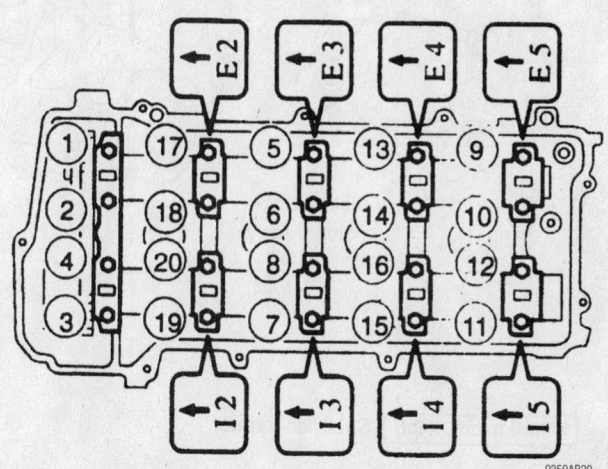

Camshaft bearing cap bolt removal sequence—2ZZ-GE engine

camshaft timing sprockets and VVT timing sprockets are in a straight line as shown. If not, turn the crankshaft 1 complete revolution (360°) and align the marks.

7. Remove the drive belt tensioner.

✳✳ WARNING

Do not turn the crankshaft without the tensioner installed.

8. Make sure the No. 1 cylinder is at TDC of the compression stroke.

9. Matchmark the timing chain and camshaft sprockets

10. Remove the 2 nuts and chain tensioner.

11. Hold the camshafts with a wrench and loosen the camshaft set bolt.

12. Using several passes, gradually remove the bearing cap bolts from the No. 2 camshaft, in the proper sequence.

13. Remove the camshaft and timing gear as shown.

14. Using several passes, gradually remove the bearing cap bolts from the other camshaft, in the proper sequence.

15. Remove the camshaft while holding the timing chain.

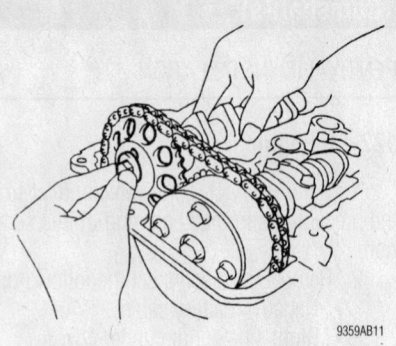

Carefully remove the cam and timing gear

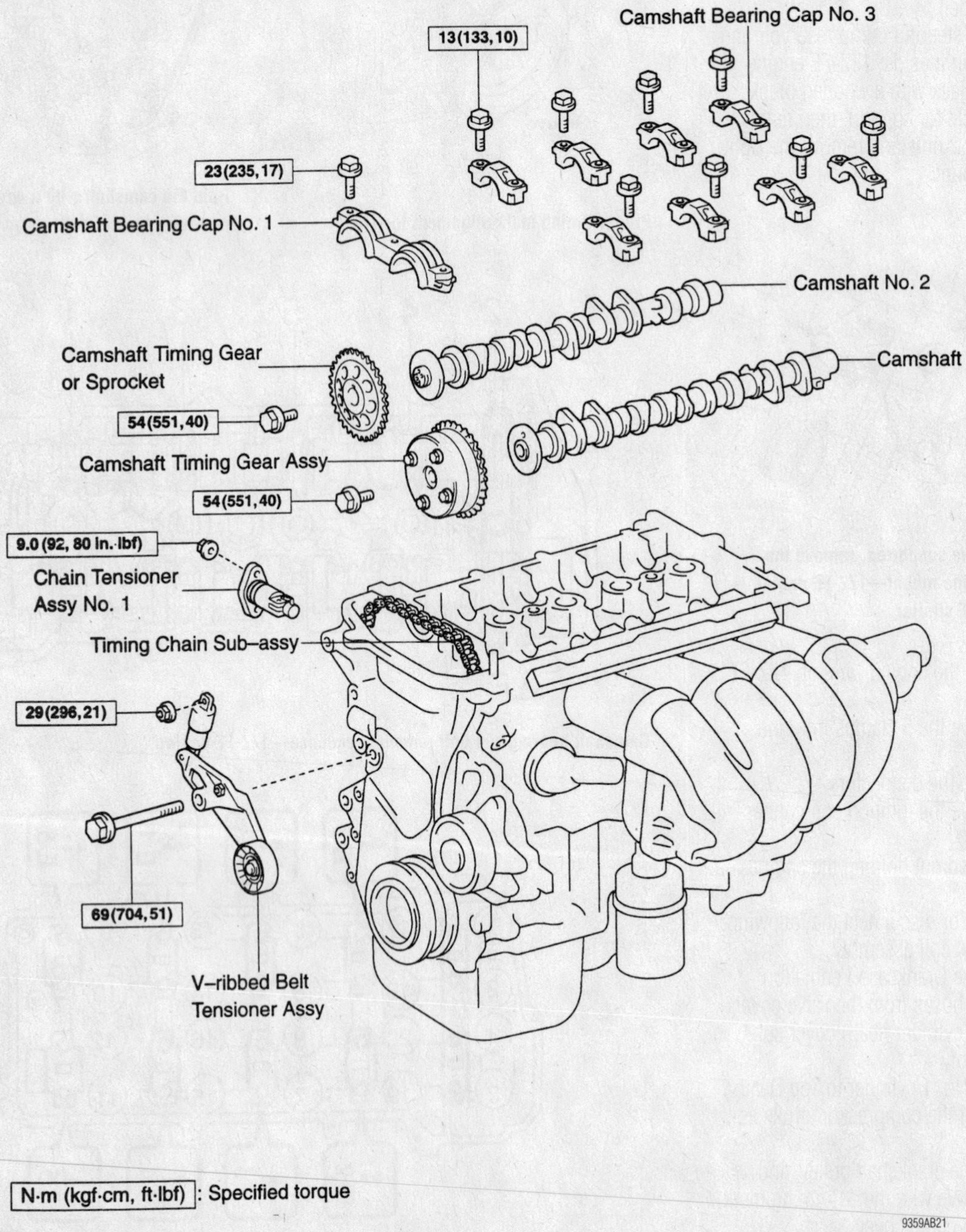

Camshaft Bearing Cap No. 3

13(133,10)

23(235,17)

Camshaft Bearing Cap No. 1

Camshaft No. 2

Camshaft Timing Gear or Sprocket

Camshaft

54(551,40)

Camshaft Timing Gear Assy

54(551,40)

9.0(92, 80 in.·lbf)

Chain Tensioner Assy No. 1

Timing Chain Sub-assy

29(296,21)

69(704,51)

V-ribbed Belt Tensioner Assy

N·m (kgf·cm, ft·lbf) : Specified torque

Exploded view of the camshafts and related components—1ZZ-FE engine

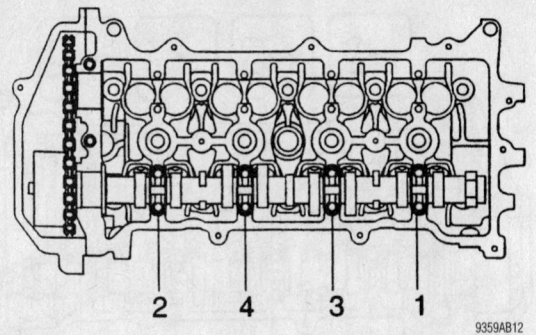

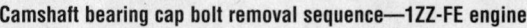

2   4   3   1

9359AB12

Camshaft bearing cap bolt removal sequence—1ZZ-FE engine

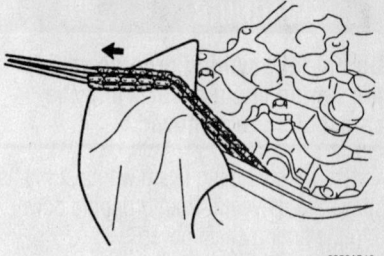

9359AB13

Secure the timing chain with string to prevent it from slipping down into the timing chain cover

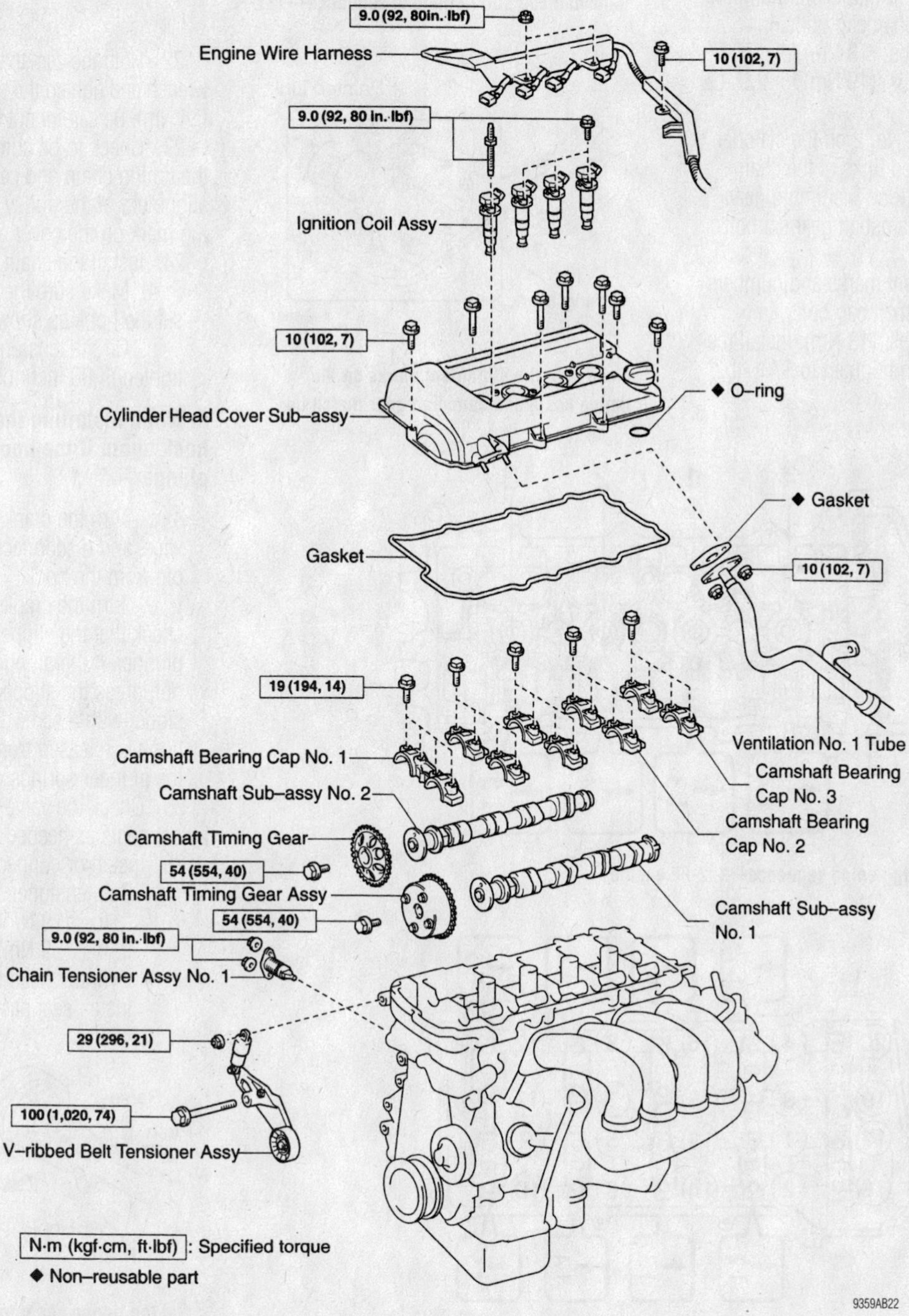

Engine Wire Harness — 9.0 (92, 80 in.·lbf)

10 (102, 7)

9.0 (92, 80 in.·lbf)

Ignition Coil Assy

10 (102, 7)

Cylinder Head Cover Sub–assy — ◆ O-ring

◆ Gasket

Gasket — ◆ Gasket

10 (102, 7)

19 (194, 14)

Ventilation No. 1 Tube

Camshaft Bearing Cap No. 1

Camshaft Sub–assy No. 2

Camshaft Bearing Cap No. 3

Camshaft Timing Gear

Camshaft Bearing Cap No. 2

54 (554, 40)

Camshaft Timing Gear Assy

54 (554, 40)

Camshaft Sub–assy No. 1

9.0 (92, 80 in.·lbf)

Chain Tensioner Assy No. 1

29 (296, 21)

100 (1,020, 74)

V–ribbed Belt Tensioner Assy

N·m (kgf·cm, ft·lbf): Specified torque
◆ Non–reusable part

9359AB22

Exploded view of the camshafts and related components—2ZZ-GE engine

✳✳ WARNING

Do not let anything drop down into the timing chain cover while the camshafts are removed.

16. Tie the timing chain with a string as shown, to prevent it from dropping down into the timing chain cover.

To install:

17. Position the camshaft on the cylinder head, then install the timing chain on the cam timing gear, with the painted links aligned with the marks on the timing gear.

18. Check the front marks and numbers and torque the camshaft cap bolts, in sequence, to 10 ft. lbs. (13 Nm) for 1ZZ-FE engine, or to 14 ft. lbs. (19 Nm) for 2ZZ-GE engines.

19. Put camshaft No. 2 on the cylinder head, with the painted links of the chain aligned with the mark on the timing gear.

20. Tighten the camshaft gear set bolt temporarily.

21. Check the front marks and numbers and torque the camshaft cap bolts, in sequence, to 10 ft. lbs. (13 Nm). Install the No. 1 bearing cap and tighten to 17 ft. lbs. (23 Nm).

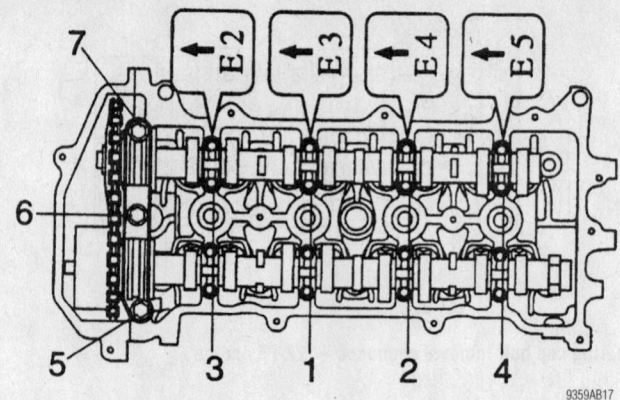

Camshaft cap bolt tightening sequence—1ZZ-FE

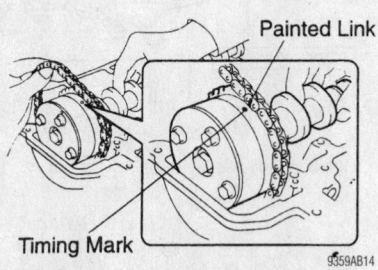
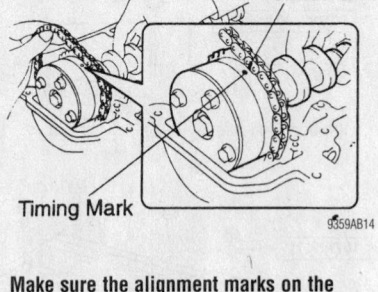

Make sure the alignment marks on the timing chain and camshaft gear match up

22. Hold the camshaft secure with a wrench and tighten the set bolt to 40 ft. lbs. (54 Nm). Be careful not the damage the lifters.

23. Check to be sure the matchmarks on the timing chain and cam sprockets, and the alignment of the pulley groove with the timing mark on the cover are still aligned.

24. Install the chain tensioner:

a. Make sure the O-ring is clean, then set the hook as shown.

b. Oil the tensioner, then install and tighten to 80 inch lbs. (9 Nm).

➡**When installing the tensioner, set the hook again if the hook releases the plunger.**

c. Turn the crankshaft counterclockwise, and disconnect the plunger knock pin from the hook.

d. Turn the crankshaft clockwise and check that the slipper is pushed by the plunger. If the plunger does not spring out, press the slipper into the chain tensioner with a screwdriver so that the hook is released from the knock pin and the plunger springs out.

25. Check the valve clearance and make adjustments as needed.

26. Install or connect the following:

- Belt tensioner. Tighten the nut to 21 ft. lbs. (29 Nm) and the bolt to 51 ft. lbs. (69 Nm).
- Cylinder head sub-assembly cover. Install seal packing into the loca-

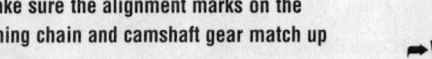
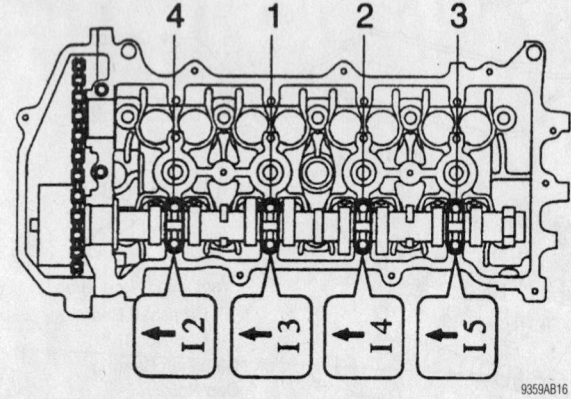

Camshaft cap bolt tightening sequence—1ZZ-FE engine

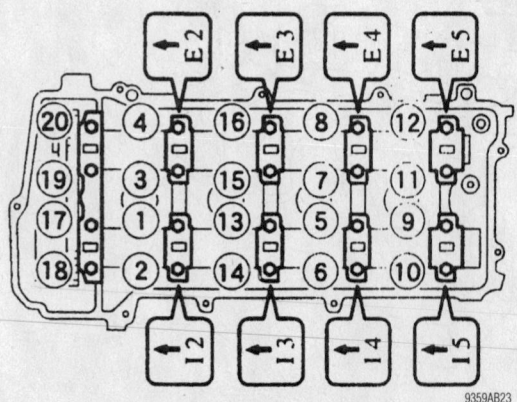

Camshaft cap bolt tightening sequence—2ZZ-GE engine

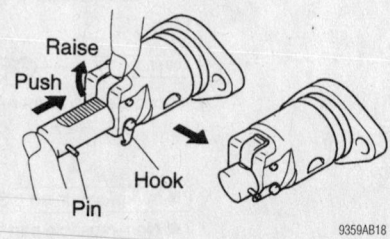

Set the timing chain tensioner hook properly

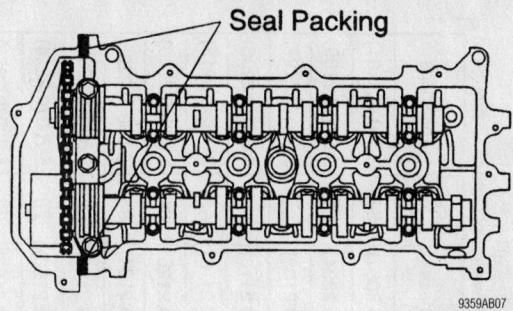

Seal packing installation locations

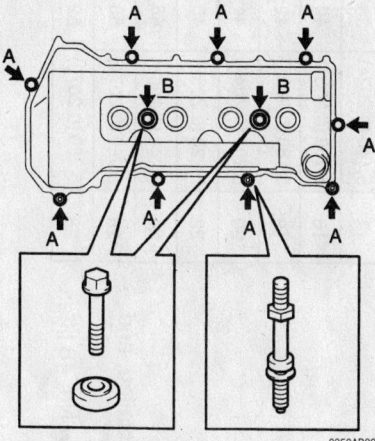

Cylinder head (valve) cover bolt locations—1ZZ-FE engine

tions shown and install within 3 minutes. Tighten the "A" bolts to 8 ft. lbs. (11 Nm) and the "B" bolts to 80 inch lbs. (9 Nm) for 1ZZ-FE engine and to 7 ft. lbs. (10 Nm) for 2ZZ-GE engines.

- Ignition coil assembly. Torque the bolts to 80 inch lbs. (9 Nm).
- Engine wire and tighten to 80 inch lbs. (9 Nm)
- Right side engine mount. Tighten to 38 ft. lbs. (52 Nm).
- Cylinder head (valve) cover
- Negative battery cable

Valve Lash

ADJUSTMENT

1ZZ-FE Engine

➡ **Adjust the valve clearance when the engine is cold.**

1. Before servicing the vehicle, refer to the precautions in the beginning of this section.

2. Remove or disconnect the following:
- Negative battery cable.

- Cylinder head covers
- Engine wire
- Ignition coil
- Positive Crankcase Ventilation (PCV) hoses
- Cylinder head cover sub-assembly

3. Set the No. 1 cylinder to Top Dead Center (TDC) of the compressor stroke as follows:

a. Turn the crankshaft pulley, and align its groove with the "0" timing mark of the timing chain cover.

b. Make sure the point marks of the

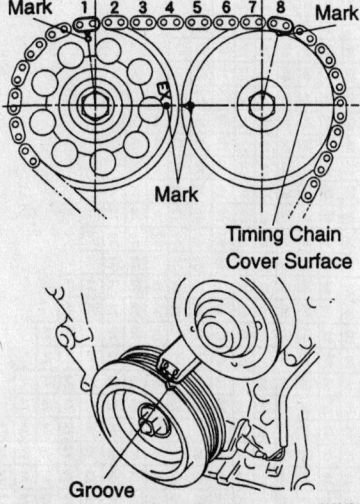

Proper timing mark alignment for TDC— 1ZZ-FE and 2ZZ-GE engines

camshaft timing sprockets and VVT timing sprockets are in a straight line as shown. If not, turn the crankshaft 1 complete revolution (360°) and align the marks.

4. Check the valve clearance of the first set of the valves shown:

a. Use a feeler gauge to measure the clearance between the valve lifter and camshaft. The clearance of the intake valves should be 0.0059–0.0098 in. (0.15–0.25mm). The clearance of the exhaust valves should be 0.0098–0.0138 in. (0.25–0.35mm).

b. Note the out-of-specification valve clearance measurements. You will need them later to determine the required replacement valve lifter.

c. Turn the crankshaft 1 revolution (360°) to set the No. 4 cylinder to TDC.

5. Check the valve clearance of the second set of the valves shown:

a. Use a feeler gauge to measure the clearance between the valve lifter and camshaft. The clearance of the intake valves should be 0.0059–0.0098 in. (0.15–0.25mm). The clearance of the exhaust valves should be 0.0098–0.0138 in. (0.25–0.35mm).

b. Note the out-of-specification valve clearance measurements. You will need them later to determine the required replacement valve lifter.

6. Remove or disconnect the following:
- Drive belt
- Right side engine mount
- Drive belt tensioner

❊❊ WARNING

DO NOT turn the crankshaft while the tensioner is removed!

7. Set the No. 1 cylinder to TDC of the compression stroke.
- Camshafts
- Valve lifters.

8. Use a micrometer to measure the thickness of the used lifter. Calculate the

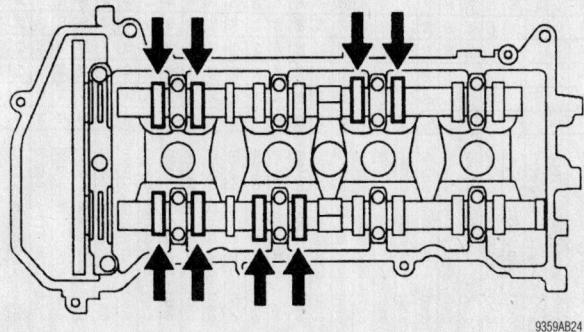

Check the clearance of the 1st set of valves–1ZZ-FE engine

1ZZ–FE: Valve Lifter Selection Chart (Intake)

New lifter thickness   mm (in.)

Lifter No.	Thickness	Lifter No.	Thickness	Lifter No.	Thickness
06	5.060 (0.1992)	30	5.300 (0.2087)	54	5.540 (0.2181)
08	5.080 (0.2000)	32	5.320 (0.2094)	56	5.560 (0.2189)
10	5.100 (0.2008)	34	5.340 (0.2102)	58	5.580 (0.2197)
12	5.120 (0.2016)	36	5.360 (0.2110)	60	5.600 (0.2205)
14	5.140 (0.2024)	38	5.380 (0.2118)	62	5.620 (0.2213)
16	5.160 (0.2031)	40	5.400 (0.2126)	64	5.640 (0.2220)
18	5.180 (0.2039)	42	5.420 (0.2134)	66	5.660 (0.2228)
20	5.200 (0.2047)	44	5.440 (0.2142)	68	5.680 (0.2236)
22	5.220 (0.2055)	46	5.460 (0.2150)	70	5.700 (0.2244)
24	5.240 (0.2063)	48	5.480 (0.2157)	72	5.720 (0.2252)
26	5.260 (0.2071)	50	5.500 (0.2165)	74	5.740 (0.2260)
28	5.280 (0.2079)	52	5.520 (0.2173)		

9307WG70

Intake valve clearance (Cold):
0.15 – 0.25 mm (0.006 – 0.010 in.)

EXAMPLE: The 5.250 mm (0.2067 in.) lifter is installed, and the measured clearance is 0.400 mm (0.0157 in.).
Replace the 5.250 mm (0.2067 in.) lifter with a new No. 48 lifter.

Adjusting shim chart (intake)—1ZZ-FE engine

Valve Lifter Selection Chart (Intake) — installed lifter thickness across the top (mm (in.)): 5.060 (0.1992), 5.100 (0.2008), 5.120 (0.2016), 5.140 (0.2024), 5.160 (0.2031), 5.180 (0.2039), 5.200 (0.2047), 5.210 (0.2051), 5.220 (0.2055), 5.230 (0.2059), 5.240 (0.2063), 5.250 (0.2067), 5.260 (0.2071), 5.270 (0.2075), 5.280 (0.2079), 5.290 (0.2083), 5.300 (0.2087), 5.310 (0.2091), 5.320 (0.2094), 5.330 (0.2098), 5.340 (0.2102), 5.350 (0.2106), 5.360 (0.2110), 5.370 (0.2114), 5.380 (0.2118), 5.390 (0.2122), 5.400 (0.2126), 5.410 (0.2130), 5.420 (0.2134), 5.430 (0.2138), 5.440 (0.2142), 5.450 (0.2146), 5.460 (0.2150), 5.470 (0.2154), 5.480 (0.2157), 5.490 (0.2161), 5.500 (0.2165), 5.510 (0.2169), 5.520 (0.2173), 5.530 (0.2177), 5.540 (0.2181), 5.550 (0.2185), 5.560 (0.2189), 5.570 (0.2193), 5.580 (0.2197), 5.590 (0.2201), 5.600 (0.2205), 5.620 (0.2213), 5.640 (0.2220), 5.660 (0.2228), 5.680 (0.2236), 5.700 (0.2244), 5.720 (0.2252), 5.740 (0.2260).

Measured clearance mm (in.) down the left side:

Measured clearance mm (in.)
0.000 – 0.030 (0.0000 – 0.0012)
0.031 – 0.050 (0.0012 – 0.0020)
0.051 – 0.070 (0.0020 – 0.0028)
0.071 – 0.090 (0.0028 – 0.0035)
0.091 – 0.110 (0.0036 – 0.0043)
0.111 – 0.130 (0.0044 – 0.0051)
0.131 – 0.149 (0.0052 – 0.0059)
0.150 – 0.250 (0.0059 – 0.0098)
0.251 – 0.270 (0.0099 – 0.0105)
0.271 – 0.290 (0.0107 – 0.0114)
0.291 – 0.310 (0.0115 – 0.0122)
0.311 – 0.330 (0.0122 – 0.0130)
0.331 – 0.350 (0.0130 – 0.0138)
0.351 – 0.370 (0.0138 – 0.0146)
0.371 – 0.390 (0.0146 – 0.0154)
0.391 – 0.410 (0.0154 – 0.0161)
0.411 – 0.430 (0.0162 – 0.0169)
0.431 – 0.450 (0.0170 – 0.0177)
0.451 – 0.470 (0.0178 – 0.0185)
0.471 – 0.490 (0.0185 – 0.0193)
0.491 – 0.510 (0.0193 – 0.0201)
0.511 – 0.530 (0.0201 – 0.0209)
0.531 – 0.550 (0.0209 – 0.0217)
0.551 – 0.570 (0.0217 – 0.0224)
0.571 – 0.590 (0.0225 – 0.0232)
0.591 – 0.610 (0.0233 – 0.0240)
0.611 – 0.630 (0.0241 – 0.0248)
0.631 – 0.650 (0.0248 – 0.0256)
0.651 – 0.670 (0.0256 – 0.0264)
0.671 – 0.690 (0.0264 – 0.0272)
0.691 – 0.710 (0.0272 – 0.0280)
0.711 – 0.730 (0.0280 – 0.0287)
0.731 – 0.750 (0.0287 – 0.0295)
0.751 – 0.770 (0.0296 – 0.0303)
0.771 – 0.790 (0.0304 – 0.0311)
0.791 – 0.810 (0.0311 – 0.0319)
0.811 – 0.830 (0.0319 – 0.0327)
0.831 – 0.850 (0.0327 – 0.0335)
0.851 – 0.870 (0.0335 – 0.0343)
0.871 – 0.890 (0.0343 – 0.0350)
0.891 – 0.910 (0.0351 – 0.0358)
0.911 – 0.930 (0.0359 – 0.0366)

1ZZ–FE: Valve Lifter Selection Chart (Exhaust)

New lifter thickness — mm (in.)

Lifter No.	Thickness	Lifter No.	Thickness	Lifter No.	Thickness
06	5.060 (0.1992)	30	5.300 (0.2087)	54	5.540 (0.2181)
08	5.080 (0.2000)	32	5.320 (0.2094)	56	5.560 (0.2189)
10	5.100 (0.2008)	34	5.340 (0.2102)	58	5.580 (0.2197)
12	5.120 (0.2016)	36	5.360 (0.2110)	60	5.600 (0.2205)
14	5.140 (0.2024)	38	5.380 (0.2118)	62	5.620 (0.2213)
16	5.160 (0.2031)	40	5.400 (0.2126)	64	5.640 (0.2220)
18	5.180 (0.2039)	42	5.420 (0.2134)	66	5.660 (0.2228)
20	5.200 (0.2047)	44	5.440 (0.2142)	68	5.680 (0.2236)
22	5.220 (0.2055)	46	5.460 (0.2150)	70	5.700 (0.2244)
24	5.240 (0.2063)	48	5.480 (0.2157)	72	5.720 (0.2252)
26	5.260 (0.2071)	50	5.500 (0.2165)	74	5.740 (0.2260)
28	5.280 (0.2079)	52	5.520 (0.2173)		

Exhaust valve clearance (Cold):
0.25 – 0.35 mm (0.010 – 0.014 in.)

EXAMPLE: The 5.340 mm (0.2102 in.) lifter is installed, and the measured clearance is 0.440 mm (0.0173 in.).
Replace the 5.340 mm (0.2102 in.) lifter with a new No. 48 lifter.

Adjusting shim chart (exhaust)—1ZZ-FE engine

Valve Lifter Selection Chart (Exhaust) — matrix of installed lifter thickness (columns, mm/in. from 5.060 (0.1992) to 5.740 (0.2260)) versus measured clearance (rows, mm/in.), giving the required new lifter number.

Measured clearance mm (in.) column:

Measured clearance mm	(in.)
0.000 – 0.030	(0.0000 – 0.0012)
0.031 – 0.050	(0.0012 – 0.0020)
0.051 – 0.070	(0.0020 – 0.0028)
0.071 – 0.090	(0.0028 – 0.0035)
0.091 – 0.110	(0.0036 – 0.0043)
0.111 – 0.130	(0.0044 – 0.0051)
0.131 – 0.150	(0.0052 – 0.0059)
0.151 – 0.170	(0.0059 – 0.0067)
0.171 – 0.190	(0.0067 – 0.0075)
0.191 – 0.210	(0.0075 – 0.0083)
0.211 – 0.230	(0.0083 – 0.0091)
0.231 – 0.249	(0.0091 – 0.0098)
0.250 – 0.350	(0.0098 – 0.0138)
0.351 – 0.370	(0.0138 – 0.0146)
0.371 – 0.390	(0.0146 – 0.0154)
0.391 – 0.410	(0.0154 – 0.0161)
0.411 – 0.430	(0.0162 – 0.0169)
0.431 – 0.450	(0.0170 – 0.0177)
0.451 – 0.470	(0.0178 – 0.0185)
0.471 – 0.490	(0.0185 – 0.0193)
0.491 – 0.510	(0.0193 – 0.0201)
0.511 – 0.530	(0.0201 – 0.0209)
0.531 – 0.550	(0.0209 – 0.0217)
0.551 – 0.570	(0.0217 – 0.0224)
0.571 – 0.590	(0.0225 – 0.0232)
0.591 – 0.610	(0.0233 – 0.0240)
0.611 – 0.630	(0.0241 – 0.0248)
0.631 – 0.650	(0.0248 – 0.0256)
0.651 – 0.670	(0.0256 – 0.0264)
0.671 – 0.690	(0.0264 – 0.0272)
0.691 – 0.710	(0.0272 – 0.0280)
0.711 – 0.730	(0.0280 – 0.0287)
0.731 – 0.750	(0.0288 – 0.0295)
0.751 – 0.770	(0.0296 – 0.0303)
0.771 – 0.790	(0.0304 – 0.0311)
0.791 – 0.810	(0.0311 – 0.0319)
0.811 – 0.830	(0.0319 – 0.0327)
0.831 – 0.850	(0.0327 – 0.0335)
0.851 – 0.870	(0.0335 – 0.0343)
0.871 – 0.890	(0.0343 – 0.0350)
0.891 – 0.910	(0.0351 – 0.0358)
0.911 – 0.930	(0.0359 – 0.0365)
0.931 – 0.950	(0.0365 – 0.0374)
0.951 – 0.970	(0.0374 – 0.0382)
0.971 – 0.990	(0.0382 – 0.0390)
0.991 – 1.010	(0.0390 – 0.0398)
1.011 – 1.030	(0.0398 – 0.0406)

9307WG71

Adjusting shim chart (intake)—2ZZ-GE engine

New Shim thickness   mm (in.)

Shim No.	Thickness	Shim No.	Thickness	Shim No.	Thickness
00	2.000 (0.0787)	28	2.280 (0.0898)	56	2.560 (0.1008)
02	2.020 (0.0795)	30	2.300 (0.0906)	58	2.580 (0.1016)
04	2.040 (0.0803)	32	2.320 (0.0913)	60	2.600 (0.1024)
06	2.060 (0.0811)	34	2.340 (0.0921)	62	2.620 (0.1031)
08	2.080 (0.0819)	36	2.360 (0.0929)	64	2.640 (0.1039)
10	2.100 (0.0827)	38	2.380 (0.0937)	66	2.660 (0.1047)
12	2.120 (0.0835)	40	2.400 (0.0945)	68	2.680 (0.1055)
14	2.140 (0.0843)	42	2.420 (0.0953)	70	2.700 (0.1063)
16	2.160 (0.0850)	44	2.440 (0.0961)	72	2.720 (0.1071)
18	2.180 (0.0858)	46	2.460 (0.0969)	74	2.740 (0.1079)
20	2.200 (0.0866)	48	2.480 (0.0976)	76	2.760 (0.1087)
22	2.220 (0.0874)	50	2.500 (0.0984)	78	2.780 (0.1094)
24	2.240 (0.0882)	52	2.520 (0.0992)	80	2.800 (0.1102)
26	2.260 (0.0890)	54	2.540 (0.1000)		

Intake valve clearance (Cold):
0.08 – 0.18 mm (0.0031 – 0.0071 in.)

EXAMPLE: The 2.200 mm (0.0826 in.) shim is installed, and
the measured clearance is 0.400 mm (0.0157 in.).
Replace the 2.600 mm (0.1024 in.) shim with a new No. 60 shim.

Adjusting shim chart (intake)—2ZZ-GE engine

The large adjusting shim chart cross-references "Installed shim thickness mm(in.)" (columns, from 2.000 (0.0787) through 2.800 (0.1102)) against "Measure clearance mm(in.)" (rows):

Measure clearance mm (in.)
0.000 – 0.030 (0.0000 – 0.0012)
0.031 – 0.050 (0.0012 – 0.0020)
0.051 – 0.070 (0.0020 – 0.0028)
0.071 – 0.099 (0.0028 – 0.0035)
0.100 – 0.160 (0.0036 – 0.0039)
0.161 – 0.180 (0.0063 – 0.0071)
0.181 – 0.200 (0.0071 – 0.0079)
0.201 – 0.220 (0.0079 – 0.0087)
0.221 – 0.240 (0.0087 – 0.0094)
0.241 – 0.260 (0.0095 – 0.0102)
0.261 – 0.280 (0.0103 – 0.0110)
0.281 – 0.300 (0.0111 – 0.0118)
0.301 – 0.320 (0.0119 – 0.0126)
0.321 – 0.340 (0.0126 – 0.0134)
0.341 – 0.360 (0.0134 – 0.0142)
0.361 – 0.380 (0.0142 – 0.0150)
0.381 – 0.400 (0.0150 – 0.0157)
0.401 – 0.420 (0.0158 – 0.0165)
0.421 – 0.440 (0.0166 – 0.0173)
0.441 – 0.460 (0.0174 – 0.0181)
0.461 – 0.480 (0.0181 – 0.0189)
0.481 – 0.500 (0.0189 – 0.0197)
0.501 – 0.520 (0.0197 – 0.0205)
0.521 – 0.540 (0.0205 – 0.0213)
0.541 – 0.560 (0.0213 – 0.0220)
0.561 – 0.580 (0.0221 – 0.0228)
0.581 – 0.600 (0.0229 – 0.0236)
0.601 – 0.620 (0.0237 – 0.0244)
0.621 – 0.640 (0.0244 – 0.0252)
0.641 – 0.660 (0.0252 – 0.0260)
0.661 – 0.680 (0.0260 – 0.0268)

9359AB26

New Shim thickness mm (in.)

Shim No.	Thickness	Shim No.	Thickness	Shim No.	Thickness
00	2.000 (0.0787)	28	2.280 (0.0898)	56	2.560 (0.1008)
02	2.020 (0.0795)	30	2.300 (0.0906)	58	2.580 (0.1016)
04	2.040 (0.0803)	32	2.320 (0.0913)	60	2.600 (0.1024)
06	2.060 (0.0811)	34	2.340 (0.0921)	62	2.620 (0.1031)
08	2.080 (0.0819)	36	2.360 (0.0929)	64	2.640 (0.1039)
10	2.100 (0.0827)	38	2.380 (0.0937)	66	2.660 (0.1047)
12	2.120 (0.0835)	40	2.400 (0.0945)	68	2.680 (0.1055)
14	2.140 (0.0843)	42	2.420 (0.0953)	70	2.700 (0.1063)
16	2.160 (0.0850)	44	2.440 (0.0961)	72	2.720 (0.1071)
18	2.180 (0.0858)	46	2.460 (0.0969)	74	2.740 (0.1079)
20	2.200 (0.0866)	48	2.480 (0.0976)	76	2.760 (0.1087)
22	2.220 (0.0874)	50	2.500 (0.0984)	78	2.780 (0.1094)
24	2.240 (0.0882)	52	2.520 (0.0992)	80	2.800 (0.1102)
26	2.260 (0.0890)	54	2.540 (0.1000)		

Exhaust valve clearance (Cold):
0.22 – 0.32 mm (0.0087 – 0.0126 in.)
EXAMPLE: The 2.200 mm (0.0862 in.) shim is installed, and
the measured clearance is 0.500 mm (0.0197 in.).
Replace the 2.540 mm (0.1000 in.) shim with a new No. 54 shim.

Adjusting shim chart (exhaust)—2ZZ-GE engine

9359AB27

Adjusting shim selection chart (exhaust)

Installed shim thickness mm (in.) — column headers (top row, with shim No.):

2.000 (0.0787), 2.020 (0.0795), 2.040 (0.0803), 2.060 (0.0811), 2.080 (0.0819), 2.100 (0.0827), 2.120 (0.0835), 2.140 (0.0843), 2.160 (0.0850), 2.180 (0.0858), 2.190 (0.0862), 2.200 (0.0866), 2.210 (0.0870), 2.220 (0.0874), 2.230 (0.0878), 2.240 (0.0882), 2.250 (0.0886), 2.260 (0.0890), 2.270 (0.0894), 2.280 (0.0898), 2.290 (0.0902), 2.300 (0.0906), 2.310 (0.0909), 2.320 (0.0913), 2.330 (0.0917), 2.340 (0.0921), 2.340 (0.0921), 2.350 (0.0925), 2.360 (0.0929), 2.370 (0.0933), 2.380 (0.0937), 2.390 (0.0941), 2.400 (0.0945), 2.410 (0.0949), 2.420 (0.0953), 2.430 (0.0957), 2.440 (0.0961), 2.450 (0.0965), 2.460 (0.0969), 2.470 (0.0972), 2.480 (0.0976), 2.490 (0.0980), 2.500 (0.0984), 2.510 (0.0988), 2.520 (0.0992), 2.530 (0.0996), 2.540 (0.1000), 2.550 (0.1004), 2.560 (0.1008), 2.580 (0.1016), 2.600 (0.1024), 2.620 (0.1031), 2.640 (0.1039), 2.660 (0.1047), 2.680 (0.1055), 2.700 (0.1063), 2.720 (0.1071), 2.740 (0.1079), 2.760 (0.1087), 2.780 (0.1094), 2.800 (0.1102)

Measure clearance mm (in.) — row headers (left):

Measure clearance mm (in.)	
0.000 – 0.030	(0.0000 – 0.0012)
0.031 – 0.050	(0.0012 – 0.0020)
0.051 – 0.070	(0.0020 – 0.0028)
0.071 – 0.090	(0.0028 – 0.0035)
0.091 – 0.110	(0.0036 – 0.0043)
0.111 – 0.130	(0.0044 – 0.0051)
0.131 – 0.150	(0.0052 – 0.0059)
0.151 – 0.170	(0.0059 – 0.0067)
0.171 – 0.190	(0.0067 – 0.0075)
0.191 – 0.210	(0.0075 – 0.0083)
0.211 – 0.230	(0.0083 – 0.0091)
0.231 – 0.239	(0.0091 – 0.0094)
0.321 – 0.340	(0.0126 – 0.0134)
0.341 – 0.360	(0.0134 – 0.0142)
0.361 – 0.380	(0.0142 – 0.0150)
0.381 – 0.400	(0.0150 – 0.0157)
0.401 – 0.420	(0.0157 – 0.0165)
0.421 – 0.440	(0.0165 – 0.0173)
0.441 – 0.460	(0.0174 – 0.0181)
0.461 – 0.480	(0.0181 – 0.0189)
0.481 – 0.500	(0.0189 – 0.0197)
0.501 – 0.520	(0.0197 – 0.0205)
0.521 – 0.540	(0.0205 – 0.0213)
0.541 – 0.560	(0.0213 – 0.0220)
0.561 – 0.580	(0.0221 – 0.0228)
0.581 – 0.600	(0.0228 – 0.0236)
0.601 – 0.620	(0.0236 – 0.0244)
0.621 – 0.640	(0.0244 – 0.0252)
0.641 – 0.660	(0.0252 – 0.0260)
0.661 – 0.680	(0.0260 – 0.0268)
0.681 – 0.700	(0.0268 – 0.0276)
0.701 – 0.720	(0.0276 – 0.0283)
0.721 – 0.740	(0.0284 – 0.0291)
0.741 – 0.760	(0.0292 – 0.0299)
0.761 – 0.780	(0.0299 – 0.0307)
0.781 – 0.800	(0.0307 – 0.0315)
0.801 – 0.820	(0.0315 – 0.0323)

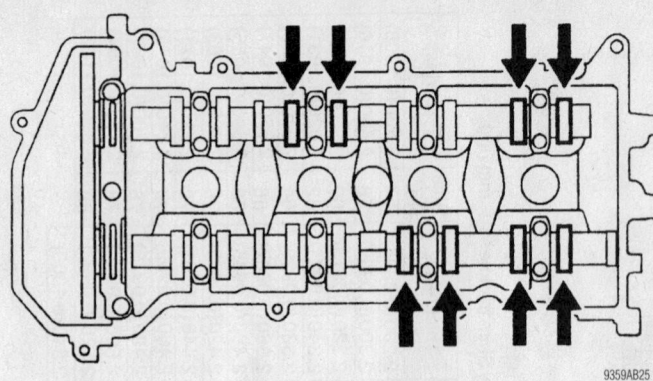

Check the clearance of the 2nd set of valves–1ZZ-FE engine

9359AB25

thickness of a new lifter. so the valve clearance comes within the specified value:

 a. A: Thickness of new lifter.

 b. B: Thickness of used lifter.

 c. C: Measured valve clearance.

 d. Intake valve clearance: A = B + (C—0.0079 in. (0.20mm).

 e. Exhaust valve clearance: A = B + (C—0.0118 in. (0.30mm).

 f. Select a new lifter with a thickness as close as possible to the calculated values. Lifters come in 35 sizes in increments of 0.0008 in. (0.020mm) from 0.1992–0.2260 in (5.060–5.740mm).

9. Install or connect the following:

- Camshafts
- Drive belt tensioner
- Right hand engine mount
- Cylinder head (valve) cover sub-assembly
- Ignition coil
- Engine wire
- Cylinder head (valve) cover
- Negative battery cable

2ZZ-GE Engine

➡**Adjust the valve clearance when the engine is cold.**

 1. Before servicing the vehicle, refer to the precautions in the beginning of this section.

 2. Remove or disconnect the following:

- Negative battery cable.
- Right side engine under cover
- Cylinder head cover
- Ignition coil assembly
- Wire harness clamp
- Suction hose sub-assembly
- Cylinder head cover sub-assembly
- Drive belt
- Right side engine mount

 3. Set the No. 1 cylinder to Top Dead Center (TDC) of the compressor stroke as follows:

 a. Turn the crankshaft pulley, and align its groove with the "0" timing mark of the timing chain cover.

 b. Make sure the point marks of the camshaft timing sprockets and VVT timing sprockets are in a straight line as shown. If not, turn the crankshaft 1 complete revolution (360°) and align the marks.

 4. Check the valve clearance of the first set of the valves shown:

 a. Use a feeler gauge to measure the clearance between the valve lifter and camshaft. The clearance of the intake valves should be 0.0031–0.0071 in.

(0.08–0.18mm). The clearance of the exhaust valves should be 0.0087–0.0126 in. (0.22–0.32mm).

 b. Note the out-of-specification valve clearance measurements. You will need them later to determine the required replacement valve lifter.

 c. Turn the crankshaft 1 revolution (360°) to set the No. 4 cylinder to TDC.

 5. Check the valve clearance of the second set of the valves shown:

 a. Use a feeler gauge to measure the clearance between the valve lifter and camshaft. The clearance of the intake valves should be 0.0031–0.0071 in. (0.08–0.18mm). The clearance of the exhaust valves should be 0.0087–0.0126 in. (0.22–0.32mm).

 b. Note the out-of-specification valve clearance measurements. You will need them later to determine the required replacement valve lifter.

 6. To adjust the intake valve clearance:

 a. Set the SST. Turn the crankshaft so the related rocker arm, where the valve clearance is adjusted, is fully pushed down.

➡**Remove the spark plug and take off the compression.**

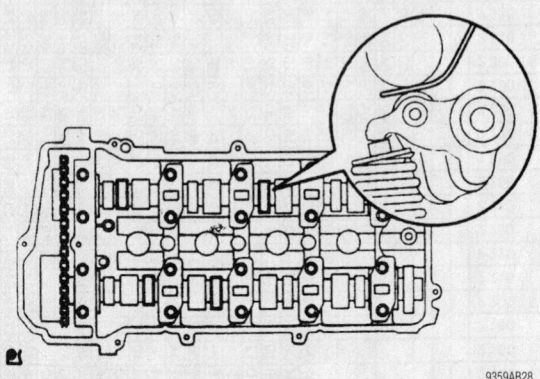

Check the clearance of the 1st set of valves–2ZZ-GE engine

9359AB28

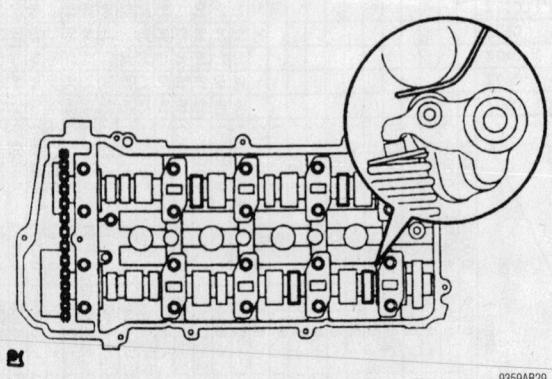

Check the clearance of the 2nd set of valves–2ZZ-GE engine

9359AB29

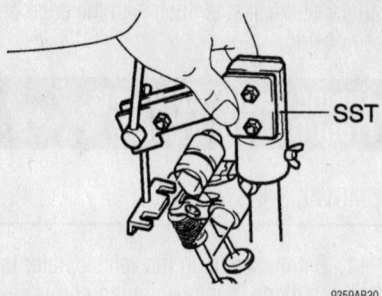

Insert the special tool into the plug tube—2ZZ-GE

b. Insert SST 09248-77010 into the plug tube. The tool cannot be inserted unless the set screw is loosened.

c. Operate the lever so that the SST's seat surface comes to contact with the valve retainer and lock them with the set screw. Clearance between the valve retainer and SST's set surface is not allowed. Be careful not to make clearance when inserting the SST, since clearance may unlock the keeper.

d. Lock the set screw on the tube side of the SST.

e. Rotate the crankshaft so that the camshaft is position as shown. During rotation, pay attention to the direction, to prevent the nose of the camshaft from interfering with the SST's shaft. Do not rotate the crankshaft excessively.

f. Lift the rocker arm to make room and remove the adjusting shim using SST 09248-77010.

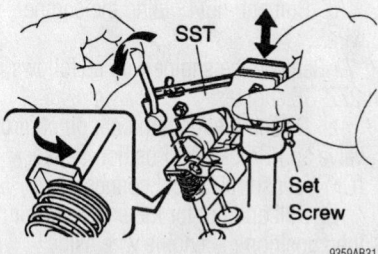

Operate the lever so that the SST's seat surface comes to contact with the valve retainer and lock them with the set screw

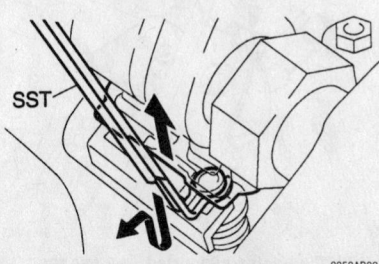

Setting the tool from the right side, makes shim removal easier—2ZZ-GE

7. Determine the size of the replaced shim according to the chart or the following formula:

a. Use a dial indicator to measure the thickness of the removed shim.

b. Calculate the thickness of a new shim so that the valve clearance comes within the specified value.

c. A: Thickness of new shim.

d. B: Thickness of used shim.

e. C: Measured valve clearance.

f. Intake: $A = B + (C—0.005$ in. [0.13mm])

g. Exhaust: $A = B + (C—0.011$ in. [0.27mm])

h. Select a new shim with a thickness as close as possible to the calculated values. Shims come in 41 sizes in increments of 0.0008 in. (0.020mm) from 0.0787–0.1102 in (2.0–2.8mm).

8. Lift the rocker arm to make room, then install the adjusting shim using the SST. To remove the tool from the shim, push down on the rocker arm.

9. Turn the crankshaft so the related rocker arm, where the valve clearance is adjusted, is fully pushed down.

10. Loosen the 2 set-screws, then remove the SST.

11. Install all components in the reverse of the removal procedure.

Starter

REMOVAL & INSTALLATION

1. Before servicing the vehicle, refer to the precautions in the beginning of this section.

2. Remove or disconnect the following:
- Negative battery cable
- Right side engine undercover
- Starter wiring
- Starter

3. Installation is the reverse of removal. Torque the bolts to 27 ft. lbs. (37 Nm) and the nut to 7 ft. lbs. (10 Nm).

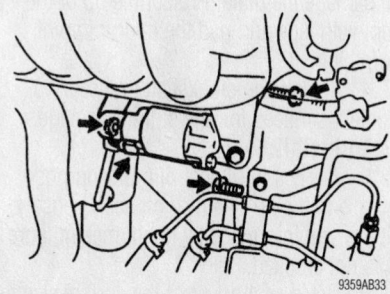

Starter mounting—Matrix

Oil Pan

REMOVAL & INSTALLATION

1ZZ-FE Engine

1. Before servicing the vehicle, refer to the precautions in the beginning of this section.

2. Drain the engine oil.

3. Remove or disconnect the following:
- Negative battery cable
- Undercovers
- Front exhaust pipe
- Oil pan mounting bolts and nuts
- Oil pan, cutting off the applied sealer.

To install:

4. Remove any old sealant from the oil pan flange and thoroughly clean the sealing surface.

5. Install or connect the following:
- Oil pan. Tighten the bolts and nuts in several passes to 80 inch lbs. (9 Nm).
- Front exhaust pipe
- Negative battery cable
- Undercovers

6. Fill the engine with clean oil.

7. Start the vehicle, check for leaks and repair if necessary.

2ZZ-GE Engine

1. Before servicing the vehicle, refer to the precautions in the beginning of this section.

2. Drain the engine oil.

3. Remove or disconnect the following:
- Negative battery cable. On vehicles

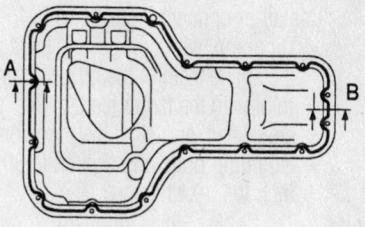

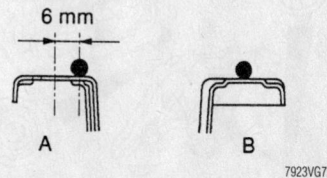

Seal Width
4 – 5 mm

Apply sealant to the oil pan as shown—1.8L (1ZZ-FE) engine

equipped with an air bag, wait at least 90 seconds before proceeding.
- Undercovers
- Front exhaust pipe
- Oil pan mounting bolts and nuts
- Oil pan, cutting off the applied sealer

To install:

4. Remove any old sealant from the oil pan flange and thoroughly clean the sealing surface.

5. Install or connect the following:
- Oil pan. Tighten the bolts and nuts in several passes to 80 inch lbs. (9 Nm).
- Front exhaust pipe
- Negative battery cable
- Undercovers

6. Fill the engine with clean oil.

7. Start the vehicle, check for leaks and repair if necessary.

Oil Pump

REMOVAL & INSTALLATION

1ZZ-FE Engine

1. Before servicing the vehicle, refer to the precautions in the beginning of this section.

2. Drain the engine oil.

3. Remove or disconnect the following:
- Negative battery cable
- Timing chain and crankshaft sprocket
- Timing chain vibration damper
- Oil pump bolts, pump and gasket

To install:

4. Clean the mounting surface.

5. Install or connect the following:
- Oil pump, with new gasket. Engage the spline teeth of the oil pump drive rotor with the larger teeth of the crankshaft, and slide the pump on.
- Oil pump bolts and tighten to 80 inch lbs. (9 Nm)

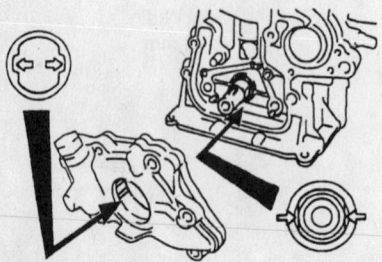

Oil pump mounting—1ZZ-FE and 2ZZ-GE engines

- Crankshaft vibration damper and tighten to 80 inch lbs. (9 Nm)
- Crankshaft sprocket and timing chain
- Negative battery cable

6. Fill the engine with clean oil.

7. Start the vehicle, check for leaks and repair if necessary.

2ZZ-GE Engine

1. Before servicing the vehicle, refer to the precautions in the beginning of this section.

2. Drain the engine oil.

3. Remove or disconnect the following:
- Negative battery cable
- Timing chain and crankshaft sprocket
- Oil pump and gasket

To install:

4. Clean the mounting surface.

5. Install or connect the following:
- Oil pump, with new gasket. Engage the spline teeth of the oil pump drive rotor with the larger teeth of the crankshaft, and slide the pump on.
- Oil pump bolts and tighten to 80 inch lbs. (9 Nm)
- Crankshaft sprocket and timing chain
- Negative battery cable

6. Fill the engine with clean oil.

7. Start the vehicle, check for leaks and repair if necessary.

Rear Main Seal

REMOVAL & INSTALLATION

1. Remove or disconnect the following:
- Transaxle
- Clutch assembly
- Flywheel or flexplate

2. Use a small sharp knife to cut off the lip of the oil seal. Take great care not to score any metal with the knife.

3. Use a small prytool to pry the old seal from the retaining plate. Be careful not to damage the plate. Protect the tip of the tool with tape and pad the fulcrum point with cloth.

4. Inspect the crankshaft and seal lip contact surfaces for any sign of damage.

To install:

5. Apply a light coat of multi-purpose grease to the lip of a new oil seal. Loosely fit the seal into place by hand, making sure it is not crooked.

6. Use a seal driver of the correct size to install the seal. Tap it into place until the

surface of the seal is flush with the edge of the housing.

Timing Chain, Sprockets, Front Cover and Seal

REMOVAL & INSTALLATION

1. Before servicing the vehicle, refer to the precautions in the beginning of this section.

2. Drain the cooling system.

3. Remove or disconnect the following:
- Right side engine under cover
- Right front wheel and tire
- Cylinder head cover
- Wire harness clamp and suction hose assembly, 2ZZ-GE engine
- Drive belt

4. Separate the vane pipe assembly, but do not disconnect the hose, 1ZZ-FE engine.
- Alternator bracket, 2ZZ-GE
- Alternator
- Power steering pump reservoir and position it aside, 1ZZ-FE engine

5. Place a jack with a wooden block under the vehicle for support, then remove the 4 bolts and 2 nuts and remove the right side engine mount.

6. Remove the engine wire as follows, on 1ZZ-FE engines:

a. Remove the 5 clamps from the brackets.

b. Detach the connectors.

c. Remove the ignition coil connectors.

d. Bolt and nut holding the engine wire.

7. Remove the engine wire as follows, on 2ZZ-GE engines:

a. Detach the ignition coil, oil control valve and Crankshaft Position Sensor (CKP) sensor electrical connectors.

b. Bolt and nut for the engine ground, then position the engine wire aside

8. Remove or disconnect the following:
- Ignition coil assembly

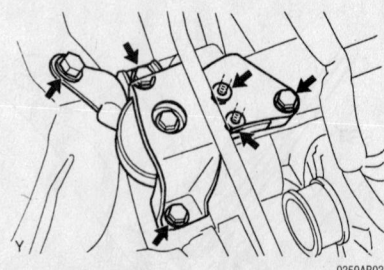

With the engine supported, remove the right side engine mount—1ZZ-FE engine shown, 2ZZ-GE similar

- Positive Crankcase Ventilation (PCV) hoses from the cylinder head cover, if necessary
- Cylinder head (valve) cover sub-assembly

9. Set the No. 1 cylinder to Top Dead Center (TDC) of the compressor stroke as follows:

a. Turn the crankshaft pulley, and align its groove with the "0" timing mark of the timing chain cover.

b. Make sure the point marks of the camshaft timing sprockets and VVT timing sprockets are in a straight line as shown. If not, turn the crankshaft 1 complete revolution (360°) and align the marks.

- Crankshaft pulley, using SST 09960-10010
- Belt tensioner
- Water pump pulley, if equipped, and pump

- Transverse engine mounting bracket
- Crankshaft Position (CKP) sensor
- No. 1 chain tensioner assembly, making sure not to revolve the crankshaft without the tensioner
- Timing chain cover. The cover is retained with 11 bolts and nuts and a Torx® stud bolt. Pry the cover between the cylinder head and block to remove it.
- Timing gear cover oil seal
- CKP sensor plate No. 1
- Timing chain tensioner slipper

➡ **In case you turn the camshafts with the timing chain removed, turn the crankshaft ¼ turn for the valve to avoid contact with the pistons.**

- Timing chain sub-assembly. Remove the chain with the crankshaft gear, using screwdrivers as shown

To install:

10. Set the No. 1 cylinder to TDC of the compression stroke:

a. Turn the hexagonal wrench head part of the camshafts, and align the point marks of the cam sprockets.

b. Using the crankshaft pulley bolt, turn the crankshaft and position the crankshaft set key upward.

11. Install or connect the following:

- Timing chain on the crank sprocket with the yellow link aligned with the mark on the crank sprocket. There are 3 yellow links on the timing chain.
- Crankshaft sprocket, using SST 09223-22010
- Timing chain on the camshaft sprockets with the yellow links aligned with the marks on the cam sprockets
- Timing chain tensioner slipper and tighten the bolt to 14 ft. lbs. (19 Nm)
- Crankshaft position sensor plate, with the "F" mark facing forward
- Timing gear cover oil seal
- Timing cover. For 1ZZ-FE engine, tighten the "A" bolts to 10 ft. lbs. (13 Nm), the "B" bolts to 14 ft. lbs. (19 Nm) and the stud bolt to 84

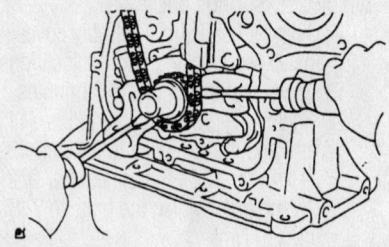
Remove the timing chain with the crankshaft gear

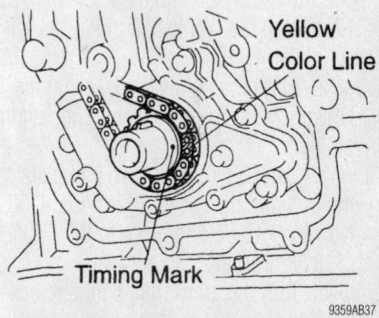
Make sure the yellow link is aligned with the crankshaft sprocket timing mark—1ZZ-FE and 2ZZ-GE engines

Proper timing mark alignment for TDC

Proper alignment of the camshaft sprockets—1ZZ-FE engine

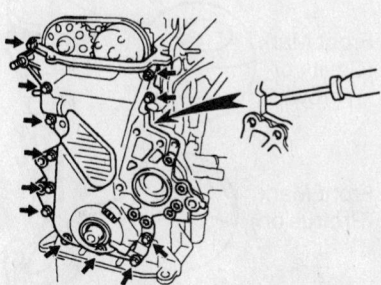
Timing chain cover mounting—1ZZ-FE engine shown, 2ZZ-GE similar

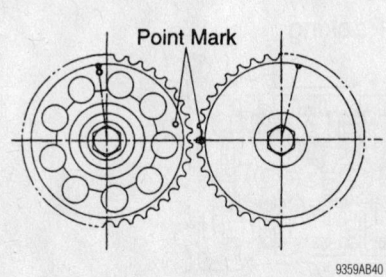
Proper alignment of the camshaft sprockets—2ZZ-GE engine

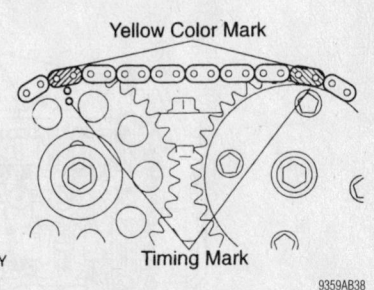
The yellow links of the timing chain must align with the camshaft sprocket timing marks—1ZZ-FE and 2ZZ-GE engines

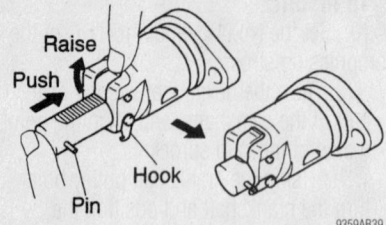

Timing chain tensioner—1ZZ-FE engine

inch lbs. (9.5 Nm), using a Torx® wrench. For 2ZZ-GE engines, tighten the M8 bolts to 15 ft. lbs. (21 m), the M6 bolts to 8 ft. lbs. (11 Nm) and the stud bolt to 84 inch lbs. (9.5 Nm).

➡ **When installing the tensioner, make sure to set the hook again if the hook releases the plunger.**

- Timing chain tensioner. Torque the nuts to 80 inch lbs. (9 Nm).
- CKP sensor and tighten the bolts to 80 inch lbs. (9 Nm)
- Transverse engine mounting bracket. Tighten the bolts to 35 ft. lbs. (47 Nm).
- Water pump and pulley
- Drive belt tensioner. Tighten the nut to 21 ft. lbs. (29 Nm) and the bolt to 51 ft. lbs. (69 Nm) on 1ZZ-FE engines or to 74 ft. lbs. (100 Nm) on 2ZZ-GE engines.

12. Install the crankshaft pulley, as follows:

   a. Align the pulley set key with the key groove of the pulley and slide on the pulley.

   b. Use SST 09960-11010 to install the bolt and tighten to 102 ft. lbs. (138 Nm) for 1ZZ-FE engine or to 87 ft. lbs. (118 Nm) on 2ZZ-GE engines.

   c. Turn the crankshaft counterclockwise and disconnect the plunger knock pin from the hook.

   d. Turn the crankshaft clockwise and check that the slipper is pushed by the plunger. If the plunger does not spring

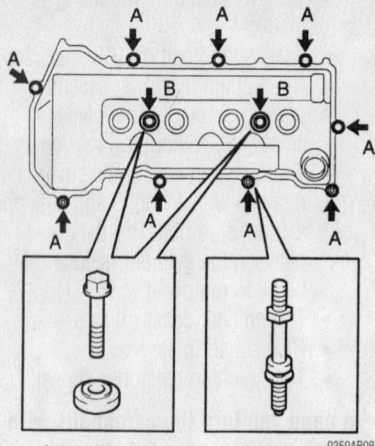

Cylinder head (valve) cover bolt locations—1ZZ-FE engine

out, press the slipper into the chain tensioner with a screwdriver so that the hook is released from the knock pin and the plunger springs out.

- Cylinder head sub-assembly cover. Install seal packing into the locations shown and install within 3 minutes. Tighten the "A" bolts to 8 ft. lbs. (11 Nm) and the "B" bolts to 80 inch lbs. (9 Nm) for 1ZZ-FE engines. For 2ZZ-GE engines, tighten the bolts to 7 ft. lbs. (10 Nm).
- Ignition coil assembly. Torque the bolts to 80 inch lbs. (9 Nm).
- Engine wire and tighten to 80 inch lbs. (9 Nm)
- Right side engine mount. Tighten to 38 ft. lbs. (52 Nm).
- Alternator bracket, 2ZZ-GE engine
- Alternator
- Vane pump, 1ZZ-FE
- Main cylinder head cover and tighten to 62 inch lbs. (7 Nm)
- Right front wheel and tire. Tighten the lug nuts to 76 ft. lbs. (103 Nm).

13. Fill the cooling system to the proper level.

14. Start the vehicle, check for leaks and repair if necessary.

Piston and Ring Positioning

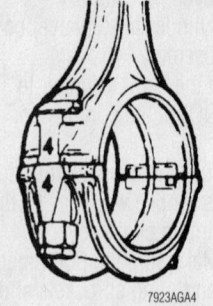

Before removing the caps from the connecting rods, be sure to matchmark them as shown

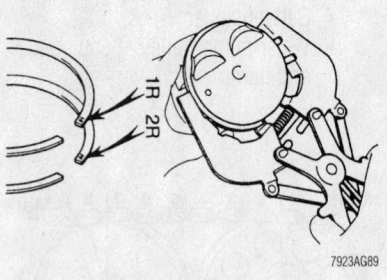

Piston ring identification mark locations—1ZZ-FE and 2ZZ-GE engines

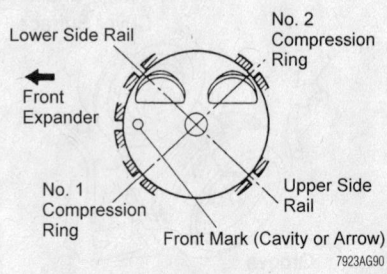

Piston ring end-gap spacing —1ZZ-FE and 2ZZ-GE engines

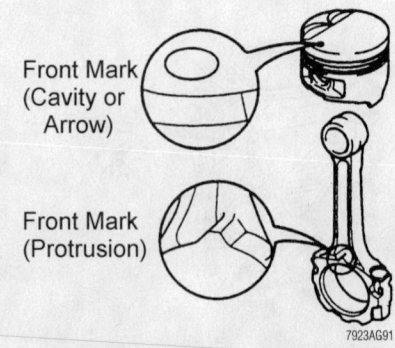

Piston-to-connecting rod assembly —1ZZ-FE and 2ZZ-GE engines

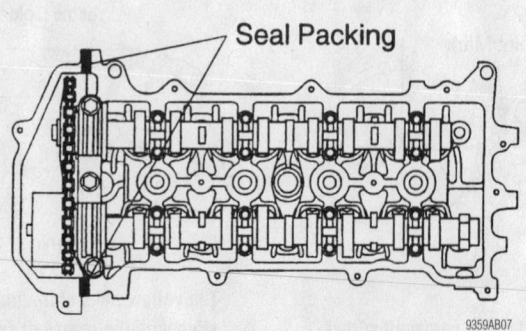

Seal packing installation locations

FUEL SYSTEM

Fuel System Service Precautions

Safety is the most important factor when performing not only fuel system maintenance, but any type of maintenance. Failure to conduct maintenance and repairs in a safe manner may result in serious personal injury or death. Work on a vehicle's fuel system components can be accomplished safely and effectively by adhering to the following rules and guidelines.

• To avoid the possibility of fire and personal injury, always disconnect the negative battery cable unless the repair or test procedure requires that battery voltage by applied.

• Always relieve the fuel system pressure prior to disconnecting any fuel system component (injector, fuel rail, pressure regulator, etc.) fitting or fuel line connection. Exercise extreme caution whenever relieving fuel system pressure, to avoid exposing skin, face and eyes to fuel spray. Please be advised that fuel under pressure may penetrate the skin or any part of the body that it contacts.

• Always place a shop towel or rag around the fitting or connection prior to loosening to absorb any excess fuel due to spillage. Ensure that all fuel spillage is quickly remove from engine surfaces. Ensure that all fuel-soaked cloths or towels are deposited into a flame-proof waste container with a lid.

• Always keep a dry chemical (Class B) fire extinguisher near the work area.

• Do not allow fuel spray or fuel vapors to come into contact with a light bulb, spark or open flame.

• Always use a second wrench when loosening or tightening fuel line connections fittings. This will prevent unnecessary stress and torsion to fuel piping. Always follow the proper torque specifications.

• Always replace worn fuel fitting O-rings with new ones. Do not substitute fuel hose where rigid pipe is installed.

Fuel System Pressure

RELIEVING

✳✳ CAUTION

Failure to relieve fuel pressure before repairs or disassembly can cause serious personal injury and/or property damage. Fuel pressure is maintained within the fuel lines, even if the engine is OFF or has not been run in a period of time. This pressure must be safely relieved before any fuel-bearing line or component is loosened or removed. On vehicles equipped with inflatable restraints or air bag systems, wait at least 90 seconds after disconnecting the battery cable before performing any other work. The back-up power will keep the restraint system energized for a period of time after the battery is disconnected.

1. Before servicing the vehicle, refer to the precautions in the beginning of this section.
2. Perform the following:
   a. Remove the rear seat cushion.
   b. Remove the rear floor service hole cover.
   c. Disconnect the fuel pump connector.
   d. Start and run the engine, until it stalls.
   e. Turn the ignition key to the **LOCK** position.
   f. Disconnect the negative battery cable.
   g. Connect the fuel pump connector.
   h. Install the service hole cover and rear seat cushion.
   i. Place a catch-pan under the joint to be disconnected. A large quantity of fuel may be released when the joint is opened.
   j. Wear eye or full face protection.
   k. Place a shop towel over the area and slowly release the joint using a wrench of the correct size.
   l. Allow the any fuel left in the line to bleed off slowly before fully disconnecting the joint.
   m. Plug the opened lines.

Fuel Filter

REMOVAL & INSTALLATION

1. Before servicing the vehicle, refer to the precautions in the beginning of this section.
2. Relieve the fuel system pressure.
3. Remove or disconnect the following:
   • Negative battery cable
   • Protective shield for the fuel filter
   • Air cleaner hose and cap, if necessary

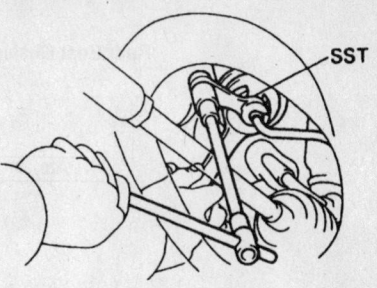

7923VG85

A line wrench with an extension may be needed to loosen the inlet line at the filter

   • Charcoal canister, if necessary
   • Slowly loosen the lower flare nut fitting until all the pressure is relieved
   • Banjo fitting and 2 metal gaskets. Discard the gaskets.
   • Fuel line with the flared nut from the filter
   • Filter from the mounting bracket

To install:
4. Install or connect the following:
   • New fuel filter
   • Banjo fitting with a new metal gasket on each side and install the union bolt. Bolt: 22 ft. lbs. (30 Nm).
   • Flare nut to the lower connection. Nut: 22 ft. lbs. (30 Nm).
   • Charcoal canister
   • Air cleaner hose and cap
   • Protective shield
   • Negative battery cable

Fuel Pump

REMOVAL & INSTALLATION

1. Before servicing the vehicle, refer to the precautions in the beginning of this section.
2. Remove or disconnect the following:
   • Negative battery cable
   • Rear seat cushion and floor service hole cover
   • Fuel pump and vapor pressure sensor connectors
   • Start and run the engine, until it stalls
3. Turn the ignition key to the **LOCK** position.
   • Negative battery cable
4. Connect the fuel pump connector.
   • Fuel tank protector, AWD vehicles
   • Fuel tank main tube sub-assembly
   • Fuel emission tube sub-assembly No. 1, FWD vehicles

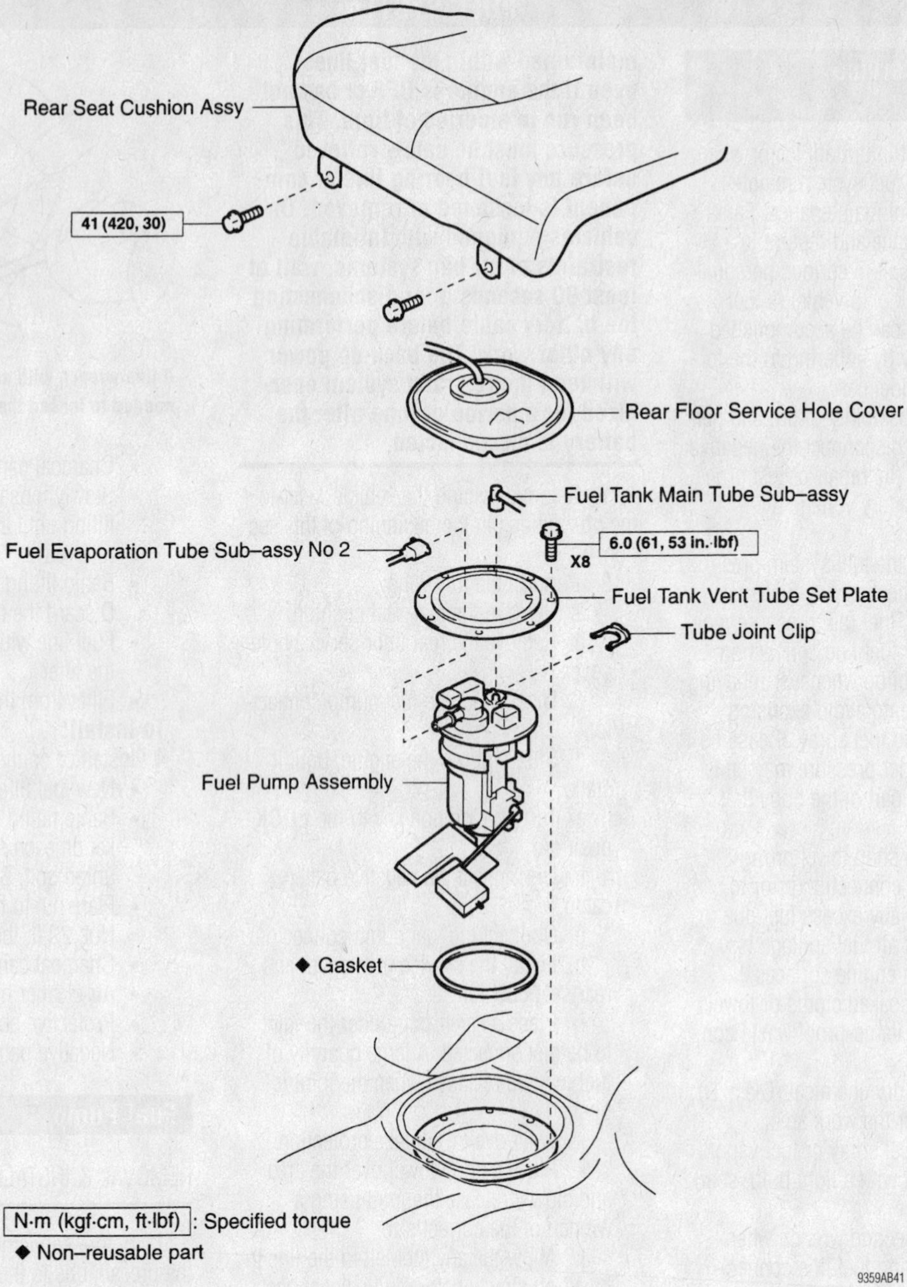

Rear Seat Cushion Assy

41 (420, 30)

Rear Floor Service Hole Cover

Fuel Tank Main Tube Sub–assy

Fuel Evaporation Tube Sub–assy No 2

6.0 (61, 53 in. lbf)
X8

Fuel Tank Vent Tube Set Plate

Tube Joint Clip

Fuel Pump Assembly

◆ Gasket

N·m (kgf·cm, ft·lbf) : Specified torque
🔧 ◆ Non–reusable part

9359AB41

Exploded view of the fuel pump mounting—FWD shown, AWD similar

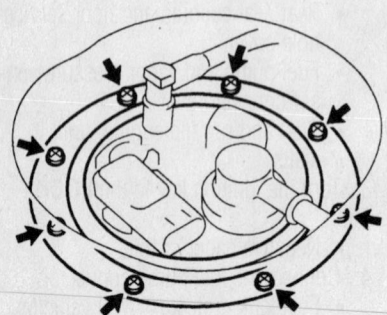

9359AB42

The fuel tank vent tube set plate is secured with 8 bolts on FWD vehicles

- Fuel tank vent tube set plate. The plate is secured with 8 bolts on FWD vehicles, or 5 bolts on AWD vehicles.
- Fuel pump assembly, being careful not to damage the filter or bend the arm of the fuel sender gauge
- Fuel suction tube set gasket
- Fuel suction support No. 2
- Fuel pump rubber cushion
- Fuel sender gauge assembly. Unplug the connector, then use a screwdriver to unlock the gauge and slide it to remove.

- Fuel section plate sub-assembly
- Vapor pressure sensor
- Fuel pump harness
- Fuel pump
- Fuel pump filter
- Fuel pressure regulator and O-ring

To install:
5. Install or connect the following:
- New regulator O-ring and regulator
- Fuel pump filter
- Fuel pump
- Vapor pressure sensor
- Fuel suction tube set gasket
- Fuel pump assembly

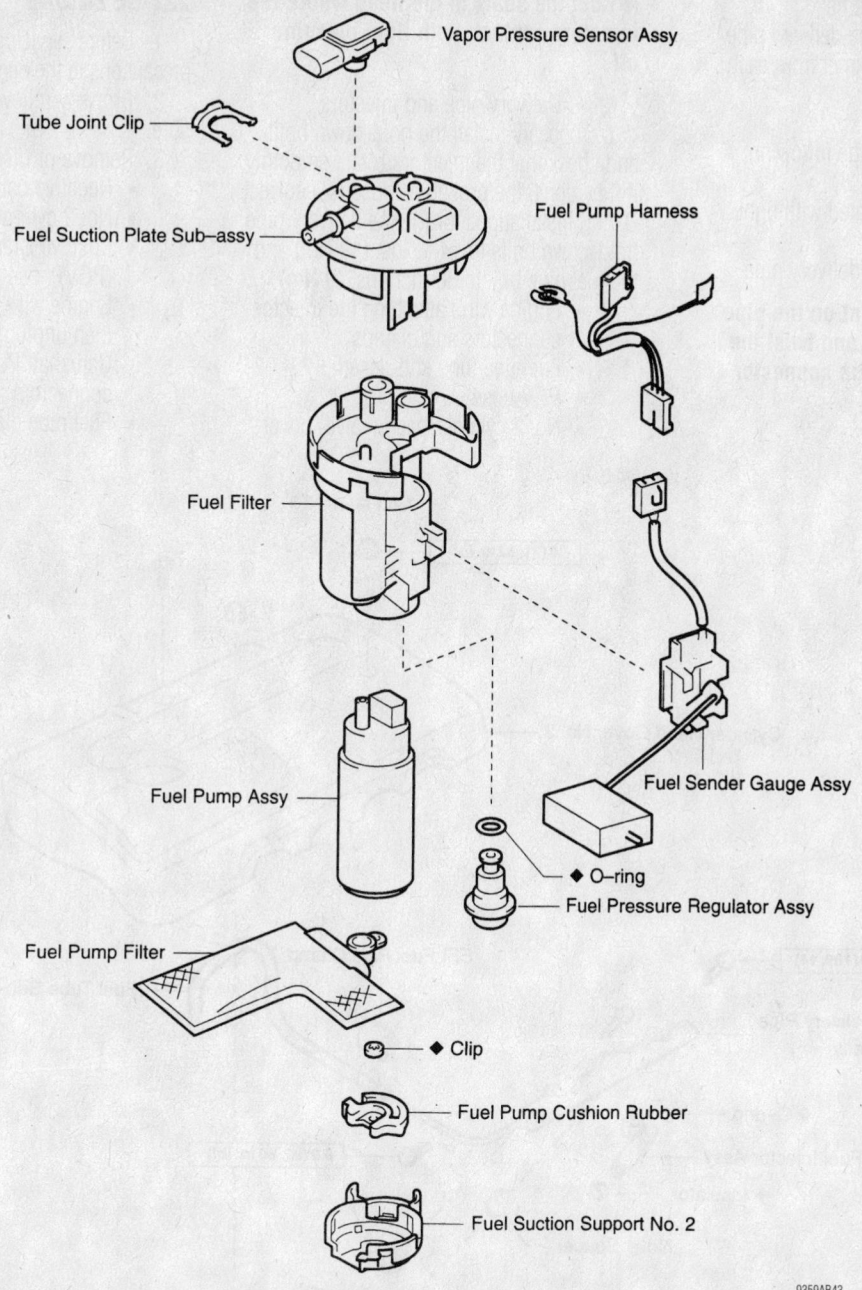

Vapor Pressure Sensor Assy

Tube Joint Clip

Fuel Suction Plate Sub–assy

Fuel Pump Harness

Fuel Filter

Fuel Pump Assy

Fuel Sender Gauge Assy

◆ O–ring

Fuel Pressure Regulator Assy

Fuel Pump Filter

◆ Clip

Fuel Pump Cushion Rubber

Fuel Suction Support No. 2

9359AB43

Fuel pump assembly components—FWD vehicles shown, AWD similar

- Fuel tank vent tube set plate. Tighten the bolts to 53 inch lbs. (6 Nm).
- Connect the fuel emission tube sub-assembly
- Fuel tank main tube sub-assembly
- Fuel tank protector No. 2, AWD vehicles
- Negative battery cable. Check for fuel leaks.
- Floor service hole cover. Use butyl tape to seal the cover.
- Rear seat cushion

Fuel Injectors

REMOVAL & INSTALLATION

1ZZ-FE Engine

1. Before servicing the vehicle, refer to the precautions in the beginning of this section.
2. Properly relieve the fuel system pressure.
3. Remove or disconnect the following:
   - Negative battery cable.
   - No. 2 cylinder head cover

- Positive Crankcase Ventilation (PCV) hose
- Engine wire, unplugging the injector connectors and clamps
- Fuel pipe clamp
- Fuel line/tube sub-assembly

✳✳ WARNING

Be careful not to drop the fuel injectors when removing the delivery pipe.

- Fuel delivery pipe sub-assembly with the injectors attached
- Delivery pipe and injectors

- Spacers from the head
- Injectors from the delivery pipe
- O-ring and grommet from each injector

To install:

4. Install or connect the following:
- New grommets
- New O-rings coated with light machine oil
- Injectors on the delivery pipe

➡ **Coat the contact point on the pipe with light machine oil and twist the injectors into place. The connector should face outward.**

- Spacers

➡ **Coat the seats in the head where the injectors contact, with light machine oil.**

- Delivery pipe and injectors

5. Loosely install the hold-down bolts and check that the injectors rotate smoothly. If they don't, the probable cause is incorrect O-ring installation. Torque the delivery pipe hold-down bolts to 14 ft. lbs. (19 Nm) and the fuel pipe bolt to 80 inch lbs. (9 Nm).

- Engine wire, attaching the injector connectors and clamps
- Fuel line/tube sub-assembly
- PCV hose
- No. 2 cylinder head (valve) cover

2ZZ-GE ENGINE

1. Before servicing the vehicle, refer to the precautions in the beginning of this section.
2. Properly relieve the fuel system pressure.
3. Remove or disconnect the following:
- Negative battery cable.
- No. 2 cylinder head cover
- Positive Crankcase Ventilation (PCV) hose
- Engine wire, by removing the bolt, then unplugging the injector and Camshaft Position (CMP) sensor connectors
- Fuel pipe clamp

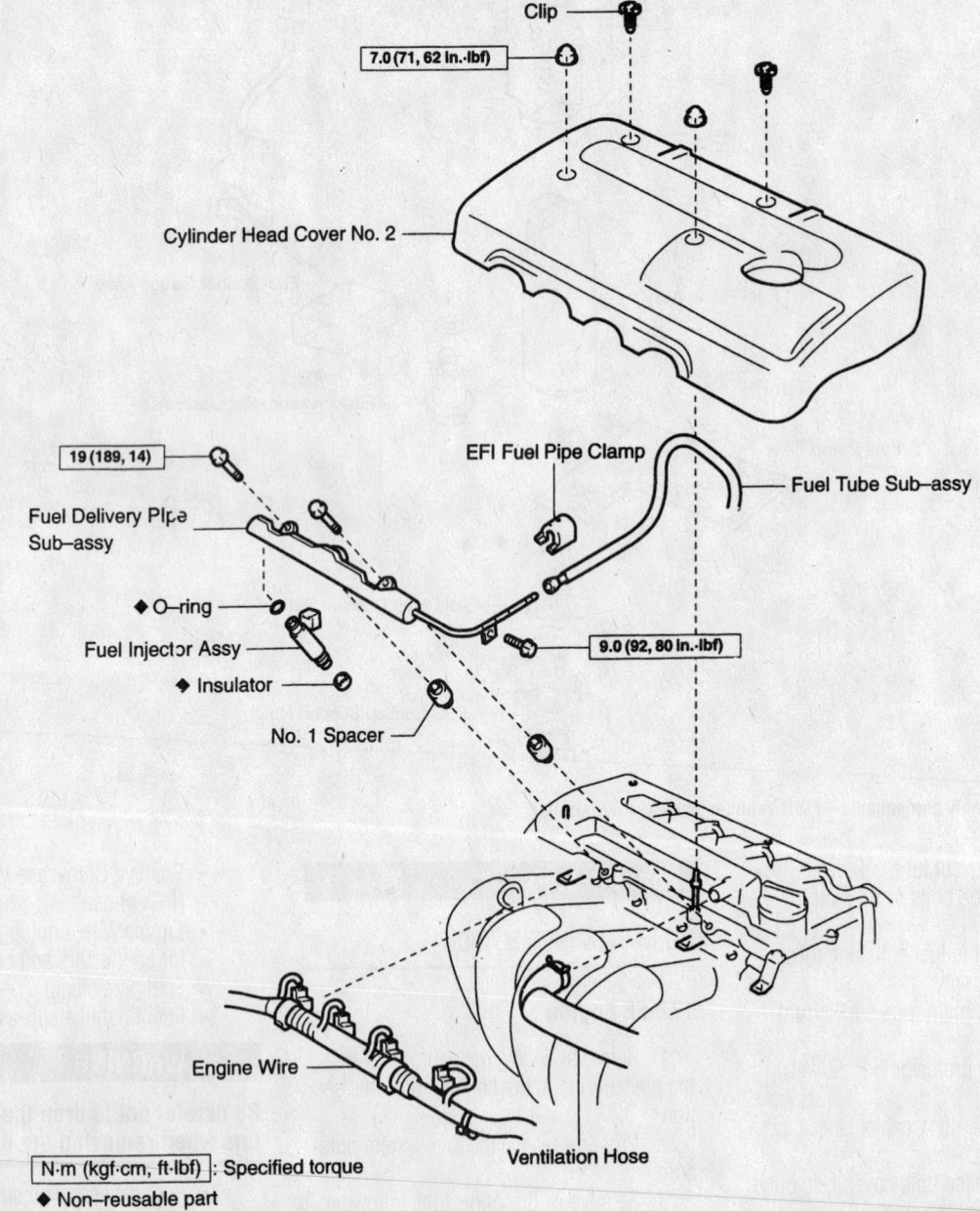

Clip

7.0 (71, 62 in.·lbf)

Cylinder Head Cover No. 2

EFI Fuel Pipe Clamp

Fuel Tube Sub–assy

19 (189, 14)

Fuel Delivery Pipe Sub–assy

◆ O–ring

Fuel Injector Assy

◆ Insulator

No. 1 Spacer

9.0 (92, 80 in.·lbf)

Engine Wire

Ventilation Hose

N·m (kgf·cm, ft·lbf): Specified torque

◆ Non–reusable part

9359AB44

Fuel injector removal and installation—1ZZ-FE engine

❊❊ WARNING

Be careful not to drop the fuel injectors when removing the delivery pipe.

- Fuel delivery pipe sub-assembly with the injectors attached
- Delivery pipe and injectors
- Spacers from the head
- Injectors from the delivery pipe
- O-ring and grommet from each injector

To install:

4. Install or connect the following:
   - New grommets
   - New O-rings coated with light machine oil
   - Injectors on the delivery pipe

➡ **Coat the contact point on the pipe with light machine oil and twist the injectors into place. The connector should face outward.**

   - Spacers

➡ **Coat the seats in the head where the injectors contact, with light machine oil.**

- Delivery pipe and injectors

5. Loosely install the hold-down bolts and check that the injectors rotate smoothly. If they don't, the probable cause is incorrect O-ring installation. Torque the delivery pipe hold-down bolts to 14 ft. lbs. (19 Nm) and the fuel pipe bolt to 80 inch lbs. (9 Nm).

- Fuel line/tube sub-assembly
- PCV hose
- Engine wire, by connecting the CMP sensor and injector connectors and installing the bolt. Tighten the bolt to 7 ft. lbs. (10 Nm).
- No. 2 cylinder head (valve) cover

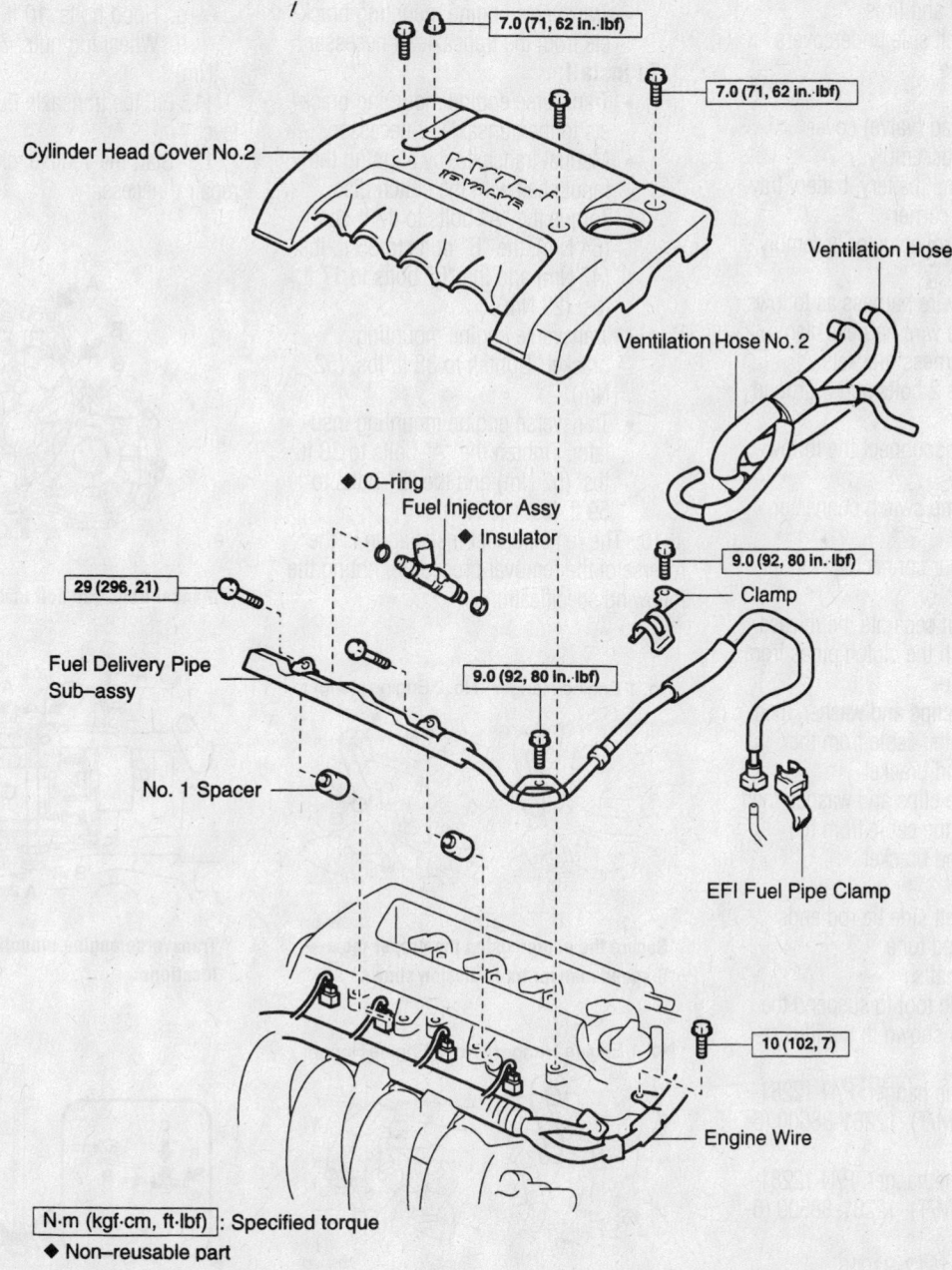

7.0 (71, 62 in. lbf)

7.0 (71, 62 in. lbf)

Cylinder Head Cover No.2

Ventilation Hose

Ventilation Hose No. 2

◆ O–ring
Fuel Injector Assy
◆ Insulator

29 (296, 21)

9.0 (92, 80 in. lbf)

Clamp

Fuel Delivery Pipe Sub–assy

9.0 (92, 80 in. lbf)

No. 1 Spacer

EFI Fuel Pipe Clamp

10 (102, 7)

Engine Wire

N·m (kgf·cm, ft·lbf) : Specified torque

◆ Non–reusable part

9359AB45

Fuel injector removal and installation—2ZZ-GE engine

DRIVE TRAIN

Manual Transaxle Assembly

REMOVAL & INSTALLATION

1. Before servicing the vehicle, refer to the precautions in the beginning of this section.
2. Drain the transaxle fluid.
3. Place the front wheels in the straight-ahead position.
4. Remove or disconnect the following:

- Steering intermediate shaft
- Front wheel and tires
- Right and left side undercovers
- Exhaust pipe
- Hood
- Cylinder head (valve) cover
- Air cleaner assembly
- Battery clamp, battery, battery tray and battery carrier
- Cruise control actuator assembly, if equipped

5. Remove the wire harness as follows:

a. Remove the wire harness clamp, 2 bolts and wire harness brackets.

b. Remove the 2 bolts and 2 ground cables.

6. Remove or disconnect the following:

- Back-up lamp switch connector, with ABS
- Speed sensor connector, without ABS
- 5 bolts, then separate the release cylinder with the clutch pipes from the transaxle
- Shift cable clips and washer, then disconnect the cable from the transaxle and bracket
- Select cable clips and washer, then disconnect the cable from the transaxle and bracket
- Starter
- Right and left side tie rod ends
- Pressure feed tube
- Front halfshafts

7. Use a suitable tool to suspend the engine assembly, as shown in the illustration:

a. No. 1 engine hanger: P/N 12281-22021 (5-speed M/T), 12281-88600 (6-speed M/T).

b. No. 2 engine hanger: P/N 12281-15040 (5-speed M/T), 12281-88600 (6-speed M/T).

c. Bolt: P/N 91512-B1016.

d. Torque the bolts to 28 ft. lbs. (38 Nm).

✱✱ CAUTION

Do not try to suspend the engine by hooking the chain to any other part.

e. Attach an engine chain hoist to the hangers.

8. Remove or disconnect the following:

- Front suspension crossmember

9. Support the transaxle with a floor jack.

- Transverse engine mounting insulator and brackets
- Manual transaxle assembly
- Transverse engine mounting brackets from the transaxle, if necessary

To install:

- Transverse engine mounting brackets to the transaxle, if necessary
- Manual transaxle, by aligning the input shaft with the clutch disc. Torque the "A" bolts to 47 ft. lbs. (64 Nm), the "B" bolts to 35 ft. lbs. (47 Nm) and the "C" bolts to 17 ft. lbs. (23 Nm).
- Transverse engine mounting bracket. Tighten to 38 ft. lbs. (52 Nm).
- Transverse engine mounting insulator. Tighten the "A" bolts to 38 ft. lbs. (52 Nm) and the "B" bolts to 59 ft. lbs. (80 Nm).

10. The remainder of installation is the reverse of the removal procedure, noting the following specifications:

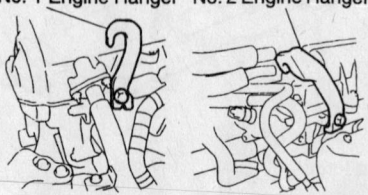

Secure the engine using the proper tools—5-speed manual transmission shown

9359AB46

Secure the engine using the proper tools—6-speed manual transmission shown

9359AB49

a. Starter mounting bolts: 27 ft. lbs. (37 Nm).

b. Clutch release cylinder bolts: "A" bolts 19 ft. lbs. (25 Nm), "B" bolts 9 ft. lbs. (12 Nm) and "C" bolts 44 inch lbs. (5 Nm).

c. Battery carrier bolts: 10 ft. lbs. (13 Nm).

d. Battery clamp bolt: 44 inch lbs. (5 Nm).

e. Battery clamp nut: 31 inch lbs. (3.5 Nm).

f. Cylinder head cover bolts: 62 inch lbs. (7 Nm).

g. Hood bolts: 10 ft. lbs. (13 Nm).

h. Wheel lug nuts: 76 ft. lbs. (103 Nm).

11. Fill the transaxle fluid to the proper level.

12. Start the vehicle, check for leaks and repair if necessary.

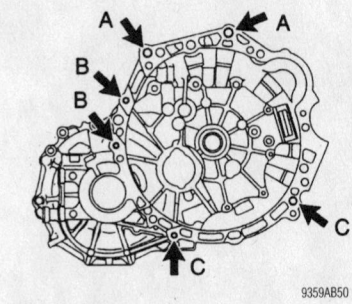

Manual transaxle bolt installation locations

9359AB50

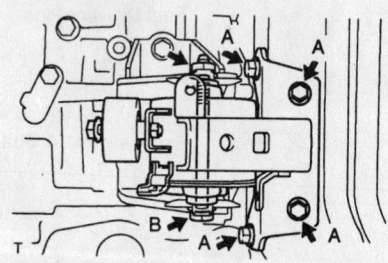

Transverse engine mounting insulator bolt locations

9359AB51

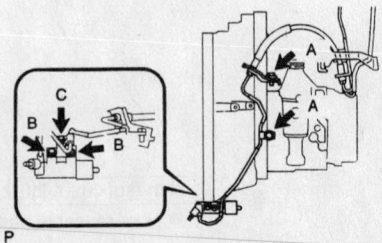

Clutch release cylinder bolt locations

9359AB52

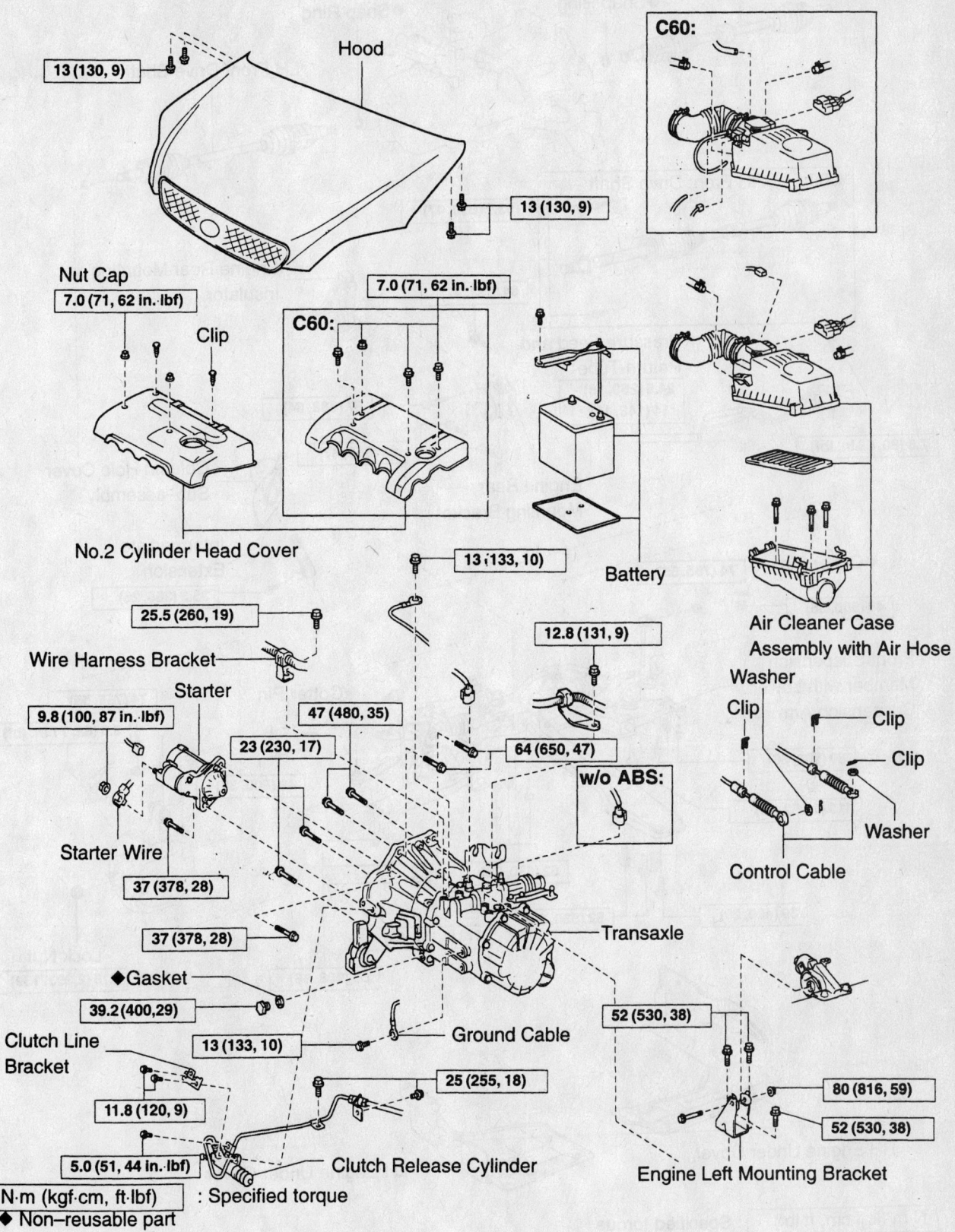

Hood

13 (130, 9)

C60:

13 (130, 9)

7.0 (71, 62 in.·lbf)

Nut Cap
7.0 (71, 62 in.·lbf)

Clip

C60:

7.0 (71, 62 in.·lbf)

Battery

No.2 Cylinder Head Cover

13 (133, 10)

Air Cleaner Case Assembly with Air Hose

25.5 (260, 19)

12.8 (131, 9)

Washer

Wire Harness Bracket

Clip

Clip

Starter

Clip

9.8 (100, 87 in.·lbf)

47 (480, 35)

64 (650, 47)

23 (230, 17)

w/o ABS:

Washer

Control Cable

Starter Wire

37 (378, 28)

Transaxle

37 (378, 28)

◆**Gasket**

52 (530, 38)

39.2 (400, 29)

Ground Cable

Clutch Line Bracket

13 (133, 10)

11.8 (120, 9)

25 (255, 18)

80 (816, 59)

5.0 (51, 44 in.·lbf)

52 (530, 38)

Clutch Release Cylinder

Engine Left Mounting Bracket

N·m (kgf·cm, ft·lbf) : Specified torque
◆ Non–reusable part

9359AB47

Exploded view of the manual transaxle (1 of 2)

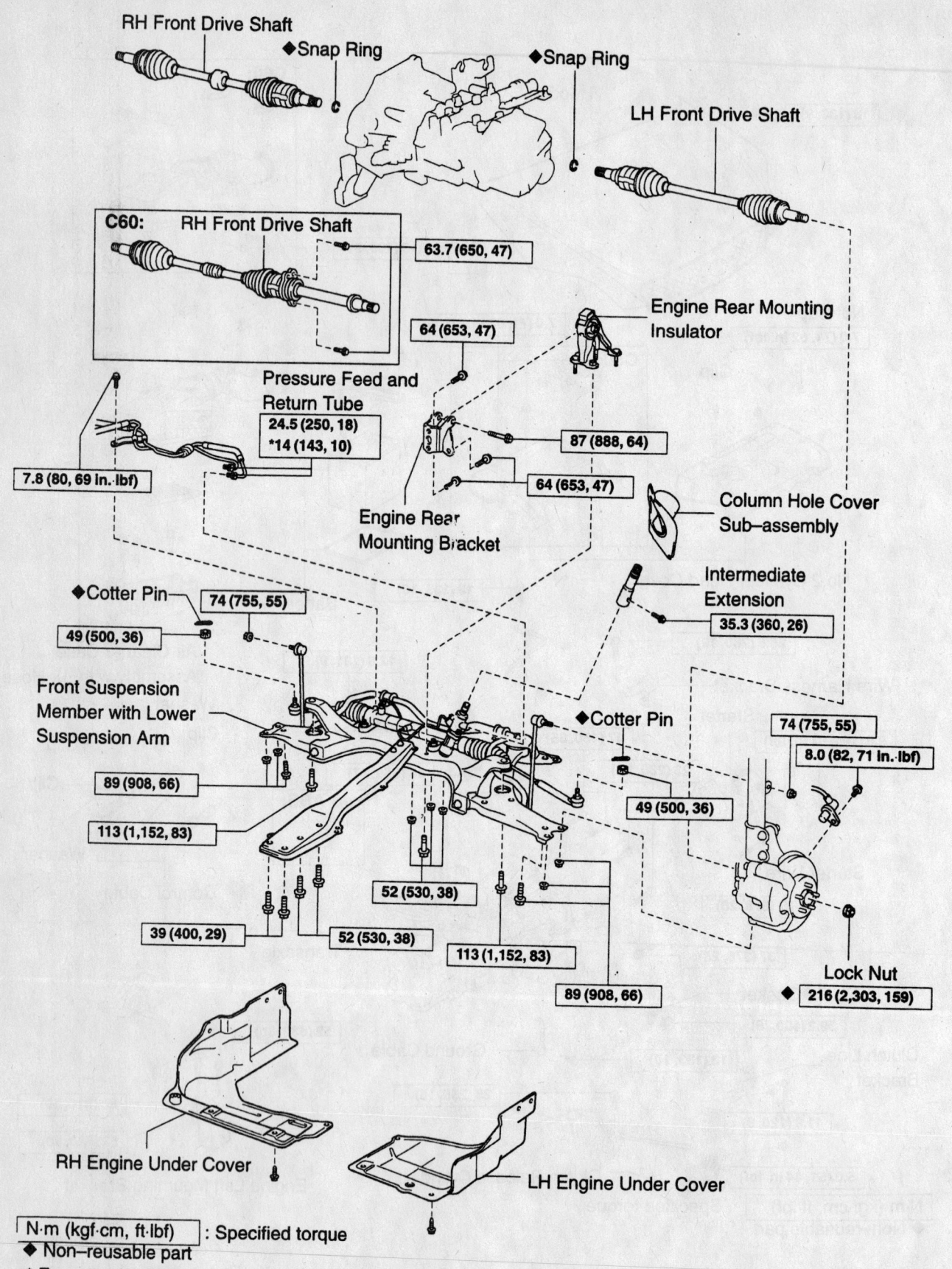

RH Front Drive Shaft

◆ Snap Ring

◆ Snap Ring

LH Front Drive Shaft

C60: RH Front Drive Shaft

63.7 (650, 47)

64 (653, 47)

Engine Rear Mounting Insulator

Pressure Feed and Return Tube

24.5 (250, 18)
*14 (143, 10)

87 (888, 64)

64 (653, 47)

7.8 (80, 69 In. lbf)

Engine Rear Mounting Bracket

Column Hole Cover Sub–assembly

Intermediate Extension

◆ Cotter Pin

74 (755, 55)

49 (500, 36)

35.3 (360, 26)

Front Suspension Member with Lower Suspension Arm

◆ Cotter Pin

74 (755, 55)

8.0 (82, 71 In. lbf)

89 (908, 66)

113 (1,152, 83)

49 (500, 36)

52 (530, 38)

39 (400, 29)

52 (530, 38)

113 (1,152, 83)

89 (908, 66)

Lock Nut

◆ 216 (2,303, 159)

RH Engine Under Cover

LH Engine Under Cover

N·m (kgf·cm, ft·lbf) : Specified torque
◆ Non–reusable part
* For use with SST

9359AB48

Exploded view of the manual transaxle (2 of 2)

Automatic Transaxle

REMOVAL & INSTALLATION

FWD—A246E & U240E Transaxles

1. Before servicing the vehicle, refer to the precautions in the beginning of this section.
2. Drain the transaxle fluid.
3. Remove or disconnect the following:
   - Negative battery cable
   - Hood
   - No. 2 cylinder head cover
   - Battery and battery carrier
   - Air cleaner assembly with hose
   - Floor shift cable transmission control shift
   - Transmission control cable support
   - No. 1 transmission control cable bracket
   - Wiring harness and brackets
   - Transmission wire connector
   - Park/neutral position switch connector, with Anti-lock Brake System (ABS)
   - Speedometer sensor connector, without ABS
   - Transmission revolution sensor connectors, if equipped
   - Transmission fluid filler tube
   - No. 1 oil cooler inlet and outlet tubes
   - Foot rest
   - Floor carpet
   - Oxygen (O$_2$) sensor connector
4. Suspend the engine as follows:
   a. Disconnect the 2 Positive Crankcase Ventilation (PCV) hoses.
   b. Install the No. 1 and No. 2 engine hangers in the correct direction.
   c. No. 1 engine hanger: P/N 12281-

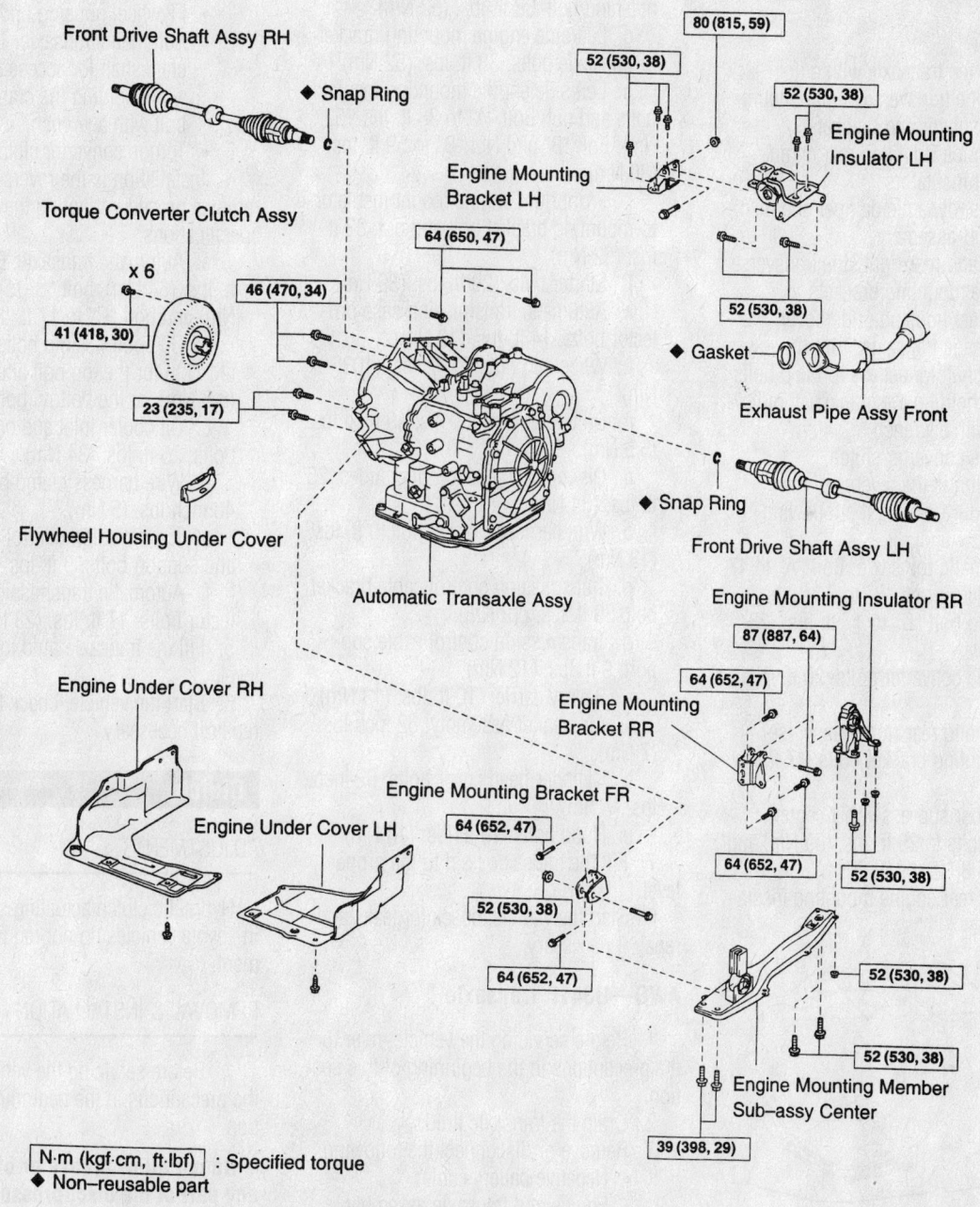

N·m (kgf·cm, ft·lbf) : Specified torque
◆ Non–reusable part

Automatic transaxle and related components—U240E transaxle shown, A246E similar

9355AB55

22021 (A246E) or 12281-88600 (U240E).

d. No. 2 engine hanger: P/N 12281-15040 (A246E) or 12281-88600 (U240E)

e. Bolt: P/N 91512-B1016.

f. Torque the bolt to 28 ft. lbs. (38 Nm).

g. Attach an engine chain hoist to the engine hangers.

- Front wheels
- Right and left engine undercovers
- Front floor panel brace, U240E transaxle
- Front exhaust pipe
- Front halfshafts
- Automatic transmission case protector
- Starter

5. Support the transaxle with a floor jack

- Left side transverse engine mounting insulator and bracket
- Right side front and rear engine mount insulators
- 4 bolts, dynamic damper and member sub-assembly
- Front and rear right side transverse engine mounting brackets
- Flywheel housing undercover
- Automatic transaxle. Turn the crankshaft for access to the 6 bolts while holding the crankshaft pulley bolt with a wrench.
- Torque converter clutch

6. Installation is the reverse of the removal procedure, noting the following specifications:

a. Automatic transaxle: Bolt "A" to 47 ft. lbs. (64 Nm), bolt "B" to 34 ft. lbs. (47 Nm) and bolt "C" to 17 ft. lbs. (23 Nm).

b. Torque converter bolts: 20 ft. lbs. (28 Nm).

c. Front and rear right transverse engine mounting bracket bolts: 47 ft. lbs. (64 Nm).

d. Member sub-assembly center bolts: "A" bolts to 29 ft. lbs. (39 Nm) and "B" bolts to 38 ft. lbs. (52 Nm).

e. Right rear engine mounting insula-

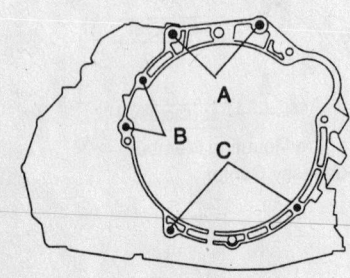

P

Automatic transaxle bolt locations

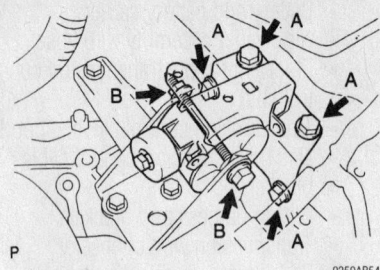

P

Left side engine mount insulator bolt and nut locations

tor-to-engine mounting bracket bolt: 64 ft. lbs. (87 Nm).

f. Right rear engine mount insulator nuts and bolt: 38 ft. lbs. (52 Nm).

g. Left side engine mounting bracket-to-transaxle bolts: 38 ft. lbs. (52 Nm).

h. Left side engine mounting insulator bolts and nut: Bolt "A" to 38 ft. lbs. (52 Nm), Bolt "B" and Nut "B" to 59 ft. lbs. (80 Nm).

i. Front right engine mount insulator-to-mounting bracket bolt and nut: 38 ft. lbs. (52 Nm).

j. Starter bolts: 29 ft. lbs. (39 Nm).

k. Automatic transmission case protector bolts: 14 ft. lbs. (18 Nm).

l. Wheel lug nuts: 76 ft. lbs. (103 Nm).

m. Oil cooler clamp bolts: 49 inch lbs. (5.5 Nm).

n. Oil cooler inlet and outlet tubes: 25 ft. lbs. (34 Nm).

o. Wire harness bracket bolt: 9 ft. lbs. (13 Nm).

p. Transmission control cable bracket bolts: 9 ft. lbs. (12 Nm).

q. Transmission control cable support: 9 ft. lbs. (12 Nm).

r. Battery carrier: 10 ft. lbs. (13 Nm).

s. Air cleaner assembly: 62 inch lbs. (7 Nm).

t. Cylinder head cover bolts: 62 inch lbs. (7 Nm).

u. Hood bolts: 10 ft. lbs. (13 Nm).

7. Fill the transaxle fluid to the proper level.

8. Start the vehicle, check for leaks and repair if necessary.

AWD—U341F Transaxle

1. Before servicing the vehicle, refer to the precautions in the beginning of this section.

2. Drain the transaxle fluid.

3. Remove or disconnect the following:

- Negative battery cable
- Engine and transaxle assembly
- Transfer case

- Automatic transmission case protector
- Front left side halfshaft
- Transmission control cable support and bracket
- Wire harness clamp bracket, bolts and 2 wire harnesses
- Transmission wire connector
- Park/neutral position switch connector
- Transmission revolution sensor connectors, if equipped
- Transmission fluid filler tube
- Oil cooler inlet and outlet tubes
- Transverse engine mounting brackets
- Flywheel housing undercover
- Automatic transaxle. Turn the crankshaft for access to the 6 bolts while holding the crankshaft pulley bolt with a wrench.
- Torque converter clutch

4. Installation is the reverse of the removal procedure, noting the following specifications:

a. Automatic transaxle: Bolt "A" to 47 ft. lbs. (64 Nm), bolt "B" to 34 ft. lbs. (47 Nm) and bolt "C" to 17 ft. lbs. (23 Nm).

b. Oil cooler clamp bolts: 8 ft. lbs. (11 Nm) for the top bolt and 49 inch lbs. (5.5 Nm) for the bottom bolt.

c. Oil cooler inlet and outlet tube bolts: 25 ft. lbs. (34 Nm).

d. Wire harness clamp bracket bolt: 48 inch lbs. (5 Nm).

e. Transmission control cable bracket and support bolts: 9 ft. lbs. (12 Nm).

f. Automatic transmission case protector bolts: 17 ft. lbs. (23 Nm).

5. Fill the transaxle fluid to the proper level.

6. Start the vehicle, check for leaks and repair if necessary.

Clutch

ADJUSTMENTS

Hydraulic clutch actuating systems used in Toyota vehicles do not require adjustment.

REMOVAL & INSTALLATION

1. Before servicing the vehicle, refer to the precautions in the beginning of this section.

➡**Do not allow grease or oil to get on any part of the disc, pressure plate, or flywheel surfaces.**

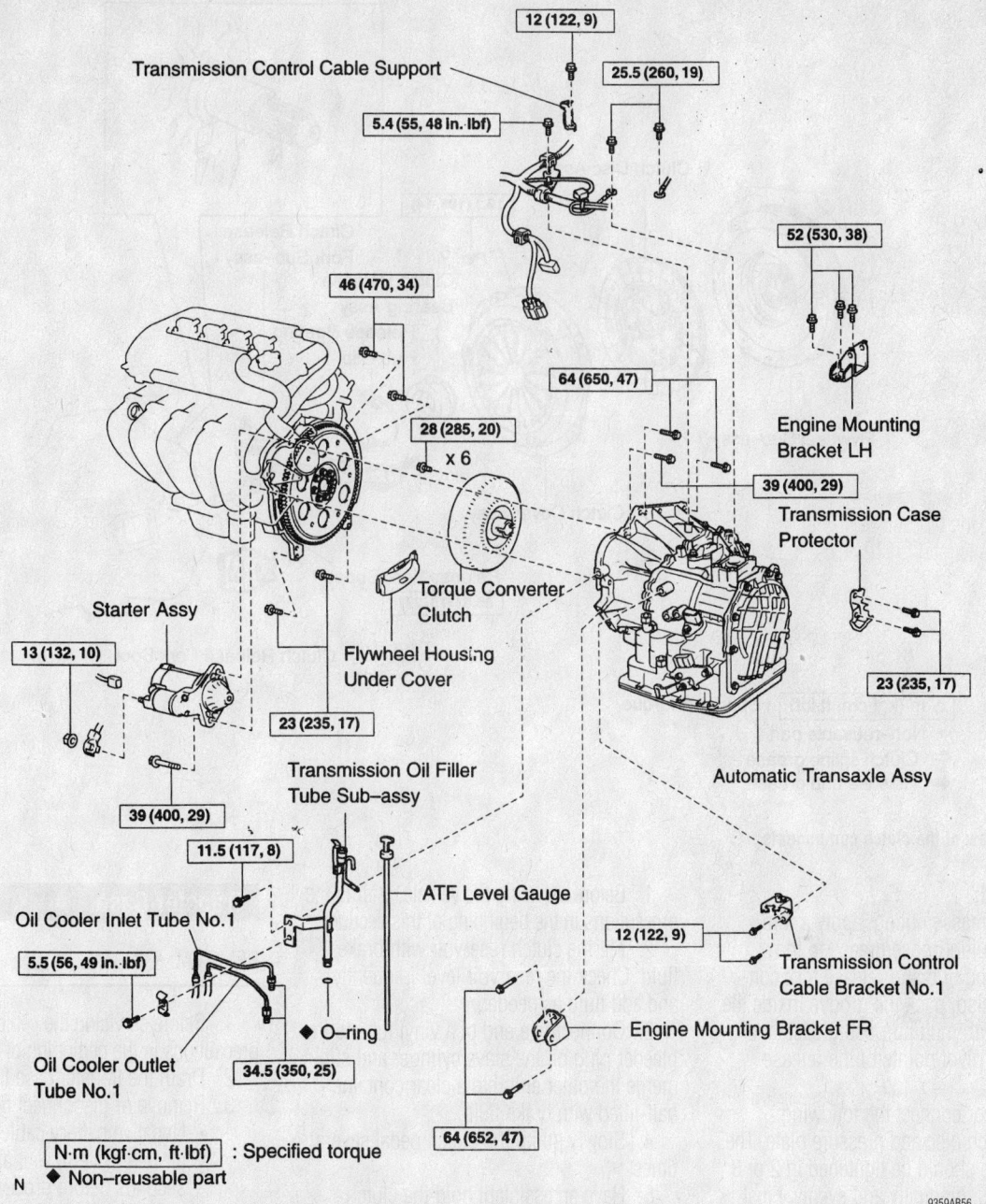

Transmission Control Cable Support

12 (122, 9)

25.5 (260, 19)

5.4 (55, 48 in. lbf)

52 (530, 38)

46 (470, 34)

64 (650, 47)

Engine Mounting Bracket LH

28 (285, 20) x 6

39 (400, 29)

Transmission Case Protector

Starter Assy

Torque Converter Clutch

13 (132, 10)

Flywheel Housing Under Cover

23 (235, 17)

23 (235, 17)

39 (400, 29)

Transmission Oil Filler Tube Sub–assy

Automatic Transaxle Assy

11.5 (117, 8)

ATF Level Gauge

Oil Cooler Inlet Tube No.1

12 (122, 9)

Transmission Control Cable Bracket No.1

5.5 (56, 49 in. lbf)

◆ O–ring

Engine Mounting Bracket FR

Oil Cooler Outlet Tube No.1

34.5 (350, 25)

64 (652, 47)

N·m (kgf·cm, ft·lbf) : Specified torque

N ◆ Non–reusable part

9359AB56

Automatic transaxle and related components—U341F transaxle

2. Remove or disconnect the following:
- Negative battery cable. On vehicles equipped with an air bag, wait at least 90 seconds before proceeding
- Transaxle assembly

3. Make matchmarks on the clutch cover (pressure plate) and flywheel so that the pressure plate can be returned to its original position during installation.

4. Remove or disconnect the following:
- Release fork bearing clips
- Release bearing hub, complete with the release bearing
- Release fork and support

❋❋ CAUTION

Slowly unfasten the bolts which attach the pressure plate. Loosen each bolt 1 turn at a time until the spring tension is released. If the bolts are released improperly the clutch assembly could fly apart, causing possible injury.

- Pressure plate from the clutch cover/spring assembly

5. Inspect the disc, pressure plate and flywheel for damage and wear using a

caliper to measure depth and width and a dial indicator to measure runout.

a. The minimum clutch disc rivet head depth is 0.012 in. (0.3mm).

b. The maximum clutch disc runout is 0.031 in. (0.8mm).

c. The maximum pressure plate spring depth is 0.024 in. (0.6mm).

d. The maximum pressure plate spring width is 0.197 in. (5.0mm).

e. The maximum flywheel runout is 0.004 in. (0.1mm).

6. Replace or machine parts as necessary.

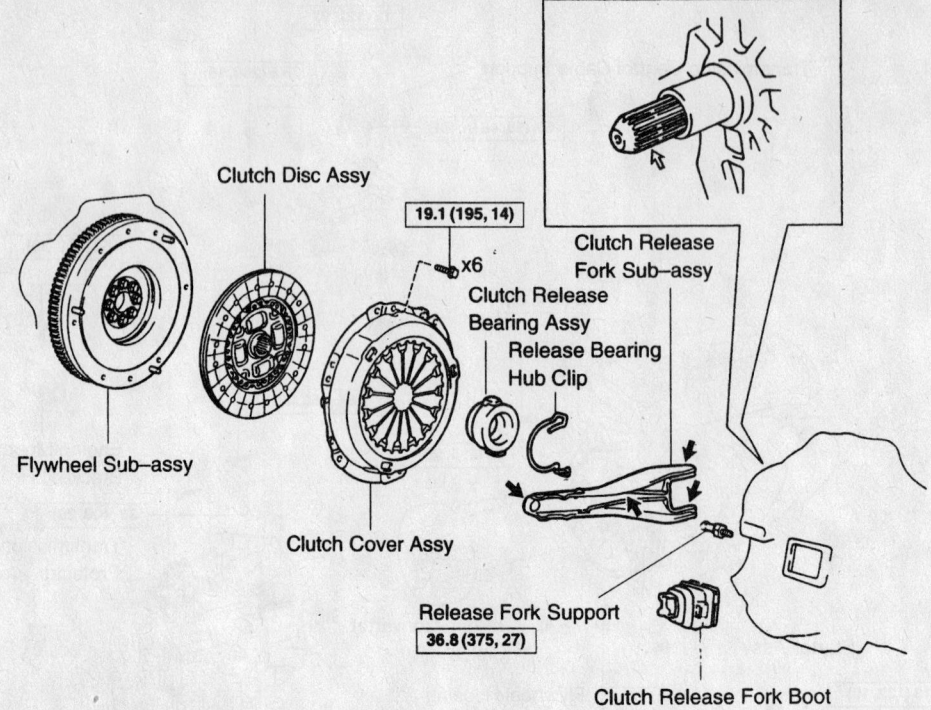

19.1 (195, 14)

Clutch Disc Assy

x6

Clutch Release
Fork Sub—assy

Clutch Release
Bearing Assy

Release Bearing
Hub Clip

Flywheel Sub—assy

Clutch Cover Assy

Release Fork Support
36.8 (375, 27)

Clutch Release Fork Boot

N·m (kgf·cm, ft·lbf) : Specified torque
◆ Non—reusable part
⇐ Clutch spline grease
⇐ Release hub grease

9359AB57

Exploded view of the clutch components

To install:

7. When reassembling, apply a thin coating of multipurpose grease to the release bearing hub and release fork contact points. Also, pack the groove inside the clutch hub with multipurpose grease and lubricate the pivot points of the release fork.

8. Install or connect the following:
- Clutch disc and pressure plate. The bolts should be tightened in 2 or 3 steps, gradually and evenly. Final bolt torque is 14 ft. lbs. (19 Nm).
- Release bearing, fork and boot
- Transaxle assembly
- Negative battery cable

Hydraulic Clutch System

BLEEDING

➡ **If any maintenance on the clutch system was performed or the system is suspected of containing air, bleed the system. Use care; brake fluid will remove the paint from any surface. If the brake fluid spills onto any painted surface, wash it off immediately with soap and water.**

1. Before servicing the vehicle, refer to the precautions in the beginning of this section.
2. Fill the clutch reservoir with brake fluid. Check the reservoir level frequently and add fluid as needed.
3. Connect one end of a vinyl tube to the bleeder plug on the slave cylinder and submerge the other end into a clear container half-filled with brake fluid.
4. Slowly pump the clutch pedal several times.
5. Have an assistant hold the clutch pedal down and loosen the bleeder plug until fluid and/or air starts to run out of the bleeder plug. Close the bleeder plug while the pedal is held to the floor.

➡ **Do not allow the pedal to rise back-up while the bleeder is still open. If this happens, it will allow air to re-enter the slave cylinder and cause the clutch system not to work properly.**

6. Repeat Steps 2 and 3 until all the air bubbles are removed from the system.
7. Tighten the bleeder plug when all the air is gone.
8. Refill the master cylinder to the proper level as required.
9. Check the system for leaks.

Transfer Case

REMOVAL & INSTALLATION

1. Before servicing the vehicle, refer to the precautions in the beginning of this section.
2. Drain the transfer case fluid.
3. Remove or disconnect the following:
- Negative battery cable. Due to the air bag system, wait at least 90 seconds before proceeding
- Engine and transaxle assembly
- Separate vane pump
- Steering gear
- Crossmember
- Manifold stay
- Oxygen (O₂) sensor
- Exhaust manifold heat shield
- Exhaust manifold
- Starter
- Right side halfshaft
- Transverse engine mounting bracket
- Center and right side transfer stiffener plates

✶✶ WARNING

When removing the transfer case, DO NOT touch the oil seal.

- Transfer case bolts, and transfer assembly, using a mallet to dislodge it from the transaxle

4. Installation is the reverse of the removal procedure, noting the following specifications:

   a. Transfer case stiffener case bolts: 25 ft. lbs. (34 Nm).

   b. Engine mounting bracket bolts: 47 ft. lbs. (64 Nm).

5. Add fluid to the transfer case, and check for leaks.

Halfshaft

REMOVAL & INSTALLATION

➡The hub bearing could be damaged if subjected to the full weight of the vehicle, such as if the vehicle is moved without the halfshafts. If it is absolutely necessary to place the full vehicle weight on the hub bearing, first support the bearing with SST No. 09608–16041.

1. Before servicing the vehicle, refer to the precautions in the beginning of this section.

2. Drain the transaxle fluid.

3. Remove or disconnect the following:

- Negative battery cable. Due to the air bag system, wait at least 90 seconds before proceeding.
- Both front wheels
- Cotter pin, locknut cap, and the hub nut
- Undercovers

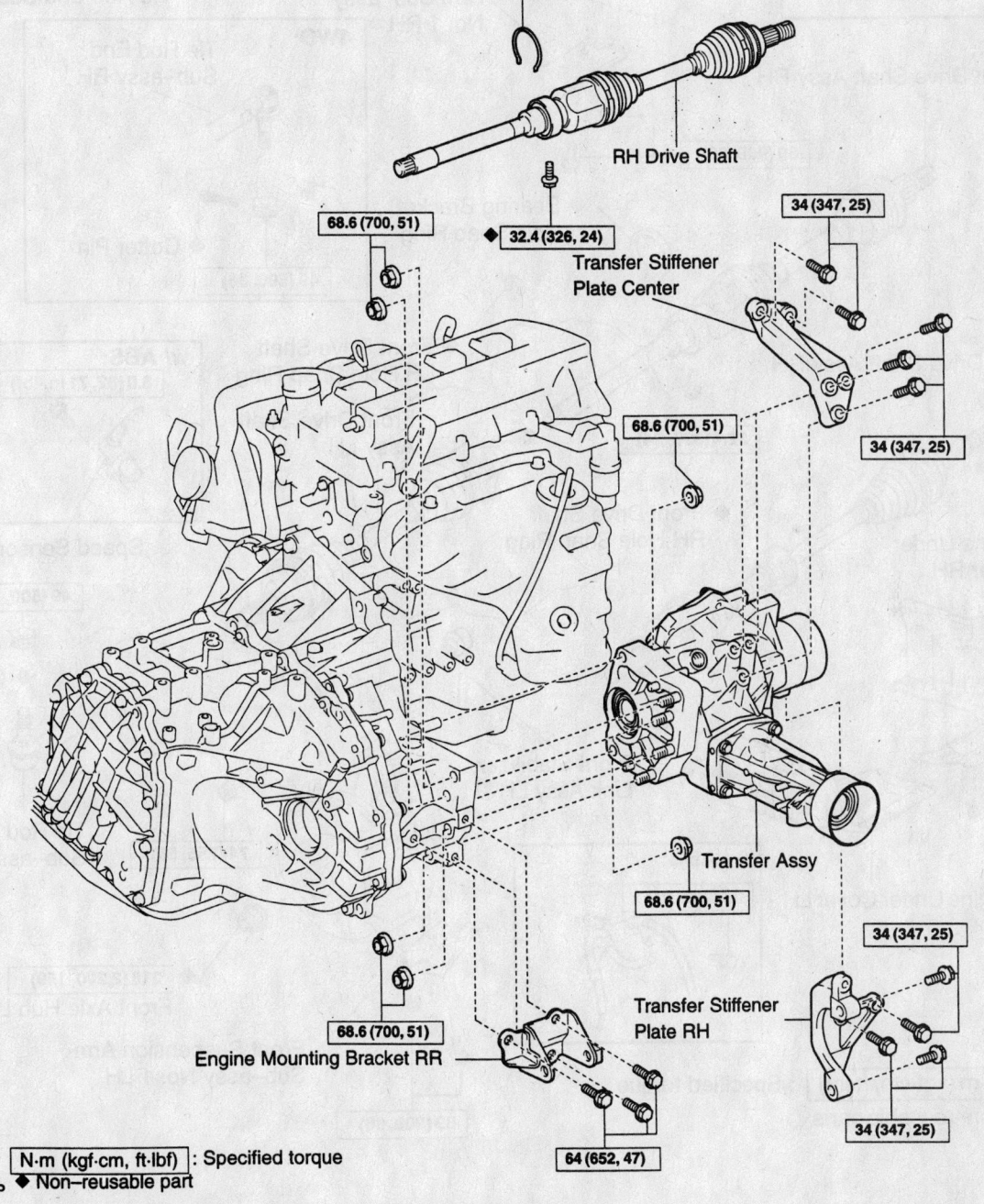

◆ Snap Ring

RH Drive Shaft

68.6 (700, 51)

◆ 32.4 (326, 24)

Transfer Stiffener Plate Center

34 (347, 25)

68.6 (700, 51)

34 (347, 25)

Transfer Assy

68.6 (700, 51)

34 (347, 25)

Transfer Stiffener Plate RH

Engine Mounting Bracket RR

68.6 (700, 51)

64 (652, 47)

34 (347, 25)

N·m (kgf·cm, ft·lbf) : Specified torque
◆ Non–reusable part

Exploded view of the transfer case mounting

9359AB65

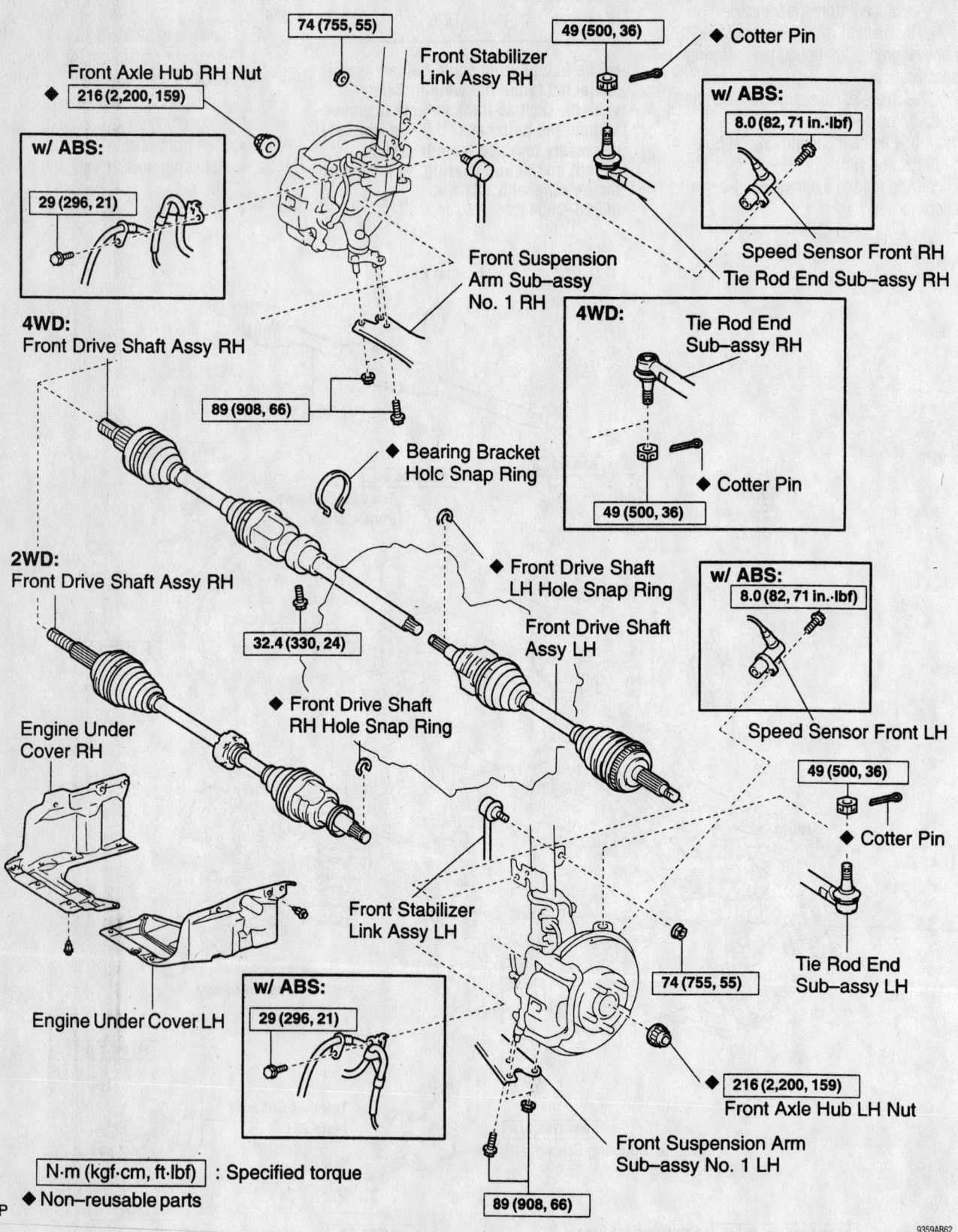

Front Axle Hub RH Nut
◆ 216 (2,200, 159)

74 (755, 55)

Front Stabilizer Link Assy RH

49 (500, 36)

◆ Cotter Pin

w/ ABS:
8.0 (82, 71 in.·lbf)

Speed Sensor Front RH
Tie Rod End Sub–assy RH

w/ ABS:
29 (296, 21)

Front Suspension Arm Sub–assy No. 1 RH

4WD:
Front Drive Shaft Assy RH

4WD:
Tie Rod End Sub–assy RH

◆ Cotter Pin

49 (500, 36)

89 (908, 66)

◆ Bearing Bracket Hole Snap Ring

2WD:
Front Drive Shaft Assy RH

◆ Front Drive Shaft LH Hole Snap Ring

Front Drive Shaft Assy LH

w/ ABS:
8.0 (82, 71 in.·lbf)

32.4 (330, 24)

◆ Front Drive Shaft RH Hole Snap Ring

Speed Sensor Front LH

49 (500, 36)

◆ Cotter Pin

Engine Under Cover RH

Front Stabilizer Link Assy LH

Tie Rod End Sub–assy LH

Engine Under Cover LH

w/ ABS:
29 (296, 21)

74 (755, 55)

216 (2,200, 159)
Front Axle Hub LH Nut

Front Suspension Arm Sub–assy No. 1 LH

N·m (kgf·cm, ft·lbf) : Specified torque

P

◆ Non–reusable parts

89 (908, 66)

9359AB62

Halfshafts and related components

- Speed sensors
- Tie rod ball joint from the steering knuckle
- Stabilizer bar link from the lower suspension arm
- Lower ball joint from the lower suspension arm
- Halfshaft from the knuckle

➡ **Be careful not to damage the inner oil seal or the ABS sensor rotor on the halfshaft.**

4. To remove the left side halfshaft, separate the halfshaft from the transaxle.

5. To remove the right side halfshaft perform the following steps:
- Remove the 2 bolts of the center bearing bracket
- Pull the halfshaft out together with the center bearing case and the center halfshaft.
- Remove the center shaft with the right-hand halfshaft from the transaxle through the bearing bracket.

➡ **Do not damage the oil seal lip.**

To install:

6. Install or connect the following:
- Snapring opening side facing downward, on the oiled inboard joint tulip
- Left side halfshaft into the transaxle
- Right side halfshaft, with the bearing case and center shaft, into the transaxle
- Center bearing case (right side).

7. After installing either halfshaft, check that there is 0.08–0.12 in. (2–3mm) of axial play. Check that the halfshaft is making contact with the pinion shaft and that the halfshaft cannot be pulled out.

8. Install or connect the following:
- Halfshaft into the knuckle
- Lower suspension arm to the lower ball joint. Torque the bolt and nuts to 66 ft. lbs. (89 Nm).
- Tie rod end to the steering knuckle. Tighten the nut to 36 ft. lbs. (49 Nm).
- Stabilizer bar link to the lower suspension arm. Torque the nuts to 55 ft. lbs. (74 Nm).
- Front wheels
- Hub nut and washer and tighten to 159 ft. lbs. (216 Nm)
- Negative battery cable
- Locknut cap and a new cotter pin.
- Speed sensors
- Undercover

9. Fill the transaxle fluid to the proper level

10. Start the vehicle, check for leaks and repair if necessary.

CV-Joints

OVERHAUL

1. Before servicing the vehicle, refer to the precautions in the beginning of this section.

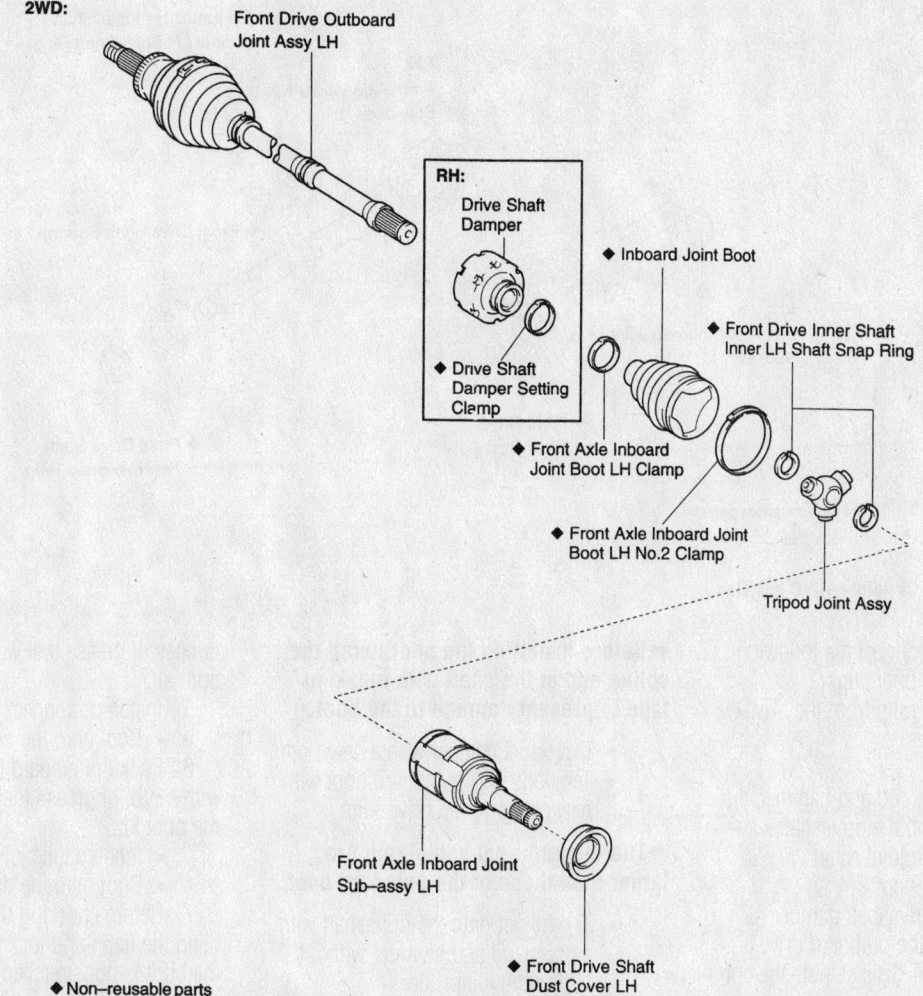

2WD:
Front Drive Outboard Joint Assy LH

RH:
Drive Shaft Damper

◆ Drive Shaft Damper Setting Clamp

◆ Inboard Joint Boot

◆ Front Drive Inner Shaft Inner LH Shaft Snap Ring

◆ Front Axle Inboard Joint Boot LH Clamp

◆ Front Axle Inboard Joint Boot LH No.2 Clamp

Tripod Joint Assy

Front Axle Inboard Joint Sub–assy LH

◆ Front Drive Shaft Dust Cover LH

◆ Non–reusable parts

P

9359AB63

Exploded view of the CV-joint—FWD vehicles

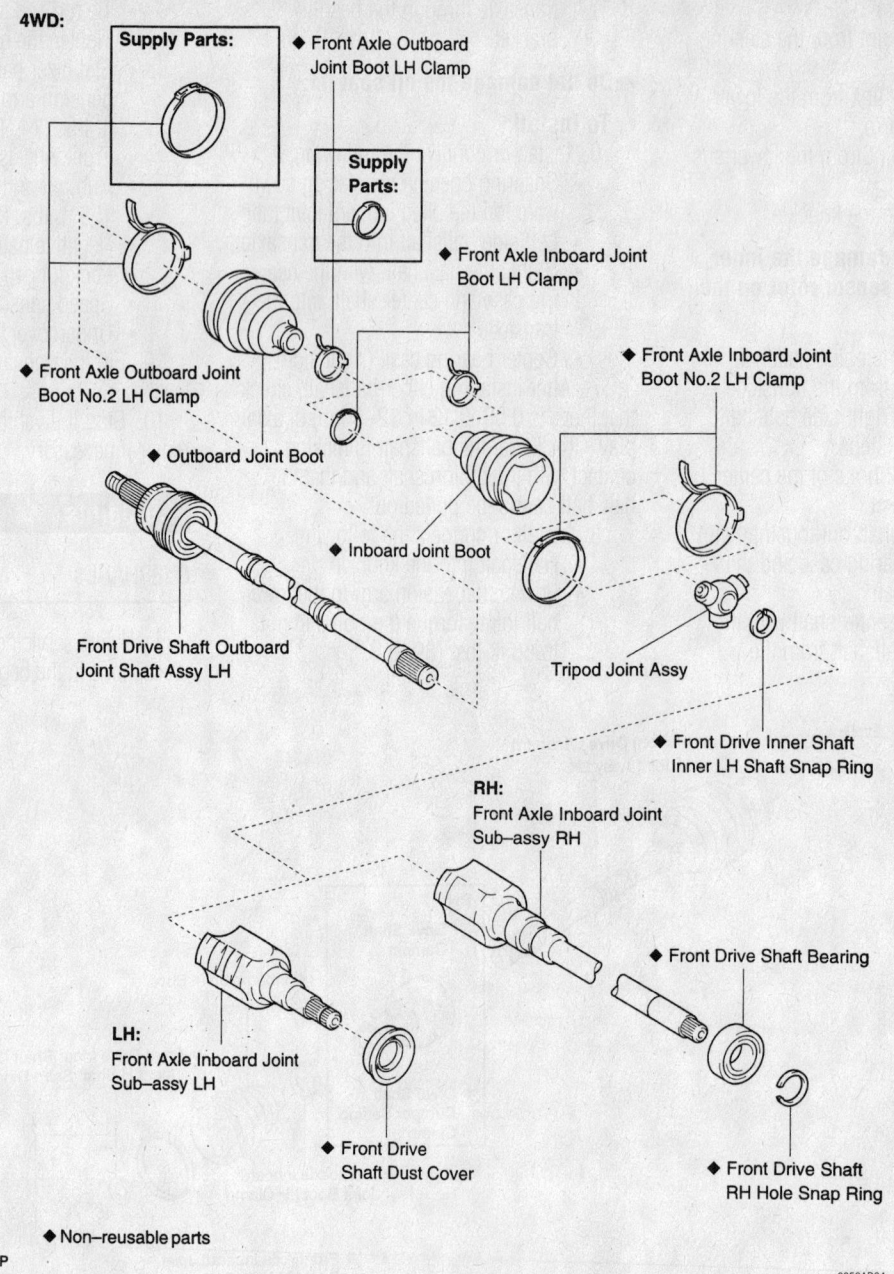

4WD:

Supply Parts:

◆ Front Axle Outboard Joint Boot LH Clamp

Supply Parts:

◆ Front Axle Inboard Joint Boot LH Clamp

◆ Front Axle Outboard Joint Boot No.2 LH Clamp

◆ Front Axle Inboard Joint Boot No.2 LH Clamp

◆ Outboard Joint Boot

◆ Inboard Joint Boot

Front Drive Shaft Outboard Joint Shaft Assy LH

Tripod Joint Assy

◆ Front Drive Inner Shaft Inner LH Shaft Snap Ring

RH:
Front Axle Inboard Joint Sub–assy RH

◆ Front Drive Shaft Bearing

LH:
Front Axle Inboard Joint Sub–assy LH

◆ Front Drive Shaft Dust Cover

◆ Front Drive Shaft RH Hole Snap Ring

◆ Non–reusable parts

P

9359AB64

Exploded view of the CV-joint—AWD vehicles

2. Remove or disconnect the following:
- Inboard joint boot clips
- Inboard joint tulip from the drive-shaft
- Snapring
- Using a brass rod and hammer, the tri-pot joint off the driveshaft without hitting the joint roller
- Inboard joint boot
- Clamp and driveshaft damper
- Clamps and the outboard drive boot. DO NOT disassemble the outboard joint.

To assemble:
3. Install or connect the following:

➡**Before installing the boot, wrap the spline end of the shaft with masking tape to prevent damage to the boot.**

- Driveshaft damper with a new clamp
- Temporarily, the inboard boot with new clamp to the drive joint

➡**The inboard boot and clamp are larger than those of the outboard boot.**

- The tri-pot onto the driveshaft with a brass rod and hammer without hitting the joint roller
- The snapring

4. Pack the outboard tulip joint and the outboard boot with about 0.26–0.33 lbs.

ounces of grease that was supplied with the boot kit.

5. Install or connect the following:
- Boot onto the outboard joint

6. Pack the inboard tulip joint and boot with ½ lb. of grease that was supplied with the boot kit.

- Inboard tulip joint onto the driveshaft
- Boot onto the driveshaft

7. Before checking the standard length, bend the band and lock it. Make sure that the boot is not stretched or squashed when the driveshaft is at standard length. Standard driveshaft length: LH: 540.2 mm (21.268 in.); RH: 857.4 mm (33.756 in.)

STEERING AND SUSPENSION

Air Bag

PRECAUTIONS

Several precautions must be observed when handling the inflator module to avoid accidental deployment and possible personal injury.

• Never carry the inflator module by the wires or connector on the underside of the module.

• When carrying a live inflator module, hold securely with both hands, and ensure that the bag and trim cover are pointed away.

• Place the inflator module on a bench or other surface with the bag and trim cover facing up.

• With the inflator module on the bench, never place anything on or close to the module that may be thrown in the event of an accidental deployment.

DISARMING

To avoid personal injury when working on vehicles equipped with an air bag, the negative battery cable must be disconnected and at least 90 seconds must elapse before working on the system. Failure to do so may result in deployment of the air bag.

REARMING

After vehicle service is completed, reattach the battery cables (positive cable first!) to rearm the air bag system.

Rack and Pinion Steering Gear

REMOVAL & INSTALLATION

1. Before servicing the vehicle, refer to the precautions in the beginning of this section.

2. Position the front wheels straight ahead.

3. Remove or disconnect the following:

• Negative battery cable. Because these vehicles are equipped with air bags, wait at least 90 seconds before proceeding.
• Horn button
• Steering wheel
• Front wheels
• Left and right engine undercovers
• Left and right tie rod ends
• Column hose cover silencer sheet

• Steering intermediate shaft
• Pressure feed and return tubes
• Left and right side front stabilizer links
• Right and left front lower control arms from the ball joints
• Hood
• No. 2 cylinder head (valve) cover

4. Install an engine support and tension it to support the engine without raising it.

    a. No. 1 engine hanger: P/N 12281-22021 1ZZ-FE, 12281-88600 2ZZ-GE.

    b. No. 2 engine hanger: P/N 12281-15040 1ZZ-FE, 12281-88600 2ZZ-GE.

    c. Bolt: P/N 91512-B1016.

    d. Torque the bolts to 28 ft. lbs. (38 Nm).

✳✳ CAUTION

Do not try to suspend the engine by hooking the chain to any other part.

    e. Attach an engine chain hoist to the hangers.

✳✳ CAUTION

The engine hoist is now in place and under tension. Use care when repositioning the vehicle and make necessary adjustments to the engine support.

5. Remove or disconnect the following:

• Bolt and nuts holding in the middle of the crossmember and support the crossmember with a jack
• Bolts from the outer side of the suspension crossmember

1ZZ–FE:

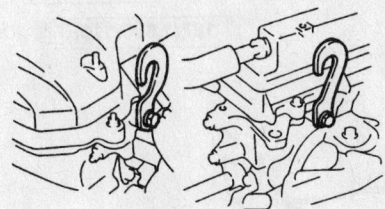

2ZZ–GE:

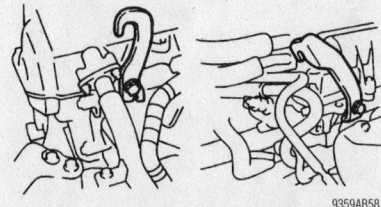

9359AB58

Proper installation of engine hangers

• Suspension crossmember with the steering gear assembly
• Steering intermediate shaft, after matchmarking it
• Rack and pinion steering gear from the crossmember

6. Installation is the reverse of the removal procedure, noting the following specifications:

    a. Steering gear bolts and nuts: 43 ft. lbs. (58 Nm) FWD, 60 ft. lbs. (82 Nm) AWD.

    b. Steering intermediate shaft: 26 ft. lbs. (35 Nm).

    c. Suspension crossmember bolts: 116 ft. lbs. (157 Nm).

    d. Engine mount insulator bolts: 38 ft. lbs. (52 Nm).

    e. Center member-to-frame bolts: 29 ft. lbs. (39 Nm).

    f. Stabilizer bar link-to-the lower control arms nuts: 55 ft. lbs. (74 Nm).

    g. Fluid return and pressure tubes: 17 ft. lbs. (23 Nm).

    h. Tie rod ends: 36 ft. lbs. (49 Nm).

    i. Wheel lug nuts: 76 ft. lbs. (103 Nm).

7. Check and top off the power steering fluid.

8. Check and adjust the alignment, if needed.

Strut and Coil Spring

REMOVAL & INSTALLATION

Front

1. Before servicing the vehicle, refer to the precautions in the beginning of this section.

2. Remove or disconnect the following:

• Negative battery cable. Because of the air bag system, wait at least 90 seconds before proceeding

✳✳ WARNING

Do not support the weight of the vehicle on the suspension arm; the arm will deform under its weight.

• Wheel
• Stabilizer link from the strut
• Bolt, and disconnect the brake hose from the strut
• With ABS brakes, speed sensor wiring harness from the strut
• Lower strut bolts and nuts
• Upper strut nuts

2WD:

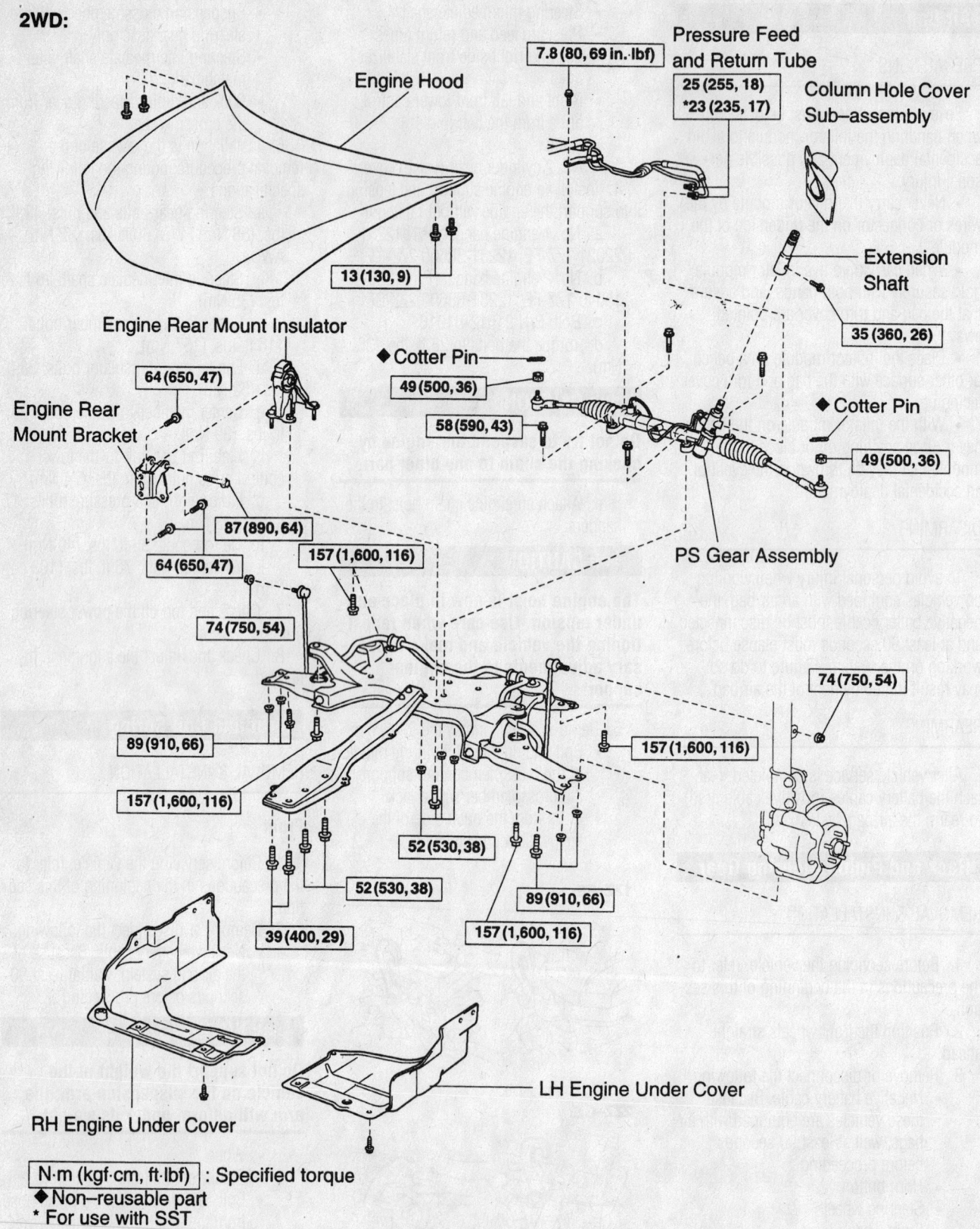

Engine Hood

7.8 (80, 69 in.·lbf)

Pressure Feed
and Return Tube

25 (255, 18)
*23 (235, 17)

Column Hole Cover
Sub–assembly

13 (130, 9)

Extension
Shaft

35 (360, 26)

Engine Rear Mount Insulator

64 (650, 47)

◆Cotter Pin

49 (500, 36)

58 (590, 43)

◆Cotter Pin

49 (500, 36)

Engine Rear
Mount Bracket

87 (890, 64)

64 (650, 47)

157 (1,600, 116)

74 (750, 54)

PS Gear Assembly

74 (750, 54)

89 (910, 66)

157 (1,600, 116)

157 (1,600, 116)

52 (530, 38)

52 (530, 38)

89 (910, 66)

39 (400, 29)

157 (1,600, 116)

LH Engine Under Cover

RH Engine Under Cover

N·m (kgf·cm, ft·lbf) : Specified torque
◆Non–reusable part
* For use with SST

9359AB59

Exploded view of a typical power rack and pinion steering gear unit—FWD shown

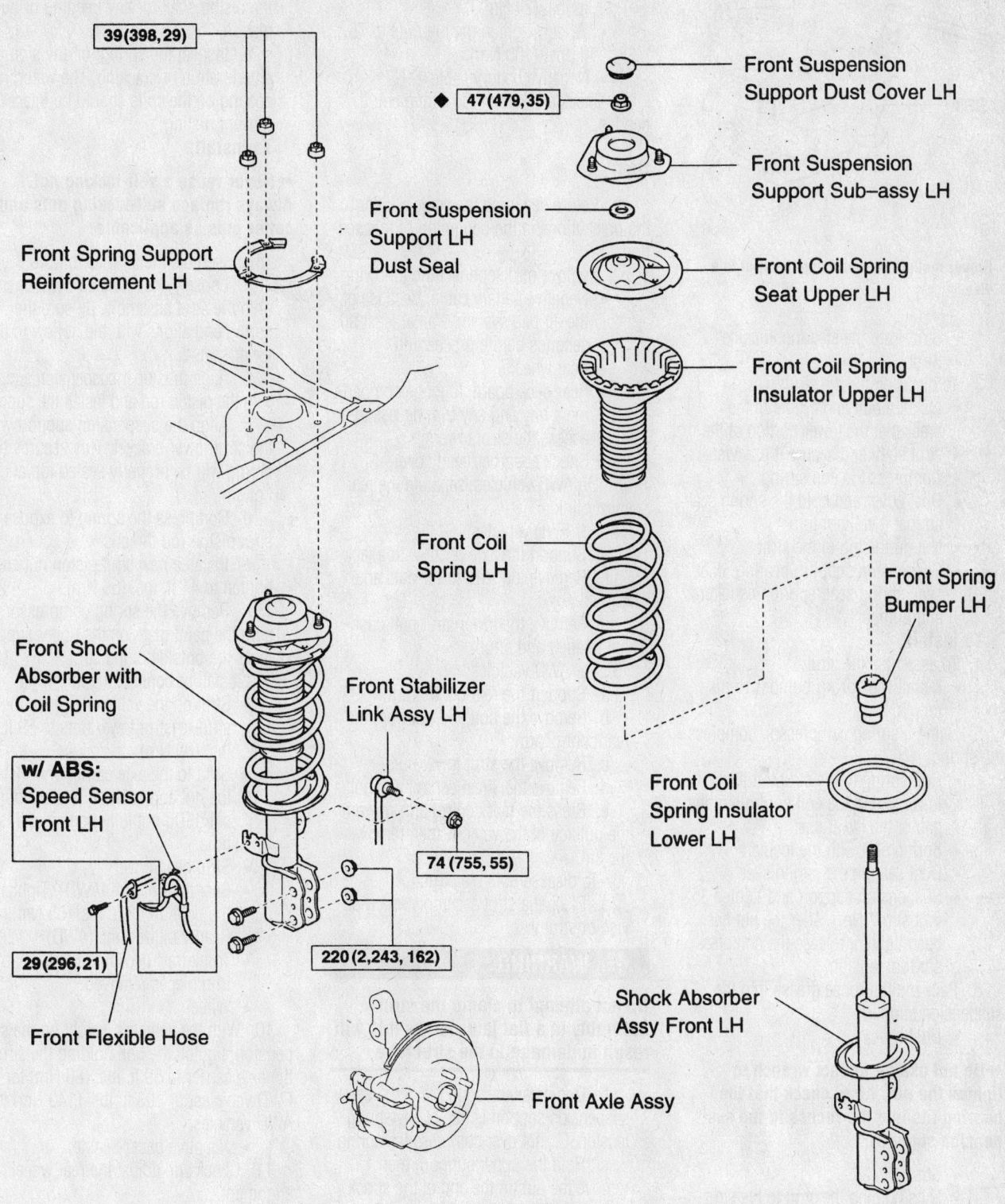

39(398, 29)

◆ 47(479, 35)

Front Suspension
Support Dust Cover LH

Front Suspension
Support Sub–assy LH

Front Spring Support
Reinforcement LH

Front Suspension
Support LH
Dust Seal

Front Coil Spring
Seat Upper LH

Front Coil Spring
Insulator Upper LH

Front Coil
Spring LH

Front Spring
Bumper LH

Front Shock
Absorber with
Coil Spring

Front Stabilizer
Link Assy LH

Front Coil
Spring Insulator
Lower LH

w/ ABS:
Speed Sensor
Front LH

74 (755, 55)

29 (296, 21)

220 (2,243, 162)

Shock Absorber
Assy Front LH

Front Flexible Hose

Front Axle Assy

N·m (kgf·cm, ft·lbf) : Specified torque

P   ◆ Non–reusable part

9359AB60

Common coil spring and strut component assembly

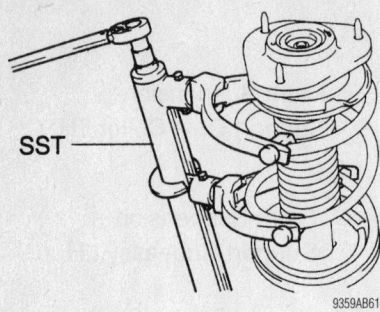

Proper method of supporting the strut in a vise

- Strut from the steering knuckle
- Strut
3. To disassemble the strut:
- Install a bolt and 2 nuts to the bracket at the lower portion of the strut shell and secure it in a vise
- Compress the coil spring
- Dust cover and hold the spring seat so that it will not turn
- Nut on the top of the strut
- Suspension support, bearing, dust seal, spring seat, spring, insulators and bumper

To install:
4. To assemble the strut:
- Install the spring bumper to piston
5. Using a spring compressor, compress the spring.
- Coil spring to the strut. Fit the lower end of the coil spring into the gap of the lower seat.
- Spring seat with the insulator
- Dust seal on the spring seat
- Suspension support and tighten 35 ft. lbs. (47 Nm). After the nut has been tighten, release the compressor tool tension.
6. Pack multipurpose grease into the suspension support.
- Dust cover.

➡Do not use an impact wrench to tighten the nut. Also, check that the bearing fits into the recess in the suspension support.

- Strut
- Nuts holding the strut to the strut tower. Tighten the nuts to 29 ft. lbs. (39 Nm).
- 2 lower strut bolts and nuts. Tighten to 162 ft. lbs. (220 Nm).
- Brake line to the steering knuckle. Tighten the line bolt to 21 ft. lbs. (29 Nm).
- Secure the wiring harness, if equipped with ABS

- Stabilizer link. Tighten the nut to 55 ft. lbs. (74 Nm).
- Wheel. Tighten the lug nuts to 76 ft. lbs. (103 Nm).
- Negative battery cable
7. Check and adjust the alignment, if needed.

Rear

1. Before servicing the vehicle, refer to the precautions in the beginning of this section.
2. Remove or disconnect the following:
- Negative battery cable. Because of the air bag system, wait at least 90 seconds before proceeding.
- Rear wheel
- Rear deck board, luggage compartment tray and any trim necessary to access the strut towers
- Shock absorber head cover
3. On AWD vehicles, separate the rear stabilizer link.
4. For FWD vehicles:
a. Support the axle beam with a jack.
b. Remove the strut tower nuts and bolt.
c. Remove the lower strut nut, cushion retainer and strut .
5. For AWD vehicles:
a. Support the rear control arm.
b. Remove the bolt and nut from the rear control arm.
c. Remove the strut tower nuts.
d. Remove the 3 rear control arm bolts.
e. Press the rear control arm down to the outside of the vehicle, then remove the strut.
6. To disassemble the strut:
a. Place the strut assembly in a pipe vise or strut vise.

✱✱ WARNING

Do not attempt to clamp the strut assembly in a flat jaw vise as this will result in damage to the strut tube.

b. Compress the spring until the upper suspension support is free of any spring tension. Do not over-compress the spring.
c. Hold the upper support, then remove the nut on the end of the shock piston rod.
d. Remove the support, coil spring, insulator, and bumper.
7. Inspect the strut as follows:
a. Check the shock absorber by moving the piston shaft through its full range of travel. It should move smoothly and evenly throughout its entire travel without any trace of binding or notching.

b. Use a small straightedge to check the piston shaft for any bending or deformation.
c. Inspect the spring for any sign of deterioration or cracking. The waterproof coating on the coils should be intact to prevent rusting.
To install:

➡**Never reuse a self-locking nut. Always replace self-locking nuts and cotter pins as applicable.**

8. Assemble the strut as follows:
a. Loosely assemble all components onto the strut assembly. Be sure the spring end aligns with the hollow in the lower seat.
b. Align the upper suspension support with the piston rod and install the support.
c. Align the suspension support with the strut lower bracket. This assures the spring will be properly seated top and bottom.
d. Compress the spring to expose the strut piston rod threads.
e. Install a new strut piston nut and tighten to 41 ft. lbs. (56 Nm).
f. Remove the spring compressor. Be sure the paint mark on the upper support faces the outside of the strut.
9. Install or connect the following:
- Strut on the vehicle. Tighten the strut-to-strut tower nuts to 59 ft. lbs. (80 Nm).
- Strut to the axle carrier and install the nut and cushion retainer/bolt snug. Do not fully tighten at this time.
- Strut head cover
- Rear control arm (AWD). Tighten the bolts to 48 ft. lbs. (65 Nm).
- Rear stabilizer link (AWD)
- Trunk tray, deckboard and any other trim pieces removed
- Wheel
10. With the vehicle's weight on the suspension, tighten the bolt holding the strut to the axle carrier to 59 ft. lbs. (80 Nm) for FWD vehicles, or 103 ft. lbs. (140 Nm) for AWD vehicles.
- Negative battery cable
11. Check and adjust the rear wheel alignment.

Lower Ball Joint

REMOVAL & INSTALLATION

1. Before servicing the vehicle, refer to the precautions in the beginning of this section.

2. Remove or disconnect the following:
- Negative battery cable. Wait at least 90 seconds before proceeding.
- Front wheel

3. Depress the brake pedal and loosen the hub nut
- ABS speed sensor, if equipped
- Cotter pin and nut from the tie rod end. Using a tie rod end removal tool, separate the tie rod end from the steering knuckle.
- Lower control arm ball joint, using a suitable puller
- Separate the front halfshaft
- Lower ball joint cotter pin and castle nut
- Lower ball joint from the steering knuckle using a puller

SST

9359AB71

Removing the ball joint from the knuckle

To install:

4. Install or connect the following:
- Lower ball joint to the lower arm. Tighten the castle nut to 76 ft. lbs. (103 Nm).
- New cotter pin
- Front halfshaft
- Lower control arm
- Tie rod end to the knuckle
- ABS speed sensor
- Hub nut
- Wheel
- Negative battery cable

5. Check and adjust the alignment, if needed.

Upper Control Arm

REMOVAL & INSTALLATION

Rear—AWD Only

1. Before servicing the vehicle, refer to the precautions in the beginning of this section.
2. Remove or disconnect the following:
- Negative battery cable. Wait at least 90 seconds before proceeding
- Rear wheel
- Exhaust pipe

- Propeller shaft with center bearing shaft
- Rear stabilizer links
- Rear hub nuts
- Rear brake drum
- Speed sensor
- Front brake shoe
- Parking brake shoe strut set
- Rear brake shoe
- Parking brake cables
- Rear brake hoses
- Separate the rear suspension arms
- Separate the upper control arm
- Rear drive axle assembly
- Rear strut nut and bolt
- Rear strut
- Rear suspension arm
- Rear suspension member
- Upper control arm assembly. Matchmark the camber adjust cams and rear suspension member prior to removal.

3. Installation is the reverse of the removal procedure.

Lower Control Arm

REMOVAL & INSTALLATION

1. Before servicing the vehicle, refer to the precautions in the beginning of this section.
2. Remove or disconnect the following:
- Negative battery cable. Wait at least 90 seconds before proceeding..
- Front wheel
- Stabilizer link
- Bolt and nuts and separate the lower control arm from the lower ball joint
- Bolts and nuts, then separate the steering gear. Loosen the bolt, since the nut cannot be rotated, then suspend the steering gear.

3. Support the engine, using the engine lifting hooks and the procedure under Engine Removal & Installation.
- Crossmember
- Lower control arm from the crossmember

4. Installation is the reverse of the removal procedure.

Wheel Bearings

REMOVAL & INSTALLATION

Front

1. Before servicing the vehicle, refer to the precautions in the beginning of this section.

2. Remove or disconnect the following:
- Negative battery cable. On vehicles equipped with an air bag, wait at least 90 seconds before proceeding.
- Wheels
- Hub nut
- Front stabilizer link
- Anti-lock Brake System (ABS) speed sensor
- Brake caliper
- Rotor
- Tie rod end from the steering knuckle
- Lower control arm ball joint
- Front halfshaft from the hub, using a mallet to tap it out. Be careful not to damage the boot or speed sensor.

3. Loosen the nuts on the lower side of the strut assembly. Do not remove at this time.
- Lower ball joint using a puller
- Tie rod end from the steering knuckle
- Steering knuckle from the lower control arm
- Knuckle from the strut assembly
- Hub

➡**Cover the halfshaft boot with a shop rag to protect it from any damage.**

4. Clamp the steering knuckle in a vise and remove the dust deflector. Remove the nut holding the steering knuckle to the ball joint. Press the ball joint out of the steering knuckle.

5. Remove the inner axle seal.

6. Using a Torx® wrench, remove the bolts securing the dust cover.

7. Using hub puller, remove the hub and backing plate from the steering knuckle.

8. Using a proper sized driver and a press, remove the inner hub race from the axle hub.

9. Using seal removal tool, remove the outer axle seal.

10. Using snapring pliers, remove the snapring from the inner side of the steering knuckle.

11. Using a proper sized driver and a press, remove the bearing from the steering knuckle. The bearing is pressed from the front of the steering knuckle and is removed through the back of the steering knuckle.

To install:

12. Perform the following:

13. Using a proper sized driver and a press, install a new bearing to the steering knuckle.

14. Install the snapring to the steering knuckle using snapring pliers.

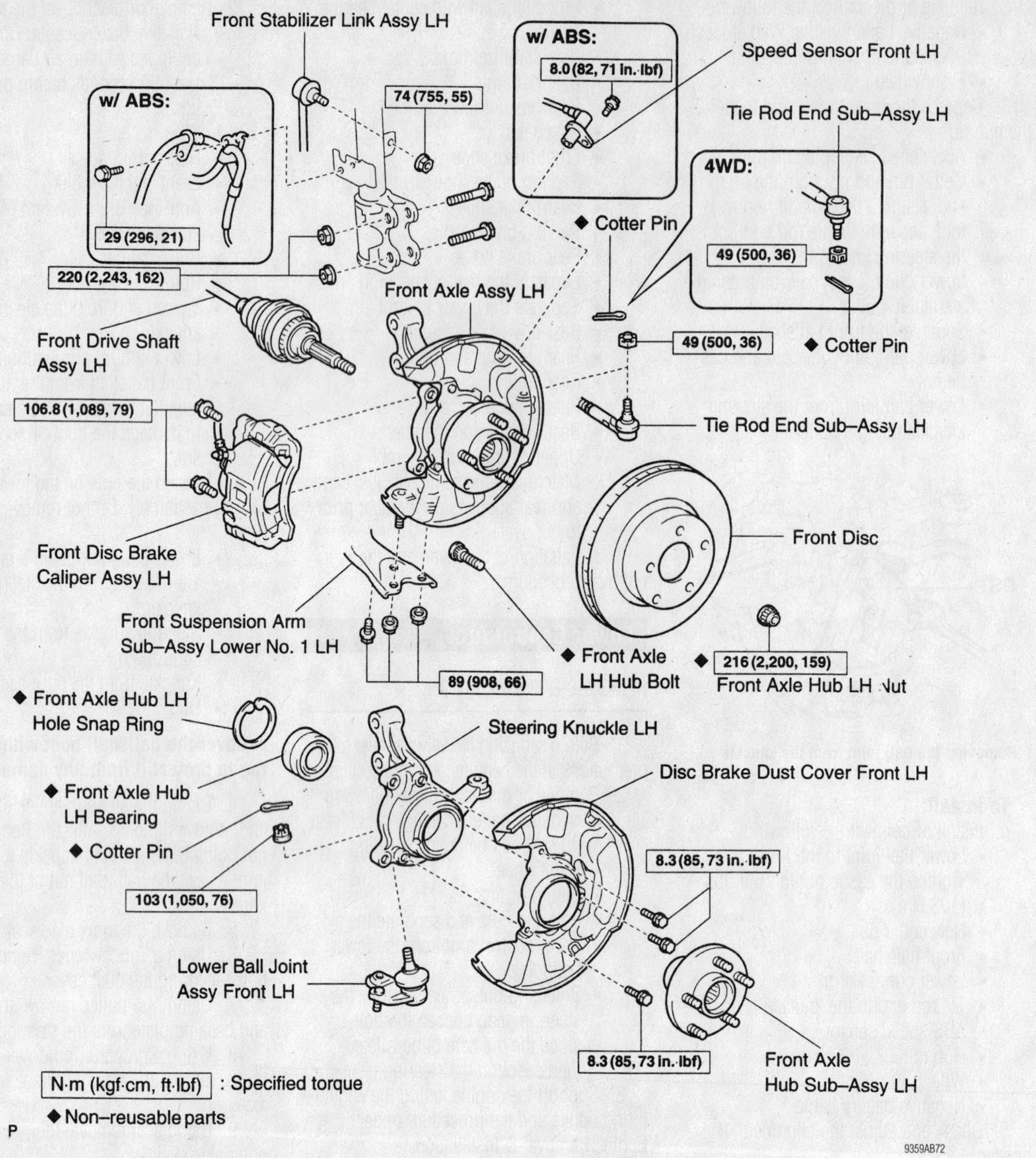

Front Stabilizer Link Assy LH

w/ ABS:

w/ ABS:

Speed Sensor Front LH

8.0 (82, 71 in.·lbf)

Tie Rod End Sub–Assy LH

74 (755, 55)

4WD:

29 (296, 21)

◆ Cotter Pin

49 (500, 36)

◆ Cotter Pin

220 (2,243, 162)

Front Axle Assy LH

49 (500, 36)

Front Drive Shaft
Assy LH

Tie Rod End Sub–Assy LH

106.8 (1,089, 79)

Front Disc

Front Disc Brake
Caliper Assy LH

Front Suspension Arm
Sub–Assy Lower No. 1 LH

216 (2,200, 159)

◆ Front Axle
LH Hub Bolt

Front Axle Hub LH Nut

89 (908, 66)

◆ Front Axle Hub LH
Hole Snap Ring

Steering Knuckle LH

Disc Brake Dust Cover Front LH

◆ Front Axle Hub
LH Bearing

8.3 (85, 73 in.·lbf)

◆ Cotter Pin

103 (1,050, 76)

Lower Ball Joint
Assy Front LH

8.3 (85, 73 in.·lbf)

Front Axle
Hub Sub–Assy LH

N·m (kgf·cm, ft·lbf) : Specified torque

◆ Non–reusable parts

P

9359AB72

Exploded view of the front hub and bearing, and related components

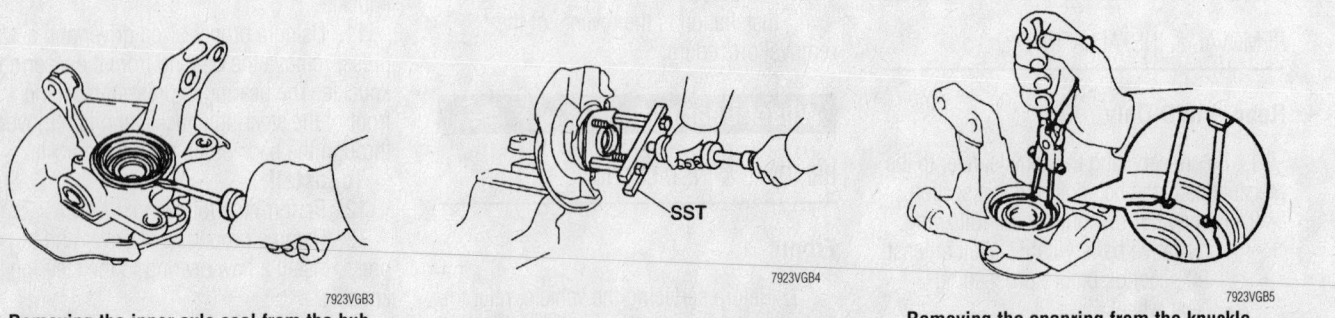

7923VGB3

7923VGB4

SST

7923VGB5

**Removing the inner axle seal from the hub
assembly**

Removing the axle hub from the knuckle

**Removing the snapring from the knuckle
before pressing out the bearing**

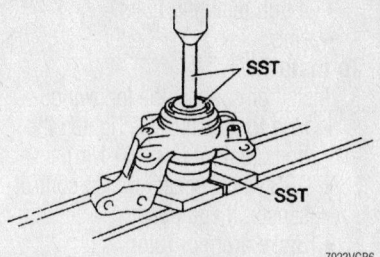

Removing the bearing from the steering knuckle using a press

15. Using a seal driver and a hammer, install a new outer oil seal. Apply multipurpose grease to the oil seal lip.

16. Place the dust cover on the steering knuckle. Tighten the bolts: 78 inch lbs. (9 Nm).

17. Using a press and a proper sized driver, install the axle hub to the steering knuckle.

18. Attach the ball joint to the steering knuckle. Install a new cotter pin.

19. Using a seal driver and a hammer, install a new inner oil seal. Apply multipurpose grease to the oil seal lip.

20. Install the knuckle and hub assembly to the axle and temporarily tighten the axle nut.

21. Connect the knuckle assembly to the lower strut bracket. Temporarily insert the mounting bolts from the rear and install the nuts making sure the matchmarks made earlier are in alignment.

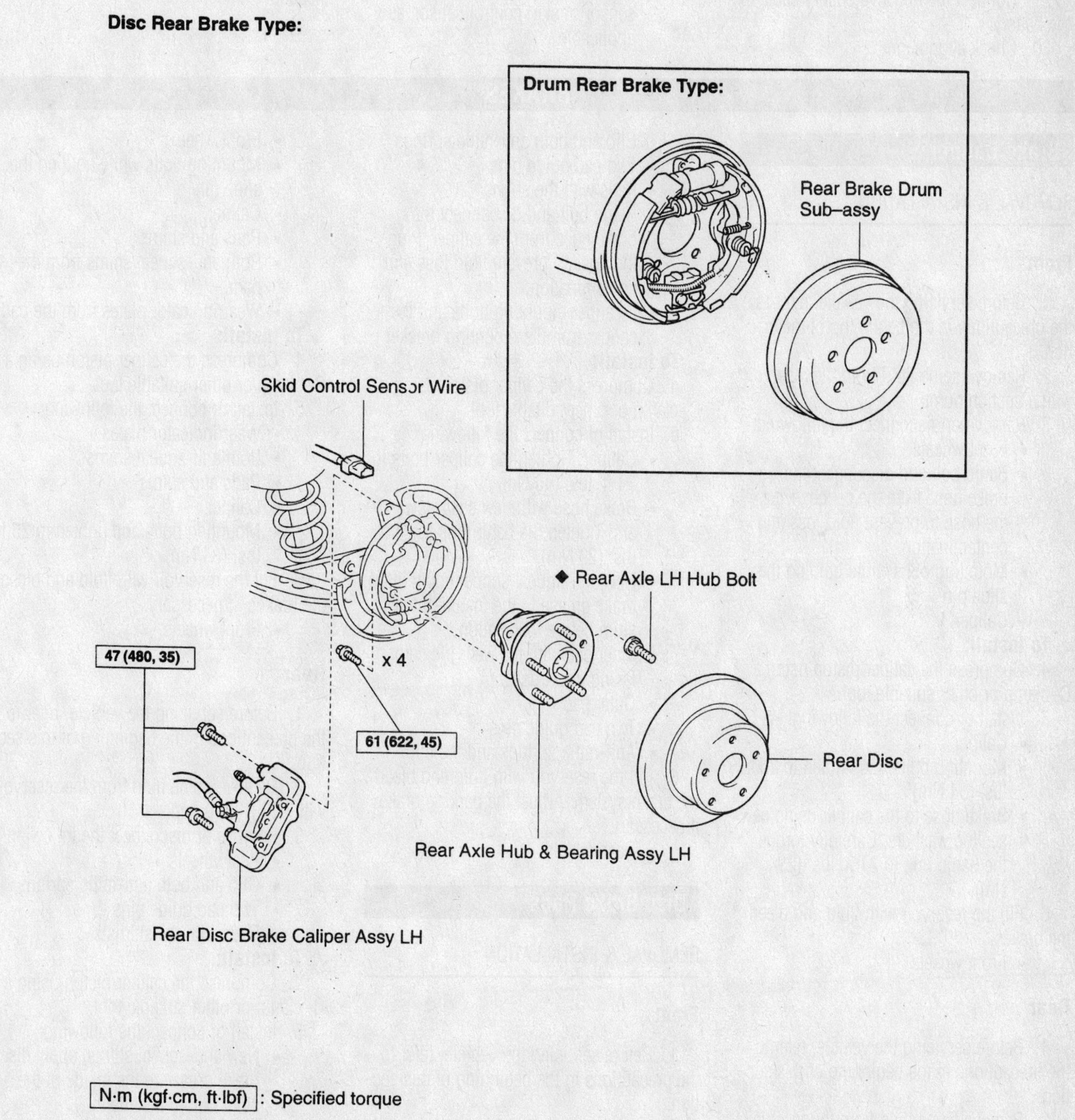

Disc Rear Brake Type:

Drum Rear Brake Type:

Rear Brake Drum Sub–assy

Skid Control Sensor Wire

◆ Rear Axle LH Hub Bolt

47 (480, 35)

x 4

61 (622, 45)

Rear Disc

Rear Axle Hub & Bearing Assy LH

Rear Disc Brake Caliper Assy LH

N·m (kgf·cm, ft·lbf) : Specified torque

◆ Non–reusable part

Exploded view of the hub and wheel bearing assembly

22. Connect the lower ball joint to lower arm.

23. Connect the tie rod end to the knuckle.

24. Tighten the bolts on the lower side of the strut assembly.

25. If equipped, install the ABS speed sensor.

26. Install the brake disc and the caliper.

27. Tighten the axle nut while someone depresses the brake pedal.

28. Install the wheels to the vehicle. Verify that the wheel turns freely.

29. Connect the negative battery cable to the battery.

30. Check alignment.

Brake Caliper

REMOVAL & INSTALLATION

Front

1. Before servicing the vehicle, refer to the precautions in the beginning of this section.

2. Remove some fluid from the reservoir with a suction pump.

3. Remove or disconnect the following:
   - Front wheels
   - Banjo bolt and disconnect the brake hose from the caliper. Plug the hose to prevent fluid loss and contamination.
   - Mounting bolts while holding the slide pin
   - Caliper

To Install:

4. Compress the caliper piston using a C–clamp or other suitable tool.

5. Install or connect the following:
   - Caliper
   - Mounting bolts and tighten to 25 ft. lbs. (34 Nm)
   - Brake hose to the caliper using new sealing washers. Carefully torque the banjo bolt to 21 ft. lbs. (29 Nm).

6. Fill the reservoir with fluid and bleed the brakes.
   - Front wheels

Rear

1. Before servicing the vehicle, refer to the precautions in the beginning of this section.

2. Remove some fluid from the reservoir with a suction pump.

3. Remove or disconnect the following:
   - Rear wheels

Rear

1. Before servicing the vehicle, refer to the precautions in the beginning of this section.

2. Remove or disconnect the following:
   - Negative battery cable. On vehicles equipped with an air bag, wait at least 90 seconds before proceeding.
   - Wheel
   - Brake drum or rotor
   - With ABS brakes, ABS wheel speed sensor or skid control sensor, as applicable

BRAKES

- Clip and both anti-rattle springs
- Two pad guide pins
- Pads with the shims
- Banjo bolt and disconnect the brake hose from the caliper. Plug the hose to prevent fluid loss and contamination.
- 2 caliper mounting bolts and the caliper from its mounting bracket

To Install:

4. Compress the caliper piston using a C–clamp or other suitable tool.

5. Install or connect the following:
   - Caliper. Tighten the caliper bolts to 34 ft. lbs. (46 Nm).
   - Brake hose with new sealing washers. Tighten the banjo bolt to 21 ft. lbs. (29 Nm).
   - New anti-squeal shims, apply disc brake grease to the inside of the shim before installation
   - Inner pad with the wear indicator facing upwards
   - Outer pad
   - Two pad guide pins
   - Anti-rattle springs and the clip

6. Fill the reservoir with fluid and bleed the brake system. Adjust the parking brake if necessary.
   - Rear wheels

Disc Brake Pads

REMOVAL & INSTALLATION

Front

1. Before servicing the vehicle, refer to the precautions in the beginning of this section.

2. Remove some fluid from the reservoir with a suction pump.

3. Remove or disconnect the following:

- 4 hub retaining bolts
- Hub

To install:

3. Install or connect the following:
   - Hub to the knuckle. Tighten the bolts to 45 ft. lbs. (61 Nm).
   - ABS wheel speed or skid control sensor, if equipped
   - Brake drum or rotor
   - Wheel
   - Negative battery cable

4. Check and adjust the alignment, if needed.

- Front wheels
- Mounting bolts while holding the slide pin
- Caliper
- Pads and shims
- Both anti-squeal shims from the pads
- Wear indicator plates from the pads

To Install:

4. Compress the caliper piston using a C–clamp or other suitable tool.

5. Install or connect the following:
   - Wear indicator plates
   - Both anti-squeal shims
   - Pads and shims
   - Caliper
   - Mounting bolts and tighten to 25 ft. lbs. (34 Nm)

6. Fill the reservoir with fluid and bleed the brakes, if necessary.
   - Front wheels

Rear

1. Before servicing the vehicle, refer to the precautions in the beginning of this section.

2. Remove some fluid from the reservoir with a suction pump.

3. Remove or disconnect the following:
   - Rear wheels
   - Clip and both anti-rattle springs
   - Two pad guide pins
   - Pads with the shims

To Install:

4. Compress the caliper piston using a C–clamp or other suitable tool.

5. Install or connect the following:
   - New anti-squeal shims, apply disc brake grease to the inside of the shim before installation
   - Inner pad with the wear indicator facing upwards
   - Outer pad

1ZZ–FE(FF) Type:

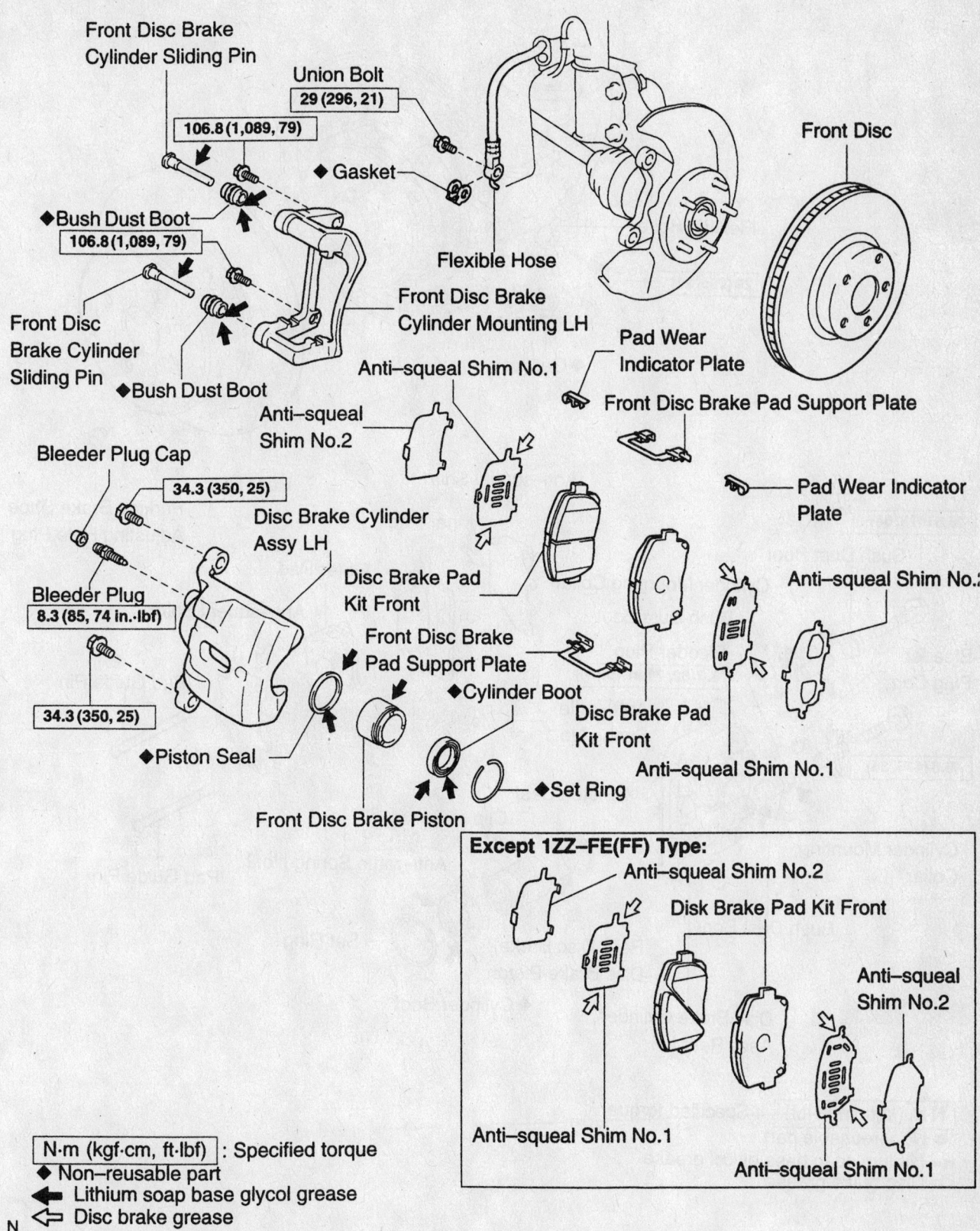

N·m (kgf·cm, ft·lbf) : Specified torque
◆ Non–reusable part
◀ Lithium soap base glycol grease
◁ Disc brake grease

Exploded view of the front caliper components

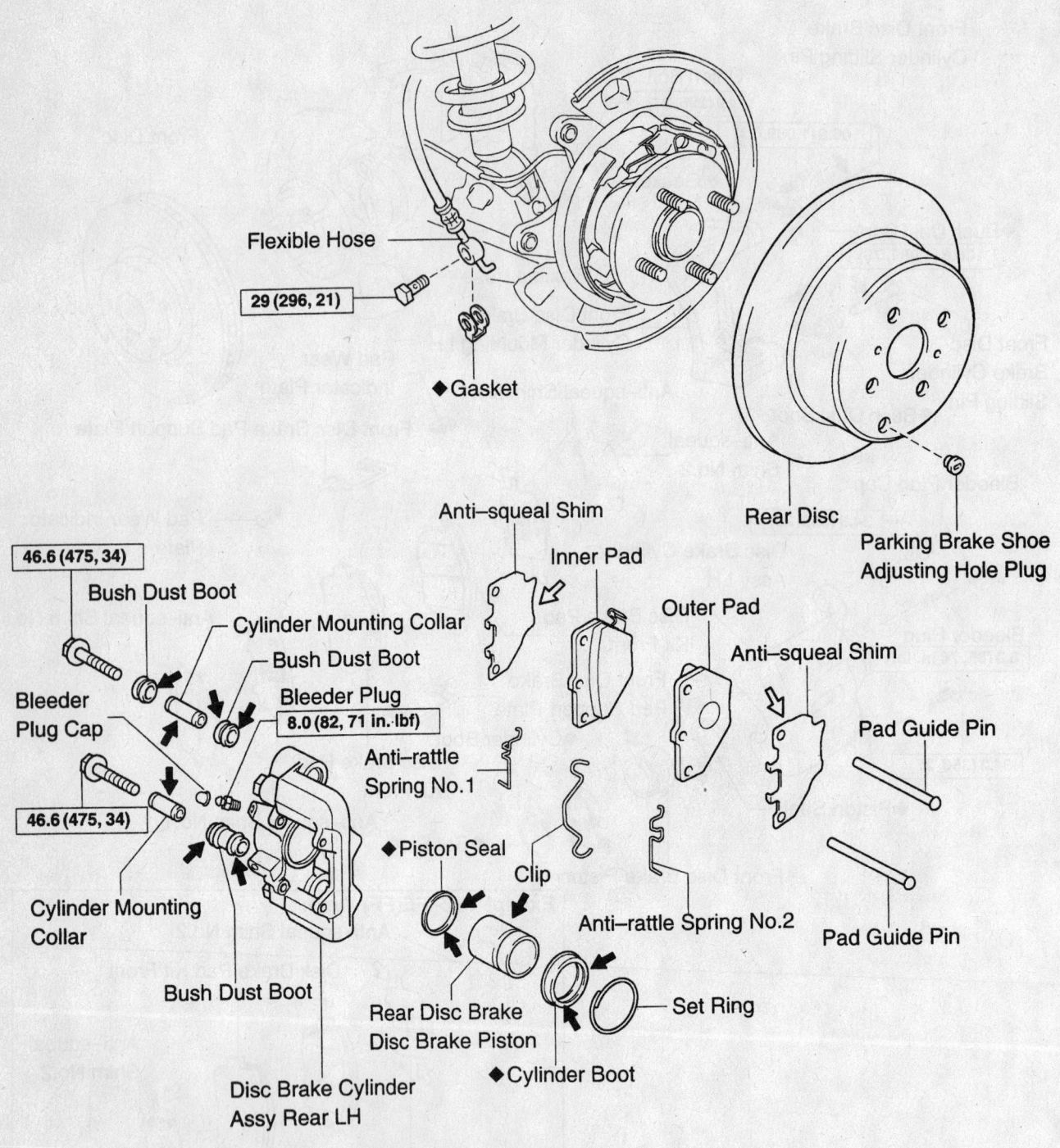

Flexible Hose

29 (296, 21)

◆Gasket

Anti–squeal Shim

Inner Pad

Rear Disc

Parking Brake Shoe Adjusting Hole Plug

Outer Pad

Anti–squeal Shim

46.6 (475, 34)

Bush Dust Boot

Cylinder Mounting Collar

Bush Dust Boot

Bleeder Plug

8.0 (82, 71 in. lbf)

Bleeder Plug Cap

Pad Guide Pin

Anti–rattle Spring No.1

46.6 (475, 34)

Cylinder Mounting Collar

◆Piston Seal

Clip

Anti–rattle Spring No.2

Pad Guide Pin

Bush Dust Boot

Rear Disc Brake Disc Brake Piston

Set Ring

◆Cylinder Boot

Disc Brake Cylinder Assy Rear LH

N·m (kgf·cm, ft·lbf) : Specified torque
◆ Non–reusable part
◀ Lithium soap base glycol grease
◁ Disc brake grease

42356-TMAT-G02

Exploded view of the rear caliper components

1ZZ–FE(FF) Type:

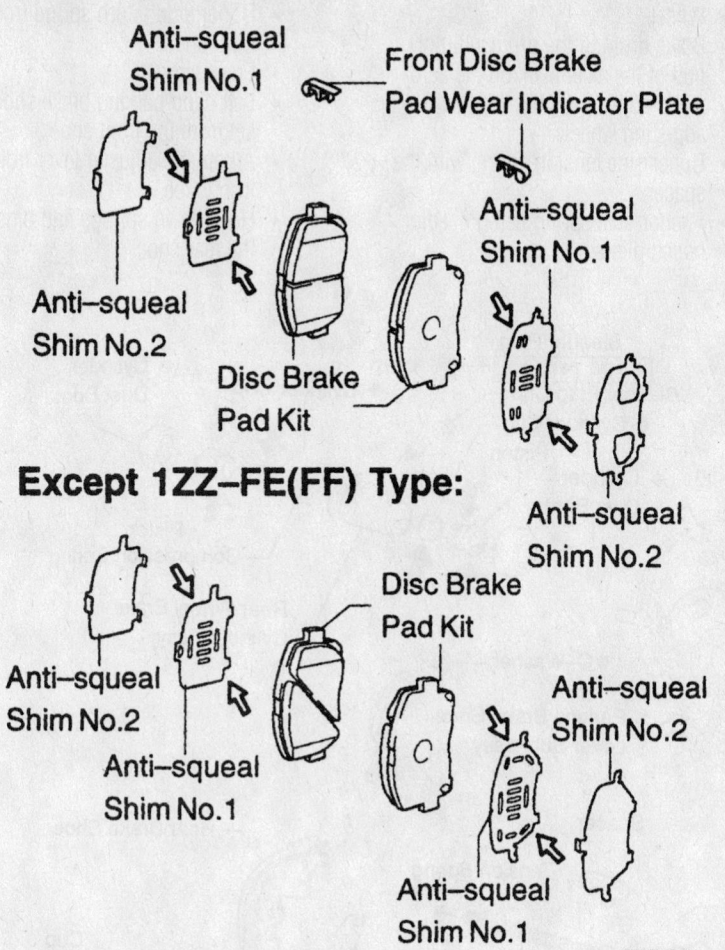

Anti–squeal
Shim No.1

Front Disc Brake
Pad Wear Indicator Plate

Anti–squeal
Shim No.1

Anti–squeal
Shim No.2

Disc Brake
Pad Kit

Anti–squeal
Shim No.2

Except 1ZZ–FE(FF) Type:

Disc Brake
Pad Kit

Anti–squeal
Shim No.2

Anti–squeal
Shim No.1

Anti–squeal
Shim No.2

Anti–squeal
Shim No.1

42356-TMAT-G03

Exploded view of the front pads and related components

- Two pad guide pins
- Anti-rattle springs and the clip

6. Fill the reservoir with fluid and bleed the brake system. Adjust the parking brake if necessary.
- Rear wheels

Brake Drums

REMOVAL & INSTALLATION

1. Before servicing the vehicle, refer to the precautions in the beginning of this section.
2. Remove or disconnect the following:
- Wheel
- Brake drum. If the drum will not pull of the axle, back off the automatic adjuster by turning the adjusting wheel.

To install:
3. Install or connect the following:
- Drum on the axle
- Wheel

4. Refill the master cylinder and pump pedal to attain full brake pedal before road-testing the vehicle.

Brake Shoes

REMOVAL & INSTALLATION

FWD Models

1. Before servicing the vehicle, refer to the precautions in the beginning of this section.

LH

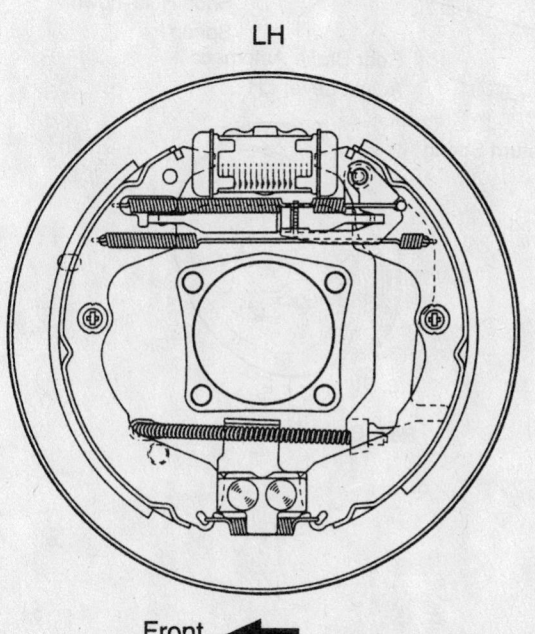

Front ◀

RH

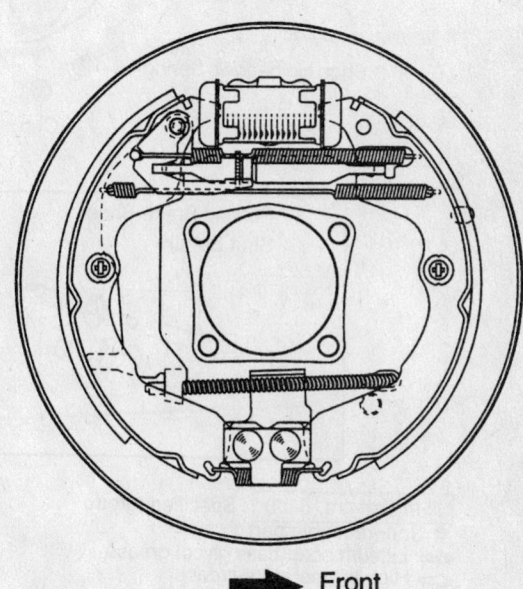

▶ Front

42356-TMAT-G04

View of the properly installed rear drum brake assembly

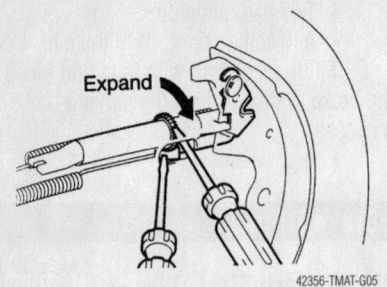

Expand

42356-TMAT-G05

Turn the adjuster to expand the shoe until the drum locks

2. Remove or disconnect the following:
- Wheel
- Brake drum. If the drum will not pull of the axle, back off the automatic adjuster by turning the adjusting wheel.
- Upper side tension spring with the spacer
- Anchor side spring using needle nosed pliers
- Hold-down springs and pins from the front shoe
- Upper side return spring from the front shoe
- Front shoe
- Left hand parking brake shoe strut set from the front shoe
- Automatic adjuster lever from the front shoe
- Hold-down springs and pins from the rear shoe

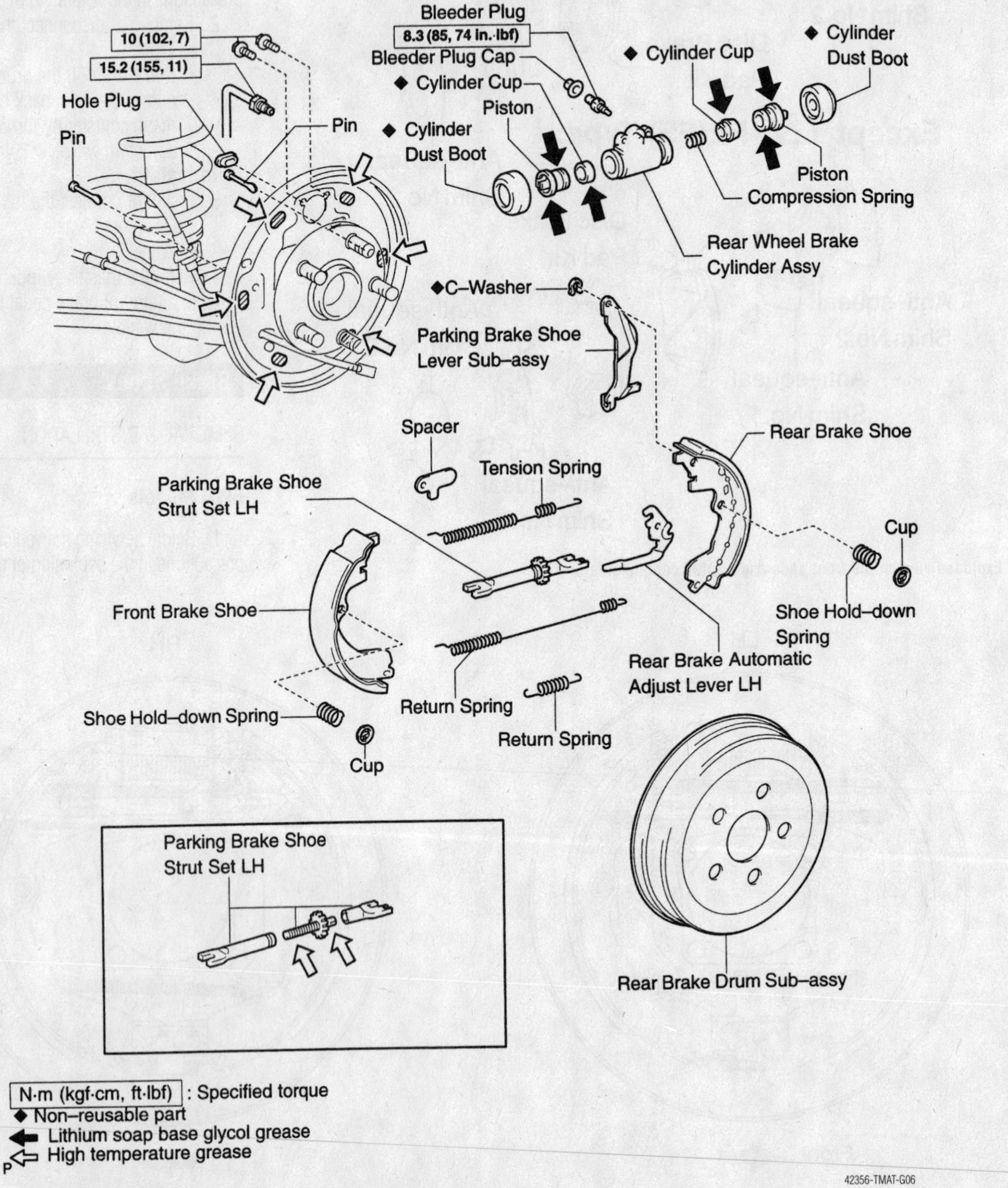

10 (102, 7)
15.2 (155, 11)

Hole Plug

Pin

Pin

Bleeder Plug
8.3 (85, 74 in.·lbf)
Bleeder Plug Cap
◆ Cylinder Cup
Piston
◆ Cylinder Dust Boot

◆ Cylinder Cup

◆ Cylinder Dust Boot

◆ Cylinder Cup

Piston
Compression Spring

Rear Wheel Brake Cylinder Assy

◆ C–Washer

Parking Brake Shoe Lever Sub–assy

Spacer

Tension Spring

Rear Brake Shoe

Cup

Parking Brake Shoe Strut Set LH

Front Brake Shoe

Shoe Hold–down Spring

Rear Brake Automatic Adjust Lever LH

Shoe Hold–down Spring

Return Spring

Return Spring

Cup

Parking Brake Shoe Strut Set LH

Rear Brake Drum Sub–assy

N·m (kgf·cm, ft·lbf) : Specified torque
◆ Non–reusable part
◀ Lithium soap base glycol grease
P◁ High temperature grease

42356-TMAT-G06

Exploded view of the drum brake components—1ZZ–FE (FWD) models

- Upper side return spring from the rear shoe
- Parking brake cable from the rear shoe using needle nosed pliers
- Rear brake shoe
- C–washer using a suitable pry tool from the shoe
- Parking brake lever from the shoe

To install:

3. Lubricate the contact points on the backing plate and the adjuster with lithium grease.

4. Install or connect the following:
- Parking brake lever, and attach using a new C–washer
- Parking brake lever to the shoe lever
- Upper return spring to the rear shoe
- Shoe assembly onto the backing plate
- Hold-down springs and pins to retain the rear shoe
- Automatic adjuster lever

5. Apply lithium grease to the adjuster bolt.
- Left hand parking brake shoe strut set
- Upper side return spring to the front shoe
- Hold-down springs and pins to the front shoe
- Anchor side spring using needle nosed pliers
- Upper side tension spring with the spacer

6. Adjust the rear brakes as follows:
  a. Temporarily install the drum and hub nuts.
  b. Remove the hole plug from the backing plate.
  c. Turn the adjuster to expand the shoe until the drum locks.
  d. Back off the adjuster eight notches using a suitable adjustment tool.
  e. Install the hole plug. into the backing plate to prevent dirt and moisture from entering.
  f. Readjust the parking brake cable as necessary.
7. Install the wheels.
8. Refill the master cylinder and pump pedal to attain full brake pedal before Road-testing the vehicle.

AWD Models

1. Before servicing the vehicle, refer to the precautions in the beginning of this section.

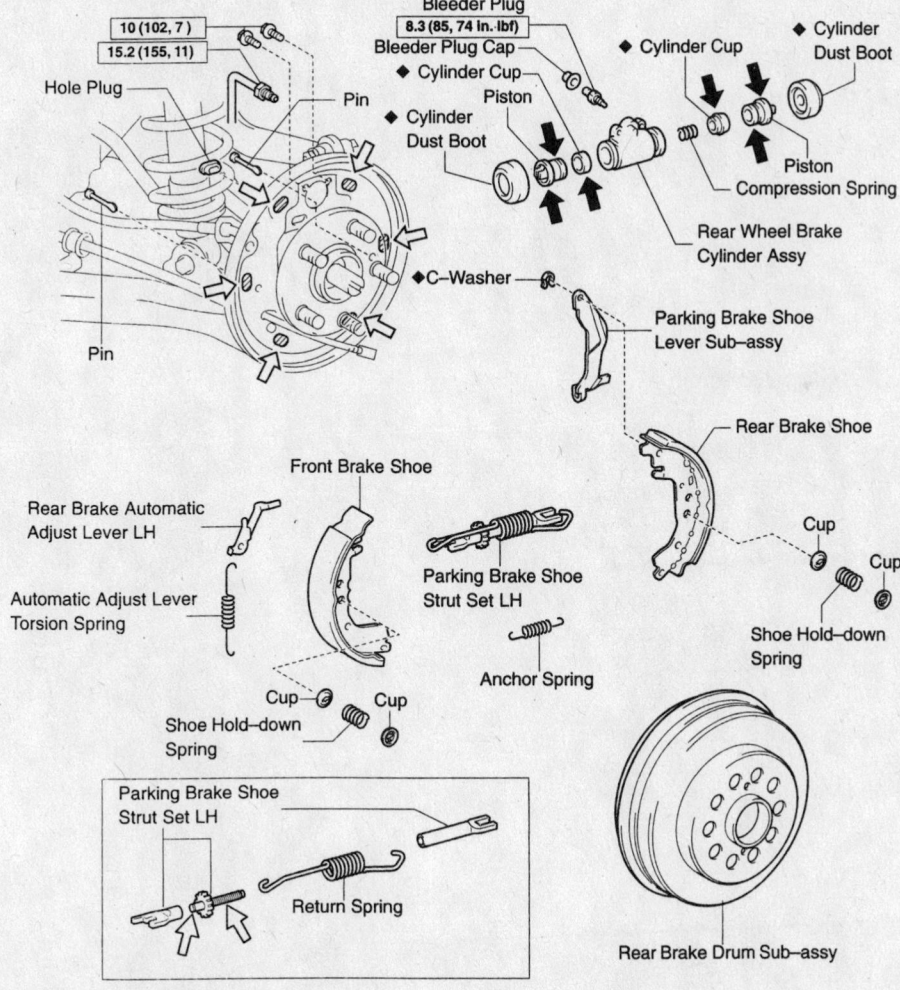

N·m (kgf·cm, ft·lbf) : Specified torque
◆ Non–reusable part
◀ Lithium soap base glycol grease
◁ High temperature grease

42356-TMAT-G07

Exploded view of the drum brake components—1ZZ–FE (AWD) models

2. Remove or disconnect the following:
- Wheel
- Brake drum. If the drum will not pull of the axle, back off the automatic adjuster by turning the adjusting wheel.
- Return spring from the front brake shoe
- Anchor spring using needle nosed pliers
- Hold-down springs and pins from the front shoe
- Front shoe
- Automatic adjuster lever spring and the lever
- Hold-down springs and pins from the rear shoe
- Parking brake cable from the rear shoe using needle nosed pliers
- Rear brake shoe
- Anchor spring from the rear shoe

- C-washer using a suitable pry tool from the shoe
- Parking brake lever from the shoe

To install:

3. Lubricate the contact points on the backing plate and the adjuster with lithium grease.

4. Install or connect the following:
- Parking brake lever, and attach using a new C-washer
- Parking brake cable to the lever
- Shoe assembly onto the backing plate
- Hold-down springs and pins to retain the rear shoe
- Automatic adjuster lever

5. Apply lithium grease to the adjuster bolt.
- Left hand parking brake shoe strut set
- Hold-down springs and pins to the front shoe

- Anchor spring to each shoe using needle nosed pliers
- Return spring to each shoe

6. Adjust the rear brakes as follows:

a. Temporarily install the drum and hub nuts.

b. Remove the hole plug from the backing plate.

c. Turn the adjuster to expand the shoe until the drum locks.

d. Back off the adjuster eight notches using a suitable adjustment tool.

e. Install the hole plug. into the backing plate to prevent dirt and moisture from entering.

f. Readjust the parking brake cable as necessary.

7. Install the wheels.

8. Refill the master cylinder and pump pedal to attain full brake pedal before Road-testing the vehicle.

TOYOTA AND LEXUS

Land Cruiser • Sequoia • LX470

38

SPECIFICATION CHARTS

ENGINE AND VEHICLE IDENTIFICATION

Code ①	Liters (cc)	Cu. In.	Cyl.	Fuel Sys.	Engine Type	Eng. Mfg.	Code ②	Year
2UZ-FE	4.7 (4664)	285	8	SFI	DOHC	Toyota	Y	2000
							1	2001
							2	2002
							3	2003
							4	2004

SFI: Sequential Fuel Injection

DOHC: Double Overhead Camshaft

① Stamped on the left side of the engine block

② 10th digit of the Vehicle Identification Number (VIN)

42356-SEQU-C01

GENERAL ENGINE SPECIFICATIONS

Year	Model	Engine Displacement Liters (cc)	Engine Series (ID/VIN)	Fuel System	Net Horsepower @ rpm	Net Torque @ rpm (ft. lbs.)	Bore x Stroke (in.)	Compression Ratio	Oil Pressure @ rpm
2000	LX470	4.7 (4665)	2UZ-FE	SFI	230@4800	320@3400	3.70x3.31	9.6:1	43-85@3000
	Land Cruiser	4.7 (4664)	2UZ-FE	SFI	230@4800	320@3400	3.70x3.30	9.6:1	45-65@3000
2001	LX470	4.7 (4665)	2UZ-FE	SFI	245@4800	315@3400	3.70x3.31	9.6:1	45-65@3000
	Land Cruiser	4.7 (4664)	2UZ-FE	SFI	245@4800	315@3400	3.70x3.30	9.6:1	45-65@3000
	Sequoia	4.7 (4664)	2UZ-FE	SFI	245@4800	315@3400	3.70x3.30	9.6:1	45-65@3000
2002	LX470	4.7 (4665)	2UZ-FE	SFI	245@4800	315@3400	3.70x3.31	9.6:1	45-65@3000
	Land Cruiser	4.7 (4664)	2UZ-FE	SFI	245@4800	315@3400	3.70x3.30	9.6:1	45-65@3000
	Sequoia	4.7 (4664)	2UZ-FE	SFI	245@4800	315@3400	3.70x3.30	9.6:1	45-65@3000
2003	LX470	4.7 (4665)	2UZ-FE	SFI	245@4800	315@3400	3.70x3.31	9.6:1	45-65@3000
	Land Cruiser	4.7 (4664)	2UZ-FE	SFI	245@4800	315@3400	3.70x3.30	9.6:1	45-65@3000
	Sequoia	4.7 (4664)	2UZ-FE	SFI	245@4800	315@3400	3.70x3.30	9.6:1	45-65@3000

SFI: Sequential Fuel Injection

42356-SEQU-C02

ENGINE TUNE-UP SPECIFICATIONS

Year	Engine Displacement Liters (cc)	Engine ID/VIN	Spark Plug Gap (in.)	Ignition Timing (deg.)*	Fuel Pump (psi)	Idle Speed (rpm) MT	Idle Speed (rpm) AT	Valve Clearance Intake	Valve Clearance Exhaust
2000	4.7 (4664)	2UZ-FE	0.043	5-15B	38-44	—	650-750	0.006-0.009	0.011-0.014
2001	4.7 (4664)	2UZ-FE	0.031	5-15B	38-44	—	650-750	0.006-0.010	0.010-0.014
2002	4.7 (4664)	2UZ-FE	0.031	5-15B	38-44	—	650-750	0.006-0.010	0.010-0.014
2003	4.7 (4664)	2UZ-FE	0.031	5-15B	38-44	—	650-750	0.006-0.010	0.010-0.014

NOTE: The Vehicle Emission Control Information label often reflects specification changes made during production.

The label figures must be used if they differ from those in this chart.

B: Before top dead center

* With terminals TC and E1 connected to DLC1

42356-SEQU-C03

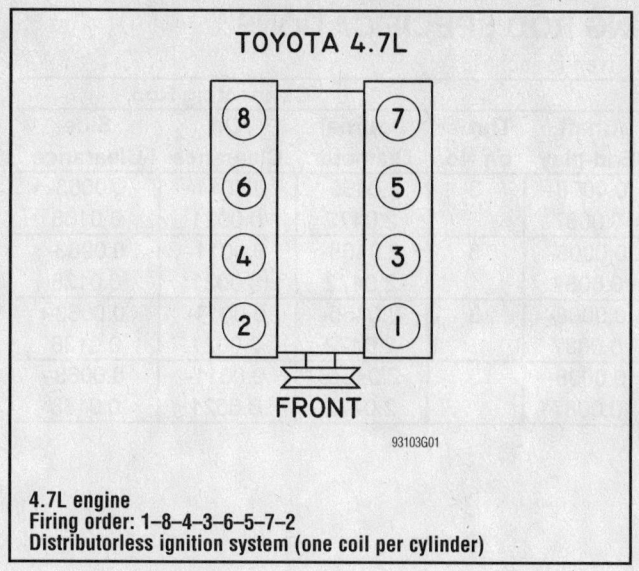

TOYOTA 4.7L

FRONT

93103G01

4.7L engine
Firing order: 1–8–4–3–6–5–7–2
Distributorless ignition system (one coil per cylinder)

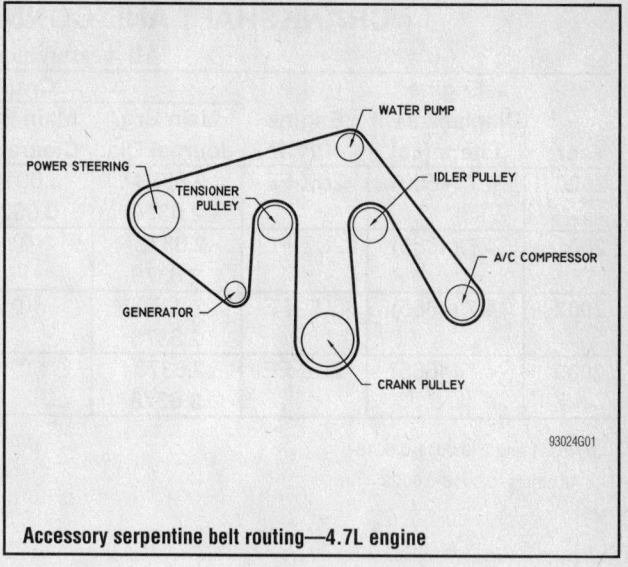

WATER PUMP
POWER STEERING
TENSIONER PULLEY
IDLER PULLEY
A/C COMPRESSOR
GENERATOR
CRANK PULLEY

93024G01

Accessory serpentine belt routing—4.7L engine

CAPACITIES

Year	Model	Engine Displacement Liters (cc)	Engine ID/VIN	Engine Oil with Filter (qts.)	Transmission (pts.) 5-Spd	Transmission (pts.) Auto.*	Transfer Case (pts.)	Drive Axle Front (pts.)	Drive Axle Rear (pts.)	Fuel Tank (gal.)	Cooling System (qts.)
2000	Land Cruiser	4.7 (4664)	2UZ-FE	7.2	—	3.6	3.6	3.6	6.8	25.1	①
	LX470	4.7 (4664)	2UZ-FE	7.2	—	3.6	2.8	3.6	6.8	25.4	①
2001	Land Cruiser	4.7 (4664)	2UZ-FE	7.2	—	3.6	3.6	3.6	6.8	25.1	①
	Sequoia	4.7 (4664)	2UZ-FE	5.5	—	3.6	2.4	3.6	6.6	21.6	②
	LX470	4.7 (4664)	2UZ-FE	7.2	—	3.6	2.8	3.6	6.8	25.4	①
2002	Land Cruiser	4.7 (4664)	2UZ-FE	7.2	—	7.4	2.8	3.6	③	25.4	②
	Sequoia	4.7 (4664)	2UZ-FE	6.6	—	4.2	2.6	2.4	7.7	26.1	12.3
	LX470	4.7 (4664)	2UZ-FE	6.6	—	4.2	2.6	2.4	7.7	25.4	12.3
2003	Land Cruiser	4.7 (4664)	2UZ-FE	7.2	—	7.4	2.8	3.6	③	25.4	②
	Sequoia	4.7 (4664)	2UZ-FE	6.6	—	4.2	2.6	2.4	7.7	26.1	12.3
	LX470	4.7 (4664)	2UZ-FE	6.6	—	4.2	2.6	2.4	7.7	25.4	12.3

*After draining, add the following amounts, then fill to the cold full line

① With rear heater: 9.5
Without rear heater: 8.5

② With rear heater: 16.2
Without rear heater: 15.6

③ w/o diff. Lock: 6.98
w/diff lock: 6.76

42356-SEQU-C04

CRANKSHAFT AND CONNECTING ROD SPECIFICATIONS

All measurements are given in inches.

Year	Engine Displacement Liters (cc)	Engine ID/VIN	Crankshaft				Connecting Rod		
			Main Brg. Journal Dia.	Main Brg. Clearance	Shaft End-play	Thrust on No.	Journal Diameter	Oil Clearance	Side Clearance
2000	4.7 (4665)	2UZ-FE	2.6373-2.6378	0.0016-0.0023	0.0008-0.0087	3	2.0465-2.0472	0.0011-0.0021	0.0063-0.0138
2001	4.7 (4665)	2UZ-FE	2.6373-2.6378	①	0.0008-0.0087	3	2.0465-2.0472	0.0011-0.0021	0.0063-0.0138
2002	4.7 (4665)	2UZ-FE	2.6373-2.6378	①	0.0008-0.0087	3	2.0465-2.0472	0.0011-0.0021	0.0063-0.0138
2003	4.7 (4665)	2UZ-FE	2.6373-2.6378	①	0.0008-0.0087	3	2.0465-2.0472	0.0011-0.0021	0.0063-0.0138

① Nos. 1 and 2: 0.0011-0.0018
All others: 0.0016-0.0023

42356-SEQU-C05

VALVE SPECIFICATIONS

Year	Engine Displacement Liters (cc)	Engine ID/VIN	Seat Angle (deg.)	Face Angle (deg.)	Spring Test Pressure (lbs. @ in.)	Spring Installed Height (in.)	Stem-to-Guide Clearance (in.)		Stem Diameter (in.)	
							Intake	Exhaust	Intake	Exhaust
2000	4.7 (4664)	2UZ-FE	45	44.5	45.9-50.7@1.378	1.380	0.0010-0.0024	0.0012-0.0026	0.2154-0.2159	0.2152-0.2157
2001	4.7 (4664)	2UZ-FE	45	44.5	45.9-50.7@1.378	1.380	0.0010-0.0024	0.0012-0.0026	0.2154-0.2159	0.2152-0.2157
2002	4.7 (4664)	2UZ-FE	45	44.5	45.9-50.7@1.378	1.380	0.0010-0.0024	0.0012-0.0026	0.2154-0.2159	0.2152-0.2157
2003	4.7 (4664)	2UZ-FE	45	44.5	45.9-50.7@1.378	1.380	0.0010-0.0024	0.0012-0.0026	0.2154-0.2159	0.2152-0.2157

42356-SEQU-C06

PISTON AND RING SPECIFICATIONS

All measurements are given in inches.

Year	Engine Displacement Liters (cc)	Engine ID/VIN	Piston Clearance	Ring Gap			Ring Side Clearance		
				Top Comp.	Bottom Comp.	Oil Control	Top Comp.	Bottom Comp.	Oil Control
2000	4.7 (4664)	2UZ-FE	0.0035-0.0044	0.0118-0.0197	0.0157-0.0256	0.0051-0.0189	0.0012-0.0031	0.0012-0.0028	SNUG
2001	4.7 (4664)	2UZ-FE	0.0035-0.0044	0.0118-0.0157	0.0157-0.0217	0.0051-0.0150	0.0012-0.0031	0.0012-0.0028	SNUG
2002	4.7 (4664)	2UZ-FE	0.0035-0.0044	0.0118-0.0157	0.0157-0.0217	0.0051-0.0150	0.0012-0.0031	0.0012-0.0028	SNUG
2003	4.7 (4664)	2UZ-FE	0.0035-0.0044	0.0118-0.0157	0.0157-0.0217	0.0051-0.0150	0.0012-0.0031	0.0012-0.0028	SNUG

42356-SEQU-C07

TORQUE SPECIFICATIONS
All readings in ft. lbs.

Year	Engine Displacement Liters (cc)	Engine ID/VIN	Cylinder Head Bolts	Main Bearing Bolts	Rod Bearing Bolts	Crankshaft Damper Bolts	Flywheel Bolts	Manifold Intake	Manifold Exhaust	Spark Plugs	Lug Nuts
2000	4.7 (4664)	2UZ-FE	①	②	③	181	④	13	33	13	97
2001	4.7 (4664)	2UZ-FE	⑤	②	③	181	④	13	33	13	97
2002	4.7 (4664)	2UZ-FE	⑤	②	③	181	④	13	33	13	97
2003	4.7 (4664)	2UZ-FE	⑤	②	③	181	④	13	33	13	97

① Step 1: 24 ft. lbs.
Step 2: Plus 180 degrees

② Step 1: 20 ft. lbs.
Step 2: Plus 90 degrees

③ Step 1: 18 ft. lbs.
Step 2: Plus 90 degrees

④ Step 1: 35 ft. lbs.
Step 2: Plus 90 degrees

⑤ Step 1: 24
Step 2: Plus 90 degrees
Step 3: Plus 90 degrees

42356-SEQU-C08

WHEEL ALIGNMENT

Year	Model	Caster Range (+/-Deg.)	Caster Preferred Setting (Deg.)	Camber Range (+/-Deg.)	Camber Preferred Setting (Deg.)	Toe-in (in.)	Steering Axis Inclination (Deg.)
2000	Land Cruiser	0.75	+2.50	0.75	+0.08	0.04+/-0.08	12.17+/-0.75
	LX470	0.75	+3.08	0.75	0	0+/-0.08	12.25+/-0.75
2001	Land Cruiser	0.75	+2.50	0.75	+0.08	0.04+/-0.08	12.17+/-0.75
	Sequoia	0.75	+2.95	0.75	+0.13	0.05+/-0.08	10.65+/-0.75
	LX470	0.75	+3.08	0.75	0	0+/-0.08	12.25+/-0.75
2002	Land Cruiser	0.75	+2.50	0.75	+0.08	0.04+/-0.08	12.17+/-0.75
	Sequoia	0.75	①	0.75	+0.13	0.05+/-0.08	10.635+/-0.75
	LX470	0.75	①	0.75	+0.13	0.05+/-0.08	10.635+/-0.75
2003	Land Cruiser	0.75	+2.50	0.75	+0.08	0.04+/-0.08	12.17+/-0.75
	Sequoia	0.75	①	0.75	+0.13	0.05+/-0.08	10.635+/-0.75
	LX470	0.75	①	0.75	+0.13	0.05+/-0.08	10.635+/-0.75

Note: All alignment specifications are based on nominal ride height and standard tires

① P245/70R16: +2.95
P265/70R16: +3.00

42356-SEQU-C09

TIRE, WHEEL AND BALL JOINT SPECIFICATIONS

Year	Model	OEM Tires Standard	OEM Tires Optional	Tire Pressures (psi) Front	Tire Pressures (psi) Rear	Wheel Size	Ball Joint Inspection
2000	Land Cruiser	P275/70R15	None	32	32	8-JJ	①
	LX470	P275/70HR16	None	32	32	8-JJ	①
2001	Land Cruiser	P275/70R15	None	32	32	8-JJ	①
	Sequoia	P245/70R16	P265/70R16	32	Std: 35; Opt: 32	7-JJ	②
	LX470	P275/70HR16	None	32	32	8-JJ	①
2002	Land Cruiser	P275/70R16	None	29③	32③	8-JJ	①
	Sequoia	P245/70R16	P265/70R16	32	Std: 35; Opt: 32	7-JJ	②
	LX470	P275/70HR16	None	32	32	8-JJ	①
2003	Land Cruiser	P275/65R17	P275/60R18	NA	NA	8-JJ	①
	Sequoia	P265/70R16	P265/6R17	32	Std: 35; Opt: 32	7-JJ	②
	LX470	P275/60HR18	None	NA	NA	8-JJ	②

OEM: Original Equipment Manufacturer

PSI: Pounds Per Square Inch

STD: Standard

OPT: Optional

NA: Not Available

① Replace if any measurable movement is found.

② Upper ball joint turning torque: 6-39 inch lbs.
   Lower ball joint turning torque: 1-22 inch lbs.
   Lower ball joint excessive play: 0.020 in.

③ Trailer towing: front 32; rear 35

42356-SEQU-C10

BRAKE SPECIFICATIONS
All measurements in inches unless noted

Year	Model		Brake Disc Original Thickness	Brake Disc Minimum Thickness	Brake Disc Maximum Runout	Brake Drum Diameter Original Inside Diameter	Max. Wear Limit	Maximum Machine Diameter	Minimum Lining Thickness	Brake Caliper Bracket Bolts (ft. lbs.)	Brake Caliper Mounting Bolts (ft. lbs.)
2000	Land Cruiser	F	1.260	1.181	0.0059	—	—	—	0.039	—	90
		R	0.709	0.611	—	—	—	—	0.039	65	76
	LX470	F	1.260	1.181	0.0028	—	—	—	0.039	76	90
		R	0.709	0.611	0.0040	—	—	—	0.039	—	65
2001	Land Cruiser	F	1.260	1.181	0.0059	—	—	—	0.039	—	90
		R	0.709	0.611	—	—	—	—	0.039	65	76
	Sequoia	F	1.102	1.024	0.0028	—	—	—	0.039	—	90
		R	0.709	0.611	0.0039	—	—	—	0.039	—	65
	LX470	F	1.260	1.181	0.0028	—	—	—	0.039	76	90
		R	0.709	0.611	0.0040	—	—	—	0.039	—	65
2002	Land Cruiser	F	1.260	1.181	0.0028	—	—	—	0.039	—	90
		R	0.709	0.611	—	—	—	—	0.039	—	20
	Sequoia	F	1.102	1.024	0.0028	—	—	—	0.039	—	90
		R	0.709	0.611	0.0039	—	—	—	0.039	—	65
	LX470	F	1.260	1.181	0.0028	—	—	—	0.039	—	90
		R	0.709	0.611	0.0040	—	—	—	0.039	—	20
2003	Land Cruiser	F	1.260	1.181	0.0028	—	—	—	0.039	—	90
		R	0.709	0.611	—	—	—	—	0.039	—	20
	Sequoia	F	1.102	1.024	0.0028	—	—	—	0.039	—	90
		R	0.709	0.611	0.0039	—	—	—	0.039	—	65
	LX470	F	1.260	1.181	0.0028	—	—	—	0.039	—	90
		R	0.709	0.611	0.0040	—	—	—	0.039	—	20

F: Front

R: Rear

42356-SEQU-C11

SCHEDULED MAINTENANCE INTERVALS
Lexus LX470

TO BE SERVICED	TYPE OF SERVICE	7.5	15	22.5	30	37.5	45	52.5	60	67.5	75	82.5	90	97.5
Engine oil & filter	R	✓	✓	✓	✓	✓	✓	✓	✓	✓	✓	✓	✓	✓
Automatic transmission fluid & filter	S/I		✓		✓		✓		✓		✓		✓	
Ball joints & dust covers	S/I		✓		✓		✓		✓		✓		✓	
Bolts & nuts on chassis & body	S/I		✓		✓		✓		✓		✓		✓	
Brake line pipes & hoses	S/I		✓		✓		✓		✓		✓		✓	
Brake pads & discs	S/I		✓		✓		✓		✓		✓		✓	
Propeller shaft grease	S/I		✓		✓		✓		✓		✓		✓	
Steering knuckle & chassis grease	S/I		✓		✓		✓		✓		✓		✓	
Steering linkage	S/I		✓		✓		✓		✓		✓		✓	
Transfer and differential oil	S/I	✓			✓		✓		✓		✓		✓	
Air cleaner filter	R				✓				✓				✓	
Spark plugs ①	R				✓				✓				✓	
Drive belts	S/I				✓				✓				✓	
Exhaust pipes & mountings	S/I				✓				✓				✓	
Fuel lines & connections	S/I				✓				✓				✓	
Engine coolant	R					✓					✓			
Charcoal canister	R								✓					
Fuel tank cap gasket	R								✓					
Heated oxygen sensors (exc. Cal.) ②	R													

R: Replace   S/I: Service or Inspect
① Platinum plugs, replace every 100,000 miles.
② Heated oxygen sensors (except Calif.): replace every 80,000 miles.

FREQUENT OPERATION MAINTENANCE (SEVERE SERVICE)
If a vehicle is operated under any of the following conditions it is considered severe service:
- Extremely dusty areas.
- 50% or more of the vehicle operation is in 32°C (90°F) or higher temperatures, or constant operation in temperatures below 0°C (32°F).
- Prolonged idling (vehicle operation in stop and go traffic).
- Frequent short running periods (engine does not warm to normal operating temperatures).
- Police, taxi, delivery usage or trailer towing usage.

Air cleaner filter: service or inspect every 3750 miles
Engine oil & filter: replace every 3750 miles.
Ball joints & dust covers: service or inspect every 7500 miles.
Bolts & nuts on chassis & body: service or inspect every 7500 miles.
Brake pads & discs (front & rear): service or inspect every 7500 miles.
Steering knuckle & chassis grease: service or inspect every 7500 miles.
Steering linkage: service or inspect every 7500 miles.
Propeller shaft grease: service or inspect every 7500 miles.
Exhaust pipes & mountings: service or inspect every 15,000 miles.

42356-SEQU-C12

SCHEDULED MAINTENANCE INTERVALS
Toyota Sequoia

TO BE SERVICED	TYPE OF SERVICE	VEHICLE MILEAGE INTERVAL (x1000)												
		7.5	15	22.5	30	37.5	45	52.5	60	67.5	75	82.5	90	97.5
Engine oil & filter	R	✓	✓	✓	✓	✓	✓	✓	✓	✓	✓	✓	✓	✓
Automatic transmission fluid & filter	S/I		✓		✓		✓		✓		✓		✓	
Ball joints & dust covers	S/I		✓		✓		✓		✓		✓		✓	
Bolts & nuts on chassis & body	S/I		✓		✓		✓		✓		✓		✓	
Brake line pipes & hoses	S/I		✓		✓		✓		✓		✓		✓	
Brake pads & discs	S/I		✓		✓		✓		✓		✓		✓	
Propeller shaft grease	S/I		✓		✓		✓		✓		✓		✓	
Steering knuckle & chassis grease	S/I		✓		✓		✓		✓		✓		✓	
Steering linkage	S/I		✓		✓		✓		✓		✓		✓	
Transfer and differential oil	S/I		✓		✓		✓		✓		✓		✓	
Air cleaner filter	R				✓				✓				✓	
Spark plugs ①	R				✓				✓				✓	
Drive belts	S/I				✓				✓				✓	
Exhaust pipes & mountings	S/I				✓				✓				✓	
Fuel lines & connections	S/I				✓				✓				✓	
Engine coolant	R						✓				✓			
Charcoal canister	R								✓					
Fuel tank cap gasket	R								✓					
Heated oxygen sensors (except Calif.)	R										✓			

R: Replace   S/I: Service or Inspect

① Platinum plugs, replace every 100,000 miles

② Heated oxygen sensors (except Calif.): replace every 80,000 miles.

FREQUENT OPERATION MAINTENANCE (SEVERE SERVICE)

If a vehicle is operated under any of the following conditions it is considered severe service:

- Extremely dusty areas.

- 50% or more of the vehicle operation is in 32°C (90°F) or higher temperatures, or constant operation in temperatures below 0°C (32°F).

- Prolonged idling (vehicle operation in stop and go traffic).

- Frequent short running periods (engine does not warm to normal operating temperatures).

- Police, taxi, delivery usage or trailer towing usage.

Air cleaner filter: service or inspect every 3750 miles

Engine oil & filter: replace every 3750 miles.

Ball joints & dust covers: service or inspect every 7500 miles.

Bolts & nuts on chassis & body: service or inspect every 7500 miles.

Brake pads & discs (front & rear): service or inspect every 7500 miles.

Steering knuckle & chassis grease: service or inspect every 7500 miles.

Steering linkage: service or inspect every 7500 miles.

Propeller shaft grease: service or inspect every 7500 miles.

Exhaust pipes & mountings: service or inspect every 15,000 miles.

42356-SEQU-C13

SCHEDULED MAINTENANCE INTERVALS
Toyota Land Cruiser

TO BE SERVICED	TYPE OF SERVICE	VEHICLE MILEAGE INTERVAL (x1000)																		
		5	10	15	20	25	30	35	40	45	50	55	60	65	70	75	80	85	90	95
Automatic transmission and differential fluid	S/I			✓			✓			✓			✓			✓			✓	
Ball joints and boots	S/I			✓			✓			✓			✓			✓			✓	
Brake system	S/I			✓			✓			✓			✓			✓			✓	
Charcoal canister	S/I												✓							
Drive belts	S/I						✓						✓						✓	
Driveshaft bushing (4WD)	L						✓						✓						✓	
Engine coolant	R						✓						✓						✓	
Engine oil & filter	R	✓	✓	✓	✓	✓	✓	✓	✓	✓	✓	✓	✓	✓	✓	✓	✓	✓	✓	✓
Exhaust pipes & mounts	S/I			✓			✓			✓			✓			✓			✓	
Fuel lines	S/I						✓						✓						✓	
Fuel tank cap gasket	S/I						✓						✓						✓	
Halfshaft boots & flange bolts	S/I			✓			✓			✓			✓			✓			✓	
Limited slip differential fluid	R						✓						✓						✓	
Non-platinum spark plugs	R						✓						✓						✓	
Platinum spark plugs	R												✓							
Propeller shaft (4WD)	L			✓			✓			✓			✓			✓			✓	
Propeller shaft bolts	S/I			✓			✓			✓			✓			✓			✓	
Rack and pinion assembly	S/I			✓			✓			✓			✓			✓			✓	
Rear wheel bearing	L						✓						✓							
Steering Knuckle	L			✓			✓			✓			✓			✓			✓	
Steering linkage	S/I			✓			✓			✓			✓			✓			✓	
Valves	S/I												✓							

R: Replace   S/I: Service or Inspect   L: Lubricate

FREQUENT OPERATION MAINTENANCE (SEVERE SERVICE)

If a vehicle is operated under any of the following conditions it is considered severe service:

- Towing a trailer or using a camper or car-top carrier.
- Repeated short trips of less than 5 miles in temperatures below freezing.
- Excessive idling or low-speed driving for long distances as in heavy commercial use, such as delivery, taxi or police cars.
- Operating on rough, muddy or salt-covered roads.
- Operating on unpaved or dusty roads.

Oil filter: service or inspect every 5000 miles or 4 months, whichever occurs first.

Brake linings and discs or drums: service or inspect every 5000 miles or 4 months, whichever occurs first.

Steering linkage: service or inspect every 5000 miles or 4 months, whichever occurs first.

Ball joints and boots: service or inspect every 5000 miles or 4 months, whichever occurs first.

Brake discs & pads (front): service or inspect every 6000 miles.

Halfshaft boots: service or inspect every 5000 miles or 4 months. Retighten the flange bolts, whichever occurs first.

Body chassis bolts and nuts: service or inspect every 5000 miles or 4 months, whichever occurs first.

Transmission and differential fluid: replace every 15,000 miles or 12 months, whichever occurs first.

Transfer case and differential fluid: replace every 15,000 miles or 12 months, whichever occurs first.

Timing belt: replace every 60,000 miles or 48 months, whichever occurs first.

42356-SEQU-C14

PRECAUTIONS

Before servicing any vehicle, please be sure to read all of the following precautions, which deal with personal safety, prevention of component damage, and important points to take into consideration when servicing a motor vehicle:

• Never open, service or drain the radiator or cooling system when the engine is hot; serious burns can occur from the steam and hot coolant.

• Observe all applicable safety precautions when working around fuel. Whenever servicing the fuel system, always work in a well-ventilated area. Do not allow fuel spray or vapors to come in contact with a spark, open flame, or excessive heat (a hot drop light, for example). Keep a dry chemical fire extinguisher near the work area. Always keep fuel in a container specifically designed for fuel storage; also, always properly seal fuel containers to avoid the possibility of fire or explosion. Refer to the additional fuel system precautions later in this section.

• Fuel injection systems often remain pressurized, even after the engine has been turned **OFF**. The fuel system pressure must be relieved before disconnecting any fuel lines. Failure to do so may result in fire and/or personal injury.

• Brake fluid often contains polyglycol ethers and polyglycols. Avoid contact with the eyes and wash your hands thoroughly after handling brake fluid. If you do get brake fluid in your eyes, flush your eyes with clean, running water for 15 minutes. If eye irritation persists, or if you have taken brake fluid internally, IMMEDIATELY seek medical assistance.

• The EPA warns that prolonged contact with used engine oil may cause a number of skin disorders, including cancer. You should make every effort to minimize your exposure to used engine oil. Protective gloves should be worn when changing oil. Wash your hands and any other exposed skin areas as soon as possible after exposure to used engine oil. Soap and water, or waterless hand cleaner should be used.

• All new vehicles are now equipped with an air bag system. The system must be disabled before performing service on or around system components, steering column, instrument panel components, wiring and sensors. Failure to follow safety and disabling procedures could result in accidental air bag deployment, possible personal injury and unnecessary system repairs.

• Always wear safety goggles when working with, or around, the air bag system. When carrying a non-deployed air bag, be sure the bag and trim cover are pointed away from your body. When placing a non-deployed air bag on a work surface, always face the bag and trim cover upward, away from the surface. This will reduce the motion of the module if it is accidentally deployed. Refer to the additional air bag system precautions later in this section.

• NEVER disconnect the negative battery cable with the ignition **ON** or the engine running. Removing power from the computer control module with the ignition **ON** may destroy the module.

• Clean, high quality brake fluid from a sealed container is essential to the safe and proper operation of the brake system.

You should always buy the correct type of brake fluid for your vehicle. If the brake fluid becomes contaminated, completely flush the system with new fluid. Never reuse any brake fluid. Any brake fluid that is removed from the system should be discarded. Also, do not allow any brake fluid to come in contact with a painted surface; it will damage the paint.

• Never operate the engine without the proper amount and type of engine oil; doing so WILL result in severe engine damage.

• Timing belt maintenance is extremely important. Many models utilize an interference-type, non-freewheeling engine. If the timing belt breaks, the valves in the cylinder head may strike the pistons, causing potentially serious (also time-consuming and expensive) engine damage. Refer to the maintenance interval charts in the front of this section for the recommended replacement interval for the timing belt, and to the timing belt section for belt replacement and inspection.

• Disconnecting the negative battery cable on some vehicles may interfere with the functions of the on-board computer system(s) and may require the computer to undergo a relearning process once the negative battery cable is reconnected.

• When servicing drum brakes, only disassemble and assemble one side at a time, leaving the remaining side intact for reference.

• Only an MVAC-trained, EPA-certified automotive technician should service the air conditioning system or its components.

ENGINE REPAIR

➡**Disconnecting the negative battery cable on some vehicles may interfere with the functions of the on board computer system. The computer may undergo a relearning process once the negative battery cable is reconnected.**

Distributor

All models are equipped with a distibutorless ignition system.

Alternator

REMOVAL

1. Before servicing the vehicle, refer to the precautions in the beginning of this section.
2. Drain the cooling system.
3. Remove or disconnect the following:
   • Negative battery cable
   • Accessory drive belt
   • Engine under cover

• Radiator
• Power steering pump pulley
• Alternator harness connectors
• Alternator

INSTALLATION

1. Install or connect the following:
   • Alternator. Tighten the fasteners to 29 ft. lbs. (39 Nm).
   • Alternator harness connectors
   • Power steering pump pulley

- Radiator
- Engine under cover
- Accessory drive belt
- Negative battery cable

2. Fill the cooling system.
3. Start the engine and check for leaks.

Ignition Timing

ADJUSTMENT

All engines are equipped with a Distributorless Ignition System (DIS). No timing adjustment is possible.

Engine Assembly

REMOVAL & INSTALLATION

Sequoia

1. Before servicing the vehicle, refer to the precautions in the beginning of this section.
2. Relieve the fuel system pressure.
3. Drain the cooling system.
4. Drain the engine oil.
5. Remove or disconnect the following:
   - Battery and tray
   - Hood
   - Engine appearance cover
   - Air intake pipe
   - Engine under covers
   - Coolant recovery tank
   - Radiator hoses
   - Radiator and fan shroud
   - Accessory drive belt
   - Cooling fan and pulley
   - Powertrain Control Module (PCM) harness connectors and pass the wiring harness through the firewall
   - Accelerator cable
   - Power steering vacuum hoses
   - Alternator harness connectors
   - Heater hoses
   - Engine control wiring harness and grommet at the firewall
   - Ground cable connector
   - Fuel lines
   - Evaporative Emissions (EVAP) canister hoses
   - Wire clamp at right inner fender
   - Negative battery cable at the relay box and right inner fender
   - Positive battery cable
   - Center console
   - Transmission shift lever assembly
   - Transfer case shift lever and rod
   - Exhaust front pipes
   - Stabilizer bar

- Front and rear driveshafts
- A/C compressor
- Power steering pump

6. Attach a hoist to the engine lifting eyes.
7. Remove or disconnect the following:
   - Transfer case skid plate
   - Left and right motor mounts
   - Transmission mount crossmember
8. Attach a hoist to the engine lifting eyes and raise the powertrain out of the vehicle.

To install:

9. Lower the powertrain into the vehicle.
10. Install or connect the following:
    - Transmission mount crossmember. Tighten the bolts to 37 ft. lbs. (50 Nm) and the nuts to 55 ft. lbs. (74 Nm).
    - Transfer case skid plate
    - Left and right motor mounts. Tighten the fasteners to 22 ft. lbs. (30 Nm).
    - Power steering pump. Tighten the bolts to 13 ft. lbs. (17 Nm).
    - A/C compressor. Tighten the bolts to 36 ft. lbs. (49 Nm).
    - Front driveshaft. Tighten the fasteners to 59 ft. lbs. (80 Nm).
    - Rear driveshaft. Tighten the fasteners to 78 ft. lbs. (106 Nm).
    - Stabilizer bar. Tighten the bracket bolts to 13 ft. lbs. (18 Nm) and the link nuts to 18 ft. lbs. (25 Nm).
    - Exhaust front pipes
    - Transfer case shift lever and rod
    - Transmission shift lever assembly
    - Center console
    - Positive battery cable
    - Negative battery cable at the relay box and right inner fender
    - Wire clamp at right inner fender
    - EVAP canister hoses
    - Fuel lines
    - Ground cable connector
    - Engine control wiring harness and grommet at the firewall
    - Heater hoses
    - Alternator harness connectors
    - Power steering vacuum hoses
    - Accelerator cable
    - PCM harness connectors
    - Cooling fan and pulley
    - Accessory drive belt
    - Radiator and fan shroud
    - Radiator hoses
    - Coolant recovery tank
    - Engine under covers
    - Air intake pipe
    - Engine appearance cover

- Hood
- Battery and tray

11. Fill the crankcase to the correct level.
12. Fill the cooling system.
13. Start the engine and check for leaks.

Land Cruiser and LX470

1. Before servicing the vehicle, refer to the precautions in the beginning of this section.
2. Relieve the fuel system pressure.
3. Drain the cooling system.
4. Drain the engine oil.
5. Remove or disconnect the following:
   - Battery and tray
   - Hood
   - Engine appearance cover
   - Air intake pipe
   - Engine under covers
   - Coolant recovery tank
   - Radiator hoses
   - Radiator and fan shroud
   - Accessory drive belt
   - Cooling fan and pulley
   - Powertrain Control Module (PCM) harness connectors and pass the wiring harness through the firewall
   - Accelerator cable
   - Power steering vacuum hoses
   - Alternator harness connectors
   - Heater hoses
   - Engine control wiring harness and grommet at the firewall
   - Ground cable connector
   - Fuel lines
   - Evaporative Emissions (EVAP) canister hoses
   - Wire clamp at right inner fender
   - Negative battery cable at the relay box and right inner fender
   - Positive battery cable
   - Center console
   - Transmission shift lever assembly
   - Transfer case shift lever and rod
   - Exhaust front pipes
   - Stabilizer bar
   - Front and rear driveshafts
   - A/C compressor
   - Power steering pump
6. Attach a hoist to the engine lifting eyes.
7. Remove or disconnect the following:
   - Transfer case skid plate
   - Left and right motor mounts
   - Transmission mount crossmember
8. Attach a hoist to the engine lifting eyes and raise the powertrain out of the vehicle.

To install:

9. Lower the powertrain into the vehicle.

10. Install or connect the following:
    - Transmission mount crossmember. Tighten the bolts to 37 ft. lbs. (50 Nm) and the nuts to 55 ft. lbs. (74 Nm).
    - Transfer case skid plate
    - Left and right motor mounts. Tighten the fasteners to 22 ft. lbs. (30 Nm).
    - Power steering pump. Tighten the bolts to 13 ft. lbs. (17 Nm).
    - A/C compressor. Tighten the bolts to 36 ft. lbs. (49 Nm).
    - Front driveshaft. Tighten the fasteners to 59 ft. lbs. (80 Nm).
    - Rear driveshaft. Tighten the fasteners to 78 ft. lbs. (106 Nm).
    - Stabilizer bar. Tighten the bracket bolts to 13 ft. lbs. (18 Nm) and the link nuts to 18 ft. lbs. (25 Nm).
    - Exhaust front pipes
    - Transfer case shift lever and rod
    - Transmission shift lever assembly
    - Center console
    - Positive battery cable
    - Negative battery cable at the relay box and right inner fender
    - Wire clamp at right inner fender
    - EVAP canister hoses
    - Fuel lines
    - Ground cable connector
    - Engine control wiring harness and grommet at the firewall
    - Heater hoses
    - Alternator harness connectors
    - Power steering vacuum hoses
    - Accelerator cable
    - PCM harness connectors
    - Cooling fan and pulley
    - Accessory drive belt
    - Radiator and fan shroud
    - Radiator hoses
    - Coolant recovery tank
    - Engine under covers
    - Air intake pipe
    - Engine appearance cover
    - Hood
    - Battery and tray
11. Fill the crankcase to the correct level.
12. Fill the cooling system.
13. Start the engine and check for leaks.

Heater Core

REMOVAL & INSTALLATION

Land Cruiser

FRONT HEATER

1. Disconnect the negative battery cable.

2. Drain the cooling system into a clean container for reuse.

3. Disconnect the heater hoses from the heater core.

4. Remove the steering wheel by performing the following procedure:

a. Position the front wheels facing straight-ahead.

b. Remove the steering wheel side covers.

c. Using a Torx® wrench, loosen the 2 screws located at each side of the steering wheel until the screw's circumference groove catches on the screw case.

d. Pull the air bag module from the steering wheel and disconnect the electrical connector.

⁂ CAUTION

Place the air bag module in a safe place with the front side facing upward.

e. Remove the steering wheel nut.

f. Place alignment marks on the steering wheel and the main shaft.

g. Using a steering wheel puller, press the steering wheel from the steering column.

5. Remove the instrument panel and reinforcement by performing the following procedure:

a. Remove the front door scuff plates, the cowl side trim and the front door opening trim.

b. At the driver's side, remove the 2 assist grip plugs, the 2 screws and assist grip and the front pillar garnish.

c. At the passenger's side, remove the 4 assist grip plugs, the 4 screws, the 2 assist grips and the front pillar garnish.

d. Remove the instrument cluster finish panel.

e. Remove the 2 screws and the hood lock control cable.

f. Remove the 2 screws and the fuel lid control cable lever.

g. Remove the lower No. 1 panel screw and the panel.

h. Remove the lower left side panel.

i. Remove the 3 steering column cover screws and the covers.

j. At the steering column, disconnect the electrical connectors; then, remove the clamp, the 3 screws and the combination switch.

k. Remove the No. 2 heater-to-register duct screw and the duct.

l. Remove the steering column-to-instrument panel bolts and the steering column.

m. At the combination meter, disconnect the electrical connectors; then, remove the 4 screws and the combination meter.

n. Remove the glove compartment door stoppers, the 2 screws and the glove box door.

o. At the passenger's side air bag module, remove the No. 1 undercover, pull the air bag connector up from the undercover and disconnect it; then, remove the air bag.

⁂ CAUTION

Place the air bag module in a safe place with the front side facing upward.

p. Remove the 3 lower No. 2 panel screws and the panel.

q. Remove the center cluster; then, pry the center cluster from the dash by prying the 8 clips in the following order:

- Left side
- Right side
- Top left side
- Top right side

r. Remove the 4 radio screws, pull the radio outward, disconnect the electrical connectors and remove the radio.

s. At the rear console panel, remove the transfer shift lever knob. Pry the panel upward disengaging the 4 clips (2 on each side) and remove the panel.

t. At the rear of the console, remove the 2 rear end panel-to-console screws; then, pry the end panel rearward disengaging the 2 clips and remove the panel.

u. If not equipped with a rear air conditioning system, disconnect the connector and control cable; then, remove the 3 rear heater control panel screws and the panel.

v. Remove the 4 rear console box-to-chassis screws/bolts and the console box.

w. Remove the center lower cluster finish panel by prying panel rearward disengaging the 5 clips; then, disconnect the electrical connector.

x. Remove the 2 front console-to-chassis bolts/screws, disengage the 2 clips and remove the console.

y. At the instrument panel, disconnect the junction connectors (the connectors can be disconnected by loosening the bolts), the instrument panel-to-chassis 8 bolts and 2 nuts. Using an assistant, remove the instrument panel.

z. Disconnect the electrical connector and remove the ECM.

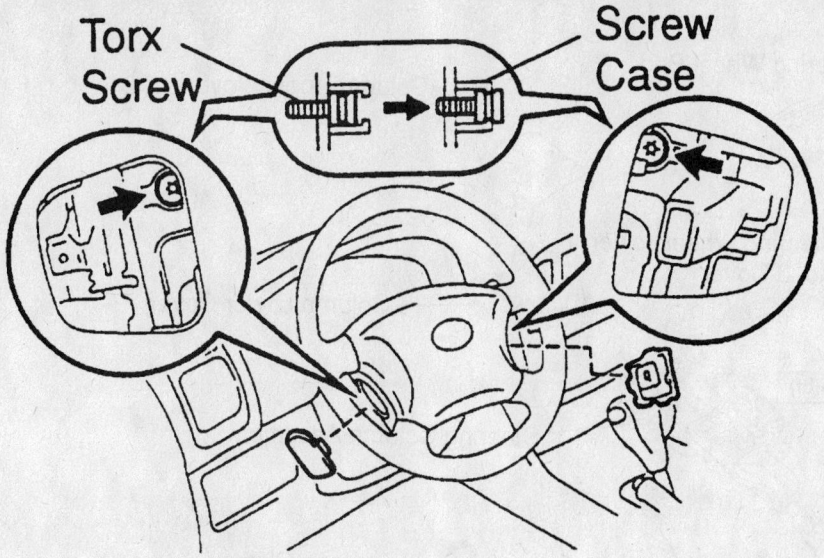

93113GG4

View the steering wheel's Torx® bolts—Toyota Land Cruiser

aa. Remove the No. 3 and No. 4 heater-to-register ducts.

bb. Remove the floor brace, the No. 1 brace and the reinforcement.

6. Remove the evaporator housing by performing the following procedure:

a. Discharge and recover the air conditioning system refrigerant.

b. Remove the air conditioning liquid line clamp.

c. Remove the air conditioning suction line clamp.

d. Disconnect both air conditioning lines and plug the openings to prevent contamination. Discard the 4 O-rings.

e. Remove the antenna relay electrical connector, the 2 screws and the relay.

f. Remove the evaporator housing-to-chassis 4 screws/2 nuts and the housing.

7. Remove the heater housing by performing the following procedure:

a. Remove the defroster nozzle.

b. Disconnect the electrical connector.

c. Remove the 4 nuts and the heater housing.

8. Remove the heater core-to-heater housing packing, the screw, the bracket, the clamp and the heater core.

To install:

9. Install the heater core, the clamp, the bracket, the screw and the heater core-to-heater housing packing.

10. Install the heater housing by performing the following procedure:

a. Install the heater housing and the 4 nuts.

b. Connect the electrical connector.

c. Install the defroster nozzle.

11. Install the evaporator housing by performing the following procedure:

a. Install the evaporator housing and the housing-to-chassis 4 screws and 2 nuts.

b. Install the antenna relay, the 2 screws and the electrical connector.

c. Using new O-rings, connect both air conditioning lines.

d. Install the air conditioning liquid line and suction line clamp.

12. Install the instrument panel and reinforcement by performing the following procedure:

a. Install the reinforcement, the No. 1 brace and the floor brace.

b. Install the No. 3 and No. 4 heater-to-register ducts.

c. Install the ECM and connect the electrical connector.

d. Using an assistant, install the instrument panel, connect the junction connectors, the instrument panel-to-chassis 8 bolts and 2 nuts.

e. Install the front the console, engage the 2 clips and install the 2 console-to-chassis bolts/screws.

f. Connect the electrical connector; then, install the center lower cluster finish panel by engaging the 5 clips.

g. Install the console box and the 4 rear console box-to-chassis screws/bolts.

h. If not equipped with a rear air conditioning system, install rear heater control panel, the 3 panel screws; then, connect the connector and control cable.

i. Install the rear of the console and engage the 2 clips; then, install the 2 rear end panel-to-console screws.

j. Install the rear console panel and engage the 4 clips (2 on each side); then, install the transfer shift lever knob.

k. Install the radio, connect the electrical connectors and the 4 radio screws.

l. Install the center cluster and engage the 8 center cluster clips.

m. Install the lower No. 2 panel and the 3 panel screws.

n. Install the passenger's side air bag module, connect it and install the No. 1 undercover.

o. Install the glove box door, the 2 screws and the glove compartment door stoppers.

p. Install the combination meter and the 4 screws; then, connect the electrical connectors.

q. Install the steering column and the steering column-to-instrument panel bolts.

r. Install the No. 2 heater-to-register duct and the duct screw.

s. At the steering column, install the combination switch, the 3 screws and the clamp; then, connect the electrical connectors.

t. Install the steering column covers and the 3 covers screws.

u. Install the lower left side panel.

v. Install the lower No. 1 panel and the panel screw.

w. Install the fuel lid control cable lever and the 2 screws.

x. Install the hood lock control cable and the 2 screws.

y. Install the instrument cluster finish panel.

z. At the passenger's side, install the front pillar garnish, the 2 assist grips, the 4 screws and the 4 assist grip plugs.

aa. At the driver's side, install the front pillar garnish, assist grip, the 2 screws and the 2 assist grip plugs.

bb. Install the front door scuff plates, the cowl side trim and the front door opening trim.

13. Install the steering wheel by performing the following procedure:

a. Install the steering wheel to the steering column.

b. Align the steering wheel-to-main shaft marks.

c. Install the steering wheel nut and torque to 25 ft. lbs. (34 Nm).

d. Install the air bag module to the steering wheel and connect the electrical connector.

e. Using a Torx® wrench, tighten the 2

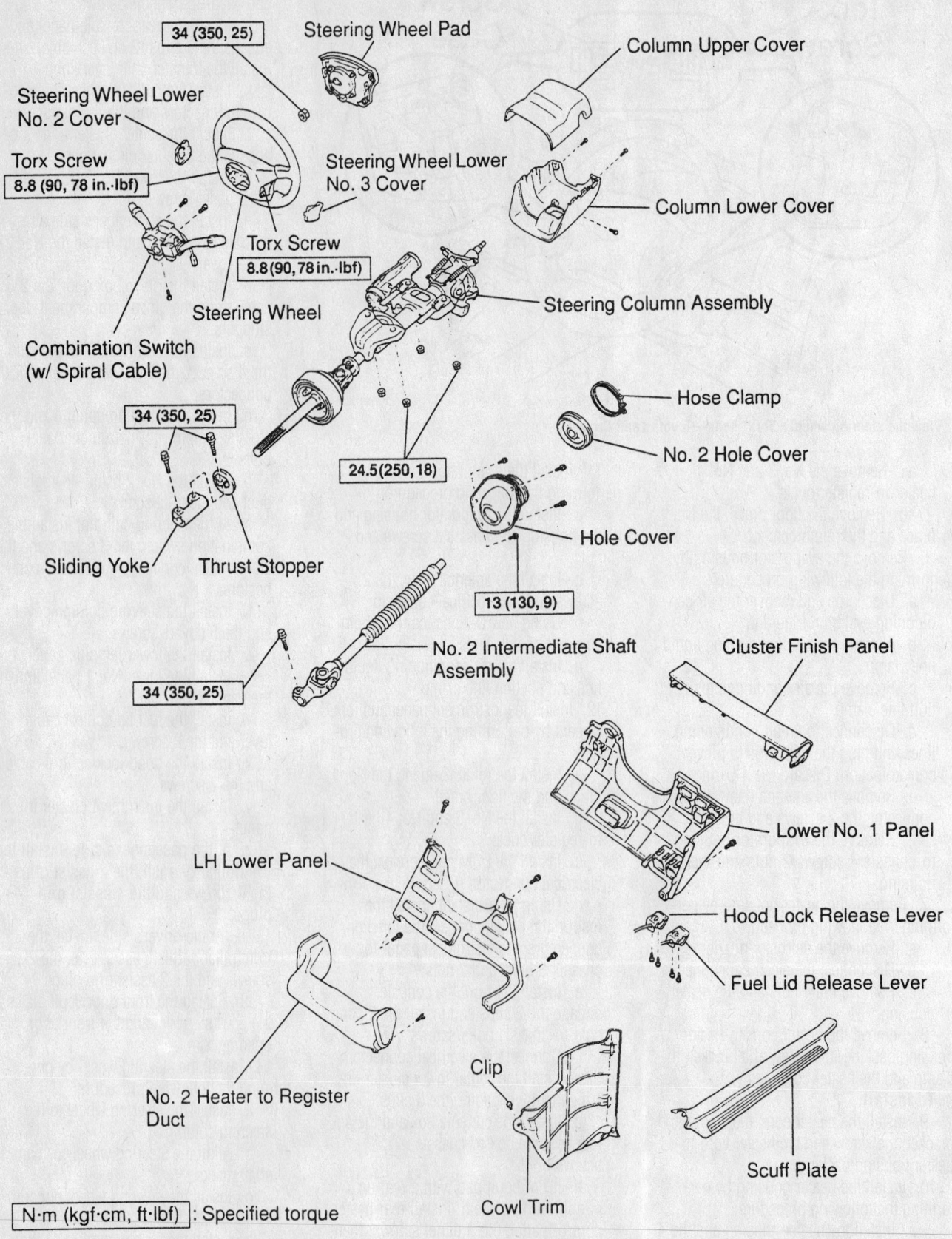

34 (350, 25)

Steering Wheel Pad

Column Upper Cover

Steering Wheel Lower
No. 2 Cover

Torx Screw
8.8 (90, 78 in.·lbf)

Column Lower Cover

Steering Wheel Lower
No. 3 Cover

Torx Screw
8.8 (90, 78 in.·lbf)

Steering Column Assembly

Steering Wheel

Combination Switch
(w/ Spiral Cable)

Hose Clamp

34 (350, 25)

No. 2 Hole Cover

24.5 (250, 18)

Hole Cover

Sliding Yoke   Thrust Stopper

13 (130, 9)

No. 2 Intermediate Shaft
Assembly

Cluster Finish Panel

34 (350, 25)

Lower No. 1 Panel

LH Lower Panel

Hood Lock Release Lever

Fuel Lid Release Lever

No. 2 Heater to Register
Duct

Clip

Scuff Plate

Cowl Trim

N·m (kgf·cm, ft·lbf) : Specified torque

Exploded view the steering column—Toyota Land Cruiser (Part 1 of 2)

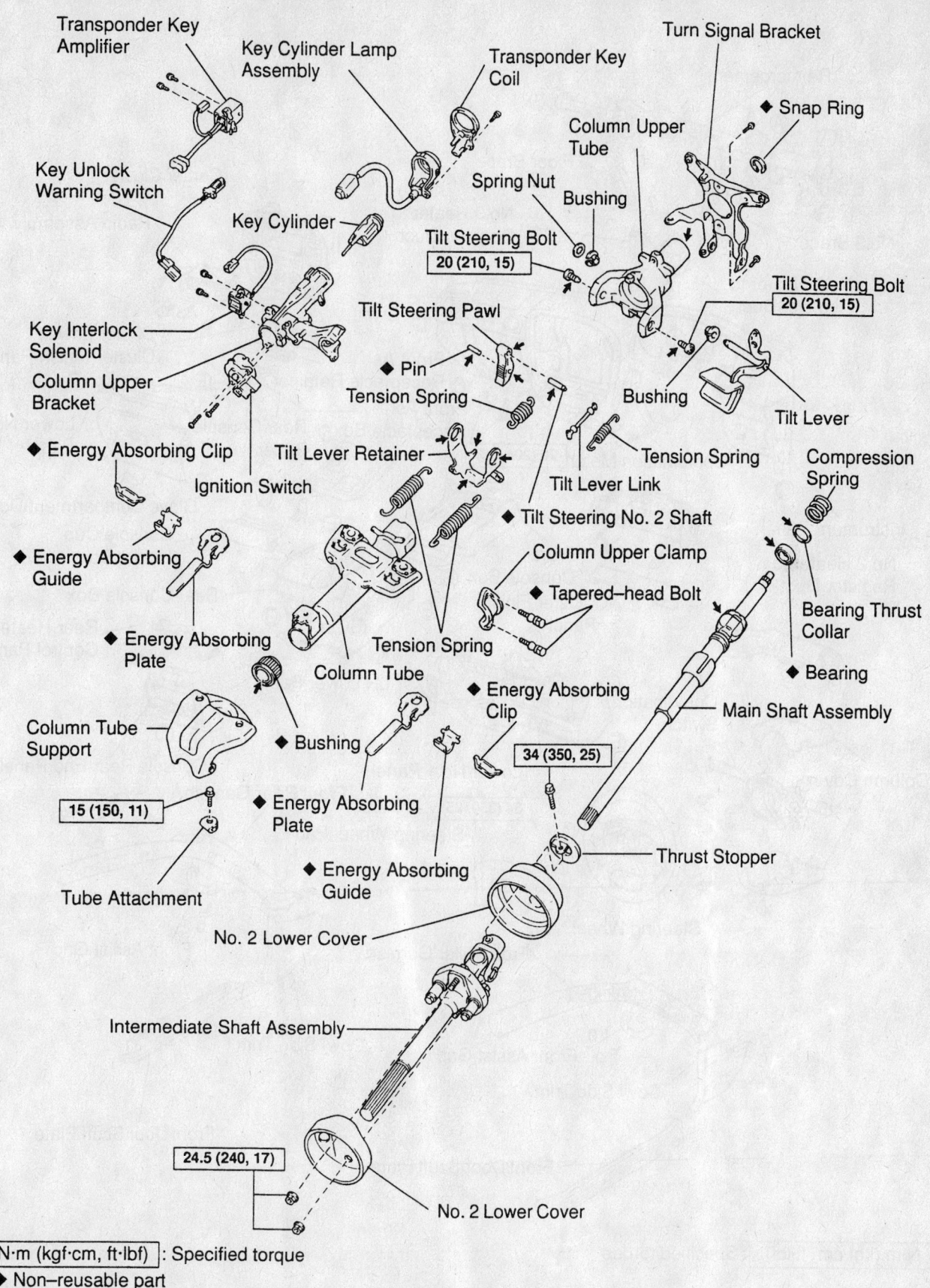

Transponder Key Amplifier

Key Cylinder Lamp Assembly

Transponder Key Coil

Turn Signal Bracket

◆ Snap Ring

Column Upper Tube

Key Unlock Warning Switch

Key Cylinder

Spring Nut

Bushing

Tilt Steering Bolt
20 (210, 15)

Tilt Steering Bolt
20 (210, 15)

Tilt Steering Pawl

Key Interlock Solenoid

◆ Pin

Tension Spring

Bushing

Tilt Lever

Column Upper Bracket

◆ Energy Absorbing Clip

Tilt Lever Retainer

Tension Spring

Compression Spring

Ignition Switch

Tilt Lever Link

◆ Energy Absorbing Guide

◆ Tilt Steering No. 2 Shaft

Bearing Thrust Collar

Column Upper Clamp

◆ Energy Absorbing Plate

◆ Tapered-head Bolt

◆ Bearing

Tension Spring

Main Shaft Assembly

Column Tube

Column Tube Support

◆ Bushing

◆ Energy Absorbing Clip

34 (350, 25)

15 (150, 11)

◆ Energy Absorbing Plate

Thrust Stopper

Tube Attachment

◆ Energy Absorbing Guide

No. 2 Lower Cover

Intermediate Shaft Assembly

24.5 (240, 17)

No. 2 Lower Cover

N·m (kgf·cm, ft·lbf) : Specified torque

◆ Non–reusable part

Molybdenum disulfide lithium base grease
N

93113GG6

Exploded view the steering column—Toyota Land Cruiser (Part 2 of 2)

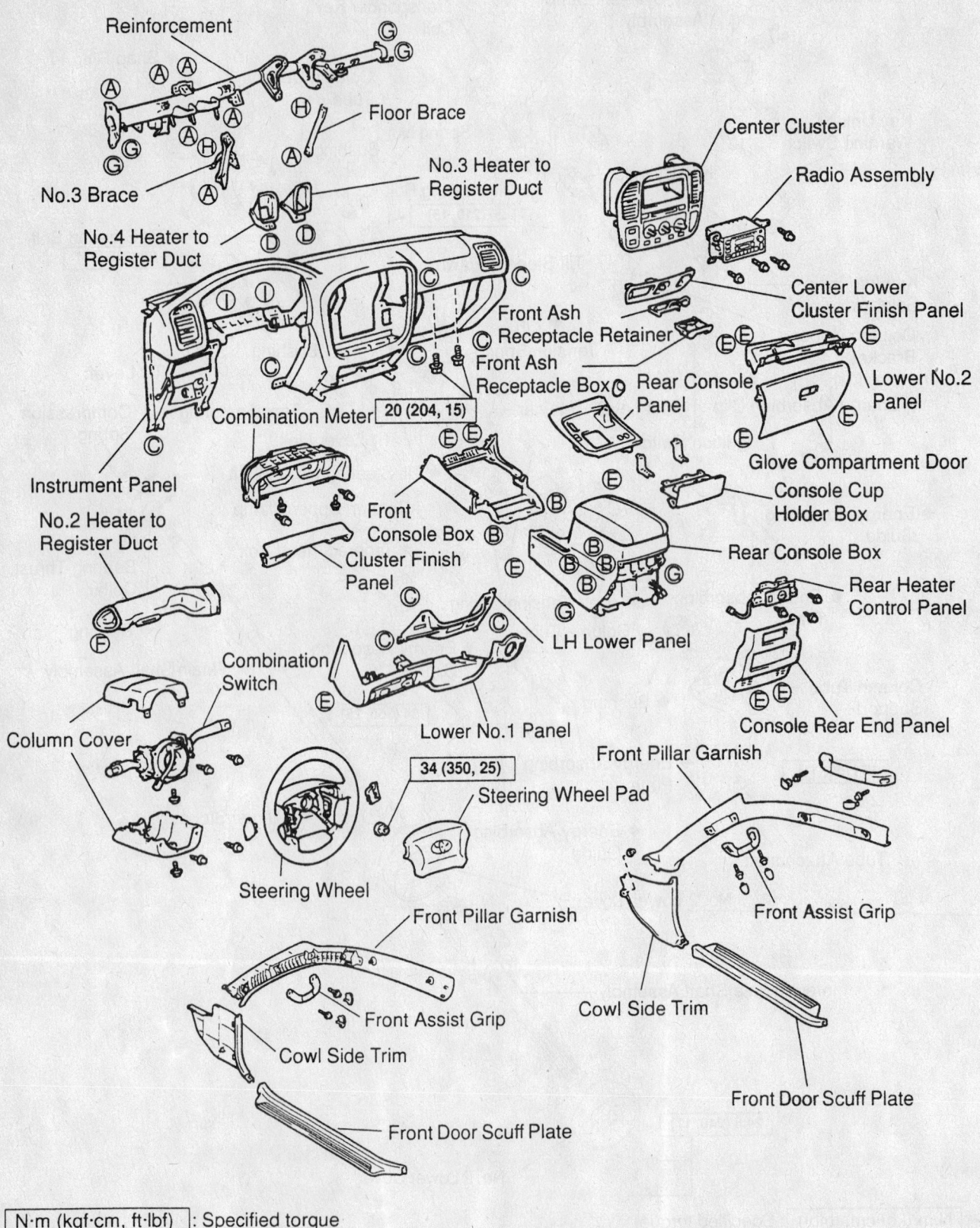

Reinforcement

Floor Brace

No.3 Brace

No.4 Heater to
Register Duct

No.3 Heater to
Register Duct

Center Cluster

Radio Assembly

Center Lower
Cluster Finish Panel

Front Ash
Receptacle Retainer

Front Ash
Receptacle Box

Rear Console
Panel

Lower No.2
Panel

Combination Meter   20 (204, 15)

Glove Compartment Door

Console Cup
Holder Box

Instrument Panel

No.2 Heater to
Register Duct

Front
Console Box

Cluster Finish
Panel

Rear Console Box

Rear Heater
Control Panel

LH Lower Panel

Combination
Switch

Column Cover

Lower No.1 Panel

Console Rear End Panel

Front Pillar Garnish

34 (350, 25)

Steering Wheel Pad

Front Assist Grip

Steering Wheel

Front Pillar Garnish

Front Assist Grip

Cowl Side Trim

Cowl Side Trim

Front Door Scuff Plate

Front Door Scuff Plate

N·m (kgf·cm, ft·lbf) : Specified torque

93113GG7

Exploded view the instrument panel and related components—Toyota Land Cruiser

screws located at each side of the steering wheel to 78 inch lbs. (8.8 Nm).

f. Install the steering wheel side covers.

14. Connect the heater hoses to the heater core.

15. Refill the cooling system.

16. Connect the negative battery cable.

a. Evacuate and charge the air conditioning system refrigerant.

17. Run the engine to normal operating temperatures. Check the climate control operation and check for leaks.

REAR AUXILIARY HEATER

1. Disconnect the negative battery cable.

2. Drain the cooling system into a clean container for reuse.

3. Disconnect the heater hoses from the rear heater core.

4. Remove the front seats.

5. Remove the rear heater control assembly.

6. Remove the rear console box.

7. Remove the front console box cover.

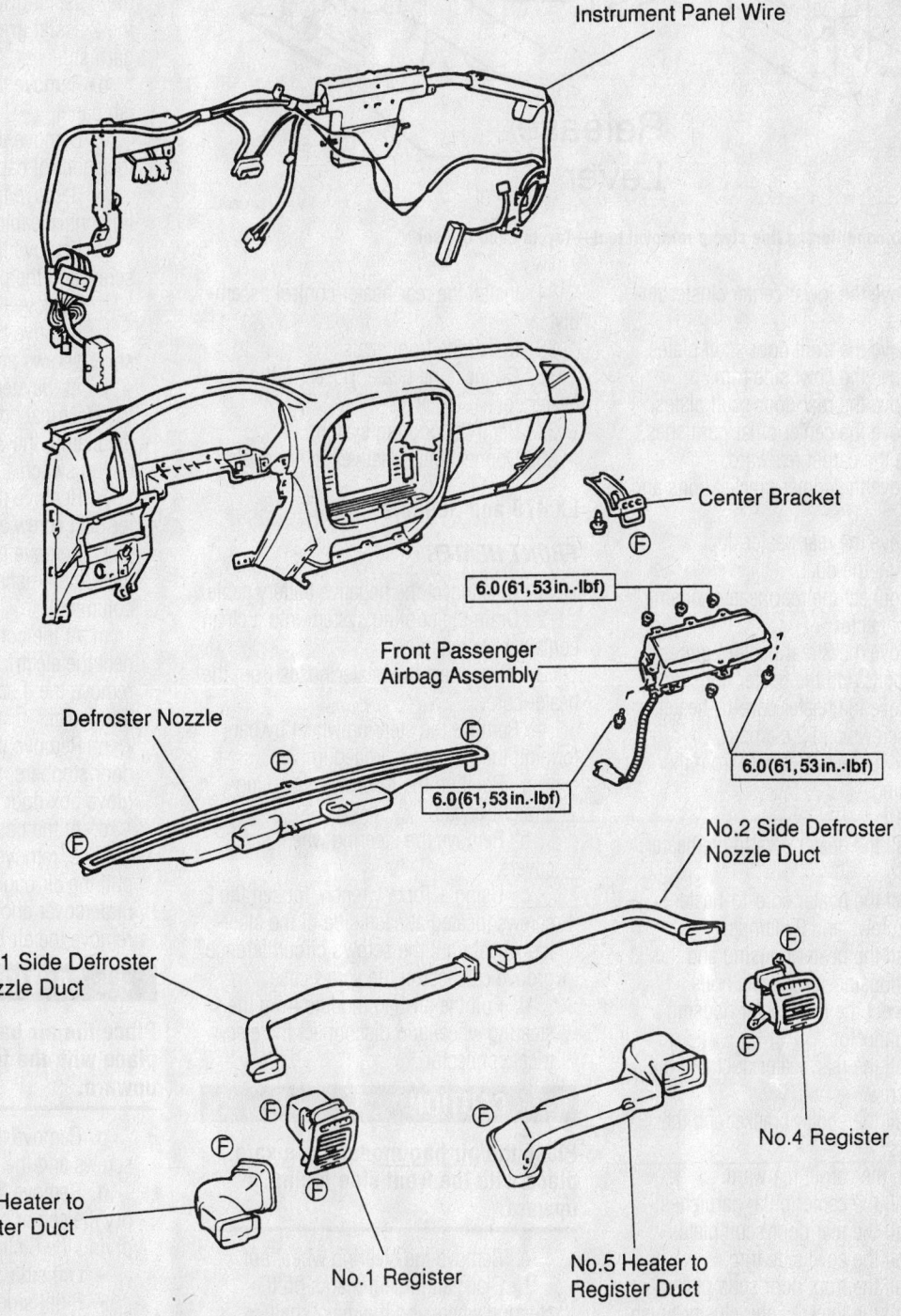

Instrument Panel Wire

Center Bracket

6.0 (61, 53 in.·lbf)

Front Passenger Airbag Assembly

6.0 (61, 53 in.·lbf)

Defroster Nozzle

6.0 (61, 53 in.·lbf)

No.2 Side Defroster Nozzle Duct

No.1 Side Defroster Nozzle Duct

No.4 Register

No.1 Heater to Register Duct

No.1 Register

No.5 Heater to Register Duct

N·m (kgf·cm, ft·lbf) : Specified torque

93113GG8

Exploded view the front ventilation ducts and related components—Toyota Land Cruiser

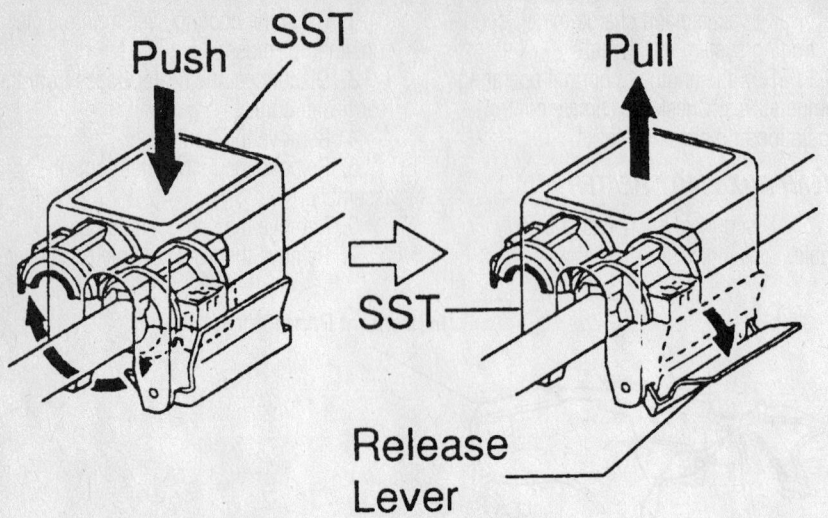

Push SST   Pull

SST

Release
Lever

93113GG9

View the air conditioning line clamp removal tool—Toyota Land Cruiser

8. Remove the lower center cluster finish panel.

9. Remove the front door scuff plates.

10. Remove the cowl side trim.

11. Remove the rear door scuff plates.

12. Remove the center pillar garnishes.

13. Slide the carpet rearward.

14. Remove the cooler bracket bolts and the bracket.

15. Remove the rear heater duct bolt/screw and the duct.

16. Disconnect the rear heater housing electrical connector.

17. Remove the 3 rear heater housing-to-chassis bolts and the heater housing.

18. Remove the heater core-to-heater housing 3 screws and 2 clamps.

19. Remove the heater core from the heater housing.

To install:

20. Install the heater core to the heater housing.

21. Install the heater core-to-heater housing 3 screws and 2 clamps.

22. Install the heater housing and the 3 rear heater housing-to-chassis bolts.

23. Connect the rear heater housing electrical connector.

24. Install the rear heater duct and the duct bolt/screw.

25. Install the cooler bracket and the bracket bolts.

26. Slide the carpet rearward.

27. Install the center pillar garnishes.

28. Install the rear door scuff plates.

29. Install the cowl side trim.

30. Install the front door scuff plates.

31. Install the lower center cluster finish panel.

32. Install the front console box cover.

33. Install the rear console box.

34. Install the rear heater control assembly.

35. Install the front seats.

36. Connect the heater hoses to the rear heater core.

37. Refill the cooling system.

38. Connect the negative battery cable.

LX 470 and Sequoia

FRONT HEATER

1. Disconnect the negative battery cable.

2. Drain the cooling system into a clean container for reuse.

3. Disconnect the heater hoses from the heater core.

4. Remove the steering wheel by performing the following procedure:

   a. Position the front wheels facing straight-ahead.

   b. Remove the steering wheel side covers.

   c. Using a Torx® wrench, loosen the 2 screws located at each side of the steering wheel until the screw's circumference groove catches on the screw case.

   d. Pull the air bag module from the steering wheel and disconnect the electrical connector.

✳✳ **CAUTION**

Place the air bag module in a safe place with the front side facing upward.

   e. Remove the steering wheel nut.

   f. Place alignment marks on the steering wheel and the main shaft.

   g. Using a steering wheel puller, press the steering wheel from the steering column.

5. Remove the instrument panel and reinforcement by performing the following procedure:

   a. Remove the front door scuff plates, the cowl side trim and the front door opening trim.

   b. At the driver's side, remove the 2 assist grip plugs, the 2 screws and assist grip and the front pillar garnish.

   c. At the passenger's side, remove the 4 assist grip plugs, the 4 screws, the 2 assist grips and the front pillar garnish.

   d. Remove the instrument cluster finish panel.

   e. Remove the 2 screws and the hood lock control cable.

   f. Remove the 2 screws and the fuel lid control cable lever.

   g. Remove the lower No. 1 panel screw and the panel.

   h. Remove the lower left side panel.

   i. Remove the 3 steering column cover screws and the covers.

   j. At the steering column, disconnect the electrical connectors; then, remove the clamp, the 3 screws and the combination switch.

   k. Remove the No. 2 heater-to-register duct screw and the duct.

   l. Remove the steering column-to-instrument panel bolts and the steering column.

   m. At the combination meter, disconnect the electrical connectors; then, remove the 4 screws and the combination meter.

   n. Remove the glove compartment door stoppers, the 2 screws and the glove box door.

   o. At the passenger's side air bag module, remove the No. 1 undercover, pull the air bag connector up from the undercover and disconnect it; then, remove the air bag.

✳✳ **CAUTION**

Place the air bag module in a safe place with the front side facing upward.

   p. Remove the 3 lower No. 2 panel screws and the panel.

   q. Remove the center cluster; then, pry the center cluster from the dash by prying the 8 clips in the following order:

   • Left side
   • Right side
   • Top left side
   • Top right side

   r. Remove the 4 radio screws, pull

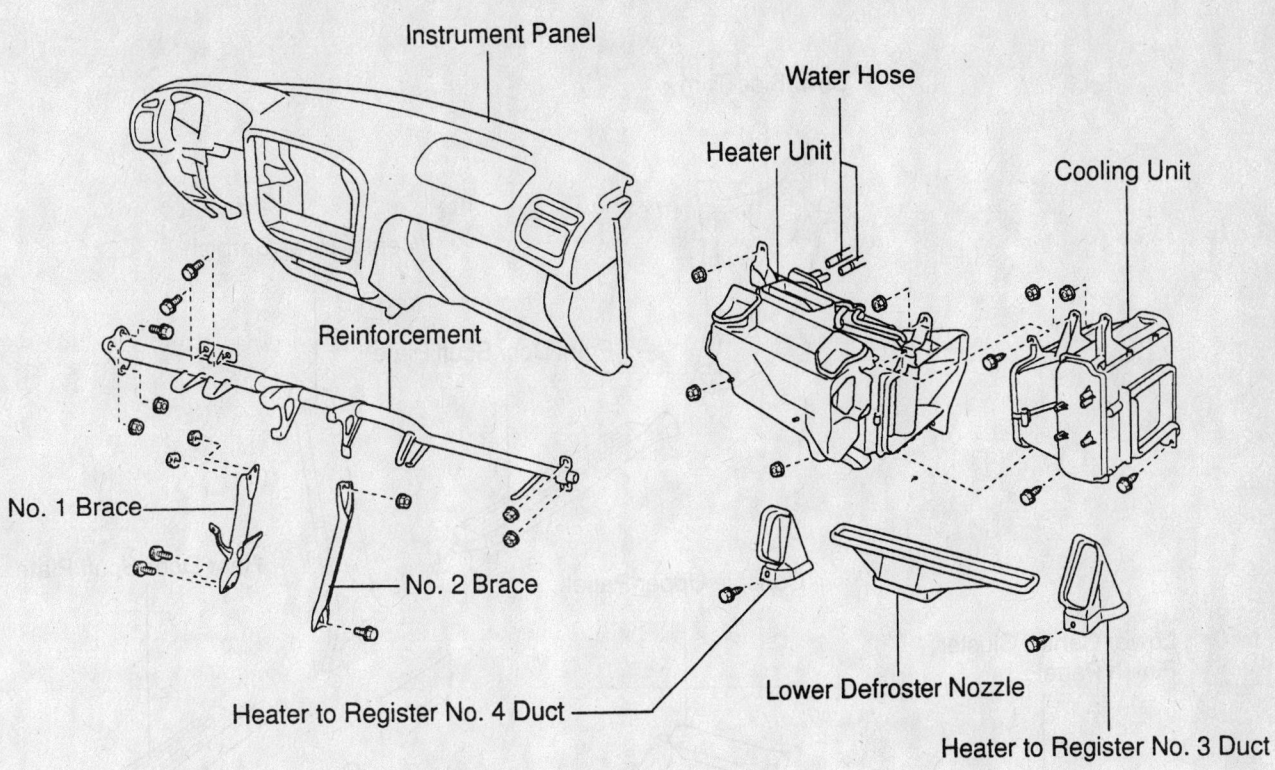

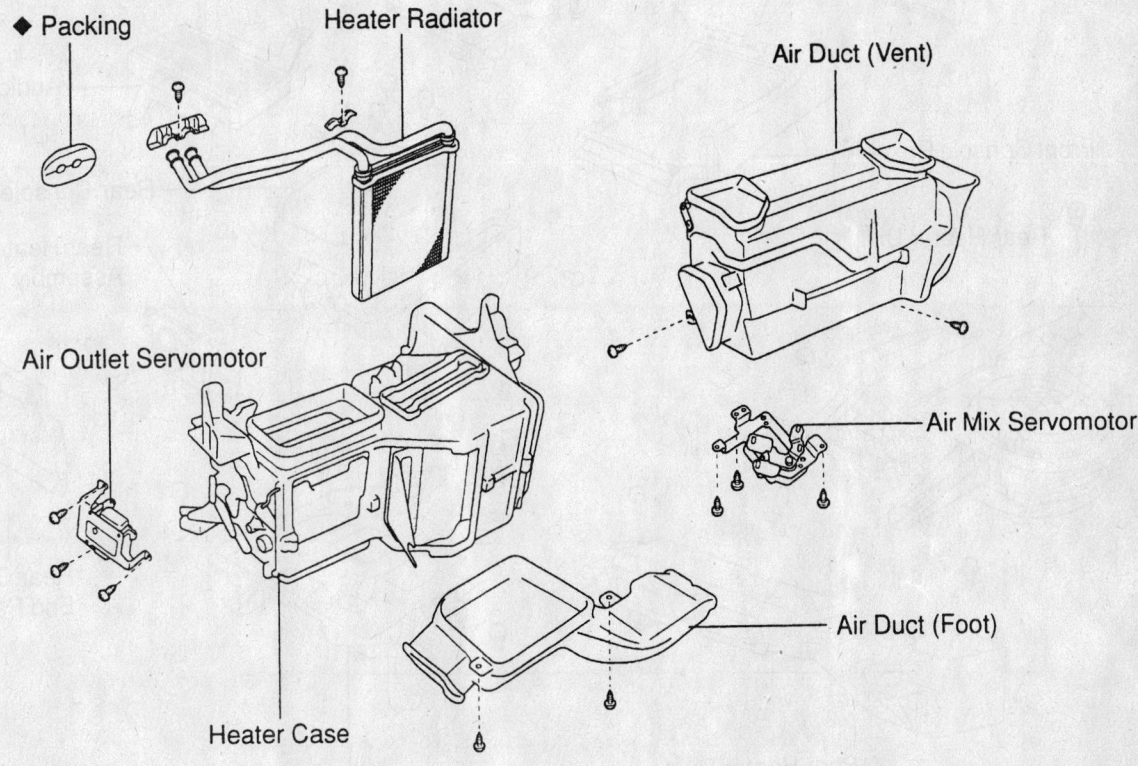

Exploded view of the front heater core, heater housing, evaporator housing and related components—Toyota Land Cruiser

93113GG0

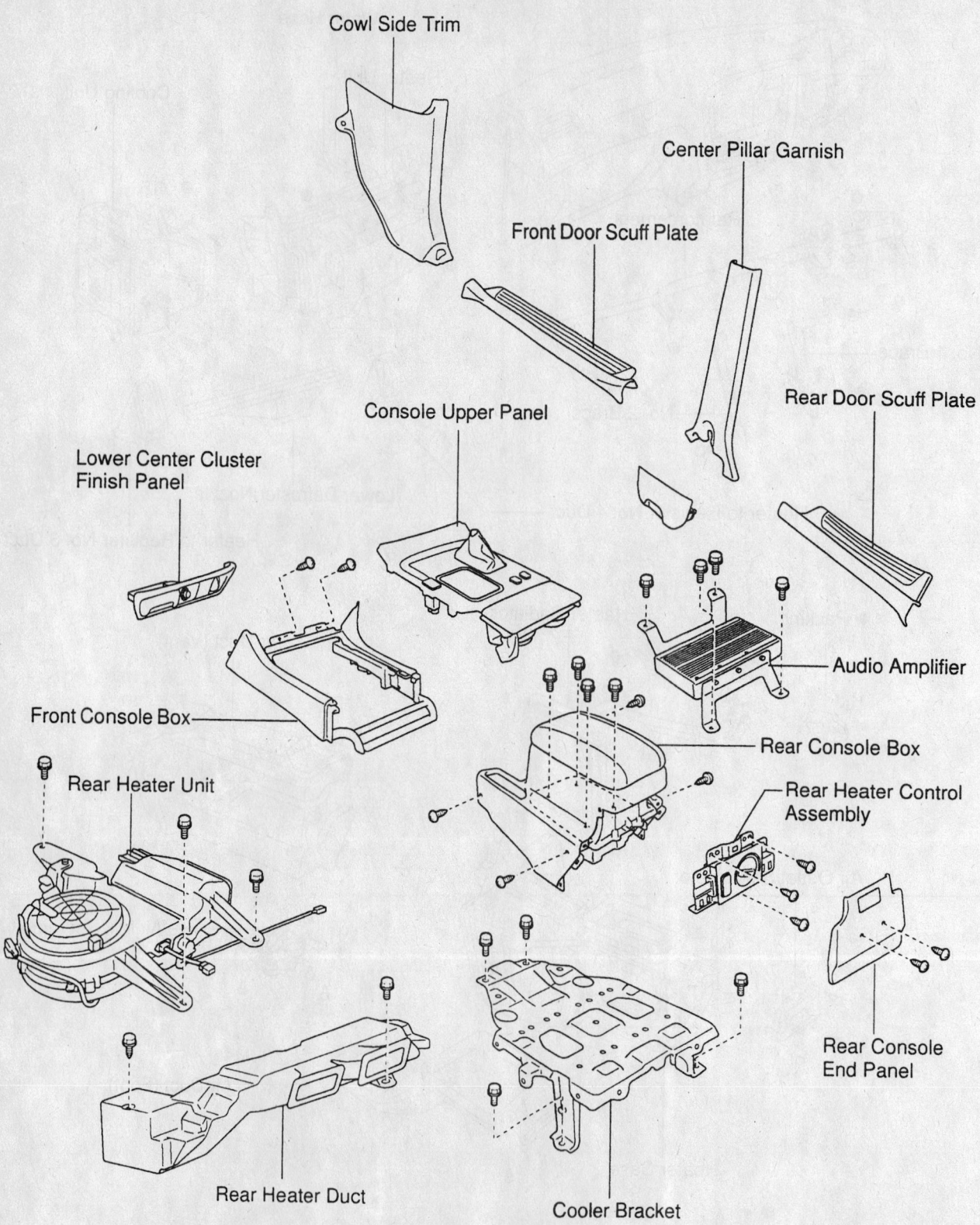

Cowl Side Trim

Center Pillar Garnish

Front Door Scuff Plate

Console Upper Panel

Rear Door Scuff Plate

Lower Center Cluster
Finish Panel

Audio Amplifier

Front Console Box

Rear Console Box

Rear Heater Unit

Rear Heater Control
Assembly

Rear Console
End Panel

Rear Heater Duct

Cooler Bracket

93113GH1

Exploded view of the rear heater housing and related components—Toyota Land Cruiser

the radio outward, disconnect the electrical connectors and remove the radio.

s. At the rear console panel, remove the transfer shift lever knob; then, pry the panel upward disengaging the 4 clips (2 on each side) and remove the panel.

t. At the rear of the console, remove the 2 rear end panel-to-console screws; then, pry the end panel rearward disengaging the 2 clips and remove the panel.

u. If not equipped with a rear air conditioning system, disconnect the connector and control cable; then, remove the 3 rear heater control panel screws and the panel.

v. Remove the 4 rear console box-to-

chassis screws/bolts and the console box.

w. Remove the center lower cluster finish panel by prying panel rearward disengaging the 5 clips; then, disconnect the electrical connector.

x. Remove the 2 front console-to-chassis bolts/screws, disengage the 2 clips and remove the console.

y. At the instrument panel, disconnect the junction connectors (the connectors can be disconnected by loosening the bolts), the instrument panel-to-chassis 8 bolts and 2 nuts. Using an assistant, remove the instrument panel.

z. Disconnect the electrical connector and remove the ECM.

aa. Remove the No. 3 and No. 4 heater-to-register ducts.

bb. Remove the floor brace, the No. 1 brace and the reinforcement.

6. Remove the evaporator housing by performing the following procedure:

a. Discharge and recover the air conditioning system refrigerant.

b. Remove the air conditioning liquid line clamp.

c. Remove the air conditioning suction line clamp.

d. Disconnect both air conditioning

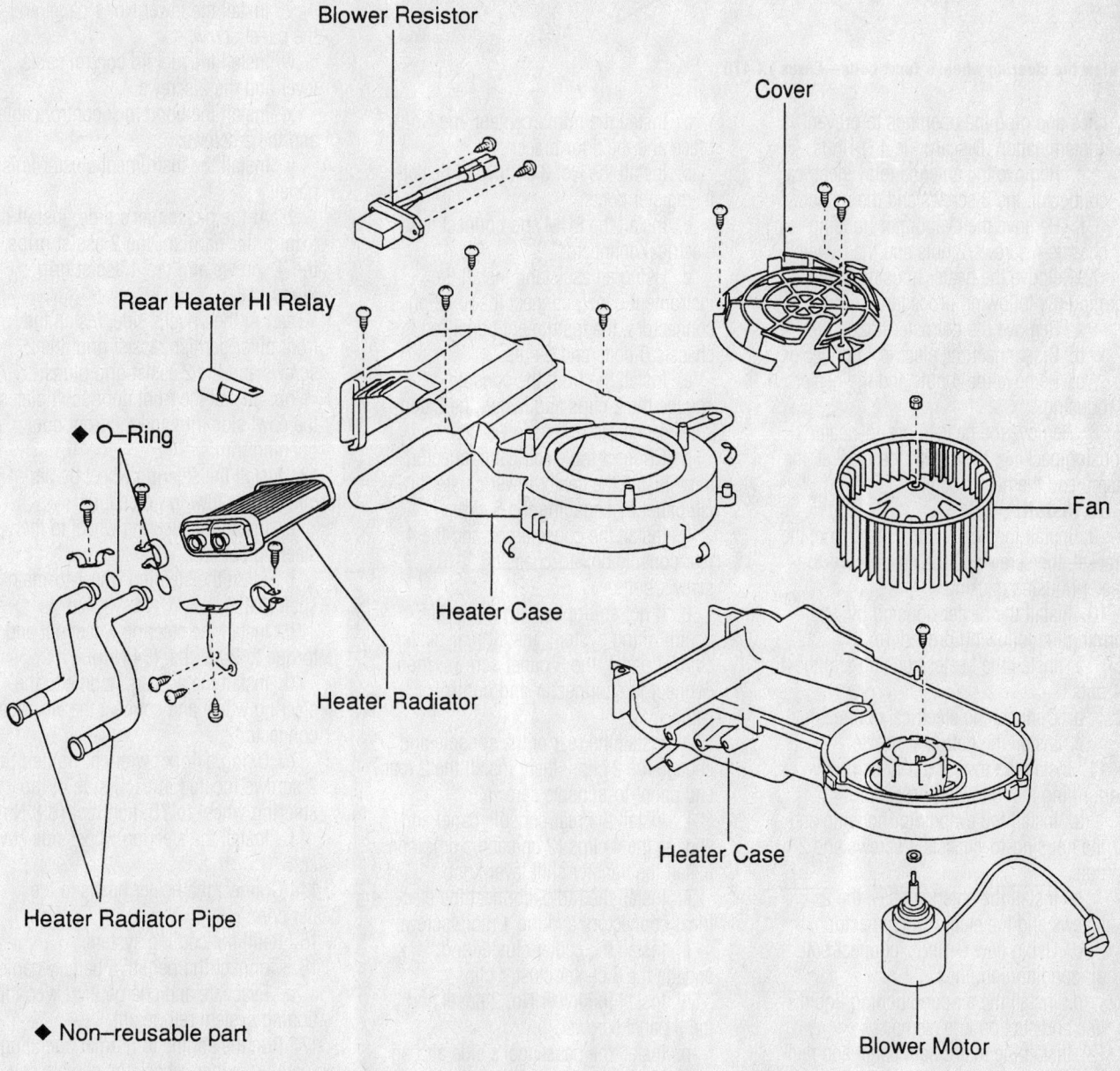

Blower Resistor

Cover

Rear Heater HI Relay

◆ O–Ring

Fan

Heater Case

Heater Radiator

Heater Case

Heater Radiator Pipe

Blower Motor

◆ Non–reusable part

93113GH2

Exploded view of the rear heater core, heater housing and related components—Toyota Land Cruiser

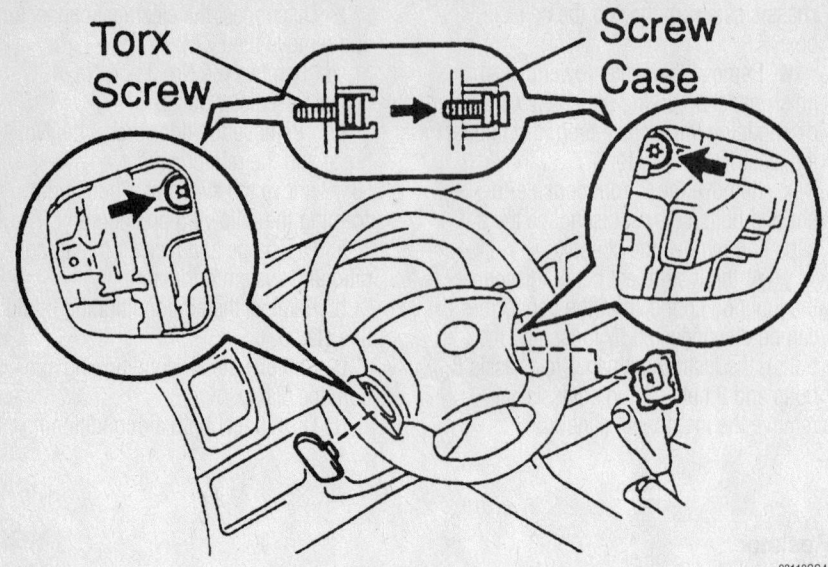

93113GG4

View the steering wheel's Torx® bolts—Lexus LX 470

lines and plug the openings to prevent contamination. Discard the 4 O-rings.

e. Remove the antenna relay electrical connector, the 2 screws and the relay.

f. Remove the evaporator housing-to-chassis 4 screws/2 nuts and the housing.

7. Remove the heater housing by performing the following procedure:

a. Remove the defroster nozzle.

b. Disconnect the electrical connector.

c. Remove the 4 nuts and the heater housing.

8. Remove the heater core-to-heater housing packing, the screw, the bracket, the clamp and the heater core.

To install:

9. Install the heater core, the clamp, the bracket, the screw and the heater core-to-heater housing packing.

10. Install the heater housing by performing the following procedure:

a. Install the heater housing and the 4 nuts.

b. Connect the electrical connector.

c. Install the defroster nozzle.

11. Install the evaporator housing by performing the following procedure:

a. Install the evaporator housing and the housing-to-chassis 4 screws and 2 nuts.

b. Install the antenna relay, the 2 screws and the electrical connector.

c. Using new O-rings, connect both air conditioning lines.

d. Install the air conditioning liquid line and suction line clamp.

12. Install the instrument panel and reinforcement by performing the following procedure:

a. Install the reinforcement, the No. 1 brace and the floor brace.

b. Install the No. 3 and No. 4 heater-to-register ducts.

c. Install the ECM and connect the electrical connector.

d. Using an assistant, install the instrument panel, connect the junction connectors, the instrument panel-to-chassis 8 bolts and 2 nuts.

e. Install the front the console, engage the 2 clips and install the 2 console-to-chassis bolts/screws.

f. Connect the electrical connector; then, install the center lower cluster finish panel by engaging the 5 clips.

g. Install the console box and the 4 rear console box-to-chassis screws/bolts.

h. If not equipped with a rear air conditioning system, install rear heater control panel, the 3 panel screws; then, connect the connector and control cable.

i. Install the rear of the console and engage the 2 clips; then, install the 2 rear end panel-to-console screws.

j. Install the rear console panel and engage the 4 clips (2 on each side); then, install the transfer shift lever knob.

k. Install the radio, connect the electrical connectors and the 4 radio screws.

l. Install the center cluster and engage the 8 center cluster clips.

m. Install the lower No. 2 panel and the 3 panel screws.

n. Install the passenger's side air bag module, connect it and install the No. 1 undercover.

o. Install the glove box door, the 2 screws and the glove compartment door stoppers.

p. Install the combination meter and the 4 screws; then, connect the electrical connectors.

q. Install the steering column and the steering column-to-instrument panel bolts.

r. Install the No. 2 heater-to-register duct and the duct screw.

s. At the steering column, install the combination switch, the 3 screws and the clamp; then, connect the electrical connectors.

t. Install the steering column covers and the 3 covers screws.

u. Install the lower left side panel.

v. Install the lower No. 1 panel and the panel screw.

w. Install the fuel lid control cable lever and the 2 screws.

x. Install the hood lock control cable and the 2 screws.

y. Install the instrument cluster finish panel.

z. At the passenger's side, install the front pillar garnish, the 2 assist grips, the 4 screws and the 4 assist grip plugs.

aa. At the driver's side, install the front pillar garnish, assist grip, the 2 screws and the 2 assist grip plugs.

bb. Install the front door scuff plates, the cowl side trim and the front door opening trim.

13. Install the steering wheel by performing the following procedure:

a. Install the steering wheel to the steering column.

b. Align the steering wheel-to-main shaft marks.

c. Install the steering wheel nut and torque to 25 ft. lbs. (34 Nm).

d. Install the air bag module to the steering wheel and connect the electrical connector.

e. Using a Torx® wrench, tighten the 2 screws located at each side of the steering wheel to 78 inch lbs. (8.8 Nm).

f. Install the steering wheel side covers.

14. Connect the heater hoses to the heater core.

15. Refill the cooling system.

16. Connect the negative battery cable.

a. Evacuate and charge the air conditioning system refrigerant.

17. Run the engine to normal operating temperatures; then, check the climate control operation and check for leaks.

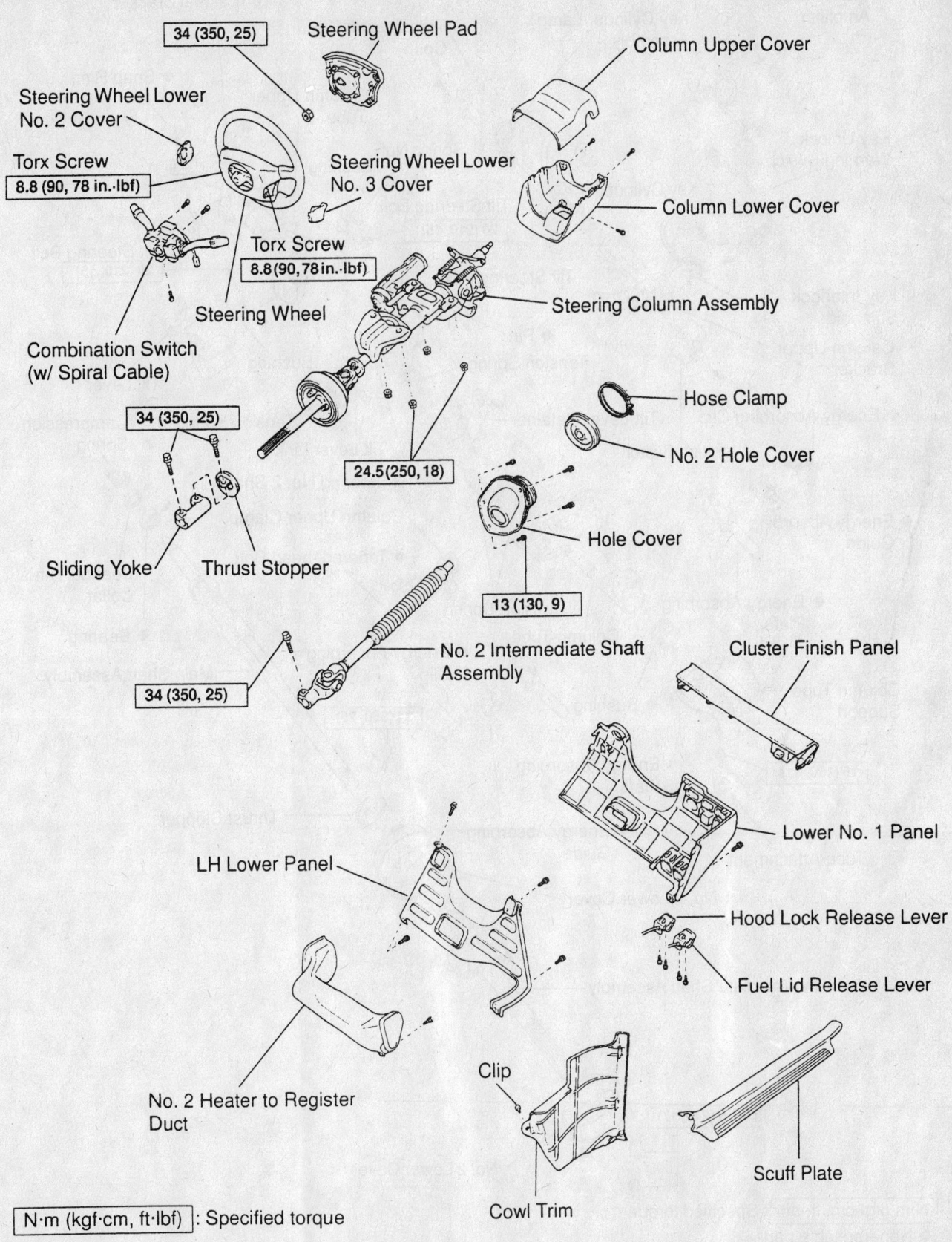

34 (350, 25)

Steering Wheel Pad

Column Upper Cover

Steering Wheel Lower No. 2 Cover

Torx Screw
8.8 (90, 78 in.·lbf)

Steering Wheel Lower No. 3 Cover

Column Lower Cover

Torx Screw
8.8 (90, 78 in.·lbf)

Steering Wheel

Steering Column Assembly

Combination Switch (w/ Spiral Cable)

Hose Clamp

No. 2 Hole Cover

34 (350, 25)

24.5 (250, 18)

Hole Cover

Sliding Yoke

Thrust Stopper

13 (130, 9)

34 (350, 25)

No. 2 Intermediate Shaft Assembly

Cluster Finish Panel

Lower No. 1 Panel

LH Lower Panel

Hood Lock Release Lever

Fuel Lid Release Lever

No. 2 Heater to Register Duct

Clip

Scuff Plate

Cowl Trim

N·m (kgf·cm, ft·lbf) : Specified torque

93113GG5

Exploded view the steering column—Lexus LX 470 (Part 1 of 2)

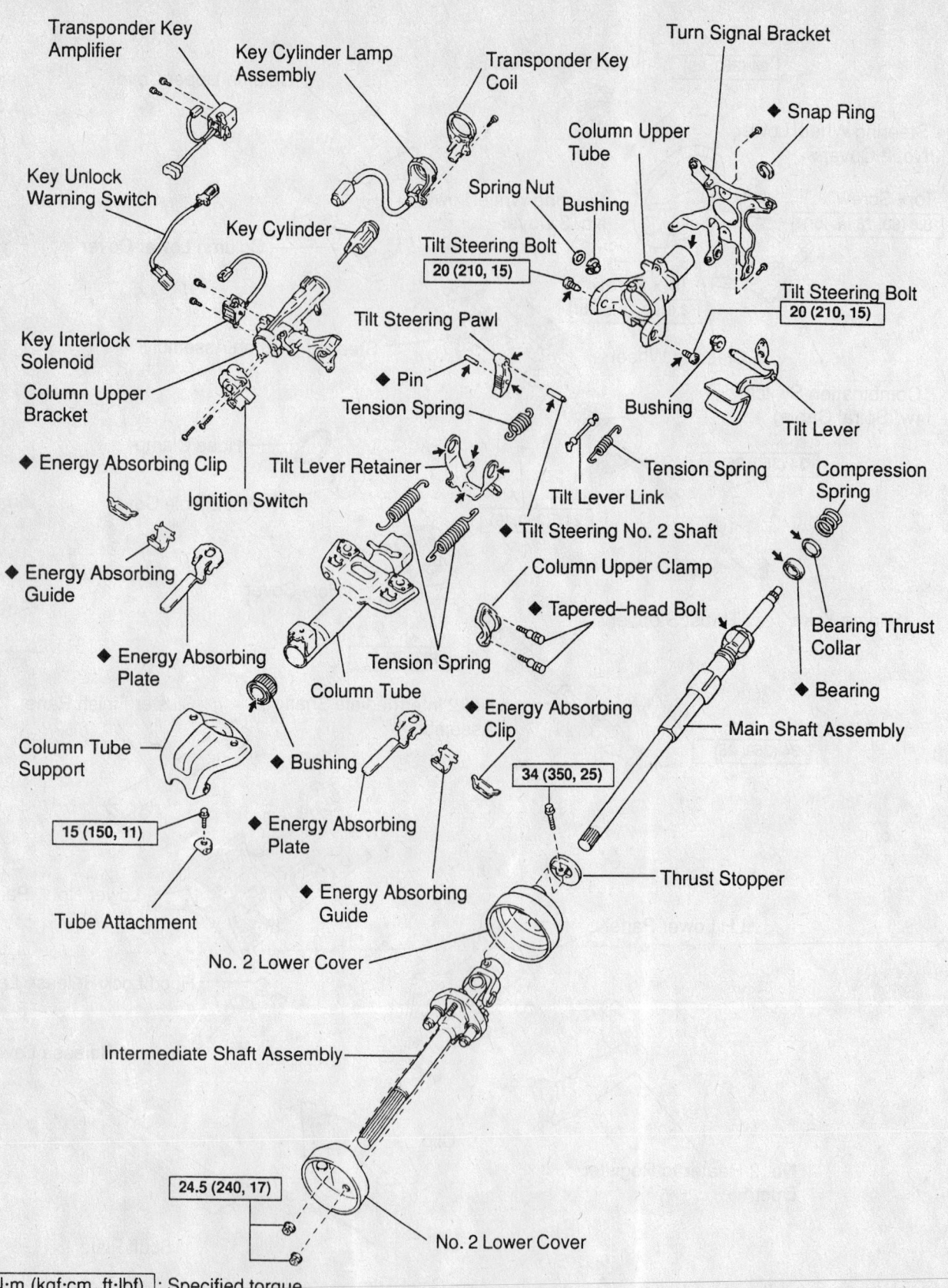

Transponder Key Amplifier

Key Cylinder Lamp Assembly

Transponder Key Coil

Turn Signal Bracket

Column Upper Tube

◆ Snap Ring

Key Unlock Warning Switch

Spring Nut

Bushing

Key Cylinder

Tilt Steering Bolt
20 (210, 15)

Tilt Steering Bolt
20 (210, 15)

Key Interlock Solenoid

Tilt Steering Pawl

◆ Pin

Tension Spring

Bushing

Tilt Lever

Column Upper Bracket

Tension Spring

Compression Spring

◆ Energy Absorbing Clip

Tilt Lever Retainer

Tilt Lever Link

Ignition Switch

◆ Tilt Steering No. 2 Shaft

◆ Energy Absorbing Guide

Column Upper Clamp

Bearing Thrust Collar

◆ Tapered–head Bolt

◆ Bearing

◆ Energy Absorbing Plate

Tension Spring

Column Tube

Main Shaft Assembly

Column Tube Support

◆ Bushing

◆ Energy Absorbing Clip

◆ Energy Absorbing Plate

34 (350, 25)

15 (150, 11)

◆ Energy Absorbing Guide

Thrust Stopper

Tube Attachment

No. 2 Lower Cover

Intermediate Shaft Assembly

24.5 (240, 17)

No. 2 Lower Cover

N·m (kgf·cm, ft·lbf) : Specified torque

◆ Non–reusable part

N Molybdenum disulfide lithium base grease

93113GG6

Exploded view the steering column—Lexus LX 470 (Part 2 of 2)

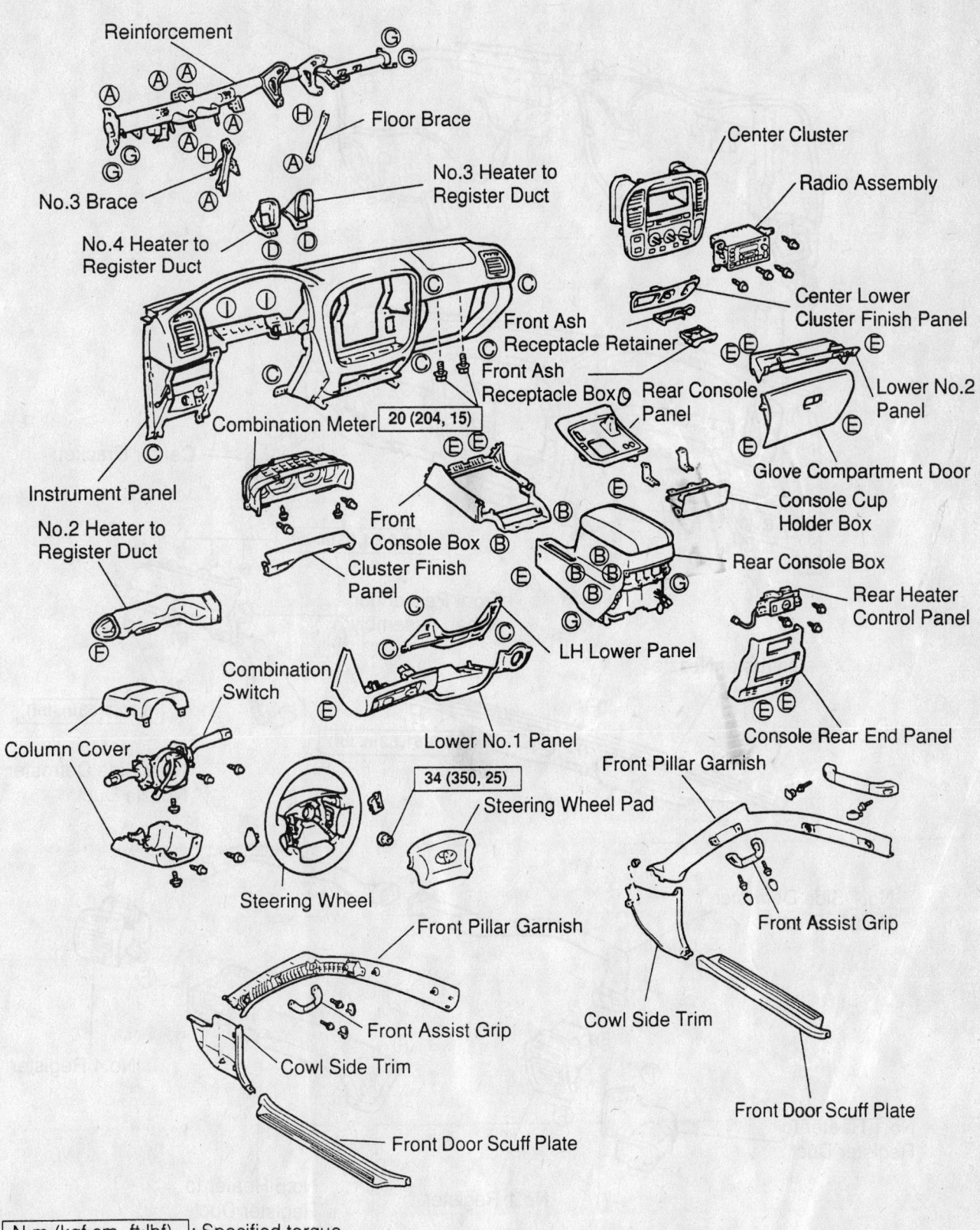

Reinforcement

Floor Brace

No.3 Brace

No.4 Heater to Register Duct

No.3 Heater to Register Duct

Center Cluster

Radio Assembly

Center Lower Cluster Finish Panel

Front Ash Receptacle Retainer

Front Ash Receptacle Box

Lower No.2 Panel

Combination Meter

Rear Console Panel

20 (204, 15)

Glove Compartment Door

Console Cup Holder Box

Instrument Panel

No.2 Heater to Register Duct

Front Console Box

Cluster Finish Panel

Rear Console Box

Rear Heater Control Panel

LH Lower Panel

Combination Switch

Column Cover

Lower No.1 Panel

Console Rear End Panel

34 (350, 25)

Steering Wheel Pad

Front Pillar Garnish

Front Assist Grip

Steering Wheel

Front Pillar Garnish

Cowl Side Trim

Front Assist Grip

Front Assist Grip

Cowl Side Trim

Front Door Scuff Plate

Front Door Scuff Plate

N·m (kgf·cm, ft·lbf) : Specified torque

Exploded view the instrument panel and related components—LX 470

93113GG7

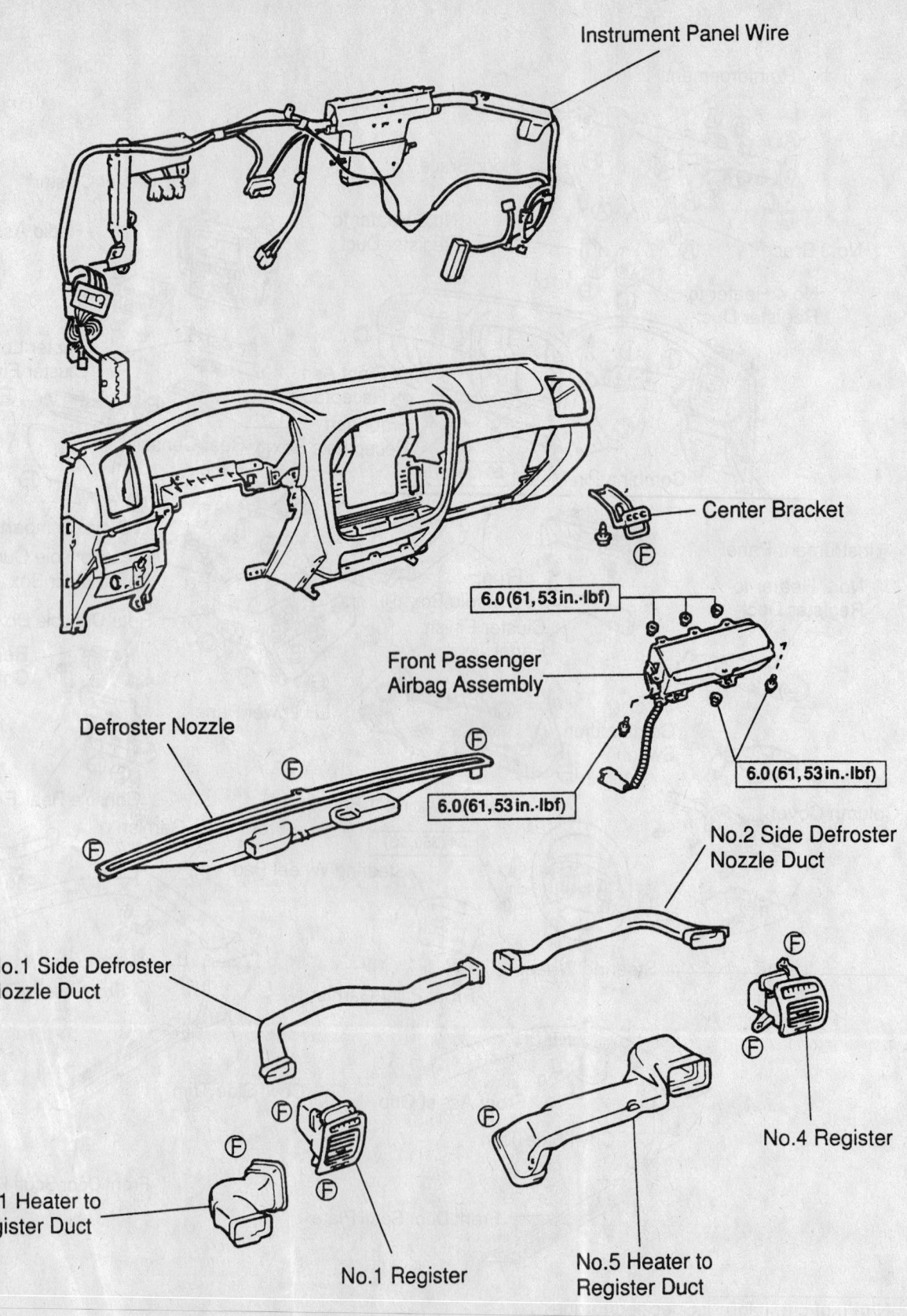

Instrument Panel Wire

Center Bracket

6.0 (61, 53 in.·lbf)

Front Passenger Airbag Assembly

6.0 (61, 53 in.·lbf)

Defroster Nozzle

6.0 (61, 53 in.·lbf)

No.2 Side Defroster Nozzle Duct

No.1 Side Defroster Nozzle Duct

No.4 Register

No.1 Heater to Register Duct

No.1 Register

No.5 Heater to Register Duct

N·m (kgf·cm, ft·lbf) : Specified torque

93113GG8

Exploded view the front ventilation ducts and related components—Lexus LX 470

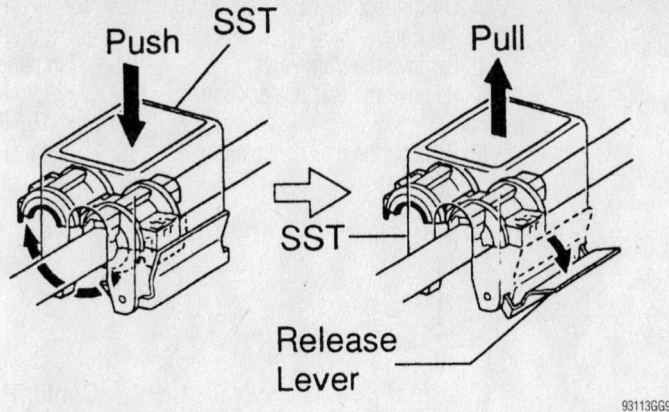

Push SST   Pull

SST

Release
Lever

93113GG9

View the air conditioning line clamp removal tool—Lexus LX 470

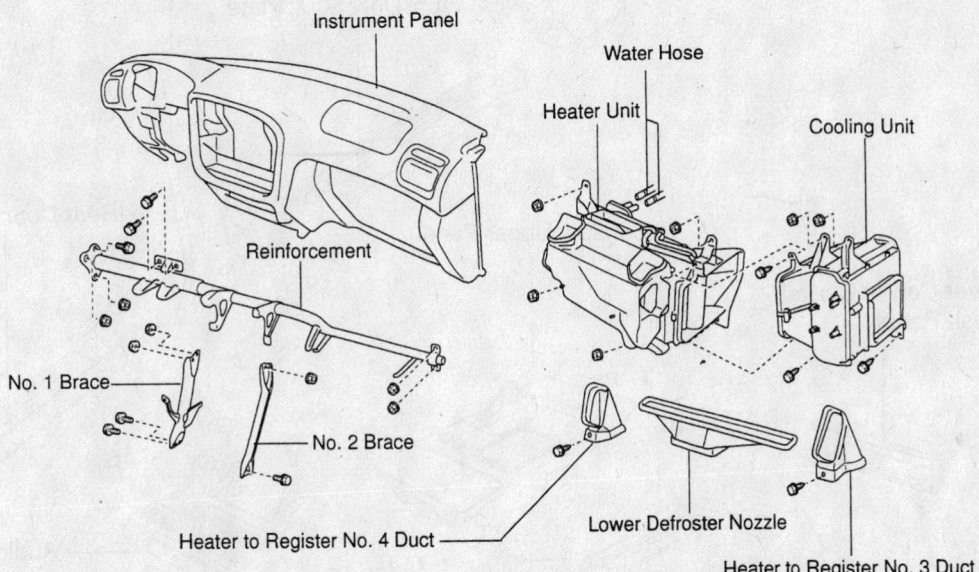

Instrument Panel

Water Hose

Heater Unit

Cooling Unit

Reinforcement

No. 1 Brace

No. 2 Brace

Heater to Register No. 4 Duct

Lower Defroster Nozzle

Heater to Register No. 3 Duct

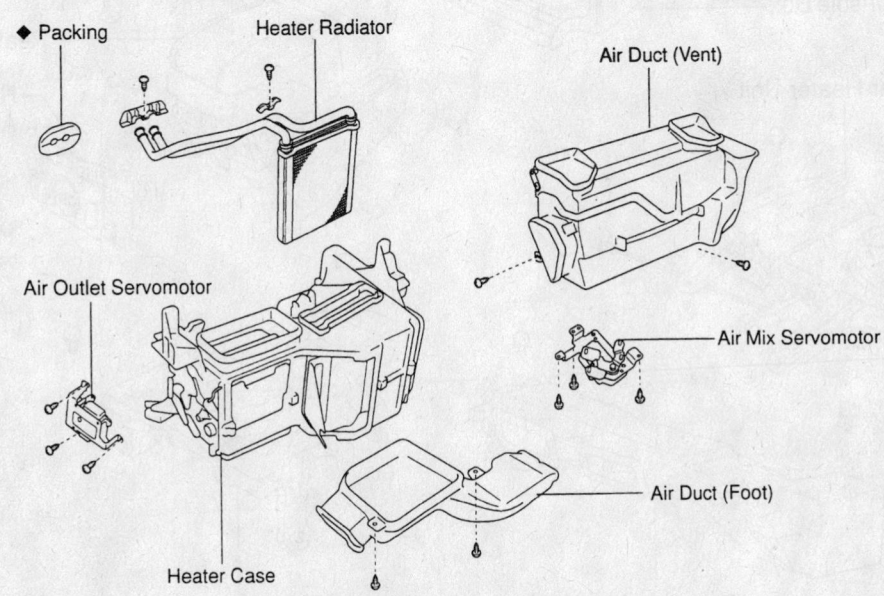

◆ Packing

Heater Radiator

Air Duct (Vent)

Air Outlet Servomotor

Air Mix Servomotor

Air Duct (Foot)

Heater Case

◆ Non–reusable part

93113GG0

Exploded view the front heater core, heater housing, evaporator housing and related components—Lexus LX 470

REAR AUXILIARY HEATER

1. Disconnect the negative battery cable.
2. Drain the cooling system into a clean container for reuse.

3. Disconnect the heater hoses from the rear heater core.
4. Remove the front seats.
5. Remove the rear heater control assembly.
6. Remove the rear console box.

7. Remove the front console box cover.
8. Remove the lower center cluster finish panel.
9. Remove the front door scuff plates.
10. Remove the cowl side trim.
11. Remove the rear door scuff plates.

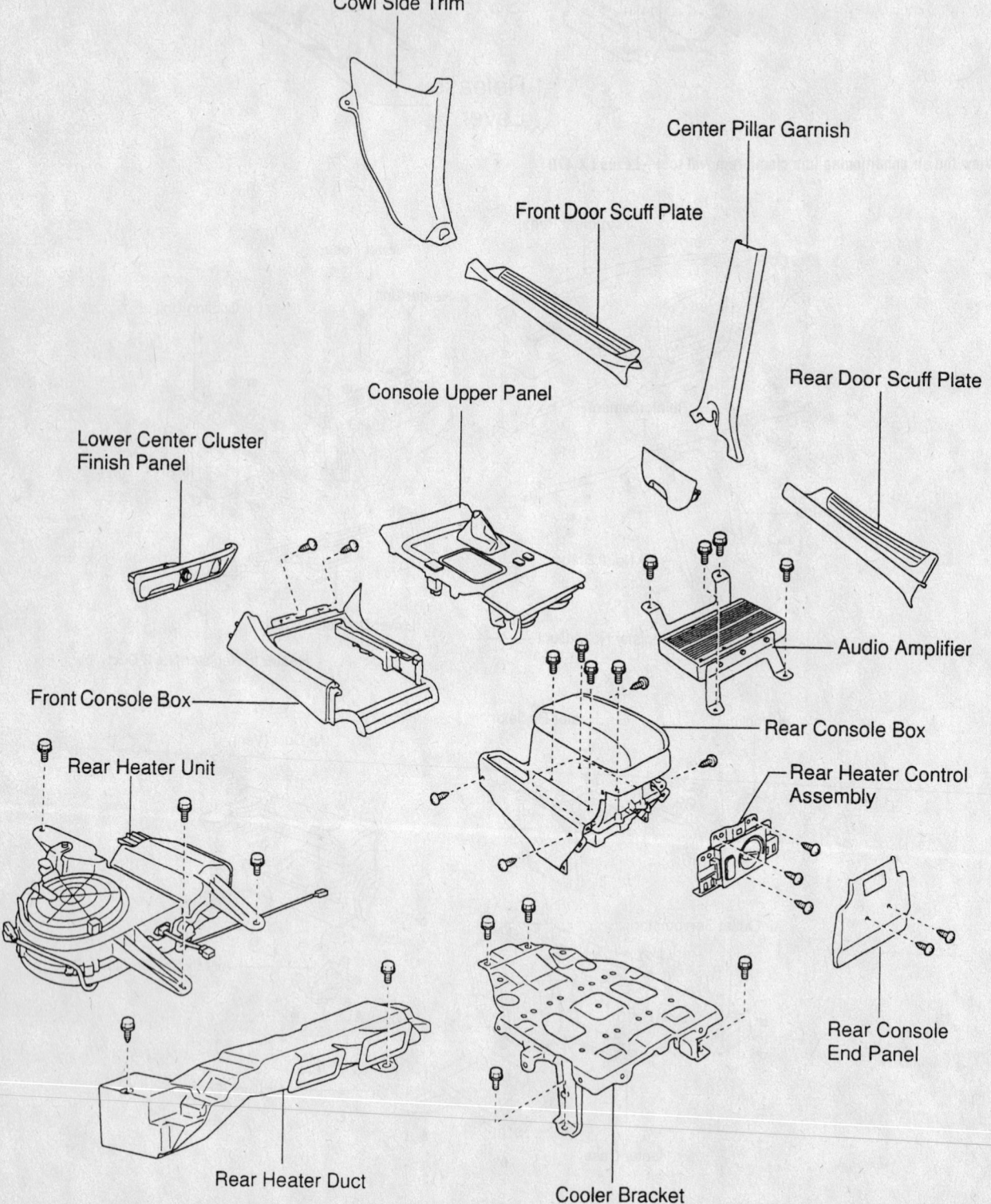

Cowl Side Trim

Center Pillar Garnish

Front Door Scuff Plate

Rear Door Scuff Plate

Console Upper Panel

Lower Center Cluster Finish Panel

Audio Amplifier

Front Console Box

Rear Console Box

Rear Heater Unit

Rear Heater Control Assembly

Rear Console End Panel

Rear Heater Duct

Cooler Bracket

93113GH1

Exploded view of the rear heater housing and related components—Lexus LX 470

12. Remove the center pillar garnishes.

13. Slide the carpet rearward.

14. Remove the cooler bracket bolts and the bracket.

15. Remove the rear heater duct bolt/screw and the duct.

16. Disconnect the rear heater housing electrical connector.

17. Remove the 3 rear heater housing-to-chassis bolts and the heater housing.

18. Remove the heater core-to-heater housing 3 screws and 2 clamps.

19. Remove the heater core from the heater housing.

To install:

20. Install the heater core to the heater housing.

21. Install the heater core-to-heater housing 3 screws and 2 clamps.

22. Install the heater housing and the 3 rear heater housing-to-chassis bolts.

23. Connect the rear heater housing electrical connector.

24. Install the rear heater duct and the duct bolt/screw.

25. Install the cooler bracket and the bracket bolts.

26. Slide the carpet rearward.

27. Install the center pillar garnishes.

28. Install the rear door scuff plates.

29. Install the cowl side trim.

30. Install the front door scuff plates.

31. Install the lower center cluster finish panel.

32. Install the front console box cover.

33. Install the rear console box.

34. Install the rear heater control assembly.

35. Install the front seats.

36. Connect the heater hoses to the rear heater core.

37. Refill the cooling system.

38. Connect the negative battery cable.

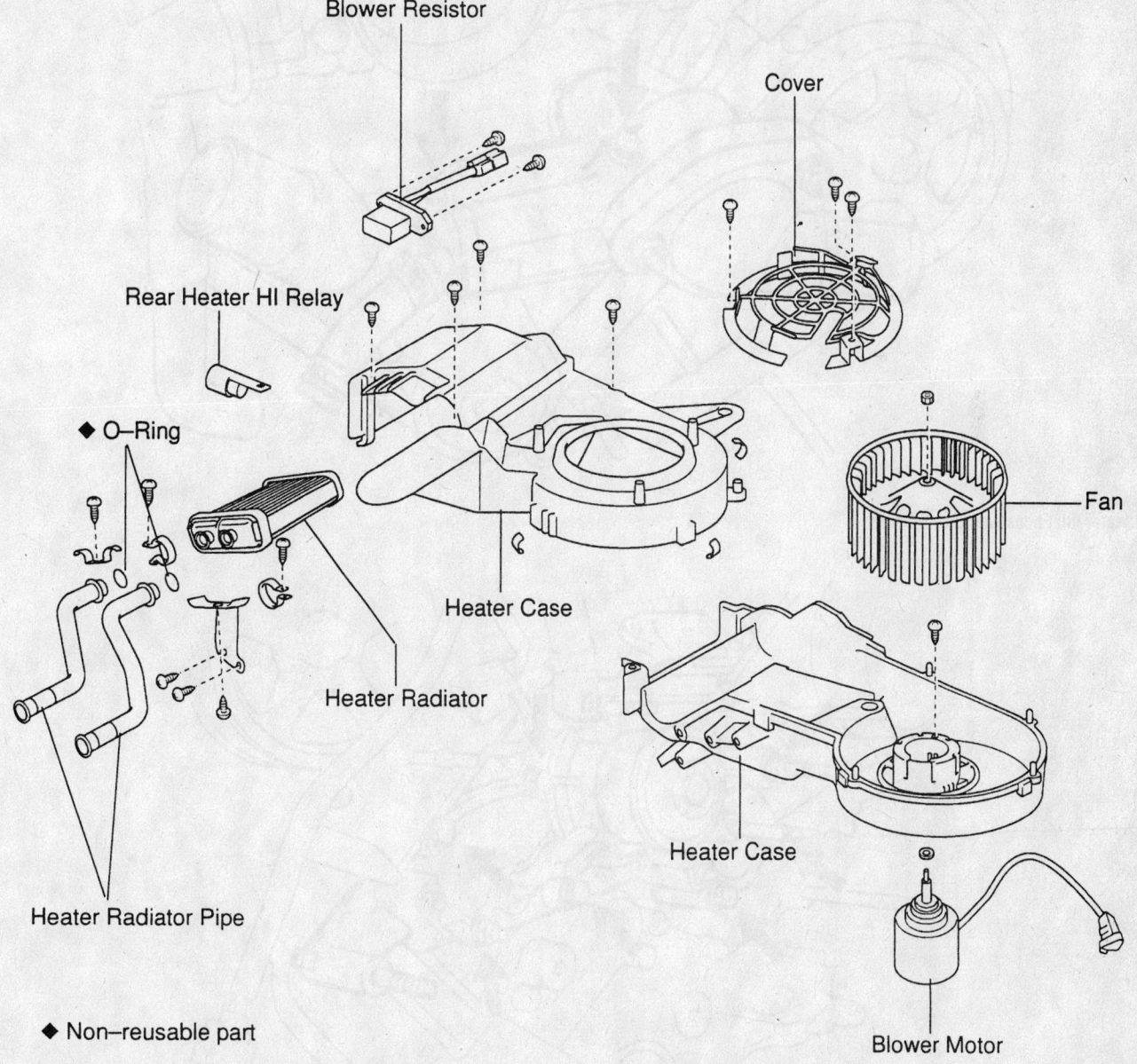

Exploded view of the rear heater core, heater housing and related components—Lexus LX 470

93113GH2

Water Pump

REMOVAL & INSTALLATION

1. Before servicing the vehicle, refer to the precautions in the beginning of this section.
2. Drain the cooling system.
3. Remove or disconnect the following:
   - Negative battery cable
   - Timing belt
   - No. 2 idler pulley
   - Radiator hose
   - Bypass hose
   - Water inlet housing assembly
   - Water pump

To install:

4. Install or connect the following:
   - Water pump. Use a new gasket and tighten the bolts to 15 ft. lbs. (21 Nm). Tighten the stud bolt and nut to 13 ft. lbs. (18 Nm).
   - Water inlet housing assembly. Use a new O-ring and apply sealant as shown. Tighten the bolts to 13 ft. lbs. (18 Nm).
   - Bypass hose
   - Radiator hose
   - No. 2 idler pulley
   - Timing belt
   - Negative battery cable
5. Fill the cooling system.
6. Start the engine and check for leaks.

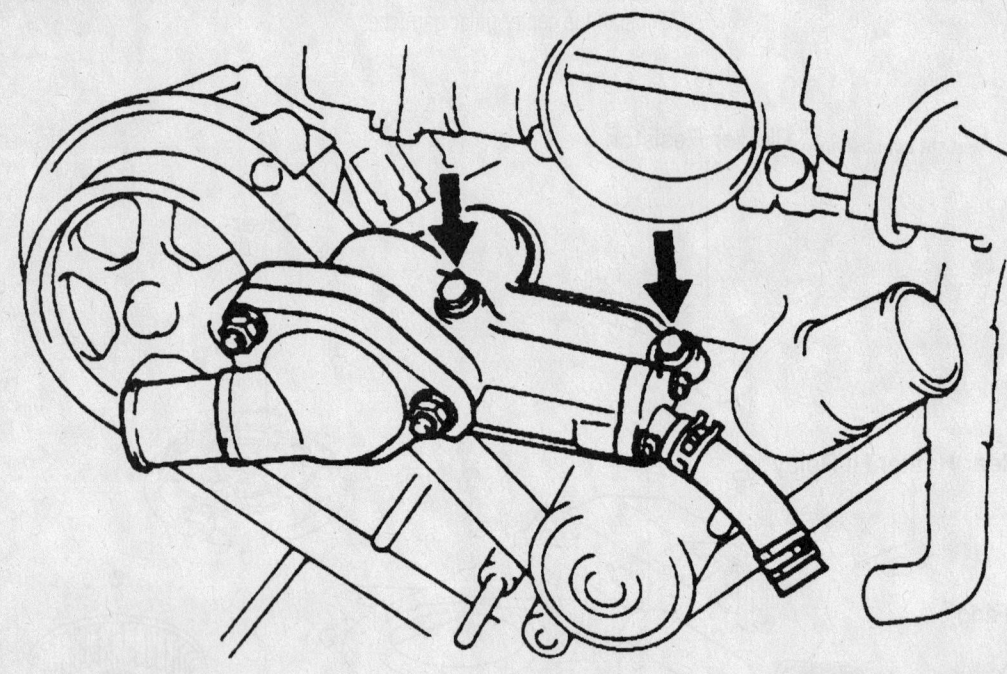

Water inlet housing attaching bolts

7924SG40

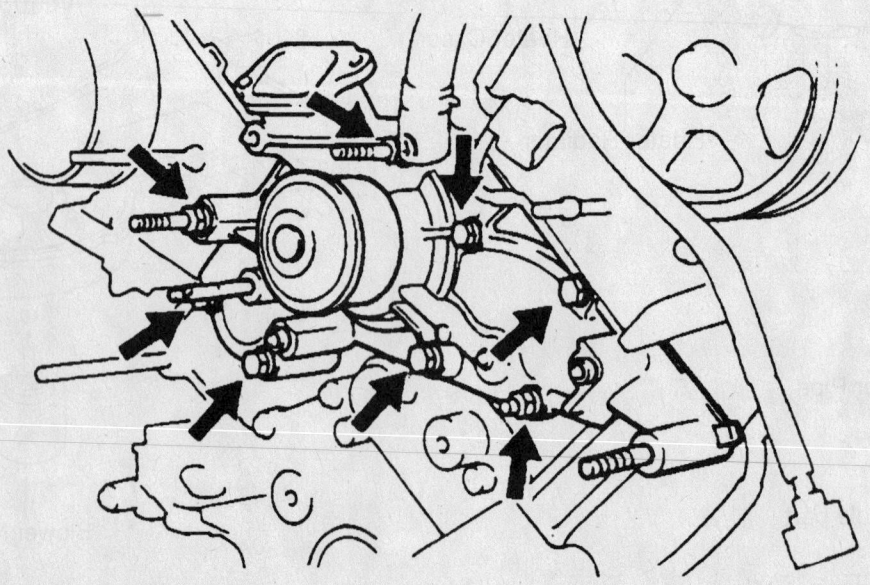

Water pump mounting bolts, stud bolts and nut locations

7924SG41

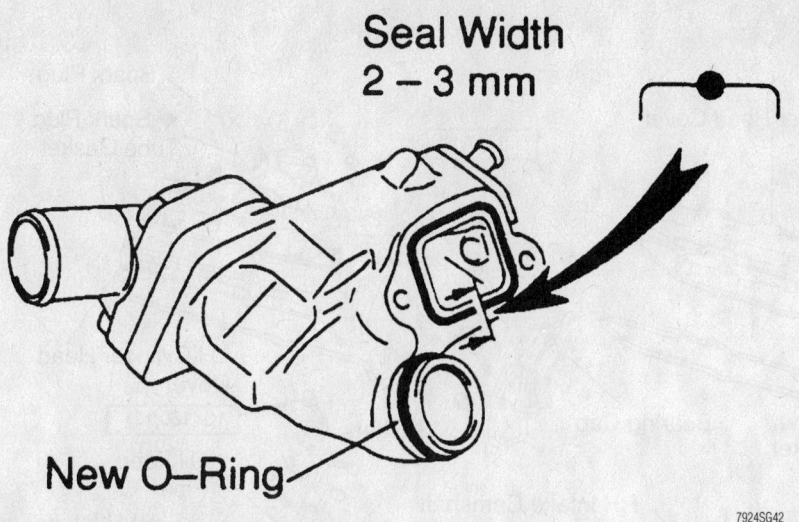

Seal Width
2 – 3 mm

New O–Ring

Water inlet housing sealant application

Cylinder Head

REMOVAL & INSTALLATION

1. Before servicing the vehicle, refer to the precautions in the beginning of this section.
2. Drain the cooling system.
3. Relieve the fuel system pressure.
4. Remove or disconnect the following:
   - Battery and tray
   - Engine appearance cover
   - Engine under covers
   - Air intake assembly
   - Accessory drive belt
   - A/C compressor and bracket
   - Cooling fan and bracket
   - Radiator
   - Idler pulley
   - Front covers
   - Timing belt
   - Camshaft sprockets
   - Camshaft Position (CMP) sensor
   - Power steering pump
   - Exhaust front pipes
   - Transmission dipstick tube
   - Ignition coils
   - Rear timing belt covers
   - Fuel lines
   - Intake manifold
   - Water inlet housing assembly
   - Front and rear water bypass joints
   - Engine lifting eyes
   - Oil dipstick tube
   - Valve covers
   - Camshafts

- Cylinder heads with the exhaust manifolds attached. Loosen the bolts in the sequence shown.

To install:
5. Install the cylinder heads with new gaskets. Tighten the bolts in sequence as follows:
   a. Step 1: 24 ft. lbs. (32 Nm)
   b. Step 2: Plus 180 degrees
6. Install or connect the following:
   - Camshafts
   - Valve covers
   - Oil dipstick tube
   - Engine lifting eyes
   - Front and rear water bypass joints
   - Water inlet housing assembly
   - Intake manifold
   - Fuel lines
   - Rear timing belt covers
   - Ignition coils
   - Transmission dipstick tube
   - Exhaust front pipes
   - Power steering pump
   - CMP sensor
   - Camshaft sprockets
   - Timing belt
   - Front covers
   - Idler pulley
   - Radiator
   - Cooling fan and bracket
   - A/C compressor and bracket
   - Accessory drive belt
   - Air intake assembly
   - Engine under covers
   - Engine appearance cover
   - Battery and tray
7. Fill the cooling system.
8. Start the engine and check for leaks.

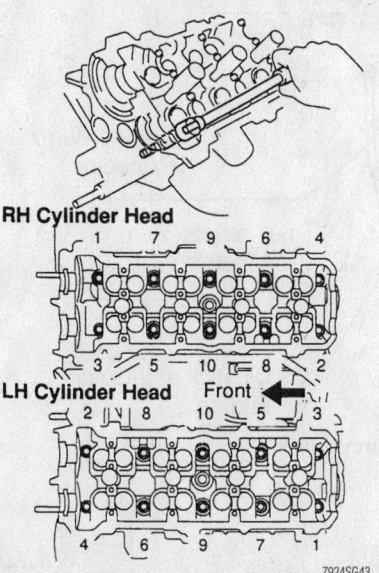

Cylinder head loosening sequence

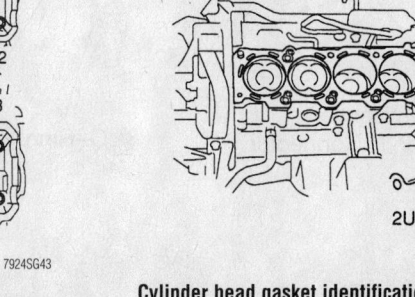

Cylinder head gasket identification

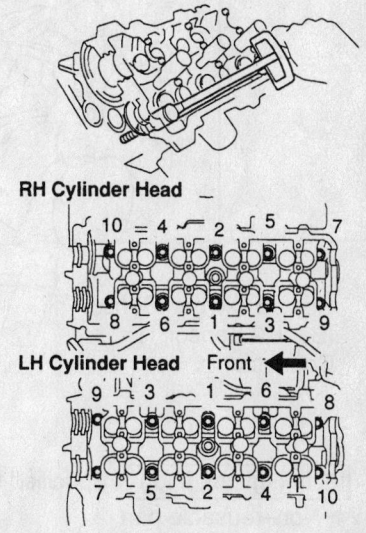

Cylinder head torque sequence

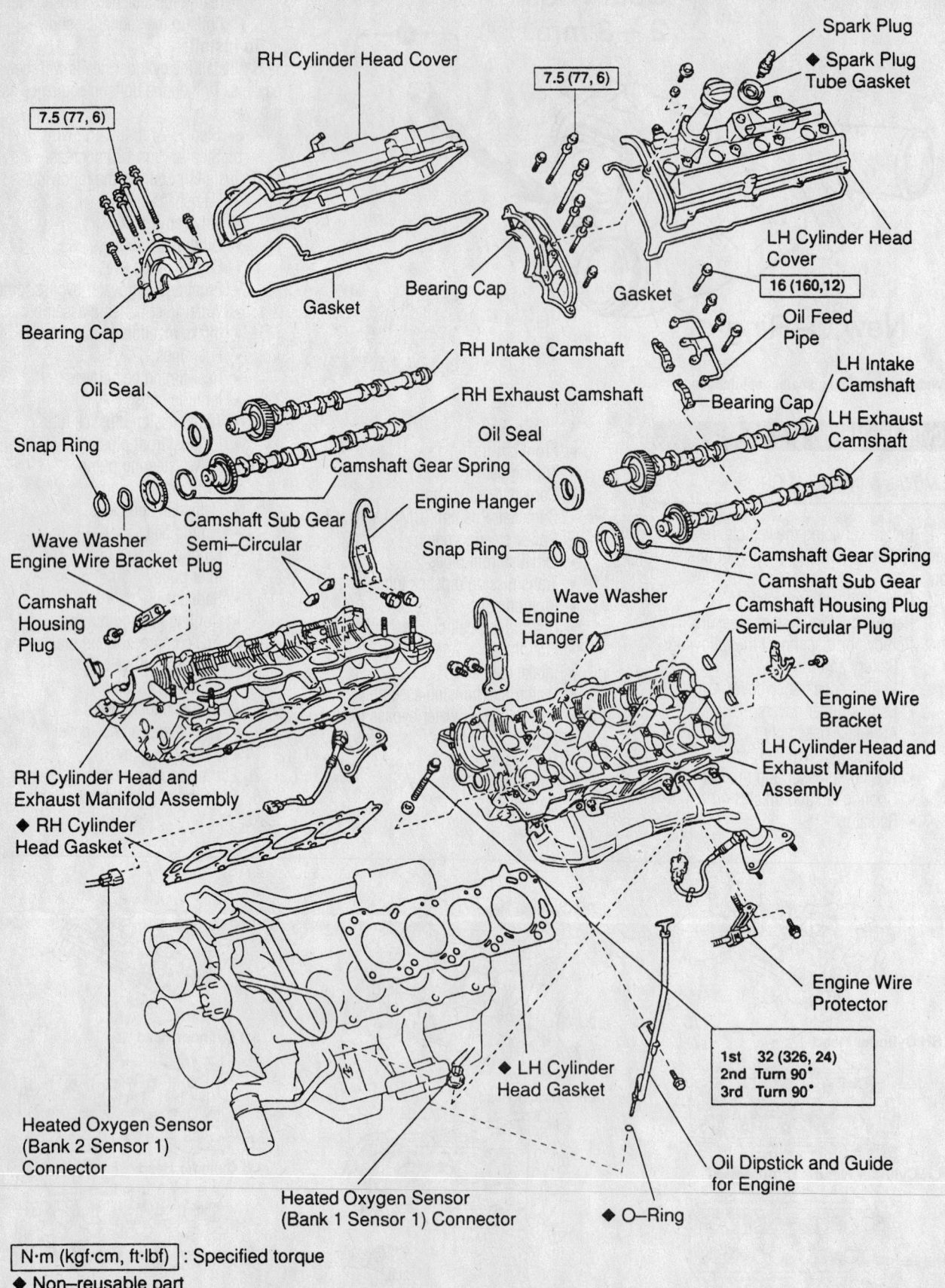

RH Cylinder Head Cover

7.5 (77, 6)

Spark Plug

◆ Spark Plug Tube Gasket

7.5 (77, 6)

LH Cylinder Head Cover

7.5 (77, 6)

Bearing Cap

Gasket

Bearing Cap

Gasket

16 (160,12)

Oil Feed Pipe

Bearing Cap

Bearing Cap

Oil Seal

Snap Ring

RH Intake Camshaft

RH Exhaust Camshaft

Oil Seal

Camshaft Gear Spring

Engine Hanger

Snap Ring

LH Intake Camshaft

LH Exhaust Camshaft

Camshaft Gear Spring

Camshaft Sub Gear

Camshaft Housing Plug

Semi–Circular Plug

Wave Washer

Engine Wire Bracket

Camshaft Sub Gear

Semi–Circular Plug

Wave Washer

Engine Hanger

Engine Wire Bracket

Camshaft Housing Plug

RH Cylinder Head and Exhaust Manifold Assembly

LH Cylinder Head and Exhaust Manifold Assembly

◆ RH Cylinder Head Gasket

◆ LH Cylinder Head Gasket

Engine Wire Protector

1st 32 (326, 24)
2nd Turn 90°
3rd Turn 90°

Heated Oxygen Sensor (Bank 2 Sensor 1) Connector

Heated Oxygen Sensor (Bank 1 Sensor 1) Connector

Oil Dipstick and Guide for Engine

◆ O–Ring

N·m (kgf·cm, ft·lbf) : Specified torque

◆ Non–reusable part

7924SG49

Exploded view of the cylinder head mounting—4.7L Land Cruiser/LX470

Intake Manifold

REMOVAL & INSTALLATION

1. Before servicing the vehicle, refer to the precautions in the beginning of this section.

2. Drain the cooling system.
3. Relieve the fuel system pressure.
4. Remove or disconnect the following:
   • Negative battery cable
   • Engine appearance cover
   • Accelerator cable
   • Throttle Position (TP) sensor connector

   • Accelerator pedal position sensor
   • Throttle motor connector
   • Evaporative Emissions (EVAP) vacuum switching valve connector
   • Fuel injector connectors
   • Engine Coolant Temperature (ECT) sensor connector
   • ETC gauge sender connector

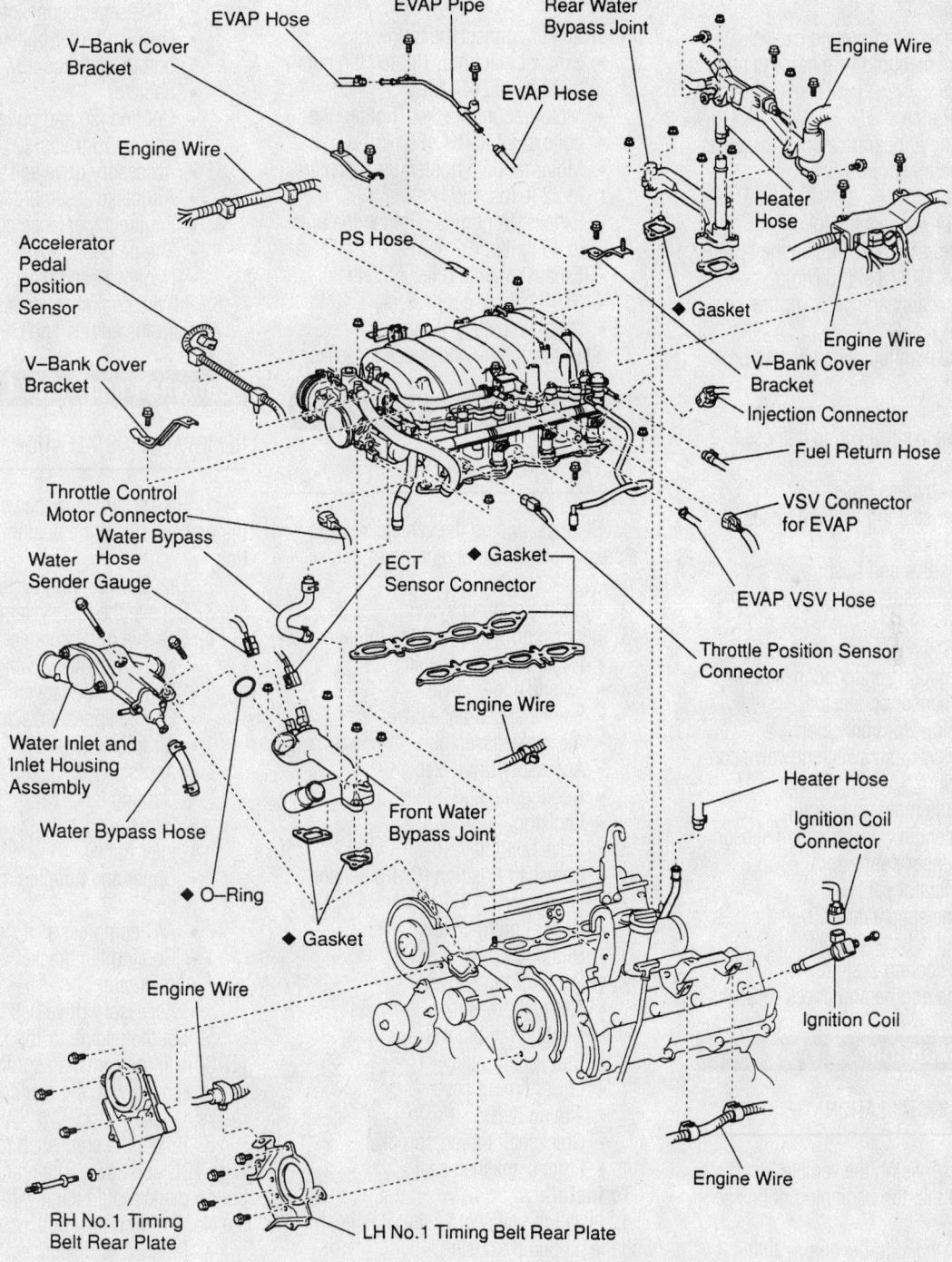

◆ Non–reusable part

Exploded of the intake manifold mounting

7924SG50

- Heated Oxygen (HO2S) sensor connectors
- Fuel pressure regulator vacuum hose
- Positive Crankcase Ventilation (PCV) valve and hose
- EVAP hoses
- Power steering vacuum hoses
- Water bypass hose
- Engine control wiring harness clamps
- Cylinder head ground cables
- Intake manifold wire harness protector
- EVAP pipe
- Engine appearance cover brackets
- Intake manifold

To install:

5. Install or connect the following:
- Intake manifold. Tighten the fasteners to 13 ft. lbs. (18 Nm).
- Engine appearance cover brackets
- EVAP pipe
- Intake manifold wire harness protector
- Cylinder head ground cables
- Engine control wiring harness clamps
- Water bypass hose
- Power steering vacuum hoses
- EVAP hoses
- PCV valve and hose
- Fuel pressure regulator vacuum hose
- HO2S sensor connectors
- ETC gauge sender connector
- ECT sensor connector
- Fuel injector connectors
- EVAP vacuum switching valve connector
- Throttle motor connector
- Accelerator pedal position sensor
- TP sensor connector
- Accelerator cable
- Engine appearance cover
- Negative battery cable
6. Fill the cooling system.
7. Start the engine and check for leaks.

Exhaust Manifold

REMOVAL & INSTALLATION

1. Before servicing the vehicle, refer to the precautions in the beginning of this section.
2. Attach a hoist to the engine lifting eyes.
3. Remove or disconnect the following:
- Negative battery cable

- Heated Oxygen (HO2S) sensor connectors
- Exhaust manifold heat shield
- Exhaust front pipe
- Motor mount
- Motor mount bracket
- Exhaust manifold

To install:

➡ **Use new exhaust manifold nuts for assembly.**

4. Install or connect the following:
- Exhaust manifold. Tighten the nuts to 32 ft. lbs. (44 Nm).
- Motor mount bracket. Tighten the bolts to 27 ft. lbs. (36 Nm).
- Motor mount. Tighten the fasteners to 22 ft. lbs. (30 Nm).
- Exhaust front pipe. Tighten the nuts to 46 ft. lbs. (62 Nm).
- Exhaust manifold heat shield
- HO2S sensor connectors
- Negative battery cable
5. Start the engine and check for leaks.

Front Crankshaft Seal

REMOVAL & INSTALLATION

1. Before servicing the vehicle, refer to the precautions in the beginning of this section.
2. Drain the cooling system.
3. Remove or disconnect the following:
- Negative battery cable
- Engine under cover
- Engine appearance cover
- Air intake assembly
- Accessory drive belt
- Cooling fan and pulley
- Radiator
- Drive belt idler pulley
- Camshaft Position (CMP) sensor connector
- Upper timing covers
- Oil cooler pipe
- Center timing cover
- A/C compressor
- Cooling fan bracket
- Crankshaft pulley
- Lower timing cover
- Timing belt
- Crankshaft timing sprocket
- Front crankshaft seal

To install:

4. Install the oil seal so that it is flush with the oil pump housing.
5. Install or connect the following:
- Crankshaft timing sprocket
- Timing belt
- Lower timing cover

- Crankshaft pulley. Tighten the bolt to 181 ft. lbs. (245 Nm).
- Cooling fan bracket. Tighten the 12mm bolts to 12 ft. lbs. (16 Nm) and the 14mm bolts to 24 ft. lbs. (32 Nm).
- A/C compressor
- Center timing cover
- Oil cooler pipe
- Upper timing covers
- CMP sensor connector
- Drive belt idler pulley. Tighten the bolt to 27 ft. lbs. (37 Nm).
- Radiator
- Cooling fan and pulley. Tighten the nuts to 16 ft. lbs. (21 Nm).
- Accessory drive belt
- Air intake assembly
- Engine appearance cover
- Engine under cover
- Negative battery cable
6. Fill the cooling system.
7. Start the engine and check for leaks.

Camshaft and Valve Lifters

REMOVAL & INSTALLATION

1. Before servicing the vehicle, refer to the precautions in the beginning of this section.
2. Drain the cooling system.
3. Relieve the fuel system pressure.
4. Remove or disconnect the following:
- Negative battery cable
- Engine under covers
- Engine appearance cover
- Air intake hose
- Accessory drive belt
- Cooling fan
- Radiator
- Idler pulley
- Upper and middle timing belt covers
- A/C compressor
- Cooling fan bracket
- Alternator
- Accessory drive belt tensioner
5. Set the engine to Top Dead Center (TDC) with the camshaft sprocket timing marks aligned with the rear cover timing marks.
6. Rotate the crankshaft to 50 degrees After TDC as shown. The crankshaft pulley timing mark should align with the center of the No. 2 idler pulley bolt.
7. Remove or disconnect the following:
- Crankshaft pulley
- Lower timing cover
- Timing belt
- Camshaft timing sprockets

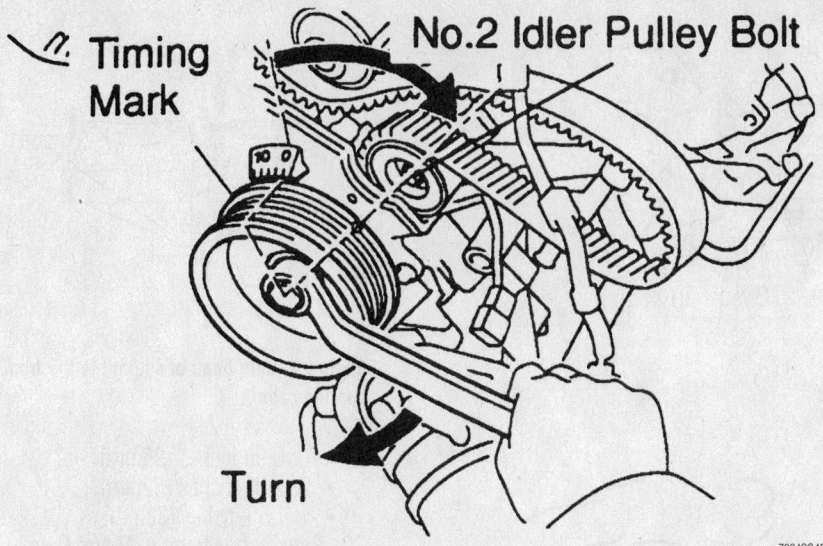

Setting the crankshaft to 50 degrees ATDC

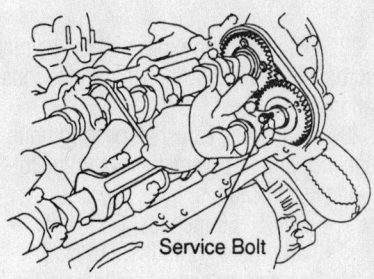

Camshaft service bolt installation

timing marks to a **10** degree angle as shown.

10. Loosen the bearing cap bolts in sequence and in several passes.

11. Remove the right bank camshafts.

12. Rotate the left bank camshafts as necessary to access the exhaust camshaft sub-gear service bolt hole and install a 6mm x 1.0mm bolt.

13. Align the left bank camshaft 2 dot timing marks as shown.

14. Loosen the bearing cap bolts in sequence and in several passes.

15. Remove the left bank camshafts.

16. Remove the valve lifters and shims.

To install:

17. Ensure that the crankshaft is at 50 degrees After TDC.

18. Install or connect the following:
  • Valve lifters and shims in their original positions
  • Right bank camshafts with the 1 dot timing marks at 10 degrees
  • Left bank camshafts with the 2 dot timing marks aligned
  • Left and right bank camshaft bearing caps in their original positions. Apply sealant to the front bearing caps as shown.
  • Camshaft oil seals

19. The bearing cap bolts vary in length and are identified as follows:
  • A: 3.70 inches (94mm)
  • B: 2.83 inches (72mm)

  • Camshaft Position (CMP) sensor
  • Ignition coils
  • Valve cover
  • Timing belt rear covers

8. Rotate the right bank camshafts as necessary to access the exhaust camshaft sub-gear service bolt hole and install a 6mm x 1.0mm bolt.

➡**Keep all valvetrain components in order for assembly.**

9. Align the right bank camshaft 1 dot

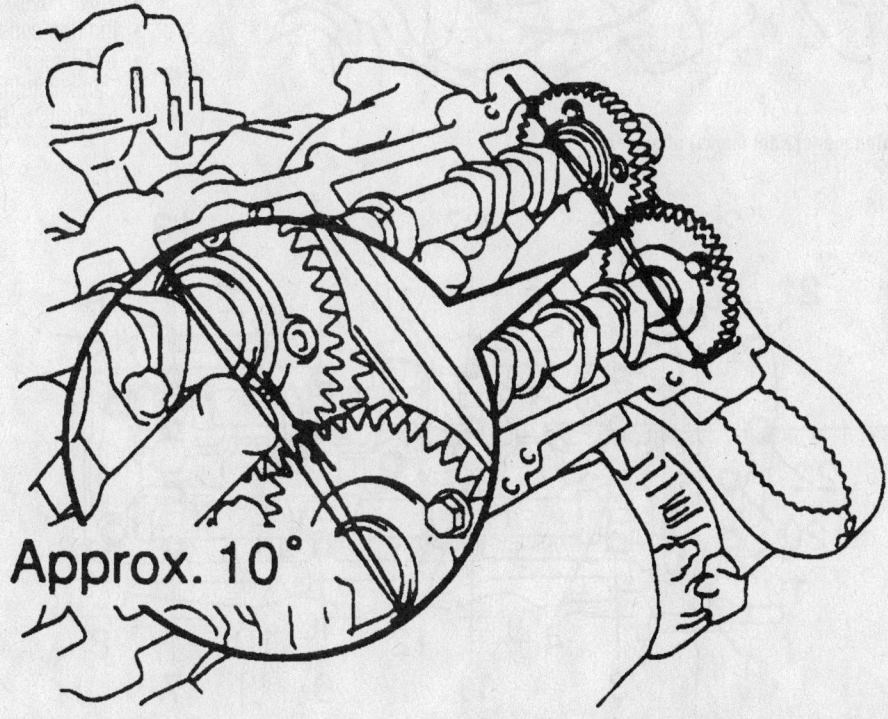

Right bank camshaft timing mark (1 dot marks) alignment

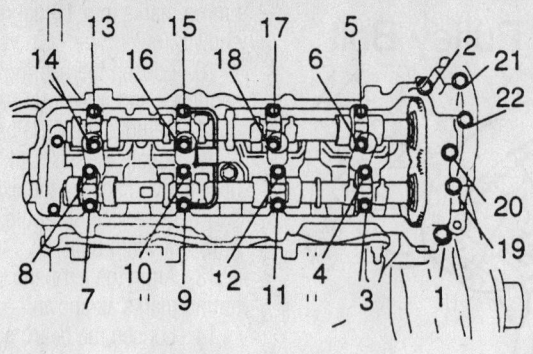

Right bank camshaft bearing cap loosening sequence

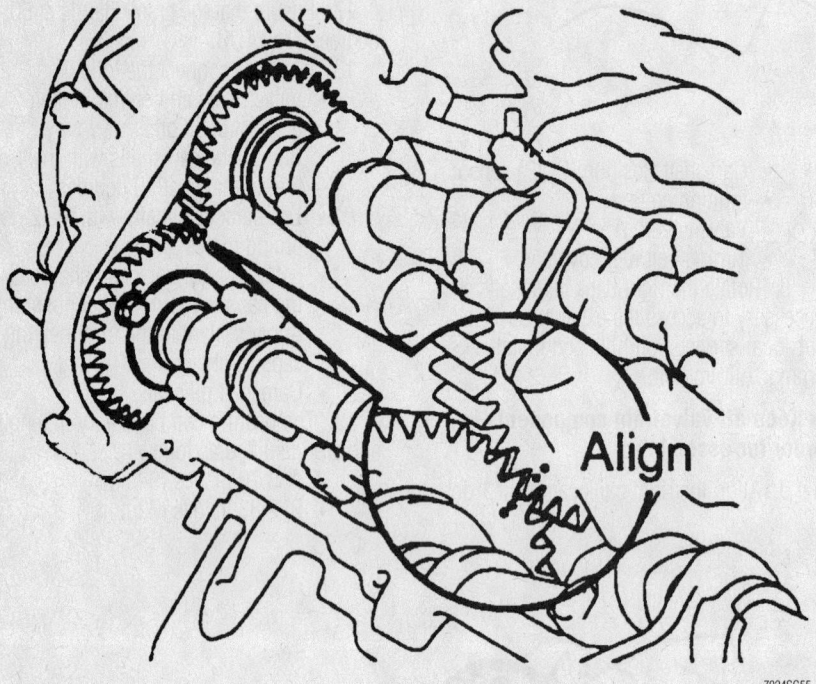

Left bank camshaft timing mark (2 dot marks) alignment

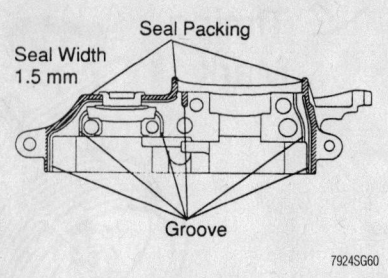

Apply a 1.5mm bead of sealant to the front bearing caps

- C: 0.98 inches (25mm)
- D: 2.05 inches (52mm)
- E: 1.50 inches (38mm)

20. Bolts in positions **A**, **B** and **C** are installed dry.

21. Lubricate the threads and under the contact flange for bolts in positions **D** and **E**.

22. Install oil feed pipes and the bearing cap bolts according to position in the illustrations.

23. Tighten the camshaft bearing bolts in sequence and in several passes to the following specifications:

- Bolt C: 66 inch lbs. (7.5 Nm)
- All others: 12 ft. lbs. (16 Nm)

24. Remove the service bolts from the exhaust camshaft gears.

25. Install or connect the following:

- Timing belt rear covers
- Valve cover
- Ignition coils
- CMP sensor
- Camshaft timing sprockets. Tighten the bolts to 80 ft. lbs. (108 Nm).

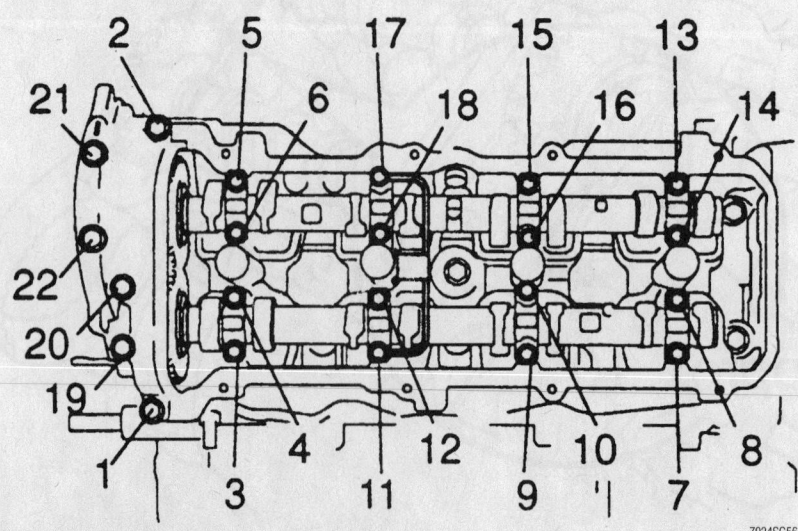

Left bank camshaft bearing cap loosening sequence

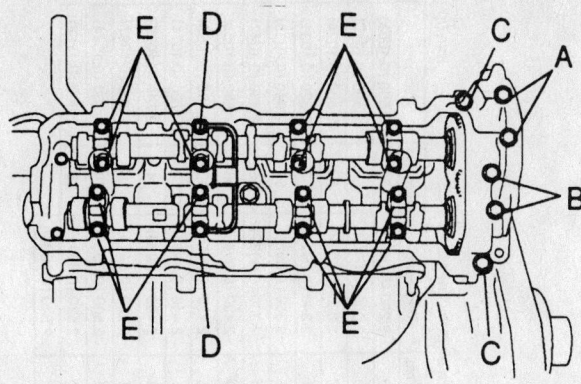

Right bank bearing cap bolt location

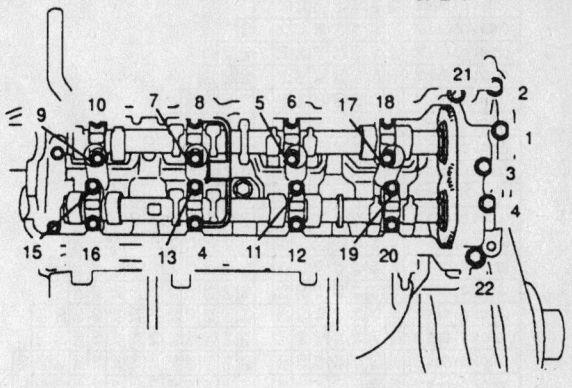

Right bank camshaft bearing cap bolt torque sequence

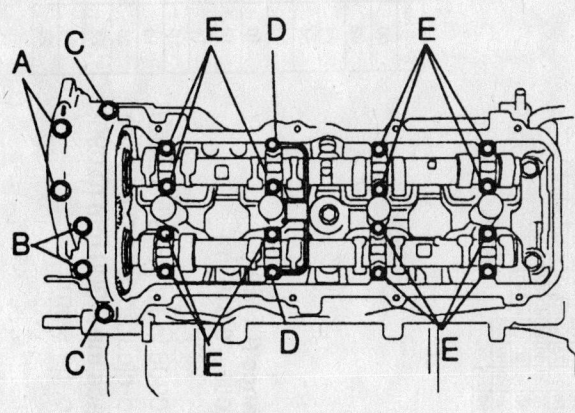

Left camshaft bearing cap bolt locations

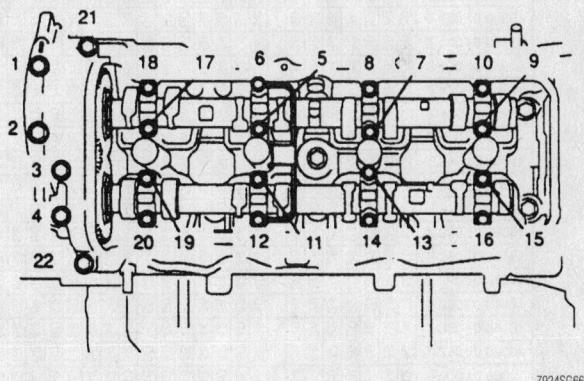

Left bank camshaft bearing cap bolt torque sequence

- Timing belt
- Lower timing cover
- Crankshaft pulley. Tighten the bolt to 181 ft. lbs. (245 Nm).
- Accessory drive belt tensioner
- Alternator
- Cooling fan bracket
- A/C compressor
- Upper and middle timing belt covers
- Idler pulley. Tighten the bolt to 27 ft. lbs. (37 Nm).
- Radiator
- Cooling fan
- Accessory drive belt
- Air intake hose
- Engine appearance cover
- Engine under covers
- Negative battery cable
26. Fill the cooling system.
27. Start the engine and check for leaks.

Valve Lash

ADJUSTMENT

➡**Measure valve clearance with the engine cold.**

1. Before servicing the vehicle, refer to the precautions in the beginning of this section.
2. Drain the cooling system.
3. Remove or disconnect the following:
   - Negative battery cable
   - Ignition coils
   - Valve covers
4. Set the engine to the top of the compression stroke with the valves closed for the cylinder to be measured.
5. Check the valve clearance. The valve clearance specifications are as follows:
   - Intake: 0.006–0.010 in. (0.15–0.25mm)
   - Exhaust: 0.010–0.014 in. (0.25–0.35mm)
6. Record the measurements for each valve.
7. When all valve clearances have been measured, remove the camshafts.
8. Remove the valve shims and measure them. Note this measurement along with the clearance measurement recorded earlier.
9. Using the valve clearance and shim thickness measurements, find replacement shims in the Adjusting Shim Selection charts.
10. Install or connect the following:
    - Replacement valve shims
    - Camshafts
    - Valve covers
    - Ignition coils
    - Negative battery cable
11. Fill the cooling system.
12. Start the engine and check for leaks.

Measured clearance mm (in.) \ Installed shim thickness mm (in.)	2.000 (0.0787)	2.020 (0.0795)	2.040 (0.0803)	2.060 (0.0811)	2.080 (0.0819)	2.100 (0.0827)	2.120 (0.0835)	2.140 (0.0843)	2.160 (0.0850)	2.180 (0.0858)	2.200 (0.0866)	2.210 (0.0870)	2.220 (0.0874)	2.230 (0.0878)	2.240 (0.0882)	2.250 (0.0886)	2.260 (0.0890)	2.270 (0.0894)	2.280 (0.0898)	2.290 (0.0902)	2.300 (0.0906)	2.310 (0.0909)	2.320 (0.0913)	2.330 (0.0917)	2.340 (0.0921)	2.350 (0.0925)	2.360 (0.0929)	2.370 (0.0933)	2.380 (0.0937)	2.390 (0.0941)	2.400 (0.0945)	2.410 (0.0949)	2.420 (0.0953)	2.430 (0.0957)	2.440 (0.0961)	2.450 (0.0965)	2.460 (0.0969)	2.470 (0.0972)	2.480 (0.0976)	2.490 (0.0980)	2.500 (0.0984)	2.510 (0.0988)	2.520 (0.0992)	2.530 (0.0996)	2.540 (0.1000)	2.550 (0.1004)	2.560 (0.1008)	2.570 (0.1012)	2.580 (0.1016)	2.590 (0.1020)	2.600 (0.1024)	2.620 (0.1031)	2.640 (0.1039)	2.660 (0.1047)	2.680 (0.1055)	2.700 (0.1063)	2.720 (0.1071)	2.740 (0.1079)	2.760 (0.1087)	2.780 (0.1094)	2.800 (0.1102)
0.000-0.030 (0.0000-0.0012)							00	00	00	00	02	02	04	04	06	06	08	08	10	10	12	12	14	14	16	16	18	18	20	20	22	22	24	24	26	26	28	28	30	30	32	32	34	34	36	36	38	38	40	40	42	44	46	48	50	52	54	56	58	60	62
0.031-0.050 (0.0012-0.0020)						00	00	00	00	02	04	06	06	08	08	10	10	12	12	14	14	16	16	18	18	20	20	22	22	24	24	26	26	28	28	30	30	32	32	34	34	36	36	38	38	40	40	42	42	44	44	46	48	50	52	54	56	58	60	62	64
0.051-0.070 (0.0020-0.0028)					00	00	00	00	02	04	06	08	08	10	10	12	12	14	14	16	16	18	18	20	20	22	22	24	24	26	26	28	28	30	30	32	32	34	34	36	36	38	38	40	40	42	42	44	44	46	46	48	50	52	54	56	58	60	62	64	66
0.071-0.090 (0.0028-0.0035)				00	00	00	00	02	04	06	08	10	10	12	12	14	14	16	16	18	18	20	20	22	22	24	24	26	26	28	28	30	30	32	32	34	34	36	36	38	38	40	40	42	42	44	44	46	46	48	48	50	52	54	56	58	60	62	64	66	68
0.091-0.110 (0.0036-0.0043)			00	00	00	00	02	04	06	08	10	12	12	14	14	16	16	18	18	20	20	22	22	24	24	26	26	28	28	30	30	32	32	34	34	36	36	38	38	40	40	42	42	44	44	46	46	48	48	50	50	52	54	56	58	60	62	64	66	68	70
0.111-0.130 (0.0044-0.0051)		00	00	00	00	02	04	06	08	10	12	14	14	16	16	18	18	20	20	22	22	24	24	26	26	28	28	30	30	32	32	34	34	36	36	38	38	40	40	42	42	44	44	46	46	48	48	50	50	52	52	54	56	58	60	62	64	66	68	70	72
0.131-0.149 (0.0052-0.0059)		00	00	00	02	04	06	08	10	12	14	16	16	18	18	20	20	22	22	24	24	26	26	28	28	30	30	32	32	34	34	36	36	38	38	40	40	42	42	44	44	46	46	48	48	50	50	52	52	54	54	56	58	60	62	64	66	68	70	72	74
0.150-0.250 (0.0059-0.0098)																																																													
0.251-0.270 (0.0099-0.0106)	06	08	10	12	14	16	18	20	22	24	26	28	28	30	30	32	32	34	34	36	36	38	38	40	40	42	42	44	44	46	46	48	48	50	50	52	52	54	54	56	56	58	58	60	60	62	62	64	64	66	66	68	70	72	74	76	78	80	80	80	
0.271-0.290 (0.0107-0.0114)	08	10	12	14	16	18	20	22	24	26	28	30	30	32	32	34	34	36	36	38	38	40	40	42	42	44	44	46	46	48	48	50	50	52	52	54	54	56	56	58	58	60	60	62	62	64	64	66	66	68	68	70	72	74	76	78	80	80	80		
0.291-0.310 (0.0115-0.0122)	10	12	14	16	18	20	22	24	26	28	30	32	32	34	34	36	36	38	38	40	40	42	42	44	44	46	46	48	48	50	50	52	52	54	54	56	56	58	58	60	60	62	62	64	64	66	66	68	68	70	70	72	74	76	78	80	80	80			
0.311-0.330 (0.0122-0.0130)	12	14	16	18	20	22	24	26	28	30	32	34	34	36	36	38	38	40	40	42	42	44	44	46	46	48	48	50	50	52	52	54	54	56	56	58	58	60	60	62	62	64	64	66	66	68	68	70	70	72	72	74	76	78	80	80	80				
0.331-0.350 (0.0130-0.0138)	14	16	18	20	22	24	26	28	30	32	34	36	36	38	38	40	40	42	42	44	44	46	46	48	48	50	50	52	52	54	54	56	56	58	58	60	60	62	62	64	64	66	66	68	68	70	70	72	72	74	74	76	78	80	80	80					
0.351-0.370 (0.0138-0.0146)	16	18	20	22	24	26	28	30	32	34	36	38	38	40	40	42	42	44	44	46	46	48	48	50	50	52	52	54	54	56	56	58	58	60	60	62	62	64	64	66	66	68	68	70	70	72	72	74	74	76	76	78	80	80	80						
0.371-0.390 (0.0146-0.0154)	18	20	22	24	26	28	30	32	34	36	38	40	40	42	42	44	44	46	46	48	48	50	50	52	52	54	54	56	56	58	58	60	60	62	62	64	64	66	66	68	68	70	70	72	72	74	74	76	76	78	78	80	80	80							
0.391-0.410 (0.0154-0.0161)	20	22	24	26	28	30	32	34	36	38	40	42	42	44	44	46	46	48	48	50	50	52	52	54	54	56	56	58	58	60	60	62	62	64	64	66	66	68	68	70	70	72	72	74	74	76	76	78	78	80	80	80	80								
0.411-0.430 (0.0162-0.0169)	22	24	26	28	30	32	34	36	38	40	42	44	44	46	46	48	48	50	50	52	52	54	54	56	56	58	58	60	60	62	62	64	64	66	66	68	68	70	70	72	72	74	74	76	76	78	78	80	80	80	80	80									
0.431-0.450 (0.0170-0.0177)	24	26	28	30	32	34	36	38	40	42	44	46	46	48	48	50	50	52	52	54	54	56	56	58	58	60	60	62	62	64	64	66	66	68	68	70	70	72	72	74	74	76	76	78	78	80	80	80	80	80	80										
0.451-0.470 (0.0178-0.0185)	26	28	30	32	34	36	38	40	42	44	46	48	48	50	50	52	52	54	54	56	56	58	58	60	60	62	62	64	64	66	66	68	68	70	70	72	72	74	74	76	76	78	78	80	80	80	80	80	80	80											
0.471-0.490 (0.0185-0.0193)	28	30	32	34	36	38	40	42	44	46	48	50	50	52	52	54	54	56	56	58	58	60	60	62	62	64	64	66	66	68	68	70	70	72	72	74	74	76	76	78	78	80	80	80	80	80	80	80													
0.491-0.510 (0.0193-0.0201)	30	32	34	36	38	40	42	44	46	48	50	52	52	54	54	56	56	58	58	60	60	62	62	64	64	66	66	68	68	70	70	72	72	74	74	76	76	78	78	80	80	80	80	80	80	80															
0.511-0.530 (0.0201-0.0209)	32	34	36	38	40	42	44	46	48	50	52	54	54	56	56	58	58	60	60	62	62	64	64	66	66	68	68	70	70	72	72	74	74	76	76	78	78	80	80	80	80	80	80	80																	
0.531-0.550 (0.0209-0.0217)	34	36	38	40	42	44	46	48	50	52	54	56	56	58	58	60	60	62	62	64	64	66	66	68	68	70	70	72	72	74	74	76	76	78	78	80	80	80	80	80	80	80																			
0.551-0.570 (0.0217-0.0224)	36	38	40	42	44	46	48	50	52	54	56	58	58	60	60	62	62	64	64	66	66	68	68	70	70	72	72	74	74	76	76	78	78	80	80	80	80	80	80	80																					
0.571-0.590 (0.0225-0.0232)	38	40	42	44	46	48	50	52	54	56	58	60	60	62	62	64	64	66	66	68	68	70	70	72	72	74	74	76	76	78	78	80	80	80	80	80	80	80																							
0.591-0.610 (0.0233-0.0240)	40	42	44	46	48	50	52	54	56	58	60	62	62	64	64	66	66	68	68	70	70	72	72	74	74	76	76	78	78	80	80	80	80	80	80	80																									
0.611-0.630 (0.0241-0.0248)	42	44	46	48	50	52	54	56	58	60	62	64	64	66	66	68	68	70	70	72	72	74	74	76	76	78	78	80	80	80	80	80	80	80																											
0.631-0.650 (0.0248-0.0256)	44	46	48	50	52	54	56	58	60	62	64	66	66	68	68	70	70	72	72	74	74	76	76	78	78	80	80	80	80	80	80	80																													
0.651-0.670 (0.0256-0.0264)	46	48	50	52	54	56	58	60	62	64	66	68	68	70	70	72	72	74	74	76	76	78	78	80	80	80	80	80	80	80																															
0.671-0.690 (0.0264-0.0272)	48	50	52	54	56	58	60	62	64	66	68	70	70	72	72	74	74	76	76	78	78	80	80	80	80	80	80	80																																	
0.691-0.710 (0.0272-0.0280)	50	52	54	56	58	60	62	64	66	68	70	72	72	74	74	76	76	78	78	80	80	80	80	80	80	80																																			
0.711-0.730 (0.0280-0.0287)	52	54	56	58	60	62	64	66	68	70	72	74	74	76	76	78	78	80	80	80	80	80	80	80																																					
0.731-0.750 (0.0288-0.0295)	54	56	58	60	62	64	66	68	70	72	74	76	76	78	78	80	80	80	80	80	80	80																																							
0.751-0.770 (0.0296-0.0303)	56	58	60	62	64	66	68	70	72	74	76	78	78	80	80	80	80	80	80	80																																									
0.771-0.790 (0.0304-0.0311)	58	60	62	64	66	68	70	72	74	76	78	80	80	80	80	80	80	80																																											
0.791-0.810 (0.0311-0.0319)	60	62	64	66	68	70	72	74	76	78	80	80	80	80	80	80																																													
0.811-0.830 (0.0319-0.0327)	62	64	66	68	70	72	74	76	78	80	80	80	80	80																																															
0.831-0.850 (0.0327-0.0335)	64	66	68	70	72	74	76	78	80	80	80	80																																																	
0.851-0.870 (0.0335-0.0343)	66	68	70	72	74	76	78	80	80	80																																																				
0.871-0.890 (0.0343-0.0350)	68	70	72	74	76	78	80	80	80																																																					
0.891-0.910 (0.0351-0.0358)	70	72	74	76	78	80	80	80																																																					
0.911-0.930 (0.0359-0.0366)	72	74	76	78	80	80	80																																																						
0.931-0.950 (0.0367-0.0374)	74	76	78	80	80	80																																																							
0.951-0.970 (0.0374-0.0382)	76	78	80	80	80																																																								
0.971-0.990 (0.0382-0.0390)	78	80	80	80																																																									
0.991-1.010 (0.0390-0.0398)	80	80	80																																																									
1.011-1.030 (0.0398-0.0406)	80	80																																																											
1.031-1.050 (0.0406-0.0413)	80																																																												

Intake valve clearance (Cold):
0.15 – 0.25 mm (0.006 – 0.010 in.)

EXAMPLE:
The 2.300 mm (0.0906 in.) shim is installed, and the measured clearance is 0.440 mm (0.0173 in.). Replace the 2.300 mm (0.0906 in.) shim with a No. 54 shim.

New shim thickness mm (in.)

Shim No.	Thickness	Shim No.	Thickness	Shim No.	Thickness
00	2.000 (0.0787)	28	2.280 (0.0898)	56	2.560 (0.1008)
02	2.020 (0.0795)	30	2.300 (0.0906)	58	2.580 (0.1016)
04	2.040 (0.0803)	32	2.320 (0.0913)	60	2.600 (0.1024)
06	2.060 (0.0811)	34	2.340 (0.0921)	62	2.620 (0.1031)
08	2.080 (0.0819)	36	2.360 (0.0929)	64	2.640 (0.1039)
10	2.100 (0.0827)	38	2.380 (0.0937)	66	2.660 (0.1047)
12	2.120 (0.0835)	40	2.400 (0.0945)	68	2.680 (0.1055)
14	2.140 (0.0843)	42	2.420 (0.0953)	70	2.700 (0.1063)
16	2.160 (0.0850)	44	2.440 (0.0961)	72	2.720 (0.1071)
18	2.180 (0.0858)	46	2.460 (0.0969)	74	2.740 (0.1079)
20	2.200 (0.0866)	48	2.480 (0.0976)	76	2.760 (0.1087)
22	2.220 (0.0874)	50	2.500 (0.0984)	78	2.780 (0.1094)
24	2.240 (0.0882)	52	2.520 (0.0992)	80	2.800 (0.1102)
26	2.260 (0.0890)	54	2.540 (0.1000)		

7924SG71

Intake valve clearance shim selection chart

New shim thickness

mm (in.)

Shim No.	Thickness	Shim No.	Thickness	Shim No.	Thickness
00	2.000 (0.0787)	28	2.280 (0.0898)	56	2.560 (0.1008)
02	2.020 (0.0795)	30	2.300 (0.0906)	58	2.580 (0.1016)
04	2.040 (0.0803)	32	2.320 (0.0913)	60	2.600 (0.1024)
06	2.060 (0.0811)	34	2.340 (0.0921)	62	2.620 (0.1031)
08	2.080 (0.0819)	36	2.360 (0.0929)	64	2.640 (0.1039)
10	2.100 (0.0827)	38	2.380 (0.0937)	66	2.660 (0.1047)
12	2.120 (0.0835)	40	2.400 (0.0945)	68	2.680 (0.1055)
14	2.140 (0.0843)	42	2.420 (0.0953)	70	2.700 (0.1063)
16	2.160 (0.0850)	44	2.440 (0.0961)	72	2.720 (0.1071)
18	2.180 (0.0858)	46	2.460 (0.0969)	74	2.740 (0.1079)
20	2.200 (0.0866)	48	2.480 (0.0976)	76	2.760 (0.1087)
22	2.220 (0.0874)	50	2.500 (0.0984)	78	2.780 (0.1094)
24	2.240 (0.0882)	52	2.520 (0.0992)	80	2.800 (0.1102)
26	2.260 (0.0890)	54	2.540 (0.1000)		

Exhaust valve clearance (Cold):
0.25 – 0.35 mm (0.010 – 0.014 in.)

EXAMPLE:
The 2.300 mm (0.0906 in.) shim is installed, and the measured clearance is 0.440 mm (0.0173 in.). Replace the 2.300 mm (0.0906 in.) shim with a No. 44 shim.

Exhaust valve clearance shim selection chart

7924SG72

Starter Motor

REMOVAL & INSTALLATION

1. Before servicing the vehicle, refer to the precautions in the beginning of this section.
2. Drain the cooling system.
3. Relieve the fuel system pressure.
4. Remove or disconnect the following:
   - Negative battery cable
   - Engine appearance cover
   - Air intake tube
   - Intake manifold
   - Starter motor mounting bolts
   - Starter wiring connectors
   - Starter motor

To install:

5. Install or connect the following:
   - Starter motor
   - Starter wiring connectors. Tighten the cable nut to 86 inch lbs. (10 Nm).
   - Starter motor mounting bolts. Tighten the bolts to 29 ft. lbs. (39 Nm).
   - Intake manifold
   - Air intake tube
   - Engine appearance cover
   - Negative battery cable
6. Fill the cooling system.
7. Start the engine and check for leaks.

Timing Belt

REMOVAL & INSTALLATION

1. Disconnect the negative battery cable.
2. Raise and safely support the vehicle.
3. Remove the oil pan protector and the engine under cover.
4. Drain the cooling system and store the coolant for refilling purposes.
5. Lower the vehicle and remove the battery clamp cover.
6. From the top of the engine, remove the fuel return hose, the engine cover nuts/bolts and the cover.
7. Remove the air cleaner and the intake air connector assembly.
8. Remove the cooling fan pulley by performing the following procedures:
   a. Loosen the 4 fan clutch-to-fan pulley nuts.
   b. Using a box-end wrench on the serpentine drive belt tensioner bolt, rotate the tensioner counterclockwise and remove the drive belt.

➡ **The serpentine drive belt tensioner bolt is a left-hand thread.**

   c. Remove the fan clutch-to-fan pulley nuts, the fan, the clutch assembly and the fan pulley.
9. Remove the radiator by performing the following procedures:
   a. Disconnect the upper, lower and reservoir hoses from the radiator.
   b. Disconnect and plug the automatic transmission oil cooler at the radiator. Disconnect the automatic transmission oil cooler hoses from the fan shroud clamp.
   c. Remove the radiator reservoir tank.
   d. Remove the fan shroud-to-radiator bolts and the shroud.
   e. Remove the 2 upper radiator-to-chassis nuts.
   f. Remove the middle radiator-to-chassis nut/bolts and brackets.
   g. Carefully, lift the radiator from the vehicle.
10. Remove the serpentine drive belt idler pulley bolt, cover plate and pulley.
11. Remove the right side (No. 3) timing belt cover.
12. Remove the left side (No. 3) timing belt cover by performing the following procedures:
   a. Disconnect the engine wire from both wire clamps.
   b. Disconnect the camshaft position sensor wire from the wire clamp on the left-side (No.3) timing belt cover.
   c. Disconnect the sensor connector from the connector bracket.
   d. Disconnect the sensor connector.
   e. Remove the wire grommet from the left-side (No. 3) timing belt cover.
   f. Remove the oil cooler tube bolts and tube.
13. Remove the middle (No. 2) timing belt cover bolts and cover.
14. Remove the cooling fan bracket nuts/bolts and bracket.

➡ **If reusing the timing belt, make sure that there are 3 installation marks on the belt; if there are none, install them.**

15. Using the Crankshaft Pulley Holding tool 09213-70010, Bolt tool 90105-08076 and Companion Flange Holding tool 09330-00021, or equivalent, loosen the crankshaft pulley bolt.
16. Position the No. 1 cylinder to approximately 50 degrees After Top Dead Center (ATDC) of the compression stroke by performing the following procedures:
   a. Rotate the crankshaft pulley (CLOCKWISE) to align its groove with the timing mark "0" on the lower (No. 1) timing belt cover.
   b. Check that the camshaft sprocket timing marks are aligned with the rear timing belt plate marks; if not, rotate the crankshaft 1 revolution (360 degrees).
   c. Rotate the crankshaft pulley approximately 50 degrees (CLOCKWISE) and align the crankshaft pulley timing mark between the centers of the crankshaft pulley bolt and the idler pulley bolt.

✳✳ WARNING

If the timing belt is disengaged, having the crankshaft pulley in the wrong angle can cause the valve to come into contact with the piston when removing the camshaft pulley.

17. Remove the crankshaft pulley bolt.

➡ **If reusing the timing belt and the installation marks have disappeared, place new installation marks on the timing belt to match the camshaft timing sprocket marks.**

➡ **To avoid meshing the timing sprocket and the timing belt, secure one with a string; then, place matchmarks on the timing belt and the right-side camshaft timing sprocket.**

18. Remove the timing belt tensioner bolts and the tensioner.
19. Using the Camshaft Holding tool 09960-10010, or equivalent, slightly turn the left-side camshaft sprocket clockwise to loosen the tension spring. Then, disconnect the timing belt from the camshaft sprockets.
20. Remove the alternator by performing the following procedures:
   a. Disconnect the electrical connector from the alternator.
   b. Remove the rubber cap/nut and disconnect the battery wire from the alternator.
   c. Disconnect the wire clamp from the alternator cord clip.
   d. Remove the alternator-to-engine nuts/bolts and the alternator.
21. Remove the serpentine drive belt tensioner nuts/bolts and the tensioner.
22. Using the Crankshaft Puller Assembly tool 09950-50012, or equivalent, press the crankshaft pulley from the crankshaft.

✳✳ WARNING

DO NOT rotate the crankshaft pulley.

23. Remove the lower (No. 1) timing belt cover bolts and the cover.

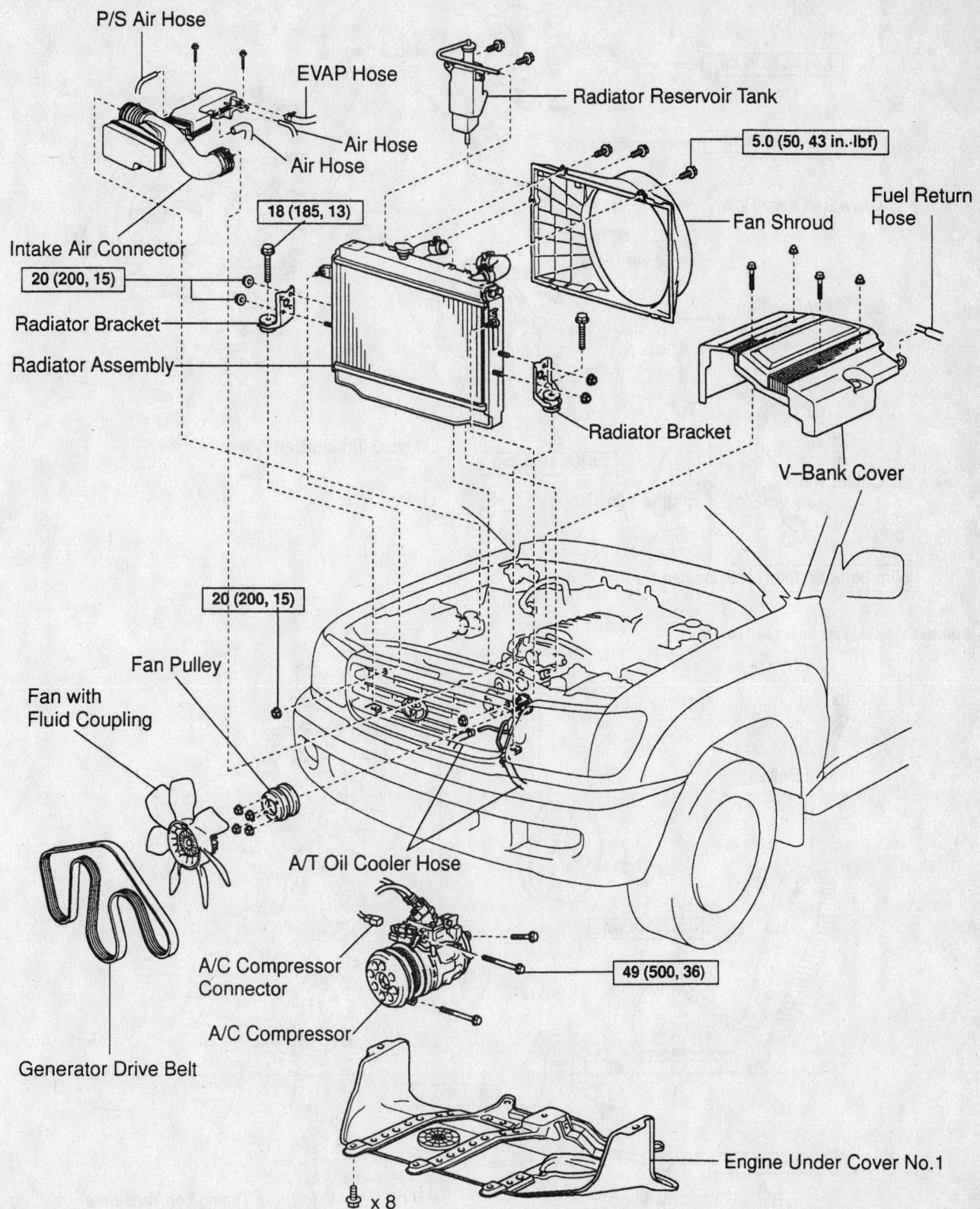

P/S Air Hose

EVAP Hose

Air Hose

Air Hose

Radiator Reservoir Tank

5.0 (50, 43 in.·lbf)

Fan Shroud

Fuel Return Hose

18 (185, 13)

Intake Air Connector

20 (200, 15)

Radiator Bracket

Radiator Assembly

Radiator Bracket

V–Bank Cover

20 (200, 15)

Fan Pulley

Fan with Fluid Coupling

A/T Oil Cooler Hose

A/C Compressor Connector

A/C Compressor

49 (500, 36)

Generator Drive Belt

x 8

Engine Under Cover No.1

93025G24

Exploded view of vehicle components for timing belt replacement

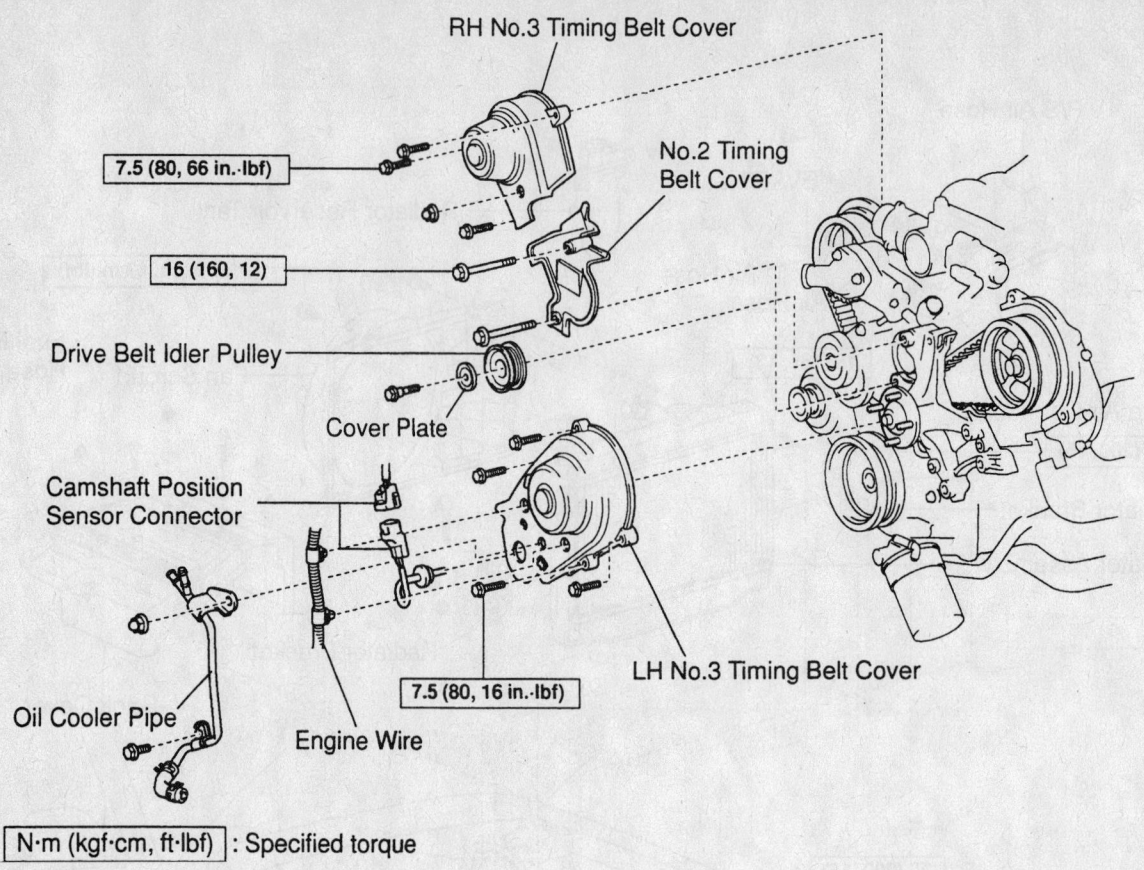

7.5 (80, 66 in.·lbf)

16 (160, 12)

RH No.3 Timing Belt Cover

No.2 Timing Belt Cover

Drive Belt Idler Pulley

Cover Plate

Camshaft Position Sensor Connector

Oil Cooler Pipe

Engine Wire

7.5 (80, 16 in.·lbf)

LH No.3 Timing Belt Cover

N·m (kgf·cm, ft·lbf) : Specified torque

93025G25

Exploded view of upper timing belt covers

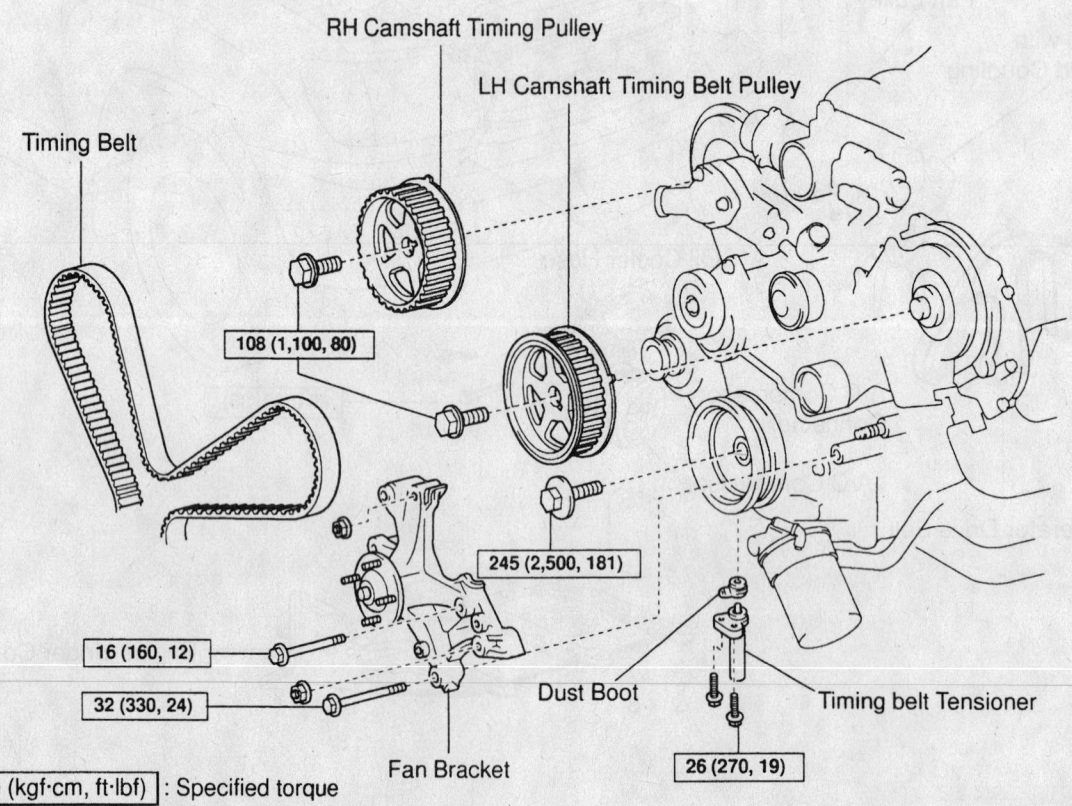

Timing Belt

RH Camshaft Timing Pulley

LH Camshaft Timing Belt Pulley

108 (1,100, 80)

245 (2,500, 181)

16 (160, 12)

32 (330, 24)

Fan Bracket

Dust Boot

Timing belt Tensioner

26 (270, 19)

N·m (kgf·cm, ft·lbf) : Specified torque

93025G26

Exploded view of upper timing sprockets and components

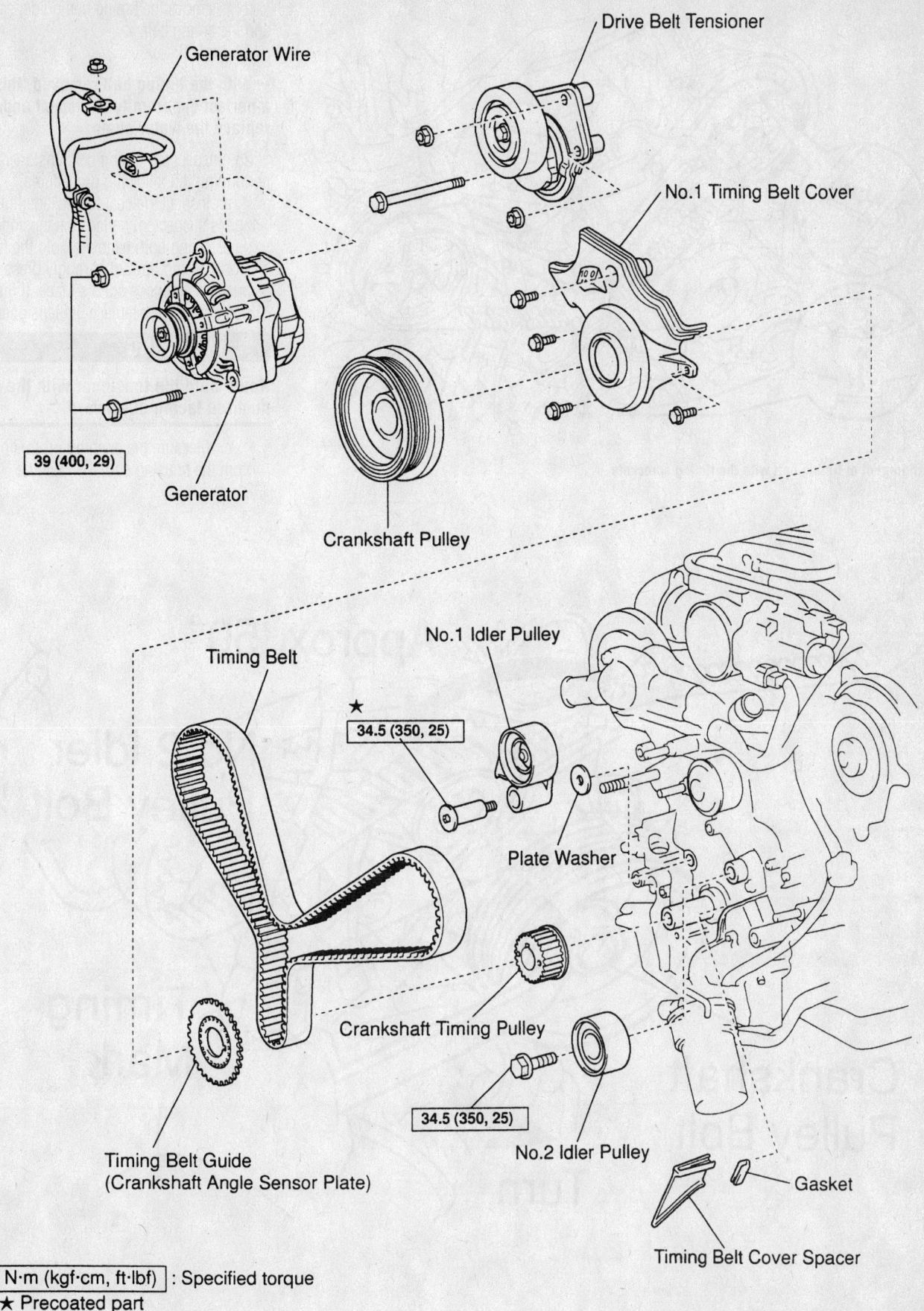

Generator Wire

Drive Belt Tensioner

No.1 Timing Belt Cover

39 (400, 29)

Generator

Crankshaft Pulley

Timing Belt

No.1 Idler Pulley

★ 34.5 (350, 25)

Plate Washer

Crankshaft Timing Pulley

Timing Belt Guide
(Crankshaft Angle Sensor Plate)

34.5 (350, 25)

No.2 Idler Pulley

Gasket

Timing Belt Cover Spacer

N·m (kgf·cm, ft·lbf) : Specified torque
★ Precoated part

Exploded view of lower timing belt cover, sprockets and components

93025G27

Alignment of timing belt with the timing sprockets

93025G28

24. Remove the timing belt guide, spacer and the timing belt.

To install:

➡ **With the timing belt removed, this is a perfect opportunity to inspect and/or replace the water pump.**

25. Inspect the timing belt tensioner by performing the following procedures:

    a. Inspect the seal for leakage; if leakage is suspected, replace the tensioner.

    b. Using both hands to hold the tensioner facing upward, strongly press the pushrod against a solid surface. If the pushrod moves, replace the tensioner.

✳✳ WARNING

Never hold the tensioner with the pushrod facing downward.

    c. Measure the pushrod protrusion from the housing end, it should be

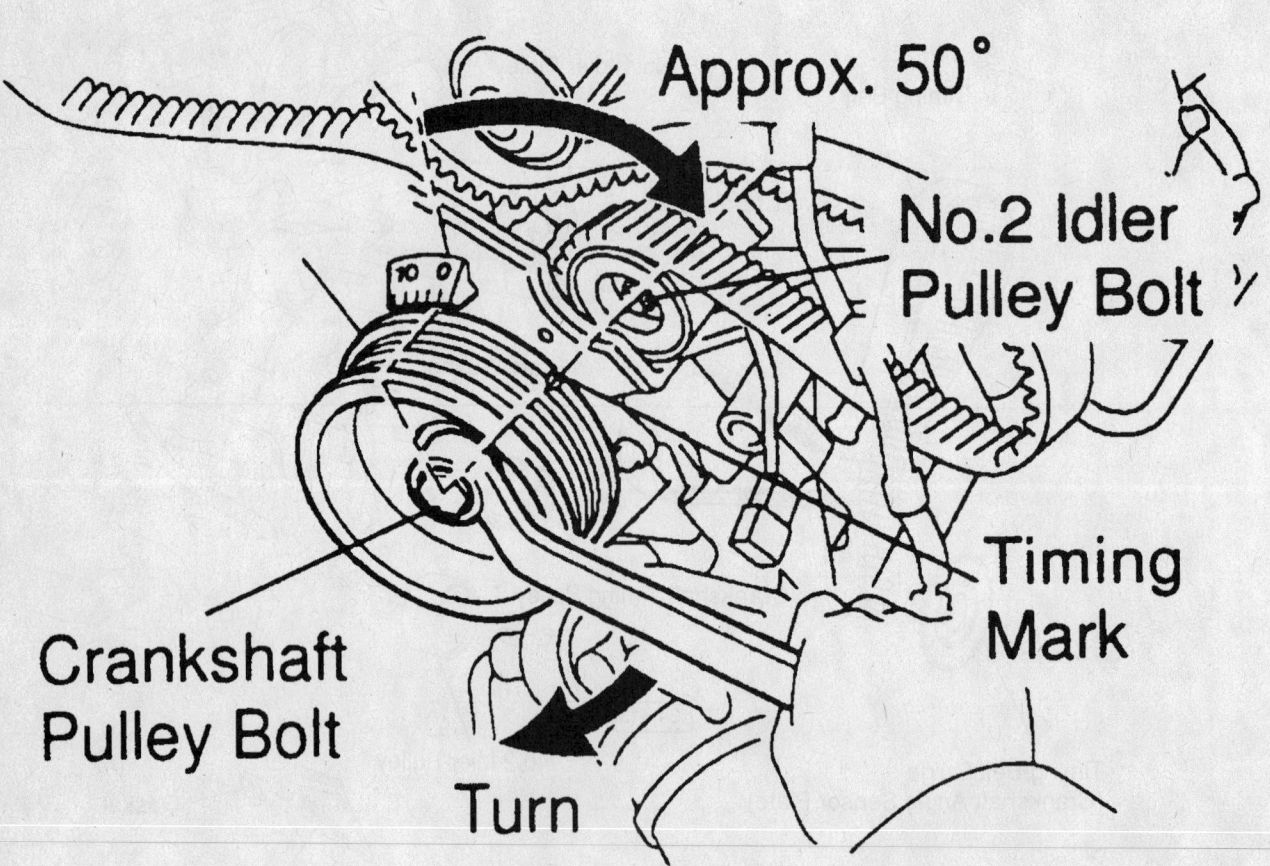

93025G29

Aligning of crankshaft pulley timing mark with the center line of the crankshaft pulley bolt and the idler pulley bolt

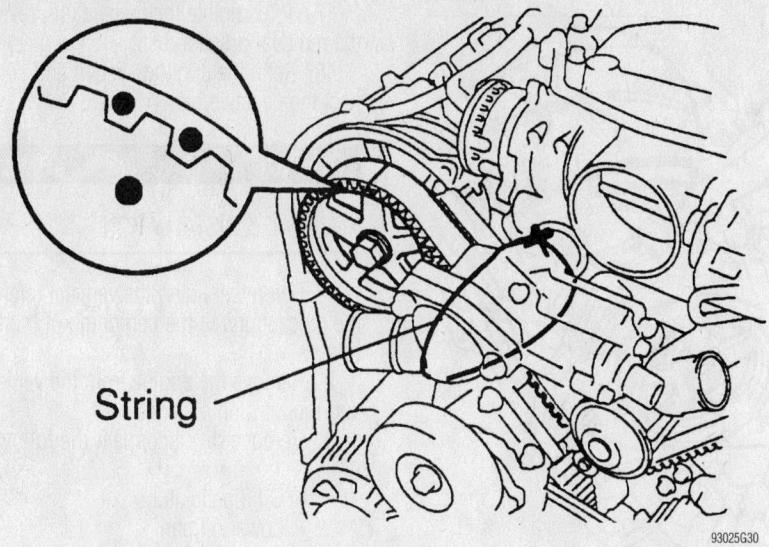

Securing the timing belt with string and matchmarking the camshaft with the timing belt

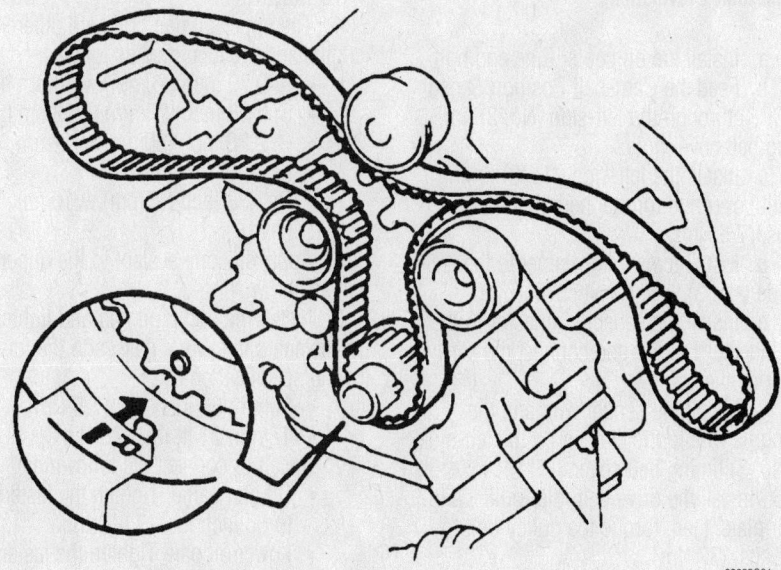

Installing the timing belt on the crankshaft sprocket

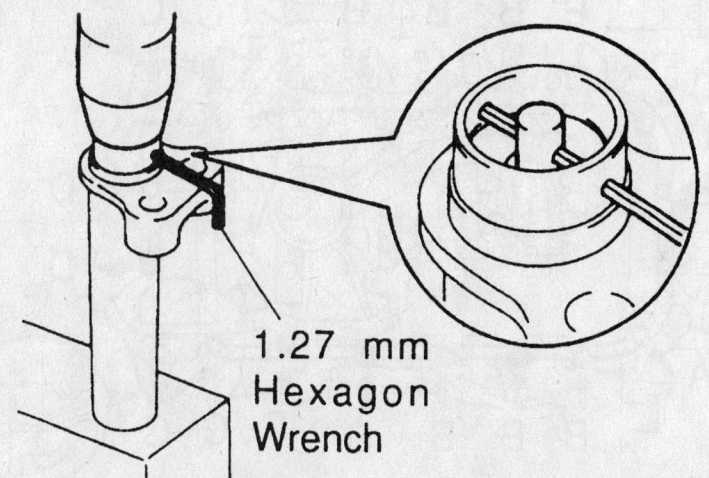

Securing the timing belt tensioner pushrod

0.413–0.453 in. (10.5–11.5mm). If the protrusion is not as specified, replace the tensioner.

26. Temporarily install the timing belt by performing the following procedures:

    a. Align the timing belt's installation mark with the crankshaft timing sprocket.

    b. Install the timing belt on the crankshaft timing sprocket, the No. 1 idler pulley and the No. 2 idler pulley.

27. Install the gasket to the timing belt cover spacer and install the cover spacer.

28. Install the timing belt guide with the cup side facing outward.

29. Install the lower (No. 1) timing belt cover.

30. Install the crankshaft pulley by performing the following procedures:

    a. Align the crankshaft pulley with the crankshaft key.

    b. Using the Crankshaft Installer tool 09223-46011, or equivalent, and a hammer, tap the crankshaft pulley into position.

31. Install the serpentine drive belt tensioner and torque the tensioner-to-engine bolts to 12 ft. lbs. (16 Nm).

➡**To install the serpentine drive belt tensioner, use a bolt 4.18 in. (106mm) in length.**

32. Check that the crankshaft pulley's timing mark is aligned with the centers of the idler pulley and crankshaft pulley bolts.

33. Install the alternator and torque the alternator-to-engine nuts/bolts to 29 ft. lbs. (39 Nm). Connect the alternator's electrical connectors and clip.

34. Install the timing belt to the left-side camshaft by performing the following procedures:

    a. Rotate the left-side camshaft pulley to align the timing belt installation mark with the camshaft sprocket's timing mark and slide the belt onto the camshaft timing sprocket.

    b. Using the Camshaft Holding tool 09960-10010, or equivalent, slightly turn the left-side camshaft sprocket counterclockwise to place tension on the timing belt between the crankshaft sprocket and the camshaft sprocket.

35. Rotate the right-side camshaft pulley to align the timing belt installation mark with the camshaft sprocket's timing mark and slide the belt onto the camshaft timing sprocket.

36. Using a vertical press, slowly press the pushrod into the housing using 200–2205 lbs. (981–9807 N) until the holes align, then, install a 1.27mm Allen® wrench to secure the pushrod and release

93025G33

Checking the TDC alignment marks after rotating the crankshaft 2 revolutions

the press. Install the dust boot on the tensioner housing.

37. Install the timing belt tensioner and torque the bolts to 19 ft. lbs. (26 Nm).

38. Using a pair of pliers, remove the Allen® wrench from the tensioner housing.

39. Check the valve timing by performing the following procedure:

a. Temporarily install the crankshaft pulley bolt.

b. Slowly, rotate the crankshaft pulley 2 revolutions (CLOCKWISE) and realign the TDC marks.

➡ **If the pulley/sprocket timing marks do not realign, remove the timing belt and reinstall it.**

40. Using the Crankshaft Pulley Holding tool 09213-70010, Bolt tool 90105-08076 and Companion Flange Holding tool 09330-00021, or equivalent, torque the crankshaft pulley bolt to 181 ft. lbs. (245 Nm).

41. Install the cooling fan bracket and torque the 12mm (head size) bolt to 12 ft. lbs. (16 Nm) and the 14mm (head size) bolt to 24 ft. lbs. (32 Nm).

42. Install the air conditioning compressor.

43. Install the middle (No. 2) timing belt cover and torque the bolts to 12 ft. lbs. (16 Nm).

44. Install the upper right-side (No. 3) timing belt cover and torque the bolts to 66 inch lbs. (7.5 Nm).

45. Install the upper left-side (No. 3) timing belt cover by performing the following procedures:

a. Install the oil cooler tube and bolt.

b. Feed the Camshaft Position Sensor (CPS) through the left-side (No. 3) timing belt cover hole.

c. Install the left-side (No. 3) timing belt cover and torque the bolts to 66 inch lbs. (7.5 Nm).

d. Install the wire grommet to the left-side (No. 3) timing belt cover.

e. Install the sensor connector to the connector bracket and connect the sensor connector.

f. Install the sensor wire and the engine wire to the clamps on the left-side (No. 3) timing belt cover.

46. Install the drive belt idler pulley and cover plate; then, torque the pulley bolt to 27 ft. lbs. (37 Nm).

47. To complete the installation, reverse the removal procedures.

48. Refill the cooling system and connect the negative battery cable.

Oil Pan

REMOVAL & INSTALLATION

1. Before servicing the vehicle, refer to the precautions in the beginning of this section.

2. Remove the engine from the vehicle and mount it on a stand.

3. Remove or disconnect the following:

- Oil dipstick tube
- Lower oil pan
- Oil pan baffle
- Upper oil pan

To install:

4. The upper oil pan bolts are different lengths and are identified as follows:

- A: 0.79 inch (20mm) w/10mm head
- B: 0.98 inch (25mm) w/12mm head
- C: 2.36 inch (60mm) w/12mm head
- D: 1.38 inch (35mm) w/10mm head

5. Apply silicone sealant to the upper oil pan as shown.

6. Install the upper oil pan and tighten the fasteners in several passes to the following specifications:

- 10mm: 66 inch lbs. (7.5 Nm)
- 12mm: 21 ft. lbs. (28 Nm)

7. Install or connect the following:

- Oil pan baffle. Tighten the fasteners to 66 inch lbs. (7.5 Nm).
- Lower oil pan. Tighten the fasteners

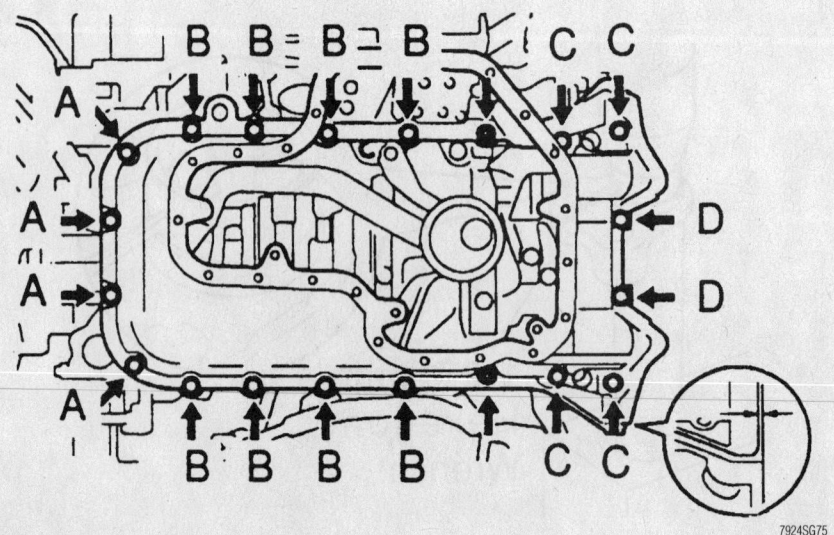

7924SG75

Upper oil pan bolt location

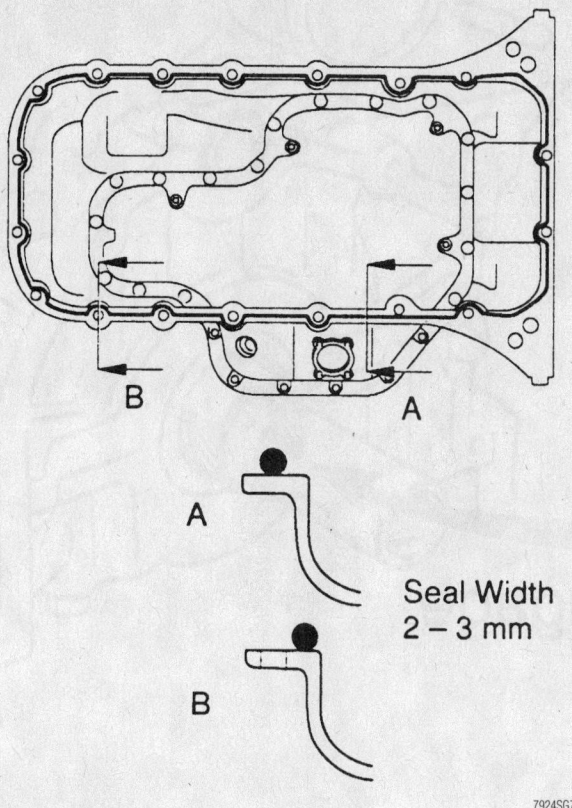

Seal Width
2 – 3 mm

7924SG74

Upper oil pan sealant application

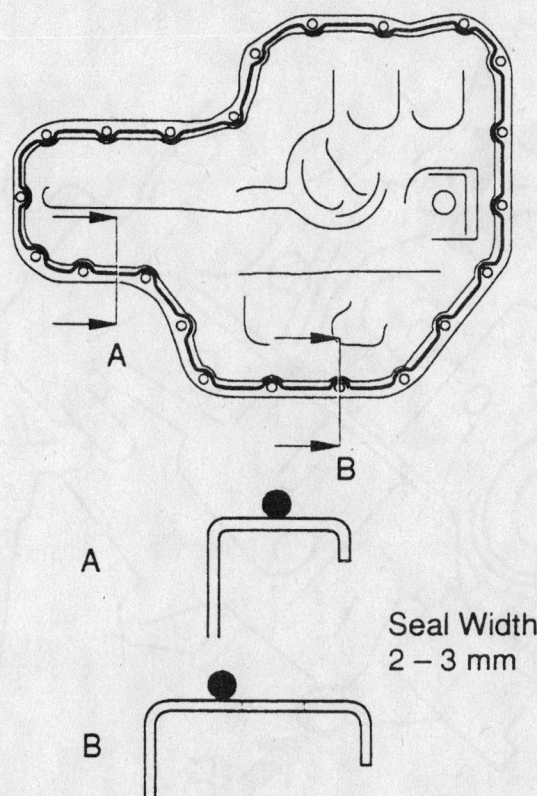

Seal Width
2 – 3 mm

7924SG76

Lower oil pan sealant application

in several passes to 66 inch lbs.
(7.5 Nm).
- Oil dipstick tube
8. Install the engine.

Oil Pump

REMOVAL & INSTALLATION

1. Before servicing the vehicle, refer to the precautions in the beginning of this section.

2. Remove the engine from the vehicle and mount it on a stand.

3. Remove or disconnect the following:

- Front cover
- Timing belt
- Timing belt idler pulleys
- Crankshaft timing sprocket
- Oil dipstick tube
- Oil filter and bracket
- Crankshaft Position (CKP) sensor
- Oil pan and baffle
- Oil pump pickup tube
- Oil pump

To install:

4. The upper oil pan bolts are different lengths and are identified as follows:

- A: 1.38 inch (35mm) w/12mm head
- B: 1.97 inch (50mm) w/12mm head
- C: 4.17 inch (106mm) w/12mm head
- D: 1.57 inch (40mm) w/14mm head
- E: 1.18 inch (30mm) w/6mm hex head

5. Install a new O-ring on the engine block.

6. Apply silicone sealant to the oil pump housing as shown.

7. Install the oil pump. Tighten the bolts in several passes to the following specifications:

- 12mm: 11 ft. lbs. (15.5 Nm)
- 14mm: 22 ft. lbs. (30.5 Nm)
- 6mm Hex: 11 ft. lbs. (15.5 Nm)

8. Install or connect the following:

- Oil pump pickup tube. Tighten the bolts to 66 inch lbs. (7.5 Nm).
- Oil pan and baffle
- CKP sensor
- Oil filter and bracket. Tighten the bolts to 13 ft. lbs. (18 Nm).
- Oil dipstick tube
- Crankshaft timing sprocket
- Timing belt idler pulleys
- Timing belt
- Front cover

9. Install the engine.

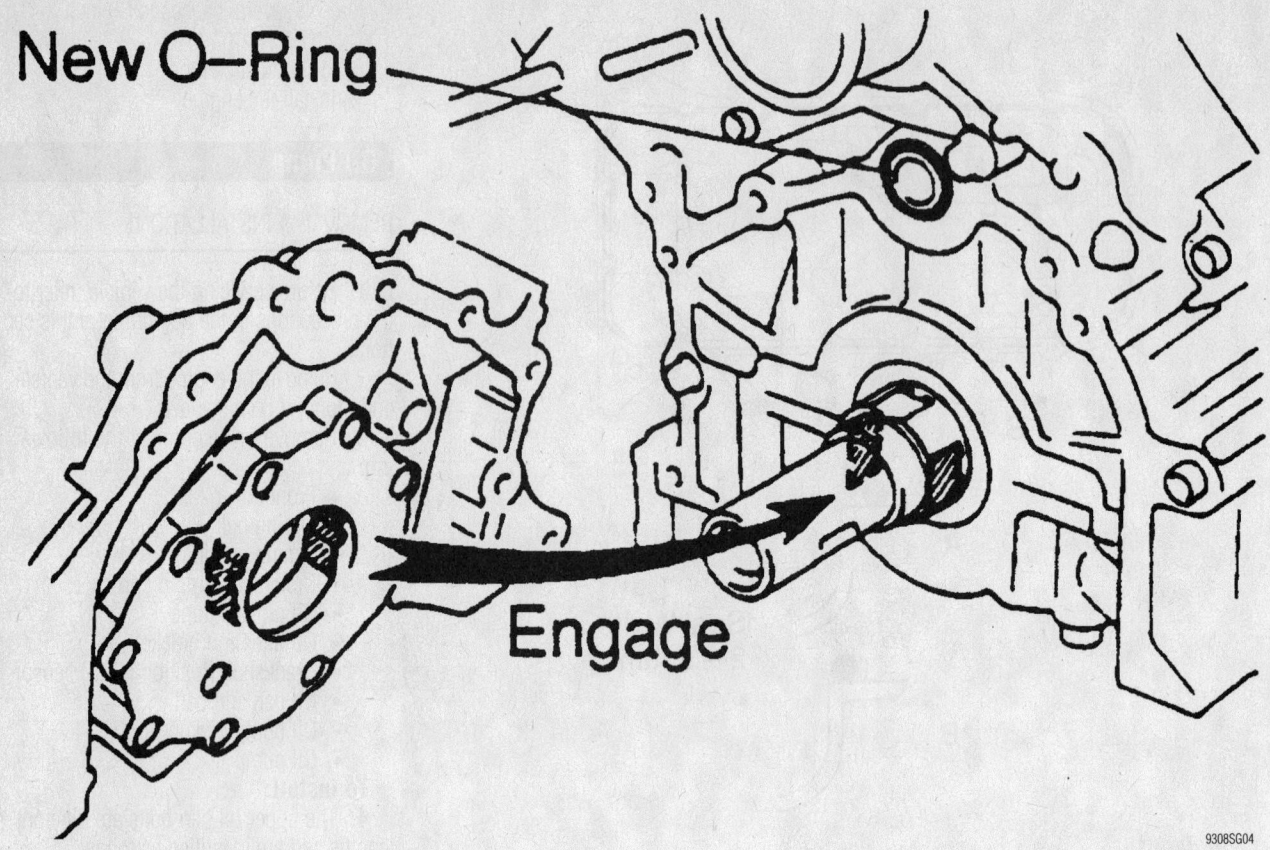

New O-Ring

Engage

9308SG04

Location of the O-ring seal

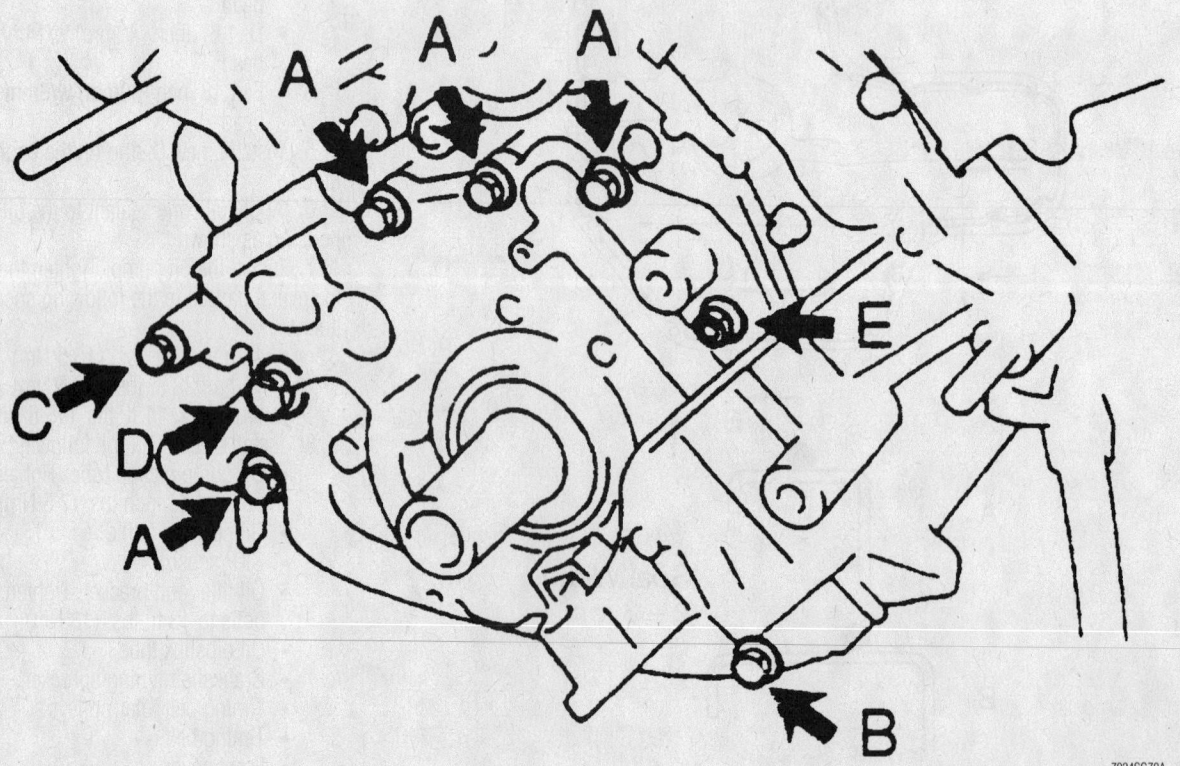

7924SG79A

Oil pump bolt location

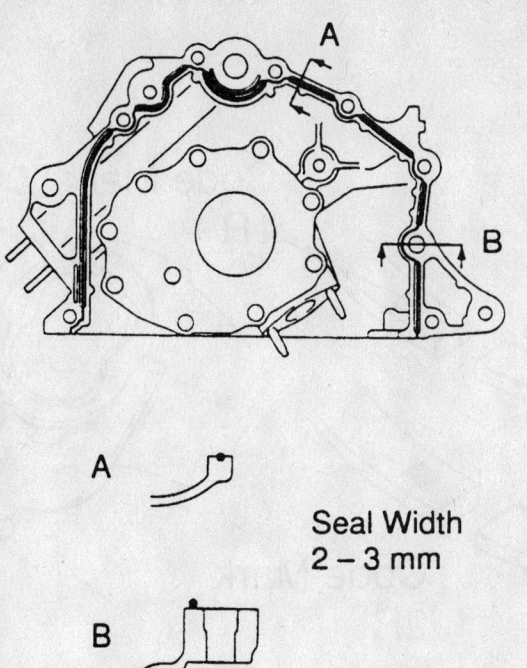

Seal Width
2 – 3 mm

7924SG78

Oil pump housing sealant application

REMOVAL & INSTALLATION

1. Before servicing the vehicle, refer to the precautions in the beginning of this section.
2. Remove the transmission and flywheel from the vehicle.
3. Cut off the rubber lip portion of the seal with a sharp knife.
4. Pry out the oil seal.
To install:
5. Install the rear main seal so that it is flush with the seal retainer housing.
6. Install or connect the following:
   • Flywheel/driveplate. Tighten the bolts to 35 ft. lbs. (48 Nm) plus a 90 degree turn.
   • Transmission

Piston and Ring

POSITIONING

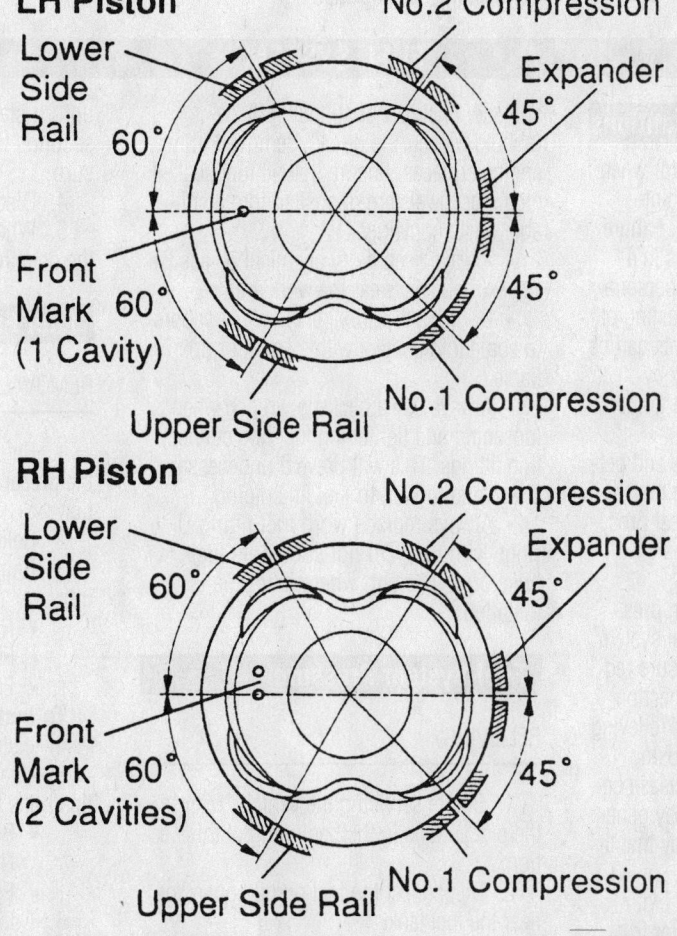

Piston ring positioning

9302AG07

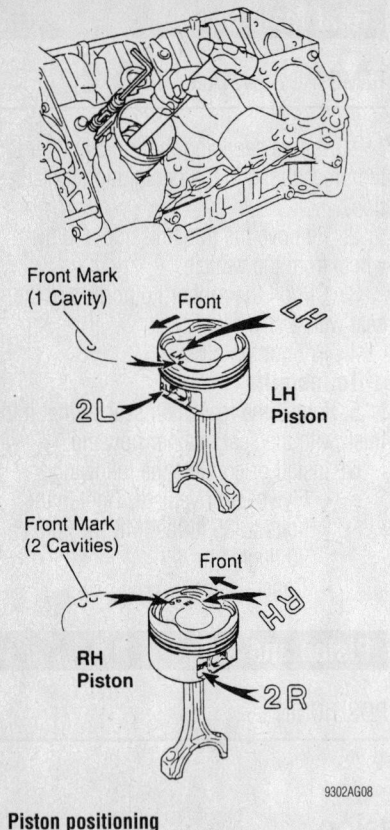

Piston positioning

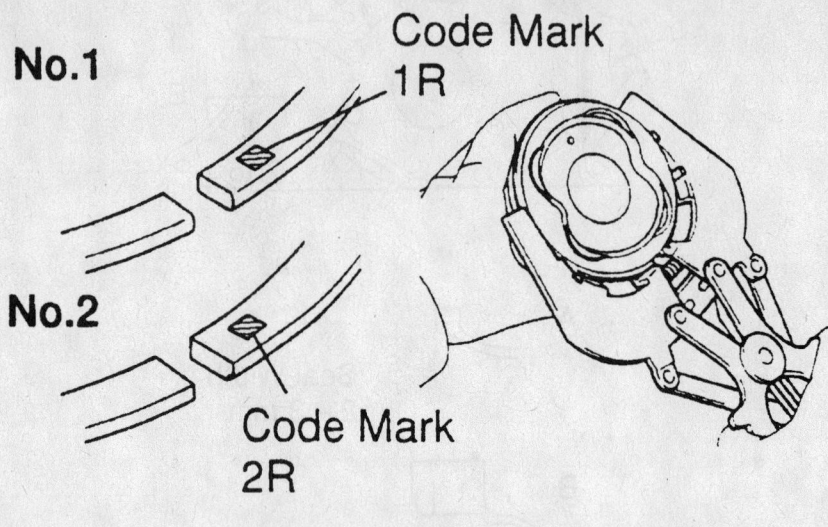

No.1

No.2

Code Mark 1R

Code Mark 2R

Piston ring identification

FUEL SYSTEM

Fuel System Service Precautions

Safety is the most important factor when performing not only fuel system maintenance but any type of maintenance. Failure to conduct maintenance and repairs in a safe manner may result in serious personal injury or death. Maintenance and testing of the vehicle's fuel system components can be accomplished safely and effectively by adhering to the following rules and guidelines.

• To avoid the possibility of fire and personal injury, always disconnect the negative battery cable unless the repair or test procedure requires that battery voltage be applied.

• Always relieve the fuel system pressure prior to disconnecting any fuel system component (injector, fuel rail, pressure regulator, etc.), fitting or fuel line connection. Exercise extreme caution whenever relieving fuel system pressure, to avoid exposing skin, face and eyes to fuel spray. Please be advised that fuel under pressure may penetrate the skin or any part of the body that it contacts.

• Always place a shop towel or cloth around the fitting or connection prior to loosening to absorb any excess fuel due to spillage. Ensure that all fuel spillage (should it occur) is quickly removed from engine surfaces. Ensure that all fuel soaked cloths or towels are deposited into a suitable waste container.

• Always keep a dry chemical (Class B) fire extinguisher near the work area.

• Do not allow fuel spray or fuel vapors to come into contact with a spark or open flame.

• Always use a back-up wrench when loosening and tightening fuel line connection fittings. This will prevent unnecessary stress and torsion to fuel line piping.

• Always replace worn fuel fitting O-rings with new. Do not substitute fuel hose or equivalent, where fuel pipe is installed.

Fuel System Pressure

RELIEVING

1. Before servicing the vehicle, refer to the precautions in the beginning of this section.
2. Disconnect the fuel pump connector near the fuel tank.
3. Start the engine and allow it to run until it stalls. Crank the engine for a few seconds to relieve additional fuel pressure.
4. Disconnect the negative battery cable.
5. When repairs are complete, connect the negative battery cable.

Fuel Filter

REMOVAL & INSTALLATION

1. Before servicing the vehicle, refer to the precautions in the beginning of this section.
2. Relieve the fuel system pressure.
3. Remove or disconnect the following:

• Negative battery cable
• Fuel lines
• Fuel filter

To install:
4. Install the fuel filter.
5. Use new washers and tighten the fuel line bolts to the following specifications:

• Banjo bolt fittings: 21 ft. lbs. (29 Nm)
• Flare nut fitting: 28 ft. lbs. (38 Nm)
6. Connect the negative battery cable.
7. Start the engine and check for leaks.

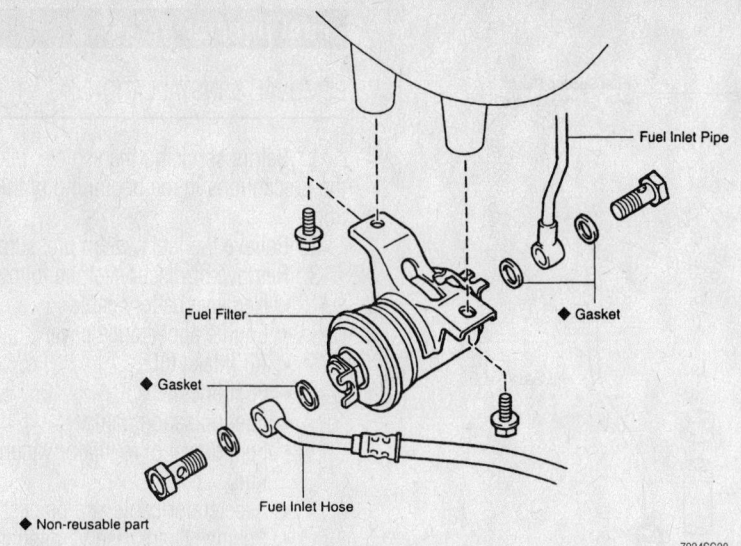

7924SG28

Always use new gaskets when replacing the fuel filter

Fuel Pump

REMOVAL & INSTALLATION

Sequoia

1. Before servicing the vehicle, refer to the precautions in the beginning of this section.
2. Relieve the fuel system pressure.
3. Remove or disconnect the following:
   - Negative battery cable
   - Fuel tank
   - Fuel pump harness connector
   - Fuel lines
   - Fuel pump module

To install:

4. Install or connect the following:
   - Fuel pump module. Tighten the bolts to 35 inch lbs. (4 Nm).

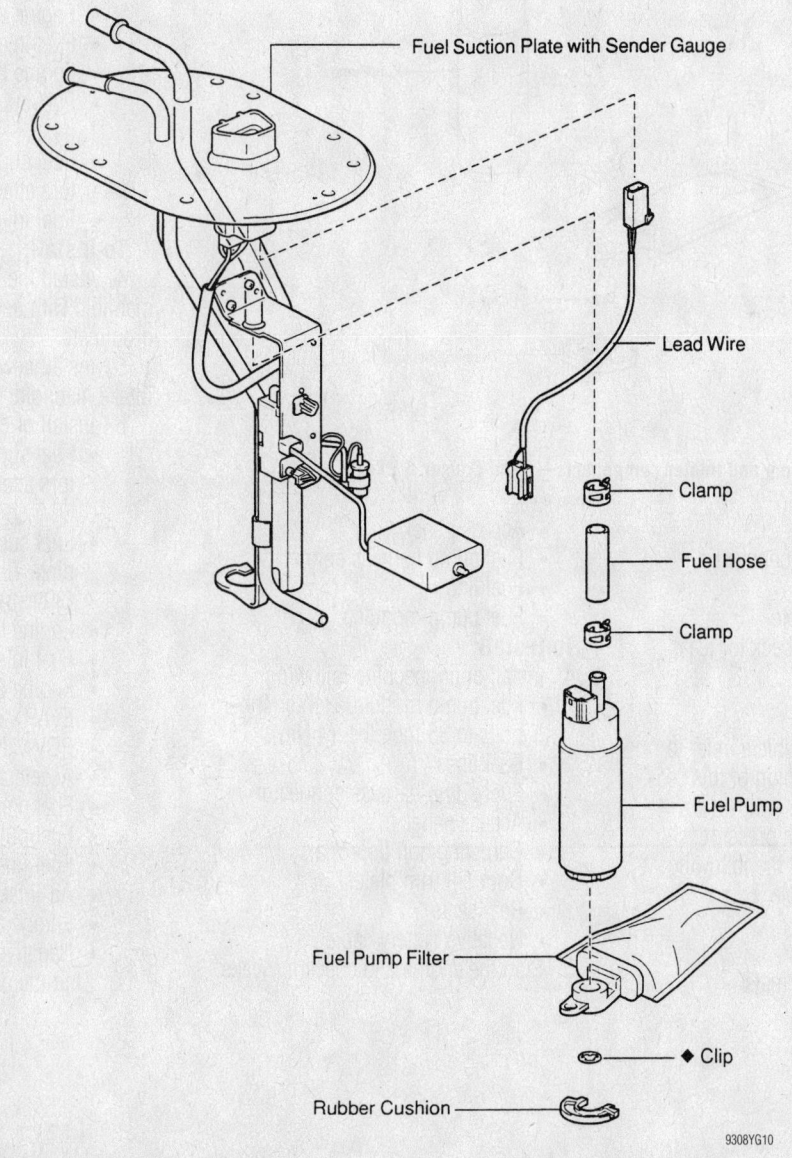

9308YG10

Exploded view of the fuel pump and related components—Sequoia

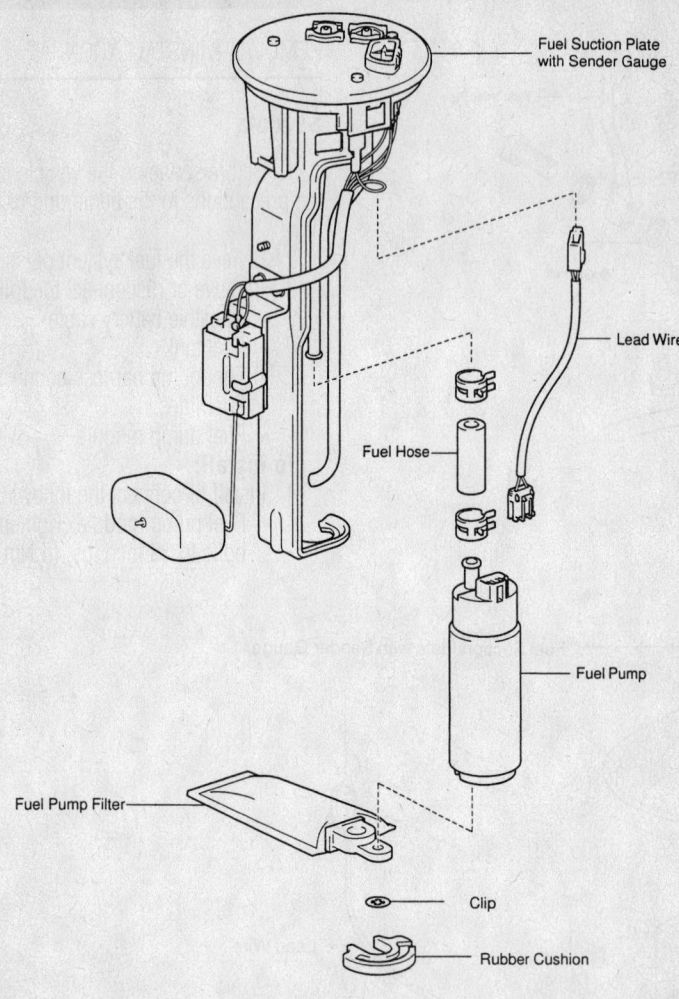

Exploded view of the fuel pump and related components— Land Cruiser & LX470

- Fuel lines
- Fuel pump harness connector
- Fuel tank
- Negative battery cable

5. Start the engine and check for leaks.

Land Cruiser & LX470

1. Before servicing the vehicle, refer to the precautions in the beginning of this section.

2. Relieve the fuel system pressure.

3. Remove or disconnect the following:
   - Negative battery cable
   - Rear seats
   - Door sill trim plates
   - Carpeting and floor mats

- Access panel
- Fuel pump harness connector
- Fuel lines
- Fuel pump module

To install:

4. Install or connect the following:
   - Fuel pump module. Tighten the bolts to 35 inch lbs. (4 Nm).
   - Fuel lines
   - Fuel pump harness connector
   - Access panel
   - Carpeting and floor mats
   - Door sill trim plates
   - Rear seats
   - Negative battery cable

5. Start the engine and check for leaks.

Fuel Injector

REMOVAL & INSTALLATION

1. Before servicing the vehicle, refer to the precautions in the beginning of this section.

2. Relieve the fuel system pressure.

3. Remove or disconnect the following:
   - Negative battery cable
   - Engine appearance cover
   - Air intake tube
   - Fuel lines
   - Fuel pulsation damper
   - Fuel pressure regulator vacuum line
   - Accelerator cable and bracket
   - Positive Crankcase Ventilation (PCV) valve and hose
   - Evaporative Emissions (EVAP) vacuum switching valve
   - Engine appearance cover brackets
   - Fuel injector harness connectors
   - Engine harness protector
   - Fuel supply manifold crossover pipe
   - Fuel supply manifolds with injectors attached
   - Fuel injectors

To install:

4. Install the fuel injectors to the supply manifold with new O-ring seals and new grommets.

5. Install new injector insulators to the intake manifold.

6. Install or connect the following:
   - Fuel supply manifolds with injectors attached. Tighten the bolts to 66 inch lbs. (7.5 Nm).
   - Fuel supply manifold crossover pipe. Tighten the bolts to 29 ft. lbs. (39 Nm).
   - Engine harness protector
   - Fuel injector harness connectors
   - Engine appearance cover brackets
   - EVAP vacuum switching valve
   - PCV valve and hose
   - Accelerator cable and bracket
   - Fuel pressure regulator vacuum line
   - Fuel pulsation damper
   - Fuel lines
   - Air intake tube
   - Engine appearance cover
   - Negative battery cable

7. Start the engine and check for leaks.

DRIVE TRAIN

Automatic Transmission Assembly

REMOVAL & INSTALLATION

Sequoia

1. Before servicing the vehicle, refer to the precautions in the beginning of this section.
2. Remove or disconnect the following:
   - Oil filler pipe
   - No. 1 engine undercover
   - Exhaust pipes
   - Front and rear driveshafts
   - Nos. 1&2 vehicle speed sensors
   - Solenoid connector
   - Shift cable at the transmission
   - Overdrive sensor connector
   - Oil cooler lines
   - ATF temperature sensor
   - Park/Neutral switch
   - End plate and converter clutch mounting bolts
3. Raise the transmission slightly.
4. Remove or disconnect the following:
   - Crossmember
   - Rear mount insulator
   - Transmission and transfer case as a unit
5. Installation is the reverse of removal. Note the following torques:
   - Transmission-to-transfer case bolts: 53 ft. lbs. (71Nm)
   - Rear mount insulator-to-transmission: 48 ft. lbs. (65Nm)
   - Rear mount insulator-to-crossmember: 13 ft. lbs. (18Nm)
   - Crossmember-to-frame: 53 ft. lbs. (71Nm)
   - Torque converter clutch: 30 ft. lbs. (41Nm)

➡**Install the green bolt first.**

   - Rear end plate: 13 ft. lbs. (18Nm)
   - Oil cooler lines: 25 ft. lbs. (34Nm)
   - Shift control bracket: 13 ft. lbs. (18Nm)

Land Cruiser & LX470

1. Before servicing the vehicle, refer to the precautions in the beginning of this section.
2. Remove or disconnect the following:
   - Battery and tray
   - Air intake assembly
   - Cooling fan and shroud
   - Coolant recovery reservoir
   - Transmission dipstick tube
   - Center console

   - Transmission gear select lever and rod
   - Transfer case shift lever and rod
   - Engine under covers
   - Exhaust front pipes
   - Front and rear driveshafts
   - Vehicle Speed (VSS) sensor connectors
   - Overdrive clutch speed sensor connector
   - Solenoid harness connector
   - Transmission fluid temperature sensor connector
   - Park/Neutral Position (PNP) switch connector
   - Center differential lock indicator switch connector
   - L4 solenoid valve position switch connector
   - Motor actuator connector
   - Torque converter
   - Transmission oil cooler lines
   - Transmission mount crossmember. Support the transmission with a jack.
   - Transmission flange bolts
   - Transmission

To install:
3. Install or connect the following:
   - Transmission. Tighten the flange bolts to 53 ft. lbs. (72 Nm).
   - Transmission mount crossmember. Tighten the bolts to 37 ft. lbs. (50 Nm) and the nuts to 54 ft. lbs. (74 Nm).
   - Transmission oil cooler lines
   - Torque converter. Tighten the bolts to 35 ft. lbs. (48 Nm).
   - Motor actuator connector
   - L4 solenoid valve position switch connector
   - Center differential lock indicator switch connector
   - PNP switch connector
   - Transmission fluid temperature sensor connector
   - Solenoid harness connector
   - Overdrive clutch speed sensor connector
   - VSS sensor connectors
   - Front driveshaft. Tighten the fasteners to 59 ft. lbs. (80 Nm).
   - Rear driveshaft. Tighten the fasteners to 78 ft. lbs. (106 Nm).
   - Exhaust front pipes
   - Engine under covers
   - Transfer case shift lever and rod
   - Transmission gear select lever and rod

   - Center console
   - Transmission dipstick tube
   - Coolant recovery reservoir
   - Cooling fan and shroud
   - Air intake assembly
   - Battery and tray
4. Check the transmission and transfer case fluid levels and adjust as necessary.

Transfer Case Assembly

REMOVAL & INSTALLATION

Sequoia

1. Before servicing the vehicle, refer to the precautions in the beginning of this section.
2. Drain the transfer case oil.
3. Place the shift lever in the **H** position and the one-touch 2-4 switch **OFF**.
4. Remove or disconnect the following:
   - Shift lever
   - Skid plate
   - Oil from the transfer case
   - Exhaust pipes
   - Front and rear driveshafts
   - Crossmember
   - Rear engine mount
   - All wiring connectors
5. Support the transfer case, remove the case-to-adapter bolts and lower the transfer case.
6. Installation is the reverse of removal. Note the following torques:
   - Transfer case-to-adapter bolts: 17 ft. lbs. (24Nm)
   - Engine rear mount-to-adapter: 48 ft. lbs. (65Nm)
   - Crossmember: 53 ft. lbs. (72Nm)
   - Engine rear mount set bolts: 13 ft. lbs. (18Nm)
   - Skid plate: 13 ft. lbs. (18Nm)

Land Cruiser & LX470

1. Before servicing the vehicle, refer to the precautions in the beginning of this section.
2. Drain the transfer case oil.
3. Remove or disconnect the following:
   - Transfer case protector
   - Front and rear driveshafts
   - Transfer case shift lever rod
   - Ground cable
   - Transmission mount crossmember. Support the transmission with a jack.
   - Transfer case vent hose

- Vehicle Speed (VSS) sensor connector
- Center differential lock indicator switch connector
- Motor actuator connectors
- Transfer case adapter bolts
- Transfer case

To install:

4. Install or connect the following:
- Transfer case. Tighten the adapter bolts to 51 ft. lbs. (69 Nm).
- Motor actuator connectors
- Center differential lock indicator switch connector
- VSS sensor connector

- Transfer case vent hose
- Transmission mount crossmember. Tighten the bolts to 37 ft. lbs. (50 Nm) and the nuts to 54 ft. lbs. (74 Nm).
- Ground cable
- Transfer case shift lever rod
- Front driveshaft. Tighten the fasteners to 59 ft. lbs. (80 Nm).
- Rear driveshaft. Tighten the fasteners to 78 ft. lbs. (106 Nm).
- Transfer case protector

5. Fill the transfer case to the correct level.

Halfshaft

REMOVAL & INSTALLATION

Sequoia

1. Before servicing the vehicle, refer to the precautions in the beginning of this section.
2. Remove or disconnect the following:
- Front wheel
- Under cover
3. Drain the differential oil.
4. Remove or disconnect the following:

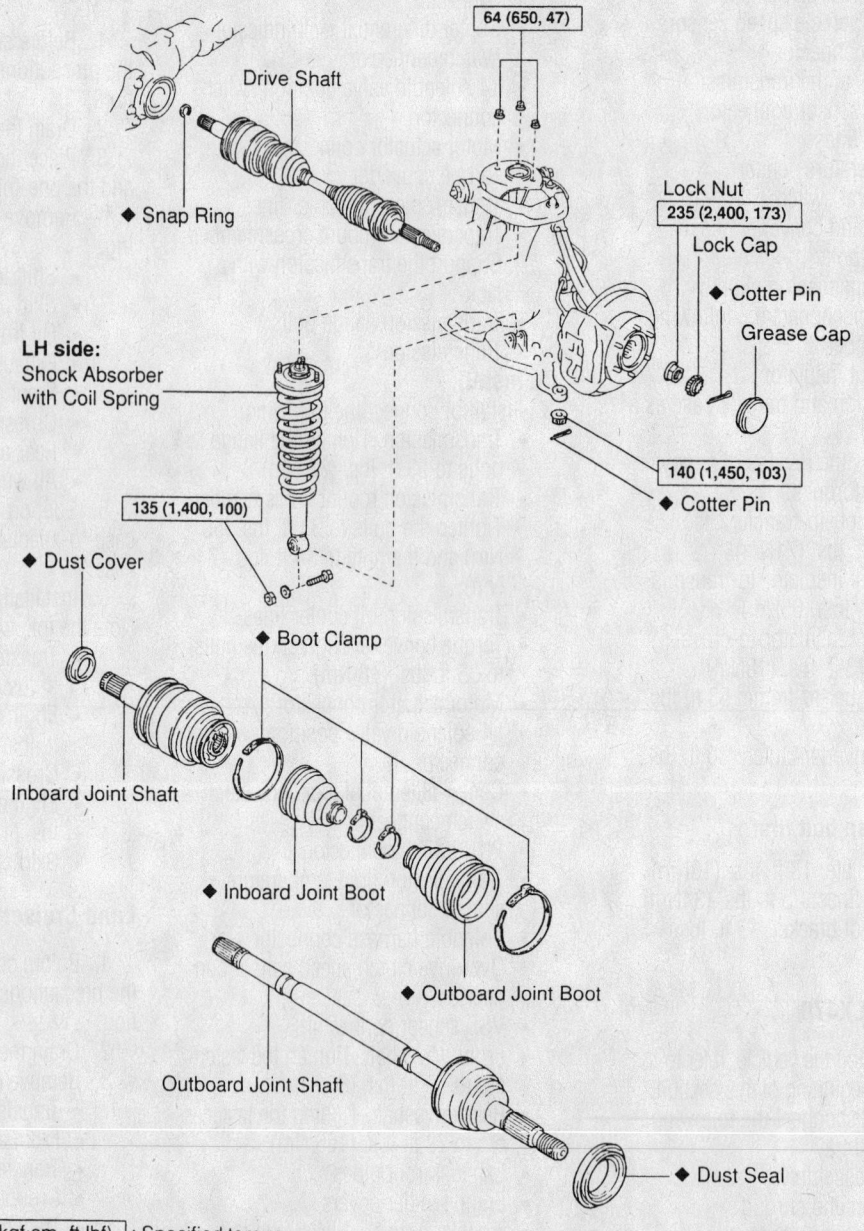

N·m (kgf·cm, ft·lbf) : Specified torque
◆ Non-reusable part

View of the halfshaft and related components—Sequoia

9308YG12

- Grease cap
- Cotter pin and lock cap
- Halfshaft locknut by applying the brakes
- Lower control arm from the lower ball joint
- Halfshaft from the steering knuckle, using a plastic hammer
- Left strut, for the left halfshaft
- Right halfshaft, using a brass bar and a hammer
- Left halfshaft, using tools 09520-01010 and 09520-24010
- Snapring from the inboard joint shaft

To install:

5. Install or connect the following:
   - New snapring, onto the inboard joint shaft with the opening facing downward

LH side:

RH side:

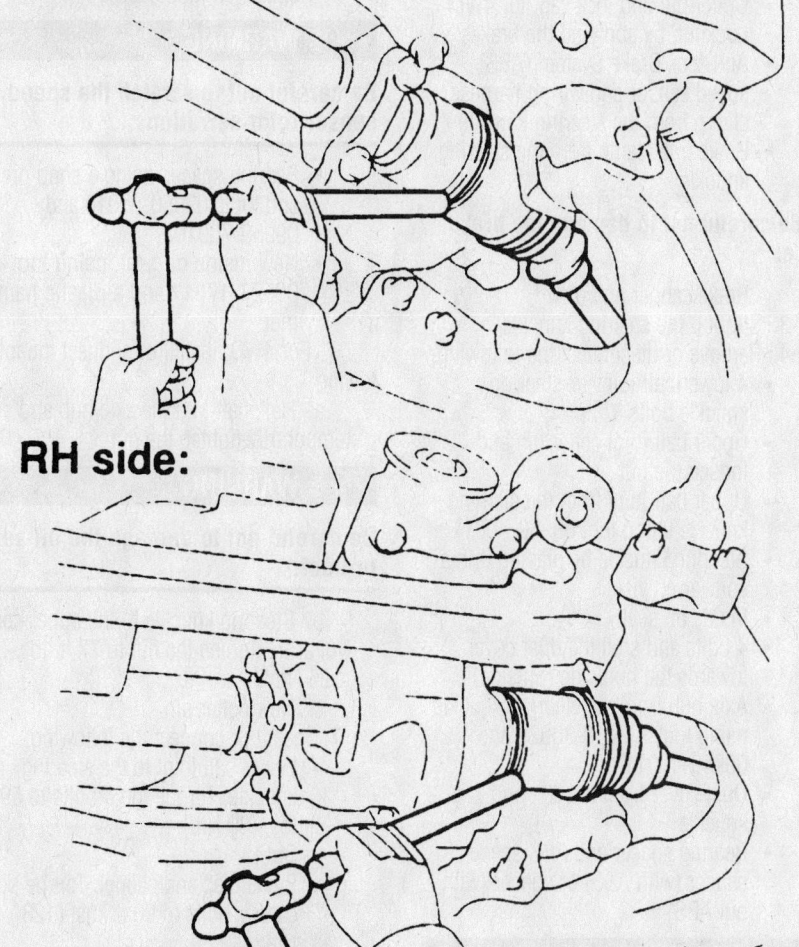

Axle halfshaft removal—Land Cruiser/LX470

- Halfshafts to the differential using a brass bar and a hammer
- Halfshafts to the steering knuckles

❉❉ WARNING

Be careful not to damage the oil seal, boot or dust seal.

- Lower control arm to the lower ball joint using a new cotter pin. Torque the ball joint nut to 103 ft. lbs. (140 Nm).
- Halfshaft locknut by applying the brakes. Torque the nut to 173 ft. lbs. (235 Nm).
- Lock cap and a new cotter pin
- Grease cap

6. Refill the differential with oil.
7. Install or connect the following:

- Under cover
- Front wheel

Land Cruiser & LX470

1. Before servicing the vehicle, refer to the precautions in the beginning of this section.
2. Remove or disconnect the following:
   - Front wheel
   - Brake caliper
   - Grease cap
   - Snapring
   - Wheel speed sensor and wire harness
   - Steering knuckle arm
   - Lower ball joint
   - Upper ball joint
   - Steering knuckle
   - Axle halfshaft

To install:

➡**Use new split pins, snaprings and circlips for assembly.**

3. Install or connect the following:
   - Axle halfshaft
   - Steering knuckle
   - Upper ball joint. Tighten the nut to 81 ft. lbs. (110 Nm).
   - Lower ball joint. Tighten the nut to 117 ft. lbs. (159 Nm).
   - Steering knuckle arm. Tighten the bolts to 108 ft. lbs. (147 Nm).
   - Wheel speed sensor and wire harness
   - Snapring
   - Grease cap
   - Brake caliper
   - Front wheel

CV-Joints

OVERHAUL

Land Cruiser & LX470

OUTER CV-JOINT

The outer CV-joint is serviced with the axle shaft as an assembly. The outer CV-joint boot can be serviced by removing the inner CV-joint.

INNER CV-JOINT

1. Before servicing the vehicle, refer to the precautions in the beginning of this section.
2. Remove or disconnect the following:
   - Halfshaft from the vehicle
   - Grease boot clamps
   - Outer race snapring
   - Outer race

9308SG07

- Shaft snapring
- Inner race, cage and balls

To install:

3. Install or connect the following:
   - Inner race, cage and balls
   - Shaft snapring
   - Outer race
   - Outer race snapring

4. Fill the outer race and the grease boot with CV-joint grease and tighten the boot clamps.

5. Install the axle halfshaft.

Sequoia

OUTER CV-JOINT

The outer CV-joint is serviced with the axle shaft as an assembly. The outer CV-joint boot can be serviced by removing the inner CV-joint.

INNER CV-JOINT

1. Before servicing the vehicle, refer to the precautions in the beginning of this section.

2. Remove or disconnect the following:
   - Halfshaft from the vehicle
   - Large boot clamps
   - Small boot clamps

3. Matchmark inboard CV-joint to the shaft

4. Remove or disconnect the following:
   - Inboard CV-joint from the shaft by expanding the snapring
   - Both CV-joint boots
   - Outer dust seal, using a shop press and tool 09950-00020
   - Outer dust cover, using a shop press and tool 09950-00020

To install:

5. Install or connect the following:
   - Outer dust cover, using a screwdriver and a hammer
   - Outer dust seal, using a screwdriver and a hammer

6. Wrap the shaft splines with tape to protect the boot from damage.

7. Install or connect the following:
   - Both CV-joint boots with clamps, temporarily
   - Inboard CV-joint to the shaft by aligning the matchmarks and expanding the snapring

8. Lubricate the outboard joint with 7.23–7.94 oz. (205–225g) grease, provided in the boot kit.

9. Lubricate the inboard joint with 6.70–7.41 oz. (190–210g) grease, provided in the boot kit.

10. Install or connect the following:
    - Both joint boots making sure the boots are in the shaft groove

- Standard halfshaft length is 20.531–20.689 in. (521.5–525.5mm) when the shaft is not expanded or contracted
- Large inboard boot clamp
- All other boot clamps using tool 09521-24010. Tighten the crimping tool until the clamp clearance is 0.039–0.059 in. (1.0–1.5mm)
- Halfshaft

Spindle Bearings

REMOVAL, PACKING AND INSTALLATION

Sequoia

1. Before servicing the vehicle, refer to the precautions in the beginning of this section.

2. Remove or disconnect the following:
   - Front wheel
   - Grease cap, for 2WD
   - Cotter pin and lock cap, for 4WD
   - Locknut, by applying the brakes
   - Anti-lock Brake System (ABS) speed sensor and wiring harness clamp from the steering knuckle
   - Brake line clamp from the steering knuckle

➡Be careful not to damage the brake tube.

   - Brake caliper and disc

3. Support the steering knuckle.

4. Remove or disconnect the following:
   - 4 lower ball joint-to-steering knuckle bolts
   - Upper ball joint cotter pin and loosen the nut
   - Upper ball joint from the steering knuckle using tool 09950-40011
   - Steering knuckle by placing it in a soft-jawed vise
   - Inside oil seal
   - 4 bolts and shift the dust cover towards the hub side (outside)
   - Axle hub from the steering knuckle using tools 09710-30021 and 09950-40011
   - Dust cover from the steering knuckle
   - Bearing spacer and ABS speed sensor (with ABS) or spacer (without ABS)

❋❋ WARNING

Be careful not to scratch the speed sensor rotor serrations

   - Outside oil seal from the steering knuckle

- Bearing snapring from the steering knuckle
- Bearing from the steering knuckle, using a shop press and tools 09950-60020 and 09950-70010

To install:

5. Install or connect the following:
   - Bearing to the steering knuckle, using a shop press and tools 09527-17011 and 09950-60020
   - Bearing snapring to the steering knuckle
   - New outside oil seal to the steering knuckle, using tools 09223-15030 and 09527-17011
   - Dust cover to the steering knuckle. Torque the 4 bolts to 13 ft. lbs. (18 Nm).
   - Axle hub to the steering knuckle using a shop press and tool 09649-17010
   - ABS speed sensor (with ABS) or spacer (without ABS)

❋❋ WARNING

Be careful not to scratch the speed sensor rotor serrations

   - Bearing spacer, using a shop press and tools 09950-60010 and 09950-70010
   - New inside oil seal, using tool 09527-17011 and a plastic hammer

6. For 4WD, install or connect the following:

   a. Halfshaft into the axle hub and temporarily tighten the nut

❋❋ WARNING

Be careful not to damage the oil seal or boot.

   b. Steering knuckle to the upper control arm. Tighten the nut to 77 ft. lbs. (105 Nm).
   c. New cotter pin.

7. Install or connect the following:
   - Lower ball joint to the steering knuckle. Torque the 4 bolts to 59 ft. lbs. (80 Nm).
   - Strut
   - Brake disc and caliper. Torque both caliper bolts to 90 ft. lbs. (123 Nm).
   - Brake line clamp to the steering knuckle. Torque the bolt to 21 ft. lbs. (28 Nm).
   - ABS speed sensor. Torque both bolts to 7.1 ft. lbs. (8.2 Nm).
   - Halfshaft locknut. Torque the nut to

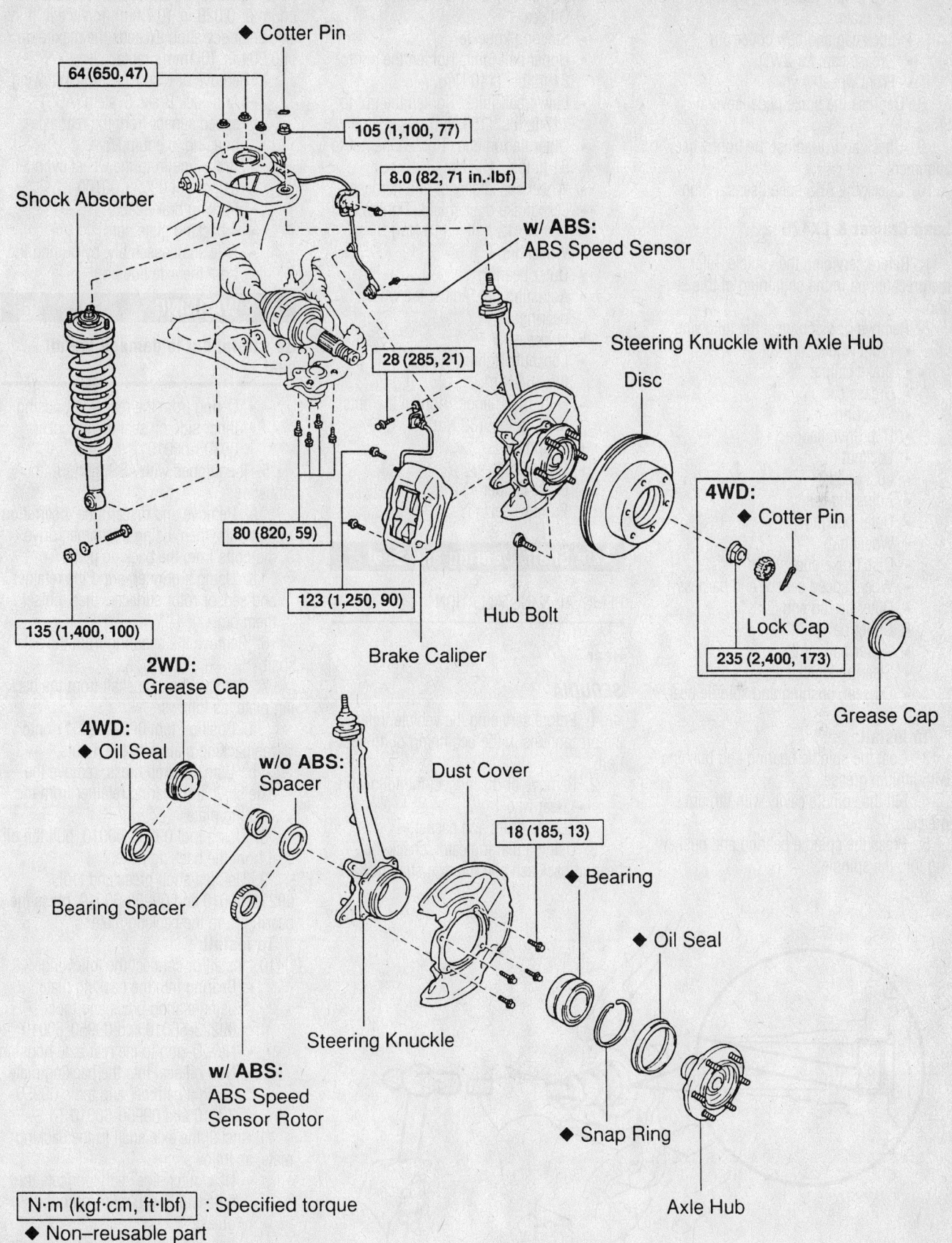

◆ Cotter Pin

64 (650, 47)

105 (1,100, 77)

8.0 (82, 71 in.·lbf)

Shock Absorber

w/ ABS:
ABS Speed Sensor

Steering Knuckle with Axle Hub

Disc

28 (285, 21)

4WD:
◆ Cotter Pin

80 (820, 59)

Lock Cap

135 (1,400, 100)

123 (1,250, 90)

Hub Bolt

235 (2,400, 173)

Brake Caliper

Grease Cap

2WD:
Grease Cap

4WD:
◆ Oil Seal

w/o ABS:
Spacer

Dust Cover

18 (185, 13)

◆ Bearing

Bearing Spacer

◆ Oil Seal

Steering Knuckle

w/ ABS:
ABS Speed
Sensor Rotor

◆ Snap Ring

Axle Hub

N·m (kgf·cm, ft·lbf) : Specified torque

◆ Non–reusable part

Exploded view of the front axle hub and related components—Sequoia

9308YG13

173 ft. lbs. (235 Nm), by applying the brakes.
- Lock cap and new cotter pin
- Grease cap, for 2WD
- Front wheel

8. Depress the brake pedal several times.

9. Check and/or adjust the front wheel alignment.

10. Check the ABS speed sensor signal.

Land Cruiser & LX470

1. Before servicing the vehicle, refer to the precautions in the beginning of this section.

2. Remove or disconnect the following:
- Front wheel
- Brake caliper
- Grease cap
- Snapring
- Hub drive flange
- Locknut
- Lockwasher
- Adjusting nut
- Outer bearing
- Wheel hub
- Disc brake dust shield
- Wheel speed sensor and harness
- Outer tie rod end
- Upper ball joint
- Lower ball joint
- Steering knuckle
- Oil seal, bushing and spindle bearing

To install:

3. Coat the spindle bearing and bushing with lithium grease.

4. Fill the spindle cavity with lithium grease.

5. Press the spindle bearing and bushing into the spindle.

6. Install or connect the following:
- Oil seal
- Steering knuckle
- Upper ball joint. Tighten the nut to 81 ft. lbs. (110 Nm).
- Lower ball joint. Tighten the nut to 117 ft. lbs. (159 Nm).
- Outer tie rod end. Tighten the nut to 91 ft. lbs. (122 Nm).
- Wheel speed sensor and harness
- Disc brake dust shield. Tighten the bolts to 13 ft. lbs. (18 Nm).
- Wheel hub
- Outer bearing
- Adjusting nut. Adjust the wheel bearings.
- Lockwasher
- Locknut. Tighten the nut to 47 ft. lbs. (64 Nm).
- Hub drive flange. Tighten the nuts to 26 ft. lbs. (35 Nm).
- Snapring
- Grease cap
- Brake caliper
- Front wheel

Axle Shaft, Bearing and Seal

REMOVAL & INSTALLATION

Rear

SEQUOIA

1. Before servicing the vehicle, refer to the precautions in the beginning of this section.

2. Remove or disconnect the following:
- Rear wheel
- Brake drum and gasket

3. Using a dial indicator, check the bearing backlash and the axle shaft devia-

tion. If the bearing backlash exceeds a maximum or 0.028 in. (0.7mm), replace it. If the axle shaft deviation exceeds the maximum of 0.004 in. (0.1mm), replace it.

4. Remove or disconnect the following:
- Anti-lock Brake System (ABS) speed sensor from the rear axle housing, if equipped
- Brake line from the wheel cylinder, using tool 09023-00100
- Parking brake cable
- 4 backing plate nuts
- Axle shaft assembly, by pulling it from the axle housing

✱✱ WARNING

Be careful not to damage the oil seal.

- O-ring from the rear axle housing
- Inner side oil seal using tool 09308-00010

5. If equipped with ABS, perform the following:

a. Remove and discard the 4 serration bolt nuts; then, using a hammer, drive the bolts from the backing plate.

b. Using a grinder, grind the retainer and sensor rotor surfaces; then, chisel them out.

6. Remove the snapring from the axle shaft.

7. Remove the axle shaft from the backing plate, as follows:

a. Position tool 09521-25011 onto the backing plate with the 4 nuts.

b. Using a shop press, remove the axle shaft and bearing retainer from the backing plate.

8. Using tool 09308-00010, pull the oil seal from the backing plate.

9. Using a shop press and tools 09223-56010 and 09950-60010, press the bearing from the backing plate.

To install:

10. Install or connect the following:
- Bearing into the backing plate, using a shop press and tools 09223-56010 and 09950-60010
- New O-ring to the rear axle housing
- New oil seal into the backing plate, using a hammer and tools 09950-70010 and 09950-60010

11. Install the axle shaft to the backing plate, as follows:
- New outer side seal, lubricate the oil seal lip with multi-purpose grease
- Backing plate and bearing retainer onto the rear axle shaft
- Axle shaft onto the backing plate,

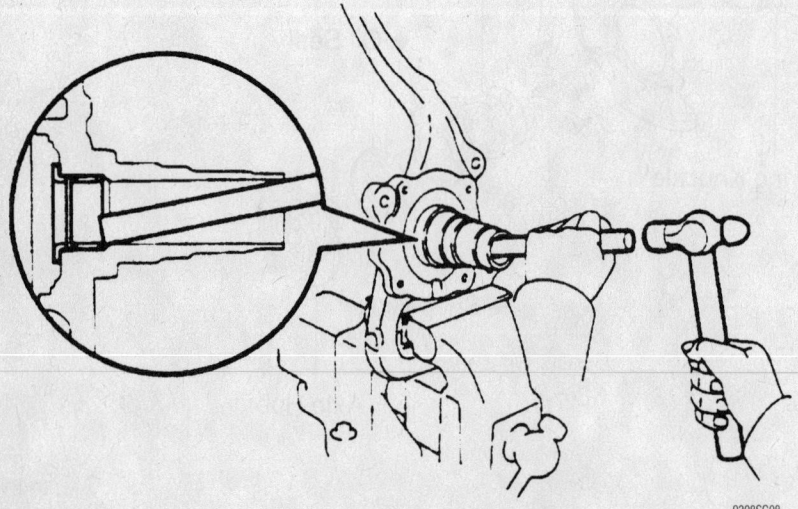

9308SG08

Removing the oil seal, bushing and spindle bearing—Land Cruiser/LX470

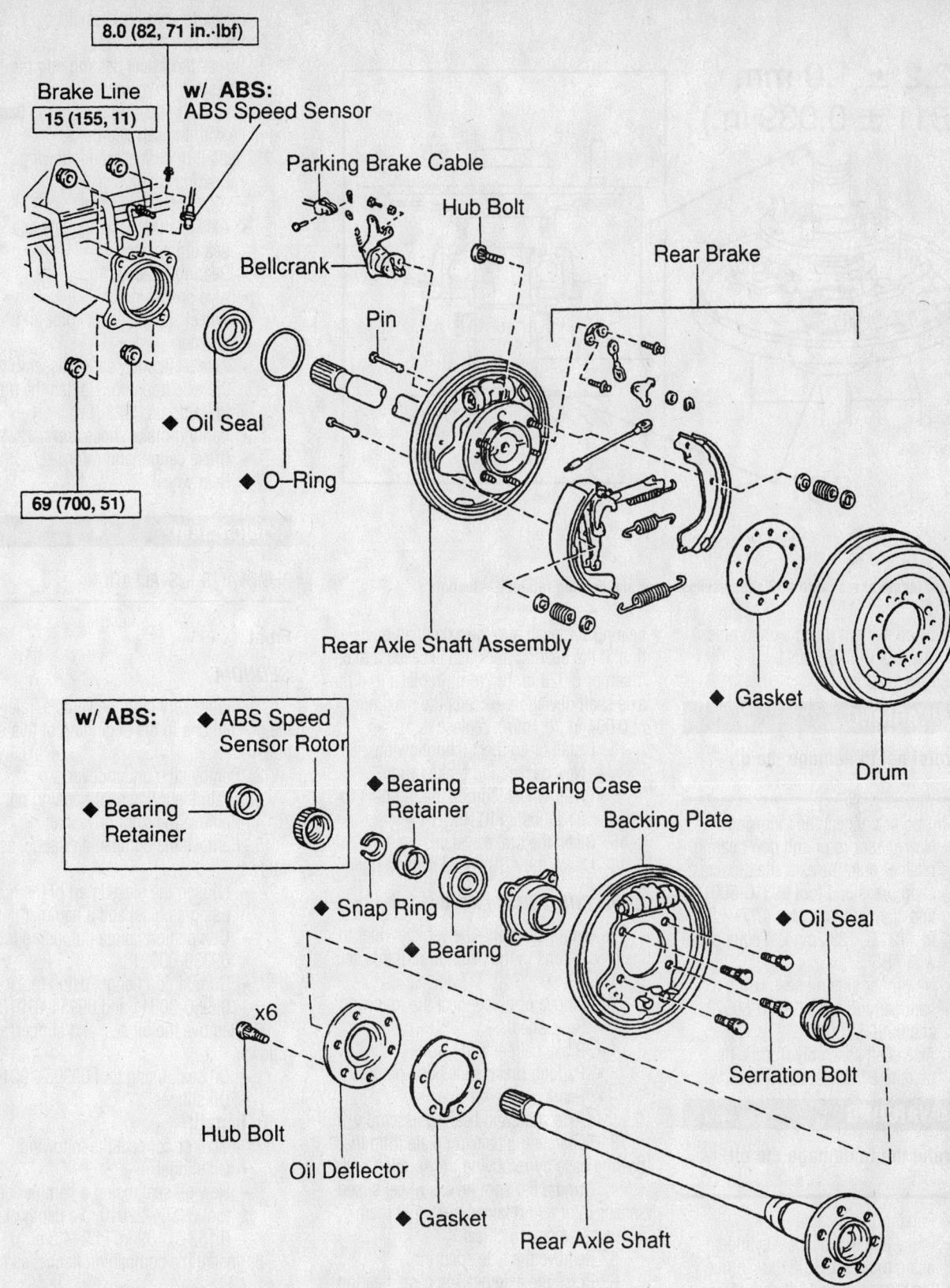

8.0 (82, 71 in.·lbf)

Brake Line
15 (155, 11)

w/ **ABS:**
ABS Speed Sensor

Parking Brake Cable

Hub Bolt

Rear Brake

Bellcrank

Pin

◆ Oil Seal

◆ O–Ring

69 (700, 51)

Rear Axle Shaft Assembly

◆ Gasket

Drum

w/ **ABS:** ◆ ABS Speed Sensor Rotor

◆ Bearing Retainer

◆ Bearing Retainer

Bearing Case

Backing Plate

◆ Snap Ring

◆ Bearing

◆ Oil Seal

x6

Serration Bolt

Hub Bolt

Oil Deflector

◆ Gasket

Rear Axle Shaft

N·m (kgf·cm, ft·lbf) : Specified torque

◆ Non–reusable part

Exploded view of the rear axle—Sequoia

9308YG14

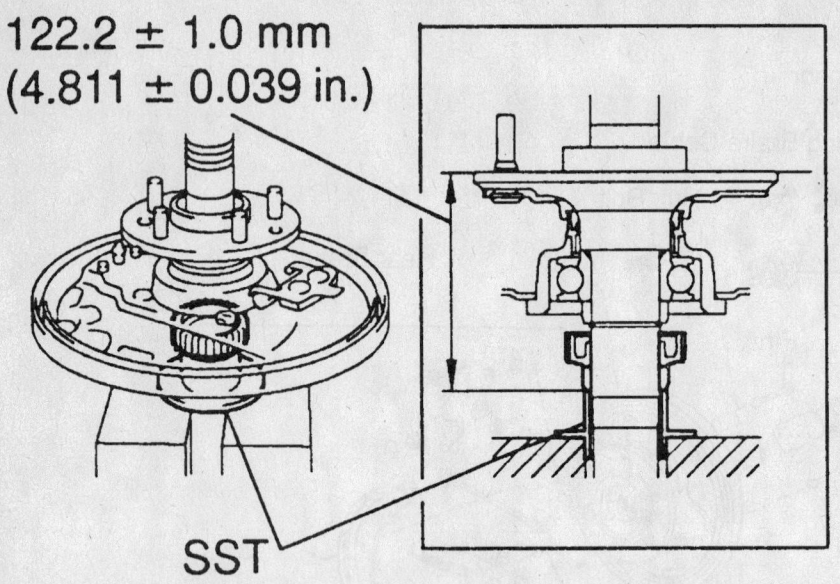

122.2 ± 1.0 mm (4.811 ± 0.039 in.)

SST

9308YG15

Standard length of rear axle ABS speed sensor rotor and bearing retainer—Sequoia

by pressing it using a shop press and tool 09316-60011
- New snapring

❋❋ WARNING

Be careful not to damage the oil seal.

12. Install or connect the following:
- New sensor rotor and new bearing retainer onto the axle shaft, using a shop press and tool 09316-60011 to a standard length of 4.77–4.85 in. (121.2–123.2mm), if equipped with ABS
- New inner side oil seal, using a hammer and tools 09950-60020 and 09950-70010
- Axle shaft assembly. Torque the bolts to 51 ft. lbs. (69 Nm).

❋❋ WARNING

Be careful not to damage the oil seal.

- Parking brake cable
- Brake line to the wheel cylinder, using tool 09023-00100. Torque the brake line to 11 ft. lbs. (15 Nm).
- Rear brake assembly
- ABS speed sensor to the rear axle housing. Torque it to 7.1 ft. lbs. (8.0 Nm).
13. Using a dial indicator, check the

bearing backlash and the axle shaft deviation. If the bearing backlash exceeds a maximum or 0.028 in. (0.7mm), replace it. If the axle shaft deviation exceeds the maximum of 0.004 in. (0.1mm), replace it.
14. Install or connect the following:
- New gasket and brake drum
- Rear wheel. Torque the lug nuts to 81 ft. lbs. (110 Nm).
15. Bleed the brake system.
16. Check the ABS speed sensor signal.

LAND CRUISER & LX470

1. Before servicing the vehicle, refer to the precautions in the beginning of this section.
2. Remove or disconnect the following:
- Rear wheel
- Brake caliper and rotor
- Parking brake shoes and hardware
- Bearing case nuts
- Axle shaft and bearing assembly
3. Separate the backing plate from the bearing case by removing the serrated bolts.
4. Grind a flat spot on the wheel speed sensor rotor and retainer, then split them with a hammer and chisel.
5. Remove the axle snapring.
6. Press the axle bearing case, bearing and retainer off of the axle.
7. Press the axle bearing from the bearing case.
8. Remove or disconnect the following:
- Backing plate
- Axle housing oil seal

- Bearing case oil seal

To install:
9. Press the wheel bearing into the bearing case.
10. Install the bearing case to the backing plate with the serrated bolts.
11. Install or connect the following:
- Bearing case oil seal
- Axle housing oil seal
- Axle shaft to backing plate and bearing assembly
- Bearing retainer
- Axle snapring
- Wheel speed sensor rotor and retainer
- Axle shaft and bearing assembly to the axle housing. Tighten the nuts to 91 ft. lbs. (123 Nm).
- Parking brake shoes and hardware
- Brake caliper and rotor
- Rear wheel

Pinion Seal

REMOVAL & INSTALLATION

Front

SEQUOIA

1. Before servicing the vehicle, refer to the precautions in the beginning of this section.
2. Remove the under cover.
3. Drain the differential housing oil.
4. Remove the front driveshaft.
5. Remove the companion flange, as follows:
- Loosen the staked part of the nut, using a chisel and a hammer
- Companion flange nut, using tool 09330-00021
- Companion flange, using tools 09950-30011 and 09954-03010
6. Remove the oil seal and slinger, as follows:
- Oil seal, using tool 09308-10010
- Oil slinger

To install:
7. Install or connect the following:
- Oil slinger
- New oil seal, using a hammer and tool 09554-22010 to a depth of 0.153–0.189 in. (4.2–4.8mm).
8. Install the companion flange, as follows:
- Companion flange
- New nut, lubricated with hypoid gear oil
- Torque the nut to 80 ft. lbs. (108 Nm), using tool 09330-00021.
9. Adjust the drive pinion preload

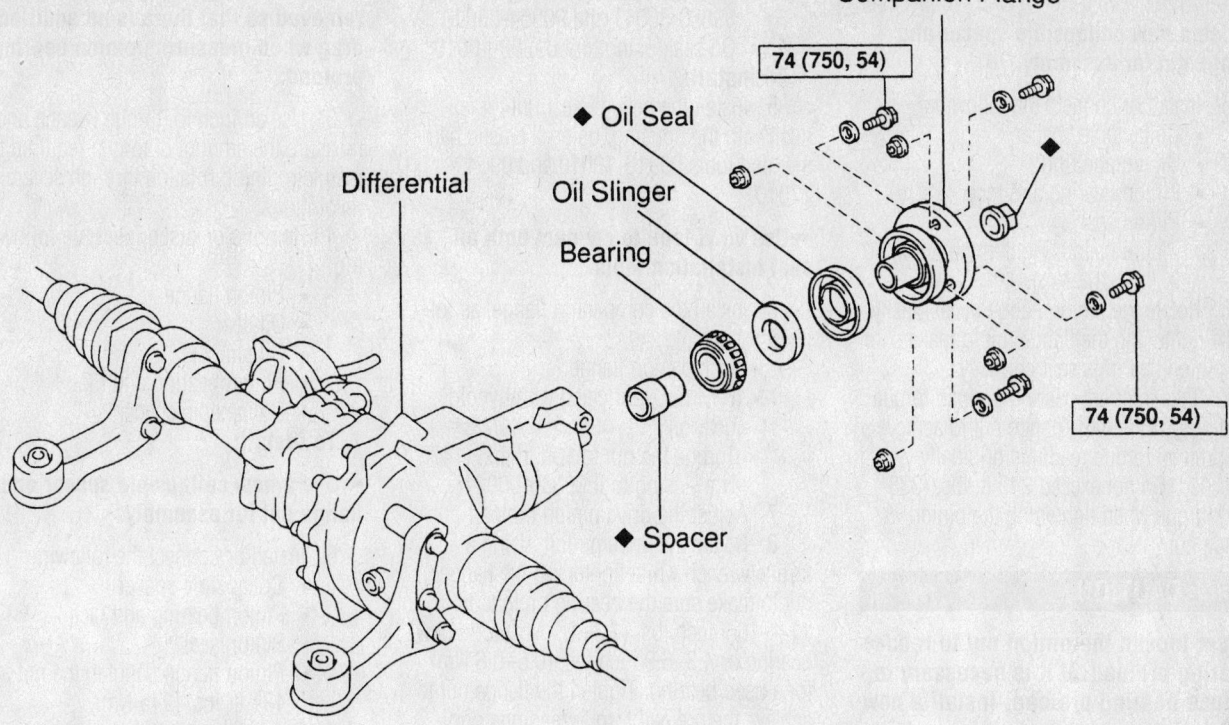

Companion Flange

◆ Oil Seal

Differential

Oil Slinger

Bearing

◆ Spacer

74 (750, 54)

74 (750, 54)

N·m (kgf·cm, ft·lbf) : Specified torque

◆ Non–reusable part

9308YG16

Exploded view of the Sequoia front differential assembly—Rear differential assembly is similar

10. Rotate the drive pinion, using a torque wrench while tightening the flange nut to make sure the bearing preload is 10.4–16.5 inch lbs. (1.2–1.9 Nm) for a new bearing or 5.2–8.7 inch lbs. (0.6–1.0 Nm) for a used bearing. Tighten the flange nut to achieve the preload torque readings originally recorded.

✳✳ CAUTION

Never loosen the pinion nut to reduce bearing preload.

11. Install or connect the following:
• Drive pinion nut, stake it
• Front driveshaft. Tighten the fasteners to 54 ft. lbs. (74 Nm).

• Under cover

12. Fill the differential with gear lubricant and check for leaks.

LAND CRUISER & LX470

1. Before servicing the vehicle, refer to the precautions in the beginning of this section.

2. Remove or disconnect the following:
• Driveshaft
• Front wheels
• Front brake calipers

➡The front brake calipers must be removed so that there is no additional drag when measuring pinion bearing preload.

3. Use an inch lb. torque wrench and measure the amount of torque required to maintain pinion rotation through several revolutions.

4. Remove or disconnect the following:
• Pinion flange
• Oil seal
• Oil slinger
• Pinion bearing and race
• Oil storage ring
• Collapsible spacer

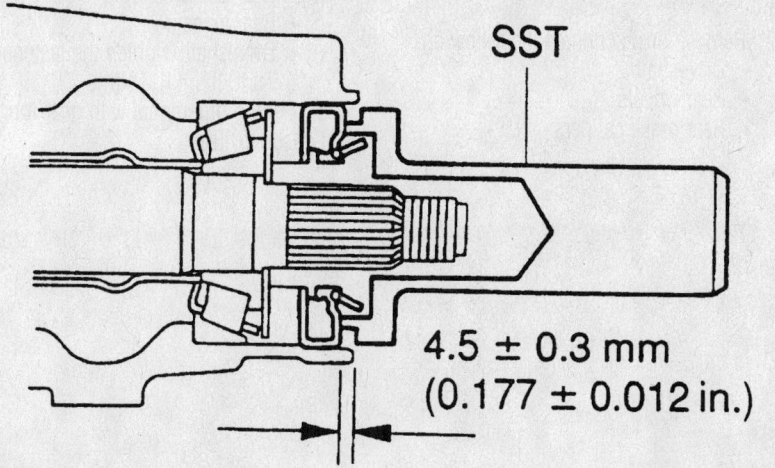

SST

4.5 ± 0.3 mm
(0.177 ± 0.012 in.)

9308YG17

Positioning the Sequoia front pinion seal in the differential housing—Rear differential assembly is similar

To install:

→**Use a new collapsible spacer and flange nut for assembly.**

5. Install or connect the following:
  • Collapsible spacer
  • Oil storage ring
  • Pinion bearing and race
  • Pinion seal
  • Pinion flange. Tighten the nut to 80 ft. lbs. (108 Nm).

6. Rotate the pinion flange occasionally while tightening the flange nut to make sure the pinion bearings seat correctly.

7. Take frequent bearing preload torque readings. Tighten the flange nut to achieve the preload torque readings originally recorded. Do not exceed 249 ft. lbs. (338 Nm) torque when tightening the pinion flange nut.

❋❋ CAUTION

Never loosen the pinion nut to reduce bearing preload. If it is necessary to reduce bearing preload, install a new collapsible spacer and pinion nut.

8. Install or connect the following:
  • Front brake calipers
  • Front wheels
  • Driveshaft. Tighten the fasteners to 59 ft. lbs. (80 Nm).

9. Fill the differential with gear lubricant and check for leaks.

Rear

SEQUOIA

1. Before servicing the vehicle, refer to the precautions in the beginning of this section.

2. Drain the differential housing oil.

3. Remove the rear driveshaft.

4. Remove the companion flange, as follows:
  • Loosen the staked part of the nut, using a chisel and a hammer
  • Companion flange nut, using tool 09330-00021

• Companion flange, using tools 09950-30011 and 09954-03010
• Oil seal, using tool 09308-10010

To install:

5. Install the new oil seal until it is flush with the housing, using a plastic hammer and tools 09316-12010 and 09649-17010

→**Use vinyl tape to connect both oil seal installation tools.**

6. Install the companion flange, as follows:
  • Companion flange
  • New nut, lubricated with hypoid gear oil
  • Torque the nut to 109 ft. lbs. (147 Nm), using tool 09330-00021

7. Adjust the drive pinion preload

8. Rotate the drive pinion, using a torque wrench while tightening the flange nut to make sure the bearing preload is 11.4–16.7 inch lbs. (1.3–1.9 Nm) for a new bearing or 4.3–6.9 inch lbs. (0.5–0.8 Nm) for a used bearing. Tighten the flange nut to achieve the preload torque readings originally recorded.

❋❋ CAUTION

Never loosen the pinion nut to reduce bearing preload.

9. Install or connect the following:
  • Drive pinion nut, stake it
  • Rear driveshaft. Tighten the fasteners to 54 ft. lbs. (74 Nm).

10. Refill the differential with gear lubricant and check for leaks; 3.33 qts. for 2WD or 3.12 qts. for 4WD.

LAND CRUISER & LX470

1. Before servicing the vehicle, refer to the precautions in the beginning of this section.

2. Remove or disconnect the following:
  • Driveshaft
  • Rear wheels
  • Rear brake calipers

→**The rear brake calipers must be removed so that there is no additional drag when measuring pinion bearing preload.**

3. Use an inch lb. torque wrench and measure the amount of torque required to maintain pinion rotation through several revolutions.

4. Remove or disconnect the following:
  • Pinion flange
  • Oil seal
  • Oil slinger
  • Pinion bearing and race
  • Collapsible spacer

To install:

→**Use a new collapsible spacer and flange nut for assembly.**

5. Install or connect the following:
  • Collapsible spacer
  • Pinion bearing and race
  • Pinion seal
  • Pinion flange. Tighten the nut to 181 ft. lbs. (245 Nm).

6. Rotate the pinion flange occasionally while tightening the flange nut to make sure the pinion bearings seat correctly.

7. Take frequent bearing preload torque readings. Tighten the flange nut to achieve the preload torque readings originally recorded. Do not exceed 326 ft. lbs. (441 Nm) torque when tightening the pinion flange nut.

❋❋ CAUTION

Never loosen the pinion nut to reduce bearing preload. If it is necessary to reduce bearing preload, install a new collapsible spacer and pinion nut.

8. Install or connect the following:
  • Rear brake calipers
  • Rear wheels
  • Driveshaft. Tighten the fasteners to 78 ft. lbs. (106 Nm).

9. Fill the differential with gear lubricant and check for leaks.

STEERING AND SUSPENSION

Air Bag

☀☀ CAUTION

Some vehicles are equipped with an air bag system. The system must be disarmed before performing service on, or around, system components, the steering column, instrument panel components, wiring and sensors. Failure to follow the safety precautions and the disarming procedure could result in accidental air bag deployment, possible injury and unnecessary system repairs.

PRECAUTIONS

Several precautions must be observed when handling the inflator module to avoid accidental deployment and possible personal injury.

• Never carry the inflator module by the wires or connector on the underside of the module.

• When carrying a live inflator module, hold securely with both hands and ensure that the bag and trim cover are pointed away.

• Place the inflator module on a bench or other surface with the bag and trim cover facing up.

• With the inflator module on the bench, never place anything on or close to the module which may be thrown in the event of an accidental deployment.

DISARMING

To avoid personal injury when working on vehicles equipped with an air bag, the negative battery cable must be disconnected and at least 90 seconds must elapse before working on the system. Failure to do so may result in deployment of the air bag.

Power Rack And Pinion Steering Gear

REMOVAL & INSTALLATION

Sequoia

1. Before servicing the vehicle, refer to the precautions in the beginning of this section.

2. Position the front wheels in the straight-ahead position.

3. Remove or disconnect the following:

• Engine under cover
• Steering wheel pad
• Steering wheel
• Left and right outer tie-rod ends from the steering knuckles

4. Matchmark the No. 2 intermediate shaft to the steering gear input shaft.

5. Remove or disconnect the following:

• Clamp plate
• Pressure feed and return tubes from the power steering gear, using tool 09631-22020
• Power steering gear assembly

To install:

6. Install or connect the following:

• Power steering gear assembly.

Torque the set bolt to 123 ft. lbs. (165 Nm) and the set nut/bolt to 96 ft. lbs. (91 Nm).

• Pressure feed and return tubes to the power steering gear. Torque them to 27 ft. lbs. (32 Nm), using tool 09631-22020.

• Clamp plate. Torque the bolt to 21 ft. lbs. (29 Nm).

• No. 2 intermediate shaft to the steering gear input shaft

• Left and right outer tie-rod ends to the steering knuckles. Torque the nuts to 67 ft. lbs. (91 Nm).

• Steering wheel. Torque the nut to 26 ft. lbs. (35 Nm).

• Steering wheel pad
• Engine under cover

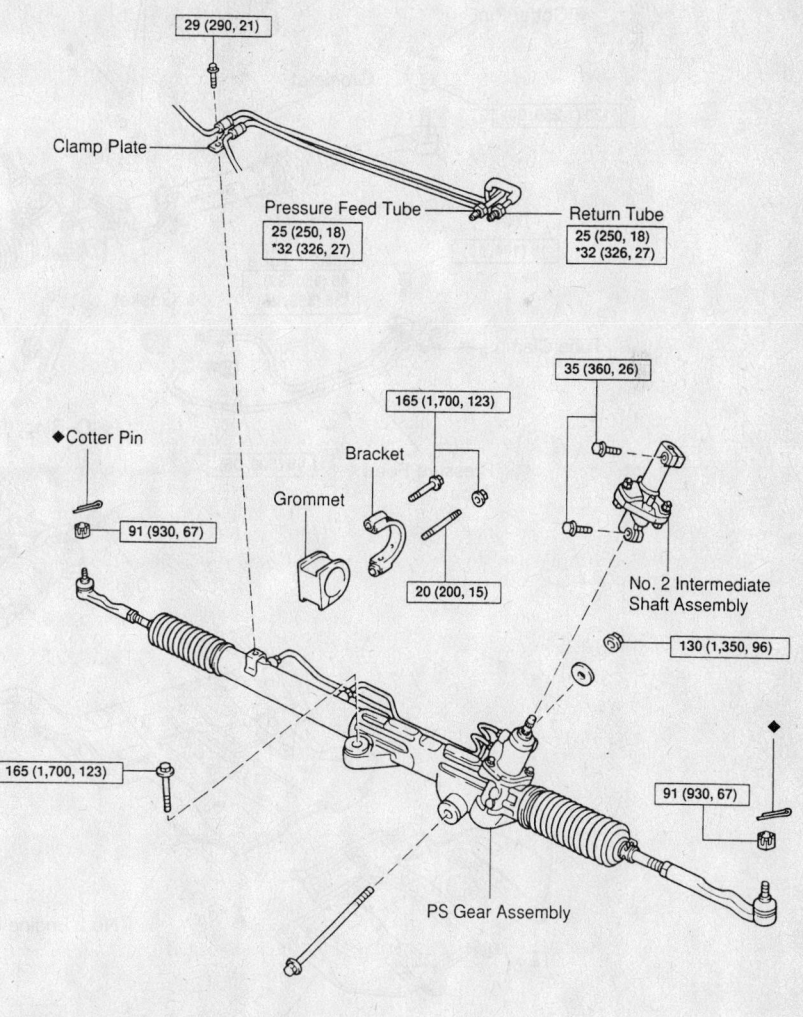

29 (290, 21)

Clamp Plate

Pressure Feed Tube
25 (250, 18)
*32 (326, 27)

Return Tube
25 (250, 18)
*32 (326, 27)

35 (360, 26)

◆Cotter Pin

165 (1,700, 123)

Bracket

Grommet

91 (930, 67)

20 (200, 15)

No. 2 Intermediate
Shaft Assembly

130 (1,350, 96)

165 (1,700, 123)

91 (930, 67)

PS Gear Assembly

N·m (kgf·cm, ft·lbf) : Specified torque
◆Non-reusable part
* For use with SST

9308YG18

Exploded view of the power rack and pinion steering gear mounting—Sequoia

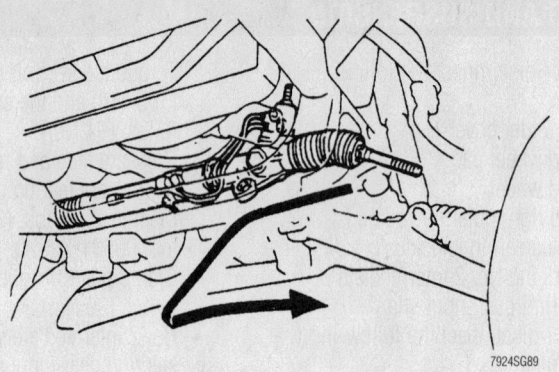

Power rack and pinion steering gear removal—Land Cruiser/LX470

7924SG89

7. Fill and bleed the power steering system.

8. Check and/or adjust the wheel alignment, as necessary.

Land Cruiser & LX470

1. Before servicing the vehicle, refer to the precautions in the beginning of this section.

2. Matchmark the intermediate shaft to the steering gear input shaft.

3. Remove or disconnect the following:

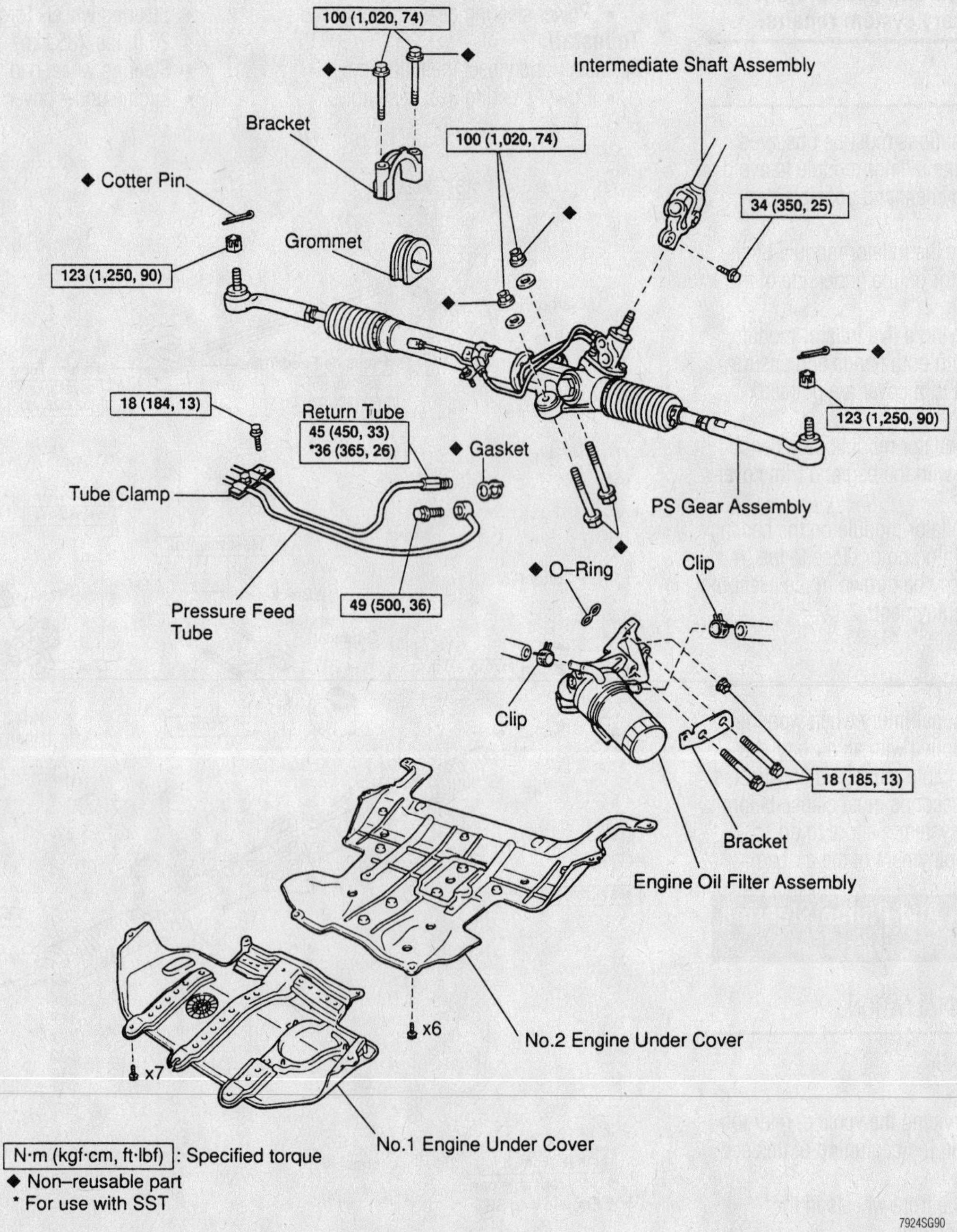

Bracket

100 (1,020, 74)

Intermediate Shaft Assembly

100 (1,020, 74)

34 (350, 25)

Cotter Pin

Grommet

123 (1,250, 90)

123 (1,250, 90)

18 (184, 13)

Return Tube
45 (450, 33)
*36 (365, 26)

Gasket

Tube Clamp

PS Gear Assembly

Pressure Feed Tube

49 (500, 36)

O–Ring

Clip

Clip

18 (185, 13)

Bracket

Engine Oil Filter Assembly

x6

No.2 Engine Under Cover

x7

No.1 Engine Under Cover

N·m (kgf·cm, ft·lbf) : Specified torque
◆ Non–reusable part
* For use with SST

7924SG90

Exploded view of the rack and pinion steering gear mounting—Land Cruiser/LX470

- Negative battery cable
- Engine under covers
- Outer tie rod ends
- Engine oil filter adapter
- Intermediate steering shaft
- Power steering hoses and bracket
- Power steering gear

To install:

4. Install or connect the following:
- Power steering gear. Tighten the fasteners to 74 ft. lbs. (100 Nm).
- Power steering hoses and bracket
- Intermediate steering shaft. Tighten the bolts to 25 ft. lbs. (34 Nm).
- Engine oil filter adapter. Tighten the bolts to 13 ft. lbs. (18 Nm).
- Outer tie rod ends. Tighten the nuts to 90 ft. lbs. (122 Nm).
- Engine under covers
- Negative battery cable

5. Fill the power steering fluid reservoir.

6. Check the wheel alignment and adjust as necessary.

Shock Absorber

REMOVAL & INSTALLATION

Land Cruiser & LX470 Without Active Height Control

FRONT

1. Before servicing the vehicle, refer to the precautions in the beginning of this section.
2. Support the axle with a jackstand.
3. Remove or disconnect the following:
- Front wheel
- Shock absorber

To install:

4. Install or connect the following:
- Shock absorber. Tighten the nut to 51 ft. lbs. (69 Nm) and the bolt to 100 ft. lbs. (135 Nm).
- Front wheel

REAR

1. Before servicing the vehicle, refer to the precautions in the beginning of this section.
2. Support the axle with a jackstand.
3. Remove or disconnect the following:
- Rear wheel
- Shock absorber

To install:

4. Install or connect the following:
- Shock absorber. Tighten the nut to

51 ft. lbs. (69 Nm) and the bolt to 72 ft. lbs. (98 Nm).
- Rear wheel

Land Cruiser & LX470 With Active Height Control

FRONT

✳✳ **CAUTION**

The vehicle ride height may change suddenly when relieving system pressure.

1. Before servicing the vehicle, refer to the precautions in the beginning of this section.
2. Relieve the Active Height Control (AHC) hydraulic pressure as follows:
   a. Connect a hose to the control actuator bleed screw and place the other end in a container.
   b. Open the bleed screw.
   c. When the fluid pressure has dropped and oil stops flowing, close the bleed screw.
3. Remove or disconnect the following:
- Front wheel
- Inner fender liner
- Lower shock absorber mounting bolt
- AHC pressure hose
- Upper shock absorber mounting nut
- Shock absorber

To install:

4. Install or connect the following:
- Shock absorber. Tighten the upper nut to 51 ft. lbs. (68 Nm) and the lower bolt to 101 ft. lbs. (135 Nm).
- AHC pressure hose with new O-ring seals. Tighten the bolts to 13 ft. lbs. (18 Nm).
- Inner fender liner
- Front wheel

➡**Do not let the AHC reservoir run empty during this procedure.**

5. Bleed the AHC system as follows:
   a. Fill the AHC system reservoir with AHC fluid 08886-01805.
   b. Start the engine and push **N** on the vehicle height select switch.
   c. When the AHC pump stops, turn the engine **OFF**.
   d. Open the bleed screw and allow any air in the system to escape.
   e. Repeat until no air is expelled from the bleed screw.
   f. Fill the AHC reservoir to the correct level.

REAR

✳✳ **CAUTION**

The vehicle ride height may change suddenly when relieving system pressure.

1. Before servicing the vehicle, refer to the precautions in the beginning of this section.
2. Support the rear axle with a jack or stands.
3. Relieve the Active Height Control (AHC) hydraulic pressure as follows:
   a. Connect a hose to the control actuator bleed screw and place the other end in a container.
   b. Open the bleed screw.
   c. When the fluid pressure has dropped and oil stops flowing, close the bleed screw.
4. Remove or disconnect the following:
- Rear wheel
- Lower shock absorber mounting bolt
- AHC pressure hose
- Upper shock absorber mounting nut
- Shock absorber

To install:

5. Install or connect the following:
- Shock absorber. Tighten the upper nut to 51 ft. lbs. (68 Nm) and the lower bolt to 72 ft. lbs. (98 Nm).
- AHC pressure hose with new O-ring seals. Tighten the bolts to 13 ft. lbs. (18 Nm).
- Rear wheel

➡**Do not let the AHC reservoir run empty during this procedure.**

6. Bleed the AHC system as follows:
   a. Fill the AHC system reservoir with AHC fluid 08886-01805.
   b. Start the engine and push **N** on the vehicle height select switch.
   c. When the AHC pump stops, turn the engine **OFF**.
   d. Open the bleed screw and allow any air in the system to escape.
   e. Repeat until no air is expelled from the bleed screw.
   f. Fill the AHC reservoir to the correct level.

Sequoia

REAR

1. Before servicing the vehicle, refer to the precautions in the beginning of this section.

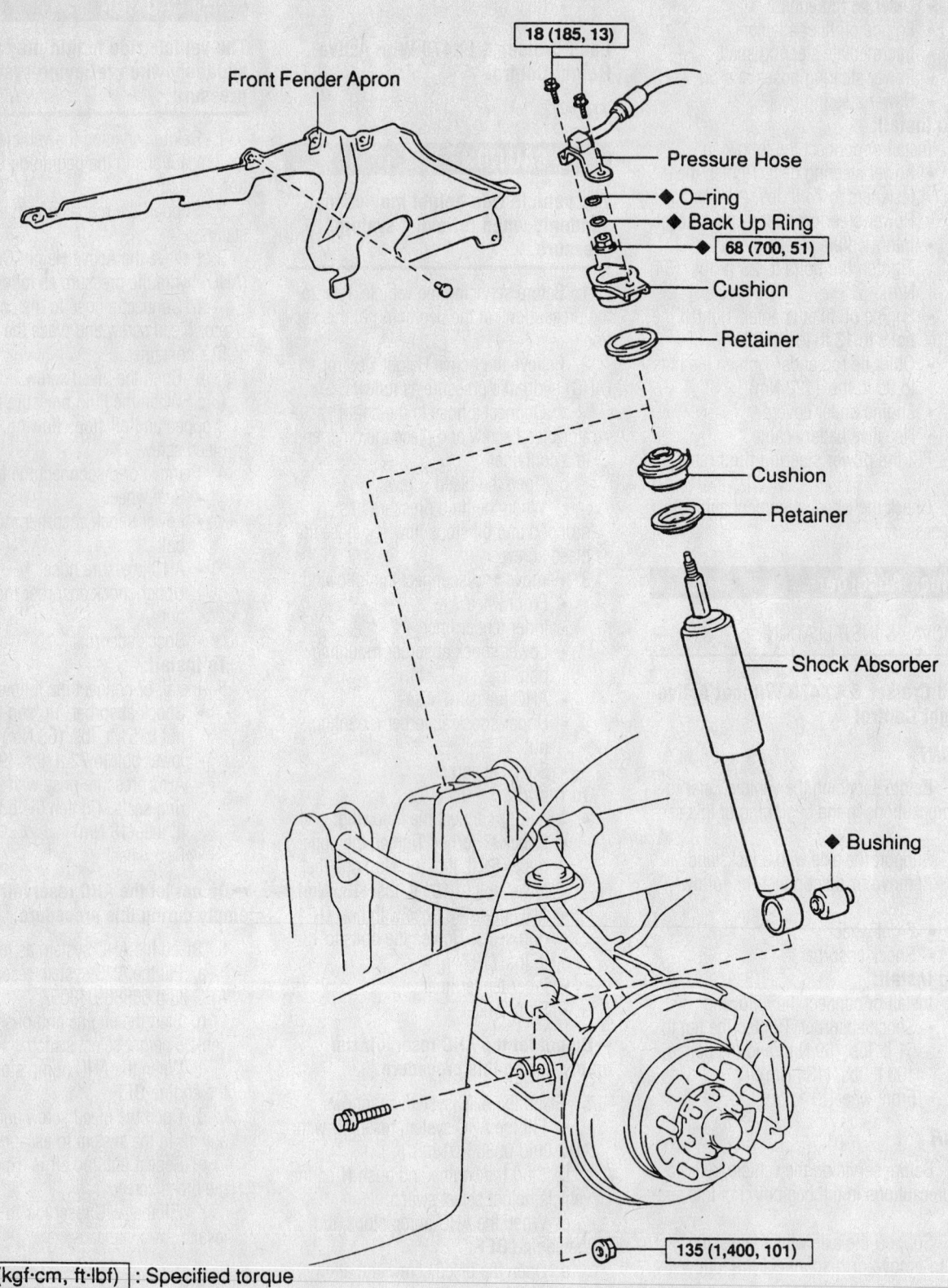

Front Fender Apron

18 (185, 13)

Pressure Hose

◆ O—ring

◆ Back Up Ring

◆ 68 (700, 51)

Cushion

Retainer

Cushion

Retainer

Shock Absorber

◆ Bushing

135 (1,400, 101)

N·m (kgf·cm, ft·lbf) : Specified torque

◆ Non—reusable part

7924SG86

Exploded view of the front shock absorber mounting—Land Cruiser/LX470 models with Active Height Control (AHC)

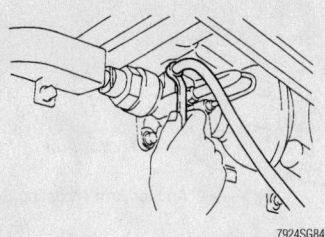

Relieving system pressure—Land Cruiser/LX470 with Active Height Control (AHC)

7924SG84

2. Remove or disconnect the following:
   - Rear wheel
   - Shock absorber

To install:

3. Install or connect the following:
   - Shock absorber. Tighten the upper nut to 15 ft. lbs. (20 Nm) and the lower nut/bolt to 64 ft. lbs. (87 Nm).
   - Rear wheel. Torque the lug nuts to 81 ft. lbs. (110 Nm).

Strut

REMOVAL & INSTALLATION

Sequoia

FRONT

1. Before servicing the vehicle, refer to the precautions in the beginning of this section.

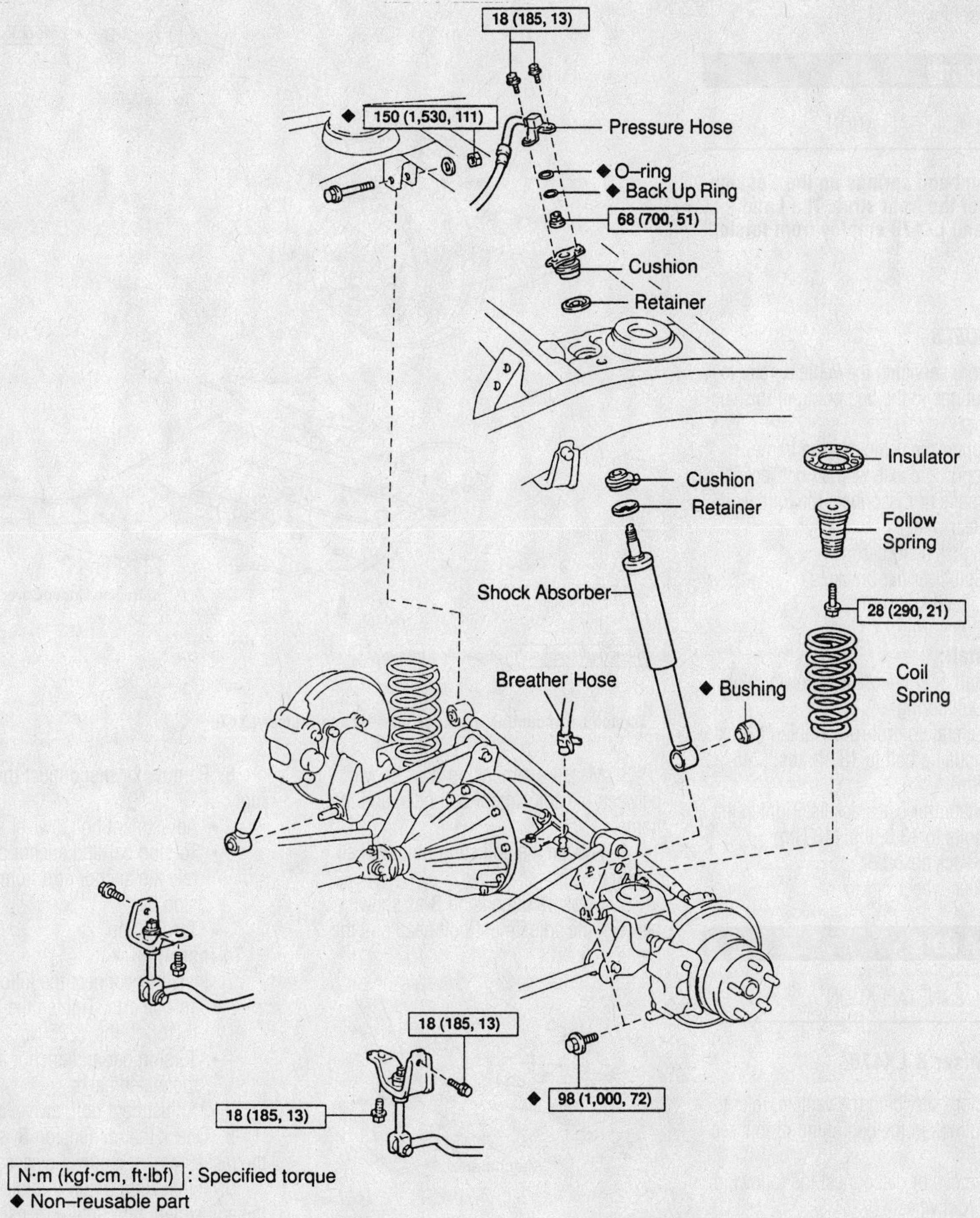

18 (185, 13)

150 (1,530, 111)

Pressure Hose

◆ O–ring

◆ Back Up Ring

68 (700, 51)

Cushion

Retainer

Insulator

Cushion

Retainer

Follow Spring

Shock Absorber

28 (290, 21)

Breather Hose

◆ Bushing

Coil Spring

18 (185, 13)

18 (185, 13)

◆ 98 (1,000, 72)

N·m (kgf·cm, ft·lbf) : Specified torque
◆ Non–reusable part

7924SG87

Exploded view of the rear shock absorber mounting—Land Cruiser/LX470 with Active Height Control (AHC)

2. Remove or disconnect the following:
- Front wheel
- Strut-to-lower control arm nut/bolt and the strut
- Strut-to-chassis 3 nuts/bolts and the strut

To install:

3. Install or connect the following:
- Strut to the chassis. Torque the 3 nuts/bolts to 47 ft. lbs. (64 Nm).
- Strut to the lower control arm. Torque the nut/bolt to 100 ft. lbs. (135 Nm).
- Front wheel

Coil Spring

REMOVAL & INSTALLATION

➡ The front coil springs on the Sequoia are part of the front strut. The Land Cruiser and LX470 employ front torsion bars.

Rear

ALL MODELS

1. Before servicing the vehicle, refer to the precautions in the beginning of this section.
2. Support the vehicle at the frame.
3. Support the axle with a floor jack.
4. Remove or disconnect the following:
- Rear wheel
- Shock absorber
- Stabilizer bar brackets
- Lateral control rod
- Coil spring

To install:

5. Install or connect the following:
- Coil spring
- Lateral control rod. Tighten the axle housing bolt to 181 ft. lbs. (245 Nm).
- Stabilizer bar brackets. Tighten the bolts to 13 ft. lbs. (18 Nm)
- Shock absorber
- Rear wheel

Torsion Bars

REMOVAL & INSTALLATION

Land Cruiser & LX470

1. Before servicing the vehicle, refer to the precautions in the beginning of this section.
2. Remove or disconnect the following:
- Front wheel
- Engine under cover

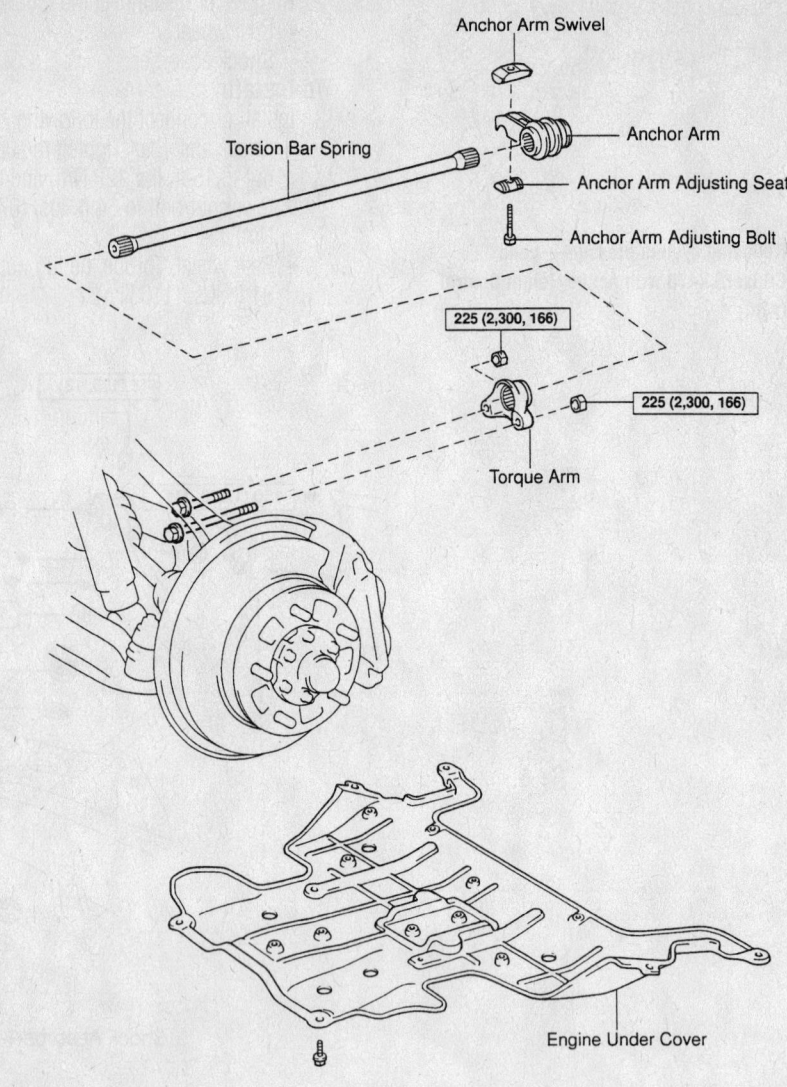

N·m (kgf·cm, ft·lbf) : Specified torque

9308SG10

Torsion bar mounting exploded view—Land Cruiser/LX470

3. Measure dimension **A** as shown between the adjustment bolt head and the frame.
4. Loosen the adjusting bolt until all spring tension is relieved.
5. Measure dimension **B** as shown between the adjustment bolt head and the frame.

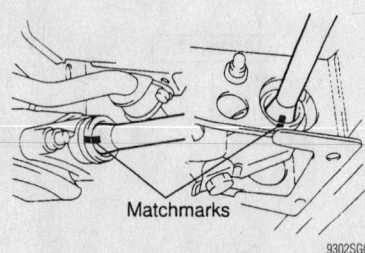

9302SG03

Matchmark the torsion bar to the anchor arm and torque arm—Land Cruiser/LX470

6. Remove or disconnect the following:
- Adjustment bolt, swivel and seat
- Torsion bar and anchor arm. Separate the anchor arm from the torsion bar.
- Torque arm

To install:

7. Install or connect the following:
- Torque arm. Tighten the fasteners to 166 ft. lbs. (225 Nm).
- Torsion bar and anchor arm. Align the matchmarks.
- Adjustment bolt, swivel and seat

8. Check that dimension **B** is close to the measurement made at disassembly.
9. If installing a new torsion bar, tighten the adjustment bolt until dimension **A** is as follows:

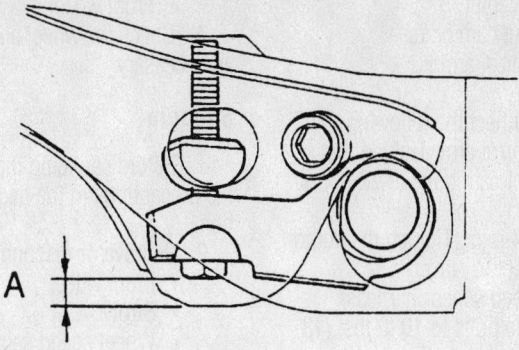

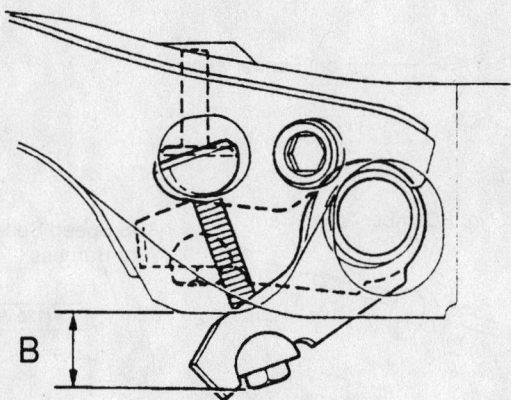

9302SG04

Reference measurements A and B—Land Cruiser/LX470

- Left torsion bar: 0.315–0.984 inches (8–25mm)
- Right torsion bar: 0.079–0.709 inches (2–18mm)

10. If installing the original torsion bar, tighten the adjustment bolt until dimension **A** is close to the measurement made at disassembly.

11. Install or connect the following:
- Engine under cover
- Front wheel

12. Place the vehicle on a flat, level surface and check the vehicle curb height as follows:
- a. Step 1: Measure dimension **A**

Front:

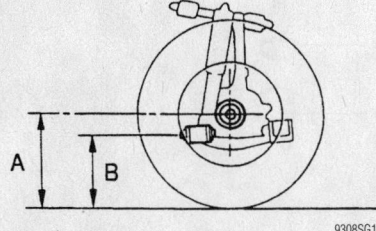

9308SG12

Ride height measurements A and B—Land Cruiser/LX470

between the spindle center and the ground.

b. Step 2: Measure dimension **B** between the lower control arm front bolt center and the ground.

c. Step 3: Turn the adjusting bolt so that **A** minus **B** is equal to 2.795 inches (71mm).

Upper Ball Joint

REMOVAL & INSTALLATION

Land Cruiser & LX470

The upper ball joint is serviced with the upper control arm as an assembly.

Sequoia

1. Before servicing the vehicle, refer to the precautions in the beginning of this section.

2. Remove or disconnect the following:
- Front wheel
- Steering knuckle with the axle hub
- Wire and boot
- Snapring
- Upper ball joint from the steering

knuckle, using a deep socket wrench and tool 09050-40011

To install:

3. Install or connect the following:
- New upper ball joint to the steering knuckle, using a deep socket and tool 09309-37010
- New snapring

4. Using a torque wrench, inspect the upper ball joint rotation, as follows:

a. Flip the ball joint back-and-forth 5 times.

b. Using a torque wrench, continuously turn the nut 1 turn in 2–4 seconds.

c. Take the reading on the 5th turn; it should be 6–39 inch lbs. (0.7–4.4 Nm). If not, replace the upper ball joint.

5. Install or connect the following:
- New boot secured with a wire
- Front wheel. Torque the lug nuts to 81 ft. lbs. (110 Nm).

6. Check and/or adjust the front wheel alignment.

Lower Ball Joint

REMOVAL & INSTALLATION

Land Cruiser & LX470

The lower ball joint is serviced with the lower control arm as an assembly.

Sequoia

1. Before servicing the vehicle, refer to the precautions in the beginning of this section.

2. Remove or disconnect the following:
- Front wheel
- 4 lower ball joint set bolts
- Tie-rod end from the lower ball joint, using tool 09610-20012
- Lower ball joint nut.
- Lower ball joint from the lower control arm, using tool 09628-62011

To install:

3. Install or connect the following:
- New lower ball joint to the lower control. Torque the bolts to 103 ft. lbs. (140 Nm).
- New cotter pin
- Tie-rod end to the lower ball joint. Torque the nut to 67 ft. lbs. (91 Nm).
- Lower ball joint set bolts. Torque the 4 bolts to 59 ft. lbs. (80 Nm).
- Front wheel. Torque the lug nuts to 81 ft. lbs. (110 Nm).

4. Check and/or adjust the front wheel alignment.

Upper Control Arm

REMOVAL & INSTALLATION

Land Cruiser & LX470

1. Before servicing the vehicle, refer to the precautions in the beginning of this section.

2. Remove or disconnect the following:
   - Front wheel
   - Inner fender liner
   - Wheel speed sensor harness
   - Upper ball joint
   - Adjustment cam bolts
   - Upper control arm

To install:

3. Install or connect the following:
   - Upper control arm. Tighten the adjustment cam bolts to 72 ft. lbs. (98 Nm).
   - Upper ball joint. Tighten the nut to 81 ft. lbs. (110 Nm).
   - Wheel speed sensor harness. Tighten the bolts to 10 ft. lbs. (13 Nm).
   - Inner fender liner
   - Front wheel

4. Check the wheel alignment and adjust as necessary.

Sequoia

1. Before servicing the vehicle, refer to the precautions in the beginning of this section.

2. Remove or disconnect the following:
   - Front wheel
   - Strut
   - Wheel speed sensor harness, if equipped with Anti-lock Brake System (ABS)

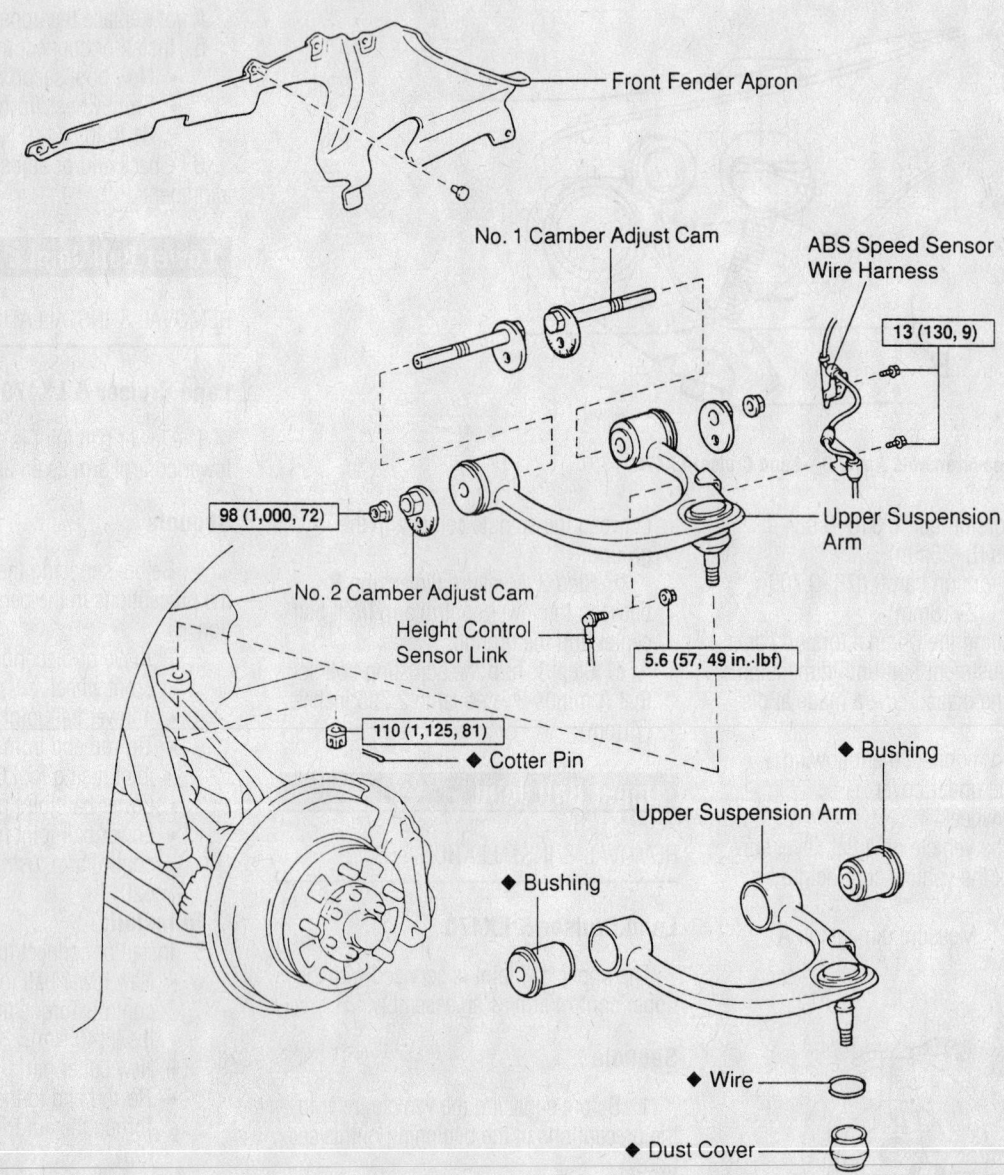

N·m (kgf·cm, ft·lbf) : Specified torque

◆ Non–reusable part

9302SG01

Exploded view of the upper control arm and related components—Land Cruiser/LX470

3. Upper ball joint, as follows:
   - Cotter pin and loosen the nut
   - Upper ball joint from the upper control arm, using tool 09950-40011
   - Steering knuckle, support it securely
   - Upper ball joint nut
4. Remove or disconnect the following:
   - 4 clips and the fender apron seal
   - Brake/fuel line clamp nut and clamp

- Both upper control arm-to-chassis nuts/bolts
- Upper control arm

To install:

5. Install or connect the following:
   - Upper control arm. Torque both upper control arm-to-chassis nuts/bolts to 72 ft. lbs. (98 Nm).
   - Brake/fuel line clamp nut and clamp. Torque the clamp nut to 49 inch lbs. (5.5 Nm).
   - Fender apron seal

- Upper ball joint. Torque the nut to 77 ft. lbs. (105 Nm).
- New cotter pin
- Steering knuckle
- Wheel speed sensor harness, if equipped with Anti-lock Brake System (ABS). Torque it to 71 inch lbs. (8.0 Nm).
- Strut
- Front wheel

6. Check and/or adjust the wheel alignment.

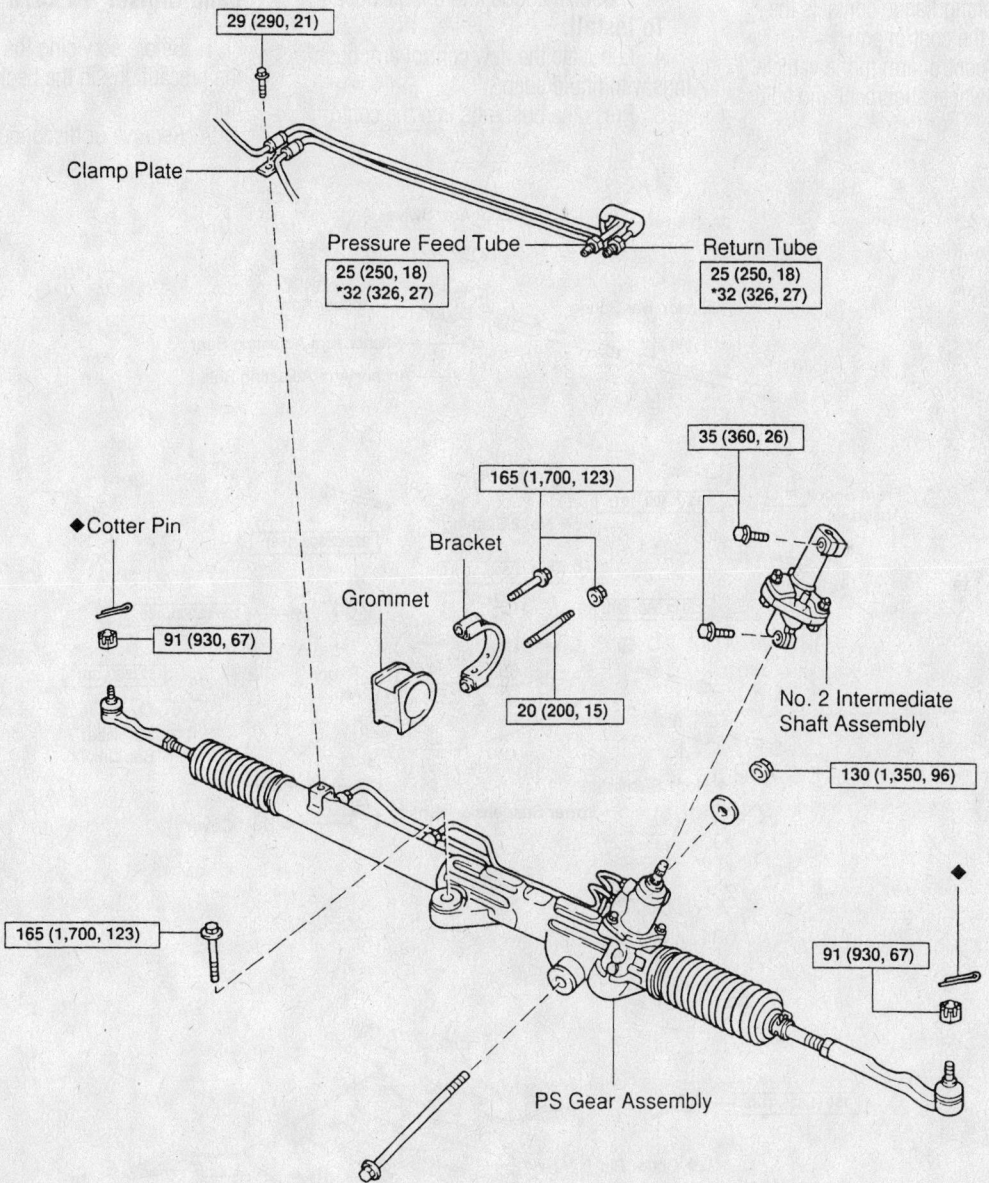

29 (290, 21)

Clamp Plate

Pressure Feed Tube
25 (250, 18)
*32 (326, 27)

Return Tube
25 (250, 18)
*32 (326, 27)

◆Cotter Pin

91 (930, 67)

Grommet

Bracket

165 (1,700, 123)

20 (200, 15)

35 (360, 26)

No. 2 Intermediate Shaft Assembly

130 (1,350, 96)

165 (1,700, 123)

91 (930, 67)

PS Gear Assembly

N·m (kgf·cm, ft·lbf) : Specified torque
◆Non–reusable part
* For use with SST

Exploded view of the front suspension and related components—Sequoia

9308YG19

CONTROL ARM BUSHING REPLACEMENT

Land Cruiser & LX470

1. Before servicing the vehicle, refer to the precautions in the beginning of this section.
2. Remove the control arm from the vehicle.
3. Remove the control arm bushings with a hydraulic press.

To install:

4. Lubricate the control arm bushings with liquid soap.
5. Press the bushings into the control arm until the bushing flange contacts the housing edge of the control arm.
6. Install the control arm to the vehicle.
7. Check the wheel alignment and adjust as necessary.

Sequoia

1. Before servicing the vehicle, refer to the precautions in the beginning of this section.
2. Remove the upper control arm from the vehicle.
3. Remove the control arm bushings, as follows:

- Pry up the bushing flange, using a chisel and a hammer
- Press the bushing(s) from the upper control arm, using a shop press and tools 09613-26010, 09631-20060 and 09950-00020

To install:

4. Lubricate the new control arm bushings with liquid soap.
5. Press the bushings into the control arm until the bushing flange contacts the housing edge of the control arm, using a shop press, a steel plate and tools 09631-12090 and 09710-30021
6. Install the upper control arm to the vehicle.
7. Check and/or adjust the wheel alignment.

Lower Control Arm

REMOVAL & INSTALLATION

Land Cruiser & LX470

1. Before servicing the vehicle, refer to the precautions in the beginning of this section.
2. Remove or disconnect the following:

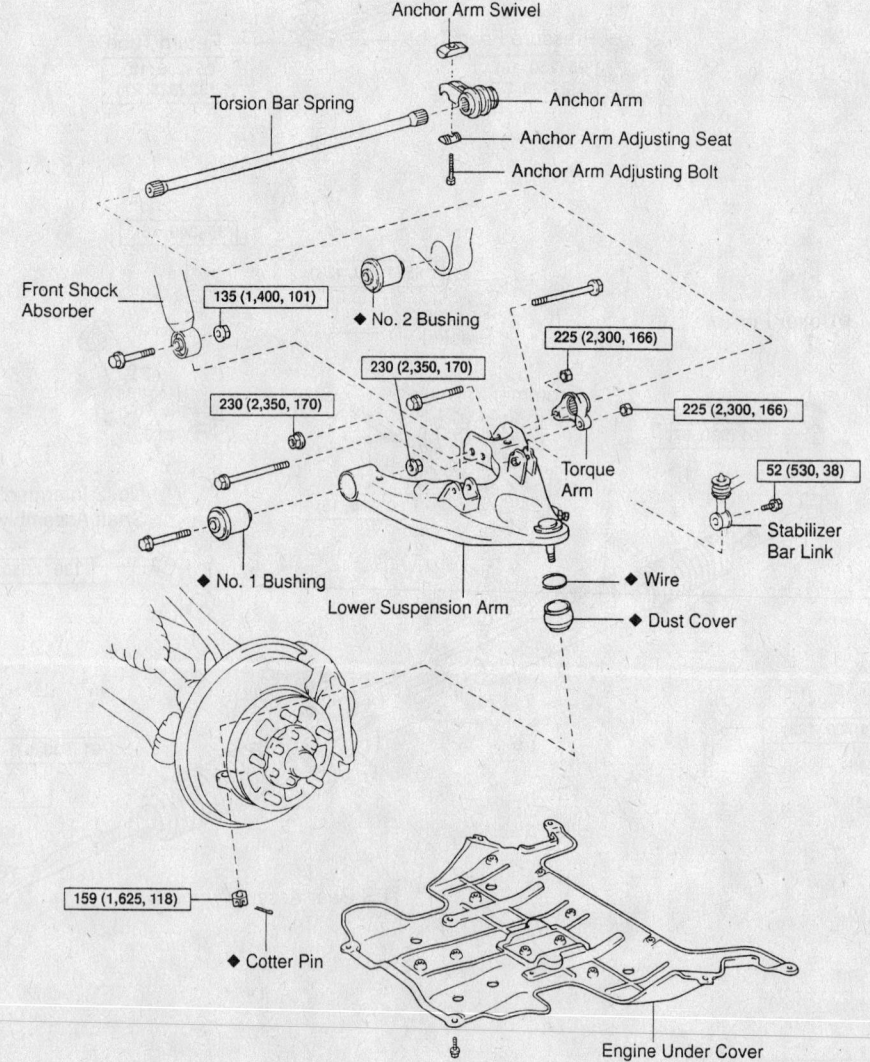

N·m (kgf·cm, ft·lbf) : Specified torque
◆ Non–reusable part

9302SG02

Exploded view of the lower control arm and related components—Land Cruiser/LX470

- Front wheel
- Engine under cover
- Torsion bar
- Stabilizer bar link
- Shock absorber
- Lower ball joint
- Lower control arm

To install:
3. Install or connect the following:
- Lower control arm. Tighten the bolts to 170 ft. lbs. (230 Nm).
- Lower ball joint. Tighten the nut to 117 ft. lbs. (159 Nm).
- Shock absorber
- Stabilizer bar link. Tighten the bolt to 38 ft. lbs. (52 Nm).
- Torsion bar
- Engine under cover
- Front wheel

4. Check the wheel alignment and adjust as necessary.

Sequoia

1. Before servicing the vehicle, refer to the precautions in the beginning of this section.
2. Remove front wheel.
3. Disconnect the tie-rod end, as follows:
- Cotter pin and nut
- Tie-rod end from the lower ball joint, using tool 09610-20012
4. Remove or disconnect the following:
- Power steering gear set bolts and nuts
- Stabilizer bar link from the lower control arm
- Strut from the lower control arm
5. Disconnect the lower ball joint, as follows:
- Cotter pin and nut
- Lower ball joint from the lower control arm
6. Matchmark both front and rear cam plates and chassis frame.
7. Remove the lower control arm while slightly shifting the power steering gear rearward.

To install:
8. Install or connect the following:
- Lower control arm while slightly shifting the power steering gear rearward
- Align both front and rear cam plates and chassis frame matchmarks. Torque both bolts to 96 ft. lbs. (130 Nm).
9. Connect the lower ball joint, as follows:
- Lower ball joint to the lower control arm. Torque the nut to 103 ft. lbs. (140 Nm).

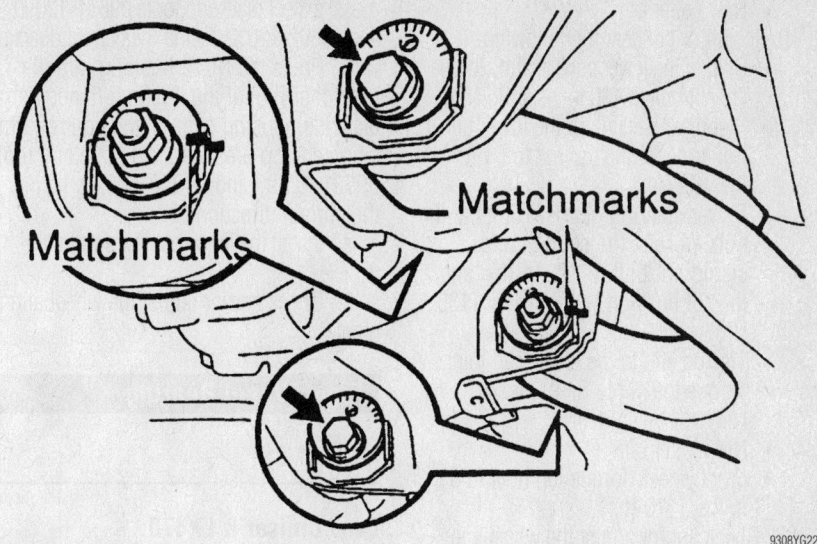

9308YG22

View of the lower control arm's cam plate alignment—Sequoia

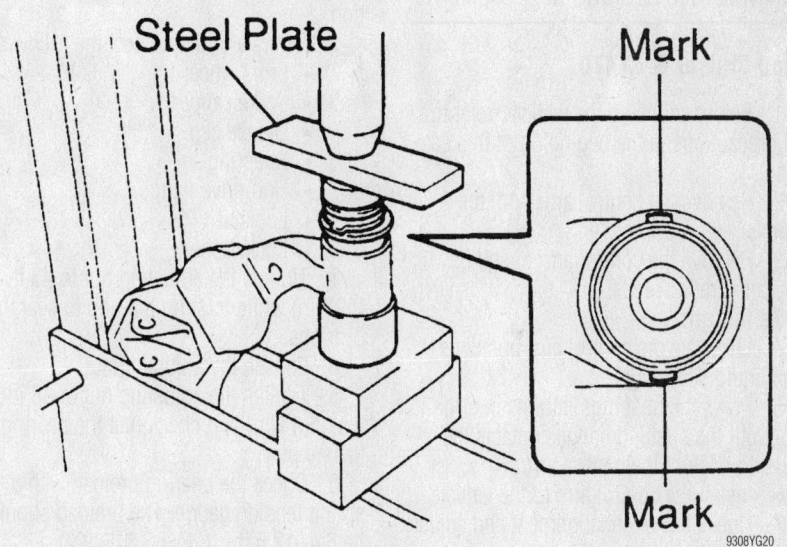

9308YG20

View of the No. 1 bushing's installed direction—Sequoia

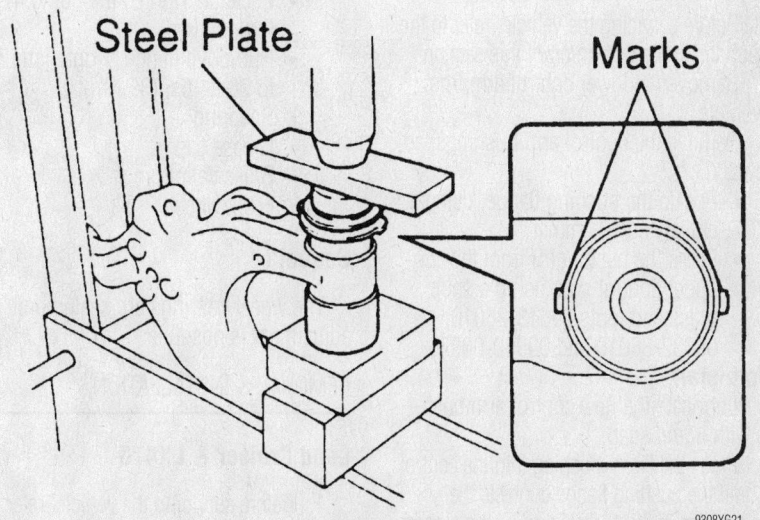

9308YG21

View of the No. 2 bushing's installed direction—Sequoia

- New cotter pin
10. Install or connect the following:
  - Strut to the lower control arm. Torque the nut/bolt to 100 ft. lbs. (135 Nm).
  - Stabilizer bar link to the lower control arm. Torque the nut to 51 ft. lbs. (69 Nm).
  - Power steering gear set bolts and nuts. Torque the set bolt and clamp nut/bolt to 122 ft. lbs. and the set nut/bolt to 96 ft. lbs. (130 Nm)
  - Tie-rod end to the lower ball joint. Torque the nut to 67 ft. lbs. (91 Nm).
  - New cotter pin
  - Front wheel. Torque lug nuts to 81 ft. lbs. (110 Nm).
11. Check and/or adjust the wheel alignment.

CONTROL ARM BUSHING REPLACEMENT

Land Cruiser & LX470

1. Before servicing the vehicle, refer to the precautions in the beginning of this section.
2. Remove the control arm from the vehicle.
3. Remove the control arm bushings with a hydraulic press.
To install:
4. Lubricate the control arm bushings with liquid soap.
5. Press the bushings into the control arm until the bushing flange contacts the housing edge of the control arm.
6. Install the control arm to the vehicle.
7. Check the wheel alignment and adjust as necessary.

Sequoia

1. Before servicing the vehicle, refer to the precautions in the beginning of this section.
2. Remove the lower control arm from the vehicle.
3. Remove the control arm bushings, as follows:
  - Pry up the bushing flange, using a chisel and a hammer
  - Press the bushing(s) from the upper control arm, using a shop press and tools 09613-26010, 09632-36010 and 09950-00020
To install:
4. Lubricate the new control arm bushings with liquid soap.
5. Press the No. 1 bushing into the control arm until the bushing flange contacts the housing edge of the control arm, using a shop

press, a steel plate and tools 09631-12090 and 09502-12010, facing the correct direction.
6. Press the No. 2 bushing into the control arm until the bushing flange contacts the housing edge of the control arm, using a shop press, a steel plate and tools 09631-12090 and 09950-60020, facing the correct direction.
7. Install the lower control arm to the vehicle.
8. Check and/or adjust the wheel alignment.

Front Wheel Bearing

ADJUSTMENT

Land Cruiser & LX470

1. Before servicing the vehicle, refer to the precautions in the beginning of this section.
2. Remove or disconnect the following:
  - Front wheel
  - Brake caliper
  - Grease cap
  - Snapring
  - Hub drive flange
  - Locknut
  - Lockwasher
3. Tighten the adjusting nut to 43 ft. lbs. (59 Nm) while rotating the hub to seat the bearings.
4. Loosen the adjusting nut.
5. Tighten the adjusting nut to 48 inch lbs. (5.4 Nm) and check that the bearing has no play.
6. Check the bearing preload with a spring tension gauge. The preload should be 6.4–12.6 lbs. (28–56 N).
7. Install or connect the following:
  - Lockwasher
  - Locknut. Tighten the nut to 47 ft. lbs. (64 Nm).
  - Hub drive flange. Tighten the nuts to 26 ft. lbs. (35 Nm).
  - Snapring
  - Grease cap
  - Brake caliper
  - Front wheel

Sequoia

The wheel bearings are sealed unit; no adjustment is possible.

REMOVAL & INSTALLATION

Land Cruiser & LX470

1. Before servicing the vehicle, refer to the precautions in the beginning of this section.

2. Remove or disconnect the following:
  - Front wheel
  - Brake caliper
  - Grease cap
  - Snapring
  - Hub drive flange
  - Locknut
  - Lockwasher
  - Adjusting nut
  - Outer bearing
  - Wheel hub
  - Inner grease seal
  - Inner bearing
To install:
3. Install or connect the following:
  - Inner bearing
  - Inner grease seal
  - Wheel hub
  - Outer bearing
  - Adjusting nut. Adjust the wheel bearings.
  - Lockwasher
  - Locknut. Tighten the nut to 47 ft. lbs. (64 Nm).
  - Hub drive flange. Tighten the nuts to 26 ft. lbs. (35 Nm).
  - Snapring
  - Grease cap
  - Brake caliper
  - Front wheel

Sequoia

1. Before servicing the vehicle, refer to the precautions in the beginning of this section.
2. Remove or disconnect the following:
  - Front wheel
  - Axle hub/steering knuckle assembly and place it in a vise
  - Grease cap, for 2WD
  - Inner grease seal, for 4WD
3. Remove the axle hub from the steering knuckle
  - 4 bolts and shift the dust cover towards the outside (hub side)
  - Axle hub from the steering knuckle, using tools 09710-30021 and 09950-40011
  - Dust cover from the steering knuckle
  - Bearing spacer and Anti-lock Brake System (ABS) speed sensor, if equipped with ABS
  - Spacer, if not equipped with ABS

✳✳ WARNING

Be careful not to scratch the speed sensor rotor serrations.

4. Remove the outside oil seal from steering knuckle, using a small prybar

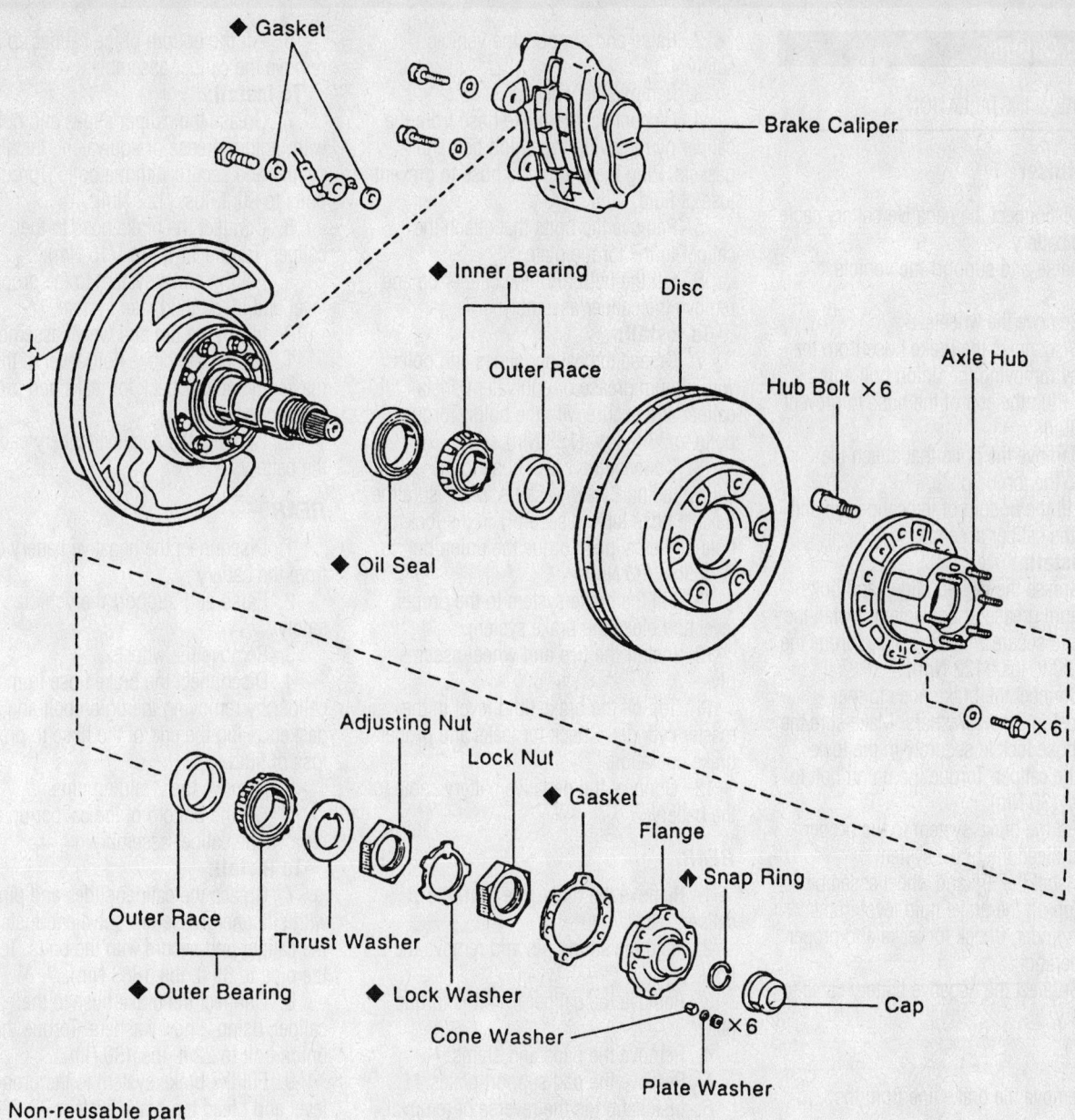

- Gasket
- Brake Caliper
- Inner Bearing
- Outer Race
- Disc
- Hub Bolt ×6
- Axle Hub
- Oil Seal
- ×6
- Adjusting Nut
- Lock Nut
- Gasket
- Flange
- Snap Ring
- Outer Race
- Thrust Washer
- Outer Bearing
- Lock Washer
- Cone Washer
- Cap
- Plate Washer

◆ Non-reusable part

7924SG31

Exploded view of the front hub and related components—Land Cruiser/LX470

5. Remove the bearing from the steering knuckle, as follows:
- Snapring
- Bearing from the steering knuckle, using tools 09950-60020 and 09950-70010

To install:
6. Install the bearing to the steering knuckle, as follows:
- Bearing to the steering knuckle, using tools 09950-60020 and 09527-17011
- Snapring
7. Install the new outside oil seal to

steering knuckle, using a plastic hammer and tools 09223-15030 and 09527-17011
8. Install the axle hub to the steering knuckle, as follows:
- Dust cover to the steering knuckle. Torque the 4 bolts to 13 ft. lbs. (18 Nm).
- Axle hub to the steering knuckle, using a shop press and tool 09649-17010
9. Install or connect the following:
- Bearing spacer and Anti-lock Brake System (ABS) speed sensor, if equipped with ABS

※※ **WARNING**

Be careful not to scratch the speed sensor rotor serrations.

- Bearing spacer, if not equipped with ABS, using a shop press and tools 09950-60010 and 09950-70010
- Grease cap, for 2WD
- Inner grease seal, for 4WD, using a plastic hammer and tool 09527-17011
- Axle hub/steering knuckle assembly
- Front wheel

BRAKES

Brake Caliper

REMOVAL & INSTALLATION

Land Cruiser

1. Disconnect the negative battery cable from the battery.
2. Raise and support the vehicle safely.
3. Remove the wheels.
4. Disconnect the brake hose from the caliper by removing the union bolt and 2 gaskets. Plug the end of the hose to prevent loss of fluid.
5. Remove the bolts that attach the caliper to the torque plate.
6. Lift the bottom of the caliper up and remove the caliper assembly.
To install:
7. Grease the caliper slides and bolts with lithium grease or equivalent. Install the caliper and secure with the bolts. Torque the bolts to 90 ft. lbs. (123 Nm).
8. Connect the brake hose to the caliper, using 2 new washers. Make sure the flexible hose lock is securely in the lock hole of the caliper. Torque the union bolt to 22 ft. lbs. (30 Nm).
9. Fill the brake system to the proper level and bleed the brake system.
10. Install the tire and wheel assembly.
11. Top off the brake fluid level in the master cylinder. Check for leaks and proper brake operation.
12. Connect the negative battery cable to the battery.

REAR

1. Remove the brake line from the caliper.
2. Hold the siding pin and remove the 2 bolts.
3. Remove the caliper from the torque plate.
4. Remove the pads and shims.
5. Remove the pad support plates.
6. Installation is the reverse of removal. Torque the caliper bolts to 20 ft. lbs. (26 Nm). Torque the brake line union bolt to 22 ft. lbs. (30 Nm).

LX470

FRONT

1. Disconnect the negative battery cable from the battery.

2. Raise and support the vehicle safely.
3. Remove the wheels.
4. Disconnect the brake hose from the caliper by removing the union bolt and 2 gaskets. Plug the end of the hose to prevent loss of fluid.
5. Remove the bolts that attach the caliper to the torque plate.
6. Lift the bottom of the caliper up and remove the caliper assembly.
To install:
7. Grease the caliper slides and bolts with lithium grease or equivalent. Install the caliper and secure with the bolts. Torque the bolts to 90 ft. lbs. (123 Nm).
8. Connect the brake hose to the caliper, using 2 new washers. Make sure the flexible hose lock is securely in the lock hole of the caliper. Torque the union bolt to 22 ft. lbs. (30 Nm).
9. Fill the brake system to the proper level and bleed the brake system.
10. Install the tire and wheel assembly.
11. Top off the brake fluid level in the master cylinder. Check for leaks and proper brake operation.
12. Connect the negative battery cable to the battery.

REAR

1. Remove the brake line from the caliper.
2. Hold the siding pin and remove the 2 bolts.
3. Remove the caliper from the torque plate.
4. Remove the pads and shims.
5. Remove the pad support plates.
6. Installation is the reverse of removal. Torque the caliper bolts to 20 ft. lbs. (26 Nm). Torque the brake line union bolt to 22 ft. lbs. (30 Nm).

Sequoia

FRONT

1. Disconnect the negative battery cable from the battery.
2. Raise and support the vehicle safely.
3. Remove the wheels.
4. Disconnect the brake hose from the caliper. Plug the end of the hose to prevent loss of fluid.
5. Remove the bolts that attach the caliper to the torque plate.

6. Lift the bottom of the caliper up and remove the caliper assembly.
To install:
7. Grease the caliper slides and bolts with lithium grease or equivalent. Install the caliper and secure with the bolts. Torque the bolts to 90 ft. lbs. (123 Nm).
8. Connect the brake hose to the caliper. Torque 11 ft. lbs. (15 Nm).
9. Fill the brake system to the proper level and bleed the brake system.
10. Install the tire and wheel assembly.
11. Top off the brake fluid level in the master cylinder. Check for leaks and proper brake operation.
12. Connect the negative battery cable to the battery.

REAR

1. Disconnect the negative battery cable from the battery.
2. Raise and support the vehicle safely.
3. Remove the wheels.
4. Disconnect the brake hose from the caliper by removing the union bolt and 2 gaskets. Plug the end of the hose to prevent loss of fluid.
5. Remove the 2 sliding pins.
6. Lift the bottom of the caliper up and remove the caliper assembly.
To install:
7. Grease the caliper slides and pins with silicone grease or equivalent. Install the caliper and secure with the bolts. Torque the pins to 65 ft. lbs. (883 Nm).
8. Connect the brake hose to the caliper, using 2 new washers. Torque the union bolt to 22 ft. lbs. (30 Nm).
9. Fill the brake system to the proper level and bleed the brake system.
10. Install the tire and wheel assembly.
11. Top off the brake fluid level in the master cylinder. Check for leaks and proper brake operation.
12. Connect the negative battery cable to the battery.

Disc Brake Pads

REMOVAL & INSTALLATION

Land Cruiser and Sequoia

FRONT

1. Raise the vehicle and support it safely.

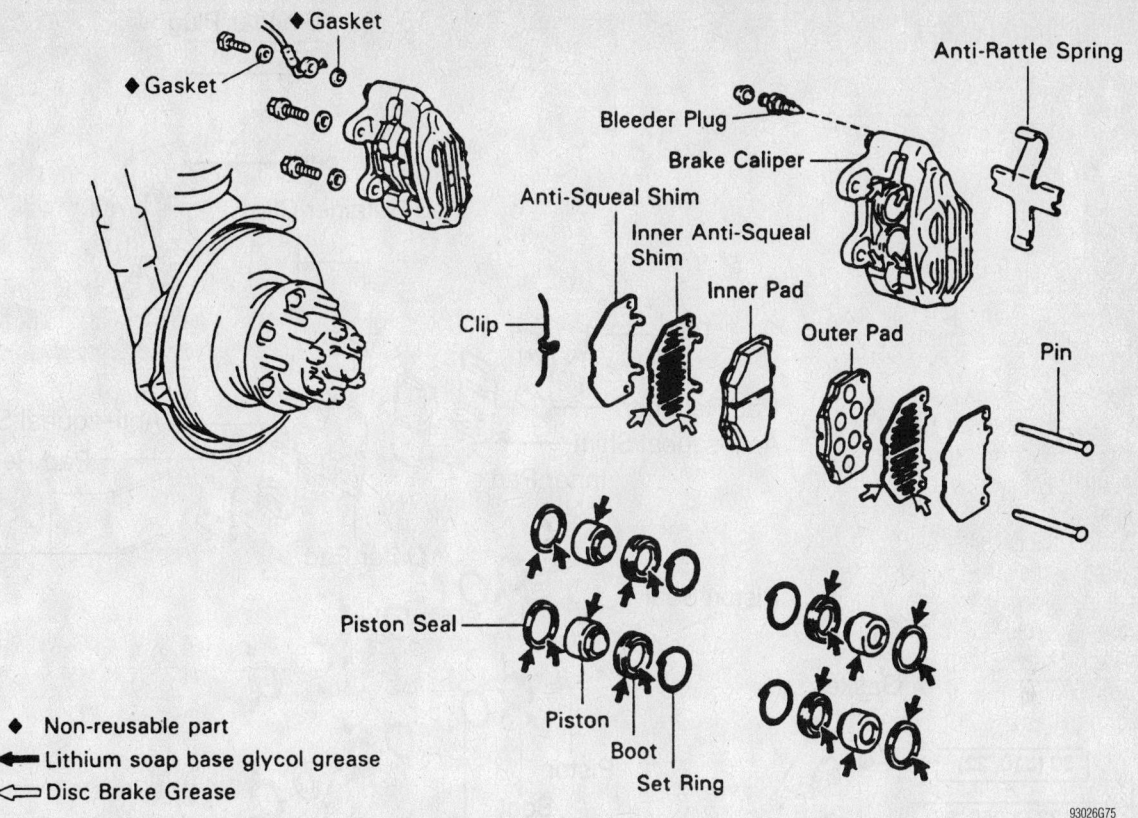

◆ Non-reusable part

◀ Lithium soap base glycol grease

◁ Disc Brake Grease

Exploded view of the front disc brake components—Land Cruiser

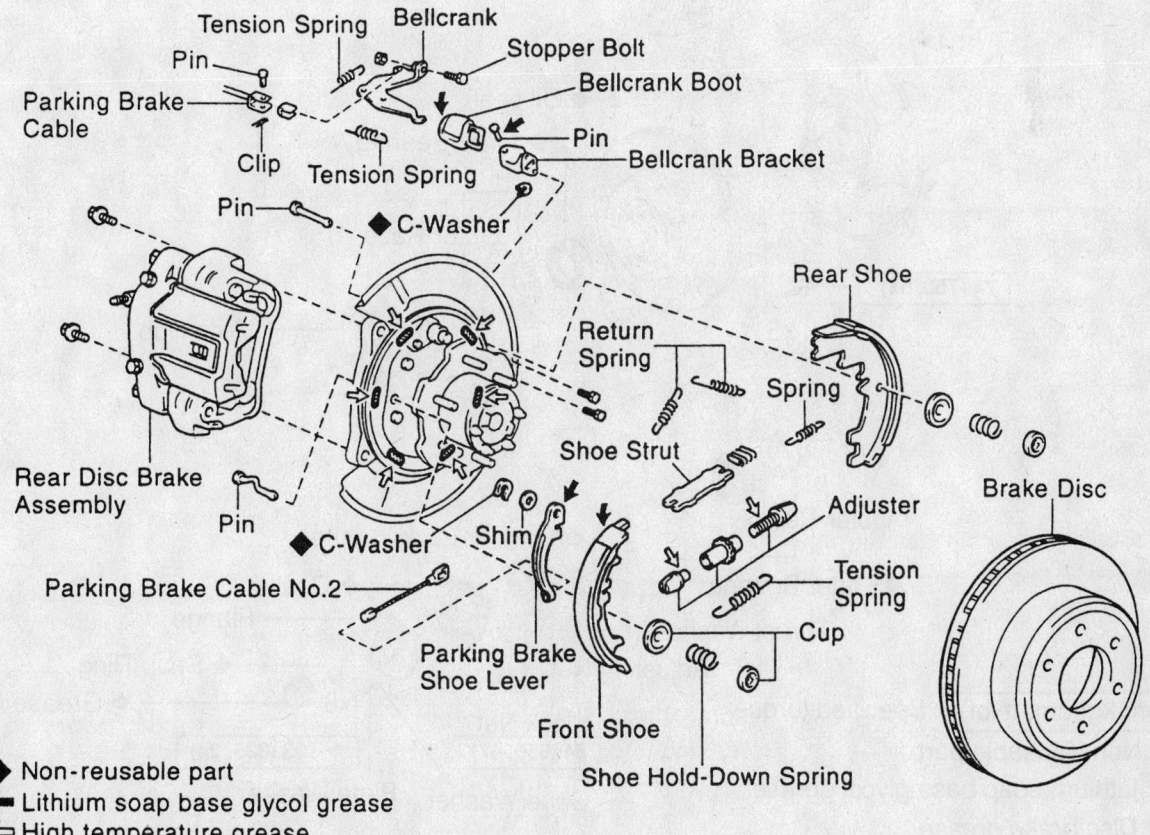

◆ Non-reusable part

◀ Lithium soap base glycol grease

◁ High temperature grease

Exploded view of the rear disc brake components—Land Cruiser

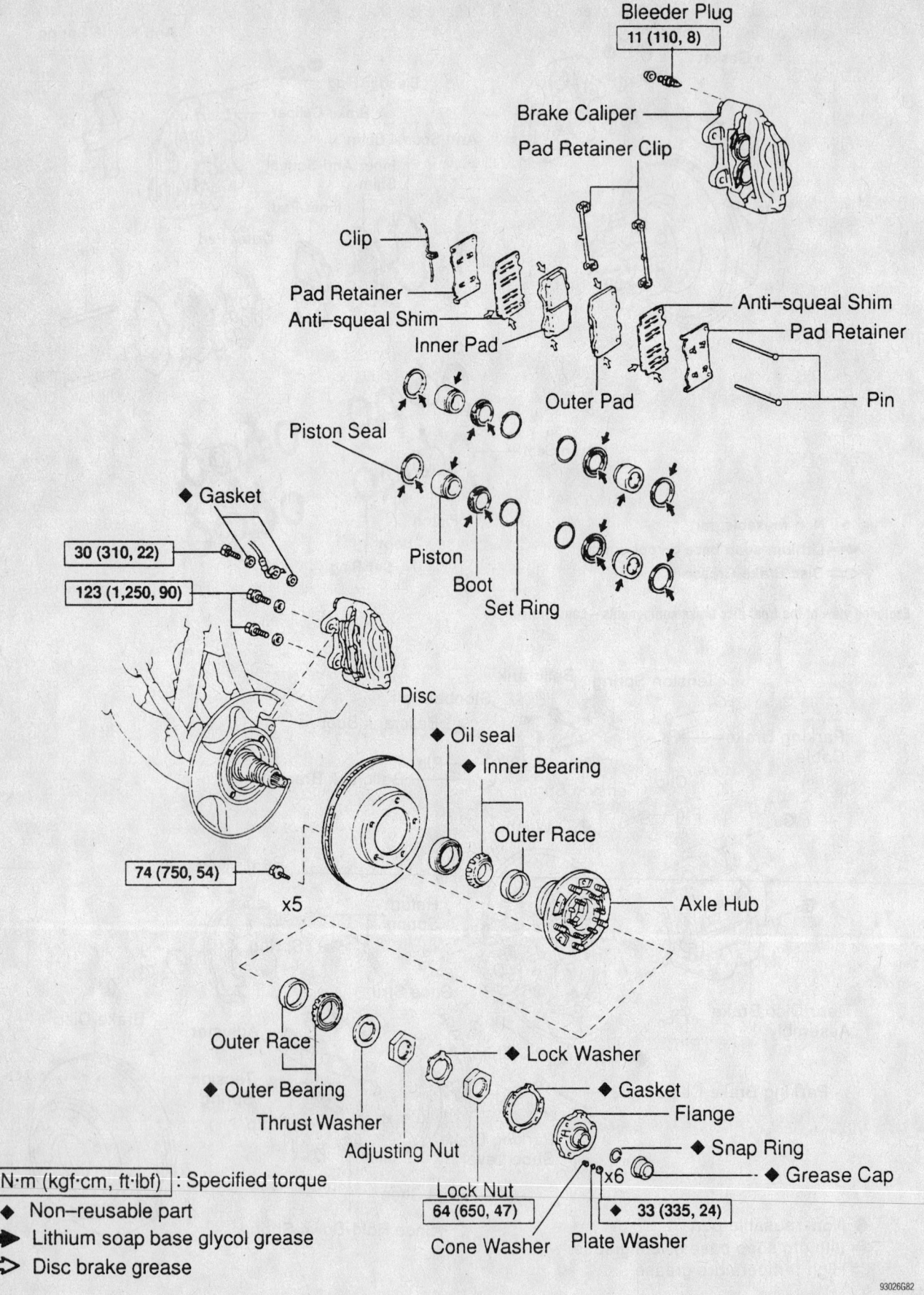

Bleeder Plug
11 (110, 8)

Brake Caliper

Pad Retainer Clip

Clip

Pad Retainer

Anti-squeal Shim

Inner Pad

Outer Pad

Anti-squeal Shim

Pad Retainer

Pin

Piston Seal

◆ Gasket

30 (310, 22)

123 (1,250, 90)

Piston

Boot

Set Ring

Disc

◆ Oil seal

◆ Inner Bearing

Outer Race

Axle Hub

74 (750, 54)

x5

Outer Race

◆ Outer Bearing

Thrust Washer

Adjusting Nut

Lock Nut
64 (650, 47)

Cone Washer

Plate Washer

◆ Lock Washer

◆ Gasket

Flange

◆ Snap Ring

◆ Grease Cap

x6

33 (335, 24)

N·m (kgf·cm, ft·lbf) : Specified torque

◆ Non-reusable part

➤ Lithium soap base glycol grease

⇨ Disc brake grease

93026G82

Exploded view of the front disc brake components—LX470

2. Remove the wheels.

3. Remove the clip, pins and anti-rattle spring.

4. Withdraw the pads and remove the anti-squeal shims.

To install:

5. Before installing the new pads, check the disc thickness and disc runout.

6. Siphon out a small amount of brake fluid from the reservoir.

7. Press in the pistons with a hammer handle or equivalent.

8. Apply disc brake grease to both sides of the inner anti-squeal shim. Install the anti-squeal shims to the new pads.

9. Install the pads.

10. Install the anti-rattle springs and pins. Install the clip.

11. Install the wheels.

12. Check and adjust the fluid level. Apply the brake pedal several times.

13. Road-test the vehicle for proper operation.

REAR

1. Raise the vehicle and support it safely.

2. Remove the wheels.

3. Remove the brake caliper and suspend it so the hose is not stretched.

4. Remove the brake pads, anti-squeal shim, pad support plates and wear indicators.

To install:

5. Before installing the new pads, check the disc thickness and disc runout.

6. Install the pad support plates.

7. Install the pad wear indicator plates on each pad.

8. Install the anti-squeal shim to the outer pad. Install the pads so the wear indicator plate is facing upward.

9. Install the brake caliper.

10. Install the wheels.

11. Apply the brake pedal several times.

12. Road-test the vehicle for proper operation.

LX470

FRONT

1. Raise the vehicle and support it safely.

2. Remove the wheels.

3. Remove the clip, pins and anti-rattle spring.

4. Withdraw the pads and remove the anti-squeal shims.

To install:

5. Before installing the new pads, check the disc thickness and disc runout.

6. Siphon out a small amount of brake fluid from the reservoir.

7. Press in the pistons with a hammer handle or equivalent.

8. Apply disc brake grease to both sides of the inner anti-squeal shim. Install the anti-squeal shims to the new pads.

9. Install the pads.

10. Install the anti-rattle springs and pins. Install the clip.

11. Install the wheels.

12. Check and adjust the fluid level. Apply the brake pedal several times.

13. Road-test the vehicle for proper operation.

REAR

1. Raise the vehicle and support it safely.

2. Remove the wheels.

3. Remove the brake caliper and suspend it so the hose is not stretched.

4. Remove the brake pads, anti-squeal

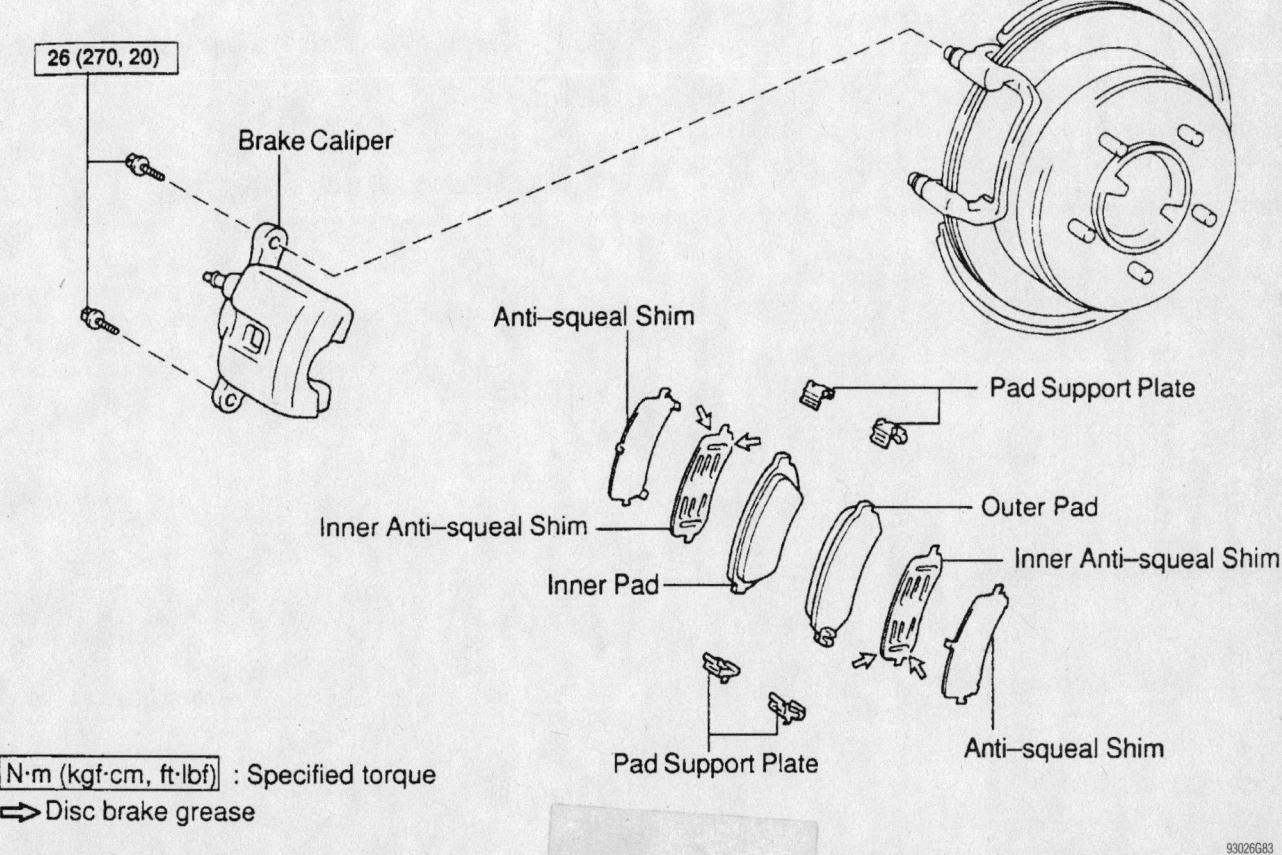

26 (270, 20)

Brake Caliper

Anti-squeal Shim

Pad Support Plate

Inner Anti-squeal Shim

Inner Pad

Outer Pad

Inner Anti-squeal Shim

Pad Support Plate

Anti-squeal Shim

N·m (kgf·cm, ft·lbf) : Specified torque

⇨ Disc brake grease

93026G83

Exploded view of the rear disc brake components—LX470

shim, pad support plates and wear indicators.

To install:

5. Before installing the new pads, check the disc thickness and disc runout.

6. Install the pad support plates.

7. Install the pad wear indicator plate to each pads.

8. Install the anti-squeal shim to the outer pad. Install the pads so the wear indicator plate is facing upward.

9. Install the brake caliper.

10. Install the wheels.

11. Apply the brake pedal several times.

12. Road-test the vehicle for proper operation.